CHILTON'S

ELECTRONIC ENGINE CONTROLS MANUAL

AF894578

Publisher Kerry A. Freeman, S.A.E. □ **Editor-In-Chief** Dean F. Morgantini, S.A.E.
Senior Editor Richard J. Rivele, S.A.E. □ **Senior Editor** Nick D'Andrea □ **Senior Editor** Ron Webb
Project Manager Peter M. Conti, Jr. □ **Project Manager** Ken Grabowski, A.S.E., S.A.E.
Project Manager Martin J. Gunther □ **Project Manager** Richard T. Smith
Service Editors Lawrence C. Braun, S.A.E., A.S.C., Robert E. Doughten, Jeff H. Fisher, A.S.E., Jacques Gordon, Michael L. Grady, Ben Greisler, S.A.E., Jeffrey M. Hoffman, Steve Horner, Joseph T. Lafferty, Neil Leonard, A.S.E., Kevin Maher, James R. Marotta, Robert McAnally, Steven Morgan, Michael W. Parks, John H. Rutter, Don Schnell, A.S.E., S.A.E., James B. Steele, Larry E. Stiles, Jim Taylor, Anthony Tortorici, A.S.E., S.A.E.
Editorial Consultants Edward K. Shea, S.A.E., Stan Stephenson

Director of Manufacturing Mike D'Imperio
Production Manager W. Calvin Settle, Jr., S.A.E
Assistant Production Manager Andrea Steiger
Production Assistants Marsha Park Herman, Margaret Stoner
Mechanical Artists Lisa Gressen,Lorraine Martinelli, Kim Tansey

National Sales Manager Lawrence R. Rufo
Manager, National Account Sales Donald A. Wright
National Account Sales Coordinator Jacquelyn T. Powers
Regional Sales Managers Herbert Marshall, Larry W. Marshall, Bruce McCorkle

OFFICERS
Senior Vice President Ronald A. Hoxter

CHILTON *PROFESSIONAL AUTOMOTIVE*
ONE OF THE ***DIVERSIFIED PUBLISHING COMPANIES,***
A PART OF ***CAPITAL CITIES/ABC, INC.***
Manufactured in USA • Chilton Way, Radnor, Pa. 19089
ISBN 0-8019-8273-1 1234567890 1098765432 ISSN 1050-1207

HOW TO USE THIS MANUAL

For ease of use, this manual is divided into sections:

The **CONTENTS**, inside the front cover, summarizes the subjects covered in each section

To quickly locate the proper service section, use the chart on the following pages. It references applicable **MODELS** and **SERVICE SECTIONS** for major engine performance control systems.

It is recommended that the service technician be familiar with the applicable **GENERAL INFORMATION** and **SERVICE PRECAUTIONS** (if any) before testing or servicing the system. Major service sections are grouped by individual vehicle or component. Each sub-section contains:

- **GENERAL INFORMATION** Information pertaining to the operation of the system, individual components and the overall logic by which components work together.

- **SERVICE PRECAUTIONS (if any)** Precautions of which the service technician should be aware to prevent injury or damage to the vehicle or components.

- **TESTS, ADJUSTMENTS AND COMPONENT REMOVAL AND INSTALLATION** Performance tests and specifications, adjustments and component replacement procedures.

- **FAULT DIAGNOSIS** Complete troubleshooting for the entire system.

SAFETY NOTICE

Proper service and repair procedures are vital to the safe, reliable operation of all motor vehicles, as well as the personal safety of those performing repairs. This manual outlines procedures for servicing and repairing vehicles using safe, effective methods. The procedures contain many NOTES and CAUTIONS which should be followed along with standard safety procedures to eliminate the possibilty of personal injury or improper service which could damage the vehicle or compromise its safety.

It is important to note that the repair procedures and techniques, tools and parts for servicing motor vehicles, as well as the skill and experience of the individual performing the work vary widely. It is not possible to anticipate all of the conceivable ways or conditions under which vehicles may be serviced, or to provide cautions as to all of the possible hazards that may result. Standard and accepted safety precautions and equipment should be used when handling toxic or flammable fluids, and safety glasses or other protection should be used during cutting, grinding, chiseling, prying, or any other process that can cause material removal or projectiles.

Some procedures require the use of tools specially designed for a specific purpose. Before substituting another tool or procedure, you must be completely satisfied that neither your personal safety, nor the performance of the vehicle will be endangered

PART NUMBERS

Part numbers listed in this reference are not recommendations by Chilton Book Company for any product by brand name. They are references that can be used with interchange manuals and aftermarket supplier catalogs to locate each brand supplier's discrete part number.

Although information in this manual is based on industry sources and is as complete as possible at the time of publication, the possibilty exists that the manufacturer made later changes which could not be included here. While striving for total accuracy, Chilton Book Company cannot assume responsibility for any errors, changes or omissions that may occur in the compilation of this data.

ACURA

Year	Model	Body VIN	Engine Displacement Liter (cc)	Engine ① ID	Ignition System	Fuel ② System
1990	Integra Hatchbk	DA9	1.8 (1834)	B18A1	PGM-IG	Port
	Sedan	DB1	1.8 (1834)	B18A1	PGM-IG	Port
	Legend Sedan	KA4	2.7 (2675)	C27A1	PGM-IG	Port
	Coupe	KA3	2.7 (2675)	C27A1	PGM-IG	Port
1991-92	Integra Hatchbk	DA9	1.8 (1834)	B18A1	PGM-IG	Port
	Sedan	DB1	1.8 (1834)	B18A1	PGM-IG	Port
	Legend Sedan	KA8	3.2 (3206)	C32A1	PGM-IG	Port
	Coupe	KA7	3.2 (3206)	C32A1	PGM-IG	Port
	NSX	NA1	3.0 (2977)	C30A1	PGM-IG	Port
	Vigor Sedan	CC2	2.5 (2451)	G25A1	PGM-IG	Port

Port—Multi-port Fuel Injection
PGM-IG—Programmed Ignition System
① First 5 characters of engine number; not in VIN
② Electronic control via PGM-FI system

CHRYSLER IMPORTS

Year	Model	Body VIN	Engine Liter (cc)	Engine ID	Ignition System	Fuel System
1990	Colt	V	1.5 (1468)	X-G15B	ESC	MPI
	Colt GL	V	1.5 (1468)	X-G15B	ESC	MPI
	Colt GT	V	1.5 (1468)	X-G15B	ESC	MPI
			1.6 (1595)	Y-4G61	ESC	MPI
	Colt DL (2WD)	V	1.5 (1468)	X-4G15	ESC	MPI
	Colt DL (4WD)	W	1.8 (1755)	T-4G37	ESC	MPI
	Colt Vista (2WD, 4WD)	V,W	2.0 (1997)	V-G63B	ESC	MPI
1991	Colt	V	1.5 (1468)	X-G15B	ESC	MPI
	Colt GL	V	1.5 (1468)	X-G15B	ESC	MPI
	Colt GT	V	1.5 (1468)	X-G15B	ESC	MPI
			1.6 (1595)	Y-4G61	ESC	MPI
	Colt DL (2WD)	V	1.5 (1468)	X-4G15	ESC	MPI
	Colt DL (4WD)	W	1.8 (1755)	T-4G37	ESC	MPI
	Colt Vista (2WD, 4WD)	V,W	2.0 (1997)	V-G63B	ESC	MPI

CHRYSLER IMPORTS

Year	Model	Body VIN	Engine Liter (cc)	Engine ID	Ignition System	Fuel System
1992	Colt	V	1.5 (1468)	X-G15B	ESC	MPI
	Colt GL	V	1.5 (1468)	X-G15B	ESC	MPI
	Colt GT	V	1.5 (1468)	X-G15B	ESC	MPI
			1.6 (1595)	Y-4G61	ESC	MPI
	Colt DL (2WD)	V	1.5 (1468)	X-4G15	ESC	MPI
	Colt DL (4WD)	W	1.8 (1755)	T-4G37	ESC	MPI
	Colt Vista (2WD, 4WD)	V,W	2.0 (1997)	V-G63B	ESC	MPI

MPI—Multiport Injection
VIN Engine Code is 8th digit of VIN number
ECI—Electronically Controlled Fuel Injection
(Throttle Body Injection)
ESC—Electronic Spark Control
FBC—Feedback Carburetor

DAIHATSU

Year	Model	Body VIN	Engine Displacement Liter (cc)	Engine ID (VIN)	Ignition System	Fuel System
1990	Charade ①	G	1.0 (993)	CB (0)	Electronic	EFI
			1.3 (1295)	HC (2)	Electronic	EFI
	Rocky	F	1.6 (1589)	HD (0)	Electronic	EFI
1991	Charade ①	G	1.0 (993)	CB (0)	Electronic	EFI
			1.3 (1295)	HC (2)	Electronic	EFI
	Rocky	F	1.6 (1589)	HD (0)	Electronic	EFI
1992	Charade ①	G	1.0 (993)	CB (0)	Electronic	EFI
			1.3 (1295)	HC (2)	Electronic	EFI
	Rocky	F	1.6 (1589)	HD (0)	Electronic	EFI

① With 1.0L 3 cylinder engine—G100 model
With 1.3L 4 cylinder engine—G102 model

HONDA

Year	Model		Body VIN	Engine Displacement Liter (cc)	Engine ① ID (VIN)	Ignition System	Fuel ② System
1990	Accord	DX	CB7	2.2 (2156)	F22A1	PGM-IG	Port
		LX	CB7	2.2 (2156)	F22A1	PGM-IG	Port
		EX	CB7	2.2 (2156)	F22A4 ③	PGM-IG	Port
	Civic	1500 SDN	ED3	1.5 (1493)	D15B1	PGM-IG	Dual Pt.
		1600 SDN	ED4	1.6 (1590)	D16A6	PGM-IG	Port
		1500 Htchbk	ED6	1.5 (1493)	D15B2	PGM-IG	Dual Pt.
		1600 Htchbk	ED7	1.6 (1590)	D16A6	PGM-IG	Port
		1500 CRX	ED8	1.5 (1493)	D15B2	PGM-IG	Dual Pt.
		1500 CRX HF	ED8	1.5 (1493)	D15B6	PGM-IG	Dual Pt.
		1600 CRX-Si	ED9	1.6 (1590)	D16A6	PGM-IG	Port
		1600 WGN	EE4	1.6 (1590)	D16A6	PGM-IG	Port
		1500 WGN	EE2	1.5 (1493)	D15B2	PGM-IG	Dual Pt.
	Prelude	S	BA4	2.0 (1958)	B20A3	Electronic ⑤	Carb.
		2.0 Si	BA4	2.0 (1958)	B20A5	PGM-IG	Port
		Si	BA4	2.1 (2056)	B21A1	PGM-IG	Port
1991-92	Accord	DX	CB7	2.2 (2156)	F22A1	PGM-IG	Port
		LX	CB7	2.2 (2156)	F22A1	PGM-IG	Port
		EX	CB7	2.2 (2156)	F22A4 ③	PGM-IG	Port
		SE	CB7	2.2 (2156)	F22A6 ④	PGM-IG	Port
	Civic	1500 SDN	ED3	1.5 (1493)	D15B1	PGM-IG	Dual Pt.
		1600 SDN	ED4	1.6 (1590)	D16A6	PGM-IG	Port
		1500 Htchbk	ED6	1.5 (1493)	D15B2	PGM-IG	Dual Pt.
		1600 Htchbk	ED7	1.6 (1590)	D16A6	PGM-IG	Port
		1500 CRX	ED8	1.5 (1493)	D15B2	PGM-IG	Dual Pt.
		1500 CRX HF	ED8	1.5 (1493)	D15B6	PGM-IG	Dual Pt.
		1600 CRX-Si	ED9	1.6 (1590)	D16A6	PGM-IG	Port
		1600 WGN	EE4	1.6 (1590)	D16A6	PGM-IG	Port
		1500 WGN	EE2	1.5 (1493)	D15B2	PGM-IG	Dual Pt.
	Prelude	S	BA4	2.0 (1958)	B20A5	PGM-IG	Port
		Si	BA4	2.1 (2056)	B21A1	PGM-IG	Port

PGM-IG—Programmed ignition system
Dual Pt.—Dual point throttle body injection
Carb.—Prelude uses 2 1bbl carburetor
Port—Ported fuel injection
① First 5 digits of engine number; not in VIN
② Programmed fuel injection system (PGM-F1)
③ Dual exhaust manifold
④ Dual intake manifold
⑤ Temperature controlled vacuum advance

HYUNDAI

Year	Model	Body VIN	Engine Liter (cc)	Engine VIN	Ignition Type	Fuel Type
1990	Excel	V	1.5 (1468)	J	Electronic	MPI
	Sonata	B	2.4 (2351)	S	Electronic	MPI
1991	Scoupe	V	1.5 (1468)	J	Electronic	MPI
	Sonata	B	3.0 (2972)	T	Electronic	MPI
1992	Elantra	J	1.6 (1595)	R	Electronic	MPI
	Scoupe	V	1.5 (1468)	J	Electronic	MPI
	Sonata	B	2.0 (1997)	F	Electronic	MPI
			3.0 (2972)	T	Electronic	MPI

MPI—Multi-point Fuel Injection

INFINITI

Year	Model	Body VIN ①	Engine Displacement Liter (cc)	Engine ID (VIN) ②	Ignition System	Fuel System
1990	M30	F	3.0 (2960)	VG30E (H)	Electronic	Port
	Q45	G	4.5 (4494)	VH45DE (N)	DIS	Port
1991	G20	P	2.0 (1998)	SR20DE (C)	Electronic	Port
	M30	F	3.0 (2960)	VG30E (H)	Electronic	Port
	Q45	G	4.5 (4494)	VH45DE (N)	DIS	Port
1992	G20	P	2.0 (1998)	SR20DE (C)	Electronic	Port
	M30	F	3.0 (2960)	VG30E (H)	Electronic	Port
	Q45	G	4.5 (4494)	VH45DE (N)	DIS	Port

DIS—Distributorless Ignition System
Port—ECCS—Electronic Concentrated Control System
① Body VIN is 5th digit
② Engine VIN is 4th digit

ISUZU

Year	Model	Body VIN	Engine Displacement Liter (cc)	Engine ID (VIN)	Ignition System	Fuel System
1990	Impulse	JI	1.6 (1588)	4XE1-W (5)	HEI	MPI
	Amigo	TC	2.3 (2254)	4ZD1 (L)	Electronic	2bbl
			2.6 (2559)	4ZE1 (E)	Electronic	MPI
	Pick-Up	TF	2.3 (2254)	4ZD1 (L)	Electronic	2bbl
			2.6 (2559)	4ZE1 (E)	Electronic	MPI
	Trooper/Trooper II	KT	2.6 (2559)	4ZE1 (E)	Electronic	MPI
			2.8 (2835)	V6 (E)	HEI	TBI
1991	Impulse	JI	1.6 (1588)	4XE1-W (5)	HEI	MPI
			1.6 (1588)	4XE1-WT (4)	DIS	MPI
	Stylus	JS	1.6 (1588)	4XE1-V (6)	HEI	MPI
			1.6 (1588)	4XE1-W (5)	HEI	MPI
	Amigo	TC	2.3 (2254)	4ZD1 (L)	Electronic	2bbl
			2.6 (2559)	4ZE1 (E)	Electronic	MPI
	Pick-Up	TF	2.3 (2254)	4ZD1 (L)	Electronic	2bbl
			2.6 (2559)	4ZE1 (E)	Electronic	MPI
			3.1 (3130)	V6 (Z)	HEI	TBI
	Rodeo	UC	2.6 (2559)	4ZE1 (E)	Electronic	MPI
			3.1 (3130)	V6 (Z)	HEI	TBI
	Trooper/Trooper II	KT	2.6 (2559)	4ZE1 (E)	Electronic	MPI
			2.8 (2835)	V6 (R)	HEI	TBI
1992	Impulse	JI	1.6 (1588)	4XE1-W (5)	HEI	MPI
			1.6 (1588)	4XE1-WT (4)	DIS	MPI
			1.8 (1809)	4XF1-W (8)	HEI	MPI
	Stylus	JS	1.6 (1588)	4XE1-V (6)	HEI	MPI
			1.6 (1588)	4XE1-W (5)	HEI	MPI
			1.8 (1809)	4XF1-W (8)	HEI	MPI
	Amigo	TC	2.3 (2254)	4ZD1 (L)	Electronic	2bbl
			2.6 (2559)	4ZE1 (E)	Electronic	MPI
	Pick-Up	TF	2.3 (2254)	4ZD1 (L)	Electronic	2bbl
			2.6 (2559)	4ZE1 (E)	Electronic	MPI
			3.1 (3130)	V6 (Z)	HEI	TBI
	Rodeo	UC	2.6 (2559)	4ZE1 (E)	Electronic	MPI
			3.1 (3130)	V6 (Z)	HEI	TBI
	Trooper/Trooper II	KT	2.6 (2559)	4ZE1 (E)	Electronic	MPI
			2.8 (2835)	V6 (R)	HEI	TBI

HEI—High Energy Ignition
DIS—Direct Ignition System
MPI—Multi-Port Ignition
2bbl—2 barrel carburetor
TBI—Throttle Body Injection

LEXUS

Year	Model	Body VIN ①	Engine Displacement Liter (cc)	Engine ID (VIN) ②	Ignition System	Fuel System
1990	ES250	C	2.5 (2508)	2VZ-FE	Electronic	EFI
	LS400	A	4.0 (3969)	IUZ-FE	Electronic	EFI
1991	ES250	C	2.5 (2508)	2VZ-FE	Electronic	EFI
	LS400	A	4.0 (3969)	IUZ-FE	Electronic	EFI
1992	LS400	A	4.0 (3969)	IUZ-FE	Electronic	EFI
	ES300	B	3.0 (2959)	3VZ-FE	Electronic	EFI
	SC300	A	3.0 (2997)	2JZ-GE	Electronic	EFI
	SC400	A	4.0 (3969)	IUZ-FE	Electronic	EFI

EFI—Electronic Fuel Injection
① 7th digit of VIN
② First 5 digits of VIN; VZV21-2VZ-FE engine; UCF10—IV2-FE

MAZDA

Year	Model	Body VIN ④	Engine Displacement Liter (cc)	Engine ID (VIN) ⑥	Ignition System	Fuel System
1990	323	BG	1.6 (1597)	B6① (2)	Electronic	EGI
	Protege	BG	1.8 (1839)	BP① (4)	Electronic	EGI
			1.8 (1839)	BP② (6)⑨	Electronic	EGI
	626	GD	2.2 (2184)	F2 (③)	Electronic	EGI
	MX6	GE	2.2 (2184)	F2 (③)	Electronic	EGI
	929	HC	3.0 (2954)	JE① (2)	Electronic	EGI
			3.0 (2954)	JE② (4)⑩	Electronic	EGI
	Miata	NA	1.6 (1597)	B6② (⑦)	Electronic	EGI
	RX7	FC	1.3 (1308)	13B (⑦)	Electronic	EGI
			1.3 (1308)	13B (⑦)	Electronic	Turbo
	MPV	LV	2.6 (2606)	G6 (1)	Electronic	EGI
			3.0 (2954)	JE (⑤)	Electronic	EGI
	B2200	UF	2.2 (2184)	F2 (3)	Electronic	Carb.
			2.2 (2184)	F2 (3)	Electronic	EGI
	B2600i	UF	2.6 (2606)	G6 (4)	Electronic	EGI

MAZDA

Year	Model	Body VIN ④	Engine Displacement Liter (cc)	Engine ID (VIN) ⑥	Ignition System	Fuel System
1991	323	BG	1.6 (1597)	B6① (2)	Electronic	EGI
	Protege	BG	1.8 (1839)	BP① (4)	Electronic	EGI
			1.8 (1839)	BP② (6)⑨	Electronic	EGI
	626	GD	2.2 (2184)	F2 (③)	Electronic	EGI
	MX6	GE	2.2 (2184)	F2 (③)	Electronic	EGI
	929	HC	3.0 (2954)	JE① (2)	Electronic	EGI
			3.0 (2954)	JE② (4)⑩	Electronic	EGI
	Miata	NA	1.6 (1597)	B6② (⑦)	Electronic	EGI
	RX7	FC	1.3 (1308)	13B (⑦)	Electronic	EGI
			1.3 (1308)	13B (⑦)	Electronic	Turbo
	MPV	LV	2.6 (2606)	G6 (1)	Electronic	EGI
			3.0 (2954)	JE (⑤)	Electronic	EGI
	Navajo	V44	4.0 (4000)	⑧ (X)	EDIS	EEC IV
	B2200	UF	2.2 (2184)	F2 (3)	Electronic	Carb.
			2.2 (2184)	F2 (3)	Electronic	EGI
	B2600i	UF	2.6 (2606)	G6 (4)	Electronic	EGI
1992	323	BG	1.6 (1597)	B6① (2)	Electronic	EGI
	Protege	BG	1.8 (1839)	BP① (4)	Electronic	EGI
			1.8 (1839)	BP② (6)⑨	Electronic	EGI
	626	GD	2.2 (2184)	F2 (③)	Electronic	EGI
	MX6	GD	2.2 (2184)	F2 (③)	Electronic	EGI
	929	HD	3.0 (2954)	JE (⑦)	Electronic	EGI
	MX3	EC	1.6 (1597)	B6① (1)	Electronic	EGI
			1.8 (1845)	K8 (2)	Electronic	EGI
	Miata	NA	1.6 (1597)	B6② (⑦)	Electronic	EGI
	MPV	LV	2.6 (2606)	G6 (1)	Electronic	EGI
			3.0 (2954)	JE (⑤)	Electronic	EGI
	Navajo	V44	4.0 (4000)	⑧ (X)	EDIS	EEC IV
	B2200	UF	2.2 (2184)	F2 (3)	Electronic	Carb.
			2.2 (2184)	F2 (3)	Electronic	EGI
	B2600i	UF	2.6 (2606)	G6 (4)	Electronic	EGI

EGI—Electronic Gasoline Injection
EDIS—Electronic Distributorless Ignition System
EEC IV—Electronic Engine Control IV
① SOHC
② DOHC
③ VIN 1, 2, A or B—Nonturbo, VIN 3, 4, C or D—Turbo
④ 4th and 5th digits except Navajo—5th, 6th and 7th digits
⑤ VIN 2 is 3.0L engine with 2WD, VIN 3 is 3.0L engine with 4WD
⑥ 8th digit
⑦ Information not provided in VIN
⑧ Information not provided on Navajo
⑨ VIN 8 with 4WD
⑩ VIN 3 in Canada

MITSUBISHI

Year	Model	Body VIN	Engine Displacement Liter (cc)	Engine ID (VIN)	Ignition System	Fuel System
1990	Eclipse	S/T	1.8 (1795)	4G37 (T)	Distributor	MPI
			2.0 (1997)	4G63 (R) ①	DIS	MPI
			2.0 (1997)	4G63 (U) ②	DIS	MPI
	Galant	R/X	2.0 (1997)	4G63 (R) ①	DIS	MPI
			2.0 (1997)	4G63 (U) ②	DIS	MPI
			2.0 (1997)	4G63 (V) ③	Distributor	MPI
	Mirage	U	1.5 (1468)	4G15 (A)	Distributor	MPI
			1.6 (1597)	4G61 (Y)	DIS	MPI
	Montero	J	2.6 (2555)	G54B (E)	Distributor	Carb
	Montero	J	3.0 (2972)	6G72 (S)	Distributor	MPI
	Precis	V/U	1.5 (1468)	64DJ (J)	Distributor	MPI
	Sigma	B	3.0 (2972)	6G72 (S)	Distributor	MPI
	3000GT	D/E	3.0 (2972)	6G72 (B)	DIS	MPI
			3.0 (2972)	6G72 (C)	DIS	MPI
	Truck	M/L	2.4 (2350)	4G64 (W)	Distributor	MPI
			3.0 (2972)	6G72 (S)	Distributor	MPI
1991	Eclipse	S/T	1.8 (1795)	4G37 (T)	Distributor	MPI
			2.0 (1997)	4G63 (R) ①	DIS	MPI
			2.0 (1997)	4G63 (U) ②	DIS	MPI
	Galant	R/X	2.0 (1997)	4G63 (R) ①	DIS	MPI
			2.0 (1997)	4G63 (U) ②	DIS	MPI
			2.0 (1997)	4G63 (V) ③	Distributor	MPI
	Mirage	U	1.5 (1468)	4G15 (A)	Distributor	MPI
			1.6 (1597)	4G61 (Y)	DIS	MPI
	Montero	J	3.0 (2972)	6G72 (S)	Distributor	MPI
	Precis	V/U	1.5 (1468)	64DJ (J)	Distributor	MPI
	Sigma	B	3.0 (2972)	6G72 (S)	Distributor	MPI
	3000GT	D/E	3.0 (2972)	6G72 (B)	DIS	MPI
			3.0 (2972)	6G72 (C)	DIS	MPI
	Truck	M/L	2.4 (2350)	4G64 (W)	Distributor	MPI
			3.0 (2972)	6G72 (S)	Distributor	MPI

MITSUBISHI

Year	Model	Body VIN	Engine Displacement Liter (cc)	Engine ID (VIN)	Ignition System	Fuel System
1992	Diamante	C	3.0 (2972)	6G72 (B)	DIS	MPI
			3.0 (2972)	6G72 (S)	Distributor	MPI
	Eclipse	S/T	1.8 (1795)	4G37 (T)	Distributor	MPI
			2.0 (1997)	4G63 (R) ①	DIS	MPI
			2.0 (1997)	4G63 (U) ②	DIS	MPI
	Expo/Expo LRV	V/W/Y/Z	1.8 (1795)	4G93 (D)	Distributor	MPI
			2.4 (2350)	4G64 (W)	Distributor	MPI
	Galant	R/X	2.0 (1997)	4G63 (R) ①	DIS	MPI
			2.0 (1997)	4G63 (U) ②	DIS	MPI
			2.0 (1997)	4G63 (V) ③	Distributor	MPI
	Mirage	U	1.5 (1468)	4G15 (A)	Distributor	MPI
			1.6 (1597)	4G61 (Y)	DIS	MPI
	Montero	K	3.0 (2972)	6G72 (S)	Distributor	MPI
	Precis	V/U	1.5 (1468)	G4DJ (J)	Distributor	MPI
	3000GT	D/E	3.0 (2972)	6G72 (B)	DIS	MPI
			3.0 (2972)	6G72 (C)	DIS	MPI
	Truck	M/L	2.4 (2350)	4G64 (W)	Distributor	MPI
			3.0 (2972)	6G72 (S)	Distributor	MPI

DIS—Distributorless Ignition System
MPI—Multi-Point Fuel Injection
① 2.0L (VIN R) engine is dual overhead camshaft
② 2.0L (VIN U) engine is turbocharged
③ 2.0L (VIN V) engine is single overhead camshaft

Basic Electricity

SECTION 1

FUNDAMENTALS OF ELECTRICITY

In order to be able to successfully troubleshoot and repair any electrical problem, one must first understand the theory behind how electricity works. A clear understanding of how electrical current behaves in a circuit and how components such as solenoids and relays work is essential for correct interpretation of test results. The following section covers basic knowledge necessary to perform all service and testing procedures outlined in this manual. This section should be read before attempting any repair or diagnosis procedures.

All matter is made up of tiny particles called molecules. Each molecule is made up of two or more atoms. Atoms may be divided into even smaller particles called protons, neutrons and electrons. These particles are the same in all matter and differences in materials (hard or soft, conductive or non-conductive) occur only because of the number and arrangement of these particles. In other words, the protons, neutrons and electrons in a drop of water are the same as those in an ounce of lead, there are just more of them (arranged differently) in a lead molecule than in a water molecule. Protons and neutrons packed together form the nucleus of the atom, while electrons orbit around the nucleus much the same way as the planets of the solar system orbit around the sun.

The proton is a small positive natural charge of electricity, while the neutron has no electrical charge. The electron carries a negative charge equal to the positive charge of the proton. Every electrically neutral atom contains the same number of protons and electrons, the exact number of which determines the element. The only difference between a conductor and an insulator is that a conductor possesses free electrons in large quantities, while an insulator has only a few. An element must have very few free electrons to be a good insulator, and vice-versa. When we speak of electricity, we're talking about these free electrons.

In a conductor, the movement of the free electrons is hindered by collisions with the adjoining atoms of the element (matter). This hindrance to movement is called RESISTANCE and it varies with different materials and temperatures. As temperature increases, the movement of the free electrons increases, causing more frequent collisions and therefore increasing resistance to the movement of the electrons. The number of collisions (resistance) also increases with the number of electrons flowing (current). Current is defined as the movement of electrons through a conductor such as a wire. In a conductor (such as copper) electrons can be caused to leave their atoms and move to other atoms. This flow is continuous in that every time an atom gives up an electron, it collects another one to take its place. This movement of electrons is called electric current and is measured in amperes. When 6.28 billion, billion electrons pass a certain point in the circuit in one second, the amount of current flow is called one ampere.

The force or pressure which causes electrons to flow in any conductor (such as a wire) is called voltage. It is measured in volts and is similar to the pressure that causes water to flow in a pipe. Voltage is the difference in electrical pressure measured between two different points in a circuit. In a 12 volt system, for example, the force measured between the two battery posts is 12 volts. Two important concepts are voltage potential and polarity. Voltage potential is the amount of voltage or electrical pressure at a certain point in the circuit with respect to another point. For example, if the voltage potential at one post of the 12 volt battery is zero, the voltage potential at the other post is 12 volts with respect to the first post. One post of the battery is said to be positive (+); the other post is negative (-) and the conventional direction of current flow is from positive to negative in an electrical circuit. It should be noted that the electron flow in the wire is opposite the current flow. In other words, when the circuit is energized, the current flows from positive to negative, but the electrons actually move from negative to positive. The voltage or pressure needed to produce a current flow in a circuit must be greater than the resistance present in the circuit. In other words, if the voltage drop across the resistance is greater than or equal to the voltage input, the voltage potential will be zero—no voltage will flow through the circuit. Resistance to the flow of electrons is measured in ohms. One volt will cause one ampere to flow through a resistance of one ohm.

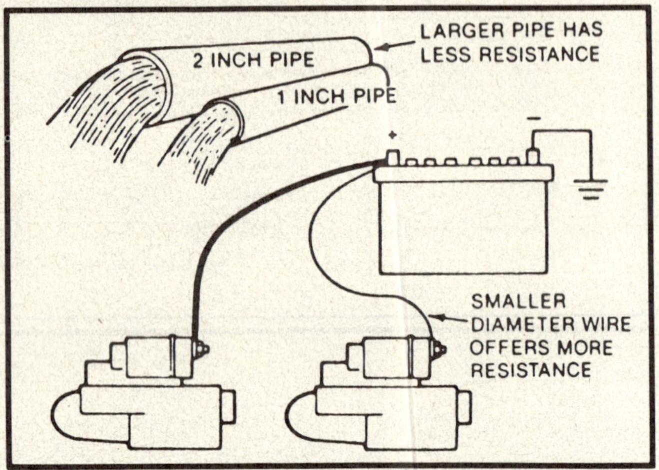

Electrical resistance can be compared to water flow through a pipe. The smaller the wire (pipe), the more resistance to the flow of electrons (water)

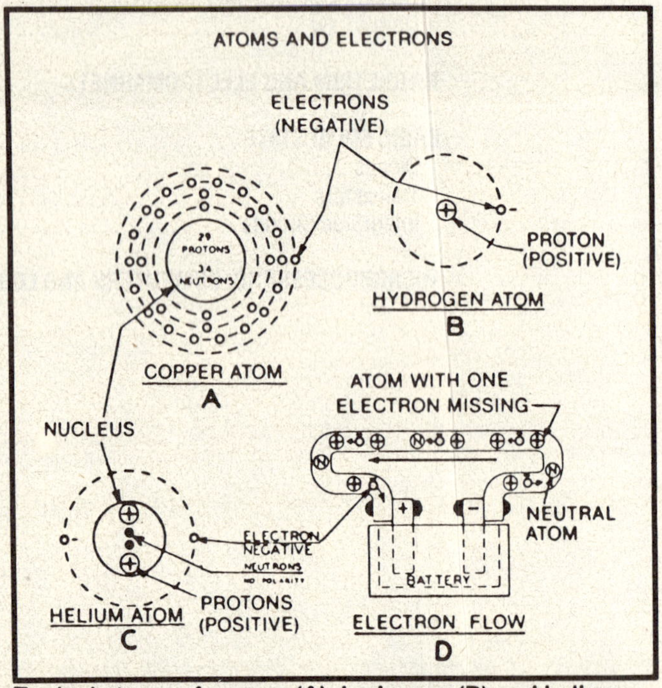

Typical atoms of copper (A), hydrogen (B) and helium (C). Electron flow in battery circuit (D)

Units Of Electrical Measurement

There are three fundamental characteristics of a direct-current electrical circuit: volts, amperes and ohms.

VOLTAGE

The voltage in a circuit controls the intensity with which the loads in the circuit operate. The brightness of a lamp, the heat of an electrical defroster, the speed of a motor are all directly proportional to the voltage, if the resistance in the circuit and/or mechanical load on electric motors remains constant. Voltage available from the battery is constant (normally 12 volts), but as it operates the various loads in the circuit, voltage decreases (drops).

AMPERE

An ampere is the unit of measurement of current in an electrical circuit. One ampere is the quantity of current that will flow through a resistance of one ohm at a pressure of one volt. The amount of current that flows in a circuit is controlled by the voltage and the resistance in the circuit. Current flow is directly proportional to resistance. Thus, as voltage is increased or decreased, current is increased or decreased accordingly. Current is decreased as resistance is increased, however, and is increased as resistance is decreased. With little or no resistance in a circuit, current is high.

OHM

The ohm is the unit of measurement of resistance, represented by the Greek letter Omega (Ω). One ohm is the resistance of a conductor through which a current of one ampere will flow at a pressure of one volt. Electrical resistance can be measured on an instrument called an ohmmeter. The loads (electrical devices) are the primary resistances in a circuit. Loads such as lamps, solenoids, and electric heaters have a resistance that is essentially fixed; at a normal fixed voltage, they will draw a fixed current. Motors, on the other hand, do not have a fixed resistance. Increasing the mechanical load on a motor (such as might be caused by a misadjusted track in a power window system) will decrease the motor speed. The drop in motor rpm has the effect of reducing the internal resistance of the motor because the current draw of the motor varies directly with the mechanical load on the motor, although its actual resistance is unchanged. Thus, as the motor load increases, the current draw of the motor increases, and may increase up to the point where the motor stalls (cannot move the mechanical load).

Circuits are designed with the total resistance of the circuit taken into account. Troubles can arise when unwanted resistances enter into a circuit. If corrosion, dirt, grease, or any other contaminant occurs in places like switches, connectors, and grounds, or if loose connections occur, resistances will develop in these areas. These resistances act like additional loads in the circuit and cause problems.

OHMS LAW

Ohms law is a statement of the relationship between the three fundamental characteristics of an electrical circuit. These rules apply to direct current (DC) only.

Ohms law provides a means to make an accurate circuit analysis without actually seeing the circuit. If, for example, one wanted to check the condition of the rotor winding in a alternator whose specifications indicate that the field (rotor) current draw is normally 2.5 amperes at 12 volts, simply connect the rotor to a 12 volt battery and measure the current with an ammeter. If it measures about 2.5 amperes, the rotor winding can be assumed good.

An ohmmeter can be used to test components that have been removed from the vehicle in much the same manner as an ammeter. Since the voltage and the current of the rotor windings used as an earlier example are known, the resistance can be calculated using Ohms law. The formula would be:

If the rotor resistance measures about 4.8 ohms when checked with an ohmmeter, the winding can be assumed good. By plugging in different specifications, additional circuit information can be determined such as current draw, etc.

$$I = \frac{E}{R} \quad \text{or} \quad \text{AMPERES} = \frac{\text{VOLTS}}{\text{OHMS}}$$

$$R = \frac{E}{I} \quad \text{or} \quad \text{OHMS} = \frac{\text{VOLTS}}{\text{AMPERES}}$$

$$E = I \times R \quad \text{or} \quad \text{VOLTS} = \text{AMPERES} \times \text{OHMS}$$

$$R = \frac{E}{I} \quad \text{Where: } E = 12 \text{ volts}, \; I = 2.5 \text{ amperes}$$

$$R = \frac{12 \text{ volts}}{2.5 \text{ amps}} = 4.8 \text{ ohms}$$

Electrical Circuits

An electrical circuit must start from a source of electrical supply and return to that source through a continuous path. Circuits are designed to handle a certain maximum current flow. The maximum allowable current flow is designed higher than the normal current requirements of all the loads in the circuit. Wire size, connections, insulation, etc., are designed to prevent undesirable voltage drop, overheating of conductors, arcing of contacts, and other adverse effects. If the safe maximum current flow level is exceeded, damage to the circuit components will result; it is this condition that circuit protection devices are designed to prevent.

Protection devices are fuses, fusible links or circuit breakers designed to open or break the circuit quickly whenever an overload, such as a short circuit, occurs. By opening the circuit quickly, the circuit protection device prevents damage to the wiring, battery, and other circuit components. Fuses and fusible links are designed to carry a preset maximum amount of current and to melt when that maximum is exceeded, while circuit breakers merely break the connection and may be manually reset. The maximum amperage rating of each fuse is marked on the fuse body and all contain a see-through portion that shows the break in the fuse element when blown. Fusible link maximum amperage rating is indicated by gauge or thickness of the wire. Never replace a blown fuse or fusible link with one of a higher amperage rating.

CAUTION

Resistance wires, like fusible links, are also spliced into conductors in some areas. Do not make the mistake of replacing a fusible link with a resistance wire. Resistance wires are longer than fusible links and are stamped "RESISTOR-DO NOT CUT OR SPLICE."

Circuit breakers consist of two strips of metal which have different coefficients of expansion. As an overload or current flows through the bimetallic strip, the high-expansion metal will elongate due to heat and break the contact. With the circuit open, the bimetal strip cools and shrinks, drawing the strip down until contact is re-established and current flows once again. In actual operation, the contact is broken very quickly if the overload is

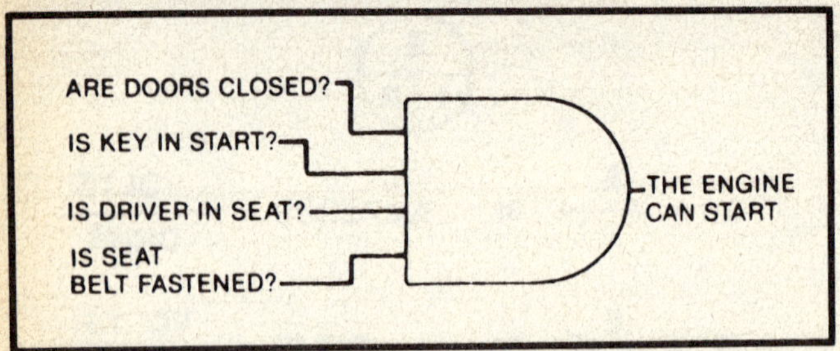

Multiple input AND operation in a typical automotive starting circuit

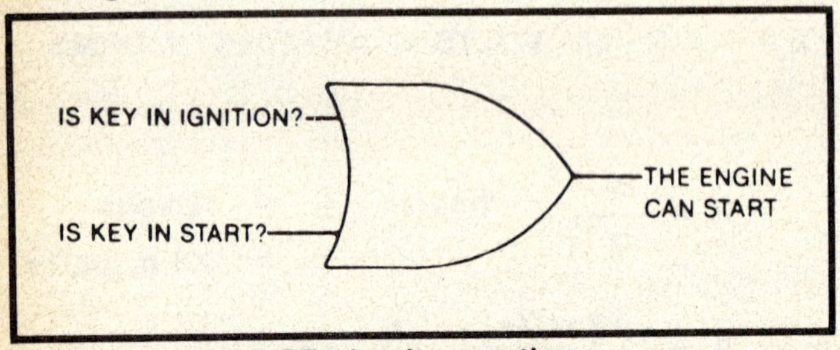

Typical two-input OR circuit operation

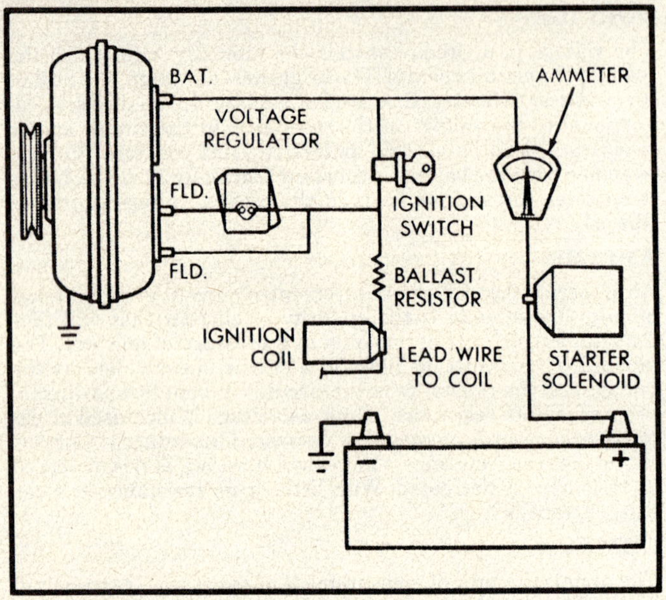

Typical alternator circuit

continuous and the circuit will be repeatedly broken and remade until the source of the overload is corrected.

The self-resetting type of circuit breaker is the one most generally used in automotive electrical systems. On manually reset circuit breakers, a button will pop up on the circuit breaker case. This button must be pushed in to reset the circuit breaker and restore power to the circuit. Always repair the source of the overload before resetting a circuit breaker or replacing a fuse or fusible link. When searching for overloads, keep in mind that the circuit protection devices protect only against overloads between the protection device and ground.

There are two basic types of circuit; Series and Parallel. In a series circuit, all of the elements are connected in chain fashion with the same amount of current passing through each element or load. No matter where an ammeter is connected in a series circuit, it will always read the same. The most important fact to remember about a series circuit is that the sum of the voltages across each element equals the source voltage. The total resistance of a series circuit is equal to the sum of the individual resistances within each element of the circuit. Using ohms law, one can determine the voltage drop across each element in the circuit. If the total resistance and source voltage is known, the amount of current can be calculated. Once the amount of current (amperes) is known, values can be substituted in the Ohms law formula to calculate the voltage drop across each individual element in the series circuit. The individual voltage drops must add up to the same value as the source voltage.

A parallel circuit, unlike a series circuit, contains two or more branches, each branch a separate path independent of the others. The total current draw from the voltage source is the sum of all the currents drawn by each branch. Each branch of a parallel circuit can be analyzed separately. The individual branches can be either simple circuits, series circuits or combinations of series-parallel circuits. Ohms law applies to parallel circuits just as it applies to series circuits, by considering each branch independently of the others. The most important thing to remember is that the voltage across each branch is the same as the source voltage. The current in any branch is that voltage divided by the resistance of the branch. A practical method of determining the resistance of a parallel circuit is to divide the product of the two resistances by the sum of two resistances at a time. Amperes through a parallel circuit is the sum of the amperes through the separate branches. Voltage across a parallel circuit is the same as the voltage across each branch.

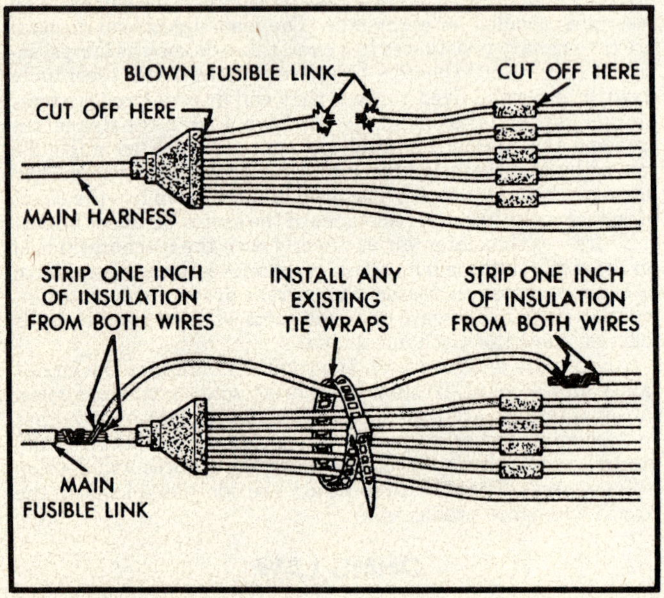

Typical fusible link repair

By measuring the voltage drops, you are in effect measuring the resistance of each element within the circuit. The greater the voltage drop, the greater the resistance. Voltage drop measurements are a common way of checking circuit resistances in automotive electrical systems. When part of a circuit developes excessive resistance (due to a bad connection) the element will show a higher than normal voltage drop. Normally, automotive wiring is selected to limit voltage drops to a few tenths of a volt. In parallel circuits, the total resistance is less than the sum of the individual resistances; because the current has two paths to take, the total resistance is lower.

Magnetism and Electromagnets

Electricity and magnetism are very closely associated because when electric current passes through a wire, a magnetic field is

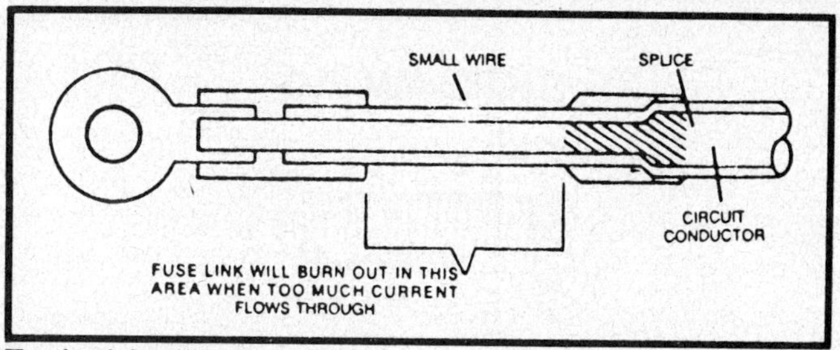

Typical fusible link wire

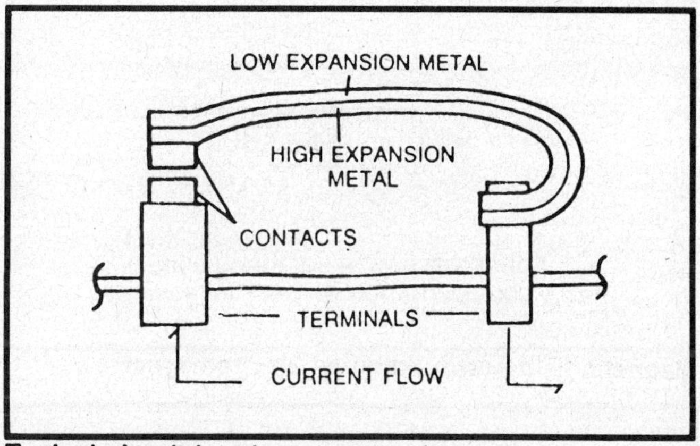

Typical circuit breaker construction

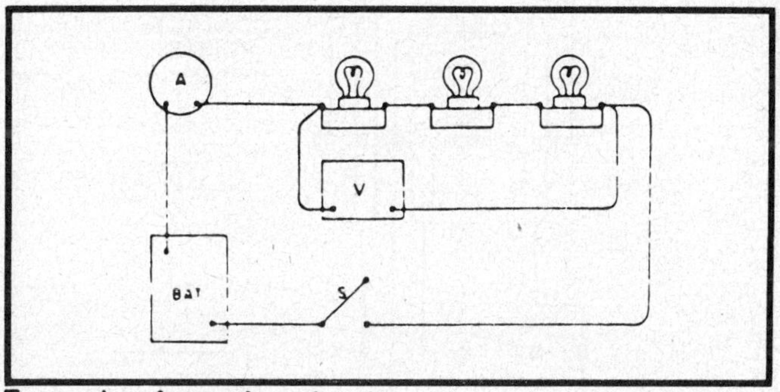

Example of a series circuit

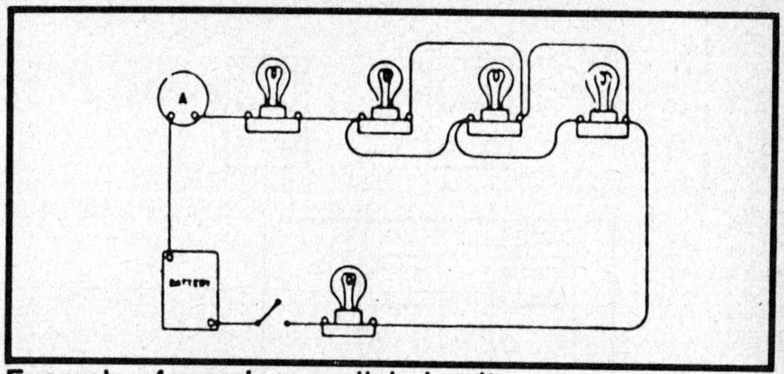

Example of a series-parallel circuit

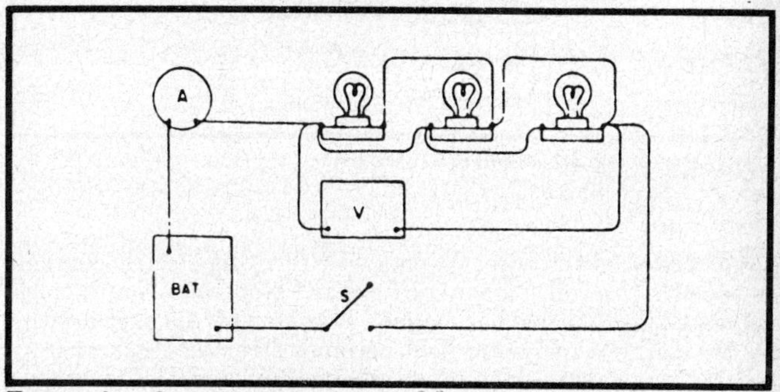

Example of a parallel circuit

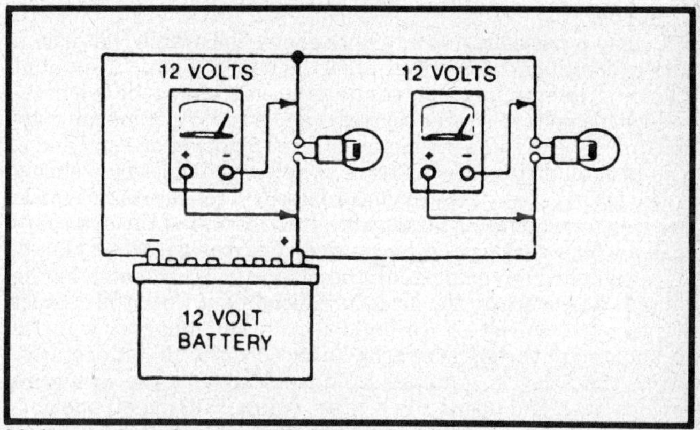

Voltage drop in a parallel circuit. Voltage drop across each lamp is 12 volts

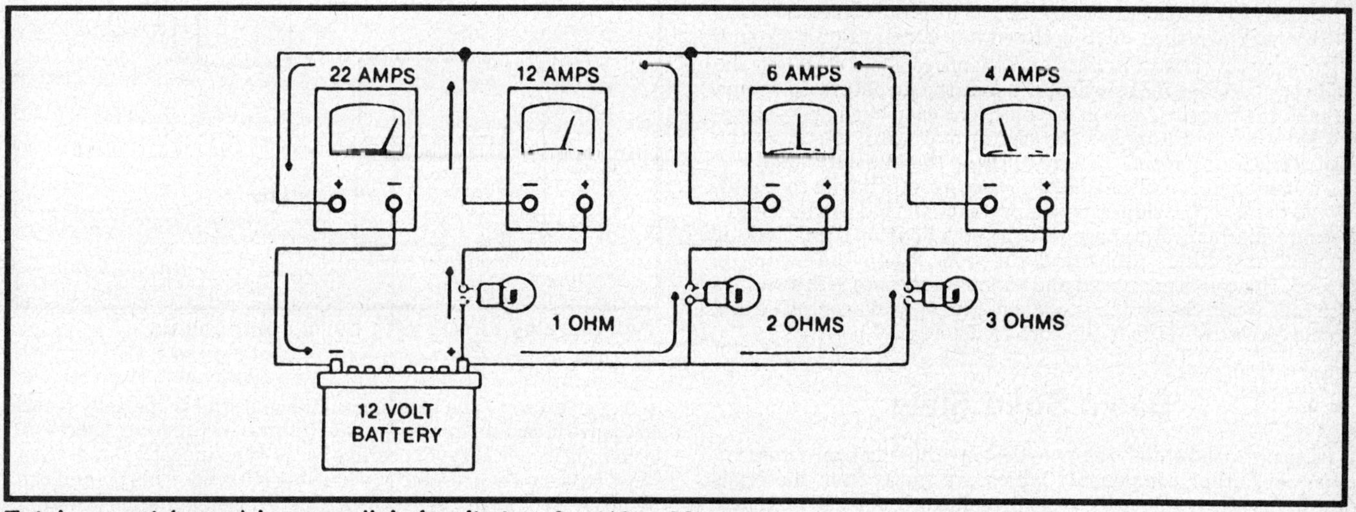

Total current (amps) in a parallel circuit. 4 + 6 + 12 = 22 amps

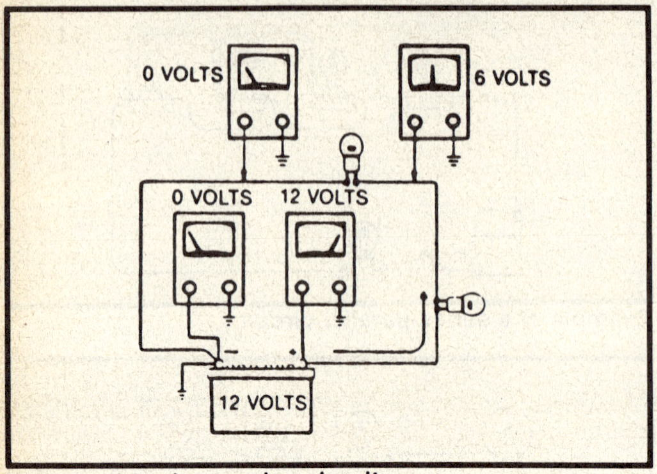

Voltage drop in a series circuit

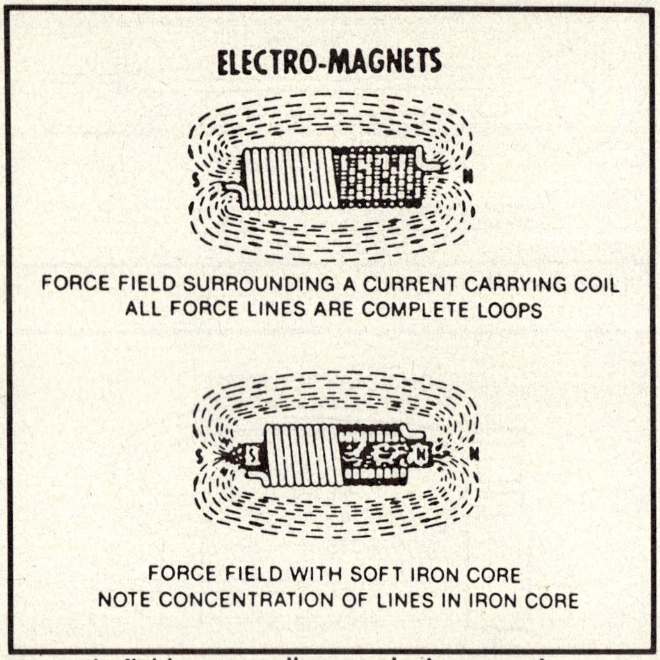

Magnetic field surrounding an electromagnet

created around the wire. When a wire carrying electric current is wound into a coil, a magnetic field with North and South poles is created just like in a bar magnet. If an iron core is placed within the coil, the magnetic field becomes stronger because iron conducts magnetic lines much easier than air. This arrangement is called an electromagnet and is the basic principle behind the operation of such components as relays, buzzers and solenoids.

A relay is basically just a remote-controlled switch that uses a small amount of current to control the flow of a large amount of current. The simplest relay contains an electromagetic coil in series with a voltage source (battery) and a switch. A movable armature made of some magnetic material pivots at one end and is held a small distance away from the electromagnet by a spring or the spring steel of the armature itself. A contact point, made of a good conductor, is attached to the free end of the armature with another contact point a small distance away. When the relay is switched on (energized), the magnetic field created by the current flow attracts the armature, bending it until the contact points meet, closing a circuit and allowing current to flow in the second circuit through the relay to the load the circuit operates. When the relay is switched off (de-energized), the armature springs back and opens the contact points, cutting off the current flow in the secondary, or controlled, circuit. Relays can be designed to be either open or closed when energized, depending on the type of circuit control a manufacturer requires.

A buzzer is similar to a relay, but its internal connections are different. When the switch is closed, the current flows through the normally closed contacts and energizes the coil. When the coil core becomes magnetized, it bends the armature down and breaks the circuit. As soon as the circuit is broken, the spring-loaded armature remakes the circuit and again energizes the coil. This cycle repeats rapidly to cause the buzzing sound.

A solenoid is constructed like a relay, except that its core is allowed to move, providing mechanical motion that can be used to actuate mechanical linkage to operate a door or trunk lock or control any other mechanical function. When the switch is closed, the coil is energized and the movable core is drawn into the coil. When the switch is opened, the coil is de-energized and spring pressure returns the core to its original position.

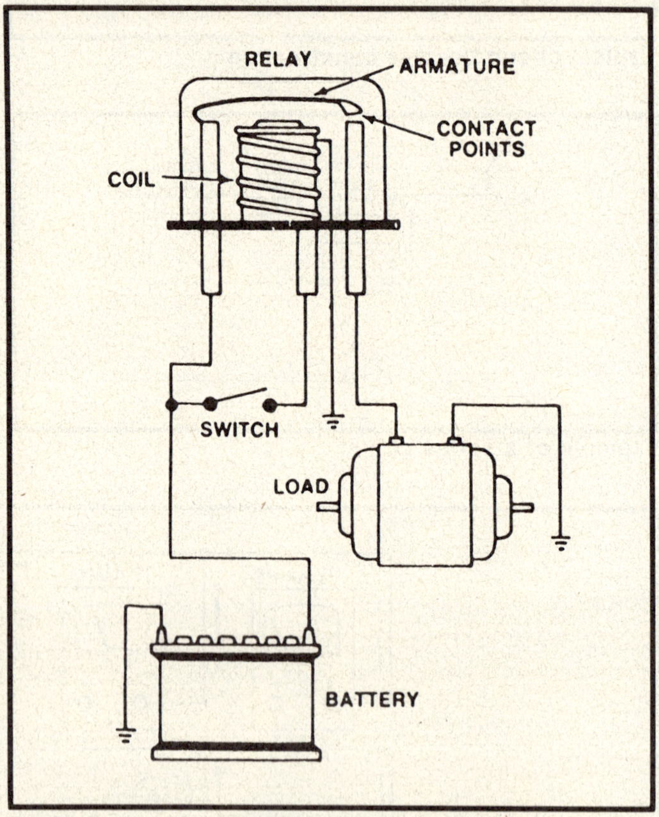

Typical relay circuit with basic components

Basic Solid State

The term "solid state" refers to devices utilizing transistors, diodes and other components which are made from materials known as semiconductors. A semiconductor is a material that is neither a good insulator or a good conductor; principally silicon and germanium. The semiconductor material is specially treated to give it certain qualities that enhance its function, therefore becoming either P-type (positive) or N-type (negative) material. Most modern semiconductors are constructed of silicon and can be made to change their characteristics depending on whether its function calls for an insulator or conductor.

DIODES

The simplest semiconductor function is that of the diode or rectifier (the two terms mean the same thing). A diode will pass current in one direction only, like a one-way valve, because it has low resistance in one direction and high resistance on the other. Whether the diode conducts or not depends on the polarity of the voltage applied to it. A diode has two electrodes, an anode and a cathode. When the anode receives positive (+) voltage and the cathode receives negative (–) voltage, current can flow easily through the diode. When the voltage is reversed, the diode becomes non-conducting and only allows a very slight amount of current to flow in the circuit. Because the semiconductor is not a perfect insulator, a small amount of reverse current leakage will occur, but the amount is usually too small to consider. The application of voltage to maintain the current flow described is called "forward bias."

A light-emitting diode (LED) is made of a particular type of crystal that glows when current is passed through it. LED's are used in display faces of many digital or electronic instrument clusters. LED's are usually arranged to display numbers (digital readout), but can be used to illuminate a variety of electronic graphic displays.

Like any other electrical device, diodes have certain ratings that must be observed and should not be exceeded. The forward current rating (or bias) indicates how much current can safely pass through the diode without causing damage or destroying it. Forward current rating is usually given in either amperes or milliamperes. The voltage drop across a diode remains constant regardless of the current flowing through it. Small diodes designed to carry low amounts of current need no special provision for dissipating the heat generated in any electrical device, but large current carrying diodes are usually mounted on heat sinks to keep the internal temperature from rising to the point where the silicon will melt and destroy the diode. When diodes are operated in a high ambient temperature environment, they must be de-rated to prevent failure.

Another diode specification is its peak inverse voltage rating. This value is the maximum amount of voltage the diode can safely handle when operating in the blocking mode. This value can be anywhere from 50–1000 volts, depending on the diode, and if exceeded can damage the diode just as too much forward current will. Most semiconductor failures are caused by excessive voltage or internal heat.

One can test a diode with a small battery and a lamp with the same voltage rating. With this arrangement one can find a bad diode and determine the polarity of a good one. A diode can fail and cause either a short or open circuit, but in either case it fails to function as a diode. Testing is simply a matter of connecting the test bulb first in one direction and then the other and making sure that current flows in one direction only. If the diode is shorted, the test bulb will remain on no matter how the light is connected.

TRANSISTORS

The transistor is an electrical device used to control voltage within a circuit. A transistor can be considered a "controllable diode" in that, in addition to passing or blocking current, the transistor can control the amount of current passing through it. Simple transistors are composed of three pieces of semiconductor material, P and N type, joined together and enclosed in a container. If two sections of P material and one section of N material are used, it is known as a PNP transistor; if the reverse is true, then it is known as an NPN transistor. The two types cannot be interchanged.

Most modern transistors are made from silicon (earlier transistors were made from germanium) and contain three elements; the emitter, the collector and the base. In addition to passing or blocking current, the transistor can control the

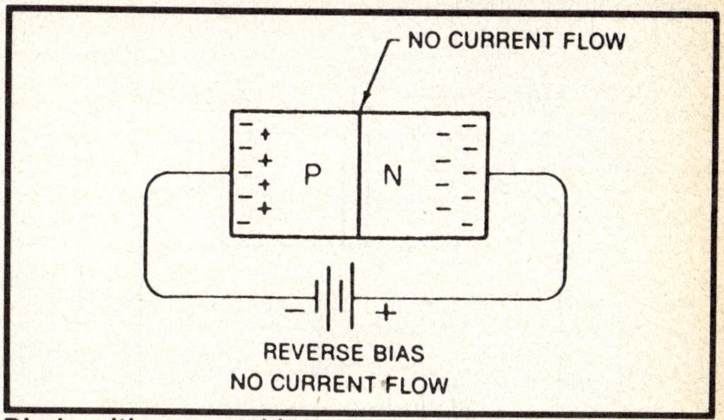

Diode with reverse bias

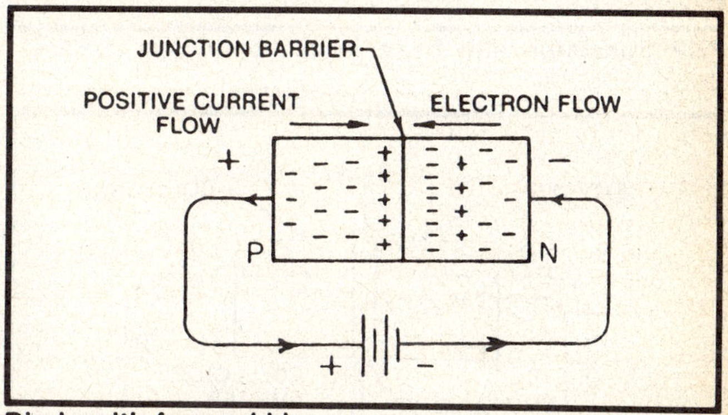

Diode with forward bias

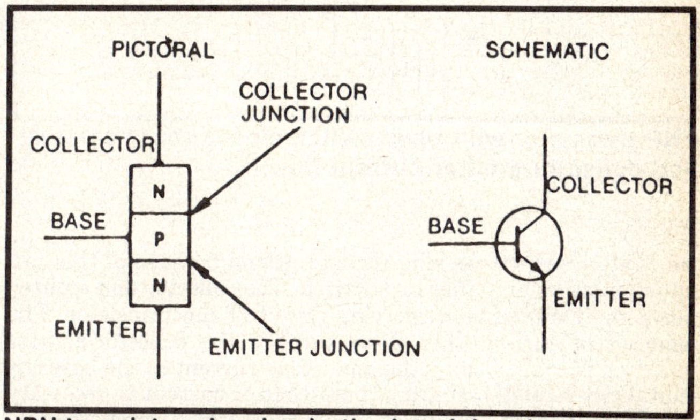

NPN transistor showing both pictorial and schematic illustrations

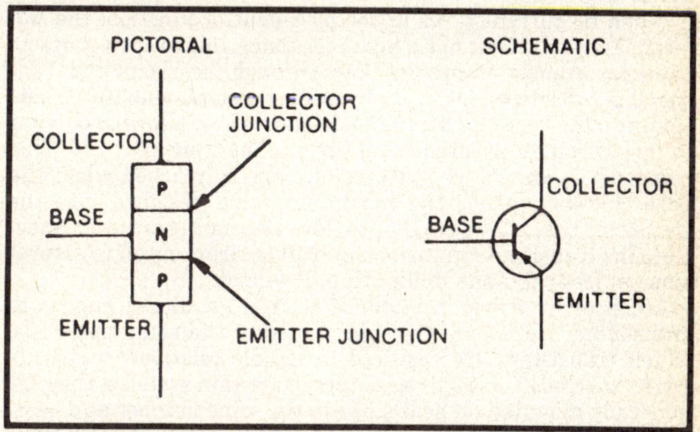

PNP transistor showing both pictorial and schematic illustrations

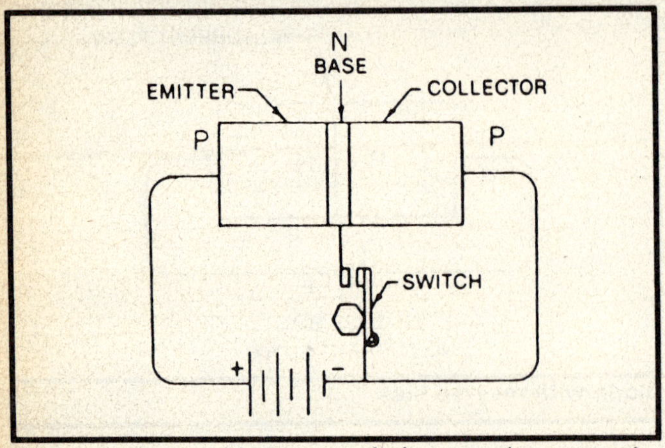

PNP transistor with base switch open (no current flow)

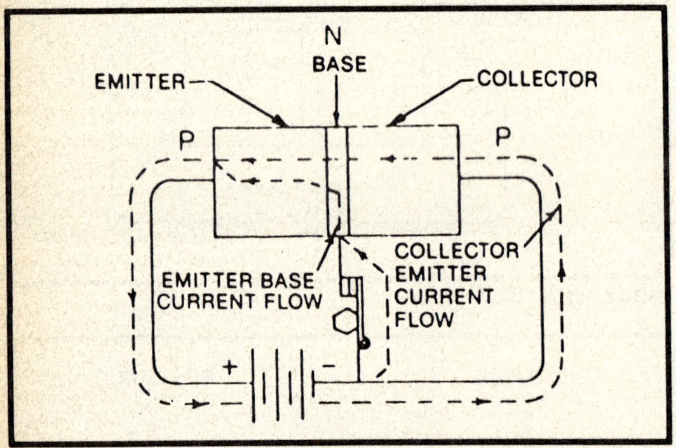

PNP transistor with base switch closed (base emitter and collector emitter current flow)

amount of current passing through it and because of this can function as an amplifier or a switch. The collector and emitter form the main current-carrying circuit of the transistor. The amount of current that flows through the collector-emitter junction is controlled by the amount of current in the base circuit. Only a small amount of base-emitter current is necessary to control a large amount of collector-emitter current (the amplifier effect). In automotive applications, however, the transistor is used primarily as a switch.

When no current flows in the base-emitter junction, the collector-emitter circuit has a high resistance, like to open contacts of a relay. Almost no current flows through the circuit and transistor is considered OFF. By bypassing a small amount of current into the base circuit, the resistance is low, allowing current to flow through the circuit and turning the transistor ON. This condition is known as "saturation" and is reached when the base current reaches the maximum value designed into the transistor that allows current to flow. Depending on various factors, the transistor can turn on and off (go from cutoff to saturation) in less than one millionth of a second.

Much of what was said about ratings for diodes applies to transistors, since they are constructed of the same materials. When transistors are required to handle relatively high currents, such as in voltage regulators or ignition systems, they are generally mounted on heat sinks in the same manner as diodes. They can be damaged or destroyed in the same manner if their voltage ratings are exceeded. A transistor can be checked for proper operation by measuring the resistance with an ohmmeter between the base-emitter terminals and then between the base-collector terminals. The forward resistance should be small, while the reverse resistance should be large. Compare the readings with those from a known good transistor. As a final check, measure the forward and reverse resistance between the collector and emitter terminals.

INTEGRATED CIRCUITS

The integrated circuit (IC) is an extremely sophisticated solid state device that consists of a silicone wafer (or chip) which has been doped, insulated and etched many times so that it contains an entire electrical circuit with transistors, diodes, conductors and capacitors miniaturized within each tiny chip. Integrated circuits are often referred to as "computers on a chip" and are largely responsible for the current boom in electronic control technology.

Microprocessors, Computers and Logic Systems

Mechanical or electromechanical control devices lack the precision necessary to meet the requirements of modern control standards. Since they generally function in an ON/OFF cycle, they do not have the ability to respond to the variety of input conditions common to antilock brakes, climate control and electronic suspension operation. To meet these requirements, manufacturers have gone to solid state logic systems and microprocessors to control the basic functions of suspension, brake and temperature control, as well as other systems and accessories.

One of the vital roles of microprocessor-based systems is their ability to perform logic functions and make decisions. Logic designers use a shorthand notation to indicate whether a voltage is present in a circuit (the number 1) or not present (the number 0). Their systems are designed to respond in different ways depending on the output signal (or the lack of it) from various control devices.

There are 3 basic logic functions or "gates" used to construct a microprocessor control system: the AND gate, the OR gate or the NOT gate. Stated simply, the AND gate works when voltage is present in 2 or more circuits which then energize a third (A and B energize C). The OR gate works when voltage is present at either circuit A or circuit B which then energizes circuit C. The NOT function is performed by a solid state device called an "inverter" which reverses the input from a circuit so that, if voltage is going in, no voltage comes out and vice versa. With these three basic building blocks, a logic designer can create complex systems easily. In actual use, a logic or decision making system may employ many logic gates and receive inputs from a number of sources (sensors), but for the most part, all utilize the basic logic gates discussed above.

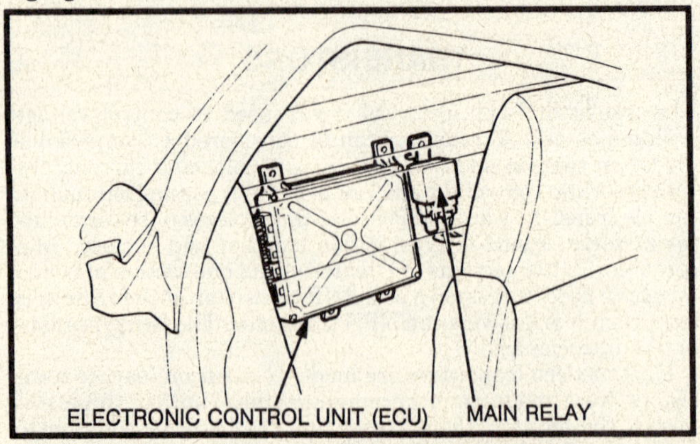

Location of the Electronic control Unit (ECU)—Acura

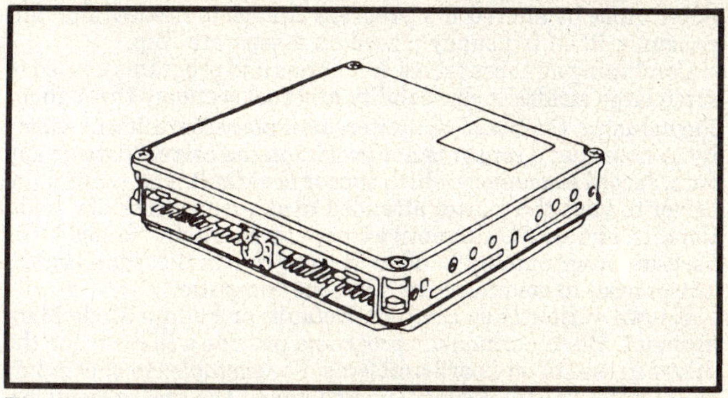

Electronic control Unit (ECU) – Infiniti

Stripped to its bare essentials, a computerized decision-making system is made up of three subsystems:

a. Input devices (sensors or switches)
b. Logic circuits (computer control unit)
c. Output devices (actuators or controls)

The input devices are usually nothing more than switches or sensors that provide a voltage signal to the control unit logic circuits that is read as a 1 or 0 (on or off) by the logic circuits. The output devices are anything from a warning light to solenoid-operated valves, motors, linkage, etc. In most cases, the logic circuits themselves lack sufficient output power to operate these devices directly. Instead, they operate some intermediate device such as a relay or power transistor which in turn operates the appropriate device or control. Many problems diagnosed as computer failures are really the result of a malfunctioning intermediate device like a relay. This must be kept in mind whenever troubleshooting any microprocessor-based control system.

As computer capacity is improved by the manufacturers, so does sensor technology. A few years ago, the on-board computer would receive a message from an engine sensor in a "go or no-go" form; for example, engine coolant temperature is either below or above 150°F. The sensor would or would not pass electric current depending on the temperature. Today's systems allow the same sensor to be much more communicative; it will pass progressively more voltage as the engine warms up. The computer can react to the changing voltage instantly and control other engine functions during the warm-up period.

The logic systems discussed above are called "hardware" systems, because they consist only of the physical electronic components (gates, resistors, transistors, etc.). Hardware systems do not contain a program and are designed to perform specific or "dedicated" functions which cannot readily be changed. For many simple automotive control requirements, such dedicated logic systems are perfectly adequate. When more complex logic functions are required, or where it may be desirable to alter these functions (e.g. from one model vehicle to another) a true computer system is used. A computer can be programmed through its software to perform many different functions and, if that program is stored on a separate integrated circuit chip called a ROM (Read Only Memory), it can be easily changed simply by plugging in a different ROM with the desired program. Most on-board automotive computers are designed with this capability. The on-board computer method of engine control offers the manufacturer a flexible method of responding to data from a variety of input devices and of controlling an equally large variety of output controls. The computer response can be changed quickly and easily by simply modifying its software program.

The microprocessor is the heart of the microcomputer. It is the thinking part of the computer system through which all the data from the various sensors passes. Within the microprocessor, data is acted upon, compared, manipulated or stored for future use. A microprocessor is not necessarily a microcomputer, but the differences between the 2 are becoming very minor. Originally, a microprocessor was a major part of a microcomputer, but today microprocessors are being called "single-chip microcomputers". They contain all the essential elements to make them behave as a computer, including the most important ingredient–the program.

All computers require a program. In a general purpose computer, the program can be easily changed to allow different tasks to be performed. In a "dedicated" computer, such as most on-board automotive computers, the program isn't quite so easily altered. These automotive computers are designed to perform one or more specific tasks, such as maintaining the passenger compartment temperature at a specific, predetermined level. A program is what makes a computer smart; without a program a computer can do absolutely nothing. The term "software" refers to the computer's program that makes the hardware perform the functions needed.

The software program is simply a listing in sequential order of the steps or commands necessary to make a computer perform the desired task. Before the computer can do anything at all, the program must be fed into it by one of several possible methods. A computer can never be "smarter" than the person programming it, but it is a lot faster. Although it cannot perform any calculation or operation that the programmer himself cannot perform, its processing time is measured in millionths of a second.

Because a computer is limited to performing only those operations (instructions) programmed into its memory, the program must be broken down into a large number of very simple steps. Two programmers can come up with 2 different programs, since there is usually more than one way to perform any task or solve a problem. In any computer, however, there is only so much memory space available, so an overly long or inefficient program may not fit into the memory. In addition to performing arithmetic functions (such as with a trip computer), a computer can also store data, look up data in a table and perform the logic functions previously discussed. A Random Access Memory (RAM) al-

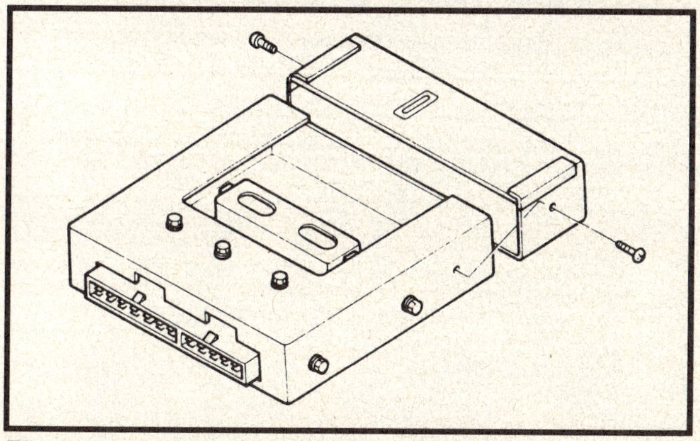

Electronic control Unit (ECU) – Isuzu

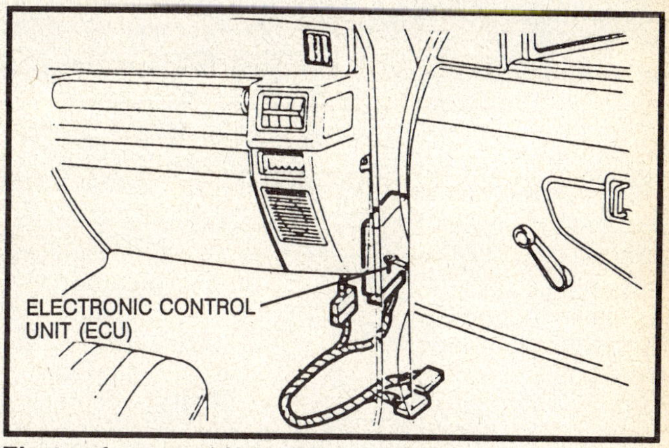

Electronic control Unit (ECU) – Daihatsu

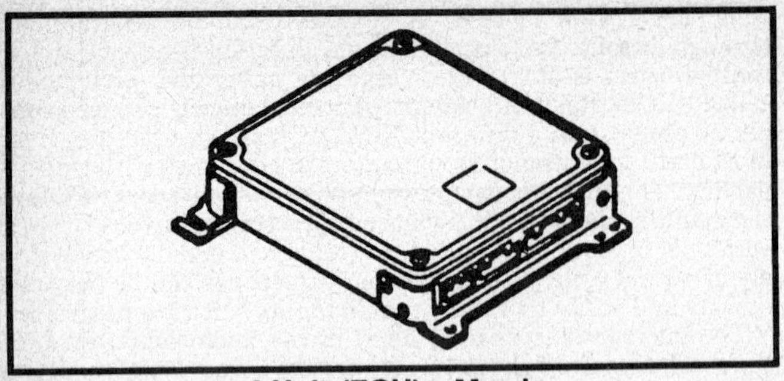

Electronic control Unit (ECU)—Mazda

lows the computer to store bits of data temporarily while waiting to be acted upon by the program. It may also be used to store output data that is to be sent to an output device. Whatever data is stored in a RAM is lost when power is removed from the system by turning **OFF** the ignition key, for example.

Computers have another type of memory called a Read Only Memory (ROM) which is permanent. This memory is not lost when the power is removed from the system. Most programs for automotive computers are stored on a ROM memory chip. Data is usually in the form of a look-up table that saves computing time and program steps. For example, a computer designed to control the amount of distributor advance can have this information stored in a table. The information that determines distributor advance (engine rpm, manifold vacuum and temperature) is coded to produce the correct amount of distributor advance over a wide range of engine operating conditions. Instead of the computer computing the required advance, it simply looks it up in a pre-programmed table.

Not all electronic control functions can be handled in this manner; some must be computed. On an antilock brake system, for example, the computer must measure the rotation of each separate wheel and then calculate how much brake pressure to apply in order to prevent one wheel from locking up and causing a loss of control.

There are several ways of programming a ROM, but once programmed the ROM cannot be changed. If the ROM is made on the same chip that contains the microprocessor, the whole computer must be altered if a program change is needed. For this reason, a ROM is usually placed on a separate chip.

Combining all these tables, memories and programs has led to some large strides in driveability and fuel economy throughout the industry. On-board computers now possess sufficient capacity to note that a sensor is not providing the correct data or any data, record in memory which sensor is at fault, and remind the driver to have the matter attended to as soon as possible. Additionally, almost all automotive computers contain "fail-safe" or back-up programming to allow the car to run if certain critical sensors fail to communicate with the computer.

Known variously as Limp-in, Default, or Failure Mode Management, these permanent programs provide a lifeboat for the driver in case of on-board problems. For example, most computers monitor the intake air flow and adjust the fuel to the injectors accordingly. The air flow varies drastically between conditions of idle, hard acceleration and highway cruising. If, for some reason–perhaps a loose wiring connector–the sensor message no longer reaches the computer, the computer notes the failure. Instead of shutting the engine off due to lack of necessary data, the computer simply "looks" to its back-up programming which provides it with a fixed value for the air flow. The computer adjusts the fuel flow to meet this assumed value. Engine performance will be compromised (because the metering is fixed, not changing with the conditions) but the engine keeps running and the driver can "limp home" instead of being stranded.

Development is also leading to the use of more than one computer or microprocessor within the vehicle. It is now quite possible to have separate control units for engine function, emission control components, climate control, cruise control, anti-lock brakes and/or passive restraints and seat belts.

As complicated as all these systems may appear, keep in mind that any computer simply processes information. In the automotive world, the computer itself is rarely at fault in a problem car. The network of sensors and wires sending messages to the computer or the network of components and wires receiving the output messages from the computer are generally much more prone to failure or impaired function. Too many people are intimidated by the presence of a computer and immediately blame it instead of hunting down the true cause. Careful, thorough investigation coupled with a firm understanding of how a system functions will lead to the cause of almost any problem.

Diagnostic Equipment and Special Tools

SECTION 2

DIAGNOSTIC EQUIPMENT AND SPECIAL TOOLS

General Information

This section outlines basic diagnostic techniques for dealing with computerized control systems. Along with a general explanation of the various types of test equipment available to aid in servicing modern electronic automotive systems, basic repair techniques for wiring harnesses and connectors are given. Read the basic information before attempting any repairs or testing on any computerized system. This will provide the background information necessary to avoid the most common and obvious mistakes. Likewise, the individual system sections should be read from the beginning to the end before any repair or diagnosis is attempted.

Although the replacement and testing procedures are simple, the systems are not. Unless one has a thorough understanding of all components and their function, the logical test sequences cannot be followed. Minor malfunctions can make a big difference, so it is important to know how each component affects the operation of the system to find the cause of a problem. It is not enough to use the correct test equipment; the test equipment must be used correctly.

SERVICE PRECAUTIONS

CAUTION

Whenever working on or around any computer-based microprocessor control system, always observe these general precautions to prevent the possibility of personal injury or damage to electronic components:

- Never install or remove battery cables with the ignition key **ON** or the engine running. Jumper cables should be connected with the ignition key **OFF** to avoid power surges that can damage electronic control units.
- Always remove the battery cables before charging the battery. Never use a high-output charger on an installed battery. Never attempt to use any type of "hot shot" (24 volt) starting aid.
- Always use a breakout or pin-out box if possible. If probing a connector cannot be avoided, exercise great care when inserting test probes to insure good connections without damaging the connector or spreading the pins. Avoid accidental shorting of terminals during test procedures by probing from the rear or wire side of the connector.
- Never remove or attach wiring harness connectors with the ignition switch **ON**. The resulting voltage spike may destroy an electronic control unit.
- Avoid dropping a component or subjecting it to rough handling. Never apply 12 volts directly to any component unless instructed specifically to do so. Some components are designed to safely handle only 4–5 volts and can be destroyed in seconds if 12 volts are applied directly to the connector.
- Remove the electronic control unit if the vehicle is to be placed in an environment where temperatures exceed approximately 176°F (80°C), such as a paint spray booth.
- If welding is to be done on the vehicle, remove or disconnect all control units on the vehicle. Connect the welder ground as close as possible to the work area.
- Avoid touching connector pins with the fingers.
- Leave new components and modules in the shipping package until ready to install them.
- Avoid static discharge at or near any solid-state component. Always touch a vehicle ground after sliding across the seat or walking over carpeted or vinyl floors. If possible, wear a grounded wrist strap during all testing and component replacement.
- Never allow welding cables to lie on, near or across any vehicle wiring.
- Avoid draping extension cords or droplight cables on or around vehicle wiring.
- If the vehicle is equipped with air bag or inflatable restraints, always properly disable the system before working on or around system components.
- Always use a digital, high-impedance volt-ohmmeter (DVOM) for testing unless specified otherwise. Minimum impedance should be 10 kilo-ohms per volt.

Organized Troubleshooting

When diagnosing a specific problem, organized troubleshooting is a must. The complexity of modern vehicle systems demands that the approach to any problem be logical and organized. There are certain troubleshooting techniques that are standard:

1. Establish when the problem occurs. Does the problem appear only under certain conditions? Were there any noises, odors, or other unusual symptoms? Make notes on any symptoms found, including warning lights and trouble codes, if applicable.
2. Isolate the problem area. To do this, make some simple tests and observations; then eliminate the systems that are working properly. Experience has shown that many problems in computer controlled systems are the result of fairly simple and obvious causes such as loose or corroded connectors or air leaks in the intake system. Always check the obvious before assuming something complicated is the cause.
3. Once the problem is isolated, test for problems systematically. Are all the components functioning properly? Is there power going to electrical switches and motors? Is there vacuum at vacuum switches and/or actuators? Is there a mechanical problem such as bent linkage or loose screws? Doing careful, systematic checks will often turn up causes on the first inspection without wasting time checking components that have little or no relationship to the problem.
4. Test all repairs after the work is done to make sure that the problem is fixed. Some causes can be traced to more than one component. A careful verification of repair work is important to pick up additional or remaining faults. For example, a blown fuse is a simple problem that may require more than just replacing a fuse. If the problem that caused a fuse to blow isn't searched for, a shorted wire may go undetected.

The diagnostic tree charts are designed to help solve problems by leading the user through closely defined conditions and tests; only the most likely components, vacuum and electrical circuits are checked for proper operation when troubleshooting a particular malfunction. By using the trouble trees to eliminate those items which normally will not cause the fault, a problem can be isolated within one system or circuit without wasting time on unnecessary testing.

Frequent references to special test equipment will be found in the text and in the diagnosis charts. These devices or their equivalents are necessary to perform some of the test procedures; however, many components can be functionally tested with common shop tools such as volt or ohm meters, test lights, etc. Special test equipment may be purchased from either an aftermarket supplier or the vehicle manufacturer. Care must be taken to insure that the equipment is correct for the vehicle or system on which it will be used.

NOTE: Pinpointing the exact cause of trouble in an electrical system can sometimes only be performed with special equipment. The following describes commonly used test equipment and explains how to put it to best use in diagnosis. Additionally, the tool manufacturer's instruction booklet should be read and clearly understood before attempting any test procedures.

Equipment and Tools

Almost every vehicle manufactured or sold today contains at least one solid-state processor or on-board computer. These units can control many systems and subsystems efficiently, leading to improved driveability, lowered emissions and safer operation. Control modules are currently used to oversee such diverse applications as fuel and ignition management, cruise control, anti-lock brakes, security systems, lighting and suspension.

It is easy to think that a small box containing no moving parts will last the life of the vehicle. In the real world, malfunctions occur due to either electrical problems or external causes. Any computer-based system is extremely sensitive to incorrect voltage; they cannot tolerate careless procedures or improper testing. An inexperienced person can cause extensive and expensive damage in pursuit of a minor problem. Since electricity travels at just under the speed of light, this damage occurs instantly and permanently.

Before beginning any testing or diagnosis on a vehicle, review the safety precautions as well as the procedures to be followed. Make certain that the test equipment to be used is compatible with the vehicle. Read all instructions carefully and double-check test points before connecting probes or other test equipment.

The following section outlines basic diagnostic techniques for dealing with computerized engine control systems. Along with a general explanation of the various types of test equipment available, basic repair techniques for wiring harnesses and connectors are given. Read the basic information before attempting any repairs or testing on any computerized system. This will provide the background information necessary to avoid the most common and obvious mistakes.

The individual system sections for engine controls or fuel injection should be read from the beginning to the end before any repairs or diagnosis is attempted. Although the replacement and testing procedures are simple in themselves, the systems are not. Unless the technician has a thorough understanding of all components and their function within a particular system, the logical test sequence cannot be followed. A minor problem, such as a loose connector, can affect an entire system or several components; it is important to know how each component relates to the operation of the complete system. It is not enough to use the correct test equipment; the test equipment must be used correctly.

JUMPER WIRES

Jumper wires are simple, yet extremely valuable pieces of test equipment. Jumper wires are merely wires that are used to bypass sections of a circuit. The simplest type of jumper wire is a length of multi-strand wire with an alligator clip at each end. Jumper wires are usually fabricated from lengths of standard automotive wire and whatever type of connector (alligator clip, spade connector or pin connector) is required for the particular circuit being tested. The well-equipped tool box will have several different styles of jumper wires in several different lengths. Some jumper wires are made with 3 or more terminals coming from a common splice for special-purpose testing. In cramped, hard-to-reach areas it is advisable to have insulated boots over the jumper wire terminals to prevent accidental grounding or sparks, especially when testing fuel system components.

Jumper wires are used primarily to locate open electrical circuits, on either the ground (−) side of the circuit or on the hot (+) side. If an electrical component fails to operate, connect the jumper wire between the component and a good ground. If the component operates only with the jumper installed, the ground circuit is open. If the ground circuit is known to be good but the component does not operate, the circuit between the power feed and component is open.

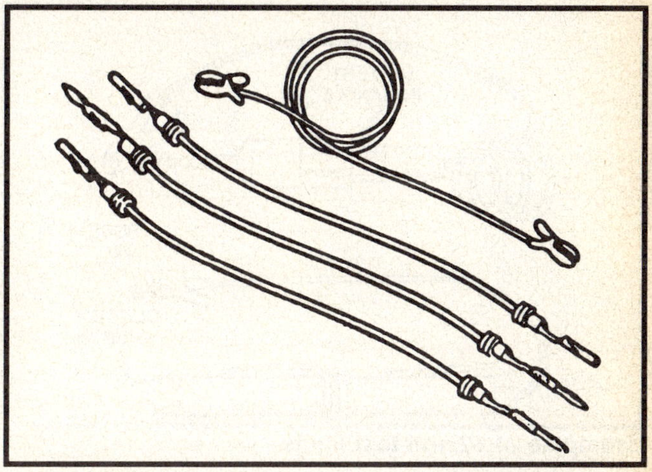

Jumper wires should have assorted connectors

Sometimes a jumper wire connected directly from the battery to the hot terminal of the component can be used, but first make sure the component uses 12 volts in operation. Some electrical components, such as fuel injectors or certain sensors, are designed to operate on low voltage; running 12 volts to the terminals can burn out the wiring or the component.

NOTE: Inserting an in-line fuseholder into a set of test leads provides a fused jumper which is highly recommended for use in testing. This will protect both the component and wiring harnesses in the event of a connection error. A 5 amp fuse will provide protection for most circuits.

Never use jumpers made from wire of lighter gauge (thickness) than used in the circuit being tested. If the jumper wire is of too small gauge, it may overheat or melt; in extreme cases, the heat may cause fire. Never use jumpers to bypass high-resistance loads (such as motors) in a circuit. Bypassing resistances in effect creates a short circuit which may cause damage or fire. Never use a jumper for anything other than temporary bypassing of components in a circuit.

12 VOLT TEST LIGHT

The 12 volt test light is used to check circuits and components while electrical current is flowing through them. It is used for voltage and ground tests. Twelve volt test lights come in different styles but all have 3 main parts; a ground clip, a probe, and a light. The most commonly used 12 volt test lights have pick-type probes. To use a 12 volt test light, connect the ground clip to a good ground and probe wherever necessary with the pick. The pick should be sharp so that it can penetrate wire insulation to make contact with the wire, without making a large hole in the insulation. The wrap-around light is handy in hard to reach areas or where it is difficult to support a wire to push a probe pick into it. To use the wrap around light, hook the wire to be probed with the hook and pull the trigger. A small pick will be forced through the wire insulation into the wire core.

NOTE: Do not use a test light to probe electronic ignition, spark plug or coil wires. Never use a pick-type test light to probe wiring on computer controlled systems unless specifically instructed to do so. Never pierce any shielded cable, such as that used on anti-lock brake systems or data lines between computers. Any wire insulation that is pierced by the test light probe should be taped and sealed with silicone after testing.

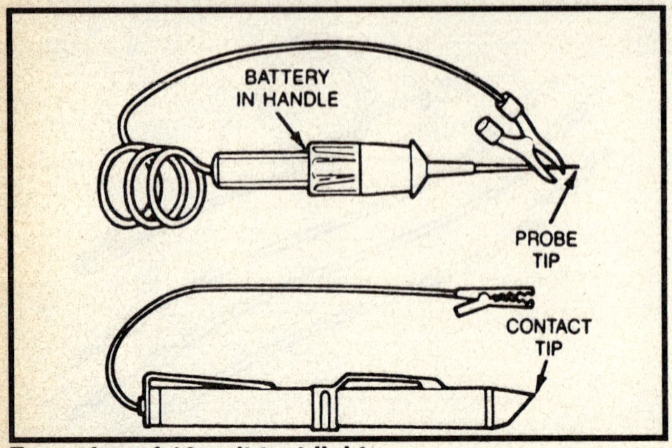

Examples of 12 volt test lights

Like the jumper wire, the 12 volt test light is used to isolate breaks in circuits. While the jumper wire is used to bypass the open to operate the load, the 12 volt test light is used to locate the presence of voltage in a circuit. If the test light glows, there is power up to that point; if the 12 volt test light does not glow when its probe is inserted into the wire or connector, there is an open circuit (no power). Move the test light in successive steps back toward the power source until the light in the handle illuminates. When it lights, the open is between the probe location and the point previously probed.

NOTE: The test light does not detect that 12 volts (or any particular voltage) is present; it only indicates that some voltage is present. It is advisable before using the test light to touch its terminals across the battery posts to make sure the light is operating properly. Do not attempt to determine voltage by the brightness of the lamp; a voltmeter must be used to determine voltage.

Self-Powered Test Light

The self-powered test light usually contains a 1.5 volt penlight battery. One type of self-powered test light is similar in design to the 12 volt test light. This type has both the battery and the light in the handle and pick-type probe tip. The second type has the light toward the open tip, so that the light illuminates the contact point.

The self-powered test light is dual-purpose piece of test equipment. It can be used to test for either open or short circuits when power is isolated from the circuit (continuity test). Do not use a self-powered test light on any computer controlled system or component unless specifically instructed to do so. Many engine sensors can be destroyed by even this small amount of voltage.

OPEN CIRCUIT TESTING

To use the self-powered test light to check for open circuits, first isolate the circuit from the power supply by removing a fuse, disconnecting a wiring harness connector or even turning a dashboard switch **OFF**. Avoid disconnecting the battery cables if at all possible; doing so will erase the electronic memories in the control modules. Besides the obvious effects on the clock and radio, driveability may be affected until the vehicle systems relearn the settings.

Connect the test light ground clip to a good ground and probe sections of the circuit sequentially with the test light. (start from either end of the circuit). If the light is out, the open is between the probe and the circuit ground. If the light is on, the open is between the probe and end of the circuit toward the power source.

SHORT CIRCUIT TESTING

After isolating the circuit both from power and from ground, a self-powered test light, may be used to find shorts within a circuit. Connect the test light ground clip to a good ground and probe any easy-to-reach test point in the circuit. If the light comes on, there is a short somewhere in the circuit. To isolate the short, probe a test point at either end of the isolated circuit (the light should be on). Leave the test light probe connected and open connectors, switches, remove parts, etc., sequentially, until the light goes out. When the light goes out, the short is between the last circuit component opened and the previous circuit opened.

NOTE: The 1.5 volt battery in the test light does not provide much current. A weak battery may not provide enough power to illuminate the test light even when a complete circuit is made. Always make sure that the test battery is strong. To check the battery, briefly touch the ground clip to the probe; if the light glows brightly the battery is strong enough for testing.

Never use a self-powered test light to perform checks for opens or shorts when power is applied to the electrical system under test. The 12-volt vehicle power will quickly burn out the 1.5 volt light bulb in the test light.

VOLTMETER

A voltmeter is used to measure voltage at any point in a circuit, or to measure the voltage drop across any part of a circuit. Voltmeters usually have various scales on the meter dial and a selector switch to allow the selection of different test ranges. The voltmeter has a positive and a negative lead. To avoid damage to the meter, connect the negative lead, usually black, to the negative (−) side of circuit or to ground. Connect the positive lead, usually red, to the positive (+) or power side of the circuit.

A voltmeter can be connected either in parallel or in series with a circuit and has a very high resistance to current flow. When connected in parallel, only a small amount of current will flow through the voltmeter current path; the rest will flow through the normal current path and the circuit will work normally. When the voltmeter is connected in series with a circuit, only a small amount of current can flow through the circuit. The circuit will not work properly, but the voltmeter reading will show if the circuit is complete or not.

Available Voltage Measurement

Set the voltmeter selector switch to the 20V position and connect the meter negative lead to the negative post of the battery and connect the positive meter lead to the positive post of the battery. Read the voltage on the meter or digital display. A well-charged battery should register over 12 volts. If the meter reads below 11.5 volts, the battery power may be insufficient to operate the electrical system properly. This test determines voltage available from the battery and should be the first step in any electrical trouble diagnosis procedure. Many electrical problems, especially on computer controlled systems, can be caused by a low state of charge in the battery. Excessive corrosion at the battery cable terminals can cause a poor contact that will prevent proper charging and full battery current flow.

Nominal battery voltage is 12 volts but, when fully charged, should be about 13.2 volts. When the battery is supplying current to 1 or more circuits it is said to be under load. When everything is **OFF** the electrical system is under a no-load condition. A fully charged battery showing about 12.5 volts at no load may drop to 12 volts under medium load and will drop even lower under heavy load. If the battery is partially discharged, the voltage decrease under heavy load may be excessive, even though the battery shows 12 volts or more at no load. For this reason, it is important that the battery be fully charged during all testing

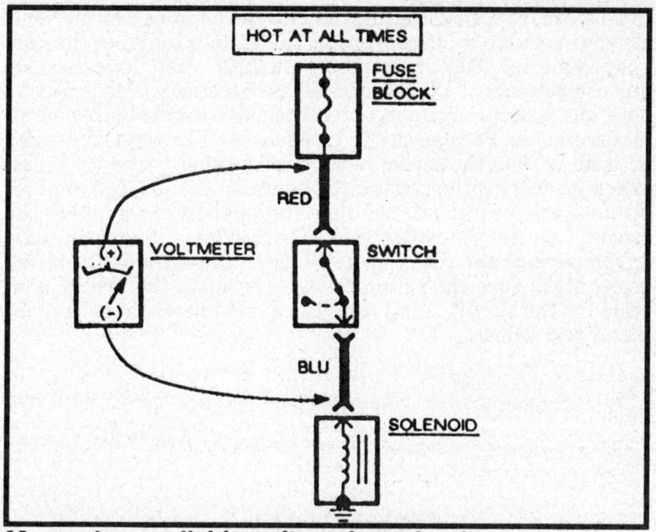

Measuring available voltage in a circuit

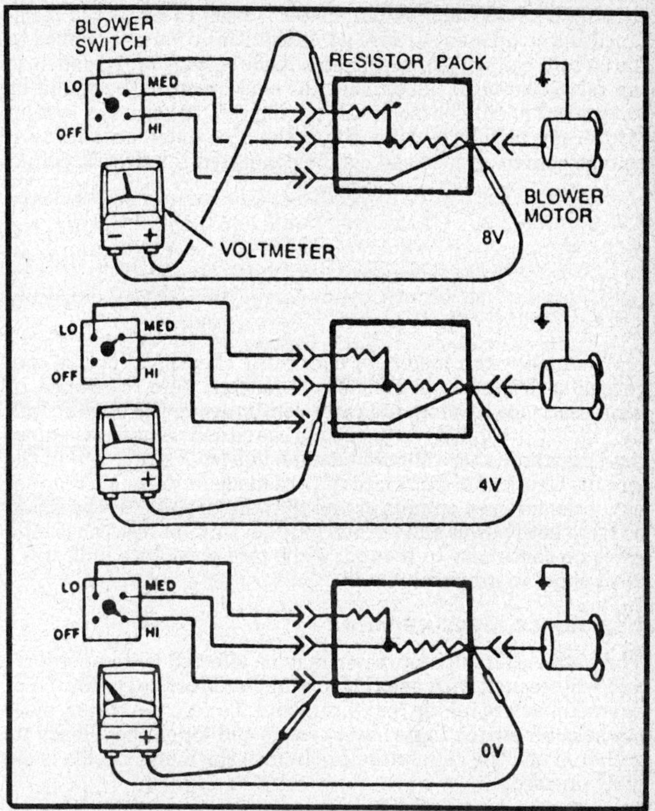

Direct measurement of voltage drops in a circuit

procedures to avoid errors in diagnosis and incorrect test results.

Voltage Drop

When current flows through a resistance, the voltage beyond the resistance is reduced. The larger the current, the greater the voltage reduction. When the circuit is off, there is no voltage drop because there is no current. All points in the circuit which are connected to the power source are at the same voltage as the power source. In a long circuit with many connectors, a series of small, unwanted voltage drops due to corrosion at the connectors can add up to a total loss of voltage which impairs the operation of the loads in the circuit.

INDIRECT COMPUTATION OF VOLTAGE DROPS

1. Set the voltmeter selector switch to the 20 volt position.
2. Connect the meter negative lead to a good ground.
3. Probe all resistances in the circuit with the positive meter lead.
4. Operate the circuit in all modes and observe the voltage readings.

DIRECT MEASUREMENT OF VOLTAGE DROPS

1. Set the voltmeter switch to the 20 volt position.
2. Connect the voltmeter negative lead to the ground side of the resistance load to be measured.
3. Connect the positive lead to the positive side of the resistance or load to be measured.
4. Read the voltage drop directly on the 20 volt scale.

Too high a voltage indicates too high a resistance. If, for example, a blower motor runs too slowly, there may be too high a resistance in the resistor pack. By taking voltage drop readings in all parts of the circuit, the problem can be isolated. Too low a voltage drop indicates too low a resistance. If, for example, a blower motor runs too fast in the **MED** and/or **LOW** position, the problem can be isolated to the resistor pack by taking voltage drop readings in all parts of the circuit to locate a possibly shorted resistor. The maximum allowable voltage drop under load is critical, especially if there is more than one high resistance problem in a circuit; all voltage drops are cumulative. A small drop is normal due to the resistance of the conductors.

High Resistance Testing

1. Set the voltmeter selector switch to the 2 volt position.
2. Connect the voltmeter positive lead to the positive post of the battery.
3. Turn **ON** the headlights and heater blower to provide a load.
4. Probe various points in the circuit with the negative voltmeter lead.
5. Read the voltage drop. Some average maximum allowable voltage drops are:
   - Fuse panel – 0.7 volts
   - Ignition switch – 0.5volts
   - Headlight switch – 0.7 volts
   - Ignition coil (+) – 0.5 volts
   - Any other load – 0.5–1.3 volts

NOTE: Voltage drops are all measured while a load is operating; without current flow, there will be no voltage drop.

OHMMETER

The ohmmeter is designed to read resistance (ohms) in a circuit or component. Although there are several different styles of ohmmeters, all will usually have a selector switch which permits the measurement of different ranges of resistance. Usually the selector switch allows the multiplication of the meter reading by 10, 100, 1000, 10,000, etc. A calibration knob allows the meter to be set at zero for accurate measurement. Since all ohmmeters are powered by an internal battery (usually 9 volts), the ohmmeter can be used as a self-powered test light. When the ohmmeter is connected, current from the ohmmeter flows through the circuit or component being tested. Since the ohmmeter's internal resistance and voltage are known values, the amount of current flow through the meter depends on the resistance of the circuit or component being tested.

The ohmmeter can be used to perform continuity tests for opens or shorts and to read actual resistance in a circuit. It should be noted that the ohmmeter is used to check the resistance of a component or wire while there is no voltage applied to the circuit. Current flow from an outside voltage source (such as the vehicle battery) can damage the ohmmeter, so the circuit or component should be isolated from the vehicle electrical system before any testing is done. Since the ohmmeter uses its own voltage source, either lead can be connected to any test point.

NOTE: When checking diodes or other solid state components, the ohmmeter leads can only be connected one way in order to measure current flow in a single direction. Make sure the positive (+) and negative (−) terminal connections are as described in the test procedures to verify the one-way diode operation.

When using the meter for continuity checks, do not be concerned with the actual resistance readings. Zero resistance, or any resistance reading, indicates continuity in the circuit. Infinite resistance indicates an open in the circuit. A high resistance reading where there should be none indicates a problem in the circuit. Checks for short circuits are made in the same manner as checks for open circuits except that the circuit must be isolated from both power and normal ground. Infinite resistance indicates no continuity to ground, while zero resistance indicates a dead short to ground.

Resistance Measurement

The batteries in an ohmmeter may be affected by temperature and will weaken with age. The ohmmeter must be calibrated or "zeroed" before taking measurements. To zero the meter, place the selector switch in its lowest range and touch the 2 leads together. Turn the calibration knob until the meter needle is exactly on zero.

NOTE: All analog (needle) type ohmmeters must be zeroed before use, but some digital ohmmeter models are automatically calibrated when the switch is turned ON. Self-calibrating digital ohmmeters do not have an adjusting knob, but it's a good idea to check for a zero readout before use by touching the leads together. All computer controlled systems require the use of a digital ohmmeter with at least 10 megohms impedance for testing. Before any test procedures are attempted, make sure the ohmmeter used is compatible with the electrical system, or damage to the on-board computer could result.

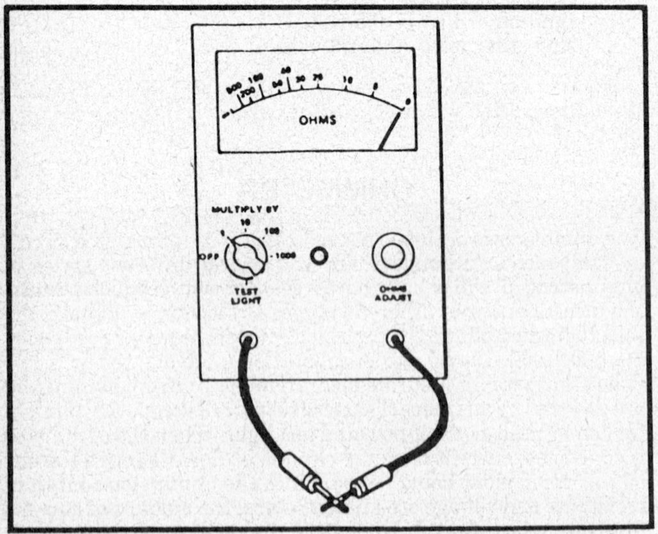

Analog ohmmeters must be calibrated before use by touching the probes together and adjusting the zero-set knob

To measure resistance, first isolate the circuit from the vehicle power source by disconnecting the battery cables or the harness connector. Make sure the key is **OFF** when disconnecting any components or the battery. Where necessary, also isolate at least one side of the circuit to be checked to avoid reading parallel resistances. Parallel circuit resistances will always give a lower reading than the actual resistance of either of the branches. When measuring the resistance of parallel circuits, the total resistance will always be lower than the smallest resistance in the circuit. Connect the meter leads to both sides of the circuit (wire or component) and read the actual measured ohms on the meter scale. Make sure the selector switch is set to the proper ohm scale for the circuit being tested to avoid misreading the ohmmeter test value.

NOTE: Never use an ohmmeter with power applied to the circuit. Like the self-powered test light, the ohmmeter is designed to operate on its own power supply. The normal 12 volt automotive system could damage the meter.

AMMETERS

An ammeter measures the amount of current flowing through a circuit in units called amperes or amps. Amperes are units of electron flow which indicate how fast the electrons are flowing through the circuit. Since Ohm's Law dictates that current flow in a circuit is equal to the circuit voltage divided by the total circuit resistance, increasing voltage also increases the current level (amps). Likewise, any decrease in resistance will increase the amount of amps in a circuit. At normal operating voltage, most circuits have a characteristic amount of amperes, called "current draw" which can be measured using an ammeter. By referring to a specified current draw rating, measuring the amperes, and comparing the 2 values, one can determine what is happening within the circuit to aid in diagnosis. An open circuit, for example, will not allow any current to flow so the ammeter reading will be zero. More current flows through a heavily loaded circuit or when the charging system is operating.

An ammeter is always connected in series with the circuit being tested. All of the current that normally flows through the circuit must also flow through the ammeter; if there is any other path for the current to follow, the ammeter reading will not be accurate. The ammeter itself has very little resistance to current flow and therefore will not affect the circuit, but it will measure current draw only when the circuit is closed and electricity is flowing. Excessive current draw can blow fuses and/or drain the battery; a reduced current draw can cause motors to

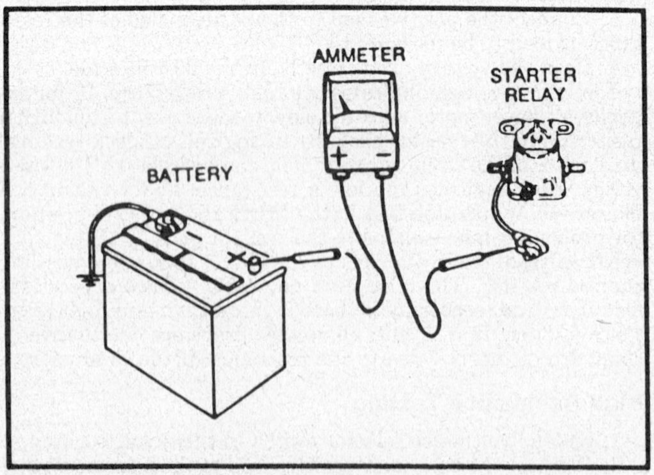

The ammeter is useful in tracking current drain

run slowly, lights to dim and other components to operate improperly.

MULTIMETERS

Different combinations of test meters can be built into a single unit designed for specific tests. Some of the more common combination test devices are known as Volt-Amp testers, Tach-Dwell meters, or Digital Multimeters. The Volt-Amp tester is used for charging system, starting system or battery tests and consists of a voltmeter, an ammeter and a variable resistance carbon pile. The voltmeter will usually have at least 2 ranges for use with 6, 12 and 24 volt systems. The ammeter also has more than 1 range for testing various levels of battery loads and starter current draw and the carbon pile can be adjusted to offer different amounts of resistance. The Volt-Amp tester has heavy leads to carry large amounts of current and many later models have an inductive ammeter pickup that clamps around the wire to simplify test connections. On some models, the ammeter also has a zero-center scale to allow testing of charging and starting systems without switching leads or polarity.

A digital multimeter is a voltmeter, ammeter and ohmmeter combined in an instrument which gives a digital readout. These are often used when testing solid state circuits because of their high input impedance (usually 10 megohms or more). It is important to note that the ammeter portion of this meter may only be useful on extremely low current circuits. Refer to the instructions provided with the meter before attempting to use the ammeter on automotive circuits.

The tach-dwell meter combines a tachometer with a dwell (cam angle) meter and is a specialized kind of voltmeter. The tachometer scale is marked to show engine speed in rpm and the dwell scale is marked to show degrees of distributor shaft rotation. In most electronic ignition systems, dwell is determined by the control unit, but the dwell meter can also be used to check the duty cycle (operation) of some electronic engine control systems.

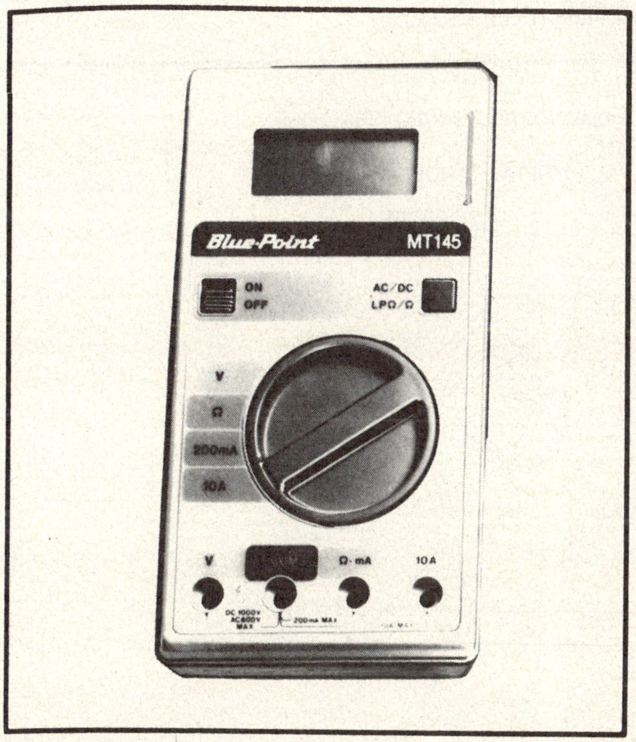

A digital multi-meter is required for testing most solid state circuits

VACUUM LEAK DETECTOR

Although electronically or computer controlled engine management systems have greatly reduced the number of vacuum circuits within the engine compartment, the remaining circuits have become more important. In many cases, the vacuum signal is translated into an electrical signal sent to the engine controller. For this reason, the integrity of the vacuum system must be maintained at all times.

A hand-held electronic vacuum leak detector can be of great value in tracing leaks affecting the system but not easily seen or heard. These battery powered tools employ a microphone and beeper system. The unit is calibrated to respond to ultrasonic signals, well above the level of human hearing. Leaking vacuum will generate sound waves in the 40 kilo-Hz. range and the detector will sound the beeper when this signal is found. Because of the high frequency sensed by the detector, it is not generally affected by normal engine or surrounding shop noises.

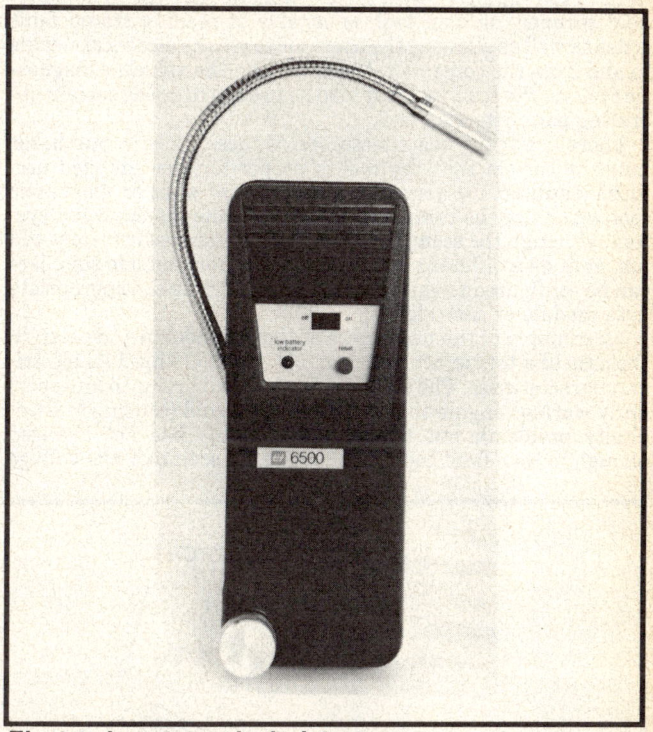

Electronic vacuum leak detector

Special Test Equipment

A variety of diagnostic tools are available to help troubleshoot and repair computerized engine control systems. Although the most sophisticated units may occupy an entire garage service bay, there are several types of aftermarket electronic testers available. These units allow quick circuit tests of the engine control system by plugging directly into a special connector located in the engine compartment, under the dashboard or in the fusebox. Several tool and equipment manufacturers offer simple, hand-held testers that measure various circuit voltage levels on command. Although these testers usually cost several hundred dollars, consider that the average computer control unit (or ECM) can cost just as much; the money saved by not replacing good sensors or components could justify the purchase price the first time it's used.

These computerized testers can allow quick and easy test measurements while the engine is operating. In addition, the on-board computer memory can be read to access any stored trouble codes, in effect allowing the computer to tell where it hurts or pinpoint which circuit is malfunctioning. In the same manner, repairs can be tested to make sure the problem has been corrected. The biggest advantage of these special testers is their relatively easy hookup, minimizing chances of making the wrong connection, getting false voltage readings or damaging the computer.

NOTE: Remember that any meter checks only the presence of electrical signals in a circuit. No meter can detect a faulty component if it is binding or disconnected internally. The presence of voltage does not necessarily mean that the related component is working properly.

SCAN TOOLS

As computers are used to control more circuits on vehicles, the need to communicate with the controllers increases. The hand-held scanner, or scan tool, is capable of reading stored fault codes as well as allowing the operator to see the electrical signals to and from the controller. Depending on the vehicle's diagnostic system, the scan tool may also be used to override certain operating parameters as well.

There are many manufacturers of these tools; a purchaser must be certain that the tool is proper for the intended use. Most manufacturers employ some form of scan or diagnostic tool which may be connected into the engine management system, although the scan tool for one vehicle almost certainly will not work on a different make. Scan tools from outside suppliers can be configured to various makes by inserting the appropriate data module or cartridge.

An example of the usefulness of the scan tool may be seen in the case of a temperature sensor which has changed its electrical characteristics. The engine controller is reacting to an apparently warmer engine that is actually still cold (causing a driveability problem), but the sensor's voltage has not changed enough to set a fault code. Connecting the scan tool, the voltage signal being sent to the ECM may be viewed; comparison to either normal values or a known good vehicle reveals the problem quickly. Additionally, some scan tools may be set for use as a volt- or ohmmeter with the push of a button. The built-in test lead allows certain electrical tests to be performed without swapping tools.

In some cases, the hand scanner may be used to test other computer-controlled vehicle systems such as Antilock Brakes (ABS) or cruise control, but many vehicle manufacturers require individual tools for the other systems. This is particularly true of ABS and/or air bag systems.

Depending on the vehicle and the amount of computer control used, systems which may be viewed or investigated with a scan tool include:

1. Engine controllers/ECUs
2. Fuel/ignition systems
3. Electronic transmission control
4. Charging system
5. Suspension control functions
6. Anti-Lock brake system.
7. Passive restraint system
8. Anti-theft system
9. Climate Control Systems
10. Body electrical systems including power locks, entertainment systems, sunroofs, and defoggers.

PIN-OUT OR TEST BOXES

Sometimes called a test or diagnostic harness, this test tool connects between the control unit and the wiring harness. A pin-out box routes each of the many wire harness conductors to a discreet, insulated terminal on the box. All testing with the volt-ohmmeter can be done safely at these terminals, eliminating the risk of damage due to backprobing at the control unit.

Additionally, some manufacturers provide plastic overlays for the pinout box. This allows the box to be used on a variety of models; different overlays identify the changes in wire use or labeling. Use of a pin-out box is highly recommended during diagnosis and circuit checking.

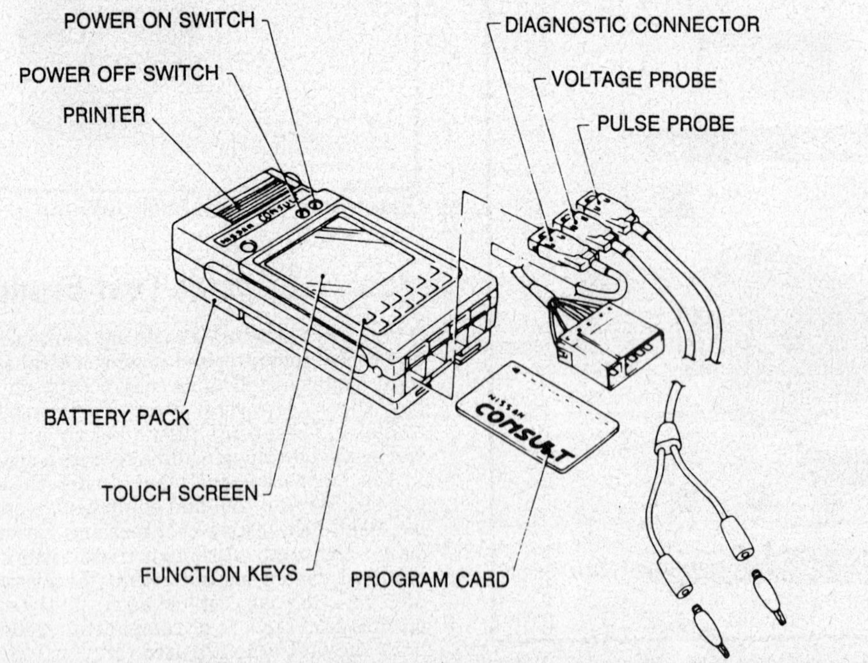

The Nissan and Infinity CONSULT tool is a good example of a multi-function hand scanner

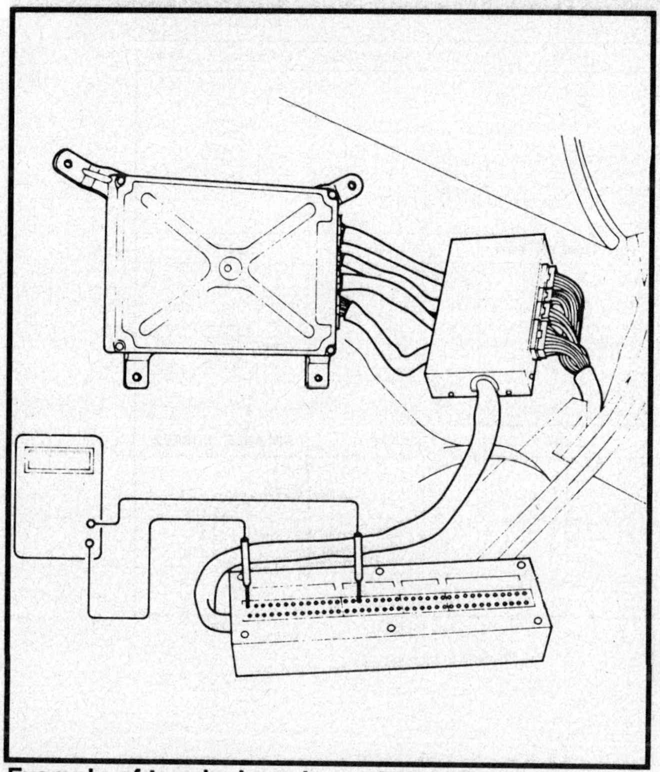

Example of terminal or pin-out box being used for system testing

Self-Diagnostics

The control units found on most systems are capable of providing diagnostic help when a malfunction occurs. The units are programmed to monitor certain input and output circuits and react to a major change in the circuit. When a fault is detected, the dashboard warning lamp is usually activated. A numbered code is assigned to the fault and stored in the memory portion of the unit. The stored code(s) become a guide to the faulty system(s).

The stored codes will usually only be output when the system is placed in the diagnostic mode. On most vehicles, this requires a specific action or series of actions, such as installing a jumper across diagnostic terminals or cycling the key ON–OFF–ON–OFF a certain number of times. In many cases, the codes may be retrieved by connecting a voltmeter to the system.

Codes will be output as a series of voltage pulses. Depending on the system, the pulses may be seen as needle sweeps on an analog voltmeter, a flashing of the dashboard warning lamp or the flashing of an LED on the ECU. In some cases, the code appear as an English message on the screen of the scan tool. Remember that these codes indicate failed circuits, not necessarily a failed component. The scan tool message screen may show FAULT 14: THROTTLE POSITION SENSOR, but the problem most likely lies in the wiring and connectors rather than in the sensor itself.

Wiring and Circuits

ELECTRICAL SCHEMATICS

An electrical schematic shows the current paths when a circuit is operating properly. It is essential to understand how a circuit

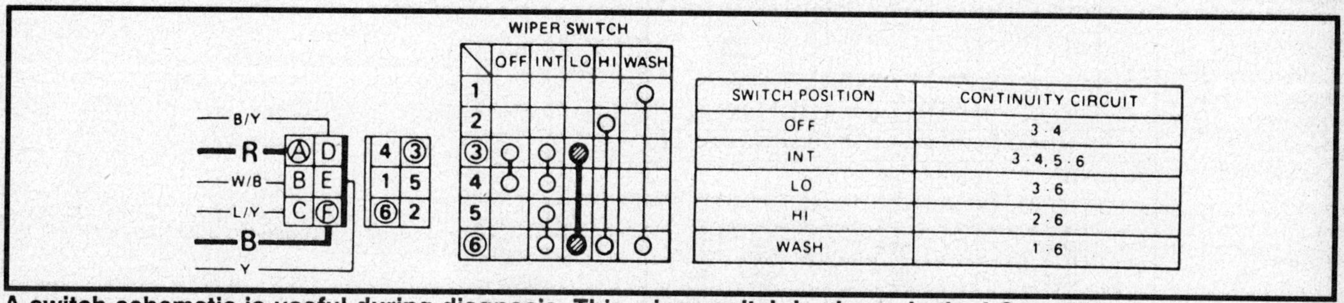

SWITCH POSITION	CONTINUITY CIRCUIT
OFF	3 - 4
INT	3 - 4, 5 - 6
LO	3 - 6
HI	2 - 6
WASH	1 - 6

A switch schematic is useful during diagnosis. This wiper switch is shown in the LO position

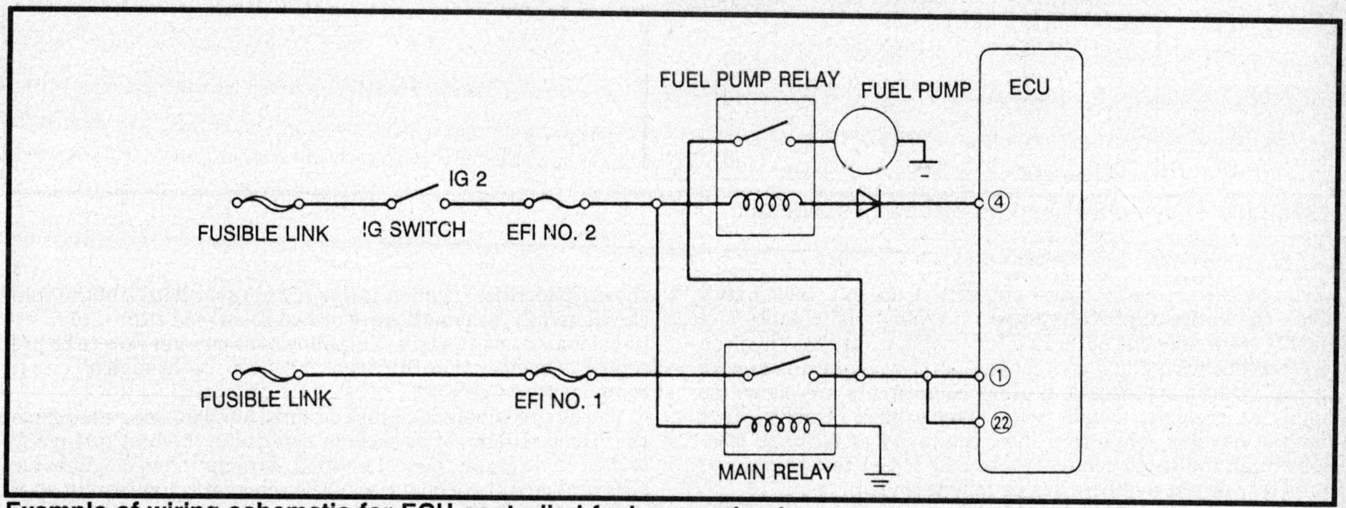

Example of wiring schematic for ECU controlled fuel pump circuit

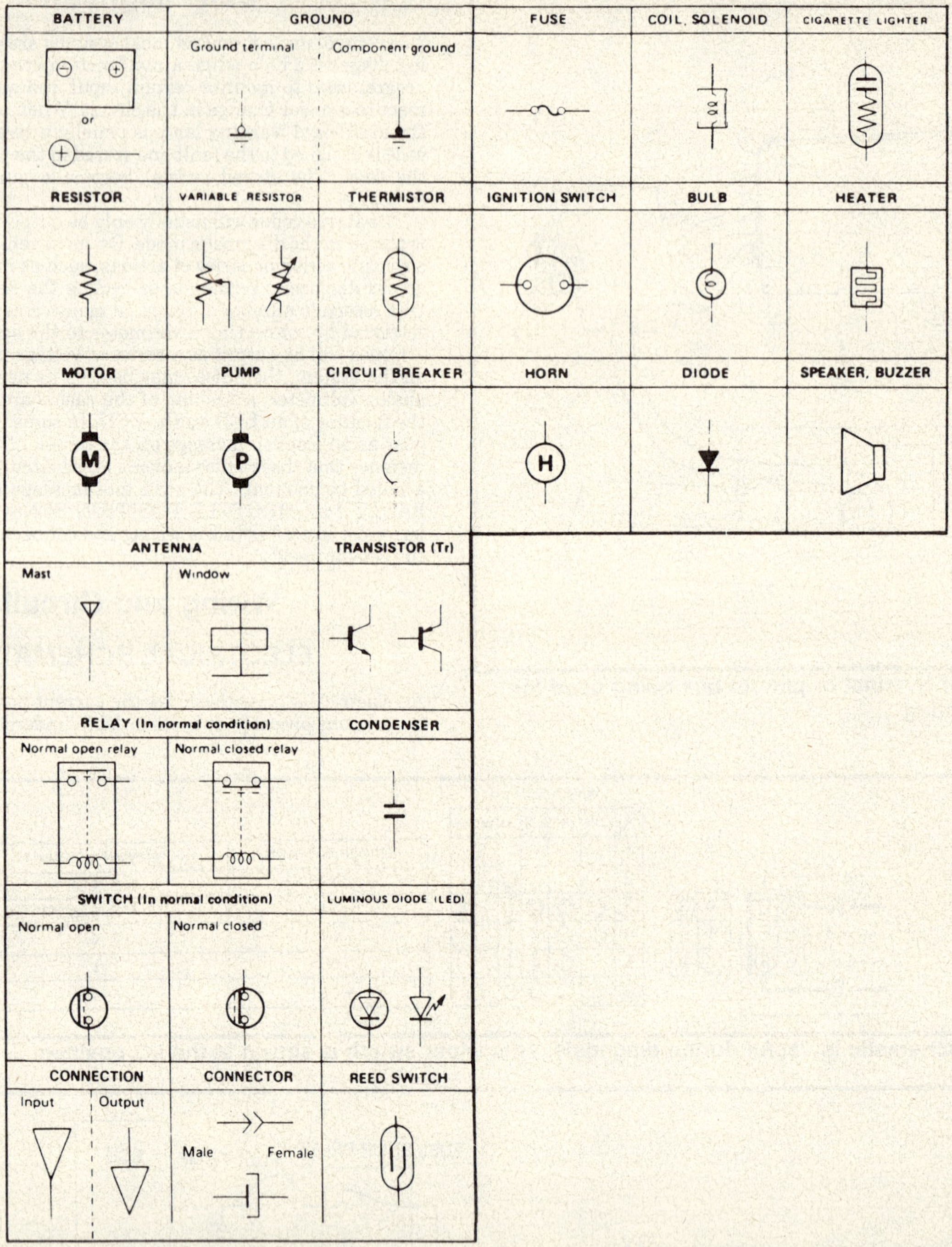

Examples of symbols used on electrical schematic

works before trying to figure out why it doesn't. Schematics break the entire electrical system into individual circuits and usually show only one path. In a schematic, no attempt is made to represent wiring and components as they physically appear on the vehicle; switches and other components are shown as simply as possible, usually with standardized symbols. Face views of harness connectors show the cavity or terminal locations in all multi-pin connectors to help locate test points and connector or pin numbers are usually provided.

Schematics are not to drawn to scale; a very short wire will be shown in identical fashion to one running the length of the vehicle. Generally, schematics are drawn to be read either from top-to-bottom or left-to-right. This allows the current flow to be presented in logical order from the power source through all circuit components to ground.

The circuit schematic must be considered an accurate map of the circuit. It should be used to determine the best test points and, in some cases, the test method. Attempting to diagnose any electrical circuit without use of the schematic may lead to wasted time and unnecessary replacement of components.

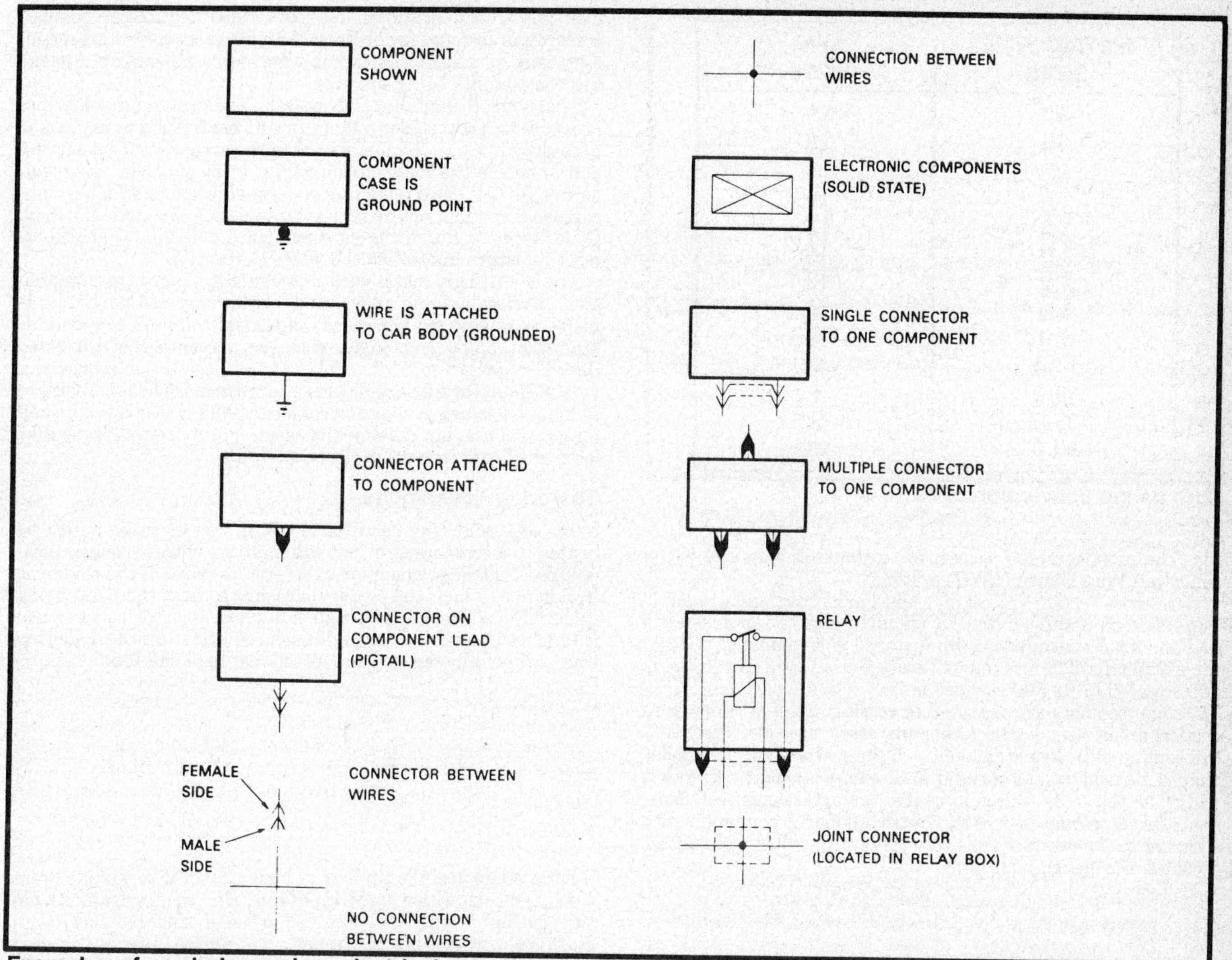

Examples of symbols used on electrical schematic

WIRE HARNESSES

The average automobile contains about ½ mile of wiring, with hundreds of individual connections. To protect the many wires from damage and to keep them from becoming a confusing tangle, they are organized into bundles, enclosed in plastic or taped together and called wire harnesses. Different wiring harnesses serve different parts of the vehicle. Individual wires are color-coded to help trace them through a harness where sections are hidden from view.

A loose, corroded connection or a replacement wire that is too small for the circuit will add extra resistance and an additional voltage drop to the circuit. A 10% voltage drop can result in slow or erratic motor operation, even though the circuit is complete. Automotive wiring or circuit conductors can be in any of 3 forms:

1. Single strand wire
2. Multi-strand wire
3. Printed circuitry

Single strand wire has a solid metal core and is usually used inside such components as alternators, motors, relays and other devices. Multi-strand wire has a core made of many small strands of wire twisted together into a single conductor. Most of the wiring in an automotive electrical system is made up of multi-strand wire, either as a single conductor or grouped together in a harness. All wiring is color-coded on the insulator, either as a solid color or as a colored wire with an identification stripe.

NOTE: While most manufacturer's wiring diagrams use common abbreviations such at R for Red or G for Green wires, certain Asian manufacturers routinely use L as a designation for Blue. Thus, a wire labeled LW is Blue with a White stripe. A lighter version of a color would appear as LT GRN or Lg.

A printed circuit is a thin film of copper or other conductor that is printed on an insulator backing. Occasionally, a printed circuit is sandwiched between 2 sheets of plastic for more protection and flexibility. A complete printed circuit, consisting of conductors, insulating material and connectors for lamps or other components is called a printed circuit board. Printed circuitry is used in place of individual wires or harnesses in places where space is limited, such as behind instrument panels.

Wire Gauge

Since computer-controlled automotive electrical systems are very sensitive to changes in resistance, the selection of properly sized wires is critical when systems are repaired. The wire gauge number is an expression of the cross section area of the conduc-

METRIC WIRE SIZES	AWG SIZES
.22	24
.35	22
.5	20
.8	18
1.0	16
2.0	14
3.0	12
5.0	10
8.0	8
13.0	6
19.0	4
32.0	2
50.0	1/0

Wire gauge conversion table

tor. The most common system for expressing wire size is the American Wire Gauge (AWG) system.

Wire cross section area is measured in circular mils. A mil is $\frac{1}{1000}$ of an in. (0.001); a circular mil is the area of a circle one mil in diameter. For example, a conductor ¼ in. in diameter is 0.250 in. or 250 mils. The circular mil cross section area of the wire is 250 squared or 62,500 circular mils.

Gauge numbers are assigned to conductors of various cross section areas. As gauge number increases, area decreases and the conductor becomes smaller. A 5 gauge conductor is smaller than a 1 gauge conductor and a 10 gauge is smaller than a 5 gauge. As the cross section area of a conductor decreases, resistance increases and so does the gauge number. A conductor with a higher gauge number will carry less current than a conductor with a lower gauge number.

NOTE: Gauge wire size refers to the size of the conductor, not the size of the complete wire. It is possible to have 2 wires of the same gauge with different outer diameters because one may have thicker insulation than the other.

Twelve volt automotive electrical systems generally use 10, 12, 14, 16 and 18 gauge wire. Main power distribution circuits and larger accessories usually use 10 and 12 gauge wire. Battery cables are usually 4 or 6 gauge, although 1 and 2 gauge wires are occasionally used. Wire length must also be considered when making repairs to a circuit. As conductor length increases, so does resistance. An 18 gauge wire, for example, can carry a 10 amp load for 10 feet without excessive voltage drop; however if a 15 foot wire is required for the same 10 amp load, a 16 gauge wire must be used.

When backprobing a connector while it is engaged, the order of the terminals must be mentally reversed. The wire color code can help in this situation, as well as any keyway, lock tab or other reference mark.

WIRING REPAIRS

Soldering is a quick, efficient method of joining metals permanently. Everyone who has to make wiring repairs should know how to solder. Electrical connections that are soldered are far less likely to come apart and will conduct electricity much better than connections that are only "pig-tailed" together. The most popular (and preferred) method of soldering is with an electric soldering tool. Soldering irons or guns are available in many sizes and wattage ratings. Irons with higher wattage ratings deliver higher temperatures and recover lost heat faster. A small soldering iron rated for no more than 50 watts is recommended, especially on electrical systems where excess heat can damage the components being soldered.

There are 3 ingredients necessary for successful soldering; proper flux, good solder and sufficient heat. A soldering flux is necessary to clean the metal of tarnish, prepare it for soldering and to enable the solder to spread into tiny crevices. When soldering, always use a resin flux or resin core solder which is non-corrosive and will not attract moisture once the job is finished. Other types of flux (acid core) will leave a residue that will attract moisture and cause the wires to corrode.

Tin is a unique metal with a low melting point. In a molten state, it dissolves and alloys easily with many metals. Solder is made by mixing tin with lead. The most common proportions are 40/60, 50/50 and 60/40, with the percentage of tin listed first. Low priced solders usually contain less tin, making them very difficult for a beginner to use because more heat is required to melt the solder. A common solder is 40/60 which is well suited for general use, but 60/40 melts easier, has more tin for a better joint and is preferred for electrical work.

Soldering Techniques

Successful soldering requires that the metals to be joined be heated to a temperature that will melt the solder (usually 360–460°F). Contrary to popular belief, the purpose of the soldering iron is not to melt the solder itself, but to heat the parts being soldered to a temperature high enough to melt the solder when it touches the work. Melting flux-cored solder on the soldering iron will usually destroy the effectiveness of the flux.

NOTE: Soldering tips are made of copper for good heat conductivity, but must be "tinned" regularly for quick transfer of heat to the project and to prevent the solder from sticking to the iron. To "tin" the iron, simply heat it and touch the flux-cored solder to the tip; the solder will flow over the hot tip. Wipe the excess off with a clean rag, but be careful as the iron will be hot.

After some use, the tip may become pitted. If so, simply dress the tip smooth with a smooth file and "tin" the tip again. An old saying that metals well cleaned are half soldered has much truth; rust, bits of insulation and oil or grease must be removed with a wire brush or emery cloth. For maximum strength in soldered parts, the joint must start off clean and mechanically tight. Weak joints will result if there are gaps.

If a separate soldering flux is used, it should be brushed or swabbed only on areas that are to be soldered. Most solders contain a core of flux and separate fluxing is unnecessary. Hold the work to be soldered firmly. It is best to solder on a wooden board, because a metal vise will only rob heat from the piece to be soldered. Hold the soldering tip with the broadest face against the work to be soldered. Apply solder under the tip close to the work, using enough solder to give a heavy film between the iron and the piece being soldered, while moving slowly and making sure the solder melts properly. Keep the work level or the solder will run to the lowest part. If the soldering tip overheats (the solder coating on the face of the tip burns), it should be retinned.

Once the soldering is completed, let the soldered joint stand until cool. Do not attempt to cool the joint with water or other artificial means. Tape and seal all soldered wire splices after the repair has cooled. The external insulation and weatherproofing must equal that of the undamaged wire.

If soldering a joint is not possible, a crimp connector may be used. Properly installed, a crimp connector will be as reliable as a soldered joint. The connector—usually called a butt connector—must be the correct diameter or gauge for the wires. The wire ends must be clean and square cut. Fit each wire end into the connector and use a crimping tool to close the connector. The most common errors in this type of repair are either not using the correct tool to get an even crimp around the connector or over-tightening the crimp to the point of distorting the connector. Once the connector is distorted, it loses clamping force and the connection becomes loose. In this case, tighter is not better. A proper terminal crimping tool has sized and color-coded notches for delivering the correct amount of pressure to the connector without damage. Test all crimps by pulling moderately on the wires; it should not be possible to pull the wires out of a good crimp.

Wires which are open, exposed or otherwise damaged are repaired by simple splicing. Where possible, if the wiring harness

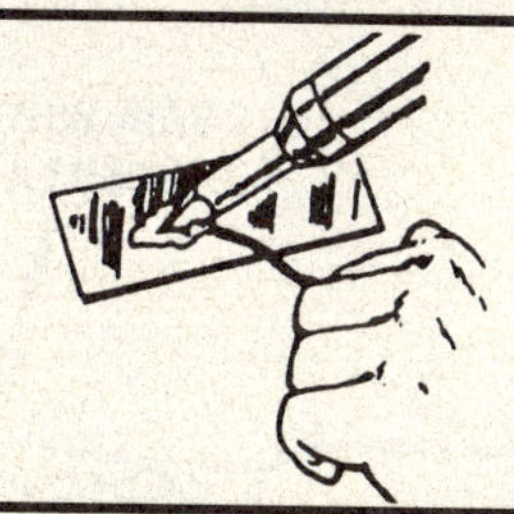

Clean and tin the soldering tool before each use

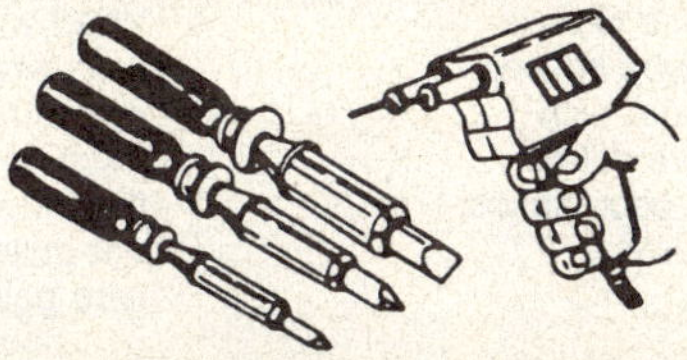

Various types of soldering tools. High wattage tools are not suitable for working on or near solid-state equipment

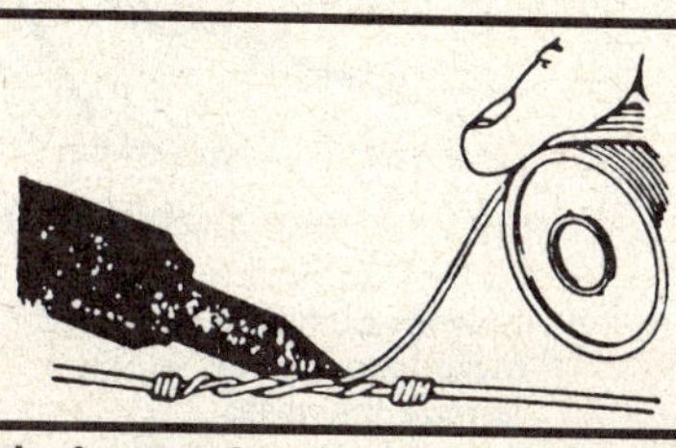

Use the soldering tool to heat the joint, then apply the solder

WIRE HARNESS REPAIR PROCEDURES

Condition	Location	Correction
Non-continuity	Using the electric wiring diagram and the wiring harness diagram as a guideline, check the continuity of the circuit in question by using a tester, and check for breaks, loose connector couplings, or loose terminal crimp contacts. 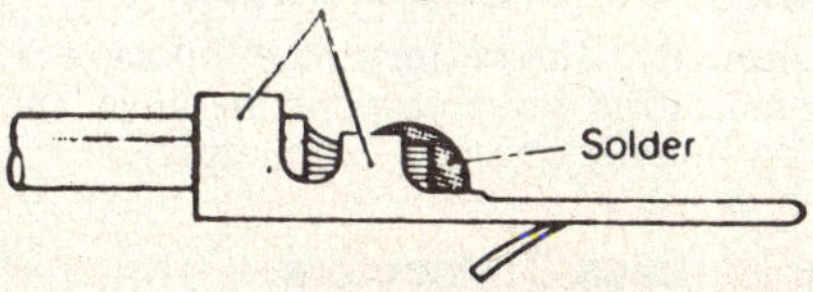	**Breaks**—Reconnect the point of the break by using solder. If the wire is too short and the connection is impossible, extend it by using a wire of the same or larger size. 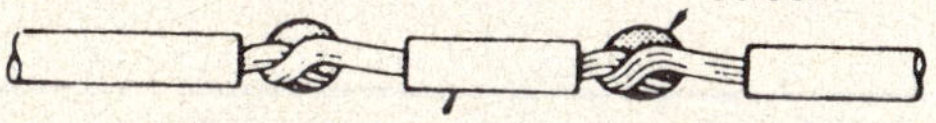 **Loose couplings**—Hold the connector securely, and insert it until there is a definite joining of the coupling. If the connector is equipped with a locking mechanism, insert the connector until it is locked securely. **Loose terminal crimp contacts**—Remove approximately 2 in. (5mm) of the insulation covering from the end of the wire, crimp the terminal contact by using a pair of pliers, and then, in addition, complete the repair by soldering.
Short-circuit	Using the electric wiring diagram and the wiring harness diagram as a guideline, check the entire circuit for pinched wires.	Remove the pinched portion, and then repair any breaks in the insulation covering with tape. Repair breaks of the wire by soldering.
Loose terminal	Pull the wiring lightly from the connector. A special terminal removal tool may be necessary for complete removal.	Raise the terminal catch pin, and then insert it until a definite clicking sound is heard. 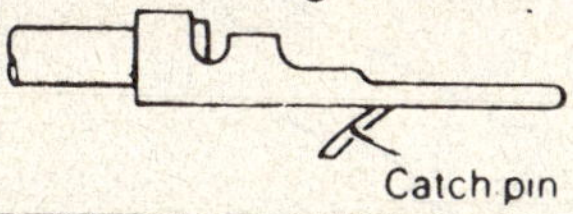

Note: There is the chance of short circuits being caused by insulation damage at soldered points. To avoid this possibility, wrap all splices with electrical tape and use a layer of silicone to seal the connection against moisture. Incorrect repairs can cause malfunctions by creating excessive resistance in a circuit.

Wire harness repair procedures

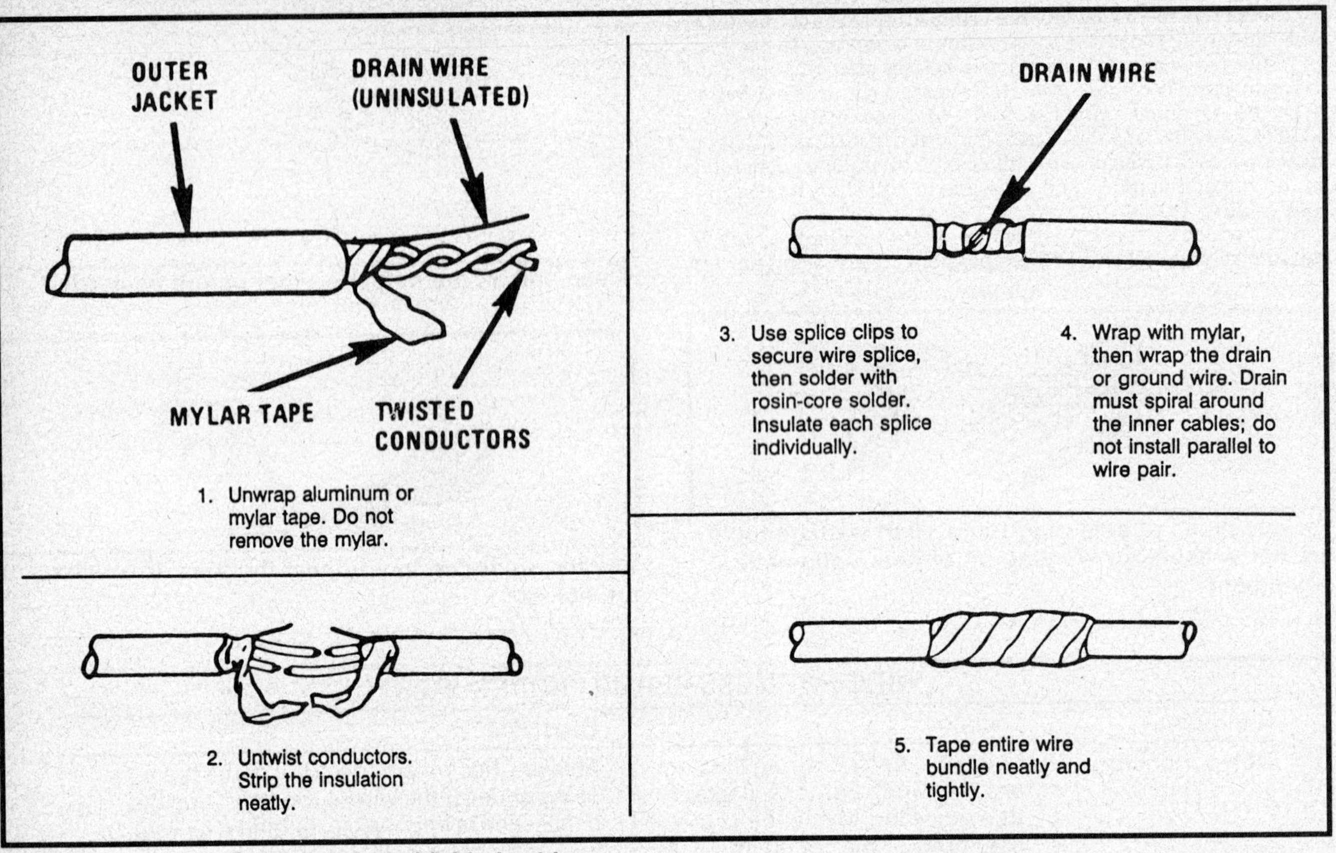

Repair procedures for twisted or shielded cable

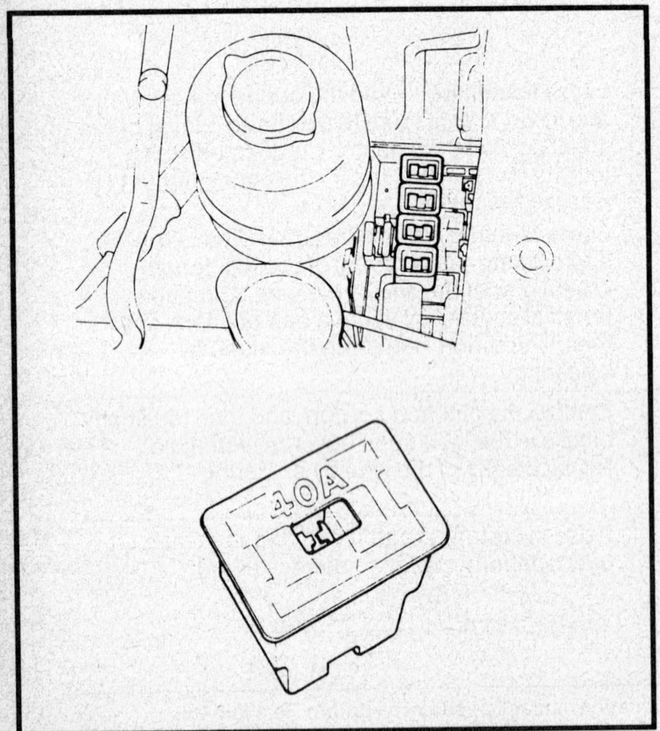

Example of main link location in the engine compartment. This type of link may be inspected visually

is accessible and the damaged place in the wire can be located, it is best to open the harness and check for all possible damage. In an inaccessible harness, the wire must be bypassed with a new insert, usually taped to the outside of the old harness.

FUSIBLE LINKS AND MAIN FUSES

The main fuses and links used on these vehicles are easily replaced but not repairable. Depending on the vehicle, these high-current fuses may be found either in the underdash fusebox or in the fuse or relay boxes in the engine compartment. Some may be mounted at the battery. The links are simply removed from their mounting and a new one installed. In most cases, the link may be visually checked for integrity.

Wire Harness Connectors

The computer wire harnesses connect the control unit to the various solenoids, switches and sensors used by the control system. Most connectors in the engine compartment or otherwise exposed to the elements are protected against moisture and dirt which could create oxidation and deposits on the terminals. This protection is important because of the very low voltage and current levels used by the computer and sensors. All connectors have a lock which secures the male and female terminals together; some use a secondary lock as well. Connectors attaching a harness to a component are also held by a retaining lock; use care not to damage the component when disconnecting the wiring.

Weatherproof and/or molded-on connectors require complete replacement of the connection. This means splicing a new connector assembly into the harness. All splices in on-board computer systems should be soldered to insure reliable contact. Use care when probing the connections or replacing terminals in them as it is possible to short between adjacent terminals. If this

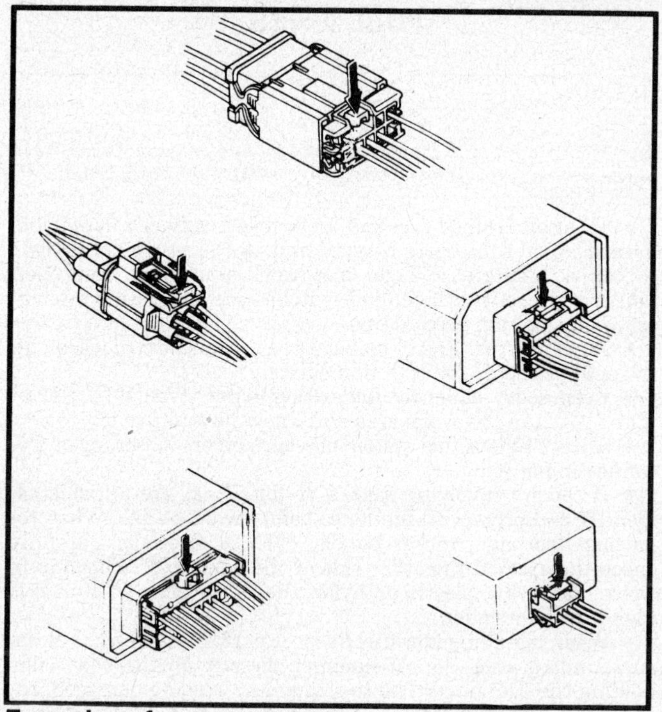

Examples of connector locks

happens to the wrong terminal pair, it is possible to damage certain components. Always use jumper wires between connectors for circuit checking and never probe through weatherproof seals.

Open circuits are often difficult to locate by sight because corrosion or terminal misalignment are hidden by the connectors. Merely wiggling a connector on a sensor or in the wiring harness may correct the open circuit. This should always be considered when an open circuit or a failed sensor is indicated. Intermittent problems may also be caused by oxidized or loose connections. When using a circuit tester for diagnosis, always probe connections from the wire side. Be careful not to damage sealed connectors with test probes; never backprobe a sealed or weatherproof connector.

NOTE: Many of the problems in the wiring harness are due to bad ground connections. Always check all vehicle ground connections for corrosion or looseness before performing any other circuit tests. This will eliminate the chance of a bad ground affecting circuit operation.

All wiring harnesses should be replaced with identical parts, using the same gauge wire and connectors. When signal or low-voltage wires are spliced into a harness, choose wire with high temperature insulation only. The low voltage and current levels found in the system require the best possible connection at all wire splices; joints must be mechanically tight and cleanly soldered.

It is seldom necessary to replace a complete harness. If replacement is necessary, pay close attention to insure proper harness routing. Secure the harness with suitable plastic wire clamps to prevent vibrations from causing the harness to wear in spots or contact any hot components.

Repairing Hard Shell Connectors

Unlike molded connectors, the terminal contacts in hard shell connectors can be replaced. Replacement usually involves the use of a special terminal removal tool that moves the locking tangs (barbs) on the connector terminal and allows the connector to be removed from the rear of the shell. The connector shell should be replaced if it shows any evidence of burning, melting, cracks, or breaks. Replace individual terminals that are burnt, corroded, distorted or loose.

NOTE: The insulation crimp must be tight to prevent the insulation from sliding back on the wire when the wire is pulled. The insulation must be visibly compressed under the crimp tabs, and the ends of the crimp should be turned in for a firm grip on the insulation.

The wire crimp must be made with all wire strands inside the crimp. The terminal must be fully compressed on the wire strands with the ends of the crimp tabs turned in to make a firm grip on the wire. Check all connections with an ohmmeter to insure a good contact. There should be no measurable resistance between the wire and the terminal when connected.

Mechanical Test Equipment

VACUUM GAUGE

Intake manifold vacuum is used to operate various systems and devices on all cars. To correctly diagnose and solve problems in vacuum control systems, a vacuum source is necessary for testing. In some cases, vacuum can be taken from the intake manifold when the engine is running, but vacuum is normally provided by a hand vacuum pump.

Most gauges are graduated in inches of Mercury (in. Hg), although a device called a manometer reads vacuum in inches of water (in. H_2O). The vacuum reading usually varies between 18 and 22 in. Hg at sea level. To test engine vacuum, the vacuum gauge must be connected to a source of manifold vacuum. Many engines have a plug in the intake manifold which can be removed and replaced with an adapter fitting. Connect the vacuum gauge to the fitting with a suitable rubber hose or, if no manifold plug is available, connect the vacuum gauge to any device using manifold vacuum, such as EGR valves, etc. The vacuum gauge can be used to determine the amount of vacuum reaching a component.

HAND VACUUM PUMP

Small, hand-held vacuum pumps come in a variety of designs and provide a source of vacuum for testing components without the engine operating. Most have a built-in vacuum gauge and allow a component to be tested without removing it from the vehicle. Operate the pump lever or plunger, applying the correct amount of vacuum required for the test. The level of vacuum in inches of Mercury (in. Hg) is indicated on the pump gauge. For some testing, an additional vacuum gauge may be necessary.

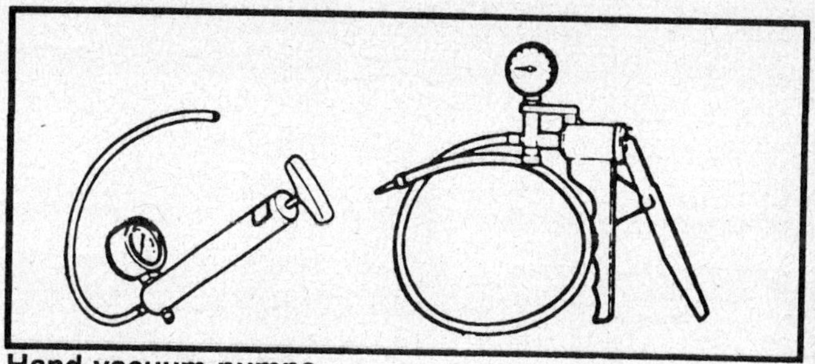

Hand vacuum pumps

COMPRESSION GAUGE

A compression gauge measures the amount of pressure in pounds per square inch (psi) that a cylinder is producing. Some gauges have a hose that screws into the spark plug hole while

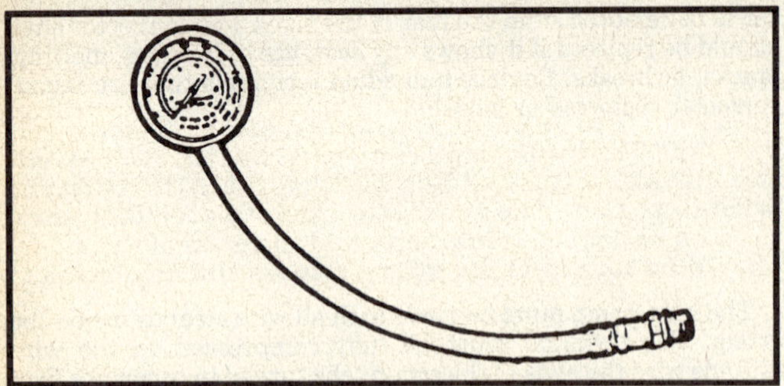

Compression gauge with threaded fitting

others have a tapered rubber tip which is held by hand in the spark plug hole. Engine compression depends on the sealing ability of the rings, valves, head gasket and spark plug gaskets. If any of these parts are not sealing properly, compression will be lost and the power output of the engine will be reduced. The compression in each cylinder should be measured and the variation between cylinders should be noted. The engine should be cranked through 5 or 6 compression strokes while warm, with all plugs removed, ignition disabled and throttle valves wide open.

FUEL PRESSURE GAUGE

A fuel pressure gauge is required to test the operation of the fuel delivery and injection systems. Some systems also need a 3-way valve to check the fuel pressure in various modes of operation. Gauges may require special adapters for making fuel connections. Always observe the cautions outlined in the Fuel System Service section when working around any pressurized fuel system.

FUEL LINES

CAUTION

Do not attempt to repair high pressure fuel lines. Any sign of scratching, kinking or deformation requires replacement of the line. Replacement lines must always be of the correct material and pressure rating. When working with or around fuel lines, observe the following precautions.

- Treat all drained fuel and fuel-soaked rags as a flammable hazard. Keep containers covered and store rags in an airtight container. Contain spillage; keep work area away from open flame, spark sources including switches and electric motors; observe no smoking precautions.
- Treat any fuel leak or spillage as a serious hazard. Clean up the spill or repair the leak immediately.
- Keep a dry chemical fire extinguisher (type B-C) within arm's reach in the work area and know how to use it.
- Always relieve fuel system pressure before loosening or disconnecting any lines.
- If repairs involving sparks or high heat are anticipated, identify and protect all fuel lines before work begins. When repairing floorpans, rocker panels, dash or firewalls, carefully check the area on the other side of the panel before beginning work. Removing plastic or nylon fuel lines prior to repair is highly recommended.
- When installing new fuel lines, or performing repairs in the surrounding area, do not hammer directly on the body clips holding the fuel lines. The fuel line may become damaged, resulting in a system leak.
- Inspect connectors for burrs, dirt, etc. Clean connectors with a soft, lint-free cloth.
- Cover all fittings to be disconnected with a rag or towel to prevent fuel spray. Some pressure may remain within the system even after the pump pressure is relieved.
- Always wear eye or full-face protection when working on any part of the fuel system.

Self-Diagnostic Systems

SECTION 3

ACURA PROGRAMMED FUEL INJECTION (PGM-FI) SYSTEM

General Information

Programmed Fuel Injection (PGM-FI) System is a fully electronic microprocessor based engine management system. The Electronic Control Unit (ECU) is given responsibility for control of injector timing and duration, intake air control, ignition timing, cold start enrichment, fuel pump control, fuel cut-off, A/C compressor operation, alternator control as well as EGR function and canister purge cycles.

The ECU receives electric signals from many sensors and sources on and around the engine. The signals are processed against pre-programmed values; correct output signals from the ECU are determined by these calculations. The ECU contains additional memories, back-up and fail-safe functions as well as self-diagnostic capabilities.

Diagnosis is generally based on symptom diagnosis and stored fault codes, if any. Testing always requires the use of the diagnostic charts in conjunction with the Acura test harness (pin-out box), a voltmeter and an ohmmeter.

ELECTRONIC CONTROL UNIT (ECU)

In order to get fuel into the cylinders at the correct instant and in correct amount, the control system must perform various separate functions. The ECU (Electronic Control Unit), the heart of the PGM-FI, uses a microcomputer and consists of a CPU (Central Processing Unit), memories, and I/O (Input/Output) ports. Basic data stored in the memories are compared to the signals sent from the various sensors to provide the correct air/fuel mixture for all engine needs.

The unit contains memories for the basic discharge duration at various engine speeds and manifold pressures. The basic discharge duration, after being read out from the memory, is further modified by signals sent from various sensors to obtain the final injector duration. Other functions also include:

Fail Safe

If the ECU notes the loss or change out-of-range of a sensor signal, the computer will ignore the faulty signal and substitute a fixed value in its place. This value may not necessarily be correct for the immediate driving situation, but will allow the engine to operate. For example, many ECU decisions are based on the engine coolant temperature. Should the coolant temperature sensor (TW sensor) signal be lost just after a cold start, the ECU will substitute the default value. This substitution value is based on an engine at normal temperature so vehicle performance will be affected until fully warmed up.

Back-up System

The ECU also tests itself while operating. If an abnormality occurs within the ECU, the system switches to a back-up circuit independent of the computing system. This system substitutes fixed values for all inputs and controls the injectors accordingly. Vehicle performance is usually reduced to minimal driveability. The back up function serves as a lifeboat to prevent the vehicle from being stranded in the event of ECU failure.

Self-Diagnosis

When a abnormality occurs, the ECU lights the engine warning light and stores a failure code in its memory. The fault code can be retrieved during diagnosis.

Self-Diagnostics

SERVICE PRECAUTIONS

- Do not operate the fuel pump when the fuel lines are empty.
- Do not operate the fuel pump when removed from the fuel tank.
- Do not reuse fuel hose clamps.
- The washer(s) below any fuel system bolt (banjo fittings, service bolt, fuel filter, etc.) must be replaced whenever the bolt is loosened. Do not reuse the washers; a high-pressure fuel leak may result.
- Make sure all ECU harness connectors are fastened securely. A poor connection can cause an extremely high voltage surge and result in damage to integrated circuits.
- Keep ECU all parts and harnesses dry during service. Protect the ECU and all solid-state components from rough handling or extremes of temperature.
- Use extreme care when working around the ECU or other components; the airbag or SRS wiring may be in the vicinity. On these vehicles, the SRS wiring and connectors are yellow; do not cut or test these circuits.
- Before attempting to remove any parts, turn the ignition switch **OFF** and disconnect the battery ground cable.
- Always use a 12 volt battery as a power source for the engine, never a booster or high-voltage charging unit.
- Do not disconnect the battery cables with the engine running.
- Do not disconnect any wiring connector with the engine running or the ignition **ON** unless specifically instructed to do so.
- Do not apply battery power directly to injectors.
- Whenever possible, use a flashlight instead of a drop light.
- Keep all open flame and smoking material out of the area.
- Use a shop cloth or similar to catch fuel when opening a fuel system. Consider the fuel-soaked rag to be a flammable solid and dispose of it in the proper manner.
- Relieve fuel system pressure before servicing any fuel system component.
- Always use eye or full-face protection when working around fuel lines, fittings or components.
- Always keep a dry chemical (class B-C) fire extinguisher near the area.

READING CODES

When a fault is noted, the ECU stores an identifying code and illuminates the CHECK ENGINE light. The code will remain in memory until cleared; the dashboard warning lamp may not illuminate during the next ignition cycle if the fault is no longer present. Not all faults noted by the ECU will trigger the dashboard warning lamp although the fault code will be set in memory. For this reason, troubleshooting should be based on the presence of stored codes, not the illumination of the warning lamp while the car is operating.

Stored codes are displayed by a flashing LED on the ECU. When the CHECK ENGINE warning lamp has been on or reported on, lift or remove the carpet from the right front passenger footwell for all vehicles except 1990 Legend. The ECU is below a protective cover; the LED may be viewed through a small window without removing the ECU cover. On 1990 Legend, the ECU is under the front passenger seat and may be viewed from the front.

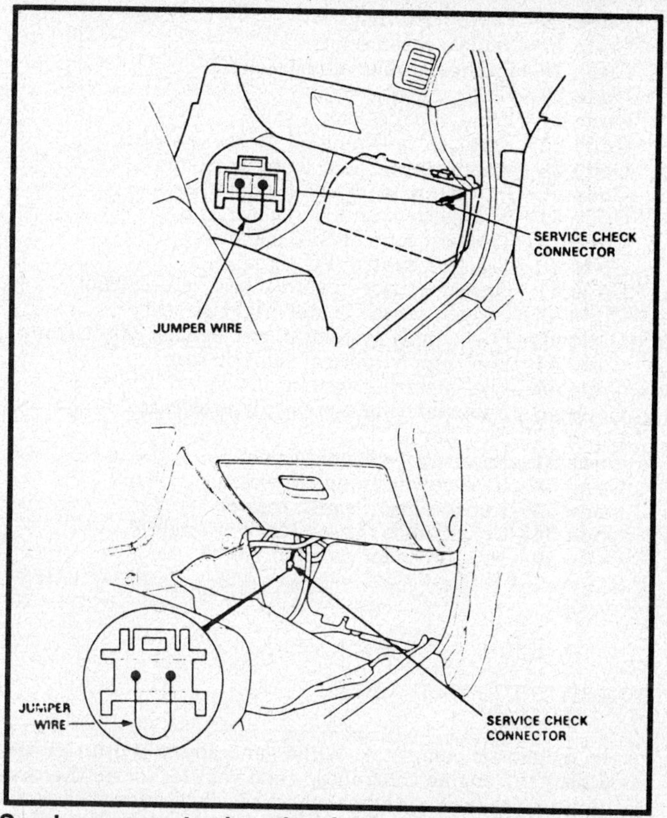

Service connector location for Legend (above) and Vigor. NSX similar to Legend

Turn the ignition switch **ON**; the LED will display any stored codes by rhythmic flashing. Note that 1990 Legends have two LEDs on the controller; one is red and one is amber. The red one will flash the fault codes; the amber one is used during idle adjustment and is not related to this procedure.

Additionally, the Vigor, NSX and 1991–92 Legend models are equipped with a service connector inside the cabin of the vehicle. If the service connector is jumpered, the CHECK ENGINE lamp will display the stored codes in the same fashion. The 2-pin service connector is located under the extreme right dashboard on Legend and NSX; on Vigor models, it is found behind the right side of the center console, well under the dashboard.

Codes 1–9 are indicated by a series of short flashes; two-digit codes use a number of long flashes for the first digit followed by the appropriate number of short flashes. For example, Code 43 would be indicated by 4 long flashes followed by 3 short flashes. Codes are separated by a longer pause between transmissions. The position of the codes during output can be helpful in diagnostic work. Multiple codes transmitted in isolated order indicate unique occurrences; a display of showing 1-1-1-pause-9-9-9 indicates two problems or problems occurring at different times. An alternating display, such as 1-9-1-9-1, indicates simultaneous occurrences of the faults.

When counting flashes to determine codes, a code not valid for the vehicle may be found. In this case, first recount the flashes to confirm an accurate count. If necessary, turn the ignition switch **OFF**, then recycle the system and begin the count again. If the Code is not valid for the vehicle, the ECU must be replaced.

NOTE: On vehicles with automatic transaxles, the S, D or D_4 lamp may flash with the CHECK ENGINE lamp if certain codes are stored. For Legend and NSX, this may occur with Codes 6, 7 or 17. On Vigor and Integra it may occur with codes 6, 7 or 13. In addition, the TCS lamp on NSX may flash with codes 3, 5, 6, 13, 15, 16, 17, 35 or 36. In all cases, proceed with the diagnosis based on the engine code shown. After repairs, recheck the lamp. If the additional warning lamp is still lit, proceed with diagnosis for that system.

CLEARING CODES

Stored codes are removed from memory by removing power to the ECU. Disconnecting the power may also clear the memories used for other solid-state equipment such as the clock and radio. For this reason, always make note of the radio presets before clearing the system. Additionally, some radios contain anti-theft programming; obtain the owner's code number before clearing the codes.

While disconnecting the battery will clear the memory, this is not the recommended procedure. The memory should be cleared after the ignition is switched **OFF** by removing the appropriate fuse for at least 10 seconds. The correct fuses and their locations are:

- 1990 Legend: ALTERNATOR SENSE, in the underhood fuse and relay panel.
- 1991–92 Legend: ACG, in the dashboard fuse panel. Removing this fuse also cancels the memory for the power seats.
- Integra, US vehicles: BACK UP, in the underhood fuse and relay panel.
- Integra, Canadian vehicles: HAZARD/BACK UP in the underhood fuse and relay panel.
- Vigor: BACK UP, located in the underhood fuse and relay panel. Removing this fuse will cancel memories for the clock and radio.

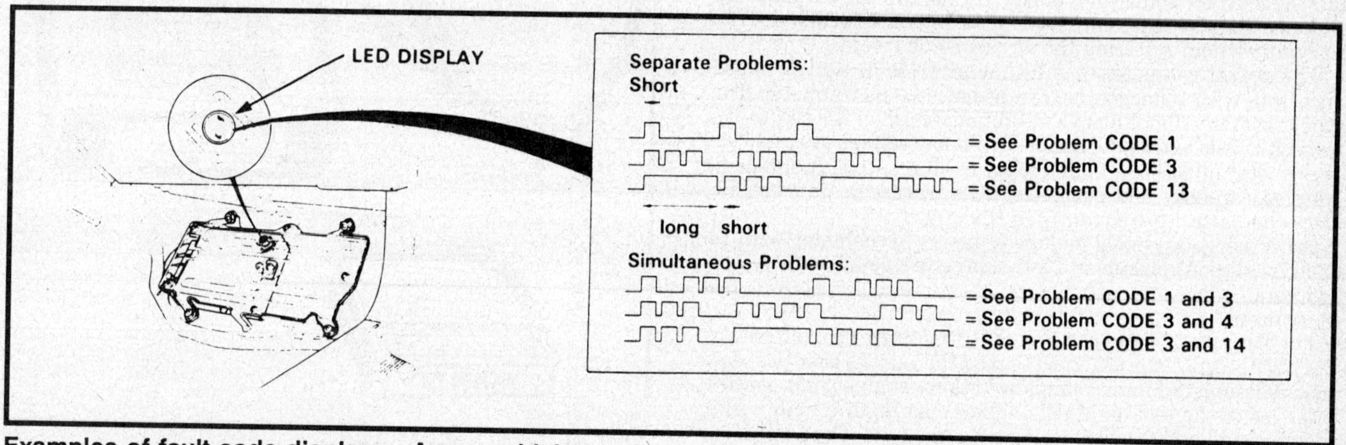

Examples of fault code displays—Acura vehicles

• NSX: CLOCK, located in the main fuse and relay panel in the front luggage compartment, right side. Removing this fuse will cancel memories for the clock and radio.

ENGINE CODES

Code 0–(Not Blinking, CHECK ENGINE lamp ON) ECU failure.
Code 1–Oxygen content or sensor. If two sensors, front or left.
Code 2–Oxygen content or sensor, rear or right
Code 3–MAP sensor, electrical signal
Code 4–Crank angle sensor. If two, No. 1 or angle A
Code 5–MAP sensor, mechanical or vacuum signal
Code 6–Coolant temperature
Code 7–Throttle angle
Code 8–TDC signal
Code 9–No. 1 cylinder signal
Code 10–Intake air temperature
Code 12–Exhaust gas recirculation
Code 13–Atmospheric pressure signal
Code 14–Electronic air control (idle control)
Code 15–Ignition output signal
Code 16–Fuel injector signal
Code 17–Vehicle speed sensor
Code 18–Ignition timing adjustment
Code 20–Electric load detector
Code 21–Front spool solenoid valve
Code 22–Front valve timing oil pressure switch
Code 23–Knock sensor; left, front or No. 1
Code 30–Automatic transaxle FI signal A
Code 31–Automatic transaxle FI signal B
Code 35–Traction control STB signal
Code 36–Traction control FC signal
Code 41–Oxygen sensor heater; if two, left or front
Code 42–Oxygen sensor heater, right or rear
Code 43–Fuel supply system; if two signals, left or front
Code 44–Fuel supply system; right or rear
Code 45–Fuel metering system; if two, left or front
Code 46–Fuel metering system; right or rear
Code 47–Fuel pump
Code 51–Rear spool solenoid valve
Code 52–Rear valve timing oil pressure switch
Code 53–Knock sensor; right, rear or No. 2
Code 54–Crank angle sensor No. 2 or angle B
Code 59–No. 1 cylinder position 2 or B

CHRYSLER IMPORTS

General Information

The engines used in these vehicles are manufactured by Mitsubishi Motor Company (MMC). The engines include single overhead cam, double overhead cam and double overhead cam turbocharged types. The vehicles which use this system are Laser, Talon, Stealth, Colt, Summit, Vista and Ram 50 pick-up truck.

All of these engines are controlled by electronic engine control units (ECUs) which oversee the multi-point fuel injection system operation. These control units may also provide control for other systems on the vehicle such as speed control. While the individual control units vary by model and are not interchangeable among engines, they share the same self-diagnostic abilities and are interrogated in the same manner. Fault codes will vary by engine or engine/chassis combination.

Self-Diagnostic System

The use of an analog (needle type) voltmeter is required if codes are to be read without a scan tool. Although the analog meter may be used for some component testing, the use of a high-impedance digital volt-ohmmeter is recommended for its accurate readout without affecting the circuit being tested.

The digital voltmeter or multimeter must be a high impedance unit, with 10 megohms of impedance in the voltmeter. This type of meter will not place an additional load on the circuit it is testing; this is extremely important in low voltage circuits. The multimeter must be of high quality in all respects. It should be handled carefully and protected from impact or damage. Replace the batteries frequently in the unit.

Other necessary tools include a quality tachometer with inductive (clip-on) pickup and a fuel pressure gauge with system adapters. A vacuum gauge may be required for certain test procedures.

The Diagnostic Read-out Box (DRB-II) Scan Tool, offered by Chrysler is especially designed to read the various electrical data from Mitsubishi Motor Company (MMC) engines and controllers. Use of the tool on MMC engines may require both wiring adaptors and additional cartridges for the unit. The DRB-II, or any scan tool to be used on Chrysler vehicles, must be connected to the diagnostic connector. While generally found under the hood near the engine controller, some vehicles locate the connector under the left or right dashboard. This connector ties directly to the computer system(s) and allows the exchange of signals between the vehicle and the scanner.

Diagnosis of a driveability problem requires attention to detail and following the diagnostic procedures in the correct order. Resist the temptation to begin extensive testing before completing the preliminary diagnostic steps. The preliminary or visual

Chrysler Corporation DRB-II tester

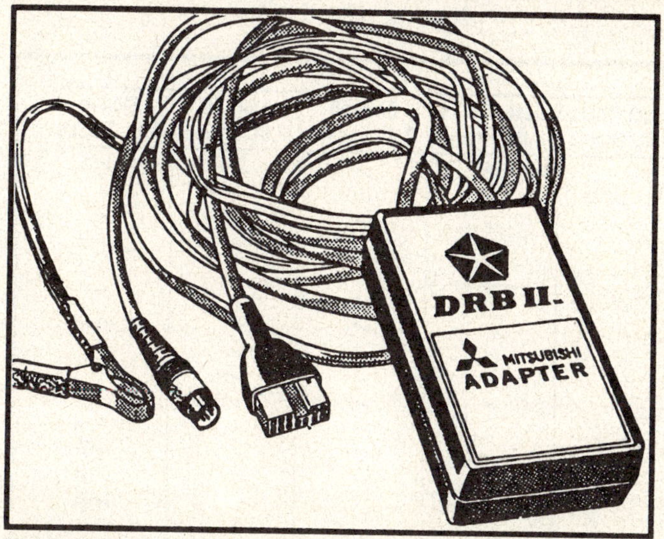

Chrysler Corporation DRB-II adapter for Mitsubishi engines

inspection must be completed in detail before diagnosis begins. In many cases this will shorten diagnostic time and often cure the problem without electronic testing.

Dashboard Warning Lamp

The CHECK ENGINE warning lamp is controlled by an output from the ECU. When an emissions related fault is detected, the warning lamp is turned on to advise the operator. Should the fault self-correct and re-establish the proper signal to the control unit, the dash warning lamp will turn off; a code may remain in the memory even though the lamp is off. The dash lamp will not light again until another fault is detected.

The CHECK ENGINE lamp should illuminate for approximately 5 seconds each time the engine is started. After this period, the lamp should extinguish and remain out during vehicle operation.

NOTE: The dash warning lamp will illuminate when the ignition timing adjustment terminal is grounded. This is a normal controller function and should not be taken to mean a fault has been detected while adjusting the engine timing.

READING CODES

All stored codes are read through connections at the diagnostic connector. The DRB-II or equivalent may be used to read stored codes; the Mitsubishi Adapter must be used as an interface between the scan tool and the vehicle. The DRB-II may be used to read codes and observe the signal values being sent to and from the ECU. Fault codes stored within the system may be cleared through the DRB-II.

In the absence of a scan tool and/or adapter, codes may be read directly from the diagnostic connector. All stored fault codes are read through an analog (needle type) voltmeter after the system enters self-diagnostics. The codes aretransmitted as pulses which cause the needle to sweep repeatedly. To enter the self-diagnostic mode:

1. The self-diagnostic connector is located either next to or below the fuse panel, under the left dash. The only exception is the Colt Wagon, which locates the connector under the left side of the glove box, behind the center console. Remove the cover if one is fitted.
2. With the ignition switch **OFF**, connect the ground probe of

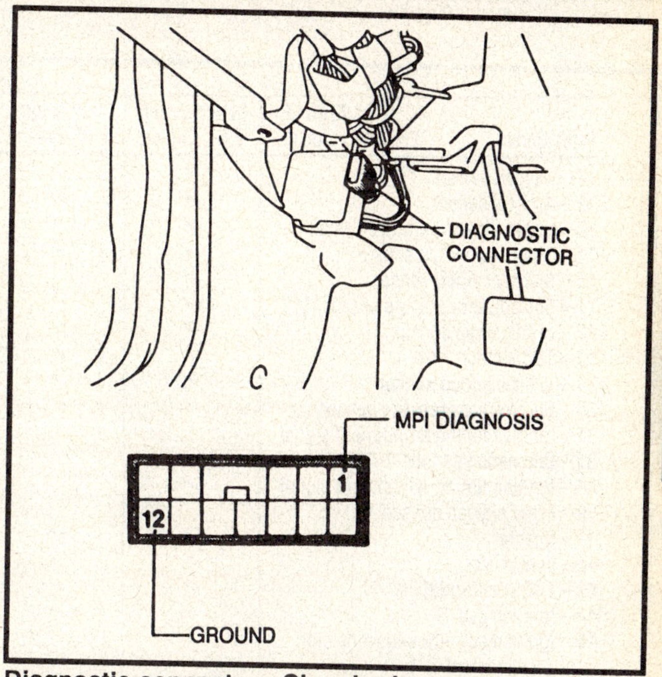

Diagnostic connector—Chrysler Imports

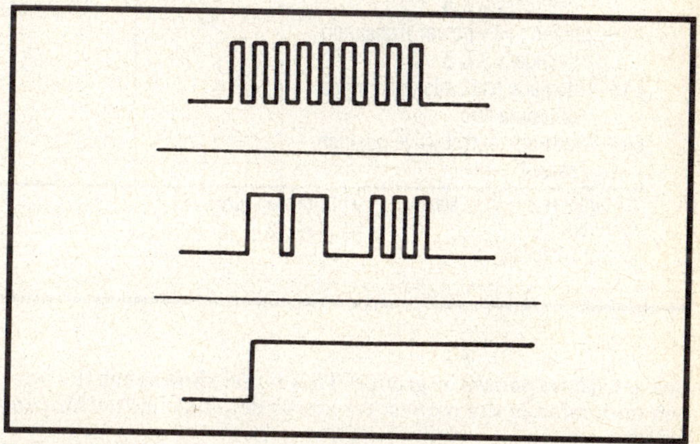

Example of fault patterns. System normal (top), Code 23 (center), and ECU failure (bottom)

the voltmeter to the ground terminal (terminal No. 12) of the diagnostic connector.

3. Connect the voltage probe of the voltmeter to the MPI diagnostic terminal (terminal No. 1) of the connector.
4. Turn the ignition switch **ON** but do not start the engine. The codes will be transmitted immediately. Each two-digit code will be sent as groups of electrical pulses; these pulses will cause needle sweeps on the voltmeter. The codes are sent in place value groups. For example, Code 23 is sent as a group of 2 long pulses followed by a group of 3 shorter pulses. There is a short pause between place value groups and a longer pause between code groups.

NOTE: If multiple codes are stored in the memory they will be transmitted in numerical order from lowest to highest, not in the order of occurrence.

5. If the ECU detects an internal error, the needle on the voltmeter will show constant voltage instead of sweeping on and off. When this signal is received, the ECU must be replaced.
6. If no faults are stored within the memory, the Normal State signal will be displayed. This is a constant, rapid needle

ENGINE FAULT CODES—CHRYLSER IMPORTS

Fault Codes / VEHICLE ENGINE (Liters)	Colt/Summit 1.5L Engine	Colt/Summit 1.6L Engine	Laser/Talon 1.8L Engine	Laser/Talon 2.0L Engine	RAM 50 2.4L Engine	RAM 50 3.0L Engine	Stealth 3.0L Engine
11—Oxygen sensor	✓	✓	✓	✓	✓	✓	✓①
12—Air flow sensor	✓	✓	✓	✓	✓	✓	✓
13—Intake air temperature sensor	✓	✓	✓	✓	✓	✓	✓
14—Throttle position sensor	✓	✓	✓	✓	✓	✓	✓
15—Motor position sensor	✓		✓		✓		
21—Coolant temperature sensor	✓	✓	✓	✓	✓	✓	✓
22—Crank angle sensor	✓	✓	✓	✓	✓	✓	✓
23—TDC sensor	✓	✓	✓	✓	✓	✓	✓
24—Vehicle speed sensor	✓	✓	✓	✓	✓	✓	✓
25—Barometric pressure sensor	✓	✓	✓	✓	✓	✓	✓
31—Ignition timing adjustment signal			✓⑤				✓③
31—Detonation sensor				✓②			✓③
36—Ignition timing adjustment signal	✓⑤				✓		✓②
39—Front oxygen sensor							✓
41—Injector	✓	✓	✓	✓	✓	✓	
42—Fuel pump	✓	✓	✓	✓	✓	✓	
43—EGR (California)	✓	✓	✓	✓	✓	✓	✓
44—Ignition coil		✓		✓			
44—Igniticn coil—power transistor, cylinders 1 & 4							✓③
52—Ignition coil—power transistor, cylinders 2 & 5							✓③
53—Ignition coil—power transistor, cylinders 3 & 6							✓③
61—Automatic transmission control unit data line							✓
62—Warm-up control valve position sensor							✓③

① Turbo: rear oxygen sensor
② Turbo
③ DOHC engines
④ DOHC non-turbo
⑤ 1991–92

sweep with no pauses or groups. This signal shows that the system understands the request for stored data but has nothing to report.

7. Record the codes transmitted for use during repairs. Once the codes are recorded, switch the ignition **OFF**. Based on the codes, perform diagnostic and test procedures on the individual components and circuits.

NOTE: If, during testing or repair, a sensor is disconnected with the ignition ON, the fault code for that sensor or circuit will be set. For this reason, having the original codes written down will eliminate confusion about additional codes appearing during testing or repair.

CLEARING CODES

After all diagnosis has been completed and the needed repairs performed, the stored fault codes must be erased from the ECU memory.

Codes may be cleared using the DRB-II. When properly connected through the Mitubishi Adapter, select ADJUSTMENTS from the Main Menu screen. Choose item 1, ERASE FAULTS and follow the directions given on the screen.

Alternately, with the ignition switch **OFF**, disconnect the negative battery terminal for at least 10 seconds. This will also clear the stored codes. When the battery is reconnected, double check the repair by road testing over a moderate distance, then checking for newly-set codes.

DAIHATSU ELECTRONIC FUEL INJECTION (EFI)

General Information

Daihatsu electronic fuel injection system consists of 3 sub-systems: air intake, electronic control and fuel supply.

ENGINE CONTROL SYSTEM

The Electronic Control Unit (ECU) controls the amount of fuel and mixture based on engine conditions and running conditions,

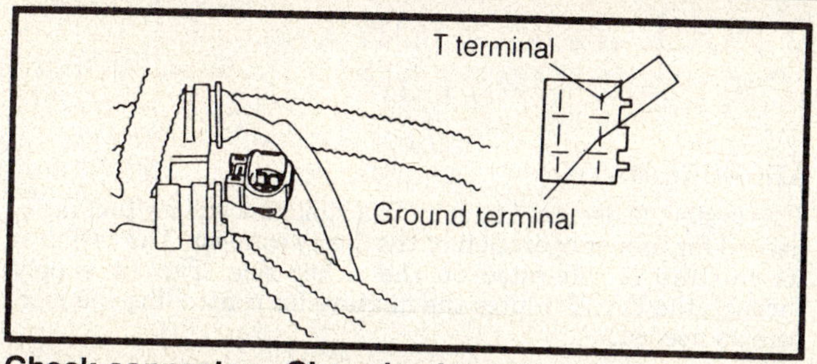

Check connector—Charade shown

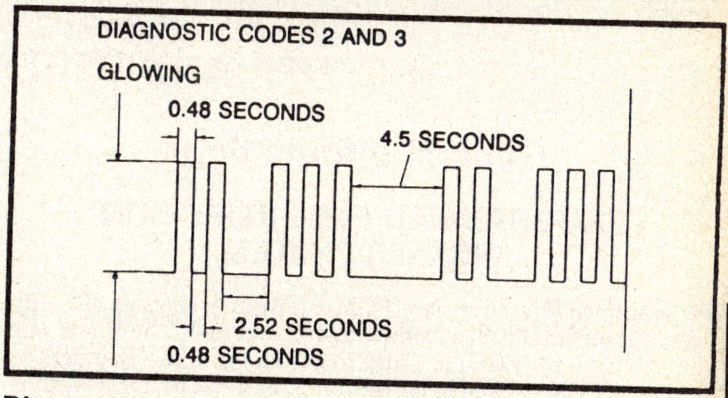

Diagnostic code patternr—Charade and Rocky

as determined by signals input by various sensors. The ECU also controls electronic ignition timing, fuel pump operation, idle-up control, a self-diagnosis function and the fail safe and back-up functions.

Self-Diagnosis

If any malfunction should occur in an input signal circuit, the ECU memorizes the fault. This malfunction is indicated during the trouble diagnosis, by codes through the checking lamps. If a major malfunction is encountered, the ECU turns ON the CHECK ENGINE lamp to warn the driver.

FAIL-SAFE FUNCTION

When a malfunction is detected, the ECU determines, based on programmed standard values, whether the engine may be allowed to continue operation or should be stopped immediately.

BACK-UP FUNCTION

In the event the ECU detects a malfunction, the back-up function makes it possible to provide fuel injection amount and ignition timing that have been predetermined by the back-up data. This is to ensure safe driving if a single or many sensors are faulty, or if the ECU malfunctions.

Self-Diagnostic System

SERVICE PRECAUTIONS

- Make sure all ECU harness connectors are fastened securely. A poor connection can cause an extremely high surge voltage and could and result in damage to integrated circuits.
- Keep ECU all parts and harnesses dry during service.
- Before attempting to remove any parts, turn **OFF** the ignition switch and disconnect the battery ground cable.
- Do not attempt to disconnect the battery cables with the engine running.
- Do not depress the accelerator pedal when starting.
- Never open the cover to the ECU.

ENTERING SELF-DIAGNOSTICS

1. Make certain battery voltage is above 11 volts. The ECU may not store codes if the battery is low.
2. Make certain the throttle is closed.
3. Turn all accessories **OFF**.
4. Locate the check terminal underhood. On the Charade it is located on the upper section of the transmission. On the Rocky it is located on the right side fender panel. Short the Brown T terminal and the Black ground terminal of the test connector.

NOTE: The Check Connector is located at the upper section of the transmission. The use of a special test harness SST No. 09991-87702-000 or equivalent, connected to the Check Connector may be necessary.

5. Turn the ignition switch to **ON**. Do not start the engine.
6. Read the diagnosis codes by counting the CHECK ENGINE lamp flashes.

CLEARING CODES

To clear the diagnosis code from the ECU, disconnect the negative terminal from the battery or disconnect the BACK-UP fuse of the relay block assembly in the engine compartment with the ignition switch **OFF** for at least 30 seconds.

NOTE: Some vehicle computers have a learning ability. If a change is noted in vehicle performance after clearing codes, it may be due to the computers learning ability. To restore performance, warm vehicle to normal operating temperature and drive at part throttle with moderate acceleration, before performing any additional diagnosis.

ENGINE CODES

Code 1—System Normal
Code 2—Pressure Sensor (MAP)
Code 3—Ignition Signal
Code 4—Water Temperature Sensor
Code 5—Oxygen Sensor Open or Short
Code 6—Revolution Sensor
Code 7—Throttle Position Sensor
Code 8—Air Intake Temperature Sensor
Code 9—Vehicle Speed Sensor
Code 10—Starter Signal
Code 11—Switch Signal
Code 12—EGR Control System
Code 15—Oxygen Sensor
Code 16—Oxygen Sensor Lean

NOTE: Code 15 and Code 16 are not available on the 1.0L engine.

HONDA PROGRAMMED FUEL INJECTION (PGM-FI) AND FEEDBACK CARBURETOR

General Information

PROGRAMMED FUEL INJECTION (PGM-FI) SYSTEM

Programmed Fuel Injection (PGM-FI) System consists of 3 subsystems: air intake, electronic control, and fuel. The Electronic Control System (ECU) in order to get fuel into the cylinders at the correct instant and in correct amount must perform various separate functions. The ECU (Electronic Control Unit), the heart of the PGM-FI, uses a microcomputer and consists of a CPU (Central Processing Unit), memories, and I/O (Input/Output) ports. Basic data stored in the memories are compensated by the signals sent from the various sensors to provide the correct air/fuel mixture for all engine needs.

The unit contains memories for the basic discharge duration at various engine speeds and manifold pressures. The basic discharge duration, after being read out from the memory, is further modified by signals sent from various sensors to obtain the final discharge duration. Other functions also include:

- Starting Control—The fuel system must vary the air/fuel ratio to suit different operating requirements. For example, the mixture must be rich for starting. The memories also contain the basic discharge durations to be read out by signals from the starter switch, and engine speed and coolant temperature sensors, thereby providing extra fuel needed for starting.
- Injector Control—The ECU controls the discharge durations at various engine speeds and loads.
- Electronic Air Control—The ECU controls the EACV to maintain correct idle speed based on engine and accessories demand.
- Ignition Timing Control—The ECU controls the basic ignition timing based on engine load, engine rpm, vehicle speed and coolant temperature.
- Fuel Pump Control—When the speed of the engine falls below the prescribed limit, electric current to the fuel pump is cut off, preventing the injectors from discharging fuel.
- Fuel Cut-Off Control—During deceleration with the throttle valve nearly closed, electric current to the injectors is cut off at speeds over 900 rpm, contributing to improved fuel economy. Fuel cut-off action also takes place when engine speed exceeds approximately 7500 rpm regardless of the position of the throttle valve.
- Safety—A fail-safe system monitors the sensors and detects any abnormality in the ECU, ensuring safe driving even if one or more sensors are faulty, or if the ECU malfunctions.
- Self-Diagnosis—When a abnormality occurs, the ECU lights the engine warning light and stores the failure code in erasable memory. The ECU LED will display the code any time the ignition is turned **ON**.

PRELUDE FEEDBACK CARBURETOR

The carburetor consists of 2 side draft carburetors, each having a variable venturi. The variable venturi carburetors allow a smooth increase of engine speed and output, due to the change in venturi area in proportion to the carburetor intake airflow rate.

Feedback control system

The feedback control system maintains the proper air/fuel mixture ratio by allowing air into the intake manifold, as is necessary, to adjust a temporarily fuel-rich condition. This system in made up of 4 subsystems: the air/fuel ratio control, shot of air control, deceleration air supply and hot engine start control.

AIR/FUEL RATIO

This system is designed to achieve a stoichimetric air/fuel ratio, needed for proper operation of the 3-way catalyst. The carburetor mixture is calibrated on the richer side. The air supply through the EACV dilutes the mixture for controlling the mixture as needed.

SHOT OF AIR CONTROL

This system provides air into the intake manifold to improve emissions performance and prevent afterburning due to over rich mixtures during deceleration.

The control unit receives signals from speed, coolant temperature, MAP and rpm sensors. The air shot is induced from the EACV when the manifold vacuum increases suddenly except when the vehicle is moving at very low speed with the coolant temperature below normal. The amount of air provides is based on the amount of manifold vacuum.

DECELERATION AIR SUPPLY

This system is designed to improve emission performance by supplying air into the intake manifold during deceleration in relatively high engine rpm.

The control unit receives signals from speed, coolant temperature, MAP, gear position and rpm sensors. The air is induced from the EACV, based on signals from the ECU.

HOT ENGINE START CONTROL

This system is designed to provide air into the intake manifold for engine starting when the coolant temperature is very high. The EACV provides the additional air flow based on signals from the ECU.

Self-Diagnostic System

SERVICE PRECAUTIONS

- Make sure all ECU harness connectors are fastened securely. A poor connection can cause an extremely high voltage surge and result in damage to integrated circuits.

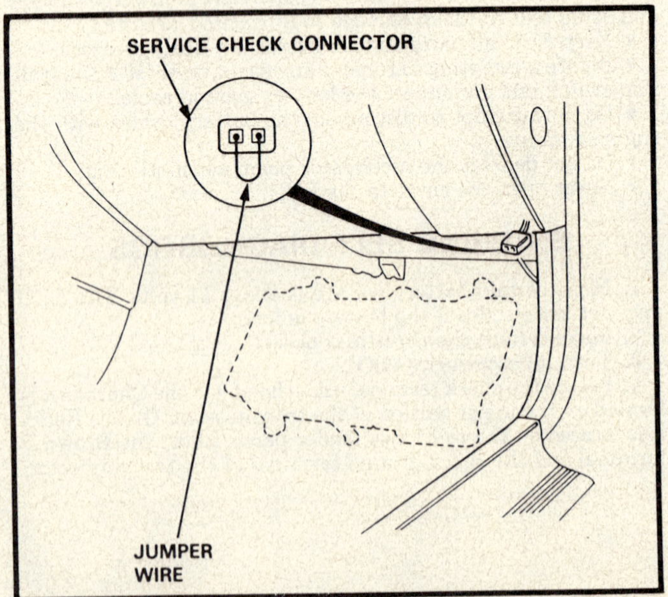

Service connector—Accord shown

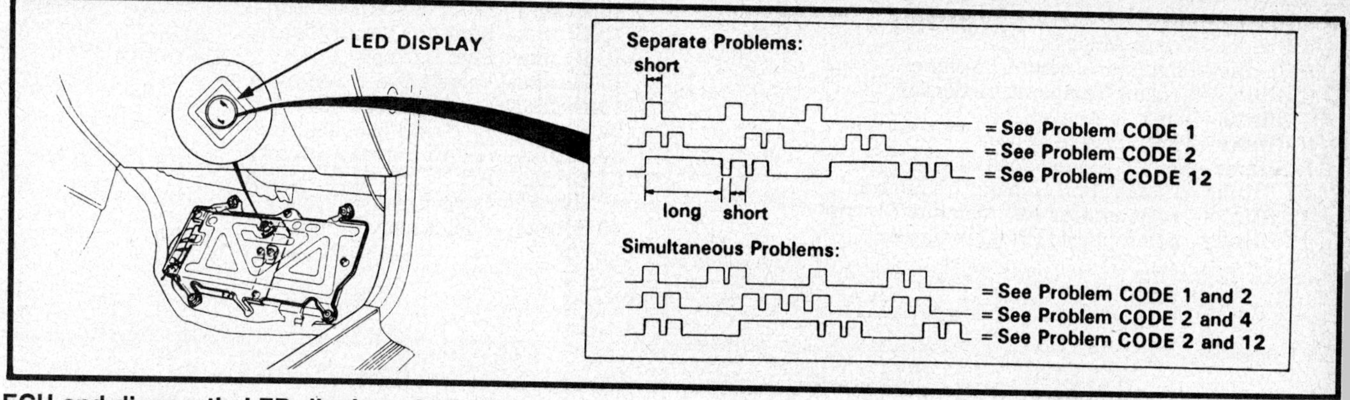

ECU and diagnostic LED display—Accord shown

- Keep ECU all parts and harnesses dry during service. Protect the ECU and all solid-state components from rough handling or extremes of temperature.
- Before attempting to remove any parts, turn the ignition switch **OFF** and disconnect the battery ground cable.
- Always use a 12 volt battery as a power source, never a booster or high-voltage charging unit.
- Do not disconnect the battery cables with the engine running.
- Do not disconnect any wiring connector with the engine running or the ignition **ON** unless specifically instructed to do so.
- Do not depress the accelerator pedal when starting.
- Do not rev up the engine immediately after starting or just prior to shutdown.
- Do not apply battery power directly to injectors.
- Whenever possible, use a flashlight instead of a drop light.
- Keep all open flame and smoking material out of the area.
- Use a shop cloth or similar to catch fuel when opening a fuel system.
- Relieve fuel system pressure before servicing.

READING CODES

When a fault is noted, the ECU stores an identifying code and illuminates the CHECK ENGINE light. The code will remain in memory until cleared; the dashboard warning lamp may not illuminate during the next ignition cycle if the fault is no longer present. Faults for either the Automatic transaxle F/I signal or the Electric Load Detector on Civic and Accord will not illuminate the warning lamp but will set a fault code in the memory.

Stored codes are displayed by a flashing LED on the ECU. When the CHECK ENGINE warning lamp has been on or reported on, lift or remove the carpet from the right front passenger footwell. The ECU is below a protective cover; the LED may be viewed through a small window without removing the ECU cover. Turn the ignition switch **ON**; the LED will display any stored codes by rhythmic flashing. Additionally, if the service connector is jumpered on Accord or 2.1L engine Preludes, the CHECK ENGINE lamp will display the stored codes in the same fashion. The 2-pin service connector is located under the extreme right dashboard on Accord or on the Prelude firewall, near the control box.

Codes 1–9 are indicated by a series of short flashes; two-digit codes use a number of long flashes for the first digit followed by the appropriate number of short flashes. For example, Code 43 would be indicated by 4 long flashes followed by 3 short flashes. Codes are separated by a pause between transmissions. The position of the codes during output can be helpful in diagnostic work. Multiple codes transmitted in isolated order indicate unique occurrences; a display of showing 1-1-1-pause-9-9-9 indicates two problems or problems occurring at different times. An alternating display, such as 1-9-1-9-1, indicates simultaneous occurrences of the faults.

When counting flashes to determine codes, a code not valid for the vehicle may be found. In this case, first recount the flashes to confirm an accurate count. If necessary, turn the ignition switch **OFF**, then recycle the system and begin the count again. If the Code is not valid for the vehicle, the ECU must be replaced.

NOTE: The S or S_4 light may illuminate with the CHECK ENGINE lamp on automatic Accord and Prelude if Code 6, 7 or 17 is stored. In this event, follow the troubleshooting procedures for the fuel injection system first. Clear the memory after repairs, then recheck the S or S_4 lamp.

CLEARING CODES

Stored codes are removed from memory by removing power to the ECU. Disconnecting the power will also clear the memories used for other solid-state equipment such as the clock and radio. For this reason, always make note of the radio presets before clearing the system.

While disconnecting the battery will clear the memory, this is not the recommended procedure. The memory should be cleared by removing the appropriate fuse for at least 10 seconds. In all cases, the correct fuse is located in the underhood fuse and relay box.

To clear the ECU trouble code memory remove the ECU memory power fuse for at least 10 seconds. The correct fuse to remove is:

Civic—HAZARD fuse at the main fuse box
Prelude with PRG-CARB—EFI-ECU fuse from the underhood relay box
Prelude with PRG-FI—CLOCK fuse from the underhood relay box
Accord—BACK UP fuse from the underhood relay box

NOTE: Some vehicle computers have a learning ability. If a change is noted in vehicle performance after clearing codes, it may be due to the computers learning ability. To restore performance, warm vehicle to normal operating temperature and drive at part throttle with moderate acceleration, before performing any additional diagnosis.

ENGINE CODES

Obtained by counting the self-diagnosis indicator blinks:
0 Blinks—ECU
1 Blink—Oxygen Content
2 Blinks—Oxygen Content (Prelude 2.0L engine)

3 Blinks—Manifold Absolute Pressure
4 Blinks—Crank Angle Sensor
5 Blinks—Manifold Absolute Pressure
6 Blinks—Coolant Temperature Sensor
7 Blinks—Throttle Angle Sensor
8 Blinks—TDC Position Sensor
9 Blinks—No. 1 Cylinder Position
10 Blinks—Intake Air Temperature Sensor
12 Blinks—Exhaust Gas Recirculation System
13 Blinks—Atmospheric Pressure Sensor
14 Blinks—Electronic Air Control
15 Blinks—Ignition Output Signal
16 Blinks—Fuel Injector
17 Blinks—Vehicle Speed Sensor
19 Blinks—Lock-up Valve (Civic)
20 Blinks—Electrical load detector
30 Blinks—A/T FI Signal A (Accord)
31 Blinks—A/T FI Signal B (Accord)
41 Blinks—Oxygen Sensor Heater
43 Blinks—Fuel Supply System

HYUNDAI MULTI-POINT FUEL INJECTION (MPI)

General Information

The fuel injection system used on Hyundai vehicles, is classified as a Multi-point Fuel Injection (MPI) system. The basic function of the system is to control the air/fuel ratio, based on input signals from various engine sensors. The air-fuel ratio is controlled by varying the injector driving time. The system is controlled by an Electronic Control Unit (ECU), which monitors the engine conditions, then calculates the injection timing and air/fuel ratio according to the signals from the sensors. The ECU consists of an 8-bit microprocessor, Random Access Memory (RAM), Read Only Memory (ROM) and input and output signal interface system.

The MPI system consists of 2 operating modes:

- Open Loop—air/fuel ratio is controlled by information programmed into the ECU by the manufacturer.
- Closed Loop—air/fuel ratio is varied by the ECU based on information supplied by the oxygen sensor.

An electric fuel pump supplies sufficient fuel to the injection system and the pressure regulator maintains a constant pressure to the injectors. These injectors inject a metered quantity of fuel into the intake manifold in accordance with signals from the Electronic Control Unit (ECU) or engine computer. After pressure regulation excess fuel is returned to the fuel tank.

The injectors have 2 modes (Injector Drive Timing) of operation:

- Non-synchronous Injection (Simultaneous Injection)
- Synchronous Injection (Sequential Injection)

Non-synchronous injection is activated during engine starting (cranking). There are 2 fuel injections, for each engine rpm, to all 4 cylinders. Also, during acceleration, fuel proportionate to the magnitude of acceleration, is injected to 2 selected cylinders during the intake and exhaust strokes.

Synchronous Injection is activated after the engine have started. The injectors are activated at the exhaust stroke of each cylinder in a sequential manner, according to crankshaft angle sensor signal. There is 1 injection per cylinder for every 2 crankshaft revolution, according to firing order.

Self-Diagnostic System

The Engine Control Unit (ECU) is capable of monitoring both the input and output signals which it performed. When the ECU noticed that an irregularity has continued for a specified time or longer, it memorizes a trouble code and outputs the signal to the self-diagnosis output terminal. There are 14 diagnosis items, including the normal state. The trouble codes can be read out with a voltmeter or multi-tester.

Amoung the self-diagnosis items, a Malfunction Indicator Light (MIL) comes ON to notify the driver of the emission control items when an irregularity is detected.

Generally, the MIL will come ON for 5 seconds when the ignition switch is turned **ON**.

NOTE: Always check that the MIL illuminates for approximately 5 seconds, when the ignition switch is turned ON. If the light does not illuminate, check for an open circuit harness, blown fuse or blown bulb.

Trouble code	Diagnosis item	Trouble code	Diagnosis item
11	Oxygen sensor	23	No.1 cylinder top dead center sensor
12	Air-flow sensor	24	Vehicle-speed reed switch
13	Air temperature sensor	25	Atmospheric pressure sensor
14	Throttle position sensor	41	Injector
15	Motor position sensor	42	Fuel pump
21	Engine coolant temperature sensor	43	EGR temperature sensor (For California Vehicles)
22	Crank angle sensor		

Diagnosis Trouble Codes—Hyundai

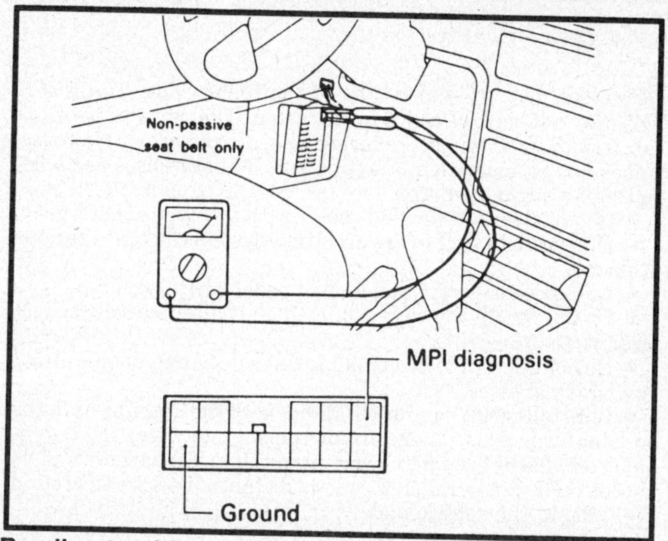

Reading trouble codes using voltmeter—Hyundai

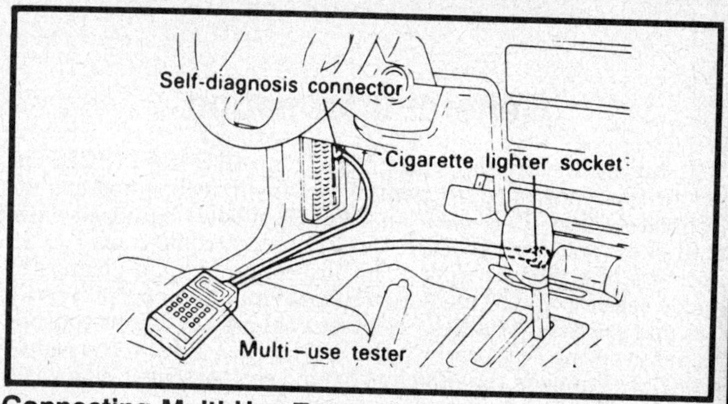

Connecting Multi-Use Tester (MUT) to vehicle—Hyundai

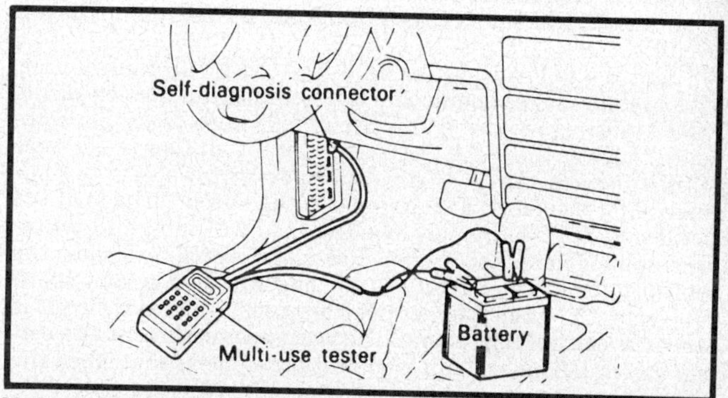

Using an auxiliary power source, if performing test during cranking—Hyundai

SERVICE PRECAUTIONS

- To avoid damage to the ECU, always turn the ignition switch **OFF** before disconnecting the battery cable.
- The control harnesses between the ECU and oxygen sensor are shielded wires with the shield grounded to the body in order to prevent the influence of ignition noises and radio interference. When the shielded wire is faulty the control harness must be replaced.
- Before removing or installing any part(s), always check the self-diagnosis system.

ENTERING SELF-DIAGNOSTICS

Using Voltmeter

1. Connect the voltmeter to the self-diagnostic connector, across the MPI diagnosis and ground terminals.
2. Turn the ignition switch **ON**. The ECU diagnostics memory will immediately start.
3. If the voltmeter displays a steady needle sweep, the system is normal and no codes are in the memory.
4. If the voltmeter displays a steady **HIGH**, signal the ECU is damaged.
5. If the ECU has detected a malfunction, the voltmeter will deflect the indicating the diagnostic code.
6. Record the codes displayed by the voltmeter needle deflections.
7. The ECU will continue to send any memorized trouble codes to the self-diagnostic for as long as the ignition is **ON**, or until the codes have been cleared from memory.

Using Multi-Unit Tester (MUT)

1. Turn the ignition switch **OFF**.
2. Connect the Multi-Unit Tester (MUT) 09391−33200 or equivalent, to the diagnosis connector in the fuse box.
3. Connect the power-source terminal of the MUT to the cigarette lighter switch.
4. Turn the ignition switch **ON** and the ECU memory contents will immediately start.
5. Read the self-diagnosis output and check output of correct code.
6. After recording the fault codes, check and repair each part according to the check items in the Diagnosis Charts.

NOTE: If the battery voltage is too low, trouble codes cannot be read. If the battery or the ECU connector is disconnected, the diagnosis memory will be erased.

When the ignition key is in the **START** position, the cigarette lighter power is OFF. If it is necessary to perform tests, using the MUT in the crank mode, an auxiliary power supply will be necessary. Use the battery clamp harness provided.

CLEARING CODES

After completion of the repair or correction of the problem:

1. Turn the ignition switch **OFF**.
2. Disconnect the negative battery cable for approximately 15 seconds.
3. Reconnect the negative battery cable.
4. Check that no trouble codes are displayed by the Multi-Use Tester (MUT).
5. Remove the Multi-Use Tester (MUT).

INFINITI FUEL INJECTION

General Information

The Infiniti Electronic Concentrated Control System (ECCS) is an air flow controlled, sequential port fuel injection and engine control system. It is used on 1990–92 models equipped with 2.0L,3.0L and 4.5L engines.The ECCS electronic control unit consists of a microcomputer, an inspection lamp, a diagnostic mode selector and connectors for signal input and output, powers and grounds. The ECU is located behind the center console panel near the accelerator pedal on the G20, behind the right-side dash panel on the M30 and behind the passenger-side dash in front of the right-side door on the Q45.

The safety relay prevents electrical damage to the ECU and injectors in case the battery terminals are accidentally connected in reverse. The safety relay is built into the fuel pump control circuit.

There is a fail-safe system built into the ECCS control unit. This system makes engine starting possible if a portion of the ECU's central processing unit circuit fails. Also, if a major component such as the crank angle sensor or the air flow meter were to malfunction, the ECU substitutes or borrows data to compensate for the fault. For example, if the output voltage of the air flow meter is extremely low, the ECU will substitute a preprogrammed value for the air flow meter signal and allow the vehicle to be driven as long as the engine speed is kept below 2000 rpm. Or, if the cylinder head temperature sensor circuit is open, the control unit clamps the warmup enrichment at a certain amount. This amount is almost the same as that when the cylinder head temperature is between 68–176°F (20–80°C).

If the fuel pump circuit malfunctions, the fuel pump relay comes on until the engine stops. This allows the fuel pump to receive power from the relay.

Self-Diagnostic System

SERVICE PRECAUTIONS

- Fuel system pressure must be relieved before disconnecting any fuel lines or attempting to remove any fuel system components.
- Do not operate the fuel pump when the fuel lines are empty.
- Do not reuse fuel hose clamps and tighten the new ones to the specified torque.
- Do not disconnect the injector harness connectors with the engine running.
- Do not apply battery power directly to the injectors.
- Always check for leaks after servicing the fuel system.
- Do not disconnect the ECCS harness connectors before the battery ground cable has been disconnected.
- Make sure all ECCS connectors are fastened securely. A poor connection can cause an extremely high surge voltage in the coil and condenser and result in damage to integrated circuits.
- When testing the ECU with a DVOM make sure that the probes of the tester never touch each other as this will result in damage to a transistor in the ECU.
- Keep the ECCS harness at least 4 in. away from adjacent harnesses to prevent an ECCS system malfunction due to external electronic "noise."
- Keep all parts and harnesses dry during service.
- Before attempting to remove any parts, turn **OFF** the ignition switch and disconnect the battery ground cable.
- Always use a 12 volt battery as a power source.
- Do not attempt to disconnect the battery cables with the engine running or the ignition key **ON**.
- If a battery terminal is disconnected (with the ignition key **OFF**), the memory will return to the ROM value. The ECCS will now self-control at its initial value. The engine operation can vary slightly when the terminal is disconnected. However, this is not an indication of a problem. Do not replace parts because of a slight variation.
- Do not clean the air flow meter with any type of detergent.
- Even a slight leak in the air intake system can cause serious problems.
- Do not depress the accelerator pedal when starting.
- Do not rev up the engine immediately after starting or just prior to shutdown.
- Do not attempt to disassemble the ECCS control unit under any circumstances.
- If installing a 2-way or CB radio, keep the antenna as far as possible away from the electronic control unit. Keep the antenna feeder line at least 8 in. away from the ECCS harness and do not let the 2 run parallel for a long distance. Be sure to ground the radio to the vehicle body.

ENTERING SELF-DIAGNOSTICS

Without Scan Tool

A check engine light has been adopted on all Infiniti models. This light blinks simultaneously with the red LED, which is located on the ECU. The ECU is located behind the center console panel near the accelerator pedal on the G20, behind the right side dash panel on the M30 and behind the front passenger-side kick-panel on the Q45. In addition to this LED, and right next to it, is a diagnostic mode selector screw that is turned to activate either of two modes.

In Mode I with the ignition switch in the **ON** position and with the engine stopped, the CHECK ENGINE LIGHT (CEL)

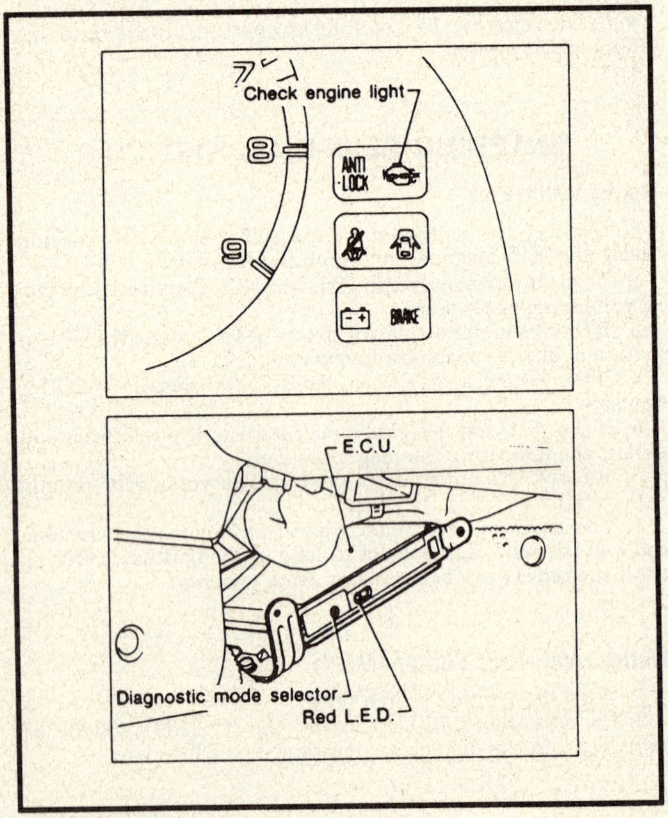

Diagnostic feature locations—G20

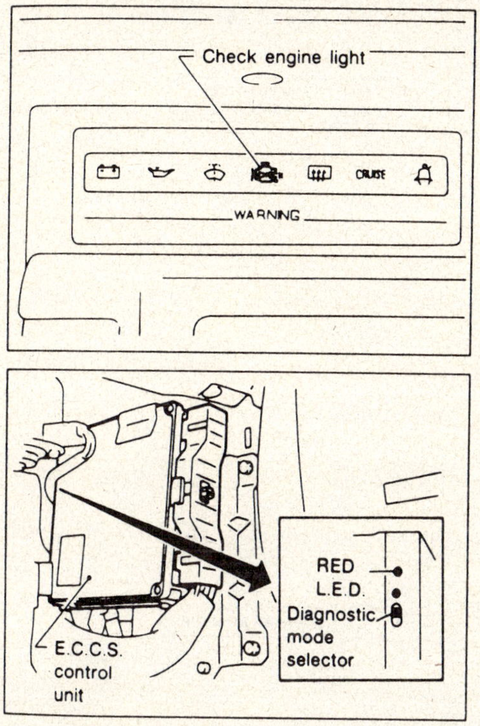

Diagnostic feature locations—M30

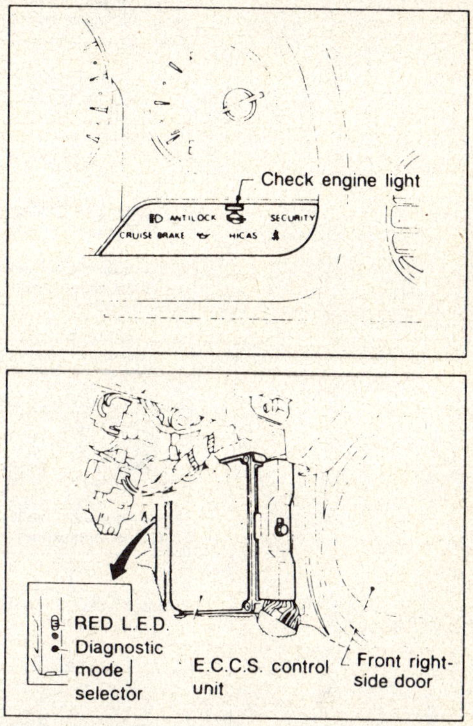

Diagnostic feature locations—Q45

and LED bulbs are checked (Both should be on.) If the engine is running in Mode I and the CEL and red LED are illuminated, then a malfunction has been detected by the ECU.

NOTE: Not all code-setting faults will cause the check-engine light to illuminate.

In Mode II with the ignition switch in the **ON** position and with the engine stopped, malfunction codes are indicated by the number of flashes from the red LED or the CEL.

When the engine is running in Mode II the exhaust gas sensor is monitored as long as the feedback system is in closed loop. The CEL and red LED are illuminated if the fuel mixture condition is lean, and off if the fuel mixture condition is rich. In open loop the bulbs remain off, regardless of the fuel mixture condition.

Condition \ Mode		Mode I	Mode II
Ignition switch in "ON" position	Engine stopped	BULB CHECK	SELF-DIAGNOSTIC RESULTS
	Engine running	MALFUNCTION WARNING	EXHAUST GAS SENSOR MONITOR

Self-diagnostic functions—Infiniti

NOTE: If crank angle sensor and ignition signal-primary codes come out at the same time, check items causing a malfunction of the crank angle sensor circuit first.

With Scan Tool

The Nissan Consult scan-tool, or equivalent, can be used in several modes to access and erase codes, read sensor data, etc. The diagnostic connector for the Consult is located either behind the fuse box cover or to the rear of the hood release handle.

DIAGNOSTIC MODES

- Self-diagnostic results mode: This is a mode in which code results can be read and erased quickly.
- Work support mode: This mode enables a technician to adjust some devices faster and more accurately by following the indications on the scan tool unit. It can also be used to relieve fuel pressure on the G20 and the Q45.
- Data monitor mode: This mode is used to read input and output data in the control unit. The technician can read the air flow meter voltage, neutral switch **ON/OFF** signal, etc.

NOTE: The data format seen on the Consult (or equivalent) is similar among the Infiniti models. The data values and codes obtained, however, may vary.

- Active test mode: In this mode the scan tool drives some actuators apart from the control units and also shifts some parameters in a specified range. For example, it enables the technician to change the amount of fuel injected, the AAC valve opening percentage, the engine coolant temperature (as seen by the ECU), etc., all by pressing buttons on the scan tool.
- ECU part number mode: It enables the technician to know what ECU part number he has, without physically gaining access to the ECU.

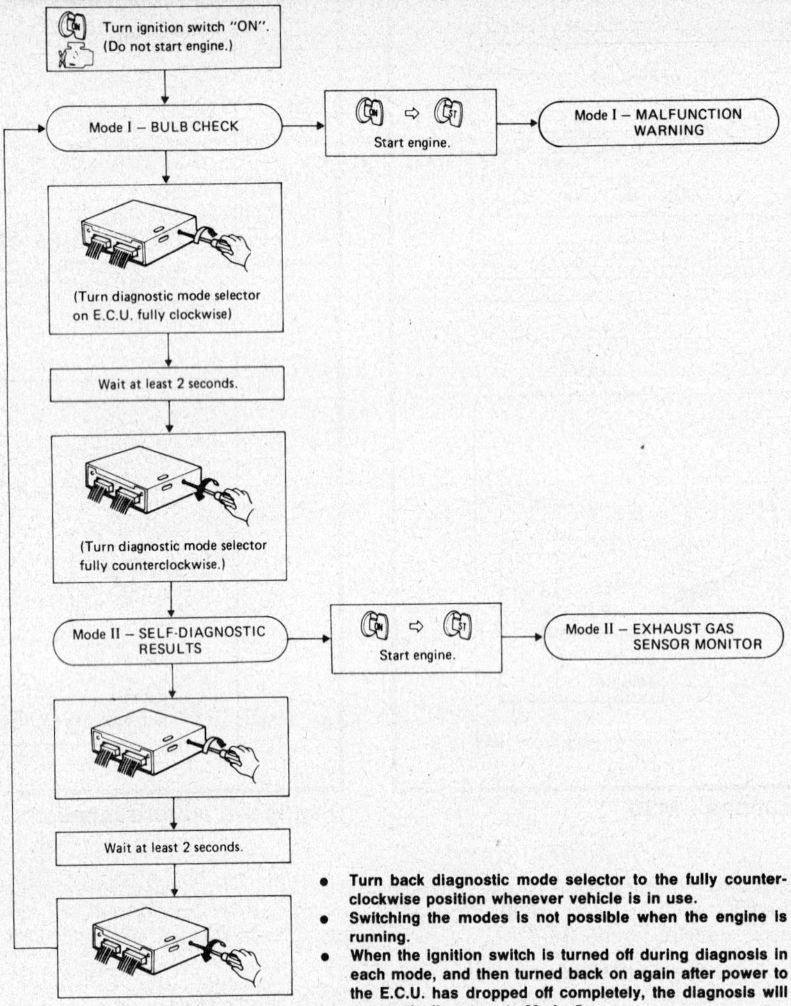

- **Turn back diagnostic mode selector to the fully counter-clockwise position whenever vehicle is in use.**
- **Switching the modes is not possible when the engine is running.**
- **When the ignition switch is turned off during diagnosis in each mode, and then turned back on again after power to the E.C.U. has dropped off completely, the diagnosis will automatically return to Mode I.**

Switching modes – Infiniti

Code No.		Detected items	California models	Non-California models
11*		Crank angle sensor circuit	X	X
12	CHECK	Air flow meter circuit	X	X
13	CHECK	Engine temperature sensor circuit	X	X
14	CHECK	Vehicle speed sensor circuit	X	X
21*		Ignition signal circuit	X	X
31	CHECK	E.C.U.	X	X
32	CHECK	E.G.R. function	X	—
33	CHECK	Exhaust gas sensor circuit	X	X
34		Detonation sensor circuit	X	X
35	CHECK	Exhaust gas temperature sensor circuit	X	—
43	CHECK	Throttle sensor circuit	X	X
45	CHECK	Injector leak	X	—
55		No malfunction in the above circuits	X	X

X: Available
—: Not available
CHECK : Check engine light item
*: Check items causing a malfunction of crank angle sensor circuit first, if both code No. 11 and 21 are displayed at the same time.

Typical code display application – Infiniti

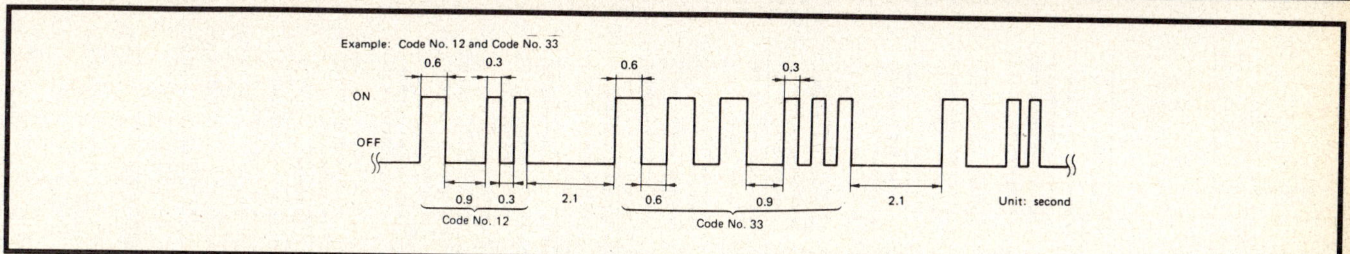

Red LED code flashing sequence – Infiniti

• Function test mode: This mode applies to the 1992 Infiniti. The Consult (or equivalent) conducts several tests on its own to determine whether a system is functioning correctly. For example, the Consult is able to conduct a power balance test by stopping the operation of each injector. It then evaluates the combustion of each cylinder based on the resultant rpm change.

CLEARING CODES

There are several ways in which a code can be erased or lost from the ECU's memory:

1. The malfunction code is erased from the backup memory on the ECU when the diagnostic mode is changed from Mode II to Mode I.
2. When the battery terminal is disconnected, the malfunction code will be lost from the backup memory within 24 hours.
3. The Nissan Consult, or equivalent, can be used to erase trouble codes quickly.
4. The diagnostic results will remain in ECU memory until the starter is operated 50 times after a diagnostic item has been judged to be malfunctioning. The diagnostic result will then be canceled automatically.

NOTE: Before starting self-diagnosis, do not erase the stored memory.

2.0L ENGINE – MODE II SELF-DIAGNOSTIC RESULTS – G20

Code No.	Detected items	Malfunction is detected when ...	Check item (remedy)
*11	Crank angle sensor circuit	• Either 1° or 180° signal is not entered for the first few seconds during engine cranking. • Either 1° or 180° signal is not input often enough while the engine speed is higher than the specified rpm.	• Harness and connector (If harness and connector are normal, replace crank angle sensor.)
12	Air flow meter circuit	• The air flow meter circuit is open or shorted. (An abnormally high or low voltage is entered.)	• Harness and connector (If harness and connector are normal, replace air flow meter.)
13	Engine temperature sensor circuit	• The engine temperature sensor circuit is open or shorted. (An abnormally high or low output voltage is entered.)	• Harness and connector • Engine temperature sensor
14	Vehicle speed sensor circuit	• The vehicle speed sensor circuit is open or shorted.	• Harness and connector • Vehicle speed sensor (reed switch)
*21	Ignition signal circuit	• The ignition signal in the primary circuit is not entered during engine cranking or running.	• Harness and connector • Power transistor unit
31	E.C.U.	• E.C.U. calculation function is malfunctioning.	(Replace E.C.C.S. control unit.)
32	E.G.R. function	• E.G.R. control valve does not operate. (E.G.R. control valve spring does not lift.)	• E.G.R. control valve • E.G.R. & canister control solenoid valve
33	Exhaust gas sensor circuit	• The exhaust gas sensor circuit is open or shorted. (An abnormally high or low output voltage is entered.)	• Harness and connector • Exhaust gas sensor • Fuel pressure • Injectors • Intake air leaks
34	Detonation sensor circuit	• The detonation circuit is open or shorted. (An abnormally high or low voltage is entered.)	• Harness and connector • Detonation sensor
35	Exhaust gas temperature sensor circuit	• The exhaust gas temperature sensor circuit is open or shorted. (An abnormally high or low voltage is entered.)	• Harness and connector • Exhaust gas temperature sensor
43	Throttle sensor circuit	• The throttle sensor circuit is open or shorted. (An abnormally high or low voltage is entered.)	• Harness and connector • Throttle sensor
45	Injector leak	• Fuel leaks from injector.	• Injector

*: Check items causing a malfunction of crank angle sensor circuit first, if both code No. 11 and 21 come out at the same time.

3.0L ENGINE—MODE II SELF-DIAGNOSTIC RESULTS—M30

Code No.	Detected items	Malfunction is detected when ...	Check item (remedy)
*11	Crank angle sensor circuit	• Either 1° or 120° signal is not entered for the first few seconds during engine cranking. • Either 1° or 120° signal is not input often enough while the engine speed is higher than the specified rpm.	• Harness and connector (If harness and connector are normal, replace crank angle sensor.)
12	Air flow meter circuit	• The air flow meter circuit is open or shorted. (An abnormally high or low voltage is entered.)	• Harness and connector (If harness and connector are normal, replace air flow meter.)
13	Engine temperature sensor circuit	• The engine temperature sensor circuit is open or shorted. (An abnormally high or low output voltage is entered.)	• Harness and connector • Engine temperature sensor
14	Vehicle speed sensor circuit	• The vehicle speed sensor circuit is open or shorted.	• Harness and connector • Vehicle speed sensor (reed switch)
*21	Ignition signal circuit	• The ignition signal in the primary circuit is not entered during engine cranking or running.	• Harness and connector • Power transistor unit
22	Fuel pump	• Abnormally high or low current supply to the fuel pump persists.	• Harness and connector • Fuel pump • Fuel pump relay
31	E.C.U.	• E.C.U. calculation function is malfunctioning.	(Replace E.C.C.S. control unit.)
32	E.G.R. function	• E.G.R. control valve does not operate. (E.G.R. control valve spring does not lift.)	• E.G.R. control valve • E.G.R. control solenoid valve
33	Exhaust gas sensor circuit	• The exhaust gas sensor circuit is open or shorted. (An abnormally high or low output voltage is entered.)	• Harness and connector • Exhaust gas sensor • Fuel pressure • Injectors • Intake air leaks
53	Exhaust gas sensor circuit (Right side)		
34	Detonation sensor circuit	• The detonation circuit is open or shorted. (An abnormally high or low voltage is entered.)	• Harness and connector • Detonation sensor
35	Exhaust gas temperature sensor circuit	• The exhaust gas temperature sensor circuit is open or shorted. (An abnormally high or low voltage is entered.)	• Harness and connector • Exhaust gas temperature sensor
42	Fuel temperature sensor circuit	• The fuel temperature sensor circuit is open or shorted. (An abnormally high or low voltage is entered.)	• Harness and connector • Fuel temperature sensor
43	Throttle sensor circuit	• The throttle sensor circuit is open or shorted. (An abnormally high or low voltage is entered.)	• Harness and connector • Throttle sensor
45	Injector leak	• Fuel leaks from injector.	• Injector
51	Injector circuit	• The injector circuit is open.	• Injector

*: Check items causing a malfunction of crank angle sensor circuit first, if both code No. 11 and 21 come out at the same time.

4.5L ENGINE—MODE II SELF-DIAGNOSTIC RESULTS—Q45

Code No.	Detected items	Malfunction is detected when ...	Check item (remedy)
11*1)	Crank angle sensor circuit	• Either 1° or 90° signal is not entered for the first few seconds during engine cranking. • Either 1° or 90° signal is not input often enough while the engine speed is higher than the specified rpm.	• Harness and connector (If harness and connector are normal, replace crank angle sensor.)
12	Air flow meter circuit	• The air flow meter circuit is open or shorted. (An abnormally high or low voltage is entered.)	• Harness and connector (If harness and connector are normal, replace air flow meter.)
13	Engine temperature sensor circuit	• The engine temperature sensor circuit is open or shorted. (An abnormally high or low output voltage is entered.)	• Harness and connector • Engine temperature sensor
14	Vehicle speed sensor circuit	• The vehicle speed sensor circuit is open or shorted.	• Harness and connector • Vehicle speed sensor (reed switch)
16*2)	TCS signal circuit	• The TCS signal circuit is open or shorted. (An abnormally high or low output voltage is entered.)	• Harness and connector • T.C.M.
21*1)	Ignition signal circuit	• The ignition signal in the primary circuit is not entered during engine cranking or running.	• Harness and connector • Power transistor unit
31	E.C.U.	• E.C.U. calculation function is malfunctioning.	(Replace E.C.C.S. control unit.)
32	E.G.R. function	• E.G.R. control valve does not operate. (E.G.R. control valve spring does not lift.)	• E.G.R. control valve • E.G.R. control solenoid valve
33	Exhaust gas sensor circuit (Left bank)	• The exhaust gas sensor circuit is open or shorted. (An abnormally high or low output voltage is entered.)	• Harness and connector • Exhaust gas sensor • Fuel pressure • Injectors • Intake air leaks
53	Exhaust gas sensor circuit (Right bank)		
34	Detonation sensor circuit	• The detonation sensor circuit is open or shorted. (An abnormally high or low voltage is entered.)	• Harness and connector • Detonation sensor
35	Exhaust gas temperature sensor circuit	• The exhaust gas temperature sensor circuit is open or shorted. (An abnormally high or low voltage is entered.)	• Harness and connector • Exhaust gas temperature sensor
43	Throttle sensor circuit	• The throttle sensor circuit is open or shorted. (An abnormally high or low voltage is entered.)	• Harness and connector • Throttle sensor
45	Injector leak	• Fuel leaks from injector.	• Injector
46*2)	Secondary throttle sensor circuit	• The secondary throttle sensor circuit is open or shorted. (An abnormally high or low output voltage is entered.)	• Harness and connector • Secondary throttle sensor • T.C.M.
51	Injector circuit	• The injector circuit is open or shorted.	• Injector
54	Signal circuit from A/T control unit to E.C.U. (A/T only)	• The A/T communication line is open or shorted.	• Harness and connector

*1): Check items causing a malfunction of crank angle sensor circuit first, if both "CRANK ANGLE SENSOR (No. 11)" and "IGN SIGNAL-PRIMARY (No. 21)" are displayed one after the other.

*2): Models with TCS only

How to Execute Self-diagnosis in Mode II

Detected items	Display Code No.	How to perform self-diagnosis judgement: Illustration	How to perform self-diagnosis judgement: Method
Crank angle sensor circuit	11	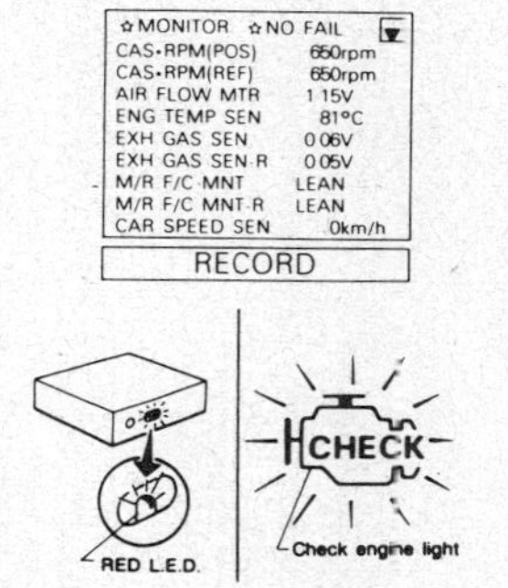	**PERFORM SELF-DIAGNOSIS.** 1) Start engine. 2) Select "DATA MONITOR" mode with CONSULT. ☆ **NO FAIL** —OR— 2) Turn ignition switch "OFF" and then "ON". 3) Perform self-diagnosis (Mode II) with E.C.U. **Check engine light and red L.E.D. display code No. 55.**
Air flow meter circuit	12	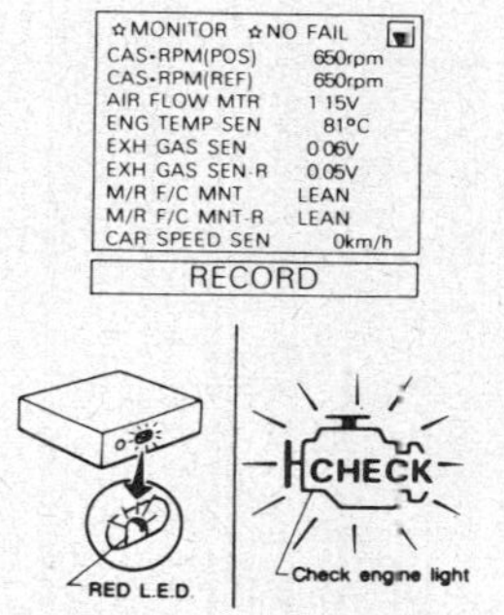	**PERFORM SELF-DIAGNOSIS.** 1) Turn ignition switch "ON" wait for at least 5 seconds and then start engine. 2) Select "DATA MONITOR" mode with CONSULT. ☆ **NO FAIL** —OR— 2) Perform self-diagnosis (Mode II) with E.C.U. **Check engine light and red L.E.D. display code No. 55.**

How to Execute Self-diagnosis in Mode II (Cont'd)

Detected items	Display Code No.	How to perform self-diagnosis judgement: Illustration	How to perform self-diagnosis judgement: Method
Engine temperature sensor circuit	13	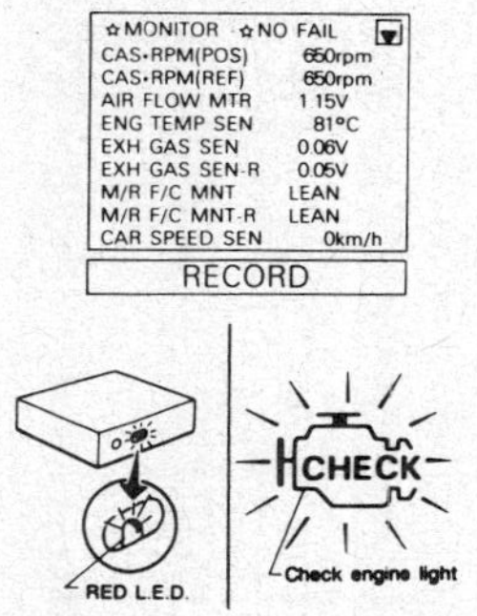	**PERFORM SELF-DIAGNOSIS.** 1) Turn ignition switch "ON" or start engine. 2) Select "DATA MONITOR" mode with CONSULT. ☆ **NO FAIL** —OR— 2) Perform self-diagnosis (Mode II) with E.C.U. **Check engine light and red L.E.D. display code No. 55.**
Vehicle speed sensor circuit*	14	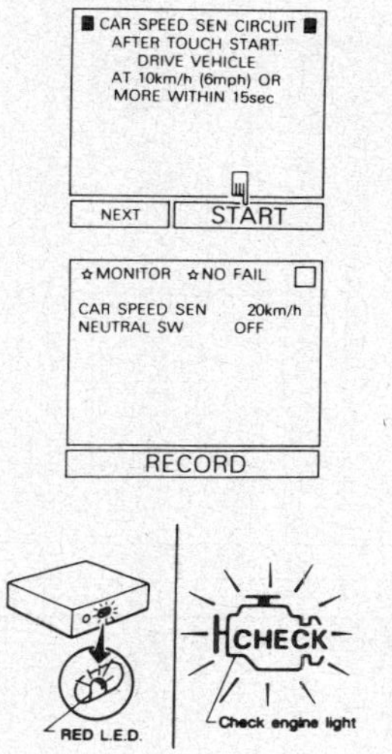	**CHECK OVERALL FUNCTION.** 1) Jack up drive wheels. 2) Start engine. 3) Perform "CAR SPEED SEN CIRCUIT" in "FUNCTION TEST" mode with CONSULT. —OR— 2) Start engine. 3) Read vehicle speed sensor signal in "DATA MONITOR" mode with CONSULT. **CONSULT value should be the same as the speedometer indication.** —OR— 1) Start engine and warm it up sufficiently. 2) Shift to a suitable gear position and maintain the following test drive conditions for at least 5 seconds. Driving conditions (1) Engine speed: 1,750 ± 150 rpm (2) Intake manifold vacuum: (−60.0 ± 6.7 kPa (−450 ± 50 mmHg, 17.72 ± 1.97 inHg) (3) Vehicle speed 5 km/h (3MPH) or more 3) If check engine light comes on during test drive, perform self-diagnosis (Mode II) with E.C.U. **Check engine light and red L.E.D. display code No. 55.**

*: Self-diagnosis is not performed but this method provides results which are equal to the self-diagnostic results.

How to Execute Self-diagnosis in Mode II (Cont'd)

Detected items	Display Code No.	How to perform self-diagnosis judgement: Illustration	How to perform self-diagnosis judgement: Method
TCS Signal circuit	16	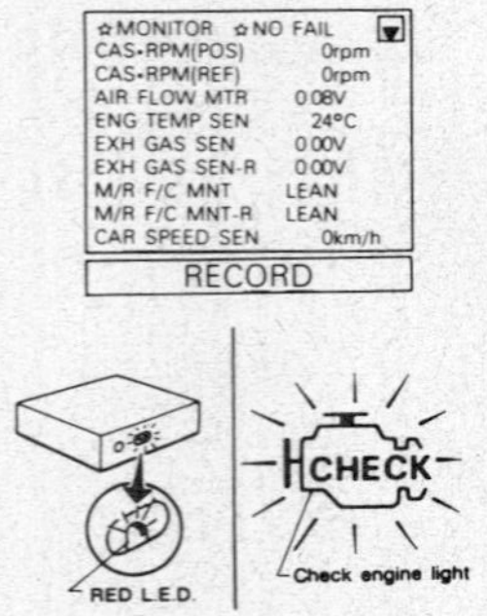	**PERFORM SELF-DIAGNOSIS.** 1) Turn ignition switch "ON". 2) Select "DATA MONITOR" mode with CONSULT ☆ **NO FAIL** — OR — 1) Turn ignition switch "ON". 2) Perform self-diagnosis (Mode II) with E.C.U. **Check engine light and red L.E.D. display code No. 55.**
Ignition signal circuit	21	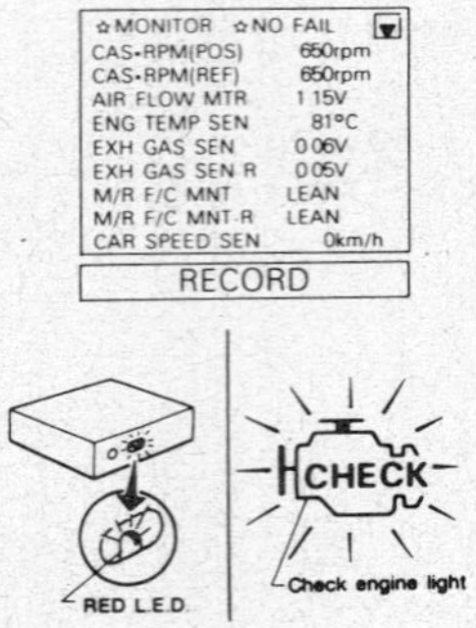	**PERFORM SELF-DIAGNOSIS.** 1) Start engine. 2) Select "DATA MONITOR" mode with CONSULT. ☆ **NO FAIL** — OR — 2) Turn ignition switch "OFF" and then "ON". 3) Perform self-diagnosis (Mode II) with E.C.U. **Check engine light and red L.E.D. display code No. 55.**

How to Execute Self-diagnosis in Mode II (Cont'd)

Detected items	Display Code No.	How to perform self-diagnosis judgement: Illustration	How to perform self-diagnosis judgement: Method
E.C.U.	31	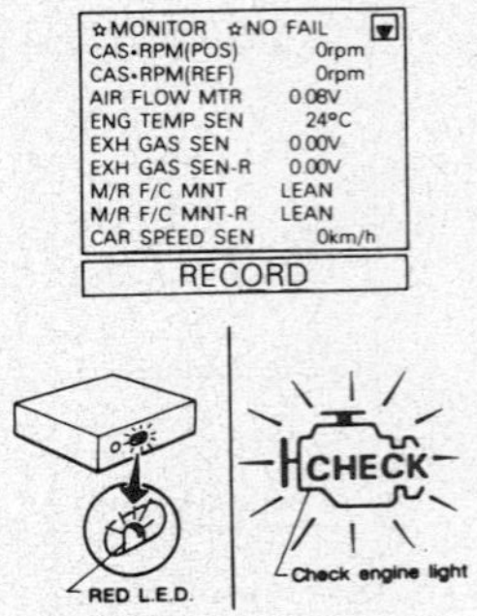	**PERFORM SELF-DIAGNOSIS.** 1) Turn ignition switch "ON". 2) Select "DATA MONITOR" mode with CONSULT. ☆ **NO FAIL** — OR — 2) Perform self-diagnosis (Mode II) with E.C.U. **Check engine light and red L.E.D. display code No. 55.**
EGR function	32	**A ROAD TEST** **Test condition** **Drive vehicle under the following conditions with a suitable shift position.** **(1) Engine speed: 1,700±200 rpm** **(2) Intake manifold vacuum: −42.7±10.7 kPa (−320±80 mmHg, −12.60±3.15 inHg)** **Driving mode** 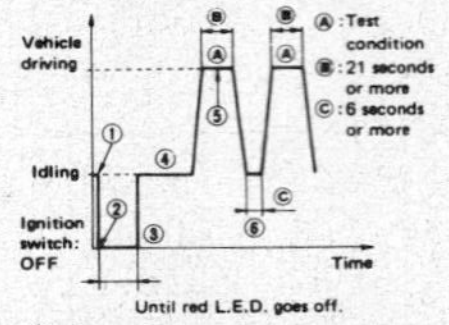 ① Start engine and warm it up sufficiently. ② Turn off ignition switch and keep it off until red L.E.D. goes off. ③ Start engine and make sure that air conditioner switch and rear defogger are turned "OFF" during test drive. ④ Keep engine running for at least 150 seconds. ⑤ Shift to suitable gear position and drive in "Test condition" for at least 21 seconds. ⑥ Decrease engine revolutions to less than 1,500 rpm for at least 6 seconds. ⑦ Repeat steps ⑤ through ⑥ at lest 1 more time. B 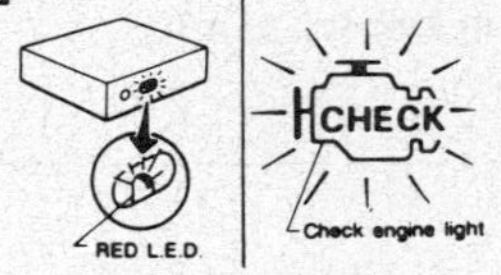	**PERFORM SELF-DIAGNOSIS.** 1) Turn ignition switch "ON". 2) Perform self-diagnosis (Mode II) with E.C.U. Make sure that code No. 11 or 12 is not displayed. 3) Perform test drive under the following conditions. (1) Warm up engine sufficiently. (2) Use test driving modes indicated in figure A. 4) If check engine light comes on during test drive, perform self-diagnosis (Mode II) with E.C.U. B **Check engine light and red L.E.D. display code No. 55.**

How to Execute Self-diagnosis in Mode II (Cont'd)

Detected items	Display Code No.	How to perform self-diagnosis judgement: Illustration	How to perform self-diagnosis judgement: Method
Exhaust gas sensor circuit (Left bank)* Exhaust gas sensor circuit (Right bank)*	33 53	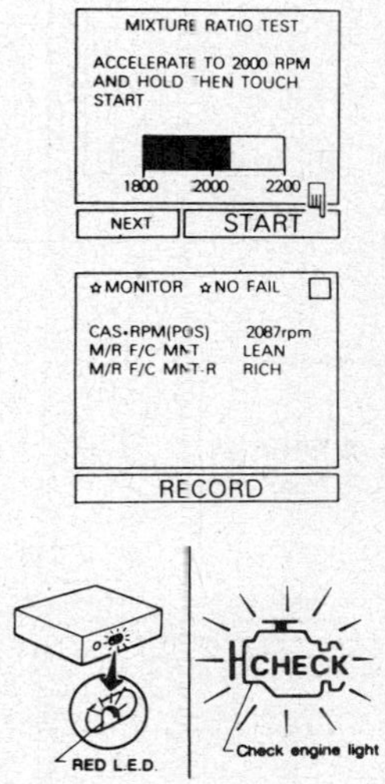 	**CHECK OVERALL FUNCTION.** 1) Start engine and warm it up sufficiently. 2) Perform "MIXTURE RATIO TEST" in "FUNCTION TEST" mode with CONSULT. — OR — 2) Make sure that "M/R F/C MNT(R)" in "DATA MONITOR" mode indicates "RICH" and "LEAN" periodically more than 5 times during 10 seconds at 2,000 rpm — OR — 2) Make sure that check engine light and red L.E.D. on E.C.U. go on and off periodically more than 5 times during 10 seconds at 2,000 rpm in self-diagnosis Mode II.
Detonation sensor circuit	34	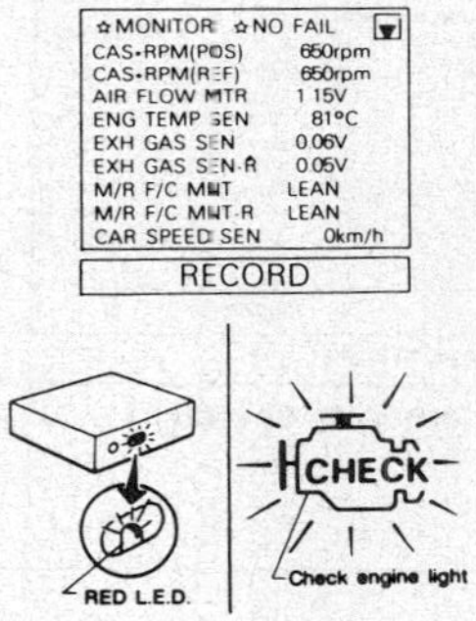 	**PERFORM SELF-DIAGNOSIS.** 1) Start engine. 2) Select "DATA MONITOR" mode with CONSULT. ☆ **NO FAIL** — OR — 2) Turn ignition switch "OFF" and then "ON". 3) Perform self-diagnosis (Mode II) with E.C.U. **Check engine light and red L.E.D. display code No. 55.**

*: Self-diagnosis is not performed but this method provides results which are equal to the self-diagnostic results.

How to Execute Self-diagnosis in Mode II (Cont'd)

Detected items	Display Code No.	How to perform self-diagnosis judgement: Illustration	How to perform self-diagnosis judgement: Method
Exhaust gas temperature sensor circuit	35	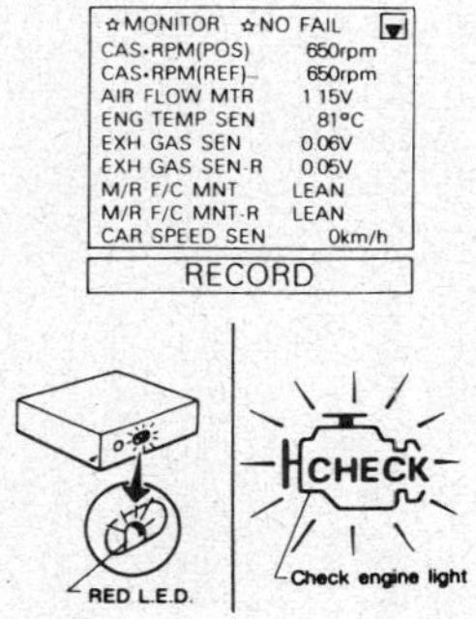 	**PERFORM SELF-DIAGNOSIS.** 1) Start engine and warm it up sufficiently. 2) Select "DATA MONITOR" mode with CONSULT. ☆ **NO FAIL** — OR — 2) Turn ignition switch "OFF" and then "ON". 3) Perform self-diagnosis (Mode II) with E.C.U. **Check engine light and red L.E.D. display code No. 55.**
Throttle sensor circuit	43	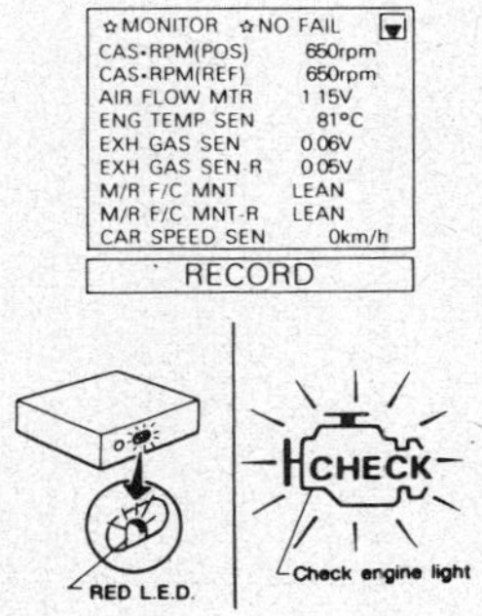 	**PERFORM SELF-DIAGNOSIS.** 1) Jack up drive wheels 2) Start engine. 3) Shift to a suitable gear position (Except "P" or "N"), and run engine at vehicle speed of 5 km/h (3 MPH) or higher for at least 10 seconds. 4) Select "DATA MONITOR" mode with CONSULT. ☆ **NO FAIL** — OR — 4) Turn ignition switch "OFF" and then "ON". 5) Perform self-diagnosis (Mode II) with E.C.U. **Check engine light and red L.E.D. display code No. 55.**

How to Execute Self-diagnosis in Mode II (Cont'd)

Detected items	Display Code No.	How to perform self-diagnosis judgement: Illustration	Method
Injector leak	45	**A ROAD TEST** **Test condition** **Drive vehicle under the following conditions with a suitable shift position.** **(1) Engine speed: 1,700±400 rpm** **(2) Intake manifold vacuum:** **−46.7±13.3 kPa** **(−350±100 mmHg, −13.78±3.94 inHg)** **Driving mode** 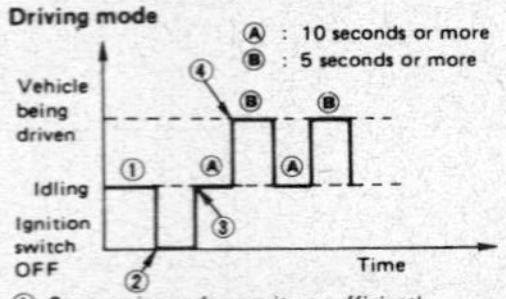 ① Start engine and warm it up sufficiently. ② Turn ignition switch OFF and wait for at least 10 seconds. ③ Start engine and keep it at idle speed for at least 10 seconds. ④ Shift to a suitable gear position and drive in "Test condition" for at least 5 seconds. ⑤ Repeat steps ③ through ④ at least 1 more times. **B** 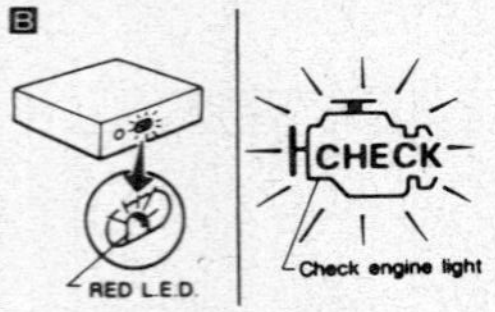 	**PERFORM SELF-DIAGNOSIS.** 1) Perform test drive as indicated in figure **A**. 2) If check engine light comes on during test drive, perform self-diagnosis (Mode II) with E.C.U. **B Check engine light and red L.E.D. display code No. 55.**
Secondary throttle sensor circuit	46	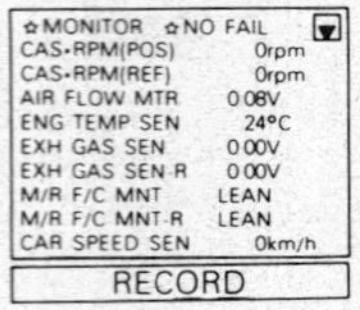 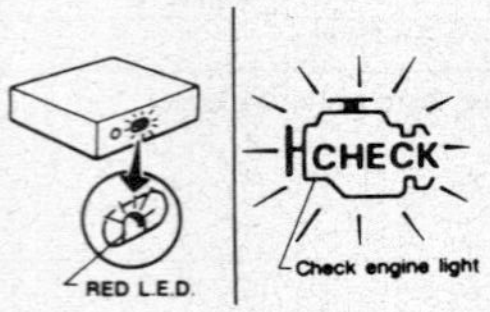 	**PERFORM SELF-DIAGNOSIS.** 1) Turn ignition switch "ON". 2) Select "DATA MONITOR" mode with CONSULT. ☆ **NO FAIL** — OR — 1) Turn ignition switch "ON". 2) Perform self-diagnosis (Mode II) with E.C.U. **Check engine light and red L.E.D. display code No. 55.**

How to Execute Self-diagnosis in Mode II (Cont'd)

Detected items	Display Code No.	How to perform self-diagnosis judgement: Illustration	Method
Injector circuit	51	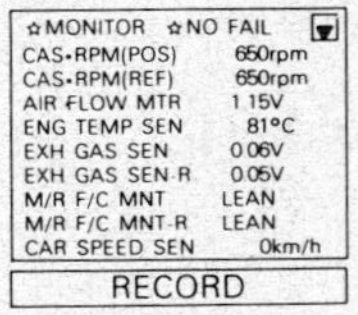 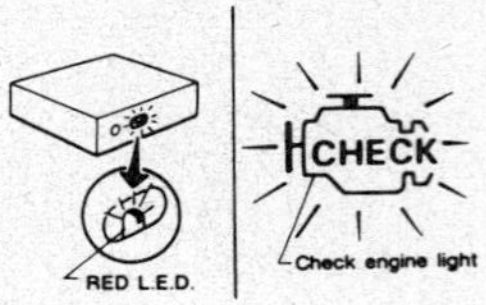 	**PERFORM SELF-DIAGNOSIS.** 1) Start engine. 2) Select "DATA MONITOR" mode with CONSULT. ☆ **NO FAIL** — OR — 2) Turn ignition switch "OFF" and then "ON". 3) Perform self-diagnosis (Mode II) with E.C.U. **Check engine light and red L.E.D. display code No. 55.**
Signal circuit from A/T control unit to E.C.U.	54	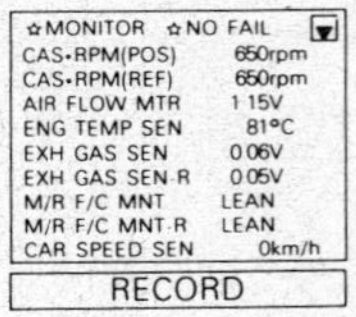 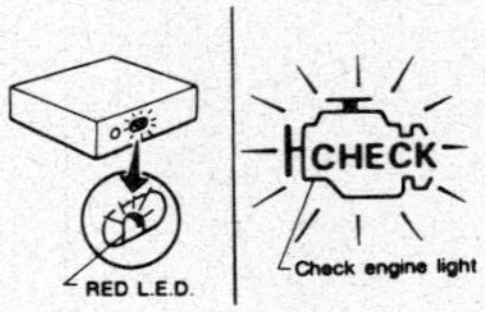 	**PERFORM SELF-DIAGNOSIS.** 1) Turn ignition switch "ON" or start engine. 2) Select "DATA MONITOR" mode with CONSULT. ☆ **NO FAIL** — OR — 2) Perform self-diagnosis (Mode II) with E.C.U. **Check engine light and red L.E.D. display code No. 55.**

ISUZU WITH FEEDBACK CARBURETOR

General Information

The FBC closed loop emission control system precisely controls the air/fuel ratio near the optimum mixture and allows the use of the 3-way catalyst to reduce the oxides of nitrogen and oxidize hydrocarbons and carbon monoxide. The essential components of the closed loop system are the coolant temperature sensor, oxygen sensor, Electronic Control Module (ECM), Feedback Carburetor (FBC), 3-way catalytic converter, idle and WOT switches, idle switch and MAP sensor, duty solenoid, fuel cut solenoid.

ELECTRONIC CONTROL MODULE

The ECM generates a control signal to the vacuum controller solenoid which controls carburetor air/fuel ratio through vacuum signals. This control signal is continuously cycling the solenoid between on and off time (duty cycle) as a function of the input voltages from the sensors. The control signal generated by the ECM is selected from 4 operational modes as follows:

1. Inhibit Mode – no signal to the vacuum controller solenoid.
2. Enrichment Mode – a fixed pre-programmed duty cycle to the vacuum controller solenoid.
3. Open Loop Mode – a fixed pre-programmed duty cycle to the vacuum controller.
4. Close Loop Mode – a calculated duty cycle is generated based on oxygen sensor and other sensor outputs.

During closed loop operation, the ECM monitors the voltage output of the oxygen sensor. As the sensor voltage increases and passes through the ECM threshold set point, the proportional gain immediately changes the duty cycle of the output signal. The duty cycle is further changed at a constant rate (integral gain) until the sensor input voltage decreases and passes through the ECM threshold point.

The selection integral and proportional gain rates by the ECM is based on the engine operating conditions (idle or off-idle condition). At idle condition different gain rates are required for optimum air/fuel ratio control than those at partial load condition. The ECM also stores in an adaptive memory the current duty cycle being used for either idle or off-idle condition (below and above the adaptive switch point). When the ECM sees a transition from idle condition (as signaled by the vacuum switch) to off-idle condition, it immediately steps to the duty cycle last recorded for stoichiometric operation. From then on while at that engine operating condition the system uses the basic proportional and integral gain controls as previously described.

The ECM outputs the signal to control the slow cut solenoid valve incorporated in the carburetor. The ECM senses the coasting condition by means of the signals from the transmission gear position switch, clutch pedal position switch and idle position switch, cutting off the flow of electric current to slow the cut solenoid valve in such a instance when the engine running speed is beyond the limit of the speed specified, thus stopping the fuel flow to the carburetor. The flow of the electric current to slow the cut solenoid valve is cut off only at a time when the following 2 conditions exist simultaneously. They are:

1. The vacuum signal of the vacuum switch is below the specified vacuum.
2. The engine speed exceeds the limit of the specified speed.

SYSTEM MALFUNCTION LAMP

This system utilizes a dashboard mounted malfunction indicator lamp which, for some failure modes will inform the driver of the need for unscheduled maintenance. In the event of a system malfunction the Legend CHECK ENGINE will light and remain ON as long as the fault function occurs and the engine is running. The electronic control unit incorporates a diagnostic program which will assist in diagnosing the closed loop control system faults. When activated the diagnostic program will flash a code through the malfunction lamp which isolates the source of the system fault.

SELF-DIAGNOSTIC SYSTEM

The diagnostic circuit check makes sure that the self-diagnostic system works, determines that the trouble codes will display and guides diagnosis to other problem areas. When the engine is running and a problem develops in the system which the self diagnosis can evaluate, the CHECK ENGINE light will come on and a trouble code will be stored in the ECM trouble code memory. The light will remain ON with the engine running as long as there is a problem. If the problem is intermittent the CHECK ENGINE light will go OFF, but the trouble codes will be stored in the ECM trouble code memory.

With the ignition turned **ON** and the engine stopped, the CHECK ENGINE lamp should be ON. This is a bulb check to indicate the light is working properly.

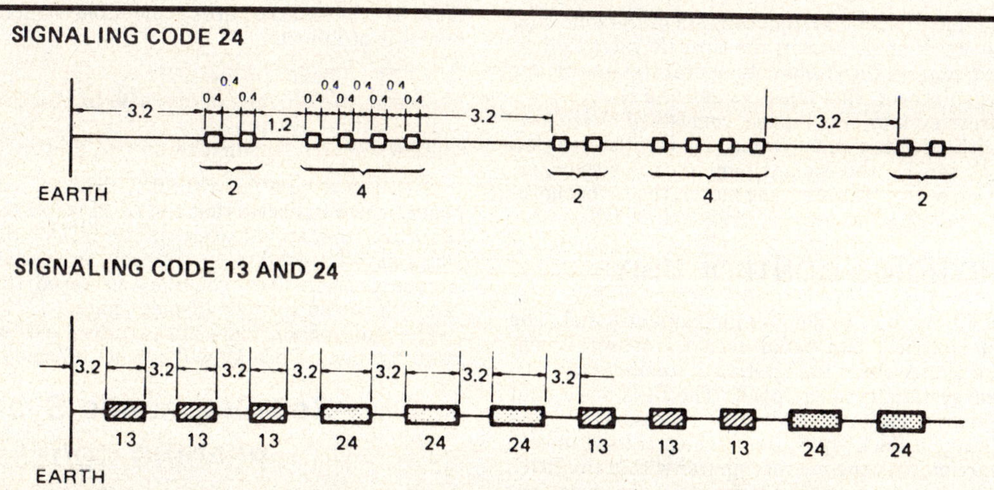

Reading diagnostic code pattern – Isuzu

On the Impulse, Amigo and Pick-Up a trouble code "TEST" lead (white cable) and a ground lead (black cable) are branched from a harness at a distance of 8 in. from the ECM connector (next to the clutch pedal).

Self-Diagnostic System

READING CODES

Entering Self-Diagnostic and Code Display

With the ignition turned **ON** and the engine stopped, the CHECK ENGINE lamp should be ON. This is a bulb check to indicate the light is working properly.

Connect the ALDL leads, on the 3 terminal connector, jumper the A and C terminals (outer terminals) or if equipped connect the trouble code "TEST" lead (white cable) and a ground lead (black cable), located 8 in. from the ECM connector (next to the clutch pedal).

The trouble code memory is activated by placing the ignition switch in the **ON** position and connecting the trouble code "TEST" lead to the ground lead.

The CHECK ENGINE light will begin to flash a trouble Code 12. Code 12 consists of 1 flash, a short pause and then 2 more flashes. There will be a longer pause and a Code 12 will repeat 2 more times. The check indicates that the self-diagnostic system is working. This cycle will repeat itself until the engine is started or the ignition switch is turned **OFF**. If more than a single fault code is stored in the memory, the lowest number code will flash 3 times followed by the next highest code number until all the codes have been flashed. The faults will then repeat in the same order. In most cases, codes will be checked with the engine running since no codes other than Codes 12 or 51 will be present on the initial "KEY ON". Remove the ground from the test terminal before starting the engine.

NOTE: The fault indicated by trouble Code 15 takes 5 minutes of engine operation before it will display. The diagnostic charts for trouble Codes 13, 31, 44 and 45 should be used if any of these trouble codes can be obtained.

CLEARING CODES

The trouble code memory is fed a continuous 12 volts even with the ignition switch in the **OFF** position. After a fault has been corrected, it will be necessary to remove the voltage for 10 seconds to clear any stored codes. The quickest way to remove the voltage is to remove the "ECM" fuse from the fuse block or the MAIN 60A fuse for 10 seconds. The voltage can also be removed by disconnecting the negative battery cable, but this will mean if the vehicle is equipped with electronic instrumentation, such as a clock and radio, they would have to be reset.

NOTE: Some vehicle computers have a learning ability. If a change is noted in vehicle performance after clearing codes, it may be due to the computers learning ability. To restore performance, warm vehicle to normal operating temperature and drive at part throttle with moderate acceleration, before performing any additional diagnosis.

ISUZU FUEL INJECTION EXCEPT 2.8L AND 3.1L ENGINES

General Information

The fuel injection control system constantly monitors and controls engine operation, which in turn helps lower emissions while maintaining the fuel economy and driveability . The Electronic Control Unit (ECU) controls the fuel injection system, the ignition system and the turbocharger control system (if so equipped).

In general, the fuel injection system consists of the following components; the crank angle sensor, the throttle valve switch, the vehicle speed sensor, the coolant temperature sensor, air flow sensor, oxygen sensor, fuel injectors and the ECU.

The turbocharger system, if equipped, consists of the following components; the throttle valve switch, the throttle position sensor, the knock sensor, the coolant temperature sensor, air flow sensor, oxygen sensor, the stepping motor, the turbocharger controller and the ECU.

ELECTRONIC CONTROL UNIT

The ECU is usually located under the instrument panel. The ECU analyzes all electrical data signals from the sensor. It controls the fuel injection system, the ignition system and the turbocharger control system (is so equipped). The ECU has a built in back-up, diagnostic and fail safe control systems.

The ECU contains a Back-Up System, which is used in case there is a malfunction with the microcomputer within the ECU, the back-up control system works to maintain the necessary functions of the control unit to permit continuous operation of the vehicle.

The sensors used in this system, provide electrical impulses to the ECM by monitoring the pressure, temperature, vacuum and other engine operating conditions.

SELF-DIAGNOSTIC SYSTEM

In the event of a failure, the ECU stores the trouble code in memory and operates the CHECK ENGINE light on the instrument panel when the nature of trouble is important, to warn the operator of failure.

NOTE: The self-diagnosis system is capable of troubleshooting the electrical circuits in the Closed Loop Emission Control system only and does not cover the trouble in the sensors, actuators or the engine itself.

With the engine not running, first, a Code 12 is flashed 3 times, which indicates that the ECM is functioning normally. If there are other trouble codes in the ECU, it displays each code 3 times at 3.2 second intervals each, in ascending order of code numbers. When all the codes are displayed, the computer will go through the entire code display again.

The malfunction codes are also displayed when the engine is running, but a Code 12 will not be indicated.

Self-Diagnostic System

READING CODES

Entering Self-Diagnostic and Code Display

When the diagnosis lead in the vicinity of the control unit is con-

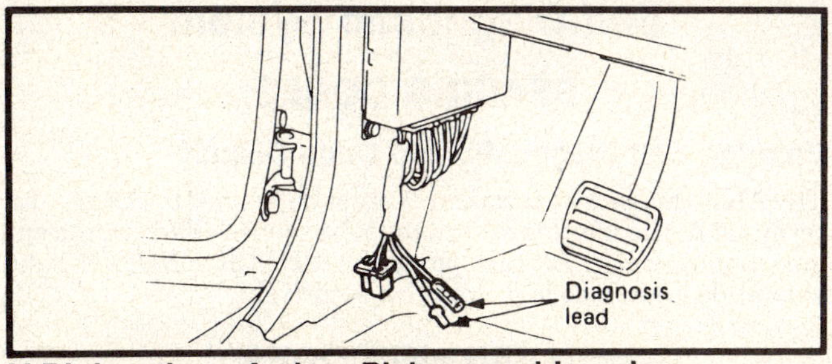

ALDL location—Amigo, Pick-up and Impulse

nected with the ignition switch in the **ON** position (engine not running), the trouble codes stored in the memory will be displayed by the CHECK ENGINE light.

To display codes on vehicles with 2 separate terminals, connect the terminals together. On vehicles with a 3 terminal connector, jumper the A and C terminals (outer terminals) together. Codes are read by the CHECK ENGINE light flashes.

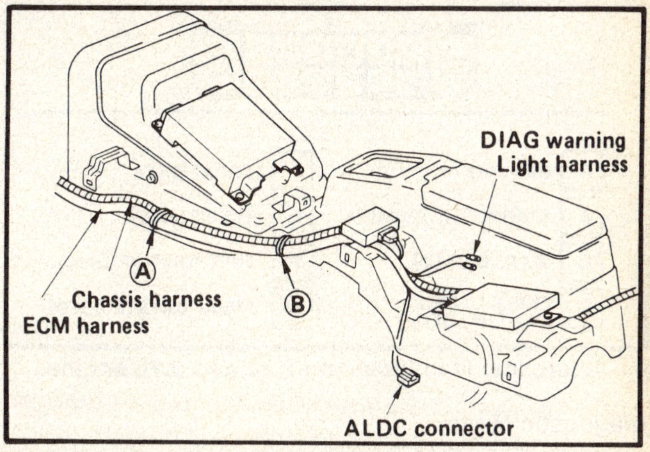

ALDL location—Trooper and Trooper II

CLEARING CODES

Clear the trouble codes stored in memory by disconnecting the memory fuse in the fuse junction block or by disconnecting the negative battery cable for at least 20 seconds, then check that only Code 12 is displayed.

Since all memory is cleared when disconnecting the battery cable or removing the fuse, the clock, radio and other electrical equipment will need to be reset.

NOTE: Some vehicle computers have a learning ability. If a change is noted in vehicle performance after clearing codes, it may be due to the computers learning ability. To restore performance, warm vehicle to normal operating temperature and drive at part throttle with moderate acceleration, before performing any additional diagnosis.

ISUZU FUEL INJECTION 2.8L AND 3.1L ENGINES

General Information

The self-diagnostic systemed used on these engines is the same type used by General Motors. Code retreival, clearing and testing is identical to 2.8L and 3.1L throttle body injected engines used by General Motors.

CHECK ENGINE LIGHT

The CHECK ENGINE light is located on the instrument panel and has 2 functions:

1. It alerts the driver that a problem has occurred in the system and that the car should be taken to the service station for evaluation as soon as possible.
2. Once in the shop, the light allows the technician to read the trouble codes to help diagnose system problems. Normally, the light will go ON with the ignition switch **ON** and the engine not running. If the ECM has detected a problem within the system, the light will remain ON. If the problem goes away, the light will go OFF in most cases but a trouble code will remain stored in the ECM.

Intermittent CHECK ENGINE light

Fault codes are generally divided into 2 categories: intermittent and hard.

An intermittent code is defined as a code that does not re-set itself, but the fault condition is not active when the technician is working on the vehicle. The most probable cause of intermittent faults are loose wiring connections.

A hard code is defined as present when the vehicle is brought in for service and that the ECM recognizes and displays to the technician.

TROUBLE CODES

The codes stored in the ECM's memory can be read either through a hand held diagnostic scanner plugged into the ALDL connector or by counting the number of flashes on the "CHECK ENGINE" light when the diagnostic terminal of the ALCL connector is grounded. The ALDL connector terminal **B** (diagnostic terminal) is the second terminal from the right of the top row in the ALDL connector. The terminal is most easily grounded by connecting to terminal **A** (internal ECM ground), the terminal to the right of the terminal **B** on the top row in the ALDL connector.

Once terminals **A** and **B** are grounded, the ignition switch must be turned **ON** with the engine not running. At this point the CHECK ENGINE light should flash a Code 12, 3 times consecutively. This would be the following flash sequence; flash, pause, flash-flash, long pause, flash, pause, flash-flash, long pause, flash, pause, flash-flash.

Code 12 indicates that the ECM's diagnostic system is operating. If Code 12 is not indicated, a problem is present within the diagnostic system itself and should by addresses by using the appropriate diagnostic chart.

Following the output of Code 12, the CHECK ENGINE light will indicate a diagnostic trouble code 3 times, if a code is present or it will simply continue to output Code 12. If more than a single code has been stored in the ECM's memory, the codes will be output from the lowest to the highest, with each code being displayed 3 times.

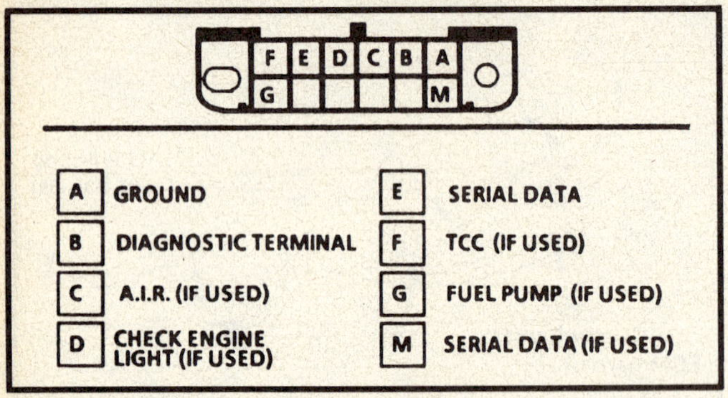

ALDL terminal identification—2.8L and 3.1L engines

Diagnostic Mode

NOTE: In the Diagnostic mode, codes can be read only with the engine stopped. Grounding the Diagnostic terminal with the engine running sends the ECM into the "FIELD SERVICE MODE" described below.

If the diagnostic terminal of the ALDL connector is grounded with the ignition switch in the **ON** position and the engine not running, the ECM will enter the Diagnostic mode. In this mode the ECM will perform the following functions:

1. Display a Code 12 by flashing the CHECK ENGINE light which indicates that the system is functioning properly. A Code 12 consists of 1 flash, followed by a short pause, then 2 quick flashes. The code will be repeated 3 times. If not other codes are stored in the system, code will continue to flash until the diagnostic terminal is ungrounded.
2. Display any stored trouble codes by flashing the CHECK ENGINE light. Each code will be code will be flashed 3 times, then Code 12 will be flashed again. If a trouble code is displayed, use the diagnostic code trouble code charts to troubleshoot the system.
3. Energize the all ECM relays and solenoids except the fuel pump relay.
4. Move the AIC valve to the fully extended position.

Field Service Mode

The Field Service Mode is entered by grounding the diagnostic terminal with the engine running and will indicate to the technician whether the ECM is in "Open" or "Closed" loop.

In "Open Loop" the CHECK ENGINE flashes 2½ time a second, in "Closed Loop" the CHECK ENGINE light once per second or will stay OFF most of the time if the system is too lean or ON most of the time if the system is too rich.

When the system is in the Field Service Mode, new trouble codes cannot be stored in the ECM and the closed loop timer is by-passed.

Self-Diagnostic System

READING CODES

Entering Self-Diagnostic and Code Display

The ALDL diagnostic terminal **B** must be jumper to the ground terminal **A**, the ignition switch must be turned **ON** with the engine not running. At this point the CHECK ENGINE light should flash each code 3 times consecutively.

CLEARING CODES

When the ECM sets a trouble code, the CHECK ENGINE light will illuminate and the code will be stored in the computer's memory. If the problem is intermittent, the light will go OFF in 10 seconds when the fault goes away. However, the code will remain in the ECU's memory until battery voltage to the ECU is removed. Removing battery voltage for 30 seconds will clear all stored trouble codes. Trouble codes will be cleared from the ECU after the fault is repaired.

NOTE: To prevent damage to the ECM, the ignition switch must be turned to the OFF position when disconnecting voltage to the ECM (battery cable, ECM pigtail, ECM fuse, jumper cables). Some vehicle computers have a learning ability. If a change is noted in vehicle performance after clearing codes, it may be due to the computers learning ability. To restore performance, warm vehicle to normal operating temperature and drive at part throttle with moderate acceleration, before performing any additional diagnosis.

ENGINE CODES

CODE 12—ECU functioning
CODE 13—Oxygen sensor
CODE 14—Coolant temperature HIGH
CODE 15—Coolant temperature LOW
CODE 21—Throttle position sensor HIGH
CODE 22—Throttle position sensor LOW
CODE 24—Vehicle speed sensor
CODE 32—EGR system
CODE 33—Manifold absolute pressure voltage HIGH
CODE 34—Manifold absolute pressure voltage LOW
CODE 42—Electronic spark timing circuit
CODE 43—Electronic spark control circuit
CODE 44—Oxygen sensor, exhaust Lean
CODE 45—Oxygen sensor, exhaust Rich
CODE 51—Prom error
CODE 52—Calpak error
CODE 54—Fuel pump circuit
CODE 55—ECM error

LEXUS FUEL INJECTION

General Information

This system is broken down into 3 major systems: the Fuel System, Air Induction System and the Electronic Control System.

Lexus engines are equipped with a computer which centrally controls the electronic fuel injection, electronic spark advance and the exhaust gas recirculation valve. The systems can be diagnosed by means of an Electronic Control Unit (ECU). The ECU receives signals from the various sensors indicating changing engine operations conditions such as:

1. Intake air flow
2. Intake air temperature
3. Coolant temperature sensor
4. Engine rpm
5. Acceleration/deceleration
6. Exhaust oxygen content

These signals are utilized by the ECU to determine the injection duration necessary for an optimum air-fuel ratio.

Fail-Safe Function

In the event of a computer malfunction, a backup circuit will take over to provide minimal driveability. Simultaneously, the "Check Engine" warning light is activated.

Data Sensors

The data sensors provide information to the ECU for processing. The information is used to adjust the controlled outputs for maximum engine control under changing situations.

Controlled Outputs

These are systems controlled by the ECU providing the adjustments to the operating parameters. The ECU adjusts these systems in response from inputs from the data sensors.

Fuel System

An electric fuel pump supplies sufficient fuel, under a constant pressure, to the EFI injectors. These injectors pass a metered quantity of fuel into the intake manifold in accordance with signals from the EFI computer.

Self-Diagnostic System

SERVICE PRECAUTIONS

- Do not operate the fuel pump when the fuel lines are empty.
- Do not reuse fuel hose clamps.
- Make sure all EFI harness connectors are fastened securely. A poor connection can cause an extremely high surge voltage and result in damage to integrated circuits.
- Keep the EFI harness at least 4 in. away from adjacent harnesses to prevent an EFI system malfunction due to external electronic "noise."
- Keep EFI all parts and harnesses dry during service.
- Before attempting to remove any parts, turn **OFF** the ignition switch and disconnect the battery ground cable.
- Always use a 12 volt battery as a power source.
- Do not attempt to disconnect the battery cables with the engine running.
- Do not depress the accelerator pedal when starting.
- Do not rev up the engine immediately after starting or just prior to shutdown.
- Do not attempt to disassemble the EFI control unit under any circumstances.
- If installing a 2-way or CB radio, keep the antenna as far as possible away from the electronic control unit. Keep the antenna feeder line at least 8 in. away from the EFI harness and do not let them run parallel for a long distance. Be sure to ground the radio to the vehicle body.
- Handle air flow meter carefully to avoid damage.
- Do not disassemble air flow meter or clean meter with any type of detergent.

TESTING PRECAUTIONS

- Before connecting or disconnecting control unit ECU harness connectors, make sure the ignition switch is **OFF** or the negative battery cable is disconnected to avoid the possibility of damage to the control unit.
- When performing ECU input/output signal diagnosis, remove the water-proofing rubber plug, if equipped, from the connectors to make it easier to insert tester probes into the connector.
- When connecting or disconnecting pin connectors from the ECU, take care not to bend or break any pin terminals. Check that there are no bends or breaks on ECU pin terminals before attempting any connections.
- Before replacing any ECU, perform the ECU input/output signal diagnosis to make sure the ECU is functioning properly or not.

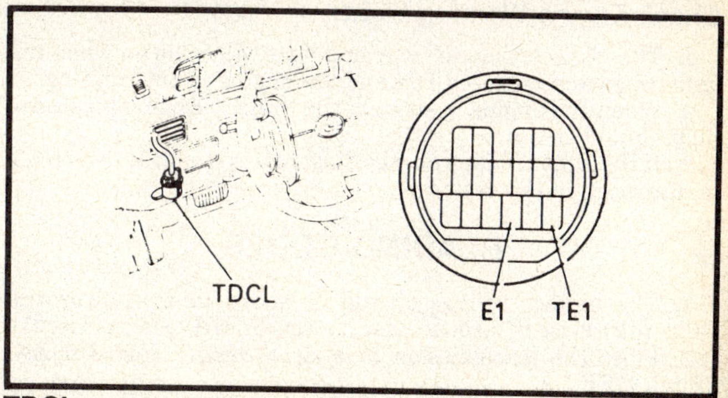

TDCL connector for retrieving codes

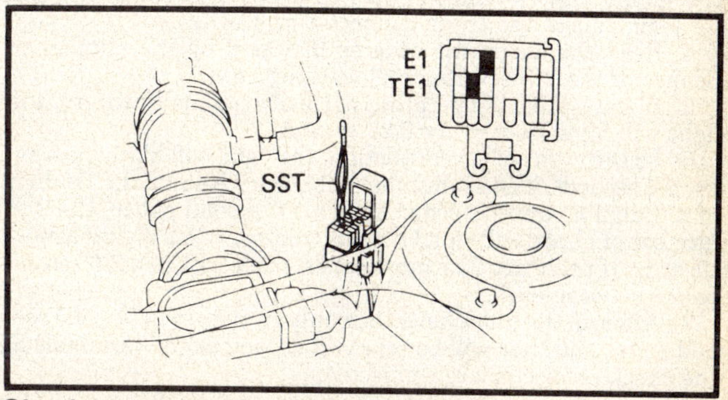

Check connector for retrieving codes

- When measuring supply voltage of ECU controlled components with a circuit tester, separate a tester probe from another. If the 2 tester probes accidentally make contact with each other during measurement, a short circuit will result and damage the ECU.

ENTERING SELF-DIAGNOSTICS

The ECU contains a built-in self diagnosis system by which troubles with the engine signal the engine signal network are detected and a "Check Engine" warning light on the instrument panel flashes Code No. 12–71 (these code numbers vary from model to model). The "Check Engine" light on the instrument panel informs the driver that a malfunction has been detected. The light goes out automatically when the malfunction has been cleared.

The diagnostic code can be read by the number of blinks of the "Check Engine" warning light when the proper terminals of the check connector are connected.

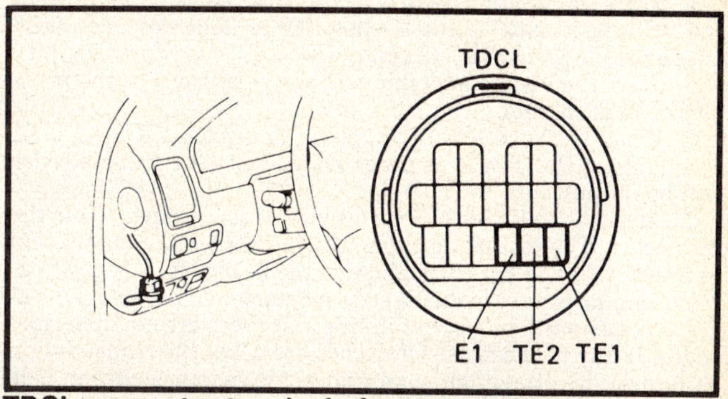

TDCL connector terminals for test mode code retrieval

Check Engine Warning Light

1. The "Check Engine" warning light will come on when the ignition switch is placed **ON** and the engine is not running.
2. When the engine is started, the "Check Engine" warning light should go out.
3. If the light remains on, the diagnosis system has detected a malfunction in the system.

READING CODES

1. The battery voltage should be above 11 volts. Throttle valve fully closed (throttle position sensor IDL points closed).
2. Place the transmission in **P** or **N** range. Turn the A/C switch **OFF**.
3. Turn the ignition switch to the **ON** position. Do not start the engine. Use a suitable jumper wire and connect the TE1 and E1 terminals of the check connector or TDCL.
4. Read the diagnostic code as indicated by the number of flashes of the "Check Engine" warning light.
5. If the system is operating normally (no malfunction), the light will blink once every 0.25 seconds.
6. In the event of a malfunction, the light will blink once every 0.5 seconds. The 1st number of blinks will equal the 1st digit of a 2 digit diagnostic code. After a 1.5 second pause, the 2nd number of blinks will equal the 2nd number of a 2 digit diagnostic code. If there are 2 or more codes, there will be a 2.5 second pause between each.
7. After all the codes have been output, there will be a 4.5 second pause and they will be repeated as long as the terminals of the check connector are connected.

NOTE: In event of multiple trouble codes, indication will begin from the smaller value and continue to the larger in order.

8. After the diagnosis check, remove the jumper wire from the check connector.

TEST MODE

The test mode is available on the ES300, SC300, SC400 and LS400. This mode allows a more comprehensive malfunction diagnosis capability than the standard output test. The test mode can detect malfunctions in the cam position sensors, the IDL contact signal from the throttle position sensor, the air conditioner signal, neutral start switch signal and the starter signal circuit. The test mode will also show the standard outputs.

1. The battery voltage should be above 11 volts. Throttle valve fully closed (throttle position sensor IDL points closed).
2. Place the transmission in **P** or **N** range. Turn the A/C switch to the **OFF** position.
3. Turn the ignition switch to the **OFF** position. Use a suitable jumper wire and connect the TE2 and E1 terminals of the check connector or TDCL.
4. Turn the ignition switch to the **ON** position. Check that the "Check Engine" light flashes. If the light does not flash check the TE2 terminal circuit.
5. Start the engine and simulate the conditions of the problem or malfunction.
6. When the road test is complete, connect the TE1 and E1 terminals of the TDCL or check connector. The trouble codes will be output.
7. After the codes are output disconnect the jumpers from the connector. The test mode cannot be initiated if the connection between TE2 and E1 are made with the key in the **ON** position. If the engine is not cranked a code for starter signal and cam position sensors will be set, but in this case these are not abnormal codes. If any of the sensed switches are used, the transmission shift lever, throttle or air conditioner, the switch condition code will be set, but this is not abnormal either.

CLEARING CODES

1. After repairing the trouble area, the diagnostic code that is retained in the ECU memory must be canceled out by removing the EFI (15A, 20A or 30A depending on model and year) fuse for 30 seconds or more, depending on the ambient temperature (the lower temperature, the longer the fuse must be left out with the ignition switch **OFF**).

NOTE: Cancellation can also be done by removing the battery negative terminal, but keep in mind, when removing the negative battery cable, the other memory systems (radio, ETR, clock, etc.) will also be canceled out.

2. If the diagnostic code is not canceled out, it will be retained by the ECU and appear along with a new code in event of future trouble. If it is necessary to work on engine components requiring removal of the battery terminal, a check must first be made to see if a diagnostic code is detected.
3. After cancellation, perform a road test, if necessary, confirm that a normal code is now read on the "Check Engine" warning light or super monitor display.
4. If the same diagnostic code is still indicated, it indicates that the trouble area has not been repaired thoroughly.

Diagnosis Indication

1. Including "Normal", the ECU is preprogrammed with diagnostic codes.
2. When more than a single code is indicated, the lowest number code will appear first. However, no other code will appear along with Code 11 which indicates normal operation.
3. All detected diagnostic codes, except Code 51 and 53, will be retained in memory by the ECU from the time of detection until canceled out.
4. Once the malfunction is cleared, the "Check Engine" warning light on the instrument panel will go out but the diagnostic code(s) remain stored in the ECU memory (except for Code 51).

Code 12—RPM signal
Code 13—RPM signal
Code 14—Ignition signal No. 1
Code 15—Ignition signal No. 2
Code 16—Electronic Controlled Transmission control signal
Code 17—No. 1 cam position sensor signal
Code 18—No. 2 cam position sensor signal
Code 21—Main oxygen sensor signal (left bank on LS400)
Code 22—Water temperature sensor signal
Code 24—Intake air temperature sensor signal
Code 25—Air-Fuel ratio lean malfunction
Code 26—Air-Fuel ratio rich malfunction
Code 27—Sub-oxygen sensor signal (left bank)
Code 28—Main oxygen sensor signal (right bank)
Code 29—Sub-oxygen sensor signal (right bank)
Code 31—Air flow meter signal
Code 32—Air flow meter signal
Code 35—High Altitude Compensator sensor signal
Code 41—Throttle position sensor signal
Code 42—Vehicle speed sensor signal
Code 43—Starter signal
Code 47—Sub-throttle position sensor signal
Code 51—Switch condition signal
Code 52—No. 1 knock sensor signal
Code 53—Knock control signal
Code 55—No. 2 knock sensor signal
Code 78—Fuel pump control signal
Code 71—EGR system malfunction

MAZDA
EXCEPT NAVAJO

General Information

ENGINE CONTROL UNIT (ECU)

The ECU, through various input signals, monitors battery voltage, engine rpm, amount of air intake, cranking signal, intake temperature, coolant temperature, oxygen concentration in the exhaust gases, throttle opening, atmospheric pressure gearshift position, clutch engagement, braking, power steering operation and A/C compressor operation.

The ECU controls the operation of the fuel injection system, idle-up system, evaporative emissions system and ignition timing. The ECU has a built in fail-safe mechanism. If a fault occurs while driving, The ECU will substitute pre-programmed values. Driving performance will be affected, but the vehicle will still be mobile.

Self-Diagnosis System

NOTE: On vehicles equipped with Audio Anti-Theft system, before performing any repairs, the Personal Code Number (P.C.N.) must be obtained from the individual owner of the vehicle, or an authorized Mazda dealer.

SERVICE PRECAUTIONS

- Before connecting or disconnecting control unit ECU harness connectors, make sure the ignition switch is OFF and the negative battery cable is disconnected to avoid the possibility of damage to the control unit.
- When performing ECU input/output signal diagnosis, remove the pin terminal retainer from the connectors to make it easier to insert tester probes into the connector.
- When connecting or disconnecting pin connectors from the ECU, take care not to bend or break any pin terminals. Check that there are no bends or breaks on ECU pin terminals before attempting any connections.
- Before replacing any ECU, perform the ECU input/output signal diagnosis to make sure the ECU is functioning properly or not.
- After checking through EGI troubleshooting, perform the EFI self-diagnosis and driving test.
- When measuring supply voltage of ECU controlled components with a circuit tester, separate 1 tester probe from another. If the 2 tester probes accidentally make contact with each other during measurement, a short circuit will result and damage to the ECU.

READING CODES

With Self-Diagnosis Checker

The self-diagnosis checker (49–H018–9A1) and System Selector (49–B019–9A0, used on Miata, MX-3, 323, Protege and 1992 929) are used to retrieve code numbers of malfunctions which have happened and were memorized or are continuing. The malfunction is indicated by the code number and a buzzer.

If there is more than 1 malfunction, the code numbers will display on the self-diagnosis checker 1 by 1 in numerical order. In the case of malfunctions, 09, 13, and 01, the code numbers are displayed in a order of 01, 09 and then 13.

The memory of malfunctions is canceled by disconnecting the negative battery cable.

The ECU has a built in fail-safe mechanism for the main input sensors. If a malfunction occurs, the emission control unit will substitute values. This will slightly effect the driving performance, but the vehicle may still be driven.

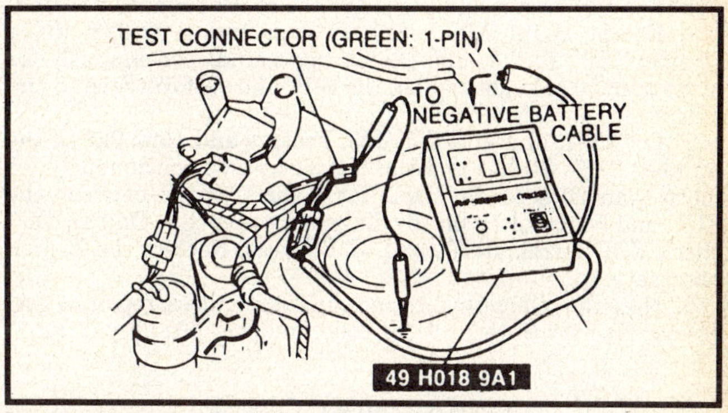

Connecting the self-diagnosis checker to vehicles without the diagnostic port

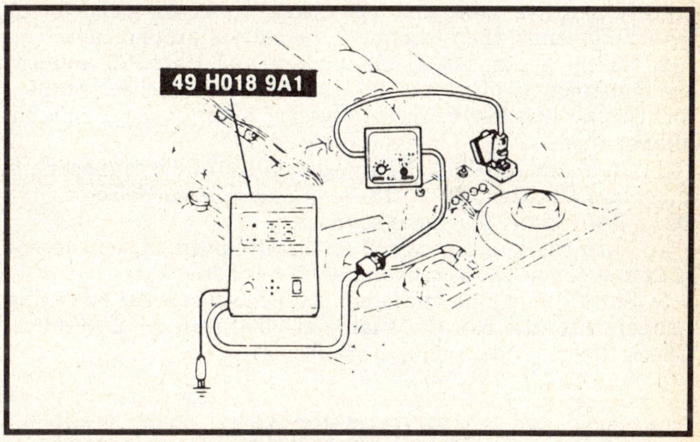

Connecting the self-diagnosis checker and system selector to vehicles with diagnostic port

The ECU continuously checks for malfunctions of the input devices. But, the ECU checks checks for malfunctions of the output devices within 3 seconds after turning the ignition switch to the **ON** position and the test connector is grounded.

The malfunction indicator light indicates a pattern the same as the buzzer of the self-diagnosis checker when the self-diagnosis check connector is grounded. When the self-diagnosis check connoctor is not grounded, the lamp illuminates steady while malfunction of the main input sensor occurs and goes out if the malfunction recovers. However, the malfunction code is memorized in the emission control unit.

Testing Procedure

1. Except Miata, MX-3, 323, Protege and 1992 929 connect the Self-Diagnosis Checker (49–H018–9A1) or equivalent to the check connector. The check connector is usually located above the right side wheel housing.
2. On Miata, MX-3, 323, Protege and 1992 929, connect Self-Diagnosis Checker (49–H018–9A1) and System Selector (49–B019–9A0) to the diagnosis connector and ground.
3. Set the select switch on the Self-Diagnosis Checker to the "A" position.
4. Except Miata, MX-3, 323, Protege and 1992 929, ground the green test connector using a suitable jumper wire.
5. On Miata, MX-3, 323, Protege and 1992 929 set the System Selector (49–B019–9A0) to position 1 and the Self-TEST switch to the up position.

6. Turn the ignition switch to the **ON** position. Check that the number "88" flashes on the digital display and the buzzer sounds for 3 seconds after turning the ignition switch **ON**.
7. If the number "88" does not flash, check the main relay, power supply circuit and the check the check connector wiring.
8. Except Miata, MX-3, 323, Protege and 1992 929, if the number "88" flashes and the buzzer sounds continuously for more than 20 seconds, replace the ECU and perform Steps 3 and 4 again.
9. On RX-7, Miata, MX-3, 323, Protege and 1992 929, if the number "88" flashes and the buzzer sounds continuously for more than 20 seconds, check for a short circuit between the ECU and terminal 1F on the diagnosis connector. On the RX-7 check ECU terminals 3X and 3Z voltages. Replace the ECU if necessary.
10. Note the code numbers and check the causes, repair as necessary. Be sure to recheck for code numbers after repairing.

CLEARING CODES

1. Cancel the memory of malfunctions by disconnecting the battery negative cable and depressing the brake pedal for at least 20 seconds, then reconnect the battery ground cable.
2. Except Miata, MX-3, 323, Protege and 1992 929, connect the Self-Diagnosis Checker 49–H018–9A1 to the check connector. Ground the test connector (green: 1 pin) using a suitable jumper wire.
3. On Miata, MX-3, 323, Protege and 1992 929, connect Self-Diagnosis Checker (49–H018–9A1) and System Selector (49–B019–9A0) to the diagnosis connector.
4. Turn the ignition switch to the **ON** position, but do not start the engine for approximately 6 seconds.
5. Start the engine and allow it to reach normal operating temperature, the run the engine at 2000 rpm for 2 minutes. Check that no code numbers are displayed.

ENGINE CODES

Except RX-7 and Carbureted B2200

Code 1—Ignition Pulse
Code 2—Crank Angle Sensor
  Ne Except MX-3 with 6 cylinder and 1992 929
  Ne2 MX-3 with 6 cylinder and 1992 929
Code 3—Crank Angle Sensor-G
Code 4—Crank Angle Sensor
  Ne1 MX-3 with 6 cylinder and 1992 929
  G2 626 Turbo, 1990–91 929
Code 5—Knock Sensor (Except 1992 929)
Code 5—Left Side Knock Sensor (1992 929)
Code 6—Vehicle Speed Sensor
Code 7—Right Side Knock Sensor (1992 929)
Code 8—Air Flow Meter
Code 9—Water Temperature Sensor
Code 10—Intake Air Temperature Sensor-at airflow meter
Code 11—Intake Air Temperature Sensor-at dynamic chamber
Code 12—Throttle Sensor
Code 14—ECA Atmospheric Pressure Sensor
Code 15—Oxygen Sensor (Left side on V6 engines)
Code 16—EGR Position Sensor
Code 17—Oxygen Sensor Feedback System
Code 23—Right Side Oxygen Sensor
Code 24—Right Side Oxygen Sensor Feedback System
Code 25—Pressure Regulator Solenoid Valve
Code 26—Purge Control Solenoid Valve
Code 28—EGR Vacuum Solenoid Valve
Code 29—EGR Vent Solenoid Valve
Code 30—Cold Start Injector Relay
Code 34—Idle Speed Control Valve
Code 36—Right Side Oxygen Sensor Heater
Code 37—Left Side Oxygen Sensor Heater
Code 40—Triple Induction Control and Oxygen Sensor Heater
Code 40—Oxygen Sensor Heater Relay (1990–91 929 DOHC)
Code 41—Variable Resonance Intake Solenoid Valve 1
Code 42—Waste Gate Solenoid Valve
Code 46—Variable Resonance Intake Solenoid Valve 2
Code 55—Pulse Generator
Code 60—1–2 Shift Solenoid Valve
Code 61—2–3 Shift Solenoid Valve
Code 62—3–4 Shift Solenoid Valve
Code 63—Lockup Solenoid Valve
Code 65—Air Conditioning Signal To EC-AT
Code 67—Electrical Cooling Fan Relay 2
Code 69—Fan Water Temperature Sensor

B2200 with carburetor

Code 1—Ignition Pulse
Code 9—Water Temperature Sensor
Code 13—Vacuum Sensor
Code 14—Atmospheric Pressure Sensor
Code 15—Oxygen Sensor Open or Short
Code 16—EGR System
Code 17—Oxygen Sensor Feedback System
Code 18—Air/Fuel Solenoid Valve
Code 22—Slow Fule Cut Solenoid Valve
Code 23—Coasting Richer Solenoid Valve
Code 26—Purge Solenoid Valve
Code 28—Duty Solenoid Vacuum Valve
Code 29—Duty Solenoid Vent Valve
Code 30—Air Control Valve Solenoid
Code 34—Idle-Up Solenoid Valve for AC
Code 35—Idle-Up Solenoid Valve for AT
Code 45—Vacuum Solenoid Valve

RX-7

Code 1—Trailing Side Ignition Pulse
Code 2—Crank Angle Sensor-Ne
Code 3—Crank Angle Sensor-G
Code 5—Knock Sensor
Code 8—Air Flow Meter
Code 9—Water Temperature Sensor
Code 10—Intake Air Temperature Sensor-at airflow meter
Code 11—Intake Air Temperature Sensor-at dynamic chamber
Code 12—Throttle Sensor
Code 13—Intake Manifold Pressure Sensor
Code 14—ECA Atmospheric Pressure Sensor
Code 15—Oxygen Sensor
Code 17—Oxygen Sensor Feedback System
Code 18—Narrow Range Throttle Sensor
Code 20—Metering Oil Pump Position Sensor
Code 25—Pressure Regulator Solenoid Valve
Code 26—Metering Oil Pump Step Motor
Code 27—Metering Oil Pump

MAZDA NAVAJO

General Information

The EEC-IV engine control system heart is a microcomputer called an Electronic Control Assembly (ECA). This EEC-IV system is the same ECC-IV system used by Ford Motor Company, testing and servicing this system is identical. The ECA receives data (system inputs) from sensors, switches, relays and other electronic components and issues command signals (system outputs) to various devices in order to control engine operation under a variety of loads and ambient conditions. The ECA is calibrated according to the powertrain, axle ratio and gross vehicle weight (GVW) to optimize fuel economy and driveability while minimizing harmful emissions.

The EEC-IV system electronically controls the fuel injectors to maintain a 14.7:1 air/fuel ratio under all driving conditions. This 14.7:1 air/fuel mixture allows the 3-way catalytic converter to operate at peak efficiency while getting the most performance and economy from the engine. Ignition spark timing, deceleration fuel cut-off, curb and fast idle speed, evaporative emissions purge, air conditioning cut-off during wide open throttle, cold engine start and enrichment, electric fuel pump and self-test engine diagnostics are also controlled by the ECA to maintain consistent driveability across a wide range of operating conditions, temperatures and altitudes. The EEC-IV system is self-adjusting for operation at high altitude elevations (over 4000 ft. above sea level) and can actually compensate for engine component wear (a worn timing chain, for example).

The ECA in the EEC-IV system contains the calibration module within the ECA assembly. The harness connectors are edge-card type contacts which provide a more positive connection. The ECA is mounted behind the right front (passenger) kick panel.

Limited Operation Strategy (LOS)

In this operation, the ECA provides the necessary output signals to allow the vehicle to "limp home" when an electronic malfunction occurs, sometimes referred to as limp-in mode. The air bypass valve goes to a fixed voltage, timing is locked at the fixed timing (depends on calibration), and the injector pulse width is constant.

Self-Diagnostic System

NOTE: On vehicles equipped with Audio Anti-Theft system, before performing any repairs, the Personal Code Number (P.C.N.) must be obtained from the individual owner of the vehicle, or an authorized Mazda dealer.

SERVICE PRECAUTIONS

- Do not operate the fuel pump when the fuel lines are empty.
- Do not reuse fuel hose clamps.
- Make sure all ECA harness connectors are fastened securely. A poor connection can cause an extremely high surge voltage in the coil and condenser and result in damage to integrated circuits.
- Keep the EGI harness at least 4 in. away from adjacent harnesses to prevent an EGI system malfunction due to external electronic "noise."
- Keep all ECA related parts and harnesses dry during service.
- Before attempting to remove any parts, turn off the ignition switch and disconnect the battery ground cable.
- Always use a 12 volt battery as a power source.
- Do not attempt to disconnect the battery cables with the engine running.
- Do not attempt to disassemble the ECA control unit under any circumstances.
- If installing a 2-way or CB radio, keep the antenna as far as possible away from the electronic control unit. Keep the antenna feeder line at least 8 in. away from the EGI harness and do not let the 2 run parallel for a long distance. Be sure to ground the radio to the vehicle body.
- Do not apply battery power directly to injectors.
- Handle air flow meter carefully to avoid damage.
- Do not disassemble air flow meter or clean meter with any type of detergent.

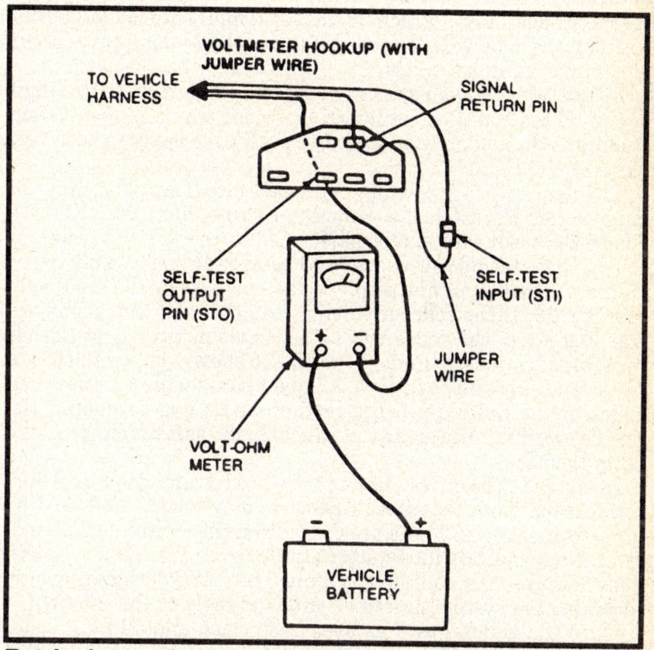

Retrieving codes using an analog voltmeter

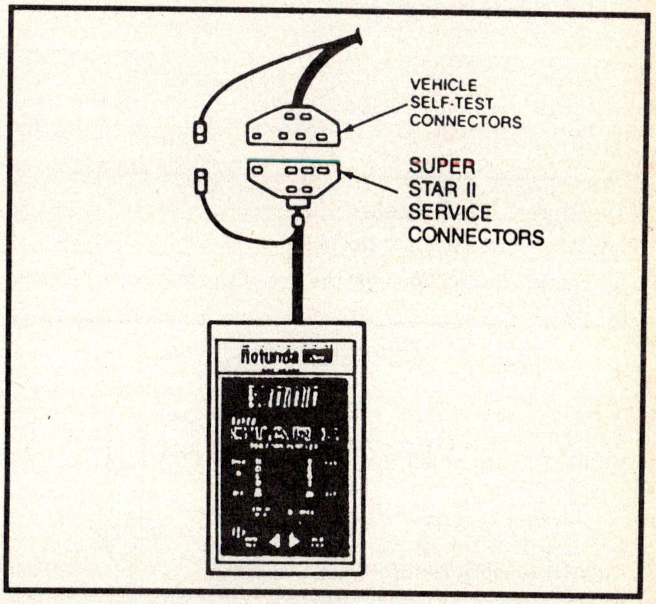

Retrieving codes using Super Star II tester

NOTE: If the battery cable(s) is disconnected for longer than 5 minutes, the adaptive fuel factor will be lost. After repair it will be necessary to drive the vehicle at least 10 miles to allow the processor to relearn the correct factors. The driving period should include steady-throttle open road driving if possible. During the drive, the vehicle may exhibit driveability symptoms not noticed before. These symptoms should clear as the ECA computes the correction factor. The ECA may also store code indicating loss of power to the controller.

READING CODES

The EEC-IV system may be interrogated for stored codes using the Quick Test procedures. These tests will reveal faults immediately present during the test as well as any intermittent codes set within the previous 80 warm up cycles. If a code was set before a problem self-corrected (such as a momentarily loose connector), the code will be erased if the problem does not reoccur within 80 warm-up cycles.

In 1991 the EEC-VI system check from 2 digit codes to 3 digit codes. These codes are obtained the same way and the system functions the same. All of the test procedures apply to both versions of the ECA.

The Quick Test procedure is divided into 2 sections, Key On Engine Off (KOEO) and Key On Engine Running (KOER). These 2 procedures must be performed correctly if the system is to run the internal self-checks and provide accurate fault codes. Codes will be output and displayed as numbers on the hand scan tool, i.e. 23. If the codes are being read through the dashboard warning lamp, the codes will be displayed as groups of flashes separated by pauses. Code 23 would be shown as two flashes, a pause and three more flashes. A longer pause will occur between codes. If the codes are being read on an analog voltmeter, the needle sweeps indicate the code digits in the same manner as the lamp flashes.

In all cases, the codes 11 or 111 are used to indicate PASS during testing. Note that the PASS code may appear, followed by other stored codes. These are codes from the continuous memory and may indicate intermittent faults, even though the system does not presently contain the fault. The PASS designation only indicates the system passes all internal tests at the moment.

Once the Quick Test has been performed and all fault codes recorded, refer to the code charts. The charts direct the use of specific pinpoint tests for the appropriate circuit and will allow complete circuit testing.

CAUTION

To prevent injury and/or property damage, always block the drive wheels, firmly apply the parking brake, place the transmission in P or N and turn all electrical loads off before performing the Quick Test procedures.

With Super Star II Tester

KEY ON ENGINE OFF (KOEO)

1. Specific instructions for the use of the tester are included with the tester.

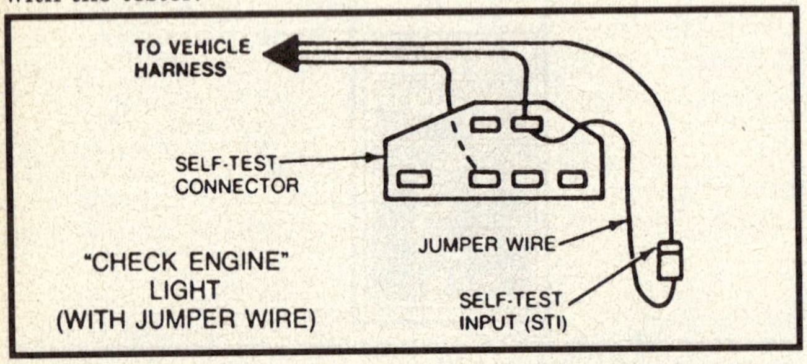

Retrieving codes using the instrument panel light

2. Plug the tester connectors into the vehicle self-test connectors.
3. Set the switches on the tester to the appropriate settings for type of EEC system and for fast or slow mode. The same information is contained in fast or slow mode, but the fast mode is 100 times faster than the slow and is for use with the Super Star Tester.
4. Turn on the tester power and depress the test button. Turn **ON** the vehicle ignition key.
5. The tester will read the codes as they exist.

KEY ON ENGINE RUNNING (KOER)

1. Connect the tester as in KOEO test and start and warm the engine to normal operating temperature.
2. Turn the engine off and depress the test button so it stays in the test position.
3. Restart the engine and the tester will read any existing codes. The tester will display an engine I.D. code, then on some vehicles a Dynamic Response code. The final display will be the service codes.

With Analog Voltmeter

KEY ON ENGINE OFF (KOEO) AND CONTINUOUS CODES

1. Turn the vehicle ignition switch **OFF**. Set the range of the analog voltmeter to 0–15 volts.
2. Connect the positive lead of an analog voltmeter to the positive terminal of the battery.
3. Connect the negative lead of the voltmeter to the Self-Test Output pin of the Self-Test connector.
4. Connect a jumper wire between the Self-Test Input connector and the Signal Return pin of the Self-Test connector.
5. The needle will sweep 3 times for a code digit of 3, for example. If the code is 123, the needle will sweep once, pause for 2 seconds, sweep twice, pause for 2 seconds, then sweep three times. There will be a 4 second pause between codes. The method of readout for 2 digit codes is similar to the three digit code readout, but only 2 sets of sweeps will occur.
4. The continuous memory codes will be displayed first, then there will be a 6 second pause, a single sweep, another 6 second pause then the KOEO codes will be displayed.

WITH MALFUNCTION INDICATOR OR CHECK ENGINE LIGHT

KEY ON ENGINE OFF (KOEO) AND CONTINUOUS CODES

1. Connect a jumper wire between the Self-Test Input (STI) connector and the Signal Return pin of the Self-Test connector.
2. This will output the codes to the MIL or Check Engine light on the instrument panel.
3. The light will flash 3 times for a code digit of 3, for example. If the code is 123, the light will flash once, pause for 2 seconds, flash twice, pause for 2 seconds, then flash three times. There will be a 4 second pause between codes. The method of readout for 2 digit codes is similar to the three digit code readout, but only 2 sets of sweeps will occur.
4. The continuous memory codes will be displayed first, then there will be a 6 second pause, a single flash, another 6 second pause then the KOEO codes will be displayed.

Other Test Modes

CONTINUOUS MONITOR OR WIGGLE TEST MODE

Once entered, this mode allows the technician to attempt to recreate intermittent faults by wiggling or tapping components, wiring or connectors. The test may be performed during either KOEO or KOER procedures. The test requires the use of either an analog voltmeter or a hand scan tool.

To enter the continuous monitor mode during KOEO testing, turn the ignition switch **ON**. Activate the test, wait 10 seconds, then deactivate and reactivate the test; the system will enter the continuous monitor mode. Tap, move or wiggle the harness, component or connector suspected of causing the problem; if a fault is detected, the code will store in the memory. When the fault occurs, the dash warning lamp will illuminate, the STAR tester will light a red indicator (and possibly beep) and the analog meter needle will sweep once.

To enter this mode in the KOER test:

1. Start the engine and run it at 2000 rpm for two minutes. This action warms up the oxygen sensor.
2. Turn the ignition switch **OFF** for 10 seconds.
3. Start the engine.
4. Activate the test, wait 10 seconds, then deactivate and reactivate the test; the system will enter the continuous monitor mode.
5. Tap, move or wiggle the harness, component or connector suspected of causing the problem; if a fault is detected, the code will store in the memory.
6. When the fault occurs, the dash warning lamp will illuminate, the STAR tester will light a red indicator (and possibly beep) and the analog meter needle will sweep once.

OUTPUT STATE CHECK

This testing mode allows the operator to energize and de-energize most of the outputs controlled by the EEC-IV system. Many of the outputs may be checked at the component by listening for a click or feeling the item move or engage by a hand placed on the case. To enter this check:

1. Enter the KOEO test mode.
2. When all codes have been transmitted, depress the accelerator all the way to the floor and release it.
3. The output actuators are now all ON. Depressing the throttle pedal to the floor again switches the all the actuator outputs OFF.
4. This test may be performed as often as necessary, switching between ON and OFF by depressing the throttle.
5. Exit the test by turning the ignition switch **OFF**, disconnecting the jumper at the diagnostic connector or releasing the test button on the scan tool.

CLEARING CODES

Continuous Codes

1. Run the Self-Test for the output system being used; Super Star II, analog voltmeter, or dash light.
2. When the service codes start to be displayed the codes will be cleared when the Test button is unlatched on the Super Star II tester. For the other retrieval methods the codes will be cleared when the jumper is removed between the Self-Test Input connector and the Signal Return pin of the Self-Test connector.
3. Disconnect the Super Star II tester or voltmeter.

Keep Alive Memory

The Keep Alive Memory will be erased when the negative battery cable is disconnected. Keep the cable off for a minimum of 5 minutes. When the cable is connected and the vehicle driven, it may exhibit some drivability problems for the first 10 miles or so. This is due to the relearning process where the ECA memorizes values needed for optimum drivability.

Code and Description

Code 11—System pass:
Code 12—Cannot control rpm during high rpm check:
Code 13—Cannot control rpm during low rpm check:
Code 14—PIP circuit failure:
Code 15—ECA keep alive only memory (KAM) test failed:
Code 15—ECA read only memory (ROM) test failed:
Code 16—IDM signal not received:
Code 18—Loss of IDM input or SAW circuit open:
Code 19—Failure in ECA internal voltage:
Code 21—ECT out of range:
Code 22—Barometric Pressure (BP) sensor:
Code 23—TP sensor out of self-test range:
Code 24—ACT sensor out of self-test range:
Code 26—MAF out of self-test range:
Code 29—Insufficient input from VSS:
Code 41—Fuel system at adaptive limits, no HEGO switch:
Code 41—Lack of HEGO switches, indicates lean:
Code 42—Lack of HEGO switches, indicates rich:
Code 45—Coil 1, 2 or 3 failure:
Code 51—ECT indicated −40°F open circuit:
Code 53—TP sensor circuit above maximum voltage
Code 54—ACT indicated −40°F open circuit:
Code 56—MAF circuit above maximum voltage:
Code 61—ECT indicated 254°F grounded circuit:
Code 63—TP sensor circuit below minimum voltage:
Code 64—ACT indicated 254°F grounded circuit:
Code 66—MAF below minimum voltage:
Code 67—Clutch switch circuit failure:
Code 67—Vehicle not in Park during KOEO:
Code 72—Insufficient MAF change during test:
Code 73—Insufficient TP change dynamic rseponse test:
Code 74—BOO circuit failure/not actuated during self-test:
Code 77—Operator error during test:
Code 79—A/C ON/ Defroster ON/ during self-test:
Code 86—Shift solenoid circuit failure
Code 87—Primary fuel pump circuit faiure:
Code 89—Clutch Converter Override (CCO) circuit failure:
Code 95—Fuel pump circuit open-ECA to motor ground:
Code 96—Fuel pump circuit open-ECA to battery:
Code 98—Hard fault present:
Code not listed:

Navajo 2 digit codes

Code and Description

Code 111—System pass:
Code 112—ACT indicated 254°F grounded circuit:
Code 113—ACT indicated −40°F open circuit:
Code 114—ACT sensor out of self-test range:
Code 116—ECT out of range:
Code 117—ECT indicated 254°F grounded circuit:
Code 118—ECT indicated −40°F open circuit:
Code 121—TP sensor out of self-test range:
Code 122—TP sensor circuit below minimum voltage:
Code 123—TP sensor circuit above maximum voltage
Code 126—Barometric Pressure (BP) sensor:
Code 128—Barometric Pressure (BP) sensor:
Code 129—Insufficient MAF change during test:
Code 136—Fuel control:
Code 137—Fuel control:
Code 139—Fuel control:
Code 144—Fuel control:
Code 157—MAF below minimum voltage:
Code 158—MAF circuit above maximum voltage:
Code 159—MAF out of self-test range:
Code 167—Insufficient TP change dynamic response test:
Code 171—Fuel system at adaptive limits, no HEGO switch:
Code 172—Lack of HEGO switches, indicates lean:
Code 173—Lack of HEGO switches, indicates rich:
Code 174—Fuel control:
Code 175—Fuel control:
Code 176—Fuel control:
Code 177—Fuel control:
Code 178—Fuel control:
Code 179—Lean adaptive limit at part throttle, system rich:
Code 181—Rich adaptive limit at part throttle, system lean:
Code 182—Fuel control:
Code 183—Fuel control:
Code 188—Fuel control:
Code 189—Fuel control:
Code 191—Fuel control:
Code 192—Fuel control:
Code 194—Fuel control:
Code 195—Fuel control:
Code 211—PIP circuit failure:
Code 212—Loss of IDM input or SAW circuit grounded:
Code 213—SAW circuit open:
Code 215—Ignition system:
Code 216—Ignition system:
Code 217—Ignition system:
Code 218—Ignition system:
Code 222—Ignition system:
Code 223—IDM signal not received:
Code 224—IDM signal not received:
Code 226—IDM signal:
Code 232—Coil 1, 2 or 3 failure:
Code 238—Ignition system:
Code 338—Temperature sensor:
Code 339—Temperature sensor:
Code 341—Octane adjust circuit open:
Code 411—Cannot control rpm during low rpm check:
Code 412—Cannot control rpm during high rpm check:
Code 452—Insufficient input from VSS:
Code 511—ECA read only memory (ROM) test failed:
Code 512—ECA keep alive only memory (KAM) test failed:
Code 513—Failure in ECA internal voltage:
Code 522—Vehicle not in P or N during KOEO:
Code 525—Vehicle not in P or N during KOEO:
Code 528—Clutch switch circuit failure:
Code 529—Check Engine light:
Code 533—Check Engine light:
Code 536—BOO circuit failure/not actuated during self-test:
Code 538—Operator error during test:
Code 539—A/C ON/ Defroster ON/ during self-test:
Code 542—Fuel pump circuit open-ECA to motor ground:
Code 542—Fuel pump circuit open-ECA to battery:
Code 556—Primary fuel pump circuit faiure:
Code 565—Canister purge (CANP) circuit failure:
Code 566—Shift solenoid circuit failure:
Code 569—Canister purge (CANP) circuit failure:
Code 629—Clutch Converter Override (CCO) circuit failure:
Code 998—Hard fault present:
Code not listed:
600 Series codes are transmission codes

Navajo 3 digit codes

MITSUBISHI FUEL INJECTION

General Information

The Electronic Fuel Injection (EFI) system, used on Mitsubishi vehicles, is classified as a Multi-Point Injection (MPI) system. The MPI system controls the fuel flow, idle speed, and ignition timing. The basic function of the MPI system is to control the air/fuel ratio in accordance with all engine operating conditions. An Electronic Control Unit (ECU) is the heart of the MPI system. Based on data from various sensors, the ECU computes the desired air/fuel ratio.

SELF-DIAGNOSIS SYSTEM

The Engine Control Unit (ECU) is capable of monitoring both the input and output signals which it performed. When the ECU noticed that an irregularity has continued for a specified time or longer, it memorizes a trouble code and outputs the signal to the self-diagnosis output terminal. There are 16 diagnosis items, including the normal state. If more than 1 item is activated, they will be displayed sequentially from the smallest code number to the largest. The trouble codes can be read out with a voltmeter or multi-tester.

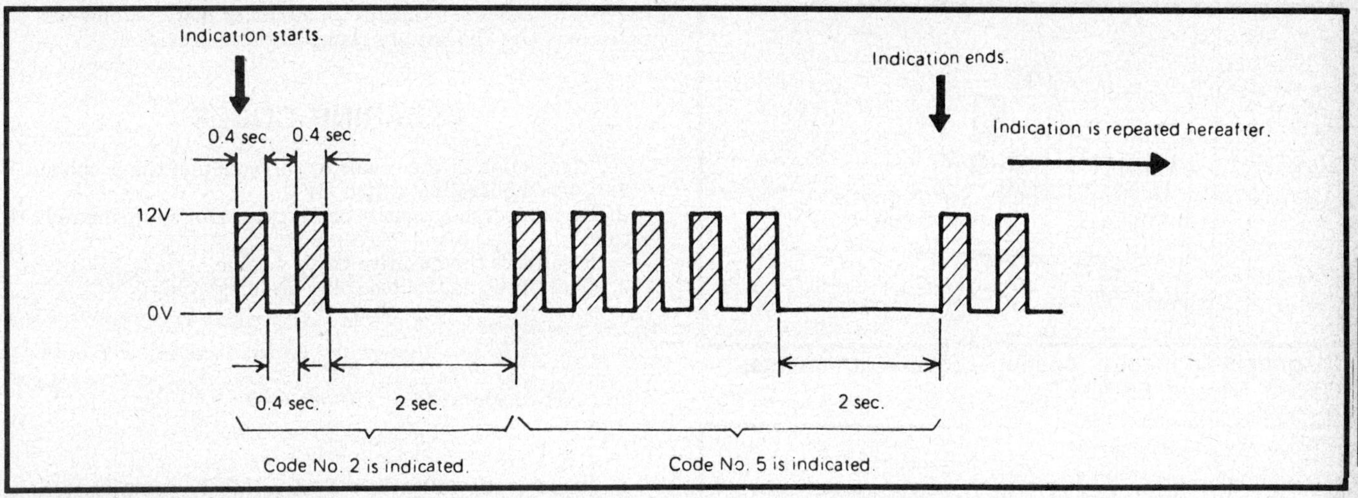

Reading fault code patterns

Among the self-diagnosis items, a Malfunction Indicator Light (MIL) comes ON to notify the driver of the emission control items when an irregularity is detected.

Generally, the MIL will come ON for 5 seconds when the ignition switch is turned **ON**.

NOTE: Always check that the MIL illuminates for approximately 5 seconds, when the ignition switch is turned ON. If the light does not illuminate, check for an open circuit harness, blown fuse or blown bulb.

Self-Diagnostic System

SERVICE PRECAUTIONS

- Do not apply battery power directly to injectors.
- To avoid damage to the ECU, always turn the ignition switch **OFF** before disconnecting the battery cable.
- Do not attempt to disconnect the battery cables with the engine running.
- Do not disconnect wiring connector with the engine running, unless instructed to do so.
- Before removing or installing any part(s), always check the self-diagnosis system.
- After completion of repairs, when the battery was disconnected, operate the engine at idle for approximately 5 minutes in order to stabilize the engine control conditions (Turn all accessories **OFF**).

READING CODES

Using Voltmeter

1. Connect an analog voltmeter to the self-diagnostic connector, across the MPI diagnosis and ground terminal.
2. Turn the ignition switch **ON**. The ECU diagnostics memory will immediately start.
3. Read the diagnosis output pattern displayed by the voltmeter needle deflections and record it.
4. Compare the readings with the diagnosis code(s) and repair the fault.
5. After repairing the fault, erase the ECU memory.

Using Multi-Unit Tester (MUT)

1. Turn the ignition switch **OFF**.
2. Connect the Multi-Unit Tester (MUT), (Tool 09391–33200 or equivalent), to the diagnosis connector in the fuse box.

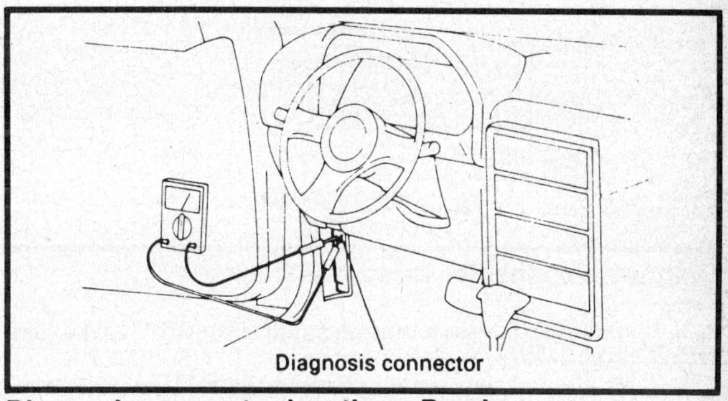

Diagnosis connector location – Precis

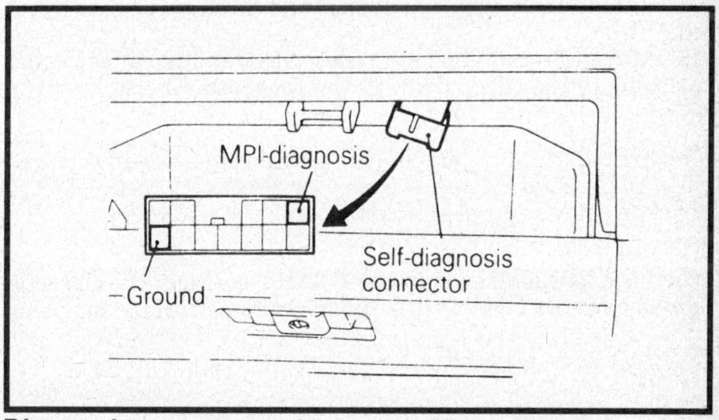

Diagnosis connector above glovebox – Sigma

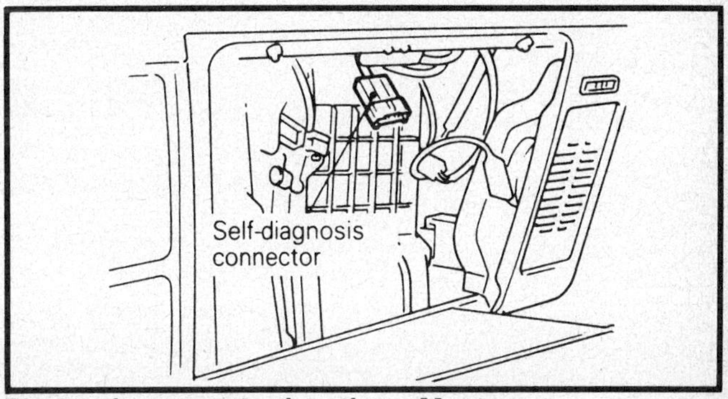

Diagnosis connector location – Montero

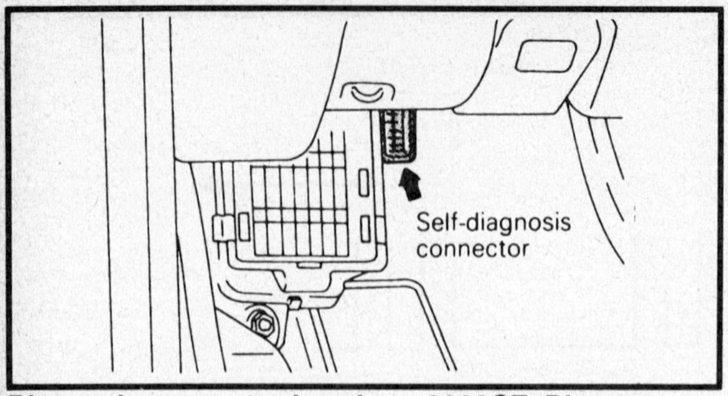

Diagnosis connector location—3000GT, Diamante, Galant, Mirage, EXPO/LRV

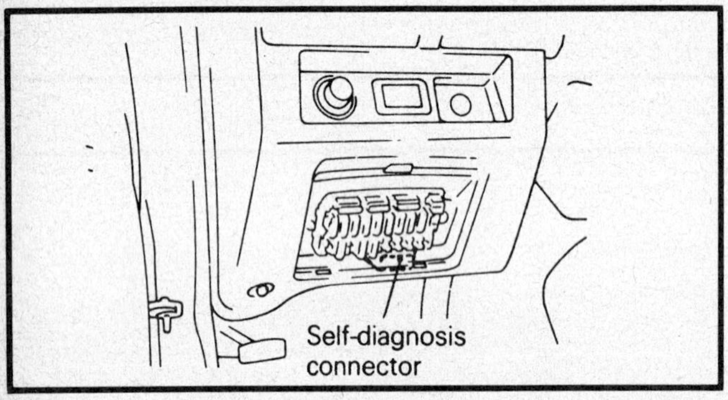

Diagnosis connector location—Truck

3. Connect the power-source terminal of the MUT to the cigarette lighter switch.
4. Turn the ignition switch **ON** and the ECU memory contents will immediately start.
5. Read the self-diagnosis output and check output of correct code.
6. After recording the fault codes, check and repair each part according to the check items in the Diagnosis Charts.

NOTE: If the battery voltage is too low, trouble codes cannot be read. If the battery or the ECU connector is disconnected, the diagnosis memory will be erased.

When the ignition key is in the **START** position, the cigarette lighter power is OFF. If it is necessary to perform tests, using the MUT in the crank mode, an auxiliary power supply will be necessary. Use the battery clamp harness provided.

CLEARING CODES

After completion of the repair or correction of the problem:

1. Turn the ignition switch **OFF**.
2. Disconnect the negative battery cable for approximately 10 seconds.
3. Reconnect the negative battery cable.
4. Check that no trouble codes are displayed.

NOTE: It is not necessary to disconnect the battery terminal, when clearing codes, if the MUT is being used. The MUT is capable of clearing the ECU self-diagnosis memory.

5. Remove the Multi-Use Tester (MUT) or voltmeter, as required.

ENGINE CODES

If the voltmeter pattern stays high; possible bad ECU. If the voltmeter pattern reveals steady pattern the system is normal and has no stored codes. Not all models are equipped with all codes.

High signal—replace ECU.
Code 11—Oxygen content or sensor.
Code 12—Air flow sensor.
Code 13—Intake air temperature.
Code 14—Throttle position sensor.
Code 15—Motor position sensor.
Code 14—Throttle position sensor.
Code 21—Coolant temperature sensor.
Code 22—Crank angle sensor.
Code 23—TDC sensor.
Code 24—Vehicle speed sensor.
Code 25—Barometric pressure sensor.
Code 31—Knock sensor.
Code 36—Ignition.
Code 41—Injector.
Code 42—Fuel pump.
Code 44—Ignition coil.
Code 43—EGR (California).

Electronic Ignition Systems

SECTION 4

ACURA PGM-IG ELECTRONIC IGNITION

General Information

In 1990 Acura continued the Programmed Ignition (PGM-IG) system in both the Integra and Legend. This ignition system is basically comprised of a distributor, an igniter, a control box, a high energy coil and related ignition wires. This ignition system is used in the 1990–92 Integra, 1990 Legend and 1992 Vigor.

The distributor contains a reluctor mounted on a rotor shaft and a magnet mounted on a base plate. The pick-up coil is located around the rotor shaft but does not rotate with the shaft. The programmed ignition employed on these vehicles, provides optimum control of ignition timing by determining the optimum timing using a microcomputer in response to engine speed and vacuum pressure in the intake manifold, which are transmitted by signals from CRANK/CYL sensor, TDC sensor, throttle angle sensor, coolant temperature sensor and MAP sensor. This system, not dependent on a governor or vacuum diaphragm, is capable of setting lead angles with complicated characteristics which cannot be provided by conventional governors or diaphragms.

On the Integra, the igniter, and ignition coil are apart of the distributor assembly. The Legend, the igniter and ignition coil are separate of the distributor and are located near the control box. The igniter contains 3 resistors, 3 diodes and 2 transistors. The transistors act as switches which are activated at a precise voltage.

When the ignition is switched **ON**, the switching of the transistors in the igniter ensures that no current can flow in the ignition coil primary windings. When the engine is cranked, the reluctor moves through the magnetic field created by the stator and, when the reluctor teeth are aligned with the stator projections, a small AC voltage is created. The igniter amplifies this voltage and uses it to switch the transistors so that an earth path is provided to the primary circuit.

As the reluctor teeth move out of alignment with the stator projections, an abrupt change occurs in the AC voltage. The transistors are switched again and the primary circuit earth path is broken. This induces a high voltage in the ignition coil secondary winding.

A time control circuit in the igniter controls the charging time for the ignition coil according to engine speed, this reduces consumption at low engine speeds and prevents secondary voltage drop at high engine speeds.

The 1991–92 Legend and NSX, are equipped with a distributorless ignition system. This system is still considered as Programmed Ignition (PGM-IG) system. The basic difference is, the distributorless system incorporates 1 ignition coil per cylinder, as opposed to the distributor type system which uses 1 ignition coil for the entire system.

System operation is accomplished when the ignition switch is in the **RUN** or **START** position. Battery current is applied through each of the ignition coils to the igniter. The igniter acts as a switch to control current through the primary windings of the ignition coils. The igniter is controlled by the PGM-FI electronic unit. When current to the ignition coil is stopped, a high voltage current flows to the spark plug. The ignition system is controlled by inputs to the PGM-FI electronic control unit. The inputs include; detonation, engine RPM, accelerator position, coolant temperature, ignition timing adjustment, manifold pressure, crankshaft position and exhaust oxygen content. Traction control and the automatic transmission also affect the ignition system.

Diagnosis and Testing

SERVICE PRECAUTIONS

- Do not leave the ignition key turned **ON** for more than 10 minutes, while the engine is not running.
- Make sure all EFI harness connectors are fastened securely. A poor connection can cause an extremely high surge voltage in the coil and condenser and result in damage to integrated circuits.
- Keep EFI all parts and harnesses dry during service.
- Before attempting to remove any parts, turn **OFF** the ignition switch and disconnect the battery ground cable.
- Always use a 12 volt battery as a power source.
- Do not attempt to disconnect the battery cables with the engine running.
- Do not depress the accelerator pedal when starting.
- Make sure the igniter is properly grounded to the body.
- Never allow the coil or tachometer terminals to touch ground or damage will result to the igniter or coil.

DISTRIBUTOR

Top End Inspection

1. Check for rough or pitted rotor and cap terminals.
2. Scrape or file off carbon deposits. Smooth rotor terminal with an oil stone or No. 600 sandpaper, if rough.
3. Check distributor cap for cracks, wear and damages. If necessary, clean or replace it.

IGNITER

Input test

NOTE: The PGM-FI ECU and tachometer should be normal. When the self diagnostic indicator blinks, perform self diagnostic checks on the fuel system. Perform an input test for the igniter unit after completing test on the ignition and fuel systems.

INTEGRA

1. Remove the distributor cap, rotor and leak cover.
2. Turn the ignition switch **ON**, check for battery voltage between the positive terminal and ground. Voltage should be present. If voltage is not present, check for an open in the BLK/YEL wire or a disconnected terminal.
3. If voltage is present, disconnect the BLK/YEL wire from the igniter. Check for battery voltage between the BLK/YEL positive wire and ground. If voltage is not present, check for an open in the BLK/YEL wire between the ignition coil and the igniter.
4. If voltage is present, disconnect the WHT/BLU wire from the igniter. Check for battery voltage between the WHT/BLU wire and ground. Voltage should be present. If voltage is not present, check for an open in the WHT/BLU wire between the ignition coil and the igniter or a disconnected terminal.
6. If voltage is present, check for continuity between the igniter body and distributor housing.
7. If all tests check good and the system still fails to function, replace the igniter.

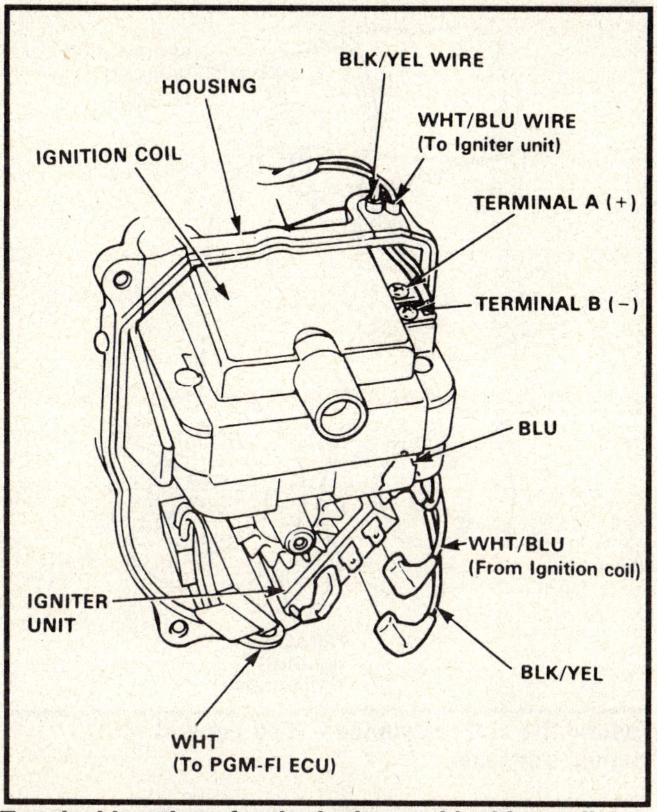

Terminal locations for the igniter and ignition coil—Integra

1990 LEGEND

1. Disconnect the 4 pin connector on from the igniter.
2. Turn the ignition switch **ON**. Check for voltage between the BLK/YEL wire and ground. Battery voltage should be present.
3. If not, check the BLK/YEL wire for voltage, between the ignition switch, ignition coil and igniter, if voltage is present, proceed to the next step. If voltage is not present, repair open ciruit.
4. Turn the ignition switch **ON**. Check for voltage between the BLK/YEL wire and ground. Voltage should be present.
5. If voltage is not present, check the BLU wire, between the ignition coil and igniter. Also check for voltage at the ignition coil.
6. If voltage is not present, proceed to next step.
7. Check the for voltage at the RED/BLU wire, between the PGM-FI ECU and igniter.
8. If all test check good, replace the igniter unit.

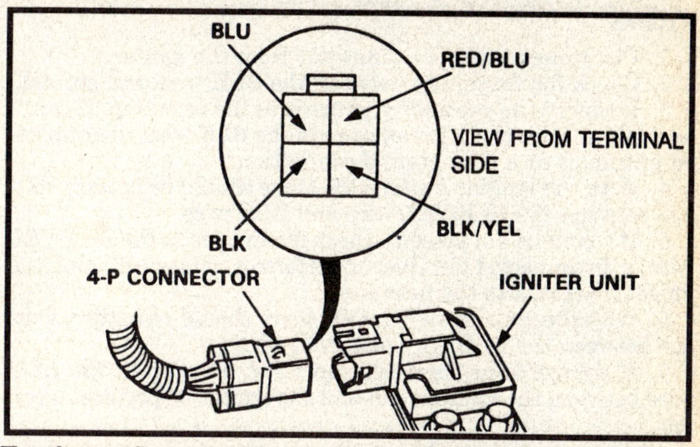

Testing voltage points at the igniter harness—1990 Legend

1991–92 LEGEND AND NSX

1. Disconnect the 8 pin connector from the igniter. Make the following tests at the harness pins.
2. Check the BLK[1] wire and BLK[2] wire for continuity to ground. Continuity should exits. If not, check for poor ground connection at G102, for Legend and G104 for NSX. If okay, check for a faulty ignition coil or and open circuit.
3. Turn the ignition switch **ON**, check for voltage at between the following circuits. Voltage should be present at each point:
   a. WHT[1] and ground
   b. WHT/GRN and ground
   c. WHT/BLK and ground
   d. WHT/BLU and ground
   e. WHT/YEL and ground
   f. WHT/RED and ground
4. If voltage was not present at each point, check for a blown fuse; No.25 fuse for Legend and No.13 fuse for NSX. Both fuses are (30 amp). If ok, check for a faulty ignition coil or an open circuit.

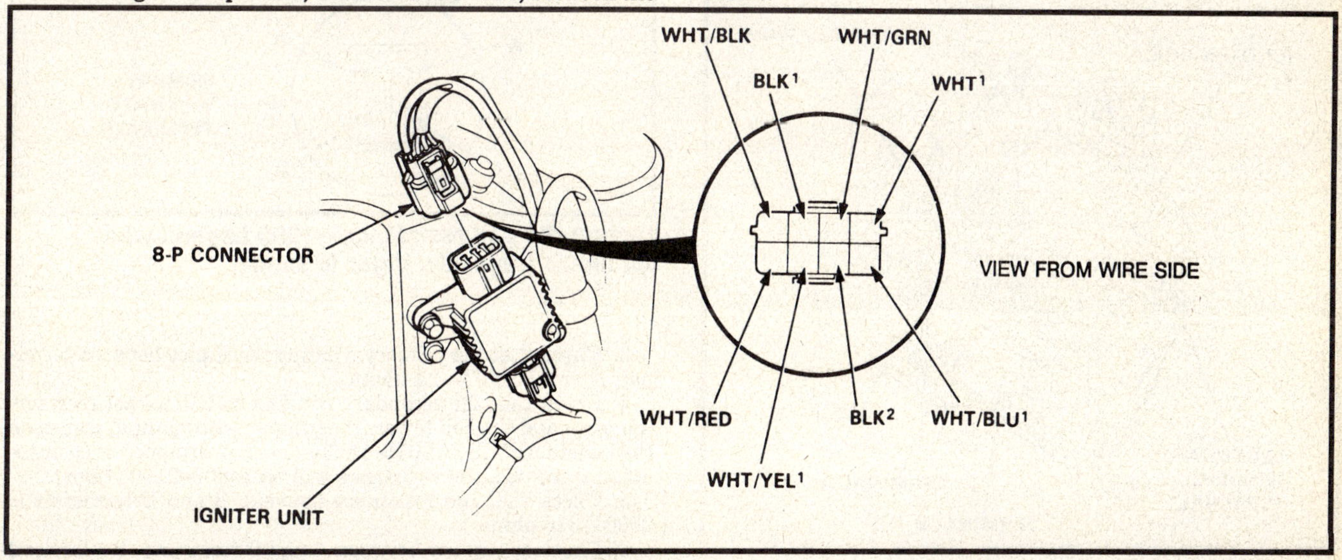

Testing the igniter—1991–92 Legend, NSX is similar

VIGOR

1. Disconnect the 4 pin connector from the igniter.
2. Check for continuity between the BLK wire and ground.
3. If there is no continuity, proceed to the next step. If continuity is present, check for an open in the BLK wire, disconnected terminals or a poor ground connection.
4. With the ignition switch **ON**, there should be battery voltage between the BLK/YEL wire and BLK wire.
5. If voltage is not present, check for an open in the BLK/YEL wire, a disconnected terminal or perform a coil test. If voltage is present, proceed to the next step.
6. With the ignition switch **ON**, there should be battery voltage between the BLU wire and BLK wire.
7. If voltage is not present, check for, an open in the BLU wire between the ignition coil and the igniter or perform a coil test.

IGNITION COIL

Testing

INTEGRA

1. With ignition switch **OFF**, remove the distributor cap.
2. Remove the 2 screws to disconnect the BLK/YEL front terminal A and WHT/BLU from terminal B of the coil.
3. Using an ohmmeter, check resistance between terminals. Replace coil if resistance is not within specification.

NOTE: Resistance will vary with coil temperature. Specifications are at 70°F (20°C)

4. Check that the primary winding resistance between terminals A and B is 0.6–0.8 ohms.
5. Check that the secondary winding resistance between terminals A and the coil tower is 9760–14640 ohms.
6. Replace coil if the resistance is not within specifications.

1990 LEGEND

1. With ignition switch **OFF**, disconnect primary connectors and coil wire.
2. Using an ohmmeter, check resistance between terminals. Replace coil if resistance is not within specification.

NOTE: Resistance will vary with coil temperature. Specifications are at 70°F (20°C)

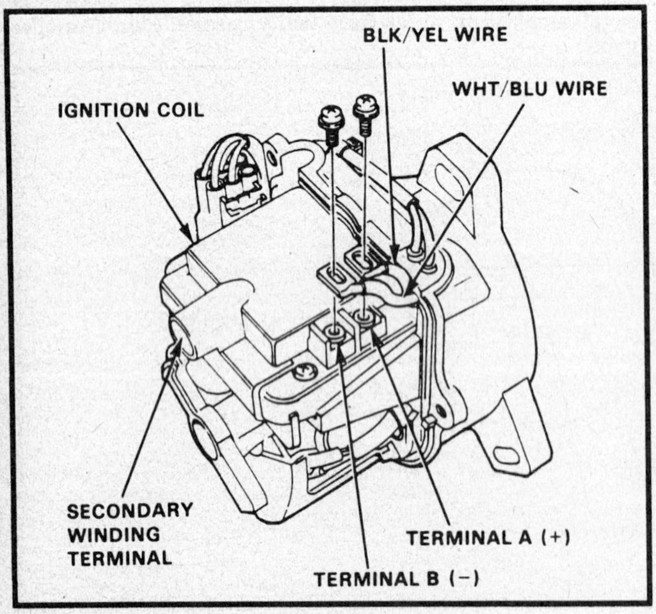

Coil resitance test—Integra

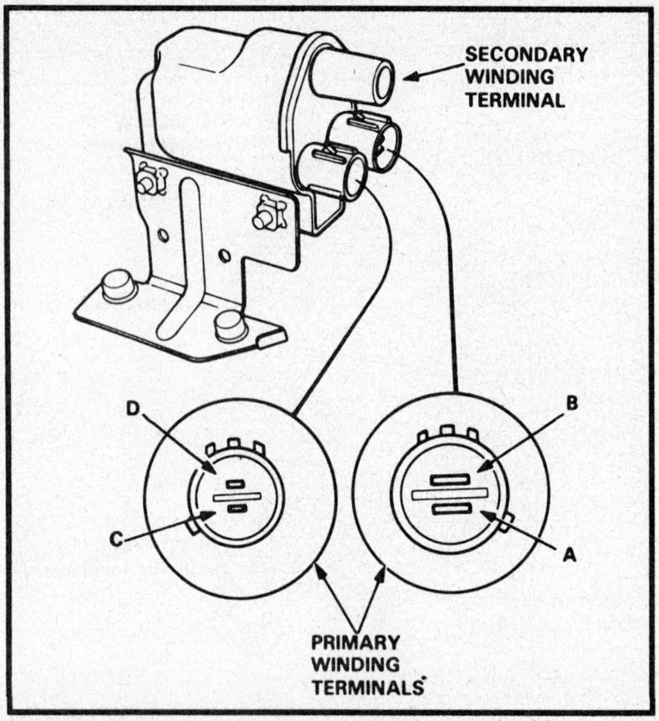

Testing the coil resistance—1990 Legend with manual transaxle

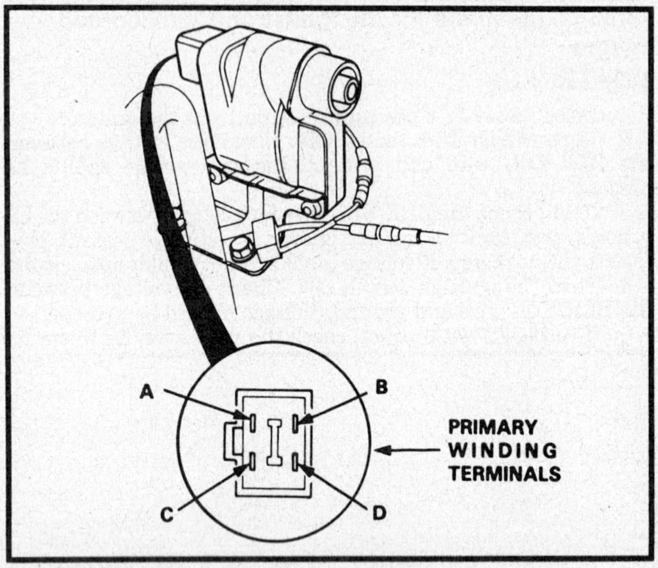

Testing the coil resistance—1990 Legend with automatic transaxle. Vigor is similar

3. Check that the primary winding resistance between terminals A and D is 0.3–0.4 ohms.
4. Check that the secondary winding resistance between terminals A and the coil tower. If equipped with manual transaxle the resistance will be 9040–13560 ohms, if equipped with automatic transaxle, the resistance will be 14400–21600 ohms.
5. Check that the resistance between B and D terminals is 2090–2310 ohms.
6. Check continuity between A and C terminals. Replace coil if no continuity.

1991–92 LEGEND AND NSX

1. With the ignition switch **OFF**. Remove the ignition coil.
2. Using an ohmmeter, check resistance between terminals. Replace coil if resistance is not within specification.

NOTE: Resistance will vary with coil temperature. Specifications are at 77°F (25°C)

3. Check that the primary winding resistance between terminals A and D is 0.9–1.1 ohms.
4. If the primary winding resistance is ok, substitute a known good ignition coil and check the system operation. If the system is normal, replace the original ignition coil.

VIGOR

1. With the ignition switch **OFF**. Disconnect the 4 pin connector and ignition coil wire.
2. Using an ohmmeter, check resistance between terminals.Replace coil if resistance is not within specification.

NOTE: Resistance will vary with coil temperature. Specifications are at 70°F (20°C)

3. Check that the primary winding resistance between terminals A and D is 0.3–0.5 ohms.
4. Check that the secondary winding resistance between terminals A and the coil tower the resistance should be between 10800–16200 ohms.
5. Check that the resistance between B and D terminals is 2090–2310 ohms.
6. Check continuity between A and C terminals. Replace coil if no continuity.

RADIO CONDENSER

Capacity Test

1. Using an appropriate condenser tester, measure the condenser capacity.
2. The condenser capacity should be 0.47 ± 0.09 microfarads.

NOTE: Although the radio noise condenser is intended to reduce ignition noise, condenser failure may can cause the engine to stop running.

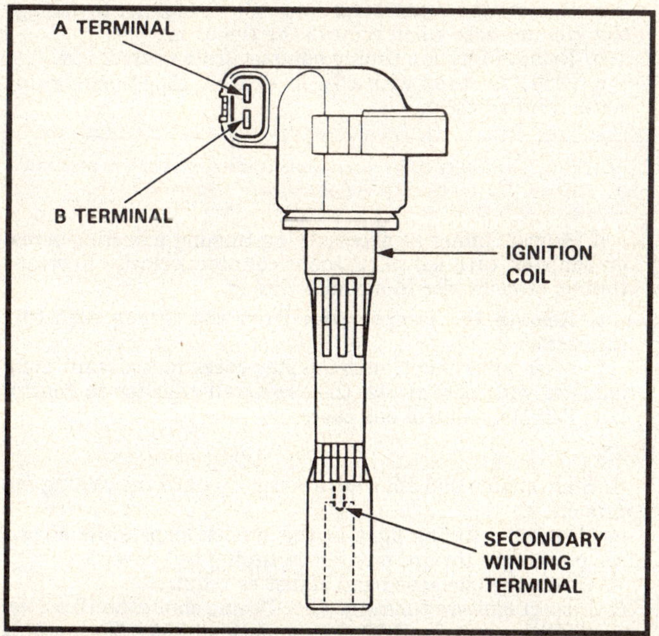

Testing the coil resistance—1991–92 Legend and NSX

IGNITION WIRES

Testing

When removing or disconnecting the ignition wires, carefully remove the wires by pulling on the rubber boots. Do not bend the wires, internal breakage may occur.

1. Check the condition of the wire terminals. If any terminal is corroded, clean it. If it is broken or distorted, replace it.
2. Using an ohmmeter, measure the resistance of the ignition wire. The resistance should not exceed 25000 ohms.
3. If resistance exceeds specification, replace the wire.

IGNITION TIMING

Adjustment

INTEGRA

1. Start engine and allow it to warm up until the cooling fan comes on.
2. Using a jumper wire, connect the GRN/WHT and BRN terminals of the ignition timing adjusting connector. The connector is gray and is located under the dash, on the right side.
3. Connect a timing light to engine; while engine idles, point light toward pointer on timing belt cover. Observe and note the timing mark position on the drive pulley.
4. The ignition timing should be 16 ± 2 degrees BTDC, indicated by the red mark, at 750 ± 50 rpm in N.
5. If timing adjustment is necessary, loosen the distributor hold-down bolts and turn the distributor as needed.
6. When the desired setting is achieved, torque the hold-down bolts to 16 ft. lbs. (22 Nm).
7. Remove the jumper wire from the timing terminal.

1990 LEGEND

1. Start engine and allow it to warm up until the cooling fan comes on.
2. Connect a timing light to engine; while engine idles, point light toward pointer on timing belt cover.
3. Inspect ignition timing at idle. Timing should be 15 ± 2 degrees BTDC, using the RED timing mark, at 680 ± 50 rpm in N.
4. If timing adjustment is necessary, complete following:
   a. Remove control box upper cover and ignition timing adjuster from control box.
   b. Drill 2 rivets off with a $^{3}/_{16}$ in. drill bit, then separate stay cover from adjuster.

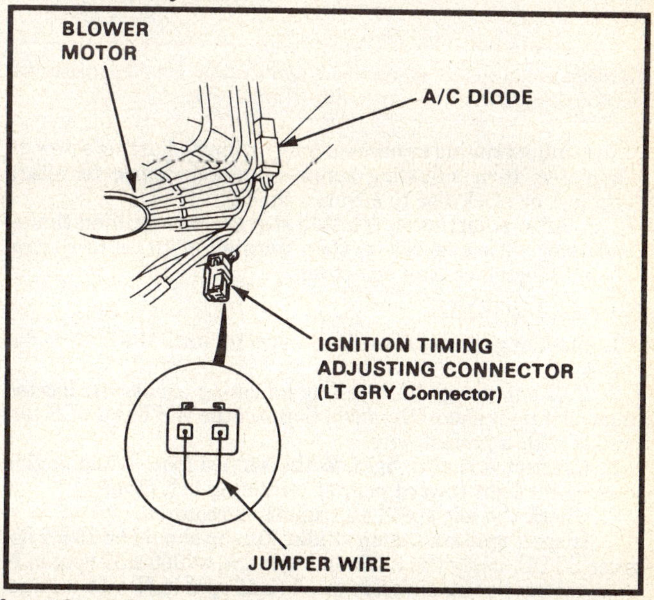

Location of the timing adjustment connector—Integra

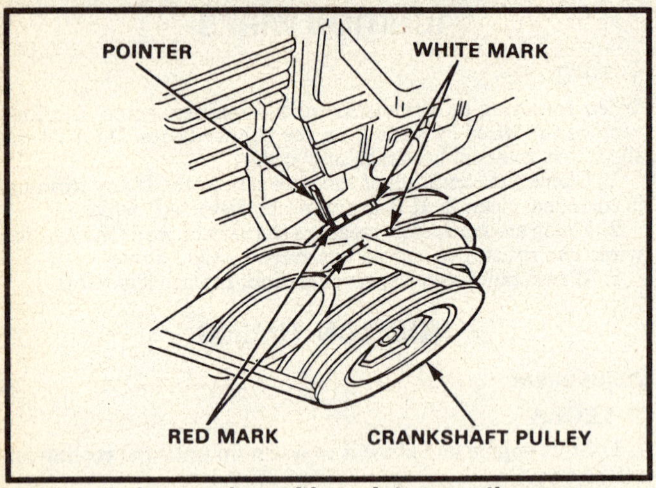

Ignition timing mark position—Integra, others are similar

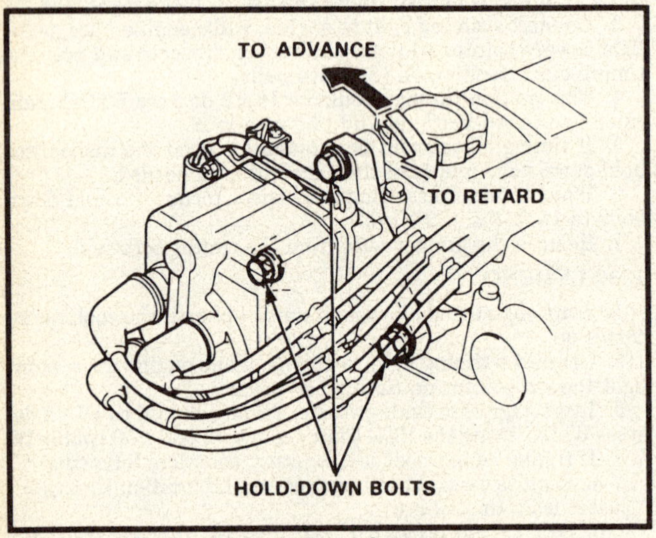

Adjusting the ignition timing—Integra

CAUTION

To avoid the possiblity of personal injury, alway wear eye protection when drilling. Do not damage adjuster when removing rivets.

c. Adjust timing as necessary by turning adjusting screw on adjuster; turn adjusting screw counterclockwise to retard timing, or clockwise to advance timing.

d. After adjustment, reinstall stay cover to ignition timing adjuster with new rivets, then reinstall adjuster to control box. Reinstall control box cover.

1991–92 LEGEND

1. Start engine and allow it to warm up until the cooling fan comes on.
2. Pull out the ignition timing adjusting connector, located under the right side of the dash. Connect the WHT and BLK terminals with a jumper wire.
3. Connect a timing light to the service loop; while engine idles, point light toward pointer on timing belt cover.
4. Check the idle speed and adjust as required.
5. Inspect ignition timing at idle. Timing should be 15±2 degrees BTDC, using the RED timing mark, at 600±50 rpm in **N** with automatic transmission or 650±50 rpm in **N** with manual transmission.

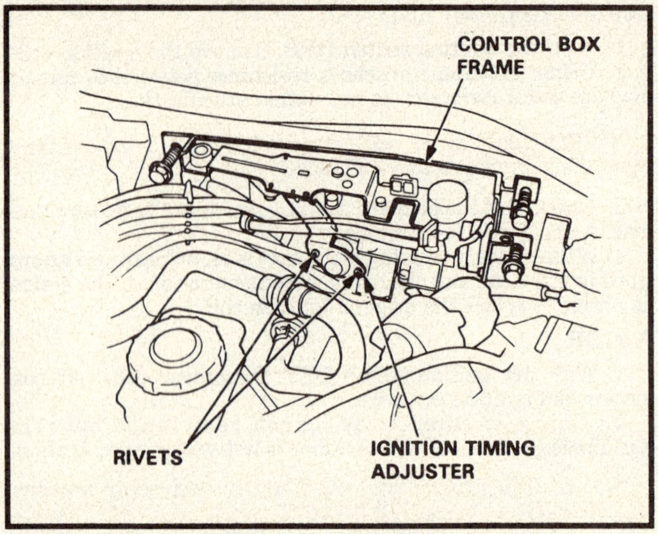

Location of the ignition timing adjuster—1990 Legend, others similar, except Integra

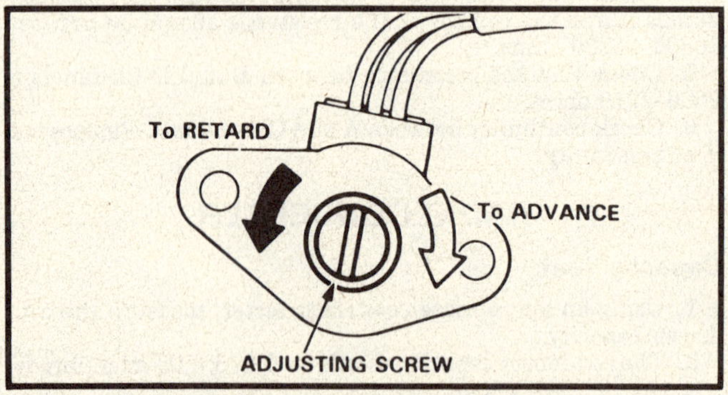

Ignition timimg adjustment—except Integra

6. If timing adjustment is necessary, complete following:

a. Remove the control box cover. Be careful not to damage the vacuum hose when remove the the cover.

b. Remove ignition timing adjuster from control box.

c. Drill 2 rivets off with a $^{3}/_{16}$ in. drill bit, then separate stay cover from adjuster.

CAUTION

To avoid the possibility of personal injury, always wear eye protection when drilling. Do not damage adjuster when removing rivets.

d. Adjust timing as necessary by turning adjusting screw on adjuster; turn adjusting screw counterclockwise to retard timing, or clockwise to advance timing.

e. Remove the jumper wire from the timing adjusting connector.

f. After adjustment, reinstall stay cover to ignition timing adjuster with new rivets, then reinstall adjuster to control box. Reinstall control box cover.

NSX

1. Start engine and allow it to warm up until the cooling fan comes on.
2. Connect a timing light to the service loop; while engine idles, point light toward pointer on timing belt cover.
4. Check the idle speed and adjust as required.
5. Inspect ignition timing at idle. Timing should be 15±2 degrees BTDC, using the RED timing mark, at 750±50 rpm in **N** with automatic transmission or 800±50 rpm in **N** with manual transmission.

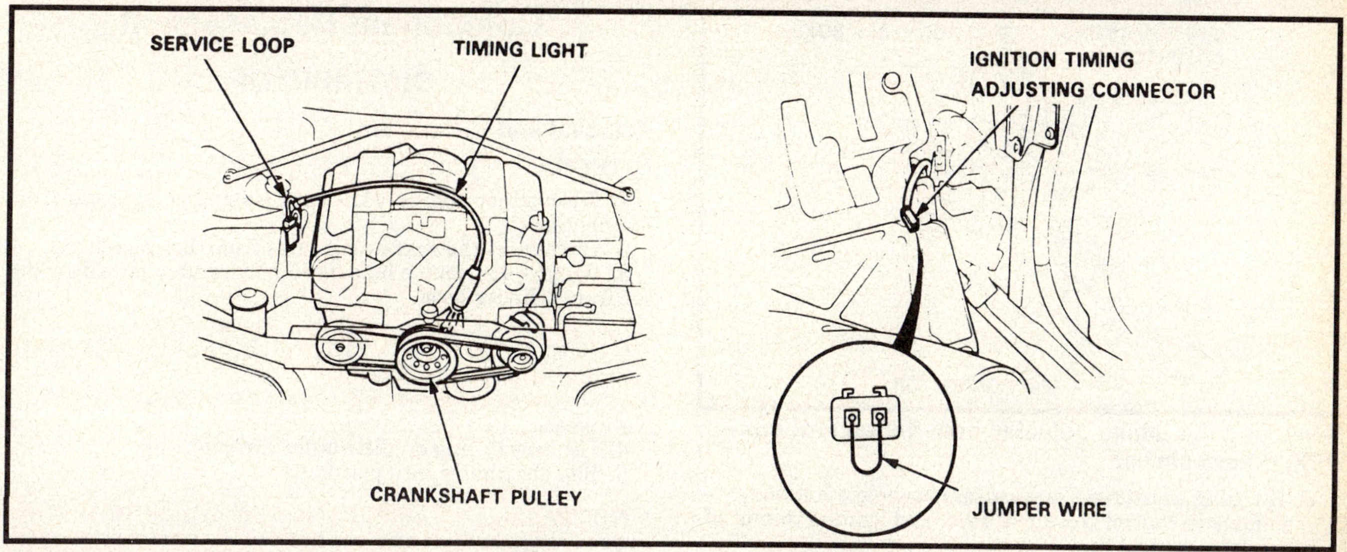

Checking the ignition timing—1991–92 Legend

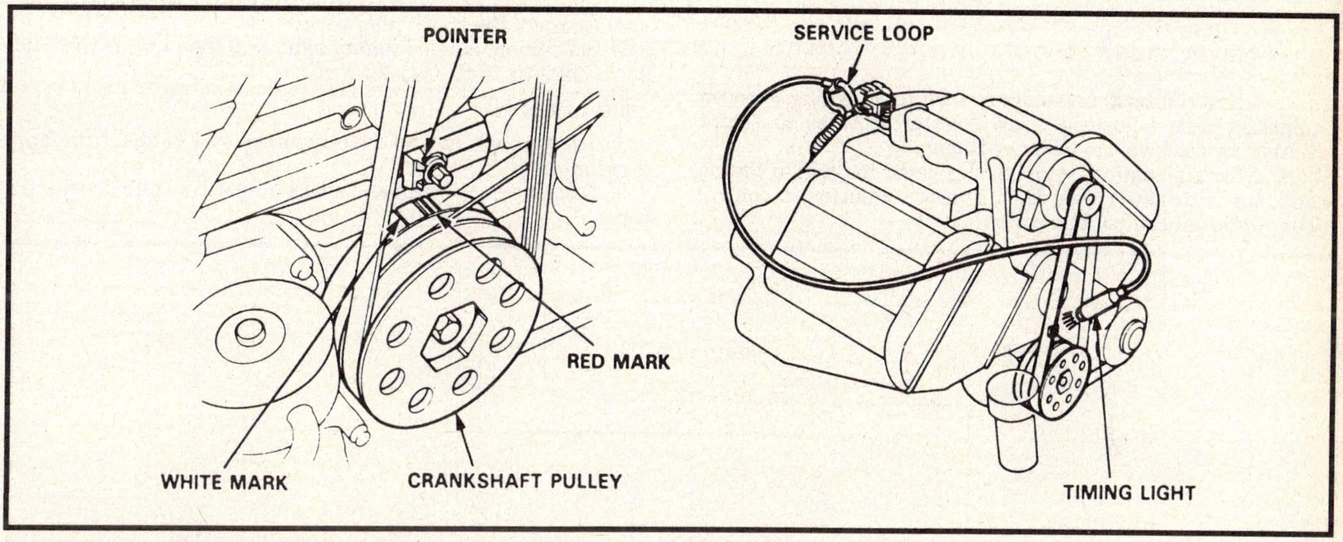

Checking the ignition timing—NSX

6. If timing adjustment is necessary, complete following:
   a. Remove the control box cover.
   b. Remove the 2 screws from the control box then disconnect the 3 pin connector from the ignition timing adjuster.
   c. Remove ignition timing adjuster from control box.
   d. Drill 2 rivets off with a $^{3}/_{16}$ in. drill bit, then separate stay cover from adjuster.

CAUTION

To avoid the possibility of personal injury, always wear eye protection when drilling. Do not damage adjuster when removing rivets.

   e. Adjust timing as necessary by turning adjusting screw on adjuster; turn adjusting screw counterclockwise to retard timing, or clockwise to advance timing.
   f. After adjustment, reinstall stay cover to ignition timing adjuster with new rivets, then reinstall adjuster to control box. Reinstall control box cover.

VIGOR

CAUTION

All SRS electrical wiring harnesses are covered with yellow outer insulation for easy identification. To avoid the possibility of personal injury, if it is necessary to disconnect the SRS harness, install the short connector on the air bag, then disconnect the wire harness. Replace the entire affected SRS harness assembly if there is an open circuit or damage to the wiring.

1. Start engine and allow it to warm up until the cooling fan comes on.
2. Pull out the service connector, located under the middle of the dash. Connect the WHT/GRN and BRN terminals with a jumper wire.
3. Check the idle speed and adjust as required.
4. Connect the timing light to the No.1 plug wire; while engine idles, point light toward pointer on timing belt cover.
5. Inspect ignition timing at idle. Timing should be 15 ± 2 degrees BTDC, using the RED timing mark, at 700 ± 50 rpm in N.

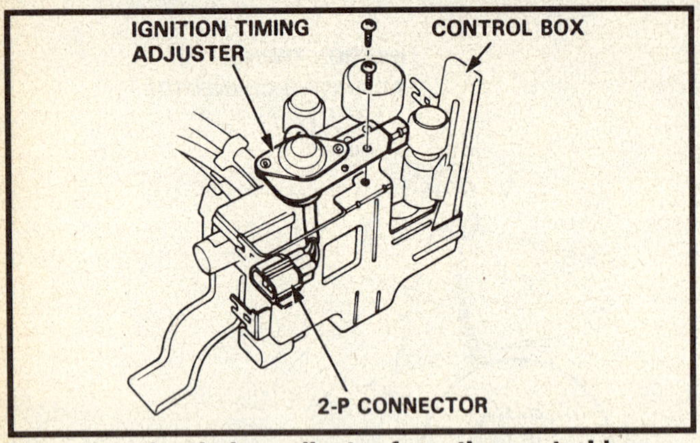

Removing the timing adjuster from the control box—NSX, others similar

6. If timing adjustment is necessary, complete following:
   a. Remove control box upper cover and ignition timing adjuster from control box.
   b. Drill 2 rivets off with a $^3/_{16}$ in. drill bit, then separate stay cover from adjuster.

CAUTION

To avoid the possibility of personal injury, always wear eye protection when drilling. Do not damage adjuster when removing rivets.

   c. Adjust timing as necessary by turning adjusting screw on adjuster; turn adjusting screw counterclockwise to retard timing, or clockwise to advance timing.
   d. After adjustment, reinstall stay cover to ignition timing adjuster with new rivets, then reinstall adjuster to control box. Reinstall control box cover.

Component Replacement

DISTRIBUTOR

Removal and Installation

INTEGRA

1. Disconnect engine wire harness and connectors from distributor.
2. Disconnect the spark plug wires from distributor cap.
3. Remove distributor hold-down bolts, and remove distributor from cylinder head.

NOTE: Coat new O-ring with engine oil before installation.

To install:

4. Use new O-ring on distributor housing.
5. Slip distributor into position.

NOTE: Lugs on end of distributor and its mating grooves in camshaft end are both offset to eliminate possibility of installing distributor 180 degrees out of time.

6. Install hold-down bolts, hand tighten.
7. Connect engine wire harness and connector to distributor.
8. Connect the spark plug wires.
9. Set timing, using a timing light, and then torque the hold-down bolts to 16 ft. lbs. (22 Nm).

1990 LEGEND

1. Disconnect engine wire harness and connectors from distributor.
2. Disconnect coil wire and spark plug wires from distributor cap.

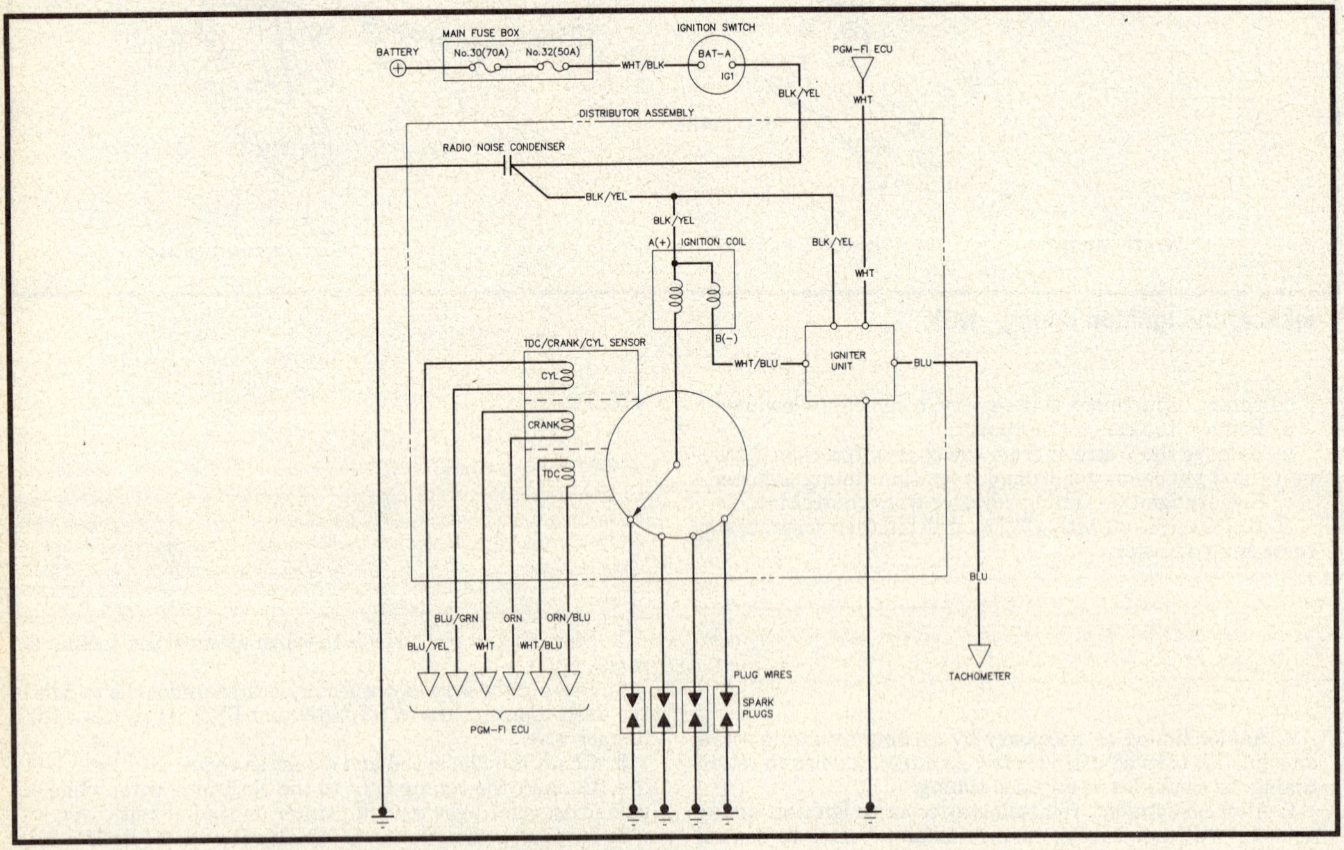

Wiring schematic of the ignition system—1990–92 Integra

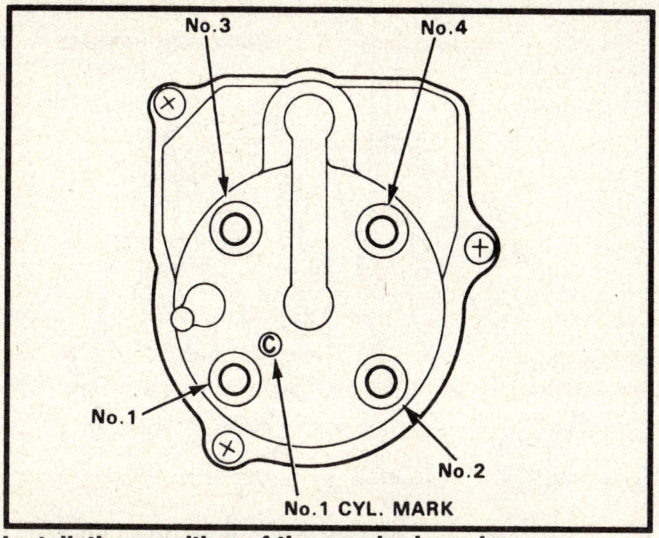

Installation position of the spark plug wires, distributor cap end—1990–91 Integra

3. Remove distributor hold-down bolt, and remove distributor from cylinder head.

NOTE: Coat new O-ring with engine oil before install.

To install:

4. Use new O-ring on distributor housing.
5. Slip distributor into position.

NOTE: Lugs on end of distributor and its mating grooves in camshaft end are both offset to eliminate possibility of installing distributor 180 degrees out of time.

6. Install hold-down bolt and tighten.
7. Connect engine wire harness and connector to distributor.
8. Connect coil wire and spark plug wires.

COIL AND SPARK PLUGS

Removal and Installation

1991–92 LEGEND AND NSX

1. Disconnect the negative battery terminal.
2. Remove the ignition coil covers.
3. Disconnect the electrical connectors from the ignition coils and carefully pull the coils from the spark plugs.

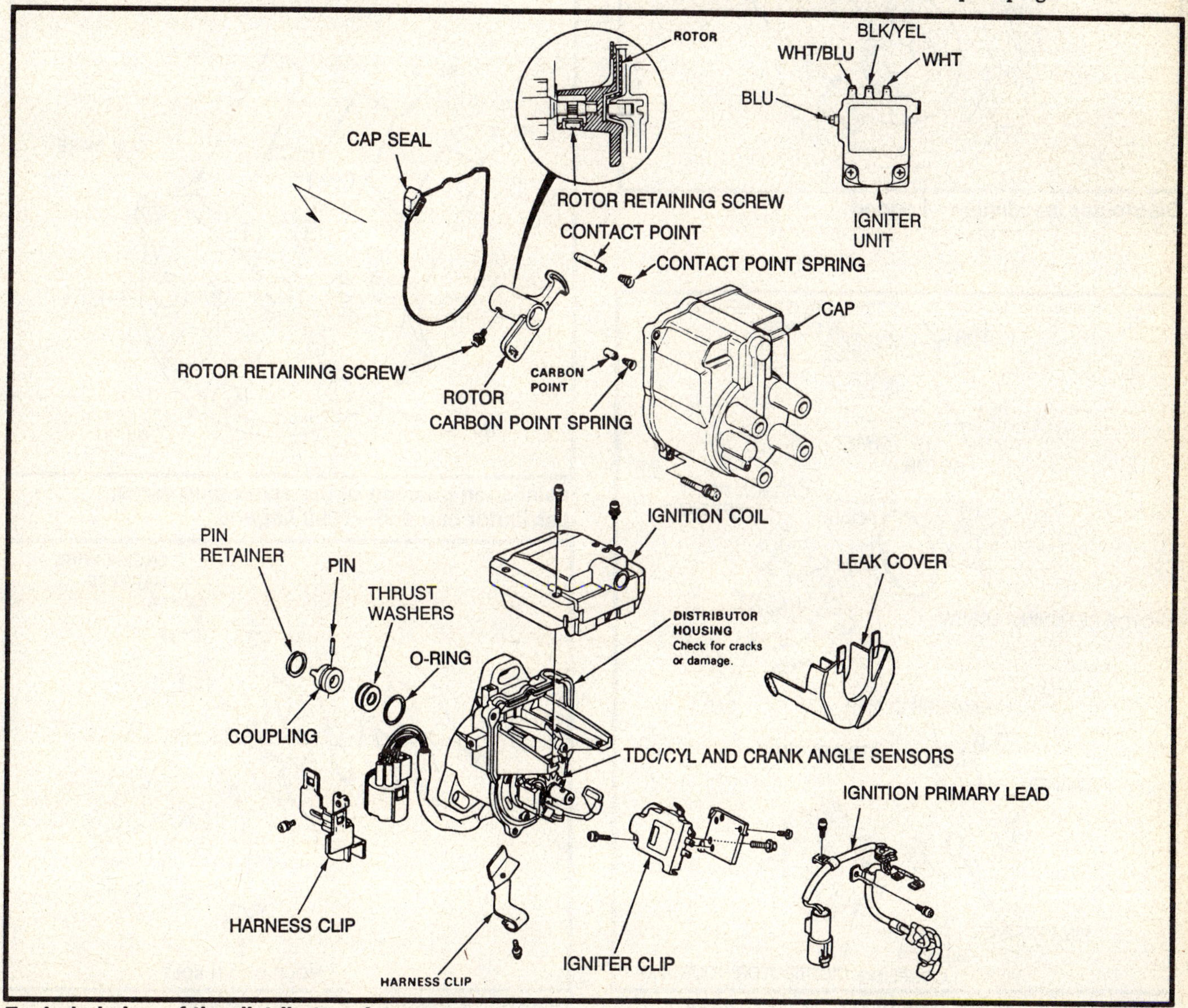

Exploded view of the distributor—Integra

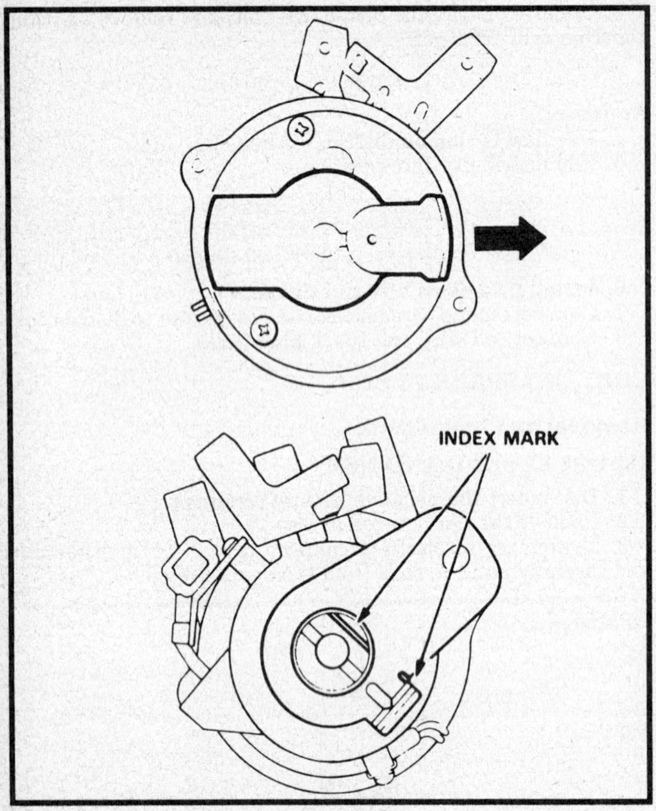

Distributor installation—Legend

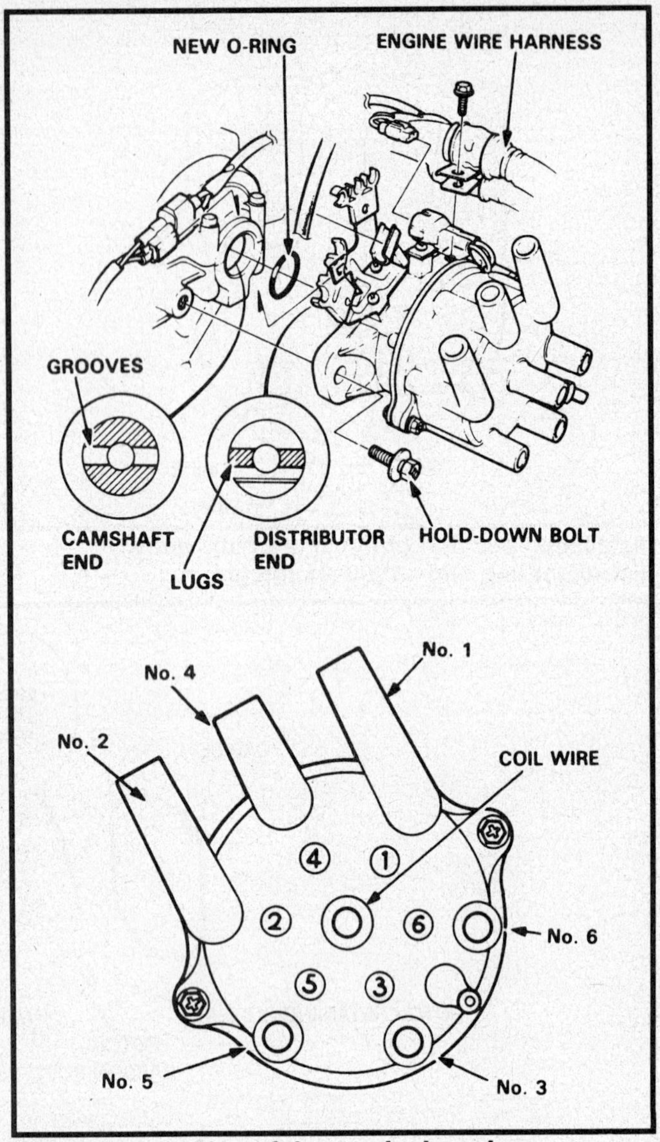

Installation position of the spark plug wires, distributor cap end—1990 Legend

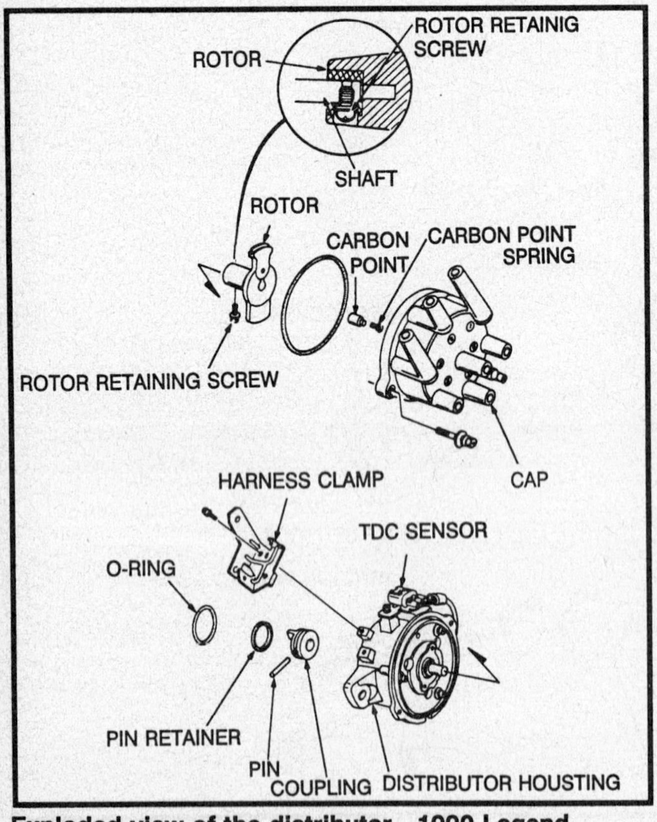

Exploded view of the distributor—1990 Legend

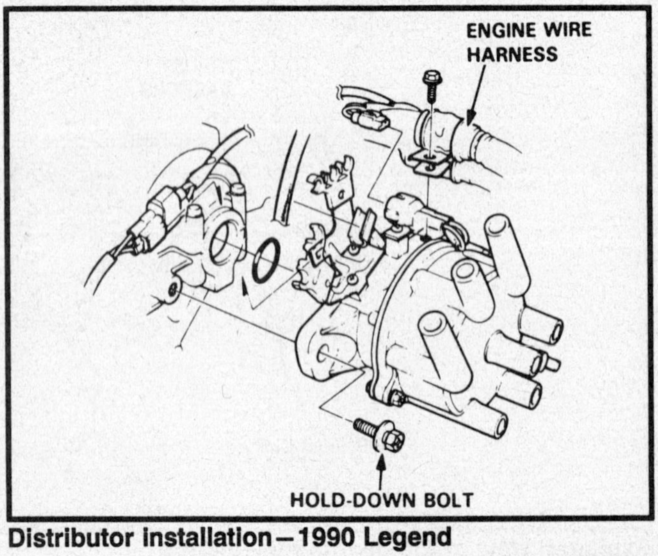

Distributor installation—1990 Legend

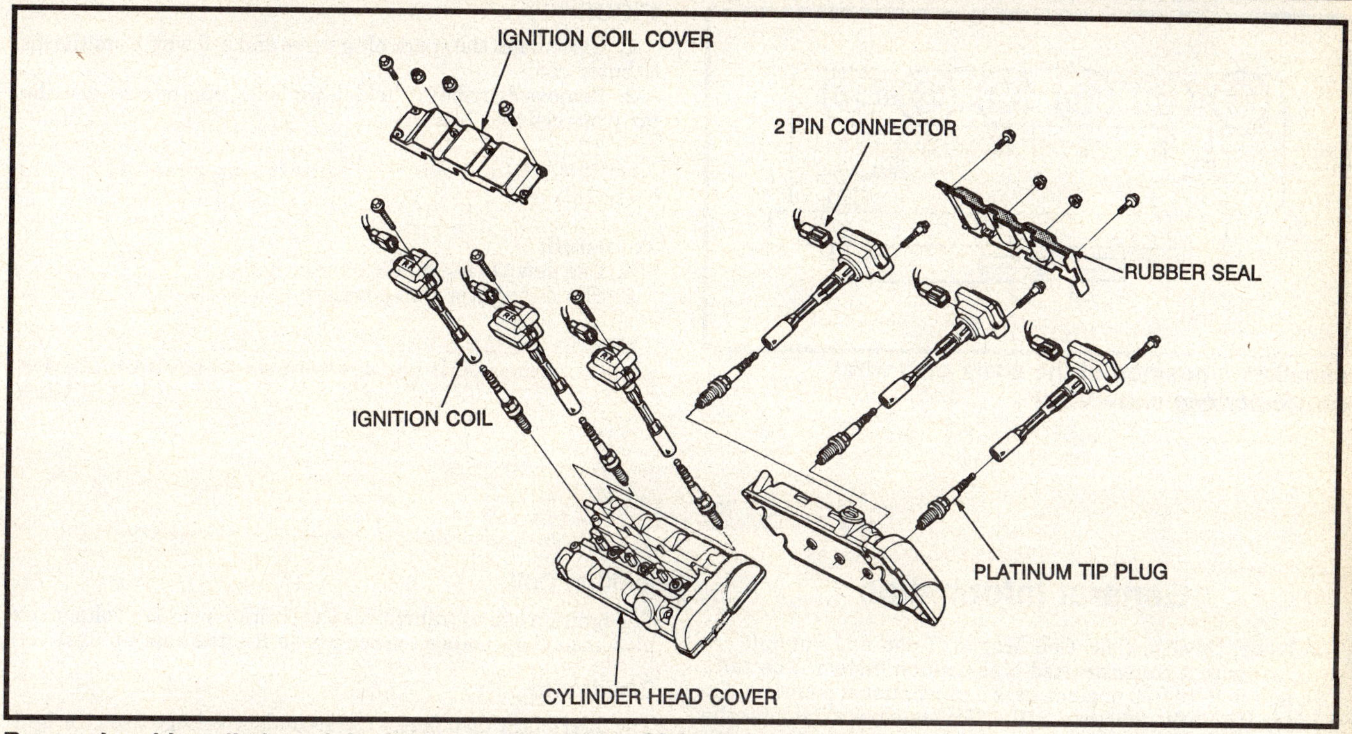

Removal and installation of the ignition coils—1991–92 Legend and NSX

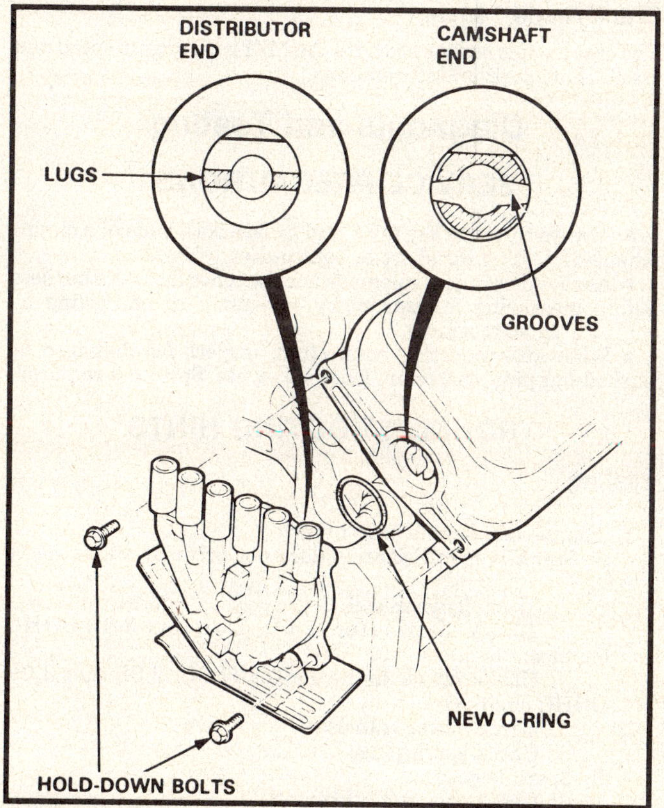

Distributor installation—Vigor

NOTE: The front and rear ignition coils and covers are different, be sure to use the correct part when installing them.

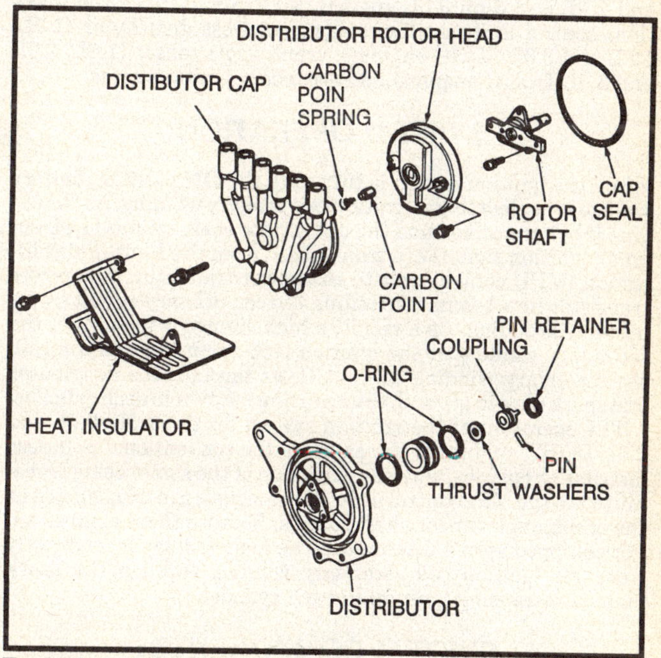

Exploded view of the distributor—Vigor

To install:

4. Position the ignition coil onto the spark plug and carefullypush it in place.
5. Connect the electrical connector on the coil.
6. Install the coil covers and secure the coils in place with the retaining screws.
7. Connect the negative battery cable.

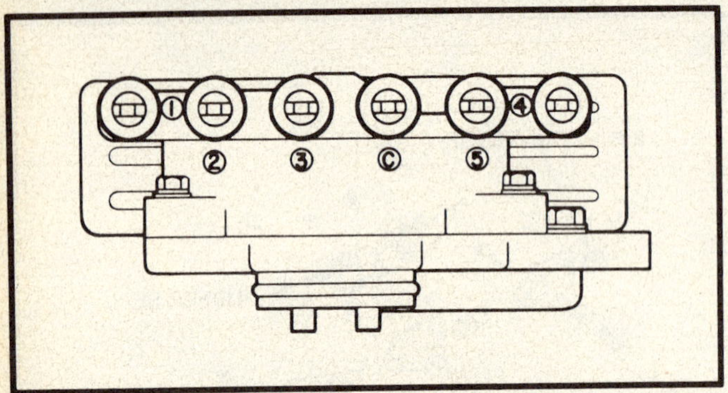

Installation position of the spark plug wires, distributor cap end—Vigor

VIGOR

1. Disconnect the spark plug wires and coil wire from the distributor cap.
3. Remove distributor hold-down bolts, and remove distributor from cylinder head.

NOTE: Coat new O-ring with engine oil before installation.

To install:

4. Use new O-ring on distributor housing.
5. Slip distributor into position.

NOTE: Lugs on end of distributor and its mating grooves in camshaft end are both offset to eliminate possibility of installing distributor 180 degrees out of time.

CHRYSLER IMPORTS

General Information

The Laser, Stealth, Colt, Colt Wagon, Talon and Summit are equipped with a transistorized type ignition system. The 1.5L, 1.8L and 3.0L SOHC engines uses a conventional pointless distributor. The 1.6L, 2.0L and 3.0L DOHC engines eliminates the conventional distributor and all its components, by using multiple ignition coils and a crank angle sensor. The ignition system consists of power transistor(s) (1 on 1.5L and 1.8L, 2 on 1.6L and 2.0L and 3 on 3.0L engine), MPI (Micro-Processor Injection) control unit, ignition coil(s), pointless distributor (1.5L, 1.8L and 3.0L SOHC engines), crank angle sensor (1.6L, 2.0L and 3.0L DOHC engines) and secondary ignition wires.

SYSTEM OPERATION

When the ignition switch is turned to the **ON** position, battery voltage is applied to the ignition coil primary winding. As the engine is being cranked and the distributor shaft rotates (1.8L engine), ignition signals are transmitted from the Multi-Point Injection (MPI) control unit to the power transistor. The power transistor now begins controlling the coil primary circuit by repeatedly grouding. As a result, a high voltage is induced in the secondary winding of the ignition coil. From the ignition coil, the secondary winding current flows through the distributor and spark plug to ground, thus causing ignition in each cylinder.

The operation of the ignition system for the 1.6L, 2.0L and 3.0L DOHC engines is the same as the conventional pointless distributor engines, with the exception of the signal being transmitted to the MPI control unit. These signals are transmitted as the crank angle sensor shaft rotates. Because these engines are not equipped with a distributor, the high voltage flows directly from the ignition coil secondary winding through the spark plug, thus causing ignition in each cylinder.

SYSTEM COMPONENTS

MPI Control Unit

The MPI (engine controller) continuously monitor various engine sensors. The computer will then electronically advance or retard the ignition timing to provide even driveability during operation.

Power Transistor

The power transistor(s) use the signal provided by the MPI to control the ignition coil primary circuit.

Ignition Coil

The ignition coil(s) transforms the ignition primary voltage to a high secondary voltage, necessary to fire the spark plugs.

Distributor

The distributor distributes the secondary voltage from the ignition coil to each spark plug. It also provides a signal to the MPI, which controls the timing of the spark at the spark plug.

Crank Angle Sensor

The crank angle sensor provides the MPI with engine speed and crank shaft position informations.

Diagnosis and Testing

SERVICE PRECAUTIONS

- Always turn the key **OFF** and isolate both ends of a circuit whenever testing for short or continuity.
- Always disconnect solenoids and switches from the harness before measuring for continuity, resistance or energizing by way of a 12 volts source.
- When disconnecting connectors, inspect for damaged or pushed-out pins, corrosion, loose wires, etc. Service if required.

TROUBLESHOOTING HINTS

Testing

1. Engine cranks, but won't start:
   A. Spark is insufficient or does not occur at all, at spark plug:
      Check ignition coil.
      Check distributor (1.5L, 1.8L and 3.0L SOHC engines).
      Check crank angle sensor (1.6L, 2.0L and 3.0L DOHC engines).
      Check power transistor.
      Check spark plugs.
   B. Spark is good:
      Check the ignition timing.
2. Engine Idles Roughly or Stalls:
   Check spark plugs.
   Check ignition timing.
   Check ignition coil.
   Check spark plug cables.

3. Poor Acceleration:
 Check ignition timing.
 Check ignition coil.
 Check spark plug cables.
4. Engine overheats or consumes excessive fuel:
 Check ignition timing.

IGNITION COIL

Testing

1.5L, 1.8L, 2.4L AND 3.0L SOHC ENGINES

1. Disconnect the negative battery cable and ignition coil harness connector.

2. Measure the primary coil resistance between the positive (+) and negative (−) terminals of the coil. Should indicate a value of 0.72–0.88 ohms, except 1991–92 1.5L engine. On 1991–92 reading should indicate a value of 0.9–1.2 ohms

3. Measure the coil secondary resistance between the positive (+) terminal and the high-voltage terminal. Should indicate a value of 10.3–13.9 kilo-ohms, except 1991–92 1.5L engine. On 1991–92 reading should indicate a value of 20–29 kilo-ohms.

4. If the readings are not within the specified value, replace the ignition coil.

1.6L AND 2.0L ENGINES

1. Disconnect the negative battery cable and ignition coil harness connector.

2. Measure the primary coil resistance.

3. On 1990 2.0L engine, measure the resistance between terminals 4 and 2 (coils at the No. 1 and No. 4 cylinder sides) of the ignition coil, and between terminals 4 and 1 (coils at the No. 2 and No. 3 cylinder sides).

4. On 1.6L engine and 1991–92 2.0L engine, measure the resistance between terminals 3 and 2 (coils at the No. 1 and No. 4 cylinder sides) of the ignition coil, and between terminals 3 and 1 (coils at the No. 2 and No. 3 cylinder sides).

5. The primary coil resistance should read 0.77–0.95 ohms.

6. Measure the coil secondary resistance.

7. Disconnect the connector of the ignition coil. Measure the resistance between the high-voltage terminals for the No. 1 and No. 4 cylinders, and between the high-voltage terminals for the No. 2 and No. 3 cylinders.

8. The secondary coil resistance should read 10.3–13.9 kilo-ohms.

9. If the readings are not within the specified value, replace the ignition coil.

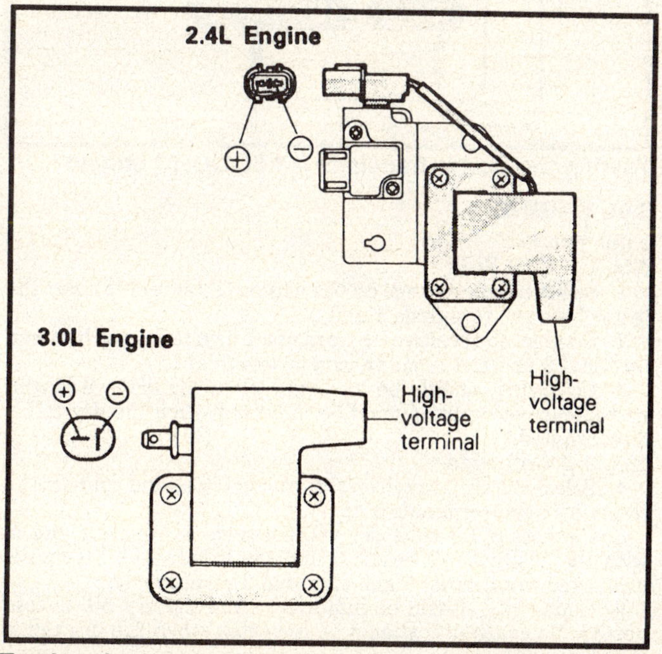

Testing the ignition coil—2.4L and 3.0L SOHC engines

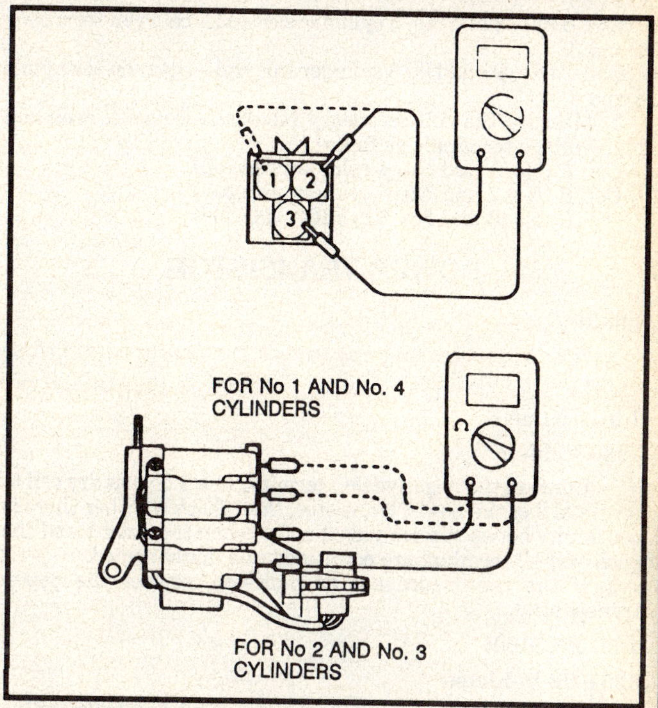

Testing the ignition coil—1.6L engine and 1991–92 2.0L engine, 1990 2.0L engine similar

3.0L DOHC ENGINE

1. Disconnect the negative battery cable and ignition coil harness connector.

2. Measure the primary coil resistance as follow:
Coil A (No. 1 and No. 4 cylinder side coil—between terminals 2 and 3.

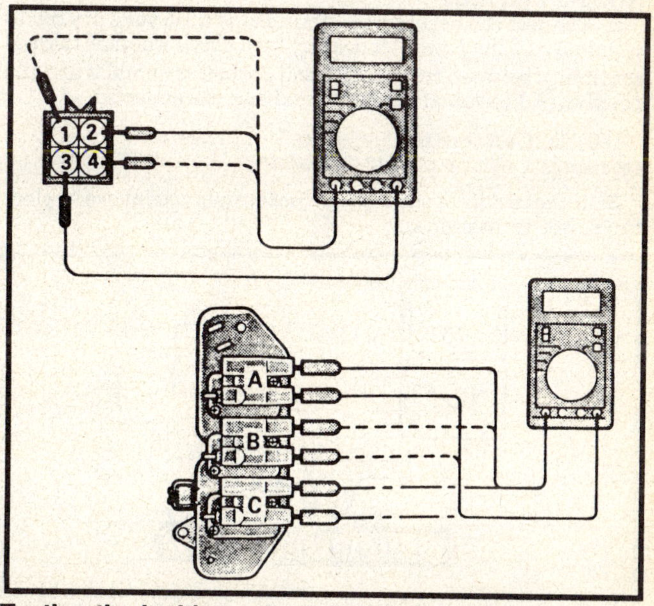

Testing the ignition coil—3.0L DOHC engine

Coil B (No. 2 and No. 5 cylinder side coil—between terminals 1 and 3.

Coil C (No. 3 and No. 6 cylinder side coil—between terminals 4 and 3.

3. Measure the coil secondary resistance between each coil high-voltage terminals as follow:

Coil A (No. 1 and No. 4 cylinder side coil)
Coil B (No. 2 and No. 5 cylinder side coil)
Coil C (No. 3 and No. 6 cylinder side coil)

POWER TRANSISTOR

Testing

NOTE: When testing the power transistor(s), an analog-type circuit tester should be used.

1.5L ENGINE

1990 Vehicle

1. Connect the negative (–) terminal of a 1.5 volts dry cell to terminal 2 of the power transistor; then check whether there is continuity between terminals 2 and 3 when terminal 1 and the positive (+) terminal are connected and disconnected.
2. If the results are not as indicated, replace the power transistor(s).

1.5L ENGINE

1991–92 Vehicles

1. Connect the negative (–) terminal of a 1.5 volts dry cell to terminal 7 of the power transistor; then check whether there is continuity between terminals 7 and 2 when terminal 8 and the positive (+) terminal are connected and disconnected.
2. If the results are not as indicated, replace the power transistor(s).

1.6L ENGINE

NO. 1–4 Coil Side

1. Connect the negative (–) terminal of a 1.5 volts dry cell to terminal 3 of the power transistor; then check whether there is continuity between terminals 7 and 3 when terminal 6 and the positive (+) terminal are connected and disconnected.

NOTE: Connect the negative (–) probe of the tester to terminal 7 of the power transistor.

NO. 2–3 Coil Side

2. Connect the negative (–) terminal of a 1.5 volts dry cell to terminal 3 of the power transistor; then check whether there is continuity between terminals 1 and 3 when terminal 2 and the positive (+) terminal are connected and disconnected.

NOTE: Connect the negative (–) probe of the tester to terminal 1 of the power transistor.

3. If the results of the tests are not as indicated above, replace the power transistor(s).

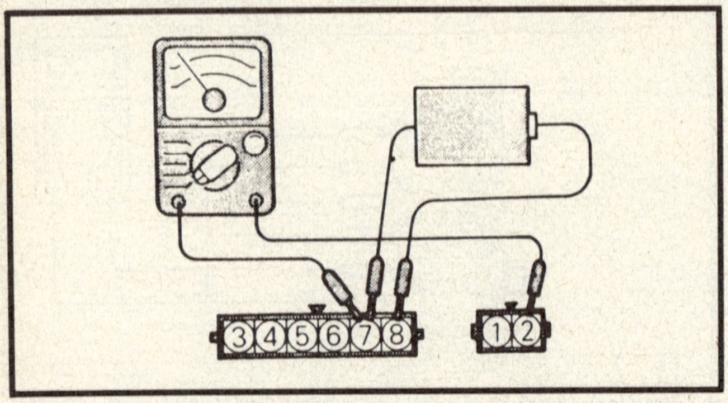

Testing the power transistor—1991–92 1.5L engine

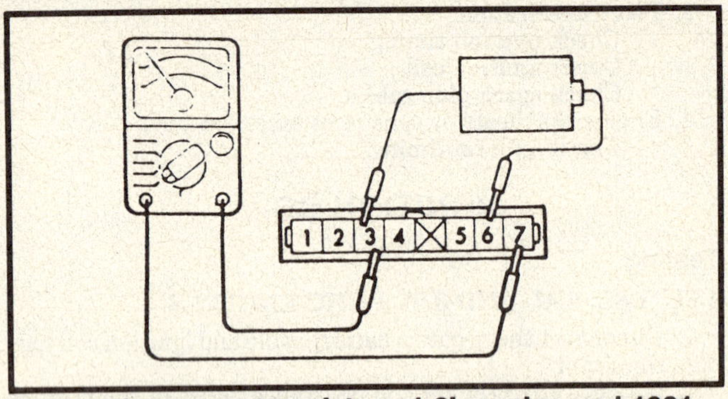

Testing the power transistor—1.6L engine and 1991–92 2.0L engine, 1990 2.0L engine similar

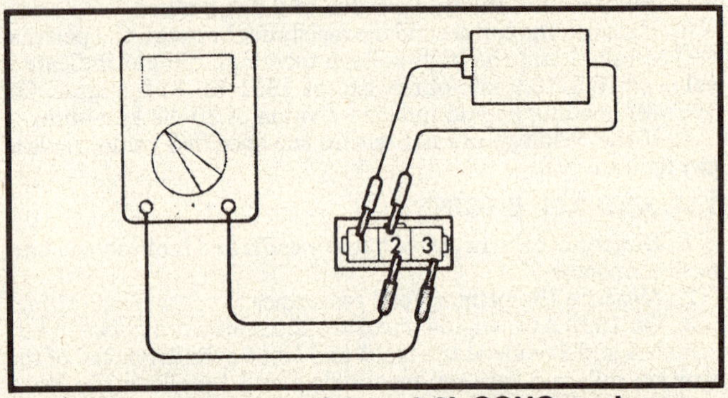

Testing the power transistor—3.0L SOHC engine

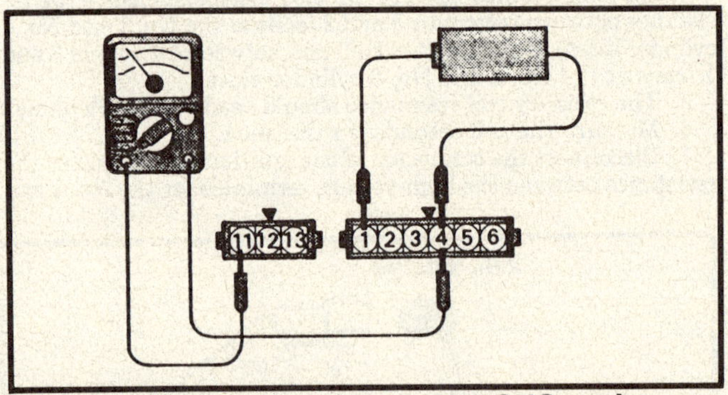

Testing the power transistor—3.0L DOHC engine

2.0L ENGINE

1990 Vehicle

NO. 1–4 Coil Side

1. Select the ohm range on the analog tester and connect the tester between terminals 1 and 3.
2. Connect the positive (+) lead of a 1.5 volts dry cell between terminal 2 and the negative lead to terminal 3.
3. Continuity should be indicated when the dry cell is connected. No continuity should be indicated when the dry cell is disconnected.

NO. 2–3 Coil Side

4. Select the ohm range on the analog tester and connect the tester between terminals 6 and 3.
5. Connect a 1.5 volts dry cell between terminals 3 and 5, with the positive (+) terminal and the negative (–) terminal connected to terminal 5 and terminal 3 respectively.
6. Continuity should be indicated when the dry cell is connected. No continuity should be indicated when the dry cell is disconnected.
7. If the results of the tests are not as indicated above, replace the power transistor(s).

1991–92 Vehicles

NO. 1–4 Coil Side

1. Connect the negative (−) terminal of a 1.5 volts dry cell to terminal 3 of the power transistor; then check whether there is continuity between terminals 7 and 3 when terminal 6 and the positive (+) terminal are connected and disconnected.

NOTE: Connect the negative (−) probe of the tester to terminal 7 of the power transistor.

NO. 2–3 Coil Side

2. Connect the negative (−) terminal of a 1.5 volts dry cell to terminal 3 of the power transistor; then check whether there is continuity between terminals 1 and 3 when terminal 2 and the positive (+) terminal are connected and disconnected.

NOTE: Connect the negative (−) probe of the tester to terminal 1 of the power transistor.

3. If the results of the tests are not as indicated above, replace the power transistor(s).

2.4L AND 3.0L SOHC ENGINE

1. Connect the negative (−) terminal of a 1.5 volts dry cell to terminal 2 of the power transistor; then check whether there is continuity between terminals 3 and 2 when terminal 1 and the positive (+) terminal are connected and disconnected.
2. If the results are not as indicated, replace the power transistor(s).

3.0L DOHC ENGINE

NO. 1–4 Coil Side

1. Connect the negative (−) terminal of a 1.5 volts dry cell to terminal 7 of the power transistor; then check whether there is continuity between terminals 3 and 7 when terminal 6 and the positive (+) terminal are connected and disconnected.

NOTE: Connect the negative (−) probe of the tester to terminal 3.

2. Replace the power transistor if the results are not as specified.

NO. 2–5 Coil Side

3. Connect the negative (−) terminal of a 1.5 volts dry cell to terminal 7 of the power transistor; then check whether there is continuity between terminals 2 and 7 when terminal 5 and the positive (+) terminal are connected and disconnected.

NOTE: Connect the negative (−) probe of the tester to terminal 2.

4. Replace the power transistor if the results are not as specified.

NO. 3–6 Coil Side

5. Connect the negative (−) terminal of a 1.5 volts dry cell to terminal 7 of the power transistor; then check whether there is continuity between terminals 1 and 7 when terminal 4 and the positive (+) terminal are connected and disconnected.

NOTE: Connect the negative (−) probe of the tester to terminal 1.

6. Replace the power transistor if the results are not as specified.

IGNITION TIMING

Adjustment

1.5L, 1.8L, 2.4L AND 3.0L SOHC ENGINES

1. Apply the parking brake and block the wheels. Run the engine until it reach operating temperature.
2. Make certain all lights, cooling fan and accessories are **OFF**.

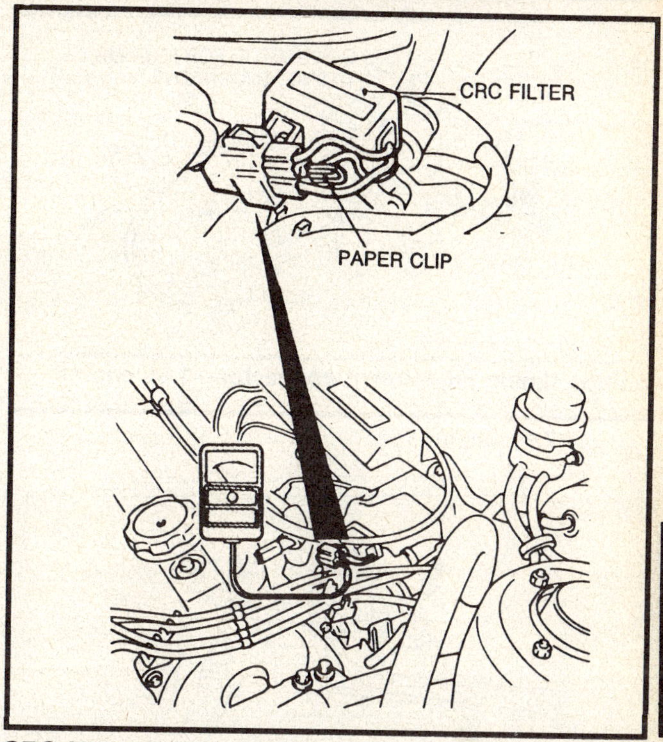

CRC filter connector—1.8L engine

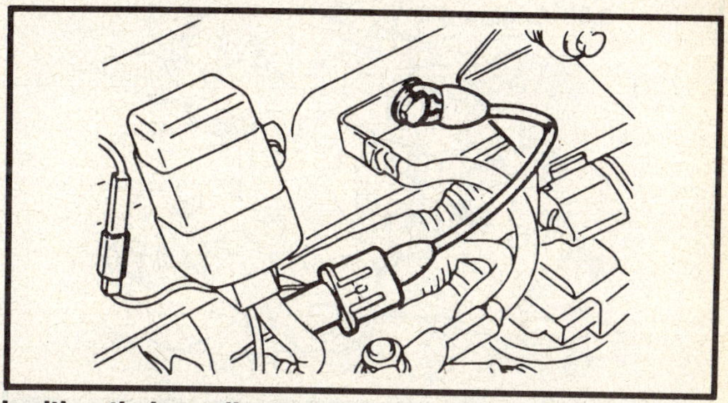

Ignition timing adjustment connector—1.8L engine

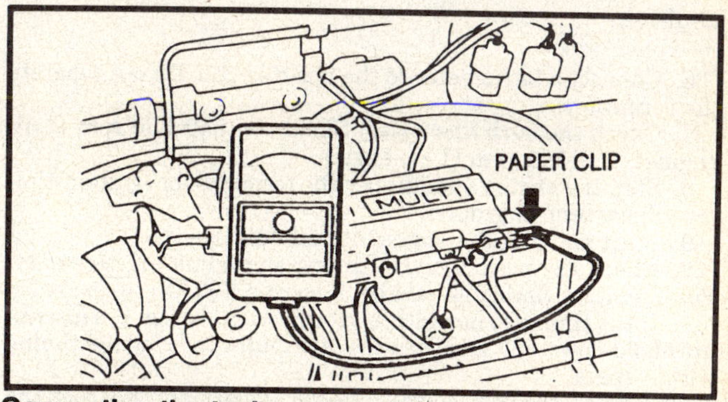

Connecting the tachometer—1.5L engine

3. Position the steering wheel in straight ahead position and the gear selector lever in **P** or **N**.
4. Connect a timing light to the engine.
5. On 1.8L engine, insert a paper clip into the CRC filter connector (3-pole connector, 0.5-W cable). On 1.5L, 2.4L and 3.0L SOHC engines, insert a paper clip from behind the connector.

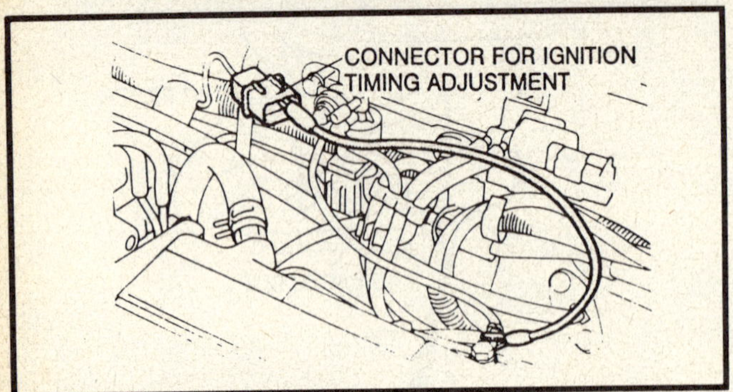

Ignition timing adjustment connector—1.5L engine

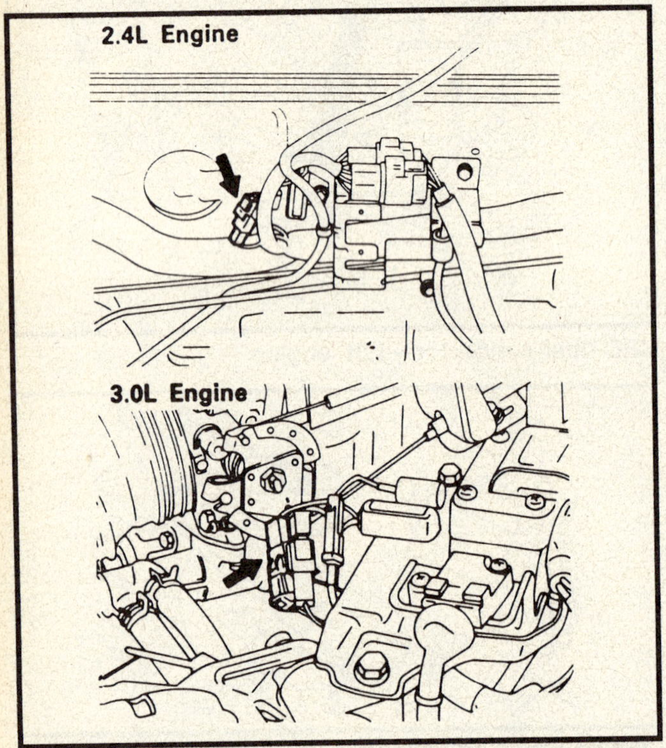

Tachometer connecter location—2.4L and 3.0L SOHC engines

6. Connect a tachometer to the inserted clip. Do not separate the connector.
7. Check the curb idle speed. Should be 650–850 rpm (1.5L engine), 600–800 rpm (1.8L engine).
8. Stop the engine and jumper the terminal for ignition-timing adjustment to ground.
9. Start and run the engine at curb idle speed.
10. Check the basic ignition timing and adjust, if necessary. Basic ignition timing should be 5 degrees BTDC.
11. If the timing is not within specification, loosen the distributor hold-down bolt and turn the distributor to bring the timing within specs.
12. Tighten the hold-down bolt after adjustment. Recheck the timing and adjust if necessary.
13. Stop the engine and unground the ignition timing connector.

NOTE: Actual ignition timing may vary, depending on the control mode of the engine control unit. In such case, recheck the basic ignition timing. If there is no deviation, the ignition timing is functioning normally.

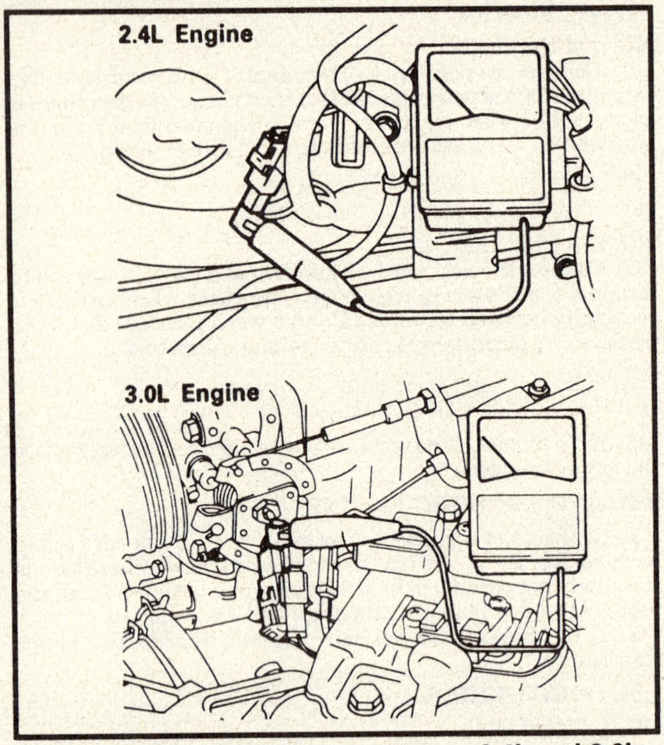

Ignition timing adjustment connector—2.4L and 3.0L SOHC engines

14. Start the engine and run at curb idle. Check the actual ignition timing. Actual ignition timing should be 10 degrees BTDC for 1.5L and 1.8L engines, 8 degrees BTDC for 2.4L engine and 15 degrees BTDC for 3.0L SOHC engine.

NOTE: At altitudes, more than approximately 2,300 ft. (701m) above sea level, the actual ignition timing is further advanced to ensure good combustion.

1.6L AND 2.0L ENGINES

1. Apply the parking brake and block the wheels. Run the engine until it reach operating temperature.
2. Make certain all lights, cooling fan and accessories are **OFF**.
3. Position the steering wheel in straight ahead position and the gear selector lever in **P** or **N**.
4. Connect a timing light to the engine.
5. On 1.6L engine, insert a paper clip from behind the connector. On 2.0L engine, insert a paper clip into the engine revolution speed detection terminal (in engine compartment) and connect a tachometer to the inserted clip.
6. Check the curb idle speed. Should be 650–850 rpm.

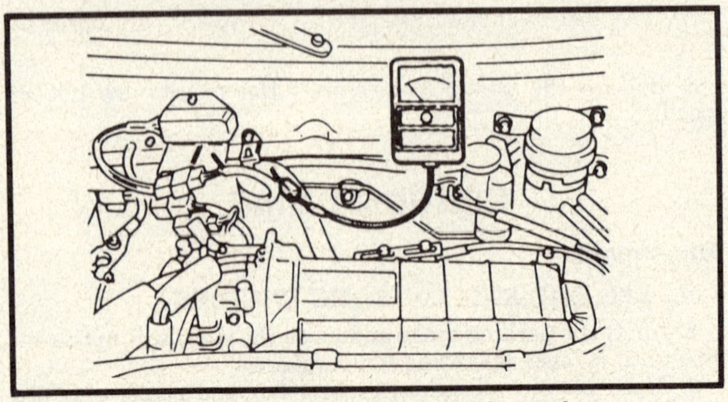

Tachometer connector location—2.0L engine

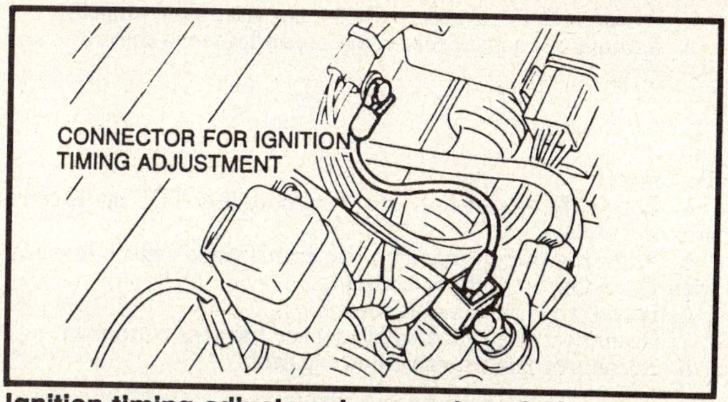

Ignition timing adjustment connector – 2.0L engine

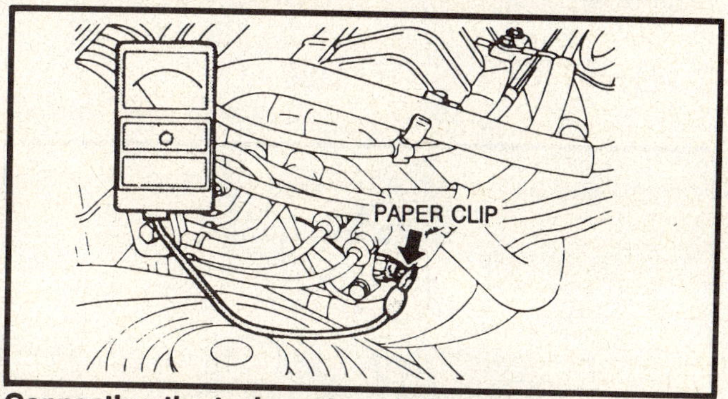

Connecting the tachometer – 1.6L engine

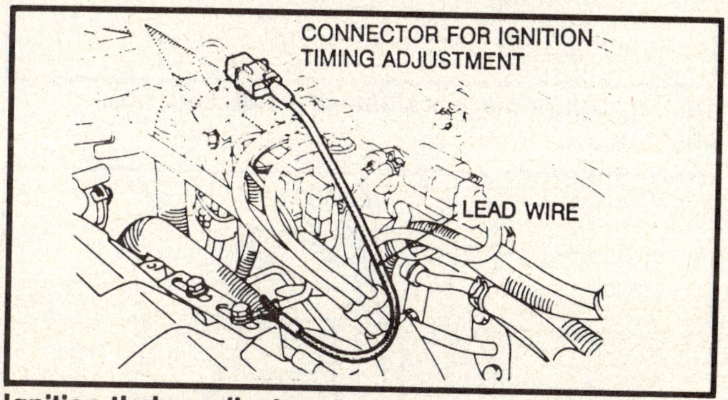

Ignition timing adjustment connector – 1.6L engine

7. Stop the engine and jumper the terminal for ignition-timing adjustment to ground.
8. Start and run the engine at curb idle speed.
9. Check the basic ignition timing and adjust, if necessary. Basic ignition timing should be 5 degrees BTDC.
10. If the timing is not within specification, loosen the crank angle sensor retaining nut and turn the crank angel sensor to bring the timing within specs.
11. Tighten the sensor retaining nut after adjustment. Recheck the timing and adjust if necessary.
12. Stop the engine and unground the ignition timing connector.
13. Start the engine and run at curb idle. Check the actual ignition timing. Actual ignition timing should be 8 degrees BTDC.

NOTE: Actual ignition timing may vary, depending on the control mode of the engine control unit. In such case, recheck the basic ignition timing. If there is no deviation, the ignition timing is functioning normally. At altitudes, more than approximately 2,300 ft. (701m) above sea level, the actual ignition timing is further advanced to ensure good combustion.

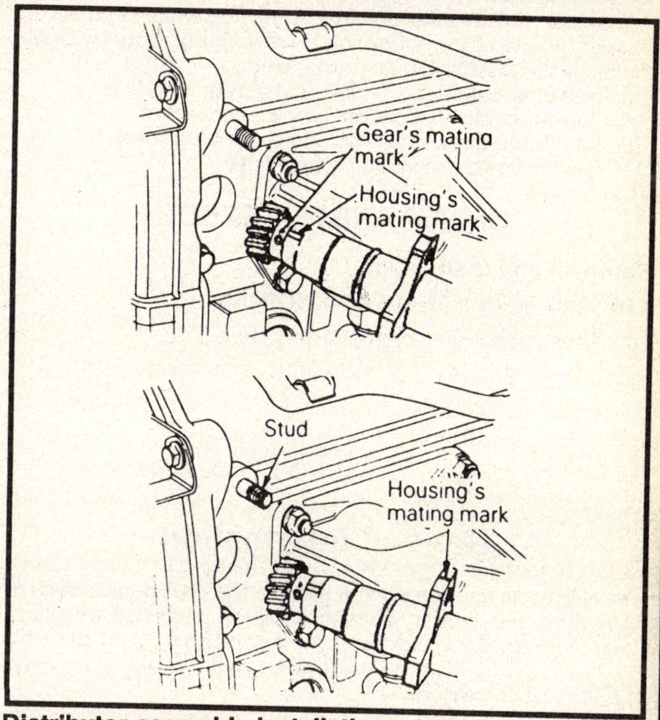

Distributor assembly installation – 1.5L, 1.8L and 3.0L SOHC engines

Crank angle sensor installation – 1.6, 2.0L and 3.0L DOHC engines

Component Replacement

DISTRIBUTOR ASSEMBLY

Removal and Installation

1.5L, 1.8L, 2.4L AND 3.0L SOHC ENGINES

1. Disconnect the negative battery cable.
2. Disconnect and tag the spark plug wires from the distributor cap.
3. Disconnect the distributor harness connector.
4. Loosen the distributor retaining bolt and remove the distributor from the engine.

To install:

5. Rotate the engine until No. 1 cylinder is at TDC on its compression stroke.

NOTE: Be careful not to turn it to the No. 4 cylinder TDC on its compression stroke by mistake.

6. Align the distributor housing and gear mating marks.

7. Install the distributor into the engine, while aligning the fine cut (groove or projection) on the distributor flange with the center of the distributor mounting stud.
8. Install and tighten the distributor retaining bolt.
9. Reconnect the distributor harness connector.
10. Install the distributor cap and spark plug wires.
11. Reconnect the negative battery cable.

CRANK ANGLE SENSOR

Removal and Installation

1.6L, 2.0L AND 3.0L DOHC ENGINES

1. Disconnect the negative battery cable.
2. Disconnect the crank angle sensor harness connector.
3. Remove the sensor retaining nut and remove the crank angle sensor.

To install:

4. Rotate the engine until No. 1 cylinder is at TDC on its compression stroke.
5. Align the punch mark on the crank angle sensor housing with the notch in plate or coupling.
6. Install the crank angle sensor.
7. Reconnect the crank angle sensor harness connector.
8. Reconnect the negative battery cable.

DAIHATSU ELECTRONIC IGNITION SYSTEM

General Information

The electronic ignition system used on the Daihatsu consists of an ignition coil, igniter, distributor, spark plugs and secondary wires. The ignition spark timing is controlled by the Electronic Control Unit (ECU), based on 2 pick-ups in the distributor and various engine sensors.

Diagnosis and Testing

SERVICE PRECAUTIONS

- Do not leave the ignition key turned **ON** for more than 10 minutes, while the engine is not running.
- Make sure all EFI harness connectors are fastened securely. A poor connection can cause an extremely high surge voltage in the coil and condenser and result in damage to integrated circuits.
- Keep EFI all parts and harnesses dry during service.
- Before attempting to remove any parts, turn **OFF** the ignition switch and disconnect the battery ground cable.
- Always use a 12 volt battery as a power source.
- Do not attempt to disconnect the battery cables with the engine running.
- Do not depress the accelerator pedal when starting.
- Make sure the igniter is properly grounded to the body.
- Never allow the coil or tachometer terminals to touch ground or damage will result to the igniter or coil.

IGNITION COIL

Testing

1.0L ENGINE

1. Turn the ignition key **OFF**.
2. Disconnect the coil secondary wire.
3. Disconnect the coil primary wires, noting their location.
4. Measure the resistance between the positive and negative terminals of the coil. Resistance reading should be as follows:

Primary resistance
0.2–1.0 ohms – 1.0L engine
1.35–1.65 ohms – 1.3L and 1.6L engines
Secondary resistance:
4.5–20.0 Kilo ohms – 1.0L engine
22–30 Kilo ohms – 1.3L and 1.6L engines

5. If the resistance is not within specifications with the coil cold, replace the coil.
6. Measure the resistance between the positive terminals and coil tower.

Testing coil primary resistance – 1.3L and 1.6L engines

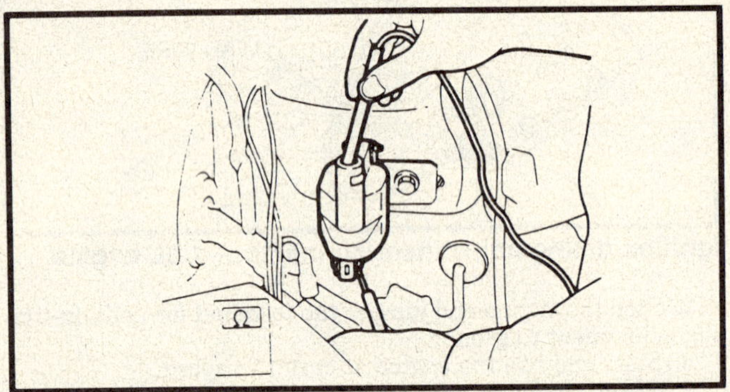

Testing coil secondary resistance – 1.3L and 1.6L engines

7. If the resistance is not within specifications with the coil cold, replace the coil.

DISTRIBUTOR PICK-UP COILS

Testing

1.0L ENGINE

1. Turn the ignition key **OFF**.
2. Disconnect the distributor connector.
3. Check the resistance between terminals 1 and 2 of the connector.

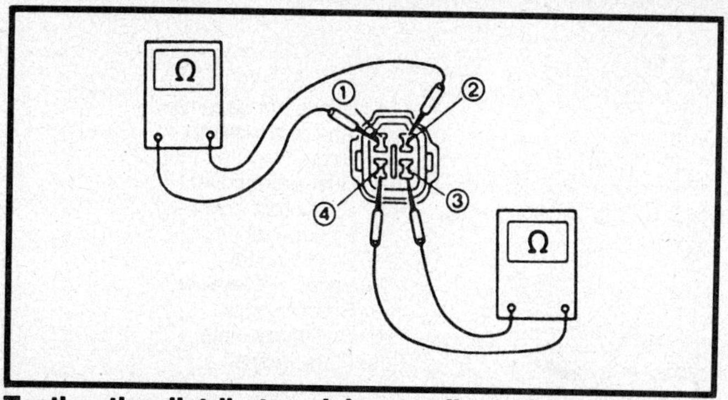

Testing the distributor pick-up coil—1.6L engine

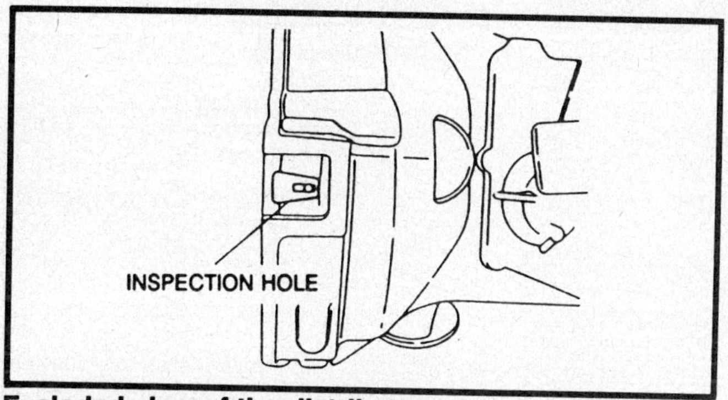

Exploded view of the distributor—1.0L engine

4. Check the resistance between terminals 3 and 4 of the connector.
5. If the resistance is not between 205–255 ohms, replace the distributor assembly.

IGNITION TIMING

Adjustment

1.0L ENGINE

1. Connect a tachometer. Never allow the tachometer terminals to touch ground or damage will result to the igniter or coil.

NOTE: If using a coil on the primary circuit that doesn't have a 3 cylinder range, set the tachometer on the 6 cylinder range and multiply the rpm by 2.

2. Allow engine to reach normal operating temperature. Cooling fan should come ON.
3. All electrical accessories should be **OFF**. Cooling fan should not be operating during adjustments.
4. Install test wire harness 09991–87702–000 or equivalent, to the service connector.
5. Jumper the Brown T terminal and the Black ground terminal of the service connector. Take care not to allow wire to touch any other terminal or ECU damage will result.
6. Remove the rubber cap from the throttle body and adjust idle speed to 750–850 rpm.
7. Adjust ignition timing by turning the distributor to obtain 5 degrees BTDC at idle with the T terminal grounded.
8. Tighten the distributor hold-down to 11–16 ft. lbs. (15–21 Nm).
9. Recheck timing.
10. Remove the jumper from the service connector.
11. Recheck and adjust idle as needed.
12. Install rubber cap to the throttle body.

1.3L AND 1.6L ENGINES

1. Connect a timing light to the No. 1 cylinder spark plug wire.
2. Connect a tachometer to the engine.

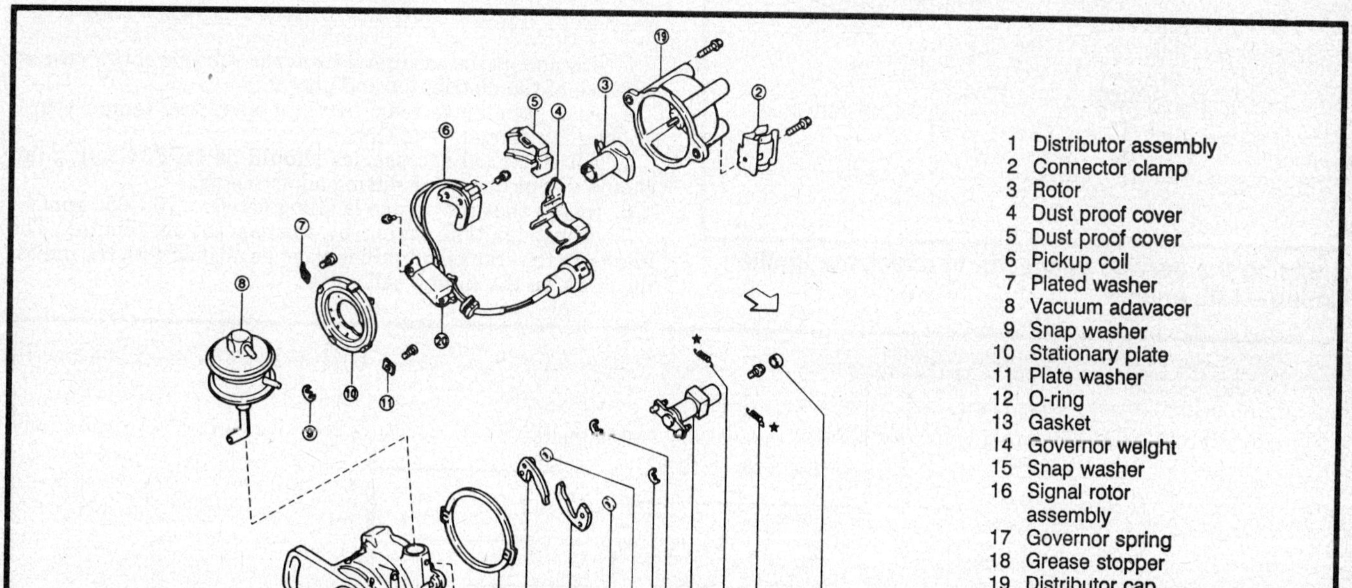

Exploded view of the distributor—1.3L engine

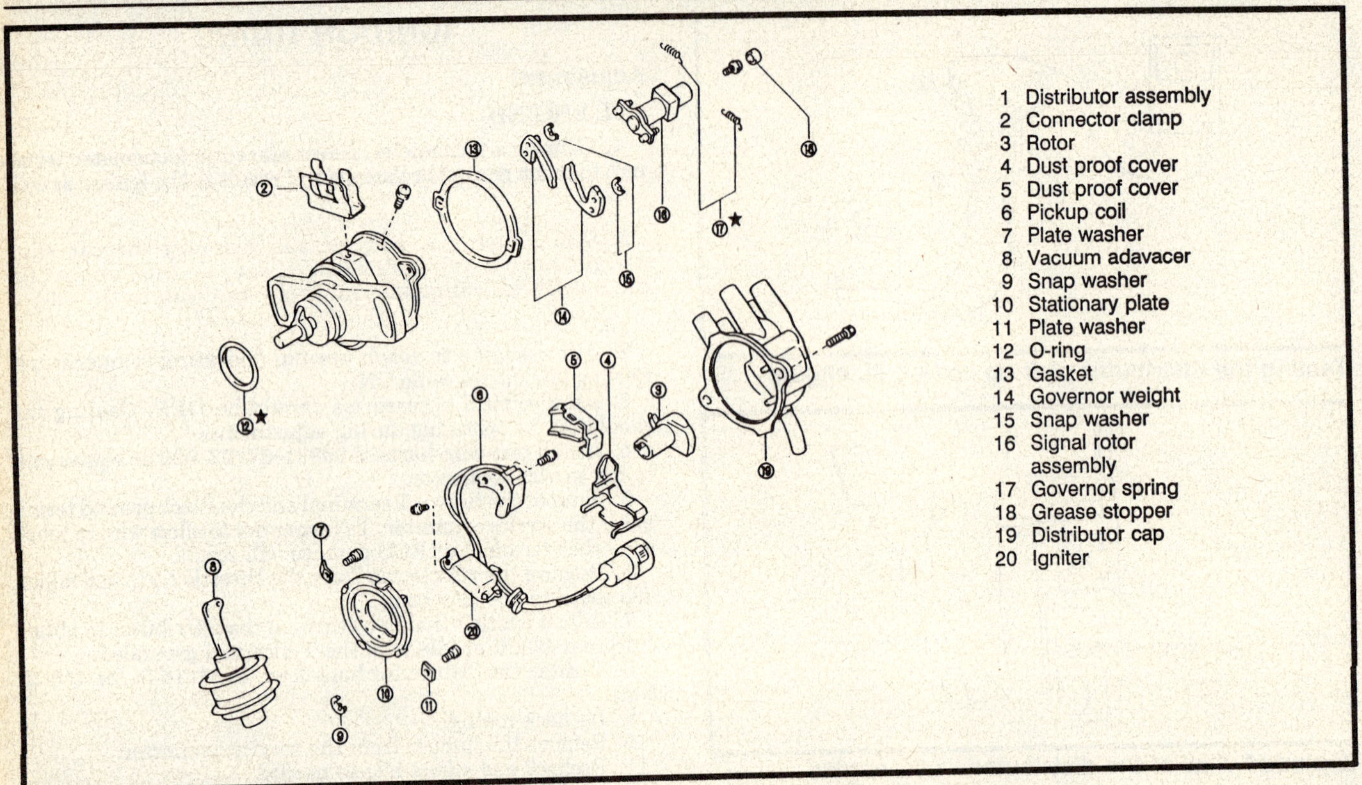

Exploded view of the distributor—1.6L engine

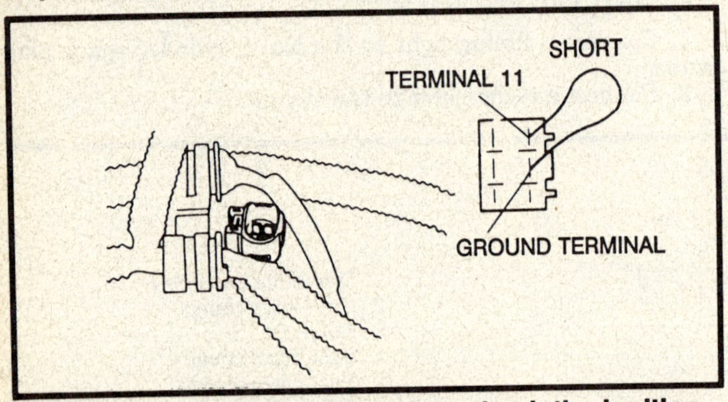

Jumping the service connector to check the ignition timing—1.0L engine

NOTE: If the tachometer being used, is to be connected to the negative terminal of the ignition coil, use a SST adapter tool or equivalent, to the distributor connector, then connect the tachometer to the SST adapter tool. Never allow the tachometer terminals to touch ground or damage will result to the igniter or coil.

3. Disconnect the vacuum hose at the sub side of the vacuum advance of the distributor and plug it.
4. Allow engine to reach normal operating temperature. Cooling fan should come ON.
5. All electrical accessories should be **OFF**. Cooling fan should not be operating during adjustments.
6. Insure that the engine is idling between 750–850 rpm.
7. Adjust ignition timing by turning the distributor. The marks on the crankshaft pulley must be aligned with the indicator mark on the timing belt cover.

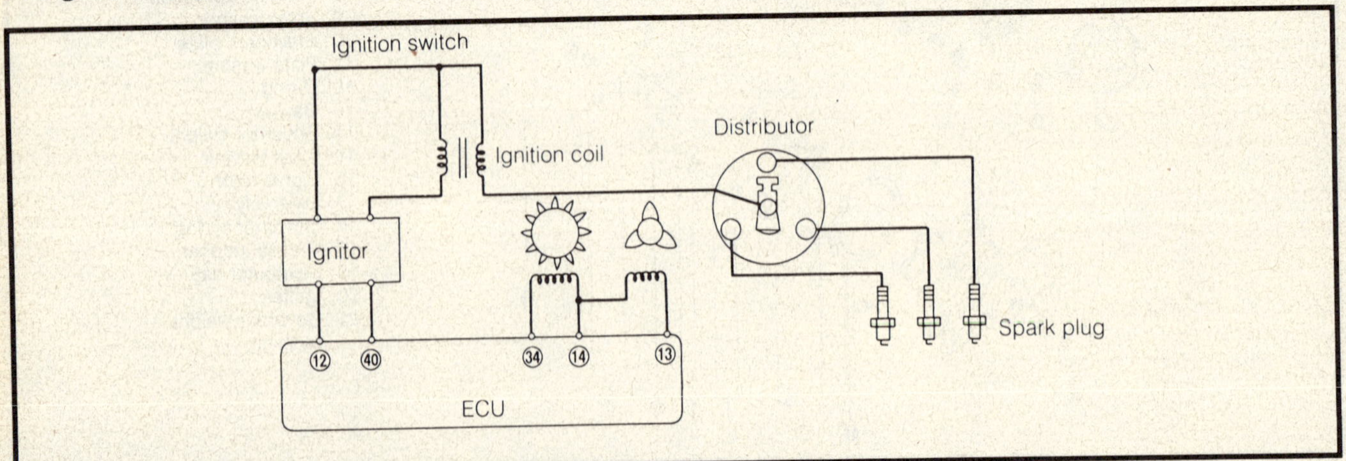

Wiring schematic of the ignition system—1.0L engine

BASIC TROUBLESHOOTING CHART

Problem	Possible causes	Remedies
Engine will not start/hard to start. (Engine crank normally.)	Incorrect ignition timing Ignition coil faulty Igniter faulty Distributor faulty Spark plug wires faulty Spark plugs faulty Ignition wiring disconnected or broken	Reset timing. Inspect coil. Inspect igniter. Inspect distributor. Inspect spark plug wires. Inspect plugs. Inspect wiring.
Rough idle or engine stalls.	Spark plugs faulty Ignition wiring faulty Incorrect ignition timing Ignition coil faulty Igniter faulty Distributor faulty Spark plug wires faulty	Inspect plugs. Inspect wiring. Reset timing. Inspect coil. Inspect igniter. Inspect distributor. Inspect spark plug wires.
Engine hesitation/poor acceleration	Spark plugs faulty Ignition wiring faulty Incorrect ignition timing	Inspect plugs. Inspect wiring. Reset timing.
Engine dieseling (Engine runs after ignition witch is turned OFF.)	Incorrect ignition timing	Reset timing.
Muffler explosion (after fire) all the time	Incorrect ignition timing	Reset timing.
Engine backfire	Incorrect ignition timing	Reset timing.
Poor fuel economy	Spark plugs faulty Incorrect ignition timing	Inspect plugs. Reset timing.
Engine overheating	Incorrect ignition timing	Reset timing.

8. Tighten the distributor hold-down to 11–16 ft. lbs. (15–21 Nm).
9. Recheck timing.
10. Connect the vacuum hose to the vacuum advance.
11. Recheck and adjust idle as needed.

Component Replacement

DISTRIBUTOR

Removal and Installation

1.0L ENGINE

1. Disconnect the negative battery cable.
2. Set the piston of the No. 1 cylinder to 5 degrees BTDC, under the compression stroke.
3. Turn the ignition key **OFF**.
4. Remove the coil wire from the distributor cap.
5. Tag and remove the spark plug wires.
6. Disconnect the distributor connector.
7. Remove the distributor hold-down bolt.
8. Pull the distributor from the housing.

To install:

9. Set the piston of the No. 1 cylinder to 5 degrees BTDC, under the compression stroke. This position can be confirmed by aligning the timing marks with each other and checking that the valve rockers for the No. 1 cylinder are in a free state, by removing the oil filler cap.
10. Remove the distributor cap.
11. Align the drilled mark at the forward end of the distributor drive gear with the drilled mark on the distributor body.
12. Install the distributor assembly into the distributor housing in such a way that the split line of the distributor body may be aligned with the embossed line of the distributor housing.
13. Install the distributor hold-down bolt.
14. Install the distributor cap.
15. Connect the distributor connector.

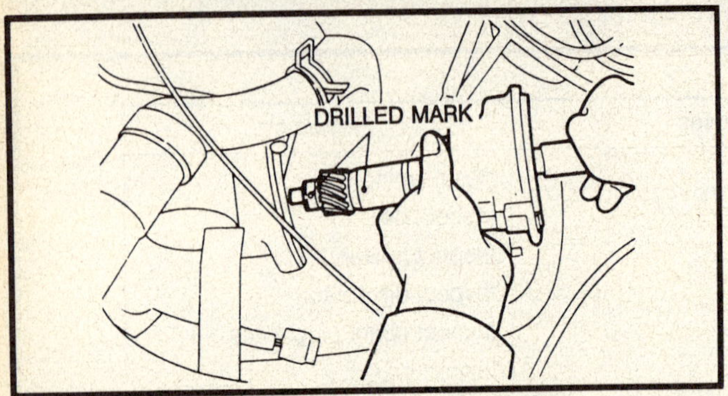

Aligning the distributor shaft—1.0L engine

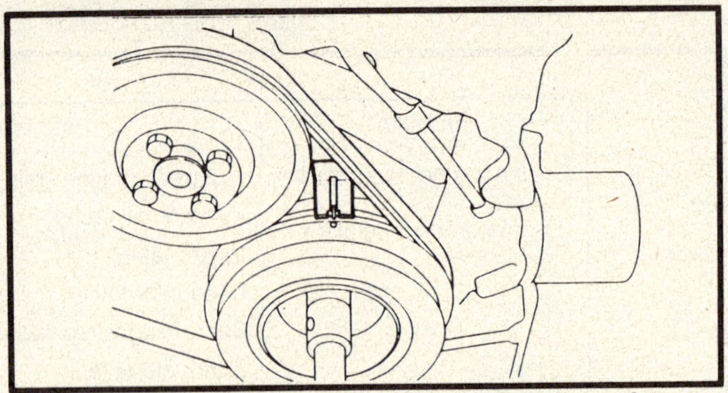

Crankshaft position at TDC—1.3L and 1.6L engines

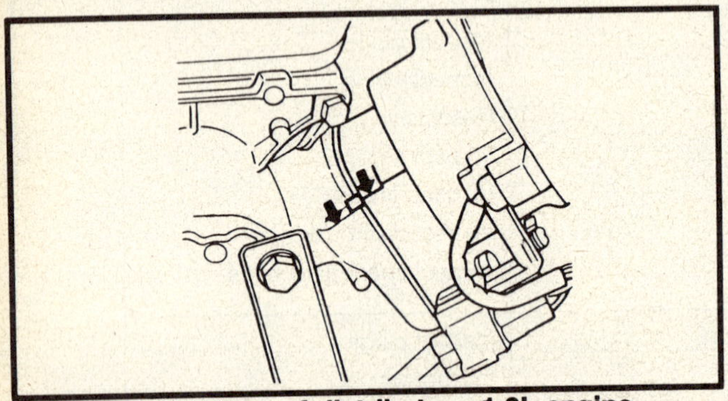

Installation position of distributor—1.0L engine

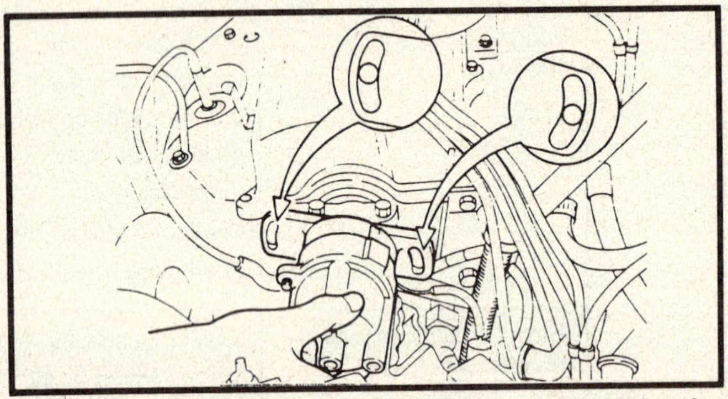

Installing the distributor—1.3L engine, 1.6L engine is similar

16. Install the spark plug and coil wires in the correct locations.
17. Adjust the timing and tighten the

1.3L AND 1.6L ENGINE

1. Disconnect the negative battery cable.
2. Set the piston of the No. 1 cylinder to 5 degrees BTDC, under the compression stroke.
3. Turn the ignition key **OFF**.
4. Remove the coil wire from the distributor cap.
5. Tag and remove the spark plug wires from the distributor cap.
6. Disconnect the distributor connector.
7. Tag and remove the vacuum hoses from the vacuum advance.
8. Remove the distributor hold-down bolts.

NOTE: Engine oil will pour out as the distributor is removed from the engine, place a cloth under the distributor while removing it.

9. Pull the distributor from the housing.

To install:

10. Set the piston of the No. 1 cylinder to 5 degrees BTDC, under the compression stroke. This position can be confirmed by aligning the timing marks with each other and checking that the valve rockers for the No. 1 cylinder are in a free state, by removing the oil filler cap.
11. Align the cut out section of the distributor housing with the distributor shaft coupling tangs.
12. Install the distributor into the cylinder head. At this time, ensure that the distributor attaching hole of the cylinder head comes to the center of the elongaged hole for the distributor bolt.
13. Install the distributor hold-down bolts and tighten snug at this time.
14. Connect the vacuum advance hoses.
15. Connect the distributor connector.
16. Install the distributor cap.
17. Install the spark plug and coil wires in the correct locations.
18. Adjust the timing and tighten the distributor hold-down bolts.

Air Gap Adjustment

1. Disconnect the negative battery cable.
2. Tag and remove the spark plug wires from the distributor cap.
3. Remove the distributor connector from the connector clamp.
4. Remove the distributor cap.
5. Check the vacuum advance. Tag and remove the vacuum hoses from the advance.
6. Apply vacuum to the vacuum advance of 150 mmHg. or higher. Check and make sure that the vacuum advance operates. If it does not operate, repair or replace it.
7. Reconnect the vacuum hoses to the vacuum advance.
8. Check the governor advance. Turn the rotor counterclock-

Checking vacuum advance operation—1.3L and 1.6L engines

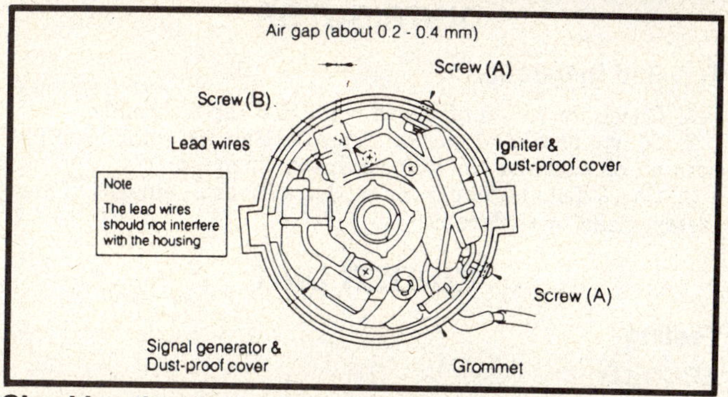

Checking the air gap between the signal generator and signal rotor—1.3L engine, 1.6L engine is similar

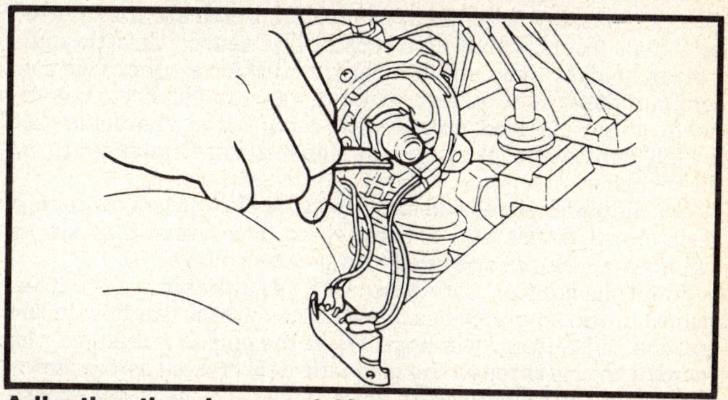

Adjusting the air gap—1.3L engine, 1.6L engine is similar

wise and release it. The rotor should return rapidly. Repair or replace the governor advance, as required.

9. Check the rotor for excessive play. If it is excessive, replace the governor advance.
10. Remove the rotor. Turn the crankshaft until the signal rotor faces toward the signal generator.
11. Inspect the air gap. Using a brass feeler gauge tool, check the gap at all points between the signal generator and the signal rotor. The gap should be 0.0079–0.0015 in. (0.2–0.4mm). If not adjust the air gap accordingly.
12. If it is necessary to adjust the air gap, proceed as follows:

NOTE: The procedure below indicates igniter removal, this applies to the 1.3L engine. It is not necessary to remove the igniter on the 1.6L engine, all other procedures apply.

a. Remove the distributor gasket. Remove the screws retaining the igniter. Carefully pull the igniter from the distributor housing, making sure not to damage the wire connected to it.
b. Loosen the screws retaining the signal generator.
c. Place the brass feeler gauge between the signal generator and the signal rotor and adjust the gap to 0.0079–0.0015 in. (0.2–0.4mm).
d. When the gap is correct, tighten the retaining screws of the signal generator.
e. Check the air gap again. If the gap has changed, adjust as required.

13. Intall the igniter to the distributor, it must be installed so that the 2 leads are twisted 360 degrees in clockwise direction.
14. Hold the lead wires using a clamp mounted on the side of the housing.

NOTE: The mounting surface of the igniter and distributor housing must be kept clean. It should be noted that if any foreign object is present between the igniter and the housing, heat developed by the igniter will not radiate properly.

15. Tighten the igniter mounting screws to 14–23 inch lbs. (1.7–2.5 Nm).

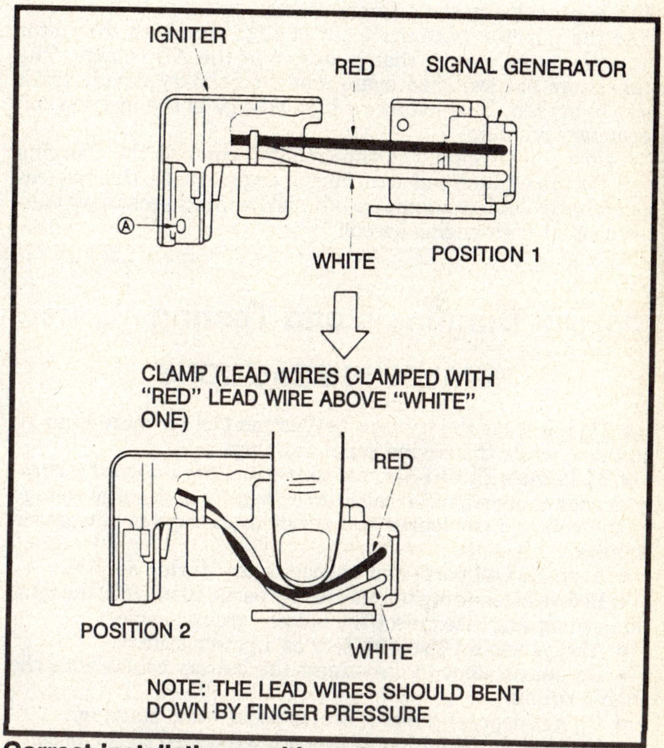

Correct installation position of the igniter—1.3L engine

16. Install the rotor. Check the distributor cap gasket for cracks or damage. Replace it, as required.
17. Install the distributor cap, connect the distributor connector and install it to the retaining clamp.
18. Install the spark plug and coil wires in the correct locations.
19. Check and adjust the ignition timing, as required.

HONDA PGM-IG ELECTRONIC IGNITION

General Information

The fully transistorized programmed ignition (PGM-IG) system comprises of a distributor driven directly from the front camshaft, an amplifier mounted on the side of the distributor and a high energy coil.

The distributor contains a reluctor mounted on a rotor shaft and a magnet mounted on a base plate. The pick-up coil is located around the rotor shaft but does not rotate with the shaft. The programmed ignition employed on these vehicles, provides optimum control of ignition timing by determining the optimum timing using a microcomputer in response to engine speed and vacuum pressure in the intake manifold, which are transmitted by signals from CRANK/CYL sensor, TDC sensor, throttle angle sensor, coolant temperature sensor and MAP sensor. This system, not dependent on a governor or vacuum diaphragm, is capable of setting lead angles with complicated characteristics which cannot be provided by conventional governors or diaphragms.

The amplifier connected directly to the pick-up coil contains 3 resistors, 3 diodes and 2 transistors. The transistors act as switches which are activated at a precise voltage.

When the ignition is switched **ON**, the switching of the transistors in the amplifier ensures that no current can flow in the ignition coil primary windings. When the engine is cranked, the reluctor moves through the magnetic field created by the stator and, when the reluctor teeth are aligned with the stator projections, a small AC voltage is created. The amplifier amplifies this voltage and uses it to switch the transistors so that an earth path is provided to the primary circuit.

As the reluctor teeth move out of alignment with the stator projections, an abrupt change occurs in the AC voltage. The transistors are switched again and the primary circuit earth path is broken. This induces a high voltage in the ignition coil secondary winding.

A time control circuit in the amplifier controls the charging time for the ignition coil according to engine speed, this reduces consumption at low engine speeds and prevents secondary voltage drop at high engine speeds.

Diagnosis and Testing

SERVICE PRECAUTIONS

- Do not leave the ignition key turned **ON** for more than 10 minutes, while the engine is not running.
- Make sure all EFI harness connectors are fastened securely. A poor connection can cause an extremely high surge voltage in the coil and condenser and result in damage to integrated circuits.
- Keep EFI all parts and harnesses dry during service.
- Before attempting to remove any parts, turn **OFF** the ignition switch and disconnect the battery ground cable.
- Always use a 12 volt battery as a power source.
- Do not attempt to disconnect the battery cables with the engine running.
- Do not depress the accelerator pedal when starting.
- Make sure the igniter is properly grounded to the body.
- Never allow the coil or tachometer terminals to touch ground or damage will result to the igniter or coil.

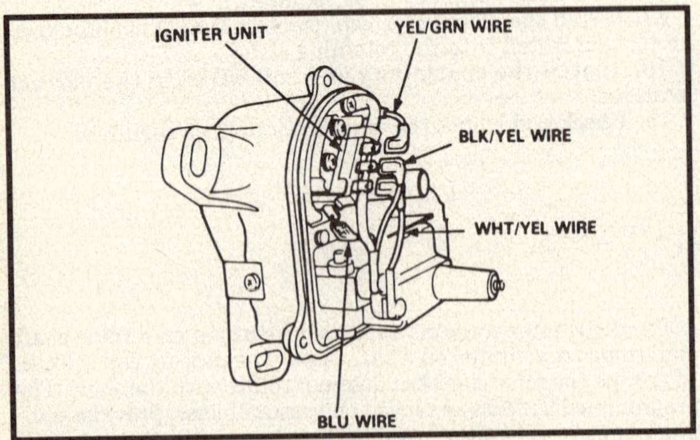

Testing the igniter unit—Accord and Civic

DISTRIBUTOR

Top End Inspection

1. Check for rough or pitted rotor and cap terminals.
2. Scrape or file off carbon deposits. Smooth rotor terminal with an oil stone or No. 600 sandpaper if rough.
3. Check distributor cap for cracks, wear and damages. If necessary, clean or replace it.

IGNITER UNIT

Testing

ACCORD

1. Check and service any engine codes before proceeding.
2. Remove the distributor cap, rotor and the leak cover.
3. Disconnect the BLK/YEL, WHT/BLU, YEL/GRN and BLU wires from the igniter.
4. Turn the ignition switch **ON**.
5. Check for voltage between the BLK/YEL wire and ground. If there is no voltage, check the BLK/YEL wire and ignition switch.
6. Check for voltage between the WHT/BLU wire and ground. If there is no voltage, check the WHT/BLU wire and the ignition coil.
7. Check the YEL/GRN wire between the PGM-FI ECU and the igniter.
8. Check the BLU wire between the tachometer and the igniter.
9. If all tests are normal, replace the igniter unit.

CIVIC

1. Check and service any engine codes before proceeding.
2. Check the No. 32 fuse (50A). in the dash fuse box before testing.
3. Remove the distributor cap, rotor and the leak cover.
4. Disconnect the wires from the igniter.
5. Turn the ignition switch **ON**.
6. Check for voltage between the BLK/YEL wire and ground. If there is not voltage check the BLK/YEL wire and ignition switch.
7. Check for voltage between the WHT/BLU wire and ground. If there is not voltage check the WHT/BLU wire and the ignition coil.
8. Check the continuity between the WHT wire and body ground. Service the wire or connectors as needed.
9. Check the continuity between the BLU wire and body ground. If no continuity check the wire and connectors between the igniter unit and the tachometer or the A/T control unit.

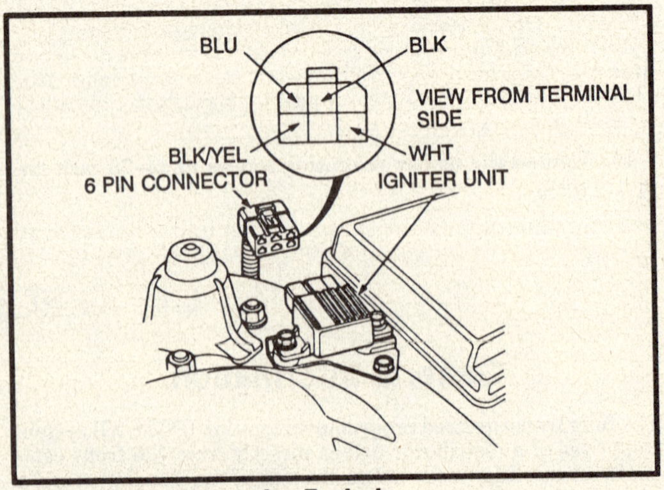

Testing the igniter unit—Prelude

10. If all tests are normal, replace the igniter unit.

IGNITION COIL

Testing

ACCORD

1. With ignition switch **OFF**, remove the distributor cap.
2. Remove the 2 screws to disconnect the BLK/YEL front terminal A and WHT/BLU from terminal B of the coil.
3. Using an ohmmeter, check resistance between terminals. Replace coil if resistance is not within specification.

NOTE: Resistance will vary with coil temperature. Specifications are at 70°F (20°C)

4. Check that the primary winding resistance between terminals A and B is 0.6–0.8 ohms.
5. Check that the secondary winding resistance between terminals A and the coil tower is 12880–19320 ohms.
6. If not within specifications, replace the coil.

CIVIC

1. With ignition switch **OFF**, remove the distributor cap.
2. Remove the 2 screws to disconnect the BLK/YEL front terminal A and WHT/BLU from terminal B of the coil.

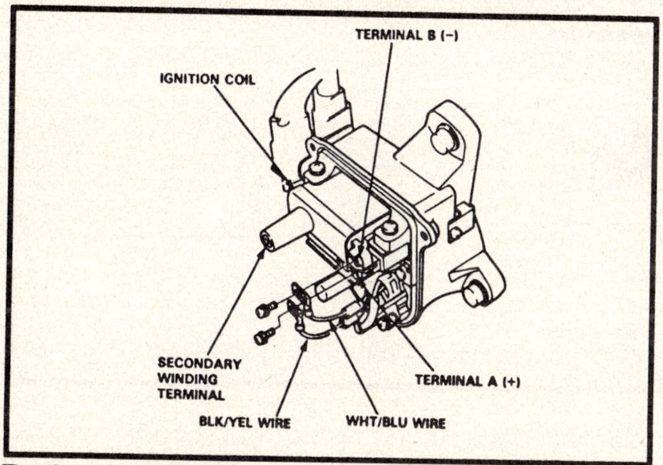

Testing the ignition coil—Accord and Civic

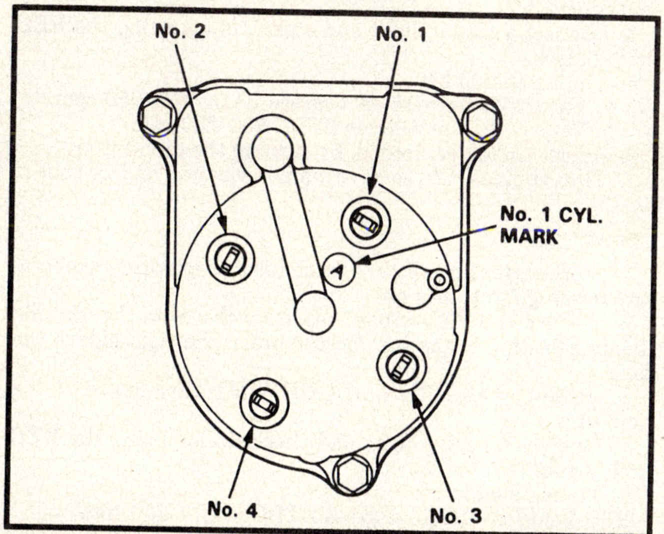

Spark plug location in the distributor cap towers—Accord

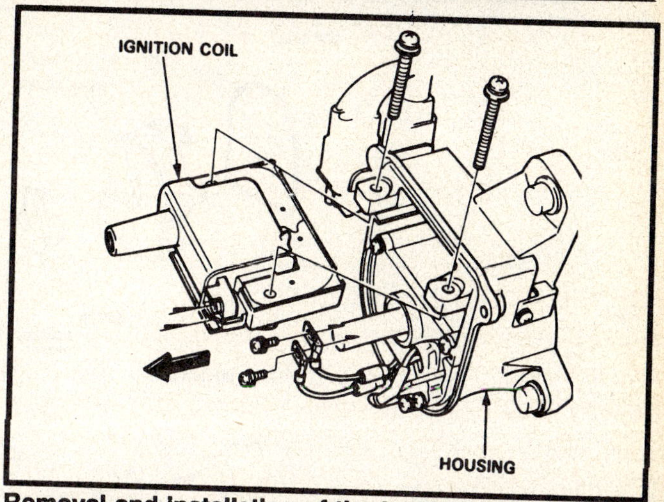

Removal and installation of the ignition coil—Prelude

3. Using an ohmmeter, check resistance between terminals. Replace coil if resistance is not within specification.

NOTE: Resistance will vary with coil temperature. Specifications are at 70°F (20°C)

4. Check that the primary winding resistance between terminals A and B is 0.6–0.8 ohms.
5. Check that the secondary winding resistance between terminals A and the coil tower is 9760–14640 ohms.
6. If not within specifications, replace the coil.

PRELUDE

1. With ignition switch **OFF**, disconnect primary connectors and coil wire.
2. Using an ohmmeter, check resistance between terminals. Replace coil if resistance is not within specification.

NOTE: Resistance will vary with coil temperature.

3. Primary winding resistance between terminals A and D should be 0.3–0.4 ohms at 70°F (20°C).
4. Secondary winding resistance between terminals A and secondary winding terminals should be 9040–13560 ohms at 70°F (20°C).
5. Resistance between B and D terminals should be 2090–2310 ohms at 70°F (20°C).

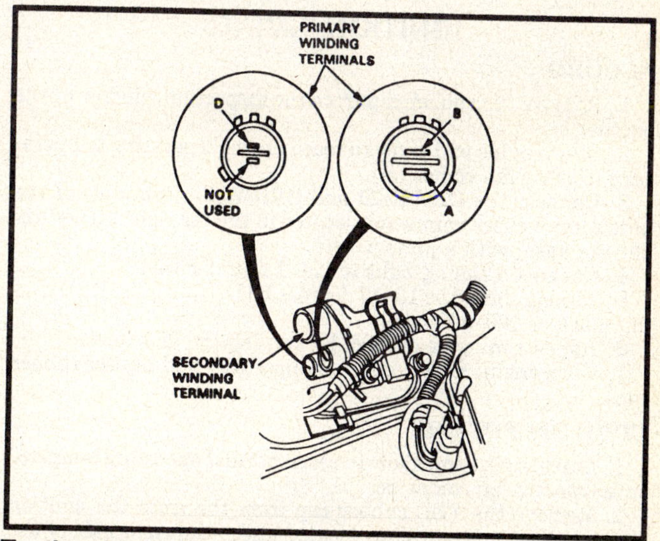

Testing the ignition coil—Prelude

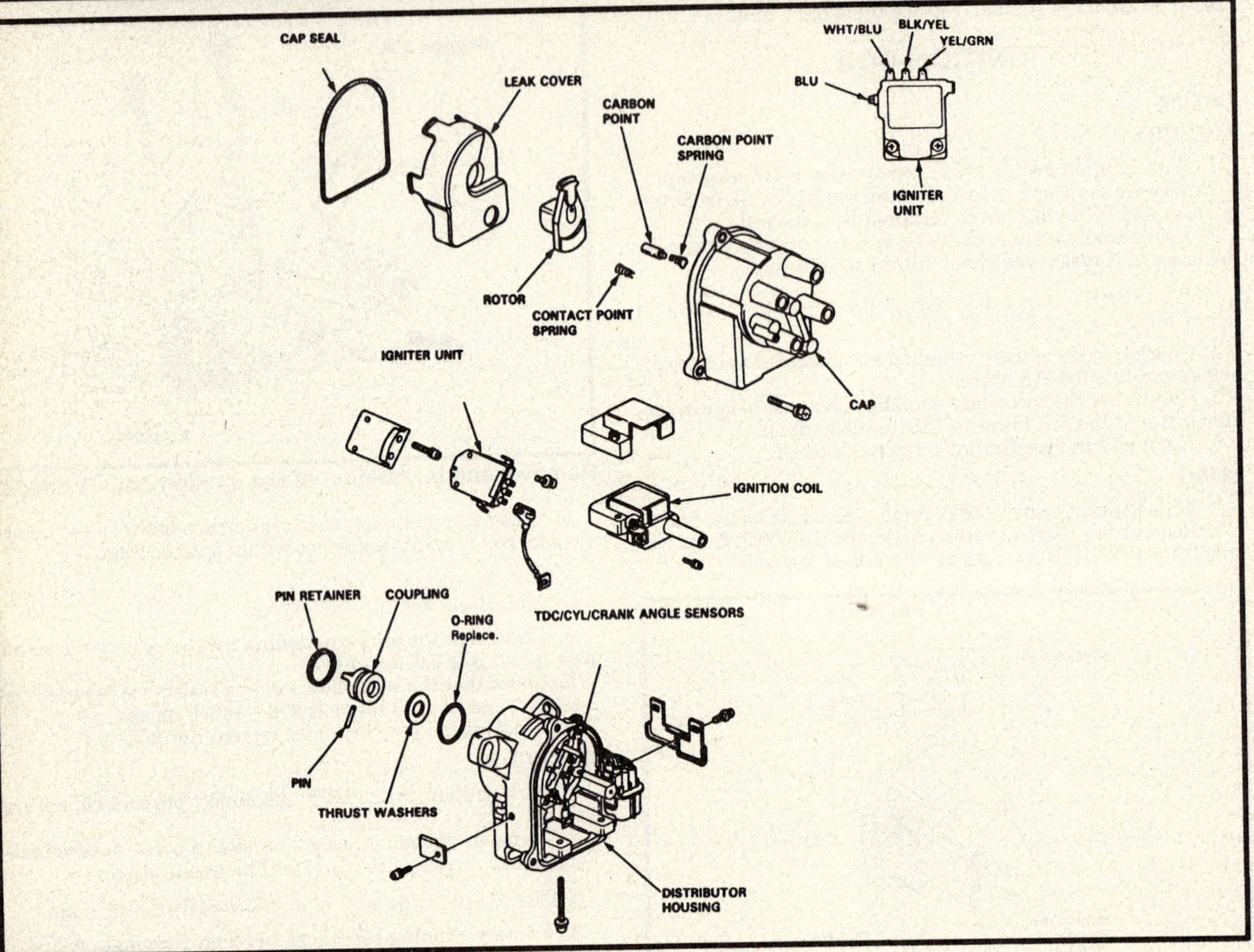

Exploded view of the distributor—Accord

6. Check continuity between A and C terminals. Replace coil if no continuity.

IGNITION TIMING

ACCORD

1. Remove the rubber cap from the inspection window of the engine.
2. Start engine and allow to reach normal operating temperature (cooling fan comes on).
3. Connect the ORN/RED and GRN/WHT terminals of the blue service check connector, located in the far right corner under the dash, with a jumper wire.
4. Connect a timing light to No. 1 cylinder wire.
5. Timing should be 13–17 degrees BTDC using the RED timing mark at 700–800 rpm.
6. Adjust timing as needed, by turning the distributor.
7. Turn engine **OFF**, remove jumper wire and replace rubber caps.

CIVIC EXCEPT CRX

1. Start engine and allow to reach normal operating temperature (cooling fan comes on).
2. Remove the YEL rubber cap from the from the ignition timing adjusting connector, located under the right side of the dash.
3. Jumper the BRN and GRN/WHT terminals of the connector.
4. Connect a timing light and check timing using the RED timing mark.
5. Timing should be:
   a. Except Wagon—16–20 degrees BTDC 675–750 rpm.
   b. Wagon—16–20 degrees BTDC 700–800 rpm.
6. Adjust timing as needed, by turning the distributor.
7. Turn engine **OFF**, remove jumper wire and replace rubber caps.

CIVIC CRX

1. Start engine and allow to reach normal operating temperature (cooling fan comes on).
2. Remove the YEL rubber cap from the from the ignition timing adjusting connector, located under the right side of the dash.
3. Jumper the BRN and GRN/WHT terminals of the connector.
4. Connect a timing light and check timing using the RED timing mark.
5. Timing should be:
   STD Vehicles—16–20 degrees BTDC at 675–750 rpm
   SI Vehicles—16–20 degrees BTDC at 700–800 rpm
   HF Vehicles (Except California)—12–16 degrees BTDC at 550–650 rpm

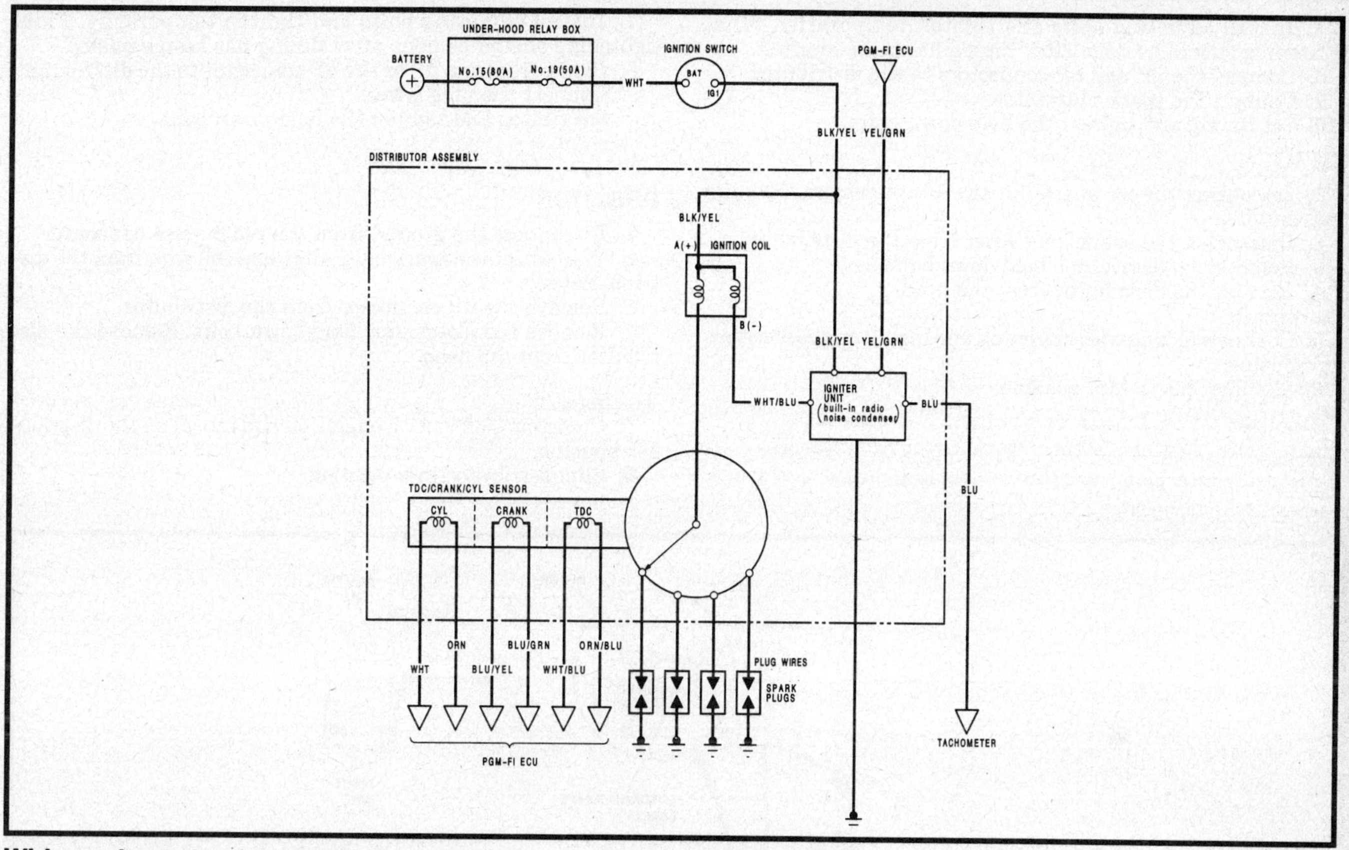

Wiring schematic of the ignition system—Accord shown, others similar

HF Vehicles (California)—12–16 degrees BTDC at 600–700 rpm

6. Adjust timing as needed, by turning the distributor.
7. Turn engine **OFF**, remove jumper wire and replace rubber caps.

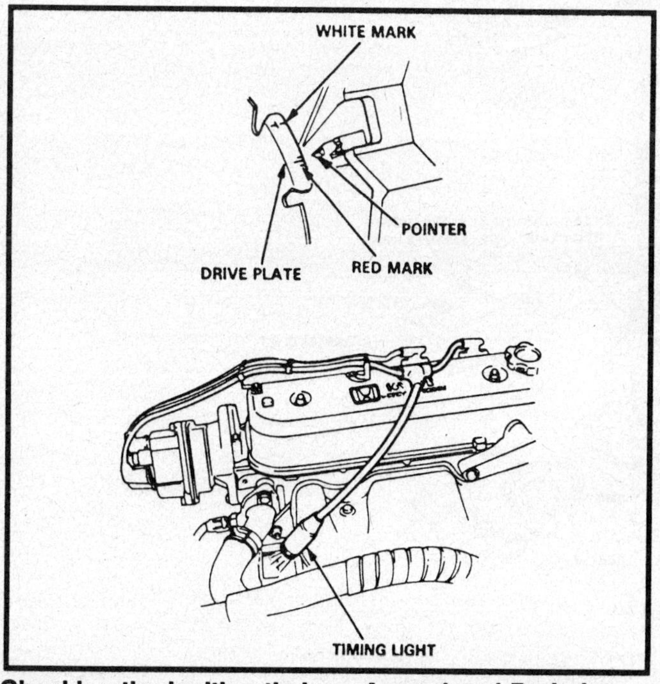

Checking the ignition timing—Accord and Prelude

PRELUDE

1. Start engine and allow to reach normal operating temperature (cooling fan comes on).
2. Remove the YEL rubber cap from the from the ignition timing adjusting connector, located behind the ignition coil.
3. Jumper the BRN and GRN/WHT terminals of the connector.
4. Connect a timing light and check timing using the RED timing mark.
5. Timing should be 13–17 degrees BTDC at 700–800 rpm.
6. Adjust timing as needed, by turning the distributor.
7. Turn engine **OFF**, remove jumper wire and replace rubber caps.

Component Replacement

DISTRIBUTOR

Removal and Installation

ACCORD

1. Disconnect the 2P and 8P connectors from the distributor.
2. Disconnect the spark plug wires from the distributor.
3. Remove the distributor hold-down bolts.
4. Remove the distributor from the head.

To install:

5. Coat new O-ring with engine oil and install on the distributor housing.
6. Slip distributor into position.

NOTE: Lugs on end of distributor and its mating grooves in camshaft end are both offset to eliminate possibility of installing distributor 180 degrees out of time.

7. Install adjusting bolts and tighten temporarily. Final tightening should be done after timing has been adjusted.
8. Connect the 2P and 8P connectors to the distributor.
9. Connect the spark plug wires.
10. Set timing and tighten the hold-down bolts.

CIVIC

1. Disconnect the 2P and 6P or the 8P connectors from the distributor.
2. Disconnect the spark plug wires from the distributor.
3. Remove the distributor hold-down bolts.
4. Remove the distributor from the head.

To install:

5. Coat new O-ring with engine oil and install on the distributor housing.
6. Slip distributor into position.

NOTE: Lugs on end of distributor and its mating grooves in camshaft end are both offset to eliminate possibility of installing distributor 180 degrees out of time.

7. Install adjusting bolts and tighten temporarily. Final tightening should be done after timing has been adjusted.
8. Connect 2P and 6P or the 8P connector to the distributor.
9. Connect the plug wires.
10. Set timing and tighten the hold-down bolts.

PRELUDE

1. Disconnect the ground from the radio noise condenser.
2. Disconnect the spark plug wires and coil wire from the distributor cap.
3. Remove the 4P connector from the distributor.
4. Remove the distributor hold-down bolts. Remove the distributor from the head.

To install:

5. Coat new O-ring with engine oil and install on the distributor housing.
6. Slip distributor into position.

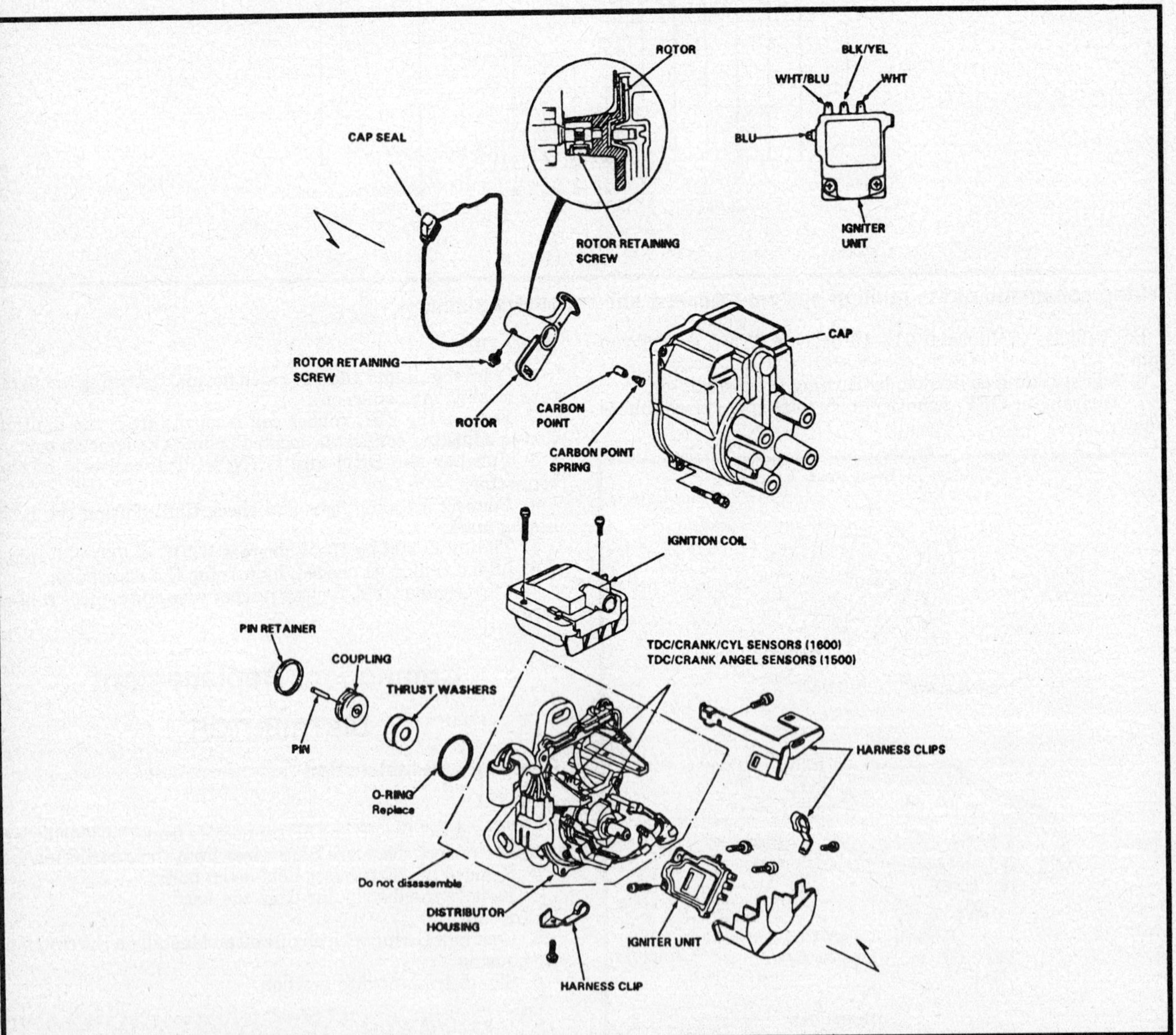

Exploded view of the distributor – Civic

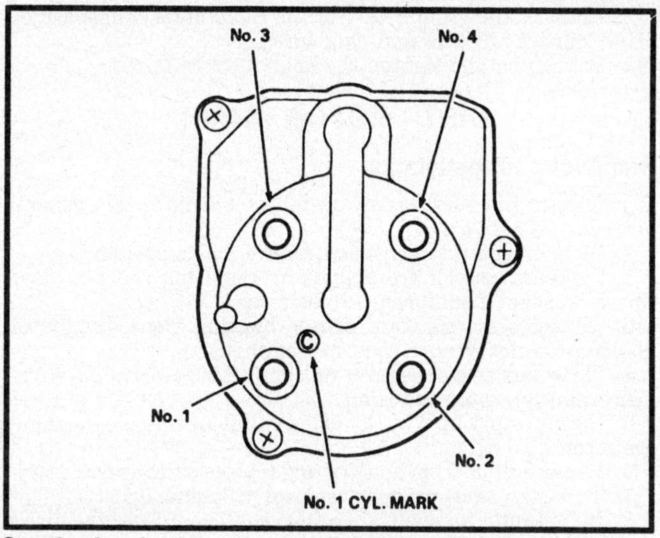

Spark plug location in the distributor cap towers—Civic

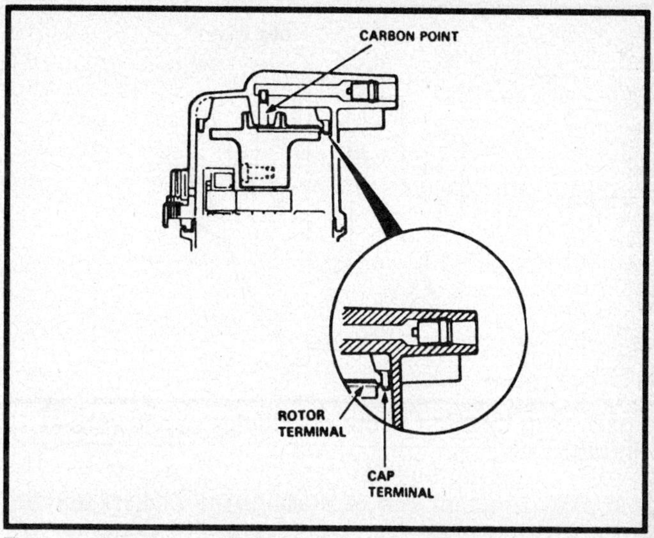

Removal and installation of the distributor—Prelude

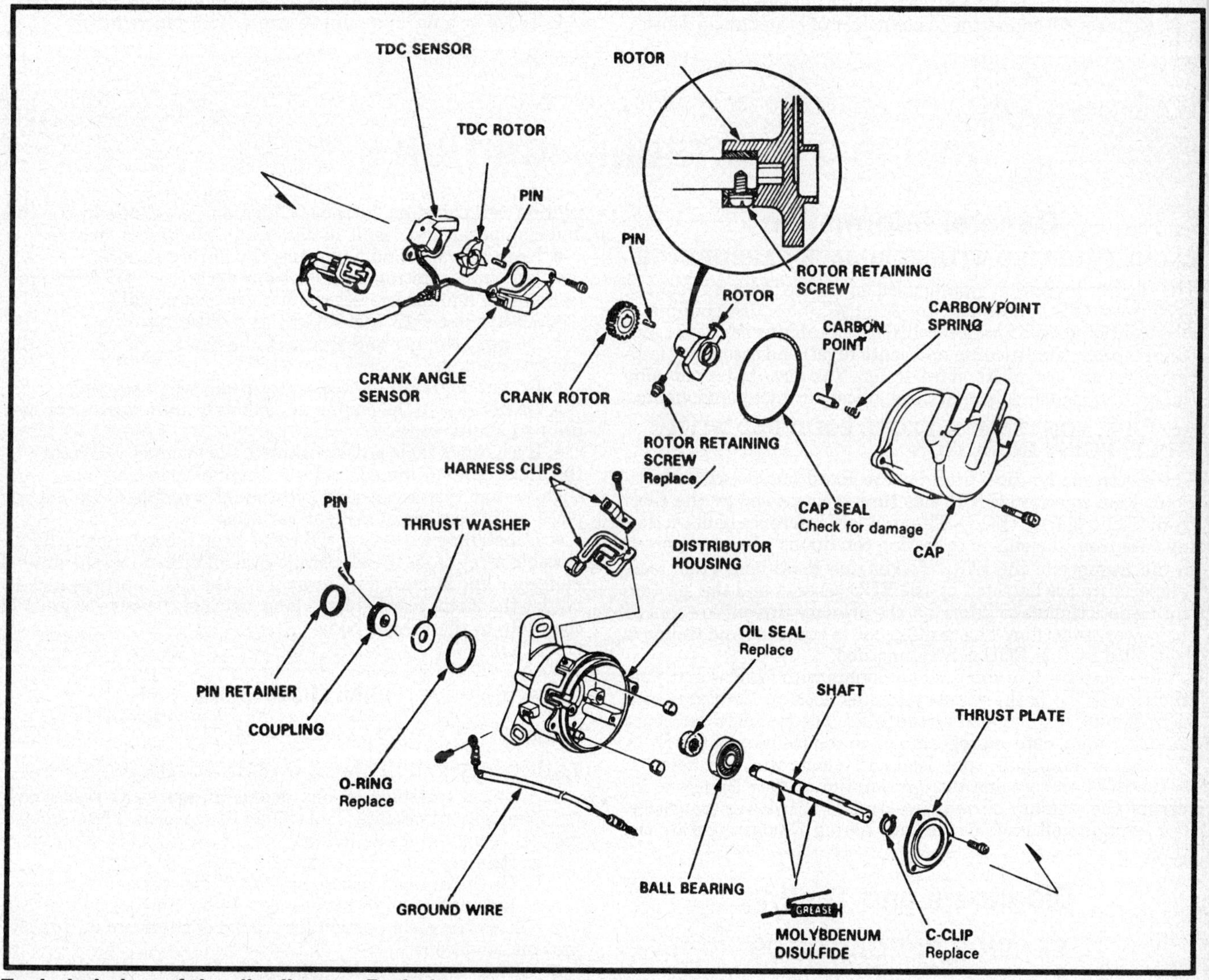

Exploded view of the distributor—Prelude

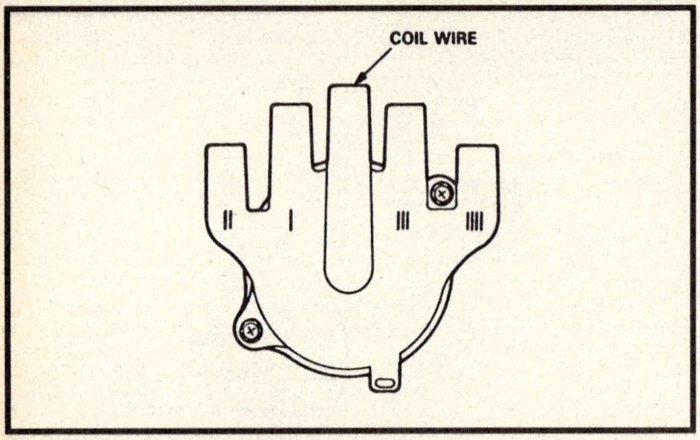

Spark plug location in the distributor cap towers—Prelude

NOTE: Lugs on end of distributor and its mating grooves in camshaft end are both offset to eliminate possibility of installing distributor 180 degrees out of time.

7. Install adjusting bolts and tighten temporarily. Final tightening should be done after timing has been adjusted.
8. Connect 4P connector to connector of crank angle sensor.
9. Connect the ground wire to the radio noise condenser.
10. Connect coil wire and plug wires.
11. Set timing and tighten the hold-down bolts.

DISTRIBUTOR SHAFT

Removal and Installation

Distributors may vary slightly on different vehicles. The general procedure is similar.

1. Slide off pin retainer being careful not to stretch it.
2. Separate coupler from shaft by removing roll pin. Save thrust washer (if equipped) for later use.
3. If necessary, separate sensor housing from distributor housing. Do not remove sensors from housing.
4. Drive out sensor rotor's roll pin after scribing an alignment mark on rotor and shaft.
5. Carefully pry up rotor by using suitable tools. Do not damage rotor.
6. Remove shaft from housing and replace parts as necessary.
7. To install shaft, reverse removal procedure.
8. Install rotor on shaft, then turn shaft so it faces the direction of the of No. 1 cylinder mark.
9. Set thrust washer and coupling on the shaft.
10. Check that rotor is still pointing toward No. 1 cylinder, then align index mark on housing with index mark on coupling.
11. Drive in a new pin and secure it with pin retainer.

HYUNDAI ELECTRONIC IGNITION SYSTEM

General Information

EXCEL EQUIPPED WITH FEEDBACK CARBURETOR

The electronic ignition system used on engines with a Feedback Carburetor (FBC), uses a magnetic pulse/igniter distributor and a conventional type ignition coil. The distributor cap, rotor, advance mechanism (vacuum and centrifugal) and secondary ignition wires are also of standard design. The distributor contains the igniter, signal rotor and mechanical advance components.

SCOUPE, SONATA AND EXCEL EQUIPPED WITH MULTI-POINT INJECTION

The electronic ignition used on the Excel engine with Multi-Point Fuel Injection (MPI), has timing controlled by the Electronic Control Unit (ECU). The standard reference ignition timing data from the engine operating conditions are programmed in the memory of the ECU. The engine conditions, rpm, load, temperature are detected by the ECU sensors and the ignition timing data signals to interrupt the primary current are sent to the power transistor. The ignition coil is activated and timing is controlled by this ECU control method.

The electronic ignition used on Sonata with Multi-Point Fuel Injection (MPI) is the same system as used on the Excel with MPI. Timing is controlled by the ECU. The standard reference ignition timing data is programmed in the memory of the ECU. The engine conditions, rpm, load and temperature are detected by the ECU sensors and the ignition timing data signals to interrupt the primary current are sent to the power transistor. The ignition coil is activated and timing is controlled by the ECU.

Diagnosis and Testing

SERVICE PRECAUTIONS

- Make sure all connectors are fastened securely. A poor connection can cause an extremely high surge voltage in the coil and condenser and result in damage to integrated circuits.
- Keep all parts and harnesses dry during service.
- Before attempting to remove any parts, turn **OFF** the ignition switch and disconnect the battery ground cable.
- Always use a 12 volt battery as a power source.
- Do not attempt to disconnect the battery cables with the engine running.
- Do not depress the accelerator pedal when starting.
- Do not rev up the engine immediately after starting or just prior to shutdown.
- If a battery cable is disconnected, the memory will return to the ROM (programmed) values. Engine operation may vary slightly, but this is not an indication of a problem. Do not replace parts because of a slight variation.
- If installing a 2-way or CB radio, keep the antenna as far as possible away from the electronic control unit. Keep the antenna feeder line at least 8 in. away from the ECU harness and do not let the 2 run parallel for a long distance. Be sure to ground the radio to the vehicle body.

IGNITION COIL

Testing

EXCEL WITH FEEDBACK CARBURETOR

1. Using a suitable volt/ohmmeter, measure the resistance. An open or short circuited coil should be replaced. The standard resistance values are as follows:
   a. Primary coil resistance—1.2 ohms
   b. Secondary coil resistance—13.7 kilo-ohms
   c. External resistor resistance—1.35 ohms
2. Check the resin portion for cracks. If there are any cracks present, replace it.
3. Check for any fluid (oil) leaks. If there are any leaks present, replace it.

SCOUPE, SONATA AND EXCEL WITH MPI

1. Using a suitable volt/ohmmeter, measure the resistance. An open or short circuited coil should be replaced. The standard resistance values are as follows:
   a. Primary coil resistance – 0.72–0.88 ohms
   b. Secondary coil resistance – 10.9–13.3 kilo-ohms
2. Check the resin portion for cracks. If there are any cracks present, replace it.
3. Check for any fluid (oil) leaks. If there are any leaks present, replace it.

ELANTRA

1. Using a suitable volt/ohmmeter, measure the primary and secondary resistances. An open or short circuited coil should be replaced. The standard resistance values are as follows:
   a. Primary coil resistance – 0.77–0.95 ohms. Measure resistance between connector terminals 3 and 1 (the coils at the No.1 and No.4 cylinder sides) of the ignition coil and between terminals 3 and 2, the coils at the No.2 and No.3 cylinder sides.
   b. Secondary coil resistance – 10.3–13.9 kilo-ohms. Measure the resistance between the high voltage terminal for the No.1 and No.4 cylinders and between the high voltage terminals for the No.2 and No.3 cylinders.

NOTE: When measuring the resistance of the secondary coil, disconnect the ignition coil connector.

2. Check the resin portion for cracks. If there are any cracks present, replace it.
3. Check for any fluid (oil) leaks. If there are any leaks present, replace it.

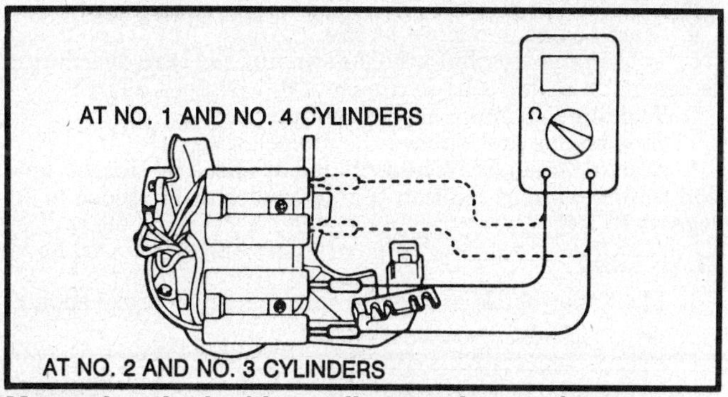

Measuring the ignition coil secondary resistance – Elantra

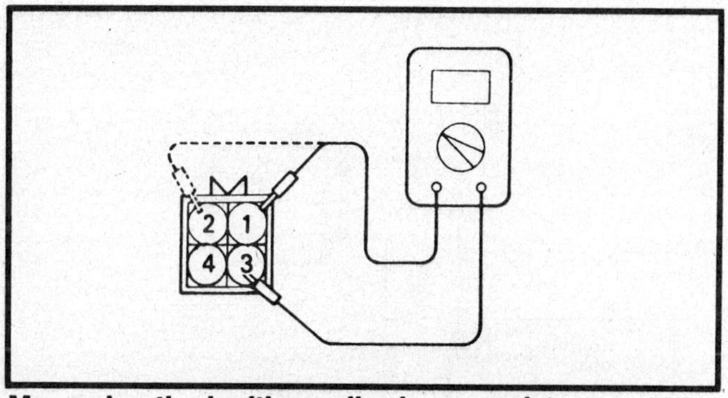

Measuring the ignition coil primary resistance – Elantra

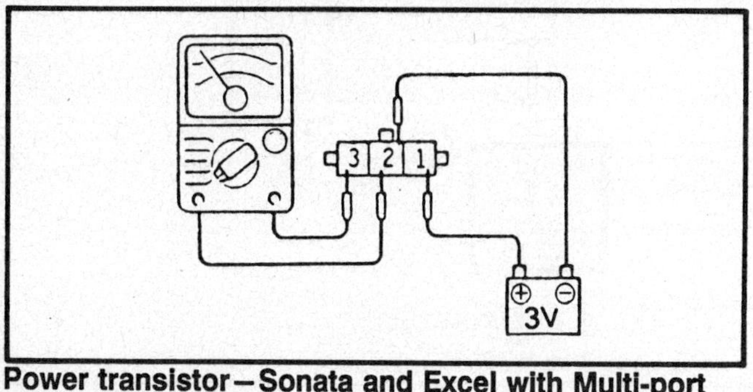

Power transistor – Sonata and Excel with Multi-port fuel injection

POWER TRANSISTOR

Testing

SCOUPE, SONATA AND EXCEL WITH MPI

1. Connect an analog type ohmmeter across terminals 2 and 3 of the power transistor.
2. There should be no continuity.
3. Connect the negative terminal of a 3 volt battery to terminal 2 and the positive terminal to terminal 1 of the power transistor. The ohmmeter should show continuity.
4. Replace the power transistor, if there is a malfunction.

ELANTRA

1. Check the power transistor for coil No.1 and No.4 as follows:
   a. Connect the negative lead of a 1.5 volt battery source to terminal 3 of the power transistor; then check for continuity between terminal 5 and terminal 3 when the terminal 4 and the positive terminal are connected.
   b. Continuity should exist when terminal 4 is connected. Continuity should not exist when terminal 4 is disconnected.
3. Check the power transistor for coil No.2 and No.3 as follows:
   a. Connect the negative lead of a 1.5 volt battery source to terminal 3 of the power transistor; then check for continuity between terminal 1 and terminal 3 when the terminal 2 and the positive terminal are connected.
   b. Continuity should exist when terminal 2 is connected. Continuity should not exist when terminal 2 is disconnected.
4. If the problem is still evident after test are completed, replace the power transistor.

IGNITION TIMING

Adjustment

EXCEL WITH FEEDBACK CARBURETOR

1. Place the vehicle in **P** or **N**, the emergency brake applied and the drive wheels blocked. Start the engine and let it reach normal operating temperature.
2. Connect a suitable tachometer and timing light to the engine. Turn **OFF** all accessories. Check the curb idle and adjust as necessary. The curb idle should be as follows:
   a. Manual Transmissions – 600–800 rpm
   b. Automatic Transmission – 650–850 rpm
3. If the timing marks are difficult to see, use chalk or a dab of paint to make them more visible.
4. On the Canadian engines, disconnect the vacuum line from the distributor vacuum advance and plug the line.
5. Aim the timing light onto the timing marks and check the engine timing. If the timing is out of specifications, loosen the distributor mounting bolt and advance or retard the timing as necessary.
6. Once the proper timing has been reached, tighten the distributor mounting bolt and then recheck the timing again. The correct timing specifications are as follows:
   a. California – 1–5 degrees BTDC
   b. Federal – 3–7 degrees BTDC
   c. Canada – 2–6 degrees BTDC

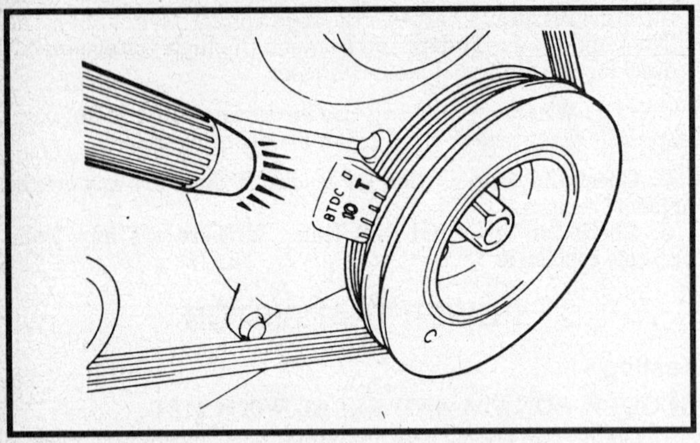

Checking the ignition timing—All vehicles

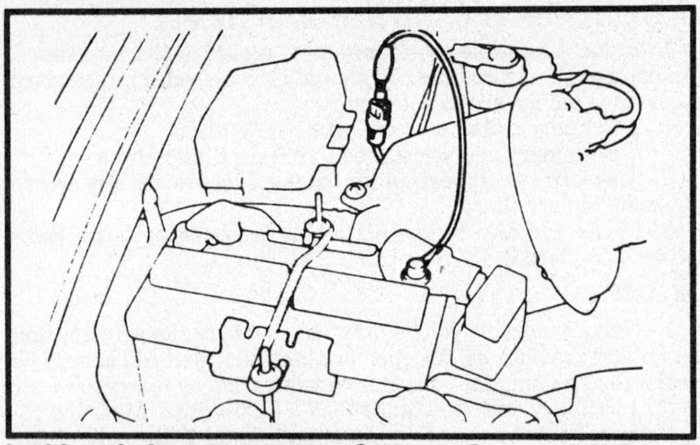

Ignition timing connector—Scoupe, Sonata and Excel with Multi-port fuel injection

7. When the timing has be set properly, recheck the curb idle speed and adjust as necessary.

8. When all final adjustments have been made, shut **OFF** the engine, reconnect all vacuum lines and remove all test equipment.

SCOUPE, SONATA AND EXCEL WITH MPI

1. Place the vehicle in **P** or **N**, the emergency brake applied and the drive wheels blocked. Start the engine and let it reach normal operating temperature.

2. Connect a suitable tachometer and timing light to the engine. Turn **OFF** all accessories. Check the curb idle and adjust as necessary to 650–850 rpm.

3. Stop engine and connect a jumper wire from the ignition timing adjustment connector (located at the rear of the engine compartment) to ground.

4. Start engine and allow to idle.

5. Loosen the distributor hold-down nut and turn distributor as needed to obtain a basic timing of 3–7 degrees BTDC.

6. Tighten distributor hold-down nut.

7. Stop engine and remove the jumper wire.

8. Start engine and recheck timing at idle. The correct ignition timing without ignition timing connector grounded is 10 degrees BTDC.

ELANTRA

1. Place the vehicle in **P** or **N**, the emergency brake applied

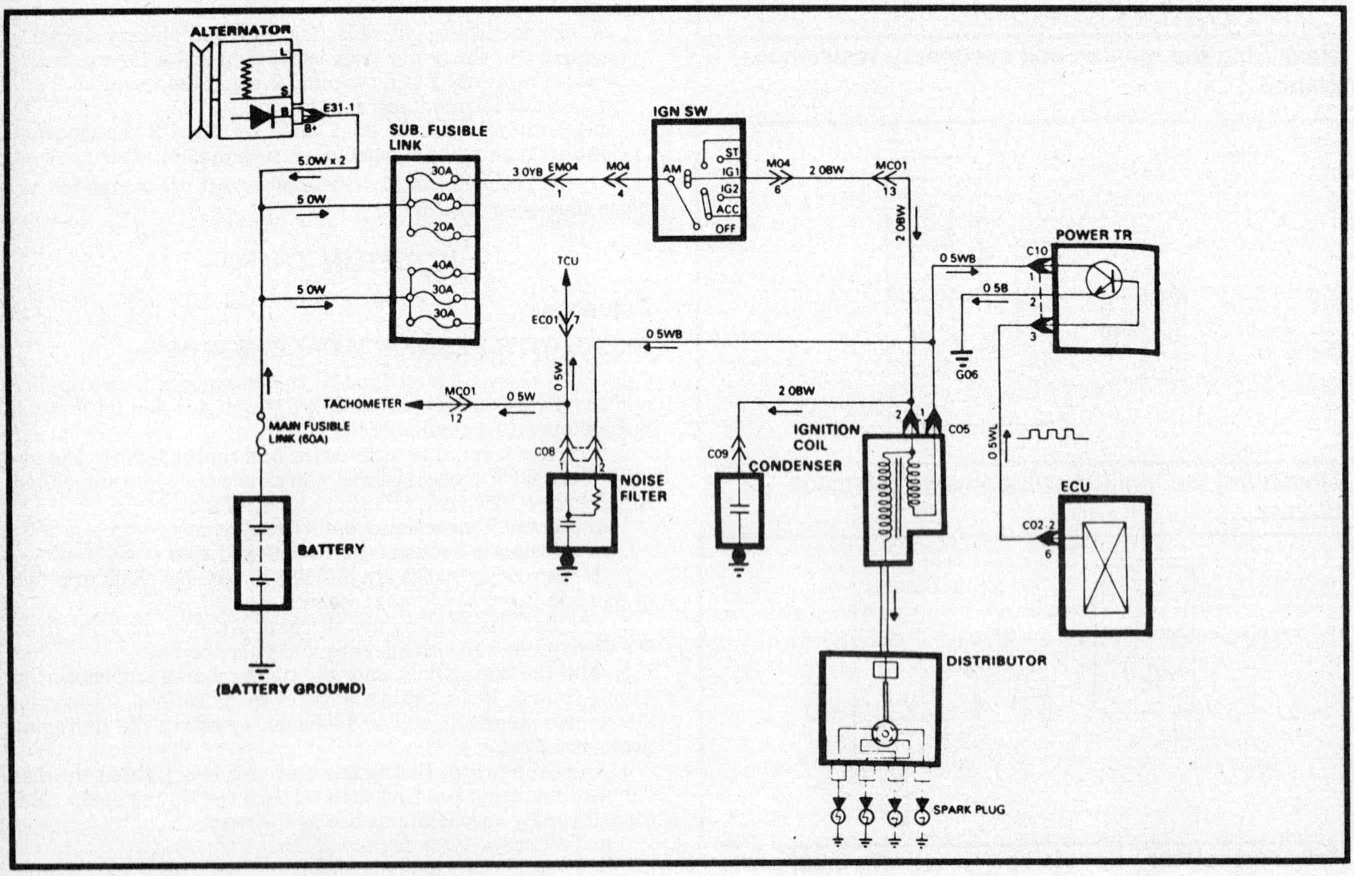

Ignition system wiring diagram—Sonata

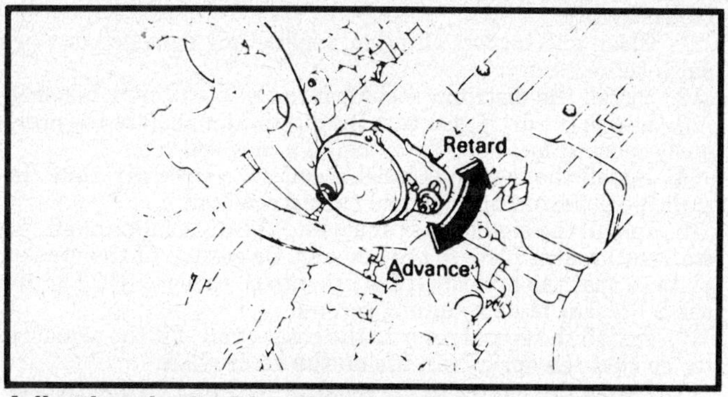

Adjusting the ignition timing with the crank angle sensor – Elantra

and the drive wheels blocked. Start the engine and let it reach normal operating temperature.

2. Connect a suitable tachometer and timing light to the engine. Turn **OFF** all accessories. Check the curb idle and adjust as necessary to 750–800 rpm.

3. Stop engine and connect a jumper wire from the ignition timing adjustment connector (located at the rear of the engine compartment) to ground.

4. Start engine and allow to idle.

5. Check the basic ignition timing and adjust as required. The basic timing should be 3–7 degrees BTDC.

6. If the timing is incorrect, loosen the crank angle sensor mounting nut and rotate the sensor until the correct timing is obtained.

6. Tighten the crank angle mounting nut to 7–9 ft. lbs. (10–13 Nm).

7. Stop engine and remove the jumper wire.

8. Start engine and recheck timing at idle. The correct ignition timing without ignition timing connector grounded is 8degrees BTDC.

Component Replacement

DISTRIBUTOR

Removal and Installation

EXCEL WITH FEEDBACK CARBURETOR

1. Rotate the engine and bring up the No. 1 cylinder to top dead center of its compression stroke.

2. Disconnect the negative battery cable. Disconnect and tag (if necessary) the electrical connectors along with the spark plug wires from the distributor.

3. Disconnect the wiring harness from the distributor lead wire. Disconnect the vacuum hoses from the vacuum controller.

4. Using a scribe tool or a marker, scribe or mark an alignment mark on the base of the distributor and the cylinder head. This can be used for an easier installation procedure.

5. Remove the distributor mounting nut and remove the distributor assembly from the engine cylinder head.

To install:

6. Install the distributor assembly into the cylinder block, be sure to align the scribe marks made previously. Lightly tighten the distributor mounting bolt.

7. Connect the wiring harness to the distributor lead wire. Connect the vacuum hoses to the vacuum controller.

8. Connect all the electrical connectors along with the spark plug wires to the distributor.

9. Connect the negative battery cable. Start the engine and set the timing. Then tighten the distributor mounting bolt.

SCOUPE, SONATA AND EXCEL WITH MPI

1. Rotate the engine and bring up the No.1 cylinder to TDC of its compression stroke.

2. Disconnect the negative battery cable. Disconnect and tag the electrical connectors and spark plug wires from the distributor.

3. Unclip distributor cap and set aside.

4. Disconnect the wiring harness from the distributor.

5. Remove the distributor mounting nut and remove the distributor assembly.

To install:

6. If necessary, turn crankshaft so that No. 1 cylinder is at TDC on compression stroke.

7. Align the distributor housing and gear mating marks.

8. Install the distributor to the engine while aligning the fine cut of the distributor flange with the center of the distributor stud.

Overhaul

EXCEL WITH FEEDBACK CARBURETOR

1. Lightly clamp the distributor assembly into a suitable soft jaw vise.

2. Remove the retaining clips and lift off the distributor cap and seal ring.

3. Pull off the rotor from the rotor shaft. Remove the 2 vacuum controller retaining screws.

4. Remove the link of the controller from the pin on the breaker base, then remove the vacuum controller.

5. Remove the 2 igniter retaining screws and remove the igniter. Be sure to remove the lead (black) wire.

6. Remove the signal rotor shaft tightening screw and 2 breaker plate retaining screws. Remove the signal rotor shaft and breaker plate assembly.

7. Remove the signal rotor shaft from the signal rotor.

8. Remove the 2 spring retainers with a suitable pair of pliers

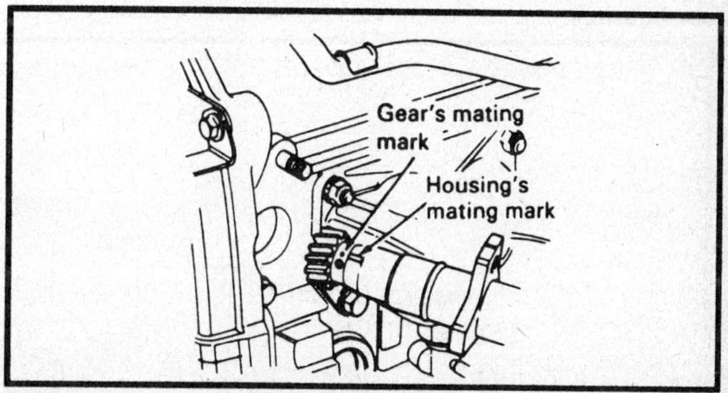

Distributor installation gear marks – Sonata and Excel with Multi-port fuel injection

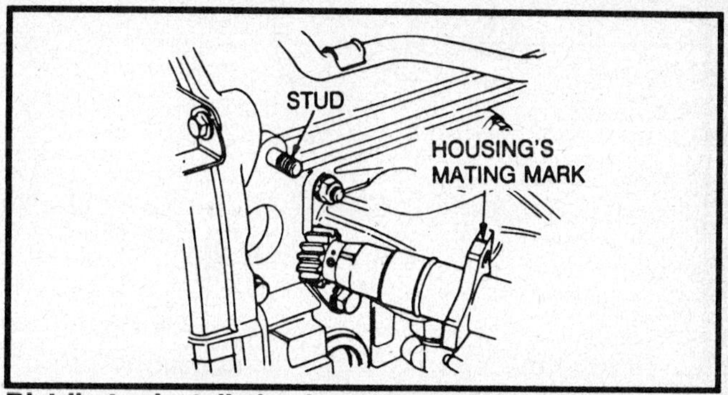

Distributor installation housing marks – Scoupe, Sonata and Excel with Multi-port fuel injection

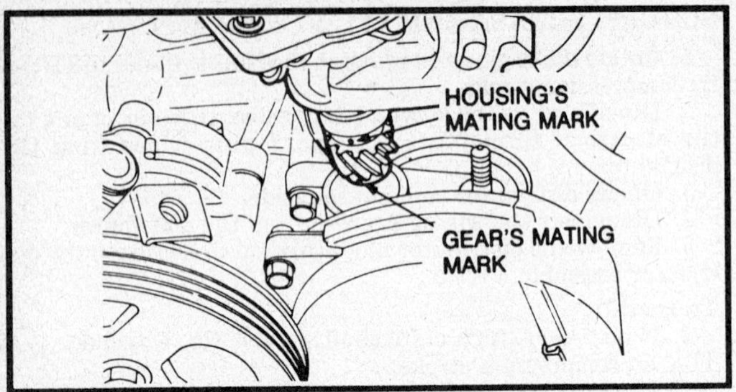

Installing 6 cylinder distributor—Sonata

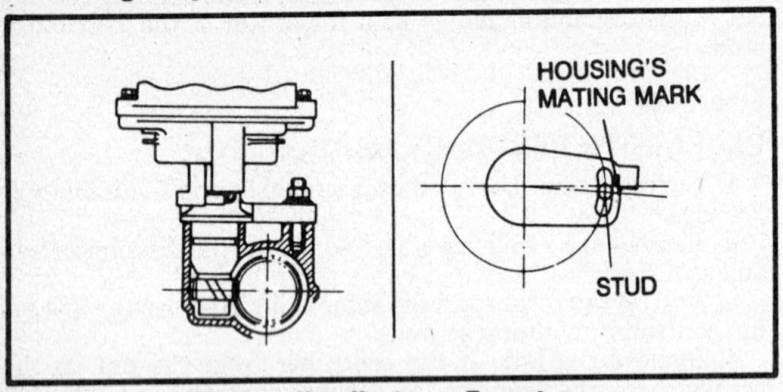

Aligning 6 cylinder distributor—Sonata

and then remove the 2 governor springs. Remove the 2 governor weights.

9. Mark the location of the drive gear on the distributor shaft. Place the drive gear on a soft base (wooden block) so that the roll pin can be removed.

10. Using a suitable pin punch, remove the roll pin. Remove the drive gear and washer. Remove the distributor shaft from the housing.

Reassembly:

11. Clean and inspect all parts, replace any damaged or worn parts as necessary.

12. Install the distributor shaft into the distributor housing.

13. Install the drive gear into the distributor shaft at the previously marked location. Then install a new roll pin.

14. Install the governor weights onto the governor plate. Install the governor springs and spring retainer.

15. Install the signal rotor shaft into the distributor shaft. Install the breaker plate to the housing. Be sure to fit the breaker plate so that the projection (A) fits into the groove (B). Tighten the 2 breaker plate retaining screws.

16. Install the signal rotor to the rotor shaft. Fit the signal rotor so that the spring pin fits on the rotor shaft.

17. Install the igniter to the breaker plate and install the 2 retaining screws. Connect one end of the lead (black) wire to the igniter attaching screw and the other end to the breaker plate.

18. Adjust the air gap between the signal rotor and the pick up of the igniter. The air gap should be 0.03 in.

19. Connect the vacuum control link to the breaker plate and tighten the 2 vacuum controller retaining screws.

20. Install the rotor to the rotor shaft. Install the seal ring and distributor cap to the housing and set the retaining clips.

SCOUPE, SONATA AND EXCEL WITH MPI

1. Lightly clamp the distributor assembly into a suitable soft jaw vise.
2. Pull off the rotor from the rotor shaft.
3. Remove the packing rubber.
4. Lift off the cover.
5. Remove the rotor shaft using a driver.
6. Remove the disc and spacer.
7. Disconnect the wire connector.
8. Remove the mounting screws and lift off the plate and unit.
9. Remove the shaft bearing screws.
10. Scribe a position identification mark for the drive gear on the shaft.

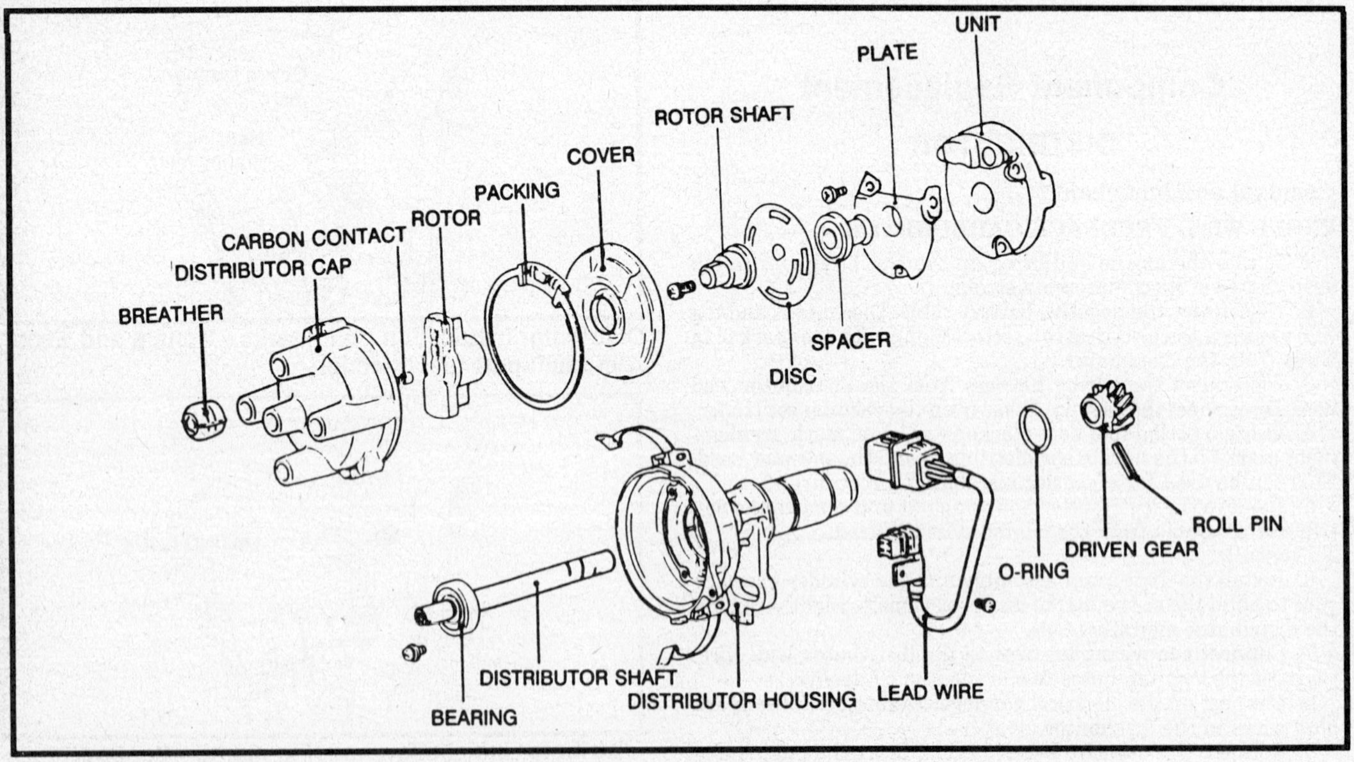

Distributor exploded view—Scoupe, Sonata 4 cylinder and Excel with Multi-port fuel injection

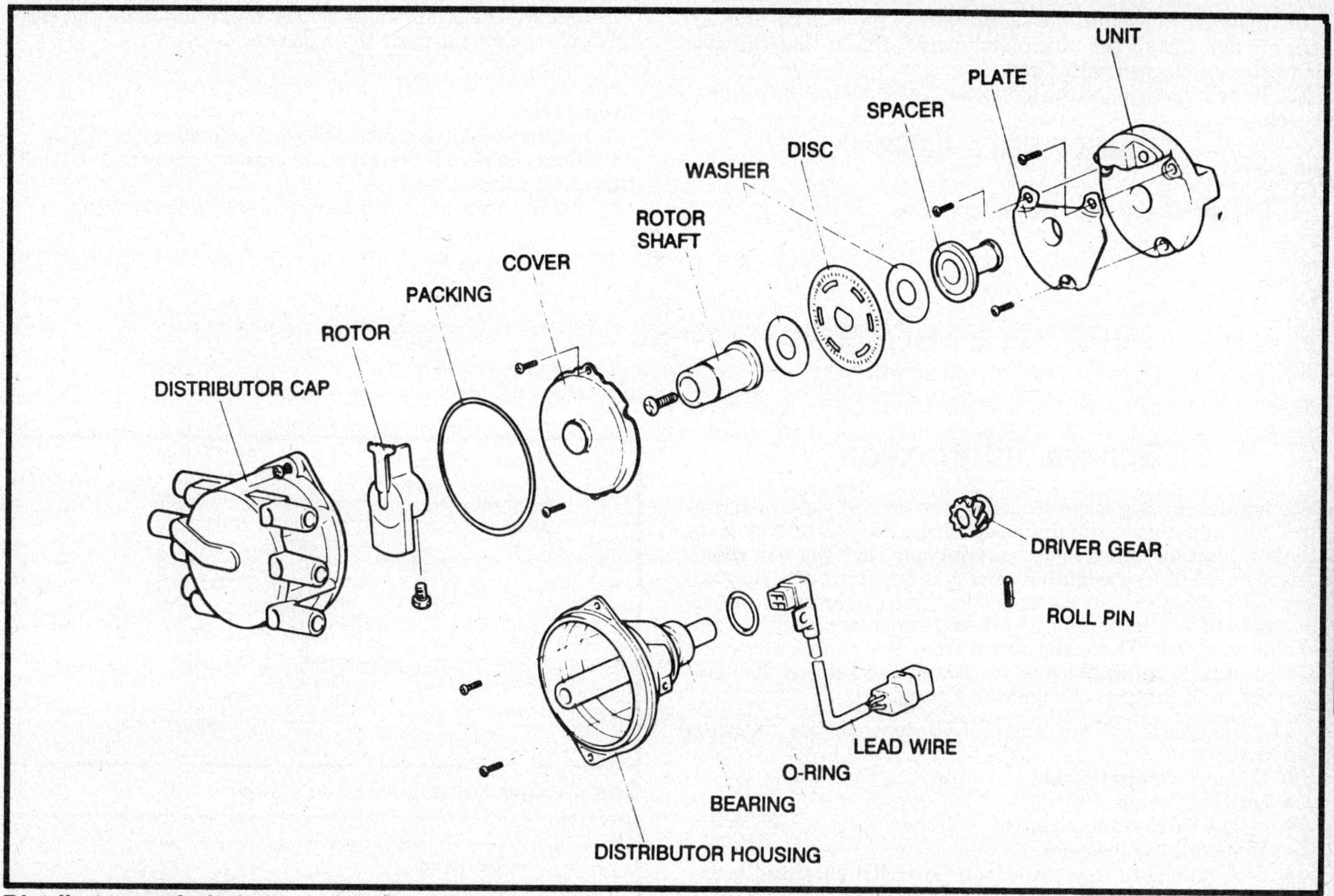

Distributor exploded view—Sonata with 3.0L engine

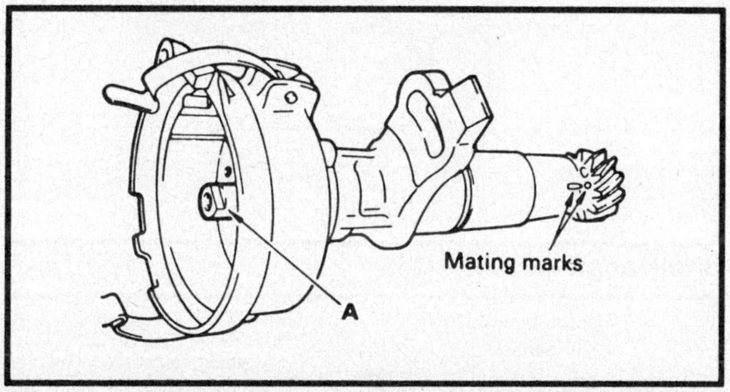

Aligning notch and gear marks—Scoupe, Sonata and Excel with Multi-port fuel injection

11. Place the drive gear on a wooden block and drive out the roll pin.

Reassembly:

12. Clean and inspect all parts, replace any damaged or worn parts as necessary.
13. Coat shaft with engine oil and insert into the housing.
14. Align the drive gear with mark made during disassembly.
15. When aligning the drive gear mating mark, make certain the the flat notch "A" at the shaft end is correctly aligned and then align the roll pin hole and drive in a new roll pin.

NOTE: Drive the roll pin so that the slits are at right angle relative to the shaft.

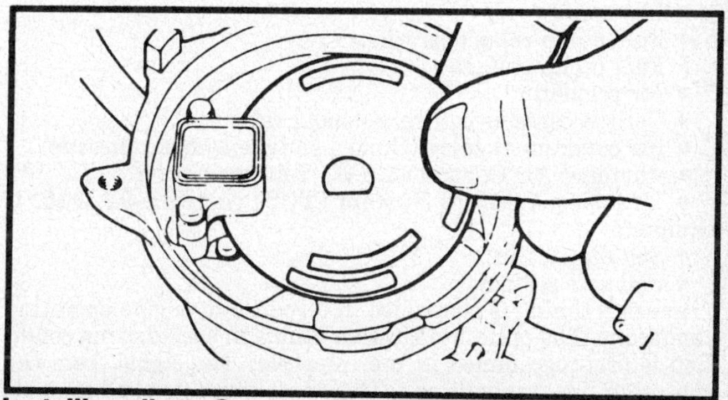

Installing disc—Scoupe, Sonata and Excel with Multi-port fuel injection

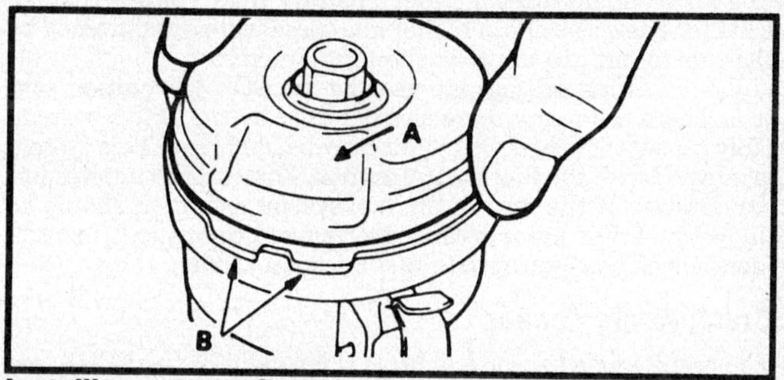

Installing cover—Scoupe, Sonata and Excel with Multi-port fuel injection

16. Insert the disc into the sensor part of the pick-up unit assembly and install the alignment spacer. Check that the disc slits are not clogged with dirt.
17. Install the cover so the indentation aligns with the housing notch.

CRANK ANGLE SENSOR

ELANTRA

1. Disconnect the negative battery cable.
2. Remove the crank angle sensor mounting bolts and carefully pull the sensor from the engine.

To install:

3. Rotate the crankshaft until No.1 cylinder is at TDC.
4. Align the punch mark on the crank angle sensor housing with the notch in the plate.
5. Install the crank angle sensor on the cylinder head.

INFINITI ELECTRONIC CONCENTRATED CONTROL (ECCS) IGNITION SYSTEM

General Information

The ignition timing is controlled by the central electronic control unit adjusting to the engine operating conditions: that is, as the best ignition timing in each driving condition has been memorized in the unit, the ignition timing is determined by the electric signal calculated in the unit. The signals used for the determination of ignition timing are water temperature, engine rpm, engine load, etc. Then, the signal from the central electronic control unit is transmitted to the power transistor of the ignition coil, and controls the ignition timing.

The electronic control unit (ECU) controls the following functions:

- Amount of injected fuel
- Ignition timing
- Intake valve timing control
- Mixture ratio feedback
- Exhaust Gas Recirculation (E.G.R.) solenoid valve operation
- Exhaust gas sensor heater
- Idle speed
- Fuel pump relay operation
- Fuel pump voltage
- Air regulator
- Carbon canister control solenoid valve
- Air conditioner relay (During early wide-open throttle)
- Auxiliary Air Control (A.A.C.) Valve operation
- Traction Control System (TCS) in 1991–92 Q45, if equipped
- Self-diagnostics
- Fail safe system

Ignition timing is controlled in response to engine operating conditions. The optimum ignition timing in each driving condition is preprogrammed in the computer. The signal from the control unit is transmitted to the power transistor and controls ignition timing. The idle speed is also controlled according to engine operating conditions, temperature and gear position. Vehicles with manual transmission, if battery voltage is less than 12 volts for a few seconds, a higher idle speed will be maintained by the control unit to improve charging function.

The following sensors are used by the ECU for ignition control. There are many more sensors used by the ECCS system. Only sensors involved with ignition are listed here. This system is integral with the fuel control system. For a more complete understanding of the system the fuel system operation should be understood. For more detailed testing and component replacement check the appropriate fuel injection section.

Crank Angle Sensor

The crank angle sensor is a basic component of the ECCS system. It monitors engine speed and piston position, as well as sending signals which the ECU uses to control fuel injection, ignition timing and other functions. The crank angle sensor has a rotor plate and a wave forming circuit. The rotor plate has 360 slits for 1 degree signals (crank angle), and 6 slits for 120 degree signal (engine speed).

The light emitting diodes (LED's) and photo diodes are built into the wave forming circuit. When the rotor plate passes the

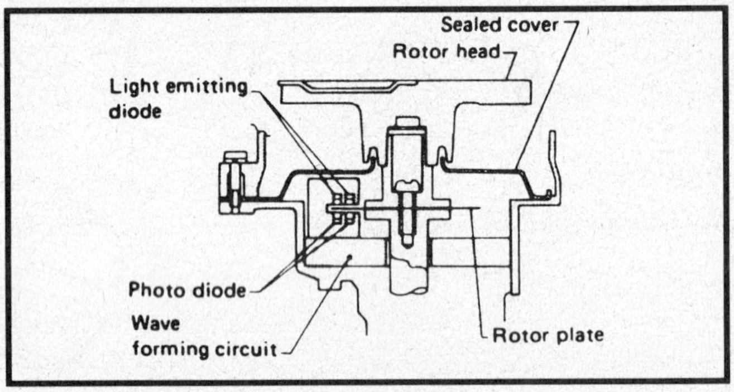

Crank angle sensor—G20 and M30

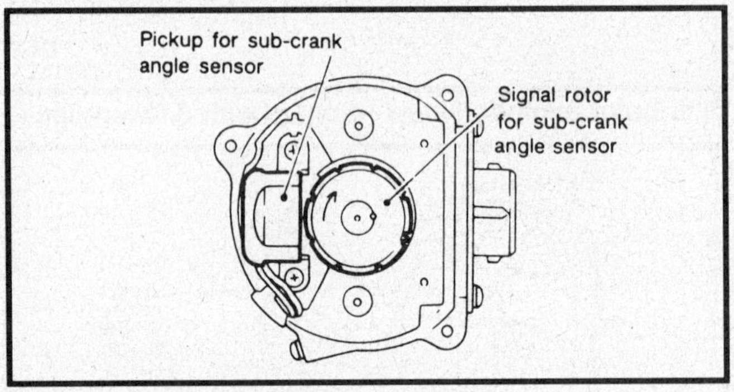

Crank angle sensor—Q45

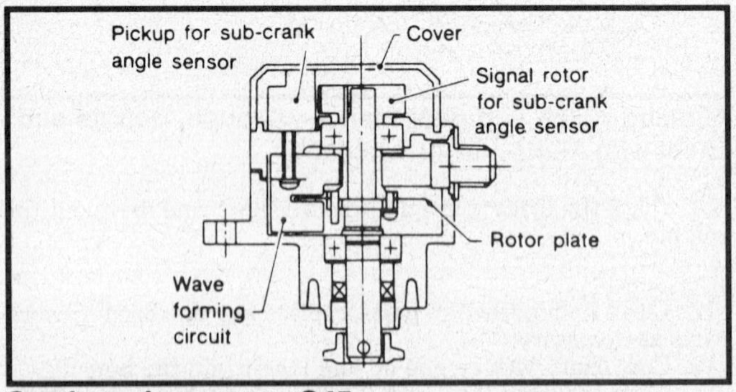

Crank angle sensor—Q45

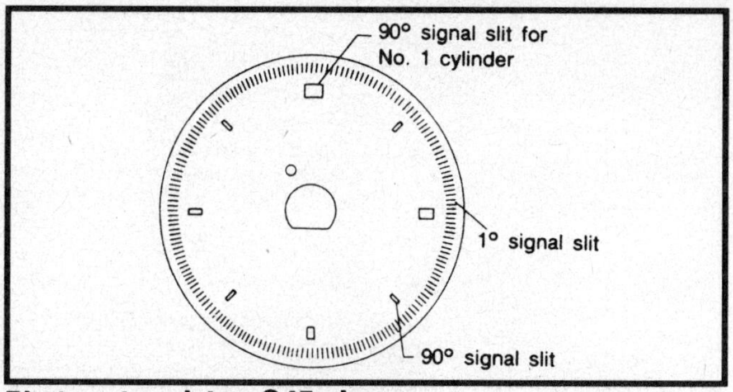

Photo rotor plate – Q45 shown

space between the LED and the photo diode, the slits of the rotor plate continually cut the light which is sent to the photo diode from the LED. This generates rough haped pulses which are converted into ON/OFF pulses by the wave forming circuit and then sent to the ECU.

Engine Temperature Sensor

The engine temperature sensor monitors changes in engine coolant temperature and transmits a signal to the ECU. The temperature sensing unit employs a thermistor which is sensitive to the change in temperature, with electrical resistance decreasing as temperature rises.

Detonation Sensor

The detonation sensor is attached to the cylinder block and senses engine knocking conditions. A knocking vibration from the cylinder block is applied as pressure to the piezoelectric element. This vibrational pressure is then converted into a voltage signal which is delivered as output.

Power Transistor and Ignition Coil

The ignition signal from the ECU is amplified by the power transistor, which turns the ignition coil primary circuit on and off, inducing the necessary high voltage in the secondary circuit to fire the spark plugs. The ignition coil is a small, molded type.

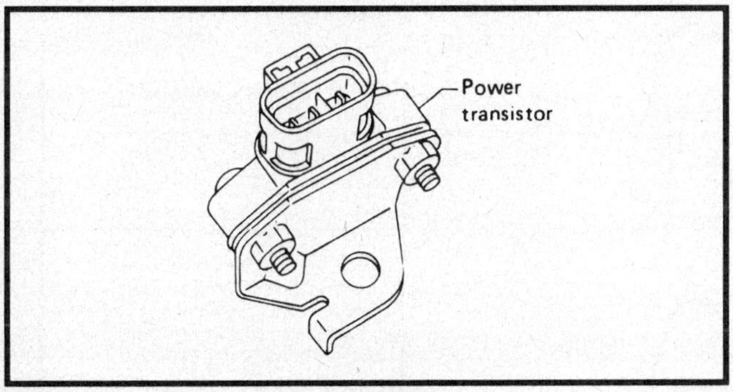

Power transistor – G20 and M30

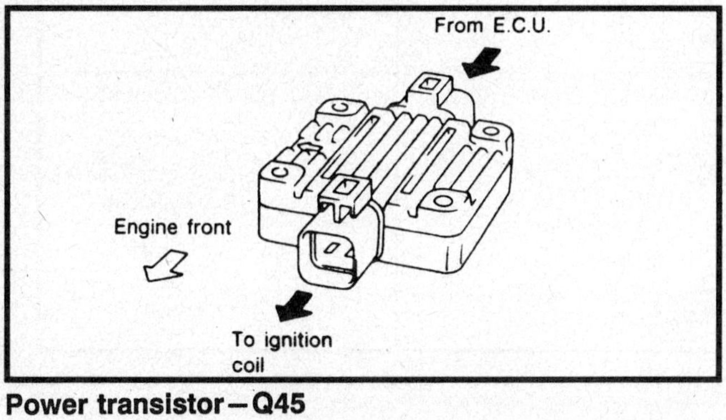

Power transistor – Q45

Diagnosis and Testing

SERVICE PRECAUTIONS

- Make sure all ECCS connectors are fastened securely. A poor connection can cause an extremely high surge voltage in the coil and condenser and result in damage to integrated circuits.
- Keep the ECCS harness at least 4 in. away from adjacent harnesses to prevent an ECCS system malfunction due to external electronic "noise."
- Keep all parts and harnesses dry during service.
- Before attempting to remove any parts, turn **OFF** the ignition switch and disconnect the battery ground cable.
- Always use a 12 volt battery as a power source.
- Do not attempt to disconnect the battery cables with the engine running.
- Do not depress the accelerator pedal when starting.
- Do not rev up the engine immediately after starting or just prior to shutdown.
- Do not attempt to disassemble the ECCS control unit under any circumstances.
- If a battery cable is disconnected, the memory will return to the ROM (programmed) values. Engine operation may vary slightly, but this is not an indication of a problem. Do not replace parts because of a slight variation.
- If installing a 2-way or CB radio, keep the antenna as far as possible away from the electronic control unit. Keep the antenna feeder line at least 8 in. away from the ECCS harness and do not let the 2 run parallel for a long distance. Be sure to ground the radio to the vehicle body.

NOTE: For Diagnosis and Testing of the ECCS Ignition System, refer to the appropriate "Fuel Injection Section." Component replacement and adjustments are also covered in the fuel section.

POWER TRANSISTOR

Testing

1. Disconnect the power transistor harness connector.
2. Check the power transistor continuity between terminals with a digital tester. If it is not within specifications, replace the power transistor.

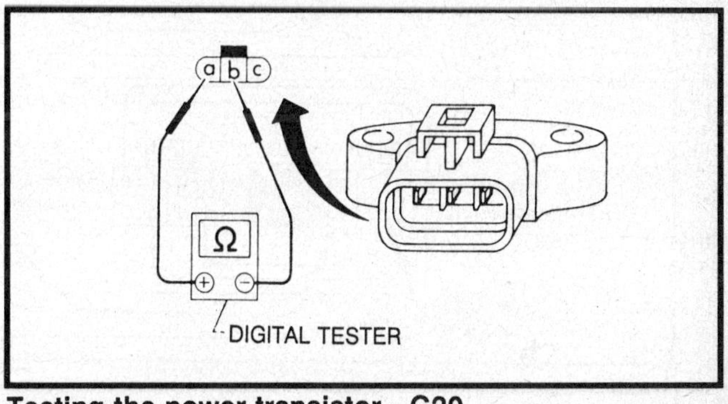

Testing the power transistor – G20

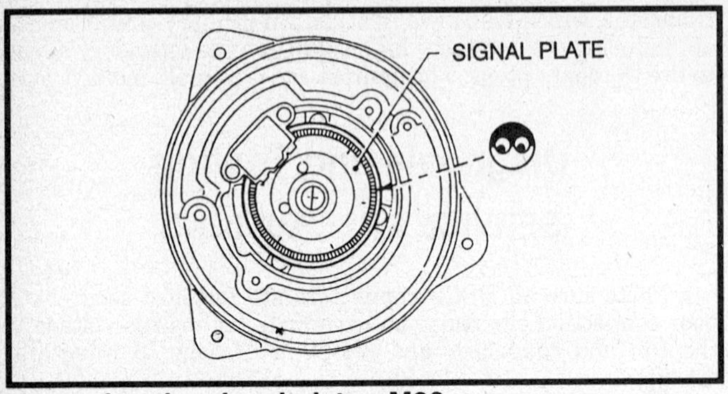

Inspecting the signal plate—M30

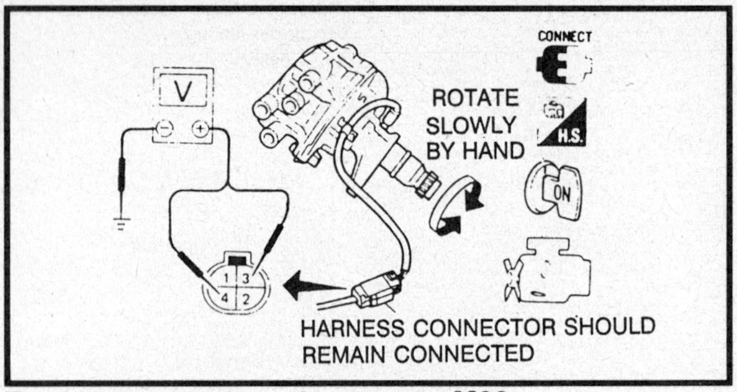

Testing the crank angle sensor—M30

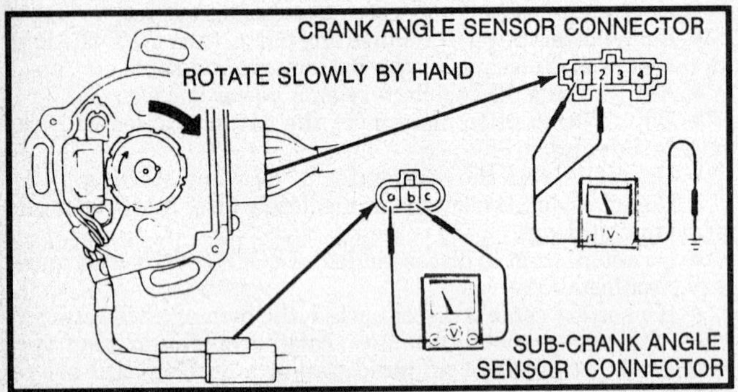

Testing the crank angle sensor and sub-crank angle sensor—Q45

CRANK ANGLE SENSOR

Component Testing

G20 AND M30

1. Visually check the signal plate for damage or dust.
2. Remove the distributor from the engine without disconnecting the crank angle sensor harness.
3. Disconnect the ignition wires.
4. Turn the ignition switch **ON**.
5. Rotate the distributor shaft slowly by hand and check the voltage between terminals a, d and ground on the G20, and terminals 3, 4 and ground on the M30.
6. If the tester's pointer does not fluctuate between 5–0 volts, replace the distributor assembly with crank angle sensor.

Connected terminal connections and measured terminals	Voltage
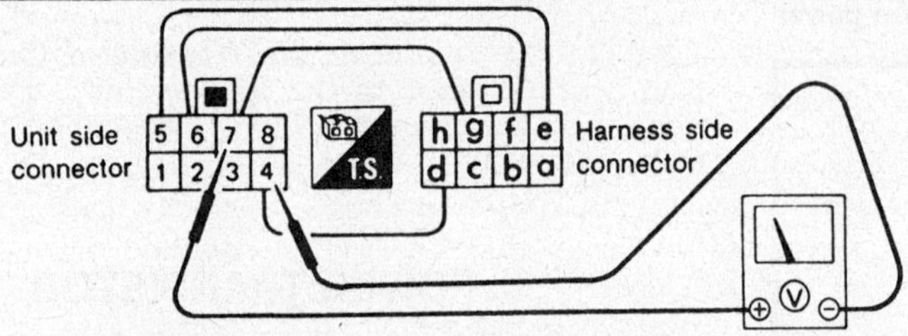	Battery voltage
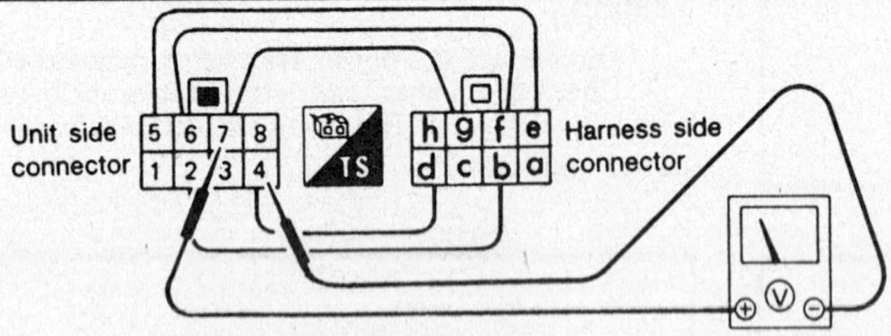	7.5 - 8.1V
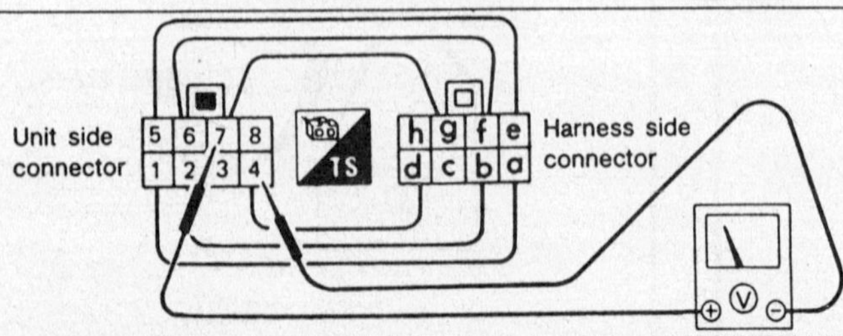	5.4 - 5.8V

Fuel pump control unit testing—Q45

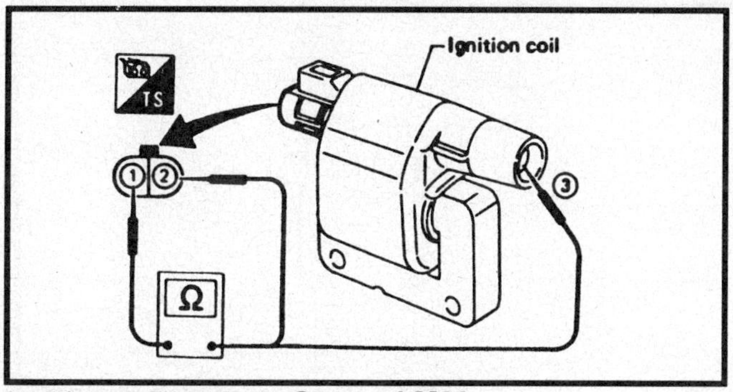

Ignition coil testing—G20 and M30

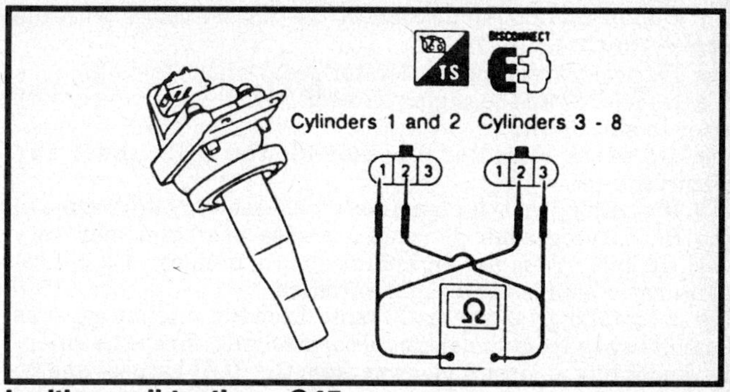

Ignition coil testing—Q45

Q45

This procedure includes testing the crank angle sensor and the sub-crank angle sensor on the Q45.

1. Remove the crank angle sensor from the engine without disconnecting the sensor harness.
2. Turn the ignition switch **ON**.
3. Rotate the crank angle sensor shaft slowly by hand and check the voltage between terminals 1, 2 and ground. If the voltage does not fluctuate between 5–0 volts, replace the crank angle sensor.
4. Rotate the crank angle sensor shaft slowly by hand and check the voltage between terminals a and the b ground, using a circuit tester set in the 100 millivolt range, AC. If the tester pointer does not deflect, replace the crank angle sensor.
5. After this inspection, a malfunction Code 11 might be displayed even though the crank angle sensor is functioning properly. In this case erase the stored memory.

ISUZU ELECTRONIC IGNITION SYSTEM

General Information

There are 2 types of ignition systems used on Izuzu engines: one is the Direct Ignition system (DIS). The other, is the computer controlled Electronic Spark Timing (EST) ignition system.

The Direct Ignition System (DIS) is use only on the turbocharged 1.6L engine. This ignition system is considered to be a distributorless ignition. Distributorless igniton systems use a waste spark method of spark distribution. Each cylinder is paired with its opposing cylinder in the firing order, so that when 1 cylinder on conpression, it fires simultaneouly with the opposing cylinder on exhaust. Since the cylinder on exhaust requires very little of the available voltage to fire its plug, most of the voltage is used to fire the cylinder on compression. The process reverses when the cylinders reverse roles. There are 2 coils used in the Direct Ignition System (DIS).

The Direct Ignition System (DIS) is comprised of a coil pack, an ignition module, cam angle sensor and Electronic Control Module (ECM). The coil pack consists of 2 separate, interchangeable, ignition coils. It is necessary to use 2 coils because each coil only fires per 2 cylinders. The ignition module is located under the coil pack and is connected to the Electronic Control Module (ECM) by a 6 pin connector. The ignition module controls the primary circuit to the coils, turning them on and off. It also controls spark timing below 400 rpm and if the ECM bypass circuit becomes open or grounded.

The magnetic pickup sensor inserts through the cam angle sensor body, near the camshaft reluctor ring. Notches in the camshaft reluctor ring trigger the magnnetic pickup sensor to provide timing information to the ECM. The magnetic pickup sensor provides a cam signal to identify correct firing sequence and crank signals to trigger each coil at the proper time. This system also uses Electronic Spark Timing (EST) and ignition wires from the ECM, as with the distibutor systems. The ECM controls timing using crankshaft position, engine rpm, engine temperature and manifold absolute pressure sensing.

The computer controlled Electronic Spark Timing (EST) ignition system used on the 1.6L non-turbocharged engine, the 2.3L, 2.6L, 2.8L and 3.1L engines. This system consists of the battery, distributor, engine control switch, spark plugs, primary and secondary wiring.

The computer, or Electronic Control Module (ECM), monitors the information from the engine sensor network. The ECM uses this information to calculate the proper spark timing and tell the distributor when to make timing modifications.

The distributor does not have centrifugal advance weights, springs or a vacuum advance unit.

This ignition sytem also ultilizes an Electronic Spark Control (ESC). At the heart of the ESC system is the knock sensor which is mounted to the engine block. The knock sensor is connected to the electronic spark control module which is located on the right fender panel in the engine compartment or on a bracket mounted to the block. In response to engine knock, the sensor transmits a signal to the electronic spark module ("half function box"). The spark control module sends the signal to the ECM which in turn tells the distributor to retard the spark timing up to 20 degrees to reduce spark knock in the engine.

Isuzu identifies each engine by a code designation. The following information identifies each engine according to code ID and Liter:

The code identification is as follows: 4XE1–W–1.6L engine (non-turbocharged), 4XE1–WT–1.6L engine (turbocharged), 4ZD1–2.3L engine, 4ZE1–2.6L engine, No code designation–2.8L engine, CPC–3.1L engine.

Diagnosis and Testing

SERVICE PRECAUTIONS

- Make sure all connectors are fastened securely. A poor connection can cause an extremely high surge voltage in the coil and condenser and result in damage to integrated circuits.
- Keep all parts and harnesses dry during service.
- Before attempting to remove any parts, turn **OFF** the ignition switch and disconnect the battery ground cable.
- Always use a 12 volt battery as a power source.

• Do not attempt to disconnect the battery cables with the engine running.
• Do not depress the accelerator pedal when starting.
• Do not rev up the engine immediately after starting or just prior to shutdown.
• Do not attempt to disassemble the ECU under any circumstances.
• If a battery cable is disconnected, the memory will return to the ROM (programmed) values. Engine operation may vary slightly, but this is not an indication of a problem. Do not replace parts because of a slight variation.
• If installing a 2-way or CB radio, keep the antenna as far as possible away from the electronic control unit. Keep the antenna feeder line at least 8 in. away from the ECU harness and do not let the 2 run parallel for a long distance. Be sure to ground the radio to the vehicle body.
• Before attempting to remove any parts, turn **OFF** the ignition switch and disconnect the battery ground cable.
• Always use a 12 volt battery as a power source.
• Make sure the igniter is properly grounded to the body.
• Never allow the coil or tachometer terminals to touch ground or damage will result to the igniter or coil.

NOTE: For further diagnosis and testing of the ignition, refer to the fuel injection section.

IGNITION COIL

Testing

1.6L ENGINE

Non-Turbocharged

1. Disconnect the wiring along with the ignition coil and high tension cable at the connector.
2. Using a suitable ohmeter, measure the resistance of the coil. The resistance should be as follows:
   a. Primary coil resistance (1–2)—1.130–1.529 ohms
   b. Secondary coil resistance (1–3)—10.20–13.80 kilo-ohms
   c. Insulation resistance (2–4)—more than 10 megohms

Turbocharged

1. With the ignition switch **OFF**, install a spark tester.
2. Crank the engine and observe the spark tester. Spark should be evident, if not measure the resistance of each spark plug wire. The resistance should be less than 30000 ohms.

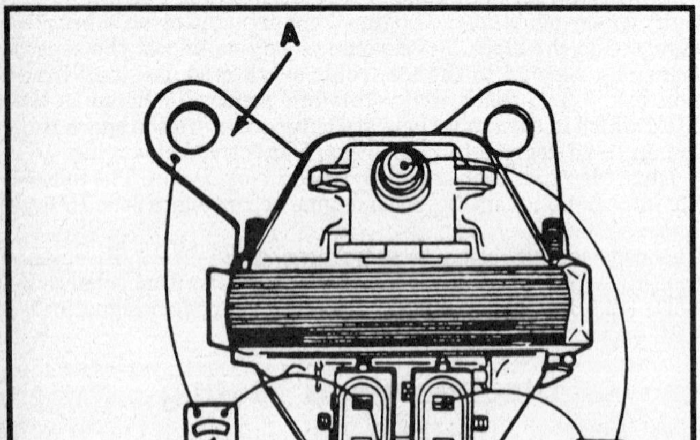

Ignition coil test—1.6L, 2.8L, 3.1L engines

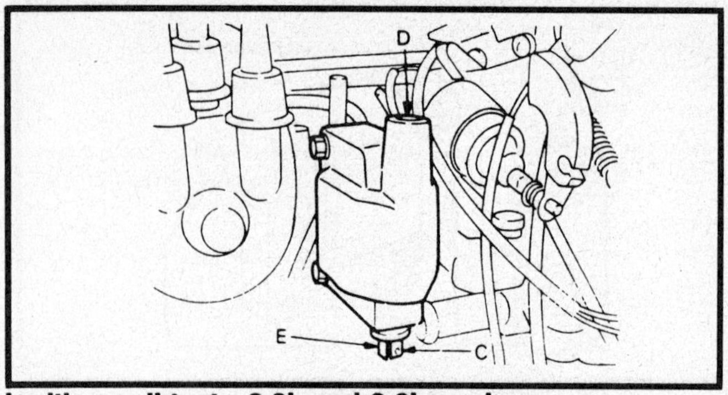

Ignition coil test—2.3L and 2.6L engines

3. If the wires a check good, inspect the ignition coils. If not, replace the defective wire and retest spark.
4. Remove the ignition coils from the DIS module and visually inspect for carbon tracking. If carbon tracking is evident, replace the affected coil.
5. If the ignition coil looks good externaly, but is still suspected, install a known good coil. Retest for spark, if spark is now evident the original coil was defective.

2.3L AND 2.6L ENGINES

Resistance Tests

1. Disconnect the wiring along with the ignition coil and high tension cable at the connector.
2. Using a suitable ohmeter, measure the resistance of the coil across the appropriate coil terminals. The resistance should be as follows:
   a. Primary coil resistance:
      2.3L engines (C–E)—1.2–1.4 ohms
      2.6L engines (C–E)—0.81–0.99 ohms
   b. Secondary coil resistance:
      2.3L engines (C–D)—8.6–13.0 kilo-ohms
      2.6L engines (C–D)—7.5–13.5 kilo-ohms
   c. Insulation resistance:
      2.3L and 2.6L engines (C-Body)—more than 10 megaohms.

2.8L AND 3.1L ENGINE

1. Disconnect the distributor wire and wiring from the coil.
2. Make sure the metal ground surface is clean.
3. Connect an ohmmeter as shown in position 1. Set the ohmmeter to the high ohms scale and measure the resistance. It should read infinite resistnace.
4. Connect an ohmmeter as shown in position 2. Set the ohmmeter to the low ohms scale and measure the resistance. It should read very low or zero resistance.
5. Connect an ohmmeter as shown in position 3. Set the ohmmeter to the high ohms scale and measure the resistance. It should not read infinite resistance.
6. If the resistance is not as described, replace the ignition coil.
7. Connect the coil wiring.

DISTRIBUTOR AIR GAP, GOVERNOR AND VACUUM ADVANCE

Inspection

1.6L ENGINE

Non-Turbocharged

1. Inspect the air gap, by using a non-magnetic feeler gauge. Measure the gap between the signal rotor and the pick-up coil projection. You should be able to insert the feeler gauge with a slight drag. The air gap should be 0.008–0.0016 in.

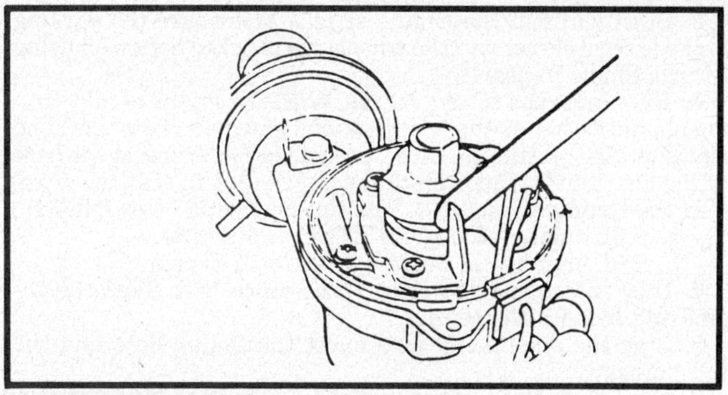

Checking the air gap

2. If the measurements are not within specification, replace the signal generator.

2.3L AND 2.6L ENGINES

1. Inspect the air gap, by using a non-magnetic feeler gauge. Measure the gap between the signal rotor and the pick-up coil projection. You should be able to insert the feeler gauge with a slight drag. The air gap should be 0.008–0.0016 in.
2. If the gap in not within specification, adjust the gap as required.
3. To adjust the air gap, loosen the igniter retaining screws and move igniter until the gap is correct. Tighten the retaining screws.
4. Check the governor function by turning the rotor both to the right and the left. The rotor should spin freely in both directions.
5. To check the vacuum advance unit, apply 400mm of vacuum to both the main diaphragm and the sub diaphragm ports on the unit. Check to see that both diaphragms move when vacuum is applied. If the unit does not function as described, replace it.

2.8L AND 3.1L ENGINES

1. Inspect the air gap, by using a non-magnetic feeler gauge. Measure the gap between the signal rotor and the pick-up coil projection. You should be able to insert the feeler gauge with a slight drag. The air gap should be 0.008–0.0016 in.
2. If the measurements are not within specification, replace the signal generator.

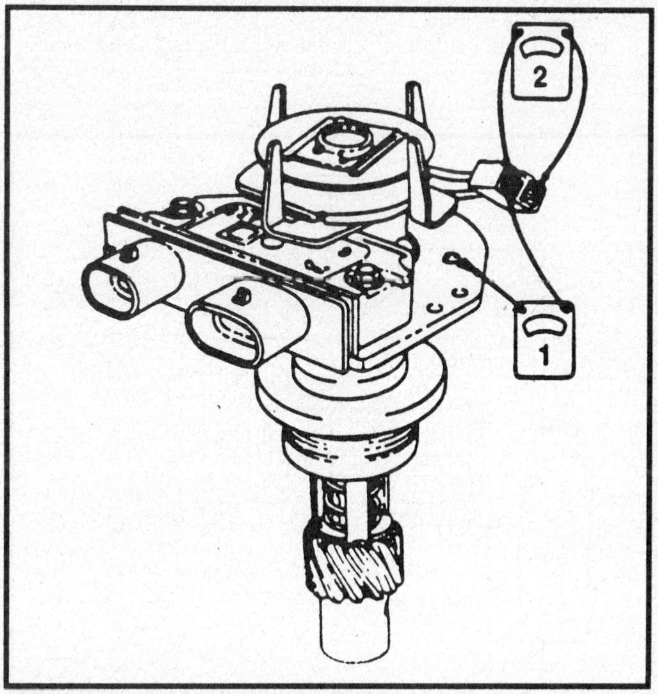

Checking pick-up coil resistance—1.6L, 2.8L, 3.1L engines

PICK-UP COIL RESISTANCE

Testing

This test is not applicable to the 1.6L turbocharged engine, the 2.3L and 2.6L engines. The ignition system in these engines are not equipped with a pick-up coil.

1. Disconnect the negative battery cable.
2. Remove the distributor cap.
3. Disconnect the pick-up coil connector from the module.
4. Measure the coil resistance between either of the pick-up coil leads and the housing. Set the ohmmeter to the high ohms range. Infinite resistance should be read.
5. Connect the ohmmeter probes across the coil leads and measure the resistance. It should read a constant, steady reading in the 500–1500 ohms range. Flex the lead wires at the coil and connector by hand to check for intermittent opens in the wiring.
6. If the resistance readings are not within specifications, replace the coil.

IGNITER

Testing

This test is only applicable to the 2.3L and 2.6L engines. All other engines not equipped with an igniter.

2.3L AND 2.6L ENGINES

1. Remove the distributor cap.
2. Disconnect the ignition coil high tension cable at the distributor side. Move the high tension cable end to the coil fixing screw and maintain a 5mm clearance.
3. Place the ignition switch to the **ON** position.
4. Connect a 1.5 volt dry cell to the red igniter wiring terminal at the positive (+) side and the white wiring terminal at the negative (−) side.

NOTE: Do not apply voltage to the igniter for more than three seconds at a time to avoid destroying the power transistor in the igniter.

5. The igniter is normal when sparks are generated between the high tension cable and ground when one connected wiring (or probe) is disconnected.
6. If the igniter fails this test (no sparks), replace it.
7. Reconnect all disconnected wiring and install all removed parts.

IGNITION TIMING

Adjustment

1.6L ENGINE

Non-Turbocharged

The ignition timing adjustment is performed under the following conditions:

a. Parking brake firmly applied.
b. Front and rear wheels blocked.
c. Transmission in **P** or **N**.
d. All accessories in the **OFF** position.
e. Front wheels straight ahead (if equipped with power steering).

1. Using a jumper wire, connect the both end terminals of the ALDL connector, located under the instrument panel.
2. Connect a tachometer to the engine.
3. Connect a timing light to the engine according to the man-

ufacturer's instructions. If the timing marks are difficult to see, use chalk or a dab of paint to make them more visible.

4. Start the engine and allow it to reach normal operating temperature.

5. Illuminate the timing marks. With the engine at idle, timing should be set at the specifications given on the underhood emission control sticker. If not, loosen the pinch bolt at the base of the distributor, then rotate the distributor to advance or retard the timing as required. The timing should be 9–11 degrees BTDC at 1500–2000 rpm.

6. Stop the engine and tighten the pinch bolt. Start the engine and recheck the timing.

7. Stop the engine and disconnect the timing light and the tachometer.

8. Remove the jumper wire from the ALDL connector.

Turbocharged

The ignition timing adjustment is performed under the following conditions:

a. Parking brake firmly applied.
b. Front and rear wheels blocked.
c. Transmission in **P** or **N**.
d. All accessories in the **OFF** position.
e. Front wheels straight ahead (if equipped with power steering).

1. Using a jumper wire, connect the both end terminals of the ALDL connector, located under the instrument panel.
2. Connect a tachometer to the engine.
3. Connect a timing light to the engine according to the manufacturer's instructions. If the timing marks are difficult to see, use chalk or a dab of paint to make them more visible.
4. Start the engine and allow it to reach normal operating temperature.
5. Illuminate the timing marks. With the engine at idle, timing should be set at the specifications given on the underhood emission control sticker. If not, loosen the mounting bolt on the cam angle sensor,then rotate the sensor clockwise or counterclockwise to advance or retard the timing as required. The timing should be 9–11 degrees BTDC at 1500–2000 rpm.
6. Stop the engine and tighten the mounting bolt. Start the engine and recheck the timing.
7. Stop the engine and disconnect the timing light and the tachometer.

NOTE: Idle speed is controlled be the ECM, therefore idle adjustment is not required.

8. Remove the jumper wire from the ALDL connector.

2.3L AND 2.6L ENGINES

The ignition timing adjustment is performed under the following conditions:

a. Parking brake firmly applied.
b. Front and rear wheels blocked.
c. Transmission in **P** or **N**.
d. All accessories in the **OFF** position.
e. Front wheels straight ahead (if equipped with power steering).

1. Start the engine and allow it to reach normal operating temperature. Connect a tachometer and check that the idle speed is within specifications. If not, adjust the idle speed to the following specifications:
   a. Amigo and Pickup—900 rpm
   b. Trooper and Trooper II—850–950 rpm
2. Stop the engine.
3. Connect a timing light to the engine according to the manufacturer's instructions either to the No. 1 or No. 4 cylinder spark plug wire. If the timing marks are difficult to see, use chalk or a dab of paint to make them more visible.
4. On 2.3L engines, disconnect the vacuum line(s) from the distributor vacuum advance and plug the line.
5. Start and run the engine at idle. Make sure the parking brake is securely set and the wheels are chocked before allowing the car to idle in gear.
6. Illuminate the timing marks. With the engine at idle, timing should be set at the specifications given on the underhood emission control sticker. If not, loosen the pinch bolt at the base of the distributor, then rotate the distributor to advance or retard the timing as required. The timing should be as follows:
   a. 2.3L engine—6 degrees BTDC at 900 rpm
   b. 2.6L engine—12 degrees BTDC at 900 rpm
7. Stop the engine and tighten the pinch bolt. Start the engine and recheck the timing.
8. Stop the engine and disconnect the timing light and the tachometer.
9. On 2.3L engine, reconnect the vacuum line to the distributor advance.

2.8L AND 3.1L ENGINE

For ignition timing procedures refer to the "Vehicle Emissions Control Information" label affixed to the hood. Follow the directions on the label.

Component Replacement

IGNITION COIL

Removal and Installation

1.6L ENGINE

Non-Tuborcharged

1. Disconnect the negative battery cable.
2. Disconnect the high tension wire from the coil.
3. Disconnect electrical connectors from the coil.
4. Remove coil bracket mounting bolts and remove the coil.
5. Installation is the reverse order of the removal procedure.
6. Connect the negative battery cable.

Tuborcharged

1. Disconnect the negative battery cable.
2. Remove the inner cooler mounting bolts and remove the cooler from the engine.
3. Remove the ignition coil cover retaining bolts and remove the cover.

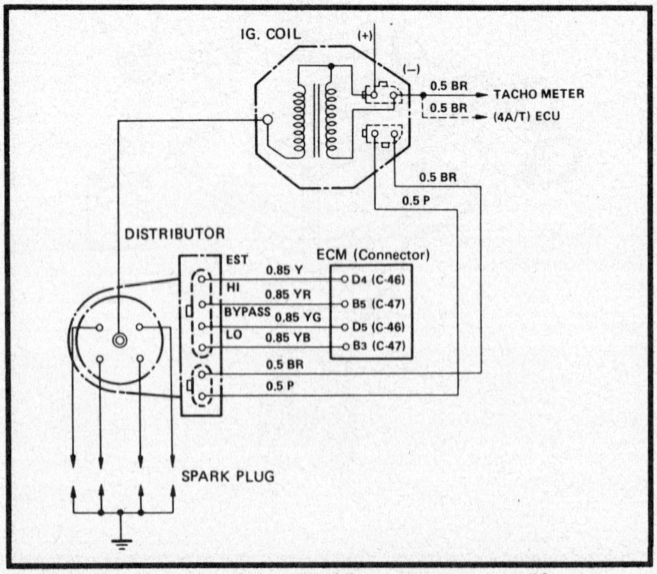

Wiring schematic of the ignition system—1.6L non-turbochargedengine

1.6L ENGINE—IGNITION SYSTEM TESTS—IMPULSE WITH TURBOCHARGER

CHART C-4D-3
"DIS" NO SPARK

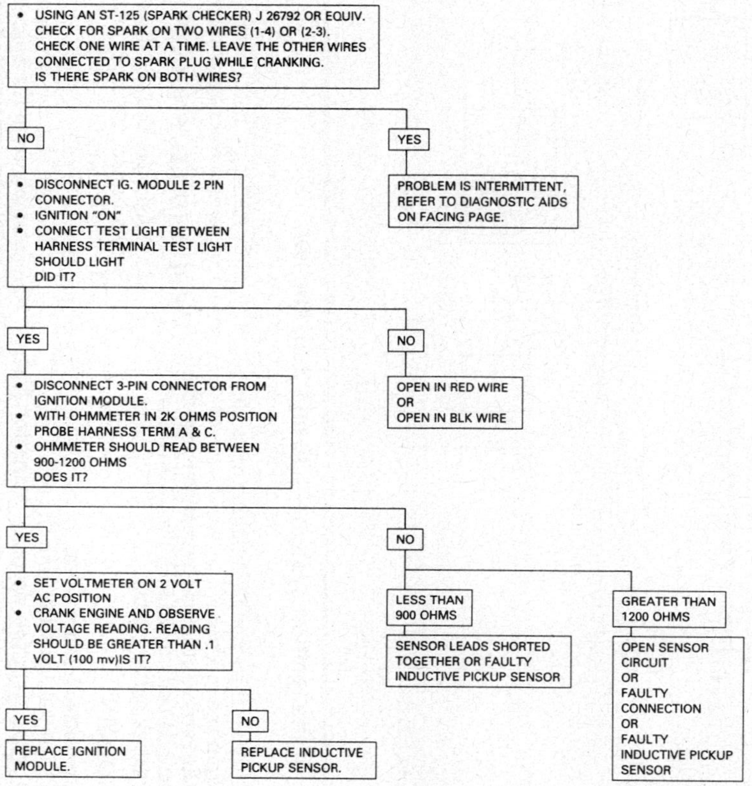

CLEAR CODES AND CONFIRM "CLOSED LOOP" OPERATION AND NORMAL "CHECK ENGINE" LIGHT OPERATION.

CHART C-4D-2
DIS MISFIRE UNDER LOAD

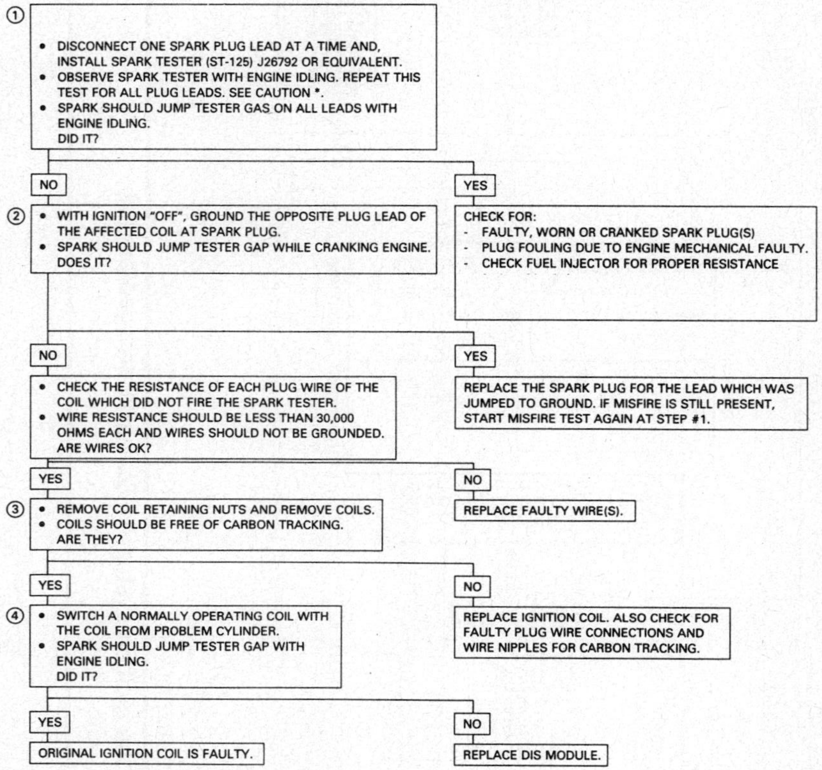

***CAUTION:** When handling secondary spark plug leads with engine running, insulted pliers must be used and care exercised to prevent a possible electrical shock.

CLEAR CODES AND CONFIRM "CLOSED LOOP" OPERATION AND NORMAL "CHECK ENGINE" LIGHT OPERATION.

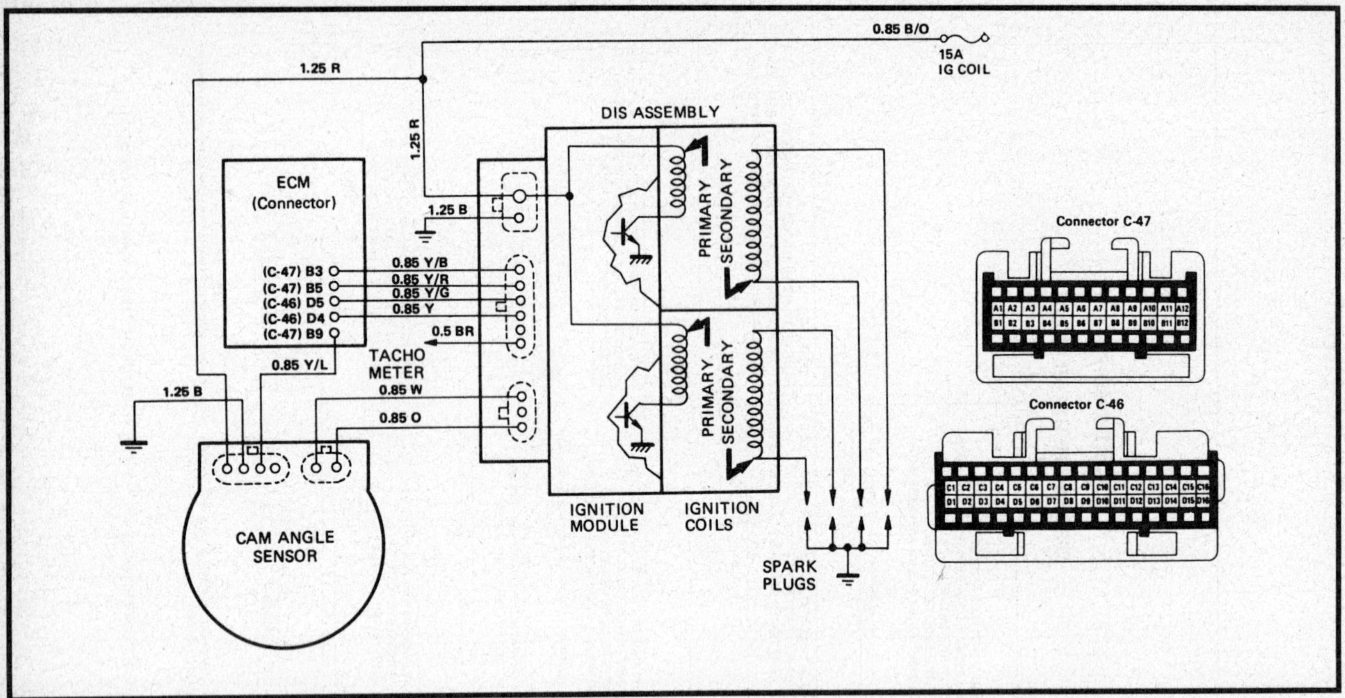

Wiring schematic of the ignition system—1.6L turbo charged engine

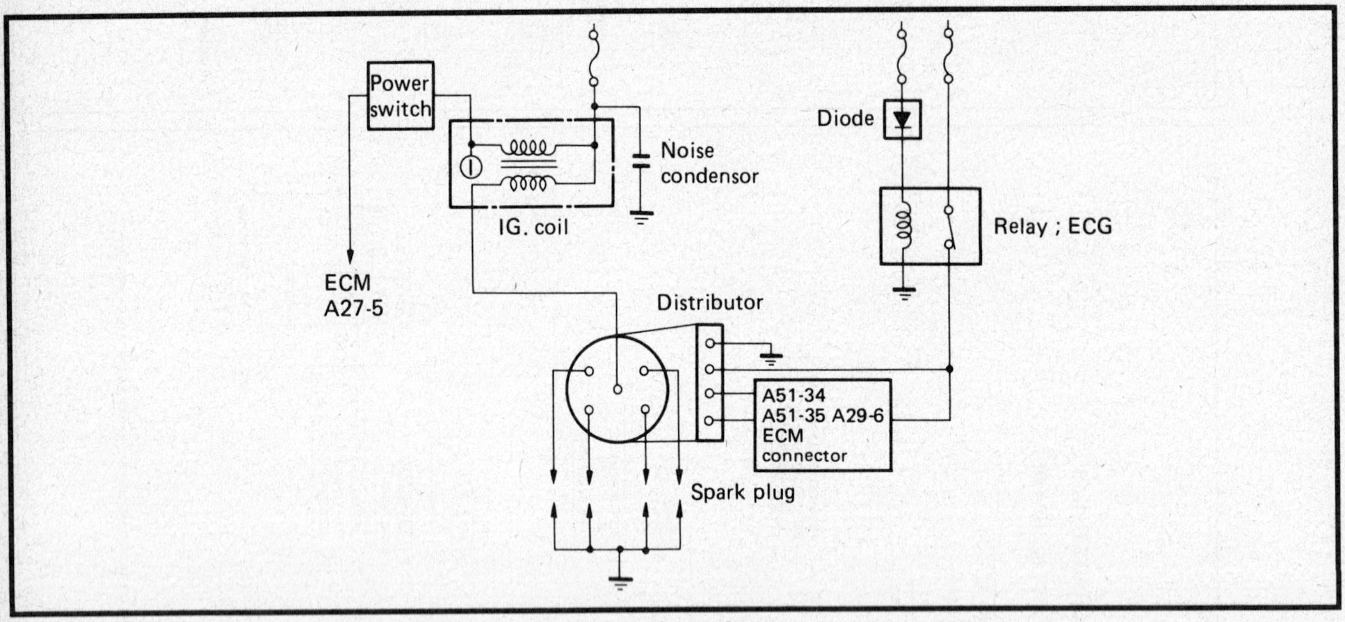

Wiring schematic of the ignition system—2.3L and 2.6L engines

4. Remove the ignition coil-to-module retaining bolts and remove the coil pack from the module.
5. Installation is the reverse order of the removal procedure. Torque the coil-to-module bolts to 40 inch lbs. (4.5 Nm).
6. Connect the negative battery cable.

2.3L AND 2.6L ENGINES

1. Disconnect the negative battery cable.
2. Disconnect the high tension wire from the coil.
3. Disconnect the wires from the electrical connector from the coil.
4. Remove the coil mounting bolts and remove the coil.
5. Installation is the reverse of the removal procedure.
6. Connect the negative battery cable.

2.8L AND 3.1L ENGINES

1. Disconnect the negative battery cable.
2. Disconnect the high tension wire from the coil.
3. Disconnect the electrical connector and tachometer terminal from the coil.
4. Remove the coil and bracket mounting bolts and remove the assembly from the engine.
6. Drill and punch out the rivets holding the coil to the bracket. Remove the coil from the bracket.

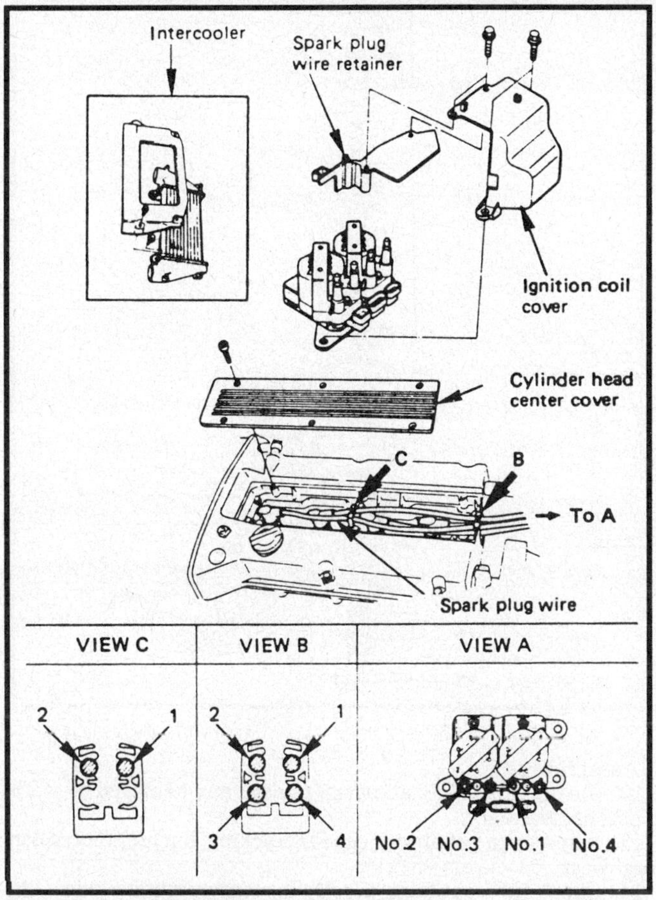

Removal and installation of the ignition coil—1.6L turbochargedengine

7. Installation is the reverse of the removal procedure. Assemble the coil and bracket assembly, using retaining bolts in place of the rivets.
6. Connect the negative battery cable.

DISTRIBUTOR

Removal and Installation

1. Rotate the engine and bring up the No. 1 cylinder to top dead center of its compression stroke.
2. Disconnect the negative battery cable. Disconnect and tag (if necessary) all the electrical connectors along with the spark plug wires from the distributor.
3. Disconnect the wiring harness from the distributor lead wire. Disconnect the vacuum hoses from the vacuum advance if so equipped.
4. Using a scribe tool or a marker, scribe or mark an alignment mark on the base of the distributor and the cylinder head. This can be used for an easier installation procedure.
5. Remove the distributor clamp bolt and bracket and lift out the distributor assembly from the engine cylinder head. Remove the distributor housing seal and discard it. Lubricate the new seal with clean engine oil prior to installation.

To install:

6. Install the distributor assembly into the cylinder block, be sure to align the scribe marks made previously. Lightly tighten the distributor mounting bolt.
7. Connect the wiring harness to the distributor lead wire. Connect the vacuum hoses to the vacuum controller.
8. Connect all the electrical connectors along with the spark plug wires to the distributor.
9. Connect the negative battery cable. Start the engine, set the timing and check the idle speed. Tighten the distributor mounting bolt.

Disassembly

2.3L ENGINE

1. Remove the distributor cap retaining screw.
2. Remove the distributor cap and rotor.
3. Remove the carbon points from the cap.
4. Remove distributor packing gasket and packing.
5. Remove the distributor lead wire.
6. Remove the ground terminal and vacuum control mounting screw. Remove the vacuum control unit.
7. Remove the reluctor as follows:
   a. Pry loose the reluctor outer cover.
   b. Using 2 small pry bars or equivalent, slide the pry bars into the lower side of the reluctor and pull it free.

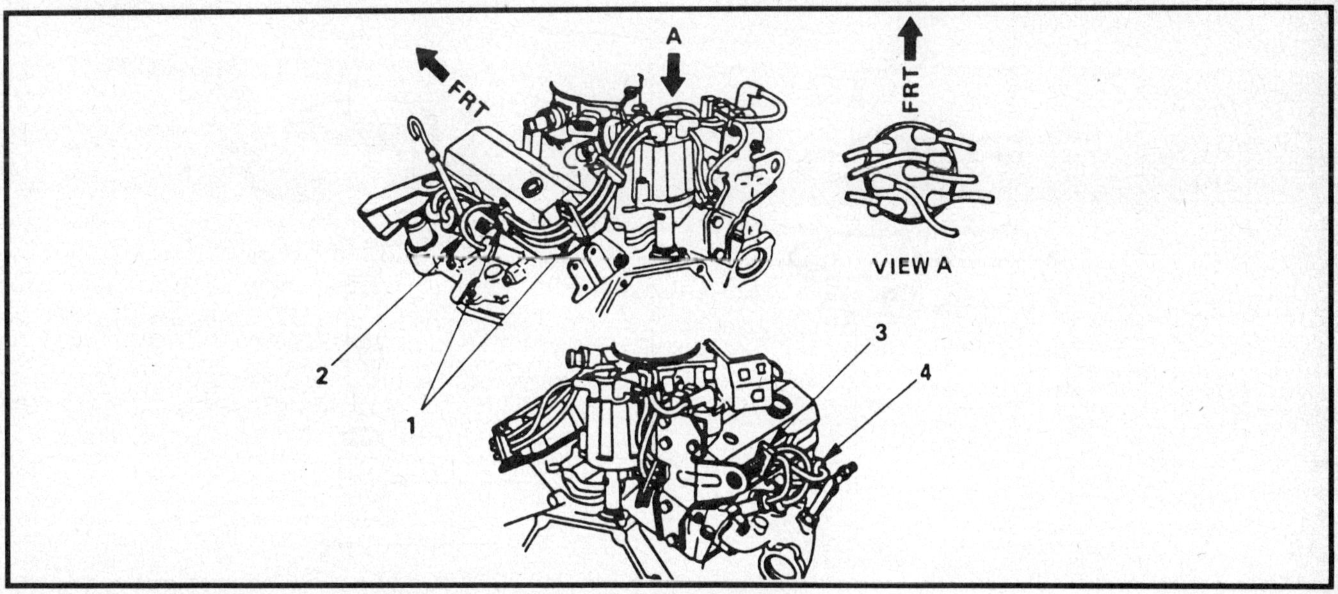

Removal and installation of the distributor—2.8L and 3.0L engines

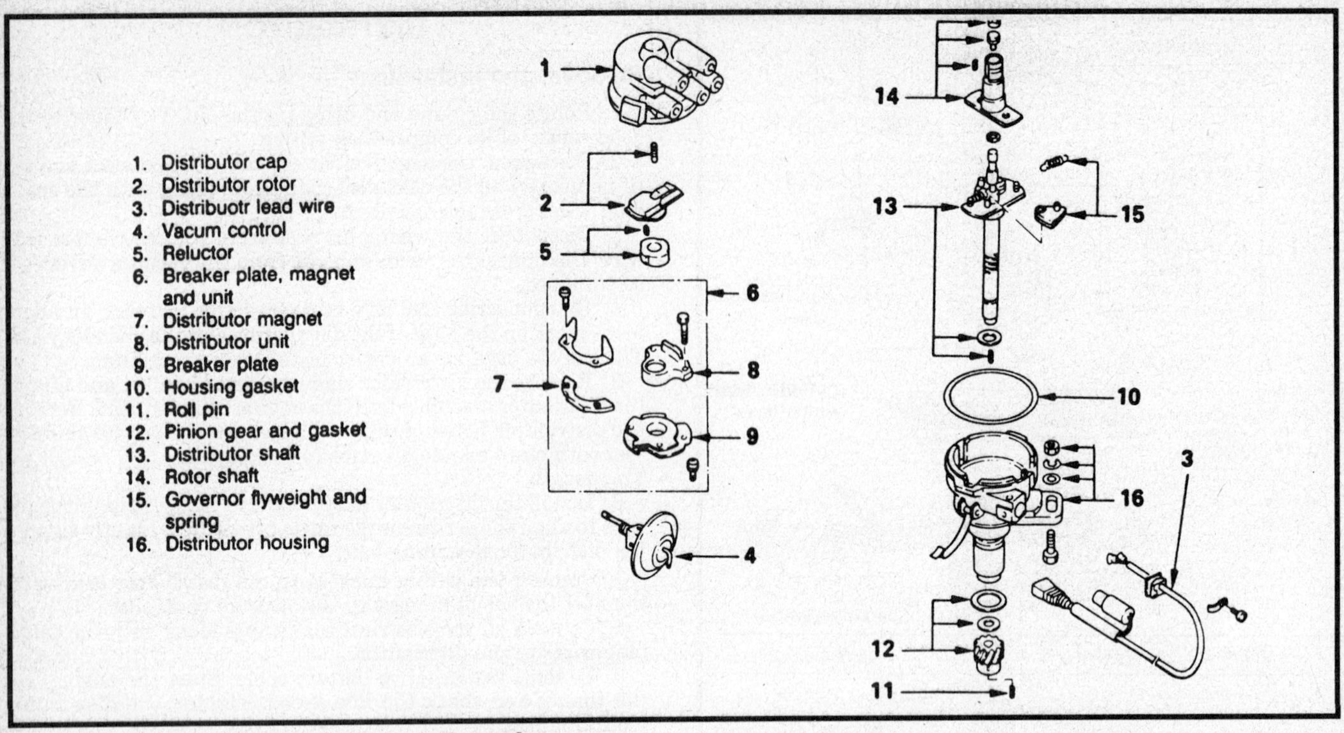

Exploded view of the distributor–2.3L engine

8. Remove the breaker plate, magnet and distributor unit assembly.

9. Remove the roll pin, pinion gear and gasket as follows:

a. File off the staked end of the roll pin.

b. From the opposite side, drive the roll pin from the gear set using a suitable drill press.

c. Pull the pinion gear and gasket from the distributor shaft.

10. Remove the distributor shaft and rotor shaft.

11. Remove the governor springs. Release the snap ring and remove the governor weights.

Reassembly

12. Clean and inspect all parts, replace any damaged or worn parts as necessary.

13. Inspect the distributor cap for cracking, burning, corrosion and wear.

14. Check the ventilator, electrode, center contact peice and spring.

15. The carbon length limit is 0.087 in. (2.2mm) or less.

16. Apply a vacuum to the vacuum advance controller and check to see that it is operating smoothly.

17. To reassemble, follow the disassembly procedure in the re-

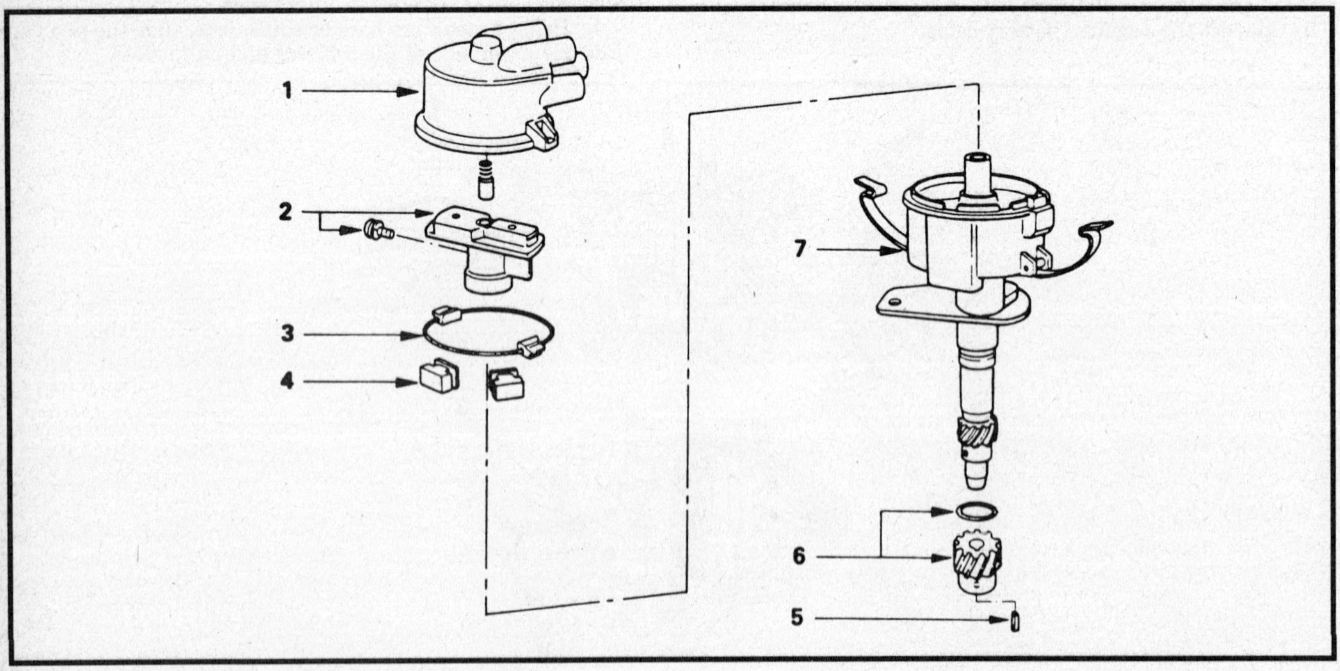

Exploded view of the distributor–2.6L engine

verse order. When installing the pinion gear roll pin make sure to stake it on both sides after it is in place. Be sure to carefully align the rotor shaft notches and the shaft notches when installing the rotor shaft. Also when installing the igniter and breaker assembly, align the scribe mark on the breaker plate and the housing. When installing the reluctor, the roll pin notch and the reluctor notch must be parallel when the roll pin is inserted into the reluctor.

18. Inspect the air gap, by using a feeler gauge. Measure the gap between the signal rotor and the pick-up coil projection. The air gap should be 0.008–0.0016 in.

19. If a air gap adjustment is needed, loosen the 2 screws and move the igniter until the proper specified gap (0.012–0.020 in.) is reached. Then tighten the 2 screws and recheck the air gap.

20. Also check the governor function by turning the rotor both to the right and the left. The rotor should spin freely in both directions.

2.6L ENGINE

1. Remove the distributor cap assembly.
2. Remove the rotor set screw and remove the rotor.
3. Remove the distributor housing gasket.
4. Remove the distributor dust bushing from the housing gasket.
5. Remove the roll pin, pinion gear and gasket as follows:
   a. File off one of the staked ends of the roll pin.
   b. From the opposite side, drive the roll pin from the gear set using a suitable drill press.
   c. Pull the pinion gear and gasket from the distributor shaft.
6. Remove the distributor base plate O-ring and discard it.

To assemble:

7. Clean and inspect all parts, replace any damaged or worn parts as necessary.
8. Inspect the distributor cap for cracking, burning, corrosion and wear.
9. Remove the 2 screws of the fixing rotor plate cover, then remove it. Using a soft brush, clean the rotor plate in the angle signal port area while blowing some air over it.

NOTE: The distributor body assembly is of an integral design and cannot be disassembled. The entire assembly must be replaced if service becomes necessary.

10. Install the pinion gear, gasket and roll pin as follows:
    a. Align the pinion gear and distributor shaft roll pin holes.
    b. Slide the pinion gear and gasket onto the distributor shaft.
    c. Insert the new roll pin into the pin hole.
    d. Drive the pin into place using a small pin punch.
    e. Stake the roll pin on both sides.
11. Assemble the dust bushing onto the housing gasket and install the gasket and dust bushings onto the distributor housing.
12. Install the rotor.
13. Install the distributor cap.

Air Gap Adjustments

Air gap adjustment is not required on the 2.6L engine.

2.3L ENGINE

1. Insert the proper size feeler gage between the pickup coil projections to measure the air gap. The air gap should be 0.012–0.020 in.
2. If the air gap is not within specs, loosen the 2 adjustings screws and move the igniter until the air gap is correct.
3. Tighten the 2 adjusting screws and re-check the air gap.

1.6L, 2.8L AND 3.1L ENGINES

1. Remove the distributor cap and rotor.
2. Remove the ignition module retaining screws and remove module.
3. Remove the roll pin retaining spring and remove the roll pin from the base of the drive coupling.
4. Remove the drive coupling from the distributor shaft.
5. Remove the flat washer, tang washer and thrust spring from the distributor shaft.
6. Slide the distributor shaft assembly from the distributor housing.
7. Remove the pick-up coil retainer and remove the pick-up coil and pole piece.

To install:

8. Clean and inspect all parts, replace any damaged or worn parts as necessary.
9. Inspect the distributor cap for cracking, burning, corrosion and wear.
10. Install the pole piece and pick-up coil to the distributor housing and secure in place with the retainer.
11. Slide the distributor shaft into the housing. Spin the shaft to ensure there is no binding.
12. Install the flat washer, tang washer and thrust spring on the distributor shaft.
13. Install the drive coupling and align the role pin hole in the distributor shaft.
14. Insert a new role pin and the retaining spring.
15. Install the ignition module and retaining screws.
16. Inspect the air gap, by using a feeler gauge. Measure the gap between the signal generator and the pick-up coil projection. The air gap should be 0.008–0.0016 in. If the gap is not within specifications, replace the signal generator.
17. Install the distributor rotor and cap.

LEXUS ELECTRONIC SPARK IGNITION SYSTEM

General Information

The electronic control unit is programmed with data for optimum ignition timing under any and all operating conditions. Using data provided by the sensors which monitor various engine functions (rpm, intake air volume, engine temperature, etc.) the electronic control unit (microcomputer triggers the spark at precisely the right instant.

SYSTEM COMPONENTS

The major components of the ignition system include the igniter, ignition coil, distributor and engine speed sensors.

Igniter

The LS 400 and SC 400 use a dual ignition system with 2 separate igniters. The igniter temporarily interrupts the primary current with the ignition signal (IGT signal) from the ECU and generates sparks at the spark plug. As a fail safe, when ignition occurs, a confirmation signal (IGF signal) is sent to the ECU.

Ignition Coil

The ignition coil uses a closed core coil with the primary coil wrapped around the core and the secondary coil wrapped around the primary coil. This allows the generation of a high voltage secondary voltage, sufficient to cause a spark to jump the plug gap.

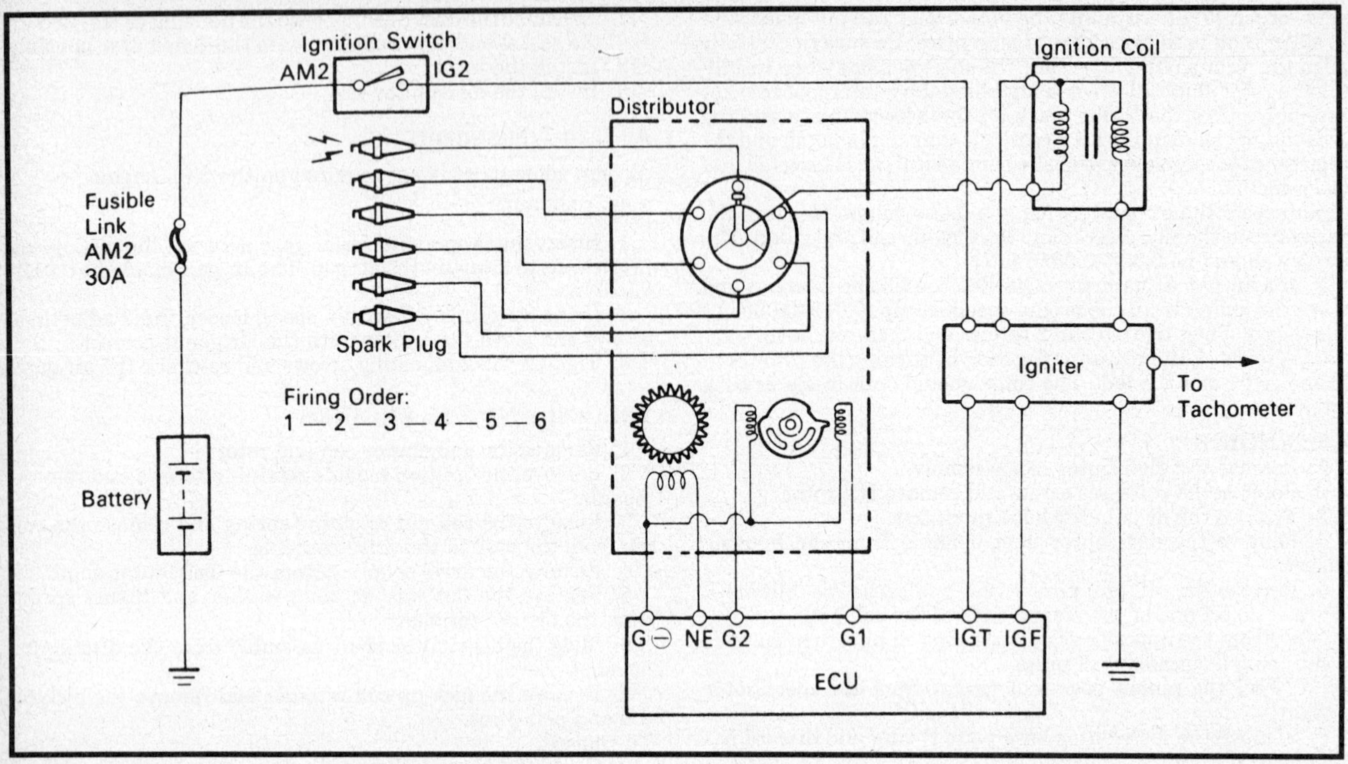

Ignition system diagram—ES 250 and ES 300

Distributor

The distributor distributes high voltage spark to each spark plug through a cap and rotor. The LS 400 and SC 400 use a dual distributor system.

Sensors

Lexus engines use two different ways of timing ignition spark: a pickup coil, located in the distributor or an engine speed sensor and camshaft angle sensor combination, located externally of the distributor.

PICKUP COILS

On the ES 250, SC 300 and ES 300 the NE coil detects crank angle and the G1 and G2 coils detect the cam angle.

ENGINE SPEED SENSOR

On the LS 400 AND SC 400, the engine speed sensor detects the crankshaft angle.

CAMSHAFT ANGLE SENSOR

On the LS 400 and SC 400, cam position sensors, one for each bank of cylinders, detect the camshaft angle.

Diagnosis and Testing

SERVICE PRECAUTIONS

- Do not leave the ignition swtich **ON** for more than 10 minutes if the engine will not start.
- Make sure all electronic controller connectors are fastened securely. A poor connection can cause an extremely high surge voltage in the coil and condenser and result in damage to integrated circuits.
- Keep the electronic control unit wire harness at least 4 in. away from adjacent harnesses to prevent an electronic system malfunction due to external electronic "noise."
- Keep all parts and harnesses dry during service.
- Before attempting to remove any parts, turn **OFF** the ignition switch and disconnect the battery ground cable.
- Always use a 12 volt battery as a power source.
- Do not attempt to disconnect the battery cables with the engine running.
- Do not depress the accelerator pedal when starting.
- Do not rev up the engine immediately after starting or just prior to shutdown.
- Do not attempt to disassemble the electronic control units, they are not serviceable.
- If a battery cable is disconnected, the memory will return to the ROM (programmed) values. Engine operation may vary slightly, but this is not an indication of a problem. Do not replace parts because of a slight variation.
- If installing a 2-way or CB radio, keep the antenna as far as possible away from the electronic control unit. Keep the antenna feeder line at least 8 in. away from the electronic control unit wiring harness and do not let the 2 run parallel for a long distance. Be sure to ground the radio to the vehicle body.

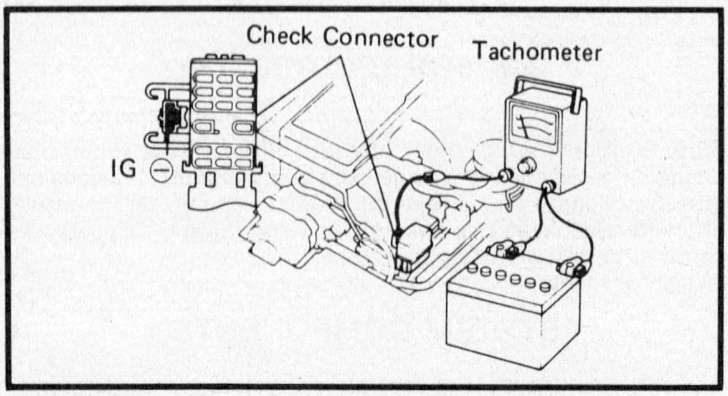

Connecting a tachometer—ES 250 and LS 400

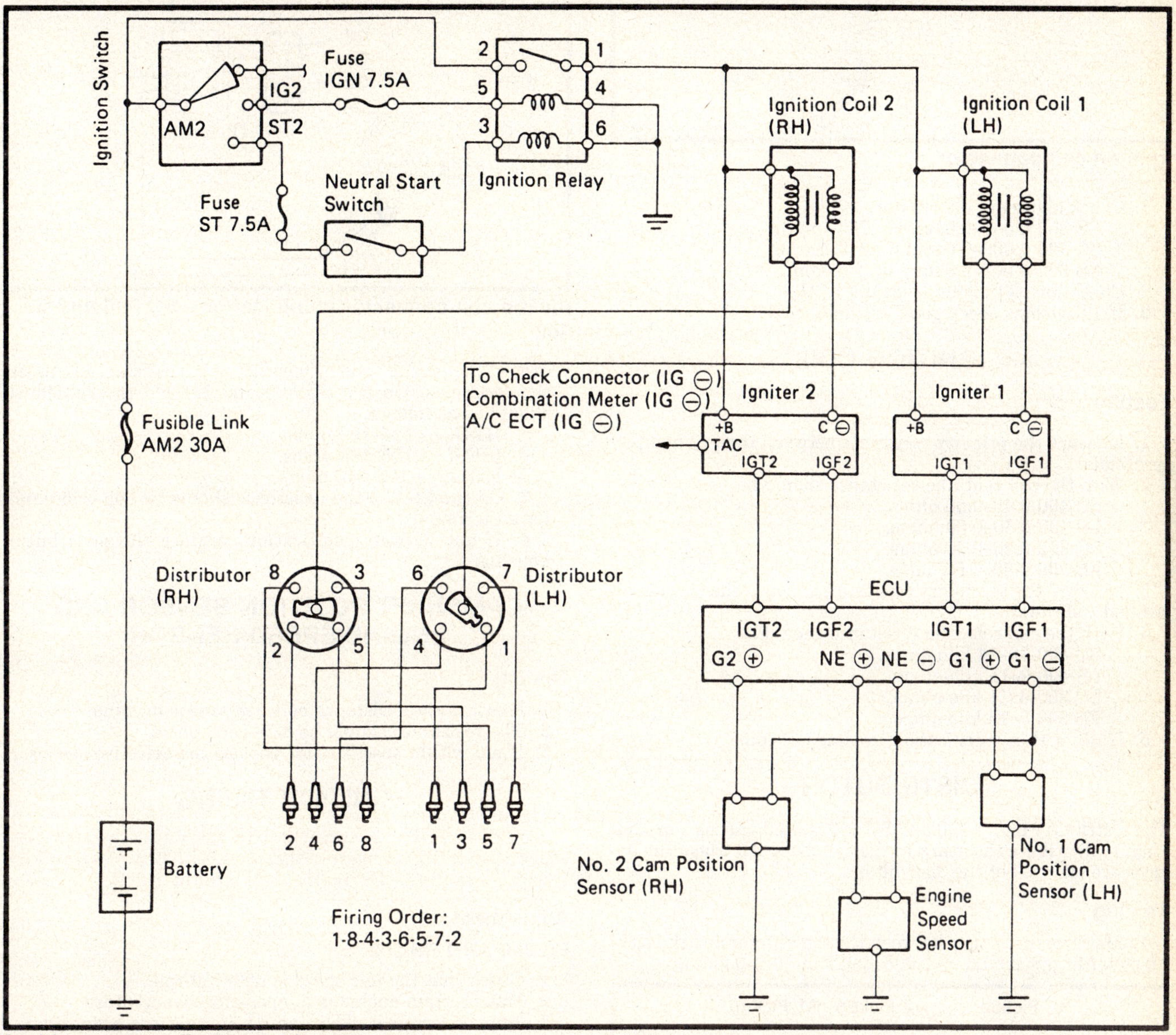

Ignition system diagram—LS 400

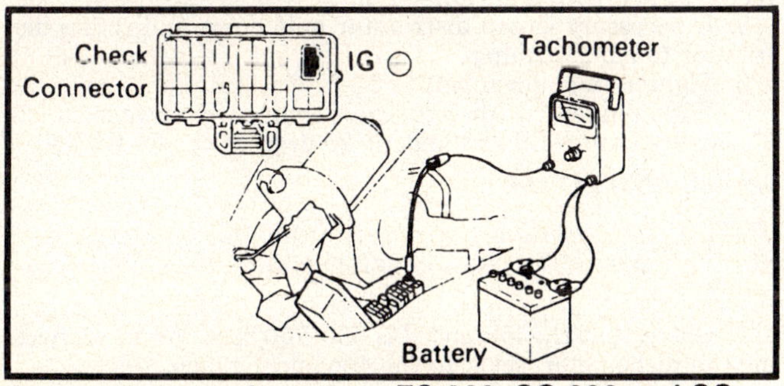

Connecting a tachometer—ES 300, SC 300 and SC 400

NO SPARK—NO START

Test

1. Check the connection at coil, igniter and distributor.
2. Check that the resistance of spark plug wires in less than 25 kilo-ohms.
3. Check for power to the positive side of coil(s).
4. Check the resistance of coil primary winding.
   SC 400 0.40–0.50 ohms.
   LS 400 0.30–0.60 ohms.
   SC 300 0.20–0.30 ohms.
   ES 300 0.30–0.60 ohms.
5. Check the resistance of the coil secondary winding.
   SC 400 10–14 kilo-ohms.
   LS 400 9–15 kilo-ohms.
   SC 300 6–11 kilo-ohms.
   ES 300 9–15 kilo-ohms.
6. Check pickup coil resistance.
   ES 250 205–255 ohms. SC 300 125–190 ohms between G1/G(−) and G2/G(−).
   SC 300 155–240 ohms between NE and G(−).
7. Check camshaft/crankshaft position sensor resistance.
   LS 400 835–1350 ohms.
   SC 400 835–1350 ohms.

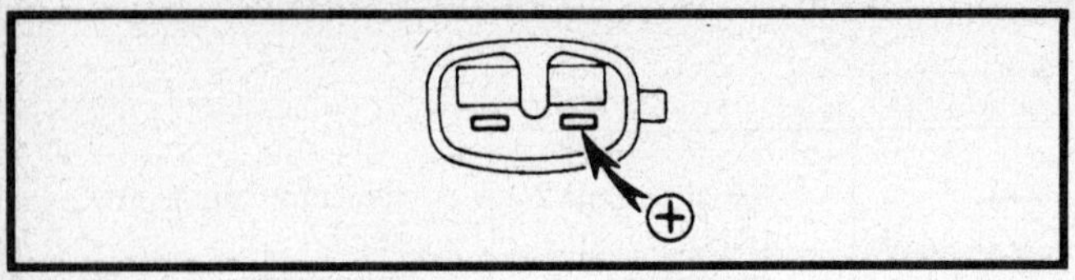

Ignition coil connector

8. Check air gap of distributor.
   ES 250 0.008–0.016in. (0.2–0.4mm).
   SC 300 0.008–0.016in. (0.2–0.4mm).
   ES 300 0.008–0.016in. (0.2–0.4mm).
8. Check for IGT signal from the ECU.
9. If all systems check good, try another igniter.

IGNITION COIL

Testing

1. Measure the primary resistance between the (+) and (−) terminals.
2. With the coil cold, the resistance should be:
   SC 400 0.40–0.50 ohms.
   LS 400 0.30–0.60 ohms.
   SC 300 0.20–0.30 ohms.
   ES 300 0.30–0.60 ohms.
3. Measure the secondary resistance between the (+) terminal and coil tower.
4. With the coil cold, the resistance should be:
   SC 400 10–14 kilo-ohms.
   LS 400 9–15 kilo-ohms.
   SC 300 6–11 kilo-ohms.
   ES 300 9–15 kilo-ohms.
5. If not within specification, replace the coil.

DISTRIBUTOR

The distributor on LS 400 and SC 400 is used only as a mechanism to distribute the spark to the cylinders. All triggering devices are external of the distributor.

Testing

1. Measure the air gap between the signal rotor and the pickup coil. Air gap should be 0.008–0.016 in. (0.2–0.4mm).

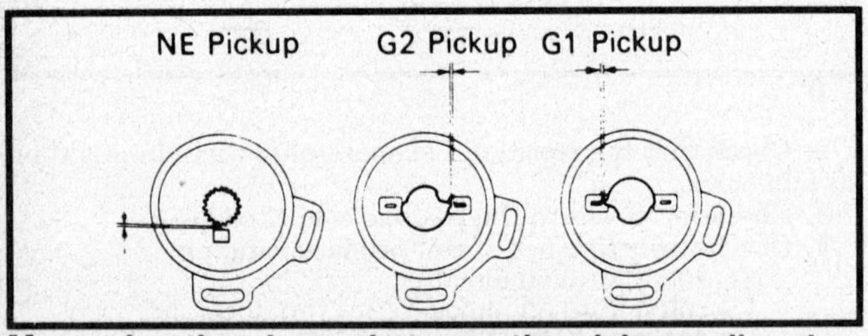

Measuring the air gap between the pickup coil and the signal rotor—ES 250, ES 300 and SC 300

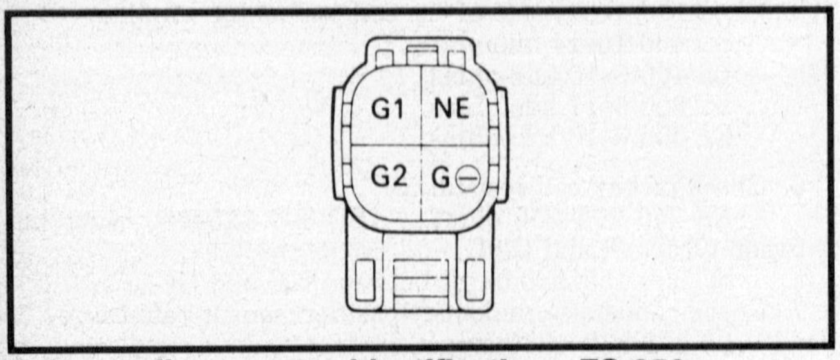

Pickup coil connector identification—ES 250

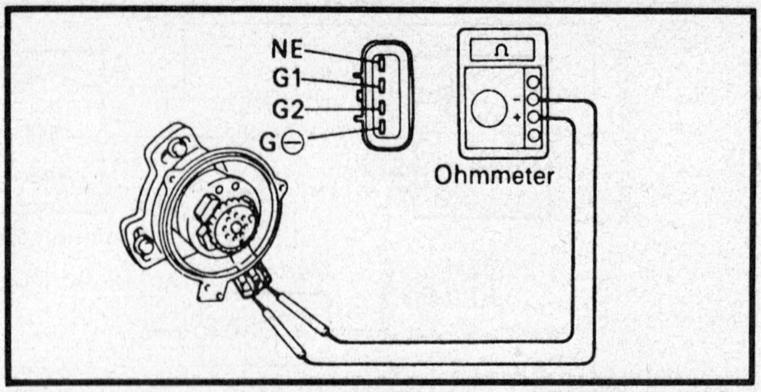

Pickup coil connector identification—SC 300 and ES 300

2. Inspect the pickup coils by measuring the resistance between the terminals. On the SC 300 and ES 300 the resistance should be as follows:
   G1/G(−) 125–190 ohms.
   G2/G(−) 125–190 ohms.
   NE/G(−) 155–240 ohms.
3. On the ES 250 the resistance should be 205–255 ohms for all terminals.
4. If not within specification, replace the distributor housing.

CAMSHAFT POSITION SENSOR AND ENGINE SPEED SENSOR

Testing

1. Measure the resistance between sensor terminals.
2. The resistance should be 835–1350 ohms.
3. If not within specification, replace the defective sensor.

IGNITION TIMING

NOTE: Failure to connect the tachometer correctly will cause ignition system damage. The use of an inductive type tachometer is highly recommended.

Adjustment

ES 250 AND ES 300

1. Check that the idle speed is 650–750 rpm.
2. Make certain engine is at operating temperature.
3. Jumper terminals TE1 and E1 of the check connector.
4. Using a timing light check that the timing is 10 degrees BTDC at idle with transmission in **N**.
5. If necessary loosen distributor hold down and rotate distributor to adjust timing.
6. Tighten hold down bolt to 13 ft. lbs. (18 Nm).
7. With jumper wire disconnected, timing will move from 13–27 degrees on the ES 250 and 10–20 degrees on the ES 300.

LS 400 AND SC 400

NOTE: The timing on this vehicle is not adjustable, if not within specification check the engine and fuel control systems.

1. Check that the idle speed is 600–700 rpm. When connecting the tachometer, set it to the 4-cylinder range.
2. Make certain engine is at operating temperature.
3. Jumper terminals TE1 and E1 of the check connector.
4. Connect a timing light to the number 6 spark plug wire.
5. Check that the timing is 8–12 degrees BTDC at idle with transmission in **N**.
5. If not within specification check that the throttle valve is fully closed. Check for continuity between terminals IDL and E2 of the throttle position sensor and check valve timing.

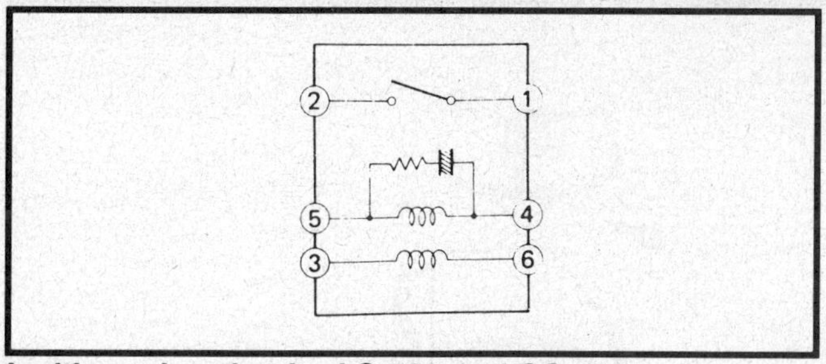

Ignition relay circuit—LS 400 and SC 400

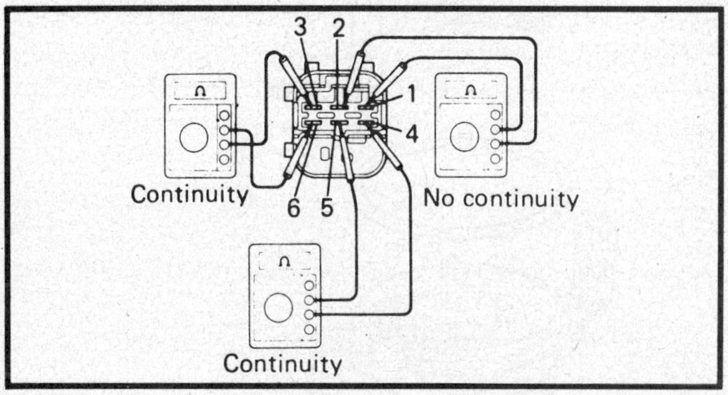

Ignition relay testing—LS 400 and SC 400

SC 300

1. Jumper terminals TE1 and E1 of the check connector.
2. Check that the idle speed is 700 rpm for manual transmission vehicles or 650 rpm for automatic transmission vehicles with transmission in **N**.
3. Ensure all accessories are **OFF**.
4. Using a timing light check that the timing is 10 degrees BTDC at idle with transmission in **N**.
5. If necessary loosen distributor hold down and rotate distributor to adjust timing.
6. Tighten hold down bolt to 10 ft. lbs. (14 Nm).
7. With jumper wire disconnected, timing will move from 9–11 degrees.

IGNITION RELAY

Testing

LS 400 AND SC 400

1. Using an ohmmeter check for continuity between terminals 3 and 6.
2. Check for continuity between terminals 4 and 5.
3. Check for no continuity between terminals 1 and 2.
4. Apply battery voltage across terminals 3 and 6.
5. Check for continuity between terminals 1 and 2.
6. Apply battery voltage across terminals 4 and 5.
7. Check for continuity between terminals 1 and 2.
8. If not as specified replace the relay.

Component Replacement

DISTRIBUTOR

Removal and Installation

ES 250

1. Disconnect the negative battery cable.
2. Remove the upper brace.
3. Disconnect the air inlet, air flow meter and hoses.

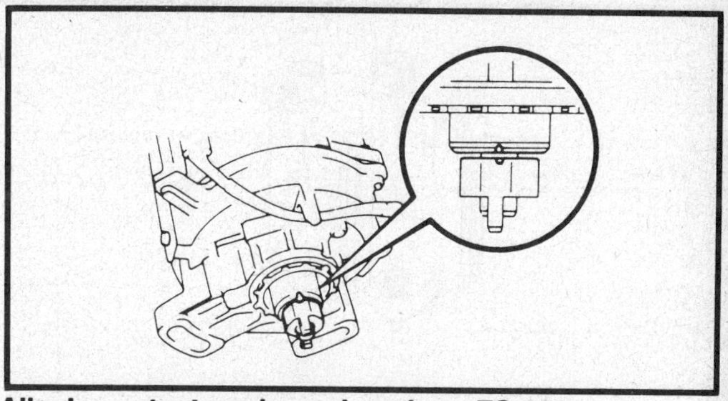
Aligning cutout marks on housing—ES 250

4. Label and disconnect the spark plug wires.
5. Disconnect the electrical connector.
6. Remove the distributor.

To install:

7. Turn crankshaft clockwise and position slit of the intake camshaft, right head, so slit aligns with center of engine.
8. Use new O-ring coating with engine oil.
9. Align the cutout marks of the coupling with the housing.
10. Insert the distributor hand tighten the hold down until timing is adjusted.

ES 300

1. Disconnect the negative battery cable and wait for 30 seconds prior to servicing.
2. Disconnect the air flow meter electrical connection.
3. Disconnect the air hoses and loosen the air cleaner clamp bolt. Disconnect the air cleaner cap clips.
4. Remove the air cleaner cap and air flow meter together with the air cleaner hose.
5. Label and disconnect the spark plug wires from the distributor cap. Note that the spark plug wires use a locking cover. Pry the lock tab carefully to remove.
6. Disconnect the distributor electrical connection.
7. Remove the 2 hold-down bolts and pull out the distributor. Discard the O-ring.

To install:

6. Set the No. 1 cylinder to TDC on the compression stroke. Turn the crankshaft clockwise and position the groove of the intake camshaft on the right side cylinder head at approximately the 2 o'clock position with the wider section of the camshaft end facing upwards.
7. Install a new O-ring on the distributor body and apply a light coat of engine oil. Locate the small cut-out marks on the distributor coupling and the housing and align.
8. Install the distributor aligning the line of the housing with the cut-out of the distributor attachment bearing cap. Lightly tighten the 2 hold-down bolts.

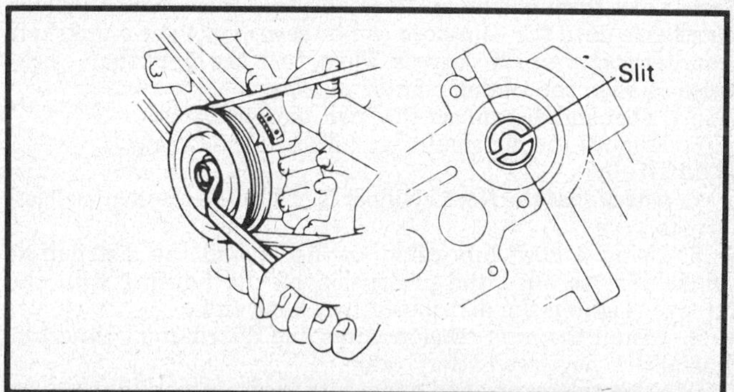

Distributor installation—ES 250 and ES 300

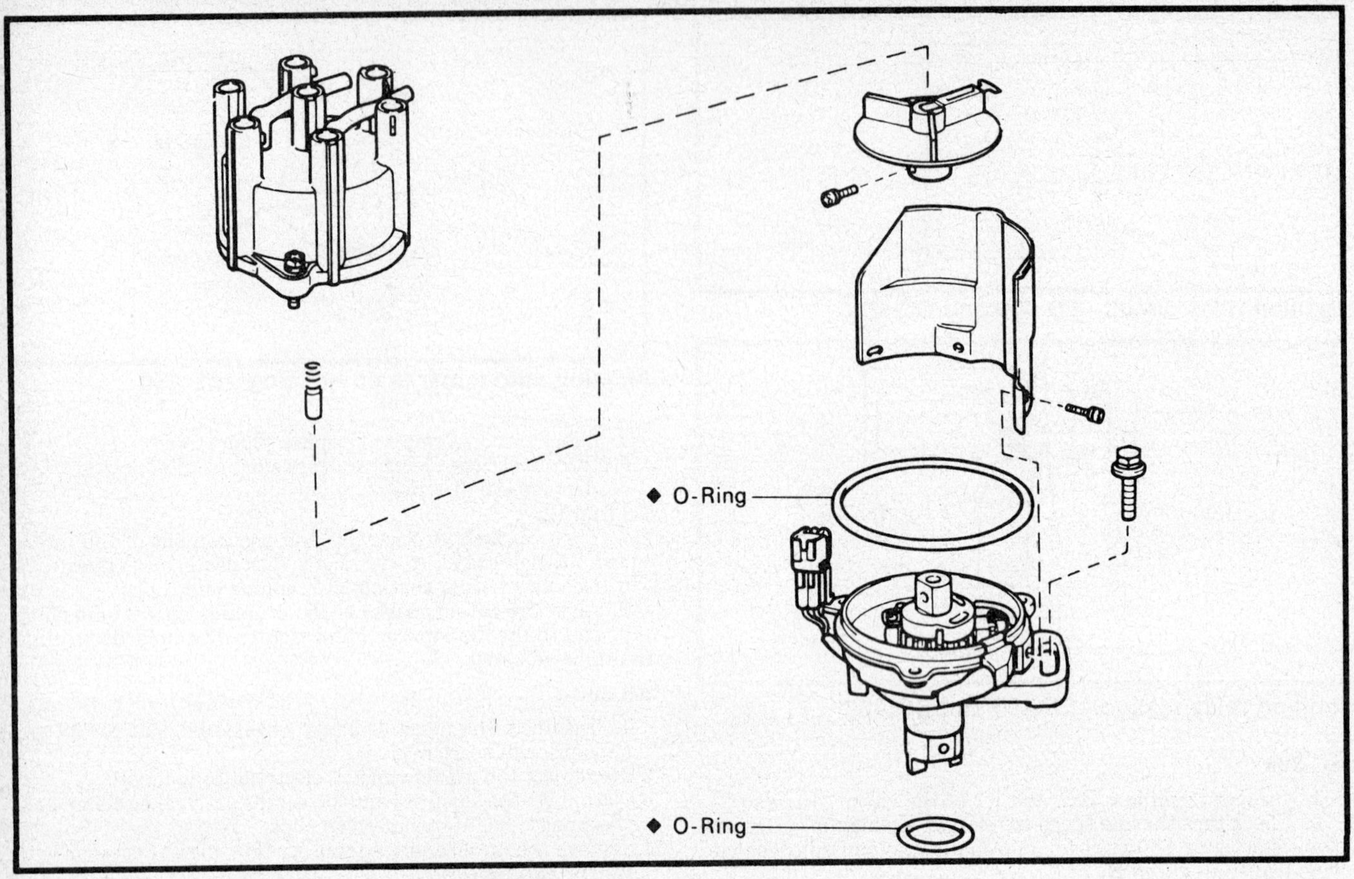

ES 250 distributor and components – ES 300 similar

9.Install the spark plug wires. Install the grommet part first, then align the spline of the distributor cap with the spline groove of the holder into place. Check that the claw of the holder is engaged by pulling lightly on the holder.

10. Install the distributor electrical connector.

11. Connect the air cleaner hose and install the air cleaner cap and air flow meter with the 4 clips.

12. Tighten the air cleaner clamp bolt, connect the air hoses and connect the air flow meter connection.

13. Start the engine and allow it to reach operating temperature. Adjust the timing.

SC 300

1. Disconnect the negative battery cable.
2. Disconnect the distributor connector.
3. Crank the engine so that the No. 1 piston is at TDC on the compression stroke.
4. Look through the oil filler hole and turn the crankshaft clockwise until the cam nose can be seen. Turn the crankshaft counterclockwise 120 degrees. Then, turn the crankshaft clockwise to align the timing pointer to zero.
5. Label and disconnect the high tension wires.
6. Remove the distributor and discard the O-ring.

To install:

7. Check that the No. 1 cylinder is at TDC on the compression stroke.
8. Using a new, lubricated O-ring, install the distributor. Make sure to align the protrusion on the housing with the groove. Tighten the distributor bolt hand tight.
9. Install the high tension wires and distributor connector. Install the negative battery cable.
10. Start the engine and allow it to reach operating temperature. Adjust the ignition timing.

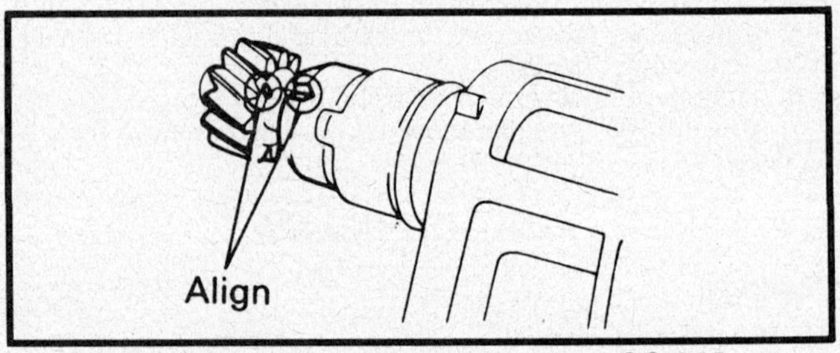

Aligning distributor gear with housing – SC 300

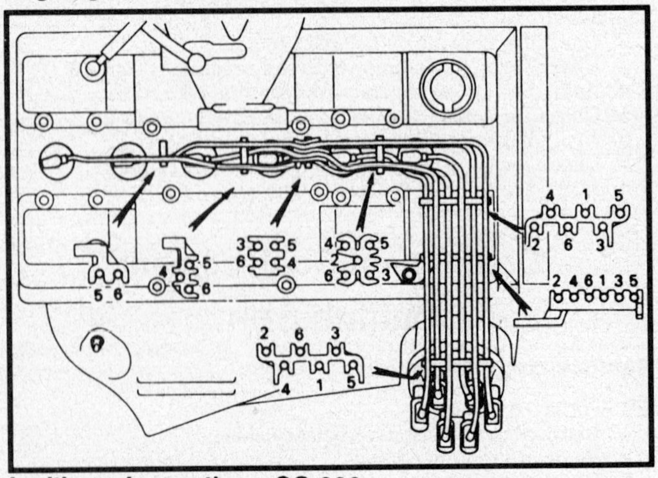

Ignition wire routing – SC 300

LS 400

Right Side

1. Remove air duct.
2. Remove air flow meter assembly.
3. Remove throttle body cover.
4. Remove the No. 1 air hose.
5. Remove the spark plug wire cover.
6. Remove the engine right side wire cover.
7. Remove the No. 3 timing belt cover.
8. Remove the No. 2 timing belt cover.
9. Label and disconnect the spark plug wires.
10. Remove the distributor cap and rotor.
11. Remove the distributor housing assembly.
12. Remove sensor as needed.

To install:

13. Align crankshaft pulley with "0" mark for No. 1 cylinder.
14. Install sensor if removed.
15. Install the distributor.
16. Install rotor, cap and spark plug wires.
17. Install Install timing belt and spark plug wire covers.
18. Install air hoses, throttle body cover and air flow meter assembly.

Left Side

1. Drain coolant.
2. Remove engine wire left cover.
3. Remove the No. 3 timing belt cover.
4. Remove the No. 2 timing belt cover.
5. Label and disconnect the spark plug wires.

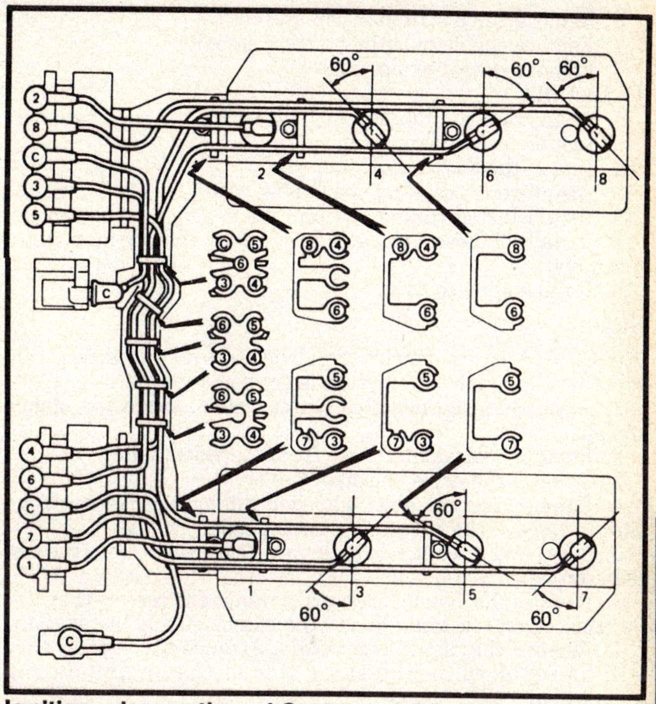

Ignition wire routing—LS 400 and SC 400

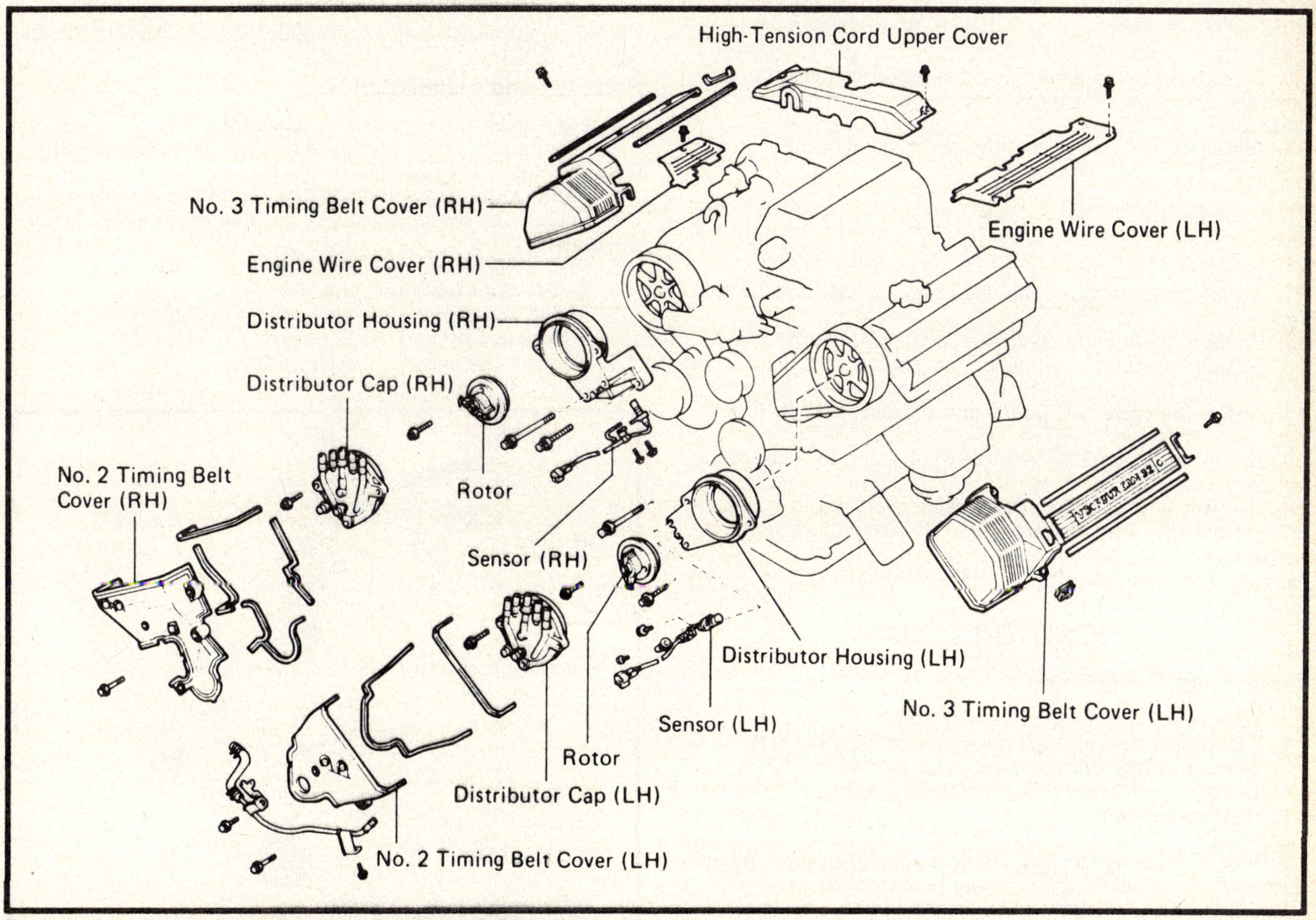

Distributor and components—LS 400 and SC 400

6. Remove the distributor cap and rotor.
7. Remove the distributor housing assembly.
8. Remove sensor as needed.

To install:

9. Align crankshaft pulley with "0" mark for No. 1 cylinder.
10. Install sensor if removed.
11. Install the distributor.
12. Install rotor, cap and spark plug wires.
13. Install Install timing belt and spark plug wire covers.
14. Install air hoses, throttle body cover and air flow meter assembly.
15. Refill engine coolant.

SC400

The SC 400 uses 2 distributors. The removal and installation procedure for both distributors is the same.

1. Disconnect the negative battery cable. Drain the engine coolant.
2. Remove the radiator hose from the water inlet.
3. Remove the drive belt idler pulley.
4. Remove the left and right side timing belt covers. Take care to remove the camshaft position sensor wires.
5. Label and disconnect the high tension wires from the distributor caps.
6. Remove the distributor caps. Arrange the caps so they will be replaced on the same side of the engine during installation.
7. Remove the distributor rotors. Arrange the caps so they will be replaced on the same side of the engine during installation.

NOTE: The distributor attaching bolts are of varying lengths and must be replaced in their original positions. Take care to note the positions of the bolts during removal.

8. Disconnect the camshaft position sensor connector and remove the distributor from the engine

To install:

9. Install the distributor and tighten the hold down bolt to 13 ft. lbs. (18 Nm).

NOTE: The distributors are marked with an "R" for right and an "L" for left.

10. Install the rotor and tighten to 34 inch lbs. (3.8 Nm).
11. Install the distributor cap and tighten to 34 inch lbs. (3.8 Nm).
12. Connect the high tension wire to the distributor cap.
13. Install the left and right side timing belt covers. Tighten to 12 ft. lbs. (16 Nm). Install the camshaft position sensor wires.
14. Install the drive belt pulley and tighten to 27 ft. lbs. (37 Nm).
15. Install the radiator hose to the water inlet. Fill the cooling system.
16. Connect the negative battery cable.
17. Start the engine and allow it to reach operating temperature.
18. Check the ignition timing and idle speed.

IGNITION COIL

Removal and Installation

1. Disconnect the negative battery cable.
2. Disconnect the coil high tension wire from the coil.
3. Disconnect the coil electrical connector.
4. Loosen and remove the coil attaching bolts.
5. Remove the coil.

To install:

6. When replacing the coil, check the high tension wire and electrical connector for damage and replace as necessary.
7. Install the coil and tighten the attaching bolts securely.
8. Connect the coil electrical connector.
9. Connect the coil high tension wire.

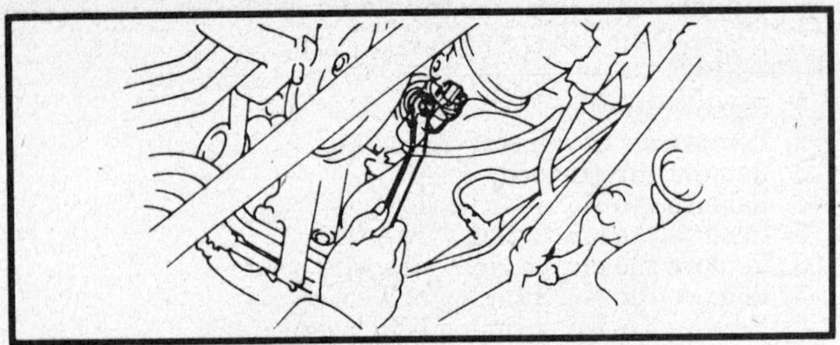

Engine speed sensor removal and installation

10. Connect the negative battery cable.

ENGINE SPEED SENSOR

Removal and Installation

1. Disconnect the negative battery cable and wait 20 seconds prior to starting service.
2. Remove the engine under cover.
3. Disconnect the engine speed sensor connector.
4. Remove the engine speed sensor from the front of the engine.

To install:

5. Install the engine speed sensor and tighten the attaching bolt to 56 inch lbs. (6.4Nm).
6. Connect the engine speed sensor connector.
7. Install the engine under cover.
8. Connect the negative battery cable.

CAMSHAFT POSITION SENSOR

The camshaft position sensor is mounted on the side of the distributor housing.

Removal and Installation

1. Disconnect the negative battery cable.
2. Disconnect the camshaft position sensor electrical connector and remove the distributor.
3. Remove the screw and wire stay from the distributor.
4. Loosen the sensor attaching nut and remove the sensor.

To install:

5. Install the sensor and tighten the nut to 13 ft. lbs. (18 Nm).
6. Install the wire stay and screw.
7. Install the distributor and connect the camshaft position sensor electrical connector.
8. Connect the negative battery cable.
9. Start the engine and check the ignition timing.

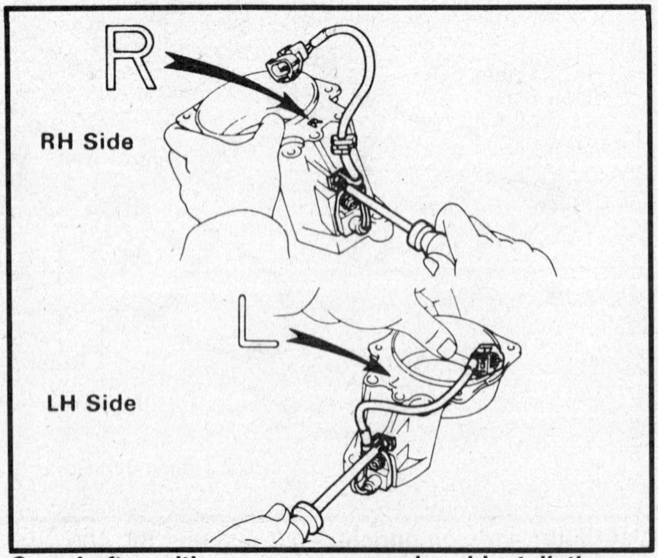

Camshaft position sensor removal and installation

MAZDA 323 AND PROTEGE ELECTRONIC IGINTION SYSTEM

General Information

The ignition system in the 323 is a basic electronic system. It is comprised of the ignition switch, an ignition coil, an electronic distributor driven off off the overhead camshaft, an igniter, a Electronic Control Unit (ECU) and related ignition wires and spark plugs.

Operation of the system is quite simple. When the ignition switch is in **RUN** or **START** position, voltage is applied to the ignition coil. As the distributor shaft turns, the igniter acts as a switch to control current flow through the primary winding of the ignition coil. When current flow through the coil is stopped, a high voltage current is induced in the secondary winding of the ignition coil. The high voltage current flows through the distributor cap and rotor to the proper spark plug.

Diagnosis and Testing

SERVICE PRECAUTIONS

- Make sure all connectors are fastened securely. A poor connection can cause an extremely high surge voltage in the coil and condenser and result in damage to integrated circuits.
- Keep all parts and harnesses dry during service.
- Before attempting to remove any parts, turn **OFF** the ignition switch and disconnect the battery ground cable.
- Always use a 12 volt battery as a power source.
- Do not attempt to disconnect the battery cables with the engine running.
- Do not depress the accelerator pedal when starting.
- Do not rev up the engine immediately after starting or just prior to shutdown.
- Do not attempt to disassemble the ECU under any circumstances.
- If a battery cable is disconnected, the memory will return to the ROM (programmed) values. Engine operation may vary slightly, but this is not an indication of a problem. Do not replace parts because of a slight variation.
- If installing a 2-way or CB radio, keep the antenna as far as possible away from the electronic control unit. Keep the antenna feeder line at least 8 in. away from the ECU harness and do not let the 2 run parallel for a long distance. Be sure to ground the radio to the vehicle body.

IGNITION COIL

Testing

Before checking the coil, run the engine so as to get the coil to normal operating temperature.

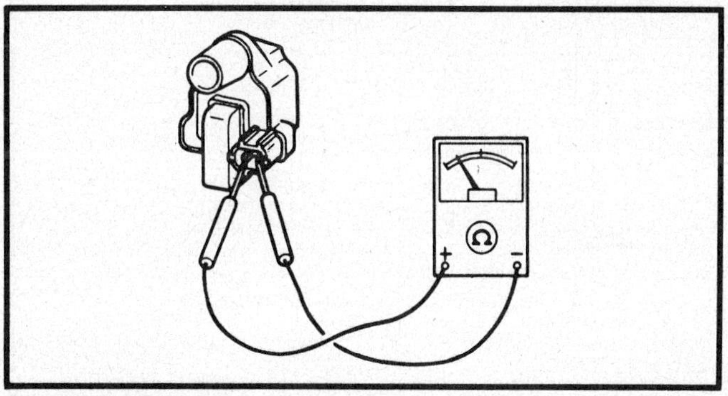

Checking the primary resistance of the ignition coil—323 and Protege

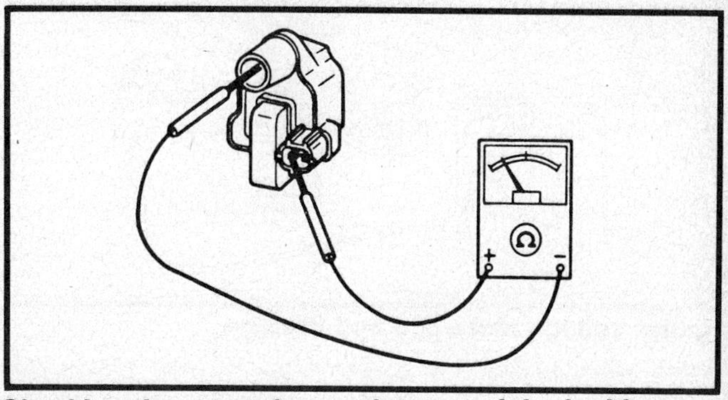

Checking the secondary resistance of the ignition coil—323 and Protege

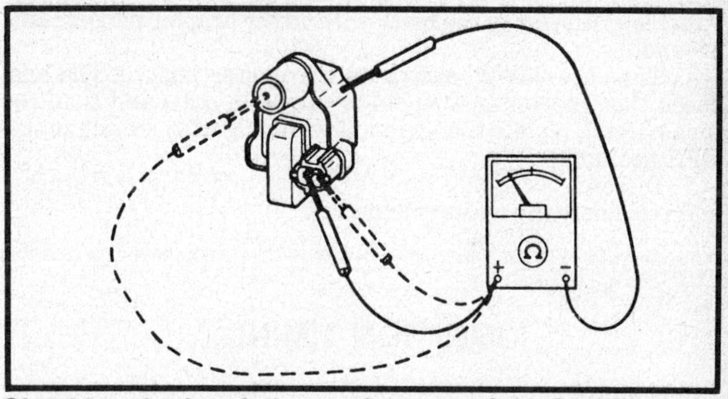

Checking the insulation resistance of the ignition coil—323 and Protege

1. Check the primary resistance by using a suitable ohmmeter. Primary resistance should be 0.81–0.99 ohms.
2. Measure the secondary resistance by using a suitable ohmmeter. Secondary resistance should be 10–16 kilo-ohms.
3. Measure the insulation resistance between the primary terminal and the casing. Be sure to use a 500 volt megger resistance range for this measurement. Insulation resistance should be 10 megaohms or more.
4. If the coil fails any of these tests, replace it.

IGNITER

Testing

1. Disconnect the electrical connector from the igniter. Connect SST 49–N018–011 (Adapter Harness) between the igniter and the harness connector.

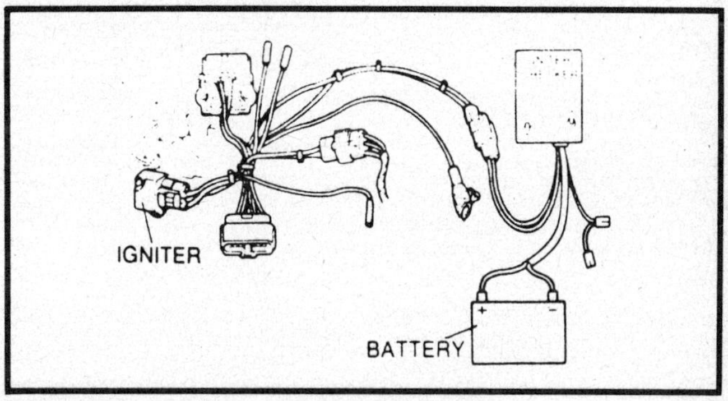

Testing the igniter—323 and Protege

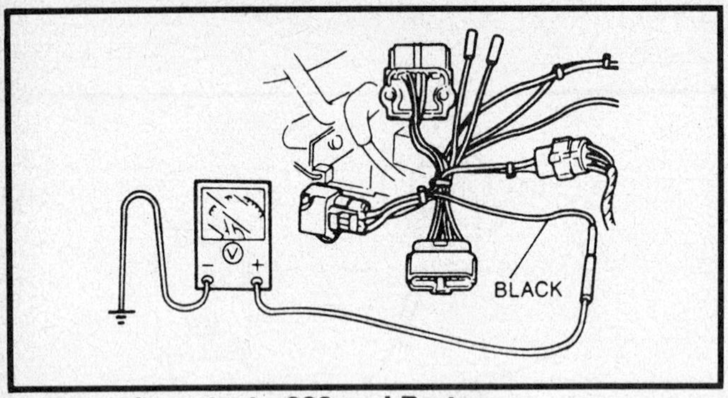

Igniter voltage test—323 and Protege

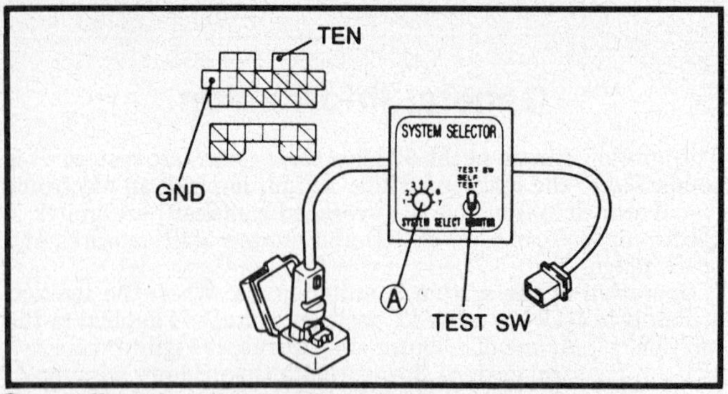

Grounding the diagnostic connector for ignition timing check

2. Connect SST 49–F018–002 (Igniter Checker) to the adapter harness and to the battery.

3. Turn the ignition switch to the **ON** position. Connect a suitable voltmeter to the black wire on the adapter harness and ground.

4. Turn the "SW2" switch on the Igniter Checker **ON** and check that the voltmeter pointer jumps ½ scale and then returns. If not, replace the igniter. Turn the ignition switch to the **OFF** position.

5. Disconnect the SST Igniter Checker and Adapter Harness, and reconnect the igniter connector.

IGNITION TIMING

Adjustment

1. Place the vehicle in **N**, block the drive wheels and apply the emergency brake. Make sure that all the accessories are in the **OFF** position.

2. Start the engine and allow it to reach normal operating temperature. Connect SST 49–B019–9A0 (System Selector) to the diagnosis connector.

3. Set switch (A) on the System Selector to position 1 and the Test Switch to Self-Test.

NOTE: To check ignition timing without using the System Selector, connect a jumper wire between terminal No. 10 and ground on the diagnosis connector.

4. Connect a timing light to the engine according to manufacturer's instructions. Connect a tachometer to the IG terminal on the diagnosis connector. If the timing marks (usually a yellow mark) are difficult to see, use chalk or a dab of paint to make them more visible.

5. Point the timing light at the timing marks. Timing marks are on the crankshaft pulley and the timing belt cover. Ignition timing should be as follows:

a. Vehicles with 1.6L SOHC engine—7±1 degrees BTDC at 700–800 rpm

b. Vehicles with 1.8L SOHC engine—5±1 degrees BTDC at 700–800 rpm

c. Vehicles with 1.8L DOHC engine—10±1 degrees BTDC at 700–800 rpm

6. If specifications are not as specified, loosen the distributor hold-down bolt, then rotate the distributor to advance or retard the timing as necessary.

7. Stop the engine and tighten the hold-down bolt. Start the engine and recheck the timing.

8. Stop the engine and disconnect the SST Tester and/or jumper wire, tachometer and timing light.

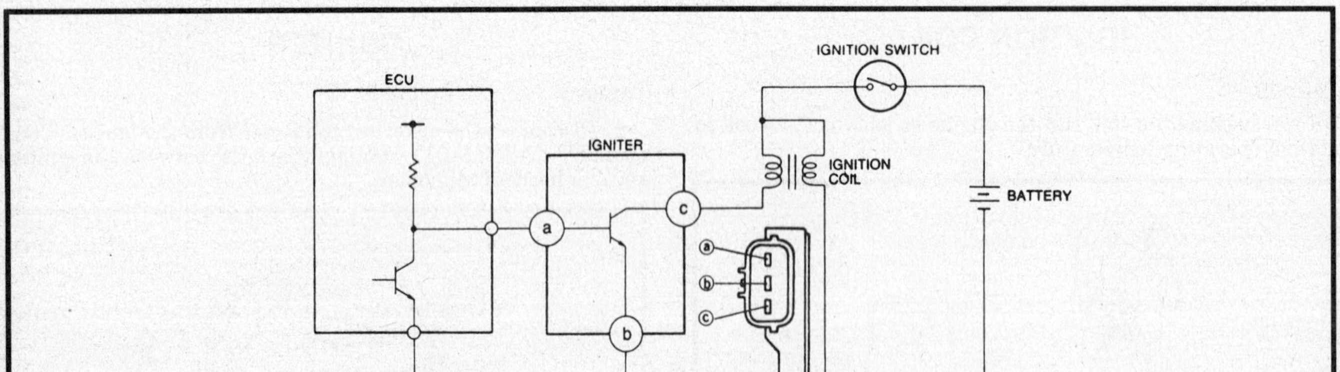

Wiring schematic of the ignition system—323 and Protege

IGNITION SYSTEM TESTS—323 AND PROTEGE

STEP	INSPECTION		ACTION
1	Check if MIL illuminates when engine is running 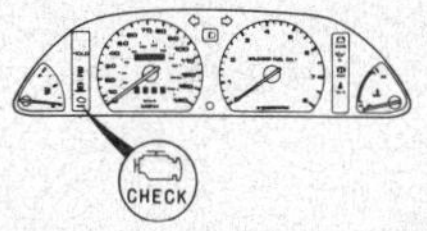	Yes	Check for malfunction code number and perform troubleshooting (TEN terminal grounded)
		No	Go to next step
2	Disconnect high-tension lead from distributor and check for strong blue spark while cranking engine 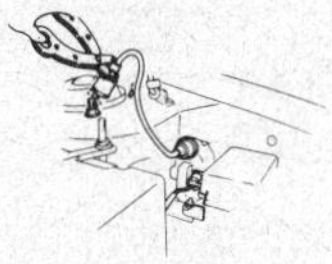	Yes	• Check distributor cap and rotor • Check high-tension lead ⇨ Ignition system normal
		No	Go to next step
3	Check if resistance of ignition coil is OK **Specification (at 20°C [68°F]):** **Primary coil winding: 0.81—0.99Ω** **Secondary coil winding: 10—16 kΩ**	Yes	Go to next step
		No	Replace ignition coil
4	Check if resistance of high-tension leads is OK **Specification: 16 kΩ per 1 m (3.28 ft)**	Yes	Go to next step
		No	Replace high-tension lead
5	Check if igniter is OK 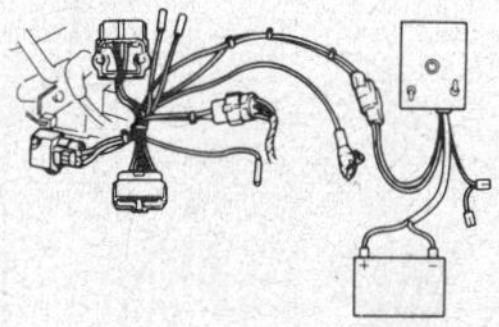	Yes	Replace ECU
		No	Replace igniter

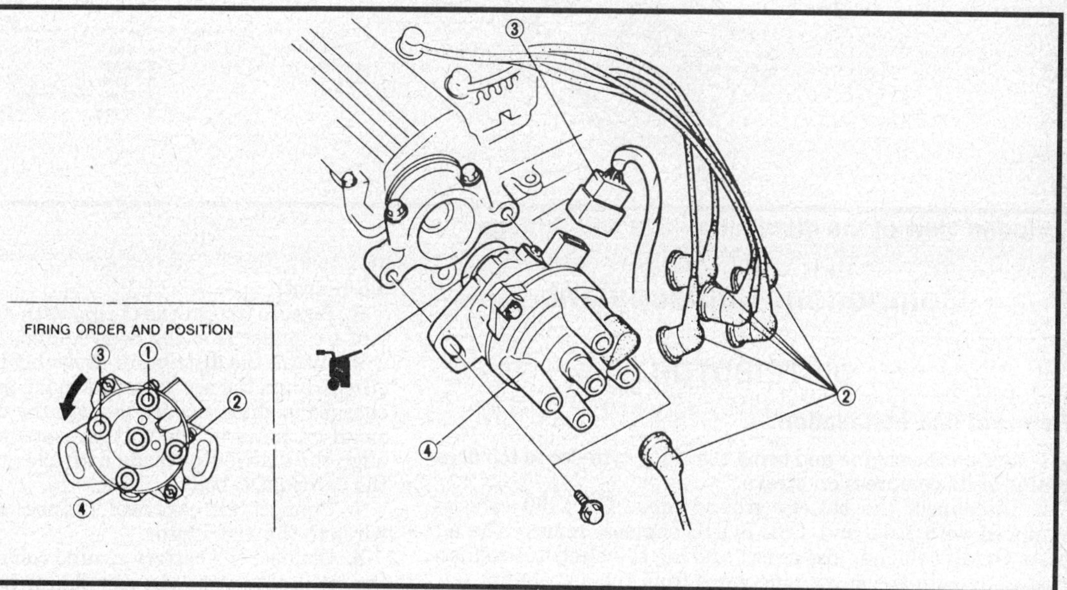

Removal and installation of the distributor—1.6L and 1.8L engines

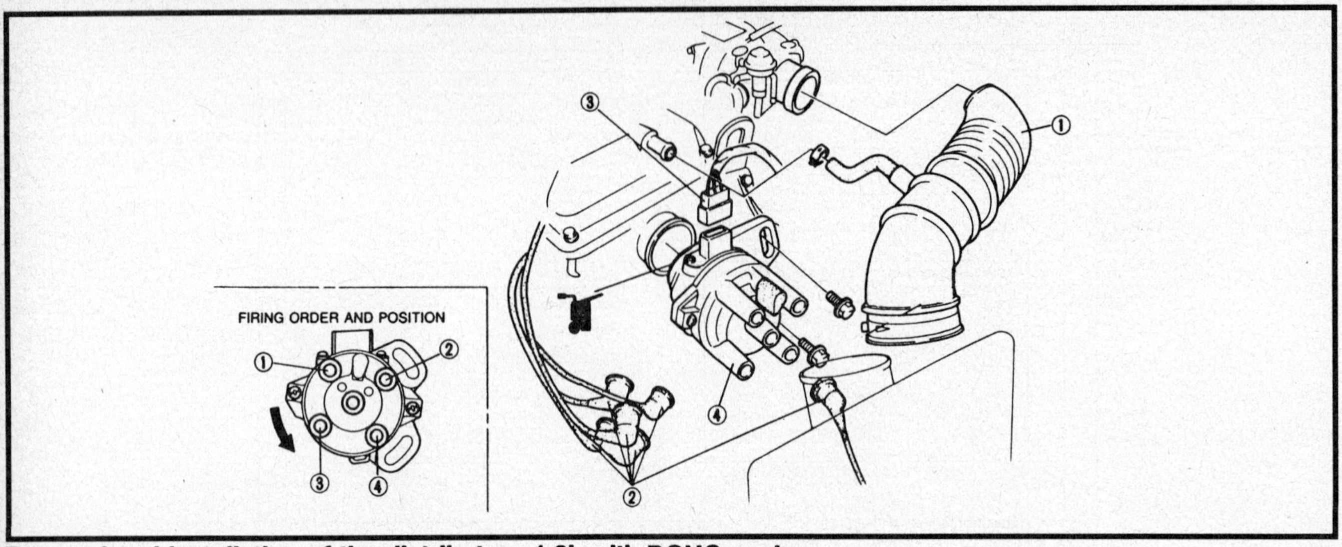

Removal and installation of the distributor—1.8L with DOHC engine

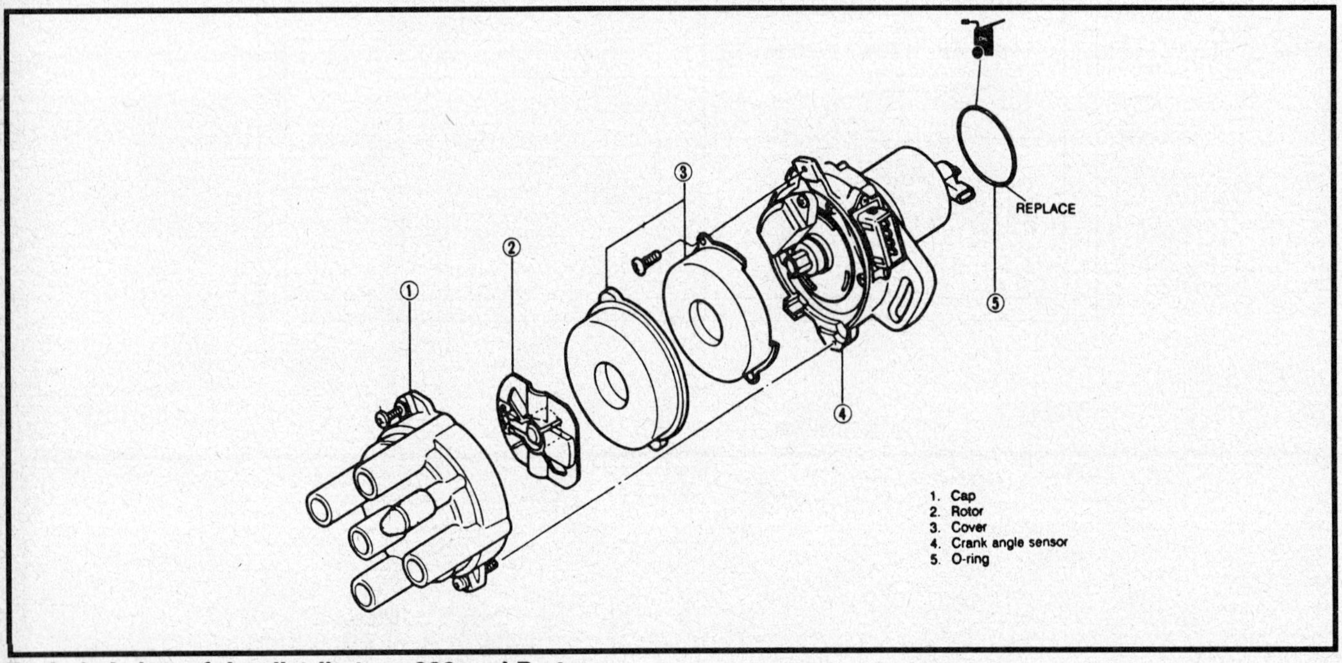

Exploded view of the distributor—323 and Protege

Component Replacement

DISTRIBUTOR

Removal and Installation

1. Rotate the engine and bring the No. 1 cylinder to top dead center of its compression stroke.
2. Disconnect the battery ground cable. On 1990 vehicles equipped with 1.6L and 1.8L SOHC engines, remove the air hose. On all vehicles, disconnect and tag the electrical connectors along with the spark plug wires from the distributor.
3. Mark the rotor position on the distributor body. Mark the distributor body and cap using a scribe or a marker. This can be used for an easier installation procedure.
4. Remove the distributor hold-down bolt, then pull the distributor assembly out of the engine cylinder head.

To install:

5. Be sure to coat the O-ring with engine oil. Ensure that the No. 1 cylinder is at top dead center.
6. Install the distributor assembly into the cylinder block (be sure to align the scribe marks made previously). On non-turbocharged engines, ensure to align the distributor blade with the small oil holes at the bottom portion of the shaft. Ensure to align the distributor blade with the grooved matching mark on the distributor body.
7. Connect all electrical connectors along with the spark wires to the distributor.
8. Connect the battery ground cable. Start the engine and set the timing. Tighten the distributor mounting bolt.

Disassembly

1. Remove the distributor cap and rotor.
2. Remove the 2 dust covers.

Reassembly:

3. Clean and inspect all parts, replace any damaged or worn parts as necessary.

4. Inspect the distributor cap for cracking, burning, corrosion and wear. Replace all O-rings as necessary.

5. Install the 2 dust covers.

6. Install the distributor rotor and cap.

MAZDA 626 AND MX-6 ELECTRONIC IGNITION SYSTEM

General Information

The ignition system is a basic electronic system. It is comprised of the ignition switch, an electronic distributor driven off off the overhead camshaft, an igniter, an ignition coil, an Electronic Control Unit (ECU) and related ignition wires and spark plugs. There are 2 different distributor applications for the non-turbocharged and the turbocharged engines. The distributor for the non-turbocharged engine incorporates a vacuum advance and a pick-up coil with igniter combined. The distributor for the turbocharged engine is an integrated type and does not use a vacuum advance. Ignition timing is controlled by the ECU base on signals from the various sensors and switches.

Operation of the system is quite simple. When the ignition switch is in **RUN** or **START** position, voltage is applied to the ignition coil. As the distributor shaft turns, the igniter acts as a switch to control current flow through the primary winding of the ignition coil. When current flow through the coil is stopped, a high voltage current is induced in the secondary winding of the ignition coil. The high voltage current flows through the distributor cap and rotor to the proper spark plug.

Diagnosis and Testing

SERVICE PRECAUTIONS

- Make sure all connectors are fastened securely. A poor connection can cause an extremely high surge voltage in the coil and condenser and result in damage to integrated circuits.
- Keep all parts and harnesses dry during service.
- Before attempting to remove any parts, turn **OFF** the ignition switch and disconnect the battery ground cable.
- Always use a 12 volt battery as a power source.
- Do not attempt to disconnect the battery cables with the engine running.
- Do not depress the accelerator pedal when starting.
- Do not rev up the engine immediately after starting or just prior to shutdown.
- Do not attempt to disassemble the ECU control unit under any circumstances.
- If a battery cable is disconnected, the memory will return to the ROM (programmed) values. Engine operation may vary slightly, but this is not an indication of a problem. Do not replace parts because of a slight variation.
- If installing a 2-way or CB radio, keep the antenna as far as possible away from the electronic control unit. Keep the antenna feeder line at least 8 in. away from the ECU harness and do not let the 2 run parallel for a long distance. Be sure to ground the radio to the vehicle body.

IGNITION COIL

Testing

Before checking the coil, run the engine as to get the coil to normal operating temperature.

1. Using a suitable ohmmeter, measure the primary resistance of the coil. Primary resistance should be as follows:
   Turbocharged – 0.72–0.88 ohms
   Non-Turbocharged – 0.77–0.95 ohms
2. Using a suitable ohmmeter, measure the secondary resistance of the coil. Secondary resistance should be 10.3–13.9 kilo-ohms.
3. Measure the insulation resistance between the primary terminal and the casing. Be sure to use a 500 volt megger resistance range for this measurement. Insulation resistance should be 10 Megaohms.

SPARK ADVANCE CONTROL

Testing

CENTRIFUGAL

1. Start the engine and allow it to reach normal operating temperature.
2. Connect a timing light and tachometer to the engine. Check that the idle speed and ignition timing are within specification. Remove the tachometer.
3. Disconnect and plug the vacuum hoses from the vacuum controller.
4. While gradually increasing engine rpm, check the advance angle on the crankshaft pulley using the timing light.
   Excess advance – Indicates a weak governor spring
   Insufficient advance – Governor weight or cam defective

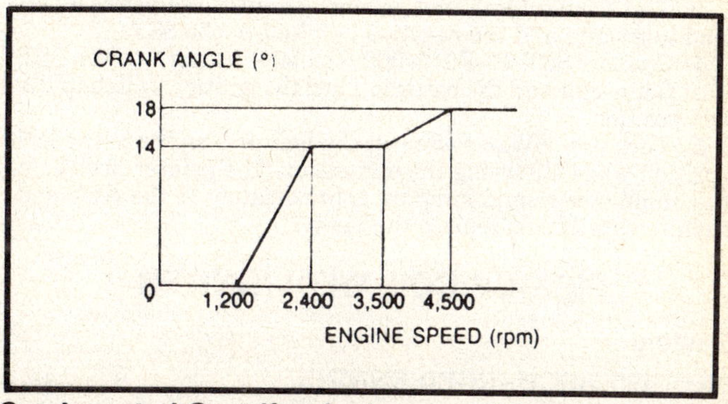

Spark control-Centrifugal advance – 626 and Mx-6

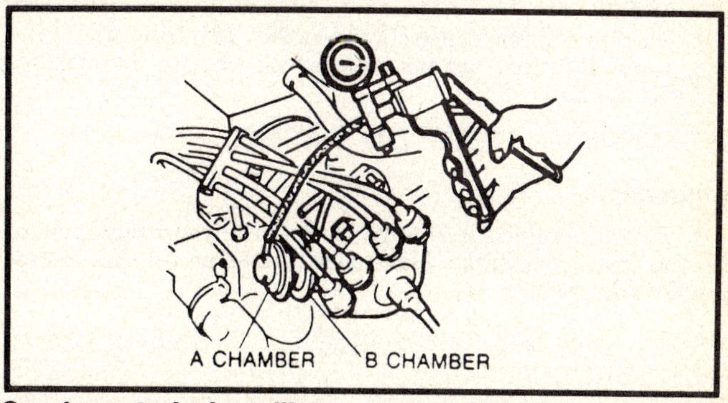

Spark control – Installing vacuum pump for vacuum advance – 626 and MX-6

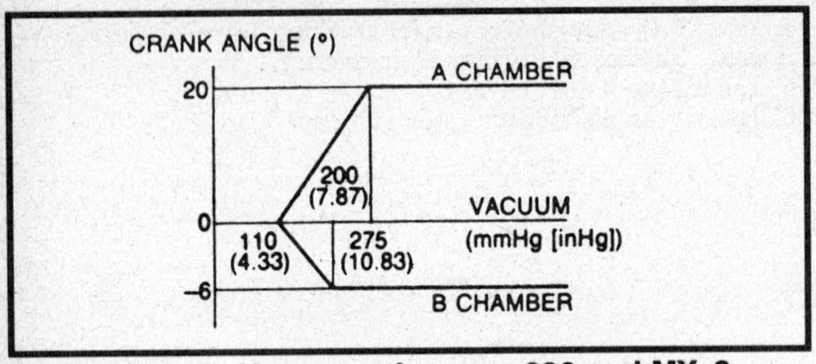

Spark control—Vacuum advance—626 and MX-6

VACUUM

1. Start the engine and allow to reach normal operating temperature.
2. Connect a timing light and tachometer to the engine. Check that the idle speed and ignition timing are within specification. Remove the tachometer.
3. Disconnect and plug the vacuum hoses from the vacuum controller. Run the engine at idle.
4. Connect a suitable hand operated vacuum pump to the vacuum controller (on non-turbocharged engines, chamber A, then B). Using the timing light, check the crank angle advance on the crankshaft pulley while applying vacuum.

IGNITER

Testing

TURBOCHARGED ENGINE

Before checking the igniter, disconnect the electrical connector from the coil.

1. Disconnect the igniter electrical connector.
2. Connect SST 49-H018-910 (Test Connectors), between the igniter and the igniter connector.
3. Connect a suitable ohmmeter between the SST 49-H018-910 (Test Connectors), and ground. Connect the positive (+) lead to ground and the negative (−) lead to the SST.
4. Connect SST 49-F018-002 (Igniter Checker), to the SST Test Connector and the battery. Turn the ignition switch to the **ON** position.
5. Turn the "SW2" switch on the Igniter Checker to the **ON** position while observing the ohmmeter. The pointer needle on the ohmmeter should jump up approximately ½ the scale and then return. If not, replace the igniter.

PICK-UP COIL WITH IGNITER

Testing

NON-TURBOCHARGED ENGINE

1. Using a suitable ohmmeter measure the resistance of the pick-up coil.
2. Pick-up coil resistance should be 900–1200 ohms.
3. If specifications are not as specified, replace the pick-up coil with igniter.

IGNITION TIMING

Adjustment

1. Place the vehicle in **P** or **N**, block the drive wheels and apply the emergency brake. Make sure that all accessories are in the **OFF** position.

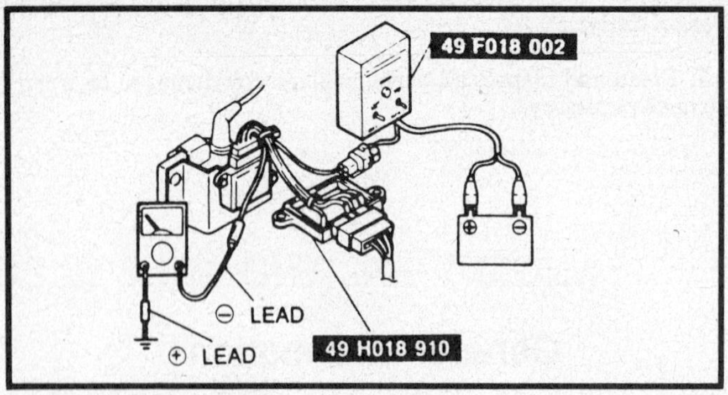

Igniter checker tool

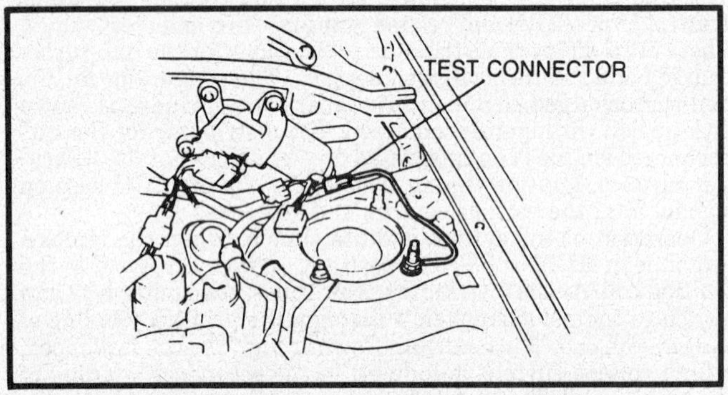

Grounding the test connector—626 and MX-6

2. On turbocharged engines, connect a jumper wire between the test connector and ground.
3. Start the engine and allow it to reach normal operating temperature. Connect a tachometer and check that the idle speed is within specification. If not, adjust the idle speed. Stop the engine.
4. Connect a timing light to the engine according to the manufacturer's instructions. If the timing marks (usually a yellow mark) are difficult to see, use chalk or a dap of paint to make them more visible.
5. On non-turbocharged engines, disconnect and plug the vacuum hoses from the distributor vacuum controller.
6. Start and run the engine at idle with the transmission in **N** on manual vehicles, or **D** on automatic vehicles. Ensure that the parking brake is securely set and the wheels are chocked.
7. Point the timing light at the timing marks. Ignition timing should be as follows:
   Turbocharged engines—9±1 degrees BTDC at 725–775 rpm
   Non-turbocharged engines—6±1 degrees BTDC at 725–775 rpm
8. If specifications are not as specified, loosen the distributor hold-down bolts and rotate the distributor to advance or retard the timing as necessary.
9. Stop the engine and tighten the hold-down bolts. Start the engine and recheck the timing.
10. Stop the engine and disconnect the timing light and tachometer. Reconnect the vacuum hoses (non-turbocharged) and test connector (turbocharged).

IGNITION SYSTEM TESTS—626 AND MX-6—TURBOCHARGED ENGINE

Code No.01 (Ignition pulse)

- Are there poor connections in igniter circuit?
  - YES → Repair or replace connector
  - NO ↓
- Is there battery voltage at igniter terminal WR?
  - NO → PC: Open circuit in wiring harness from main relay to igniter
  - YES ↓
- Is igniter OK?
  - NO → Replace igniter
  - YES ↓
- Is engine control unit 1G terminal OK?
  - NO → PC: • Malfunction of IGt signal or engine control unit • Open circuit in related wiring harness
  - YES ↓
- Is same Code No. present after performing after-repair procedure?
  - NO → Ignition pulse and circuit OK
  - YES ↓
- Is engine control unit 1V terminal OK?
  - NO → PC: Open circuit in wiring harness between engine control unit 1V terminal and igniter
  - YES → PC: Engine control unit malfunction

Code No.02 (Ne signal)

- Are there poor connections in distributor circuit?
  - YES → Repair or replace connector
  - NO ↓
- Is resistance between A and B terminal OK?
  - NO → Relace distributor
  - YES ↓
- Is there continuity between distributor and engine control unit?

Distributor	Control unit
A	3E
B	3F

  - NO → PC: Open circuit in wiring harness from distributor to engine control unit
  - YES ↓
- Is same Code No. present after performing after-repair procedure?
  - NO → Ne signal and circuit OK
  - YES ↓
- Are control unit 3E and 3F terminal voltages OK?
  - NO → PC: Open or short circuit in wiring harness
  - YES → PC: Engine control unit malfunction

Code No.03 (G1 signal)

- Are there poor connections in distributor circuit?
  - YES → Repair or replace connector
  - NO ↓
- Is resistance between B and D terminal OK?
  - NO → Replace distributor
  - YES ↓
- Is there continuity between distributor and engine control unit?

Distributor	Control unit
B	3F
D	3G

  - NO → PC: Open circuit in wiring harness from distributor to engine control unit
  - YES ↓
- Is same Code No. present after performing after-repair procedure?
  - NO → G1 signal and circuit OK
  - YES ↓
- Are engine control unit 3F and 3G terminal voltages OK?
  - NO → PC: Short circuit in wiring harness
  - YES → PC: Engine control unit malfunction

Code No.04 (G2 signal)

- Are there poor connections in distributor circuit?
  - YES → Repair or replace connector
  - NO ↓
- Is resistance between B and C terminal OK?
  - NO → Replace distributor
  - YES ↓
- Is there continuity between distributor and engine control unit?

Distributor	Control unit
B	3F
C	3H

  - NO → PC: Open circuit in wiring harness from distributor to engine control unit
  - YES ↓
- Is same Code No. present after performing after-repair procedure?
  - NO → G2 signal and circuit OK
  - YES ↓
- Are engine control unit 3F and 3H terminal voltages OK?
  - NO → PC: Short circuit in wiring harness
  - YES → PC: Engine control unit malfunction

IGNITION SYSTEM TESTS—626 AND MX-6 NON-TURBOCHARGED ENGINE

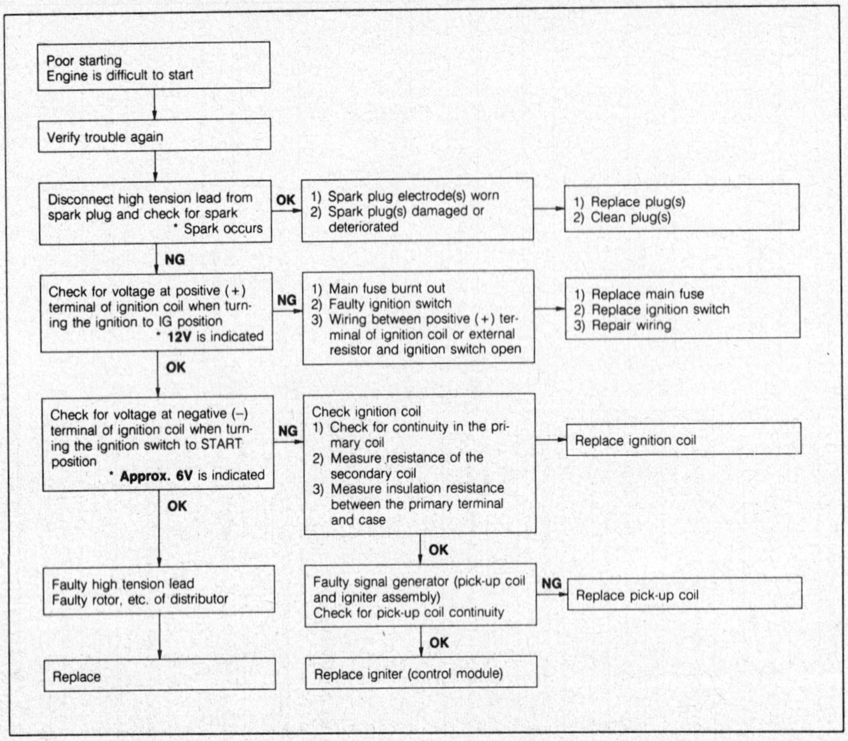

Component Replacement

DISTRIBUTOR

Removal and Installation

1. Rotate the engine and bring the No. 1 cylinder to top dead center of its compression stroke.
2. Disconnect the battery ground cable. Disconnect and tag (if necessary) the electrical connectors along with the spark plug wires from the distributor.
3. Disconnect the vacuum hoses (non-turbocharged engines) and remove the distributor cap.
4. Mark the rotor position on the distributor body. Mark the distributor body and cap using a scribe or marker. This is done for an easier installation procedure.
5. Remove the distributor hold-down bolt(s) and pull the distributor assembly out of the cylinder head.

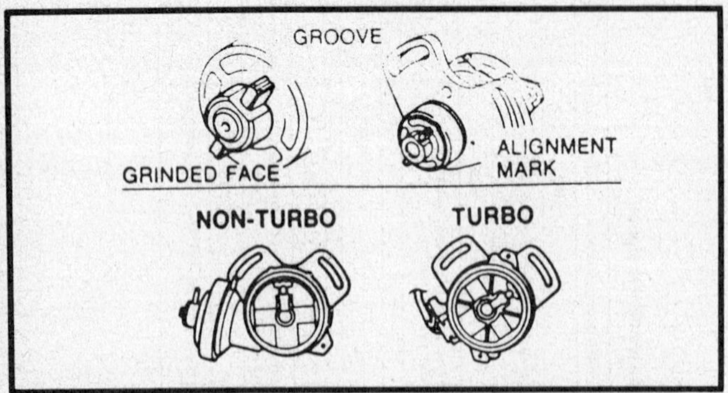

Aligning the distributor blade with alignment mark—626 and MX-6

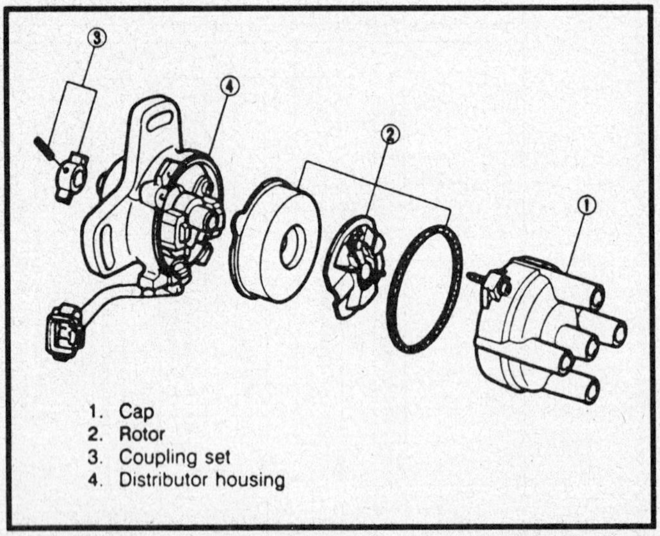

Exploded view of the distributor—626 and MX-6, with turbocharger

To install:

6. Be sure to coat the O-ring with oil. Install the distributor assembly into the cylinder head, be sure to align scribe marks made previously. Lightly tighten the distributor hold-down bolt(s).
7. Connect all electrical connectors and vacuum hoses (non-turbocharged).
8. Install the distributor cap with spark plug wires.
9. Connect the battery ground cable. Start the engine and set the timing. Tighten the distributor hold-down bolt(s).

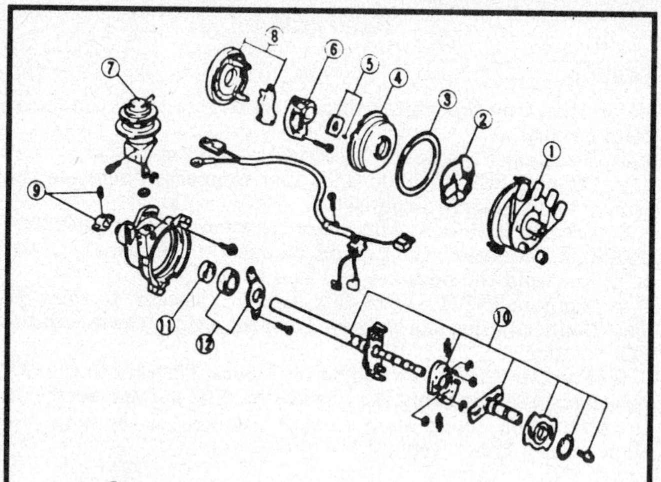

Exploded view of the distributor—626 and MX-6, without turbocharger—626 and MX-6

Disassembly

NON-TURBOCHARGED ENGINE

1. Remove the cap, rotor, gasket and dust cover.
2. Place the distributor assembly into a suitable holding fixture.
3. Using a suitable drift, remove the signal rotor pin. Using 2 suitable small pry bars or equivalent, remove the signal rotor.
4. Remove the screw(s) retaining the pick-up coil and remove the pick-up coil with igniter from the distributor.
5. Remove the screw(s) attaching the vacuum controller, then remove the vacuum controller.
6. Remove the breaker plate assembly. Using a suitable drift, remove the coupling pin, then the coupling.

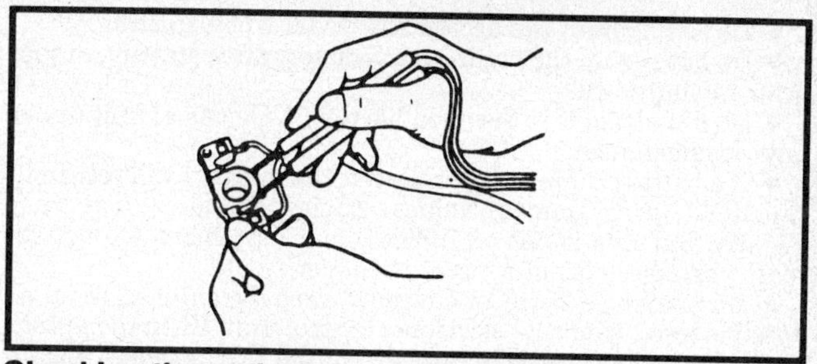

Checking the resistance of the pick-up coil with the igniter—626 and MX-6

7. Remove the governor set and shaft, then the oil seal and bearing.

TURBOCHARGED ENGINE

1. Remove the cap, rotor, gasket and dust cover.
2. Place the distributor assembly into a suitable holding fixture.
3. Using a suitable drift, remove coupling pin, then the coupling from the distributor.

NOTE: Do not disassemble the distributor set (housing).

4. Check the pick-up coil and igniter assembly resistance. The resistance should be 900–1200 ohms.
5. If specifications are not as specified, replace the pick-up coil with igniter.

Reassembly

1. Clean and inspect all parts. Replace any damaged or worn parts as necessary.
2. Inspect the distributor cap for cracking, burning, corrosion and wear. Replace O-rings as necessary.
3. To reassemble, follow the disassembly procedure in the reverse order.
4. Align the coupling set blade with the alignment mark on the distributor, then check that the rotor is aligned.

MAZDA 929 ELECTRONIC IGNITION SYSTEM

General Information

The ignition system a basic electronic system. It is comprised of the ignition switch, an electronic distributor driven off off the overhead camshaft, an igniter, an ignition coil, a Electronic Control Unit (ECU), ignition wires and spark plugs.

Operation of the system is quite simple. When the ignition switch is in **RUN** or **START** position, voltage is applied to the ignition coil. As the distributor shaft turns, the igniter acts as a switch to control current flow through the primary winding of the ignition coil. When current flow through the coil is stopped, a high voltage current is induced in the secondary winding of the ignition coil. The high voltage current flows through the distributor cap and rotor to the proper spark plug.

Diagnosis and Testing

SERVICE PRECAUTIONS

- Make sure all connectors are fastened securely. A poor connection can cause an extremely high surge voltage in the coil and condenser and result in damage to integrated circuits.
- Keep all parts and harnesses dry during service.
- Before attempting to remove any parts, turn **OFF** the ignition switch and disconnect the battery ground cable.
- Always use a 12 volt battery as a power source.
- Do not attempt to disconnect the battery cables with the engine running.

- Do not depress the accelerator pedal when starting.
- Do not rev up the engine immediately after starting or just prior to shutdown.
- Do not attempt to disassemble the ECU control unit under any circumstances.
- If a battery cable is disconnected, the memory will return to the ROM (programmed) values. Engine operation may vary slightly, but this is not an indication of a problem. Do not replace parts because of a slight variation.
- If installing a 2-way or CB radio, keep the antenna as far as possible away from the electronic control unit. Keep the antenna feeder line at least 8 in. away from the ECU harness and do not let the 2 run parallel for a long distance. Be sure to ground the radio to the vehicle body.

IGNITION COIL

Testing

Before checking the coil, run the engine so as to get the coil to normal operating temperature.

1. Check the primary resistance using a suitable ohmmeter. Primary resistance should be 0.72–0.88 ohms for 1990–91.
2. On 1992 engines, measure primary resistance at 2 points:
   Primary side 1 – should be 0.7–1.0 ohms
   Primary side 2 – should be 0.9–1.1 ohms
2. Measure the secondary resistance using a suitable ohmmeter. Secondary resistance should be 10–30 kilo-ohms on 1990–91 and 9–17 kilo-ohms for 1992 engines.
3. Measure the insulation resistance between the primary terminal and the casing. Be sure to use a 500 volt megger resistance range for this measurement. Insulation resistance should be 10 Megaohms or more.
4. If the coil fails any of these test, replace it.

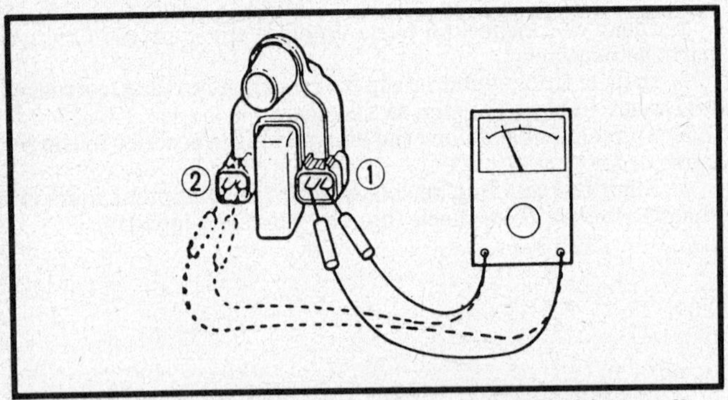

Checking the primary resistance of the ignition coil – 1992 929

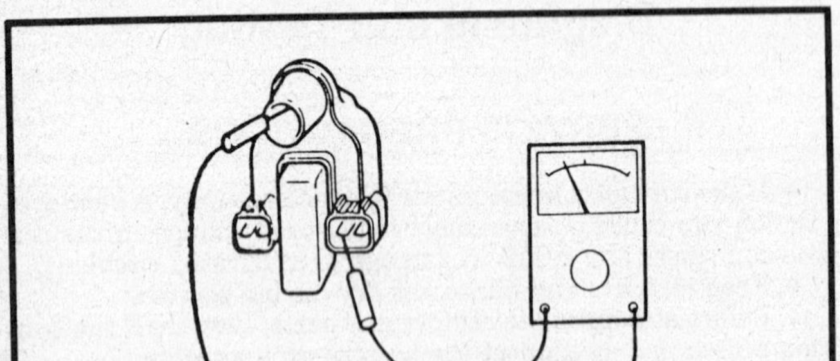

Checking the secondary resistance of the ignition coil – 1992 929

IGNITER

Testing

Before checking the igniter, disconnect the electrical connector from the coil.

1. Disconnect the igniter electrical connector.
2. Connect SST 49–9200–443, Test Connector, between the igniter and the igniter connector.
3. Connect a suitable ohmmeter between the SST 49–9200–443, Test Connector, and ground. Connect the positive (+) lead to ground and the negative (−) lead to the SST.
4. Connect SST 49–F018–002, Igniter Checker, to the SST Test Connector and the battery. Turn the ignition switch to the **ON** position.
5. Turn the "SW2" switch on the Igniter Checker to the **ON** position while observing the ohmmeter. The pointer needle on the ohmmeter should jump up approximately ½ the scale and then return. If not, replace the igniter.

IGNITION TIMING

Adjustment

1. Place the vehicle in **P** or **N**, block the drive wheels and apply the emergency brake. Make sure that all accessories are in the OFF position. On 1990–91 vehicles, connect a jumper wire to test connector (green: 1-pin) and ground.
2. On 1992 vehicles, connect a jumper wire to terminals 10 and ground of the diagnosis connector.
3. Start the engine and it to reach normal operating temperature. Connect a tachometer and check that the idle speed is within specification. Stop the engine.
4. Connect a timing light to the engine according to manufacturer's instructions. If timing marks (usually a yellow mark) are difficult to see, use chalk or a dap of paint to make them more visible.
5. Start and run the engine at idle with the transmission in **N** on manual transmissions, or **P** on automatic transmissions.
6. Point the timing timing light at the timing marks. With the engine at idle, ignition timing should be (15±1) degrees BTDC at 650 rpm for 1990–91 and 11–13 degrees BTDC at 680–720 rpm for 1992 vehicles.
6. If specifications are not as specified, loosen the distributor hold-down bolt and rotate the distributor to advance or retard the timing as necessary.
7. Stop the engine and tighten the hold-down bolt. Start the engine and recheck the timing. Disconnect the timing light, tachometer and the jumper wire.

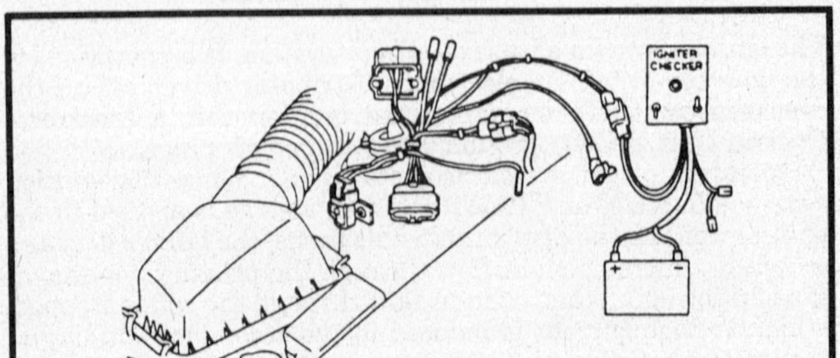

Testing the igniter – 929

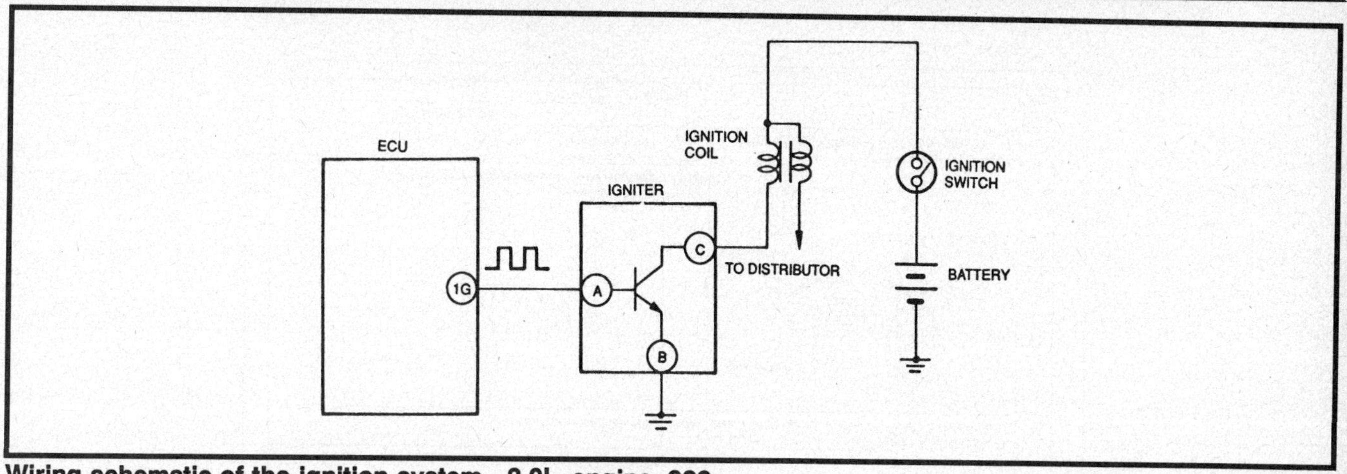

Wiring schematic of the ignition system—3.0L engine, 929

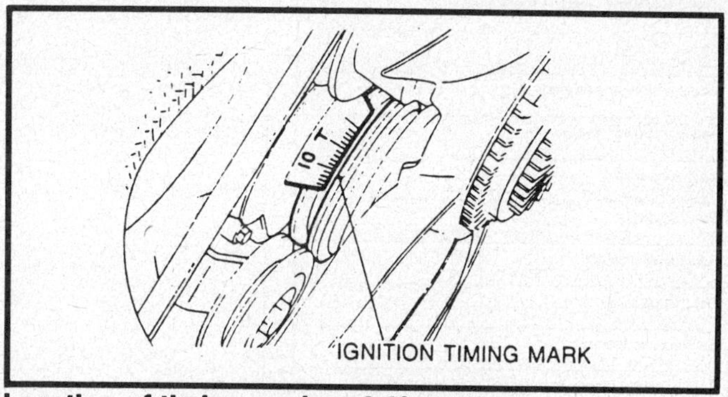

Location of timing marks—3.0L engine

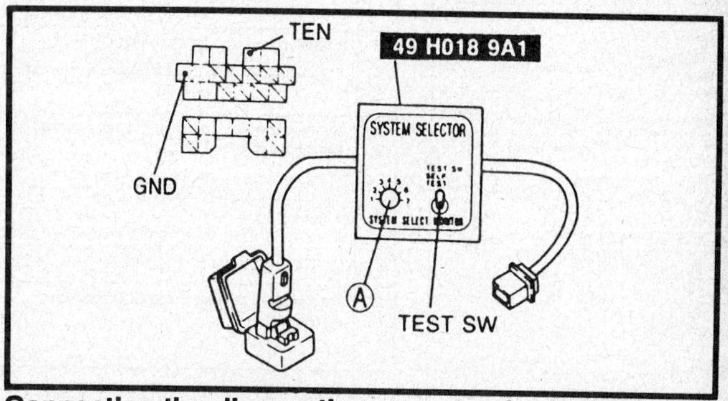

Connecting the diagnostic connector for ignition timing check—1992 929

Component Replacement

DISTRIBUTOR

Removal and Installation

1. Rotate the engine and bring the No. 1 cylinder to top dead center of its compression stroke.
2. Disconnect the battery ground cable. Disconnect and tag (if necessary) the electrical connector along with the spark plug wires for the distributor.
3. Remove the distributor cap. Mark the rotor position on the distributor body. Mark the distributor body and the cap using a scribe or a marker. This can be used for an easier installation procedure.
4. Remove the distributor hold-down bolt, then remove the distributor assembly out from the engine.

To install:

5. Be sure to coat the O-ring and drive gear with engine oil. Check that the No. 1 cylinder is at top dead center.
6. Align the matching marks on the distributor housing and the drive gear.

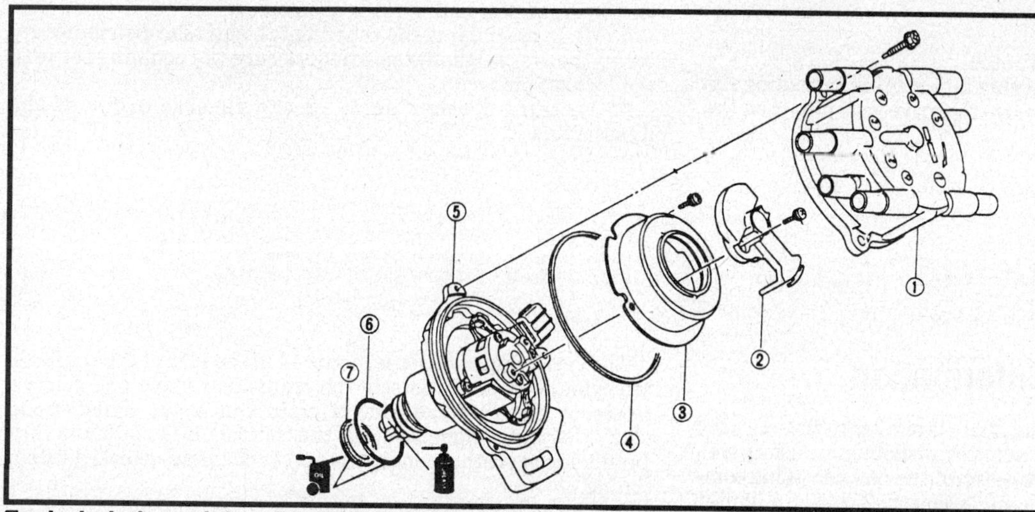

1. Cap
2. Rotor
3. Cover
4. Cover retaining ring
5. Distributor assembly
6. O-ring seal
7. Shim

Exploded view of the distributor—3.0L engine, 929

IGNITION SYSTEM TESTS—929

06	Misfire		
STEP	INSPECTION		ACTION
1	Is "00" displayed on Self-Diagnosis Checker with ignition switch ON? 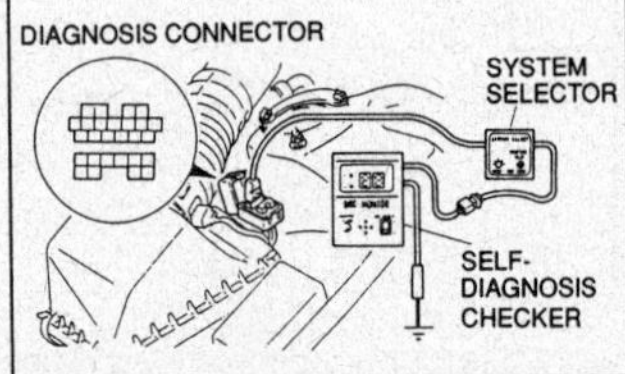	Yes	Go to next step
		No	**Service code number displayed** Check for cause
2	Remove spark plug; is there a strong blue spark while engine is cranking? 	Yes	Ignition system normal
		No	Check spark plug If OK, go to next step If not OK, clean or replace spark plug
3	Is resistance of high-tension leads OK? **Specification: 9.6—22.4 kΩ per 1m (3.28 ft)**	Yes	Go to next step
		No	Replace high-tension leads
4	Are there corrosion, damage or cracks on the distributor cap and rotor?	Yes	Replace cap or rotor
		No	Go to next step
5	Is ignition coil normal?	Yes	Go to next step
		No	Replace ignition coil
6	Are distributor connector and ECU connector connected?	Yes	Got to next step
		No	Connect connectors
7	Is igniter OK?	Yes	Check wiring harness from igniter to ECU terminal 1G If OK, replace ECU If not OK, repair wiring harness
		No	Replace igniter

7. Install the distributor assembly into the cylinder block, be sure to align the scribe marks made previously.
8. Connect the electrical connector along with the spark plug wires to the distributor.
9. Connect the battery ground cable. Start the engine and set the timing. Tighten the distributor hold-down bolt.

Disassembly

1. Remove the cap, seal and rotor.
2. Place the distributor assembly into a suitable holding fixture. Using a suitable drift, remove the drive gear pin, then the drive gear (coupling set).

NOTE: Do not disassemble the distributor set.

Reassembly:

3. Clean and inspect all parts, replace any damaged or worn parts as necessary.
4. Inspect the distributor cap for cracking, burning, corrosion and wear. Replace all O-rings as necessary.
5. When assembling the coupling set, align the distributor rotor to the No. 1 cylinder position. Secure the coupling set with the locking pin.
6. Install all other parts in the reverse order of the disassembly.

MAZDA MIATA ELECTRONIC IGNITION SYSTEM

General Information

The ignition system used on the Miata is similar to the one used on the RX-7. This system does not use a distributor, pick-up coil, reluctor or a ignition module to control the ignition. This computerized system uses a crank angle sensor, 2 coils, an igniter and an electronic control unit.

The crank angle sensor is mounted in the rear of the engine in the cylinder head. It detects the crankshaft angle and sends a reference signal to the control unit. This input signal, along with other input signals, allows the control unit to monitor various engine operating conditions and to compute vehicle ignition timing.

IGNITION SYSTEM TESTS—MIATA

Troubleshooting
If a malfunction code number is shown on the **SST**, check for the cause by using the chart related to the code number shown.

CODE No.	01 (IGf-SIGNAL)		
STEP	INSPECTION		ACTION
1	Are there any poor connections at ignition coil connectors and igniter connectors?	Yes	Repair or replace connector
		No	Go to next step
2	Does tachometer operate?	Yes	Go to next step
		No	Check for open circuit in wiring from igniter to ECU terminal 2I
3	Is resistance of ignition coil OK? Resistance: Primary 0.78—0.94Ω Secondary 11.2—15.2 kΩ	Yes	Go to next step
		No	Replace ignition coil
4	Is there continuity between ignition coil and igniter? (Ignition coil / Igniter: A (W) / A (W); B (Y) / H (Y))	Yes	Go to next step
		No	Check for open circuit in wiring from ignition coil to igniter
5	Is ignition coil terminal wire (L) voltage OK?	Yes	Go to next step
		No	Check for open circuit in wiring from ignition coil to ignition switch
6	Is igniter terminal-wire (L) voltage OK?	Yes	Go to next step
		No	Check for open circuit in wiring from igniter to ignition switch
7	Is there continuity between igniter and ground?	Yes	Go to next step
		No	Check for open circuit in wiring from igniter to ground
8	Check igniter	Yes	Replace ECU
		No	Replace igniter

Circuit Diagram

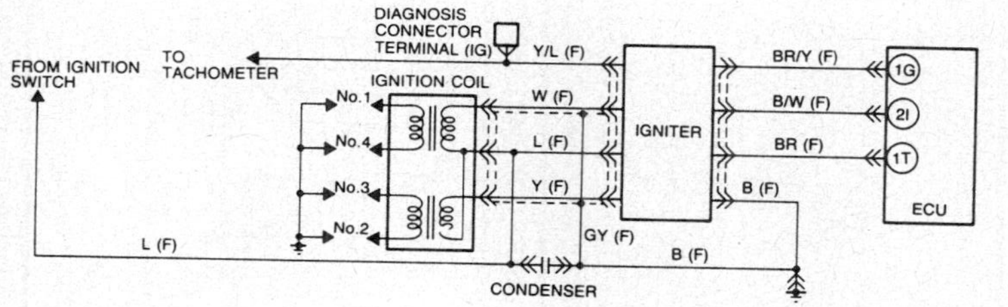

CODE No.	02 (Ne-SIGNAL)		
STEP	INSPECTION		ACTION
1	Are there any poor connections in crank angle sensor circuit?	Yes	Repair or replace connector
		No	Go to next step
2	Is Code No.03 present at same time?	Yes	Go to next step
		No	Go to Step 5
3	Is there continuity between crank angle sensor terminal-wire (B/LG) and ground?	Yes	Go to next step
		No	Check for open circuit in wiring from crank angle sensor to ground
4	Is there battery voltage at crank angle sensor terminal-wire (W/R)?	Yes	Go to next step
		No	Check for open circuit in wiring from crank angle sensor to main relay
5	Is there continuity between crank angle sensor and ECU? (Crank angle sensor / ECU: C (W) / 2E)	Yes	Go to next step
		No	Check for open circuit in wiring from crank angle sensor to ECU
6	Is there approx. 5V at ECU terminal 2E? (With crank angle sensor connector disconnected)	Yes	Go to next step
		No	Replace ECU
7	Is there approx. 5V at crank angle sensor terminal-wire (W)? (At harness-side connector with connector disconnected)	Yes	Replace crank angle sensor
		No	Check for short circuit in wiring from crank angle sensor to ECU

CODE No.	03 (G-SIGNAL)		
STEP	INSPECTION		ACTION
1	Are there any poor connections in crank angle sensor circuit?	Yes	Repair or replace connector
		No	Go to next step
2	Is Code No.02 also present?	Yes	Go to next step
		No	Go to Step 5
3	Is there continuity between crank angle sensor terminal-wire (B/LG) and ground?	Yes	Go to next step
		No	Check for open circuit in wiring from crank angle sensor to ground
4	Is there battery voltage at crank angle sensor terminal-wire (W/R)?	Yes	Go to next step
		No	Check for open circuit in wiring from crank angle sensor to main relay
5	Is there continuity between crank angle sensor and ECU? (Crank angle sensor / ECU: D (Y/L) / 2G)	Yes	Go to next step
		No	Check for open circuit in wiring from crank angle sensor to ECU
6	Is there approx. 5V at ECU terminal 2E? (With crank angle sensor connector disconnected)	Yes	Go to next step
		No	Replace ECU
7	Is there approx. 5V at crank angle sensor terminal-wire (Y/L)? (At harness-side connector with connector disconnected)	Yes	Replace crank angle sensor
		No	Check for short circuit in wiring from crank angle sensor to ECU

Diagnosis and Testing

SERVICE PRECAUTIONS

- Make sure all connectors are fastened securely. A poor connection can cause an extremely high surge voltage in the coil and condenser and result in damage to integrated circuits.
- Keep all parts and harnesses dry during service.
- Before attempting to remove any parts, turn **OFF** the ignition switch and disconnect the battery ground cable.
- Always use a 12 volt battery as a power source.
- Do not attempt to disconnect the battery cables with the engine running.
- Do not depress the accelerator pedal when starting.
- Do not rev up the engine immediately after starting or just prior to shutdown.
- Do not attempt to disassemble the ECU control unit under any circumstances.
- If a battery cable is disconnected, the memory will return to the ROM (programmed) values. Engine operation may vary slightly, but this is not an indication of a problem. Do not replace parts because of a slight variation.
- If installing a 2-way or CB radio, keep the antenna as far as possible away from the electronic control unit. Keep the antenna feeder line at least 8 in. away from the ECU harness and do not let the 2 run parallel for a long distance. Be sure to ground the radio to the vehicle body.

IGNITION COIL

Testing

Before checking the coils, run the engine as to get the the coils to normal operating temperature.

1. Disconnect the ignition coil electrical connector. Turn the ignition switch to the **ON** position.
2. Using a suitable voltmeter, check for voltage at the positive (+) terminal of the ignition control connector. Voltage should be approximately 12 volts. If no voltage exists, check the main fuse, ignition switch and harness.

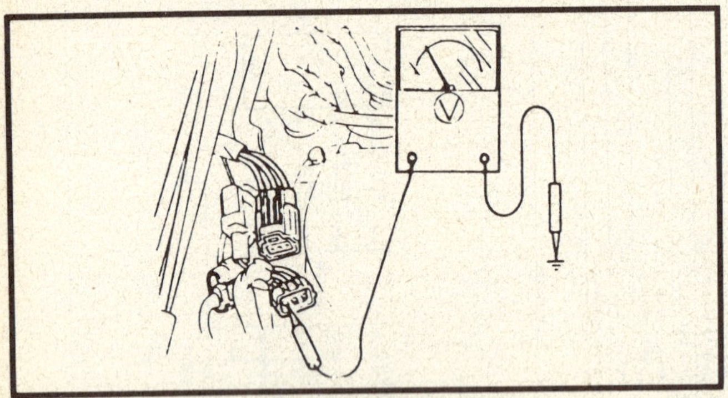

Checking ignition coil connector

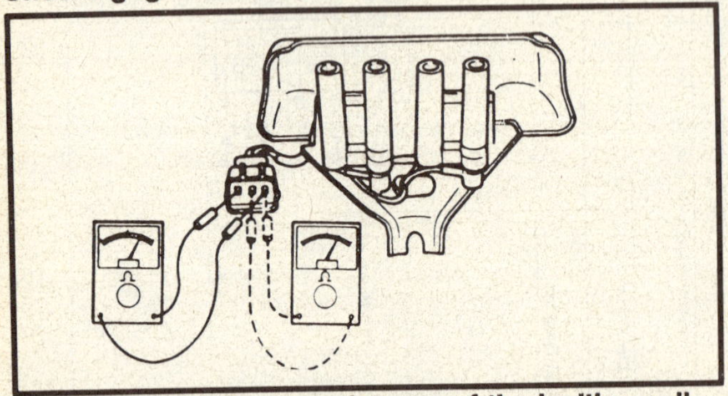

Checking the primary resistance of the ignition coil—Miata

3. Using a suitable ohmmeter, measure the primary resistance of the coils. Primary resistance should be 0.78–0.94 ohms.
4. Using a suitable ohmmeter, measure the secondary resistance of the coils. Secondary resistance should be 11.2–15.2 kilo-ohms.
5. Measure the insulation resistance between the primary terminal and the case. Be sure to use a 500 Volt megger resistance range for this measurement. Insulation resistance should be 10 Megaohms or more.
6. If the coil fails any of these tests, replace it.

IGNITER

Testing

1. Disconnect the igniter electrical connector.
2. Connect SST 49–N018–001 (Adapter Harness) between the igniter and the igniter connector.
3. Connect SST 49–F018–002 (Igniter Checker) to the power connector (blue: 1 pin), to ground and to the adapter harness.
4. Connect voltmeter between the SST adapter harness (brown wire) and ground.
5. Turn the ignition switch to the **ON** position. Check that 12 volts exists at the terminal.
6. Push the "SW2" switch, on the SST igniter checker, while observing the voltmeter. Voltmeter should read approximately 12 volts. If not, replace the igniter.
7. Disconnect the voltmeter from the SST adapter harness (brown wire) and connect it to the SST adapter harness (gray wire) and ground. Repeat Steps 5 and 6.

IGNITION TIMING

Adjustment

1. Place the vehicle in **N**, block the drive wheels and apply the emergency brake. Make sure that all the accessories are in the **OFF** position.
2. Start the engine and allow it to reach normal operating temperature. Connect a timing light and tachometer to the diagnosis connector terminal (IG−).

NOTE: When using a externally powered timing light and/or tachometer, connect it to the power connector (blue: 1 − pin).

3. Connect SST System Selector 49–B019–9A0 to the diagnosis connector and set the "TEST SW" to the "SELF TEST" position.

NOTE: If the System Selector is not available, connect a jumper wire between terminals No. 10 and ground on the diagnosis connector. Take care when making connections to the diagnosis connector, as a mistaken connection will cause a malfunction or damage to the ECU.

4. Check that the idle speed is within specification. If not, adjust the idle speed to specifications.
5. Point the timing light at the timing marks, on the crankshaft pulley and timing belt cover. If the timing marks (usually a yellow mark) are difficult to see, use chalk or a dab of paint to make them more visible.
6. With the engine at idle, ignition timing should be (10 ± 1) degrees BTDC at 800–900 rpm.
7. If ignition timing is not as specified, loosen the crank angle sensor hold-down bolt, and rotate the crank angle sensor to advance or retard the timing as necessary.
8. Disconnect the SST System Selector and Adapter Harness and reconnect the electrical connectors. Disconnect the timing light and tachometer.

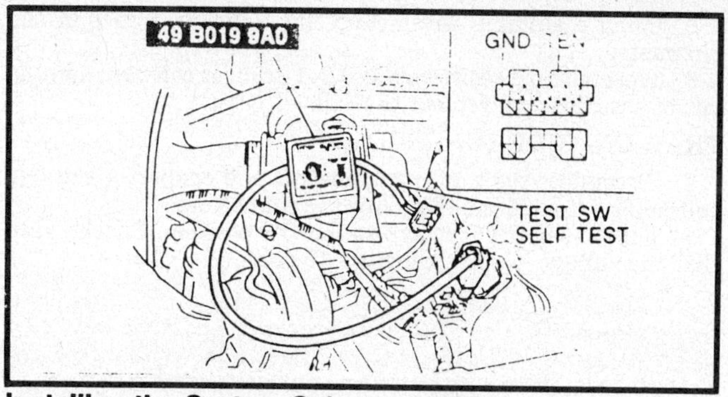

Installing the System Selector—Miata

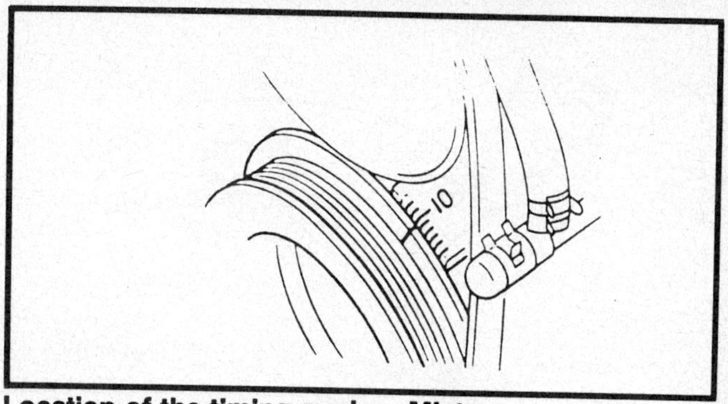

Location of the timing marks—Miata

MAZDA RX-7 ELECTRONIC IGNITION SYSTEM

General Information

The RX-7 uses a distributorless ignition system. This system does not use a distributor, pick-up coil, reluctor or ignition module to control the ignition. This computerized system uses a crank angle sensor, 2 coil and igniter assemblies and an electronic control unit.

The crank angle sensor is mounted in the top of the front crankshaft housing, extending down toward the eccentric shaft. The sensor detects the angle of the eccentric shaft and sends a reference signal to the control unit. This input signal, along with other input signals, allows the control unit to monitor various engine operating conditions and to compute vehicle ignition timing. The control unit varies the ignition timing by controlling the interruption of the current flow in the primary windings of the coil/igniter units.

Diagnosis and Testing

SERVICE PRECAUTIONS

- Make sure all connectors are fastened securely. A poor connection can cause an extremely high surge voltage in the coil and condenser and result in damage to integrated circuits.
- Keep all parts and harnesses dry during service.
- Before attempting to remove any parts, turn **OFF** the ignition switch and disconnect the battery ground cable.
- Always use a 12 volt battery as a power source.
- Do not attempt to disconnect the battery cables with the engine running.
- Do not depress the accelerator pedal when starting.
- Do not rev up the engine immediately after starting or just prior to shutdown.
- Do not attempt to disassemble the ECU control unit under any circumstances.
- If a battery cable is disconnected, the memory will return to the ROM (programmed) values. Engine operation may vary slightly, but this is not an indication of a problem. Do not replace parts because of a slight variation.
- If installing a 2-way or CB radio, keep the antenna as far as possible away from the electronic control unit. Keep the antenna feeder line at least 8 in. away from the ECU harness and do not let the 2 run parallel for a long distance. Be sure to ground the radio to the vehicle body.

IGNITION COIL WITH IGNITER

Testing

Before making this inspection, check to ensure that the battery is at or near a full charge.

LEADING SIDE IGNITER

1. Disconnect the 2 prong connector, then connect a suitable voltmeter between the (BY) wire terminals.
2. Turn the ignition switch to the **ON** position. Check to see that the voltage is approximately 12 volts.

NOTE: Do not misconnect the voltmeter leads. They should be connected as follows: positive lead (+) to ground and negative lead (−) to the "B" wire. Disconnect the battery ground cable before removing the "B" wire from the coil with igniter, or damage will result.

3. Turn the ignition switch to the **OFF** position and disconnect the voltmeter and battery ground cable.
4. Connect Igniter Checker Tool 49-FO18-002 or equivalent between the 2 prong connectors.
5. Disconnect the "B" wire from the coil with igniter. Reconnect the battery ground cable.

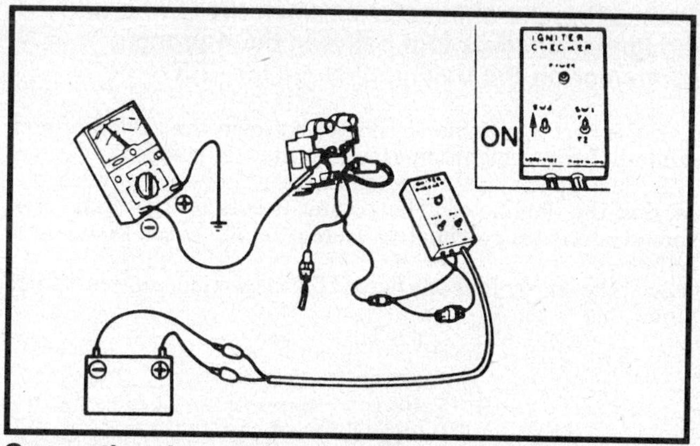

Connecting the ohmmeter between the B wire an ground and the igniter checker tool between the 2 prong connectors

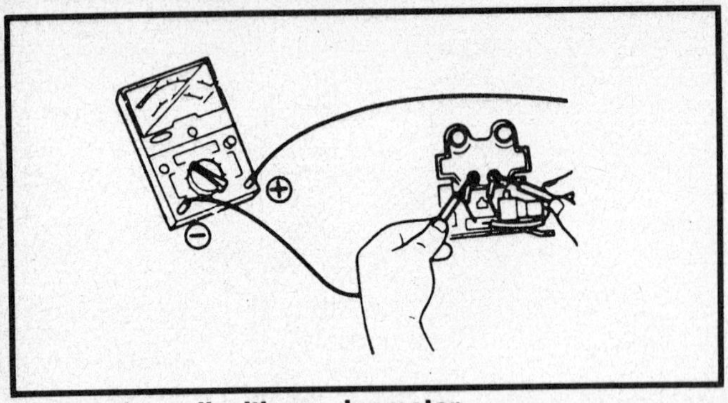

Testing the coil with an ohmmeter

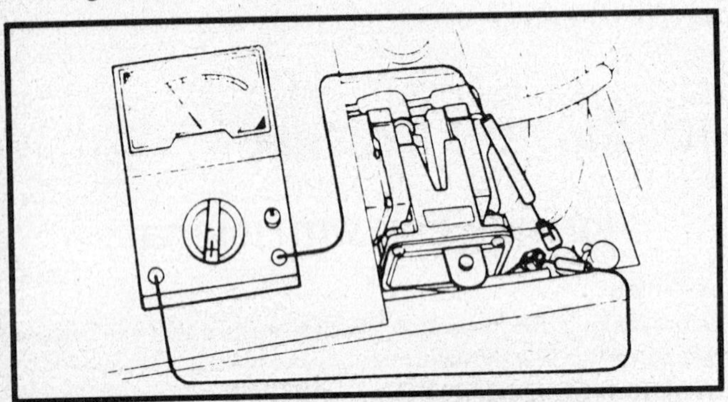

Connecting the voltmeter between the 2 prong connectors on the trailing side igniter—RX7

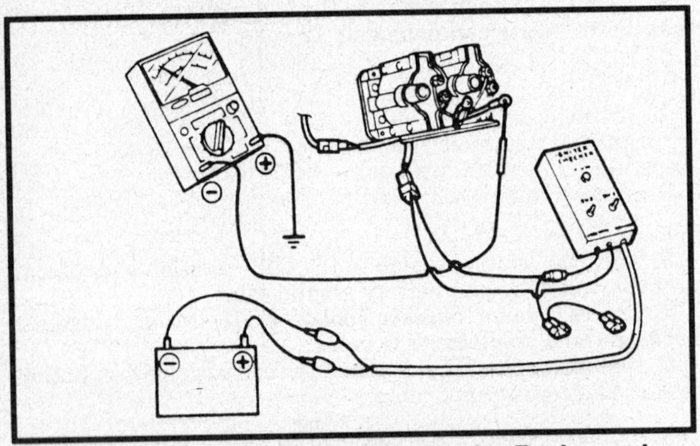

Connecting the ohmmeter between the B wire and the igniter checker tool between the 4 prong connector on the trailing side igniter—RX7

6. Connect a suitable ohmmeter between the "B" wire and ground. Turn the ignition switch to the **ON** position.

7. Push up the "SW2" switch on the igniter checker while observing the ohmmeter. The pointer needle on the ohmmeter should jump to approximately ⅓ scale (on the XI scale) and then return.

8. If the igniter fails any part of the inspection, replace it with a new one.

NOTE: While checking the coil with igniter as outlined above, the high tension leads must remain connected to the coil. If the high tension leads are disconnected, a high voltage ignition spark(s) may occur.

COIL

1. Disconnect the battery ground cable.

2. Using a suitable ohmmeter, check the resistance of the ohmmeter.

3. The standard resistance is 0.2–1 ohm. If coil resistance is out of specification, replace the coil.

TRAILING SIDE IGNITER

1. Disconnect the 2 prong connector and connect a suitable voltmeter between the (BY) 2 prong connectors.

2. Turn the ignition switch to the **ON** position. Check to see that the voltage is approximately 12 volts.

NOTE: When connecting the voltmeter leads, the leads should be connected as follows; positive lead (+) to ground and negative lead (−) to the "B" wire. Disconnect the negative battery cable before removing the "B" or "L" wires from the coil with igniter, or damage will result.

3. Turn the ignition switch to the **OFF** position.

4. Disconnect the 4 prong connector and connect Igniter Tool 49–FO18–002 or equivalent to the 4 prong connector of the coil with igniter.

5. Disconnect the "B" and "L" wires from the coil with ignitor. Reconnect the negative battery cable.

6. Connect a suitable ohmmeter between the "B" wire and ground. Turn the ignition switch to the **ON** position.

7. Set the "SW1" switch on the igniter checker to the "T1" position. While observing the ohmmeter, push up the "SW2" switch and the ohmmeter pointer needle should jump up to approximately ⅓ scale (on the XI scale) and then return.

8. Turn the ignition switch to the **OFF** position and disconnect the ohmmeter. Reconnect the ohmmeter between the "L" wire and ground.

NOTE: When connecting the voltmeter leads, the leads should be connected as follows; positive lead (+) to ground and negative lead (−) to the "L" wire.

9. Set the "SW1" switch on the igniter checker to the "T2" position. While observing the ohmmeter, push up the "SW2" switch and the ohmmeter pointer needle should jump to approximately ⅓ scale (on the X1) scale and then return.

10. If the igniter fails any part of this inspection, replace it with a new one.

NOTE: While checking the coil with igniter as outlined above, the high tension leads must remain connected to the coil. If the high tension leads are disconnected, a high voltage ignition spark(s) may occur.

CRANK ANGLE SENSOR

Testing

1. Disconnect the crank angle sensor connector.

2. Connect a suitable ohmmeter to the terminals of the crank angle sensor connector. The resistance should be as follows:

a. Between terminals G-1 (black) and G-2 (green) should be 110–210 ohms.

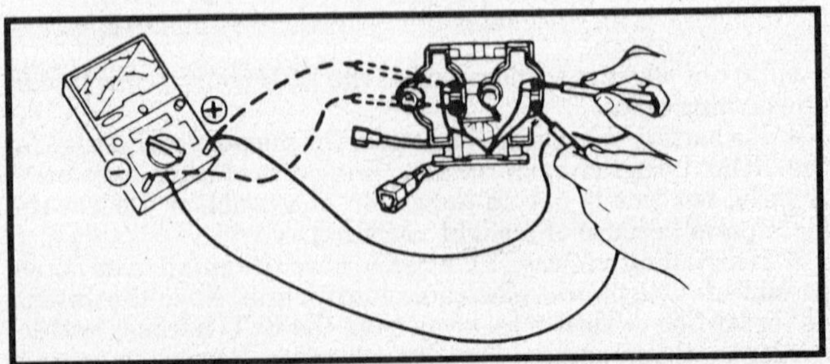

Testing the trailing side coil with an ohmmeter—RX7

IGNITION SYSTEM TESTS—RX-7

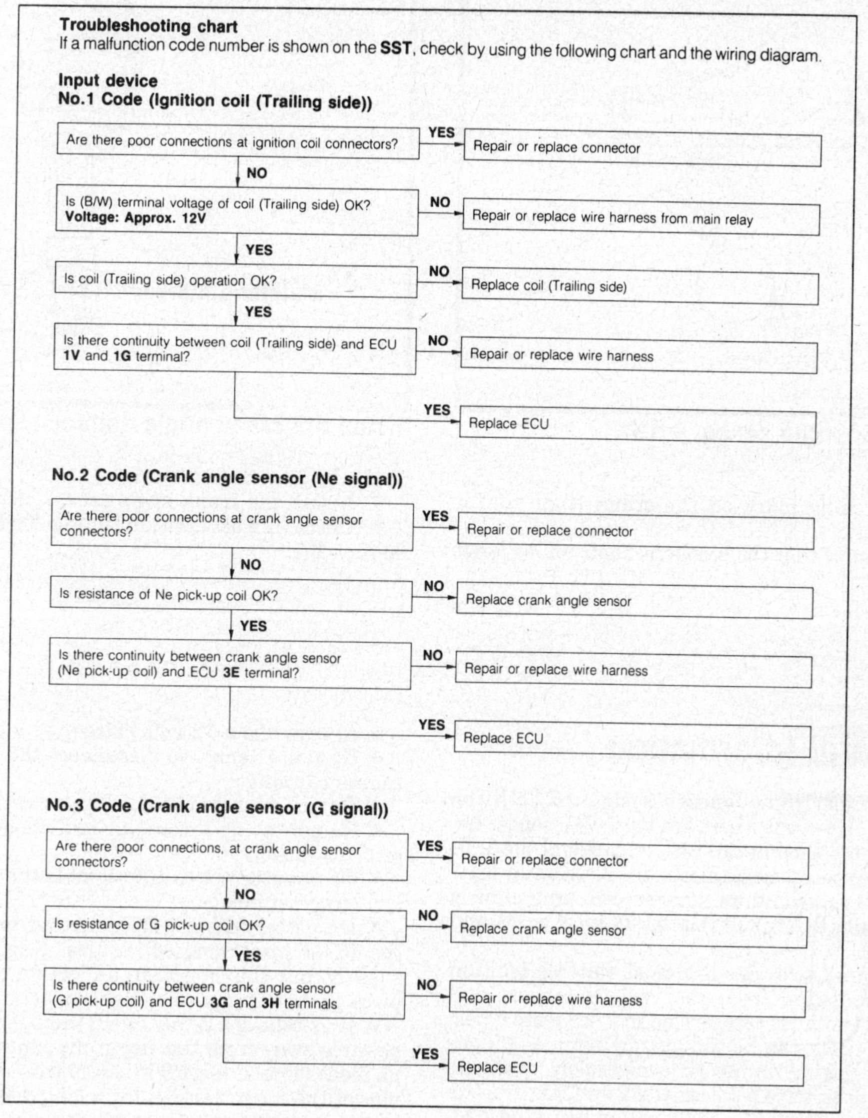

b. Between terminals NE-1 (white) and NE-2 (red) should be 110–210 ohms.

3. If the crank angle sensor resistance does not fall into specification, check the wiring and the connectors involved first and replace the crank angle sensor only if necessary.

IGNITION TIMING

Adjustment

1. Start the engine and allow it to reach normal operating temperature. Stop the engine. Connect a tachometer to the engine. Connect a timing light to the "L" high tension wire on the leading spark plug.
2. Start the engine, check the idle speed and adjust if necessary.
3. Point the timing light at the timing indicator (yellow pointer) located to the rear of the crankshaft pulley.
4. Check to see that the mark on the pulley and the indicator are aligned. If they are not aligned and the timing is out of specification, loosen the crank angle sensor lock bolt and rotate the sensor housing until the correct leading timing is obtained.

Timing specifications are as follows:
Leading timing (yellow mark) – 5 ± 1 degrees ATDC
Trailing timing (red mark) – 20 ± 2 degrees ATDC

NOTE: Check the trailing side with the same procedure as the leading side.

5. Tighten the sensor lock bolt, recheck the timing and idle speed, and adjust if necessary.

Component Replacement

CRANK ANGLE SENSOR

Removal and Installation

1. Set the eccentric shaft pulley to the leading timing mark (usually painted yellow) by turning the pulley.
2. Disconnect the crank angle sensor connector. Remove the crank angle sensor lock nut.
3. Slowly pull up on the crank angle sensor and remove it from the vehicle.

Inspecting the crank angle sensor—RX7

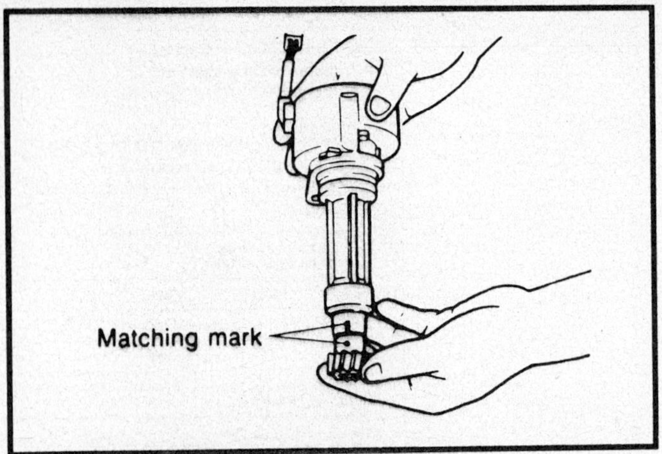

Install the crank angle sensor—RX7

To Install:

4. Aligning the matching mark on the crank angle sensor housing and driven gear.
5. Check and make sure that the eccentric shaft pulley is set to the leading timing mark.
6. Install the crank angle sensor lock nut.
7. Check and set ignition timing. Tighten the lock bolt to 70–96 inch lbs.

MAZDA MX-3 ELECTRONIC IGNITION SYSTEM

General Information

The MX-3 ultilizes an electronic ignition system in both the 1.6L and 1.8L engines. However, there are basic differences between the ignition system used in the 1.6L engine and the 1.8L engine. Both ignition systems use an electronic distributor driven off off the overhead camshaft, an ignition coil, an igniter, a Electronic Control Unit (ECU) and related ignition wires and spark plugs.

The engine control unit provides the most suitable ignition timing based on information from the various sensors and switches. The ignition timing is categorized in 2 separate types; base ignition timing, which can be changed by manual adjustment and changeable ignition timing, governed solely by engine conditions. The main difference between the 2 systems are, the 1.6L engine uses a separate igniter and ignition coil. The 1.8L engine incorporates the ignition coil, igniter and a crank angle sensor within the distributor assembly and is completed controlled by the Electronic Control Unit (ECU).

Operation of the system is conventional. When the ignition switch is in **RUN** or **START** position, voltage is applied to the ignition coil. As the distributor shaft turns, the igniter acts as a switch to control current flow through the primary winding of the ignition coil. When current flow through the coil is stopped, a high voltage current is induced in the secondary winding of the ignition coil. The high voltage current flows through the distributor cap and rotor to the proper spark plug.

Diagnosis and Testing

SERVICE PRECAUTIONS

- Make sure all connectors are fastened securely. A poor connection can cause an extremely high surge voltage in the coil and condenser and result in damage to integrated circuits.
- Keep all parts and harnesses dry during service.
- Before attempting to remove any parts, turn **OFF** the ignition switch and disconnect the battery ground cable.
- Always use a 12 volt battery as a power source.
- Do not attempt to disconnect the battery cables with the engine running.
- Do not depress the accelerator pedal when starting.
- Do not rev up the engine immediately after starting or just prior to shutdown.
- Do not attempt to disassemble the ECU control unit under any circumstances.
- If a battery cable is disconnected, the memory will return to the ROM (programmed) values. Engine operation may vary slightly, but this is not an indication of a problem. Do not replace parts because of a slight variation.
- If installing a 2-way or CB radio, keep the antenna as far as possible away from the electronic control unit. Keep the antenna feeder line at least 8 in. away from the ECU harness and do not let the 2 run parallel for a long distance. Be sure to ground the radio to the vehicle body.

IGNTION COIL

Testing

Before checking the coil, run the engine so as to get the coil to normal operating temperature.

1.6L ENGINE

1. Check the primary resistance by using a suitable ohmmeter. Primary resistance should be 0.81–0.99 ohms.
2. Measure the secondary resistance by using a suitable ohmmeter. Secondary resistance should be 10–16 kilo-ohms.
3. Measure the insulation resistance between the primary terminal and the casing. Be sure to use a 500 volt megger resistance range for this measurement. Insulation resistance should be 10 Megaohms or more.
4. If the coil fails any of these tests, replace it.

1.8L ENGINE

1. The ignition coil is apart of the distributor assembly. At the distributor, disconnect the electrical connector from the ignition coil.

IGNITION SYSTEM TESTS—RX-7

Troubleshooting chart

If a malfunction code number is shown on the **SST**, check by using the following chart and the wiring diagram.

Input device

No.1 Code (Ignition coil (Trailing side))

- Are there poor connections at ignition coil connectors? — **YES** → Repair or replace connector
  - **NO** ↓
- Is (B/W) terminal voltage of coil (Trailing side) OK? **Voltage: Approx. 12V** — **NO** → Repair or replace wire harness from main relay
  - **YES** ↓
- Is coil (Trailing side) operation OK? — **NO** → Replace coil (Trailing side)
  - **YES** ↓
- Is there continuity between coil (Trailing side) and ECU **1V** and **1G** terminal? — **NO** → Repair or replace wire harness
  - **YES** → Replace ECU

No.2 Code (Crank angle sensor (Ne signal))

- Are there poor connections at crank angle sensor connectors? — **YES** → Repair or replace connector
  - **NO** ↓
- Is resistance of Ne pick-up coil OK? — **NO** → Replace crank angle sensor
  - **YES** ↓
- Is there continuity between crank angle sensor (Ne pick-up coil) and ECU **3E** terminal? — **NO** → Repair or replace wire harness
  - **YES** → Replace ECU

No.3 Code (Crank angle sensor (G signal))

- Are there poor connections, at crank angle sensor connectors? — **YES** → Repair or replace connector
  - **NO** ↓
- Is resistance of G pick-up coil OK? — **NO** → Replace crank angle sensor
  - **YES** ↓
- Is there continuity between crank angle sensor (G pick-up coil) and ECU **3G** and **3H** terminals — **NO** → Repair or replace wire harness
  - **YES** → Replace ECU

b. Between terminals NE-1 (white) and NE-2 (red) should be 110–210 ohms.

3. If the crank angle sensor resistance does not fall into specification, check the wiring and the connectors involved first and replace the crank angle sensor only if necessary.

IGNITION TIMING

Adjustment

1. Start the engine and allow it to reach normal operating temperature. Stop the engine. Connect a tachometer to the engine. Connect a timing light to the "L" high tension wire on the leading spark plug.
2. Start the engine, check the idle speed and adjust if necessary.
3. Point the timing light at the timing indicator (yellow pointer) located to the rear of the crankshaft pulley.
4. Check to see that the mark on the pulley and the indicator are aligned. If they are not aligned and the timing is out of specification, loosen the crank angle sensor lock bolt and rotate the sensor housing until the correct leading timing is obtained.

Timing specifications are as follows:

Leading timing (yellow mark)—5±1 degrees ATDC
Trailing timing (red mark)—20±2 degrees ATDC

NOTE: Check the trailing side with the same procedure as the leading side.

5. Tighten the sensor lock bolt, recheck the timing and idle speed, and adjust if necessary.

Component Replacement

CRANK ANGLE SENSOR

Removal and Installation

1. Set the eccentric shaft pulley to the leading timing mark (usually painted yellow) by turning the pulley.
2. Disconnect the crank angle sensor connector. Remove the crank angle sensor lock nut.
3. Slowly pull up on the crank angle sensor and remove it from the vehicle.

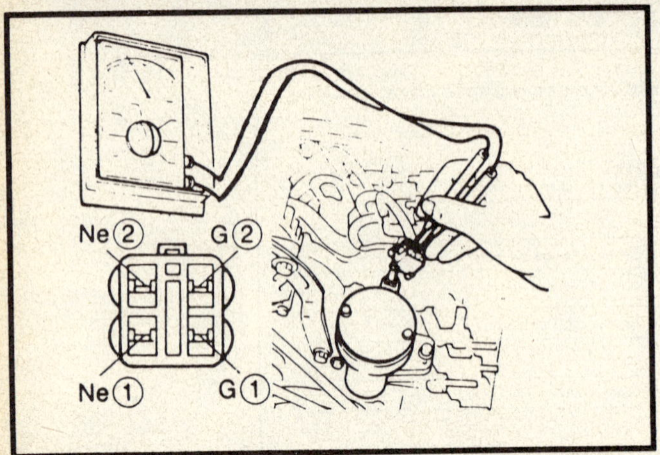

Inspecting the crank angle sensor—RX7

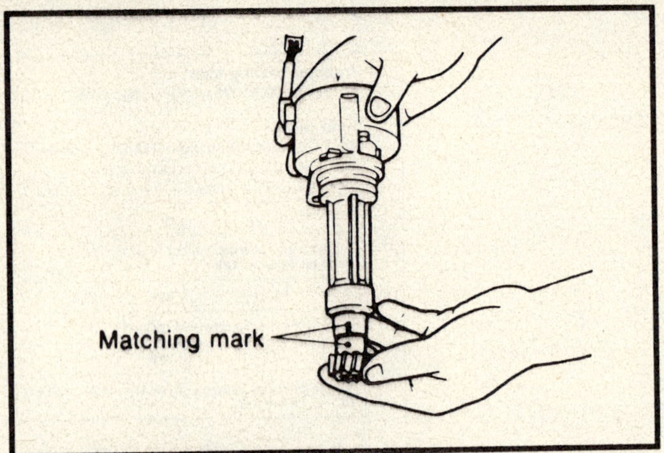

Install the crank angle sensor—RX7

To Install:

4. Aligning the matching mark on the crank angle sensor housing and driven gear.
5. Check and make sure that the eccentric shaft pulley is set to the leading timing mark.
6. Install the crank angle sensor lock nut.
7. Check and set ignition timing. Tighten the lock bolt to 70–96 inch lbs.

MAZDA MX-3 ELECTRONIC IGNITION SYSTEM

General Information

The MX-3 ultilizes an electronic ignition system in both the 1.6L and 1.8L engines. However, there are basic differences between the ignition system used in the 1.6L engine and the 1.8L engine. Both ignition systems use an electronic distributor driven off off the overhead camshaft, an ignition coil, an igniter, a Electronic Control Unit (ECU) and related ignition wires and spark plugs.

The engine control unit provides the most suitable ignition timing based on information from the various sensors and switches. The ignition timing is categorized in 2 separate types; base ignition timing, which can be changed by manual adjustment and changeable ignition timing, governed solely by engine conditions. The main difference between the 2 systems are, the 1.6L engine uses a separate igniter and ignition coil. The 1.8L engine incorporates the ignition coil, igniter and a crank angle sensor within the distributor assembly and is completed controlled by the Electronic Control Unit (ECU).

Operation of the system is conventional. When the ignition switch is in **RUN** or **START** position, voltage is applied to the ignition coil. As the distributor shaft turns, the igniter acts as a switch to control current flow through the primary winding of the ignition coil. When current flow through the coil is stopped, a high voltage current is induced in the secondary winding of the ignition coil. The high voltage current flows through the distributor cap and rotor to the proper spark plug.

Diagnosis and Testing

SERVICE PRECAUTIONS

- Make sure all connectors are fastened securely. A poor connection can cause an extremely high surge voltage in the coil and condenser and result in damage to integrated circuits.
- Keep all parts and harnesses dry during service.
- Before attempting to remove any parts, turn **OFF** the ignition switch and disconnect the battery ground cable.
- Always use a 12 volt battery as a power source.
- Do not attempt to disconnect the battery cables with the engine running.
- Do not depress the accelerator pedal when starting.
- Do not rev up the engine immediately after starting or just prior to shutdown.
- Do not attempt to disassemble the ECU control unit under any circumstances.
- If a battery cable is disconnected, the memory will return to the ROM (programmed) values. Engine operation may vary slightly, but this is not an indication of a problem. Do not replace parts because of a slight variation.
- If installing a 2-way or CB radio, keep the antenna as far as possible away from the electronic control unit. Keep the antenna feeder line at least 8 in. away from the ECU harness and do not let the 2 run parallel for a long distance. Be sure to ground the radio to the vehicle body.

IGNTION COIL

Testing

Before checking the coil, run the engine so as to get the coil to normal operating temperature.

1.6L ENGINE

1. Check the primary resistance by using a suitable ohmmeter. Primary resistance should be 0.81–0.99 ohms.
2. Measure the secondary resistance by using a suitable ohmmeter. Secondary resistance should be 10–16 kilo-ohms.
3. Measure the insulation resistance between the primary terminal and the casing. Be sure to use a 500 volt megger resistance range for this measurement. Insulation resistance should be 10 Megaohms or more.
4. If the coil fails any of these tests, replace it.

1.8L ENGINE

1. The ignition coil is apart of the distributor assembly. At the distributor, disconnect the electrical connector from the ignition coil.

2. Using a suitable ohmmeter, check the primary resistance. Primary resistance should be 0.6–0.9 ohms.

3. Remove the distributor cap and rotor. Using a suitable ohmmeter, check the secondary resistance. Secondary resistance should be 11.5–18.5 ohms.

4. If the coil is does not meet specification, the distributor and coil must be replaced as an assembly.

IGNITER

Testing

1.6L ENGINE

1. Disconnect the electrical connector from the igniter. Connect SST 49-N018-011 (Adapter Harness) between the igniter and the harness connector.

2. Connect SST 49-F018-002 (Igniter Checker) to the adapter harness and to the battery.

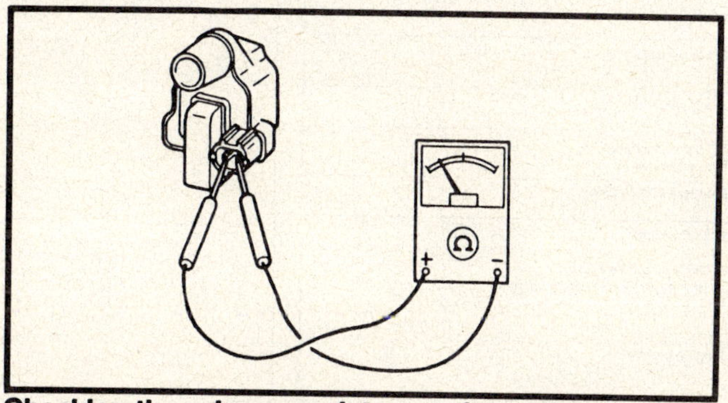

Checking the primary resistance of the ignition coil—MX3 with 1.6L engine

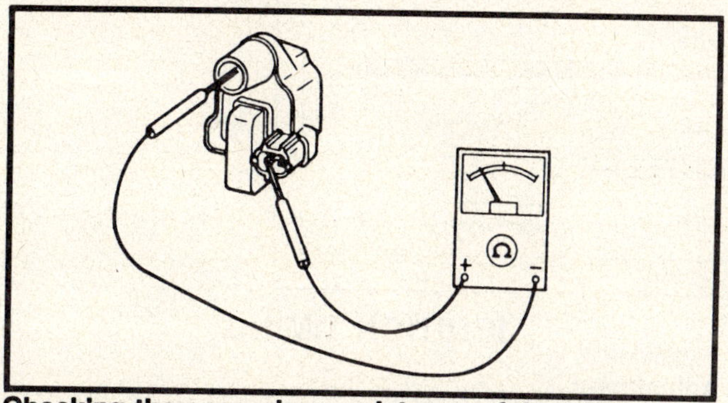

Checking the secondary resistance of the ignition coil—MX3 with 1.6L engine

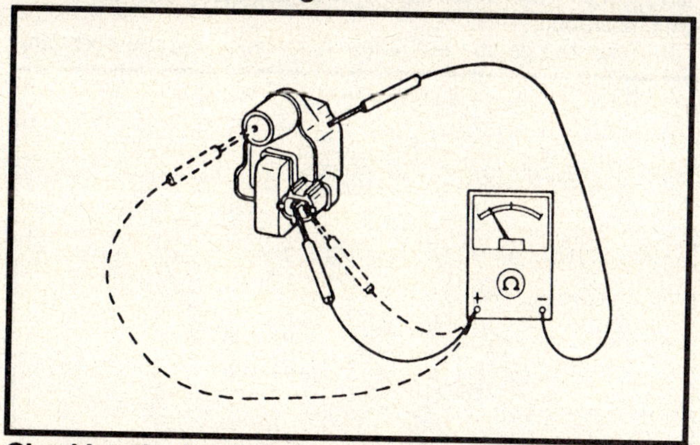

Checking the insulation resistance of the ignition coil—MX3 with 1.6L engine

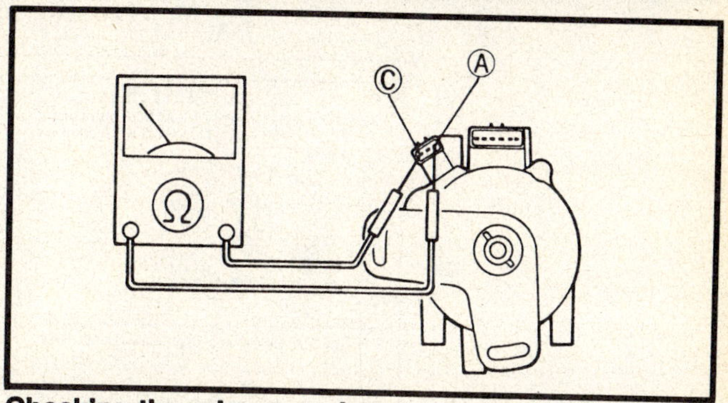

Checking the primary resistance of the ignition coil—MX3 with 1.8L engine

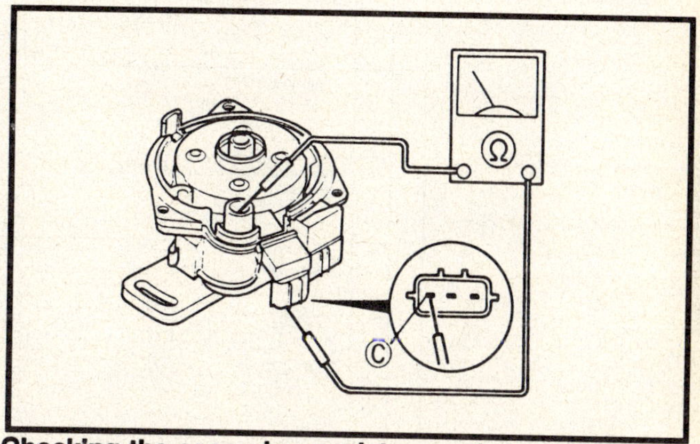

Checking the secondary resistance of the ignition coil—MX3 with 1.8L engine

3. Turn the ignition switch to the **ON** position.

4. Disconnect the high tension lead for the distributor and hold it ¼ inch from the ground.

5. Flip the SW2 switch ON and OFF and verify that a strong blue sparks discharge for the high tension lead. If not, replace the igniter.

NOTE: Do not hold the SW2 switch for more than 1 second in the ON or OFF position. Damage could occur to the coil or igniter.

6. Turn the ignition switch to the **OFF** position.

7. Disconnect the SST Igniter Checker and Adapter Harness, and reconnect the igniter connector.

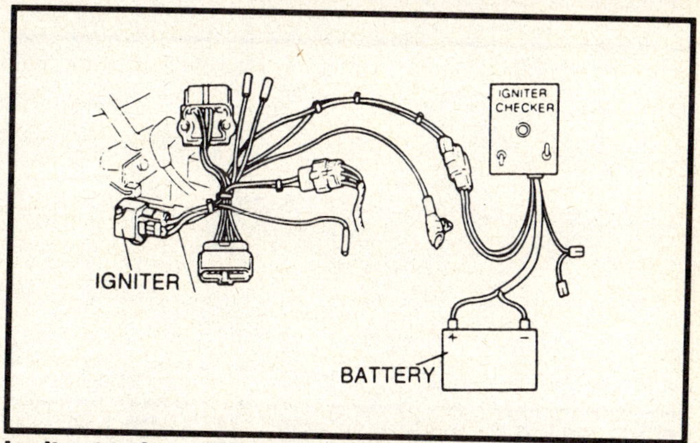

Igniter testing—MX3 with 1.6L engine

IGNITION SYSTEM TESTS—MX-3

STEP	INSPECTION		ACTION
1	Disconnect high-tension lead from distributor and check for strong blue spark while cranking engine 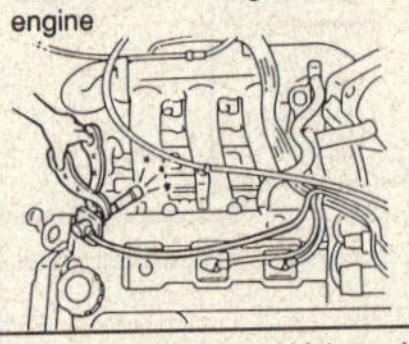	Yes	• Check distributor cap and rotor • Check high-tension lead ➪ Ignition system normal
		No	Go to next step
2	Check if resistance of high-tension leads is OK **Specification: 16 kΩ per 1 m (3.28 ft)**	Yes	Go to next step
		No	Replace high-tension lead
3	Are there corrosion, damage or cracks on the distributor cap and rotor?	Yes	Replace cap and rotor
		No	Go to next step
4	Is ignition coil OK?	Yes	Go to next step
		No	Replace ignition coil
5	Are distributor connector and ECU connector connected?	Yes	B6 go to next step K8 go to step 7
		No	Connect connectors
6	B6 Is ignitor OK? 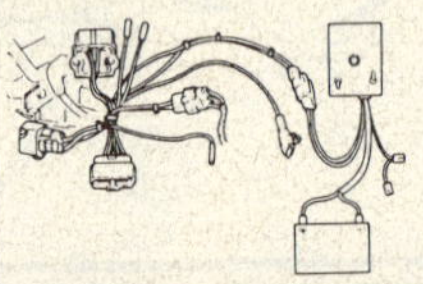	Yes	Check wiring ➪ If OK, replace ECU ➪ If not OK, repair wiring harness
		No	Replace igniter
7	Measure voltage of C terminal (Distribution connector). Is it correct? **Voltage: V_B** 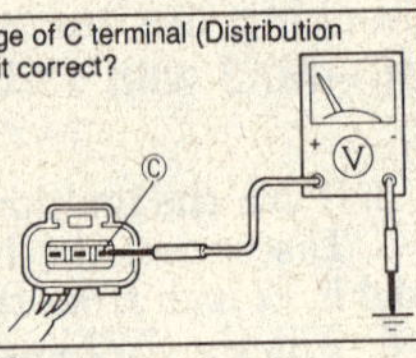	Yes	Go to next step
		No	Check main fuse, ignition switch, wiring harness, and ECU.
8	Is ECU normal?	Yes	Replace distributor
		No	Replace ECU

1.8L ENGINE

1. Test the Electronic Control Unit (ECU) for proper operation.
2. Check the condition of the ECU connector, the distributor, ignition coil and high tension leads.
3. If all check good, replace the igniter.

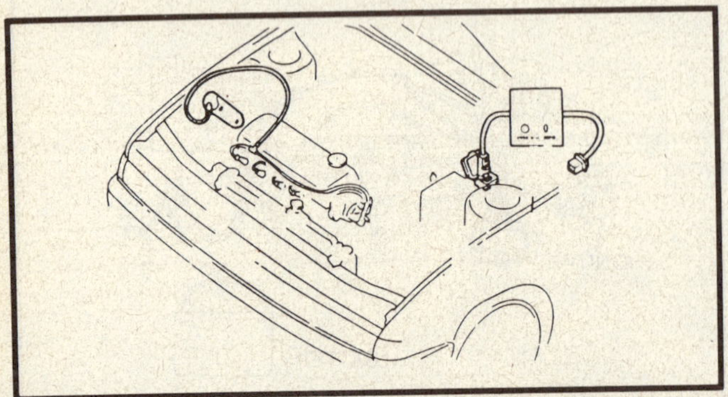

Setting switch on system selector—MX3 with 1.6L and 1.8L engines

IGNITION TIMING

Adjustment

1. Place the vehicle in **N**, block the drive wheels and apply the emergency brake. Make sure that all the accessories are in the **OFF** position.
2. Start the engine and allow it to reach normal operating

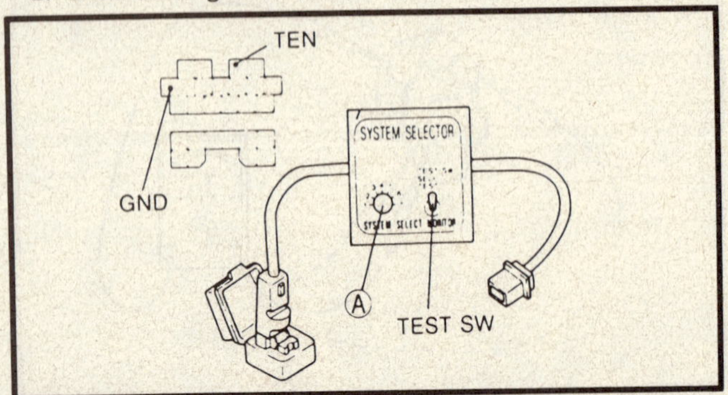

Ignition timing mark position—MX3 with 1.6L and 1.8L engines

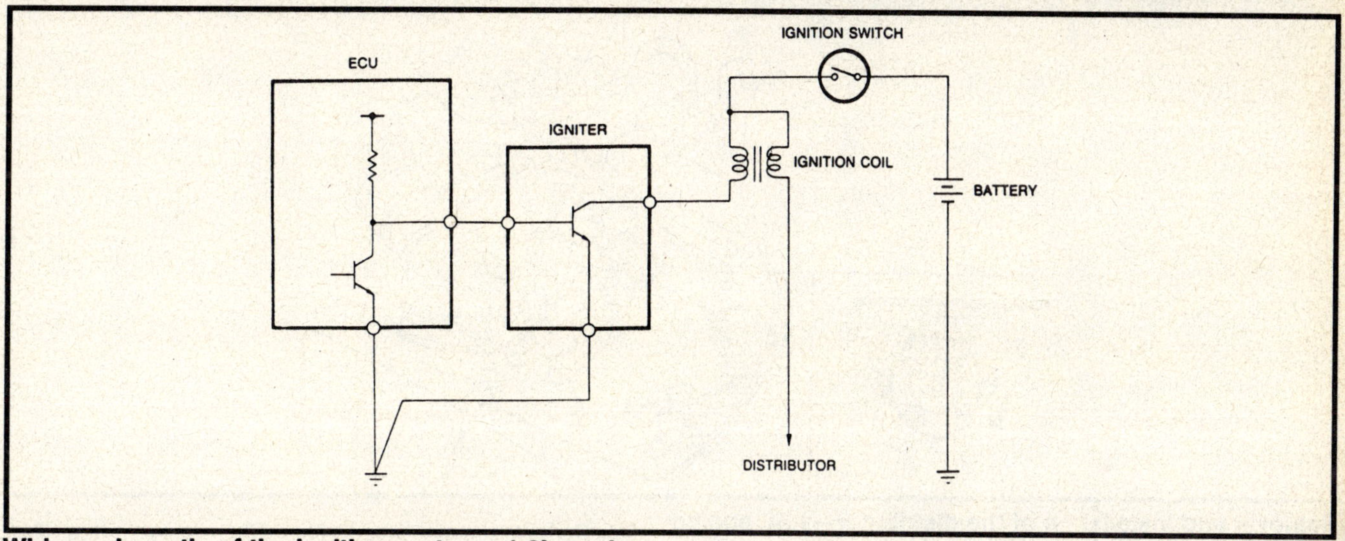

Wiring schematic of the ignition system—1.8L engine

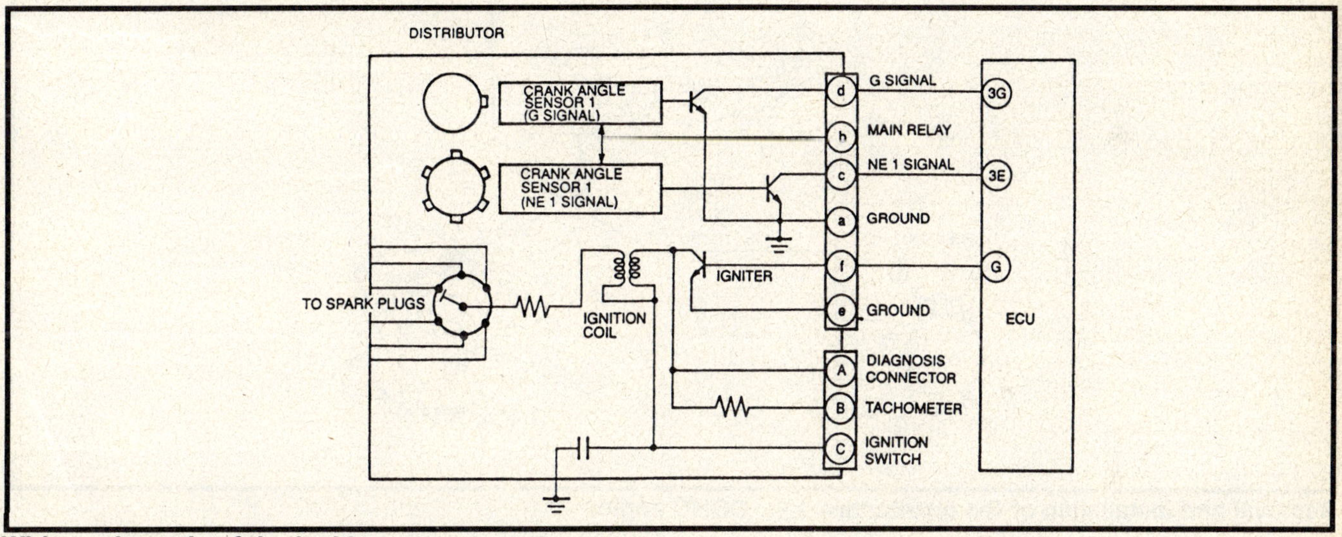

Wiring schematic of the ignition system—1.6L engine

temperature. Connect SST 49–B019–9A0 (System Selector) to the diagnosis connector.

3. Set switch (A) on the System Selector to position 1 and the Test Switch to Self-Test.

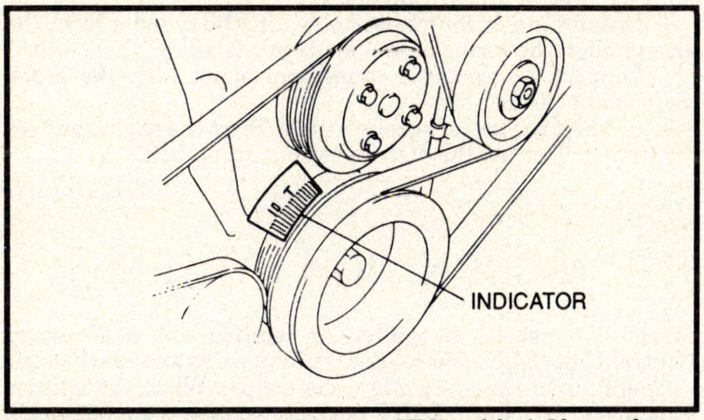

Adjusting the ignition timing—MX3 with 1.8L engine

NOTE: To check ignition timing without using the System Selector, connect a jumper wire between terminal No. 10 and ground on the diagnosis connector.

4. Connect a timing light to the engine according to manufacturer's instructions. Connect a tachometer to the IG terminal on the diagnosis connector. If the timing marks (usually a yellow mark) are difficult to see, use chalk or a dab of paint to make them more visible.

5. Point the timing light at the timing marks. Timing marks are on the crankshaft pulley and the timing belt cover. Ignition timing should be 10±1 degrees BTDC at 650–700 rpm.

6. If specifications are not as specified, loosen the distributor hold-down bolt, then rotate the distributor to advance or retard the timing as necessary.

7. Stop the engine and tighten the hold-down bolt. Start the engine and recheck the timing.

8. Stop the engine and disconnect the SST Tester and/or jumper wire, tachometer and timing light.

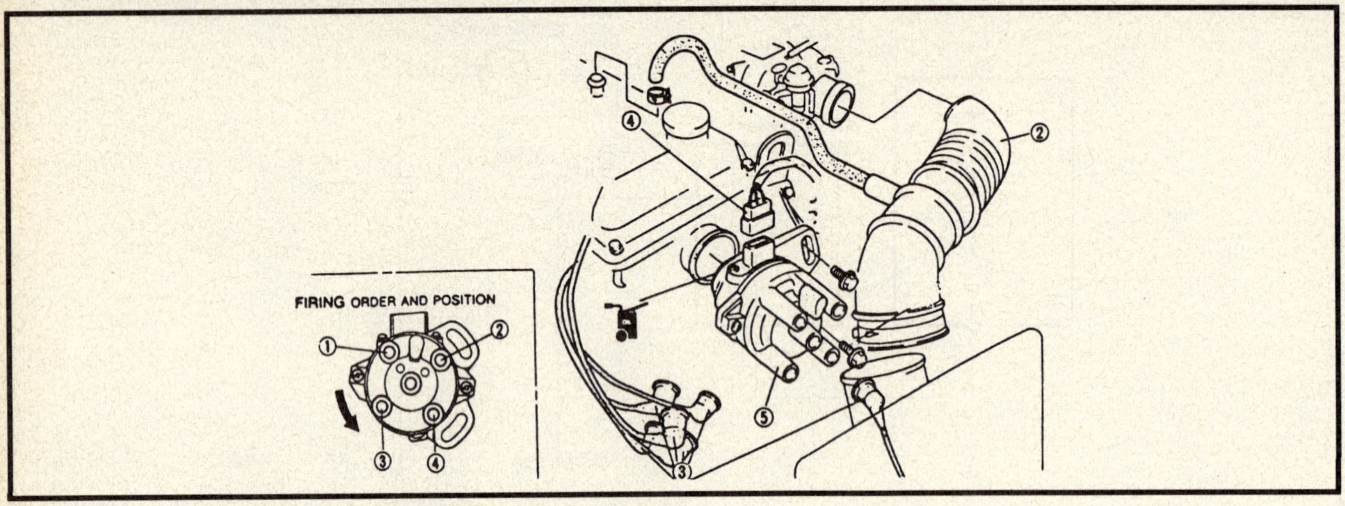

Removal and installation of the distributor—1.6L engine

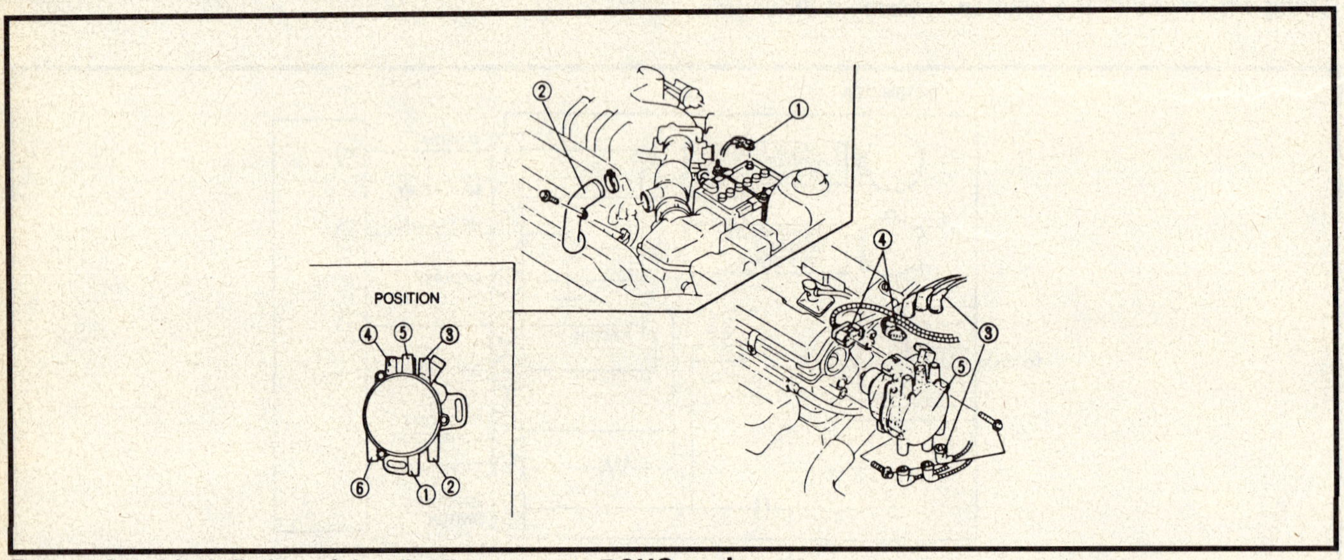

Removal and installation of the distributor—1.8L DOHC engine

Component Replacement

DISTRIBUTOR

Removal and Installation

1. Rotate the engine and bring the No. 1 cylinder to top dead center of its compression stroke.
2. Disconnect the battery ground cable. If so equipped, remove the air hose. Disconnect and tag (if necessary) the electrical connectors along with the spark plug wires from the distributor.
3. Mark the rotor position on the distributor body. Mark the distributor body and cap using a scribe or a marker. This can be used for an easier installation procedure.
4. Remove the distributor hold-down bolt, then pull the distributor assembly out of the engine cylinder head.

To install:

5. Be sure to coat the O-ring with engine oil. Ensure that the No. 1 cylinder is at top dead center.
6. Install the distributor assembly into the cylinder block (be sure to align the scribe marks made previously).
7. Connect all electrical connectors along with the spark wires to the distributor.
8. Connect the battery ground cable. Start the engine and set the timing. Tighten the distributor mounting bolt.

MAZDA B2200 AND B2600 ELECTRONIC IGNITION SYSTEM

General Information

The ignition system a basic electronic system. It is comprised of the ignition switch, an electronic distributor driven off off the overhead camshaft, an igniter, an ignition coil, a Electronic Control Unit (ECU) and related ignition wires and spark plugs.

Operation of the system is quite simple. When the ignition switch is in **RUN** or **START** position, voltage is applied to the ignition coil. As the distributor shaft turns, the igniter acts as a switch to control current flow through the primary winding of the ignition coil. When current flow through the coil is stopped, a high voltage current is induced in the secondary winding of the ignition coil. The high voltage current flows through the distributor cap and rotor to the proper spark plug.

Diagnosis and Testing

SERVICE PRECAUTIONS

- Make sure all connectors are fastened securely. A poor connection can cause an extremely high surge voltage in the coil and condenser and result in damage to integrated circuits.
- Keep all parts and harnesses dry during service.
- Before attempting to remove any parts, turn **OFF** the ignition switch and disconnect the battery ground cable.
- Always use a 12 volt battery as a power source.
- Do not attempt to disconnect the battery cables with the engine running.
- Do not depress the accelerator pedal when starting.
- Do not rev up the engine immediately after starting or just prior to shutdown.
- Do not attempt to disassemble the ECU control unit under any circumstances.
- If a battery cable is disconnected, the memory will return to the ROM (programmed) values. Engine operation may vary slightly, but this is not an indication of a problem. Do not replace parts because of a slight variation.
- If installing a 2-way or CB radio, keep the antenna as far as possible away from the electronic control unit. Keep the antenna feeder line at least 8 in. away from the ECU harness and do not let the 2 run parallel for a long distance. Be sure to ground the radio to the vehicle body.

IGNITION COIL

Testing

Before checking the coil, start and run the engine as to get the coil to normal operating temperature.

1. Using a suitable ohmmeter, check the primary resistance. Primary resistance should be as follows:
   1.0–1.3 ohms – B2200
   0.8–0.99 ohms – B2600
2. Using a suitable ohmmeter, check the secondary resistance. Secondary resistance should be 6–30 kilo-ohms for all engines.
3. Measure the insulation resistance between the primary terminal and the casing. Be sure to use a 500 volt megger resistance range for this measurement. Insulation resistance should be 10 Megaohms or more.

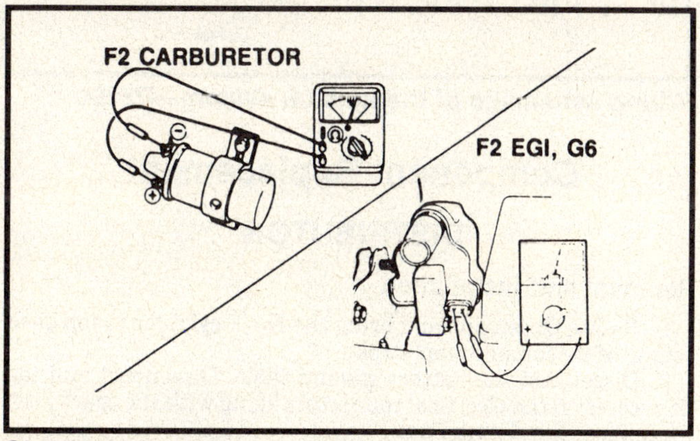

Checking the primary resistance of the ignition coil – B2200 and B2600

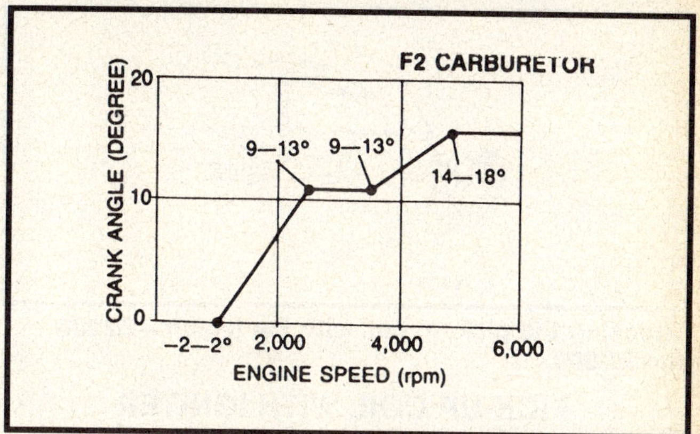

Spark control centrifugal advance – B2200

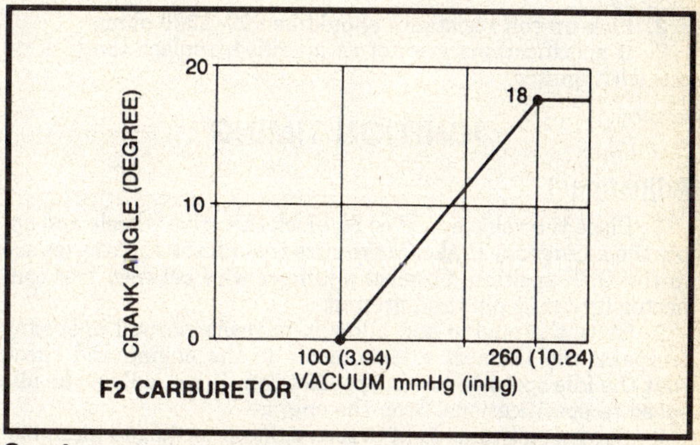

Spark control centrifugal advance – B2600

SPARK ADVANCE CONTROL

Testing

CENTRIFUGAL

1. Start the engine and allow it to reach normal operating temperature.
2. Connect a timing light and tachometer to the engine. Check that the idle speed and ignition timing are within specification.
3. Disconnect and plug the vacuum hoses from the vacuum controller.
4. While gradually increasing engine rpm, check the advance angle on the crankshaft pulley using the timing light.
   Excess advance – Indicates a weak governor spring
   Insufficient advance – Governor weight or cam defective

VACUUM

1. Start the engine and allow to reach normal operating temperature.
2. Connect a timing light and tachometer to the engine. Check that the idle speed and ignition timing are within specification. Remove the tachometer.
3. Disconnect and plug the vacuum hoses from the vacuum controller. Run the engine at idle.
4. Connect a suitable hand operated vacuum pump to the vacuum controller. Using the timing light, check the advance on the crankshaft pulley while applying vacuum.

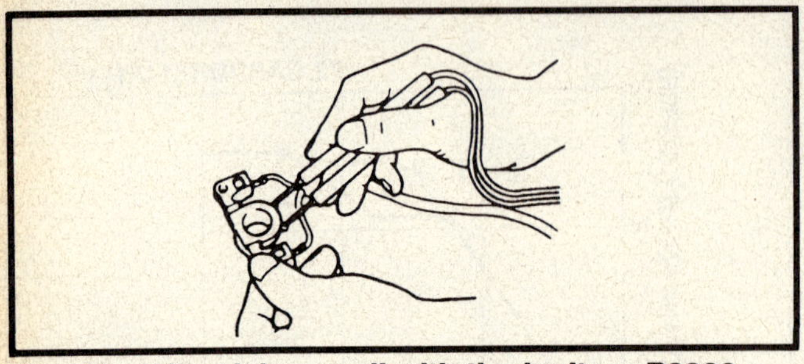

Checking the pick-up coil with the igniter—B2200 and B2600

PICK-UP COIL WITH IGNITER

Testing

1. Using a suitable ohmmeter measure the resistance of the pick-up coil.
2. Pick-up coil resistance should be 900–1200 ohms.
3. If specifications are not as specified, replace the pick-up coil with igniter.

IGNITION TIMING

Adjustment

1. Place the vehicle in **P** or **N**, block the drive wheels and apply the emergency brake. Make sure that all the accessories are in the OFF position. Connect a jumper wire between test connector (green, 1 pin) and ground.
2. Start the engine and allow it to reach normal operating temperature. Connect a tachometer to the engine and check that the idle speed is within specification. If not, adjust the idle speed to specifications. Stop the engine.
3. Connect a timing light to the engine according to manufacturer's instructions. If the timing marks (usually yellow) are difficult to see, use chalk or a dab of paint to make them more visible.
4. Disconnect and plug the vacuum hose(s) from the vacuum controller.
5. Start and run the engine at idle with the transmission in **N** on manual transmissions, or **D** on automatics. Make sure that the parking brake is set and the wheels are blocked.
6. Point the timing light at the timing marks (the timing marks are on the crankshaft pulley and the timing belt cover). With the engine at idle, ignition timing should be as follows:
   B2200—6 degrees BTDC at 800–850 rpm
   B2600—7 degrees BTDC at 800–850 rpm
7. If specifications are not as specified, loosen the distributor hold-down bolt, and rotate the distributor to advance or retard the timing a required.
8. Stop the engine and tighten the hold-down bolt. Start the engine and recheck the timing.
9. Stop the engine and disconnect the timing light and tachometer. Reconnect the vacuum lines and electrical connectors.

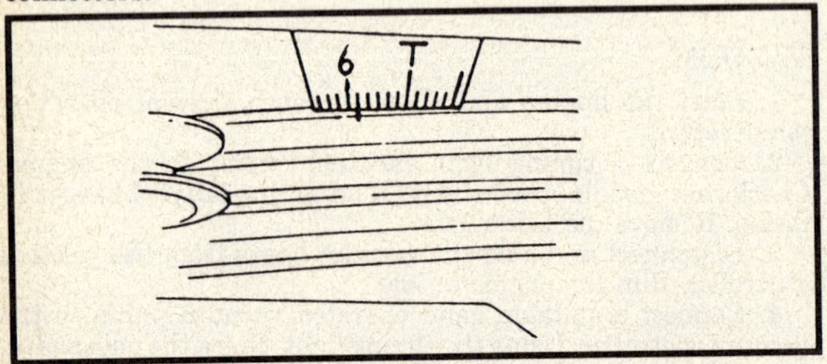

Location of the timing marks—B2200 and B2600

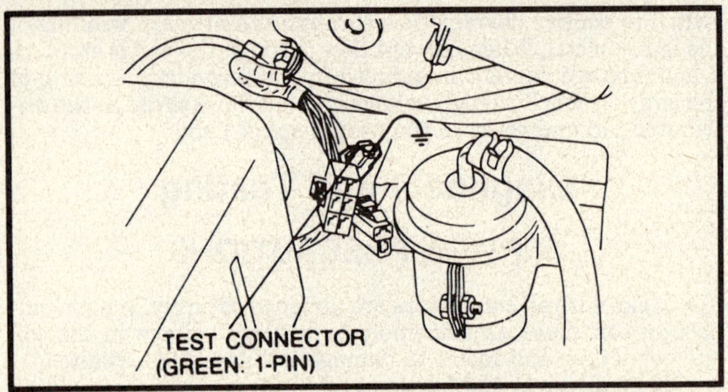

Location of the 1 pin green test connector—B2200 andB2600

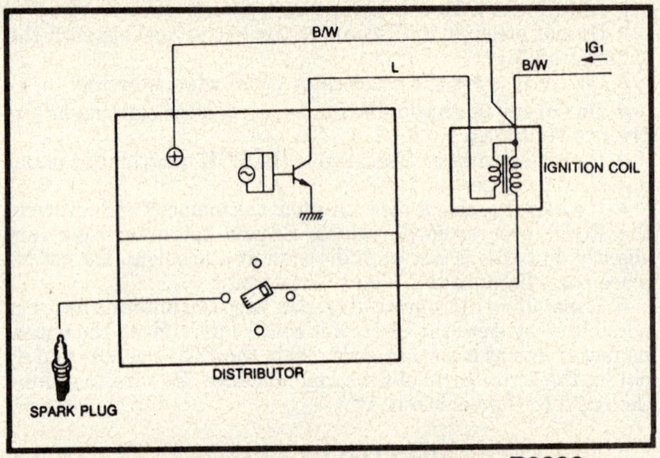

Wiring schematic of the ignition system—B2200

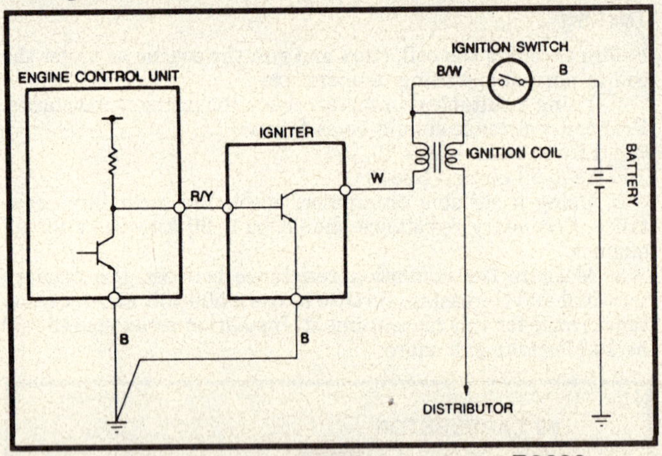

Wiring schematic of the ignition system—B2600

Component Replacement

DISTRIBUTOR

Removal and Installation

1. Rotate the engine and bring the No. 1 cylinder to top dead center of its compression stroke.
2. Disconnect the battery ground cable. Disconnect and tag (if necessary) the electrical connectors along with the spark plug wires from the distributor.
3. Disconnect the vacuum hose(s) and remove distributor cap.

IGNITION SYSTEM TESTS—2200 and 2600

TROUBLESHOOTING

MISFIRE (NO SPARK, WEAK SPARK)
All cylinders
Note: When check spark test, hold lead with insulated pliers approx. 5—10mm (0.20—0.39 in) from a ground or cap.

- Ignition coil spark test
  - NG → Distributor pick up coil resistance (F2 Carb.); Igniter (F2 EGI, G6); Ignition coil resistance; High-tension lead resistance
  - OK ↓
- Distributor spark test
  - NG → Distributor cap or rotor
  - OK ↓
- Spark plug spark test
  - NG → High-tension lead resistance
  - OK ↓
- Spark plug gap, dirty **Plug gap: 0.75—0.85mm (0.028—0.033 in)...F2 (Carb.), 1.0—1.1mm (0.039—0.043 in)...F2 (EGI), G6**

Some cylinder(s)
Spark plug gap, dirty **Plug gap: 0.75—0.85mm (0.028—0.033 in)...F2 (Carb.), 1.0—1.1mm (0.039—0.043 in)...F2 (EGI), G6**
High-tension lead resistance
Distributor cap

Ignition coil spark test

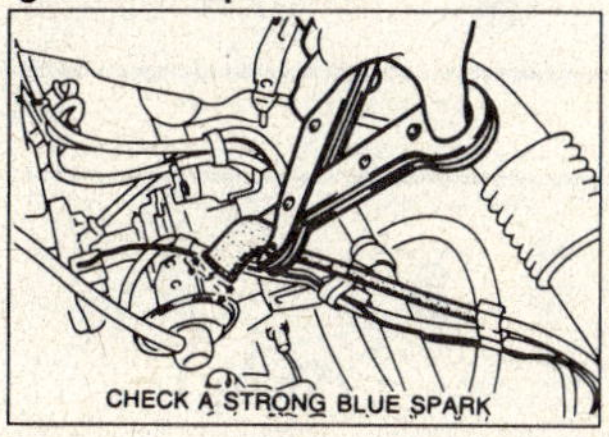

Distributor pickup coil resistance

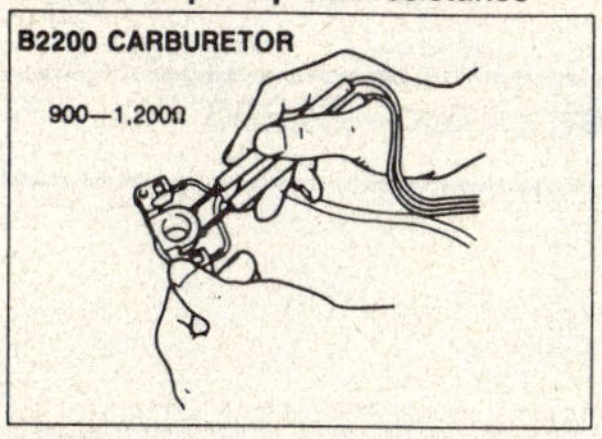

Distributor spark test

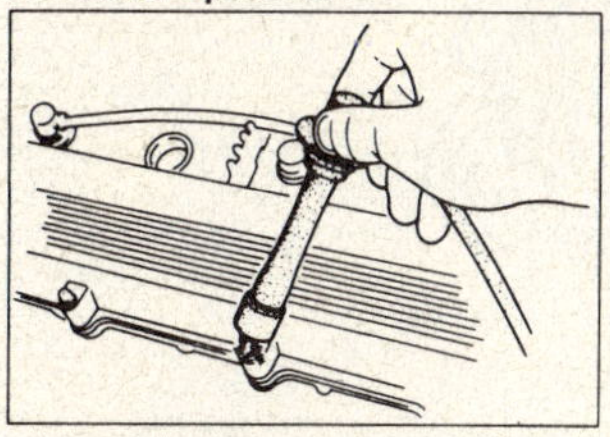

Ignition coil resistance

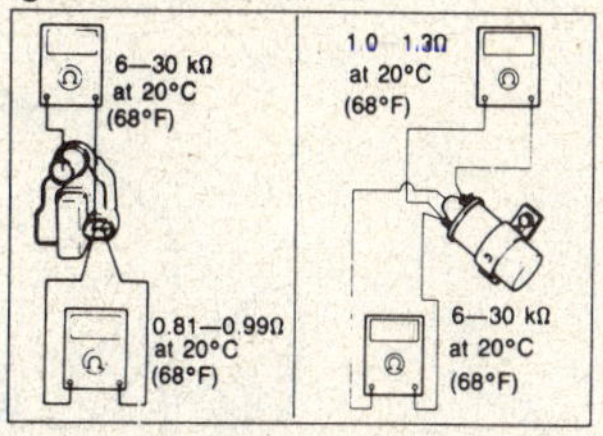

Spark plug spark test

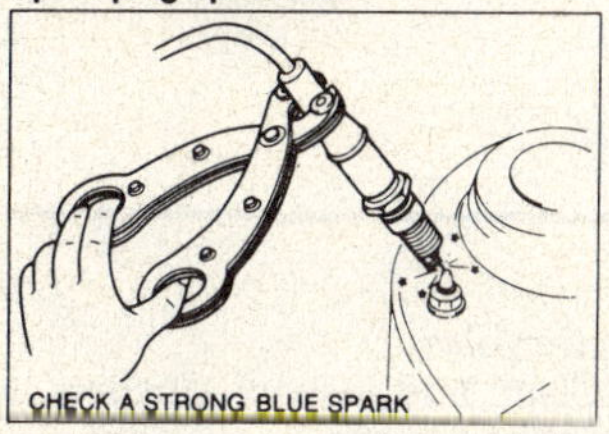

High-tension lead resistance

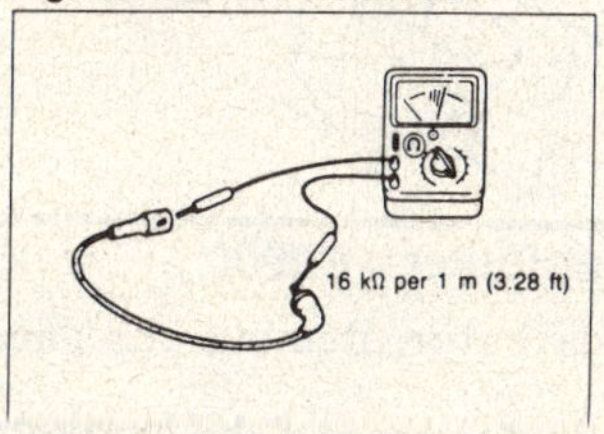

4. Mark the rotor position on the distributor body. Mark the distributor body and cap using a scribe or a marker. This can be used for an easier installation procedure.
5. Remove the distributor hold-down bolt, and pull the distributor assembly out of from the engine.

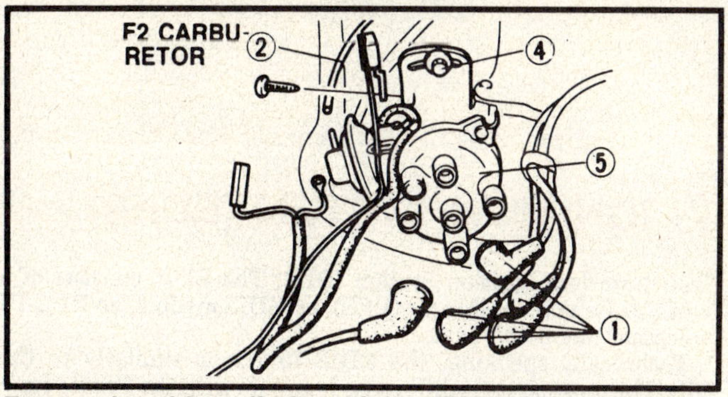

Removal and installation of the distributor—B2200, B2600 is simular

To install:

6. Be sure to coat the O-ring(s) with engine oil. Ensure that the No. 1 cylinder is at top dead center of its compression stroke.
7. Align the matching marks on the drive gear with the mark on the distributor shaft. Install the distributor assembly into the engine, be sure to align the scribe marks made previously. Lightly tighten the distributor hold-down bolt.
8. Connect the wiring harness to the distributor lead wire. Connect the vacuum hoses to the vacuum controller.
9. Connect the electrical connectors along with the spark plug wires to the distributor.
10. Connect the battery ground cable. Start the engine and set the timing. Tighten the distributor hold-down bolt.

Disassembly

1. Place the distributor assembly into a suitable holding fixture. Remove the cap and rotor.
2. Using 2 small pry bars or equivalent, remove the reluctor and/or signal rotor and pin.
3. Remove the pick-up coil attaching screws, then remove the

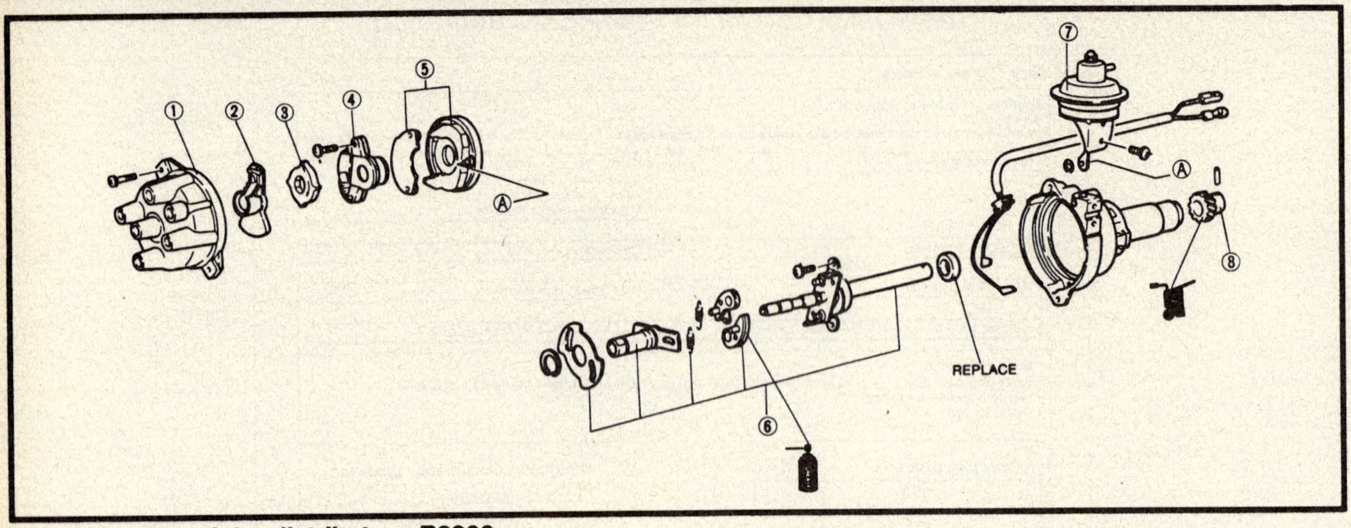

Exploded view of the distributor—B2200

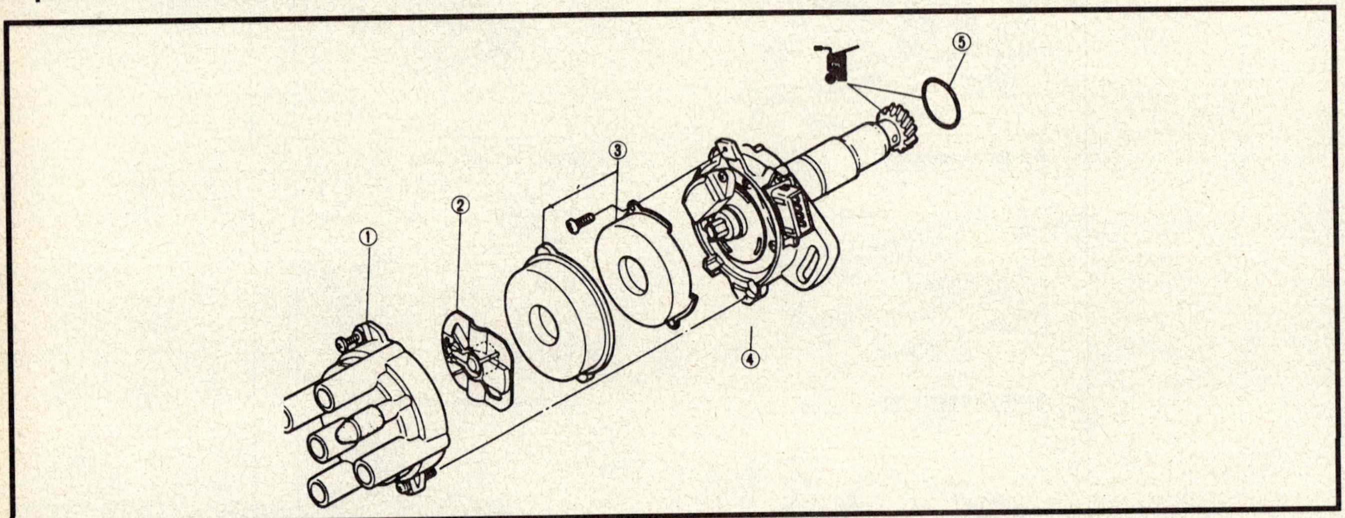

Exploded view of the distributor—B2600

pick-up coil from the distributor. Remove the governor assembly.

4. Remove the vacuum controller and the breaker plate assembly.

5. Using a suitable drift, remove the distributor drive gear pin, then the drive gear. Remove the shaft from the distributor housing.

Reassembly:

5. Clean and inspect all parts, replace any damaged or worn parts as necessary.

6. Inspect the distributor cap for cracking, burning, corrosion and wear.

7. Complete the assembly in the reverse order of the disassembly procedure.

MAZDA NAVAJO ELECTRONIC DISTRIBUTORLESS IGNITION SYSTEM (EDIS)

General Information

The new high data rate EDIS system is controlled by an EDIS module and the EEC-IV module operating in union. The system is designed to deliver a full energy spark at a crank angle selected by the EEC-IV. This system is a modified version of the Distributorless Ignition System (DIS). The EDIS consists of a Variable Reluctance Sensor (VRS), an EDIS module, an EEC-IV processor and a coil pack.

Technically speaking, the EDIS functions similarly to the DIS. The most obvious difference has to do with the Profile Ignition Pickup (PIP) signal. In the DIS, the PIP is a signal generated by a hall-effect device and passed through the DIS module to

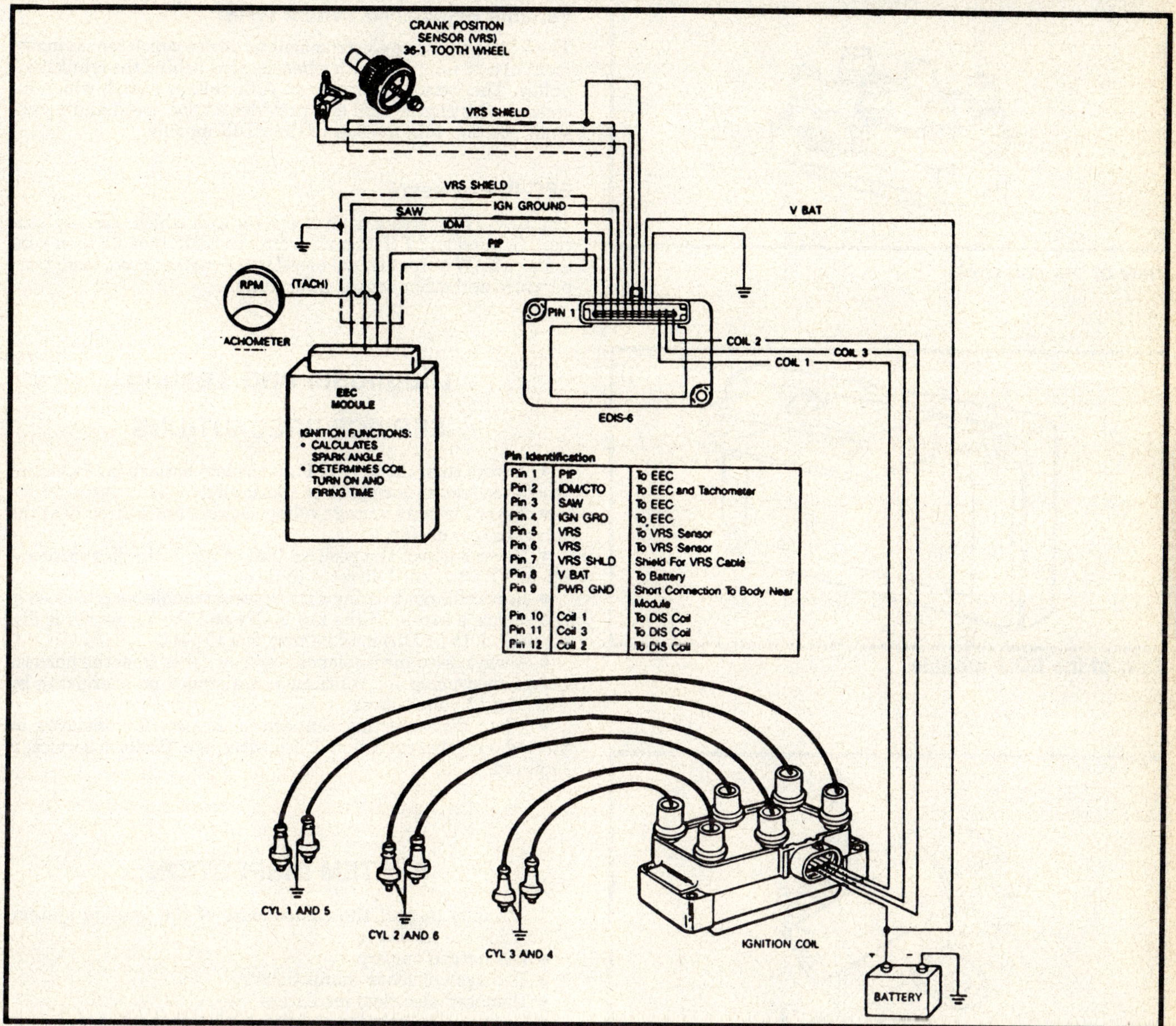

Pin Identification

Pin 1	PIP	To EEC
Pin 2	IDM/CTO	To EEC and Tachometer
Pin 3	SAW	To EEC
Pin 4	IGN GRD	To EEC
Pin 5	VRS	To VRS Sensor
Pin 6	VRS	To VRS Sensor
Pin 7	VRS SHLD	Shield For VRS Cable
Pin 8	V BAT	To Battery
Pin 9	PWR GND	Short Connection To Body Near Module
Pin 10	Coil 1	To DIS Coil
Pin 11	Coil 3	To DIS Coil
Pin 12	Coil 2	To DIS Coil

Component layout view of the EDIS system

the EEC-IV. PIP provided the EEC-IV with base timing information. In the EDIS, the crankshaft position sensing and rpm is accomplished by the use of a trigger wheel and a Variable Reluctance Sensor (VRS). The position is determined by a missing tooth on the sensor wheel.

SYSTEM OPERATION

The EDIS operates by sending crankshaft position information from the VRS to the EDIS module. The VRS senses the passage of teeth from the 36 minus 1 tooth trigger wheel, creating an input signal for the EDIS module. The module process the signal indicated by the missing tooth, generates a PIP signal and sends it to the EEC-IV processor. The processor responds with a Spark Angle Word (SAW) signal containing advance or retard timing information back to the EDIS module. The ignition coil pack, turned ON and then OFF by the EDIS module. This condition fires 2 spark plugs at once, one for the cylinder on its compression stroke and the other for the mating cylinder which is on its exhaust stroke. The next time the coil is fired, the situation is reversed.

The module generates an Ignition Diagnostic Monitor (IDM) signal to the processor which is used to indicate a failure mode and also provide a tach output signal. In addition, the IDM line indicates Key ON/Engine OFF (KOEO) status. By providing a pulse, the KOEO signal communicates module power on status.

SYSTEM COMPONENTS

Coil Pack

The coil pack is controlled by the EDIS module, firing 2 spark plugs at once. One is for the cylinder being fired, which is on its compression stroke and the other is for the cylinder on its exhaust stroke.

EDIS Module

The EDIS module is a microprocessor based device with coil drivers which make decisions about spark timing and coil firing. The module controls the coil, based on information from the VRS signal and a pulse width modulated signal (SAW) generated from the EEC-IV processor.

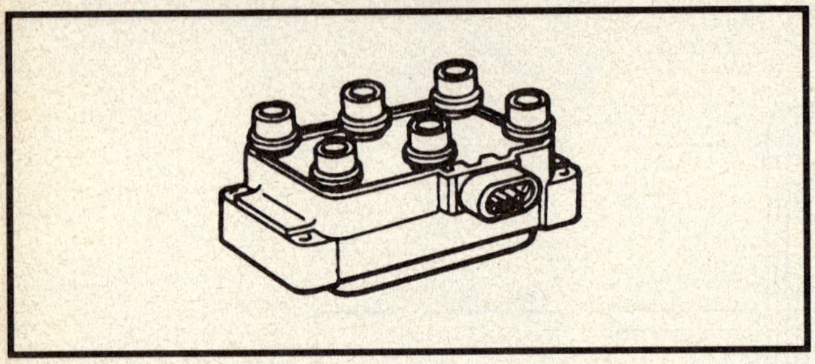

View of the coil pack

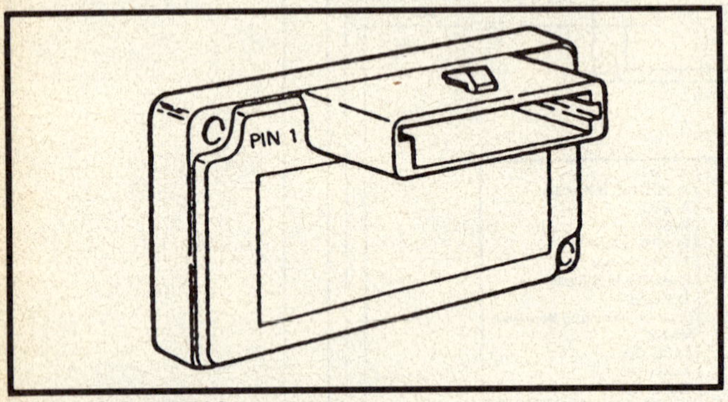

View of the EDIS module

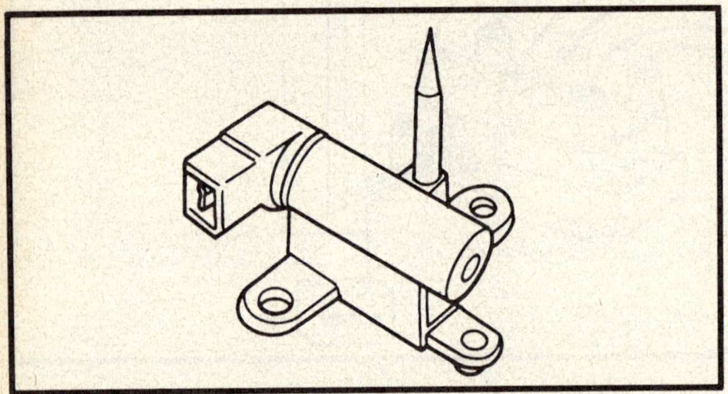

View of the Variable Reluctance Sensor (VRS)

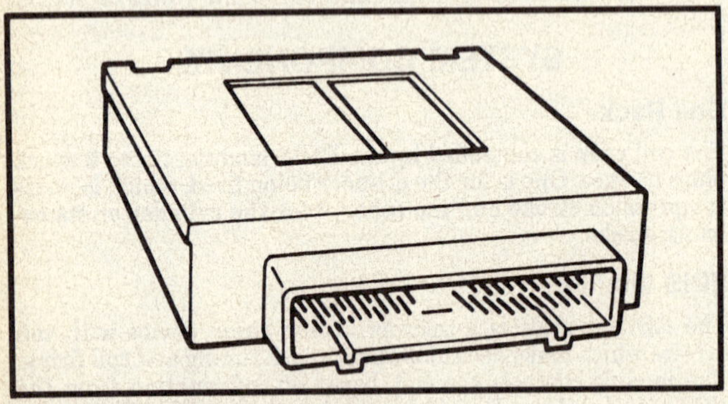

View of the EEC-IV Processor

Variable Reluctance Sensor (VRS)

The VRS is a passive electromagnetic device which senses movement of a 36 minus 1 tooth wheel located behind the crankshaft pulley. The sensor generates an A.C. voltage signal, which increases with engine rpm and provides engine speed and crankshaft position information to the EDIS module.

EEC-IV Processor

The EEC-IV processor is an electronic unit which receives Ignition, Ground and PIP signals from the EDIS module then generates a SAW output signal based upon engine speed, load, temperature and other sensor information.

Diagnosis and Testing

SERVICE PRECAUTIONS

- Always turn the key **OFF** and isolate both ends of a circuit whenever testing for short or continuity.
- Never measure voltage voltage or resistance directly at the processor connector.
- Never connect the positive lead of the EDIS diagnostic cable or VRS tee until directed to do so.
- Be careful not to bring a fluorescent trouble lamp to close to the vehicle's wiring. If the key is **ON** and the VR sensor is disconnected, the EDIS module may fire the coil.
- Always disconnect solenoids and switches from the harness before measuring for continuity, resistance or energizing by way of a 12 volts source.
- When disconnecting connectors, inspect for damaged or pushed-out pins, corrosion, loose wires, etc. Perform service, if required.

SYSTEM INSPECTION

1. Visually inspect the components of the ignition system. Check for the following:
   - Discharged battery
   - Damaged or loose connections
   - Damaged electrical insulation
   - Poor coil, distributor and spark plug connections
   - Ignition module connections
   - Blown fuses
   - Damaged vacuum hoses
   - Damaged or worn rotor and distributor cap
   - Damaged spark plugs
   - Distributor cap, rotor and spark plug wires are properly seated
2. Check the vehicle's maintenance schedule to ensure spark plugs and wire have been properly maintained.
3. Check spark plug wires and boots for signs of poor insulation that could cause cross firing.
4. A damaged or worn timing belt can cause symptoms that appear to be ignition timing related.
5. Make certain the engine idle speed is within specification.
6. Be certain that the battery is fully charged and that all accessories should be **OFF** during the diagnosis.

IGNITION TIMING

Spark timing adjustment is not possible on the EDIS system.

4.0L ENGINE—IGNITION SYSTEM TESTS—NAVAJO

EDIS No Start — Pinpoint Test A

TEST STEP	RESULT ▶	ACTION TO TAKE
A1 PERFORM EEC-IV QUICK TEST		
• **Has Quick Test been performed?**	Yes ▶	GO to A2.
	No ▶	first and perform Quick Test.
A2 CHECK FOR SPARK DURING CRANK		
• Using a Neon Bulb Spark Tester (OTC D 89P-6666-A), check for spark at all spark plug wires while cranking. • **Was spark present on ALL spark plug wires and consistent (one spark per crankshaft revolution)?**	Yes ▶	GO to A3.
	No ▶	GO to A10.
A3 CHECK PIP AT EDIS MODULE		
• Key off. • Install EDIS diagnostic cable to breakout box. • Use EDIS 6 overlay. • Connect LED test lamp between J43 (PIP E) and J7 (GND). • **Does the LED test lamp blink continuously during crank?**	Yes ▶	GO to A4.
	No ▶	GO to A7.
A4 CHECK PIP CONTINUITY TO EEC		
• Key off. • Disconnect the EEC processor. • **Is the resistance between J43 (PIP E) and Pin 56 (PIP) of the EEC connector less than 5 ohms?**	Yes ▶	GO to A5.
	No ▶	SERVICE open circuit. RECONNECT the EEC processor and RERUN Quick Test.
A5 CHECK IGND AT EDIS		
• Key off. • **Is the resistance between J47 (IGND E) and J7 (BAT−) less than 5 ohms?**	Yes ▶	GO to A6.
	No ▶	REPLACE EDIS module and RERUN Quick Test.

EDIS No Start — Pinpoint Test A

TEST STEP	RESULT ▶	ACTION TO TAKE
A6 CHECK IGND CONTINUITY TO EEC		
• Key off. • **Is the resistance between J47 (IGND E) and Pin 16 (IGN GND) of the EEC connector less than 5 ohms?**	Yes ▶	REPLACE EDIS module. RUN Quick Test.
	No ▶	SERVICE open circuit. RECONNECT EEC and RERUN Quick Test.
A7 CHECK PIP AT EDIS MODULE/ISOLATE EEC		
• Key off. • Disconnect the EEC processor. • Crank the engine. • **Does the LED test lamp blink continuously during crank?**	Yes ▶	REPLACE EEC processor. RERUN Quick Test.
	No ▶	GO to A8.
A8 CHECK PIP CIRCUIT FOR SHORT TO GROUND		
• Key off. • Disconnect the EDIS module from the EDIS diagnostic cable. • Disconnect the HEGO sensor. • **Is the resistance between J43 (PIP E) and J7 (BAT−) less than 5 ohms?**	Yes ▶	PIP is shorted to ground in harness. CHECK connectors, SERVICE or REPLACE harness. RERUN Quick Test.
	No ▶	GO to A9.
A9 CHECK PIP CIRCUIT FOR SHORT TO POWER		
• **Is the resistance between J43 (PIP E) and J51 (VBAT E) less than 5 ohms?**	Yes ▶	PIP is shorted to power in harness. CHECK connectors, SERVICE or REPLACE harness. RUN Quick Test.
	No ▶	REPLACE EDIS module. RERUN Quick Test.

EDIS No Start | **Pinpoint Test** | **A**

TEST STEP	RESULT ▶	ACTION TO TAKE
A10 DETERMINE MISSING SPARK PATTERN		
• **Is spark missing from BOTH cylinder 1 and 5 plug wires?** or • **Is spark missing from BOTH cylinder 2 and 6 plug wires?** or • **Is spark missing from BOTH cylinder 3 and 4 plug wires?**	Yes ▶ No ▶	GO to A12. GO to A11.
A11 CHECK PLUGS AND WIRES		
• Check spark plug wires for insulation damage, looseness, shorting or other damage. • Remove and check spark plugs for damage, wear, carbon deposits and proper plug gap. • **Are spark plugs and wires OK?**	Yes ▶ No ▶	REINSTALL plugs and wires. GO to A12. SERVICE or REPLACE damaged component. RERUN Quick Test.
A12 CHECK VBAT C AT COIL		
• Key off. • Install EDIS diagnostic cable to breakout box. • Use EDIS 6 overlay. • Connect the LED test lamp between J5 (VBAT C) and J7 (BAT-). • Key On. • **Is the LED test lamp on and bright?**	Yes ▶ No ▶	GO to A13. SERVICE open circuit. RECONNECT the EEC processor and RERUN Quick Test.
A13 CHECK COIL FOR FAILURE		
• Connect LED test lamp between J7 (BAT-) to each coil - J3, J6 and J10, one at a time. • Key on. • **Is the LED test lamp on and bright?**	Yes ▶ No ▶	GO to A14. REPLACE Coil pack. RERUN Quick Test.
A14 CHECK GND CONTINUITY TO EDIS MODULE		
• Key off. • **Is the resistance between J27 (GND E) and J7 (BAT-) less than 5 ohms?**	Yes ▶ No ▶	GO to A15. SERVICE open circuit. RECONNECT EEC and RERUN Quick Test.

EDIS No Start | **Pinpoint Test** | **A**

TEST STEP	RESULT ▶	ACTION TO TAKE
A15 CHECK GND IN EDIS MODULE		
• Key off. • **Is the resistance between J27 (GND E) and J47 (IGND E) less than 5 ohms?**	Yes ▶ No ▶	GO to A16. REPLACE EDIS module. RERUN Quick Test.
A16 CHECK VBAT E AT EDIS MODULE		
• Connect the LED test lamp between J51 (VBAT E) and J7 (BAT-). • Key on. • **Is the LED test lamp on and bright?**	Yes ▶ No ▶	GO to A17. SERVICE open circuit. RECONNECT the EEC processor and RERUN Quick Test.
A17 CHECK VRS AMPLITUDE AT EDIS MODULE		
• **Is the A/C voltage between J35 (VRS+ E) and J48 (VRS- E) more than 1 volt when the engine is cranked?**	Yes ▶ No ▶	GO to A18. GO to A19.
A18 CHECK VRS BIAS AT EDIS MODULE		
• Key on. • **Is the DC voltage between J35 (VRS+ E) and J27 (GND E) between 1.0 and 2.0 volts?**	Yes ▶ No ▶	REPLACE EDIS module. RERUN Quick Test. GO to A28.
A19 CHECK VRS CRANKING VOLTAGE AT EDIS		
• Key off. • Disconnect the EDIS module from the EDIS diagnostic cable. • **Is the A/C voltage more than 1 volt when the engine is cranked?**	Yes ▶ No ▶	REPLACE EDIS module. RERUN Quick Test. GO to A20.
A20 CHECK VRS CRANKING VOLTAGE AT SENSOR		
• Key off. • Disconnect the VRS engine harness from the EDIS diagnostic cable. • **Is the A/C voltage between J32 (VRS- S) and J31 (VRS+ S) more than 1 volt when the engine is cranked?**	Yes ▶ No ▶	GO to A22. GO to A21.

4.0L ENGINE—IGNITION SYSTEM TESTS—NAVAJO

EDIS No Start | **Pinpoint Test** | **A**

TEST STEP	RESULT ▶	ACTION TO TAKE
A21 CHECK VR SENSOR AND CRANKSHAFT WHEEL		
• Check the crankshaft data wheel for damage. • Check the air gap. • **Is the air gap correct and is the crankshaft data wheel OK?**	Yes ▶	REPLACE VR sensor. RERUN Quick Test.
	No ▶	REPAIR or REPLACE bad parts. SET air gap.
A22 CHECK VRS+ CIRCUIT FOR OPEN		
• Key off. • Reconnect the EDIS diagnostic cable. • **Is the resistance between J35 (VRS+ E) and J31 (VRS+ S) less than 5 ohms?**	Yes ▶	GO to A23 .
	No ▶	SERVICE open circuit in VRS+ harness. RERUN Quick Test.
A23 CHECK VRS- CIRCUIT FOR OPEN		
• Key off. • **Is the resistance between J32 (VRS- S) and J48 (VRS– E) less than 5 ohms?**	Yes ▶	GO to A24 .
	No ▶	SERVICE open circuit in VRS- harness. RERUN Quick Test.
A24 CHECK VRS- CIRCUIT FOR SHORT TO POWER		
• Key off. • Disconnect the HEGO sensor. • **Is the resistance between J48 (VRS- E) and J51 (VBAT E) greater than 10K ohms?**	Yes ▶	GO to A25 .
	No ▶	SERVICE short circuit in VRS- harness. RERUN Quick Test.
A25 CHECK VRS- CIRCUIT FOR SHORT TO GROUND		
• **Is the resistance between J48 (VRS- E) and J7 (BAT-) greater than 10K ohms?**	Yes ▶	GO to A26 .
	No ▶	SERVICE short circuit in VRS- harness. RERUN Quick Test.

EDIS No Start | **Pinpoint Test** | **A**

TEST STEP	RESULT ▶	ACTION TO TAKE
A26 CHECK VRS+ CIRCUIT FOR SHORT TO GROUND		
• **Is the resistance between J35 (VRS+ E) and J7 (BAT-) greater than 10K ohms?**	Yes ▶	GO to A27 .
	No ▶	SERVICE short circuit in VRS+ harness. RERUN Quick Test.
A27 CHECK VRS+ CIRCUIT FOR SHORT TO POWER		
• **Is the resistance between J35 (VRS+ E) and J51 (VBAT E) greater than 10K ohms?**	Yes ▶	VRS+ is shorted to VRS-. SERVICE short in harness. RERUN Quick Test.
	No ▶	SERVICE short circuit in VRS+ harness. RERUN Quick Test.
A28 CHECK VRS BIAS AT EDIS MODULE		
• Key off. • Disconnect the VR sensor from the EDIS diagnostic cable. • Key on. • **Is the DC voltage between J35 (VRS+ E) and J7 (BAT-) between 1.0 and 2.0 volts?**	Yes ▶	REPLACE VR sensor. RERUN Quick Test.
	No ▶	GO to A29 .
A29 CHECK VRS+ CIRCUIT FOR SHORT TO POWER		
• Key off. • Disconnect the EDIS module from the EDIS diagnostic cable. • Disconnect the HEGO sensor. • **Is the resistance between J35 (VRS+ E) and J51 (VBAT E) greater than 10K ohms?**	Yes ▶	GO to A30 .
	No ▶	VRS+ is shorted to power in harness. CHECK connectors, SERVICE or REPLACE harness. RERUN Quick Test.
A30 CHECK VRS+ CIRCUIT FOR SHORT TO GROUND		
• **Is the resistance between J35 (VRS+ E) and J7 (BAT-) greater than 10K ohms?**	Yes ▶	GO to A31 .
	No ▶	SERVICE short circuit in VRS+ harness. RERUN Quick Test.

4.0L ENGINE—IGNITION SYSTEM TESTS—NAVAJO

EDIS No Start	Pinpoint Test	A

TEST STEP	RESULT ▶	ACTION TO TAKE
A31 CHECK VRS– CIRCUIT FOR SHORT TO POWER		
• **Is the resistance between J48 (VRS– E) and J51 (VBAT E) greater than 10K ohms?**	Yes ▶	GO to A32.
	No ▶	SERVICE short circuit in VRS– harness. RERUN Quick Test.
A32 CHECK VRS– CIRCUIT FOR SHORT TO GROUND		
• **Is the resistance between J48 (VRS– E) and J7 (BAT–) less than 5 ohms?**	Yes ▶	SERVICE short circuit in VRS– harness. RERUN Quick Test.
	No ▶	REPLACE EDIS module. RERUN Quick Test.

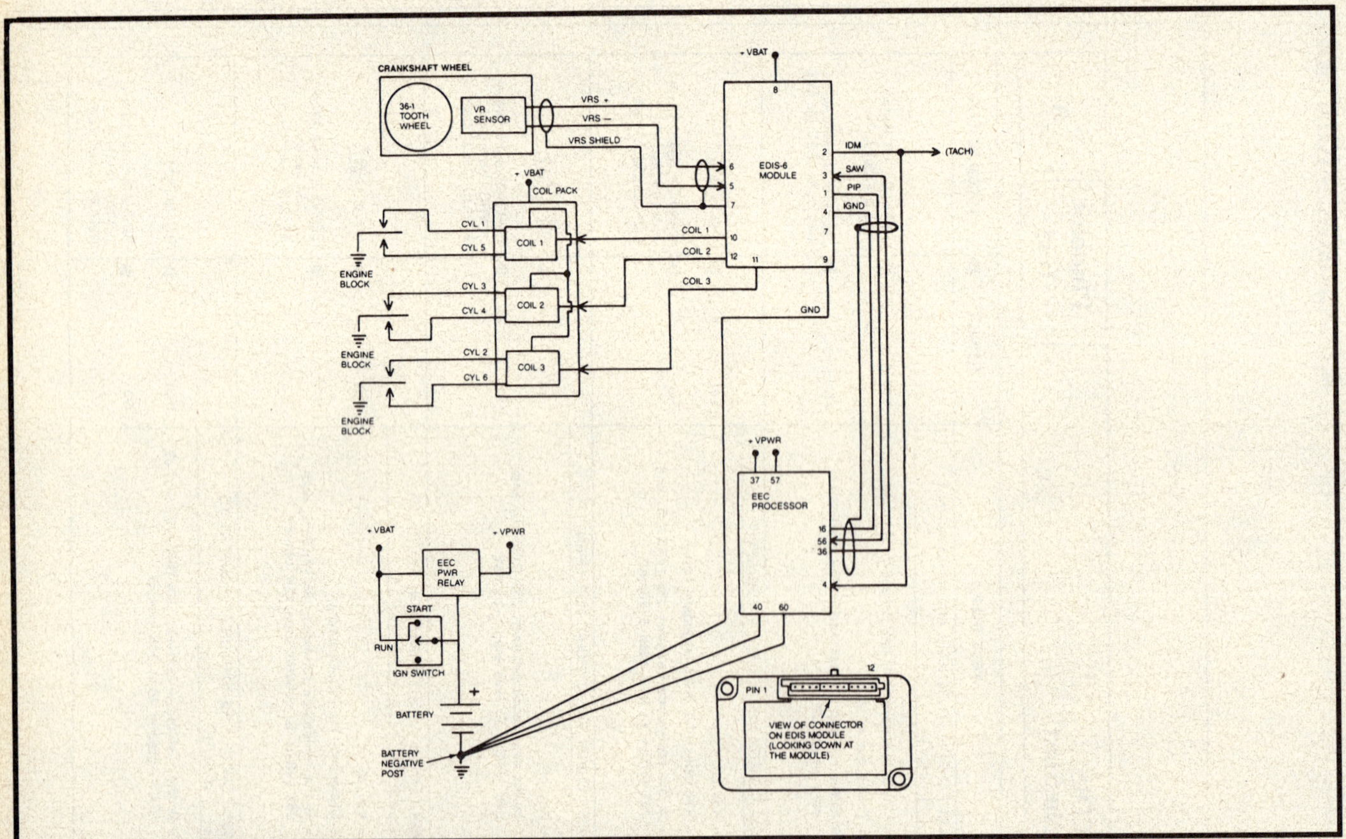

Wiring schematic of the EDIS ignition system

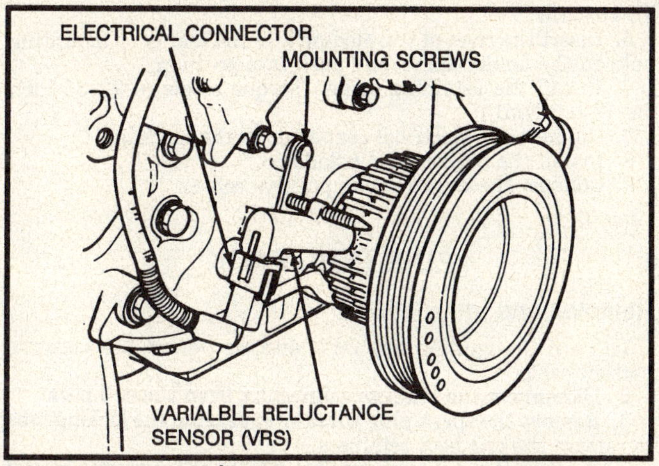

Removal and installation of the crankshaft timing sensor

Component Replacement

CRANKSHAFT TIMING SENSOR

Removal and Installation

1. Disconnect the negative battery cable.
2. Disconnct the electrical harness connector from the sensor.
3. If necessary, remove the serpintine drive belt.
4. Remove the sensor retaining bolts and remove the sensor from the engine.

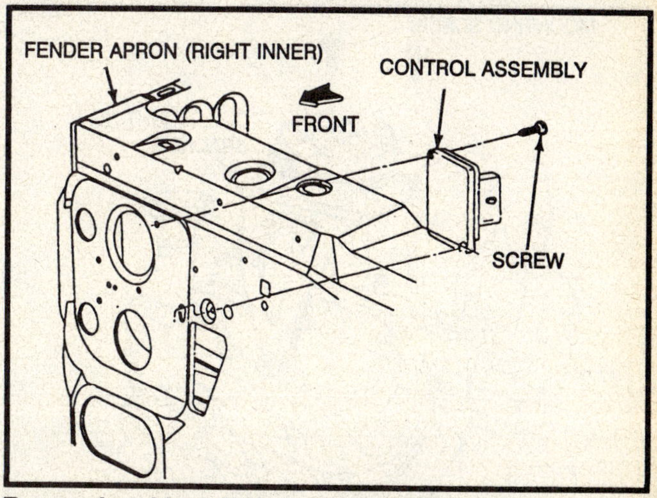

Removal and installation of the EDIS module

To install:

5. Position the crankshaft sensor assembly on the engine and install the retaining bolts. Torque the bolts 75–106 inch lbs. (9–12 Nm).
6. Route sensor wiring harness and connect the electrical connector.
7. Connect the negative battery cable.

EDIS IGNITION MODULE

Removal and Installation

1. Disconnect the negative and positive battery cables.

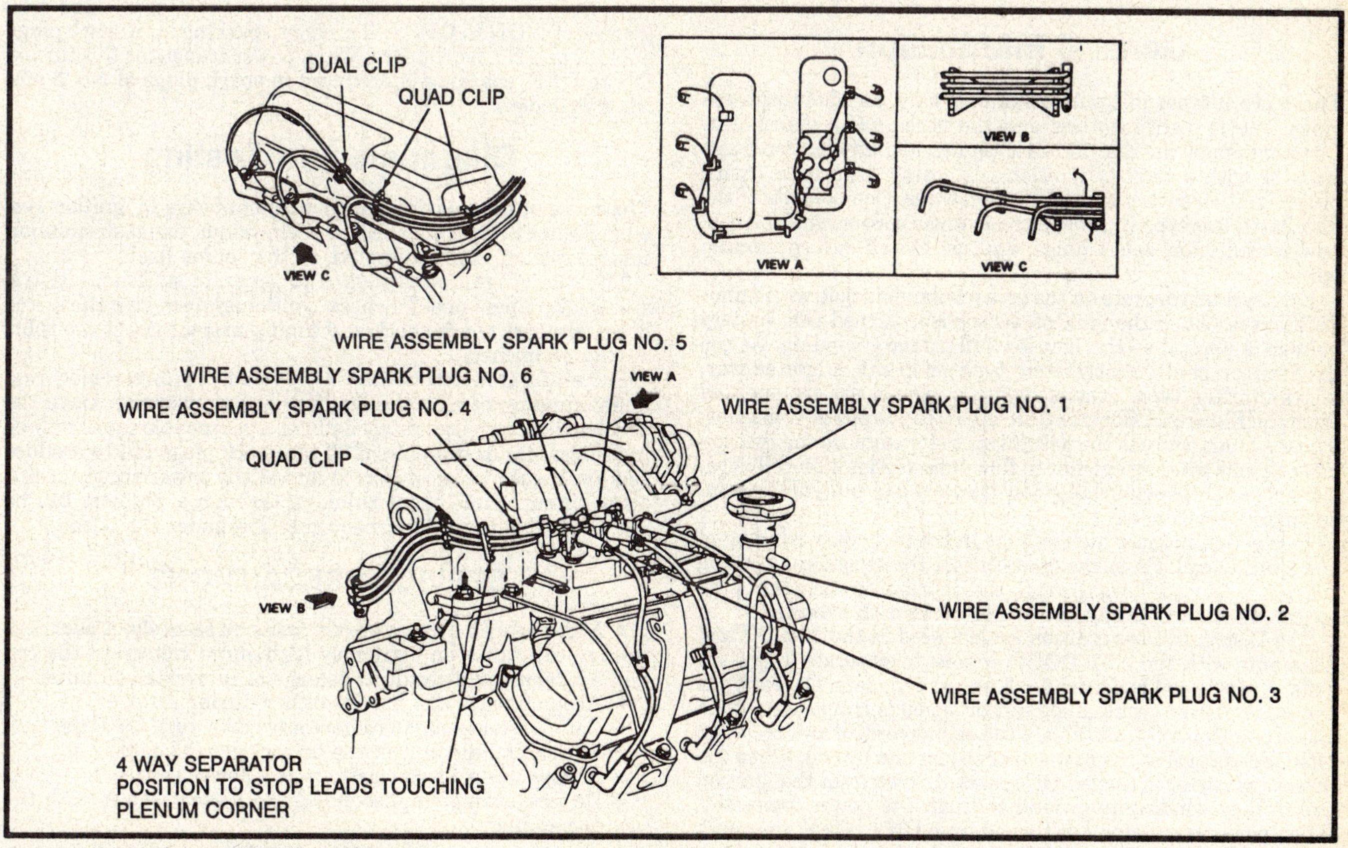

Spark plug wire routing locations

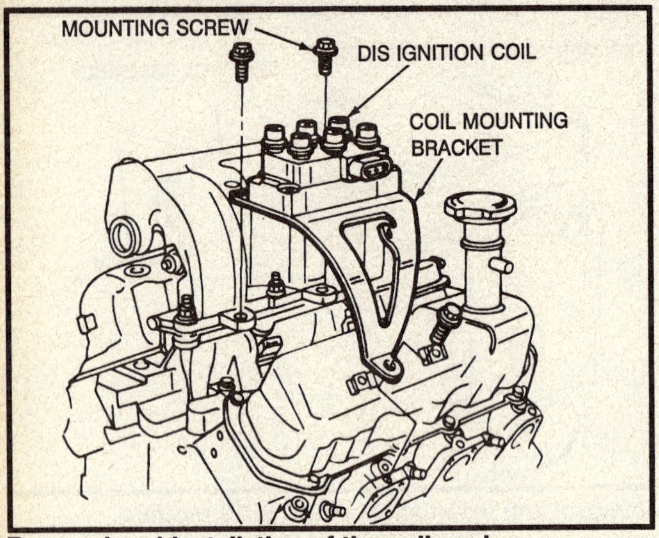

Removal and installation of the coil pack

2. Remove the battery holddown and remove the battery.
3. Disconnect electrical harness connector from the module.
4. Remove the retaining screw and remove the module from the fender apron.

To install:

5. Insert the rivet of the module into the tear drop mounting hole on the fender apron and push into position.
6. Install the retaining screw. Torque screw to 22–31 inch lbs. (2–3.5 Nm).
7. Connect the electrical connector to the module.
8. Install the battery and holddown.
9. Connect the negative and positive cables.

EDIS COIL PACK

Removal and Installation

1. Turn the ignition key **OFF** and disconnect the negative battery cable.
2. Disconnect the harness connectors from the coil pack.
3. Remove the spark plug wires by squeezing the locking tabs to release the coil boot retainers.
4. Remove the coil pack retaining bolts and remove the coil pack.

To install:

5. Position the coil pack in place on the engine and install the retaining bolts. Tighten to 44–62 inch. lbs. (5–7 Nm).
6. Install the spark plug wires. Be certain the boot retainers are firmly seated and latched on the coil pack.
7. Reconnect the coil pack electrical connector.
8. Reconnect the negative battery cable.

MITSUBISHI ELECTRONIC CONTROL IGNITION TIMING (ECIT) SYSTEM

General Information

There are 2 types of ignition systems used on Mitsubishi engines: one is distributorless and the other has a distributor. Components of the distributor type include; an electronic distributor with a built-in ignition coil, power transistor, spark plugs, ignition switch and connection wires. Components of the distributorless type include; a crank angle sensor, ignition coil, power transistor, spark plugs, ignition switch and connection wires.

Both systems operate on the on a similar principle with minor differences. When the ignition switch is is turned **ON**, battery voltage is applied to the ignition coil primary winding. As the distributor shaft, distributor type or crank angle sensor, distributorless type, rotates, ignition signals are transmitted from the Electronic Control Unit (ECU) to the power transistor. These signals activate the power transistor to cause the ignition coil primary winding current to flow from the ignition coil negative terminal through the power transistor to ground or be interrupted, repeatedly.

This action induces high voltage in the secondary winding of the ignition coil. From the ignition coil, the secondary winding current produced flows through the distributor and spark plug to ground, thus creating ignition spark in each cylinder.

The Distributorless ignition system used in the 3000GT and Diamante with the 3.0L DOHC engine, incorporate 3 ignition coils. A single coil is paired for 2 paired cylinders. Operation is the same. As the crank angle sensor signal is transmitted to to Electronic Control Unit (ECU), the engine control unit creates a ON-OFF control to the power transistors one by one. When the power transistor is turned **ON**, current flows from the ignition coil primay winding to ground through the power transistor. When power transistor A is turned from **ON** to **OFF**, the spark plugs of No. 1 and No. 4 cylinders spark. Turning of the power transistor B **ON** to **OFF** will produce sparking in in spark plugs of No. 2 and No. 5 cylinders. When power transistor C is turned **ON** to **OFF**, sparking is produced in spark plugs of No. 3 and No. 6 cylinders.

Diagnosis and Testing

The cause of hard engine starting is not always in ignition system. Faulty parts may exist in fuel system, exhaust emission control system, starting system or the engine itself.

The role of ignition system is to generate sufficient electric sparks at proper time. To check ignition system, therefore, it is necessary that spark check and timing measurement are thoroughly performed.

Successful troubleshooting on the vehicle requires a moderate understanding of how the ignition system operates. Once the problem has been accurately defined, a reasonable cause may be investigated. For example, if all the spark plugs fail to produce sparks, a reasonable cause is failure of the power supply or primary circuit. Time spent thinking through a problem can be more valuable than time spent guessing under the hood.

SERVICE PRECAUTIONS

- Make sure all connectors are fastened securely. A poor connection can cause an extremely high surge voltage in the coil and condenser and result in damage to integrated circuits.
- Keep all parts and harnesses dry during service.
- Before attempting to remove any parts, turn **OFF** the ignition switch and disconnect the battery ground cable.
- Always use a 12 volt battery as a power source.
- Do not attempt to disconnect the battery cables with the engine running.
- Do not depress the accelerator pedal when starting.

• Do not rev up the engine immediately after starting or just prior to shutdown.

• Do not attempt to disassemble the ECU control unit under any circumstances.

• If a battery cable is disconnected, the memory will return to the ROM (programmed) values. Engine operation may vary slightly, but this is not an indication of a problem. Do not replace parts because of a slight variation.

• If installing a 2-way or CB radio, keep the antenna as far as possible away from the electronic control unit. Keep the antenna feeder line at least 8 in. away from the ECU harness and do not let the 2 run parallel for a long distance. Be sure to ground the radio to the vehicle body.

• Before attempting to remove any parts, turn **OFF** the ignition switch and disconnect the battery ground cable.

• Always use a 12 volt battery as a power source.

• Make sure the igniter is properly grounded to the body.

• Never allow the coil or tachometer terminals to touch ground or damage will result to the igniter or coil.

NO-SPARK, NO-START TEST

1. If the engine cranks but will not start, check ignition system for insufficient spark or no spark at all. Perform test in the following areas:
   a. Visually inspect the ignition coil and perform a coil resistance test.
   b. Inspect the distributor, cap and wires and test the power transistor.
   c. Check the crank angle sensor, harness and wiring.
   d. Inspect the condition of the spark plugs.
2. If the engine idles rough or stalls, check the following:
   a. Inspect the spark plugs, replace as required.
   b. Check the ignition timing.
   c. Check the ignition coil.
3. If the vehicle accelerates poorly, overheats or consumes excessive fuel, check the ignition timing and fuel system.

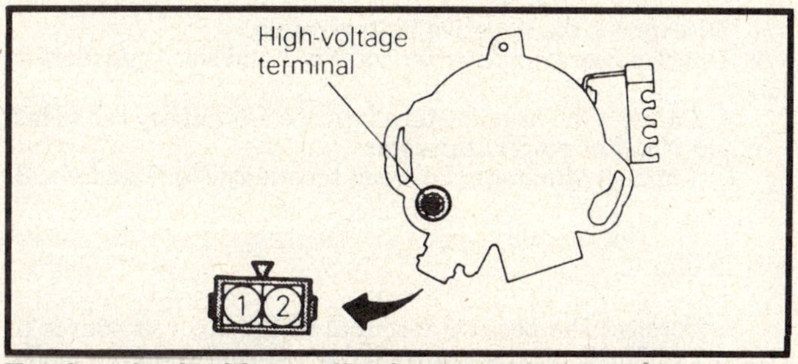

Testing the ignition coil primary and secondary resistances—coil and distributor system

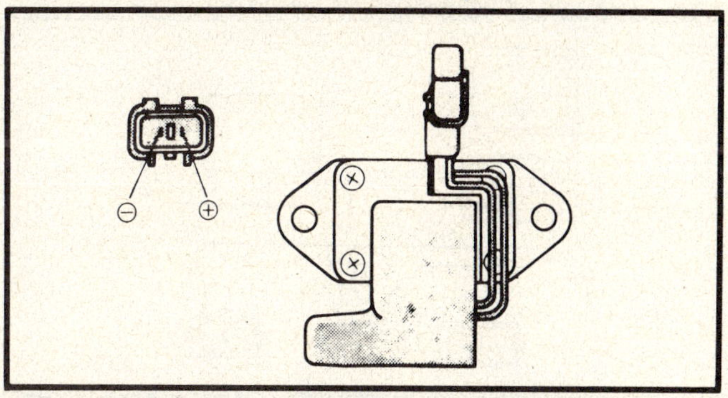

Testing the ignition coil primary and secondary resistances—single coil system

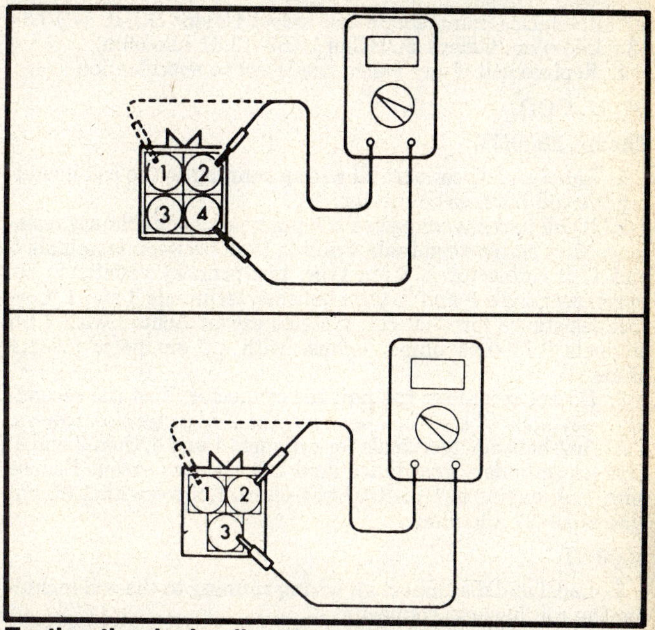

Testing the dual coil system with either 4 pin or 3 pin connector

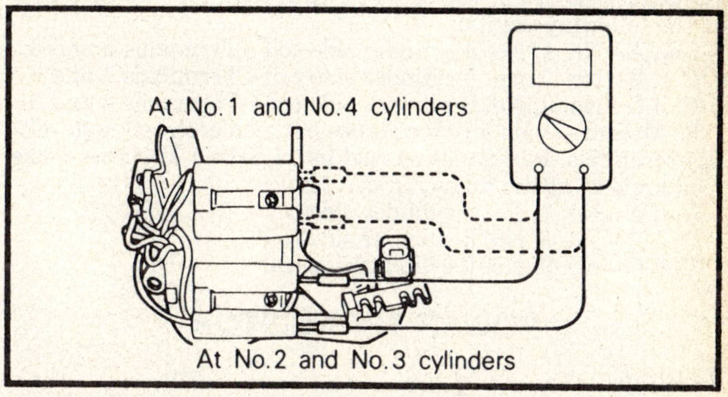

Testing the secondary coil resistance—dual coil system

IGNITION COIL

Testing

COIL AND DISTRIBUTOR

1. Disconnect the negative battery cable.
2. Label and disconnect the coil electrical connector at the distributor, including high tension cable.
2. Measure primary resistance between terminal 1 and 2 of the ignition coil. Resistace should be 0.9–1.2 ohms.
3. Measure secondary resistance between terminal 1 and high voltage terminal of the ignition coil. Resistace should be 20–29 kilo-ohms.
4. If resistance reading is not correct, replace the coil. Connect the distributor wiring and hight tension cable. Connect negative battery cable.

SINGLE COIL

1. Label and disconnect all wiring to coil, including high tension cable.
2. Measure primary coil resistance across positive and negative terminals of ignition coil. Resistance should be 0.72–0.88 ohms.
3. Measure the secondary coil resistance between ignition coil positive terminal and high voltage terminal. Correct resistance

for all vehicles using single coil except Galant SOHC is 10.3–13.9 kilo-ohm. Galant SOHC is 10.89–13.31 kilo-ohm.

4. Replace coil if any resistance is not to specification.

DUAL COIL

Except 3000GT

1. Label and disconnect all wiring running to the coil including the coil harness connector.
2. If coil harness connector is 4-pin type, check primary resistance first across terminals 4 and 2; then between terminals 4 and 1. If connector is 3-pin type, test primary resistance between terminals 3 and 2, then between terminals 3 and 1. Correct resistance for dual coil systems except Eclipse with 1.8L engineis 0.77–0.95 ohms. Eclipse with 1.8 engine is 0.9–1.1 ohms.
3. Do not reconnect the harness connector. Test the secondary resistance of each coil between the 2 high tension towers. Test first between terminals for cylinders 1 and 4, then 2 and 3. Correct secondary resistance dual coil systems except Eclipse with 1.8L engine is 10.3–13.9 kilo-ohms. Eclipse with 1.8L engine is 19–27 kilo-ohm.

3000GT

1. Label and disconnect all wiring running to the coil including the coil harness connector.
2. Measure the resistance between connector terminal No. 3 and each coil terminal. Resistance should be 0.67–0.81 ohms. Resistance should be the same for each coil. Take meaurement at the following points:
   Coil A–No. 1, No. 4 cylinder side coil–Terminals 2 and 3
   Coil B–No. 2, No. 5 cylinder side coil–Terminals 1 and 3
   Coil C–No. 3, No. 6 cylinder side coil–Terminals 4 and 3
3. Measure secondary resistance between each coil high voltage terminals. Resistance should be 11.3–15.3 kil-ohms. Take meaurement at the following points:
   Coil A–No. 1, No. 4 cylinder side coil
   Coil B–No. 2, No. 5 cylinder side coil
   Coil C–No. 3, No. 6 cylinder side coil

POWER TRANSISTOR

Testing

COIL AND DISTRIBUTOR SYSTEM

1. Disconnect the negative battery cable.
2. Label and disconnect electrical connectors at the distributor.
3. Connect the negative terminal of a 1.5 volt dry cell to terminal No. 7 of power transistor.
4. Connect an ohmmeter between terminals 7 and 2. Connect the positive terminal of the 1.5 volt source to terminal No. 8 on the connector; when connection is made, the ohmmeter should show continuity. No continuity should exist when voltage is removed from terminal No. 8.

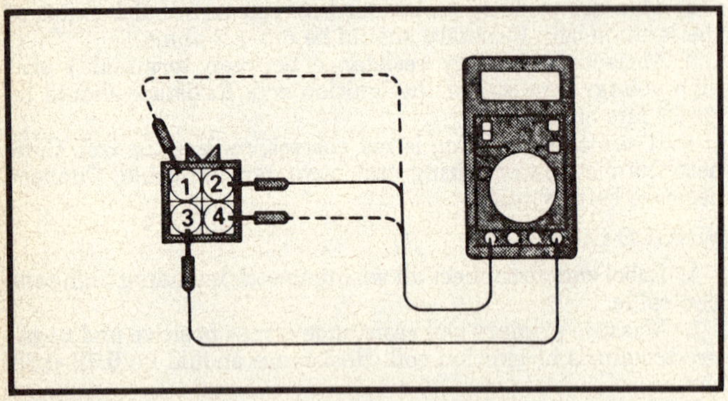

Testing coil primary resistance–3000GT and Diamante with DOHC engine

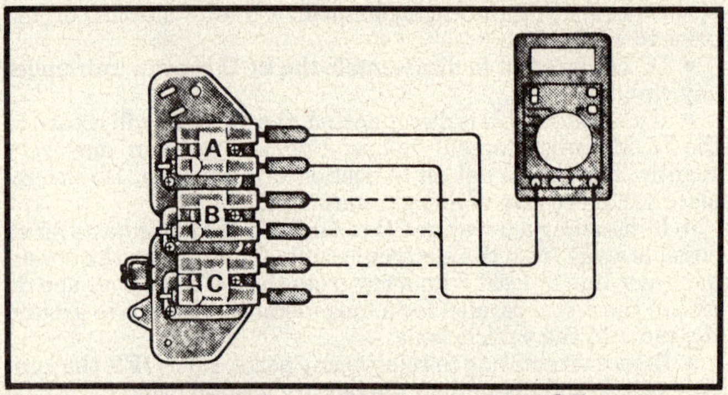

Testing coil secondary resistance–3000GT and Diamante with DOHC engine

5. If any test condition is not met, replace the power transistor. Connect the negative battery cable.

SINGLE COIL SYSTEM

1. Connect the negative terminal of a 1.5 volt dry cell to terminal No. 2 of power transistor.
2. Connect an ohmmeter between terminals 2 and 3. Connect the positive terminal of the 1.5 volt source to terminal No. 1 on the connector; when connection is made, the ohmmeter should show continuity. No continuity should exist when voltage is removed from terminal No. 1.
3. If any test condition is not met, replace the power transistor.

DUAL COIL

EXCEPT 3000GT

NOTE: The use of an accurate analog ohmmeter is recommended.

1. Disconnect the negative battery cable.
2. Check power transistor for coil No. 1 and No. 4 cylinders as follows:
   a. Connect the negative terminal of a 1.5 volt dry cell to terminal No. 3 of power transistor.
   b. Connect ohmmeter between terminals No. 7 and No. 3.

NOTE: The negative probe of the tester must be to terminal No. 7.

   c. Connect the positive terminal of the 1.5 volt source to terminal No. 6 on the connector; when connection is made, the ohmmeter should show continuity. No continuity should exist when voltage is removed from terminal No. 6.

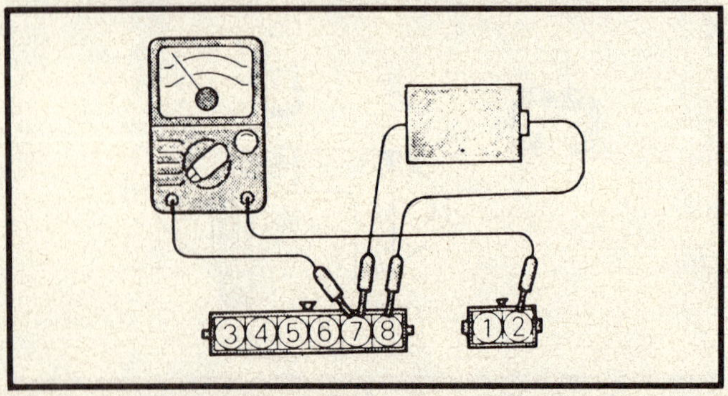

Testing the power transistor, using a analog type tester–coil and distributor system

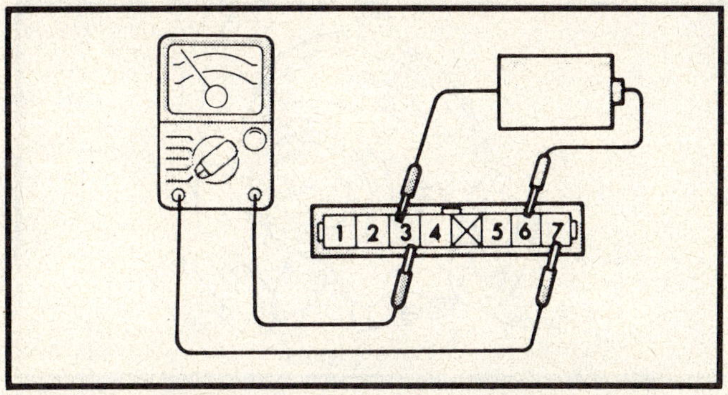

Testing the power transistor for the coil of the No. 1 and No. 4 cylinders—Mirage 1990–92 1.6L engine

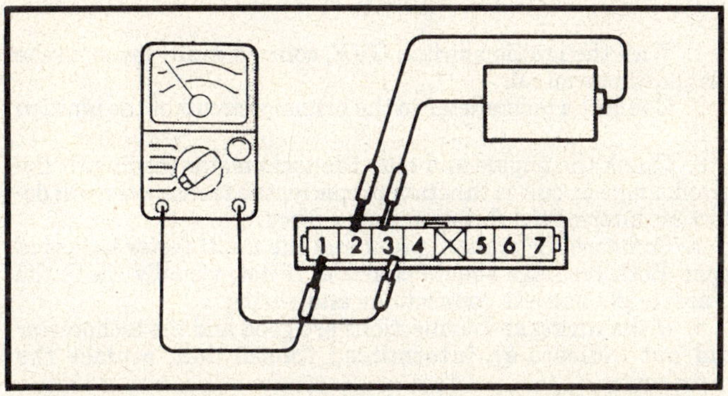

Testing the power transistor for the coil of the No. 2 and No. 3 cylinders—Mirage 1990–92 1.6L engine

3. Check power transistor for coil No. 2 and No. 4 cylinders as follows:
   a. Connect the negative terminal of a 1.5 volt dry cell to terminal No. 3 of power transistor.
   b. Connect ohmmeter between terminals 1 and 3.

NOTE: The negative probe of the tester must be to terminal No. 1.

   c. Connect the positive terminal of the 1.5 volt source to terminal No. 2 on the connector; when connection is made, the ohmmeter should show continuity. No continuity should exist when voltage is removed from terminal No. 2. If any test condition is not met, replace the powertransistor.

3000GT

NOTE: The use of an accurate analog ohmmeter is recommended.

1. Disconnect the negative battery cable.
2. Check power transistor for coil No. 1, No. 4 coil side as follows:
   a. Connect the negative terminal of a 1.5 volt dry cell to terminal No. 7 of power transistor.
   b. Connect ohmmeter between terminals No. 3 and No. 7.

NOTE: The negative probe of the tester must be to terminal No. 3.

   c. Connect the positive terminal of the 1.5 volt source to terminal No. 6 on the connector; when connection is made, the ohmmeter should show continuity. No continuity should exist when voltage is removed from terminal No. 6.
3. Check power transistor for coil No. 2 and No. 5 coil side as follows:
   a. Connect the negative terminal of a 1.5 volt dry cell to terminal No. 7 of power transistor.
   b. Connect an ohmmeter between terminals 2 and 7.

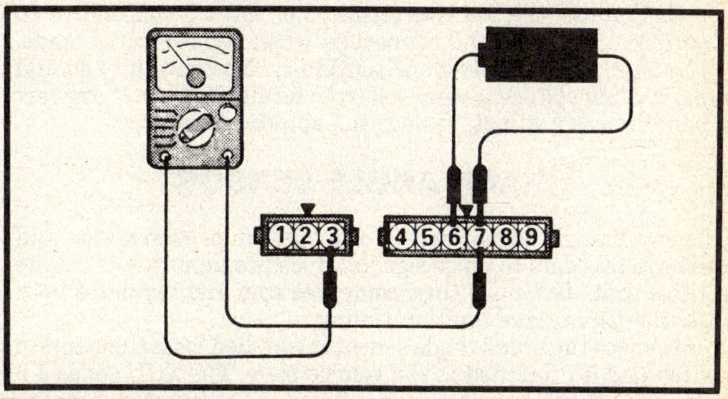

Testing the power transistor for the coil of the No. 1 and No. 4 cylinders—3000GT and Diamante with DOHC engine

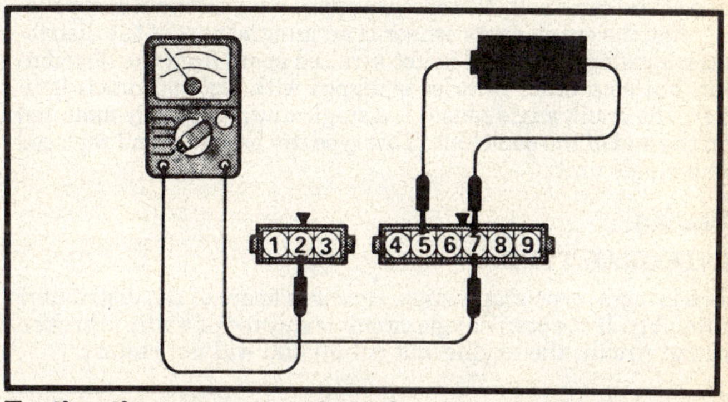

Testing the power transistor for the coil of the No. 2 and No. 5 cylinders—3000GT and Diamante with DOHC engine

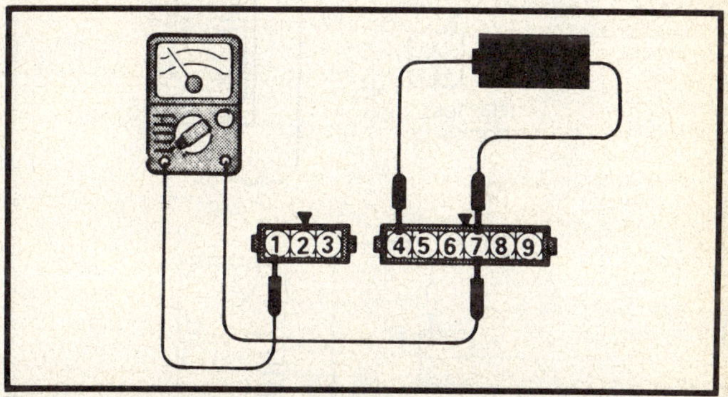

Testing the power transistor for the coil of the No. 3 and No. 6 cylinders—3000GT and Diamante with DOHC engine

NOTE: The negative probe of the tester must be to terminal No. 2.

   c. Connect the positive terminal of the 1.5 volt source to terminal No. 5 on the connector; when connection is made, the ohmmeter should show continuity. No continuity should exist when voltage is removed from terminal No. 5. If any test condition is not met, replace the powertransistor.
4. Check power transistor for coil No. 3 and No. 6 coil side as follows:
   a. Connect the negative terminal of a 1.5 volt dry cell to terminal No. 7 of power transistor.
   b. Connect an ohmmeter between terminals 1 and 7.

NOTE: The negative probe of the tester must be to terminal No. 1.

c. Connect the positive terminal of the 1.5 volt source to terminal No. 4 on the connector; when connection is made, the ohmmeter should show continuity. No continuity should exist when voltage is removed from terminal No. 5. If any test condition is not met, replace the powertransistor.

CRANK ANGLE SENSOR

The crank angle sensor detects the position of each piston and converts this data to pulse signals which are input to the engine control unit. The ECU then computes rpm and regulates both injection timing and ignition timing.

Power for the crank angle sensor is supplied from the ignition switch and is grounded to the vehicle body. The ECU sends a 5 volt signal to the sensor, which returns an interrupted signal to the computer. The pattern of interruption or pulsing is read by the ECU.

The crank angle sensor, if not within the distributor, is on the end of the camshaft. Vehicles equipped with distributor type ignitions, the crank angle sensor is an integral part of the distributor housing. This type is not serviced apart from the distributor, but as a unit. Vehicles equipped with distributorless ignition, the crank angle sensor is a single unit, externally mounted on the end of the camshaft. This type can be tested and replaced as a single unit.

Testing

INTEGRAL TYPE

The integral type crank angle sensor is apart of the distributor assembly. If the crank angle circuit is shorted or a wire is broken in the circuit, the engine will cut off and will not restart.

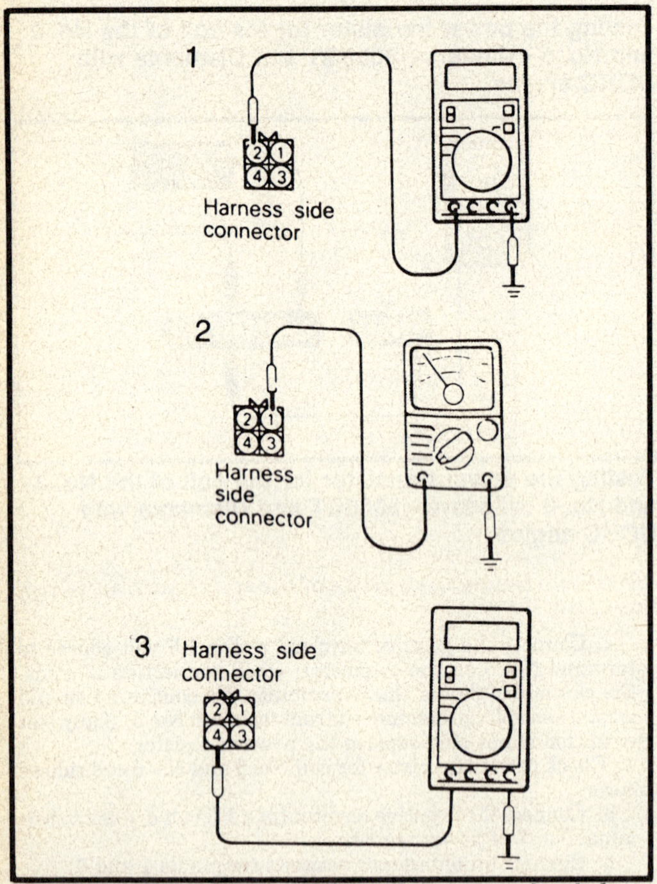

Checking the external crank angle sensor circuit for supply voltage

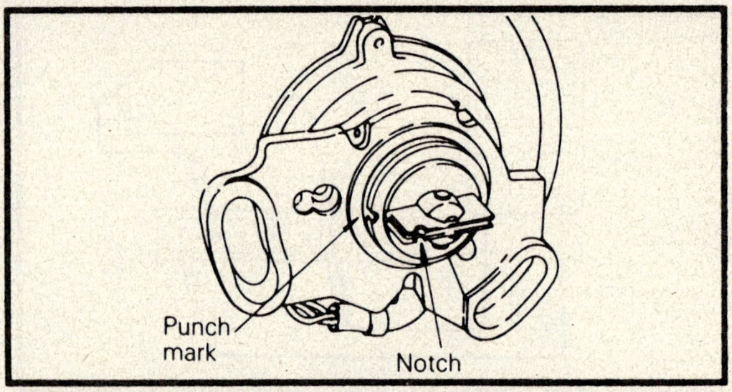

Alignment of the crank angle sensor before installation

1. With the ignition switch **OFF**, connect multi-tester to the diagnostic terminal.
2. Connect a tachometer to the primary circuit of the ignition coil.
3. Crank the engine and note the tachometer reading. If the crank angle circuit is function properly, the tachometer will detect an intermitted flutation in the circuit.
4. Compare the cranking rpm and the multi-tester indicated rpm. Both readings should coinside. If not, visually check the crank angle harness connections and wiring.
5. If the wiring and connections are good and the tachometer did not indicate an intermitted fluctuation, replace the distributor.

EXTERNAL TYPE

1. With engine **OFF**, disconnect external sensor lead and connect voltmeter between terminal 2 of wire harness (not sensor lead) and ground.
2. With ignition **ON**, meter should show supply voltage, approx. 12 volt.
3. Connect voltmeter between terminal 4 of wire harness and ground.
4. With ignition **ON**, meter should show ECU control voltage, 4.8–5.2 volts.
5. Connect ohmmeter between terminal 1 of wire harness and ground. Continuity should exist.
6. If any voltage or ground condition is not within specifications, the wiring and circuitry must be repaired.

IGNITION TIMING

Adjustment

NOTE: Engine must be at operating temperature, all lights and accessories OFF, transaxle in N or P, steering wheel centered and parking brake applied.

1. Connect timing light.
2. Insert a paper clip (from harness side) to the connector located between noise filter and primary side of ignition coil. Connect the tachometer lead to this clip.

NOTE: Wiring connector should not be separated. Insert paper clip along terminal surface. Incorrect insertion of paper clip may cause clip to become caught and not able to be removed. For male pin, insert on lock tab side. For female pin, insert on side opposite lock tab.

3. Inspect curb idle speed. Refer to the underhood emissions label for correct idle speed.
4. With engine stopped, connect a jumper wire with alligator clips to terminal for ignition-timing adjustment (located in engine compartment), and ground it.
5. Start and run engine at curb idle speed.
6. Check basic ignition timing and adjust as required.
7. If there is a deviation from standard value, adjust timing by either:
   a. Loosening distributor holding nut, and then turning distributor to make necessary adjustment or
   b. Loosening retaining nut on crank angle sensor and turning sensor to adjust timing.
8. Once timing is correctly set, tighten retaining nut. Make certain distributor or crank angle sensor does not move out of place during tightening.
9. Stop engine.
10. Disconnect lead wire connected at ignition timing adjustment connector.
11. Start and run engine at curb idle speed.
12. Check that actual timing is correct. Refer underhood emission label.

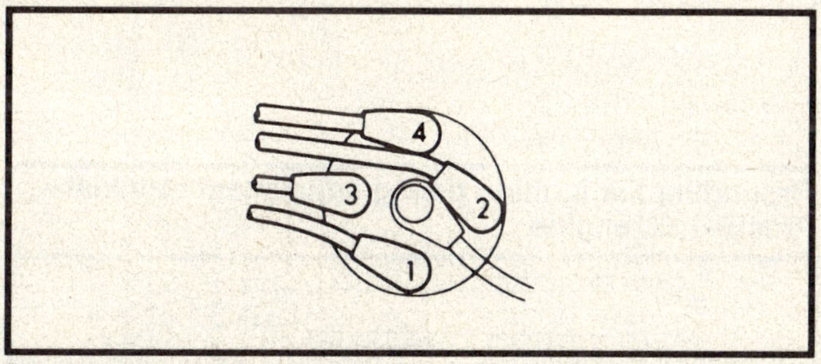

Spark plug wiring locations—1.5L, 1.8L, 2.0L and 2.4L engines are all similar

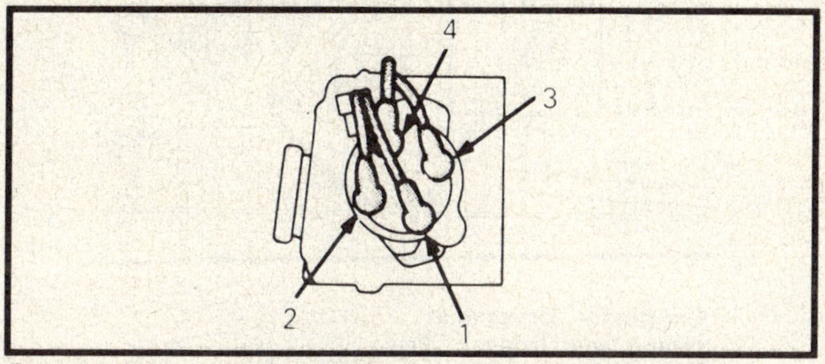

Spark plug wiring locations—1991–92 with 1.5L engine

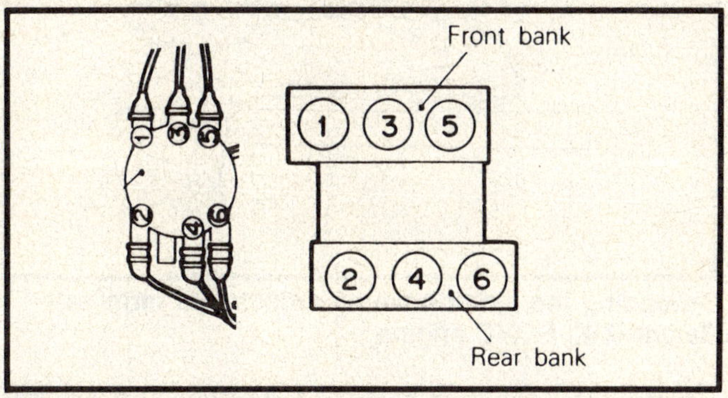

Spark plug wiring locations—Sigma, Diamante with 3.0L SOHC engine is similar

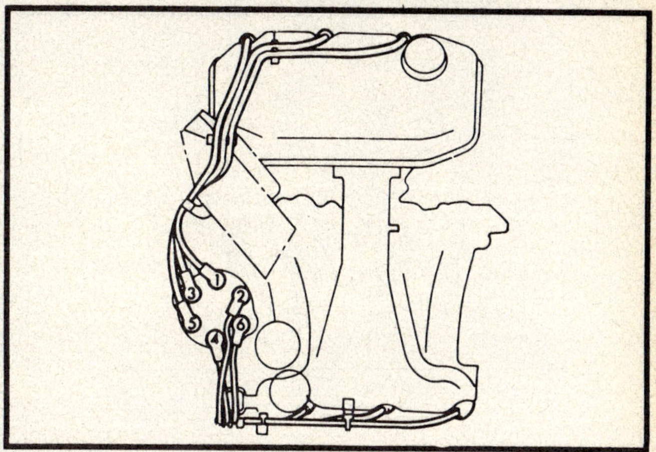

Spark plug wiring locations—Montero and Truck with 3.0L SOHC engine

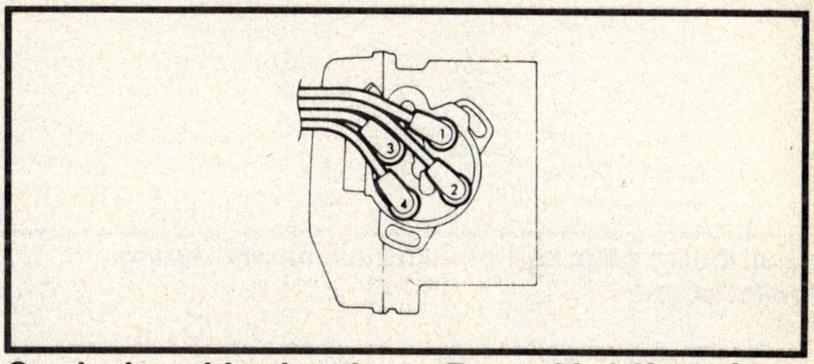

Spark plug wiring locations—Expo with 1.8L engine

NOTE: Timing at idle will fluctuate above and below ideal value; this variation is normal due to computer controlling idle quality.

Component Replacement

CRANK ANGLE SENSOR

Removal and Installation

INTEGRAL TYPE

The integral type crank angle sensor is apart of the distributor assembly. If service is required, the distributor must be replace as unit.

EXTERNAL TYPE

1. Disconnect the negative battery cable.
2. Prior to removing the crank sensor, turn the crankshaft so that cylinder No. 1 is at TDC on compression.
3. Disconnect the wiring harness to sensor. Remove retaining bolts and remove the sensor.
4. When installing, align punch mark on sensor housing with notch in plate., then install sensor to head.

DISTRIBUTOR ASSEMBLY

Removal and Installation

1. Rotate the crankshaft and bring up No. 1 piston to top dead center of its compression stroke.
2. Disconnect the negative battery cable.
3. Remove the retaining clips; remove distributor cap and seal ring from distributor.
4. Disconnect the wiring harness from distributor lead wire.
5. Disconnect the vacuum hoses from vacuum controller.

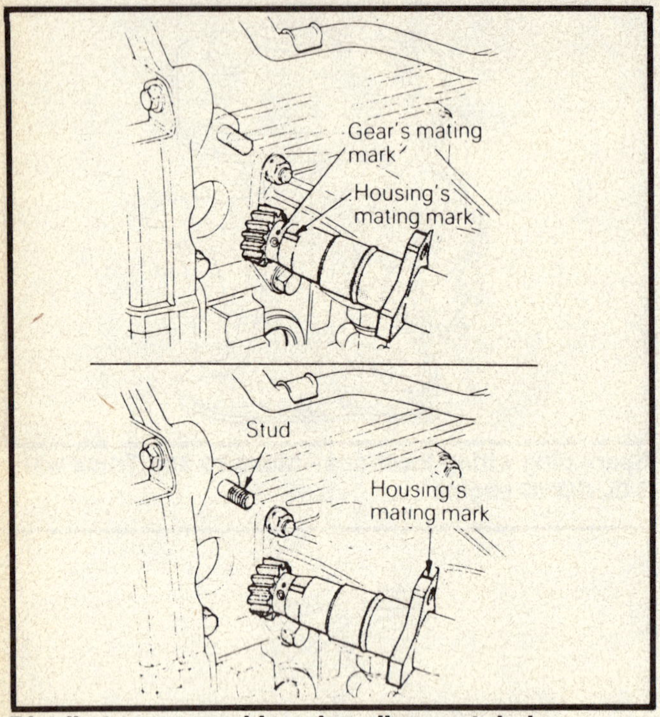

Distributor gear and housing alignment during installation

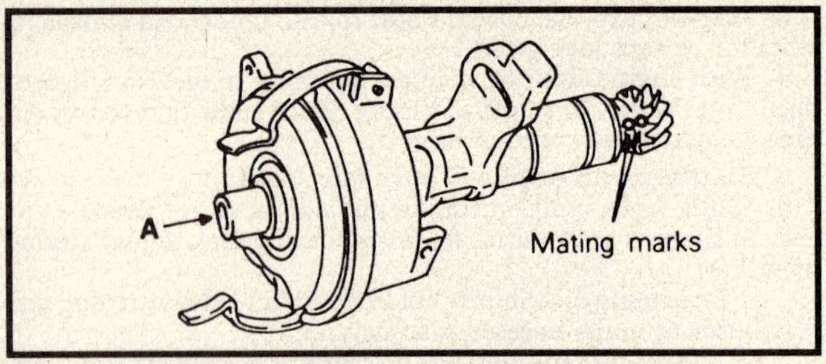

Aligning the driven gears mating mark and the distributor housing mating mark

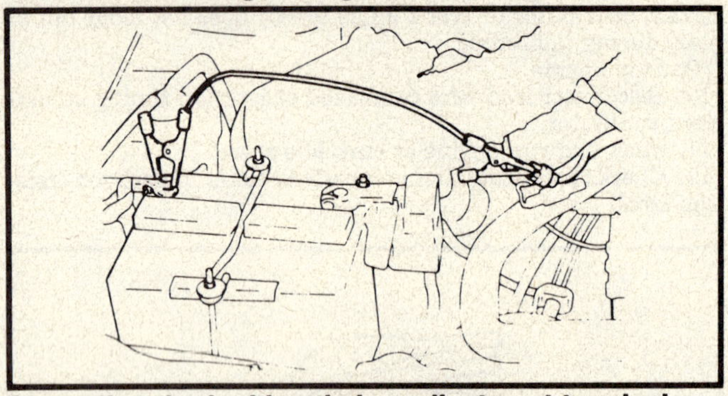
Grounding the ignition timing adjustment terminal—Precis–1.5L engine

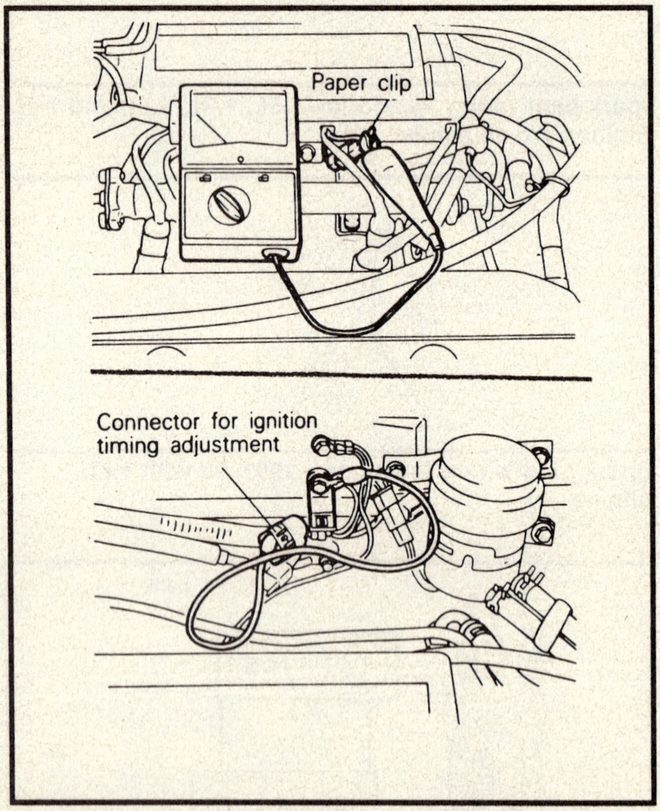

Grounding the ignition timing adjustment terminal—Galant–2.0L SOHC engine

6. Remove distributor mounting nut and remove distributor assembly from engine.

To Install:

7. Align the mating mark on distributor housing with mating mark (punch) on distributor driven gear.
8. Install distributor assembly into cylinder block while aligning mating mark (on distributor attaching flange) with center of distributor installing stud. Lightly tighten distributor mounting bolt.
9. Connect the wiring harness to distributor lead wire. Connect the vacuum hoses to vacuum controller.
10. Install seal ring and distributor cap to housing and engage retaining clips.
11. Connect the negative battery cable. Start engine and set timing. Tighten the distributor mounting bolt.

Disassembly

Distributors are similar but not identical. Procedure below is general.

1. Remove distributor from the engine.
2. Remove distributor cap, O–ring and rotor.
3. Disconnect lead wire from housing. Not all distributors allow wiring disconnection.
4. Make position identifying mark for driven gear on distributor shaft.
5. Place gear on soft base (wood block) and drive out spring pin with pin punch.
6. Remove O–ring and distributor housing assembly.
7. After cleaning or inspection, install O–ring on housing assembly.

Reassembly:

8. Align marks made at time of disassembly and install the gear to distributor shaft. When aligning the driven gear's mating mark and the housing's mating mark, set the shaft so that the rotor locating notch (top of the shaft) is positioned as shown. Align the pin hole and drive a new spring pin. Install the pin so that the slot is at a right angle relative to the shaft.
9. Connect the wire harness to the housing if it was disconnected.
10. Install the rotor, O-ring and cap. Reinstall the distributor.

Acura Fuel Injection

SECTION 5

ACURA PROGRAMMED FUEL INJECTION (PGM-FI) SYSTEM

General Information

Programmed Fuel Injection (PGM-FI) System is a fully electronic microprocessor based engine management system. The Electronic Control Unit (ECU) is given responsibility for control of injector timing and duration, intake air control, ignition timing, cold start enrichment, fuel pump control, fuel cut-off, A/C compressor operation, alternator control as well as EGR function and canister purge cycles.

The ECU receives electric signals from many sensors and sources on and around the engine. The signals are processed against pre-programmed values; correct output signals from the ECU are determined by these calculations. The ECU contains additional memories, back-up and fail-safe functions as well as self diagnostic capabilities.

Diagnosis is generally based on symptom diagnosis and stored fault codes, if any. Testing always requires the use of the diagnostic charts in conjunction with the Acura test harness (pinout box), a voltmeter and an ohmmeter.

ELECTRONIC CONTROL UNIT (ECU)

In order to get fuel into the cylinders at the correct instant and in correct amount, the control system must perform various separate functions. The ECU (Electronic Control Unit), the heart of the PGM-FI, uses a microcomputer and consists of a CPU (Central Processing Unit), memories, and I/O (Input/Output) ports. Basic data stored in the memories are compared to the signals sent from the various sensors to provide the correct air/fuel mixture for all engine needs.

The unit contains memories for the basic discharge duration at various engine speeds and manifold pressures. The basic discharge duration, after being read out from the memory, is further modified by signals sent from various sensors to obtain the final injector duration. Other functions also include those listed below.

Starting Control

The fuel system must vary the air/fuel ratio to suit different operating requirements. For example, the mixture must be rich for starting. The memories also contain the injector duration values to be triggered by signals from the starter switch, engine speed and coolant temperature sensors, thereby providing extra fuel needed for starting.

Electronic Air Control

The ECU controls the EACV to maintain correct idle speed based on engine and accessories demand.

Ignition Timing Control

The ECU controls the basic ignition timing based on engine load, engine rpm, vehicle speed and coolant temperature.

Fuel Pump Control

When the engine is not running but the ignition is **ON**, electric current to the fuel pump is cut off, preventing the injectors from discharging fuel. Additionally, the fuel pump is energized for 2 seconds when the ignition is first switched **ON**. Once the engine is running, the pump is constantly energized.

Fuel Cut-Off Control

During deceleration with the throttle valve nearly closed, electric current to the injectors is cut off at engine speeds of approximately 1000–1500 rpm, depending on the engine. This contributes to improved fuel economy. Fuel cut-off also takes place when engine speed exceeds the red line or safe operation rpm limit, regardless of throttle position.

VTEC Control

The NSX employs Variable Valve Timing and Lift Electronic Control (VTEC). The ECU is responsible for the control of this system as well as control of the intake manifold chamber and fuel pump discharge pressure.

Fail Safe

If the ECU notes the loss or change out-of-range of a sensor signal, the computer will ignore the faulty signal and substitute a fixed value in its place. This value may not necessarily be correct for the immediate driving situation, but will allow the engine to operate. For example, many ECU decisions are based on the engine coolant temperature. Should the coolant temperature sensor (TW sensor) signal be lost just after a cold start, the ECU will substitute the default value. This substitution value is based on an engine at normal temperature so vehicle performance will be affected until fully warmed up.

Back-up System

The ECU also tests itself while operating. If an abnormality occurs within the ECU, the system switches to a back-up circuit independent of the computing system. This system substitutes fixed values for all inputs and controls the injectors accordingly. Vehicle performance is usually reduced to minimal driveability. The back up function serves as a lifeboat to prevent the vehicle from being stranded in the event of ECU failure.

Self-Diagnosis

When a abnormality occurs, the ECU lights the engine warning light and stores a failure code in its memory. The fault code can be retrieved during diagnosis.

SENSORS AND INPUTS

CRANK, TDC AND CYL Sensors

The CRANK or crankshaft angle sensor signal is used to determine the timing for fuel injection and ignition timing. It is also used to generate the engine RPM signal. The TDC sensor signal, found only on Integra, Vigor and 1990 Legend, is used to determine ignition timing during start-up or when the CRANK signal is abnormal. The CYL sensor is used to detect the position of No. 1 cylinder; the signal is used to trigger the sequential fuel injection.

On Integra, Vigor and 1990 Legend, these sensors are contained within the distributor. The signals are generated by the rotation of toothed wheels passing through pick-up coils. In some cases, the CYL sensor is incorporated with the other two. The function of these components does not change, but if any single sensor is defective, the entire distributor must be replaced as an assembly. The 1990 Legend uses only a TDC sensor within the distributor.

For all Legend and NSX, the CRANK and CYL sensors are located behind camshaft pulleys on the left or front bank of cylinders.

Manifold Air Pressure (MAP) Sensor

The sensor converts manifold air pressure readings into electrical voltage signals and sends them to the ECU. This information is used along with signals from the crank angle sensor to compute the basic injector duration.

For Legend and NSX, the MAP sensor is located within the emission control box. On Integras, the sensor is locate on a bracket on the firewall. The Vigor sensor is located on the left fender inner apron, just behind the air cleaner housing.

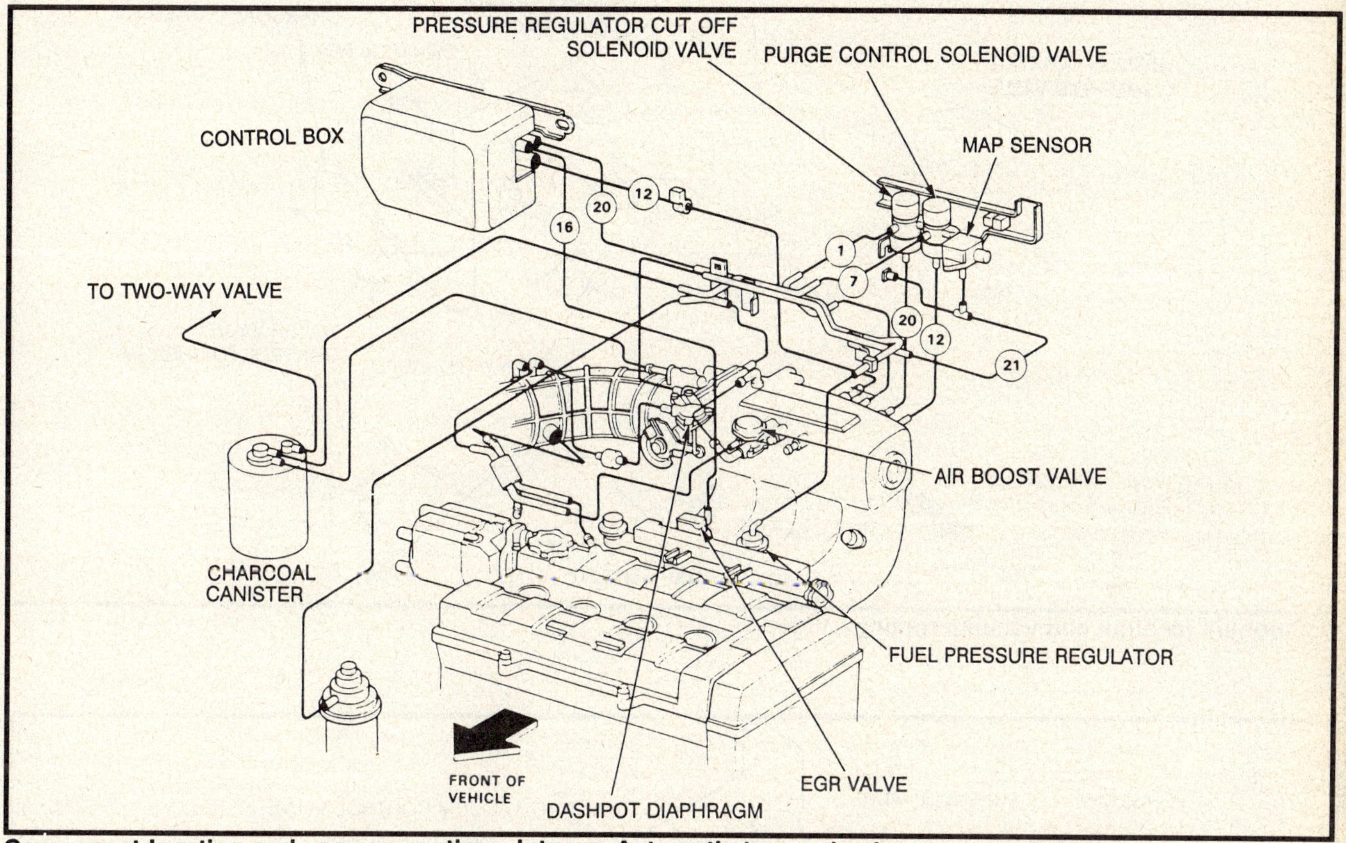

Component location and vacuum routing—Integra. Automatic transaxle shown, manual similar

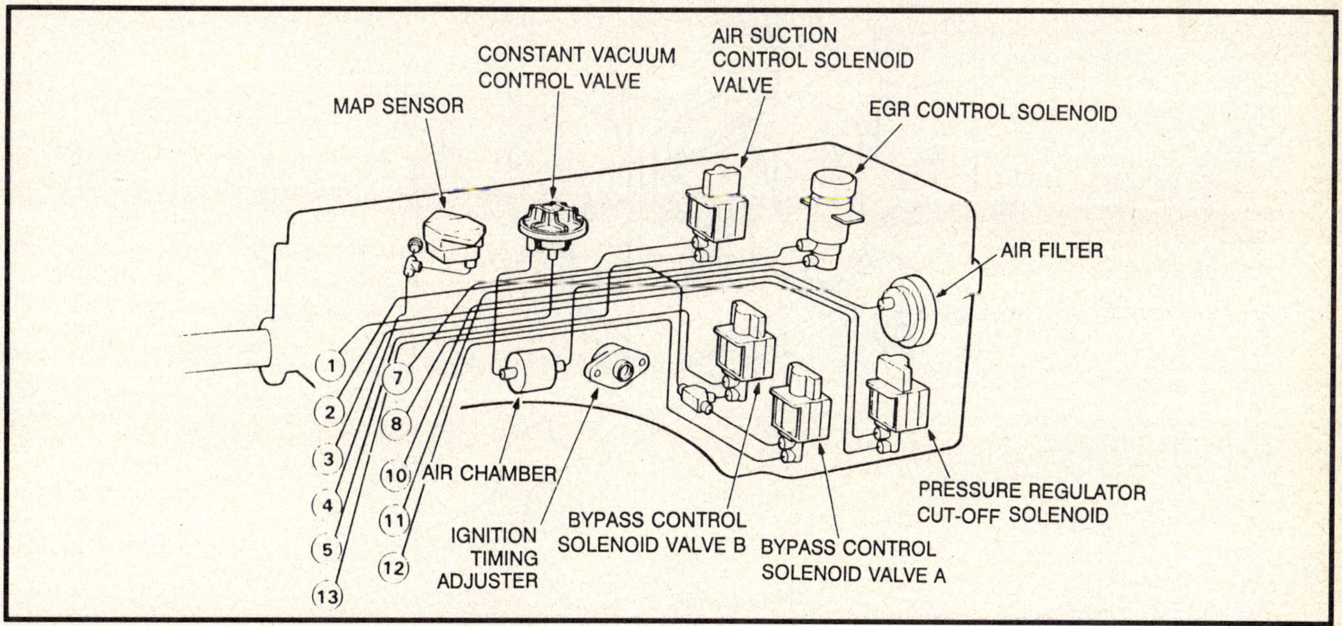

Emission control box components—1990 Legend and Coupe

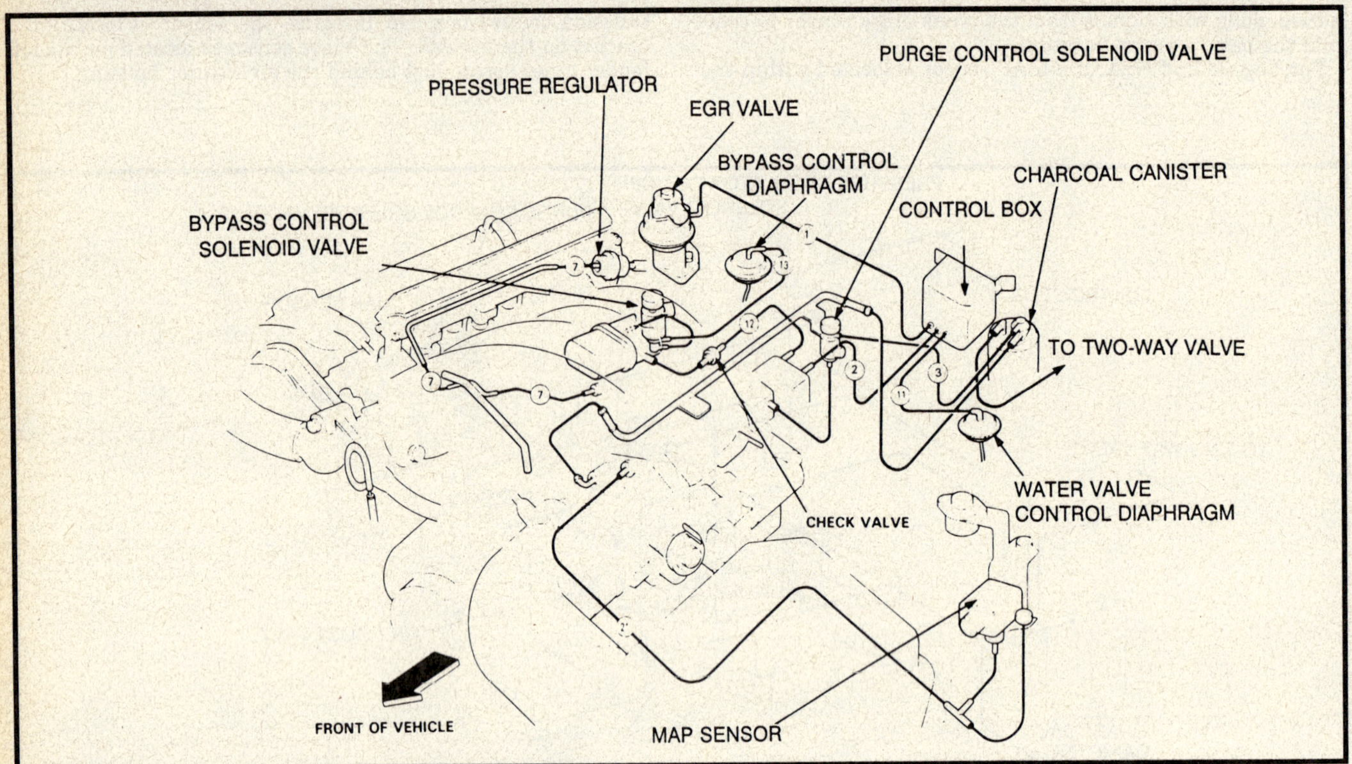

Component location and vacuum routing—Vigor

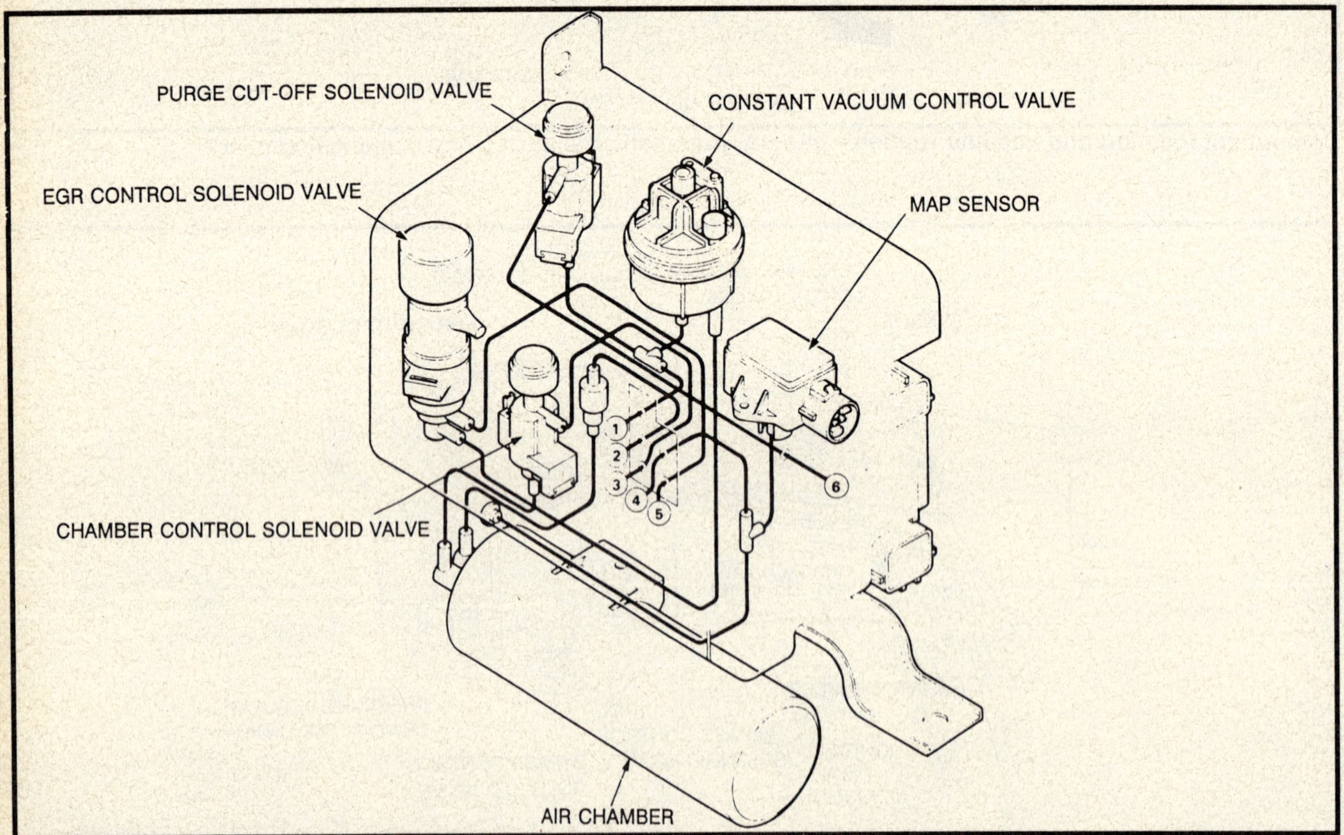

Emission control box components—NSX

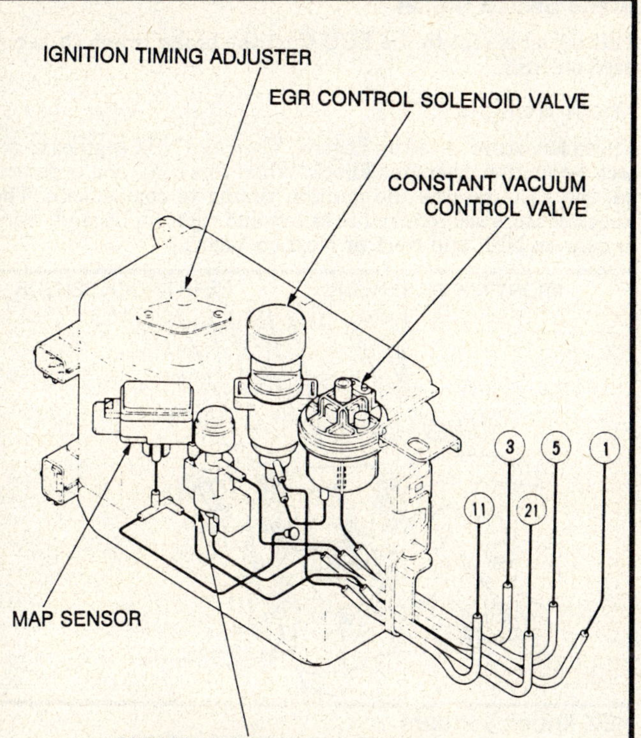

Emission control box components—1991–92 Legend and Coupe

Coolant Temperature (TW) Sensor

The sensor uses a temperature-dependent diode (thermistor) to measure differences in the coolant temperature. The basic injector duration is partially based on the signals sent from this sensor through the ECU. The resistance of the thermistor decreases with a rise in coolant temperature.

Intake Air Temperature (TA) Sensor

This device is also a thermistor and is placed in the intake manifold. It acts much like the water temperature sensor but with a reduced thermal capacity for for quicker response. The injector duration determined by the ECU is altered for different operating conditions by the signals sent from this sensor.

Throttle Angle Sensor

This sensor is a potentiometer which translates the position of the throttle plate to an electrical signal. The signal is near zero at idle and increases to just under 5 volts at wide-open throttle. The sensor is mounted to the side of the throttle body.

Oxygen Sensor

The oxygen sensor detects the oxygen content in the exhaust gas and sends an electrical signal to the ECU. The ECU uses this signal to control the fuel injector duration, maintaining the stoichiometric air/fuel ratio of 14.7:1. The sensor is located in the exhaust manifold; 6 cylinder engines use two sensors, one in each manifold.

This sensor is a hollow shaft of zirconia with a closed end. The inner and outer surfaces are plated with platinum, thus forming a platinum electrode. The inner surface or chamber is open to the atmosphere; the outer surface is exposed to the exhaust gas flow through the manifold.

Voltage is induced at the platinum electrode when there is any difference in oxygen concentration on the outside and inside. Operation of the device is dependent upon the changes in the induced voltage as the exhaust stream content changes.

Oxygen sensors only become efficient when operating at the proper temperature. All models except Integra use a heated oxygen sensor. The heater stabilizes the sensor's output and allows the sensor to heat quicker after the engine as been started.

EGR Valve Lift Sensor

The EGR lift sensor is found on all Acura vehicles except Integra with manual transaxle. The ECU controls the function of the EGR valve through the EGR control solenoid. The lift sensor translates the movement of the EGR shaft into an electric signal which is sent back to the ECU. The control unit compares this signal to pre-programmed values for optimum function. The sensor is located on the EGR valve.

Atmospheric Pressure (PA) Sensor

Used only on Integra and the 1990 Legend, this unit converts atmospheric pressures into voltage signals and sends them to the ECU. The signals then modify the basic injector duration to compensate for changes in the atmospheric pressure.

The PA sensor is located inside the passenger compartment of the vehicle. On 1990 Legend, it is located under the left dash, immediately behind the main relay. The Integra sensor is built into the ECU and is not accessible.

Vehicle Speed Sensor (VSS)

The signal from the vehicle speed sensor is used by the ECU as well as the cruise control and/or automatic transmission control units. For 1990 Legends and all Integra, the signal is generated by a pulser unit incorporated into the back of the speedometer/odometer unit.

Vigor and 1991–92 Legend use an external sensor mounted on top of the power steering speed sensor in the engine compartment. The power steering speed sensor is gear-driven by the engine and controls the steering boost in relation to the speed of the vehicle. An electric pulser unit is mounted on top of this unit and is coupled to it by a small shaft. The motion from the lower unit is transferred to the upper unit; the resultant electrical signal is sent to the appropriate control units.

The NSX, having no hydraulic power steering system, uses speed sensors mounted in the transmission case to communicate with the ECU.

Starter Signal

The **START** position on the ignition switch causes a signal to be sent to the ECU. During cranking, the ECU will increase the amount of fuel injected into the manifold according to the engine temperature. The amount of fuel injected is gradually reduced when the starter switch is released.

Alternator FR Signal

The ECU receives a signal from the alternator when the system is charging.

Air Conditioning Signal

The load on the engine increases when the compressor is engaged; the ECU must increase the idle as necessary. In some cases, the ECU will limit A/C operation, cutting it off through the A/C control relay in cases of wide-open throttle or preventing it from engaging in cold engine conditions. If the vehicle has a separate control unit for the air conditioning, the two controllers exchange information regarding system status.

Battery Voltage (IGN 1)

A 12 volt signal controlled by the IGN 1 (ON or RUN) position of the ignition switch is sent to the ECU.

Automatic Transaxle Shift Position Signal
Manual Transaxle Neutral and Clutch Signals

This signal is sent to the ECU through the neutral safety switch. The control unit will not allow the engine to start if the shift selector is not in **N** or **PARK**.

1991–92 Legend and NSX vehicles with manual transaxle employ switches at the clutch pedal and the shifter. The signals from the switches are used by the ECU to determine that the vehicle is in N and/or the clutch is disengaged.

Power Steering Fluid Pressure Switch

This switch is used all but NSX vehicles. When the power steering fluid pressure exceeds a predetermined level, the ECU is signaled by a pressure operated switch. If the idle is below a predetermined rpm, as during a parking maneuver, the control unit will increase the idle speed of the engine to compensate for the additional load.

Electric Load Detector (ELD)

This sensor is used on Vigor vehicles; it is located in the underhood fuse and relay box. It detects the present of moderate to large electrical current in the system and sends a signal to the ECU.

The ECU will control the idle speed to compensate for the additional load of headlights, rear defroster, heater fan or similar heavy electrical loads.

The ELD unit is integral with the fuse box in which it is mounted; should the sensor fail, the fuse box assembly must be replaced.

Brake Switch Signal

This signal is sent to the ECU from the brake switch. It is not used on NSX.

Knock Sensors

These sensors are used on Legend, Vigor and NSX engines to detect detonation or engine knock. When this condition is detected, the ECU retards the ignition timing to compensate. The knock sensors are referred to as left and right on Legend, front or rear on NSX and No.1 or No.2 on Vigor.

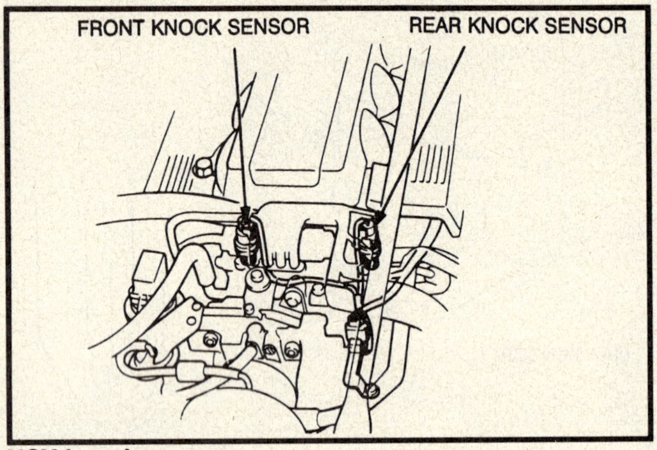

NSX knock sensors

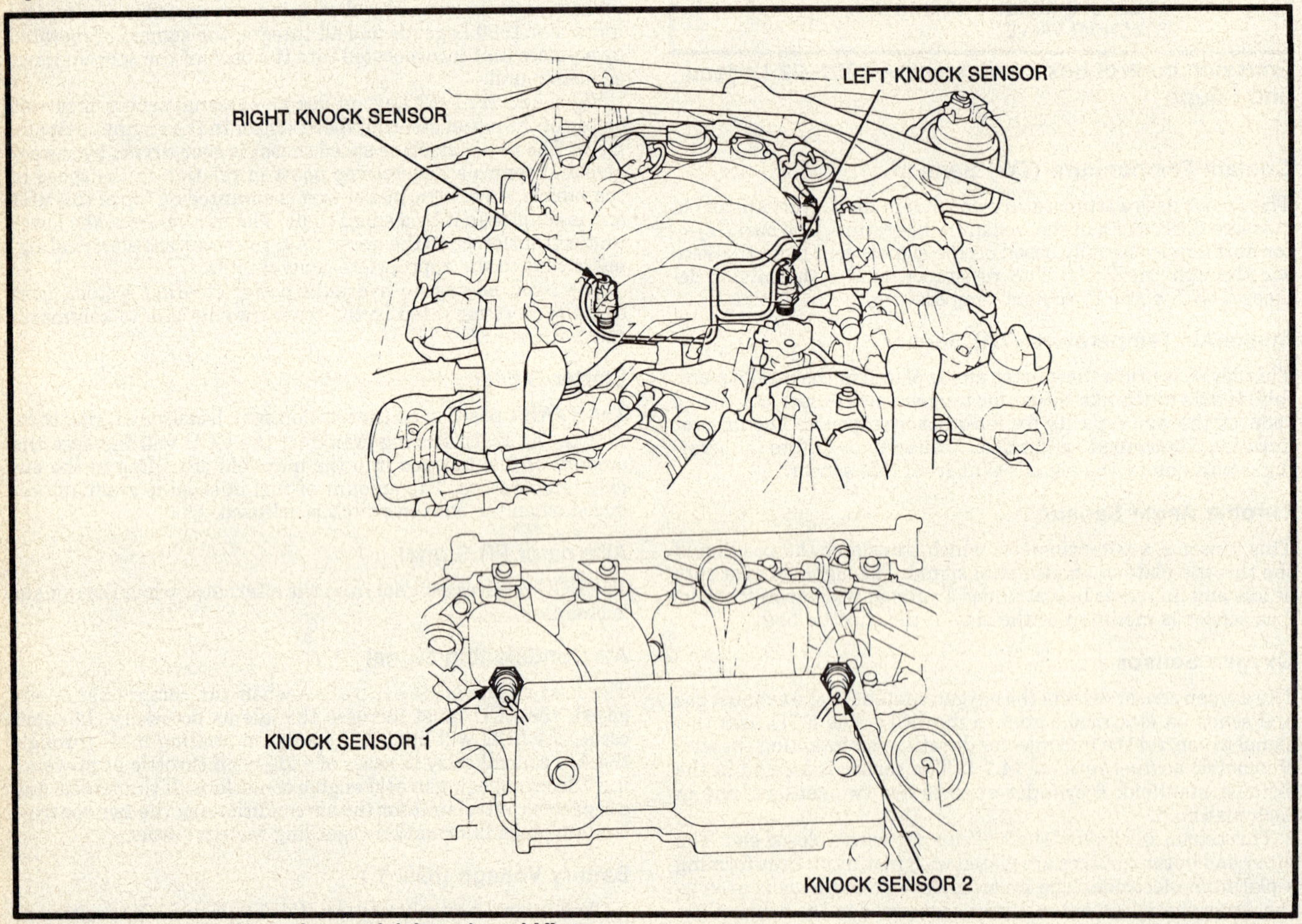

Knock sensor locations, Legend (above) and Vigor

Ignition Timing Adjuster

This component allows the electronic ignition advance to be set to 15 degrees BTDC. The sensor provides signals to the ECU; it is a solid-state electrical component having no mechanical parts. On Legend and NSX the sensor is located within the emission control box; on Vigor, it is located on the left fender apron, behind the air filter housing. The adjuster is covered or concealed by a cover held by 2 rivets.

Automatic Transaxle FI Signals

Used only on the NSX, the automatic transaxle control unit communicates with the engine ECU. The signal lines between the two computers are designated signal line 1 and 2; a fault or loss of signal on either line will cause the engine ECU to set a fault code; when this code is set, the dash warning lamp will not illuminate.

TCS Signals

The NSX engine controller receives data from the Traction Control System (TCS) control unit. Based on these signals, the engine controller alters engine function to reduce wheelspin during acceleration or cornering.

VTEC Oil Pressure Switch Signals

For each set of two valves—intake or exhaust—in each cylinder, there are three corresponding lobes on the camshaft. The 2 outboard lobes have a significantly different profile from the center lobe. The center or third lobe is not used during part-throttle or low load conditions.

During high speed or high load operation, the ECU signals a spool valve which delivers engine oil to small pistons in the rocker arms. The pressure causes the pistons to move outward, locking all 3 rockers together. Once locked, the rockers are forced to follow the center cam lobe, thus changing the performance of the engine. The change to high lift occurs in approximately 0.1 second and is virtually undetectable by the operator.

Once engine operation returns to part throttle or low load, the additional oil pressure is cut off; the pistons retract and the rockers are controlled by the outer lobes of the camshaft.

The ECU receives signals from both spool valves and both oil pressure switches indicating system status. The ECU will set one of 4 possible fault codes if any out-of-range signals occur.

ECU CONTROLLED OUTPUTS

The prime mission of the ECU is to maintain good driveability and proper emission levels under all operating conditions. The ECU interprets and compares the varied input signals from the sensors and uses the resulting calculations to control electrical outputs. In many cases, the circuit control provided by the ECU closes the ground rather than the power side of the circuit.

Fuel Injectors

All models use fuel injectors in the intake manifold (port injection) controlled individually by the ECU.

The ECU controls the system electrically, energizing the magnet within each injector according to engine speed, manifold pressure and other variables. The ECU can recognize an injector which is not functioning electrically and set a fault code. Physical problems, such as dirt lodged within the injector or polluted fuel cannot be detected because they are not electrical in nature. When tracing an apparently fuel-related driveability problem, the electrical and physical condition of the injector and condition of the fuel must be considered.

Main Relay

Although part of the main relay is energized by the ignition switch, the ECU controls ground to another part of the relay. When the ECU receives information that the the ignition switch is **ON** and a either a cranking or engine rpm signal is present, the ECU closes the relay circuit for the fuel pump system.

CHECK ENGINE Warning Lamp

The dashboard warning lamp will be lit by the ECU for 2 seconds during the initial ignition **ON** sequence. After this test sequence has ended, the ECU will illuminate the lamp anytime an electrical fault is noted within the system.

When a fault is detected, the dash warning lamp will stay on for the duration of the ignition **ON** cycle. Once the ignition is switched **OFF**, then recycled, the dashboard warning lamp will not light unless the original fault or some other (newer) fault is still present. Although the lamp does not re-light, the stored code for the original fault will remain in memory with the ignition **OFF**. For this reason, diagnosis should be based on the presence of fault codes, not the illumination of the warning lamp.

Electronic Air Control Valve (EACV)

The EACV controls the amount of air bypassing the throttle plate in response to signals from the ECU. During conditions of cold engine operation, A/C compressor engaged, automatic transaxle in gear, or alternator charging, the EACV is energized to control the idle speed correctly. Function of the EACV is controlled by the ECU proportional to engine coolant temperature. Integra engines incorporate an additional vacuum operated air boost valve to provide more intake air during cranking.

All vehicles except Integra with manual transaxle also employ a fast idle valve. The Integra with automatic transaxle controls this valve electrically through the ECU; all others are controlled by a temperature-sensitive thermowax valve reacting to the coolant being circulated through the valve body. On a cold engine, more air is admitted into the engine, raising the idle.

Note that in all cases, the external idle adjustment screw acts on an air-bleed passage, not on the throttle mechanism.

Air Conditioning Compressor Clutch Relay

When the engine control unit receives a request for air conditioning from either the A/C switch or the compressor control unit, the signal to the compressor clutch relay is momentarily delayed until the ECU can enrichen the air/fuel mixture. This allows a smoother engagement of the compressor with less effect on driveability.

Purge Cut-off Solenoid Valve

When the coolant temperature is below a predetermined level, the ECU controls the function of the purge cut-off solenoid or purge control solenoid. This eliminates vacuum to the purge control valve. The exact coolant temperature limit or engagement point is different for each engine but is approximately 150–160 °F.

Igniter

Although part of the ignition system, the igniter is triggered by a signal from the ECU. The timing of the spark, including advance, is handled by the engine controller.

EGR Control Solenoid Valve (EGR CSV)

When EGR function is necessary to provide control of NOx emissions, the ECU grounds the control solenoid valve. The solenoid allows regulated vacuum to pass to the EGR, thus precisely controlling EGR function. The system is used on all vehicles except Integra with manual transxaxle.

Pressure Regulator Cut-off Solenoid Valve

Used on Integra and Legend, this valve is activated by the ECU during hot restarts if the intake air and/or coolant temperature are above preset limits. When energized, the valve cuts off vacuum to the fuel pressure regulator for 60–80 seconds, allowing the pressure to build within the fuel rail. This reduces the occurrence of fuel boiling and improves hot-restart characteristics.

Bypass Control Solenoid Valves (BPCSV) Chamber Volume Control Solenoid Valve

Used only on the Legend , Vigor and NSX engines, this system provides 2 or even 3 air intake paths within the manifold. By selectively switching between them, both high torque at low rpm and high power at high rpm can be maintained.

On 1990 Legend, the BPCS valves are designated A and B. When engine speed is below 3200 rpm, the valves are activated by the ECU and intake air is flowed through the longer intake path. Above 3800 rpm, the valves are de-energized and the shorter port is opened.

1991 and 1992 Legends identify the valves as Bypass Low Control Solenoid and Bypass High Control Solenoid valves. When the engine speed is below 3100 rpm, both valves are energized by the ECU, causing the airflow to travel through an additional chamber. Between 3200 and 3800 rpm, the Bypass Low valve is deactivated, allowing air to flow through a shorter intake path. Above 3900 rpm, both valves are turned off, allowing air to flow through the shortest intake path.

The 5-cylinder Vigor engines use one BPCSV directing air into the longer intake path below 4900 rpm and through the shorter intakes above 4900 rpm. The NSX uses a chamber control system similar to the 1991–92 Legend. When the engine rpm is below 4800, the chamber is opened, creating a longer intake path for the air. Above this rpm, the control solenoid is de-energized by the ECU and the air is forced through a shorter intake path.

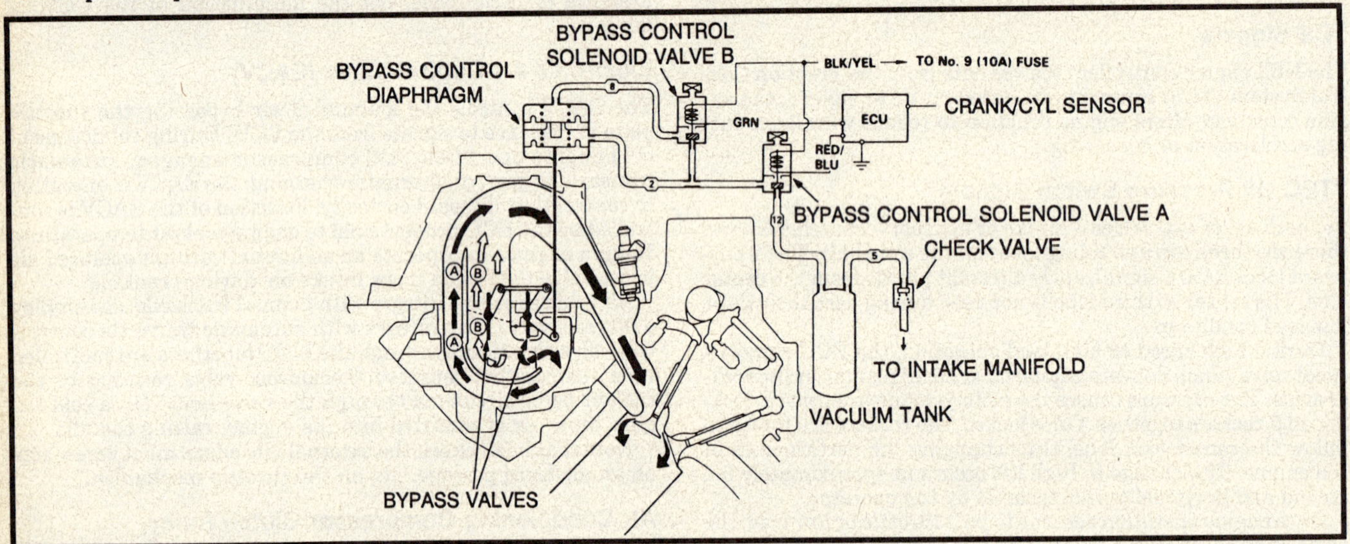

Bypass control system—1990 Legend

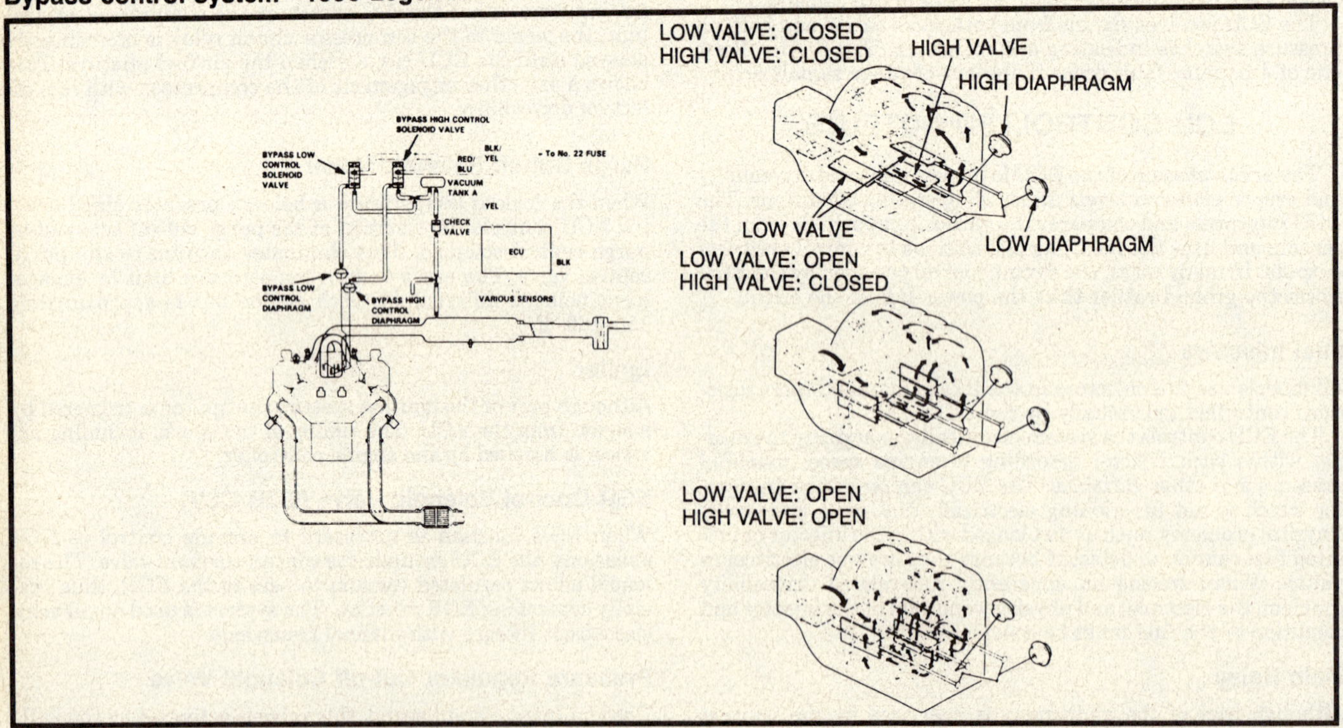

Chamber control system—1991–92 Legend

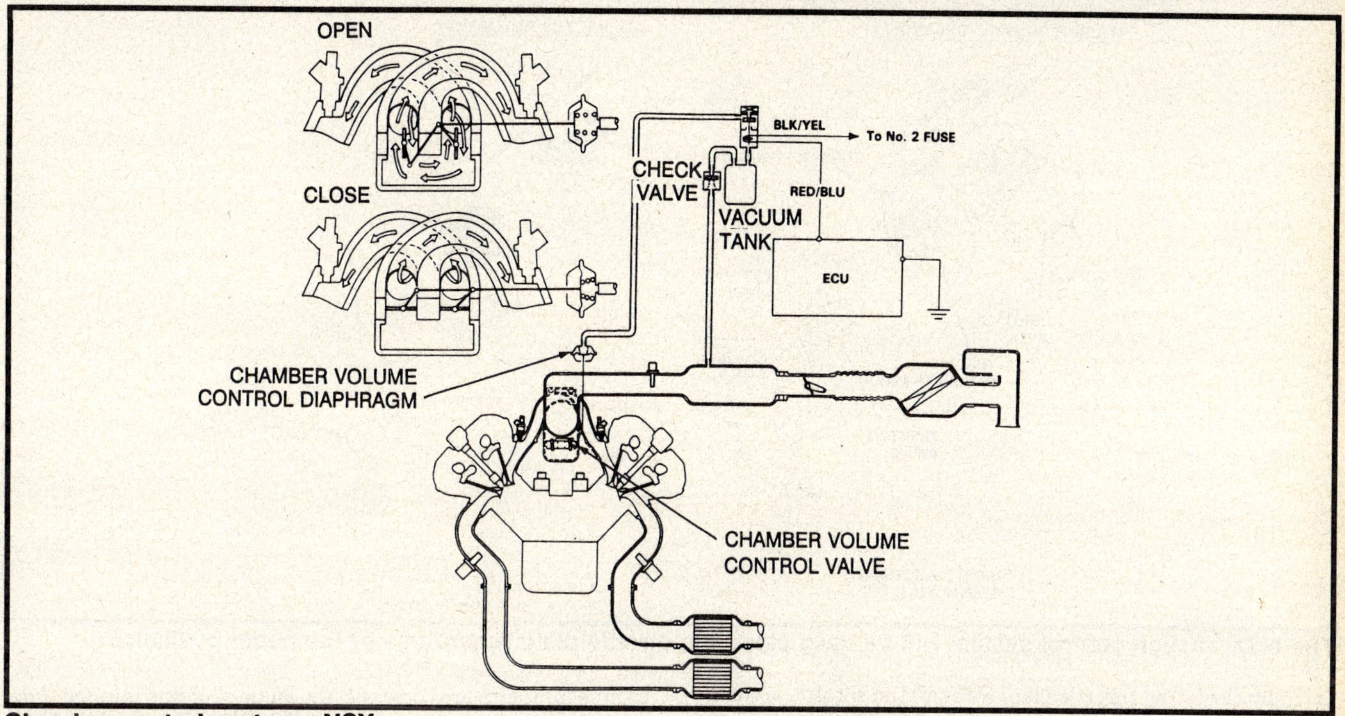

Chamber control system—NSX

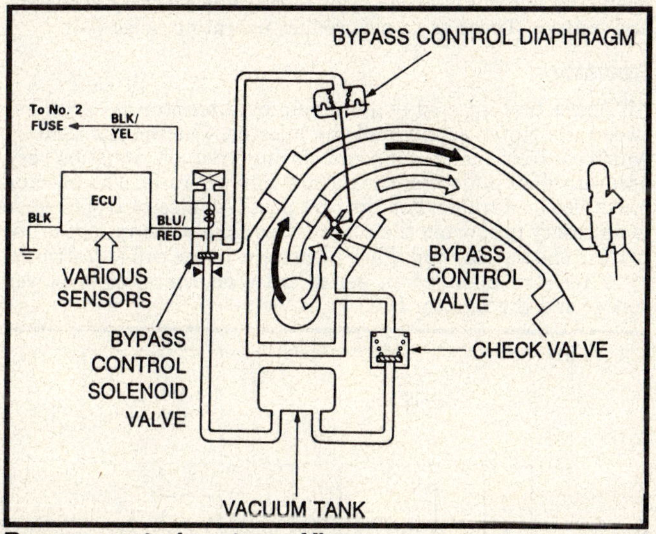

Bypass control system—Vigor

Fan Operation

On all vehicles except Integra and 1990 Legend, the ECU either communicates with the fan control unit or timer or controls the function of the cooling fan relays. On NSX, the additional engine compartment cooling fan is also controlled by the ECU.

Resonator Control Solenoid Valve

Used only on the 1990 Legend, this ECU-operated valve allows selective control of airflow into the intake air resonator chamber. The solenoid is placed in the vacuum line controlling the resonator diaphragm. The system decreases intake noise during operation below 3800 rpm.

Fuel Cut-Off Function

All ECUs are designed to shut off current to the fuel injectors when the vehicle is decelerating with the throttle closed. The cut off point varies with each engine , generally in the vicinity of 1000 rpm.

Additionally, the ECU will shut off current to the fuel injectors when the safe rpm or red-line is exceeded. Again, the cut-off point varies with each engine.

Air Suction Control Solenoid Valve

This valve is used on 1990–92 Legends. The system is designed to reduce exhaust emissions by introducing fresh air from the air cleaner into the exhaust manifold. When the solenoid valve is activated by the ECU, manifold vacuum raises the diaphragm valve of the air suction valve. The pulses in the exhaust stream cause a reed valve—within the air suction valve—to admit fresh air.

Ne Signal

The output signal to the tachometer connector on 1991–92 Legends is controlled by the ECU.

Oxygen Sensor Heaters

Operation of the heating elements used in the oxygen sensors for 1991–92 Legend and Vigor is controlled by the ECU.

Alternator Control

The Vigor ECU controls the voltage output of the alternator. Output is varied according to electrical load and drive mode, reducing engine load and improving fuel economy.

VTEC Spool Valves

The NSX engine control unit oversees the operation of the spool valves; they are energized or released depending on the engine rpm.

Traction Control System (TCS)

The TCS operates to reduce wheel spin during periods of limited traction. The TCS control unit exchanges signals with the engine ECU. When traction control is required, the TCS controller

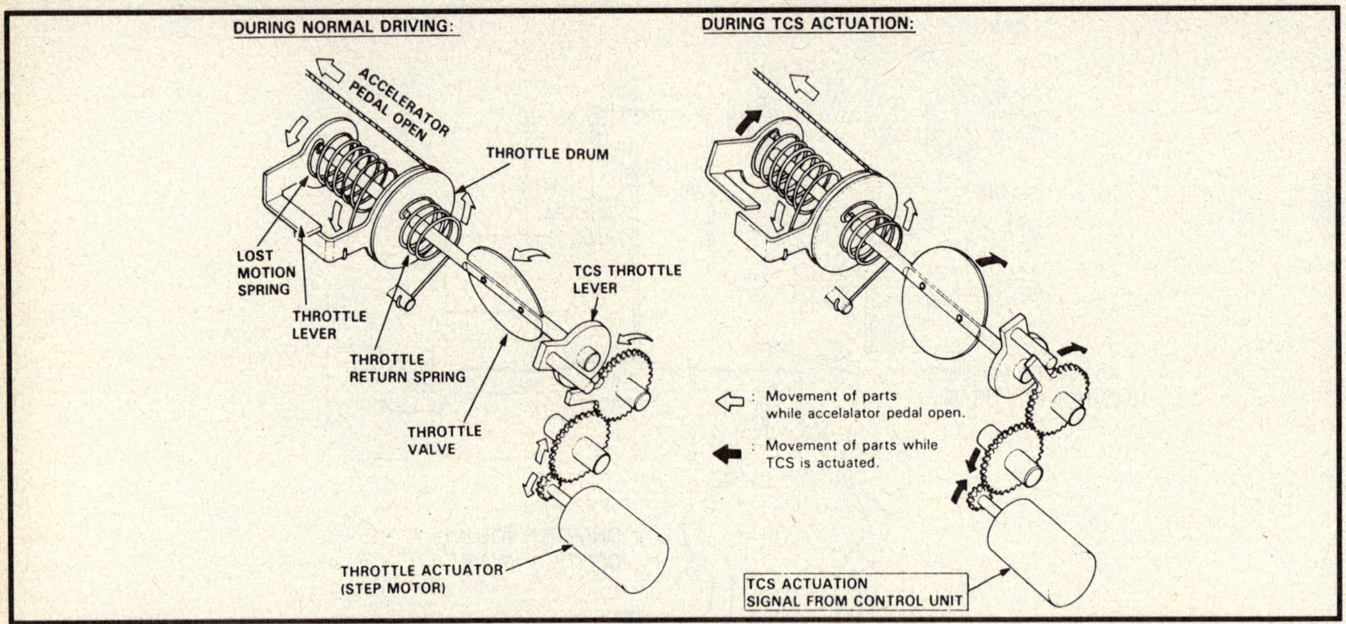

The NSX traction control system will partially close the throttle plate regardless of the pedal position

partially closes the throttle plate through the throttle actuator stepper motor. The stepper motor moves the throttle plate into a less open position, reducing engine power. Note that this closing of the throttle plate occurs regardless of the accelerator position; even with the accelerator calling for wide-open throttle the stepper motor will partially close the throttle is sensors indicate wheelspin is present.

FUEL SYSTEM

Fuel Pump

The fuel pump is installed inside the fuel tank, thereby saving space and simplifying the fuel line system.

The fuel pump is comprised of a DC motor, a circumference flow pump, a relief valve for protecting the fuel line systems, a check valve for retaining residual line pressure after the pump stops, an inlet port, and a discharge port. If the flow of fuel is obstructed at the discharge side of the pump, an internal relief valve will open to bypass the fuel to the inlet port.

The pump assembly consists of the impeller (driven by the motor), the pump casing (which forms the pumping chamber), and cover of the pump.

Pressure Regulator

The fuel pressure regulator maintains a constant fuel pressure at the injectors. The regulator is located on one end of the fuel rail. The spring chamber of the pressure regulator is connected to intake manifold vacuum to constantly maintain the fuel pressure at a specified pressure higher than the pressure in the manifold. When the difference between the fuel pressure and manifold pressure exceeds 36 psi (2.55 kg/cm^2) on Integra, Legend and Vigor or 50 psi (3.5 kg/cm^2), the diaphragm is pushed upward, and the excess fuel is fed back into the fuel tank through the return line.

Injector

The injector is of the solenoid-actuated constant-stroke pintle type consisting of a solenoid, plunger, needle valve and housing. When current is applied to the solenoid coil, the valve lifts up and pressurized fuel fills the inside of the injector and is injected near the intake valve. Because the needle valve lift height and the fuel pressure are constant, the amount of fuel injected is determined by the length of time (duration) that the injector is open. The injector is sealed by an O-ring and seal ring at the top and bottom. These seals also reduce operating noise.

Resistor

All Acura fuel injected engines employ a resistor in series between the power supply and the injector. The injector timing, which controls the opening and closing intervals, must be very accurate since it dictates the air/fuel mixture ratio. The injector must also be durable. For the best possible injector response, it is necessary to shorten the current rise time when voltage is applied to the injector coil. Flow of current in the coil is restricted by a resistor installed in series between the electric power source and the injector coil.

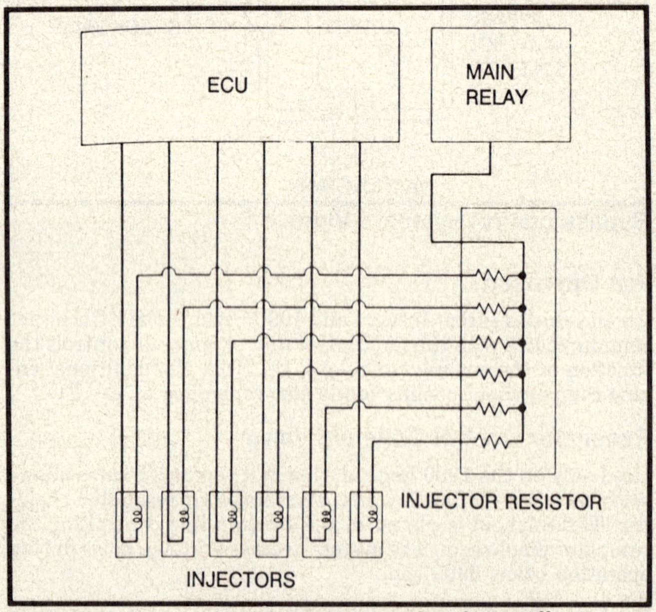

A resistor is used in the injector circuit on all models. 6 cylinder shown

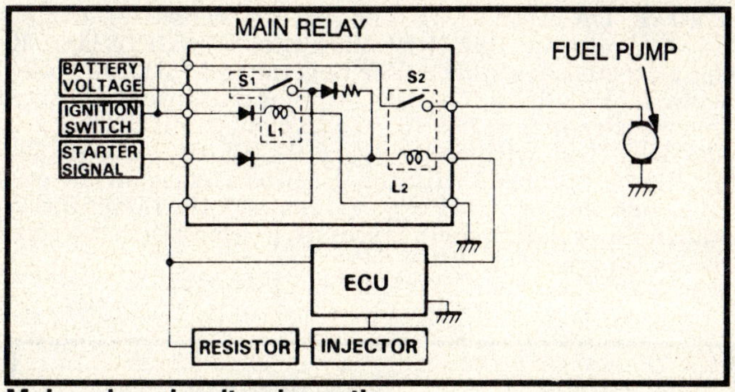

Main relay circuit schematic

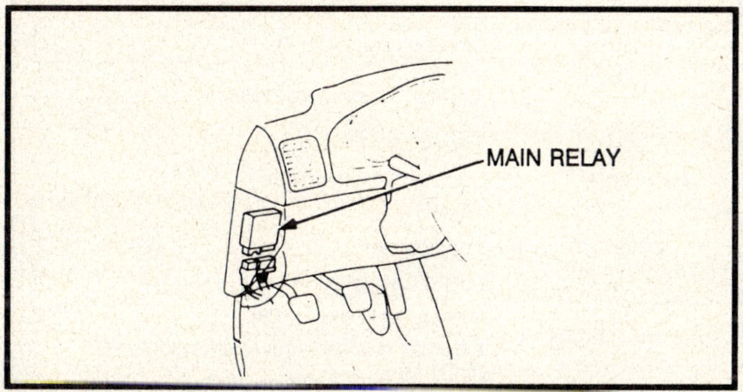

Main relay location, Legend and Integra. Vigor location is behind the left kick panel

Main relay location—NSX

Main Relay

The relay is found at the extreme left side of the dashboard, above and behind the fusebox on Integra and Legend or behind the left front kick panel on Vigor. On NSX, it is located behind the right seat on the bulkhead or firewall.

The main relay unit contains two separate relays within one case.The first relay is energized whenever the ignition is switched **ON**. Power is supplied to ECU, the injectors and to the second relay. The second relay circuit supplies power to the fuel pump. It is engaged for 2 seconds any time the ignition is switched **ON** and is always engaged when the engine is running.

Diagnosis and Testing

SERVICE PRECAUTIONS

- Do not operate the fuel pump when the fuel lines are empty.
- Do not operate the fuel pump when removed from the fuel tank.
- Do not reuse fuel hose clamps.
- The washer(s) below any fuel system bolt (banjo fittings, service bolt, fuel filter, etc.) must be replaced whenever the bolt is loosened. Do not reuse the washers; a high-pressure fuel leak may result.
- Make sure all ECU harness connectors are fastened securely. A poor connection can cause an extremely high voltage surge and result in damage to integrated circuits.
- Keep ECU all parts and harnesses dry during service. Protect the ECU and all solid-state components from rough handling or extremes of temperature.
- Use extreme care when working around the ECU or other components; the airbag or SRS wiring may be in the vicinity. On these vehicles, the SRS wiring and connectors are yellow; do not cut or test these circuits.
- Before attempting to remove any parts, turn the ignition switch **OFF** and disconnect the battery ground cable.
- Always use a 12 volt battery as a power source for the engine, never a booster or high-voltage charging unit.
- Do not disconnect the battery cables with the engine running.
- Do not disconnect any wiring connector with the engine running or the ignition **ON** unless specifically instructed to do so.
- Do not apply battery power directly to injectors.
- Whenever possible, use a flashlight instead of a drop light.
- Keep all open flame and smoking material out of the area.
- Use a shop cloth or similar to catch fuel when opening a fuel system. Consider the fuel-soaked rag to be a flammable solid and dispose of it in the proper manner.
- Relieve fuel system pressure before servicing any fuel system component.
- Always use eye or full-face protection when working around fuel lines, fittings or components.
- Always keep a dry chemical (class B-C) fire extinguisher near the area.

READING CODES

When a fault is noted, the ECU stores an identifying code and illuminates the CHECK ENGINE light. The code will remain in memory until cleared; the dashboard warning lamp may not illuminate during the next ignition cycle if the fault is no longer present. Not all faults noted by the ECU will trigger the dashboard warning lamp although the fault code will be set in memory. For this reason, troubleshooting should be based on the presence of stored codes, not the illumination of the warning lamp while the car is operating.

Stored codes are displayed by a flashing LED on the ECU. When the CHECK ENGINE warning lamp has been on or reported on, lift or remove the carpet from the right front passenger footwell. The ECU is below a protective cover; the LED may be viewed through a small window without removing the ECU cover. Turn the ignition switch **ON**; the LED will display any stored codes by rhythmic flashing. Note that 1990 Legends have two LEDs on the controller; one is red and one is amber. The red one will flash the fault codes; the amber one is used during idle adjustment and is not related to this procedure.

Additionally, the Vigor, NSX and 1991–92 Legend models are equipped with a service connector in side the cabin of the vehicle. If the service connector is jumpered, the CHECK ENGINE lamp will display the stored codes in the same fashion. The 2-pin service connector is located under the extreme right dashboard on Legend and NSX; on Vigor models, it is found behind the right side of the center console, well under the dashboard.

Codes 1–9 are indicated by a series of short flashes; two-digit codes use a number of long flashes for the first digit followed by the appropriate number of short flashes. For example, Code 43 would be indicated by 4 long flashes followed by 3 short flashes. Codes are separated by a longer pause between transmissions. The position of the codes during output can be helpful in diagnostic work. Multiple codes transmitted in isolated order indicate unique occurrences; a display of showing 1-1-1-pause-9-9-9 indicates two problems or problems occurring at different times. An alternating display, such as 1-9-1-9-1, indicates simultaneous occurrences of the faults.

When counting flashes to determine codes, a code not valid for the vehicle may be found. In this case, first recount the flashes to confirm an accurate count. If necessary, turn the ignition switch **OFF**, then recycle the system and begin the count again. If the Code is not valid for the vehicle, the ECU must be replaced.

NOTE: On vehicles with automatic transaxles, the S, D or D_4 lamp may flash with the CHECK ENGINE lamp if certain codes are stored. For Legend and NSX, this may occur with Codes 6, 7 or 17. On Vigor and Integra it may occur with codes 6, 7 or 13. In addition, the TCS lamp on NSX may flash with codes 3, 5, 6, 13, 15, 16, 17, 35 or 36. In all cases, proceed with the diagnosis based on the engine code shown. After repairs, recheck the lamp. If the additional warning lamp is still lit, proceed with diagnosis for that system.

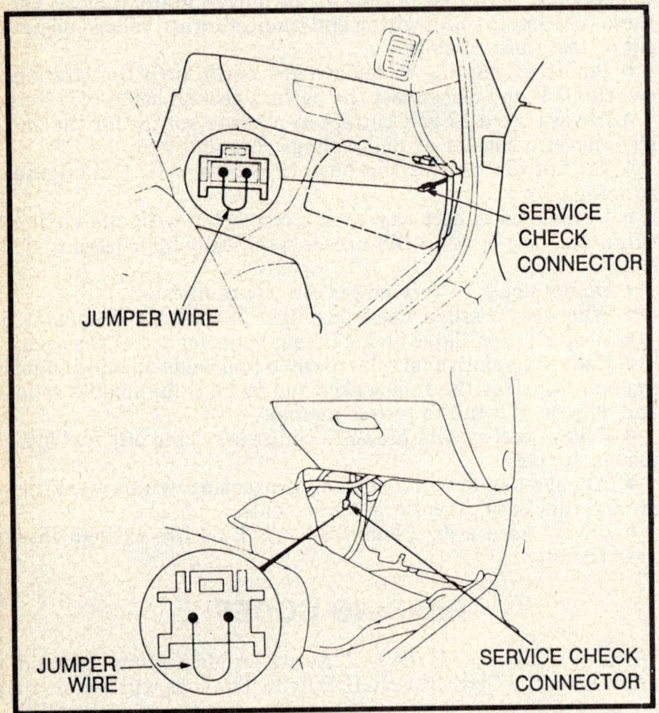

Service connector location for Legend (above) and Vigor. NSX similar to Legend

SELF-DIAGNOSIS INDICATOR BLINKS	SYSTEM INDICATED
0	ECU
1	FRONT OXYGEN CONTENT
2	REAR OXYGEN CONTENT
3 5	MANIFOLD ABSOLUTE PRESSURE
4	CRANK ANGLE
6	COOLANT TEMPERATURE
7	THROTTLE ANGLE
8	TDC POSITION
9	No. 1 CYLINDER POSITION
10	INTAKE AIR TEMPERATURE
12	EXHAUST GAS RECIRCULATION SYSTEM
13	ATMOSPHERIC PRESSURE
14	ELECTRONIC IDLE CONTROL
15	IGNITION OUTPUT SIGNAL
17	VEHICLE SPEED PULSER
18	IGNITION TIMING ADJUSTMENT
30	A/T FI SAIGNAL A
31	A/T FI SIGNAL B

Fault codes—1990 Legend

SELF-DIAGNOSIS INDICATOR BLINKS	SYSTEM INDICATED
0	ECU
1	OXYGEN CONTENT
3 5	MANIFOLD ABSOLUTE PRESSURE
4	CRANK ANGLE
6	COOLANT TEMPERATURE
7	THROTTLE ANGLE
8	TDC POSITION
9	No. 1 CYLINDER POSITION
10	INTAKE AIR TEMPERATURE
12	EXHAUST GAS RECIRCULATION SYSTEM (A T)
13	ATMOSPHERIC PRESSURE
14	ELECTRONIC AIR CONTROL
15	IGNITION OUTPUT SIGNAL
16	FUEL INJECTOR
17	VEHICLE SPEED SENSOR
43	FUEL SUPPLY SYSTEM

Fault codes—Integra

SELF DIAGNOSIS INDICATOR BLINKS	SYSTEM INDICATED
0	ECU
1	LEFT OXYGEN SENSOR
2	RIGHT OXYGEN SENSOR
3 5	MANIFOLD ABSOLUTE PRESSURE (MAP SENSOR)
4	CRANK ANGLE 1
6	COOLANT TEMPERATURE (TW SENSOR)
7	THROTTLE ANGLE
9	No. 1 CYLINDER POSITION 1 (CYL SENSOR)
10	INTAKE AIR TEMPERATURE (TA SENSOR)
12	EXHAUST GAS RECIRCULATION SYSTEM (EGR)
13	ATMOSPHERIC PRESSURE (PA SENSOR)
14	ELECTRONIC AIR CONTROL (EACV)
15	IGNITION OUTPUT SIGNAL
16	FUEL INJECTOR
17	VEHICLE SPEED SENSOR
18	IGNITION TIMING ADJUSTMENT
23	LEFT KNOCK SENSOR
41	LEFT OXYGEN SENSOR HEATER
42	RIGHT OXYGEN SENSOR HEATER
43	LEFT FUEL SUPPLY SYSTEM
44	RIGHT FUEL SUPPLY SYSTEM
45	LEFT FUEL METERING
46	RIGHT FUEL METERING
53	RIGHT KNOCK SENSOR
54	CRANK ANGLE 2
59	No. 1 CYLINDER POSITION 2 (CYL SENSOR)

Fault codes—1991–92 Legend

SELF-DIAGNOSIS INDICATOR BLINKS	SYSTEM INDICATED
0	ECU
1	OXYGEN SENSOR
3 5	MANIFOLD ABSOLUTE PRESSURE (MAP SENSOR)
4	CRANK ANGLE (CRANK SENSOR)
6	COOLANT TEMPERATURE (TW SENSOR)
7	THROTTLE ANGLE
8	TDC POSITION (TDC SENSOR)
9	No. 1 CYLINDER POSITION (CYL SENSOR)
10	INTAKE AIR TEMPERATURE (TA SENSOR)
12	EXHAUST GAS RECIRCULATION SYSTEM (EGR)
13	ATMOSPHERIC PRESSURE (PA SENSOR)
14	ELECTRONIC AIR CONTROL (EACV)
15	IGNITION OUTPUT SIGNAL
16	FUEL INJECTOR
17	VEHICLE SPEED SENSOR
18	IGNITION TIMING ADJUSTMENT
20	ELECTRIC LOAD DETECTOR (ELD)
23	KNOCK SENSOR 1
30	A/T FI SIGNAL A
31	A/T FI SIGNAL B
41	OXYGEN SENSOR HEATER
43	FUEL SUPPLY SYSTEM
45	FUEL METERING
53	KNOCK SENSOR 2

Fault codes—Vigor

SELF DIAGNOSIS INDICATOR BLINKS	SYSTEM INDICATED
0	ECU
1	FRONT OXYGEN SENSOR
2	REAR OXYGEN SENSOR
3 5	MANIFOLD ABSOLUTE PRESSURE (MAP SENSOR)
4	CRANK ANGLE A
6	COOLANT TEMPERATURE (TW SENSOR)
7	THROTTLE ANGLE
9	No. 1 CYLINDER POSITION A (CYL SENSOR)
10	INTAKE AIR TEMPERATURE (TA SENSOR)
12	EXHAUST GAS RECIRCULATION SYSTEM (EGR)
13	ATMOSPHERIC PRESSURE (PA SENSOR)
14	ELECTRONIC AIR CONTROL (EACV)
15	IGNITION OUTPUT SIGNAL
16	FUEL INJECTOR
17	VEHICLE SPEED SENSOR
18	IGNITION TIMING ADJUSTMENT
21	FRONT SPOOL SOLENOID VALVE
22	FRONT VALVE TIMING OIL PRESSURE SWITCH
23	FRONT KNOCK SENSOR
30	A/T FI SIGNAL A
31	A/T FI SIGNAL B
35	TC STB SIGNAL
36	TCFC SIGNAL
41	FRONT OXYGEN SENSOR HEATER
42	REAR OXYGEN SENSOR HEATER
43	FRONT FUEL SUPPLY SYSTEM
44	REAR FUEL SUPPLY SYSTEM
45	FRONT FUEL METERING
46	REAR FUEL METERING
47	FUEL PUMP
51	REAR SPOOL SOLENOID VALVE
52	REAR VALVE TIMING OIL PRESSURE SWITCH
53	REAR KNOCK SENSOR
54	CRANK ANGLE B
59	No. 1 CYLINDER POSITION B (CYL SENSOR)

Fault codes—NSX

CLEARING CODES

Stored codes are removed from memory by removing power to the ECU. Disconnecting the power may also clear the memories used for other solid-state equipment such as the clock and radio. For this reason, always make note of the radio presets before clearing the system. Additionally, some radios contain anti-theft programming; obtain the owner's code number before clearing the codes.

While disconnecting the battery will clear the memory, this is not the recommended procedure. The memory should be cleared after the ignition is switched **OFF** by removing the appropriate fuse for at least 10 seconds. The correct fuses and their locations are:

1990 Legend—ALTERNATOR SENSE, in the underhood fuse and relay panel.

1991–92 Legend—ACG, in the dashboard fuse panel. Removing this fuse also cancels the memory for the power seats.

Integra, U.S. vehicles—BACK UP, in the underhood fuse and relay panel.

Integra, Canadian vehicles—HAZARD/BACK UP in the underhood fuse and relay panel.

Vigor—BACK UP, located in the underhood fuse and relay panel. Removing this fuse will cancel memories for the clock and radio.

NSX—CLOCK, located in the main fuse and relay panel in the front luggage compartment, right side. Removing this fuse will cancel memories for the clock and radio.

FUEL PRESSURE

Relief

CAUTION

The fuel system is under constant pressure, even when the engine has not been run for a period of time. To avoid injury and/or property damage, always safely relieve the fuel system pressure before servicing any fuel or return line component.

1. Remove the fuel filler cap.
2. With the ignition **OFF**, disconnect the negative battery cable.
3. Locate the 6mm service bolt at the top of the fuel filter.
4. Use a wrench to hold the larger bolt below the service bolt. Use a box-end wrench on the service bolt. Place a rag or cloth over the wrenches and service bolt.
5. Slowly loosen the service bolt 1 full turn. The pressurized fuel will escape into the rag.

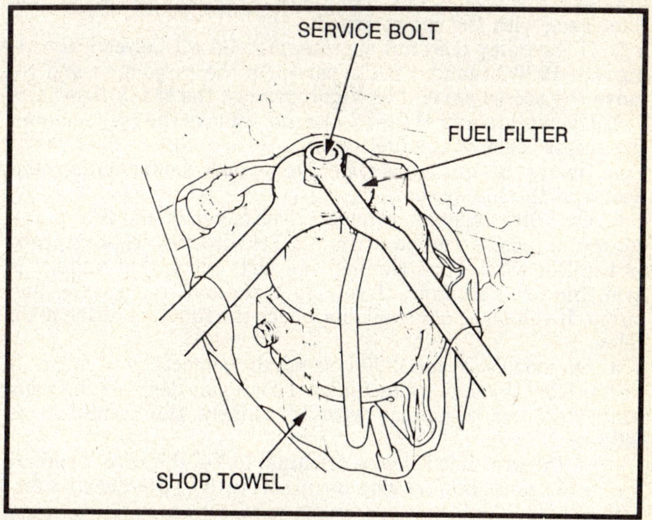

Relieving the fuel pressure

6. Remove the service bolt, replace the washer and reinstall the bolt. Tighten the bolt to 9 ft. lbs. (12 Nm). The washer must be replaced any time the service bolt is loosened or removed. Failure to replace the washer may result in high-pressure fuel leaks.
7. Dispose of the fuel-soaked rag in a manner correct for flammable items. Check the surrounding area for any spilled fuel.
8. Reconnect the battery cable. The pressure within the system will stay reduced until the ignition switch is turned **ON** or the fuel pump is operated by other means.

Testing

1. Relieve the fuel pressure using the correct procedure.
2. Remove the service bolt. Install a fuel pressure gauge in the service bolt port.
3. Start the engine.
4. Disconnect the vacuum hose at the pressure regulator. Measure the fuel pressure with the engine at idle. Correct pressure should be:
   - 1990 Legend: 36–41 psi.
   - 1991–92 Legend: 38–46 psi.
   - Integra: 37–44 psi.
   - Vigor: 43–50 psi.
   - NSX: 46–53 psi.
5. Reconnect the vacuum hose to the regulator on 1991 Legend, Vigor and NSX. Regulated fuel pressure should drop to:
   - 1991–92 Legend: 31–37 psi.
   - Vigor: 33–41 psi.
   - NSX: 36–44 psi.
6. Reconnect the vacuum hose to the regulator on 1990 Legend and Integra.
7. If fuel pressure is above specification, inspect for a pinched or clogged fuel return hose or faulty regulator.
8. If fuel pressure is below specification, causes may include fuel line leaks, a faulty regulator or a clogged fuel filter.
9. Shut the engine **OFF**.
10. Remove the test equipment. Install a new washer on the service bolt; reinstall the bolt, tightening it to 9 ft. lbs.(12 Nm).

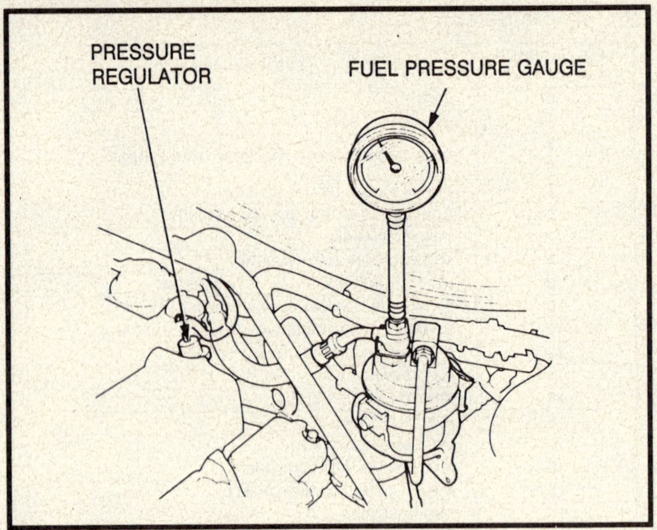

Testing the fuel pressure. Legend shown; procedure is the same for all

FUEL PUMP

Testing

EXCEPT NSX

1. The simplest test for the fuel pump is listening to it. The pump should be heard to operate for 2 seconds when the ignition switch is turned **ON**. The noise is easily heard at the fuel filler neck with the cap removed.
2. If the pump does not operate, proceed with electrical testing. On 1990 Legend, lift the carpet in the cargo area and remove the access panel. For Vigor, remove the trunk floor.
3. On Integra and 1991–92 Legend, remove the rear seat and the access cover to the fuel tank.
4. Insure the ignition is **OFF**. Disconnect the fuel pump connector at the fuel pump.
5. On 1990 Legends, connect a voltmeter at the fuel pump connector, positive probe to the Blk/Yel wire and negative probe to the Blk wire. Turn the ignition **ON**; check that voltage is available for 2 seconds. If voltage is present, replace the fuel pump. If voltage is not available, check the function of the main relay.
6. On models except 1990 Legend, disconnect the wiring connector from the main relay under the left side dash. At the relay connector, use a jumper wire to connect the terminals as follows:
   - Integra: Blk/Yel wire terminal to Yel/Blk wire terminal
   - Legend: Blk/Yel wire terminal No. 5 to Blk/Yel wire terminal No. 7
   - Vigor: Blk/Yel wire terminal to Yel wire terminal

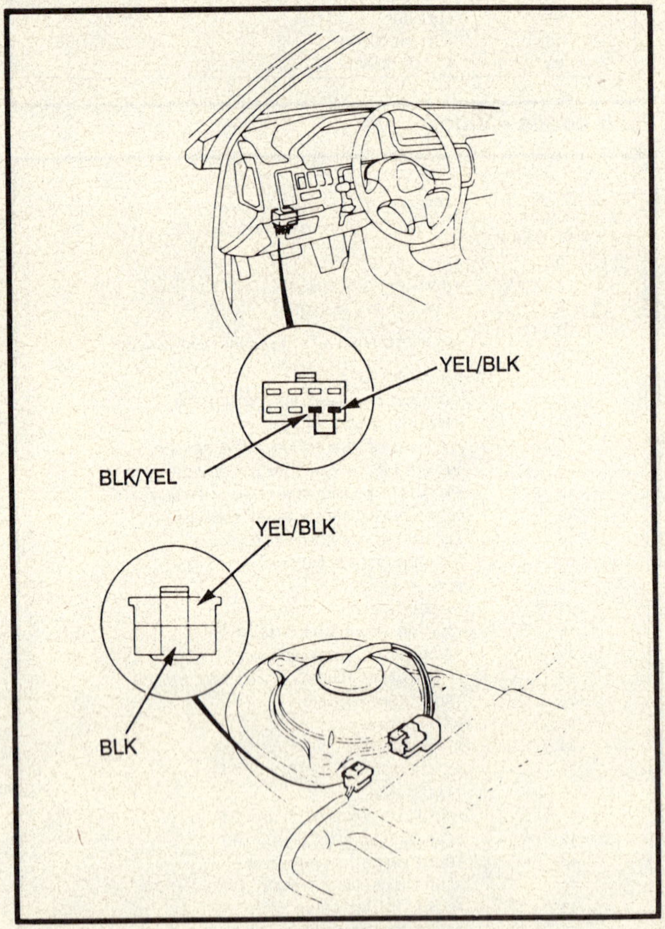

Testing fuel pump circuit—Integra

7. Turn the ignition switch **ON**. Battery voltage should be available at the fuel pump connector. Connect the negative probe of the meter to the Blk terminal and the positive probe to the Integra Yel/Blk terminal, the Legend Blk/Yel terminal or the Vigor Yellow wire terminal.
8. If battery voltage is available, the fuel pump must be replaced.

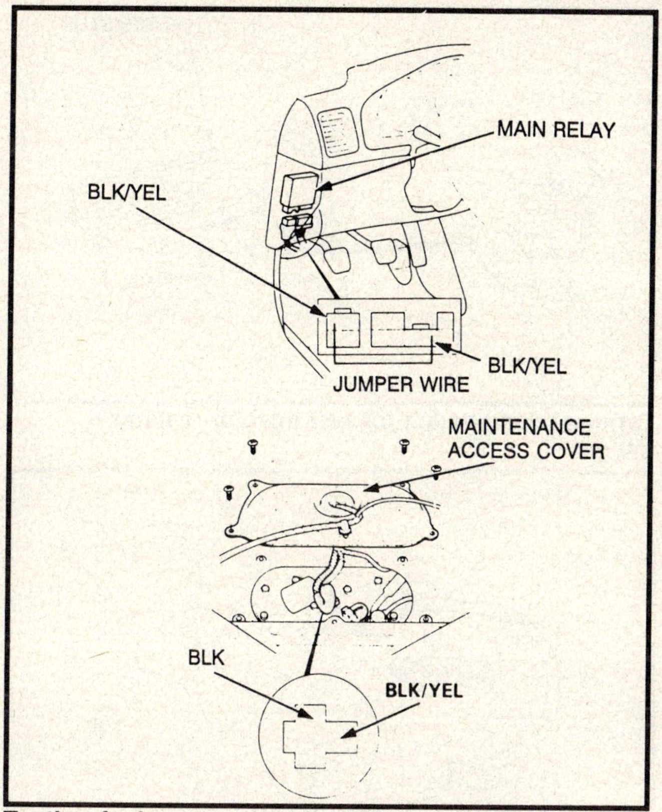

Testing fuel pump circuit—1991–92 Legend

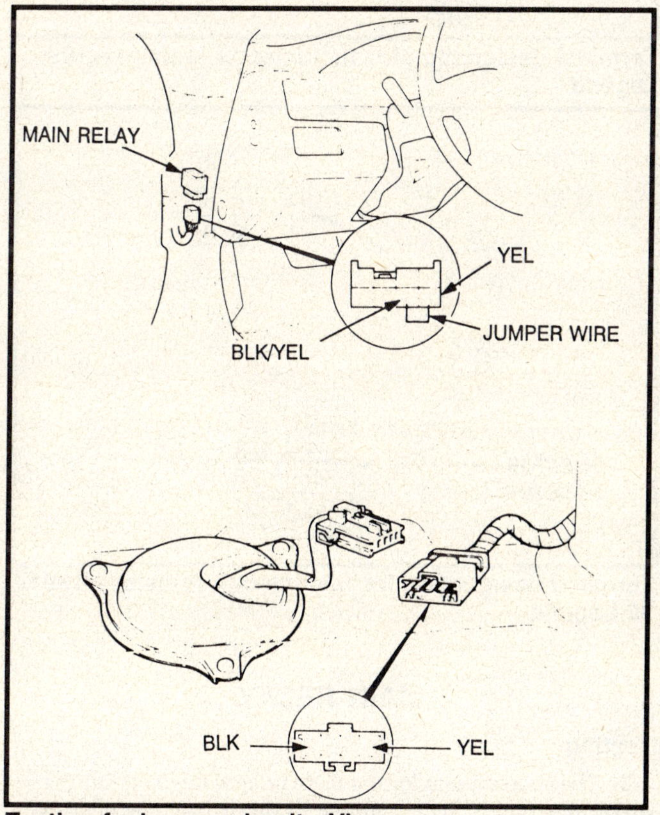

Testing fuel pump circuit—Vigor

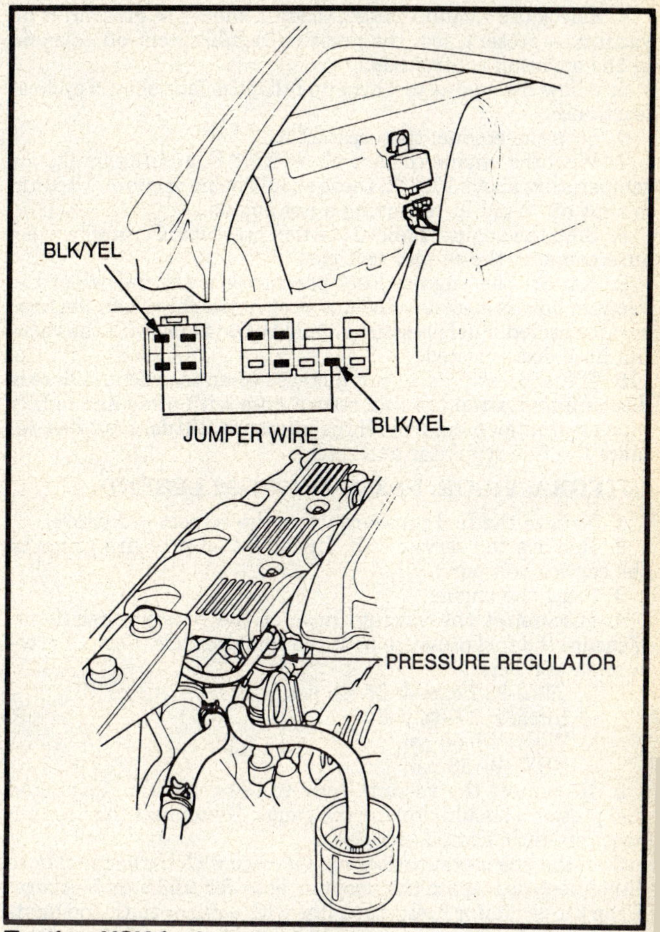

Testing NSX fuel pump delivery

9. If no voltage is available, check the main relay and/or wire harness.

NSX

1. With the ignition switch **OFF**, disconnect the wiring connector at the main relay.
2. Use a jumper to connect the Blk/Yel wire terminal No. 5 to Blk/Yel wire terminal No. 7.
3. Relieve the fuel pressure following the correct procedure. Once pressure is bled off, retighten the service bolt.
4. Disconnect the fuel return hose from the fuel pressure regulator.
5. Place the disconnected hose end into a large glass container, preferably one marked for fluid ounces.
6. Turn the ignition **ON**d for 10 seconds; measure the volume of fuel delivered into the container. Correct fuel delivery is 11.2 oz (333cc) in 10 seconds at 12 volts.
7. If fuel volume is less than specified, check for clogged filter or lines or a failed pressure regulator.

FUEL PRESSURE REGULATOR

Testing

1990 Legend

1. Inspect the vacuum line running to the pressure regulator for any signs of cracking or splitting.
2. Disconnect the vacuum hose from the pressure regulator; connect a vacuum gauge to the hose.
3. Start the engine, allowing it to idle.

4. Check the vacuum gauge; vacuum should be present. If no vacuum is present, test the pressure regulator cut-off solenoid in the emission control box.
5. Allow the engine to warm up fully; the fans should cycle at least once.
6. Stop the engine, then restart it.
7. With the engine coolant above 221° F. and the intake air temperature above 176° F., there should be no vacuum. Vacuum is shut off to aid in restarting a hot engine.
8. Stop the engine. Relieve the fuel pressure. Connect a pressure gauge at the service bolt port.
9. Restart the engine. Fuel pressure should rise when the vacuum hose is disconnected and drop noticeably when the hose is reconnected. Fuel pressure should be 36–41 psi with the vacuum hose disconnected.
10. If fuel pressure does not increase when the vacuum hose is disconnected, wrap the fuel return hose with a rag and lightly pinch it shut momentarily. If fuel pressure still does not rise, replace the fuel regulator and retest.

INTEGRA, VIGOR, NSX AND 1991–92 LEGEND

1. Relieve the fuel pressure using the correct procedure.
2. Remove the service bolt. Install a fuel pressure gauge in the service bolt port.
3. Start the engine.
4. Disconnect the vacuum hose at the pressure regulator. Measure the fuel pressure with the engine at idle. Correct pressure should be:
   - 1991–92 Legend: 38–46 psi.
   - Integra: 37–44 psi.
   - Vigor: 43–50 psi.
   - NSX: 46–53 psi.
5. Reconnect the vacuum hose to the regulator. Regulated fuel pressure should drop on the gauge. Disconnect the vacuum hose; pressure should increase.
6. If the line pressure does not increase with the vacuum hose disconnected, inspect the vacuum hose for kinks or blockage. Wrap the regulator fuel return line with a rag or cloth and lightly pinch off the line. Fuel pressure should increase. If the pressure still does not rise, the regulator is faulty and must be replaced.

FUEL INJECTOR

Electrical Testing

1. Turn the ignition switch **OFF**.
2. Disconnect the wiring harness from each injector.
3. Measure the resistance at the 2 terminals of the injector. Correct resistance for all engines is 1.5–2.5 ohms. Any injector not showing the proper resistance must be replaced.
4. Further testing of the injector circuits and wiring are included in the troubleshooting charts.

INJECTOR RESISTOR

Testing

A resistor is used on all Acura vehicles.
1. With the ignition **OFF**, disconnect the harness connector from the resistor.
2. Measure the resistance between the power terminal of the resistor connector and each other connector. On all but 1991–92 Legend, the power terminal is A; on the 1991–92 Legend, it's terminal E.
3. Correct resistance between the power terminal and each other connector is 5–7 ohms. If any circuit is out of range, the resistor must be replaced.

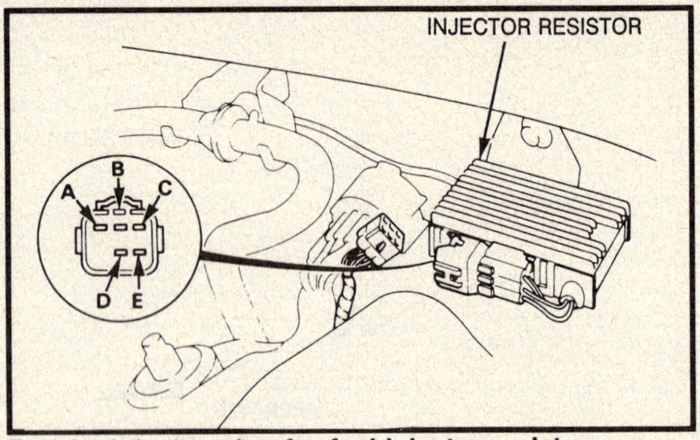

Terminal designation for fuel injector resistor—Integra

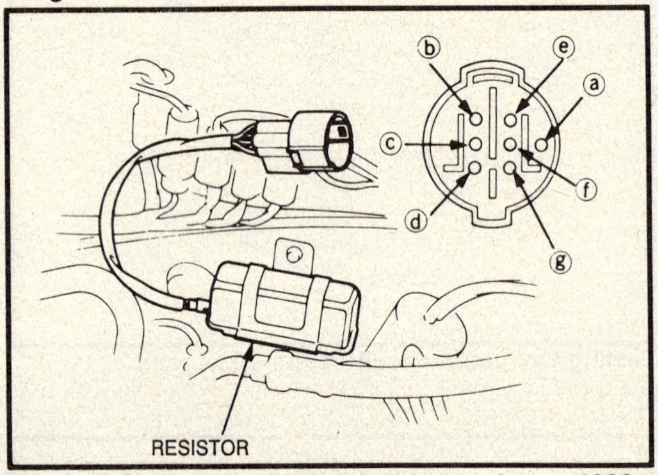

Terminal designation for fuel injector resistor—1990 Legend

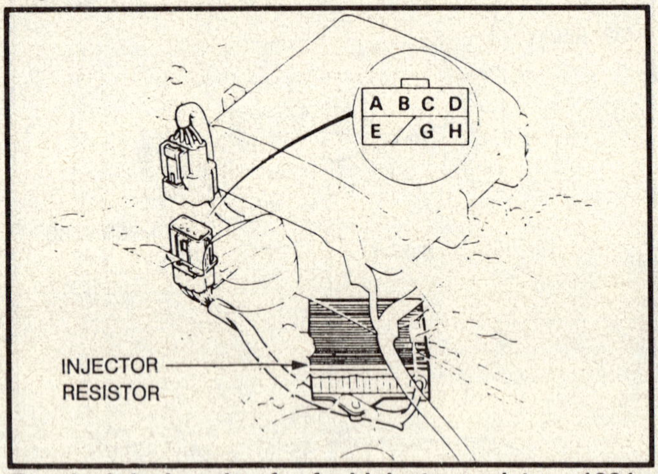

Terminal designation for fuel injector resistor—1991-92 Legend

MAIN RELAY

Testing

1. The simplest test for the main relay is the fact that the vehicle starts and remains running. If this is true, the main relay is functioning properly.

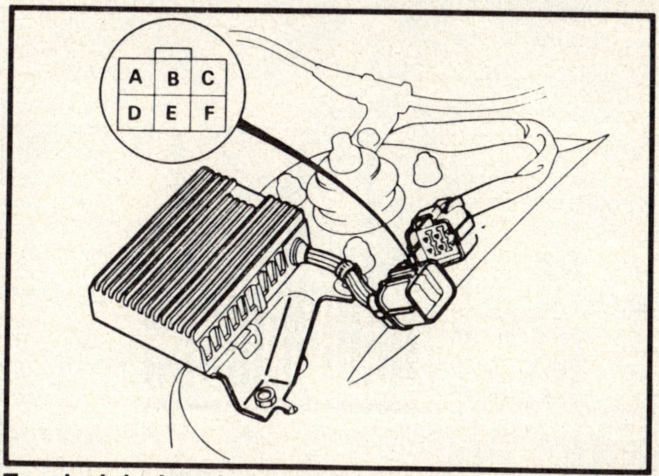

Terminal designation for fuel injector resistor—Vigor

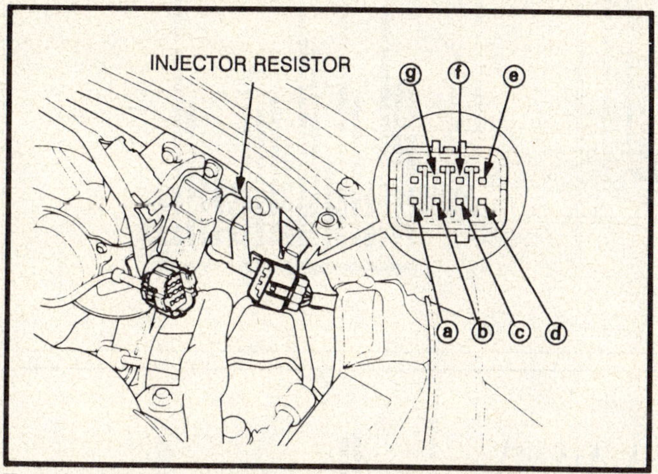

Terminal designation for fuel injector resistor—NSX

2. If the relay must be tested, remove it from its mount and disconnect it from the harness.
3. Using jumper wires, connect battery + to terminal 4 for all vehicles except Vigor; on Vigor, connect 12 volt + to terminal 6. Connect the − battery terminal to relay terminal 8.
4. Check for continuity between relay terminals 5 and 7. If no continuity is present replace the relay.
5. On Integra, Legend and 1990 Legend, connect the + battery jumper to relay terminal 5; for NSX and 1991–92 Legend, connect the jumper to terminal 6. Connect the negative jumper to relay terminal 2. Check for continuity between relay terminals 1 and 3. If there is no continuity, replace the relay.
6. Connect the + battery jumper to relay terminal 3 and the negative jumper to relay terminal 8. Check for continuity between relay terminals 5 and 7. If there is no continuity, replace the relay. If continuity is present, the relay has not failed.
7. Reinstall the relay.

CHECK ENGINE Light

The function of the CHECK ENGINE lamp on the dashboard is checked by the ECU every time the ignition is turned **ON**. The lamp should illuminate for 2 seconds and then extinguish. In normal operation, the lamp should not come on while driving.

If the warning lamp lights while the vehicle is operating, the ECU has sensed a system fault. The fault may be either loss of signal or signal out-of-range, either high or low. Additionally, when the CHECK ENGINE lamp is lit, the ECU has entered the default or fail-safe mode in which fixed values are substituted for the faulty or missing signal.

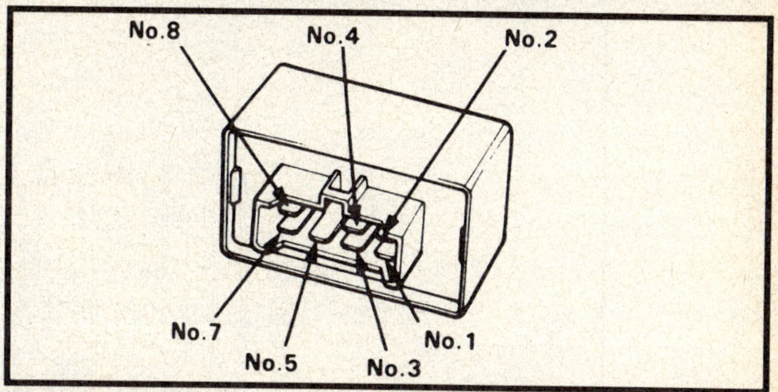

Main relay terminal designation—Integra and 1990 Legend

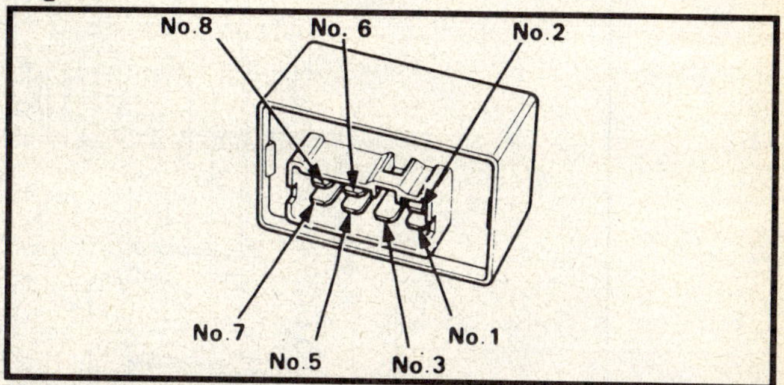

Main relay terminal designation—Vigor

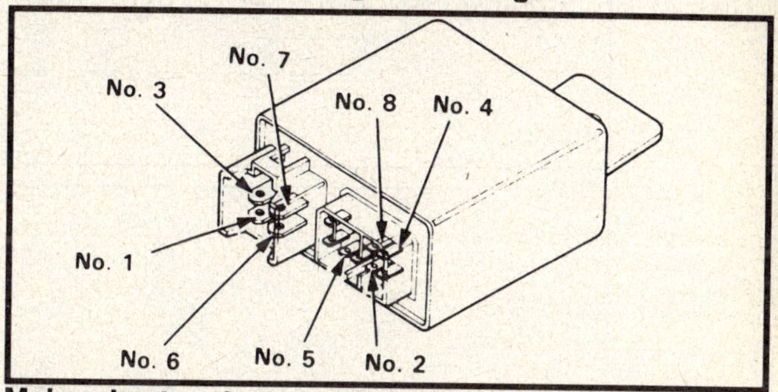

Main relay terminal designation—NSX and 1991–92 Legend

If, during diagnosis, the service connector is jumpered, the CHECK ENGINE lamp will light, even if no faults are present. When repairs or diagnosis are completed, the memory should be checked for any stored codes and cleared if necessary. Recording and clearing the codes before a diagnostic test drive can be helpful in separating intermittent faults from current or hard faults.

USING TROUBLESHOOTING CHARTS

The symptom to system charts are divided into sub-groups for the PGM-FI Control, Idle Control, Air Intake Control, Fuel Supply and Emission Control systems. They also provide a list of items to be checked in order of likelihood; the most common cause is numbered 1, and so on. Note also that in some cases the same troubleshooting chart is used for multiple codes. The codes may not be sequential; for example, the NSX knock sensor chart covers fault Codes 23 and 53 with the same chart.

When beginning diagnosis, refer to the symptom-to-system troubleshooting charts; this chart will direct the use of other troubleshooting charts to test circuits and components related to the problem at hand. Both the symptom and troubleshooting charts include the applicable self-diagnostic or fault codes when necessary.

VACUUM COMPONENT AND HOSE ROUTING 1990 LEGEND

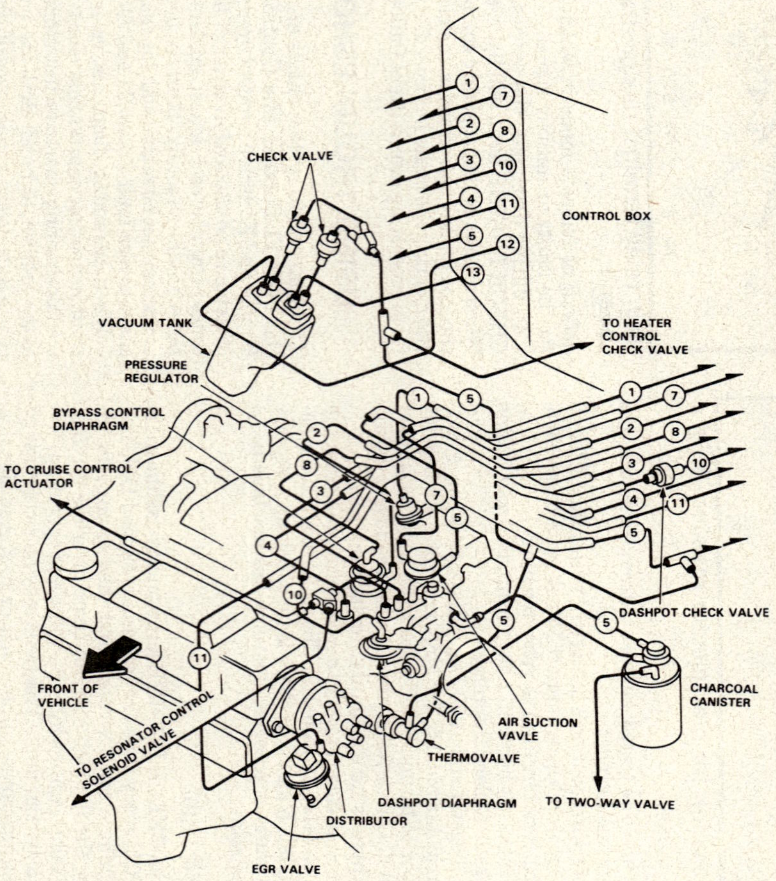

VACUUM SYSTEM SCHEMATIC—1990 LEGEND

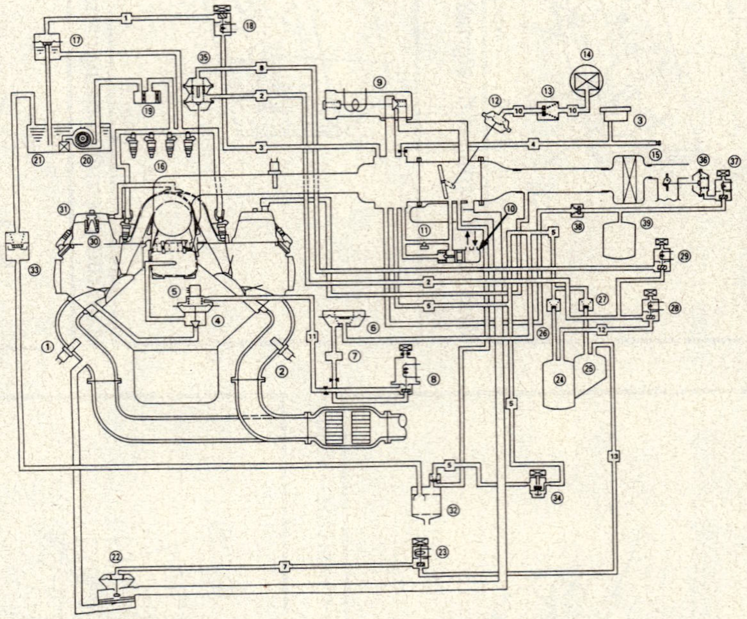

1. FRONT OXYGEN (O_2) SENSOR
2. REAR OXYGEN (O_2) SENSOR
3. MANIFOLD ABSOLUTE PRESSURE (MAP) SENSOR
4. EGR VALVE
5. EGR VALVE LIFT SENSOR
6. CONSTANT VACUUM CONTROL (CVC) VALVE
7. AIR CHAMBER
8. EGR CONTROL SOLENOID VALVE
9. ELECTRONIC AIR CONTROL VALVE (EACV)
10. FAST IDLE VALVE
11. IDLE ADJUSTING SCREW
12. DASHPOT DIAPHRAGM
13. DASHPOT CHECK VALVE
14. AIR FILTER
15. AIR CLEANER
16. FUEL INJECTOR
17. PRESSURE REGULATOR
18. PRESSURE REGULATOR CUT-OFF SOLENOID VALVE
19. FUEL FILTER
20. FUEL PUMP
21. FUEL TANK
22. AIR SUCTION VALVE
23. AIR SUCTION CONTROL SOLENOID VALVE
24. VACUUM TANK A
25. VACUUM TANK B
26. CHECK VALVE A
27. CHECK VALVE B
28. BYPASS CONTROL SOLENOID VALVE A
29. BYPASS CONTROL SOLENOID VALVE B
30. PCV VALVE
31. BREATHER CHAMBER
32. CHARCOAL CANISTER
33. TWO-WAY VALVE
34. THERMOVALVE
35. BYPASS CONTROL DIAPHRAGM
36. VACUUM CONTROL DIAPHRAGM
37. RESONATOR CONTROL SOLENOID VALVE
38. CHECK VALVE C
39. SURGE TANK

ECU INPUT/OUTPUT WIRING SCHEMATIC
1990 LEGEND

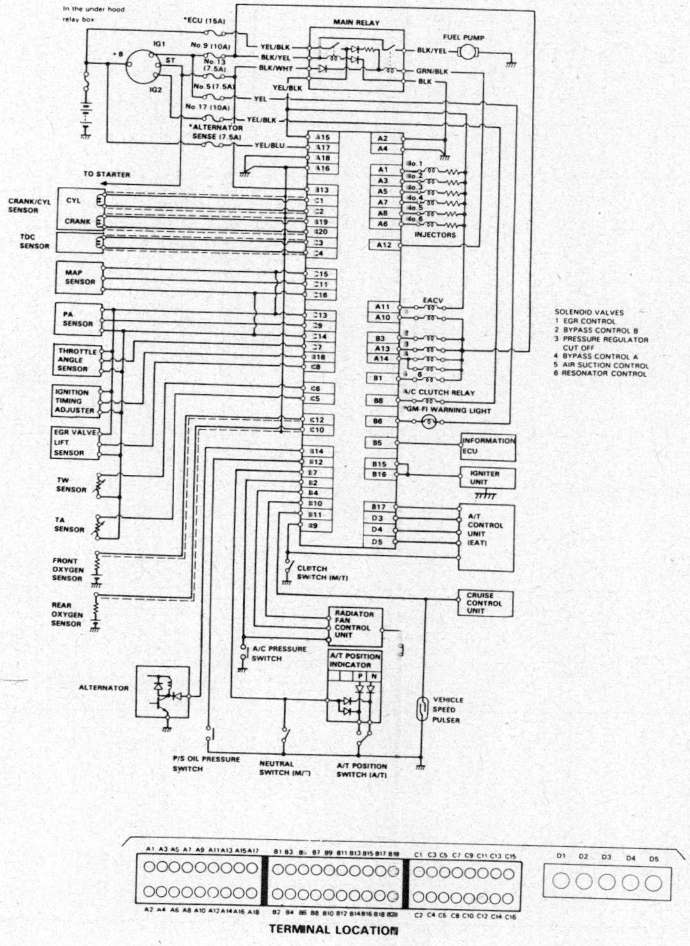

SYMPTOM TO SYSTEM DIAGNOSIS
1990 LEGEND

When the Check Engine warning light has been reported on, turn the ignition on, pull down the passenger side carpet from under the dashboard and observe the LED on the top of the ECU. The Red LED indicates a system failure code by blinking frequency.
(The Yellow LED is for idle speed adjustment and is not related to the following troubleshooting codes.)

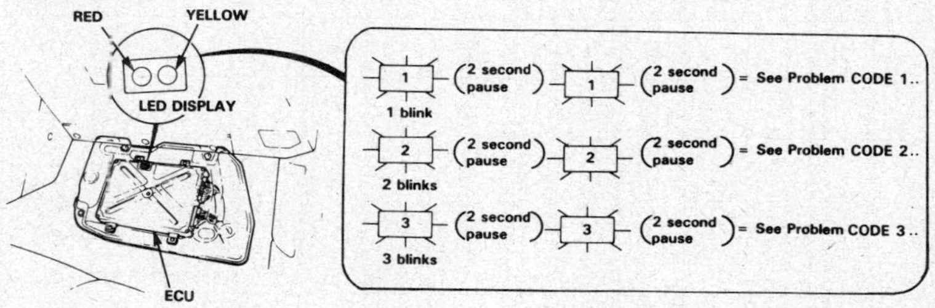

SELF-DIAGNOSIS INDICATOR BLINKS	SYSTEM INDICATED	PAGE
0	ECU	11-17, 18
1	FRONT OXYGEN CONTENT	11-20
2	REAR OXYGEN CONTENT	11-20
3	MANIFOLD ABSOLUTE PRESSURE	11-22
5		1124
4	CRANK ANGLE	11-26
6	COOLANT TEMPERATURE	11-30
7	THROTTLE ANGLE	11-32
8	TDC POSITION	11-34
9	No. 1 CYLINDER POSITION	11-28
10	INTAKE AIR TEMPERATURE	11-38
12	EXHAUST GAS RECIRCULATION SYSTEM	11-103
13	ATMOSPHERIC PRESSURE	11-40
14	ELECTRONIC IDLE CONTROL	11-54
15	IGNITION OUTPUT SIGNAL	11-42
17	VEHICLE SPEED PULSER	11-44
18	IGNITION TIMING ADJUSTMENT	11-46
30	A/T FI SAIGNAL A	11-48
31	A/T FI SIGNAL B	11-50

- If codes other than those listed above are indicated, count the number of blinks again, if the indicator is in fact blinking these codes, substitute a known-good ECU and recheck. If the indication goes away, replace the original ECU.
  The Check Engine dash warning light and ECU Red LED may come on, indicating a system problem, when, in fact, there is a poor or intermittent electrical connection. First, check the electrical connections, clean or repair connections if necessary.
- The Check Engine warning light and S warning light may light simultaneously when the self-diagnosis indicator blinks 6,7 and 17. Check the PGM-FI system according to the PGM-FI control system troubleshooting, then recheck the S warning light.
- The Check Engine warning light does not come on when there is a malfunction in the A/T FI signal. However, the ECU LED will indicate the codes.

SYMPTOM TO SYSTEM DIAGNOSIS—1990 LEGEND

PGM-FI						IDLE CONTROL				
ATMOSPHERIC PRESSURE SENSOR	IGNITION OUTPUT SIGNAL	VEHICLE SPEED PULSER	IGNITION TIMING ADJUSTER	A/T FI SIGNAL A	A/T FI SIGNAL B	ELECTRONIC AIR CONTROL VALVE	OTHER IDLE CONTROLS	FUEL SUPPLY	AIR INTAKE	EGR CONTROL SYSTEM
13	15	17	18	30	31	14				12
3	3							1		
								2		
						2	1			
						2	3			1
						1	2			
						1				2
						1		2		3
						2		3		1
						2		1		3
								1	3	

SYMPTOM \ SUB SYSTEM		IDLE ADJUSTING SCREW	ELECTRONIC AIR CONTROL VALVE	STARTER SIGNAL	ALTERNATOR FR SIGNAL	A/T SHIFT POSITION SIGNAL	P/S OIL PRESSURE SIGNAL	AIR CONDITIONING SIGNAL	FAST IDLE VALVE	M/T NEUTRAL SWITCH SIGNAL	CLUTCH SWITCH SIGNAL	HOSES AND CONNECTIONS
												•
DIFFICULT TO START ENGINE WHEN COLD			2						1			
WHEN COLD	Fast idle speed is not as specified (1,100–1,900 rpm).		2						1			
WHEN WARM RPM TOO HIGH	Idle speed is above specified rpm.		3				3		2			1
WHEN WARM RPM TOO LOW	Idle speed is below specified rpm.	2	1						3			
	Idle speed does not increase after initial start up.		1									
	On models with automatic transmission, the idle speed drops in gear.		2			1						
	Idle speed drops when steering wheel is turning.		2				1					
	Idle speed drops when air conditioner is ON.		2					1				
FREQUENT STALLING WHILE WARMING UP		3	2						1			

• – If by-pass passages are blocked, a low idle speed will result.
– If hoses or by-pass passages are leaking, a high idle speed will result.

SYMPTOM TO SYSTEM DIAGNOSIS—1990 LEGEND

NOTE: Across each row in the chart, the sub systems that could be sources of a symptom are ranked in the order they should be inspected, starting with ①. Find the symptom in the left column, read across to the most likely source, then refer to the page listed at the top of that column. If inspection shows the system is OK, try the next system ②, etc.

SYMPTOM \ SUB SYSTEM		FUEL INJECTOR	INJECTOR RESISTOR	PRESSURE REGULATOR	FUEL FILTER	FUEL PUMP	MAIN RELAY	FUEL TANK	CONTAMINATED FUEL
									*
ENGINE WON'T START			③		③	①	②		
FREQUENT STALLING					③	①			②
POOR PERFORMANS	MISFIRE	①	②						③
	LOSS OF POWER				①				②

* Fuel with dirt, water or a high percentage of alcohol is considered contaminated.

SYMPTOM \ SUB SYSTEM		THROTTLE CABLE	THROTTLE BODY	DASHPOT SYSTEM	BYPASS CONTROL SYSTEM
IRREGULAR IDLING	WHEN WARM IDLE SPEED OUT OF SPECIFIED RPM	②	①		
FREQUENT STALLING WHILE WARMING UP			①		
POOR PERFORMANCE	LOSS OF POWER	②	①		③

NOTE: Across each row in the chart, the sub systems that could be sources of a symptom are ranked in the order they should be inspected, starting with ①. Find the system in the left column, read across to the most likely source, then refer to the page listed at the top of that column. If inspection shows the system is OK, try the next system ②, etc.

SYMPTOM \ SUB SYSTEM		CATALYTIC CONVERTER	AIR INJECTION SYSTEM	EGR SYSTEM	POSITIVE CRANKCASE VENTILATION SYSTEM	EVAPORATIVE EMISSION CONTROLS
IRREGULAR IDLING	ROUGH IDLE			①		
	WHEN WARM RPM TOO HIGH				①	
FREQUENT STALLING				①		
POOR PERFORMANCE	FAILS EMISSION TEST	①				②
	LOSS OF POWER	①		②		

TROUBLESHOOTING CHARTS—1990 LEGEND

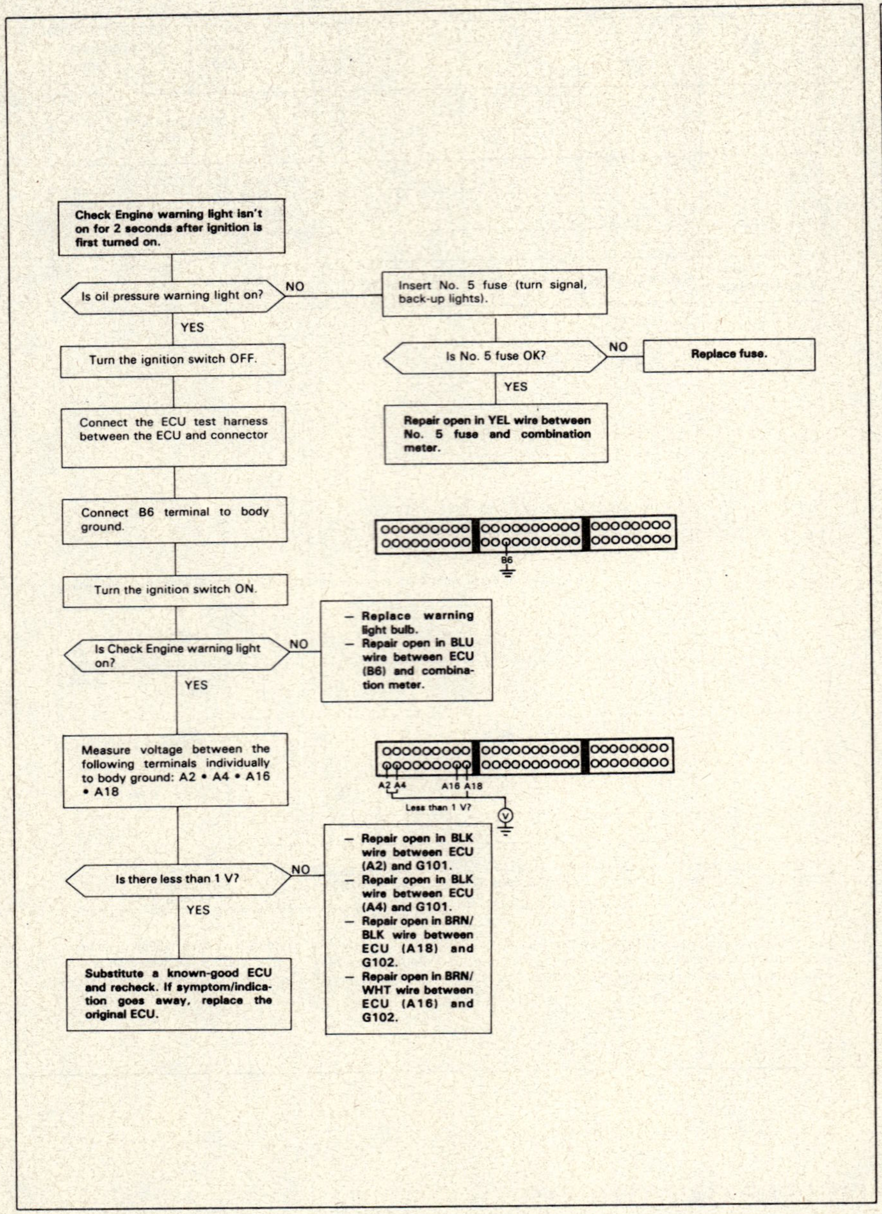

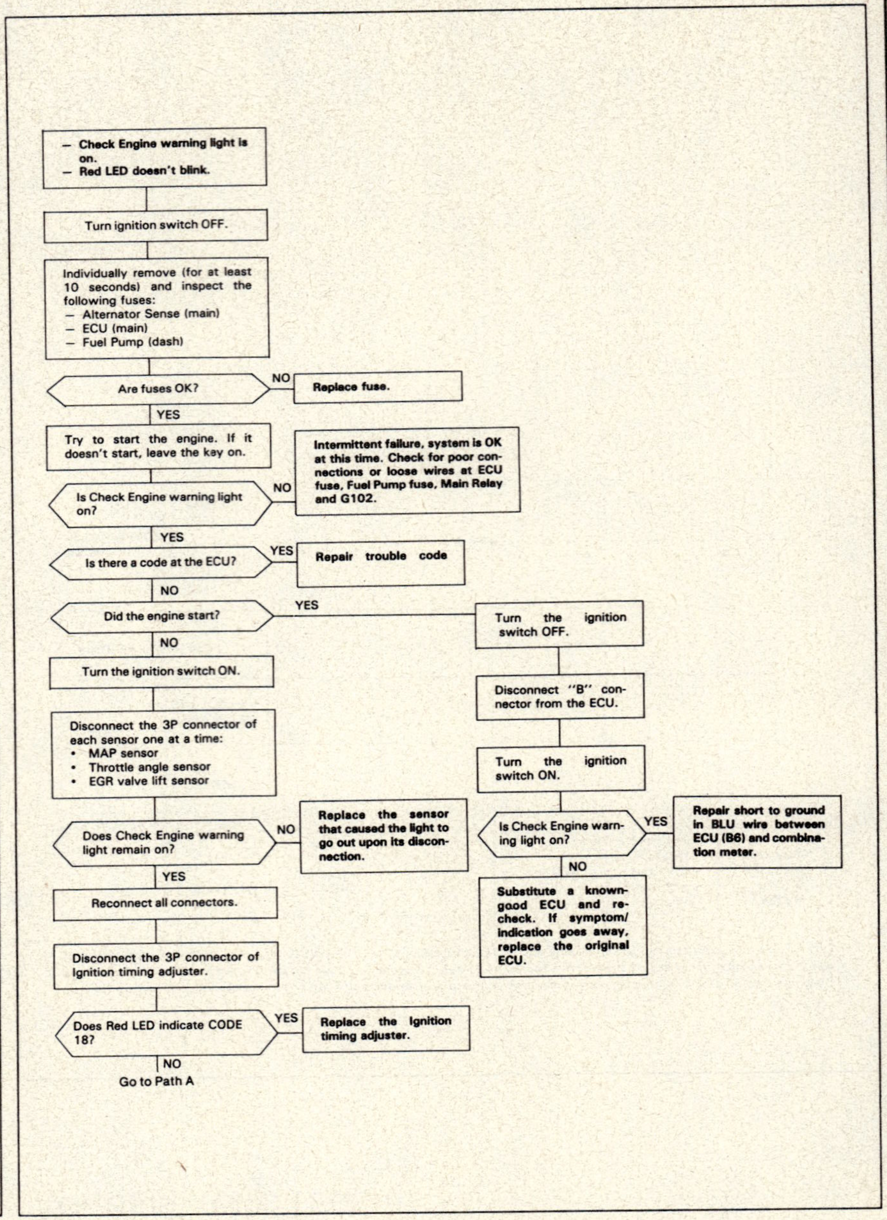

TROUBLESHOOTING CHARTS—1990 LEGEND

Path A

Reconnect the connector.

Turn the ignition switch OFF.

Connect the ECU test harness between the ECU and connector. Disconnect "C" connector from ECU only, not the main wire harness.

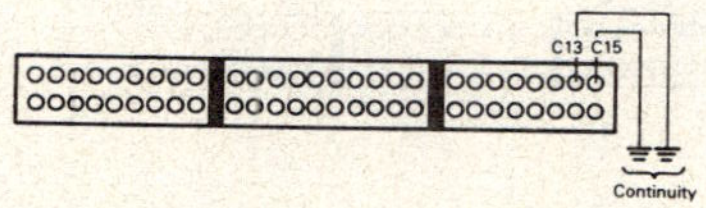

Check for continuity between the body ground and the following terminals individually: • C13, • C15

Does continuity exist?

- YES → Repair short to ground in YEL/WHT wire between ECU (C15) and MAP sensor, or in YEL/WHT wire between ECU (C13) and the PA sensor, Ignition timing adjuster, throttle angle sensor or EGR valve lift sensor.
- NO ↓

Reconnect "C" connector to ECU and disconnect "B" connector from the ECU only, not the main wire harness.

Turn the ignition switch ON.

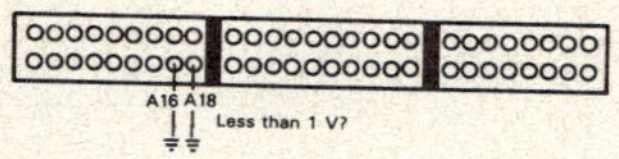

Measure voltage between the following terminals individually to body ground: • A16, • A18

Is there less than 1 V?

- NO → – Repair open in BRN/BLK wire between ECU (A18) and G102. – Repair open in BRN/WHT wire between ECU (A16) and G102.
- YES ↓

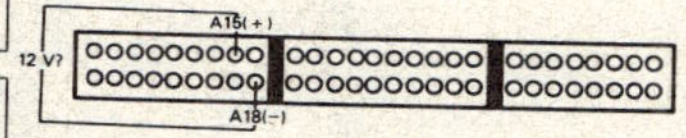

Measure voltage between A15 (+) terminal and A18 (−) terminal.

Is there battery voltage?

- NO → – Repair open in YEL/BLK wire between ECU (A15) and main relay. – Check main relay and wiring connectors at main relay
- YES ↓

Turn the ignition switch OFF.

Substitute a known-good ECU.

Turn the ignition switch ON.

Does the Check Engine warning light remain on after 2 sec.?

- NO → Replace the original ECU.
- YES → Replace the PA sensor.

Self-diagnosis Red LED indicates code 1: A problem in the Front Oxygen (O_2) Sensor circuit.

Self-diagnosis Red LED indicates code: A problem in the Rear Oxygen (O_2) Sensor circuit.

The oxygen sensor, detects the oxygen content in the exhaust gas, and inputs the ECU. In operation, the ECU receives the signals from the sensor and varies the duration during which fuel is injected. The oxygen sensors are installed on the exhaust manifolds.

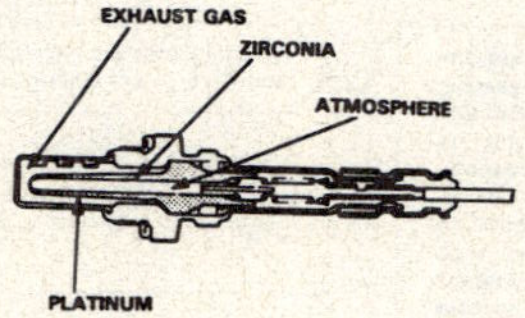

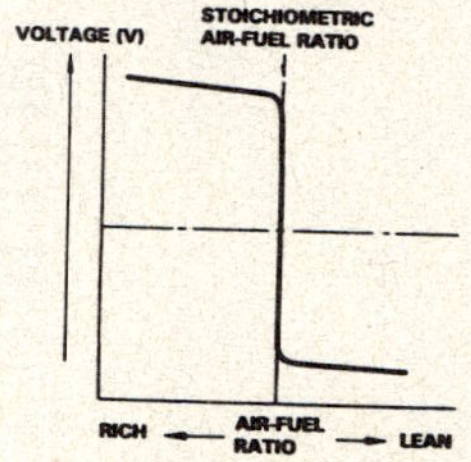

① or ②

- Check Engine warning light has been reported on.
- Red LED indicates CODE 1 (Front O_2 sensor) or CODE 2 (Rear O_2 sensor).

Turn the ignition switch OFF.

Remove ALTERNATOR SENSE fuse in the under-hood relay box for 10 seconds to reset ECU.

Inspect fuel pressure at the fuel filter

Is the fuel pressure above 34 psi with the pressure regulator vacuum hose connected?

- YES → Replace pressure regulator
- NO ↓

Warm up engine to normal operating temperature (cooling fan comes on).

Road test necessary. With the transmission in second gear, accelerate from 1000 RPM to 4000 RPM (A/T) or 4500 RPM (M/T) using wide open throttle, then decelerate to 1000 RPM with the throttle completely closed.

Is Check Engine warning light on and does LED indicates CODE 1 or CODE 2?

- NO → Intermittent failure; system is OK at this time. Check for poor connections or loose wires at the indicated O_2 sensor and C102–C305 near the right shock tower, and at the ECU.
- YES → Go to Path B

Path B

Disconnect engine wire harness from indicated O_2 sensor.

Warm up engine to normal operating temperature again.

Measure voltage between the connector terminal and body ground.

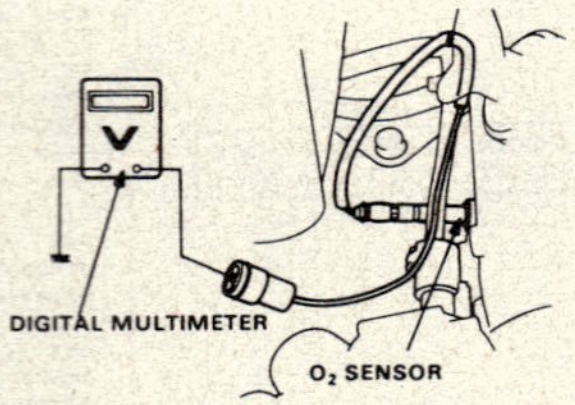

Is voltage above 0.6 V at 5,000 rpm and below 0.4 V when the throttle is quickly released from 5,000 rpm?

- NO → Replace O_2 sensor.
- YES ↓

Stop engine.

Reconnect the O_2 sensor.

Connect the PGM-FI test harness between the ECU and connector

Restart and warm up engine to normal operating temperature.

Measure voltage between:
- Front O_2 sensor C12 (+) and A18 (−) terminals
- Rear O_2 sensor C10 (+) and A18 (−) terminals

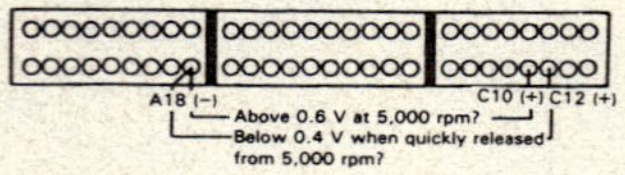

Is voltage above 0.6 V at 5,000 rpm and below 0.4 V when the throttle is quickly released from 5,000 rpm?

- NO → Repair wire between ECU and O_2 sensor for short or open:
  - Front O_2 sensor WHT wire to C12.
  - Rear O_2 sensor RED/BLU wire to C10.
- YES → Substitute a known-good ECU and recheck. If symptom/indication goes away, replace the original ECU.

Self-diagnosis Red LED indicates code 3: Most likely an electrical problem in the Manifold Absolute Pressure (MAP) Sensor system.

The MAP sensor converts manifold absolute pressure into electrical signals and inputs the ECU.

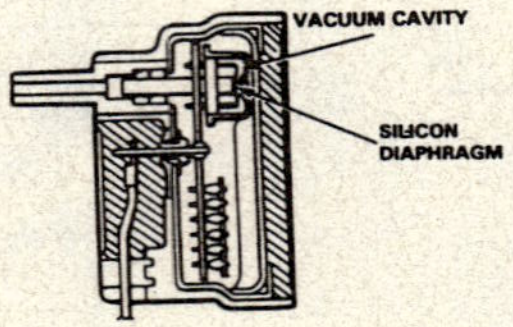

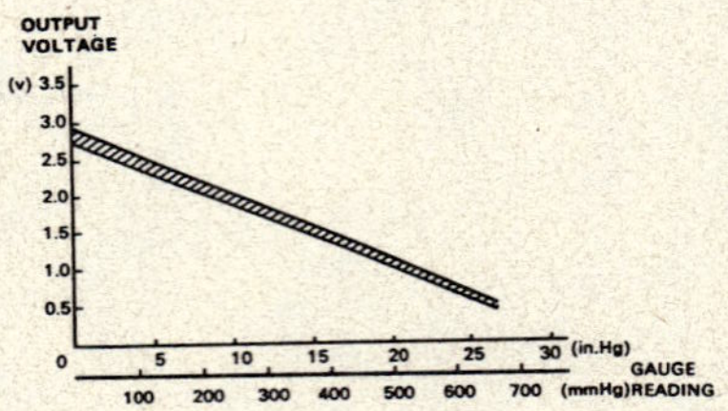

- Engine is warm and running.
- Check Engine warning light has been reported on.
- Red LED indicates CODE 3.

Turn the ignition switch OFF.

Remove ALTERNATOR SENSE fuse in the under-hood relay box for 10 seconds to reset ECU.

Start the engine and allow it to idle.

Is Check Engine warning light on and does LED indicate CODE 3?

- NO → Intermittent failure, system is OK at this time. (Test drive may be necessary.) Check for poor connections or loose wires at the MAP sensor connector on the right firewall shown in the illustration below.
- YES ↓

Turn the ignition switch OFF.

Connect the test harness between the MAP sensor and wire harness.

Turn the ignition switch ON.

Go to Path C

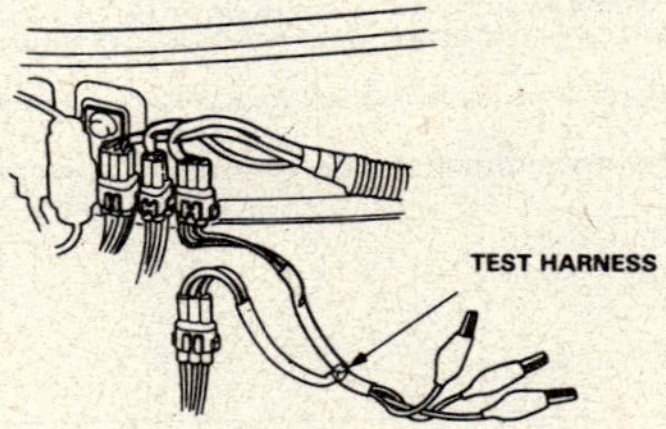

TROUBLESHOOTING CHARTS—1990 LEGEND

Path C

Measure voltage between RED (+) terminal and GRN (−) terminal.

Is there approx. 5 V?

- NO → Measure voltage between RED (+) terminal and body ground.
  - Is there approx. 5 V?
    - YES → Repair open in GRN/WHT wire between ECU (C16) and MAP sensor.
    - NO → Turn the ignition switch OFF.
      - Connect the ECU test harness between the ECU and connector
      - Turn the ignition switch ON.
      - Measure voltage between C15 (+) terminal and C16 (−) terminal with the ignition switch on.

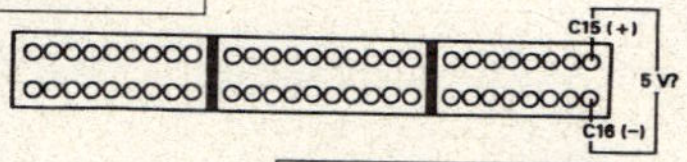

      - Is there approx. 5 V?
        - YES → Repair open in YEL/WHT wire between ECU (C15) and MAP sensor.
        - NO → Substitute a known-good ECU and recheck. If prescribed voltage is now available, replace the original ECU.
- YES → Measure voltage between WHT (+) terminal and GRN (−) terminal.
  - Is there approx. 3 V?
    - NO →
      - Replace MAP sensor.
      - Repair open or short in WHT wire between ECU (C11) and MAF sensor.
    - YES → Turn the ignition switch OFF.
      - Connect the ECU test harness between the ECU and connector
      - Turn the ignition switch ON.
      - Measure voltage between C11 (+) terminal and C16 (−) terminal.

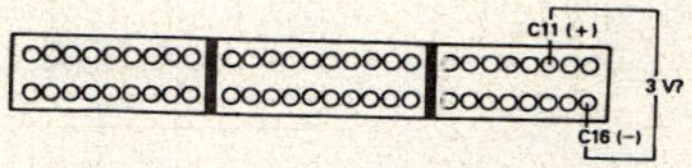

      - Is there approx. 3 V?
        - NO → Repair open in WHT wire between ECU (C11) and MAP sensor.
        - YES → Substitute a known-good ECU and recheck. If symptom/indication goes away, replace the original ECU.

Self-diagnosis Red LED indicates code 4: A problem in the CRANK circuit of the CRANK/CYL Sensor.

The CRANK sensor determines timing for fuel injection and ignition of each cylinder and also detects engine RPM. The CYL sensor detects the position of No.1 cylinder for sequential fuel injection to each cylinder.

- Check Engine warning light has been reported on.
- Red LED indicates CODE 4.

Turn the ignition switch OFF.

Remove ALTERNATOR SENSE fuse in the under-hood relay box for 10 seconds to reset ECU.

Start engine.

Is Check Engine warning light on and does LED indicate CODE 4?

- NO → Intermittent failure, system is OK at this time. (Test drive may be necessary.) Check for poor connections or loose wires at CRANK/CYL sensor connector and C102—C305 at the right shock tower.
- YES → Stop engine.
  - Disconnect the 4P connector near the CRANK/CYL sensor.
  - Go to Path D

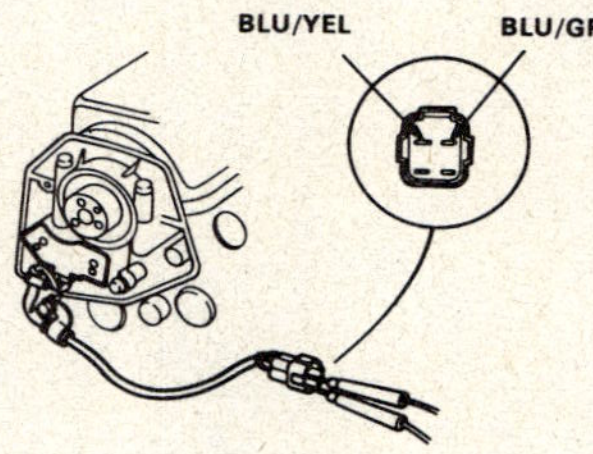

TROUBLESHOOTING CHARTS – 1990 LEGEND

Path D

Measure resistance between BLU/YEL terminal and BLU/GRN terminal.

Is there 500–1,000 Ω?
- NO → **Replace CRANK/CYL sensor.**
- YES ↓

Check for continuity to body ground on BLU/YEL terminal and BLU/GRN terminal individually.

Does continuity exist?
- YES → **Replace CRANK/CYL sensor**
- NO ↓

Reconnect the connector.

Connect the ECU test harness only to the main wire harness, not to the ECU

Measure resistance between B19 terminal and B20 terminal.

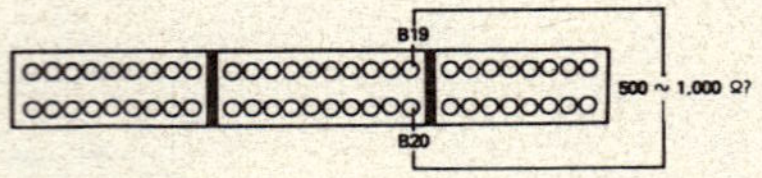

Is there 500–1,000 Ω?.
- NO → **Repair open in BLU/YEL and/or BLU/GRN wires.**
- YES ↓

Check continuity to body ground on B19 terminal.

Does continuity exist?
- YES → **Repair short to ground in BLU/GRN wire between ECU (B19) and 4P connector.**
- NO ↓

Substitute a known-good ECU and recheck. If symptom/indication goes away, replace the original ECU.

– Check Engine warning light has been reported on.
– Red LED indicates CODE 5.

Turn the ignition switch OFF.

Remove ALTERNATOR SENSE fuse in the under-hood relay box for 10 seconds to reset ECU.

Start engine and keep engine rpm at idle.

Is Check Engine warning light on and and does LED indicate CODE 5?
- NO → **Intermittent failure, system is OK at this time. (Test drive may be necessary.) Check #4 vacuum hose for cracks, leaks or a pinched hose.**
- YES ↓

Stop engine.

Remove #4 hose from the vacuum hose manifold and connect a T-fitting from a vacuum gauge between the intake manifold and the MAP sensor.

Start engine.

Is there manifold vacuum?
- NO → **Repair as necessary.**
- YES ↓

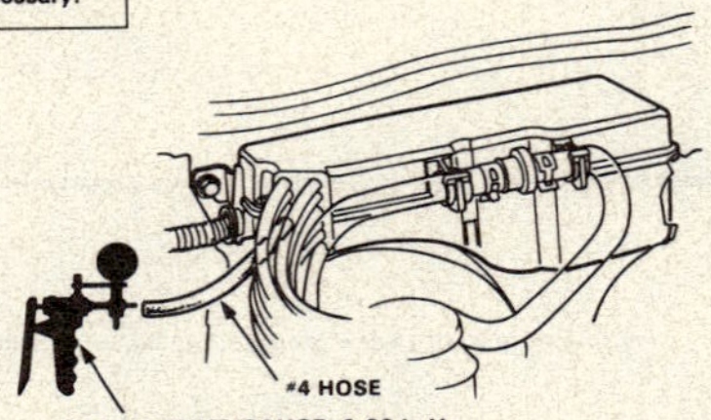

Connect vacuum pump to #4 hose and apply vacuum.

Does it hold vacuum?
- NO → **• Replace #4 hose • Replace MAP sensor.**
- YES ↓

Go to Path E

Path E

Stop engine.

Connect the test harness between the MAP sensor and wire harness.

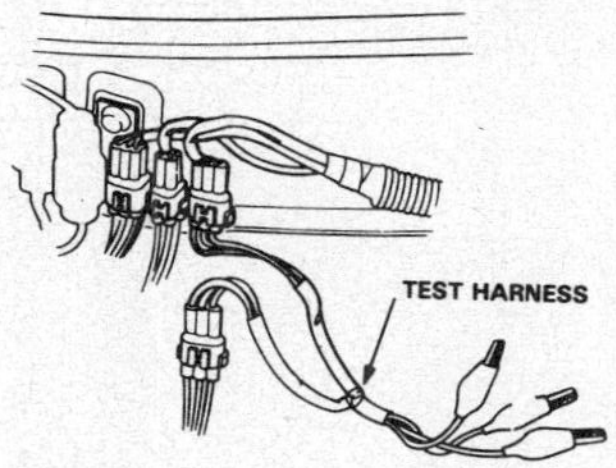

Turn the ignition switch ON.

Measure voltage between WHT (+) terminal and GRN (−) terminal.

Is there approx. 3V?

- NO → **Replace MAP sensor.**
- YES → **Substitute a known-good ECU and recheck. If symptom/indication goes away, replace the original ECU.**

Self-diagnosis Red LED indicates code 6: Most likely a problem in the Coolant Temperature (TW) Sensor circuit.

The TW sensor is a temperature dependant resistor (thermistor). The resistance of the thermistor decreases as the coolant temperature increases as shown below.

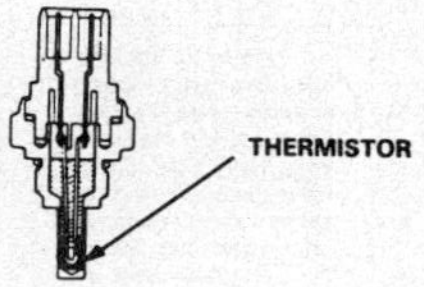

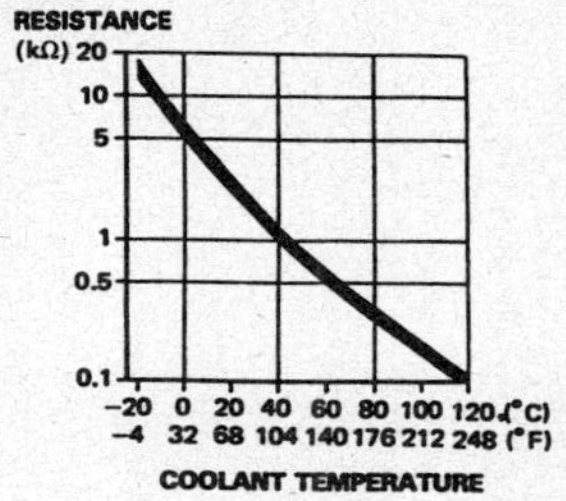

- Check Engine warning light has been reported on.
- Red LED indicates CODE 6.

Turn the ignition switch OFF.

Remove ALTERNATOR SENSE fuse in the under-hood relay box for 10 seconds to reset ECU.

Turn the ignition switch ON.

Is Check Engine warning light on and does LED indicate CODE 6?

- NO → **Intermittent failure, system is OK at this time. (Test drive may be necessary.) Check for poor connections or loose wires at TW sensor and RED/WHT wire at C303—C103 and GRN/WHT at C305—C102 at the right shock tower.**
- YES ↓

Turn the ignition switch OFF.

Connect the ECU test harness between the ECU and connector

Warm up engine to normal operating temperature (cooling fan comes on).

Turn the ignition switch OFF.

Disconnect "C" connector from ECU only, not the main wire harness.

Measure resistasnce between C6 terminal and C14 terminal.

Is there 200—400 Ω?

- NO → **Inspect for an open circuit in GRN/WHT wire between ECU (C14) and TW sensor or open circuit in RED/WHT wire between ECU (C6) and TW sensor. If wires are OK, replace TW sensor.**
- YES ↓

Reconnect "C" connector to ECU and disconnect "C" connector from main wire harness.

Go to Path F

TROUBLESHOOTING CHARTS—1990 LEGEND

Path F

- Turn the ignition switch ON.
- Measure voltage between C6 (+) terminal and C14 (−) terminal.
- Is there approx. 5 V?
  - NO → Substitute a known-good ECU and recheck. If prescribed voltage is now available, replace the original ECU.
  - YES → Turn the ignition switch OFF.
- Reconnect "C" connector to main wire harness.
- Turn the ignition switch ON.
- Measure voltage between C6 (+) terminal and C14 (−) terminal.
- Is there less than 0.50 V?
  - NO → Substitute a known-good ECU and recheck. If symptom/indication goes away, replace the original ECU.
  - YES → AT ONLY:
    - Disconnect "A" connector from Automatic Transmission control unit.
    - Is there less than 0.50 V?
      - NO → Replace Automatic Transmission control unit.
      - YES → Repair short in RED/WHT wire between ECU (C6) and TW sensor.

Self-diagnosis Red LED indicates code 7: Most likely a problem in the Throttle Angle Sensor circuit.

The throttle angle sensor is a potentiometer. It is connected to the primary throttle valve shaft. As the throttle angle changes, the throttle angle sensor varies the voltage signal to the ECU.

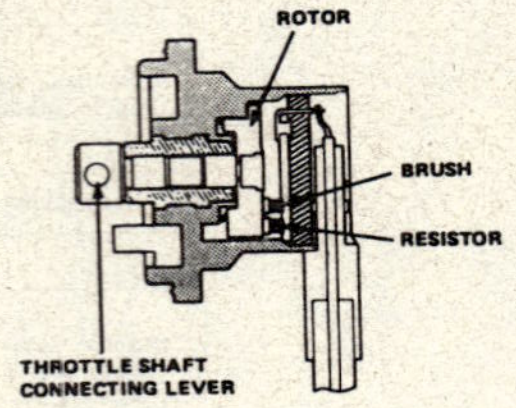

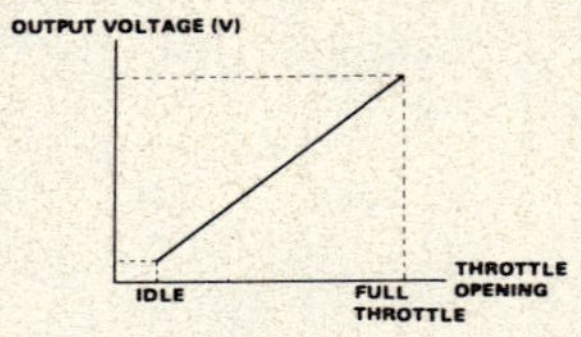

- Engine is running.
- Check Engine warning light has been reported on.
- Red LED indicates CODE 7.

1. Turn the ignition switch OFF.
2. Remove ALTERNATOR SENSE fuse in the under-hood relay box for 10 seconds to reset ECU.
3. Start engine.
4. Is Check Engine warning light on and does LED indicate CODE 7?
   - NO → Intermittent failure, system is OK at this time. (Test drive may be necessary.) Check for poor connections or loose wires at the throttle angle sensor and GRN/WHT wire at C305—C102 and YEL/WHT wires at C303—C103 at the right shock tower.
   - YES → Turn the ignition switch OFF.
5. Connect the test harness between the throttle angle sensor and wire harness.
6. Turn the ignition switch ON.

Go to Path G

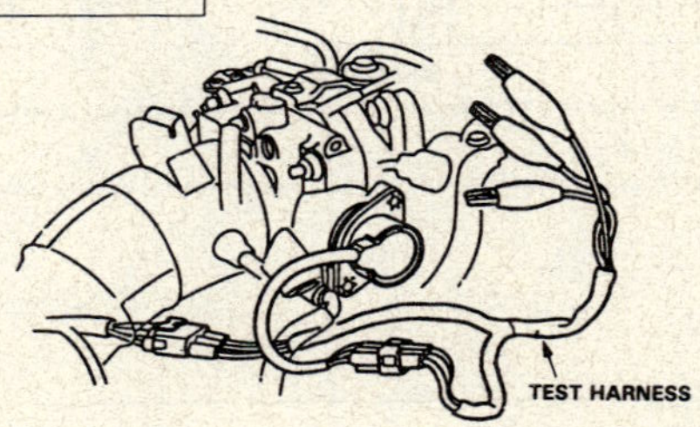

TROUBLESHOOTING CHARTS – 1990 LEGEND

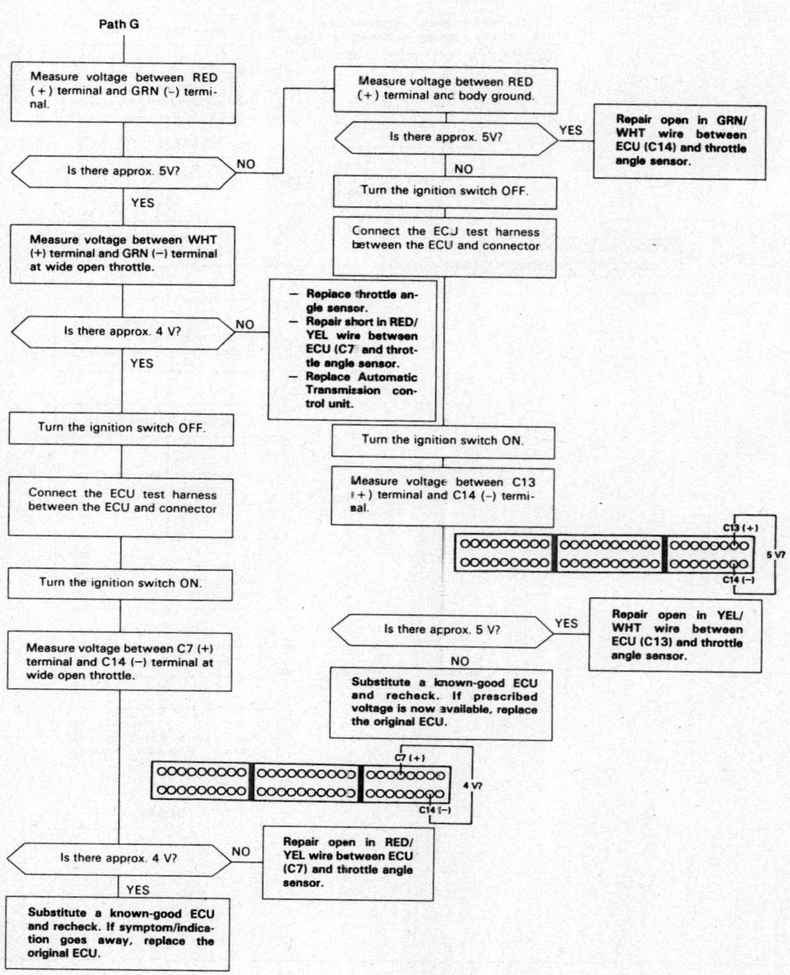

Self-diagnosis LED indicates code 8: A problem in the TDC sensor.

The TDC sensor determines ignition timing at start-up (cranking) and when crank angle signal is abnormal.

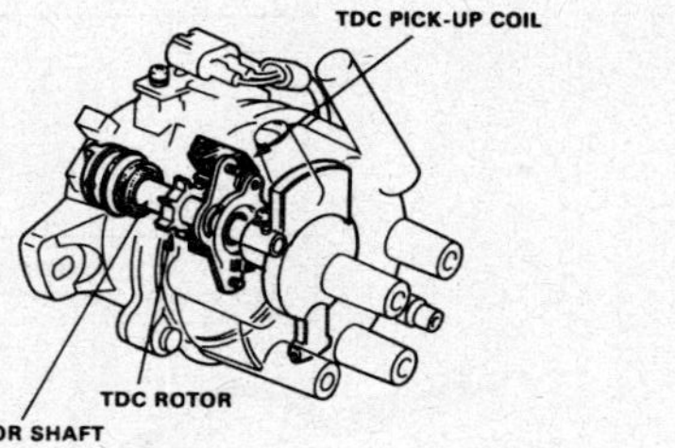

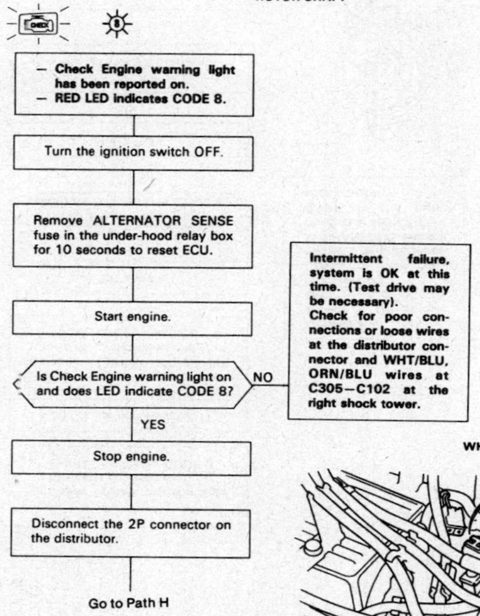

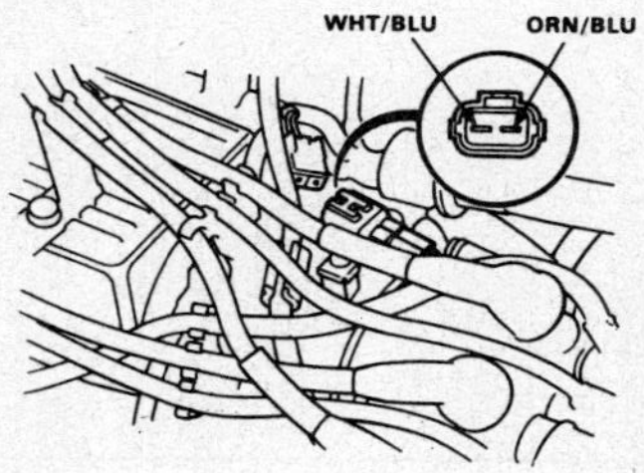

Path H

- Measure resistance between WHT/BLU terminal and ORN/BLU terminal.
- Is there 500 — 1,000 Ω?
  - NO → Replace TDC sensor
  - YES ↓
- Check for continuity to body ground on WHT/BLU terminal and ORN/BLU terminal individually.
- Does continuity exist?
  - YES → Replace TDC sensor
  - NO ↓
- Reconnect the connector.
- Connect the ECU test harness only to the main wire harness, not to the ECU
- Measure resistance between C3 terminal and C4 terminal.

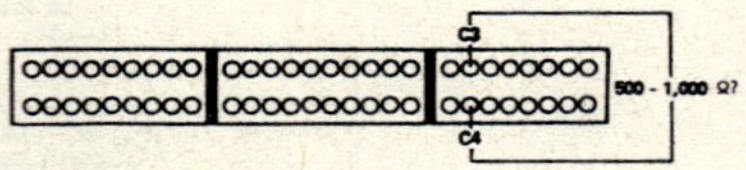

- Is there 500 — 1,000 Ω?
  - NO → Repair open is ORN/BLU and/or WHT/BLU wires.
  - YES ↓
- Check for continuity to body ground on C3 terminal.
- Does continuity exist?
  - YES → Repair short in ORN/BLU wire between ECU (C3) and 29 connector.
  - NO ↓
- Substitute a known-good ECU and recheck. If symptom/indication goes away, replace the original ECU.

- – Check Engine warning light has been reported on.
  – Red LED indicates CODE 9.
- Turn the ignition switch OFF.
- Remove ALTERNATOR SENSE fuse in the under-hood relay box for 10 seconds to reset ECU.
- Start engine.
- Is Check Engine warning light on and does LED indicate CODE 9?
  - NO → Intermittent failure, system is OK at this time. (Test drive may be necessary.) Check for poor connections or loose wires at CRANK/CYL sensor connector and C102—C305 at the right shock tower.
  - YES ↓
- Stop engine.
- Disconnect the 4P connector near the CRANK/CYL sensor.
- Measure resistance between ORN terminal and WHT terminal.

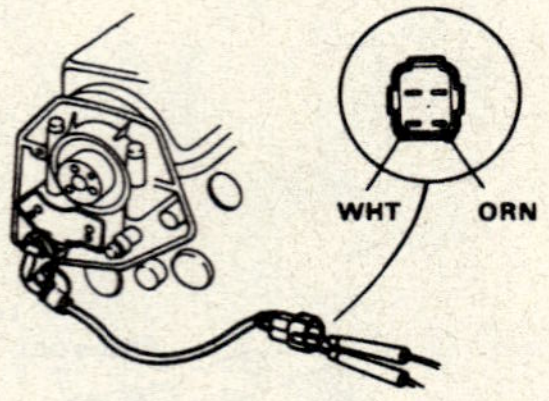

- Is there 500—1,000 Ω?
  - NO → Replace CRANK/CYL sensor
  - YES ↓
- Check for continuity to body ground on ORN terminal and WHT terminal individually.
- Does continuity exist?
  - YES → Replace CRANK/CYL sensor
  - NO ↓
- Reconnect the connector.
- Connect the ECU test harness only to the main harness, not to the ECU (page 11-13).
- Go to Path I

Path I

Measure resistance between C1 terminal and C2 terminal.

Is there 500–1,000Ω?
- NO → **Repair open in ORN and/or WHT wires.**
- YES → Check for continuity to body, ground on C1 terminal.

Does continuity exist?
- YES → **Repair short to ground in ORN wire between ECU (C1) and 4P connector.**
- NO → **Substitute a known-good ECU and recheck. If symptom/indication goes away, replace the original ECU.**

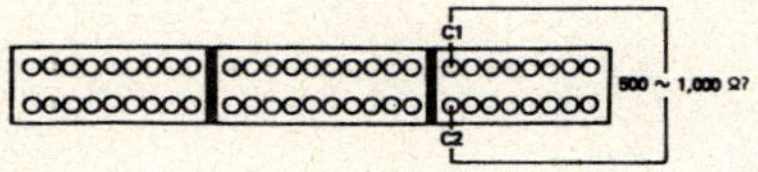

Removal:
1. Remove the cruise control actuator.
2. Remove the front upper timing belt cover.
3. Remove the timing belt
4. Remove 3 bolts and detach the pulley.
5. Remove 4 bolts and remove the front upper cover back plate.

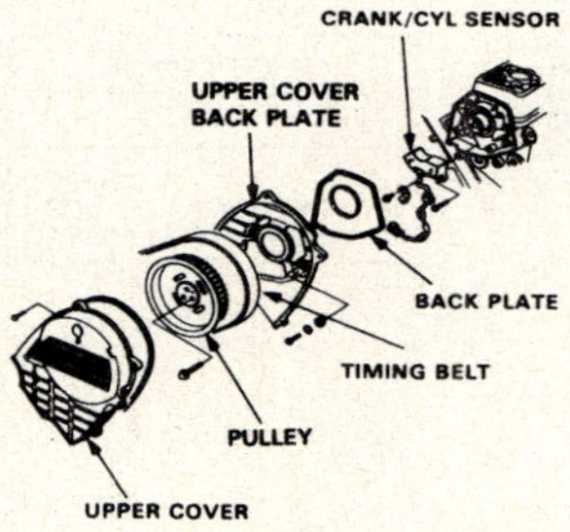

Installation:
1. Install the CRANK/CYL sensor on the cylinder head.
2. Align the cam pulley pin with the camshaft hole to install the cam pulley.

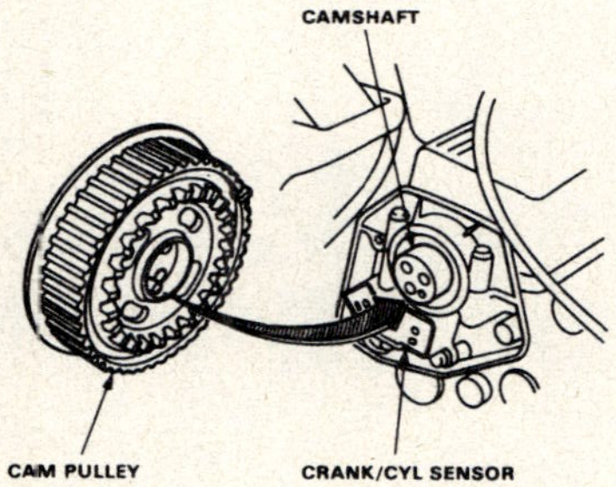

Self-diagnosis Red LED indicates code 10: Most likely a problem in the Intake Air Temperature (TA) Sensor circuit.

The TA sensor is a temperature dependant resistor (thermistor). The resistance of the thermistor decreases as the intake air temperature increases as shown below.

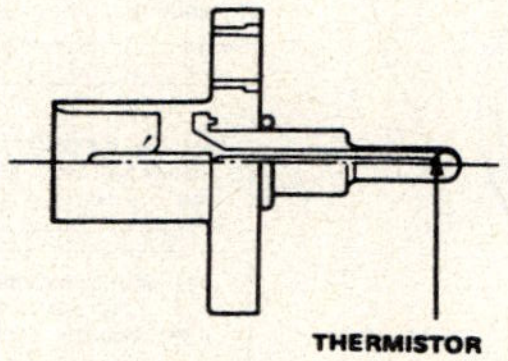

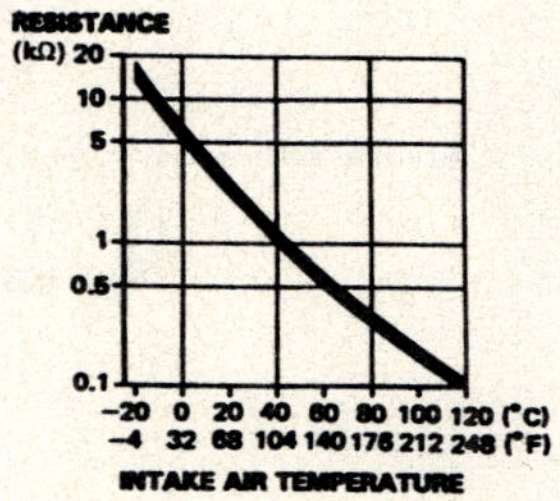

- **Check Engine warning light has been reported on.**
- **Red LED indicates CODE 10.**

Turn the ignition switch OFF.

Remove ALTERNATOR SENSE fuse in the under-hood relay box for 10 seconds to reset ECU.

Turn the ignition switch ON.

Is Check Engine warning light on and does LED indicate CODE 10?
- NO → **Intermittent failure, system is OK at this time. (Test drive may be necessary.) Check for poor connections or loose wires at the TA sensor, GRN/WHT wire at C-305—C-102 and RED/BLK wire at C-303—C-103.**
- YES → Disconnect the 2P connector on the TA sensor.

Measure resistance between the 2 terminals on the TA sensor.

Is there 1–4 kΩ?
- NO → **Replace TA sensor.**
- YES → Go to Path J

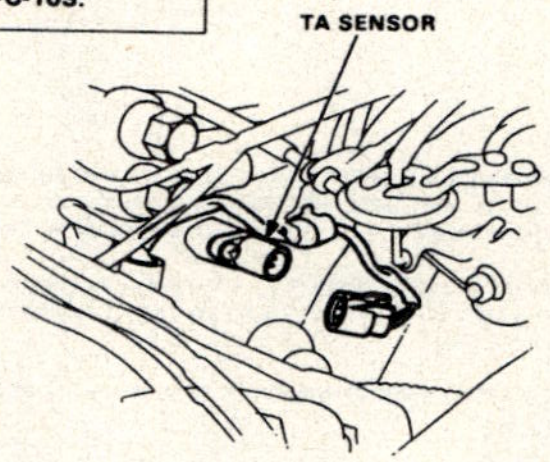

Path J

Measure voltage between RED/BLK (+) terminal and body ground.

Is there approx. 5 V?

NO → **Repair open or short in RED/BLK wire between ECU (C5) and TA sensor. If wires is OK, substitute a known-good ECU and recheck. If prescribed voltage is now a available, replace the original ECU.**

YES ↓

Measure voltage between RED/BLK (+) terminal and GRN/WHT (-) terminal.

Is there approx. 5 V?

NO → **Repair open in GRN/WHT wire between ECU (C14) and TA sensor.**

YES ↓

Substitute a known-good ECU and recheck. If symptom/indication goes away, replace the original ECU.

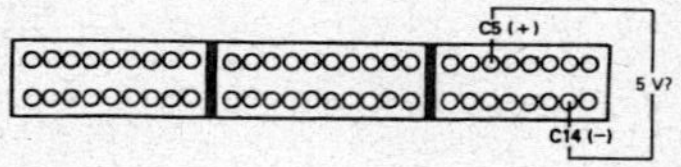

Self-diagnosis Red LED indicates code 12: Most likely a problem in the Exhaust Gas Recirculation (EGR) system.

The EGR System is designed to reduce oxides of nitrogen emissions (NOx), by recirculating exhaust gas through the EGR valve and the intake manifold into the combustion chambers. It is composed of the EGR valve, CVC valve, EGR control solenoid valve, PGM-FI ECU and various sensors.
The ECU contains memories for ideal EGR valve lifts for varying operating conditions. The EGR valve lift sensor detects the amount of EGR valve lift and sends the information to the ECU. The ECU then compares it with the ideal EGR valve lift which is determined by signals sent from the other sensors. If there is any difference between the two, the ECU cuts current to the EGR control solenoid valve to reduce vacuum applied to the EGR valve.

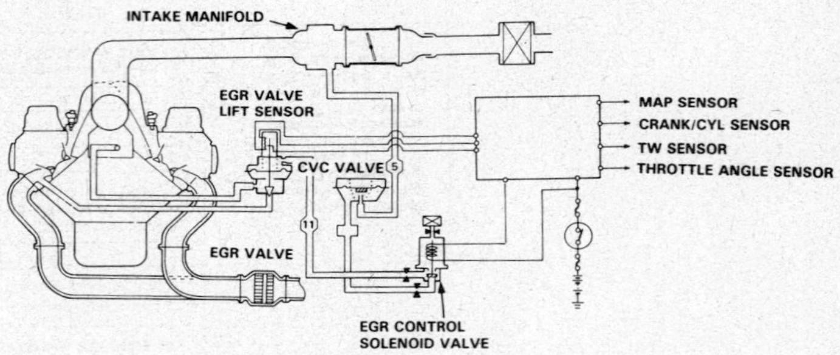

– **Check Engine warning light has been reported on.**
– **Red LED indicates CODE 12.**

Verify that all connectors between the EGR valve lift sensor and the PGM-FI ECU are securely connected.

Warm up engine to normal operating temperature (cooling fan comes on).

Disconnect #11 hose from EGR valve.

Connect vacuum pump to EGR valve and apply vacuum.

Does EGR valve hold vacuum and did engine stall?

NO → **Replace EGR valve.**

YES → Go to Path K

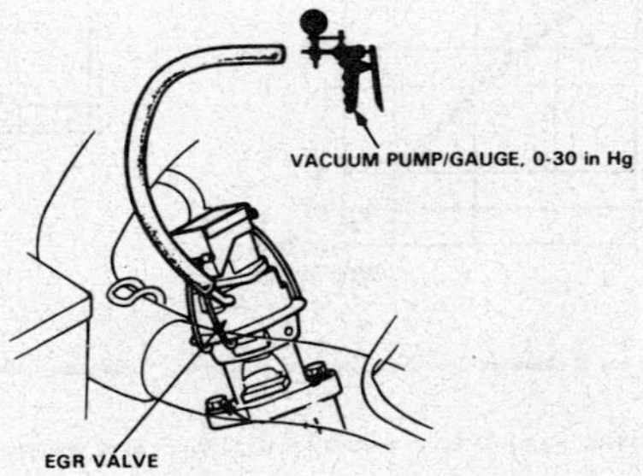

TROUBLESHOOTING CHARTS—1990 LEGEND

Path K

Disconnect the #5 hose (between the vacuum hose manifold and the 3-way joint) from the 3-way joint and connect vacuum gauge to the hose.

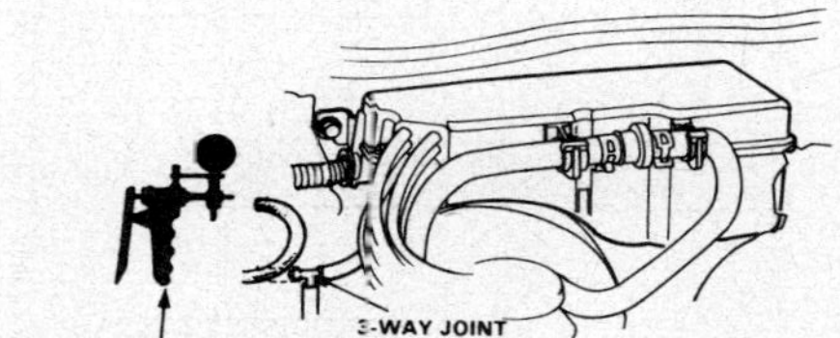

Restart engine and check for manifold vacuum.

Is there manifold vacuum?
- NO → Repair blockage at port or pinch in the #5 hose.
- YES → Reconnect the #5 hose.

Disconnect the #11 hose from the EGR control solenoid valve, connect a vacuum pump/gauge to the #11 hose and check for vacuum.

Is there vacuum?
- YES →
  - Inspect hose routing.
  - Replace EGR control solenoid valve.
  - Repair short in WHT wire between ECU (A10) and EGR control solenoid valve.
- NO → Disconnect 6P connector at emission control box.

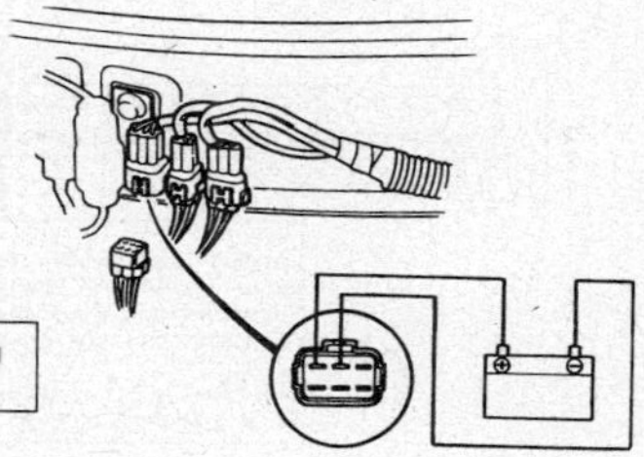

Connect the battery positive terminal to the BLK/YEL terminal and the negative terminal to the WHT terminal.

Is there approx. 8'' of vacuum within one second?
- NO → Replace EGR control solenoid valve.
- YES → Go to Path L

Path L

Measure voltage between BLK/YEL (+) terminal and body ground.

Is there battery voltage?
- NO → Repair open in BLK/YEL wire between the solenoid valve and No. 9 (10A) fuel pump fuse.
- YES → Turn the ignition switch OFF.

Connect the test harness between the EGR valve lift sensor and engine wire harness.

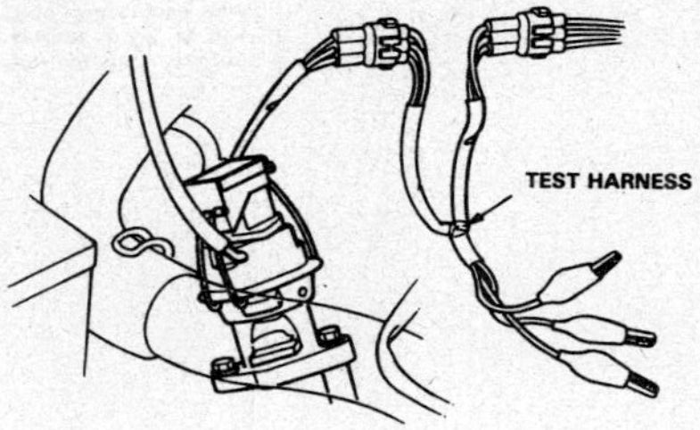

Reconnect the #11 hose.

Turn the ignition switch ON.

Measure voltage between RED (+) terminal and GRN (–) terminal.

Is there approx. 5V?
- NO → Measure voltage between RED (+) terminal and body ground.
  - Is there approx. 5V?
    - YES → Repair open in GRN/WHT wire between ECU (C14) and the sensor.
    - NO → Go to Path N
- YES → Measure voltage between WHT (+) terminal and GRN (–) terminal.

Is there approx. 1V?
- NO →
  - Replace EGR valve.
  - Repair short in WHT/GRN wire between ECU (C8) and EGR valve lift sensor.
- YES → Go to Path M

Path M

Turn the ignition switch OFF.

Connect the ECU test harness between the ECU and connector

Turn the ignition switch ON.

Measure voltage between C8 (+) terminal and C14 (−) terminal.

Is there approx. 1V?

- NO: **Repair open in WHT/GRN wire between ECU (C8) and the sensor.**
- YES: Disconnect "A" connector from the ECU only, not the main wire harness.

Measure voltage between A10 (+) terminal and A18 (−) terminal.

Is there battery voltage?

- NO: **Repair open in WHT wire between ECU (A10) and EGR control solenoid valve.**
- YES: **Substitute a known-good ECU and recheck. If symptom/indication goes away, replace the original ECU.**

Path N

Repair open in YEL/WHT wire between ECU (C13) and the sensor.
If wire is OK, substitute a known-good ECU and recheck. If prescribed voltage is now available replace the original ECU.

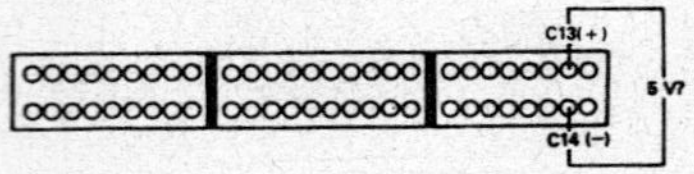

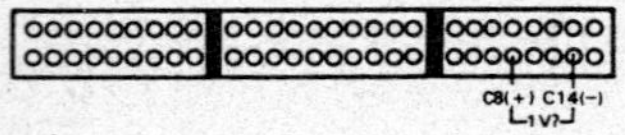

Self-diagnosis Red LED indicates code 13: A problem in the Atmospheric Pressure (PA) Sensor circuit.

The PA sensor converts atmospheric pressure into electrical signals and inputs the ECU.

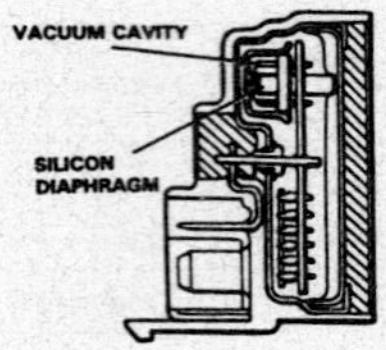

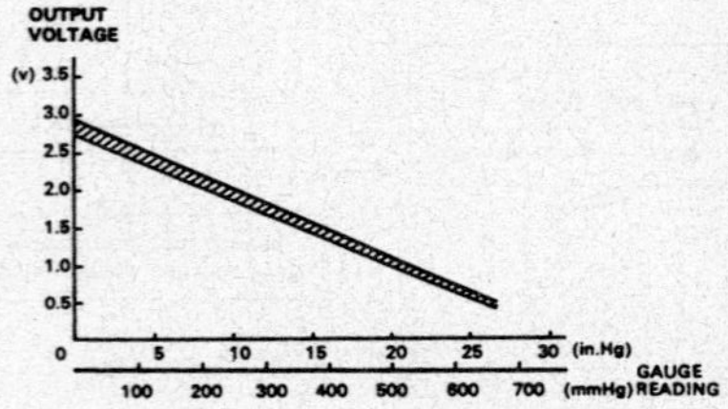

- Check Engine warning light has been reported on.
- Red LED indicates CODE 13.

Turn the ignition switch OFF.

Remove ALTERNATOR SENSE fuse in the under-hood relay box for 10 seconds to reset ECU.

Turn the ignition switch ON.

Is Check Engine warning light on and does LED indicate CODE 13?

- NO: **Intermittent failure, system is OK at this time. (Test drive may be necessary.) Check for poor connections or loose wires at PA sensor, GRN/WHT, YEL/WHT and RED wires at C-403—C-312.**
- YES: Turn the ignition switch OFF.

Connect the ECU test harness between the ECU and connector

Turn the ignition switch ON.

Measure voltage between C13 (+) terminal and C14 (−) terminal.

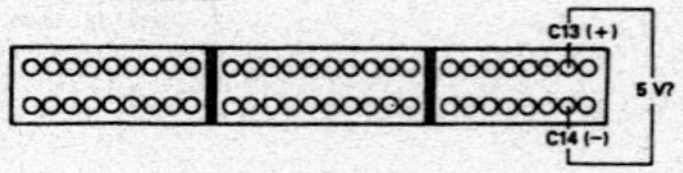

Is there approx. 5V?

- NO: **Substitute a known-good ECU and recheck. If prescribed voltage is now available replace the original ECU.**
- YES: Go to Path O

TROUBLESHOOTING CHARTS — 1990 LEGEND

Path O

Measure voltage between C9 (+) terminal and C14 (−) terminal.

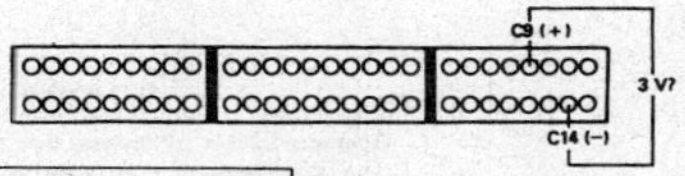

Is there approx. 3 V?
- YES: Substitute a known-good ECU and recheck. If symptom/indication goes away, replace the original ECU.
- NO:

Is there approx. 5 V?
- YES: Repair open RED wire between ECU (C9) and PA sensor.
- NO:

Turn the ignition switch OFF.

Disconnect the 3P connector of the PA sensor.

Turn the ignition switch ON.

Measure voltage between YEL/WHT (+) terminal and body ground.

Is there approx. 5 V?
- NO: Repair open in YEL/WHT wire between ECU (C13) and PA sensor.
- YES:

Measure voltage between YEL/WHT (+) terminal and GRN/WHT (−) terminal.

Is there approx. 5 V?
- NO: Repair open in GRN/WHT wire between ECU (C14) and the sensor.
- YES:

Measure voltage between RED (+) terminal and GRN/WHT (−) terminal.

Is there approx. 5 V?
- NO:
  - Repair short in RED wire between ECU (C9) and PA sensor.
  - Replace ECU.
- YES: Replace PA sensor.

Self-diagnosis Red LED indicates code 14: A problem in the Electronic Air Control Valve (EACV) circuit.

The EACV changes the amount of air bypassing the throttle body in response to a current signal from the ECU in order to maintain the proper idle speed.

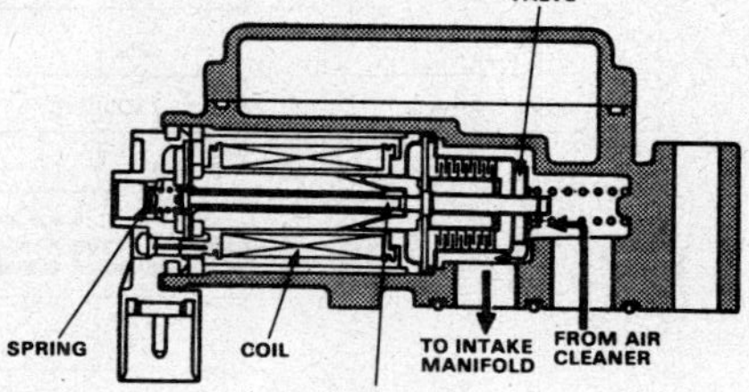

- Engine is running.
- Check Engine warning light has been reported on.
- Red LED indicates CODE 14.

Turn the ignition switch OFF.

Remove ALTERNATOR SENSE fuse in the under-hood relay box for 10 seconds to reset ECU.

Start engine.

Is Check Engine warning light on and does LED indicate CODE 14?
- NO: Intermittent failure, system is OK at this time. (Test driving may be necessary). Check for poor connections or loose wires at the EACV, C-303—C-103, C-228—C-127 and C-439—C-257.
- YES:

Stop engine.

Disconnect the 2P connector on the EACV.

Measure resistance between the 2 terminals on the EACV.

Go to Path P

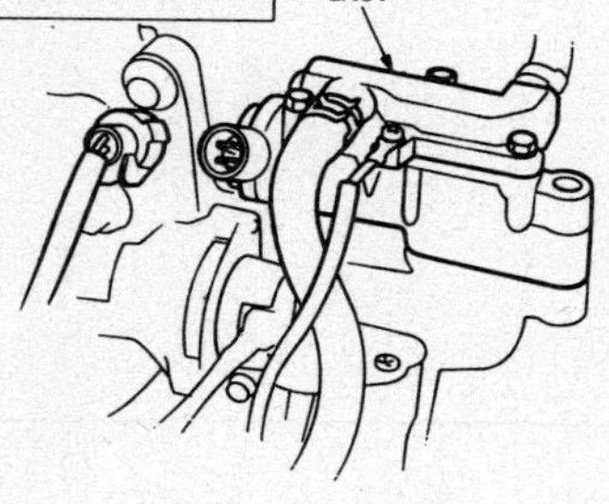

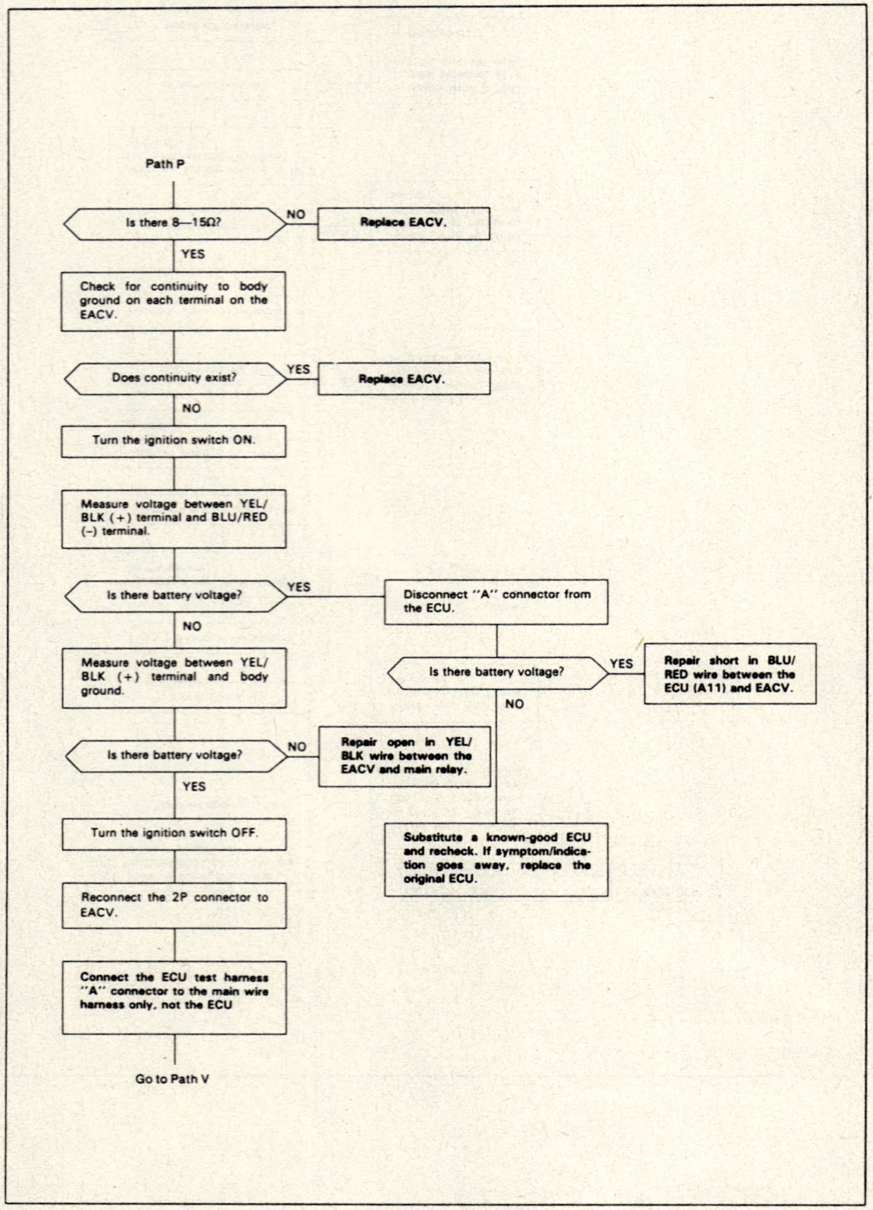
Path P
Is there 8—15Ω?
NO
Replace EACV.
YES
Check for continuity to body ground on each terminal on the EACV.
Does continuity exist?
YES
Replace EACV.
NO
Turn the ignition switch ON.
Measure voltage between YEL/BLK (+) terminal and BLU/RED (–) terminal.
Is there battery voltage?
YES
Disconnect "A" connector from the ECU.
Is there battery voltage?
YES
Repair short in BLU/RED wire between the ECU (A11) and EACV.
NO
Substitute a known-good ECU and recheck. If symptom/indication goes away, replace the original ECU.
NO
Measure voltage between YEL/BLK (+) terminal and body ground.
Is there battery voltage?
NO
Repair open in YEL/BLK wire between the EACV and main relay.
YES
Turn the ignition switch OFF.
Reconnect the 2P connector to EACV.
Connect the ECU test harness "A" connector to the main wire harness only, not the ECU
Go to Path V

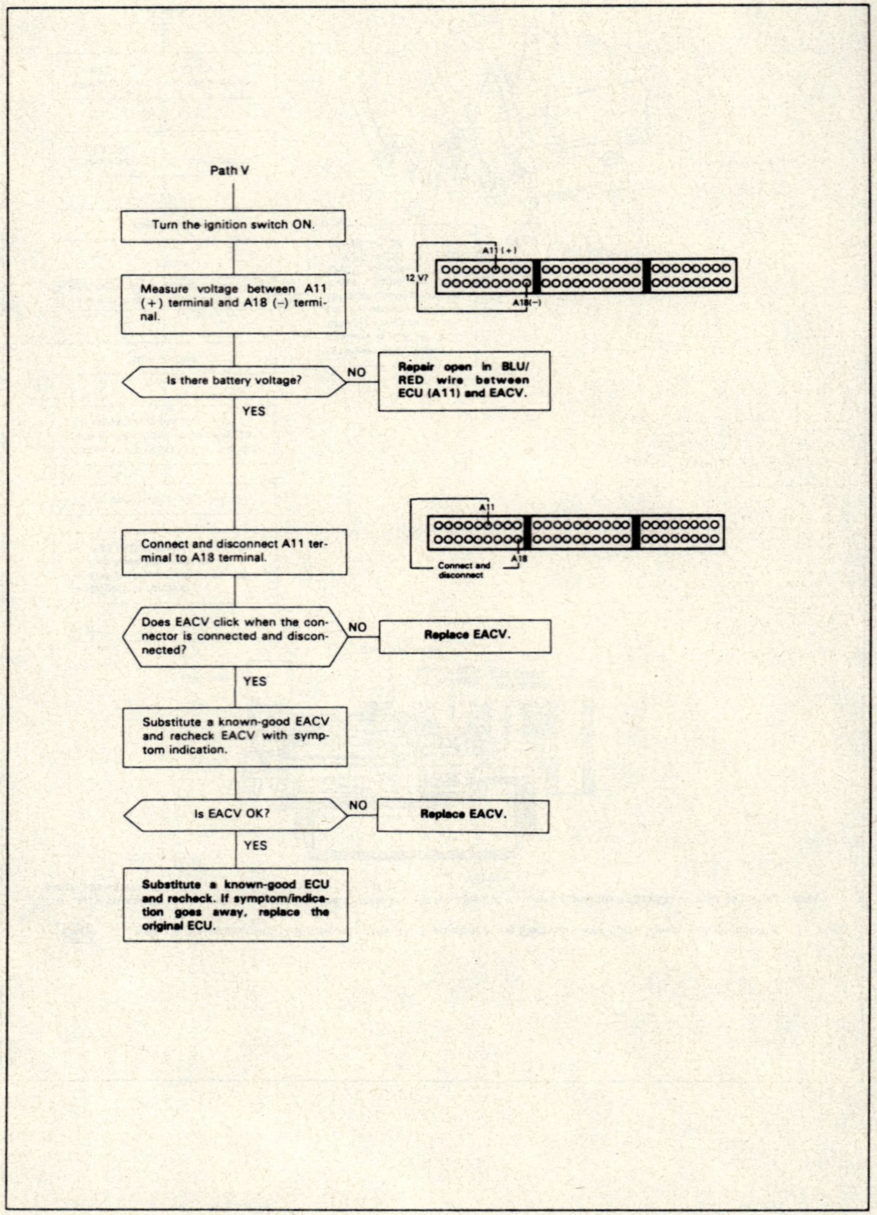
Path V
Turn the ignition switch ON.
Measure voltage between A11 (+) terminal and A18 (–) terminal.
12 V?
A11 (+)
A18 (–)
Is there battery voltage?
NO
Repair open in BLU/RED wire between ECU (A11) and EACV.
YES
Connect and disconnect A11 terminal to A18 terminal.
A11
A18
Connect and disconnect
Does EACV click when the connector is connected and disconnected?
NO
Replace EACV.
YES
Substitute a known-good EACV and recheck EACV with symptom indication.
Is EACV OK?
NO
Replace EACV.
YES
Substitute a known-good ECU and recheck. If symptom/indication goes away, replace the original ECU.

TROUBLESHOOTING CHARTS—1990 LEGEND

Self-diagnosis Red LED indicates code 15: A problem in the Ignition Output Signal circuit.

- Check Engine warning light has been reported on.
- Red LED indicates CODE 15.

Turn the ignition switch OFF.

Remove ALTERNATOR SENSE fuse in the under-hood relay box for 10 seconds to reset ECU.

Turn the ignition switch ON.

Is Check Engine warning light on and does LED indicate CODE 15?

- NO → Intermittent failure, system is OK at this time. (Test drive may be necessary.) Check for poor connections or loose wires at igniter and RED/BLU wire at C-439—C-257 and C-403—C-312.
- YES ↓

Turn the ignition switch OFF.

Disconnect the 4P connector on the igniter unit.

Turn the ignition switch ON.

Measure voltage between the BLK/YEL (+) terminal and body ground.

Is there battery voltage?

- NO → Repair open in BLK/YEL wire between the igniter unit and ignition switch.
- YES ↓

Turn the ignition switch OFF.

Go to Path Q

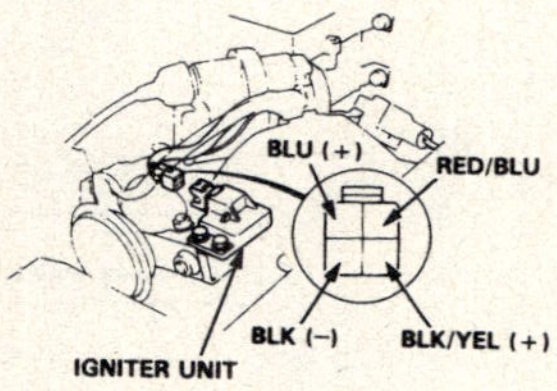

Path Q

Connect the ECU test harness between the ECU and connector

Reconnect the 4P connector on the igniter unit.

Turn the ignition switch ON.

Measure voltage individually between B15(+), B16(+) terminals and A18(-) terminal.

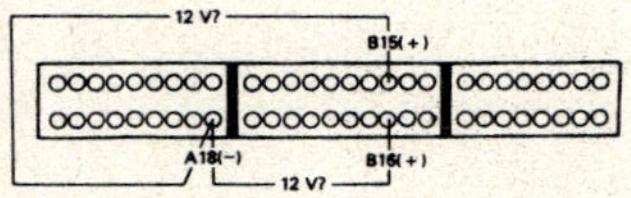

Is there battery voltage?

- YES → Substitute a known-good ECU and recheck. If symptom/indication goes away, replace the original ECU.
- NO ↓

Turn the ignition switch OFF.

Disconnect the 4P connector on the igniter unit and "B" connector from the ECU.

Check for continuity of RED/BLU wire between the ECU (B15, B16) and the igniter unit.

Does continuity exist?

- NO → Repair open in RED/BLU wire between ECU (B15, B16) and the igniter unit.
- YES ↓

Check for continuity to body ground on RED/BLU wire on the 4P connector.

Does continuity exist?

- NO → Replace the igniter unit.
- YES ↓

Repair short in the RED/BLU wire. Then replace the igniter unit (Igniter unit will be broken in this case).

Self-diagnosis Red LED indicates code 17: A problem in the Vehicle Speed Pulser circuit.

The signal generated by the speed pulser, which produces 4 pulses (switch closures to ground) for each revolution of the speedometer cable.

- **Check Engine warning light has been reported on.**
- **Red LED indicates CODE 17.**

Drive vehicle and operate cruise control.

Does cruise control operate properly?

- NO → **Cruise Control System Inspection**
- YES ↓

Turn the ignition switch OFF.

Connect the ECU test harness between the ECU and connector. Disconnect "B" connector from the main wire harness.

Turn the ignition switch ON.

Measure voltage between B11 (+) terminal and A18 (−) terminal.

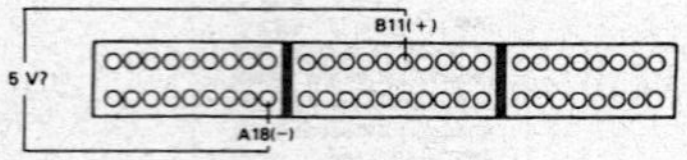

Is there approx. 5V?

- NO → **Substitute a known-good ECU and recheck. If prescribed voltage is now available, replace the original ECU.**
- YES ↓

Turn the ignition switch OFF.

Reconnect "B" connector to the main wire harness.

Block rear wheels and set the parking brake. Jack up the front of the car and support with safety stands.

⚠WARNING Block rear wheels before jacking up front of car.

Turn the ignition switch ON.

Slowly rotate left front wheel and measure voltage between B11 (+) terminal and A18 (−) terminal.

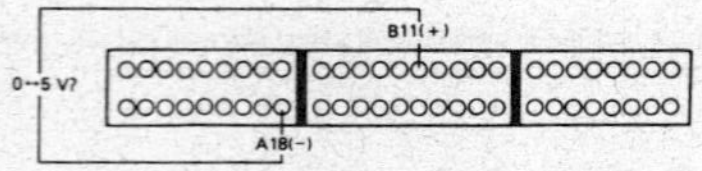

Does voltage pulse 0V and 5V?

- YES → **Intermittent failure, system is OK at this time. Check for poor connections or loose wires at the speed pulser and ECU.**
- NO → Go to Path R

Path R

Does voltage hold 0 V or 5 V?

- 5 V →
  - **Repair open in YEL/RED wire between ECU (B11) and the speed pulser.**
  - **Faulty speed pulser.**
- 0 V →
  - **Repair short in YEL/RED wire between ECU (B11) and the speed pulser.**
  - **Faulty speed pulser.**
  - **If OK, substitute a known-good ECU and recheck. If prescribed voltage is now available, replace the original ECU.**

Self-diagnosis Red LED indicates code 18: A problem in the Ignition Timing Adjuster circuit.

The ignition timing adjuster allows the electronic ignition advance to be set to 15° BTDC at idle.

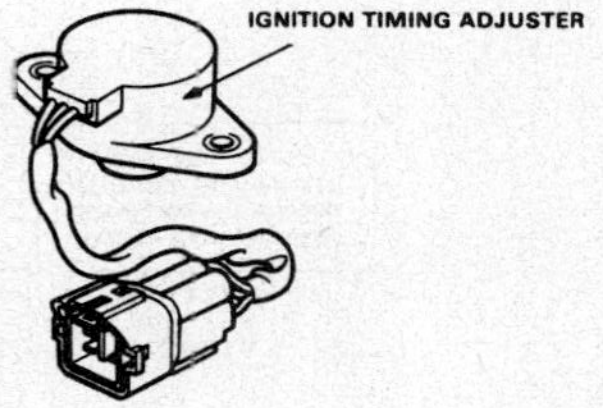

- **– Check Engine warning light has been reported on.**
  – Red LED indicates CODE 18.
- Turn the ignition switch OFF.
- Remove ALTERNATOR SENSE fuse in the under-hood relay box for 10 seconds to reset ECU.
- Turn the ignition switch ON.
- Is Check Engine warning light on and does LED indicate CODE18?
  - NO → **Intermittent failure, system is OK at this time. (Test drive may be necessary). Check for poor connections or loose wires at the Ignition Timing adjuster connecter shown below.**
  - YES ↓
- Turn the ignition switch OFF.
- Disconnect the 4P connector on the right side of the firewall that has only 3 wires.
- Measure resistance between YEL/WHT terminal and GRN/WHT terminal on Ignition timing adjuster harness.
- Is there 3—6 kΩ?
  - NO → **Replace Ignition timing adjuster.**
  - YES → Go to Path S

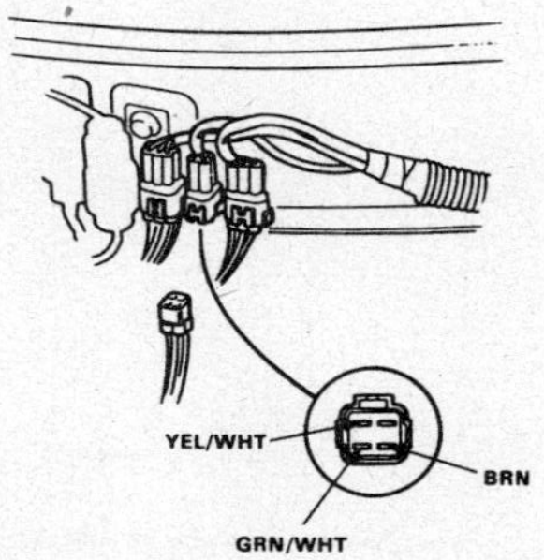

Path S

- Measure resistance between YEL/WHT and BRN terminals and between GRN/WHT and BRN terminals.
- Does the sum of the two resistance checks equal 3.5–6.5 kΩ?
  - NO → **Replace Ignition timing adjuster.**
  - YES ↓
- Turn the ignition switch ON.
- **Measure voltage between YEL/WHT (+) terminal and GRN/WHT (–) terminal on the main wire harness.**

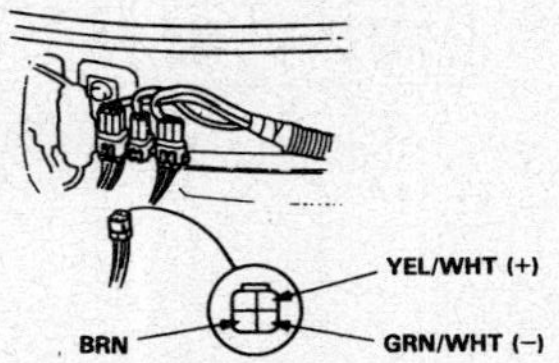

- Is there approx. 5V?
  - NO → Measure voltage between YEL/WHT (+) terminal and body ground.
    - Is there approx. 5V?
      - YES → **Repair open in GRN/WHT wire between ECU (C14) and Ignition timing adjuster.**
      - NO ↓
    - Turn the ignition switch OFF.
    - Connect the ECU test harness between the ECU and connector
    - Turn the ignition switch ON.
    - Measure voltage between C13 (+) terminal and C14 (–) terminal.

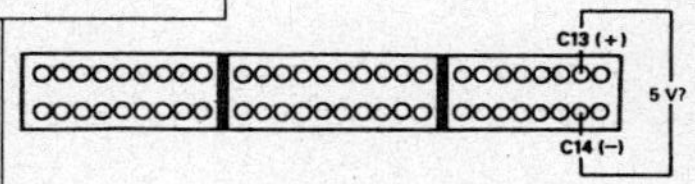

    - Is there approx. 5V?
      - YES → **Repair open in YEL/WHT wire between ECU (C13) and Ignition timing adjuster.**
      - NO → **Substitute a known-good ECU and recheck. If prescribed voltage is now available, replace the original ECU.**
  - YES ↓
- Turn the ignition switch OFF.
- **Reconnect 4P connector.**
- Connect the ECU test harness between the ECU and connector
- Turn the ignition switch ON.
- Measure voltage between B18 (+) terminal and C14 (–) terminal.
- Is there 0.5—4.5V?
  - NO → **Repair open or short in BRN wire between ECU (B18) and Ignition timing adjuster.**
  - YES → **Substitute a known-good ECU and recheck. If symptom/indication goes away, replace the original ECU.**

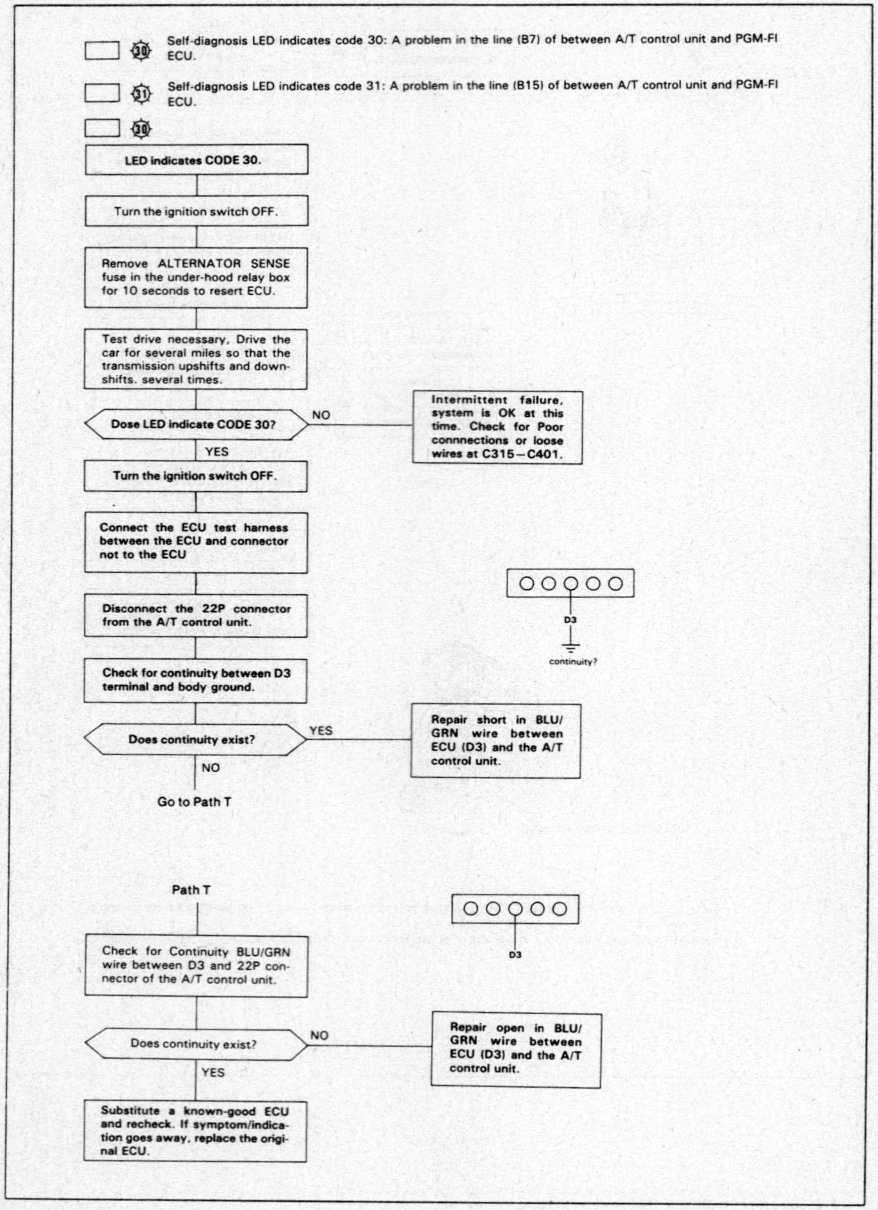
Self-diagnosis LED indicates code 30: A problem in the line (B7) of between A/T control unit and PGM-FI ECU.
Self-diagnosis LED indicates code 31: A problem in the line (B15) of between A/T control unit and PGM-FI ECU.
30
LED indicates CODE 30.
Turn the ignition switch OFF.
Remove ALTERNATOR SENSE fuse in the under-hood relay box for 10 seconds to resert ECU.
Test drive necessary. Drive the car for several miles so that the transmission upshifts and downshifts. several times.
Dose LED indicate CODE 30?
NO
Intermittent failure, system is OK at this time. Check for Poor connnections or loose wires at C315—C401.
YES
Turn the ignition switch OFF.
Connect the ECU test harness between the ECU and connector not to the ECU
Disconnect the 22P connector from the A/T control unit.
D3
continuity?
Check for continuity between D3 terminal and body ground.
Does continuity exist?
YES
Repair short in BLU/GRN wire between ECU (D3) and the A/T control unit.
NO
Go to Path T
Path T
D3
Check for Continuity BLU/GRN wire between D3 and 22P connector of the A/T control unit.
Does continuity exist?
NO
Repair open in BLU/GRN wire between ECU (D3) and the A/T control unit.
YES
Substitute a known-good ECU and recheck. If symptom/indication goes away, replace the original ECU.

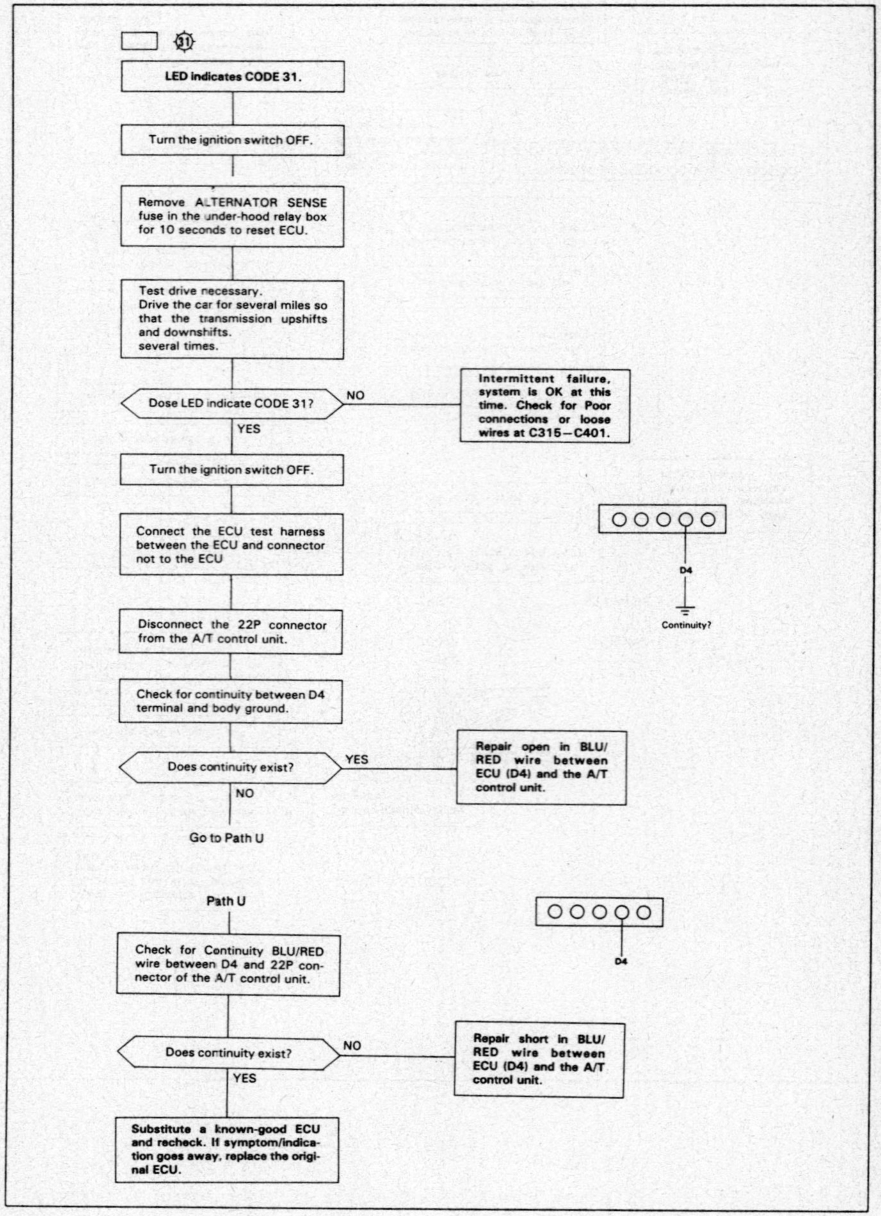
31
LED indicates CODE 31.
Turn the ignition switch OFF.
Remove ALTERNATOR SENSE fuse in the under-hood relay box for 10 seconds to reset ECU.
Test drive necessary.
Drive the car for several miles so that the transmission upshifts and downshifts.
several times.
Dose LED indicate CODE 31?
NO
Intermittent failure, system is OK at this time. Check for Poor connections or loose wires at C315—C401.
YES
Turn the ignition switch OFF.
D4
Continuity?
Connect the ECU test harness between the ECU and connector not to the ECU
Disconnect the 22P connector from the A/T control unit.
Check for continuity between D4 terminal and body ground.
Does continuity exist?
YES
Repair open in BLU/RED wire between ECU (D4) and the A/T control unit.
NO
Go to Path U
Path U
D4
Check for Continuity BLU/RED wire between D4 and 22P connector of the A/T control unit.
Does continuity exist?
NO
Repair short in BLU/RED wire between ECU (D4) and the A/T control unit.
YES
Substitute a known-good ECU and recheck. If symptom/indication goes away, replace the original ECU.

TROUBLESHOOTING CHARTS—1990 LEGEND

This signals the PGM-FI ECU when the engine is cranking. The ECU controls the current to the EACV to promote easy starting.

Inspection of Starter Switch Signal

Connect the ECU test harness between the ECU and connector

Measure voltage B13 (+) terminal and A18 (-) terminal with ignition switch in the start position.

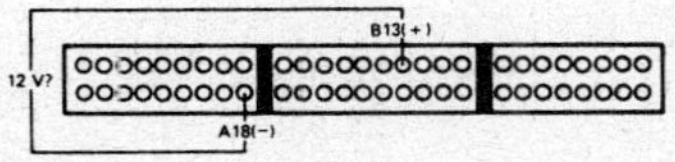

Is there battery voltage?

- YES → **Starter signal is OK.**
- NO → Inspect No. 13 fuse.
  - Is No. 13 fuse OK?
    - YES → **Repair open in BLK/WHT wire between ECU (B13) and No. 13 fuse.**
    - NO → **Replace fuse.**

This signals the PGM-FI ECU when the alternator is charging.

Note: Make sure the battery is fully charged before performing inspection of the FR terminal.

Inspection of Alternator FR signal

Connect the ECU test harness between the ECU and connector. Disconnect "B" connector from the main wire harness only, not the ECU.

Turn the ignition switch ON.

Measure voltage between B14 (+) terminal and A18 (-) terminal.

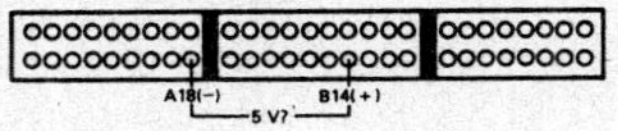

Is there approx. 5V?

- NO → **Substitute a known-good ECU and recheck. If prescribed voltage is now available, replace the original ECU.**
- YES → Turn the ignition switch OFF.

Reconnect "B" connector to the main wire harness.

Warm up engine to normal operating temperature (cooling fan comes on).

Measure voltage between B14 (+) terminal and A18 (-) terminal.

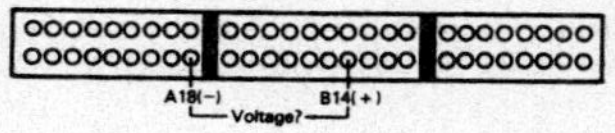

Does the voltage decrease when headlight and rear defogger are turned on?

- YES → **Alternator FR signal is OK. Turn the ignition switch off and remove ALTERNATOR SENSE fuse for at least 10 seconds to reset ECU.**
- NO → Stop engine.

Go to Path W

Path W

Disconnect "B" connector from ECU only, not the main wire harness.

Disconnect the negative battery cable from the battery.

Check for continuity between B14 terminal and body ground.

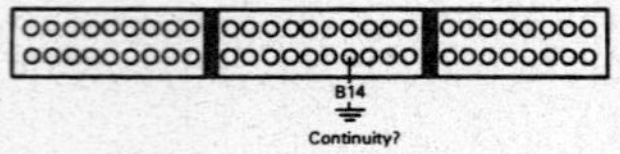

Does continuity exist?

- YES → Disconnect GRN connector from the alternator.
  - Check for continuity between B14 terminal and body ground.
  - Does continuity exist?
    - NO → Alternator Inspection
    - YES → Repair short in WHT/RED wire between ECU (B14) and alternator.
- NO → Disconnect GRN connector from the alternator.
  - Connect WHT/RED wire to body ground.
  - Check for continuity between B14 terminal and body ground.
  - Does continuity exist?
    - YES → Alternator Inspection
    - NO → Repair open in WHT/RED wire between ECU (B14) and alternator.

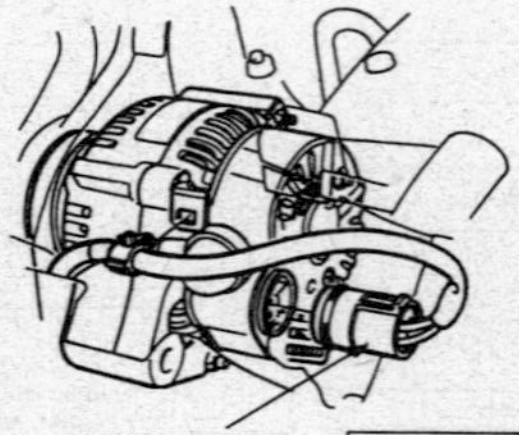

GRN CONNECTOR

This signals the PGM-FI ECU when the transmission is in Neutral or Park.

Inspection of A/T Shift Position Signal

Turn the ignition switch ON.

Observe the A/T shift indicator and select each position separately.

Does the indicator light properly?

- NO → **A/T shift position Indicator Inspection**
- YES → Turn the ignition switch OFF.

Connect the ECU test harness between the ECU and connector. Disconnect "B" connector from the main wire harness only, not the ECU.

Turn the ignition switch ON.

Measure voltage between B7 (+) terminal and A18 (–) terminal.

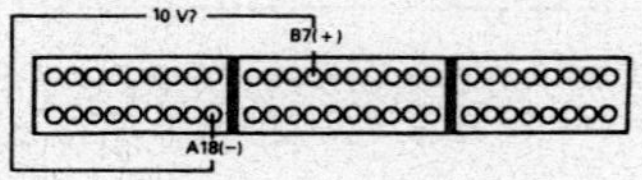

Is there approx. 10 V?

- NO → **Substitute a known-good ECU and recheck. If prescribed voltage is now available, replace the original ECU.**
- YES → Reconnect "B" connector to the main wire harness.

Measure voltage between B7 (+) terminal and A18 (–) terminal in Neutral position.

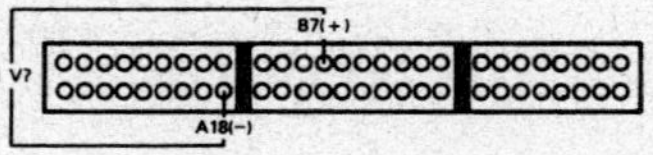

Go to Path X

Path X

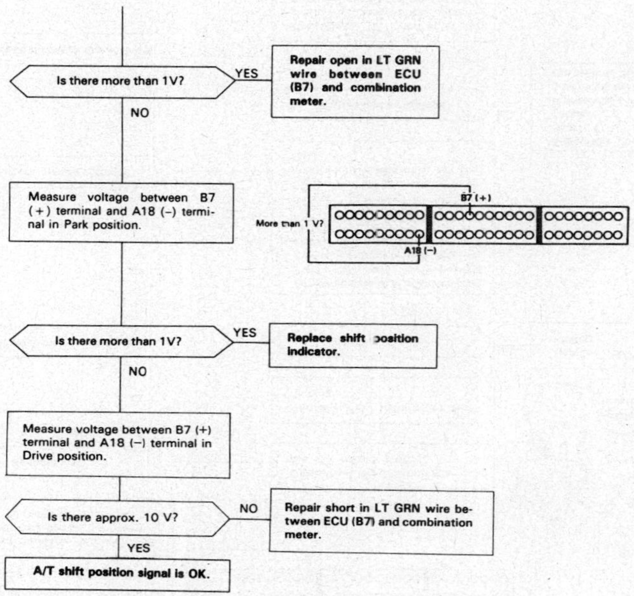

This signals the PGM-FI ECU when the power steering load is high.

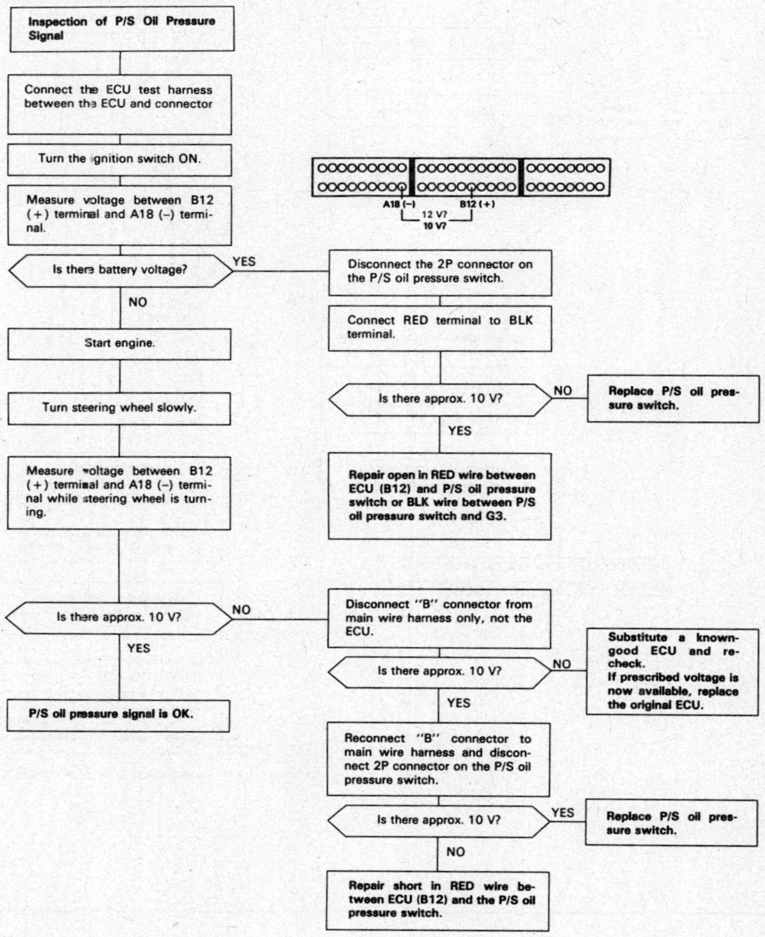

This signals the PGM-FI ECU when the transmission is in Neutral.

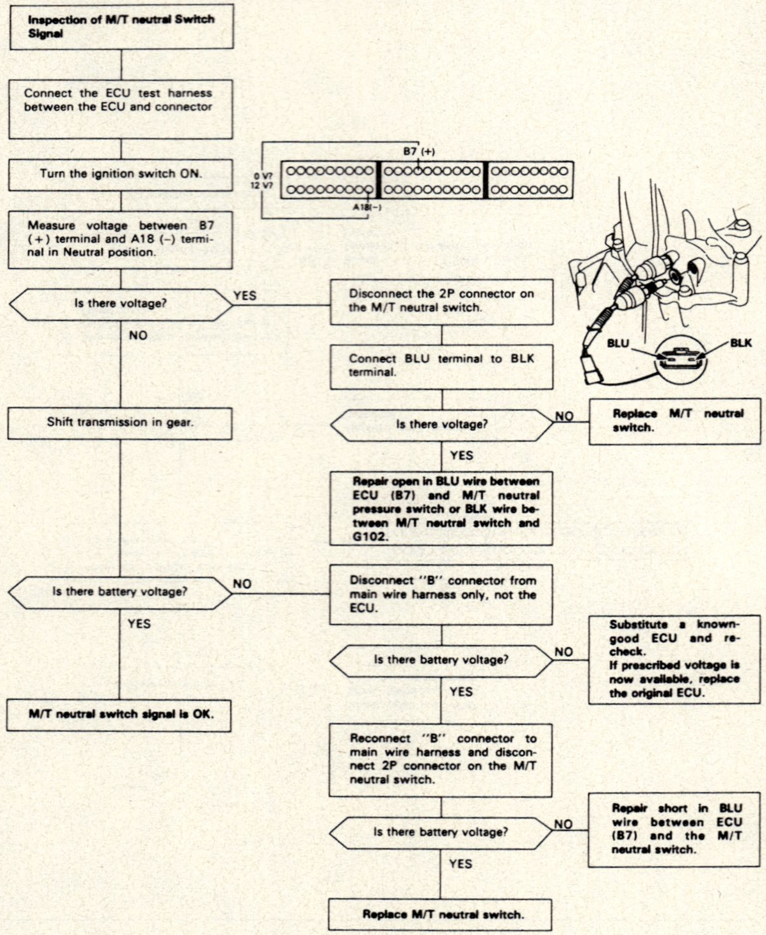

This signals the PGM-FI ECU when the clutch is engaged.

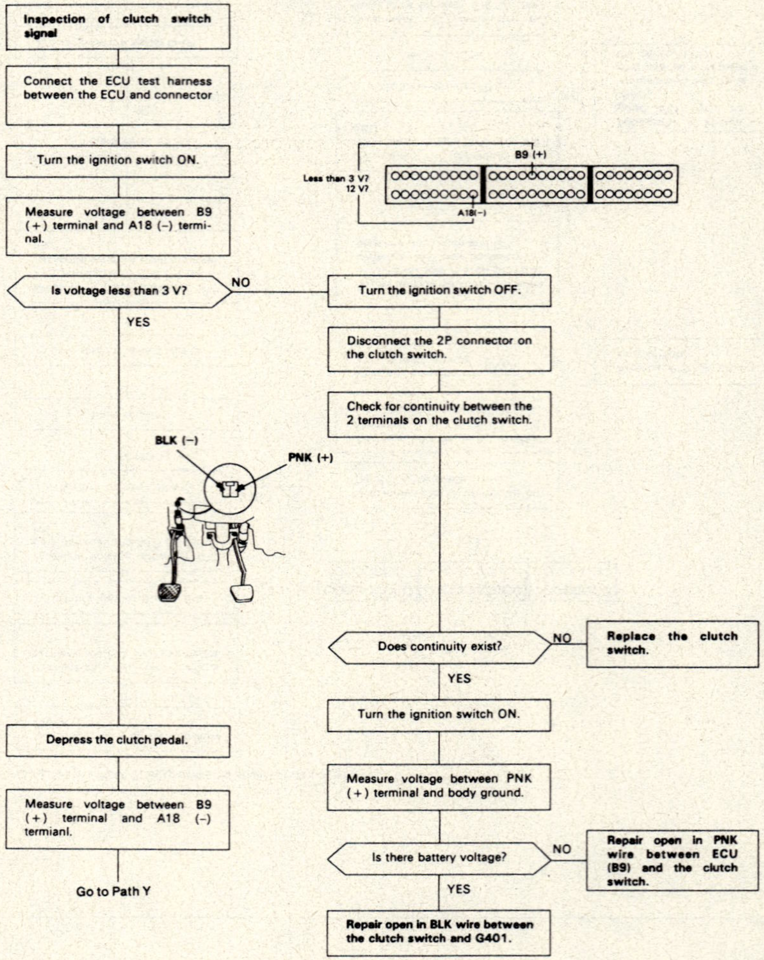

TROUBLESHOOTING CHARTS – 1990 LEGEND

Path Y

- Is there battery voltage?
  - YES: Clutch switch signal is OK.
  - NO: Turn the ignition switch OFF.
    - Disconnect the 2P connector on the clutch switch.
    - Turn the ignition switch ON.
    - Measure voltage between B9 (+) terminal and A18 (–) terminal.
    - Is there battery voltage?
      - YES: Replace the clutch switch.
      - NO: Disconnect "B" connector from main wire harness only, not the ECU.
        - Is there battery voltage?
          - YES: Repair short in PNK wire between ECU (B9) and the clutch switch.
          - NO: Substitute a known-good ECU and recheck. If prescribed voltage is now available, replace the original ECU.

Inspection of Bypass Control System

NOTE: Check for normal operation of the hypass valve prior to checking the system

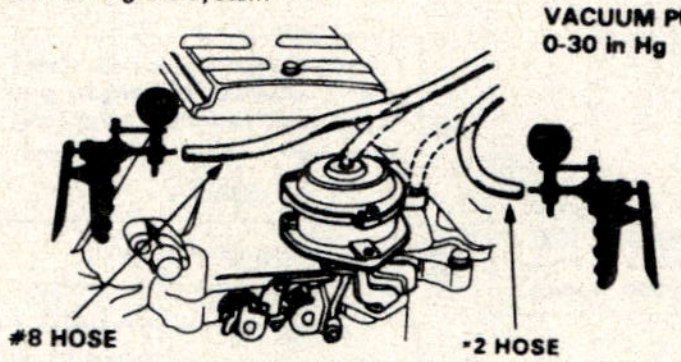

- Start engine and allow to idle.
- Remove #2 vacuum hose from the bypass control diaphragm and connect vacuum gauge to the hose.
- Is there vacuum?
  - YES: (continued)
  - NO: Remove #12 vacuum hose from the vacuum tank A, then check for vacuum at the tank.
    - Is there vacuum?
      - NO: Check the vacuum line between the vacuum tank A and the intake manifold.
      - YES: Disconnect 6P connector.
        - Measure voltage between BLK/YEL (+) terminal and RED/BLU (–) terminal.

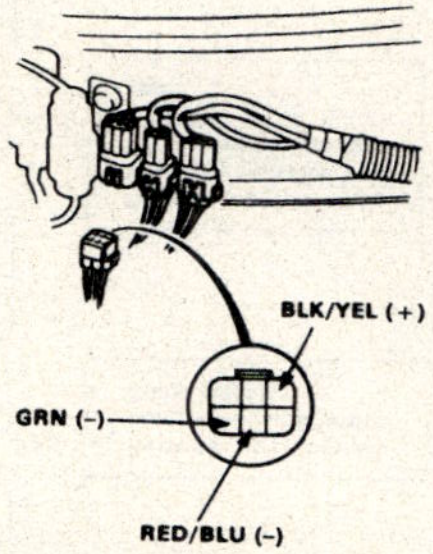

- Is there battery voltage?
  - YES: Replace the bypass control solenoid valve A and retest.
  - NO: Measure voltage between BLK/YEL (+) terminal and body ground.

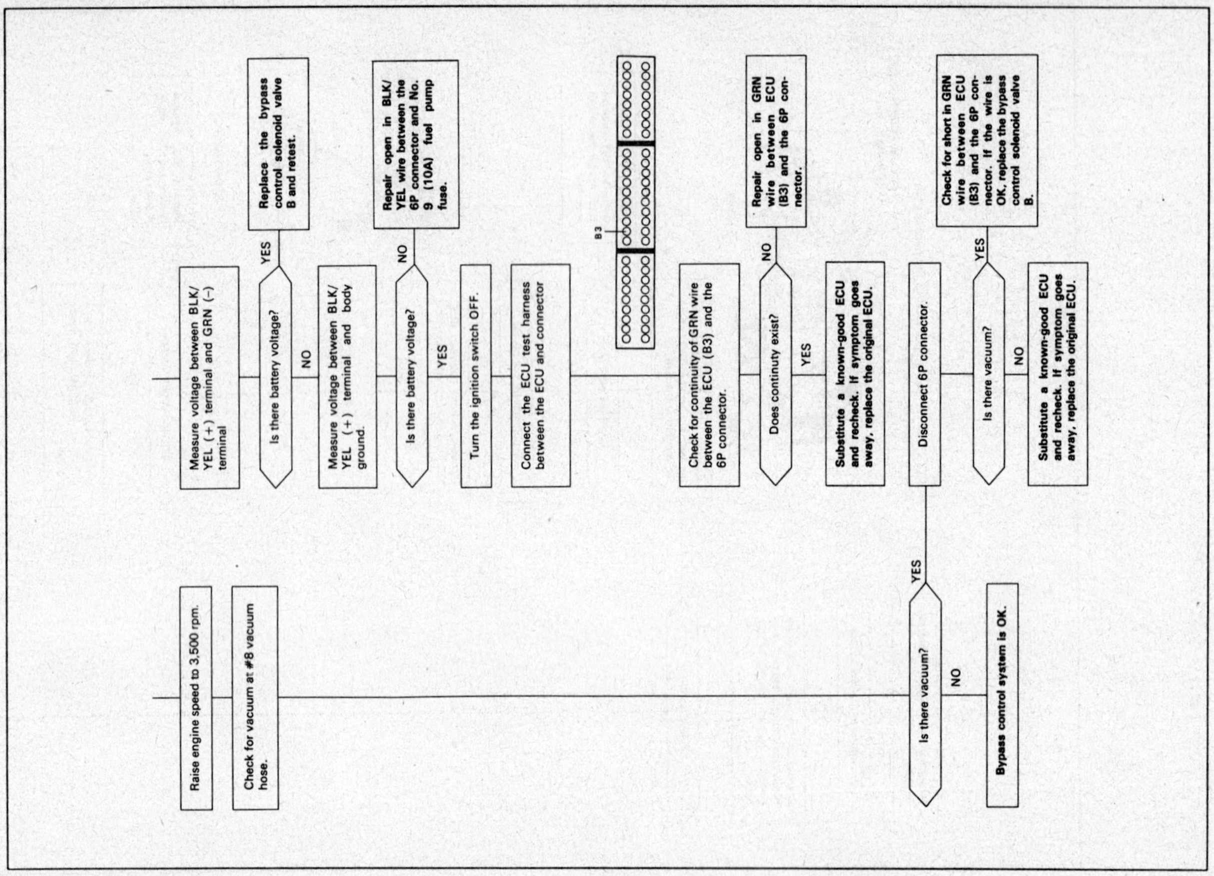
Raise engine speed to 3,500 rpm.
Check for vacuum at #8 vacuum hose.
Is there vacuum?
NO
Bypass control system is OK.
YES
Disconnect 6P connector.
Is there vacuum?
YES
Check for short in GRN wire between ECU (B3) and the 6P connector. If the wire is OK, replace the bypass control solenoid valve B.
NO
Substitute a known-good ECU and recheck. If symptom goes away, replace the original ECU.
Measure voltage between BLK/YEL (+) terminal and GRN (–) terminal
Is there battery voltage?
YES
Replace the bypass control solenoid valve B and retest.
NO
Measure voltage between BLK/YEL (+) terminal and body ground.
Is there battery voltage?
NO
Repair open in BLK/YEL wire between the 6P connector and No. 9 (10A) fuel pump fuse.
YES
Turn the ignition switch OFF.
Connect the ECU test harness between the ECU and connector
B3
Check for continuity of GRN wire between the ECU (B3) and the 6P connector.
Does continuity exist?
NO
Repair open in GRN wire between ECU (B3) and the 6P connector.
YES
Substitute a known-good ECU and recheck. If symptom goes away, replace the original ECU.

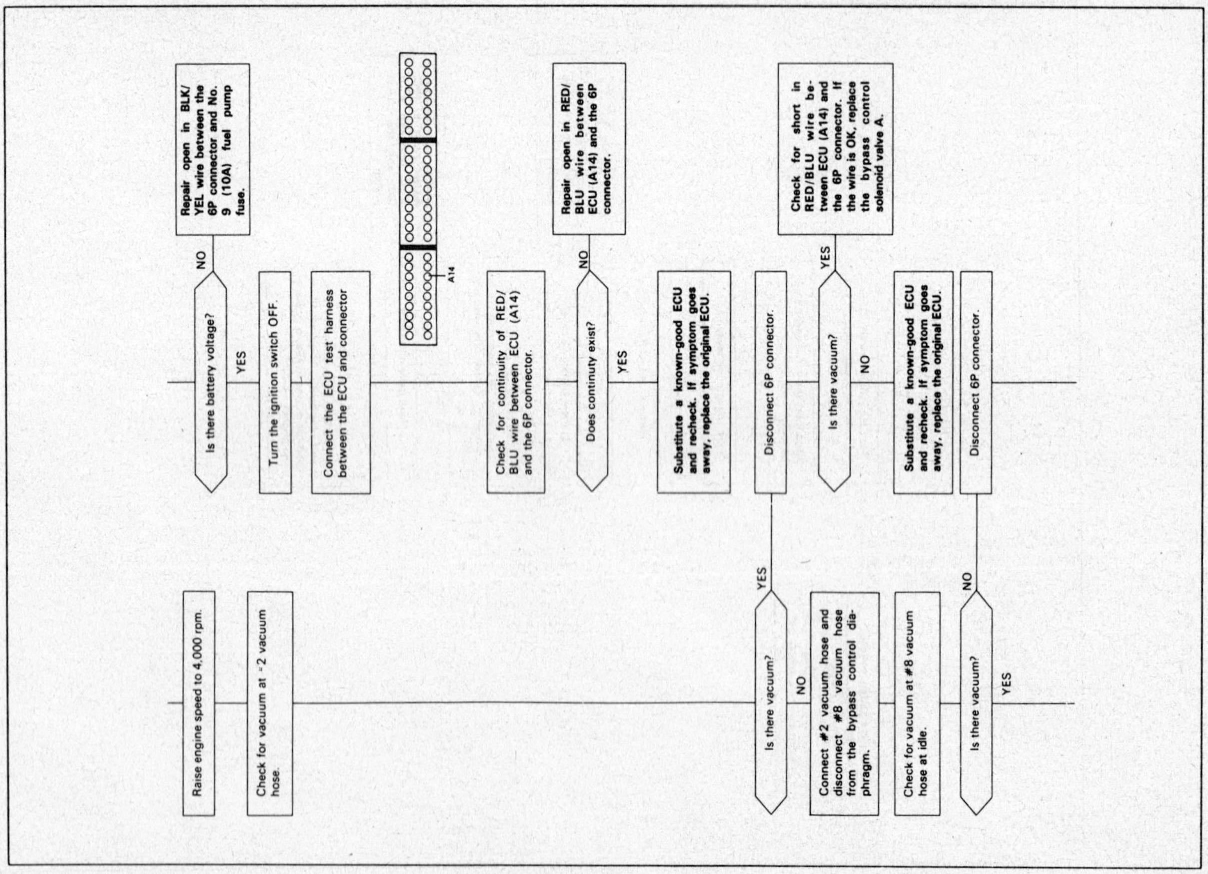
Raise engine speed to 4,000 rpm.
Check for vacuum at #2 vacuum hose.
Is there vacuum?
YES
Disconnect 6P connector.
Is there vacuum?
YES
Check for short in RED/BLU wire between ECU (A14) and the 6P connector. If the wire is OK, replace the bypass control solenoid valve A.
NO
Substitute a known-good ECU and recheck. If symptom goes away, replace the original ECU.
NO
Connect #2 vacuum hose and disconnect #8 vacuum hose from the bypass control diaphragm.
Check for vacuum at #8 vacuum hose at idle.
Is there vacuum?
NO
Disconnect 6P connector.
YES
Is there battery voltage?
NO
Repair open in BLK/YEL wire between the 6P connector and No. 9 (10A) fuel pump fuse.
YES
Turn the ignition switch OFF.
Connect the ECU test harness between the ECU and connector
A14
Check for continuity of RED/BLU wire between ECU (A14) and the 6P connector.
Does continuity exist?
NO
Repair open in RED/BLU wire between ECU (A14) and the 6P connector.
YES
Substitute a known-good ECU and recheck. If symptom goes away, replace the original ECU.

TROUBLESHOOTING CHARTS—1990 LEGEND

Inspection of Resonator Control System

Start engine and allow it to idle.

Remove the vacuum hose from the resonator diaphragm and connect a vacuum gauge to the hose.

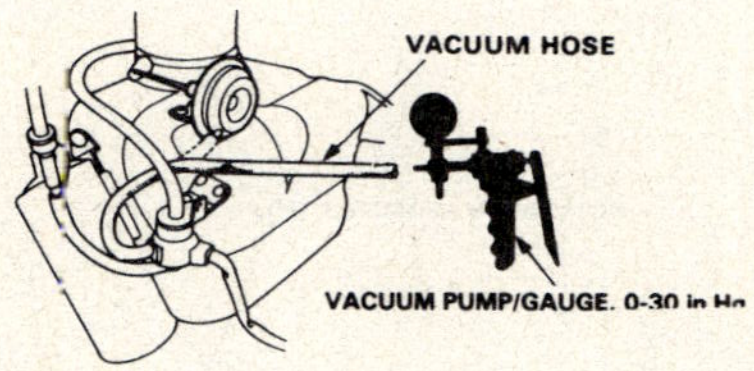

Is there vacuum?
- NO → Remove resonator control vacuum hose from surge tank, and check for vacuum at the tank.
  - Is there vacuum?
    - NO → **Repair the blockage or vacuum leak between surge tank and intake manifold.**
    - YES → Disconnect the 2P connector from the resonator control solenoid valve.
      - Measure voltage between BLK/YEL (+) terminal and RED/BLU (–) terminal.

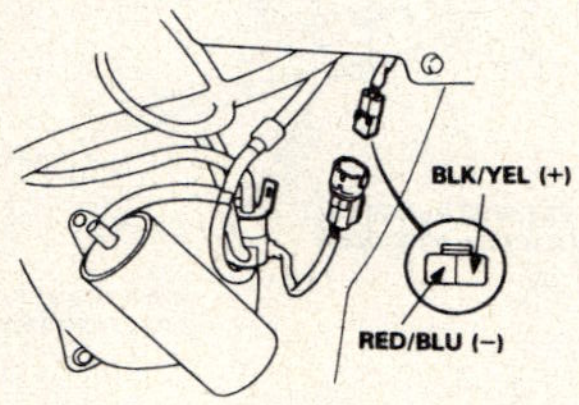

      - Is there battery voltage?
        - YES → **Replace the resonator control solenoid valve.**
        - NO → Measure voltage between BLK/YEL (+) terminal and body ground.
- YES → Raise engine speed to 3,800 rpm.
  - Check for vacuum at resonator diaphragm vacuum hose.
  - Is there vacuum?
    - NO → **Resonator control system is OK.**
    - YES → Disconnect the 2P connector
      - Is there vacuum?
        - YES → **Replace resonator control solenoid valve.**
        - NO → Turn the ignition switch OFF.
          - Disconnect "B" connector from ECU only, not the main wire harness.
          - Check for continuity to ground on the wire.
          - Is there continuity to ground?
            - YES → **Repair short to ground in RED/BLU wire between ECU (A14) and the 2P connector.**
            - NO → **Substitute a known-good ECU and check. If symptom goes away, replace the original ECU.**

(Continued from BLK/YEL (+) terminal and body ground measurement:)

Is there battery voltage?
- NO → **Repair open in wire between the 2P connector and No. 9 (10A) fuse.**
- YES → Turn the ignition switch OFF.
  - Connect the PGM-FI test harness between the ECU and connector
  - Check for continuity of wire between ECU (B3) and the 2P connector.
  - Does continuity exist?
    - NO → **Repair open in RED/BLU wire between ECU (A14) and the 2P connector.**
    - YES → **Substitute a known-good ECU and recheck. If symptom/indication goes away, replace the original ECU.**

VACUUM COMPONENT AND HOSE ROUTING INTEGRA WITH MANUAL

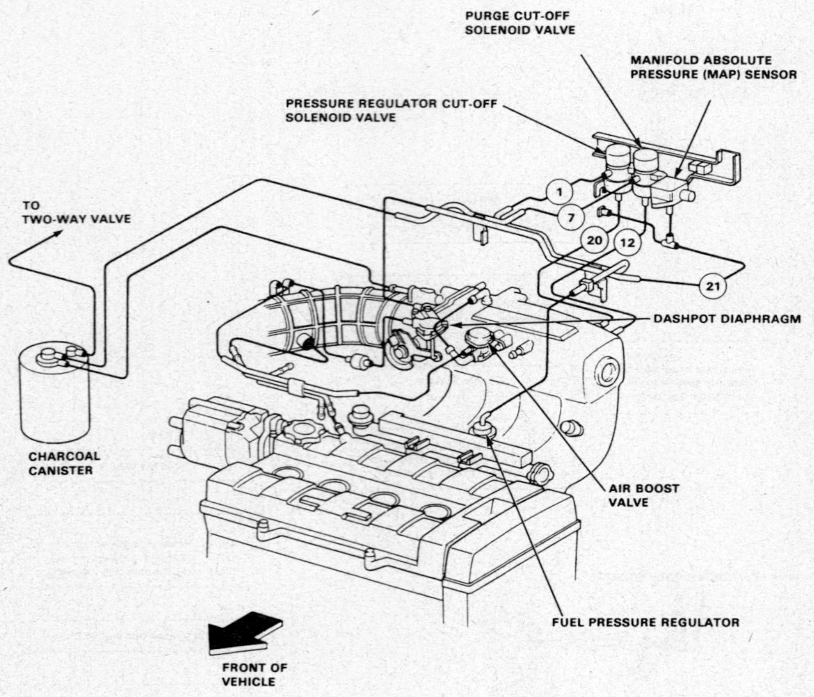

VACUUM COMPONENT AND HOSE ROUTING INTEGRA WITH AUTOMATIC

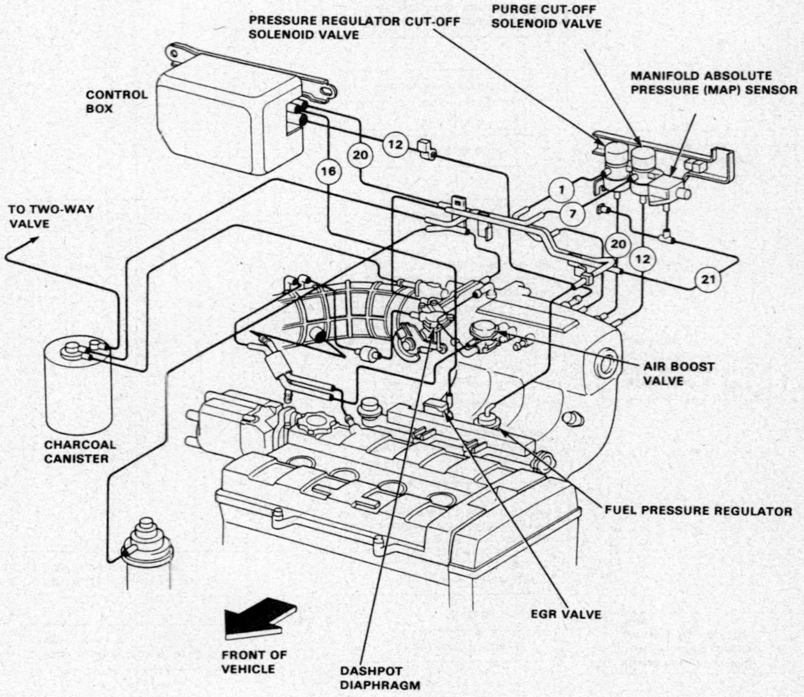

CONTROL BOX COMPONENTS—INTEGRA

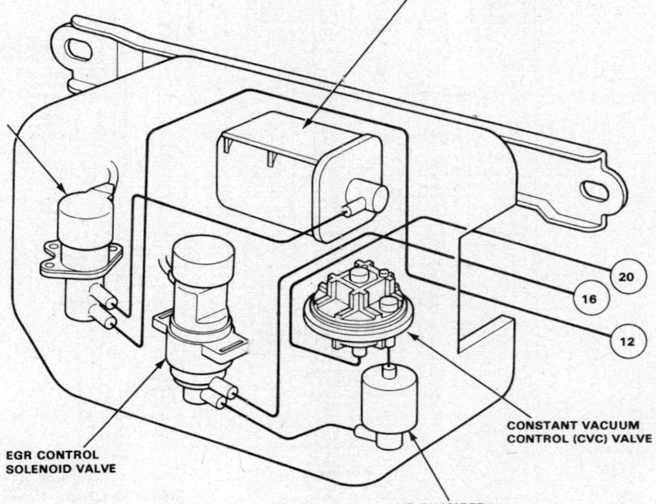

VACUUM SYSTEM SCHEMATIC—INTEGRA

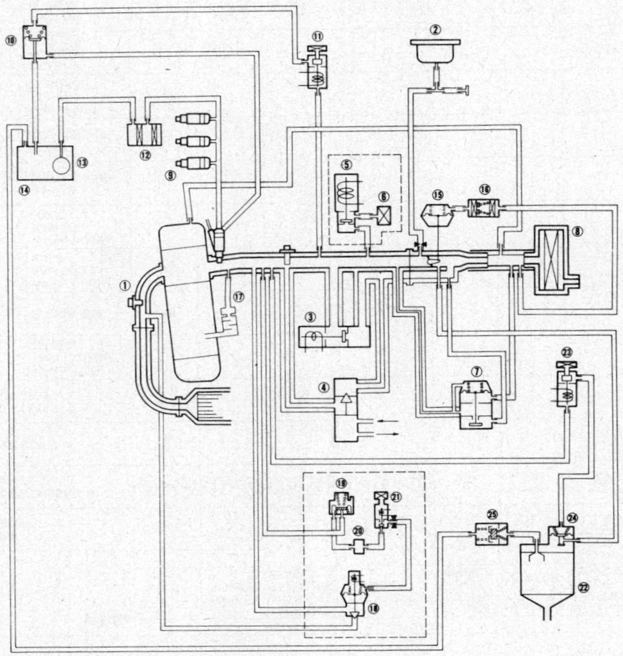

1. OXYGEN (O_2) SENSOR
2. MANIFOLD ABSOLUTE PRESSURE (MAP) SENSOR
3. ELECTRONIC AIR CONTROL VALVE (EACV)
4. FAST IDLE VALVE
5. FAST IDLE CONTROL SOLENOID VALVE (A/T ONLY)
6. AIR FILTER
7. AIR BOOST VALVE
8. AIR CLEANER
9. FUEL INJECTOR
10. PRESSURE REGULATOR
11. PRESSURE REGULATOR CUT-OFF SOLENOID VALVE
12. FUEL FILTER
13. FUEL PUMP
14. FUEL TANK
15. DASHPOT DIAPHRAGM
16. DASHPOT CHECK VALVE
17. PCV VALVE
18. EGR VALVE (A/T ONLY)
19. CONSTANT VACUUM CONTROL (CVC) VALVE (A/T ONLY)
20. AIR CHAMBER (A/T ONLY)
21. EGR CONTROL SOLENOID VALVE (A/T ONLY)
22. CHARCOAL CANISTER
23. PURGE CUT-OFF SOLENOID VALVE
24. PURGE CONTROL DIAPHRAGM VALVE
25. TWO-WAY VALVE

ECU INPUT/OUTPUT WIRING SCHEMATIC—INTEGRA

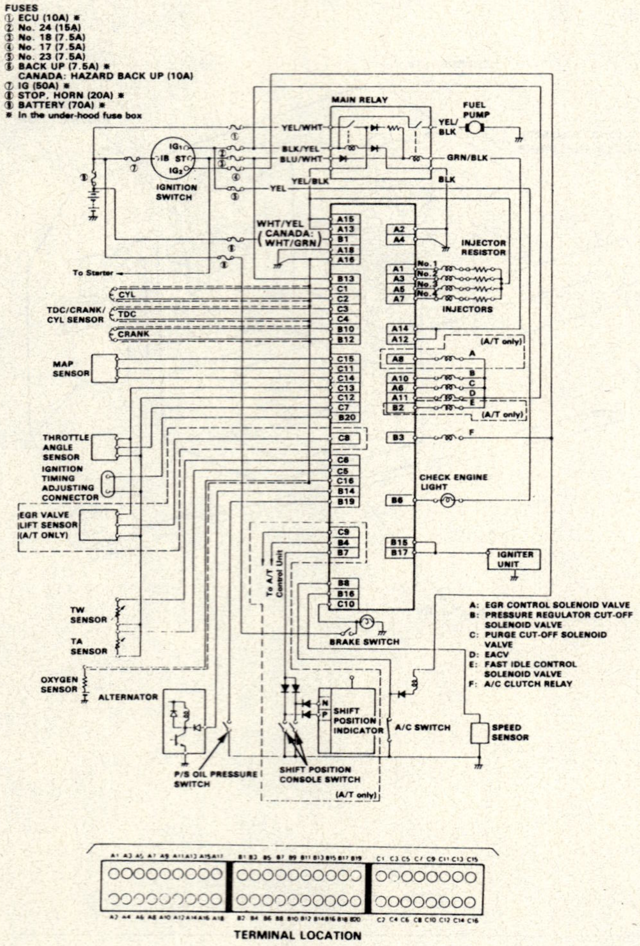

SYMPTOM TO SYSTEM DIAGNOSIS—INTEGRA

NOTE: Across each row in the chart, the systems that could be sources of a symptom are ranked in the order they should be inspected starting with ①. Find the symptom in the left column, read across to the most likely source, then refer to the page listed at the top of that column. If inspection shows the system is OK, try the next most likely system ②, etc.

SYMPTOM		SYSTEM: ECU	PGM-FI: OXYGEN SENSOR	MANIFOLD ABSOLUTE PRESSURE SENSOR	TDC/CRANK/CYL SENSOR	COOLANT TEMPERATURE SENSOR	THROTTLE ANGLE SENSOR	INTAKE AIR TEMPERATURE SENSOR	ATMOSPHERIC PRESSURE SENSOR
CHECK ENGINE LIGHT TURNS ON		▭ or ▭	▭	▭	▭	▭	▭	▭	▭
SELF-DIAGNOSIS INDICATOR (LED) BLINKS		0 or *	1 or 43	3 or 5	4 or 8 or 9	6	7	10	13
ENGINE WON'T START		3							
DIFFICULT TO START ENGINE WHEN COLD		BU		3		1			
IRREGULAR IDLING	WHEN COLD FAST IDLE OUT OF SPEC	BU				3			
	ROUGH IDLE	BU		3					
	WHEN WARM RPM TOO HIGH	BU							
	WHEN WARM RPM TOO LOW	BU							
FREQUENT STALLING	WHILE WARMING UP	BU							
	AFTER WARMING UP	BU							
POOR PERFORMANCE	MISFIRE OR ROUGH RUNNING	BU							
	FAILS EMISSION TEST	BU	3	2					
	LOSS OF POWER	BU		3			2		

If codes other than those listed above are indicated, count the number of blinks again. If the indicator is in fact blinking these codes, replace the original ECU.

(BU): When the Check Engine light and the self-diagnosis indicator are on, the back-up system is in operation. Substitute a known-good ECU and recheck. If the indication goes away, replace the original ECU.

▭ : CHECK (USA), ENGINE MOTEUR (CANADA)

SYMPTOM TO SYSTEM DIAGNOSIS—INTEGRA

PGM-FI		IDLE CONTROL		FUEL SUPPLY			EMISSION CONTROL	
IGNITION OUTPUT SIGNAL	VEHICLE SPEED SENSOR	ELECTRONIC AIR CONTROL VALVE	OTHER IDLE CONTROLS	FUEL INJECTOR	OTHER FUEL SUPPLY	AIR INTAKE	EGR CONTROL SYSTEM (A/T only)	OTHER EMISSION CONTROLS
⑮	⑰	⑭		⑯			⑫	
				②	①			
			②					
		①	②					
		①		②			③	
		②	①					
		①		②				
		①	②		③			
		①	②		③		③	
		②		①			③	
				①				
					①			

NOTE:
- Across each row in the chart, the sub systems that could be sources of a symptom are ranked in the order they should be inspected, starting with ①. Find the symptom in the left column, read across to the most likely source, then refer to the page listed at the top of that column. If inspection shows the system is OK, try the next system ②, etc.
- If the idle speed is out of specification and LED does not blink CODE 14, go to inspection.

SYMPTOM	(SUB SYSTEM)	IDLE ADJUSTING SCREW	EACV	AIR CONDITIONING SIGNAL	ALTERNATOR FR SIGNAL	A/T SHIFT POSITION SIGNAL (A/T ONLY)	BRAKE SWITCH SIGNAL	STARTER SWITCH SIGNAL	P/S OIL PRESSURE SWITCH SIGNAL	FAST IDLE VALVE	AIR BOOST VALVE	FAST IDLE CONTROL SOLENOID VALVE (A/T ONLY)	HOSES AND CONNECTIONS
DIFFICULT TO START ENGINE WHEN COLD		③	②							①	③		
WHEN COLD FAST IDLE OUT OF SPEC (1,000–2,000 rpm)		③	②							①		③	
ROUGH IDLE			②										①
WHEN WARM RPM TOO HIGH		③	①						③	②		③	
WHEN WARM RPM TOO LOW	Idle speed is below specified rpm (no load)	②	①		③								
	Idle speed does not increase after initial start up.		①					②					
	On models with automatic transmission, the idle speed drops in gear		②			①							
	Idle speeds drops when air conditioner in ON		②	①									
	Idle speed drops when steering wheel is turning		②						①				
	Idle speed fluctuates with electrical local	③	②		①								
FREQUENT STALLING	WHILE WARMING UP		①										
	AFTER WARMING UP	②	①										
FAILS EMISSION TEST													①

SYMPTOM TO SYSTEM DIAGNOSIS—INTEGRA

NOTE: Across each row in the chart, the systems that could be sources of a symptom are ranked in the order they should be inspected starting with ①. Find the symptom in the left column, read across to the most likely source, then refer to the page listed at the top of that column. If inspection shows the system is OK, try the next most likely system ②, etc.

SYMPTOM \ SUB SYSTEM		FUEL INJECTOR	INJECTOR RESISTOR	PRESSURE REGULATOR (PRESSURE REGULATOR CUT-OFF SOLENOID VALVE)	FUEL FILTER	FUEL PUMP	MAIN RELAY	CONTAMINATED FUEL *
ENGINE WON'T START			③		③	①	②	
DIFFICULT TO START ENGINE WHEN COLD OR HOT		③		③ (WHEN HOT)	②	①		
ROUGH IDLE		①		②				③
FREQUENT STALLING	WHILE WARMING UP	①			②			
	AFTER WARMING UP	①			③	②		
POOR PERFORMANCE	MISFIRE OR ROUGH RUNNING	①		②				③
	FAILS EMISSION TEST	①		②				
	LOSS OF POWER				①	③		②

* Fuel with dirt, water or a high percentage of alcohol is considered contaminated.

NOTE: Across each row in the chart, the sub systems that could be sources of a symptom are ranked in the order they should be inspected starting with ①. Find the symptom in the left column, read across to the most likely source, then refer to the page listed at the top of that column. If inspection shows the system is OK, try the next system ②, etc.

SYMPTOM \ SUB SYSTEM	THROTTLE CABLE	THROTTLE BODY	THROTTLE CONTROL SYSTEM
WHEN COLD FAST IDLE OUT OF SPEC	③	②	①
WHEN WARM RPM TOO HIGH	③	②	①
WHEN WARM RPM TOO LOW		①	
FREQUENT STALLING WHILE WARMING UP		①	
LOSS OF POWER	①	②	

NOTE: Across each row in the chart, the systems that could be sources of a symptom are ranked in the order they should be inspected starting with ①. Find the symptom in the left column, read across to the most likely source, then refer to the page listed at the top of that column. If inspection shows the system is OK, try the next most likely system ②, etc.

M/T:

SYMPTOM \ SUB SYSTEM		CATALYTIC CONVERTER	POSITIVE CRANKCASE VENTILATION SYSTEM	EVAPORATIVE EMISSION CONTROLS
ROUGH IDLE			①	
POOR PERFORMANCE	FAILS EMISSION TEST	①		②
	LOSS OF POWER	①		

A/T:

SYMPTOM \ SUB SYSTEM		CATALYTIC CONVERTER	EGR SYSTEM	POSITIVE CRANKCASE VENTILATION SYSTEM	EVAPORATIVE EMISSION CONTROLS
ROUGH IDLE			①	②	
FREQUENT STALLING	WHILE WARMING UP		①		
	AFTER WARMING UP		①		
POOR PERFORMANCE	FAILS EMISSION TEST	①		②	
	LOSS OF POWER	①			

TROUBLESHOOTING CHARTS—INTEGRA

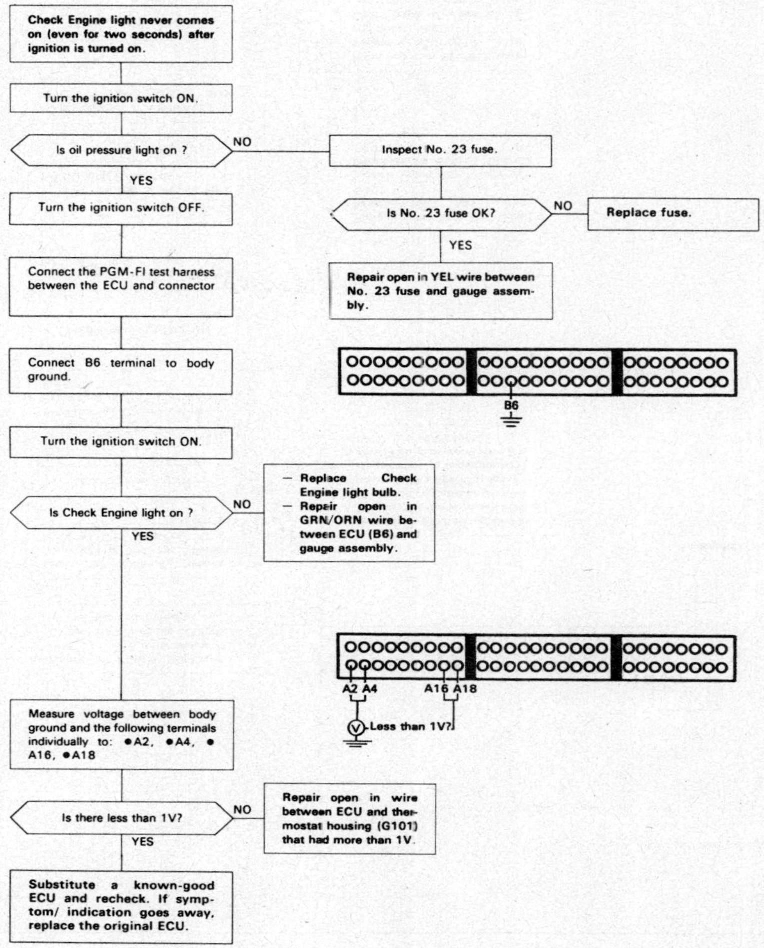

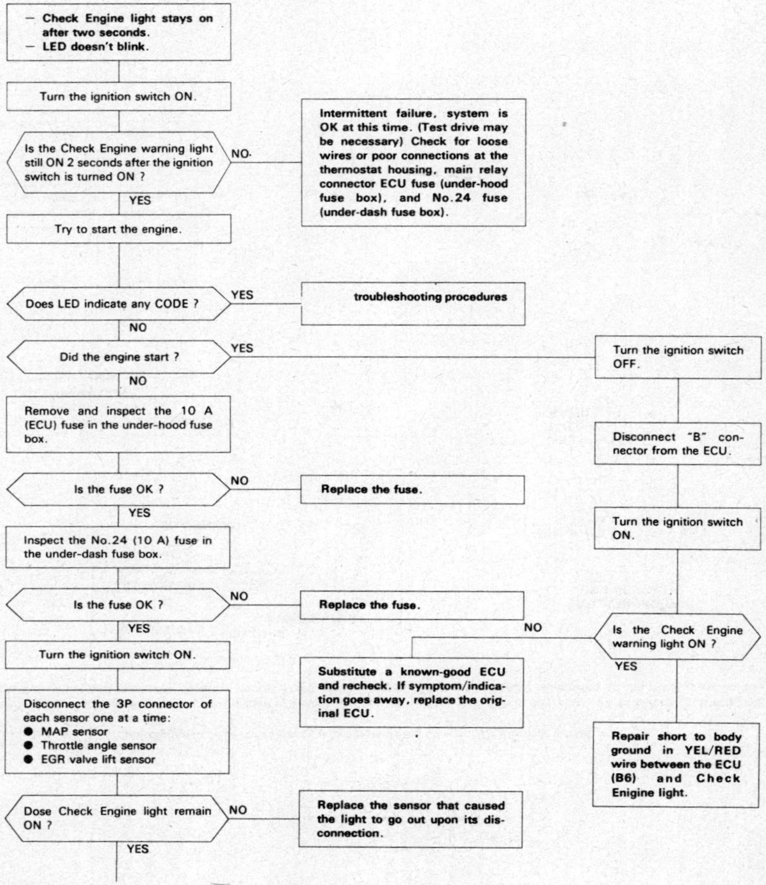

Self-diagnosis LED blinks once: A problem in the Oxygen (O_2) Sensor circuit.

The oxygen sensor, detects the oxygen content in the exhaust gas, and inputs the ECU. In operation, the ECU receives the signals from the sensor and varies the duration during which fuel is injected. The oxygen sensor is installed on the exhaust manifold.

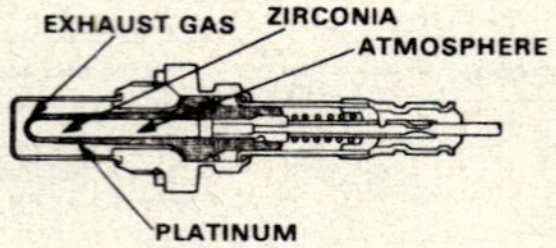

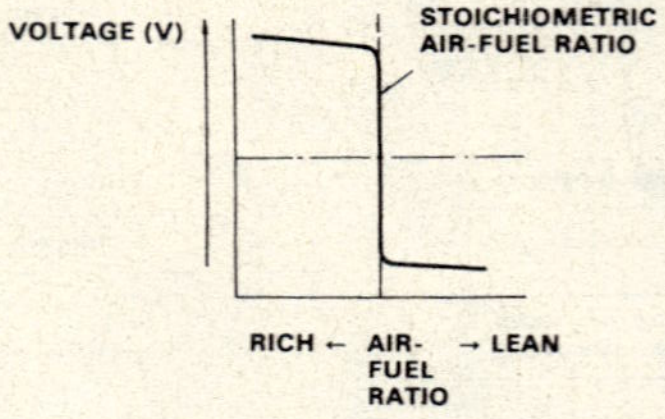

—Check Engine light has been reported on.
—LED indicates CODE 1.

Do the ECU Reset Procedure

Warm up engine to normal operating temperature (cooling fan comes on).

Run engine for 60 seconds.

Turn the ignition switch OFF.

Connect the PGM-FI test harness But disconnect the "C" connector from the ECU only, not the main wire harness.

Check for continuity between body ground and the following terminals: C13, C15.

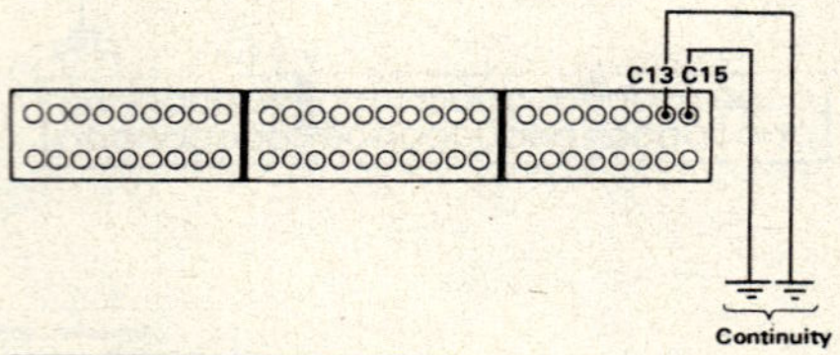

Does continuity exist ?

YES:
- Repair short to body ground in YEL/WHT wire between ECU (C13) and throttle angle sensor or EGR valve lift sensor.
- Repair short to body ground in RED/YEL wire between ECU (C15) and MAP sensor.

NO:

Reconnect all the connectors. Reconnect the "C" connector to the ECU.

Turn the ignition switch ON.

Individually connect the following terminal to Body Ground. A16 · A18

Is the Check Engine warning light still ON after 2 seconds ?

NO:
- Repair open in BLK/RED wire between ECU (A18) and G101.
- Repair open in BRN/BLK wire between ECU (A16) and G101.

YES:

Measure voltage between A18 (−) and the following: A13 (+) and A15 (+).

Is there battery voltage ?

NO:
- Repair open in YEL/BLK wire between ECU (A13, A15) and main relay.
- Check main relay and wiring connectors at main relay

YES:

Substitute a known-good ECU and recheck. If symptom/indication goes away, replace the original ECU.

Road test necessary with the Transmission in 2nd gear, accelerate using wide open throttle for at least 5 seconds. Then decelerate for at least 5 seconds with the throttle completely closed.

Is Check Engine light on and does LED indicate CODE 1 ?

- NO → Intermittent failure, system is OK at this time. Check for poor connections or loose wires at C216 (round connector located at right shock tower).
- YES ↓

Disconnect the O_2 sensor connector and measure voltage between the O_2 sensor connector and body ground.

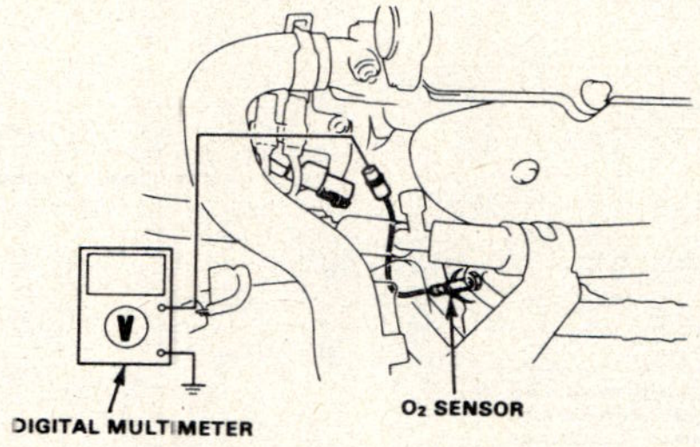

Is the voltage above 0.6 V at 4,500 rpm and below 0.4 V when the throttle is quickly released from 4,500 rpm ?

- NO → Replace O_2 sensor.
- YES → perform test for CODE 43.

Self-diagnosis LED indicator blinks three times: An electrical problem in the Manifold Absolute Pressure (MAP) Sensor system.

Self-diagnosis LED indicator blinks five times: A mechanical problem (broken hose) in the Manifold Absolute Pressure (MAP) Sensor system.

The MAP sensor converts manifold absolute pressure into electrical signals and inputs the ECU.

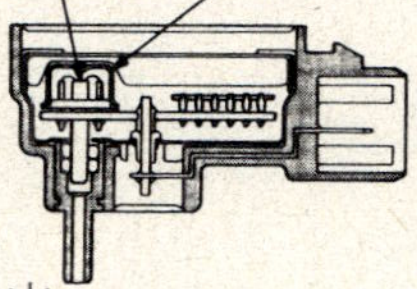

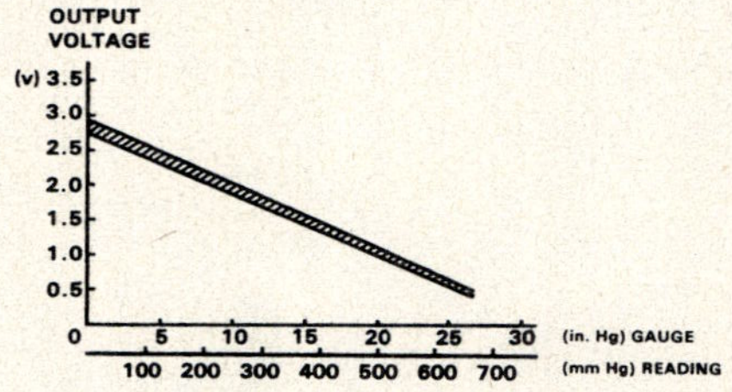

- Engine is warm and running.
- Check Engine light is on.
- LED indicates CODE 3.

Do the ECU Reset Procedure

Warm up engine to normal operating temperature (cooling fan comes on).

Is Check Engine light on and does LED indicate CODE 3?

- NO → Intermittent failure, system is OK at this time (test drive may be necessary). Check for poor connection or loose wires at MAP sensor connector and ECU.
- YES ↓

Turn the ignition switch OFF.

Disconnect the 3P connector from the MAP sensor.

Turn the ignition switch ON.

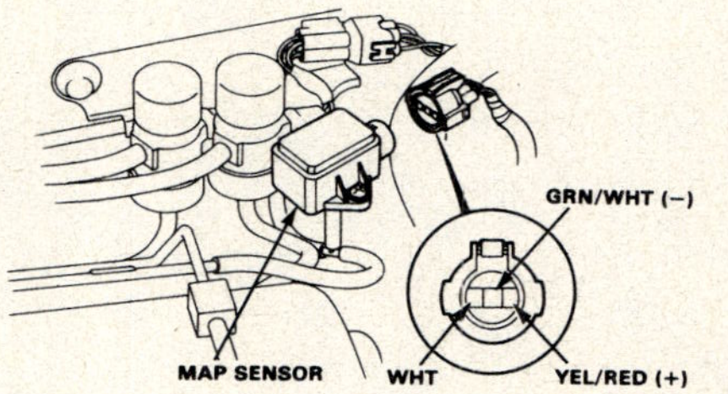

TROUBLESHOOTING CHARTS—INTEGRA

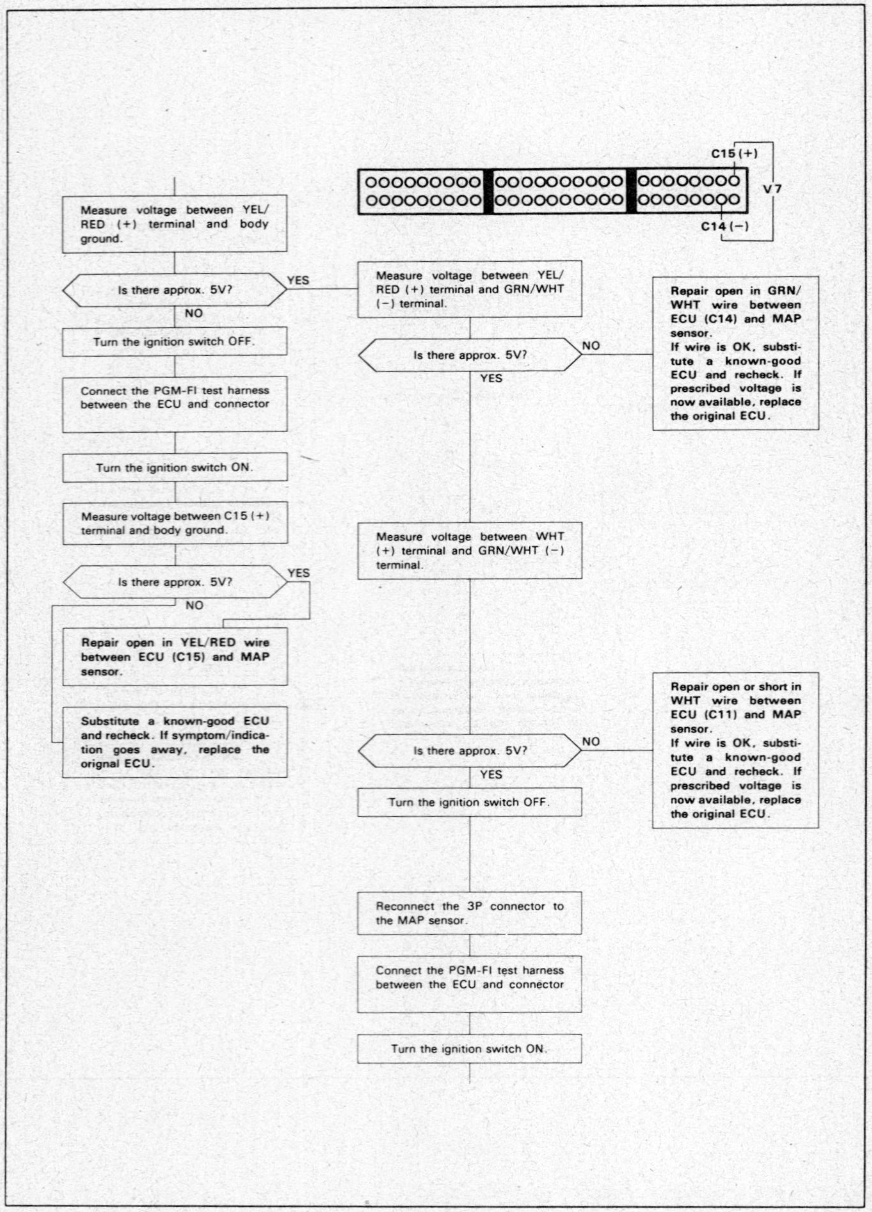

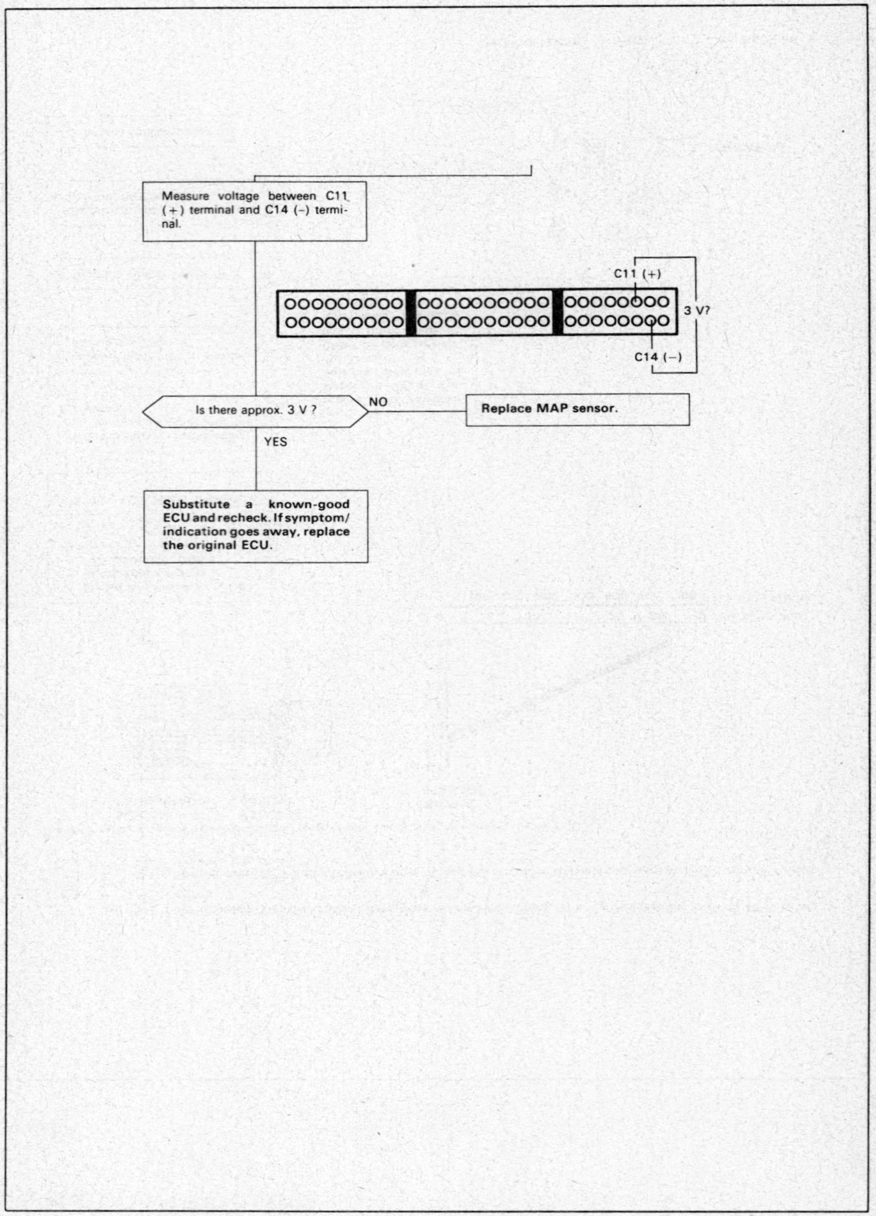

TROUBLESHOOTING CHARTS—INTEGRA

Self-diagnosis LED indicator blinks four times: A problem in the circuit of the CRANK Sensor.

Self-diagnosis LED indicator blinks eight times: A problem in the circuit of the TDC Sensor.

Self-diagnosis LED indicator blinks nine times: A problem in the circuit of the CYL Sensor.

The CRANK sensor determines timing for fuel injection and ignition of each cylinder and also detects engine RPM. The TDC sensor determines ignition timing at start-up (cranking) and when crank angle is abnormal. The CYL sensor detects the position of No.1 cylinder for sequential fuel injection to each cylinder.

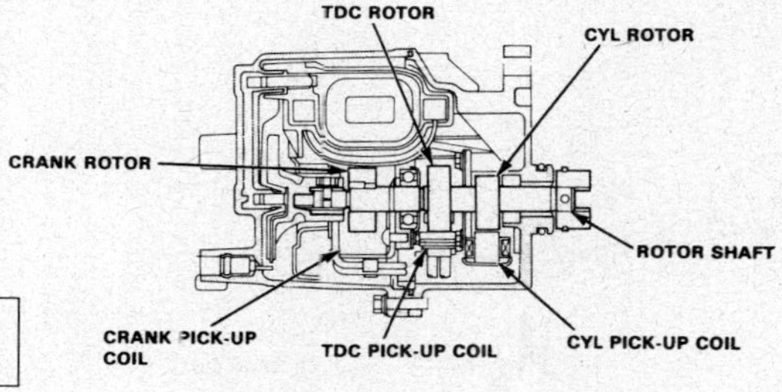

- Check Engine light has been reported on.
- LED indicates CODE 4.

Do the ECU Reset Procedure

Start engine.

Is Check Engine light on and does LED indicate CODE 4?

NO → **Intermittent failure, system is OK at this time (test drive may be necessary). Check for poor connections or loose wires at the distributor connector and C216 (round connector located at the right shock tower).**

YES ↓

Stop engine.

Disconnect the 8P connector from the TDC/CRANK/CYL sensor.

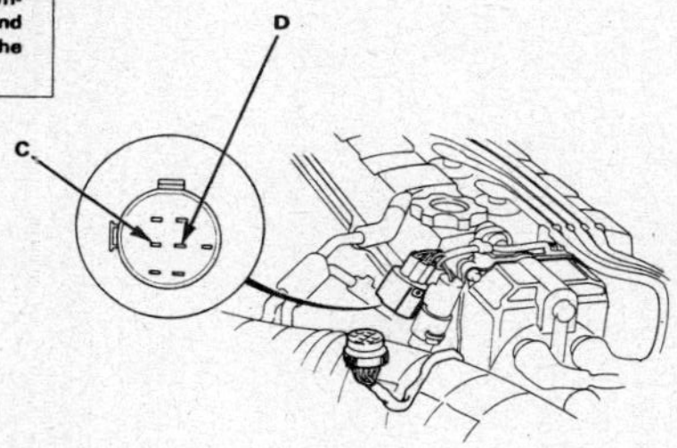

Measure resistance between C terminal and D terminal.

Is there 350–700 Ω ?

NO → **Replace the distributor assembly**

YES ↓

Check for continuity to body ground on C terminal and D terminal individually.

Does continuity exist ?

YES → **Replace the distributor assembly**

NO ↓

Reconnect the connector.

Connect the PGM-FI test harness only to the main wire harness, not to the ECU

Measure resistance between B10 terminal and B12 terminal.

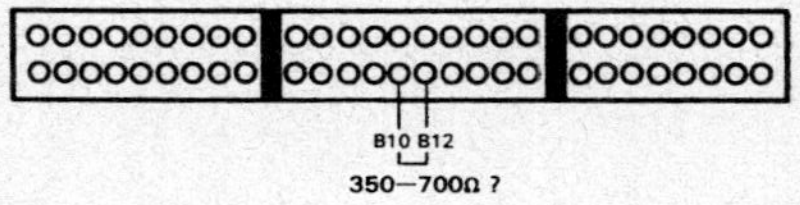

Is there 350–700 Ω ?

NO → **Repair open in ORN and/or WHT wires.**

YES ↓

Check for continuity to body ground on B10 terminal.

Does continuity exist?

YES → **Repair short in ORN wire between ECU (B10) and distributor connector.**

NO ↓

Substitute a known-good ECU and recheck. If symptom/indication goes away, replace the original ECU.

TROUBLESHOOTING CHARTS—INTEGRA

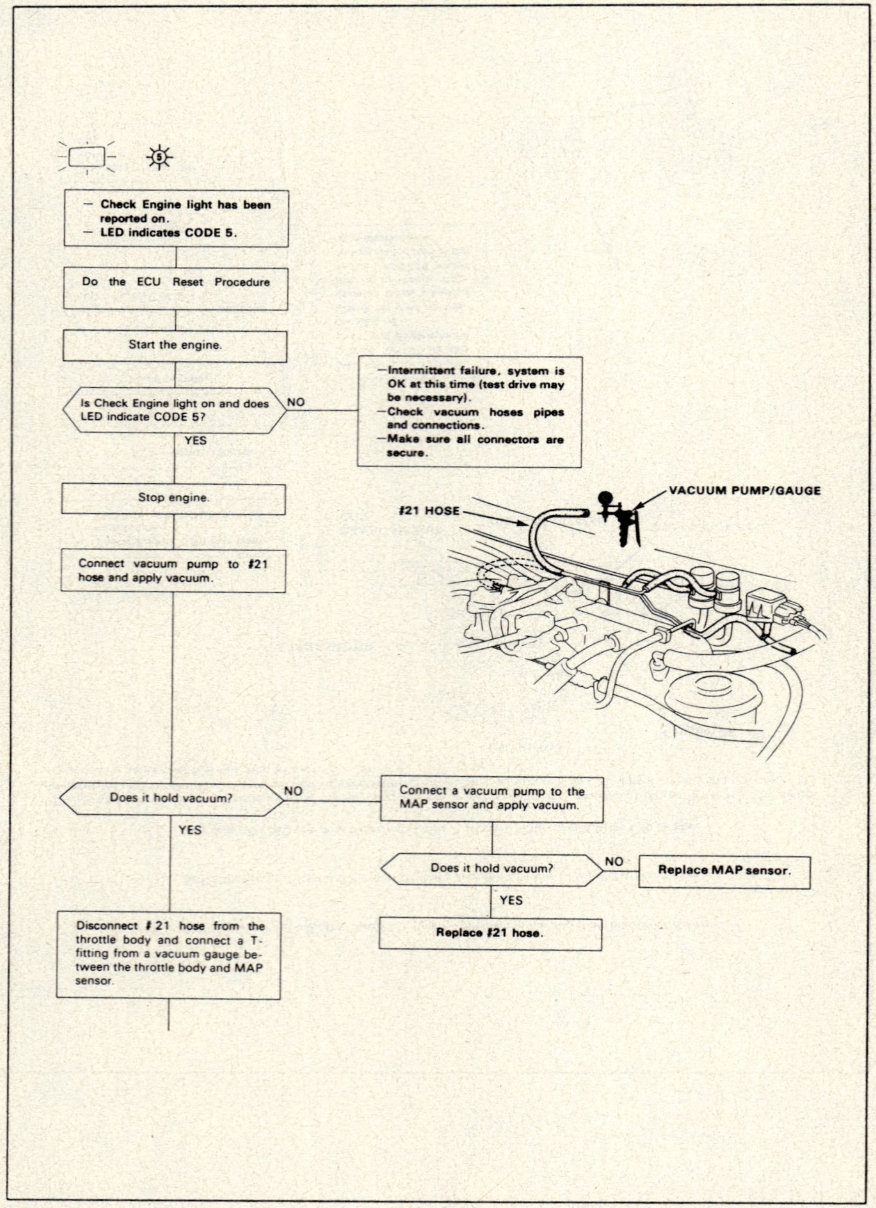

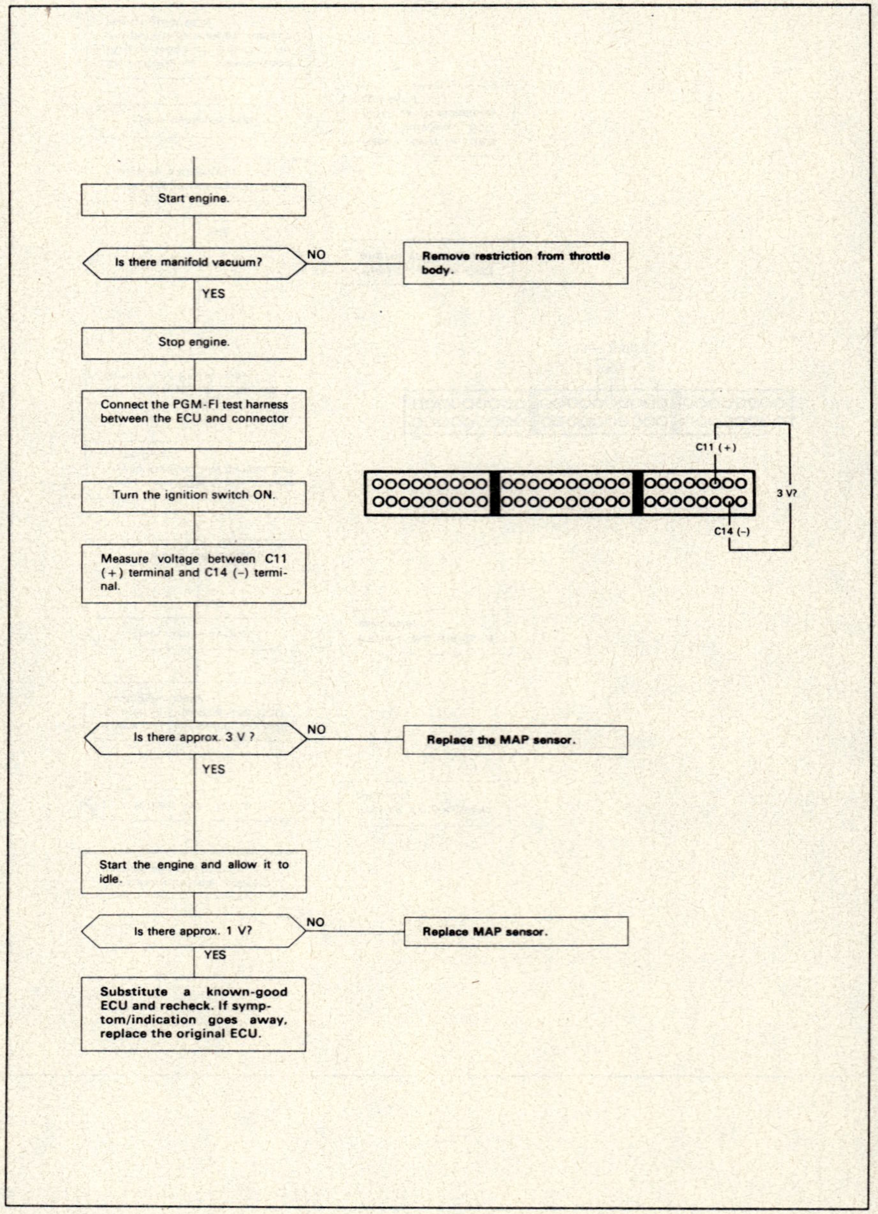

TROUBLESHOOTING CHARTS – INTEGRA

Self-diagnosis LED indicator blinks six times: A problem in the Coolant Temperature (TW) Sensor circuit.

The TW sensor is a temperature dependant resistor (thermistor). The resistance of the thermistor decreases as the coolant temperature increases as shown below.

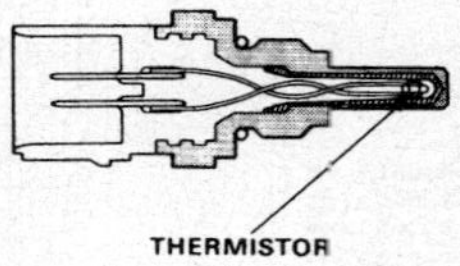

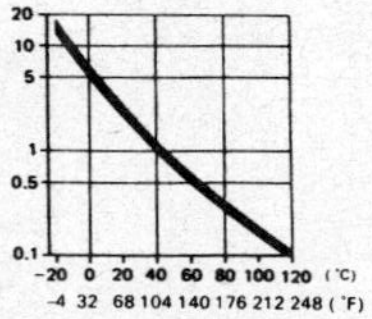

- Check Engine light is on.
- LED indicates CODE 6.

↓

Do the ECU Reset Procedure

↓

Turn the ignition switch ON.

↓

Is Check Engine light on and does LED indicate CODE 6?

- NO → **Intermittent failure, system is OK at this time (test drive may be necessary). Check for poor connections or loose wires at TW sensor, C216 and C318 (round connector located at the right left shock tower).**
- YES ↓

Warm up engine to normal operating temperature (cooling fan comes on).

↓

Turn the ignition switch OFF.

↓

Disconnect the 2P connector from the TW sensor.

↓

Measure resistance between the 2 terminals on the TW sensor.

↓

Is there 200—400 Ω ?

- NO → **Replace TW sensor.**
- YES ↓

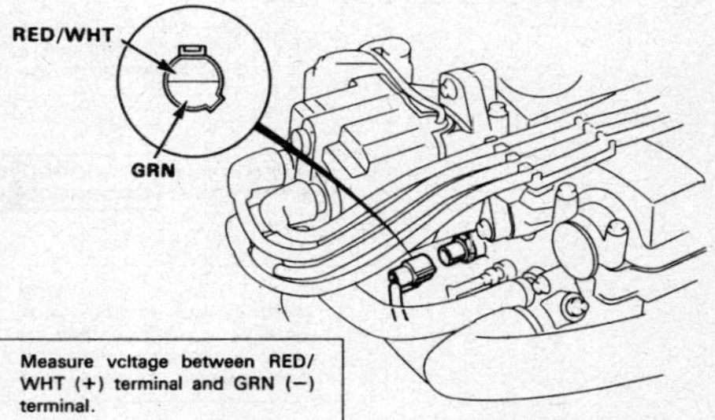

Turn the ignition switch ON.

↓

Measure voltage between RED/WHT (+) and body ground.

↓

Is there apporx. 5V ?

- YES → Measure voltage between RED/WHT (+) terminal and GRN (–) terminal.
  - Is there approx. 5V ?
    - NO → **Repair open in GRN, GRN/WHT wire between ECU (C12) and TW sensor.**
    - YES → **Substitute a known-good ECU and recheck. If symptom/indication goes away, replace the original ECU.**
- NO ↓

A/T only

Disconnect the 18P connector from the A/T control unit.

↓

Is there apporox. 5V ?

- YES → **Replace the A/T control unit.**
- NO ↓

Connect the PGM-FI test harness to the ECU only, not to the main wire harness (page 11-18).

↓

Measure voltage between C6 (+) terminal and C12 (–) terminal.

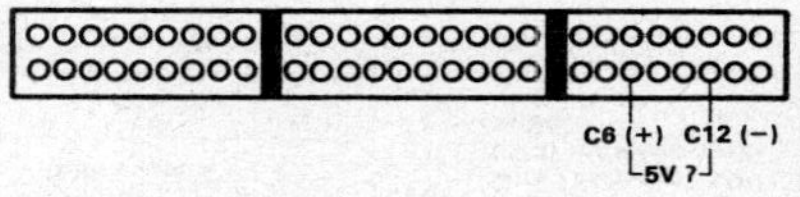

↓

Is there approx. 5V ?

- YES → **Repair open in RED/WHT wire between ECU (C6), A/T control unit and sensor.**
- NO → **Substitute a known-good ECU and recheck. If symptom/indication goes away, replace the original ECU.**

TROUBLESHOOTING CHARTS—INTEGRA

Self-diagnosis LED indicator blinks seven times: A problem in the Throttle Angle Sensor circuit.

The throttle angle sensor is a potentiometer. It is connected to the throttle valve shaft. As the throttle angle changes, the throttle angle sensor varies the voltage signal to the ECU.

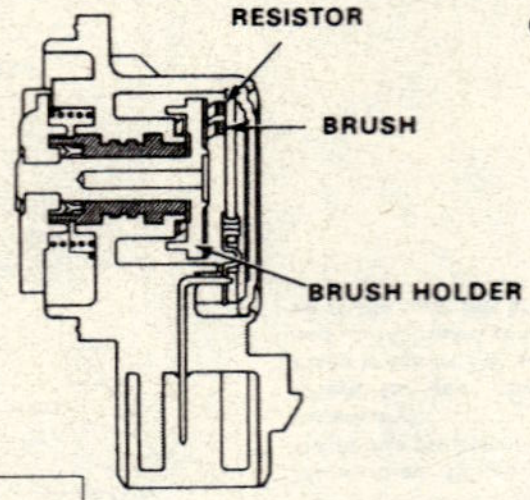

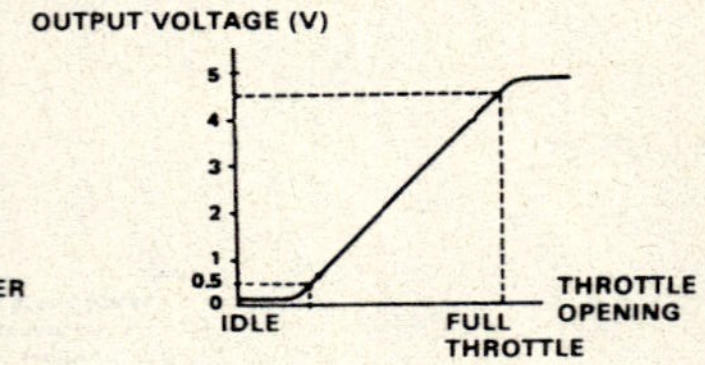

- Engine is running.
- Check Engine light is on.
- LED indicates CODE 7.

Do the ECU Reset Procedure

Start engine.

Is Check Engine light on and does LED indicate CODE 7?

NO → Intermittent failure, system is OK at this time (test drive may be necessary). Check for poor connections or loose wires at throttle angle sensor, C216 (round connector located at the right shock tower) and C318 (round connector located at the left shock tower).

YES ↓

Turn the ignition switch OFF.

Disconnect the 3P connector from the throttle angle sensor.

Turn the ignition switch ON.

Measure voltage between YEL/WHT (+) terminal and GRN (−) terminal.

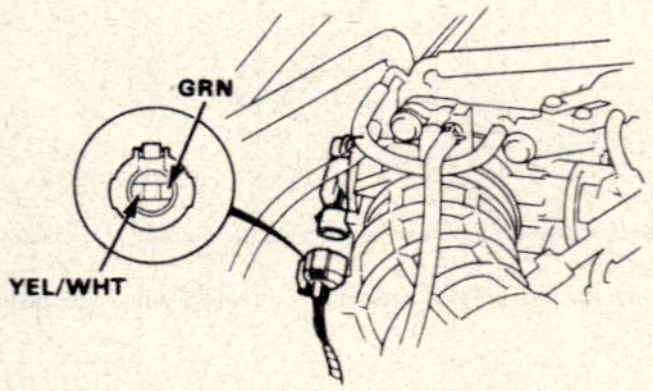

Is there approx. 5V ?

NO → Measure voltage between YEL/WHT (+) terminal and body ground.

YES ↓

Turn the ignition switch OFF.

Reconnect the 3P connector.

Connect the PGM-FI test harness between the ECU and connector

Turn the ignition switch ON.

Measure voltage between C7(+) terminal and C12(−) terminal.

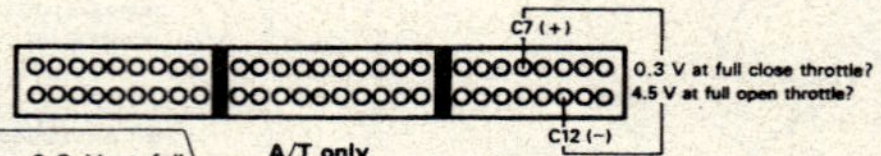

Is voltage approx. 0.3 V at full close throttle, and approx. 4.5 V at full open throttle ?
NOTE: There should be a smooth transition from 0.3 V to 4.5 V as the throttle is depressed.

YES → Substitute a known-good ECU and recheck. If symptom/indication goes away, replace the original ECU.

NO → A/T only:

Disconnect the 18 P connector from the A/T control unit.

Is voltage approx. 0.3 V at full close throttle, and approx. 4.5 V at full open throttle ?
NOTE: There should be a smooth transition from 0.3 V to 4.5 V as the throttle is depressed.

YES → Replace the A/T control unit.

NO →
- Replace throttle angle sensor.
- Repair open or short in RED/BLU wire between ECU (C7), A/T control unit and throttle angle sensor.

(From "Measure voltage between YEL/WHT (+) terminal and body ground":)

Is there approx. 5V ?

YES → Repair open in GRN wire between ECU (C12) and throttle angle sensor.

NO ↓

Turn the ignition switch OFF.

Connect the PGM-FI test harness between the ECU and connector

Turn the ignition switch ON.

Measure voltage between C13 (+) terminal and C12 (−) terminal.

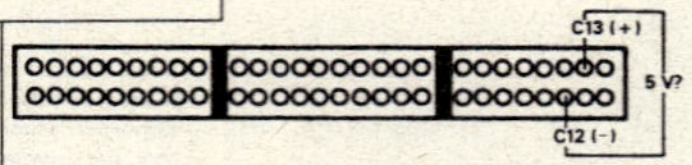

Is there approx. 5V ?

YES → Repair open in YEL/WHT wire between ECU(C13) and throttle angle sensor.

NO → Substitute a known-good ECU and recheck. If prescribed voltage is now available, replace the original ECU.

TROUBLESHOOTING CHARTS—INTEGRA

- Check Engine light has been reported on.
- LED indicates CODE 8.

Do the ECU Reset Procedure

Start engine.

Is Check Engine light on and does LED indicate CODE 8?

- NO → Intermittent failure, system is OK at this time (test drive may be necessary). Check for poor connections or loose wires at distributor connector and C216 (round connector located at the right shock tower).
- YES → Stop engine.

Disconnect the 8P connector from the TDC/CRANK/CYL sensor.

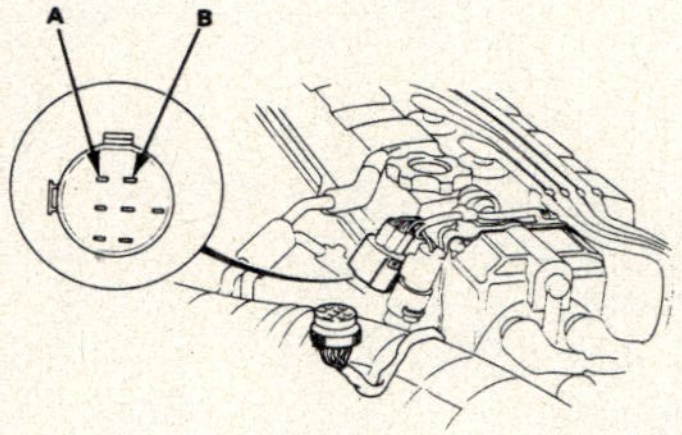

Measure resistance between A terminal and B terminal.

Is there 350–700 Ω ?

- NO → Replace the distributor assembly
- YES → Check for continuity to body ground on A terminal and B terminal individually.

Does continuity exist ?

- YES → Replace the distributor assembly
- NO → Reconnect the connector.

Connect the PGM-FI test harness only to the main wire harness, not to the ECU

Measure resistance between C3 terminal and C4 terminal.

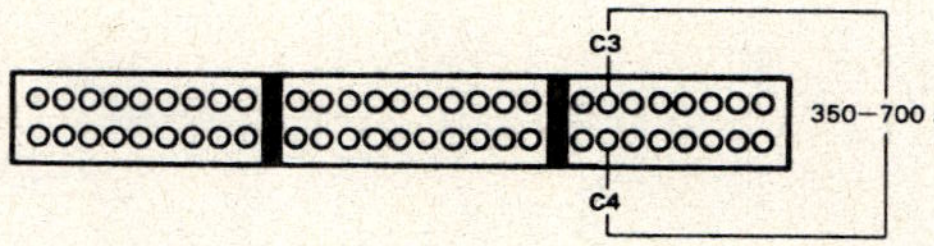

Is there 350–700 Ω ?

- NO → Repair open in ORN/BLU and/or WHT/BLU wires.
- YES → Check for continuity to body ground on C3 terminal.

Does continuity exist?

- YES → Repair short in ORN/BLU wire between ECU (C3) and distributor connector.
- NO → Substitute a known-good ECU and recheck. If symptom/indication goes away, replace the original ECU.

TROUBLESHOOTING CHARTS – INTEGRA

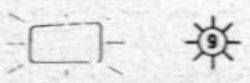

- Check Engine light has been reported on.
- LED indicates CODE 9.

Do the ECU Reset Procedure

Start engine.

Is Check Engine light on and does LED indicate CODE 9?

- NO → Intermittent failure, system is OK at this time (test drive may be necessary). Check for poor connection or loose wires at the distributor connector and C216 (round connector located at the right shock tower).
- YES ↓

Stop engine.

Disconnect the 8P connector from the TDC/CRANK/CYL sensor.

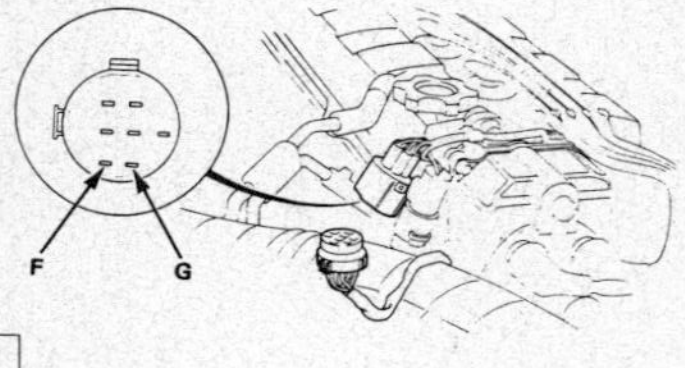

Measure resistance between F terminal and G terminal.

Is there 350–700 Ω ?

- NO → Replace the distributor assembly
- YES ↓

Check for continuity to body ground on F terminal and G terminal individually.

Does continuity exist ?

- YES → Replace the distributor assembly
- NO ↓

Reconnect the connector.

Connect the PGM-FI test harness only to the main wire harness, not to the ECU

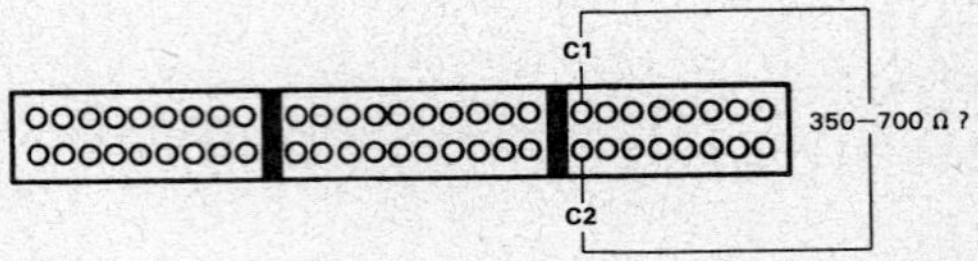

Measure resistance between C1 terminal and C2 terminal.

Is there 350–700 Ω ?

- NO → Repair open in BLU/GRN and/or BLU/YEL wires.
- YES ↓

Check for continuity to body ground on C1 terminal.

Does continuity exist?

- YES → Repair short in BLU/GRN wire between ECU (C1) and distributor connector.
- NO ↓

Substitute a known-good ECU and recheck. If symptom/indication goes away, replace the original ECU.

TROUBLESHOOTING CHARTS—INTEGRA

Self-diagnosis LED indicates code 10: A problem in the Intake Air Temperature (TA) Sensor circuit.

The TA sensor is a temperature dependant resistor (thermistor). The resistance of the thermistor decreases as the intake air temperature increases as shown below.

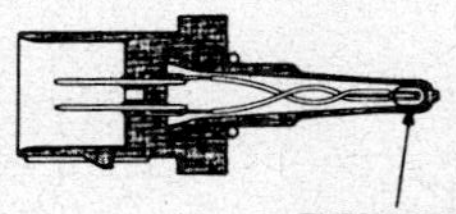

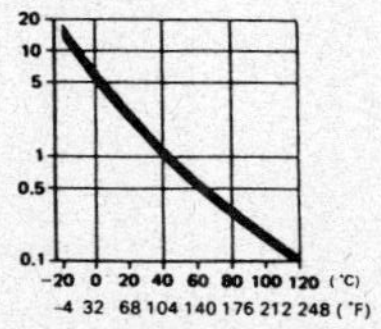

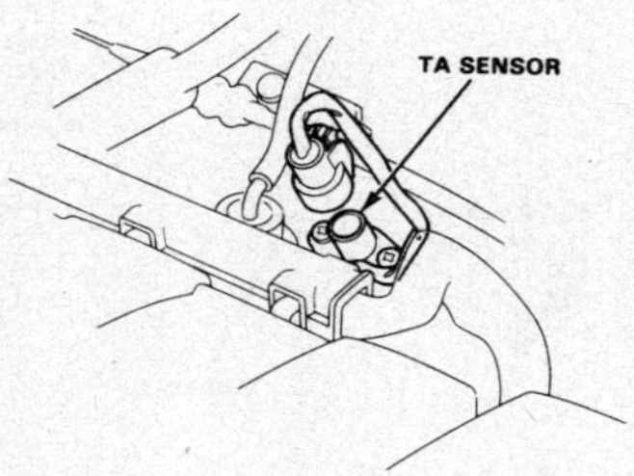

- Check Engine light is on.
- LED indicates CODE 10.

Do the ECU Reset Procedure

Turn the ignition switch ON.

Is Check Engine light on and does LED indicate CODE 10?

- NO → Intermittent failure, system is OK at this time (test drive may be necessary). Check for poor connections or loose wires at TA sensor and C318 (round connector located at left shock tower).
- YES ↓

Turn the ignition switch OFF.

Disconnect the 2P connector from the TA sensor.

Measure resistance between the 2 terminals on the TA sensor.

Is there 1–4 kΩ ?

- NO → Replace TA sensor.
- YES ↓

Turn the ignition switch ON.

Measure voltage between RED/WHT (+) terminal and body ground.

Is there apporx. 5V ?

- YES → Measure voltage between RED/YEL (+) terminal and GRN (−) terminal.
  - Is there approx. 5V ?
    - NO → Repair open in GRN, GRN/WHT wire between ECU (C12), and TA sensor.
    - YES → Substitute a known-good ECU and recheck. If symptom/indication goes away, replace the original ECU.
- NO ↓

Connect the PGM-FI test harness to the ECU only, not to the main wire harness

Measure voltage between C5 (+) terminal and C12 (−) terminal.

Is there approx. 5V ?

- YES → Repair open in RED/YEL wire between ECU (C5) and TA sensor.
- NO → Substitute a known-good ECU and recheck. If symptom/indication gose away, replace the original ECU.

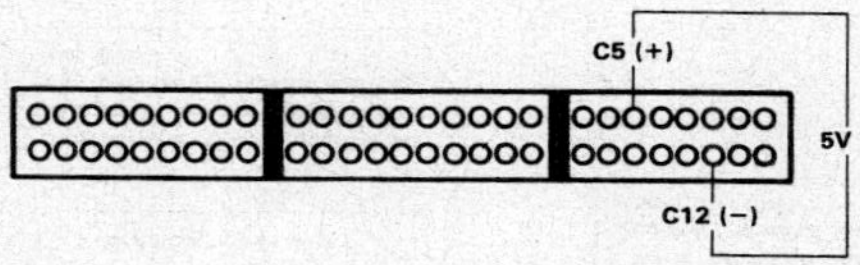

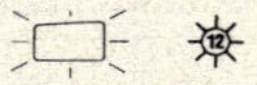

- Check Engine light has been reported on.
- LED indicates CODE 12.

Do the ECU Reset Procedure

Road test necessary: Warm up the engine to normal operating temperature (cooling fan comes on). Drive the car on the road for approx. 10 minutes. Try to keep the engine speed in the 1700— 2500 range.

Is Check Engine light on and does LED indicate CODE 12?

NO → Intermittent failure, system is OK at this time (test drive may be necessary). Check for poor connections or loose wires at EGR (valve, C216 at the right shock tower, C318 at the left shock tower, control box and ECU.

YES → With the engine at idle, disconnect the #16 hose from the EGR valve and connect a vacuum pump/gauge to the hose.

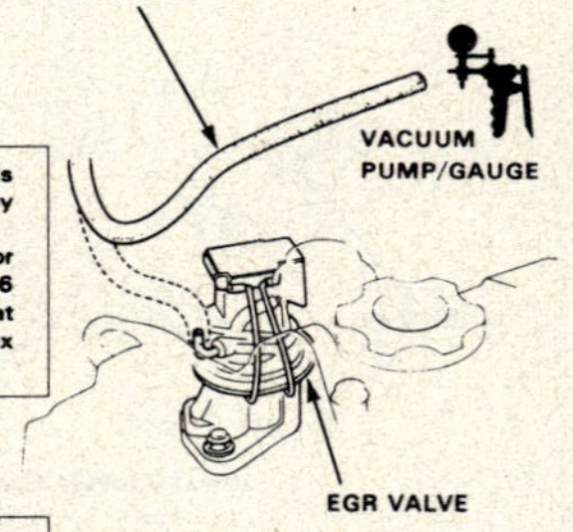

Is there any vacuum ?

YES → Disconnect 4P connector from the control box and check the #16 hose for vacuum again.

NO → Move the vacuum pump/gauge to the EGR valve.

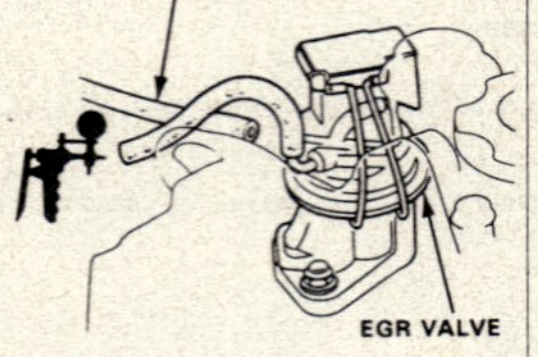

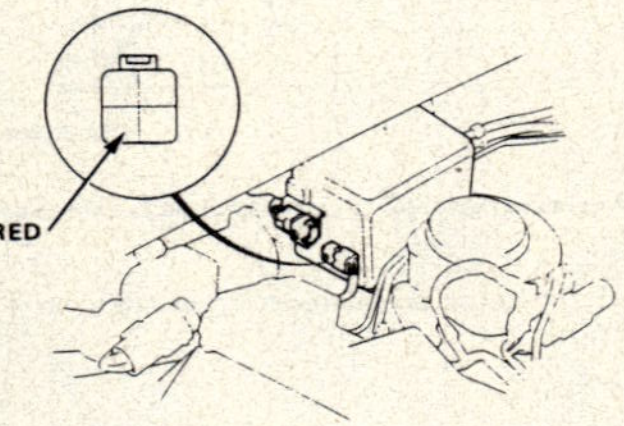

With the engine at idle, apply 8" of vacuum to the EGR valve.

Is there any vacuum ?

YES → Check vacuum hose routing of the entire EGR system. If hose routing is OK, replace EGR control solenoid valve.

NO → Check for continuity ground on RED wire of 4P connector.

Does continuity exist ?

YES → Repair open in RED wire between EGR control solenoid valve and ECU (A8).

NO → Substitute a known-good ECU and retest. If symptom/indication goes away. replace the original ECU.

Does the engine stall or run rough and does the EGR valve hold vacuum ?

NO → Replace EGR valve.

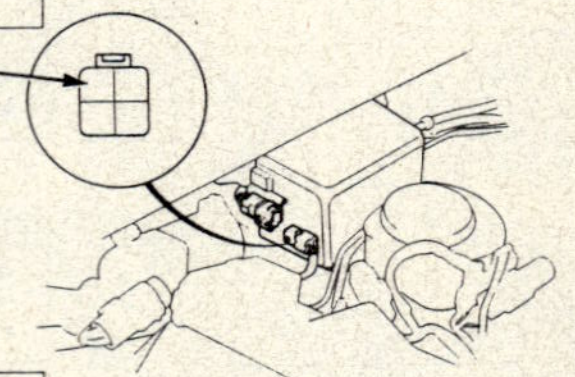

YES → Disconnect the 4P connector from the control box.

Measure voltage between BLK/YEL (+) terminal on the main wire harness and body ground.

Is there battery voltage ?

NO → Repair open in BLK/YEL wire between the solenoid valve and No.24 fuse.

YES → Reconnect the vacuum pump/gauge to the #16 hose.

Start the engine and allow it to idle.

TROUBLESHOOTING CHARTS—INTEGRA

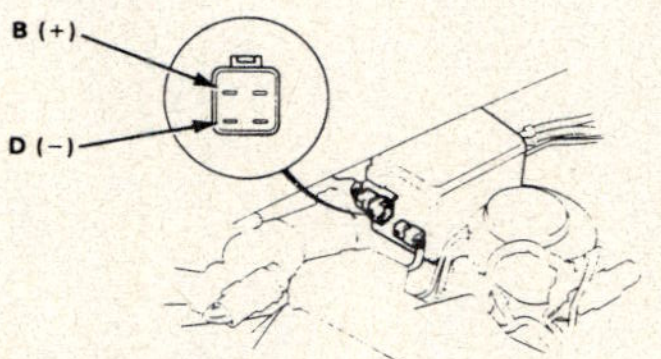

- Connect the battery positive terminal to the B terminal of the 4P connector. While watching the vacuum gauge, connect the battery negative terminal to the D terminal.
- Is there approx. 6" within 1 second?
  - NO → Turn the ignition switch OFF and inspect the #16 and #20 hoses for leaks, restrictions, or mis-routing.
    - Are the hose OK ?
      - NO → **Connect as necessary.**
      - YES → Disconnect the lower hose on EGR control solenoid valve and connect a vacuum gauge to the hose.
        - Start the engine and allow it to idle.
        - Is there 6"— 10" of vacuum ?
          - NO → **Replace CVC valve.**
          - YES → **Replace the EGR control solenoid valve.**
  - YES → Turn the ignition switch OFF and reconnect the 4P connector.
    - Connect the test harness between the EGR valve lift sensor and engine wire harness.
    - Turn the ignition switch ON.
    - Measure voltage between RED (+) terminal and GRN (−) terminal.
    - Is there approx. 5 V ?
      - NO → **— Repair open in YEL/WHT wire (C216) between EGR valve and ECU (C13).** **— Repair open in GRN, GRN/WHT wire (C318) between EGR valve and ECU (C12).**
      - YES → (continued)

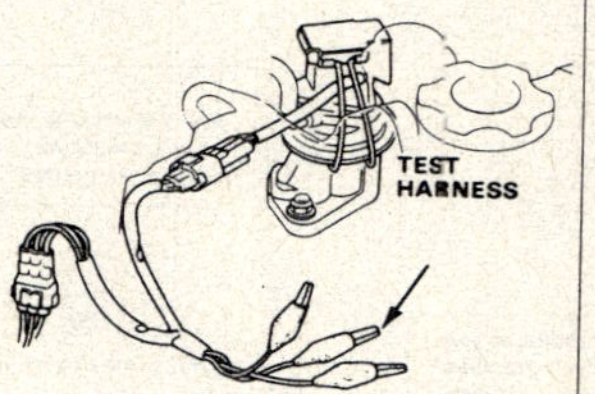

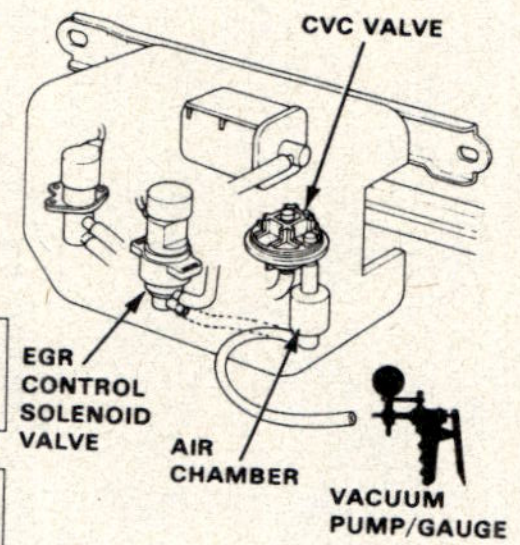

- Measure voltage between WHT (+) terminal and GRN (−) terminal.
- Is there approx 1.2 V ?
  - NO → **— Replace EGR valve.** **— Repair short in GRN/RED wire between ECU (C8) and EGR valve lift sensor.** **— Substitute a known-good ECU and recheck. If prescribed voltage is now available, replace the original ECU.**
  - YES → While watching the voltmeter, slowly apply a continuous 8" of vacuum to the EGR valve. Repeat several times, completely releasing vacuum between.
- Is the maximum voltage approx. 4 V ?
  - NO → **Replace EGR valve.**
  - YES ↓
- Does the voltage consistently increase/decrease as the vacuum increases/decreases ?
  - NO → **Replace EGR valve.**
  - YES → Reconnect the #16 hose to the EGR valve.
- Turn the ignition switch OFF.
- Connect the PGM-FI test harness between the ECU and connector (page 11-18).
- Start the engine and allow it to idle.

(13) Self-diagnosis LED indicates code: A problem in the Atmospheric Pressure (PA) Sensor circuit.

The PA sensor is built into the ECU.

(13)

- Check Engine light is on.
- LED indicates CODE 13.

Do the ECU Reset Procedure

Turn the ignition switch ON.

Is Check Engine light on and does LED indicate CODE 13?

- NO → Intermittent failure, system is OK at this time (test drive may be necessary).
- YES → Turn the ignition switch OFF.

Connect the PGM-FI test harness to the main wire harness not to the ECU.

Disconnect the 18P connector from the A/T control unit.

Check for continuity between C9 terminal and body ground.

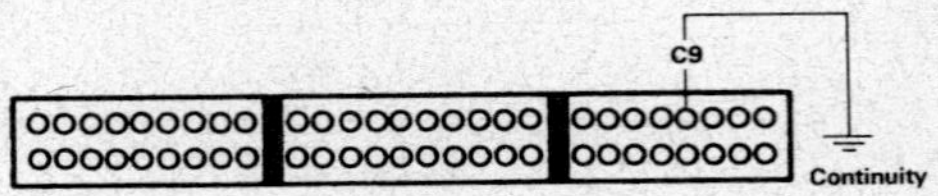

Dose continuity exist ?

- YES → Repair short in RED/WHT wire between ECU (C9) and the A/T control unit.
- NO → Substitute known-good A/T control unit and recheck.

Is Check Enginge warning light on and does LED indicate CODE 13?

- NO → Replace the original A/T control unit.
- YES → Substitute a known-good ECU and recheck. If precribed voltage is now available replace the original ECU.

Measure voltage between C8 (+) terminal and C12 (−) terminal.

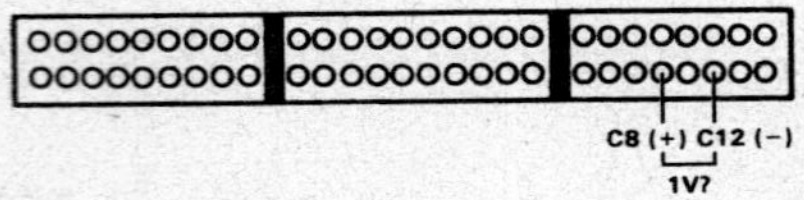

Is there approx. 1.2 V ?

- NO → Repair open in GRN/RED wire between ECU (C8) and the sensor.
- YES → Connect A8 terminal to A18 terminal with a jumper wire.

Did the engine stall or run rough?

- NO → Repair open in RED wire between ECU (A8) and EGR control solenoid valve.
- YES → Substitute a known-good ECU and recheck. If symptom/indication goes away, replace the original ECU.

TROUBLESHOOTING CHARTS—INTEGRA

Self-diagnosis LED indicates code 14: A problem in the Electronic Air Control Valve (EACV) circuit.

The EACV changes the amount of air bypassing the throttle body in response to a current signal from the ECU in order to maintain the proper idle speed.

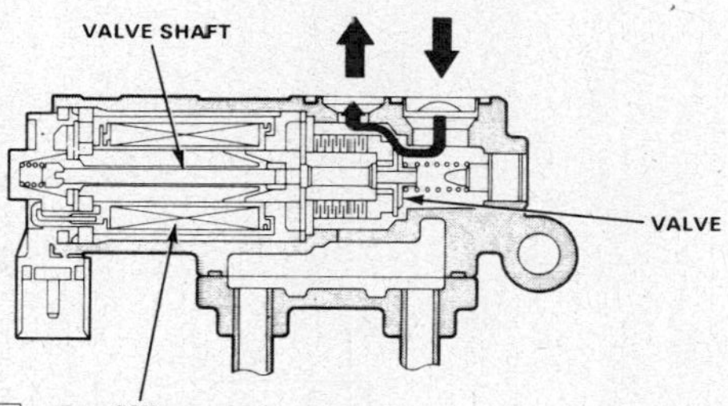

- Engine is running.
- Check Engine light is on.
- LED indicates CODE 14.

Do the ECU Reset Procedure

Start engine.

Is Check Engine light on and does LED indicate CODE 14?
- NO → Intermittent failure, system is OK at this time (test driving may be necessary). Check for poor connections or loose wies at EACV and at C317, C318.
- YES ↓

Stop engine.

Disconnect the 2P connector from the EACV.

Measure resistance between the 2 terminals on the EACV.

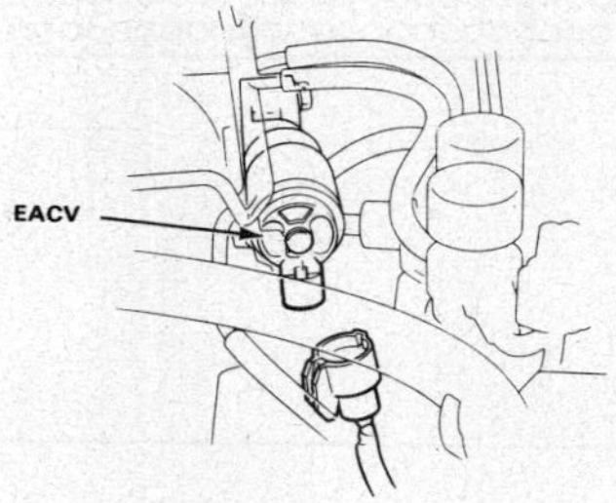

Is there 8—15Ω ?
- NO → Replace EACV.
- YES ↓

Check for continuity to body ground on each terminal on the EACV.

Does continuity exist ?
- YES → Replace EACV.
- NO ↓

Turn the ignition switch ON.

Measure voltage between BLK/YEL (+) terminal and BLU/YEL (-) terminal.

Is there battery voltage ?
- YES → Disconnect the "A" connector from the ECU. → Measure voltage between BLK/YEL (+) terminal and BLU/YEL (−) terminal. → Is there battery voltage ?
  - YES → Repair short in BLU/YEL wire between ECU (A11) and EACV.
  - NO → Substitute a known-good ECU and recheck. If symptom/indication goes away, replace the original ECU.
- NO ↓

Measure voltage between BLK/YEL (+) terminal and body ground.

Is there battery voltage ?
- NO → Repair open in BLK/YEL wire between the EACV and No.24 fuse.
- YES ↓

Turn the ignition switch OFF.

Reconnect the 2P connector to EACV.

Connect the PGM-FI test harness "A" connector to the main wire harness only, not the ECU (page 11-18).

Turn the ignition switch ON.

Measure voltage between A11 (+) terminal and A2 (−) terminal.

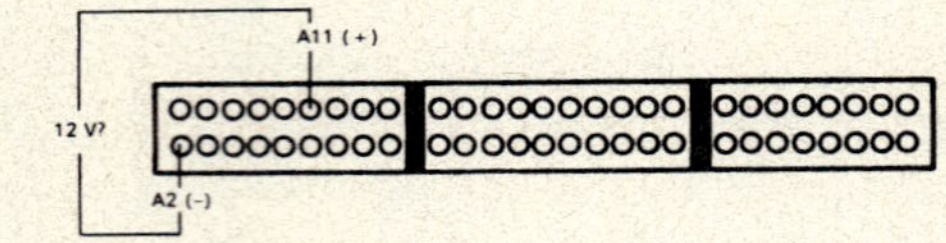

Is there battery voltage ?
- NO → Repair open in BLU/YEL wire between ECU (A11) and EACV.
- YES ↓

Connect and disconnect A11 terminal to A2 terminal.

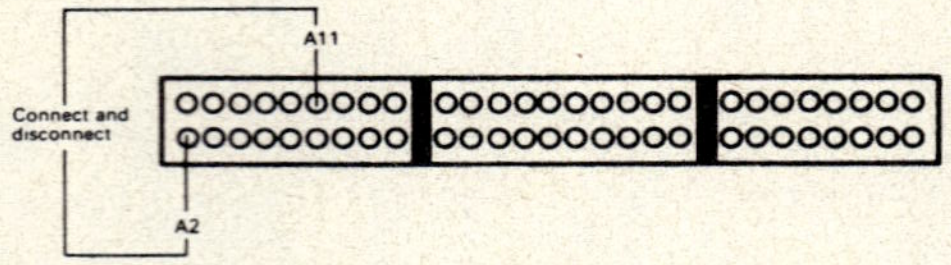

Does EACV click when the connector is connected and disconnected ?
- NO → Replace EACV.
- YES ↓

Substitute a known-good ECU and recheck. If symptom/indication goes away, replace the original ECU.

Self-diagnosis LED indicates code 15: A problem in the Ignition Output Signal circuit.

– Check Engine light is on.
– LED indicates CODE 10.

Do the ECU Reset Procedure

Start engine.

NOTE: If the engine won't start, it may take 20 seconds of cranking to set the code.

Is Check Engine light on and does LED indicate CODE 15?
- NO → Intermittent failure, system is OK at this time (test drive may be necessary). Check for poor connections or loose wires at the distributor connector and C216 (round connector located at the right shock tower).
- YES ↓

Turn the ignition switch OFF.

Disconnect the 2P connector from the distributor.

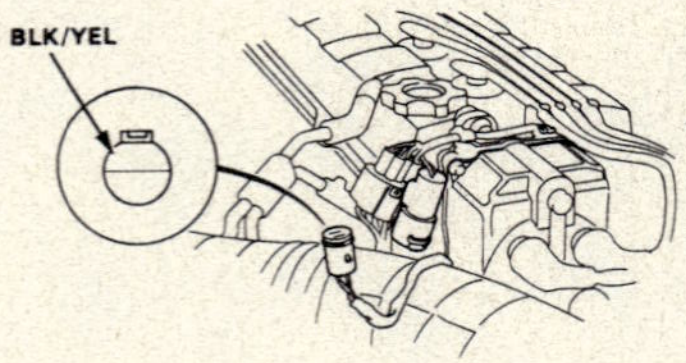

Turn the ignition switch ON.

Measure voltage between BLK/YEL (+) terminal and body ground.

Is there battery voltage ?
- NO → Repair open in BLK/YEL wire between the 2P connector and ignition switch.
- YES ↓

Turn the ignition switch OFF.

Reconnect the 2P connector.

TROUBLESHOOTING CHARTS—INTEGRA

Troubleshooting Flowchart

Self-diagnosis LED indicates code 16: A problem in the fuel injector circuit.

The injectors are the solenoid-actuated constant-stroke pintle type consisting of a solenoid, plunger needle valve and housing. When current is applied to the solenoid coil, the valve lifts up and pressurized fuel is injected close to the intake valve. Because the needle valve lift and the fuel pressure are constant, the injection quantity is determined by the length of time that the valve is open (i.e., the duration the current is supplied to the solenoid coil). The injector is sealed by an O-ring and seal ring at the top and bottom. These seals also reduce operating noise.

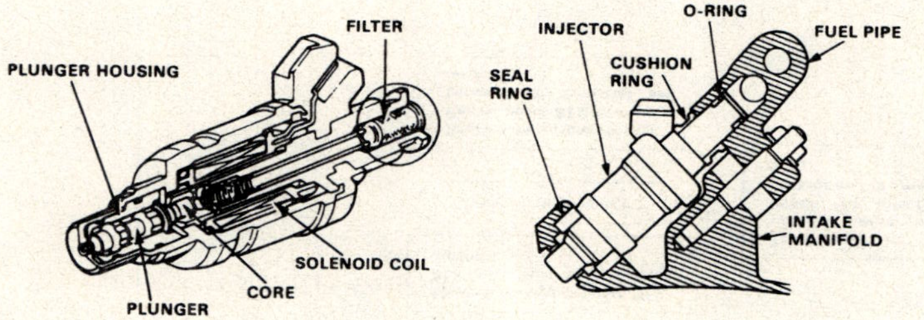

- **– Check Engine light is on.**
- **– LED indicates CODE 16.**

Do the ECU Reset Procedure

Turn the ignition switch to START position.

Does the engine start ?
- NO → **Intermittent failure, system is OK at this time (test drive may be necessary). Check for poor connections or loose wires at injectors, injector resistor and C318 (round connector on left shock tower).**
- YES → Is Check Engine light on and does LED indicate CODE 16?
  - NO → **Intermittent failure, system is OK at this time (test drive may be necessary). Check for poor connections or loose wires at injectors, injector resistor and C318 (round connector on left shock tower).**
  - YES → (continued)

Connect the PGM-FI test harness between the ECU and connector

Turn the ignition switch ON.

Measure voltage individually between B15 (+), B17 (+) terminals and A18 (–) terminal.

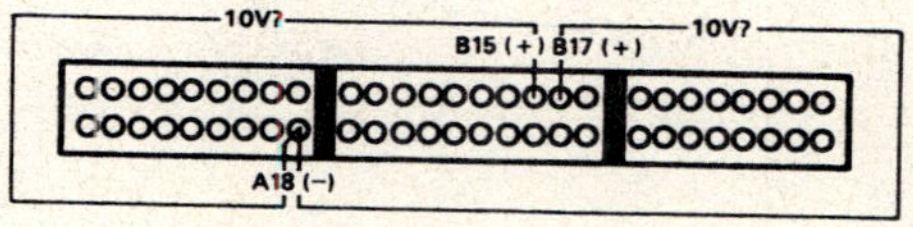

Is there approx. 10 V?
- NO → **–Replace the igniter unit.**
  –Repair open or short WHT wires between distributor and ECU (B15 or B17).

  NOTE: If the WHT wire was shorted, the igniter may be damaged.
- YES → **Substitute a known-good ECU and recheck. If symptom/indication goes away, replace the original ECU.**

Check the clicking sound of each injector by means of a stethoscope when the engine is idling.

Do the injectors click ?

YES → **Substitute a known-good ECU and recheck. If symptom/indication goes away, replace the original ECU.**

NO → Turn the ignition switch OFF.

Disconnect the 2P connector from the injector that does not click.

Measure resistance between the 2 terminals of the injector.

Turn the ignition switch OFF.

Disconnect the 2P connector from each injector.

Measure resistance between the 2 terminals of the injector.

Is there 1.5—2.5Ω ?

NO → **Replace the injector.**

YES → Turn the ignition switch ON.

Measure voltage between RED/BLK (+) terminal on the 2P connector and body ground.

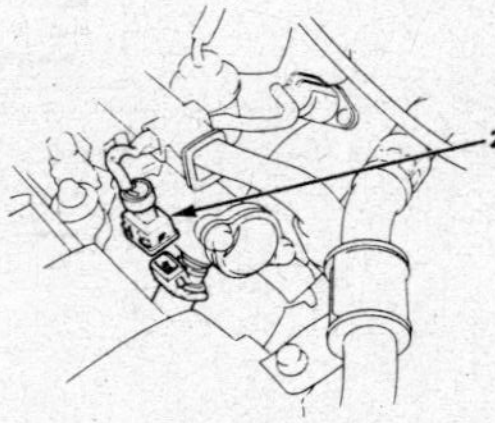

Is there battery voltage ?

NO → Turn the ignition switch OFF.

Disconnect 6P connector on the injector resistor.

Turn the ignition switch ON.

Measure voltage between YEL/BLK (+) terminal and body ground.

Is there battery voltage ?

NO → **Repair open in the YEL/BLK wire between the injector resistor and the main relay.**

YES →
- **Replace the injector resistor.**
- **Repair open RED/BLK wire between 2P connector and resistor.**

YES → Reconnect the 2P connector to the injector.

Turn the ignition switch OFF.

TROUBLESHOOTING CHARTS—INTEGRA

Connect the PGM-FI test harness between the ECU and connector (page 11-18).

Turn the ignition switch ON.

Measure voltage between A2 (−) terminal and following terminals.
- No.1 injector: A1 (+) terminal.
- No.2 injector: A3 (+) terminal.
- No.3 injector: A5 (+) terminal.
- No.4 injector: A7 (+) terminal.

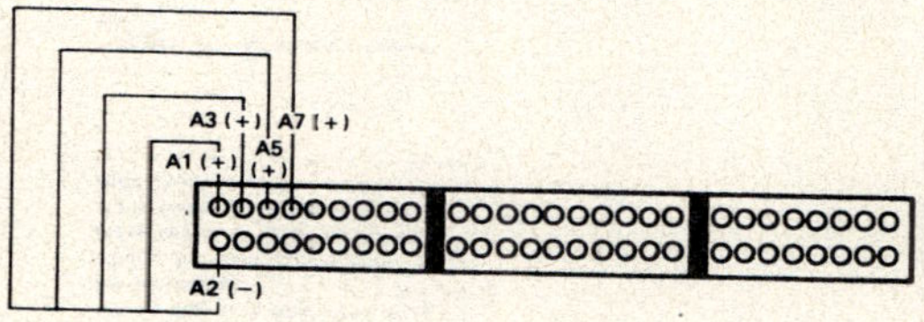

Is there battery voltage?

NO → Repair open in the wire between the ECU (A1, A3, A5 or A7) and the injector.

YES → **Substitute a known-good ECU and recheck. If symptom/indication goes away, replace the original ECU.**

Self-diagnosis LED indicates code 17: A problem in the Vehicle Speed Sensor circuit.

The signal generated by the speed sensor, which produces 4 pulses (switch closures to ground) for each revolution of the speedometer cable.

— Check Engine light is on.
— LED indicates CODE 17.

Do the ECU Reset Procedure

Road test necessary.
In 2nd gear accelerate to 3,500 rpm and decelerate to 1,500 rpm with throttle fully closed.

Is Check Engine light on and does LED indicate CODE 15?

NO → **Intermittent failure, system is OK at this time. Check for poor connections or loose wires at C401 and C714.**

YES → Block rear wheels and set the parking brake. Jack up the front of the car and support with safety stands.

⚠WARNING Block rear wheels before jacking up front of car.

Connect the PGM-FI test harness between the ECU and connector

Turn the ignition switch ON.

Slowly rotate left front wheel and measure voltage between B16 (+) terminal and A18 (−) terminal.

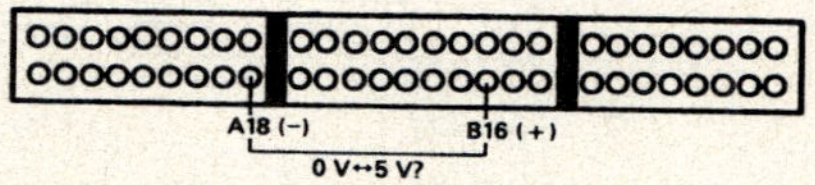

Does voltage pulse 0V and 5V?

NO →
— Repair open or short in YEL/RED wire between ECU (B16) and the speed sensor.
— Faulty speed sensor.
— Substitute a known-good ECU and recheck. If symptom/indication goes away, replace the original ECU.

YES → **Substitute a known-good ECU and recheck. If symptom/indication goes away, replace the original ECU.**

TROUBLESHOOTING CHARTS—INTEGRA

Self-diagnosis LED indicates code 43: A problem in the Oxygen (O_2) Sensor circuit or a problem in the Fuel Supply System.

NOTE: If 43 code is accompanied by poor driveability, go to Fuel Supply System.

- Check Engine light has been reported on.
- LED indicates CODE 43.
- or continued from CODE 1.

Do the ECU Reset Procedure

Warm up engine to normal operating temperature (cooling fan comes on).

Hold engine at 3,000 rpm for 2 minutes.
(A/T: Transmission is N or P.)

Is the Check Engine light on and does LED indicate CODE 43 ?

- NO → **Intermittent failure, system is OK at this time (test drive may be necessary). Check for poor connections or loose wires at O_2 sensor C122, C216 (round connector located at right shock tower), and ECU.**
- YES ↓

Inspect the fuel pressure

Is it normal ?

- NO → **Replace the fuel pressure regulator**
- YES ↓

Disconnect the O_2 sensor connector and measure voltage between the O_2 sensor connector and body ground.

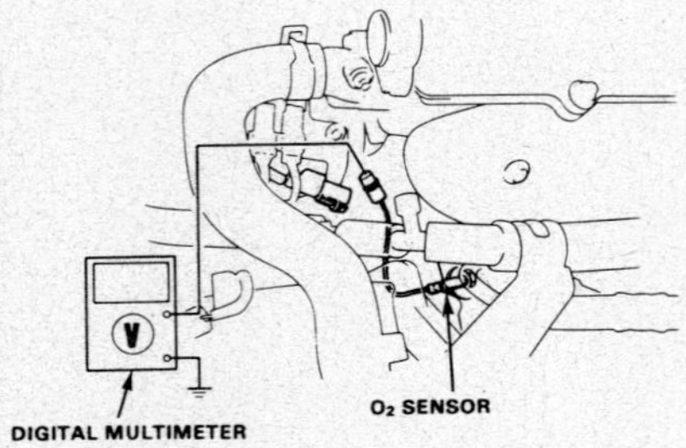

Is the voltage above 0.6 V at 4,500 rpm and below 0.4 V when the throttle is quickly released from 4,500 rpm ?

- NO → **Replace O_2 sensor.**
- YES ↓

Stop engine.

Connect the O_2 sensor connector to engine wire harness.

Connect the PGM-FI test harness between the ECU and connector

Restart and warm up engine to normal operating temperature (cooling fan comes on).

Measure voltage between C16 (+) and A18 (−) terminal.

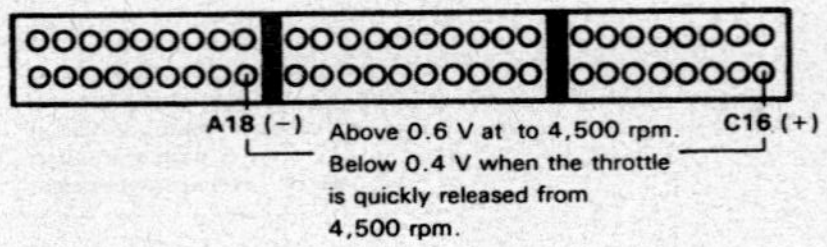

Is the voltage above 0.6 V at 4,500 rpm and 0.4 V when the throttle is quickly released from 4,500 rpm ?

- NO → **Repair short or open in white wire between ECU (C16) and O_2 sensor.**
- YES ↓

Substitute a known-good ECU and recheck. If symptom/indication goes away, replace the original ECU.

TROUBLESHOOTING CHARTS – INTEGRA

This signals the PGM-FI ECU when there is a demand for cooling from the air conditioning circuits.

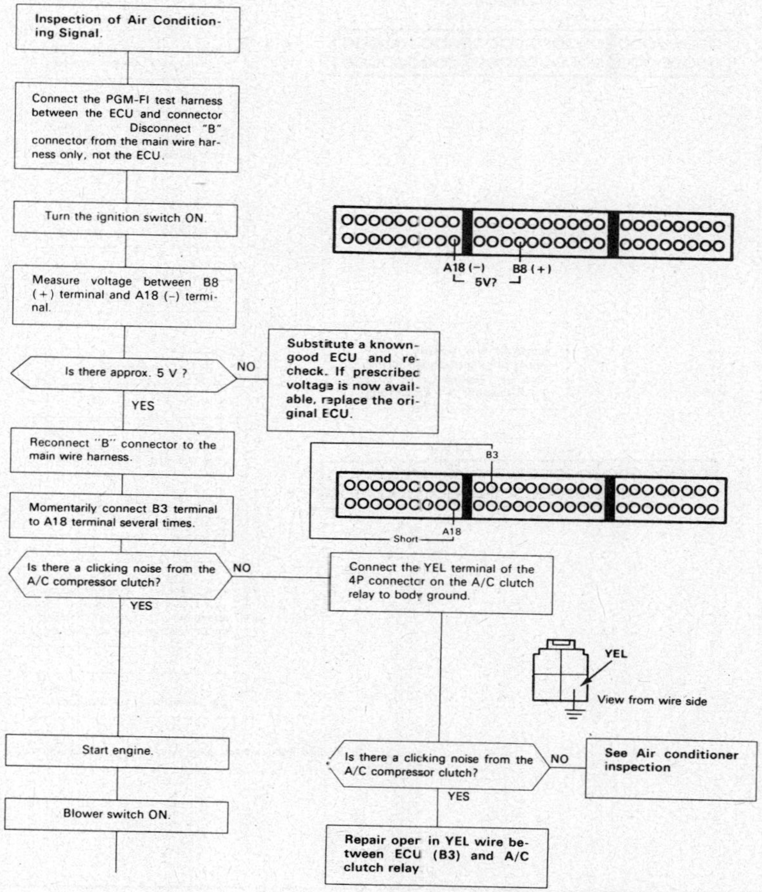

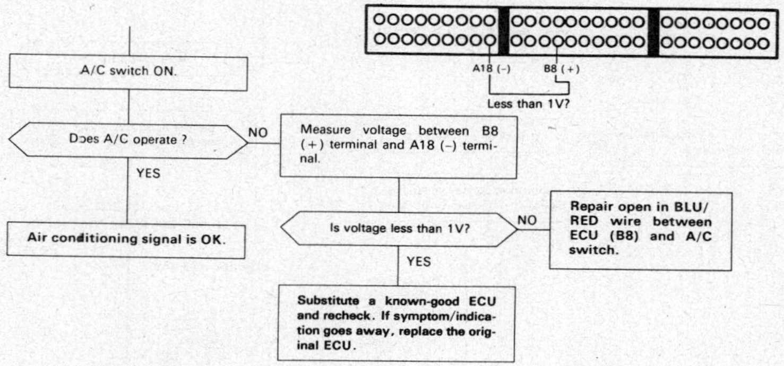

This signals the PGM-FI ECU when the alternator is charging.

Inspection of Alternator FR signal.

Connect the PGM-FI test harness between the ECU and connector Disconnect "B" connector from the main wire harness only, not the ECU.

Turn the ignition switch ON.

Measure voltage between B14 (+) terminal and A18 (−) terminal.

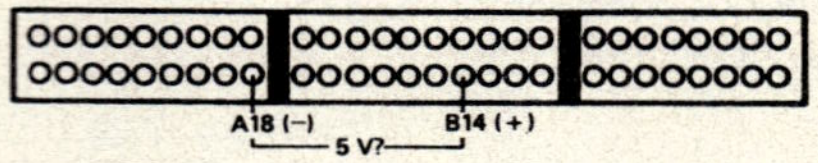

Is there approx. 5V ?

- NO → **Substitute a known-good ECU and recheck. If prescribed voltage is now available, replace the original ECU.**
- YES ↓

Turn the ignition switch OFF.

Reconnect "B" connector to the main wire harness.

Warm up engine to normal operating temperature (cooling fan comes on).

Measure voltage between B14 (+) terminal and A18 (−) terminal.

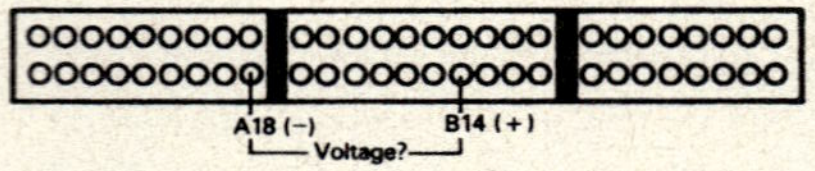

Does the voltage decrease when headlights and rear defogger are turned on ?

- YES → **Alternator FR signal is OK.**
- NO → Stop engine. ↓

Disconnect "B" connector from ECU only, not the main wire harness.

Disconnect the negative battery cable from the battery.

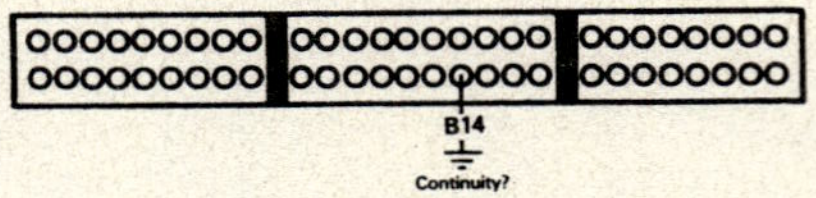

Check for continuity between B14 terminal and body ground.

Does continuity exist ?

- YES → Disconnect GRN connector from the alternator.
  - Check for continuity between B14 terminal and body ground.
  - Does continuity exist ?
    - NO → **Alternator Inspection**
    - YES → **Repair short in BLU wire between ECU (B14) and alternator.**
- NO → Disconnect GRN connector from the alternator.

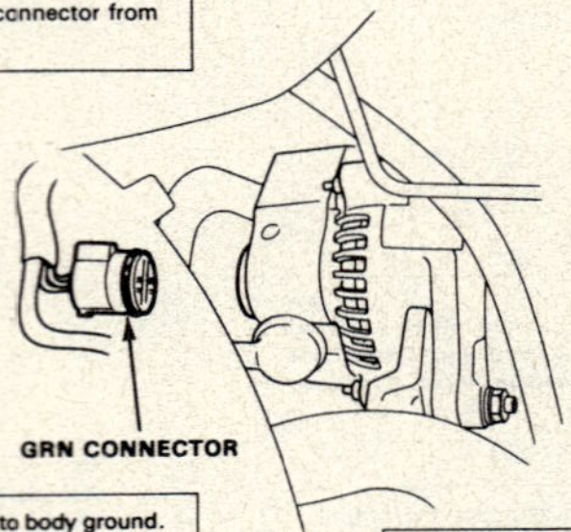

Connect BLU wire to body ground.

Check for continuity between B14 terminal and body ground.

Does continuity exist ?

- YES → **See Alternator Inspection**
- NO → **Repair open in BLU wire between ECU (B14) and alternator.**

TROUBLESHOOTING CHARTS—INTEGRA

This signals the PGM-FI ECU when the transmission is in Neutral Park.

Inspection of A/T Shift Position Signal

Turn the ignition switch ON.

Observe the A/T shift indicator and select each position separately.

Does the indicator light properly?
- NO → **A/T shift position Indicator Inspection**
- YES ↓

Turn the ignition switch OFF.

Connect the PGM-FI test harness between the ECU and connector Disconnect "B" connector from the main wire harness only, not the ECU.

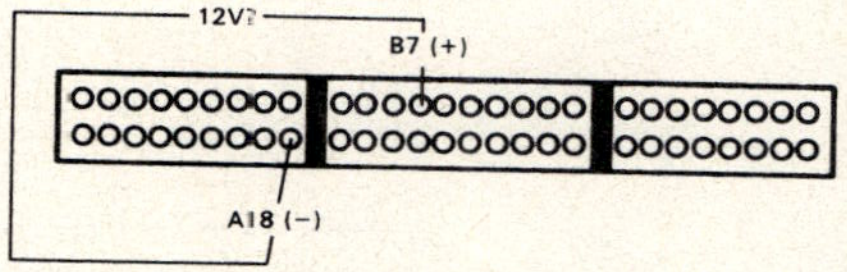

Turn the ignition switch ON.

Measure voltage between B7 (+) terminal and A18 (−) terminal.

Is there battery voltage?
- NO → **Substitute a known-good ECU and recheck If prescribed voltage is now available, replace the original ECU.**
- YES ↓

Turn the ignition switch OFF.

Reconnect "B" connector to the main wire harness.

Turn the ignition switch ON.

Measure voltage between B7 (+) terminal and A18 (−) terminal in Neutral positon.

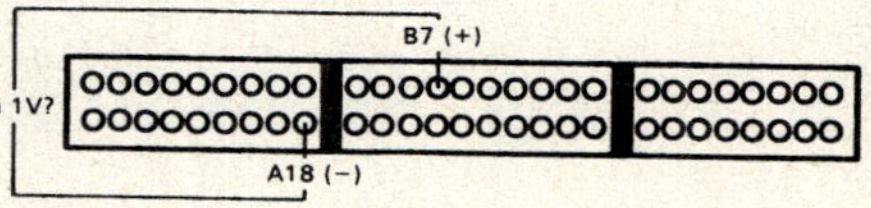

Is voltage less than 1V?
- NO → **–Repair open in GRN wire between ECU (B7) and gauge assembly. –Repair open in GRN wire between the gauge assembly and shift position cosole switch.**
- YES ↓

Measure voltage between B7 (+) terminal and A18 (−) terminal in Park position.

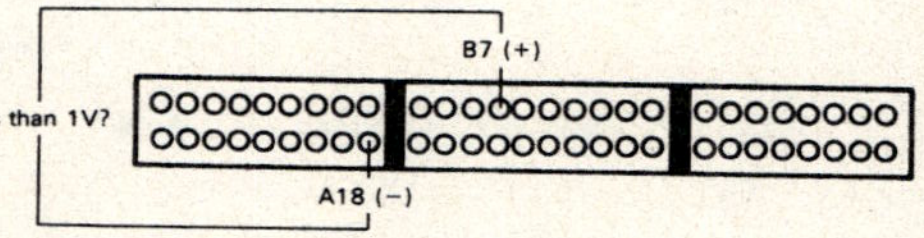

Is voltage less than 1V?
- NO → **Repair open in GRN/WHT wire between gauge assembly and shift position console switch.**
- YES ↓

Measure voltage between B7 (+) terminal and A18 (−) except in Park and Neutral.

Is there battery voltage?
- NO → **Repair short in GRN wire between ECU (B7) and gauge assembly.**
- YES ↓

A/T shift position signal is OK.

This signals the PGM-FI ECU when the brake pedal is depressed.

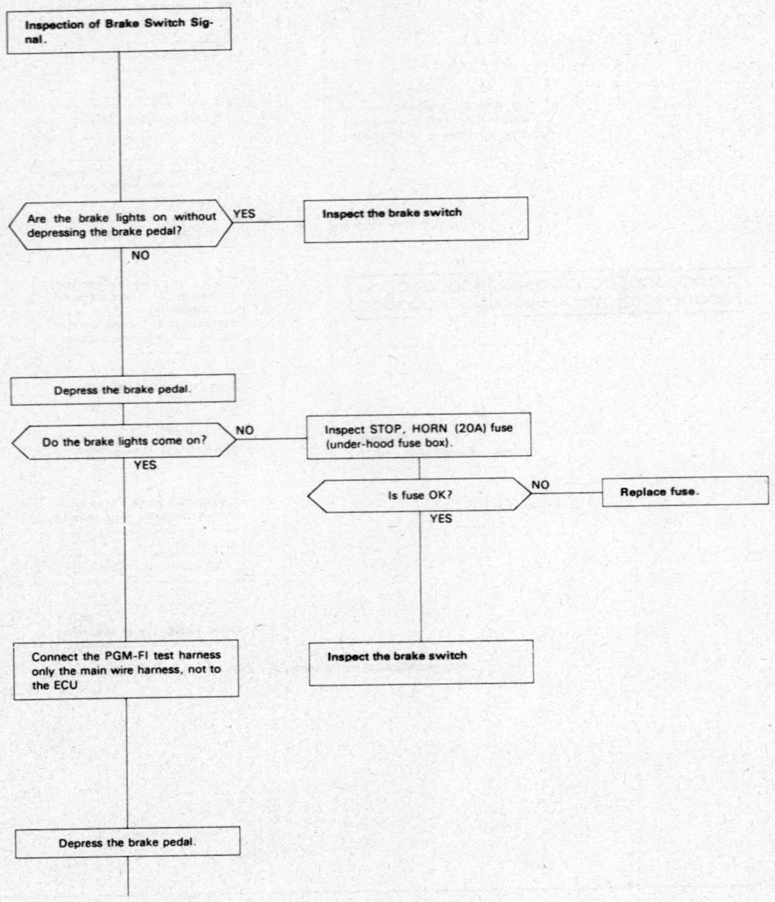

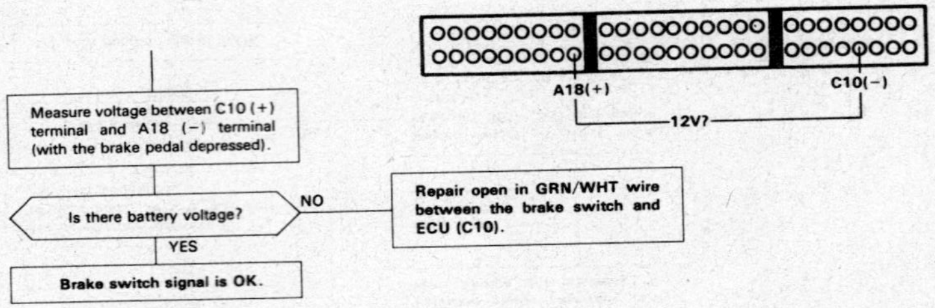

This signals the PGM-FI ECU when the engine is cranking.

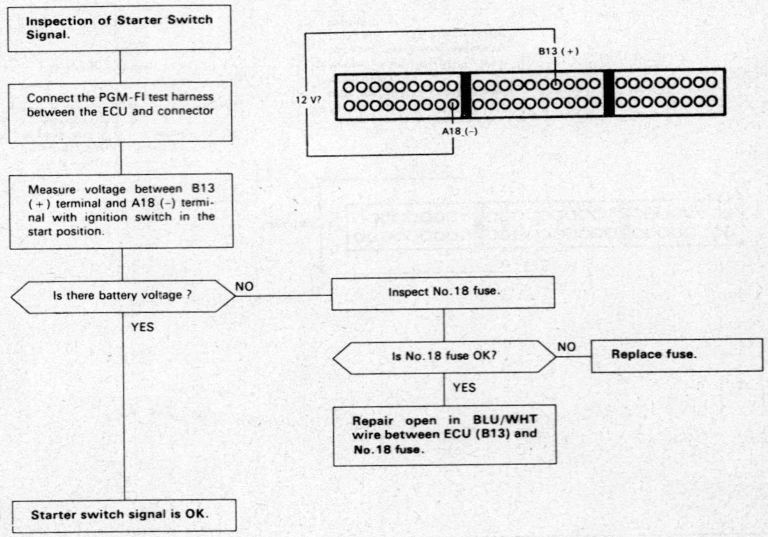

This signals the PGM-FI ECU when the power steering load is high.

Inspection of P/S Oil Pressure Signal

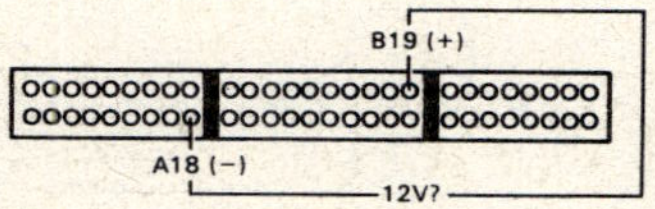

- Connect the PGM-FI test harness between the ECU and connector
- Start engine.
- Measure voltage between B19 (+) terminal and A18 (−) terminal.
- Is there less than 1V ?
  - YES → Disconnect the 2P connector on the P/S oil pressure switch.
    - Connect RED terminal to BLK terminal.
    - Is there less than 1V?
      - YES → **Replace P/S oil pressure switch.**
      - NO → **Repair open in RED wire between ECU (B19) and P/S oil pressure switch or BLK wire between P/S oil pressure switch and G301.**
  - NO → Turn steering wheel slowly.
    - Measure voltage between B19 (+) terminal and A18 (−) terminal while steering wheel is turning.
    - Is there battery voltage ?
      - YES → **P/S oil pressure signal is OK.**
      - NO → Turn the ignition switch OFF.
        - Disconnect "B" connector from main wire harness only, not the ECU.
        - Turn the ignition switch ON.
        - Is there battery voltage ?
          - NO → **Substitute a known-good ECU and recheck. If prescribed voltage is now available, replace the original ECU.**
          - YES → Reconnect "B" connector to main wire harness and disconnect 2P connector on the P/S oil pressure switch.
            - Is there battery voltage ?
              - YES → **Replace P/S oil pressure switch.**
              - NO → **Repair short in RED wire between ECU (B19) and the P/S oil pressure switch.**

Troubleshooting Flowchart

Inspection of Fast Idle Control Solenoid Valve.

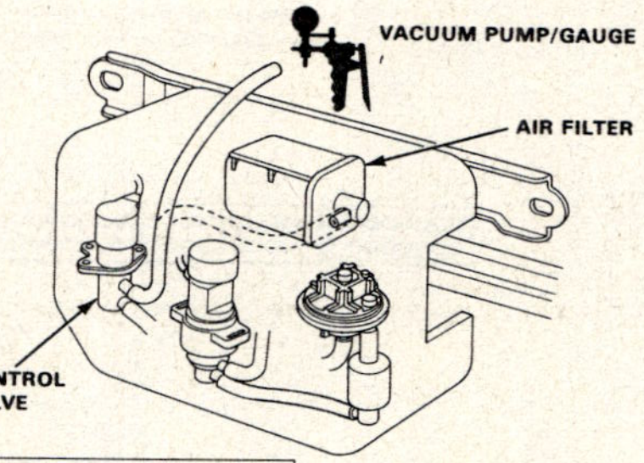

- Warm up engine to normal operating temperature (cooling fan comes on).
- Remove the control box cover, disconnect the vacuum hose from the air filter and connect a vacuum gauge to the hose.
- Is there any vacuum?
  - YES → Disconnect the 4P connector from the control box.
    - Is there any vacuum?
      - YES → **Replace the fast idle control solenoid valve.**
      - NO → **Repair short to ground at BLU wire between ECU (B2) and the 4P connector. If wire is OK, substitute a known-good ECU and recheck. If symptom goes away, replace the original ECU.**
  - NO → Turn the ignition switch OFF.
    - Disconnect the 4P connector from the control box.
    - Connect battery positive to terminal A and battery negative to terminal C of the control box.
    - Start the engine.
    - Is there manifold vacuum?
      - NO → **Replace the fast idle control solenoid valve.**
      - YES → **Fast Idle Control Solenoid Valve is OK.**

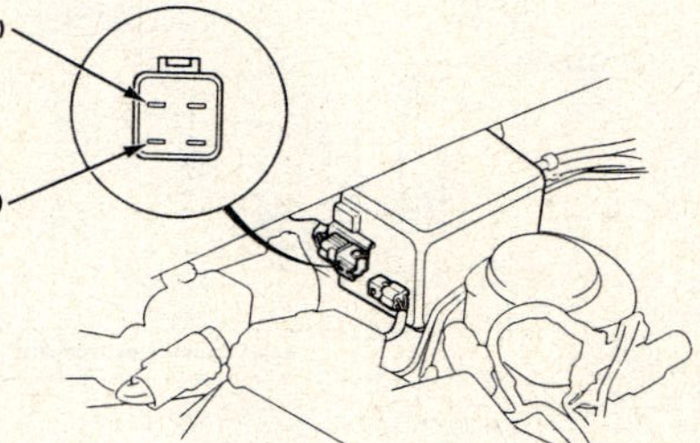

Pressure Regulator Cut-off Solenoid Valve Troubleshooting Flowchart

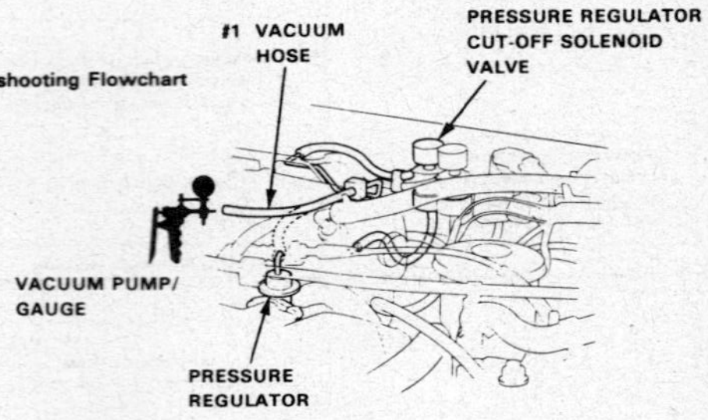

Inspection of Pressure Regulator Cut-off Solenoid Valve.

Start the engine and warm up to normal operating temperature (the cooling fan comes on).

Disconnect the #1 vacuum hose from the pressure regulator and connect a vacuum gauge to the hose.

Is there manifold vacuum ?

- NO → Disconnect the 4P connector from the solenoid valve
  - Is there vacuum ?
    - NO → **Check the #1 and #20 vacuum hose. If hoses are OK, replace the solenoid valve.**
    - YES → **Repair short to ground in GRN/YEL wire between ECU (A10) and 4P connector. If wire is OK, substitute a known-good ECU and recheck. If symptom goes away, replace the original ECU.**
- YES ↓

Turn the ignition switch OFF.

Disconnect the 4P connector from the solenoid valve.

Connect battery positive to terminal C and battery negative to terminal D of the solenoid valve.

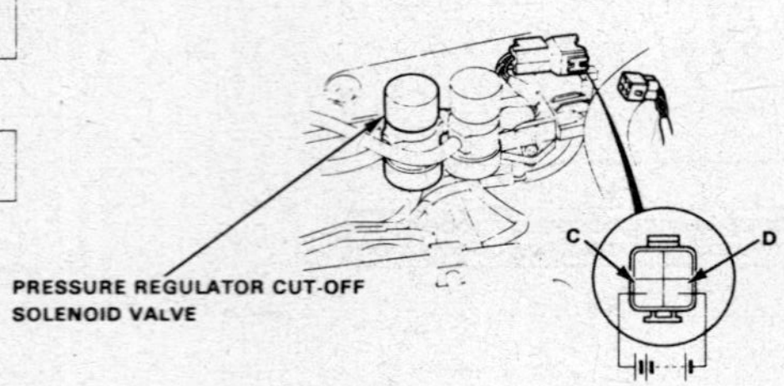

Start the engine and allow it to idle.

Is there manifold vacuum ?

- YES → **Replace the solenoid valve.**
- NO ↓

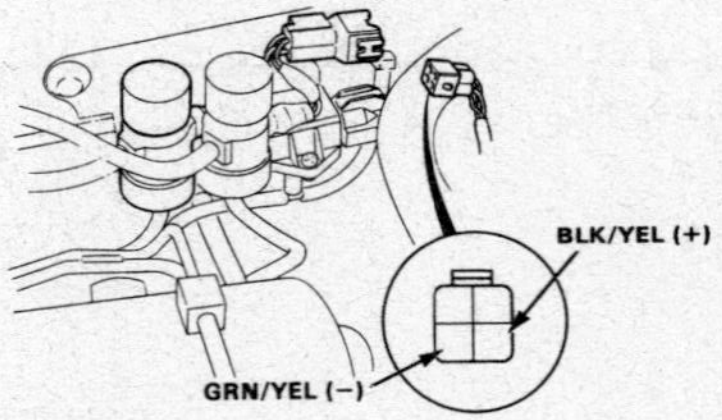

Measure voltage between BLK/YEL (+) terminal and body ground.

Is there battery voltage ?

- NO → **Repair in BLK/YEL wire between No. 24 fuse and 4P connector.**
- YES ↓

Turn the ignition switch OFF.

Reconnect the 4P connector to the solenoid valve.

Connect the PGM-FI test harness between the ECU and connector

Start the engine and allow it to idle.

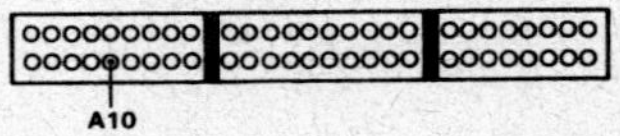

Connect A10 terminal to body ground with a jumper wire.

Is there manifold vacuum ?

- YES → **Repair open in GRN/YEL wire between ECU (A10) and the 4P connector.**
- NO ↓

Pressure Regulator Cut-off Solenoid Valve is OK.

Inspection of Evaporative Emission Controls

Disconnect vacuum hose from the purge control diaphragm valve (on the charcoal canister) and connect a vacuum gauge to the hose.

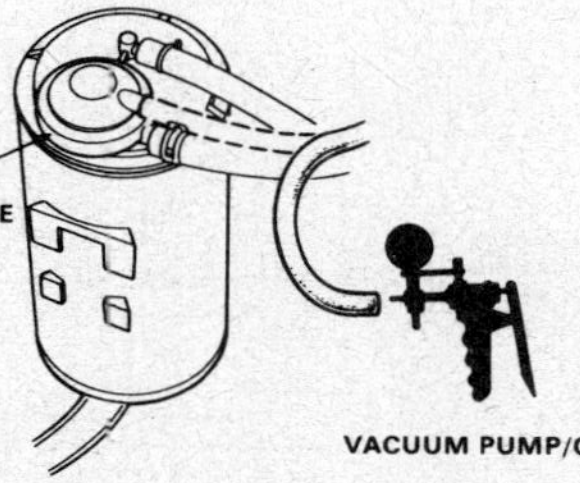

Start the engine and allow to idle.
NOTE: Engine coolant temperature must be below 73°C (163°F).

Is there vacuum ?

- YES → Disconnect the 4P connector.

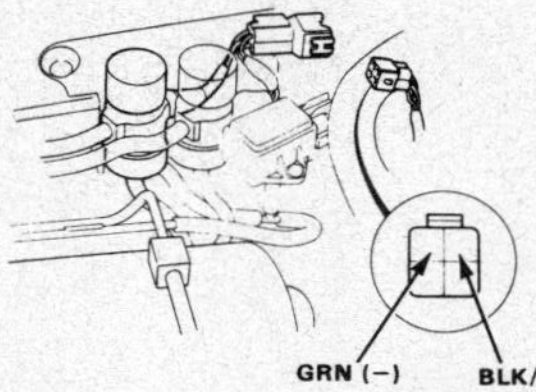

  Measure voltage between BLK/YEL (+) terminal and GRN (–) terminal.

  Is there battery voltage ?
  - YES → **Inspect vacuum hose routing. If OK, replace purge cut-off solenoid valve.**
  - NO → Measure voltage between BLK/YEL (+) terminal and body ground.

    Is there battery voltage ?
    - NO → **Repair open in BLK/YEL wire between No. 24 fuse and the 4P connector.**
    - YES → **Inspect GRN wire for a short to body ground between ECU (A6) and the connector. If wire is OK, substitute a known-good ECU and recheck. If symptom/indication goes away, replace the original ECU.**

- NO → Warm up the engine to normal operating temperature (cooling fan comes on).

  Check for vacuum at vacuum hose 10 seconds after starting the engine.

  Is there manifold vacuum?
  - NO → Disconnect the 4P connector.

    Is there manifold vacuum?
    - NO → **Inspect vacuum hose routing. If OK, purge cut-off solenoid valve.**
    - YES → **Inspect for an open in GRN wire between ECU (A6) and the connector. If wire is OK, substitute a known-good ECU and recheck. If prescribed voltage is now available, replace the original ECU.**
  - YES → Reconnect the hose.

    Remove fuel filler cap.

TROUBLESHOOTING CHARTS – INTEGRA

Connect a vacuum gauge to canister purge air hose.

Start the engine and raise speed to 3,500 rpm.

Does vacuum appear on gauge within 1 minute?

- YES → two-way valve test to complete. Evaporative emission control are OK.
- NO → Connect a vacuum gauge to the canister purge hose and raise the engine speed to 3,500 rpm.
  - Does vacuum appear on the gauge?
    - NO → Inspect the purge hose and throttle body port for pinch or blockage.
    - YES → Replace the canister.

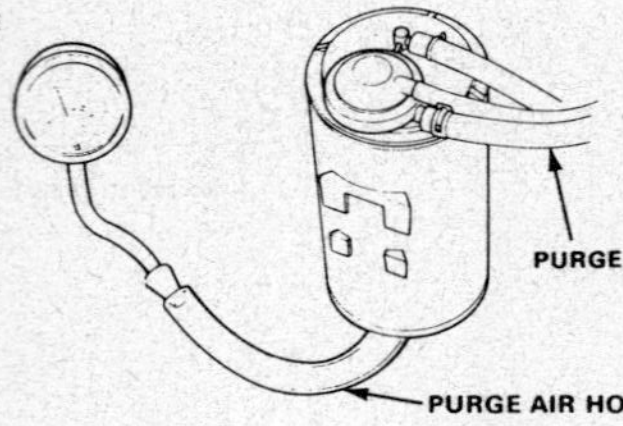

VACUUM COMPONENT AND HOSE ROUTING 1991–92 LEGEND

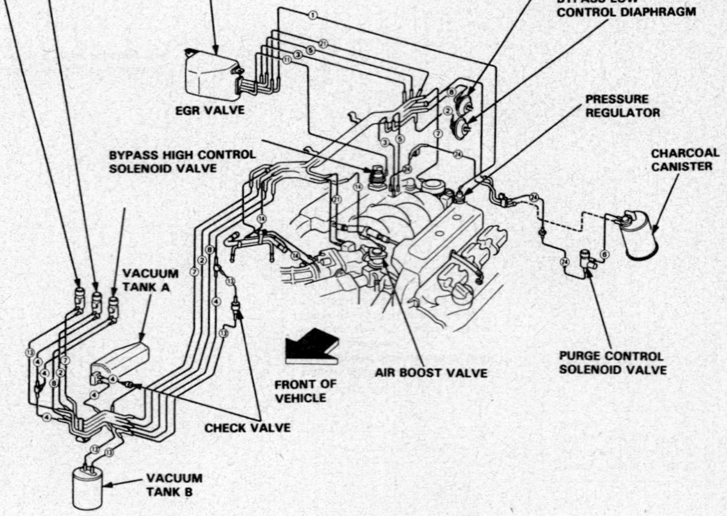

CONTROL BOX COMPONENTS—1991–92 LEGEND

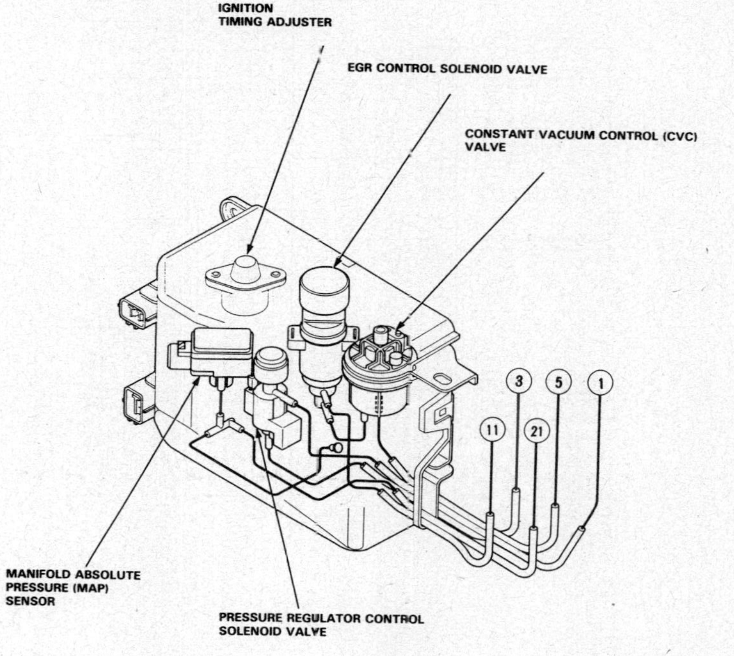

ECU INPUT/OUTPUT WIRING SCHEMATIC 1991–92 LEGEND

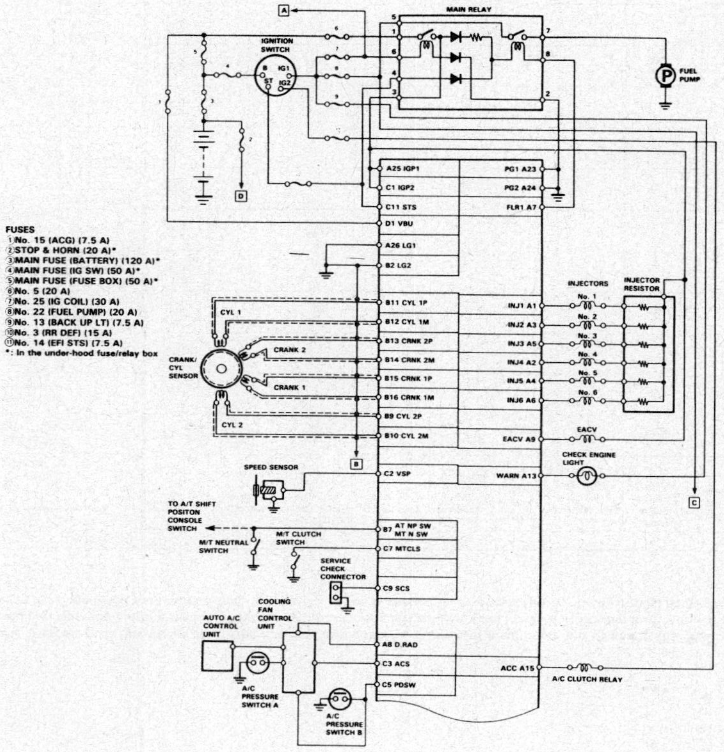

ECU INPUT/OUTPUT WIRING SCHEMATIC 1991–92 LEGEND

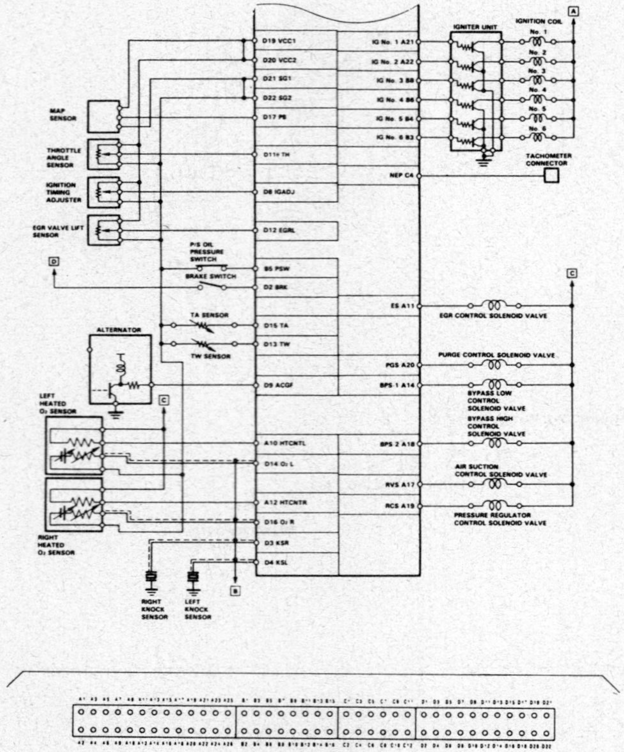

TERMINAL LOCATION

SYMPTOM TO SYSTEM DIAGNOSIS 1991–92 LEGEND

NOTE: Across each row in the chart, the systems that could be sources of a symptom are ranked in the order they should be inspected starting with ①. Find the symptom in the left column, read across to the most likely source, then refer to the page listed at the top of that column. If inspection shows the system is OK, try the next most likely system ②, etc.

SYMPTOM \ SYSTEM		ECU	OXYGEN SENSOR	MANIFOLD ABSOLUTE PRESSURE SENSOR	CRANK/CYL SENSOR	COOLANT TEMPERATURE SENSOR	THROTTLE ANGLE SENSOR	INTAKE AIR TEMPERATURE SENSOR	ATMOSPHERIC PRESSURE SENSOR	IGNITION OUTPUT SIGNAL
			PGM-FI							
CHECK ENGINE LIGHT** TURNS ON		▭ or (lit)	(blinking)	(blinking)	(blinking)	(blinking)	(blinking)	(blinking)	(blinking)	(blinking)
CHECK ENGINE LIGHT BLINKS		0 or *	1 or 2 / 41 or 42	3 or 5	4 or 54 / 9 or 59	6	7	10	13	15
ENGINE WON'T START		①				③				③
DIFFICULT TO START ENGINE WHEN COLD		BU		③	③	①			③	
IRREGULAR IDLING	WHEN COLD FAST IDLE OUT OF SPEC	BU				②				
	ROUGH IDLE	BU		③						
	WHEN WARM RPM TOO HIGH	BU				②				
	WHEN WARM RPM TOO LOW	BU								
FREQUENT STALLING	WHILE WARMING UP	BU				③				
	AFTER WARMING UP	BU							③	
POOR PERFORMANCE	MISFIRE OR ROUGH RUNNING	BU		②	③					
	FAILS EMISSION TEST	BU	③	②						
	LOSS OF POWER	BU		③			②			

* If codes other than those listed above are indicated, count the number of blinks again. If the indicator is in fact blinking these codes, substitute a known-good ECU and recheck. If the indication goes away, replace the original ECU.

(BU): When the Check Engine light and the self-diagnosis indicator are on, the back-up system is in operation. Substitute a known-good ECU and recheck. If the indication goes away, replace the original ECU.

** Check Engine light

USA: [CHECK] CANADA: [engine symbol]

SYMPTOM TO SYSTEM DIAGNOSIS
1991–92 LEGEND

PGM-FI			IDLE CONTROL		FUEL SUPPLY			AIR INTAKE	EMISSION CONTROL	
VEHICLE SPEED SENSOR	IGNITION TIMING ADJUSTER	KNOCK SENSOR	ELEC-TRONIC AIR CONTROL VALVE	OTHER IDLE CONTROLS	FUEL INJECTOR	FUEL METERING	OTHER FUEL SUPPLY		EGR CONTROL SYSTEM	OTHER EMISSION CONTROLS
CHECK	CHECK	CHECK	CHECK		CHECK	CHECK			CHECK	
17	18	23 or 53	14		16	43 or 44 45 or 46			12	
					②		③			
				①			②			
				①						
				①	②					
				1						
				1	②					
				1			②			
				1			②		③	
			③		1				③	
					②	③	③			1
		③			③		1	③		③

SYMPTOM TO SYSTEM DIAGNOSIS
1991–92 LEGEND

NOTE:
- Across each row in the chart, the sub systems that could be sources of a symptom are ranked in the order they should be inspected, starting with ①. Find the symptom in the left column, read across to the most likely source, then refer to the page listed at the top of that column. If inspection shows the system is OK, try the next system ②, etc.
- If the idle speed is out of specification and the Check Engine light does not blink CODE 14, go to inspection

SYMPTOM \ SUB SYSTEM		IDLE ADJUST-ING SCREW	EACV	AIR CONDI-TIONING SIGNAL	ALTER-NATOR FR SIGNAL	A/T SHIFT POSITION SIGNAL	M/T NEUTRAL SWITCH SIGNAL	M/T CLUTCH SWITCH SIGNAL	STARTER SWITCH SIGNAL	BRAKE SWITCH SIGNAL	P/S OIL PRESSURE SWITCH SIGNAL	FAST IDLE VALVE	AIR BOOST VALVE	HOSES AND CONNEC-TIONS
DIFFICULT TO START ENGINE WHEN COLD												①	②	
WHEN COLD FAST IDLE OUT OF SPEC (1,000–2,000 rpm)		③	②									①		
ROUGH IDLE			②											①
WHEN WARM RPM TOO HIGH		③	①								③	②		③
WHEN WARM RPM TOO LOW	Idle speed is below specified rpm (no load)	②	①											
	Idle speed does not increase after initial start up.		①											
	On models with automatic transmission, the idle speed drops in gear		②			①								
	Idle speeds drops when air conditioner in ON		②	①										
	Idle speed drops when steering wheel is turning		②								①			
	Idle speed fluctuates with electrical load		②		③									①
FREQUENT STALLING	WHILE WARMING UP	②	①											
	AFTER WARMING UP	①	②											
FAILS EMISSION TEST														①

SYMPTOM TO SYSTEM DIAGNOSIS – 1991–92 LEGEND

NOTE: Across each row in the chart, the systems that could be sources of a symptom are ranked in the order they should be inspected starting with ①. Find the symptom in the left column, read across to the most likely source, then refer to the page listed at the top of that column. If inspection shows the system is OK, try the next most likely system ②, etc.

SYMPTOM \ SUB SYSTEM		FUEL INJECTOR	INJECTOR RESISTOR	PRESSURE REGULATOR	FUEL FILTER	FUEL PUMP	MAIN RELAY	CONTAMINATED FUEL
ENGINE WON'T START		③	③		③	①	②	③
DIFFICULT TO START ENGINE WHEN COLD OR HOT								①
ROUGH IDLE		①	②					③
POOR PERFORMANCE	MISFIRE OR ROUGH RUNNING	①	②	③		③		③
	FAILS EMISSION TEST	②	③	①				
	LOSS OF POWER	③	③		①	③		②
FREQUENT STALLING	WHILE WARMING UP			①				
	AFTER WARMING UP			①				

NOTE: Across each row in the chart, the sub systems that could be sources of a symptom are ranked in the order they should be inspected starting with ①. Find the symptom in the left column, read across to the most likely source, then refer to the page listed at the top of that column. If inspection shows the system is OK, try the next system ②, etc.

SYMPTOM \ SUB SYSTEM	THROTTLE CABLE	THROTTLE BODY	CHAMBER VOLUME CONTROL SYSTEM
WHEN COLD FAST IDLE OUT OF SPEC	③	②	①
WHEN WARM RPM TOO HIGH	②	①	
LOSS OF POWER		①	②

NOTE: Across each row in the chart, the systems that could be sources of a symptom are ranked in the order they should be inspected starting with ①. Find the symptom in the left column, read across to the most likely source, then refer to the page listed at the top of that column. If inspection shows the system is OK, try the next most likely system ②, etc.

SYMPTOM \ SUB SYSTEM		CATALYTIC CONVERTER	EGR SYSTEM	AIR INJECTION SYSTEM	POSITIVE CRANKCASE VENTILATION SYSTEM	EVAPORATIVE EMISSION CONTROLS
ROUGH IDLE			①		②	
FREQUENT STALLING	AFTER WARMING UP		①			
POOR PERFORMANCE	MISFIRE OR ROUGH RUNNING		①			
	FAILS EMISSION TEST	①	③			②
	LOSS OF POWER	①	②	③		

TROUBLESHOOTING CHARTS—1991–92 LEGEND

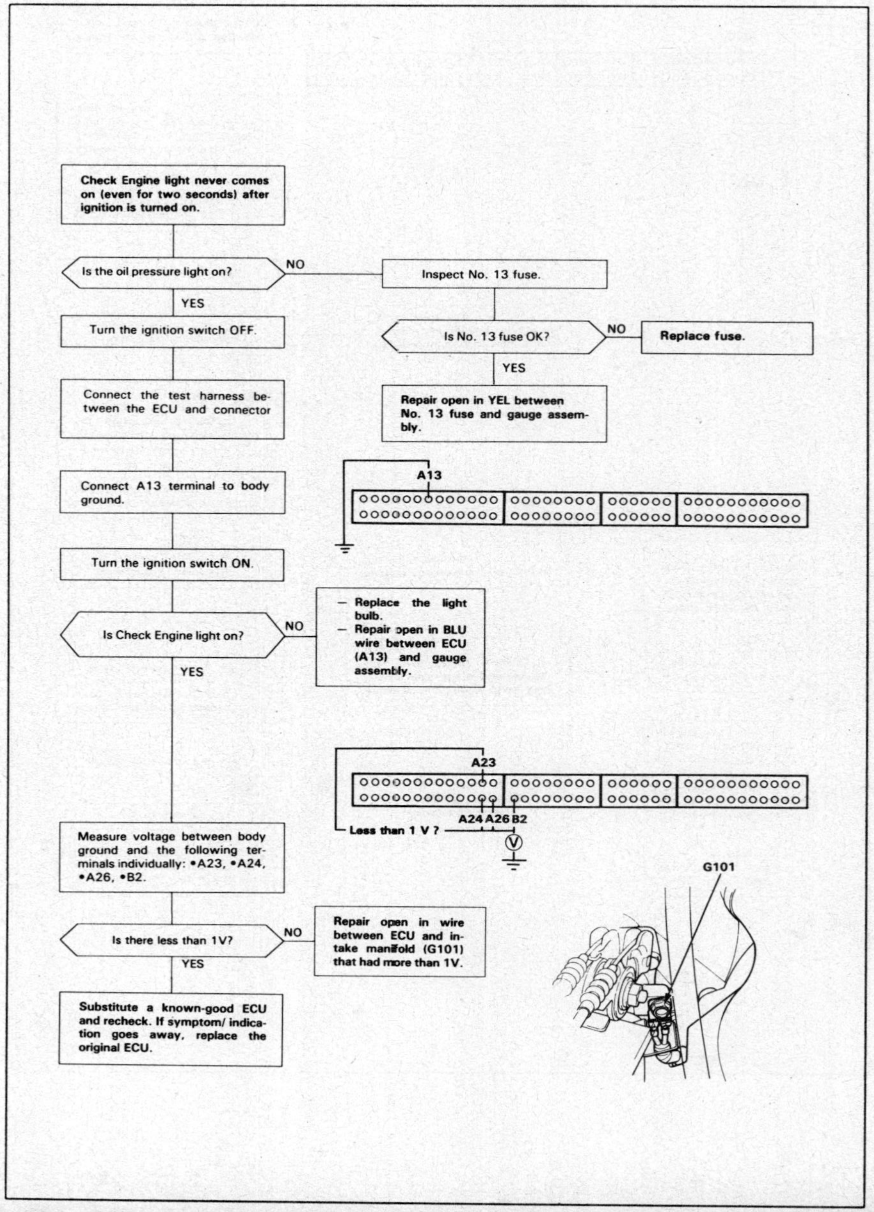

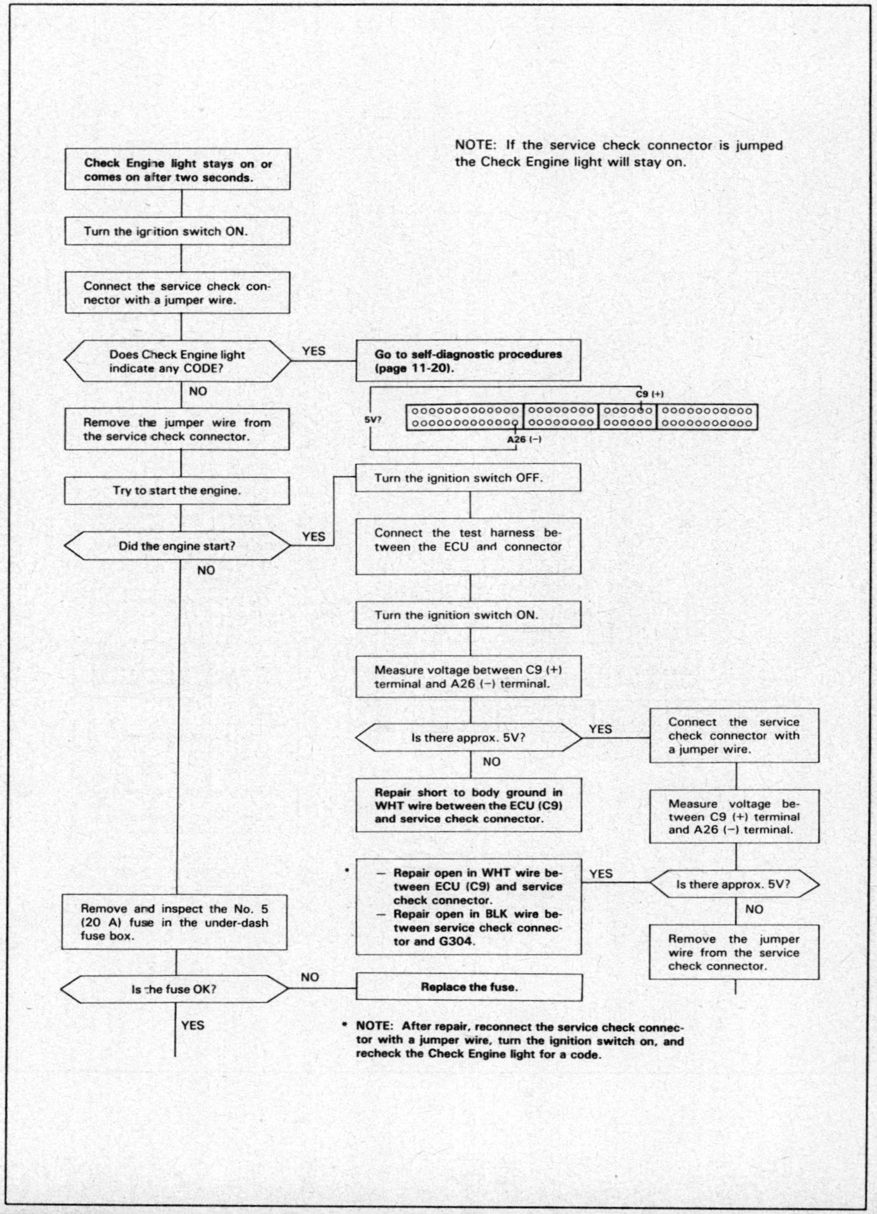

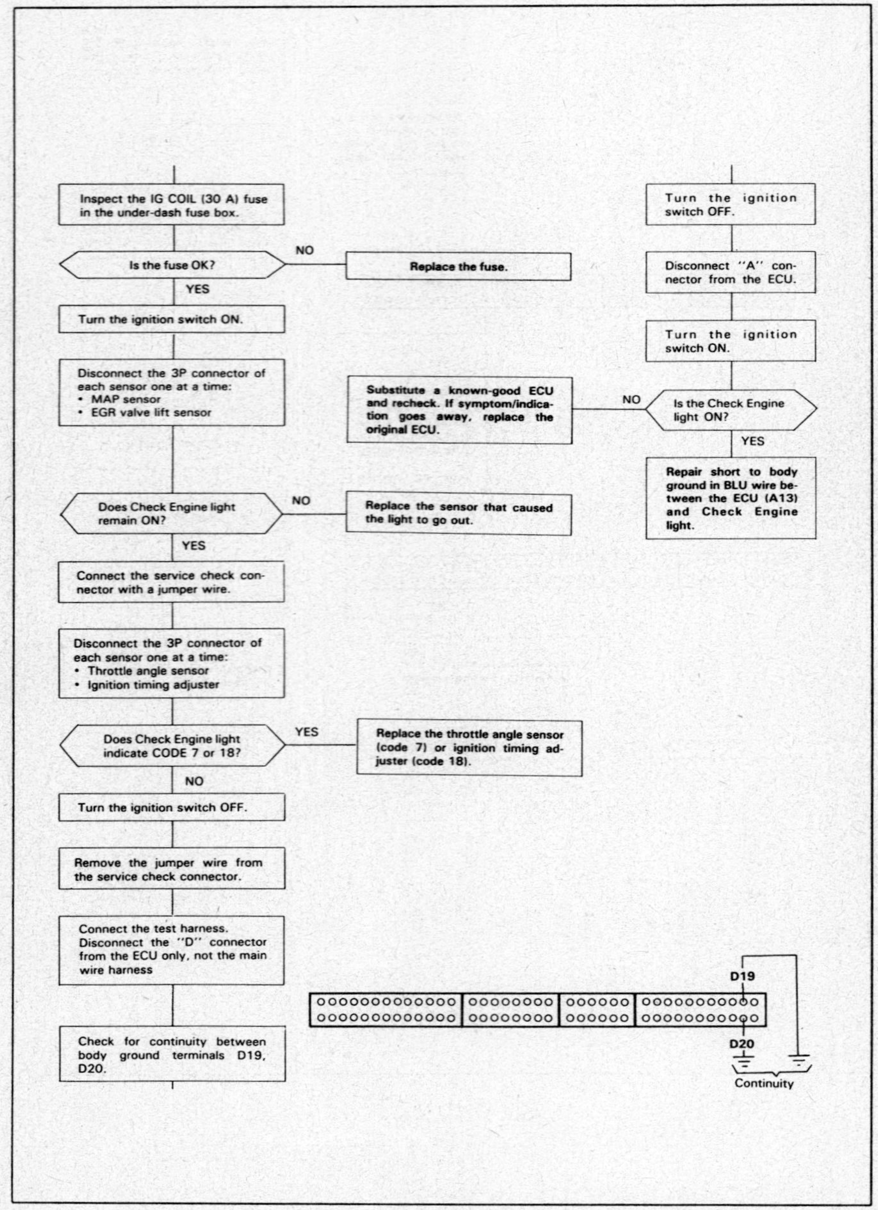
Inspect the IG COIL (30 A) fuse in the under-dash fuse box.
Is the fuse OK?
NO
Replace the fuse.
YES
Turn the ignition switch ON.
Disconnect the 3P connector of each sensor one at a time:
• MAP sensor
• EGR valve lift sensor
Does Check Engine light remain ON?
NO
Replace the sensor that caused the light to go out.
YES
Connect the service check connector with a jumper wire.
Disconnect the 3P connector of each sensor one at a time:
• Throttle angle sensor
• Ignition timing adjuster
Does Check Engine light indicate CODE 7 or 18?
YES
Replace the throttle angle sensor (code 7) or ignition timing adjuster (code 18).
NO
Turn the ignition switch OFF.
Remove the jumper wire from the service check connector.
Connect the test harness. Disconnect the "D" connector from the ECU only, not the main wire harness
Check for continuity between body ground terminals D19, D20.
D19
D20
Continuity
Turn the ignition switch OFF.
Disconnect "A" connector from the ECU.
Turn the ignition switch ON.
Is the Check Engine light ON?
NO
Substitute a known-good ECU and recheck. If symptom/indication goes away, replace the original ECU.
YES
Repair short to body ground in BLU wire between the ECU (A13) and Check Engine light.

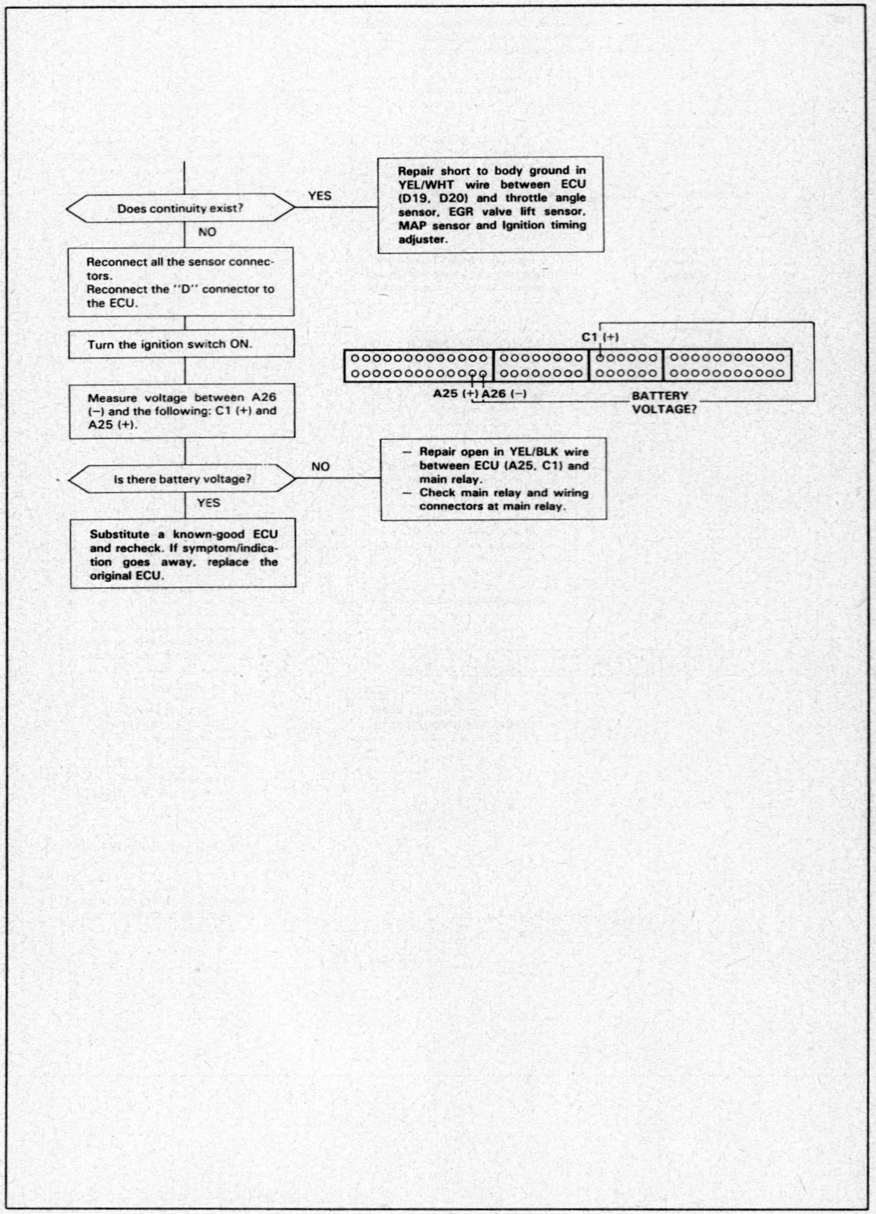
Does continuity exist?
YES
Repair short to body ground in YEL/WHT wire between ECU (D19, D20) and throttle angle sensor, EGR valve lift sensor, MAP sensor and Ignition timing adjuster.
NO
Reconnect all the sensor connectors.
Reconnect the "D" connector to the ECU.
Turn the ignition switch ON.
Measure voltage between A26 (−) and the following: C1 (+) and A25 (+).
C1 (+)
A25 (+) A26 (−)
BATTERY VOLTAGE?
Is there battery voltage?
NO
– Repair open in YEL/BLK wire between ECU (A25, C1) and main relay.
– Check main relay and wiring connectors at main relay.
YES
Substitute a known-good ECU and recheck. If symptom/indication goes away, replace the original ECU.

1 Self-diagnosis Check Engine light indicates code 1: A problem in the Left Oxygen (O_2) Sensor circuit.

2 Self-diagnosis Check Engine light indicates code 2: A problem in the Right Oxygen (O_2) Sensor circuit.

The oxygen sensors detect the oxygen content in the exhaust gas and signal the ECU. In operation, the ECU receives the signals from the sensors and varies the duration during which fuel is injected. The oxygen sensors have internal heaters. The heaters stabilize the sensors' output. The oxygen sensors are installed in the exhaust manifolds.

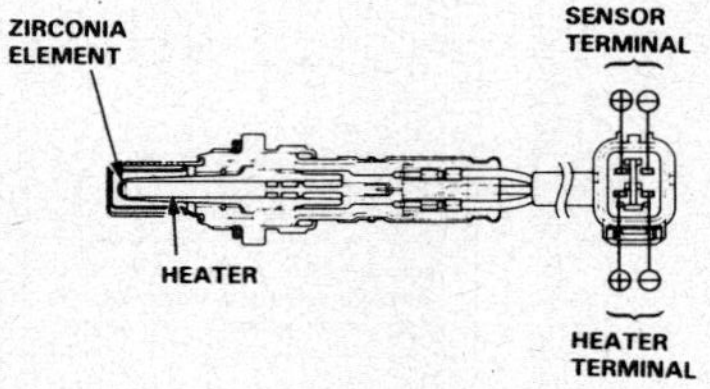

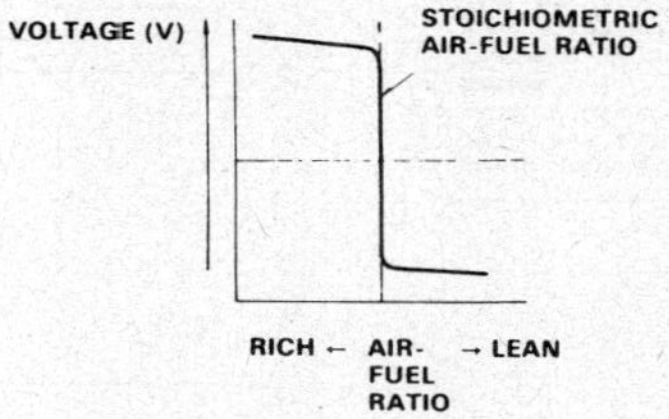

1 or 2

- Check Engine light has been reported on.
- With service check connector jumped (page 11-18), CODE 1 (Left O_2 sensor) and/or 2 (Right O_2 sensor) are indicated.

Do the ECU Reset Procedures

Warm up engine to normal operating temperature (cooling fan comes on).

Run engine for 60 seconds.

Road test with the transmission in 2nd gear (M/T: 4th gear). Starting at 1200 rpm accelerate using wide open throttle for at least 5 seconds. Then decelerate for at least 5 seconds with the throttle completely closed.

Is Check Engine light on and does it indicate CODE 1 or CODE 2?

NO → Intermittent failure, system is OK at this time. Check for poor connections or loose wires at the C353, (located at right shock tower), C111, C123 (O_2 sensors) and ECU.

YES → Go to test for CODE 43 or CODE 44.

3 Self-diagnosis Check Engine light indicates code 3: An electrical problem in the Manifold Absolute Pressure (MAP) Sensor system.

5 Self-diagnosis Check Engine light indicates code 5: Most likely mechanical problem (broken hose) in the Manifold Absolute Pressure (MAP) Sensor System.

The MAP sensor converts manifold absolute pressure into electrical signals and inputs the ECU.

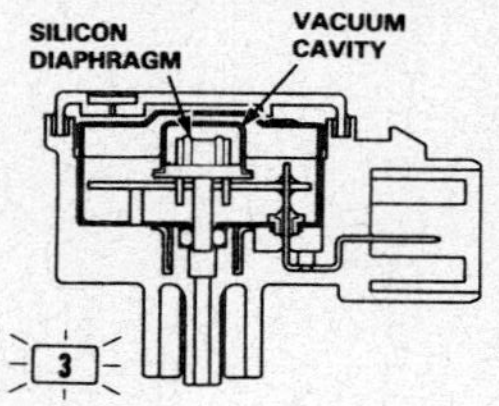

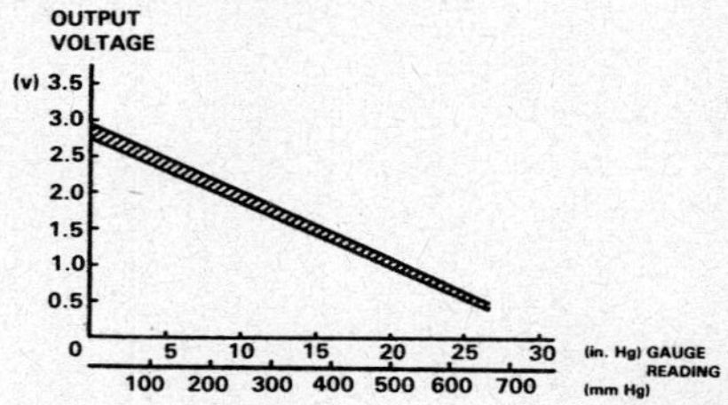

3

- Engine is warm and running.
- Check Engine light has been reported on.
- With service check connector jumped CODE 3 is indicated.

Do the ECU Reset Procedures

Start the engine and allow it to idle.

Is Check Engine light on and does it indicate CODE 3?

NO → Intermittent failure, system is OK at this time (test drive may be necessary). Check for poor connection or loose wires at C360 (MAP sensor) and ECU.

YES ↓

Turn the ignition switch OFF.

Disconnect the 3P connector from the MAP sensor.

Turn the ignition switch ON.

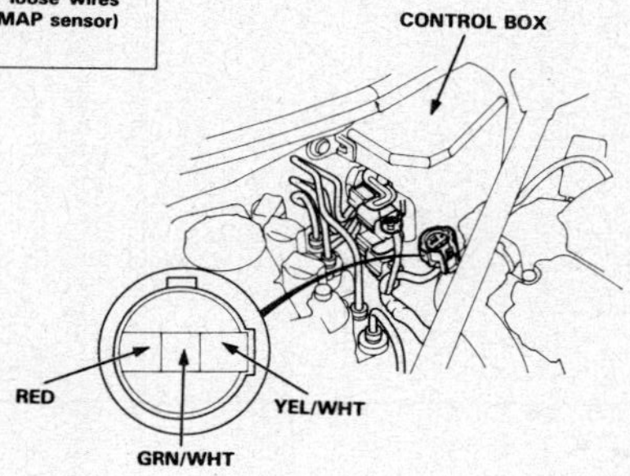

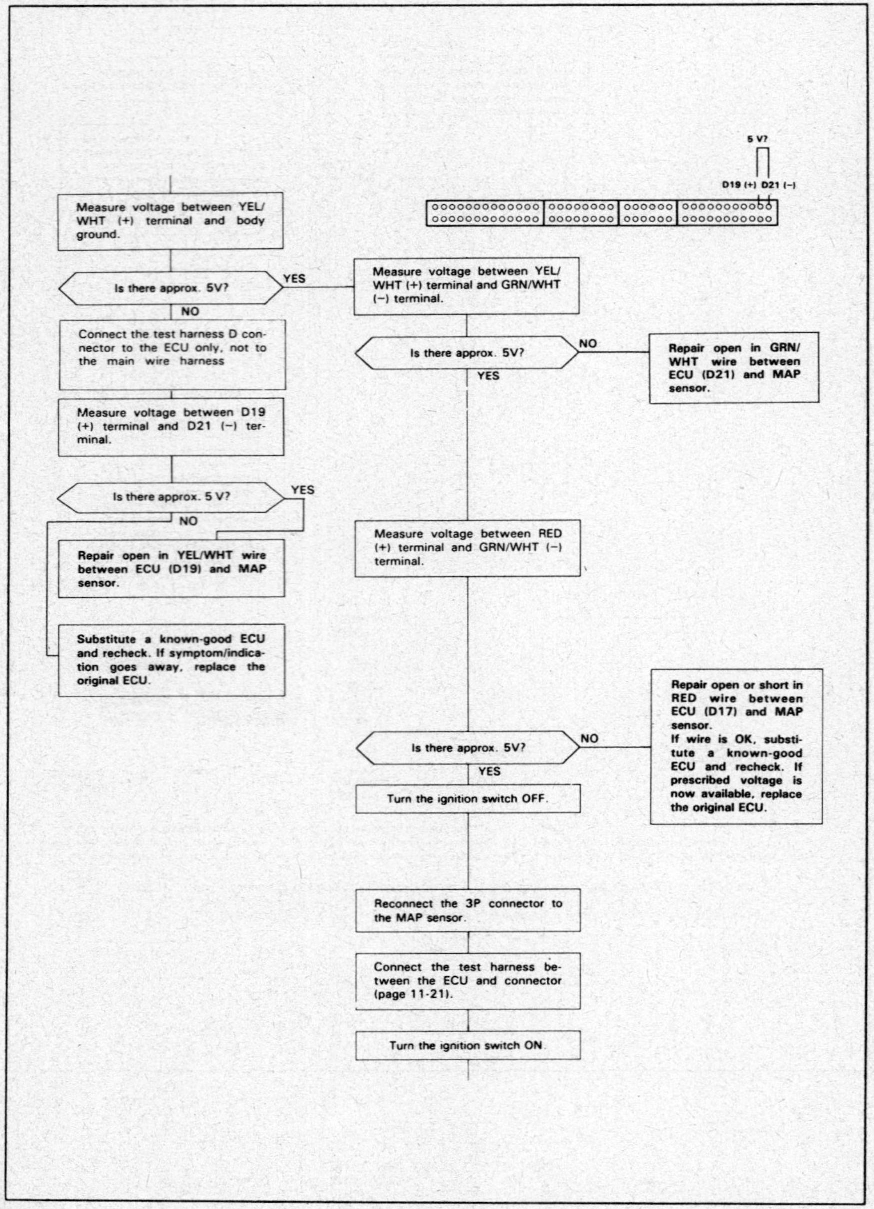
Measure voltage between YEL/WHT (+) terminal and body ground.
Is there approx. 5V?
YES
NO
Connect the test harness D connector to the ECU only, not to the main wire harness
Measure voltage between D19 (+) terminal and D21 (−) terminal.
Is there approx. 5 V?
YES
NO
Repair open in YEL/WHT wire between ECU (D19) and MAP sensor.
Substitute a known-good ECU and recheck. If symptom/indication goes away, replace the original ECU.
5 V?
D19 (+) D21 (−)
Measure voltage between YEL/WHT (+) terminal and GRN/WHT (−) terminal.
Is there approx. 5V?
NO
YES
Repair open in GRN/WHT wire between ECU (D21) and MAP sensor.
Measure voltage between RED (+) terminal and GRN/WHT (−) terminal.
Is there approx. 5V?
NO
YES
Repair open or short in RED wire between ECU (D17) and MAP sensor.
If wire is OK, substitute a known-good ECU and recheck. If prescribed voltage is now available, replace the original ECU.
Turn the ignition switch OFF.
Reconnect the 3P connector to the MAP sensor.
Connect the test harness between the ECU and connector (page 11-21).
Turn the ignition switch ON.

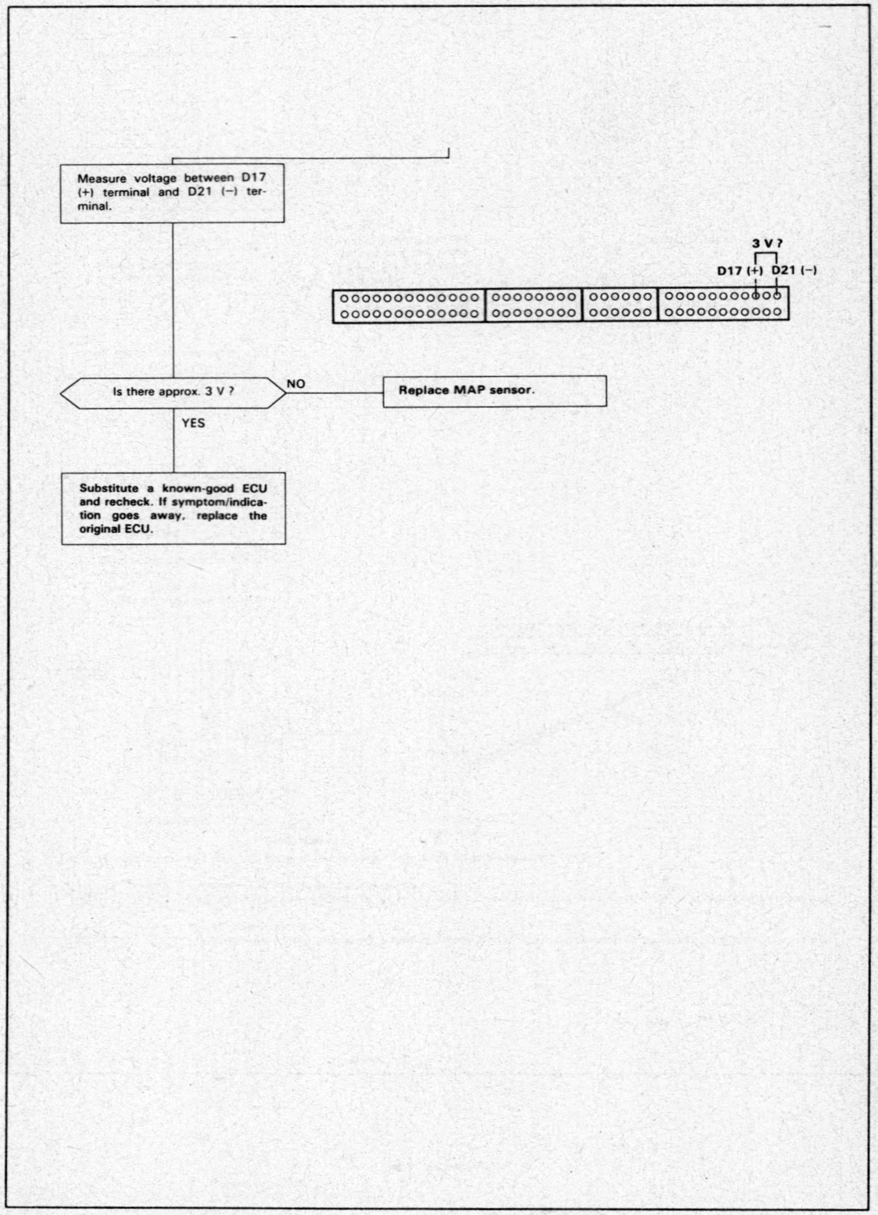
Measure voltage between D17 (+) terminal and D21 (−) terminal.
3 V ?
D17 (+) D21 (−)
Is there approx. 3 V ?
NO
YES
Replace MAP sensor.
Substitute a known-good ECU and recheck. If symptom/indication goes away, replace the original ECU.

CHECK – 4 – Self-diagnosis Check Engine light indicates code 4: A problem in the circuit of the CRANK 1 Sensor.

CHECK – 54 – Self-diagnosis Check Engine light indicates code 54: A problem in the circuit of the CRANK 2 Sensor.

CHECK – 9 – Self-diagnosis Check Engine light indicates code 9: A problem in the circuit of the CYL 1 Sensor.

CHECK – 59 – Self-diagnosis Check Engine light indicates code 59: A problem in the circuit of the CYL 2 Sensor.

The CRANK sensor determines timing for fuel injection and ignition of each cylinder and also detects engine RPM. The CYL sensor detects the position of No. 1 cylinder for sequential fuel injection to each cylinder and ignition timing.

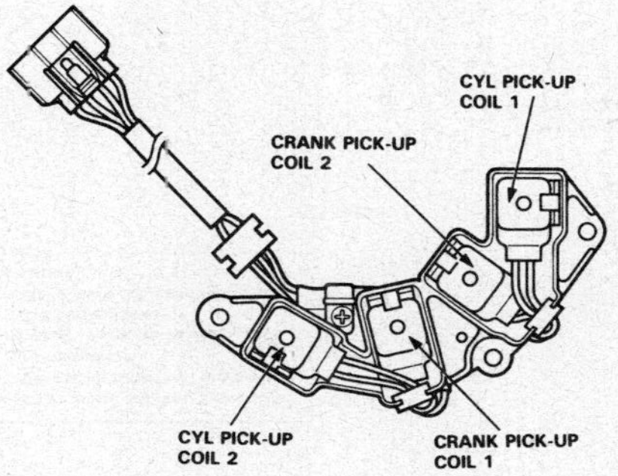

CHECK – 4 – 54 – 9 – or – 59

- Check Engine light has been reported on.
- With service check connector jumped CODE 4, 54, 9 and/or 59 are indicated.

Do the ECU Reset Procedures

Start the engine.

Is Check Engine light on and does it indicate CODE 4, 54, 9 or 59?

NO → Intermittent failure, system is OK at this time (test drive may be necessary). Check for poor connections or loose wires at C353 (located at right shock tower), C119 (CRANK/CYL sensor) and ECU.

YES ↓

Stop the engine.

Remove the engine harness cover.

Disconnect the 8P connector from the CRANK/CYL sensor.

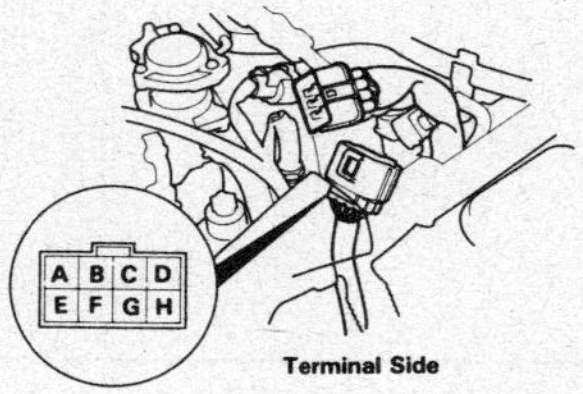

Measure resistance between terminals of the indicated sensor.
*see table

Is there 650–900 Ω?

NO → Replace the CRANK/CYL sensor

YES ↓

Check for continuity to body ground on all terminals individually.

Does continuity exist?

YES → Replace the CRANK/CYL sensor

NO ↓

Reconnect the connector.

Connect the test harness to the main wire harness only, not to the ECU

Measure resistance between terminals of the indicated sensor on test harness.
*see table

Is there 650–900 Ω?

NO → Repair open in the indicated sensor wires.

YES ↓

Check for continuity to body ground on B15, B13, B11 or B9 terminals.

Does continuity exist?

YES → Repair short in the indicated sensor wires.

NO ↓

Substitute a known-good ECU and recheck. If symptom/indication goes away, replace the original ECU.

SENSOR	CODE	SENSOR TERMINAL	ECU TERMINAL	WIRE COLOR
CRANK 1	4	A	B15	PNK
		B	B16	GRY
CRANK 2	54	C	B14	GRN/WHT
		D	B13	BLK/WHT
CYL 1	9	G	B12	BLU/YEL
		H	B11	YEL/GRN
CYL 2	59	E	B9	ORN
		F	B10	WHT

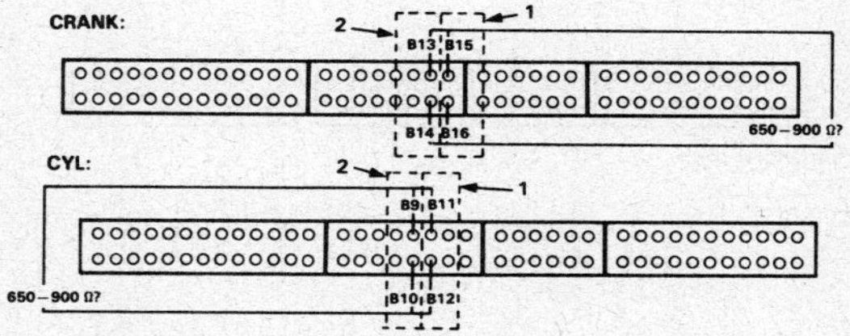

TROUBLESHOOTING CHARTS – 1991–92 LEGEND

- Check Engine light has been reported on.
- With service check connector jumped CODE 5 is indicated.

Do the ECU Reset Procedures

Start the engine and keep engine rpm at idle.

Is Check Engine light on and does it indicate CODE 5?

NO → —Intermittent failure, system is OK at this time (test drive may be necessary).
—Check vacuum hoses, pipes and connections.
—Make sure all connectors are secure.

YES → Stop the engine.

Disconnect #21 hose from the throttle body base, connect vacuum pump to the hose and apply vacuum.

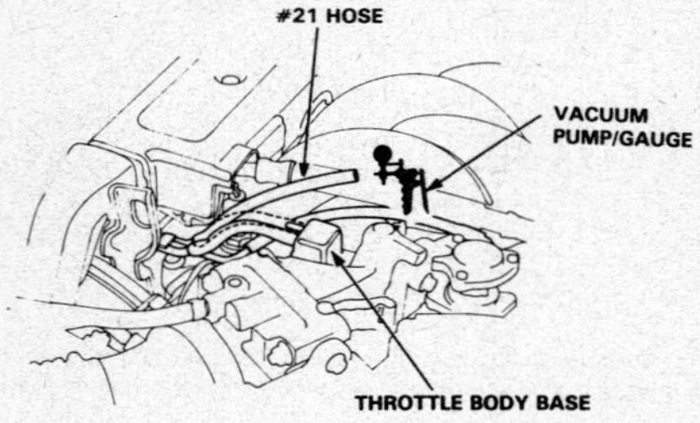

Does it hold vacuum?

NO → Connect a vacuum pump to the MAP sensor and apply vacuum.

Does it hold vacuum?

NO → Replace MAP sensor.

YES → Repair vacuum leak in hose routing between MAP sensor and intake manifold.

YES → Connect a T-fitting from a vacuum gauge between the throttle body base and #21 hose.

Start the engine.

Is there manifold vacuum?

NO → Remove restriction from throttle body base.

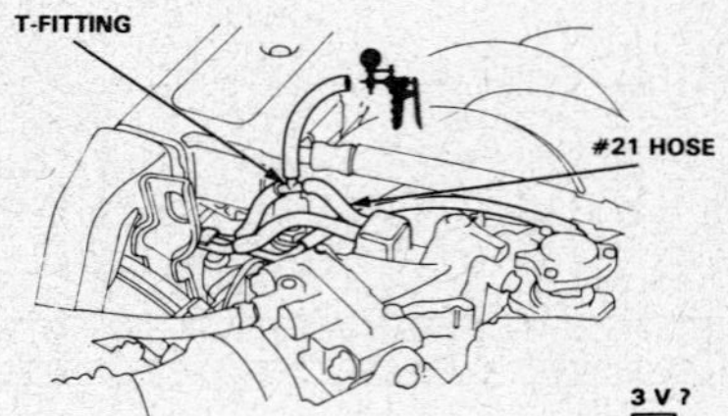

YES → Stop the engine.

Connect the test harness between the ECU and connector

Turn the ignition switch ON.

Measure voltage between D17 (+) terminal and D21 (–) terminal.

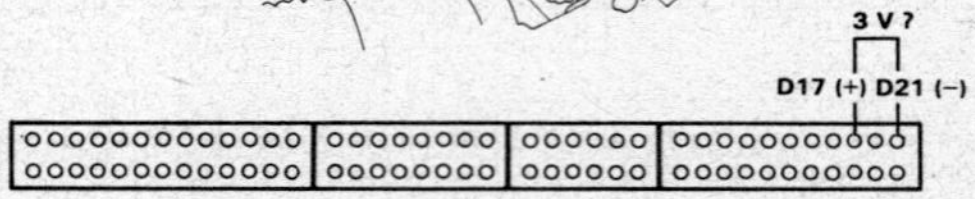

Is there approx. 3 V ?

NO → Replace MAP sensor.

YES → Start the engine and allow it to idle.

Is there approx. 1 V?

NO → Replace MAP sensor.

YES → Substitute a known-good ECU and recheck. If symptom/indication goes away, replace the original ECU.

TROUBLESHOOTING CHARTS – 1991–92 LEGEND

6 Self-diagnosis Check Engine light indicates code 6: A problem in the Coolant Temperature (TW) Sensor circuit.

The TW sensor is a temperature dependant resistor (thermistor). The resistance of the thermistor decreases as the coolant temperature increases as shown below.

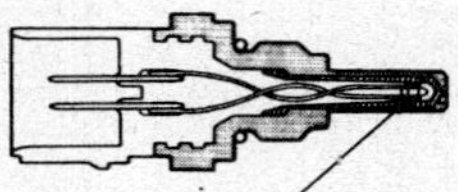

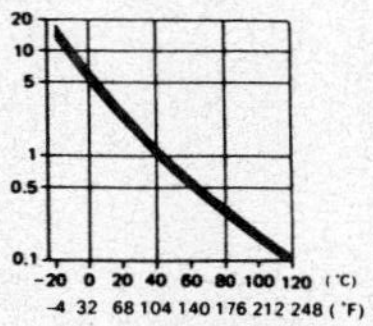

6

- Check Engine light has been reported on.
- With service check connector jumped CODE 6 is indicated.

Do the ECU Reset Procedures

Turn the ignition switch ON.

Is Check Engine light on and does it indicate CODE 6?

NO → Intermittent failure, system is OK at this time (test drive may be necessary). Check for poor connections or loose wires at C355 (located at right shock tower), C107 (TW sensor) and ECU.

YES ↓

Warm up engine to normal operating temperature (cooling fan comes on).

Turn the ignition switch OFF.

Remove the engine harness cover.

Disconnect the 2P connector from the TW sensor.

Measure resistance between the 2 terminals on the TW sensor.

Is there 200—400 Ω ?

NO → Replace TW sensor.

YES ↓

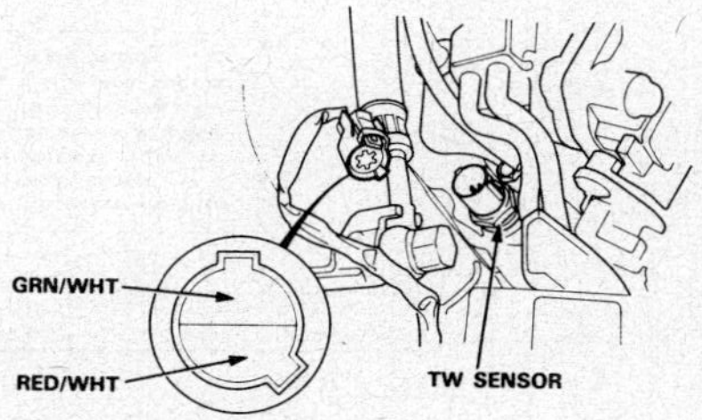

Turn the ignition switch ON.

Measure voltage between RED/WHT and body ground.

Is there approx. 5V ?

YES → Measure voltage between RED/WHT (+) terminal and GRN/WHT (–) terminal.

Is there approx. 5V ?

NO → Repair open in GRN/WHT wire between ECU (D22) and TW sensor.

YES → Substitute a known-good ECU and recheck. If symptom/indication goes away, replace the original ECU.

NO ↓

Connect the test harness D connector to the ECU only, not to the main wire harness.

Measure voltage between D13 (+) terminal and D22 (–) terminal.

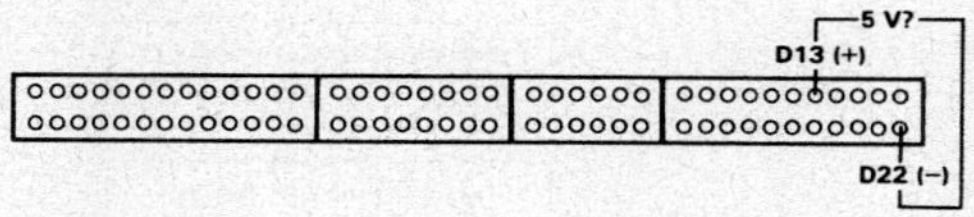

Is there approx. 5V?

YES → Repair open or short in RED/WHT wire between ECU (D13) and sensor.

NO → Substitute a known-good ECU and recheck. If symptom/indication goes away, replace the original ECU.

[CHECK] – 7 – Self-diagnosis Check Engine light indicates code 7: A problem in the Throttle Angle Sensor circuit.

The throttle angle sensor is a potentiometer. It is connected to the throttle valve shaft. As the throttle angle changes, the throttle angle sensor varies the voltage signal to the ECU.

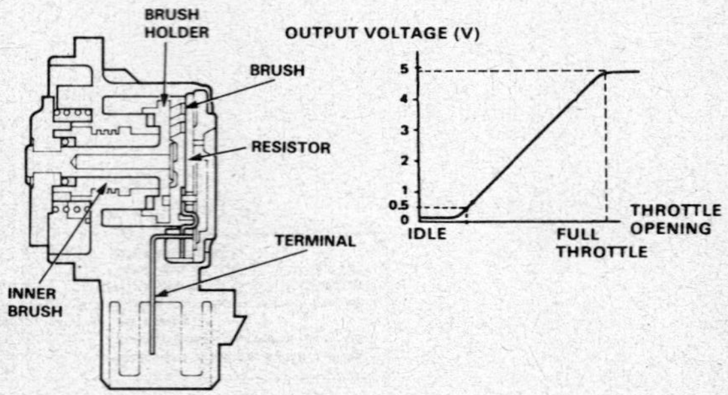

[CHECK] 7

- Engine is running.
- Check Engine light has been reported on.
- With service check connector jumped CODE 7 is indicated.

Do the ECU Reset Procedures

Turn the ignition switch ON.

Is Check Engine light on and does it indicate CODE 7?

NO → Intermittent failure, system is OK at this time (test drive may be necessary). Check for poor connections or loose wires at C355 (located at right shock tower), C108 (throttle angle sensor) and ECU.

YES ↓

Turn the ignition switch OFF.

Disconnect the 3P connector from the throttle angle sensor.

Turn the ignition switch ON.

Measure voltage between YEL/WHT (+) terminal and GRN/WHT (–) terminal.

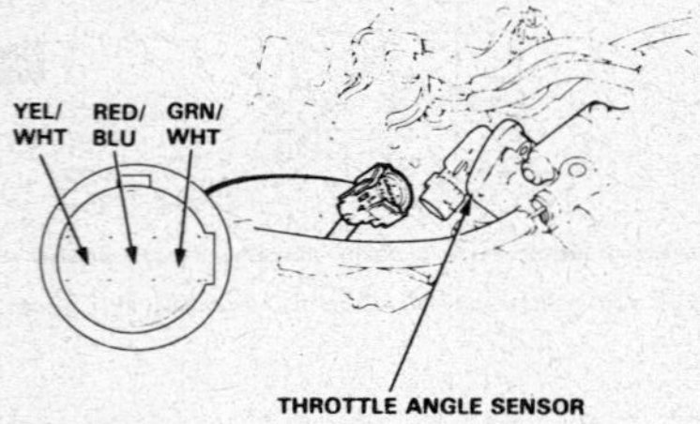

Is there approx. 5V ?

NO → Measure voltage between YEL/WHT (+) terminal and body ground.

Is there approx. 5V ?

YES → Repair open in GRN/WHT wire between ECU (D22) and throttle angle sensor.

NO ↓

Turn the ignition switch OFF.

Connect the test harness between the ECU and connector

Turn the ignition switch ON.

Measure voltage between D20 (+) terinal and D22 (–) terminal.

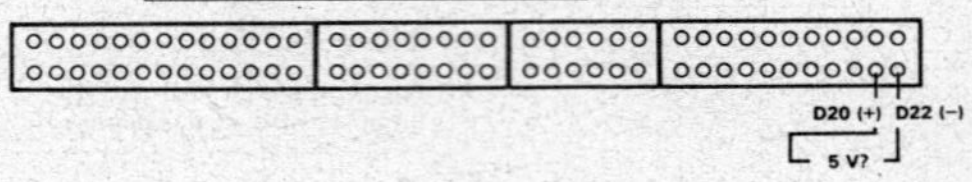

Is there approx. 5V ?

YES → Repair open in YEL/WHT wire between ECU (D20) and throttle angle sensor.

NO ↓

Substitute a known-good ECU and recheck. If prescribed voltage is now available, replace the original ECU.

YES (from first "Is there approx. 5V ?") ↓

Turn the ignition switch OFF.

Reconnect the 3P connector.

Connect the test harness between the ECU and connector

Turn the ignition switch ON.

Measure voltage between D11(+) terminal and D22 (–) terminal.

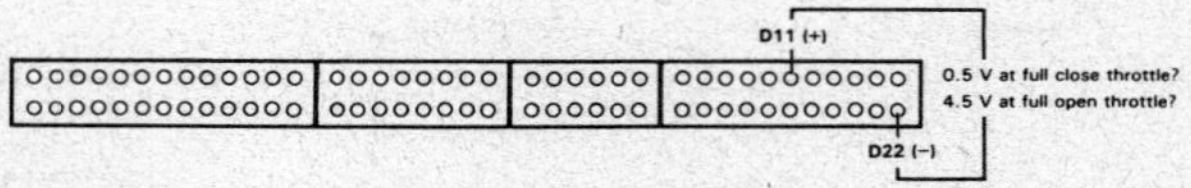

Is voltage 0.5 V at full close throttle, and approx. 4.5 V at full open throttle?
NOTE: There should be a smooth transition from 0.5 V to approx. 4.5 V as the throttle is depressed

NO →
- Replace throttle angle sensor.
- Repair open or short in RED/BLU wire between ECU (D11) and throttle angle sensor.

YES ↓

Substitute a known-good ECU and recheck. If symptom/indication goes away, replace the original ECU.

TROUBLESHOOTING CHARTS – 1991–92 LEGEND

10 — Self-diagnosis Check Engine light indicates code 10: A problem in the Intake Air Temperature (TA) Sensor circuit.

The TA sensor is a temperature dependant resistor (thermistor). The resistance of the thermistor decreases as the intake air temperature increases as shown below.

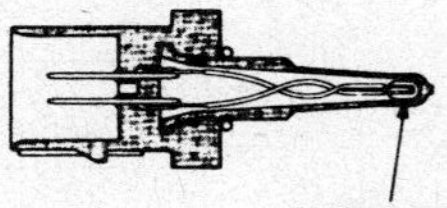

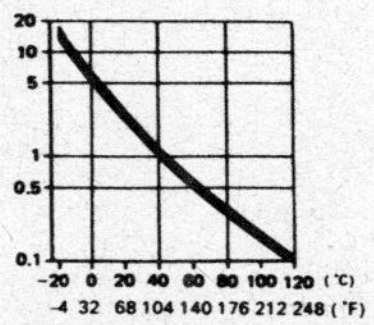

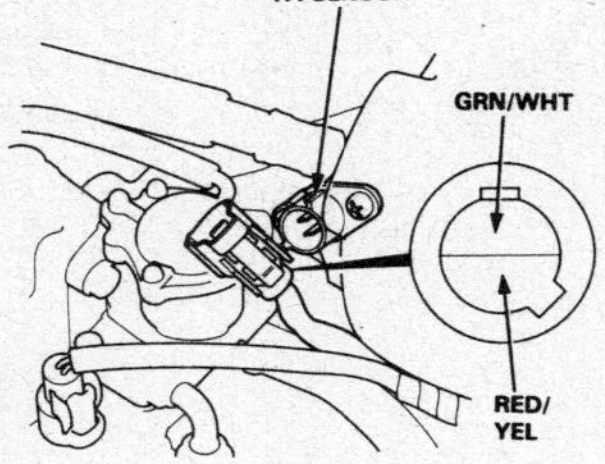

10

- Check Engine light has been reported on.
- With service check connector jumped CODE 10 is indicated.

Do the ECU Reset Procedures

Turn the ignition switch ON.

Is Check Engine light on and does it indicate CODE 10?

- NO → Intermittent failure, system is OK at this time (test drive may be necessary) Check for poor connections or loose wires at C355 (located at right shock tower), C118 (TA sensor) and ECU.
- YES ↓

Turn the ignition switch OFF.

Disconnect the 2P connector from the TA sensor.

Measure resistance between the 2 terminals on the TA sensor.

Is there 0.4–4.0 kΩ?

- NO → Replace TA sensor.
- YES ↓

Turn the ignition switch ON.

Measure voltage between RED/YEL (−) terminal and body ground.

Is there approx. 5 V ?

- YES → Measure voltage between RED/YEL (+) terminal and GRN/WHT (−) terminal.
  - Is there approx. 5 V ?
    - NO → Repair open in GRN/WHT wire between ECU (D22) and TA sensor.
    - YES → Substitute a known-good ECU and recheck. If symptom/indication goes away, replace the original ECU.
- NO ↓

Connect the test harness D connector to the ECU only, not to the main wire harness

Measure voltage between D15 (+) terminal and D22 (−) terminal.

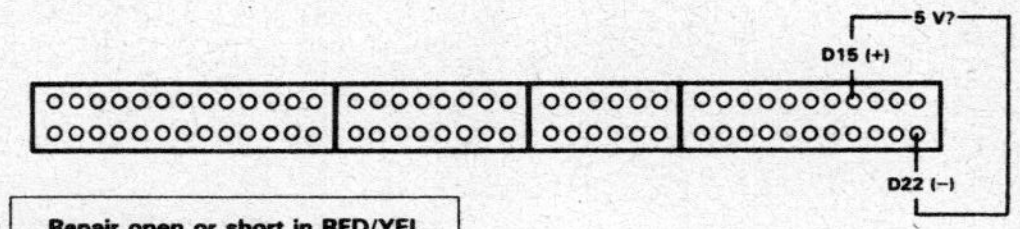

Is there approx. 5V?

- YES → Repair open or short in RED/YEL wire between ECU (D15) and TA sensor.
- NO → Substitute a known-good ECU and recheck. If symptom/indication goes away, replace the original ECU.

[CHECK] 12

- Check Engine light has been reported on.
- With service check connector jumped , CODE 12 is indicated.

Do the ECU Reset Procedures

Road test necessary: Warm up the engine to normal operating temperature (cooling fan comes on). Drive the car on the road for approx. 10 minutes. Keep the engine speed in the 1700–2500 range.

Is Check Engine light on and does it indicate CODE 12?

- NO → Intermittent failure, system is OK at this time. Check for poor connections or loose wires at C355 (located at right shock tower), C106 (EGR valve), C359 (control box) and ECU.
- YES → With the engine at idle, disconnect the #11 hose from the EGR valve and connect a vacuum pump/gauge to the hose.

Is there any vacuum?

- YES → Disconnect 4P connector from the control box and check the #11 hose for vacuum again.
- NO → Move the vacuum pump/gauge to the EGR valve.

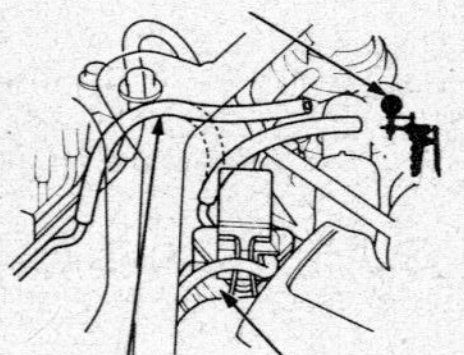

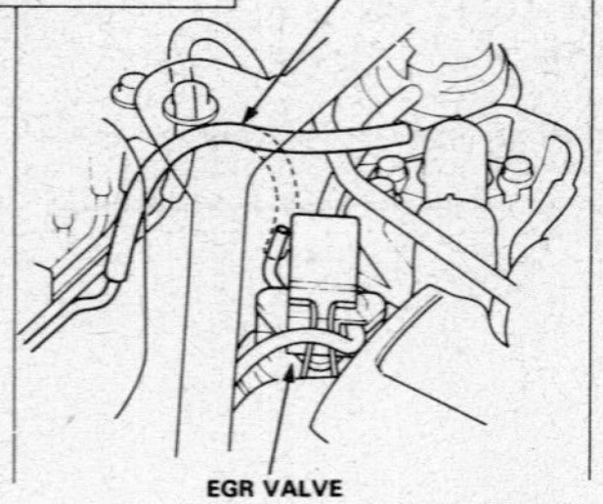

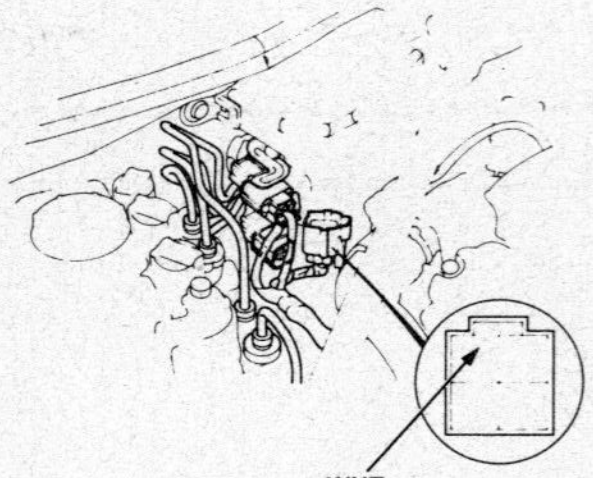

With the engine at idle, apply 8" of vacuum to the EGR valve.

Is there any vacuum ?

- YES → Check vacuum hose routing of the entire EGR system. If hose routing is OK, replace EGR control solenoid valve.
- NO → Check for continuity to ground on WHT wire of 4P connector.

Does continuity exist?

- YES → Repair short in WHT wire between EGR control solenoid valve and ECU (A11).
- NO → Substitute a known-good ECU and retest. If symptom/indication goes away, replace the original ECU.

Does the engine stall and does the EGR valve hold vacuum ?

- NO → Replace EGR valve.
- YES → Disconnect the 4P connector from the control box.

Measure voltage between BLK/YEL (+) terminal on the main wire harness and body ground.

Is there battery voltage ?

- NO → Repair open in BLK/YEL wire between the solenoid valve and No. 22 fuse.
- YES → Reconnect the vacuum pump/gauge to the #11 hose.

Start the engine and allow it to idle.

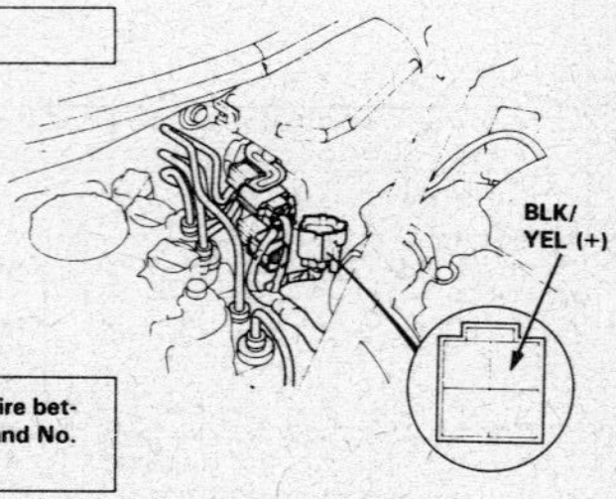

Connect the battery positive terminal to the A terminal of the 4P connector. While watching the vacuum gauge, connect the battery negative terminal to the B terminal.

Is there approx. 8" within 1 second?
- NO → Turn the ignition switch OFF and inspect the #11 and #5 hoses for leaks, restrictions, or misrouting.
  - Are the hoses OK?
    - NO → Connect as necessary
    - YES → Disconnect the lower hose on EGR control solenoid valve and connect a vacuum gauge to the hose.
      - Start the engine and allow it to idle.
      - Is there 6"— 10" of vacuum ?
        - NO → Replace CVC valve.
        - YES → Replace the EGR control solenoid valve.
- YES → Turn the ignition switch OFF and reconnect the 4P connector.

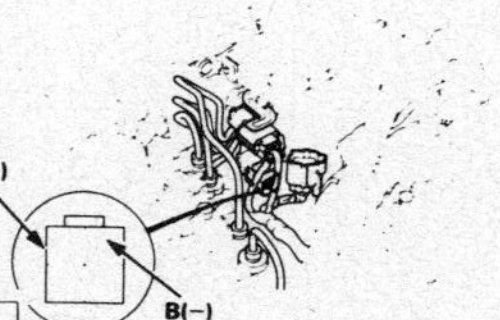

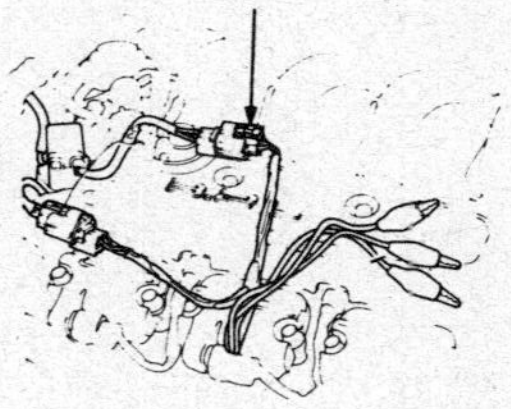

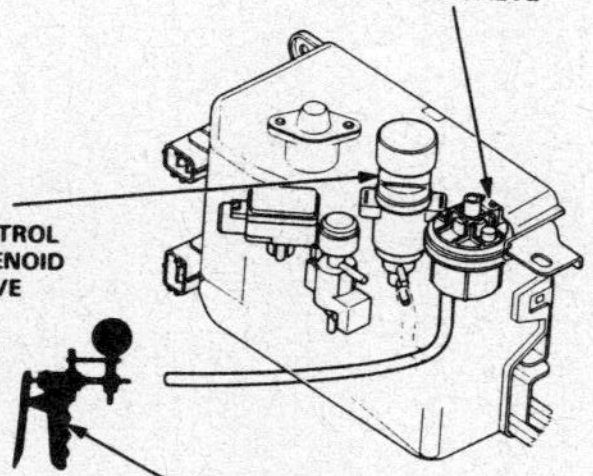

Connect the test harness between the EGR valve lift sensor and engine wire harness.

Turn the ignition switch ON.

Measure voltage between RED (+) terminal and GRN (−) terminal.

Is there approx. 5 V ?
- NO →
  - Repair open in YEL/WHT wire between EGR valve and ECU (D20).
  - Repair open in GRN/WHT wire between EGR valve and ECU (D22).
- YES

Measure voltage between WHT (+) terminal and GRN (−) terminal.

Is there approx 1.2 V ?
- NO →
  - Replace EGR valve.
  - Repair short in BLK/WHT wire between ECU (D12) and EGR valve lift sensor.
  - Substitute a known-good ECU and recheck. If prescribed voltage is now available, replace the original ECU.
- YES

While watching the voltmeter, slowly apply a continuous 8" of vacuum to the EGR valve. Repeat several times, completely releasing vacuum between.

Is the maximum voltage approx. 4 V ?
- NO → Replace EGR valve.
- YES

Does the voltage consistently increase/decrease as the vacuum increases/decreases ?
- NO → Replace EGR valve.
- YES

Reconnect the #11 hose to the EGR valve.

Turn the ignition switch OFF.

Connect the test harness between the ECU and connector

Start the engine and allow it to idle.

13 Self-diagnosis Check Engine light indicates code 13: A problem in the Atmospheric Pressure (PA) Sensor.

The PA sensor is built into the ECU.

13

- **Check Engine light has been reported on.**
- **With service check connector jumped (page 11-18), CODE 13 is indicated.**

Do the ECU Reset Procedures

Turn the ignition switch ON.

Is Check Engine light on and does it indicate CODE 13?

NO → **Intermittent failure, system is OK at this time (test drive may be necessary).**

YES → **Substitute a known-good ECU and recheck. If symptom/indication goes away, replace the original ECU.**

Measure voltage between D12 (+) terminal and D22 (−) terminal.

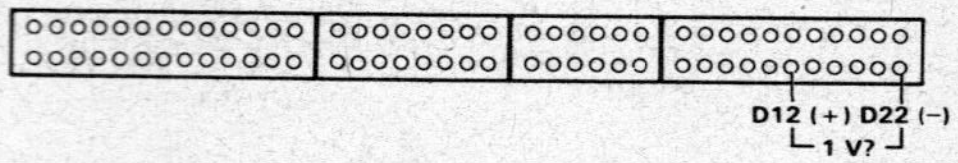

Is there approx. 1.2 V ?

NO → **Repair open in BLK/WHT wire between ECU (D12) and the sensor.**

YES → Connect A11 terminal to A26 terminal with a jumper wire.

Did the engine stall ?

NO → **Repair open in WHT wire between ECU (A11) and EGR control solenoid valve.**

YES → **Substitute a known-good ECU and recheck. If symptom/indication goes away, replace the original ECU.**

TROUBLESHOOTING CHARTS—1991–92 LEGEND

[Check Engine light] [14] Self-diagnosis Check Engine light indicates code 14: A problem in the Electronic Air Control Valve (EACV) circuit.

The EACV changes the amount of air bypassing the throttle body in response to a current signal from the ECU in order to maintain the proper idle speed.

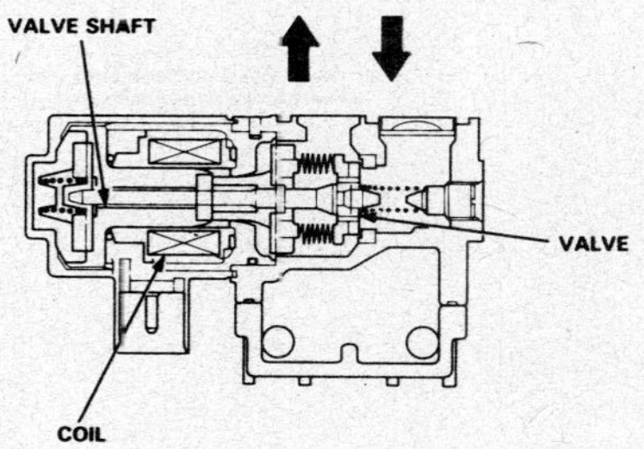

[Check Engine light] [14]

- Check Engine light has been reported on.
- With service check connector jumped CODE 14 is indicated.

Do the ECU Reset Procedures

Start the engine.

Is Check Engine light on and does it indicate CODE 14?

- YES: Remove the 2P connector from the EACV.
- NO: With the engine running and the accelerator pedal released disconnect the 2P connector on the EACV.
  - Is there a reduction in engine rpm?
    - YES: **Intermittent failure, system is OK at this time (test driving may be necessary). Check for poor connection or loose wires at C356 (located at right shock tower), C116 (EACV) and ECU.**
    - NO: **Substitute a known-good EACV and retest.**

Measure voltage between the YEL/BLK wire and body ground.

Is there battery voltage?

- NO: **Repair open in YEL/BLK wire between EACV and main relay.**
- YES: Turn the ignition switch off and reconnect the 2P connector the EACV.

Connect the test harness "A" connector to the main wire harness only, not the ECU (page

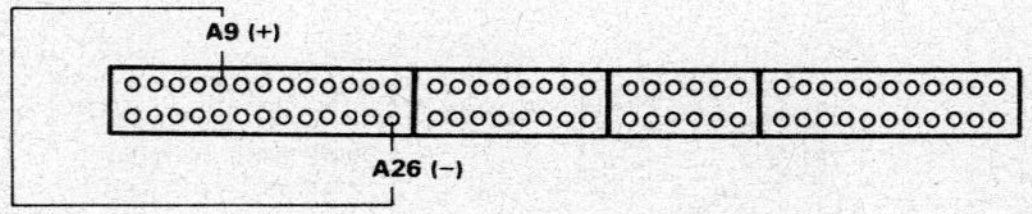

Turn the ignition switch ON.

Momentarily connect A9 terminal to A26 terminal several times.

Does the EACV click?

- YES: **Substitute a known-good ECU and retest. If symptom/indication goes away, replace the original ECU.**
- NO: **Repair open or short in BLU/RED wire between EACV and ECU. If the wire is OK, replace the EACV.**

15 Self-diagnosis Check Engine light indicates code 15: A problem in the Ignition Output Signal circuit.

- Check Engine light has been reported on.
- With service check connector jumped CODE 15 is indicated.

Do the ECU Reset Procedures

Turn the ignition switch ON.

Is Check Engine light on and does it indicate CODE 15?

NO → **Intermittent failure, system is OK at this time. Check for poor connections or loose wires at C380,C382 (located under the dash on the passenger side of the car) C155, C113 (igniter unit), ECU and G102.**

YES ↓

Turn the ignition switch OFF.

Disconnect the 8P connector from the igniter unit.

Check for continuity to body ground on C and F terminals individually.

Does continuity exist?

NO → **Repair open in BLK wire between 8P connector and G102.**

YES ↓

Disconnect the 6P connector from the igniter unit.

Connect the test harness between the ECU and connector (page 11-21).

NOTE:
Inspection of igniter unit secondary circuit

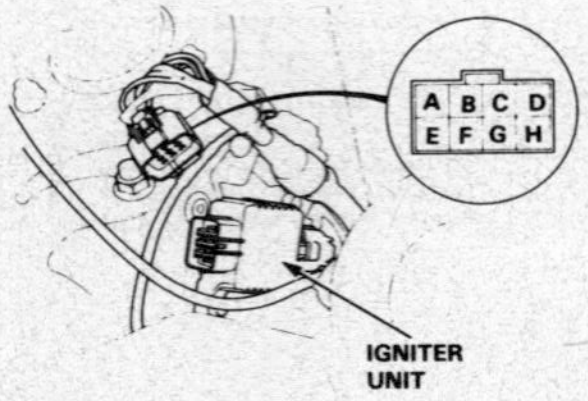

*1:

IGNITER UNIT SECONDARY TERMINAL (8P)	IGNITION COIL (2P)	WIRE COLOR
A	No. 1	WHT
B	No. 2	WHT/GRN
C	GROUND	BLK
D	No. 3	WHT/BLK
E	No. 4	WHT/BLU
F	GROUND	BLK
G	No. 5	WHT/YEL
H	No. 6	WHT/RED

Check for continuity between body ground and the following terminals: A21, A22, B8, B6, B4, B3.

Does continuity exist?

YES → **Repair short in each terminal wire.**

NO ↓

Check for continuity between terminals of 6P connector and ECU individually. *See table 2.

Does continuity exist?

NO → **Repair open in each terminal wire.**

YES ↓

Substitute a known-good igniter unit and recheck.

Does Check Engine light indicate CODE 15?

NO → **Replace the original igniter unit.**

YES ↓

Substitute a known-good ECU and recheck. If symptom/indication goes away, replace the original ECU.

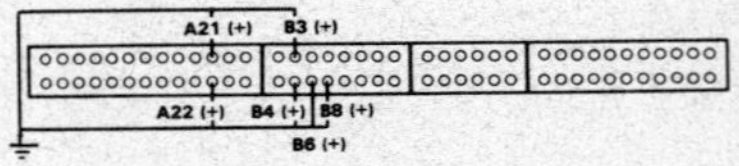

*2:

IGNITER UNIT PRIMARY TERMINAL (6P)	ECU TERMINAL	WIRE COLOR
A	B8 (No. 3)	BLU
B	A22 (No. 2)	BRN
C	A21 (No. 1)	PNK
D	B3 (No. 6)	RED
E	B4 (No. 5)	GRY
F	B6 (No. 4)	GRN

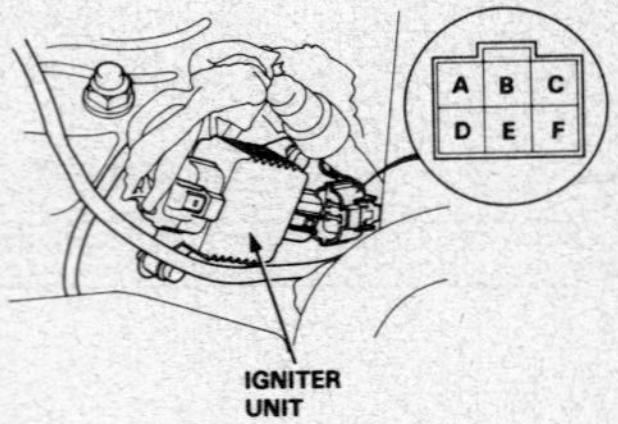

TROUBLESHOOTING CHARTS—1991–92 LEGEND

[CHECK] 16 Self-diagnosis Check Engine light indicates code 16: A problem in the fuel injector circuit.

The injectors are a solenoid-actuated constant-stroke pintle type consisting of a solenoid, plunger needle valve and housing. When current is applied to the solenoid coil, the valve lifts up and pressurized fuel is injected. Because the needle valve lift and the fuel pressure are constant, the injection quantity is determined by the length of time that the valve is open (i.e., the duration the current is supplied to the solenoid coil). The injector is sealed by an O-ring and seal ring at the top and bottom. These seals also reduce operating noise.

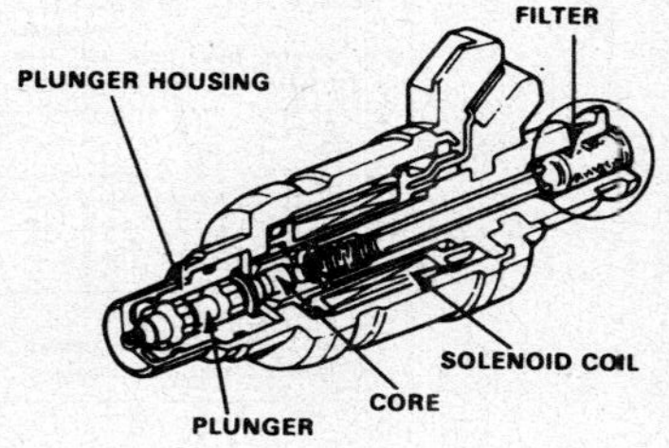

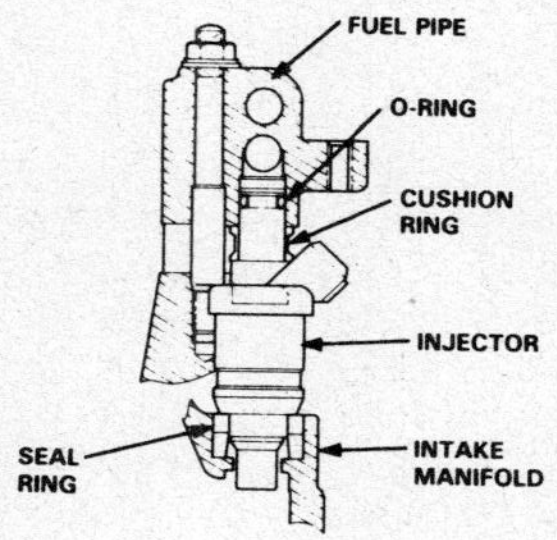

[CHECK] 16

- Check Engine light has been reported on.
- With service check connector jumped CODE 16 is indicated.

Do the ECU Reset Procedures

Turn the ignition switch to ON position.

Is Check Engine light on and does it indicate CODE 16?

- NO → Intermittent failure, system is OK at this time (test drive may be necessary). Check for poor connections or loose wires at C354, C356 (located at right shock tower), C136, C137, C138, C139, C140, C141 (injector), C125 (injector resistor) and ECU.
- YES ↓

Start engine and listen at each injector for a clicking sound.

Turn the ignition switch OFF.

Disconnect the 2P connector from the injector that does not click.

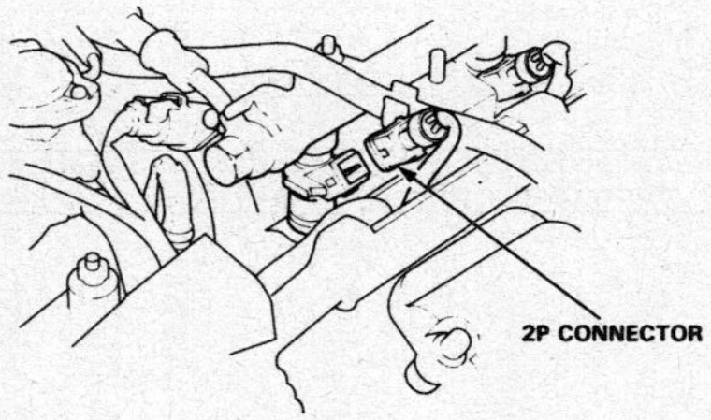

Measure resistance between the 2 terminals of the injector.

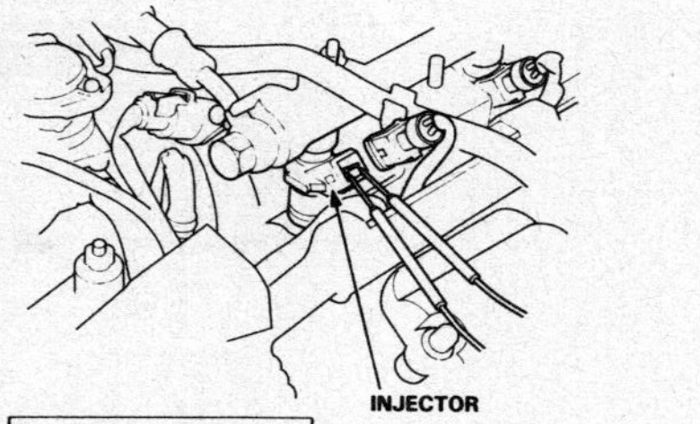

Is there 1.5—2.5Ω ?

- NO → Replace the injector/injectors that are not 1.5—2.5Ω.
- YES ↓

Turn the ignition switch ON.

Measure voltage between RED/BLK (+) terminal on the 2P connector and body ground.

Is there battery voltage ?

- NO → Turn the ignition switch OFF. → Disconnect 8P connector on the injector resistor. → Turn the ignition switch ON. → Measure voltage between YEL/BLK (+) terminal and body ground.
- YES ↓

Reconnect the 2P connector to the injector.

Turn the ignition switch OFF.

Connect the test harness between the ECU and connector

Turn the ignition switch ON.

Measure voltage between A23 (−) terminal and following terminal:
- No. 1 injector: A1 (+) terminal.
- No. 2 injector: A3 (+) terminal.
- No. 3 injector: A5 (+) terminal.
- No. 4 injector: A2 (+) terminal.
- No. 5 injector: A4 (+) terminal.
- No. 6 injector: A6 (+) terminal.

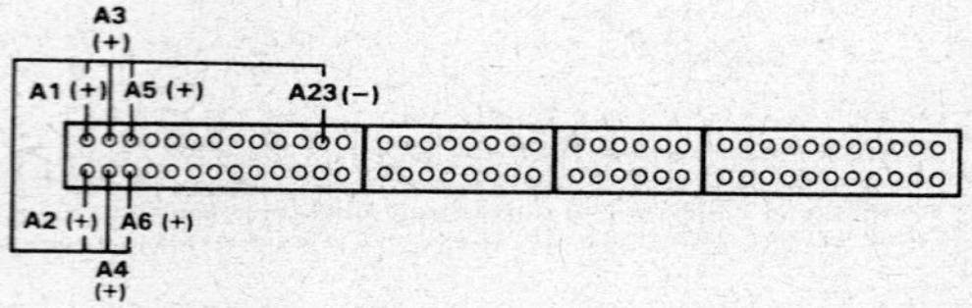

Is there battery voltage ?
- NO → Repair open in the wire between the ECU (A1, A3, A5, A2, A4 or A6) and the injector.
- YES → Substitute a known-good ECU and recheck. If symptom/indication goes away, replace the original ECU.

Is there battery voltage?
- NO → Repair open in the YEL/BLK wire between the injector resistor and the main relay.
- YES → Test the injector resistor

Does the injector resistor test OK?
- NO → Replace the injector resistor.
- YES → Repair open RED/BLK wire between 2P connector and resistor.

---

17 — Self-diagnosis Check Engine light indicates code 17: A problem in the Vehicle Speed Sensor circuit.

The signal generated by the speed sensor, produces pulses when the front wheels turn.

- Check Engine light has been reported on.
- With service check connector jumped CODE 17 is indicated.

Do the ECU Reset Procedures

Road test necessary.
In 2nd gear accelerate to 4,000 rpm, then decelerate to 1,500 rpm with throttle fully closed.

Is Check Engine light on and does it indicate CODE 17?
- NO → Intermittent failure, system is OK at this time. Check for poor connections or loose wires at C355 (located at right shock tower), C109 (speed sensor) and ECU.
- YES → Block rear wheels and set the parking brake. Jack up the front of the car and support with safety stands.

WARNING Block rear wheels before jacking up front of car.

Connect the test harness between the ECU and connector

Turn the ignition switch ON.

Slowly rotate left front wheel and measure voltage between C2 (+) terminal and A26 (−) terminal.

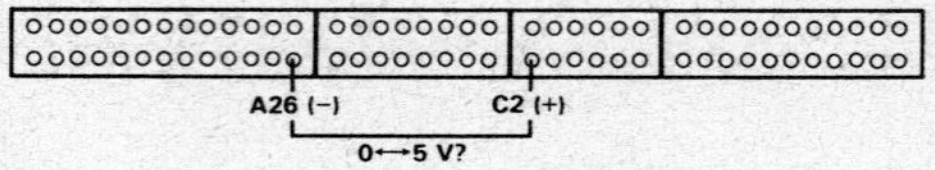

Does voltage pulse 0 V and 5 V?
- NO →
  - Repair open or short in YEL/RED wire between ECU (C2) and the speed sensor.
  - Faulty speed sensor.
  - Substitute a known-good ECU and recheck. If symptom/indication goes away, replace the original ECU.
- YES → Substitute a known-good ECU and recheck. If symptom/indication goes away, replace the original ECU.

TROUBLESHOOTING CHARTS – 1991–92 LEGEND

CHECK – 18 – Self-diagnosis Check Engine light indicates code 18: A problem in the Ignition Timing Adjuster circuit.

The ignition timing adjuster allows the electronic ignition advance to be set to 15° BTDC at idle.

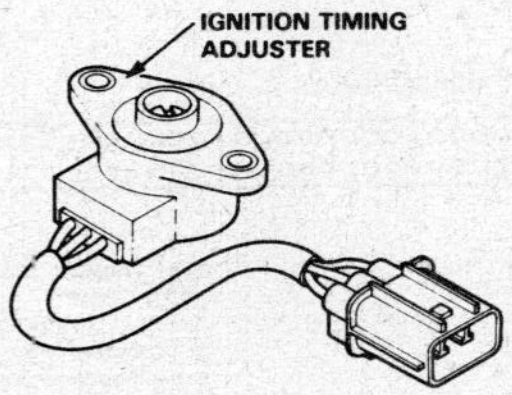

CHECK – 18

- – Check Engine light has been reported on.
  – With service check connector jumped CODE 18 is indicated.
- Do the ECU Reset Procedures
- Turn the ignition switch ON.
- Is Check Engine light on and does it indicate CODE 18?
  - NO → Intermittent failure system is OK at this time (Test drive may be necessary). Check for poor connections or loose wires at C361 (ignition timing adjuster) and ECU.
  - YES ↓
- Turn the ignition switch OFF.
- Disconnect the 3P connector from the control box.
- Measure resistance between A terminal and C terminal on Ignition timing adjuster.
- Is there 3.5–6.5 kΩ?
  - NO → Replace ignition timing adjuster (Section 23).
  - YES ↓

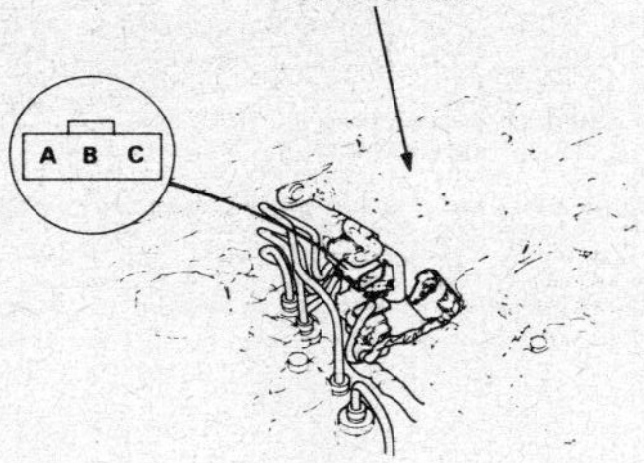

- Measure resistance between A and B terminals and between C and B terminals.
- Does the sum of the two resistance checks equal 3.5–6.5 kΩ?
  - NO → Replace Ignition timing adjuster
  - YES ↓
- Turn the ignition switch ON.
- Measure voltage between YEL/WHT (+) terminal and GRN/WHT (–) terminal on the main wire harness.
- Is there approx. 5 V?
  - NO → Measure voltage between YEL/WHT (+) terminal and body ground.
    - Is there approx. 5 V?
      - YES → Repair open in GRN/WHT wire between ECU (D22) and Ignition timing adjuster.
      - NO ↓
    - Turn the ignition switch OFF.
    - Connect the test harness between the ECU and connector
    - Turn the ignition switch ON.
    - Measure voltage between D20 (+) terminal and D22 (–) terminal.
    - Is there approx. 5 V?
      - YES → Repair open in YEL/WHT wire between ECU (D20) and Ignition timing adjuster.
      - NO → Substitute a known-good ECU and recheck. If prescribed voltage is now available, replace the original ECU.
  - YES ↓
- Turn the ignition switch OFF.
- Reconnect 3P connector.
- Connect the test harness between the ECU and connector
- Turn the ignition switch ON.
- Measure voltage between D8 (+) terminal and D22 (–) terminal.
- Is there 0.5–4.5 V?
  - NO → Repair open or short in BLU/YEL wire between ECU (D8) and Ignition timing adjuster.
  - YES → Substitute a known-good ECU and recheck. If symptom/indication goes away, replace the original ECU.

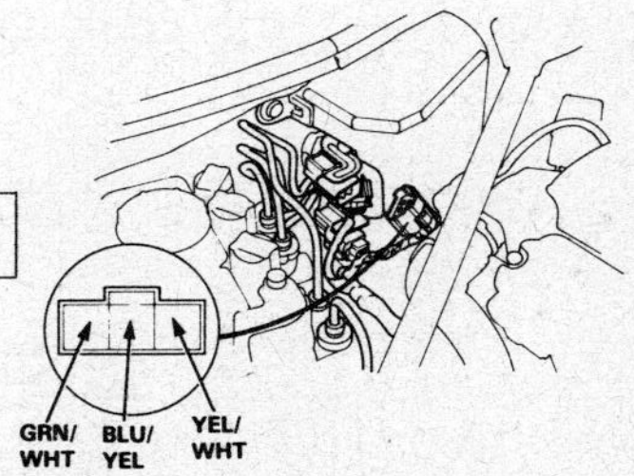

D20 (+) D22 (–) 5 V?

[23] Self-diagnosis Check Engine light indicates code 23: A problem in the circuit of the Left Knock Sensor.

[53] Self-diagnosis Check Engine light indicates code 53: A problem in the circuit of the Right Knock Sensor.

[23] or [53]

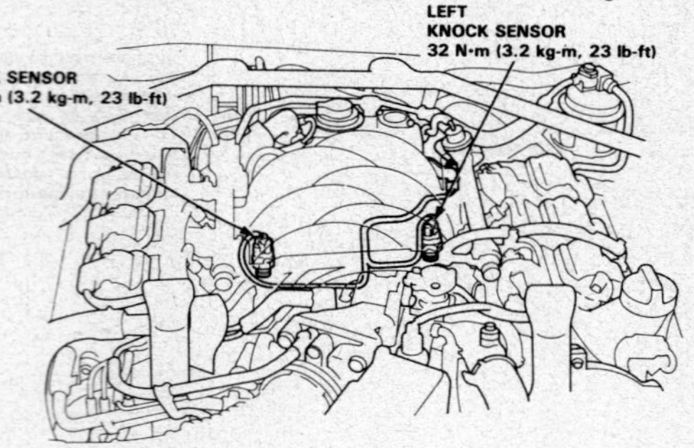

- Check Engine light has been reported on.
- With service check connector jumped CODE 23 and/or 53 are indicated.

Do the ECU Reset Procedures

Warm up the engine to normal operating temperature (cooling fan comes on).

Hold engine at 3000–4000 rpm for 10 seconds. (A/T: Transmission in [N] or [P])

Is Check Engine light on and does it indicate CODE 23 or 53?

- NO → Intermittent failure, system is OK at this time (test drive may be necessary). Check for poor connection or loose wires at C353 (located at right shock tower), C120, C232, C233 (knock sensor) and ECU.
- YES ↓

Turn the ignition switch OFF.

Connect the test harness to the main wire harness only, not to the ECU

Disconnect the knock sensor connector from the engine wire harness.

Check for continuity between D4 or D3*, terminal and body ground.

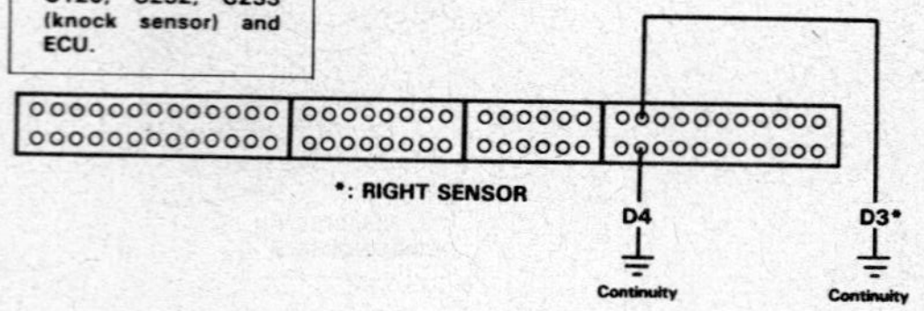

Does continuity exist?

- YES → Repair short in RED/BLU or WHT* wire between ECU (D4 or D3*) and the knock sensor.
  *: RIGHT SENSOR
- NO ↓

Check for continuity on RED/BLU or WHT* wire between D4 or D3* and 3P connector of engine wire harness.

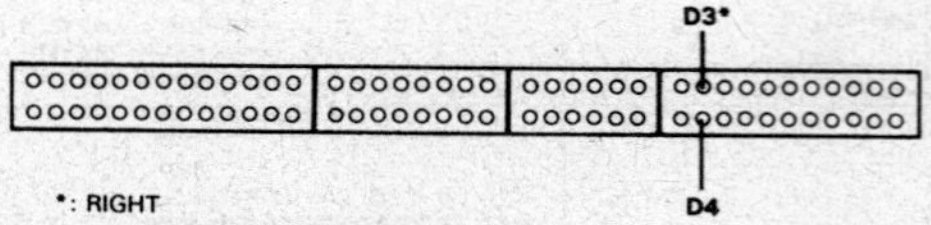

Does continuity exist?

- NO → Repair open in RED/BLU or WHT* wire between ECU (D4 or D3*) and the indicated knock sensor.
  *: RIGHT
- YES ↓

Substitute a known-good ECU and then hold the engine rpm at 3,000–4,000 rpm for 10 seconds and recheck.

Is Check Engine light on and does it indicate CODE 23 or 53?

- NO → Replace the original ECU.
- YES → Replace left or right knock sensor and recheck.

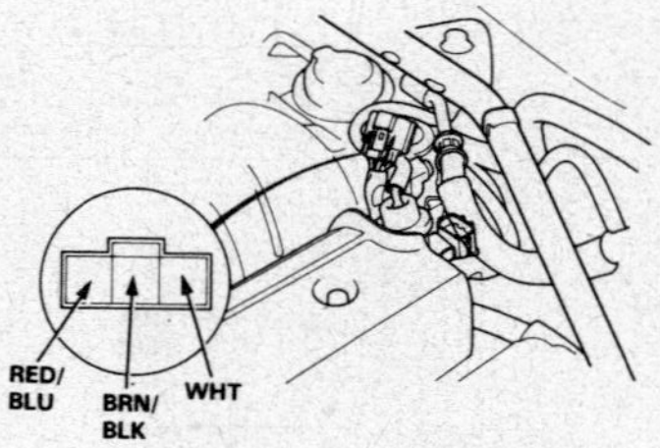

[CHECK ENGINE] 41 — Self-diagnosis Check Engine light indicates code 41: A problem in the Left Oxygen (O_2) Sensor Heater circuit.

[CHECK ENGINE] 42 — Self-diagnosis Check Engine light indicates code 42: A problem in the Right Oxygen (O_2) Sensor Heater circuit.

[CHECK ENGINE] 41 or 42

- Engine is running.
- Check Engine light has been reported on. With service check connector jumped CODE 41 and/or 42 are indicated.

Do the ECU Reset Procedures

Start the engine.

Is Check Engine light on and does it indicate CODE 41 or 42 ?

- NO → Intermittent failure, system is OK at this time (test drive may be necessary). Check for poor connections or loose wires at C355, C356 (located at right shock tower), C111, C123 (O_2 sensors) and ECU.
- YES ↓

Stop the engine

Disconnect the 4P connector from the O_2 sensor.

Measure resistance between terminals C and D on the O_2 sensor.

Is there 10–20 Ω?

- NO → Replace O_2 sensor.
- YES ↓

LEFT:

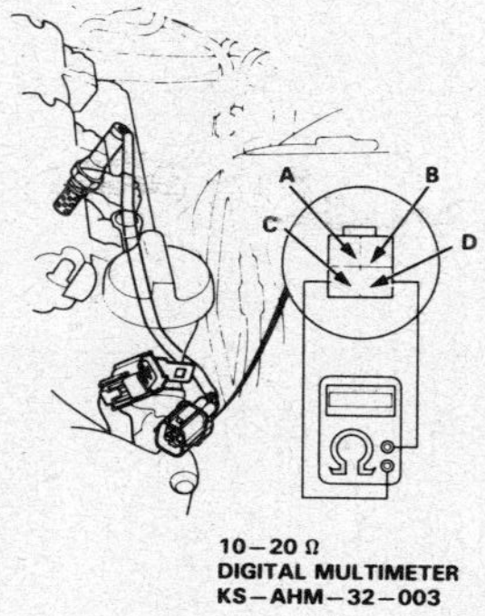

RIGHT:

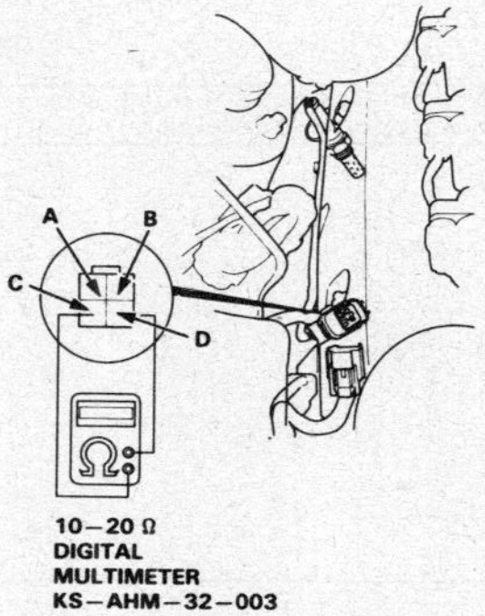

Check for continuity to body ground on each terminal on the O_2 sensor.

Does continuity exist?

- YES → Replace O_2 sensor.
- NO ↓

Check for continuity between terminal A and terminals C and D individually.

Does continuity exist?

- YES → Replace O_2 sensor.
- NO ↓

Turn the ignition switch ON.

At O_2 sensor harness measure voltage between BLK/YEL (+) terminal and GRN/BLU (–) or GRN/RED (–)* terminal.

*: RIGHT SENSOR

Is there battery voltage?

- YES → Disconnect the "A" connector from the ECU.
  - At O_2 sensor harness measure voltage between BLK/YEL (+) terminal and GRN/BLU (–) or GRN/RED (–)* terminal.
  - *: RIGHT SENSOR
  - Is there battery voltage?
    - YES → Repair short in GRN/BLU or GRN/RED* wire between ECU (A10 or A12*) and O_2 sensor.
    - NO → Substitute a known-good ECU and recheck. If symptom/indication goes away, replace the original ECU.
- NO ↓

Measure voltage between BLK/YEL (+) terminal and body ground.

Is there battery voltage?

- NO →
  - Repair open in BLK/YEL wire between O_2 sensors and No. 22 fuse.
  - Replace blown No. 22 fuse (20 A) in the under-dash fuse/relay box.
- YES ↓

Turn the ignition switch OFF.

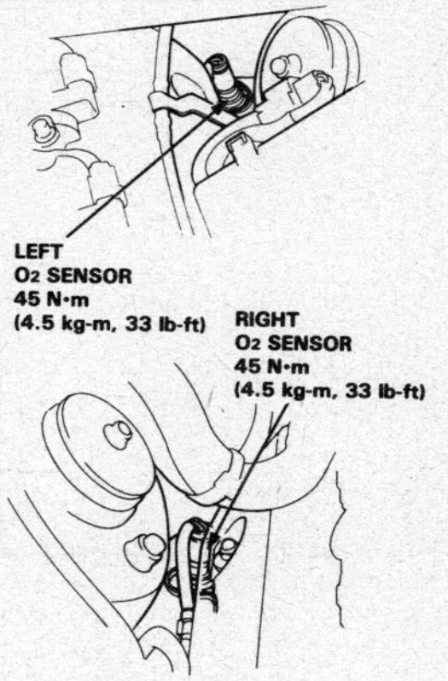

(cont'd)

TROUBLESHOOTING CHARTS—1991–92 LEGEND

[CHECK] 43: Self-diagnosis Check Engine light indicates code 43: Most likely a problem in the Left Oxygen (O_2) Sensor circuit or a problem in the Left Fuel Supply System.

[CHECK] 44: Self-diagnosis Check Engine light indicates code 44: Most likely a problem in the Right Oxygen (O_2) Sensor circuit or a problem in the Right Fuel Supply System.

[CHECK] 43 or 44

- Check Engine light has been reported on.
- With service check connector jumped (page 11-18), CODE 43 and/or 44 are indicated.
- or continued from code 1 or 2.

Do the ECU Reset Procedures

Warm up engine to normal operating temperature (cooling fan comes on).

Hold engine at 3,000 rpm for 2 minutes.
(A/T: Transmission in [N] or [P].)

Is the Check Engine light on and does it indicate CODE 43 or 44?

NO → Intermittent failure, system is OK at this time (test drive may be necessary). Check for poor connections or loose wires at C353, C355, C356 (located at right shock tower), C111, C123 (O_2 sensors) and ECU.

YES ↓

Turn the ignition switch OFF.

Connect the test harness between the ECU and connector

With the ignition switch OFF, wait for at least two minutes.

Install a jumper wire on the test harness between A10 and A26 (left): or A12 and A26 (right).

Turn the ignition switch ON.

Measure voltage between D14 (+) (left) or D16 (+) (right) terminal and A26 (−) terminal as soon as the ignition switch is turned on.

NOTE:
- Use DIGITAL MULTIMETER (KS–AHM–32–003) or equivalent.
- Use 2 Volt range.

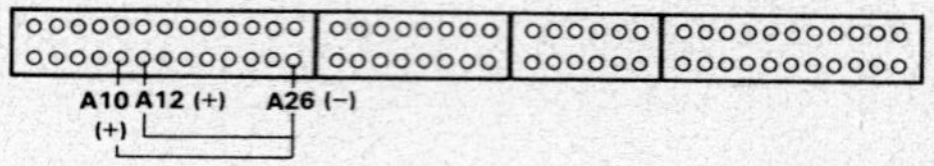

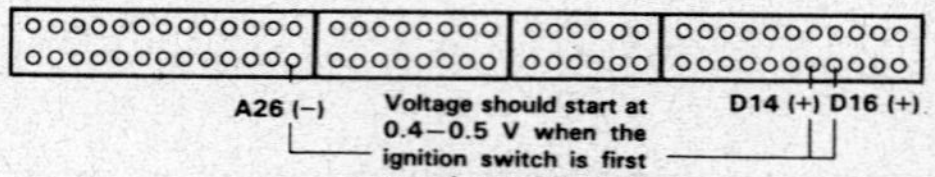

Voltage should start at 0.4–0.5 V when the ignition switch is first turned on, and decrease to below 0.1 V in less than two minutes.

Reconnect the 4P connector to O_2 sensor.

Connect the test harness "A" connector to the main wire harness only, not the ECU (page 11-21).

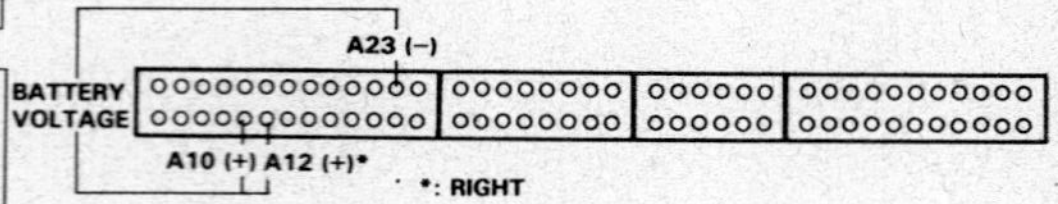

Turn the ignition switch ON.

Measure voltage between A10 (+) or A12 (+)* terminal and A23 (−) terminal.

*: RIGHT SENSOR

Is there battery voltage?

NO → Repair open in GRN/BLU or GRN/RED* wire between ECU (A10 or A12*) and O_2 sensor.

YES ↓

Substitute a known-good ECU and recheck. If symptom/indication goes away, replace the original ECU.

(From page 11-36)

Is there 0.1 V or less when the ignition switch is first turned on?

YES → Disconnect the 4P connector from the O2 sensor.

→ Measure voltage between D14 (+) (left) or D16 (+) (right) terminal and A26 (–) terminal.

→ Is there more than 0.1 V?

- YES → Replace O2 sensor.
- NO → Disconnect the "D" connector from the main wire harness.

→ Measure voltage between D14 (+) (left) or D16 (+) (right) terminal and A26 (–) terminal.

→ Is there more than 0.1 V?

- YES → Repair short in WHT (left) or RED/BLU (right) wire between ECU (D14 or D16) and O2 sensor.
- NO → Substitute a known-good ECU and recheck. If symptom/indication goes away, replace the original ECU.

NO → Disconnect the 4P connector from the O2 sensor.

→ Measure voltage at the engine wire harness side of the O2 sensor connector between A (+) and B (–).

→ Is there more than 0.1 V?

- YES → Replace O2 sensor.
- NO → Repair open in WHT (left) or RED/BLU (right) wire between ECU (D14 or D16) and O2 sensor.

RIGHT:

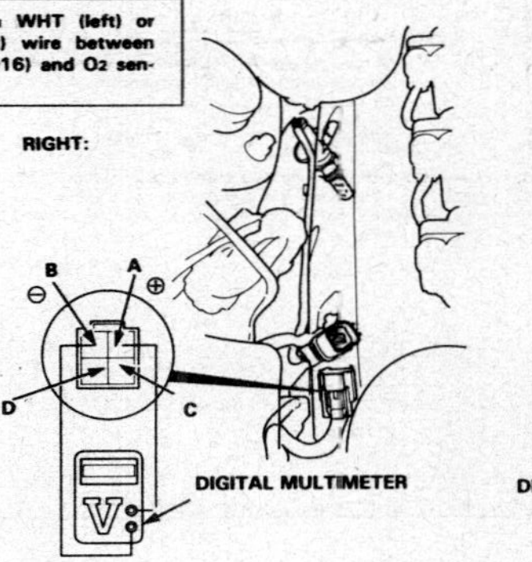

LEFT:

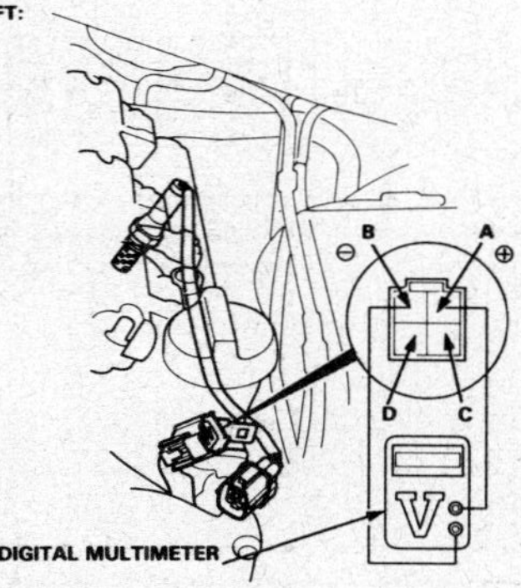

45 — Self-diagnosis Check Engine light indicates code 45: Most likely a problem in the Left Fuel Metering System.

46 — Self-diagnosis Check Engine light indicates code 46: Most likely a problem in the Right Fuel Metering System.

45 or 46

- Check Engine light has been reported on.
- With service check connector jumped CODE 45 and/or 46 are indicated.

→ Do the ECU Reset Procedures

→ Start the engine

→ Warm up engine to normal operating temperature (cooling fan comes on).

→ Do the Road Test.*

→ Does Check Engine light indicate CODE 45 and/or 46?

- NO → Intermittent failure, system is OK at this time.
- YES → Turn the ignition switch OFF.

→ Relieve fuel pressure (page 11-90).

→ Attach the fuel pressure gauge and measure the fuel pressure at idle (page 11-91).

→ Is the pressure less than 195 kpa?

- YES → fuel filter replacement / fuel pump testing.
- NO → Substitute a known-good MAP sensor.

→ Do the ECU Reset Procedures (page 11-19).

→ Do the Road Test.*

* Road Test
- Place the gear shift or selector lever in 3rd gear (M/T) or D3 position (A/T), cruise at exactly 35 mph for at least 10 seconds.
- Remove your foot from the throttle until the vehicle decels below 30 mph.
- Accelerate to 35 mph and cruise at exactly 35 mph for at least 10 seconds.
- The cruise/decel procedure must be repeated 10 times.
- If the above does not recreate the code, do the following:
- Stop the vehicle, and turn off the ignition switch.
- Start the engine.
- Repeat the cruise/decel procedure 5 times.

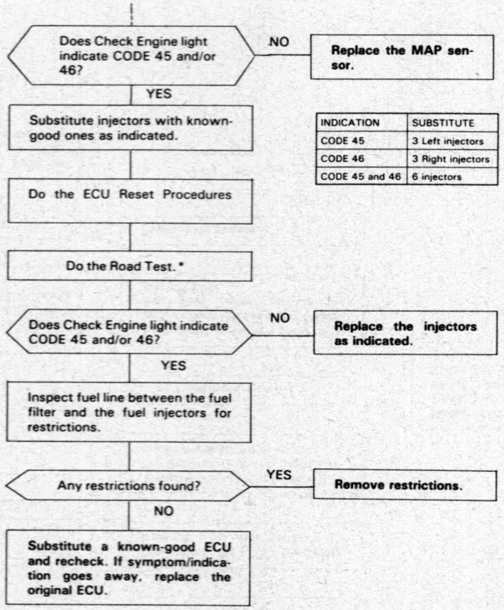

INDICATION	SUBSTITUTE
CODE 45	3 Left injectors
CODE 46	3 Right injectors
CODE 45 and 46	6 injectors

This signals the ECU when there is a demand for cooling from the air conditioning system.

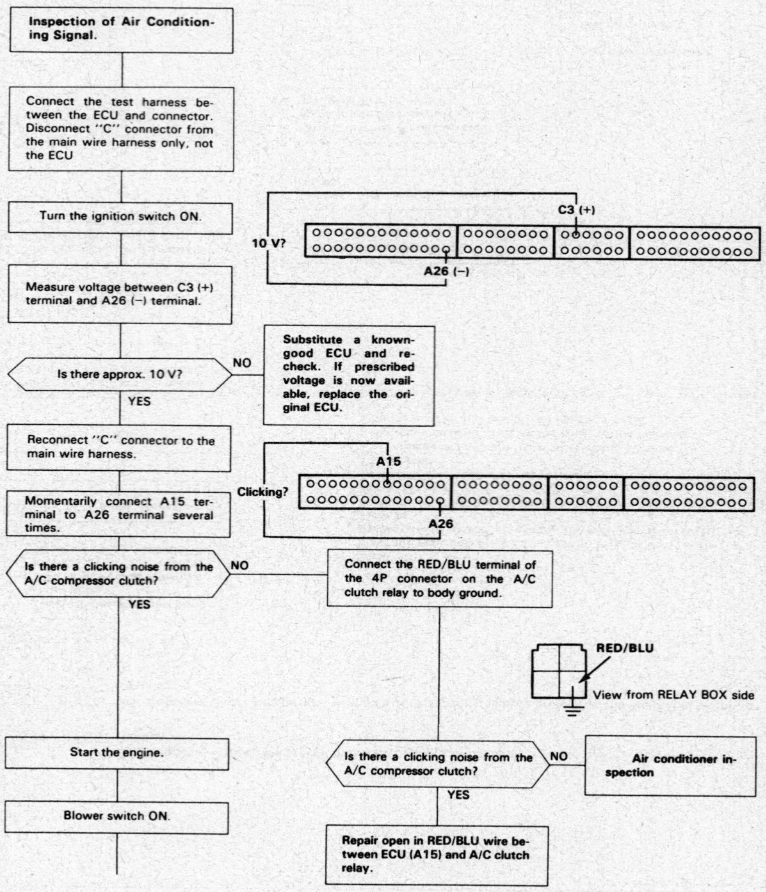

TROUBLESHOOTING CHARTS—1991–92 LEGEND

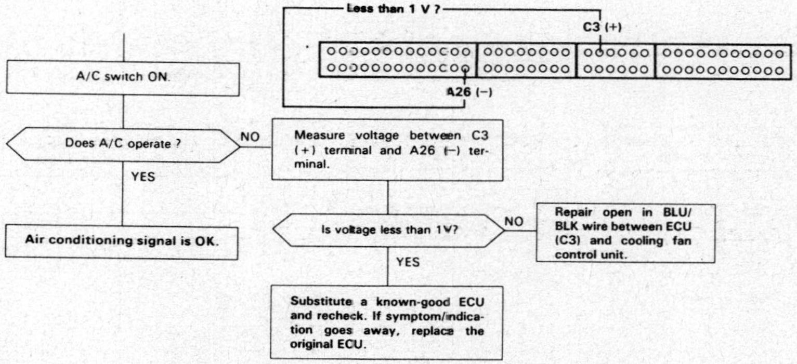

This signals the ECU when the alternator is charging.

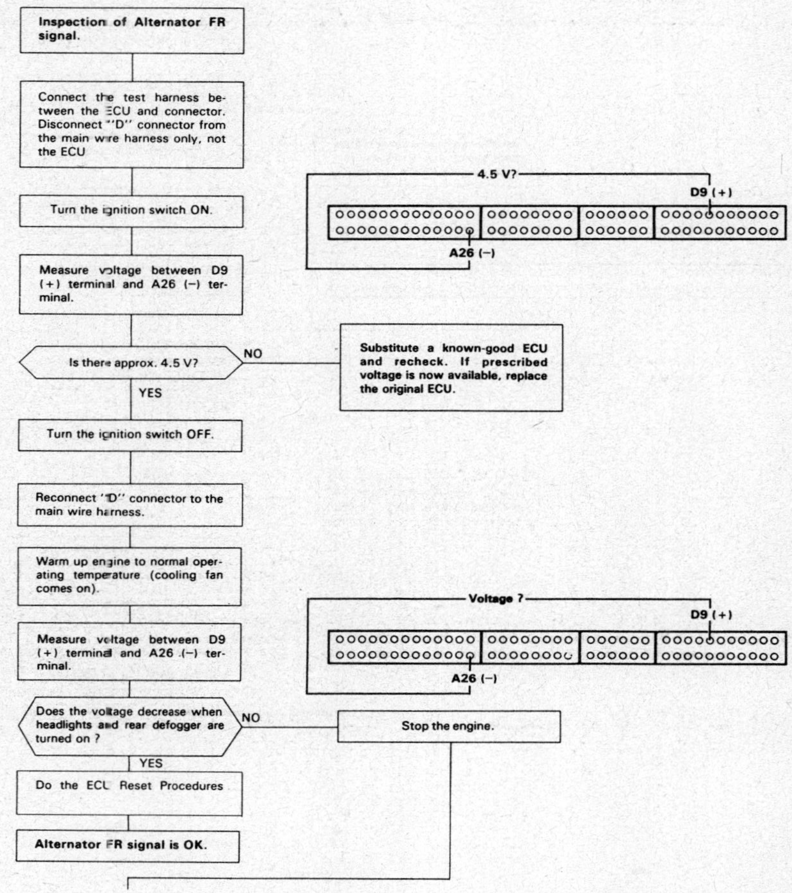

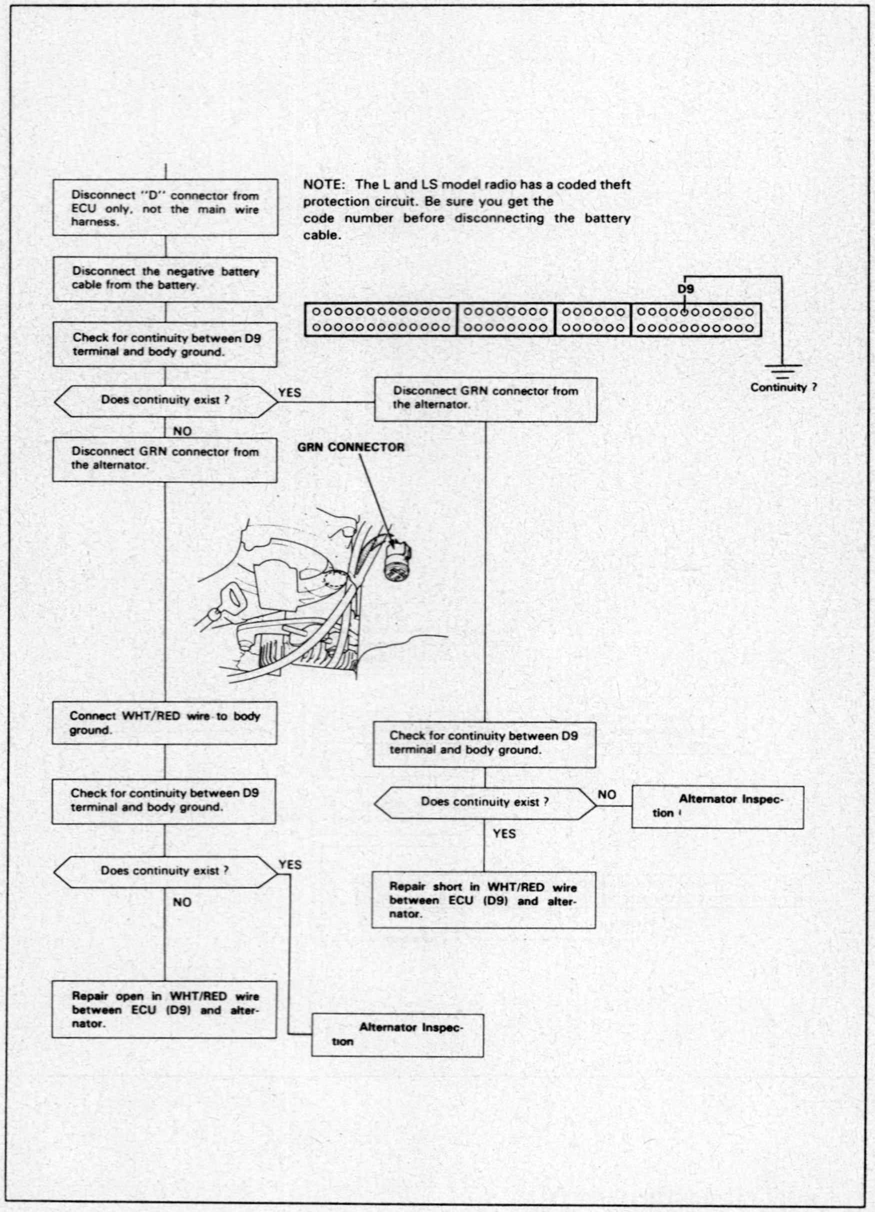
Disconnect "D" connector from ECU only, not the main wire harness.
NOTE: The L and LS model radio has a coded theft protection circuit. Be sure you get the code number before disconnecting the battery cable.
Disconnect the negative battery cable from the battery.
D9
Continuity ?
Check for continuity between D9 terminal and body ground.
Does continuity exist ?
YES
Disconnect GRN connector from the alternator.
NO
Disconnect GRN connector from the alternator.
GRN CONNECTOR
Connect WHT/RED wire to body ground.
Check for continuity between D9 terminal and body ground.
Check for continuity between D9 terminal and body ground.
Does continuity exist ?
NO
Alternator Inspection
YES
Does continuity exist ?
YES
NO
Repair short in WHT/RED wire between ECU (D9) and alternator.
Repair open in WHT/RED wire between ECU (D9) and alternator.
Alternator Inspection

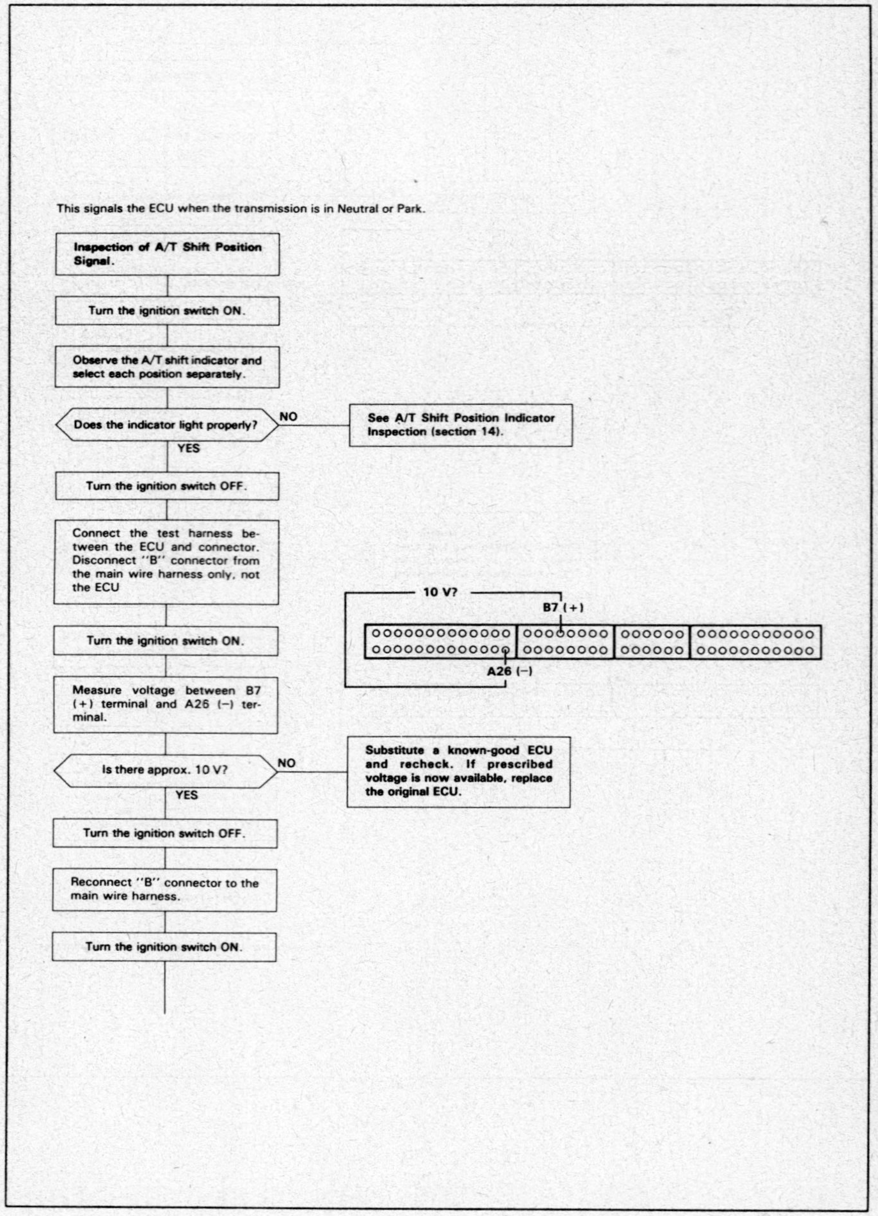
This signals the ECU when the transmission is in Neutral or Park.
Inspection of A/T Shift Position Signal.
Turn the ignition switch ON.
Observe the A/T shift indicator and select each position separately.
Does the indicator light properly?
NO
See A/T Shift Position Indicator Inspection (section 14).
YES
Turn the ignition switch OFF.
Connect the test harness between the ECU and connector. Disconnect "B" connector from the main wire harness only, not the ECU
10 V?
B7 (+)
A26 (−)
Turn the ignition switch ON.
Measure voltage between B7 (+) terminal and A26 (−) terminal.
Is there approx. 10 V?
NO
Substitute a known-good ECU and recheck. If prescribed voltage is now available, replace the original ECU.
YES
Turn the ignition switch OFF.
Reconnect "B" connector to the main wire harness.
Turn the ignition switch ON.

TROUBLESHOOTING CHARTS – 1991–92 LEGEND

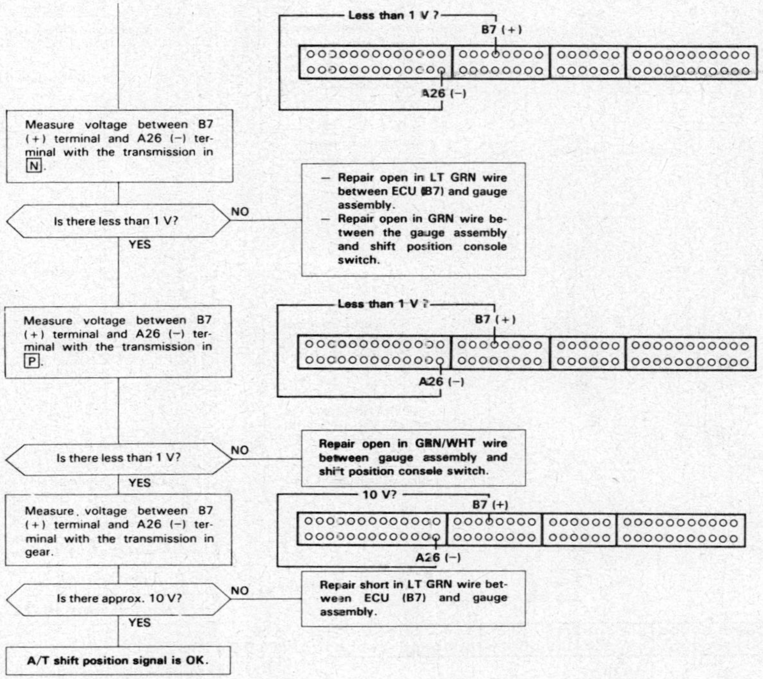

This signals the ECU when the transmission is in Neutral.

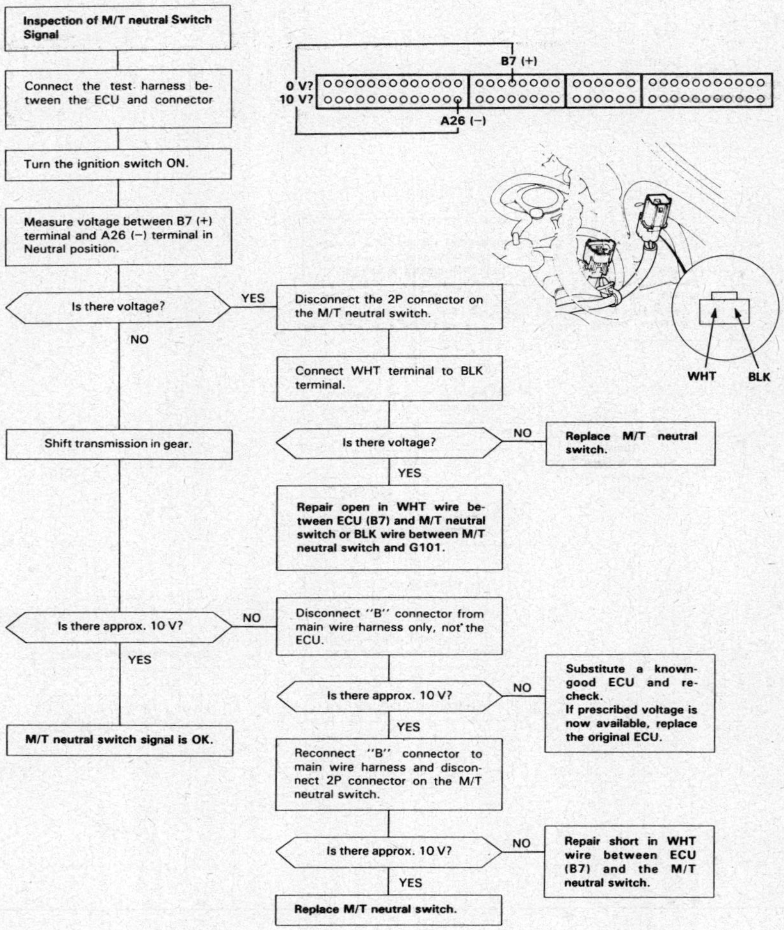

This signals the ECU when the clutch is engaged.

Inspection of clutch switch signal.

Connect the test harness between the ECU and connector

Turn the ignition switch ON.

Measure voltage between C7 (+) terminal and A26 (−) terminal.

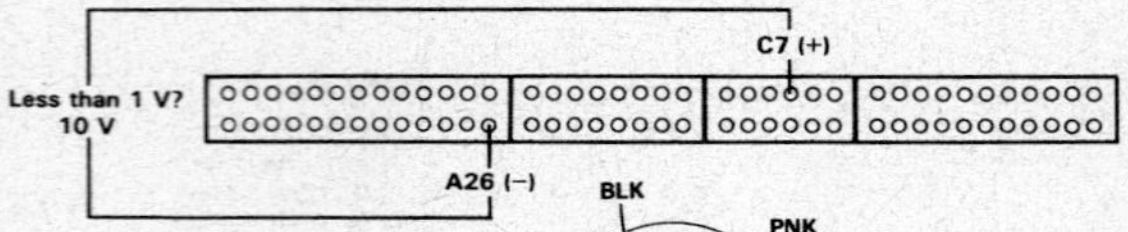

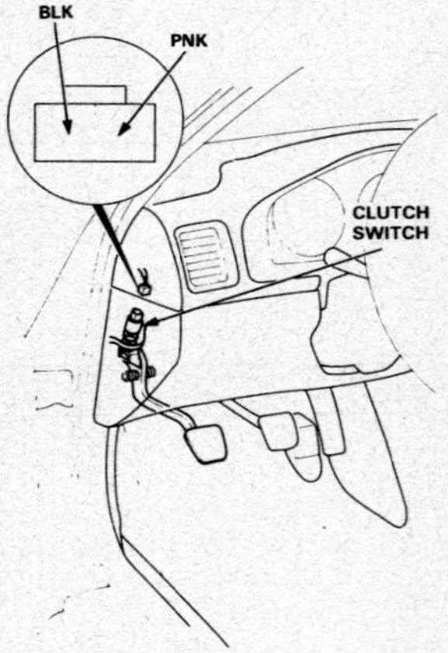

Is voltage less than 1 V?

- NO → Turn the ignition switch OFF.
  - Disconnect the 2P connector from the clutch switch.
  - Check for continuity between the A terminal and B terminal on the clutch switch.
  - Does continuity exist?
    - NO → **Replace the clutch switch.**
    - YES → Turn the ignition switch ON.
      - Measure voltage between PNK (+) terminal and body ground.
      - Is there approx. 10 V?
        - NO → **Repair open in PNK wire between ECU (C7) and the clutch switch.**
        - YES → **Repair open in BLK wire between the clutch switch and G301.**
- YES → Depress the clutch pedal.
  - Measure voltage between C7 (+) terminal and A26 (−) terminal.
  - Is there approx. 10 V?
    - YES → **Clutch switch signal is OK.**
    - NO → Turn the ignition switch OFF.
      - Disconnect the 2P connector from the clutch switch.
      - Turn the ignition switch ON.
      - Measure voltage between C7 (+) terminal and A26 (−) terminal.
      - Is there approx. 10 V?
        - YES → **Replace the clutch switch.**
        - NO → Disconnect "C" connector from main wire harness only, not the ECU.
          - Is there approx. 10 V?
            - YES → **Repair short in PNK wire between ECU (C7) and the clutch switch.**
            - NO → **Substitute a known-good ECU and recheck. If prescribed voltage is now available, replace the original ECU.**

TROUBLESHOOTING CHARTS – 1991–92 LEGEND

This signals the ECU when the engine is cranking.

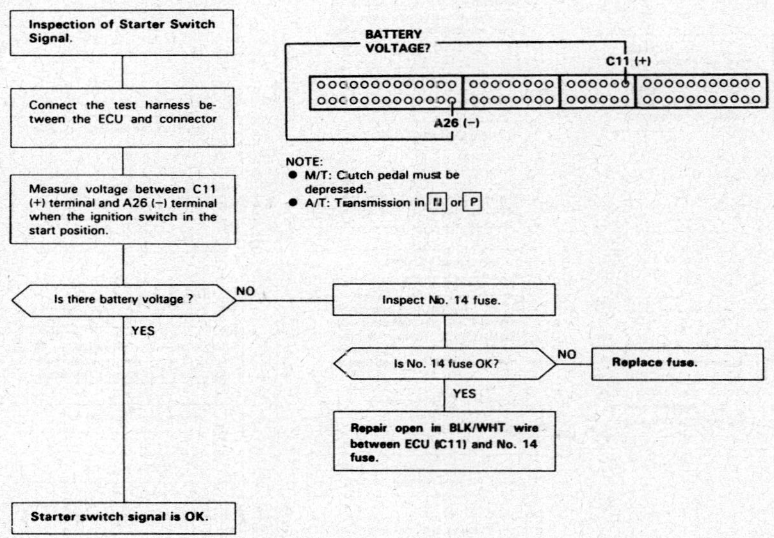

This signals the ECU when the brake pedal is depressed.

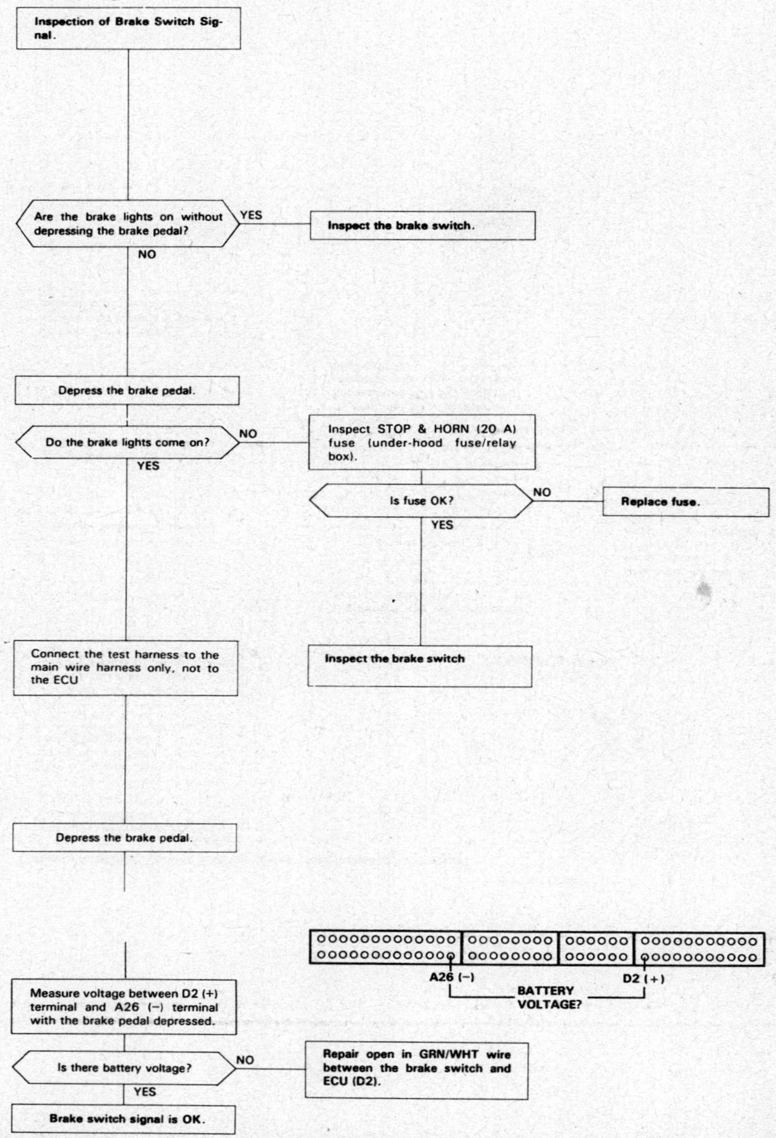

This signals the ECU when the power steering load is high.

Inspection of P/S Oil Pressure Signal

Connect the test harness between the ECU and connector

Turn the ignition switch ON.

Measure voltage between B5 (+) terminal and D22 (−) terminal.

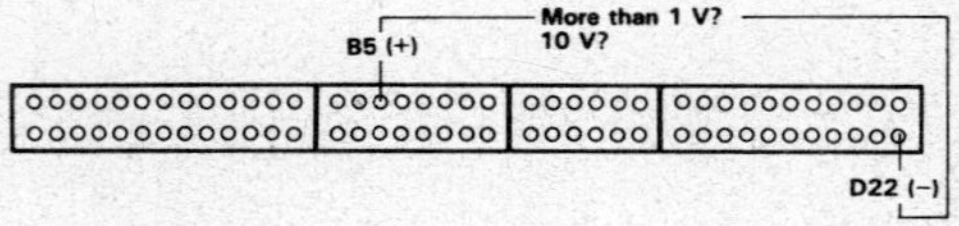

Is there more than 1V?

- YES → Disconnect the 2P connector on the P/S oil pressure switch.
  - Connect GRN terminal to GRN/WHT terminal.
  - Is there more than 1V?
    - NO → **Replace P/S oil pressure switch.**
    - YES → **Repair open in GRN wire between ECU (B5) and P/S oil pressure switch or GRN/WHT wire between P/S oil pressure switch and ECU (D22).**
- NO → Start the engine.
  - Turn steering wheel slowly.
  - Measure voltage between B5 (+) terminal and D22 (−) terminal while steering wheel is turning.
  - Is there approx. 10V?
    - YES → **P/S oil pressure signal is OK.**
    - NO → Turn the ignition switch OFF.
      - Disconnect "B" connector from main wire harness only, not the ECU.
      - Turn the ignition switch ON.
      - Is there approx. 10V?
        - NO → **Substitute a known-good ECU and recheck. If prescribed voltage is now available, replace the original ECU.**
        - YES → Reconnect "B" connector to main wire harness and disconnect 2P connector on the P/S oil pressure switch.
          - Is there approx. 10V?
            - YES → **Replace P/S oil pressure switch.**
            - NO → **Repair short in GRN wire between ECU (B5) and the P/S oil pressure switch.**

Pressure Regulator Control Solenoid Valve Troubleshooting Flowchart

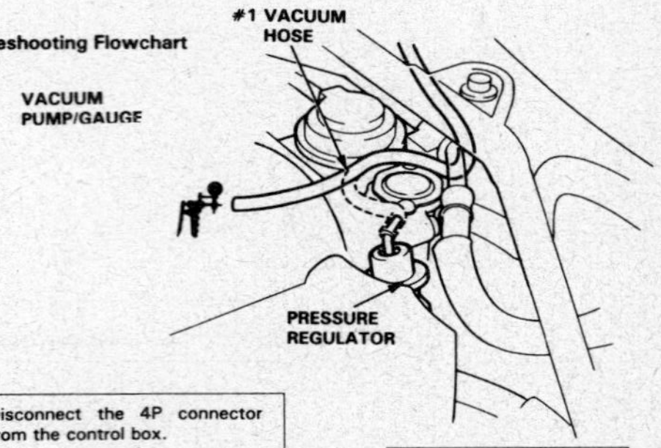

Inspection of Pressure Regulator Control Solenoid Valve.

Start the engine and warm up to normal operating temperature (the cooling fan comes on).

Disconnect the #1 vacuum hose from the pressure regulator and connect a vacuum gauge to the hose.

Is there manifold vacuum ?

- NO → Disconnect the 4P connector from the control box.
  - Is there vacuum ?
    - NO → **Check the #1 and #3 vacuum hose. If hoses are OK, replace the solenoid valve.**
    - YES → **Repair short to ground in LT GRN wire between ECU (A19) and 4P connector. If wire is OK, substitute a known-good ECU and recheck. If symptom goes away, replace the original ECU.**
- YES → Turn the ignition switch OFF.
  - Disconnect the 4P connector from the control box
  - Connect battery positive to terminal D and battery negative to terminal C of the solenoid valve.
  - Start the engine and allow it to idle.

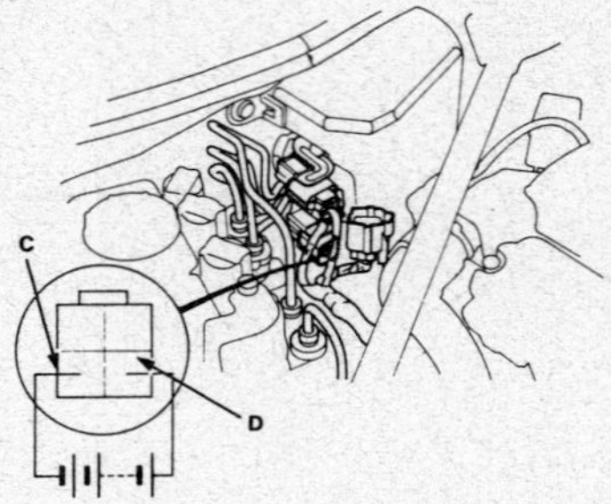

TROUBLESHOOTING CHARTS – 1991–92 LEGEND

Is there manifold vacuum ?
- YES → Replace the solenoid valve.
- NO ↓

Measure voltage between BLK/YEL (+) terminal and body ground.

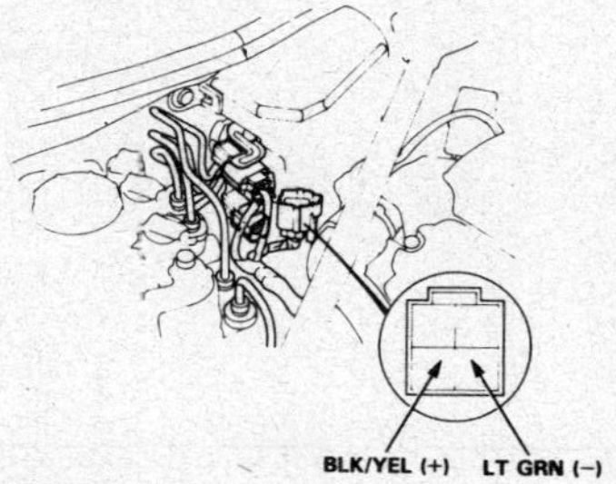

Is there battery voltage ?
- NO → Repair open in BLK/YEL wire between No. 22 fuse and 4P connector.
- YES ↓

Turn the ignition switch OFF.

Reconnect the 4P connector to the control box.

Connect the test harness between the ECU and connector

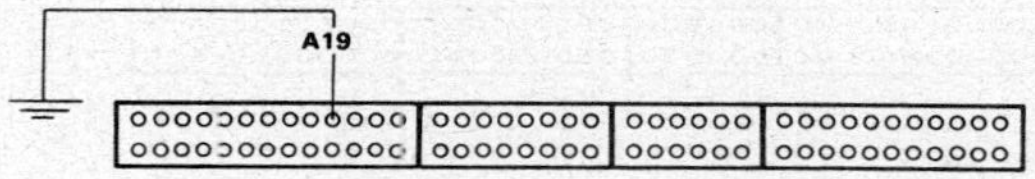

Start the engine and allow it to idle.

Connect A19 terminal to body ground with a jumper wire.

Is there manifold vacuum ?
- YES → Repair open in LT GRN wire between ECU (A19) and the 4P connector.
- NO ↓

Pressure Regulator Control Solenoid Valve is OK.

– Engine will not start.
– Inspection of main relay.

Disconnect the main relay connectors.

Measure the voltage between YEL/BLK terminal ① and body ground.

Is there battery voltage?
- NO → Replace No. 5 fuse.
- YES ↓

Turn the ignition switch ON.

Measure the voltage between BLK/YEL terminal ⑤ and body ground.

Is there battery voltage?
- NO → Replace No. 22 fuse.
- YES ↓

Measure the voltage between BLK/YEL terminal ⑥ and body ground.

Is there battery voltage?
- NO → Replace No. 25 fuse.
- YES ↓

Turn the ignition switch to the START position.

Measure the voltage between BLK/WHT terminal ④ and body ground.

Is there battery voltage?
- NO → Replace No. 14 fuse.
- YES ↓

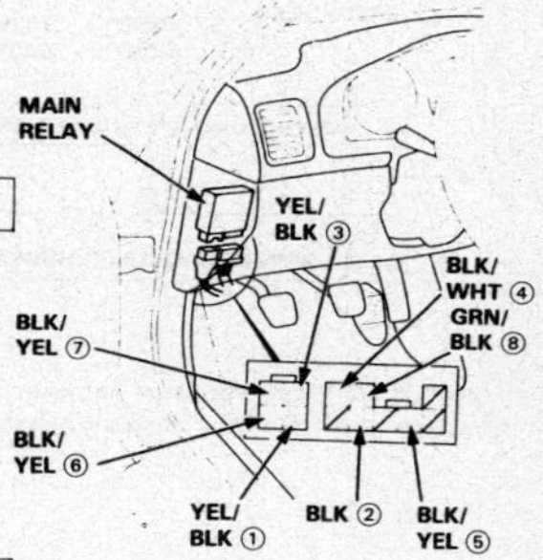

1. Turn the ignition switch off.
2. Connect the test harness between the ECU and connector. Disconnect "A" connector from the main wire harness only, not the ECU

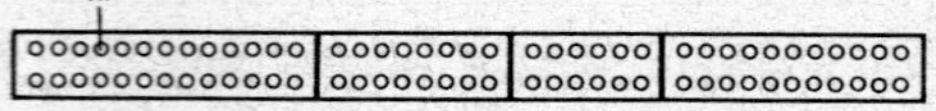

3. Check for continuity between GRN/BLK terminal ⑧ and A7 terminal.
4. Does continuity exist?
   - NO → Repair open in GRN/BLK wire between ECU (A7) and Main Relay.
   - YES → continue.
5. Reconnect "A" connector to the main wire harness.

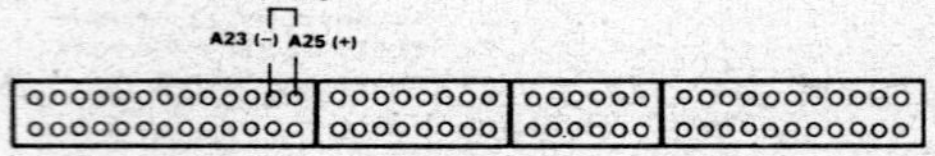

6. Connect the main relay connectors.
7. Turn the ignition switch ON.
8. Measure the voltage between A25 (+) terminal and A23 (−) terminal.
9. Is there battery voltage?
   - NO → Replace main relay.
   - YES → continue.
10. Measure the voltage between A7 (+) terminal and A23 (−) terminal.

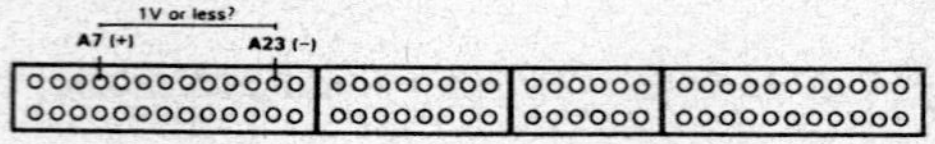

11. Is there 1V or less? (for 2 seconds)
    - NO → Substitute a known-good ECU and recheck. If prescribed voltage is now available, replace the original ECU.
    - YES → Check the main relay

Description

Satisfactory power performance is achieved by closing and opening the bypass control valves. High torque at low RPM is achieved when the valves are closed, whereas high power at high RPM is achieved by when the valves are opened.

ENGINE RPM IS ABOVE 3,300 rpm ⇒ BYPASS LOW CONTROL SOLENOID VALVE OFF

ENGINE RPM IS ABOVE 4,000 rpm ⇒ BYPASS HIGH CONTROL SOLENOID VALVE OFF

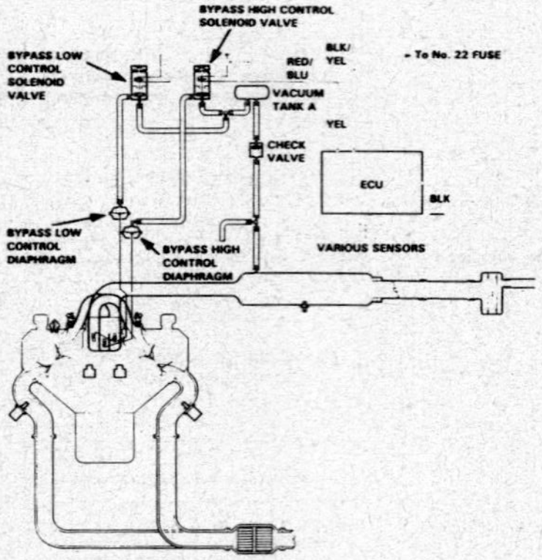

LOW VALVE : CLOSED
HIGH VALVE : CLOSED

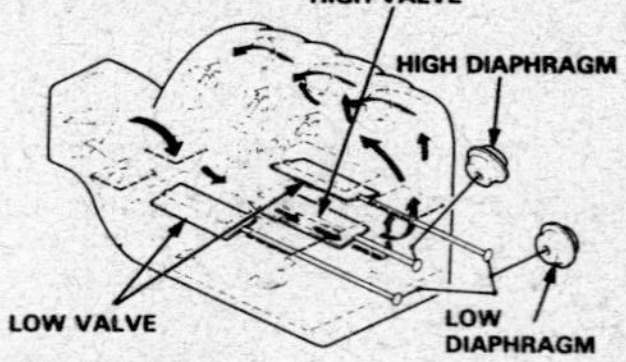

LOW VALVE : OPEN
HIGH VALVE : CLOSED

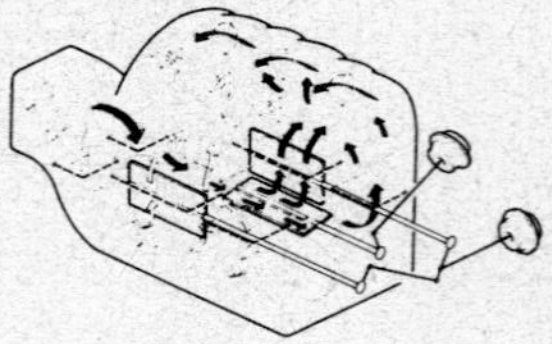

LOW VALVE: OPEN
HIGH VALVE: OPEN

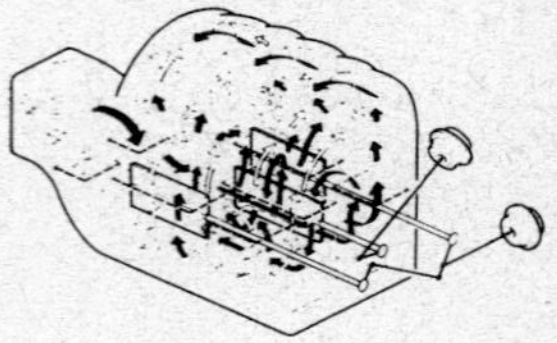

Inspection of Chamber Volume Control System.

- Start engine and allow it to idle.
- Remove # 2 vacuum hose from the bypass low control diaphragm and connect vacuum gauge to the hose.
- Is there vacuum?
  - YES
  - NO → Disconnect the #4 hose on bypass low control solenoid valve and connect a vacuum gauge to the hose.
    - Is there vacuum?
      - NO → **Repair the blockage or vacuum leak between the vacuum tank and the intake manifold.**
      - YES → Disconnect the 6P connector.
        - Measure voltage between BLK/YEL (+) terminal and RED/BLU (–) terminal.
        - Is there battery voltage?
          - YES → Replace the bypass low control solenoid valve.
          - NO → Measure voltage between BLK/YEL (+) terminal and body ground.
            - Is there battery voltage?
              - NO → **Repair open in BLK/YEL wire between the 6P connector and No. 22 fuse.**
              - YES → Turn the ignition switch OFF.
                - Connect the test harness between the ECU and connector.
                - Check for continuity of RED/BLU wire between ECU (A14) and the 6P connector.
                - Does continuity exist?
                  - NO → **Repair open in RED/BLU wire between ECU (A14) and the 6P connector.**
                  - YES → **Substitute a known-good ECU and recheck. If symptom goes away, replace the original ECU.**

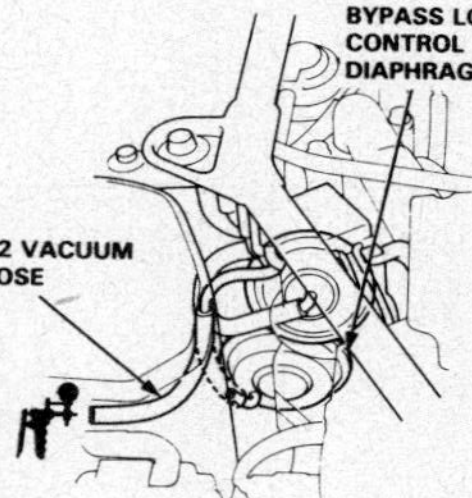

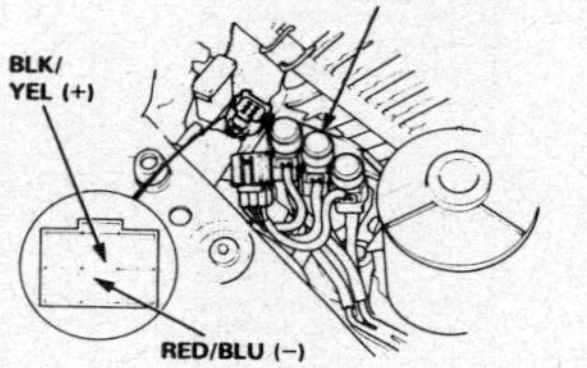

- YES (from "Is there vacuum?" above) → Remove #8 vacuum hose from the bypass high control diaphragm and connect vacuum gauge to the hose.
- Is there vacuum?
  - YES
  - NO → Disconnect the #4 hose on bypass high control solenoid valve and connect a vacuum gauge to the hose.
    - Is there vacuum?
      - NO → **Repair the blockage or vacuum leak between the vacuum tank and the intake manifold.**
      - YES → Disconnect the 6P connector.
        - Measure voltage between BLK/YEL (+) terminal and YEL (–) terminal.

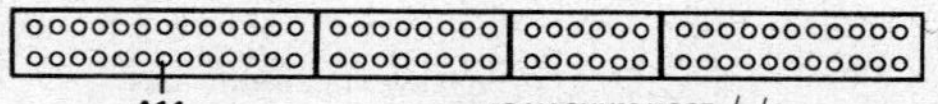

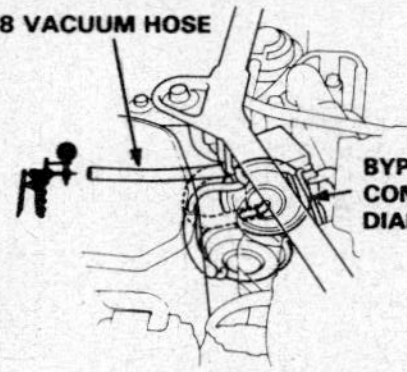

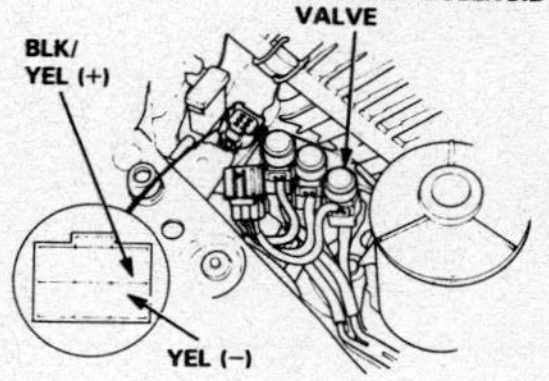

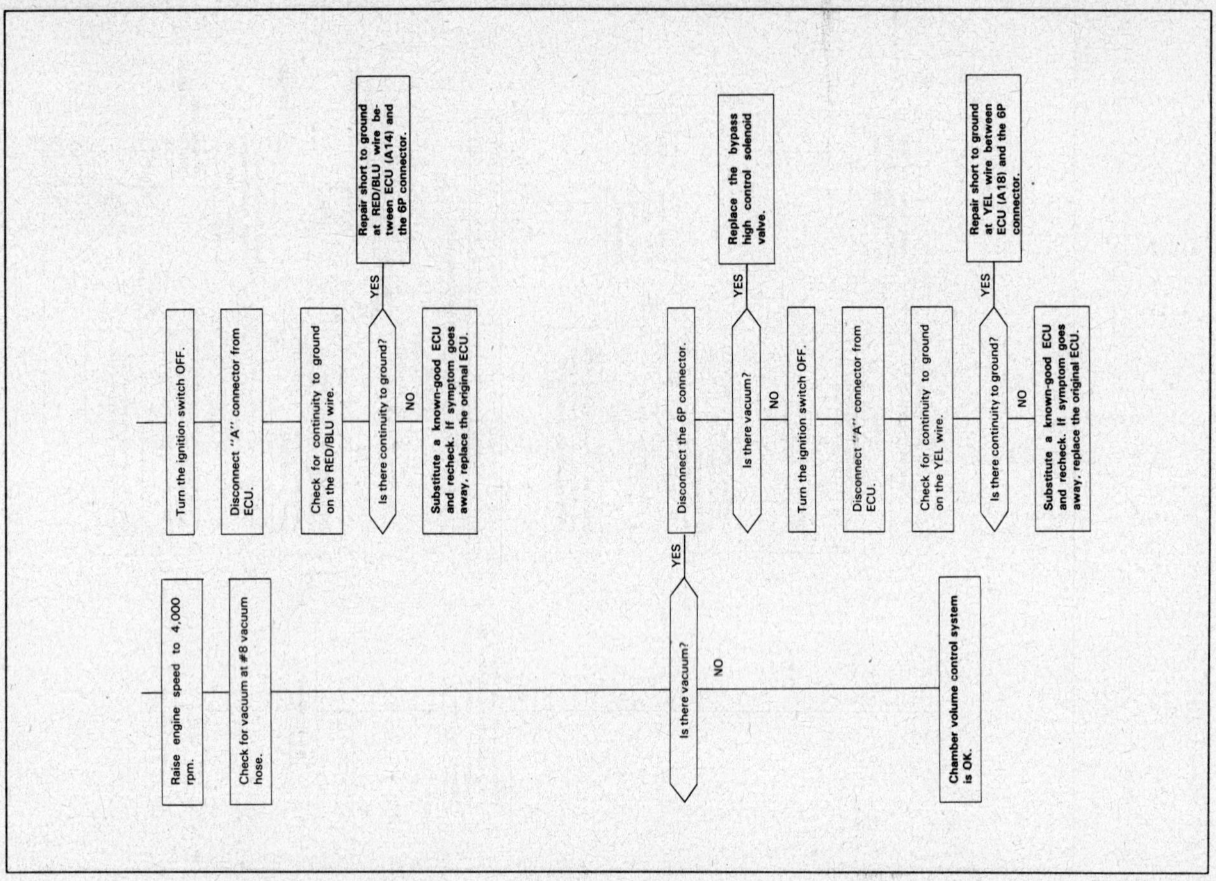
Raise engine speed to 4,000 rpm.
Check for vacuum at #8 vacuum hose.
Is there vacuum?
YES
NO
Chamber volume control system is OK.
Disconnect the 6P connector.
Is there vacuum?
YES
Replace the bypass high control solenoid valve.
NO
Turn the ignition switch OFF.
Disconnect "A" connector from ECU.
Check for continuity to ground on the YEL wire.
Is there continuity to ground?
YES
Repair short to ground at YEL wire between ECU (A18) and the 6P connector.
NO
Substitute a known-good ECU and recheck. If symptom goes away, replace the original ECU.
Turn the ignition switch OFF.
Disconnect "A" connector from ECU.
Check for continuity to ground on the RED/BLU wire.
Is there continuity to ground?
YES
Repair short to ground at RED/BLU wire between ECU (A14) and the 6P connector.
NO
Substitute a known-good ECU and recheck. If symptom goes away, replace the original ECU.

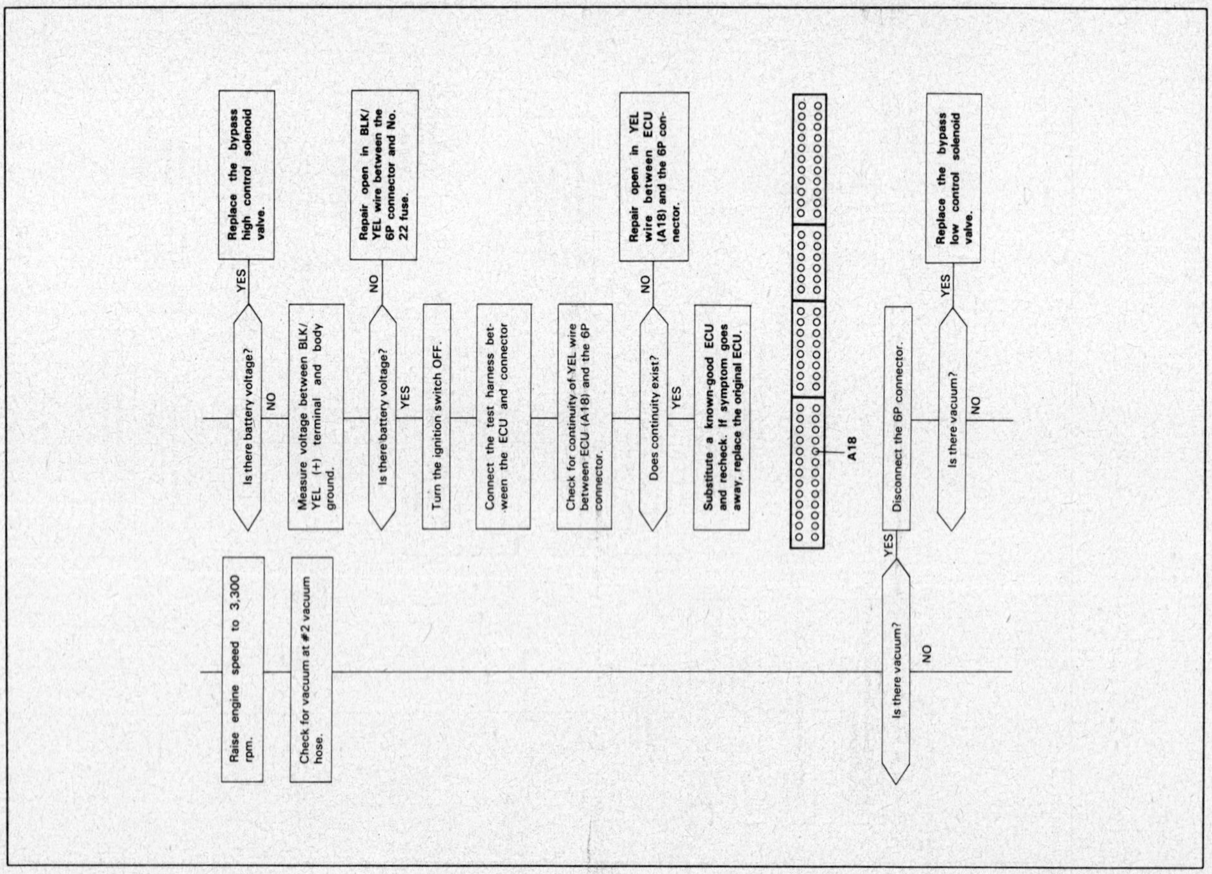
Raise engine speed to 3,300 rpm.
Check for vacuum at #2 vacuum hose.
Is there vacuum?
YES
NO
Disconnect the 6P connector.
Is there vacuum?
YES
Replace the bypass low control solenoid valve.
NO
Is there battery voltage?
YES
Replace the bypass high control solenoid valve.
NO
Measure voltage between BLK/YEL (+) terminal and body ground.
Is there battery voltage?
NO
Repair open in BLK/YEL wire between the 6P connector and No. 22 fuse.
YES
Turn the ignition switch OFF.
Connect the test harness between the ECU and connector
Check for continuity of YEL wire between ECU (A18) and the 6P connector.
Does continuity exist?
NO
Repair open in YEL wire between ECU (A18) and the 6P connector.
YES
Substitute a known-good ECU and recheck. If symptom goes away, replace the original ECU.
A18

Inspection of Air Injection System.

Start the engine and warm up to normal operating temperature (the cooling fan comes on).

Disconnect the #7 vacuum hose from the air suction valve and connect a vacuum gauge to the hose.

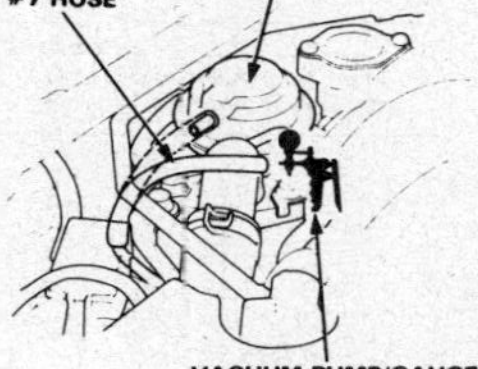

Is there vacuum?

- YES → Disconnect the 6P connector.
  - Is there vacuum?
    - YES → Check vacuum hose routing of entire air injection system. If hose routing is OK, replace air suction control solenoid valve.
    - NO → Inspect GRY wire for a short to body around between ECU (A17) and the 6P connector. If wire is OK, substitute a known-good ECU and recheck. If symptom goes away, replace the original ECU.
- NO ↓

Block rear wheels and set the parking brake. Jack up the front of the car and support with safety stands.

⚠WARNING Block rear wheels before jacking up front of car.

Place the transmission in second or "2."

Check for vacuum during deceleration above 9 mph and 1,500 rpm.

Is there vacuum?

- NO → Turn the ignition switch OFF.
  - Disconnect the 6P connector.
  - Restart the engine.
  - Measure voltage between BLK/YEL (+) terminal and GRY (−) terminal during deceleration above 9 mph and 1,500 rpm.
  - Is there battery voltage?
    - YES → Inspect vacuum hose routing. If OK, replace the air suction control solenoid value.
    - NO → Measure voltage between BLK/YEL (+) terminal and body ground.
      - Is there battery voltage?
        - NO → Repair open in BLK/YEL wire between No. 22 fuse and the 6P connector.
        - YES → Inspect for an open in GRY wire between ECU (A17) and the 6P connector. If wire is OK, substitute a known-good ECU and recheck. If symptom goes away, replace the original ECU.
- YES ↓

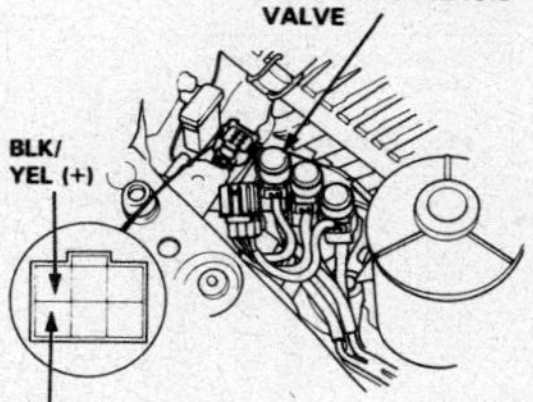

Reconnect the #7 hose.

Check for air suction noise (bubbling noise) from the air suction valve during operation above 9 mph and 1,500 rpm.

Can air suction noise be heard?

- NO → Replace the air suction valve.
- YES → Air Injection System is OK.

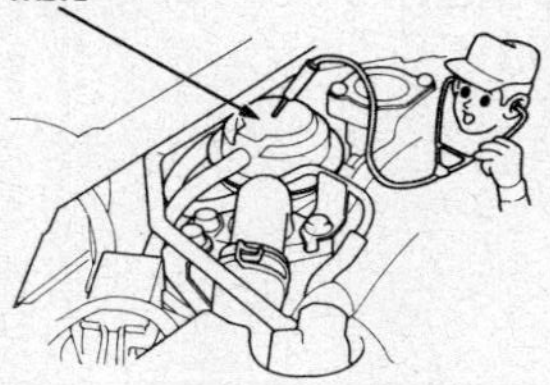

Inspection of Evaporative Emission Controls

Disconnect vacuum hose from the purge control diaphragm valve (on the charcoal canister) and connect a vacuum gauge to the hose.

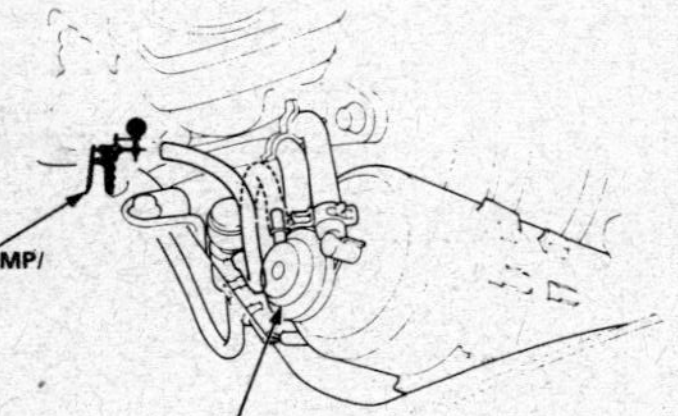

Start the engine and allow it to idle.
NOTE: Engine coolant temperature must be below 70°C (158°F).

Is there vacuum ?

- YES → Disconnect the 2P connector.

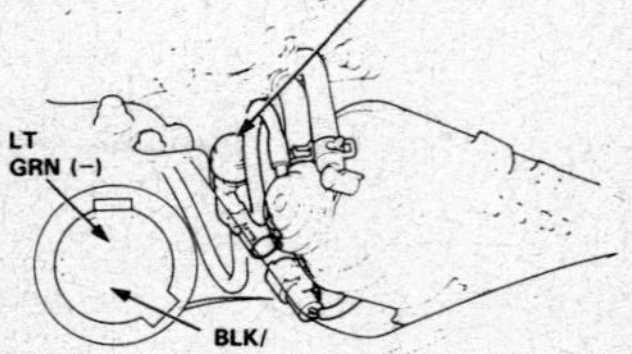

Measure voltage between BLK/YEL (+) terminal and LT GRN (–) terminal.

Is there battery voltage ?

- YES → Inspect vacuum hose routing. If OK, replace purge control solenoid valve.
- NO → Measure voltage between BLK/YEL (+) terminal and body ground.

Is there battery voltage ?

- NO → Repair open in BLK/YEL wire between No. 22 fuse and the 2P connector.
- YES → Inspect LT GRN wire for an open to body ground between ECU (A20) and the connector. If wire is OK, substitute a known-good ECU and recheck. If symptom goes away, replace the original ECU.

- NO (Is there vacuum ?) →

Warm up the engine to normal operating temperature (cooling fan comes on).

Hold engine at 2,000 rpm.

Check for vacuum at #6 vacuum hose after starting the engine.

Is there manifold vacuum?

- NO → Disconnect the 2P connector.

  Is there manifold vacuum?

  - NO → Inspect vacuum hose routing. If OK, replace purge control solenoid valve.
  - YES → Inspect for a short in LT GRN wire between ECU (A20) and the connector. If wire is OK, substitute a known-good ECU and recheck. If symptom goes away, replace the original ECU.

- YES →

Reconnect the hose.

Remove fuel filler cap.

TROUBLESHOOTING CHARTS—1991–92 LEGEND

Connect a vacuum gauge to canister purge air hose.

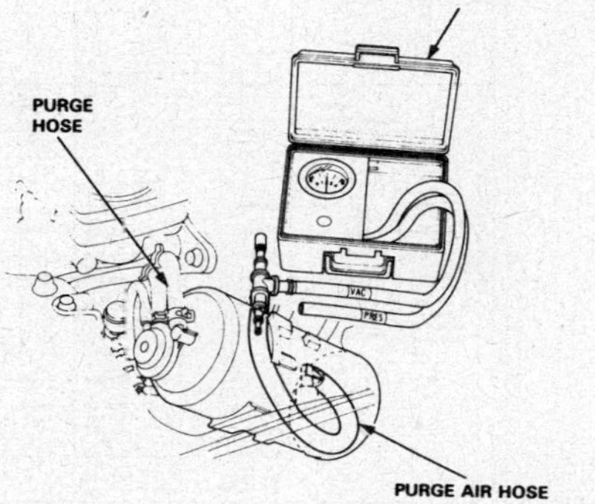

Start the engine and raise speed to 3,500 rpm.

Does vacuum appear on gauge within 1 minute?

- YES: **See two-way valve test to complete. Evaporative emission controls are OK.**
- NO: Connect a vacuum gauge to the canister purge hose and raise the engine speed to 3,500 rpm.

Does vacuum appear on the gauge?

- NO: **Inspect the purge hose and throttle body port for pinch or blockage.**
- YES: **Replace the canister.**

VACUUM COMPONENT AND HOSE ROUTING—VIGOR

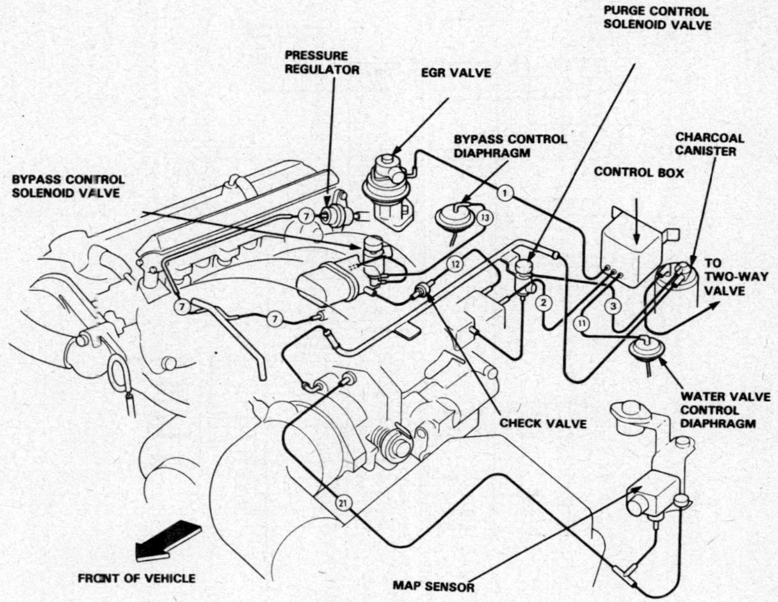

ECU INPUT/OUTPUT WIRING SCHEMATIC—VIGOR

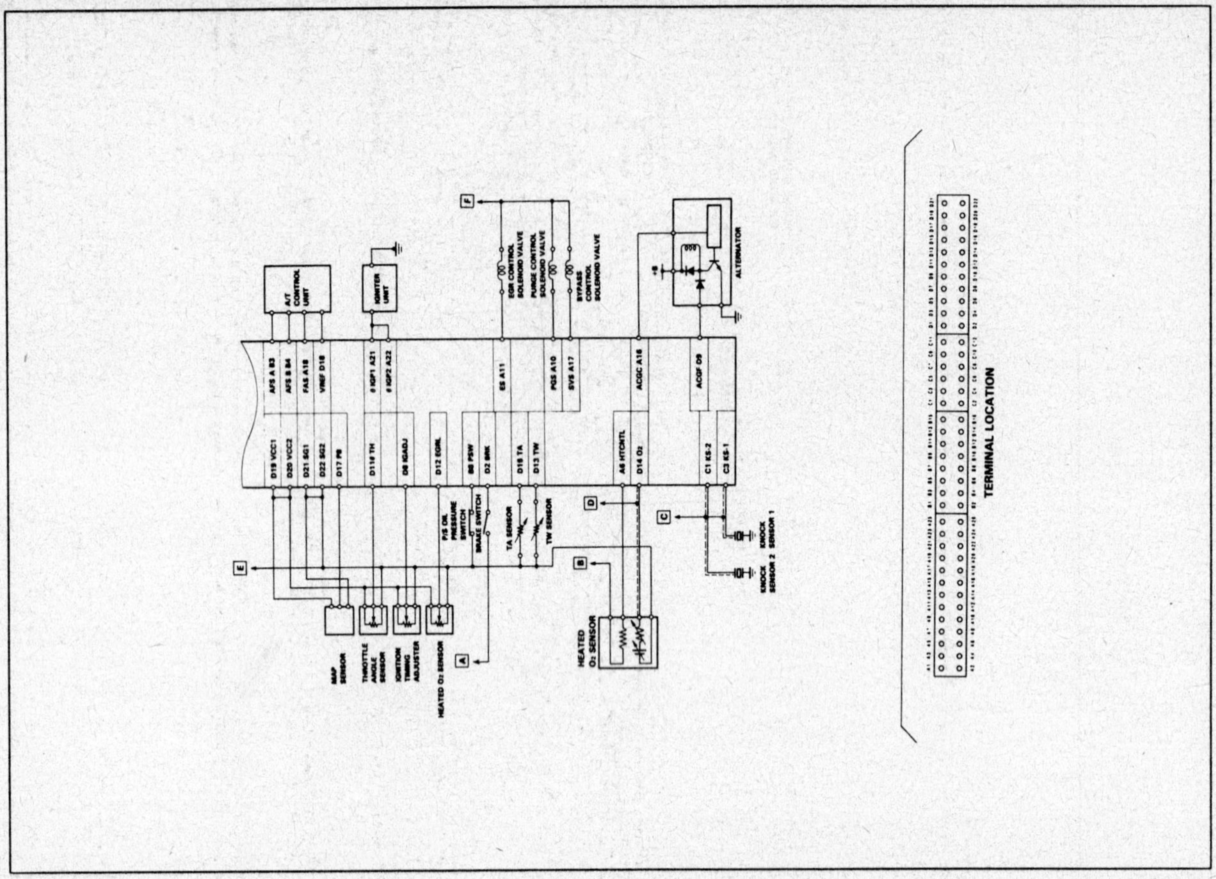

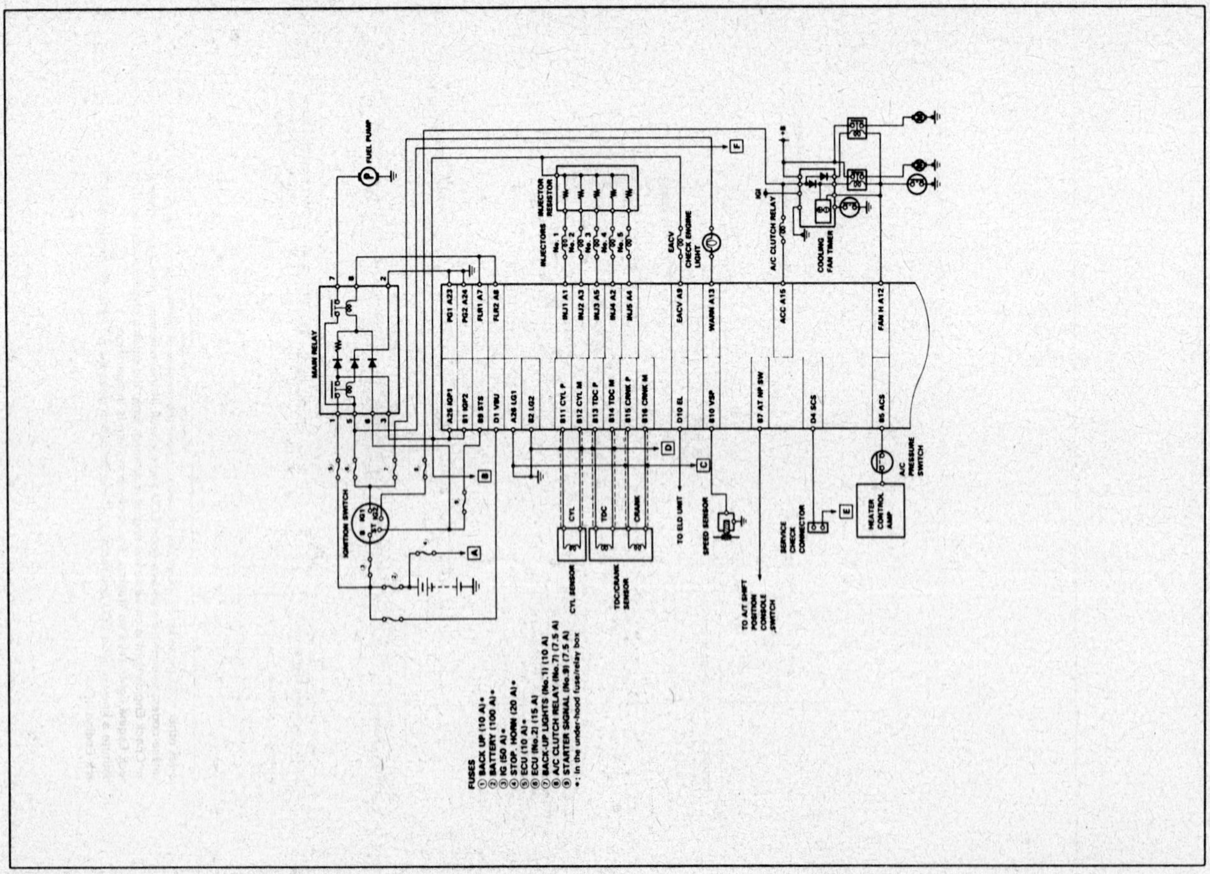

SYMPTOM TO SYSTEM DIAGNOSIS

NOTE: Across each row in the chart, the systems that could be sources of a symptom are ranked in the order they should be inspected starting with ①. Find the symptom in the left column, read across to the most likely source, then refer to the page listed at the top of that column. If inspection shows the system is OK, try the next most likely system ②, etc.

SYMPTOM		ECU	OXYGEN SENSOR	MANIFOLD ABSOLUTE PRESSURE SENSOR	TDC/CRANK/CYL SENSOR	COOLANT TEMPERATURE SENSOR	THROTTLE ANGLE SENSOR	INTAKE AIR TEMPERATURE SENSOR	ATMOSPHERIC PRESSURE SENSOR	IGNITION OUTPUT SIGNAL	VEHICLE SPEED SENSOR
		PGM-FI									
CHECK ENGINE LIGHT** TURNS ON		□ or □	(CHECK)	(CHECK)	(CHECK)	(CHECK)	(CHECK)	(CHECK)	(CHECK)	(CHECK)	(CHECK)
CHECK ENGINE LIGHT BLINKS		0 or [illegible]	1 or 41 or 43	3 or 5	4 or 8 or 9	6	7	10	13	15	17
ENGINE WON'T START		③			③					③	
DIFFICULT TO START ENGINE WHEN COLD		(BU)		③	②	①					
IRREGULAR IDLING	WHEN COLD FAST IDLE OUT OF SPEC	(BU)				③					
	ROUGH IDLE	(BU)		③							
	WHEN WARM RPM TOO HIGH	(BU)									
	WHEN WARM RPM TOO LOW	(BU)									
FREQUENT STALLING	WHILE WARMING UP	(BU)				③					
	AFTER WARMING UP	(BU)									
POOR PERFORMANCE	MISFIRE OR ROUGH RUNNING	(BU)		②	③						
	FAILS EMISSION TEST	(BU)	③	②							
	LOSS OF POWER	(BU)		③			②				

SYMPTOM		IGNITION TIMING ADJUSTER	ELECTRIC LOAD DETECTOR	KNOCK SENSOR	A/T FI SIGNAL A	A/T FI SIGNAL B	ELECTRONIC AIR CONTROL VALVE	OTHER IDLE CONTROLS	FUEL INJECTOR	FUEL METERING	OTHER FUEL SUPPLY	AIR INTAKE	EGR CONTROL SYSTEM	OTHER EMISSION CONTROLS
		PGM-FI					IDLE CONTROL		FUEL SUPPLY				EMISSION CONTROL	
CHECK ENGINE LIGHT** TURNS ON		(CHECK)	□	(CHECK)	□	□	(CHECK)		(CHECK)	(CHECK)			(CHECK)	
CHECK ENGINE LIGHT BLINKS		18	20	23 or 53	30	31	14		16	45			12	
ENGINE WON'T START									②		①			
DIFFICULT TO START ENGINE WHEN COLD														
IRREGULAR IDLING	WHEN COLD FAST IDLE OUT OF SPEC						①	②						
	ROUGH IDLE						①		②				③	
	WHEN WARM RPM TOO HIGH						①	②						
	WHEN WARM RPM TOO LOW		③				①		②					
FREQUENT STALLING	WHILE WARMING UP						①	②			③			
	AFTER WARMING UP						③				①		②	
POOR PERFORMANCE	MISFIRE OR ROUGH RUNNING								①				③	
	FAILS EMISSION TEST									③				①
	LOSS OF POWER								③		①	③		

- * If codes other than those listed above are indicated, count the number of blinks again. If the indicator is in fact blinking these codes, substitute a known-good ECU and recheck. If the indication goes away, replace the original ECU.

(BU) If the Check Engine light is on while the engine is running, jump the service check connector. If no code is displayed (Check Engine light stays on steady), the back-up system is in operation. Substitute a known-good ECU and recheck. If the indication goes away, replace the original ECU.

** Check Engine light
USA: (CHECK) CANADA: (engine symbol)

SYMPTOM TO SYSTEM DIAGNOSIS

NOTE:
- Across each row in the chart, the sub systems that could be sources of a symptom are ranked in the order they should be inspected, starting with ①. Find the symptom in the left column, read across to the most likely source, then refer to the page listed at the top of that column. If inspection shows the system is OK, try the next system ②, etc.

SYMPTOM \ SUB SYSTEM		IDLE ADJUSTING SCREW	EACV	AIR CONDI-TIONING SIGNAL	ALTER-NATOR FR SIGNAL	A/T SHIFT POSITION SIGNAL	STARTER SWITCH SIGNAL	BRAKE SWITCH SIGNAL	P/S OIL PRESSURE SWITCH SIGNAL	FAST IDLE VALVE	HOSES AND CONNEC-TIONS
PAGE		88	74	76	78	80	82	84	86	87	•
DIFFICULT TO START ENGINE WHEN COLD										①	
WHEN COLD FAST IDLE OUT OF SPEC (1,000–2,000 rpm)		③	②							①	
ROUGH IDLE			②								①
WHEN WARM RPM TOO HIGH		③	①						③	②	③
WHEN WARM RPM TOO LOW	Idle speed is below specified rpm (no load)	②	①								
	Idle speed does not increase after initial start up.		①								
	On models with automatic transmission, the idle speed drops in gear		②			①					
	Idle speeds drops when air conditioner in ON		②	①							
	Idle speed drops when steering wheel is turning		②						①		
	Idle speed fluctuates with electrical load		②		③						①
FREQUENT STALLING	WHILE WARMING UP	②	①								
	AFTER WARMING UP	①	②								
FAILS EMISSION TEST											①

NOTE: Across each row in the chart, the systems that could be sources of a symptom are ranked in the order they should be inspected starting with ①. Find the symptom in the left column, read across to the most likely source, then refer to the page listed at the top of that column. If inspection shows the system is OK, try the next most likely system ②, etc.

SYMPTOM \ SUB SYSTEM		FUEL INJECTOR	PRESSURE REGULATOR	FUEL FILTER	FUEL PUMP	MAIN RELAY	CONTAMI-NATED FUEL
ENGINE WON'T START				③	①	②	
DIFFICULT TO START ENGINE WHEN COLD OR HOT				①			
ROUGH IDLE		①					②
POOR PERFORMANCE	MISFIRE OR ROUGH RUNNING	①	③				②
	FAILS EMISSION TEST	②	①				
	LOSS OF POWER	③		②	③		
FREQUENT STALLING	WHILE WARMING UP		①				
	AFTER WARMING UP		①				

SYMPTOM TO SYSTEM DIAGNOSIS

NOTE: Across each row in the chart, the sub systems that could be sources of a symptom are ranked in the order they should be inspected starting with ①. Find the symptom in the left column, read across to the most likely source, then refer to the page listed at the top of that column. If inspection shows the system is OK, try the next system ②, etc.

SUB SYSTEM / SYMPTOM	THROTTLE CABLE	THROTTLE BODY	BYPASS CONTROL SYSTEM
PAGE			
WHEN COLD FAST IDLE OUT OF SPEC	③	②	①
WHEN WARM RPM TOO HIGH	②	①	
LOSS OF POWER	①		②

NOTE: Across each row in the chart, the systems that could be sources of a symptom are ranked in the order they should be inspected starting with ①. Find the symptom in the left column, read across to the most likely source, then refer to the page listed at the top of that column. If inspection shows the system is OK, try the next most likely system ②, etc.

SYMPTOM	SUB SYSTEM	CATALYTIC CONVERTER	EGR SYSTEM	POSITIVE CRANKCASE VENTILATION SYSTEM	EVAPORATIVE EMISSION CONTROLS
ROUGH IDLE			①	②	
FREQUENT STALLING	AFTER WARMING UP		①		
POOR PERFORMANCE	MISFIRE OR ROUGH RUNNING		①		
	FAILS EMISSION TEST	①	③		②
	LOSS OF POWER	①	②		

TROUBLESHOOTING CHARTS—VIGOR

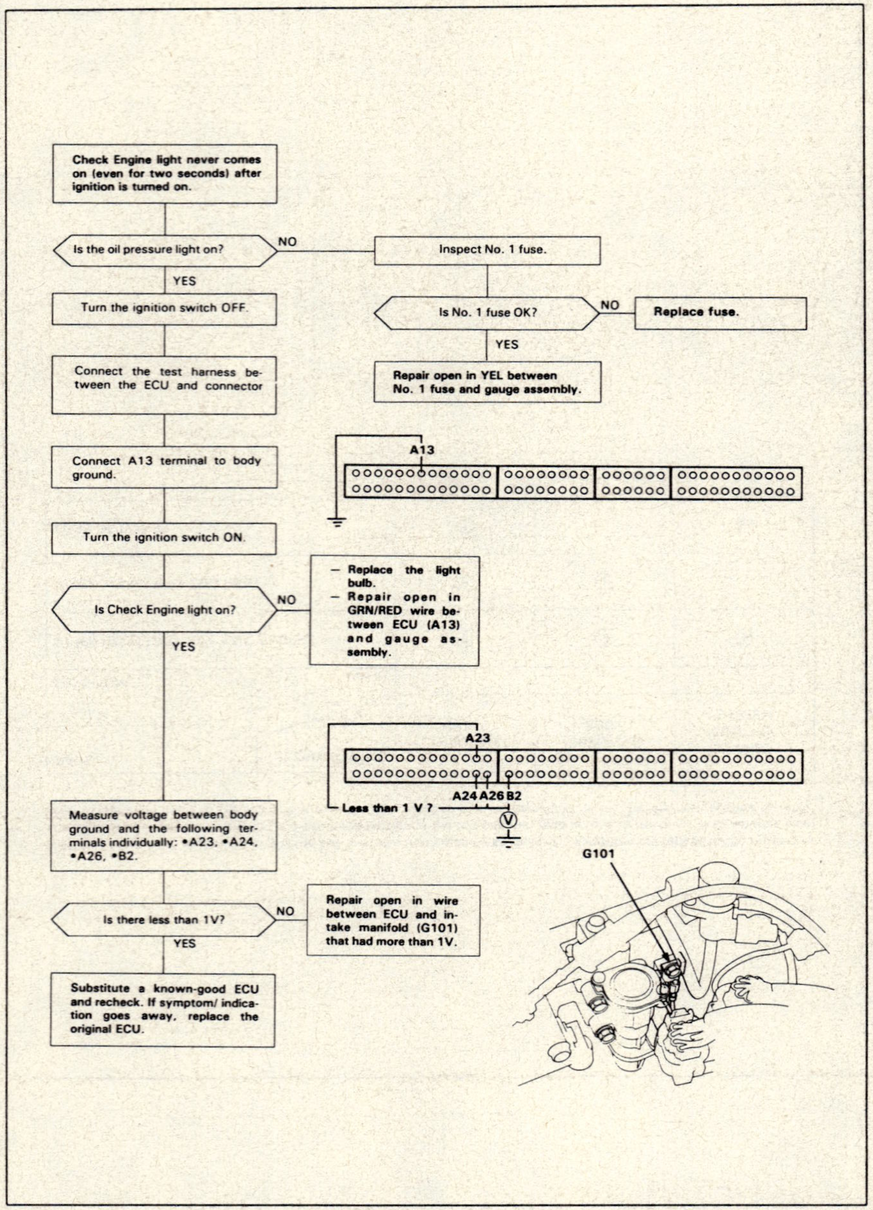

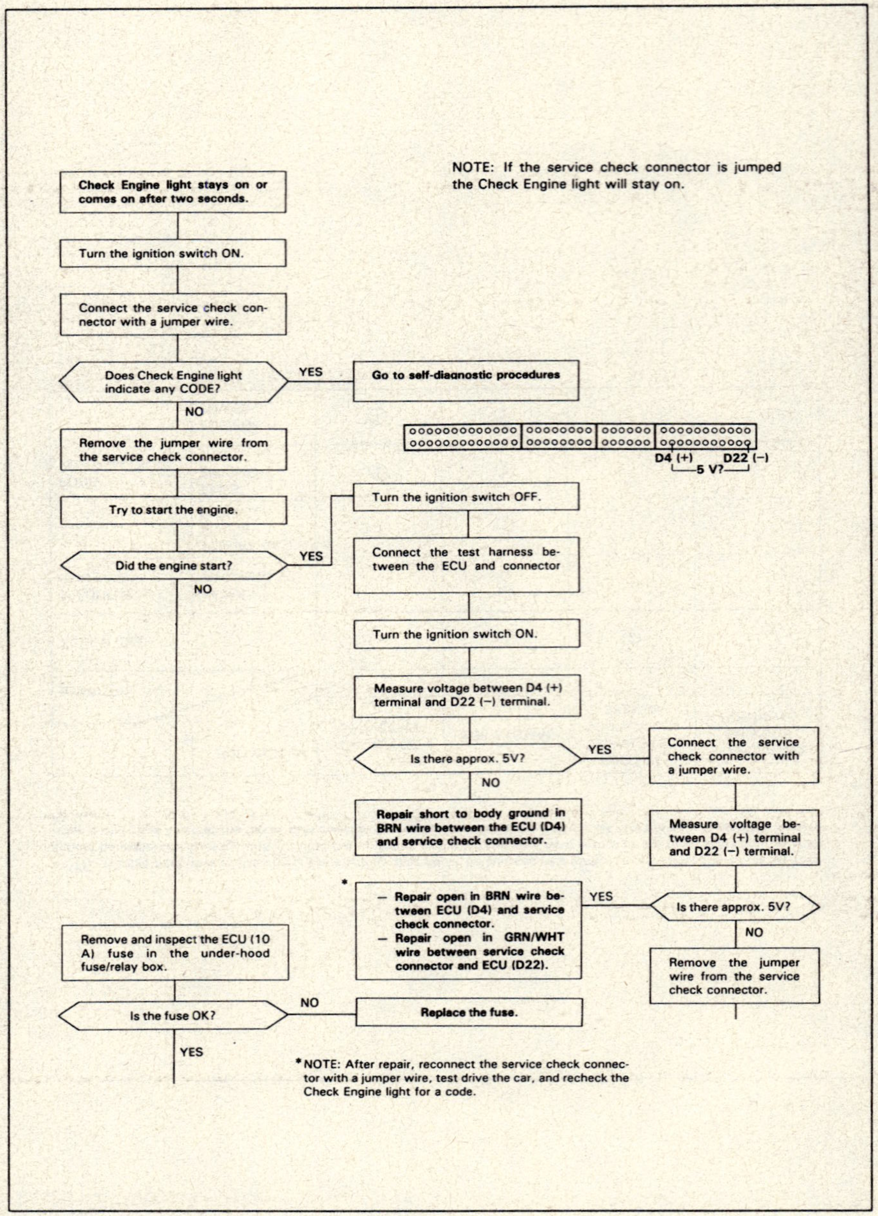

TROUBLESHOOTING CHARTS—VIGOR

1. Inspect the ECU (No. 2) (15 A) fuse in the under-dash fuse box.
2. Is the fuse OK?
   - NO → Replace the fuse.
   - YES ↓
3. Turn the ignition switch ON.
4. Disconnect the 3P connector of each sensor one at a time:
   - MAP sensor
   - EGR valve lift sensor
   - Throttle angle sensor
5. Does Check Engine light remain ON?
   - NO → Replace the sensor that caused the light to go out.
   - YES ↓
6. Connect the service check connector with a jumper wire.
7. Disconnect the 3P connector of the Ignition timing adjuster.
8. Does Check Engine light indicate CODE 18?
   - YES → Replace the ignition timing adjuster.
   - NO ↓
9. Turn the ignition switch OFF.
10. Remove the jumper wire from the service check connector.
11. Connect the test harness. Disconnect the "D" connector from the ECU only, not the main wire harness
12. Check for continuity between body ground terminals D19, D20.

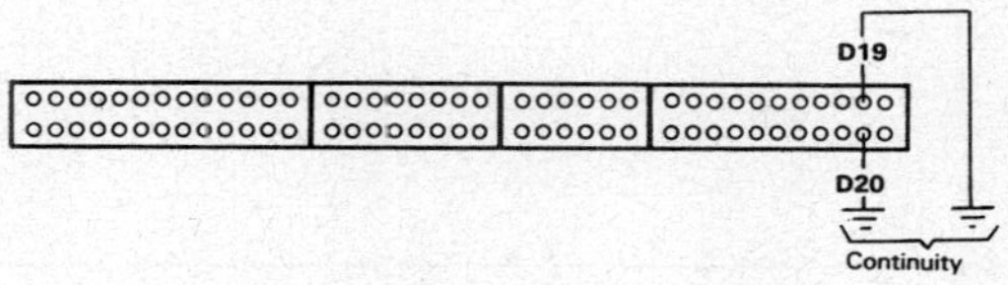

---

1. Turn the ignition switch OFF.
2. Disconnect "A" connector from the ECU.
3. Turn the ignition switch ON.
4. Is the Check Engine light ON?
   - NO → Substitute a known-good ECU and recheck. If symptom/indication goes away, replace the original ECU.
   - YES → Repair short to body ground in GRN/RED wire between the ECU (A13) and Check Engine light.

---

1. Does continuity exist?
   - YES → Repair short to body ground in YEL/WHT wire between ECU (D19, D20) and throttle angle sensor, EGR valve lift sensor, MAP sensor and Ignition timing adjuster.
   - NO ↓
2. Reconnect all the sensor connectors. Reconnect the "D" connector to the ECU.
3. Turn the ignition switch ON.
4. Measure voltage between A26 (−) and the following: B1 (+) and A25 (+).

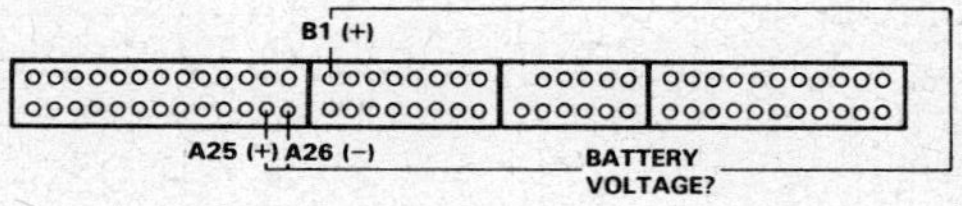

5. Is there battery voltage?
   - NO →
     - Repair open in YEL/BLK wire between ECU (A25, B1) and main relay.
     - Check main relay and wiring connectors at main relay.
   - YES → Substitute a known-good ECU and recheck. If symptom/indication goes away, replace the original ECU.

Self-diagnosis Check Engine light indicates code 1: A problem in the Oxygen (O_2) Sensor circuit.

The Oxygen sensor detects the oxygen content in the exhaust gas and signals the ECU. In operation, the ECU receives the signals from the sensor and varies the duration during which fuel is injected. The oxygen sensor has an internal heater. The heater stabilizes the sensor's output. The oxygen sensor is installed in the exhaust manifold.

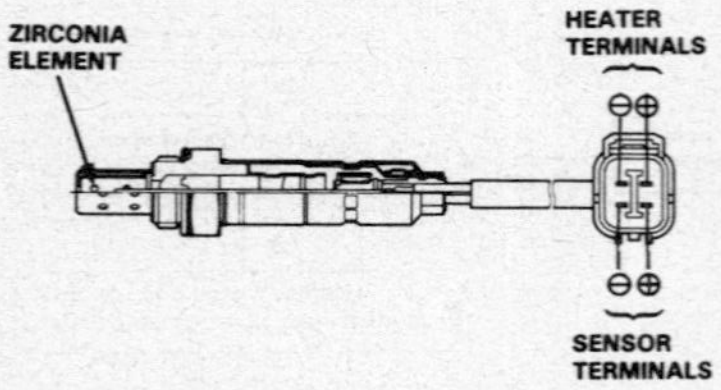

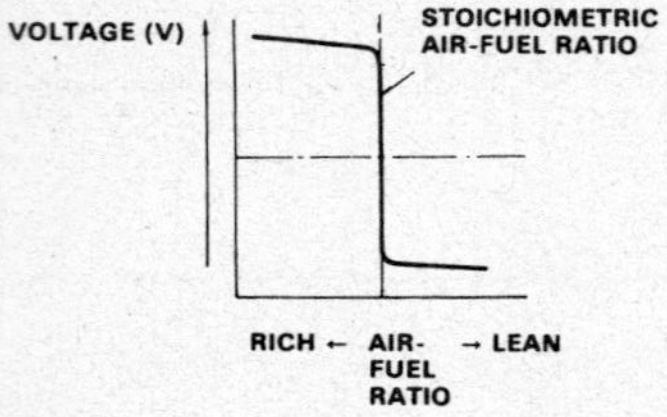

Code 1

- Check Engine light has been reported on.
- With service check connector jumped CODE 1 is indicated.

Do the ECU Reset Procedures

Warm up engine to normal operating temperature (cooling fan comes on).

Run engine for 60 seconds.

Road test with the transmission in 2nd gear (M/T: 4th gear). Starting at 1200 rpm, accelerate using wide open throttle for at least 5 seconds. Then decelerate for at least 5 seconds with the throttle completely closed.

Is Check Engine light on and does it indicate CODE 1?

NO → Intermittent failure, system is OK at this time. Check for poor connections or loose wires at the C202 (O_2 sensor) and ECU.

YES → test for CODE 43.

Self-diagnosis Check Engine light indicates code 3: An electrical problem in the Manifold Absolute Pressure (MAP) Sensor system.

Self-diagnosis Check Engine light indicates code 5: A mechanical problem (broken hose) in the Manifold Absolute Pressure (MAP) sensor System.

The MAP sensor converts manifold absolute pressure into electrical signals and inputs the ECU.

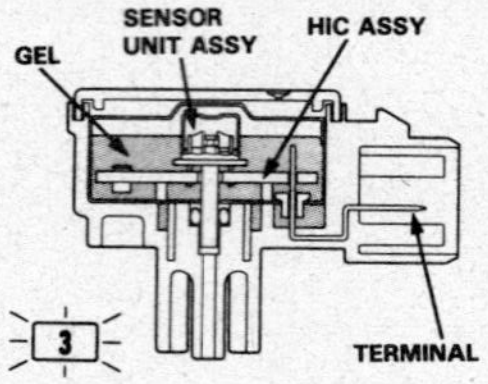

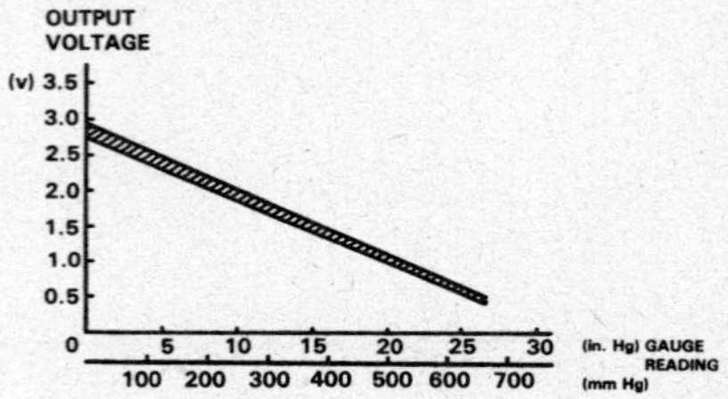

Code 3

- Engine is warm and running.
- Check Engine light has been reported on.
- With service check connector jumped CODE 3 is indicated.

Do the ECU Reset Procedures

Start the engine and allow it to idle.

Is Check Engine light on and does it indicate CODE 3?

NO → Intermittent failure, system is OK at this time (test drive may be necessary). Check for poor connection or loose wires at 329 (MAP sensor) and ECU.

YES

Turn the ignition switch OFF.

Disconnect the 3P connector from the MAP sensor.

Turn the ignition switch ON.

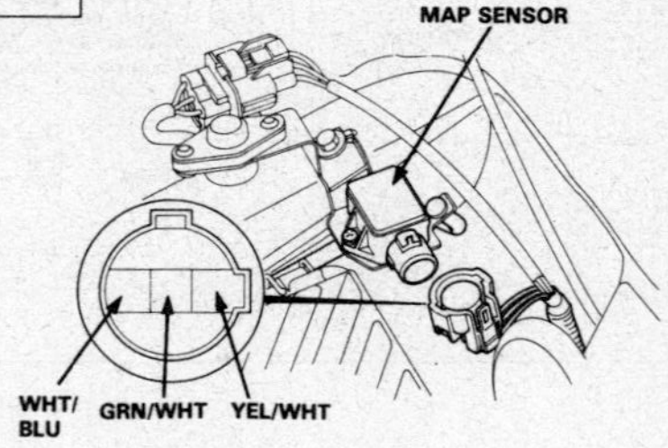

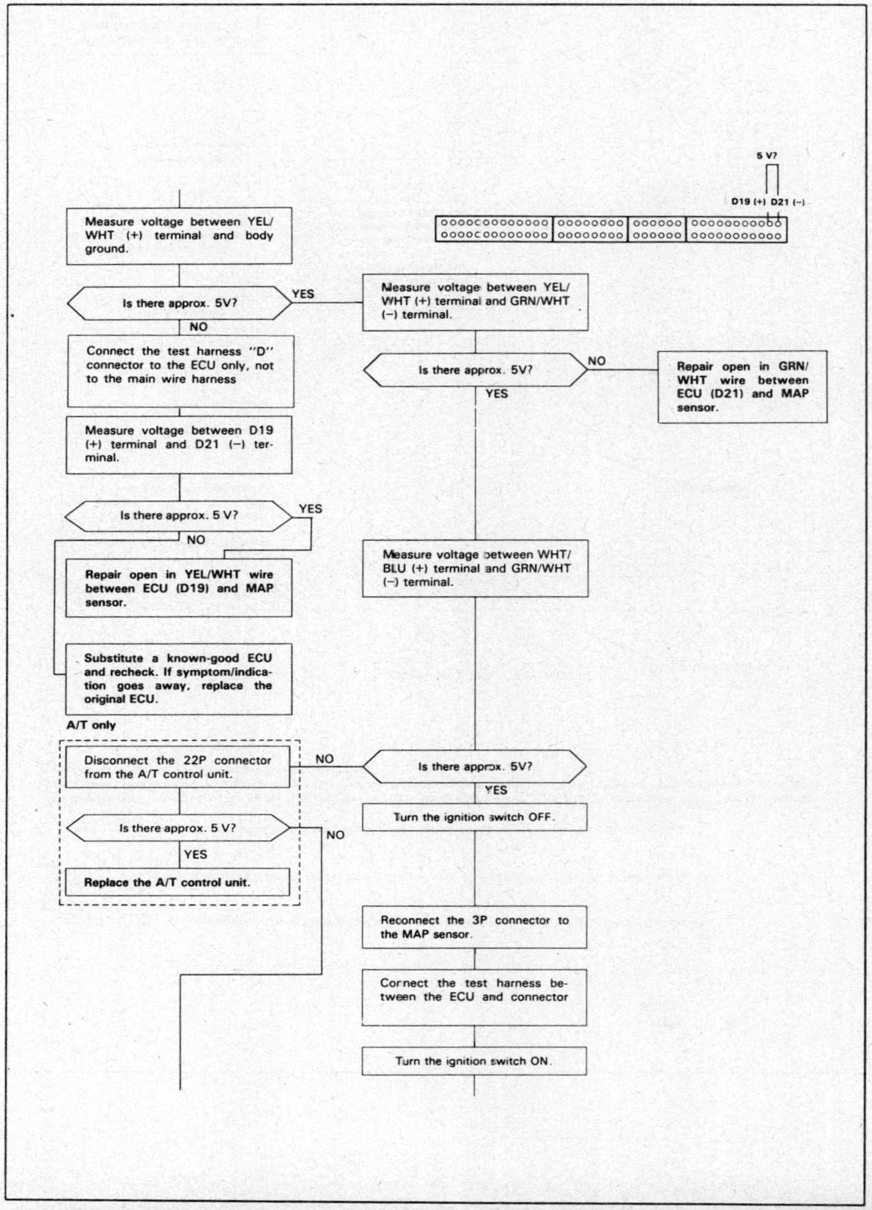
Measure voltage between YEL/WHT (+) terminal and body ground.
Is there approx. 5V?
YES
NO
Connect the test harness "D" connector to the ECU only, not to the main wire harness
Measure voltage between D19 (+) terminal and D21 (−) terminal.
Is there approx. 5 V?
YES
NO
Repair open in YEL/WHT wire between ECU (D19) and MAP sensor.
Substitute a known-good ECU and recheck. If symptom/indication goes away, replace the original ECU.
A/T only
Disconnect the 22P connector from the A/T control unit.
Is there approx. 5 V?
NO
YES
Replace the A/T control unit.
5 V?
D19 (+) D21 (−)
Measure voltage between YEL/WHT (+) terminal and GRN/WHT (−) terminal.
Is there approx. 5V?
NO
YES
Repair open in GRN/WHT wire between ECU (D21) and MAP sensor.
Measure voltage between WHT/BLU (+) terminal and GRN/WHT (−) terminal.
NO
Is there approx. 5V?
YES
Turn the ignition switch OFF.
Reconnect the 3P connector to the MAP sensor.
Connect the test harness between the ECU and connector
Turn the ignition switch ON.

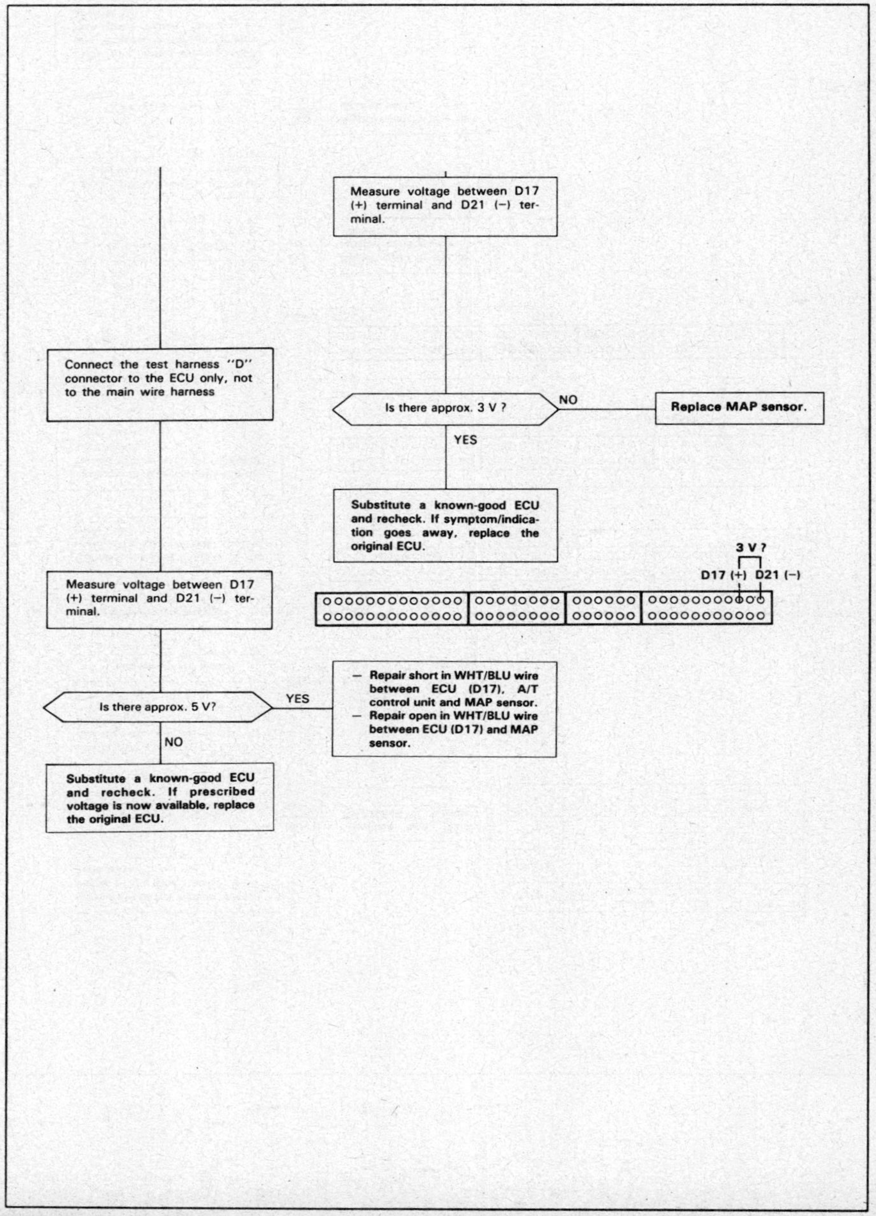
Connect the test harness "D" connector to the ECU only, not to the main wire harness
Measure voltage between D17 (+) terminal and D21 (−) terminal.
Is there approx. 5 V?
YES
NO
Substitute a known-good ECU and recheck. If prescribed voltage is now available, replace the original ECU.
Measure voltage between D17 (+) terminal and D21 (−) terminal.
Is there approx. 3 V ?
NO
Replace MAP sensor.
YES
Substitute a known-good ECU and recheck. If symptom/indication goes away, replace the original ECU.
3 V ?
D17 (+) D21 (−)
– Repair short in WHT/BLU wire between ECU (D17), A/T control unit and MAP sensor.
– Repair open in WHT/BLU wire between ECU (D17) and MAP sensor.

TROUBLESHOOTING CHARTS—VIGOR

4 — Self-diagnosis Check Engine light indicates code 4: A problem in the CRANK Sensor circuit.

8 — Self-diagnosis Check Engine light indicated code 8: A problem in the TDC Sensor circuit.

9 — Self-diagnosis Check Engine light indicates code 9: A problem in the CYL Sensor circuit.

The CRANK sensor determines timing for fuel injection and ignition of each cylinder and also detects engine RPM. The TDC sensor determines ignition timing at start-up (cranking) and when crank angle is abnormal. The CYL sensor detects the position of No.1 cylinder for sequential fuel injection to each cylinder.

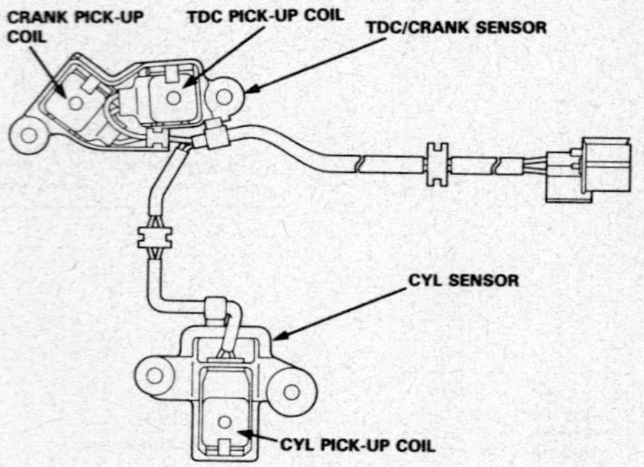

4 — 8 — or — 9

- Check Engine light has been reported on.
- With service check connector jumped CODE 4, 8 and/or 9 are indicated.

Do the ECU Reset Procedures

Start the engine.

Is Check Engine light on and does it indicate CODE 4, 8 or 9?

NO → Intermittent failure, system is OK at this time (test drive may be necessary). Check for poor connections or loose wires at C308 (located at left shock tower), C117 (TDC/CRANK/CYL sensor) and ECU.

YES ↓

Stop the engine.

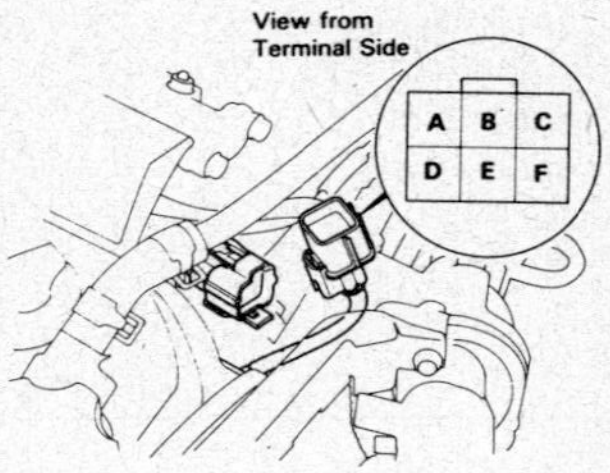

Disconnect the 6P connector from the TDC/CRANK/CYL sensor.

Measure resistance between terminals of the indicated sensor.
*see table

Is there 650–850 Ω?

NO → Replace the TDC/CRANK/CYL sensor

YES ↓

Check for continuity to body ground on all terminals individually.

Does continuity exist?

YES → Replace the TDC/CRANK/CYL sensor

NO ↓

Reconnect the connector.

Connect the test harness to the main wire harness only, not to the ECU

Measure resistance between terminals of the indicated sensor on test harness.
*see table

Is there 650–850 Ω?

NO → Repair open in the indicated sensor wires.
*see table

YES ↓

Check for continuity to body ground on B13 or B11 terminals.

Does continuity exist?

YES → Repair short in the indicated sensor wires.

NO ↓

Substitute a known-good ECU and recheck. If symptom/indication goes away, replace the original ECU.

*:

SENSOR	CODE	SENSOR TERMINAL	ECU TERMINAL	WIRE COLOR
CRANK	4	B	B16	BLU/YEL
		E	B15	YEL/GRN
TDC	8	A	B14	WHT/BLU
		D	B13	ORN/BLU
CYL	9	C	B12	WHT
		F	B11	ORN

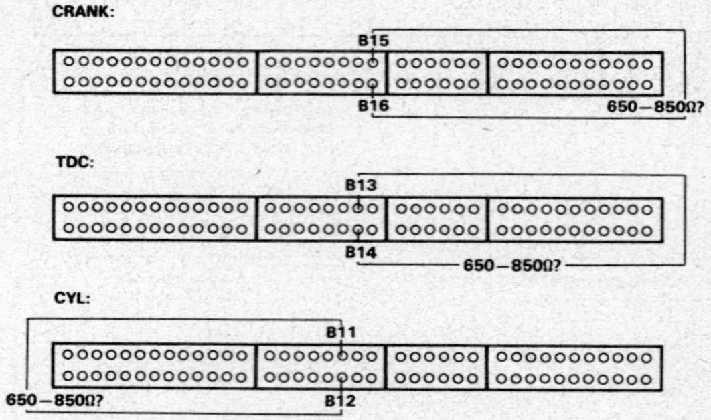

TROUBLESHOOTING CHARTS – VIGOR

CHECK 5

- Check Engine light has been reported on.
- With service check connector jumped CODE 5 is indicated.

Do the ECU Reset Procedures

Start the engine and keep engine rpm at idle.

Is Check Engine light on and does it indicate CODE 5?

NO →
- Intermittent failure, system is OK at this time (test drive may be necessary).
- Check vacuum hoses, pipes and connections.
- Make sure all connectors are secure.

YES ↓

Stop the engine.

Disconnect #21 hose from the throttle body, connect vacuum pump to the hose and apply vacuum.

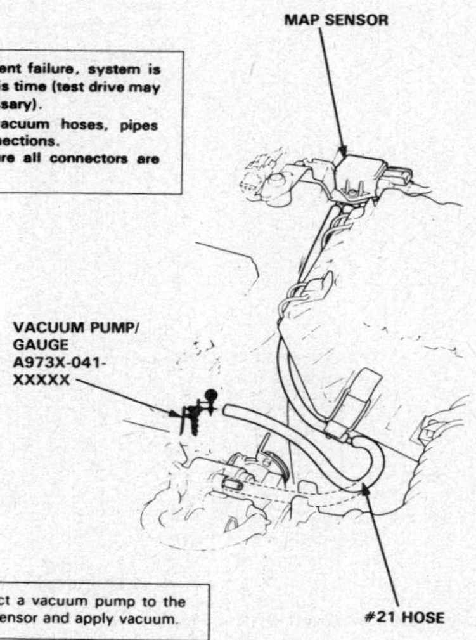

Does it hold vacuum?

NO → Connect a vacuum pump to the MAP sensor and apply vacuum.

Does it hold vacuum?

NO → Replace MAP sensor.

YES → Repair vacuum leak in hose routing between MAP sensor and intake manifold.

YES ↓

Connect a T-fitting from a vacuum gauge between the throttle body base and #21 hose.

Start the engine.

Is there manifold vacuum?

NO → Remove restriction from throttle body.

YES ↓

Stop the engine.

Connect the test harness between the ECU and connector

Turn the ignition switch ON.

Measure voltage between D17 (+) terminal and D21 (−) terminal.

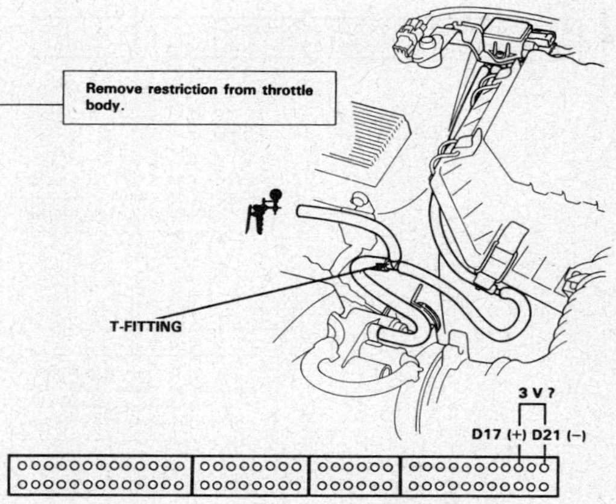

Is there approx. 3 V ?

NO → Replace MAP sensor.

YES ↓

Start the engine and allow it to idle.

Is there approx. 1 V?

NO → Replace MAP sensor.

YES ↓

Substitute a known-good ECU and recheck. If symptom/indication goes away, replace the original ECU.

6 Self-diagnosis Check Engine light indicates code 6: A problem in the Coolant Temperature (TW) Sensor circuit.

The TW sensor is a temperature dependant resistor (thermistor). The resistance of the thermistor decreases as the coolant temperature increases as shown below.

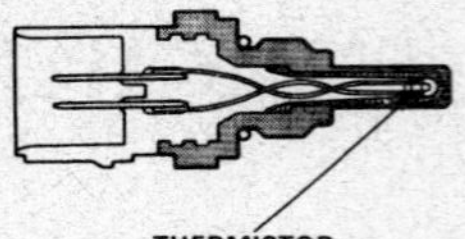

THERMISTOR

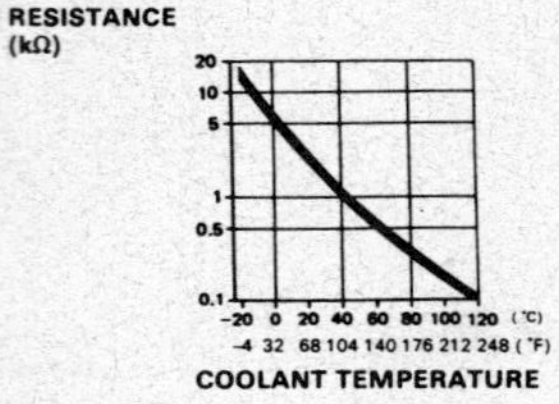

6

- Check Engine light has been reported on.
- With service check connector jumped CODE 6 is indicated.

Do the ECU Reset Procedures

Turn the ignition switch ON.

Is Check Engine light on and does it indicate CODE 6?

- NO → Intermittent failure, system is OK at this time (test drive may be necessary). Check for poor connections or loose wires at C219 (located under the hood on the front right side of the car), C127 (TW sensor) and ECU.
- YES ↓

Warm up engine to normal operating temperature (cooling fan comes on).

Turn the ignition switch OFF.

Disconnect the 2P connector from the TW sensor.

Measure resistance between the 2 terminals on the TW sensor.

Is there 20—400 Ω?

- NO → Replace TW sensor.
- YES ↓

Turn the ignition switch ON.

Measure voltage between YEL/GRN and body ground.

Is there approx. 5V ?

- YES → Measure voltage between YEL/GRN (+) terminal and GRN/WHT (−) terminal.
  - Is there approx. 5V ?
    - NO → Repair open in GRN/WHT wire between ECU (D22) and TW sensor.
    - YES → Substitute a known-good ECU and recheck. If symptom/indication goes away, replace the original ECU.
- NO ↓

A/T only

Disconnect the 22P connector from the A/T control unit.

Is there approx. 5 V?

- YES → Replace the A/T control unit.
- NO ↓

Connect the test harness "D" connector to the ECU only, not to the main wire harness

Measure voltage between D13 (+) terminal and D22 (−) terminal.

Is there approx. 5V?

- YES → Repair open or short in YEL/GRN wire between ECU (D13), A/T control unit and sensor.
- NO → Substitute a known-good ECU and recheck. If symptom/indication goes away, replace the original ECU.

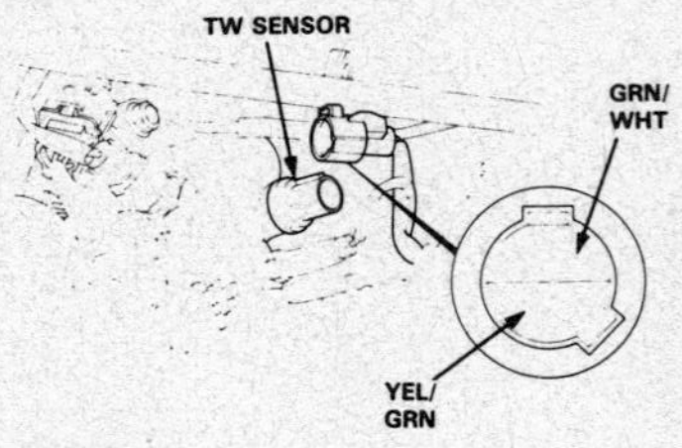

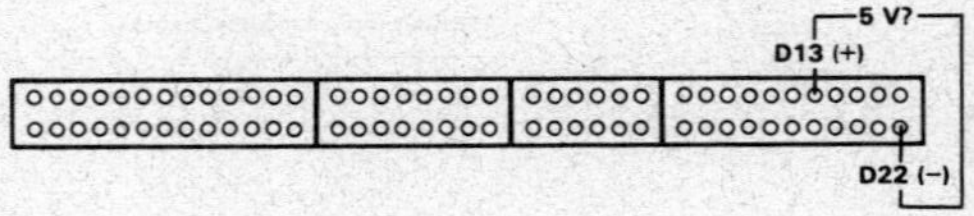

7 — Self-diagnosis Check Engine light indicates code 7: A problem in the Throttle Angle Sensor circuit.

The throttle angle sensor is a potentiometer. It is connected to the throttle valve shaft. As the throttle angle changes, the throttle angle sensor varies the voltage signal to the ECU.

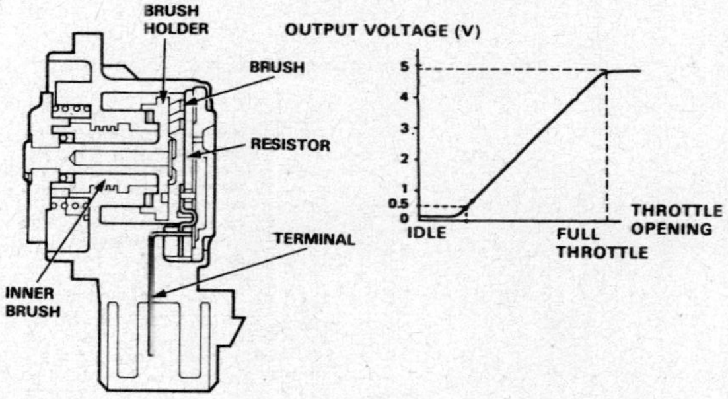

7

- Engine is running.
- Check Engine light has been reported on.
- With service check connector jumped CODE 7 is indicated.

Do the ECU Reset Procedures

Start the engine.

Is Check Engine light on and does it indicate CODE 7?

NO → Intermittent failure, system is OK at this time (test drive may be necessary). Check for poor connections or loose wires at C219 (located under the hood on the front right side of the car), C124 (throttle angle sensor) and ECU.

YES ↓

Turn the ignition switch OFF.

Disconnect the 3P connector from the throttle angle sensor.

Turn the ignition switch ON.

Measure voltage between YEL/WHT (+) terminal and GRN/WHT (−) terminal.

Is there approx. 5V ?

NO → Measure voltage between YEL/WHT (+) terminal and body ground.

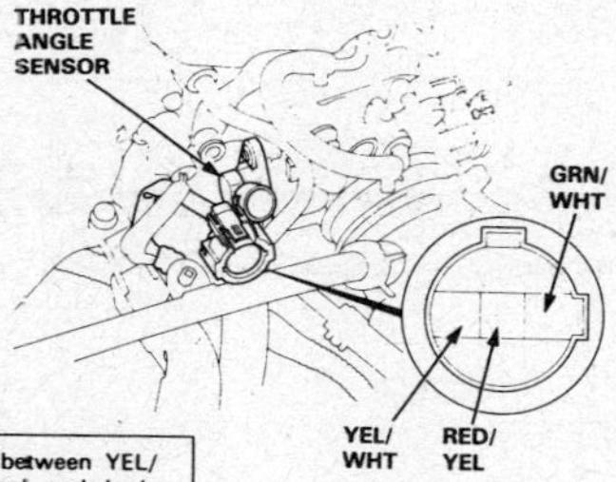

YES ↓

Turn the ignition switch OFF.

Reconnect the 3P connector.

Connect the test harness between the ECU and connector

Turn the ignition switch ON.

Measure voltage between D11(+) terminal and D22 (−) terminal.

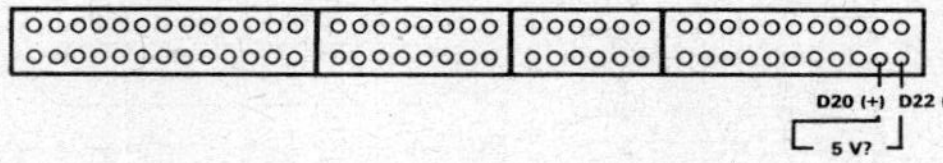

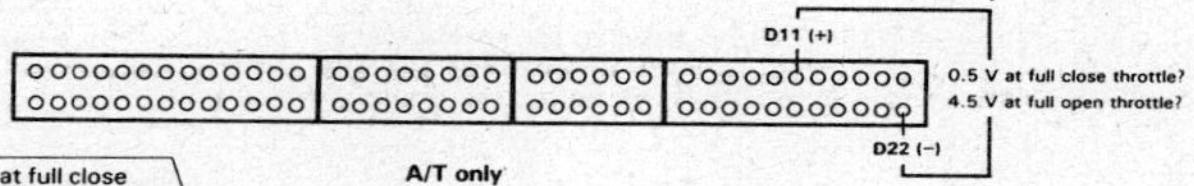

Is voltage 0.5 V at full close throttle, and approx. 4.5 V at full open throttle?
NOTE: There should be a smooth transition from 0.5 V to approx. 4.5 V as the throttle is depressed.

YES → Substitute a known-good ECU and recheck. If symptom/indication goes away, replace the original ECU.

NO ↓

A/T only

Disconnect the 22P connector from the A/T control unit.

Is voltage 0.5 V at full close throttle, and 4.8 V at full open throttle?
NOTE: There should be a smooth transition from 0.5 V to 4.8 V as the throttle is depressed.

YES → Replace the A/T control unit.

NO →
- Replace throttle angle sensor.
- Repair open or short in RED/YEL wire between ECU (D11), A/T control unit and throttle angle sensor.

(From "Measure voltage between YEL/WHT (+) terminal and body ground.")

Is there approx. 5V ?

YES → Repair open in GRN/WHT wire between ECU (D22) and throttle angle sensor.

NO ↓

Turn the ignition switch OFF.

Connect the test harness between the ECU and connector

Turn the ignition switch ON.

Measure voltage between D20 (+) terinal and D22 (−) terminal.

Is there approx. 5V ?

YES → Repair open in YEL/WHT wire between ECU (D20) and throttle angle sensor.

NO ↓

Substitute a known-good ECU and recheck. If prescribed voltage is now available, replace the original ECU.

10 Self-diagnosis Check Engine light indicates code 10: A problem in the Intake Air Temperature (TA) Sensor circuit.

The TA sensor is a temperature dependant resistor (thermistor). The resistance of the thermistor decreases as the intake air temperature increases as shown below.

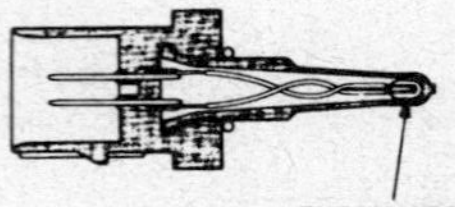

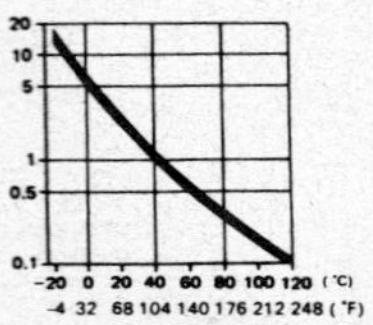

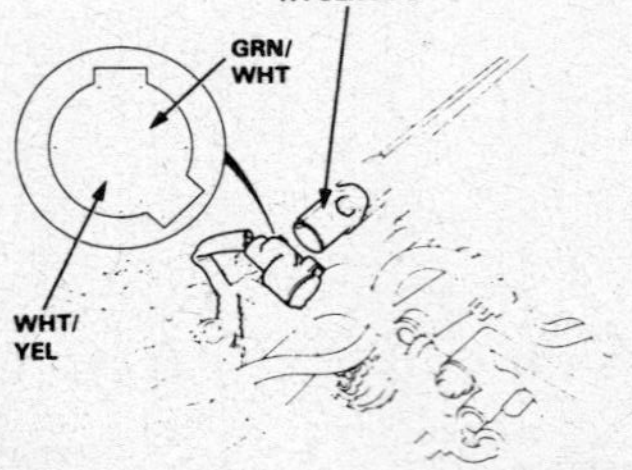

10

- Check Engine light has been reported on.
- With service check connector jumped CODE 10 is indicated.

Do the ECU Reset Procedures

Turn the ignition switch ON.

Is Check Engine light on and does it indicate CODE 10?
- NO → Intermittent failure, system is OK at this time (test drive may be necessary). Check for poor connections or loose wires at C219 (located under the hood on the front right side of the car), C125 (TA sensor) and ECU.
- YES ↓

Turn the ignition switch OFF.

Disconnect the 2P connector from the TA sensor.

Measure resistance between the 2 terminals on the TA sensor.

Is there 0.4–4.0 kΩ?
- NO → Replace TA sensor.
- YES ↓

Turn the ignition switch ON.

Measure voltage between WHT/YEL (+) terminal and body ground.

Is there approx. 5 V ?
- YES → Measure voltage between WHT/YEL (+) terminal and GRN/WHT (–) terminal.
  - Is there approx. 5 V ?
    - NO → Repair open in GRN/WHT wire between ECU (D22) and TA sensor.
    - YES → Substitute a known-good ECU and recheck. If symptom/indication goes away, replace the original ECU.
- NO ↓

Connect the test harness "D" connector to the ECU only, not to the main wire harness

Measure voltage between D15 (+) terminal and D22 (–) terminal.

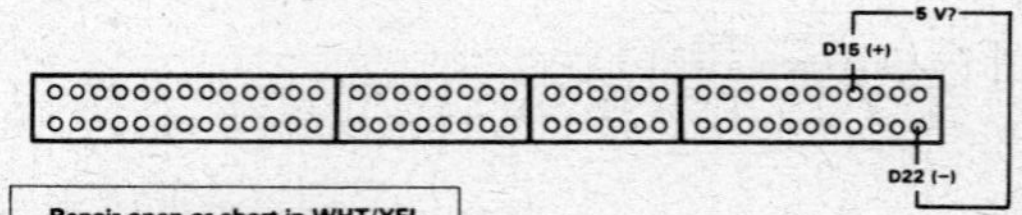

Is there approx. 5V?
- YES → Repair open or short in WHT/YEL wire between ECU (D15) and TA sensor.
- NO → Substitute a known-good ECU and recheck. If symptom/indication goes away, replace the original ECU.

[CHECK ENGINE] [12]

- Check Engine light has been reported on.
- With service check connector jumped CODE 12 is indicated.

Do the ECU Reset Procedures

Road test necessary: Warm up the engine to normal operating temperature (cooling fan comes on). Drive the car on the road for approx. 10 minutes. Keep the engine speed in the 1700–2500 range.

Is Check Engine light on and does it indicate CODE 12?

- NO → Intermittent failure, system is OK at this time. Check for poor connections or loose wires at C219 (located under the hood on the front right side of the car), C110 (EGR valve), C305 (control box) and ECU.
- YES → With the engine at idle, disconnect the #1 hose from the EGR valve and connect a vacuum pump/gauge to the hose.

Is there any vacuum?

- YES → Disconnect 4P connector from the control box and check the #1 hose for vacuum again.
- NO → Move the vacuum pump/gauge to the EGR valve.

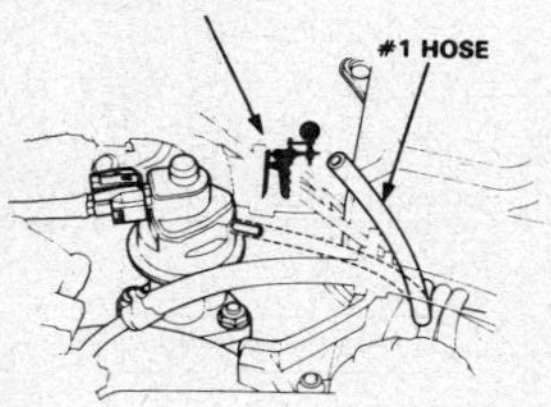

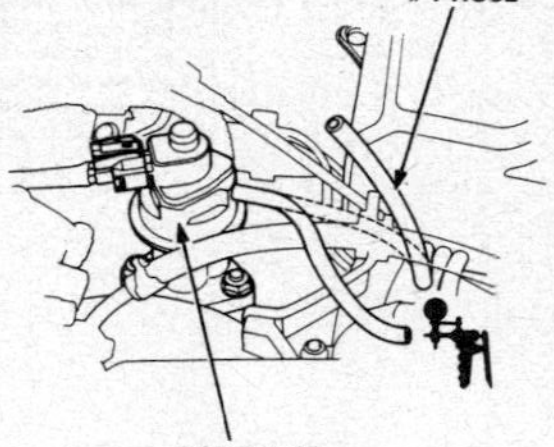

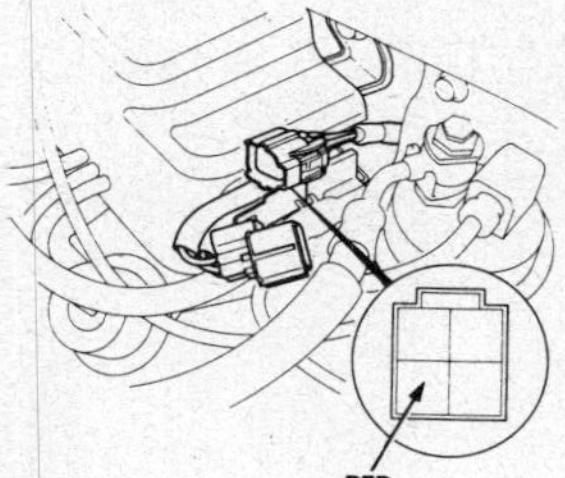

With the engine at idle, apply 8" of vacuum to the EGR valve.

Is there any vacuum ?

- YES → Check vacuum hose routing of the entire EGR system. If hose routing is OK, replace EGR control solenoid valve.
- NO → Turn the ignition switch OFF and disconnect the "A" connector from the ECU.

Check for continuity to ground on RED wire of 4P connector.

Does continuity exist?

- YES → Repair short in RED wire between EGR control solenoid valve and ECU (A11).
- NO → Substitute a known-good ECU and retest. If symptom/indication goes away, replace the original ECU.

Does the engine stall and does the EGR valve hold vacuum ?

- NO → Replace EGR valve.
- YES → Disconnect the 4P connector from the control box.

Measure voltage between BLK/YEL (+) terminal on the main wire harness and body ground.

Is there battery voltage ?

- NO → Repair open in BLK/YEL wire between the solenoid valve and No. 2 fuse.
- YES → Reconnect the vacuum pump/gauge to the #1 hose.

Start the engine and allow it to idle.

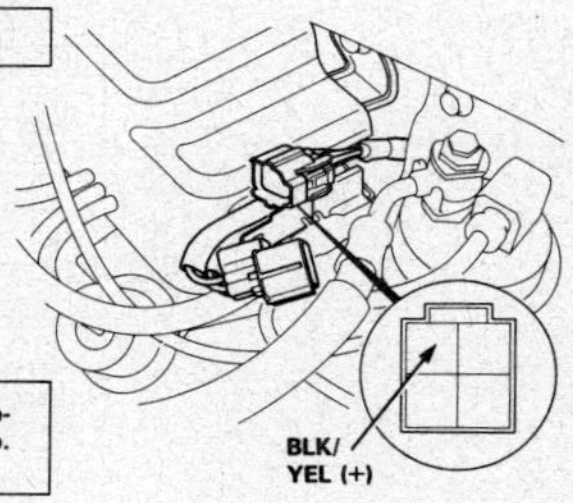

Connect the battery positive terminal to the B terminal of the 4P connector. While watching the vacuum gauge, connect the battery negative terminal to the D terminal.

Is there approx. 8" within 1 second?

- NO → Turn the ignition switch OFF and inspect the #1 and #2 hoses for leaks, restrictions, or misrouting.
  - Are the hoses OK?
    - NO → Correct as necessary.
    - YES → Disconnect the lower hose on EGR control solenoid valve and connect a vacuum gauge to the hose.
      - Start the engine and allow it to idle.
      - Is there 6"— 10" of vacuum ?
        - NO → Replace CVC valve.
        - YES → Replace the EGR control solenoid valve.
- YES → Turn the ignition switch OFF and reconnect the 4P connector.

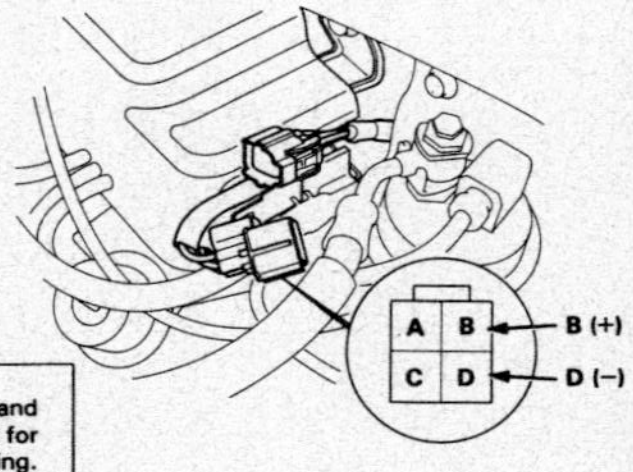

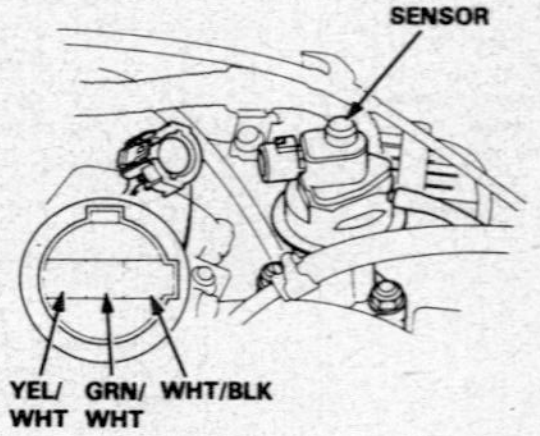

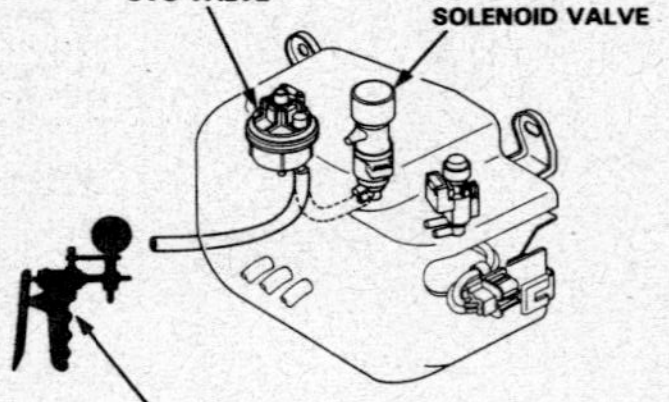

Disconnect 3P connector from the EGR valve.

Turn the ignition switch ON.

Measure voltage between YEL/WHT (+) terminal and GRN/WHT (−) terminal.

Is there approx. 5 V ?

- NO → Measure voltage between YEL/WHT (+) terminal and body ground.
- YES ↓

Is there approx. 5 V?

- YES → Repair open in GRN/WHT wire between EGR valve and ECU (D22).
- NO → Connect the test harness "D" connector to the ECU only, not to the main wire harness
  - Measure voltage between D20 (+) terminal and D22 (−) terminal.
  - Is there approx. 5 V?
    - NO → Repair open in YEL/WHT wire between EGR valve and ECU (D20).
    - YES → Substitute a known-good ECU and recheck. If symptom/indication goes away, replace the original ECU.

Turn the ignition switch OFF.

Reconnect the 3P connector to the EGR valve.

Connect the test harness between the ECU and connector

Turn the ignition switch ON.

Measure voltage between D12 (+) terminal and D22 (−) terminal.

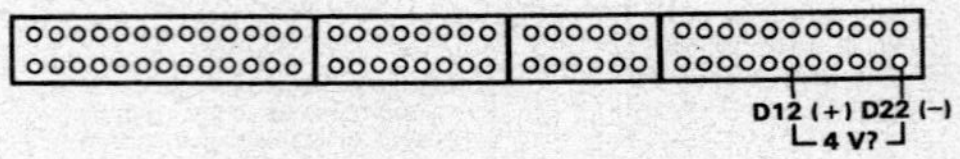

Is the voltage approx. 1.2 V with no vacuum applied and approx. 4.3 V with 8" of vacuum applied?

- NO →
  - Repair open or short in WHT/BLK wire between EGR valve and ECU (D12).
  - If wire is OK, replace the EGR valve.
- YES

TROUBLESHOOTING CHARTS—VIGOR

Does the voltage consistently increase/decrease as the vacuum increases/decreases?
- NO → Replace the EGR valve.
- YES → Reconnect the #1 hose to the EGR valve.

Start the engine and allow it to idle.

Connect A11 terminal to A26 terminal with a jumper wire.

Did the engine stall ?
- NO → Repair open in RED wire between ECU (A11) and EGR control solenoid valve.
- YES → Substitute a known-good ECU and recheck. If symptom/indication goes away, replace the original ECU.

13 — Self-diagnosis Check Engine light indicates code 13: A problem in the Atmospheric Pressure (PA) Sensor.

The PA sensor is built into the ECU.

13

- Check Engine light has been reported on.
- With service check connector jumped CODE 13 is indicated.

Do the ECU Reset Procedures

Turn the ignition switch ON.

Is Check Engine light on and does it indicate CODE 13?
- NO → Intermittent failure, system is OK at this time (test drive may be necessary).
- YES → Substitute a known-good ECU and recheck. If symptom/indication goes away, replace the original ECU.

14 — Self-diagnosis Check Engine light indicates code 14: A problem in the Electric Air Control Valve (EACV) circuit.

The EACV changes the amount of air bypassing the throttle body in response to a current signal from the ECU in order to maintain the proper idle speed.

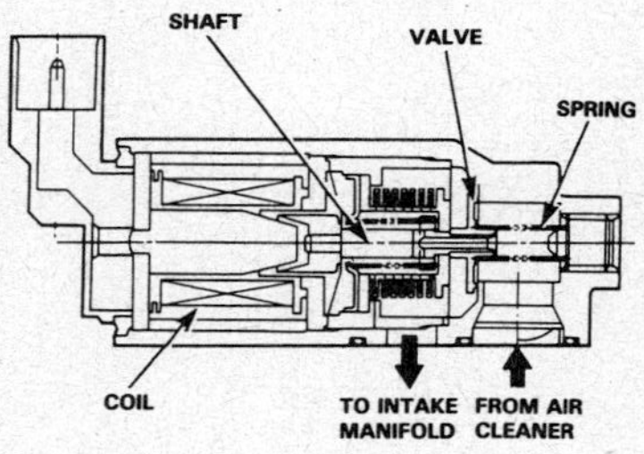

14

- Check Engine light has been reported on.
- With service check connector jumped CODE 14 is indicated.

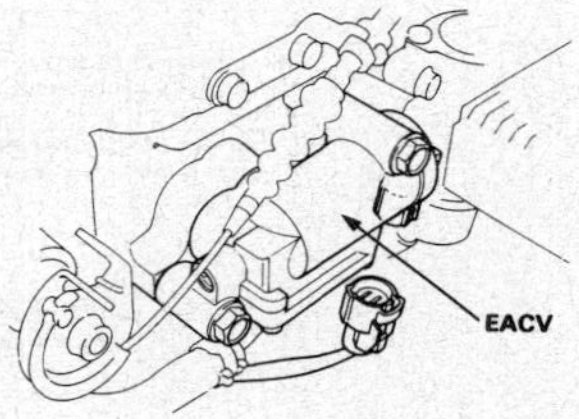

Do the ECU Reset Procedures

Start the engine.

Is Check Engine light on and does it indicate CODE 14?
- YES → Remove the 2P connector from the EACV.
- NO → With the engine running and the accelerator pedal released disconnect the 2P connector on the EACV.

Is there a reduction in engine rpm?
- YES → Intermittent failure, system is OK at this time (test driving may be necessary). Check for poor connection or loose wires at C307 (located at left shock tower), C122 (EACV) and ECU.
- NO → Substitute a known-good EACV and retest.

Measure voltage between the YEL/BLK wire and body ground.

Is there battery voltage?
- NO → Repair open in YEL/BLK wire between EACV and main relay.
- YES ↓

Turn the ignition switch off and reconnect the 2P connector the EACV.

Connect the test harness "A" connector to the main wire harness only, not the ECU.

Turn the ignition switch ON.

Momentarily connect A9 terminal to A26 terminal several times.

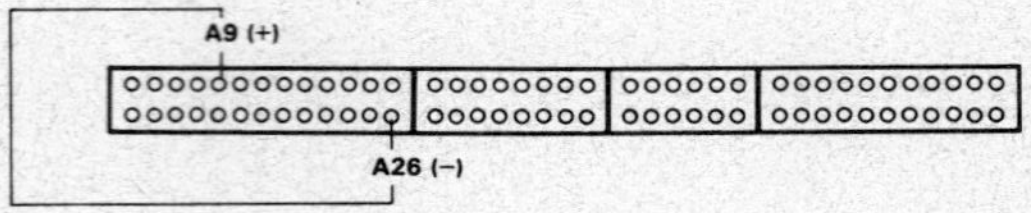

Does the EACV click?
- YES → Substitute a known-good ECU and retest. If symptom/indication goes away, replace the original ECU.
- NO → Repair open or short in BLK/BLU wire between EACV and ECU (A9). If the wire is OK, replace the EACV.

---

15 — Self-diagnosis Check Engine light indicates code 15: A problem in the Ignition Output Signal circuit.

- Check Engine light has been reported on.
- With service check connector jumped CODE 15 is indicated.

Do the ECU Reset Procedure

Start the engine.

NOTE: If the engine won't start, it may take 20 seconds of cranking to set the code.

Is Check Engine light on and does it indicate CODE 15?
- NO → Intermittent failure, system is OK at this time (test drive may be necessary). Check for poor connections or loose wires at C203 (igniter unit) and ECU.
- YES ↓

Turn the ignition switch OFF.

Disconnect the 4P connector from the igniter unit.

Turn the ignition switch ON.

Measure voltage between BLK/YEL (+) terminal and body ground.

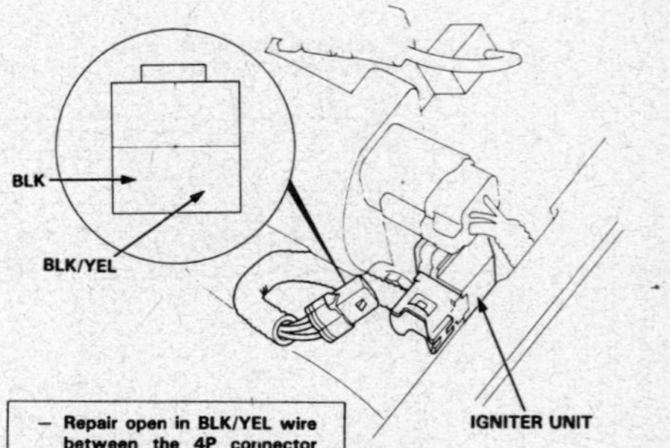

Is there battery voltage ?
- NO →
  - Repair open in BLK/YEL wire between the 4P connector and ignition coil.
  - Repair open in BLK/YEL wire between the ignition coil and ignition switch.
- YES ↓

Measure voltage between BLK/YEL (+) terminal and BLK (–) terminal.

Is there battery voltage?
- NO → Repair open in BLK wire between igniter unit and G101.
- YES

Turn the ignition switch OFF.

Reconnect the 4P connector.

Connect the test harness between the ECU and connector

Turn the ignition switch ON.

Measure voltage individually between A21 (+), A22 (+) terminals and A26 (−) terminal.

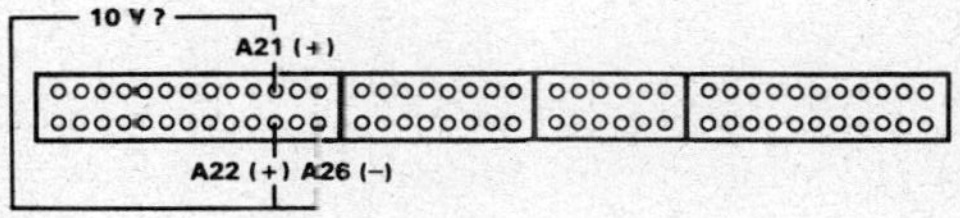

Is there approx. 10 V?

NO:
- Replace the igniter unit.
- Repair open or short YEL/GRN wires between igniter unit and ECU (A21 or A22).

NOTE: If the YEL/GRN wire was shorted, the igniter may be damaged.

YES:

Substitute a known-good ECU and recheck. If symptom/indication goes away, replace the original ECU.

16 — Self-diagnosis Check Engine light indicates code 16: A problem in the fuel injector circuit.

The injectors are a solenoid-actuated constant-stroke pintle type consisting of a solenoid, plunger needle valve and housing. When current is applied to the solenoid coil, the valve lifts up and pressurized fuel is injected. Because the needle valve lift and the fuel pressure are constant, the injection quantity is determined by the length of time that the valve is open (i.e., the duration the current is supplied to the solenoid coil). The injector is sealed by an O-ring and seal ring at the top and bottom. These seals also reduce operating noise.

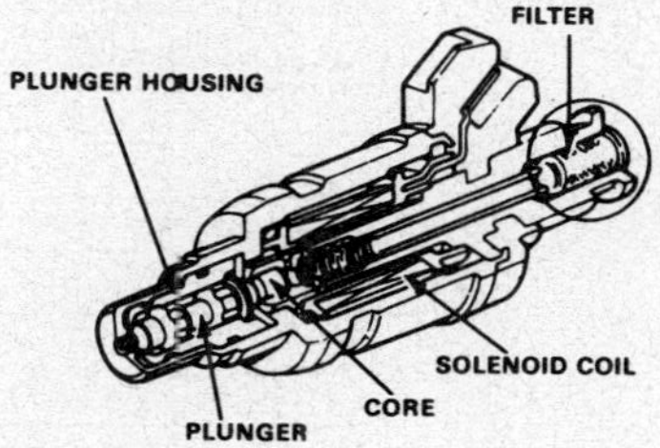

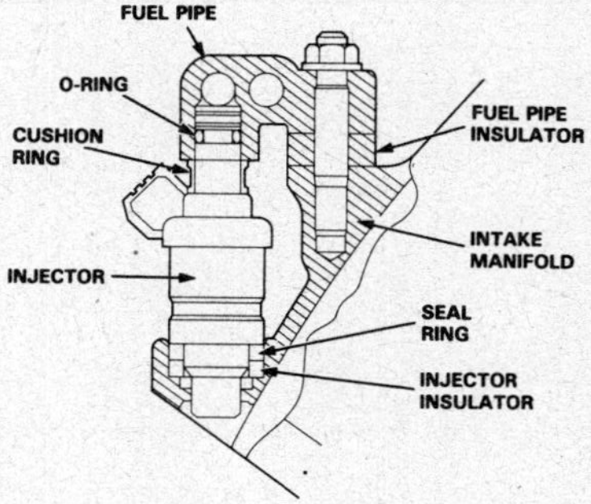

16

- Check Engine light has been reported on.
- With service check connector jumped CODE 16 is indicated.

Do the ECU Reset Procedures

Start the engine and allow it to idle.

NOTE: If engine will not start, it may take 10 seconds of cranking to set the code.

Is Check Engine light on and does it indicate CODE 16?

NO:

Intermittent failure, system is OK at this time (test drive may be necessary).
Check for poor connections or loose wires at C307 (located at left shock tower), C111, C112, C113, C114, C115 (injector), C104 (injector resistor) and ECU.

YES

TROUBLESHOOTING CHARTS—VIGOR

Start engine and listen at each injector for a clicking sound.

Turn the ignition switch OFF.

Disconnect the 2P connector from the injector that does not click.

Measure resistance between the 2 terminals of the injector.

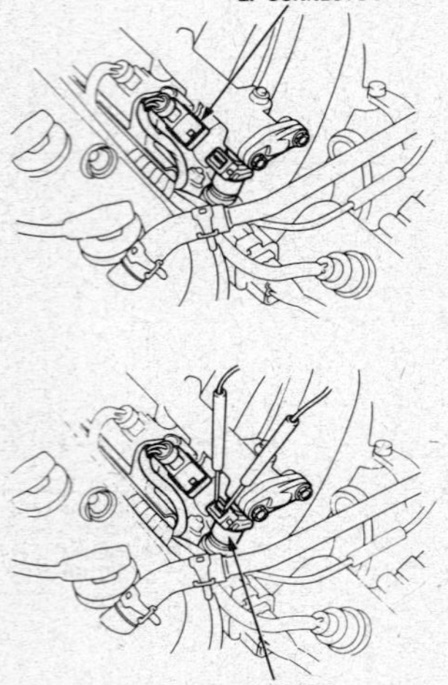

Is there 1.5—2.5Ω ?

- NO → Replace the injector/injectors that are not 1.5–2.5 Ω.
- YES ↓

Turn the ignition switch ON.

Measure voltage between RED/BLK (+) terminal in the 2P connector and body ground.

Is there battery voltage ?

- NO → Turn the ignition switch OFF. → Disconnect 8P connector from the injector resistor. → Turn the ignition switch ON. → Measure voltage between YEL/BLK (+) terminal and body ground.
  - Is there battery voltage?
    - NO → Repair open in the YEL/BLK wire between the injector resistor and the main relay.
    - YES → Test the injector resistor
  - Does the injector resistor test OK?
    - NO → Replace the injector resistor.
    - YES → Repair open RED/BLK wire between 2P connector and resistor.
- YES ↓

Reconnect the 2P connector to the injector.

Turn the ignition switch OFF.

Connect the test harness between the ECU and connector

Turn the ignition switch ON.

Measure voltage between A23 (−) terminal and following terminal:
- No. 1 injector: A1 (+) terminal.
- No. 2 injector: A3 (+) terminal.
- No. 3 injector: A5 (+) terminal.
- No. 4 injector: A2 (+) terminal.
- No. 5 injector: A4 (+) terminal.

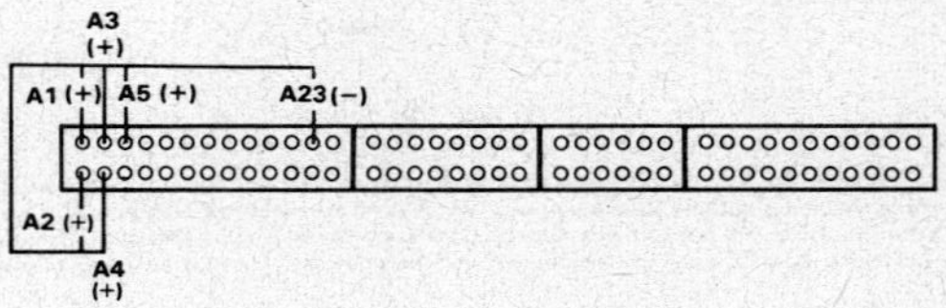

Is there battery voltage ?

- NO → Repair open in the wire between the ECU (A1, A3, A5, A2 or A4) and the injector.
- YES → Substitute a known-good ECU and recheck. If symptom/indication goes away, replace the original ECU.

TROUBLESHOOTING CHARTS—VIGOR

17 Self-diagnosis Check Engine light indicates code 17. A problem in the Vehicle Speed Sensor circuit.

The signal generated by the speed sensor, produces pulses when the front wheels turn.

- Check Engine light has been reported on.
- With service check connector jumped CODE 17 is indicated.

Do the ECU Reset Procedures

Road test necessary.
In 2nd gear accelerate to 4,000 rpm, then decelerate to 1,500 rpm with throttle fully closed.

Is Check Engine light on and does it indicate CODE 17?

NO → Intermittent failure, system is OK at this time.
Check for poor connections or loose wires at C308 (located at left shock tower), C120 (speed sensor) and ECU.

YES ↓

Block rear wheels and set the parking brake. Jack up the front of the car and support with safety stands.

WARNING Block rear wheels before jacking up front of car.

Connect the test harness between the ECU and connector

Turn the ignition switch ON.

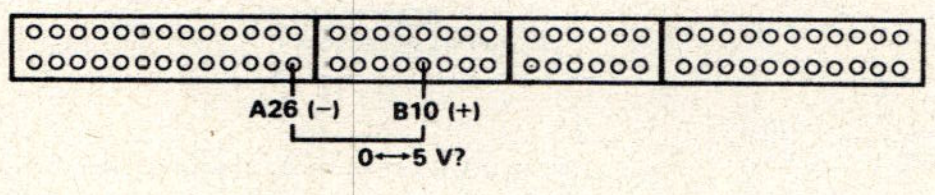

Slowly rotate left front wheel and measure voltage between B10 (+) terminal and A26 (−) terminal.

Does voltage pulse 0 V and 5 V?

NO → Turn the ignition switch OFF.

YES ↓

Substitute a known-good ECU and recheck. If symptom/indication goes away, replace the original ECU.

Disconnect the B connector from the ECU only, not the main wire harness.

Turn the ignition switch ON.

Slowly rotate left front wheel and measure voltage between B10 (+) terminal and A26 (−) terminal.

Does voltage pulse 0 V and 5 V?

NO →
- Repair short in ORN wire between ECU (B10) and the speed sensor, A/T control unit, speedometer, or cruise control unit.
- Repair open in ORN wire between ECU (B10) and speed sensor.
- If wire is OK, test the speed sensor

YES ↓

Substitute a known-good ECU and recheck. If prescribed voltage is now available, replace the original ECU.

TROUBLESHOOTING CHARTS—VIGOR

CHECK 18 — Self-diagnosis Check Engine light indicates code 18: A problem in the Ignition Timing Adjuster circuit.

The ignition timing adjuster allows the electronic ignition advance to be set to 15° BTDC at idle.

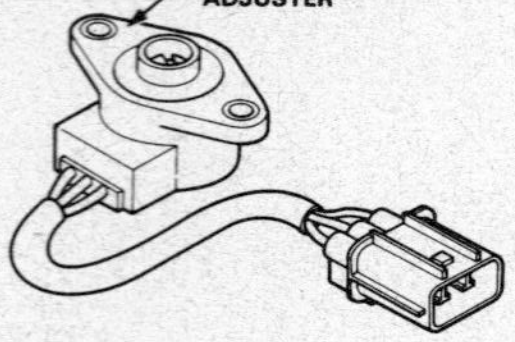

CHECK 18

- Check Engine light has been reported on.
- With service check connector jumped CODE 18 is indicated.

Do the ECU Reset Procedures

Turn the ignition switch ON.

Is Check Engine light on and does it indicate CODE 18?
- NO → Intermittent failure system is OK, at this time (test drive may be necessary). Check for poor connections or loose wires at C328 (ignition timing adjuster) and ECU.
- YES ↓

Turn the ignition switch OFF.

Disconnect the 3P connector from the control box.

Measure resistance between A terminal and C terminal on Ignition timing adjuster.

Is there 3.5–6.5 kΩ?
- NO → Replace Ignition timing adjuster
- YES ↓

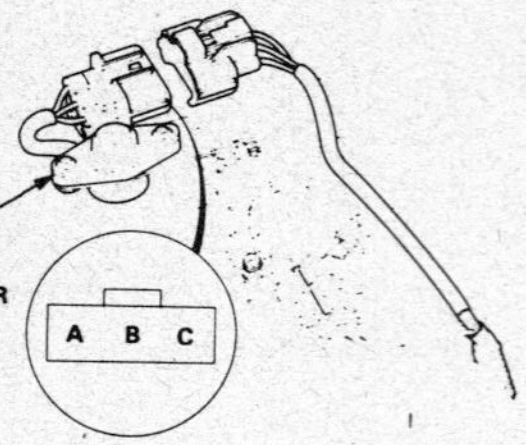

Measure resistance between A and B terminals and between C and B terminals.

Does the sum of the two resistance checks equal 3.5–6.5 kΩ?
- NO → Replace Ignition timing adjuster
- YES ↓

Turn the ignition switch ON.

Measure voltage between YEL/WHT (+) terminal and GRN/WHT (−) terminal on the main wire harness.

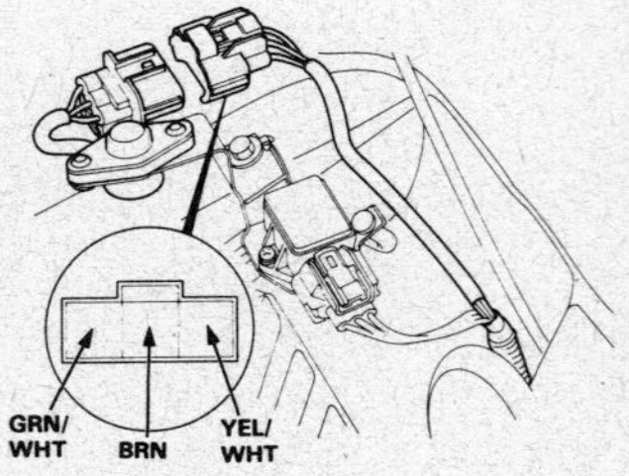

Is there approx. 5 V?
- NO → Measure voltage between YEL/WHT (+) terminal and body ground.
  - Is there approx. 5 V?
    - YES → Repair open in GRN/WHT wire between ECU (D22) and Ignition timing adjuster.
    - NO ↓
  - Turn the ignition switch OFF.
  - Connect the test harness between the ECU and connector
  - Turn the ignition switch ON.
  - Measure voltage between D20 (+) terminal and D22 (−) terminal.
  - Is there approx. 5 V?
    - YES → Repair open in YEL/WHT wire between ECU (D20) and Ignition timing adjuster.
    - NO → Substitute a known-good ECU and recheck. If prescribed voltage is now available, replace the original ECU.
- YES ↓

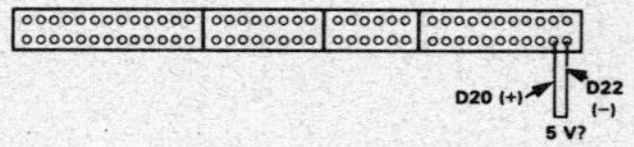

Turn the ignition switch OFF.

Reconnect 3P connector.

Connect the test harness between the ECU and connector

Turn the ignition switch ON.

Measure voltage between D8 (+) terminal and D22 (−) terminal.

Is there 0.5–4.5 V?
- NO → Repair open or short in BRN wire between ECU (D8) and Ignition timing adjuster.
- YES → Substitute a known-good ECU and recheck. If symptom/indication goes away, replace the original ECU.

TROUBLESHOOTING CHARTS—VIGOR

20 Self-diagnosis Check Engine light indicates code 20: A problem in the Electric Load Detector circuit.

With service check connector jumped CODE 20 is indicated.

Do the ECU Reset Procedure

Start engine and keep engine rpm at idle.
Turn on headlights.

Does Check Engine light indicate CODE 20?

- NO → **Intermittent failure, system is OK at this time (test drive may be necessary). Check for poor connections or loose wires at C218 (ELD) and ECU.**
- YES ↓

Turn the ignition switch OFF.

Open the under-hood fuse/relay box lid.

Disconnect the 3P connector from the electric load detector.

Turn the ignition switch ON.

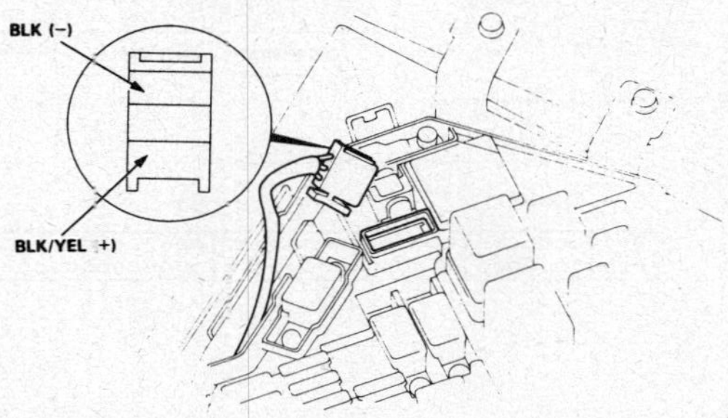

Measure voltage between BLK/YEL (+) terminal and BLK (−) terminal.

Is there battery voltage?

- NO → Measure voltage between BLK/YEL (+) terminal and body ground.
  - Is there battery voltage?
    - YES → **Repair open in BLK wire between the 3P connector and G201.**
    - NO → **Repair open in BLK/YEL wire between No. 2 fuse and the 3P connector.**
- YES ↓

Measure voltage between GRN/RED terminal and body ground.

Is there 4.5–5 V?

- NO → **Repair open or short in GRN/RED wire between ECU (D10) and the 3P connector. If wire is OK, substitute a known-good ECU and recheck.**
- YES ↓

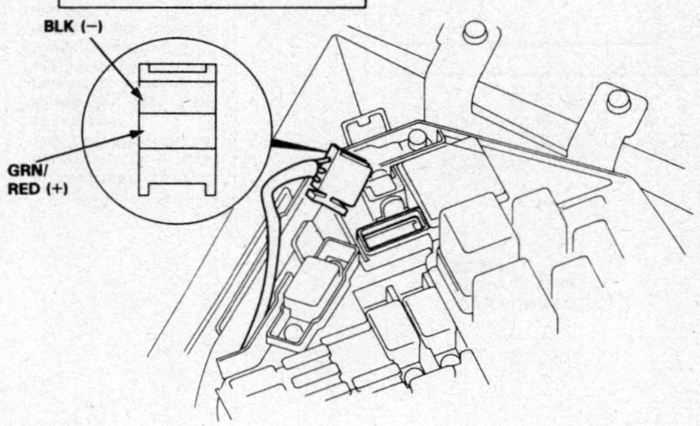

Turn the ignition switch OFF.

Connect the 3P connector to the electric load detector.

Connect the test harness between the ECU and connector

Turn the ignition switch on.

Under the conditions listed in the chart to the right, measure voltage between D10 (+) terminal and A26 (−) terminal.

Condition	Voltage
Headlight switch, first position (•)	2.5–3.5 V
Headlight switch, second position (●)	1.5–2.5 V

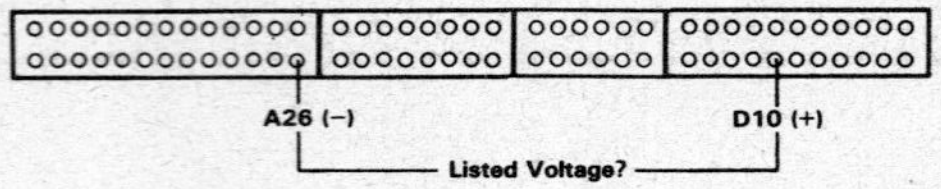

Is the voltage listed in the chart available?

- NO → Faulty electric load detector.
- YES → Substitute a known-good ECU and recheck. If symptom/indication goes away, replace the original ECU.

[23] Self-diagnosis Check Engine light indicates code 23: A problem in the circuit of the Knock Sensor 1.

[53] Self-diagnosis Check Engine light indicates code 53: A problem in the circuit of the Knock Sensor 2.

[23] or [53]

- Check Engine light has been reported on.
- With service check connector jumped , CODE 23 and/or 53 are indicated.

Do the ECU Reset Procedures

Warm up the engine to normal operating temperature (cooling fan comes on).

Hold engine at 3000–4000 rpm for 10 seconds. (A/T: Transmission in [N] or [P])

Is Check Engine light on and does it indicate CODE 23 or 53?

- NO → Intermittent failure, system is OK at this time (test drive may be necessary). Check for poor connection or loose wires at C308 (located at left shock tower), C123, C119 (knock sensor) and ECU.
- YES ↓

Turn the ignition switch OFF.

Connect the test harness to the main wire harness only, not to the ECU

Disconnect the C308 connector.

Check for continuity between C1 or C3* terminal and body ground.

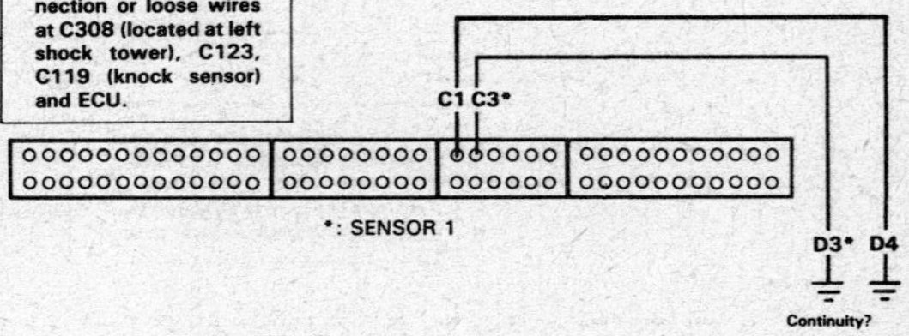

Does continuity exist?

- YES → Repair short in ORN/WHT or RED/BLU* wire between ECU (C1 or C3*) and the knock sensor.
- NO

*: SENSOR 1

TROUBLESHOOTING CHARTS – VIGOR

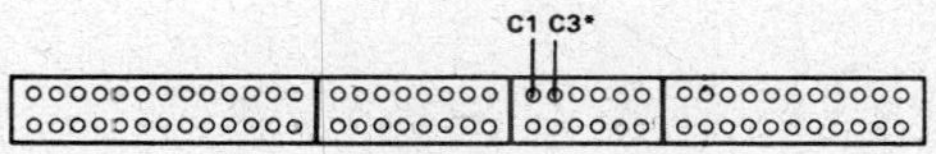

*: SENSOR 1

Check for continuity on ORN/WHT or RED/BLU* wire between C1 or C3* and C308 (14P) connector of main wire harness.

Does continuity exist?

- NO → Repair open in ORN/WHT or RED/BLU* wire between ECU (C1 or C3*) and the indicated knock sensor.
- YES ↓

Substitute a known-good ECU and then hold the engine rpm at 3,000 – 4,000 rpm for 10 seconds and recheck.

Is Check Engine light on and does it indicate CODE 23 or 53?

- NO → Replace the original ECU
- YES ↓

Replace knock sensor 1 or 2 and recheck.

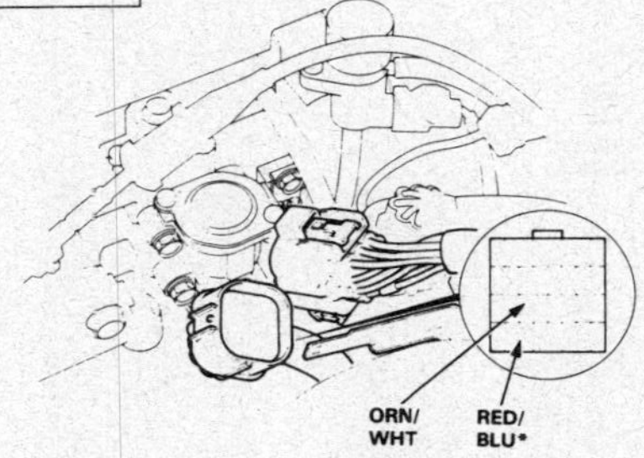

---

30 — Self-diagnosis Check Engine light indicates code 30: A problem in the circuit (B3) between A/T control unit and ECU (A/T FI Signal A).

31 — Self-diagnosis Check Engine light indicates code 31: A problem in the circuit (B4) between A/T control unit and ECU (A/T FI Signal B).

30 or 31

With service check connector jumped CODE 30 and/or 31 are indicated.

Do the ECU Reset Procedure

Test drive necessary.
Drive the car for several miles so that the transmission upshifts and downshifts several times.

Does Check Engine light indicate CODE 30 or 31?

- NO → Intermittent failure, system is OK at this time. Check for poor connections or loose wires at C441 (A/T control unit) and ECU.
- YES ↓

Turn the ignition switch OFF.

Connect the test harness to the main harness only, not to the ECU

Disconnect the 22P connector from the A/T control unit.

Check for continuity between B3 or B4* terminal and body ground.

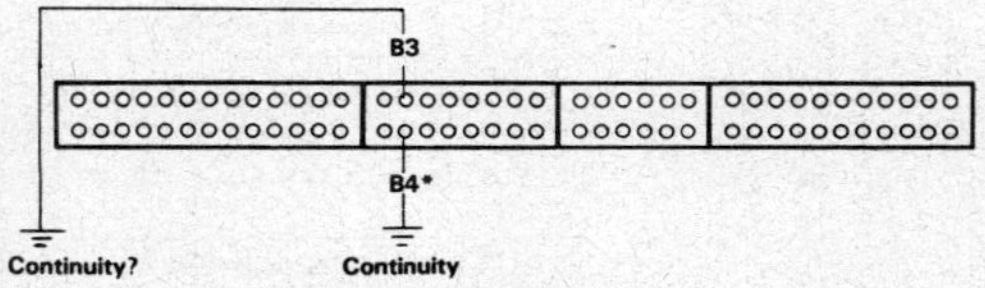

*: CODE 31 (A/T FI Signal B)

Does continuity exist?

- YES → Repair short in WHT/GRN wire between ECU (B3 or B4*) and the A/T control unit.
- NO ↓

Check for continuity on WHT/GRN or WHT/RED* wire between B3 or B4* and 22P connector of the A/T control unit.

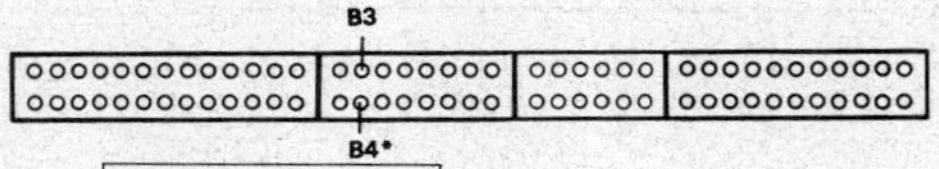

Does continuity exist?

- NO → Repair open in WHT/GRN or WHT/RED* wire between ECU (B3 or B4*) and the A/T control unit.
- YES ↓

Substitute a known-good ECU and recheck. If symptom/indication goes away, replace the original ECU.

CHECK 41 — Self-diagnosis Check Engine light indicates code 41: A problem in the Oxygen (O_2) Sensor Heater circuit.

CHECK 41

- Engine is running.
- Check Engine light has been reported on. With service check connector jumped CODE 41 is indicated.

Do the ECU Reset Procedures

Start the engine.

Is Check Engine light on and does it indicate CODE 41?
- NO → Intermittent failure, system is OK at this time (test drive may be necessary). Check for poor connections or loose wires at C202 (O_2 sensor) and ECU.
- YES ↓

Stop the engine

Remove the O_2 sensor covers.

Disconnect the 4P connector from the O_2 sensor.

Measure resistance between terminals C and D on the O_2 sensor.

Is there 10–40 Ω
- NO → Replace O_2 sensor.
- YES ↓

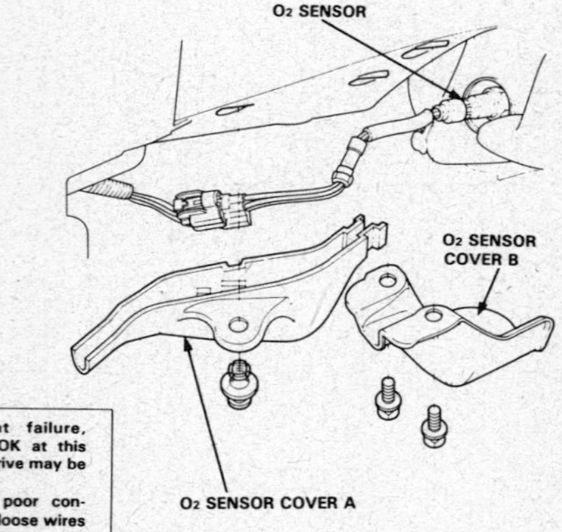

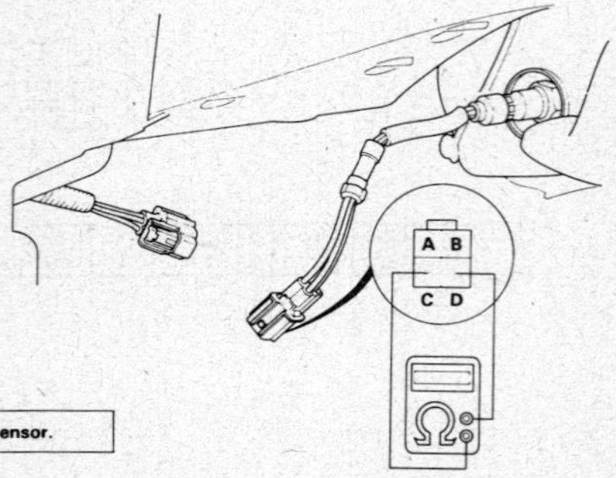

Check for continuity to body ground on each terminal on the O_2 sensor.

Does continuity exist?
- YES → Replace O_2 sensor.
- NO ↓

Check for continuity between terminal D and terminals A and B individually.

Does continuity exist?
- YES → Replace O_2 sensor.
- NO ↓

Turn the ignition switch ON.

At O_2 sensor harness, measure voltage between YEL/BLK (+) terminal and PNK/WHT (−) terminal.

Is there battery voltage?
- YES → Disconnect the "A" connector from the ECU.
  - At O_2 sensor harness, measure voltage between YEL/BLK (+) terminal and PNK/WHT (−) terminal.
  - Is there battery voltage?
    - YES → Repair short in PNK/WHT wire between ECU (A6) and O_2 sensor.
    - NO → Reconnect the O_2 sensor connector.
- NO ↓

Measure voltage between YEL/BLK (+) terminal and body ground.

Is there battery voltage?
- NO → Repair open in YEL/BLK wire between O_2 sensor and main relay.
- YES ↓

Turn the ignition switch OFF.

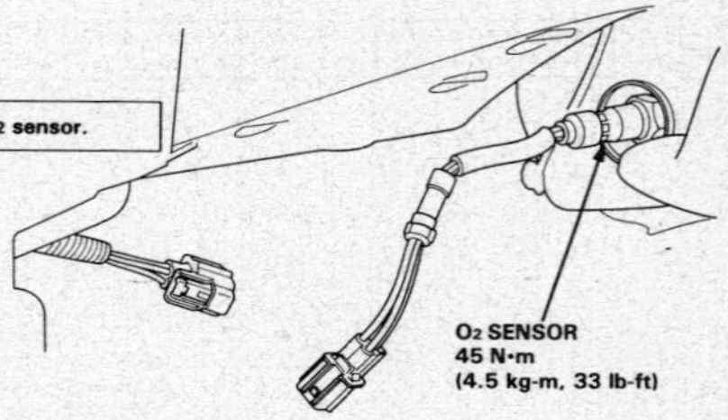

TROUBLESHOOTING CHARTS—VIGOR

Self-diagnosis Check Engine light indicates code 43: Most likely a problem in the Oxygen (O_2) Sensor circuit or a problem in the Fuel Supply System.

- Check Engine light has been reported on.
- With service check connector jumped (page 11-18), CODE 43 is indicated.
- or continued from code 1.

Is the 43 code accompanied by the Check Engine light and poor driveability?

YES → Fuel Supply System

NO ↓

Do the ECU Reset Procedures

Warm up engine to normal operating temperature (cooling fan comes on).

Hold engine at 3,000 rpm for 2 minutes. (A/T: Transmission in [N] or [P].)

Is the Check Engine light on and does it indicate CODE 43?

NO → Intermittent failure, system is OK at this time (test drive may be necessary). Check for poor connections or loose wires at C202 (O_2 sensor) and ECU.

YES ↓

Turn the ignition switch OFF.

Connect the test harness between the ECU and connector

With the ignition switch OFF, wait for at least two minutes.

Install a jumper wire on the test harness between A6 and A26.

Turn the ignition switch ON.

Measure voltage between D14 (+) terminal and A26 (−) terminal as soon as the ignition switch is turned on.

NOTE:
- Use DIGITAL MULTIMETER (KS—AHM—32—003) or equivalent.
- Use 2 Volt range.

A6 A26

A26 (−) D14 (+)

Voltage should start at 0.4—0.5 V when the ignition switch is first turned on, and decrease to below 0.1 V in less than two minutes.

Reconnect the 4P connector to O_2 sensor.

Connect the test harness "A" connector to the main wire harness only, not the ECU

BATTERY VOLTAGE A23 (−) A6 (+)

Turn the ignition switch ON.

Measure voltage between A6 (+) terminal and A23 (−) terminal.

Is there battery voltage?

NO → Repair open in PNK/WHT wire between ECU (A6) and O_2 sensor.

YES ↓

Substitute a known-good ECU and recheck. If symptom/indication goes away, replace the original ECU.

Connect the test harness "A" connector to the main wire harness only, not the ECU

Connect an ammeter between terminals A6 (+) and A26 (−).

Is the current less than 0.1 A?*

YES → Replace O_2 sensor.

NO ↓

Substitute a known-good ECU and recheck. If symptom/indication goes away, replace the original ECU.

* Monitor over a 5 minutes period.

- Is there 0.1 V or less when the ignition switch is first turned on?
  - YES:
    - Remove the O₂ sensor covers.
    - Disconnect the 4P connector from the O₂ sensor.
    - Measure voltage between D14 (+) terminal and A26 (–) terminal.
    - Is there more than 0.1 V?
      - YES: **Replace O₂ sensor.**
      - NO:
        - Disconnect the "D" connector from the main wire harness.
        - Measure voltage between D14 (+) terminal and A26 (–) terminal.
        - Is there more than 0.1 V?
          - YES: **Repair short in WHT wire between ECU (D14) and O₂ sensor.**
          - NO: **Substitute a known-good ECU and recheck. If symptom/indication goes away, replace the original ECU.**
  - NO:
    - Remove the O₂ sensor covers.
    - Disconnect the 4P connector from the O₂ sensor.
    - Measure voltage at the engine wire harness side of the O₂ sensor connector between A (–) and B (+).
    - Is there more than 0.1 V?
      - YES: **Replace O₂ sensor.**
      - NO: **Repair open in WHT wire between ECU (D14) and O₂ sensor.**

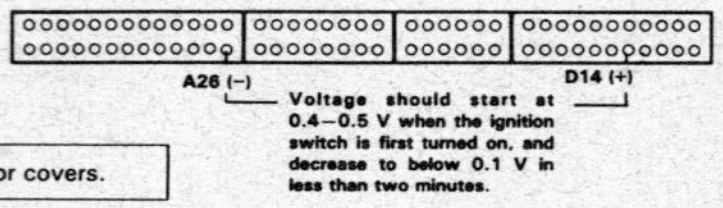

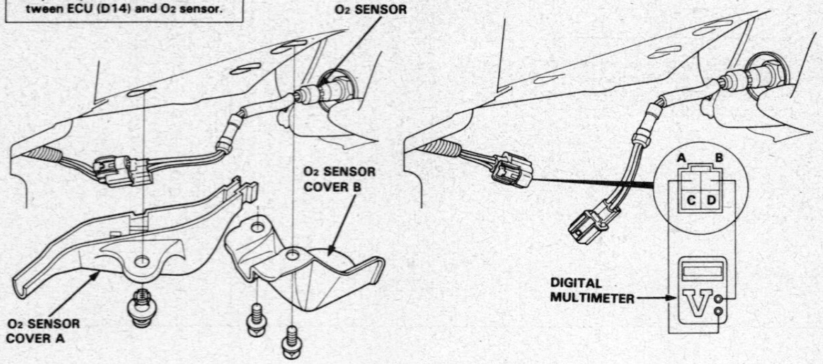

CHECK — 45 — Self-diagnosis Check Engine light indicates code 45: Most likely a problem in the Fuel Metering System.

- **Check Engine light has been reported on.**
- **With service check connector jumped CODE 45 is indicated.**
  - Do the ECU Reset Procedures
  - Start the engine
  - Warm up engine to normal operating temperature (cooling fan comes on).
  - Do the Road Test.*
  - Does Check Engine light indicate CODE 45?
    - NO: **Intermittent failure, system is OK at this time.**
    - YES:
      - Turn the ignition switch OFF.
      - Relieve fuel pressure
      - Attach the fuel pressure gauge and measure the fuel pressure at idle
      - Is the pressure less than 215 kpa (31 psi)?
        - YES: **fuel filter replacement** / **fuel pump testing.**
        - NO:
          - Substitute a known-good MAP sensor.
          - Do the ECU Reset Procedures
          - Do the Road Test.*

* Road Test
- Place the gear shift or selector lever in 3rd gear (M/T) or D_3 position (A/T), cruise at exactly 35 mph for at least 10 seconds.
- Remove your foot from the throttle until the vehicle decels below 30 mph.
- Accelerate to 35 mph and cruise at exactly 35 mph for at least 10 seconds.
- The cruise/decel procedure must be repeated 10 times.
- If the above does not recreate the code, do the following:
- Stop the vehicle, and turn off the ignition switch.
- Start the engine.
- Repeat the cruise/decel procedure 5 times.

TROUBLESHOOTING CHARTS—VIGOR

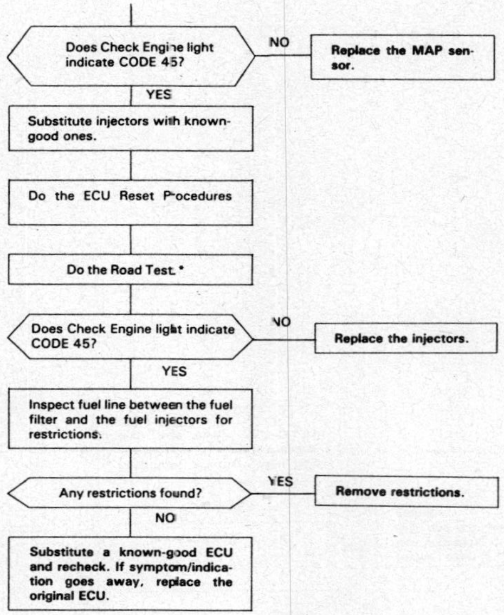

This signals the ECU when there is a demand for cooling from the air conditioning system.

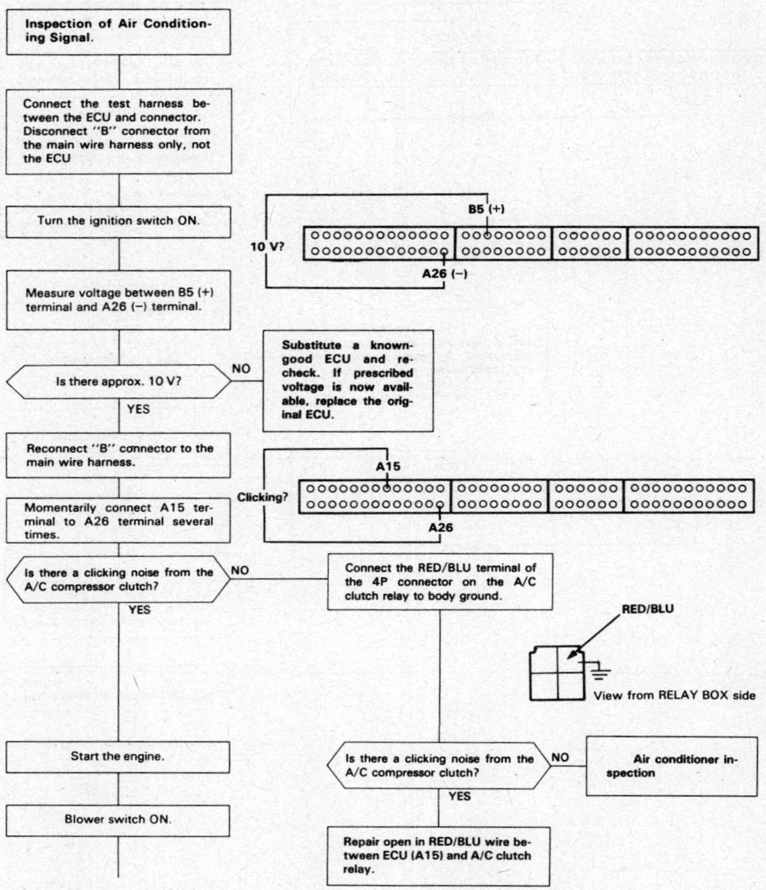

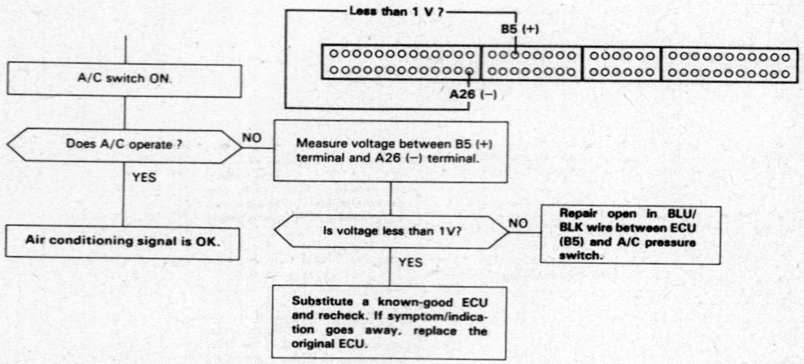

This signals the ECU when the alternator is charging.

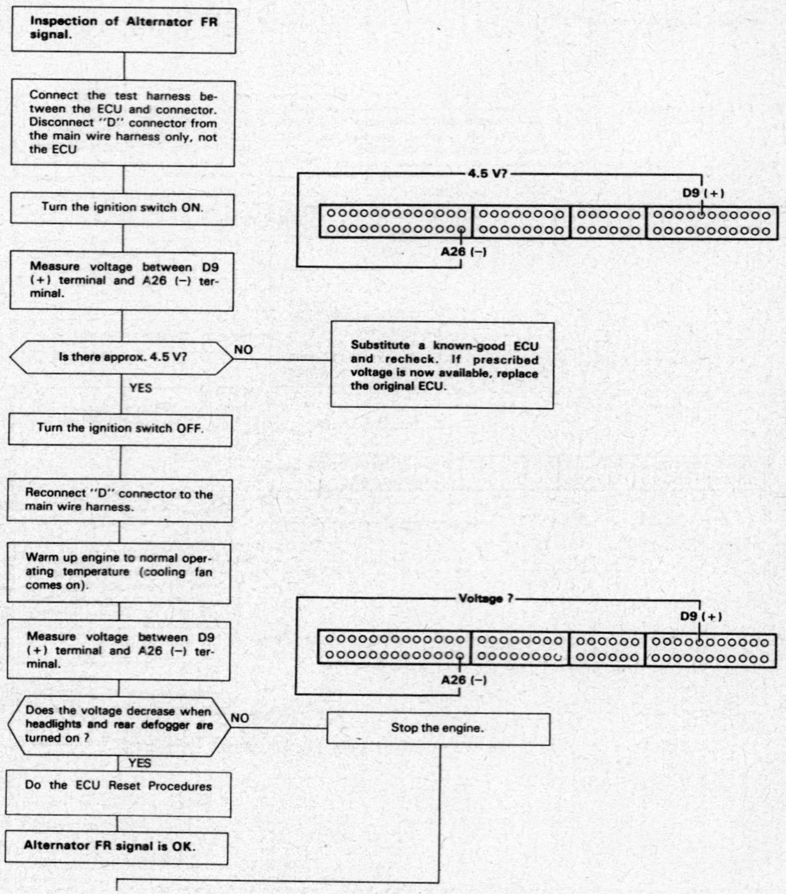

TROUBLESHOOTING CHARTS – VIGOR

NOTE: The radio has a coded theft protection circuit. Be sure you get the customer's code number before disconnecting the battery cable.

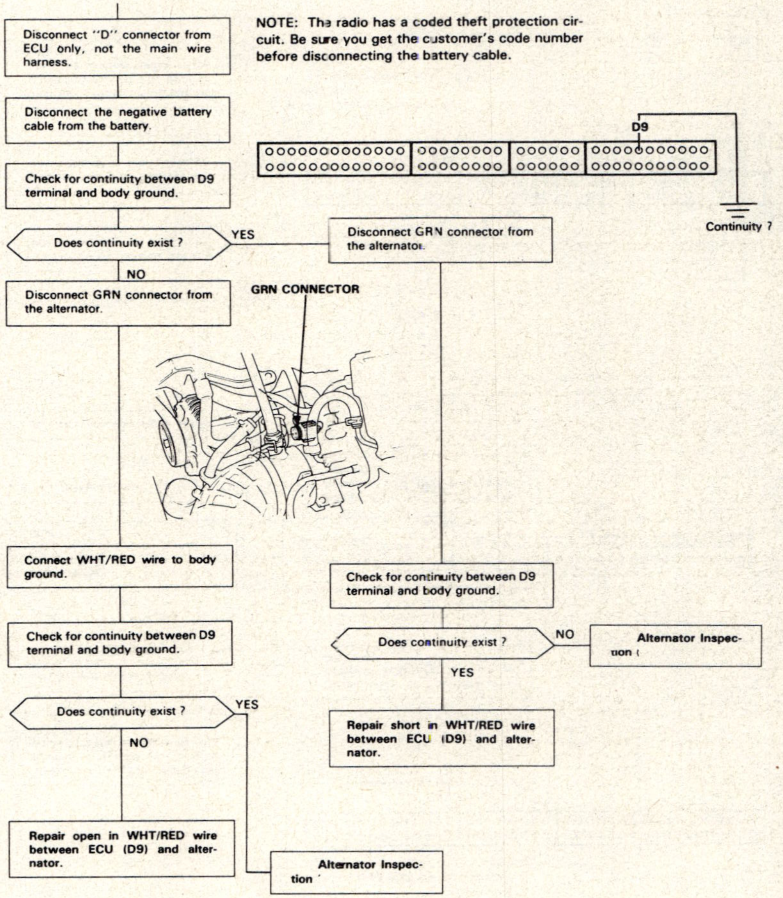

This signals the ECU when the transmission is in Neutral or Park.

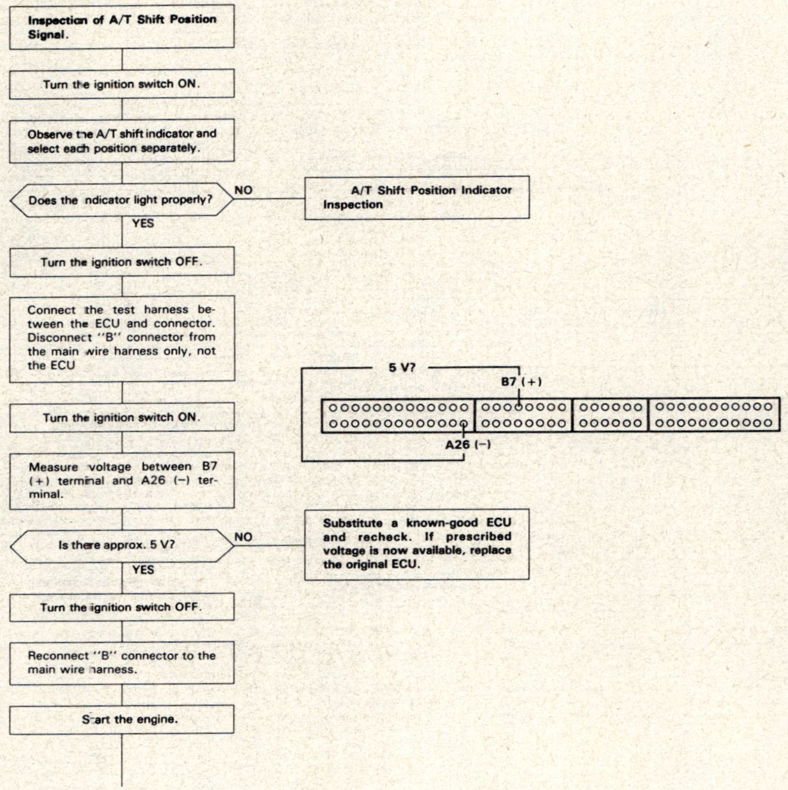

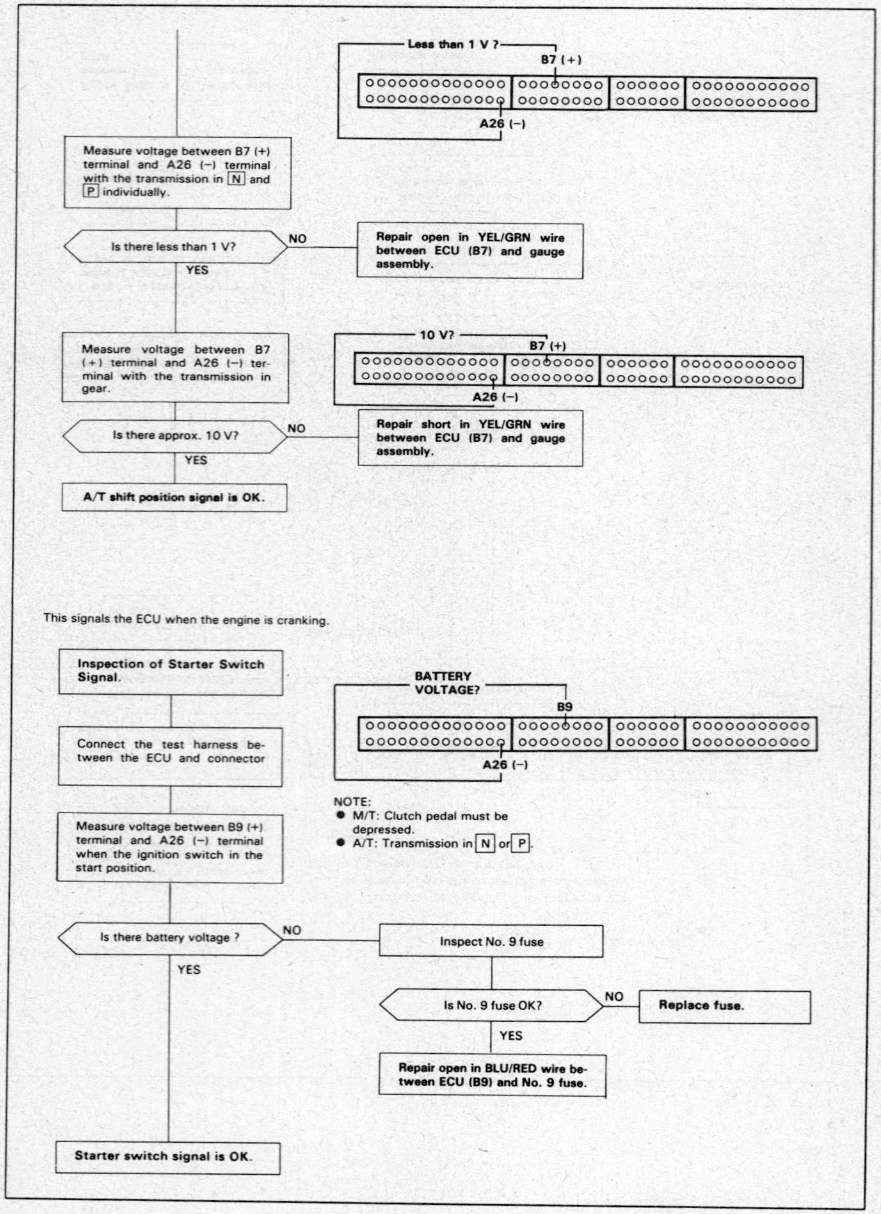
Measure voltage between B7 (+) terminal and A26 (−) terminal with the transmission in N and P individually.
Less than 1 V ?
B7 (+)
A26 (−)
Is there less than 1 V?
NO
Repair open in YEL/GRN wire between ECU (B7) and gauge assembly.
YES
Measure voltage between B7 (+) terminal and A26 (−) terminal with the transmission in gear.
10 V?
B7 (+)
A26 (−)
Is there approx. 10 V?
NO
Repair short in YEL/GRN wire between ECU (B7) and gauge assembly.
YES
A/T shift position signal is OK.
This signals the ECU when the engine is cranking.
Inspection of Starter Switch Signal.
Connect the test harness between the ECU and connector
BATTERY VOLTAGE?
B9
A26 (−)
NOTE:
● M/T: Clutch pedal must be depressed.
● A/T: Transmission in N or P.
Measure voltage between B9 (+) terminal and A26 (−) terminal when the ignition switch in the start position.
Is there battery voltage ?
NO
Inspect No. 9 fuse
Is No. 9 fuse OK?
NO
Replace fuse.
YES
Repair open in BLU/RED wire between ECU (B9) and No. 9 fuse.
YES
Starter switch signal is OK.

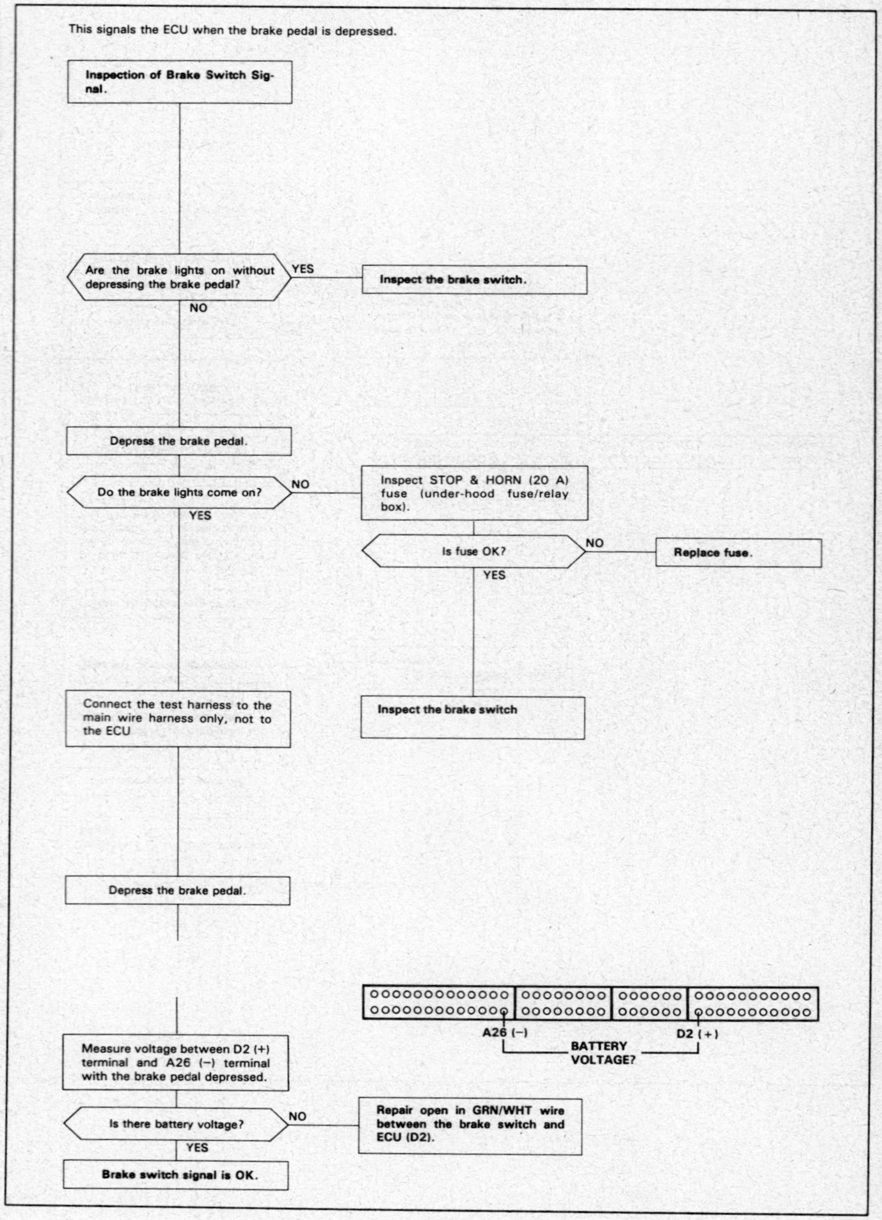
This signals the ECU when the brake pedal is depressed.
Inspection of Brake Switch Signal.
Are the brake lights on without depressing the brake pedal?
YES
Inspect the brake switch.
NO
Depress the brake pedal.
Do the brake lights come on?
NO
Inspect STOP & HORN (20 A) fuse (under-hood fuse/relay box).
Is fuse OK?
NO
Replace fuse.
YES
Inspect the brake switch
YES
Connect the test harness to the main wire harness only, not to the ECU
Depress the brake pedal.
Measure voltage between D2 (+) terminal and A26 (−) terminal with the brake pedal depressed.
A26 (−)
D2 (+)
BATTERY VOLTAGE?
Is there battery voltage?
NO
Repair open in GRN/WHT wire between the brake switch and ECU (D2).
YES
Brake switch signal is OK.

TROUBLESHOOTING CHARTS – VIGOR

This signals the ECU when the power steering load is high.

Inspection of P/S Oil Pressure Signal

1. Connect the test harness between the ECU and connector.
2. Turn the ignition switch ON.
3. Measure voltage between B8 (+) terminal and A26 (–) terminal.
4. Is there more than 1V?
   - YES → Disconnect the 2P connector on the P/S oil pressure switch.
     - Connect RED terminal to BLK terminal.
     - Is there more than 1V?
       - NO → **Replace P/S oil pressure switch.**
       - YES → **Repair open in RED wire between ECU (B8) and P/S oil pressure switch or BLK wire between P/S oil pressure switch and G302.**
   - NO → Start the engine.
5. Turn steering wheel slowly.
6. Measure voltage between B8 (+) terminal and A26 (–) terminal while steering wheel is turning.
7. Is there approx. 10V?
   - YES → **P/S oil pressure signal is OK.**
   - NO → Turn the ignition switch OFF.
     - Disconnect "B" connector from main wire harness only, not the ECU.
     - Turn the ignition switch ON.
     - Is there approx. 10V?
       - NO → **Substitute a known-good ECU and recheck. If prescribed voltage is now available, replace the original ECU.**
       - YES → Reconnect "B" connector to main wire harness and disconnect 2P connector on the P/S oil pressure switch.
         - Is there approx. 10V?
           - YES → **Replace P/S oil pressure switch.**
           - NO → **Repair short in RED wire between ECU (B8) and P/S oil pressure switch.**

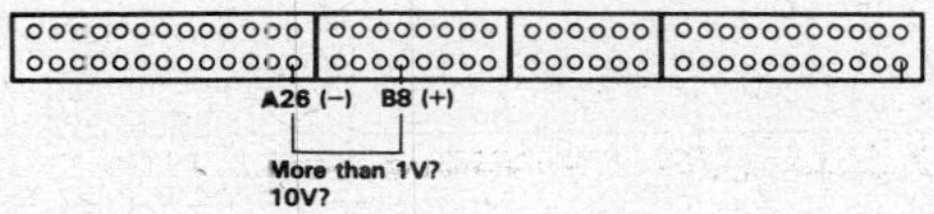

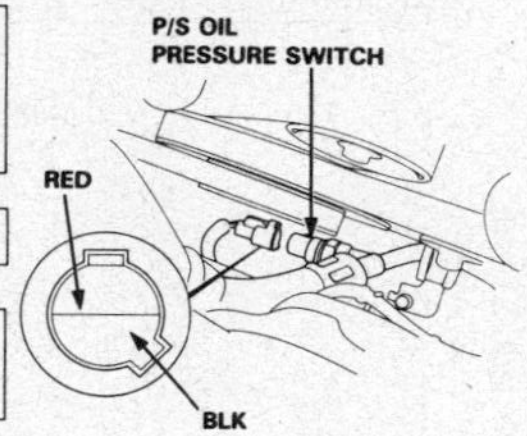

Troubleshooting Flowchart

– **Engine will not start.**
– **Inspection of main relay.**

1. Disconnect the main relay connectors.
2. Check for continuity between BLK terminal ② and body ground.
3. Does continuity exist?
   - NO → **Repair open in BLK wire between main relay and G101.**
   - YES ↓
4. Measure the voltage between YEL/BLU terminal ① and body ground.
5. Is there battery voltage?
   - NO → **– Replace ECU fuse. – Repair open in the YEL/BLU wire between the main relay and the ECU fuse.**
   - YES ↓
6. Turn the ignition switch ON.
7. Measure the voltage between BLK/YEL terminal ⑤ and body ground.
8. Is there battery voltage?
   - NO → **– Replace No. 2 fuse. – Repair open in the BLK/YEL wire between the main relay and the No. 2 fuse.**
   - YES ↓
9. Turn the ignition switch to the START position.
10. Measure the voltage between BLK/WHT terminal ⑥ and body ground.
11. Is there battery voltage?
    - NO → **Repair open in the BLK/WHT wire between the main relay and the ignition switch.**
    - YES ↓

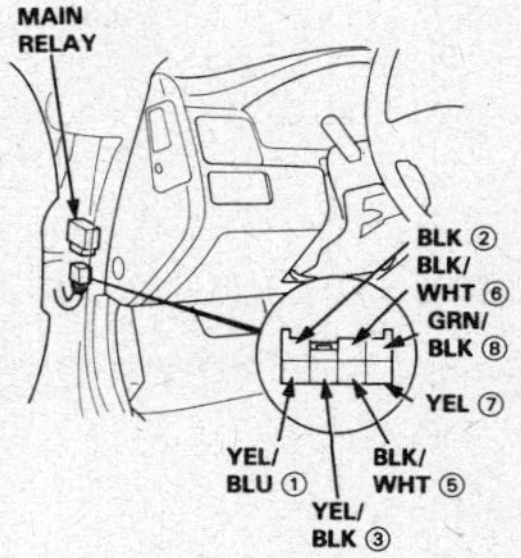

Turn the ignition switch off.

Connect the test harness between the ECU and connector. Disconnect "A" connector from the ECU only, not the main wire harness

Check for continuity between GRN/BLK terminal ⑧ and the following terminals; A7, A8.

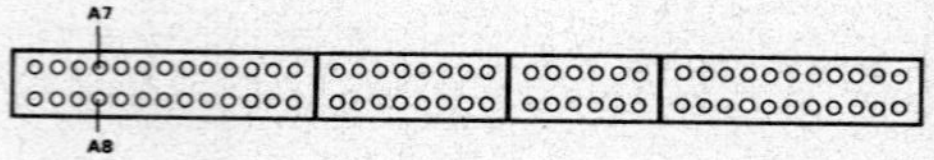

Does continuity exist?

- NO → Repair open in GRN/BLK wire between ECU (A7, A8) and main relay.
- YES → Reconnect "A" connector to the main wire harness.

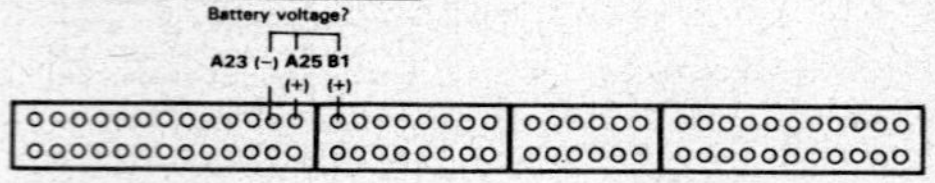

Connect the main relay connector.

Turn the ignition switch ON.

Measure the voltage between A23 (−) terminal and the following terminals: A25 (+) B1 (+).

Is there battery voltage?

- NO →
  - Repair open in the YEL/BLK wire between the ECU (A25, B1) and main relay.
  - Replace main relay.
- YES → Turn the ignition switch OFF.

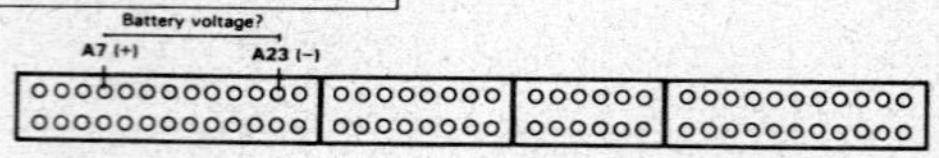

Connect a voltmeter between A7 (+) terminal and A23 (−) terminal.

Is there battery voltage for two seconds when the ignition switch is first turned on?

- YES → Substitute a known-good ECU and recheck. If prescribed voltage is now available, replace the original ECU.
- NO → Check the main relay

Troubleshooting Flowchart

Inspection of Bypass Control System

Start engine and allow to idle.

Remove #13 vacuum hose from the bypass control diaphragm and connect vacuum gauge to the hose.

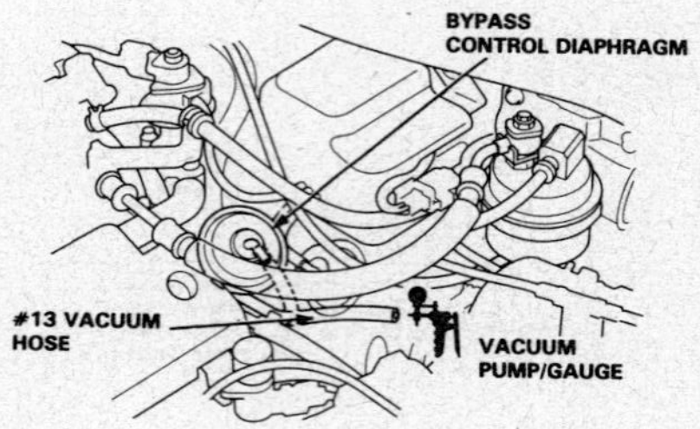

Is there vacuum ?

- YES
- NO → Disconnect the 2P connector from the bypass control solenoid valve.

Measure voltage between BLK/YEL (+) terminal and BLU/RED (−) terminal.

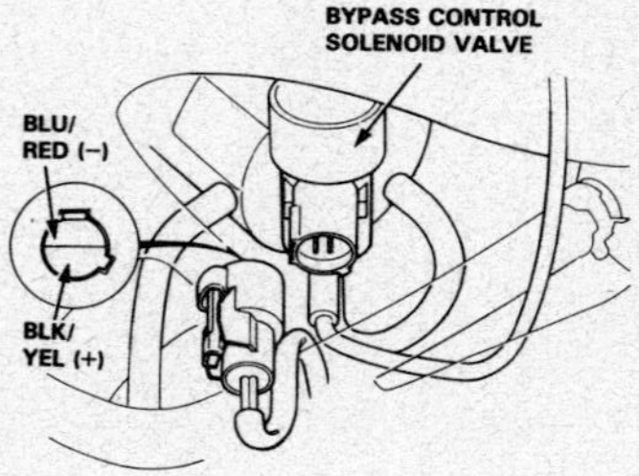

Is there battery voltage ?

- YES → Repair the blockage or vacuum leak between the intake manifold and diaphragm, if vacuum is OK, replace the bypass control solenoid valve.
- NO → Measure voltage between BLK/YEL (+) terminal and body ground.

Is there battery voltage ?

- NO → Repair open in BLK/YEL wire between the 2P connector and No. 2 (10 A) fuse.
- YES → Turn the ignition switch OFF.

Raise engine speed to 4,900 rpm.

Check for vacuum at #13 vacuum hose.

Is there vacuum ?

- NO → Bypass control system is OK.
- YES → Disconnect the 2P connector from the bypass control solenoid valve.

Is there vacuum ?

- YES → Replace bypass control solenoid valve.
- NO → Turn the ignition switch OFF.

Disconnect "A" connector from ECU.

Check for continuity to ground on the BLU/RED wire.

Is there continuity to ground ?

- YES → Repair short to ground at BLU/RED wire between ECU (A17) and the 2P connector.
- NO → Substitute a known-good ECU and recheck. If symptom goes away, replace the original ECU.

Reconnect the 2P connector to the bypass control solenoid valve.

Connect the test harness to the main wire harness only

Turn the ignition switch ON and connect a jumper wire between ECU (A17) and ECU (A26).

Does the solenoid valve click when the jumper is connected?

- NO → Repair open in BLU/RED wire between ECU (A17) and the 2P connector. If the wire is OK, replace the bypass control solenoid valve.
- YES → Substitute a known-good ECU and recheck. If prescribed voltage is now available replace the original ECU.

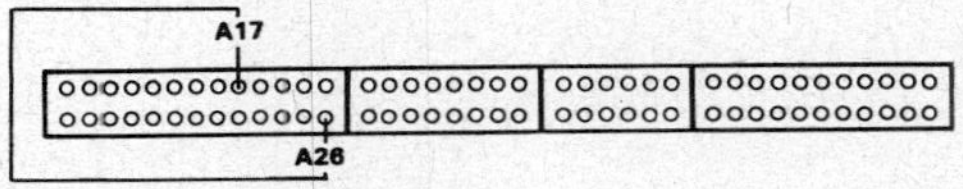

Inspection of Evaporative Emission Controls

Disconnect #3 vacuum hose from the purge control diaphragm valve (on the charcoal canister) and connect a vacuum gauge to the hose.

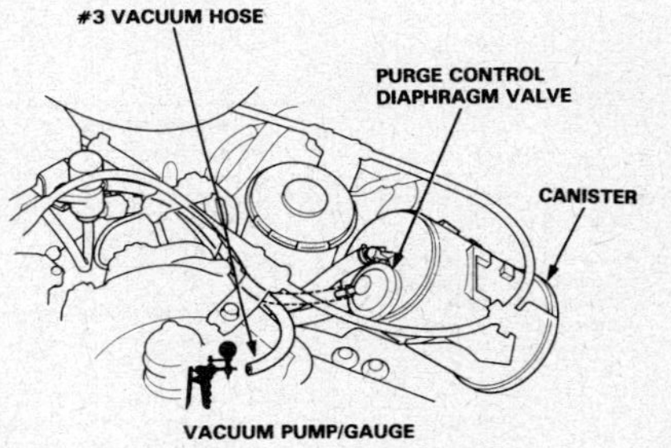

Start the engine and allow it to idle.
NOTE: Engine coolant temperature must be below 65°C (149°F).

Is there vacuum ?

- NO
- YES → Disconnect the 2P connector.

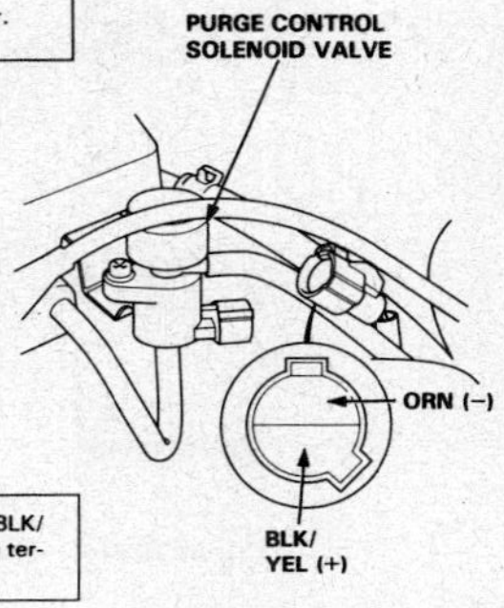

Measure voltage between BLK/YEL (+) terminal and ORN (−) terminal.

Is there battery voltage ?

- YES → Inspect vacuum hose routing. If OK, replace purge control solenoid valve.
- NO → Measure voltage between BLK/YEL (+) terminal and body ground.

Warm up the engine to normal operating temperature (cooling fan comes on)

Hold engine at 2,000 rpm.

Check for vacuum at #3 vacuum hose after staring the engine.

Is there manifold vacuum?
- NO → Disconnect the 2P connector.
  - Is there manifold vacuum?
    - NO → Inspect vacuum hose routing. If OK, replace purge control solenoid valve.
    - YES → Inspect for a short in ORN wire between ECU (A10) and the connector. If wire is OK, substitute a known-good ECU and recheck. If symptom goes away, replace the original ECU.
- YES → Reconnect the hose.

Is there battery voltage ?
- NO → Repair open in BLK/YEL wire between No. 2 fuse and the 2P connector.
- YES → Inspect ORN wire for an open to body ground between ECU (A10) and the connector. If wire is OK, substitute a known-good ECU and recheck. If symptom goes away, replace the original ECU.

Remove fuel filler cap.

Connect a vacuum gauge to canister purge air hose.

Start the engine and raise speed to 3,500 rpm.

Does vacuum appear on gauge within 1 minute ?
- YES → See two-way valve test to complete. Evaporative emission controls are OK.
- NO → Connect a vacuum gauge to the canister purge hose and raise the engine speed to 3,500 rpm.
  - Does vacuum appear on the gauge?
    - NO → Inspect the purge hose and throttle body port for pinch or blockage.
    - YES → Replace the canister.

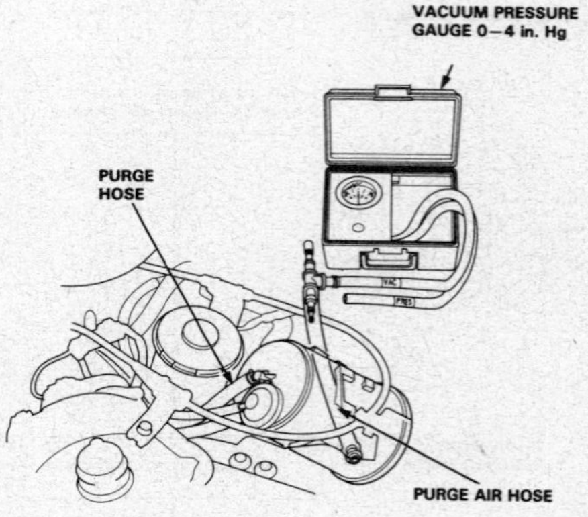

VACUUM COMPONENT AND HOSE ROUTING—NSX

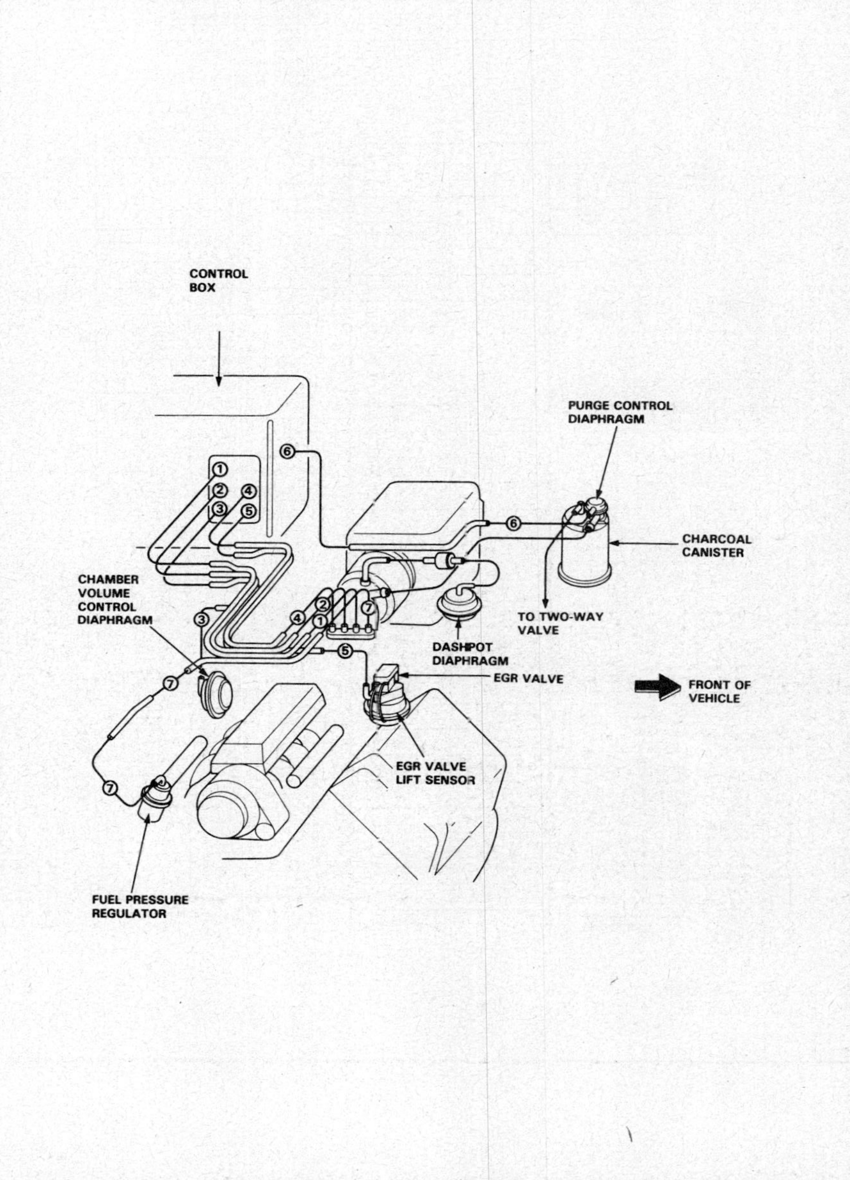

ECU INPUT/OUTPUT WIRING SCHEMATIC—NSX

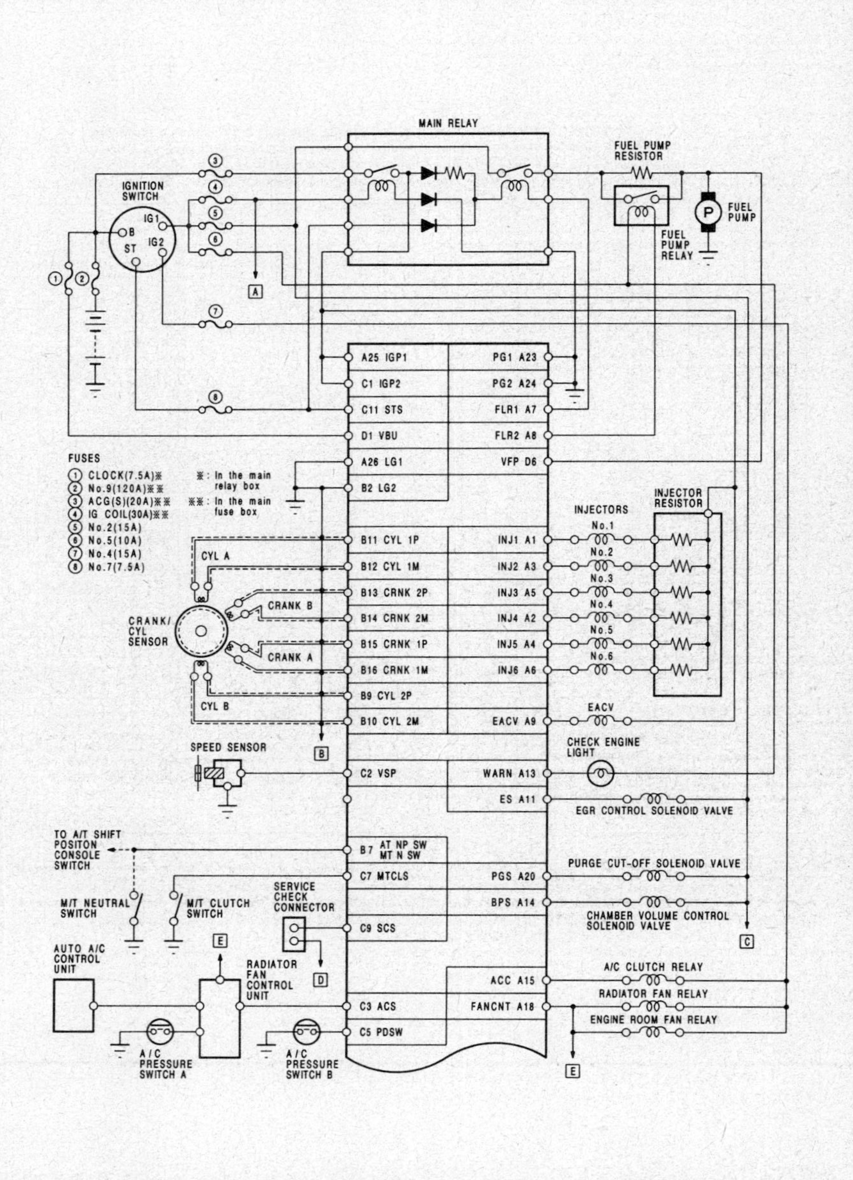

ECU INPUT/OUTPUT WIRING SCHEMATIC—NSX

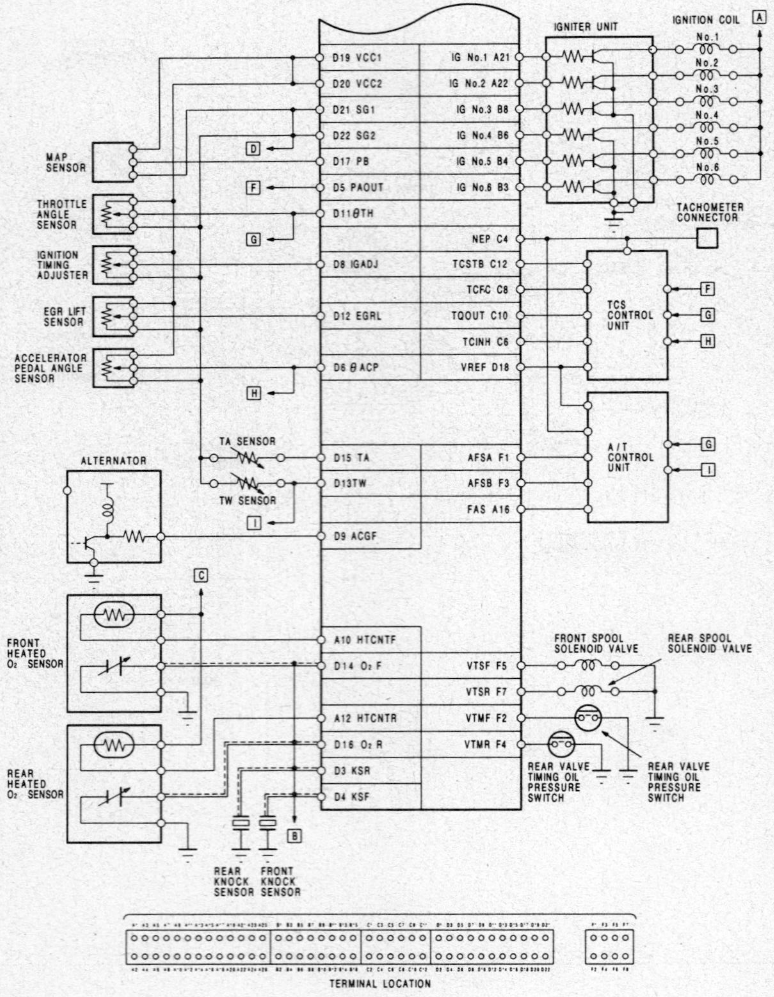

SYMPTOM TO SYSTEM DIAGNOSIS—NSX

NOTE: Across each row in the chart, the systems that could be sources of a symptom are ranked in the order they should be inspected starting with ①. Find the symptom in the left column, read across to the most likely source, then refer to the page listed at the top of that column. If inspection shows the system is OK, try the next most likely system ②, etc.

SYMPTOM \ SYSTEM		PGM-FI: ECU	OXYGEN SENSOR	MANIFOLD ABSOLUTE PRESSURE SENSOR	CRANK/CYL SENSOR	COOLANT TEMPERATURE SENSOR	THROTTLE ANGLE SENSOR	INTAKE AIR TEMPERATURE SENSOR	ATMOSPHERIC PRESSURE SENSOR	IGNITION OUTPUT SIGNAL	SPOOL SOLENOID VALVE
CHECK ENGINE LIGHT TURNS ON											
CHECK ENGINE LIGHT BLINKS		8 or *	1 or 2, 41 or 42	3 or 5	4 or 54, 9 or 59	6	7	10	13	15	21 or 51
ENGINE WON'T START		③			③					①	
DIFFICULT TO START ENGINE WHEN COLD		(BU)		③	③	①			③		
IRREGULAR IDLING	WHEN COLD FAST IDLE OUT OF SPEC	(BU)				③					
	ROUGH IDLE	(BU)		③							
	WHEN WARM RPM TOO HIGH	(BU)									
	WHEN WARM RPM TOO LOW	(BU)									
FREQUENT STALLING	WHILE WARMING UP	(BU)				③					
	AFTER WARMING UP	(BU)							③		
POOR PERFORMANCE	MISFIRE OR ROUGH RUNNING	(BU)			③						
	FAILS EMISSION TEST	(BU)	③	②							
	LOSS OF POWER	(BU)		③			②				③

* If codes other than those listed above are indicated, count the number of blinks again. If the indicator is in fact blinking these codes, substitute a known-good ECU and recheck. If the indication goes away, replace the original ECU.

(BU): When the Check Engine light and the self-diagnosis indicator are on, the back-up system is in operation. Substitute a known-good ECU and recheck. If the indication goes away, replace the original ECU.

SYMPTOM TO SYSTEM DIAGNOSIS—NSX

PGM-FI						IDLE CONTROL		FUEL SUPPLY					EMISSION CONTROL	
VALVE TIMING OIL PRESSURE SWITCH	VEHICLE SPEED SENSOR	IGNITION TIMING ADJUSTER	KNOCK SENSOR	A/T FI SIGNAL A	A/T FI SIGNAL B	ELECTRONIC AIR CONTROL VALVE	OTHER IDLE CONTROLS	FUEL INJECTOR	FUEL METERING	FUEL PUMP	OTHER FUEL SUPPLY	AIR INTAKE	EGR CONTROL SYSTEM	OTHER EMISSION CONTROLS
CHECK	CHECK	CHECK	CHECK			CHECK		CHECK	CHECK				CHECK	
22 or 52	17	18	23 or 53	30	31	14		16	43 or 44 / 45 or 46	47			12	
								②			③			
							②			②				
						①	②							
②						①		②					③	
						①	②							
						①		②						
						①	②				③			
						③	①				②		③	
						③		①					③	
								②	③		③			①
③			③					③		②	①	③		③

SYMPTOM TO SYSTEM DIAGNOSIS—NSX

NOTE:
- Across each row in the chart, the sub systems that could be sources of a symptom are ranked in the order they should be inspected, starting with ①. Find the symptom in the left column, read across to the most likely source, then refer to the page listed at the top of that column. If inspection shows the system is OK, try the next system ②, etc.
- If the idle speed is out of specification and the Check Engine light does not blink CODE 14, go to inspection

SYMPTOM \ SUB SYSTEM		IDLE ADJUSTING SCREW	EACV	AIR CONDITIONING SIGNAL	ALTERNATOR FR SIGNAL	A/T SHIFT POSITION SIGNAL	M/T NEUTRAL SWITCH SIGNAL	M/T CLUTCH SWITCH SIGNAL	STARTER SWITCH SIGNAL	FAST IDLE VALVE	HOSES AND CONNECTIONS
DIFFICULT TO START ENGINE WHEN COLD										①	
WHEN COLD FAST IDLE OUT OF SPEC (1,000–2,000 rpm)		③	②							①	
ROUGH IDLE			②								①
WHEN WARM RPM TOO HIGH		③	①							②	③
WHEN WARM RPM TOO LOW	Idle speed is below specified rpm (no load)	②	①								
	Idle speed does not increase after initial start up.		①								
	On models with automatic transmission, the idle speed drops in gear		②			①					
	Idle speeds drops when air conditioner in ON		②	①							
	Idle speed fluctuates with electrical load		②								①
FREQUENT STALLING	WHILE WARMING UP		①								
	AFTER WARMING UP	①									
FAILS EMISSION TEST											①

SYMPTOM TO SYSTEM DIAGNOSIS—NSX

NOTE: Across each row in the chart, the systems that could be sources of a symptom are ranked in the order they should be inspected starting with ①. Find the symptom in the left column, read across to the most likely source, then refer to the page listed at the top of that column. If inspection shows the system is OK, try the next most likely system ②, etc.

SYMPTOM \ SUB SYSTEM		FUEL INJECTOR	INJECTOR RESISTOR	PRESSURE REGULATOR	FUEL FILTER	FUEL PUMP	MAIN RELAY	FUEL PUMP RELAY FUEL PUMP RESISTOR	CONTAMI-NATED FUEL
ENGINE WON'T START		③	③		③	①	②		③
DIFFICULT TO START ENGINE WHEN COLD OR HOT									①
ROUGH IDLE		①	②						③
POOR PERFORMANCE	MISFIRE OR ROUGH RUNNING	①	②	③		③		③	③
	FAILS EMISSION TEST	②	③	①					
	LOSS OF POWER	③	③		①	③		③	②

NOTE: Across each row in the chart, the sub systems that could be sources of a symptom are ranked in the order they should be inspected starting with ①. Find the symptom in the left column, read across to the most likely source, then refer to the page listed at the top of that column. If inspection shows the system is OK, try the next system ②, etc.

SYMPTOM \ SUB SYSTEM	THROTTLE CABLE	THROTTLE BODY	DASHPOT SYSTEM	CHAMBER VOLUME CONTROL SYSTEM
WHEN COLD FAST IDLE OUT OF SPEC	③	②		①
WHEN WARM RPM TOO HIGH	②	①	③	
LOSS OF POWER		①		②

NOTE: Across each row in the chart, the systems that could be sources of a symptom are ranked in the order they should be inspected starting with ①. Find the symptom in the left column, read across to the most likely source, then refer to the page listed at the top of that column. If inspection shows the system is OK, try the next most likely system ②, etc.

SYMPTOM \ SUB SYSTEM		CATALYTIC CONVERTER	EGR SYSTEM	POSITIVE CRANKCASE VENTILATION SYSTEM	EVAPORATIVE EMISSION CONTROLS
ROUGH IDLE			①	②	
FREQUENT STALLING AFTER WARM-ING UP	AFTER WARM-ING UP		①		
POOR PERFORMANCE	FAILS EMISSION TEST	①	③		②
	LOSS OF POWER	①	②		

TROUBLESHOOTING CHARTS—NSX

Check Engine light never comes on (even for two seconds) after ignition is turned on.

Is the oil pressure light on?
- NO → Inspect No. 5 fuse. Is No. 5 fuse OK?
  - NO → **Replace fuse.**
  - YES → **Repair open in YEL between No. 5 fuse and gauge assembly.**
- YES → Turn the ignition switch OFF.

Connect the ECU test harness between the ECU and connector

Connect A13 terminal to body ground.

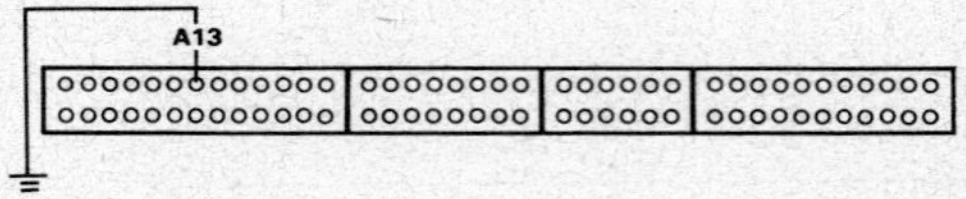

Turn the ignition switch ON.

Is Check Engine light on?
- NO → – **Replace the light bulb.** – **Repair open in BLU wire between ECU (A13) and gauge assembly.**
- YES → Measure voltage between body ground and the following terminals individually: •A23, •A24, •A26, •B2.

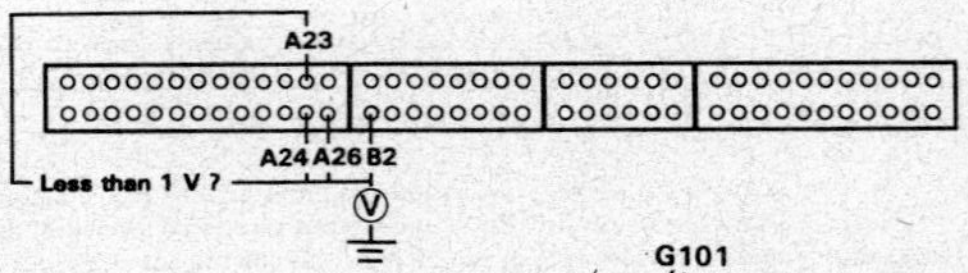

Is there less than 1V?
- NO → **Repair open in wire between ECU and intake manifold (G101) that had more than 1V.**
- YES → **Substitute a known-good ECU and recheck. If symptom/ indication goes away, replace the original ECU.**

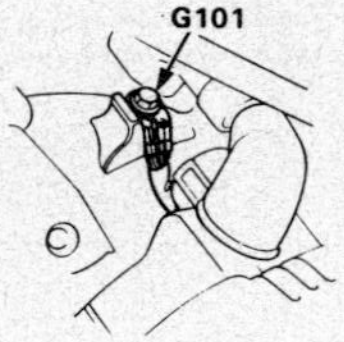

TROUBLESHOOTING CHARTS—NSX

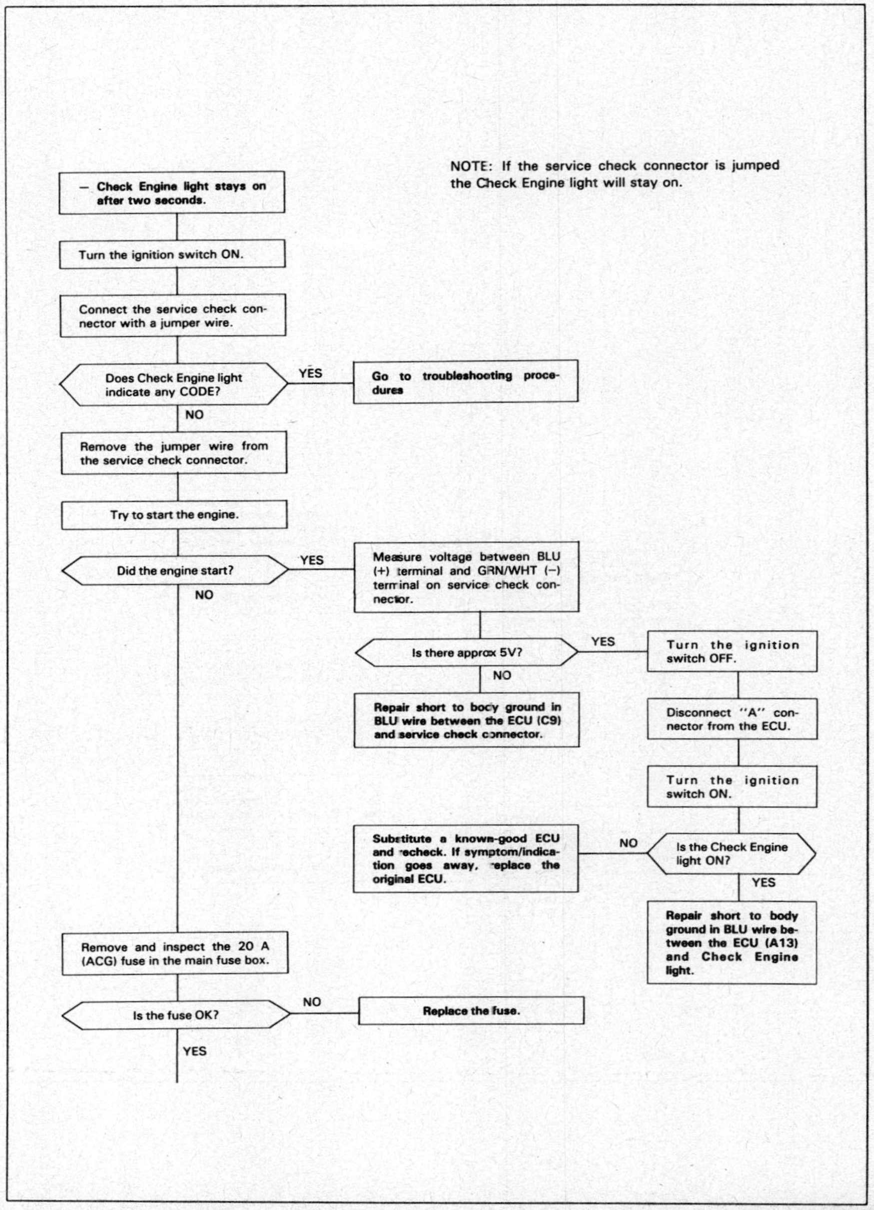

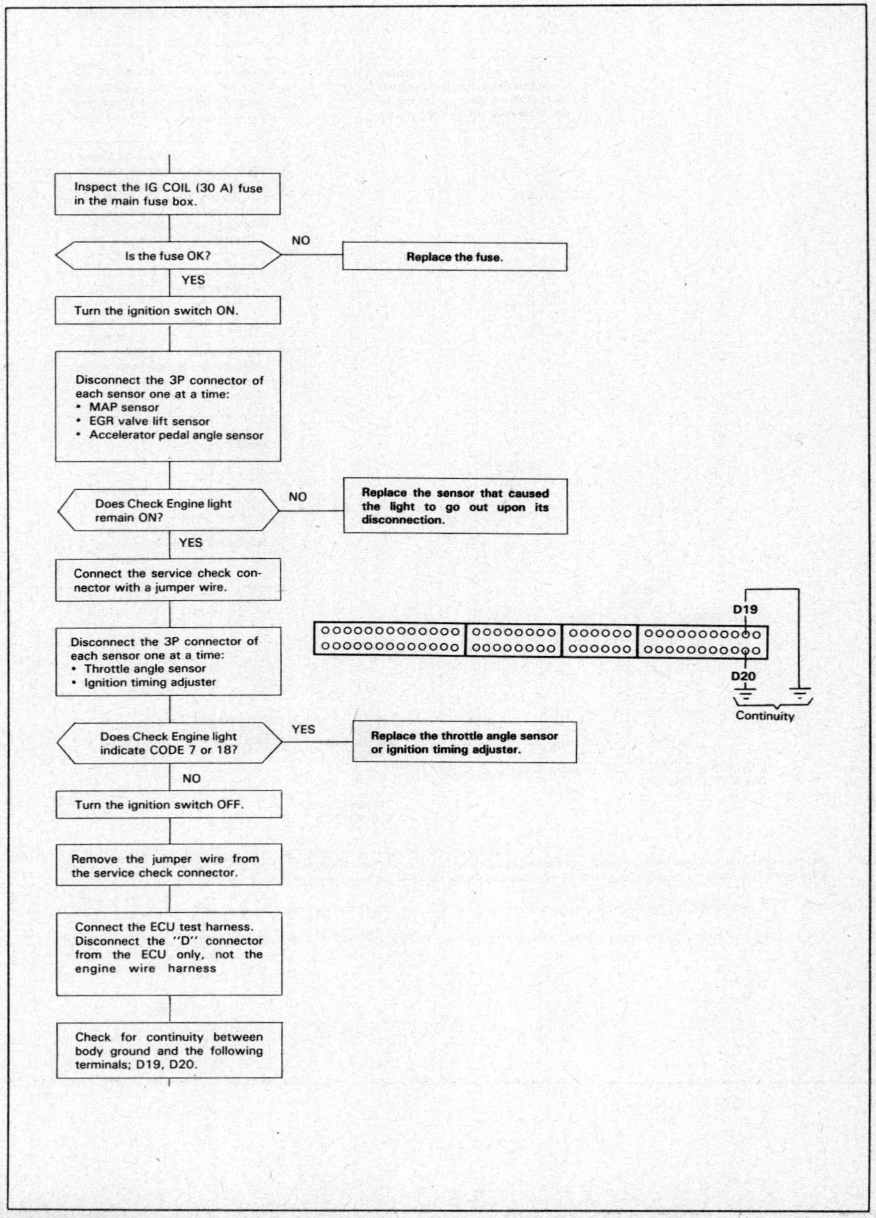

Does continuity exist?

- YES → Repair short to body ground in YEL/WHT wire between ECU (D19, D20) and throttle angle sensor, EGR valve lift sensor MAP sensor, Ignition timing Adjuster and Accelerator Pedal angle sensor.
- NO ↓

Reconnect all the connectors. Reconnect the "D" connector to the ECU.

Turn the ignition switch ON.

Measure voltage between A26 (−) and the following: C1 (+) and A25 (+).

Is there battery voltage?

- NO → – Repair open in YEL/BLK wire between ECU (A25, C1) and main relay.
  – Check main relay and wiring connectors at main relay.
- YES ↓

Substitute a known-good ECU and recheck. If symptom/indication goes away, replace the original ECU.

---

[1] Self-diagnosis Check Engine light indicates code 1: A problem in the Front Oxygen (O_2) Sensor circuit.

[2] Self-diagnosis Check Engine light indicates code 2: A problem in the Rear Oxygen (O_2) Sensor circuit.

The oxygen sensors, detects the oxygen content in the exhaust gas, and input the ECU. In operation, the ECU receives the signals from the sensors and varies the duration during which fuel is injected. The oxygen sensors are combined with a heaters. The heaters stabilize the sensors output. The oxygen sensors are installed in the exhaust manifolds.

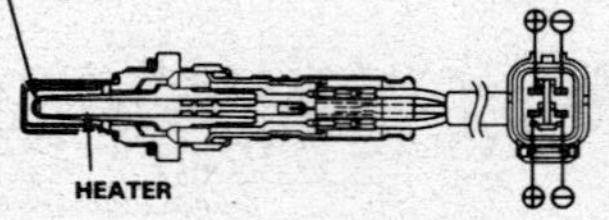

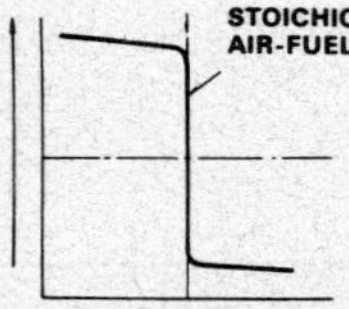

[1] or [2]

– Check Engine light has been reported on.
– With service check connector jumped (page 11-18), CODE 1 (Front O_2 sensor) and/or 2 (Rear O_2 sensor) are indicated.

Do the ECU Reset Procedures

Warm up engine to normal operating temperature (cooling fan comes on).

Run engine for 60 seconds.

Road test with the transmission in 4th gear, starting at 1200 rpm accelerate using wide open throttle for at least 5 seconds. Then decelerate for at least 5 seconds with the throttle completely closed.

Is Check Engine light on and does it indicate CODE 1 or CODE 2?

- NO → Intermittent failure, system is OK at this time. Check for poor connections or loose wires at the O_2 sensors C144, C148.
- YES ↓

test for CODE 43 or CODE 44.

3 — Self-diagnosis check Engine light indicates code 3: Most likely an elektrical problem in the Manifold Absolute Pressure (MAP) Sensor system.

5 — Self-diagnosis check Engine light indicates code 5: Most likely mechanical problem (broken hose) in the Manifold Absolute Pressure (MAP) Sensor System.

The MAP sensor converts manifold absolute pressure into electrical signals and inputs the ECU.

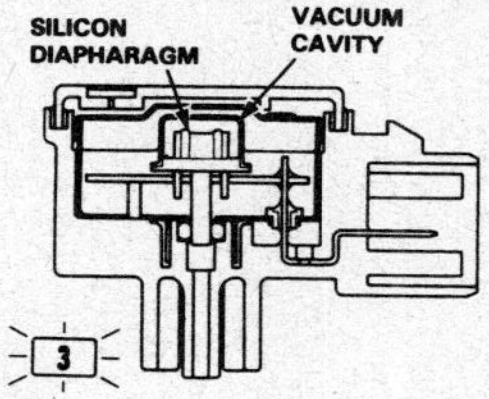

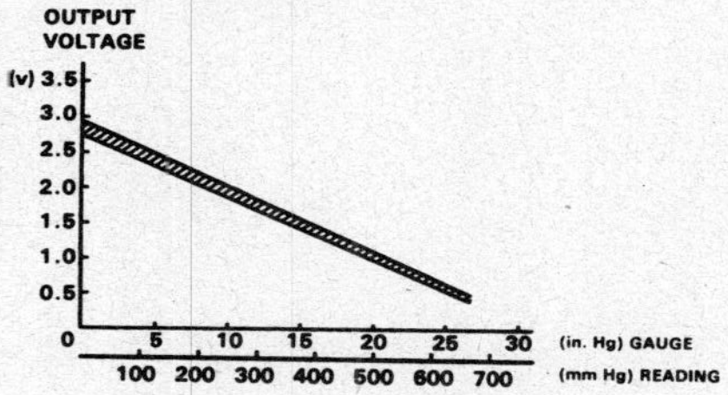

3

- Engine is warm and running.
- Check Engine light has been reported on.
- With service check connector jumped CODE 3 is indicated.

Do the ECU Reset Procedures

Start the engine and allow it to idle.

Is Check Engine light on and does it indicate CODE 3?

NO → Intermittent failure, system is OK at this time (test drive may be necessary). Check for poor connection or loose wires at MAP sensor C534 and ECU.

YES ↓

Turn the ignition switch OFF.

Disconnect the 3P connector from the MAP sensor.

Turn the ignition switch ON.

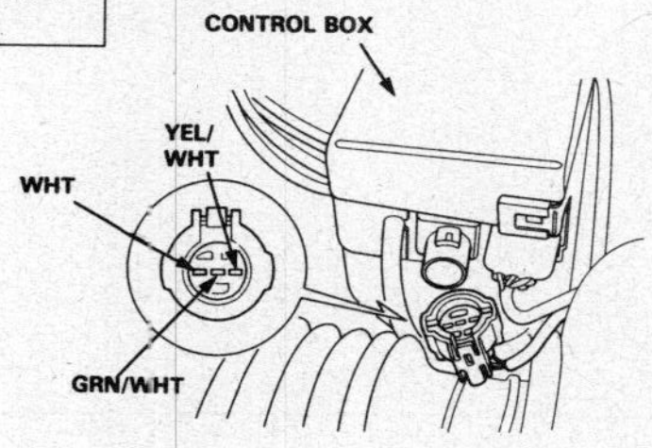

Measure voltage between YEL/WHT (+) terminal and body ground.

Is there approx. 5V?

NO ↓

Connect the ECU test harness to the ECU only, not to the engine wire harness

Measure voltage between D19 (+) terminal and D21 (−) terminal.

Is there approx. 5 V?

NO → Repair open in YEL/WHT wire between ECU (D19) and MAP sensor.

YES → Substitute a known-good ECU and recheck. If symptom/indication goes away, replace the orignal ECU.

YES (from first "Is there approx. 5V?") ↓

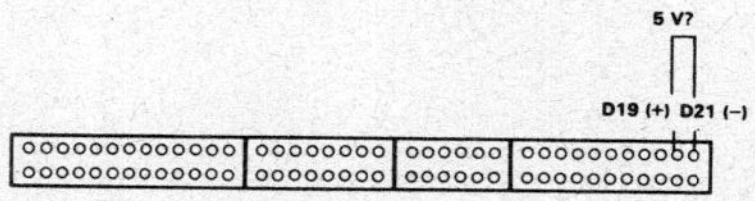

Measure voltage between YEL/WHT (+) terminal and GRN/WHT (−) terminal.

Is there approx. 5V?

NO → Repair open in GRN/WHT wire between ECU (D21) and MAP sensor.

YES ↓

Measure voltage between WHT (+) terminal and GRN/WHT (−) terminal.

Is there approx. 5V?

NO → Repair open or short in WHT wire between ECU (D17) and MAP sensor. If wire is OK, substitute a known-good ECU and recheck. If prescribed voltage is now available, replace the original ECU.

YES ↓

Turn the ignition switch OFF.

Reconnect the 3P connector to the MAP sensor.

Connect the ECU test harness between the ECU and connector

Turn the ignition switch ON.

Measure voltage between D17 (+) terminal and D21 (−) terminal.

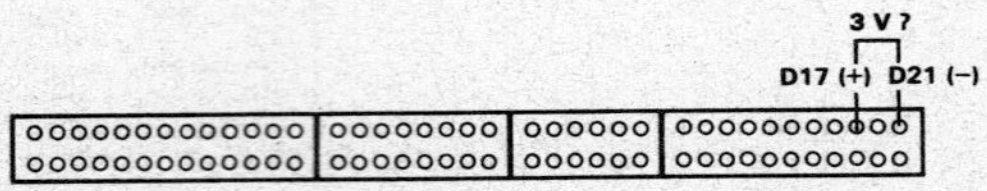

Is there approx. 3 V ?

NO → Replace MAP sensor.

YES → Substitute a known-good ECU and recheck. If symptom/indication goes away, replace the original ECU.

---

4 — Self-diagnosis Check Engine light indicates code 4: A problem in the circuit of the CRANK A Sensor.

54 — Self-diagnosis Check Engine light indicates code 54: A problem in the circuit of the CRANK B Sensor.

9 — Self-diagnosis Check Engine light indicates code 9: A problem in the circuit of the CYL A Sensor.

59 — Self-diagnosis Check Engine light indicates code 59: A problem in the circuit of the CYL B Sensor.

The CRANK sensor determines timing for fuel injection and ignition of each cylinder and also detects engine RPM. The CYL sensor detects the position of No. 1 cylinder for sequential fuel injection to each cylinder.

4 — 54 — 9 or 59

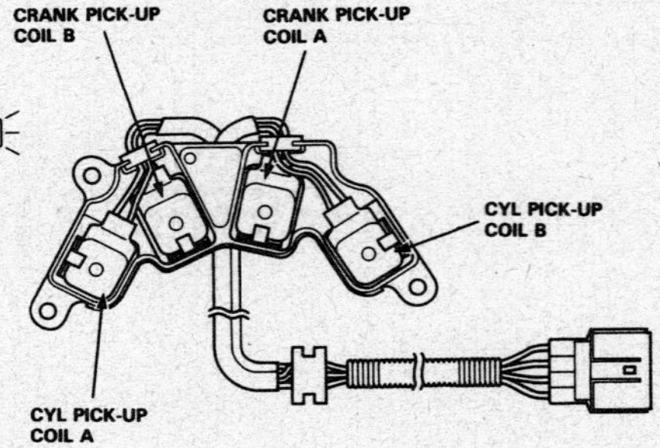

- Check Engine light has been reported on.
- With service check connector jumped CODE 4, 54, 9 and/or 59 are indicated.

Do the ECU Reset Procedures

Start the engine.

Is Check Engine light on and does it indicate CODE 4, 54, 9 or 59?

NO → Intermittent failure, system is OK at this time (test drive may be necessary). Check for poor connections or loose wires at CRANK/CYL sensor connector C146 and ECU.

YES → Stop the engine.

Disconnect the 8P connector from the CRANK/CYL sensor.

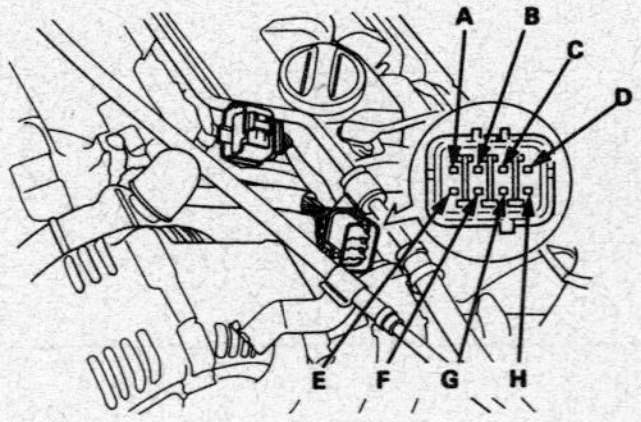

TROUBLESHOOTING CHARTS—NSX

Measure resistance between terminals of the indicated sensor.
*see table

Is there 650–850 Ω — NO → Replace the CRANK/CYL sensor

YES

Check for continuity to body ground on A—H terminals individually.

Does continuity exist? — YES → Replace the CRANK/CYL sensor

NO

Reconnect the connector.

Connect the ECU test harness to the engine wire harness only, not to the ECU

Measure resistance between terminals of the indicated sensor on ECU test harness.
*see table

Is there 650–850 Ω? — NO → Repair open in the indicated sensor wires.
*see table

YES

Check for continuity to body ground on B15, B13, B11 or B9 terminals.

Does continuity exist? — YES → Repair short in the indicated sensor wires.
*see table

NO

Substitute a known-good ECU and recheck. If symptom/indication goes away, replace the original ECU.

SENSOR	CODE	SENSOR TERMINAL	ECU TERMINAL	WIRE COLOR
CRANK A	4	A	B15	ORN/BLU
		B	B16	WHT/BLU
CRANK B	54	C	B14	WHT/BLU
		D	B13	ORN/BLU
CYL A	9	G	B12	BLU/YEL
		H	B11	BLU/GRN
CYL B	59	E	B9	ORN
		F	B10	WHT

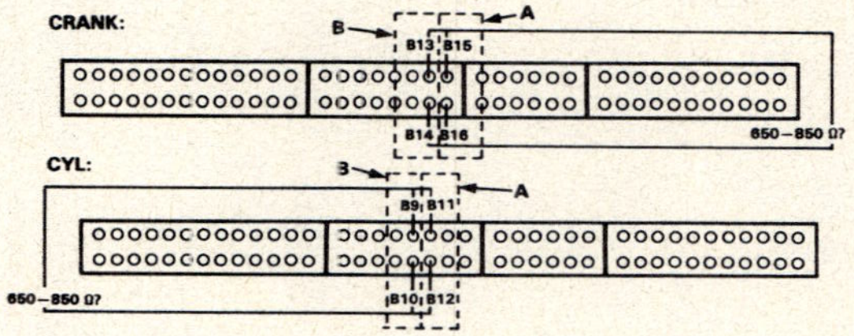

- Check Engine light has been reported on.
- With service check connector jumped CODE 5 is indicated.

Do the ECU Reset Procedures

Start the engine and keep engine rpm at idle.

Is Check Engine light on and does it indicate CODE 5? — NO →
- Intermittent failure, system is OK at this time (test drive may be necessary).
- Check vacuum hoses, pipes and connections.
- Make sure all connectors are secure.

YES

Stop the engine.

Disconnect #4 hose from the throttle body, connect vacuum pump to the hose and apply vacuum.

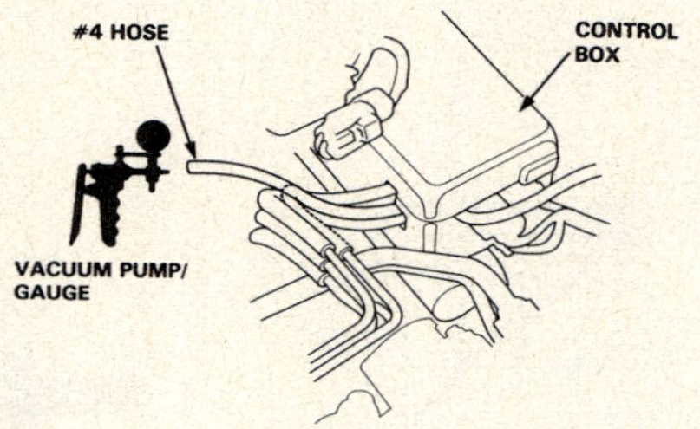

Does it hold vacuum? — NO → Connect a vacuum pump to the MAP sensor and apply vacuum.

Does it hold vacuum? — NO → Replace MAP sensor.

YES → Replace #4 hose.

YES

Connect a T-fitting from a vacuum gauge between the throttle body and MAP sensor.

- Start the engine.
- Is there manifold vacuum?
  - NO: – Remove restriction from throttle body.
  - YES: Stop the engine.
- Connect the ECU test harness between the ECU and connector
- Turn the ignition switch ON.
- Measure voltage between D17 (+) terminal and D21 (–) terminal.

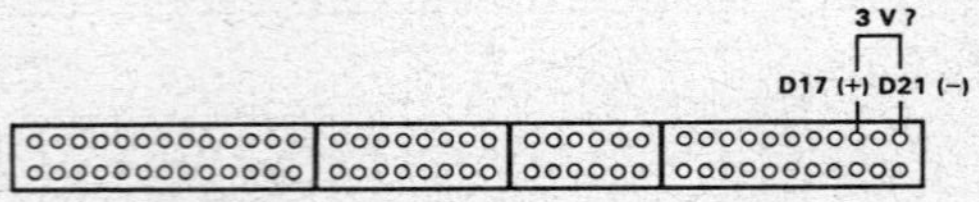

- Is there approx. 3 V ?
  - NO: Replace the MAP sensor.
  - YES: Start the engine and allow it to idle.
- Is there approx. 1 V?
  - NO: Replace MAP sensor.
  - YES: Substitute a known-good ECU and recheck. If symptom/indication goes away, replace the original ECU.

[6] Self-diagnosis Check Engine light indicates code 6: Most likely a problem in the Coolant Temperature (TW) Sensor circuit.

The TW sensor is a temperature dependant resistor (thermistor). The resistance of the thermistor decreases as the coolant temperature increases as shown below.

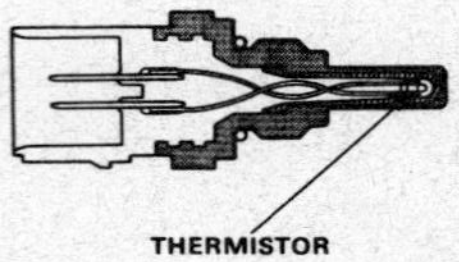

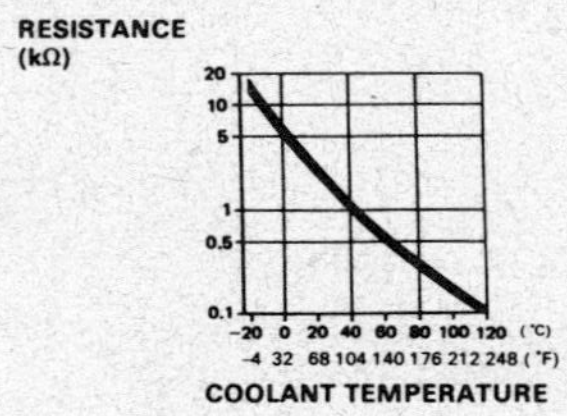

[6]

- – Check Engine light has been reported on.
  – With service check connector jumped CODE 6 is indicated.
- Do the ECU Reset Procedures
- Turn the ignition switch ON.
- Is Check Engine light on and does it indicate CODE 6?
  - NO: Intermittent failure, system is OK at this time (test drive may be necessary). Check for poor connections or loose wires at TW sensor C108 and ECU.
  - YES: Warm up engine to normal operating temperature (cooling fan comes on).
- Turn the ignition switch OFF.
- Disconnect the 2P connector from the TW sensor.
- Measure resistance between the 2 terminals on the TW sensor.
- Is there 200—400 Ω ?
  - NO: Replace TW sensor.
  - YES

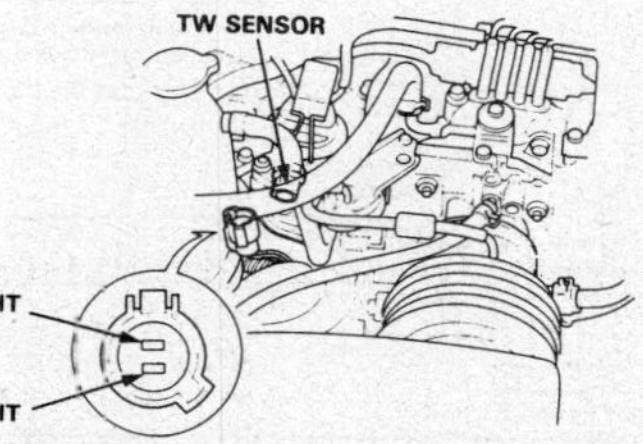

Turn the ignition switch ON.

Measure voltage between RED/WHT and body ground.

Is there approx. 5V ?

- YES: Measure voltage between RED/WHT (+) terminal and GRN/WHT (–) terminal.
  - Is there approx. 5V ?
    - NO: Repair open in GRN/WHT wire between ECU (D22) and TW sensor.
    - YES: Substitute a known-good ECU and recheck. If symptom/indication goes away, replace the original ECU.
- NO:

A/T only

Disconnect the 22P connector from the A/T control unit.

Is there approx. 5V?

- YES: Replace the A/T control unit.
- NO:

Connect the ECU test harness to the ECU only, not to the engine wire harness (page 11-21).

Measure voltage between D13 (+) terminal and D22 (–) terminal.

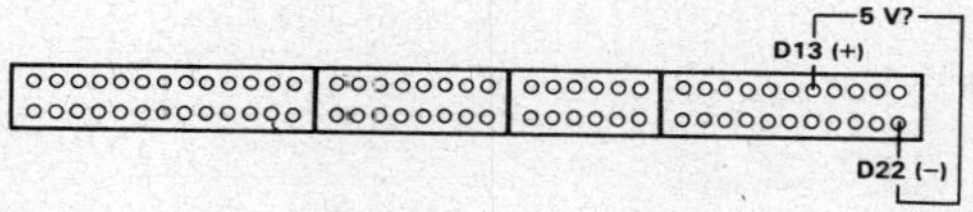

Is there approx. 5V?

- YES: Repair open in RED/WHT wire between ECU (D13), A/T control unit and sensor.
- NO: Substitute a known-good ECU and recheck. If symptom/indication goes away, replace the original ECU.

[CHECK] [7] Self-diagnosis Check Engine light indicates code 7: Most likely a problem in the Throttle Angle Sensor circuit.

The throttle angle sensor is a potentiometer. It is connected to the throttle valve shaft. As the throttle angle changes, the throttle angle sensor varies the voltage signal to the ECU.

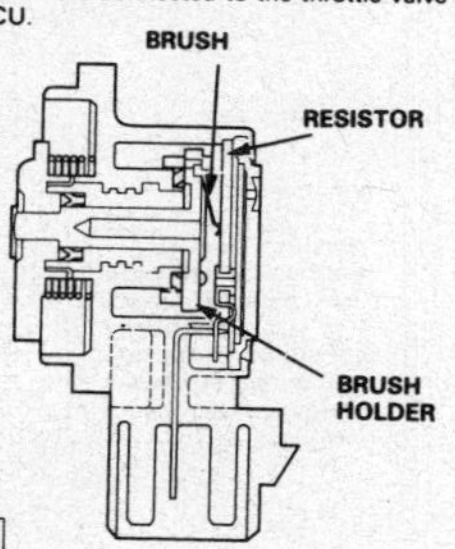

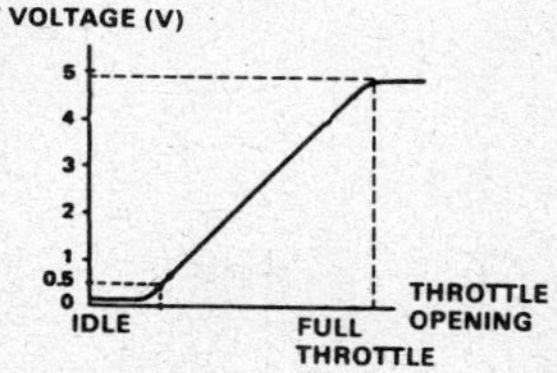

[CHECK] [7]

- Engine is running.
- Check Engine light has been reported on.
- With service check connector jumped CODE 7 is indicated.

Do the ECU Reset Procedures

Trun the ignition switch ON.

Is Check Engine light on and does it indicate CODE 7?

- NO: Intermittent failure, system is OK at this time (test drive may be necessary). Check for poor connections or loose wires at throttle angle sensor C130 and ECU.
- YES:

Turn the ignition switch OFF.

Disconnect the 3P connector from the throttle angle sensor.

Turn the ignition switch ON.

Measure voltage between YEL/WHT (+) terminal and GRN/WHT (–) terminal.

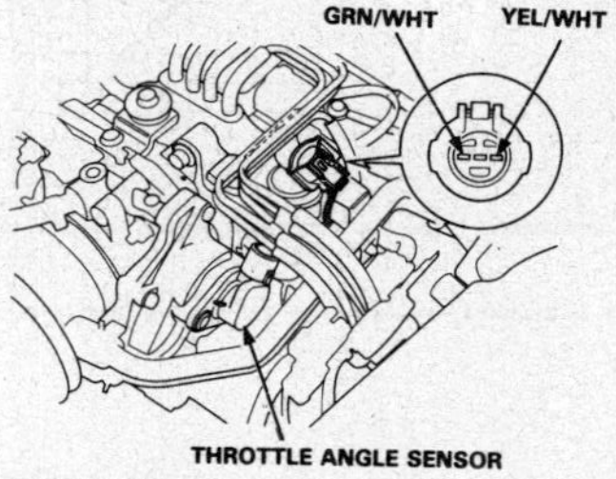

Is there approx. 5V ?

- NO: Measure voltage between YEL/WHT (+) terminal and body ground.
- YES:

Turn the ignition switch OFF.

Reconnect the 3P connector.

Connect the ECU test harness between the ECU and connector

Turn the ignition switch ON.

Is there approx. 5V ?

YES → Repair open in GRN/WHT wire between ECU (D22) and throttle angle sensor.

NO → Turn the ignition switch OFF.

Connect the ECU test harness between the ECU and connector

Turn the ignition switch ON.

Measure voltage between D20 (+) terinal and D22 (−) terminal.

Is there approx. 5V ?

YES → Repair open in YEL/WHT wire between ECU (D20) and throttle angle sensor.

NO → Substitute a known-good ECU and recheck. If prescribed voltage is now available, replace the original ECU.

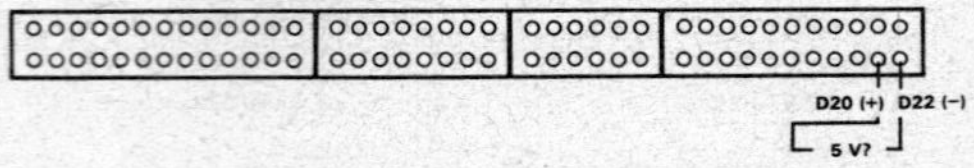

Measure voltage between D11(+) terminal and D22 (−) terminal.

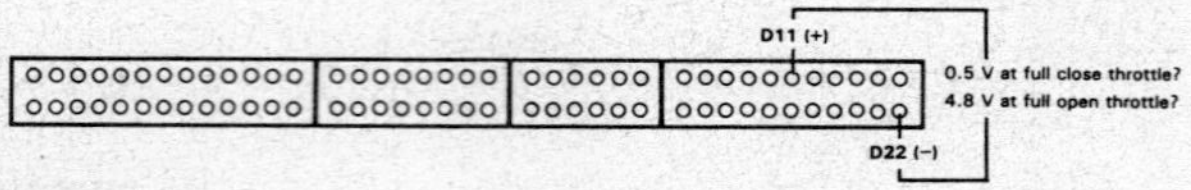

Is voltage 0.5 V at full close throttle, and 4.8 V at full open throttle?
NOTE: There should be a smooth transition from 0.5 V to 4.8 V as the throttle is depressed.

YES → Substitute a known-good ECU and recheck. If symptom/indication goes away, replace the original ECU.

NO → **A/T only**

Disconnect the 22P connector from the A/T control unit.

Is voltage 0.5 V at full close throttle, and 4.8 V at full open throttle?
NOTE: There should be a smooth transition from 0.5 V to 4.8 V as the throttle is depressed.

YES → Replace the A/T control unit.

NO →
- Replace throttle angle sensor.
- Repair open or short in RED/YEL wire between ECU (D11), A/T control unit and throttle angle sensor.

CHECK 10 Self-diagnosis Check Engine light indicates code 10: Most likely a problem in the Intake Air Temperature (TA) Sensor ciucuit.

The TA sensor is a temperature dependant resistor (thermistor). The resistance of the thermistor decreases as the intake air temperature increases as shown below.

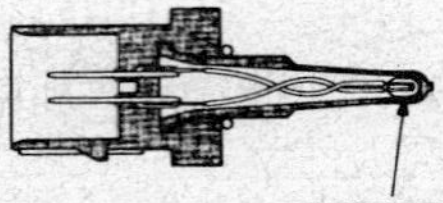

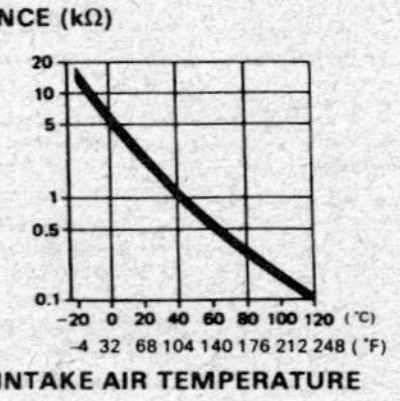

CHECK 10

- Check Engine light has been reported on.
- With service check connector jumped CODE 10 is indicated.

Do the ECU Reset Procedures

Turn the ignition switch ON.

Is Check Engine light on and does it indicate CODE 10?

NO → Intermittent failure, system is OK at this time (test drive may be necessary). Check for poor connections or loose wires at TA sensor C140 and ECU.

YES → Turn the ignition switch OFF.

Disconnect the 2P connector from the TA sensor.

Measure resistance between the 2 terminals on the TA sensor.

Is there 1–4 kΩ ?

NO → Replace TA sensor.

YES

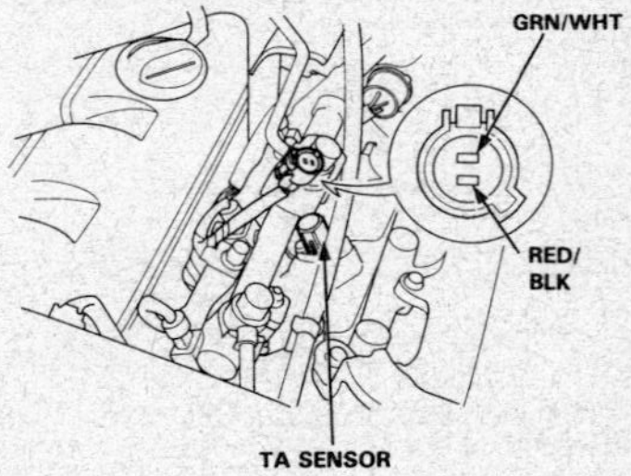

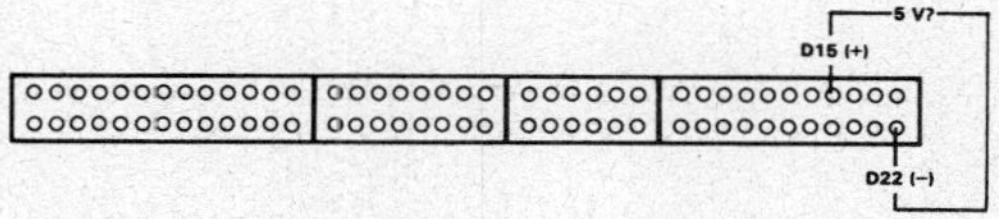

Turn the ignition switch ON.

Measure voltage between RED/BLK (+) terminal and body ground.

Is there approx. 5 V ?

- YES → Measure voltage between RED/BLK (+) terminal and GRN/WHT (−) terminal.
  - Is there approx. 5 V ?
    - NO → Repair open in GRN/WHT wire between ECU (D22) and TA sensor.
    - YES → Substitute a known-good ECU and recheck. If symptom/indication goes away, replace the original ECU.
- NO → Connect the ECU test harness to the ECU only, not to the engine wire harness

Measure voltage between D15 (+) terminal and D22 (−) terminal.

Is there approx. 5V?

- YES → Repair open in RED/BLK wire between ECU (D15) and TA sensor.
- NO → Substitute a known-good ECU and recheck. If symptom/indication goes away, replace the original ECU.

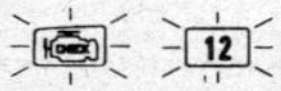

- Check Engine light has been reported on.
- With service check connector jumped CODE 12 is indicated.

Do the ECU Reset Procedures

Road test necessary: Warm up the engine to normal operating temperature (cooling fan comes on). Drive the car on the road for approx. 10 minutes. Try to keep the engine speed in the 1700— 2500 range.

Is Check Engine light on and does it indicate CODE 12?

- NO → Intermittent failure, system is OK at this time. Check for poor connections or loose wires at EGR valve C141, and control box C532.

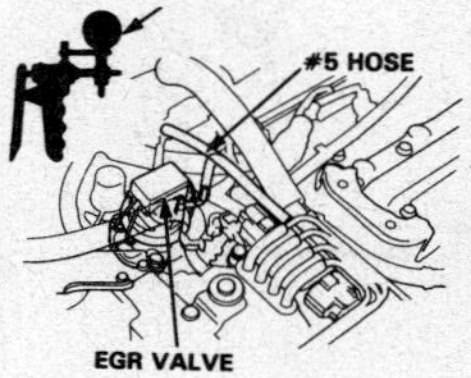

- YES → With the engine at idle, disconnect the #5 hose from the EGR valve and connect a vacuum pump/gauge to the hose.

Is there any vacuum?

- YES → Disconnect 6P connector from the control box and check the #5 hose for vacuum again.

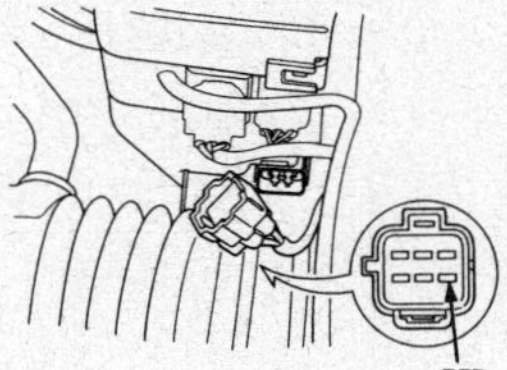

- NO → Move the vacuum pump/gauge to the EGR valve.

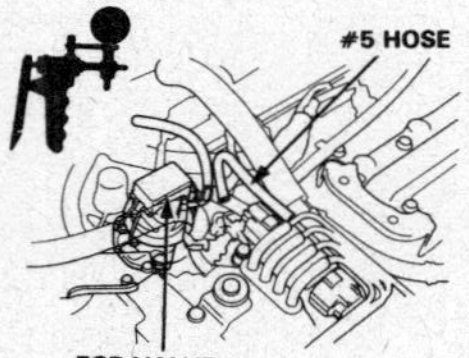

With the engine at idle, apply 8" of vacuum to the EGR valve.

Is there any vacuum ?
- YES → Check vacuum hose routing of the entire EGR system. If hose routing is OK, replace EGR control solenoid valve.
- NO → Check for continuity ground on RED wire of 6P connector.

Does continuity exist?
- YES → Repair open in RED wire between EGR control solenoid valve and ECU (A11).
- NO → Substitute a known-good ECU and retest. If symptom/indication goes away, replace the original ECU.

Does the engine stall and does the EGR valve hold vacuum ?
- NO → Replace EGR valve.
- YES → Disconnect the 6P connector from the control box.

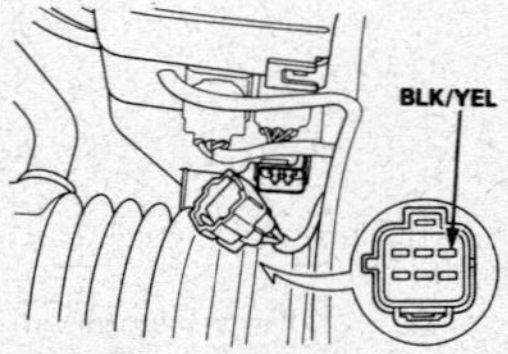

Measure voltage between BLK/YEL (+) terminal on the main wire harness and body ground.

Is there battery voltage ?
- NO → Repair open in BLK/YEL wire between the solenoid valve and No.2 fuse.
- YES → Reconnect the vacuum pump/gauge to the #5 hose.

Start the engine and allow it to idle.

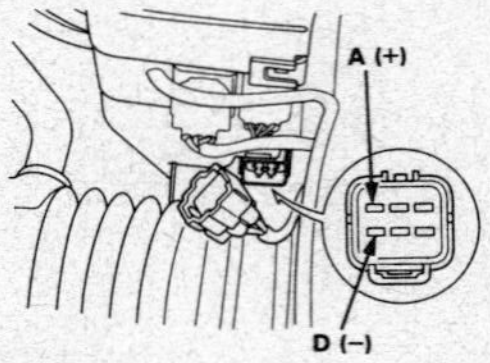

Connect the battery positive terminal to the A terminal of the 6P connector. While watching the vacuum gauge, connect the battery negative terminal to the D terminal.

Is there approx. 8" within 1 second?
- NO → Turn the ignition switch OFF and inspect the #1 and #5 hoses for leaks, restrictions, or misrouting.
  - Are the hoses OK?
    - NO → Connect as necessary.
    - YES → Disconnect the lower hose on EGR control solenoid valve and connect a vacuum gauge to the hose.
      - Start the engine and allow it to idle.
      - Is there 6"— 10" of vacuum ?
        - NO → Replace CVC valve.
        - YES → Replace the EGR control solenoid valve.
- YES → Turn the ignition switch OFF and reconnect the 6P connector.

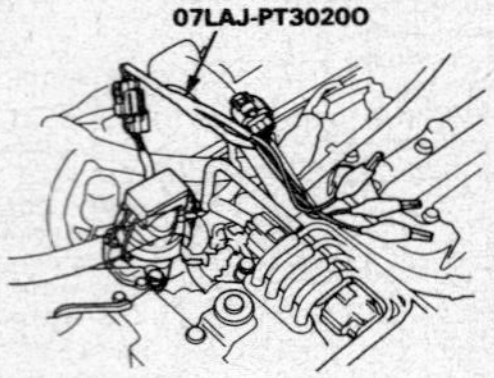

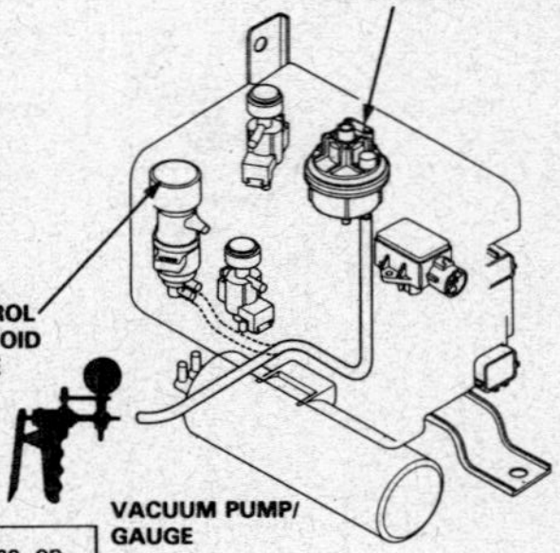

Connect the test harness between the EGR valve lift sensor and engine wire harness.

Turn the ignition switch ON.

Measure voltage between GRN (+) teminal and RED (–) terminal.

Is there approx. 5 V ?
- NO → – Repair open in YEL/WHT wire between EGR valve and ECU (D20).
  – Repair open in GRN/WHT wire between EGR valve and ECU (D22).
- YES

TROUBLESHOOTING CHARTS—NSX

- Measure voltage between WHT (+) terminal and RED (–) terminal.
- Is there approx 1.2 V ?
  - NO:
    - Replace EGR valve.
    - Repair short in WHT/GRN wire between ECU (D12) and EGR valve lift sensor.
    - Substitute a known-good ECU and recheck. If prescribed voltage is now available, replace the original ECU.
  - YES:
- While watching the voltmeter, slowly apply a continuous 8" of vacuum to the EGR valve. Repeat several times, completely releasing vacuum between.
- Is the maximum voltage approx. 4 V ?
  - NO: Replace EGR valve.
  - YES:
- Does the voltage consistently increase/decrease as the vacuum increases/decreases ?
  - NO: Replace EGR valve.
  - YES:
- Reconnect the #5 hose to the EGR valve.
- Turn the ignition switch OFF.
- Connect the ECU test harness between the ECU and connector
- Start the engine and allow it to idle.

- Measure voltage between D12 (+) terminal and D22 (–) terminal.

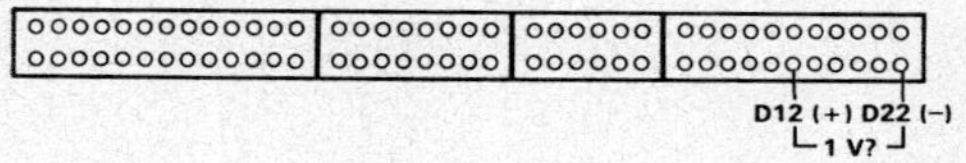

- Is there approx. 1.2 V ?
  - NO: Repair open in WHT/BLK wire between ECU (D12) and the sensor.
  - YES:
- Connect A11 terminal to A26 terminal with a jumper wire.
- Did the engine stall ?
  - NO: Repair open in RED wire between ECU (A11) and EGR control solenoid valve.
  - YES:
- Substitute a known-good ECU and recheck. If symptom/indication goes away, replace the original ECU.

13 Self-diagnosis Check Engine light indicates code 13: A problem in the Atmospheric Pressure (PA) Sensor.

The PA sensor is built into the ECU.

13

- Check Engine light has been reported on.
- With service check connector jumped CODE 13 is indicated.

Do the ECU Reset Procedures

Turn the ignition switch ON.

Is Check Engine light on and does it indicate CODE 13?

NO → Intermittent failure, system is OK at this time (test drive may be necessary.)

YES ↓

Turn the ignition switch OFF.

Connect the ECU test harness to the engine wire harness not to the ECU

Disconnect the 20P connector from the TCS control unit.

Check for continuity between D5, terminal and body ground.

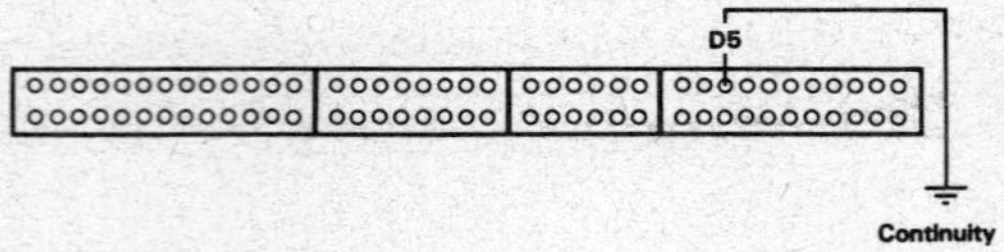

Does continuity exist?

YES → Repair short in RED/ORN wire between ECU (D5) and the TCS control unit.

NO ↓

Substitute known-good TCS control unit and recheck.

Turn the ignition switch ON.

Is Check Engine light on and does it CODE 13?

NO → Replace the original TCS control unit.

YES ↓

Substitute a known-good ECU and recheck. If symptom/indication goes away, replace the original ECU.

14 Self-diagnosis check Engine light indicates code 14: A problem in the Electronic Air Control Valve (EACV) circuit.

The EACV changes the amount of air bypassing the throttle body in response to a current signal from the ECU in order to maintain the proper idle speed.

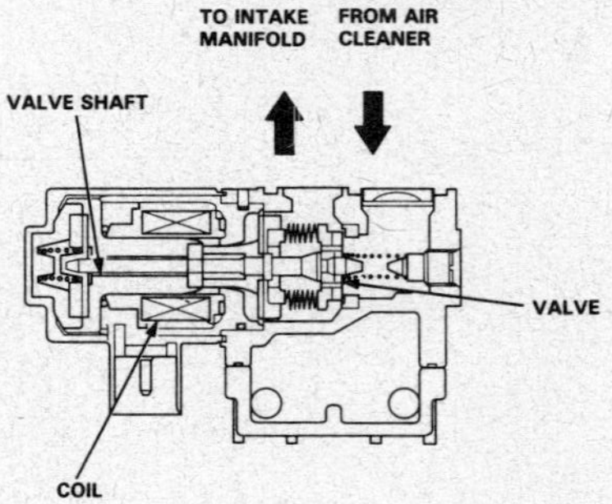

14

- Check Engine light has been reported on.
- With service check connector jumped CODE 14 is inaicated.

Do the ECU Reset Procedures

Start the engine.

Is Check Engine light on and does it indicate CODE 14?

NO → Intermittent failure, system is OK at this time (test driving may be necessary). Check for poor connection or loose wires at EACV C133 and ECU.

YES ↓

Stop the engine.

Connect the ECU test harness "A" connector to the engine wire harness only, not the ECU

Turn the ignition switch ON.

Measure voltage between A9 (+) terminal and A23 (–) terminal.

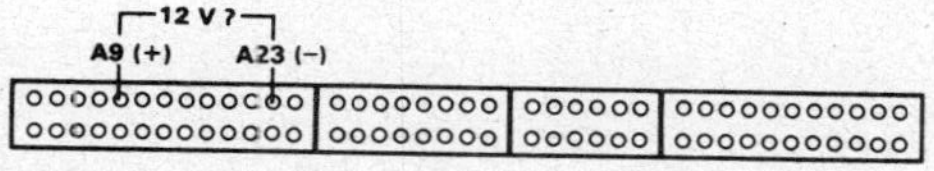

Is there battery voltage? — YES → Replace EACV.

NO ↓

Remove the C133 connector from the EACV and connect a jumper wire* to the two pins on the wire harness connector.

CAUTION:
*** the "A" connector must be disconnected for this test or the ECU will be damaged.**

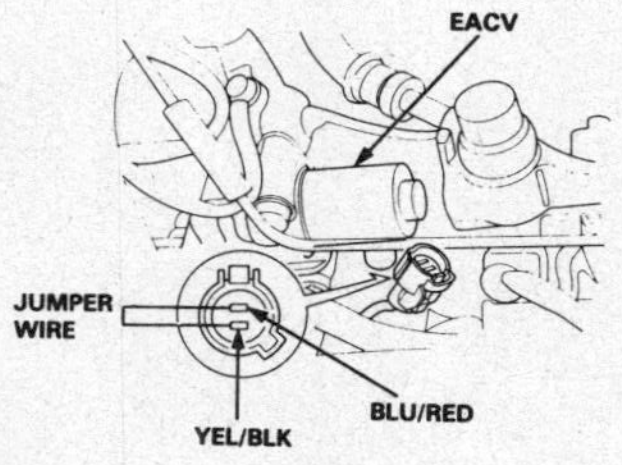

Measure voltage between A9 (+) terminal and A23 (–) terminal.

Is there battery voltage? — YES → Replace EACV.

NO ↓

Measure voltage between the YEL/BLK wire on C133 and body ground.

Is there battery voltage? — NO → Repair open in YEL/BLK wire between the EACV and Main Relay.

YES ↓

Repair open in BLU/RED wire between ECU (A9) and EACV.

– 15 – Self-diagnosis Check Engine light indicates code 15: A problem in the Ignition Output Signal circuit.

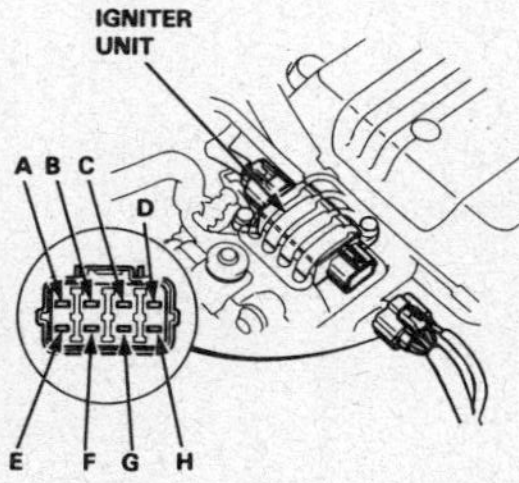

- Check Engine light has been reported on.
- With service check connector jumped CODE 15 is indicated.

Do the ECU Reset Procedures

Turn the ignition switch ON.

Is Check Engine light on and does it indicate CODE 15? — NO → Intermittent failure, system is OK at this time. Check for poor connections or loose wires at igniter unit C143, C123 and ECU.

YES ↓

Turn the ignition switch OFF.

Disconnect the 8P connector from the igniter unit.

Check for continuity to body ground on C and F terminals individually.

Does continuity exist? — NO → Repair open in BLK wire between 8P connector and G103.

YES ↓

Disconnect the 6P connector from the igniter unit.

Connect the ECU test harness between the ECU and connector

NOTE:
Inspection of igniter unit secondary circuit

*1:

IGNITER UNIT SECONDARY TERMINAL (8P)	IGNITION COIL (2P)	WIRE COLOR
A	No. 1	WHT
B	No. 2	WHT/GRN
C	GROUND	BLK
D	No. 3	WHT/BLK
E	No. 4	WHT/BLU
F	GROUND	BLK
G	No. 5	WHT/YEL
H	No. 6	WHT/RED

Check for continuity between body ground and the following terminals: A21, A22, B8, B6, B4, B3.

Does continuity exist?
- YES → Repair short in each terminal wire.
- NO → Check for continuity between terminals of 6P connector and ECU individually. *See table 2.

Does continuity exist?
- NO → Repair open in each terminal wire.
- YES → Substitute a known-good igniter unit and recheck.

Does Check Engine warning light indicate CODE 15?
- NO → Replace the original igniter unit.
- YES → Substitute a known-good ECU and recheck. If symptom/indication goes away, replace the original ECU.

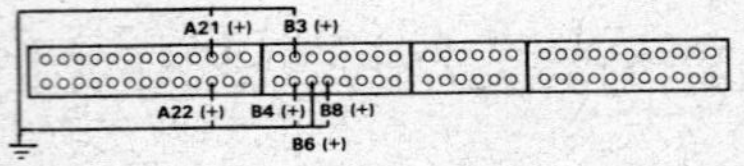

*2:

IGNITER UNIT PRIMARY TERMINAL (6P)	ECU TERMINAL	WIRE COLOR
A	B8 (No. 3)	WHT/BLK
B	A22 (No. 2)	WHT/GRN
C	A21 (No. 1)	WHT
D	B3 (No. 6)	WHT/RED
E	B4 (No. 5)	WHT/YEL
F	B6 (No. 4)	WHT/BLU

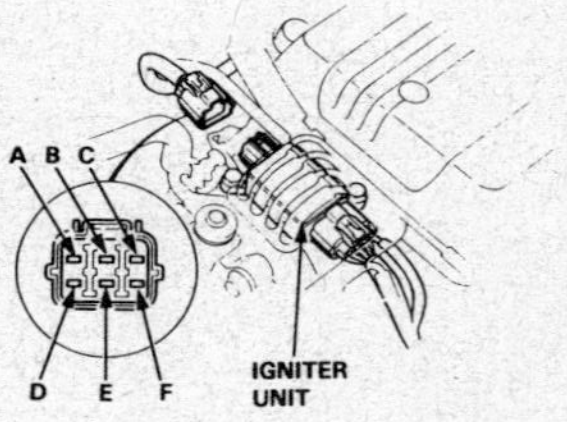

16 Self-diagnosis Check Engine light indicates code 16: A problem in the fuel injector circuit.

The injectors are the solenoid-actuated constant-stroke pintle type consisting of a solenoid, plunger needle valve and housing. When current is applied to the solenoid coil, the valve lifts up and pressurized fuel is injected close to the intake valve. Because the needle valve lift and the fuel pressure are constant, the injection quantity is determined by the length of time that the valve is open (i.e., the duration the current is supplied to the solenoid coil). The injector is sealed by an O-ring and seal ring at the top and bottom. These seals also reduce operating noise.

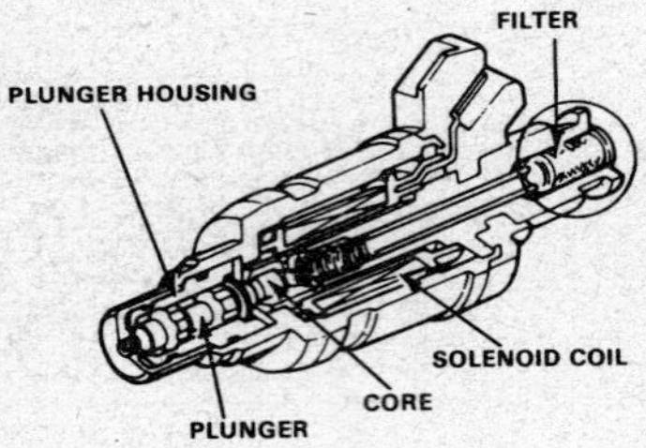

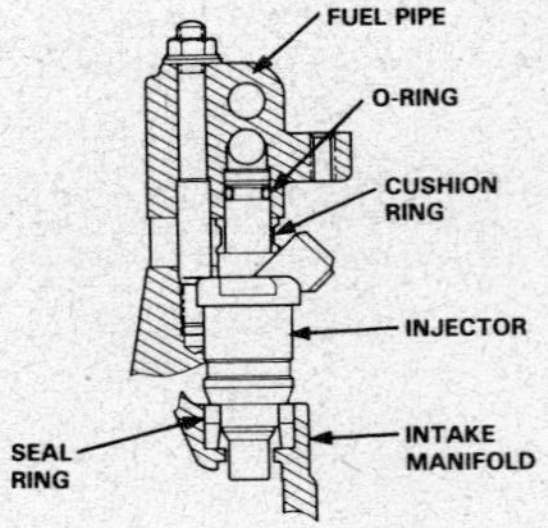

16

- Check Engine light has been reported on.
- With service check connector jumped CODE 16 is indicated.

Do the ECU Reset Procedures

Turn the ignition switch to START position.

Is Check Engine light on and does it indicate CODE 16?
- NO → Intermittent failure, system is OK at this time (test drive may be necessary). Check for poor connections or loose wires at injectors, injector resistor and ECU.
- YES

TROUBLESHOOTING CHARTS – NSX

Turn the ignition switch OFF.

Disconnect the 2P connector from the injector that does not click.

Measure resistance between the 2 terminals of the injector.

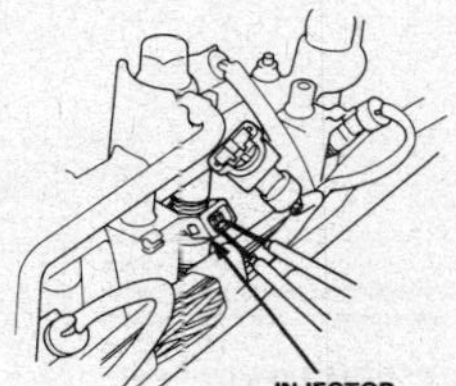

Is there 1.5—2.5Ω ?
- NO → Replace the injector/injectors that are not 1.5–2.5Ω.
- YES ↓

Turn the ignition switch ON.

Measure voltage between RED/BLK (+) terminal on the 2P connector and body ground.

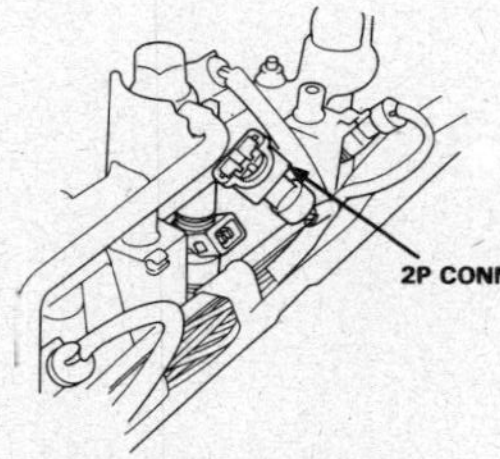

Is there battery voltage ?
- NO → Turn the ignition switch OFF. → Disconnect 8P connector on the injector resistor. → Turn the ignition switch ON. → Measure voltage between YEL/BLK (+) terminal and body ground.
  - Is there battery voltage?
    - NO → Repair open in the YEL/BLK wire between the injector resistor and the main relay.
    - YES → Test the of injector resistor (page 11-96)
      - Is the injector resistor testing OK?
        - NO → Replace injector resistor.
        - YES → – Replace the injector resistor. – Repair open RED/BLK wire between 2P connector and resistor.
- YES ↓

Reconnect the 2P connector to the injector.

Turn the ignition switch OFF.

Connect the ECU test harness between the ECU and connector

Turn the ignition switch ON.

Measure voltage between A23 (–) terminal and following terminals.
- No. 1 injector: A1 (+) terminal.
- No. 2 injector: A3 (+) terminal.
- No. 3 injector: A5 (+) terminal.
- No. 4 injector: A2 (+) terminal.
- No. 5 injector: A4 (+) terminal.
- No. 6 injector: A6 (+) terminal.

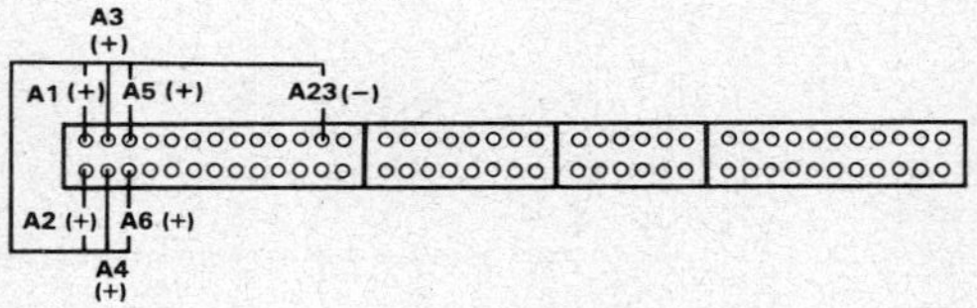

Is there battery voltage ?
- NO → Repair open in the wire between the ECU (A1, A3, A5, A2, A4 or A6) and the injector.
- YES → Substitute a known-good ECU and recheck. If symptom/indication goes away, replace the original ECU.

TROUBLESHOOTING CHARTS – NSX

Code 17

Self-diagnosis Check Engine light indicates code 17: A problem in the Vehicle Speed Sensor circuit.

The signal generated by the speed sensor, produces pulses when the front wheels turn.

- Check Engine light has been reported on.
- With service check connector jumped CODE 17 is indicated.

Do the ECU Reset Procedures

Road test necessary.
In 2nd gear accelerate to 3,500 rpm, then decelerate to 1,500 rpm with throttle fully closed.

Is Check Engine light on and does it indicate CODE 17?

NO → Intermittent failure, system is OK at this time. Check for poor connections or loose wires at and (located at left shock tower).

YES ↓

Raise the car.

WARNING Make sure lifts are placed properly.

Connect the ECU test harness between the ECU and connector

Turn the ignition switch ON.

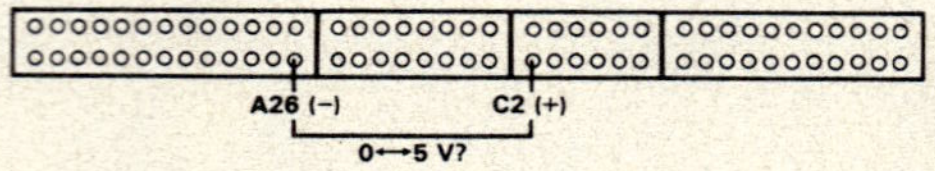

Slowly rotate left rear wheel and measure voltage between C2 (+) terminal and A26 (−) terminal.

Does voltage pulse 0 V and 5 V?

NO →
- Repair open or short in YEL/RED wire between ECU (C2) and the speed sensor.
- Faulty speed sensor.
- Substitute a known-good ECU and recheck. If symptom/indication goes away, replace the original ECU.

YES ↓

Substitute a known-good ECU and recheck. If symptom/indication goes away, replace the original ECU.

Code 18

Self-diagnosis Check Engine lights indicates code 18: A problem in the Ignition Timing Adjuster circuit.

The ignition timing adjuster allows the electronic ignition advance to be set to 15° BTDC at idle.

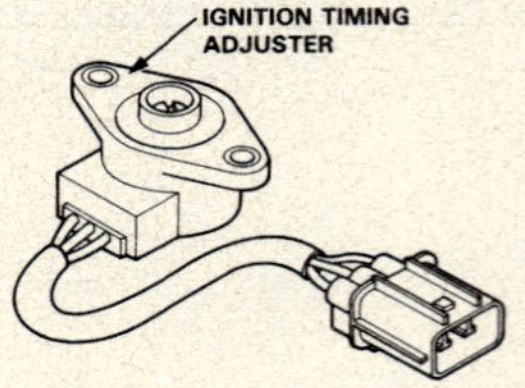

- Check Engine light is on.
- With service check connector jumped . CODE 18 is indicated.

Do the ECU Reset Procedures

Turn the ignition switch ON.

Is Check Engine light on and does it indicate CODE 18?

NO → Intermittent failure system is OK at this time (Test drive may be necessary). Check for poor connections or loose wires at the Ignition Timing adjuster connector C533 and ECU.

YES ↓

Turn the ignition switch OFF.

Disconnect the 3P connector from the control box.

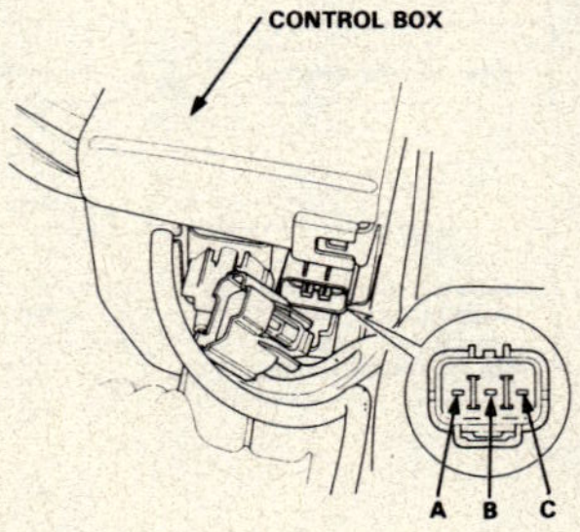

Measure resistance between A terminal and C terminal on Ignition timing adjuster.

Is there 3–6 kΩ?

NO → Replace Ianition timing adjuster

YES

- Measure resistance between A and B terminals and between C and BRN terminals.
- Does the sum of the two resistance checks equal 3.5–6.5 kΩ?
  - NO → Replace Ignition timing adjuster.
  - YES → Turn the ignition switch ON.
- Measure voltage between YEL/WHT (+) terminal and GRN/WHT (–) terminal on the main wire harness.

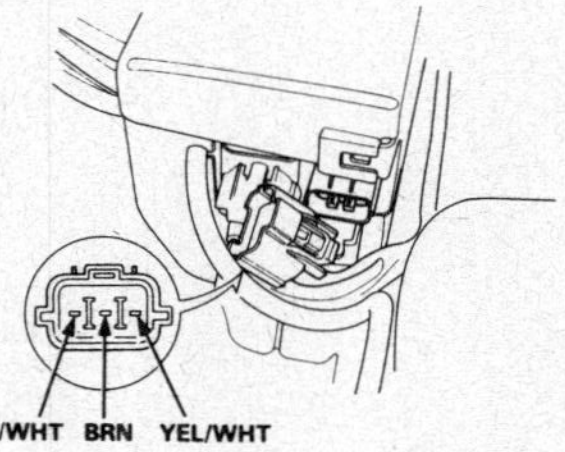

- Is there approx. 5 V?
  - NO → Measure voltage between YEL/WHT (+) terminal and body ground.
    - Is there approx. 5 V?
      - YES → Repair open in GRN/WHT wire between ECU (D22) and Ignition timing adjuster.
      - NO → Turn the ignition switch OFF.
    - Connect the ECU test harness between the ECU and connector
    - Turn the ignition switch ON.
    - Measure voltage between D20 (+) terminal and D22 (–) terminal.

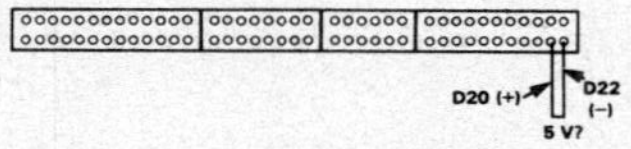

    - Is there approx. 5 V?
      - YES → Repair open in YEL/WHT wire between ECU (D20) and Ignition timing adjuster.
      - NO → Substitute a known-good ECU and recheck. If prescribed voltage is now available, replace the original ECU.
  - YES → Turn the ignition switch OFF.
- Reconnect 3P connector.
- Connect the ECU test harness between the ECU and connector
- Turn the ignition switch ON.
- Measure voltage between D8 (+) terminal and D22 (–) terminal.
- Is there 0.5–4.5 V?
  - NO → Repair open or short in BRN wire between ECU (D8) and Ignition timing adjuster.
  - YES → Substitute a known-good ECU and recheck. If symptom/indication goes away, replace the original ECU.

---

(21) (Front) or (51) (Rear) Check Engine indicates code 21 (Front) or 51 (Rear): A problem in the Spool Valve circuit.

- **Disconnect the 2P connector from the spool valve.**
- Check for continuity between YEL (Front) or BLU/YEL (Rear) terminal and body ground.

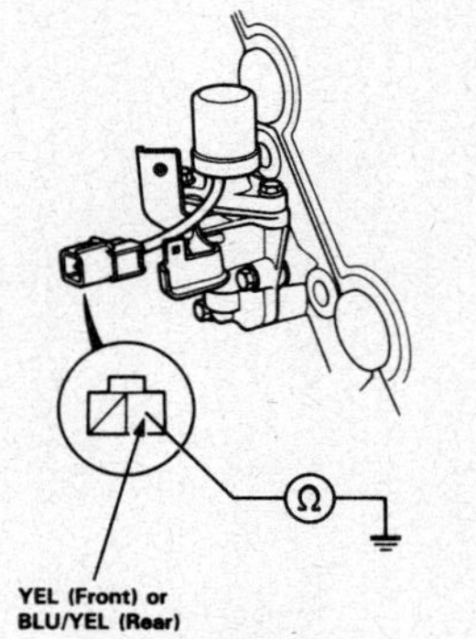

- Is there 14–30 Ω?
  - NO → **Replace the spool valve.**
  - YES → Connect the ECU test harness.
- Check for continuity between YEL (Front) or BLU/YEL (Rear) terminal and F5 (Front) or F7 (Rear) terminal.

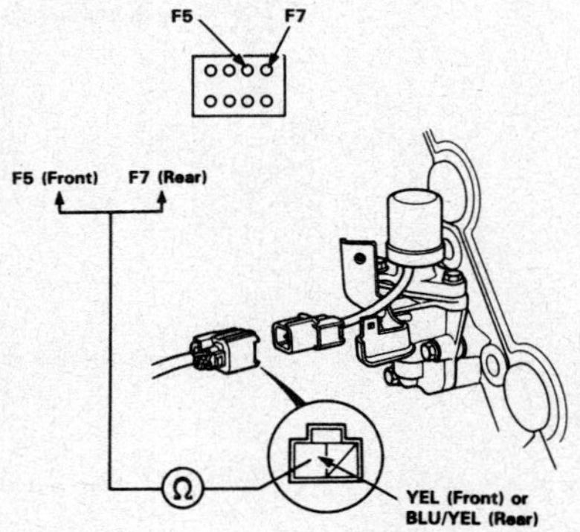

- Does continuity exist ?
  - NO → **Repair open in YEL (Front) or BLU/YEL (Rear) wire between ECU and spool valve connector.**
  - YES

Check for continuity between YEL (Front) or BLU/YEL (Rear) terminal (harness side) and body ground.

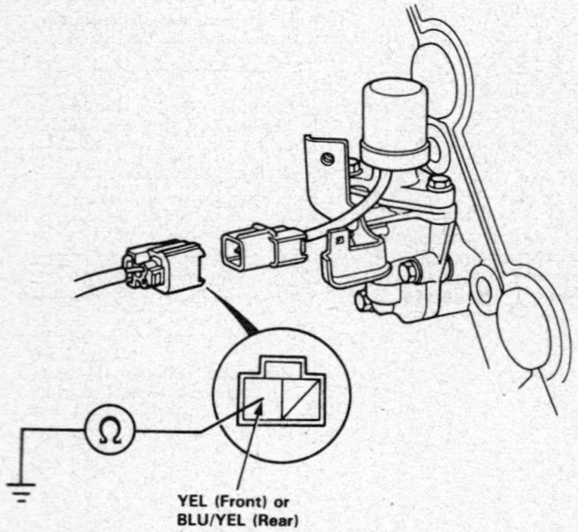

Does continuity exist ?

YES → Repair short in YEL (Front) or BLU/YEL (Rear) wire between ECU and spool valve connector.

NO → Substitute a known-good ECU and recheck If symptom/indication goes away replace the original ECU.

---

(22) (Front) or (52) (Rear) Check Engine indicates code 22 (Front), 52 (Rear): A problem in the Oil Pressure Switch circuit.

Disconnect the 2P connector from the oil pressure switch.

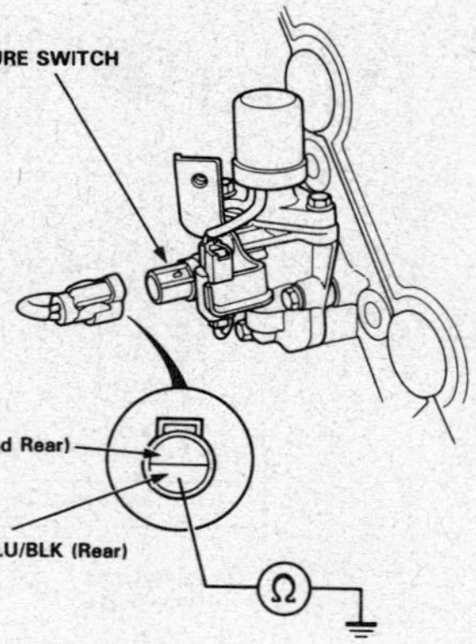

Check for continuity between BLU (Front) or BLU/BLK (Rear) terminal and body ground.

Does continuity exist ?

NO → Repair open in BLU (Front) or BLU/BLK (Rear) wire between 2P connector and body ground.

YES ↓

Connect the ECU test harness.

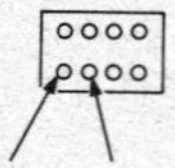

Check for continuity between BLU (Front) or BLU/BLK (Rear) terminal and F2 (Front) or F4 (Rear) terminal.

Does continuity exist ?

NO → Repair open in BLU (Front) or BLU/BLK (Rear) wire between ECU and 2P connector.

YES ↓

TROUBLESHOOTING CHARTS – NSX

Remove 10 mm sealing bolt and connect oil pressure gauge (for VTEC oil pressure test).

Connect a tachometer.

Start the engine and warm it up to normal operating temperature.

Check oil pressure at 1,000 rpm, 3,000 rpm and 5,000 rpm.

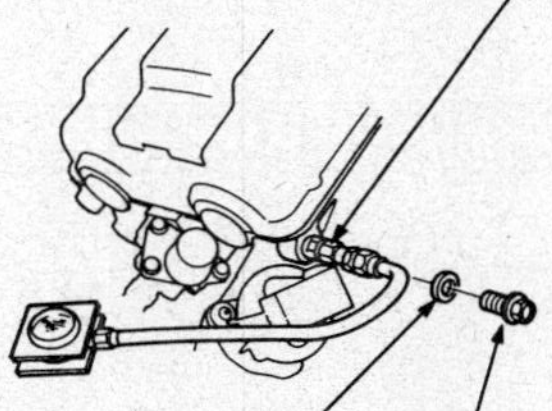

NOTE: Keep measuring time as short as possible because engine is running with no load (within one minute).

Is pressure below 49 kPa (0.5 kg/cm² 7 psi) ?
- NO → **Inspect the spool valve.**
- YES ↓

Turn off the engine. Check for continuity between the 2 terminals on the oil pressure switch.

Does continuity exist ?
- NO → **Replace oil pressure switch.**
- YES ↓

Disconnect the 2P connector from the spool valve.

Attach the battery positive terminal to the YEL (Front) or BLU/YEL (Rear) terminal.

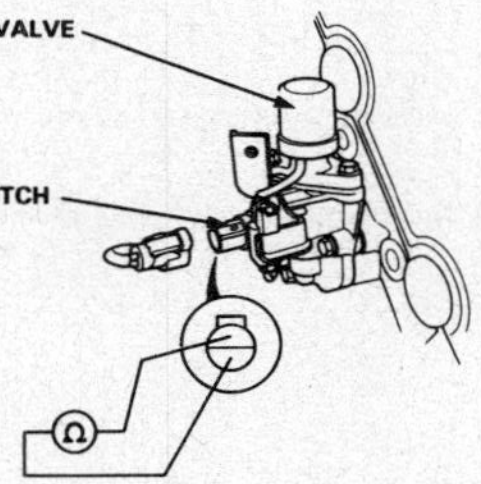

Start the engine and check oil pressure at 5,000 rpm.

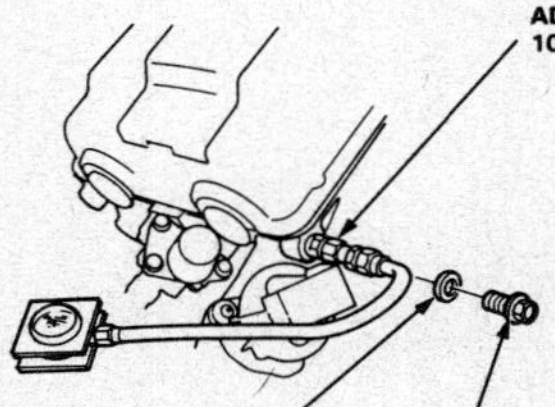

Note: Keep measuring time as short as possible because engine is running with no load (within one minute).

Is pressure above 392 kPa (4 kg/cm², 57 psi) ?
- NO → **Inspect the spool valve.**
- YES ↓

Check for continuity between the 2 terminals on the oil pressure switch under above condition.

Does continuity exist ?
- YES → **Replace oil pressure switch.**
- NO ↓

Substitute a known-good ECU and recheck. If symptom/indication goes away replace the original ECU.

(CHECK)—23 Self-diagnosis Check Engine light indicates code 23: A problem in the circuit of the Front Knock Sensor.

(CHECK)—53 Self-diagnosis Check Engine light indicates code 53: A problem in the circuit of the Rear Knock Sensor.

(CHECK)—23 or 53

- Check Engine light has been reported on.
- With service check connector jumped CODE 23 and/or 53 are indicated.

↓

Do the ECU Reset Procedures

↓

Warm up the engine to normal operating temperature (cooling fan comes on.)

↓

Hold engine at 3000–4000 rpm for 10 seconds. (A/T: Transmission is N or P)

↓

Is Check Engine light on and does it indicate CODE 23 or 53?

- NO → Intermittent failure, system is OK at this time.
- YES ↓

Turn the ignition switch OFF.

↓

Connect the ECU test harness to the engine wire harness only, not to the ECU

↓

Disconnect the knock sensor connector from the engine wire harness.

↓

Check for continuity between D4 or D3*, terminal and body ground.

↓

Does continuity exist?

- YES → Repair short in WHT or ORN* wire between ECU (D4 or D3*) and the knock sensor.
- NO ↓

*: REAR

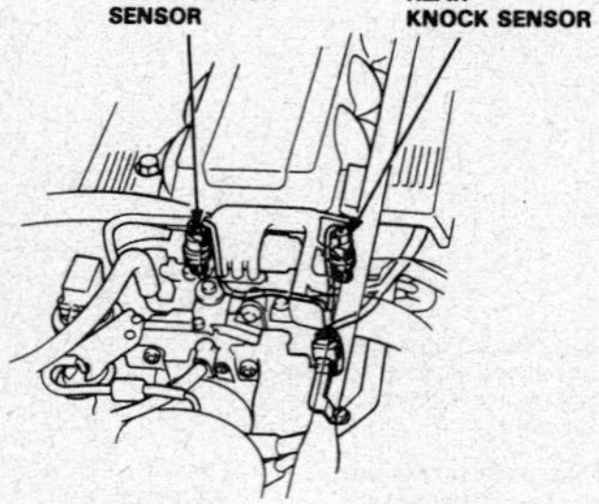

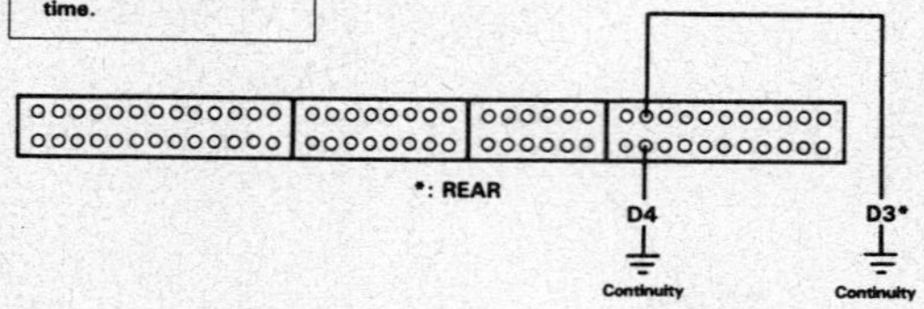

*: REAR

Check for continuity on WHT or ORN* wire between D4 or D3* and 3P connector of engine wire harness.

↓

Does continuity exist?

- NO → Repair open in WHT or ORN* wire between ECU (D4 or D3*) and the front knock sensor.
- YES ↓

Substitute a known-good front or rear knock sensor and recheck.

↓

Is Check Engine light on and does it indicate CODE 23 or 53.

- NO → Replace the original knock sensor.
- YES ↓

Substitute a known-good ECU and recheck. If symptom/indication goes away, replace the original ECU.

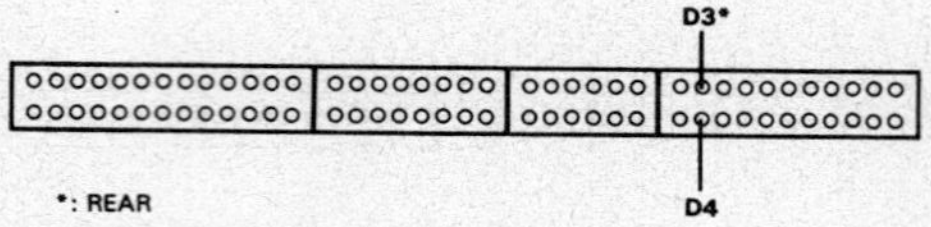

*: REAR

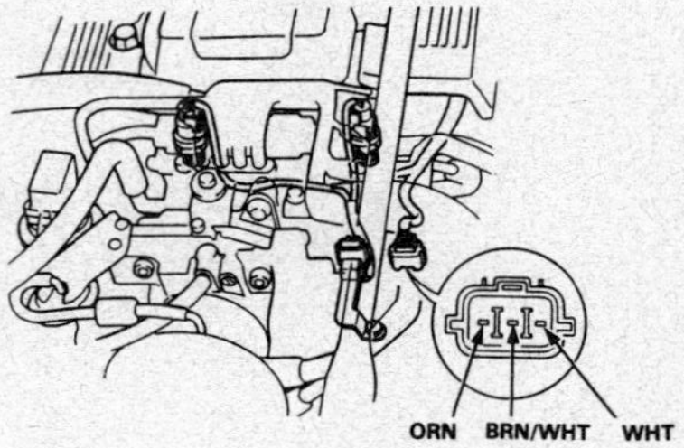

TROUBLESHOOTING CHARTS—NSX

[30] Self-diagnosis Check Engine light indicates code 30: A problem in the signal line A (F1) between A/T control unit and PGM-FI ECU.

[31] Self-diagnosis Check Engine light indicates code 31: A problem in the signal line B (F3) between A/T control unit and PGM-FI ECU.

[30] or [31]

- **Check Engine light isn't on.**
- **With service check connector jumped CODE 30 and/or 31 are indicated.**

Do the ECU Reset Procedures

Test drive necessary.
Drive the car for several miles so that the transmission upshifts and downshifts several times.

Does Check Engine light indicate CODE 30 or 31?

NO → **Intermittent failure, system is OK at this time.**

YES ↓

Turn the ignition switch OFF.

Connect the test harness to the floor wire harness only, not to the ECU

Disconnect the 22P connector from the A/T control unit.

Check for continuity between F1 or F3*, terminal and body ground.

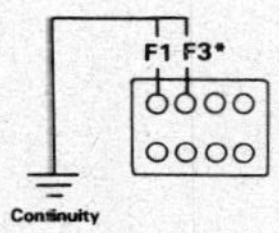

*: SIGNAL B

Does continuity exist?

YES → **Repair short in WHT/GRN or WHT/RED* wire between ECU (F1 or F3*) and the A/T control unit.**

NO ↓

Check for continuity on WHT/GRN or WHT/RED* wire between F1 or F3* and 22P connector of the A/T control unit.

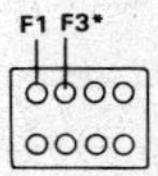

*: SIGNAL B

Does continuity exist?

NO → **Repair open in WHT/GRN or WHT/RED* wire between ECU (F1 or F3*) and the A/T control unit.**

YES ↓

Substitute a known-good A/T control unit and recheck.

Does Check Engine light indicate CODE 30 or 31?

NO → **Replace the original A/T control unit.**

YES ↓

Substitute a known-good ECU and recheck. If symptom/indication goes away, replace the original ECU.

Problem code 3-1: The PGM-FI system (TCINH, TCFS and TCFC signals).

NOTE: The PGM-FI ECU outputs the TCINH signal to the TCS control unit when the correct TCFS signal and TCFC signal have been input to it from the TCS control unit.

Remove the ALB 2.3 (20A) fuse and Clock (7.5 A) fuse for more than 10 seconds to reset the PGM-FI ECU and TCS control unit.

Start the engine.

Check for TCS problem code.

Is problem code 3-1 indicated?

- NO → Intermittant failure. Check for poor connections or loose wires.
- YES ↓

Check for a PGM-FI problem code.

Are problem codes 35 (TCFS open or short-circuit) or 36 (TCFC open or short-circuit) indicated.

- NO → **Check the TCINH signal if no problem code is indicated**
- NO → **Perform the PGM-FI troubleshooting procedure if problem codes other than 35 and 36 are indicated.**
- YES ↓

- **Check the TCFS signal when PFM-FI system problem code 35 is indicated. TCFS signal check: (see below).**
- **Check the TCFC signal when PGM-FI system problem code 36 is indicated**

Measure the voltage between the TCS ECU No. 1 terminal and body ground when the ignition switch is turned on.

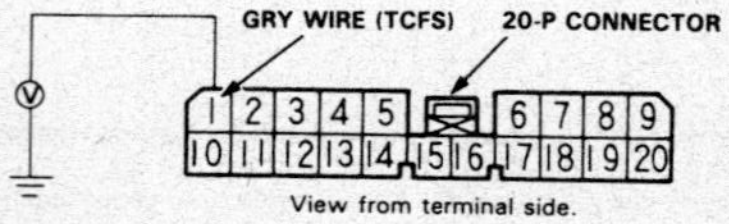

Is there approximately 4.5 V?

- NO → Turn the ignition switch OFF. → Disconnect the 20-P connector from the TCS control unit.
- YES → Turn the ignition switch OFF.

Raise the car and support it with safety stands

Start the engine

Shift the transmission into 2nd, and press the accelerator until a vehicle speed greater than 20 km/hr(12 mph) is reached.

Measure the voltage between the TCS control unit No. 1 terminal and the body ground.

Is there 0V? (TCS in operation.)

- NO → **• Substitute a known-good control unit and recheck. • Check for poor connections or loose wires.**
- YES ↓

Turn the ignition switch OFF.

Disconnect the 20-P connector from the TCS control unit and the 12-P connector from the PGM-FI ECU.

Check for continuity in the GRY wire between the TCS control unit and PGM-FI ECU.

NOTE: Use the PGM-FI test harness

Is there continuity?

- NO → **Repair open in the GRY wire between the TCS control unit and PGM-FI ECU.**
- YES → **• Substitute a known-good PGM-FI ECU and recheck. • Check for poor connections or loose wires.**

Check for continuity between the TCS control unit No. 1 terminal and the body ground.

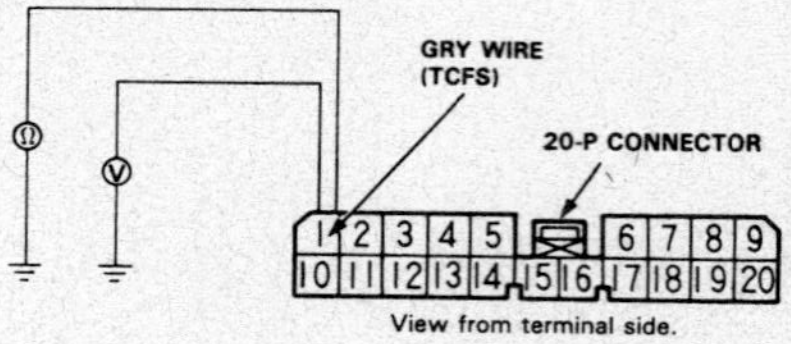

Is there continuity?

- NO → **• Substitute a known-good TCS control unit and recheck. • Check for poor connections or loose wires.**
- YES → **Repair short in the GRY wire between the TCS control unit and PGM-FI ECU.**

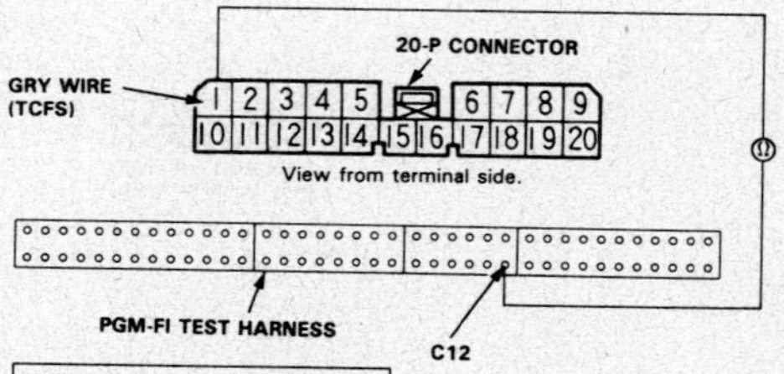

- **TCFC signal check.**

Turn the ignition switch ON.

Measure the voltage between the TCS control unit No. 32 terminal and body ground.

NOTE:
- Use the 10 V range or a similar range in the analog tester.
- Confirm that the analog tester needle is swinging at approximately 3.5V.

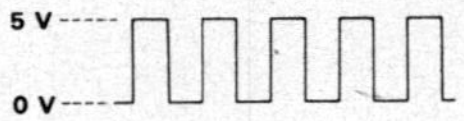

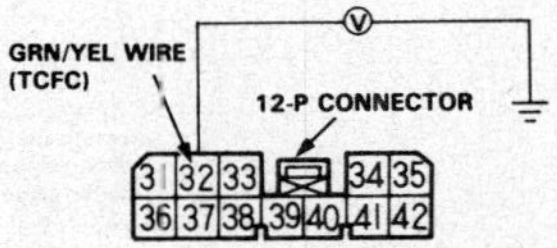

View from terminal side.

Is there approximately 3.5 V?

- NO → Turn the ignition switch OFF. → Disconnect the 12-P connector from the TCS control unit and the 12-P connector from the PGM-FI ECU. → Check for continuity between the TCS control unit No. 32 terminal and body ground. → Is there continuity?
  - NO → **• Substitute a known-good TCS control unit and recheck. • Check for poor connections or loose wires.**
  - YES → **Repair short in the GRN/YEL wire between the TCS control unit and PGM-FI ECU.**
- YES → Turn the ignition switch OFF. → Disconnect the 12-P connector from the TCS control unit and the 12-P connector from the PGM-FI ECU. → Check for continuity in the GRN/YEL wire between the TCS control unit and PGM-FI ECU.

NOTE: Use the PGM-FI test harness, connect the test harness to the PGM-FI ECU

Is there continuity?

- NO → **An open in the GRN/YEL wire between the TCS control unit and PGM-FI ECU.**
- YES → **• Substitute a known-good PGM-FI ECU and recheck. • Check for poor connections or loose wires.**

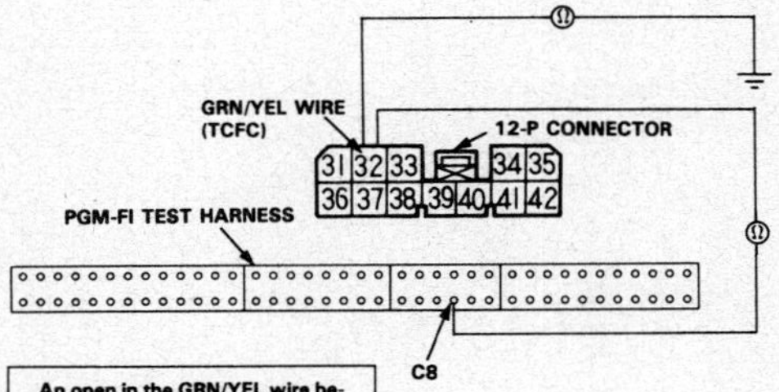

- **TCINH signal check:**

Start the engine.

Measure the voltage between the TCS control unit No. 10 terminal and body ground.

NOTE: Use the 10 V range or a similar range in the analog tester.

Check if the analog tester needle pulses in small increments in the range below 5 V.

Does the needle pulse in the range approximately 2.5V?

- NO → (continued)
- YES → **• Substitute a known-good ECU and recheck. • Check for poor connections or loose wires.**

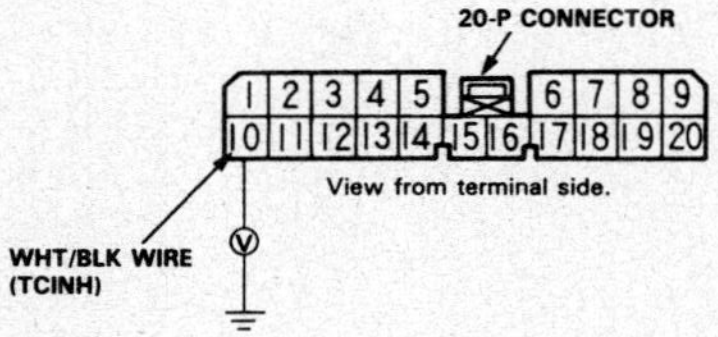

View from terminal side.

Turn the ignition switch OFF.

Disconnect the 20-P connector from the TCS control unit and the 12-P connector from the PGM-FI ECU

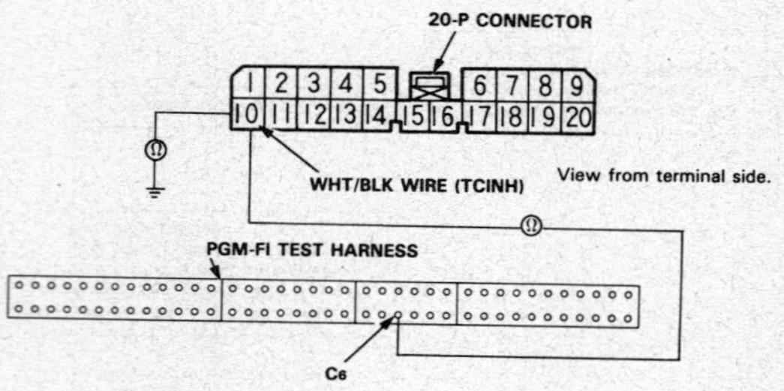

Check for open or short in the WHT/BLK wire between the TCS control unit and PGM-FI ECU.

NOTE: Use the PGM-FI test harness

Is there an open or short?

- YES → **Repair open or short in the WHT/BLK wire between the TCS control unit and PGM-FI ECU.**
- NO →
  - **Substitute a known-good PGM-FI ECU and recheck.**
  - **Check for poor connections or loose wires.**

[41] Self-diagnosis Check Engine light indicates code 41: A problem in the Front Oxygen (O_2) Sensor Heater circuit.

[42] Self-diagnosis Check Engine light indicates code 42: A problem in the Rear Oxygen (O_2) Sensor Heater circuit.

[41] or [42]

- **Engine is running.**
- **Check Engine light has been reported on, with service check connector jumped CODE 41 and/or 42 are indicated.**

Do the ECU Reset Procedures

Start the engine

Is Check Engine light on and does it indicate CODE 41 or 42 ?

- NO → **Intermittent failure, system is OK at this time (test drive may be necessary). Check for poor connections or loose wires at O_2 sensor C144, C148 and ECU.**
- YES ↓

Stop the engine

Disconnect the 4P connector from the O_2 sensor.

Measure resistance between terminals C and D on the O_2 sensor.

Is there 10–20 Ω?

- NO → **Replace O_2 sensor.**
- YES

FRONT:

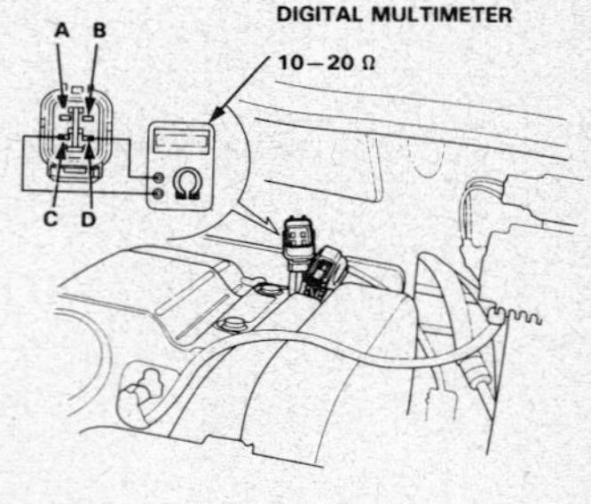

REAR:

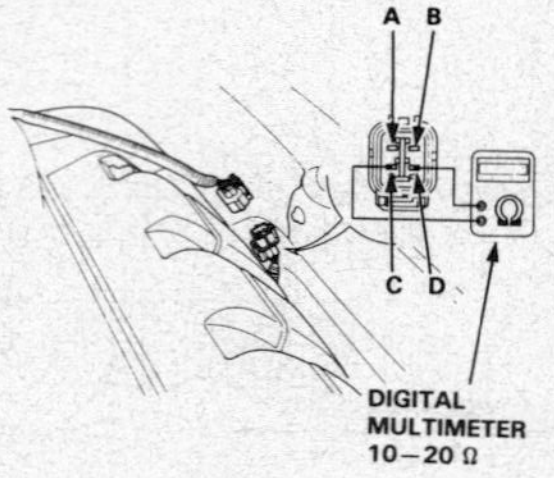

Check for continuity to body ground on each terminal on the O_2 sensor.

Does continuity exist? — YES: Replace O_2 sensor.

NO

Check for continuity between terminal A and terminals C and D individually.

Does continuity exist? — YES: Replace O_2 sensor.

NO

Turn the ignition switch ON.

Measure voltage between BLK/YEL (+) terminal and BLK (−) or BLK/BLU (−)* terminal.

*: REAR

Is there battery voltage? — YES: Disconnect the "A" connector from the ECU.

- Measure voltage between BLK/YEL (+) terminal and BLK (−) or BLK/BLU (−)* terminal. *: REAR
- Is there battery voltage? — YES: Repair short in BLK or BLK/BLU* wire between ECU (A10 or A12*) and front O_2 sensor.
- NO: Substitute a known-good ECU and recheck. If symptom/indication goes away, replace the original ECU.

NO

Measure voltage between BLK/YEL (+) terminal and body ground.

Is there battery voltage? — NO:
- Repair open in BLK/YEL wire between the O_2 sensor and Main Relay.
- Replace blown No. 2 fuse (15 A) in the dash fuse box.

YES

Turn the ignition switch OFF.

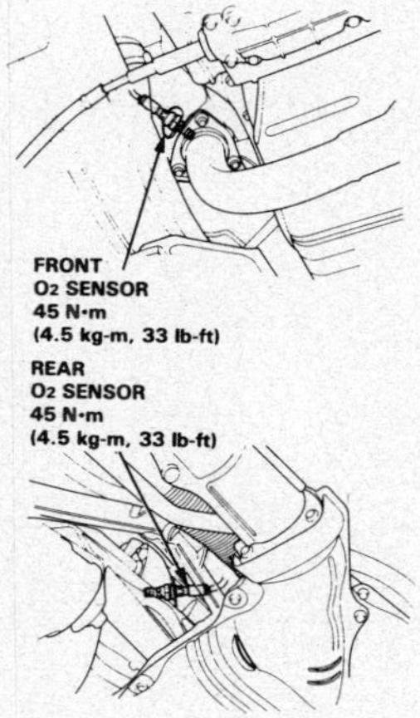

Reconnect the 4P connector to O_2 sensor.

Connect the ECU test harness "A" connector to the engine wire harness only, not the ECU.

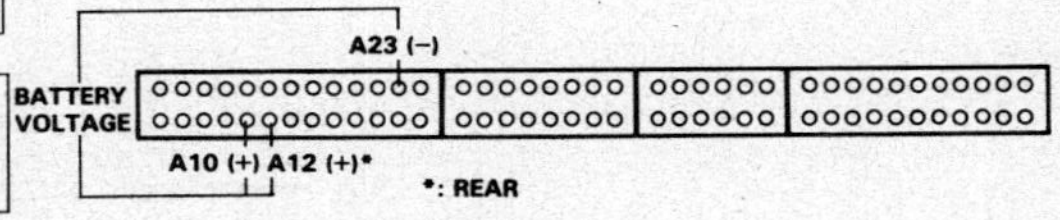

Turn the ignition switch ON.

Measure voltage between A10 (+) or A12 (+)* terminal and A23 (−) terminal.

*: REAR

Is there battery voltage? — NO: Repair open in BLK or BLK/BLU* wire between ECU (A10) and front O_2 sensor.

YES

Substitute a known-good ECU and recheck. If symptom/indication goes away, replace the original ECU.

TROUBLESHOOTING CHARTS—NSX

[CHECK] [43] Self-diagnosis Check Engine light indicates code 43: Most likely a problem in the Front Oxygen (O_2) Sensor circuit or a problem in the Front Fuel Supply System.

[CHECK] [44] Self-diagnosis Check Engine light indicates code 44: Most likely a problem in the Rear Oxygen (O_2) Sensor circuit or a problem in the Rear Fuel Supply System.

[CHECK] [43] or [44]

- Check Engine light has been reported on.
- With service check connector jumped CODE 43 and/or 44 are indicated.
- or continued from code 1 or 2.

Do the ECU Reset Procedures

Warm up engine to normal operating temperature (cooling fan comes on).

Hold engine at 3,000 rpm for 2 minutes. (A/T: Transmission is [N] or [P].)

Is the Check Engine light on and does it indicate CODE 43 or 44?

- NO → Intermittent failure, system is OK at this time (test drive may be necessary). Check for poor connections or loose wires at O_2 sensor C144, C148 and ECU.
- YES ↓

Turn the ignition switch OFF.

Connect the ECU test harness between the ECU and connector

With the ignition switch OFF, wait for at least two minutes.

Install a jumper wire on the ECU test harness between A10 and A26 (front): or A12 and A26 (rear).

Turn the ignition switch ON.

Measure voltage between D14 (+) (front) or D16 (+) (rear) terminal and A26 (−) terminal as soon as the ignition switch is turned on.

NOTE:
- Use DIGITAL MULTIMETER (KS−AHM−32−003) or equivalent.
- Use 2 Volt range.

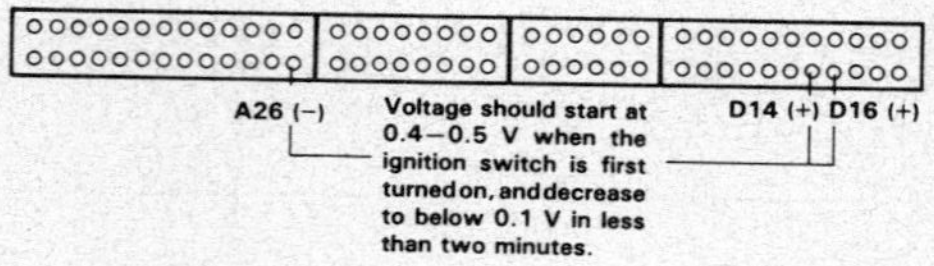

Is there 0.1 V or less when the ignition switch is first turned on?

- YES → Disconnect the 4P connector from the O_2 sensor.
  - Measure voltage between D14 (+) (front) or D16 (+) (rear) terminal and A26 (−) terminal.
  - Is there more than 0.1 V?
    - YES → Replace O_2 sensor.
    - NO → Disconnect the "D" connector from the engine wire harness.
      - Measure voltage between D14 (+) (front) or D16 (+) (rear) terminal and A26 (−) terminal.
      - Is there more than 0.1 V?
        - YES → Repair short in WHT (front) or RED/BLU (rear) wire between ECU (D14) and O_2 sensor.
        - NO → Substitute a known-good ECU and recheck. If symptom/indication goes away, replace the original ECU.
- NO → Disconnect the 4P connector from the O_2 sensor.
  - Measure voltage at the engine wire harness side of the O_2 sensor connector between A (+) and B (−).
  - Is there more than 0.1 V?
    - YES → Replace O_2 sensor.
    - NO → Repair open in WHT (front) or RED/BLU (rear) wire between ECU (D14 or D16) and O_2 sensor.

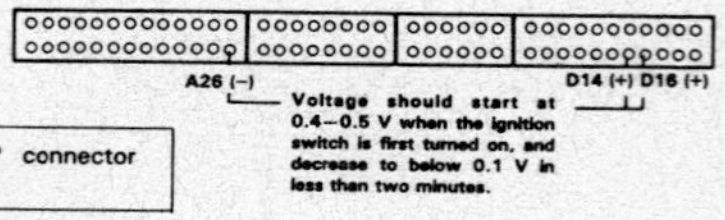

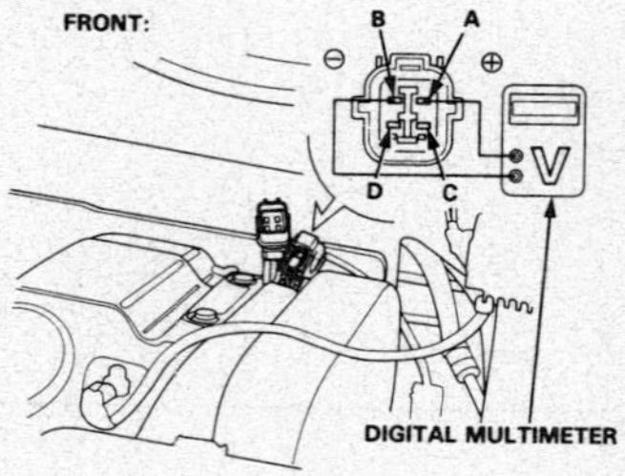

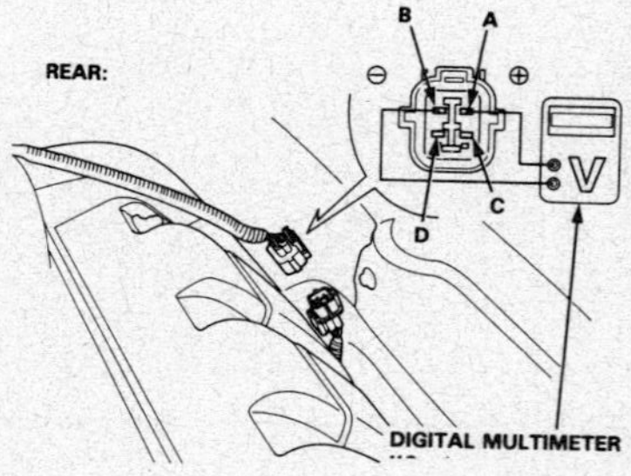

TROUBLESHOOTING CHARTS—NSX

45: Self-diagnosis Check Engine light indicates code 45: Most likely a problem in the Front Fuel Metering System.

46: Self-diagnosis Check Engine light indicates code 46: Most likely a problem in the Rear Fuel Metering System.

45 or 46

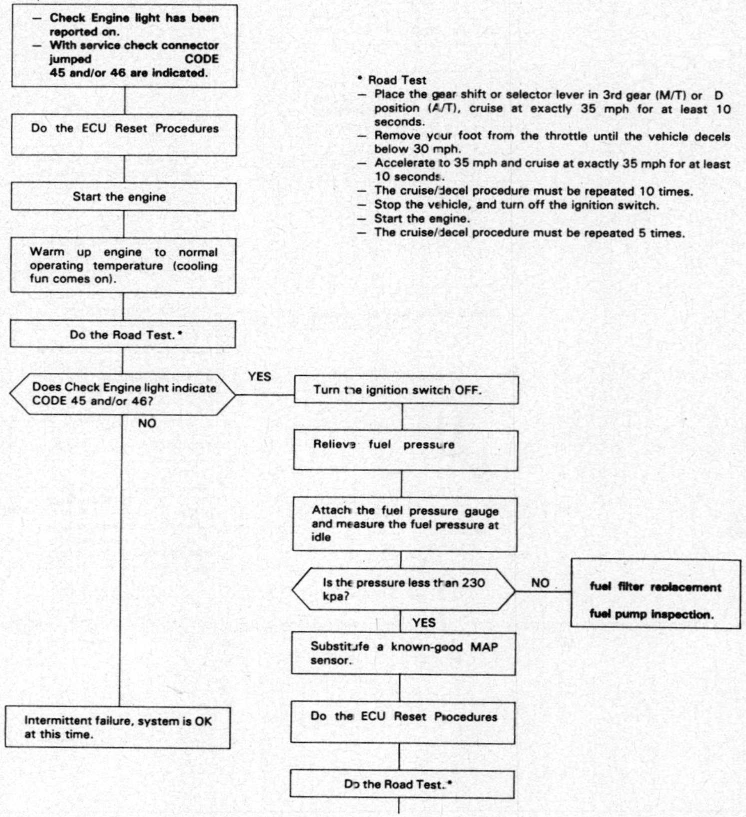

* Road Test
– Place the gear shift or selector lever in 3rd gear (M/T) or D position (A/T), cruise at exactly 35 mph for at least 10 seconds.
– Remove your foot from the throttle until the vehicle decels below 30 mph.
– Accelerate to 35 mph and cruise at exactly 35 mph for at least 10 seconds.
– The cruise/decel procedure must be repeated 10 times.
– Stop the vehicle, and turn off the ignition switch.
– Start the engine.
– The cruise/decel procedure must be repeated 5 times.

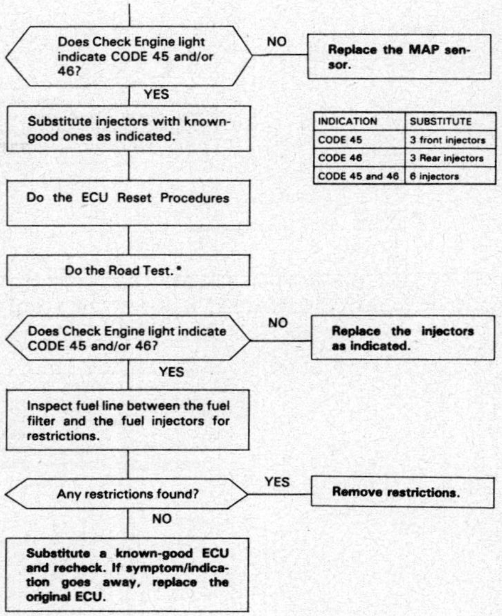

INDICATION	SUBSTITUTE
CODE 45	3 front injectors
CODE 46	3 Rear injectors
CODE 45 and 46	6 injectors

47 Self-diagnosis Check Engine light indicates code 47: A problem in the fuel Pump circuit.

- Check Engine light has been reported on.
- With service check connector jumped CODE 47 is indicated.

Do the ECU Reset Procedures

Turn the ignition switch to START position.

Does the engine start?
- NO → Inspect the fuel pump resistor
- YES ↓

Warm up engine to normal operating temperature (cooling fan comes on).

Is the check Engine light on and does it indicate CODE 47?
- YES ↓
- NO → Turn the ignition switch OFF.

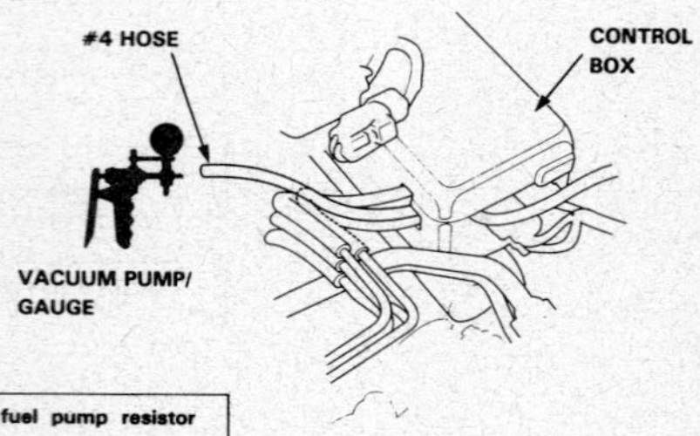

Disconnect #4 hose from the install pipe, connect vacuum pump to the hose.

Apply vacuum 300 to 400 mmHg.

Start the engine

Hold engine speed at 6500 rpm for 10 seconds.

Is the check Engine warning light on and does it indicate CODE 47?
- NO → Intermittent failure, system is OK at this time. Check for poor connections or loose wires at fuel pump, main relay, fuel pump relay and fuel pump resistor.
- YES ↓

Turn the ignition switch OFF.

Connect the ECU test harness between the ECU and connector

Start the engine.

Measure the voltage between D6 (+) terminal and A23 (−) terminal.

Is there voltage?
- NO → Repair open in BLK/RED wire between ECU (D6) and fuel pump resistor. → Connect the ECU test harness between the ECU and connector
- YES ↓

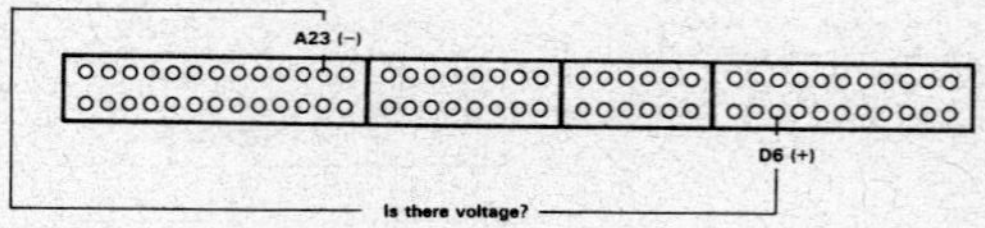

Install a jumper wire between A8 terminal and A24 terminal on the ECU test harness.

Measure the voltage between D6 (+) terminal and A23 (−) terminal.

Is there battery voltage?
- YES → Substitute a known-good ECU and recheck. If symptom/indication goes away, replace the original ECU.
- NO ↓

Turn the ignition switch OFF.

Remove a jumper wire from the ECU test harness.

Disconnect "A" connector from the ECU only, not the engine wire harness.

Turn the ignition switch ON.

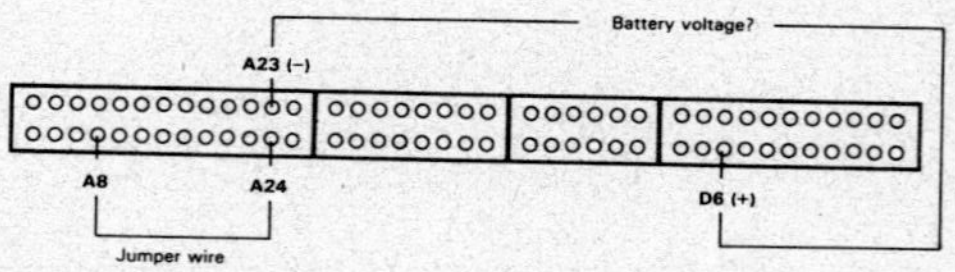

TROUBLESHOOTING CHARTS—NSX

Measure the voltage between A8 (+) terminal and A23 (−) terminal.

Is there battery voltage? — YES → **Replace Fuel Pump Relay.**

NO ↓

Turn the ignition switch OFF.

Install a jumper wire* between YEL (+) terminal and RED (−) terminal on the fuel pump relay connector.

Turn the ignition switch ON.

Measure the voltage between A8 (+) terminal and A23 (−) terminal.

Is there battery voltage? — YES → **Replace fuel pump relay.**

NO ↓

Remove a jumper wire.

Measure the voltage between YEL (+) terminal (on the fuel pump relay connector) and body ground (−).

Is there battery voltage? — NO → **Repair open in YEL wire between No. 5 fuse and fuel pump relay.**

YES ↓

Repair open in RED wire between ECU (A8) and fuel pump relay.

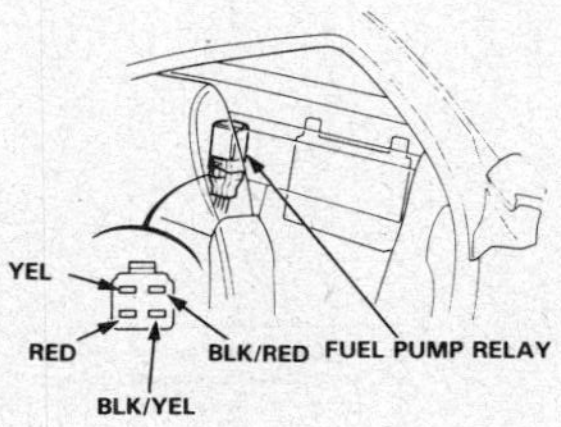

CAUTION:
***the "A" connector must be disconnected for this test or the ECU will be damaged.**

This signals the PGM-FI ECU when there is a demand for cooling from the air conditioning system.

Inspection of Air Conditioning Signal.

Connect the ECU test harness between the ECU and connector. Disconnect "C" connector from the engine wire harness only, not the ECU

Turn the ignition switch ON.

Measure voltage between C3 (+) terminal and A26 (−) terminal.

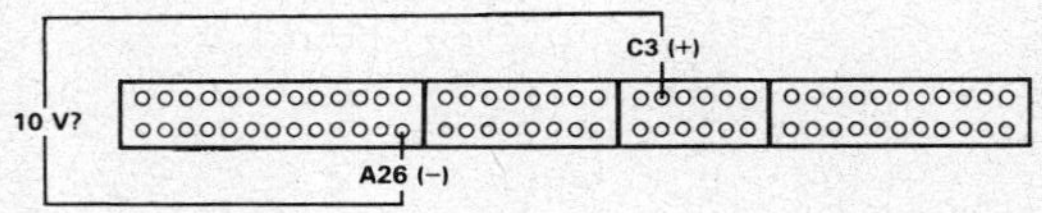

Is there approx. 10 V? — NO → **Substitute a known-good ECU and recheck. If prescribed voltage is now available, replace the original ECU.**

YES ↓

Reconnect "C" connector to the engine wire harness.

Momentarily connect A15 terminal to A26 terminal several times.

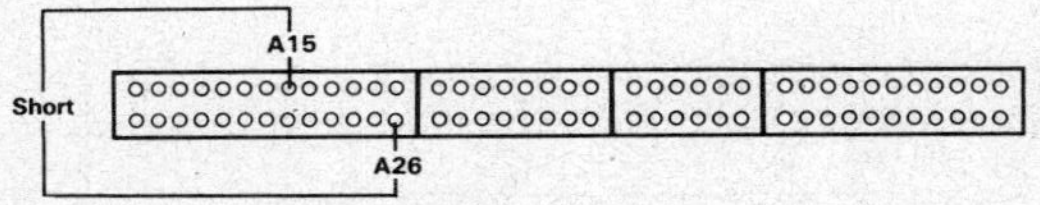

Is there a clicking noise from the A/C compressor clutch? — NO → Connect the RED/BLU terminal of the 4P connector on the A/C clutch relay to body ground.

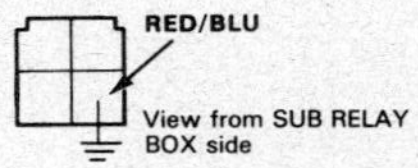

Is there a clicking noise from the A/C compressor clutch? — NO → **Air conditioner inspection**

YES → **Repair open in RED/BLU wire between ECU (A15) and A/C clutch relay.**

YES ↓ (from first clicking noise check)

Start the engine.

Blower switch ON.

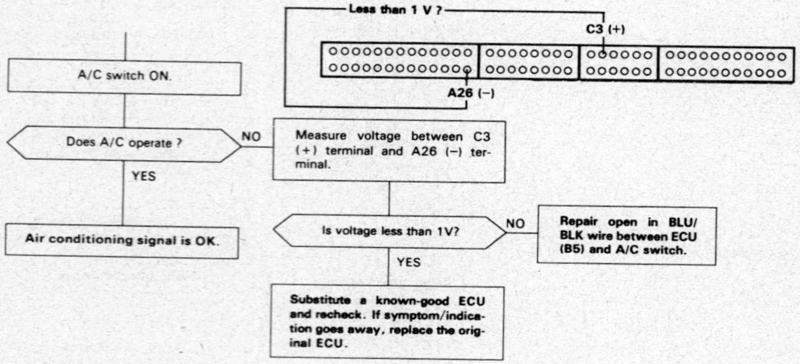

This signals the PGM-FI ECU when the alternator is charging.

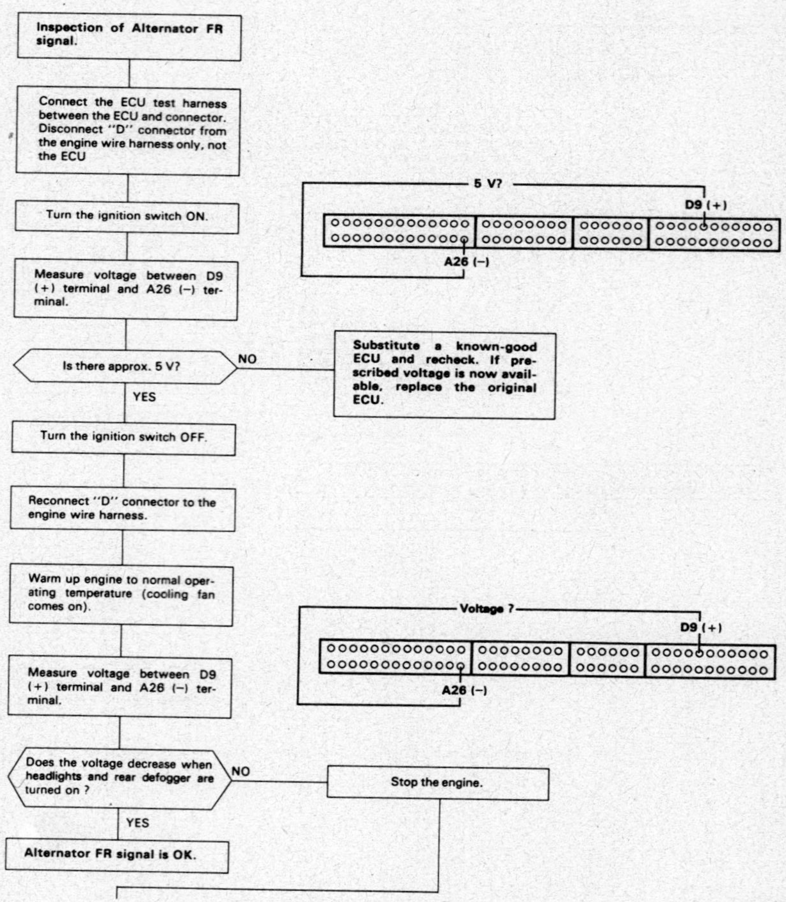

Disconnect "D" connector from ECU only, not the engine wire harness.

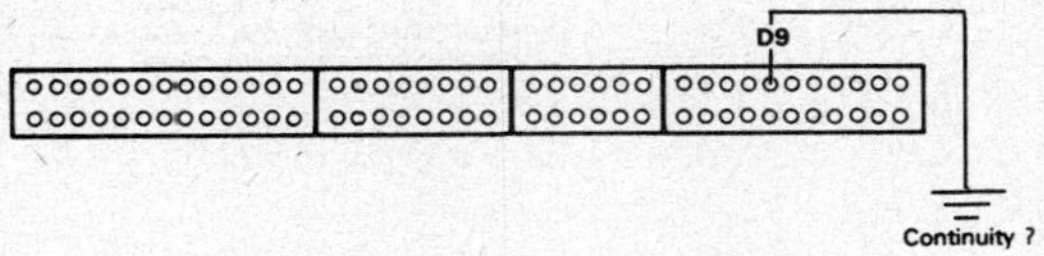

Disconnect the negative battery cable from the battery.

Check for continuity between D9 terminal and body ground.

Does continuity exist ?

- YES → Disconnect GRN connector from the alternator.
  - Check for continuity between D9 terminal and body ground.
  - Does continuity exist ?
    - NO → Alternator Inspection.
    - YES → **Repair short in WHT/RED wire between ECU (D9) and alternator.**
- NO → Disconnect GRN connector from the alternator.

GRN CONNECTOR

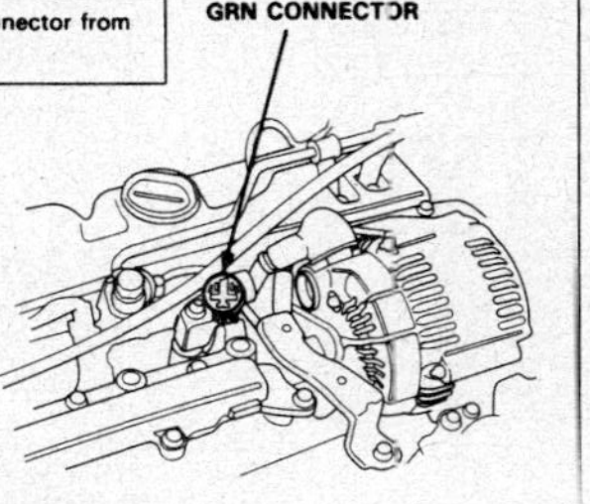

Connect WHT/RED wire to body ground.

Check for continuity between D9 terminal and body ground.

Does continuity exist ?

- YES → Alternator Inspection.
- NO → **Repair open in WHT/RED wire between ECU (D9) and alternator.**

This signals the PGM-FI ECU when the transmission is in Neutral or Park.

Inspection of A/T Shift Position Signal.

Turn the ignition switch ON.

Observe the A/T shift indicator and select each position separately.

Does the indicator light properly?

- NO → **A/T Shift Position Indicator Inspection.**
- YES

Turn the ignition switch OFF.

Connect the ECU test harness between the ECU and connector. Disconnect "B" connector from the engine wire harness only, not the ECU

Turn the ignition switch ON.

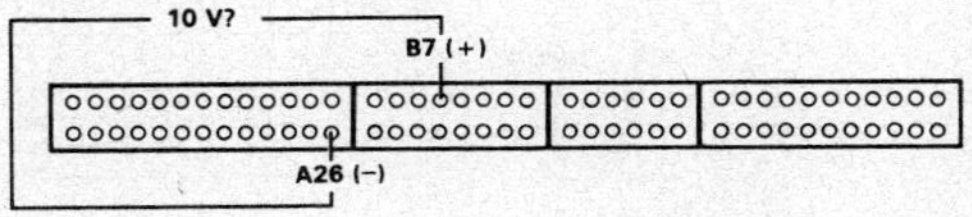

Measure voltage between B7 (+) terminal and A26 (−) terminal.

Is there approx. 10 V?

- NO → **Substitute a known-good ECU and recheck. If prescribed voltage is now available, replace the original ECU.**
- YES

Turn the ignition switch OFF.

Reconnect "B" connector to the engine wire harness.

Turn the ignition switch ON.

Measure voltage between B7 (+) terminal and A26 (–) terminal with the transmission in [N].

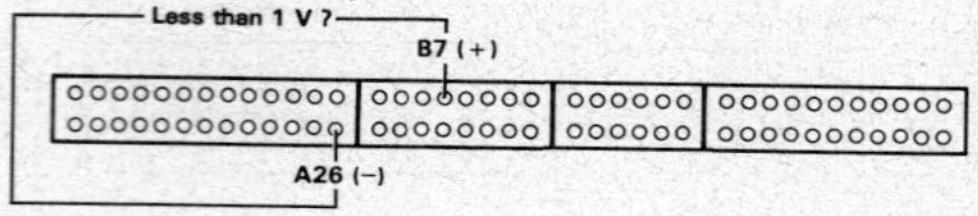

Is there less than 1 V?
- NO → – Repair open in LT GRN wire between ECU (B7) and gauge assembly. – Repair open in GRN wire between the gauge assembly and shift position console switch.
- YES ↓

Measure voltage between B7 (+) terminal and A26 (–) terminal with the transmission in [P].

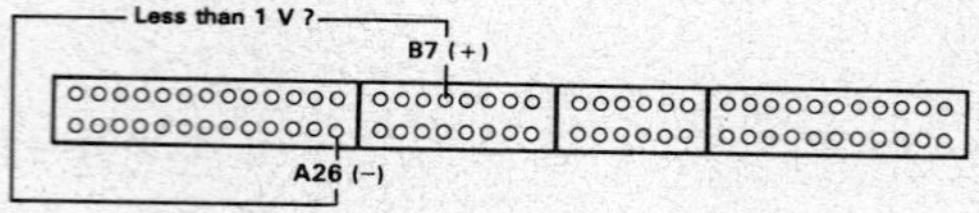

Is there less than 1 V?
- NO → Repair open in GRN/WHT wire between gauge assembly and shift position console switch.
- YES ↓

Measure voltage between B7 (+) terminal and A26 (–) terminal with the transmission in gear.

Is there approx. 10 V?
- NO → Repair short in LT GRN wire between ECU (B7) and gauge assembly.
- YES ↓

A/T shift position signal is OK.

This signals the PGM-FI ECU when the transmission is in Neutral.

Inspection of M/T neutral Switch Signal

Connect the ECU test harness between the ECU and connector

Turn the ignition switch ON.

Measure voltage between B7 (+) terminal and A26 (–) terminal in Neutral position.

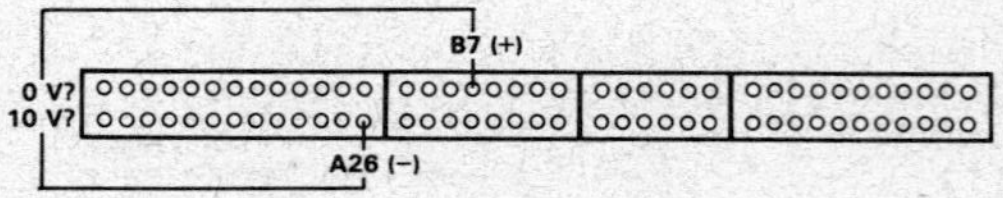

Is there voltage?
- YES → Disconnect the 3P connector on the M/T neutral switch.
  - Connect LT GRN terminal to BLK terminal.
  - Is there voltage?
    - NO → **Replace M/T neutral switch.**
    - YES → **Repair open in LT GRN wire between ECU (B7) and M/T neutral pressure switch or BLK wire between M/T neutral switch and G101.**
- NO ↓

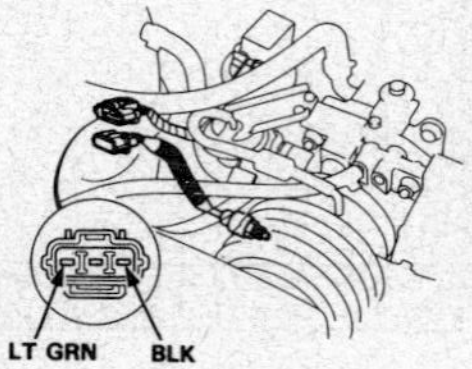

Shift transmission in gear.

Is there approx. 10 V?
- NO → Disconnect "B" connector from engine wire harness only, not the ECU.
  - Is there approx. 10 V?
    - NO → **Substitute a known-good ECU and recheck. If prescribed voltage is now available, replace the original ECU.**
    - YES → Reconnect "B" connector to main wire harness and disconnect 3P connector on the M/T neutral switch.
      - Is there approx. 10 V?
        - NO → **Repair short in LT GRN wire between ECU (B7) and the M/T neutral switch.**
        - YES → **Replace M/T neutral switch.**
- YES ↓

M/T neutral switch signal is OK.

TROUBLESHOOTING CHARTS—NSX

This signals the PGM-FI ECU when the clutch is engaged.

Inspection of clutch switch signal.

Connect the ECU test harness between the ECU and connector

Turn the ignition switch ON.

Measure voltage between C7 (+) terminal and A26 (−) terminal.

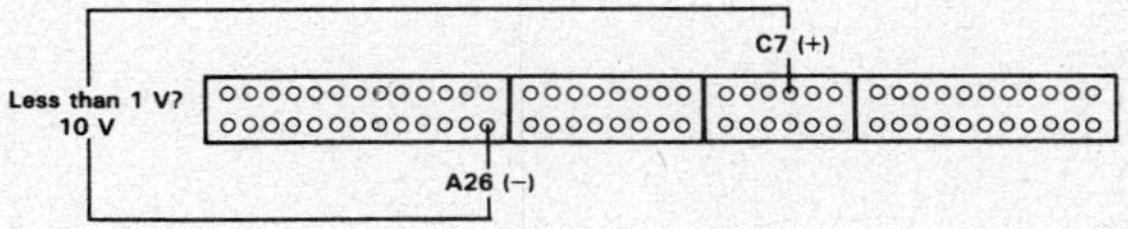

Is voltage less than 1 V?

- NO → Turn the ignition switch CFF.
  - Disconnect the 3P connector from the clutch switch.
  - Check for continuity between the A terminal and B terminal on the clutch switch.
  - Does continuity exist?
    - NO → **Replace the clutch switch.**
    - YES → Turn the ignition switch ON.
      - Measure voltage between PNK (+) terminal and body ground.
      - Is there approx. 10 V?
        - NO → **Repair open in PNK wire between ECU (C7) and the clutch switch.**
        - YES → **Repair open in BLK wire between the clutch switch and G401.**

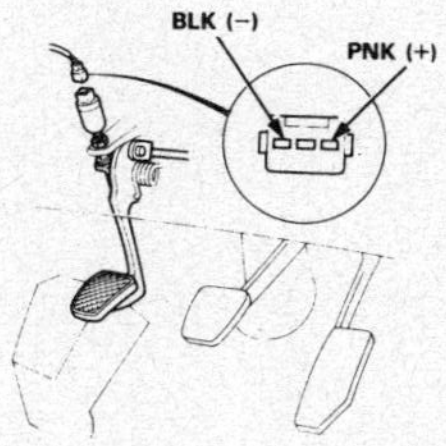

- YES → Depress the clutch pedal.
  - Measure voltage between C7 (+) terminal and A26 (−) terminal.
  - Is there approx. 10 V?
    - YES → **Clutch switch signal is OK.**
    - NO → Turn the ignition switch OFF.
      - Disconnect the 3P connector from the clutch switch.
      - Turn the ignition switch ON.
      - Measure voltage between C7 (+) terminal and A26 (−) terminal.
      - Is there approx. 10 V?
        - YES → **Replace the clutch switch.**
        - NO → Disconnect "C" connector from engine wire harness only, not the ECU.
          - Is there approx. 10 V?
            - YES → **Repair short in PNK wire between ECU (C7) and the clutch switch.**
            - NO → **Substitute a known-good ECU and recheck. If prescribed voltage is now available, replace the original ECU.**

This signals the PGM-FI ECU when the engine is cranking.

Inspection of Starter Switch Signal.

Connect the ECU test harness between the ECU and connector

Measure voltage between C11 (+) terminal and A26 (−) terminal.

Is there battery voltage ?
- NO → Inspect No. 7 fuse.
  - Is No. 7 fuse OK ?
    - NO → **Replace fuse.**
    - YES → **Repair open in BLK/WHT wire between ECU (C11) and No. 7 fuse.**
- YES → **Starter switch signal is OK.**

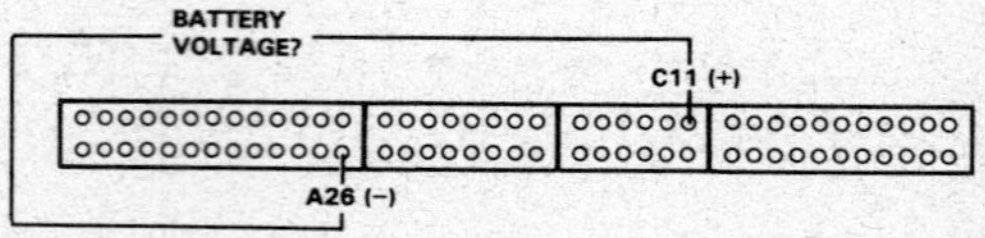

NOTE: Clutch pedal must be depressed on M/T models.

– Engine will not start.
– Inspection of main relay.

Disconnect the main relay connectors.

Measure the voltage between YEL/BLU terminal ① and body ground.

Is there battery voltage?
- NO → **Replace ACG fuse.**
- YES ↓

Turn the ignition switch ON.

Measure the voltage between BLK/YEL terminal ⑤ and body ground.

Is there battery voltage?
- NO → **Replace No. 2 fuse.**
- YES ↓

Measure the voltage between BLK/YEL terminal ⑥ and body ground.

Is there battery voltage?
- NO → **Replace IG COIL fuse.**
- YES ↓

Turn the ignition switch START position.

Measure the voltage between BLK/WHT terminal ④ and body ground.

Is there battery voltage?
- NO → **Replace No. 7 fuse.**
- YES ↓

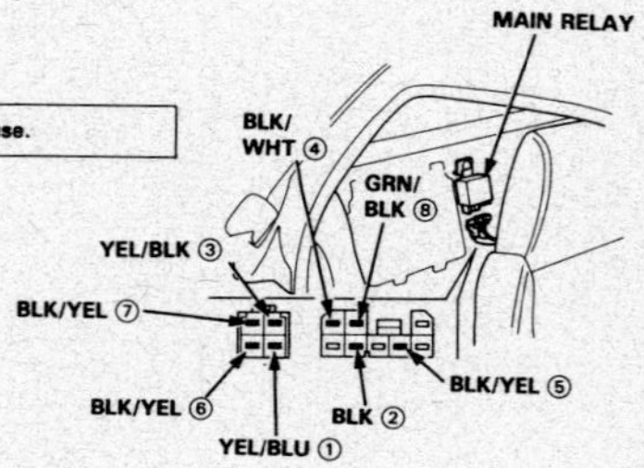

TROUBLESHOOTING CHARTS – NSX

Troubleshooting Flowchart

- **– Engine will not start.**
- **– Inspection of fuel pump resistor.**

Disconnect the fuel pump resistor connector.

Turn the ignition switch ON.

Measure the voltage between BLK/YEL (+) (on harness side) and body ground (–).

Is there battery voltage? (for 2 seconds)

- NO → **Inspect the main relay**
- YES ↓

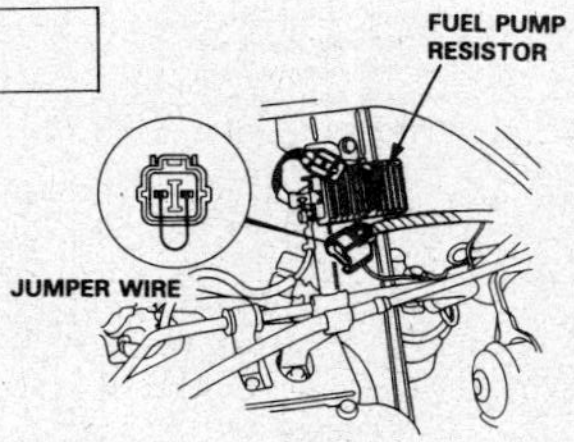

Turn the ignition switch OFF.

Install a jumper wire between two terminals on the wire harness side of fuel pump resistor.

Turn the ignition switch to START position.

Does the engine start?

- YES → **Replace fuel pump resistor.**
- NO ↓

Turn the ignition switch OFF.

Disconnect the fuel pump connector.

Turn the ignition switch ON.

Turn the ignition switch off.

Connect the ECU test harness between the ECU and connector. Disconnect "A" connector from the engine wire harness only, not the ECU

Check for continuity between GRN/BLK terminal ⑧ and A7 terminal.

Does continuity exist?

- NO → **Repair open in GRN/BLK wire between ECU (A7) and Main Relay**
- YES ↓

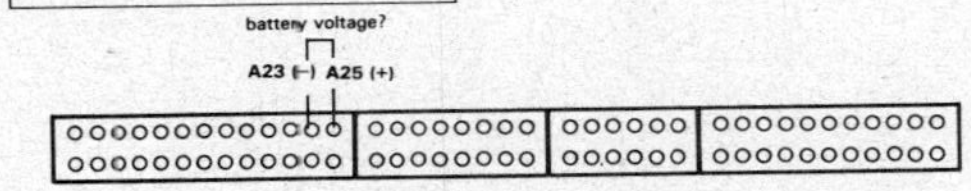

Reconnect "A" connector to the engine wire harness.

Turn the ignition switch ON.

Measure the voltage between A25 (+) terminal and A23 (–) terminal.

Is there battery voltage?

- NO → **Replace main relay.**
- YES ↓

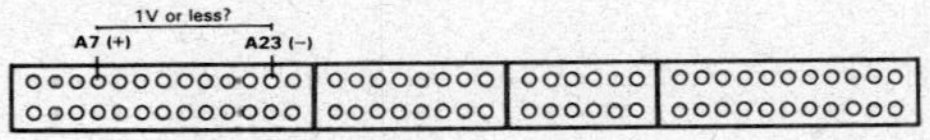

Measure the voltage between A7 (+) terminal and A23 (–) terminal.

Is there 1V or less? (for 2 seconds)

- NO → **Substitute a known-good ECU and recheck. If prescribed voltage is now available, replace the original ECU.**
- YES ↓

Replace main relay.

Measure the voltage between BLK/RED (+) (on wire harness side of fuel pump) and body ground.

Is there battery voltage? (for 2 seconds)

- YES → **Inspect the fuel pump**
- NO → **Repair open in BLK/RED wire between the fuel pump resistor and the fuel pump.**

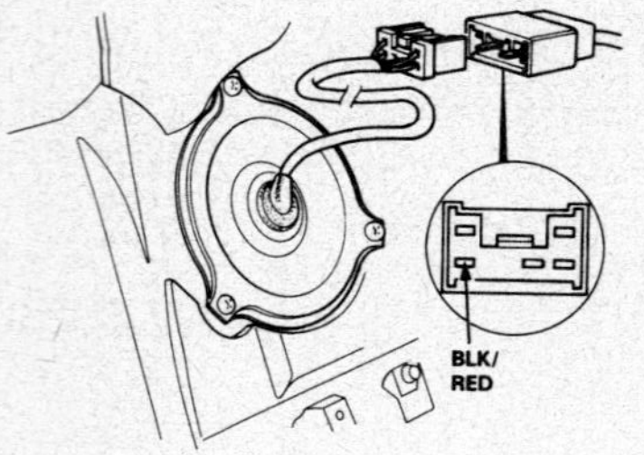

Troubleshooting Flowchart

Inspection of Chamber Volume Control System.

Start engine and allow it to idle.

Remove # 3 vacuum hose from the chamber volume control diaphragm and connect vacuum gauge to the hose.

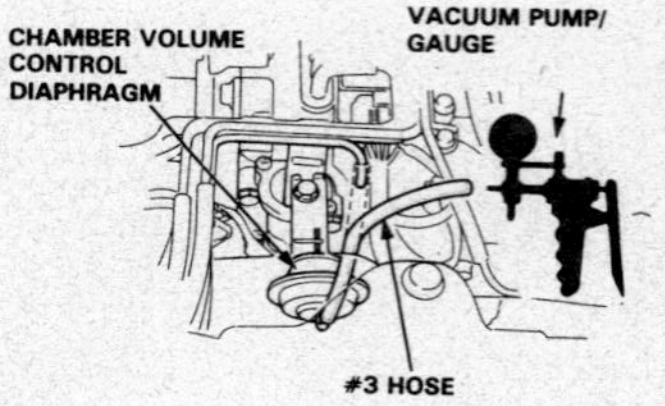

Is there vacuum?

- YES → (continued)
- NO → Disconnect the lower hose on chamber volume control solenoid valve in the control box and connect a vacuum gauge to the hose.

Is there vacuum?

- NO → **Repair the blockage or vacuum leak between the vacuum tank and the intake manifold.**
- YES → Disconnect the 6P connector from the control box.

Measure voltage between BLK/YEL (+) terminal and BLU (−) terminal.

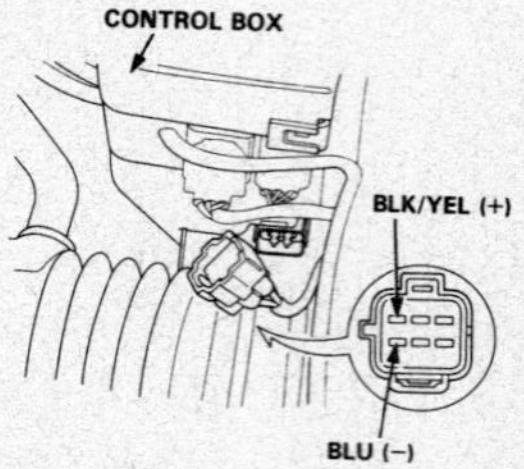

Is there battery voltage?

- YES → **Replace the chamber volume control solenoid valve.**
- NO → Measure voltage between BLK/YEL (+) terminal and body ground.

Raise engine speed to 5,000 rpm.

Check for vacuum at #3 vacuum hose.

Is there vacuum?

- NO → Chamber volume control system is OK.
- YES → Disconnect the 6P connector from the control box.

Is there vacuum?

- YES → Replace chamber volume control solenoid valve.
- NO → Turn the ignition switch OFF.

Disconnect "A" connector from ECU.

Check for continuity to ground on the RED/BLU wire.

Is there continuity to ground?

- YES → **Repair short to ground at RED/BLU wire between ECU (A14) and the 6P connector.**
- NO → **Substitute a known-good ECU and recheck. If symptom goes away, replace the original ECU.**

Is there battery voltage?

- NO → **Repair open in BLK/YEL wire between the 6P connector and No. 2 (15 A) fuse.**
- YES → Turn the ignition switch OFF.

Connect the ECU test harness between the ECU and connector

Check for continuity of RED/BLU wire between ECU (A14) and the 6P connector.

Does continuity exist?

- NO → **Repair open in RED/BLU wire between ECU (A14) and the 6P connector.**
- YES → **Substitute a known-good ECU and recheck. If symptom goes away, replace the original ECU.**

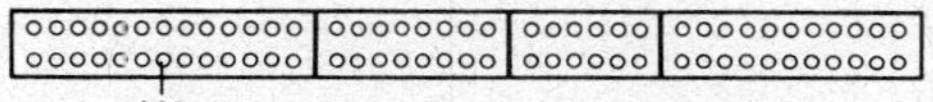

A14

Inspection of Evaporative Emission Controls

Disconnect vacuum hose from the purge control diaphragm valve (on the charcoal canister) and connect a vacuum gauge to the hose.

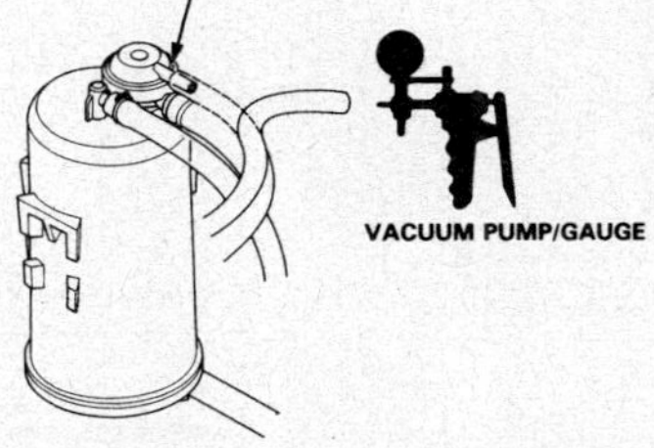

Start the engine and allow it to idle.
NOTE: Engine coolant temperature must be below 70°C (158°F).

Is there vacuum ?

- NO
- YES → Disconnect the 6P connector.

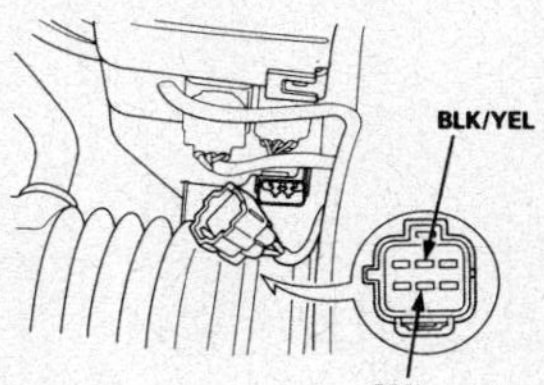

Measure voltage between BLK/YEL (+) terminal and GRN (−) terminal.

Is there battery voltage ?

- YES → **Inspect vacuum hose routing. If OK, replace purge cut-off solenoid valve.**
- NO → Measure voltage between BLK/YEL (+) terminal and body ground.

Warm up the engine to normal operating temperature (cooling fan comes on).

Is there battery voltage ?
- NO → Repair open in BLK/YEL wire between No. 2 fuse and the 6P connector.
- YES → Inspect GRN wire for a short to body ground between ECU (A20) and the connector. If wire is OK, substitute a known-good ECU and recheck. If symptom/indication goes away, replace the original ECU.

Check for vacuum at #6 vacuum hose after starting the engine.

Is there manifold vacuum?
- NO → Disconnect the 6P connector.
  - Is there manifold vacuum?
    - NO → Inspect vacuum hose routing. If OK, replace purge cut-off solenoid valve.
    - YES → Inspect for an open in GRN wire between ECU (A20) and the connector. If wire is OK, substitute a known-good ECU and recheck. If symptom/indication goes away, replace the original ECU.
- YES → Reconnect the hose.

Remove fuel filler cap.

Connect a vacuum gauge to canister purge air hose.

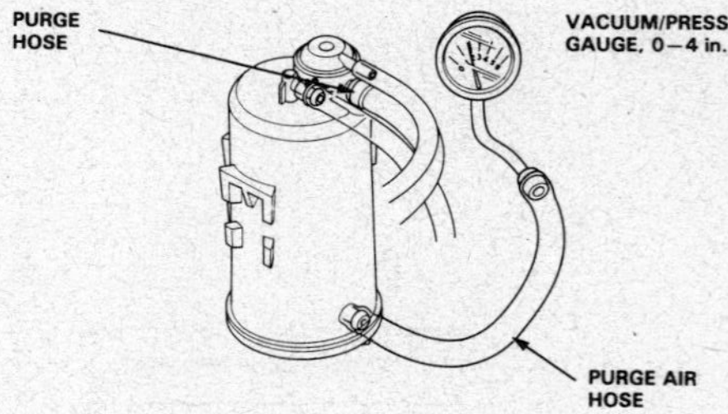

Start the engine and raise speed to 3,500 rpm.

Does vacuum appear on gauge within 1 minute ?
- YES → **See two-way valve test to complete. Evaporative emission control are OK.**
- NO → Connect a vacuum gauge to the canister purge hose and raise the engine speed to 3,500 rpm.
  - Does vacuum appear on the gauge?
    - NO → **Inspect the purge hose and throttle body port for pinch or blockage.**
    - YES → **Replace the canister.**

Component Replacement

SERVICE PRECAUTIONS

- Disconnect the negative battery terminal.
- Whenever possible, use a flashlight instead of a drop light.
- Keep all open flame and smoking material out of the area.
- Use a shop cloth or similar to catch fuel when opening a fuel system.
- Always relieve fuel system pressure before loosening any line or component.
- Always use eye or full-face protection.
- Always keep a dry chemical (class B-C) fire extinguisher near the work area.

FUEL PRESSURE RELIEF

CAUTION

The fuel system is under constant pressure, even when the engine has not been run for a period of time. To avoid injury and/or property damage, always safely relieve the fuel system pressure before servicing any fuel or return line component.

1. Remove the fuel filler cap.
2. With the ignition **OFF**, disconnect the negative battery cable.
3. Locate the 6mm service bolt; it is located at the fuel inlet line to the fuel filter.
4. Use a wrench to hold the larger bolt below the service bolt. Use a box-end wrench on the service bolt. Place a rag or cloth over the wrenches and service bolt.
5. Slowly loosen the service bolt 1 full turn. The pressurized fuel will escape into the rag.
6. Remove the service bolt, replace the washer and reinstall the bolt. Tighten the bolt to 9 ft. lbs. (12 Nm). The washer must be replaced any time the service bolt is loosened or removed. Failure to replace the washer may result in high-pressure fuel leaks.
7. Dispose of the fuel-soaked rag in a manner correct for flammable items. Check the surrounding area for any spilled fuel.
8. The pressure within the system will stay reduced until the ignition switch is turned **ON** or the fuel pump is operated by other means.

FUEL PUMP

Removal and Installation

INTEGRA, VIGOR AND NSX

1. Relieve the fuel pressure.
2. Remove the fuel tank after disconnecting all fuel lines and wiring running to the tank.
3. Remove the fuel pump mounting bolts and remove the fuel pump assembly.
4. Install the fuel pump; tighten the retaining nuts to 4 ft. lbs. (6 Nm).
5. Reinstall the fuel tank

LEGEND

1. Relieve the fuel pressure.
2. Remove the fuel tank access cover. For 1990 Legends, this may be done through the trunk. On 1991–92 Legends, the rear seat must be removed to get to the cover.
3. Disconnect the fuel lines and the electrical connector.
4. Remove the fuel pump mounting nuts and remove the pump.
5. Install the fuel pump, tightening the retaining nuts to 4 ft. lbs. (6 Nm).
6. Connect the fuel lines and wiring harnesses.

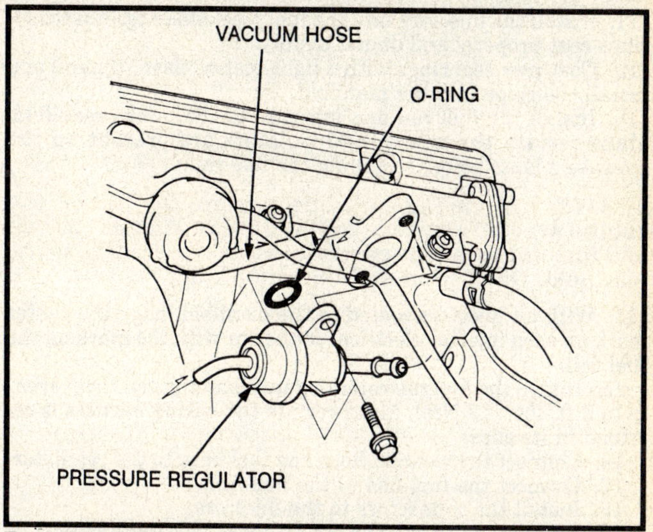

Fuel pressure regulator. Vigor shown, all are similar

7. Install the maintenance cover and reinstall the access panel. Install the rear seat if it was removed.

FUEL PRESSURE REGULATOR

Removal and Installation

1. Place a cloth or rag under the regulator. Be prepared to contain escaping fuel.
2. Relieve the fuel system pressure at the fuel filter.
3. Disconnect the vacuum hose from the regulator.
4. Disconnect the fuel line from the regulator; plug the line to prevent spillage or the entry of dirt.
5. Remove the 2 retaining bolts and remove the regulator.
6. Remove the O-ring from the regulator and discard it.

To install:

7. Install a new O-ring on the regulator; coat the ring lightly with a thin coat of clean, fresh oil.
8. Install the regulator, making sure the O-ring is not distorted or damaged. Tighten the retaining bolts to 9 ft. lbs. (12 Nm).
9. Connect the fuel line.
10. Install the vacuum line.

FUEL RAIL AND INJECTORS

Removal and Installation

1. Relieve the fuel pressure.
2. On Vigor and 1991–92 Legend remove the engine harness cover. On NSX, remove the intake manifold covers.
3. Label and disconnect the electrical harnesses from the injectors.
3. Place a rag below the fuel pressure regulator. Disconnect the vacuum hose and disconnect the fuel line. Plug the fuel line to prevent spillage and the entry of dirt.
4. Disconnect the fuel hose from the fuel rail.
5. Remove the nuts holding the fuel pipe (rail) and the wiring harness.
6. Remove the fuel rail from the injectors, leaving the injectors in the manifold.
8. Remove each injector and remove the seal ring from each manifold port.
9. Remove the cushion ring and O-ring from each injector.

To install:

10. Install new cushion rings on each injector. Install new O-rings on each injector and coat the O-rings with a light coat of clean, thin oil.

11. Install the injectors into the fuel pipe. Make certain the O-rings seat properly and do not distort.
12. Coat new seal rings with a light coat of clean, thin oil and install the rings into the manifold.
13. Install the fuel rail and injectors to the intake manifold. Make certain the mounting insulators are present on the mounting bolts before installing the fuel rail.

NOTE: Assembling each injector into the rail off the engine prevents damage to the O-rings. Handle the rail and injector assembly carefully when reinstalling to the manifold. Don't drop an injector or bang the tips.

14. With all injectors seated in the manifold, align the center mark on each injector electrical connector with the mark on the fuel rail.
15. Install the fuel rail retaining nuts and tighten them evenly to 9 ft. lbs. (12 Nm). Make certain the wiring harness is retained in its clips.
16. Connect the vacuum hose and fuel hose to the regulator.
17. Connect the fuel line to the fuel rail.
18. Install the connectors to the injectors.
19. Switch the ignition **ON** but do not engage the starter. The fuel pump should run for approximately 2 seconds, building pressure within the lines. Switch the ignition **OFF**, then **ON** 2 or 3 more times to build full system pressure. Check the work area for fuel leaks.
20. Install the engine harness covers or intake manifold covers if they were removed.

INJECTOR RESISTOR

Removal and Installation

1. Switch the ignition **OFF**.
2. The resistor is located in the engine compartment. On 1990 Legend, it is on the right side firewall; on Vigor and 1991–92 Legends, it is on the left shock tower. On Integra, look for it on the left firewall near the fender and on NSX, the resistor is located on the right side panel at the rear of the engine compartment. Disconnect the wiring connector to the resistor.
3. Remove or disconnect any wiring harness clamps or retainers holding the resistor harness.
4. Remove the retaining bolts holding the resistor; remove the resistor.
5. Reassemble in reverse order.

TDC/CRANK ANGLE/CYLINDER SENSORS

Removal and Installation

LEGEND

1. Set the engine to TDC No. 1 cylinder. On 1990 vehicles, remove the cruise control actuator. Disconnect the CRANK/CYL sensor connector.
2. Remove the upper timing belt cover from the front bank of cylinders on 1990 vehicles. On later models, remove both upper belt covers.
3. Remove the timing belt from the upper pulley(s).
4. Remove the 3 bolts holding the front or left pulley and remove it.
5. Remove the front upper cover back plate.
6. Unbolt and remove the sensor.

To install:

7. Carefully install the sensor on the head and tighten the bolts to 9 ft. lbs (12 Nm).
8. Install the upper cover back plate and gaskets.
9. Align the cam pulley pin with the camshaft hole to install the cam pulley. Tighten the bolts to 23 ft. lbs (32 Nm).
10. Install the timing belt.
11. Install the upper timing belt cover.
12. Install the cruise control actuator if it was removed.

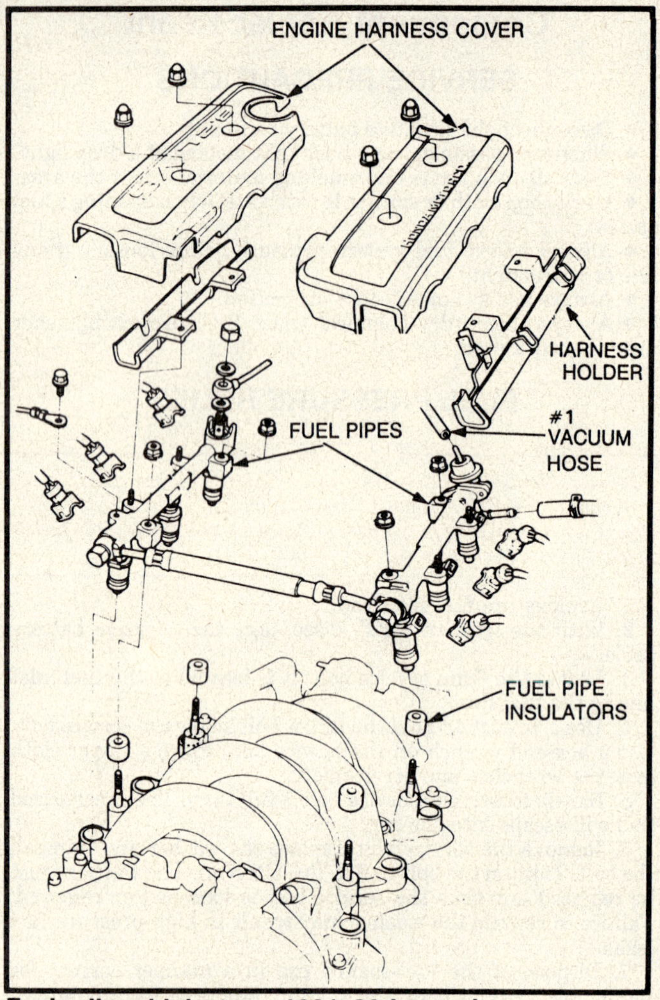

Fuel rail and injectors, 1991–92 Legend

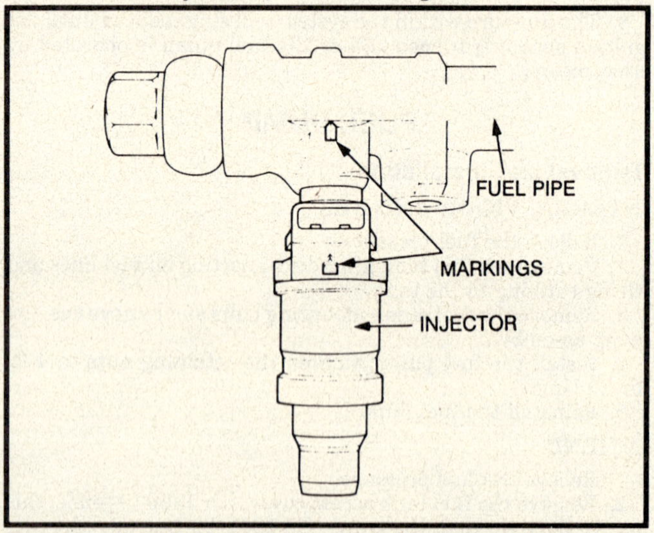

Always align the injector mark with the mark on the fuel rail

INTEGRA

The sensors are contained within the distributor housing and are not individually serviceable. Should any of the sensors become faulty, the distributor housing must be replaced. All of the

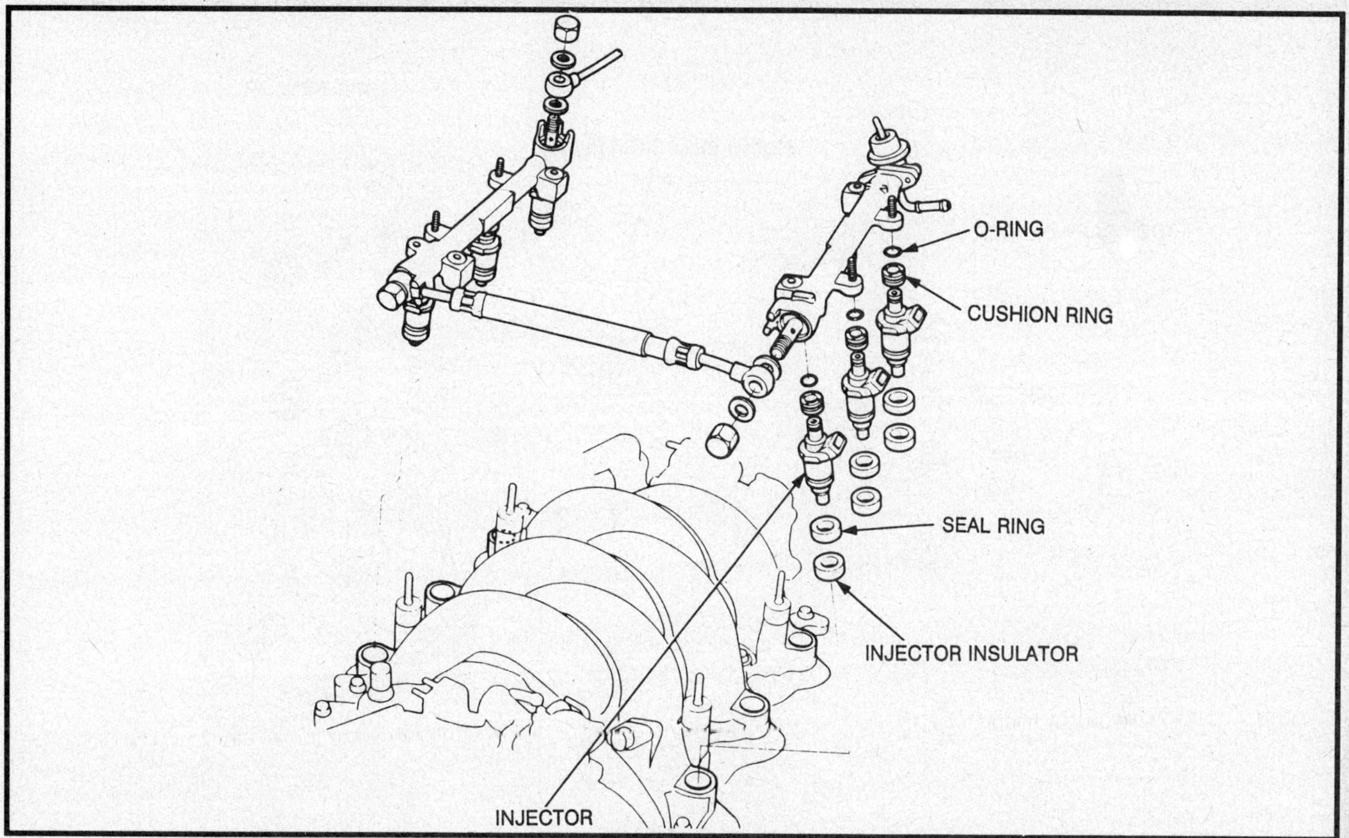

Injector and rail detail; always replace the seals and O-rings

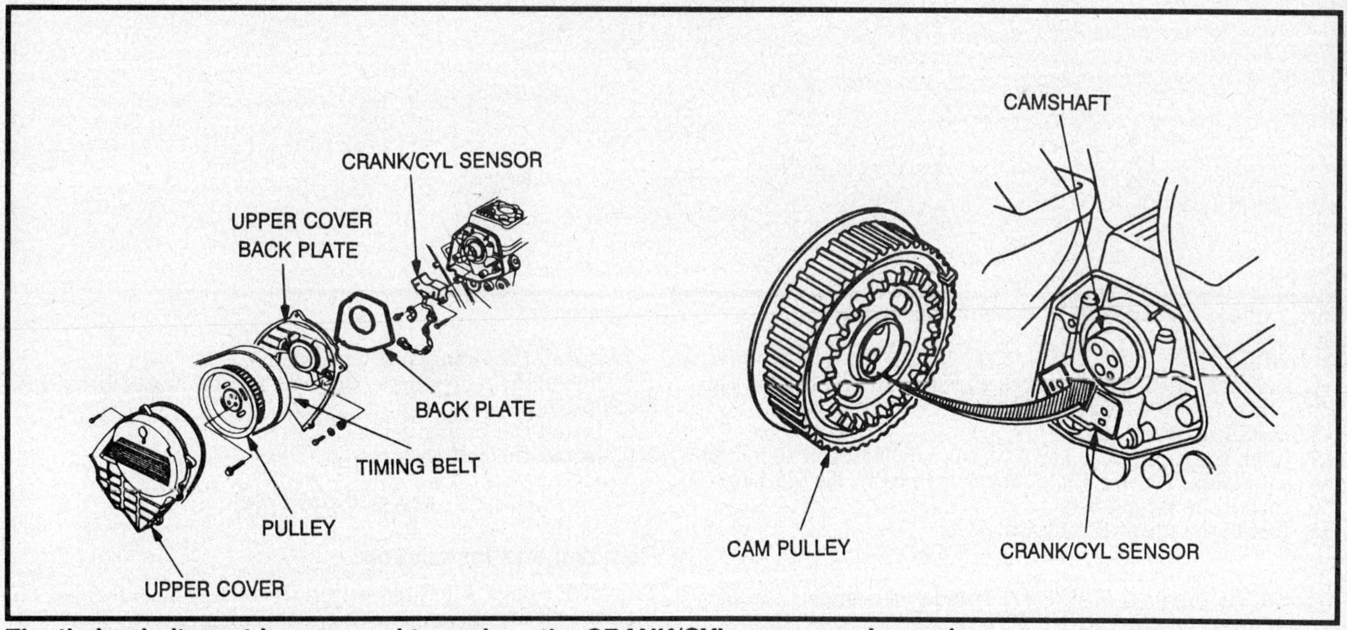

The timing belt must be removed to replace the CRANK/CYL sensor on Legends

components external to the housing, such as the cap, rotor, coil and igniter may be transferred to the new unit.

VIGOR

1. Set the engine to TDC No. 1 cylinder. Disconnect the TDC/CRANK/CYL sensor connector.
2. Remove the upper timing belt cover from the belt.
3. Remove the timing belt from the upper pulley.
4. Remove the bolt holding the pulley and remove it.
5. Remove the timing belt back plate.
6. Unbolt and remove the sensor(s). The TDC/CRANK sensor is held by two bolts; the CYL sensor is held by a nut and a bolt

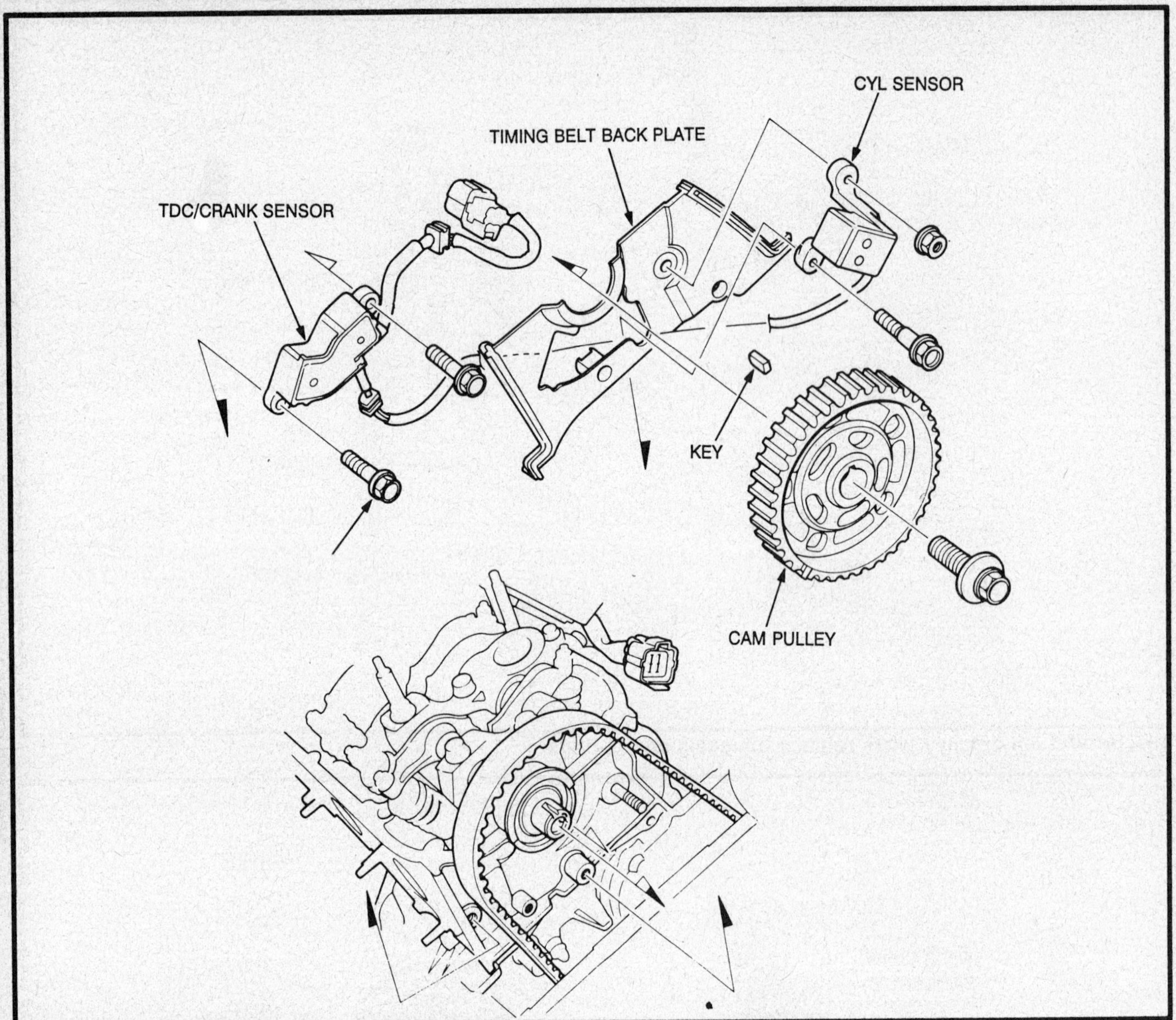

Cam pulley, timing belt and sensors – Vigor

To install:

7. Carefully install the sensors on the head and tighten the bolts and nut to 9 ft. lbs (12 Nm).
8. Install the back plate.
9. Align the cam pulley key with the camshaft slot to install the cam pulley. Tighten the pulley bolt to 54 ft. lbs (75 Nm).
10. Install the timing belt.
11. Install the upper timing belt cover.

NSX

1. Set the engine to TDC No. 1 cylinder. Disconnect the wiring connector for the CRANK/CYL sensor
2. Remove the timing belt.
3. Remove the intake and exhaust pulleys on the front bank of cylinders.
4. Remove the timing belt cover plate.
5. Remove the CRANK/CYL sensor by unbolting it from the head.

To install:

6. Install the sensor and tighten the bolts to 9 ft. lbs. (12 Nm).
7. Install the timing belt cover plate.
8. Install the cam pulleys. Tighten the retaining bolts to 51 ft. lbs. (70 Nm).
9. Install the timing belt.
10. Install the belt covers.

MAP SENSOR

Removal and Installation

The MAP sensor is located within the emission control box on the firewall for all vehicles except the Integra and Vigor. The Integra sensor is located on the firewall next to 2 solenoid valves; the Vigor sensor is located on the left fender apron.

If necessary, remove the control box from its mounts and open the box. Disconnect the wire harness connector to the sensor, then disconnect the vacuum hose(s). Remove the sensor from its mount.

When reassembling, handle the unit carefully. Install it securely in the mounting and attach the vacuum line(s) and wiring connector. Reassemble and install the emissions box.

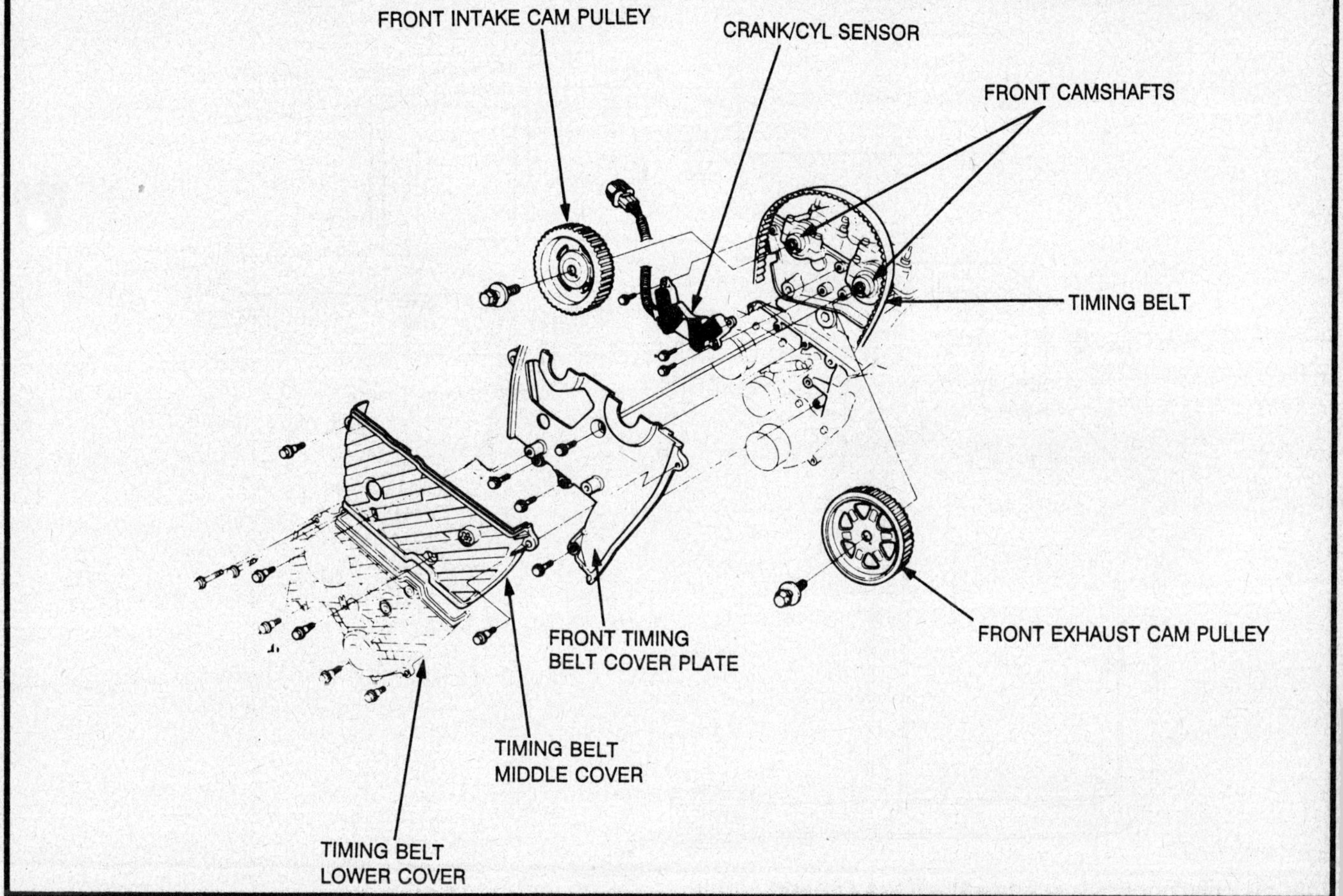

NSX CRANK/CYL sensor

INTAKE AIR TEMPERATURE (TA) SENSOR

Removal and Installation

The sensor is located on the intake manifold or air plenum. It is removed and installed in straightforward fashion; disconnect the wiring connector and remove the mounting screws. Remove the sensor.

OXYGEN SENSOR

Removal and Installation

1. Allow the exhaust system to cool before beginning the work; the pipe and sensor skin temperature may exceed 300 degrees when the engine has been operating.
2. Disconnect the wiring connector for the sensor(s). Free the wire from any retaining clips.
3. Using an oxygen sensor socket to unscrew the sensor. This tool prevents damage to the sensor or wiring harness.
4. If the sensor is to be reused, protect the tip from exposure to chemicals, oil, gasoline, etc.

To install:

5. Coat the threads of the sensor with an anti-seize compound. Take great care not to get any compound or other petroleum product on the tip of the sensor. Install the sensor to the manifold or exhaust pipe.
7. Tighten the sensor to 33 ft. lbs. (45 Nm).
8. Install the wiring into the clips and connect the wiring connector.

EGR SYSTEM CONTROLS AND PURGE CONTROL

Removal and Installation

The controls associated with control of the EGR and purge control system are located within the control box on the firewall. These components include the EGR control solenoid valve, the CVC valve and air chamber if used, as well as the purge cut-off solenoid valve.

Replacement of any of these items requires removing the control box from its mounts, disconnecting the external electrical connectors and opening the control box. Disconnect the vacuum hose(s) to the component and remove the component. Installation is in the reverse order of removal.

ELECTRONIC AIR CONTROL VALVE (EACV) FAST IDLE VALVE AIR BOOST VALVE

Removal and Installation

1. These air control valves are located at the side of the intake chamber or plenum downstream of the throttle body. On Integra vehicles, the fast idle valve is located in the emission control box.
2. Disconnect the wiring connector at the valve.
3. Label and disconnect the vacuum hoses from the unit.

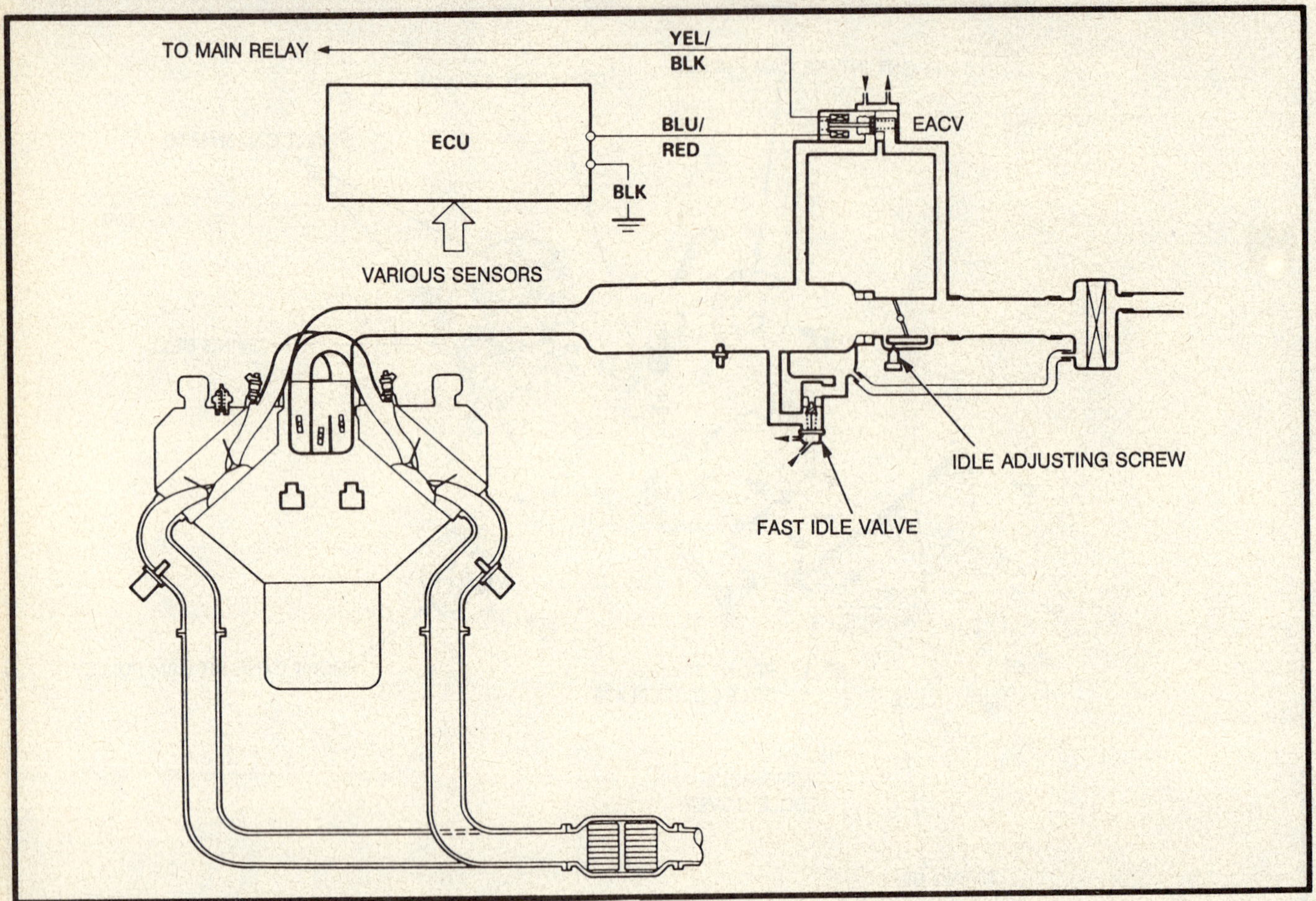

Idle air control circuit schematic—1991 Legend

4. Remove the mounting bolts and remove the valve. Recover and discard the O-rings from the vacuum ports.

NOTE: The fast idle valve is factory adjusted; it should not be disassembled or altered.

To install:

5. Replace the O-rings and make certain they are properly seated.
6. Install the valve and the retaining bolts. The pintle or control valve for the EACV must seat correctly in its port. Tighten the bolts to 9 ft. lbs. (12 Nm).
7. Connect the vacuum hoses to the valve; make certain each hose is connected to the proper port. Connect the electrical connector.

ELECTRIC LOAD DETECTOR

This sensor is used on Vigor only and is located in the underhood fuse and relay box. The ELD unit is integral with the fuse box in which it is mounted; should the sensor fail, the fuse/relay box assembly must be replaced.

IGNITER

Removal and Installation

INTEGRA

1. Remove the distributor from the engine.
2. Remove the cap, rotor and seal.
3. Remove the protective cover under the cap

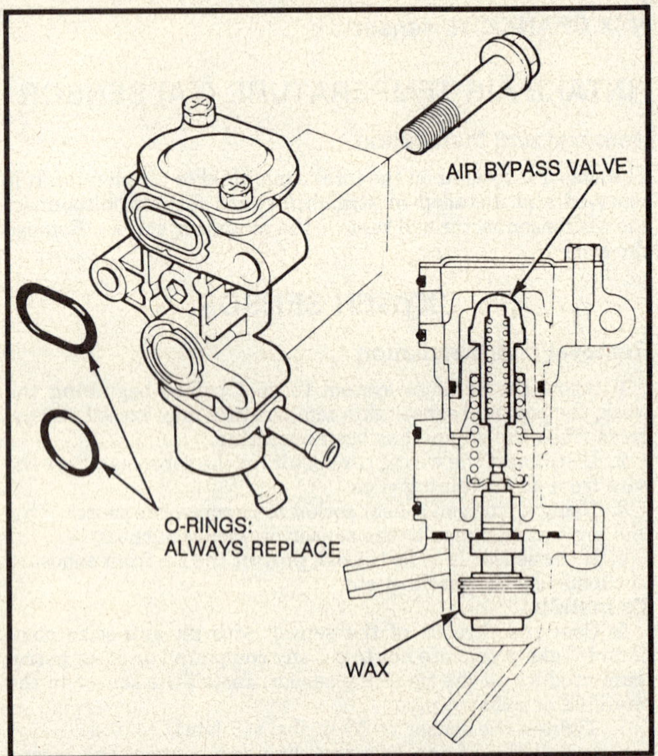

Air bypass valve mounting and cross-section

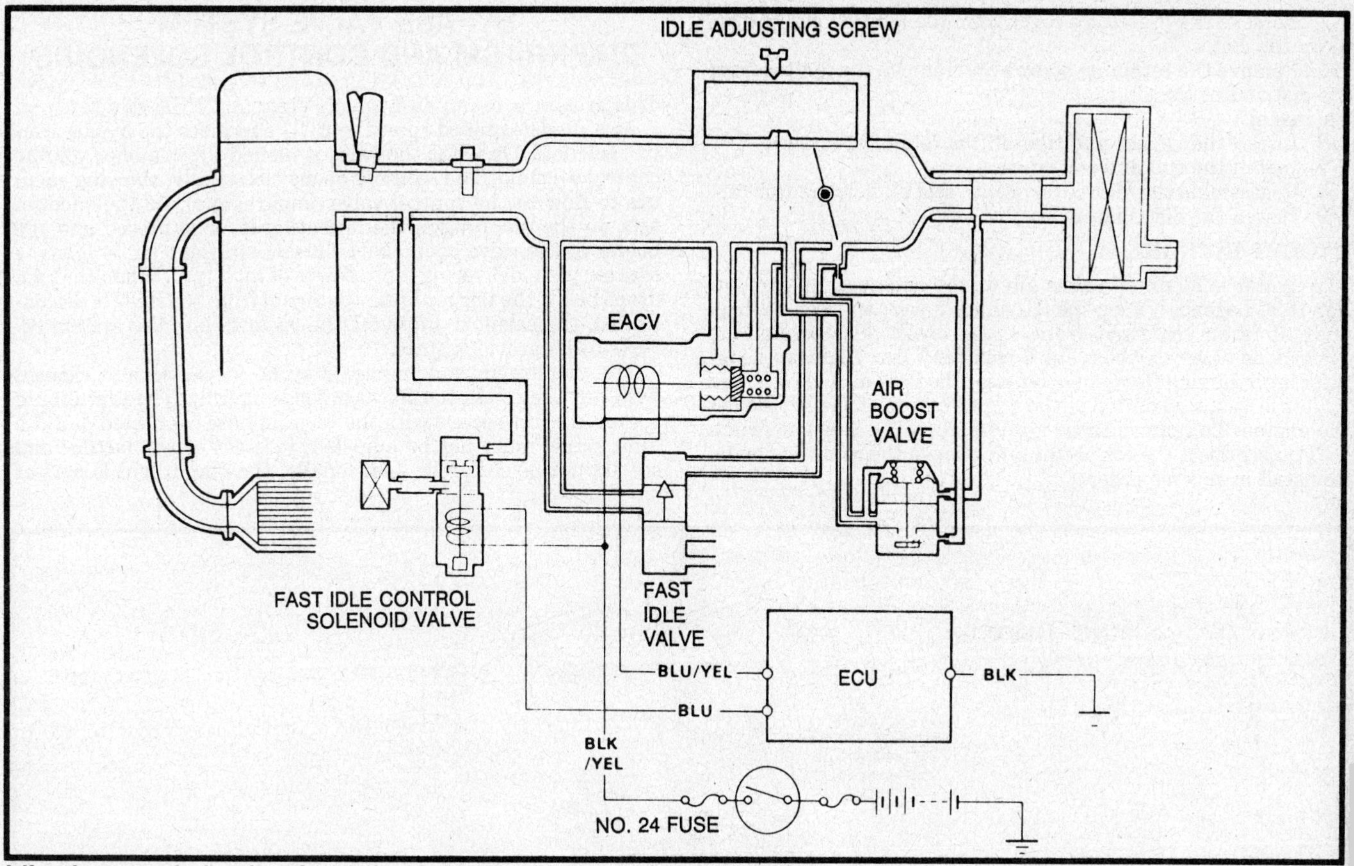

Idle air control circuit schematic—Integra

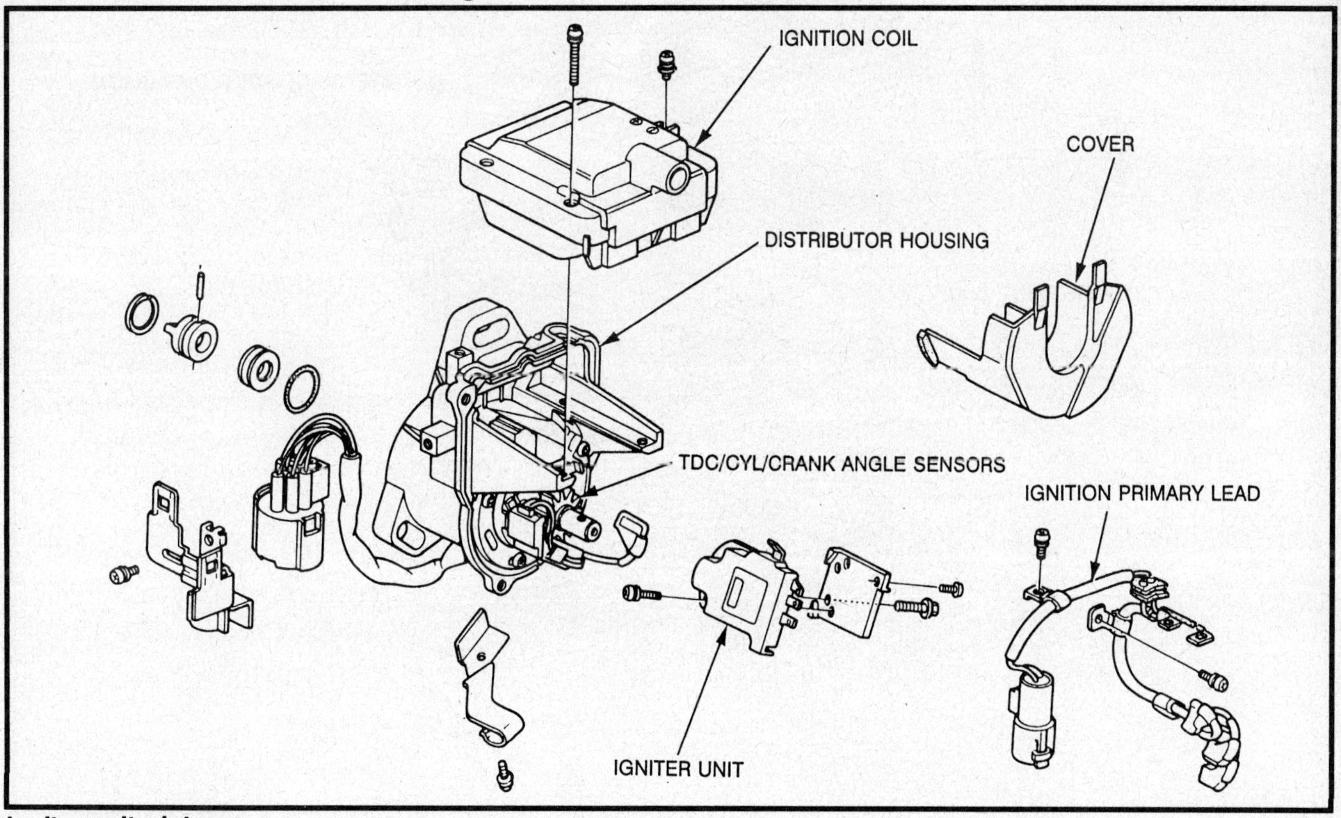

Igniter unit—Integra

4. Remove the protective cover from the ignition coil and remove the coil.
5. Remove the retaining screws and remove the igniter from the distributor housing.

To install:

6. Install the igniter and tighten the retaining screws.
7. Install the coil and coil cover.
8. Reassemble the leak cover, rotor, seal and distributor cap.
9. Install the distributor.

EXCEPT INTEGRA

The igniter is mounted externally in the engine compartment. For 1990 Legend, it is on the left shock tower; on 1991–92 Legends, it is on the right front shock tower, just behind the underhood fuse/relay box. For 5 cylinder Vigor engines, the igniter is on the right fender apron, near the firewall. On NSX, the igniter is mounted on top of the engine covers at the left side of the engine. To remove these igniters, turn the ignition switch **OFF**, disconnect the connector and remove the retaining bolts. Reinstall in reverse order.

BYPASS VALVE SYSTEM DIAPHRAGM AND CONTROL SOLENOIDS

This system is found on Legend, Vigor and NSX vehicles.

At a predetermined rpm, the ECU energizes the bypass control solenoid. On NSX, the valve is named the chamber volume control solenoid. The solenoid opens electrically, allowing vacuum to flow to the control valve diaphragm(s). As the vacuum acts on the diaphragm, the operating lever is moved and the blades of the valve open. This allows the intake air to follow a shorter path, delivering more power at high rpm. When the rpm drops below the trigger level, the signal from the ECU is discontinued, the solenoid shuts off the vacuum and the system reverts to its previous state.

The diaphragm and linkage may be inspected and cleaned with carburetor cleaner if it is not closing fully. The arm should rest on the limit screw with the vacuum hose connected, but the limit screw must not be adjusted; it is set at the factory and should not be changed. Additionally, the diaphragm is part of

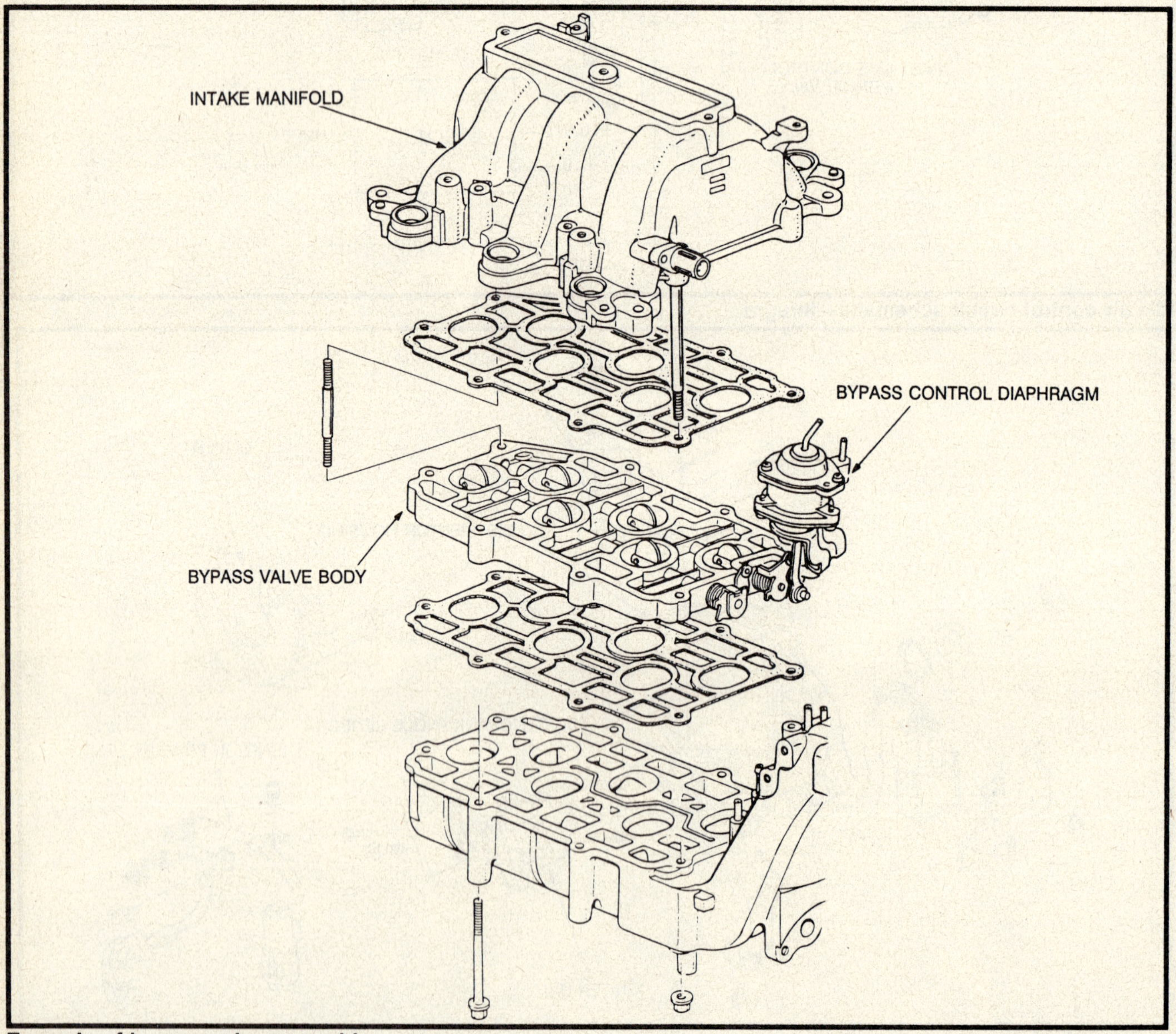

Example of bypass valve assembly

the bypass valve assembly. If the diaphragm must be replaced, the bypass valve must be removed from the intake manifold.

The control solenoids are located in the NSX and 1990 Legend emission control box, at the rear of the Vigor intake manifold or behind and below the right headlight on 1991–92 Legend. The solenoid may be removed in straightforward fashion by disconnecting the wiring connector and then the vacuum hoses. Make absolutely certain that the vacuum hoses are reinstalled to the proper ports on the valve(s).

THROTTLE BODY

Removal and Installation

1. Disconnect the air intake tube from the throttle body.
2. Drain the cooling system, at least to a level below the throttle body.
3. Disconnect the coolant hoses running to the throttle body.
4. Label and disconnect the vacuum hoses from the throttle body.
5. Disconnect the wiring connector(s) from the throttle body.
6. Label and disconnect the throttle control, cruise control and/or accelerator cable. Do not kink the cables.
7. Remove the nuts holding the throttle body to the intake chamber or plenum. Loosen each nut 1–2 turns at a time to release the pressure evenly.
8. Remove the throttle body assembly. Remove and discard the gasket. Some models use a solid insulator below the throttle body instead of a regular gasket. This base may be removed after its retaining bolts are removed. If the insulator is in good condition and free of any cracks or distortion, it may be reused.

To install:

9. Install a new gasket or insulator on the plenum. If an insulator is in use, tighten its retaining bolts to 16 ft. lbs. (22 Nm). Install the throttle body onto the studs.

NOTE: The gasket or insulator is correctly installed in only one position; make sure the grooves and passages in the gasket align with the matching openings on the air chamber.

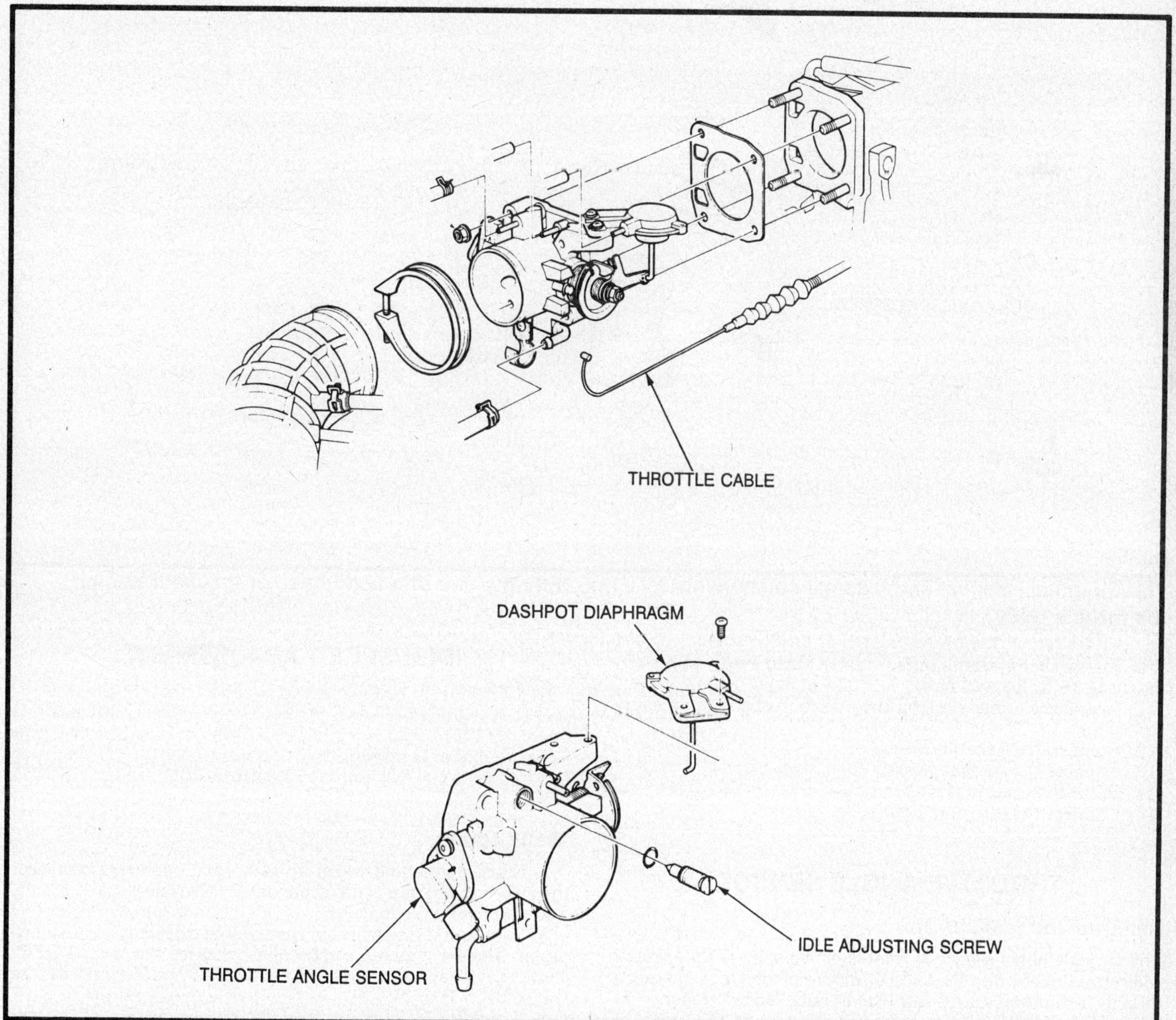

Throttle body removal and external components – Integra

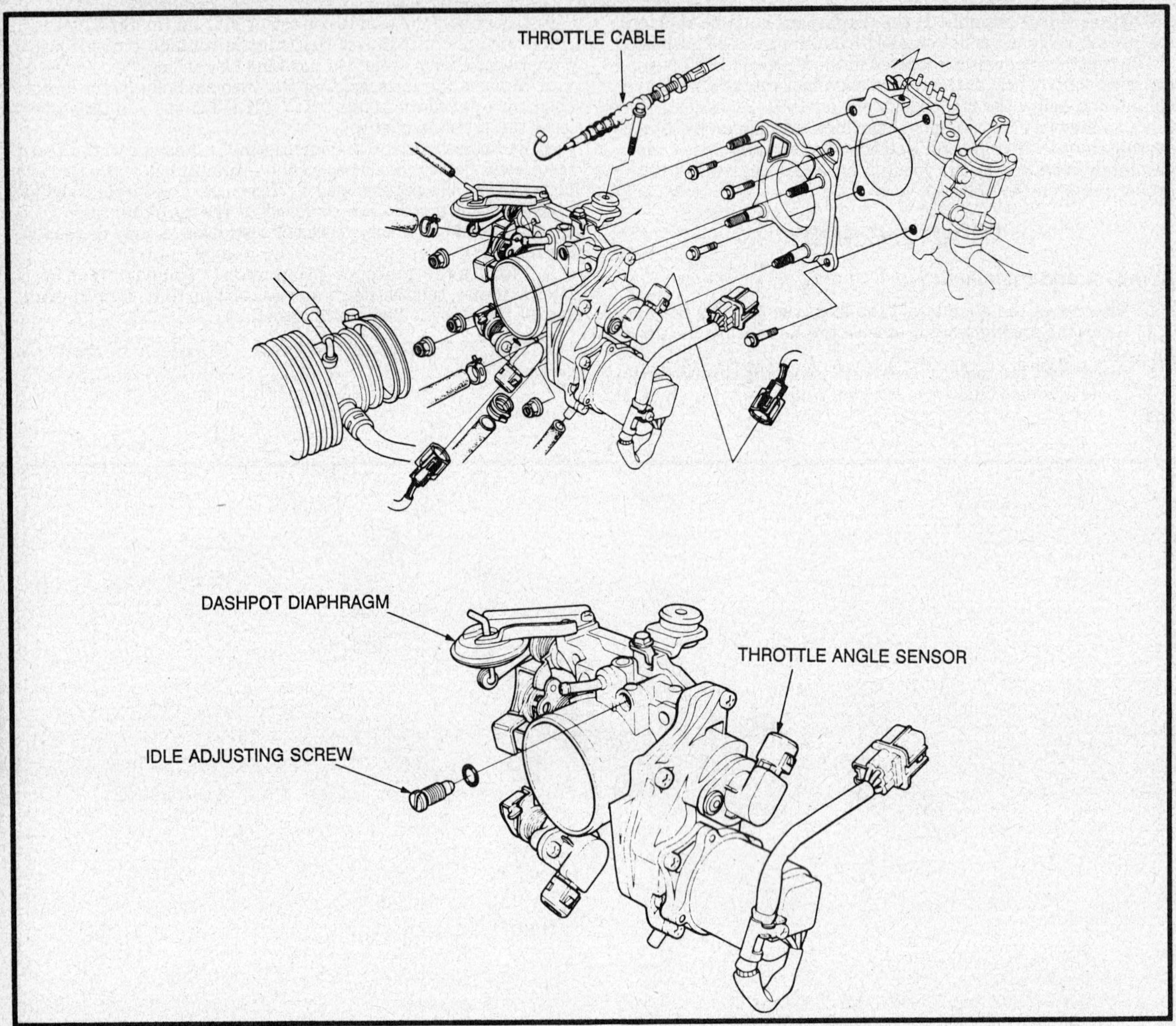

Throttle body removal and external components for NSX. Note the use of a solid base or insulator behind the throttle body

10. Install the retaining nuts. Tighten them evenly and alternately to 16 ft. lbs. (22 Nm).
11. Install the cruise control, throttle control and/or accelerator cables.
12. Connect the vacuum hoses.
13. Connect the coolant hoses.
14. Refill the cooling system.
15. Connect the air intake ducts.

THROTTLE ANGLE SENSOR

Removal and Installation

The throttle angle sensor is located on the side of the throttle body and connected to the throttle axle or pivot. Once the throttle body is removed, the sensor may be removed by removing the retaining bolts. When reinstalling, make certain the sensor is correctly mated with the throttle body shaft. Install the bolts.

IDLE SPEED ADJUSTMENT

NOTE: Idle is not adjusted at the throttle plate stop screw. That screw is set at the factory and is not adjustable. All idle adjustments are made by turning the idle adjusting screw on the throttle body. All idle tests must be met and the idle must remain stable under all test conditions.

1990 Legend

1. Start engine and warm up to normal operating temperature; the cooling fan must come on at least once.
2. Connect a tachometer.
3. Set the steering wheel in the straight-ahead or no load condition. Make certain all electrical accessories are turned **OFF**; check the idle speed. Correct idle speed is 680 ± 50 rpm with the transaxle in **P** or **N**.
4. Check the ECU under the front passenger seat. If the yellow LED is not lit, the idle speed is correct and no adjustment is

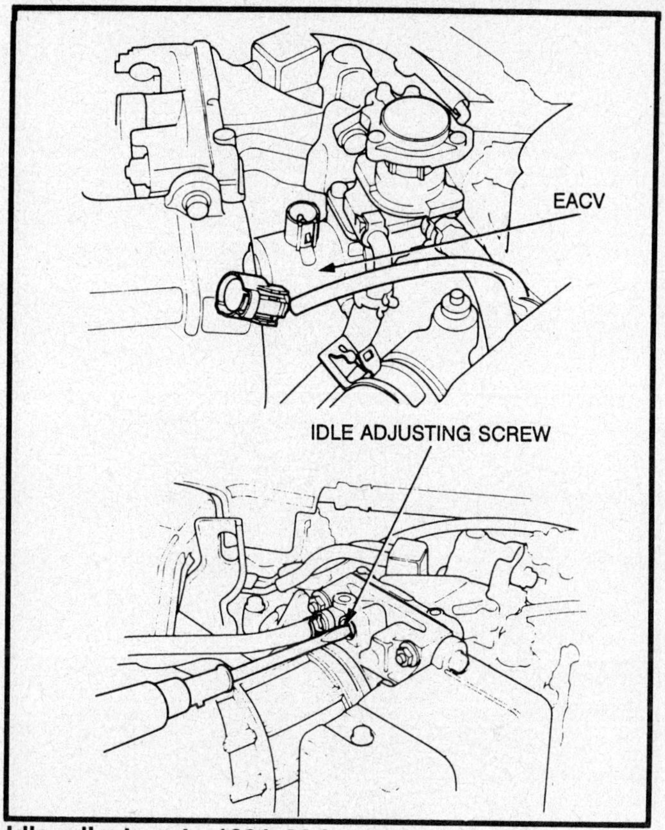

Idle adjustment—1991–92 Legend

needed. If the yellow LED is blinking, adjust the idle speed adjusting screw ¼ turn clockwise. If the yellow LED is lit constantly—not flashing—turn the idle adjusting screw ¼ turn counter-clockwise.

5. Once adjusted, the yellow LED should turn off within 30 seconds. If it does not, turn the idle adjuster another ¼ turn, continuing at 30 second intervals until the LED goes out.

6. Check the idle under each of the following conditions individually. Idle speed should remain stable at 680 ± 50 rpm under each condition: High beam headlights and rear window defogger **ON**; steering wheel turned sharply; air conditioner compressor engaged; automatic transaxle in a drive gear.

1991–92 Legend

1. Start engine and warm up to normal operating temperature; the cooling fan must come on at least once. Switch the ignition **OFF**.
2. Connect a tachometer.
3. Disconnect the 2-pin connector on the EACV.
4. Start the engine with the accelerator pedal slightly depressed. Use the pedal to stabilize the engine speed at 1000 rpm. Slowly release the throttle until the engine idles. Check the idle with air conditioning, cooling fan and all electrical loads **OFF**.
5. The idle should be 450 ± 50 rpm for manual vehicles and 480 ± 50 rpm for automatic vehicles.
6. If the rpm is not within specifications, adjust as needed by turning the idle adjusting screw on the throttle body.
7. Switch the ignition **OFF**. Reconnect the connector to the EACV, then clear the ECU by removing the No. 15 fuse in the underdash fuse box for at least 10 seconds.
8. Restart the engine, allowing it to idle in **N** or **P** for 1 minute. With all electrical loads including the air conditioning and cooling fan **OFF**, idle speed should be 650 ± 50 rpm for manual vehicles and 600 ± 50 for automatic vehicles.

Integra

1. Apply the parking brake. Start engine and warm up to normal operating temperature; the cooling fan must come on at least once.
2. Connect a tachometer.
3. Disconnect the 2-pin connector on the EACV.
4. Check the idle with the transaxle in **N** or **P** and the air conditioning, cooling fan and all electrical loads **OFF**.
5. The idle should be 650 ± 50 rpm.
6. If the rpm is not within specifications, adjust as needed by turning the idle adjusting screw on the throttle body. If the idle speed is excessively high, check the throttle control system.
7. Switch the ignition **OFF**. Reconnect the connector to the EACV, then clear the ECU by removing the BACK UP (Canadian: HAZARD, BACK-UP) fuse in the underhood relay box for at least 10 seconds.
8. Restart the engine, allowing it to idle. With all electrical loads including the air conditioning and cooling fan **OFF**, idle speed should be 750 ± 50 rpm with the transaxle in **P** or **N**.
9. Allow the engine to idle for 1 minute with the headlight high beams and rear defogger **ON**. The idle speed should be 750 ± 50 rpm.
10. Switch the air conditioner **ON** and the blower fan to **HI**. After 1 minute, the idle should be 750 ± 50 rpm.
11. If the idle speed is out of specification, refer to the system troubleshooting charts.

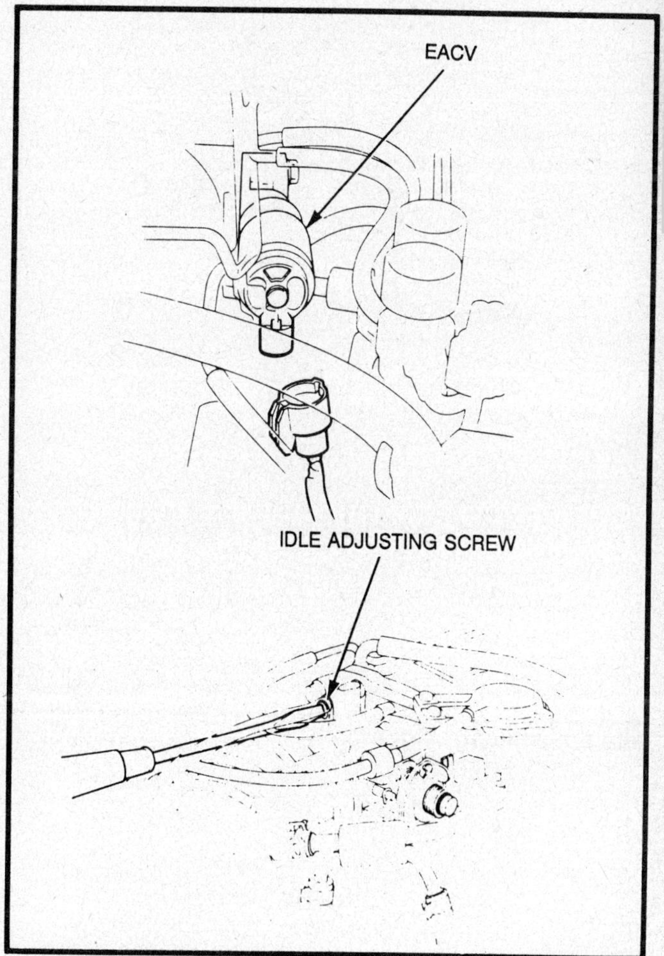

Idle adjustment—Integra

Vigor

1. Start engine and warm up to normal operating temperature; the cooling fan must come on at least once. Switch the ignition **OFF**.
2. Connect a tachometer.
3. Disconnect the 2-pin connector on the EACV.
4. Start the engine with the accelerator pedal slightly depressed. Use the pedal to stabilize the engine speed at 1000 rpm. Slowly release the throttle until the engine idles. Check the idle with air conditioning, cooling fan and all electrical loads **OFF**.
5. The idle should be 550 ± 50 rpm with the vehicle in **P** or **N**.
6. If the rpm is not within specifications, adjust as needed by turning the idle adjusting screw on the throttle body.
7. Switch the ignition **OFF**. Reconnect the connector to the EACV, then clear the ECU by removing the BACK UP fuse in the underhood fuse box for at least 10 seconds.
8. Restart the engine, allowing it to idle in **N** or **P** for 1 minute. With all electrical loads including the air conditioning and cooling fan **OFF**, idle speed should be 700 ± 50 rpm.
9. Allow the engine to idle for 1 minute with the headlight high beams and rear defogger **ON**. The idle speed should be 770 ± 50 rpm.
10. Turn the headlights and defogger OFF. Switch the air conditioner **ON** and the blower fan to **HI**. After 1 minute, the idle should be 770 ± 50 rpm.

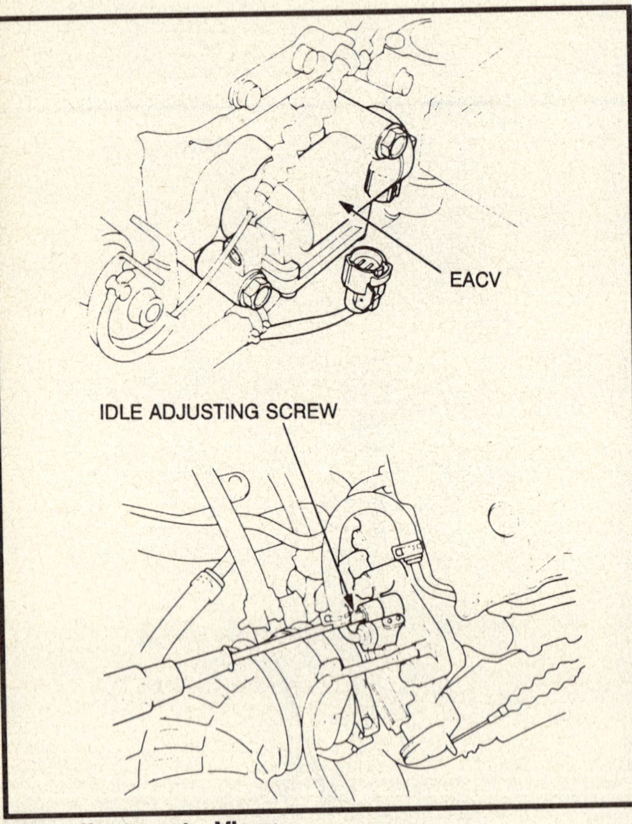

Idle adjustment—Vigor

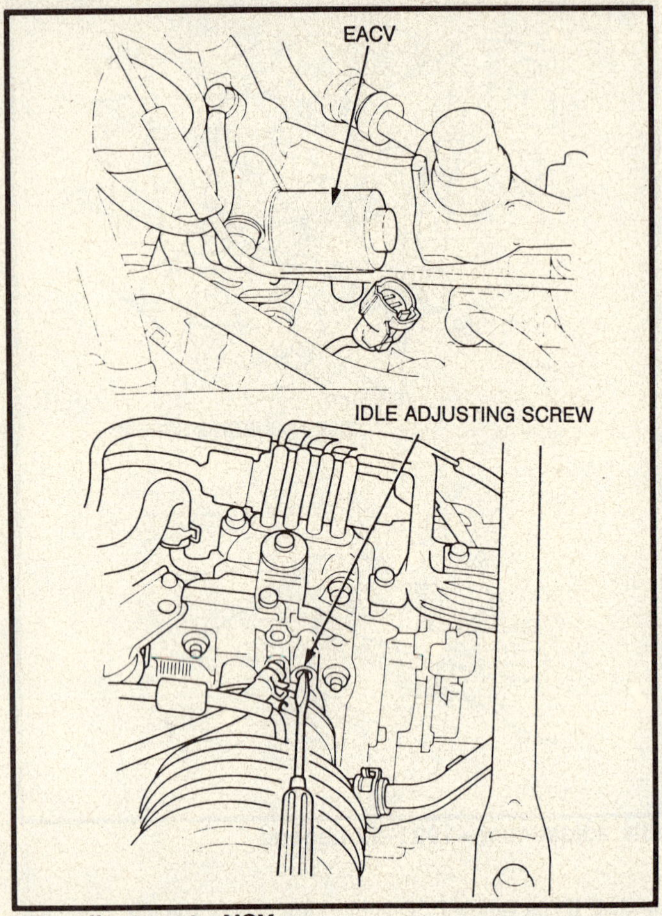

Idle adjustment—NSX

NSX

1. Start engine and warm up to normal operating temperature; the cooling fan must come on at least once. Switch the ignition **OFF**.
2. Connect a tachometer.
3. Disconnect the 2-pin connector on the EACV.
4. Start the engine with the accelerator pedal slightly depressed. Use the pedal to stabilize the engine speed at 1000 rpm. Slowly release the throttle until the engine idles. Check the idle with air conditioning, cooling fan and all electrical loads **OFF**.
5. The idle should be 650 ± 50 rpm for manual vehicles and 600 ± 50 rpm for automatic vehicles.
6. If the rpm is not within specifications, adjust as needed by turning the idle adjusting screw on the throttle body.
7. Switch the ignition **OFF**. Reconnect the connector to the EACV, then clear the ECU by removing the CLOCK fuse in the main fuse box for at least 10 seconds.
8. Restart the engine, allowing it to idle in **N** or **P** for 1 minute. With all electrical loads including the air conditioning and cooling fan **OFF**, idle speed should be 800 ± 50 rpm for manual vehicles and 750 ± 50 for automatic vehicles.

Chrysler Imports Fuel Injection

SECTION 6

CHRYSLER IMPORT FUEL SYSTEMS

General Information

The Multi-Point Injection (MPI) system controls the fuel flow, idle speed, and ignition timing. The basic function of the MPI system is to control the air/fuel ratio according to operational conditions through the Electronic Control Unit (ECU), based on data from various sensors. The MPI System is roughly divided into 4 areas: fuel system, intake system, control system and control and data sensor.

INTAKE SYSTEM

The flow rate of the air drawn through the air cleaner is measured by the Air Flow Sensor (AFS) using the Karman vortex principal. Briefly, this principal states that a triangular column placed within the airstream will generate evenly spaced vortices (whirlpools) in the air flow. The number of vortices generated is proportional to the volume of airflow in a given time. By measuring the vortices (per unit of time), volume of air can be determined accurately. The airflow signal is used by the ECU to control fuel flow.

The air is distributed to each cylinder from the air intake plenum. In the manifold, the air is mixed with the fuel from the injectors and is drawn into the cylinder. The air flow rate is controlled according to the degree of the throttle valve and the stepper motor openings.

The amount of air drawn during idling is adjusted by the Idle Speed Control (ISC) servo controlled by the ECU. Further, the amount of air drawn during warm-up and deceleration is also controlled by the ISC servo.

CONTROL SYSTEM

The control system is composed of a sensor section, which monitors engine conditions, and an Electronic Control Unit (ECU), which calculates the injection timing and rate according to the signals from the sensors. The sensors convert such conditions as the amount of intake air, amount of oxygen in the exhaust gas, coolant temperature, intake air temperature, engine revolution speed, and driving speed into electric signals, which are sent to the ECU.

Certain values and settings are built into the ECU at manufacture; these map or default settings are used under certain extreme operating conditions and/or when the ECU detects loss of signal from certain sensors.

Analyzing these signals, the ECU determines the amount of fuel to inject according to driving conditions and drives the injectors. The fuel injection is by sequential injection type, in which four injectors are sequentially driven in repeating order.

During idling, the ISC servo is driven according to the load to assure stable idling.

Injector Drive Time

Injector activation occurs 1 time per cycle for each cylinder, and the injector activation time (duration of fuel injection) is called the basic activation time.

The ECU functions calculate this basic activation time according to Air Flow Sensor (AFS) signals and crankshaft angle sensor signals as well as the oxygen sensor, ignition signal and other inputs.

When starting (cranking) the engine, the map value determined by the engine coolant temperature is used as the basic drive time. During deceleration, the basic driving time is set as zero. The injector driving time is obtained by making the corrections on the above-mentioned basic driving time.

Other modifications of basic activation time are determined by:

- Oxygen sensor feedback correction (Closed loop control correction) – In normal (warm) operation (including idling), air/fuel ratio is corrected to optimum by using oxygen sensor signals. This allows the 3-way catalytic converter to give the best degree of purification.
- Air/fuel ratio map correction (open loop control correction) – Correction to the optimum air/fuel ratio is made by the map values set by engine speed and amount of intake air.
- Engine coolant temperature correction – To maintain operability of cold engine, correction is so made that the lower the engine coolant, the greater the amount of fuel injected.
- Intake air temperature correction – Change in air/fuel ratio due to difference in intake air density.
- Barometric pressure correction – Change in air/fuel ratio due to difference in intake air density caused by change in barometric pressure is corrected.
- Acceleration/deceleration correction – In accordance with change in amount of intake air, fuel amount is corrected, improving operability.
- Dead time correction – The injectors opened by the driving signal from ECU have an operation lag which changes according to battery voltage. This means that the actual injector opening time becomes less than injector driving signal, failing to provide expected air/fuel ratio. Therefore, battery correction time corresponding to the battery voltage is added by ECU. Battery condition and state of charge can affect the performance of the injection system.

Idle Speed Control (ISC) System

The ISC system provides the following 4 modes of control:

START CONTROL

The throttle valve opening is controlled at the optimum position for start according to the engine coolant temperature and the altitude (atmospheric pressure).

FAST IDLE CONTROL

When the idle switch is on, the engine speed is controlled to a target rpm according to engine coolant temperature. When the idle switch is off, the ISC servo is actuated to move the throttle valve to a target opening position (throttle valve opening position) according to the engine coolant temperature.

IDLE LOAD CONTROL

When the air conditioner switch is turned **ON** or when the transmission is shifted from **N** to **D**, the system causes the idle speed to increase to the target rpm according to the load (rpm feedback control).

Air Conditioner Relay Control

When the air conditioner switch is turned on while the engine is at idle, the ISC servo increases the engine speed. However, there is some delay before the engine speed actually increases. To maintain the engine free from the air conditioner load during that delay period, the ECU keeps the power transistor off for a fixed time (about 0.5 seconds), disconnecting the air conditioner power relay circuit. As a result, even if the air conditioner switch is **ON**, the compressor is not driven instantly, preventing engine speed drop due to compressor load.

DATA SENSORS

Airflow Sensor

This measures the intake airflow; utilizing the Karman vortex phenomenon, it counts the number of vortices and converts this data to electric pulses which are then sent to the ECU. The ECU uses these signals to determine the basic injection time.

Barometric Pressure Sensor

This sensor detects the barometric pressure, and converts this data to voltage which is sent to the ECU. The ECU uses these signals for correct the fuel injection amount and the ignition timing.

Coolant Temperarture Sensor

This sensor detects the temperature of the engine's coolant, and converts this data to voltage which is sent to the ECU. The ECU uses these signals for correction of the fuel injection amount and the ignition spark advance.

Crankshaft Angle Sensor

This sensor detects the position of the crankshaft (and therefore, the piston). The slits in the disc are read by the light-emitting diode and the photo diode, and this data is sent from the unit assembly to the ECU as electrical pulses. This sensor will be found incorporated into the same unit as the TDC sensor.

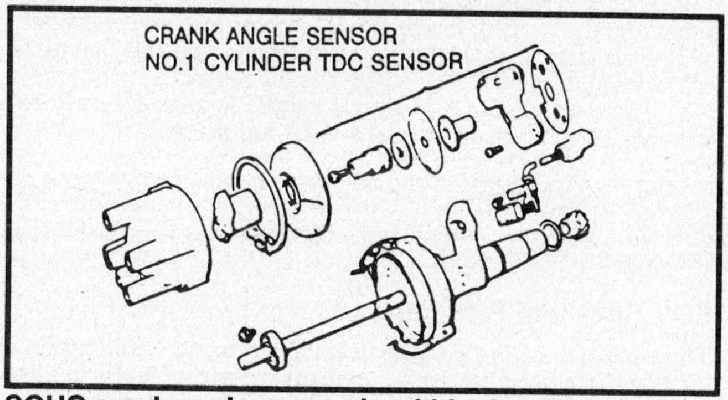

SOHC crank angle sensor is within the distributor

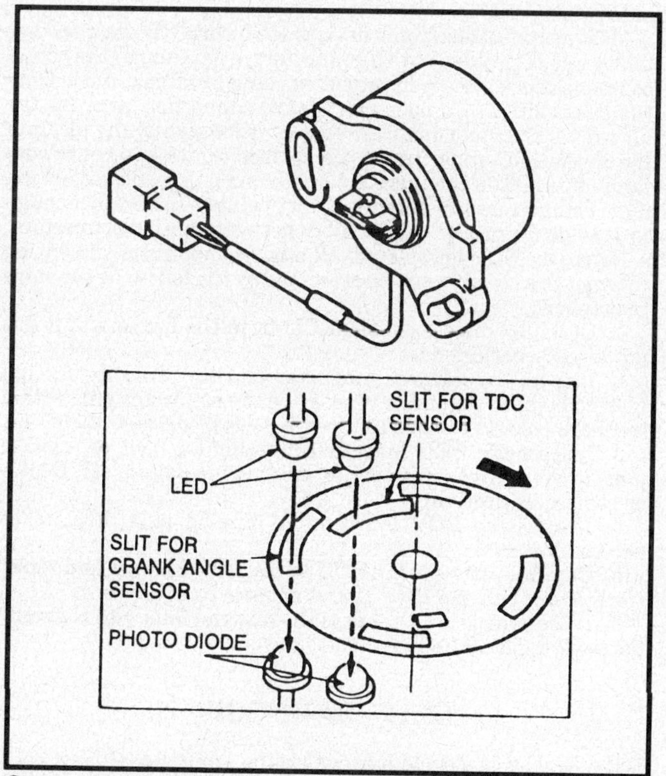

Crank angle sensor—1.6L DOHC engine

Idle Switch

This switch detects the fact that the throttle valve is in the idle position and sends this data as signals to the ECU. The ECU senses that the engine is idling and then controls the fuel injection amount, the ignition timing, the ISC servo, etc.

Inhibitor Switch

The inhibitor switch is used in vehicles with automatic transaxle. Whether the transaxle is in **N** or **D** position is detected, and the result is used as the idle speed control signal.

Intake Air Temperature (IAT) Sensor

This sensor detects the temperature of the intake air, and converts this data to voltage which is sent to the ECU. The ECU, based on these signals, performs air/fuel ratio feedback control.

Oxygen Sensor

This sensor measures the amount of oxygen in the exhaust gas, and sends these signals to the ECU. The ECU, based on these signals, performs air/fuel ratio feedback control.

Power Steering Oil Pressure Switch

Switch detects a large load on power steering system and increases idle to compensate.

Throttle Position Sensor (TPS)

The TPS detects the degree of throttle opening; this data is converted to voltage and is sent to the ECU. The ECU uses these signals to calculate the changes in the degree of throttle opening and to sense the variable speed condition.

Top Dead Center (TDC) Sensor

Located within the distributor, this sensor identifies the reference signal applicable to each cylinder. The disc slit is read by the light-emitting diode and the photo diode. This data is sent from the unit assembly to the ECU as electrical pulses. The ECU uses these signals and the signals from the crankshaft angle sensor to determine the injection timing.

Vehicle Speed Sensor

This sensor converts the rotation of the transaxle speedometer gear to electrical pulses and sends them to the ECU.

FUEL CONTROL SYSTEM

The fuel is pressure-fed through the in-tank filter by the fuel pump and is distributed to the respective injectors via the main pipe and fuel filter. The fuel pressure applied to the injector is a continuous 47.6 psi, higher than the pressure in the intake manifold. The pressure is controlled by the pressure regulator. The excess fuel, after being pressure-adjusted, is returned to the fuel tank through the return pipe.

When an electric current flows in the injector, the injector valve is fully opened to inject the fuel. Since the fuel pressure is a continuous constant, the amount of the fuel injected from the injector into the manifold is increased or decreased in proportion to the time the electric current flows. Based on ECU signals, the injectors inject fuel to the cylinder manifold ports in firing order.

Ordinarily, the intake manifold negative pressure (vacuum) acts upon the fuel-pressure regulator, the fuel pressure is maintained at a constant fixed level relative to the pressure within the intake manifold, and the amount of fuel injection pressure is regulated to a fixed amount; excess fuel is returned to the fuel tank.

If the engine coolant temperature and the intake air temperature exceed certain pre-determined values during restarting, the ECU activates the fuel pressure control valve, allowing atmospheric pressure to act on the fuel pressure regulator. This raises the fuel pressure within the system and eliminates the formation of vapor in the fuel lines. The system will return to normal fuel delivery pressures within 2 minutes of a high temperature start-up.

Fuel Pump

The fuel pump used in this system is a positive displacement, roller vane immersible pump with a permanent magnet electric motor. The fuel is drawn in through a filter sock and pushed through the electric motor to the outlet. The pump contains 2 check valves. One valve is used to relieve internal fuel pump pressure and regulate maximum pump output. The other check valve, located near the pump outlet, restricts fuel movement in either direction when the pump is not operational.

Pressure Regulator

The pressure regulator maintains the fuel volume at a fixed amount relative to fuel injection time. To improve precision, the pressure regulator controls the fuel pressure applied to the injectors at a constant pressure relative to the surge tank negative pressure.

Surge tank negative pressure is applied to the diaphragm chamber of the pressure regulator. When the fuel pressure within the pressure regulator reaches a pre-determined pressure, extra fuel is by-passed to the return hose and is returned to the fuel tank.

Injectors

Injectors are electromagnetic type injection nozzles which function to inject fuel according to injection signals calculated and provided by the ECU.

When there is continuity at the solenoid coil, the plunger and the needle valve are pulled up and fuel is sprayed from the nozzle. Because the stroke of the needle valve is always constant, the amount of fuel spray (injection) is determined by the time (duration) of the injector opening.

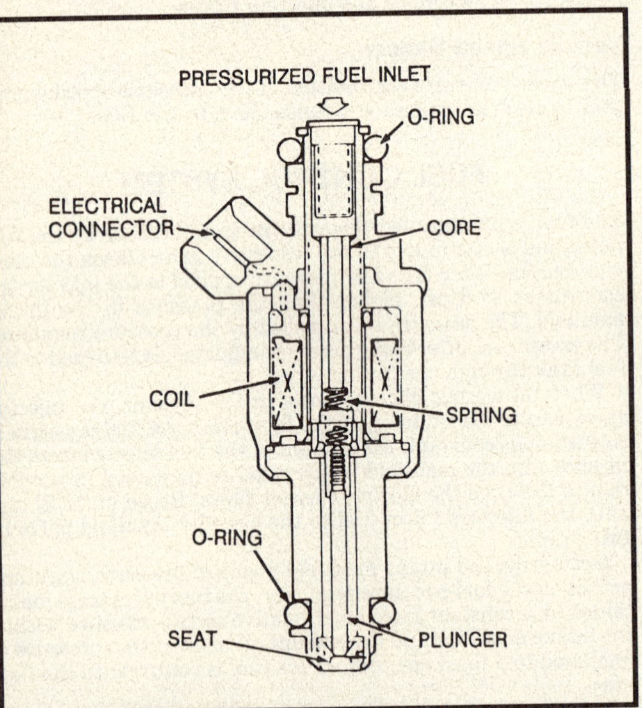

Cross sectional view of the fuel injector

Diagnosis and Testing

SERVICE PRECAUTIONS

- Before disconnecting any fuel lines, relieve the fuel system pressure by removing the fuel tank filler cap.
- Always place a shop towel under the fuel lines to catch any fuel spillage.
- Prior to disconnecting the fuel injector electrical connectors, mark the connector with the number of the corresponding injector.
- Do not disconnect the fuel tube connectors on a hot engine.
- Never allow a high voltage spark to ground on the electronic control module.

PRELIMINARY CHECKS

1. Visually inspect the engine compartment to ensure that all vacuum lines and spark plug wires are properly routed and securely connected.
2. Examine all wiring harnesses and connectors for insulation damage, burned, overheated, loose or broken conditions.
3. Be certain that the battery is fully charged and that all accessories should be **OFF** during the diagnosis.
4. The tools needed during this diagnosis procedure will be a suitable 12 volt test lamp and a volt/ohm meter.

NOTE: The connectors are protected by rubber covers. If the DRB-II diagnostic tool is to be used, please follow the manufacturer's instructions when connecting the tool to the system.

ECU Checkout Procedure

The condition of the ECU is determined by the elimination of other potentially defective components or wiring faults in a specific circuit. Even though the ECU may be determined to be faulty by process of elimination, it should not be replaced until it has been retested after some basic screening procedures have been performed.

Most electrical/electronic devices are incorrectly diagnosed as faulty due to problems with connectors, even though the actual component or device is in proper working condition. Aside from the possibilities of a poor mechanical connection between the harness and component, poor contact is frequently the result of the insulation properties of the surface oxides and other contaminants. Often the mere act of disconnecting and reconnecting a component will abrade contact points sufficiently enough to provide for proper conductivity between mating connectors.

Before an ECU is replaced, all other components and wiring relevant to a circuit must be examined by the following the steps that follow:

1. Carefully disconnect the ECU from the harness, if it has not already been done.
2. Inspect the harness connector and the ECU for foreign material on the connector pins. Wipe the connector pins with a standard contact cleaner to remove oxides, contamination, etc.
3. Inspect the ECU harness connector for bent or missing pins, broken wires, improper or missing insulation, etc. Repair or replace as necessary.
4. Inspect the ECU for bent or missing pins. Replace as necessary.
5. Carefully connect the ECU to the ECU harness and retest the system with the DRB-II and adapter or equivalent.
6. If the vehicle does not pass the test and fails with the same message, replace the ECU.

SELF-DIAGNOSIS

Self-diagnosis is a system in which the input signal from each sensor is monitored by the ECU, should any abnormality be detected, the abnormal item (or its identifying code) is memorized by the ECU. Many sensors are monitored including normal condition; these codes may be confirmed using an analog (dial) voltmeter.

The ECU checks each sensor for a signal periodically. If the signal is not present or is present but not within prescribed limits, the code for that sensor is stored in memory. If 2 or more faults are detected, the codes are stored in numerical order from lowest to highest. The order of codes in the memory does not indicate the order of failure.

READING CODES

The diagnostic connector is located either inside or behind the glove box or by the fusebox, depending on the vehicle. The codes are read using an analog (dial) voltmeter; connect the ground probe to the ground terminal and the positive probe to the test terminal within the connector.

NOTE: When battery voltage is low, fault codes may not store or transmit properly. Always check battery condition before attempting self-diagnosis procedures.

Turn the ignition switch **ON** and read the codes. The needle on the analog voltmeter will deflect periodically, indicating the codes. Count the sweeps of the needle and record each code. Once the ECU has broadcast the all the stored codes, the entire cycle will be repeated. When the codes have been read and recorded, turn the ignition **OFF**.

If the needle shows a constant 12 volts, the ECU has diagnosed itself as being the failed component. This indication may be due to a failed ECU or, more likely, to loose or faulty wiring running to or from the unit.

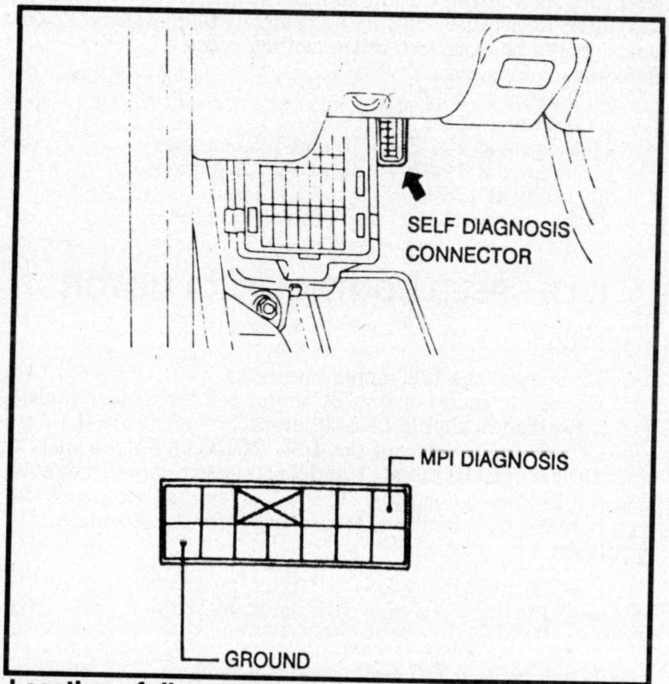

Location of diagnostic connector—Colt Summit and Colt 200

CLEARING CODES

The memory is kept by direct power, supplied from the battery. The memory is not erased by turning **OFF** the ignition switch. However, it is erased if the back-up power supply is turned off by disconnection of battery cable or ECU connector. The memory is not erased if the power supply is restored within 15 seconds.

Once repairs have been made the negative battery cable must be disconnected for 15 seconds or more. This erases the code from memory. After the engine has been operated, check again for presence of codes; if repair was successful, no code will be present.

TROUBLESHOOTING

When checking and correcting engine problems, it is important to start with inspection of the basic systems. Such conditions as failure to start, unstable idling or poor acceleration are often caused by items other than the fuel injection or engine management systems. Therefore, first check the following basic systems:

1. Power supply—check battery, fusible link and fuses.
2. Check all related body grounds.
3. Fuel supply—check fuel line, fuel filter and fuel pump.
4. Ignition system—check spark plugs, high tension cable, distributor and ignition coil.
5. Check ignition timing and idle speed.
6. Emission control system—check PCV system, EGR system, vacuum leaks

Troubles with the MPI system are often caused by poor contact of harness connectors. It is important to check harness connector contact.

System Inspection

1. Before removing or installing a part, read diagnosis code, turn the ignition **OFF** and disconnect the battery (−) terminal cable.
2. Before disconnecting the cable from battery terminal, turn the ignition switch to **OFF**. Removal or connection of battery cable during engine operation or while the ignition switch is **ON** could cause erroneous operation of the ECU or damage to semiconductors.
3. The control harness between the ECU and oxygen sensor contains shielded wires with shield grounded to the body in order to prevent ignition noise and radio interference. When the shielded wire is faulty the control harness must be replaced.
4. Pay attention to the following:

- Avoid rough operation of switches.
- Do not subject test equipment to shock and other external forces, heat, etc.
- Keep the test equipment or meter away from water and oil.

FUEL PRESSURE

Check

NOTE: Threaded adaptors and extra O-rings are required for this test. All line connections must be tight and not leak under pressure.

1. Release the pressure remaining in fuel pipe line so that fuel will not flow out.
   a. Disconnect the fuel pump connector at the fuel tank.
   b. Start the engine and after it stops by itself, turn the ignition switch to **OFF**.
   c. Disconnect the negative battery terminal.
   d. Connect the fuel pump connector.
2. Disconnect the fuel high pressure hose at the delivery pipe side.

NOTE: Cover the hose connection with rags to prevent splash of fuel that could be caused by residual pressure in the fuel line.

3. Using the adapters, install the fuel-pressure gauge to the delivery pipe. Tighten the bolt to 18–25 ft. lbs. (25–35 Nm).

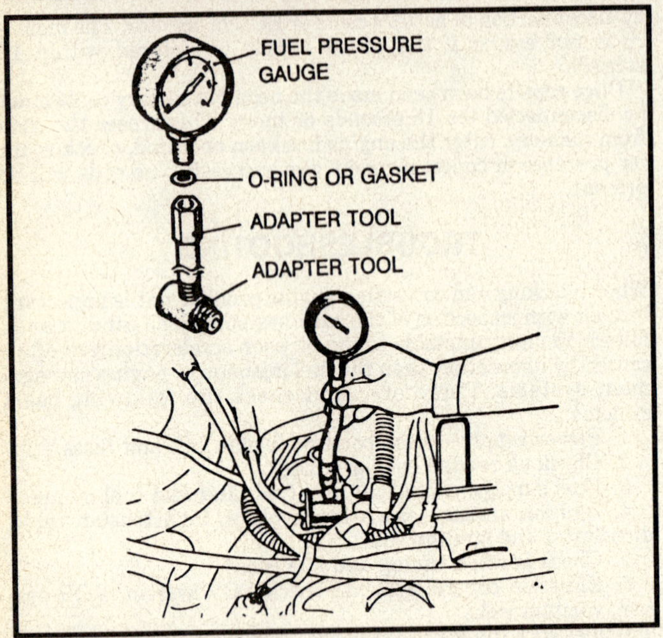

Fuel pressure testing

4. Connect the negative battery cable.
5. Apply battery voltage to the terminal for fuel pump. Check to be sure that there is no fuel leakage from the pressure gauge or the connections.
6. Disconnect the vacuum hose from the pressure regulator, and plug the hose end. Measure the fuel pressure during idling. This is the unregulated pressure.
7. Measure the fuel pressure when the vacuum hose is connected to the pressure regulator. This is regulated pressure.
8. If the results of the measurements made in Steps 6 and 7 above are not within the standard value, determine the probable cause make the necessary repairs.
9. Stop the engine and check fuel pressure gauge; pressure should not drop.
10. Release residual pressure from the fuel line.
11. Disconnect the fuel-pressure gauge from the delivery pipe.

NOTE: Cover the connection with rags to prevent splash of fuel caused by residual pressure in the fuel line.

12. Using a new gasket, connect the fuel high-pressure hose.
13. Check for fuel leaks; apply voltage to fuel pump test terminal and check connections for any leak.

NOTE: The following electrical tests require the use of the various special wiring adapters. These adapters (sometimes called break-out connectors) isolate the various conductors within a harness for ease of testing. If testing without the adapters, carefully probe from the rear of the connector. Never insert a test probe into the front of a female harness connector.

INTAKE AIR TEMPERATURE SENSOR

Inspection

1. Disconnect the sensor connector.
2. Use the ohmmeter to test the resistance value across the terminals. Record the resistance. Reference values are:
   6000 ohms at 32°F
   2700 ohms at 68°F
   400 ohms at 176°F.
3. Measure resistance while heating the sensor using a hair

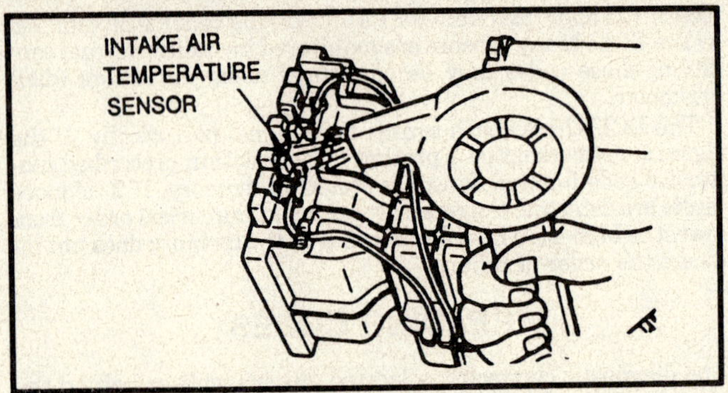

Heat the IATS during testing

drier. Resistance should become smaller as temperature becomes higher.
4. If the value deviates from the standard values or the resistance remains unchanged, replace the air flow sensor assembly.

ENGINE COOLANT TEMPERATURE SENSOR

Inspection

1. Remove engine coolant temperature sensor from the intake manifold.
2. With temperature sensing portion of engine coolant temperature sensor immersed in hot water, check resistance. Use a thermometer to test water temperature. The sensor should be held with its housing 0.12 in. (3 mm) away from the surface of the water. Resistance will vary according to temperature. Resistance should be close to 0 at the boiling point.
Reference values are:
   5900 ohms at 32°F
   2500 ohms at 68°F
   2700 ohms at 104°F – 1.5L and 1.6L engines
   1100 ohms at 104°F – 1.8L and 2.0L, engines
   300 ohms at 176°F
3. If the resistance deviates, replace the sensor.

IDLE SPEED CONTROL (ISC) MOTOR

Inspection

1. Disconnect the ISC motor connector.
2. Check continuity of the ISC motor coil across terminals 1 and 2. Resistance should be 5–28 ohms.
3. For all engines except the 1.6L DOHC engine, connect 6 volts DC between terminals 1 and 2 connector; check to be sure that the ISC servo operates. On the 1.6L engine, reconnect the wire harness, turn ignition **ON** and listen for the stepper motor operation.

NOTE: Apply only a 6 volts DC or lower voltage. Application of higher voltage could cause locking of the servo gears.

4. If not, replace ISC servo as an assembly.

IDLE SWITCH

Inspection

1. Disconnect the ISC motor connector.
2. Check continuity between the designated terminal and body ground. With accelerator depressed, no continuity should exist; with accelerator released – continuity should be present.
3. If out of specification, replace the ISC servo assembly.

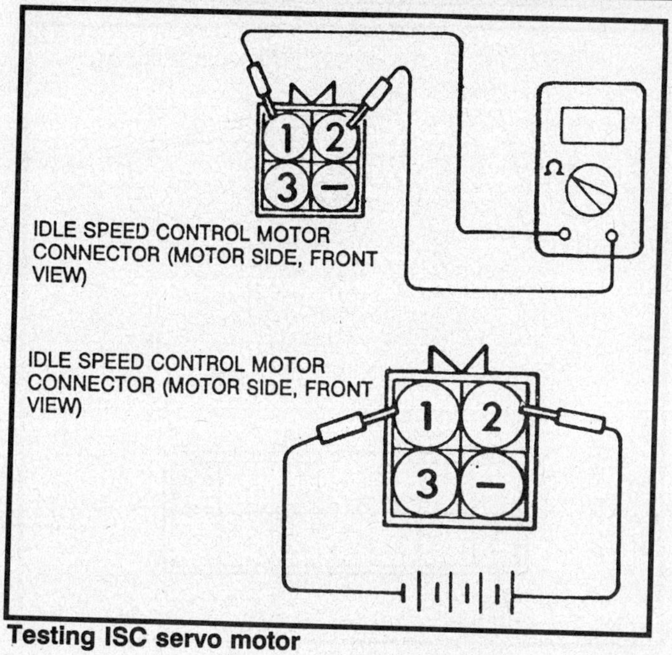

Testing ISC servo motor

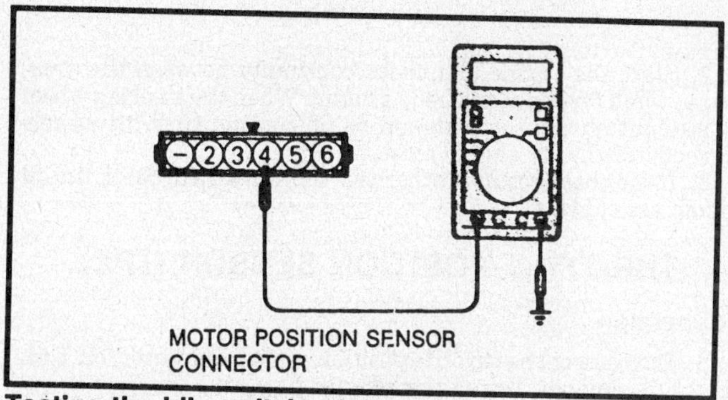

Testing the idle switch—1.5L and 1.8L engines

INJECTORS

SOUND TEST

Using a stethoscope or similar tool, check for operating sound as engine idles. Each injector gives a distinct ticking as it opens and closes. Check that the sound occurs at shorter intervals as the engine speed increases. Listen carefully; other injectors may produce similar sound even if the injector being checked is not working.

RESISTANCE TEST

With the engine off, remove the injector connector. Measure resistance between the injector terminals. Standard resistance should be between 13–16 ohms at 68°F (20°C). Install the injector connector.

MOTOR POSITION SENSOR (MPS)

1. Disconnect the motor position sensor connector.
2. For 1.5L engine, connect the adapter harness to the disconnected MPS (sensor side) connector.

NOTE: Never connect the adapter to the connector on the harness side.

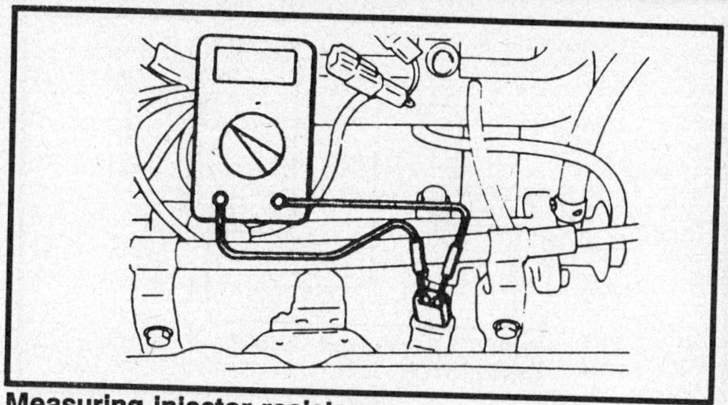

Measuring injector resistance

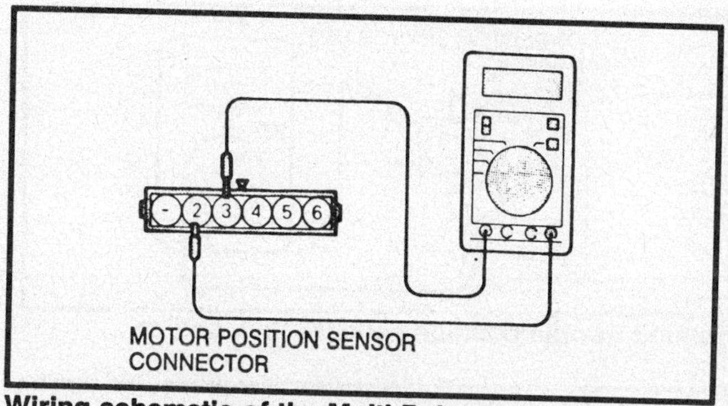

Wiring schematic of the Multi-Point Fuel Injection system—1.6L engine continued

3. Measure the resistance between terminals. Standard resistance: 4000–6000 ohms.
   1.5L and 1.8L engines—terminals 2 and 3
   2.0L engine—terminals 1 and 2
4. If the standard value is not achieved, replace the ISC servo assembly.

OXYGEN SENSOR

1.5L, 1.8L AND 2.0L ENGINES

NOTE: Before checking, warm the engine until engine coolant temperature reaches 185–205°F (85–95°C). Use an accurate digital voltmeter for testing.

1. Disconnect the oxygen sensor connector and connect a voltmeter to the oxygen sensor connector, terminal 1 and ground.

CAUTION

Avoid contact or personal injury may result. Surfaces and components will be hot.

2. With engine running, observe voltage at idle, then race engine momentarily. Voltage should increase (when raced) to approximately 1 volt, then drop as engine returns to idle.
3. If voltage is not correct, or no change occurs during rpm increases, replace oxygen sensor.

1.6L ENGINE

NOTE: Before checking, warm the engine until engine coolant temperature reaches 185–205°F (85–95°C). Use an accurate digital ohmmeter for testing.

1. Disconnect the oxygen sensor connector and connect the adapter harness to the connector on the oxygen sensor side.
2. Measure the resistance between terminals 3 and 4 of the

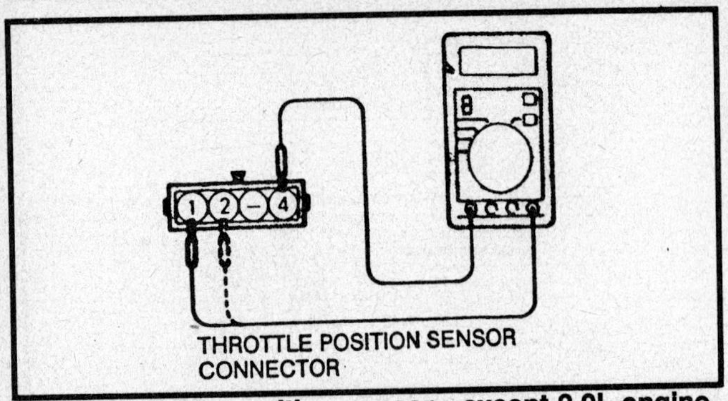

Testing throttle position sensor – except 2.0L engine

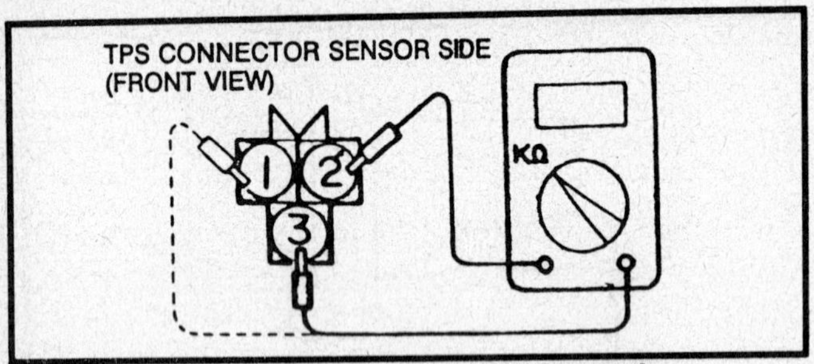

Testing throttle position sensor – 2.0L engine

Testing the oxygen sensor – 1.5, 1.8L and 2.0L engines

connector. Reference values are: 68°F (room temperature), approximately 15 ohms and 400°F, approximately 30 ohms or higher.

3. If resistance is not approximately correct, replace oxygen sensor.

POWER STEERING OIL PRESSURE SWITCH

Inspection

1. Disconnect the power steering oil pressure switch connector.

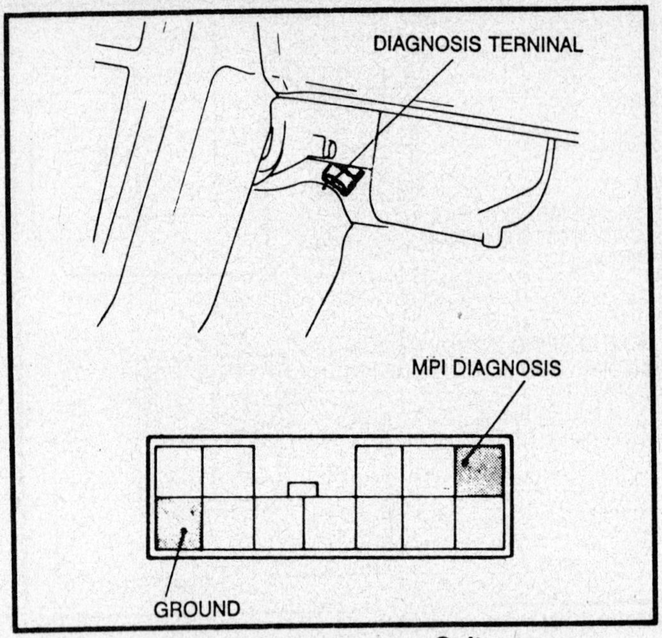

Location of diagnostic connector – Colt wagon

2. Start the engine and check continuity between the pressure switch terminal and body ground. When the steering wheel is straight ahead, there should be no continuity. With wheels turned, continuity should exist.

3. If the check result is otherwise than specified, check the oil pump assembly.

THROTTLE POSITION SENSOR (TPS)

Inspection

1. Disconnect the throttle position sensor connector; on 1.5L and 1.6L engines, install the adapter harness.

2. Measure resistance between the ground and power terminals. Standard value for all engines: 3500–6500 ohms.

3. Connect a pointer type ohmmeter between the ground terminal and the output terminal shown on the chart.

4. Operate the throttle valve slowly from the idle position to the full open position and check that the resistance changes smoothly and in proportion to the throttle valve angle.

5. If the resistance is out of specification, or fails to change smoothly, replace the TPS.

FUEL PRESSURE RELIEF

Procedure

1. Disconnect the fuel pump connector at the fuel tank.

2. Start the engine and after it stops by itself, turn the ignition switch to **OFF**.

3. Disconnect the negative battery terminal.

4. Connect the fuel pump connector.

1.5L ENGINE—DIAGNOSTIC CONNECTORS—COLT

ECU CONNECTORS AND FUNCTIONS

MULTI ECU CONNECTORS
(View from Harness End)

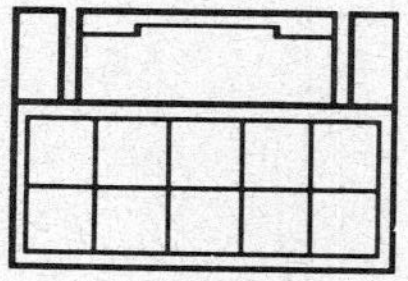

10-WAY CONNECTOR

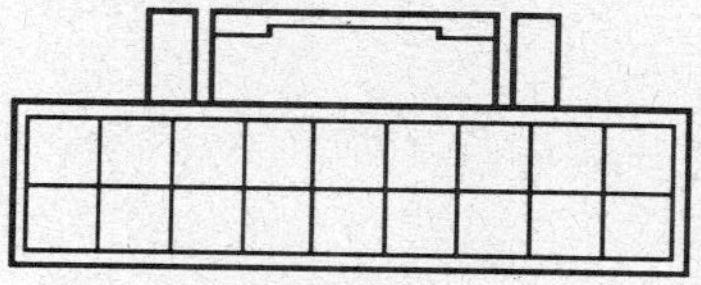

18-WAY CONNECTOR

10-WAY CONNECTOR

Cavity	Color	Function
101	BK	ECU Ground
102	RD	12 volts key "ON" from control relay to power up 5 volt regulator
103	WT/DG	Battery feed for memory (12 volts to standby 5 volt regulator)
104	BK/YL (A/T) BK (M/T)	Park/Neutral Sw. (Inhibitor) (A/T) 12 volts in gear, 0 volts in Park or Neutral (M/T) Always grounded, 0 volts
105		
106	BK	ECU ground
107	RD	12 volts key "ON" from control relay
108	BK/YL (M/T) BK/RD (A/T)	Ign. Sw. start input 12 volts during crank Ign. Sw. start input 12 volts while cranking in Park or Neutral
109	BK/WT	Control relay sense, 14 volts while running (pump feed)
110		

18-WAY CONNECTOR

Cavity	Color	Function
51	YL/DB	Injector 1 ground
52	YL/BK	Injector 2 ground
53	TN	California only, EGR solenoid ground
54	YL	Ignition coil power transistor
55		
56	WT/RD	Control relay ground (50 rpm or above, ground L1 coil)
57		
58	DB/WT	Voltage to close ISC motor
59	DG/BK	Voltage to open ISC motor
60	LG	Injector 3 ground
61	LG/WT	Injector 4 ground
62	BR/YL	Purge solenoid ground
63		
64	YL/DB	Check engine light ground (12 volts light OFF, 0 volts light ON)
65	BK/WT	A/C clutch relay ground
66		
67		
68		

ECU CONNECTORS AND FUNCTIONS

MULTI ECU CONNECTORS
(View from Harness End)

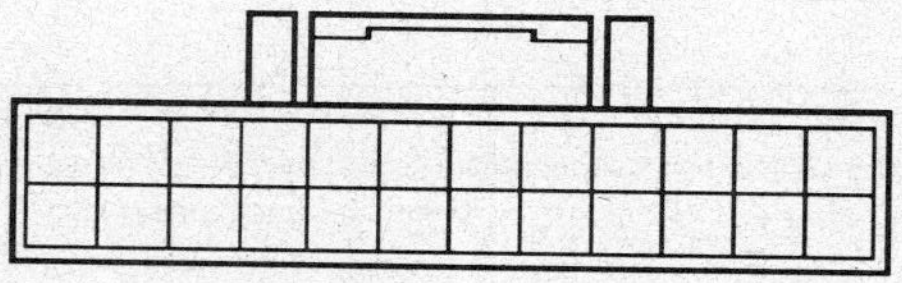

24-WAY CONNECTOR

24-WAY CONNECTOR

Cavity	Color	Function
1	YL	ECU fault code output, 12 volt pulses
2	WT	SCI change terminal. (Grounded by DRBII adapter to change pin 1 to I/O port.) Pin 1 then becomes SCI transmit & receive on one wire.
3		
4	WT	Oxygen sensor voltage 0-1 volt with digital voltmeter
5	DB/WT	12 volts to power steering switch, 0 volts with high steering pressure
6	BK	12 volts to idle switch, 0 volts with closed throttle
7	DG/YL	A/C switch sense, 12 volts when A/C
8	DG/OR	Intake air temperature sensor input
9		
10	DB/LG	5 volts out to air flow sensor .95 volts key "ON", 2.89 volts idling
11		
12	YL/RD	5 volts out ignition timing adjust terminal ground to set timing
13	DG/RD	5 volt feed to the ISC motor position sensor, TPS & barometric pressure sensor
14	DG/BK	Sensor ground
15	DB	California only, EGR temp. sensor input
16	DG/YL	Barometric pressure sensor input, 4 volts key "ON" at sea level
17	BR	ISC motor position sensor input
18	YL/WT	5 volts out to speed sensor, 0 or 5 volts
19	DG/WT	TPS voltage input
20	YL/DG	Coolant sensor input
21	BR/YL	Crank angle sensor input
22	BK/DB	TDC sensor input
23	DG/RD	Backup 5 volt feed for ISC motor position sensor, TPS & barometric pressure sensor
24	DG/BK	Sensor ground

1.8L ENGINE—DIAGNOSTIC CONNECTORS—COLT WAGON

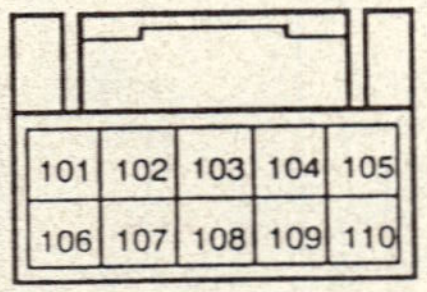

10-WAY CONNECTOR

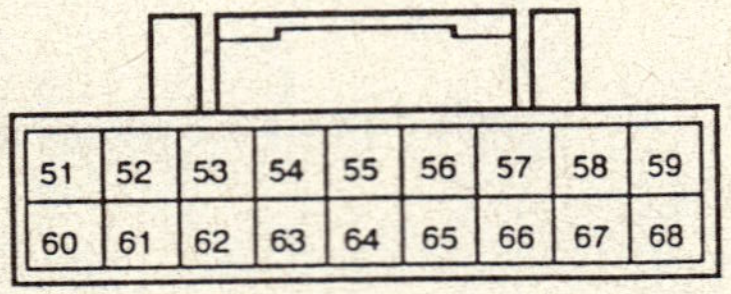

18-WAY CONNECTOR

10-WAY CONNECTOR

Cavity	Color	Function
101	BK	ECU Ground
102	RD	12 volts key "ON" from control relay to power up 5 volt regulator
103	LG/WT	Battery feed for memory (12 volts to standby 5 volt regulator)
104	BK/YL (A/T)	Park/Neutral Sw. (Inhibitor) (A/T) 12 volts in gear, 0 volts in Park or Neutral
	BK (M/T)	(M/T) Always grounded, 0 volts
105	OPEN	
106	BK	ECU ground
107	RD	12 volts battery feed
108	BK/YL (M/T)	Ign. Sw. start input 12 volts during crank
	BK/RD (A/T)	Ign. Sw. start input 12 volts while cranking in Park or Neutral
109	BK/WT	Control relay sense, 14 volts while running (pump feed)
110	OPEN	

18-WAY CONNECTOR

Cavity	Color	Function
51	YL/BK	Injector 1 ground
52	YL/RD	Injector 2 ground
53	TN	California only, EGR solenoid ground
54	WT	Ignition coil power transistor
55	OPEN	
56	WT/RD	Control relay ground (50 rpm or above, ground L1 coil)
57	OPEN	
58	DB/WT	Voltage to close ISC motor
59	DB/YL	Voltage to open ISC motor
60	YL/LG	Injector 3 ground
61	YL/WT	Injector 4 ground
62	DB/YL	Purge solenoid ground
63	OPEN	
64	LG/RD	Check engine light ground (12 volts light OFF, 0 volts light ON)
65	BK/WT	A/C clutch relay ground
66	OPEN	
67	OPEN	
68	OPEN	

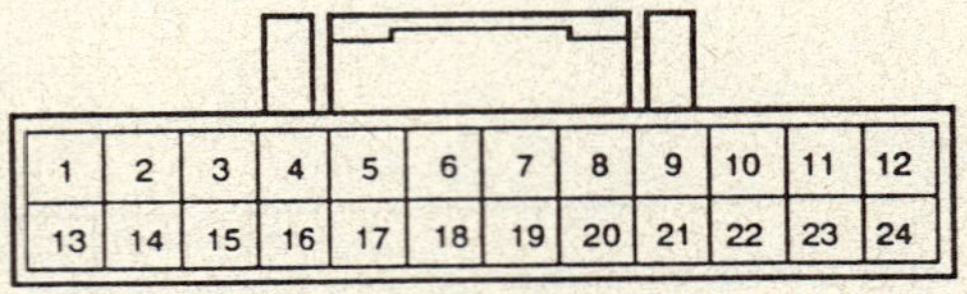

24-WAY CONNECTOR

24-WAY CONNECTOR

Cavity	Color	Function
1	LG/BK	ECU fault code output, 12 volt pulses
2	WT	SCI change terminal. (Grounded by DRBII adapter to change pin 1 to I/O port.) Pin 1 then becomes SCI transmit & receive on one wire.
3	OPEN	
4	WT	Oxygen sensor voltage 0-1 volt with digital voltmeter
5	DB/WT	12 volts to power steering switch, 0 volts with high steering pressure
6	DG	12 volts to idle switch, 0 volts with closed throttle
7	DG/RD	A/C switch sense, 12 volts when A/C
8	DG/RD	Intake air temperature sensor input
9	OPEN	
10	DG/WT	5 volts out to air flow sensor
11	OPEN	
12	YL/RD	5 volts out ignition timing adjust terminal ground to set timing
13	DG/YL	5 volt feed to the ISC motor position sensor, TPS & barometric pressure sensor
14	BK	Sensor ground
15	YL/DB	California only, EGR temp. sensor input
16	PK	Barometric pressure sensor input, 4 volts key "ON" at sea level
17	BR	ISC motor position sensor input
18	YL/WT	5 volts out to speed sensor, 0 or 5 volts
19	DG/WT	TPS voltage input
20	YL/DG	Coolant sensor input
21	BR/YL	Crank angle sensor input
22	BK/YL	TDC sensor input
23	DG/RD	Backup 5 volt feed for ISC motor position sensor, TPS & barometric pressure sensor
24	BK	Sensor ground

2.0L ENGINE—DIAGNOSTIC CONNECTORS—COLT VISTA

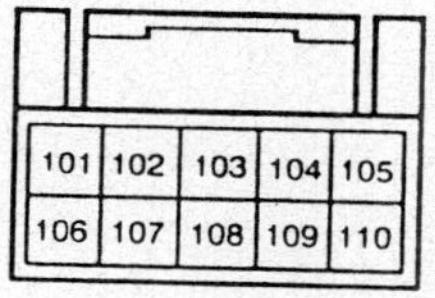

10-WAY CONNECTOR

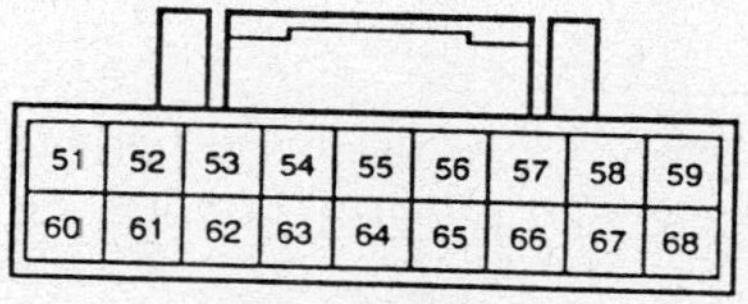

18-WAY CONNECTOR

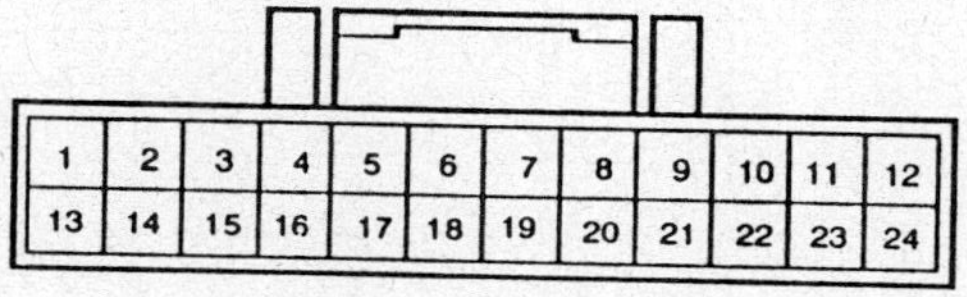

24-WAY CONNECTOR

10-WAY CONNECTOR

Cavity	Color	Function
101	BK	ECU Ground
102	RD	12 volts key "ON" from control relay to power up 5 volt regulator
103	RD/BK	Battery feed for memory (12 volts to standby 5 volt regulator)
104	BK/YL (A/T)	Park/Neutral Sw. (Inhibitor) (A/T) 12 volts in gear, 0 volts in Park or Neutral
	BK (M/T)	(M/T) Always grounded, 0 volts
105	OPEN	
106	BK	ECU ground
107	RD	12 volts key "ON" from control relay
108	BK/YL (M/T)	Ign. Sw. start input 12 volts during crank
	BR (A/T)	Ign. Sw. start input 12 volts while cranking in Park or Neutral
109	BK/WT	Control relay sense, 14 volts while running (pump feed)
110	OPEN	

18-WAY CONNECTOR

Cavity	Color	Function
51	YL/DB	Injector 1 ground
52	YL/BK	Injector 2 ground
53	BR/WT	California only, EGR solenoid ground
54	WT	Ignition coil power transistor
55	OPEN	
56	WT/RD	Control relay ground (50 rpm or above, ground L1 coil)
57	OPEN	
58	BK/YL	Voltage to close ISC motor
59	DG/BK	Voltage to open ISC motor
60	LG	Injector 3 ground
61	LG/WT	Injector 4 ground
62	BR/DB	Purge solenoid ground
63	OPEN	
64	DB/BK	Check engine light ground (12 volts light OFF, 0 volts light ON)
65	BK/RD	A/C clutch relay ground
66	OPEN	
67	OPEN	
68	OPEN	

24-WAY CONNECTOR

Cavity	Color	Function
1	YL	ECU fault code output, 12 volt pulses
2	DG/WT	SCI change terminal. (Grounded by DRBII adapter to change pin 1 to I/O port.) Pin 1 then becomes SCI transmit & receive on one wire.
3	OPEN	
4	WT	Oxygen sensor voltage 0-1 volt with digital voltmeter
5	DB/WT	12 volts to power steering switch, 0 volts with high steering pressure
6	YL/RD	12 volts to idle switch, 0 volts with closed throttle
7	DG/RD	A/C switch sense, 12 volts when A/C
8	RD/BK	Intake air temperature sensor input
9	OPEN	
10	WT/BK	5 volts out to air flow sensor
11	OPEN	
12	LB	5 volts out ignition timing adjust terminal ground to set timing
13	DG/YL	5 volt feed to the ISC motor position sensor, TPS & barometric pressure sensor
14	BK	Sensor ground
15	DB/YL	California only, EGR temp. sensor input
16	PK	Barometric pressure sensor input, 4 volts key "ON" at sea level
17	BR	ISC motor position sensor input
18	YL/WT	5 volts out to speed sensor, 0 or 5 volts
19	DG/WT	TPS voltage input
20	YL/DG	Coolant sensor input
21	BR/DG	Crank angle sensor input
22	BR/RD	TDC sensor input
23	DG/RD	Backup 5 volt feed for ISC motor position sensor, TPS & barometric pressure sensor
24	BK	Sensor ground

1.5L, 1.8L AND 2.0L ENGINES—NO START TEST

TEST NS-1 TESTING IGNITION SYSTEM - BEGIN NO START TESTS

Perform VISUAL INSPECTION Before Proceeding

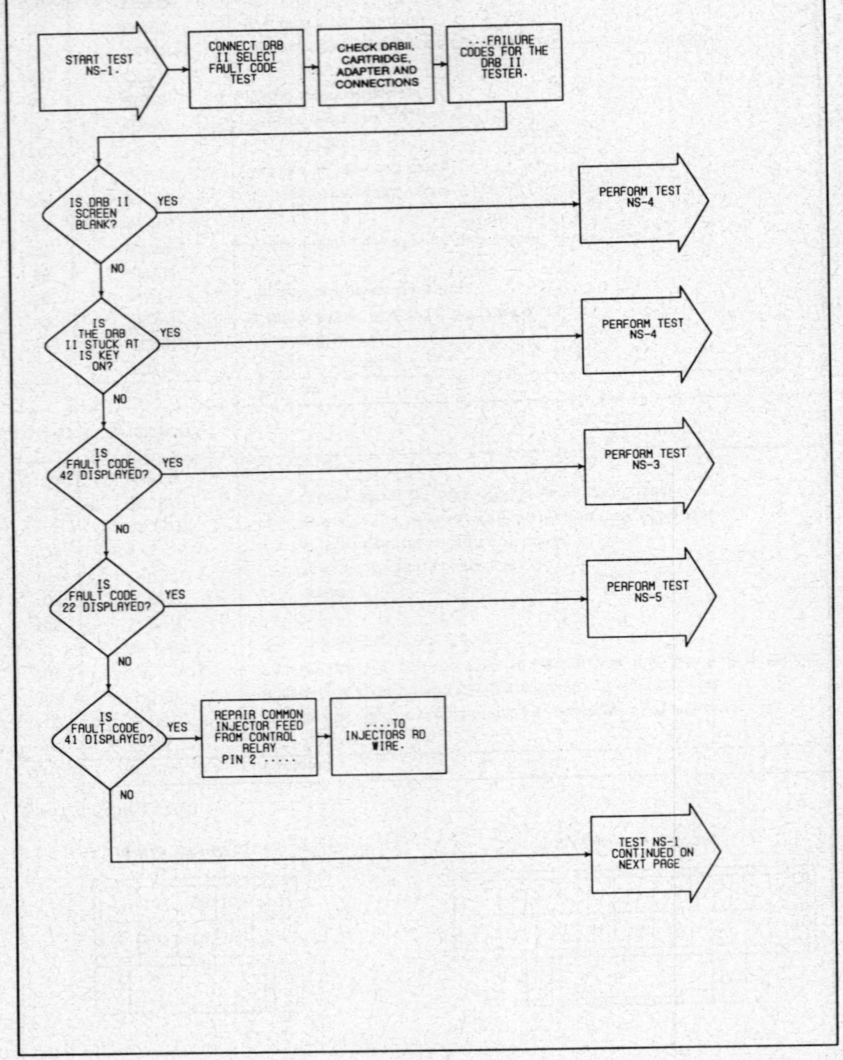

TEST NS-1 CONTINUED - CHECKING IGNITION SYSTEM

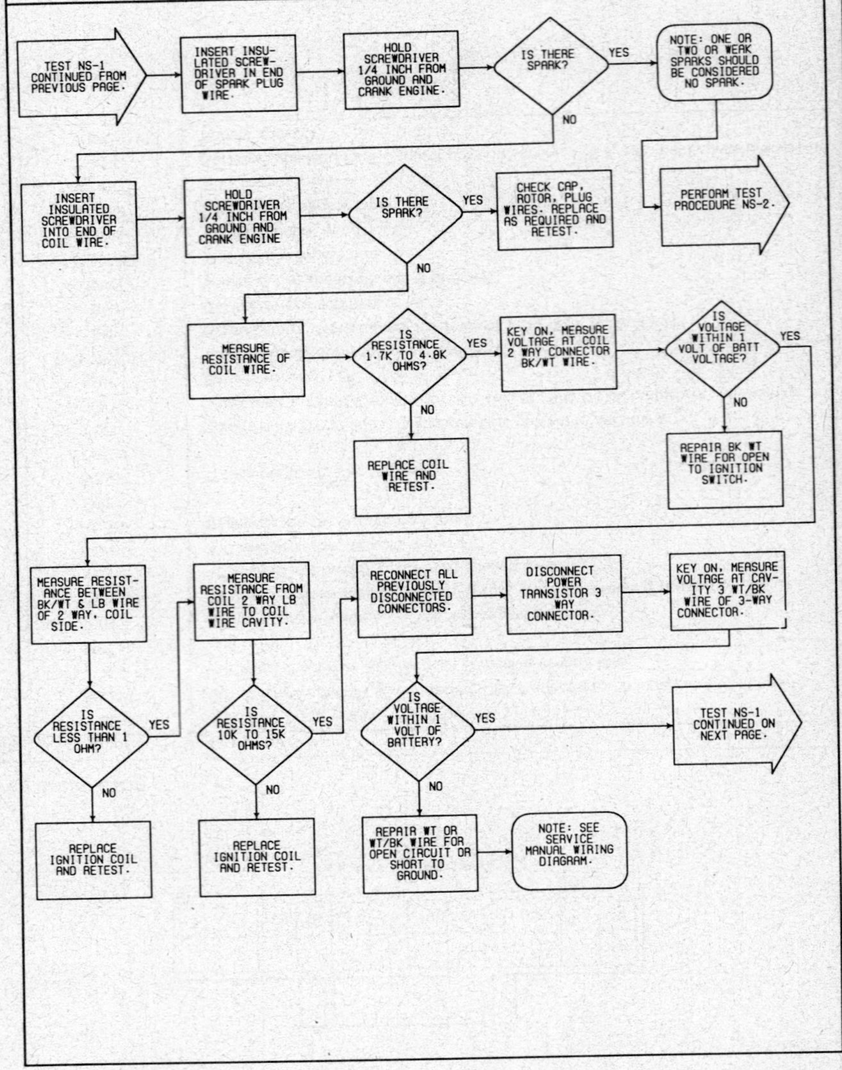

1.5L, 1.8L AND 2.0L ENGINES—NO START TEST

TEST NS-1 CONTINUED - CHECKING IGNITION SYSTEM

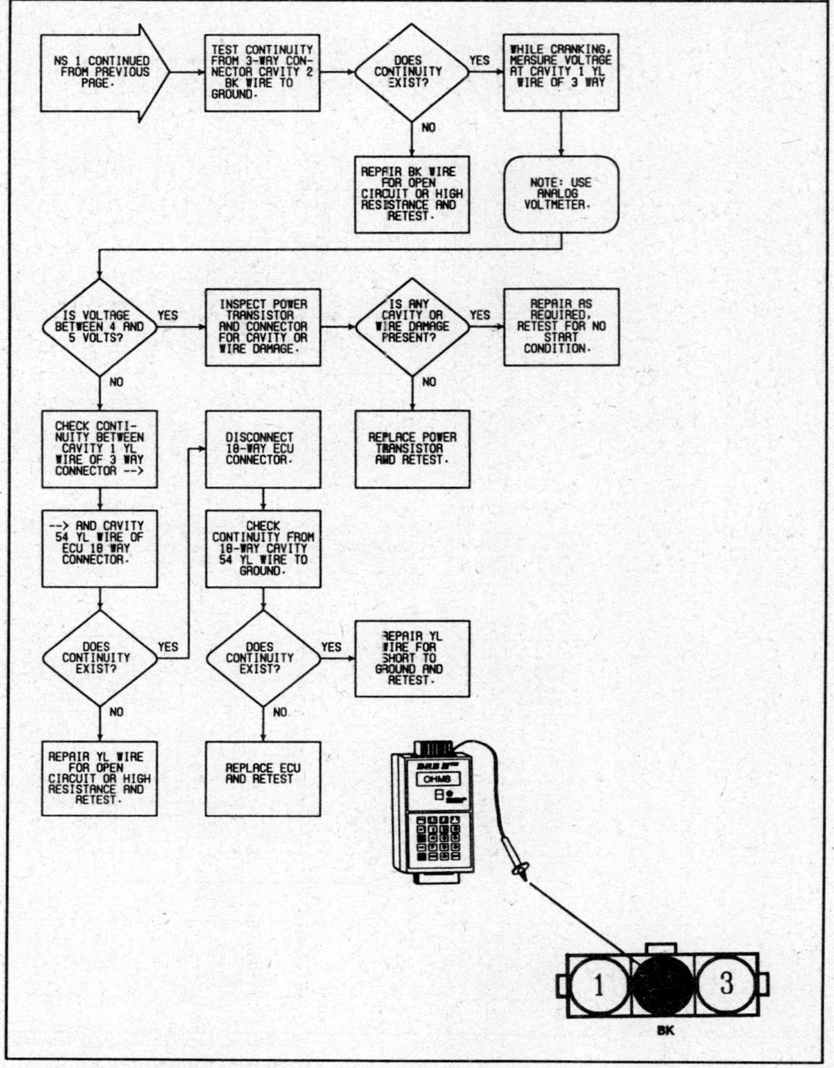

TEST NS-2 TESTING FOR FUEL

Perform TEST NS-1 Before Proceeding

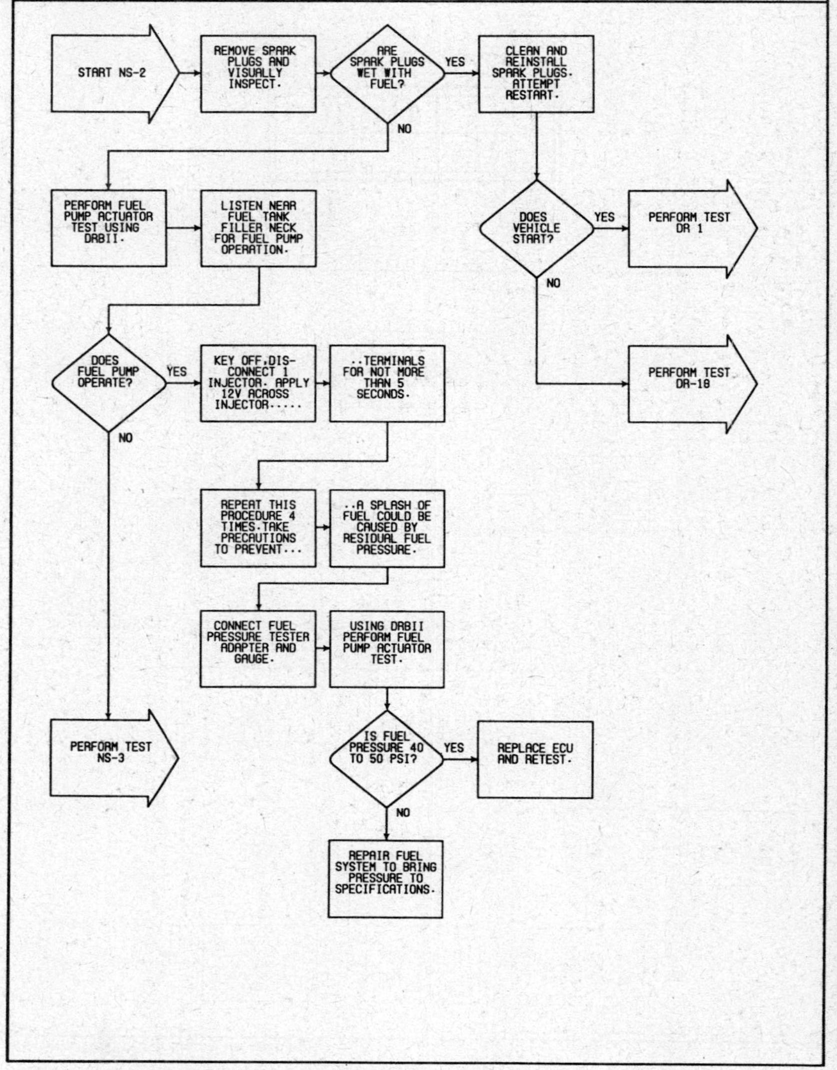

1.5L, 1.8L AND 2.0L ENGINES—NO START TEST

TEST NS-3 TESTING FUEL SYSTEM

Perform TEST NS-1 Before Proceeding

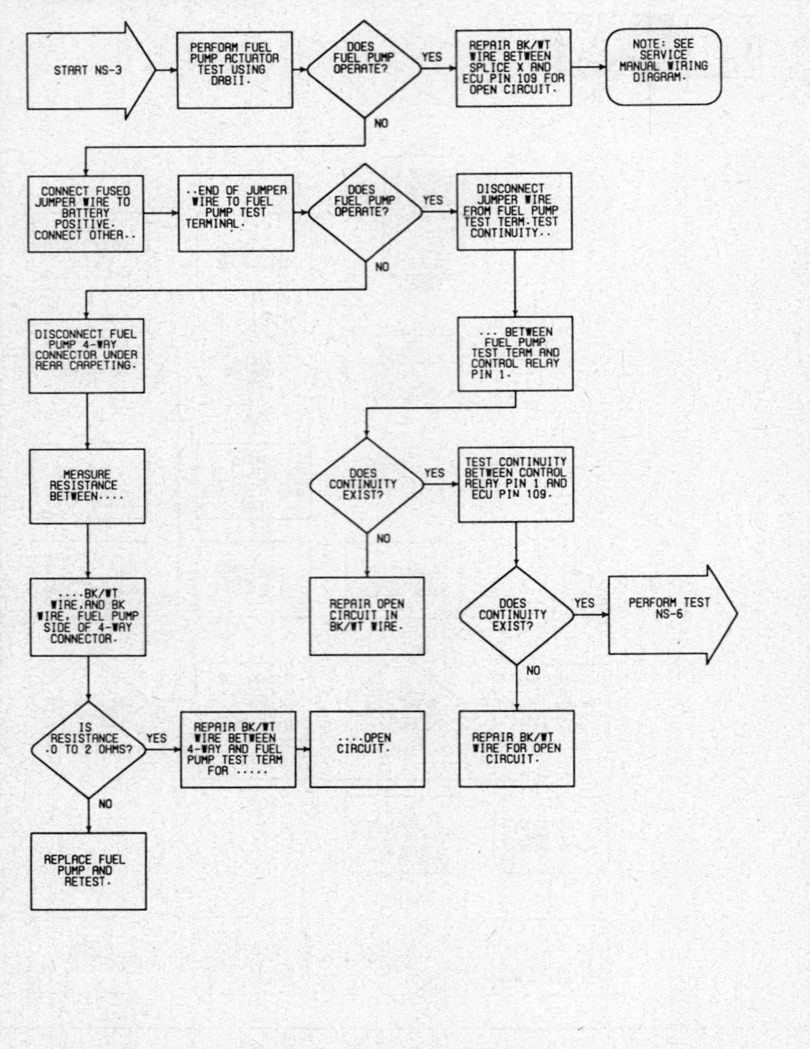

TEST NS-4 TESTING IGNITION SYSTEM - BEGIN NO START TESTS

Perform TEST NS-1 Before Proceeding

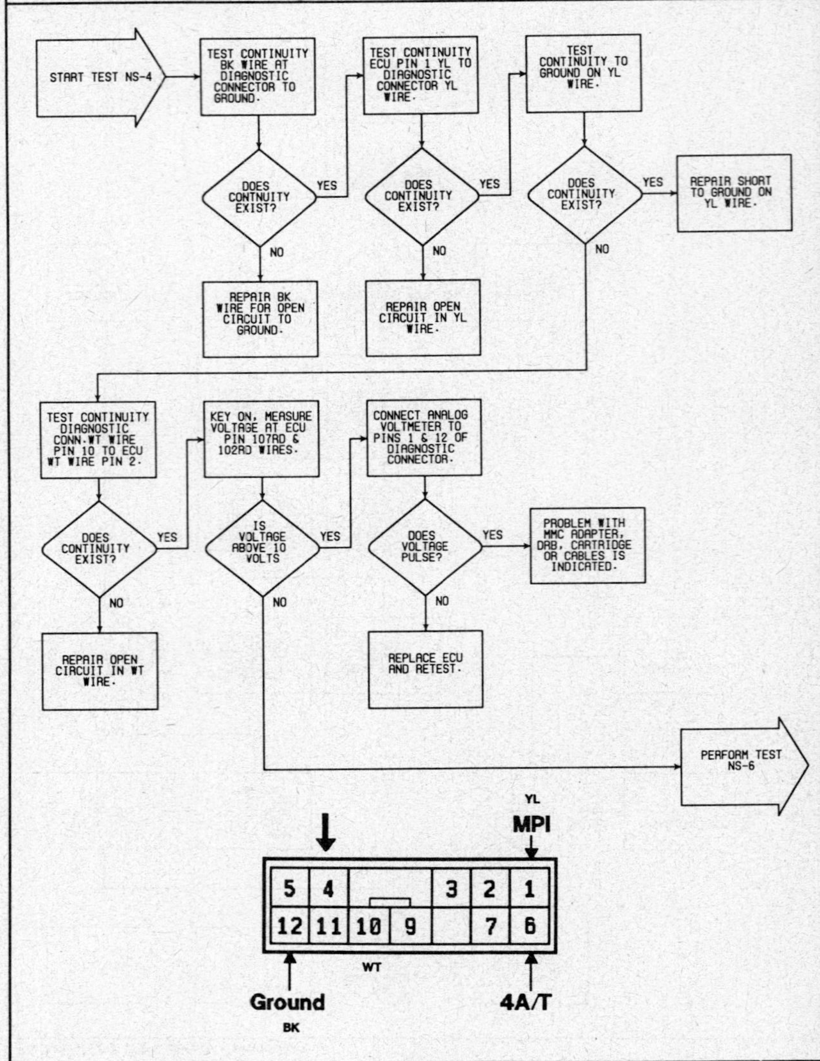

1.5L, 1.8L AND 2.0L ENGINES – NO START TEST

TEST NS-5 TESTING CRANK ANGLE SENSOR

Perform TEST NS-1 Before Proceeding

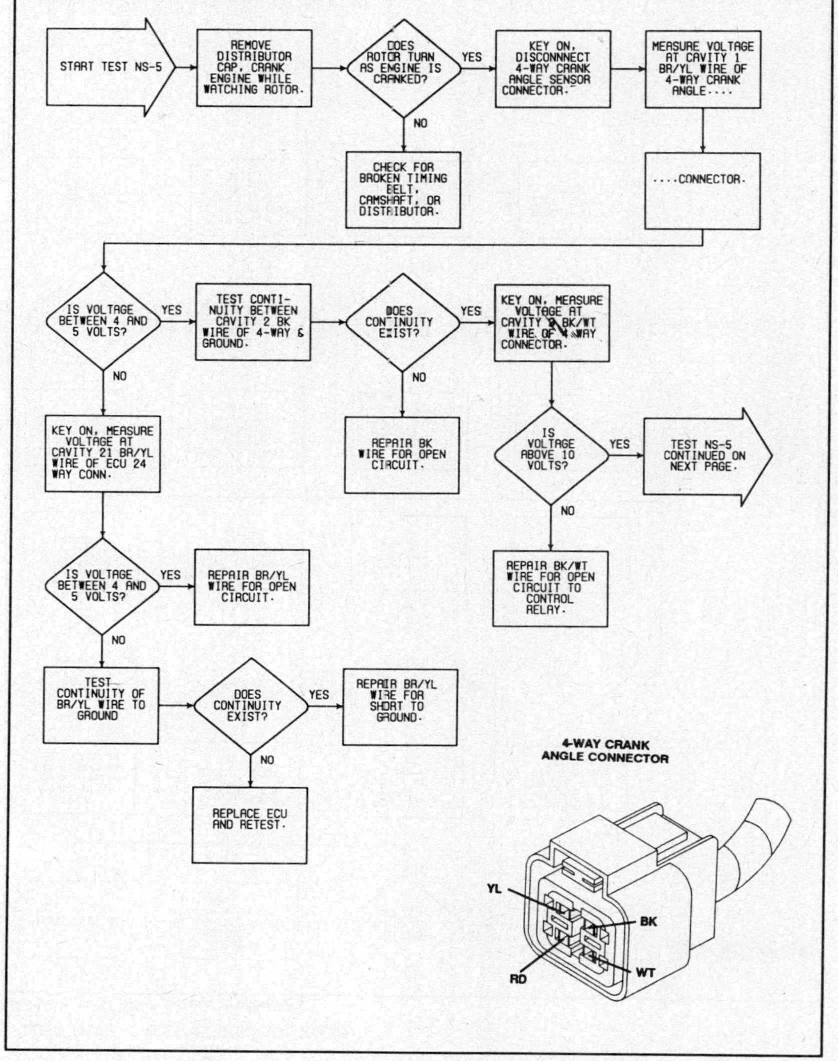

TEST NS-5 CONTINUED - TESTING CRANK ANGLE SENSOR

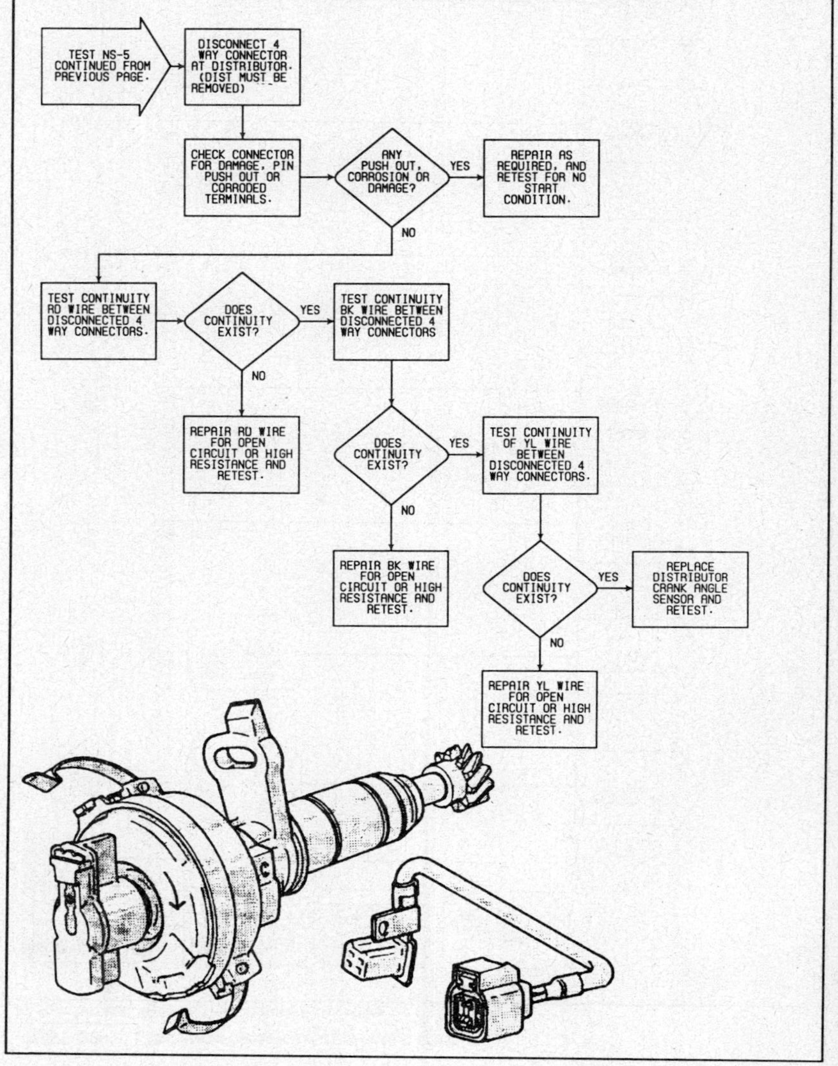

1.5L, 1.8L AND 2.0L ENGINES—NO START TEST

TEST NS-6 TESTING CONTROL RELAY

Perform TEST NS-1 Before Proceeding

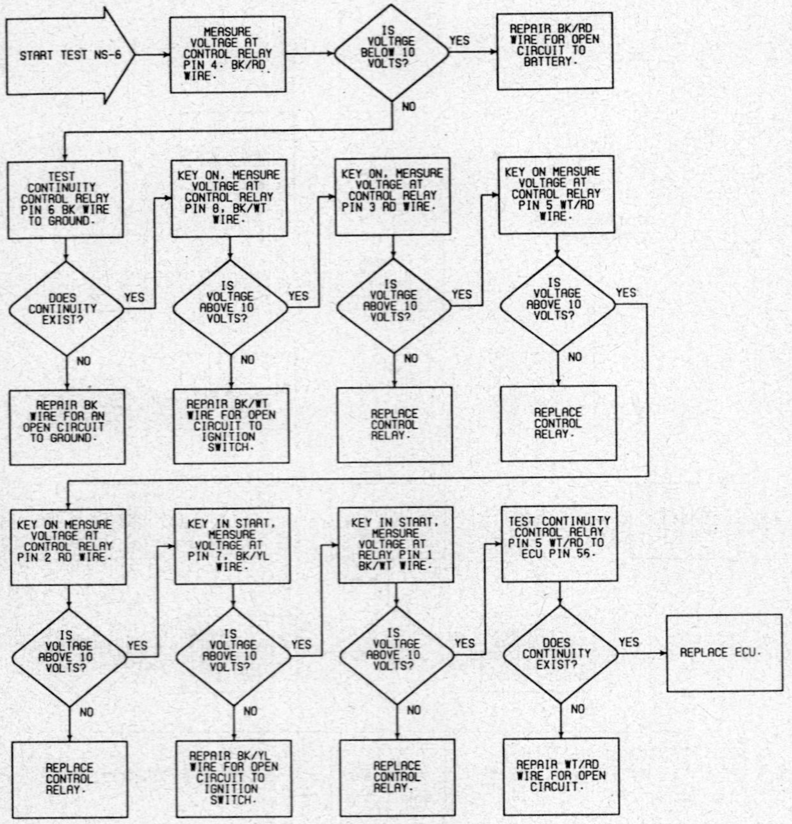

1.5L, 1.8L AND 2.0L ENGINES—DRIVEABILITY TEST

TEST DR-1 TESTING FOR MPI SYSTEM FAULT CODES

Perform VISUAL INSPECTION Before Proceeding

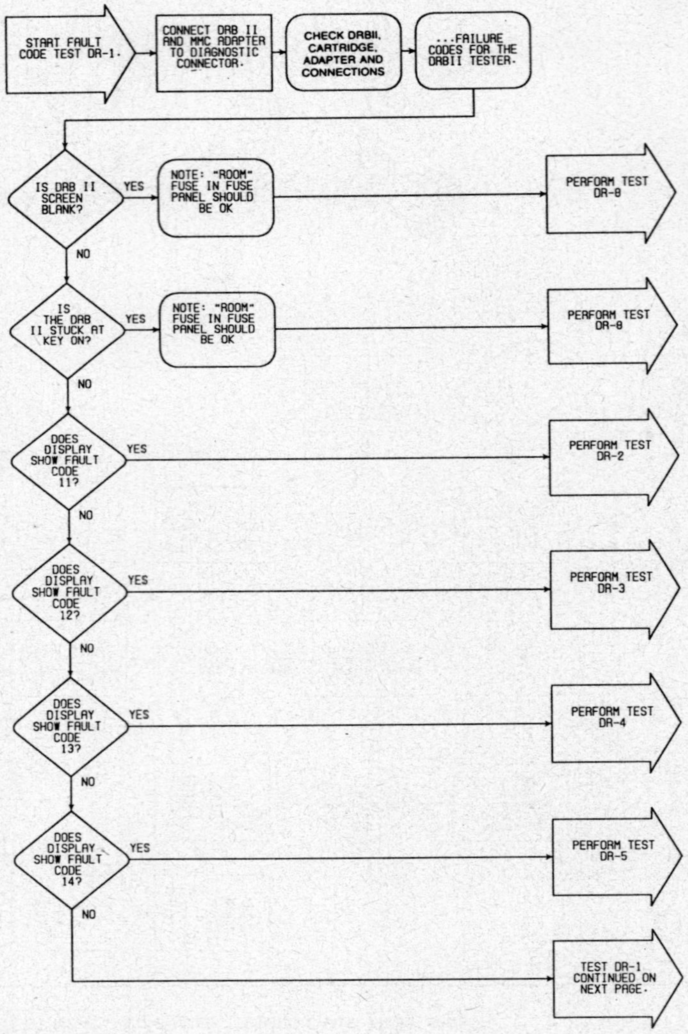

1.5L, 1.8L AND 2.0L ENGINES—DRIVEABILITY TEST

TEST DR-1 CONTINUED - TESTING FOR MPI SYSTEM FAULT CODES

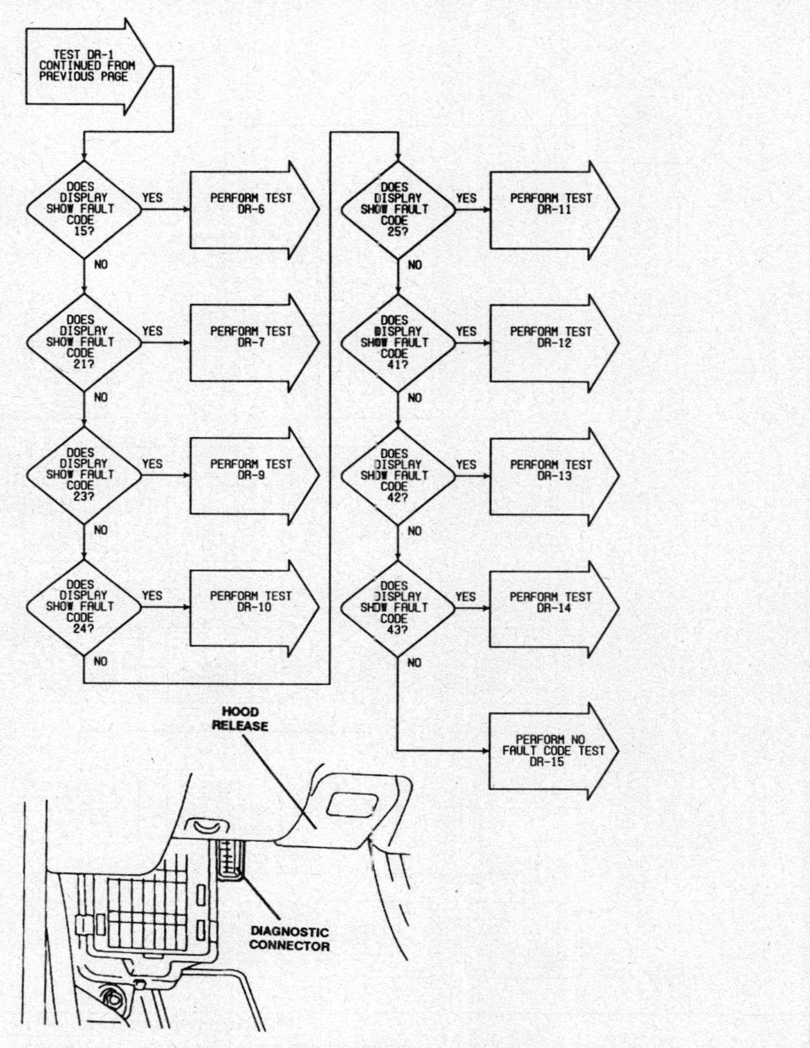

TEST DR-2 TESTING OXYGEN SENSOR CIRCUIT

Perform TEST DR-1 Before Proceeding

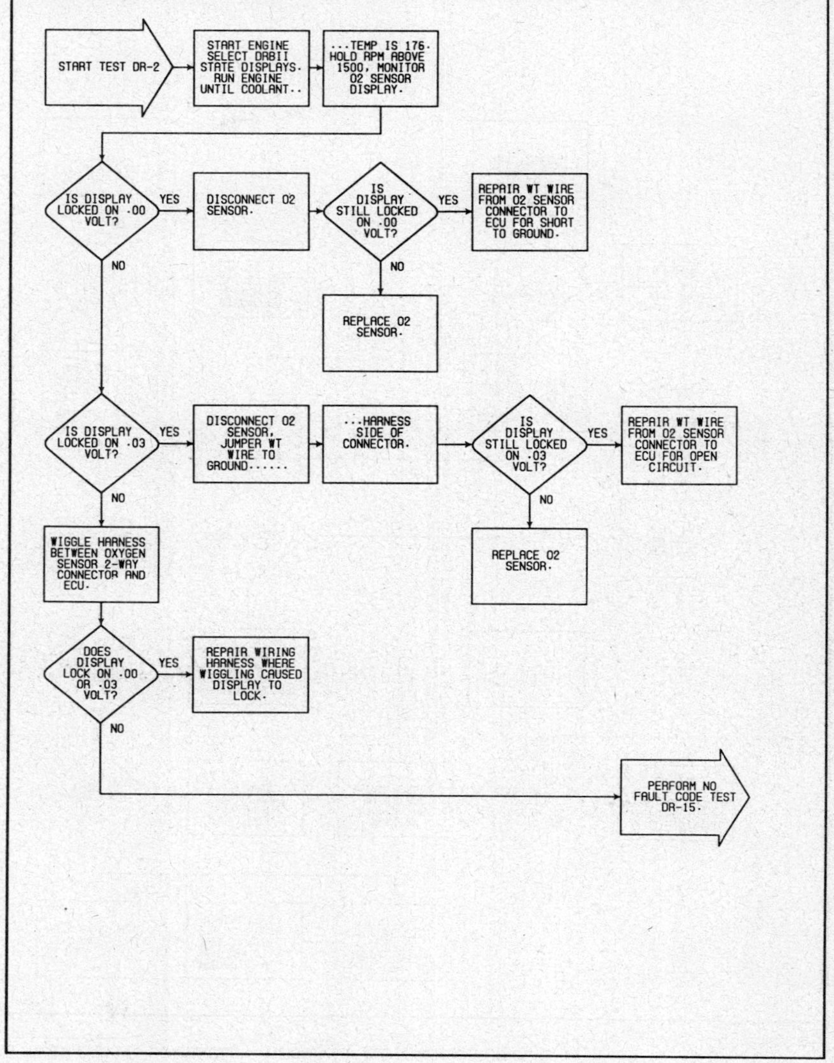

TEST DR-3 TESTING AIR FLOW SENSOR CIRCUIT

Perform TEST DR-1 Before Proceeding

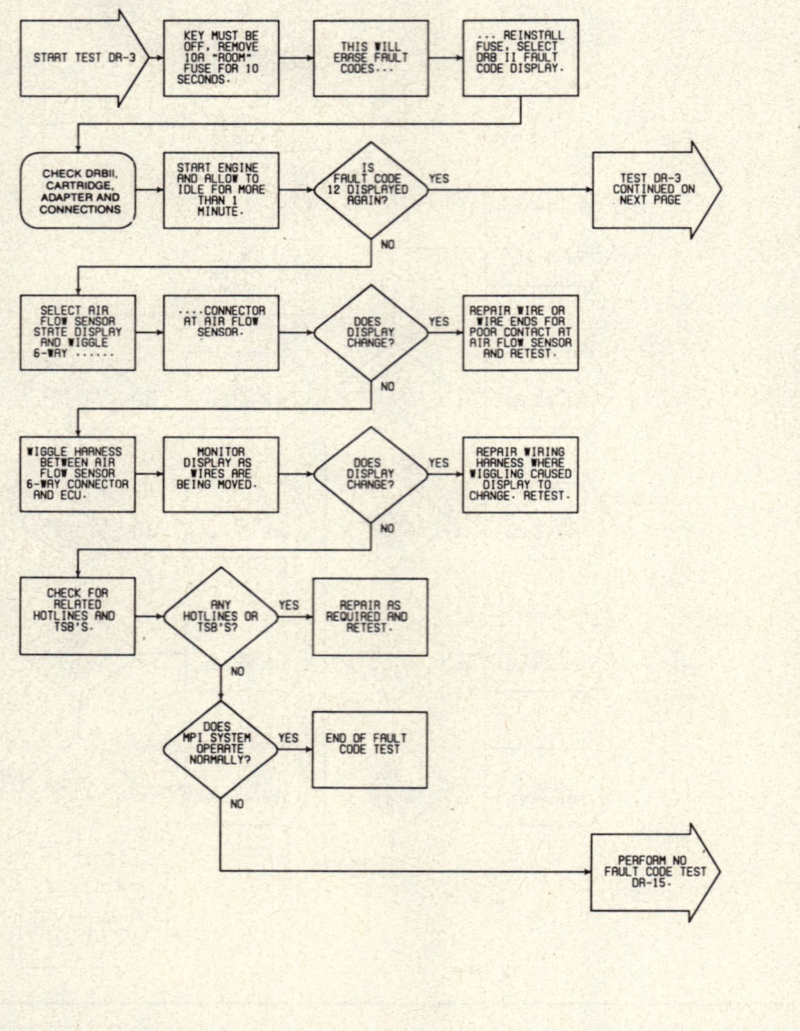

TEST DR-3 CONTINUED - TESTING AIR FLOW SENSOR CIRCUIT

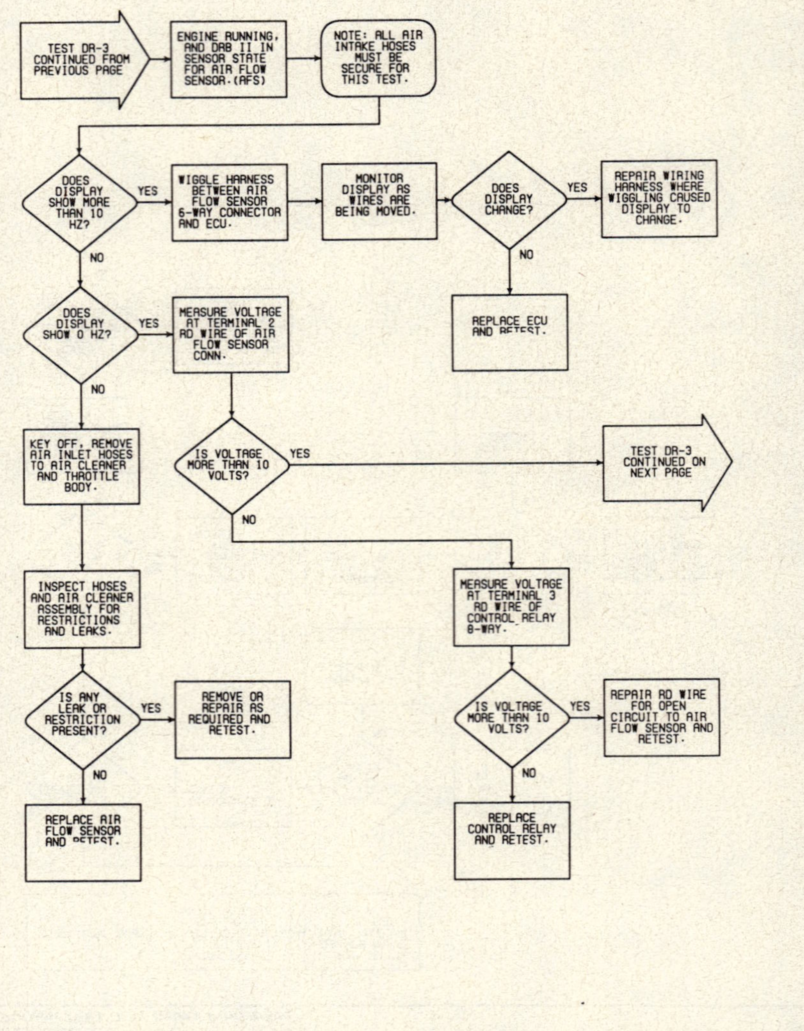

1.5L, 1.8L AND 2.0L ENGINES—DRIVEABILITY TEST

TEST DR-3 CONTINUED - TESTING AIR FLOW SENSOR CIRCUIT

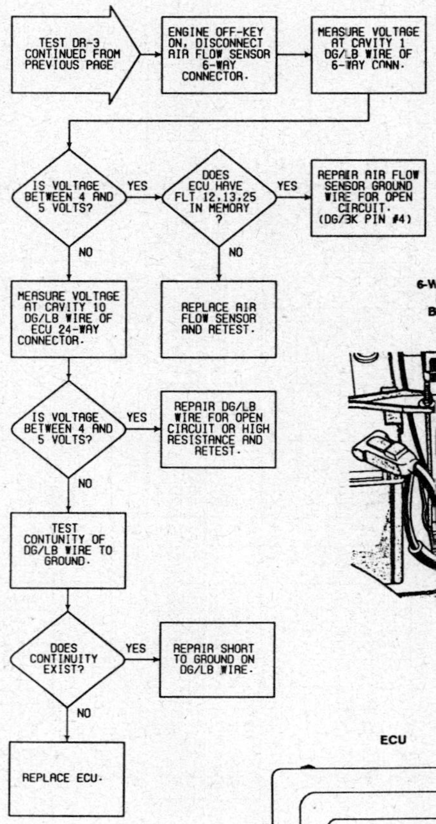

6-WAY CONNECTOR FOR:
AIR FLOW SENSOR
BARO. PRES. SENSOR
INTAKE AIR TEMP.

ECU

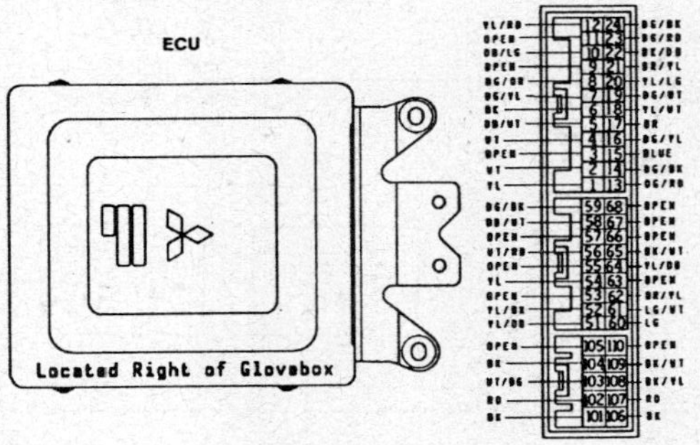

Located Right of Glovebox

TEST DR-4 TESTING INTAKE AIR TEMPERATURE SENSOR

Perform TEST DR-1 Before Proceeding

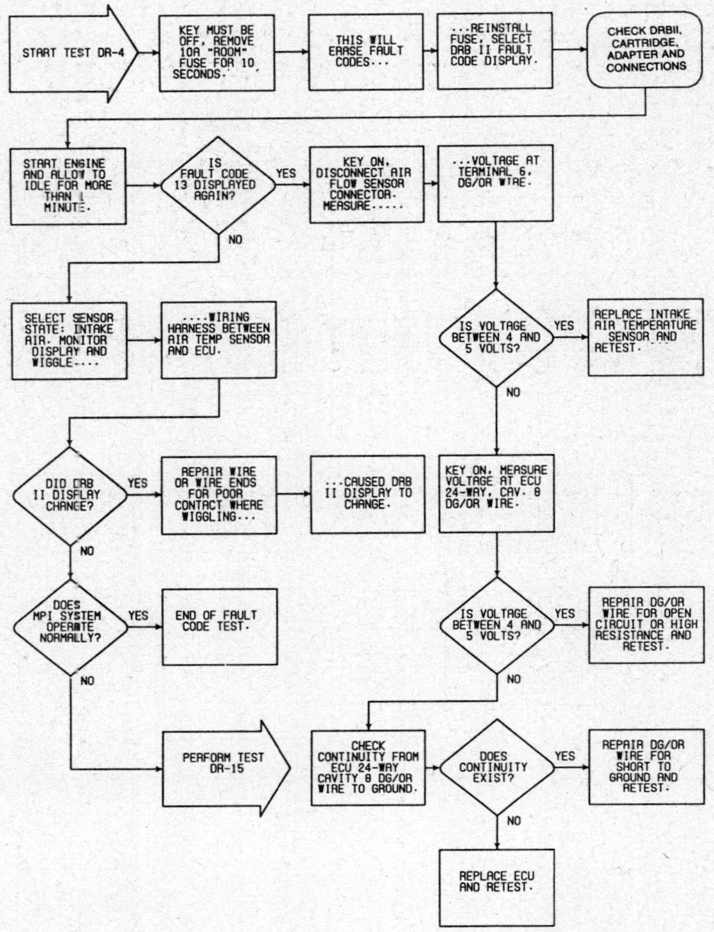

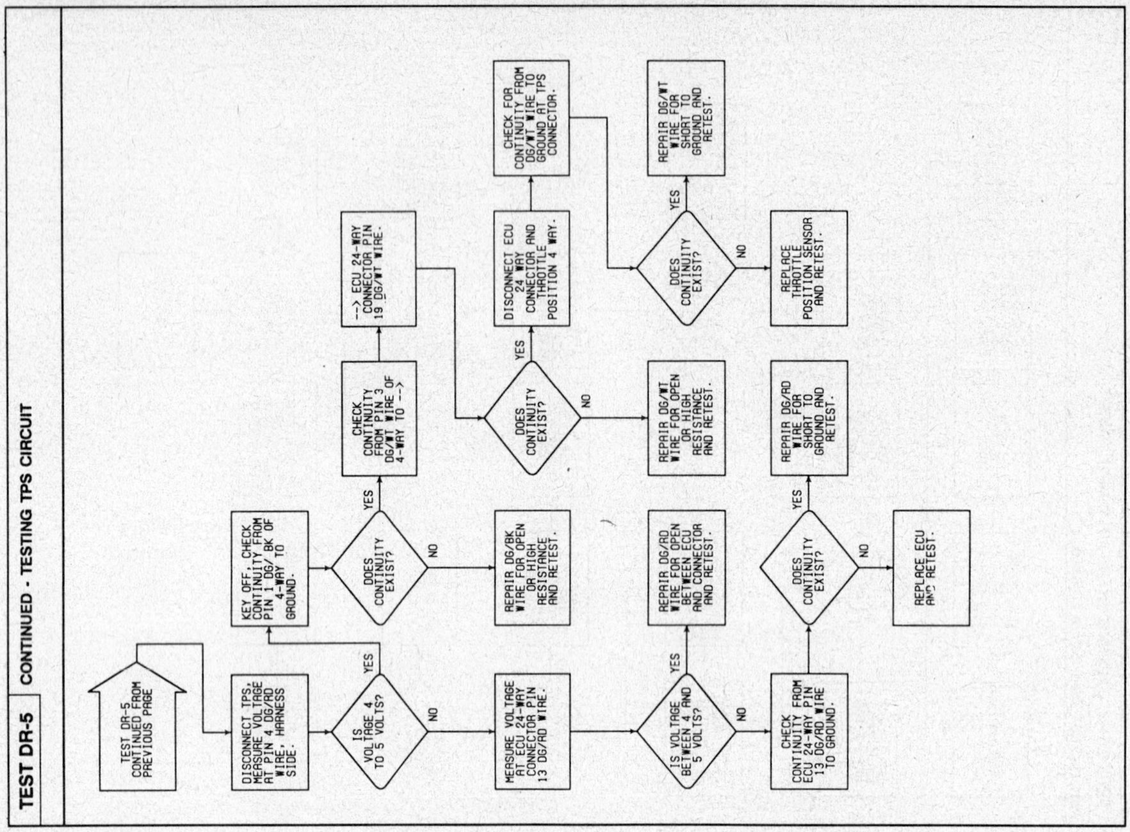
TEST DR-5
CONTINUED - TESTING TPS CIRCUIT
TEST DR-5 CONTINUED FROM PREVIOUS PAGE
DISCONNECT TPS, MEASURE VOLTAGE AT PIN 4 DG/RD WIRE, HARNESS SIDE.
IS VOLTAGE 4 TO 5 VOLTS?
YES
NO
KEY OFF, CHECK CONTINUITY FROM PIN 1 DG/BK OF 4-WAY TO GROUND.
DOES CONTINUITY EXIST?
CHECK CONTINUITY FROM PIN 3 DG/WT WIRE OF 4-WAY TO -->
--> ECU 24-WAY CONNECTOR PIN 19 DG/WT WIRE.
DOES CONTINUITY EXIST?
DISCONNECT ECU 24 WAY CONNECTOR AND THROTTLE POSITION 4 WAY.
CHECK FOR CONTINUITY FROM DG/WT WIRE TO GROUND AT TPS CONNECTOR.
DOES CONTINUITY EXIST?
REPAIR DG/WT WIRE FOR SHORT TO GROUND AND RETEST.
REPLACE THROTTLE POSITION SENSOR AND RETEST.
REPAIR DG/WT WIRE FOR OPEN OR HIGH RESISTANCE AND RETEST.
REPAIR DG/BK WIRE FOR OPEN OR HIGH RESISTANCE AND RETEST.
MEASURE VOLTAGE AT ECU 24-WAY CONNECTOR PIN 13 DG/RD WIRE.
IS VOLTAGE BETWEEN 4 AND 5 VOLTS?
REPAIR DG/RD WIRE FOR OPEN BETWEEN ECU AND CONNECTOR AND RETEST.
CHECK CONTINUITY FROM ECU 24-WAY PIN 13 DG/RD WIRE TO GROUND.
DOES CONTINUITY EXIST?
REPAIR DG/RD WIRE FOR SHORT TO GROUND AND RETEST.
REPLACE ECU AND RETEST.

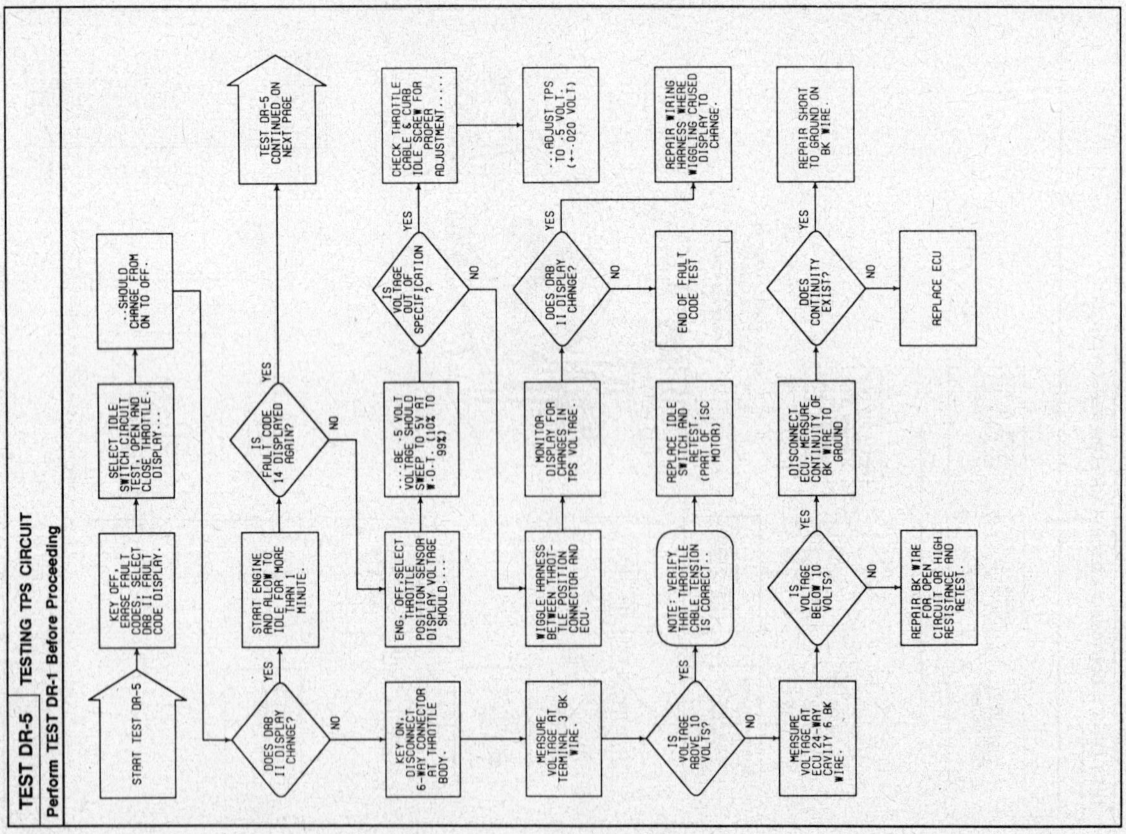
TEST DR-5
TESTING TPS CIRCUIT
Perform TEST DR-1 Before Proceeding
START TEST DR-5
KEY OFF. ERASE FAULT CODES; SELECT DRB II FAULT CODE DISPLAY.
SELECT IDLE SWITCH CIRCUIT TEST. OPEN AND CLOSE THROTTLE. DISPLAY...
...SHOULD CHANGE FROM ON TO OFF.
DOES DRB II DISPLAY CHANGE?
YES
NO
START ENGINE AND ALLOW TO IDLE FOR MORE THAN 1 MINUTE.
IS FAULT CODE 14 DISPLAYED AGAIN?
TEST DR-5 CONTINUED ON NEXT PAGE
KEY ON. DISCONNECT 6 WAY CONNECTOR AT THROTTLE BODY.
ENG. OFF; SELECT THROTTLE POSITION SENSOR DISPLAY VOLTAGE SHOULD....
....BE .5 VOLT AND SHOULD SWEEP TO 5 V AT W.O.T. (10% TO 99%)
IS VOLTAGE OUT OF SPECIFICATION ?
CHECK THROTTLE CABLE & CURB IDLE SCREW FOR PROPER ADJUSTMENT.....
...ADJUST TPS TO .5 VOLT. (+-.020 VOLT)
MEASURE VOLTAGE AT TERMINAL 3 BK WIRE.
WIGGLE HARNESS BETWEEN THROTTLE POSITION CONNECTOR AND ECU.
MONITOR DISPLAY FOR CHANGES IN TPS VOLTAGE.
DOES DRB II DISPLAY CHANGE?
REPAIR WIRING HARNESS WHERE WIGGLING CAUSED DISPLAY TO CHANGE.
END OF FAULT CODE TEST
IS VOLTAGE ABOVE 10 VOLTS?
NOTE: VERIFY THAT THROTTLE CABLE TENSION IS CORRECT.
REPLACE IDLE SWITCH AND RETEST (PART OF ISC MOTOR)
MEASURE VOLTAGE AT ECU 24-WAY CAVITY 6 BK WIRE.
IS VOLTAGE BELOW 10 VOLTS?
DISCONNECT ECU. MEASURE CONTINUITY OF BK WIRE TO GROUND
DOES CONTINUITY EXIST?
REPAIR SHORT TO GROUND ON BK WIRE.
REPLACE ECU
REPAIR BK WIRE FOR OPEN CIRCUIT OR HIGH RESISTANCE AND RETEST.

TEST DR-6 TESTING MOTOR POSITION SENSOR

Perform TEST DR-1 Before Proceeding

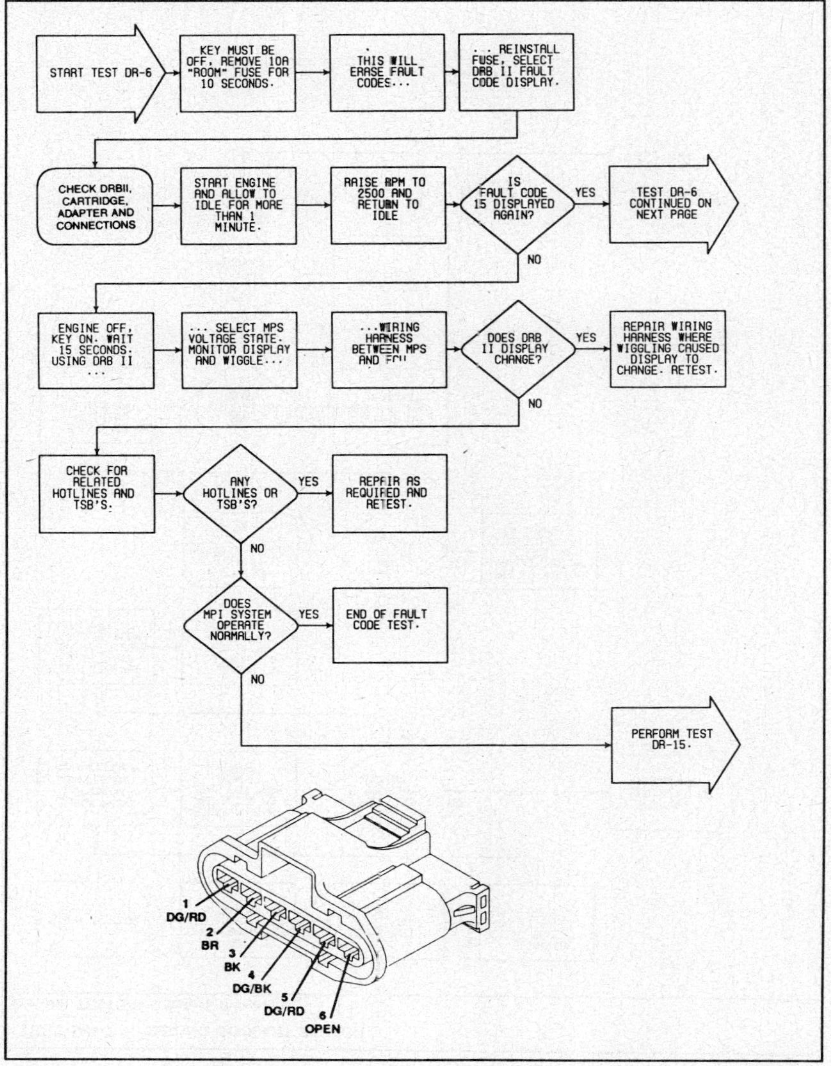

TEST DR-6 CONTINUED - TESTING MOTOR POSITION SENSOR

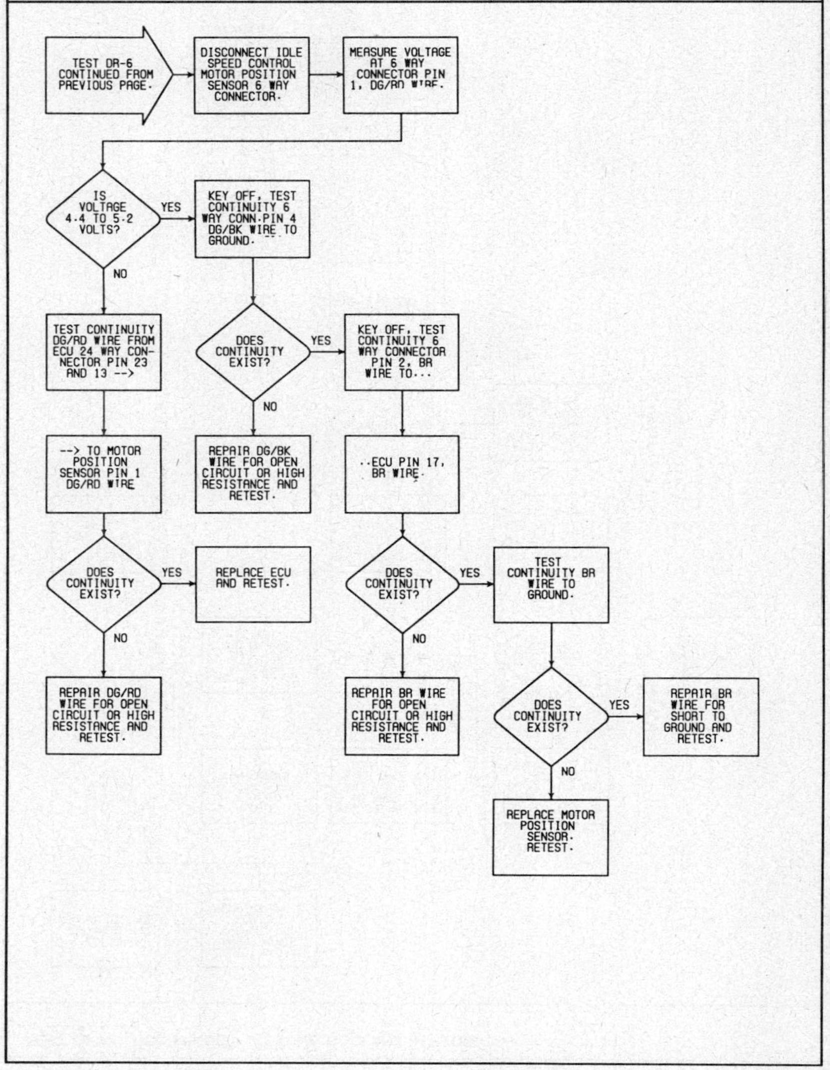

TEST DR-7 TESTING COOLANT SENSOR

Perform TEST DR-1 Before Proceeding

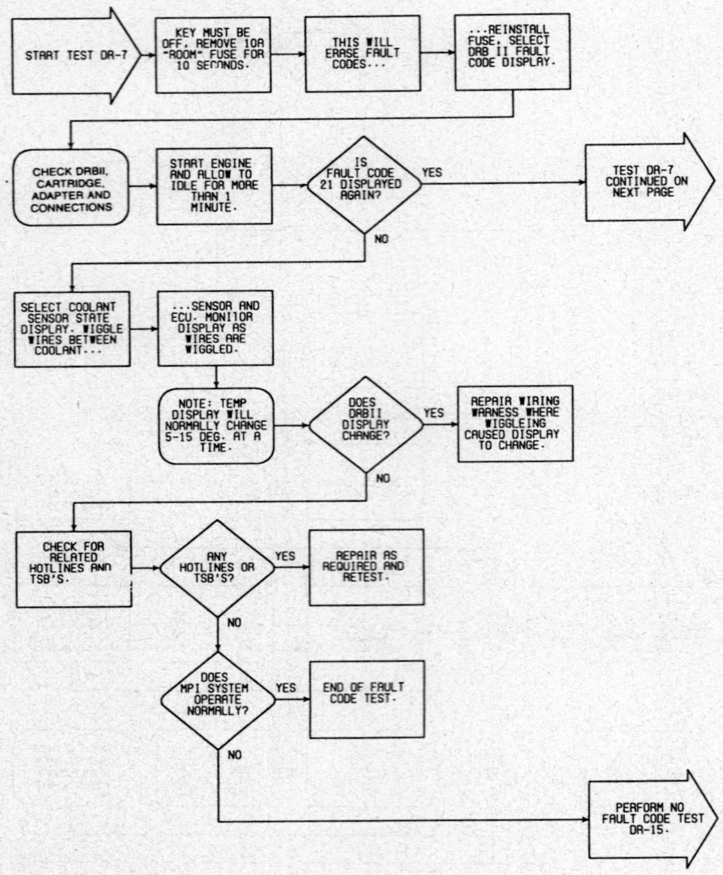

TEST DR-7 CONTINUED - TESTING COOLANT SENSOR

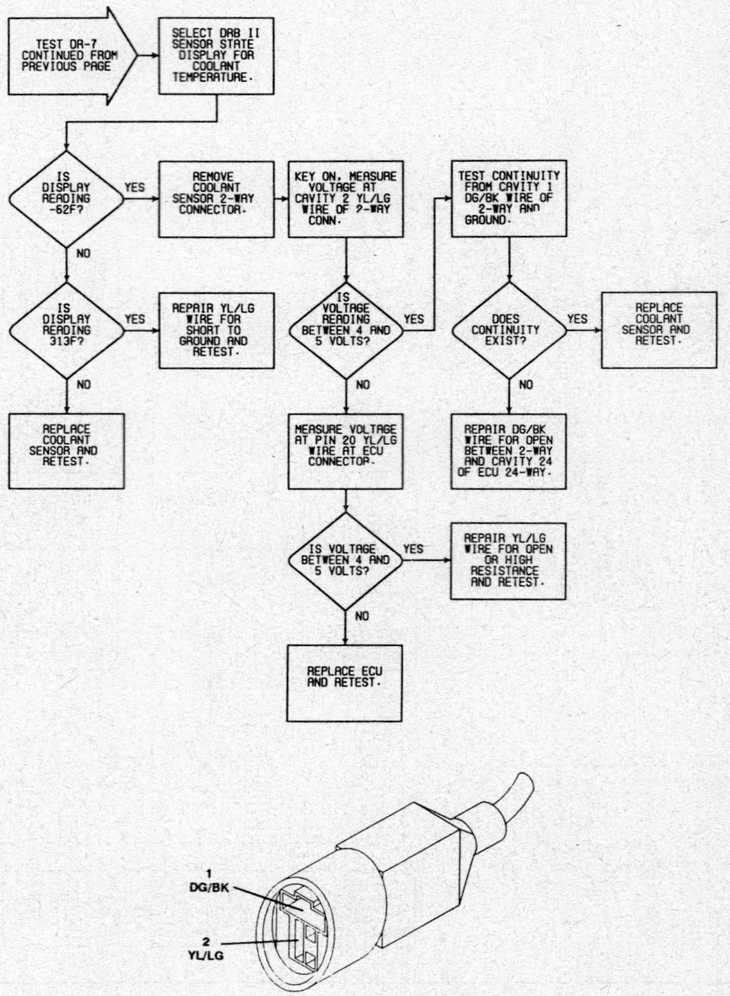

1.5L, 1.8L AND 2.0L ENGINES—DRIVEABILITY TEST

TEST DR-8 TESTING DIAGNOSTIC CONNECTOR

Perform TEST DR-1 Before Proceeding

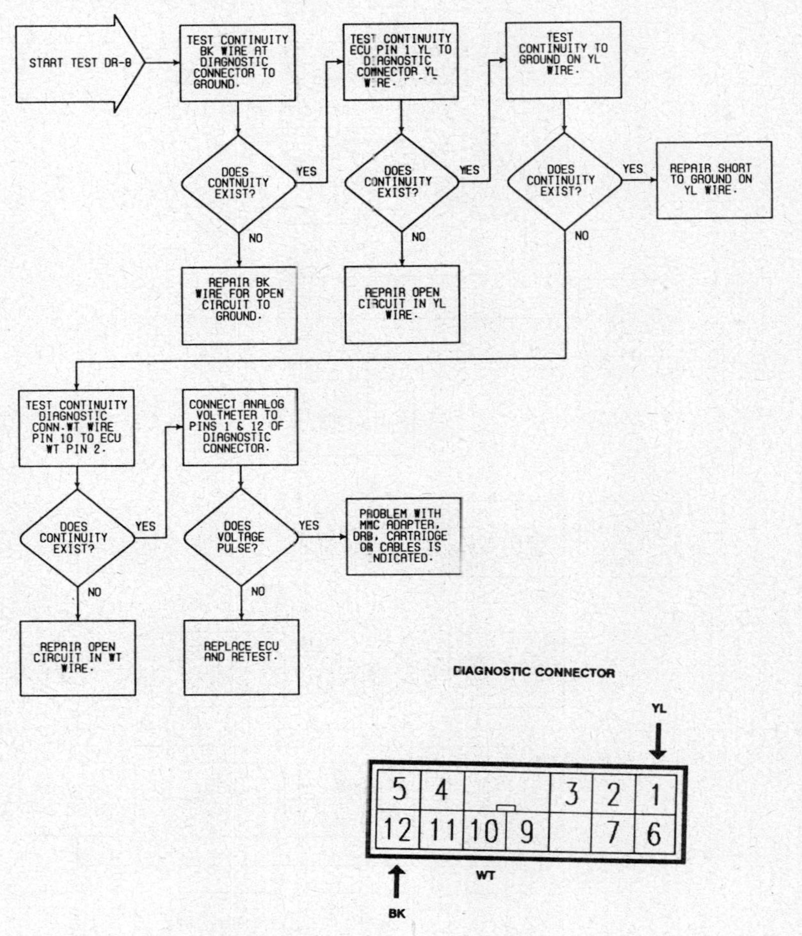

TEST DR-9 TESTING TOP DEAD CENTER SENSOR CIRCUIT

Perform TEST DR-1 Before Proceeding

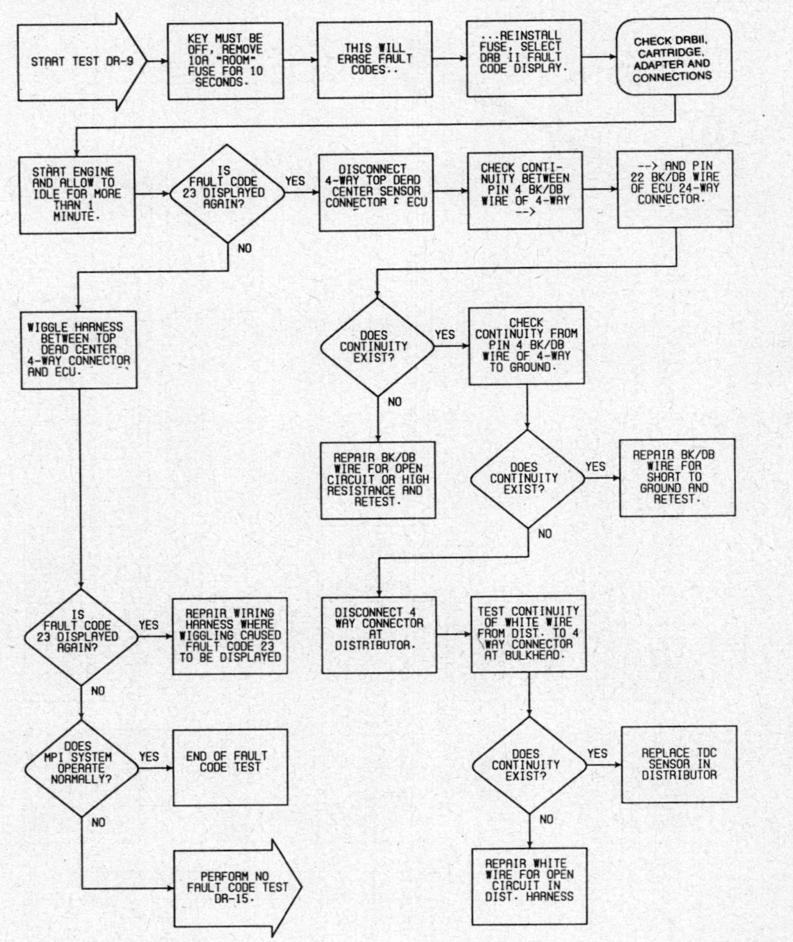

TEST DR-10 TESTING SPEED SENSOR CIRCUIT

Perform TEST DR-1 Before Proceeding

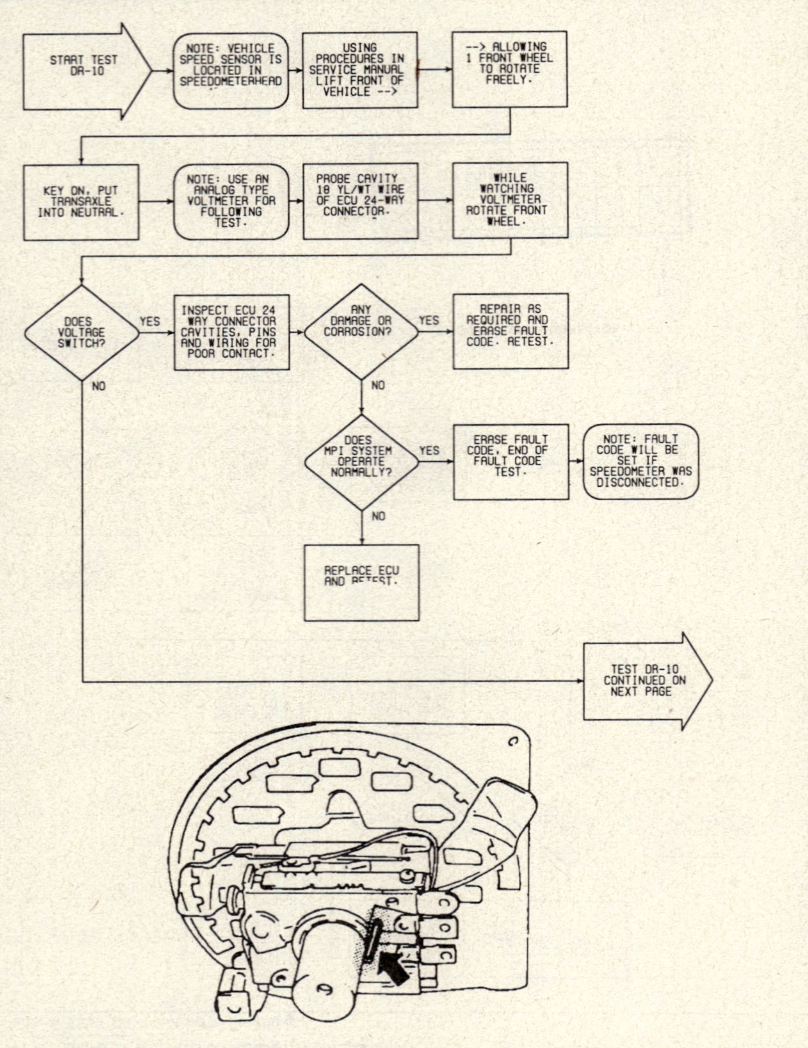

TEST DR-10 CONTINUED - TESTING SPEED SENSOR CIRCUIT

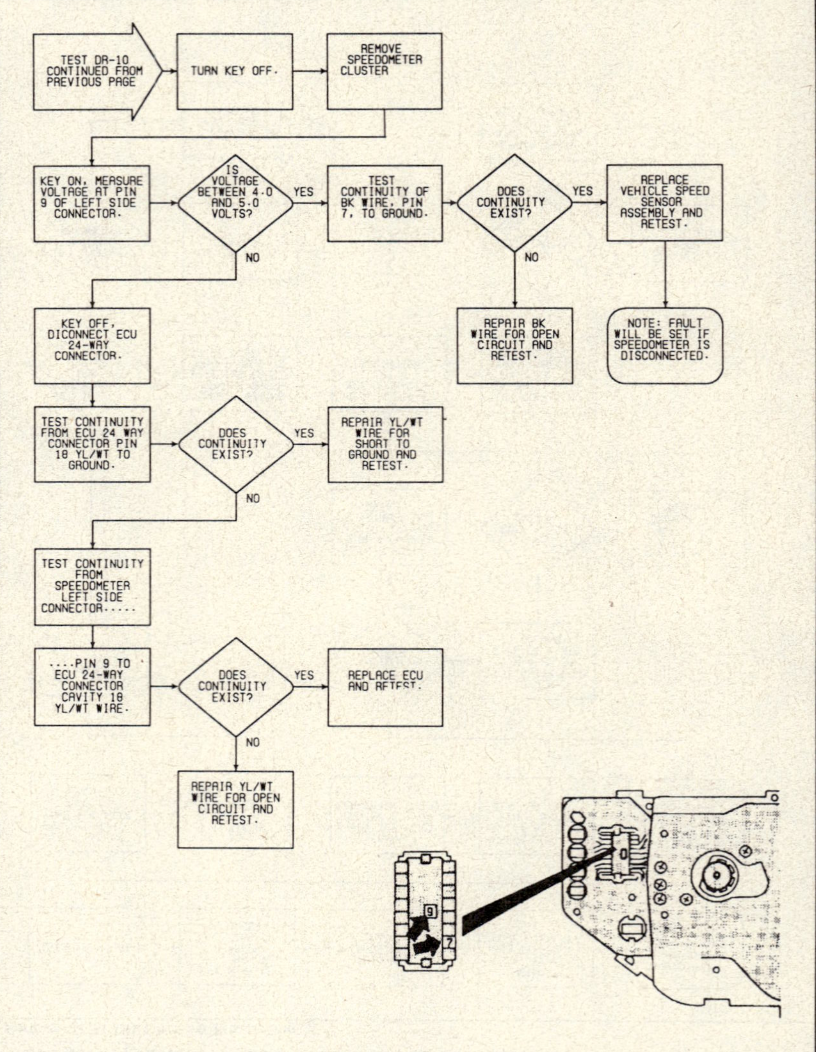

1.5L, 1.8L AND 2.0L ENGINES—DRIVEABILITY TEST

TEST DR-11 TESTING BAROMETRIC PRESSURE SENSOR

Perform TEST DR-1 Before Proceeding

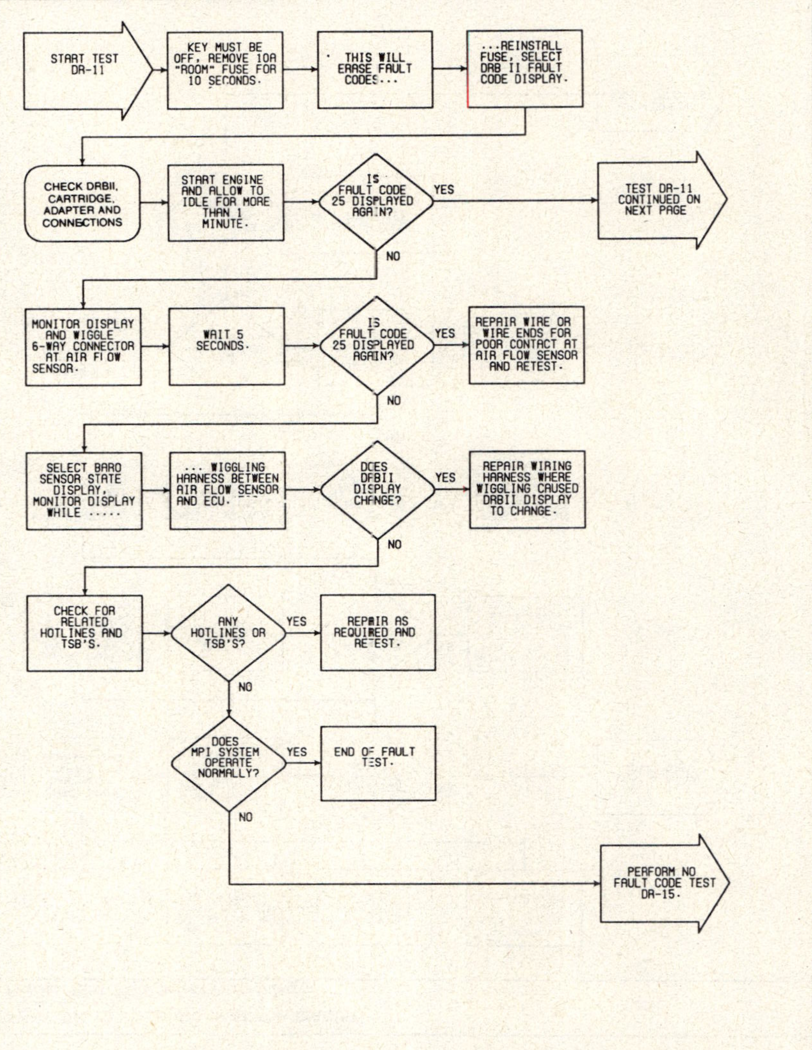

TEST DR-11 CONTINUED - TESTING BAROMETRIC PRESSURE SENSOR

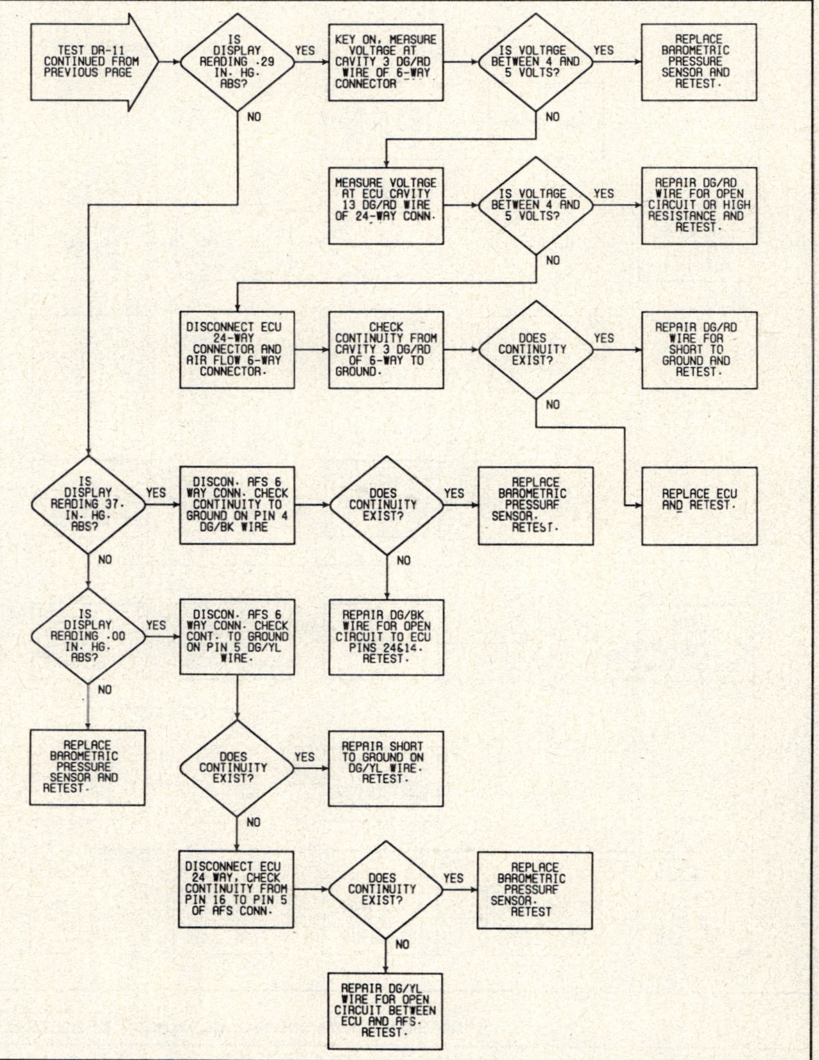

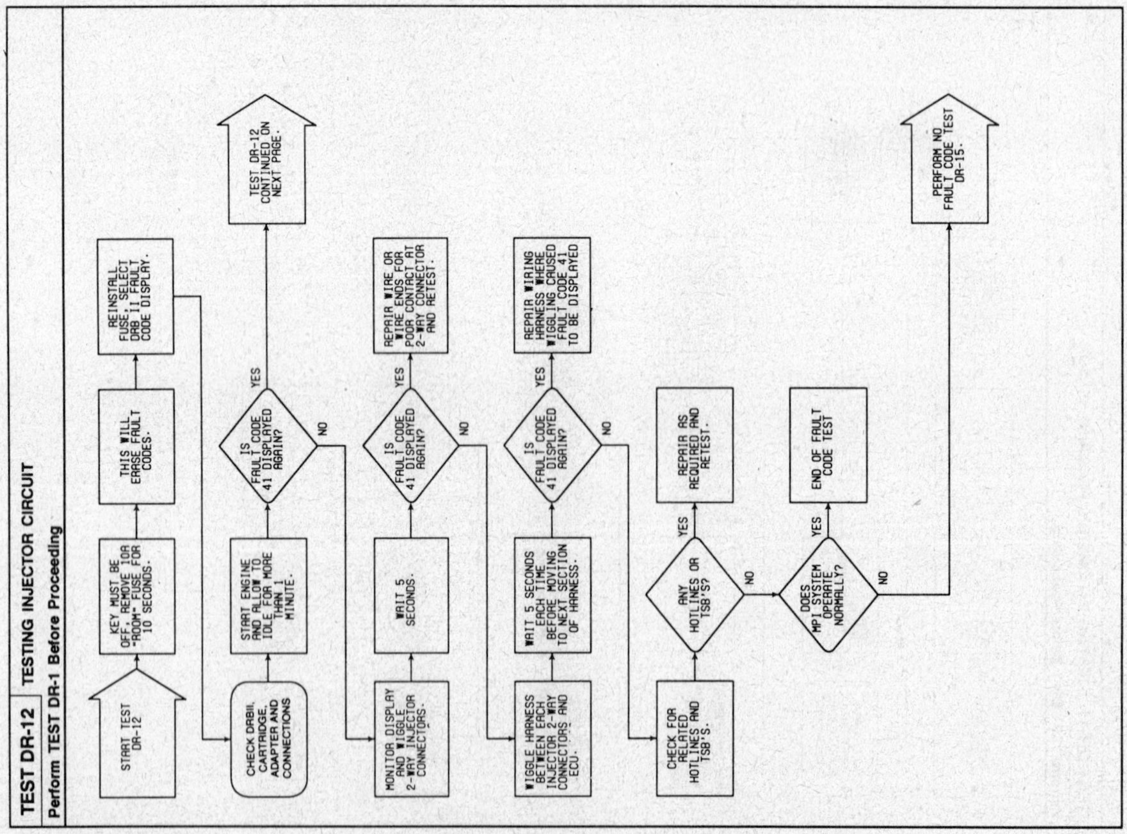
TEST DR-12 TESTING INJECTOR CIRCUIT
Perform TEST DR-1 Before Proceeding
START TEST DR-12
KEY MUST BE OFF. REMOVE 10A "ROOM" FUSE FOR 10 SECONDS.
THIS WILL ERASE FAULT CODES.
REINSTALL FUSE. SELECT DRB II FAULT CODE DISPLAY.
CHECK DRBII, CARTRIDGE, ADAPTER AND CONNECTIONS
START ENGINE AND ALLOW TO IDLE FOR MORE THAN 1 MINUTE.
IS FAULT CODE 41 DISPLAYED AGAIN?
YES
NO
TEST DR-12 CONTINUED ON NEXT PAGE.
MONITOR DISPLAY AND WIGGLE 2-WAY INJECTOR CONNECTORS.
WAIT 5 SECONDS.
IS FAULT CODE 41 DISPLAYED AGAIN?
REPAIR WIRE OR WIRE ENDS FOR POOR CONTACT AT 2-WAY CONNECTOR AND RETEST.
WIGGLE HARNESS BETWEEN EACH INJECTOR 2-WAY CONNECTORS AND ECU.
WAIT 5 SECONDS EACH TIME BEFORE MOVING TO NEXT SECTION OF HARNESS.
IS FAULT CODE 41 DISPLAYED AGAIN?
REPAIR WIRING HARNESS WHERE WIGGLING CAUSED FAULT CODE 41 TO BE DISPLAYED
CHECK FOR RELATED HOTLINES AND TSB'S.
ANY HOTLINES OR TSB'S?
REPAIR AS REQUIRED AND RETEST.
DOES MPI SYSTEM OPERATE NORMALLY?
END OF FAULT CODE TEST
PERFORM NO FAULT CODE TEST DR-15.

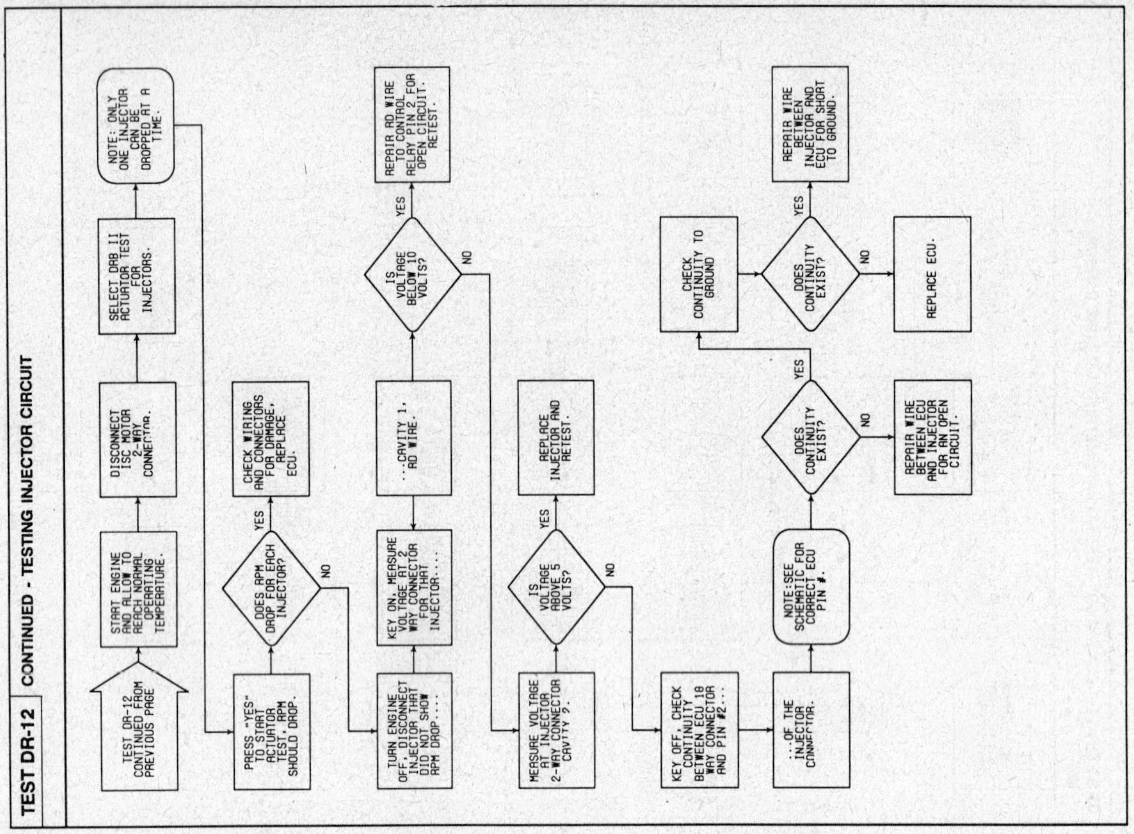
TEST DR-12 CONTINUED - TESTING INJECTOR CIRCUIT
TEST DR-12 CONTINUED FROM PREVIOUS PAGE
START ENGINE AND ALLOW TO REACH NORMAL OPERATING TEMPERATURE.
DISCONNECT ISC MOTOR 2-WAY CONNECTOR.
SELECT DRB II ACTUATOR TEST FOR INJECTORS.
NOTE: ONLY ONE INJECTOR CAN BE DROPPED AT A TIME.
PRESS "YES" TO START ACTUATOR TEST. RPM SHOULD DROP.
DOES RPM DROP FOR EACH INJECTOR?
YES
NO
CHECK WIRING AND CONNECTORS FOR DAMAGE. REPLACE ECU.
TURN ENGINE OFF. DISCONNECT INJECTOR THAT DID NOT SHOW RPM DROP...
KEY ON. MEASURE VOLTAGE AT 2-WAY CONNECTOR FOR THAT INJECTOR...
...CAVITY 1, RD WIRE.
IS VOLTAGE BELOW 10 VOLTS?
REPAIR RD WIRE TO CONTROL RELAY PIN 2 FOR OPEN CIRCUIT. RETEST.
MEASURE VOLTAGE AT INJECTOR 2-WAY CONNECTOR CAVITY 2.
IS VOLTAGE ABOVE 5 VOLTS?
REPLACE INJECTOR AND RETEST.
KEY OFF. CHECK CONTINUITY BETWEEN ECU 18 WAY CONNECTOR AND PIN #2...
...OF THE INJECTOR CONNECTOR.
NOTE: SEE SCHEMATIC FOR CORRECT ECU PIN #.
DOES CONTINUITY EXIST?
REPAIR WIRE BETWEEN ECU AND INJECTOR FOR AN OPEN CIRCUIT.
CHECK CONTINUITY TO GROUND
DOES CONTINUITY EXIST?
REPAIR WIRE BETWEEN INJECTOR AND ECU FOR SHORT TO GROUND.
REPLACE ECU.

1.5L, 1.8L AND 2.0L ENGINES—DRIVEABILITY TEST

TEST DR-13 TESTING FUEL PUMP CIRCUIT

Perform TEST DR-1 Before Proceeding

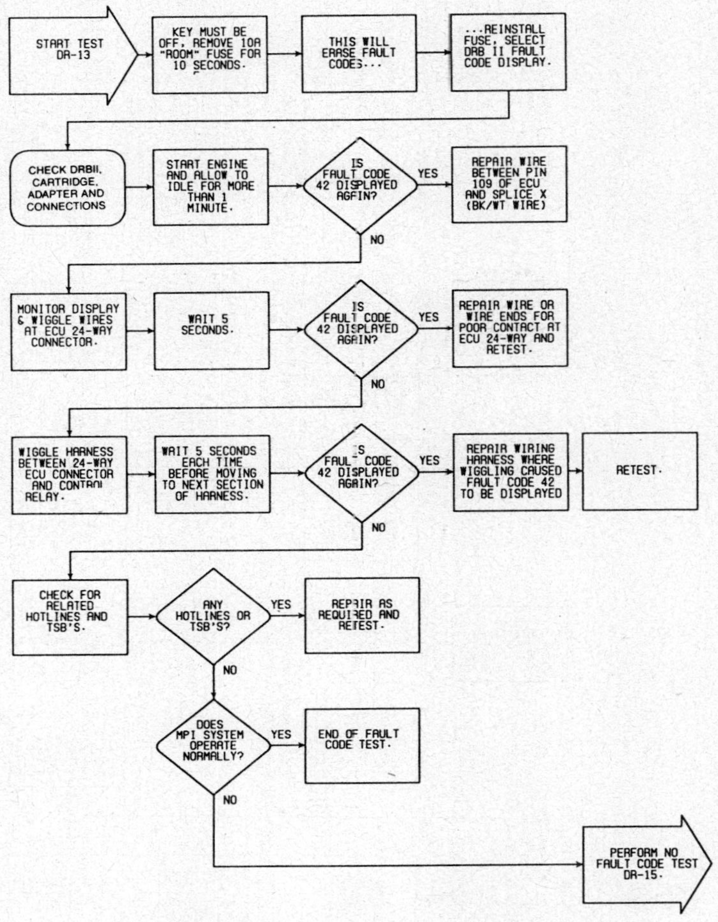

TEST DR-14 TESTING EGR TEMPERATURE SENSOR CIRCUIT

Perform TEST DR-1 Before Proceeding

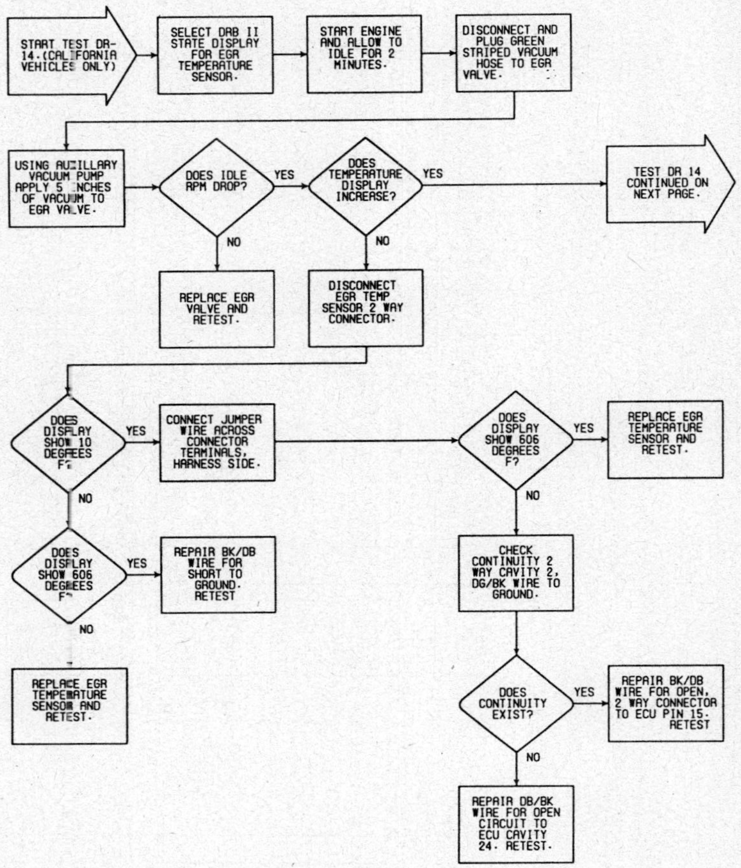

TEST DR-14 CONTINUED - TESTING EGR TEMPERATURE SENSOR CIRCUIT

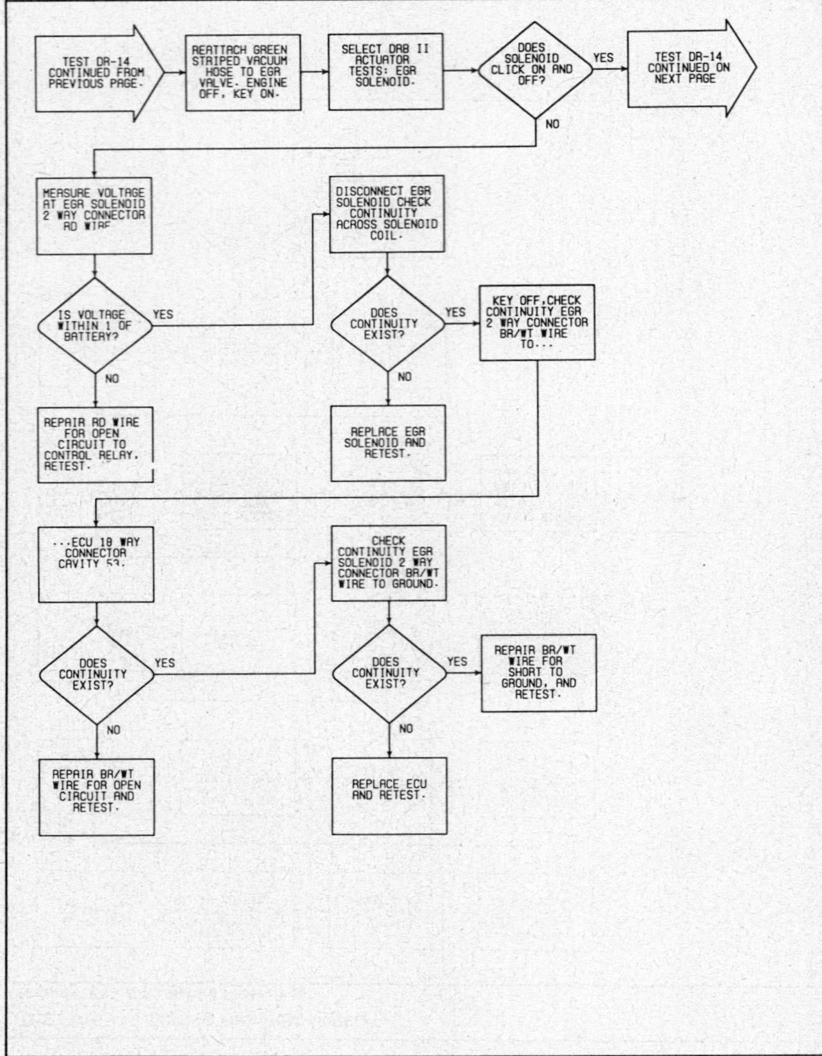

TEST DR-14 CONTINUED - TESTING EGR TEMPERATURE SENSOR CIRCUIT

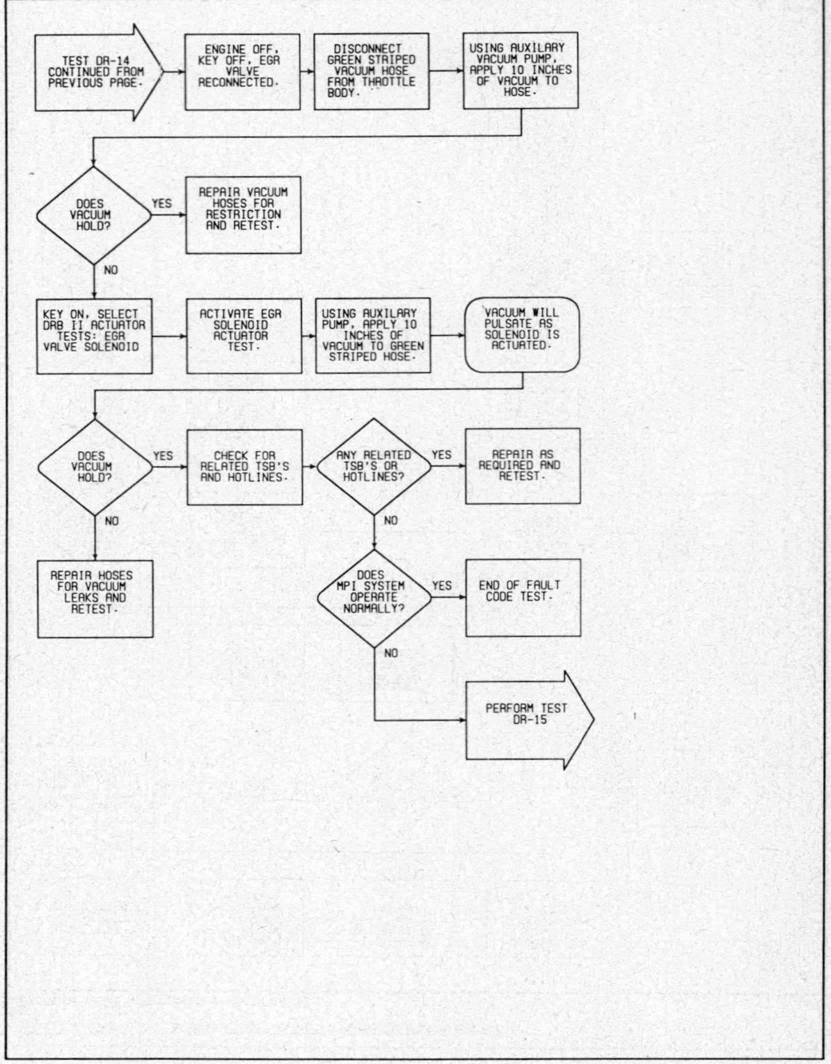

1.5L, 1.8L AND 2.0L ENGINES—DRIVEABILITY TEST

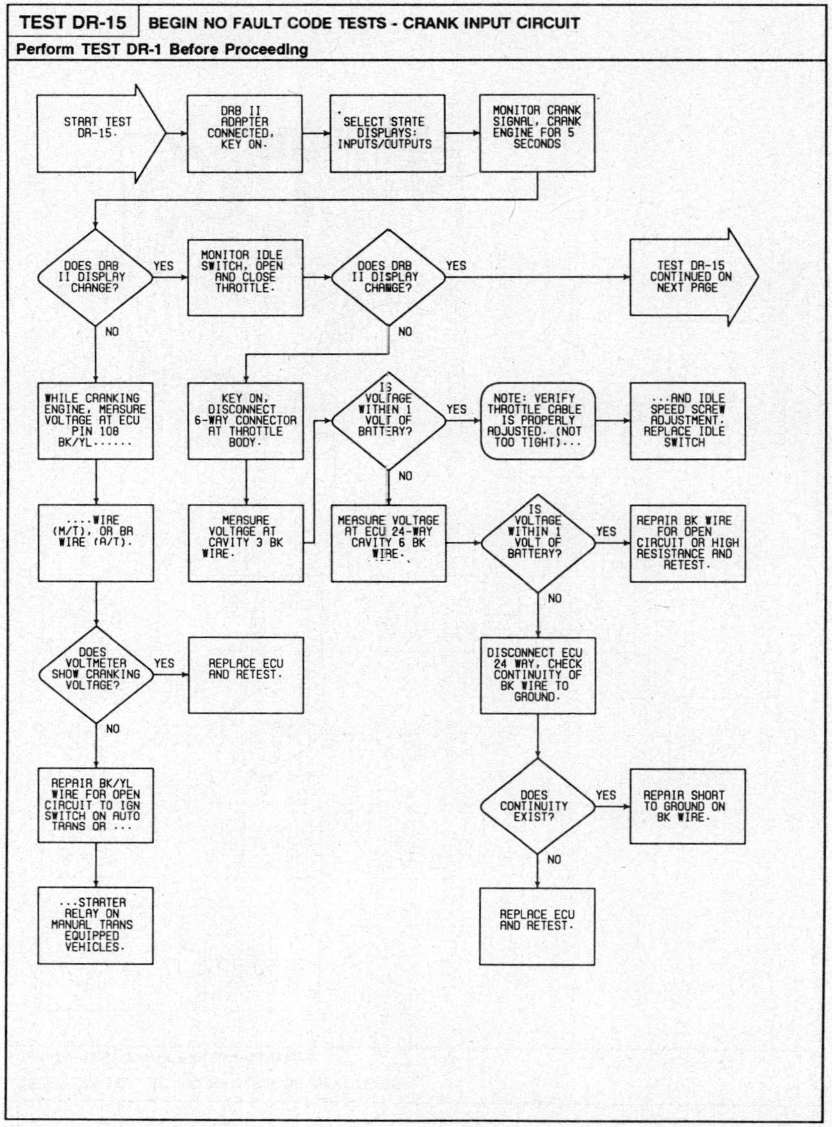

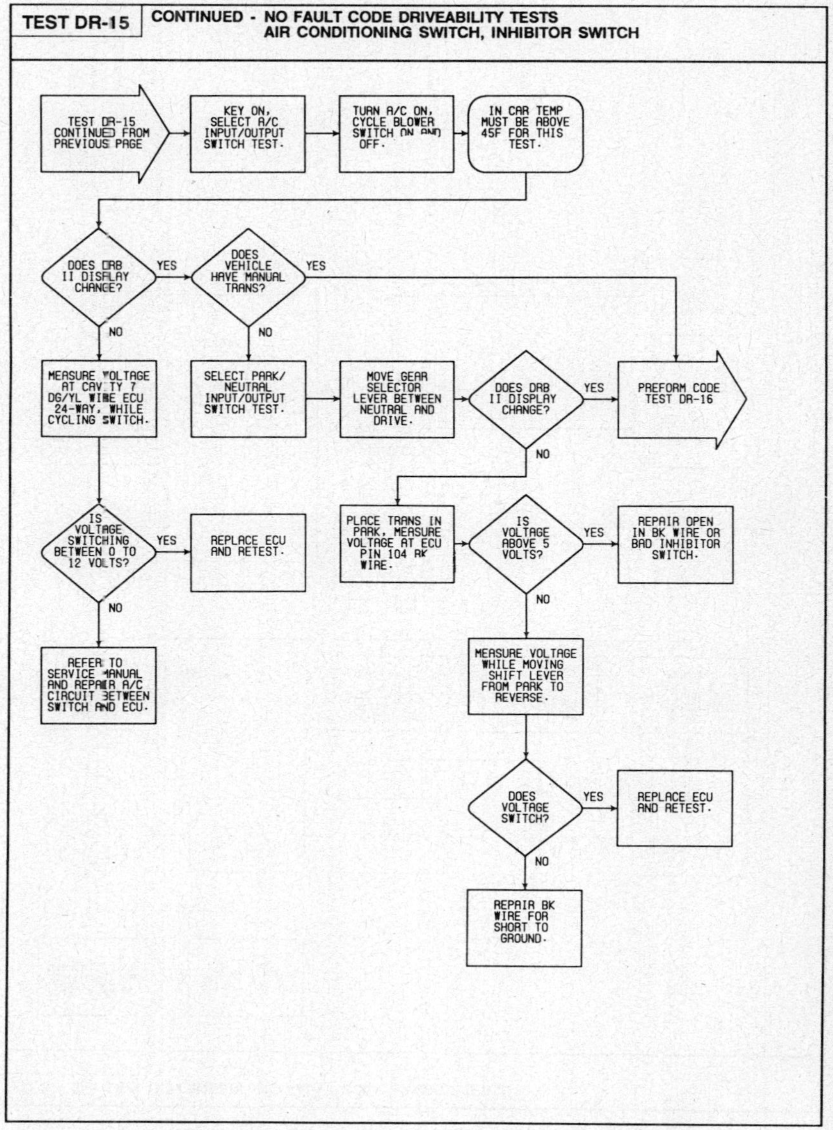

TEST DR-16 NO FAULT CODE SENSOR TESTS

Perform TEST DR-15 Before Proceeding

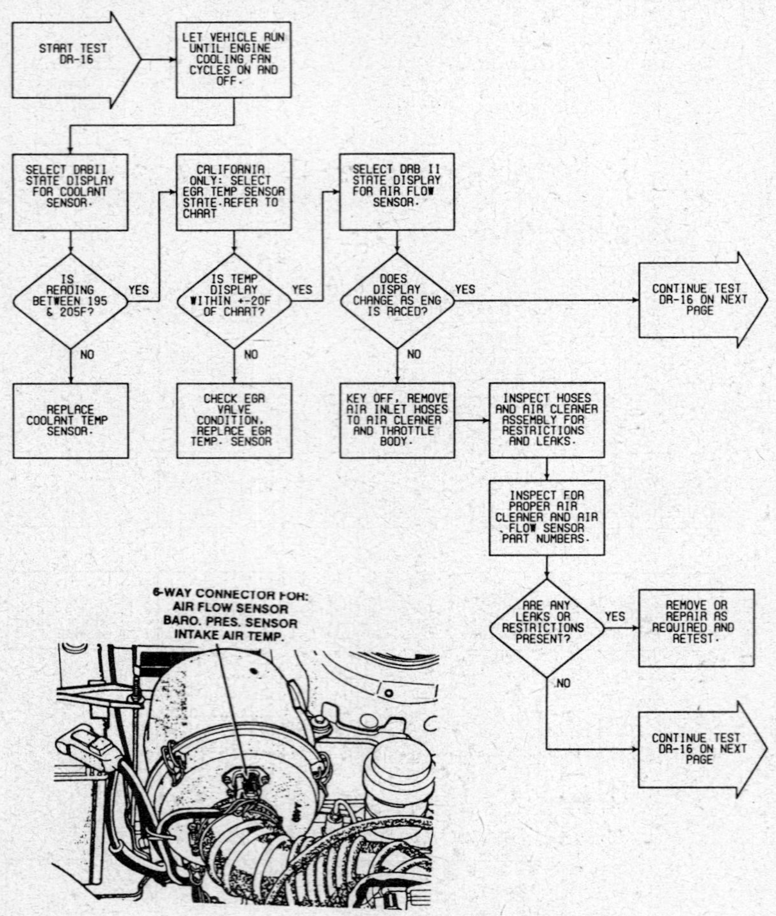

TEST DR-16 CONTINUED - NO FAULT CODE SENSOR TESTS

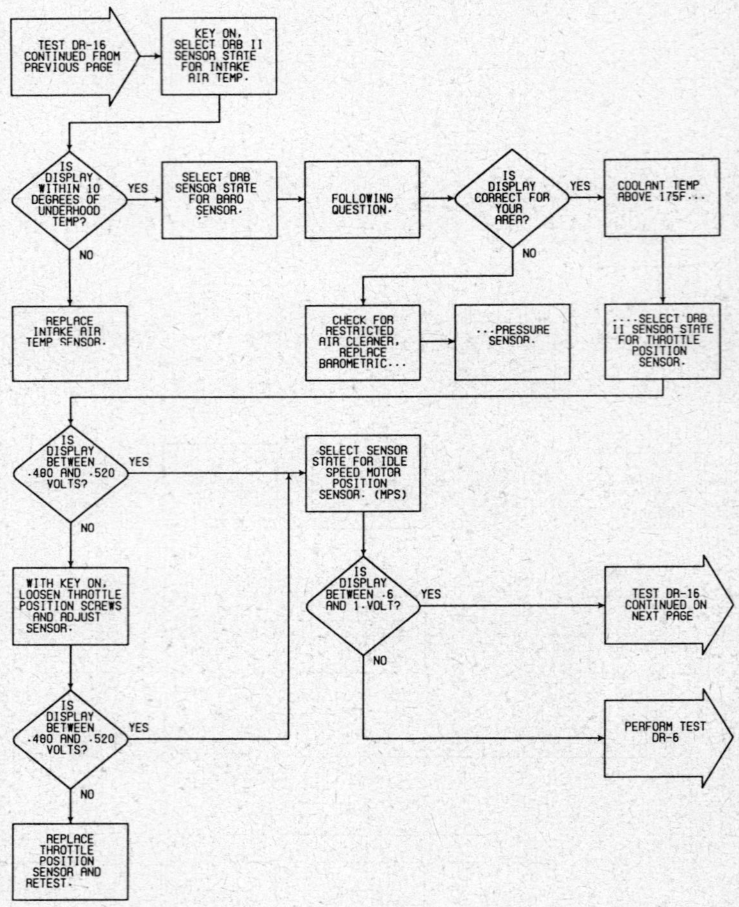

1.5L, 1.8L AND 2.0L ENGINES—DRIVEABILITY TEST

TEST DR-16 CONTINUED - NO FAULT CODE SENSOR TESTS

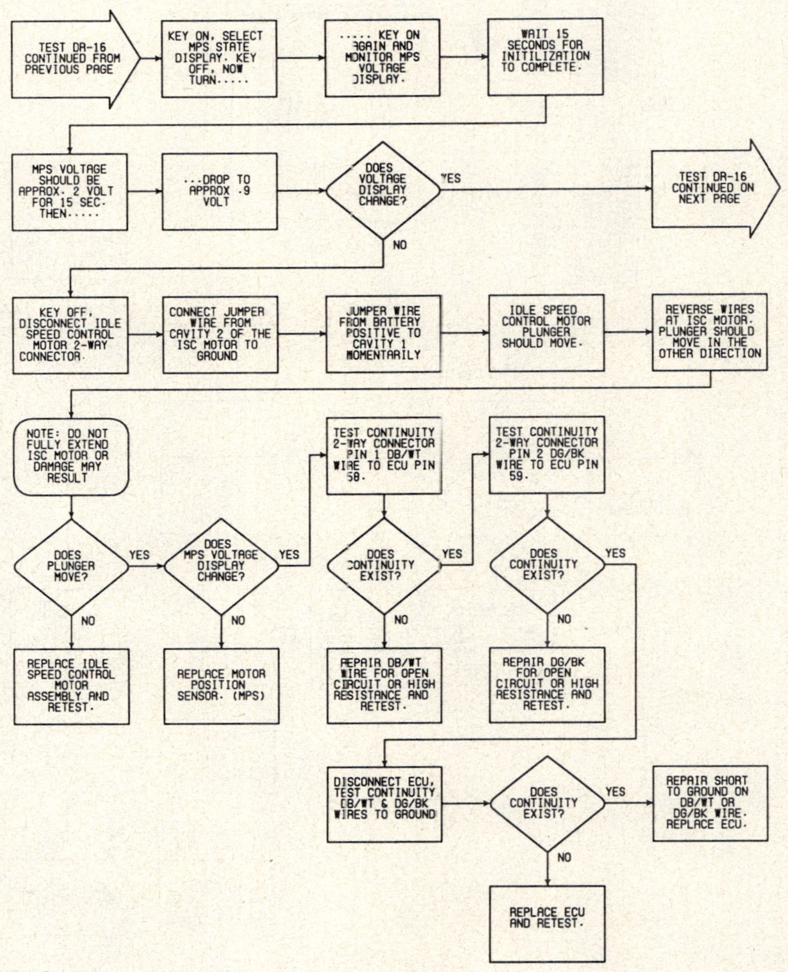

TEST DR-16 CONTINUED - NO FAULT CODE SWITCH TESTS

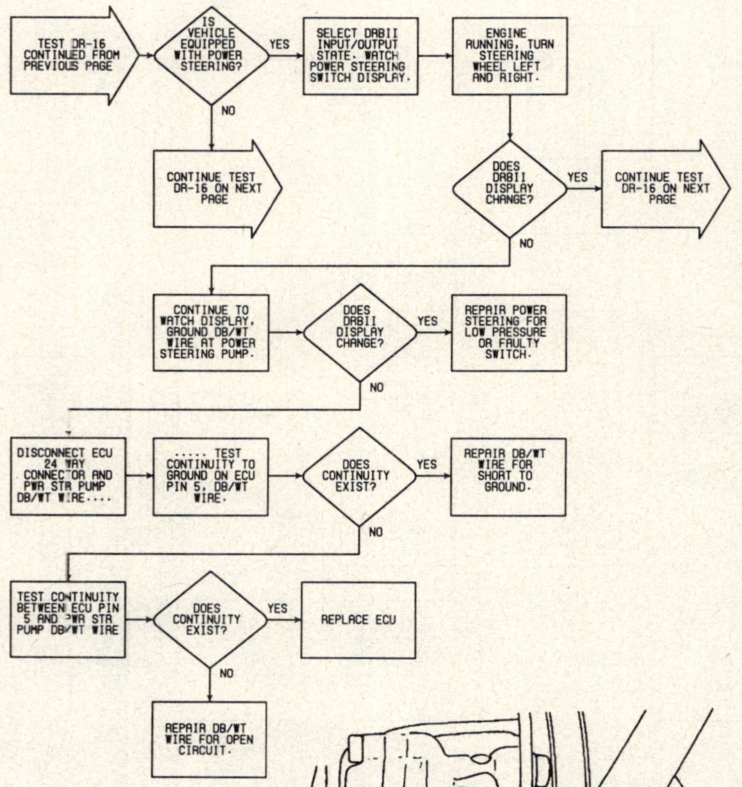

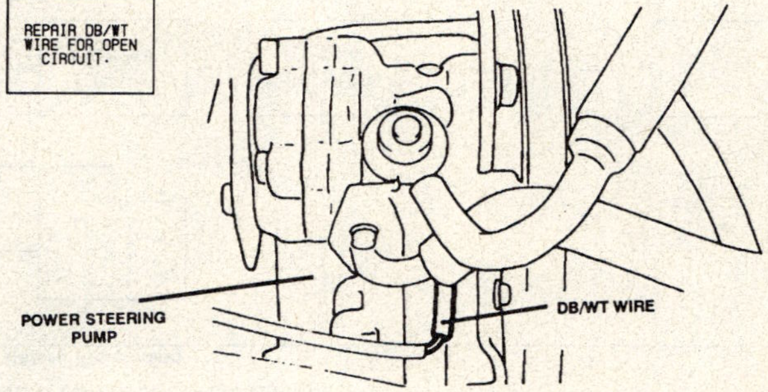

TEST DR-16 CONTINUED - NO FAULT CODE SENSOR TESTS

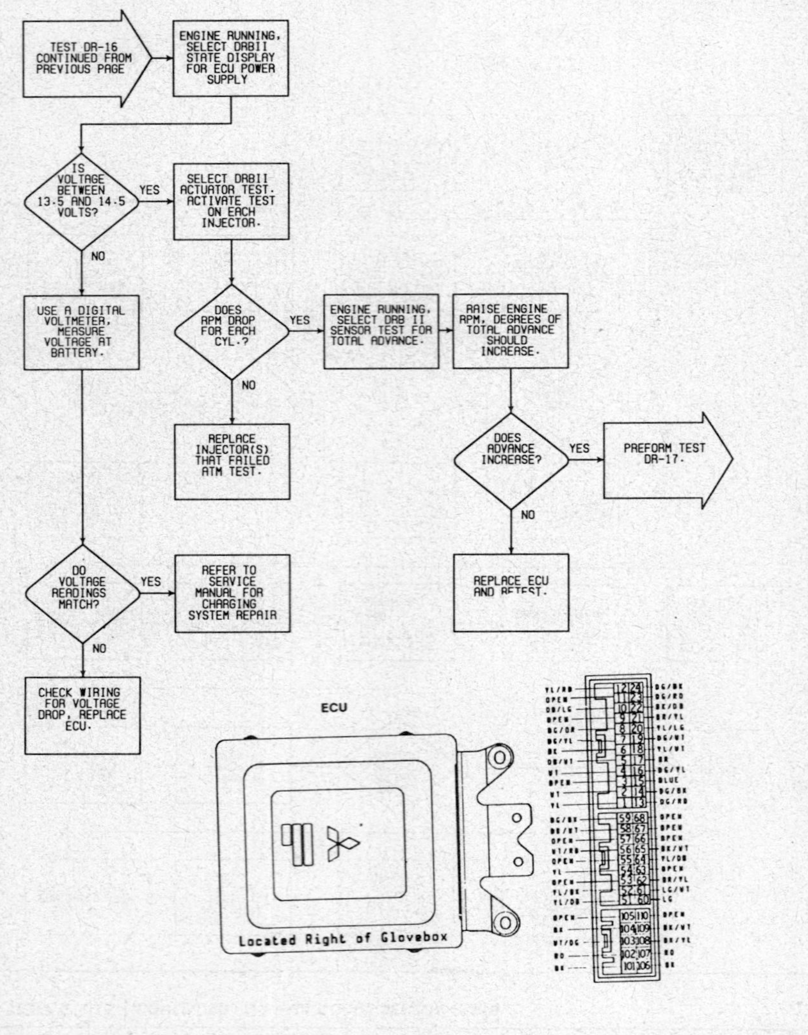

TEST DR-17 TESTING PURGE SOLENOID CIRCUIT

Perform TEST DR-16 Before Proceeding

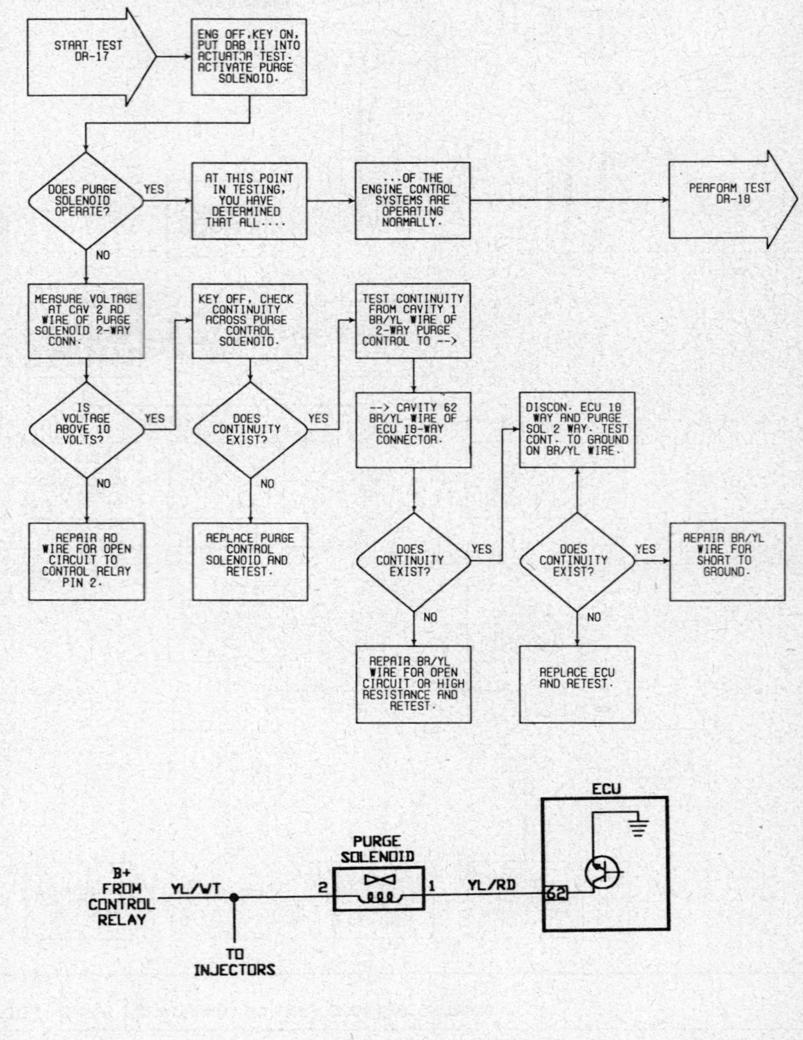

1.5L, 1.8L AND 2.0L ENGINES – DRIVEABILITY TEST

TEST DR-18 | **NO FAULT CODE MECHANICAL TEST**

Perform TEST DR-17 Before Proceeding

At this point in the driveability test procedure, you have determined that all of the engine control systems are operating as they were designed to. Therefore, they are not the cause of the driveability problem.

The following additional items can not be overlooked as possible causes of a driveability problem.

1. **THROTTLE VALVE AREA** – Dirt or ice buildup causing rough idle and stalling.
2. **ENGINE IGNITION TIMING** – Must be set with timing terminal grounded.
3. **ENGINE VACUUM** – Must be normal for your altitude.
4. **ENGINE VALVE TIMING** – To specifications.
5. **ENGINE COMPRESSION** – To specifications.
6. **ENGINE P.C.V. SYSTEM** – Must flow freely.
7. **ENGINE EXHAUST SYSTEM** – Must be free of any restrictions.
8. **POWER BRAKE BOOSTER** – No internal vacuum leaks.
9. **TORQUE CONVERTER CONDITION** – May cause very low power at breakaway or high speed. (Only 1 condition at a time)
10. **FUEL CONTAMINATION** – High alcohol or water content.
11. **FUEL INJECTORS** – Rough idle may be caused by injector wiring not connected to correct injector.
12. **ENGINE SECONDARY IGNITION CHECK** – Abnormal scope patterns.
13. **TECHNICAL SERVICE BULLETINS** – Any that apply to vehicle.
14. All air intake piping and vacuum hoses must be in place and secure. The proper air filter element must be used.
15. **FUEL PRESSURE** – Must be correct.
    Spec: With no vacuum at the regulator: 48 PSI on V6 & non-turbo 4 Cyl. engines
    36 PSI on turbo engines.

1.6L ENGINE – NO START TEST

TEST NS-1 | **START IGNITION CHECK**

Perform VISUAL INSPECTION Before Proceeding

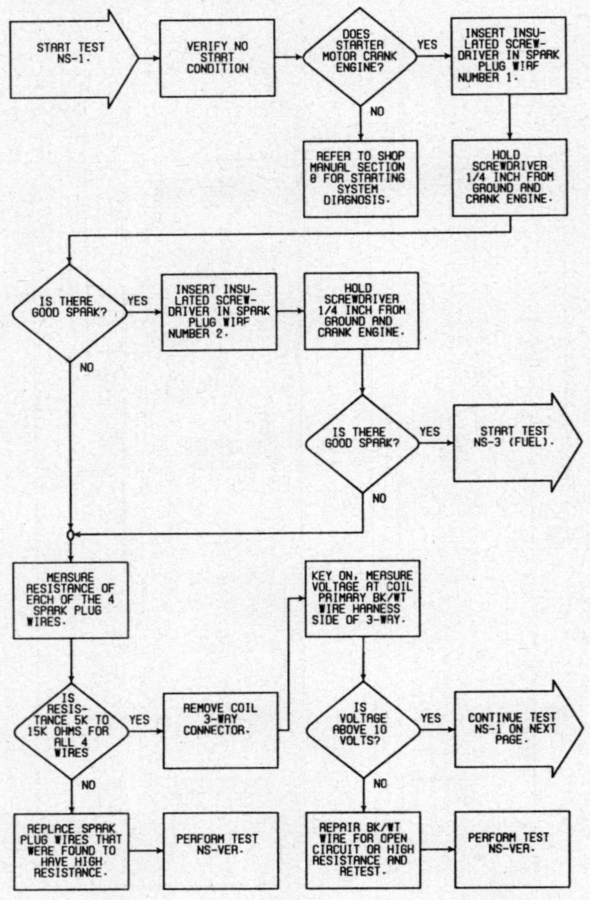

1.5L ENGINE—WIRING SCHEMATIC

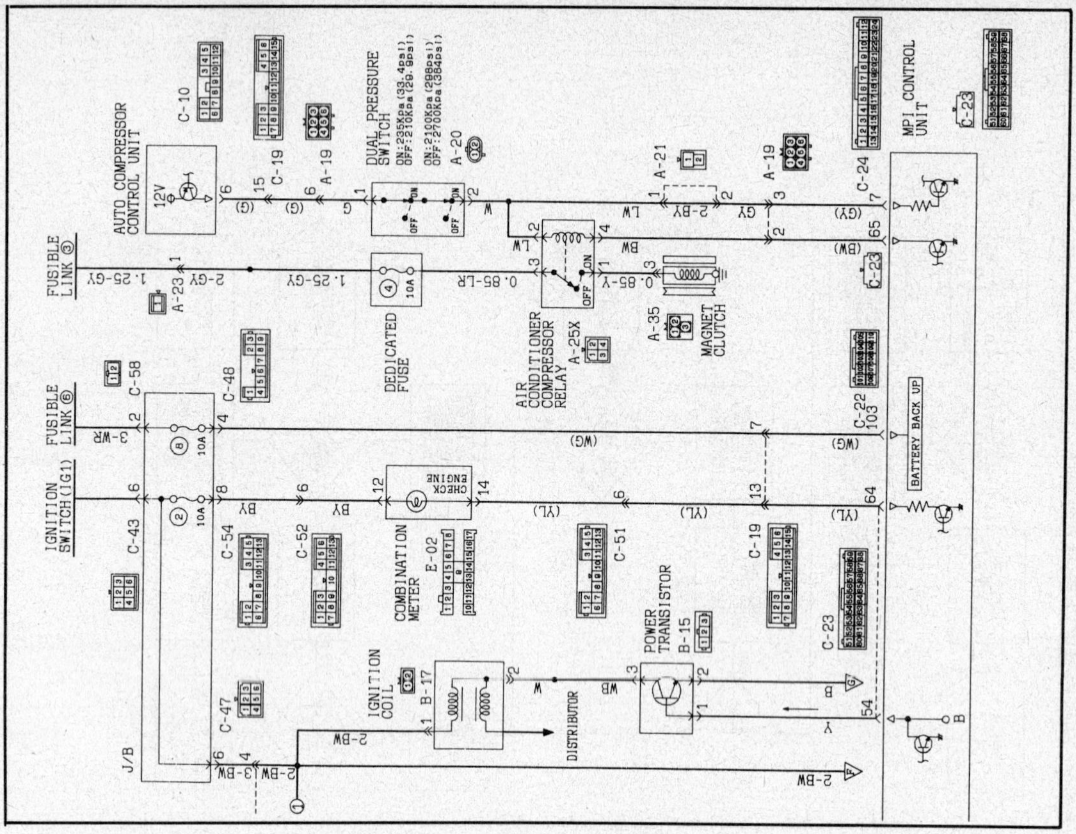

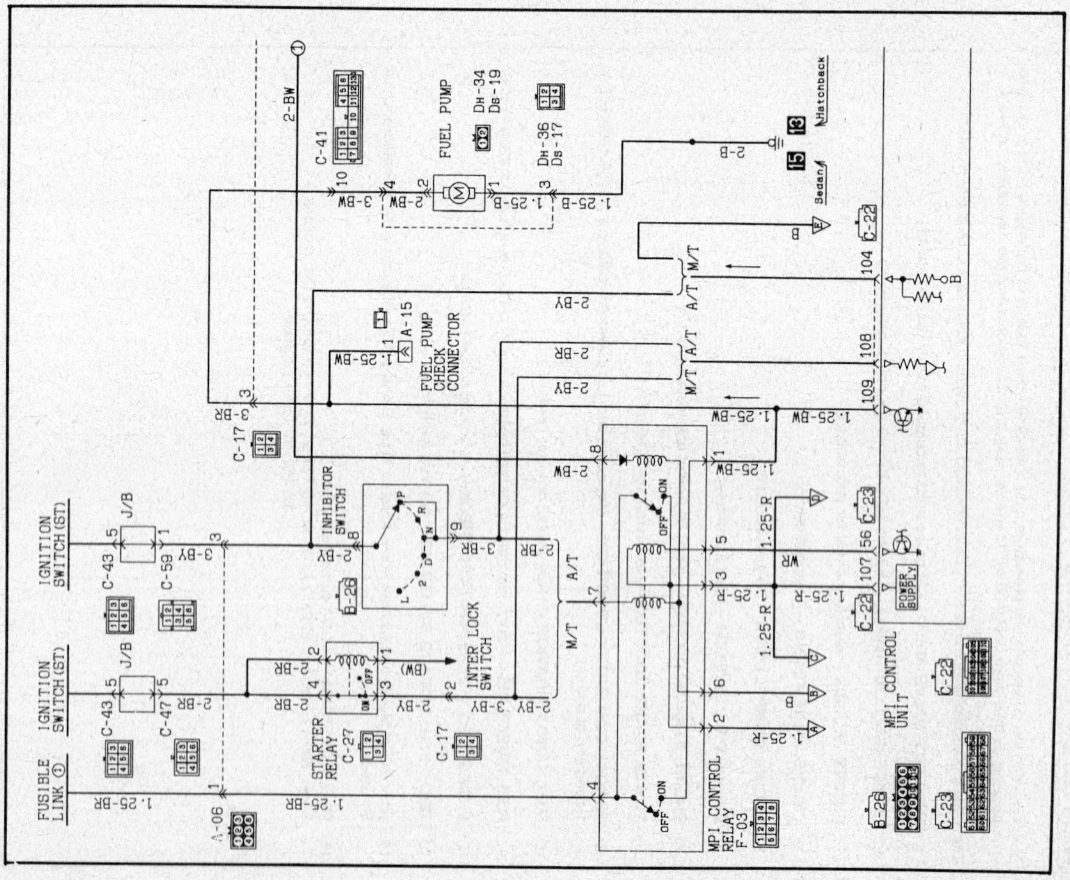

1.5L ENGINE—WIRING SCHEMATIC

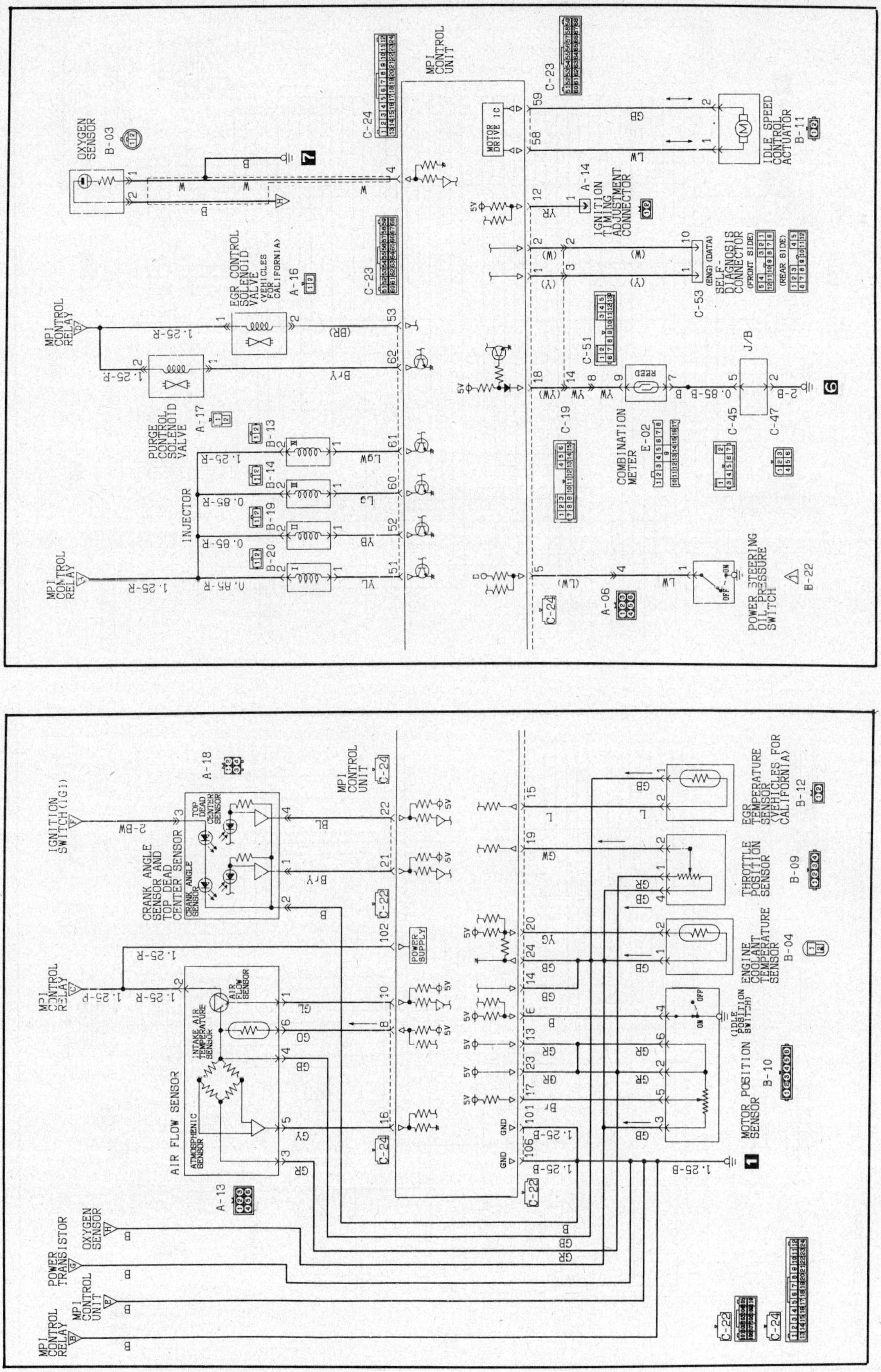

1.8L ENGINE—WIRING SCHEMATIC

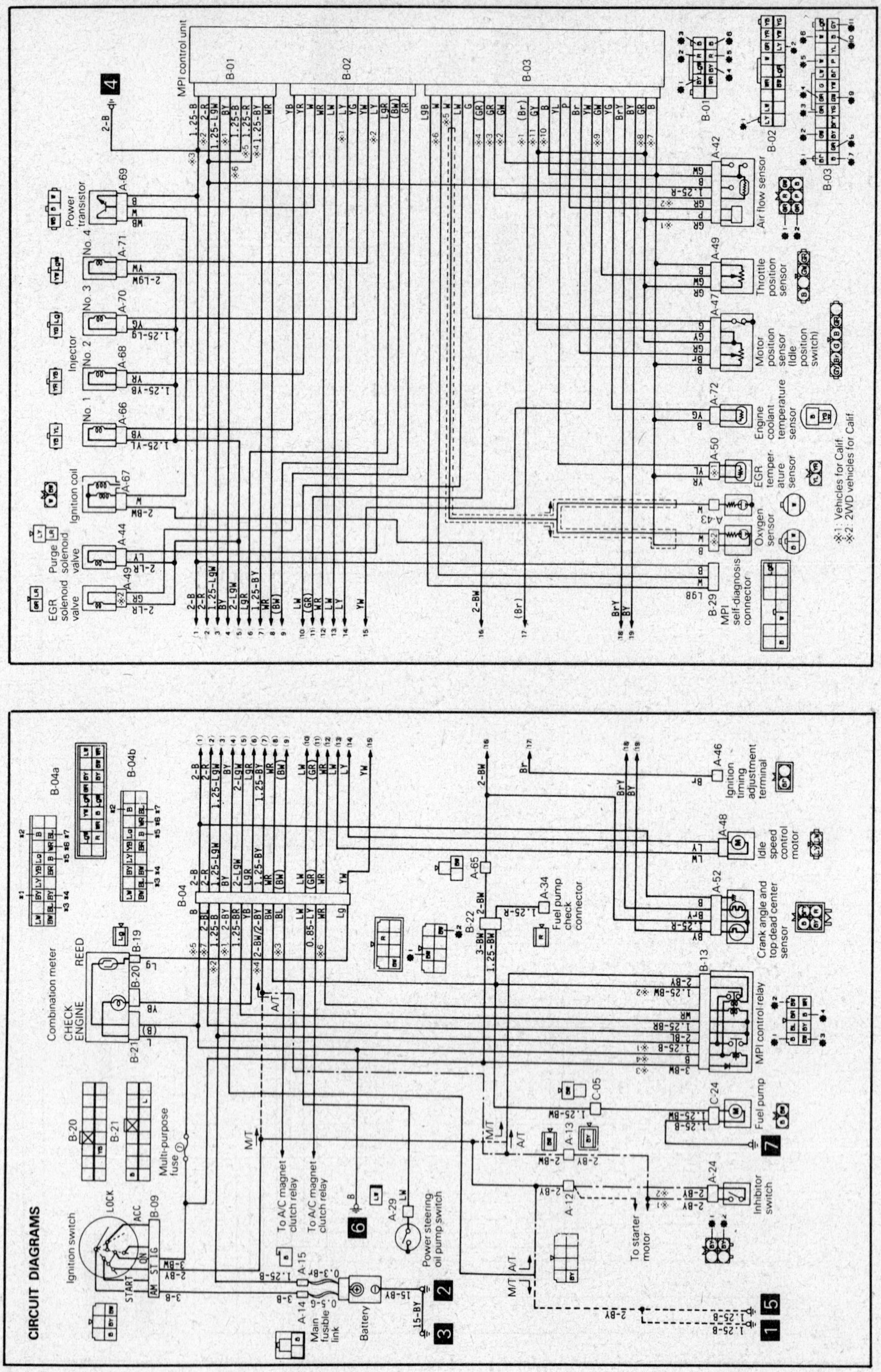

2.0L ENGINE—WIRING SCHEMATIC

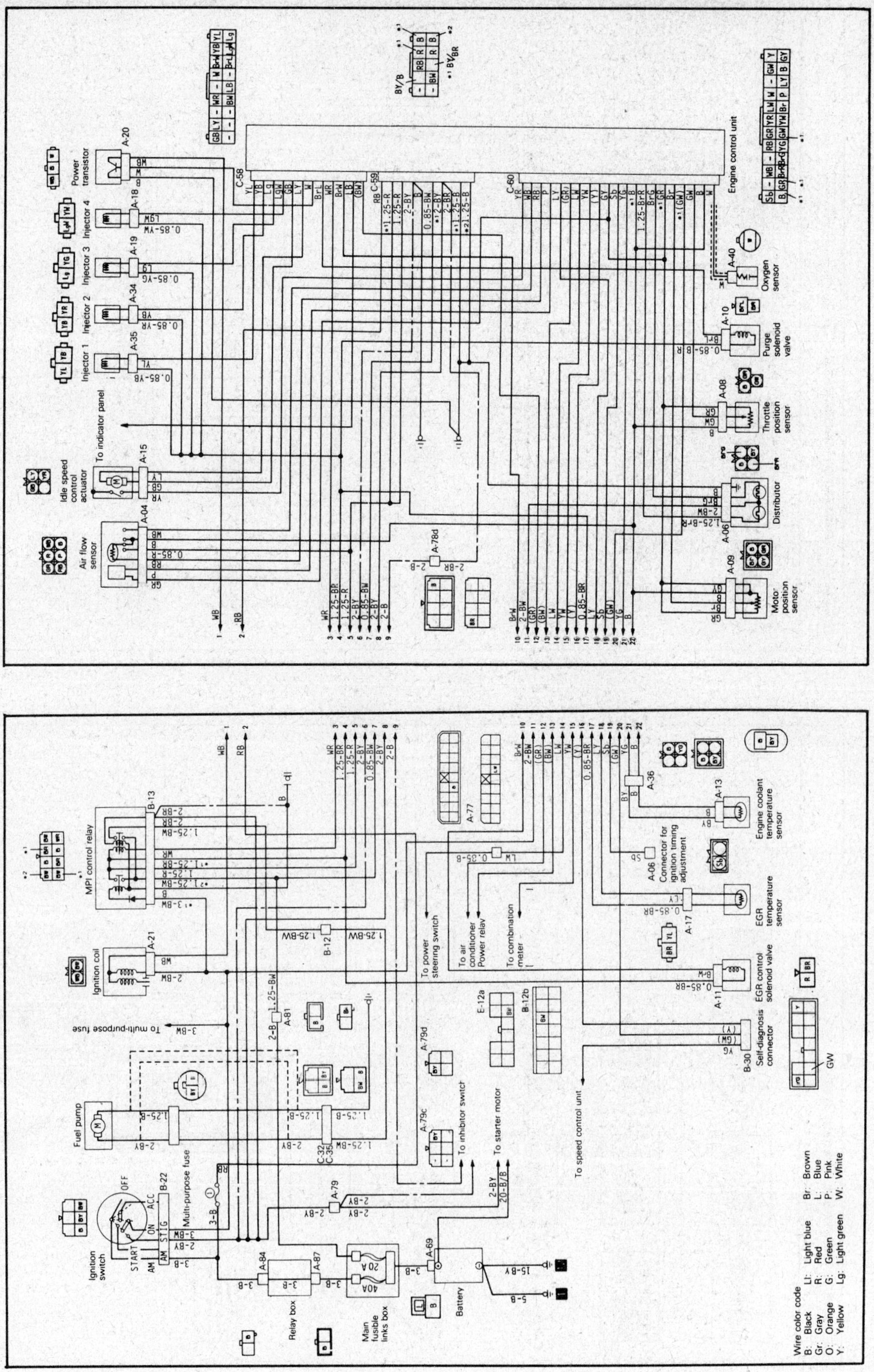

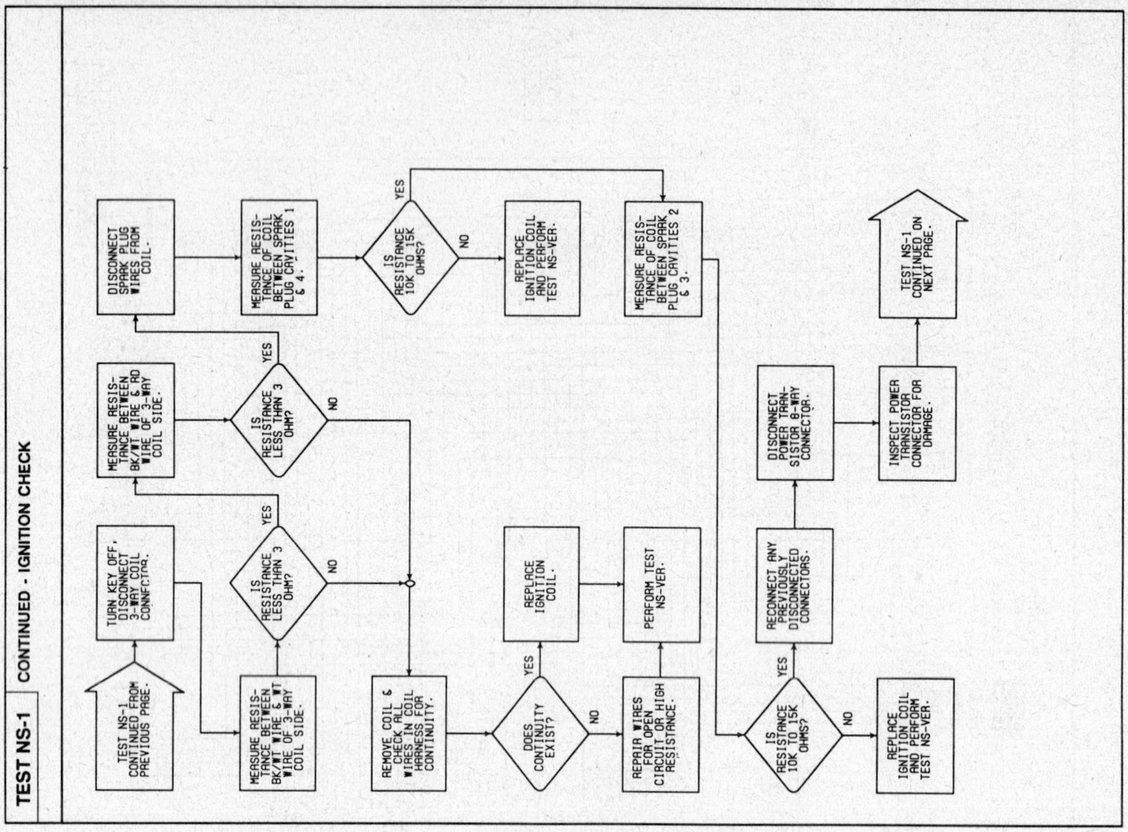
TEST NS-1
CONTINUED - IGNITION CHECK
TEST NS-1 CONTINUED FROM PREVIOUS PAGE.
TURN KEY OFF. DISCONNECT 3-WAY COIL CONNECTOR.
MEASURE RESISTANCE BETWEEN BK/WT WIRE & WT WIRE OF 3-WAY COIL SIDE.
IS RESISTANCE LESS THAN 3 OHM?
YES
NO
MEASURE RESISTANCE BETWEEN BK/WT WIRE & RD WIRE OF 3-WAY COIL SIDE.
IS RESISTANCE LESS THAN 3 OHM?
YES
NO
DISCONNECT SPARK PLUG WIRES FROM COIL.
MEASURE RESISTANCE OF COIL BETWEEN SPARK PLUG CAVITIES 1 & 4.
IS RESISTANCE 10K TO 15K OHMS?
YES
NO
REPLACE IGNITION COIL AND PERFORM TEST NS-VER.
MEASURE RESISTANCE OF COIL BETWEEN SPARK PLUG CAVITIES 2 & 3.
REMOVE COIL & CHECK ALL WIRES IN COIL HARNESS FOR CONTINUITY.
DOES CONTINUITY EXIST?
YES
NO
REPLACE IGNITION COIL.
PERFORM TEST NS-VER.
REPAIR WIRES FOR OPEN CIRCUIT OR HIGH RESISTANCE.
IS RESISTANCE 10K TO 15K OHMS?
YES
NO
REPLACE IGNITION COIL AND PERFORM TEST NS-VER.
RECONNECT ANY PREVIOUSLY DISCONNECTED CONNECTORS.
DISCONNECT POWER TRANSISTOR 8-WAY CONNECTOR.
INSPECT POWER TRANSISTOR CONNECTOR FOR DAMAGE.
TEST NS-1 CONTINUED ON NEXT PAGE.

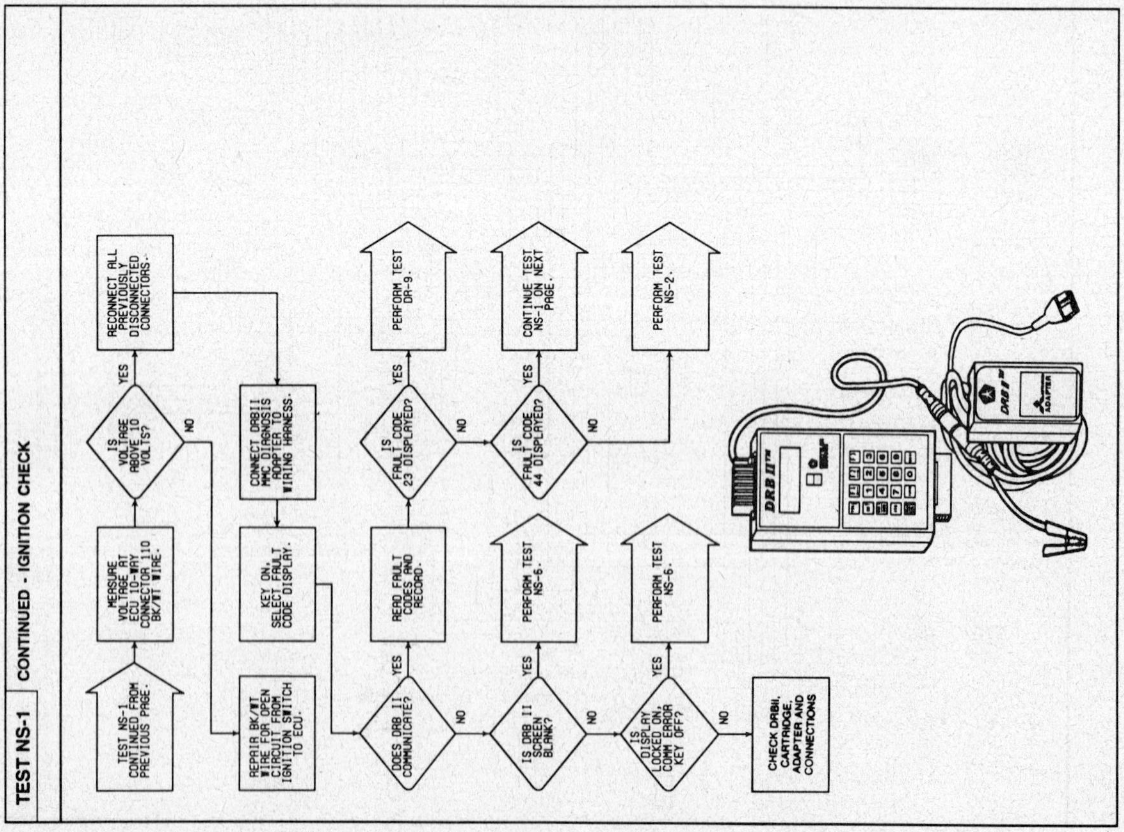
TEST NS-1
CONTINUED - IGNITION CHECK
TEST NS-1 CONTINUED FROM PREVIOUS PAGE.
MEASURE VOLTAGE AT ECU 10-WAY CONNECTOR 110 BK/WT WIRE.
IS VOLTAGE ABOVE 10 VOLTS?
YES
NO
RECONNECT ALL PREVIOUSLY DISCONNECTED CONNECTORS.
REPAIR BK/WT WIRE FOR OPEN CIRCUIT FROM IGNITION SWITCH TO ECU.
CONNECT DRBII MULTI-DIAGNOSIS ADAPTER TO WIRING HARNESS.
KEY ON. SELECT FAULT CODE DISPLAY.
DOES DRB II COMMUNICATE?
YES
NO
READ FAULT CODES AND RECORD.
IS FAULT CODE 23 DISPLAYED?
YES
NO
PERFORM TEST DR-9.
IS FAULT CODE 44 DISPLAYED?
YES
NO
CONTINUE TEST NS-1 ON NEXT PAGE.
PERFORM TEST NS-2.
IS DRB II SCREEN BLANK?
YES
NO
PERFORM TEST NS-6.
IS DISPLAY LOCKED ON COMM ERROR KEY OFF?
YES
NO
PERFORM TEST NS-6.
CHECK DRBII, CARTRIDGE, ADAPTER AND CONNECTIONS.
DRB II
ADAPTER

1.6L ENGINE—NO START TEST

TEST NS-1 CONTINUED - IGNITION CHECK

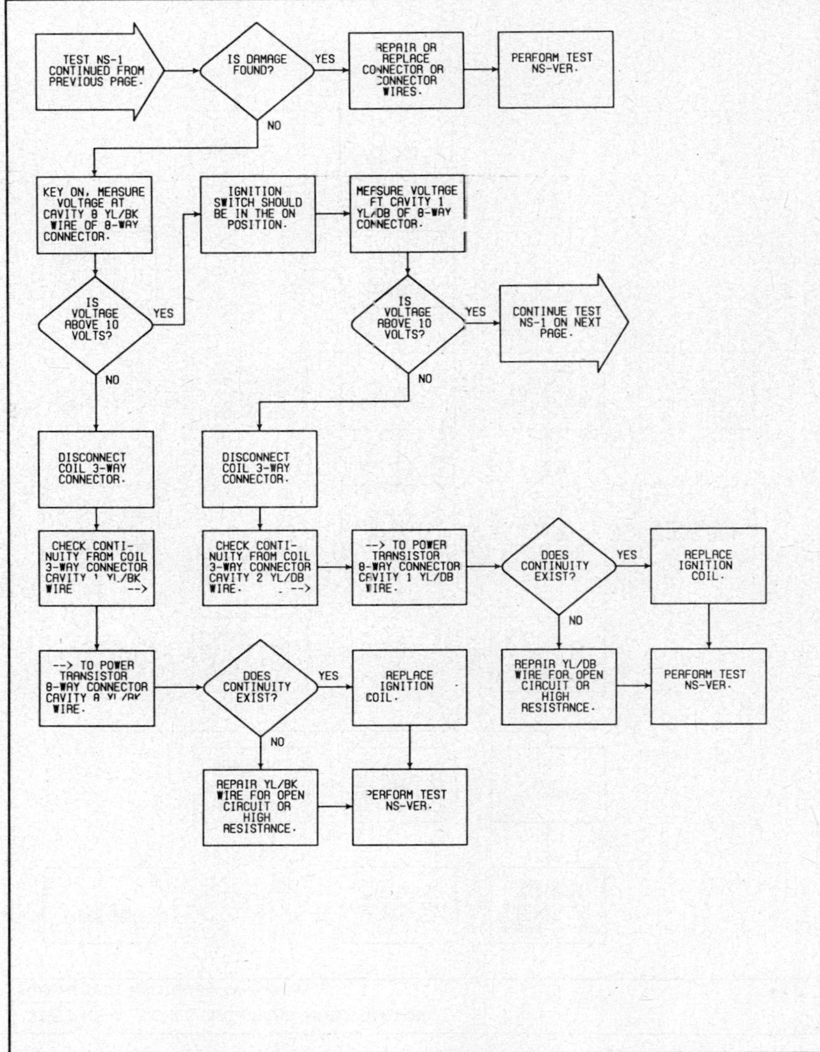

TEST NS-1 CONTINUED - IGNITION CHECK

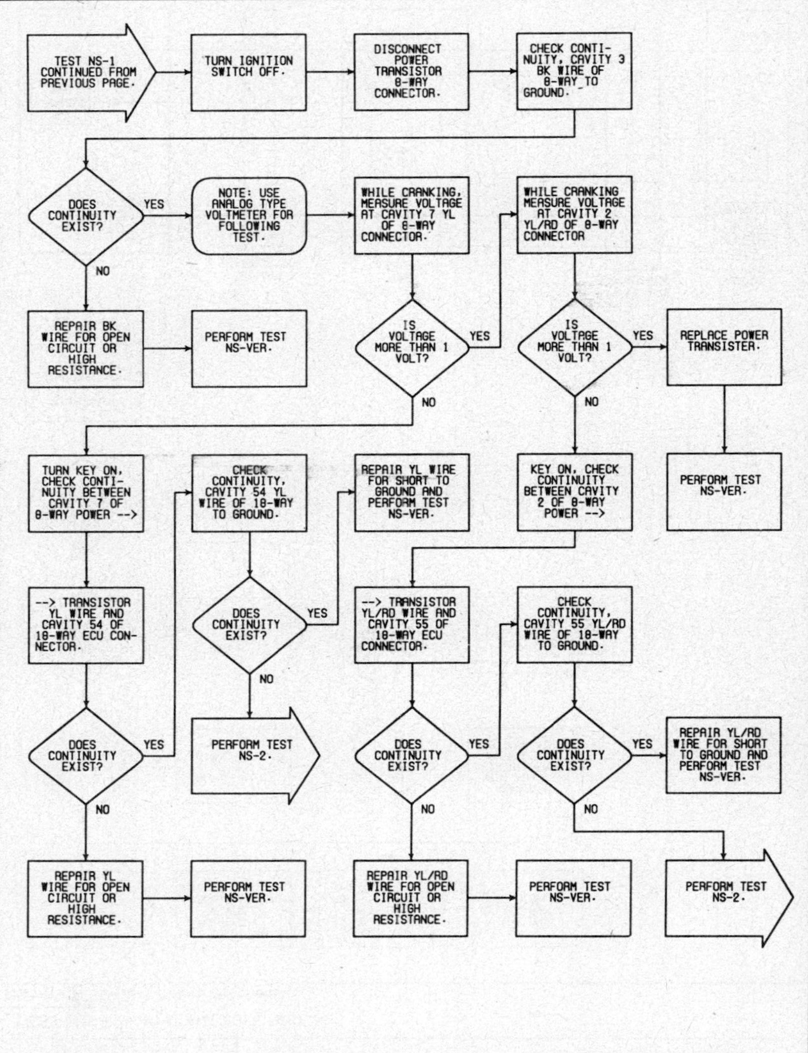

TEST NS-2 CHECKING CRANK ANGLE SENSOR

Perform TEST NS-1 Before Proceeding

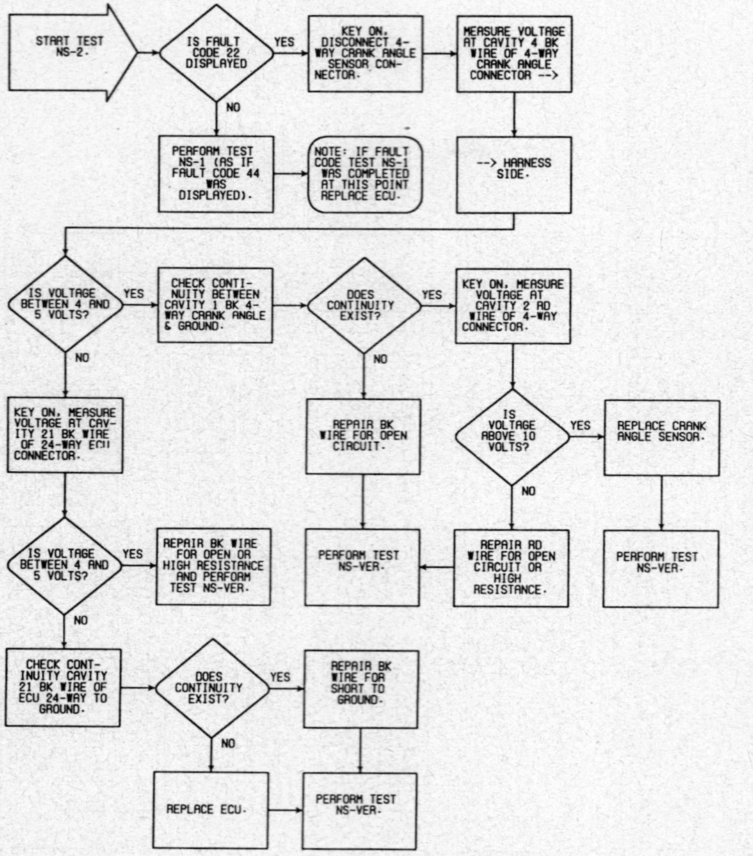

TEST NS-3 NO START FUEL TEST

Perform TEST NS-1 Before Proceeding

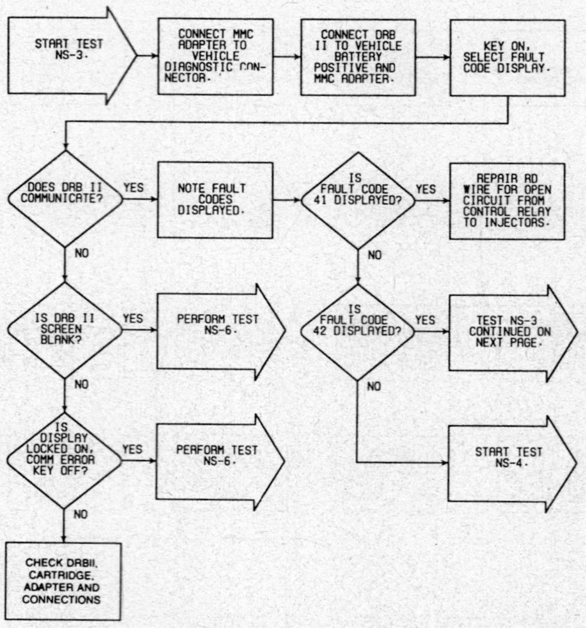

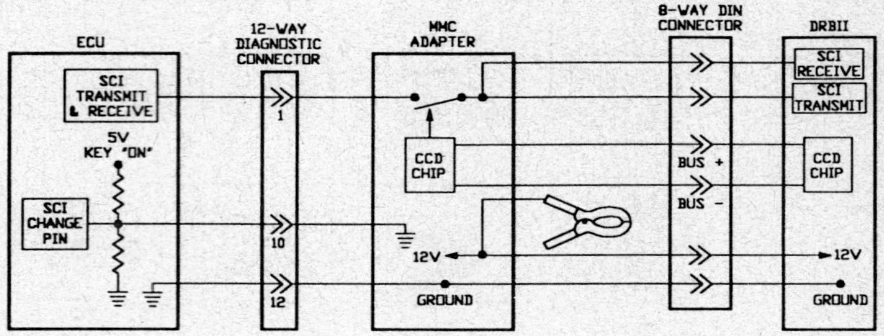

1.6L ENGINE—NO START TEST

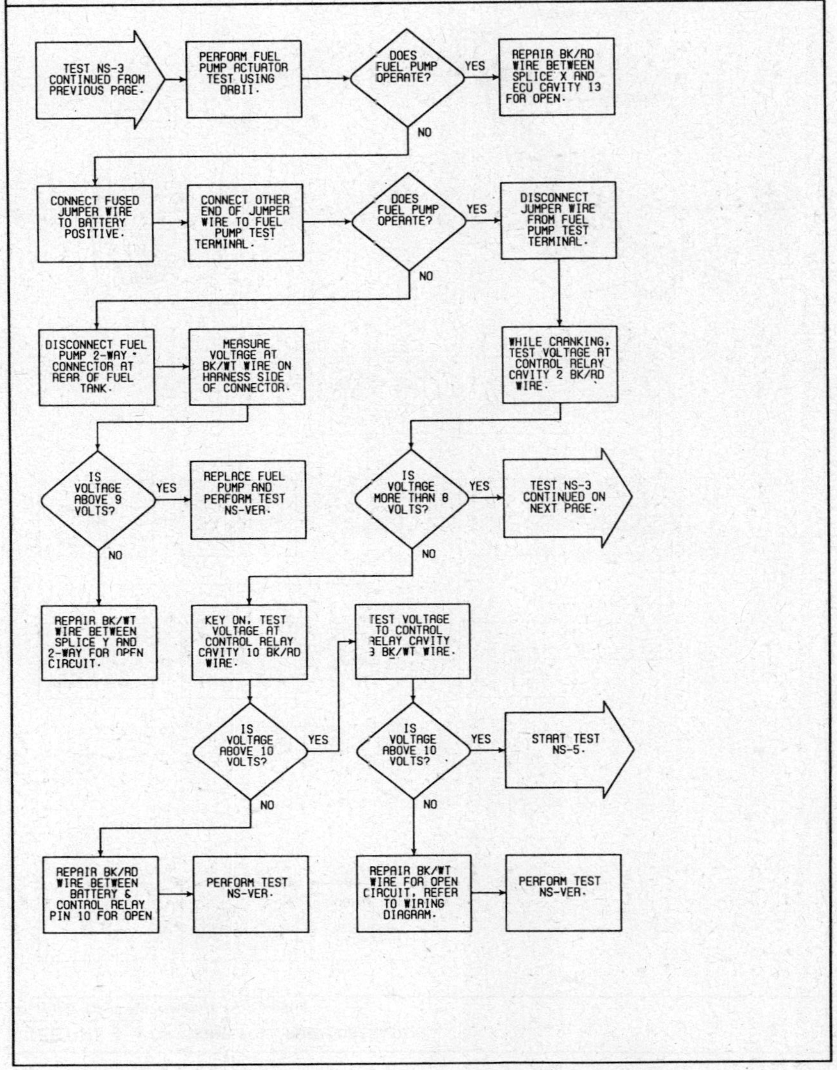

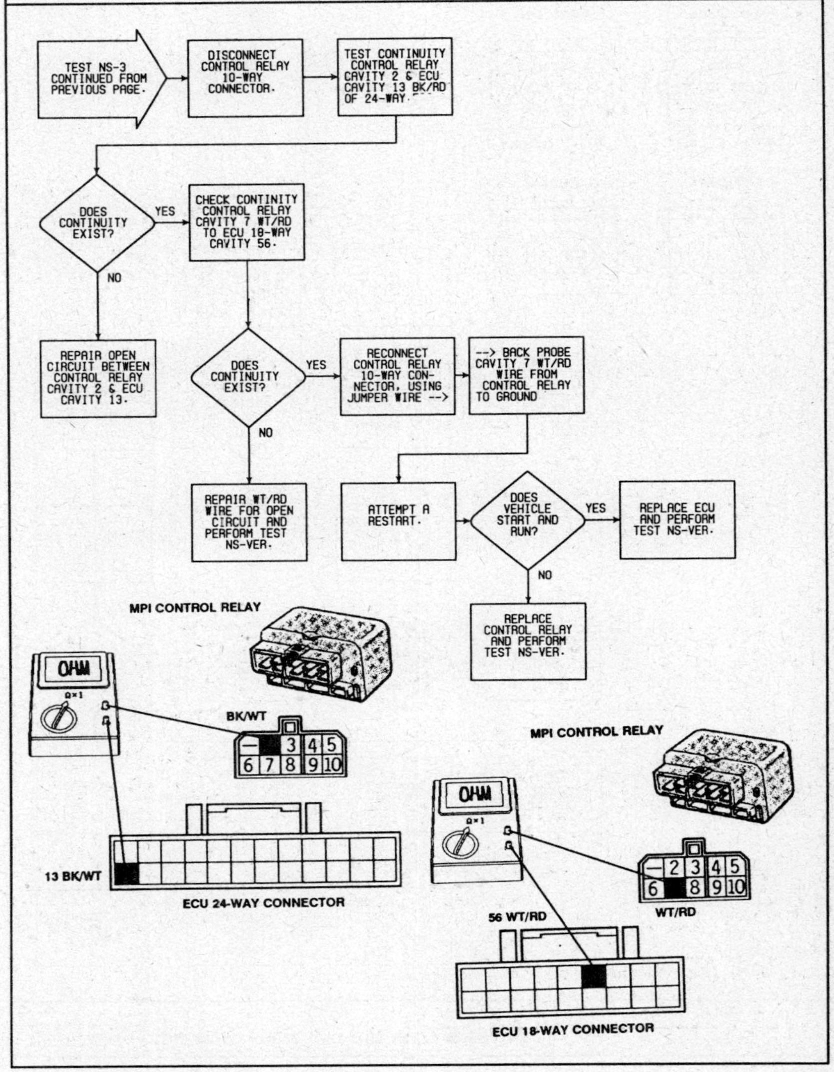

1.6L ENGINE—NO START TEST

TEST NS-4 CHECKING FUEL PUMP OPERATION

Perform TEST NS-1 Before Proceeding

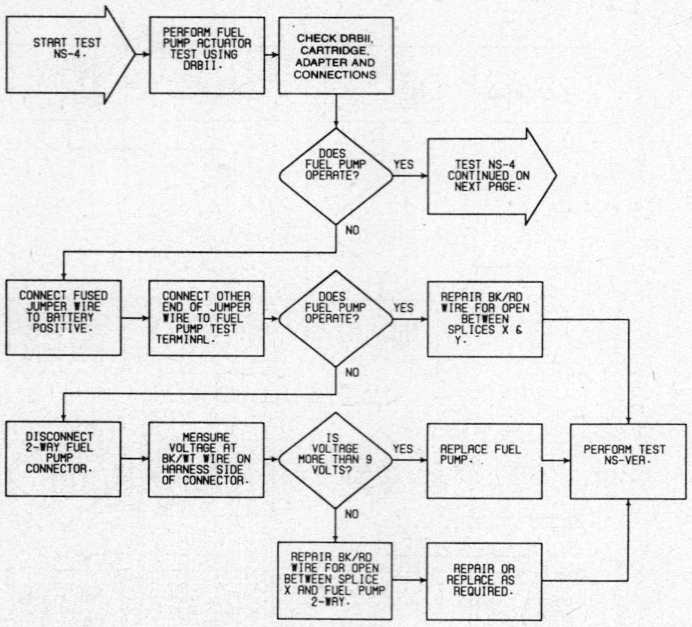

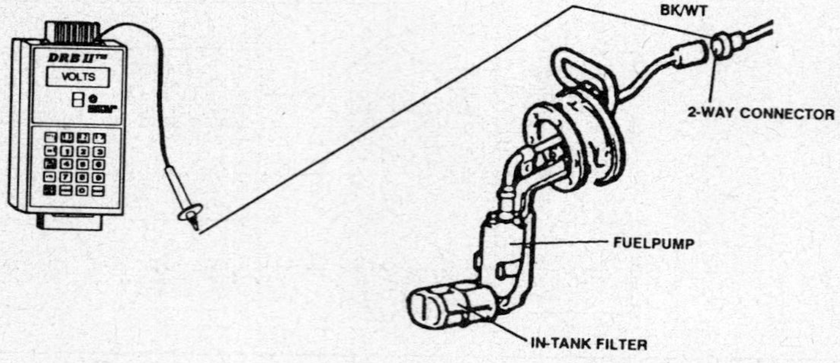

TEST NS-4 CONTINUED - CHECKING FUEL PUMP OPERATION

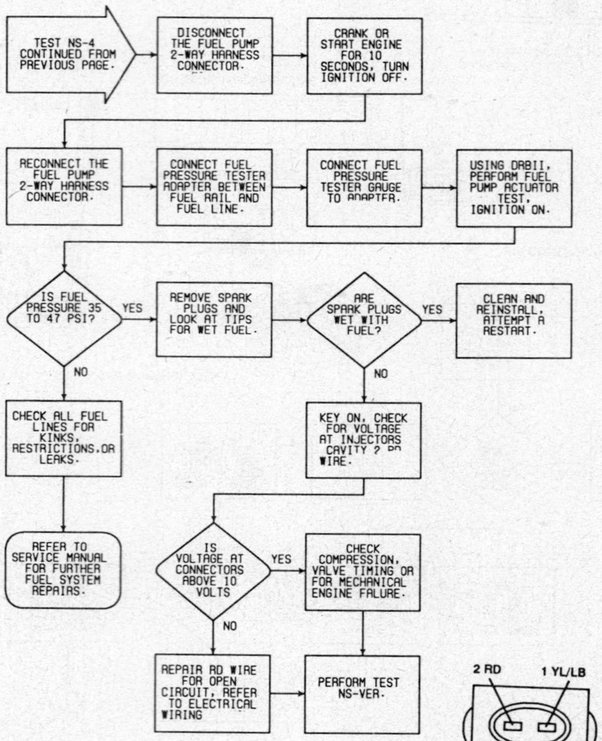

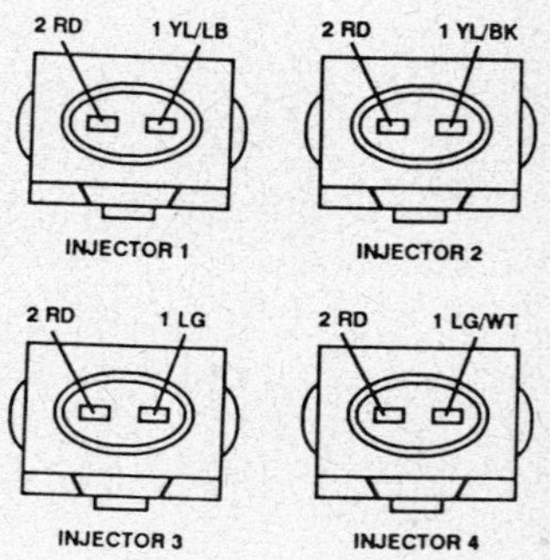

1.6L ENGINE—NO START TEST

TEST NS-5 CHECKING CONTROL RELAY CIRCUIT

Perform TEST NS-3 Before Proceeding

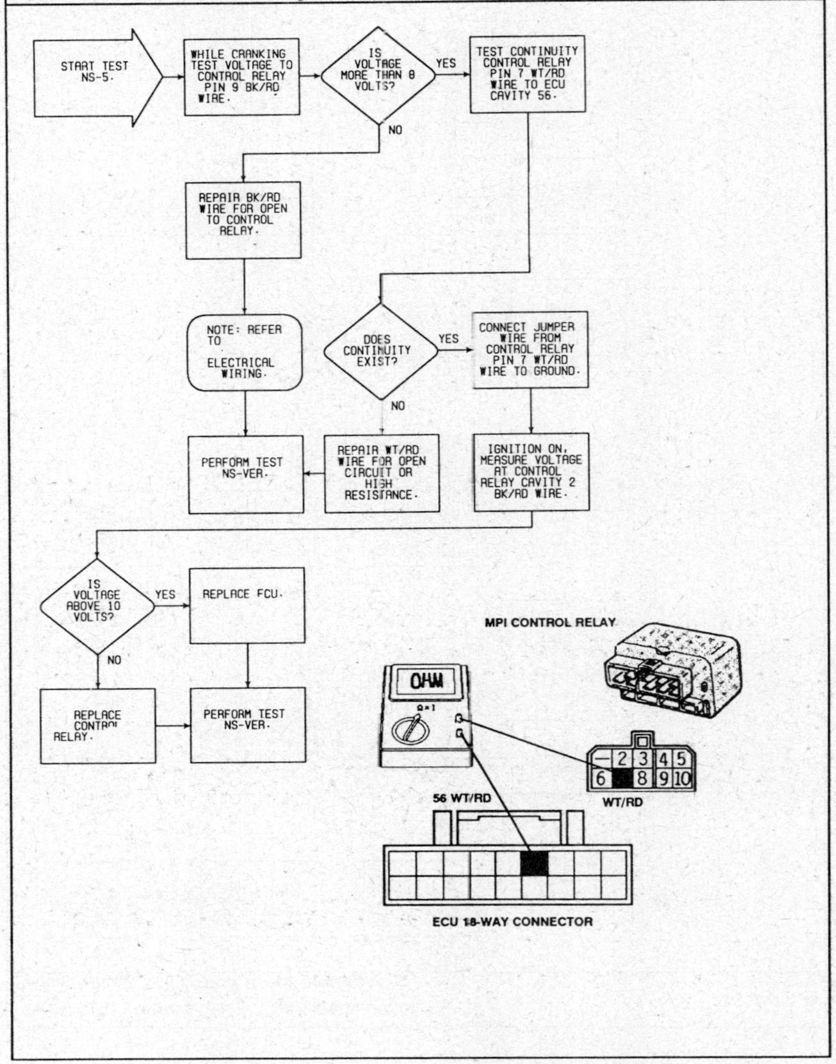

TEST NS-6 CHECKING DRBII CONNECTOR CIRCUIT

Perform TEST NS-1 Before Proceeding

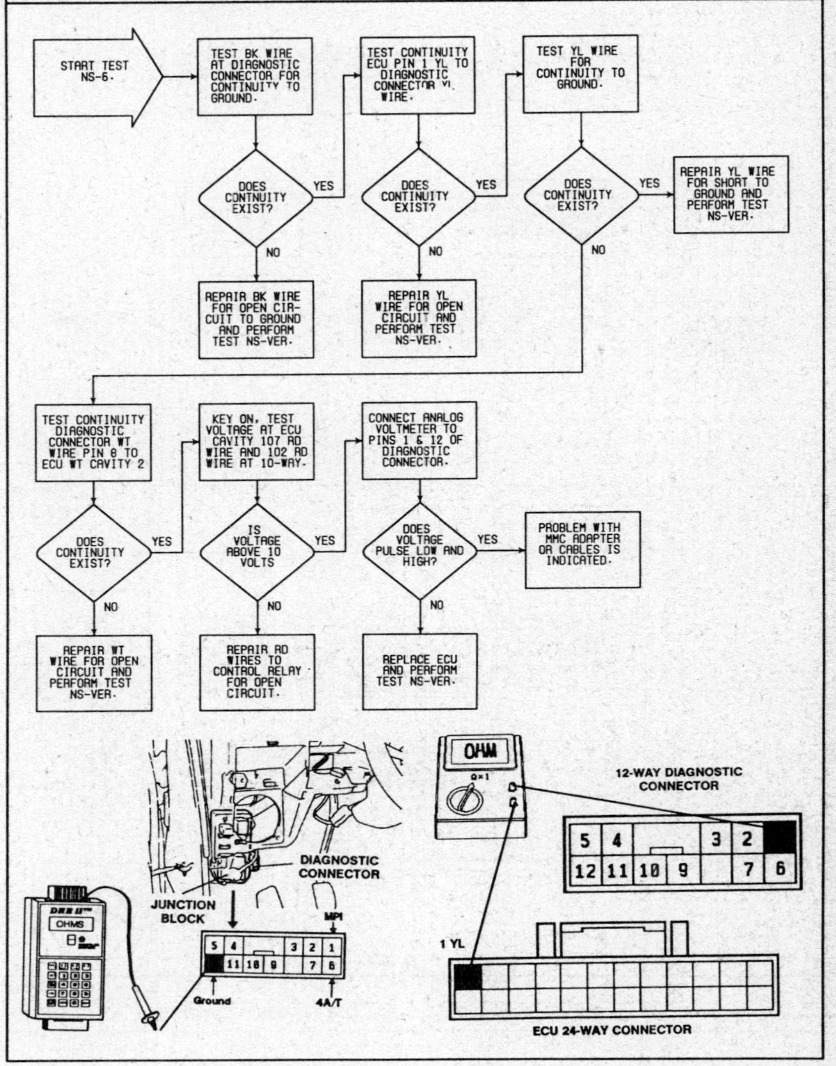

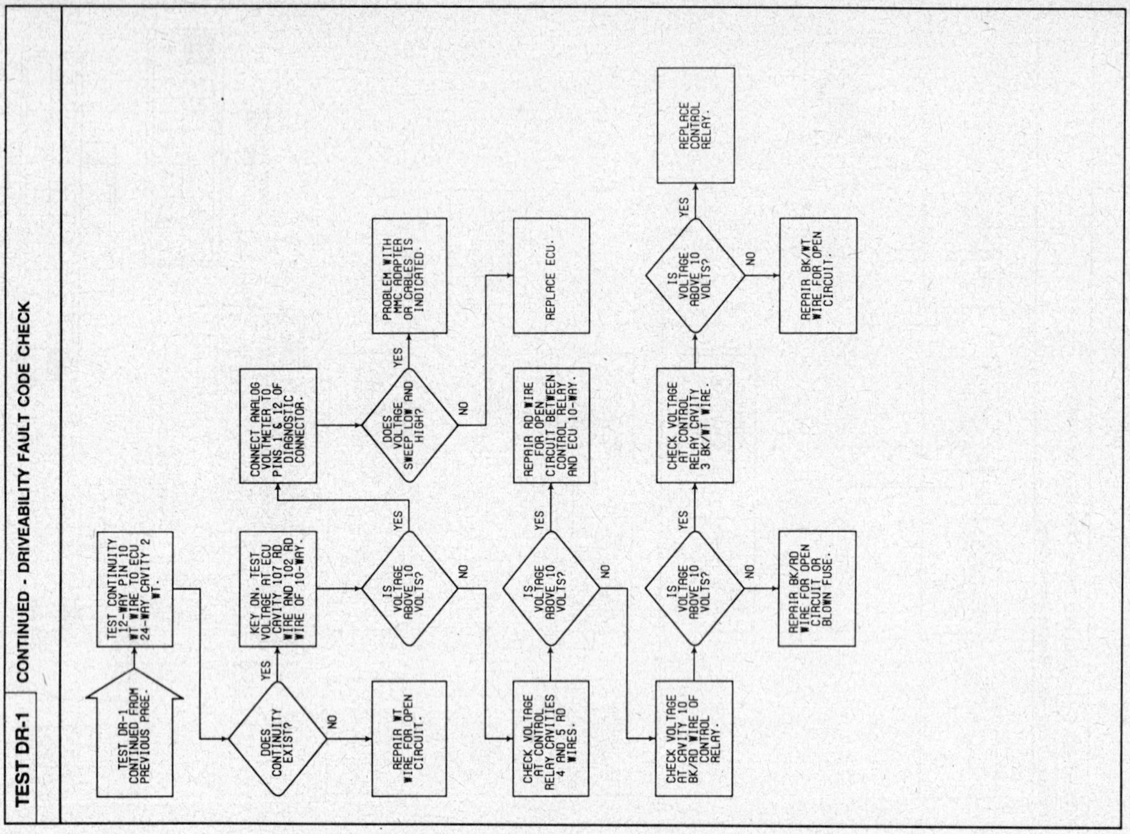
TEST DR-1
CONTINUED - DRIVEABILITY FAULT CODE CHECK
TEST DR-1 CONTINUED FROM PREVIOUS PAGE.
TEST CONTINUITY 12-WAY PIN 10 WT WIRE TO ECU 24-WAY CAVITY 2 WT.
DOES CONTINUITY EXIST?
YES
NO
REPAIR WT WIRE FOR OPEN CIRCUIT.
KEY ON, TEST VOLTAGE AT ECU CAVITY 107 RD WIRE AND 102 RD WIRE OF 10-WAY.
IS VOLTAGE ABOVE 10 VOLTS?
YES
NO
CONNECT ANALOG VOLTMETER TO PINS 1 & 12 OF DIAGNOSTIC CONNECTOR.
DOES VOLTAGE SWEEP LOW AND HIGH?
YES
NO
PROBLEM WITH MMC ADAPTER OR CABLES IS INDICATED.
REPLACE ECU.
CHECK VOLTAGE AT CONTROL RELAY CAVITIES 4 AND 5 RD WIRES.
IS VOLTAGE ABOVE 10 VOLTS?
YES
NO
REPAIR RD WIRE FOR OPEN CIRCUIT BETWEEN CONTROL RELAY AND ECU 10-WAY.
CHECK VOLTAGE AT CAVITY 10 BK/RD WIRE OF CONTROL RELAY.
IS VOLTAGE ABOVE 10 VOLTS?
YES
NO
REPAIR BK/RD WIRE FOR OPEN CIRCUIT OR BLOWN FUSE.
CHECK VOLTAGE AT CONTROL RELAY CAVITY 3 BK/WT WIRE.
IS VOLTAGE ABOVE 10 VOLTS?
YES
NO
REPLACE CONTROL RELAY.
REPAIR BK/WT WIRE FOR OPEN CIRCUIT.

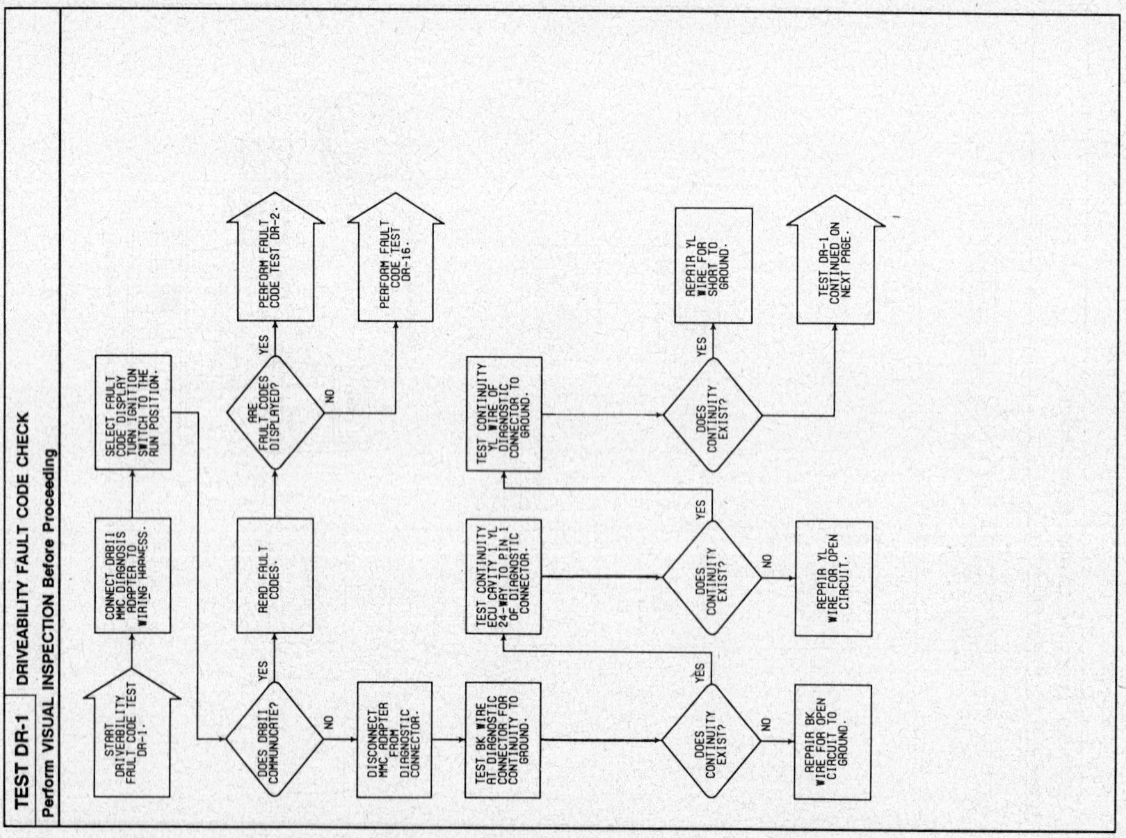
TEST DR-1
DRIVEABILITY FAULT CODE CHECK
Perform VISUAL INSPECTION Before Proceeding
START DRIVEABILITY FAULT CODE TEST DR-1.
CONNECT DRBII MMC DIAGNOSIS ADAPTER TO WIRING HARNESS.
SELECT FAULT CODE DISPLAY TURN IGNITION SWITCH TO THE RUN POSITION.
DOES DRBII COMMUNICATE?
YES
NO
READ FAULT CODES.
ARE FAULT CODES DISPLAYED?
YES
NO
PERFORM FAULT CODE TEST DR-2.
PERFORM FAULT CODE TEST DR-16.
DISCONNECT MMC ADAPTER FROM DIAGNOSTIC CONNECTOR.
TEST BK WIRE AT DIAGNOSTIC CONNECTOR FOR CONTINUITY TO GROUND.
DOES CONTINUITY EXIST?
YES
NO
REPAIR BK WIRE FOR OPEN CIRCUIT TO GROUND.
TEST CONTINUITY ECU CAVITY 21 YL 24-WAY TO PIN 1 OF DIAGNOSTIC CONNECTOR.
DOES CONTINUITY EXIST?
YES
NO
REPAIR YL WIRE FOR OPEN CIRCUIT.
TEST CONTINUITY YL WIRE OF DIAGNOSTIC CONNECTOR TO GROUND.
DOES CONTINUITY EXIST?
YES
REPAIR YL WIRE FOR SHORT TO GROUND.
TEST DR-1 CONTINUED ON NEXT PAGE.

TEST DR-2 FAULT CODE TEST

Perform TEST DR-1 Before Proceeding

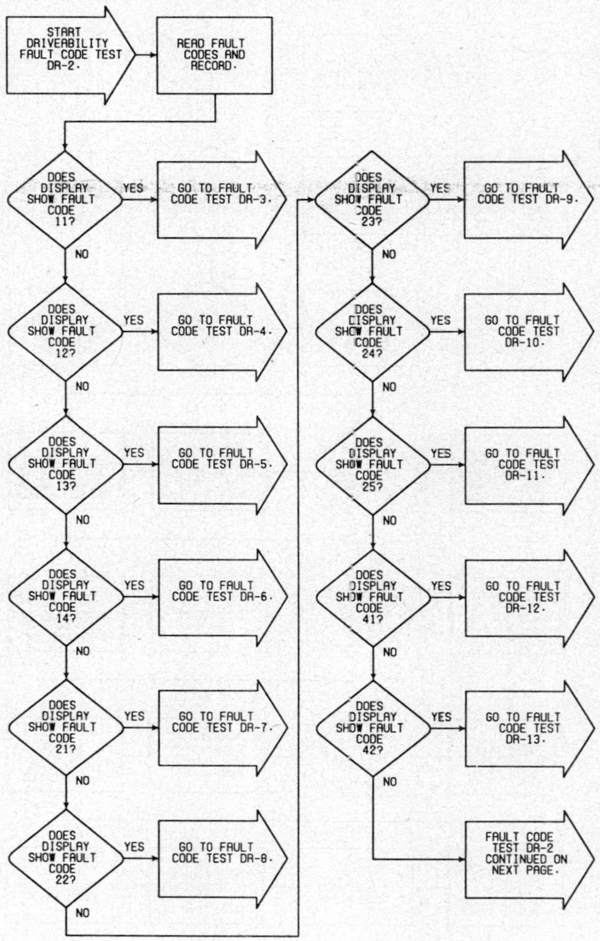

TEST DR-2 CONTINUED - FAULT CODE TEST

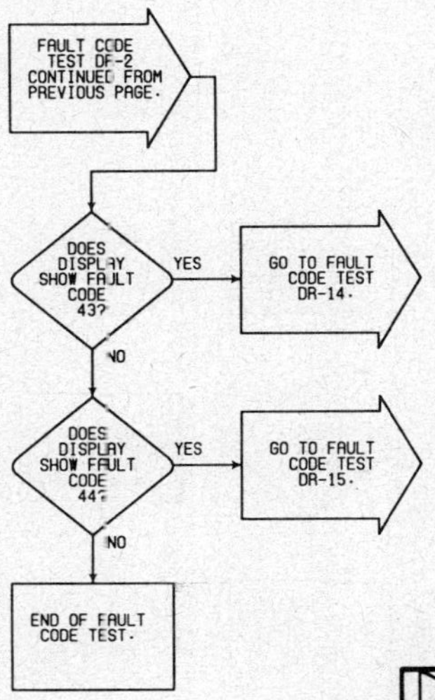

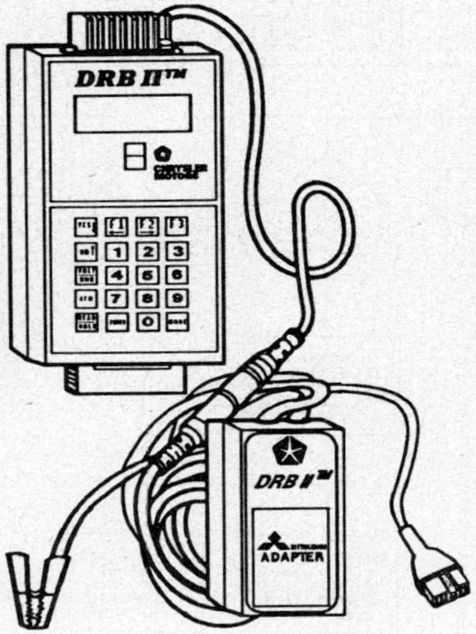

TEST DR-3 CHECKING OXYGEN SENSOR

Perform TEST DR-1 Before Proceeding

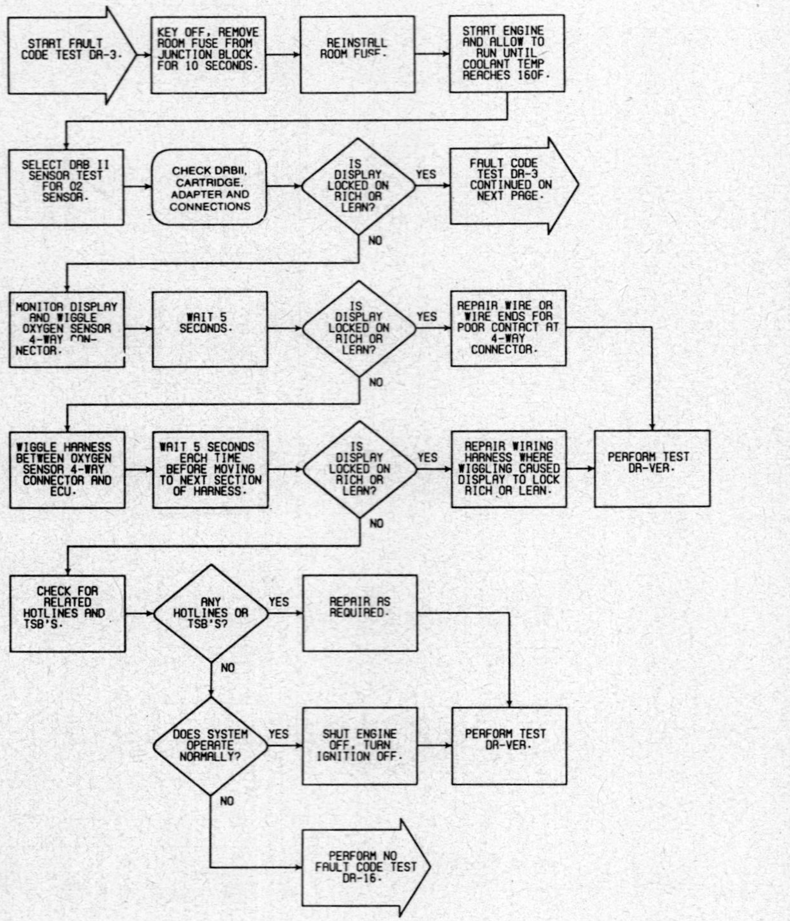

TEST DR-3 CONTINUED - CHECKING OXYGEN SENSOR

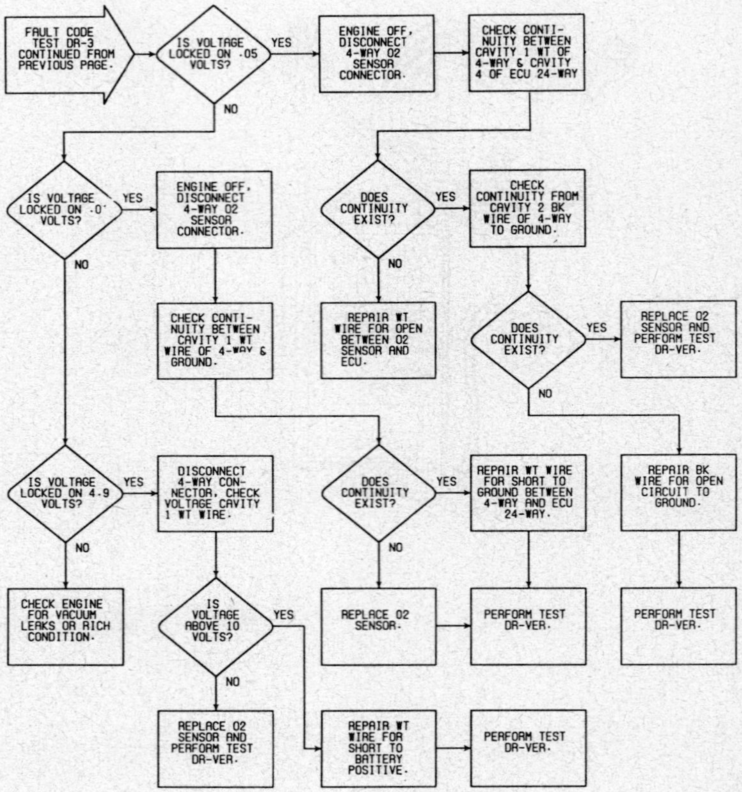

TEST DR-4 CHECKING AIR FLOW SENSOR

Perform TEST DR-1 Before Proceeding

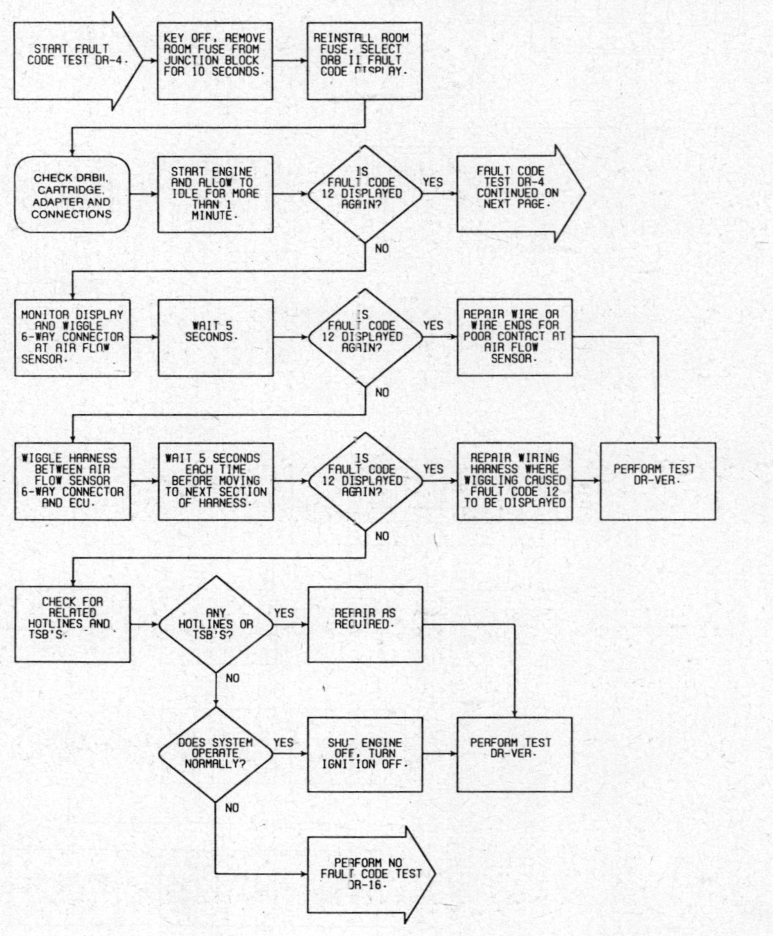

TEST DR-4 CONTINUED - CHECKING AIR FLOW SENSOR

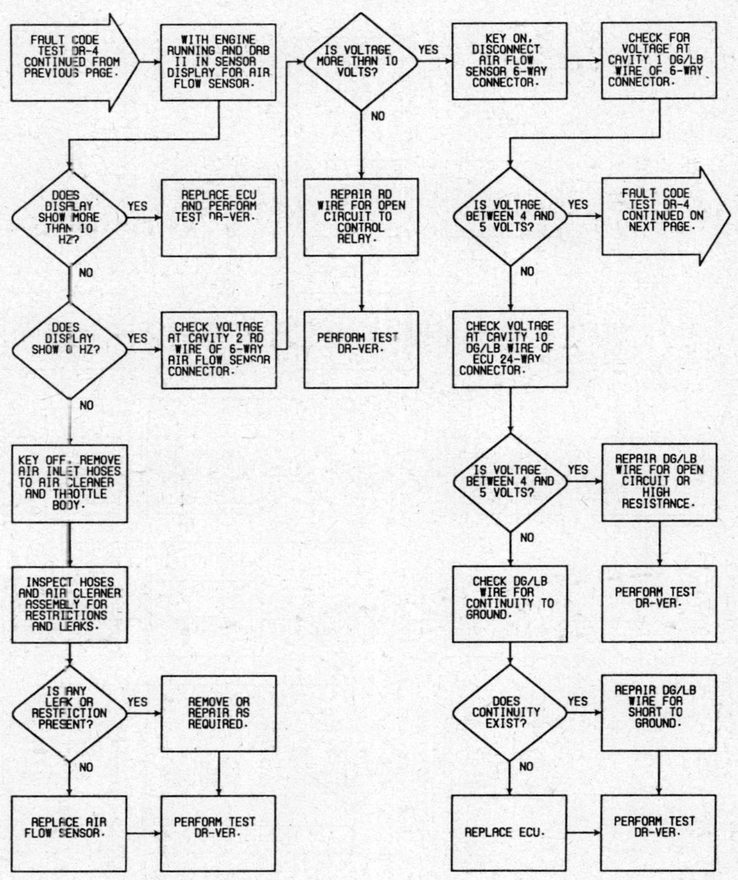

TEST DR-4 CONTINUED - CHECKING AIR FLOW SENSOR

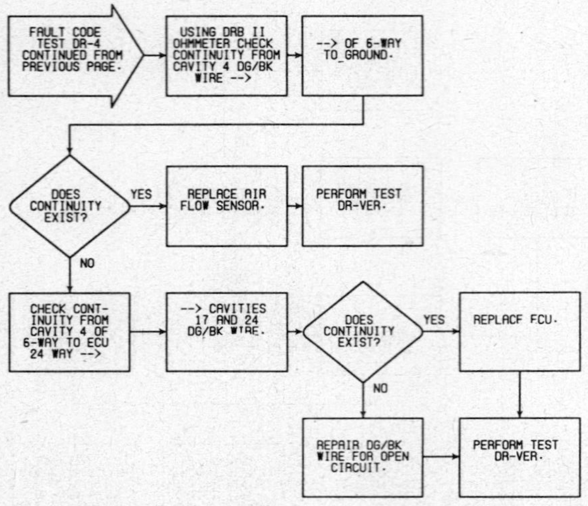

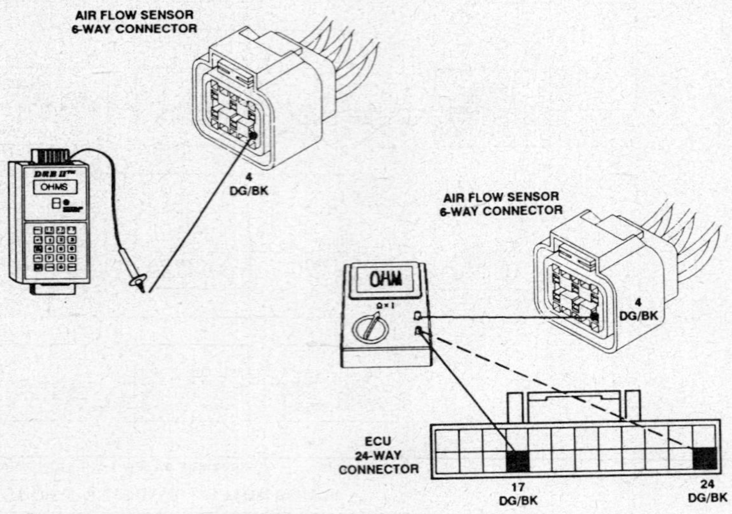

TEST DR-5 CHECKING INTAKE AIR TEMPERATURE SENSOR

Perform TEST DR-1 Before Proceeding

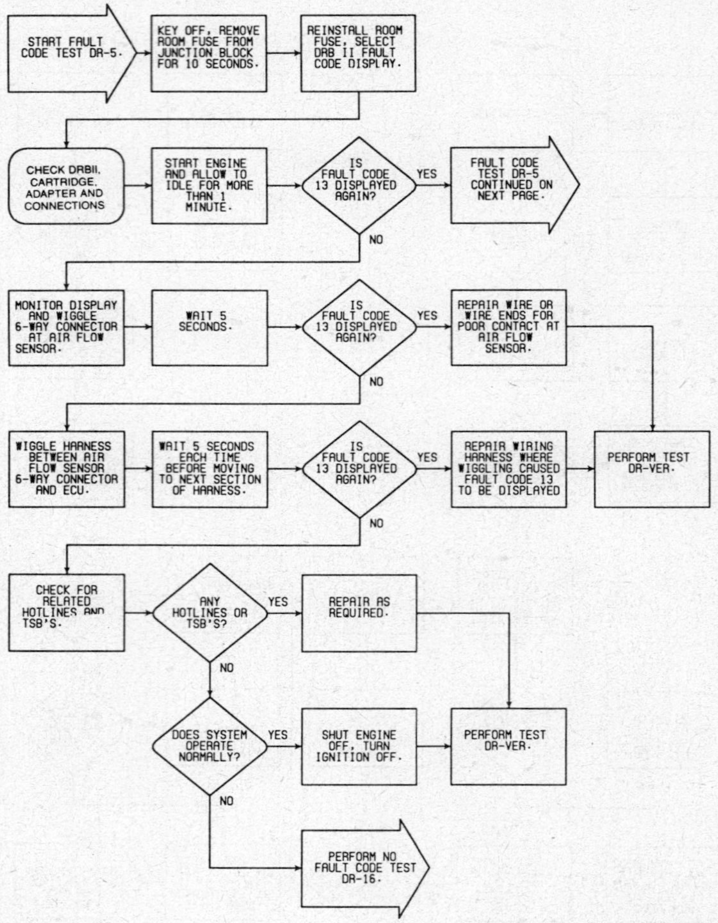

1.6L ENGINE – DRIVEABILITY TEST

TEST DR-5 CONTINUED - CHECKING INTAKE AIR TEMPERATURE SENSOR

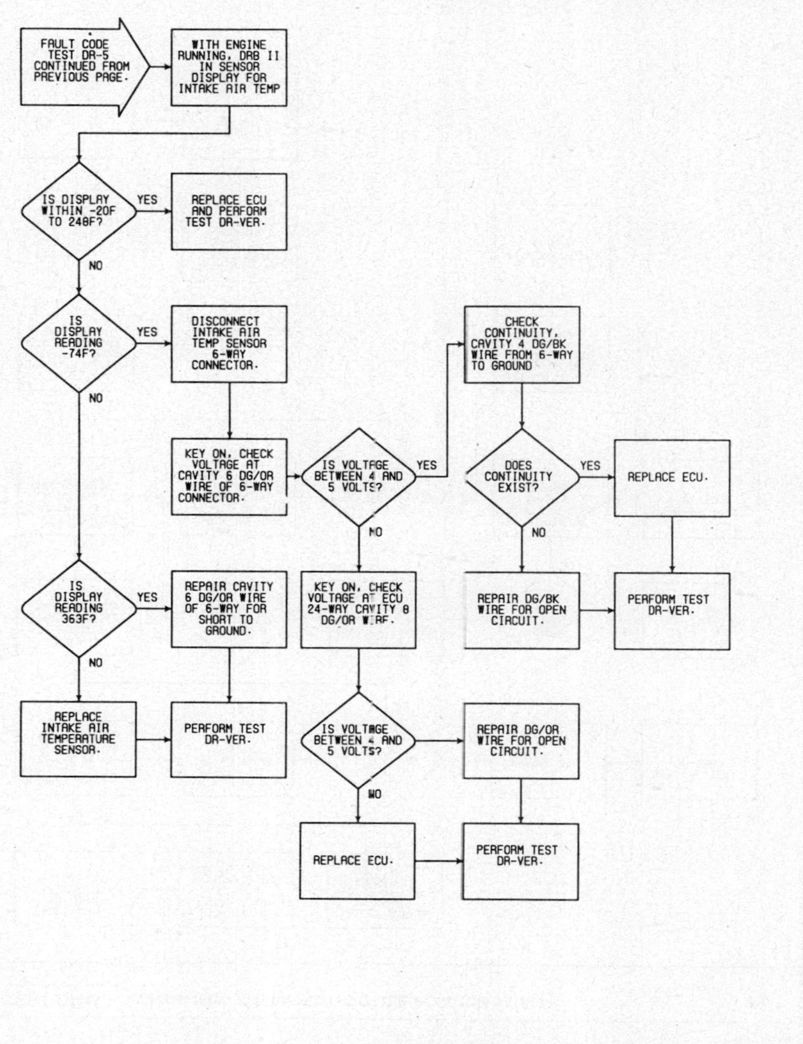

TEST DR-6 CHECKING THROTTLE POSITION SENSOR

Perform TEST DR-1 Before Proceeding

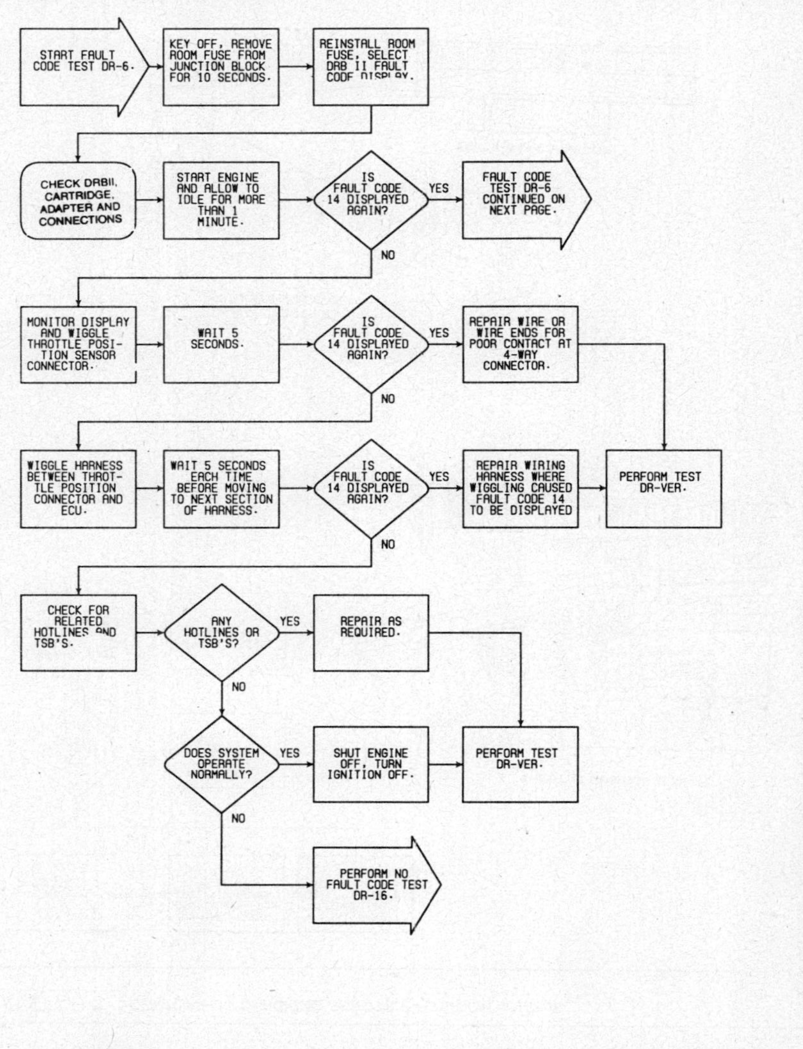

TEST DR-6 CONTINUED - CHECKING THROTTLE POSITION SENSOR

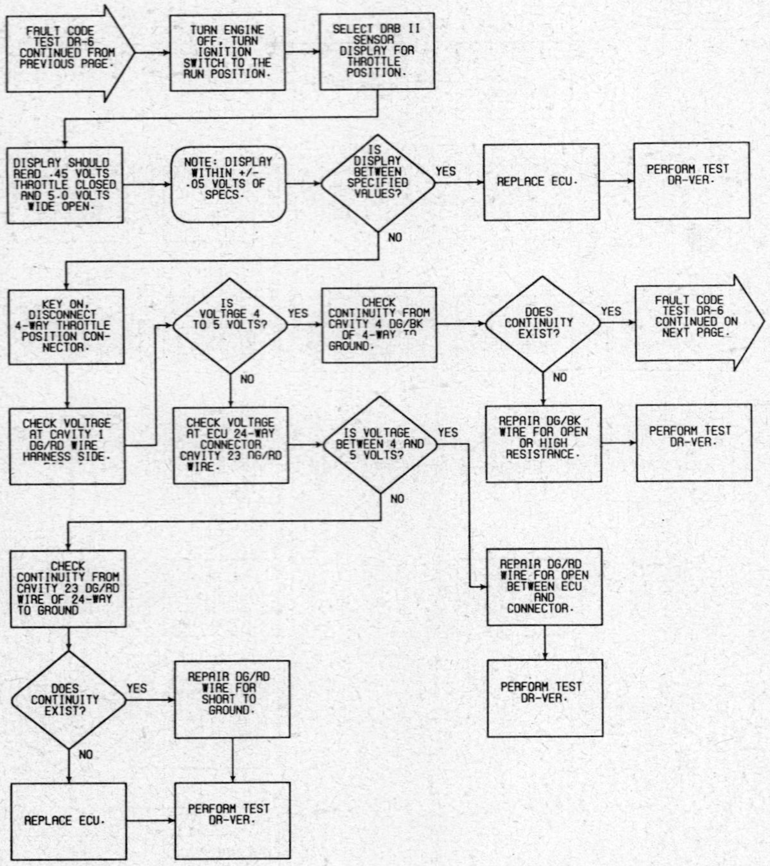

TEST DR-6 CONTINUED - CHECKING THROTTLE POSITION SENSOR

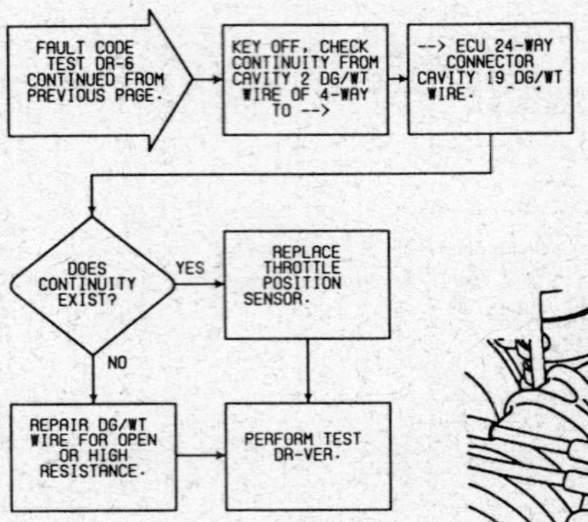

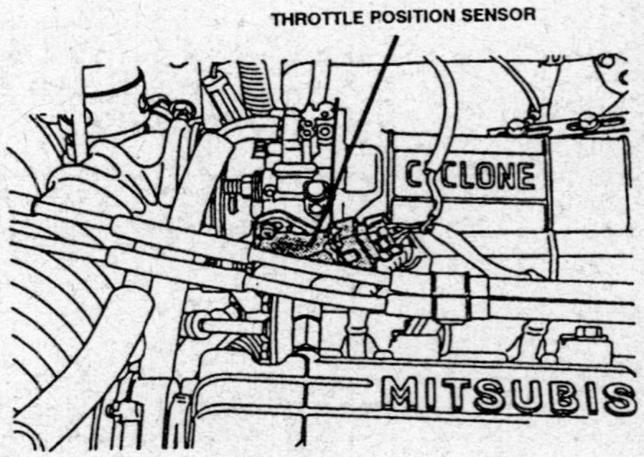

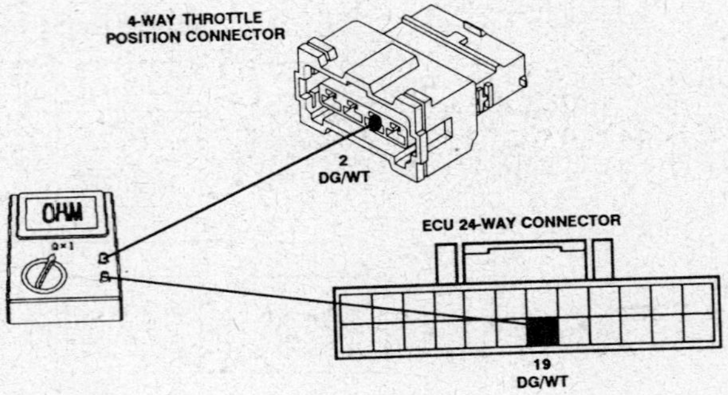

1.6L ENGINE—DRIVEABILITY TEST

TEST DR-7 CHECKING COOLANT TEMPERATURE SENSOR

Perform TEST DR-1 Before Proceeding

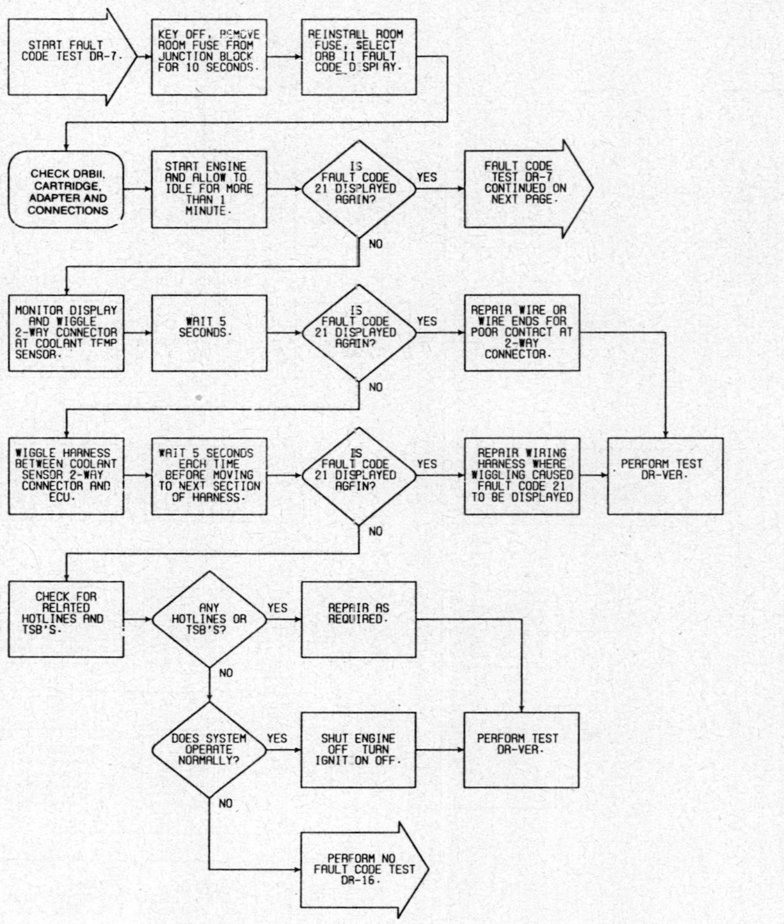

TEST DR-7 CONTINUED - CHECKING COOLANT TEMPERATURE SENSOR

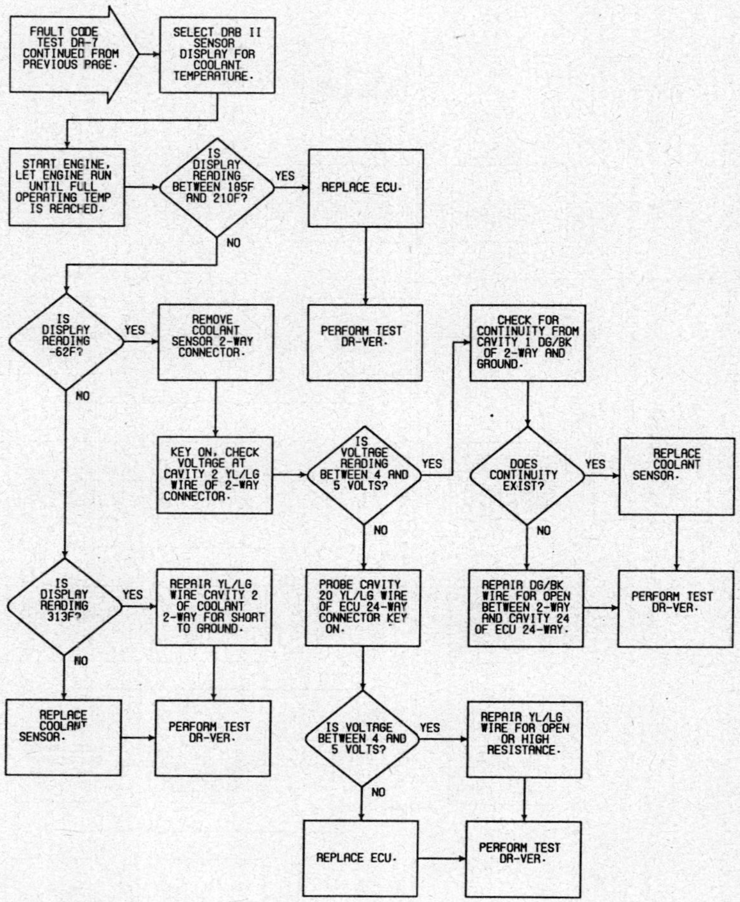

TEST DR-8 CHECKING CRANK ANGLE SENSOR

Perform TEST DR-1 Before Proceeding

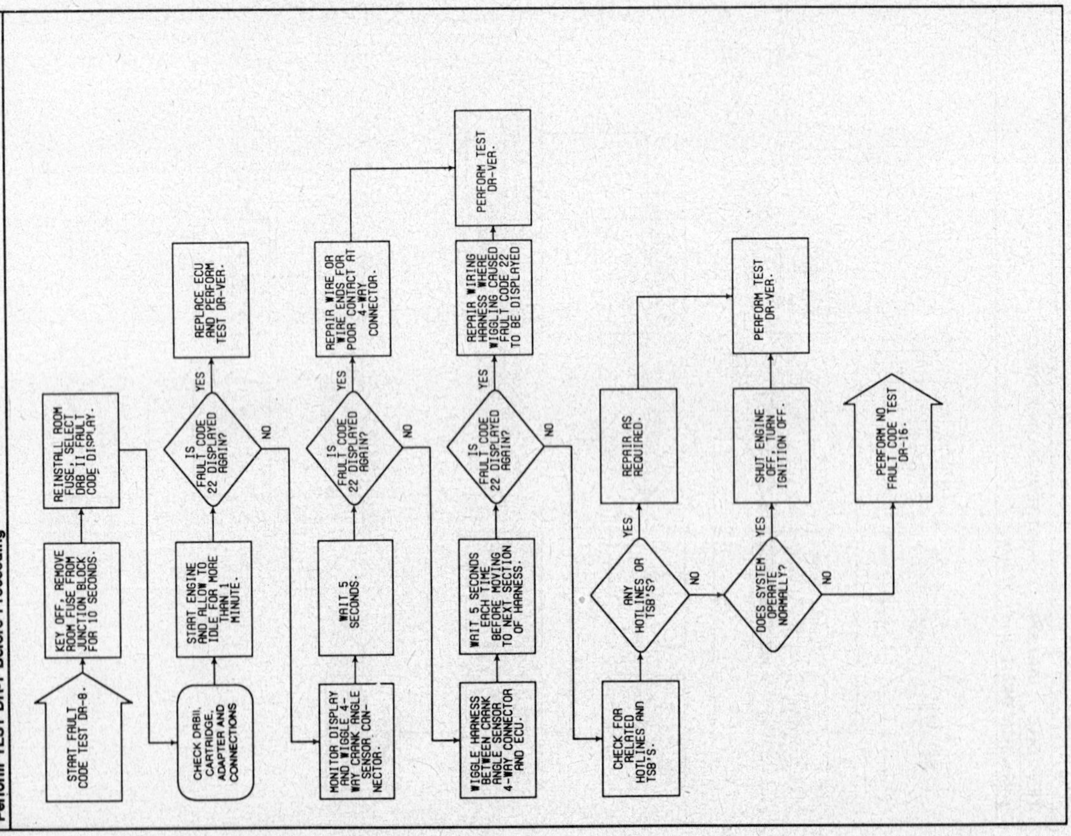

TEST DR-9 CHECKING TOP DEAD CENTER SENSOR

Perform TEST DR-1 Before Proceeding

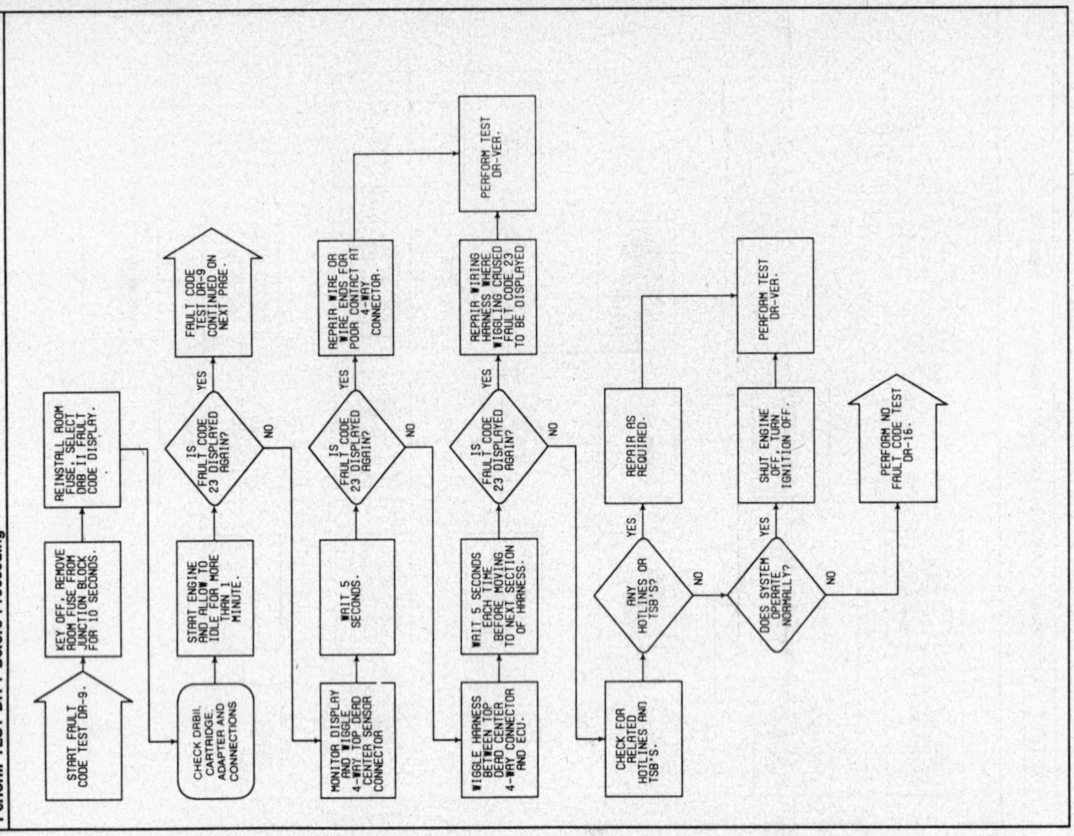

TEST DR-9 CONTINUED - CHECKING TOP DEAD CENTER SENSOR

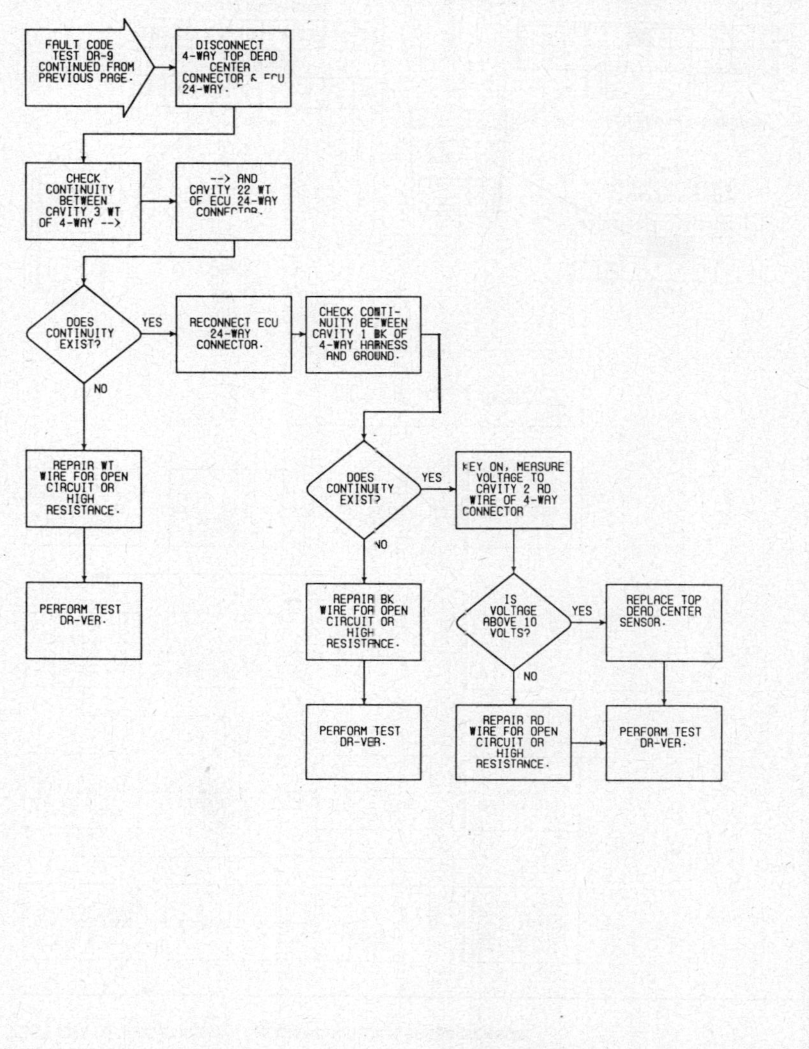

TEST DR-10 CHECKING VEHICLE SPEED SENSOR

Perform TEST DR-1 Before Proceeding

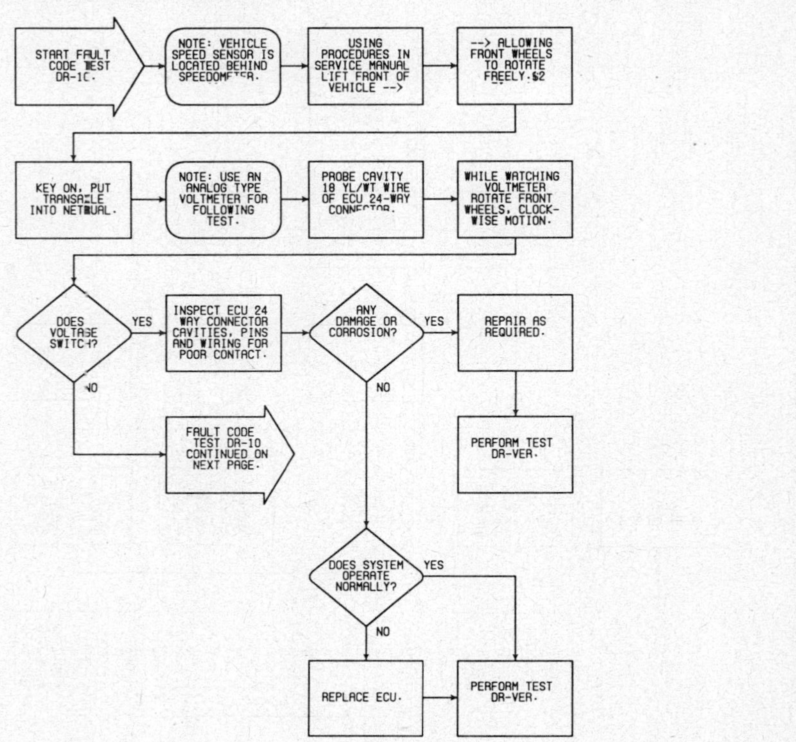

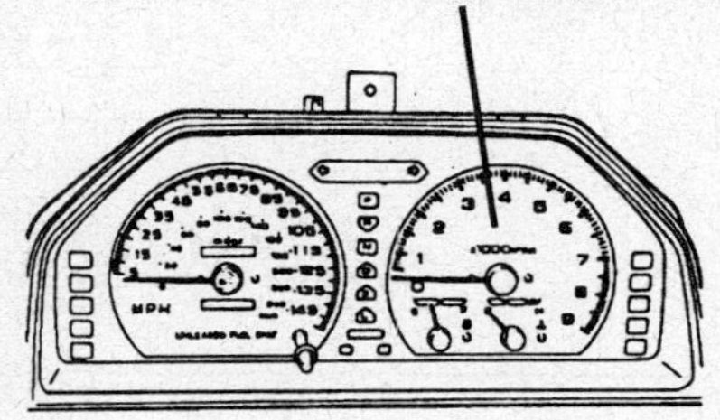

TEST DR-10 CONTINUED - CHECKING VEHICLE SPEED SENSOR

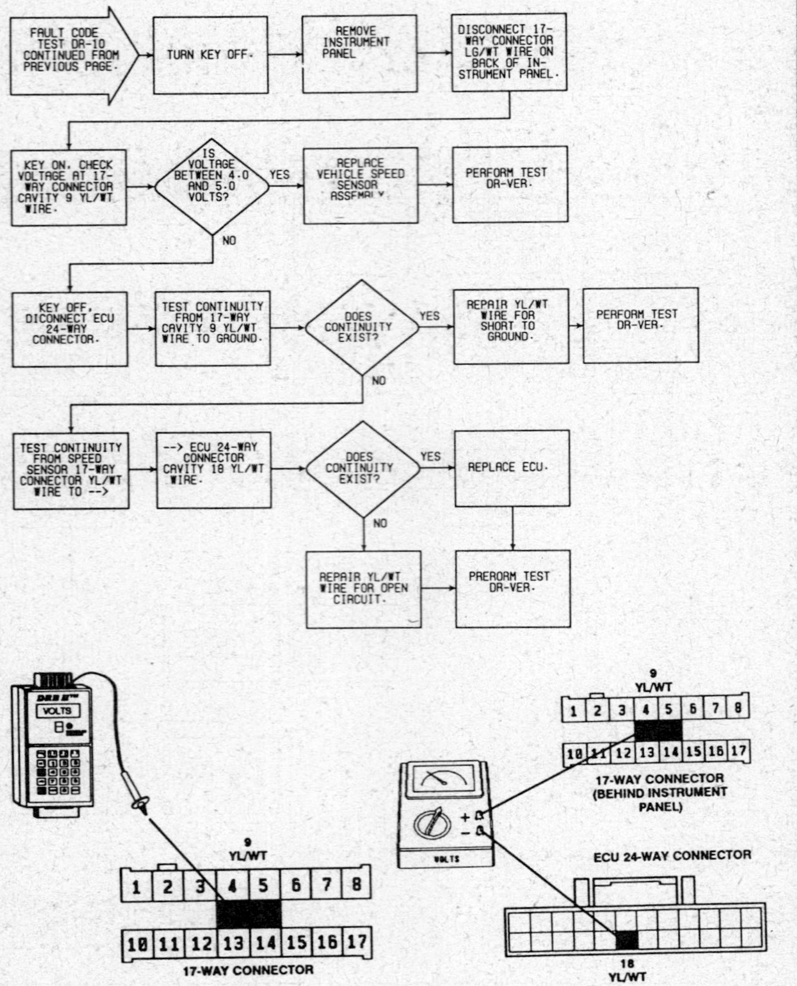

TEST DR-11 CHECKING BAROMETRIC PRESSURE SENSOR

Perform TEST DR-1 Before Proceeding

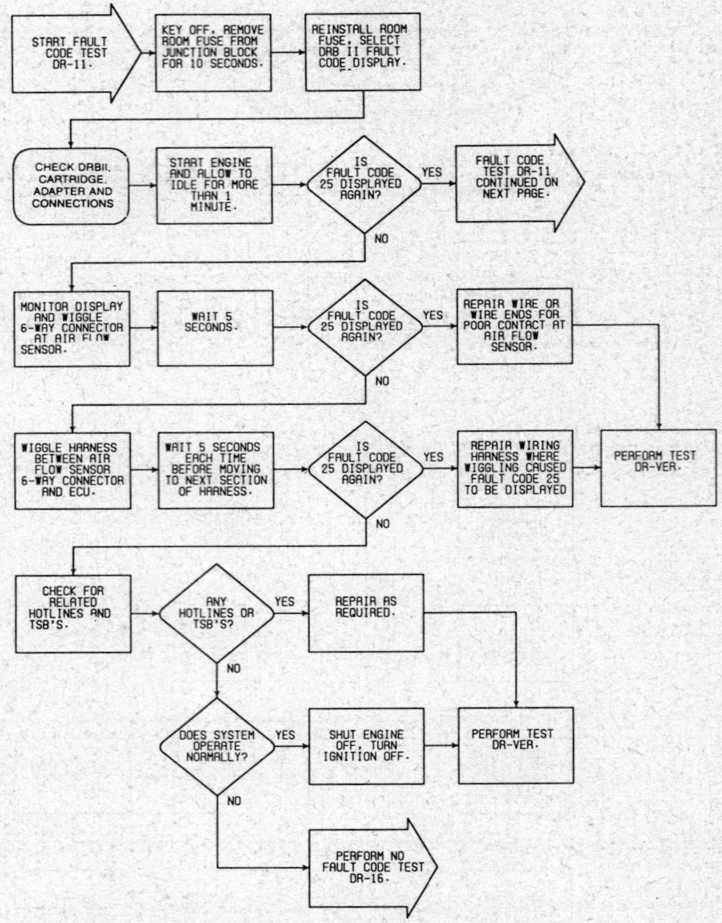

1.6L ENGINE—DRIVEABILITY TEST

TEST DR-11 CONTINUED - CHECKING BARCMETRIC PRESSURE SENSOR

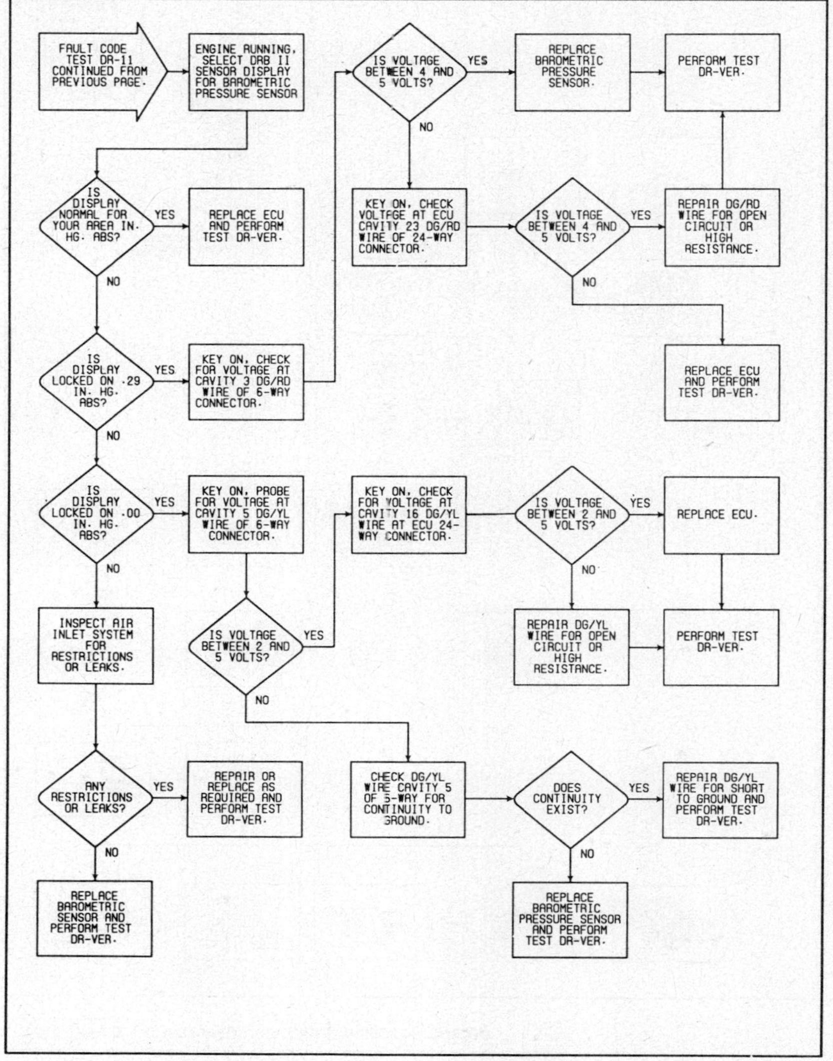

TEST DR-12 CHECKING INJECTOR CIRCUIT

Perform TEST DR-1 Before Proceeding

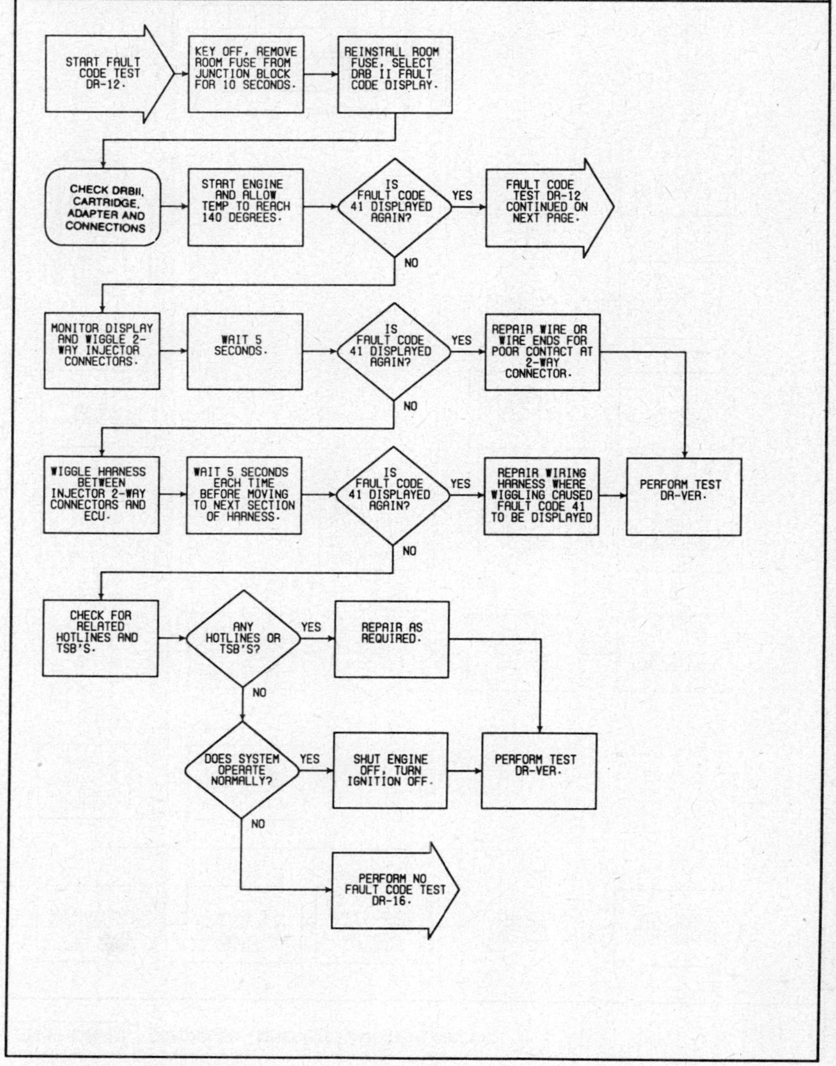

TEST DR-12 CONTINUED - CHECKING INJECTOR CIRCUIT

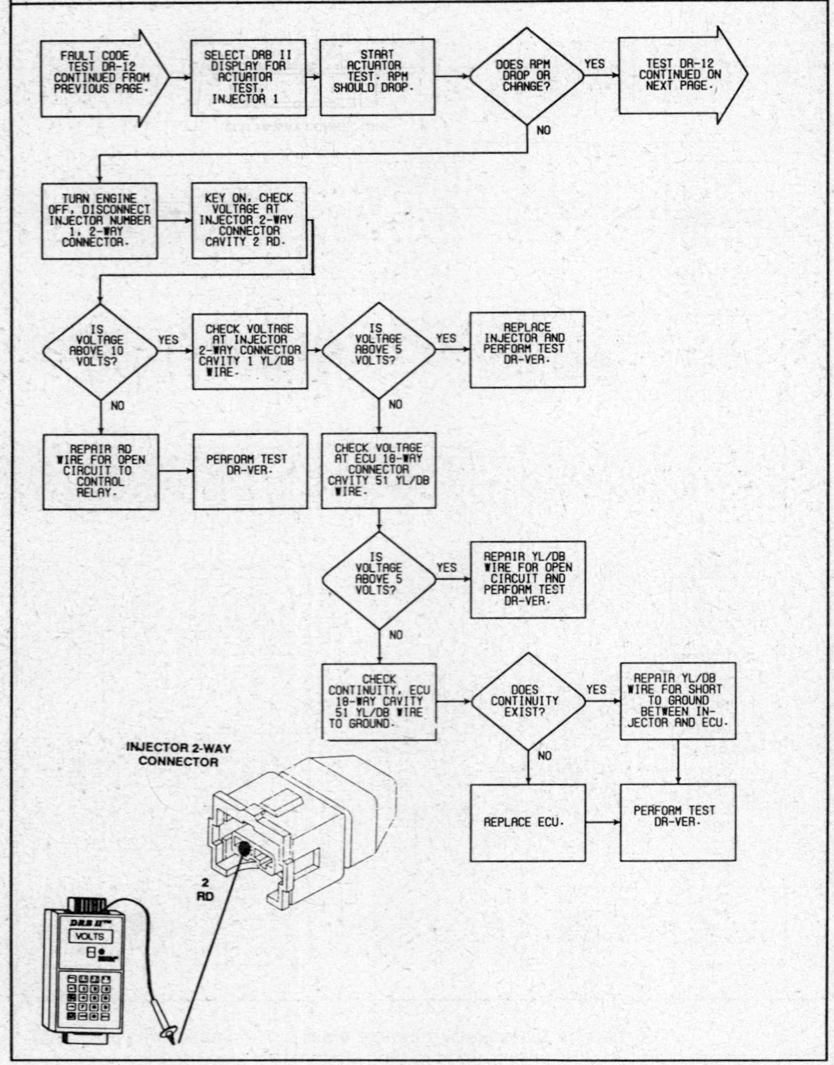

TEST DR-12 CONTINUED - CHECKING INJECTOR CIRCUIT

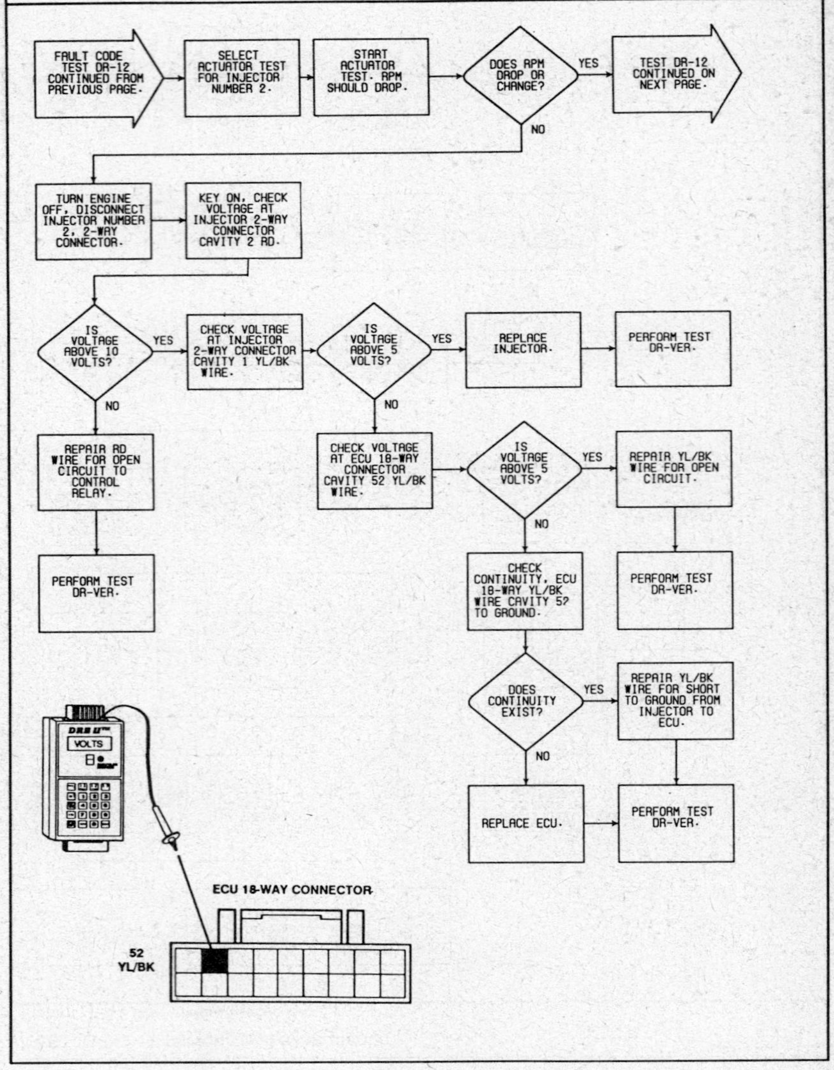

TEST DR-12 CONTINUED - CHECKING INJECTOR CIRCUIT

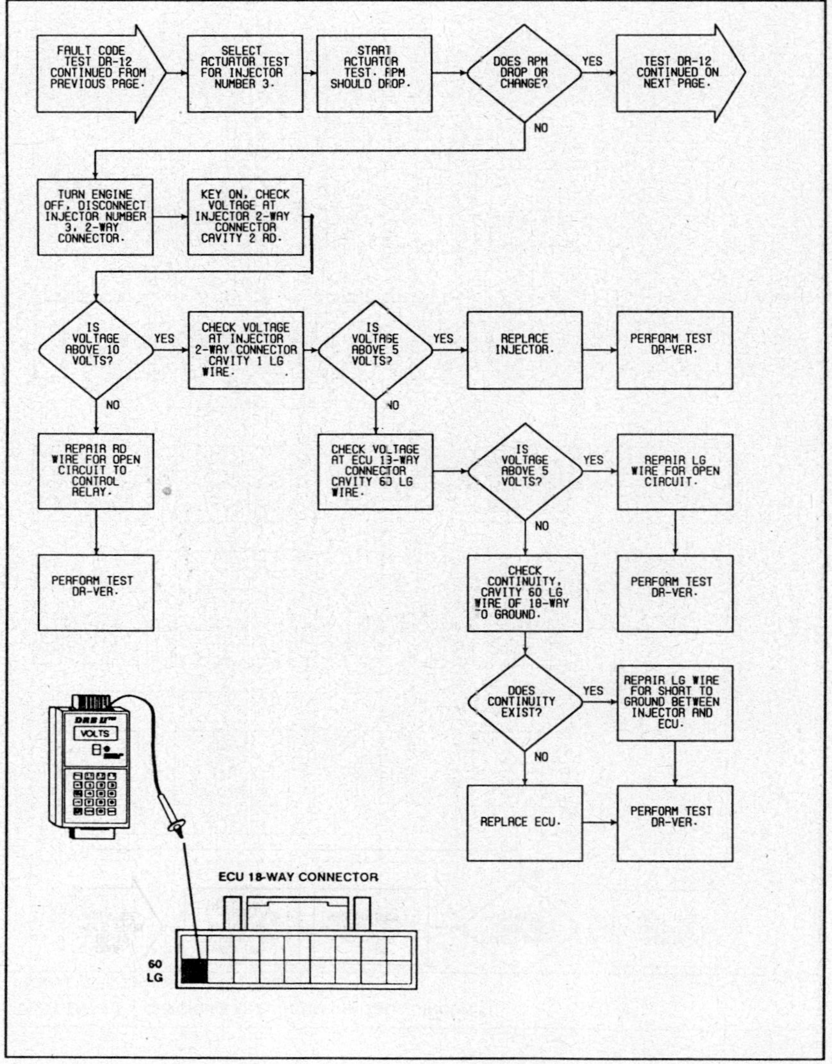

TEST DR-12 CONTINUED - CHECKING INJECTOR CIRCUIT

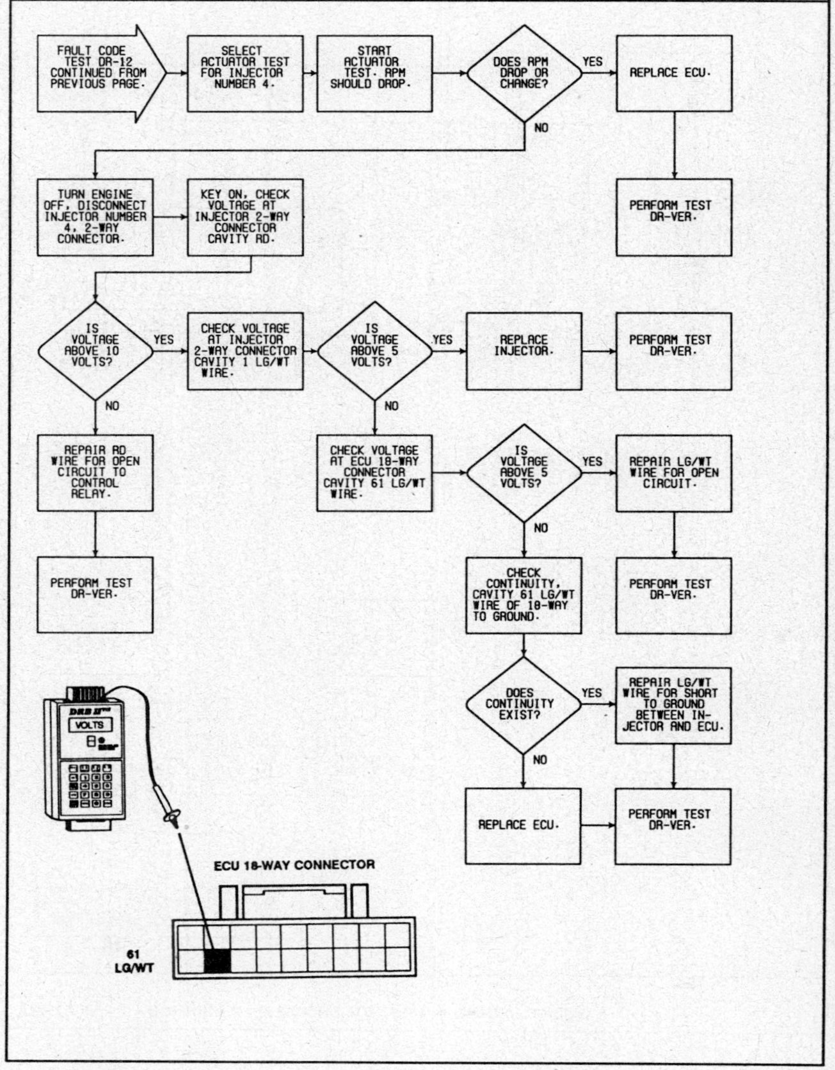

TEST DR-13 CHECKING FUEL PUMP CONTROL CIRCUIT

Perform TEST DR-1 Before Proceeding

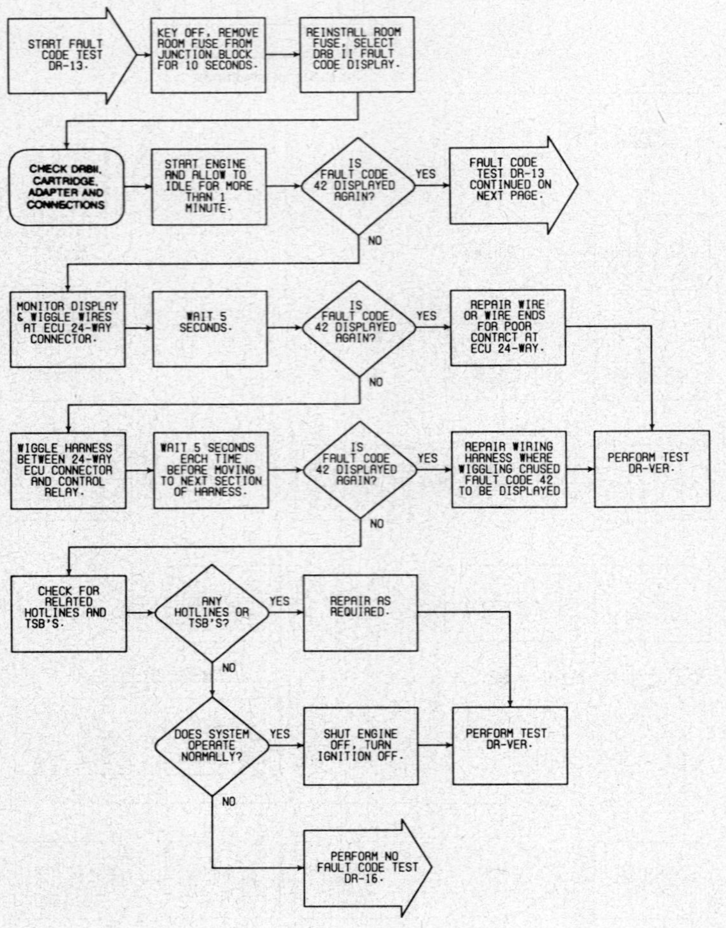

TEST DR-13 CONTINUED - CHECKING FUEL PUMP CONTROL CIRCUIT

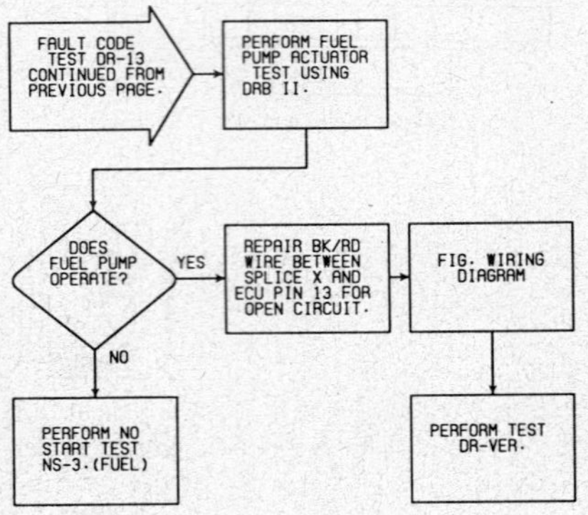

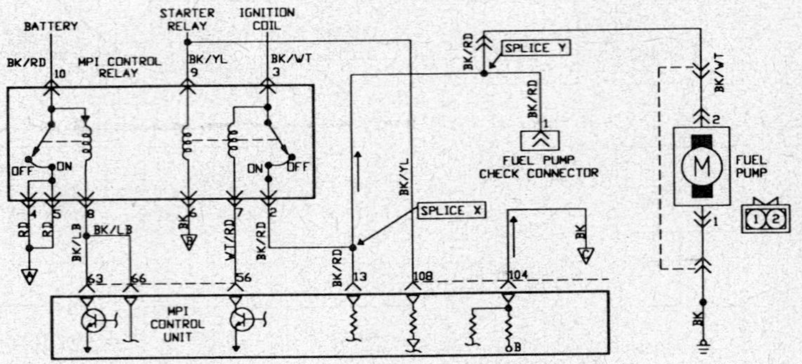

TEST DR-14 TESTING EGR TEMPERATURE SENSOR CIRCUIT (CALIFORNIA VEHICLES ONLY)

Perform TEST DR-1 Before Proceeding

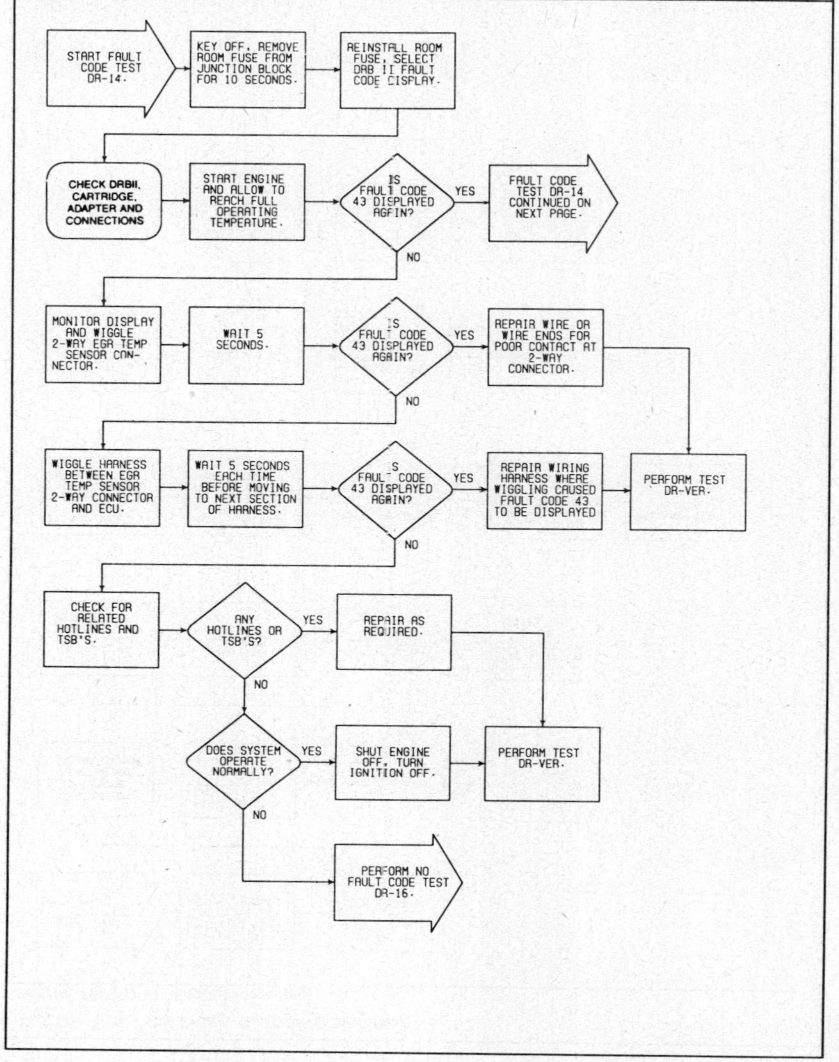

TEST DR-14 CONTINUED - TESTING EGR TEMPERATURE SENSOR CIRCUIT (CALIFORNIA VEHICLES ONLY)

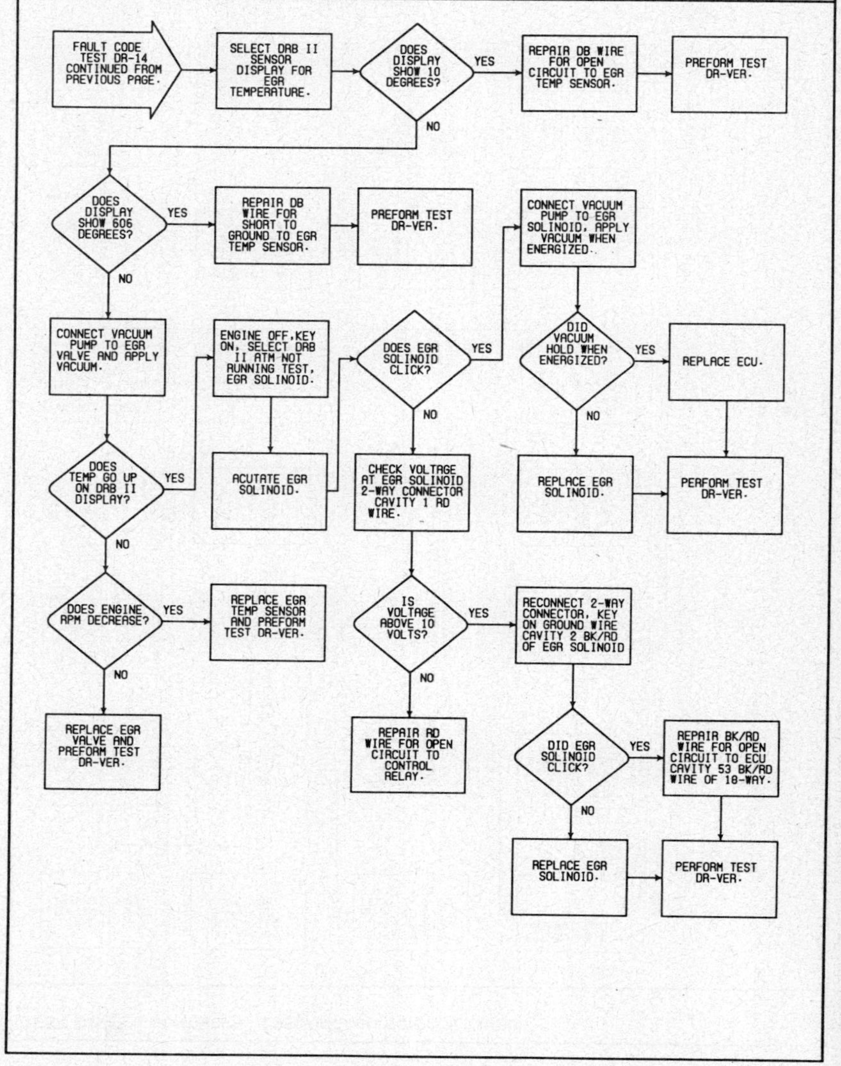

TEST DR-15 CHECKING IGNITION COIL CIRCUIT

Perform TEST DR-1 Before Proceeding

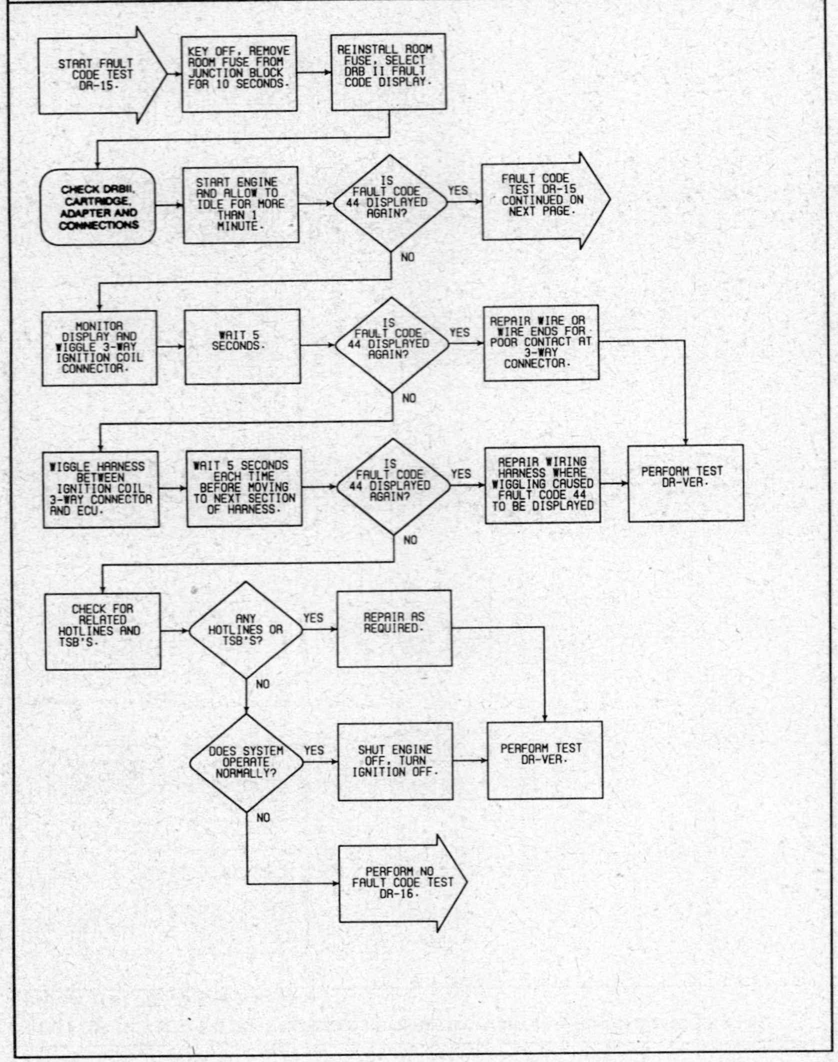

TEST DR-15 CONTINUED - CHECKING IGNITION COIL CIRCUIT

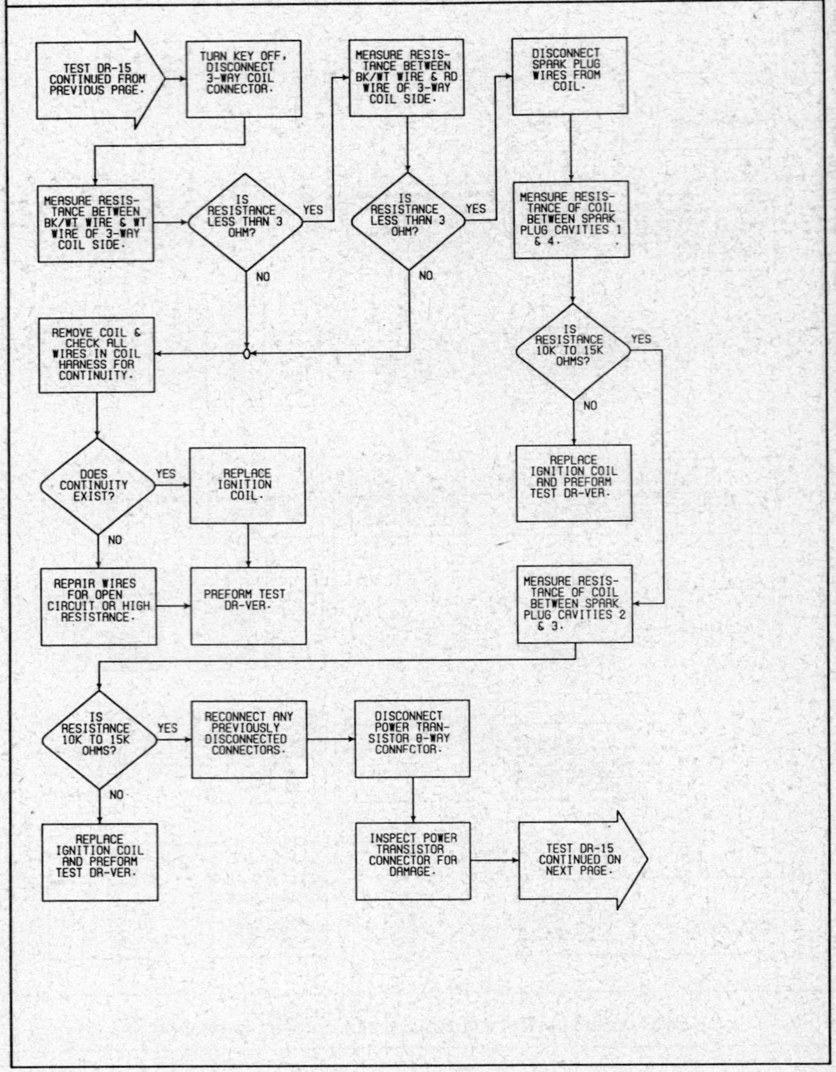

1.6L ENGINE – DRIVEABILITY TEST

TEST DR-15 CONTINUED - CHECKING IGNITION COIL CIRCUIT

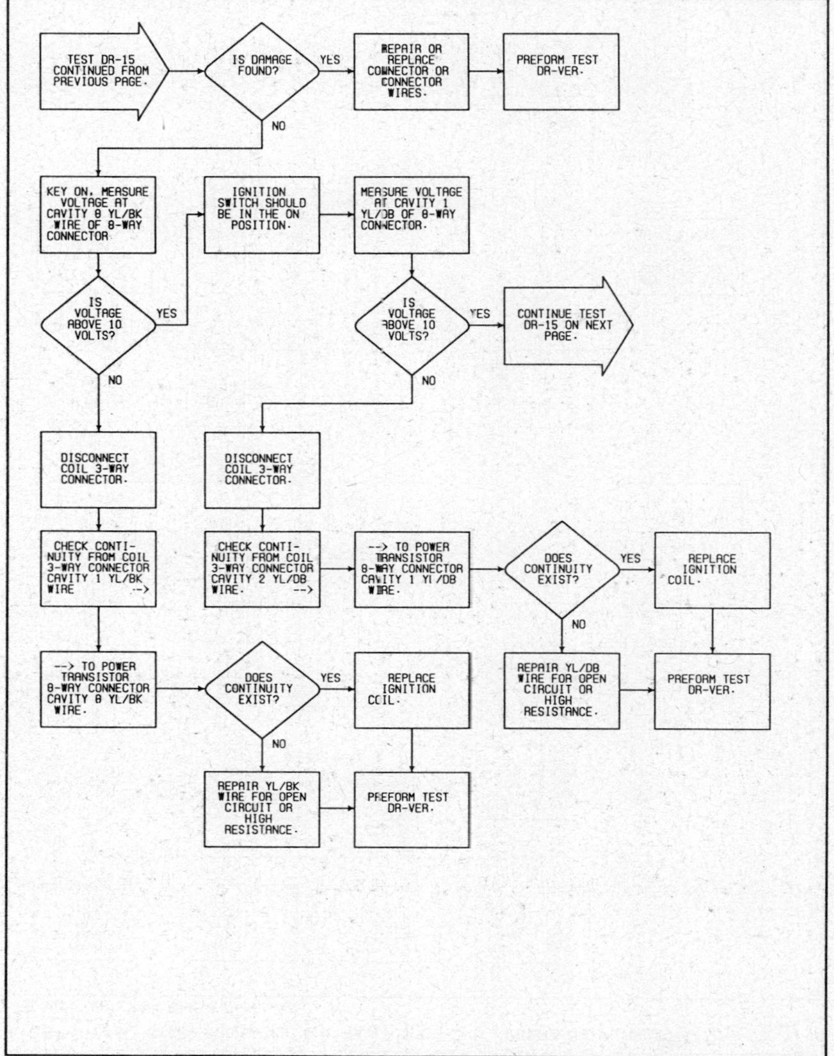

TEST DR-15 CONTINUED - CHECKING IGNITION COIL CIRCUIT

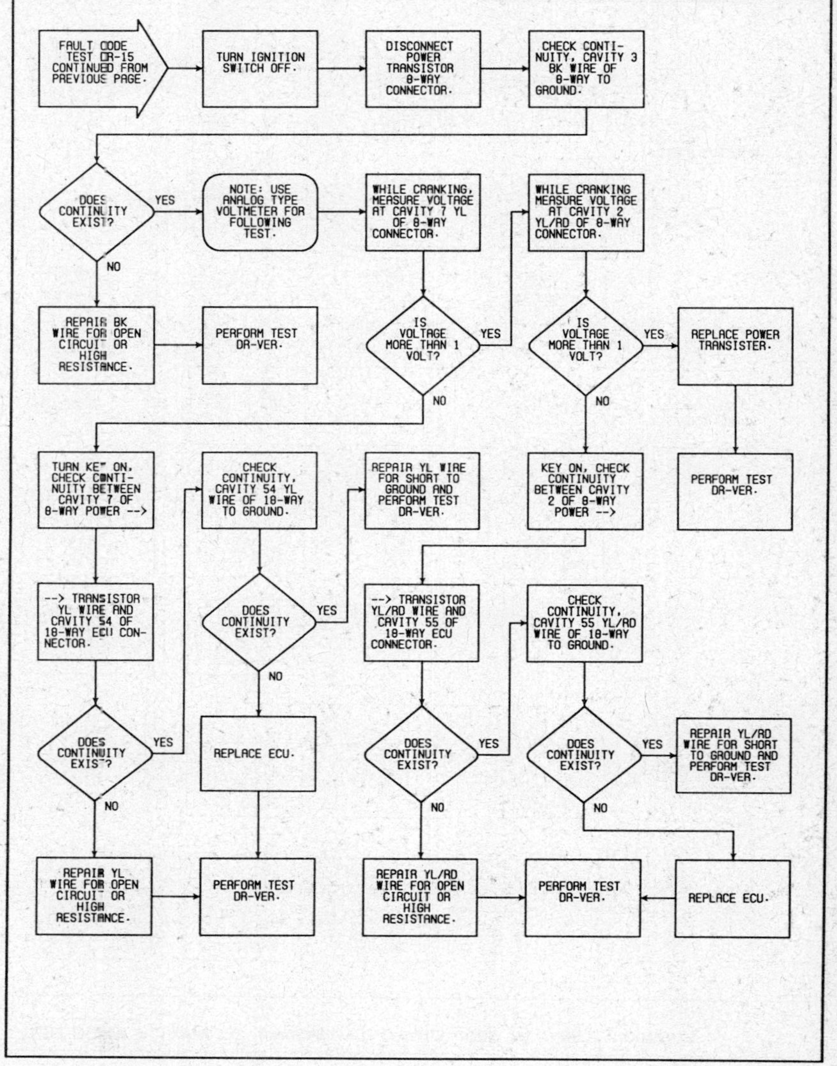

TEST DR-16 START DRIVEABILITY NO FAULT CODE TEST - INPUTS/OUTPUTS

Perform TEST DR-1 Before Proceeding

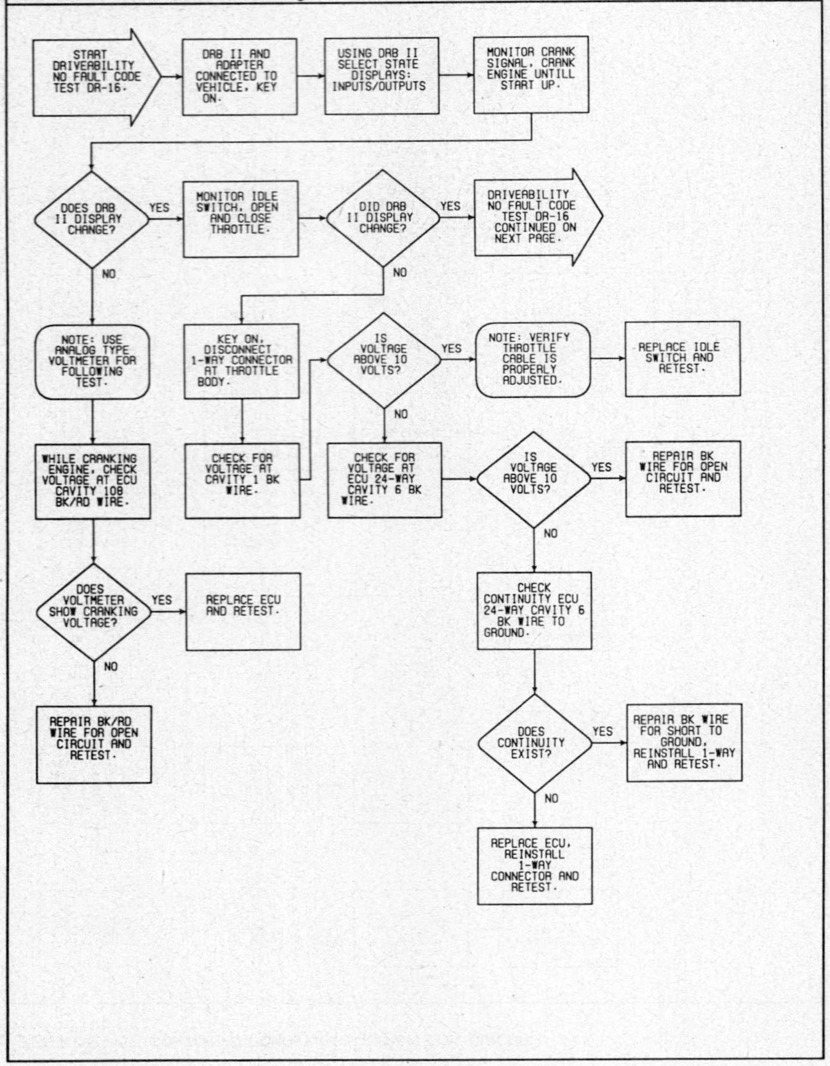

TEST DR-16 CONTINUED - DRIVEABILITY NO FAULT CODE TEST - INPUTS/OUTPUTS

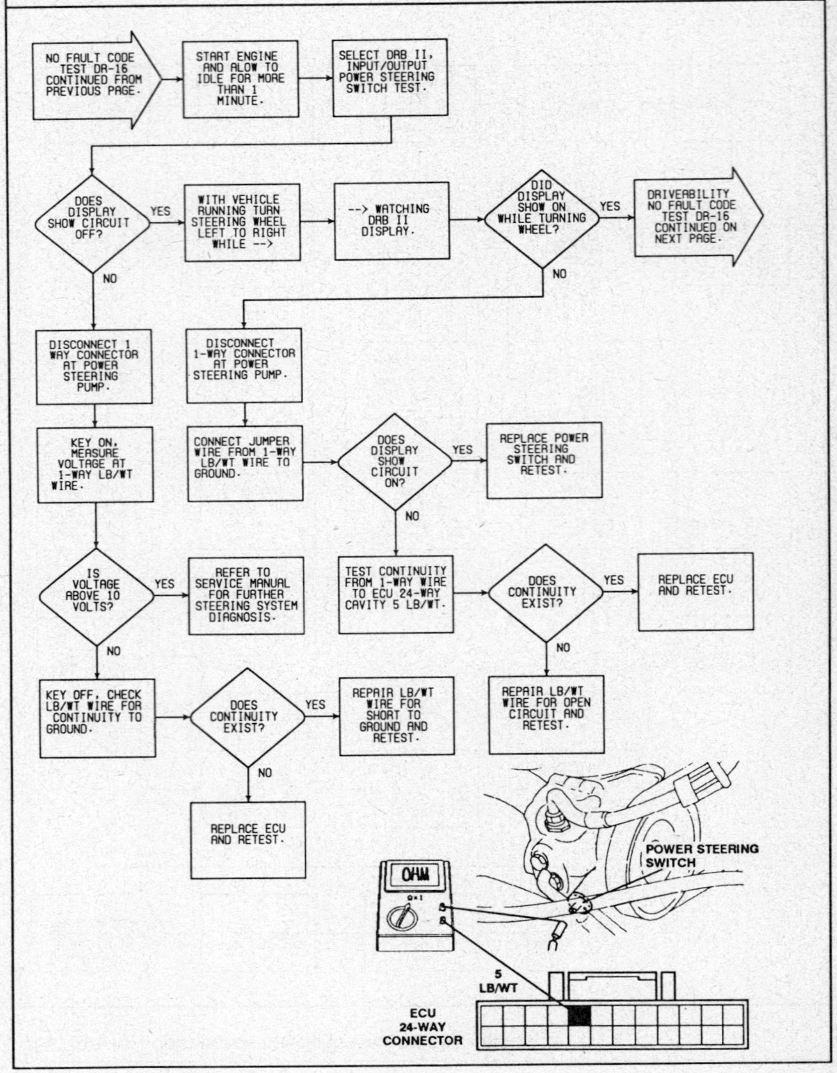

1.6L ENGINE—DRIVEABILITY TEST

TEST DR-16 CONTINUED - DRIVEABILITY NO FAULT CODE TEST - INPUTS/OUTPUTS

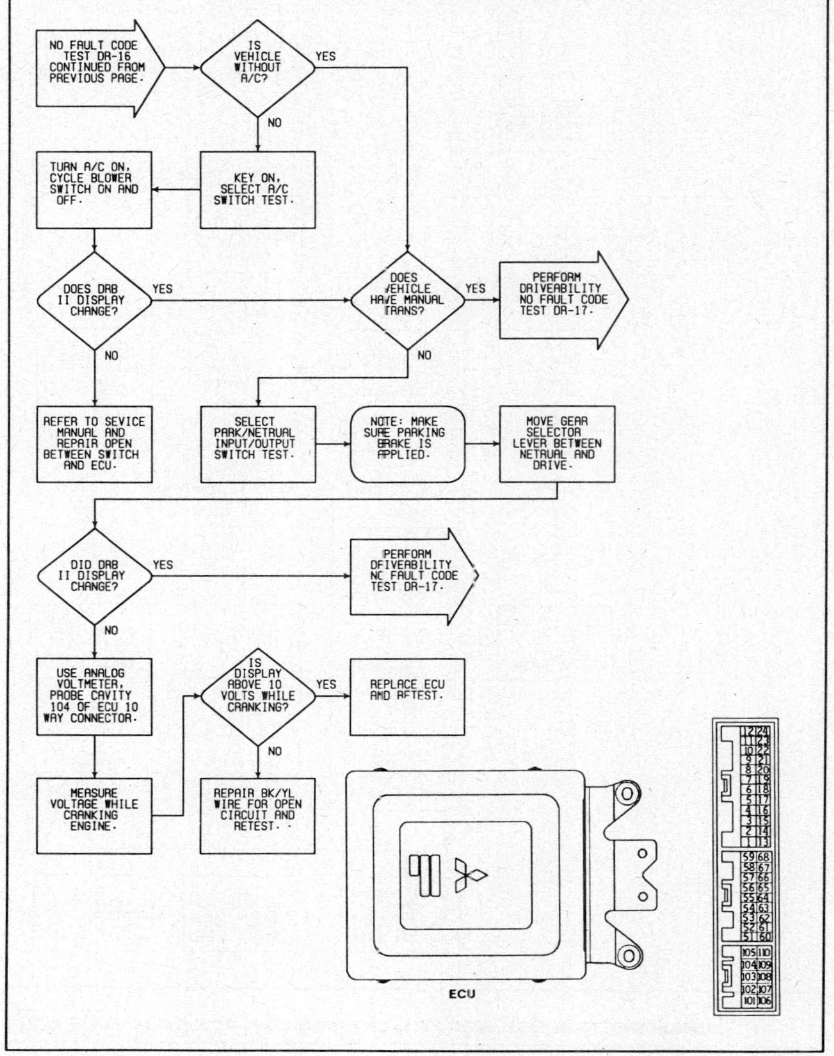

TEST DR-17 PERFORMING NO FAULT CODE TEST - DRBII SENSOR TEST

Perform TEST DR-16 Before Proceeding

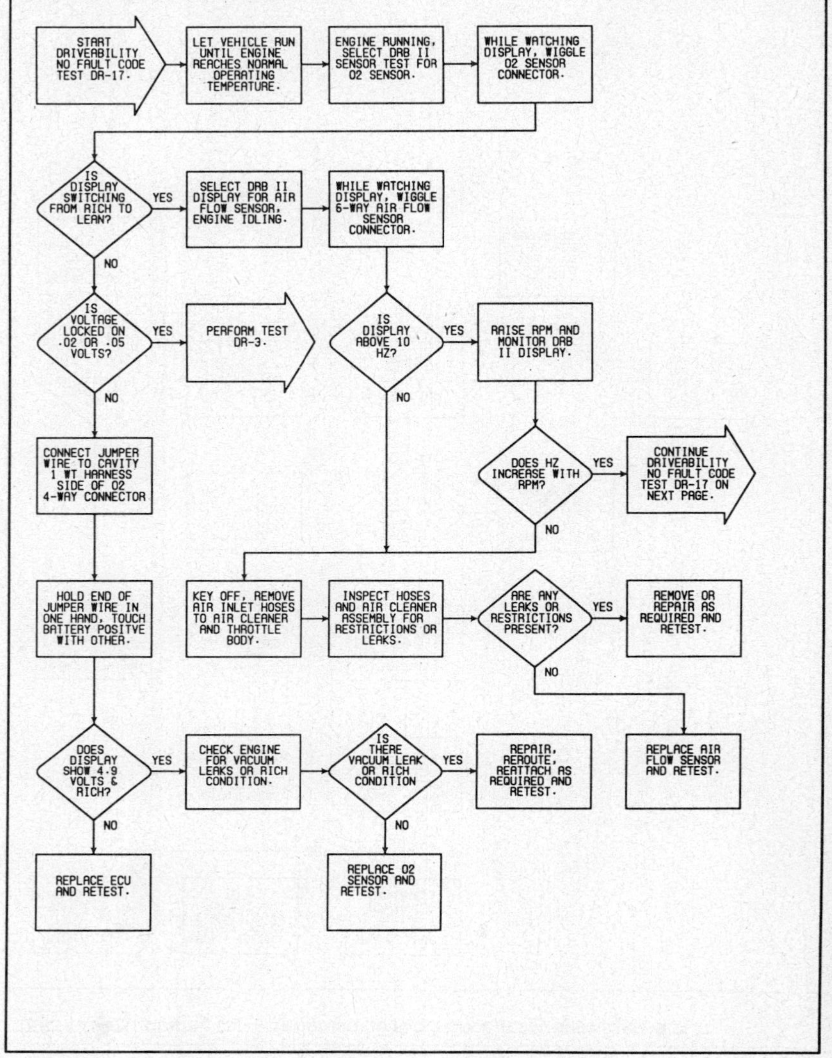

TEST DR-17 CONTINUED - PERFORMING NO FAULT CODE TEST - DRBII SENSOR TEST

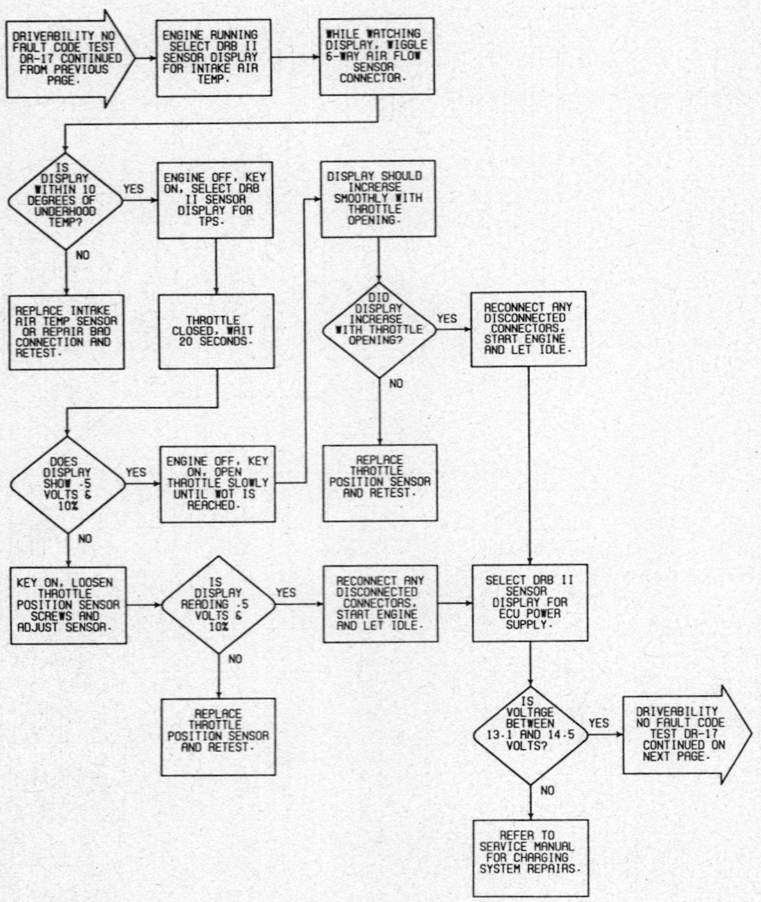

TEST DR-17 CONTINUED - PERFORMING NO FAULT CODE TEST - DRBII SENSOR TEST

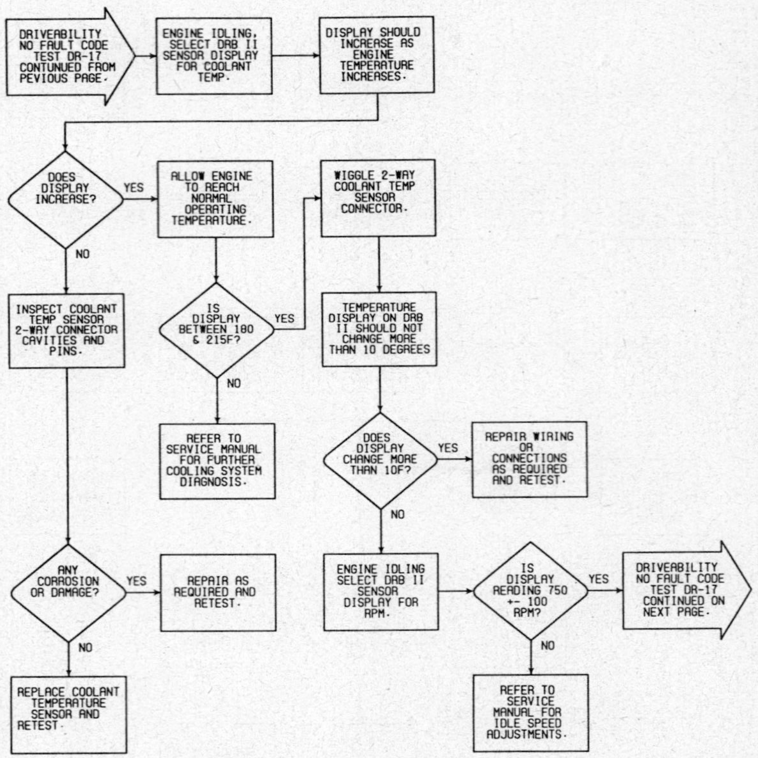

1.6L ENGINE—DRIVEABILITY TEST

TEST DR-17 CONTINUED - PERFORMING NO FAULT CODE TEST - DRBII SENSOR TEST

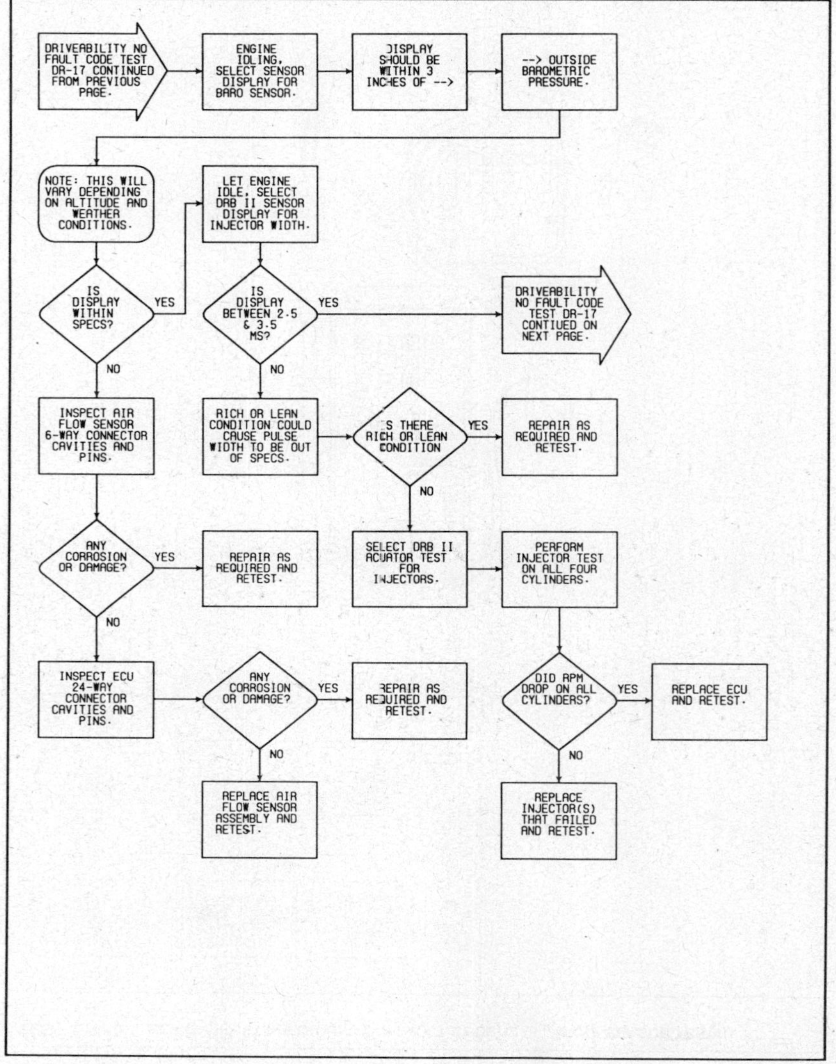

TEST DR-17 CONTINUED - PERFORMING NO FAULT CODE TEST - DRBII SENSOR TEST

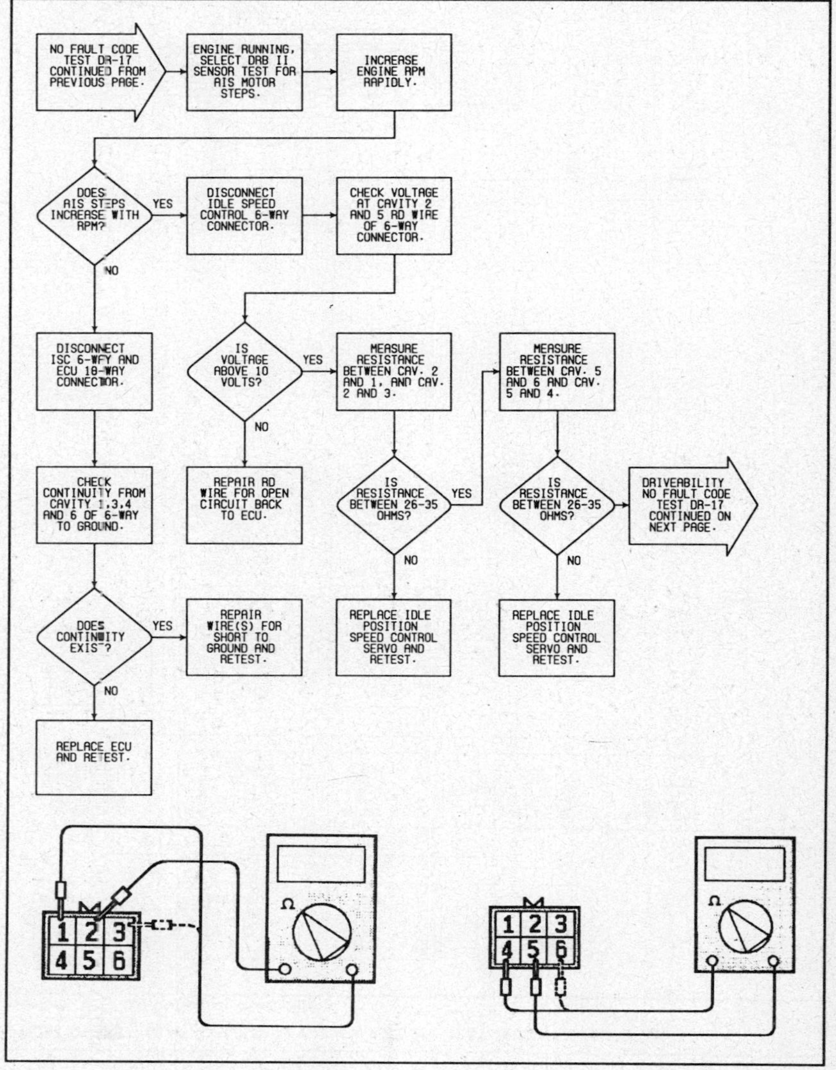

TEST DR-17 CONTINUED - PERFORMING NO FAULT CODE TEST - DRBII SENSOR TEST

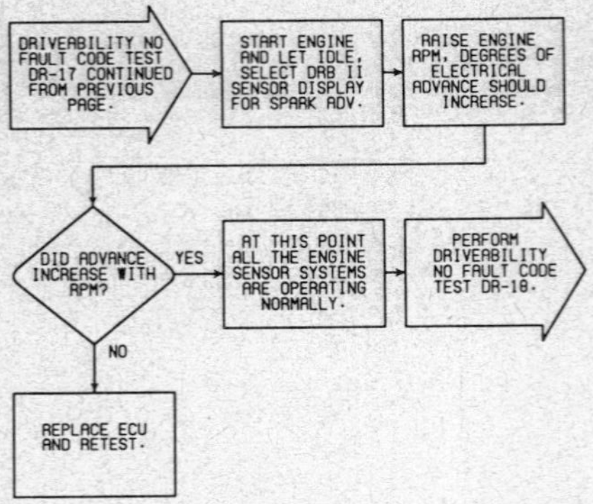

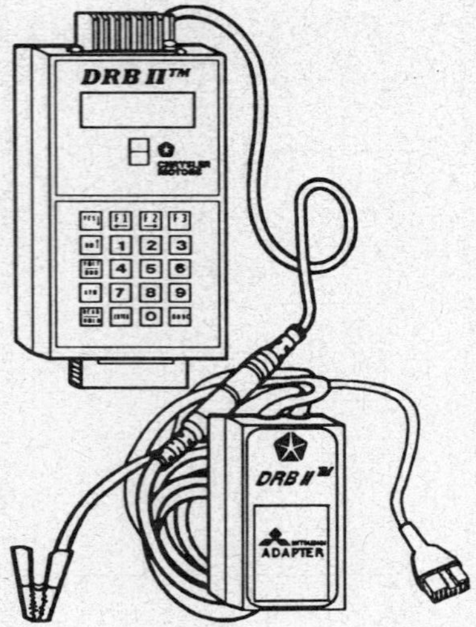

TEST DR-18 START DRIVEABILITY NO FAULT CODE TEST - FUEL PRESSURE

Perform TEST DR-16 Before Proceeding

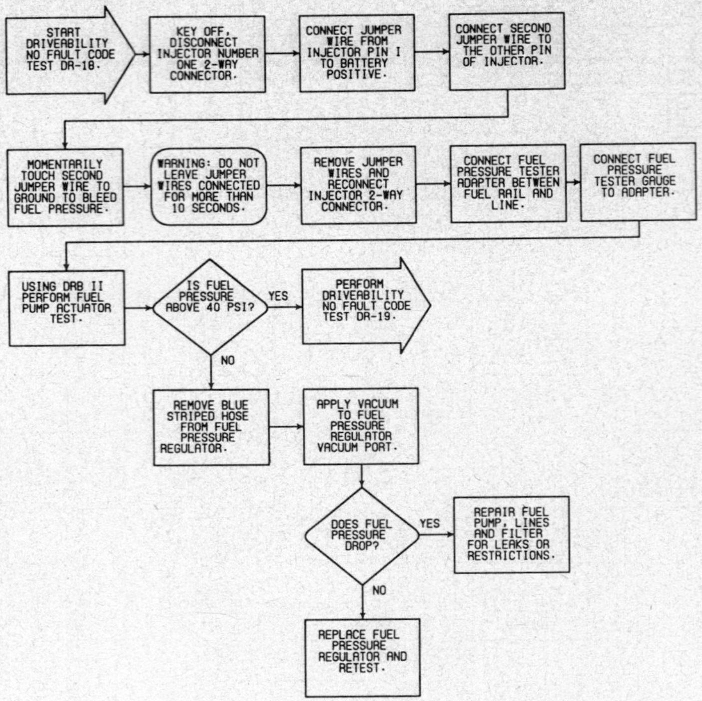

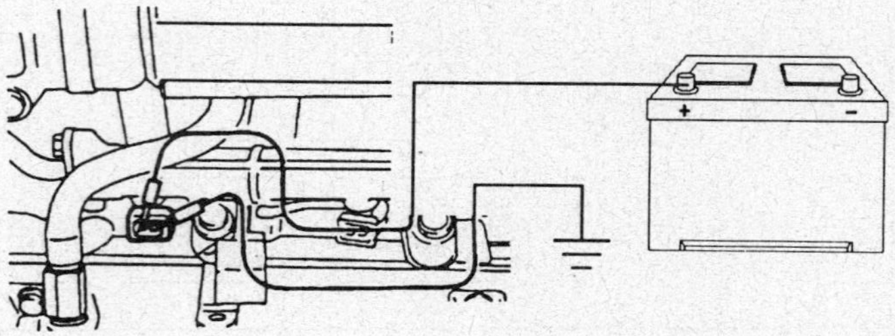

1.6L ENGINE – DRIVEABILITY TEST

TEST DR-19 START DRIVEABILITY NO FAULT CODE TEST - EGR SYSTEM

Perform TEST DR-16 Before Proceeding

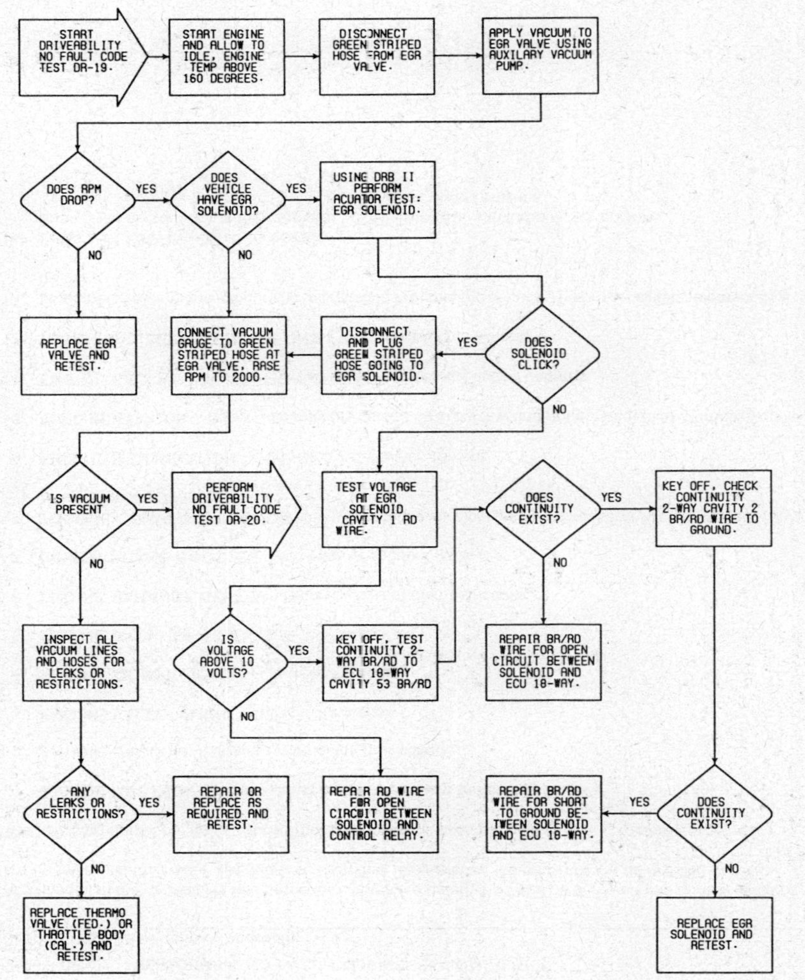

TEST DR-20 PERFORMING NO FAULT CODE TEST - PURGE CONTROL

Perform TEST DR-16 Before Proceeding

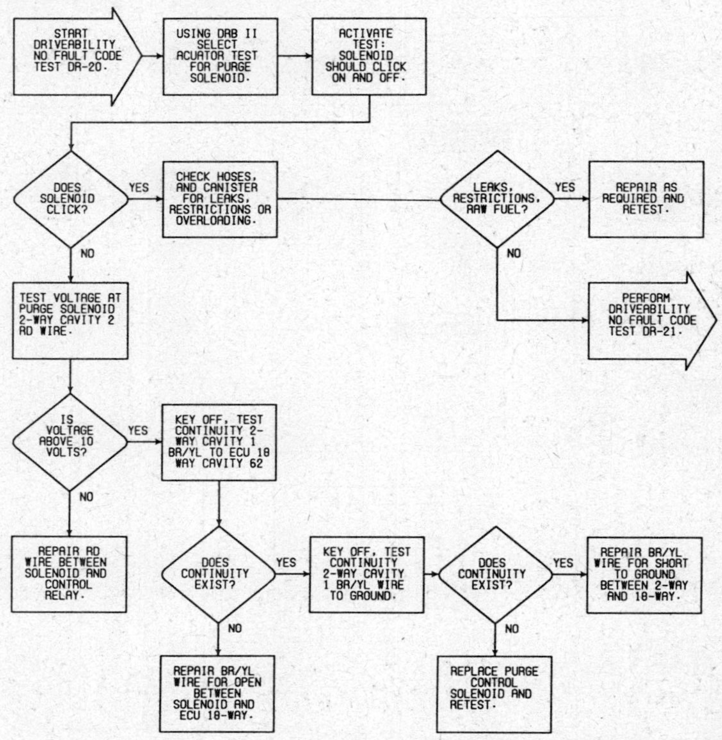

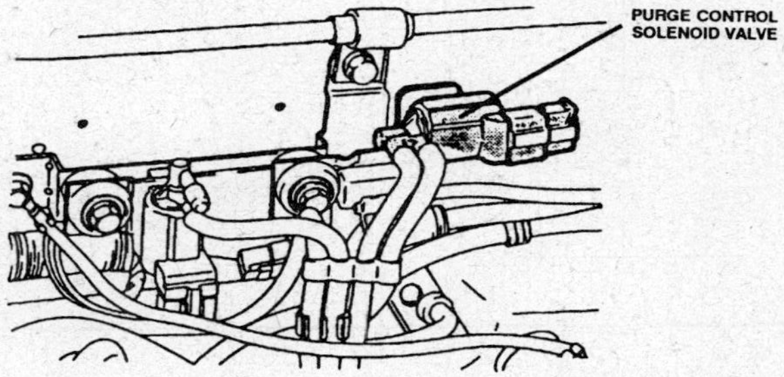

1.6L ENGINE – DRIVEABILITY TEST

TEST DR-21	PERFORMING NO FAULT CODE MECHANICAL TEST
Perform TEST DR-17 Before Proceeding	

At this point in the driveability test procedure, you have determined that all of the engine control systems are operating as they were designed to. Therefore, they are not the cause of the driveability problem.

The following additional items can not be overlooked as possible causes of a driveability problem.

1. **ENGINE IGNITION TIMING** – Must be set with timing terminal grounded.
2. **ENGINE VACUUM** – Must be normal for your altitude.
3. **ENGINE VALVE TIMING** – To specifications.
4. **ENGINE COMPRESSION** – To specifications.
5. **ENGINE P.C.V. SYSTEM** – Must flow freely.
6. **ENGINE EXHAUST SYSTEM** – Must be free of any restrictions.
7. **POWER BRAKE BOOSTER** -- No internal vacuum leaks.

9. **TORQUE CONVERTER CONDITION** – May cause very low power at breakaway or high speed. (Only 1 condition at a time)

9. **FUEL CONTAMINATION** -- High alcohol or water content.

10. **FUEL INJECTORS** – Rough idle may be caused by injector wiring not connected to correct injector.
11. **ENGINE SECONDARY IGNITION CHECK** – Abnormal scope patterns.
12. **TECHNICAL SERVICE BULLETINS** – Any that apply to vehicle.
13. All air intake piping and vacuum hoses must be in place and secure. The proper air filter element must be used.
14. **FUEL PRESSURE** – Must be correct.
    Spec: With no vacuum at the regulator: 48 PSI on V6 & non-turbo 4 Cyl. engines
    36 PSI on turbo engines.

1.6L ENGINE – WIRING SCHEMATIC

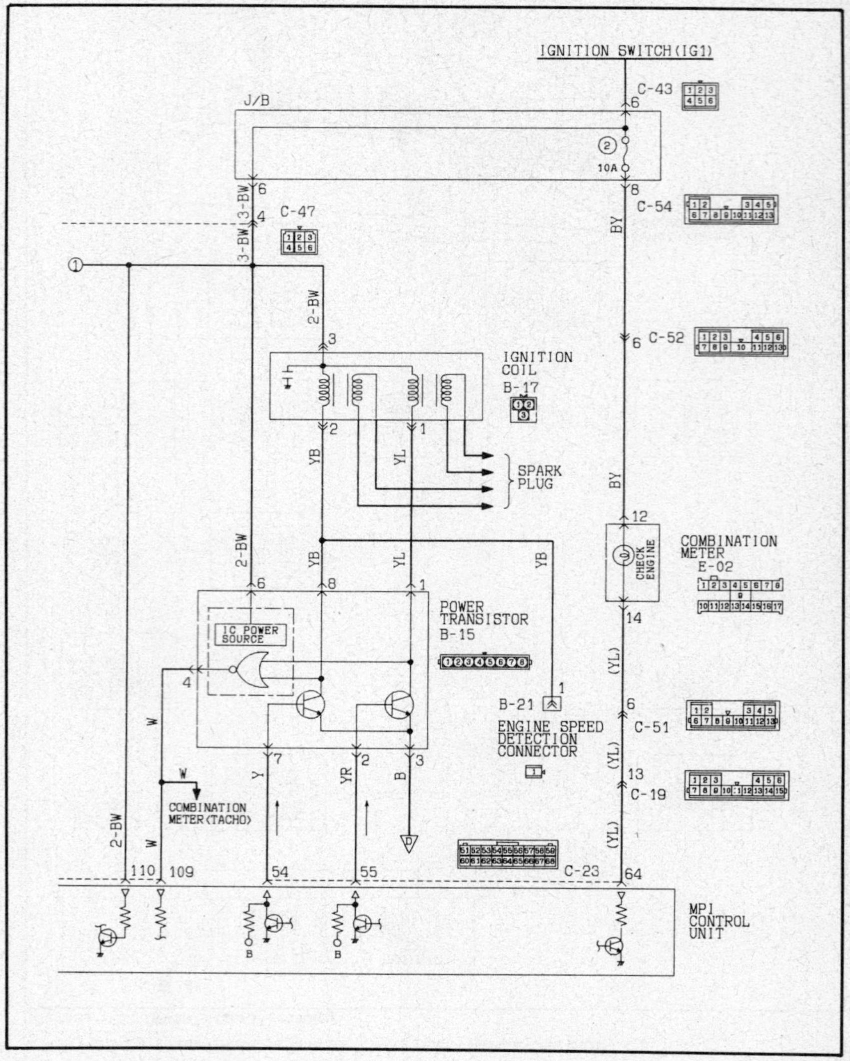

1.6L ENGINE – WIRING SCHEMATIC

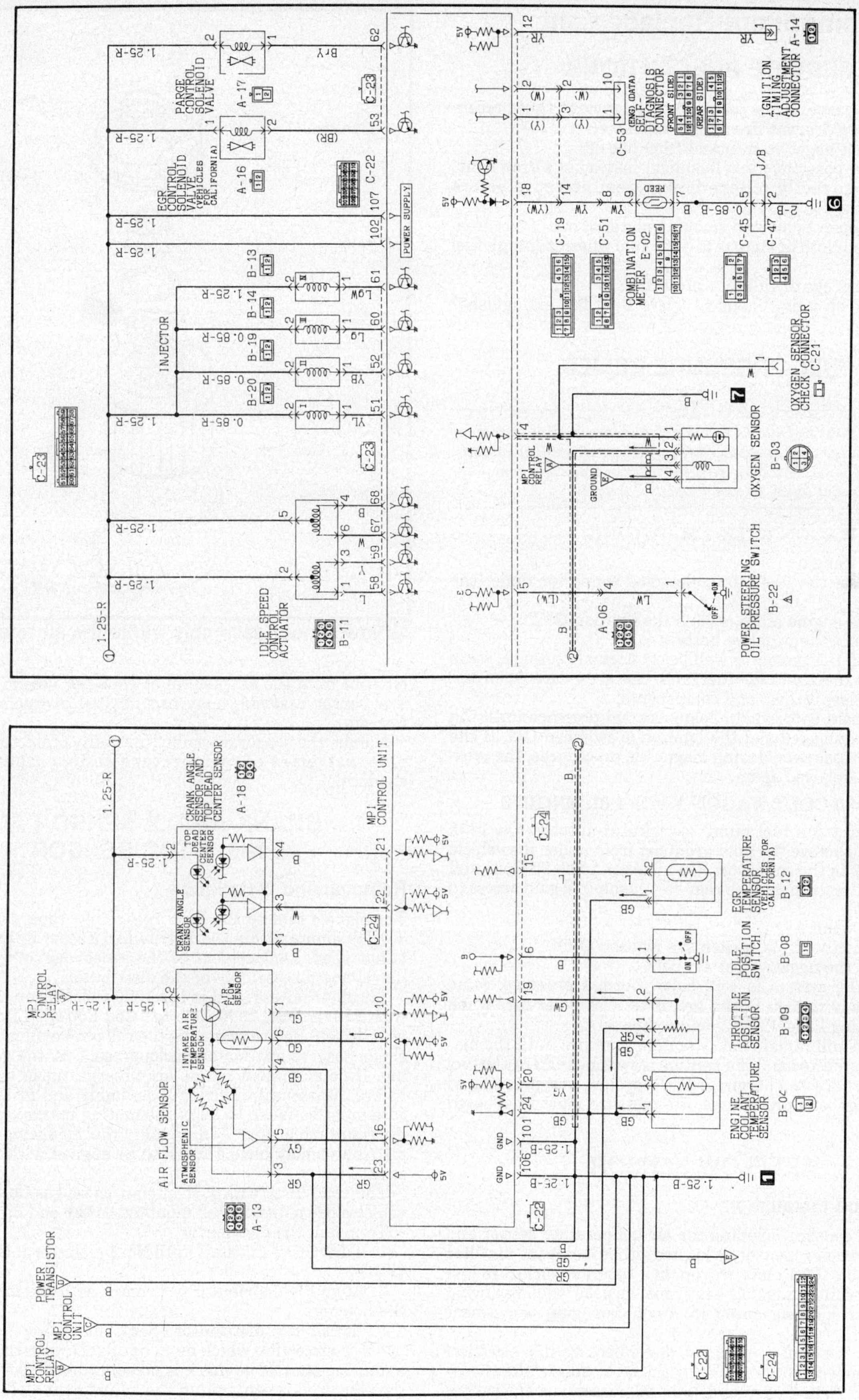

Component Replacement

SERVICE PRECAUTIONS

When working around any part of the fuel system, take precautionary steps to prevent fire and/or explosion:

- Disconnect negative terminal from battery.
- Whenever possible, use a flashlight instead of a drop light.
- Always correctly relieve fuel system pressure before servicing.
- Keep all open flame and smoking material out of the area.
- Use a shop cloth or similar to catch fuel when opening a fuel system.
- Always use eye or full-face protection.
- Always keep a dry chemical (class B-C) fire extinguisher near the area.

FUEL PRESSURE RELIEF

CAUTION

This procedure must be followed before any fuel-bearing line or component is loosened, repaired or replaced. The fuel system is under constant pressure of 38–50 psi. Failure to release this pressure properly may result in fire, personal injury and/or property damage.

1990 COLT WAGON WITH 1.8L ENGINE

1. Start the engine.
2. Disconnect the fuel pump electrical connector under the vehicle.
3. After the engine stops, switch the ignition **OFF**.
4. Disconnect the negative battery cable.
5. Although the system is well below normal pressure, some pressure may remain in the fuel lines. Use great care when disconnecting lines, fittings and components.
6. The system will remain below normal pressure until the battery is reconnected and the ignition is switched **ON**. If the system is repressurized during diagnostic procedures, the pressure must be relieved again.

EXCEPT 1990 COLT WAGON WITH 1.8L ENGINE

1. Disconnect the fuel pump electrical connector. On 1992 Vista Wagon, remove the floor grommet from under the vehicle to gain access to the connector. On all other 1990–92 Coltt and Colt 200 vehicles, remove the rear seat cushion to gain access to the connector.
2. Start the engine.
3. After the engine stops, switch the ignition **OFF**.
4. Disconnect the negative battery cable.
5. Although the system is well below normal pressure, some pressure may remain in the fuel lines. Use great care when disconnecting lines, fittings and components.
6. The system will remain below normal pressure until the battery is reconnected and the ignition is switched **ON** or the fuel pump is otherwise energized. If the system is repressurized during diagnostic procedures, the pressure must be relieved again.

AIR FLOW SENSOR

Removal and Installation

The air flow sensor, including the air temperature sensor and barometric pressure sensor, is located either inside the air filter housing, or, on 1992 Vista Wagon, in a separate enclosure just behind the air filter housing. Care must be used when removing the air filter housing cover to avoid damaging the sensor assembly.

Replacing the sensor requires disconnecting the electrical connector, then carefully removing the lid of the air filter housing. On 1992 Vista Wagon and 1991–92, disconnect the rubber air boot from the air plenum or chamber behind the air filter. The sensor assembly may be unbolted or unclipped from its mounts.

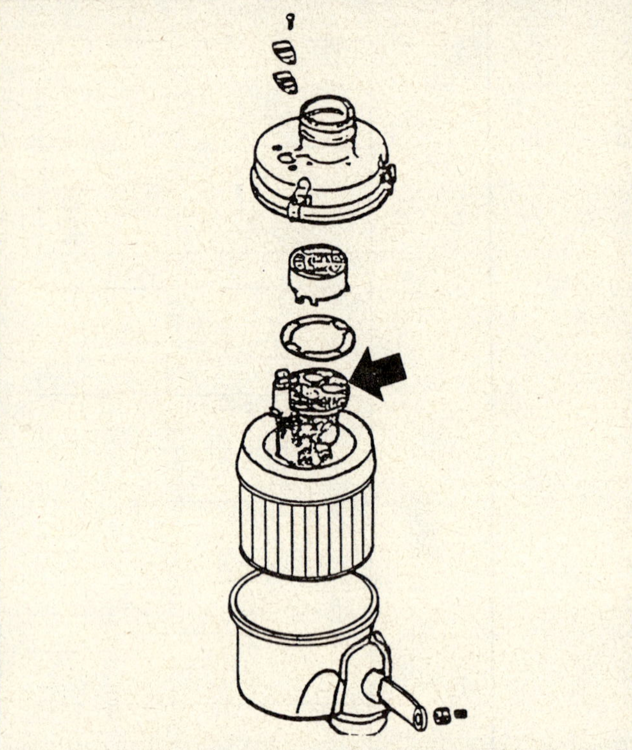

Airflow sensor assembly within the air filter housing

Handle the sensor assembly carefully, protecting it from impact, extremes of temperature and/or exposure to shop chemicals.

CRANK ANGLE SENSOR AND NO. 1 TDC SENSOR

Removal and Installation

The signals sent to the ECU regarding crank angle and No. 1 TDC originate within the distributor; a separate external sensor is not used. Replacement of the sensor assembly requires removal and disassembly of the distributor.

The distributors used in the 1990 Colt Wagon, 1990–92 Vista with 2.0L engine, 1990 Colt and Colt 200 with 1.5L engine, are gear driven. Before removing the roll pin holding the gear to the shaft, observe any matchmarks present on the gear and housing. If no alignment marks are present, make indelible marks before disassembly. During reassembly and installation, great care must be taken to align the mating marks on the gear and distributor housing. Additionally, the mounting flange of the distributor may have a mark to be aligned with the mounting stud.

The 1990 Vista with 1.8L engine, as well as Colt and Colt 200 vehicles use a distributor mounted to the end of the camshaft.

To install the distributor:

1. Turn the crankshaft until No. 1 piston is at top dead center (TDC).
2. Align the distributor gear mark with the mating mark on the housing.
3. Install the distributor. Some mounting flanges have a mark or projection which must be aligned with the center of the mounting stud. If no mark is present, install the distributor so that the stud is centered in the adjustment slot.

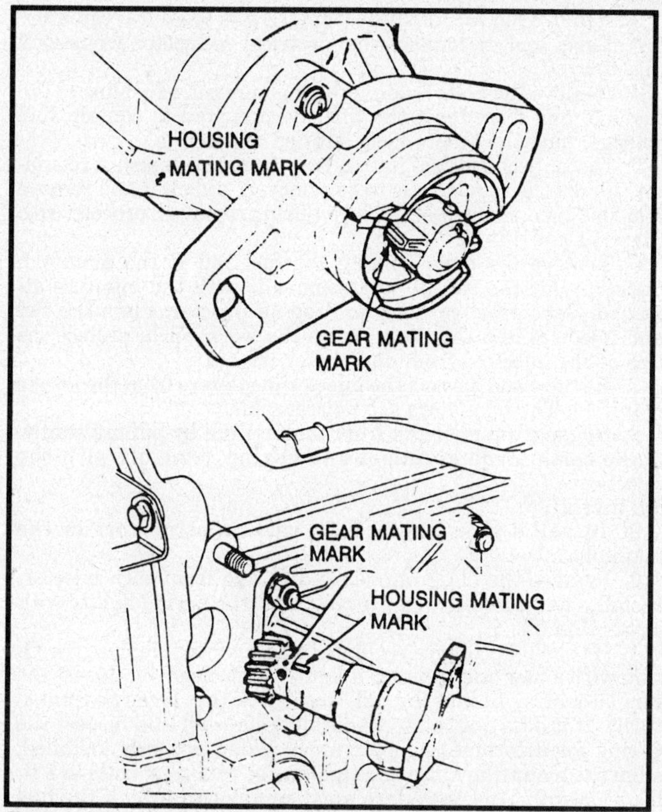

Examples of alignment marks for direct driven (above) and gear driven distributors

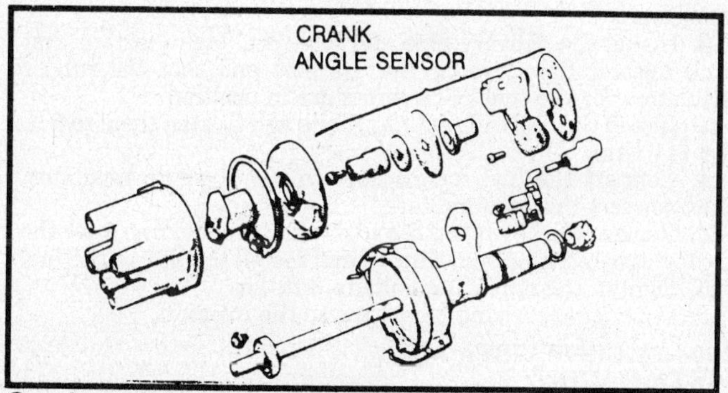

Crank angle sensor assembly

EGR CONTROL SOLENOID VALVE

Removal and Installation

The EGR control solenoid valve is used on all California specification engines except the 1990 Colt Wagon with 1.8L engine.

1. With the ignition **OFF**, disconnect the wiring harness from the solenoid.
2. Label and disconnect the vacuum hoses from the valve unit.
3. Remove the mounting bolts holding the unit.
4. Reinstall in reverse order, taking care to replace the vacuum hoses correctly.

EGR TEMPERATURE SENSOR

Removal and Installation

The EGR temperature sensor is used on all California specification engines. The sensor is mounted in the base of the EGR valve. After allowing the components to cool sufficiently, disconnect the wiring connector and unscrew the sensor.

When reinstalling, tighten the sensor to 8 ft. lbs (11 Nm) and connect the wiring harness.

ENGINE CONTROL UNIT (ECU)

Removal and Installation

On the 1990 Colt Wagon, the ECU is located on the center hump behind the center console. 1990–91 Vista Wagon locate the ECU under the driver's seat while the 1992 model place it behind the glovebox. Colt, Summit and Colt 200 vehicles with the 1.5L engine position the ECU behind the right front kick panel; 1990 vehicles with the 1.6L engine place the ECU behind the center console.

1. Make certain the ignition is **OFF**.
2. Depending on the vehicle, remove the left seat, glove box or right kick panel as necessary for access. Units mounted behind the console should be accessible without removing the console, although working space is cramped.

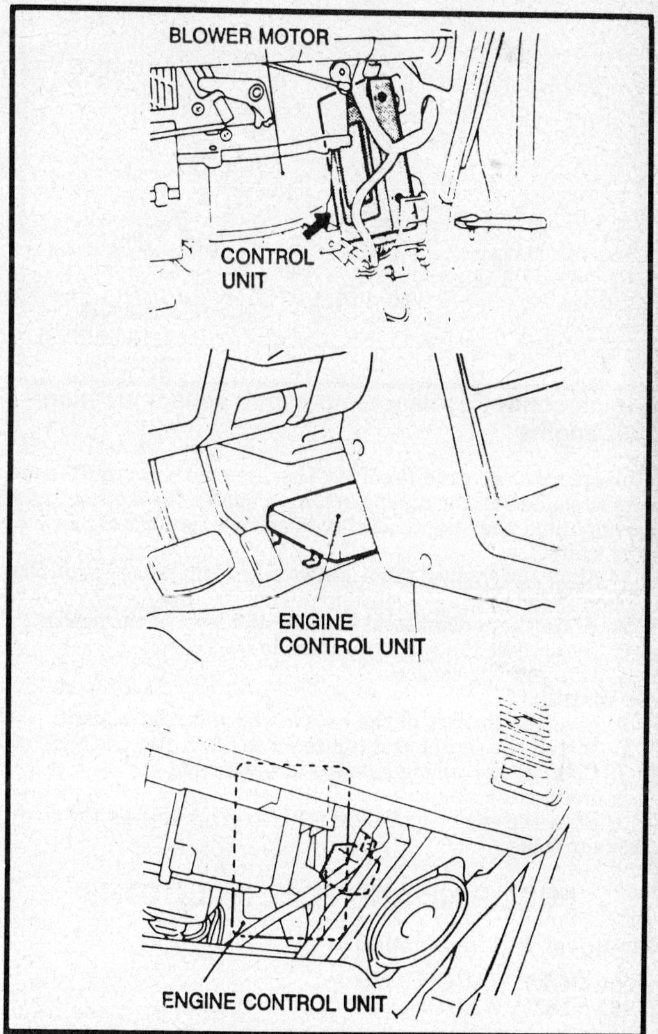

Examples of ECU placement

3. Remove the bolts holding either the ECU or its mount.
4. Free the ECU so that it may be repositioned. Carefully disconnect the wiring harness without damaging terminal pins or wiring. Remove the ECU. If the unit is to be reused, store it in an area protected from impact, static discharge or extremes of temperature.

To install:

5. Position the ECU. Connect the wiring harness, taking great care that the connector is squarely and firmly engaged. Attach any locking or retaining clips to the connector.
6. Secure the ECU in position.
7. Reinstall the left seat, glove box or right kick panel as necessary.

COOLANT TEMPERATURE SENSOR

Removal and Installation

The coolant temperature sensor is used to signal the engine control unit regarding temperature of the coolant. Do not confuse this sensor with the other switches or sensors used for a temperature gauge or air conditioning. Other systems may require sensors or senders in the cooling system. Usually, these other units are mounted near the coolant temperature sensor used for engine control.

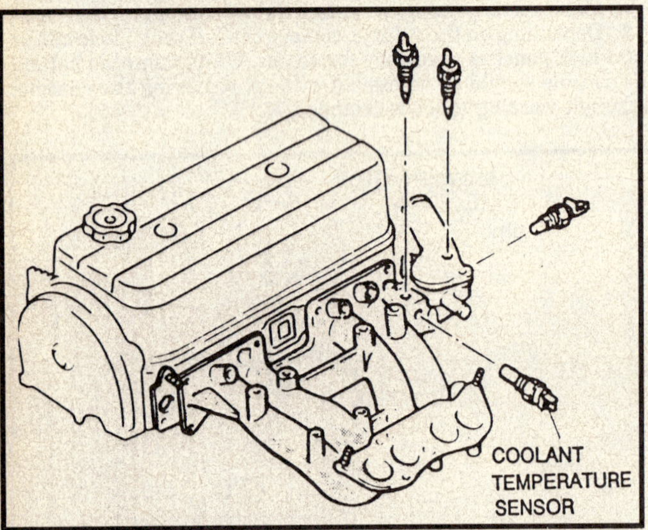

Engine control coolant temperature sensor location—2.0L engine

1. Allow the engine to cool. Safely release pressure within the cooling system.
2. Drain the coolant level to below the level of the sensor.
3. Disconnect the wiring harness to the sensor.
4. Unscrew the sensor.

To install:

5. Coat the threads of the sensor with a thread sealant.
6. Install the sensor and tighten it to 22 ft. lbs. (30 Nm).
7. Connect the wiring harness to the sensor.
8. Refill the coolant.
9. Start the engine and allow it to warm up. Inspect the work area for leaks.

FUEL RAIL AND FUEL INJECTORS

Removal and Installation

1990 COLT WAGON AND 1992 VISTA WAGON

1. Relieve the fuel pressure in the system using the procedure given in this section.
2. Remove the vacuum hose from the fuel pressure regulator.
3. Label and disconnect the electrical connector from each injector.
4. Remove the bolts holding the high pressure fuel line to the fuel rail and disconnect the line. Be prepared to contain fuel spillage; plug the line to keep out dirt and debris.
5. Disconnect the fuel return hose from the pressure regulator. If only the regulator is to be removed, unbolt it and remove it at this time. If the fuel rail is not being removed, proceed with Steps 11 and 14–16.
6. Remove the bolts holding the fuel rail to the manifold. Carefully lift the rail up and remove it with the injectors attached. Take great care not to drop an injector. Place the rail and injectors in a safe location on the workbench; protect the tips of the injectors from dirt and/or impact.
7. Remove and discard the injector insulators from the intake manifold. The insulators are not reusable.
8. Remove the injectors from the fuel rail by pulling gently. Make certain the grommet and O-ring come off with the injector.

To install:

9. Install a new insulator in each injector port in the manifold.
10. Remove the old grommet and O-ring from each injector. Install a new grommet and O-ring; coat the O-ring lightly with clean, thin oil.
11. If the fuel pressure regulator was removed, replace the O-ring with a new one and coat it lightly with clean, oil. Insert the regulator straight into the rail, then check that it can be rotated freely. If it does not rotate smoothly, remove it and inspect the O-ring for deformation or jamming. When properly installed, align the mounting holes and tighten the retaining bolts to 8 ft. lbs. (11 Nm). This procedure must be followed even if the fuel rail was not removed.
12. Install the injector into the fuel rail, constantly turning the injector left and right during installation. When fully installed, the injector should still turn freely in the rail. If it does not, remove the injector and inspect the O-ring for deformation or jamming.
13. Install the delivery pipe and injectors. Make certain that each injector fits correctly into its port and that the rubber insulators for the fuel rail mounts are in position.
14. Install the fuel rail retaining bolts and tighten them to 8 ft. lbs. (11 Nm).
14. Connect the fuel return hose to the pressure regulator, then connect the vacuum hose.
15. Replace the O-ring on the high pressure fuel line, coat the O-ring lightly with clean, thin oil and install the line to the fuel rail. Tighten the mounting bolts to 8 ft. lbs.
16. Connect the wiring harnesses to the injectors.

1991–92 COLT VISTA AND VISTA WAGON

1. Relieve the fuel pressure in the system using the procedure given in this section.
2. Disconnect the electrical harness connectors to the engine coolant temperature sender and the idle speed control servo motor. Label and disconnect the wiring to each injector.
3. Disconnect the vacuum line to the brake booster.
4. Disconnect the fuel return hose from the fuel pressure regulator. Disconnect the vacuum hose to the regulator and remove the regulator. Be prepared to contain fuel spillage. Remove and discard the O-ring from the base of the regulator.
5. Disconnect the high pressure fuel line from the fuel inlet line. Use 2 wrenches and counterhold the hose while loosening the flare fitting on the inlet line. Do not allow the hose to twist. Contain fuel spillage; plug the fuel hose immediately to eliminate entry of dirt and debris.
6. Disconnect the banjo or eye bolt holding the inlet line to the fuel rail. Discard the gasket under the bolt head. Remove the brackets holding the inlet line in place and remove the line.

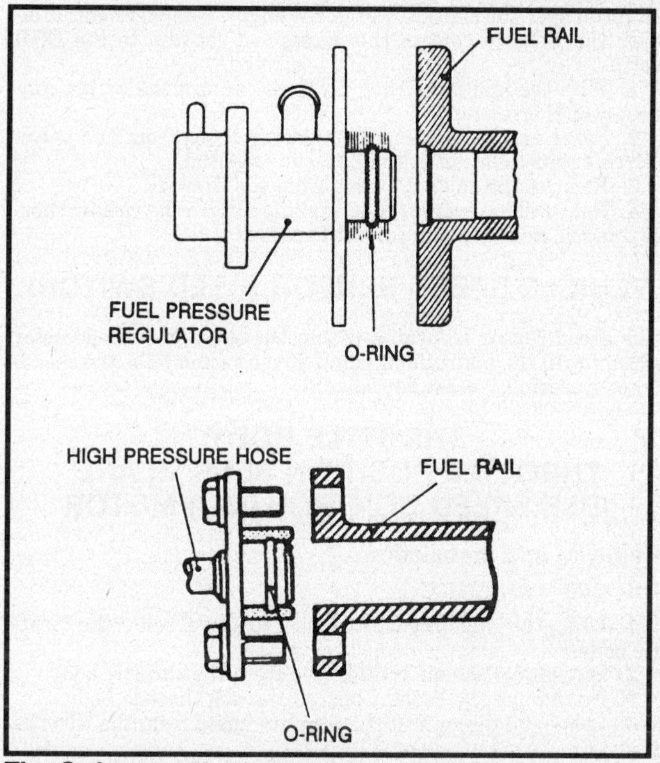

The O-rings on the fuel pressure regulator and high pressure fuel hose must be replaced whenever the units are disconnected from the fuel rail

7. Remove the bolts holding the fuel rail to the manifold. Carefully lift the rail up and remove it with the injectors attached. Take great care not to drop an injector. Place the rail and injectors in a safe location on the workbench; protect the tips of the injectors from dirt and/or impact.

8. Remove and discard the injector insulators from the intake manifold. The insulators are not reusable.

9. Remove the injectors from the fuel rail by pulling gently. Make certain the grommet and O-ring come off with the injector.

To install:

10. Install a new insulator in each injector port in the manifold.

11. Remove the old grommet and O-ring from each injector. Install a new grommet and O-ring; coat the O-ring lightly with clean, thin oil.

12. Install the injector into the fuel rail, constantly turning the injector left and right during installation. When fully installed, the injector should still turn freely in the rail. If it does not, remove the injector and inspect the O-ring for deformation or jamming.

13. Install the delivery pipe and injectors. Make certain that each injector fits correctly into its port and that the rubber insulators for the fuel rail mounts are in position.

14. Install the fuel rail retaining bolts and tighten them to 8 ft. lbs. (11 Nm).

15. Position the fuel inlet pipe and secure the bracket screws.

16. Install a new gasket under the head of the banjo or eye bolt and install the bolt, tightening it to 22 ft. lbs. (30 Nm).

17. Connect the high pressure fuel hose to the inlet line. Begin threading the flare fitting by hand. When tightening, use 2 wrenches. Do not allow the hose to turn or twist; tighten the flare fitting to 27 ft. lbs. (37 Nm).

18. Install a new O-ring on the fuel pressure regulator and coat the O-ring lightly with a clean, thin oil. Install the regulator in the same fashion as the injectors; if it does not turn freely in its mount, remove it and inspect the O-ring for deformation of jamming. Tighten the regulator mounts to 22 ft. lbs. (30 Nm).

19. Connect the vacuum hose and fuel return line to the fuel pressure regulator.

20. Connect the vacuum line for the brake booster.

21. Connect the wiring for the injectors. Reconnect the wiring harnesses to the idle servo motor and the coolant temperature sensor.

1990–92 COLT

1. Relieve the fuel pressure in the system using the procedure given in this section.

2. Disconnect the high pressure fuel line from the fuel rail. Be prepared to contain fuel spillage; plug the fuel line immediately to prevent entry of dirt and debris. Remove and discard the O-ring from the fuel line.

3. Disconnect the fuel return hose and the vacuum hose from the fuel pressure regulator.

4. Unbolt and remove the fuel pressure regulator. Remove and discard the O-ring.

5. On 1.6L engine, remove the PCV hose.

6. Label and disconnect the wiring connectors from the injectors.

7. Remove the bolts holding the fuel rail to the manifold. Carefully lift the rail up and remove it with the injectors attached. Take great care not to drop an injector. Place the rail and injectors in a safe location on the workbench; protect the tips of the injectors from dirt and/or impact.

8. Remove and discard the injector insulators from the intake manifold. The insulators are not reusable.

9. Remove the injectors from the fuel rail by pulling gently. Make certain the grommet and O-ring come off with the injector.

To install:

10. Install a new insulator in each injector port in the manifold.

11. Remove the old grommet and O-ring from each injector. Install a new grommet and O-ring; coat the O-ring lightly with clean, thin oil.

12. Install the injector into the fuel rail, constantly turning the injector left and right during installation. When fully installed, the injector should still turn freely in the rail. If it does not, remove the injector and inspect the O-ring for deformation or jamming.

13. Install the delivery pipe and injectors. Make certain that each injector fits correctly into its port and that the rubber insulators for the fuel rail mounts are in position.

14. Install the fuel rail retaining bolts and tighten them to 8 ft. lbs. (11 Nm). On 1.6L engine, make certain the throttle cable retaining brackets are reinstalled.

15. Connect the wiring harnesses to the injectors.

16. Reinstall the PCV hose if it was removed.

17. Install a new O-ring on the fuel pressure regulator and coat the O-ring lightly with a clean, thin oil. Install the regulator in the same fashion as the injectors; if it does not turn freely in its mount, remove it and inspect the O-ring for deformation of jamming. Tighten the regulator mounts to 6 ft. lbs. (8 Nm).

18. Connect the vacuum hose and fuel return line to the fuel pressure regulator.

19. Install a new O-ring on the high pressure fuel line and coat the O-ring lightly with a clean, thin oil. Install the fuel line to the fuel rail by pressing it squarely into place. Be careful not to damage or dislodge the O-ring. Tighten the mounting bolts to 27 ft. lbs. (37 Nm).

OXYGEN SENSOR

Removal and Installation

CAUTION

The oxygen sensor, exhaust system and surrounding components become very hot during engine operation. Avoid personal injury by waiting for a cooled motor or wearing protective clothing and gloves.

1. Allow the exhaust system to cool sufficiently to permit servicing.
2. Disconnect the wire connector from the oxygen sensor.
3. Remove the oxygen sensor from the exhaust system.
4. If not already done, coat the threads of the replacement sensor with anti-seize compound. New sensors are already coated with the substance. Take great care not to contaminate the oxygen sensor probe with the anti-seize compound.
5. Install the oxygen sensor into the exhaust manifold. On all vehicles except 1992 Vista Wagon with 1.8L engine, tighten the sensor to 33 ft. lbs. (45 Nm). On the 1992 1.8L engine, tighten the sensor to 26 ft. lbs. (35 Nm).
6. Reconnect the wiring to the sensor.

POWER STEERING PRESSURE SWITCH

Removal and Installation

The power steering pressure switch is located in the side of the power steering pump body. It detects a high pressure condition and signals the ECU to increase the idle speed to compensate.

CAUTION

The power steering fluid and components will be hot if the vehicle has been driven immediately before disassembly. Use caution to avoid burns from hot oil.

1. Drain the power steering fluid from the pump and hoses.
2. Disconnect the sender wire from the sender.
3. Using the correct-sized wrench, remove the terminal assembly. Use care to contain any escaping oil during removal. When the terminal assembly is free, be prepared to catch O-ring, spring, plunger and piston rod behind the terminal assembly.

NOTE: The Vista Wagon use a multi-piece terminal assembly. It should not be necessary to disassembly this assembly during or after removal. If it must be disassembled, use care to reassemble the seals and circlips in the correct order.

4. Inspect all parts for signs of binding, scoring or deterioration. The O-ring between the spring and the base of the terminal must be replaced, not reused. Clean or replace damaged seals or components as necessary.

To install:

5. Before assembly, apply a light coat of clean power steering fluid to each part.
6. Install the plunger into the pump bore.
7. Assemble the spring and piston rod. The large end of the piston rod should face into the pump bore to contact the plunger rod. The large end of the spring should face the terminal assembly or the outside of the pump case. Install the spring and piston assembly.
8. Using a new O-ring, install the terminal assembly and O-ring to the pump body. Tighten the terminal assembly to 20 ft. lbs. (27 Nm).
9. Connect the wiring to the sensor.
10. Refill the power steering fluid and bleed the system.

PURGE CONTROL SOLENOID VALVE

Removal and Installation

The purge control solenoid is located on the firewall or the fender apron near the shock tower in the engine compartment. It allows the ECU to control the passage of vacuum to the EGR valve.

1. With the ignition switch **OFF**, disconnect the wiring connector at the solenoid.
2. Label and disconnect the vacuum hoses from the valve. Correct reinstallation of the hoses is essential.
3. Remove the solenoid valve from the firewall.
4. Reinstall in reverse order, making certain the vacuum hoses are reinstalled in the correct locations.

VEHICLE SPEED SENSOR (REED SWITCH)

The speed sensor is located within the speedometer/odometer assembly in the instrument panel. If the switch fails, the speedometer assembly must be replaced.

THROTTLE BODY THROTTLE POSITION SENSOR AND IDLE SPEED CONTROL (ISC) MOTOR

Removal and Installation

EXCEPT 2.0L ENGINE

1. Drain the engine coolant at least to a level below the throttle body.
2. Disconnect the air intake hose from the throttle body.
3. Disconnect the coolant hoses from the throttle body.
4. Label and disconnect the vacuum hoses from the throttle body.
5. If equipped with an automatic transaxle, disconnect the transmission throttle cable at the throttle body assembly.
6. Disconnect the accelerator cable.

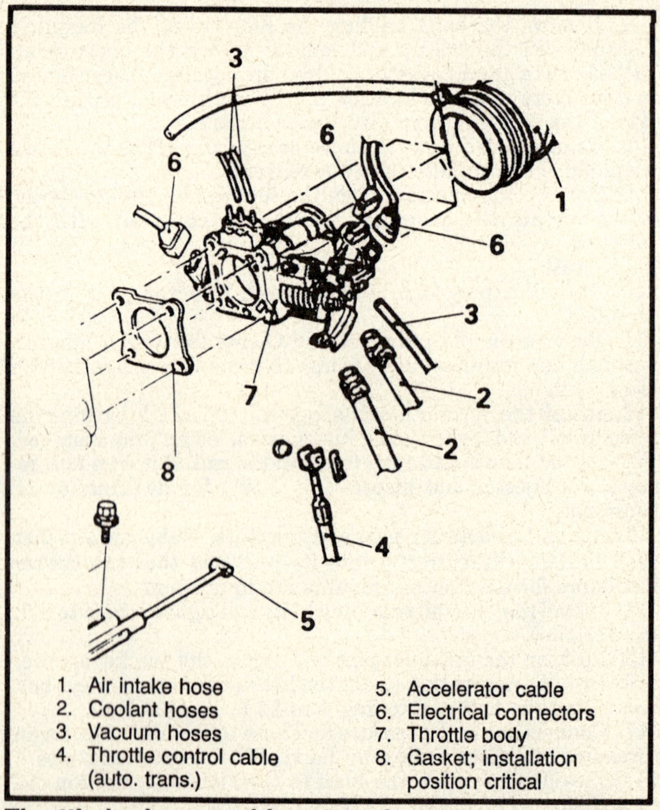

Throttle body assembly external components—1990 Colt Wagon

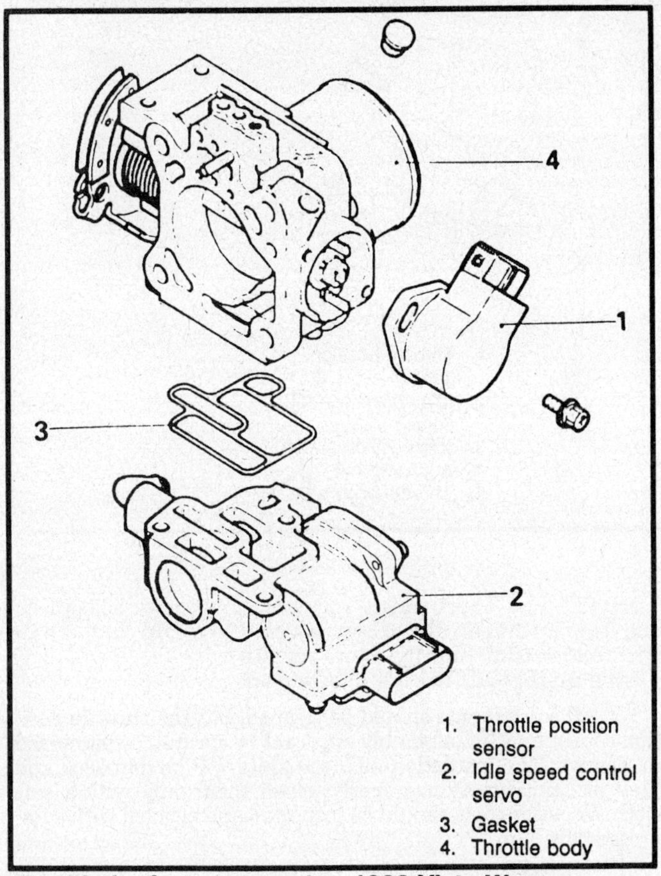

Throttle body components—1992 Vista Wagon

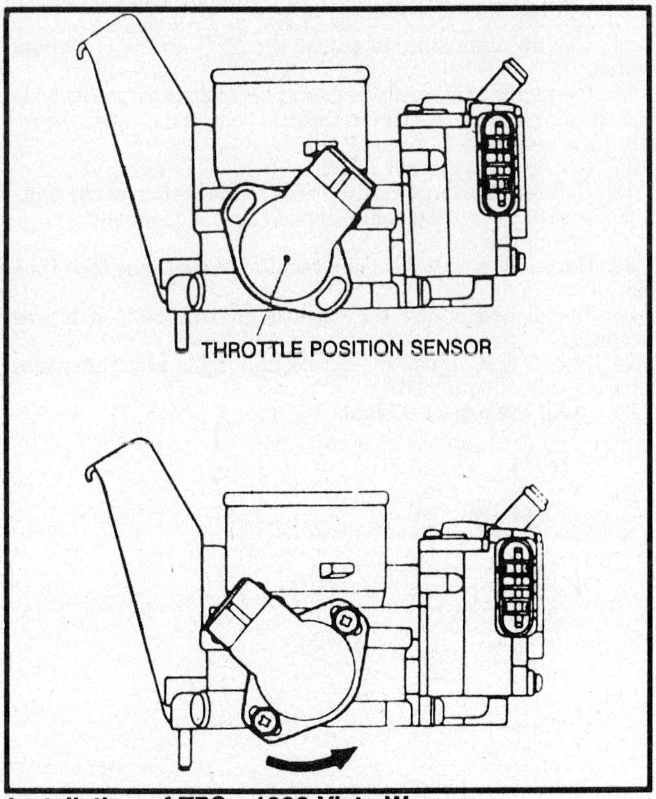

Installation of TPS—1992 Vista Wagon

7. Label and remove each wiring harness connector from the throttle body.
8. Remove the 4 bolts holding the throttle body to the air intake plenum; carefully remove the throttle body. On the 1.6L DOHC engine, it will be necessary to remove the throttle body support bracket before removing the throttle body.
9. Before removing the gasket, take careful note of its shape and position. One corner of the gasket is shaped differently from the other corners. At reassembly, the gasket must be installed in the correct position to avoid air leaks.
10. The throttle position sensor and/or the idle speed control servo assembly may be removed by unbolting them from the throttle body. Do not attempt to remove the throttle valve (blade) assembly.

NOTE: The cross-point screws are tightly installed; use a screwdriver of the correct size to avoid damage to the heads. Additionally, some screws are coated with adhesive or thread-locking compound.

11. All components should be cleaned, but the throttle position sensor and ISC assembly must not be sprayed or immersed in cleaner. The insulation of these units will be damaged and they will not function properly; clean them only with a soft cloth. Vacuum ports should be inspected and cleaned with compressed air.

To install:

12. Reassemble the TPS and ISC components to the throttle body if they were removed. On 1992 Vista Wagon, install the TPS with the connector head aligned at the 2 o'clock position as viewed from the back of the sensor. Once engaged, rotate the sensor 90 degrees counterclockwise so that the connector head is now at the 11 o'clock position. Install the retaining screws. Tighten the TPS bolts only to 1.5 ft. lbs (2 Nm or 18 inch lbs.).
13. Use an ohmmeter to adjust the TPS and idle position switch.
14. Correctly position a new gasket between the throttle body and the air plenum. Install the throttle body and tighten the retaining bolts. Tighten the bolts to 14 ft. lbs. (19 Nm). On 1.6 DOHC engines, install the throttle body support bracket.
15. Install the wiring harness connectors.
16. Connect and adjust the transmission throttle cable and the accelerator cable.
17. Connect the vacuum hoses to the throttle body.
18. Connect the coolant hoses to the throttle body.
19. Install the air intake hose.
20. Refill the coolant to the proper level.

2.0L ENGINE

1. Drain the engine coolant at least to a level below the throttle body.
2. If equipped with cruise control, disconnect the speed control cable from the throttle body.
3. Disconnect the accelerator cable.
4. If equipped with an automatic transaxle, disconnect the throttle control cable.
5. Label and disconnect the small vacuum hoses from the top of the unit.
6. Remove the air intake hose from the throttle body.
7. Label and disconnect the electrical connectors.
8. Disconnect the coolant hoses from the throttle body.
9. Disconnect the large vacuum hose from the bottom of the unit.
10. Disconnect the ground cable from the unit.
11. Remove the 4 retaining bolts. Remove the throttle body and gasket.
12. The throttle position sensor and/or idle speed control servo assembly may be removed if they are to be replaced. Do not remove these components unless replacement is required. Do not attempt to disassemble the throttle valve (blade) assembly. Never attempt to disassemble the ISC servo assembly.

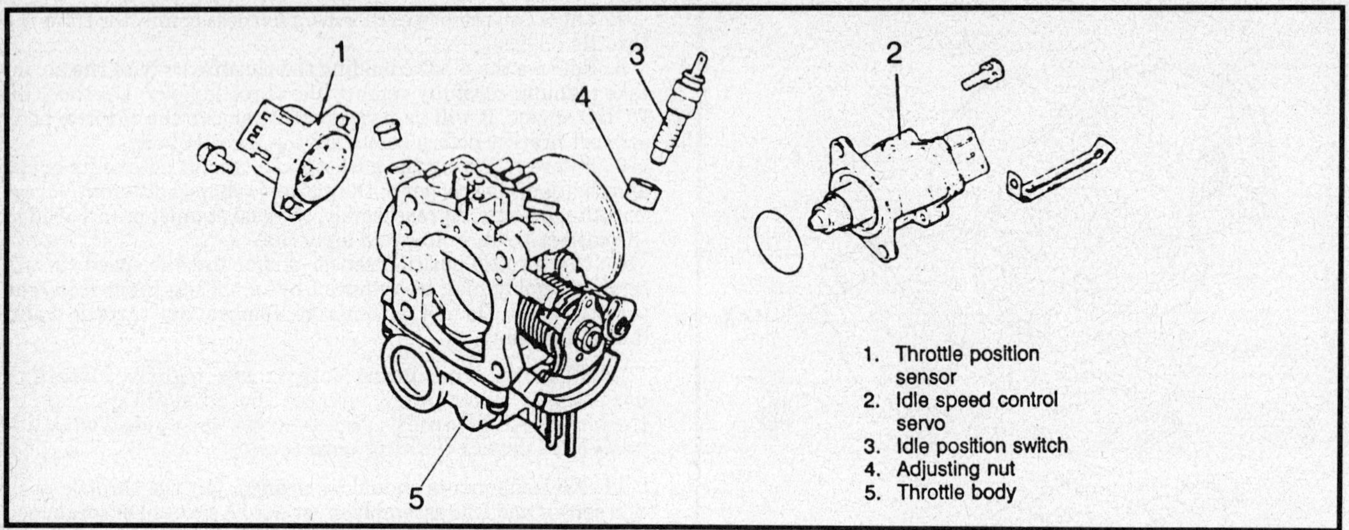

Throttle body components—1.6L DOHC engine

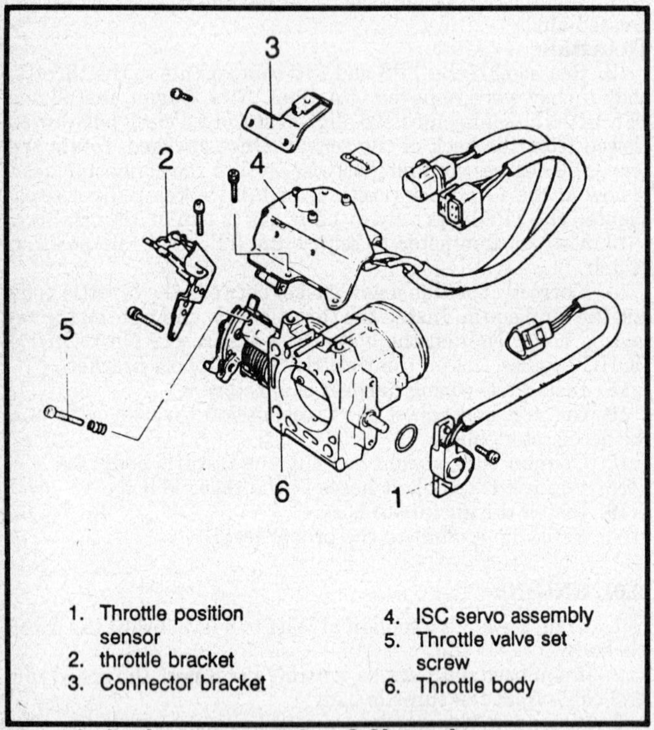

Throttle body components—2.0L engine

NOTE: The cross-point screws are tightly installed; use a screwdriver of the correct size to avoid damage to the heads. Additionally, some screws are coated with adhesive or thread-locking compound.

13. All components should be cleaned, but the throttle position sensor and ISC assembly must not be sprayed or immersed in cleaner. The insulation of these units will be damaged and they will not function properly; clean them only with a soft cloth. Vacuum ports should be inspected and cleaned with compressed air.

To install:

14. Reassembly the TPS and ISC components to the throttle body if they were removed. Install the retaining screws. Tighten the TPS bolts only to 1.5 ft. lbs (2 Nm or 18 inch lbs.).
15. Use an ohmmeter to adjust the TPS and idle position switch.
16. Correctly position a new gasket between the throttle body and the air plenum. Install the throttle body and tighten the retaining screws to 8 ft. lbs. (11 Nm).
17. Connect the ground cable.
18. Reinstall the large vacuum hose to the bottom of the unit.
20. Install the wiring connectors to each component.
21. Install the air intake hose.
22. Correctly install the small vacuum hoses to the top of the unit.
23. Install and adjust the throttle control cable if it was removed.
24. Install and adjust the accelerator cable and the cruise control cable, if equipped.
25. Refill the engine coolant.

Daihatsu Fuel Injection

SECTION 7

DAIHATSU ELECTRONIC FUEL INJECTION SYSTEM

General Information

Daihatsu electronic fuel injection system consists of 3 sub-systems: air intake, electronic control and fuel supply. The model G100 Charade is a 3 cylinder 1.0L engine while the G102 model is a 4 cylinder 1.3L engine.

AIR INTAKE SYSTEM

The system supplies air for all engine needs. It consists of the air cleaner, air intake pipe, surge tank, throttle body, auxiliary air valve, fast idle mechanism, and intake manifold.

ENGINE CONTROL SYSTEM

The Electronic Control Unit (ECU) controls the amount of fuel and mixture based on engine conditions and running conditions, as determined by signals input by various sensors. The ECU also controls electronic ignition timing, fuel pump operation, idle-up control, a self-diagnosis function and the fail safe and back-up functions.

Ignition Timing Control

Based on sensor input and engine load, the ECU controls the ignition timing. This maintains optimum engine operations.

Self Diagnosis

If any malfunction should occur in an input signal circuit, the ECU memorizes the fault. This malfunction is indicated during the trouble diagnosis, by codes through the checking lamps. If a major malfunction is encountered, the ECU turns on the check engine lamp to warn the driver.

FAIL SAFE FUNCTION

When a malfunction is detected, the ECU determines, based on programmed standard values, whether the engine may be allowed to continue operation or should be stopped immediately.

BACK-UP FUNCTION

In the event the ECU detects a malfunction, the back-up function makes it possible to provide fuel injection amount and ignition timing that have been predetermined by the back-up data. This is to ensure safe driving if a single or many sensors are faulty, or if the ECU malfunctions.

Idle-up Control

This system controls the idle-up Vacuum Switching Valve (VSV), which is necessary to increase engine idle to compensate for headlight or rear defogger operation.

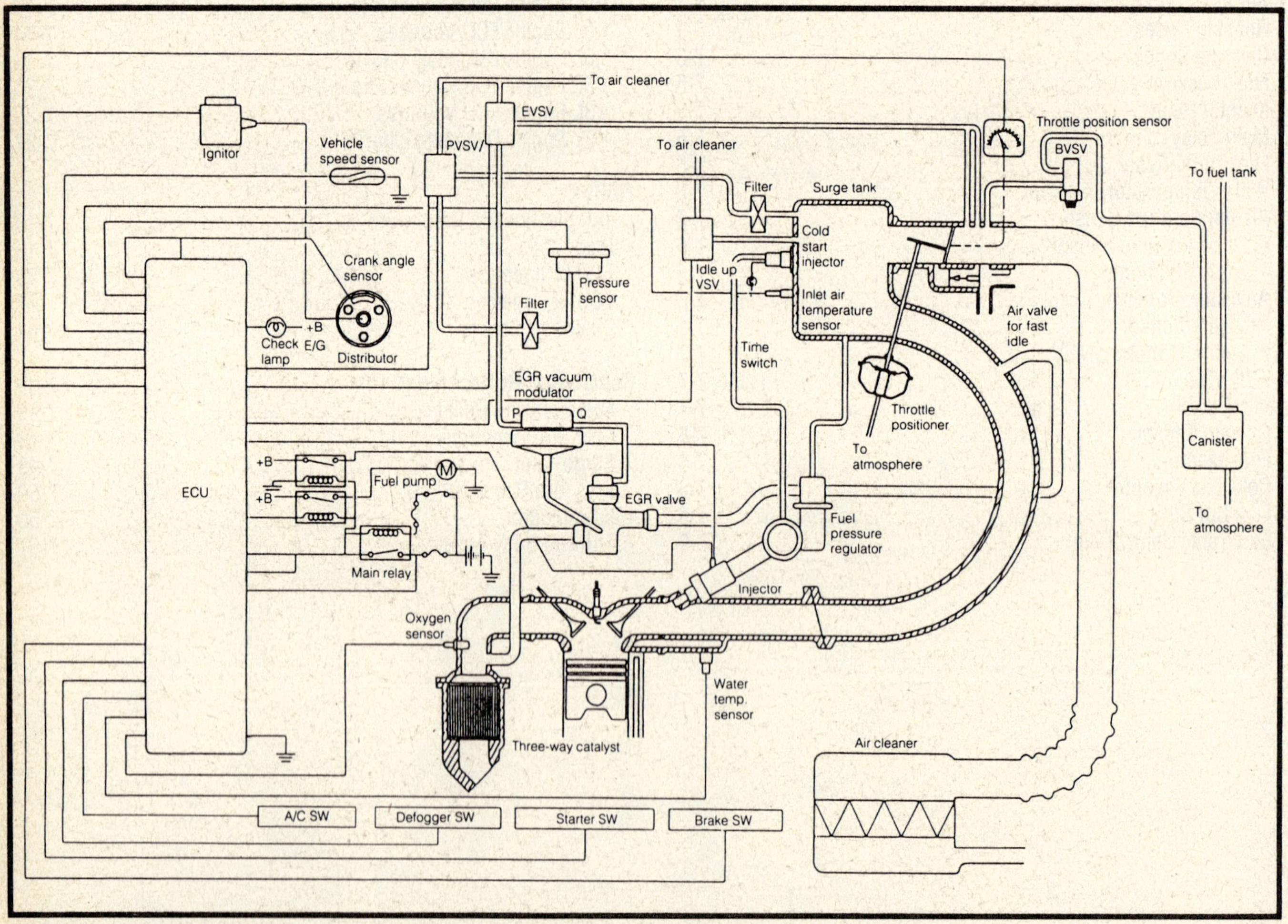

Fuel injection system diagram—Charade G100 with 1.0L engine

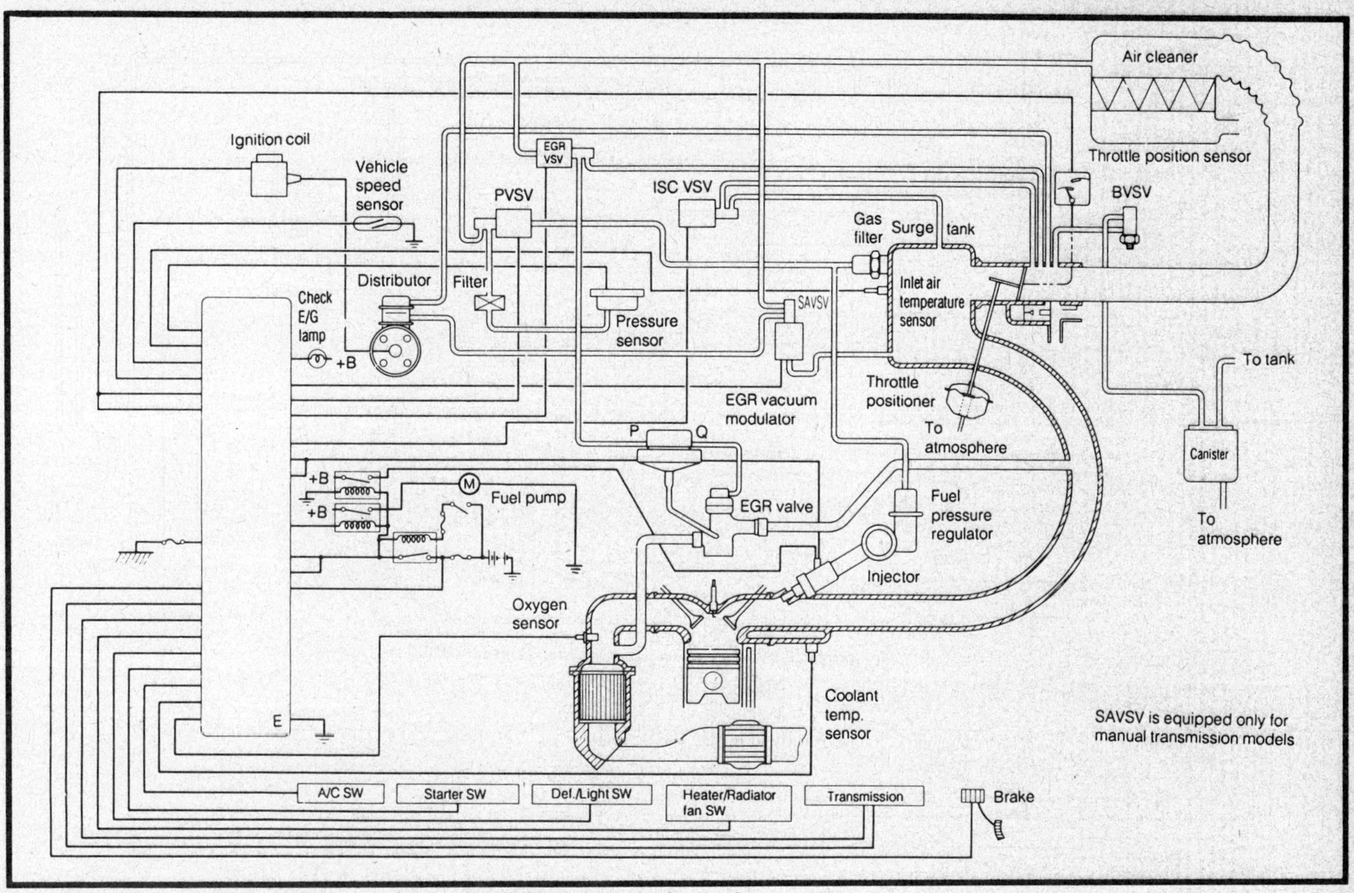

Fuel injection system diagram—Charade G102 with 1.3L engine

FUEL SYSTEM

Fuel is supplied to the engine by an in-tank fuel pump. The fuel pump delivers fuel under a pressurized condition. The pressure regulator mounted at the delivery pipe keeps the fuel pressure at 36.7 psi higher that the intake manifold inner pressure. The fuel injection amount for each injector is kept at a constant level by this method. The operation of the electric fuel pump is controlled by the Electronic Control Unit (ECU).

Fuel Pump Control

This system actuates the fuel pump by turning on the fuel pump relay when the ignition switch is turned to **START** or when the ECU detects the engine rotation.

Diagnosis and Testing

NOTE: For Self-Diagnostic System and Accessing Trouble Code Memory, see Section 3 "SELF-DIAGNOSTIC SYSTEM".

SERVICE PRECAUTIONS

- Disconnect the negative battery cable before disconnecting any fuel line.
- Do not operate the fuel pump when the fuel lines are empty.
- Do not operate the fuel pump when removed from the fuel tank.
- Make sure all ECU harness connectors are fastened securely. A poor connection can cause an extremely high surge voltage and could and result in damage to integrated circuits.
- Keep ECU all parts and harnesses dry during service.
- Before attempting to remove any parts, turn **OFF** the ignition switch and disconnect the battery ground cable.
- Do not attempt to disconnect the battery cables with the engine running.
- Do not depress the accelerator pedal when starting.
- Never open the cover to the ECU.

READING CODES

The ECU diagnostic codes are accessed through the check terminal located underhood.

1. To read the diagnostic codes the battery voltage must be at least 11 volts, the throttle fully closed and all accessory switches in the **OFF** position.
2. Locate the check terminal underhood. On the Charade it is located on the upper section of the transmission. On the Rocky it is located on the right side fender panel.

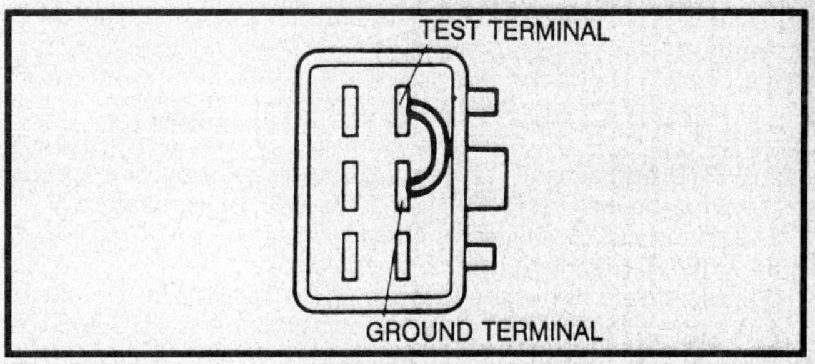

Underhood check terminal for accessing diagnostic codes

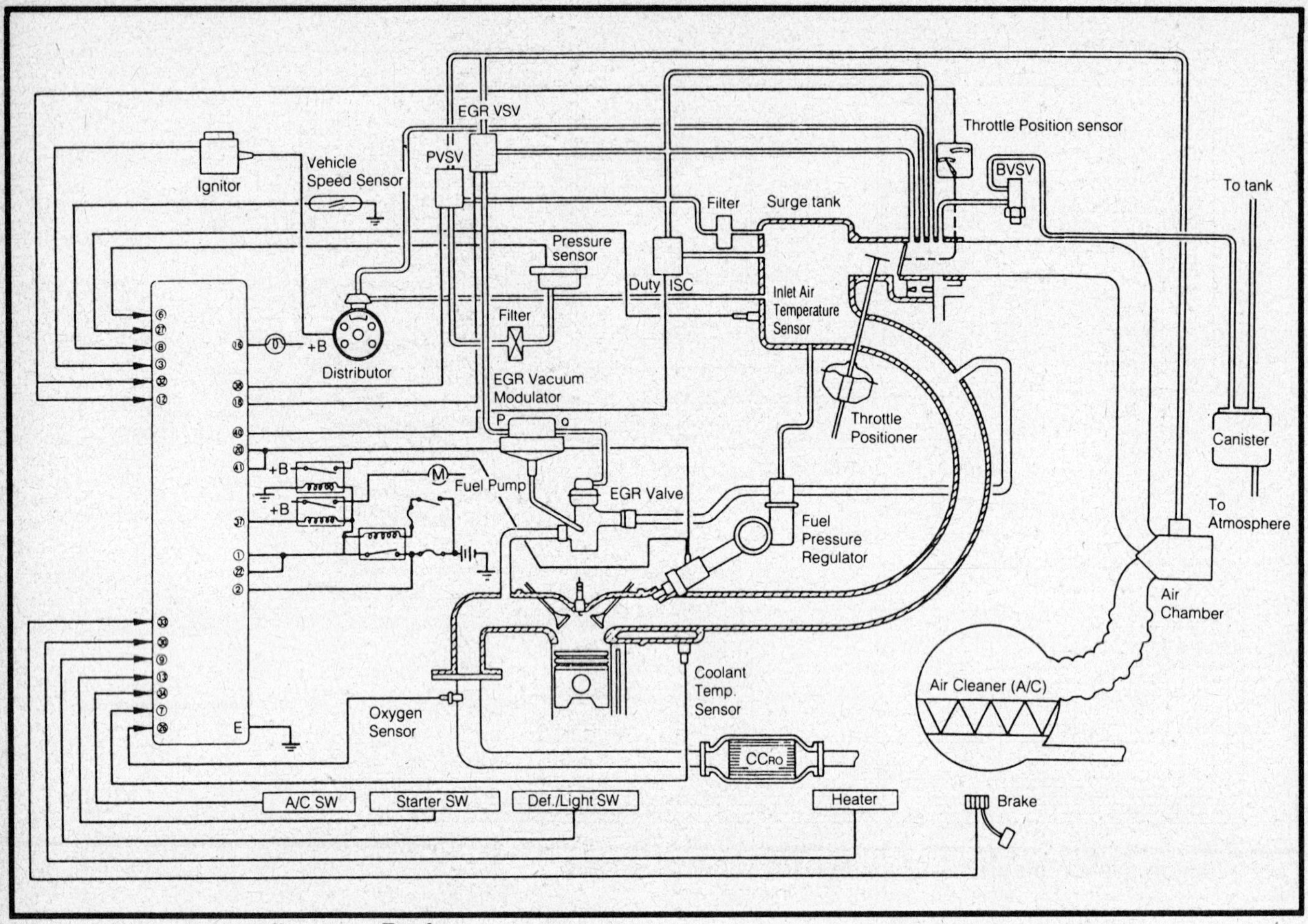

Fuel injection system diagram—Rocky

3. Short the test terminal to the ground terminal. Turn the ignition switch to the **ON** position. Be careful not to start the engine.

4. Read the pattern of flashes of the **CHECK ENGINE** lamp. If the lamp does not flash, the ECU is most likely malfunctioning. The flash pattern has the lamp glow for 0.5 seconds and pauses for 4.5 seconds. It will then repeat the pattern.

5. After the codes have been read, remove the shorting wire and cap the check terminal.

CLEARING CODES

After retrieving the diagnostic codes the memory of the ECU should be reset. Disconnect the negative battery terminal for a minimum of 10 seconds with the ignition switch in the **OFF** position.

SPECIAL SERVICE TOOL (SST)

Many ECU and sensor test require the installation of a special test wiring harness SST 09842–87704–000, or equivalent.

1. Disconnect the negative battery cable.
2. Remove the glove compartment box.
3. Disconnect the wiring harness from the ECU.
4. Connect the SST wiring harness between the ECU and the engine wiring harness. If performing resistance tests at the ECU connector terminals, never connect the SST to the ECU.
5. Reconnect the negative battery cable.

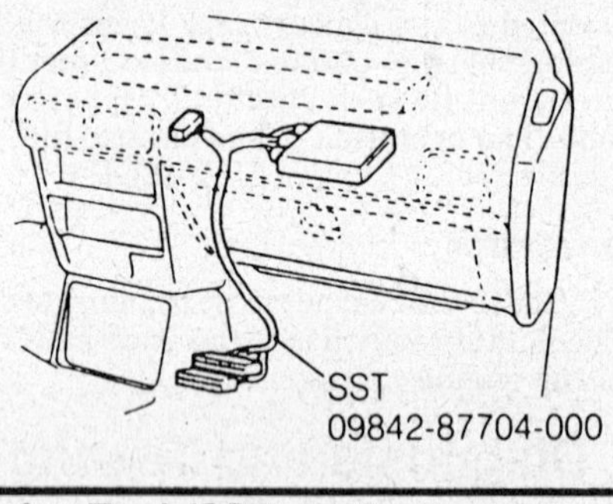

Special Service Tool (SST) testing harness

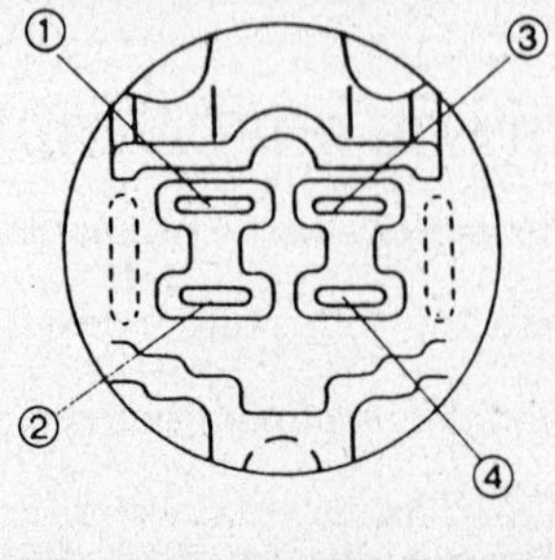

Main and injector relay terminal identification

6. After testing, disconnect the negative battery cable and replace the engine harness to the ECU.

FUEL PRESSURE

Testing

1. Disconnect the negative battery cable.
2. Slowly remove the union bolt from the fuel filter using a cloth to catch excess fuel.
3. Connect pressure gauge to the fuel pipe.
4. Install the SST No. 09991–87702–000 test harness or equivalent, to the check terminal.
5. Ground the F (white with black) terminal with the test harness.
6. Reconnect the negative battery cable. Turn **ON** the ignition switch.
7. Fuel pressure should read 33–40 psi (2.3–2.8 bar) on the Charade G100 with 1.0L engine and the Rocky. On the Charade G102 with 1.3L engine the fuel pressure reading should be 27–45.5 psi (2.6–3.2 bar).
8. Turn **OFF** the ignition switch. The fuel pressure should not drop below 25 psi (1.8 bar) after 3 or more minutes for the Charade G100 with 1.0L engine and Rocky. It should not drop below 35.5 psi (2.5 bar) for the Charade G102 with 1.3L engine.
9. Remove the SST test harness and fuel pressure tester. Start the engine and check for fuel leakage.
10. If fuel pressure is above specifications, check regulator vacuum line. If vacuum line and system check good, replace the fuel regulator.

MAIN RELAY OR INJECTOR RELAY

Testing

CAUTION

Do not touch the relay if in operation. The relay becomes very hot during operation. Personal injury could result.

1. Check that the relay clicks when the ignition switch is turned **ON**.
2. Check that there is only continuity between terminals 1 and 2.

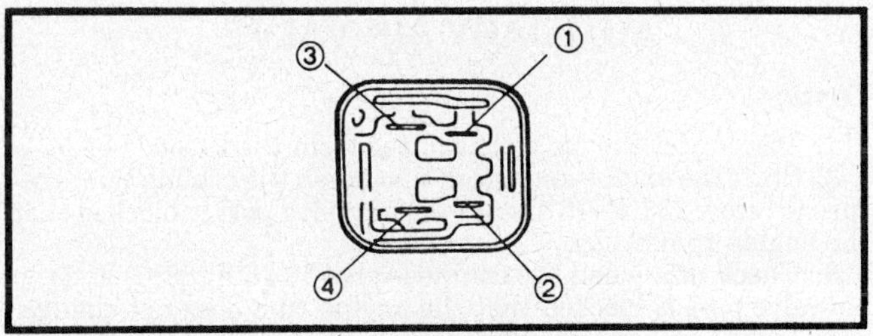

Fuel pump relay terminal identification

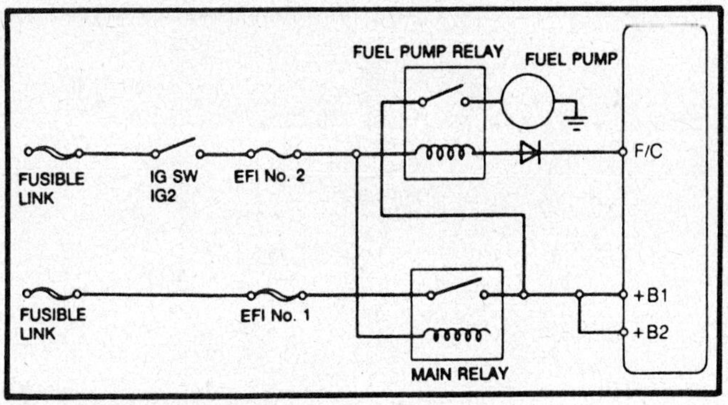

Fuel pump relay circuit

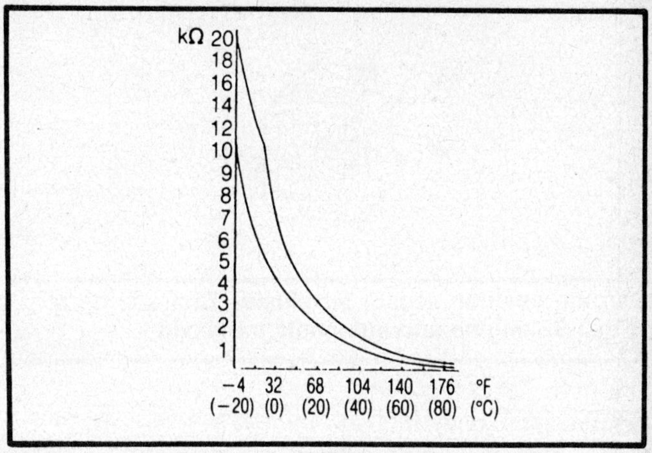

Water temperature and air temperature sensor graph

3. The resistance between terminals 1 and 2 should be 40–100 ohms.
4. Apply battery voltage across terminals 1 and 2.
5. Check that there is continuity between terminals 3 and 4.
6. If any test fails, replace the relay.

FUEL PUMP RELAY

Testing

CAUTION

Do not touch the relay if in operation. The relay becomes very hot during operation. Personal injury could result.

1. Disconnect the ST terminal of the starter.
2. Check that the relay clicks when the ignition switch is turned to the **START** position.
3. Check that there is only continuity between terminals 1 and 2 of the relay.
4. The resistance between terminals 1 and 2 should be 70–90 ohms.
5. Apply battery voltage across terminals 1 and 2.
6. Check that there is continuity between terminals 3 and 4 of the relay.
7. If any test fails, replace the relay.

WATER TEMPERATURE SENSOR

Testing

1. Disconnect sensor.
2. Measure temperature of sensor. It must be in the range of 0–180°F (−20–80°C).
3. Measure resistance between sensor terminals through temperature range and compare to graph. The test measurements must fall between the lines of the graph.
4. If not within specification, replace sensor.

AIR TEMPERATURE SENSOR

Testing

1. Disconnect sensor and check that there is no continuity between the terminals and the sensor body.
2. Measure temperature inside the surge tank. It must be in the range of 0–180°F (−20–80°C).
3. Measure resistance between sensor terminals through temperature range and compare to graph. The test measurements must fall between the lines of the graph.
4. If not within specification, replace sensor.

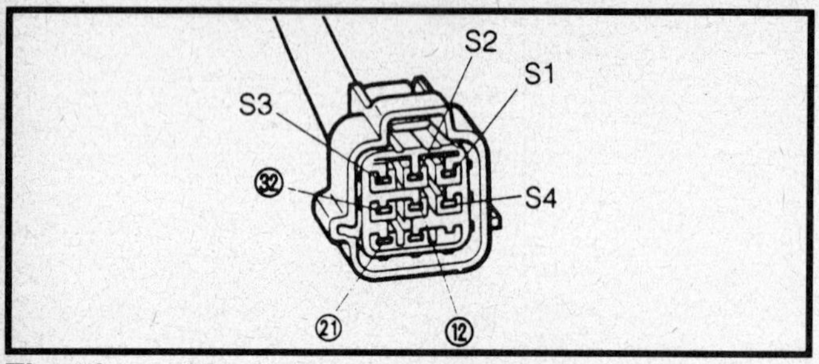

Throttle position sensor terminals—Charade G102 with 1.3L engine and automatic transaxle

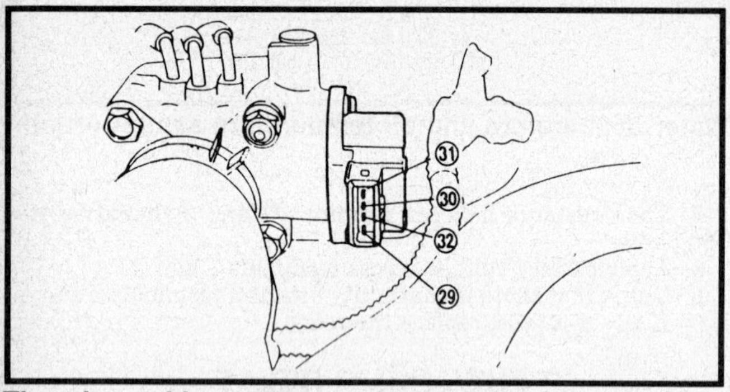

Throttle position sensor terminals—Charade G100 with 1.0L engine

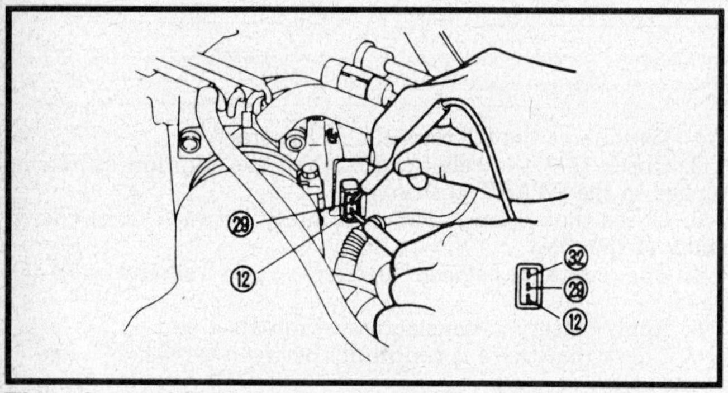

Throttle position sensor terminals—Charade G102 with 1.3L engine and manual transaxle

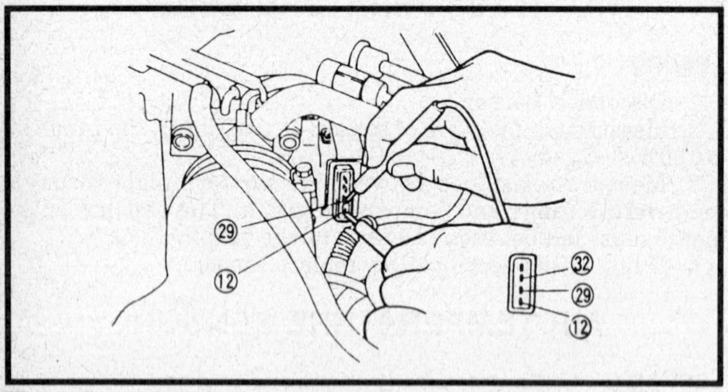

Throttle position sensor terminals—Rocky

THROTTLE POSITION SENSOR (TPS)

Testing

1. Disconnect the TPS.
2. Measure the resistance between terminals 31 and 29 on the Charade G100 with 1.0L engine.
3. The resistance should be 4.25–8.25 kilo ohms.
4. Measure the resistance between terminals 31 and 30 on the Charade G100 with 1.0L engine. Measure the resistance between terminals 12 and 29 on the Charade G102 with 1.3L engine with manual transmission and Rocky. Measure the resistance between terminals 12 and 21 on the Charade G102 with 1.3L engine with automatic transmission
5. On the Charade G100 with 1.0L engine the resistance should be 3.5–10.3 kilo ohms at closed throttle and 0.3–6.3 kilo ohms at full throttle. On the Charade G102 with 1.3L engine and Rocky the resistance should be 29 kilo ohms or less with the throttle closed and 1000 kilo ohms or more with the throttle wide open. The resistance value should change smoothly through the complete range of the throttle valve.
6. Measure the resistance between terminals 32 and 29 on Charade G100 with 1.0L engine, Charade G102 with 1.3L engine with manual transmission and Rocky. Measure the resistance between terminals 32 and 21 on the Charade G102 with 1.3L engine with automatic transmission.
7. On the Charade G100 with 1.0L engine the resistance should be less than 350 ohms at closed throttle and over 1000 kilo ohms if the throttle is opened more than 1.5 degrees. On the Charade G102 with 1.3L engine with manual transmission and the Rocky the resistance should be 1000 kilo ohms or more at closed throttle and 29 kilo ohms or less at full open throttle. On the Charade G102 with 1.3L engine with automatic transmission the resistance should be 29 kilo ohms or less at closed throttle and 1000 kilo ohms or more at full open throttle.
8. If any test fails, replace the throttle body.

THROTTLE POSITIONER (TP)

Testing

1. Allow engine to reach normal operating temperature.
2. Check that the idle speed is correct.
3. Raise engine to 3000 rpm and slowly close the throttle.
4. Observe the engine rpm at the time the dashpot lever comes in contact with the throttle lever.
5. If the engine speed is not 1500–1700 rpm for manual transmission equipped vehicles or 2700–2900 rpm for automatic transmission equipped vehicles, adjust by turning the TP adjusting screw.

AUXILIARY AIR VALVE

Testing

1. Disconnect the air cleaner hose from the throttle body.
2. Start the engine and check that when the cooling temperature is below 137°F (75°C) and the air valve port is blocked that the engine rpm drops.
3. Check that when the engine is above 146°F (80°C) and the air valve port is blocked that the engine rpm does not change.
4. If the auxiliary air valve malfunctions, replace the throttle body.

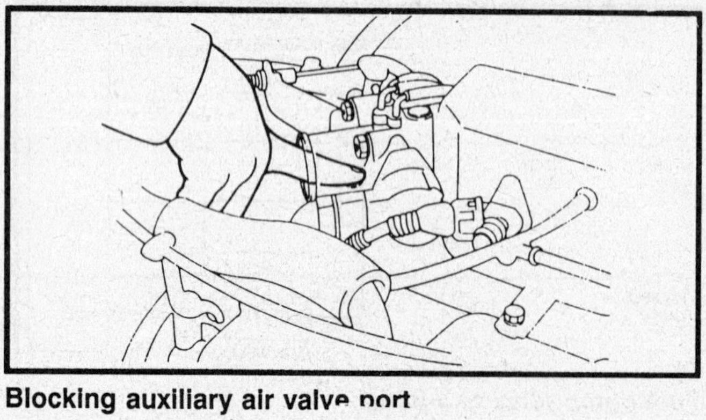

Blocking auxiliary air valve port

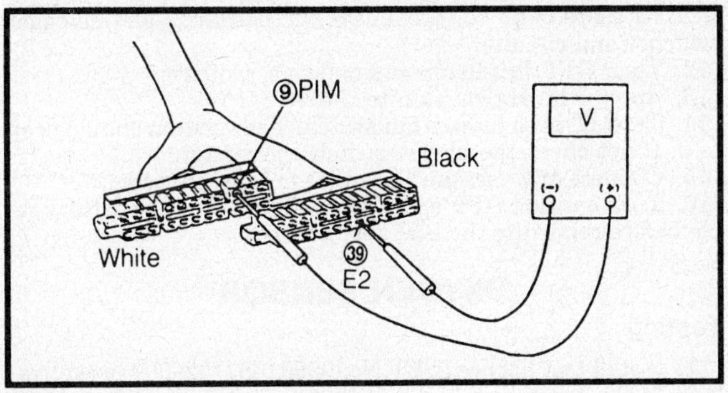

Testing the pressure sensor

Measuring point: Altitude (height above sea level) ft (m)	Atmospheric pressure inch Hg (mmHg)	Voltage V
0 (0)	29.92 (760)	2.1 - 2.7
1640 (500)	28.19 (716)	1.9 - 2.5
3280 (1,000)	26.54 (674)	1.8 - 2.4

Pressure sensor chart—Charade G100 with 1.0L engine

Measuring point: Altitude (height above sea level) m (ft)	Atmospheric pressure mmHg (inchHg)	Voltage V
0 (0)	760 (29.92)	3.2 - 4.0
500 (1640)	716 (28.19)	3.1 - 3.8
1000 (3280)	674 (26.54)	3.0 - 3.6

Pressure sensor chart—Charade G102 with 1.3L engine and Rocky

PRESSURE SENSOR

Testing

1. Disconnect the negative battery cable.
2. Connect the SST test harness to the ECU.
3. Reconnect the negative battery cable.
4. Measure the voltage between the 6 and 39 terminals (Charade G102 with 1.3L engine and Rocky) or 9 and 39 terminals (Charade G100 with 1.0L engine) of the test harness with the ignition switch **ON**.
5. If not within specification and if there is 4.5–5.5 volts between the 5 and 36 terminals (Charade G102 with 1.3L engine and Rocky) or 31 and 29 terminals (Charade G100 with 1.0L engine), replace the sensor.
6. Apply vacuum to the sensor and check that the voltage drops approximately 0.5–0.7 volts.
7. If the voltage does not drop, replace sensor. It is recommended that when the sensor is replaced that the air filter and fuel filter also be replaced.

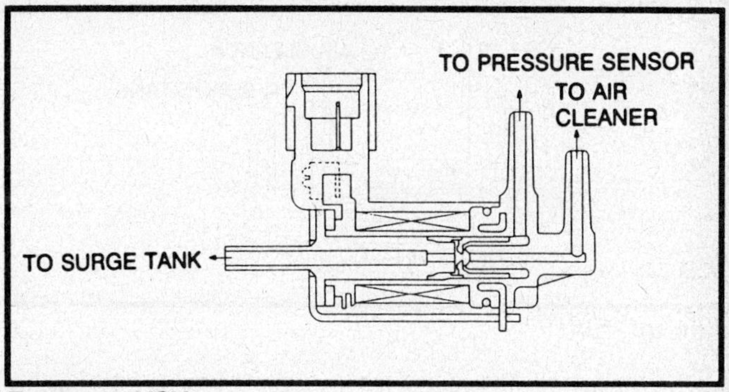

Pressure VSV

PRESSURE VACUUM SWITCHING VALVE (VSV)

Testing

1. Disconnect the negative battery cable.
2. Connect the SST test harness to the ECU.
3. Reconnect the negative battery cable.
4. Start the engine and allow it to reach normal operating temperature.
5. Measure the voltage at terminal 38–PVSV and 39–E1.
6. Raise engine to 4000 rpm and release accelerator pedal; if the voltage drops momentarily, VSV is good; if not, proceed.
7. Disconnect the vacuum hose from the VSV at the air cleaner side and apply 4 in. Hg. If VSV leaks, replace it.
8. Turn the ignition switch **ON**.
9. Short the PVSV and E1 terminals. The vacuum should drop to 0; if not, replace the VSV.
10. Turn the ignition switch **OFF**.
11. Apply vacuum to the sensor side port of the VSV. It should not hold vacuum.

EGR VACUUM SWITCHING VALVE (VSV)

Testing

1. Disconnect the negative battery cable.
2. Disconnect the EGR VSV connector and measure the resistance across the terminals. If the resistance is not approximately 44 ohms at 68°F (20°C), replace the EGR VSV.
3. Connect the SST test harness to the ECU.
4. Reconnect the negative battery cable.
5. Disconnect the hose connected to the EGR VSV.
6. Apply vacuum to the modulator side port of the VSV. It should not hold vacuum.
7. Apply vacuum to the surge tank side of the VSV and apply 4 in. Hg. It should hold vacuum.
8. EGR VSV should release this vacuum when the ignition switch is turned **ON**. If the vacuum is not released:

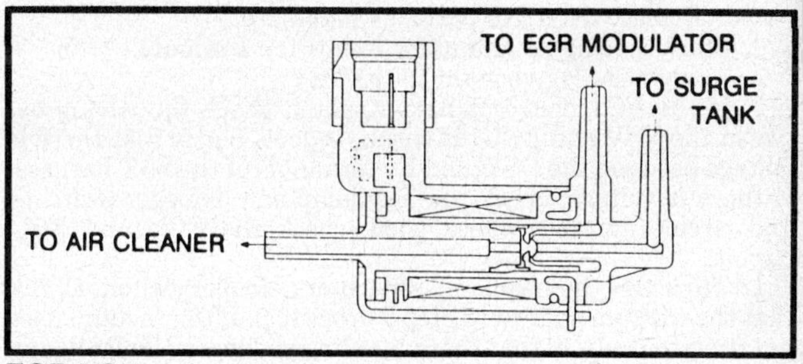

EGR VSV

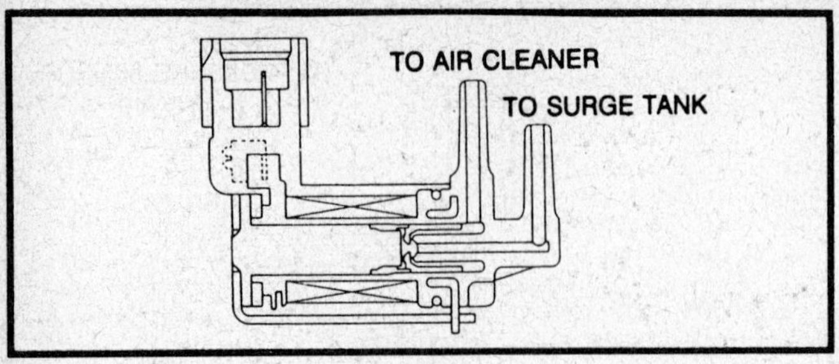

Idle-up VSV

a. Check for for voltage between the EVSV and E1 terminals of the SST harness.
b. If there is voltage, check the ECU.
c. If no voltage is applied, check the EGR VSV for power supply.
d. Measure the resistance of the EGR VSV, the resistance should be 30–50 ohms.
e. Replace defective parts, as necessary.

9. Start the engine, with the coolant temperature below 104°F (40°C). Apply vacuum to the surge tank port. If it does not hold vacuum, check the water temperature sensor.
10. With coolant temperature above 104°F (40°C), check that the engine stalls or runs poorly when vacuum is applied. If engine performance is not effected and vacuum is maintained, check the coolant temperature sensor. If engine performance is not affected without vacuum being held, check the EGR system.
11. Reconnect the vacuum hoses.
12. Start engine. While engine is warming up, measure the voltage between the EVSV and E1 terminals of the SST harness. Check that the voltage drops before the radiator cooling fan starts to rotate. If no voltage drop occurs, check the water temperature sensor. If the sensor is good, check the ECU.
13. Turn ignition **OFF** and disconnect the negative battery cable before removing the SST test harness.

IDLE-UP VACUUM SWITCHING VALVE (VSV)

Testing

1. Disconnect the negative battery cable.
2. Connect the SST test harness to the ECU.
3. Reconnect the negative battery cable.
4. Disconnect the rubber hose connected to the idle-up VSV.
5. Apply 4 in. Hg vacuum to the idle-up VSV; it should hold vacuum.
6. Turn **OFF** all accessory switches and turn **ON** the ignition switch.
7. The vacuum that was applied should drop to 0 when the ignition switch is turned **ON**. If the vacuum is not released:
   a. Check for for voltage between the VSV and E1 terminals of the SST harness.
   b. If there is voltage, check the the resistance between the VSV terminals. The resistance should be 30–50 ohms.
   c. If the resistance is correct, check the wiring between the ECU and the main relay.
   d. Replace defective parts, as necessary.
8. Start the engine and allow to idle for 1 minute.
9. Apply 4 in. Hg vacuum to the VSV.
10. If the VSV does not hold vacuum, check the wiring between the VSV and ECU. If wiring is good, check that there is voltage between the DSW and E1 terminals of the SST harness. If there is voltage, check the headlamp and defogger switches and circuits. If they check good, check the ECU and ECU circuit.
11. Turn **ON** the headlamp switch or defogger switch. Check that the vacuum applied in Step 9 drops to 0. If the vacuum does not drop to 0, check that there is voltage between the DSW and E1 terminals of the SST harness. If there is voltage, check the ECU; if there is no voltage, check the headlamp and defogger switches and circuits.
12. Turn **OFF** headlamp and defogger switches.
13. Apply 4 in. Hg vacuum to the VSV.
14. Turn **ON** the blower fan switch. The vacuum should drop to 0. If not check the blower switch and circuit.
15. Connect any vacuum lines that are disconnected.
16. Turn ignition **OFF** and disconnect the negative battery cable before removing the SST test harness.

OXYGEN SENSOR

Testing

1. Install test harness SST No. 09991-87702-000 or equivalent, to the check terminal.
2. Warm engine to normal operating temperature.
3. After warm, raise engine to 3000 rpm for 2 minutes.
4. Short the brown T terminal and the black ground terminal of the test connector.
5. Connect a voltmeter between the green VF terminal and engine ground.
6. Hold the engine speed constant at 1500 rpm.
7. The voltmeter change 8 or more times a minute.
8. If the voltmeter changes less than 8 times, repeat Steps 3 to 7; if still less than 8 changes, replace sensor.
9. If no change at all, perform diagnosis check. Check and repair and trouble codes. When trouble Code 11 is obtained continue with Step 10.
10. Repeat Steps 2 to 7 if the voltmeter reads 5 volts; the mixture is too rich. Repair condition causing rich mixture.
11. If the voltmeter reads 0 volts and engine is idling normally, replace the oxygen sensor. If the engine is not idling normally, the mixture is too lean. Check for vacuum leak or repair other condition causing a lean mixture.

FUEL INJECTOR

Testing

1. Check injector for operating sound when engine is being cranked.
2. Remove surge tank.
3. Check resistance of injector.
4. Resistance should be 11–17 ohms (Charade G102 with 1.3L engine and Rocky) or 10–15 ohms (Charade G100 with 1.0L engine).
5. If no operating sound or resistance not in range, replace injector.
6. Remove the injector and install SST 09268-87702-000 or equivalent, on the injector. Connect the fuel feed of the injector to the fuel line between the fuel filter and the pressure regulator.
7. Run the fuel pump and energize the injector with battery voltage. Measure the output of the injector with a graduated measuring cylinder. The output of the injector from the Charade G100 with 1.0L engine and G102 with 1.3L engine should be 36cc in a 15 second period and 45cc for the Rocky in a 15 second period.
8. Run the test 3 times for each injector. The measurements should be within 5cc of each other.

COLD START INJECTOR TIME SWITCH

Testing

1. Disconnect the cold start injector time switch..
2. Measure the resistance between the terminals under the following conditions:
   Terminals 1-2 – 67–79 ohms at 75°F (24°C)
   Terminals 1-3 – 67–79 ohms at 75°F (24°C)
   Terminals 2-3 – 140–155 ohms at 75°F (24°C)
   Terminals 1-2 – 33–41 ohms at 68°F (20°C)

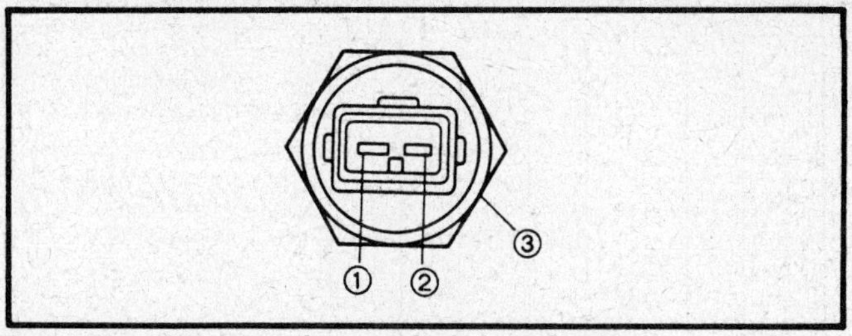

Cold start injector time switch terminals

Terminals 1–3 – 33–41 ohms at 68°F (20°C)
Terminals 2–3 – 0 ohms at 68°F (20°C)

3. If any test fails, replace the cold start injector time switch.

COLD START INJECTOR

Testing

1. Remove the injector from the intake plenum.
2. Install SST 09842-30500-000 or equivalent, on the injector.
3. Hold the end of the injector in a clear glass container. With the fuel hose still attached to the injector and battery voltage applied to the injector through the SST, run the fuel pump.
4. Observe the fuel atomization pattern. The pattern should be even without any drippage.
5. If the pattern is not satisfactory, replace the injector.

FUEL CUT RPM

Testing

1. Disconnect the negative battery cable.
2. Connect the SST test harness to the ECU.
3. Reconnect the negative battery cable.
4. Start engine and allow to reach normal operating temperature.
5. Short the IDL and E2 terminals of the SST harness (32 and 29 for Charade G100 with 1.0L engine, 12 and 39 for Charade G102 with 1.3L engine and Rocky).
6. Gradually raise the engine speed. Determine the rpm at which the engine speed ceases its increase because of fuel be cut off.
7. The fuel cut off rpm should be 1550 rpm (Charade G100 with 1.0L engine), 1600 rpm (Charade G102 with 1.3L engine with manual transmission), 1800 rpm (Charade G102 with 1.3L engine with automatic transmission), 1650 rpm (Rocky) with A/C **OFF** or 1850 rpm (Charade G100 with 1.0L engine), 2250 rpm (Charade G102 with 1.3L engine with manual transmission), 2450 rpm (Charade G102 with 1.3L engine with automatic transmission), 2300 rpm (Rocky)with the A/C **ON**.
8. The fuel return rpm should be 1200 rpm (Charade G100 with 1.0L engine), 1200 rpm (Charade G102 with 1.3L engine with manual transmission), 1400 rpm (Charade G102 with 1.3L engine with automatic transmission), 1300 rpm (Rocky) with A/C **OFF** or 1500 rpm (Charade G100 with 1.0L engine), 1500 rpm (Charade G102 with 1.3L engine with manual transmission), 1700 rpm (Charade G102 with 1.3L engine with automatic transmission), 1600 rpm (Rocky) with the A/C **ON**.

ELECTRONIC CONTROL UNIT (ECU)

Testing

1. Disconnect the negative battery cable.
2. Connect the SST test harness to the ECU.
3. Reconnect the negative battery cable.
4. Measure the voltages at the test harness.

NOTE: If measuring sensor resistances at the ECU connector, never connect the SST test harness to the ECU.

5. Check the voltage specifications chart. If voltage inputs and outputs are within specification and if the sensor resistance values are correct and the engine still malfunctions, replace the ECU.

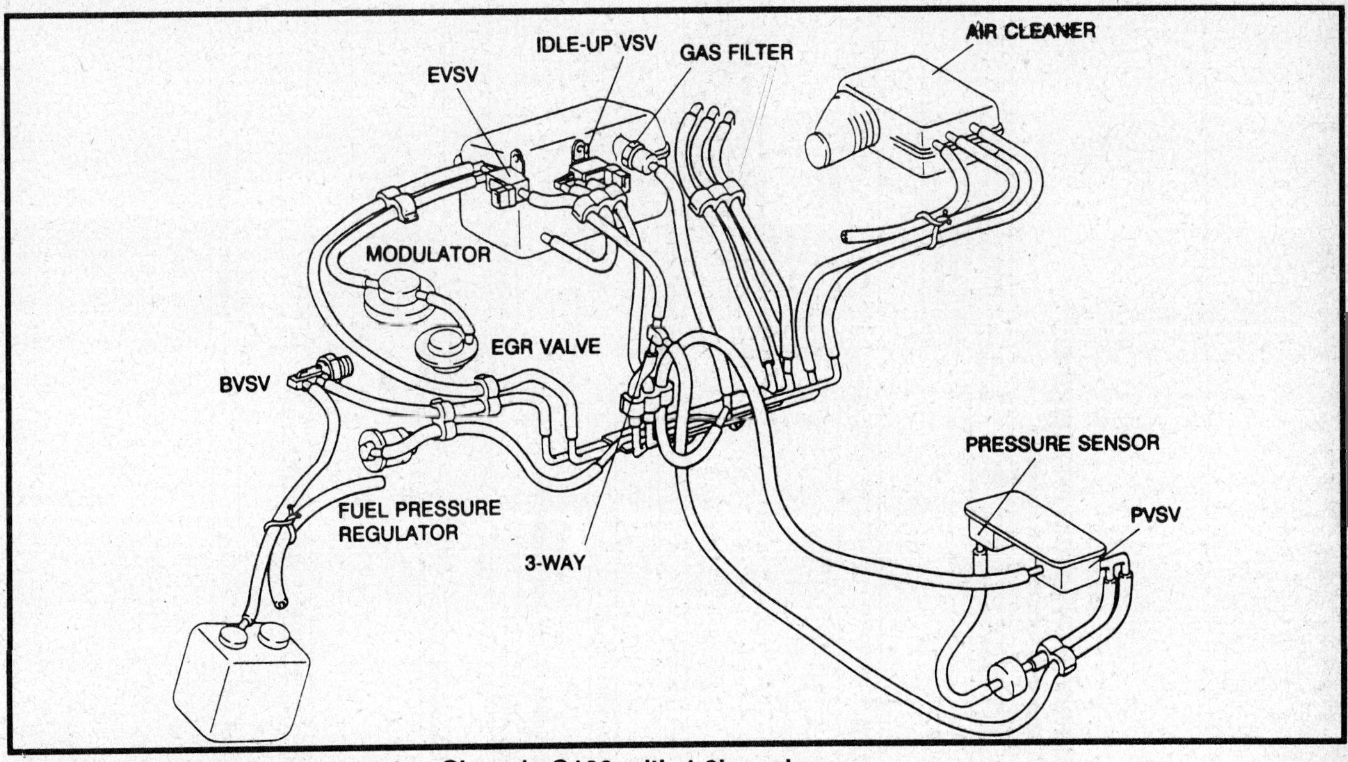

Vacuum diagram and components—Charade G100 with 1.0L engine

TROUBLE SHOOTING PROCEDURE

1 Engine will not crank or cranks slowly

Check indicator lamps in the combination meter is glows when ignition switch is turn on
- NO →
  1. Battery
     (1) Connection
     (2) Terminal voltage
  2. Fusible links or fuses
  3. Ignition switch
- OK ↓

Check starter relay sound when ignition switch is turn START
- NO →
  1. Battery
     (1) Connection
     (2) Terminal voltage
  2. Fusible links or fuses
  3. Ignition switch
  4. Starter relay
  5. Clutch switch
  6. Neutral start switch (A/T only)
  7. Starting system wiring harness/connection
- OK ↓

Check starter motor
- NO GOOD → Starter motor replace
- OK ↓

Check engine damage

2 Engine will not start (Engine cranks normally)

Check fuel gauge
- Lower position → Fuel is present in the tank
- OK ↓

Check fuel pump operating
- NO GOOD →
  1. Fusible links or fuses
  2. Wiring/connections
  3. Fuel pump relay
  4. Fuel pump
  5. ECU.
- OK ↓

Check if no water stands in the bottom of fuel tank
- → Fuel drain
- OK ↓

Check diagnosis system output code
- → Perform trouble shooting according to diagnosis codes
- OK ↓

Continued

Check injector operating sound during cranking
- →
  1. Wiring/connections for injector
  2. Injector
  3. Cold start injector (Model G100 only)
  4. Fuel pressure
- OK ↓

Check ignition spark
- →
  1. Fusible links or fuses
  2. Wiring/connections
  3. Spark plugs/gap
  4. Spark plug wires
  5. Distributor/igniter
  6. Ignition coil
- OK ↓

Check ignition timing
- →
  1. Timing
     5° BTDC at cranking
  2. ECU (G100 model)
- OK ↓

Check compression
- →
  1. Pressure
     Approx. 10.5 kg/cm² at cranking or more
  2. Valve clearances
- OK ↓

Check throttle body
- →
  1. Air valve
  2. Throttle positioner
  3. Throttle adjusting screw
  4. Throttle sensor
- OK ↓

Check PCV system
- →
  1. PCV hoses
  2. Orifice
- OK ↓

Check EGR valve
- →
  1. EGR valve
  2. EGR modulator
- OK ↓

Check ECU.

FUEL INJECTION SYSTEM TROUBLESHOOTING GUIDE

3 Engine stalls while idling

- Check fuel gauge → Fuel is present in the tank
  - OK
- Check if no water stands in the bottom of fuel tank → Fuel drain
  - OK
- Check diagnosis system output code → Perform trouble shooting according to diagnosis codes
  - OK
- Check idle speed →
  1. VSV for ISC (Model G102 only)
  2. Wiring/connections
  3. Water temperature sensor circuit
  - OK
- Check throttle body →
  1. Air valve
  2. Throttle positioner
  3. Throttle adjusting screw
  4. Throttle position sensor
  - OK
- Check EGR valve →
  1. EGR valve
  2. EGR modulator
  - OK
- Check ECU.

4 Engine idle speed will not drop

- Check diagnosis system output code → Perform trouble shooting according to diagnosis codes
  - OK
- Check throttle body →
  1. Air valve
  2. Throttle positioner
  3. Throttle valve
  4. Throttle position sensor

Continued

- Check idle speed control →
  1. Idle speed control VSV (Model G102 only)
  2. Wiring/connections
  3. Water temperature sensor
  4. Headlight switch circuit
  5. Blower fan switch circuit
  6. Defogger switch circuit
  7. Radiator fan motor circuit
  8. Air conditioner switch circuit
  9. Power steering VSV
  - OK
- Check ECU.

5 Rough idling

- Check diagnosis system output code → Perform trouble shooting according to diagnosis codes
  - OK
- Check throttle body → Throttle position sensor
  - OK
- Check idle speed control →
  1. Idle speed control circuit
  2. ISC VSV (Model G102 only)
  - OK
- Check ECU.

6 Poor driveability

- Check diagnosis system output code → Perform trouble shooting according to diagnosis codes
  - OK
- Check air cleaner element → Air cleaner element
  - OK
- Check throttle body →
  1. Throttle position sensor
  2. Wiring/connections
  - OK

Continued

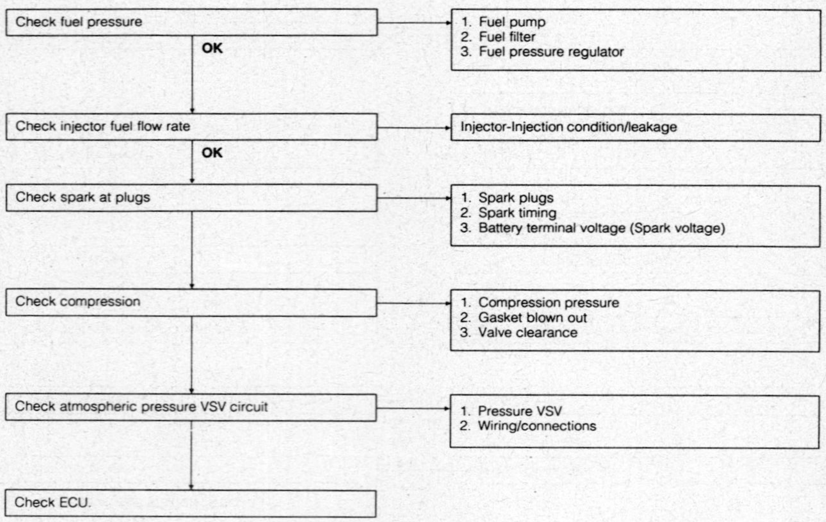

DIAGNOSIS CODE FOR G102

*When the diagnosis system detects malfunctions, the check engine lamp will go on without the ter shorted.

Code No.	Number of glowing of check engine lamp	Diagnosis item	Diagnosis contents	Trouble area
1		Normal	—	—
2*		Pressure sensor	When the input signal from the pressure sensor deviates from the specified value:	1. Pressure sensor 2. Pressure sensor circuit 3. ECU
3*		Ignition signal	When the ignition signal fails to be inputted:	1. Distributor 2. Ignitor 3. Ignition coil 4. Ignition system circuit 5. ECU
4*		Water temperature sensor	When the input signal from the water temperature sensor deviates from the specified value:	1. Water temperature sensor 2. Water temperature sensor circuit 3. ECU
5*		Oxygen sensor signal	When the input signal from the oxygen sensor fails to be inputted under the certain conditions:	1. Oxygen sensor 2. Oxygen sensor circuit 3. ECU
7*		Throttle position sensor	When both idle switch and power switch enter "ON" conditions:	1. Throttle position sensor 2. Throttle position sensor circuit 3. ECU
8*		Intake air temperature sensor	When the input signal from the intake air temperature sensor deviates from the specified value:	1. Intake air temperature sensor 2. Intake air temperature sensor circuit 3. ECU
9		Vehicle speed sensor	When the input signal from the vehicle speed sensor fails to be inputted under the certain conditions:	1. Vehicle speed sensor 2. Vehicle speed sensor circuit 3. ECU
10		Starter signal	When the input signal from the starter fails to be inputted, until the certain conditions are satisfied: However, this code may be memorized when vehicle is started by being pushed.	1. Starter 2. Starter circuit 3. ECU
11		Switch signal	When even one of the following conditions is satisfied with the test terminal shorted with the ground terminal: 1. when the air conditioner is functioning. 2. when idle switch is turned OFF. 3. when the shift lever is placed in ranges other than the neutral or parking range on A/T vehicle.	1. Air conditioner system 2. Throttle position sensor 3. Throttle position sensor circuit 4. Neutral start switch 5. ECU
12*		EGR control system	When it is judged that the EGR control system is not functioning normally under the certain conditions:	1. EGR valve 2. Modulator 3. EGR VSV 4. Water temperature sensor

DIAGNOSTIC CODES FOR ALL ENGINES

Code No.	Number of glowing of check engine lamp	Diagnosis item	Diagnosis contents	Trouble area	See page
15*		Air-to-fuel ratio (rich fail)	When the feedback function performs reduction compensation beyond the specified level:	1. Injector circuit 2. Injector 3. Fuel pressure 4. Pressure regulator 5. Pressure sensor 6. Water temperature sensor 7. Water temperature sensor circuit 8. Intake air temperature sensor 9. Intake air temperature sensor circuit	EF–90 EF–94 EF–196
16*		Air-to-fuel ratio (lean fail)	When the feedback function performs increase compensation below the specified level:	1. Injector circuit 2. Injector 3. ECU 4. Fuel pressure 5. Pressure regulator 6. Pressure sensor 7. Water temperature sensor 8. Water temperature sensor circuit 9. Intake air temperature sensor 10. Intake air temperature sensor circuit	EF–90 EF–94 EF–196

1.0L ENGINE – ECU TERMINAL CONNECTIONS
CHARADE G100 WITH 3 CYLINDER ENGINE

ECU connectors

Code	Contents of connection	Code	Contents of connection
1	Main relay (power supply)	22	Main relay (power supply)
2	Battery (back-up power supply)	23	Diagnosis lamp
3		24	
4	Fuel pump relay	25	Air conditioner amplifier
5		26	Vehicle speed sensor
6		27	
7		28	
8	Water temperature sensor	29	Sensor system ground
9	Intake manifold pressure sensor	30	Linear throttle sensor
10	Intake air temperature sensor	31	Sensor power supply output (5 V)
11	Check terminal	32	Throttle sensor full closing signal
12	Ignition monitor	33	Brake signal
13	Crank angle sensor	34	Crank angle sensor revolution speed
14	Crank angle sensor ground	35	Pressure sensor ground
15	Oxygen sensor	36	Oxygen sensor checker
16		37	Idle-up VSV
17	Electrical load signal	38	VSV for introducing atmospheric pressure
18	VSV for EGR	39	Operation system ground (engine block)
19	Starter switch	40	Igniter control output
20	Injector	41	
21	Power system ground (engine block)	42	

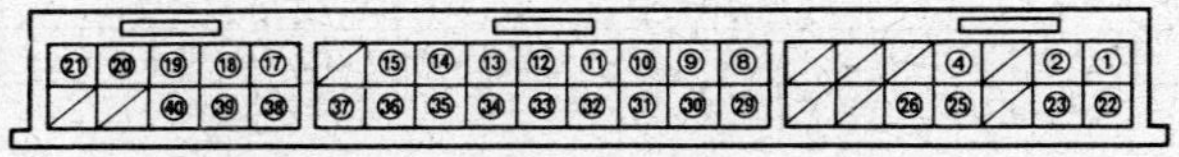

Voltage at ECU wiring connectors

Terminals	STD voltage	Condition	
① - ㊴	10 - 15.5 V	Ignition switch ON	
② - ㊴	10 - 15.5 V	At all time	
④ - ㊴	10 - 15.5 V	Ignition switch ON	When engine is stopped:
	0.1 - 1.5 V	Ignition switch ON	When engine is running:
⑧ - ㉙	0.1 - 0.7 V	Ignition switch ON	Coolant temperature 80 °C (176 °F)
⑨ - ㉙	2.1 - 2.7 V	Ignition switch ON	When atmospheric pressure is 760 mmHg (29.9 inchHg):
⑩ - ㉙	About 0.9 - 3.0 V	Ignition switch ON	Temperature of air in surge tank 20 °C (68 °F)
⑪ - ㊴	4.5 - 5.5 V	Ignition switch ON	T terminal OFF
⑫ - ㊴	0.5 - 1.5 V	Ignition switch ON	
⑬ - ㊴	About 0.6 V	Ignition switch ON	
⑭ - ㊴	About 0.6 V		
⑮ - ㊴	Change in output voltage	When engine speed is held at 3000 rpm for two minutes after engine has been fully warmed up:	
⑰ - ㊴	10 - 15.5 V	Ignition switch ON	When defogger and/or headlamp switch is turned ON:
⑱ - ㊴	10 - 15.5 V	Ignition switch ON	Coolant temperature is below 40 °C (104 °F)
	1.0 - 3.5 V		Coolant temperature is above 40 °C (104 °F)
⑲ - ㊴	0 V	At all time	
	6 - 15.5 V	When ignition switch is set to ST position:	
⑳ - ㊴	10 - 15.5 V	Ignition switch ON	
㉒ - ㊴	10 - 15.5 V	Ignition switch ON	
㉓ - ㊴	1.0 - 3.5 V	Ignition switch ON	
	10 - 15.5 V	When engine is running with ignition switch turned ON and diagnosis code is normal during diagnosis code check:	
㉕ - ㊴	10 - 15.5 V	When air conditioner switch is turned ON while engine is running:	
㉖ - ㊴	Change in voltage between 0 to 4.5–5.5 V	Ignition switch ON	When vehicle is moved:
㉚ - ㉙	About 0.55 V	Ignition switch ON	Throttle valve fully closed:
	About 3.91 V		Throttle valve fully opened:
㉛ - ㉙ or ㊱	4.5 - 5.5 V	Ignition switch ON	
㉜ - ㉙	0 V	Ignition switch ON	Throttle valve fully closed:
	4.5 - 5.5 V		Throttle valve fully opened:
㉝ - ㊴	10 - 15.5 V	Brake pedal is depressed:	
㉞ - ㊴	About 0.6 V	Ignition switch ON	
㊲ - ㊴	1.0 - 3.5 V	Ignition switch ON	
	10 - 15.5 V	After a lapse of at least 60 seconds after engine starting.	

Terminals	STD voltage	Condition	
㊳ - ㊴	10 - 15.5 V	Ignition switch ON	
㊵ - ㊴	0 V	Ignition switch ON	
	AC 0.3 - 2.0 V	When engine is running:	
㊱ - ㊴	1.5 - 3.5 V	Ignition switch ON	
	0 - 0.5 V	Ignition switch ON	When diagnosis code is memorized while throttle valve is closed fully, SST (09991-87702-000) is installed on check connector, and terminal ⑪ is shorted with ground terminal:
	4.5 - 5.5 V		When no diagnosis code is memorized while throttle valve is closed fully, SST (09991-87702-000) is installed on check connector, and terminal ⑪ is shorted with ground terminal:
	Change in voltage occurs eight times or more for 10 seconds between 0 - 5 V.		When engine speed is held at 3000 rpm for two minutes after engine has been fully warmed up, SST (09991-87702-000) is installed on check connector, and terminal ⑪ is shorted with ground terminal:

1.0L ENGINE—RESISTANCES AT ECU TERMINAL CONNECTIONS CHARADE G100 WITH 3 CYLINDER ENGINE

Resistance at ECU terminals

Terminals	Conditions	Resistance
⑧ – ㉙	Coolant temperature 80 °C (176 °F)	About 0.32 kΩ
⑩ – ㉙	Temperature of air in surge tank 20 °C (68 °F)	About 2.45 kΩ
㉛ – ㉚	Throttle valve fully closed:	3.5 - 10.3 kΩ
	Throttle valve fully opened	0.3 - 6.3 kΩ
㉜ – ㉙	Throttle valve fully closed:	350 Ω or less
	Throttle valve fully opened:	1000 kΩ or more
㉞ – ⑭	—	140 - 180 Ω
⑬ – ⑭	—	140 - 180 Ω

1.0L ENGINE—FUEL INJECTION SYSTEM TROUBLESHOOTING PROCEDURES CHARADE G100 WITH 3 CYLINDER ENGINE

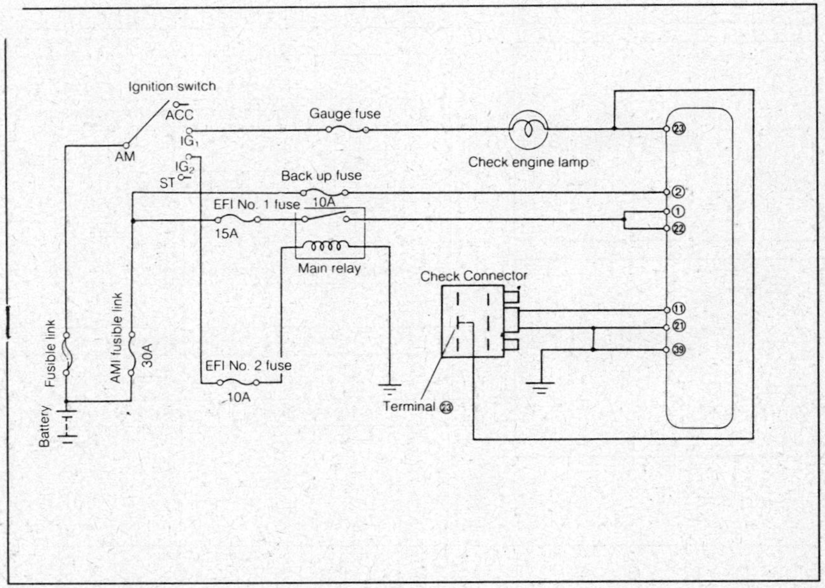

1st step

Check that check engine lamp goes on when ignition switch is turned ON.
- **YES** → System is normal.
- **NO** → Check that check engine lamp goes on when terminal ㉓ of check terminal is shorted with ground terminal using the following SST.

  SST: 09991-87702-000.
  - **YES** → Check that there is continuity between ECU terminal ㊴ and body ground and between ECU terminal ㉑ and body ground.
    - **OK** → Check ECU.
    - **BAD** → Repair or replace
  - **OK** → Check bulb, fuse and wiring between ECU and ignition switch
    - **BAD** → Repair or replace.

1.0L ENGINE—FUEL INJECTION SYSTEM TROUBLESHOOTING PROCEDURES CHARADE G100 WITH 3 CYLINDER ENGINE

2nd step

- Check that check engine lamp goes off when engine starts.
  - YES → System is normal.
  - NO ↓
- Check wiring between ECU and check engine lamp.
  - BAD → Repair or replace
  - OK ↓
- Check that diagnosis code number is indicated when terminal ⑪ of check connector is shorted with ground terminal.
  - NO → Check ECU.
  - YES ↓
- Perform trouble shooting according to diagnosis codes.
  - YES ↓
- After trouble shooting according to diagnosis codes has been performed, start engine. Check that check engine lamp goes off.
  - NO → Check diagnosis code again.
  - YES ↓
- System is normal. → Erase diagnosis code.

Terminals	Trouble	Conditions	STD voltage
② - ㊴	No voltage	At all time	10 - 15.5
① ㉒ - ㊴	No voltage	Ignition switch ON	10 - 15.5

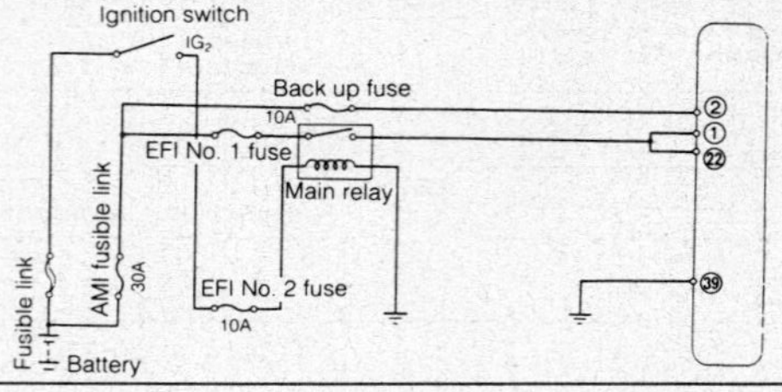

• ② – ㊴

- There is no voltage between SST terminals ② and ㊴.
  ↓
- Check that there is voltage between SST terminal ② and body ground.
  - OK ↓
    - Check wiring between SST terminal ㊴ and body ground.
      - BAD ↓
    - Repair or replace.
  - NO ↓
    - Check fuses, fusible links and wiring harness.
      - BAD ↓
    - Repair or replace.

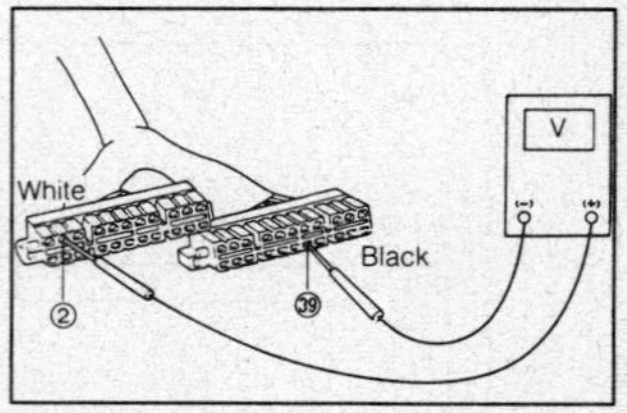

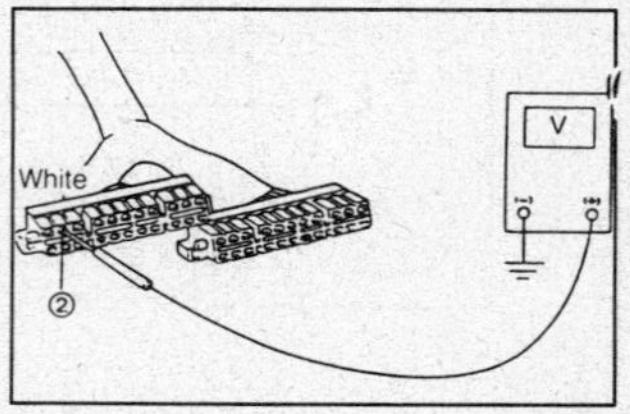

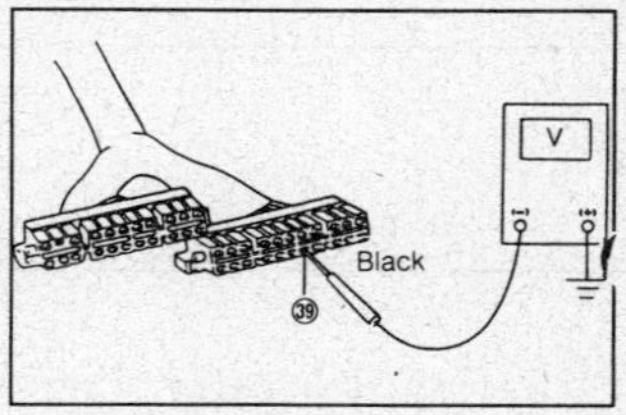

1.0L ENGINE—FUEL INJECTION SYSTEM TROUBLESHOOTING PROCEDURES CHARADE G100 WITH 3 CYLINDER ENGINE

• **① or ㉒ – ㊴**

There is no voltage between SST terminal ① or ㉒ and ㊴

OK ↓

Check that there is voltage between SST terminal ① or ㉒ and body ground when ignition switch is turned ON.

- **ON** → Check fuses, fusible links and wiring harness.
- → Check wiring between ECU terminal ㊴ and body ground.
  - **BAD** → Repair or replace.

Check fuses, fusible links and wiring harness.

- **OK** → Check EFI main relay
- **BAD** → Repair or replace.

Check EFI main relay

- **BAD** → Replace.

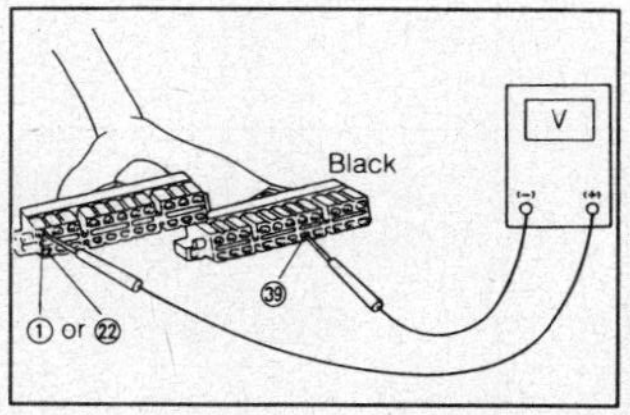

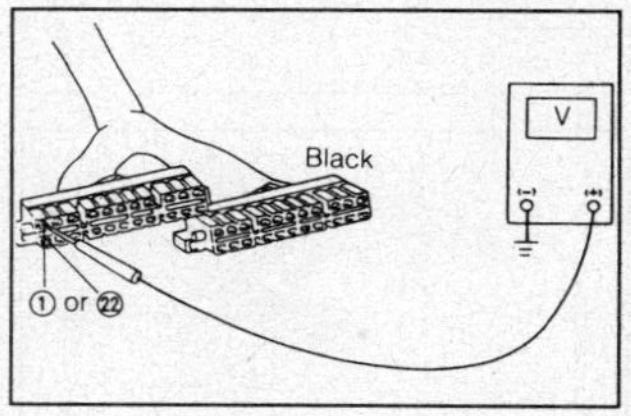

Terminals	Trouble	Conditions		STD voltage
⑧ – ㉙	No voltage	Ignition switch ON	Coolant temperature 80 °C (176 °F)	0.1 - 0.7

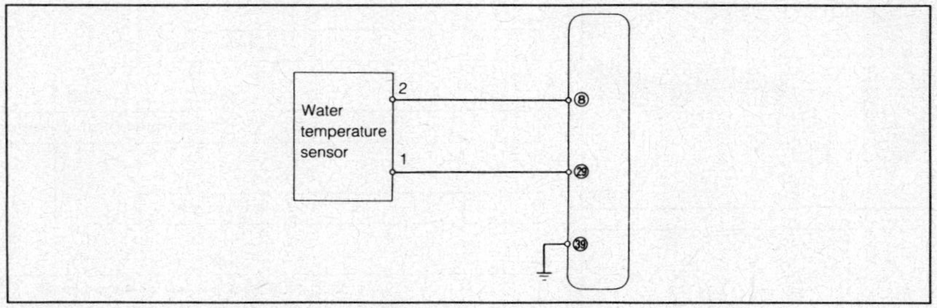

[1] There is no specified voltage between SST terminals ⑧ and ㉙ when ignition switch is turned ON.

↓

[2] Check that there is voltage between SST terminals ① or ㉒ and body ground when ignition switch is turned ON.

- **OK** → Check water temperature sensor.
- **BAD** → Check main relay circuit.

Check water temperature sensor.

- **OK** → Check wiring between ECU and water temperature sensor.
- **BAD** → Replace water temperature sensor.

Check wiring between ECU and water temperature sensor.

- **OK** → Check ECU.
- **BAD** → Repair or replace.

Check ECU.

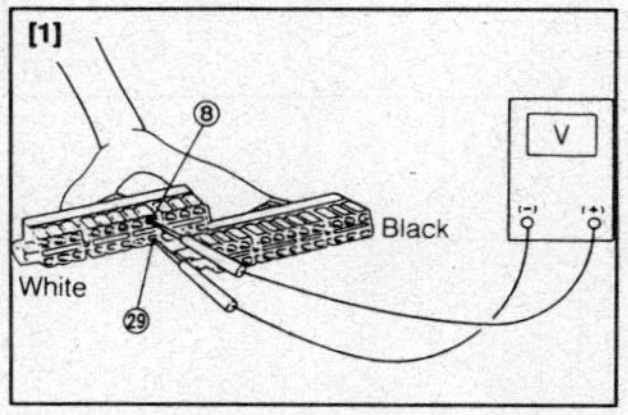

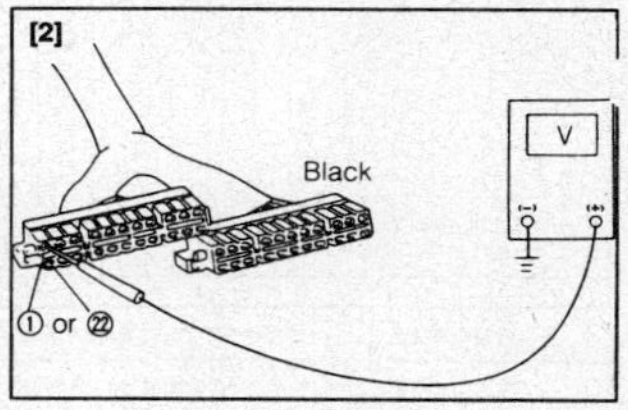

1.0L ENGINE—FUEL INJECTION SYSTEM TROUBLESHOOTING PROCEDURES
CHARADE G100 WITH 3 CYLINDER ENGINE

Terminal	Trouble	Conditions		STD voltage
⑩ – ㉙	No voltage	Ignition switch ON	Intake air temperature 20 °C (68 °F)	About 0.9 - 3.0

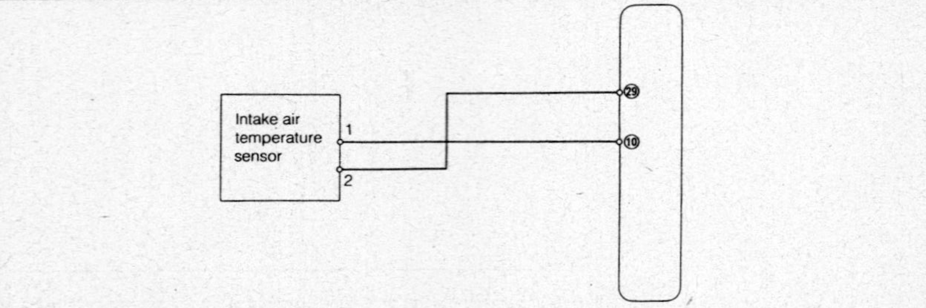

[1] There is no specified voltage between SST terminals ⑩ and ㉙ when ignition switch is turned ON.

[2] Check that there is voltage between SST terminals ① or ㉒ and ㊴.
- **OK** → Check intake air temperature sensor.
- **NO** → Check main relay circuit.

Check intake air temperature sensor.
- **OK** → Check wiring between ECU and intake air temperature sensor.
- **BAD** → Replace intake air temperature sensor.

Check wiring between ECU and intake air temperature sensor.
- **OK** → Check ECU.
- **BAD** → Repair or replace.

[1]

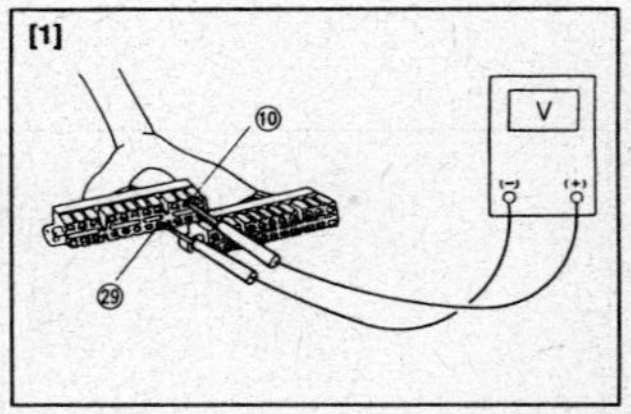

[2]

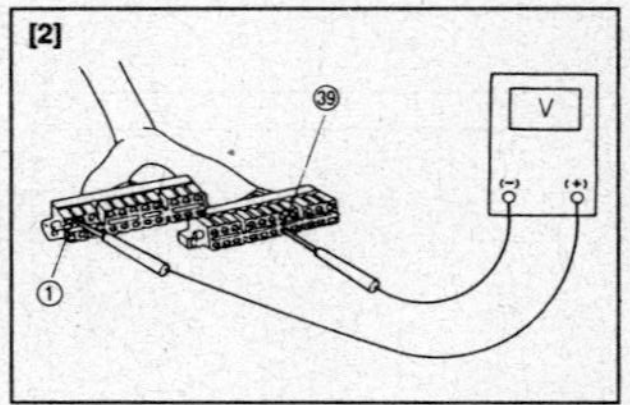

Terminal	Trouble	Conditions		STD voltage
㉛ – ㉙	No voltage	Ignition switch ON		4.5 - 5.5
㉚ – ㉙		Ignition switch ON	Throttle valve closed	About 0.55
			Throttle valve fully opened	About 3.91

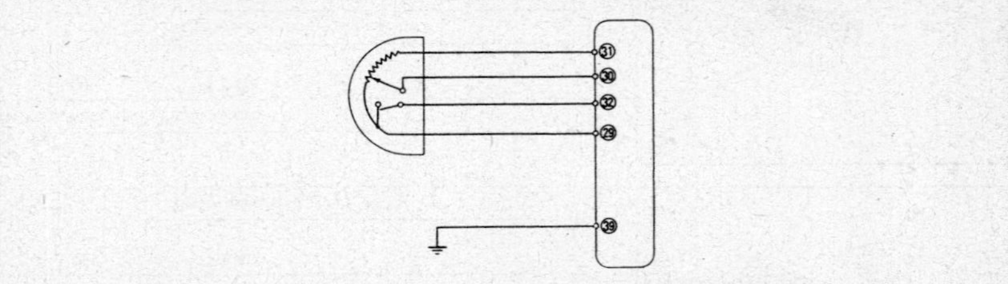

- ㉛ – ㉙

[1] There is no voltage between SST terminals ㉛ and ㉙ when ignition switch is turned ON.

[2] Check that there is voltage between SST terminal ① or ㉒ and body ground when ignition switch is turned ON.
- **OK** → Check throttle position sensor.
- **ON** → Check main relay circuit.

Check throttle position sensor.
- **OK** → Check wiring between ECU and throttle position sensor.
- **BAD** → Repair or replace.

Check wiring between ECU and throttle position sensor.
- **OK** → Check ECU.
- **BAD** → Repair or replace wiring.

Check ECU.

[1]

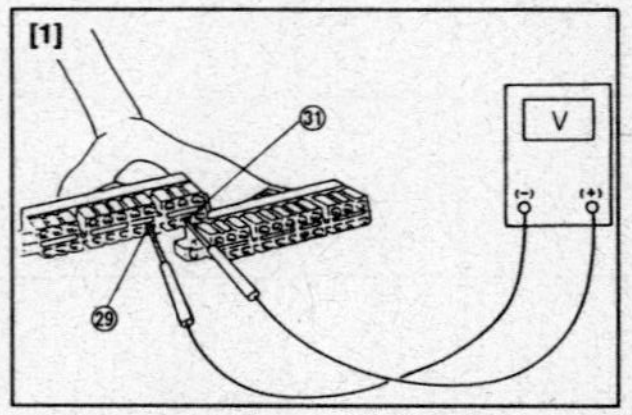

[2]

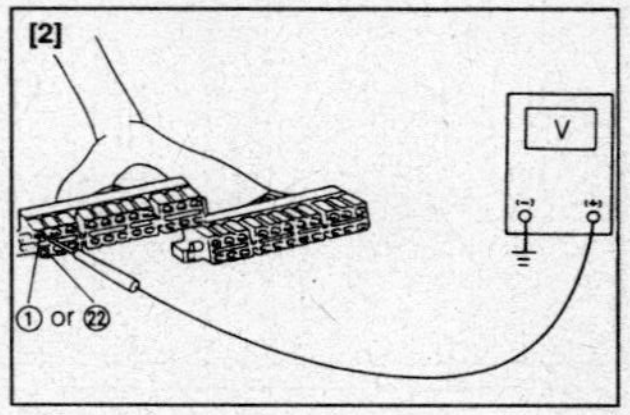

1.0L ENGINE—FUEL INJECTION SYSTEM TROUBLESHOOTING PROCEDURES
CHARADE G100 WITH 3 CYLINDER ENGINE

• ㉚ – ㉙

[1] There is no specified voltage between SST terminals ㉚ and ㉙ when ignition switch is turned ON.

[2] Check that there is voltage between SST terminals ① or ㉒ and body ground when ignition switch is turned ON.
- OK → Repair or replace.
- NO → Check main relay circuit.

Check main relay circuit.
- OK → Check throttle position sensor.

Check throttle position sensor.
- BAD → Repair or replace.
- OK → Check wiring between ECU and throttle position sensor.

Check wiring between ECU and throttle position sensor.
- BAD → Repair or replace.
- OK → Check ECU.

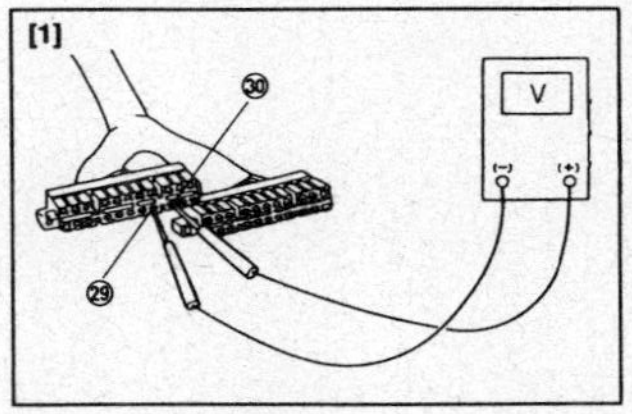

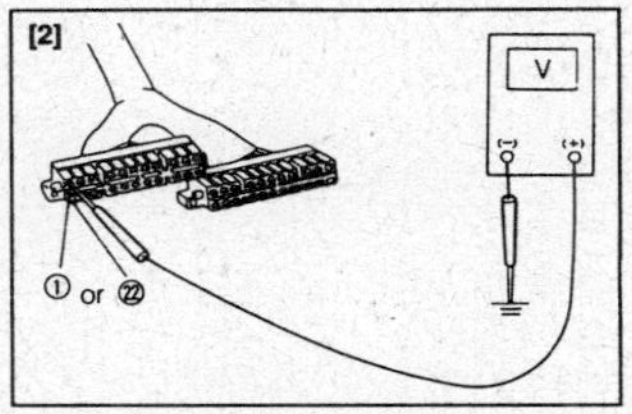

Terminals	Trouble	Conditions	STD voltage
㉛ - ㉟	No voltage	Ignition switch ON	4.5 - 5.5
⑨ - ㉟		Ignition switch ON At time of atmospheric pressure of 760 mmHg (29.9 inchHg)	2.1 - 2.7

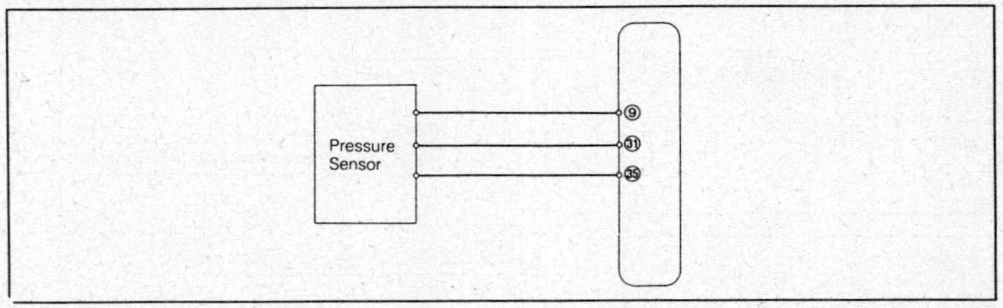

• ㉛ – ㉟

[1] There is no voltage between SST terminals ㉛ and ㉟.

[2] Check that there is voltage between SST terminal ① or ㉒ and body ground when ignition switch is turned ON.
- OK → Check wiring between SST terminal ㉟ and body ground.
- NO → Check main relay circuit.

Check wiring between SST terminal ㉟ and body ground.
- BAD → Repair or replace.

Check main relay circuit.
- BAD → Repair or replace.
- OK → Check wiring between ECU and pressure sensor and/or throttle position sensor.

Check wiring between ECU and pressure sensor and/or throttle position sensor.
- OK → Check pressure sensor.
- BAD → Repair or replace.

Check pressure sensor.
- BAD → Replace.
- OK → Check ECU.

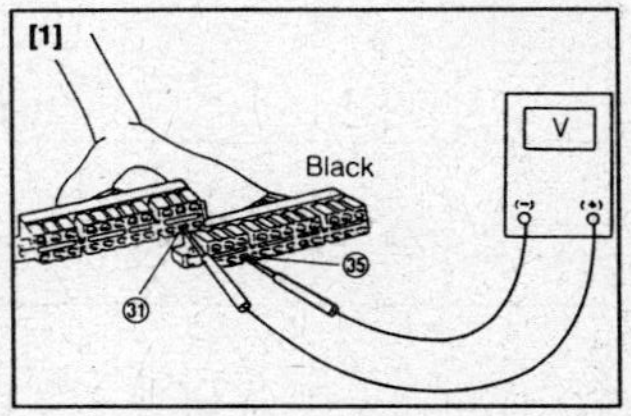

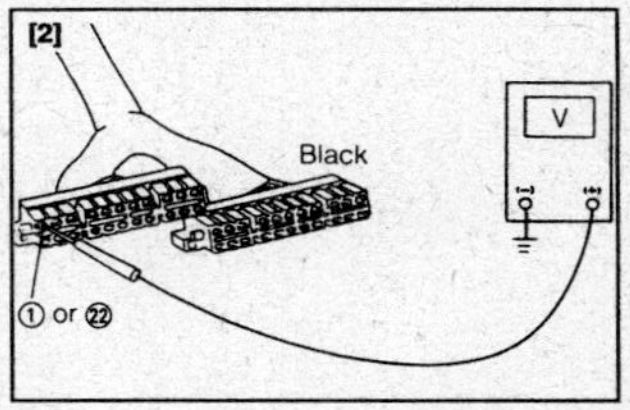

• ⑨ – ㉟

[1] There is no voltage between SST terminals ⑨ – ㉟.

↓

[2] Check that there is voltage between SST terminals ㉛ – ㉟.

- BAD → ㉛ – ㉟ trouble section
  - BAD → Repair.
- OK ↓

Check wiring between ECU and pressure sensor.

- BAD → Repair or replace.
- OK ↓

Check pressure sensor.

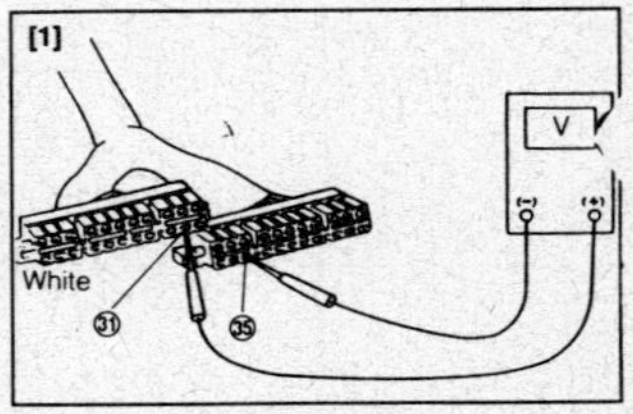

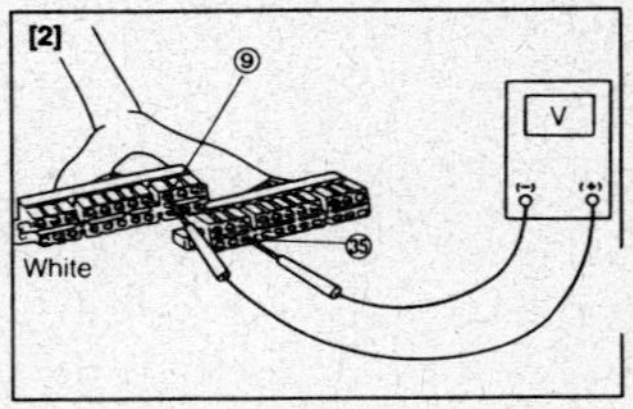

Terminal	Trouble	Conditions		STD voltage
⑱ – ㊴	No specified voltage	When engine is idling	Coolant temperature above 40 °C (104 °F)	1.0 - 3.5

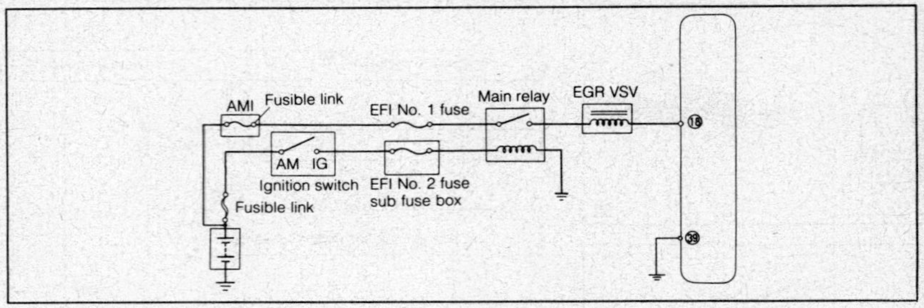

[1] There is no specified voltage between SST terminals ⑱ and ㊴ when engine is idling and coolant temperature is above 40 °C (104 °F)

↓

[2] Check that there is voltage between SST terminal ⑱ and body ground.

- OK → Check wiring between ECU terminal ㊴ and body ground.
- NG ↓

Check wiring between ECU terminal ⑱ and EGR VSV.

- BAD → Repair or replace.
- OK ↓

Check EGR VSV.

- BAD → Replace EGR VSV.
- OK ↓

Check main relay, fuse, fusible link and ignition switch.

- BAD → Repair or replace.
- OK ↓

Check wiring between battery and EGR VSV.

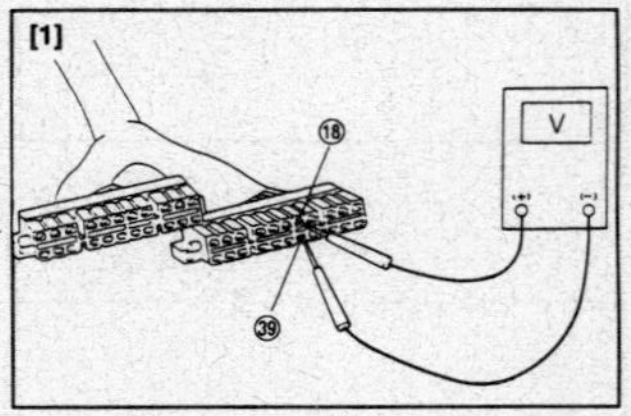

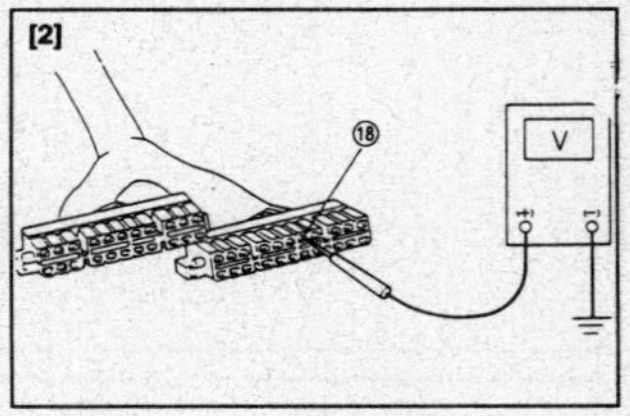

1.0L ENGINE – FUEL INJECTION SYSTEM TROUBLESHOOTING PROCEDURES CHARADE G100 WITH 3 CYLINDER ENGINE

Terminals	Trouble	Conditions		STD voltage
⑮ - ㊴	No voltage changes	Ignition switch ON	When engine speed is held at 3000 rpm for two minutes after engine has been fully warmed up:	Voltage changes more than 8 times with in 10 seconds

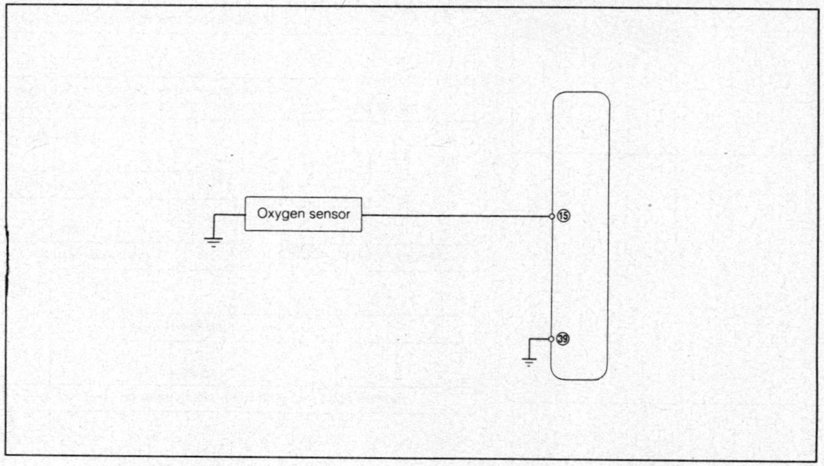

There is no change in voltage between SST terminals ⑮ and ㊴ when engine speed is held at 3000 rpm for two minutes after engine has been fully warmed up:

- Check wiring between ECU and oxygen sensor.
  - BAD → Repair or replace.
    - BAD → Check oxygen sensor.
  - OK → Check oxygen sensor.
- Check oxygen sensor.
  - OK → Check ECU.

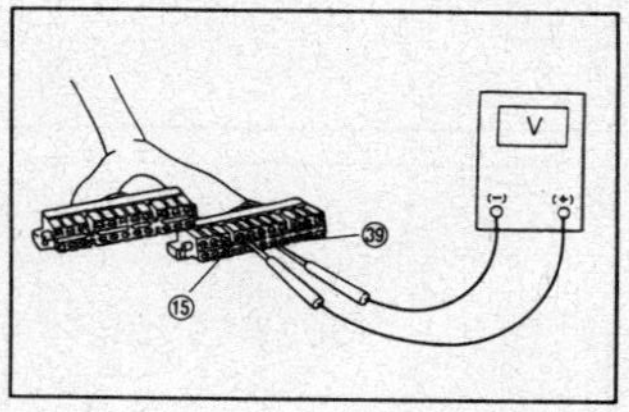

IGNITION MONITOR

Terminals	Trouble	Conditions	STD voltage
⑫ - ㊴	4.5 - 5.5 or 0	Ignition switch ON	0.5 - 1.5 (While engine is stopped)

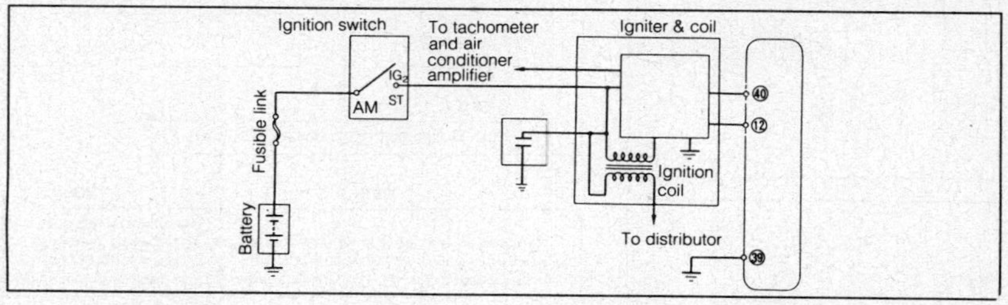

- ⑫ – ㊴ 4.5 - 5.5 V

Check that there is a voltage of 4.5 to 5.5 volts between SST terminals ⑫ and ㊴ when ignition switch is turned ON.

- Check wiring between SST terminal ㊵ and igniter.
  - BAD → Repair or replace.
  - OK → Check that voltage is applied to ignition coil.
- Check that voltage is applied to ignition coil.
  - BAD → Repair or replace.
  - OK → Check that there is continuity between igniter body and body ground.
- Check that there is continuity between igniter body and body ground.
  - BAD → Repair.
  - OK → Replace igniter.

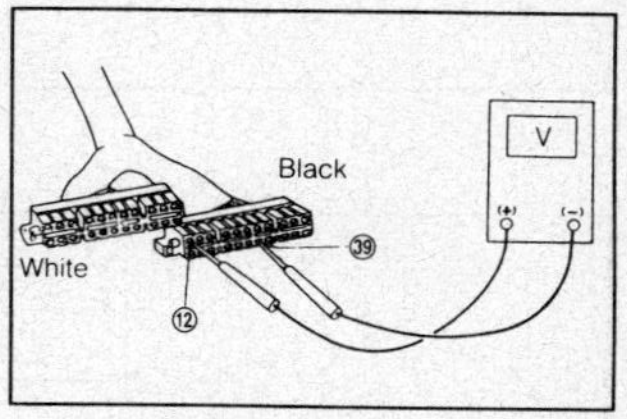

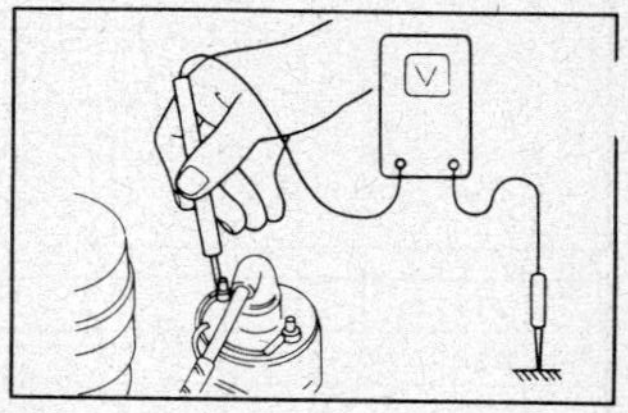

1.0L ENGINE – FUEL INJECTION SYSTEM TROUBLESHOOTING PROCEDURES CHARADE G100 WITH 3 CYLINDER ENGINE

- **(12) – (39) 0 V**

There is no voltage between SST terminals (12) and (39) when ignition key switch is turned ON.

→ Check wiring between ECU terminal (40) and igniter.
- BAD → Repair or replace.
- OK → Check insulation between igniter terminal (40) and body ground.
  - BAD → Replace igniter.
  - OK → Check insulation between igniter terminal (12) and body ground.
    - BAD → Replace ignitor.
    - → Check main relay circuit.
      - BAD → Repair or replace.
      - OK → Check ECU.

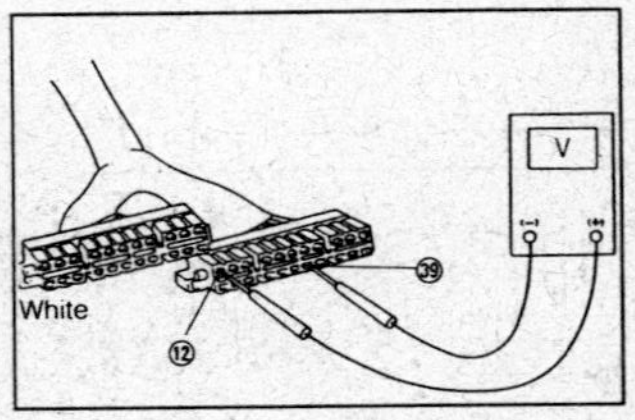

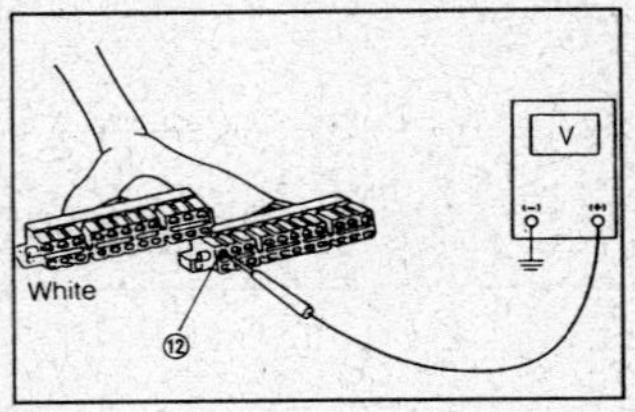

- **(40) – (39)**

Check that there is no AC voltage between SST terminals (40) and (39) while engine is being cranked or engine is running.

→ Check wiring between terminal (12) and igniter.
- NG → Repair or replace.
- OK → Check insulation between igniter terminal (12) and body ground.
  - NG → Replace igniter.
  - OK → Check main relay circuit.
    - BAD → Repair or replace.
    - OK → Check ECU.

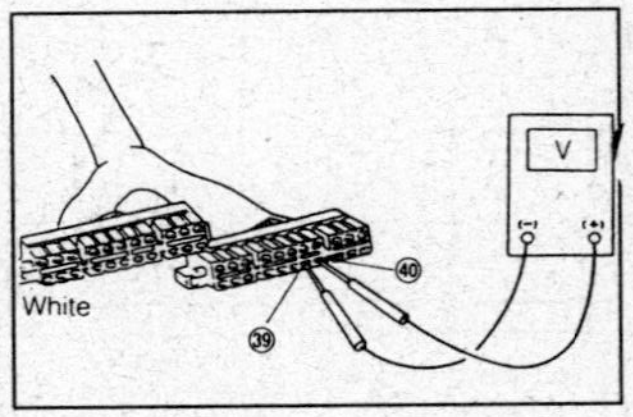

DISTRIBUTOR

Terminal	Trouble	Conditions	STD voltage
(34) – (39)	No voltage	Ignition switch ON	0.5 - 1.5
(13) – (39)	No voltage	Ignition switch ON	0.5 - 1.5

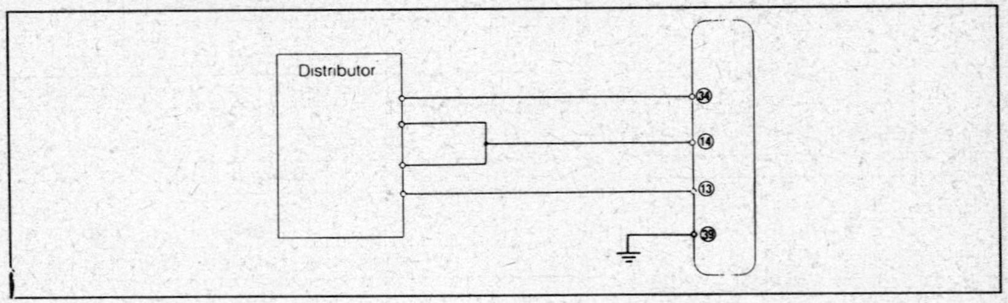

- **(34) – (14)**

[1] There is no AC voltage between SST terminals (34) and (14) when ignition key switch is turned ON.

→ [2] Check that there is a voltage of about 0.5 - 1.5 Volt between the following terminals of SST terminal when ignition switch is turned ON.: Namely, between (34) and (39), between (13) and (39), and between (14) and (39), respectively.
- OK → Check distributor.
  - BAD → Replace distributor.
- BAD → Check wiring between ECU and distributor.
  - BAD → Repair or replace.
  - OK → Check distributor.
    - OK → Check main relay circuit.
      - BAD → Repair or replace.
      - OK → Check ECU.

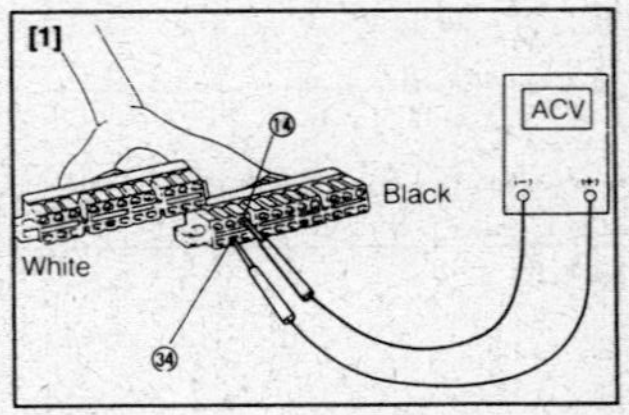

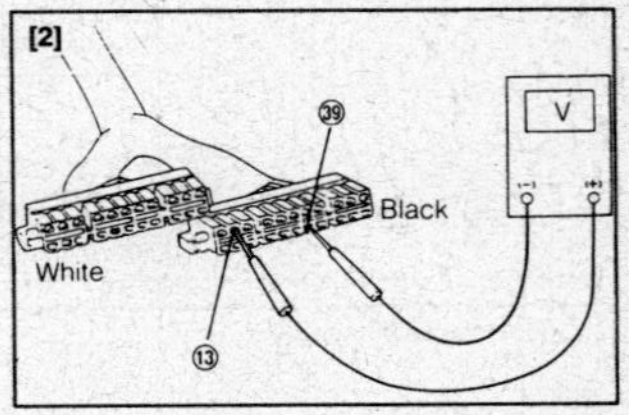

1.0L ENGINE—FUEL INJECTION SYSTEM TROUBLESHOOTING PROCEDURES
CHARADE G100 WITH 3 CYLINDER ENGINE

SPEED SENSOR

Terminal	Trouble	Conditions		STD voltage
㉖ – ㊴	No voltage changes	Ignition switch ON	When vehicle is moved slowly:	0 to 4.5 - 5.5

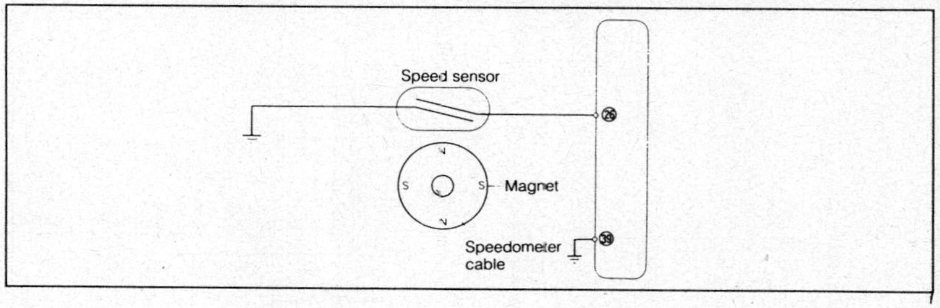

No change in voltage occurs between SST terminals ㉖ and ㊴ when vehicle is moved slowly with ignition switch turned ON.

- Check that there is voltage between SST terminal ① or ㉒ and ㊴
  - NO → Check main relay circuit.
  - OK ↓
- Check speed sensor.
  - BAD → Replace speed sensor.
  - OK ↓
- Check wiring between ECU and speed sensor.
  - BAD → Repair or replace.
  - OK ↓
- Check wiring between speed sensor and body ground.
  - BAD → Repair or replace.
  - OK → Check ECU.

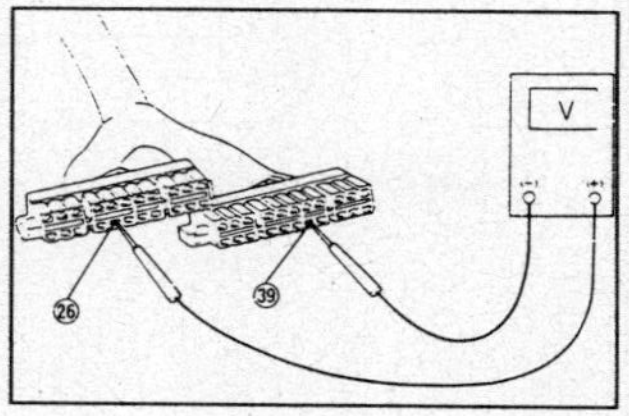

STARTER SWITCH

Terminal	Trouble	Conditions	STD voltage
⑲ – ㊴	No voltage	Ignition switch ST position	6 - 15.5

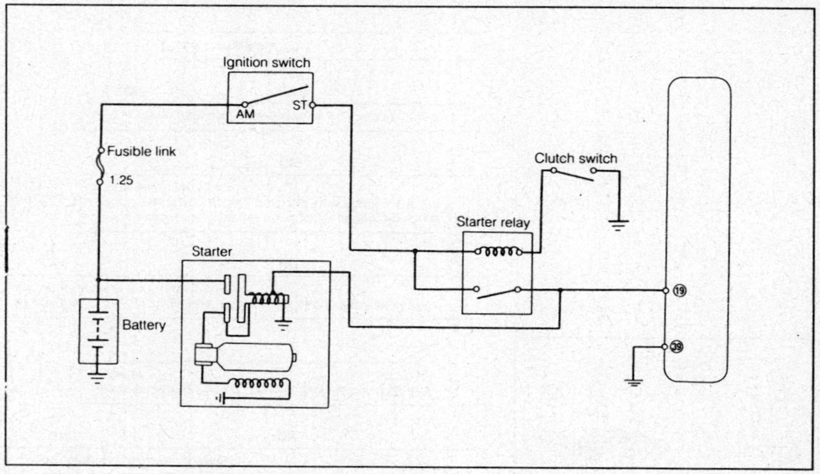

There is no voltage between SST terminals ⑲ and ㊴ when ignition switch is set to ST position.

- Check starter operation
  - OK → Check wiring between ECU and starter terminal.
    - BAD → Repair or replace.
    - OK → Check and repair wiring between ECU terminal ㊴ and body ground.
  - BAD → Check battery, fusible link, ignition switch, starter relay, clutch switch and starter.
    - BAD → Repair or replace
    - OK → Check and repair wiring between battery and ECU terminal ⑲.

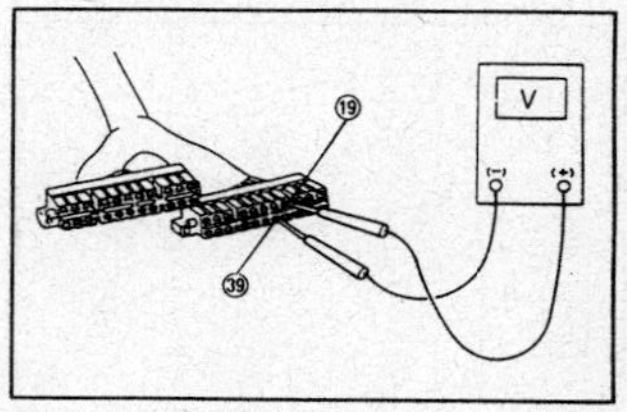

1.0L ENGINE—FUEL INJECTION SYSTEM TROUBLESHOOTING PROCEDURES CHARADE G100 WITH 3 CYLINDER ENGINE

AIR CONDITIONER SWITCH

Terminals	Trouble	Conditions		STD voltage
㉕ – ㊴	No voltage	When engine is idling	Air conditioner switch ON	10 - 15.5
㉜ – ㉙	No specified voltage	Ignition switch ON	Throttle valve fully closed	0
			Throttle valve opened	4.5 - 5.5

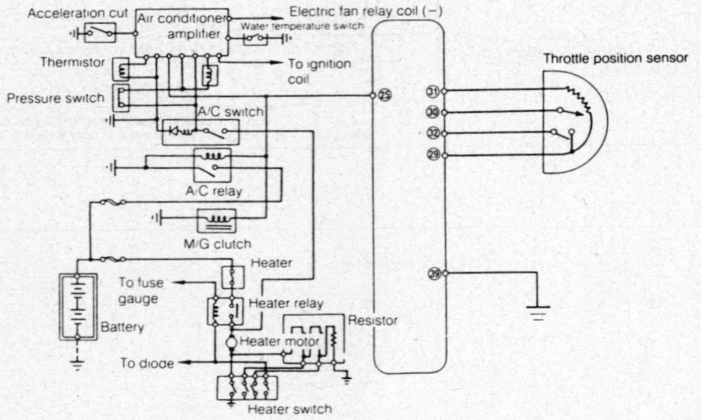

- ㉕ – ㊴

[1] There is no voltage at SST terminals ㉕ and ㊴ when air conditioner is operated.

[2] Check that there is voltage between SST terminal ㉕ and body ground.

- **OK** → Check wiring between ECU terminal ㊴ and body ground.
- **NO** → Check compressor clutch operation.

Check compressor clutch operation.

- **OK** → Check and repair wiring between ECU terminal ㉕ and amplifier.
- **BAD** → Check that there is voltage between magnet clutch terminal of the amplifier terminal and body ground.

Check that there is voltage between magnet clutch terminal of the amplifier terminal and body ground.

- **NO** → Check fusible link, fuse, air conditioner switch, pressure switch, thermistor, acceleration cut switch and water temperature switch.
  - **OK** → Check wiring between battery and air conditioner amplifier.
    - **BAD** → Repair or replace.
    - **OK** → Replace air conditioner amplifier.
- **OK** → Check and repair wiring between magnet clutch terminal of the amplifier and ECU terminal ㊴.

[1]

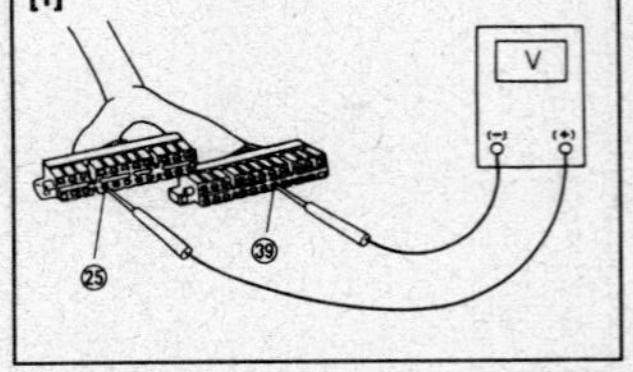

[2]

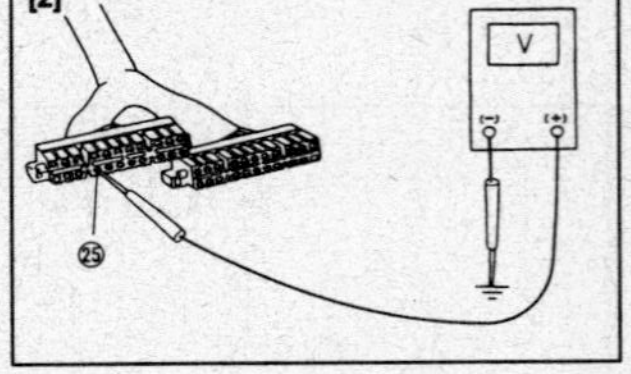

1.0L ENGINE – FUEL INJECTION SYSTEM TROUBLESHOOTING PROCEDURES CHARADE G100 WITH 3 CYLINDER ENGINE

• ㉜ – ㉙

[1] There is no voltage between SST terminals ㉜ and ㉙ when ignition switch is turned ON. (Throttle valve opened)

[2] Check that there is voltage between SST terminal ① or ㉒ and ㊴.

- **OK** → Check throttle position sensor.
- **NO** → Check main relay circuit.

Check throttle position sensor.

- **OK** → Check wiring between ECU and throttle position sensor.
- **BAD** → Replace throttle position sensor.

Check wiring between ECU and throttle position sensor.

- **OK** → Check ECU.

Check ECU.

[1]

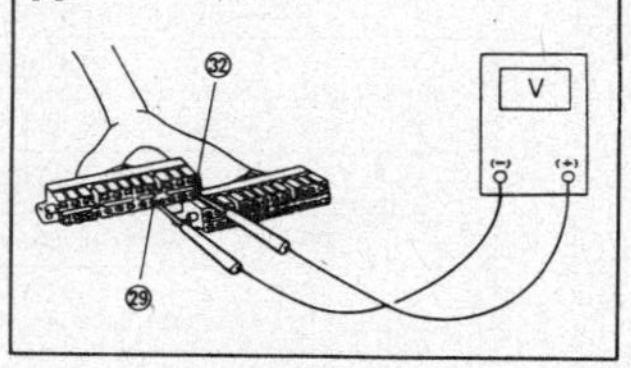

[2]

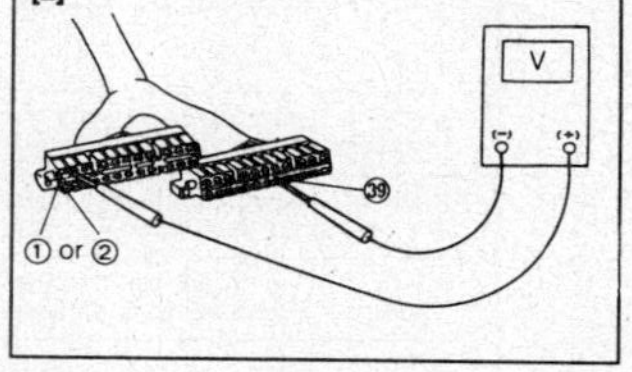

1.3L ENGINE – ECU TERMINAL CONNECTIONS CHARADE G102 WITH 4 CYLINDER ENGINE

Table showing ECU connections

Terminal code	Contents of connection	Terminal code	Contents of connection
1	Main relay (Power supply)	22	Main relay (Power supply)
2	Battery (Backup power supply)	23	
3	Ignition coil primary voltage	24	
4		25	
5	Pressure sensor power supply	26	Oxygen sensor
6	Pressure sensor	27	Intake air temperature sensor
7	Cooling water temperature sensor	28	A/T, M/T detecting terminal
8	Vehicle speed sensor	29	Operation system ground (Engine)
9	Electrical load (Headlamp and defogger)	30	Electrical load (Blower fan and radiator fan switch)
10	Neutral start switch (D, 2 and L ranges)	31	Neutral start switch (R range)
11	Check connector (Test terminal)	32	Throttle position switch (Power switch)
12	Throttle position switch (Idle switch)	33	Stop lamp
13	Starter	34	Air conditioner magnet switch
14		35	
15	Oxygen sensor feedback check terminal	36	Operation system ground
16	Check engine lamp	37	Circuit opening relay (Fuel pump relay)
17	Spark advance VSV	38	Pressure VSV
18	EGR VSV	39	System ground
19		40	Idle speed control VSV
20	Injector	41	Injector
21	Actuator drive ground (Engine)	42	Actuator drive ground (Engine)

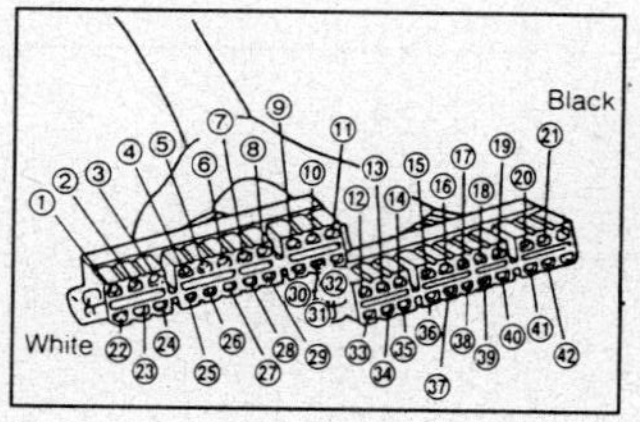

Voltages at ECU connectors

Terminals	STD voltage	Condition	
① – ㊴	Approx. battery voltage	Ignition switch ON	
② – ㊴	Approx. battery voltage	At all time	
③ – ㊴	Approx. battery voltage	Ignition switch ON	When engine is stopped:
⑤ – ㊱	4.5 - 5.5 V	Ignition switch ON	
⑥ – ㊴	3.2 - 4.0 V	Ignition switch ON	When atmospheric pressure of 760 mmHg (29.9 inchHg) exists.
⑦ – ㊴	0.4 - 0.65 V	Ignition switch ON	When cooling water temperature is 80 °C (176 °F):
⑧ – ㊴	0 - Approx. battery voltage	Ignition switch ON	Measured voltage changes when vehicle is moved 1.5 m.
⑨ – ㊴	Less than 5.0 V	Ignition switch ON	When defogger and headlamp switches are turned OFF:
	Approx. battery voltage	Ignition switch ON	When defogger and/or headlamp switches are turned ON:
In case of M/T vehicle ⑩ – ㊴	Approx. battery voltage	Ignition switch ON	
In case of A/T vehicle ⑩ – ㊴	Approx. battery voltage	Ignition switch ON	When shift lever is in "P", "N" or "R" range:
	Less than 1.0 V	Ignition switch ON	When shift lever is in "D", "2" or "L" range:
⑪ – ㊴	Approx. battery voltage	Ignition switch ON	When test terminal of check connector is not connected with ground terminal:
	Less than 1.0 V	Ignition switch ON	When test terminal of check connector is connected with ground terminal:
⑫ – ㊴	Less than 5.0 V	Ignition switch ON	Throttle valve fully closed
	Approx. battery voltage	Ignition switch ON	Throttle valve fully opened
⑬ – ㊴	0 V	Ignition switch ON	
	More than 6 V	When ignition switch is set to ST position:	
⑮ – ㊴	Measured voltage changes at a point between 0 - 5.0 V.	After warming up engine completely, connect test terminal of check connector with ground terminal. Hold engine revolution speed at 3000 rpm for two minutes.	
⑯ – ㊴	Less than 3.0 V	Ignition switch ON	• Engine is stopped. • When check engine lamp is illuminated:
	Approx. battery voltage	Ignition switch ON	• After engine starts: • When check engine lamp is extinguished:
M/T vehicle only ⑰ – ㊴	Approx. battery voltage	Ignition switch ON	
⑱ – ㊴	Approx. battery voltage	Ignition switch ON	• After engine starts: • Cooling water temperature is below 40 °C (104 °F).
	Less than 3.0 V	Ignition switch ON	• After engine starts: • Cooling water temperature is above 41 °C (106 °F).

Terminals	STD voltage	Condition	
⑳ – ㊴	Less than 1.0 V	At least 30 seconds have elapsed after turning OFF ignition switch.	
	Approx. battery voltage	Ignition switch ON	• Engine is stopped.
㉑ – ㊴	Less than 0.01 V	Ignition switch ON	
㉒ – ㊴	Approx. battery voltage	Ignition switch ON	
㉖ – ㊴	Change in output voltage	Ignition switch ON	After warming up engine completely, hold engine revolution speed at 3000 rpm for two minutes.
㉗ – ㊴	1.5 - 3.0 V	Ignition switch ON	Air temperature inside intake manifold is 20 °C (68 °F):
In case of M/T vehicle ㉘ – ㊴	4.5 - 5.5 V	Ignition switch ON	
In case of A/T vehicle ㉘ – ㊴	Less than 0.5 V	Ignition switch ON	
㉙ – ㊴	Less than 0.1 V	Ignition switch ON	
㉚ – ㊴	Approx. battery voltage	Ignition switch ON	• Blower fan switch turned OFF • -Radiator fan motor switch turned OFF
	Less than 2.0 V	Ignition switch ON	When blower fan switch and/or radiator fan motor switch are turned ON:
In case of M/T vehicle ㉛ – ㊴	0 V	At all time	
In case of A/T vehicle ㉛ – ㊴	Less than 1 V	Ignition switch ON	When shift lever is in a range other than "R" range:
	Approx. battery voltage	Ignition switch ON	When shift lever is in "R" range.
In case of M/T vehicle ㉜ – ㊴	Approx. battery voltage	Ignition switch ON	Throttle valve fully closed
	Less than 5.0 V	Ignition switch ON	Throttle valve fully opened
In case of A/T vehicle ㉜ – ㊴	Less than 5.0 V	Ignition switch ON	Throttle valve fully closed
	Approx. battery voltage	Ignition switch ON	Throttle valve fully opened
㉝ – ㊴	Less than 1 V	Ignition switch ON	When brake pedal is not depressed:
	Approx. battery voltage	At all time	When brake pedal is depressed:
㉞ – ㊴	Less than 1 V	Ignition switch ON	When compressor magnet switch of air conditioner is turned OFF:
	Approx. battery voltage	Ignition switch ON	When compressor magnet switch of air conditioner is turned ON:
㊱ – ㊴	Less than 0.1 V	Ignition switch ON	
㊲ – ㊴	Approx. battery voltage	Ignition switch ON	When fuel pump is stopped:
	Less than 2.0 V	Ignition switch ON	When fuel pump is operating:
㊳ – ㊴	Approx. battery voltage	Ignition switch ON	When pressure VSV is turned OFF:
	Less than 3.0 V	Ignition switch ON	For 0.5 second immediately after engine starts
㊴ – Engine ground	Less than 0.1 V	Ignition switch ON	

1.3L ENGINE—VOLTAGES AT ECU TERMINAL CONNECTIONS—CHARADE G102 WITH 4 CYLINDER ENGINE

Terminals	STD voltage	Condition	
㊵ – ㊴	Less than 3.0 V	Ignition switch ON	Engine is stopped.
	Approx. battery voltage	Ignition switch ON	When test terminal of check connector is connected with ground terminal:
㊶ – ㊴	Less than 1.0 V	At least 30 seconds have elapsed after turning OFF ignition switch.	
	Approx. battery voltage	Ignition switch ON	Engine is stopped.
㊷ – ㊴	Less than 0.1 V	Ignition switch ON	

1.3L ENGINE—RESISTANCES AT ECU TERMINAL CONNECTIONS—CHARADE G102 WITH 4 CYLINDER ENGINE

Terminals	STD resistance kΩ	Condition
⑦ – ㉙	0.322 ± 0.1	When cooling water temperature is F 80 °C (176 °F):
In case of M/T vehicle ⑫ – ㉙	Less than 29	Throttle valve fully closed
	More than 1000	Throttle valve fully opened
In case of A/T vehicle ⑫ – ㉑	Less than 29	Throttle valve fully closed
	More than 1000	Throttle valve fully opened
㉗ – ㉙	2.45 ± 0.5	When air temperature inside intake manifold is 20 °C (68 °F):
In case of M/T vehicle ″ ㉜ – ㉙	More than 1000	Throttle valve fully closed
	Less than 29	Throttle valve fully opened
In case of A/T vehicle ㉜ – ㉑	Less than 29	Throttle valve fully closed
	More than 1000	Throttle valve fully opened

1.3L ENGINE—FUEL INJECTION SYSTEM TROUBLESHOOTING PROCEDURES CHARADE G102 WITH 4 CYLINDER ENGINE

INSPECTION OF ECU CIRCUIT FOR G102

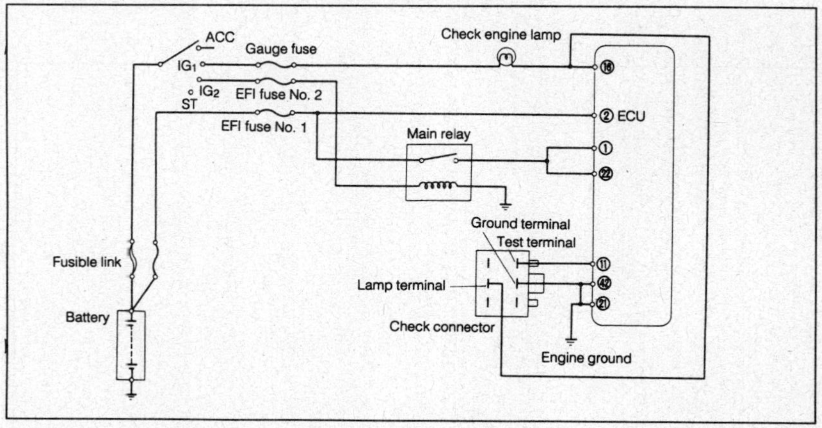

1st step

Ensure that check engine lamp goes on when ignition switch is turned ON. (Engine is in a stopped condition.)
- **YES** → System is normal.
- **NO** → Ensure that check engine lamp goes on when lamp terminal at check connector is shorted with ground terminal.
  - **YES** → Check that there is continuity between ECU terminal ㉑ and body ground, ECU terminal ㊷ and body ground, and ECU terminal ⑱ and lamp terminal in check connector.
    - **OK** → Check ECU.
    - **BAD** → Repair or replace.
  - **NO** → Check bulb, fuse and wiring between ECU and ignition switch.
    - **BAD** → Repair or replace.

1.3L ENGINE – FUEL INJECTION SYSTEM TROUBLESHOOTING PROCEDURES CHARADE G102 WITH 4 CYLINDER ENGINE

2nd step

- Ensure that check engine lamp goes off after engine starts.
  - **YES** → System is normal.
  - **NO** ↓
- Check wiring between ECU and check engine lamp.
  - **BAD** → Repair or replace.
  - **OK** ↓
- Check diagnosis code.
  - **Normal code** → Check ECU.
  - **Malfunction code** ↓
- Perform trouble shooting according to diagnosis codes.
- After trouble shooting according to diagnosis code has been performed, start engine. Check that check engine lamp goes off.
  - **NO** → Check diagnosis code again.
  - **YES** ↓
- System is normal. → Erase diagnosis code.

MAIN RELAY

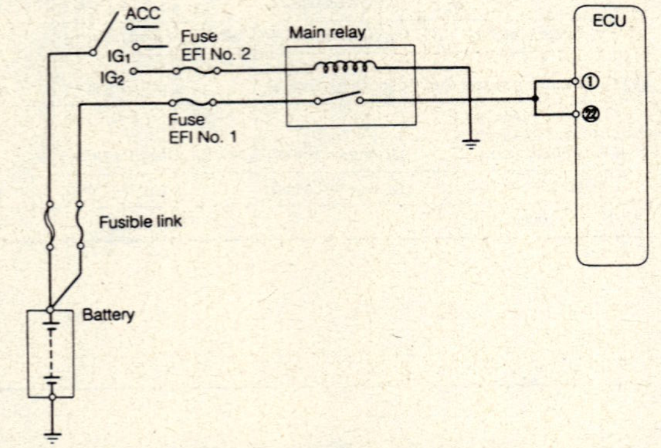

Terminals	Trouble	Condition	STD voltage
① – ㊴ Ground ㉒ – ㊴ Ground	No standard voltage	Ignition switch ON	Approx. battery voltage
② – ㊴ Ground		At all time (But, voltage drops during engine starting period)	

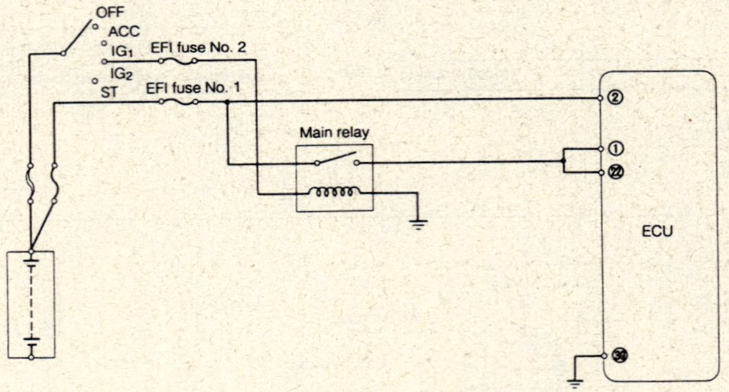

1.3L ENGINE – FUEL INJECTION SYSTEM TROUBLESHOOTING PROCEDURES
CHARADE G102 WITH 4 CYLINDER ENGINE

① or ㉒ – ㊴

[1] There is no specified voltage between SST terminals ① or ㉒ and ㊴.

[2] Check that there is specified voltage between SST terminal ① or ㉒ and body ground when ignition switch is turned ON.

- **YES** → Check between ECU terminal ㊴ and body ground.
  - **BAD** → Repair or replace.
- **NO** → Check fuses, fusible link and wiring harness.
  - **BAD** → Repair or replace.

Check ignition switch.
- **BAD** → Repair or replace.

Check EFI main relay.
- **BAD** → Repair or replace.

Check ECU.

[1]

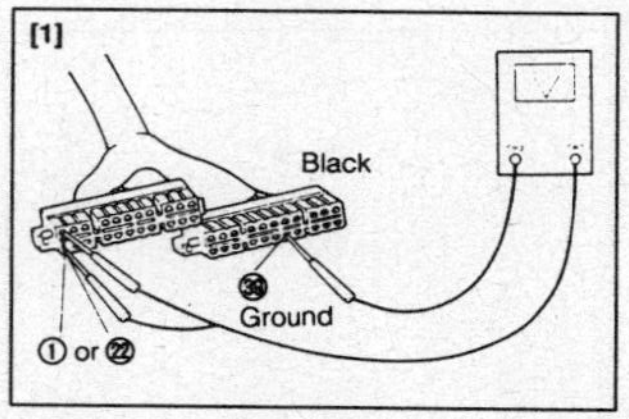

[2]

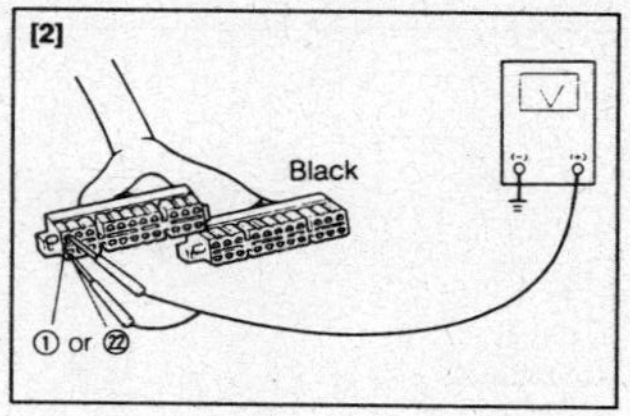

② – ㊴

[1] There is no specified voltage between SST terminals ② and ㊴.

[2] Check that there is specified voltage between terminal ② and body ground.

- **YES** → [3] Check wiring between SST terminal ㊴ and body ground.
  - **BAD** → Repair or replace.
- **NO** → Check fuses, fusible link and wiring harness.
  - **BAD** → Repair or replace.
  - **OK** → Check EFI main relay.
    - **BAD** → Repair or replace.
    - **OK** → Check ECU.

[1]

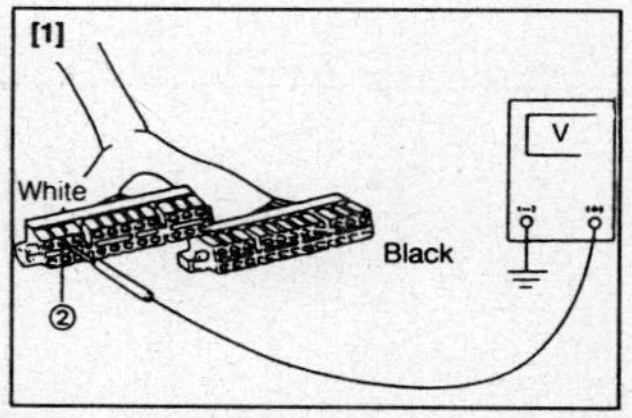

[2]

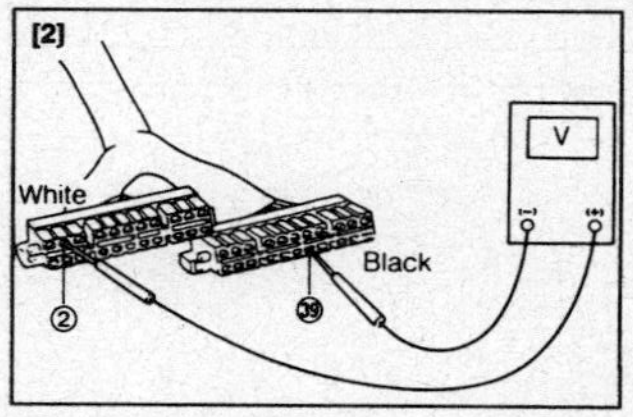

[3]

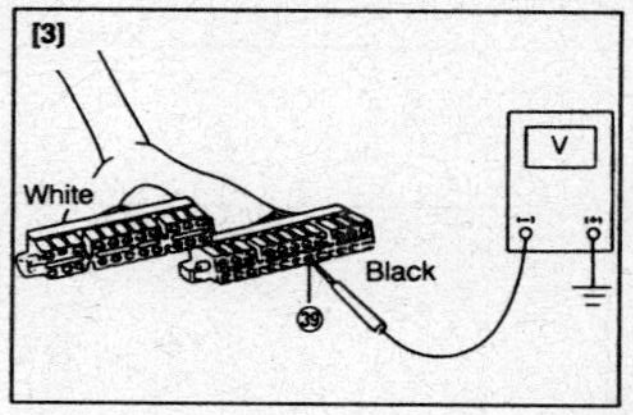

1.3L ENGINE – FUEL INJECTION SYSTEM TROUBLESHOOTING PROCEDURES CHARADE G102 WITH 4 CYLINDER ENGINE

Terminals	Trouble	Condition		STD voltage
⑦ – ㊴ Ground	No specified voltage	Ignition switch ON	When cooling water temperature is 80 °C (176 °F):	0.4 - 0.65

Water temperature sensor

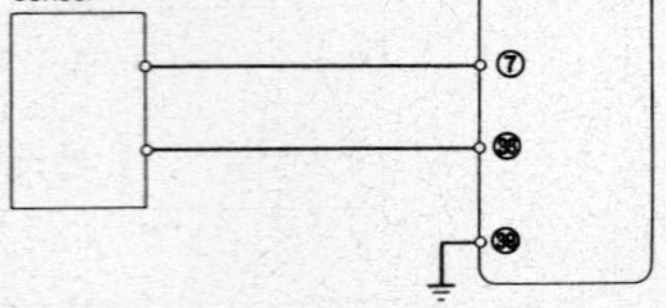

[1] There is no specified voltage between SST terminals ⑦ and ㊴.

[2] Check that there is voltage between SST terminals ① or ㉒ and ㊴ when ignition switch is turned ON.

- BAD → Check main relay circuit.
- OK → Check wiring between ECU and water temperature sensor.
  - BAD → Repair or replace.
  - OK → Check water temperature sensor.
    - BAD → Replace water temperature sensor.
    - OK → Check ECU.

[1]

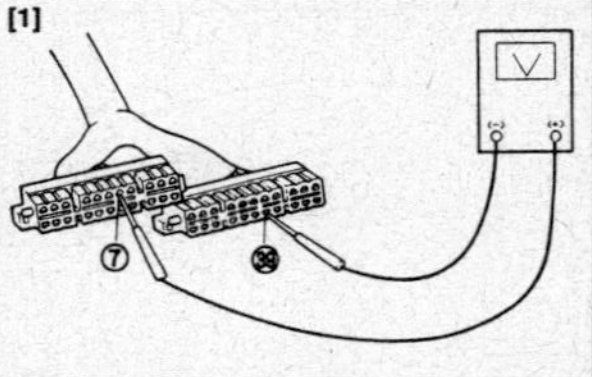

[2]

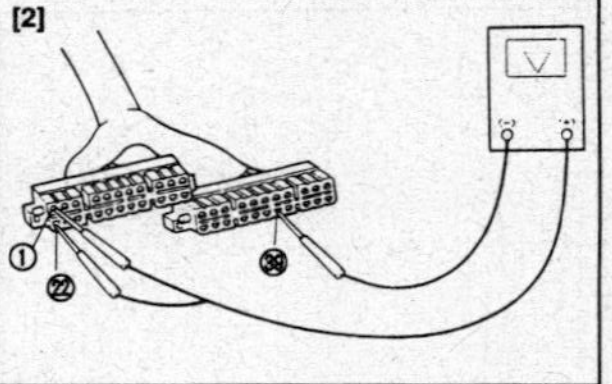

Terminals	Trouble	Condition		STD voltage
㉗ – ㊴ Ground	No specified voltage	Ignition switch ON	When air temperature inside intake manifold in 20°C (68°F):	1.5 - 3.0

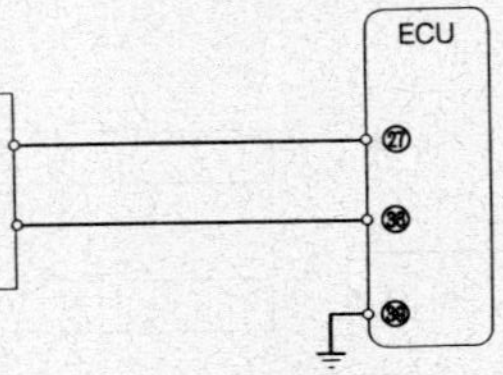

[1] There is no specified voltage between SST terminals ㉗ and ㊴.

[2] Check that there is voltage between SST terminals ① or ㉒ and ㊴ when ignition switch is turned ON.

- BAD → Check main relay circuit.
- OK → Check intake air temperature sensor.
  - BAD → Replace intake air temperature sensor.
  - OK → Check wiring between ECU and intake air temperature sensor.
    - BAD → Repair or replace.
    - OK → Check ECU.

[1]

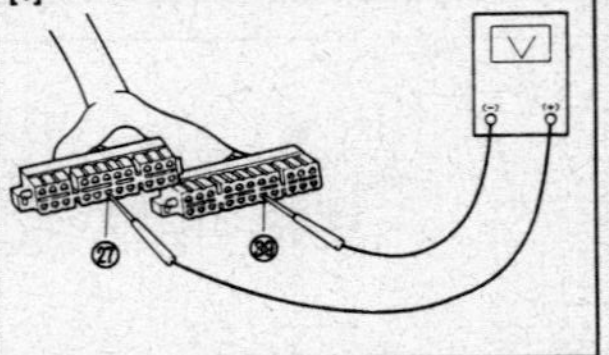

[2]

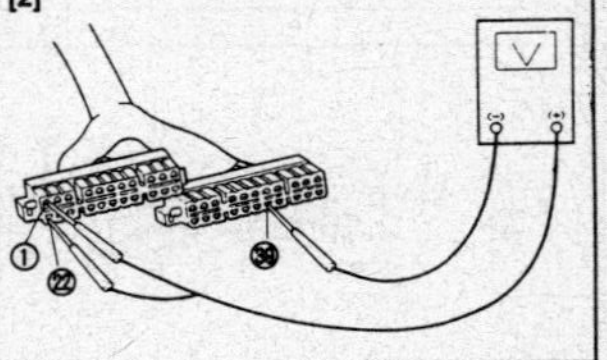

1.3L ENGINE – FUEL INJECTION SYSTEM TROUBLESHOOTING PROCEDURES
CHARADE G102 WITH 4 CYLINDER ENGINE

Terminals		Trouble	Condition		STD voltage
M/T vehicle	⑫ – ㊴ Ground	More than 5 V	Ignition switch ON	Throttle valve fully closed	Less than 5 V
		No voltage	Ignition switch ON	Throttle valve fully opened	Approx. battery voltage
	㉒ – ㊴ Ground	No voltage	Ignition switch ON	Throttle valve fully closed	Approx. battery voltage
		More than 5 V	Ignition switch ON	Throttle valve fully opened	Less than 5 V
A/T vehicle	⑫ – ㊴ Ground	More than 5 V	Ignition switch ON	Throttle valve fully closed	Less than 5 V
		No voltage	Ignition switch ON	Throttle valve fully opened	Approx. battery voltage
	㉒ – ㊴ Ground	More than 5 V	Ignition switch ON	Throttle valve fully closed	Less than 5 V
		No voltage	Ignition switch ON	Throttle valve fully opened	Approx. battery voltage

M/T vehicle
— when fully closed
--- when fully opened

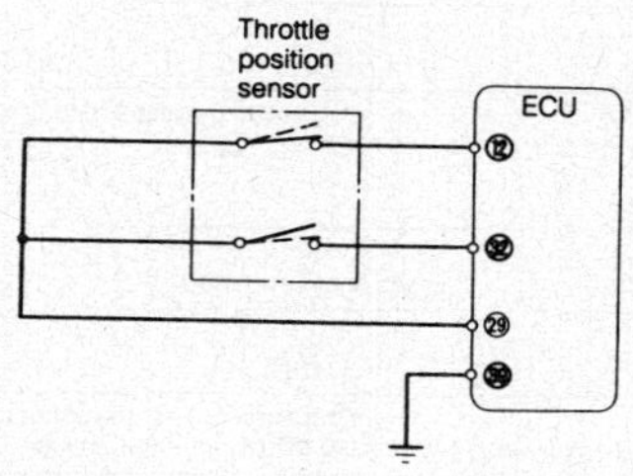

A/T vehicle
— when fully closed
--- when fully opened

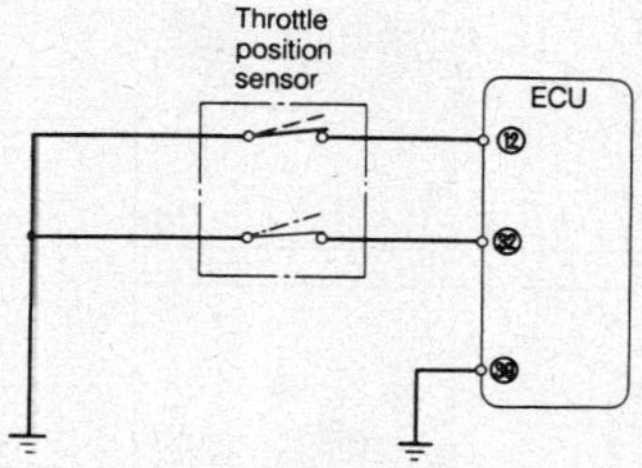

⑫ – ㊴

[1] There is no specified voltage between SST terminals ⑫ and ㊴.

↓

[2] Check that there is voltage between SST terminals ① or ㉒ and ㊴ when ignition switch is turned ON.

- BAD → Check main relay circuit.
- OK ↓

Check throttle position sensor.

- BAD → Replace.
- OK ↓

Check wiring between ECU and throttle position sensor.

- BAD → Repair or replace.
- OK ↓

(A/T vehicle only)
Check wiring between throttle position sensor and body ground.

- BAD → Repair or replace.
- OK ↓

Check ECU

[1]

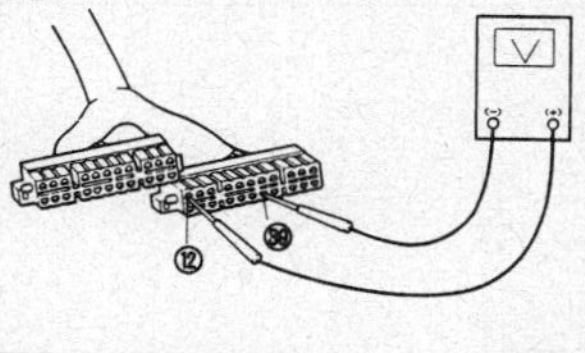

[2]

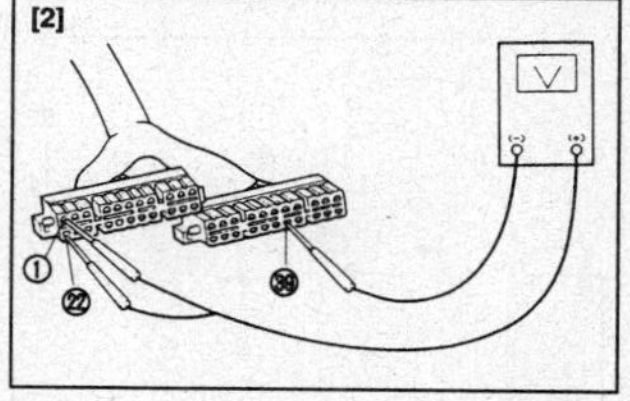

1.3L ENGINE—FUEL INJECTION SYSTEM TROUBLESHOOTING PROCEDURES
CHARADE G102 WITH 4 CYLINDER ENGINE

㉜ – ㊴

[1] There is no specified voltage between SST terminals ㉜ and ㊴.

[2] Check that there is voltage between SST terminals ① or ㉒ and ㊴ when ignition switch is turned ON.
- BAD → Check main relay circuit.
- OK ↓

Check throttle position sensor.
- BAD → Replace.
- OK ↓

Check wiring between ECU and throttle position sensor.
- BAD → Repair or replace.
- OK ↓

(A/T vehicle only)
Check wiring between throttle position sensor and body ground.
- BAD → Repair or replace.
- OK ↓

Check ECU

[1]

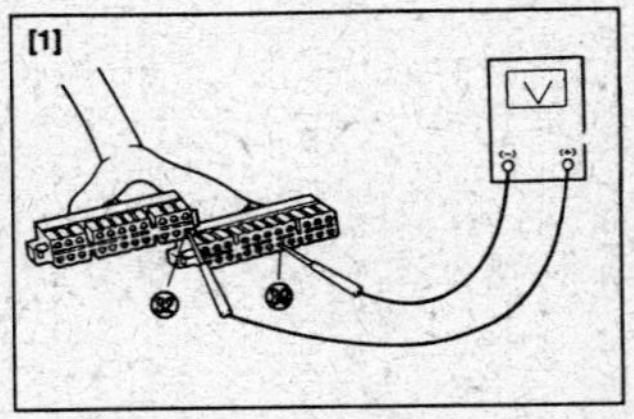

[2]

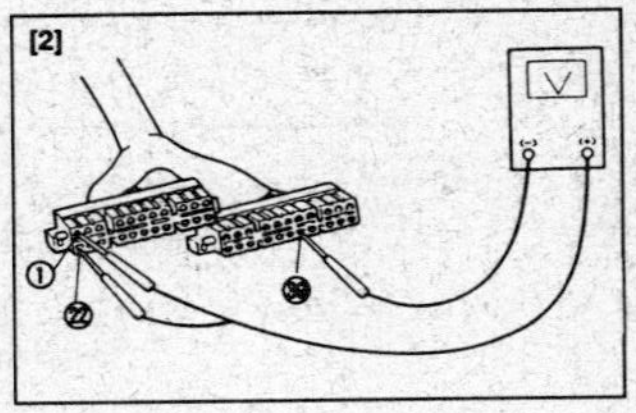

Terminals	Trouble	Condition		STD voltage
⑤ – ㊱ Ground	No voltage	Ignition switch ON		4.5 - 5.5
⑥ – ㊴ Ground	No standard voltage	Ignition switch	When atmospheric pressure of 760 mmHg (29.9 inchHg) exists:	3.2 - 4.0

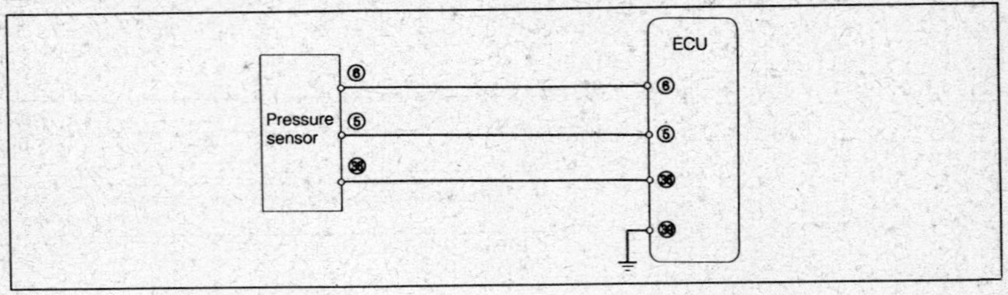

⑤ – ㊱

[1] There is no specified voltage between SST terminals ⑤ and ㊱.

[2] Check that there is specified voltage between SST terminals ① or ㉒ and ㊴.
- BAD → Check main relay circuit.
  - BAD → Repair or replace.
- OK ↓

Check wiring between ECU and pressure sensor.
- BAD → Repair or replace.
- OK ↓

Check pressure sensor.
- BAD → Repair or replace.

Check ECU.

[1]

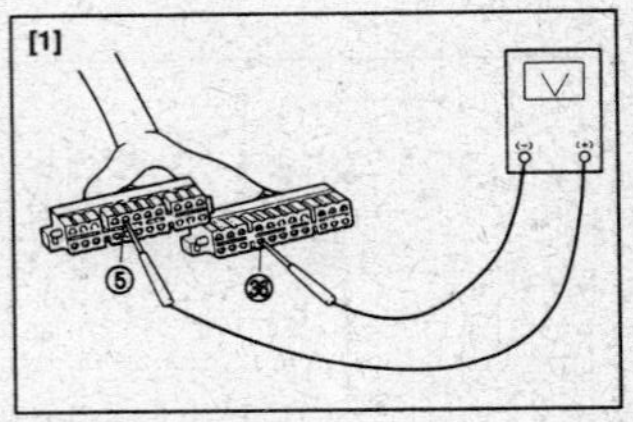

[2]

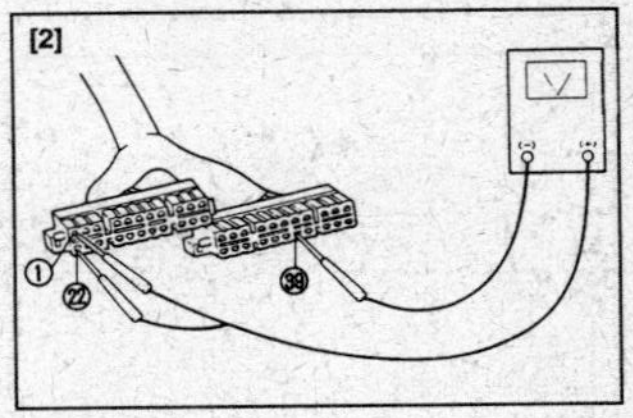

1.3L ENGINE—FUEL INJECTION SYSTEM TROUBLESHOOTING PROCEDURES
CHARADE G102 WITH 4 CYLINDER ENGINE

⑥ – ㊴

- There is no specified voltage between SST terminals ⑥ – ㊴.
- ↓
- Check that there is specified voltage between SST terminals ⑤ and ㊱.
  - **BAD** → ⑤ – ㊱ trouble section.
    - **BAD** → Repair or replace.
  - **OK** ↓
- Check wiring between ECU and pressure sensor.
  - **BAD** → Repair or replace.
  - **OK** ↓
- Check pressure sensor.
  - **OK** ↓
- Check ECU.

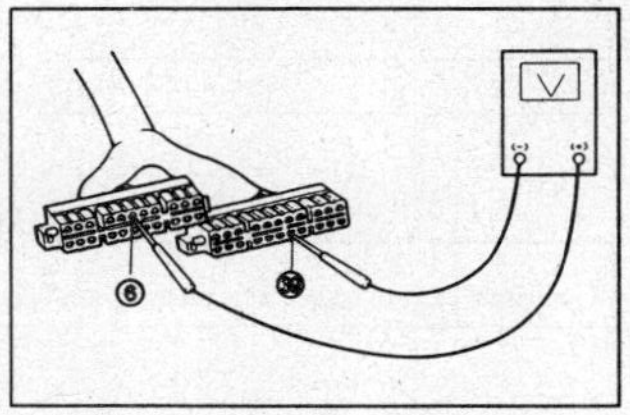

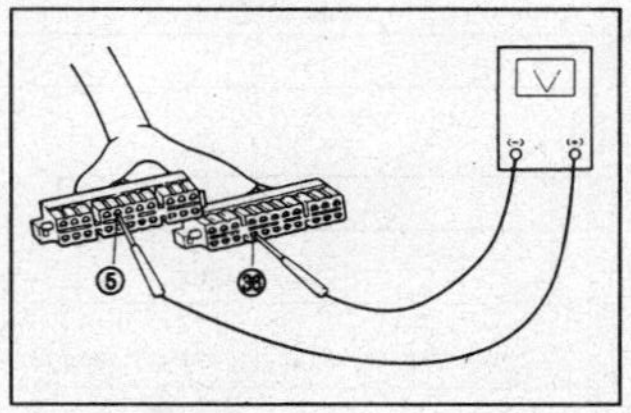

3. Perform the inspection according to the following flowchart.

- Diagnosis code check
  - **Malfunction code** → Perform trouble shooting according to diagnosis codes.
  - **Normal code** ↓
- Ensure that engine idle speed conforms to specification.
  Idle speed: M/T 800 ± 50 rpm
  A/T 850 ± 50 rpm
  - **BAD** → Check ignition timing. Adjust idle speed.
    - **OK** → (return to main flow)
    - **BAD** → Unit inspection of idle speed control VSV
      - **BAD** → Replace.
      - **OK** ↓
      - • Check that terminal T of check connector is shorted with ground terminal.
        • Check vehicle speed sensor system.
        • Check water temperature sensor system.
        • Check throttle position sensor system.
        - **BAD** → Repair or replace.
        - **OK** → Check ECU.
  - **OK** ↓
- Ensure that idle speed conforms to specification when headlamp switch is turned ON.
  - **BAD** → Check wiring from headlamp at left side to ECU terminal ⑨.
    - **BAD** → Repair or replace.
    - **OK** → Check ECU.
  - **OK** ↓
- Turn OFF headlamp switch.
- ↓

(To be continued)

1.3L ENGINE—FUEL INJECTION SYSTEM TROUBLESHOOTING PROCEDURES
CHARADE G102 WITH 4 CYLINDER ENGINE

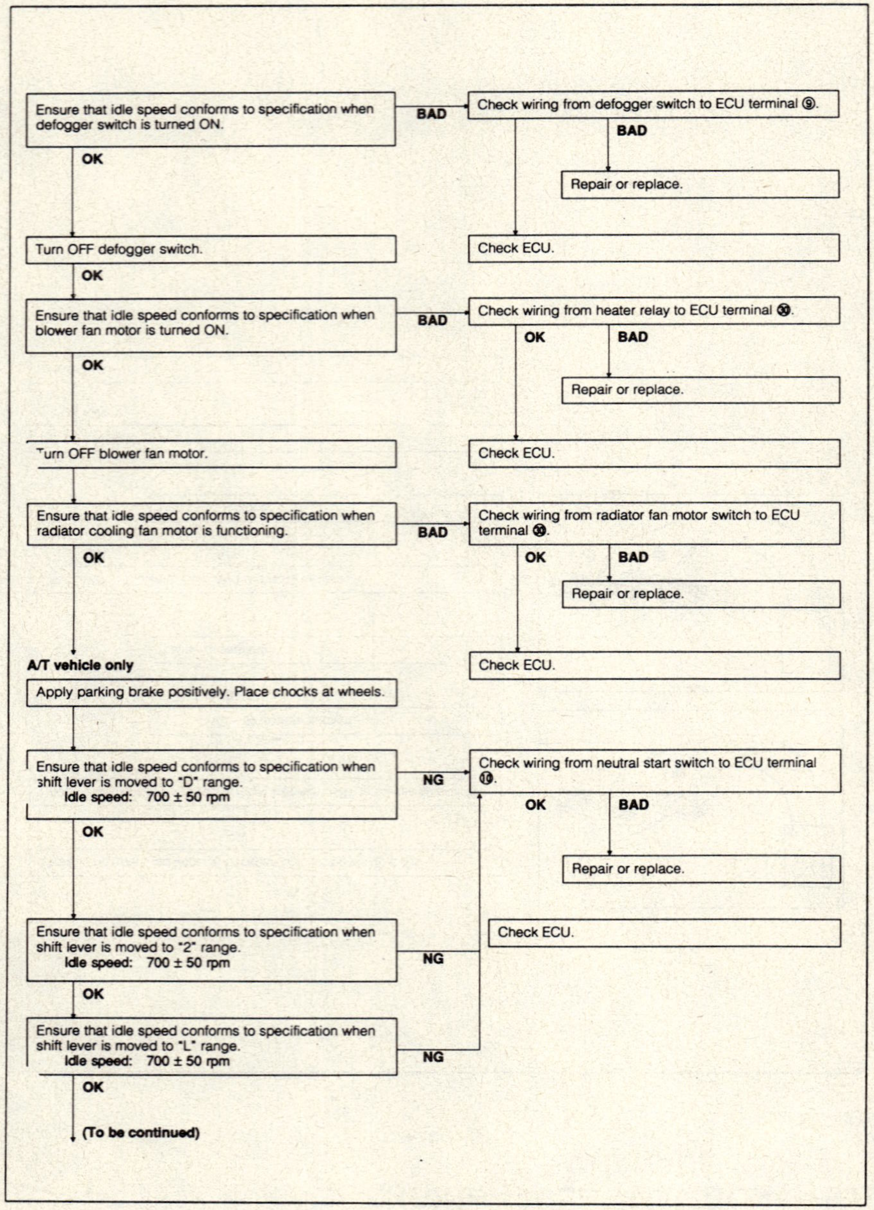

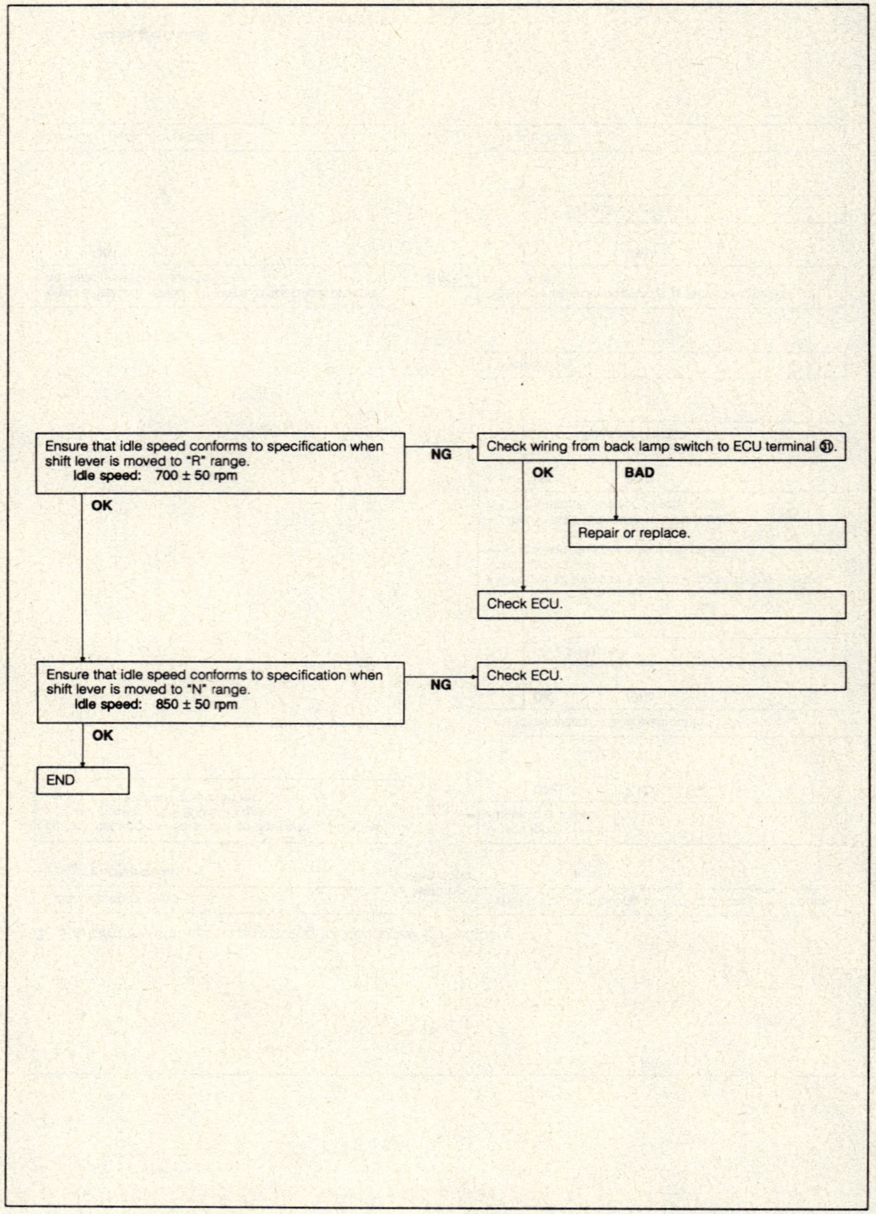

1.3L ENGINE—FUEL INJECTION SYSTEM TROUBLESHOOTING PROCEDURES CHARADE G102 WITH 4 CYLINDER ENGINE

No.	Terminals	Trouble	Condition		STD voltage
5	㉖ – ㊴ Ground	No voltage change	Ignition switch ON	When engine revolution speed is held at 3000 rpm after having warmed up engine fully:	Voltage changes more than 8 times within 10 seconds.

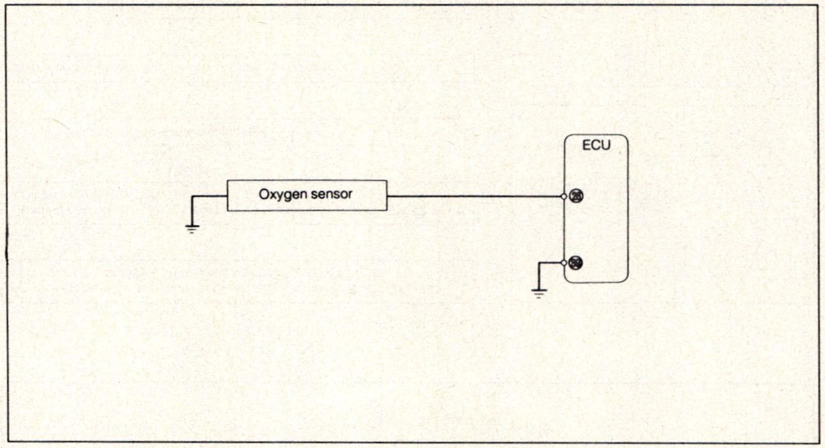

There is no change in voltage between SST terminals ㉖ and ㊴.

↓

Check wiring between ECU and oxygen sensor.

- **BAD** → Repair or replace.
- **OK** ↓

Check oxygen sensor.

- **OK** ↓

Check ECU.

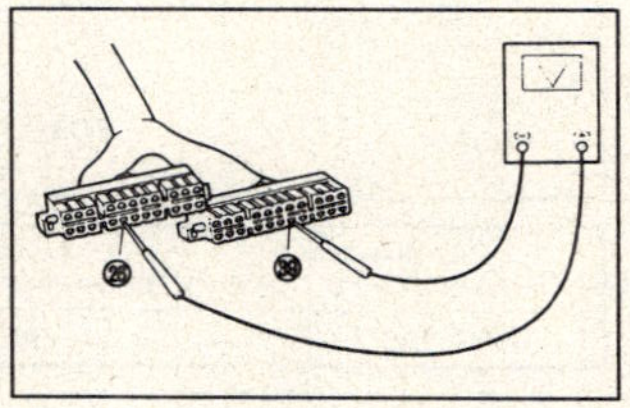

IGNITION COIL

Terminals	Trouble	Condition	STD voltage
③ – ㊴	No voltage	Ignition switch ON (Engine stopped state)	Approx. battery voltage

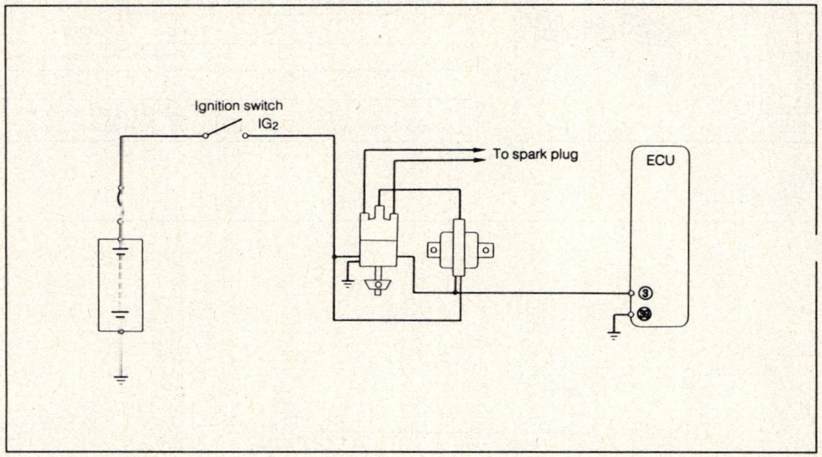

③ – ㊴

There is no voltage between SST terminals ③ and ㊴

↓

Check wiring between distribution connector positive terminals (+) and SST terminal ③.

- **BAD** → Repair or replace.
- **OK** ↓

Check ignition system.

- **BAD** → Repair or replace.
- **OK** ↓

Check ECU.

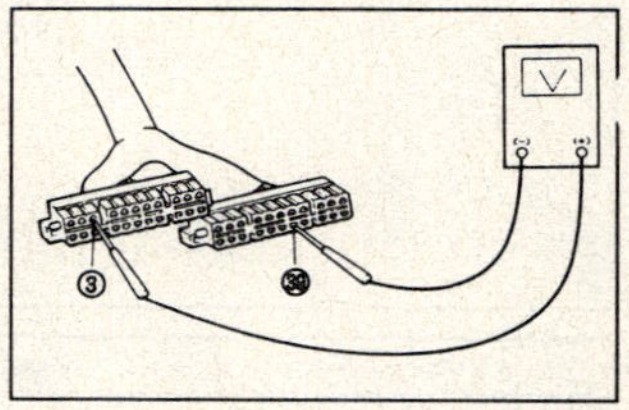

SPEED SENSOR

Terminals	Trouble	Condition		STD voltage
⑧ – ㊴ Ground	No voltage changes	Ignition switch ON	When vehicle is moved slowly:	0 - Approx. battery voltage

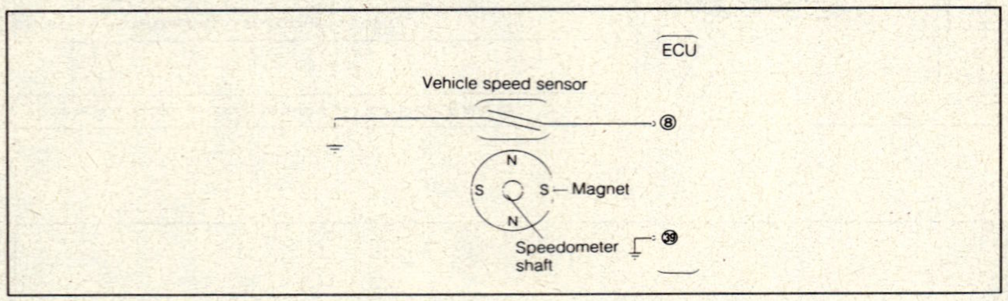

No change in voltage occurs between SST terminals ⑧ and ㊴ when vehicle is moved slowly with ignition switch turned ON.

↓

Check that there is voltage between SST terminals ① or ㉒ and ㊴ when ignition switch is turned ON.

- **BAD** → Check main relay circuit.
- **OK** ↓

Check wiring between ECU and vehicle speed sensor.

- **BAD** → Repair or replace.
- **OK** ↓

Check wiring between vehicle speed sensor and body ground.

- **BAD** → Repair or replace.
- **OK** ↓

Check vehicle speed sensor.

- **BAD** → Replace.
- ↓

Check ECU.

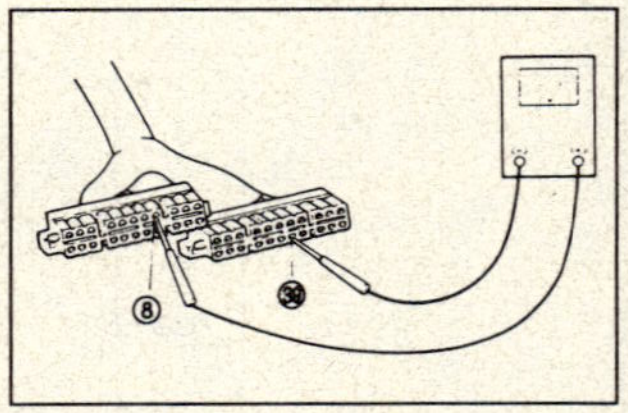

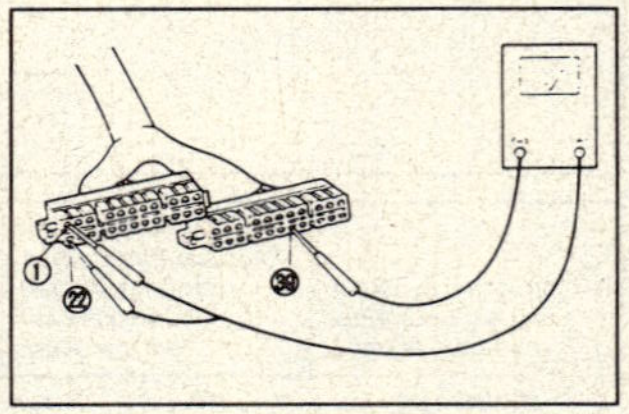

STARTER RELAY

Terminals	Trouble	Condition	STD voltage
⑬ – ㊴ Ground	No voltage	Ignition switch set to ST position	More than 5 V

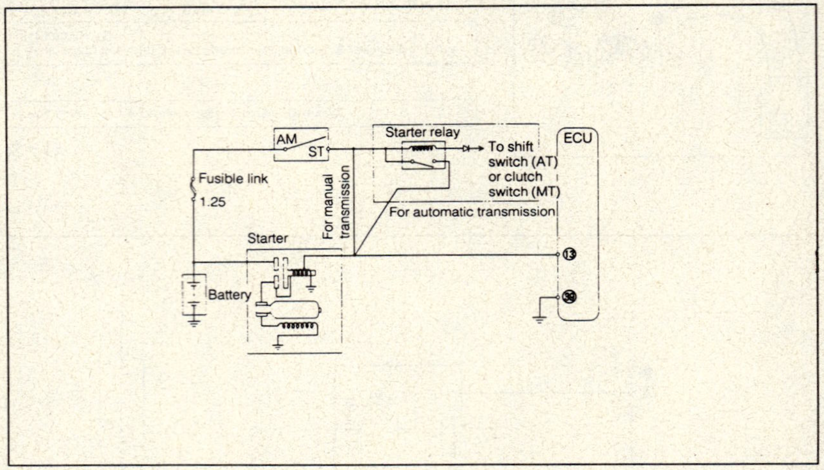

There is no voltage between SST terminals ⑬ and ㊴ when ignition switch is set to ST position.

↓

Check starter operation.

- **OK** → Check wiring between ECU and starter terminals.
  - **BAD** → Repair or replace.
- **BAD** ↓

Check battery, fusible link ignition switch, starter, starter relay and shift position switch or clutch switch.

- **BAD** → Repair or replace.
- ↓

Check ECU.

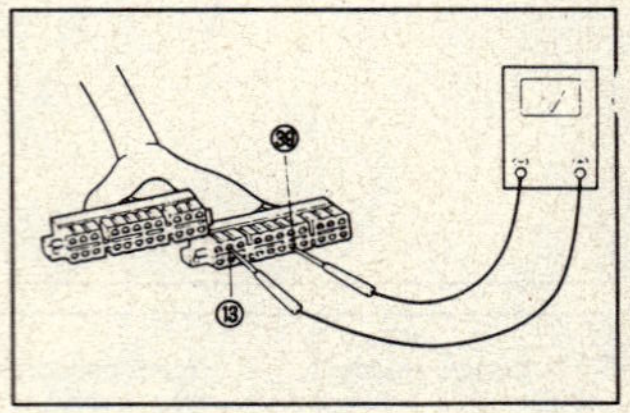

1.3L ENGINE—FUEL INJECTION SYSTEM TROUBLESHOOTING PROCEDURES
CHARADE G102 WITH 4 CYLINDER ENGINE

RELAY AND SWITCHES

Terminals	Trouble	Condition		STD voltage
(34) – (39) Ground	No voltage	When engine is operating and compressor magnet clutch of air conditioner is energized:		Approx. battery voltage
(12) – (39) Ground	More than 5 V	Ignition switch ON	Throttle valve fully closed	Less than 5 V
	No voltage		Throttle valve fully opened	Approx. battery voltage
(A/T vehicle only) (10) – (39) Ground	More than 1 V	Ignition switch ON	When shift lever is moved to "D", "2" or "L" range:	Less than 1 V
	No voltage		When shift lever is moved to "P", "N" or "R" range:	Approx. battery voltage
(A/T vehicle only) (31) – (39) Ground	More than 1 V	Ignition switch ON	When shift lever is moved to "D", "2", "L", "P" or "N" range:	Less than 1 V
	No voltage		When shift lever is moved to "R" range:	Approx. battery voltage

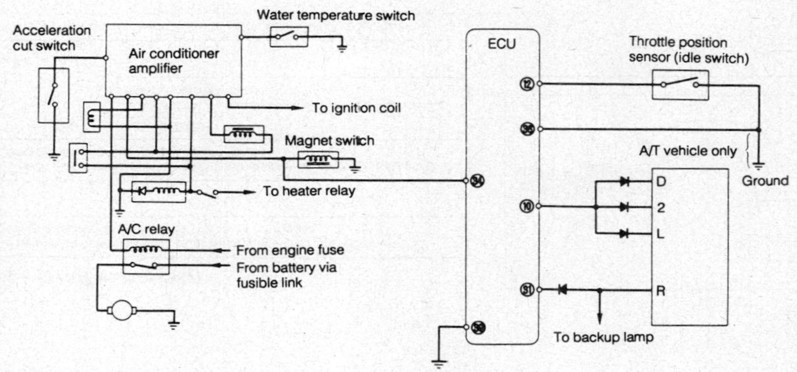

(34) – (39)

There is no specified voltage between SST terminals (34) and (39) when magnet clutch is operated.

↓

Check magnet clutch operation.

- **OK** → Check wiring between ECU terminal (34) and air conditioner amplifier terminal.
  - **BAD** → Repair or replace.
- **BAD** → Check air conditioner system.
  - **BAD** → Repair.

↓

Check ECU.

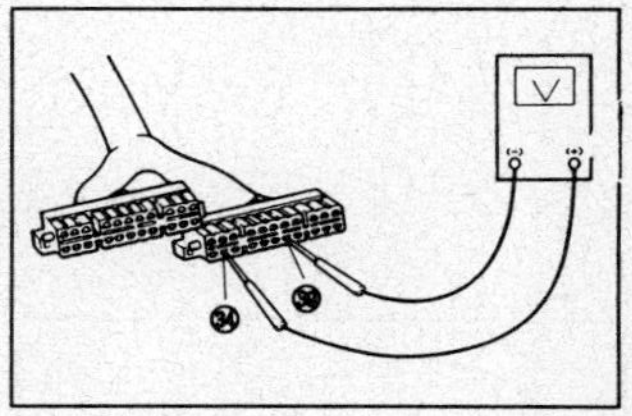

(12) – (39)

[1] There is no specified voltage between SST terminals (12) and (39).

↓

[2] Check that there is voltage between SST terminals (1) or (22) and (39) when ignition switch is turned ON.

- **BAD** → Check main relay circuit.
- **OK** ↓

Check throttle position sensor.

- **BAD** → Replace throttle position sensor.
- **OK** ↓

Check wiring between throttle position sensor and body ground.

- **BAD** → Repair or replace.
- **OK** ↓

Check ECU.

[1]

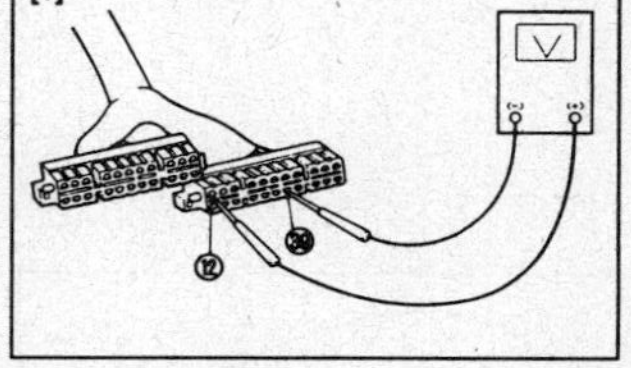

[2]

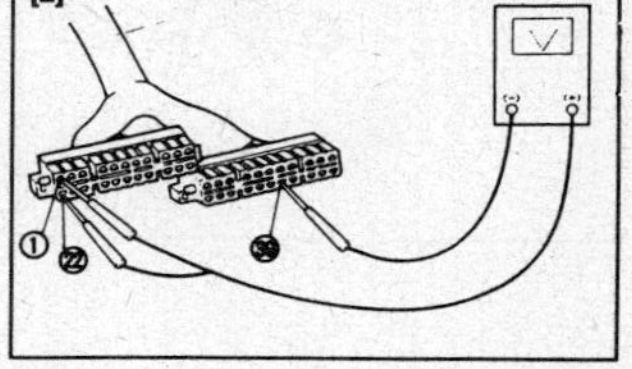

⑩ – ㊴ (A/T vehicle only)

There is no specified voltage at SST terminals ⑩ and ㊴ when shift lever is moved to each of "D", "2" and "L" ranges.

↓

Check that there is specified voltage between SST terminals ① or ㉒ and ㊴ when ignition switch is turned ON.

- BAD → Check main relay circuit.
- OK ↓

Check neutral start switch. (See section of automatic transmission.)

OK ↓

Check wiring between ECU and neutral start switch.

OK ↓

Check ECU.

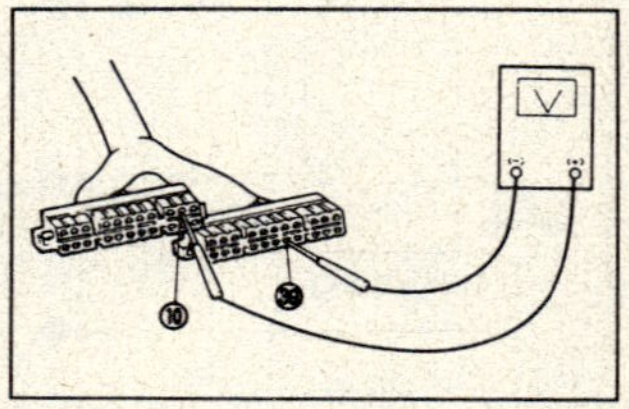

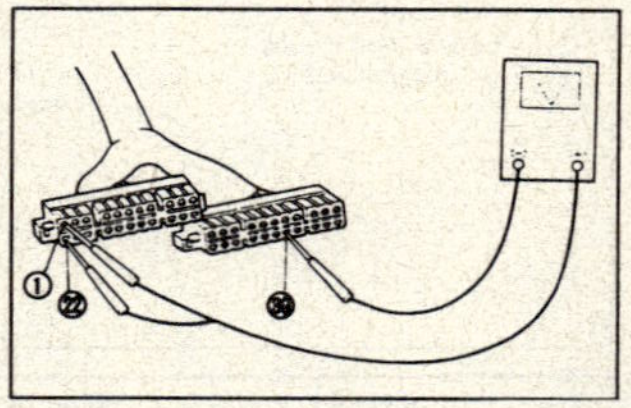

There is no specified voltage at SST terminals ⑩ and ㊴ when shift lever is moved to each of "P", "N" and "R" ranges.

↓

Check that there is specified voltage between SST terminals ① or ㉒ and ㊴ when ignition switch is turned ON.

- BAD → Check main relay circuit.
- OK ↓

Check the neutral start switch.

OK ↓

Check wiring between ECU and neutral start switch.

OK ↓

Check ECU.

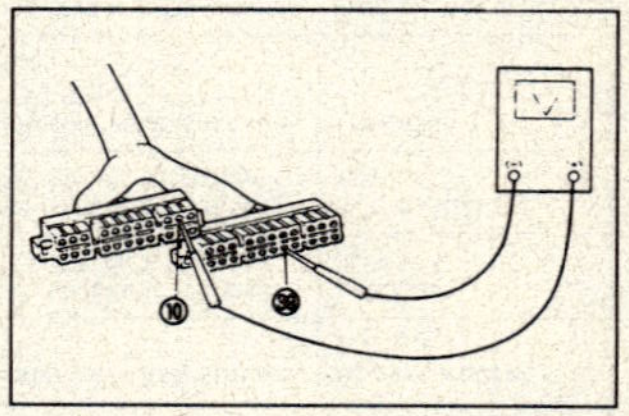

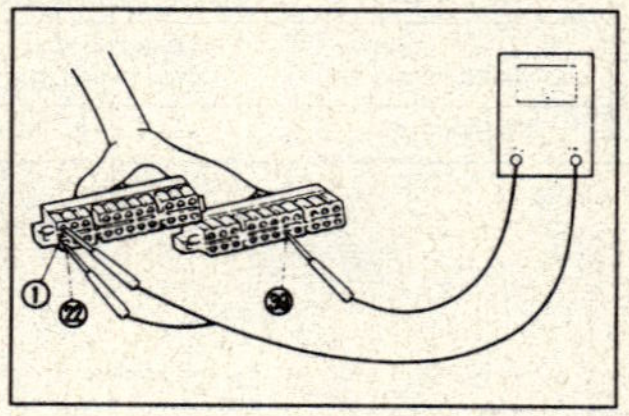

㉛ – ㊴ (AT vehicle only)

There is no specified voltage at SST terminals ㉛ and ㊴ when shift leer is moved to each of "D", "2", "L", "P" and "N" ranges.

↓

Check wiring between ECU and neutral start switch.

OK ↓

Check neutral start switch.

↓

Check ECU.

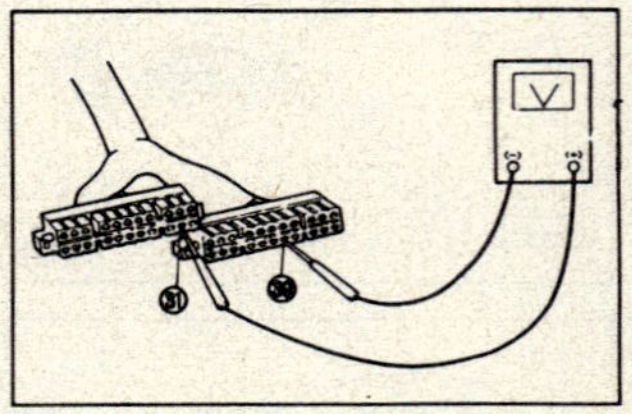

There is no specified voltage at SST terminals ㉛ and ㊴ when shift lever is moved to "R" range.

↓

Check wiring between ECU and neutral start switch.

- BAD → Repair or replace.
- OK ↓

Check neutral start switch.

- BAD → Repair or replace.
- OK ↓

Check ECU.

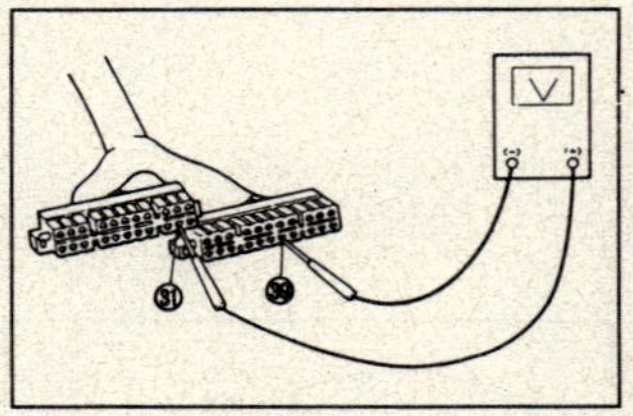

ECU TERMINAL CONNECTIONS—ROCKY

Table Showing ECU Connections

Terminal code	Contents of connection	Terminal code	Contents of connection
1	Main relay (Power supply)	22	Main relay (Power supply)
2	Battery (Backup power supply)	23	
3	Ignition coil primary voltage	24	
4		25	
5	Pressure sensor power supply	26	Oxygen sensor
6	Pressure sensor	27	Intake air temperature sensor
7	Cooling water temperature sensor	28	
8	Vehicle speed sensor	29	Operation system ground (Engine)
9	Electrical load (Headlamp and defogger)	30	Electrical load (Blower fan)
10		31	
11	Check connector (Test terminal)	32	Throttle position switch (Power switch)
12	Throttle position switch (Idle switch)	33	Stop lamp
13	Starter	34	Air conditioner magnet switch
14		35	
15	Oxygen sensor feedback check terminal	36	Operation system ground
16	Check engine lamp	37	Fuel pump relay
17		38	Pressure VSV
18	EGR VSV	39	System ground
19		40	Idle speed control VSV
20	Injector	41	Injector
21	Actuator drive ground (Engine)	42	Actuator drive ground (Engine)

21	20		18	
42	41	40	39	38

16	15		13	12	11		9	8
37	36		34	33	32		30	29

7	6	5		3	2	1
	27	26				22

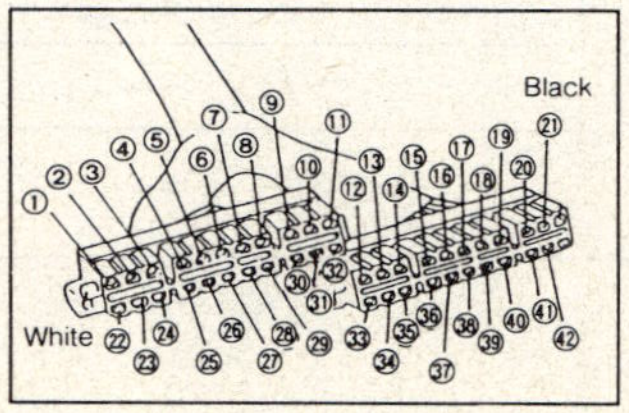

VOLTAGES AT ECU TERMINAL CONNECTIONS—ROCKY

Voltages at ECU connectors

Terminals	STD Voltage	Condition	
① — ㊴	Approx. battery voltage	Ignition switch ON	
② — ㊴	Approx. battery voltage	At all time	
③ — ㊴	Approx. battery voltage	Ignition switch ON	When engine is stopped:
⑤ — ㊴	4.5 - 5.5 V	Ignition switch ON	
⑥ — ㊴	3.2 - 4.0 V	Ignition switch ON	When atmospheric pressure of 760 mmHg (29.9 inchHg) exists.
⑦ — ㊴	0.4 - 0.65 V	Ignition switch ON	When cooling water temperature is 80°C (176°F):
⑧ — ㊴	0 - Approx. battery voltage	Ignition switch ON	Measured voltage changes when vehicle is moved 1.5 m (4.93 ft).
⑨ — ㊴	Less than 5.0 V	Ignition switch ON	When defogger and headlamp switches are turned OFF:
	Approx. battery voltage	Ignition switch ON	When defogger and/or headlamp switches are turned ON:
⑪ — ㊴	Approx. battery voltage	Ignition switch ON	When test terminal of check connector is not connected with ground terminal:
	Less than 1.0 V	Ignition switch ON	When test terminal of check connector is connected with ground terminal:
⑫ — ㊴	Less than 5.0 V	Ignition switch ON	Throttle valve fully closed
	Approx. battery voltage	Ignition switch ON	Throttle valve fully opened
⑬ — ㊴	0 V	Ignition switch ON	
	More than 6 V	When ignition switch is set to ST position:	
⑮ — ㊴	Measured voltage changes at a point between 0 - 5.0 V.	After warming up engine completely, connect test terminal of check connector with ground terminal. Hold engine revolution speed at 3000 rpm for two minutes.	
⑯ — ㊴	Less than 3.0 V	Ignition switch ON	• Engine is stopped. • When check engine lamp is illuminated:
	Approx. battery voltage	Ignition switch ON	• After engine starts: • When check engine lamp is extinguished:

VOLTAGES AT ECU TERMINAL CONNECTIONS—ROCKY

Terminals	STD Voltage	Condition	
⑱ — ㊴	Approx. battery voltage	Ignition switch ON	• After engine starts: • Cooling water temperature is below 40°C (104°F).
	Less than 3.0 V	Ignition switch ON	• After engine starts: • Cooling water temperature is above 41°C (106°F).
⑳ — ㊴	Less than 1.0 V	At least 30 seconds have elapsed after turning OFF ignition switch.	
	Approx. battery voltage	Ignition switch ON	• Engine is stopped.
㉑ — ㊴	Less than 0.01 V	Ignition switch ON	
㉒ — ㊴	Approx. battery voltage	Ignition switch ON	
㉖ — ㊴	Change in output voltage	Ignition switch ON	After warming up engine completely, hold engine revolution speed at 3000 rpm for two minutes.
㉗ — ㊴	1.5 - 3.0 V	Ignition switch ON	Air temperature inside intake manifold is 20°C(68°F):
㉙ — ㊴	Less than 0.1 V	Ignition switch ON	
㉚ — ㊴	Approx. battery voltage	Ignition switch ON	• Blower fan switch turned OFF
	Less than 2.0 V	Ignition switch ON	When blower fan switch turned ON:
㉜ — ㊴	Approx. battery voltage	Ignition switch ON	Throttle valve fully closed
	Less than 5.0 V	Ignition switch ON	Throttle valve fully opened
㉝ — ㊴	Less than 1.0 V	Ignition switch ON	When brake pedal is not depressed:
	Approx. battery voltage	At all time	When brake pedal is depressed:

Terminals	STD Voltage	Condition	
㉞ — ㊴	Less than 1.0 V	Ignition switch ON	When compressor magnet switch of air conditioner is turned OFF:
	Approx. battery voltage	Ignition switch ON	When compressor magnet switch of air conditioner is turned ON:
㊱ — ㊴	Less than 0.1 V	Ignition switch ON	
㊲ — ㊴	Approx. battery voltage	Ignition switch ON	When fuel pump is stopped:
	Less than 2.0 V	Ignition switch ON	When fuel pump is operating:
㊳ — ㊴	Approx. battery voltage	Ignition switch ON	When pressure VSV is turned OFF:
	Less than 3.0 V	Ignition switch ON	For 0.5 second immediately after engine starts
㊴ — Engine ground	Less than 0.1 V	Ignition switch ON	
㊵ — ㊴	Less than 3.0 V	Ignition switch ON	Engine is stopped.
	Approx. battery voltage	Ignition switch ON	When test terminal of check connector is connected with ground terminal:
㊶ — ㊴	Less than 1.0 V	At least 30 seconds have elapsed after turning OFF ignition switch.	
	Approx. battery voltage	Ignition switch ON	Engine is stopped.
㊷ — ㊴	Less than 0.1 V	Ignition switch ON	

RESISTANCES AT ECU TERMINAL CONNECTIONS ROCKY

No.	Terminals	Trouble	Condition	STD voltage
1	① — ㊴ Ground ㉒ — ㊴ Ground	No standard voltage	Ignition switch ON	Approx. battery voltage
	② — ㊴ Ground		At all time (But, voltage drops during engine starting period)	

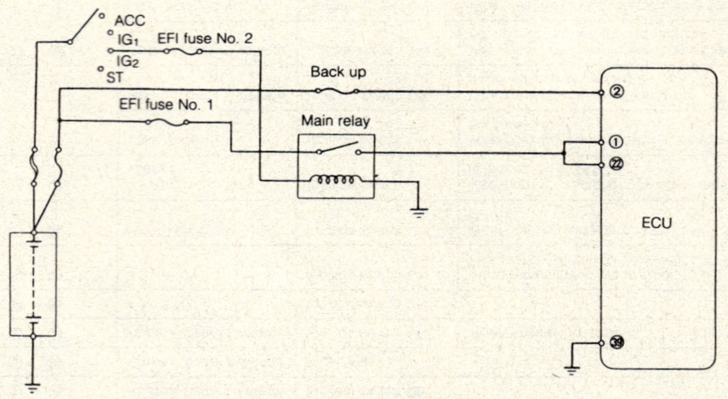

FUEL INJECTION SYSTEM TROUBLESHOOTING PROCEDURES—ROCKY

Terminals	STD Voltage	Condition
⑦ — ㉙	0.322 ± 0.1	When cooling water temperature is 80°C (176°F):
⑫ — ㊴	Less than 29	Throttle valve fully closed
	More than 1000	Throttle valve fully opened
㉗ — ㉙	2.45 ± 0.5	When air temperature inside intake manifold is 20°C (68°F):
㉜ — ㊴	More than 1000	Throttle valve fully closed
	Less than 29	Throttle valve fully opened

① or ㉒ — ㊴

[1] There is no specified voltage between SST terminals ① or ㉒ and ㊴.

[2] Check that there is specified voltage between SST terminal ① or ㉒ and body ground when ignition switch is turned ON.

- YES → Check between ECU terminal ㊴ and body ground.
  - BAD → Repair or replace.
- NO → Check fuses, fusible link and wiring harness.
  - BAD → Repair or replace.
  - OK → Check ignition switch.
    - BAD → Repair or replace.
    - OK → Check EFI main relay.
      - BAD → Repair or replace.
      - OK → Check ECU.

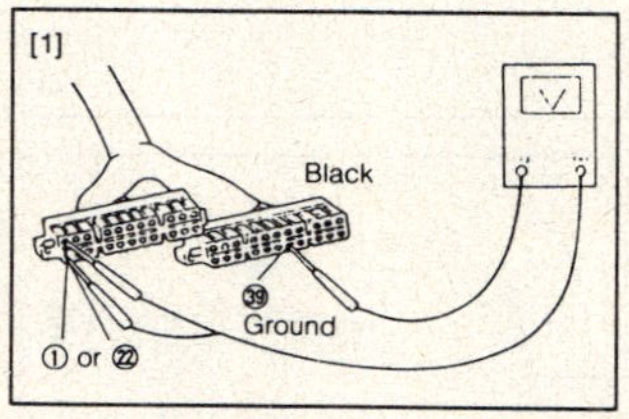

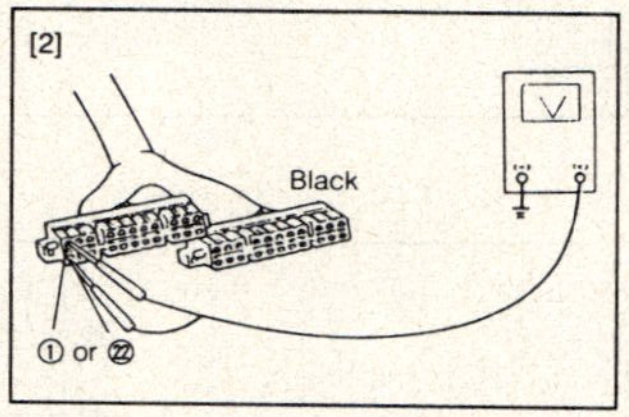

② — ㊴

[1] There is no specified voltage between SST terminals ② and ㊴.

[2] Check that there is specified voltage between terminal ② and body ground.

- YES → [3] Check wiring between SST terminal ㊴ and body ground.
  - BAD → Repair or replace.
- NO → Check fuses, fusible link and wiring harness.
  - BAD → Repair or replace.
  - OK → Check ECU.

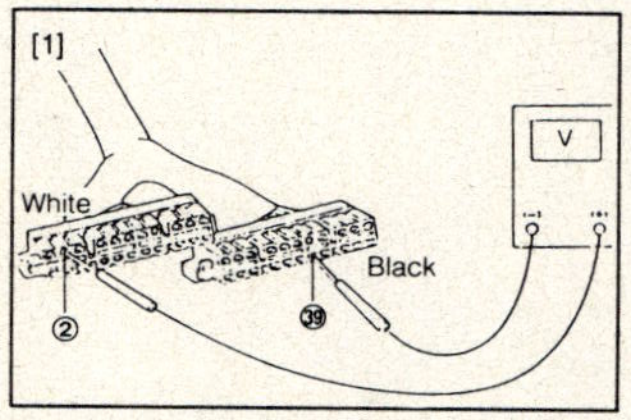

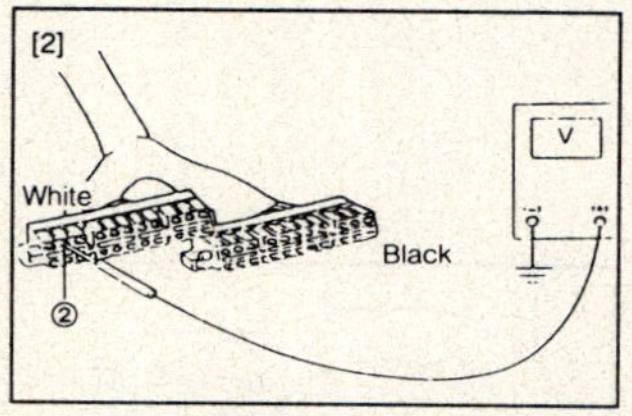

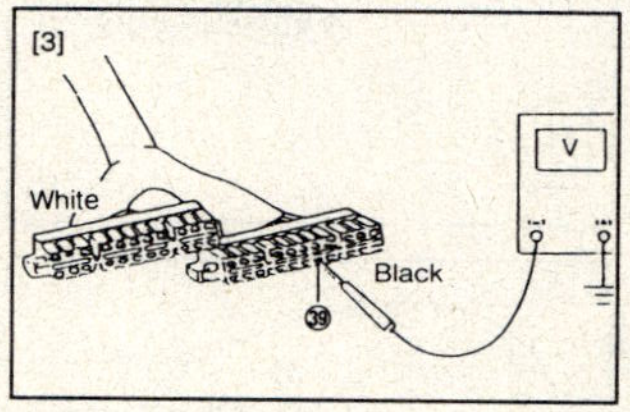

No.	Terminals	Trouble	Condition		STD voltage
2	⑤ — ㊱ Ground	No voltage	Ignition switch ON		4.5 - 5.5 V
	⑥ — ㊴ Ground	No standard voltage	Ignition switch ON	When atmospheric pressure of 760 mmHg (29.9 inchHg) exists:	3.2 - 4.0 V

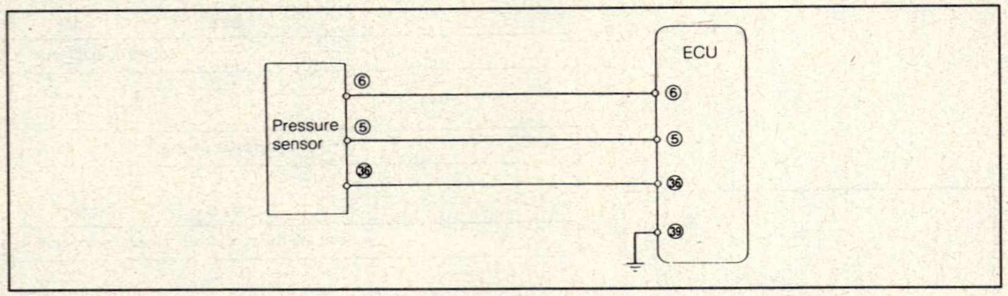

⑤ — ㊱

[1] There is no specified voltage between SST terminals ⑤ and ㊱.

[2] Check that there is specified voltage between SST terminals ① or ㉒ and ㊴.
- BAD → Check main relay circuit.
  - BAD → Repair or replace.
- OK → Check wiring between ECU and pressure sensor.
  - BAD → Repair or replace.
  - OK → Check pressure sensor.
    - BAD → Repair or replace.
    - OK → Check ECU.

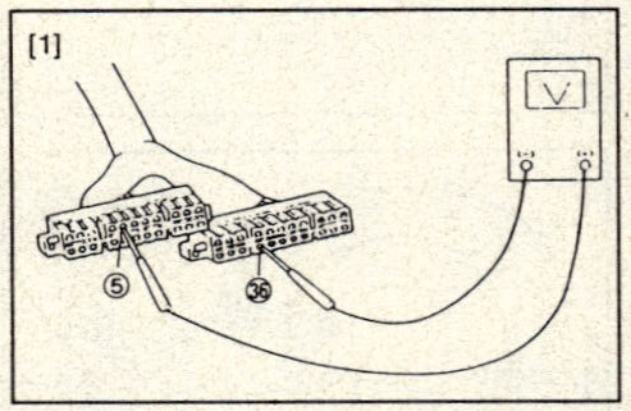

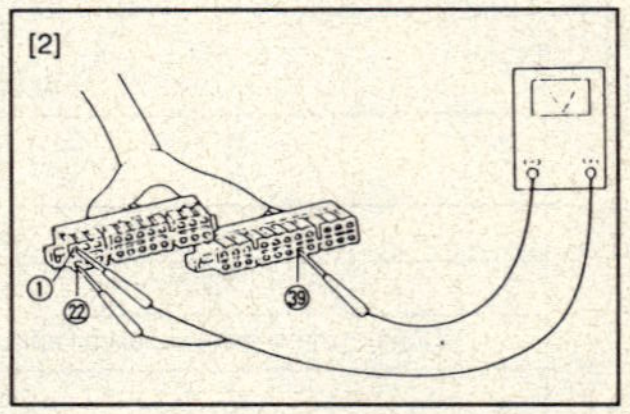

⑥ — ㊴

[1] There is no specified voltage between SST terminals ⑥ — ㊴.

[2] Check that there is specified voltage between SST terminals ⑤ and ㊱.
- BAD → ⑤ — ㊱ trouble section.
  - BAD → Repair or replace.
- OK → Check wiring between ECU and pressure sensor.
  - BAD → Repair or replace.
  - OK → Check pressure sensor.
    - OK → Check ECU.

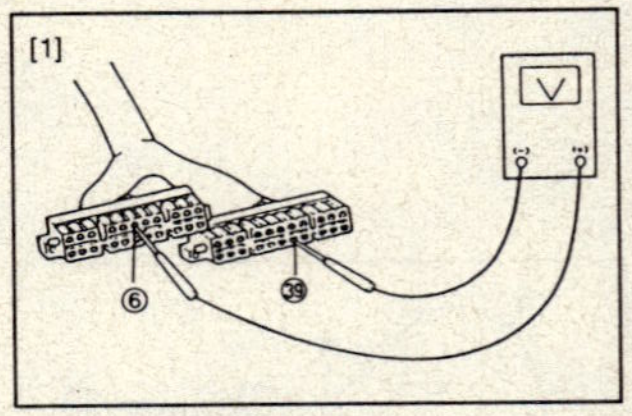

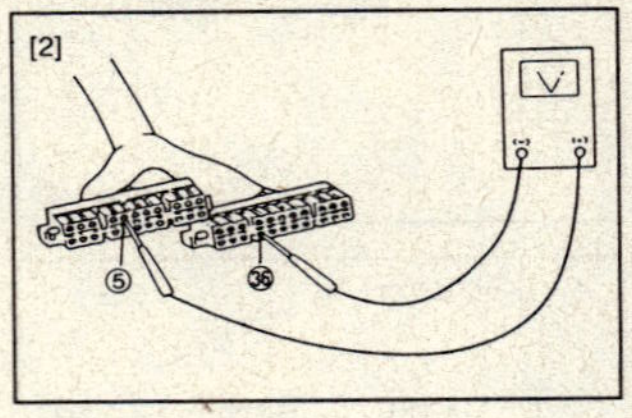

No.	Terminals	Trouble	Condition	STD voltage
3	③ — ㊴	No voltage	Ignition switch ON (Engine stopped state)	Approx. battery voltage

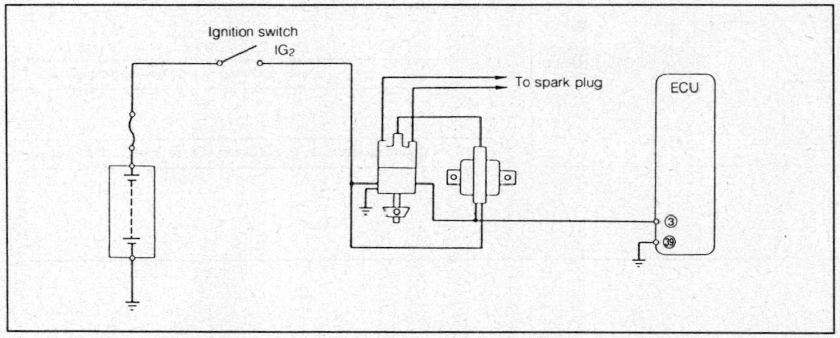

③ — ㊴

There is no voltage between SST terminals ③ and ㊴

→ Check wiring between distributor connector positive terminals (+) and SST terminal ③. (Disconnect the IG coil connector.)
- BAD → Repair or replace.
- OK → Check ignition system
  - BAD → Repair or replace.
  - OK → Check ECU.

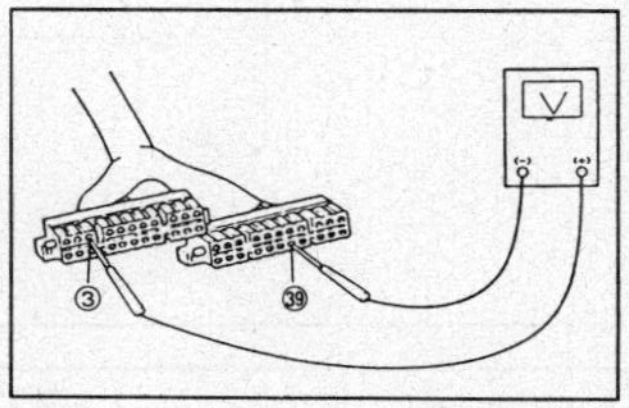

No.	Terminals	Trouble	Condition		STD voltage
4	⑦ — ㊴ Ground	No specified voltage	Ignition switch ON	When cooling water temperature is 80°C (176°F):	0.4 - 0.65 V

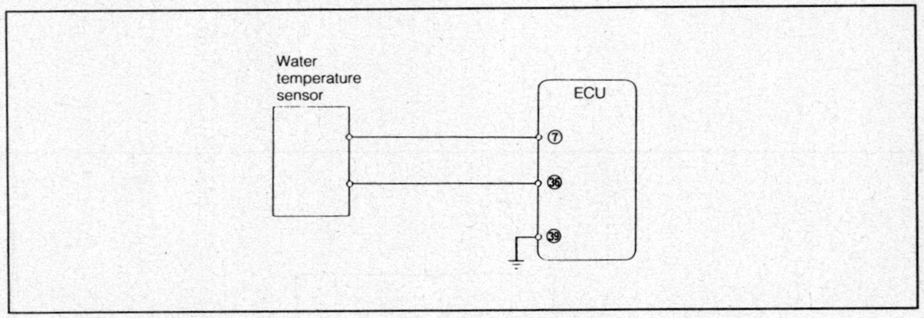

[1] There is no specified voltage between SST terminals ⑦ and ㊴.

→ [2] Check that there is voltage between SST terminals ① or ㉒ and ㊴ when ignition switch is turned ON.
- BAD → Check main relay circuit.
- OK → Check wiring between ECU and water temperature sensor.
  - BAD → Repair or replace.
  - OK → Check water temperature sensor.
    - BAD → Replace water temperature sensor.
    - OK → Check ECU.

[1]

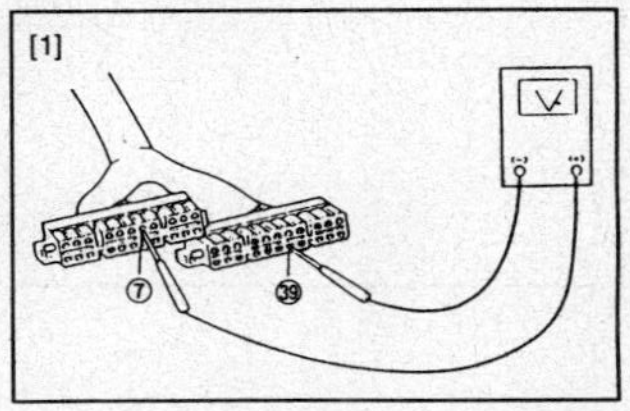

[2]

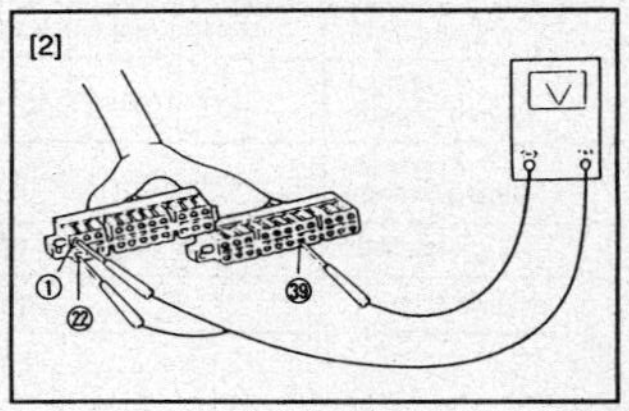

No.	Terminals	Trouble	Condition		STD voltage
5	(26) — (39) Ground	No voltage change	Ignition switch ON	When engine revolution speed is held at 3000 rpm after having warmed up engine fully:	Voltage changes more than 8 times within 10 seconds.

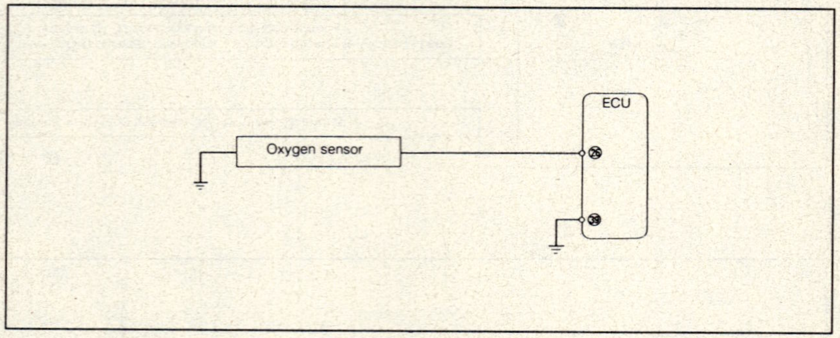

There is no change in voltage between SST terminals (26) and (39).

↓

Check wiring between ECU and oxygen sensor. — **BAD** → Repair or replace.

↓ **OK**

Check oxygen sensor.

↓ **OK**

Check ECU.

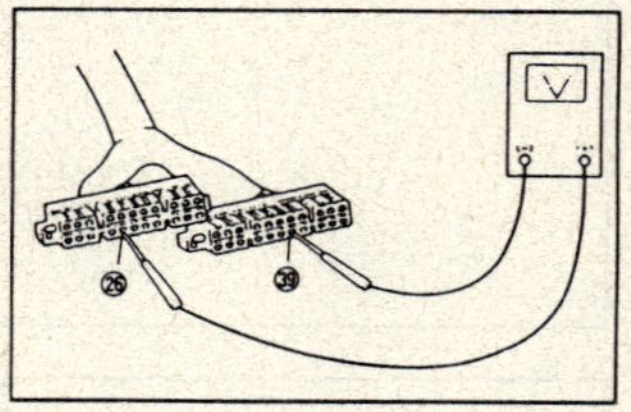

No.	Terminals	Trouble	Condition		STD voltage
7	(12) — (39) Ground	More than 5 V	Ignition switch ON	Throttle valve fully closed	Less than 5 V
		No voltage	Ignition switch ON	Throttle valve fully opened	Approx. battery voltage
	(32) — (39) Ground	No voltage	Ignition switch ON	Throttle valve fully closed	Approx. battery voltage
		More than 5 V	Ignition switch ON	Throttle valve fully opened	Less than 5 V

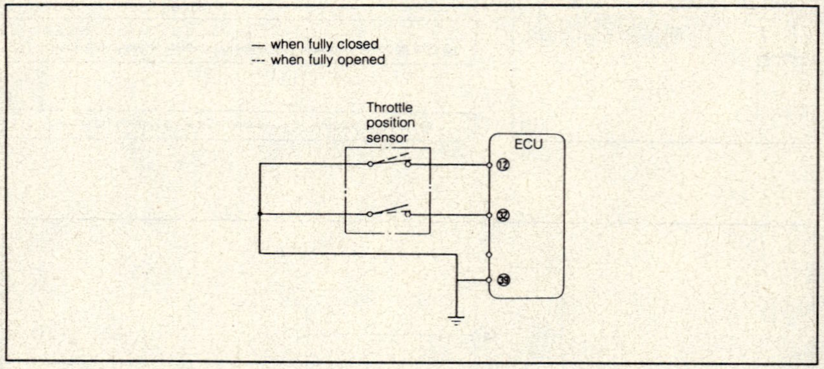

FUEL INJECTION SYSTEM TROUBLESHOOTING PROCEDURES—ROCKY

⑫ — ㊴

[1] There is no specified voltage between SST terminals ⑫ and ㊴

↓

[2] Check that there is voltage between SST terminals ○ or ㉒ and ㊴ when ignition switch is turned ON.

- BAD → Check main relay circuit.
- OK ↓

Check throttle position sensor.

- BAD → Replace.
- OK ↓

Check wiring between ECU and throttle position sensor.

- BAD → Repair or replace.
- OK ↓

Check ECU

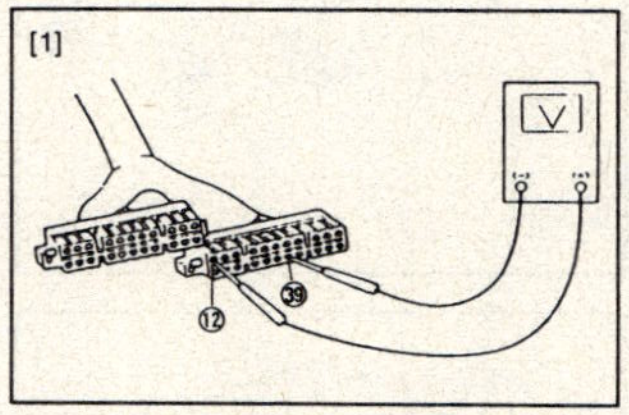

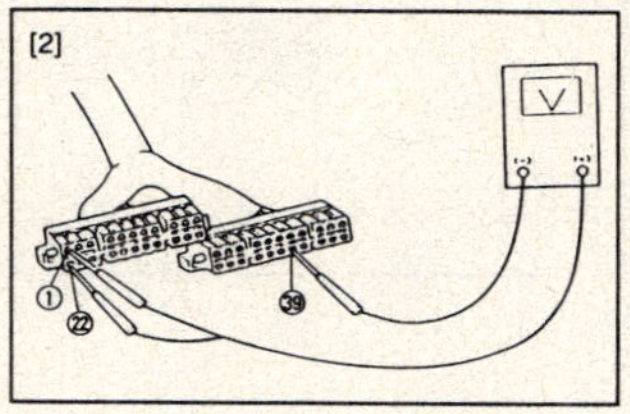

㉜ — ㊴

[1] There is no specified voltage between SST terminals ㉜ and ㊴.

↓

[2] Check that there is voltage between SST terminals ① or ㉒ and ㊴ when ignition switch is turned ON.

- BAD → Check main relay circuit.
- OK ↓

Check throttle position sensor.

- BAD → Replace.
- OK ↓

Check wiring between ECU and throttle position sensor.

- BAD → Repair or replace.
- OK ↓

Check ECU.

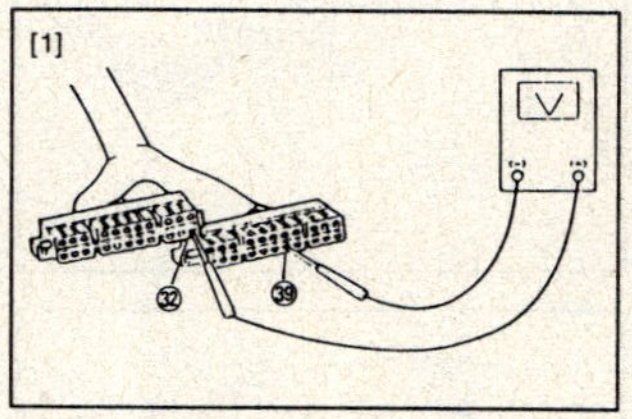

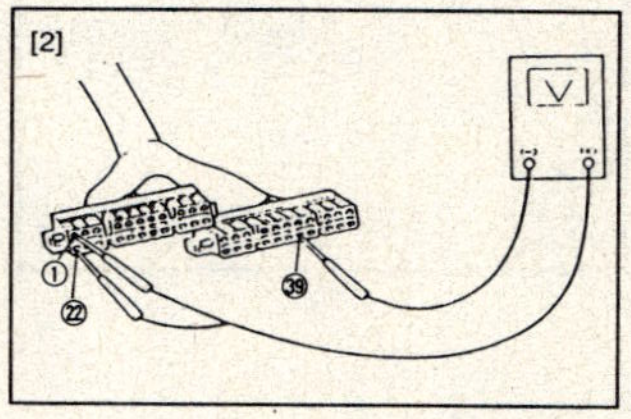

FUEL INJECTION SYSTEM TROUBLESHOOTING PROCEDURES—ROCKY

No.	Terminals	Trouble	Condition		STD voltage
8	㉗ — ㊴ Ground	No specified voltage	Ignition switch ON	When air temperature inside intake manifold is 20°C (68°F):	1.5 - 3.0 V

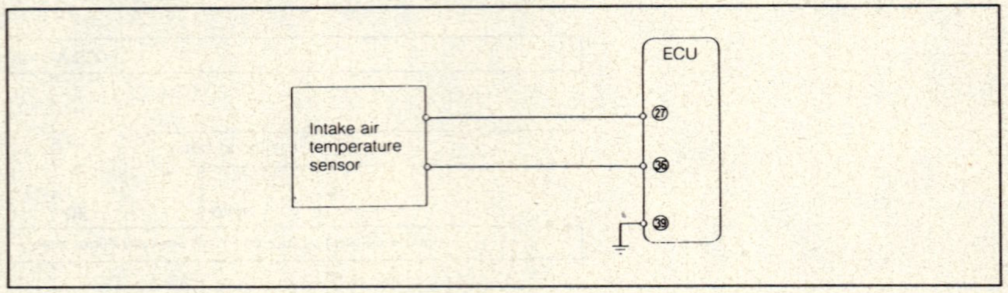

[1] There is no specified voltage between SST terminals ㉗ and ㊴.

[2] Check that there is voltage between SST terminals ① or ㉒ and ㊴ when ignition switch is turned ON.

- **OK** → Check intake air temperature sensor.
- **BAD** → Check main relay circuit.

Check intake air temperature sensor.

- **OK** → Check wiring between ECU and intake air temperature sensor.
- **BAD** → Replace intake air temperature sensor.

Check wiring between ECU and intake air temperature sensor.

- **OK** → Check ECU.
- **BAD** → Repair or replace.

Check ECU.

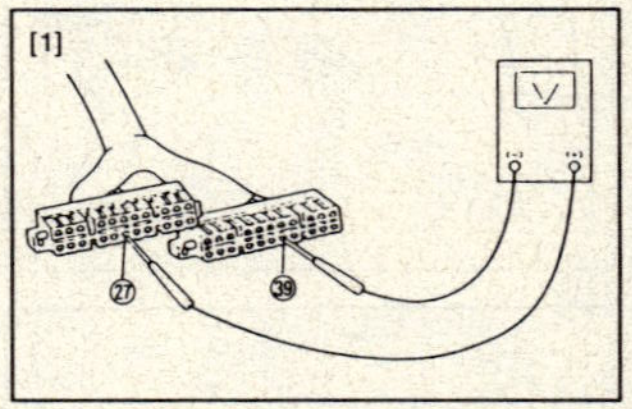

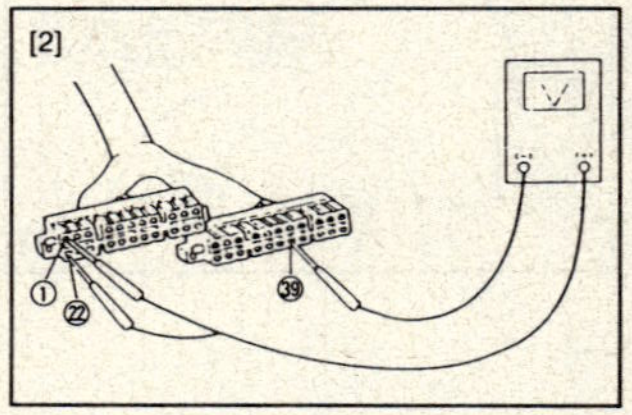

No.	Terminals	Trouble	Condition		STD voltage
9	⑧ — ㊴ Ground	No voltage changes	Ignition switch ON	When vehicle is moved slowly:	0 - Approx. battery voltage

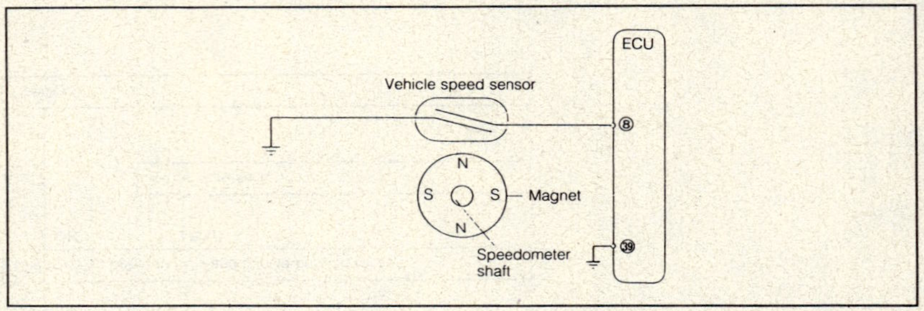

[1] No change in voltage occurs between SST terminals ⑧ and ㊴ when vehicle is moved slowly with ignition switch turned ON.

[2] Check that there is voltage between SST terminals ① or ㉒ and ㊴ when ignition switch is turned ON.

- **OK** → Check wiring between ECU and vehicle speed sensor.
- **BAD** → Check main relay circuit.

Check wiring between ECU and vehicle speed sensor.

- **OK** → Check wiring between vehicle speed sensor and body ground.
- **BAD** → Repair or replace.

Check wiring between vehicle speed sensor and body ground.

- **OK** → Check vehicle speed sensor.
- **BAD** → Repair or replace.

Check vehicle speed sensor.

- **OK** → Check ECU.
- **BAD** → Replace.

Check ECU.

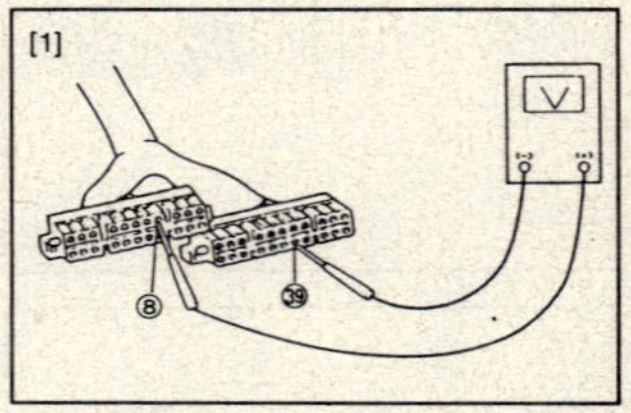

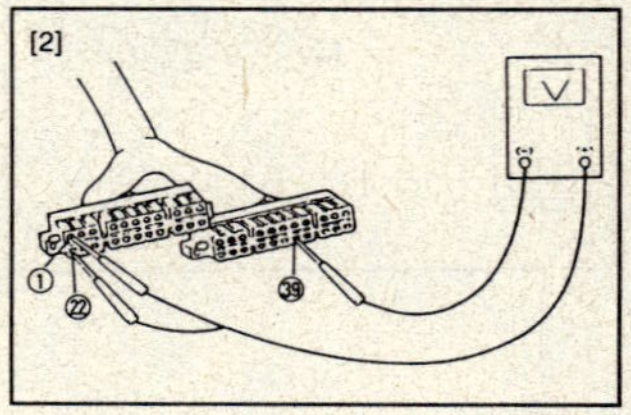

FUEL INJECTION SYSTEM TROUBLESHOOTING PROCEDURES—ROCKY

No.	Terminals	Trouble	Condition	STD voltage
10	13 — 39 Ground	No voltage	Ignition switch set to ST position	More than 6 V

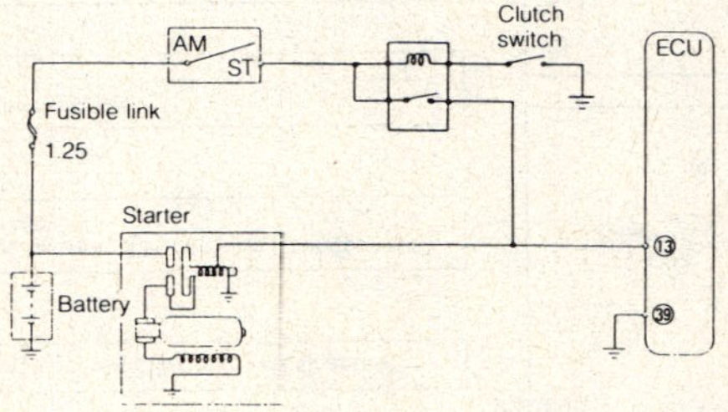

There is no voltage between SST terminals 13 and 39 when ignition switch is set to ST position.

Check starter operation.

BAD → Check battery, fusible link ignition switch, starter, starter relay and shift position switch or clutch switch.

OK → Check wiring between ECU and starter terminals.

- **BAD** → Repair or replace.

Check battery, fusible link ignition switch, starter, starter relay and shift position switch or clutch switch.

- **OK** → Check ECU.
- **BAD** → Repair or replace.

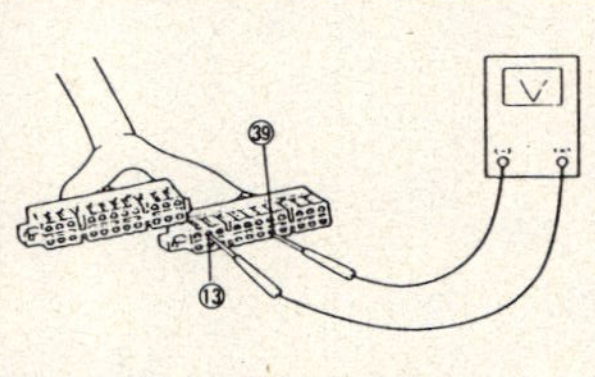

No.	Terminals	Trouble	Condition		STD voltage
11	34 — 39 Ground	No voltage	When engine is operating and compressor magnet clutch of air conditioner is energized:		Approx. battery voltage
	12 — 39 Ground	More than 5 V	Ignition switch ON	Throttle valve fully closed	Less than 5 V
		No voltage		Throttle valve fully opened	Approx. battery voltage

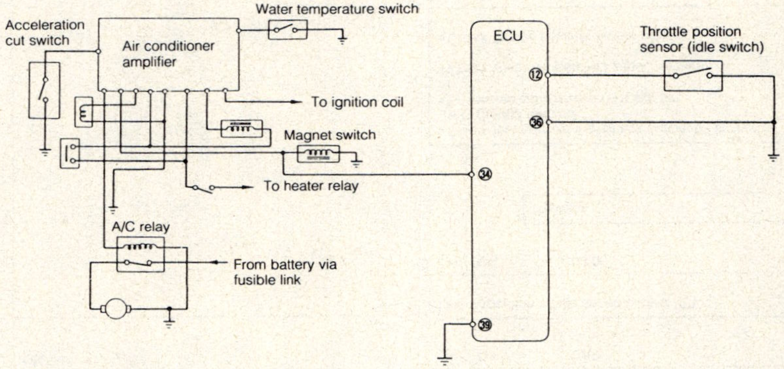

㉞ — ㊴

- There is no specified voltage between SST terminals ㉞ and ㊴ when magnet clutch is operated.
  - Check magnet clutch operation.
    - **BAD** → Check air conditioner system.
      - **OK** → Check ECU.
      - **BAD** → Repair.
    - **OK** → Check wiring between ECU terminal ㉞ and air conditioner amplifier terminal.
      - **BAD** → Repair or replace.

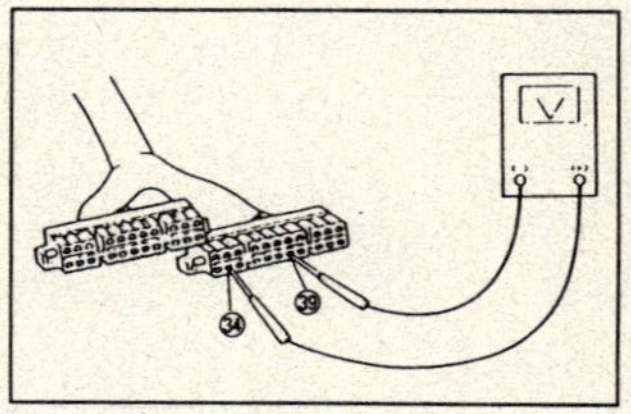

⑫ — ㊴

- [1] There is no specified voltage between SST terminals ⑫ and ㊴.
  - [2] Check that there is voltage between SST terminals ① or ㉒ and ㊴ when ignition switch is turned ON.
    - **BAD** → Check main relay circuit.
    - **OK** → Check throttle position sensor.
      - **BAD** → Replace throttle position sensor.
      - **OK** → Check wiring between throttle position sensor and body ground.
        - **BAD** → Repair or replace.
        - **OK** → Check ECU.

[1]

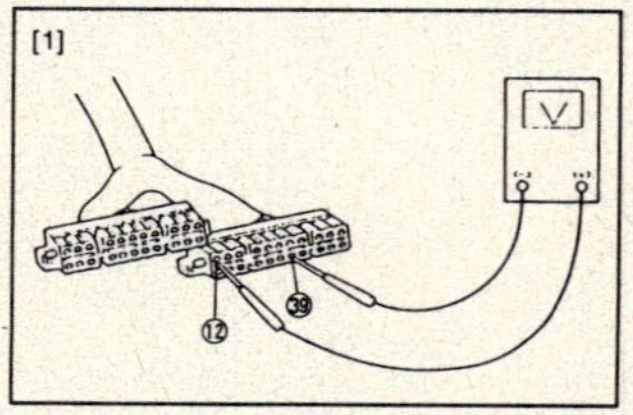

[2]

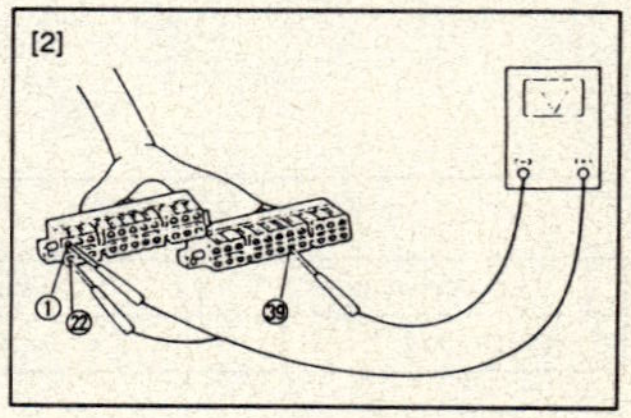

3. Perform the inspection according to the following flow chart.

- Diagnosis code check
  - **Malfunction code** → Perform trouble shooting according to diagnosis codes.
  - **Normal code** → Ensure that engine idle speed conforms to specification. Idle speed: 850 ± 50rpm
    - **BAD** → Check ignition timing. Adjust idle speed.
      - **OK** → (go to "Ensure that idle speed conforms to specification when headlamp switch is turned ON.")
      - **BAD** → Unit inspection of idle speed control VSV
        - **BAD** → Replace.
        - **OK** →
          - • Check that terminal T of check connector is shorted with ground terminal.
          - • Check vehicle speed sensor system.
          - • Check water temperature sensor system.
          - • Check throttle position sensor system.
            - **BAD** → Repair or replace.
            - **OK** → Check ECU.
    - **OK** → Ensure that idle speed conforms to specification when headlamp switch is turned ON.
      - **BAD** → Check wiring from headlamp at left side to ECU terminal ⑨.
        - **BAD** → Repair or replace.
        - **OK** → Check ECU.
      - **OK** → Turn OFF headlamp switch.

(To be continued)

FUEL INJECTION SYSTEM TROUBLESHOOTING PROCEDURES—ROCKY

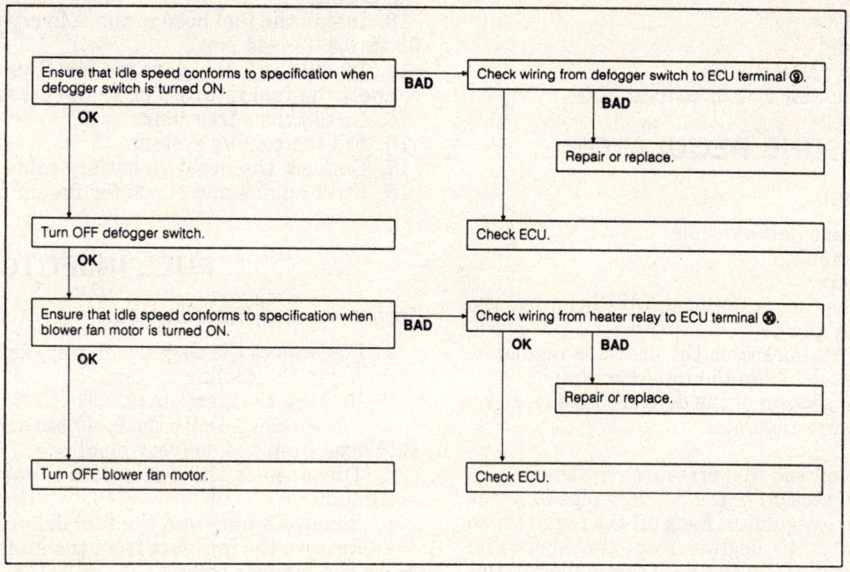

Component Replacement

SERVICE PRECAUTIONS

Do not allow sparks or open flame near during fuel system repair or testing. Wear eye protection and have a fire extinguisher at hand. Relive residual fuel system pressure before removing any fuel system parts.

FUEL PRESSURE RELIEF

1. Disconnect the negative battery terminal and remove the fuel filler cap.
2. Wrap a rag around fuel filter union bolt area.
3. Slowly unbolt the fuel filter union fitting and catch the fuel with the rag.

COLD START INJECTOR

Removal and Installation

1. Disconnect the negative battery cable.
2. Disconnect the connector of the cold start injector.
3. Place a cloth under the cold start injector and loosen the union bolt.
4. Remove the fuel hose and the cold start injector.

To Install:

5. Install the cold start injector to the surge tank using a new gasket and torque to 3–5 ft. lbs. (4–7 Nm).
6. Install the fuel hose with new gaskets to the cold start injector and torque the delivery pipe to in 3 stages to 7–11 ft. lbs. (10–15 Nm).
7. Connect the cold start injector connector.
8. Connect the negative battery cable and check for fuel leakage.

SURGE TANK

Removal and Installation

NOTE: Tag all hoses and electrical connections to use as reference during reassembly.

1. Disconnect the negative battery cable.
2. Drain the cooling system.
3. Remove the accelerator cable from the throttle body.
4. Remove the air cleaner assembly.
5. Remove oil cooler hose and water pump hose from the throttle body.
6. Remove the brake booster hose from the intake manifold.
7. Disconnect the vacuum hose connected to the air cleaner.
8. Disconnect the vacuum hose from the atmosphere pressure VSV.
9. Remove the vacuum hose from the pressure regulator.
10. Disconnect the vacuum hose connected to the intake manifold BVSV.
11. Disconnect the vacuum hose of the idle-up VSV, if equipped with air conditioning.
12. Disconnect the idle-up VSV, EGR VSV, intake air temperature sensor, throttle position sensor and cold start injector.
13. Disconnect the blow-by gas hose from the surge tank.
14. Remove the fuel hose from the cold start injector.
15. Remove the surge tank support.
16. Remove the wire retaining clamp from the surge tank.
17. Remove the EGR pipe.
18. Remove the surge tank.

To Install:

19. Install the surge tank to the intake manifold with a new gasket.
20. Torque the bolts evenly in 3 stages to 11–16 ft. lbs. (15–21 Nm).
21. Reconnect the blow-by gas, air conditioner fast idle, BVSV vacuum, pressure regulator vacuum, atmosphere VSV, air cleaner and brake booster hoses.
22. Reconnect the cold start injector, throttle position sensor, intake air temperature sensor, EGR VSV and idle-up VSV electrical connectors.
23. Install the surge tank support.
24. Install the clamp for the wiring at the back of the surge tank.
25. Install the oil cooler hose and water pump hose to the throttle body.
26. Install the air hose to the throttle body.
27. Install the accelerator cable to the throttle body and torque the locknut to 7–12 ft. lbs. (10–15 Nm).

28. Check that the accelerator cable free-play is 0.12–0.31 in. (3–8mm).
29. Install the EGR pipe.
30. Fill the cooling system.
31. Connect the negative battery cable.
32. Start engine and check for fuel or coolant leaks.

FUEL PRESSURE REGULATOR

Removal and Installation

1. Disconnect the negative battery cable.
2. Drain the cooling system.
3. Remove the surge tank.
4. Place a cloth under the fuel hose and slowly disconnect the fuel hose from the delivery pipe.
5. Remove the fuel return hose from the pressure regulator.
6. Remove the delivery pipe from the intake manifold.
7. Clamp the hexagonal section of the delivery pipe in a vise and remove the fuel pressure regulator.

To Install:

8. Install a new O-ring on the fuel pressure regulator.
9. Clamp the hexagonal section of the delivery pipe in a vise and install the fuel pressure regulator. Back off the regulator so the vacuum sensing pipe is 45 degrees from the horizontal plane, facing toward the rear of the engine. Torque the locknut to 25–32 ft. lbs. (34–44 Nm).
10. Install a new O-ring on the injector, apply gasoline to O-ring and install to the delivery pipe.
11. Torque the delivery pipe nuts in 3 stages to 7–11 ft. lbs. (10–15 Nm).
12. Check that the injector turns freely by hand; if not, replace the O-ring.
13. Install the fuel hose to the delivery pipe and torque to 25–33 ft. lbs. (34–44 Nm).
14. Install the fuel hose to the fuel filter with a new gasket and connect the fuel return pipe to the pressure regulator.
15. Install the surge tank.
16. Fill the cooling system.
17. Connect the negative battery cable.
18. Start engine and check for fuel or coolant leaks.

FUEL INJECTOR

Removal and Installation

1. Disconnect the negative battery cable.
2. Drain the cooling system.
3. Remove the surge tank.
4. Place a cloth under the fuel hose and slowly disconnect the fuel hose from the delivery pipe.
5. Disconnect the fuel return hose from the pressure regulator.
6. Remove 3 nuts and the fuel delivery pipe.
7. Remove the injectors from the intake manifold. Do not remove the injector cover.

To Install:

8. Install a new O-ring on the fuel injector.
9. Clamp the hexagonal section of the delivery pipe in a vise and install the fuel pressure regulator. Back off the regulator so the vacuum sensing pipe is 45 degrees from the horizontal

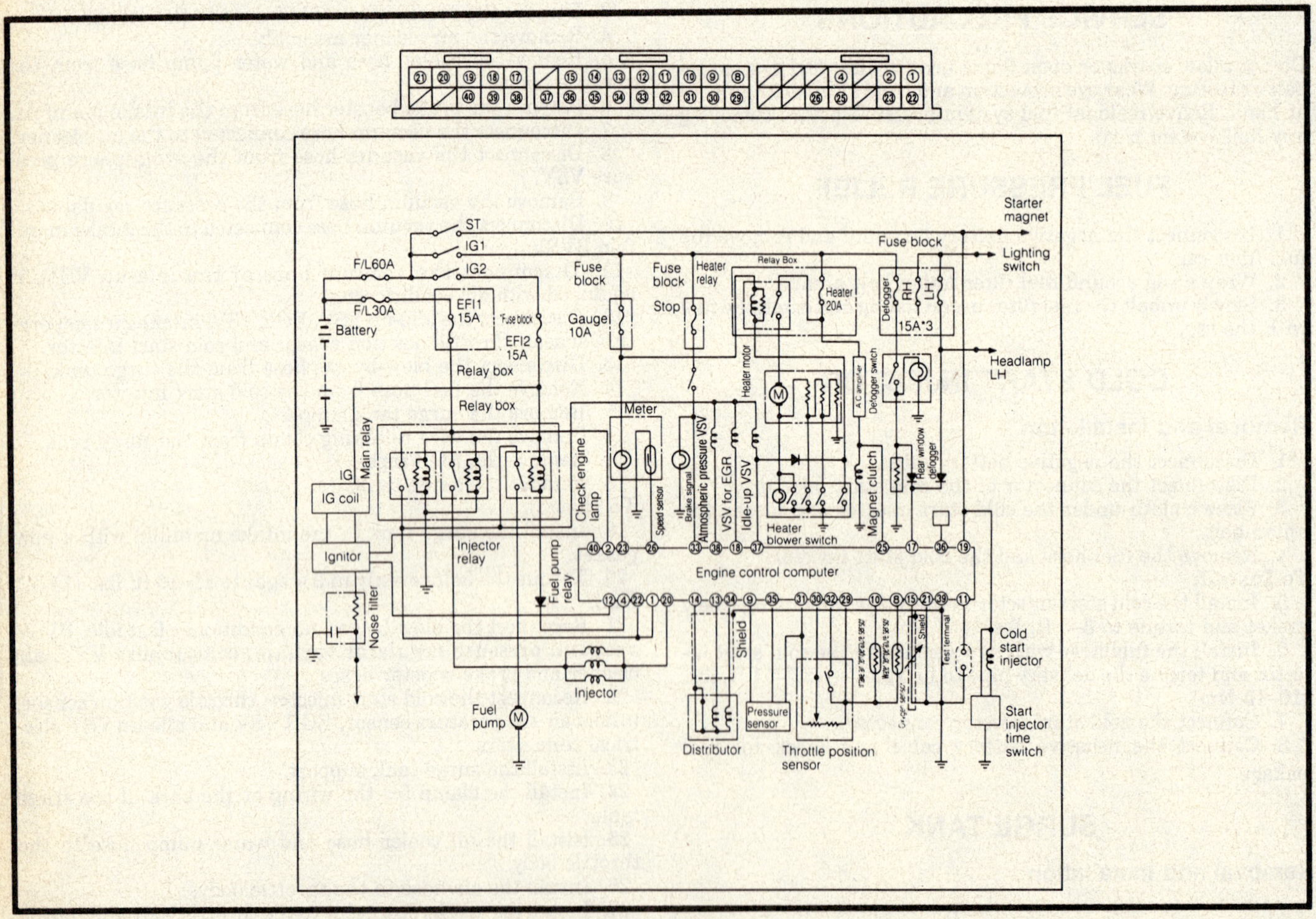

Fuel injection system wiring—Charade G100 with 1.0L engine

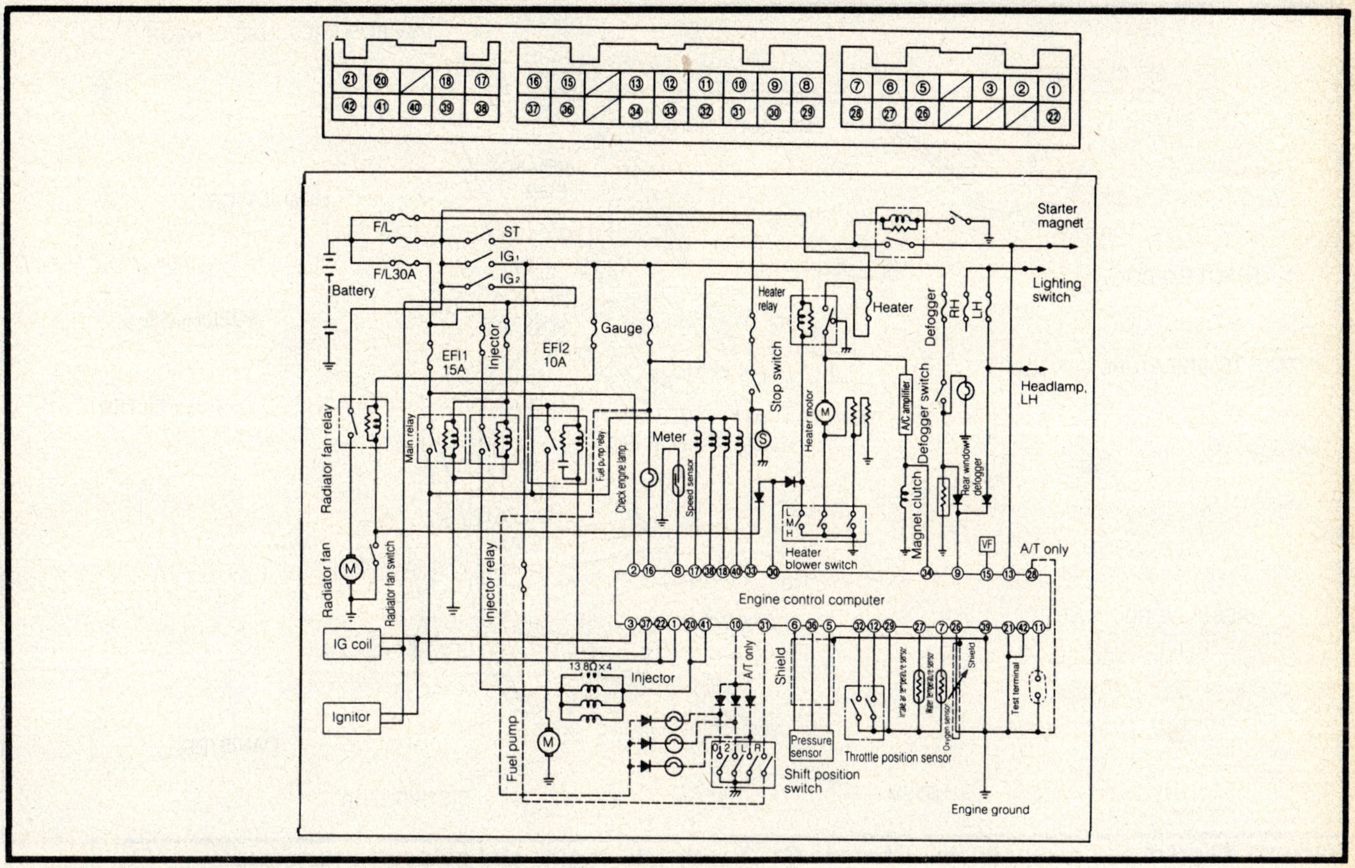

Fuel injection system wiring—Charade G102 with 1.3L engine

plane, facing toward the rear of the engine. Torque the locknut to 25–32 ft. lbs. (34–44 Nm).

10. Check the injector vibration insulator for damage. Install the injector insulator to the intake manifold.
11. Remove the O-ring from the injector, check the grommet for damaged or flatten condition. Replace if necessary.
12. Install a new O-ring onto the injector.
13. Insert the injector into the intake manifold.
14. Check the delivery pipe heat insulator for damage. Install insulator, replace with new, if necessary.
15. Apply gasoline to the O-ring of the injector and install the delivery pipe. Make certain the delivery pipe is connected to the proper injector.
16. Torque the delivery pipe nuts in 3 stages to 7–11 ft. lbs. (10–15 Nm).
17. Check that the injector turns freely by hand; if not, replace the O-ring.
18. Connect the injector connector.
19. Connect the return hose to the pressure regulator and return pipe. Attach the clip.
20. Install the fuel hose to the fuel delivery pipe with new gaskets and torque to 25–33 ft. lbs. 34–44 Nm).
21. Install the fuel hose to the fuel filter with new gaskets and torque to 25–32 ft. lbs. (34–44 Nm).
22. Install the surge tank.
23. Fill the cooling system.
24. Connect the negative battery cable.
25. Start engine and check for fuel or coolant leaks.

THROTTLE BODY

NOTE: Tag all hoses and electrical connections to use as reference during reassembly.

1. Disconnect the negative battery cable.
2. Drain the cooling system.
3. Remove the accelerator cable from the throttle body.
4. Disconnect the throttle position sensor.
5. Disconnect the air cleaner hose from the throttle body.
6. Disconnect the water hoses from the throttle body.
7. Remove 2 throttle body nuts and 2 throttle body bolts.
8. Remove the throttle body and gasket.

To Install:

9. Install the throttle body to the surge tank with a new gasket and torque the nuts and bolts to 11–16 ft. lbs. (15–21 Nm).
10. Connect the water hose to the throttle body and attach the clip.
11. Connect the air cleaner hose to the throttle body and attach the clip.
12. Connect the connector of the throttle position sensor.
13. Connect the vacuum hoses to the throttle body.
14. Install the accelerator cable to the throttle body and torque the locknut to 7–12 ft. lbs. (10–15 Nm).
15. Check that the accelerator cable free-play is 0.12–0.31 in. (3–8mm).
16. Install the EGR pipe.
17. Fill the cooling system.
18. Connect the negative battery cable.
19. Start engine and check for fuel or coolant leaks.

THROTTLE POSITION SENSOR

The throttle position sensor is not adjustable or replaceable. If a defect is found, the factory recommends the replacement of the throttle body.

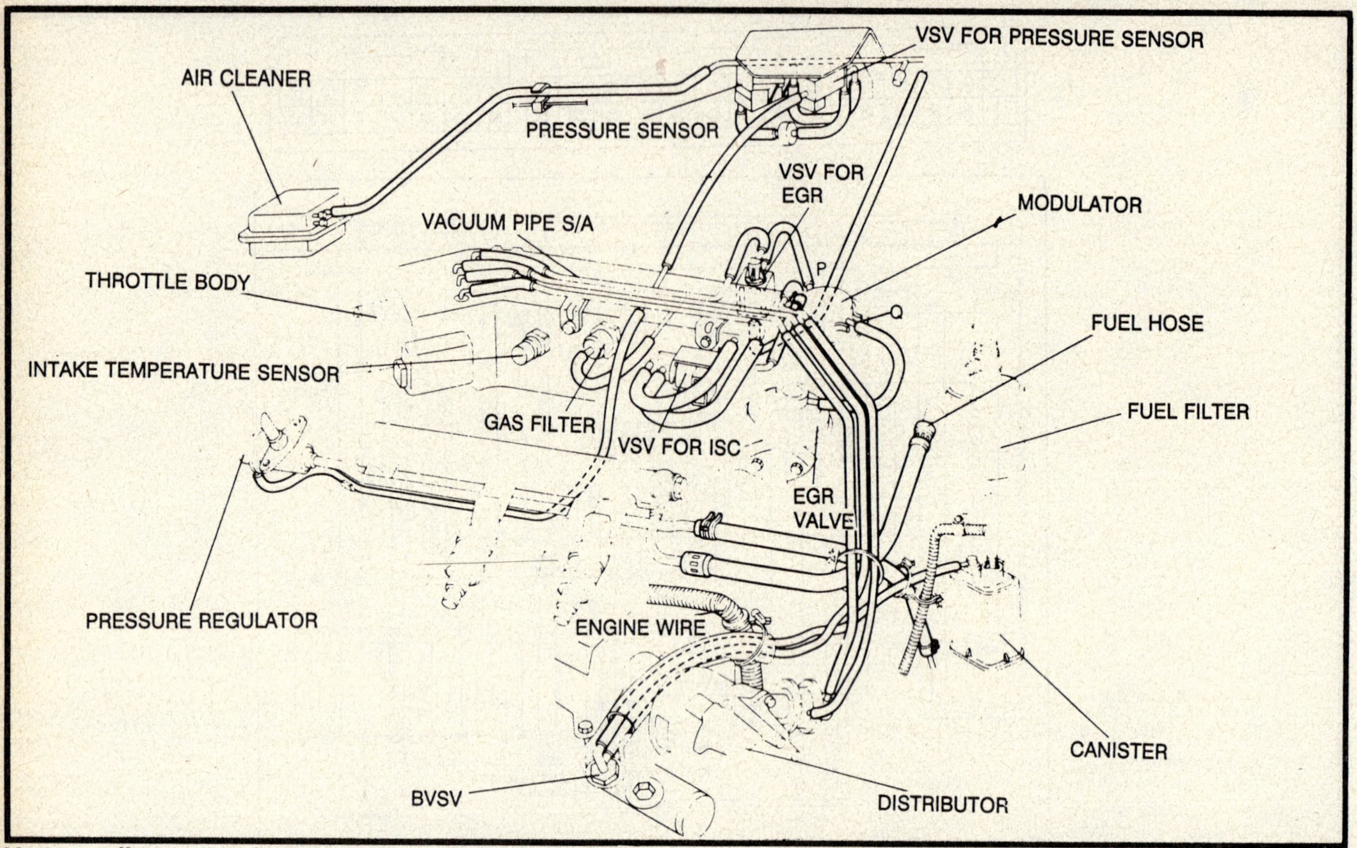

Vacuum diagram and components – Charade G102 with 1.3L engine and automatic transaxle

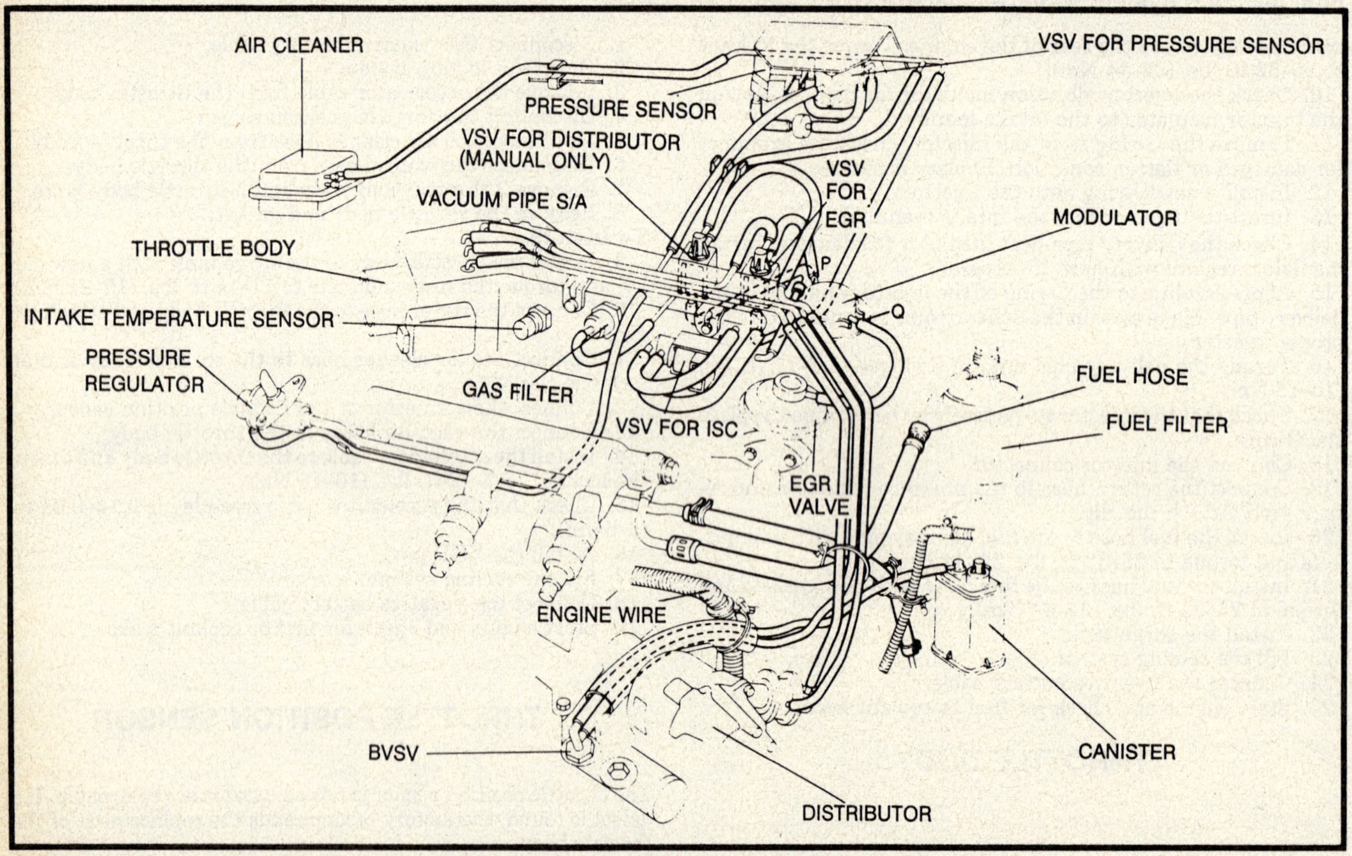

Vacuum diagram and components – Charade G102 with 1.3L engine and manual transaxle

Honda Fuel Injection

SECTION 8

HONDA PROGRAMMED FUEL INJECTION (PGM-FI) SYSTEM

General Information

Programmed Fuel Injection (PGM-FI) System is a fully electronic microprocessor based engine management system. The Electronic Control Unit (ECU) is given responsibility for control of injector timing and duration, intake air control, ignition timing, cold start enrichment, fuel pump control, fuel cut-off, A/C compressor operation, alternator control as well as EGR function and canister purge cycles.

The ECU receives electric signals from many sensors and sources on and around the engine. The signals are processed against pre-programmed values; correct output signals from the ECU are determined by these calculations. The ECU contains additional memories, back-up and fail-safe functions as well as self diagnostic capabilities.

Diagnosis is generally based on symptom diagnosis and stored fault codes, if any. Testing always requires the use of the diagnostic charts in conjunction with the Honda test harness (pinout box), a voltmeter and an ohmmeter.

ELECTRONIC CONTROL UNIT (ECU)

In order to get fuel into the cylinders at the correct instant and in correct amount, the control system must perform various separate functions. The ECU (Electronic Control Unit), the heart of the PGM-FI, uses a microcomputer and consists of a CPU (Central Processing Unit), memories, and I/O (Input/Output) ports. Basic data stored in the memories are compared to the signals sent from the various sensors to provide the correct air/fuel mixture for all engine needs.

The unit contains memories for the basic discharge duration at various engine speeds and manifold pressures. The basic discharge duration, after being read out from the memory, is further modified by signals sent from various sensors to obtain the final injector duration. Other functions include:

- Starting Control – The fuel system must vary the air/fuel ratio to suit different operating requirements. For example, the mixture must be rich for starting. The memories also contain the injector duration values to be triggered by signals from the starter switch, engine speed and coolant temperature sensors, thereby providing extra fuel needed for starting.
- Injector Control – The ECU controls the injector durations at various engine speeds and loads by precisely controlling the amount of time the injectors are open.
- Electronic Air Control – The ECU controls the EACV to maintain correct idle speed based on engine and accessories demand.
- Ignition Timing Control – The ECU controls the basic ignition timing based on engine load, engine rpm, vehicle speed and coolant temperature.
- Fuel Pump Control – When the engine is not running but the ignition is electric current to the fuel pump is cut off, preventing the injectors from discharging fuel.
- Fuel Cut-Off Control – During deceleration with the throttle valve nearly closed, electric current to the injectors is cut off at engine speeds of approximately 1300–1500 rpm, contributing to improved fuel economy. Fuel cut-off also takes place when engine speed exceeds the red line or safe operation rpm limit.
- Fail Safe – If the ECU notes the loss or change out-of-range of a sensor signal, the computer will ignore the faulty signal and substitute a fixed value in its place. This value may not necessarily be correct for the immediate driving situation, but will allow the engine to operate. For example, many ECU decisions are based on the engine coolant temperature. Should the coolant temperature sensor (TW sensor) signal be lost just after a cold start, the ECU will substitute the default value. This substitution value is based on an engine at normal temperature so vehicle performance will be affected until fully warmed up.
- Back-up System – The ECU also tests itself while operating. If an abnormality occurs within the ECU, the system switches to a back-up circuit independent of the computing system. This system substitutes fixed values for all inputs and controls the injectors accordingly. Vehicle performance is reduced to minimal driveability. The back up function serves as a lifeboat to prevent the vehicle from being stranded in the event of ECU failure.
- Self-Diagnosis – When a abnormality occurs, the ECU lights the engine warning light and stores a failure code in erasable memory. The ECU LED will display the code any time the ignition is turned **ON**.

SENSORS AND INPUTS

CRANK, TDC and CYL Sensors

The CRANK or crankshaft angle sensor signal is used to determine the timing for fuel injection and ignition timing. It is also used to generate the engine RPM signal. The TDC sensor signal is used to determine ignition timing during start-up or when the CRANK signal is abnormal. The CYL sensor is used to detect the position of No. 1 cylinder; the signal is used to trigger the sequential fuel injection. The 1.5L Civic engine does not use a CYL sensor.

These sensors are contained within the distributor. The signals are generated by the rotation of toothed wheels passing through pick-up coils. In some cases, the CYL sensor is incorporated with the other two. The function of these components does not change, but if any single sensor is defective, the entire distributor must be replaced as an assembly.

Manifold Air Pressure (MAP) Sensor

The sensor converts manifold air pressure readings into electrical voltage signals and sends them to the ECU. This information is used along with signals from the crank angle sensor to compute the basic injector duration.

Coolant Temperature (TW) Sensor

The sensor uses a temperature-dependent diode (thermistor) to measure differences in the coolant temperature. The basic injector duration is partially based on the signals sent from this sensor through the ECU. The resistance of the thermistor decreases with a rise in coolant temperature.

Intake Air Temperature (TA) Sensor

This device is also a thermistor and is placed in the intake manifold. It acts much like the water temperature sensor but with a reduced thermal capacity for for quicker response. The injector duration determined by the ECU is altered for different operating conditions by the signals sent from this sensor.

Throttle Angle Sensor

This sensor is a potentiometer which translates the position of the throttle plate to an electrical signal. The signal is near zero at idle and increases to just under 5 volts at wide-open throttle. The sensor is mounted to the side of the throttle body.

Oxygen Sensor

The oxygen sensor detects the oxygen content in the exhaust gas and sends an electrical signal to the ECU. The ECU uses this signal to control the fuel injector duration, maintaining the stoichiometric air/fuel ratio of 14.7:1. The sensor is located either in the exhaust manifold or in the exhaust piping ahead of the catalytic converter. The 2.0L carbureted 1990 Prelude uses 2 sensors.

This sensor is a hollow shaft of zirconia with a closed end. The inner and outer surfaces are plated with platinum, thus forming a platinum electrode. The inner surface or chamber is open to the atmosphere; the outer surface is exposed to the exhaust gas flow through the manifold.

Voltage is induced at the platinum electrode when there is any difference in oxygen concentration on the outside and inside. Operation of the device is dependent upon the changes in the induced voltage as the exhaust stream content changes.

Oxygen sensors only become efficient when operating at the proper temperature. Some models use a heated oxygen sensor. The heater stabilizes the sensor's output and allows the sensor to heat quicker after the engine as been started.

EGR Valve Lift Sensor

The EGR lift sensor is found on all Accords and Preludes. For Civic vehicles, it is only used on 1.5L California engines coupled to automatic transaxles. The ECU controls the function of the EGR valve through the EGR control solenoid. The lift sensor translates the movement of the EGR shaft into an electric signal which is sent back to the ECU. The control unit compares this signal to pre-programmed values for optimum function. The sensor is located on the EGR valve.

Atmospheric Pressure (PA) Sensor

Like the MAP sensor, the unit converts atmospheric pressures into voltage signals and sends them to the ECU. The signals then modify the basic injector duration to compensate for changes in the atmospheric pressure.

The PA sensor is located inside the passenger compartment of the vehicle. On Preludes, the sensor is located with, but external to, the ECU under the right front carpeting. For Civic, the sensor is behind the right front kick panel. The Accord sensor is built into the ECU and is not accessible.

Vehicle Speed Sensor (VSS)

The signal from the vehicle speed sensor is used by the ECU as well as the cruise control and/or automatic transmission control units. On Civics, the signal is generated by a pulser unit incorporated into the back of the speedometer/odometer unit.

Accord and Prelude use an external sensor mounted on top of the power steering speed sensor in the engine compartment. The power steering speed sensor is gear-driven by the engine and controls the steering boost in relation to the speed of the vehicle. An electric pulser unit is mounted on top of this unit and is coupled to it by a small shaft. The motion from the lower unit is transferred to the upper unit; the resultant electrical signal is sent to the appropriate control units.

Starter Signal

The **START** position on the ignition switch causes a signal to be sent to the ECU. During cranking, the ECU will increase the amount of fuel injected into the manifold according to the engine temperature. The amount of fuel injected is gradually reduced when the starter switch is turned **OFF**..

Alternator FR Signal

The ECU receives a signal from the alternator when the system is charging.

Air Conditioning Signal

The load on the engine increases when the compressor is engaged; the ECU must increase the idle as necessary. In some cases, the ECU will limit A/C operation, cutting it off through the A/C control relay in cases of wide-open throttle or preventing it from engaging in cold engine conditions. If the vehicle has

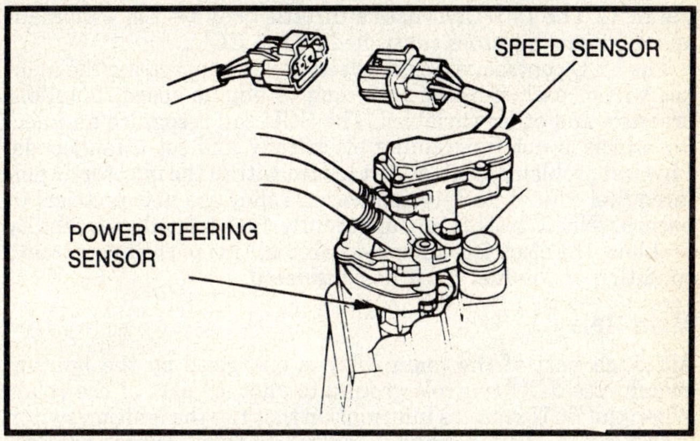

Accord and Prelude speed sensor. Civic sensor is within speedometer unit

a separate control unit for the air conditioning, the two controllers exchange information regarding system status..

Battery Voltage (IGN 1)

A 12 volt signal controlled by the IGN 1 (ON or RUN) position of the ignition switch is sent to the ECU..

Automatic Transaxle Shift Position Signal

This signal is sent to the ECU through the neutral safety switch. The control unit will not allow the engine to start if the shift selector is not in **N** or **PARK**.

Power Steering Fluid Pressure Switch

This switch is used only on Accord and Prelude. When the power steering fluid pressure exceeds a predetermined level, the ECU is signaled by a pressure operated switch. If the idle is below a predetermined rpm, as during a parking maneuver, the control unit will increase the idle speed of the engine to compensate for the additional load.

Electric Load Detector (ELD)

This sensor is used on Accord and Civic only and is located in the underhood fuse and relay box. It detects the present of moderate to large electrical current in the system and sends a signal to the ECU.

The ECU will control the idle speed to compensate for the additional load of headlights, rear defroster, heater fan or similar heavy electrical loads.

The ELD unit is integral with the fuse box in which it is mounted; should the sensor fail, the fuse box assembly must be replaced.

Brake Switch Signal

This signal is sent to the Accord and Civic ECU from the brake switch.

ECU CONTROLLED OUTPUTS

The prime mission of the ECU is to maintain good driveability and proper emission levels under all operating conditions. The ECU interprets and compares the varied input signals from the sensors and uses the resulting calculations to control electrical outputs. In many cases, the circuit control provided by the ECU closes the ground rather than the power side of the circuit.

Fuel Injectors

All models except the Civic with 1.5L engine use 4 fuel injectors in the intake manifold (port injection) controlled individually by the ECU. The 1.5L Civic uses a throttle body system with main and auxiliary injectors controlled by the ECU.

The ECU controls the system electrically, energizing the magnet within each injector according to engine speed, manifold pressure and other variables. The ECU can recognize an injector which is not functioning electrically and set a fault code. Physical problems, such as dirt lodged within the injector or polluted fuel cannot be detected because they are not electrical in nature. When tracing an apparently fuel-related driveability problem, the electrical and physical condition of the injector and condition of the fuel must be considered.

Main Relay

Although part of the main relay is energized by the ignition switch, the ECU controls ground to another part of the relay. When the ECU receives information that the the ignition switch is **ON** and a either a cranking or engine rpm signal is present, the ECU closes the relay circuit for the fuel pump system.

CHECK ENGINE Warning Lamp

The dashboard warning lamp will be lit by the ECU for 2 seconds during the initial ignition **ON** sequence. After this test sequence has ended, the ECU will illuminate the lamp anytime an electrical fault is noted within the system.

The dash warning lamp will stay on for the duration of the ignition **ON** cycle. Once the ignition is switched **OFF**, then restarted, the dashboard warning lamp will not light unless the original fault or some other (newer) fault is still present. Although the lamp does not re-light, the stored code for the original fault will remain in memory with the ignition **OFF**. For this reason, diagnosis should be based on the presence of fault codes, not the illumination of the warning lamp.

Electronic Air Control Valve (EACV)

The EACV controls the amount of air bypassing the throttle plate in response to signals from the ECU. During conditions of cold engine operation, A/C compressor engaged, automatic transaxle in gear, or alternator charging, the EACV is energized to control the idle speed correctly.

All vehicles except the Civic with 1.5L engine also employ a fast idle valve controlled by a temperature-sensitive thermowax valve. On a cold engine, more air is admitted into the engine, raising the idle. Accord and Prelude 2.2L engines incorporate an additional vacuum operated air boost valve to provide more intake air during cranking.

Note that in all cases except the Civic with 1.5L engine, the external idle adjustment screw acts on an air-bleed passage, not on the throttle mechanism.

Air Conditioning Compressor Clutch Relay

When the engine control unit receives a request for air conditioning from either the Accord and Civic A/C switch or the Prelude compressor control unit, the signal to the compressor clutch relay is momentarily delayed until the ECU can enrichen the air/fuel mixture. This allows a smoother engagement of the compressor with less effect on driveability.

Purge Cut-off Solenoid Valve

When the coolant temperature is below a predetermined level, the ECU supplies voltage to the purge cut-off solenoid which cuts vacuum to the purge control valve. The coolant temperature limit or engagement point is different for each engine

Igniter

Although part of the ignition system, the igniter is triggered by a signal from the ECU. The timing of the spark, including advance, is handled by the engine controller.

EGR Control Solenoid Valve (EGR CSV)

When EGR function is necessary to provide control of NOx emissions, the ECU grounds the control solenoid valve. The solenoid allows regulated vacuum to pass to the EGR, thus precisely controlling EGR function.

Intake Control Solenoid Valve

The Accord uses an intake control system to reduce intake or induction noise. When the engine rpm is below 3500 rpm, the ECU energizes the intake control solenoid valve. The open valve allows manifold vacuum to flow to the intake control diaphragm. As a result, intake air will traverse or bypass the resonator chamber, smoothing and quieting the airflow.

Alternator Control

On Accord engines, the ECU controls the voltage generated at

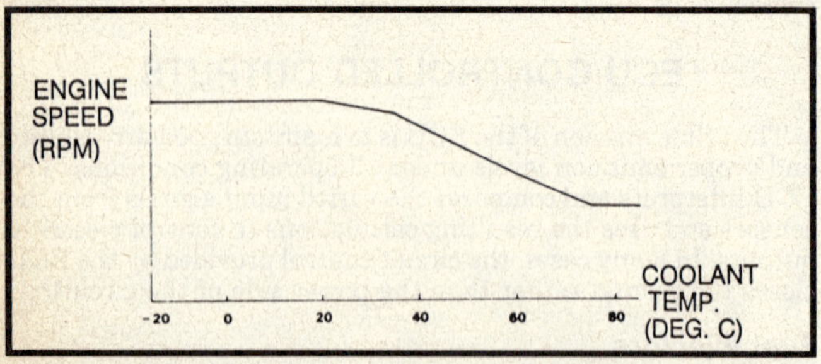

Relationship of EACV-controlled idle speed to engine temperature

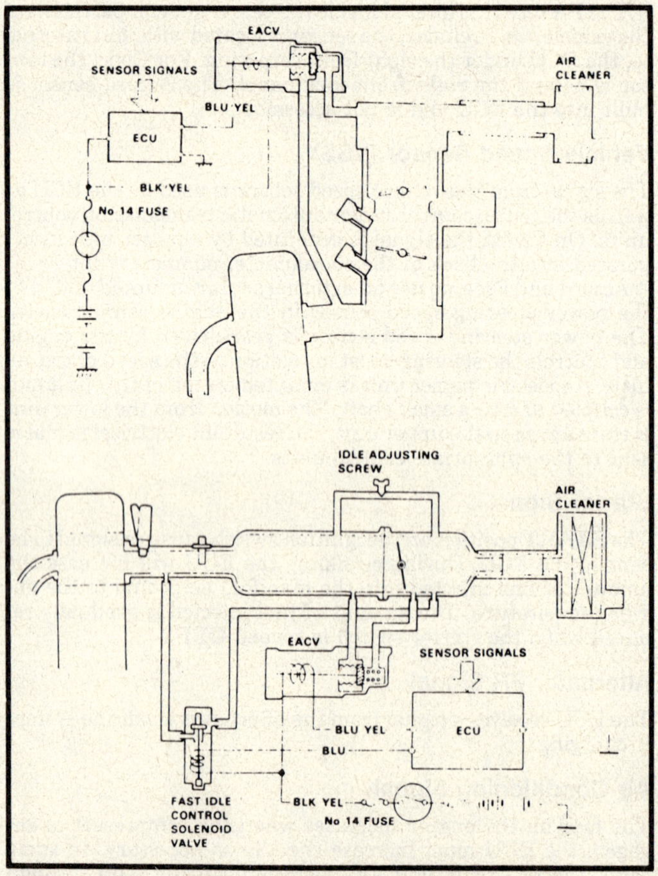

Civic EACV systems. 1.5L engine throttle body system above, 1.6L engine port injection below

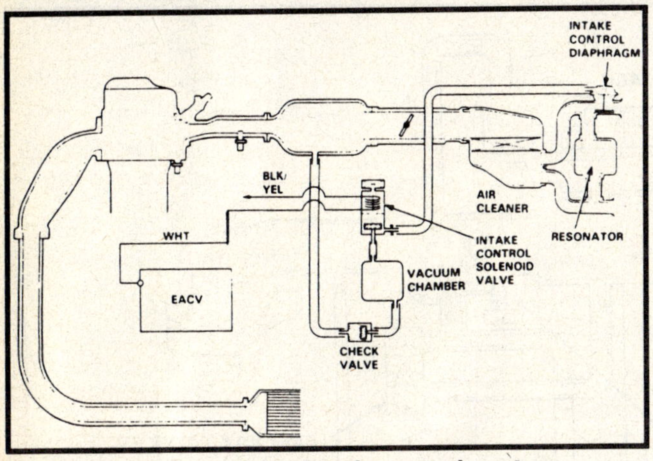

Schematic of Accord intake air control system

the alternator. By controlling power generation in proportion to electrical load and driving conditions, the load on the engine is reduced. This yields improved fuel economy.

Bypass Control Solenoid Valve (BPCSV)

Used only on the 2.0L engine in the Prelude, this system provides 2 air intake paths within the manifold. By selectively switching between them, both high torque at low rpm and high power at high rpm can be maintained.

The BPCSV is controlled by the ECU; when engine speed is below 5000 rpm, the valve is activated and intake air is flowed through the longer intake path. Above 5000 rpm, the valve is de-energized and the shorter port is opened.

Tandem Valve Control Solenoid Valve

The tandem valve is used only on 1.5L Civic engines with throttle body injection. The vacuum-operated tandem valve is used to improve the atomization of fuel injected by the main injector under changing engine conditions. By opening or closing the valve, the speed of the airflow above the injector is controlled.

The ECU controls the position of the tandem valve through the solenoid valve. The electrically-controlled solenoid valve allows venturi vacuum to pass to the tandem valve diaphragm which, in turn, moves the tandem valve.

Fast Idle Control Solenoid Valve

This system is found only on 1.6L Civic engines. It uses an ECU controlled solenoid to bypass air around the throttle plate in extremely cold conditions. The solenoid will not be signaled to engage unless the coolant temperature is below 14 degrees F and the engine speed is below 1800 rpm

Lock-up Control Solenoid Valve

The lock-up control solenoid is located on the transaxle case. This device is controlled by the ECU only on Civic models with automatic transaxle. On Accord and Prelude, the function is handled by a separate transaxle control unit. Input signals from the speed sensor and throttle angle sensor are used by the ECU to determine the precise time at which to lock the torque converter. The lock-up control solenoid valve receives only ON or OFF signals from the ECU; partial engagement is not possible.

FUEL SYSTEM

Fuel Pump

The fuel pump is installed inside the fuel tank, thereby saving space and simplifying the fuel line system.

The fuel pump is comprised of a DC motor, a circumference

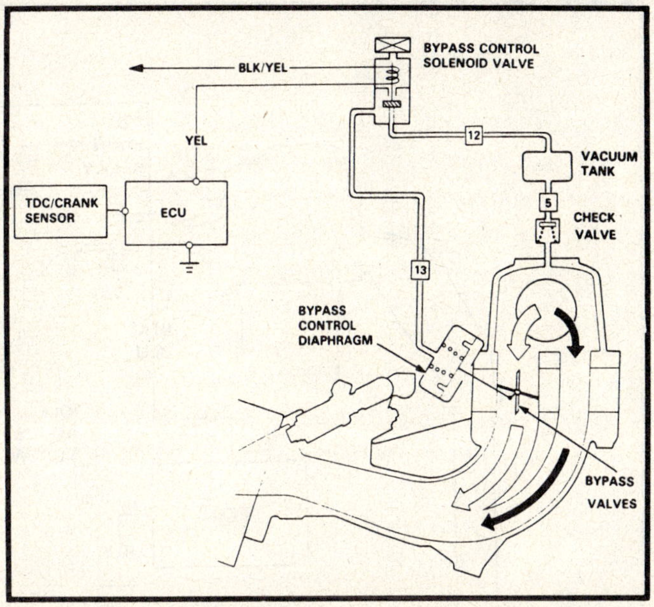

Schematic for Prelude bypass control solenoid valve. The small numbers indicate the number of the vacuum hose

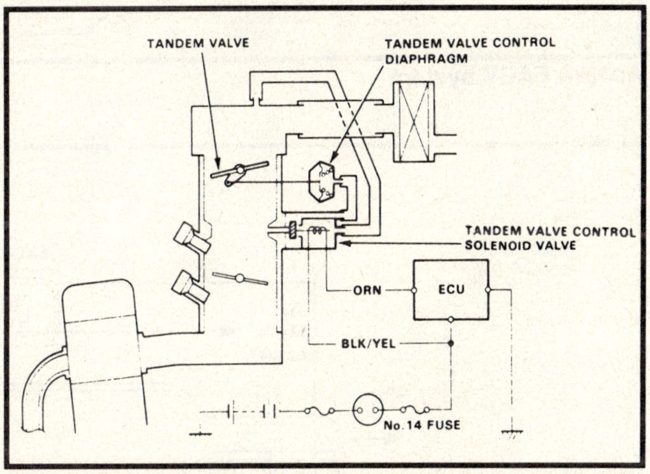

Schematic for tandem valve control system used on 1.5L Civic engine

flow pump, a relief valve for protecting the fuel line systems, a check valve for retaining residual line pressure after the pump stops, an inlet port, and a discharge port. If the flow of fuel is obstructed at the discharge side of the pump, an internal relief valve will open to bypass the fuel to the inlet port.

The pump assembly consists of the impeller (driven by the motor), the pump casing (which forms the pumping chamber), and cover of the pump.

Pressure Regulator

The fuel pressure regulator maintains a constant fuel pressure at the injectors. The spring chamber of the pressure regulator is connected to the intake manifold to constantly maintain the fuel pressure at 36 psi (2.55 kg/cm^2) higher than the pressure in the manifold. When the difference between the fuel pressure and manifold pressure exceeds 36 psi (2.55 kg/cm^2), the diaphragm is pushed upward, and the excess fuel is fed back into the fuel tank through the return line.

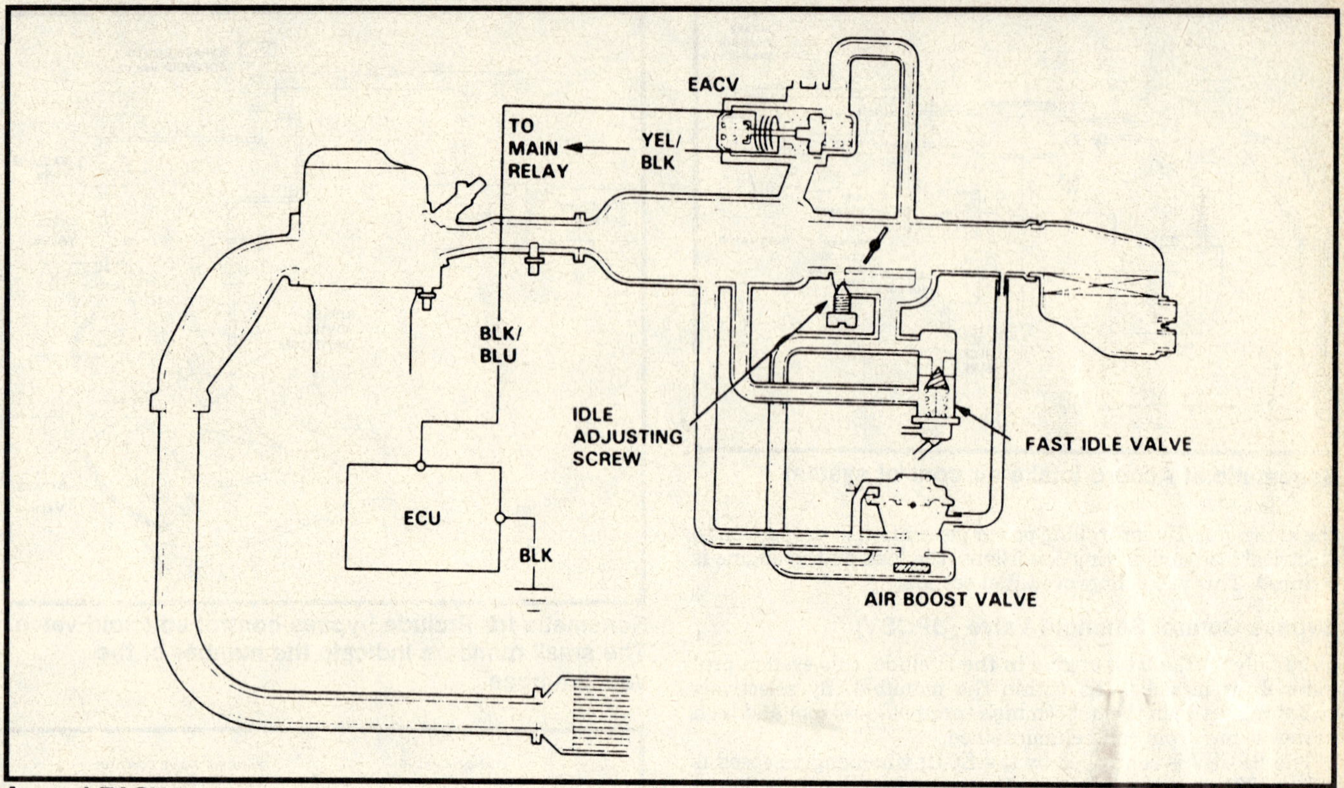

Accord EACV system

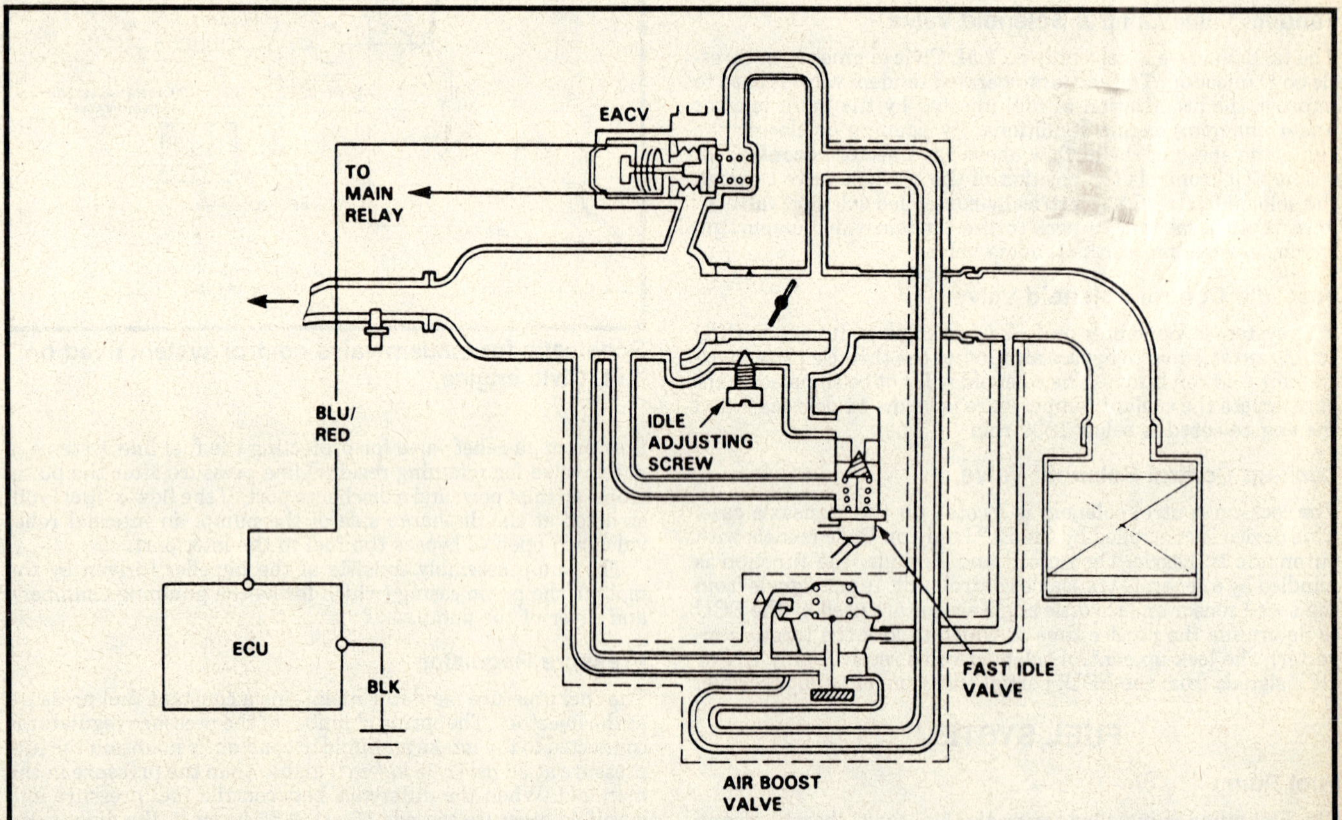

Prelude EACV system. The air boost valve is not used on 2.1L engines

Injector

The injector is of the solenoid-actuated constant-stroke pintle type consisting of a solenoid, plunger, needle valve and housing. When current is applied to the solenoid coil, the valve lifts up and pressurized fuel fills the inside of the injector and is injected near the intake valve. Because the needle valve lift height and the fuel pressure are constant, the amount of fuel injected is determined by the length of time (duration) that the injector is open. The injector is sealed by an O-ring and seal ring at the top and bottom. These seals also reduce operating noise.

Resistor

All fuel injected engines except the 1.5L Civic engine employ a resistor in series between the power supply and the injector. The injector timing, which controls the opening and closing intervals, must be very accurate since it dictates the air/fuel mixture ratio. The injector must also be durable. For the best possible injector response, it is necessary to shorten the current rise time when voltage is applied to the injector coil. Flow of current in the coil is restricted by a resistor installed in series between the electric power source and the injector coil.

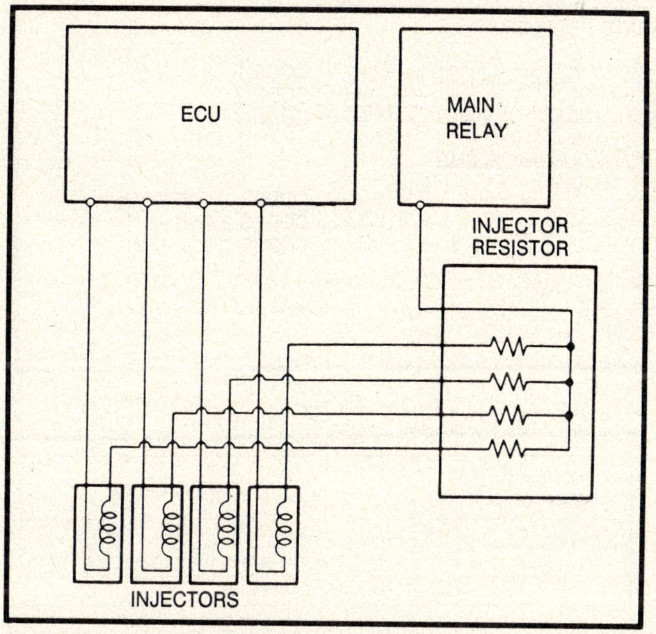

Injector resistor circuit, all fuel injected engines except Civic with 1.5L engine

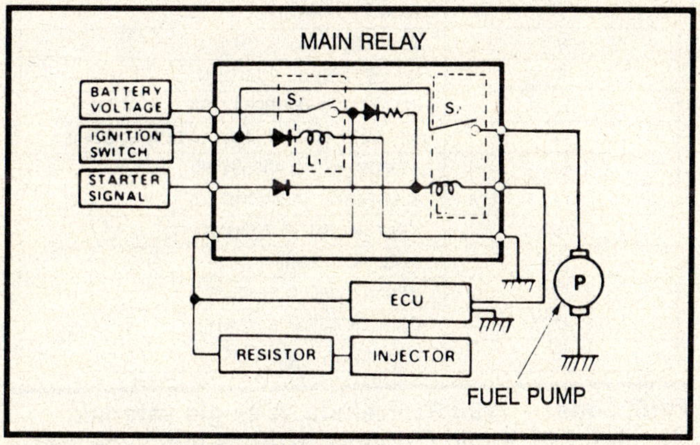

Main relay circuit schematic

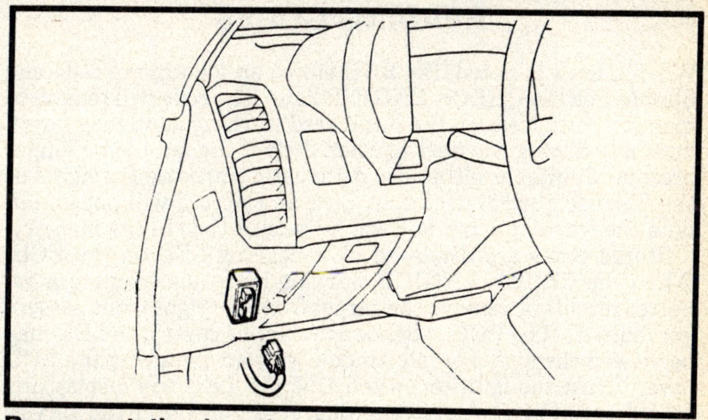

Representative location for main relay. Prelude shown, all are similar

Main Relay

The main relay unit contains two separate relays in one case. The relay is found at the extreme left side of the dashboard, above and behind the fusebox. The first relay is energized whenever the ignition is switched **ON**. Power is supplied to ECU, the injectors and to the second relay. The second relay circuit supplies power to the fuel pump. It is engaged for 2 seconds any time the ignition is switched **ON** and is always engaged when the engine is running.

Diagnosis and Testing

SERVICE PRECAUTIONS

- Do not operate the fuel pump when the fuel lines are empty.
- Do not operate the fuel pump when removed from the fuel tank.
- Do not reuse fuel hose clamps.
- The washer(s) below any fuel system bolt (banjo fittings, service bolt, fuel filter, etc.) must be replaced whenever the bolt is loosened. Do not reuse the washers; a high-pressure fuel leak may result.
- Make sure all ECU harness connectors are fastened securely. A poor connection can cause an extremely high voltage surge and result in damage to integrated circuits.
- Keep ECU all parts and harnesses dry during service. Protect the ECU and all solid-state components from rough handling or extremes of temperature.
- Before attempting to remove any parts, turn the ignition switch **OFF** and disconnect the battery ground cable.
- Always use a 12 volt battery as a power source, never a booster or high-voltage charging unit.
- Do not disconnect the battery cables with the engine running.
- Do not disconnect any wiring connector with the engine running or the ignition **ON** unless specifically instructed to do so.
- Do not depress the accelerator pedal when starting.
- Do not rev up the engine immediately after starting or just prior to shutdown.
- Do not apply battery power directly to injectors.
- Whenever possible, use a flashlight instead of a drop light.
- Keep all open flame and smoking material out of the area.
- Use a shop cloth or similar to catch fuel when opening a fuel system.
- Relieve fuel system pressure before servicing.
- Always use eye or full-face protection when working around fuel lines, fittings or components.
- Always keep a dry chemical (class B-C) fire extinguisher near the area.

READING CODES

When a fault is noted, the ECU stores an identifying code and illuminates the CHECK ENGINE light. The code will remain in memory until cleared; the dashboard warning lamp may not illuminate during the next ignition cycle if the fault is no longer present. Faults for either the Automatic transaxle F/I signal or the Electric Load Detector on Civic and Accord will not illuminate the warning lamp but will set a fault code in the memory.

Stored codes are displayed by a flashing LED on the ECU. When the CHECK ENGINE warning lamp has been on or reported on, lift or remove the carpet from the right front passenger footwell. The ECU is below a protective cover; the LED may be viewed through a small window without removing the ECU cover. Turn the ignition switch **ON**; the LED will display any stored codes by rhythmic flashing. Additionally, if the service connector is jumpered on Accord or 2.1L engine Preludes, the CHECK ENGINE lamp will display the stored codes in the same fashion. The 2-pin service connector is located under the extreme right dashboard on Accord or on the Prelude firewall, near the control box.

Codes 1–9 are indicated by a series of short flashes; two-digit codes use a number of long flashes for the first digit followed by the appropriate number of short flashes. For example, Code 43 would be indicated by 4 long flashes followed by 3 short flashes. Codes are separated by a pause between transmissions. The position of the codes during output can be helpful in diagnostic work. Multiple codes transmitted in isolated order indicate unique occurrences; a display of showing 1-1-1-pause-9-9-9 indicates two problems or problems occurring at different times. An alternating display, such as 1-9-1-9-1, indicates simultaneous occurrences of the faults.

When counting flashes to determine codes, a code not valid for the vehicle may be found. In this case, first recount the flashes to confirm an accurate count. If necessary, turn the ignition switch **OFF**, then recycle the system and begin the count again.

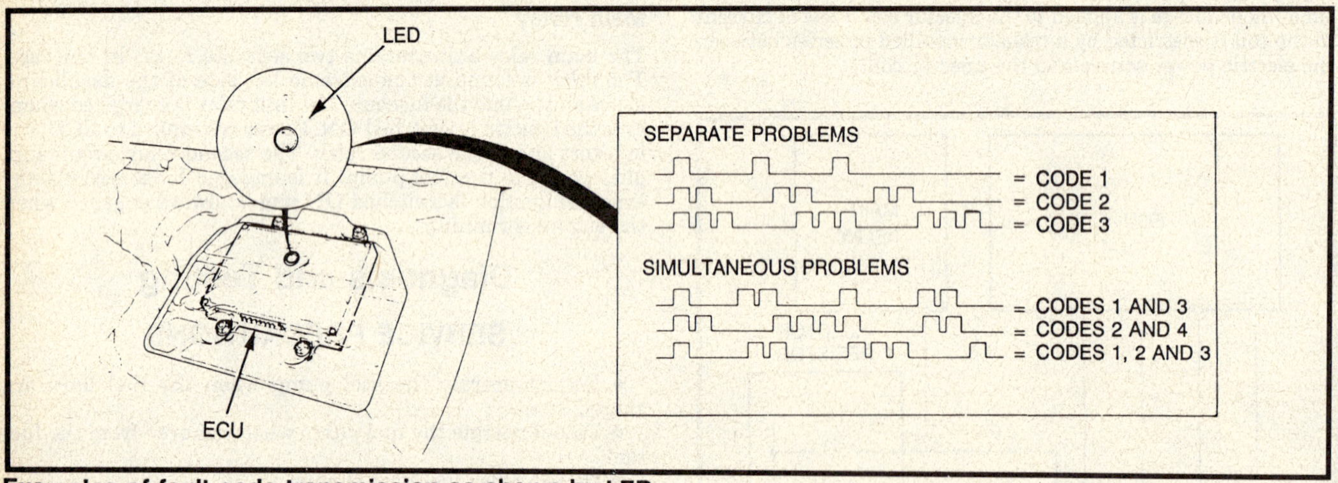

Examples of fault code transmission as shown by LED

SELF-DIAGNOSIS INDICATOR BLINKS	SYSTEM INDICATED
0	ECU
1	OXYGEN CONTENT
3	MANIFOLD ABSOLUTE PRESSURE
5	MANIFOLD ABSOLUTE PRESSURE
4	CRANK ANGLE
6	COOLANT TEMPERATURE
7	THROTTLE ANGLE
8	TDC POSITION
9	No.1 CYLINDER POSITION
10	INTAKE AIR TEMPERATURE
12	EXHAUST GAS RECIRCULATION SYSTEM
13	ATMOSPHERIC PRESSURE
14	ELECTRONIC AIR CONTROL
15	IGNITION OUTPUT SIGNAL
16	FUEL INJECTOR
17	VEHICLE SPEED SENSOR
20	ELECTRIC LOAD DETECTOR
30	A/T FI SIGNAL A
31	A/T FI SIGNAL B
41	OXYGEN SENSOR HEATER
43	FUEL SUPPLY SYSTEM

Fault codes—1990-92 Accord

SELF-DIAGNOSIS INDICATOR BLINKS	SYSTEM INDICATED
0	ECU
1	OXYGEN CONTENT A*, OXYGEN CONTENT**
2*	OXYGEN CONTENT B
3	MANIFOLD ABSOLUTE PRESSURE
5	MANIFOLD ABSOLUTE PRESSURE
4	CRANK ANGLE
6	COOLANT TEMPERATURE
7	THROTTLE ANGLE
8	TDC POSITION
9	No.1 CYLINDER POSITION
10	INTAKE AIR TEMPERATURE
12	EXHAUST GAS RECIRCULATION SYSTEM
13	ATMOSPHERIC PRESSURE
14	ELECTRONIC AIR CONTROL
15	IGNITION OUTPUT SIGNAL
16	FUEL INJECTOR
17	VEHICLE SPEED SENSOR
41**	OXYGEN SENSOR HEATER
43**	FUEL SUPPLY SYSTEM

Fault codes—1990-91 Prelude. A single asterisk denotes 2.0L engine; double asterisk indicates 2.1L engine

SELF-DIAGNOSIS INDICATOR BLINKS	SYSTEM INDICATED
0	ECU
1	OXYGEN CONTENT
3	MANIFOLD ABSOLUTE PRESSURE
5	MANIFOLD ABSOLUTE PRESSURE
4	CRANK ANGLE
6	COOLANT TEMPERATURE
7	THROTTLE ANGLE
8	TDC POSITION
9	No.1 CYLINDER POSITION (1.6 ℓ)
10	INTAKE AIR TEMPERATURE
12	EXHAUST GAS RECIRCULATION SYSTEM (1.5 ℓ CAL: A/T)
13	ATMOSPHERIC PRESSURE
14	ELECTRONIC AIR CONTROL
15	IGNITION OUTPUT SIGNAL
16	FUEL INJECTOR
17	VEHICLE SPEED SENSOR
19	LOCK-UP CONTROL SOLENOID VALVE (A/T)
20	ELECTRIC LOAD

Fault codes—1990-92 Civic

If the Code is not valid for the vehicle, the ECU must be replaced.

NOTE: The S or S_4 light may illuminate with the CHECK ENGINE lamp on automatic Accord and Prelude if Code 6, 7 or 17 is stored. In this event, follow the troubleshooting procedures for the fuel injection system first. Clear the memory after repairs, then recheck the S or S_4 lamp.

CLEARING CODES

Stored codes are removed from memory by removing power to the ECU. Disconnecting the power will also clear the memories used for other solid-state equipment such as the clock and radio. For this reason, always make note of the radio presets before clearing the system.

While disconnecting the battery will clear the memory, this is not the recommended procedure. The memory should be cleared by removing the appropriate fuse for at least 10 seconds. In all cases, the correct fuse is located in the underhood fuse and relay box. On Civic models, remove the HAZARD fuse; on Accord, remove the BACK-UP fuse and on Prelude, remove the CLOCK fuse.

FUEL PRESSURE

Relief

CAUTION

The fuel system is under constant pressure, even when the engine has not been run for a period of time. To avoid injury and/or property damage, always safely relieve the fuel system pressure before servicing any fuel or return line component.

1. Remove the fuel filler cap.
2. With the ignition **OFF**, disconnect the negative battery cable.
3. Locate the 6mm service bolt. On Accord and Civic, it is located at the fuel inlet line to the fuel filter. On Prelude, it is found on the fuel line connection to the fuel rail.
4. Use a wrench to hold the larger bolt below the service bolt. Use a box-end wrench on the service bolt. Place a rag or cloth over the wrenches and service bolt.
5. Slowly loosen the service bolt 1 full turn. The pressurized fuel will escape into the rag.
6. Remove the service bolt, replace the washer and reinstall the bolt. Tighten the bolt to 12 Nm (9 ft. lbs.). The washer must be replaced any time the service bolt is loosened or removed. Failure to replace the washer may result in high-pressure fuel leaks.
7. Dispose of the fuel-soaked rag in a manner correct for flammable items. Check the surrounding area for any spilled fuel.
8. Reconnect the battery cable. The pressure within the system will stay reduced until the ignition switch is turned **ON** or the fuel pump is operated by other means.

Testing

1. Relieve the fuel pressure using the correct procedure.
2. Remove the service bolt. Install a fuel pressure gauge in the service bolt port.
3. Start the engine.
4. Disconnect the vacuum hose at the pressure regulator. Measure the fuel pressure with the engine at idle. Correct pressure should be: Civic except CRX, 240–279 kPa (35–41 psi); all CRX, 275–324 kPa (40–50 psi), Accord except SE and Ex-R, 235–284 kPa (34–41 psi), Accord SE and Ex-R (Canada), 265–314 kPa (38–48 psi); and Prelude, 255–304 kPa (37–44 psi).
5. On Accord and Prelude, reconnect the vacuum hose to the regulator. Regulated fuel pressure should drop to: Accord, except SE and Ex-R, 176–225 kPa (26–33 psi); Accord SE and Ex-R, 196–245 kPa (28–36 psi) and Prelude, 196–245 kPa (28–36 psi).6. If fuel pressure is above specification, inspect for a pinched or clogged fuel return hose. Additional causes include a faulty regulator or clogged Prelude fuel filter.
7. If fuel pressure is below specification, causes may include fuel line leaks, a faulty regulator or, on Accord and Civic, a clogged fuel filter.
8. Shut the engine **OFF**. On Civic, reconnect the vacuum hose to the regulator.
9. Remove the test equipment. Install a new washer on the service bolt; reinstall the bolt, tightening it to 12 Nm (9 ft. lbs.).

FUEL PUMP

Testing

1. The simplest test for the fuel pump is listening to it. The pump should be heard to operate for 2 seconds when the ignition switch is turned **ON**. The noise is easily heard at the fuel filler neck with the cap removed.
2. If the pump does not operate, proceed with electrical testing. On Accord and Prelude, lift the carpet in the cargo area.
3. On Prelude, remove the left access cover in the luggage area. On Civic models, remove the rear seat.
4. Insure the ignition is **OFF**. Disconnect the fuel pump connector at the fuel pump.
5. Disconnect the wiring connector from the main relay under the left side dash.
6. At the relay connector, use a jumper wire to connect the black/yellow wire terminal to the terminal for the yellow/black (Civic, Prelude) or yellow (Accord) wire.
7. Turn the ignition switch **ON**. Battery voltage should be available at the fuel pump connector. On Civic and Prelude, connect the positive probe of the voltmeter to the yellow/black wire and the negative probe to the black wire. For Accord, test with the positive probe on the yellow wire and the negative probe on body ground.
8. If battery voltage is available, the fuel pump must be replaced.
9. If no voltage is available, check the main relay and/or wire harness.

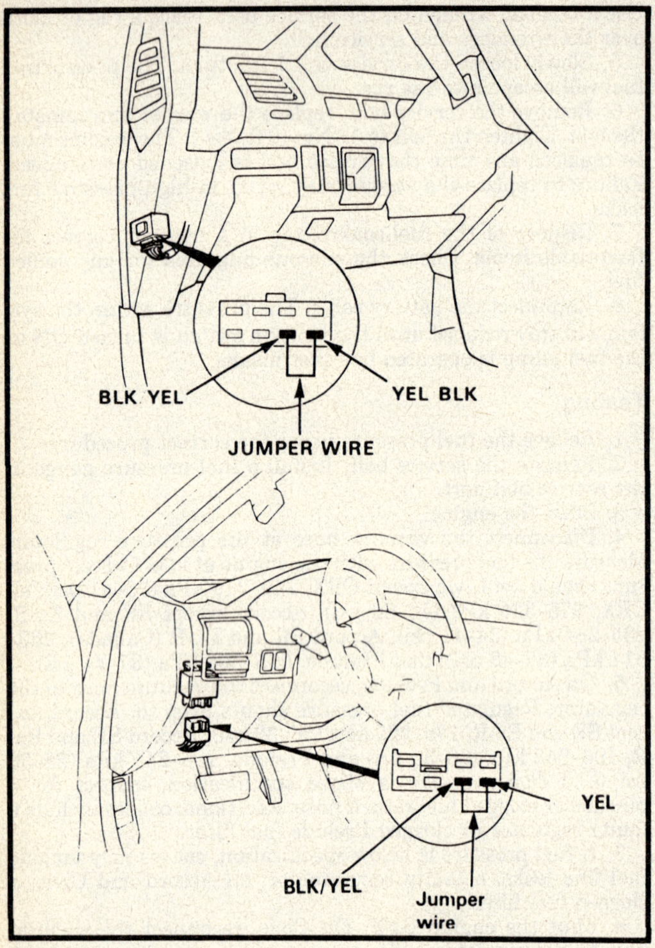

Fuel pump testing for Civic and Prelude, above, and Accord

FUEL PRESSURE REGULATOR

Testing

1. Relieve the fuel pressure using the correct procedure.
2. Remove the service bolt. Install a fuel pressure gauge in the service bolt port.
3. Start the engine.
4. Disconnect the vacuum hose at the pressure regulator. Measure the fuel pressure with the engine at idle. Correct pressure should be:
   – Civic: 275–324 kPa (40–50 psi)
   – Accord: 240–279 kPa (35–41 psi)
   – Prelude: 240–279 kPa (35–41 psi)
5. Reconnect the vacuum hose to the regulator. Regulated fuel pressure should drop on the gauge. Disconnect the vacuum hose; pressure should increase.
6. If the line pressure does not increase with the vacuum hose disconnected, inspect the vacuum hose for kinks or blockage. Wrap the regulator fuel return line with a rag or cloth and lightly pinch off the line. Fuel pressure should increase. If the pressure still does not rise, the regulator is faulty and must be replaced.

FUEL INJECTOR

Electrical Testing

1. Turn the ignition switch **OFF**.
2. Disconnect the wiring harness from each injector.
3. Measure the resistance at the 2 terminals of the injector. Correct resistance for all except 1.5L Civic engine is 1.5–2.5 ohms. For 1.5L Civic engines, injector resistance should be 0.6–1.6 ohms for the main or upper injector and 6–10 ohms for the auxiliary injector. Any injector not showing the proper resistance must be replaced.
4. Further testing of the injector circuits and wiring are included in the troubleshooting charts.

INJECTOR RESISTOR

Testing

A resistor is used on all fuel injected engines except the 1.5L throttle body injection Civic engine.

1. With the ignition **OFF**, disconnect the harness connector from the resistor.
2. Measure the resistance between terminal A of the resistor connector and each other connector. When viewing the resistor connector, terminal A is the odd or un-paired terminal within the connector housing.
3. Correct resistance between terminal A and each other connector is 5–7 ohms. If any circuit is out of range, the resistor must be replaced.

MAIN RELAY

Testing

1. The simplest test for the main relay is the fact that the vehicle starts and remains running. If this is true, the main relay is functioning properly.
2. If the relay must be tested, remove it from its mount under the dash and disconnect it from the harness.
3. Using jumper wires, connect battery + to Civic and Prelude relay terminal 4 or Accord relay terminal 6. Connect the – battery terminal to relay terminal 8.

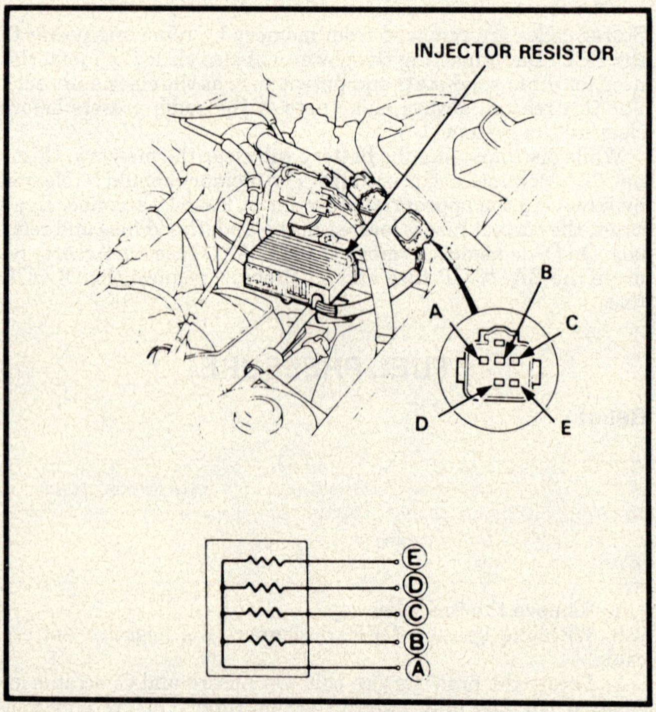

Testing the injector resistor. 1990 Accord shown; terminal pattern is the same for all

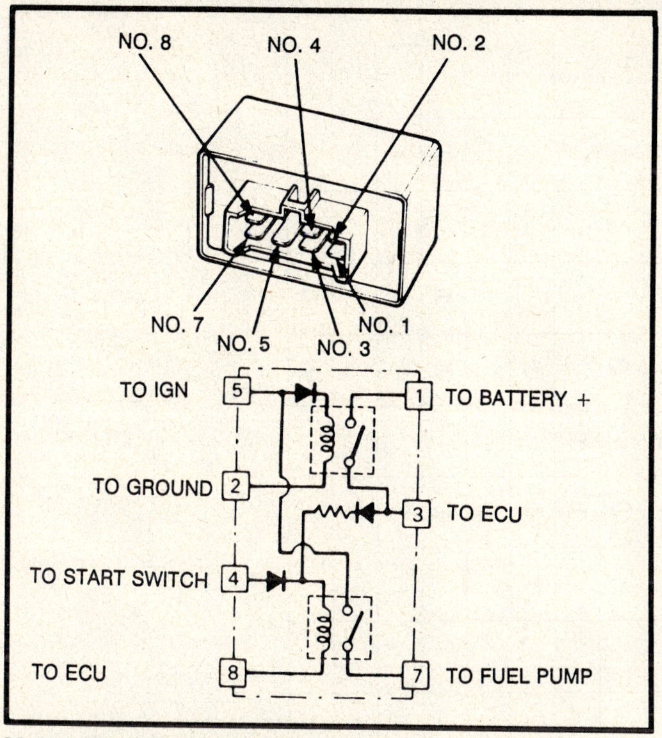

Main relay terminal designations and schematic, 1990-92 Prelude and Civic

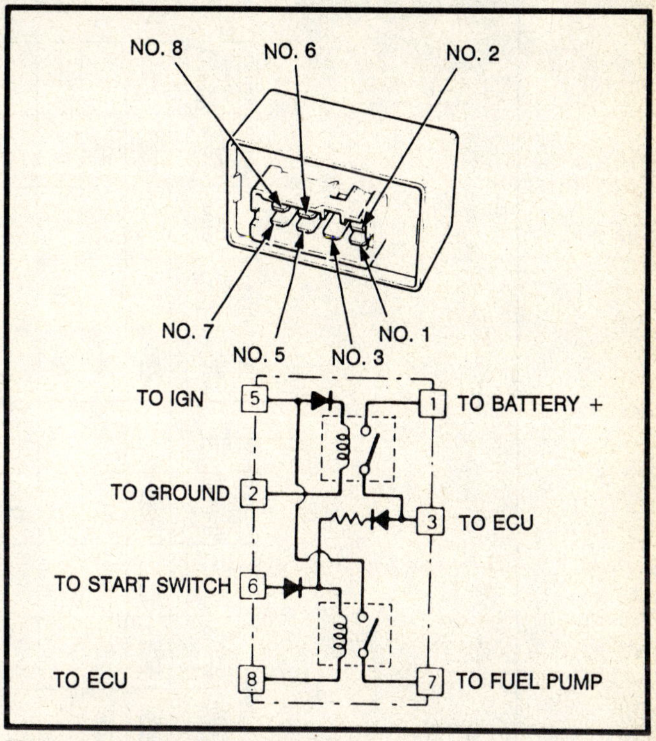

Main relay terminal designations and schematic, 1990-92 Accord

4. Check for continuity between relay terminals 5 and 7. If no continuity is present replace the relay.
5. Connect the + battery jumper to relay terminal 5 and the negative jumper to relay terminal 2. Check for continuity between relay terminals 1 and 3. If there is no continuity, replace the relay.
6. Connect the + battery jumper to relay terminal 3 and the negative jumper to relay terminal 8. Check for continuity between relay terminals 5 and 7. If there is no continuity, replace the relay. If continuity is present, the relay has not failed.
7. Reinstall the relay.

CHECK ENGINE Light

The function of the CHECK ENGINE lamp on the dashboard is checked by the ECU every time the ignition is turned **ON**. The lamp should illuminate for 2 seconds and then extinguish. In normal operation, the lamp should not come on while driving.

If the warning lamp lights while the vehicle is operating, the ECU has sensed a system fault. The fault may be either loss of signal or signal out-of-range, either high or low. Additionally, when the CHECK ENGINE lamp is lit, the ECU has entered the default or fail-safe mode in which fixed values are substituted for the faulty or missing signal.

If, during diagnosis, the Accord or Prelude service connector is jumpered, the CHECK ENGINE lamp will light, even if no faults are present. When repairs or diagnosis are completed, the memory should be checked for any stored codes and cleared if necessary.

USING TROUBLESHOOTING CHARTS

The troubleshooting charts are divided into sub-groups for the PGM-FI Control, Idle Control, Air Intake Control, Fuel Supply and Emission Control systems. Each sub-group begins with a symptom chart listing likely causes for a particular problem.

When beginning diagnosis, refer to the symptom-to-system troubleshooting charts; this chart will direct the use of other troubleshooting charts to test circuits and components related to the problem at hand. Both the symptom and troubleshooting charts include the applicable self-diagnostic or fault codes when necessary.

NOTE: Across each row in the chart, the systems that could be sources of a symptom are ranked in the order they should be inspected starting with ①. Find the symptom in the left column, read across to the most likely source.

SYMPTOM		ECU	OXYGEN SENSOR	MANIFOLD ABSOLUTE PRESSURE SENSOR	TDC/CRANK SENSOR	COOLANT TEMPERATURE SENSOR	THROTTLE ANGLE SENSOR	INTAKE AIR TEMPERATURE SENSOR	ATMOSPHERIC PRESSURE SENSOR
SYSTEM			PGM-FI						
CHECK ENGINE LIGHT TURNS ON		▭ or ☼	CHECK	CHECK	CHECK	CHECK	CHECK	CHECK	CHECK
SELF-DIAGNOSIS INDICATOR (LED) BLINKS		⓪ or ✳	①	③ or ⑤	④ or ⑧	⑥	⑦	⑩	⑬
ENGINE WON'T START		③							
DIFFICULT TO START ENGINE WHEN COLD		BU		③		①			
IRREGULAR IDLING	WHEN COLD FAST IDLE OUT OF SPEC	BU				③			
	ROUGH IDLE	BU		③					
	WHEN WARM RPM TOO HIGH	BU				③			
	WHEN WARM RPM TOO LOW	BU							
FREQUENT STALLING	WHILE WARMING UP	BU		③					
	AFTER WARMING UP	BU							
POOR PERFORMANCE	MISFIRE OR ROUGH RUNNING	BU							
	FAILS EMISSION TEST	BU	③	②					
	LOSS OF POWER	BU					②		

SYMPTOM		IGNITION OUTPUT SIGNAL	VEHICLE SPEED SENSOR	LOCK-UP CONTROL SOLENOID VALVE	ELECTRIC LOAD DETECTOR	ELECTRONIC AIR CONTROL VALVE	OTHER IDLE CONTROLS	FUEL INJECTOR	OTHER FUEL SUPPLY	AIR INTAKE	EGR CONTROL SYSTEM (CAL: A/T only)	OTHER EMISSION CONTROLS
SYSTEM		PGM-FI				IDLE CONTROL		FUEL SUPPLY			EMISSION CONTROL	
CHECK ENGINE LIGHT TURNS ON		CHECK	CHECK	CHECK	▭	CHECK		CHECK			CHECK	
SELF-DIAGNOSIS INDICATOR (LED) BLINKS		⑮	⑰	⑲	⑳	⑭		⑯			⑫	
ENGINE WON'T START								②	①			
DIFFICULT TO START ENGINE WHEN COLD						②						
IRREGULAR IDLING	WHEN COLD FAST IDLE OUT OF SPEC					①	②					
	ROUGH IDLE					①		②			③	
	WHEN WARM RPM TOO HIGH					②	①					
	WHEN WARM RPM TOO LOW				②	①		③				
FREQUENT STALLING	WHILE WARMING UP	②				①						
	AFTER WARMING UP					①		②	③		③	
POOR PERFORMANCE	MISFIRE OR ROUGH RUNNING					③		①	②		③	
	FAILS EMISSION TEST								①			
	LOSS OF POWER							③	①			

- If codes other than those listed above are indicated, count the number of blinks again. If the indicator is in fact blinking these codes, substitute a known-good ECU and recheck. If the indication goes away, replace the original ECU.

(BU): When the Check Engine light and the self-diagnosis indicator are on, the back-up system is in operation.

1.6L ENGINE—TROUBLESHOOTING—CIVIC

NOTE: Across each row in the chart, the systems that could be sources of a symptom are ranked in the order they should be inspected starting with ①. Find the symptom in the left column, read across to the most likely source

SYMPTOM \ SYSTEM		PGM-FI							
		ECU	OXYGEN SENSOR	MANIFOLD ABSOLUTE PRESSURE SENSOR	TDC/CRANK/CYL SENSOR	COOLANT TEMPERATURE SENSOR	THROTTLE ANGLE SENSOR	INTAKE AIR TEMPERATURE SENSOR	ATMOSPHERIC PRESSURE SENSOR
CHECK ENGINE LIGHT TURNS ON		□ or ☼	CHECK	CHECK	CHECK	CHECK	CHECK	CHECK	CHECK
SELF-DIAGNOSIS INDICATOR (LED) BLINKS		0 or *	1	3 or 5	4 or 8 or 9	6	7	10	13
ENGINE WON'T START		3							
DIFFICULT TO START ENGINE WHEN COLD		BU		3		1			
IRREGULAR IDLING	WHEN COLD FAST IDLE OUT OF SPEC	BU				3			
	ROUGH IDLE	BU		3					
	WHEN WARM RPM TOO HIGH	BU				3			
	WHEN WARM RPM TOO LOW	BU							
FREQUENT STALLING	WHILE WARMING UP	BU							
	AFTER WARMING UP	BU							
POOR PERFORMANCE	MISFIRE OR ROUGH RUNNING	BU							
	FAILS EMISSION TEST	BU	3	2					
	LOSS OF POWER	BU		3			2		

• If codes other than those listed above are indicated, count the number of blinks again. If the indicator is in fact blinking these codes, substitute a known-good ECU and recheck. If the indication goes away, replace the original ECU.
(BU): When the Check Engine light and the self-diagnosis indicator are on, the back-up system is in operation.

SYMPTOM		PGM-FI			IDLE CONTROL		FUEL SUPPLY			
		IGNITION OUTPUT SIGNAL	VEHICLE SPEED SENSOR	ELECTRIC LOAD DETECTOR	ELECTRONIC AIR CONTROL VALVE	OTHER IDLE CONTROLS	FUEL INJECTOR	OTHER FUEL SUPPLY	AIR INTAKE	EMISSION CONTROL
CHECK ENGINE LIGHT TURNS ON		CHECK	CHECK	□	CHECK		CHECK			
SELF-DIAGNOSIS INDICATOR (LED) BLINKS		15	17	20	14		16			
ENGINE WON'T START							2	1		
DIFFICULT TO START ENGINE WHEN COLD							2			
IRREGULAR IDLING	WHEN COLD FAST IDLE OUT OF SPEC				1	2				
	ROUGH IDLE				1		2			
	WHEN WARM RPM TOO HIGH				2	1				
	WHEN WARM RPM TOO LOW			3	1		2			
FREQUENT STALLING	WHILE WARMING UP				1	2		3		
	AFTER WARMING UP				1	2		3		
POOR PERFORMANCE	MISFIRE OR ROUGH RUNNING				2		1			
	FAILS EMISSION TEST						1			
	LOSS OF POWER							1		

IDLE CONTROL SYSTEM TROUBLESHOOTING—CIVIC

NOTE:
- Across each row in the chart, the sub systems that could be sources of a symptom are ranked in the order they should be inspected, starting with 1. Find the symptom in the left column, read across to the most likely source,

SYMPTOM \ SUB SYSTEM		IDLE ADJUSTING SCREW	EACV	AIR CONDITIONING SIGNAL	ALTERNATOR FR SIGNAL	A/T SHIFT POSITION SIGNAL (Automatic)	BRAKE SWITCH SIGNAL	STARTER SWITCH SIGNAL	FAST IDLE CONTROL (1.6 ℓ)	HOSES AND CONNECTIONS
										*
DIFFICULT TO START ENGINE WHEN COLD		②	①							
WHEN COLD FAST IDLE OUT OF SPEC (1,000–2,000 rpm)		②	①							③
ROUGH IDLE		③	②							①
WHEN WARM RPM TOO HIGH		③	②							①
WHEN WARM RPM TOO LOW	Idle speed is below specified rpm (no load)	②	①		③					
	Idle speed does not increase after initial start up.		①					②		
	On models with automatic transmission, the idle speed drops in gear	③	②			①				
	Idle speeds drops when air conditioner in ON	③	②	①						
FREQUENT STALLING	WHILE WARMING UP	②	①							
	AFTER WARMING UP	②	①							
FAILS EMISSION TEST										①

FUEL SUPPLY SYSTEM TROUBLESHOOTING—CIVIC

NOTE: Across each row in the chart, the systems that could be sources of a symptom are ranked in the order they should be inspected starting with ①. Find the symptom in the left column, read across to the most likely source,

1.5

SYMPTOM \ SUB SYSTEM		FUEL INJECTOR	PRESSURE REGULATOR	FUEL FILTER	FUEL PUMP	MAIN RELAY	CONTAMINATED FUEL
							*
ENGINE WON'T START		③		③	①	②	
DIFFICULT TO START ENGINE WHEN COLD		②	③	①			
ROUGH IDLE		①	②				
FREQUENT STALLING	WHILE WARMING UP	①		②	③		
	AFTER WARMING UP	①	③	③	②		③
POOR PERFORMANCE	MISFIRE OR ROUGH RUNNING	①	②				③
	FAILS EMISSION TEST	①	②				
	LOSS OF POWER	③		①	③		②

1.6

PAGE / SYMPTOM \ SUB SYSTEM		FUEL INJECTOR	INJECTOR RESISTOR	PRESSURE REGULATOR	FUEL FILTER	FUEL PUMP	MAIN RELAY	CONTAMINATED FUEL
								*
ENGINE WON'T START			③		③	①	②	
DIFFICULT TO START ENGINE WHEN COLD		③			②	①		
ROUGH IDLE		①		②				
FREQUENT STALLING	WHILE WARMING UP	①			②	③		
	AFTER WARMING UP	①		③	③	②		
POOR PERFORMANCE	MISFIRE OR ROUGH RUNNING	①		②				
	FAILS EMISSION TEST	①		②				
	LOSS OF POWER				①	③		

* Fuel with dirt, water or a high percentage of alcohol is considered contaminated.

AIR INTAKE SYSTEM TROUBLESHOOTING—CIVIC

NOTE: Across each row in the chart, the sub systems that could be sources of a symptom are ranked in the order they should be inspected starting with ①. Find the symptom in the left column, read across to the most likely source

SYMPTOM \ SUB SYSTEM	THROTTLE CABLE	THROTTLE BODY	TANDEM CONTROL SYSTEM (1.5)	THROTTLE CONTROL SYSTEM
DIFFICULT TO START ENGINE WHEN COLD			①	
WHEN COLD FAST IDLE OUT OF SPEC	③	②		①
WHEN WARM RPM TOO HIGH	③	②		①
WHEN WARM RPM TOO LOW		①		
FREQUENT STALLING WHILE WARMING UP	③	②	①	
LOSS OF POWER	③	②	①	

EMISSION CONTROL SYSTEM TROUBLESHOOTING—CIVIC

NOTE: Across each row in the chart, the systems that could be sources of a symptom are ranked in the order they should be inspected starting with ①. Find the symptom in the left column, read across to the most likely source,

Ex. 1.5 CAL: A/T

SYMPTOM \ SUB SYSTEM		CATALYTIC CONVERTER	POSITIVE CRANKCASE VENTILATION SYSTEM	EVAPORATIVE EMISSION CONTROLS
ROUGH IDLE			①	
POOR PERFORMANCE	FAILS EMISSION TEST	①		②
	LOSS OF POWER	①		

1.5 CAL/A/T

PAGE / SYMPTOM \ SUB SYSTEM		CATALYTIC CONVERTER	EGR SYSTEM	POSITIVE CRANKCASE VENTILATION SYSTEM	EVAPORATIVE EMISSION CONTROLS
ROUGH IDLE			①	②	
FREQUENT STALLING	WHILE WARM-ING UP		①		
	AFTER WARM-ING UP		①		
POOR PERFORMANCE	FAILS EMISSION TEST	①		②	
	LOSS OF POWER	①			

ECU—TROUBLESHOOTING—CIVIC

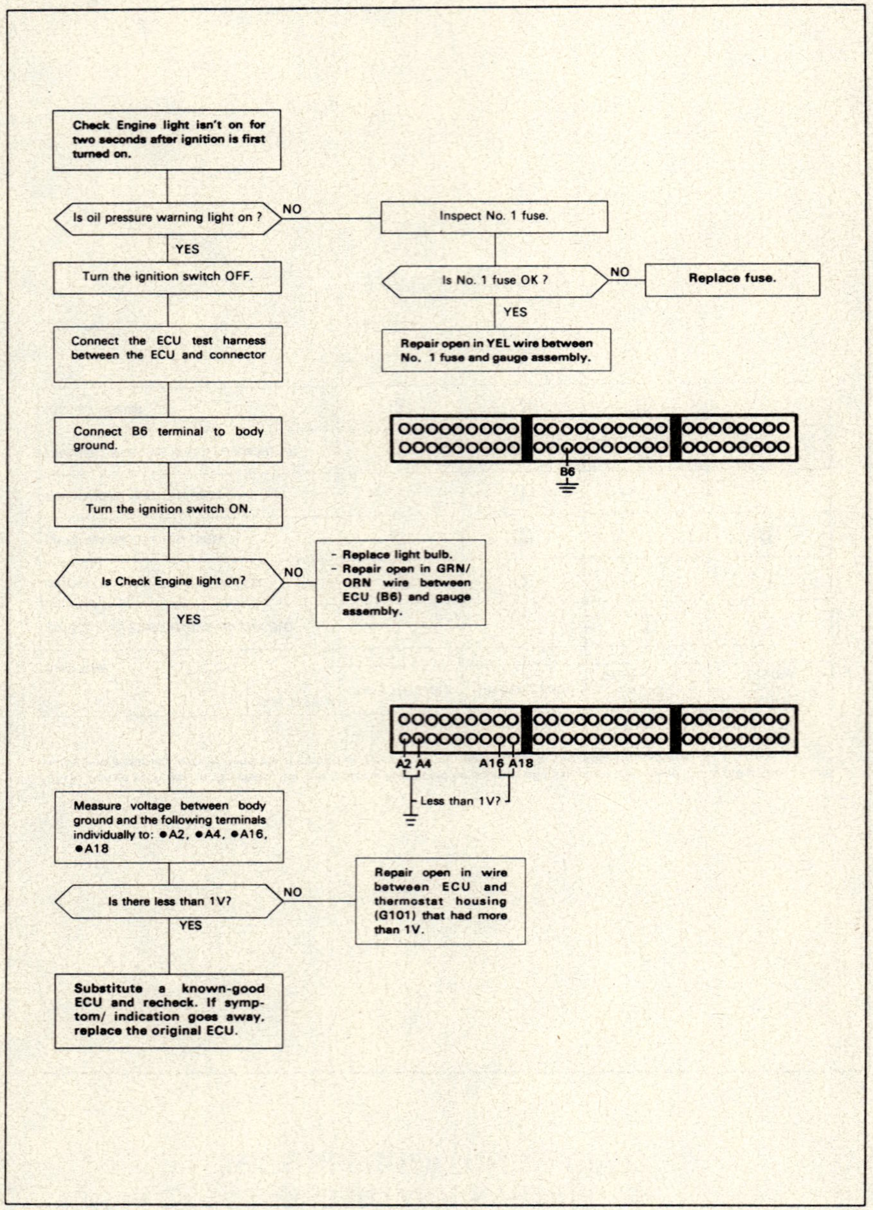

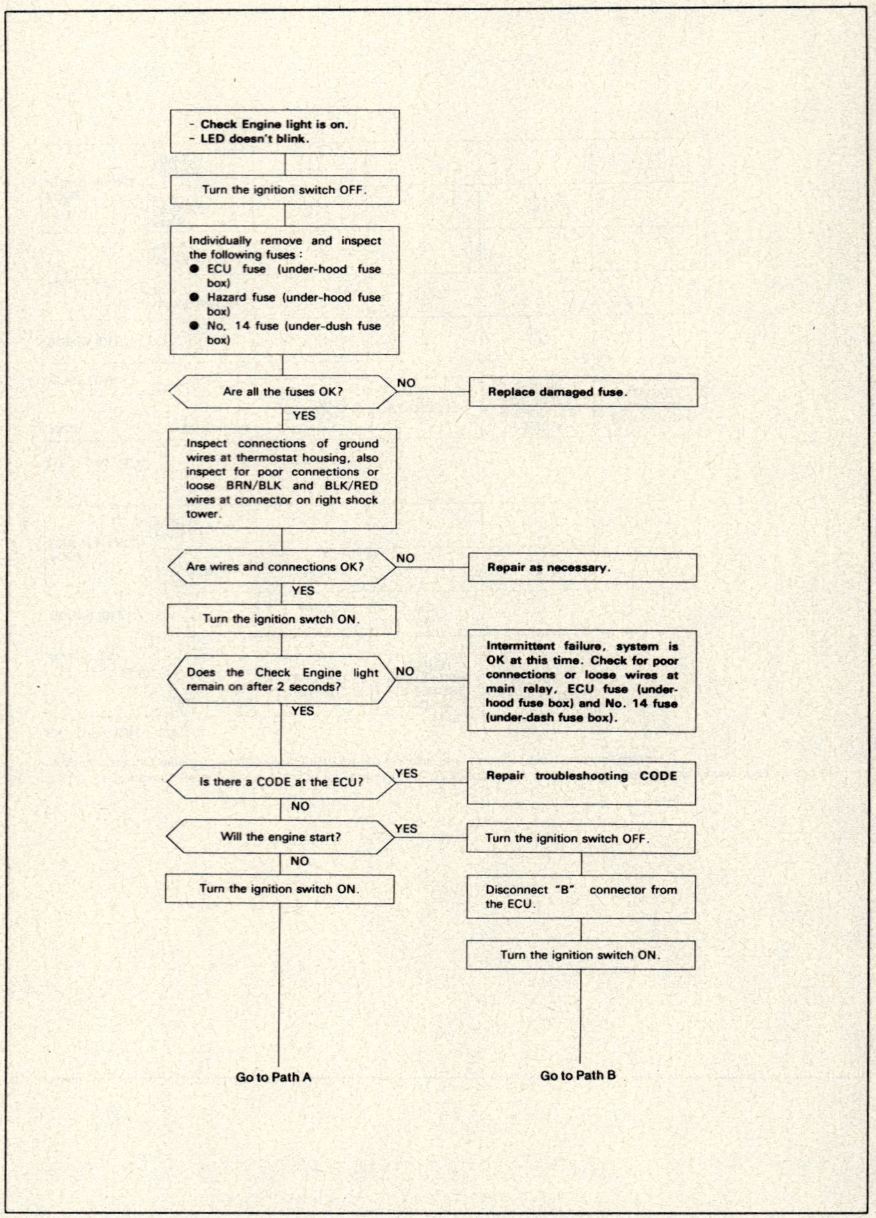

ECU—TROUBLESHOOTING—CIVIC

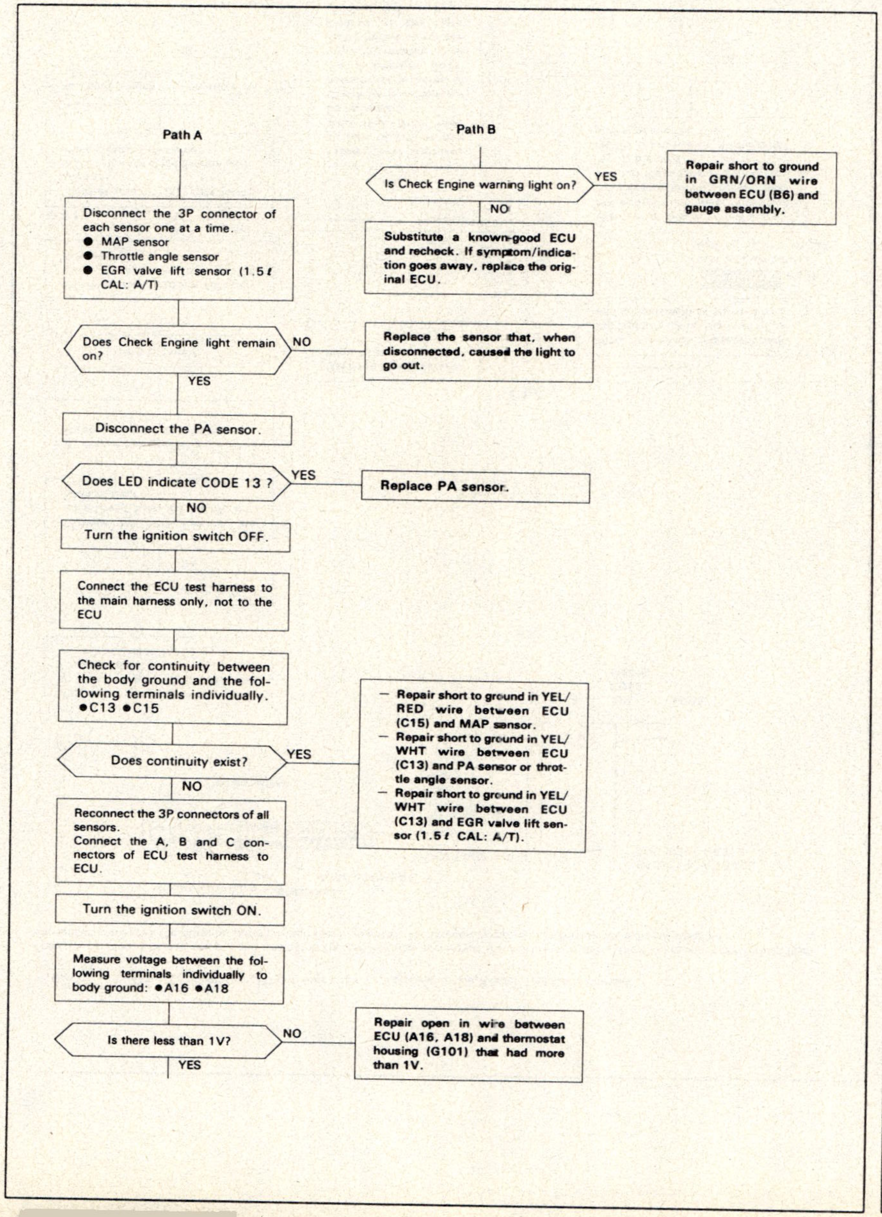

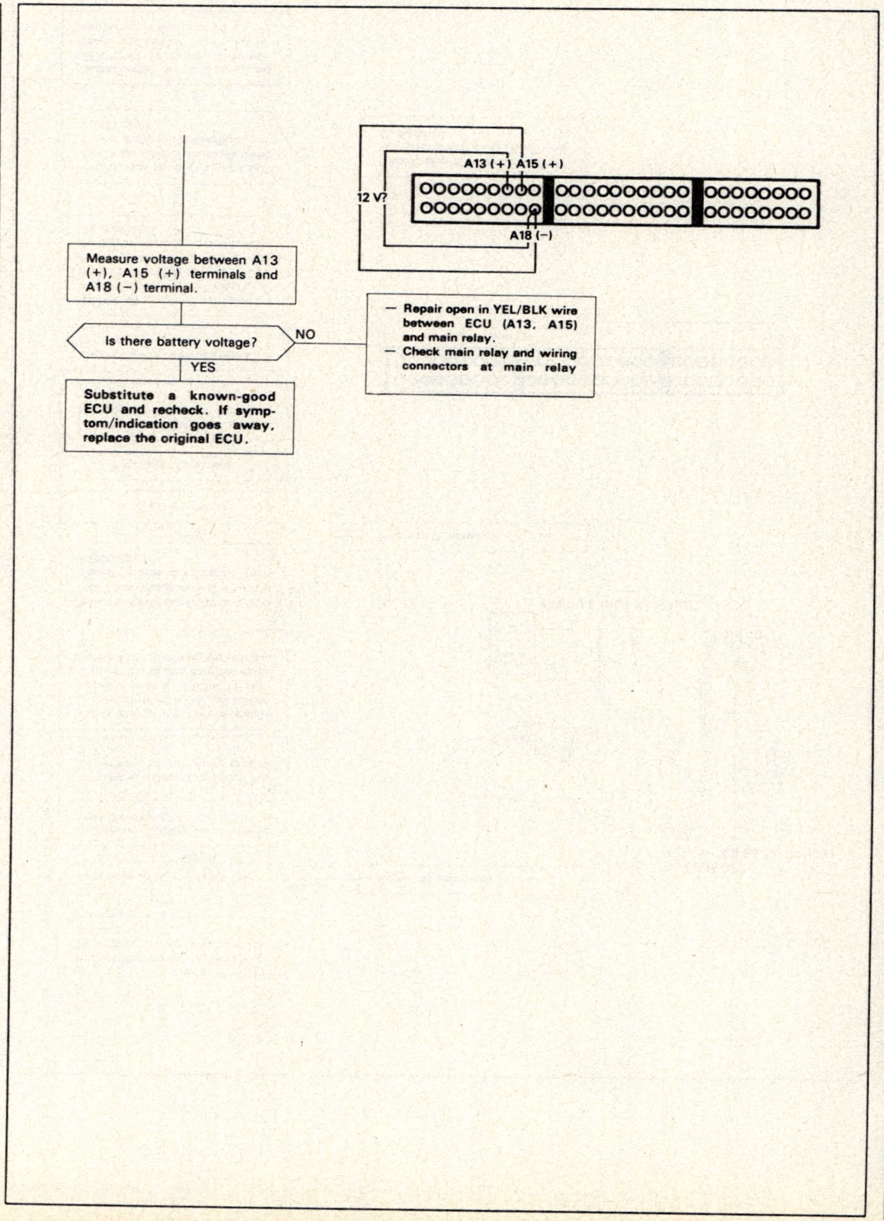

TROUBLESHOOTING—FAULT CODE 1—CIVIC

Self-diagnosis LED indicates code 1: A problem in the Oxygen (O_2) Sensor circuit.

The oxygen sensor, detects the oxygen content in the exhaust gas, and inputs the ECU. In operation, the ECU receives the signals from the sensor and varies the duration during which fuel is injected. The oxygen sensor is installed on the exhaust manifold.

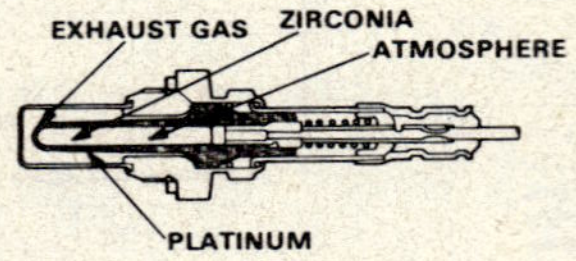

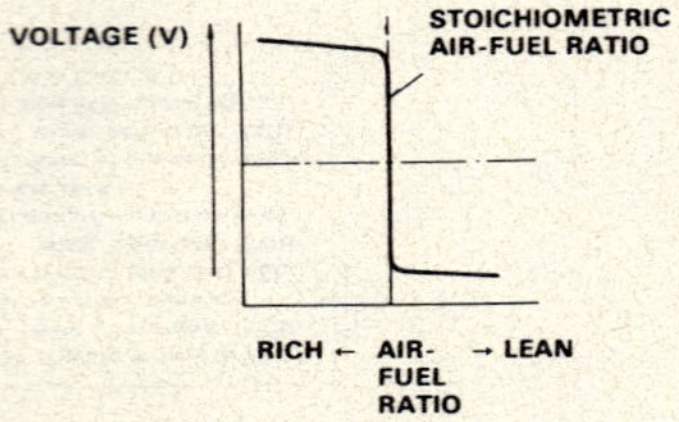

- Check Engine light has been reported on.
- LED indicates CODE 1.

Turn the ignition switch OFF.

Remove HAZARD fuse in the under-hood fuse box for 10 seconds to reset ECU.

Inspect pressure regulator

Is it normal? — NO → Replace the pressure regulator

YES

Warm up engine to normal operating temperature (cooling fan comes on).

1.6 ℓ

Hold engine at 1500 rpm for 15 minutes.
NOTE: Do not close throttle completely during this time.

1.5 ℓ

Block rear wheels and set the parking brake. Jack up the front of car and support with safety stands.

WARNING Block rear wheels before jacking up front of car.

Warm up engine to normal operating temperature again, then put the transmission into second gear and run the engine at 2,000 rpm for 15 minutes.
NOTE: Do not close throttle completely during this time.

Is Check Engine light on and does LED indicate CODE 1? — NO → Intermittent failure, system is OK at this time (test drive may be necessary). Check for poor connections or loose wires at the thermostat housing, O_2 sensor and C210 (round connector located at the right shock tower).

YES

Inspect for poor connection or loose ground wires at thermostat housing.

Are connections and wires OK? — NO → **Repair as necessary.**

YES

Disconnect engine wire harness from O_2 sensor.

Connect a voltmeter between the O_2 sensor (+) and body ground (−).

Warm up engine to normal operating temperature again, then hold engine rpm at 4,000 for 10 seconds then release throttle completely while watching the voltmeter.

Was voltage above 0.6V at 4,000 rpm? Was voltage below 0.4V during closed throttle deceleration from 4,000 rpm? — NO → **Replace O_2 sensor.**

YES

Stop engine.

Reconnect O_2 sensor.

Connect the ECU test harness between the ECU and connector

Connect a voltmeter between C16 (+) and A18 (−) terminals.

Restart and warm up engine to normal operating temperature, then hold engine rpm at 4,000 rpm for 10 seconds then release throttle completely while watching the voltmeter.

Was voltage above 0.6V at 4,000 rpm? Was voltage below 0.4V during closed throttle deceleration from 4,000 rpm? — NO → **Repair short or open in WHT wire between ECU (C16) and O_2 sensor.**

YES

Substitute a known-good ECU and recheck. If symptom/ indication goes away, replace the original ECU.

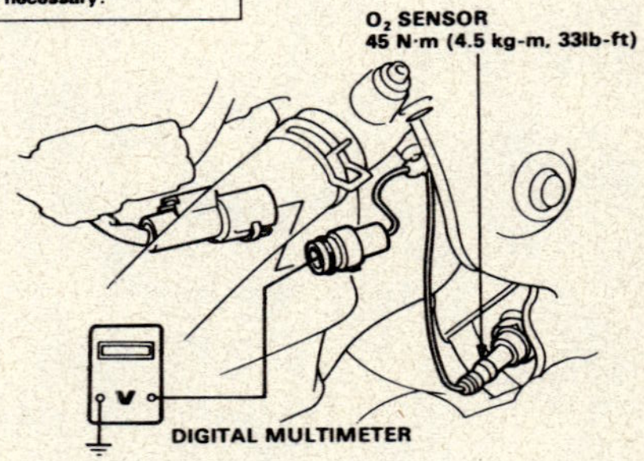

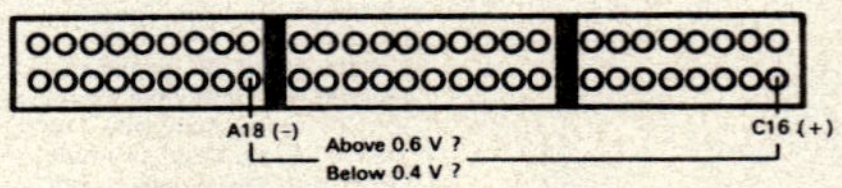

1.5L ENGINE – TROUBLESHOOTING – FAULT CODE 3 – CIVIC

Self-diagnosis LED indicates code 3: Most likely an electrical problem in the Manifold Absolute Pressure (MAP) Sensor system.

The MAP sensor converts manifold absolute pressure into electrical signals and inputs the ECU.

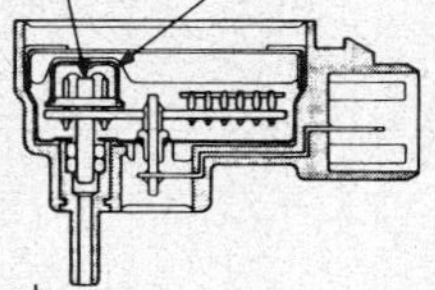

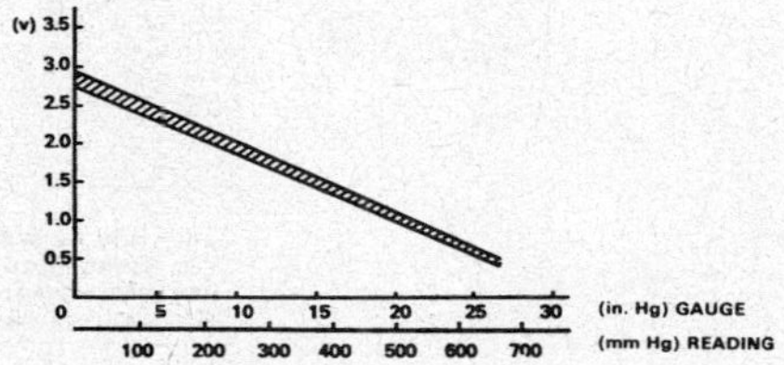

- —Engine is warm and running.
- —Check Engine light has been reported on.
- —LED indicates CODE 3.

Turn the ignition switch OFF.

Remove HAZARD fuse in the under-hood fuse box for 10 seconds to reset ECU.

Warm up engine to normal operating temperature (cooling fan comes on).

Is Check Engine light on and does LED indicate CODE 3?

NO → Intermittent failure, system is OK at this time. (test drive may be necessary). Check for poor connection or loose wires at MAP sensor connector and ECU.

YES ↓

Turn the ignition switch OFF.

Disconnect the 3P connector from the MAP sensor.

Turn the ignition switch ON.

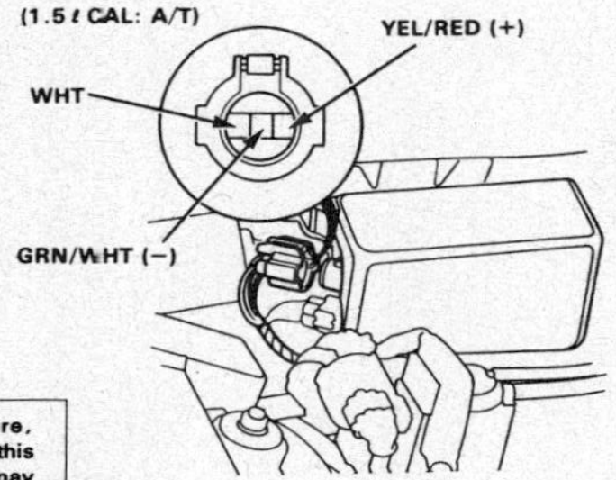

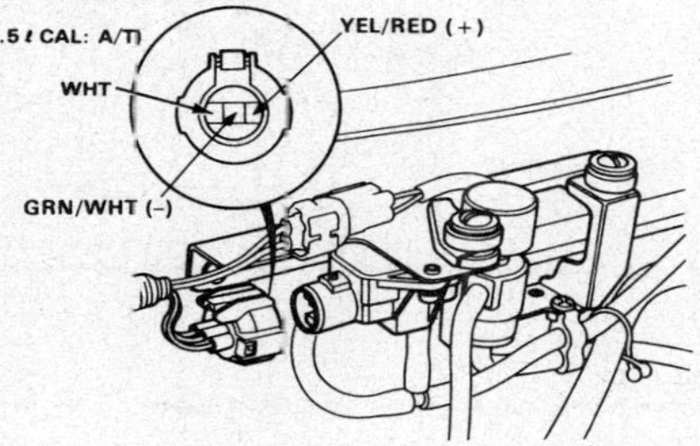

Measure voltage between YEL/RED (+) terminal and body ground.

Is there approx. 5V?

NO → Repair open in YEL/RED wire between ECU (C15) and MAP sensor. If wire is OK, substitute a known-good ECU and recheck. If prescribed voltage is now available, replace the original ECU.

YES ↓

Measure voltage between YEL/RED (+) terminal and GRN/WHT (–) terminal.

Is there approx. 5V?

NO → Repair open in GRN/WHT wire between ECU (C14) and MAP sensor. If wire is OK, substitute a known-good ECU and recheck. If prescribed voltage is now available, replace the original ECU.

YES ↓

Measure voltage between WHT (+) terminal and GRN/WHT (–) terminal.

Is there approx. 5V?

NO → Repair open or short in WHT wire between ECU (C11) and MAP sensor. If wire is OK, substitute a known-good ECU and recheck. If prescribed voltage is now available, replace the original ECU.

YES ↓

Turn the ignition switch OFF.

Reconnect the 3P connector to the MAP sensor.

Connect the ECU test harness between the ECU and connector

Turn the ignition switch ON.

Measure voltage between C11 (+) terminal and C14 (–) terminal.

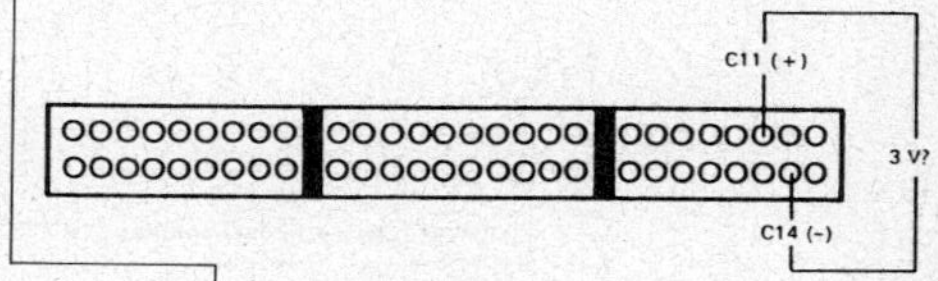

Is there approx. 3 V ?

NO → Replace MAP sensor.

YES ↓

Substitute a known-good ECU and recheck. If symptom/indication goes away, replace the original ECU.

1.5L ENGINE—TROUBLESHOOTING—FAULT CODE 4—CIVIC

Self-diagnosis LED indicates code 4: A problem in the CRANK circuit of the TDC/CRANK Sensor.

The CRANK sensor determines timing for fuel injection and ignition of each cylinder and also detects engine RPM. The TDC sensor determines ignition timing at start-up (cranking) and when crank angle signal is abnormal.

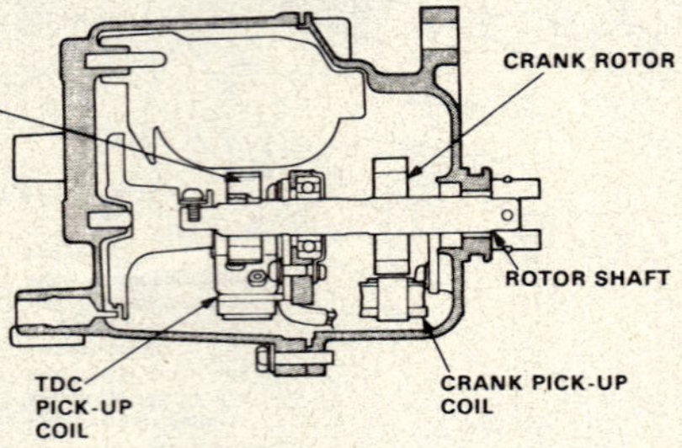

- **Check Engine light has been reported on.**
- **LED indicates CODE 4.**

Turn the ignition switch OFF.

Remove HAZARD fuse in the under-hood fuse box for 10 seconds to reset ECU.

Start engine.

Is Check Engine light on and does LED indicate CODE 4?

- NO → Intermittent failure, system is OK at this time (test drive may be necessary). Check for poor connections or loose wires at the distributor connector and C210 (round connector located at the right shock tower).
- YES ↓

Stop engine.

Disconnect 6P connector from the TDC/CRANK sensor.

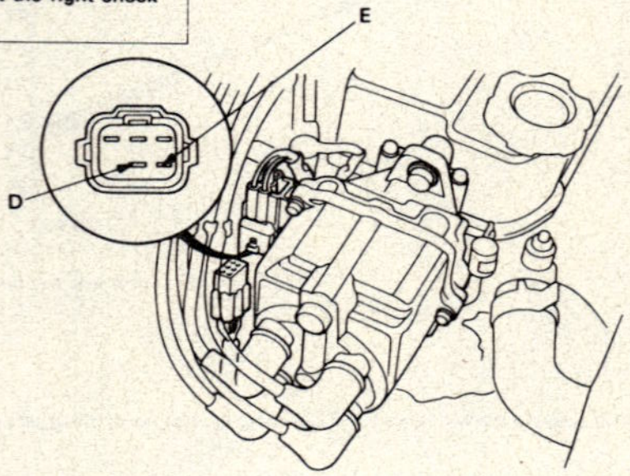

Measure resistance between D terminal and E terminal.

Is there 350–550 Ω ?

- NO → **Replace the CRANK sensor**
- YES ↓

Check for continuity to body ground on D terminal and E terminal individually.

Does continuity exist ?

- YES → **Replace the CRANK sensor**
- NO ↓

Reconnect the connector.

Connect the ECU test harness only to the main wire harness, not to the ECU.

Measure resistance between C1 terminal and C2 terminal.

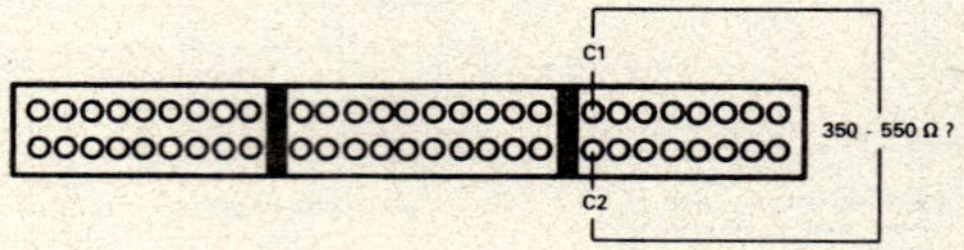

Is there 350–550 Ω ?

- NO → **Repair open in ORN and/or WHT wires.**
- YES ↓

Check for continuity to body ground on C1 terminal.

Does continuity exist?

- YES → **Repair short in ORN wire between ECU (C1) and distributor connector.**
- NO ↓

Substitute a known-good ECU and recheck. If symptom/indication goes away, replace the original ECU.

1.6L ENGINE – TROUBLESHOOTING – FAULT CODE 4 – CIVIC

Self-diagnosis LED indicates code 4: A problem in the CRANK circuit of the TDC/CRANK/CYL Sensor.

The CRANK sensor determines timing for fuel injection and ignition of each cylinder and also detects engine RPM. The TDC sensor determines ignition timing at start-up (cranking) and when crank angle is abnormal. The CYL sensor detects the position of No.1 cylinder for sequential fuel injection to each cylinder.

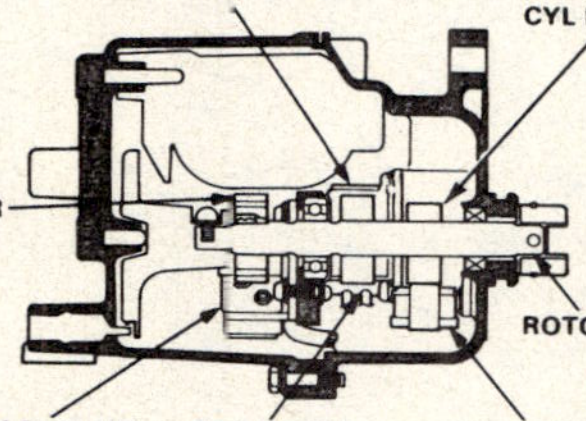

- Check Engine light has been reported on.
- LED indicates CODE 4.

Turn the ignition switch OFF.

Remove HAZARD fuse in the under-hood fuse box for 10 seconds to reset ECU.

Start engine.

Is Check Engine light on and does LED indicate CODE 4?

NO → Intermittent failure, system is OK at this time (test drive may be necessary). Check for poor connections or loose wires at the distributor connector and C210 (round connector located at the right shock tower).

YES ↓

Stop engine.

Disconnect the 8P connector from the TDC/CRANK/CYL sensor.

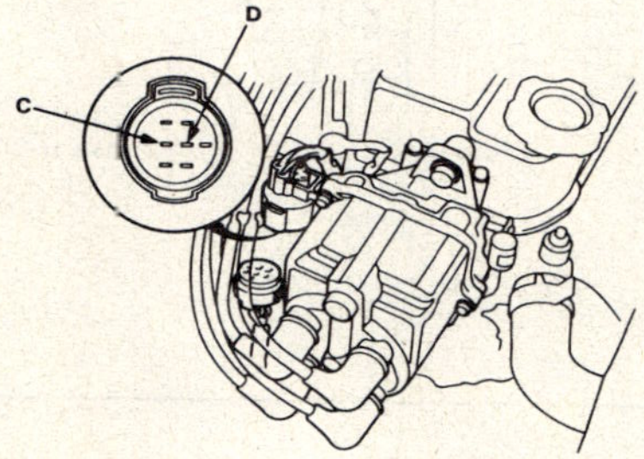

Measure resistance between C terminal and D terminal.

Is there 350-550 Ω ?

NO → Replace the distributor assembly

YES ↓

Check for continuity to body ground on C terminal and D terminal individually.

Does continuity exist ?

YES → Replace the distributor assembly

NO ↓

Reconnect the connector.

Connect the ECU test harness only to the main wire harness, not to the ECU

Measure resistance between B10 terminal and B12 terminal.

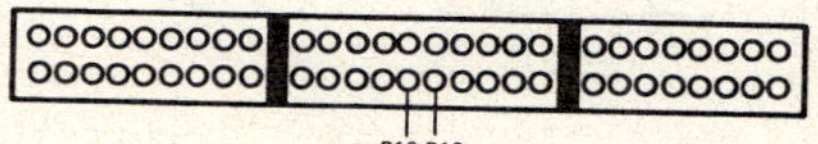

Is there 350-550 Ω ?

NO → Repair open in ORN and/or WHT wires.

YES ↓

Check for continuity to body ground on B10 terminal.

Does continuity exist?

YES → Repair short in ORN wire between ECU (B10) and distributor connector.

NO ↓

Substitute a known-good ECU and recheck. If symptom/indication goes away, replace the original ECU.

TROUBLESHOOTING—FAULT CODE 5—CIVIC

- Check Engine light has been reported on.
- LED indicates CODE 5.

Turn the ignition switch OFF.

Remove HAZARD fuse in the under-hood fuse box for 10 seconds to reset ECU.

Start the engine.

Is Check Engine light on and does LED indicate CODE 5?

NO →
- Intermittent failure, system is OK at this time (test drive may be necessary).
- Check vacuum hoses, pipes and connections.
- Make sure all connectors are secure.

YES ↓

Stop engine.

Connect vacuum pump to #21 hose and apply vacuum.

Does it hold vacuum?

NO → Connect a vacuum pump to the MAP sensor and apply vacuum.

Does it hold vacuum?

NO → Replace MAP sensor.

YES → Replace #21 hose.

YES ↓

Disconnect #21 hose from the throttle body and connect a T-fitting from a vacuum gauge between the throttle body and MAP sensor.

(1.6 ℓ)

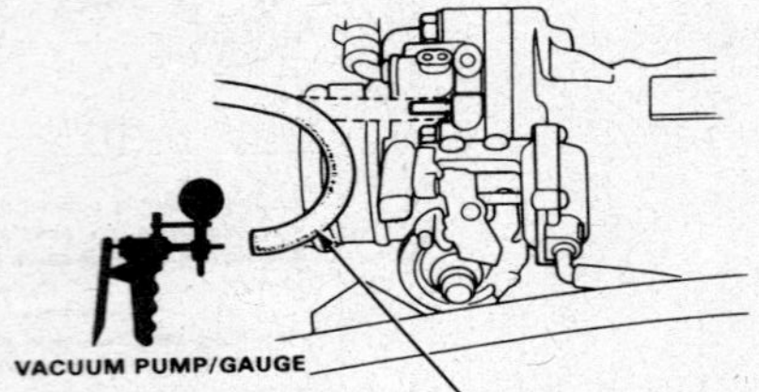

(1.5 ℓ)

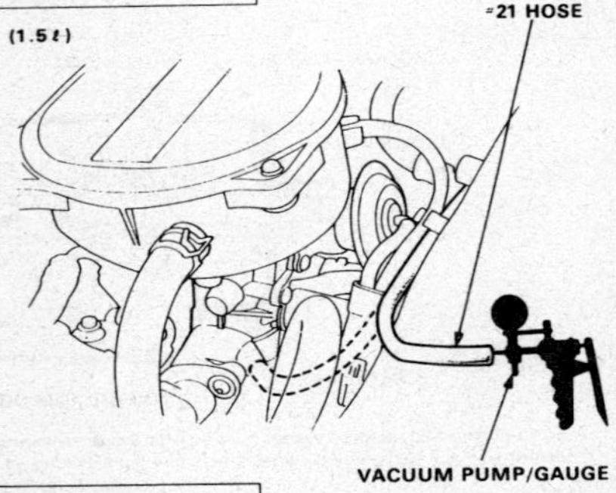

Start engine.

Is there manifold vacuum?

NO →
- Remove restriction from throttle body.
- Replace throttle body.

YES ↓

Stop engine.

Connect the ECU test harness between the ECU and connector

Turn the ignition switch ON.

Measure voltage between C11 (+) terminal and C14 (–) terminal.

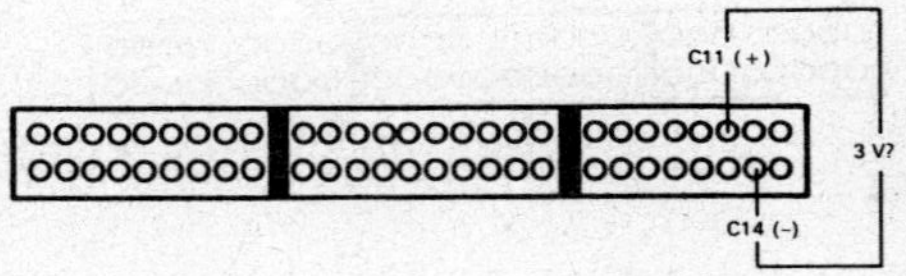

Is there approx. 3 V ?

NO → Replace the MAP sensor.

YES ↓

Start the engine and allow it to idle.

Is there approx 1V?

NO → Replace MAP sensor.

YES ↓

Substitute a known-good ECU and recheck. If symptom/indication goes away, replace the original ECU.

TROUBLESHOOTING – FAULT CODE 6 – CIVIC

Self-diagnosis LED indicates code 6: Most likely a problem in the Coolant Temperature (TW) Sensor circuit.

The TW sensor is a temperature dependant resistor (thermistor). The resistance of the thermistor decreases as the coolant temperature increases as shown below.

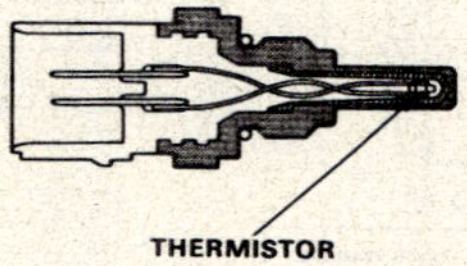

THERMISTOR

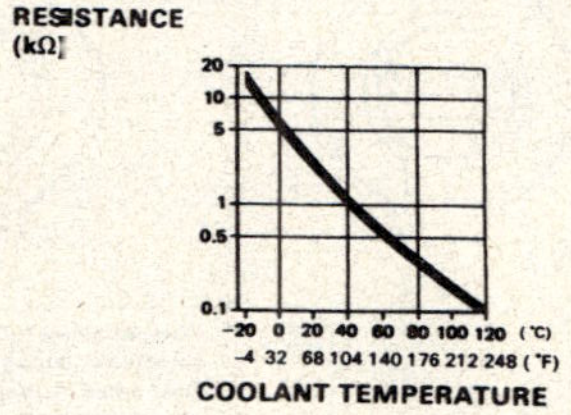

- —Check Engine light has been reported on.
  —LED indicates CODE 6.
- Turn the ignition switch OFF.
- Remove HAZARD fuse in the under-hood fuse box for 10 seconds to reset ECU.
- Turn the ignition switch ON.
- Is Check Engine light on and does LED indicate CODE 6?
  - NO → Intermittent failure, system is OK at this time (test drive may be necessary). Check for poor connections or loose wires at TW sensor, C210 and C212 (1.6 ℓ) round connector and square connector located at the right shock tower).
  - YES ↓
- Warm up engine to normal operating temperature (cooling fan comes on).
- Turn the ignition switch OFF.
- Disconnect the 2P connector from the TW sensor.
- Measure resistance between the 2 terminals on the TW sensor.
- Is there 200—400 Ω ?
  - NO → Replace TW sensor.
  - YES ↓
- Turn the ignition switch ON.
- Measure voltage between RED/WHT (+) and body ground.
- Is there approx. 5V ?
  - NO → Repair open or short in RED/WHT wire between ECU (C6) and TW sensor. If wire is OK, substitute a known-good ECU and recheck. If prescribed voltage is now available, replace original ECU.
  - YES ↓
- Measure voltage between RED/WHT (+) terminal and GRN/WHT (–) terminal.
- Is there approx. 5V ?
  - NO → Repair open in GRN/WHT wire between ECU (C12) and TW sensor.
  - YES ↓
- Substitute a known-good ECU and recheck. If symptom/indication goes away, replace the original ECU.

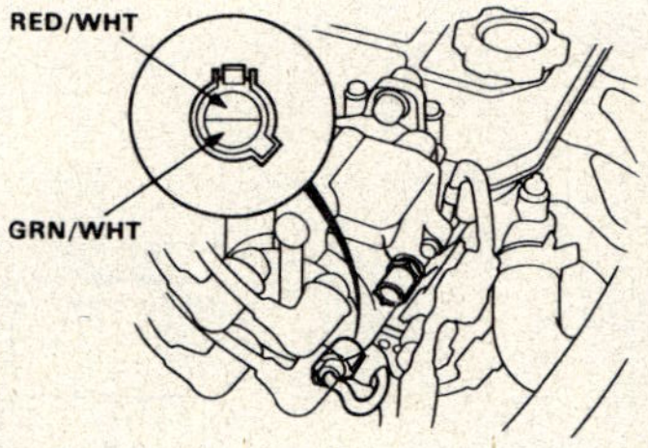

TROUBLESHOOTING—FAULT CODE 7—CIVIC

Self-diagnosis LED indicates code 7: Most likely a problem in the Throttle Angle Sensor circuit.

The throttle angle sensor is a potentiometer. It is connected to the throttle valve shaft. As the throttle angle changes, the throttle angle sensor varies the voltage signal to the ECU.

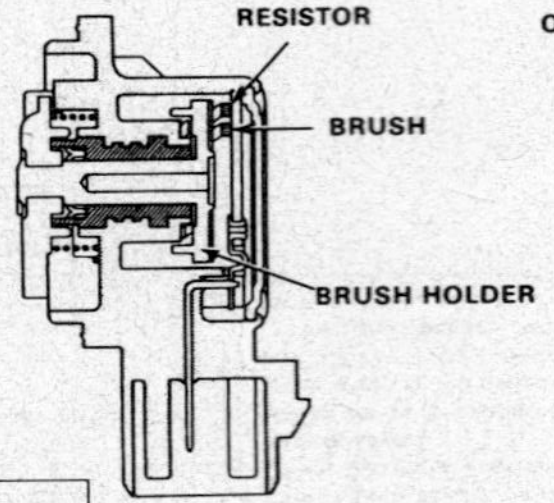

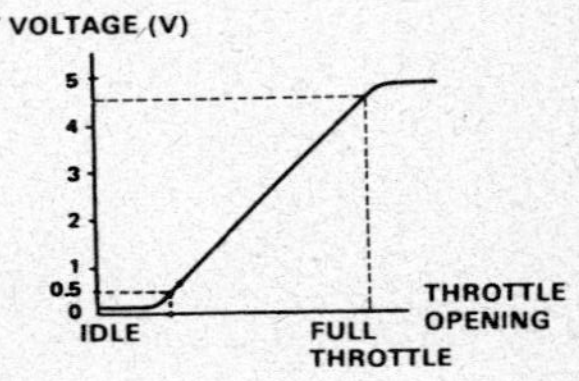

- Engine is running.
- Check Engine light has been reported on.
- LED indicates CODE 7.

Turn the ignition switch OFF.

Remove HAZARD fuse in the under-hood fuse box for 10 seconds to reset ECU.

Start engine.

Is Check Engine light on and does LED indicate CODE 7?

NO → **Intermittent failure, system is OK at this time (test drive may be necessary). Check for poor connections or loose wires at throttle angle sensor and C210 (round connector located at the right shock tower).**

YES → Turn the ignition switch OFF.

Disconnect the 3P connector from the throttle angle sensor.

Turn the ignition switch ON.

Measure voltage between YEL/WHT (+) terminal and GRN/WHT (−) terminal.

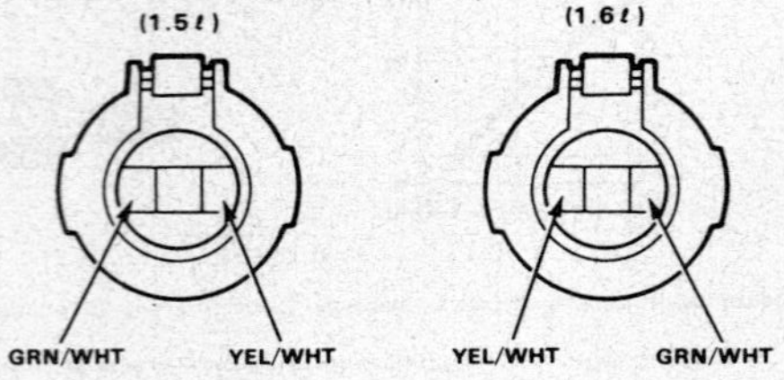

Is there approx. 5V ?

YES → Go to Path C

NO → Measure voltage between YEL/WHT (+) terminal and body ground. → Go to Path D

Path C

Turn the ignition switch OFF.

Reconnect the 3P connector.

Connect the ECU test harness between the ECU and connector

Turn the ignition switch ON.

Disconnect # 6 hose from the dashpot diaphragm and connect a vacuum pump to the diaphragm.

Apply 20 in. Hg to the diaphragm.

Measure voltage between C7(+) terminal and C12(−) terminal.

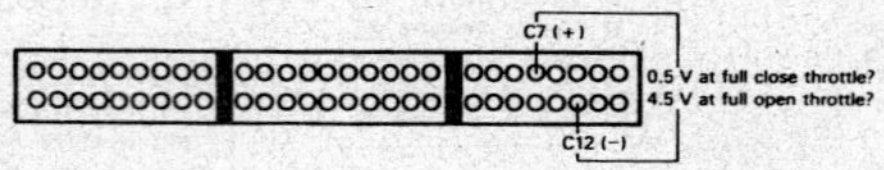

Is voltage approx. 0.5 V at full close throttle (applying vacuum to the dashpot diaphragm), and approx. 4.5 V at full open throttle? NOTE: There should be a smooth transition from 0.5 V to 4.5 V as the throttle is depressed.

NO →
- **Replace throttle angle sensor.**
- **Repair open or short in RED/BLU wire between ECU (C7) and throttle angle sensor.**

YES → **Substitute a known-good ECU and recheck. If symptom/indication goes away, replace the original ECU.**

Path D

Is there approx. 5V ?

YES → **Repair open in GRN/WHT wire between ECU (C12) and throttle angle sensor.**

NO → Turn the ignition switch OFF.

Connect the ECU test harness between the ECU and connector (page 11-23).

Turn the ignition switch ON.

Measure voltage between C13 (+) terminal and C12 (−) terminal.

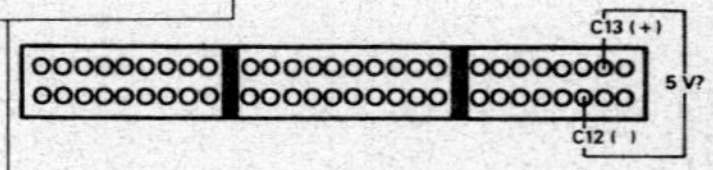

Is there approx. 5V ?

YES → **Repair open in YEL/WHT wire between ECU(C13) and throttle angle sensor.**

NO → **Substitute a known-good ECU and recheck. If prescribed voltage is now available, replace the original ECU.**

1.5L ENGINE—TROUBLESHOOTING—FAULT CODE 8—CIVIC

- Check Engine light has been reported on.
- LED indicates CODE 8.

Turn the ignition switch OFF.

Remove HAZARD fuse in the under-hood fuse box for 10 seconds to reset ECU.

Start engine.

Is Check Engine light on and does LED indicate CODE 8?

- NO → Intermittent failure, system is OK at this time (test drive may be necessary). Check for poor connections or loose wires at distributor connector and C210 (round connector located at the right shock tower).
- YES → Stop engine.

Disconnect the 6P connector from the TDC/CRANK sensor.

Measure resistance between B terminal and C terminal.

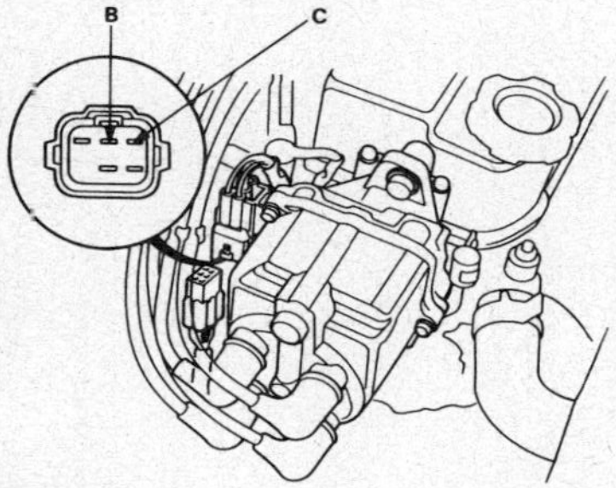

Is there 350—550 Ω ?

- NO → Replace the TDC sensor
- YES → Check for continuity to body ground on B terminal and C terminal individually.

Does continuity exist ?

- YES → Replace the TDC sensor
- NO → Reconnect the connector.

Connect the ECU test harness only to the main wire harness, not to the ECU

Measure resistance between C3 terminal and C4 terminal.

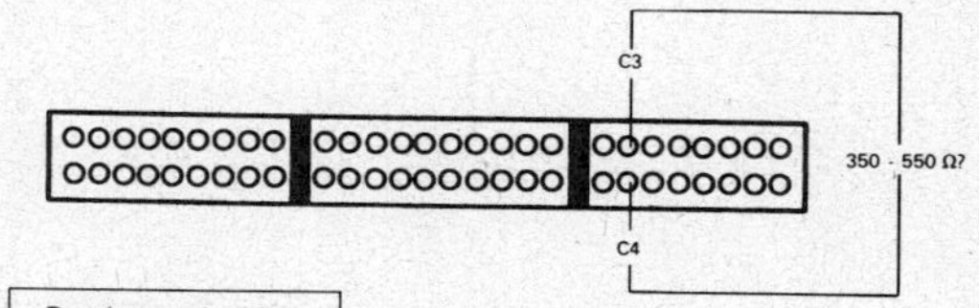

Is there 350 — 550 Ω ?

- NO → Repair open in ORN/BLU and/or WHT/BLU wires.
- YES → Check for continuity to body ground on C3 terminal.

Does continuity exist?

- YES → Repair short in ORN/BLU wire between ECU (C3) and distributor connector.
- NO → Substitute a known-good ECU and recheck. If symptom/indication goes away, replace the original ECU.

1.6L ENGINE—TROUBLESHOOTING—FAULT CODE 8—CIVIC

- Check Engine light has been reported on.
- LED indicates CODE 8.

Turn the ignition switch OFF.

Remove HAZARD fuse in the under-hood fuse box for 10 seconds to reset ECU.

Start engine.

Is Check Engine light on and does LED indicate CODE 8?

- NO → Intermittent failure, system is OK at this time (test drive may be necessary). Check for poor connections or loose wires at the distributor connector and C210 (round connector located at the right shock tower).
- YES ↓

Stop engine.

Disconnect the 8P connector from the TDC/CRANK/CYL sensor.

Measure resistance between A terminal and B terminal.

Is there 350—550 Ω ?

- NO → Replace the distributor assembly !
- YES ↓

Check for continuity to body ground on A terminal and B terminal individually.

Does continuity exist ?

- YES → Replace the distributor assembly
- NO ↓

Reconnect the connector.

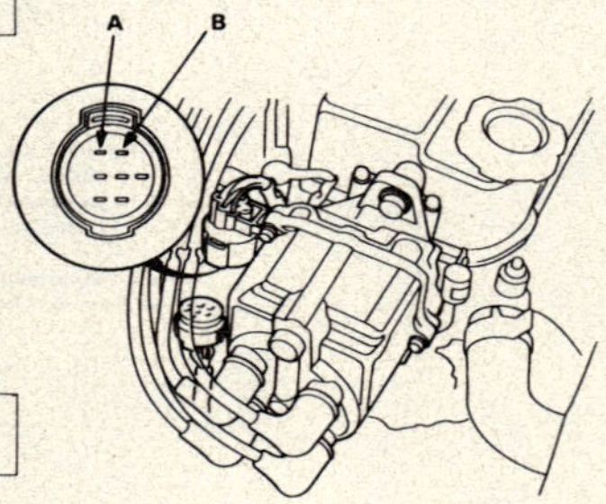

Connect the ECU test harness only to the main wire harness, not to the ECU

Measure resistance between C3 terminal and C4 terminal.

Is there 350 — 550 Ω ?

- NO → Repair open in ORN/BLU and/or WHT/BLU wires.
- YES ↓

Check for continuity to body ground on C3 terminal.

Does continuity exist?

- YES → Repair short in ORN/BLU wire between ECU (C3) and distributor connector.
- NO ↓

Substitute a known-good ECU and recheck. If symptom/indication goes away, replace the original ECU.

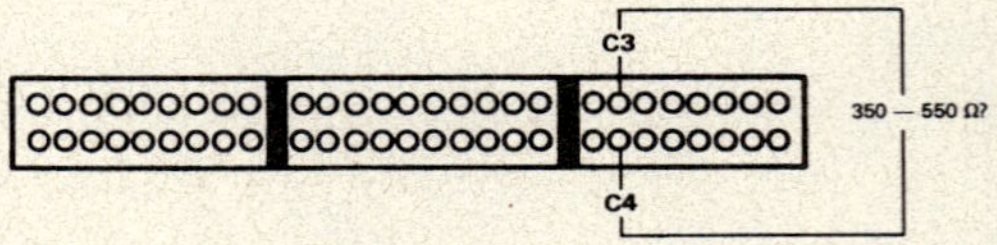

1.6L ENGINE – TROUBLESHOOTING – FAULT CODE 9 – CIVIC

- Check Engine light has been reported on.
- LED indicates CODE 9.

Turn the ignition switch OFF.

Remove HAZARD fuse in the under-hood fuse box for 10 seconds to reset ECU.

Start engine.

Is Check Engine light on and does LED indicate CODE 9?

- NO: Intermittent failure, system is OK at this time (test drive may be necessary). Check for poor connections or loose wires at the distributor connector and C210 (round connector located at the right shock tower).
- YES: Stop engine.

Disconnect the 8P connector from the TDC/CRANK/CYL sensor.

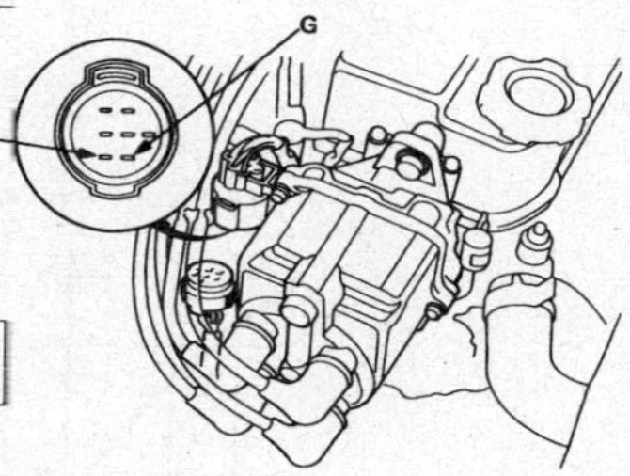

Measure resistance between F terminal and G terminal.

Is there 350—550 Ω ?

- NO: Replace the distributor assembly
- YES: Check for continuity to body ground on F terminal and G terminal individually.

Does continuity exist ?

- YES: Replace the distributor assembly
- NO: Reconnect the connector.

Connect the ECU test harness only to the main wire harness, not to the ECU

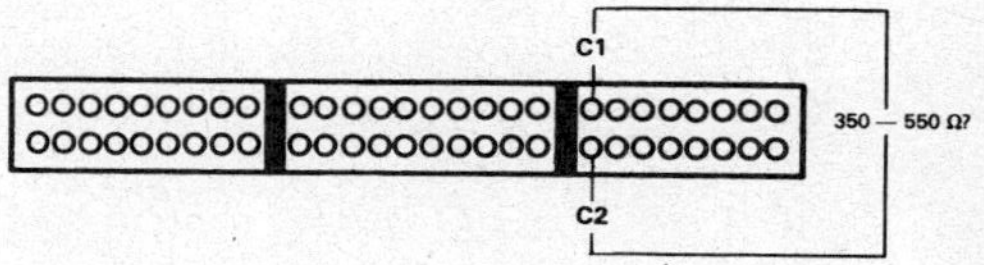

Measure resistance between C1 terminal and C2 terminal.

Is there 350 — 550 Ω ?

- NO: Repair open in BLU/GRN and/or BLU/YEL wires.
- YES: Check for continuity to body ground on C1 terminal.

Does continuity exist?

- YES: Repair short in BLU/GRN wire between ECU (C1) and distributor connector.
- NO: Substitute a known-good ECU and recheck. If symptom/indication goes away, replace the original ECU.

TROUBLESHOOTING—FAULT CODE 10—CIVIC

Self-diagnosis LED indicates code 10: Most likely a problem in the Intake Air Temperature (TA) Sensor circuit.

The TA sensor is a temperature dependant resistor (thermistor). The resistance of the thermistor decreases as the intake air temperature increases as shown below.

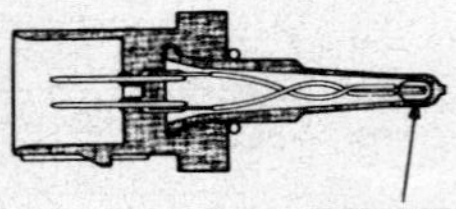

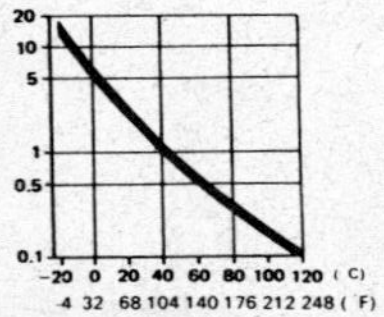

- Check Engine light has been reported on.
- LED indicates CODE 10.

Turn the ignition switch OFF.

Remove HAZARD fuse in the under-hood fuse box for 10 seconds to reset ECU.

Turn the ignition switch ON.

Is Check Engine light on and does LED indicate CODE 10?

- NO: Intermittent failure, system is OK at this time (test drive may be necessary). Check for poor connections or loose wires at TA sensor, C210 and C212 (1.6 ℓ) (round connector and square connector located at the right shock tower).
- YES: Turn the ignition switch OFF.

Disconnect the 2P connector from the TA sensor.

Measure resistance between the 2 terminals on the TA sensor.

Is there 1–4 kΩ ?

- NO: Replace TA sensor.
- YES: (continued)

(1.6 ℓ)

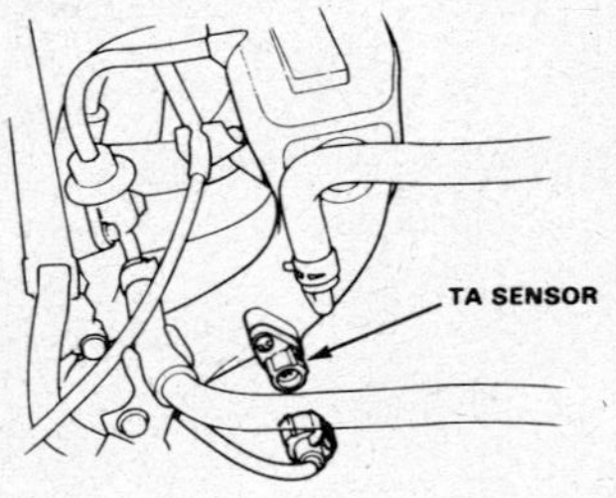

(1.5 ℓ)

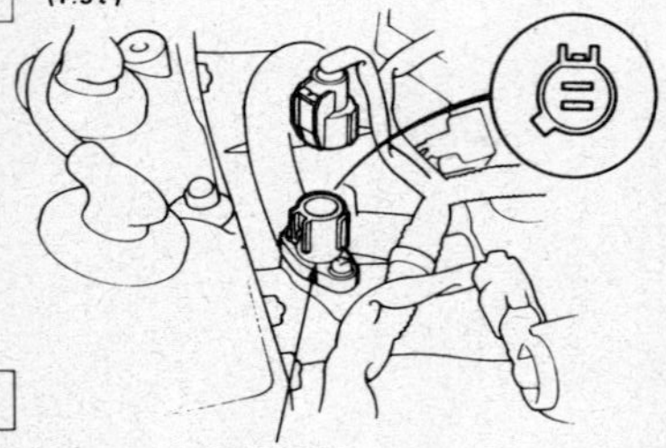

Turn the ignition switch ON.

Measure voltage between RED/YEL (+) terminal and body ground.

Is there approx. 5 V ?

- NO: Repair open or short in RED/YEL wire between ECU (C5) and TA sensor. If wire is OK, substitute a known-good ECU and recheck. If prescribed voltage is now available, replace original ECU.
- YES: Measure voltage between RED/YEL (+) terminal and GRN/WHT (−) terminal.

Is there approx. 5 V ?

- NO: Repair open in GRN/WHT wire between ECU (C12) and TA sensor.
- YES: Substitute a known-good ECU and recheck. If symptom/indication goes away, replace the original ECU.

1.5L ENGINE—TROUBLESHOOTING—FAULT CODE 12
CIVIC WITH AUTOMATIC TRANSAXLE, CALIFORNIA ONLY

Self-diagnosis indicates code 12: Most likely a problem in the Exhaust Gas Recirculation (EGR) system.

The EGR System is designed to reduce oxides of nitrogen emissions (NOx) by recirculating exhaust gas through the EGR valve and the intake manifold into the combustion chambers. It is composed of the EGR valve, CVC valve, EGR control Solenoid valve, PGM-FI ECU and various sensors.
The ECU contains memories for ideal EGR valve lifts for varying operating conditions. The EGR valve lift sensor detects the amount of EGR valve lift and sends the information to the ECU. The ECU then compares it with the ideal EGR valve lift which is determined by signals sent from the other sensors. If there is any difference between the two, the ECU cuts current to the EGR control solenoid valve to reduce vacuum applied to the EGR valve.

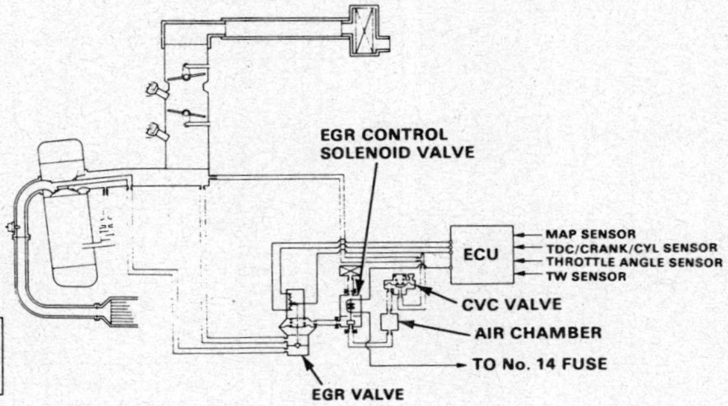

- Check Engine light has been reported on.
- LED indicates CODE 12.

Verify that all connectors between the EGR valve lift sensor and the ECU are securely connected.

Warm up engine to normal operating temperature (cooling fan comes on.)

Disconnect #16 hose from EGR valve.

Connect a vacuum pump to EGR valve and apply vacuum.

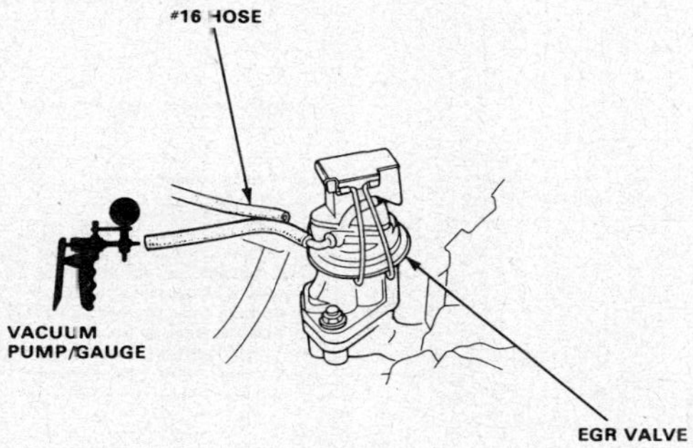

Does EGR valve hold vacuum? Did engine stall?

NO → Replace EGR valve.

YES ↓

Disconnect #12 hose from the joint and connect a vacuum gauge.

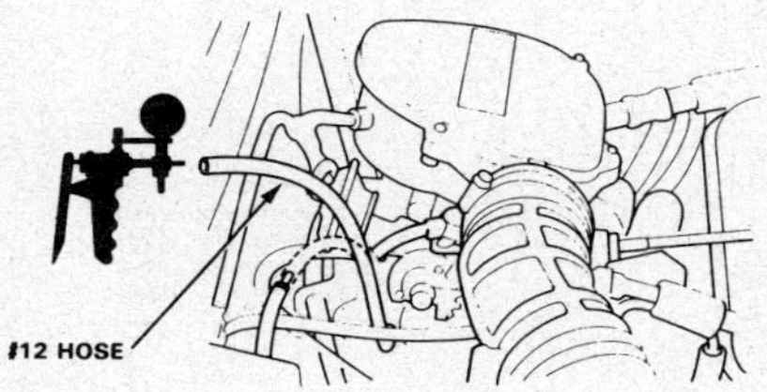

Restart engine and check for vacuum.

Is there manifold vacuum?

NO → Check vacuum hoses, pipes and connections.

YES ↓

Reconnect #12 hose.

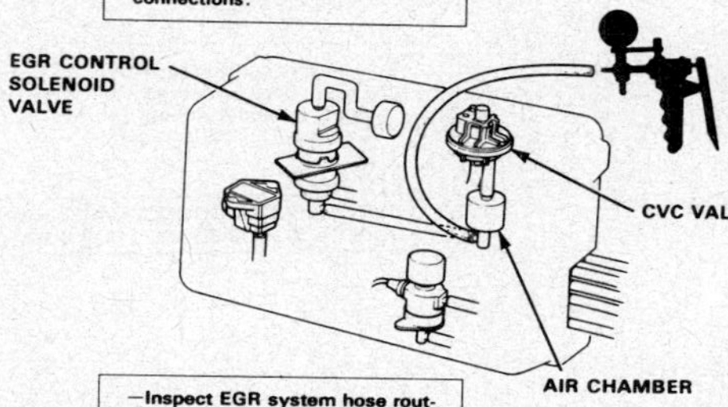

Disconnect the #16 vacuum hose from the EGR control solenoid valve, connect vacuum gauge to the #16 vacuum hose and check for vacuum.

Is there vacuum?

YES →
- Inspect EGR system hose routing.
- Check for a short in RED wire between ECU (A10) and EGR control solenoid valve. If wire is OK, replace the solenoid valve.

NO ↓

Turn the ignition switch OFF.

Disconnect the 4P connector from the control box.

1.5L ENGINE—TROUBLESHOOTING—FAULT CODE 12 CIVIC WITH AUTOMATIC TRANSAXLE, CALIFORNIA ONLY

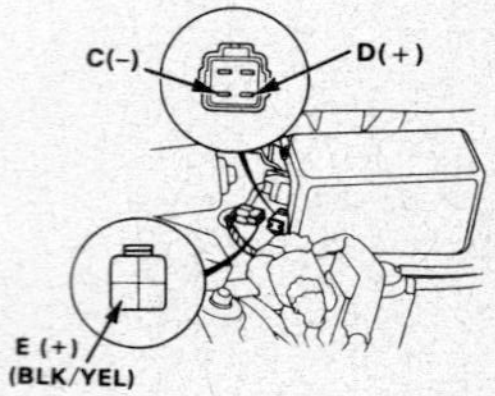

- Connect the battery positive terminal to D terminal and the negative terminal to C terminal.
- Restart engine and check for vacuum.
- Is there approx. 8"?
  - NO → Remove the control box cover, disconnect the vacuum hose between the air chamber and EGR control solenoid valve from the air chamber, and connect a vacuum gauge to the air chamber and check for vacuum.
    - Is there approx. 8" ?
      - NO → **Check vacuum hose routing to the CVC valve.**
      - YES → **Replace EGR control solenoid valve.**
  - YES → Measure voltage between E: BLK/YEL (+) terminal on the main wire harness and body ground.
- Is there battery voltage ?
  - NO → **Repair open in BLK/YEL wire between the solenoid valve and No. 14 fuse.**
  - YES → Turn the ignition switch OFF.
- Connect the test harness between the EGR valve lift sensor and engine wire harness.
- Turn the ignition switch ON.
- Measure voltage between RED (+) terminal and GRN (–) terminal.

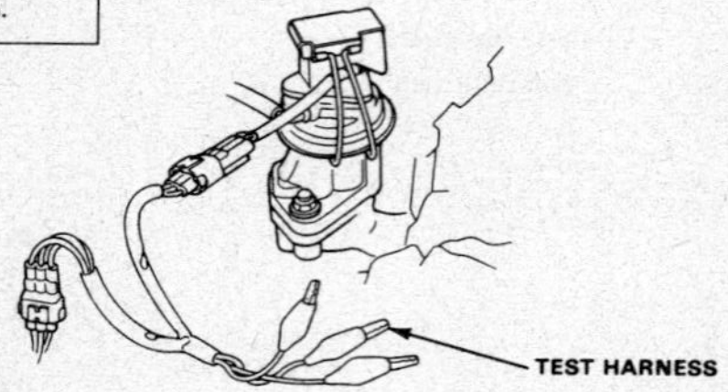

- Is there approx. 5V ?
  - NO → Measure voltage between RED (+) terminal and body ground.
    - Is there approx. 5V ?
      - YES → **Repair open in GRN/WHT wire between ECU (C12) and the sensor.**
      - NO → **Repair open in YEL/WHT wire between ECU (C13) and the sensor. If OK, substitute a known-good ECU and recheck. If prescribed voltage is now available, replace the original ECU.**
  - YES → Measure voltage between WHT (+) terminal and GRN (–) terminal.
- Is there approx. 1V ?
  - NO → **–Replace EGR valve.**
    –Repair short in YEL wire between ECU (C8) and EGR valve lift sensor.
  - YES ↓

- Turn the ignition switch OFF.
- Connect the ECU test harness between the ECU and connector
- Turn the ignition switch ON.
- Measure voltage between C8 (+) terminal and C12 (–) terminal.

- Is there approx. 1V ?
  - NO → **Repair open in YEL wire between ECU (C8) and the sensor.**
  - YES → Start the engine and allow it to idle.
- Using a jumper wire, connect A10 (+) terminal and A2 (–) terminal.

- Is there battery voltage ?
  - NO → **Repair open in RED wire between ECU (A10) and EGR control solenoid valve.**
  - YES → **Substitute a known-good ECU and recheck. If symptom/indication goes away, replace the original ECU.**

TROUBLESHOOTING—FAULT CODE 13—CIVIC

Self-diagnosis LED indicates code 13: A problem in the Atmospheric Pressure (PA) Sensor circuit.

The PA sensor converts atmospheric pressure into electrical signals and inputs the ECU.

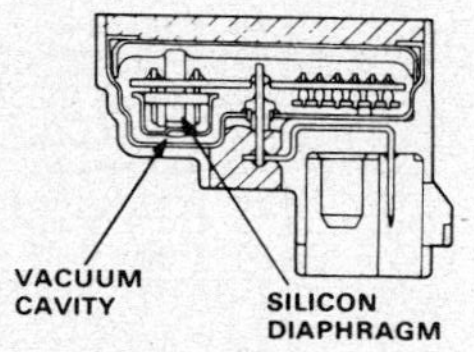

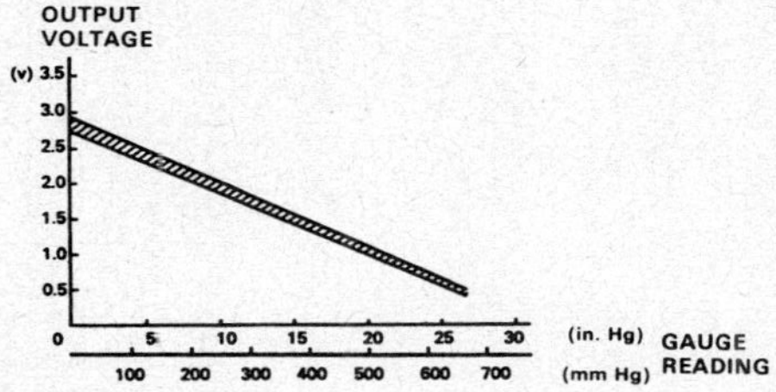

- Check Engine light has been reported on.
- LED indicates CODE 13.

Turn the ignition switch OFF.

Remove HAZARD fuse in the under-hood fuse box for 10 seconds to reset ECU.

Turn the ignition switch ON.

Is Check Engine light on and does LED indicate CODE 13?

- NO → Intermittent failure, system is OK at this time (test drive may be necessary). Check for poor connections or loose wires at the PA sensor.
- YES ↓

Turn the ignition switch OFF.

Disconnect the main wire harness from PA sensor.

Measure voltage between YEL/WHT (+) terminal and body ground.

Is there approx. 5V?

- NO → Repair open in YEL/WHT wire between ECU (C13) and the sensor. If wire is OK, substitute a known-good ECU and recheck. If prescribed voltage is now available replace the original ECU.
- YES → Measure voltage between YEL/WHT (+) terminal and GRN/WHT (−) terminal.

Is there approx. 5V?

- NO → Repair open in GRN/WHT wire between ECU (C12) and the sensor. If wire is OK, substitute a known-good ECU and recheck. If prescribed voltage is now available, replace the original ECU.
- YES ↓

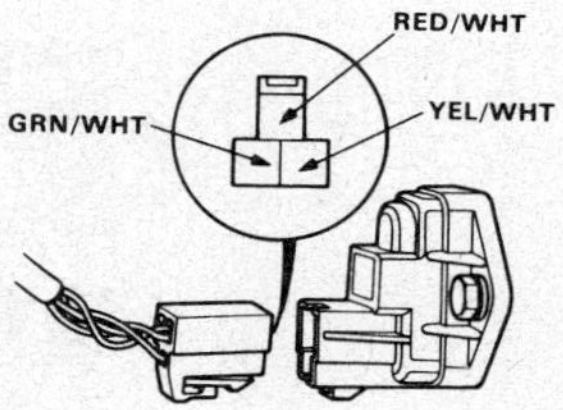

Measure voltage between RED/WHT (+) terminal and GRN/WHT (−) terminal.

Is there approx. 5V?

- NO → Repair open or short in RED/WHT wire between ECU (C9) and PA sensor. If wire is OK, substitute a known-good ECU and recheck. If prescribed voltage is now available, replace the original ECU.
- YES ↓

Turn the ignition switch OFF.

Remove HAZARD fuse in the under-hood fuse box for 10 seconds to reset ECU.

Substitute a known-good PA sensor.

Turn the ignition switch ON.

Is Check Engine light on and does LED indicate CODE 13?

- NO → Replace the original PA sensor.
- YES → Substitute a known-good ECU and recheck. If prescribed voltage is now available, replace the original ECU.

TROUBLESHOOTING—FAULT CODE 14—CIVIC

Self-diagnosis LED indicates code 14: A problem in the Electronic Air Control Valve (EACV) circuit.

The EACV changes the amount of air bypassing the throttle body in response to a current signal from the ECU in order to maintain the proper idle speed.

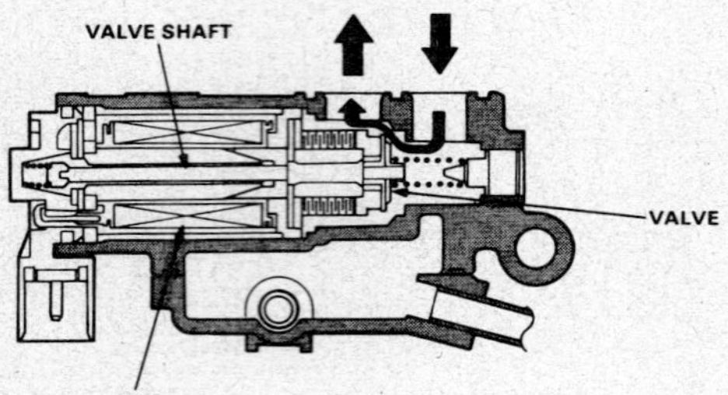

- Check Engine light has been reported on.
- LED indicates CODE 14.

Turn the ignition switch OFF.

Remove HAZARD fuse in the under-hood fuse box for 10 seconds to reset ECU.

Start engine.

Is Check Engine light on and does LED indicate CODE 14?
- NO → Intermittent failure, system is OK at this time (test driving may be necessary). Check for poor connections or loose wires at EACV and at C313.
- YES ↓

Stop engine.

Disconnect the 2P connector from the EACV.

Measure resistance between the 2 terminals on the EACV.

Is there 8—15Ω ?
- NO → Replace EACV.
- YES ↓

Check for continuity to body ground on each terminal on the EACV.

Does continuity exist ?
- YES → Replace EACV.
- NO ↓

Turn the ignition switch ON.

(1.6 ℓ)

(1.5 ℓ)

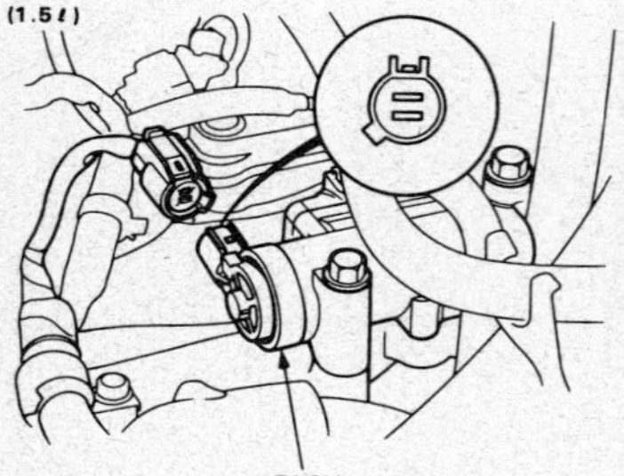

Measure voltage between BLK/YEL (+) terminal and BLU/YEL (–) terminal.

Is there battery voltage ?
- YES → Disconnect the "A" connector from the ECU. → Measure voltage between BLK/YEL (+) terminal and BLU/YEL (–) terminal. → Is there battery voltage ?
  - YES → Repair short in BLU/YEL wire between ECU (A11) and EACV.
  - NO → Substitute a known-good ECU and recheck. If symptom/indication goes away, replace the original ECU.
- NO ↓

Measure voltage between BLK/YEL (+) terminal and body ground.

Is there battery voltage ?
- NO → Repair open in BLK/YEL wire between the EACV and No.14 fuse.
- YES ↓

Turn the ignition switch OFF.

Reconnect the 2P connector to EACV.

Connect the ECU test harness "A" connector to the main wire harness only, not the ECU !

Turn the ignition switch ON.

Measure voltage between A11 (+) terminal and A2 (–) terminal.

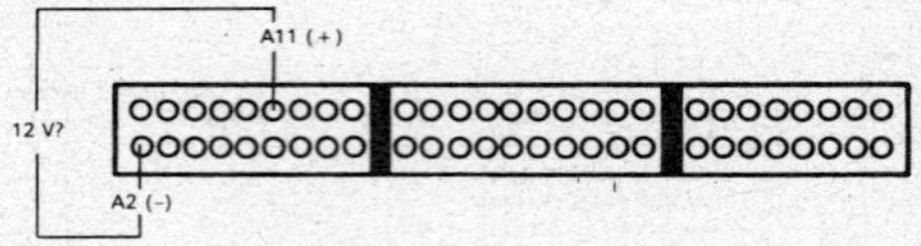

Is there battery voltage ?
- NO → Repair open in BLU/YEL wire between ECU (A11) and EACV.
- YES ↓

Connect and disconnect A11 terminal to A2 terminal.

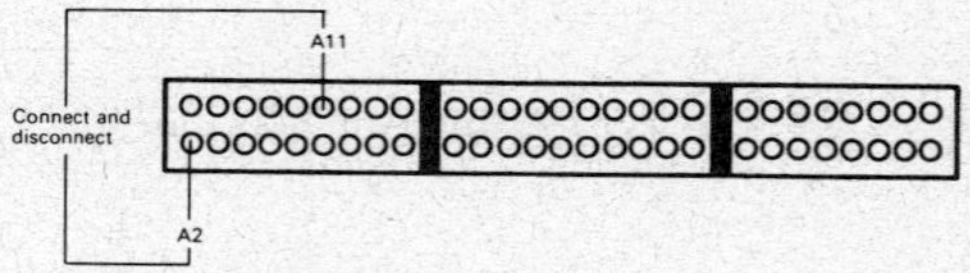

Does EACV click when the connector is connected and disconnected ?
- NO → Replace EACV.
- YES ↓

Substitute a known-good ECU and recheck. If symptom/indication goes away, replace the original ECU.

TROUBLESHOOTING—FAULT CODE 15—CIVIC

Self-diagnosis LED indicates code 15: A problem in the Ignition Output Signal circuit.

- —Check Engine light has been reported on.
  —LED indicates CODE 15.
- Turn the ignition switch OFF.
- Remove HAZARD fuse in the under-hood fuse box for 10 seconds to reset ECU.
- Start engine.
- Is Check Engine light on and does LED indicate CODE 15?
  - NO → Intermittent failure, system is OK at this time (test drive may be necessary). Check for poor connections or loose wires at the distributor connector and C211 (square connector located at the right shock tower).
  - YES ↓
- Turn the ignition switch OFF.
- Disconnect the 2P connector from the distributor.
- Turn the ignition switch ON.
- Measure voltage between BLK/YEL (+) terminal and body ground.
- Is there battery voltage ?
  - NO → Repair open in BLK/YEL wire between the 2P connector and ignition switch.
  - YES ↓
- Turn the ignition switch OFF.
- Reconnect the 2P connector.

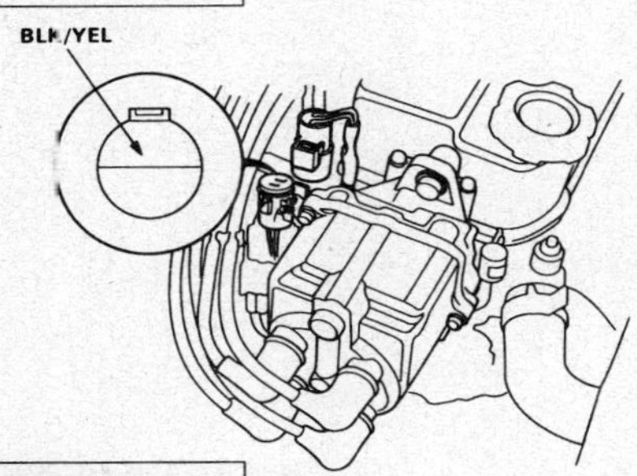

- Connect the ECU test harness to the main harness only (page 11-23).
- Turn the ignition switch ON.
- Measure voltage individually between B15 (+), B17 (+) terminals and A18 (–) terminal.
- Is there approx. 10V?
  - NO → —Replace the igniter unit.
    —Repair open or short in WHT wires between distributor and ECU (B15 or B17).
  - YES ↓
- Substitute a known-good ECU and recheck. If symptom/indication goes away, replace the original ECU.

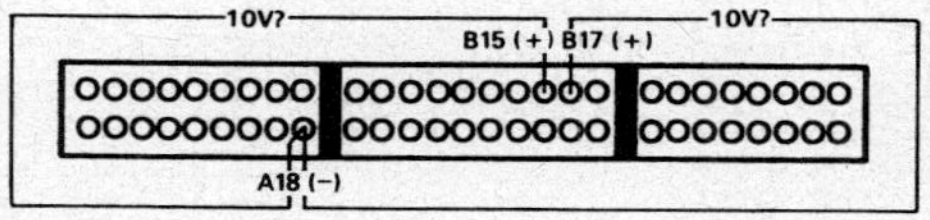

1.5L ENGINE – TROUBLESHOOTING – FAULT CODE 16 – CIVIC

Self-diagnosis LED indicates code 16: A problem in the fuel injector circuit.

The injectors are the solenoid-actuated constant-stroke pintle type consisting of a solenoid, plunger needle valve and housing. When current is applied to the solenoid coil, the valve lifts up and pressurized fuel is injected. Because the needle valve lift and the fuel pressure are constant, the injection quantity is determined by the length of time that the valve is open (i.e., the duration the current is supplied to the solenoid coil). The injector is sealed by two O-rings and a mount rubber. This also reduces operating noise.

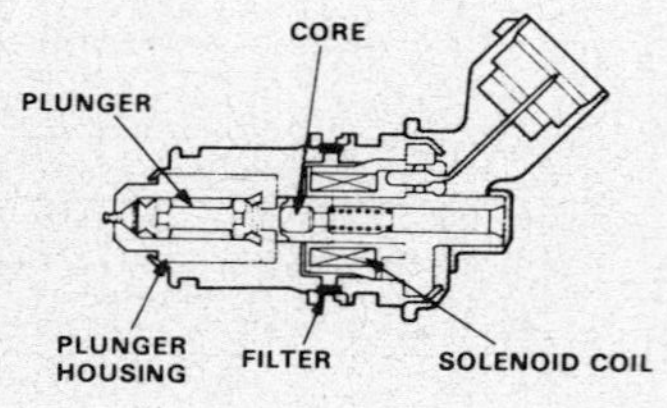

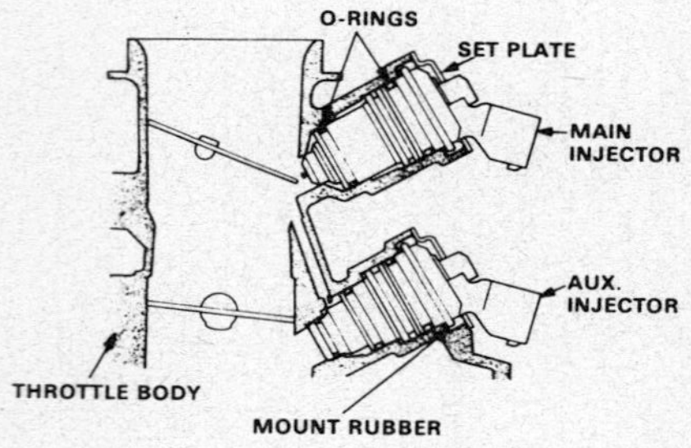

- Check Engine light has been reported on.
- LED indicates CODE 16.

Turn the ignition switch OFF.

Check for loose wires or connectors at:
- C101 and C102 (right shock tower)
- C122 (left shock tower)
- G101 (thermostat housing)

Are connections OK? — NO → **Repair as necessary.**

YES ↓

Remove HAZARD fuse in the under-hood fuse box for 10 seconds to reset ECU.

Start engine and hold at 2000 rpm for 1 minute.

NOTE: If engine will not start, continue cranking for at least 15 seconds to reproduce CODE on ECU.

Does the engine start and run at 2000 rpm for 1 minute? — NO → Is Check Engine light on and does LED indicate CODE 16? — NO → **Troubleshooting**

Is Check Engine light on and does LED indicate CODE 16? — YES → (see right column)

Left column (engine starts — YES):

Is Check Engine light on and does LED indicate CODE 16? — NO → **Intermittent failure (test drive may be necessary).**

YES ↓

Turn the ignition switch OFF.

Disconnect the 2P connector from the main injector.

Measure resistance between the 2 terminals of the injector.

Is resistance 0.6–1.6 Ω? — NO → **Replace the injector.**

YES ↓

Connect voltmeter positive to the main injector's YEL/BLK wire and the negative to body ground.

Turn the ignition switch ON.

Is there battery voltage for two seconds? — NO → **Repair open in YEL/BLK wire between main injector and main relay.**

YES ↓

Turn the ignition switch OFF.

Disconnect the 2P connector from the auxiliary injector.

Measure resistance between the 2 terminals of the injector.

Go to Path E

Right column (engine does not start — CODE 16 YES):

Turn the ignition switch OFF.

Disconnect the 2P connector from the auxiliary injector.

Measure resistance between the 2 terminals of the injector.

Is resistance 6-10 Ω? — NO → **Replace the injector.**

YES ↓

Turn the ignition switch ON.

Measure voltage between the auxiliary injectors YEL wire and body ground.

Is there approx. 10V? — NO →
- **Repair open or short in YEL wire between aux. injector and ECU (A1, A3).**
- **Substitute a known-good ECU and recheck. If prescribed voltage is now available, replace the original ECU.**

YES ↓

Go to Path F

1.5L ENGINE—TROUBLESHOOTING—FAULT CODE 16—CIVIC

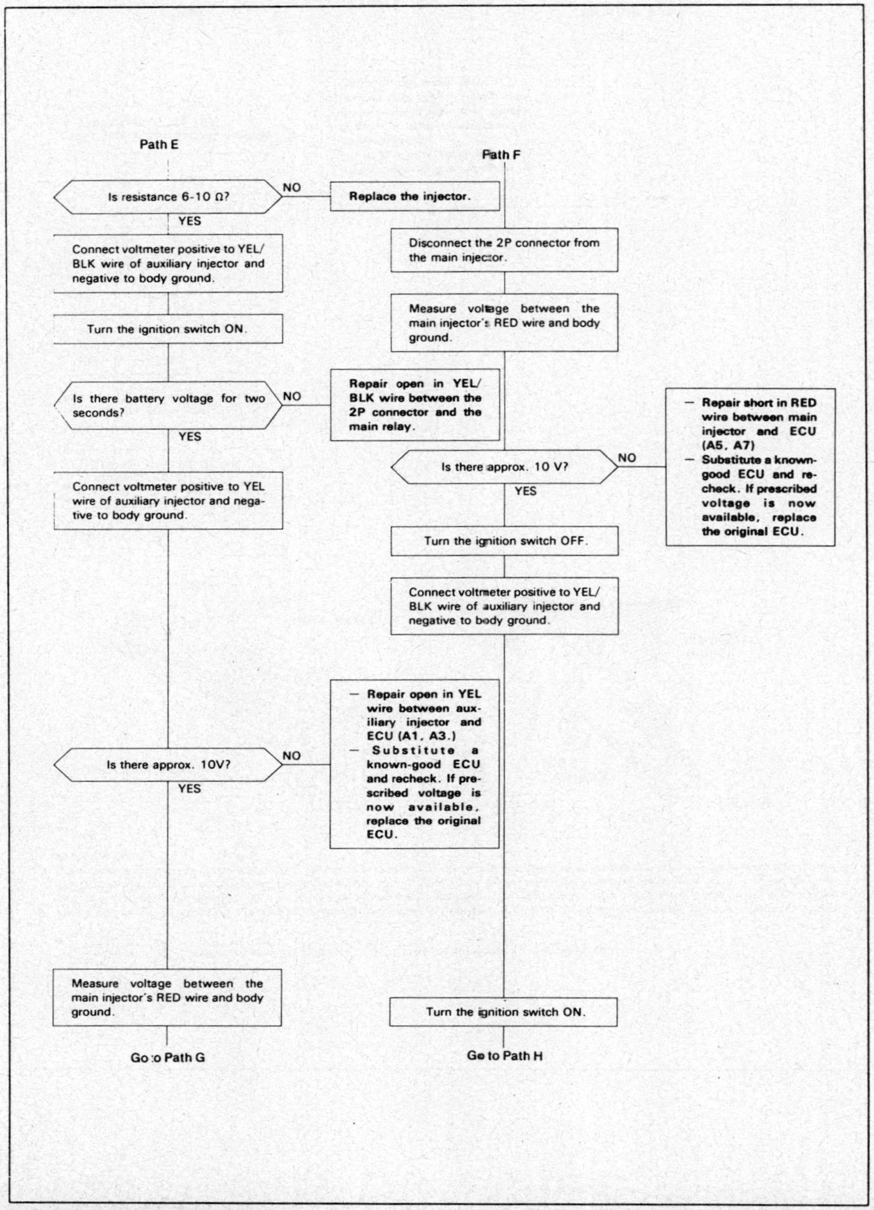

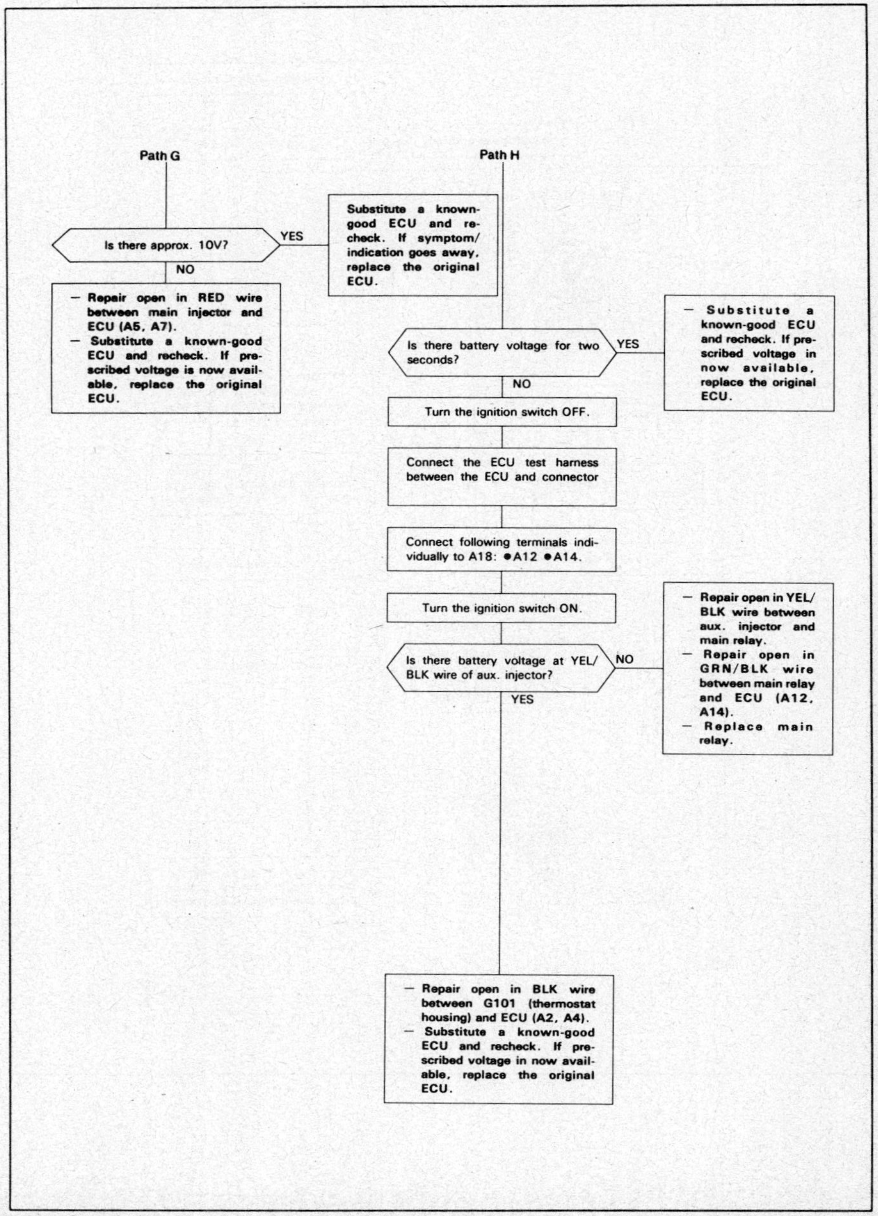

1.6L ENGINE—TROUBLESHOOTING—FAULT CODE 16—CIVIC

Self-diagnosis LED indicates code 16: A problem in the fuel injector circuit.

The injectors are the solenoid-actuated constant-stroke pintle type consisting of a solenoid, plunger needle valve and housing. When current is applied to the solenoid coil, the valve lifts up and pressurized fuel is injected close to the intake valve. Because the needle valve lift and the fuel pressure are constant, the injection quantity is determined by the length of time that the valve is open (i.e., the duration the current is supplied to the solenoid coil). The injector is sealed by an O-ring and seal ring at the top and bottom. These seals also reduce operating noise.

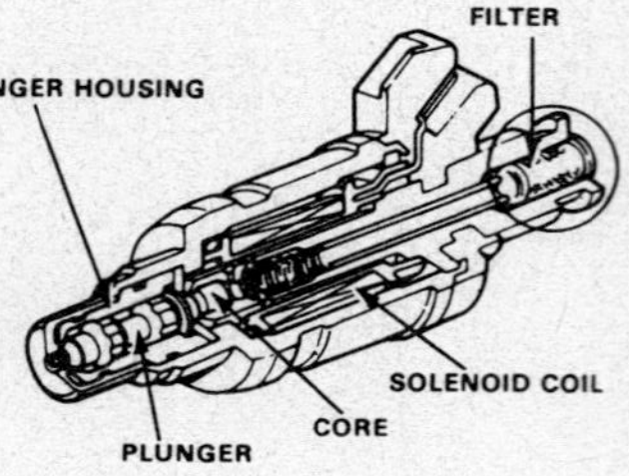

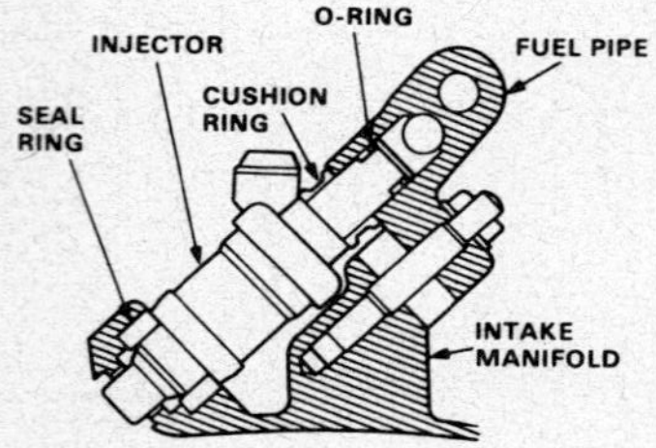

- Check Engine light has been reported on.
- LED indicates CODE 16.

Turn the ignition switch OFF.

Remove HAZARD fuse in the under-hood fuse box for 10 seconds to reset ECU.

Turn the ignition switch to START position.

Does the engine start ?
- NO: Go to Path J
- YES: Is Check Engine light on and does LED indicate CODE 16?
  - NO: Intermittent failure, system is OK at this time (test drive may be necessary). Check for poor connections or loose wires at injectors, injector resistor and C313 (round connector on left shock tower).
  - YES: Go to Path I

Path I

Check the clicking sound of each injector by means of a stethoscope when the engine is idling.

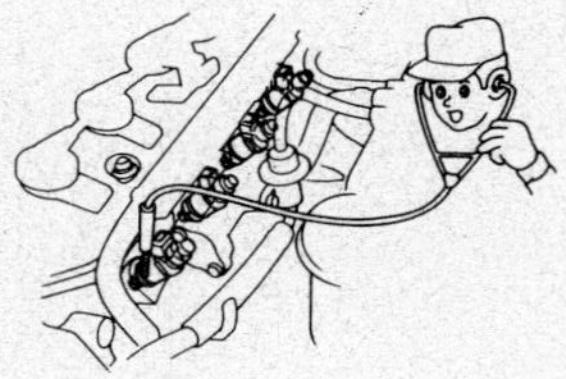

Do the injectors click ?
- YES: Substitute a known-good ECU and recheck. If symptom/indication goes away, replace the original ECU.
- NO:

Turn the ignition switch OFF.

Disconnect the 2P connector from the injector that does not click.

Measure resistance between the 2 terminals of the injector.

Path J

Turn the ignition switch OFF.

Disconnect the 2P connector from each injector.

Measure resistance between the 2 terminals of the injector.

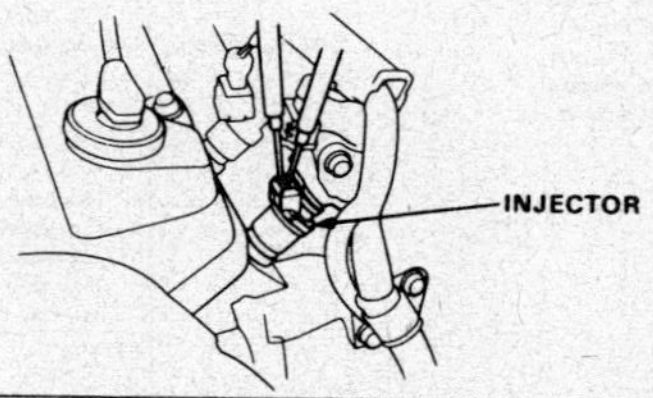

Is there 1.5—2.5Ω ?
- NO: Replace the injector.
- YES: Turn the ignition switch ON.

1.6L ENGINE—TROUBLESHOOTING—FAULT CODE 16—CIVIC

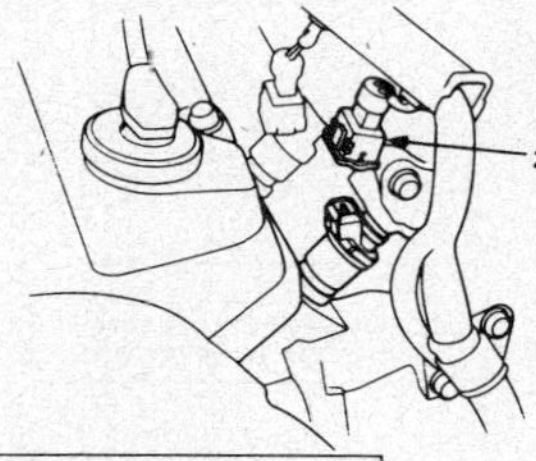

Measure voltage between RED/BLK (−) terminal on the 2P connector and body ground.

Is there battery voltage ?

- NO → Turn the ignition switch OFF. → Disconnect 6P connector on the injector resistor → Turn the ignition switch ON. → Measure voltage between YEL/BLK (+) terminal and body ground. → Is there battery voltage ?
  - NO → Repair open in the YEL/BLK wire between the injector resistor and the main relay.
  - YES →
    - Replace the injector resistor.
    - Repair open in RED/BLK wire between 2P connector and resistor.
- YES → Measure voltage between the following terminals.
  - No. 1 injector: RED/BLK (−) terminal and BRN (−) terminal.
  - No. 2 injector: RED/BLK (−) terminal and RED (−) terminal.
  - No. 3 injector: RED/BLK (−) terminal and LT BLU (−) terminal
  - No. 4 injector: RED/BLK (−) terminal and YEL (−) terminal

Is there battery voltage ?

- YES → Disconnect C451 (17P) connector from the ECU. → Is there battery voltage ?
  - YES → Repair short in the wire between the ECU (A1, A3, A5 or A7) and the injector.
  - NO → Substitute a known-good ECU and recheck. If prescribed voltage is now available, replace the original ECU.
- NO → Reconnect the 2P connector to the injector. → Turn the ignition switch OFF. → Connect the ECU test harness between the ECU and connector (page 11-23). → Turn the ignition switch ON. → Measure voltage between A2 (−) terminal and following terminals.
  - No.1 injector: A1 (−) terminal.
  - No.2 injector: A3 (−) terminal.
  - No.3 injector: A5 (−) terminal.
  - No.4 injector: A7 (−) terminal.

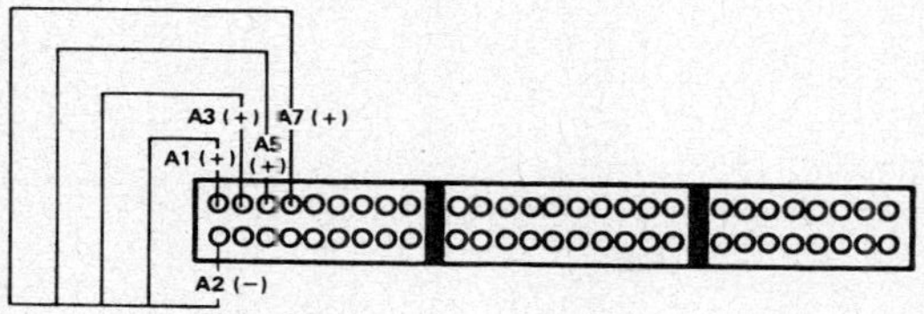

Is there battery voltage ?

- NO → Repair open in the wire between the ECU (A1, A3, A5 or A7) and the injector.
- YES → **Substitute a known-good ECU and recheck. If symptom/indication goes away, replace the original ECU.**

TROUBLESHOOTING—FAULT CODE 17—CIVIC

Self-diagnosis LED indicates code 17: A problem in the Vehicle Speed Sensor circuit.

The signal generated by the speed sensor, which produces 4 pulses (switch closures to ground) for each revolution of the speedometer cable.

- **Check Engine light has been reported on.**
- **LED indicates CODE 17.**

Turn the ignition switch OFF.

Remove HAZARD fuse in the under-hood fuse box for 10 seconds to reset ECU.

Road test necessary:
In 2nd gear accelerate to 3,500 rpm and decelerate to 1,500 rpm with throttle fully closed.

Is Check Engine light on and does LED indicate CODE 17?

- NO → **Intermittent failure, system is OK at this time. Check for poor connections or loose wires at C405 and C710.**
- YES → Block rear wheels and set the parking brake. Jack up the front of the car and support with safety stands.

WARNING Block rear wheels before jacking up front of car.

Connect the ECU test harness between the ECU and connector

Turn the ignition switch ON.

Slowly rotate left front wheel and measure voltage between B16 (+) terminal and A18 (−) terminal.

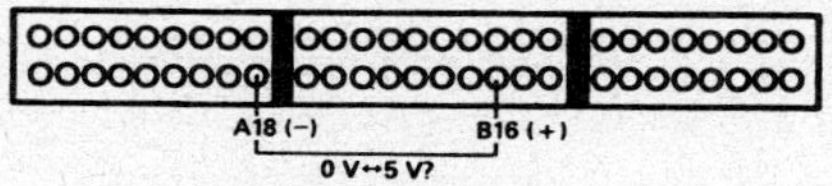

Does voltage pulse 0V and 5V ?

- NO →
  - **Repair open or short in YEL/RED wire between ECU (B16) and the speed sensor.**
  - **Faulty speed sensor.**
  - **Substitute a known-good ECU and recheck. If symptom/indication goes away, replace the original ECU.**
- YES → **Substitute a known-good ECU and recheck. If symptom/indication goes away, replace the original ECU.**

TROUBLESHOOTING—FAULT CODE 19—CIVIC WITH AUTOMATIC TRANSAXLE

Self-diagnosis LED indicates code 19: A problem in the Lock-up Control Solenoid Valve circuit.

- **—Check Engine light has been reported on.**
  —LED indicates CODE 19.
- Turn the ignition switch OFF.
- Remove HAZARD fuse in the under-hood fuse box for 10 seconds to reset ECU.
- Drive vehicle.
- Is Check Engine light on and does LED indicate CODE 19?
  - NO → **Intermittent failure, system is OK at this time. Check for poor connections or loose wires at lock-up solenoid valve and C313 (round connector located at the left shock tower).**
  - YES ↓
- Turn the ignition switch OFF.
- Connect the ECU test harness between the ECU and connector Disconnect "A" connector from the ECU only, not the main wire harness.
- Disconnect the 2P connector from the lock-up control solenoid valve.
- Check for continuity between A8 and body ground.
- Does continuity exist ?
  - YES → **Repair short in YEL wire between ECU (A8) and the 2P connector.**
  - NO ↓
- Reconnect the 2P connector.
- Measure resistance between A8 terminal and A18 terminal.

- Is there 14 — 25 Ω ?
  - YES → **Substitute a known-good ECU and recheck. If symptom/indication goes away, replace the original ECU.**
  - NO → Check for continuity in YEL wire between ECU (A8) and the 2P connector.
    - Does continuity exist ?
      - YES → **Faulty lock-up control solenoid valve.**
      - NO → **Repair open in YEL wire between ECU (A8) and the 2P connector.**

TROUBLESHOOTING—FAULT CODE 20—CIVIC

Self-diagnosis LED indicates code 20: A problem in the Electric Load Detector circuit.

- Check Engine light is off.
- LED indicates CODE 20.

Turn the ignition switch OFF.

Remove HAZARD fuse in the under-hood fuse box for 10 seconds to reset ECU.

Start engine and keep engine rpm at idle.

Does LED indicate CODE 20?

- NO → **Intermittent failure, system is OK at this time (test drive may be necessary). Check for poor connections or loose wires at main fuse box connector C214.**
- YES ↓

Turn the ignition switch OFF.

Open the under-hood fuse box lid.

Disconnect the 3P connector from the electric load detector.

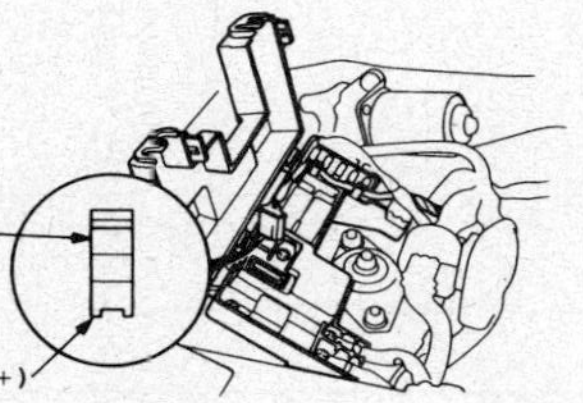

Turn the ignition switch ON.

Measure voltage between BLK/YEL (+) terminal and BLK (–) terminal.

Is there battery voltage ?

- NO → Measure voltage between BLK/YEL (+) terminal and body ground.
  - Is there battery voltage ?
    - YES → **Repair open in BLK wire between the 3P connector and G201.**
    - NO → **Repair open in BLK/YEL wire between No 14 fuse and the 3P connector.**
- YES ↓

Turn the ignition switch OFF.

Connect the ECU test harness between the ECU and connector Disconnect "B" connector from the ECU only, not the main wire harness.

Connect GRN/RED terminal of the 3P connector to body ground.

Check for continuity in GRN/RED wire between ECU (B19) and body ground.

Does continuity exist ?

- NO → **Repair open in GRN/RED wire between ECU (B19) and the 3P connector.**
- YES ↓

Disconnect GRN/RED terminal of the 3P connector from body ground.

Connect "B" connector to the ECU.

Turn the ignition switch ON.

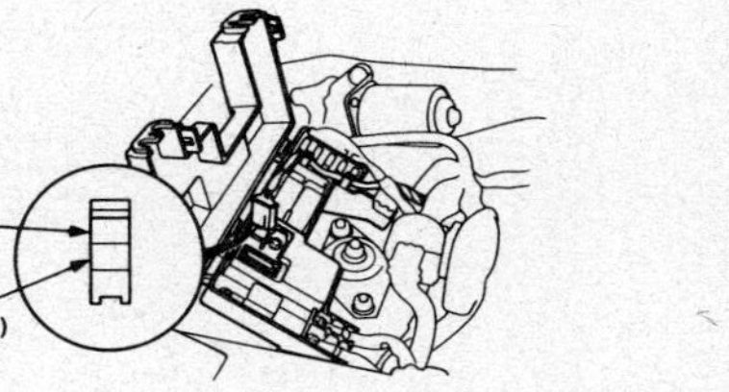

Measure voltage between GRN/RED (+) terminal and BLK (–) terminal of the 3P connector.

Is there 4.5 — 5 V ?

- NO → Disconnect "B" connector from the main wire harness only, not the ECU.
  - Measure voltage between B19 (+) terminal and A18 (–) terminal.
  - Is there 4.5 — 5 V ?
    - YES → **Repair short in GRN/RED wire between ECU (B19) and the 3P connector.**
    - NO → **Substitute a known-good ECU and recheck. If prescribed voltage is now available, replace the original ECU.**
- YES ↓

Connect the 3P connector to the electric load detector.

TROUBLESHOOTING—FAULT CODE 20—CIVIC

Condition	Voltage
Headlight switch, first position (•)	2.5 - 3.5 V
Headlight switch, second position (●)	1.5 - 2.5 V

Under the conditions listed in the chart to the right, measure voltage between B19 (+) terminal and A18 (–) terminal.

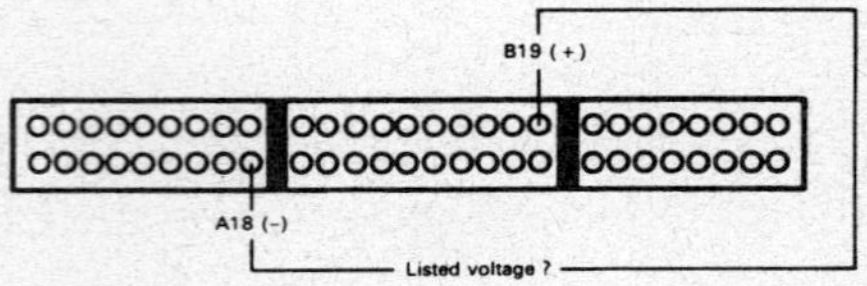

Is the voltage listed in the chart available ?

NO — Faulty electric load detector.

YES — Substitute a known-good ECU and recheck. If symptom/indication goes away, replace the original ECU.

TROUBLESHOOTING AIR CONDITIONING SIGNAL—CIVIC

This signals the PGM-FI ECU when there is a demand for cooling from the air conditioning circuits.

Inspection of Air Conditioning Signal.

Connect the ECU test harness between the ECU and connector Disconnect "B" connector from the main wire harness only, not the ECU.

Turn the ignition switch ON.

Measure voltage between B8 (+) terminal and A18 (–) terminal.

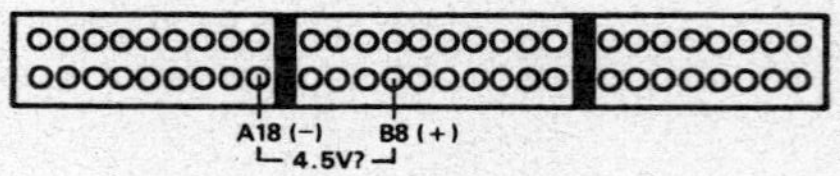

Is there approx. 4.5V?

NO — **Substitute a known-good ECU and recheck. If prescribed voltage is now available, replace the original ECU.**

YES — Reconnect "B" connector to the main wire harness.

Momentarily connect B3 terminal to A18 terminal several times.

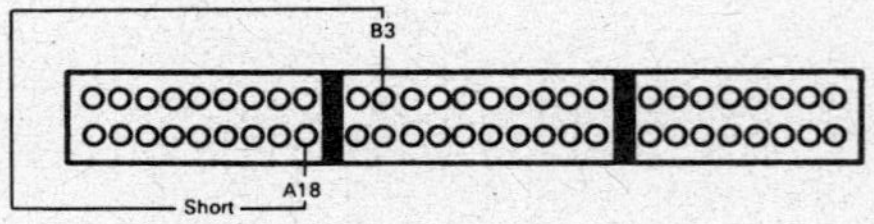

Is there a clicking noise from the A/C compressor clutch?

NO — Connect the YEL terminal of the 4P connector on the A/C clutch relay to body ground.

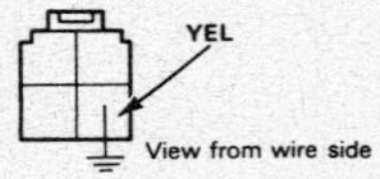

Is there a clicking noise from the A/C compressor clutch?

NO — **Air conditioner inspection**

YES — **Repair open in YEL wire between ECU (B3) and A/C clutch relay.**

YES — Start engine.

Blower switch ON.

TROUBLESHOOTING AIR CONDITIONING SIGNAL—CIVIC

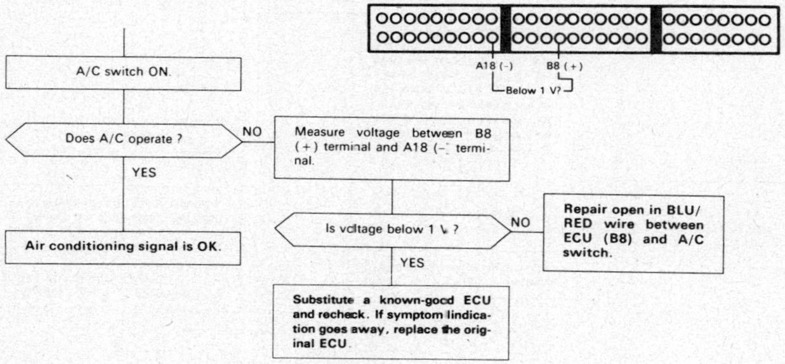

TROUBLESHOOTING ALTERNATOR FR SIGNAL—CIVIC

This signals the PGM-FI ECU when the alternator is charging.

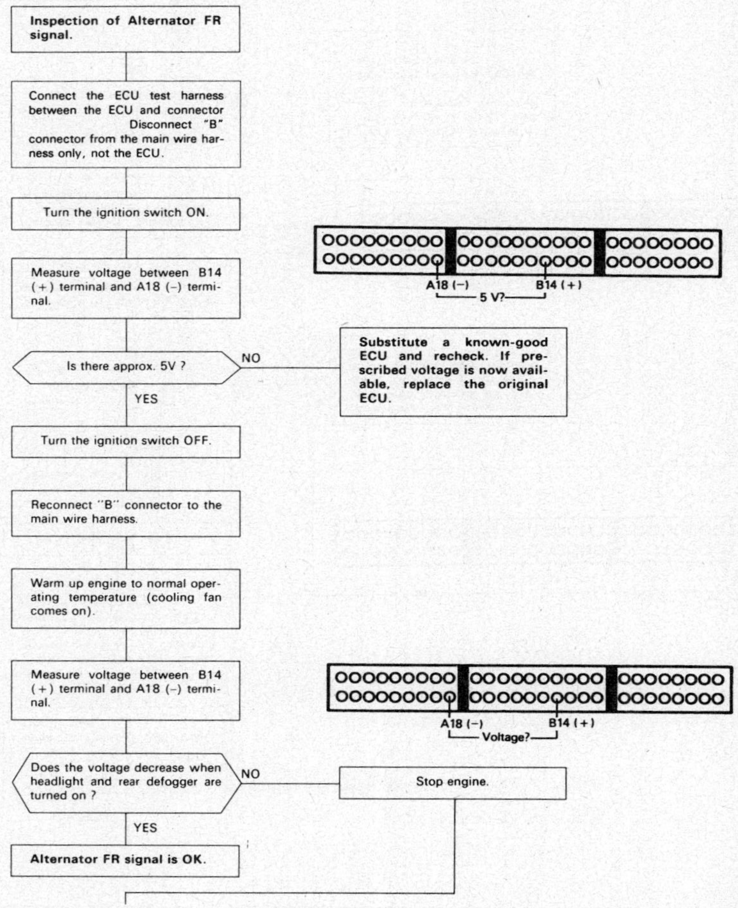

TROUBLESHOOTING ALTERNATOR FR SIGNAL—CIVIC

Disconnect "B" connector from ECU only, not the main wire harness.

Disconnect the negative battery cable from the battery.

Check for continuity between B14 terminal and body ground.

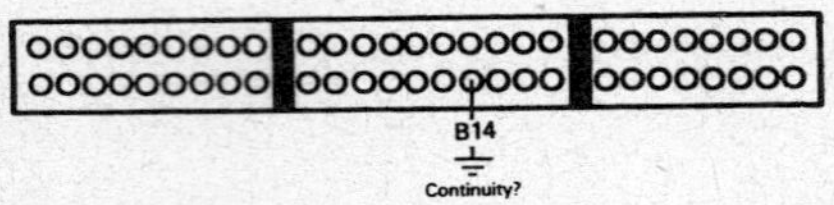

Does continuity exist ?

- YES: Disconnect GRN connector from the alternator.
  - Check for continuity between B14 terminal and body ground.
  - Does continuity exist ?
    - NO: Alternator Inspection
    - YES:
      - Repair short in BLU wire between ECU (B14) and C210 or YEL wire between C101 and alternator (1.5 ℓ).
      - Repair short in BLU wire between ECU (B14) and alternator (1.6 ℓ).
- NO: Disconnect GRN connector from the alternator.

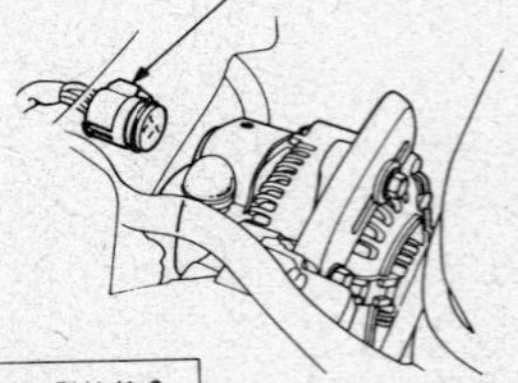

  - Connect YEL (1.5 ℓ) or BLU (1.6 ℓ) wire to body ground.
  - Check for continuity between B14 terminal and body ground.
  - Does continuity exist ?
    - YES: Alternator Inspection
    - NO:
      - Repair open in BLU wire between ECU (B14) and C210 connector, or YEL wire between C101 connector and alternator (1.5 ℓ).
      - Repair open in BLU wire between ECU (B14) and alternator (1.6 ℓ).

TROUBLESHOOTING—AUTOMATIC TRANSAXLE SHIFT POSITION SIGNAL—CIVIC

This signals the PGM-FI ECU when the transmission is in Neutral, Park or D4.

Inspection of A/T Shift Position Signal

Connect the ECU test harness between the ECU and connector Disconnect "B" connector from the main wire harness only, not the ECU.

Turn the ignition switch ON.

Measure voltage individually between B7 (+), B11 (+) terminals and A18 (−) terminal.

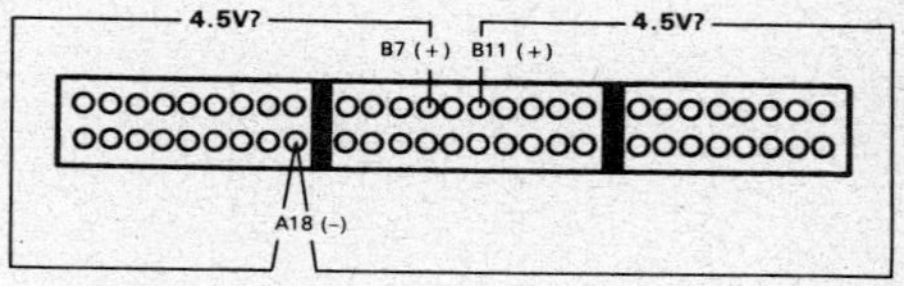

Is there approx. 4.5V?

- NO: **Substitute a known-good ECU and recheck. If prescribed voltage is now available, replace the original ECU.**
- YES: Reconnect "B" connector to the main wire harness.

Measure voltage between B7 (+) terminal and A18 (−) terminal in Neutral position.

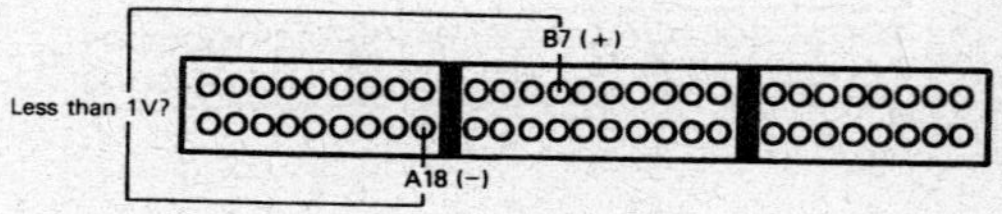

Is voltage less than 1V?

- NO:
  - **Repair open in GRN wire between ECU (B7) and shift position console switch.**
  - **See shift position console switch inspection**
- YES: Measure voltage between B7 (+) terminal and A18 (−) terminal in Park position.

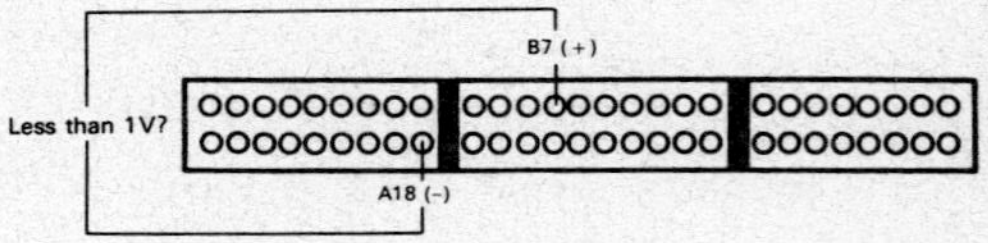

TROUBLESHOOTING—AUTOMATIC TRANSAXLE SHIFT POSITION SIGNAL—CIVIC

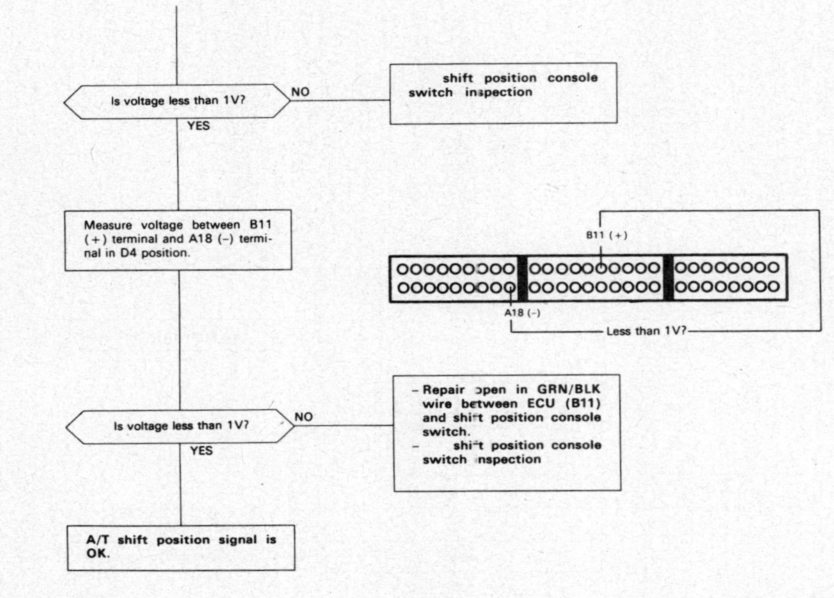

TROUBLESHOOTING BRAKE SWITCH SIGNAL—CIVIC

This signals the PGM-FI ECU when the brake pedal is depressed.

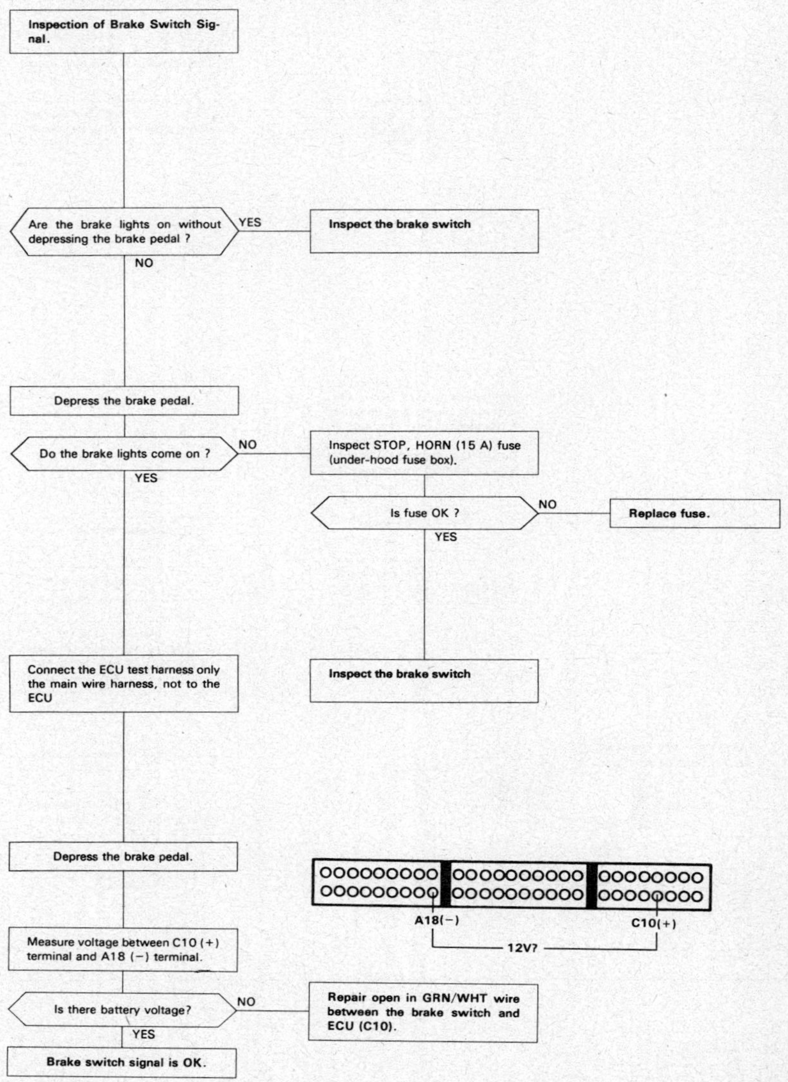

TROUBLESHOOTING—STARTER SIGNAL—CIVIC

This signals the PGM-FI ECU when the engine is cranking.

Inspection of Starter Switch Signal

Connect the ECU test harness between the ECU and connector

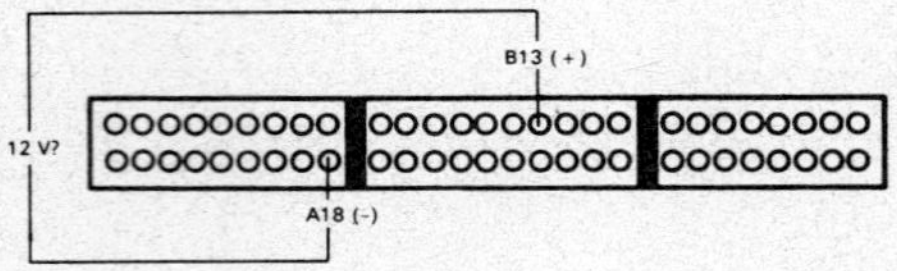

Measure voltage between B13 (+) terminal and A18 (-) terminal with ignition switch in the start position.

Is there battery voltage ?
- YES → **Starter switch signal is OK.**
- NO → Inspect No. 2 fuse.
  - Is No. 2 fuse OK ?
    - NO → **Replace fuse.**
    - YES → **Repair open in BLU/WHT wire between ECU (B13) and No. 2 fuse.**

TROUBLESHOOTING—FAST IDLE CONTROL—CIVIC

Inspection of Fast Idle Control Solenoid Valve.

Warm up engine to normal operating temperature (cooling fan comes on).

Disconnect the upper vacuum hose from the fast idle control solenoid valve and connect a vacuum gauge to the valve.

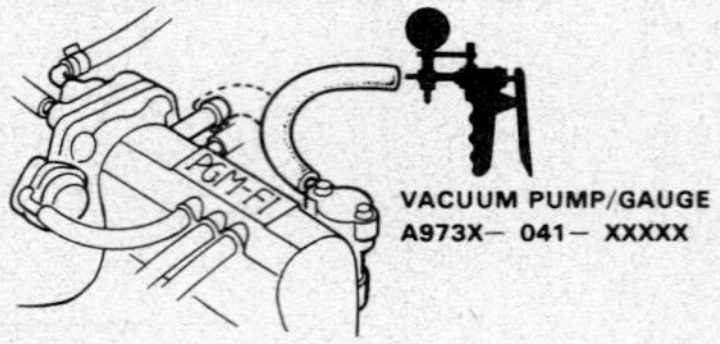

Is there any vacuum ?
- YES → Disconnect the 2P connector from the solenoid valve.
  - Is there any vacuum ?
    - YES → **Replace the fast idle control solenoid valve.**
    - NO → **Repair short to ground at BLU wire between ECU (B2) and the 2P connector. If wire is OK, substitute a known-good ECU and recheck. If symptom goes away, replace the original ECU.**
- NO → Turo the ignition switch OFF.
  - Disconnect the 2P connector from the solenoid valve.
  - Connect battery positive to terminal A and battery negative to terminal B of the solenoid valve.

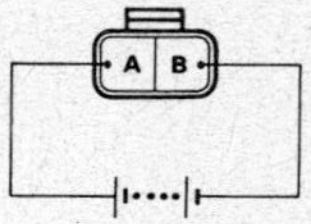

Is there manifold vacuum ?
- NO → **Replace the fast idle control solenoid valve.**
- YES → **Fast idle control Solenoid Valve is OK.**

1.5L ENGINE—TROUBLESHOOTING—TANDEM CONTROL SYSTEM—CIVIC

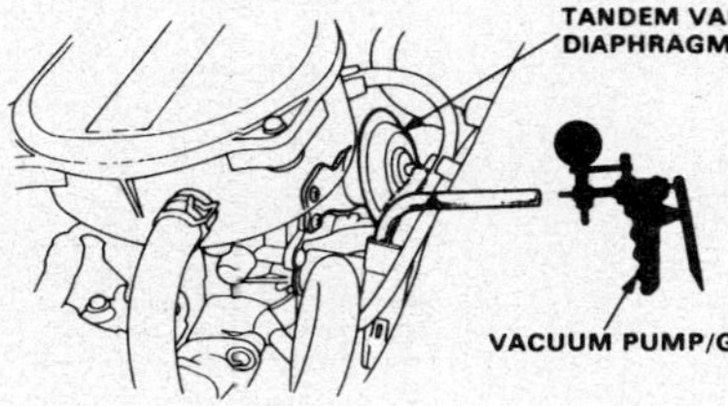

Inspection of Tandem Control System

Disconnect the vacuum hose from the tandem valve control diaphragm and connect a vacuum gauge to the hose.

Start engine and allow to idle.
NOTE: Coolant temperature must be below 70°C (160°F).

Is there vacuum ?

- YES → Disconnect the 2P connector from the tandem valve control solenoid valve.
  - Is there vacuum ?
    - YES → **Replace the solenoid valve.**
    - NO → Turn the ignition switch OFF.
      - Disconnect "B" connector from ECU.
      - Check for continuity to ground on the ORN wire.
      - Is there continuity to ground ?
        - YES → **Repair short to ground at ORN wire between ECU (B2) and the 2P connector.**
        - NO → **Substitute a known-good ECU and recheck. If symptom goes away, replace the original ECU.**
- NO → Raise engine speed to,
  - Manual: 3,000 rpm
  - Automatic: 2,000 rpm

  Is there vacuum ?
  - NO → Disconnect the 2P connector from the tandem valve control solenoid valve.
    - Go to Path L
  - YES → Go to Path K

Path K

Warm up engine to normal operating temperature (the cooling fan comes on).

Slowly open the throttle.

Is there vacuum ?

- NO → **Substitute a known-good ECU and recheck. If symptom goes away, replace the original ECU.**
- YES → **Tandem control system is OK.**

Path L

Measure voltage between BLK/YEL (+) terminal and ORN (−) terminal at 3,000 rpm.

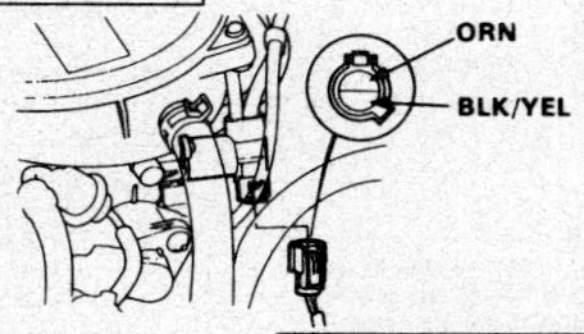

Is there battery voltage ?

- YES → **Remove the solenoid valve from the throttle body and check the port for blockage. If the port is OK, replace the solenoid valve.**
- NO → Measure voltage between BLK/YEL (+) terminal and body ground.
  - Is there battery voltage ?
    - NO → **Repair open in BLK/YEL wire between the 2P connector and No. 14 fuse.**
    - YES → Turn the ignition switch OFF.
      - Connect the ECU test harness between the ECU and connector

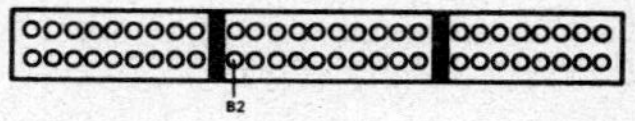

      - Check for continuity of ORN wire between ECU (B2) and the 2P connector.
      - Does continuity exist ?
        - NO → **Repair open in ORN wire between ECU (B2) and the 2P connector.**
        - YES → **Substitute a known-good ECU and recheck. If symptom goes away, replace the original ECU.**

Inspection of Evaporative Emission Controls

Disconnect #7 hose from the purge control diaphragm valve (on the charcoal canister) and connect a vacuum gauge to the hose.

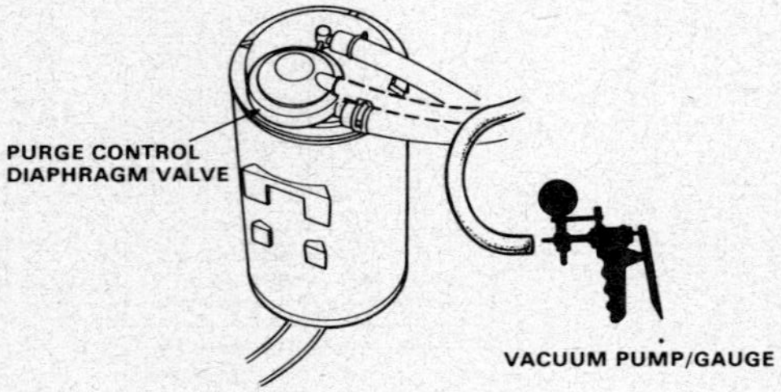

Start the engine and allow to idle. NOTE: Engine coolant temperature must be below 80°C (176°F).

Is there vacuum ?

- NO → Go to Path M
- YES → Disconnect the 2P or 4P (CAL: A/T) connector.

(Ex. CAL: A/T)

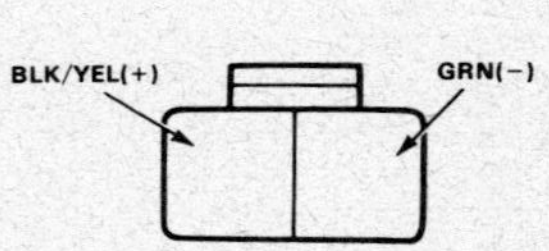

(CAL: A/T)

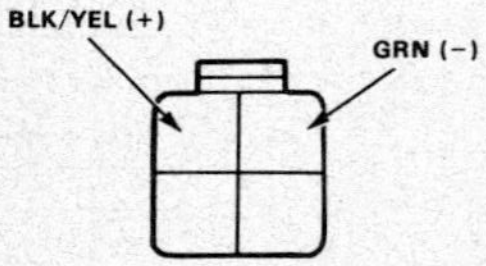

Measure voltage between BLK/YEL (+) terminal and GRN (−) terminal.

Is there battery voltage ?

- YES → Inspect #7 hose routing. If OK, replace purge cut-off solenoid valve.
- NO → Measure voltage between BLK/YEL (+) terminal and body ground. → Go to Path N

Path M

Warm up the engine to normal operating temperature (cooling fan comes on).

Check for vacuum at #7 hose 10 seconds after starting the engine. NOTE: On the 1.5 ℓ, check with the throttle valve slightly opened.

Is there manifold vacuum?

- YES → Reconnect the hose. → Remove fuel filler cap. → Go to Path O
- NO → Disconnect the 2P or 4P (CAL: A/T) connector.
  - Is there manifold vacuum?
    - NO → Inspect #7 hose routing. If OK, purge cut-off solenoid valve.
    - YES → Turn the ignition switch OFF. → Disconnect "A" connector from ECU. → Check for continuity to ground on GRN wire. → Go to Path P

Path N

Is there battery voltage ?

- NO → Repair in BLK/YEL wire between No.14 fuse and the 2P or 4P (CAL: A/T) connector
- YES → Turn the ignition switch OFF.

Connect the ECU test harness to the main wire harness only, not the ECU.

Check for continuity of GRN wire between ECU (A6) and the 2P or 4P connector.

Does continuity exist ?

- NO → Repair open in GRN wire between ECU (A6) and the connector.
- YES → Substitute a known-good ECU and recheck. If symptom goes away, replace the original ECU.

Path O

Connect a vacuum gauge to canister purge air hose.

Start the engine and raise speed to 3,500 rpm.

Does vacuum appear on gauge within 1 minute ?

- NO → **Replace the canister.**
- YES → **See two-way valve test to complete. Evaporative emission conrtols are OK.**

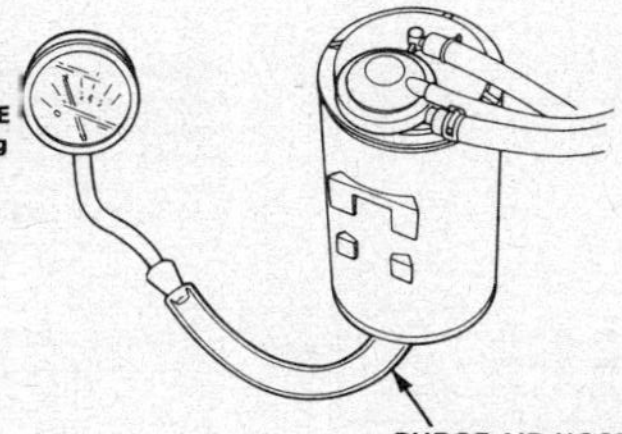

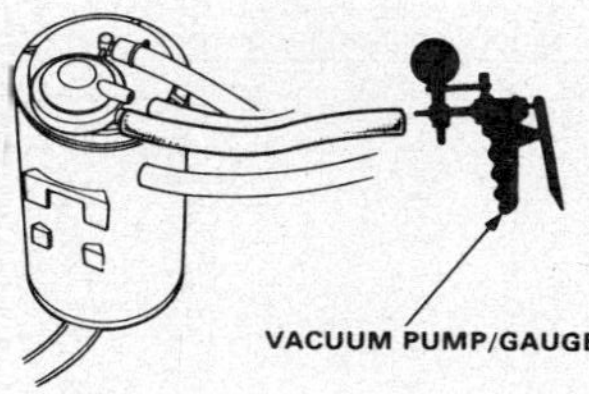

Path P

Is there continuity to ground ?

- YES → **Repair short to ground in GRN wire between ECU (A6) and the connector.**
- NO → **Substitute a known-good ECU and recheck. If symptom goes away, replace the original ECU.**

Inspection of Evaporative Emission Controls

Disconnect #7 hose from the purge control diaphragm valve (on the charcoal canister) and connect a vacuum gauge to the hose.

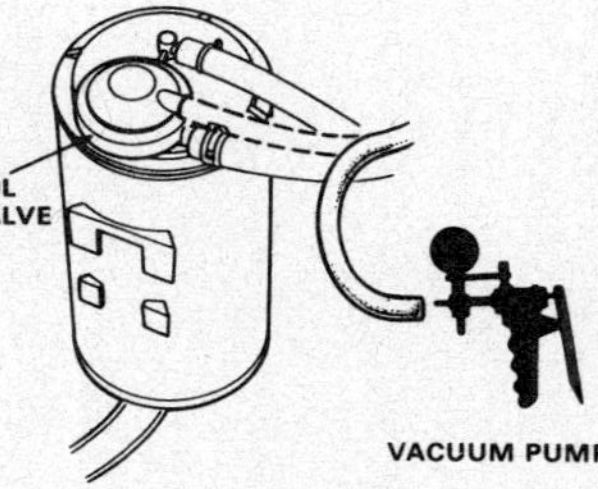

Start the engine and allow to idle.
NOTE: Engine coolant temperature must be below 57 °C (135 °F).

Is there vacuum ?

- NO → Go to Path Q
- YES → Disconnect the 2P connector.

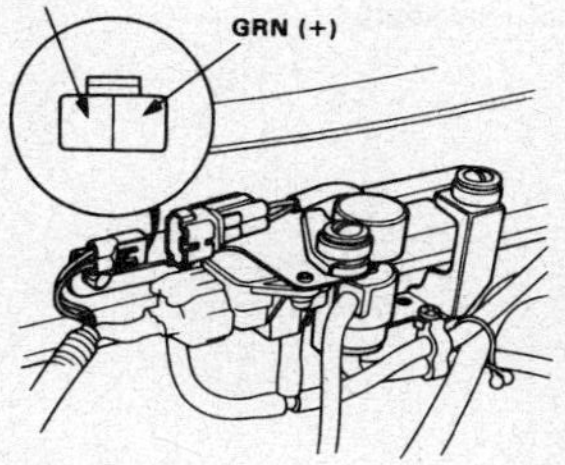

Measure voltage between GRN (+) terminal and BLK (–) terminal.

Is there battery voltage ?

- YES → **Replace purge cut-off solenoid valve.**
- NO → Measure voltage between GRN (+) terminal and body ground. → Go to Path R

1.5L ENGINE—TROUBLESHOOTING—EVAPORATIVE EMISSION CONTROL SYSTEM—CIVIC

Path Q

- Warm up the engine to normal operating temperature (cooling fan comes on).
- Check for vacuum at #7 hose 5 seconds after starting the engine.
- Is there manifold vacuum?
  - YES → (line continues down the page)
  - NO → Disconnect the 2P connector. (continues in Path R)

Path R

- Is there battery voltage ?
  - YES → **Repair open in BLK wire between the 2P connector and G201.**
  - NO ↓
- Turn the ignition switch OFF.
- Connect the ECU test harness between the ECU and connector (page 11-23).

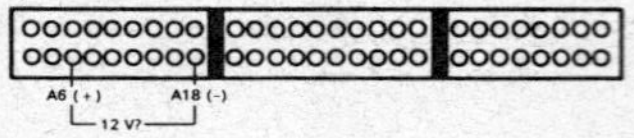

- Turn the ignition switch ON.
- Measure voltage between A6 (+) terminal and A18 (–) terminal.
- Is there battery voltage?
  - YES → **Repair open in GRN wire between ECU (A6) and the 2P connector.**
  - NO → **Inspect for a short on GRN wire between ECU (A6) and 2P connector. If wire is OK, substitute a known-good ECU and recheck. If prescribed voltage is now available, replace the original ECU.**
- Disconnect the 2P connector.
- Is there manifold vacuum?
  - NO → **Replace purge cut-off solenoid valve.**
  - YES ↓
- Turn the ignition switch OFF.
- Disconnect "A" connector from ECU.
- Check for continuity to ground on GRN wire.
- Is there continuity to ground ?
  - YES → **Repair short to ground in GRN wire between ECU (A6) and the 2P connector.**
  - NO → **Substitute a known good ECU and recheck. If symptom goes away, replace the original ECU.**

(Path Q continued)

- Remove fuel filter cap.
- Connect a vacuum gauge to canister purge air hose.
- Start the engine and raise speed to 3,500 rpm

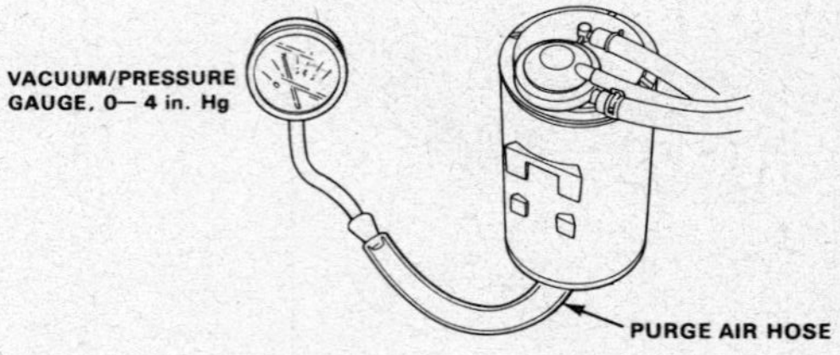

- Does vacuum appear on gauge within 1 minute ?
  - NO → **Replace the canister.**
  - YES → **Evaporative emission controls are OK.**

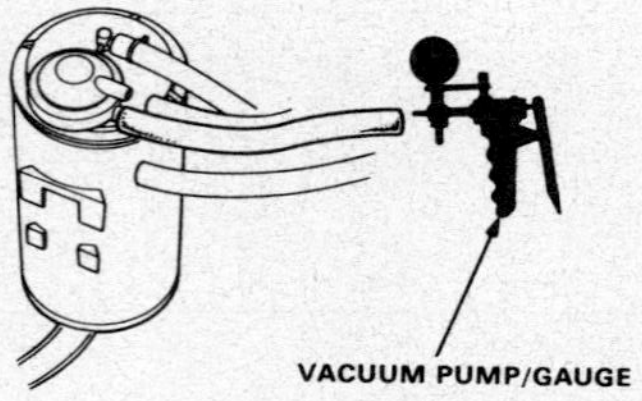

1.5L ENGINE—VACUUM DIAGRAM—CIVIC EXCEPT CALIFORNIA WITH AUTOMATIC TRANSAXLE

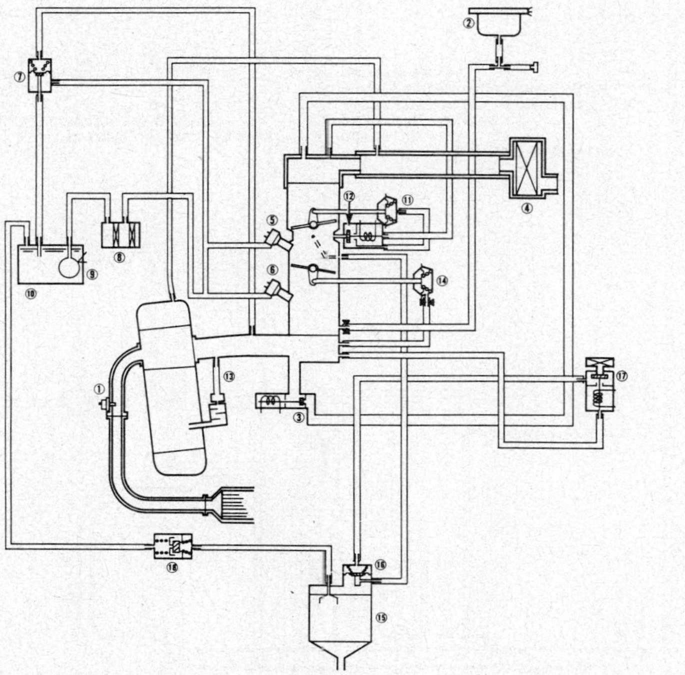

① OXYGEN (O_2) SENSOR
② MANIFOLD ABSOLUTE PRESSURE (MAP) SENSOR
③ ELECTRONIC AIR CONTROL VALVE (EACV)
④ AIR CLEANER
⑤ MAIN INJECTOR
⑥ AUX. INJECTOR
⑦ PRESSURE REGULATOR
⑧ FUEL FILTER
⑨ FUEL PUMP
⑩ FUEL TANK
⑪ TANDEM VALVE CONTROL DIAPHRAGM
⑫ TANDEM VALVE CONTROL SOLENOID VALVE
⑬ PCV VALVE
⑭ DASHPOT DIAPHRAGM
⑮ CHARCOAL CANISTER
⑯ PURGE CONTROL DIAPHRAGM VALVE
⑰ PURGE CUT-OFF SOLENOID VALVE
⑱ TWO-WAY VALVE

1.5L ENGINE—VACUUM DIAGRAM—CIVIC, CALIFORNIA WITH AUTOMATIC TRANSAXLE

CAL A/T

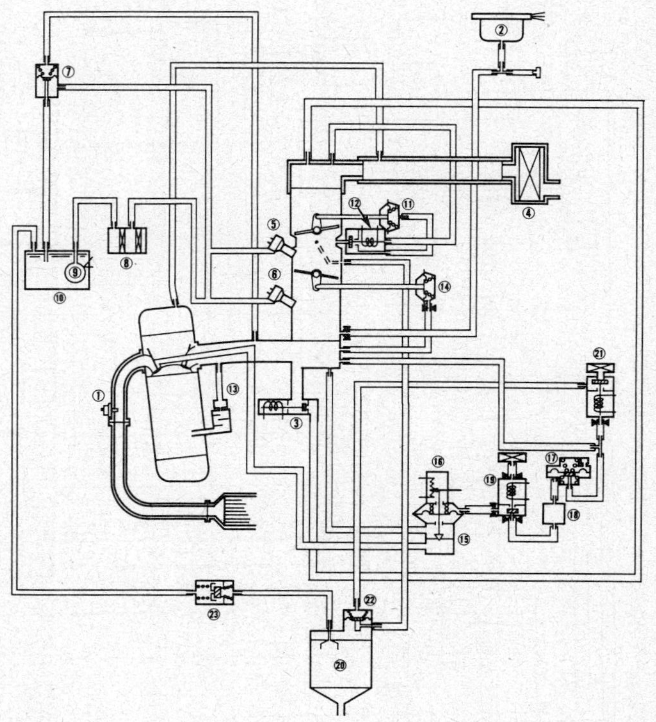

① OXYGEN (O_2) SENSOR
② MANIFOLD ABSOLUTE PRESSURE (MAP) SENSOR
③ ELECTRONIC AIR CONTROL VALVE (EACV)
④ AIR CLEANER
⑤ MAIN INJECTOR
⑥ AUX. INJECTOR
⑦ PRESSURE REGULATOR
⑧ FUEL FILTER
⑨ FUEL PUMP
⑩ FUEL TANK
⑪ TANDEM VALVE CONTROL DIAPHRAGM
⑫ TANDEM VALVE CONTROL SOLENOID VALVE
⑬ PCV VALVE
⑭ DASHPOT DIAPHRAGM
⑮ EGR VALVE
⑯ EGR VALVE LIFT SENSOR
⑰ CONSTANT VACUUM CONTROL (CVC) VALVE
⑱ AIR CHAMBER
⑲ EGR CONTROL SOLENOID VALVE
⑳ CHARCOAL CANISTER
㉑ PURGE CUT-OFF SOLENOID VALVE
㉒ PURGE CONTROL DIAPHRAGM VALVE
㉓ TWO-WAY VALVE

1.6L ENGINE—VACUUM DIAGRAM—CIVIC

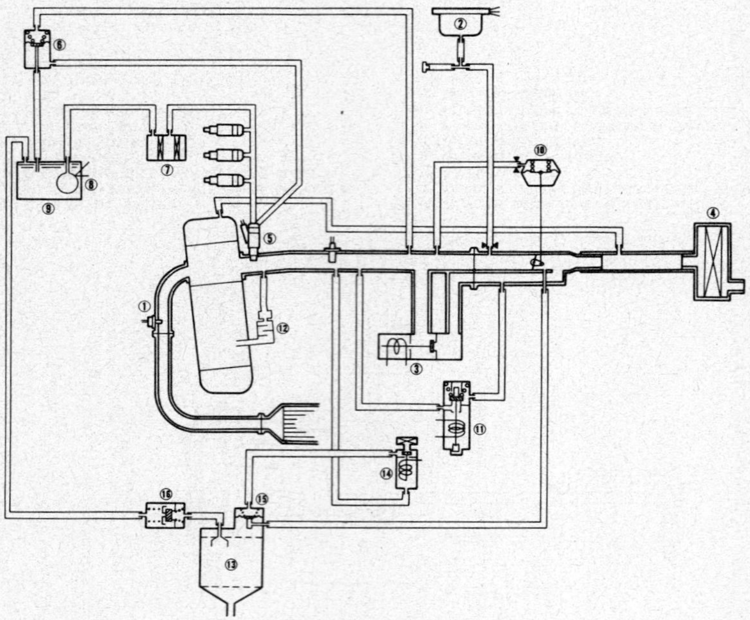

① OXYGEN (O_2) SENSOR
② MANIFOLD ABSOLUTE PRESSURE (MAP) SENSOR
③ ELECTRONIC AIR CONTROL VALVE (EACV)
④ AIR CLEANER
⑤ FUEL INJECTOR
⑥ PRESSURE REGULATOR
⑦ FUEL FILTER
⑧ FUEL PUMP
⑨ FUEL TANK
⑩ DASHPOT DIAPHRAGM
⑪ FAST IDLE CONTROL SOLENOID VALVE
⑫ PCV VALVE
⑬ CHARCOAL CANISTER
⑭ PURGE CUT-OFF SOLENOID VALVE
⑮ PURGE CONTROL DIAPHRAGM VALVE
⑯ TWO-WAY VALVE

1.5L ENGINE FUEL SYSTEM WIRING SCHEMATIC—CIVIC

FUSES
① ECU (15A)*
② No.14 (10A)
③ No.2 (10A)
④ No.1 (10A)
⑤ HAZARD (10A)*
* : In the under-hood fuse box

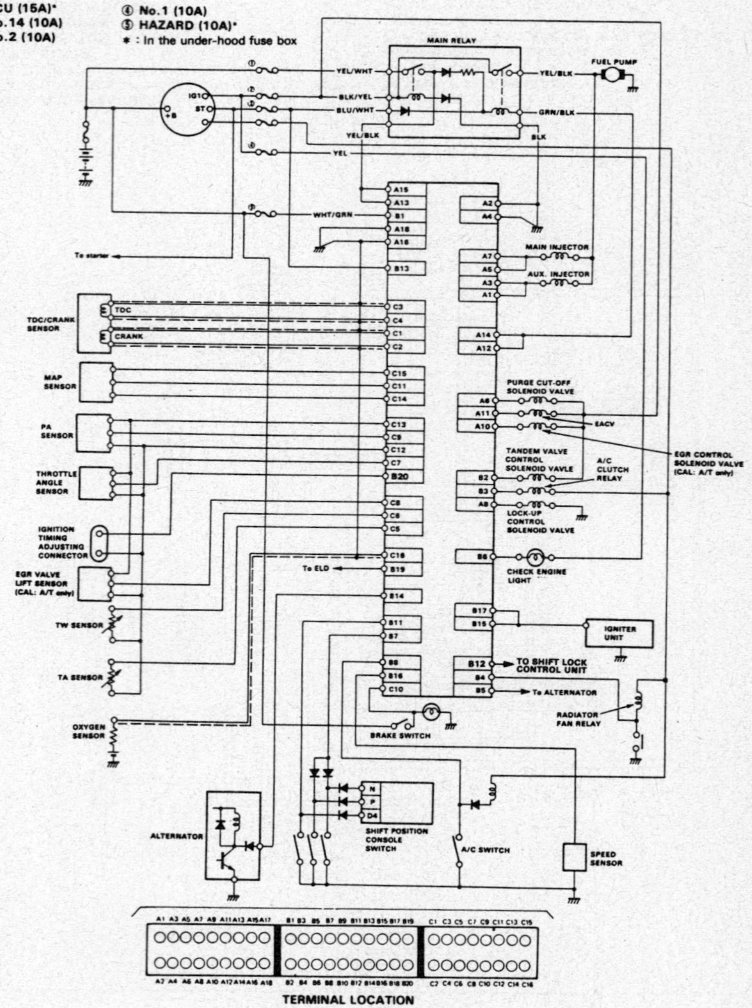

1.6L ENGINE
FUEL SYSTEM WIRING SCHEMATIC—CIVIC

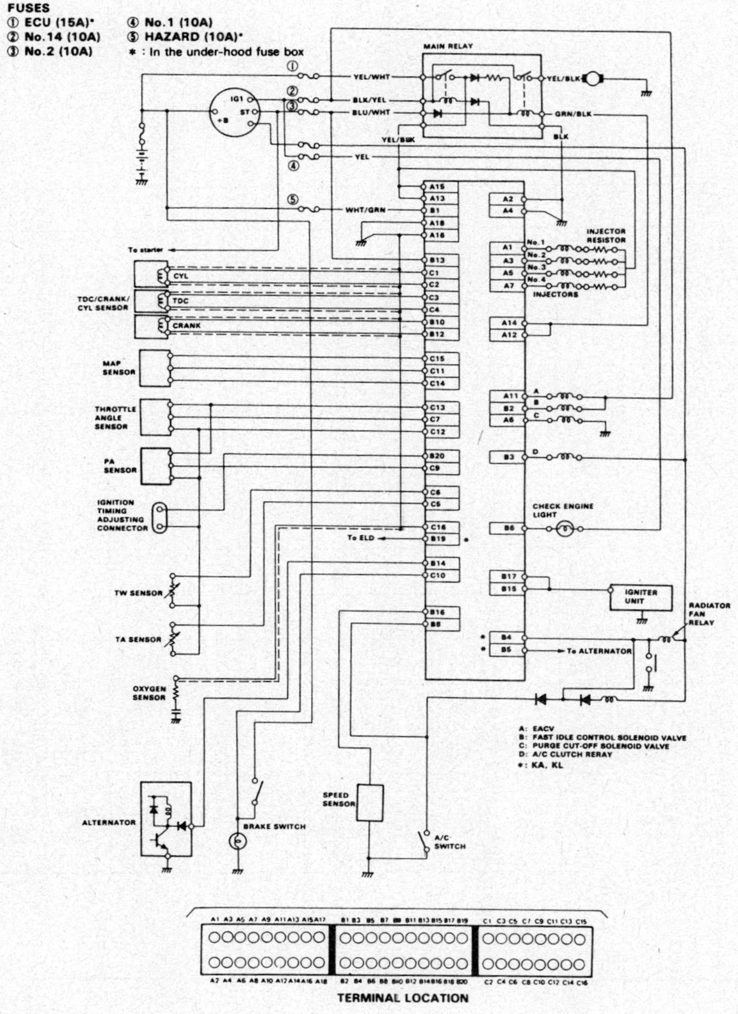

TROUBLESHOOTING—ACCORD

NOTE: Across each row in the chart, the systems that could be sources of a symptom are ranked in the order they should be inspected starting with ①. Find the symptom in the left column, read across to the most likely source, then refer to the page listed at the top of that column. If inspection shows the system is OK, try the next most likely system ②, etc.

SYMPTOM \ SYSTEM		ECU	PGM-FI: OXYGEN SENSOR	MANIFOLD ABSOLUTE PRESSURE SENSOR	TDC/CRANK/CYL SENSOR	COOLANT TEMPERA-TURE SENSOR	THROTTLE ANGLE SENSOR	INTAKE AIR TEMPERA-TURE SENSOR	ATMO-SPHERIC PRESSURE SENSOR
CHECK ENGINE WARNING LIGHT TURNS ON		□ or (lamp)	(CHECK)	(CHECK)	(CHECK)	(CHECK)	(CHECK)	(CHECK)	(CHECK)
SELF-DIAGNOSIS INDICATOR (LED) BLINKS		0 or *	1 or 41 or 43	3 or 5	4 or 8 or 9	6	7	10	13
ENGINE WON'T START		③			③				
DIFFICULT TO START ENGINE WHEN COLD		(BU)		③	③	①			③
IRREGULAR IDLING	WHEN COLD FAST IDLE OUT OF SPEC	(BU)				③			
	ROUGH IDLE	(BU)		③					
	WHEN WARM RPM TOO HIGH	(BU)							
	WHEN WARM RPM TOO LOW	(BU)							
FREQUENT STALLING	WHILE WARMING UP	(BU)				③			
	AFTER WARMING UP	(BU)							③
POOR PERFORM-ANCE	MISFIRE OR ROUGH RUNNING	(BU)			③				
	FAILS EMISSION TEST	(BU)	③	②					
	LOSS OF POWER	(BU)		③			②		

* If codes other than those listed above are indicated, count the number of blinks again. If the indicator is in fact blinking these codes, substitute a known-good ECU and recheck. If the indication goes away, replace the original ECU.

(BU): When the Check Engine warning light and the self-diagnosis indicator are on, the back-up system is in operation. Substitute a known-good ECU and recheck. If the indication goes away, replace the original ECU.

TROUBLESHOOTING—ACCORD

PGM-FI					IDLE CONTROL		FUEL SUPPLY			EMISSION CONTROL	
IGNITION OUTPUT SIGNAL	VEHICLE SPEED SENSOR	ELECTRIC LOAD DETECTOR	A/T FI SIGANL A	A/T FI SIGNAL B	ELEC-TRONIC AIR CONTROL VALVE	OTHER IDLE CONTROLS	FUEL INJECTOR	OTHER FUEL SUPPLY	AIR INTAKE	EGR CONTROL SYSTEM	OTHER EMISSION CONTROLS
CHECK	CHECK	☐	☐	☐	CHECK		CHECK			CHECK	
⑮	⑰	⑳	㉚	㉛	⑭		⑯			⑫	
①							②	③			
						②					
					①	②					
					①		②			③	
					①	②					
		③			①		②				
					①	②		③			
					③	①		②		③	
					③		①			③	
							②	③			①
							③	①	③		③

TROUBLESHOOTING IDLE CONTROL SYSTEM—ACCORD

NOTE:
- Across each row in the chart, the sub systems that could be sources of a symptom are ranked in the order they should be inspected, starting with ①
- If the idle speed is out of specification and LED does not blink CODE 14, go to inspection

SYMPTOM \ SUB SYSTEM		IDLE ADJUST-ING SCREW	EACV	AIR CONDI-TIONING SIGNAL	ALTER-NATOR FR SIGNAL	A/T SHIFT POSITION SIGNAL (A/T ONLY)	BRAKE SWITCH SIGNAL	STARTER SWITCH SIGNAL	P/S OIL PRESSURE SWITCH SIGNAL	FAST IDLE VALVE	AIR BOOST VALVE	HOSES AND CONNEC-TIONS
DIFFICULT TO START ENGINE WHEN COLD										①		
WHEN COLD FAST IDLE OUT OF SPEC (1,000–2,000 rpm)		③	②							①		
ROUGH IDLE			②									①
WHEN WARM RPM TOO HIGH		③	①						③	②		③
WHEN WARM RPM TOO LOW	Idle speed is below specified rpm (no load)	②	①									
	Idle speed does not increase after initial start up.		①									
	On models with automatic transmission, the idle speed drops in gear		②			①						
	Idle speeds drops when air conditioner in ON		②	①								
	Idle speed drops when steering wheel is turning		②						①			
	Idle speed fluctuates with electrical load		②									①
FREQUENT STALLING	WHILE WARMING UP		①									
	AFTER WARMING UP	①										
FAILS EMISSION TEST												①

TROUBLESHOOTING FUEL SUPPLY SYSTEM—ACCORD

NOTE: Across each row in the chart, the systems that could be sources of a symptom are ranked in the order they should be inspected starting with 1

SYMPTOM \ SUB SYSTEM		FUEL INJECTOR	INJECTOR RESISTOR	PRESSURE REGULATOR	FUEL FILTER	FUEL PUMP	MAIN RELAY	CONTAMINATED FUEL
ENGINE WON'T START		③	③		③	①	②	③
DIFFICULT TO START ENGINE WHEN COLD OR HOT				③	②			①
ROUGH IDLE		①	②					③
POOR PERFORMANCE	MISFIRE OR ROUGH RUNNING	①	②	③				③
	FAILS EMISSION TEST	②	③	①				
	LOSS OF POWER	③	③		①	③		②

* Fuel with dirt, water or a high percentage of alcohol is considered contaminated.

TROUBLESHOOTING AIR INTAKE SYSTEM—ACCORD

NOTE: Across each row in the chart, the sub systems that could be sources of a symptom are ranked in the order they should be inspected starting with ①. Find the symptom in the left column, read across to the most likely source.

SYMPTOM \ SUB SYSTEM	THROTTLE CABLE	THROTTLE BODY	INTAKE CONTROL SYSTEM
WHEN WARM RPM TOO HIGH	②	①	
LOSS OF POWER		①	②

TROUBLESHOOTING
EMISSION CONTROL SYSTEM—ACCORD

NOTE: Across each row in the chart, the systems that could be sources of a symptom are ranked in the order they should be inspected starting with ①.

SYMPTOM \ SUB SYSTEM		CATALYTIC CONVERTER	EGR SYSTEM	POSITIVE CRANKCASE VENTILATION SYSTEM	EVAPORATIVE EMISSION CONTROLS
ROUGH IDLE			①	②	
FREQUENT STALLING AFTER WARM-ING UP	AFTER WARM-ING UP		①		
POOR PERFORMANCE	FAILS EMISSION TEST	①			②
	LOSS OF POWER	①			

TERMINAL IDENTIFICATION
ACCORD TEST HARNESS

If the inspection for a particular failure code requires the ECU test harness, remove the right door sill molding, the small cover on the right kick panel, and pull the carpet back to expose the ECU. Unbolt the ECU bracket. Connect the ECU test harness. Then check the system according to the procedure described for the appropriate code(s) listed on the following pages.

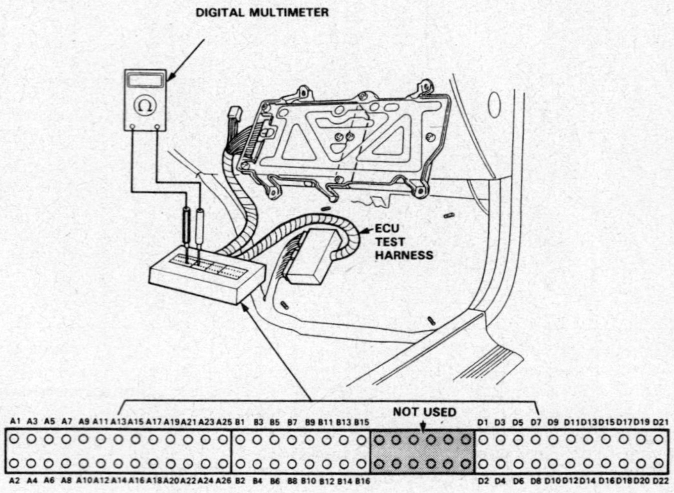

TERMINAL LOCATION

CAUTION:
- Puncturing the insulation on a wire can cause poor or intermittent electrical connections.
- For testing at connectors other than the ECU test harness, bring the tester probe into contact with the terminal from the connector side of wire harness connectors in the engine compartment. For female connectors, just touch lightly with the tester probe and do not insert the probe.

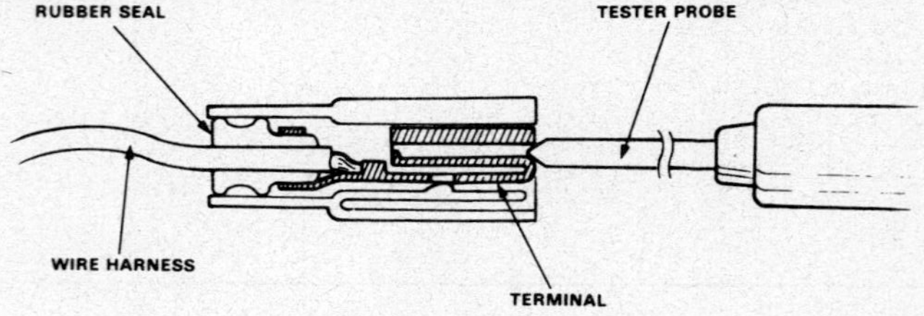

TROUBLESHOOTING—ECU—ACCORD

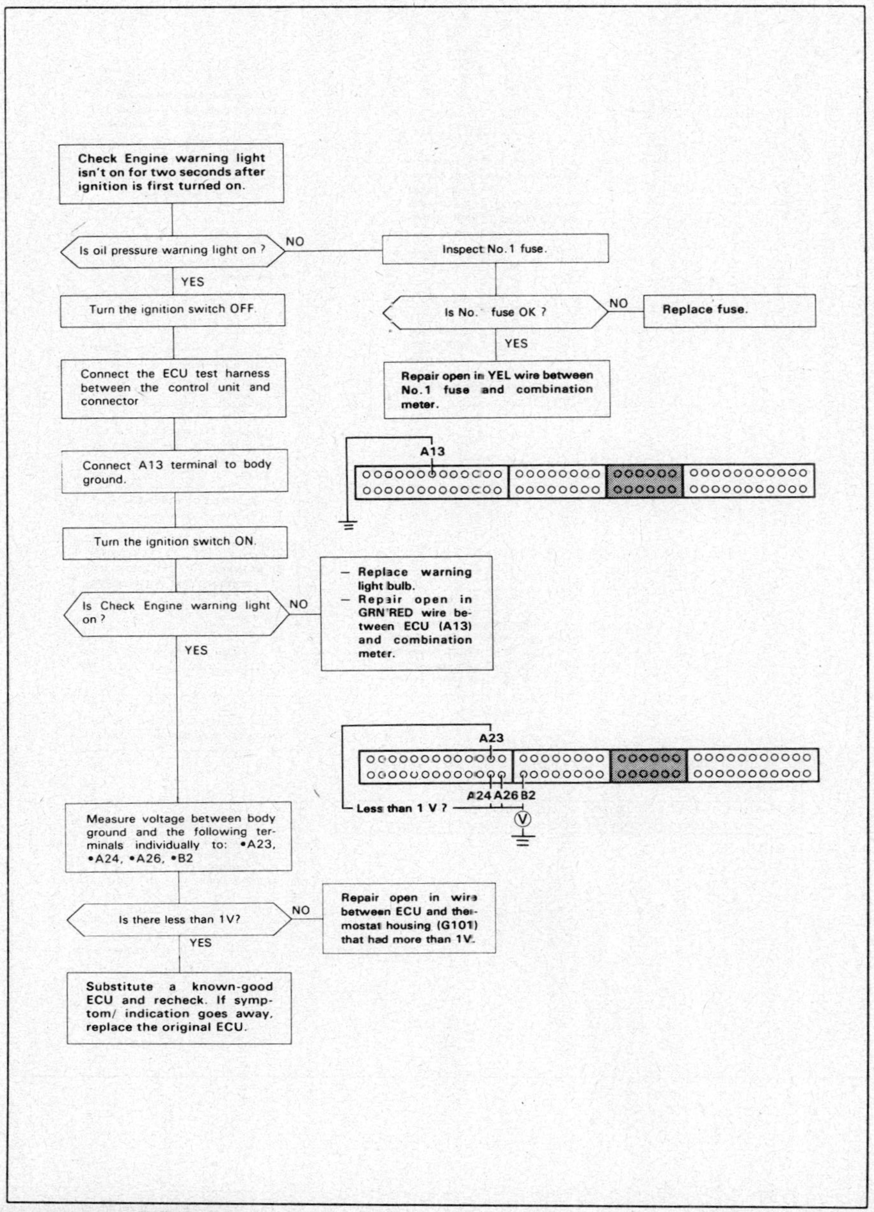

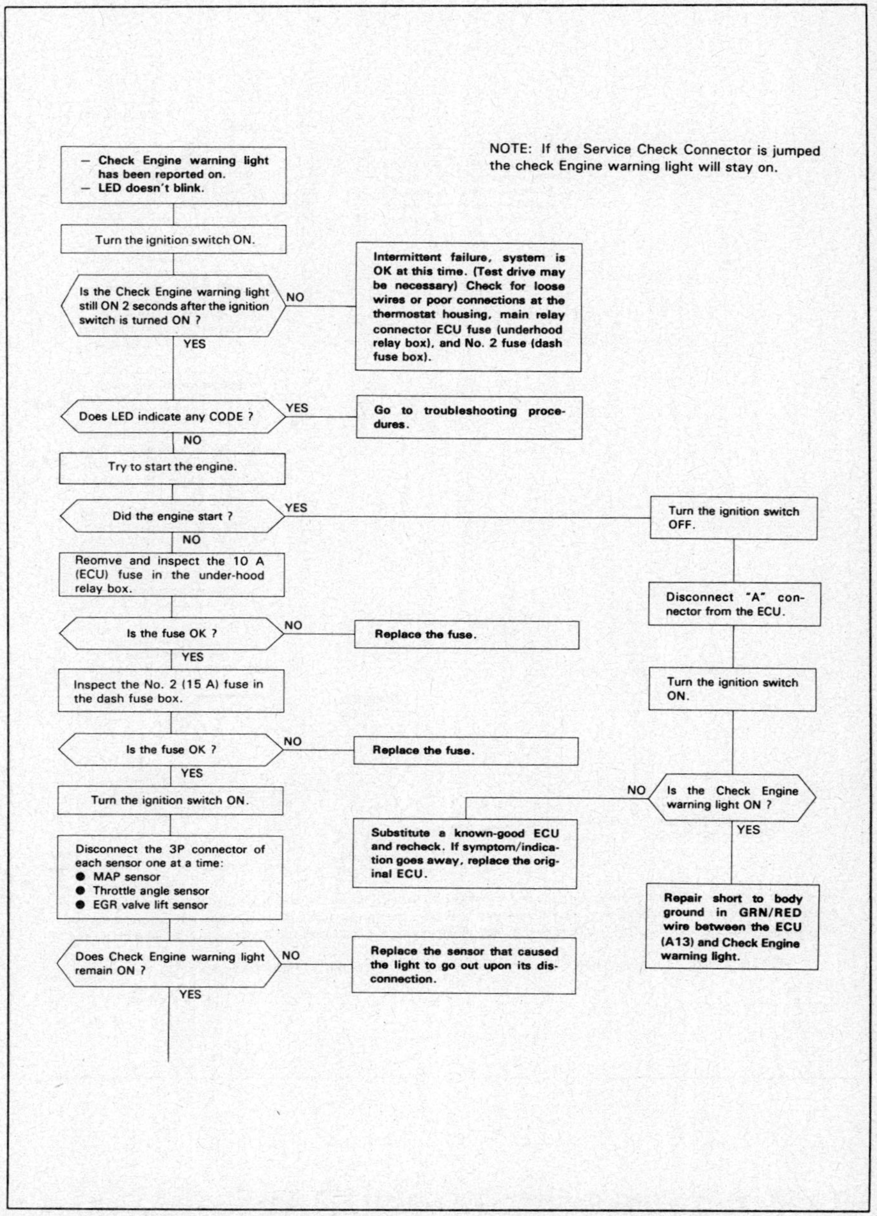

TROUBLESHOOTING—ECU—ACCORD

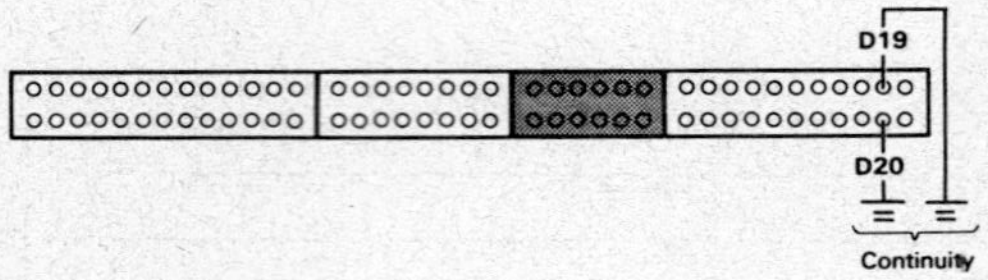

- Turn the ignition switch OFF.
- Connect the ECU test harness. Disconnect the "D" connector from the ECU only, not the main wire harness.
- Check for continuity between body ground and the following terminals; D19, D20.
- Does continuity exist?
  - YES → Repair short to body ground in RED/WHT wire, YEL/WHT wire between ECU (D19, D20) and throttle angle sensor, EGR valve lift sensor and MAP sensor.
  - NO ↓
- Reconnect all the connectors. Reconnect the "D" connector to the ECU.
- Turn the ignition switch ON.
- Individually connect the following terminal to body ground: B2 • A26
- Is the Check Engine warning light still ON after 2 seconds?
  - NO →
    - Repair open in BLK/RED wire between ECU (A26) and G101.
    - Repair open in BRN/BLK wire between ECU (B2) and G101.
  - YES ↓
- Measure voltage between A26 (−) and the following: B1 (+) and A25 (+).
- Is there battery voltage?
  - NO →
    - Repair open in YEL/BLK wire between ECU (A25, B1) and main relay.
    - Check main relay and wiring connectors at main relay
  - YES ↓
- Substitute a known-good ECU and recheck. If symptom/indication goes away, replace the original ECU.

TROUBLESHOOTING—CODE 1—ACCORD

Self-diagnosis LED indicates code 1: A problem in the Heated Oxygen (O_2) Sensor circuit.

The oxygen sensor, detects the oxygen content in the exhaust gas, and inputs the ECU. In operation, the ECU receives the signals from the sensor and varies the duration during which fuel is injected. The oxygen sensor is combined with a heater. The heater stabilizes the sensor output. The oxygen sensor is installed on the exhaust manifold (DX, LX) or in the exhaust pipe (EX).

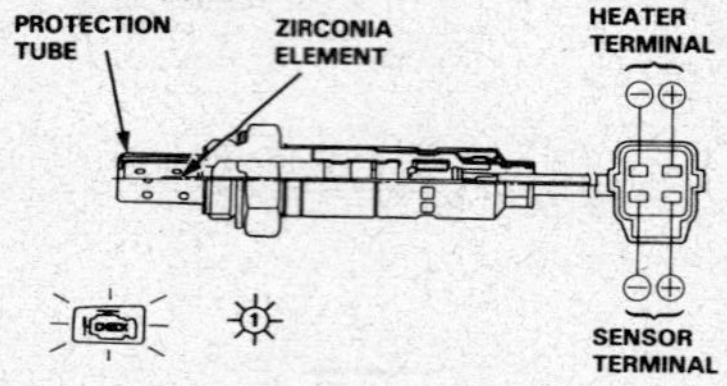

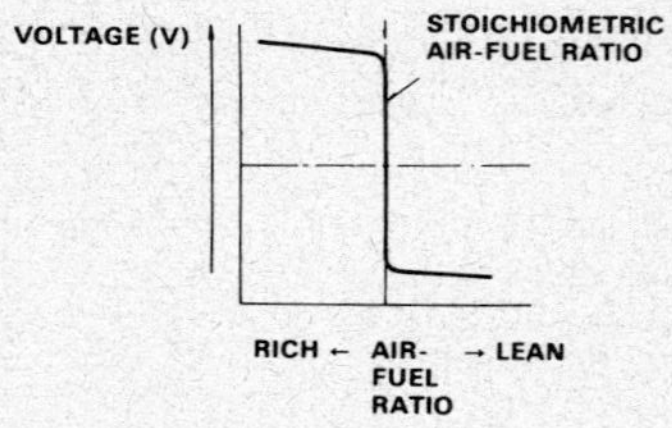

- Check Engine warning light has been reported on.
- LED indicates CODE 1.

- Turn the ignition switch OFF.
- Remove BACK UP fuse in the under-hood relay box for 10 seconds to reset ECU.
- Warm up engine to normal operating temperature (cooling fan comes on).
- Run engine for 60 seconds.
- Road test with the transmission in 2nd gear, accelerate using wide open throttle for at least 5 seconds. Then decelerate for at least 5 seconds with the throttle completely closed.
- Is Check Engine warning light on and does LED indicate CODE 1?
  - NO → Intermittent failure, system is OK at this time. Check for poor connections or loose wires at C210 (located at right shock tower) and at the O_2 sensor
  - YES → perform test for CODE 43.

TROUBLESHOOTING—CODE 3—ACCORD

Self-diagnosis LED indicates code 3: Most likely an electrical problem in the Manifold Absolute Pressure (MAP) Sensor system.

The MAP sensor converts manifold absolute pressure into electrical signals and inputs the ECU.

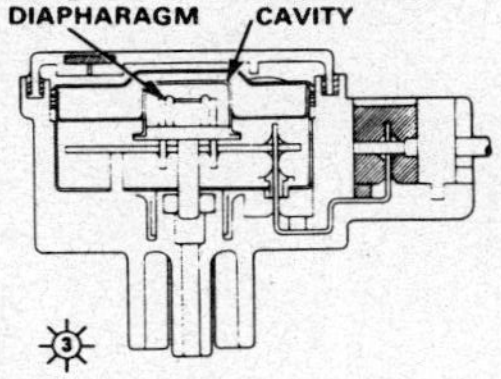

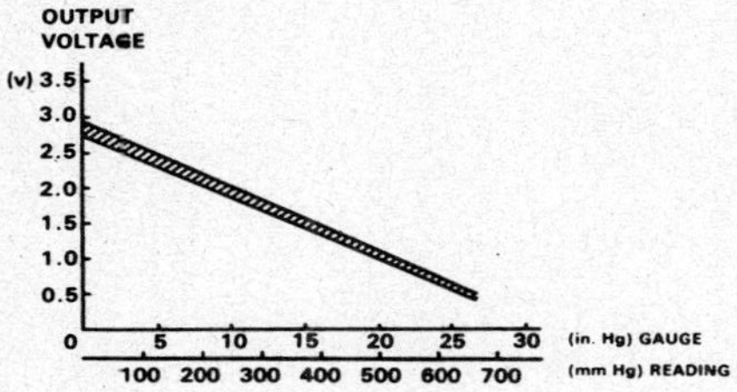

- Engine is warm and running.
- Check Engine warning light has been reported on.
- LED indicates CODE 3.

Turn the ignition switch OFF.

Remove BACK UP fuse in the under-hood relay box for 10 seconds to reset ECU.

Warm up engine to normal operating temperature (cooling fan comes on).

Is Check Engine warning light on and does LED indicate CODE 3?

NO → Intermittent failure, system is OK at this time (test drive may be necessary). Check for poor connection or loose wires at MAP sensor connector and ECU.

YES ↓

Turn the ignition switch OFF.

Disconnect the 3P connector from the MAP sensor.

Turn the ignition switch ON.

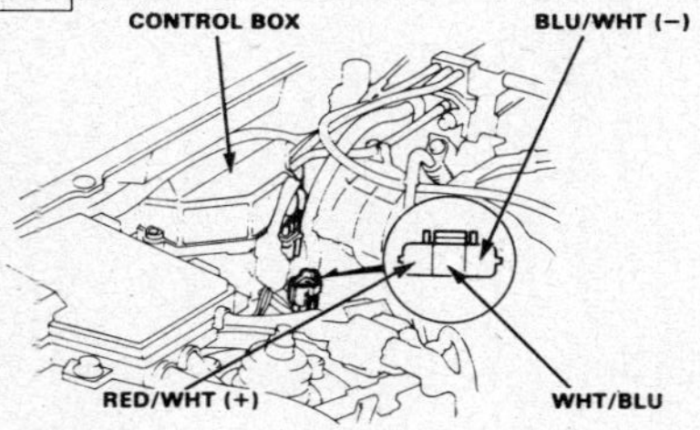

Measure voltage between RED/WHT (+) terminal and body ground.

Is there approx. 5V?

NO → Repair open in RED/WHT wire between ECU (D19) and MAP sensor. If wire is OK, substitute a known-good ECU and recheck. If prescribed voltage is now available, replace the original ECU.

YES ↓

Measure voltage between RED/WHT (+) terminal and BLU/WHT (−) terminal.

Is there approx. 5V?

NO → Repair open in BLU/WHT wire between ECU (D21) and MAP sensor. If wire is OK, substitute a known-good ECU and recheck. If prescribed voltage is now available, replace the original ECU.

YES ↓

Measure voltage between WHT/BLU (+) terminal and BLU/WHT (−) terminal.

Is there approx. 5V?

NO → Repair open or short in WHT/BLU wire between ECU (D17) and MAP sensor. If wire is OK, substitute a known-good ECU and recheck. If prescribed voltage is now available, replace the original ECU.

YES ↓

Turn the ignition switch OFF.

Reconnect the 3P connector to the MAP sensor.

Connect the ECU test harness between the ECU and connector

Turn the ignition switch ON.

TROUBLESHOOTING—CODE 3—ACCORD

Measure voltage between D17 (+) terminal and D21 (−) terminal.

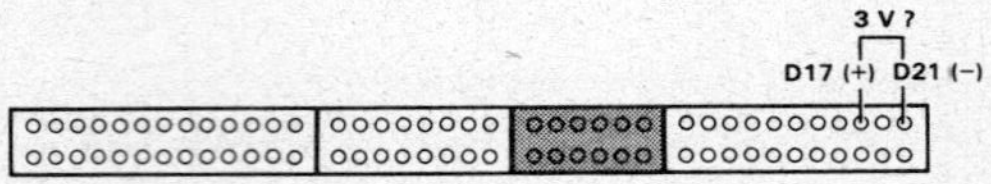

Is there approx. 3 V ?

NO → Replace MAP sensor.

YES → Substitute a known-good ECU and recheck. If symptom/indication goes away, replace the original ECU.

TROUBLESHOOTING—CODE 4—ACCORD

Self-diagnosis LED indicates code 4: A problem in the circuit of the CRANK Sensor.

The CRANK sensor determines timing for fuel injection and ignition of each cylinder and also detects engine RPM. The TDC sensor determines ignition timing at start-up (cranking) and when crank angle is abnormal. The CYL sensor detects the position of No. 1 cylinder for sequential fuel injection to each cylinder.

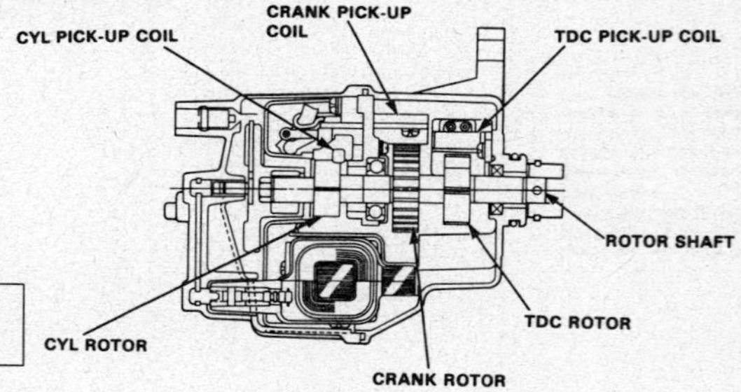

- Check Engine warning light has been reported on.
- LED indicates CODE 4.

Turn the ignition switch OFF.

Remove the BACK UP fuse in the under-hood relay box for 10 seconds to reset ECU.

Start the engine.

Is Check Engine warning light on and does LED indicate CODE 4?

NO → **Intermittent failure, system is OK at this time (test drive may be necessary). Check for poor connections or loose wires at distributor connector and C209 (located and right shock tower).**

YES → Stop the engine.

Disconnect the 8P connector from the TDC/CRANK/CYL sensor.

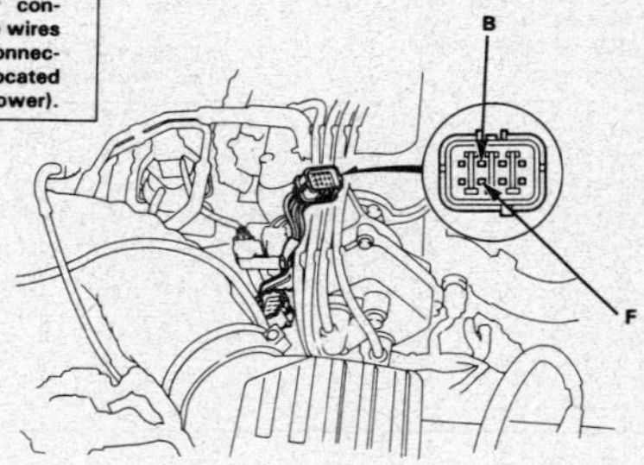

TROUBLESHOOTING—CODE 4—ACCORD

- Measure resistance between B terminal and F terminal.
- Is there 350—700 Ω ?
  - NO → Replace the distributor assembly
  - YES ↓
- Check for continuity to body ground on B terminal and F terminal individually.
- Does continuity exist on either one?
  - YES → Replace the distributor assembly
  - NO ↓
- Reconnect the connector.
- Connect the ECU test harness only to the main wire harness, not to the ECU
- Measure resistance between B15 terminal and B16 terminal.

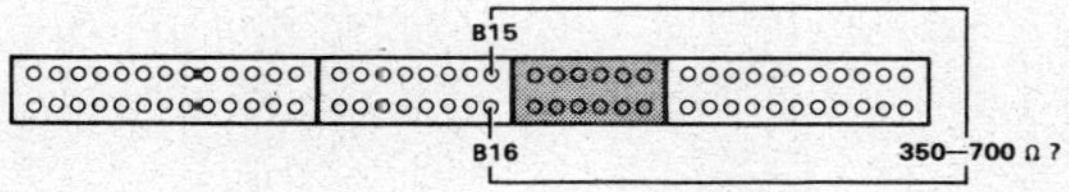

- Is there 350—700 Ω ?
  - NO → Repair open in BLU/GRN and/or BLU/YEL wires.
  - YES ↓
- Check for continuity to body ground on B15 terminal.
- Does continuity exist?
  - YES → Repair short in BLU/GRN wire between ECU (E15) and distributor connector.
  - NO ↓
- Substitute a known-good ECU and recheck. If symptom/indication goes away, replace the original ECU.

TROUBLESHOOTING—CODE 5—ACCORD

- Check Engine warning light has been reported on.
- LED indicates CODE 5.

- Turn the ignition switch OFF.
- Remove BACK UP fuse in the under-hood relay box for 10 seconds to reset ECU.
- Start the engine.
- Is Check Engine warning light on and does LED indicate CODE 5?
  - NO →
    - Intermittent failure, system is OK at this time (test drive may be necessary).
    - Check vacuum hoses, pipes and connections.
    - Make sure all connectors are secure.
  - YES ↓
- Stop the engine.
- Disconnect #21 hose from the throttle body, connect vacuum pump to the hose and apply vacuum.

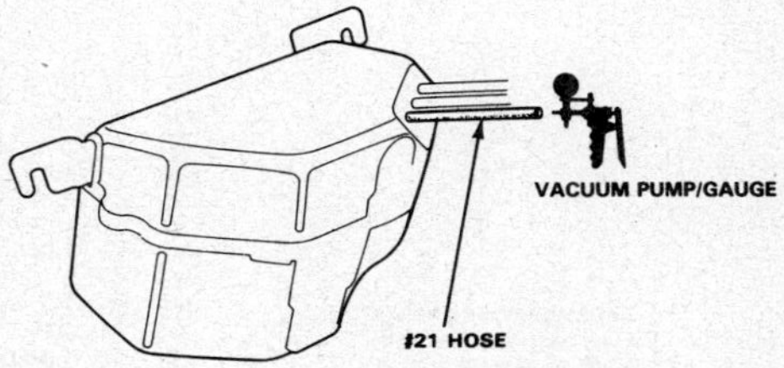

- Does it hold vacuum?
  - NO → Connect a vacuum pump to the MAP sensor and apply vacuum.
    - Does it hold vacuum?
      - NO → Replace MAP sensor.
      - YES → Replace #21 hose.
  - YES ↓
- Connect a T-fitting from a vacuum gauge between the throttle body and MAP sensor.

TROUBLESHOOTING—CODE 5—ACCORD

- Start the engine.
- Is there manifold vacuum?
  - NO → -Remove restriction from throttle body. -Replace throttle body.
  - YES ↓
- Stop the engine.
- Connect the ECU test harness between the ECU and connector
- Turn the ignition switch ON.
- Measure voltage between D17 (+) terminal and D21 (−) terminal.

- Is there approx. 3 V ?
  - NO → Replace the MAP sensor.
  - YES ↓
- Start the engine and allow it to idle.
- Is there approx. 1 V?
  - NO → Replace MAP sensor.
  - YES ↓
- Substitute a known-good ECU and recheck. If symptom/indication goes away, replace the original ECU.

TROUBLESHOOTING—CODE 6—ACCORD

Self-diagnosis LED indicates code 6: Most likely a problem in the Coolant Temperature (TW) Sensor circuit.

The TW sensor is a temperature dependant resistor (thermistor). The resistance of the thermistor decreases as the coolant temperature increases as shown below.

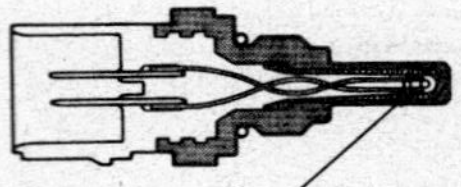

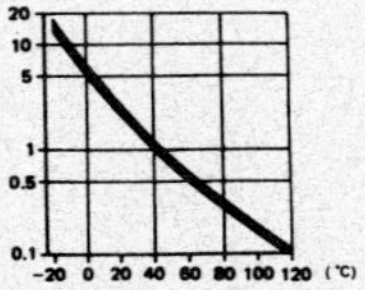

- – Check Engine warning light has been reported on. – LED indicates CODE 6.
- Turn the ignition switch OFF.
- Remove BACK UP fuse in the under-hood relay box for 10 seconds to reset ECU.
- Turn the ignition switch ON.
- Is Check Engine warning light on and does LED indicate CODE 6?
  - NO → Intermittent failure, system is OK at this time (test drive may be necessary). Check for poor connections or loose wires at TW sensor C110 and C311 (located at left shock tower).
  - YES ↓
- Warm up engine to normal operating temperature (cooling fan comes on).
- Turn the ignition switch OFF.
- Disconnect the 2P connector from the TW sensor.
- Measure resistance between the 2 terminals on the TW sensor.
- Is there 200—400 Ω ?
  - NO → Replace TW sensor.
  - YES

TROUBLESHOOTING—CODE 6—ACCORD

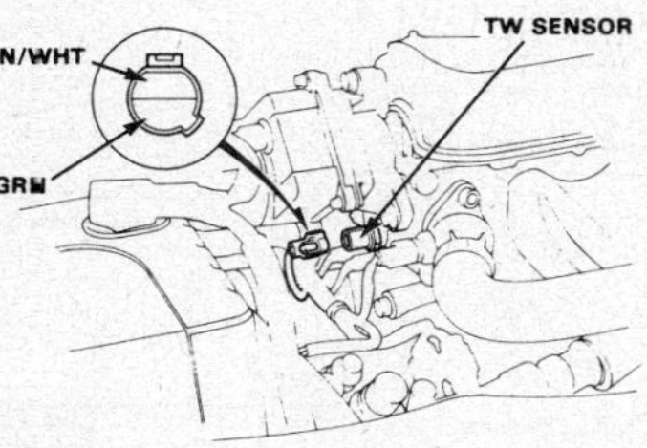

Turn the ignition switch ON.

Measure voltage between YEL/GRN and body ground.

Is there approx. 5V ?

- NO → **A/T only**
  - Disconnect the 22P connector from the A/T control unit.
  - Is there approx. 5 V ?
    - YES → **Replace the A/T control unit.**
    - NO → **Repair open or short in YEL/GRN wire between ECU (D13), A/T control unit and TW sensor. If wire is OK, substitute a known-good ECU and recheck. If prescribed voltage is now available, replace original ECU.**
- YES ↓

Measure voltage between YEL/GRN (+) terminal and GRN/WHT (−) terminal.

Is there approx. 5V ?

- NO → **Repair open in GRN/WHT wire between ECU (D22) and TW sensor.**
- YES → **Substitute a known-good ECU and recheck. If symptom/indication goes away, replace the original ECU.**

TROUBLESHOOTING—CODE 7—ACCORD

Self-diagnosis LED indicator blinks seven times: Most likely a problem in the Throttle Angle Sensor circuit.

The throttle angle sensor is a potentiometer. It is connected to the throttle valve shaft. As the throttle angle changes, the throttle angle sensor varies the voltage signal to the ECU.

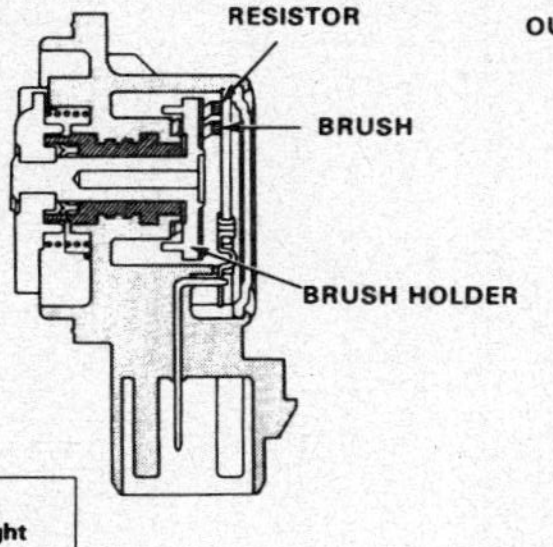

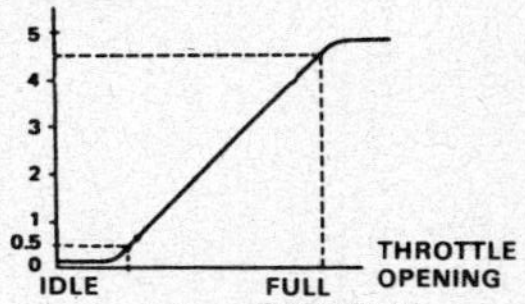

- **Engine is running.**
- **Check Engine warning light has been reported on.**
- **LED indicates CODE 7.**

Turn the ignition switch OFF.

Remove BACK UP fuse in the under-hood relay box for 10 seconds to reset ECU.

Start the engine.

Is Check Engine warning light on and does LED indicate CODE 7?

- NO → **Intermittent failure, system is OK at this time (test drive may be necessary). Check for poor connections or loose wires at throttle angle sensor C123 and C311 (located at left shock tower).**
- YES ↓

Turn the ignition switch OFF.

Disconnect the 3P connector from the throttle angle sensor.

Turn the ignition switch ON.

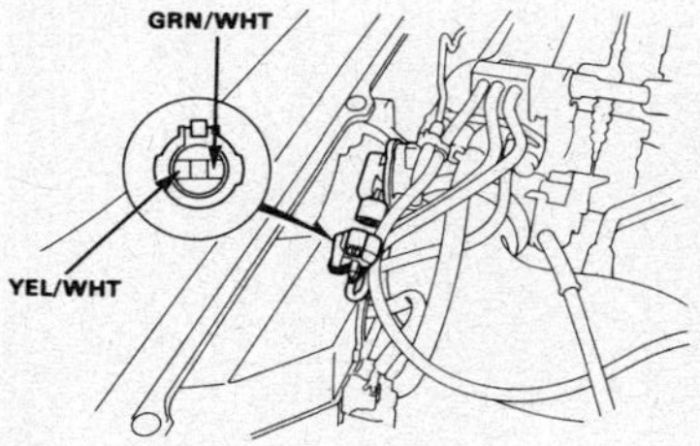

Measure voltage between YEL/WHT (+) terminal and GRN/WHT (−) terminal.

Is there approx. 5V ?

- YES → Go to Path A
- NO → Measure voltage between YEL/WHT (+) terminal and body ground. → Go to Path B

TROUBLESHOOTING—CODE 7—ACCORD

Path A

1. Turn the ignition switch OFF.
2. Reconnect the 3P connector.
3. Connect the ECU test harness between the ECU and connector
4. Turn the ignition switch ON.
5. Measure voltage between D11(+) terminal and D22 (−) terminal.

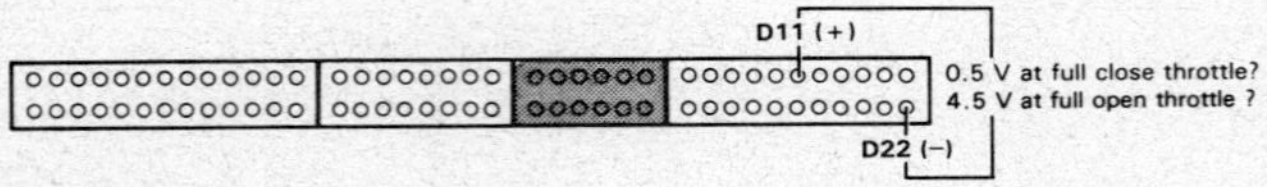

6. Is voltage approx. 0.5 V at full close throttle, and approx. 4.5 V at full open throttle? NOTE: There should be a smooth transition from 0.5 V to 4.5 V as the throttle is depressed.
   - YES → **Substitute a known-good ECU and recheck. If symptom/indication goes away, replace the original ECU.**
   - NO → **A/T only**:
     1. Disconnect the 22P connector from the A/T control unit.
     2. Is voltage approx. 0.5 V at full close throttle, and approx. 4.5 V at full open throttle? NOTE: There should be a smooth transition from 0.5 V to 4.5 V as the throttle is depressed.
        - YES → **Replace the A/T control unit.**
        - NO → **- Replace throttle angle sensor. - Repair open or short in RED/BLK wire between ECU (D11), A/T control unit and throttle angle sensor.**

Path B

1. Is there approx. 5V ?
   - YES → **Repair open in GRN/WHT wire between ECU (D22) and throttle angle sensor.**
   - NO ↓
2. Turn the ignition switch OFF.
3. Connect the ECU test harness between the ECU and connector
4. Turn the ignition switch ON.
5. Measure voltage between D20 (+) terinal and D22 (−) terminal.
6. Is there approx. 5V ?
   - YES → **Repair open in YEL/WHT wire between ECU (D20) and throttle angle sensor.**
   - NO → **Substitute a known-good ECU and recheck. If prescribed voltage is now available, replace the original ECU.**

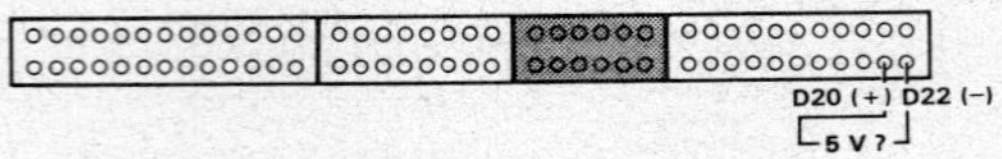

TROUBLESHOOTING—CODE 8—ACCORD

- **Check Engine warning light has been reported on.**
- **LED indicates CODE 8.**

1. Turn the ignition switch OFF.
2. Remove BACK UP fuse in the under-hood relay box for 10 seconds to reset ECU.
3. Start the engine.
4. Is Check Engine warning light on and does LED indicate CODE 8?
   - NO → **Intermittent failure, system is OK at this time (test drive may be necessary). Check for poor connections or loose wires at distributor connector and C209 (located at right shock tower).**
   - YES ↓
5. Stop the engine.
6. Disconnect the 8P connector from the TDC/CRANK/CYL sensor.
7. Measure resistance between C terminal and G terminal.
8. Is there 350—700 Ω ?
   - NO → **Replace the distributor assembly**
   - YES ↓
9. Check for continuity to body ground on C terminal and G terminal individually.
10. Does continuity exist on either one?
    - YES → **Replace the distributor assembly**
    - NO ↓
11. Reconnect the connector.

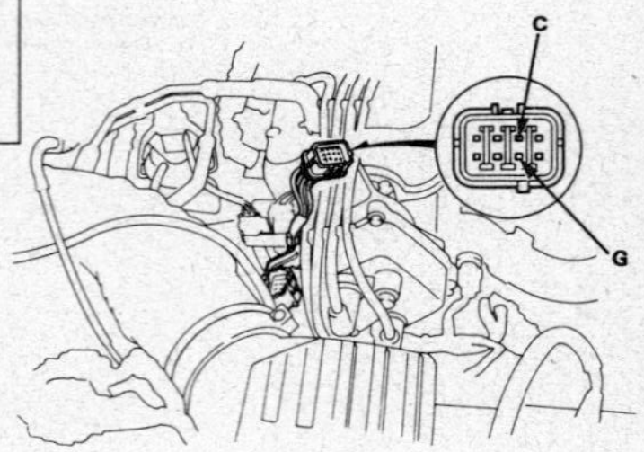

TROUBLESHOOTING—CODE 8—ACCORD

Connect the ECU test harness only to the main wire harness, not to the ECU

Measure resistance between B13 terminal and B14 terminal.

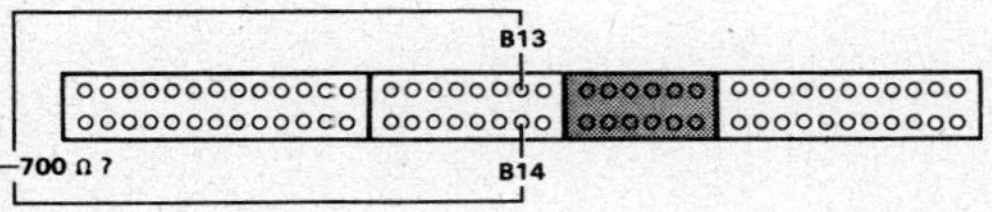

Is there 350—700 Ω ?
- NO: Repair open in ORN/BLU and/or WHT/BLU wires.
- YES: Check for continuity to body ground on B13 terminal.

Does continuity exist?
- YES: Repair short in ORN/BLU wire between ECU (B13) and distributor connector.
- NO: Substitute a known-good ECU and recheck. If symptom/indication goes away, replace the original ECU.

TROUBLESHOOTING—CODE 9—ACCORD

- Check Engine warning light has been reported on.
- LED indicates CODE 9.

Turn the ignition switch OFF.

Remove BACK UP fuse in the under-hood relay box for 10 seconds to reset ECU.

Start the engine.

Is Check Engine warning light on and does LED indicate CODE 9?
- NO: Intermittent failure, system is OK at this time (test drive may be necessary). Check for poor connections or loose wires at the distributor connector and C209 (located at right shock tower).
- YES: Stop the engine.

Disconnect the 8P connector from the TDC/CRANK/CYL sensor.

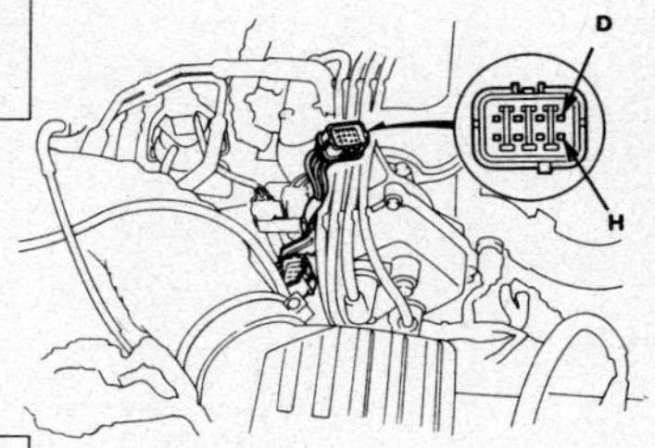

Measure resistance between D terminal and H terminal.

Is there 350—700 Ω ?
- NO: Replace the distributor assembly
- YES: Check for continuity to body ground on D terminal and H terminal individually.

Does continuity exist on either one?
- YES: Replace the distributor assembly
- NO: Reconnect the connector.

TROUBLESHOOTING—CODE 9—ACCORD

Connect the ECU test harness only to the main wire harness, not to the ECU.

Measure resistance between B11 terminal and B12 terminal.

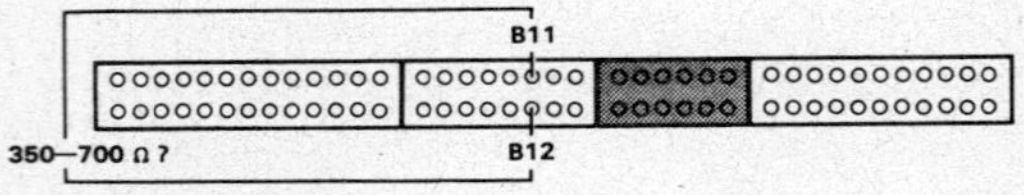

Is there 350—700 Ω ?

- NO → **Repair open in ORN and/or WHT wires.**
- YES ↓

Check for continuity to body ground on B11 terminal.

Does continuity exist?

- YES → **Repair short in ORN wire between ECU (B11) and distributor connector.**
- NO ↓

Substitute a known-good ECU and recheck. If symptom/indication goes away, replace the original ECU.

TROUBLESHOOTING—CODE 10—ACCORD

Self-diagnosis LED indicates code 10: Most likely a problem in the Intake Air Temperature (TA) Sensor circuit.

The TA sensor is a temperature dependant resistor (thermistor). The resistance of the thermistor decreases as the intake air temperature increases as shown below.

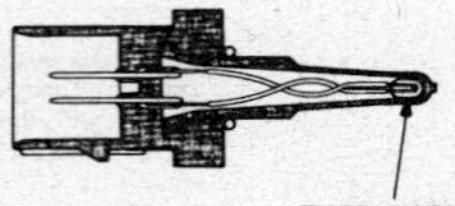

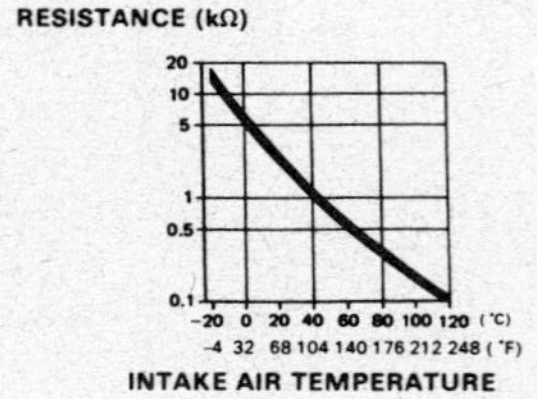

- **Check Engine warning light has been reported on.**
- **LED indicates CODE 10.**

Turn the ignition switch OFF.

Remove BACK UP fuse in the under-hood relay box for 10 seconds to reset ECU.

Turn the ignition switch ON.

Is Check Engine warning light on and does LED indicate CODE 10?

- NO → **Intermittent failure, system is OK at this time (test drive may be necessary). Check for poor connections or loose wires at TA sensor C124 and C311 (located at left shock tower).**
- YES ↓

Turn the ignition switch OFF.

Disconnect the 2P connector from the TA sensor.

Measure resistance between the 2 terminals on the TA sensor.

Is there 1–4 kΩ ?

- NO → **Replace TA sensor.**
- YES ↓

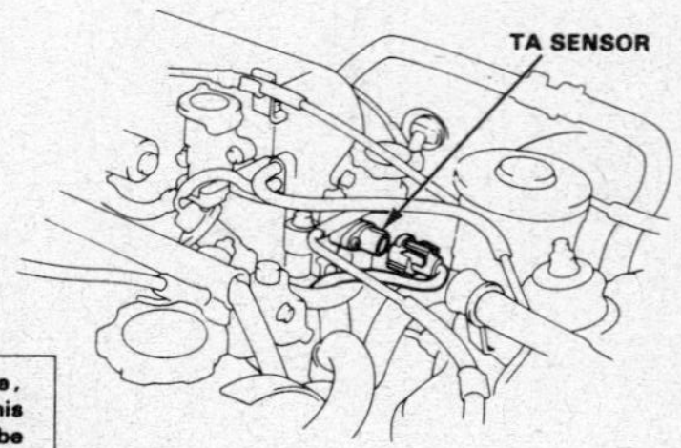

TROUBLESHOOTING—CODE 10—ACCORD

Turn the ignition switch ON.

↓

Measure voltage between RED/YEL (+) terminal and body ground.

↓

Is there approx. 5 V ?

- NO → Repair open or short in RED/YEL wire between ECU (D15) and TA sensor. **If wire is OK, substitute a known-good ECU and recheck. If prescribed voltage is now available, replace original ECU.**
- YES ↓

Measure voltage between RED/YEL (+) terminal and GRN/WHT (−) terminal.

↓

Is there approx. 5 V ?

- NO → **Repair open in GRN/WHT wire between ECU (D22) and TA sensor.**
- YES ↓

Substitute a known-good ECU and recheck. If symptom/indication goes away, replace the original ECU.

TROUBLESHOOTING—CODE 12—ACCORD

— Check Engine warning light has been reported on.
— LED indicates CODE 12.

↓

Turn the ignition switch OFF.

↓

Remove the BACK UP fuse in the under-hood relay box for 10 seconds to reset ECU.

↓

Road test necessary: Warm up the engine to normal operating temperature (cooling fan comes on). Drive the car on the road for approx. 10 minutes. Try to keep the engine speed in the 1700— 2500 range.

↓

Is Check Engine warning light on and does LED indicated CODE 12 ?

- NO → **Intermittent failure, system is OK at this time. Check for poor connections or loose wires at EGR valve C117, C210 (located at right shock tower), C311 (located at left shock tower), control box.**
- YES ↓

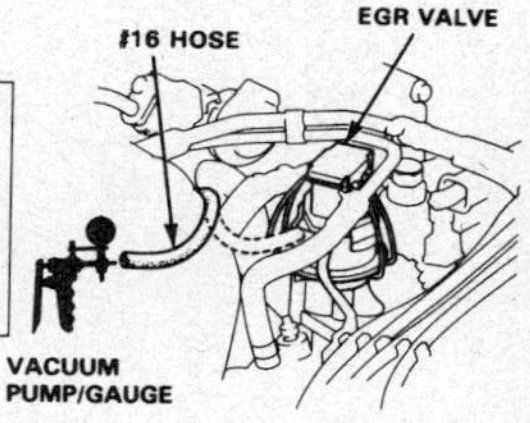

With the engine at idle, disconnect the #16 hose from the EGR valve and connect a vacuum pump/gauge to the hose.

↓

Is there any vacuum ?

- YES → Disconnect 4P connector from the control box and check the #16 hose for vacuum again. → Go to Path J
- NO ↓

Move the vacuum pump/gauge to the EGR valve. → Go to Path I

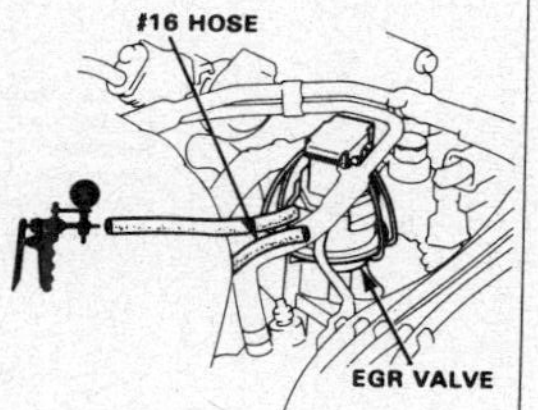

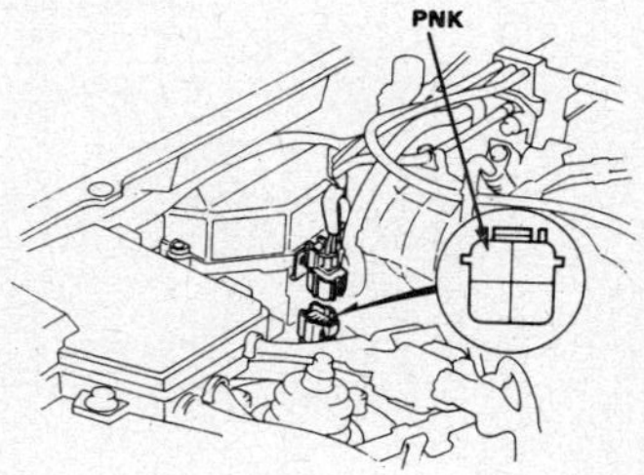

TROUBLESHOOTING—CODE 12—ACCORD

Path I

- With the engine at idle, apply 8" of vacuum to the EGR valve.
- Does the engine stall and does the EGR valve hold vacuum ?
  - NO: **Replace EGR valve.**
  - YES: Disconnect the 4P connector from the control box.
- Measure voltage between BLK/YEL (+) terminal on the main wire harness and body ground.
- Is there battery voltage ?
  - NO: **Repair open in BLK/YEL wire between the solenoid valve and No. 2 fuse.**
  - YES: Reconnect the vacuum pump/gauge to the #16 hose.
- Start the engine and allow it to idle.

Path J

- Is there any vacuum ?
  - YES: **Check vacuum hose routing of the entire EGR system. If hose routing is OK, replace EGR control solenoid valve.**
  - NO: **Inspect for short to body ground in PNK wire between EGR control solenoid valve and ECU (A11). If wire is OK, substitute a known-good ECU and retest. If symptom/indication goes away, replace the original ECU.**

- Connect the battery positive terminal to the A terminal of the 4P connector. While watching the vacuum gauge, connect the battery negative terminal to the B terminal.

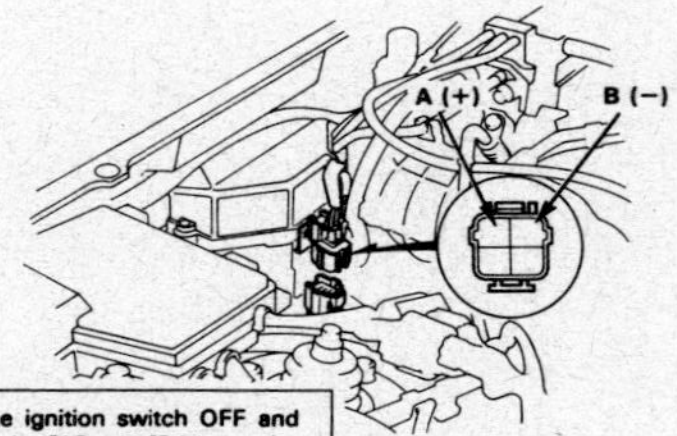

- Is there approx. 8" within 1 second?
  - NO: Turn the ignition switch OFF and inspect the #16 and #24 hoses for leaks, restrictions, or mis-routing.
    - Are the hoses OK?
      - NO: **Connect as necessary.**
      - YES: Disconnect the lower hose on EGR control solenoid valve and connect a vacuum gauge to the hose.
    - Start the engine and allow it to idle.
    - Is there 6"— 10" of vacuum ?
      - NO: **Replace CVC valve.**
      - YES: **Replace the EGR control solenoid valve.**
  - YES: Turn the ignition switch OFF and reconnect the 4P connector.

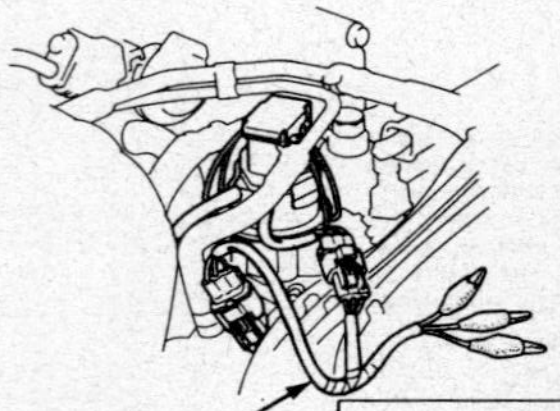

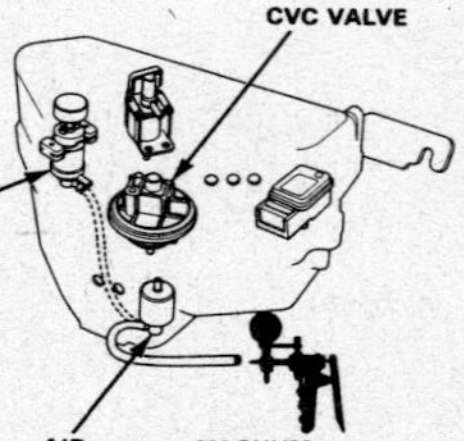

- Connect the test harness between the EGR valve lift sensor and engine wire harness.
- Turn the ignition switch ON.
- Measure voltage between RED (+) terminal and GRN (−) terminal.
- Is there approx. 5 V ?
  - NO:
    - **Repair open in YEL/WHT wire between EGR valve and ECU (D20).**
    - **Repair open in GRN/WHT wire between EGR valve and ECU (D22).**
  - YES

TROUBLESHOOTING—CODE 12—ACCORD

Measure voltage between WHT (+) terminal and GRN (−) terminal.

Is there approx 1.2 V ?

- NO →
  - Replace EGR valve.
  - Repair short in WHT/BLK* wire between ECU (D12) and EGR valve lift sensor.
  - Substitute a known-good ECU and recheck. If prescribed voltage is now available, replace the original ECU.

  *: YEL on early production.
- YES ↓

While watching the voltmeter, slowly apply a continuous 8" of vacuum to the EGR valve. Repeat several times, completely releasing vacuum between.

Is the maximum voltage approx. 4 V ?

- NO → Replace EGR valve.
- YES ↓

Does the voltage consistently increase/decrease as the vacuum increases/decreases ?

- NO → Replace EGR valve.
- YES ↓

Reconnect the #16 hose to the EGR valve.

Turn the ignition switch OFF.

Connect the ECU test harness between the ECU and connector

Start the engine and allow it to idle.

Measure voltage between D12 (+) terminal and D22 (−) terminal.

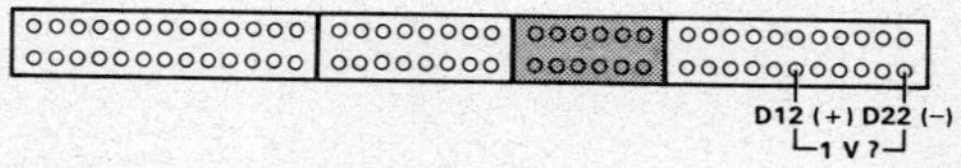

Is there approx. 1.2 V ?

- NO → Repair open in WHT/BLK* wire between ECU (D12) and the sensor.

  *: YEL on early production.
- YES ↓

Connect A11 terminal to A26 terminal with a jumper wire.

Did the engine stall ?

- NO → Repair open in PNK wire between ECU (A11) and EGR control solenoid valve.
- YES ↓

Substitute a known-good ECU and recheck. If symptom/indication goes away, replace the original ECU.

TROUBLESHOOTING—CODE 13—ACCORD

Self-diagnosis LED indicates code 13: A problem in the Atmospheric Pressure (PA) Sensor.

The PA sensor is built into the ECU.

- Check Engine warning light has been reported on.
- LED indicates CODE 13.

Turn the ignition switch OFF.

Remove BACK UP fuse in the under-hood relay box for 10 seconds to reset ECU.

Turn the ignition switch ON.

Is Check Engine warning light on and does LED indicate CODE 13 ?

NO → Intermittent failure, system is OK at this time (test drive may be necessary).

YES → Substitute a known-good ECU and recheck. If symptom/indication goes away, replace the original ECU.

TROUBLESHOOTING—CODE 14—ACCORD

Self-diagnosis LED indicates code 14: A problem in the Electronic Air Control Valve (EACV) circuit.

The EACV changes the amount of air bypassing the throttle body in response to a current signal from the ECU in order to maintain the proper idle speed.

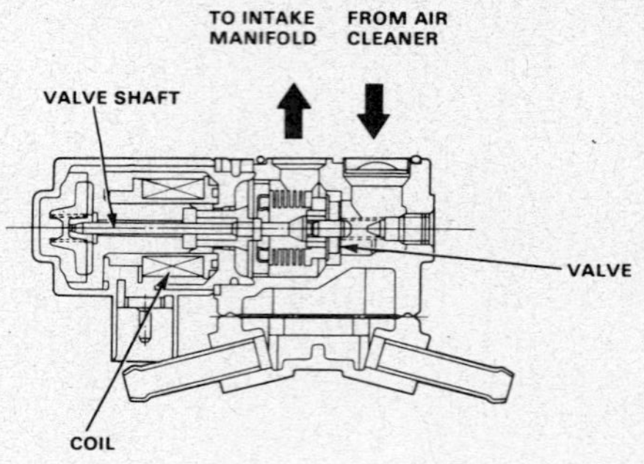

- Engine is running.
- Check Engine warning light has been reported on.
- LED indicates CODE 14.

Turn the ignition switch OFF.

Remove BACK UP fuse in the under-hood relay box for 10 seconds to reset ECU.

Start the engine.

Is Check Engine warning light on and does LED indicate CODE 14?

NO → Intermittent failure, system is OK at this time (test driving may be necessary). Check for poor connections or loose wires at EACV and at C311 (located at left shock tower).

YES → Stop the engine.

Disconnect the 2P connector from the EACV.

Measure resistance between the 2 terminals on the EACV.

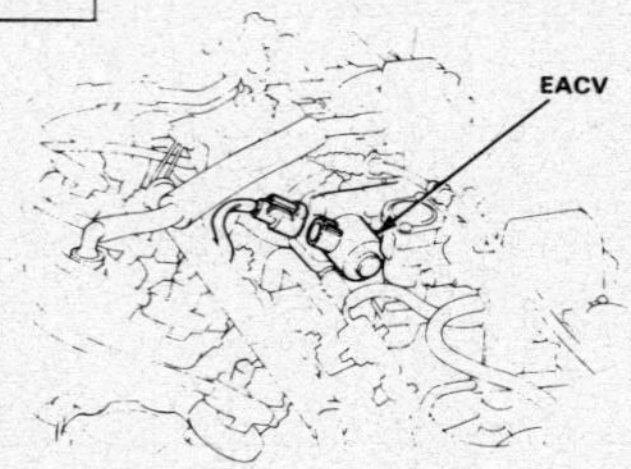

TROUBLESHOOTING—CODE 14—ACCORD

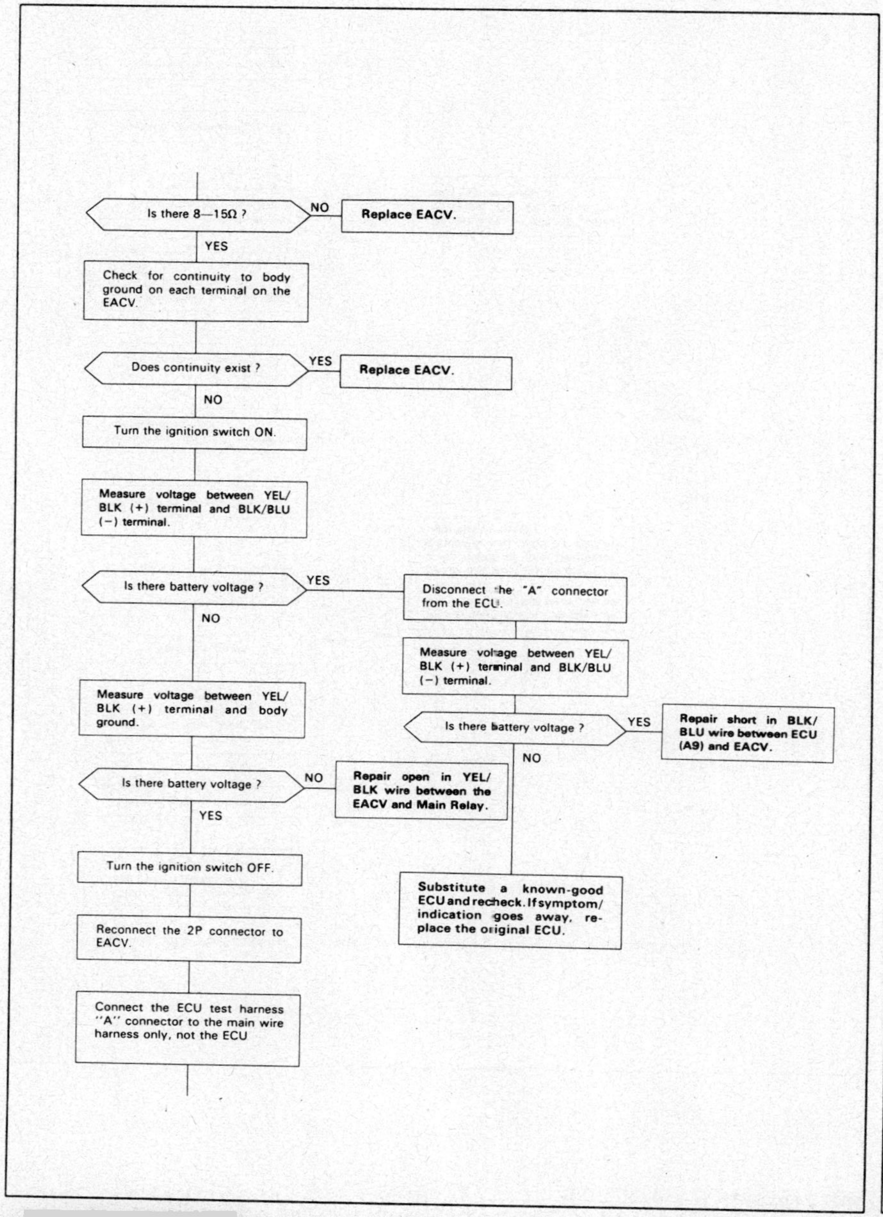

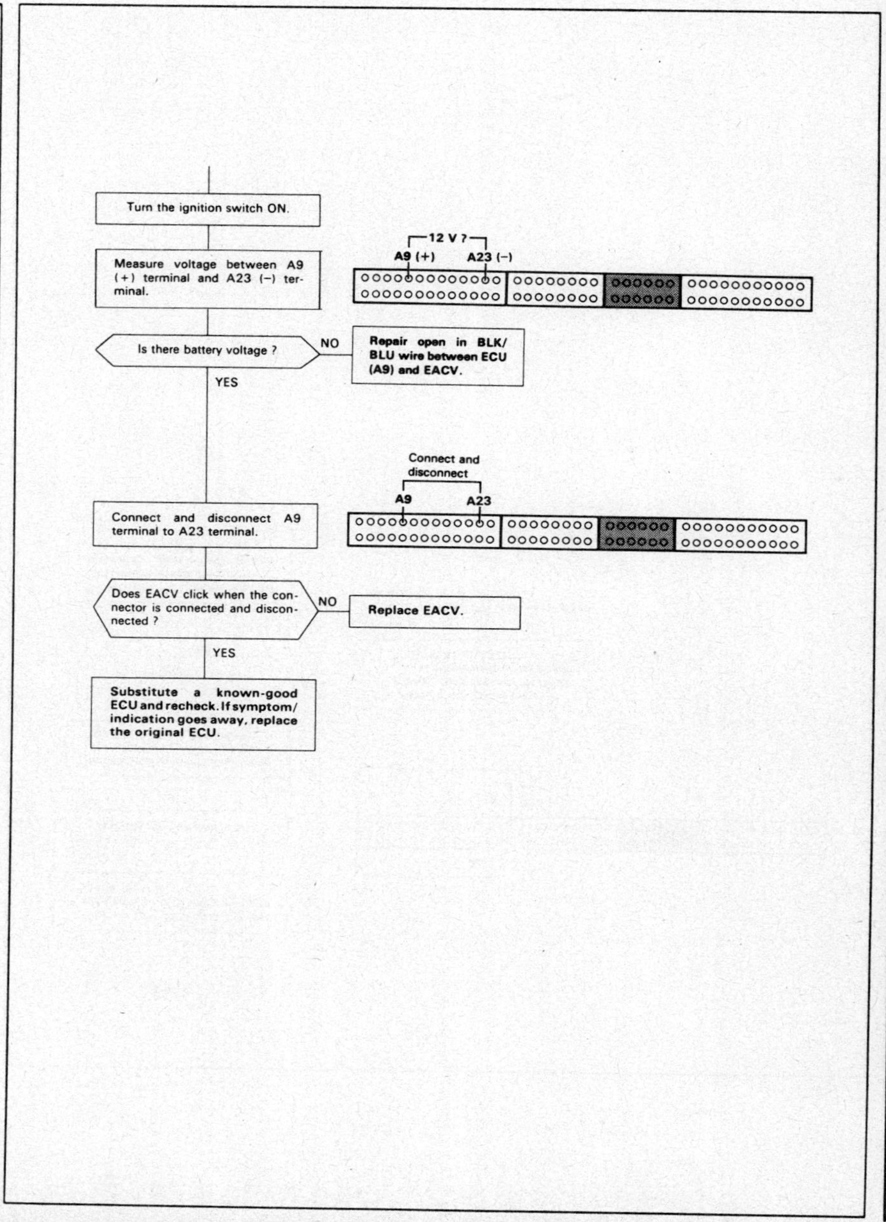

TROUBLESHOOTING—CODE 15—ACCORD

Self-diagnosis LED indicates code 15: A problem in the Ignition Output Signal circuit.

- Check Engine warning light has been reported on.
- LED indicates CODE 15.

↓

Turn the ignition switch OFF.

↓

Remove BACK UP fuse in the under-hood relay box for 10 seconds to reset ECU.

↓

Start the engine.

↓

Is Check Engine warning light on and does LED indicate CODE 15?

- NO → Intermittent failure, system is OK at this time (test drive may be necessary). Check for poor connections or loose wires at the distributor connector and C209 (located at right shock tower).
- YES ↓

Turn the ignition switch OFF.

↓

Disconnect the 2P connector from the distributor.

↓

Turn the ignition switch ON.

↓

Measure voltage between BLK/YEL (+) terminal and body ground.

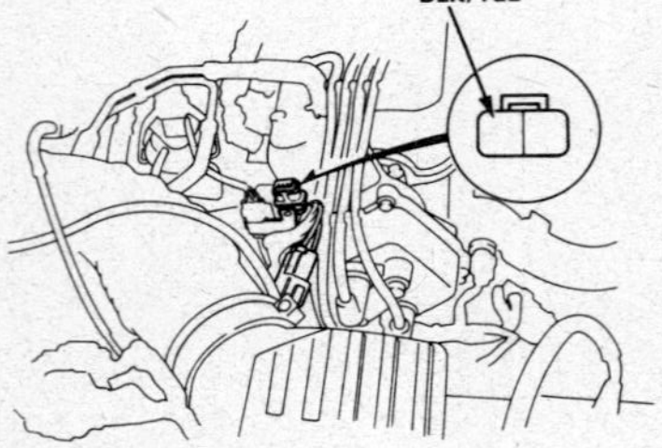

↓

Is there battery voltage ?

- NO → Repair open in BLK/YEL wire between the 2P connector and ignition switch.
- YES ↓

Turn the ignition switch OFF.

↓

Reconnect the 2P connector.

↓

Connect the ECU test harness between the ECU and connector (page 11-18).

↓

Turn the ignition switch ON.

↓

Measure voltage individually between A21 (+), A22 (+) terminals and A26 (−) terminal.

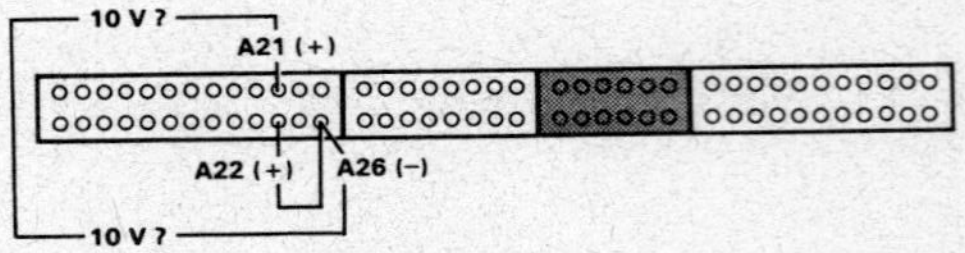

↓

Is there approx. 10 V?

- NO →
  - Replace the igniter unit.
  - Repair open or short YEL/GRN wires between distributor and ECU (A21 or A22).

  NOTE: If the YEL/GRN wire was shorted, the ignitor may be damaged.
- YES ↓

Substitute a known-good ECU and recheck. If symptom/indication goes away, replace the original ECU.

TROUBLESHOOTING—CODE 16—ACCORD

Self-diagnosis LED indicates code 16: A problem in the fuel injector circuit.

The injectors are the solenoid-actuated constant-stroke pintle type consisting of a solenoid, plunger needle valve and housing. When current is applied to the solenoid coil, the valve lifts up and pressurized fuel is injected close to the intake valve. Because the needle valve lift and the fuel pressure are constant, the injection quantity is determined by the length of time that the valve is open (i.e., the duration the current is supplied to the solenoid coil). The injector is sealed by an O-ring and seal ring at the top and bottom. These seals also reduce operating noise.

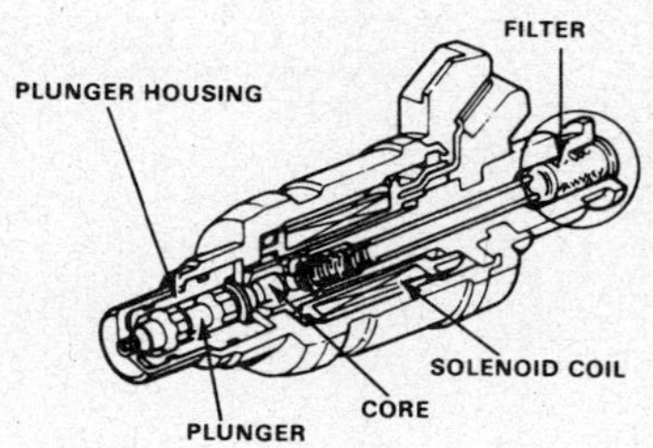

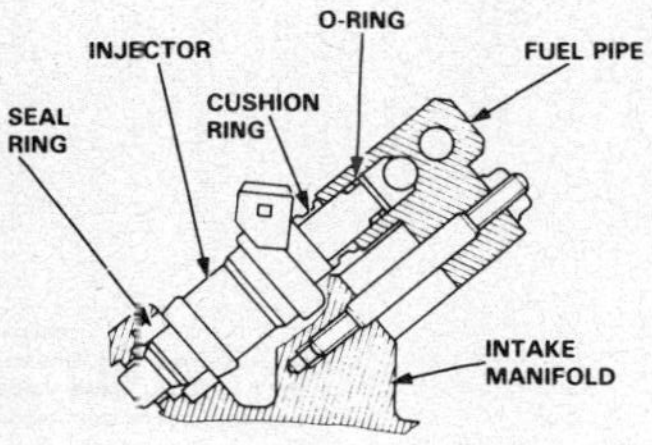

- Check Engine warning light has been reported on.
- LED indicates CODE 16.

Turn the ignition switch OFF.

Remove BACK UP fuse in the under-hood relay box for 10 seconds to reset ECU.

Turn the ignition switch to START position.

Does the engine start ?
- NO: Go to Path D
- YES: Is Check Engine warning light on and does LED indicate CODE 16?
  - NO: Intermittent failure, system is OK at this time (test drive may be necessary). Check for poor connections or loose wires at injectors, injector resistor and C311 (located at left shock tower).
  - YES: Go to Path C

Path C

Check the clicking sound of each injector by means of a stethoscope when the engine is idling.

Do the injectors click ?
- YES: Substitute a known-good ECU and recheck. If symptom/indication goes away, replace the original ECU.
- NO: Turn the ignition switch OFF.

Disconnect the 2P connector from the injector that does not click.

Measure resistance between the 2 terminals of the injector.

Path D

Turn the ignition switch OFF.

Disconnect the 2P connector from each injector.

Measure resistance between the 2 terminals of the injector.

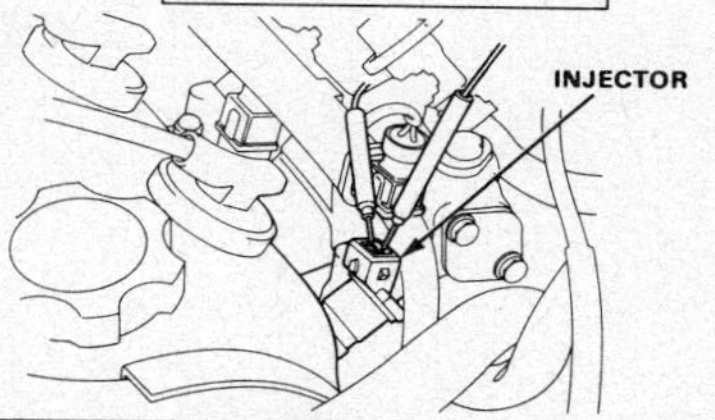

Is there 1.5—2.5Ω ?
- NO: Replace the injector.
- YES: Turn the ignition switch ON.

TROUBLESHOOTING—CODE 16—ACCORD

Measure voltage between RED/BLK (+) terminal on the 2P connector and body ground.

Is there battery voltage ?

- NO → Turn the ignition switch OFF.
  - Disconnect 6P connector on the injector resistor.
  - Turn the ignition switch ON.
  - Measure voltage between YEL/BLK (+) terminal and body ground.
  - Is there battery voltage ?
    - NO → **Repair open in the YEL/BLK wire between the injector resistor and the main relay.**
    - YES → **—Replace the injector resistor.**
      —Repair open RED/BLK wire between 2P connector and resistor.
- YES → Reconnect the 2P connector to the injector.
  - Turn the ignition switch OFF.

Connect the ECU test harness between the ECU and connector

Turn the ignition switch ON.

Measure voltage between A23(−) terminal and following terminals.
- No.1 injector: A1 (+) terminal.
- No.2 injector: A3 (+) terminal.
- No.3 injector: A5 (+) terminal.
- No.4 injector: A2 (+) terminal.

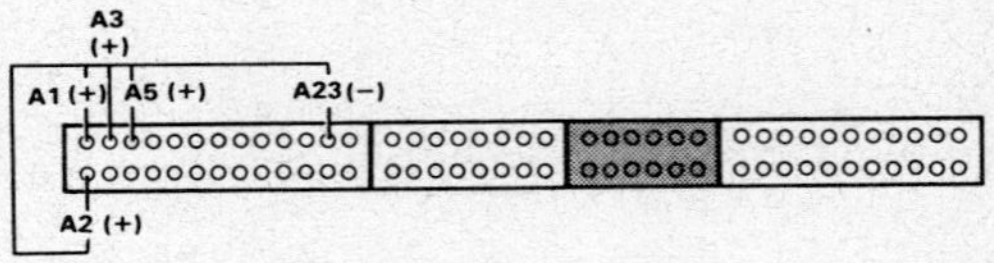

Is there battery voltage ?

- NO → **Repair open in the wire between the ECU (A1, A3, A5 or A2) and the injector.**
- YES → **Substitute a known-good ECU and recheck. If symptom/indication goes away, replace the original ECU.**

TROUBLESHOOTING – CODE 17 – ACCORD

Self-diagnosis LED indicates code 17: A problem in the Vehicle Speed Sensor circuit.

The signal generated by the speed sensor, produces pulses when the front wheels turn.

- Check Engine warning light has been reported on.
- LED indicates CODE 17.

Turn the ignition switch OFF.

Remove BACK UP fuse in the under-hood relay box for 10 seconds to reset ECU.

Road test necessary.
In 2nd gear accelerate to 3,500 rpm, then decelerate to 1,500 rpm with throttle fully closed.

Is Check Engine warning light on and does LED indicate CODE 17?

NO → Intermittent failure, system is OK at this time. Check for poor connections or loose wires at C113 and C310 (located at left shock tower).

YES ↓

Block rear wheels and set the parking brake. Jack up the front of the car and support with safety stands.

WARNING Block rear wheels before jacking up front of car.

Connect the ECU test harness between the ECU and connector

Turn the ignition switch ON.

Slowly rotate left front wheel and measure voltage between B10 (+) terminal and A26 (–) terminal.

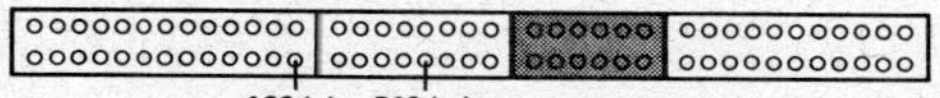

Does voltage pulse 0 V and 5 V?

NO →
- Repair open or short in ORN wire between ECU (B10) and the speed sensor.
- Faulty speed sensor.
- Substitute a known-good ECU and recheck. If symptom/indication goes away, replace the original ECU.

YES ↓

Substitute a known-good ECU and recheck. If symptom/indication goes away, replace the original ECU.

TROUBLESHOOTING – CODE 20 – ACCORD

Self-diagnosis LED indicates code 20: A problem in the Electric Load Detector circuit.

– LED indicates CODE 20.

Trun the ignition switch OFF.

Remove BACK UP fuse in the under-hood relay box for 10 seconds to reset ECU.

Start engine and keep engine rpm at idle.
Turn on headlights.

Does LED indicate CODE 20?

NO → Intermittent failure, system is ok at this time. (test drive may be necessary). Check for poor connections or loose wires.

YES ↓

Turn the ignition switch OFF.

Open the under-hood relay box lid.

Disconnect the 3P connector from the electric load detector.

Turn the ignition switch ON.

Measure voltage between BLK/YEL (+) terminal and BLK (–) terminal.

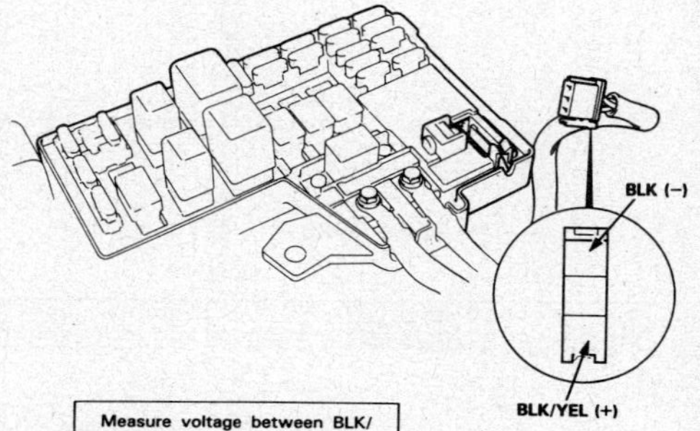

Is there battery voltage?

NO → Measure voltage between BLK/YEL (+) terminal and body ground.
Is there battery voltage?
YES → Repair open in BLK wire between the 3P connector and G251.
NO → Repair open in BLK/YEL wire between No. 2 fuse and the 3P connector.

YES ↓

Turn the ignition switch OFF.

Connect the ECU test harness between the ECU and connector. Disconnect "D" connector from the ECU only, not the main wire harness.

TROUBLESHOOTING—CODE 20—ACCORD

Connect GRN/RED terminal of the 3P connector to body ground.

Check for continuity in GRN/RED wire between ECU (D10) and body ground.

Does continuity exist?

NO → **Repair open in GRN/RED wire between ECU (D10) and the 3P connector.**

YES ↓

Connect "D" connector to the ECU.
Disconnect GRN/RED terminal from body ground.

Turn the ignition switch ON.

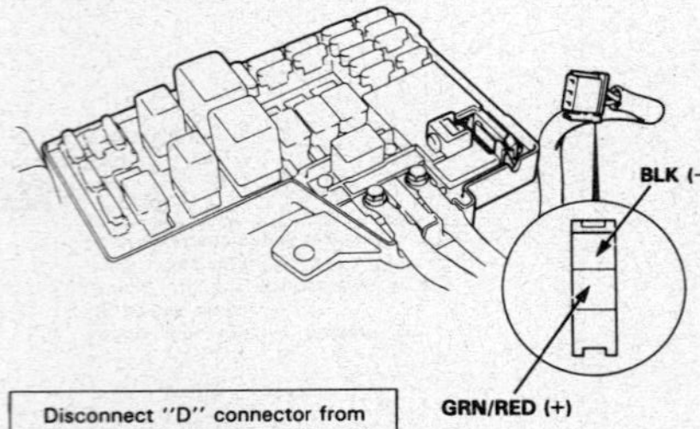

Measure voltage between GRN/RED (+) terminal and BLK (−) terminal of the 3P connector.

Is there 4.5−5 V?

NO → Disconnect "D" connector from the main wire harness only, not the ECU.

Measure voltage between D10 (+) terminal and A26 (−) terminal.

Is there 4.5−5 V?

YES → **Repair short in GRN/RED wire between ECU (D10) and the 3P connector.**

NO → **Substitute a known-good ECU and recheck. If prescribed voltage is now available, replace the original ECU.**

YES ↓

Connect the 3P connector to the electric load detector and the start the engine.

Under the conditions listed in the chart to the right, measure voltage between D10 (+) terminal and A26 (−) terminal.

Condition	Voltage
Headlight switch, first position (•)	2.5−3.5 V
Headlight switch, second position (●)	1.5−2.5 V

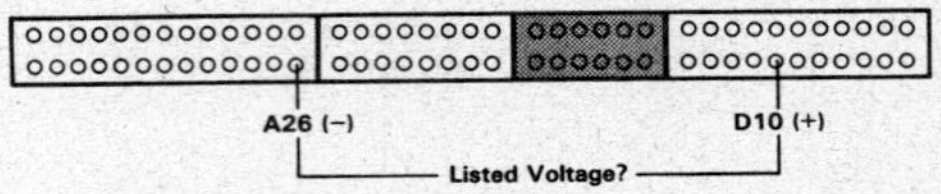

Is the voltage listed in the chart available?

NO → **Faulty electric load detector.**

YES → **Substitute a known-good ECU and recheck. If symptom/indication goes away, replace the original ECU.**

TROUBLESHOOTING – CODE 30 – ACCORD

Self-diagnosis LED indicates code 30: A problem in the line (B3) between A/T control unit and PGM-FI ECU.

LED indicates CODE 30.

Turn the ignition switch OFF.

Remove BACK UP fuse in the under-hood relay box for 10 seconds to reset ECU.

Test drive necessary.
Drive the car for several miles so that the transmission upshifts and downshifts several times.

Does LED indicate CODE 30?

NO → Intermittent failure, system is OK at this time.

YES ↓

Turn the ignition switch OFF.

Connect the ECU test harness between the ECU and connector not to the ECU

Disconnect the 22P connector from the A/T control unit.

Check for continuity between B3, terminal and body ground.

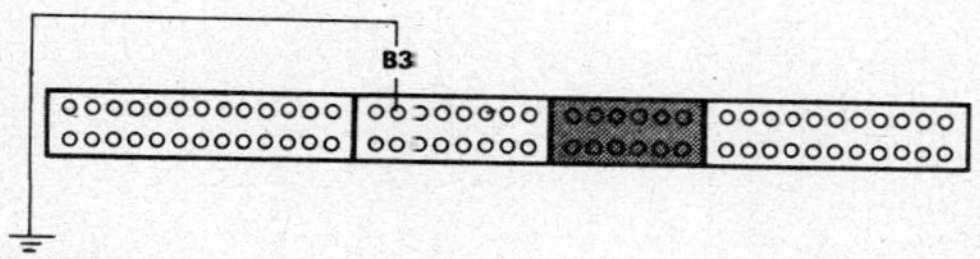

Does continuity exist?

YES → Repair short in WHT/RED wire between ECU (B3) and the A/T control unit.

NO ↓

Check for continuity on WHT/RED wire between B3 and 22P connector of the A/T control unit.

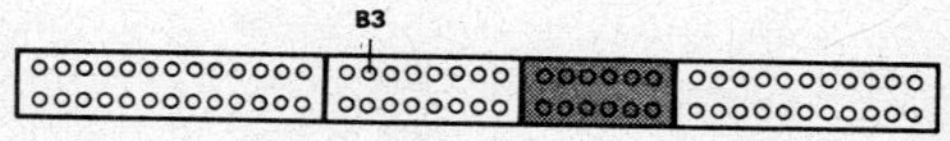

Does continuity exist?

NO → Repair open in WHT/RED wire between ECU (B3) and the A/T control unit.

YES → Substitute a known-good ECU and recheck. If symptom/indication goes away, replace the original ECU.

TROUBLESHOOTING—CODE 31—ACCORD

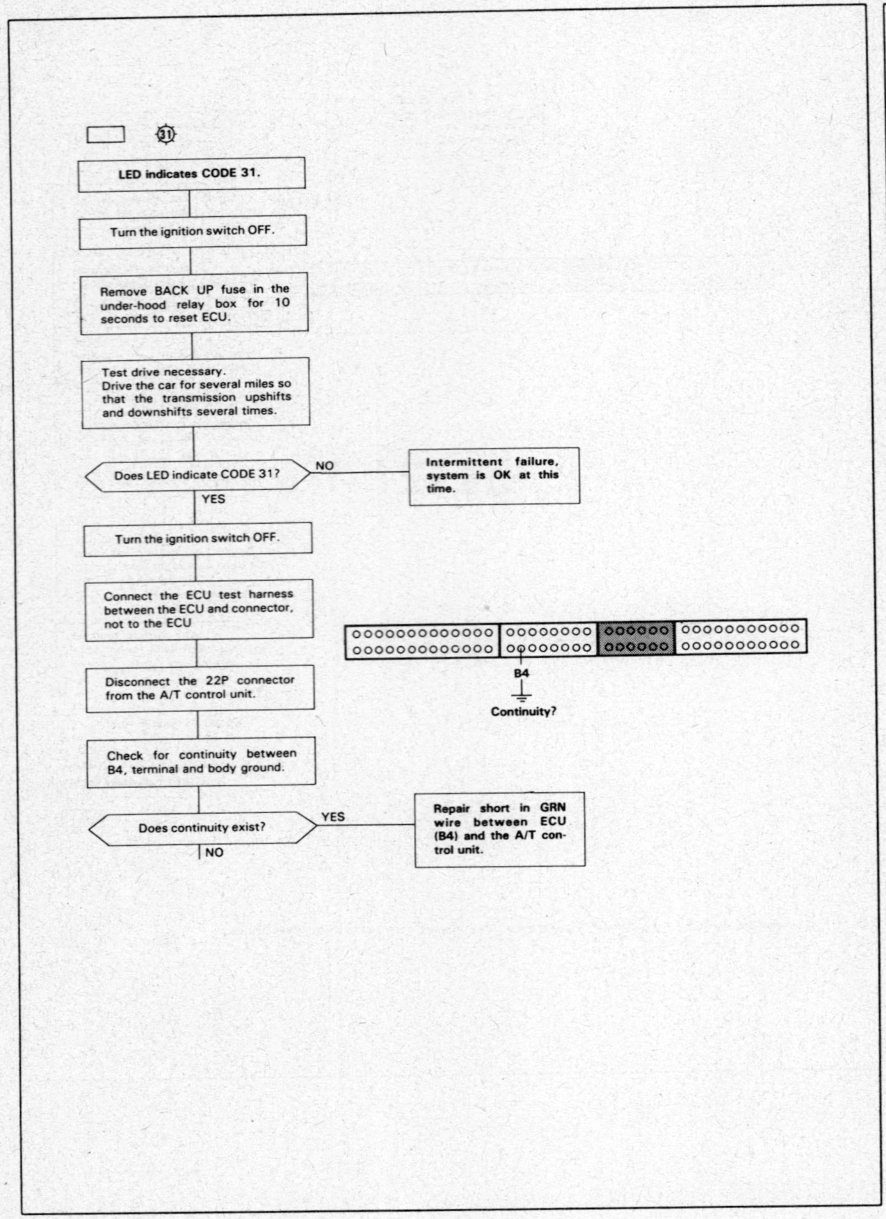

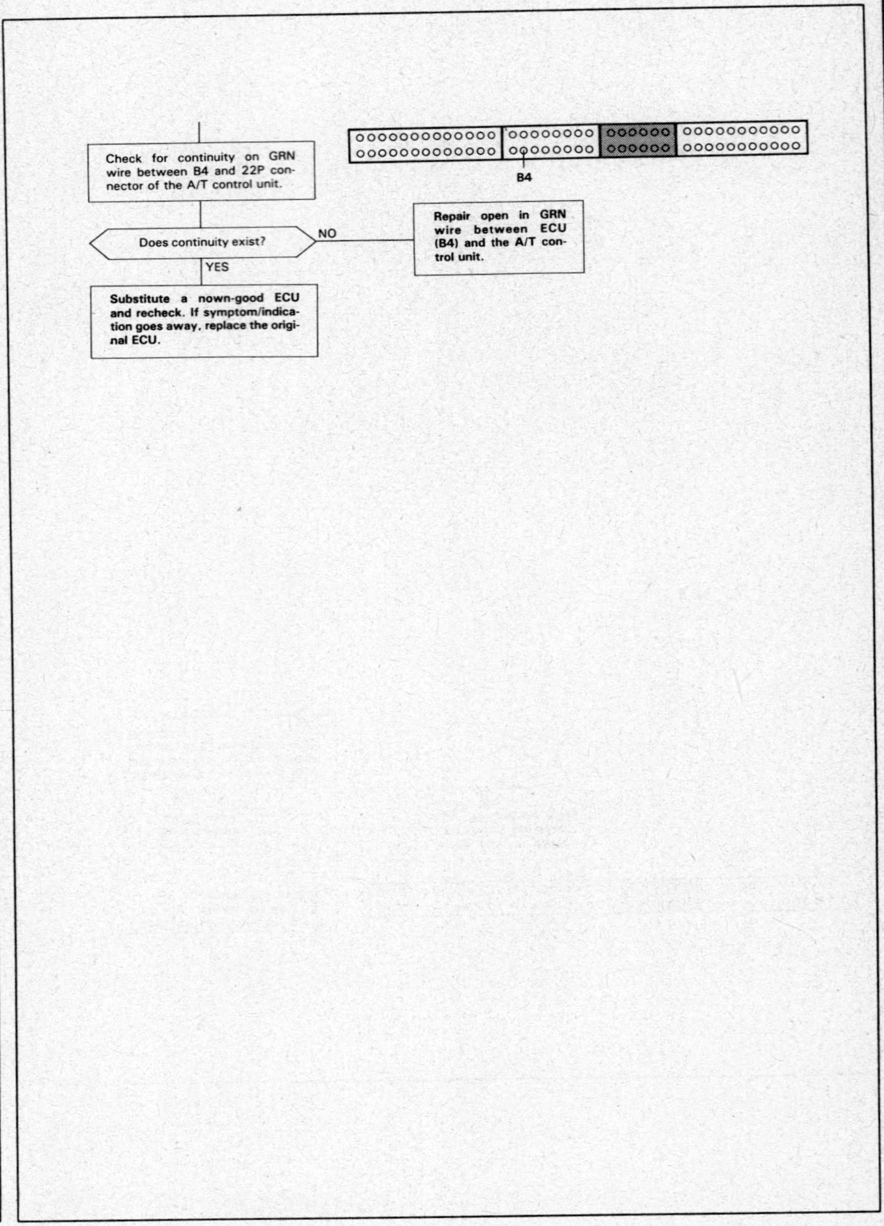

TROUBLESHOOTING—CODE 41—ACCORD

(41) Self-diagnosis LED indicates code 41: A problem in the Oxygen (O_2) Sensor Heater circuit.

(41)

- Engine is running.
- Check Engine warning light has been reported on.
- LED indicates CODE 41.

Turn the ignition switch OFF.

Remove BACK UP fuse in the under-hood relay box for 10 seconds to reset ECU.

Start the engine

Is Check Engine warning light on and does LED indicate CODE 41 ?

- NO → Intermittent failure, system is OK at this time (test drive may be necessary). Check for poor connections or loose wires at O_2 sensor C105, C210 (located at right shock tower), and ECU.
- YES ↓

Stop the engine

Disconnect the 4P connector from the O_2 sensor.

Measure resistance between terminals A and B on the O_2 sensor.

Is there 15–20 Ω ?

- NO → Replace O_2 sensor.
- YES ↓

Check for continuity to body ground on each terminal on the O_2 sensor.

Does continuity exist ?

- YES → Replace O_2 sensor.
- NO ↓

Check for continuity between terminal A and terminals C and D individually.

Does continuity exist ?

- YES → Replace O_2 sensor
- NO ↓

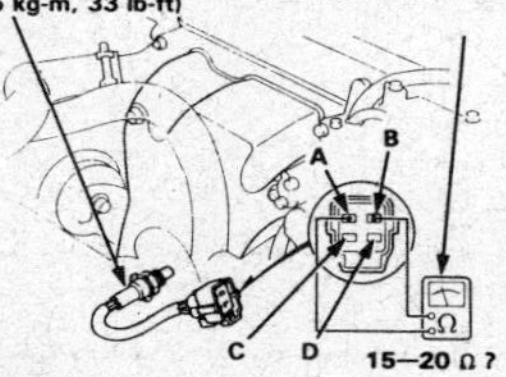

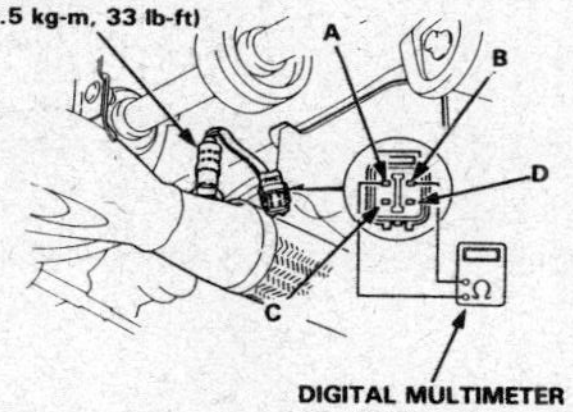

Measure voltage between YEL/BLK (+) terminal and ORN/BLK (−) terminal.

Is there battery voltage ?

- YES → Disconnect the "A" connector from the ECU.
  - Measure voltage between YEL/BLK (+) terminal and ORN/BLK (−) terminal.
  - Is there battery voltage ?
    - YES → Repair short in ORN/BLK wire between ECU (A6) and O_2 sensor.
    - NO → Substitute a known-good ECU and recheck. If symptom/indication goes away, replace the original ECU.
- NO ↓

Measure voltage between YEL/BLK (+) terminal and body ground.

Is there battery voltage ?

- NO → Repair open in YEL/BLK wire between the O_2 sensor and Main Relay.
- YES ↓

Turn the ignition switch OFF.

Reconnect the 4P connector to O_2 sensor.

Connect the ECU test harness "A" connector to the main wire harness only, not the ECU

Turn the ignition switch ON.

Measure voltage between A6 (+) terminal and A23 (−) teminal.

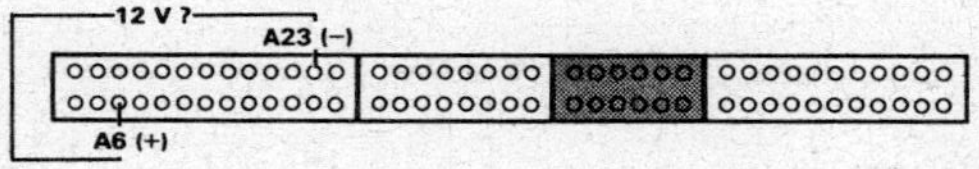

Is there battery voltage ?

- NO → Repair open in ORN/BLK wire between ECU (A6) and O_2 sensor.
- YES → Substitute a known-good ECU and recheck. If symptom/indication goes away, replace the original ECU.

TROUBLESHOOTING—CODE 43—ACCORD

(43) Self-diagnosis LED indicates code 43: Most likely a problem in the Oxygen (O_2) Sensor circuit or a problem in the Fuel Supply System.

(43) NOTE: If 43 code is accompanied by the check engine warning light and poor driveability, go to: Fuel Supply System.

- **Check Engine warning light has been reported on.**
- **LED indicates CODE 43.**
- **or continued from code 1.**

Turn the ignition switch OFF.

Remove BACK UP fuse in the under-hood relay box for 10 seconds to reset ECU.

Warm up engine to normal operating temperature (cooling fan comes on).

Hold engine at 3,000 rpm for 2 minutes.
(A/T: Transmission is [N] or [P].)

Is the Check Engine warning light on and does LED indicate CODE 43?

- NO → **Intermittent failure, system is OK at this time (test drive may be necessary). Check for poor connections or loose wires at O_2 sensor C105, C210 (located at right shock tower), and ECU.**
- YES ↓

Turn the ignition switch OFF.

Connect the ECU test harness between the ECU and connector

With the ignition switch OFF, wait for at last two minutes.

Turn the ignition switch ON.

Measure voltage between D14 (+) terminal and A26 (−) terminal as soon as the ignition switch is turned on.

NOTE
- Use DIGITAL MULTIMETER
- Use 2 Volt range.

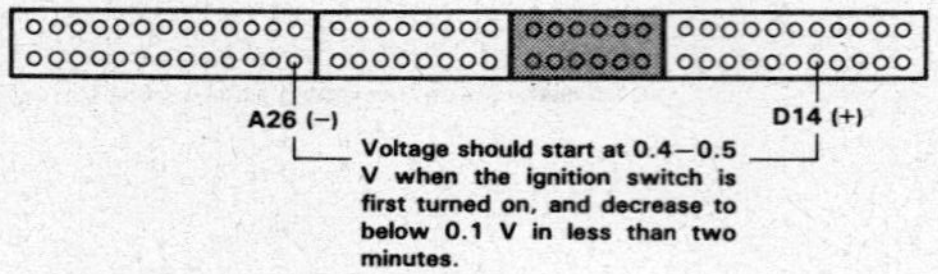

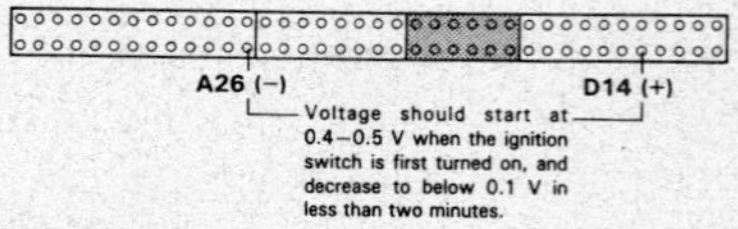

Is there 0.1 V or less when the ignition switch is first turned on?

- YES → Disconnect the 4P connector from the O_2 sensor.
  - Measure voltage between D14 (+) terminal and A26 (−) terminal.
  - Is there above 0.1 V?
    - YES → **Replace O_2 sensor.**
    - NO → Disconnect the "D" connector from the main wire harness.
      - Measure voltage between D14 (+) terminal and A26 (−) terminal.
      - Is there above 0.1 V?
        - YES → **Repair short in WHT wire between ECU (D14) and O_2 sensor.**
        - NO → **Substitute a known-good ECU and recheck. If symptom/indication goes away, replace the original ECU.**
- NO → Disconnect the 4P connector from the O_2 sensor.
  - Measure voltage at the engine wire harness side of the O_2 sensor connector between C (−) and D (+).
  - Is there above 0.1 V?
    - YES → **Replace O_2 sensor.**
    - NO → **Repair open in WHT wire between ECU (D14) and O_2 sensor.**

(DX, LX)
Engine Wire Harness side of the O_2 sensor connector

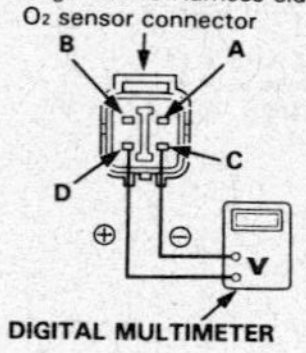

(EX)
Engine Wire Harness side of the O_2 sensor connector

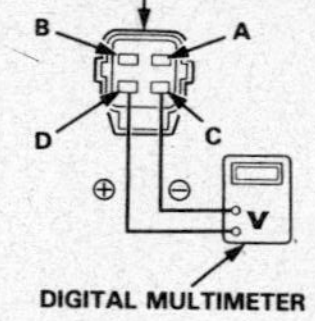

TROUBLESHOOTING—AIR CONDITIONING SIGNAL—ACCORD

This signals the PGM-FI ECU when there is a demand for cooling from the air conditioning system.

Inspection of Air Conditioning Signal.

Connect the ECU test harness between the FCU and connector Disconnect "B" connector from the main wire harness only, not the ECU.

Turn the ignition switch ON.

Measure voltage between B5 (+) terminal and A26 (−) terminal.

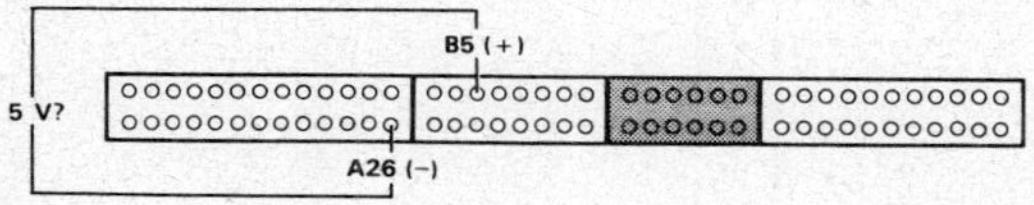

Is there approx. 5 V?

- NO → **Substitute a known-good ECU and recheck. If prescribed voltage is now available, replace the original ECU.**
- YES ↓

Reconnect "B" connector to the main wire harness.

Momentarily connect A15 terminal to A26 terminal several times.

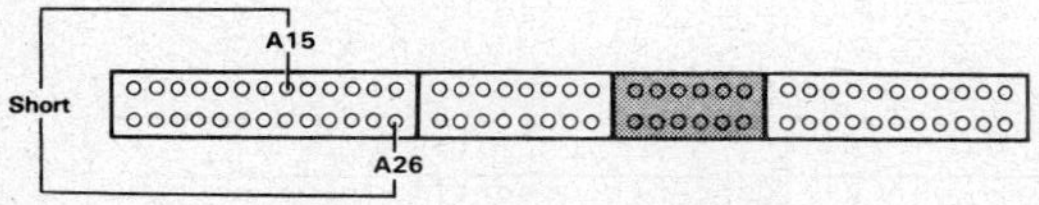

Is there a clicking noise from the A/C compressor clutch?

- NO → Connect the RED/BLU terminal of the 4P connector on the A/C clutch relay to body ground.

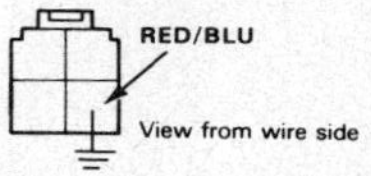

  Is there a clicking noise from the A/C compressor clutch?
  - NO → **Air conditioner inspection**
  - YES → **Repair open in RED/BLU wire between ECU (A15) and A/C clutch relay.**
- YES ↓

Start the engine.

Blower switch ON.

A/C switch ON.

Does A/C operate ?

- YES → **Air conditioning signal is OK.**
- NO → Measure voltage between B5 (+) terminal and A26 (−) terminal.

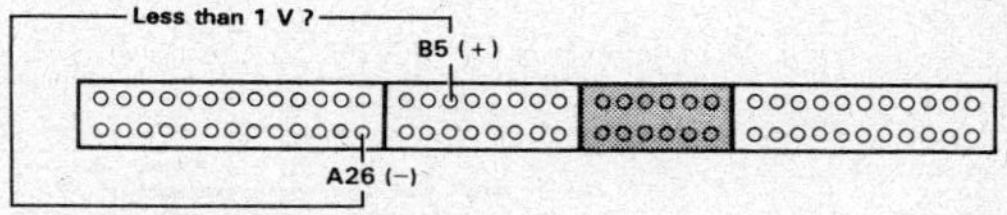

Is voltage less than 1V?

- NO → **Repair open in BLU/BLK wire between ECU (B5) and A/C switch.**
- YES → **Substitute a known-good ECU and recheck. If symptom/indication goes away, replace the original ECU.**

TROUBLESHOOTING—ALTERNATOR FR SIGNAL—ACCORD

This signals the PGM-FI ECU when the alternator is charging.

Inspection of Alternator FR signal.

Connect the ECU test harness between the ECU and connector Disconnect "D" connector from the main wire harness only, not the ECU.

Turn the ignition switch ON.

Measure voltage between D9 (+) terminal and A26 (–) terminal.

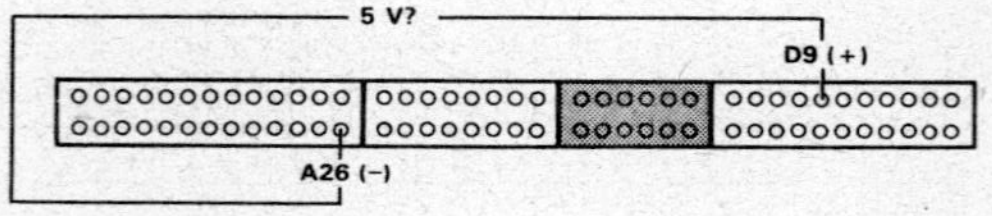

Is there approx. 5 V?

- NO → **Substitute a known-good ECU and recheck. If prescribed voltage is now available, replace the original ECU.**
- YES → Turn the ignition switch OFF.

Reconnect "D" connector to the main wire harness.

Warm up engine to normal operating temperature (cooling fan comes on).

Measure voltage between D9 (+) terminal and A26 (–) terminal.

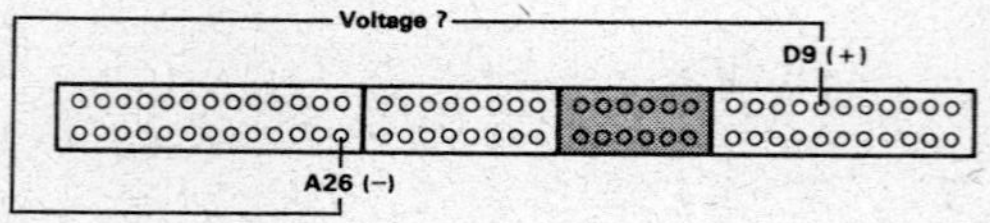

Does the voltage decrease when headlights and rear defogger are turned on ?

- YES → **Alternator FR signal is OK.**
- NO → Stop the engine.

Disconnect "D" connector from ECU only, not the main wire harness.

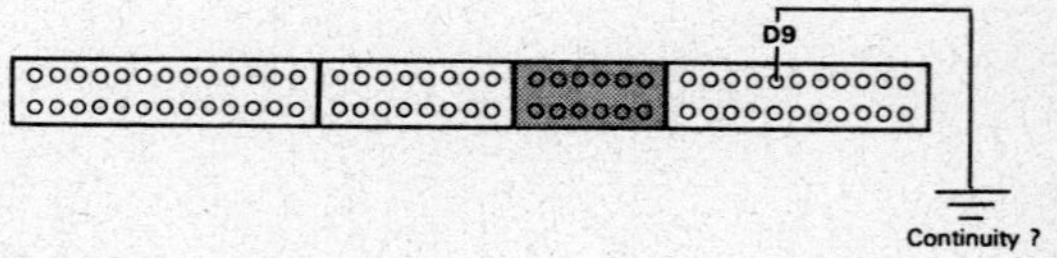

Disconnect the negative battery cable from the battery.

Check for continuity between D9 terminal and body ground.

Does continuity exist ?

- YES → Disconnect GRN connector from the alternator.
  - Check for continuity between D9 terminal and body ground.
  - Does continuity exist ?
    - NO → **Alternator Inspection**
    - YES → **Repair short in WHT/RED wire between ECU (D9) and alternator.**
- NO → Disconnect GRN connector from the alternator.

Connect WHT/RED wire to body ground.

Check for continuity between D9 terminal and body ground.

Does continuity exist ?

- YES → **Alternator Inspection**
- NO → **Repair open in WHT/RED wire between ECU (D9) and alternator.**

TROUBLESHOOTING—AUTOMATIC TRANSAXLE SHIFT POSITION SIGNAL—ACCORD

This signals the PGM-FI ECU when the transmission is in Neutral or Park.

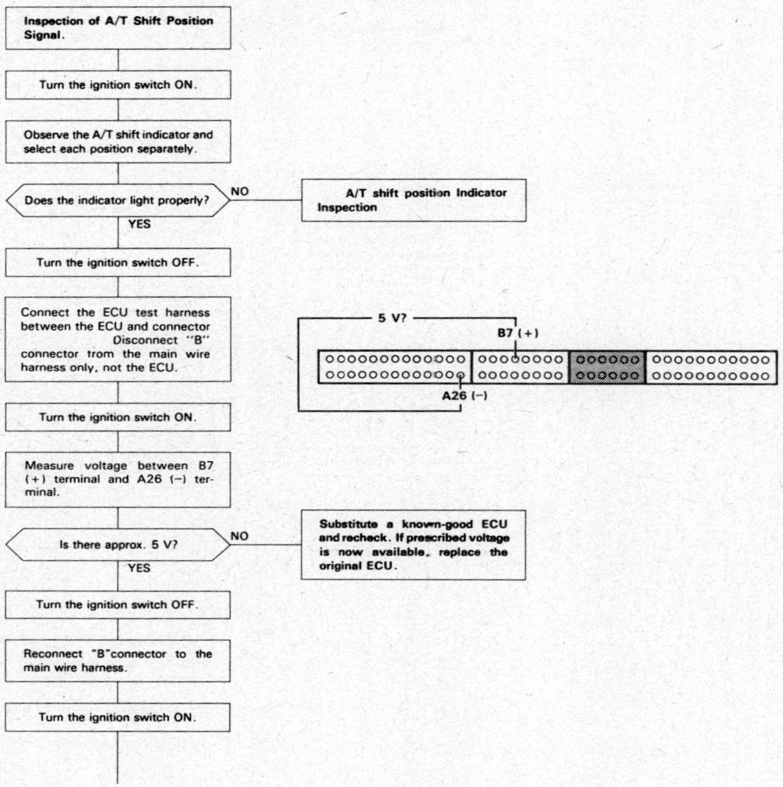

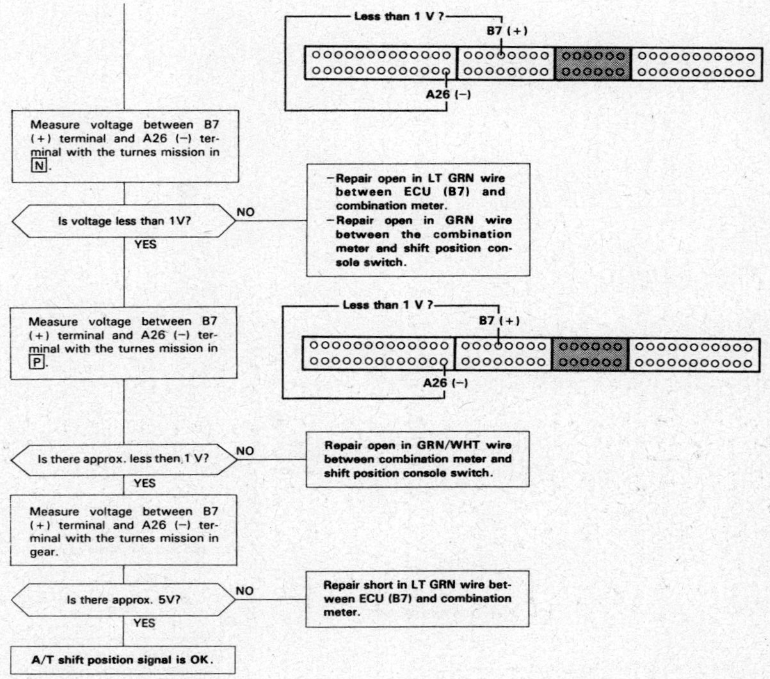

TROUBLESHOOTING—BRAKE SWITCH SIGNAL—ACCORD

This signals the PGM-FI ECU when the brake pedal is depressed.

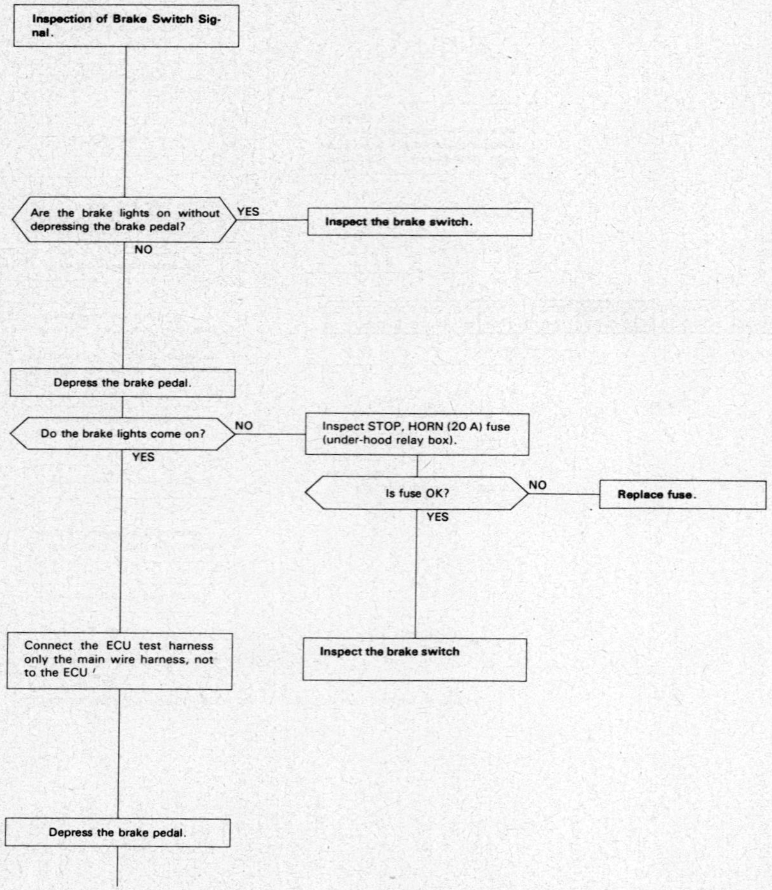

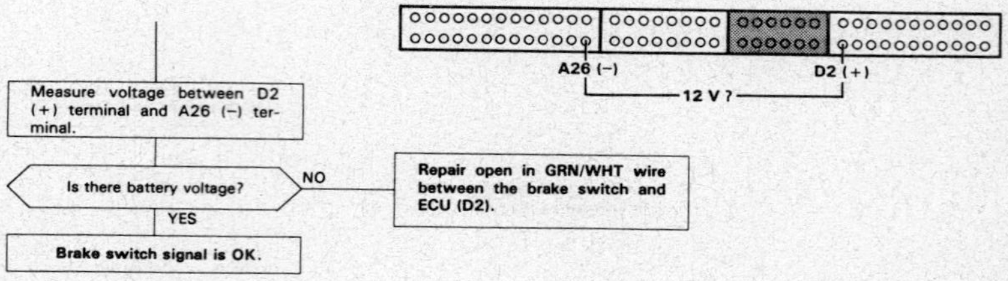

TROUBLESHOOTING—STARTER SIGNAL

This signals the PGM-FI ECU when the engine is cranking.

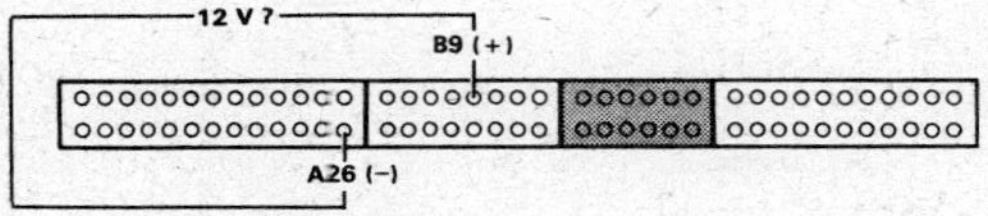

NOTE: Clutch pedal must be depressed on M/T models.

Inspection of Starter Switch Signal.

1. Connect the ECU test harness between the ECU and connector
2. Measure voltage between B9 (+) terminal and A26 (−) terminal with the ignition switch turned to the start position.
3. Is there battery voltage ?
   - YES → **Starter switch signal is OK.**
   - NO → Inspect No.9 fuse.
     - Is No.9 fuse OK ?
       - NO → **Replace fuse.**
       - YES → **Repair open in BLU/RED, BLK/GRN (M/T) wire between ECU (B9) and No. 9 fuse.**

This signals the PGM-FI ECU when the power steering load is high.

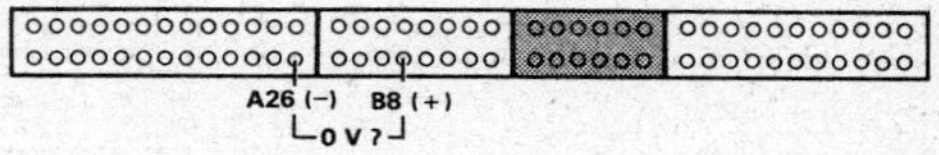

Inspection of P/S Oil Pressure Signal

1. Connect the ECU test harness between the ECU and connector
2. Turn the ignition switch ON.
3. Measure voltage between B8 (+) terminal and A26 (−) terminal.
4. Is there voltage ?
   - YES → Disconnect the 2P connector on the P/S oil pressure switch.
     - Connect RED terminal to BLK terminal.
     - Is there voltage ?
       - NO → **Replace P/S oil pressure switch.**
       - YES → **Repair open in RED wire between ECU (B8) and P/S oil pressure switch or BLK wire between P/S oil pressure switch and G301.**
   - NO → Start the engine.
5. Turn steering wheel slowly.
6. Measure voltage between B8 (+) terminal and A26 (−) terminal while steering wheel is turning.
7. Is there battery voltage ?
   - YES → **P/S oil pressure signal is OK.**
   - NO → Turn the ignition switch OFF.
     - Disconnect "B" connector from main wire harness only, not the ECU.
     - Turn the ignition switch ON.
     - Is there battery voltage ?
       - NO → **Substitute a known-good ECU and recheck. If prescribed voltage is now available, replace the original ECU.**
       - YES → Reconnect "B" connector to main wire harness and disconnect 2P connector on the P/S oil pressure switch.
         - Is there battery voltage ?
           - YES → **Replace P/S oil pressure switch.**
           - NO → **Repair short in RED wire between ECU (B8) and the P/S oil pressure switch.**

Troubleshooting Flowchart

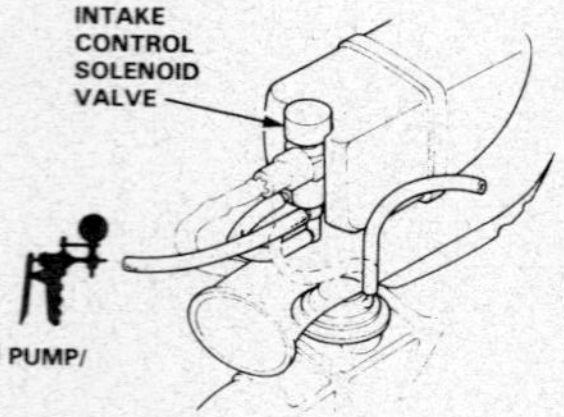

- **Inspection of Intake Control System**
- Start engine and allow to idle.
- Remove the upper vacuum hose from the intake control solenoid valve and connect a vacuum gauge to the solenoid valve.
- Is there vacuum?
  - YES → Go to Path E
  - NO → Disconnect the lower vacuum hose of the solenoid valve from the air chamber and connect a vacuum pump.
    - Is there vacuum?
      - NO → • Check the vacuum line for proper connection, cracks, blockage or disconnected hose. • Clean manifold port.
      - YES → Disconnect the 2P connector from the intake control solenoid valve.
        - Measure voltage between BLK/YEL (+) terminal and WHT (−) terminal.
        - Is there battery voltage?
          - YES → Replace the intake control solenoid valve.
          - NO → Measure voltage between BLK/YEL (+) terminal and body ground. → Go to Path F

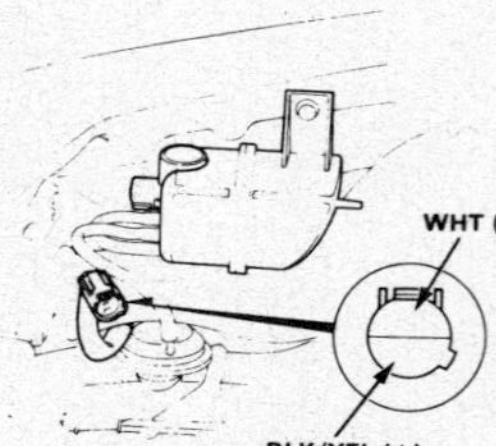

Path E

- Raise engine speed to 3,700 rpm.
- Is there vacuum?
  - NO → Go to Path G
  - YES → Disconnect the 2P connector.
    - Is there vacuum?
      - YES → Replace intake control solenoid valve.
      - NO → Turn the ignition switch OFF.
        - Disconnect "A" connector from ECU.
        - Check for continuity to ground on WHT wire.
        - Go to Path H

Path F

- Is there battery voltage?
  - NO → Repair open in BLK/YEL wire between the 2P connector and No.2 (10 A) fuse.
  - YES → Turn the ignition switch OFF.
    - Connect the ECU test harness between the ECU and connector
    - Check for continuity of WHT wire between ECU (A19) and the 2P connector.

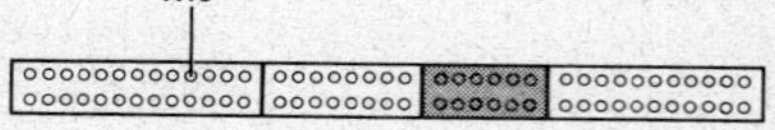

- Does continuity exist?
  - NO → Repair open in WHT wire between ECU A19 and the 2P connector.
  - YES → Substitute a known-good ECU and recheck. If symptom/indication goes away, replace the original ECU.

TROUBLESHOOTING - INTAKE CONTROL SYSTEM—ACCORD

Path G

Connect a vacuum pump to the #8 vacuum hose.

Apply vacuum.

Does it hold vacuum?

- NO: Check the vacuum line for proper connection or disconnected hose. If OK, replace the intake control diaphragm.
- YES: Intake control system is OK.

Path H

Is there continuity to ground?

- YES: Repair short to ground in WHT wire between ECU (A19) and the 2P connector.
- NO: Substitute a known good ECU and recheck. If symptom goes away, replace the original ECU.

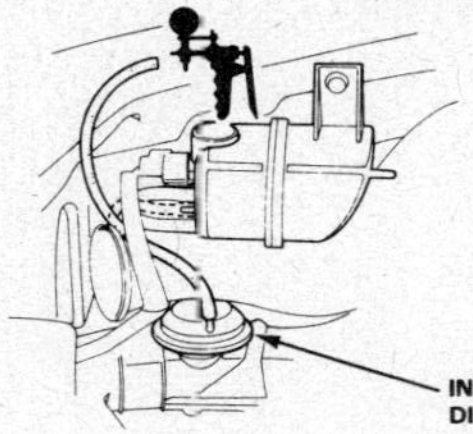

TROUBLESHOOTING—EVAPORATIVE EMISSION CONTROL SYSTEM—ACCORD

Troubleshooting Flowchart

Inspection of Evaporative Emission Controls

Disconnect vacuum hose from the purge control diaphragm valve (on the charcoal canister) and connect a vacuum gauge to the hose.

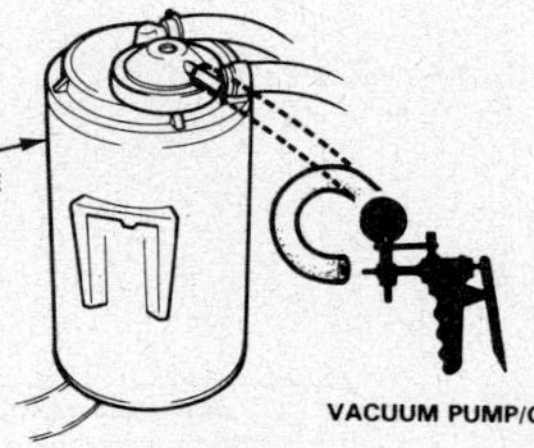

Start the engine and allow to idle. NOTE: Engine coolant temperature must be below 75°C (167°F).

Is there vacuum ?

- NO: Go to Path K
- YES: Disconnect the 4P connector.

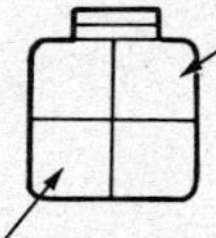

Measure voltage between BLK/YEL (+) terminal and RED/GRN (−) terminal.

Is there battery voltage ?

- YES: Inspect vacuum hose routing. If OK, replace purge cut-off solenoid valve.
- NO: Measure voltage between BLK/YEL (+) terminal and body ground. Go to Path L

TROUBLESHOOTING—EVAPORATIVE EMISSION CONTROL SYSTEM—ACCORD

Path K

- Warm up the engine to normal operating temperature (cooling fan comes on).
- Check for vacuum at #3 vacuum hose after starting the engine.
- Is there manifold vacuum?
  - NO → Disconnect the 4P connector.
    - Is there manifold vacuum?
      - NO → **Inspect vacuum hose running. If OK, replace purge cut-off solenoid valve.**
      - YES → **Inspect for a short to body ground in RED/GRN wire between ECU (A20) and the connector. If wire is OK, substitute a known-good ECU and recheck. If symptom/indication goes away, replace the original ECU.**
  - YES → Reconnect the hose.
- Remove fuel filler cap.
- Connect a vacuum gauge to canister purge air hose.
- Start the engine and raise speed to 3,500 rpm.
- Does vacuum appear on gauge within 1 minute?
  - YES → **two-way valve test to complete. Evaporative emission control are OK.**
  - NO → Connect a vacuum gauge to the canister purge hose and raise the engine speed to 3,500 rpm.
    - Does vacuum appear on the gauge?
      - NO → **Inspect the purge hose and throttle body port for pinch or blockage.**
      - YES → **Replace the canister.**

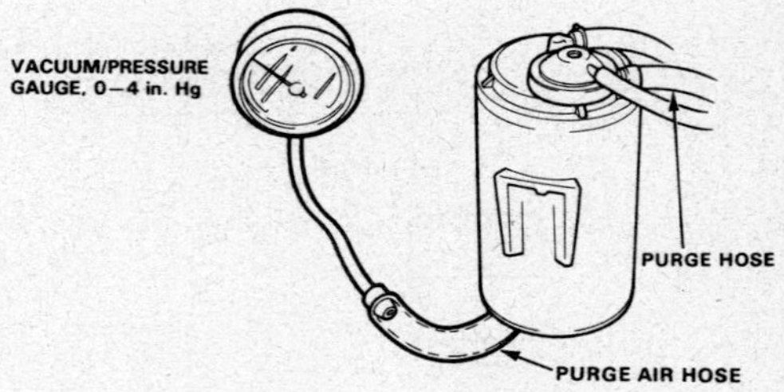

Path L

- Is there battery voltage?
  - NO → **Repair open in BLK/YEL wire between No.2 fuse and the 4P connector.**
  - YES → **Inspect RED/GRN wire for on open between ECU (A20) and the connector. If wire is OK, substitute a known-good ECU and recheck. If symptom/indication goes away, replace the original ECU.**

VACUUM DIAGRAM—ACCORD

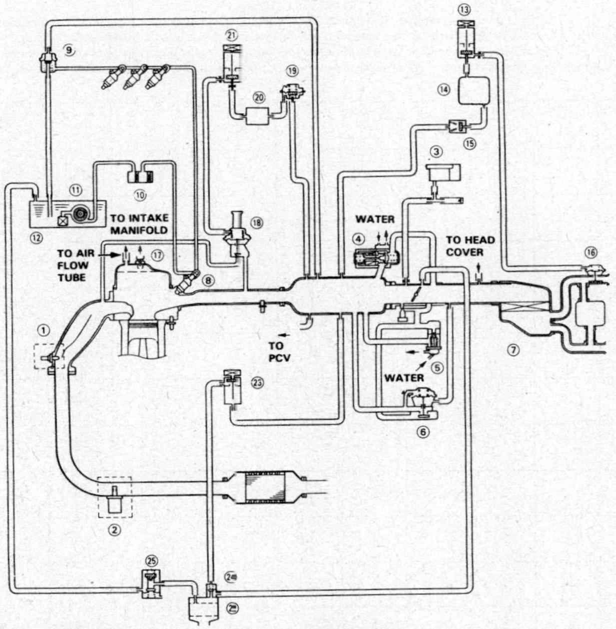

1. OXYGEN (O_2) SENSOR (DX, LX)
2. OXYGEN (O_2) SENSOR (EX)
3. MANIFOLD ABSOLUTE PRESSURE (MAP) SENSOR
4. ELECTRONIC AIR CONTROL VALVE (EACV)
5. FAST IDLE VALVE
6. AIR BOOST VALVE
7. AIR CLEANER
8. FUEL INJECTOR
9. PRESSURE REGULATOR
10. FUEL FILTER
11. FUEL PUMP
12. FUEL TANK
13. INTAKE CONTROL SOLENOID VALVE
14. AIR CHAMBER
15. CHECK VALVE
16. INTAKE CONTROL DIAPHRAGM
17. PCV VALVE
18. EGR VALVE
19. CONSTANT VACUUM CONTROL (CVC) VALVE
20. AIR CHAMBER
21. EGR CONTROL SOLENOID VALVE
22. CHARCOAL CANISTER
23. PURGE CUT-OFF SOLENOID VALVE
24. PURGE CONTROL DIAPHRAGM VALVE
25. TWO-WAY VALVE

FUEL SYSTEM WIRING SCHEMATIC—ACCORD

FUSES
1. ECU (10 A)*
2. No. 2 (15 A)
3. No. 9 (7.5 A)
4. No. 7 (7.5 A)
5. No. 1 (10 A)
6. BACK UP (7.5 A)*
7. IG (50 A)*
8. STOP, HORN (20 A)*
9. BATTERY (80 A)*

* In the under-hood relay fuse box
* BLU/RED (A/T) BLK/GRN (M/T)

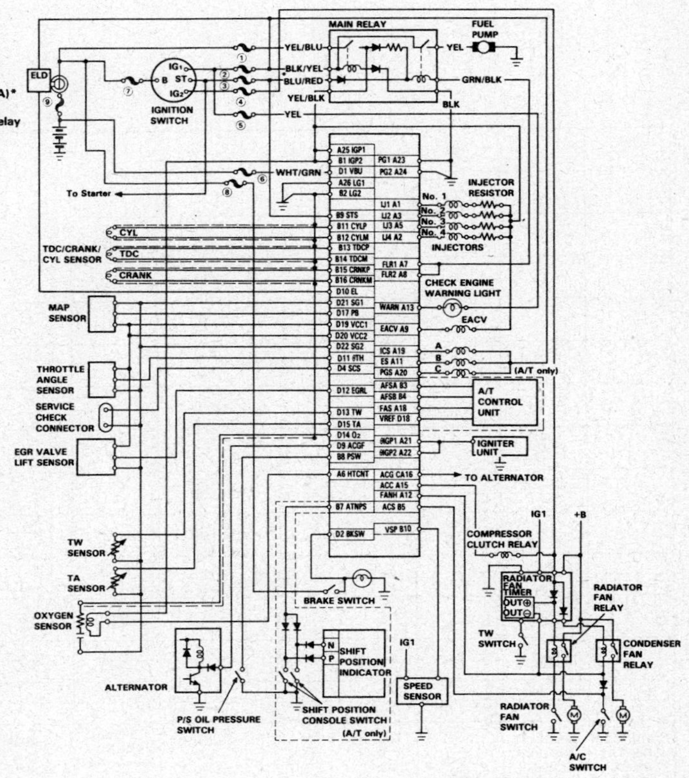

A: INTAKE CONTROL SOLENOID VALVE
B: EGR CONTROL SOLENOID VALVE
C: PURGE CUT-OFF SOLENOID VALVE

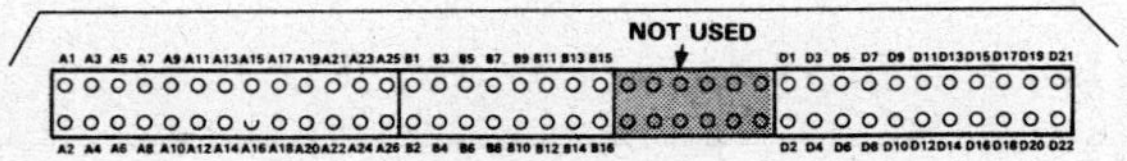

TERMINAL LOCATION

TROUBLESHOOTING – PRELUDE

SYSTEM / SYMPTOM		PGM-FI: ECU	PGM-FI: OXYGEN SENSOR	PGM-FI: MANIFOLD ABSOULUTE PRESSURE SENSOR	PGM-FI: TDC/CRANK SENSOR	PGM-FI: COOLANT TEMPERA-TURE SENSOR	PGM-FI: THROTTLE ANGLE SENSOR	PGM-FI: CYL SENSOR	PGM-FI: INTAKE AIR TEMPERA-TURE SENSOR	PGM-FI: ATMO-SPHERIC PRESSURE SENSOR
CHECK ENGINE WARNING LIGHT TURNS ON		□ or ☼	(CHECK)	(CHECK)	(CHECK)	(CHECK)	(CHECK)	(CHECK)	(CHECK)	(CHECK)
SELF-DIAGNOSIS INDICATOR (LED) BLINKS		⓪ or ✱	① or ② * / ① or ㊶ or ㊸ **	③ or ⑤	④ or ⑧	⑥	⑦	⑨	⑩	⑬
ENGINE WON'T START		③			③			③		
DIFFICULT TO START ENGINE WHEN COLD		(BU)		③	③	①		③		③
IRREGULAR IDLING	WHEN COLD FAST IDLE OUT OF SPEC	(BU)				③				
	ROUGH IDLE	(BU)		③						
	WHEN WARM RPM TOO HIGH	(BU)				③				
	WHEN WARM RPM TOO LOW	(BU)								
FREQUENT STALLING	WHILE WARMING UP	(BU)				③				
	AFTER WARMING UP	(BU)								③
POOR PERFORM-ANCE	MISFIRE OR ROUGH RUNNING	(BU)			③			③		
	FAILS EMISSION TEST	(BU)	③	②						
	LOSS OF POWER	(BU)		③			②			

- If codes other than those listed above are indicated, count the number of blinks again. If the indicator is in fact blinking these codes, substitute a known-good ECU and recheck. If the indication goes away, replace the original ECU.
  (BU): When the Check Engine warning light and the self-diagnosis indicator are on, the back-up system is in operation. Substitute a known-good ECU and recheck. If the indication goes away, replace the original ECU.

* : 2.0ℓ
** : 2.1ℓ

SYSTEM / SYMPTOM		PGM-FI: IGNITION OUTPUT SIGNAL	PGM-FI: VEHICLE SPEED SENSOR	IDLE CONTROL: ELEC-TRONIC AIR CONTROL VALVE	IDLE CONTROL: OTHER IDLE CONTROLS	FUEL SUPPLY: FUEL INJECTOR	FUEL SUPPLY: OTHER FUEL SUPPLY	AIR INTAKE	EMISSION CONTROL: EGR CONTROL SYSTEM	EMISSION CONTROL: OTHER EMISSION CONTROLS
CHECK ENGINE WARNING LIGHT TURNS ON		(CHECK)	(CHECK)	(CHECK)		(CHECK)			(CHECK)	
SELF-DIAGNOSIS INDICATOR (LED) BLINKS		⑮	⑰	⑭		⑯			⑫	
ENGINE WON'T START		①				②	③			
DIFFICULT TO START ENGINE WHEN COLD					②					
IRREGULAR IDLING	WHEN COLD FAST IDLE OUT OF SPEC			①	②					
	ROUGH IDLE			①		②			③	
	WHEN WARM RPM TOO HIGH			①	②					
	WHEN WARM RPM TOO LOW			①		②				
FREQUENT STALLING	WHILE WARMING UP			①	②		③			
	AFTER WARMING UP			③	①		②		③	
POOR PERFORM-ANCE	MISFIRE OR ROUGH RUNNING			③		①			③	
	FAILS EMISSION TEST					②	③			①
	LOSS OF POWER					③	①	③		③

TROUBLESHOOTING
IDLE CONTROL SYSTEM – PRELUDE

NOTE:
- Across each row in the chart, the sub systems that could be sources of a symptom are ranked in the order they should be inspected, starting with ①.
- If the idle speed is out of specification and LED does not blink CODE 14, go to inspection

SYMPTOM \ SUB SYSTEM		IDLE ADJUSTING SCREW	EACV	AIR CONDITIONING SIGNAL	ALTERNATOR FR SIGNAL	A/T SHIFT POSITION SIGNAL (A/T ONLY)	STARTER SWITCH SIGNAL	P/S OIL PRESSURE SWITCH SIGNAL	FAST IDLE VALVE	AIR BOOST VALVE	HOSES AND CONNECTIONS
DIFFICULT TO START ENGINE WHEN COLD									①		
WHEN COLD FAST IDLE OUT OF SPEC (1,000–2,000 rpm)		③	②						①		
ROUGH IDLE			②								①
WHEN WARM RPM TOO HIGH		③	①					③	②		③
WHEN WARM RPM TOO LOW	Idle speed is below specified rpm (no load)	②	①								
	Idle speed does not increase after initial start up.		①								
	On models with automatic transmission, the idle speed drops in gear		②			①					
	Idle speeds drops when air conditioner in ON		②	①							
	Idle speed drops when steering wheel is turning		②					①			
	Idle speed fluctuates with electrical load		①								
FREQUENT STALLING	WHILE WARMING UP		①								
	AFTER WARMING UP	②	①								
FAILS EMISSION TEST											①

TROUBLESHOOTING
FUEL CONTROL SYSTEM-PRELUDE

NOTE: Across each row in the chart, the systems that could be sources of a symptom are ranked in the order they should be inspected starting with ①.

SYMPTOM \ SUB SYSTEM		FUEL INJECTOR	INJECTOR RESISTOR	PRESSURE REGULATOR	FUEL FILTER	FUEL PUMP	MAIN RELAY	CONTAMINATED FUEL
ENGINE WON'T START		③	③		③	①	②	③
DIFFICULT TO START ENGINE WHEN COLD OR HOT				③	②			①
ROUGH IDLE		①	②					③
POOR PERFORMANCE	MISFIRE OR ROUGH RUNNING	①	②	③				③
	FAILS EMISSION TEST	②	③	①				
	LOSS OF POWER	③	③		①	③		②

- Fuel with dirt, water or a high percentage of alcohol is considered contaminated.

TROUBLESHOOTING
AIR INTAKE SYSTEM—PRELUDE

NOTE: Across each row in the chart, the sub systems that could be sources of a symptom are ranked in the order they should be inspected starting with ①.
'isted r*

SUB SYSTEM / SYMPTOM	THROTTLE CABLE	THROTTLE BODY	BYPASS CONTROL* SYSTEM
WHEN WARM RPM TOO HIGH	②	①	
LOSS OF POWER		①	②

*: 2.0 ℓ

TROUBLESHOOTING
EMISSION CONTROL SYSTEM—PRELUDE

NOTE: Across each row in the chart, the systems that could be sources of a symptom are ranked in the order they should be inspected starting with ①.

SUB SYSTEM / SYMPTOM		CATALYTIC CONVERTER	EGR SYSTEM	POSITIVE CRANKCASE VENTILATION SYSTEM	EVAPORATIVE EMISSION CONTROLS
ROUGH IDLE			①	②	
FREQUENT STALLING	AFTER WARMING UP		①		
POOR PERFORMANCE	FAILS EMISSION TEST	①			②
	LOSS OF POWER	①			

TERMINAL IDENTIFICATION PRELUDE TEST HARNESS

If the inspection for a particular failure code requires the ECU test harness, remove the right door sill molding, the small cover on the right kick panel, and pull the carpet back to expose the ECU. Unbolt the ECU bracket. Connect the ECU test harness. Then check the system according to the procedure described for the appropriate code(s) listed on the following pages.

2.0 ℓ:

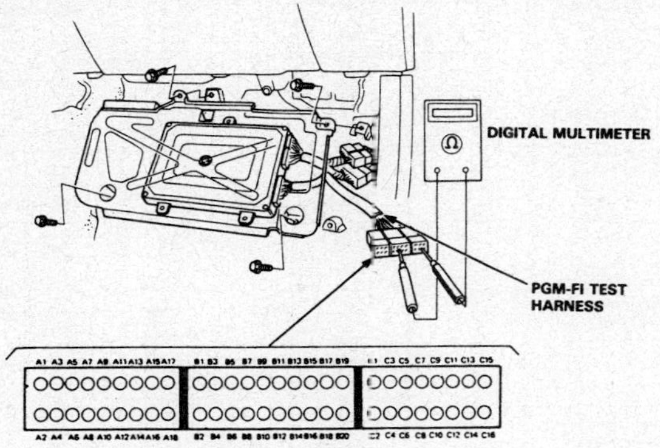

TERMINAL LOCATION

2.1 ℓ:

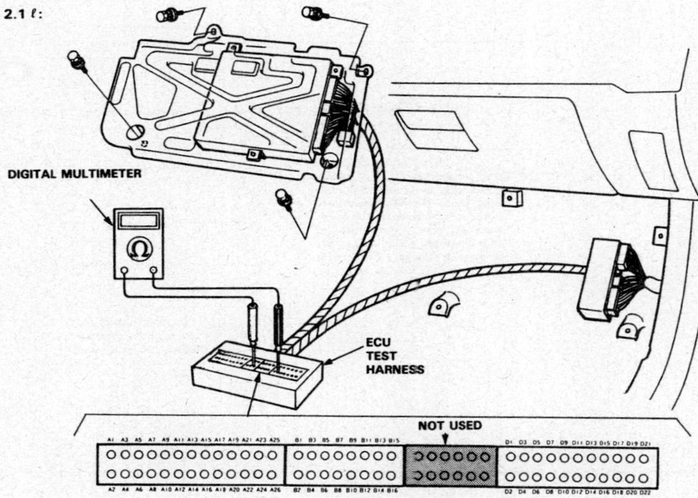

TERMINAL LOCATION

2.0L ENGINE—TROUBLESHOOTING ECU—PRELUDE

Check Engine warning light isn't on for two seconds after ignition is first turned on.

Is oil pressure warning light on?
- NO → Inspect No. 13 fuse. → Is No. 13 fuse OK?
  - NO → **Replace fuse.**
  - YES → **Repair open in YEL wire between No. 13 fuse and combination meter.**
- YES ↓

Turn the ignition switch OFF.

Connect the PGM-FI test harness between the ECU and connector

Connect B6 terminal to body ground.

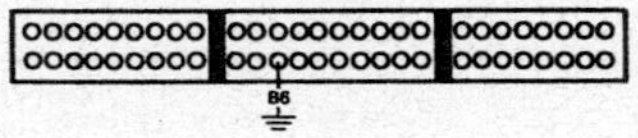

Turn the ignition switch ON.

Is Check Engine warning light on?
- NO → **– Replace warning light bulb. – Repair open in YEL/RED wire between ECU (B6) and combination meter.**
- YES ↓

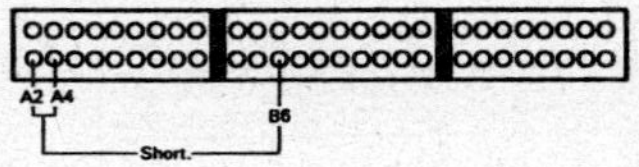

Connect the following terminals individually to B6: • A2, • A4

Is Check Engine warning light on?
- NO → **– Repair open in BLK wire between ECU (A2) and G101. – Repair open in BLK wire between ECU (A4) and G101.**
- YES ↓

Substitute a known-good ECU and recheck. If symptom/ indication goes away, replace the original ECU.

2.0L ENGINE – TROUBLESHOOTING – ECU – PRELUDE

- Check Engine warning light has been reported on.
- LED doesn't blink.

Try to start the engine.

Is the Check Engine warning light ON?
- NO → Intermittent failure. System is OK at this time. (Test drive may be necessary) Check for loose wires or poor connections at the thermostat housing, main relay ground connector at right shock tower. No.12 fuse (dash), ECU fuse (main relay box)
- YES ↓

Does LED indicate any CODE?
- YES → Go to troubleshooting
- NO ↓

Did the engine start?
- YES → Turn the ignition switch OFF. → Disconnect "B" connector from the ECU. → Turn the ignition switch ON. → Is the Check Engine warning light ON?
  - NO → Substitute a known-good ECU and recheck. If symptom/indication goes away, replace the original ECU.
  - YES → Repair short to body ground in YEL/RED wire between the ECU (B6) and check engine warning light.
- NO ↓

Remove and inspect the 10 A (ECU) fuse in the underhood fuse box.

Is the fuse OK?
- NO → Replace the fuse.
- YES ↓

Inspect the No. 12 (10 A) fuse in the dash box.

Is the fuse OK?
- NO → Replace the fuse.
- YES ↓

Turn the ignition switch ON.

Disconnect the 3P connector of each sensor one at a time:
- MAP sensor
- Throttle angle sensor
- EGR valve lift sensor

Does Check Engine warning light remain ON?
- NO → Replace the sensor that caused the light to go out upon its disconnection.
- YES ↓

Disconnect the 3P connector of the PA sensor.

Does LED indicate CODE 13?
- YES → Replace the PA sensor.
- NO ↓

Reconnect the 3P connector of the PA sensor.

Turn the ignition switch OFF.

Connect the PGM-FI test harness But disconnect the "C" connector from the ECU only, not the main wire harness.

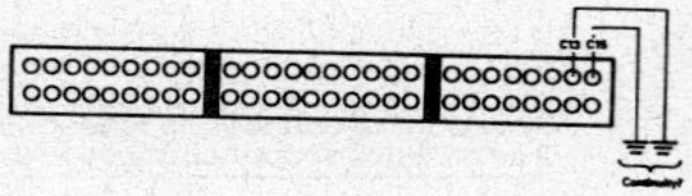

Check for continuity between body ground and the following terminals: C13, C15.

Does continuity exist?
- YES → Repair short to body ground in RED/WHT wire, YEL/WHT wire between ECU (C15, C13) and throttle angle sensor, EGR valve lift sensor, PA sensor and MAP sensor.
- NO ↓

Reconnect the all connectors. Reconnect the "C" connector to the ECU.

Turn the ignition switch ON.

Individualy connect the following terminal to Body Ground. A16 • A18

Is the Check Engine warning light ON?
- NO →
  - Repair open in BLK/RED wire between ECU (A18) and G101.
  - Repair open in BRN/BLK wire between ECU (A16) and G101.
- YES ↓

Measure voltage between A18 (−) and the following: A13 (+) and A15 (+)

Is there battery voltage?
- NO →
  - Repair open in YEL/BLK wire between ECU (A13, A15) and main relay.
  - Check main relay and wiring connectors at main relay
- YES ↓

Substitute a known-good ECU and recheck. If symptom/indication goes away, replace the original ECU.

2.0L ENGINE—TROUBLESHOOTING—CODE 1 OR CODE 2—PRELUDE

(1) Self-diagnosis LED indicates code 1: A problem in the Oxygen (O_2) Sensor A circuit.

The oxygen sensor, detects the oxygen content in the exhaust gas, and inputs the ECU. In operation, the ECU receives the signals from the sensor and varies the duration during which fuel is injected. The oxygen sensors are installed on the exhaust manifold.

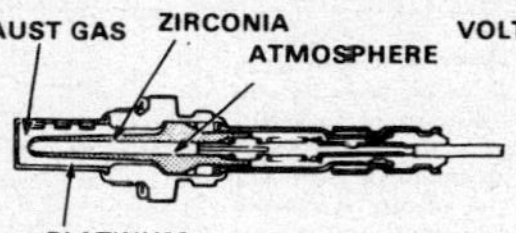

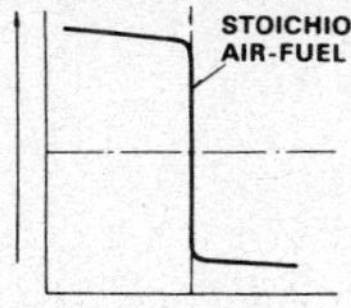

(1) or (2)

- Check Engine warning light has been reported on.
- LED indicates CODE 1 (O_2 sensor A) or CODE 2 (O_2 sensor B).

Turn the ignition switch OFF.

Remove CLOCK fuse in the under-hood relay box for 10 seconds to reset ECU.

Inspect pressure regulator

Is it normal ?
- NO: Replace the pressure regulator
- YES: continue

Warm up engine to normal operating temperature (cooling fan comes on).

Hold engine at 1500 rpm for 15 minutes.
NOTE: Do not close throttle completely during this time.

Is Check Engine warning light on and does LED indicate CODE 1 or 2 ?
- NO: Intermittent failure, system is OK at this time. Check for poor connections or loose wires at C236 (located at right shock tower) and at the O_2 sensor.
- YES: continue

Disconnect engine wire harness from indicated O_2 sensor.

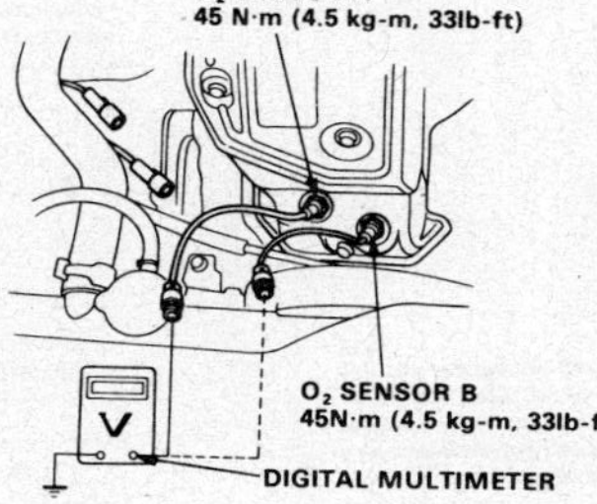

Warm up engine to normal operating temperature again, then open the throttle wide open then close it.

Measure voltage between the connector terminal and body ground.

Is voltage above 0.6 V at wide open throttle and below 0.4 V when the throttle is quickly released ?
- NO: Replace O_2 sensor.
- YES: continue

Stop engine.

Connect the PGM-FI test harness between the ECU and connector

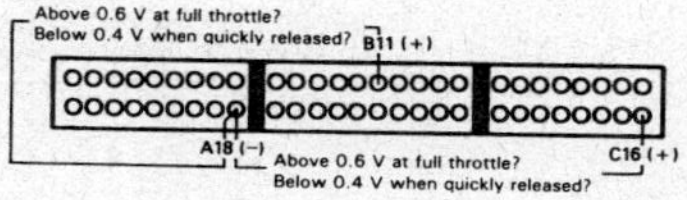

Restart and warm up engine to normal operating temperature, then open the throttle wide open then close it.

Measure voltage between:
- O_2 sensor A
  C16 (+) and A18(−) terminals
- O_2 sensor B
  B11 (+) and A18 (−) terminals

Is voltage above 0.6 V at wide open throttle and below 0.4 V when the throttle is quickly released ?
- NO: Repair wire between ECU and O_2 sensor for short or open:
  - O_2 sensor A
    WHT wire to C16
  - O_2 sensor B
    RED/BLU wire to B11.
- YES: Substitute a known-good ECU and recheck. If symptom/ indication goes away, replace the original ECU.

Self-diagnosis LED indicates code 3: Most likely an electrical problem in the Manifold Absolute Pressure (MAP) Sensor system.

The MAP sensor converts manifold absolute pressure into electrical signals and inputs the ECU.

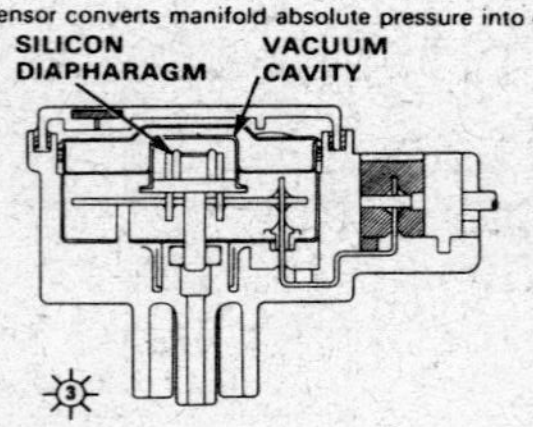

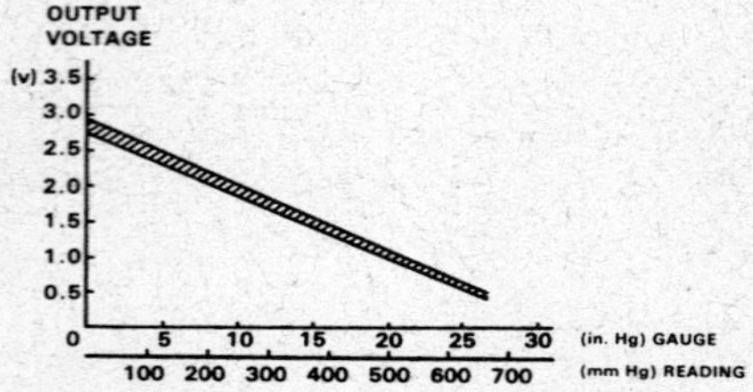

- Engine is warm and running.
- Check Engine warning light has been reported on.
- LED indicates CODE 3.

Turn the ignition switch OFF.

Remove CLOCK fuse in the under-hood relay box for 10 seconds to reset ECU.

Warm up engine to normal operating temperature (cooling fan comes on).

Is Check Engine warning light on and does LED indicate CODE 3?

- NO: Intermittent failure, system is OK at this time (test drive may be necessary). Check for poor connection or loose wires at MAP sensor connector and ECU.
- YES: Turn the ignition switch OFF.

Disconnect the 3P connector from the MAP sensor.

Turn the ignition switch ON.

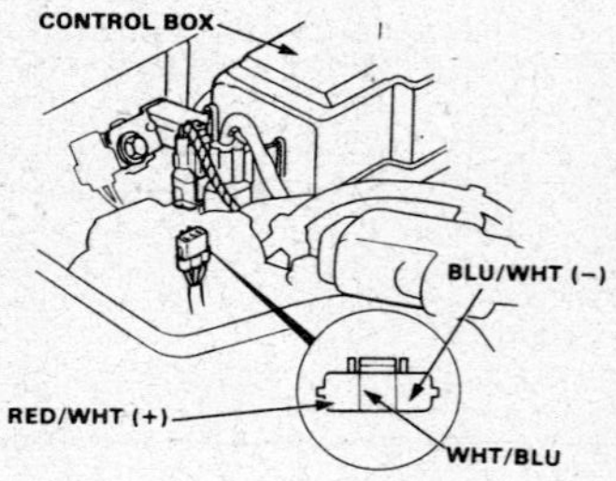

Measure voltage between RED/WHT (+) terminal and body ground.

Is there approx. 5V?

- NO: Repair open in RED/WHT wire between ECU (C15) and MAP sensor. If wire is OK, substitute a known-good ECU and recheck. If prescribed voltage is now available, replace the original ECU.
- YES: Measure voltage between RED/WHT (+) terminal and BLU/WHT (−) terminal.

Is there approx. 5V?

- NO: Repair open in BLU/WHT wire between ECU (C14) and MAP sensor. If wire is OK, substitute a known-good ECU and recheck. If prescribed voltage is now available, replace the original ECU.
- YES: Measure voltage between WHT/BLU (+) terminal and BLU/WHT (−) terminal.

Is there approx. 5V?

- NO: Repair open or short in WHT/BLU wire between ECU (C11) and MAP sensor. If wire is OK, substitute a known-good ECU and recheck. If prescribed voltage is now available, replace the original ECU.
- YES: Turn the ignition switch OFF.

Reconnect the 3P connector to the MAP sensor.

Connect the PGM-FI test harness between the ECU and connector.

Turn the ignition switch ON.

2.0L ENGINE – TROUBLESHOOTING – CODE 3 – PRELUDE

Measure voltage between C11(+) terminal and C14 (−) terminal.

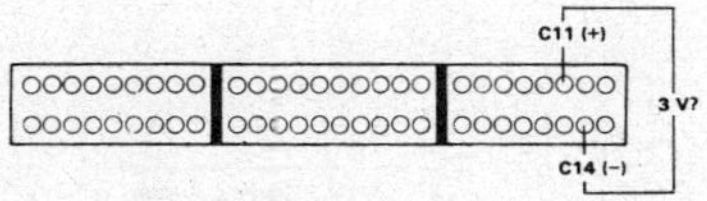

Is there approx. 3 V ?

NO → Replace MAP sensor.

YES → Substitute a known-good ECU and recheck. If symptom/indication goes away, replace the original ECU.

Self-diagnosis LED indicates code 4: A problem in the CRANK circuit of the TDC/CRANK Sensor.

The CRANK sensor determines timing for fuel injection and ignition of each cylinder and also detects engine RPM. The TDC sensor determines ignition timing at start-up (cranking) and when crank angle signal is abnormal.

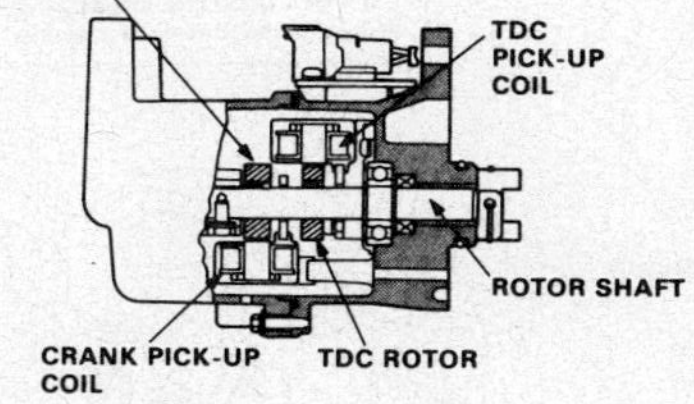

– Check Engine warning light has been reported on.
– LED indicates CODE 4.

Turn the ignition switch OFF.

Remove CLOCK fuse in the under-hood relay box for 10 seconds to reset ECU.

Start engine.

Is Check Engine warning light on and does LED indicate CODE 4 ?

NO → Intermittent failure, system is OK at this time (test drive may be necessary). Check for poor connections or loose wires at distributor connector and C236 (located at right shock tower).

YES → Stop engine.

Disconnect the 4P connector from the TDC/CRANK sensor.

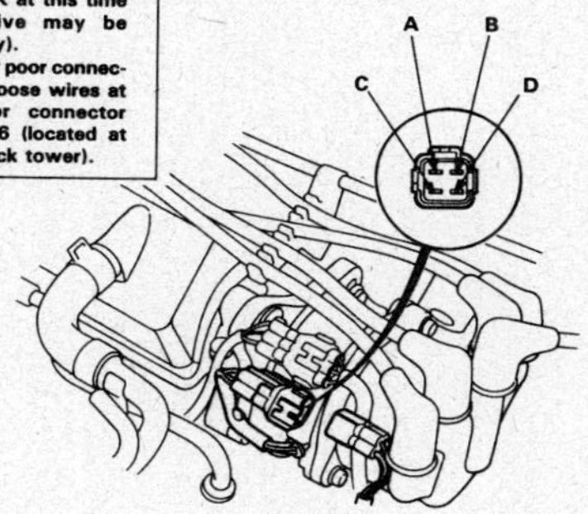

2.0L ENGINE—TROUBLESHOOTING CODE 4—PRELUDE

Measure resistance between C terminal and D terminal.

Is there 700–1,000 Ω ?
- NO → Replace pulse generator assembly
- YES ↓

Check for continuity to body ground on C terminal and D terminal individually.

Does continuity exist ?
- YES → Replace pulse generator assembly
- NO ↓

Reconnect the connector.

Connect the PGM-FI test harness only to the main wire harness, but not to the ECU

Measure resistance between B10 terminal and B12 terminal.

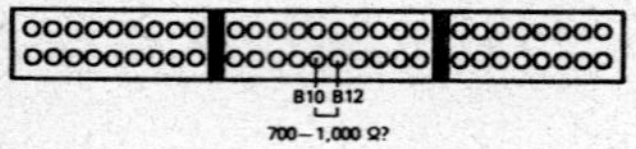

Is there 700–1,000 Ω ?
- NO → Repair open in BLU/YEL and/or BLU/GRN wires.
- YES ↓

Check for continuity to body ground on B10 terminal.

Does continuity exist?
- YES → Repair short in BLU/GRN wire between ECU (B10) and distributor connector.
- NO ↓

Substitute a known-good ECU and recheck. If symptom/indication goes away, replace the original ECU.

2.0L ENGINE—TROUBLESHOOTING CODE 5—PRELUDE

– Check Engine warning light has been reported on.
– LED indicates CODE 5.

Turn the ignition switch OFF.

Remove CLOCK fuse in the underhood relay box for 10 seconds to reset ECU.

Start engine and keep engine rpm at idle.

Is Check Engine warning light on and does LED indicate CODE 5 ?
- NO → – Intermittent failure, system is OK at this time (test drive may be necessary). – Check vacuum hoses, pipes and connections. – Make sure all connectors are secure.
- YES ↓

Stop engine.

Disconnect #21 hose from the throttle body, connect vacuum pump to the hose and apply vacuum.

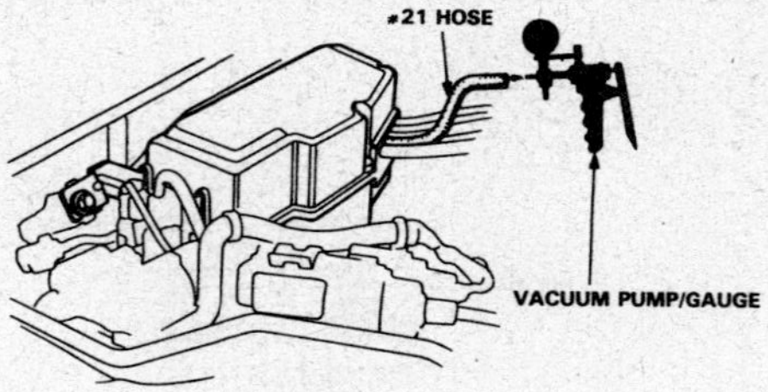

Does it hold vacuum?
- NO → Connect a vacuum pump to the MAP sensor and apply vacuum.
  - Does it hold vacuum?
    - NO → Replace MAP sensor.
    - YES → Replace #21 hose.
- YES ↓

Connect a T-fitting from a vacuum gauge between the throttle body and MAP sensor.

Start the engine.

Is there manifold vacuum?
- NO → – Remove restriction from throttle body. – Replace throttle body.
- YES

2.0L ENGINE—TROUBLESHOOTING CODE 5—PRELUDE

- Start the engine.
- Is there manifold vacuum?
  - NO → -Remove restriction from throttle body. -Replace throttle body.
  - YES ↓
- Stop the engine.
- Connect the PGM-FI test harness between the ECU and connector
- Turn the ignition switch ON.
- Measure voltage between C11 (+) terminal and C14 (−) terminal.

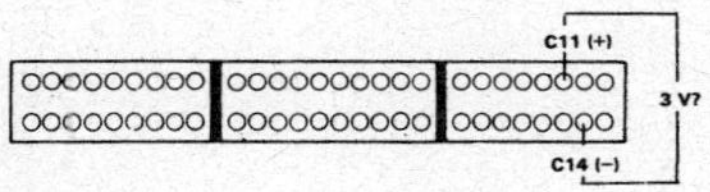

- Is there approx. 3 V ?
  - NO → **Replace the MAP sensor.**
  - YES ↓
- Start the engine and allow it to idle.
- Is there approx. 1 V?
  - NO → **Replace MAP sensor.**
  - YES ↓
- **Substitute a known-good ECU and recheck. If symptom/indication goes away, replace the original ECU.**

2.0L ENGINE—TROUBLESHOOTING CODE 6—PRELUDE

Self-diagnosis LED indicates code 6: Most likely a problem in the Coolant Temperature (TW) Sensor circuit.

The TW sensor is a temperature dependant resistor (thermistor). The resistance of the thermistor decreases as the coolant temperature increases as shown below.

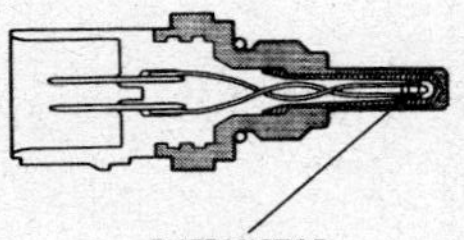

THERMISTOR

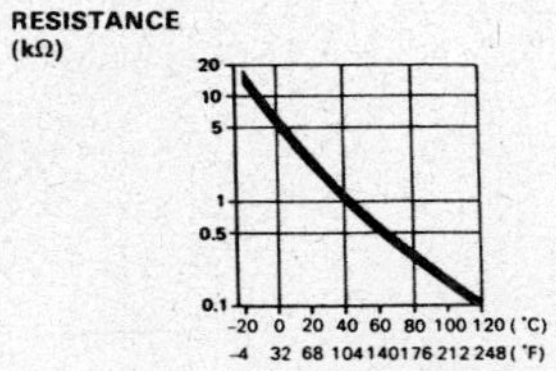

- **– Check Engine warning light has been reported on.**
  – LED indicates CODE 6.
- Turn the ignition switch OFF.
- Remove CLOCK fuse in the under-hood relay box for 10 seconds to reset ECU.
- Turn the ignition switch ON.
- Is Check Engine warning light on and does LED indicate CODE 6 ?
  - NO → **Intermittent failure, system is OK at this time (test drive may be necessary). Check for poor connections or loose wires at TW sensor and C235 (located at right shock tower).**
  - YES ↓
- Warm up engine to normal operating temperature (cooling fan comes on).
- Turn the ignition swtich OFF.
- Disconnect the 2P connector from the TW sensor.
- Measure resistance between the 2 terminals on the TW sensor.
- Is there 200–400Ω?
  - NO → **Replace TW sensor.**
  - YES

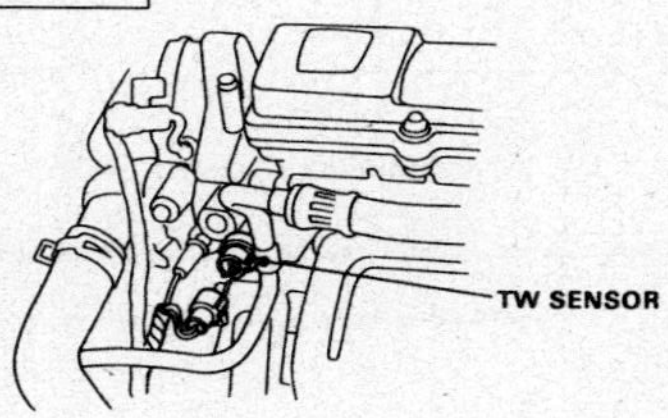

2.0L ENGINE—TROUBLESHOOTING CODE 6—PRELUDE

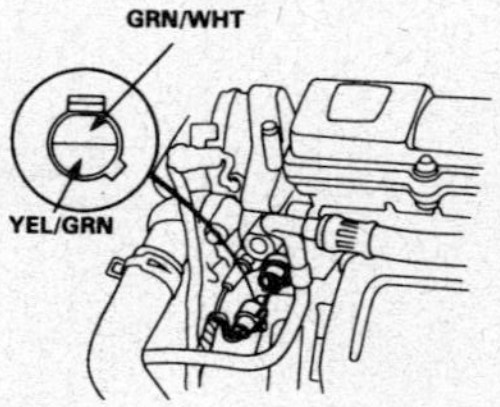

Turn the ignition switch ON.

Measure voltage between YEL/GRN and body ground.

Is there approx. 5V ?

- NO → **A/T only**: Disconnect the 18P connector from the A/T control unit.
  - Is there approx. 5 V ?
    - YES → **Replace the A/T control unit.**
    - NO → **Repair open or short in YEL/GRN wire between ECU (C6), A/T control unit and TW sensor. If wire is OK, substitute a known-good ECU and recheck. If prescribed voltage is now available, replace original ECU.**
- YES → Measure voltage between YEL/GRN (+) terminal and GRN/WHT (−) terminal.

Is there approx. 5V ?

- NO → **Repair open in GRN/WHT wire between ECU (C12) and TW sensor.**
- YES → **Substitute a known-good ECU and recheck. If symptom/indication goes away, replace the original ECU.**

2.0L ENGINE—TROUBLESHOOTING CODE 7—PRELUDE

Self-diagnosis LED indicates code 7: Most likely a problem in the Throttle Angle Sensor circuit.

The throttle angle sensor is a potentiometer. It is connected to the primary throttle valve shaft. As the throttle angle changes, the throttle angle sensor varies the voltage signal to the ECU.

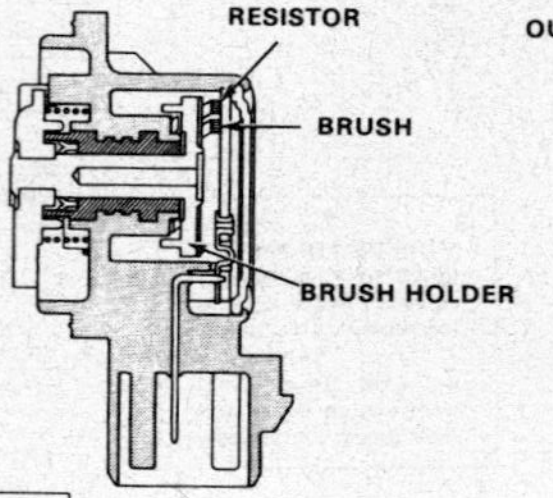

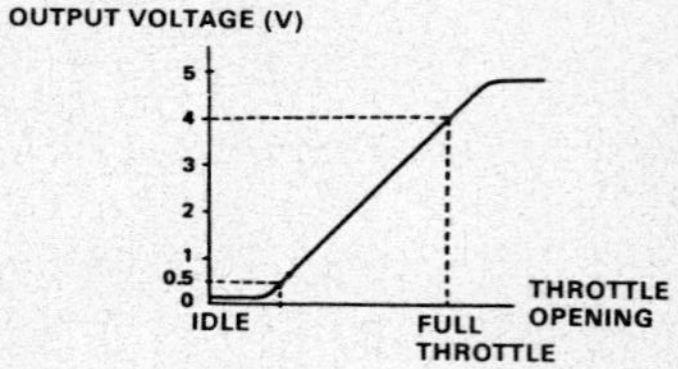

- **Engine is running.**
- **Check Engine warning light has been reported on.**
- **LED indicates CODE 7.**

Turn the ignition switch OFF.

Remove CLOCK fuse in the under-hood relay box for 10 seconds to reset ECU.

Start engine.

Is Check Engine warning light on and does LED indicate CODE 7 ?

- NO → **Intermittent failure, system is OK at this time (test drive may be necessary). Check for poor connections or loose wires at throttle angle sensor and C235 (located at right shock tower).**
- YES → Turn the ignition switch OFF.

Disconnect the 3P connector from the throttle angle sensor.

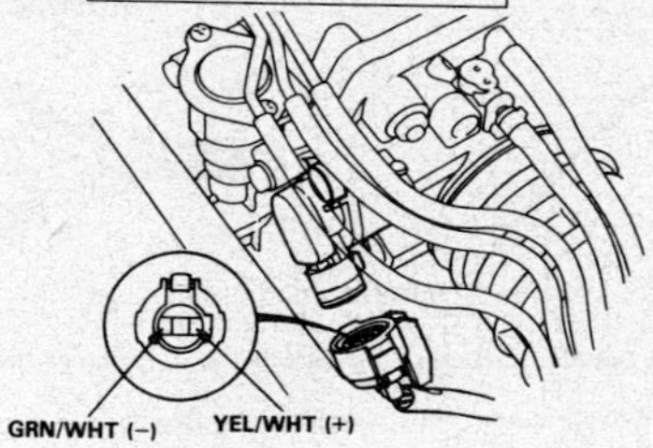

Turn the ignition switch ON.

Measure voltage between YEL/WHT (+) terminal and GRN/WHT (−) terminal.

Is there approx. 5V ?

- YES → Go to Path A
- NO → Measure voltage between YEL/WHT (+) terminal and body ground. → Go to Path B

2.0L ENGINE—TROUBLESHOOTING CODE 7—PRELUDE

Path A

- Turn the ignition switch OFF.
- Reconnect the 3P connector.
- Connect the PGM-FI test harness between the ECU and connector
- Turn the ignition switch ON.
- Measure voltage between C7(+) terminal and C12(−) terminal.

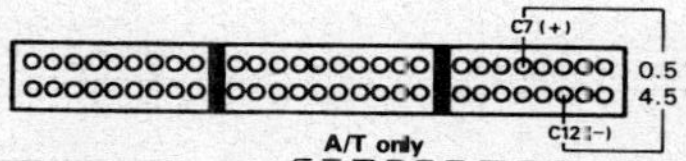

- Is voltage approx. 0.5 V at full close throttle, and approx. 4.5 V at full open throttle?
  - YES: **Substitute a known-good ECU and recheck. If symptom/indication goes away, replace the original ECU.**
  - NO: **A/T only**
    - Disconnect the 18P connector from the A/T control unit.
    - Is voltage approx. 0.5 V at full close throttle, and approx. 4.5 V at full open throttle? NOTE: There should be a smooth transition from 0.5 V to 4.5 V as the throttle is depressed.
      - YES: **Replace the A/T control unit.**
      - NO: **– Replace throttle angle sensor. – Repair open or short in RED/YEL wire between ECU (C7), A/T control unit and throttle angle sensor.**

Path B

- Is there approx. 5V ?
  - YES: **Repair open in GRN/WHT wire between ECU (C12) and throttle angle sensor.**
  - NO:
    - Turn the ignition switch OFF.
    - Connect the PGM-FI test harness between the ECU and connector
    - Turn the ignition switch ON.
    - Measure voltage between C13(+) terminal and C12(−) terminal.

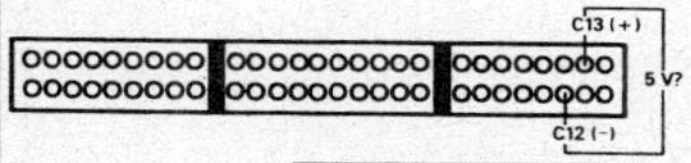

- Is there approx. 5V ?
  - YES: **Repair open in YEL/WHT wire between ECU(C13) and throttle angle sensor.**
  - NO: **Substiture a known-good ECU and recheck. If prescribed voltage is now available, replace the original ECU.**

2.0L ENGINE—TROUBLESHOOTING CODE 8—PRELUDE

- **– Check Engine warning light has been reported on.**
- **– LED indicates CODE 8.**
- Turn the ignition switch OFF.
- Remove CLOCK fuse in the under-hood relay box for 10 seconds to reset ECU.
- Start engine.
- Is Check Engine warning light on and does LED indicate CODE 8 ?
  - NO: **Intermittent failure, system is OK at this time (test drive may be necessary). Check for poor connections or loose wires at distributor connector and C236 (located at right shock tower).**
  - YES:
    - Stop engine.
    - Disconnect the 4P connector from the TDC/CRANK sensor.
    - Measure resistance between A terminal and B terminal.
    - Is there 700—1,000 Ω ?
      - NO: **Replace pulse generator assembly**
      - YES:
        - Check for continuity to body ground on A terminal and B terminal individually.
        - Does continuity exist ?
          - YES: **Replace pulse generator assembly**
          - NO: Reconnect the connector.

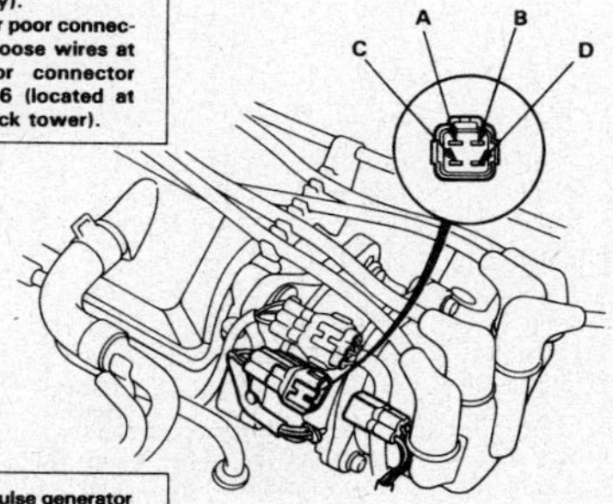

2.0L ENGINE—TROUBLESHOOTING CODE 8—PRELUDE

Connect the PGM-FI test harness only to the main wire harness, but not to the ECU

Measure resistance between C3 terminal and C4 terminal.

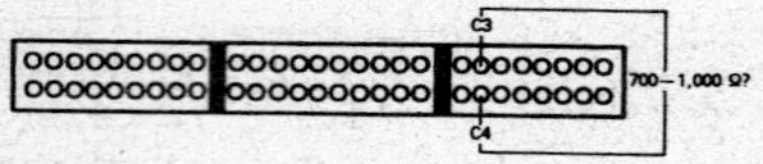

Is there 700 — 1,000 Ω ?

- NO: Repair open in ORN/BLU and/or WHT/BLU wires.
- YES: Check for continuity to body ground on C3 terminal.

Does continuity exist?

- YES: Repair short in ORN/BLU wire between ECU (C3) and distributor connector.
- NO: Substitute a known-good ECU and recheck. If symptom/indication goes away, replace the original ECU.

2.0L ENGINE—TROUBLESHOOTING CODE 9—PRELUDE

Self-diagnosis LED indicates code 9: A problem in the CYL sensor.

The CYL sensor detects the position of NO.1 cylinder for sequential fuel injection to each cylinder.

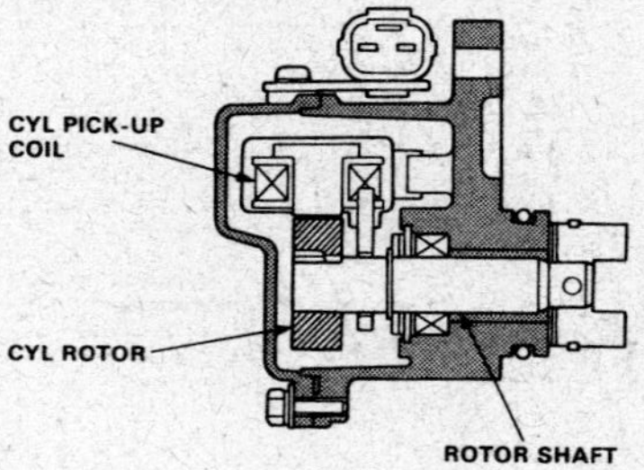

- Check Engine warning light has been reported on.
- LED indicates CODE 9.

Turn the ignition switch OFF.

Remove CLOCK fuse in the under-hood relay box for 10 seconds to reset ECU.

Start engine.

Is Check Engine warning light on and does LED indicate CODE 9 ?

- NO: Intermittent failure, system is OK at this time (test drive may be necessary). Check for poor connections or loose wires at the distributor connector and C236 (located at right shock tower).
- YES: Stop engine.

Disconnect the 2P connector from the CYL sensor.

Measure resistance between 2 terminals on the CYL sensor.

2.0L ENGINE—TROUBLESHOOTING CODE 9—PRELUDE

1. Is there 700 – 1,100 Ω?
   - NO → Replace CYL sensor assembly
   - YES ↓
2. Check for continuity to body ground on the 2 terminals individually.
3. Does continuity exist?
   - YES → Replace CYL sensor assembly
   - NO ↓
4. Reconnect the connector.
5. Connect the PGM-FI test harness only to the main wire harness, but not to the ECU
6. Measure resistance between C1 terminal and C2 terminal.

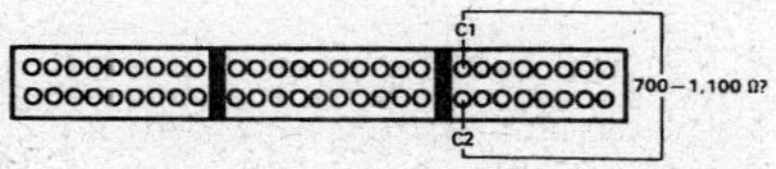

7. Is there 700 – 1,100 Ω?
   - NO → Repair open in ORN and/or WHT wires.
   - YES ↓
8. Check for continuity to body ground on C1 terminal.
9. Does continuity exist?
   - YES → Repair short in ORN wire between ECU (C1) and CYL senser connector.
   - NO ↓
10. Substitute a known-good ECU and recheck. If symptom/indication goes away, replace the original ECU

2.0L ENGINE—TROUBLESHOOTING CODE 10—PRELUDE

Self-diagnosis LED indicates code 10: Most likely a problem in the Intake Air Temperature (TA) Sensor circuit.

The TA sensor is a temperature dependant resistor (thermistor). The resistance of the thermistor decreases as the intake air temperature increases as shown below.

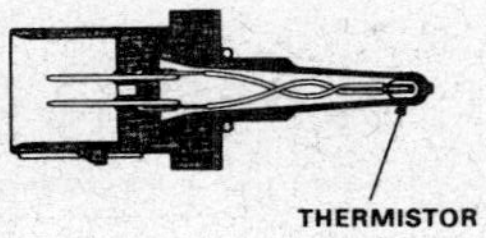

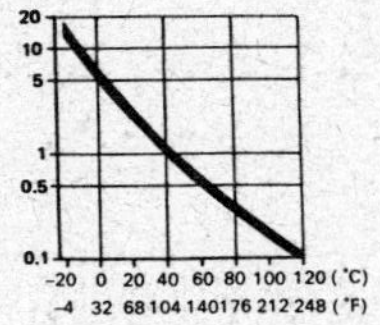

- Check Engine warning light has been reported on.
- LED indicates CODE 10.

1. Turn the ignition switch OFF.
2. Remove CLOCK fuse in the under-hood relay box for 10 seconds to reset ECU.
3. Turn the ignition switch ON.
4. Is Check Engine warning light on and does LED indicate CODE 10?
   - NO → Intermittent failure, system is OK at this time (test drive may be necessary). Check for poor connections or loose wires at TA sensor and C235 (located at right shock tower).
   - YES ↓
5. Turn the ignition switch OFF.
6. Disconnect the 2P connector from the TA sensor.
7. Measure resistance between the 2 terminals on the TA sensor.
8. Is there 1–4 kΩ?
   - NO → Replace TA sensor.
   - YES ↓

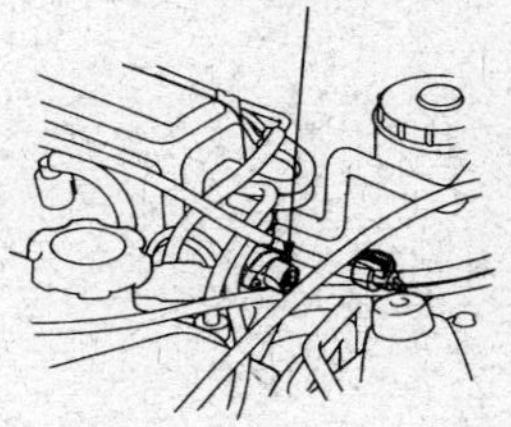

2.0L ENGINE—TROUBLESHOOTING CODE 10—PRELUDE

Turn the ignition switch ON.

Measure voltage between WHT/RED (+) terminal and body ground.

Is there approx. 5 V ?

- NO → Repair open or short in WHT/RED wire between ECU (C5) and TA sensor. If wire is OK, substitute a known-good ECU and recheck. If prescribed voltage is now available, replace original ECU.
- YES ↓

Measure voltage between WHT/RED (+) terminal and GRN/WHT (−) terminal.

Is there approx. 5 V ?

- NO → Repair open in GRN/WHT wire between ECU (C12) and TA sensor.
- YES ↓

Substitute a known-good ECU and recheck. If symptom/indication goes away, replace the original ECU.

2.0L ENGINE—TROUBLESHOOTING CODE 12—PRELUDE

- Check Engine warning light has been reported on.
- LED indicates CODE 12.

Turn the ignition switch OFF.

Remove the BACK UP fuse in the under-hood relay box for 10 seconds to reset ECU.

Road test necessary: Warm up the engine to normal operating temperature (cooling fan comes on). Drive the car on the road for approx. 10 minutes. Try to keep the engine speed in the 1700— 2500 range.

Is Check Engine warning light on and does LED indicated CODE 12 ?

- NO → Intermittent failure, system is OK at this time. Check for poor connections or loose wires at EGR valve C235, C236 (located at right shock tower), control box.
- YES ↓

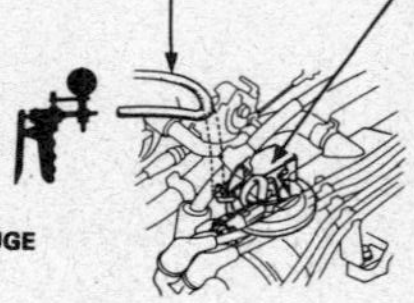

With the engine at idle, disconnect the #16 hose from the EGR valve and connect a vacuum pump/gauge to the hose.

Is there any vacuum ?

- YES → Disconnect 4P connector from the control box and check the #16 hose for vacuum again. → Go to Path M
- NO ↓

Move the vacuum pump/gauge to the EGR valve. → Go to Path L

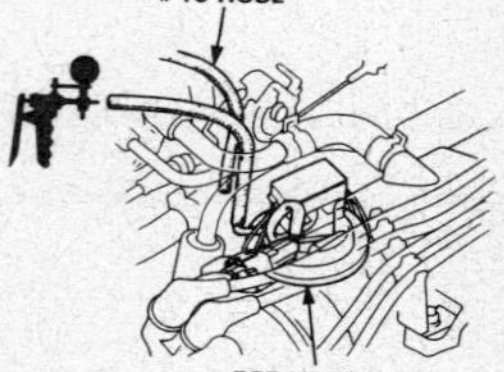

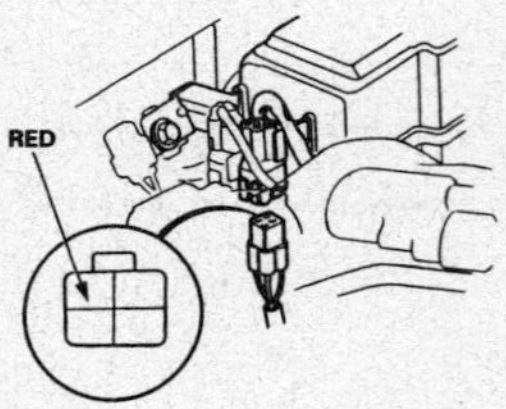

2.0L ENGINE—TROUBLESHOOTING—CODE 12—PRELUDE

Path L

With the engine at idle, apply 8" of vacuum to the EGR valve.

Does the engine stall and does the EGR valve hold vacuum ?
- NO → Replace EGR valve.
- YES ↓

Disconnect the 4P connector from the control box.

Measure voltage between BLK/YEL (+) terminal on the main wire harness and body ground.

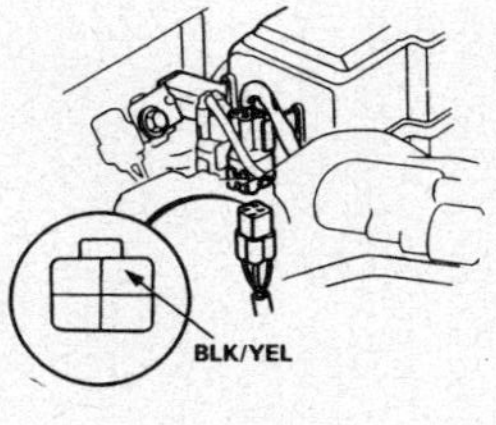

Is there battery voltage ?
- NO → Repair open in BLK/YEL wire between the solenoid valve and No.12 fuse.
- YES ↓

Reconnect the vacuum pump/gauge to the #16 hose.

Start the engine and allow it to idle.

Path M

Is there any vacuum ?
- YES → Check vacuum hose routing of the entire EGR system. If hose routing is OK, replace EGR control solenoid valve.
- NO → Inspect for short to body ground in RED wire between EGR control solenoid valve and ECU (E5). If wire is OK, substitute a known-good ECU and retest. If symptom/indication goes away, replace the original ECU.

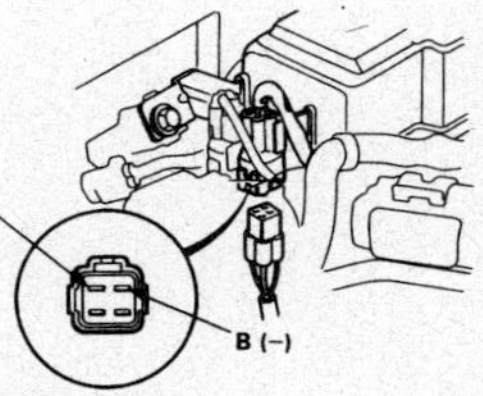

Connect the battery positive terminal to the A terminal of the 4P connector. While watching the vacuum gauge, connect the battery negative terminal to the B terminal.

Is there approx. 8" within 1 second?
- NO → Turn the ignition switch OFF and inspect the #16 and #24 hoses for leaks, restrictions, or mis-routing.
  - Are the hoses OK?
    - NO → Connect as necessary.
    - YES → Disconnect the lower hose on EGR control solenoid valve and connect a vacuum gauge to the hose.
      - Start the engine and allow it to idle.
      - Is there 6"— 10" of vacuum ?
        - NO → Replace CVC valve.
        - YES → Replace the EGR control solenoid valve.
- YES ↓

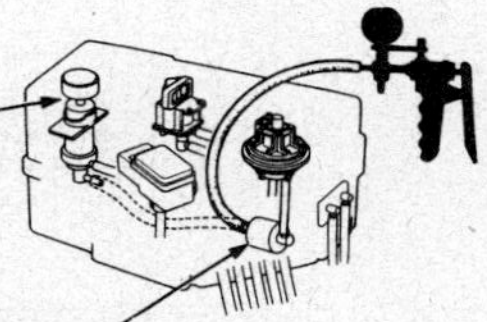

Turn the ignition switch OFF and reconnect the 4P connector.

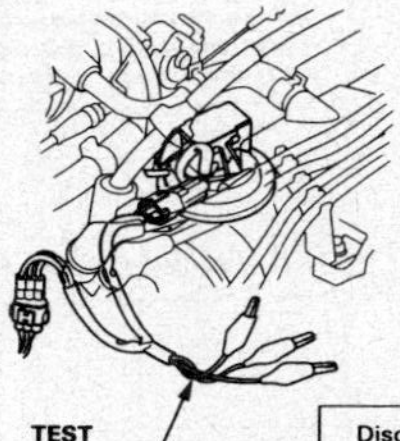

Connect the test harness between the EGR valve lift sensor and engine wire harness.

Turn the ignition switch ON:

Measure voltage between RED (+) terminal and GRN (−) terminal.

Is there approx. 5 V ?
- NO →
  - Repair open in YEL/WHT wire between EGR valve and ECU (C13).
  - Repair open in GRN/WHT wire between EGR valve and ECU (C12).
- YES

Measure voltage between WHT (+) terminal and GRN (−) terminal.

Is there approx 1.2 V ?
- NO:
  - Replace EGR valve.
  - Repair short in WHT/BLK or YEL wire between ECU (C8) and EGR valve lift sensor.
  - Substitute a known-good ECU and recheck. If prescribed voltage is now available, replace the original ECU.
- YES: While watching the voltmeter, slowly apply a continuous 8" of vacuum to the EGR valve. Repeat several times, completely releasing vacuum between.

Is the maximum voltage approx. 4 V ?
- NO: Replace EGR valve.
- YES: continue.

Does the voltage consistently increase/decrease as the vacuum increases/decreases ?
- NO: Replace EGR valve.
- YES: Reconnect the #16 hose to the EGR valve.

Turn the ignition switch OFF.

Connect the PGM-FI test harness between the ECU and connector

Start the engine and allow it to idle.

Measure voltage between C8 (+) terminal and C12 (−) terminal.

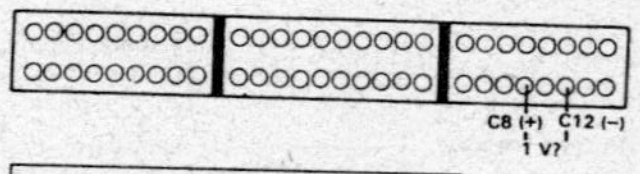

Is there approx. 1.2 V ?
- NO: Repair open in WHT/BLK or YEL wire between ECU (C8) and the sensor.
- YES: Connect A10 terminal to A18 terminal with a jumper wire.

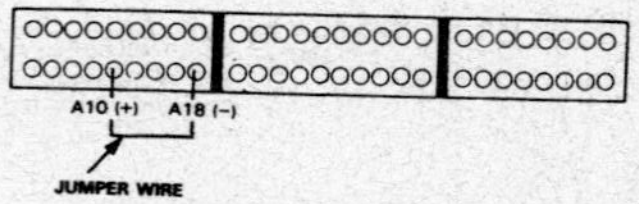

Did the engine stall ?
- NO: Repair open in RED wire between ECU (A10) and EGR control solenoid valve.
- YES: Substitute a known-good ECU and recheck. If symptom/indication goes away, replace the original ECU.

2.0L ENGINE—TROUBLESHOOTING—CODE 13—PRELUDE

Self-diagnosis LED indicates code 13: A problem in the Atmospheric Pressure (PA) Sensor circuit.

The PA sensor converts atmospheric pressure into electrical signals and inputs the ECU.

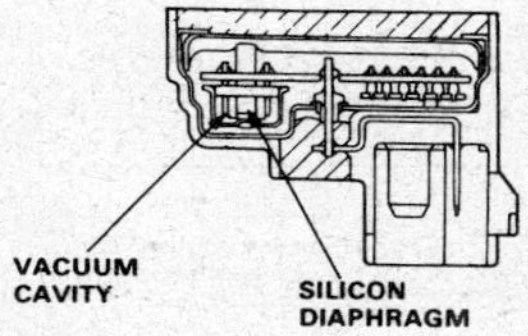

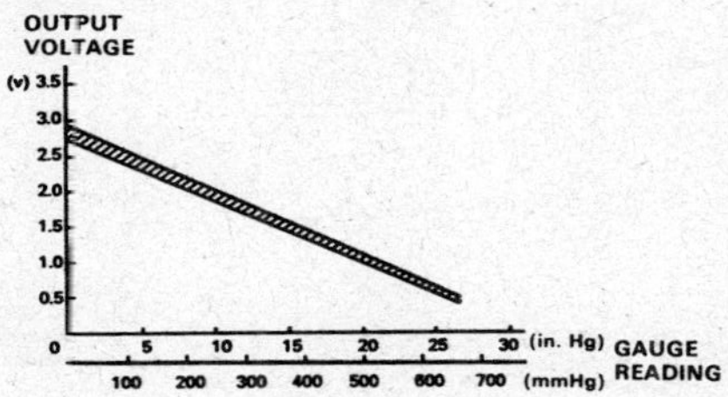

- Check Engine warning light has been reported on.
- LED indicates CODE 13.

Turn the ignition switch OFF.

Remove CLOCK fuse in the under-hood relay box for 10 seconds to reset ECU.

Turn the ignition switch ON.

Is Check Engine warning light on and does LED indicate CODE 13?
- NO: Intermittent failure, system is OK at this time (test drive may be necessary). Check for poor connections or loose wires at PA sensor connector and C263.
- YES: Turn the ignition switch OFF.

Connect the PGM-FI test harness between the ECU and connector

Turn the ignition switch ON.

Measure voltage between C13 (+) terminal and C12 (-) terminal.

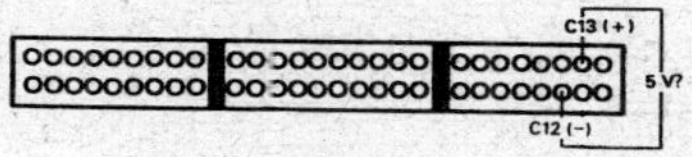

Is there approx. 5V ?
- NO: Substitute a known-good ECU and recheck. If prescribed voltage is now available replace the original ECU.
- YES: (continued)

Measure voltage between C9 (+) terminal and C12 (-) terminal.

Is there approx. 3V ?
- YES: Substitute a known-good ECU and recheck. If symptom/indication goes away, replace the original ECU.
- NO: Disconnect the main wire harness from PA sensor.

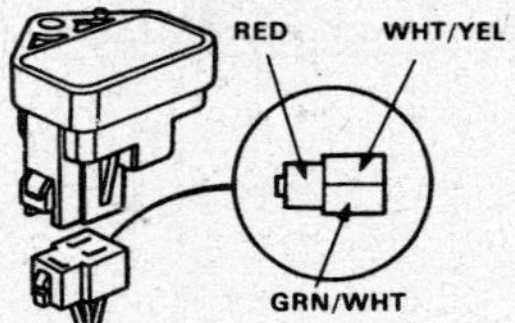

Measure voltage between C9 (+) terminal and C12 (-) terminal.

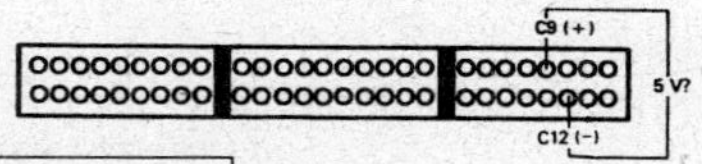

Is there approx. 5V ?
- NO: Repair short in RED wire between ECU (C9) and PA sensor.
- YES: Measure voltage between WHT/YEL (+) terminal and GRN/WHT (-) terminal.

Is there approx. 5V ?
- NO: Measure voltage between WHT/-YEL (+) terminal and body ground.
  - Is there approx. 5V ?
    - NO: Repair open in WHT/YEL wire between ECU (C13) and PA sensor.
    - YES: Repair open in GRN/WHT wire between ECU (C12) and the sensor.
- YES: Measure voltage between RED (+) terminal and GRN/WHT (-) terminal.

Is there approx. 5V ?
- NO: Repair open in RED wire between ECU (C9) and PA sensor.
- YES: **Replace PA sensor.**

Self-diagnosis LED indicates code 14: A problem in the Electronic Air Control Valve (EACV) circuit.

The EACV changes the amount of air bypassing the throttle body in response to a current signal from the ECU in order to maintain the proper idle speed.

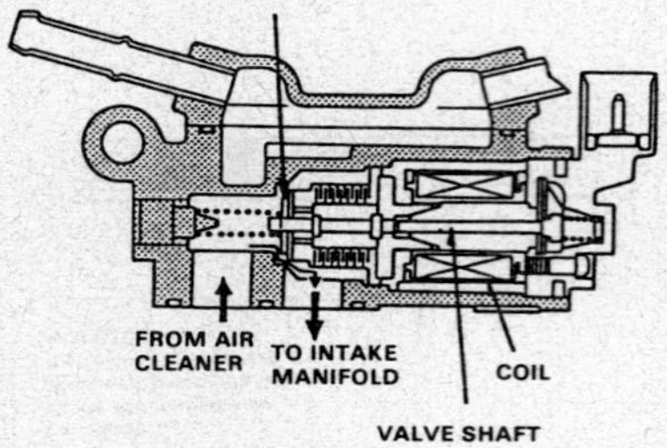

- Engine is running.
- Check Engine warning light is on.
- LED indicates CODE 14.

Turn the ignition switch OFF.

Remove CLOCK fuse in the dash fuse box for 10 seconds to reset ECU.

Start engine.

Is Check Engine warning light on and does LED indicate CODE 14?

- NO → **Intermittent failure, system is OK at this time (test driving may be necessary). Check for poor connections or loose wires at EACV and at C235 (located at right shock tower).**
- YES ↓

Stop engine.

Disconnect the 2P connector from the EACV.

Measure resistance between the 2 terminals on the EACV.

Is there 8—15Ω ?

- NO → **Replace EACV.**
- YES ↓

Check for continuity to body ground on each terminal on the EACV.

Does continuity exist ?

- YES → **Replace EACV**
- NO ↓

Turn the ignition switch ON.

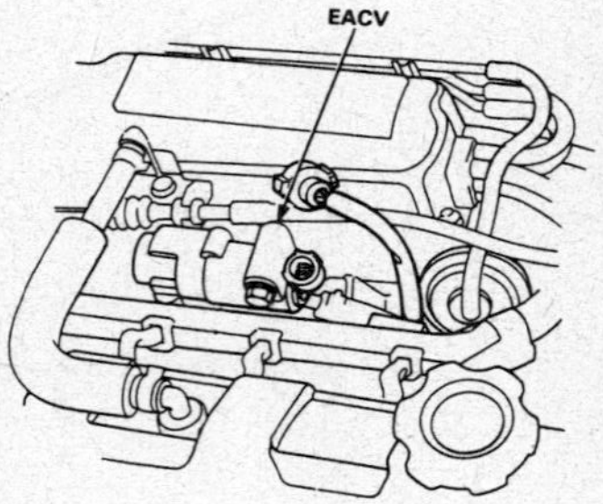

Measure voltage between BLK/YEL (+) terminal and BLU/RED (−) terminal.

Is there battery voltage ?

- YES → Disconnect the A connector from the ECU.
  - Is there battery voltage ?
    - YES → **Repair short in BLU/RED wire between ECU (A11) and EACV.**
    - NO → **Substitute a known-good ECU and recheck. If symptom/indication goes away, replace the original ECU.**
- NO ↓

Measure voltage between BLK/YEL (+) terminal and body ground.

Is there battery voltage ?

- NO → **Repair open in BLK/YEL or YEL/BLK wire between the EACV and Main Relay (wire color changes at C235).**
- YES ↓

Turn the ignition switch OFF.

Reconnect the 2P connector to EACV.

Connect the PGM-FI test harness "A" connector to the main wire harness only, not the ECU

Turn the ignition switch ON.

Measure voltage between A11 (+) terminal and A18 (−) terminal.

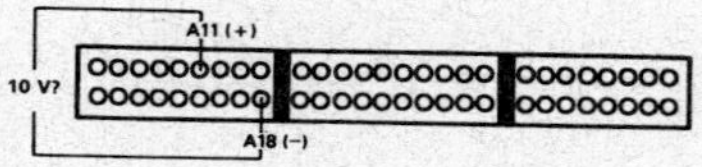

Is there approx. 10 V?

- NO → **Repair open in BLU/RED wire between ECU (A11) and EACV.**
- YES ↓

Connect and disconnect A11 terminal to A18 terminal.

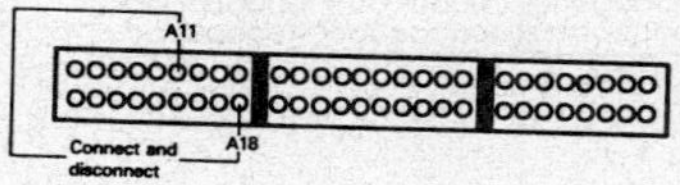

Does EACV click when the connector is connected and disconnected ?

- NO → **Replace EACV.**
- YES → **Substitute a known-good ECU and recheck. If symptom/indication goes away, replace the original ECU.**

2.0L ENGINE – TROUBLESHOOTING – CODE 15 – PRELUDE

Self-diagnosis LED indicates code 15: A problem in the Ignition Output Signal circuit.

- Check Engine warning light has been reported on.
- LED indicates CODE 15.

Turn the ignition switch OFF.

Remove CLOCK fuse in the under-hood relay box for 10 seconds to reset ECU.

Turn the ignition switch ON to the Start Position.

Is Check Engine warning light on and does LED indicate CODE 15?

- NO → **Intermittent failure, system is OK at this time (test drive may be necessary). Check for poor connections or loose wires at the igniter unit connector.**
- YES ↓

Turn the ignition switch OFF.

Disconnect the 6P connector on the igniter unit.

Turn the ignition switch ON.

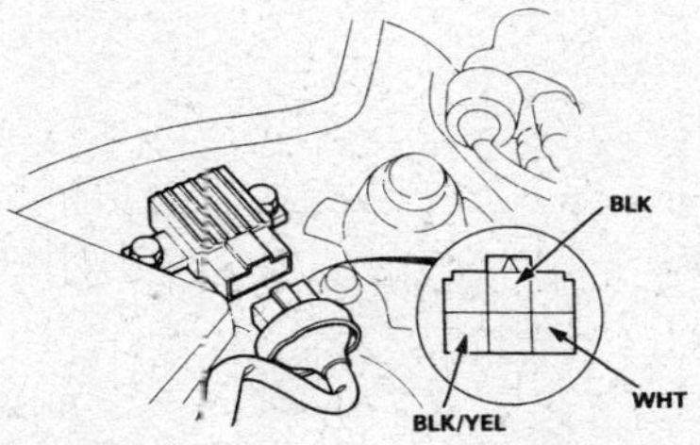

Measure voltage between the BLK/YEL (+) terminal and body ground.

Is there battery voltage ?

- NO → **Repair open in the BLK/YEL wire between the igniter unit and ignition switch.**
- YES ↓

Turn the ignition switch OFF.

Connect the PGM-FI test harness between the ECU and connector (Page 11-159)

Reconnect the 6P connector on the igniter unit.

Turn the ignition switch ON.

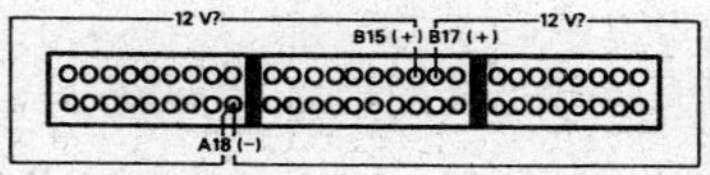

Measure voltage individually between B15 (+), B17 (+) terminals and A18 (–) terminal.

Is there battery voltage ?

- YES → **Substitute a known-good ECU and recheck. If symptom/indication goes away, replace the original ECU.**
- NO ↓

Turn the ignition switch OFF.

Disconnect the 6P connector on the igniter unit and the system checker harness from the ECU.

Check for continuity of WHT wires between the ECU (B15, B17) and the igniter unit.

Does continuity exist ?

- NO → **Repair open in WHT wires between the ECU (B15, B17) and the igniter unit.**
- YES ↓

Check for continuity between white terminal of 6P connector and body ground.

Does continuity exist ?

- YES → **Repair short in WHT wire.**
- NO ↓

Replace the igniter unit.

NOTE: If the WHT wire was shorted, the ignitor may be damaged.

Self-diagnosis LED indicates code 16: A problem in the fuel injector circuit.

The injectors are the solenoid-actuated constant-stroke pintle type consisting of a solenoid, plunger needle valve and housing. When current is applied to the solenoid coil, the valve lifts up and pressurized fuel is injected close to the intake valve. Because the needle valve lift and the fuel pressure are constant, the injection quantity is determined by the length of time that the valve is open (i.e., the duration the current is supplied to the solenoid coil). The injector is sealed by an O-ring and seal ring at the top and bottom. These seals also reduce operating noise.

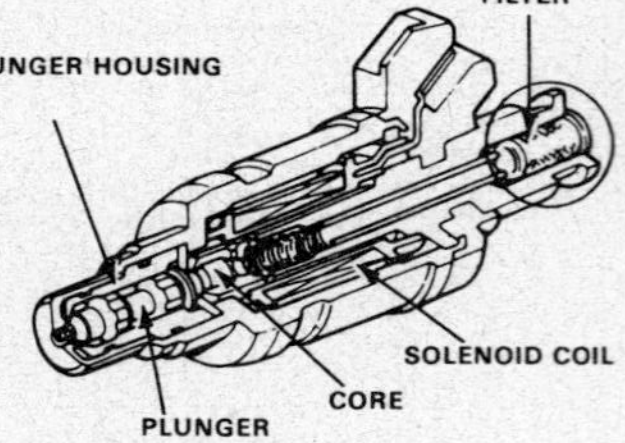

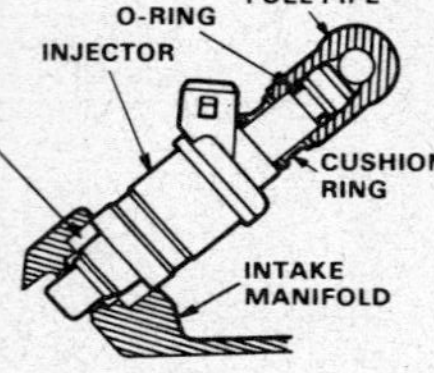

- Check Engine warning light is on.
- LED indicates CODE 16.

Turn the ignition switch OFF.

Remove CLOCK fuse in the underhood relay box for 10 seconds to reset ECU.

Turn the ignition switch to START position.
NOTE: It may be necessary to crank for a minimum of 15 seconds to reset failure CODES.

Does the engine start ? — NO: Go to Path F

YES

Is Check Engine warning light on and does LED indicate CODE 16? — NO: **Intermitten failure, system is OK at this time (test drive may be necessary). Check for poor connections or loose wires at injectors, injector resistor and C235 (located at right shock tower).**

YES

Go to Path E

Path E

Check the clicking sound of each injector by means of a stethoscope when the engine is idling.

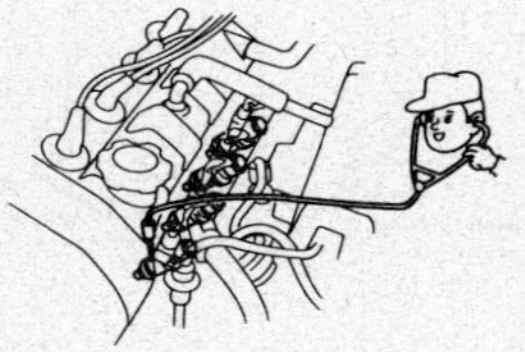

Does the injector click ? — YES: **Substitute a known-good ECU and recheck. If symptom/indication goes away, replace the original ECU.**

NO

Turn the ignition switch OFF.

Disconnect the 2P connector from the injector that does not click.

Measure resistance between the 2 terminals of the injector.

Path F

Turn the ignition switch OFF.

Disconnect the 2P connector from each injector.

Measure resistance between the 2 terminals of the injector.

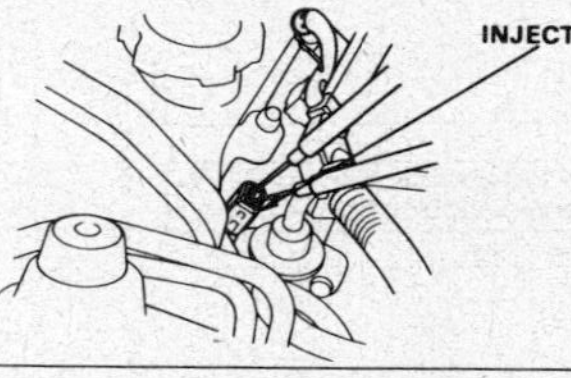

Is there 1.5—2.5Ω ? — NO: **Replace the injector.**

YES

Turn the ignition switch ON.

2.0L ENGINE—TROUBLESHOOTING—CODE 16—PRELUDE

Measure voltage between RED/BLK (+) terminal on the 2P connector and body ground.

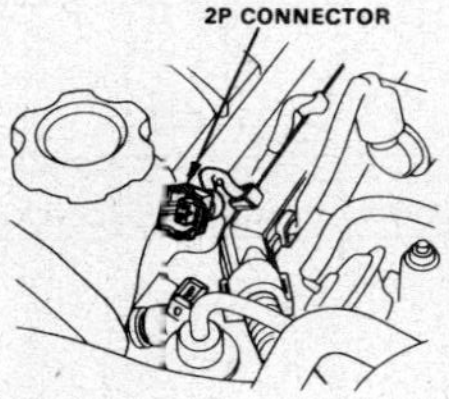

Is there battery voltage ?

- NO → Turn the ignition switch OFF.
  - Disconnect 6P connector on the injector resistor.
  - Turn the ignition switch ON.
  - Measure voltage between YEL/BLK (+) terminal and body ground.
  - Is there battery voltage ?
    - NO → **Repair open in the YEL/BLK wire between the injector resistor and the main relay.**
    - YES → **—Replace the injector resistor.**
      —Repair open RED/BLK wire between 2P connector and resistor.
- YES ↓

Reconnect the 2P connector to the injector.

Turn the ignition switch OFF.

Connect the PGM-FI test harness between the ECU and connector

Turn the ignition switch ON.

Measure voltage between A18 (−) terminal and following terminals,
- No.1 injector: A1 (+) terminal.
- No.2 injector: A3 (+) terminal.
- No.3 injector: A5 (+) terminal.
- No.4 injector: A7 (+) terminal.

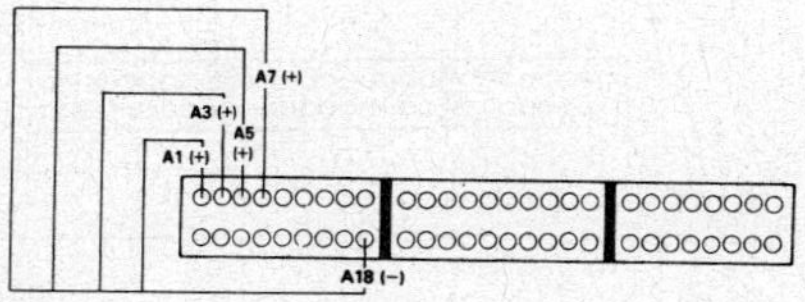

Is there battery voltage ?

- NO → **Repair open in the wire between the ECU (A1, A3, A5 or A7) and the injector.**
- YES → **Substitute a known-good ECU and recheck. If symptom/indication goes away, replace the original ECU.**

2.0L ENGINE—TROUBLESHOOTING CODE 17—PRELUDE

Self-diagnosis LED indicates code 17: A problem in the Vehicle Speed Sensor circuit.

The signal generated by the speed sensor, produces pulses when the front wheels turn.

- Check Engine warning light has been reported on.
- LED indicates CODE 17.

Turn the ignition switch OFF.

Remove CLOCK fuse in the under-hood relay box for 10 seconds to reset ECU.

Road test necessary. In 2nd gear accelerate to 3,500 rpm, then decelerate to 1,500 rpm with throttle fully closed.

Is Check Engine warning light on and does LED indicate CODE 17?

- NO: Intermittent failure, system is OK at this time. Check for poor connections or loose wires at C129 and C240 (located at right shock tower).
- YES: Block rear wheels and set the parking brake. Jack up the front of the car and support with safety stands.

WARNING Block rear wheels before jacking up front of car.

Connect the PGM-FI test harness between the ECU and connector

Turn the ignition switch ON.

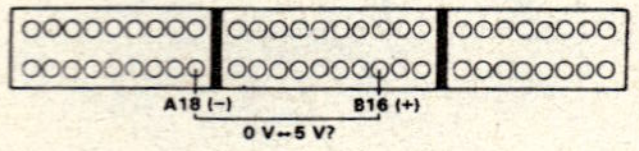

Slowly rotate left front wheel and measure voltage between B10 (+) terminal and A26 (−) terminal.

Does voltage pulse 0 V and 5 V?

- NO:
  - Repair open or short in WHT/BLU wire between ECU (B16) and the speed sensor.
  - Faulty speed sensor.
  - Substitute a known-good ECU and recheck. If symptom/indication goes away, replace the original ECU.
- YES: Substitute a known-good ECU and recheck. If symptom/indication goes away, replace the original ECU.

2.0L ENGINE—TROUBLESHOOTING AIR CONDITIONING SIGNAL—PRELUDE

This signals the PGM-FI ECU when there is a demand for cooling from the air conditioning circuits.

Inspection of Air Conditioning Signal.

Connect the PGM-FI test harness between the ECU and connector But disconnect "B" connector from the main wire harness only, not the ECU.

Turn the ignition switch ON.

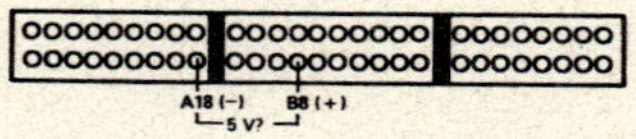

Measure voltage between B8 (+) terminal and A18 (−) terminal.

Is there approx. 5 V?

- NO: Substitute a known-good ECU and recheck. If prescribed voltage is now available, replace the original ECU.
- YES: Turn the ignition switch OFF.

Reconnect "B" connector to the main wire harness.

Turn the ignition switch ON.

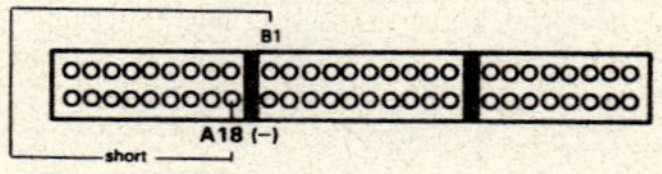

Momentarily connect B1 terminal to A18 terminal.

Is there a clicking noise from the A/C compressor clutch?

- NO: Connect the RED/BLU terminal of the 13P connector on the compressor control unit to body ground.

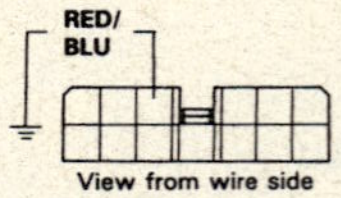

  Is there a clicking noise from the A/C compressor clutch?
  - NO: See Air conditioner inspection
  - YES: Repair open in RED/BLU wire between ECU (B1) and compressor control unit.
- YES: Start engine.

Turn the Blower switch ON.

2.0L ENGINE—TROUBLESHOOTING AIR CONDITIONING SIGNAL—PRELUDE

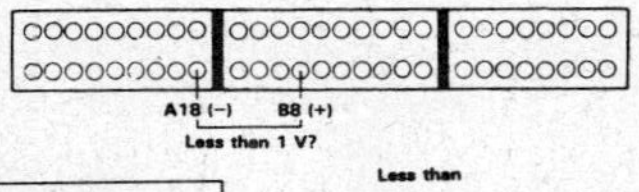

- A/C switch ON.
- Does A/C operate ?
  - YES → Air conditioning signal is OK.
  - NO → Measure voltage between B8 (+) terminal and A18 (−) terminal.
    - Is voltage less than 1V?
      - NO → **Repair open in RED/GRN wire between ECU (B8) and A/C switch.**
      - YES → **Substitute a known-good ECU and recheck. If symptom/indication goes away, replace the original ECU.**

2.0L ENGINE—TROUBLESHOOTING ALTERNATOR FR SIGNAL—PRELUDE

This signals the PGM-FI ECU when the alternator is charging.

- **Inspection of Alternator FR signal.**
- Connect the PGM-FI test harness between the ECU and connector Disconnect "B" connector from the main wire harness only, not the ECU.
- Turn the ignition switch ON.
- Measure voltage between B14 (+) terminal and A18 (−) terminal.

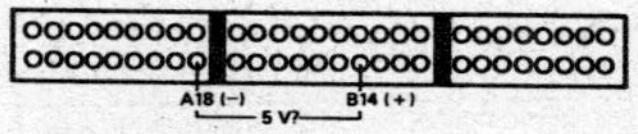

- Is there approx. 5V ?
  - NO → **Substitute a known-good ECU and recheck. If prescribed voltage is now available, replace the original ECU.**
  - YES → Turn the ignition switch OFF.
- Reconnect "B" connector to the main wire harness.
- Warm up engine to normal operating temperature (cooling fan comes on).
- Measure voltage between B14 (+) terminal and A18 (−) terminal.

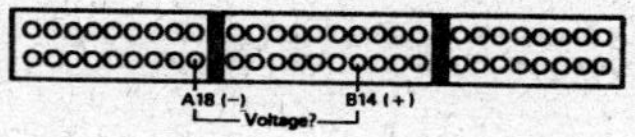

- Does the voltage decrease when headlight and rear defogger are turned on ?
  - YES → **Alternator FR signal is OK.**
  - NO → Stop engine.

2.0L ENGINE—TROUBLESHOOTING ALTERNATOR FR SIGNAL—PRELUDE

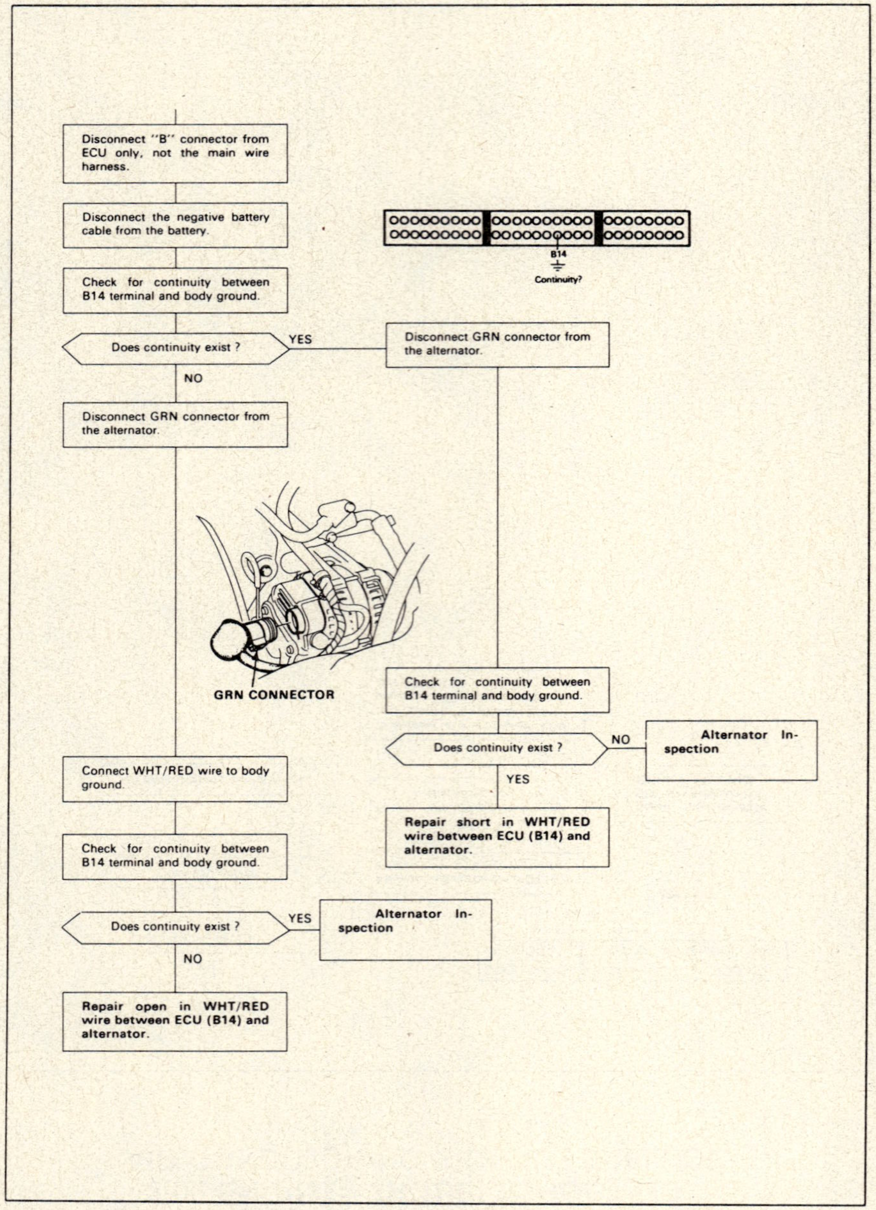

2.0L ENGINE—TROUBLESHOOTING AUTOMATIC TRANSAXLE SHIFT POSITION SIGNAL PRELUDE

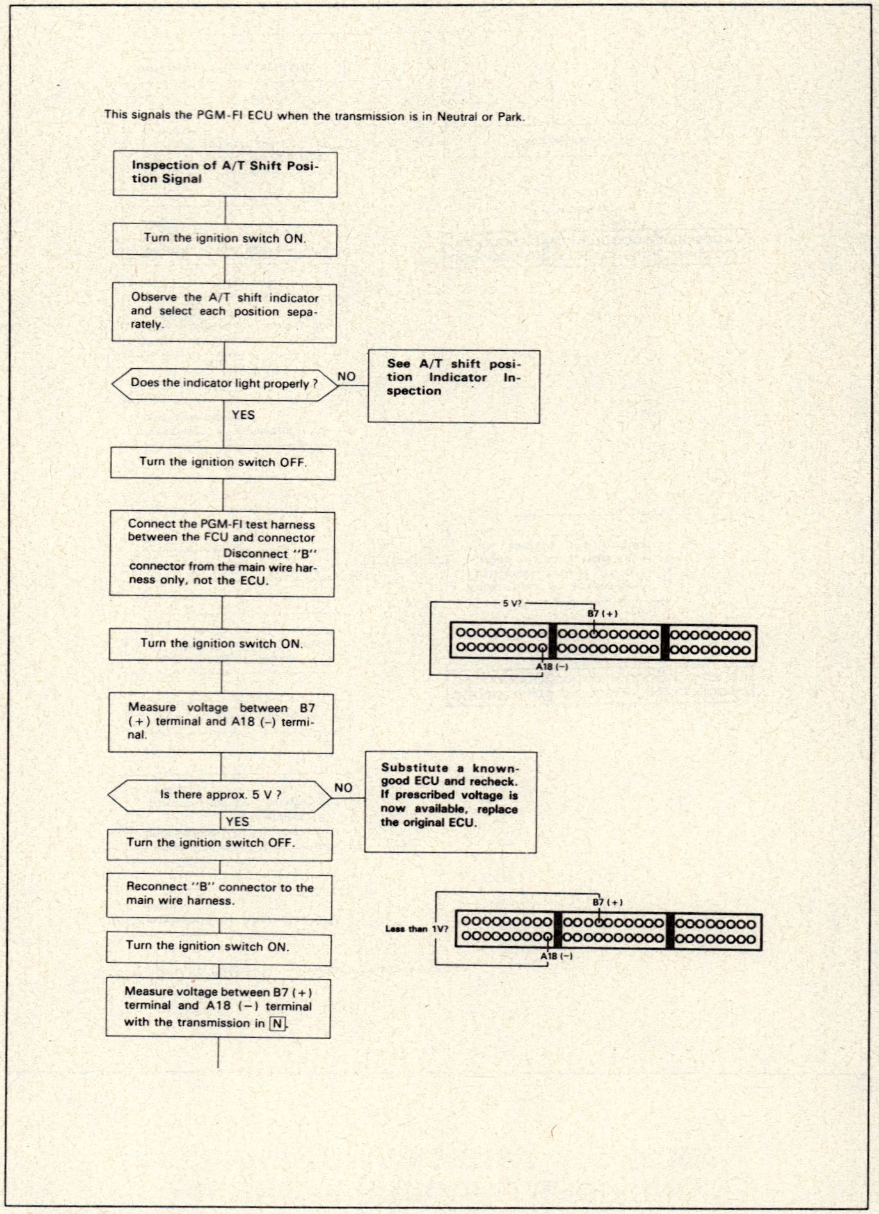

2.0L ENGINE—TROUBLESHOOTING
AUTOMATIC TRANSAXLE SHIFT POSITION SIGNAL
PRELUDE

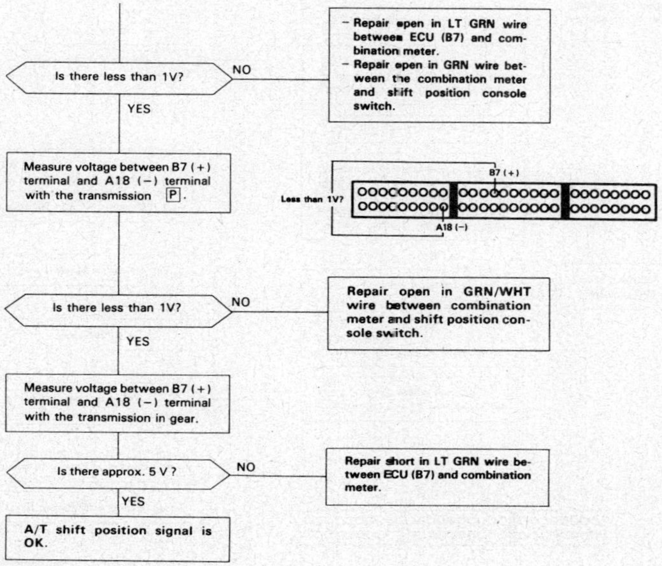

2.0L ENGINE—TROUBLESHOOTING
STARTER SIGNAL—PRELUDE

This signals the PGM-FI ECU when the engine is cranking.

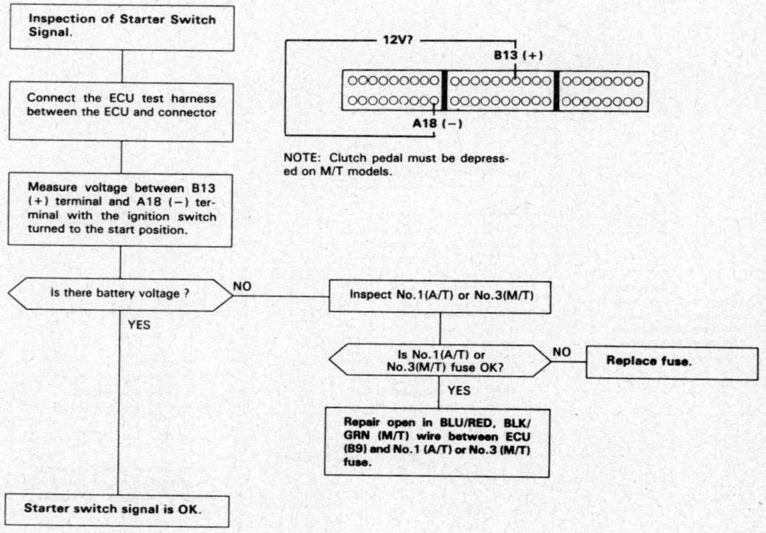

2.0L ENGINE—TROUBLESHOOTING
POWER STEERING PRESSURE SIGNAL—PRELUDE

This signals the PGM-FI ECU when the power steering load is high.

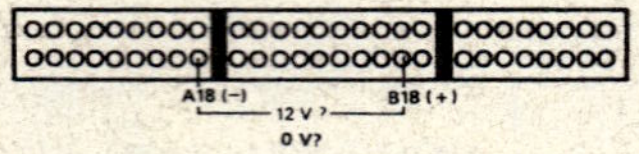

- **Inspection of P/S Oil Pressure Signal**
- Connect the PGM-FI test harness between the ECU and connector
- Turn the ignition switch ON.
- Measure voltage between B18 (+) terminal and A18 (−) terminal.
- Is there voltage?
  - YES → Disconnect the 2P connector on the P/S oil pressure switch. → Connect BLU/RED terminal to BLK terminal. → Is there voltage?
    - NO → **Replace P/S oil pressure switch.**
    - YES → **Repair open in BLU/RED wire between ECU (B18) and P/S oil pressure switch or BLK wire between P/S oil pressure switch and G301.**
  - NO → Start engine. → Turn steering wheel slowly. → Measure voltage between B18 (+) terminal and A18 (−) terminal while steering wheel is turning. → Is there battery voltage ?
    - YES → **P/S oil pressure signal is OK.**
    - NO → Turn the ignition switch OFF. → Disconnect "B" connector from main wire harness only, not the ECU. → Turn the ignition switch ON. → Is there battery voltage ?
      - NO → **Substitute a known-good ECU and recheck. If prescribed voltage is now available, replace the original ECU.**
      - YES → Reconnect "B" connector to main wire harness and disconnect 2P connector on the P/S oil pressure switch. → Is there battery voltage ?
        - YES → **Replace P/S oil pressure switch.**
        - NO → **Repair short in BLU/RED wire between ECU (B18) and the P/S oil pressure switch.**

2.0L ENGINE—TROUBLESHOOTING
BYPASS CONTROL SYSTEM—PRELUDE

- **Inspection of Bypass Control System**
- Start engine and allow to idle.
- Remove #13 vacuum hose from the bypass control diaphragm and connect vacuum gauge to the hose.

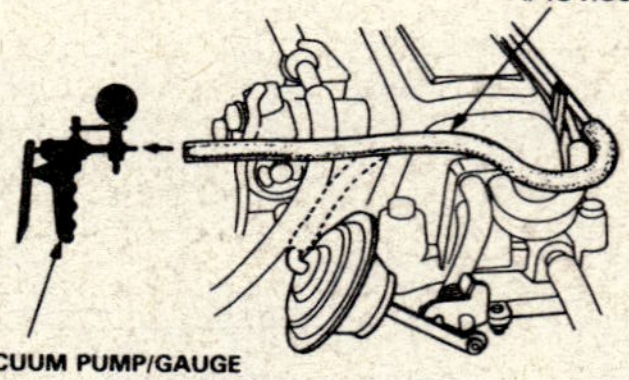

- Is there vacuum ?
  - YES → Go to Path J
  - NO → Remove #12 vacuum hose from the vacuum tank, then check for vacuum at the tank. → Is there vacuum ?
    - NO → **Repair the blockage or vacuum leak between the vacuum tank and the intake manifold.**
    - YES → Disconnect the 4P connector at the control box. → Measure voltage between BLK/YEL (+) terminal and YEL (−) terminal.

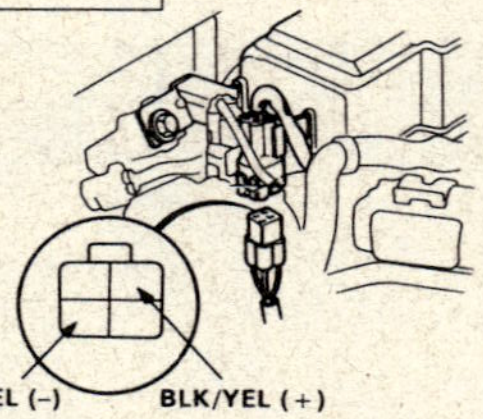

- Is there battery voltage ?
  - YES → **Replace the bypass control solenoid valve.**
  - NO → Measure voltage between BLK/YEL (+) terminal and body ground. → Go to Path K

2.0L ENGINE – TROUBLESHOOTING – BYPASS CONTROL SYSTEM – PRELUDE

Path J

- Raise engine speed to 5,000 rpm.
- Check for vacuum at #13 vacuum hose.
- Is there vacuum ?
  - NO: **Bypass control system is OK.**
  - YES: Disconnect the 4P connector. (continue below)

Path K

- Is there battery voltage?
  - NO: **Repair open in BLK/YEL wire between the 4P connector and No. 12 (10A) fuse.**
  - YES:
- Turn the ignition switch OFF.
- Connect the PGM-FI test harness between the ECU and connector

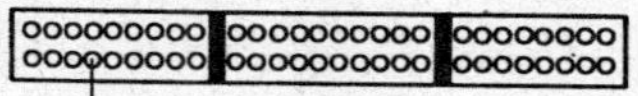

- Check for continuity of YEL wire between ECU (A8) and the 4P connector.
- Does continuity exist ?
  - NO: **Repair open in YEL wire between ECU (A8) and the 4P connector.**
  - YES: **Subsitute a known-good ECU and recheck. If prescribed voltage is now available replace the original ECU.**

- Disconnect the 4P connector.
- Is there vacuum ?
  - YES: **Replace bypass control solenoid valve.**
  - NO:
- Turn the ignition switch OFF.
- Disconnect "A" connector from ECU only, not the main wire harness.
- Check for continuity to ground on the YEL wire between ECU (A8) and 4P connector.
- Is there continuity to ground ?
  - YES: **Repair short to ground at YEL wire between ECU (A8) and the 4P connector.**
  - NO: **Substitute a known-good ECU and recheck. If symptom goes away, replace the original ECU.**

2.0L ENGINE – TROUBLESHOOTING EVAPORATIVE EMISSION CONTROL SYSTEM – PRELUDE

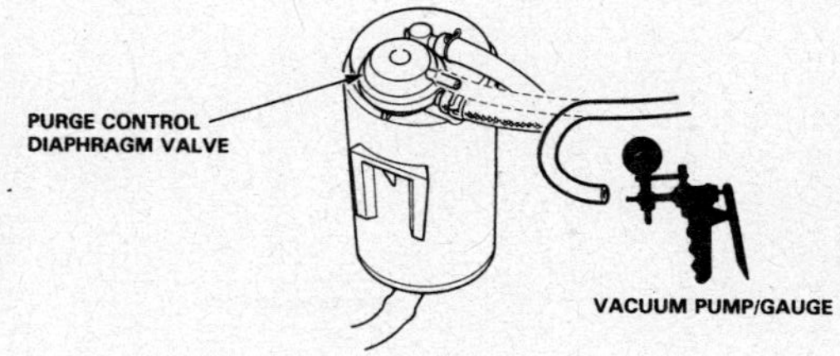

- **Inspection of Evaporative Emission Controls**
- Disconnect vacuum hose from the purge control diaphragm valve (on the charcoal canister) and connect a vacuum gauge to the hose.
- Start the engine and allow to idle. NOTE: Engine coolant temperature must be below 75°C (167°F).
- Is there vacuum ?
  - NO: Go to Path P
  - YES: Disconnect the 4P** or 2P* connector.

2.0ℓ:

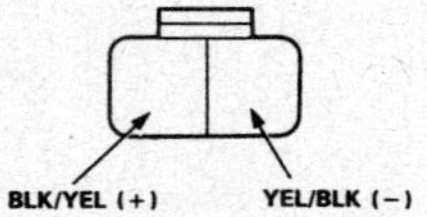

2.1ℓ:

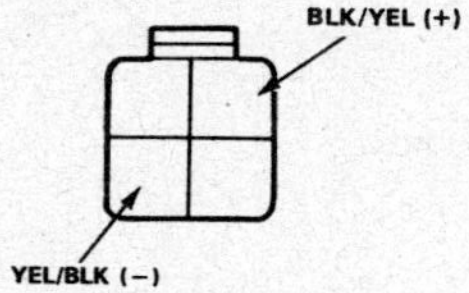

- Measure voltage between BLK/YEL (+) terminal and YEL/BLK (−) terminal.
- Is there battery voltage ?
  - YES: **Inspect vacuum hose routing. If OK, replace purge cut-off solenoid valve.**
  - NO: Measure voltage between BLK/YEL (+) terminal and body ground.
- Go to Path Q

*: 2.0 ℓ
**: 2.1 ℓ

2.0L ENGINE—TROUBLESHOOTING—EVAPORATIVE EMISSION CONTROL SYSTEM—PRELUDE

Path P

Warm up the engine to normal operating temperature (cooling fan comes on).

Check for vacuum at #3 vacuum hose after starting the engine.

Is there manifold vacuum?
- NO → Disconnect the 4P** or 2P* connector.
  - Is there manifold vacuum?
    - NO → **Inspect vacuum hose. If OK, replace purge cut-off solenoid valve.**
    - YES → **Inspect for a short to body ground in YEL/BLK wire between ECU (A20**, B2*) and the connector. If wire is OK, substitute a known-good ECU and recheck. If symptom/indication goes away, replace the original ECU.**
- YES → Reconnect the hose.

Remove fuel filler cap.

Path Q

Is there battery voltage ?
- NO → **Repair open in BLK/YEL wire between No.12 fuse and the 4P** or 2P* connector.**
- YES → **Inspect YEL/BLK wire for on open between ECU (A20**, B2*) and the connector. If wire is OK, substitute a known-good ECU and recheck. If symptom/indication goes away, replace the original ECU.**

*: 2.0 ℓ
**: 2.1 ℓ

Connect a vacuum gauge to canister purge air hose.

Start the engine and raise speed to 3,500 rpm.

Does vacuum appear on gauge within 1 minute ?
- YES → **two-way valve test to complete. Evaporative emission control are OK.**
- NO → Connect a vacuum gauge to the canister purge hose and raise the engine speed to 3,500 rpm.
  - Does vacuum appear on the gauge?
    - NO → **Inspect the purge hose and throttle body port for pinch or blockage.**
    - YES → **Replace the canister.**

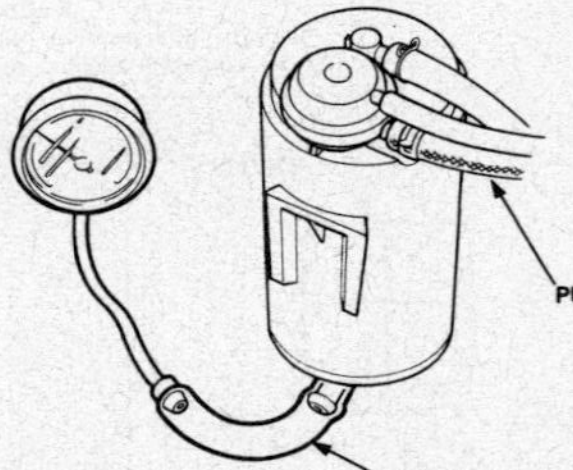

2.0L ENGINE—VACUUM DIAGRAM—PRELUDE

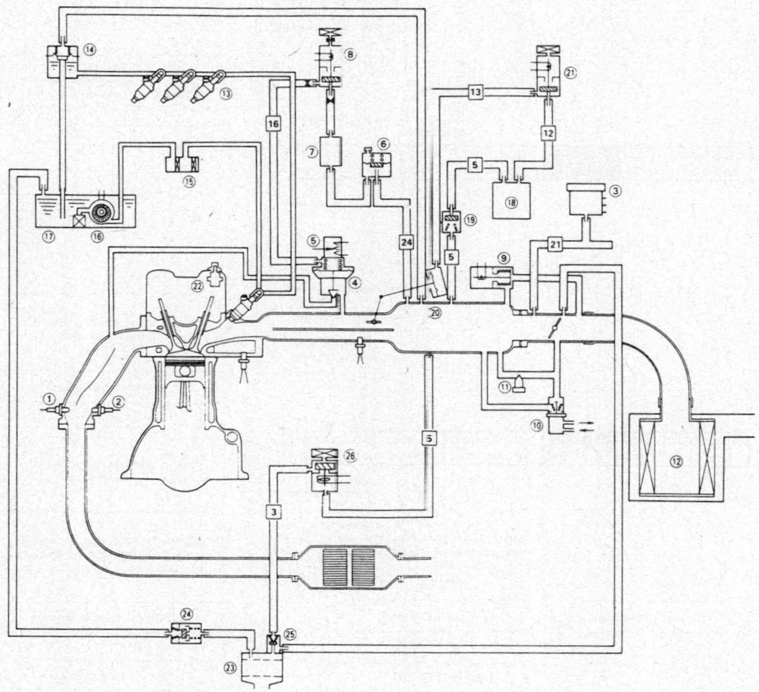

1. OXYGEN (O_2) SENSOR A
2. OXYGEN (O_2) SENSOR B
3. MANIFOLD ABSOLUTE PRESSURE (MAP) SENSOR
4. EGR VALVE
5. EGR VALVE LIFT SENSOR
6. CONSTANT VACUUM CONTROL (CVC) VALVE
7. AIR CHAMBER
8. EGR CONTROL SOLENOID VALVE
9. ELECTRONIC AIR CONTROL VALVE (EACV)
10. FAST IDLE VALVE
11. IDLE ADJUSTING SCREW
12. AIR CLEANER
13. FUEL INJECTOR
14. PRESSURE REGULATOR
15. FUEL FILTER
16. FUEL PUMP
17. FUEL TANK
18. VACUUM TANK
19. CHECK VALVE
20. BYPASS CONTROL DIAPHRAGM
21. BYPASS CONTROL SOLENOID VALVE
22. PCV VALVE
23. CHARCOAL CANISTER
24. TWO-WAY VALVE
25. PURGE CONTROL DIAPHRAGM VALVE
26. PURGE CUT-OFF SOLENOID VALVE

2.0L ENGINE—FUEL SYSTEM ELECTRICAL SCHEMATIC—PRELUDE

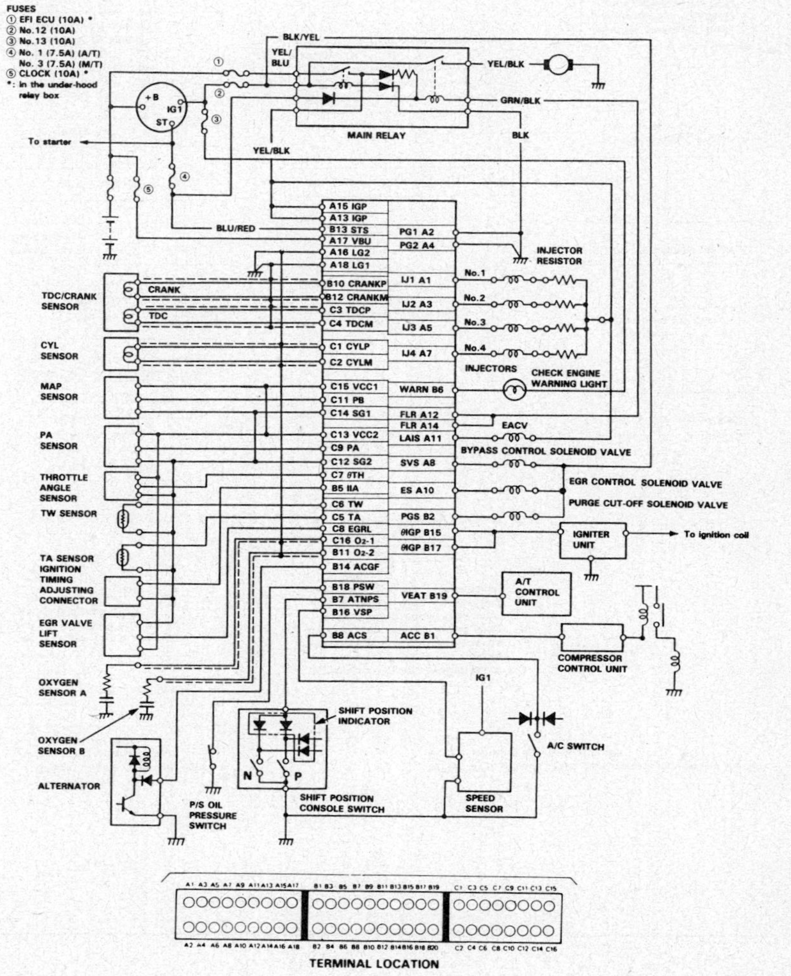

2.1L ENGINE—TROUBLESHOOTING—ECU—PRELUDE

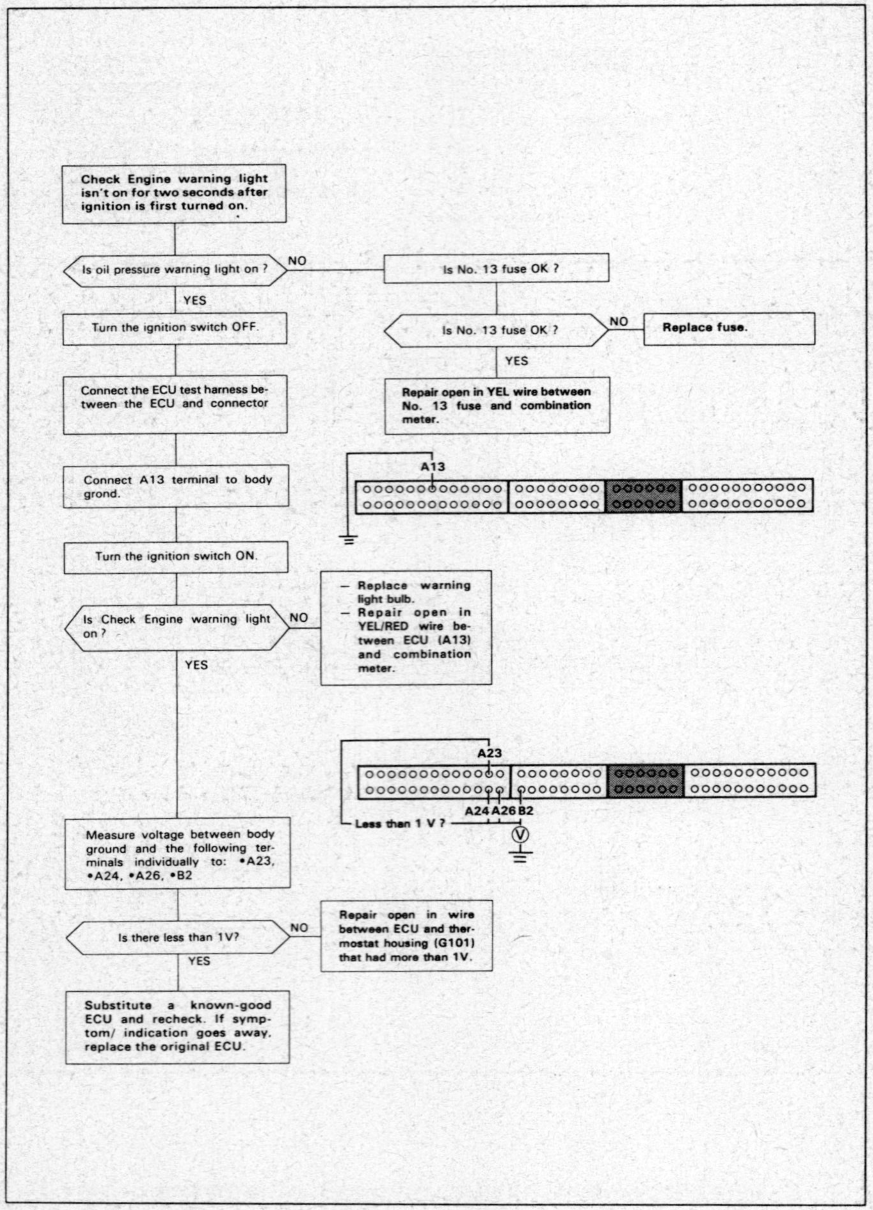

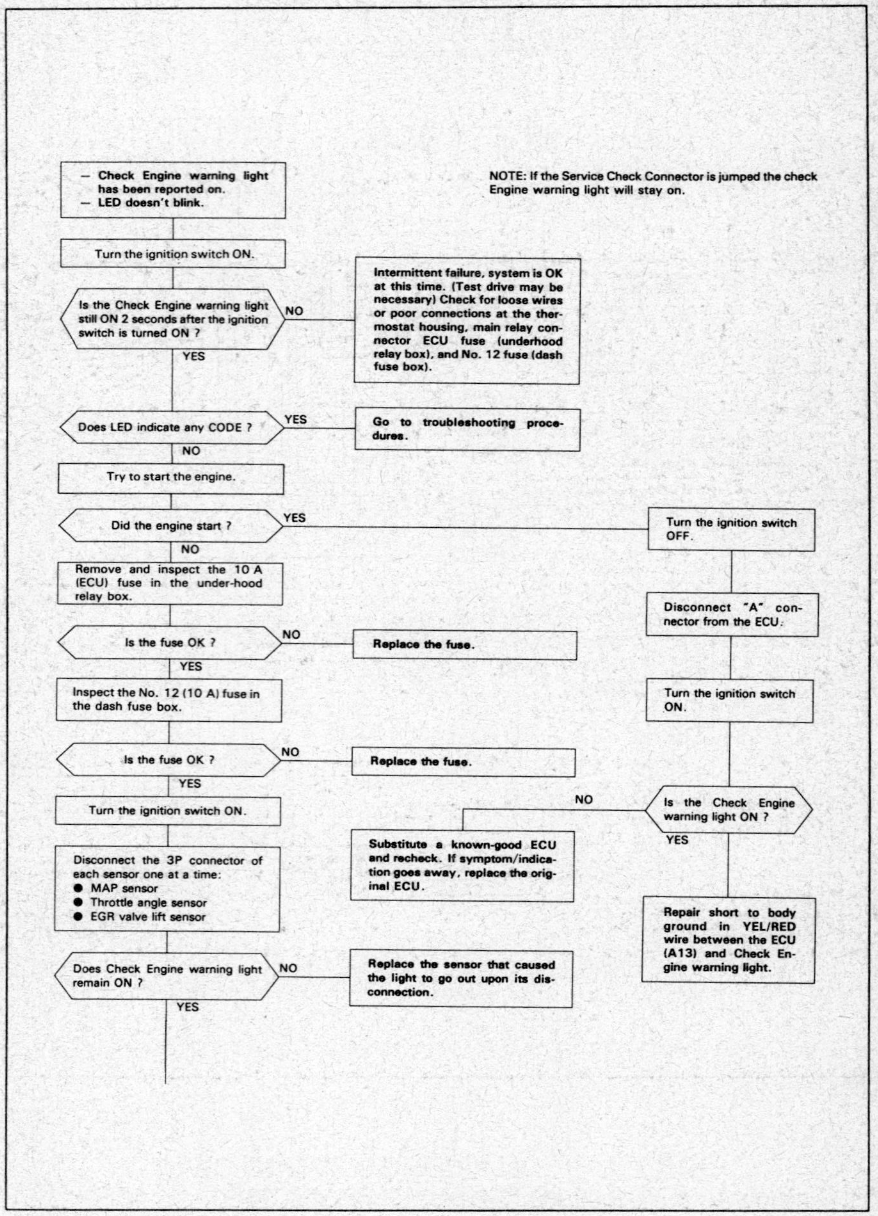

2.1L ENGINE—TROUBLESHOOTING ECU—PRELUDE

- Turn the ignition switch OFF.
- Connect the ECU test harness (page 11-159). Disconnect the "D" connector from the ECU only, not the main wire harness.

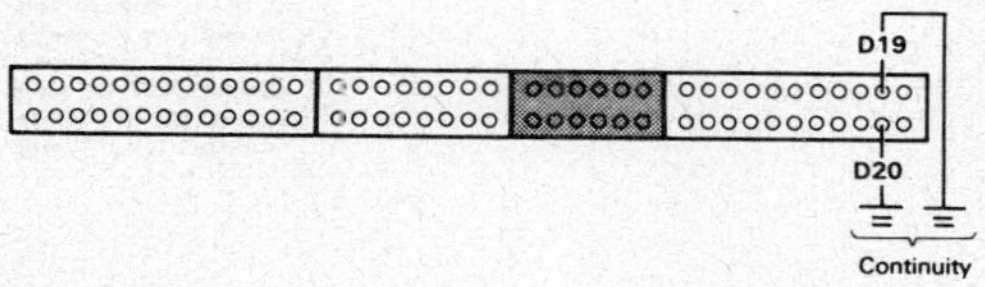

- Check for continuity between body ground and the following terminals; D19, D20.
- Does continuity exist ?
  - YES → Repair short to body ground in RED/WHT wire, YEL/WHT wire between ECU (D19, D20) and throttle angle sensor, EGR valve lift sensor and MAP sensor.
  - NO ↓
- Reconnect all the connectors. Reconnect the "D" connector to the ECU.
- Turn the ignition switch ON.
- Individually connect the following terminal to body ground: B2 • A26
- Is the Check Engine warning light still ON after 2 seconds ?
  - NO → – Repair open in BLK/RED wire between ECL (A26) and G101. – Repair open in BRN/BLK wire between ECU (B2) and G101.
  - YES ↓
- Measure voltage between A26 (–) and the following: B1 (+) and A25 (+).
- Is there battery voltage ?
  - NO → – Repair open in YEL/BLK wire between ECU (A25, B1) and main relay. – Check main relay and wiring connectors at main relay.
  - YES ↓
- Substitute a known-good ECU and recheck. If symptom/indication goes away, replace the original ECU.

2.1L ENGINE—TROUBLESHOOTING CODE 1—PRELUDE

Self-diagnosis LED indicates code 1: A problem in the Heated Oxygen (O_2) Sensor circuit.

The oxygen sensor, detects the oxygen content in the exhaust gas, and inputs the ECU. In operation, the ECU receives the signals from the sensor and varies the duration during which fuel is injected. The oxygen sensor is combined with a heater. The heater stabilizes the sensor output. The oxygen sensor is installed in the exhaust pipe.

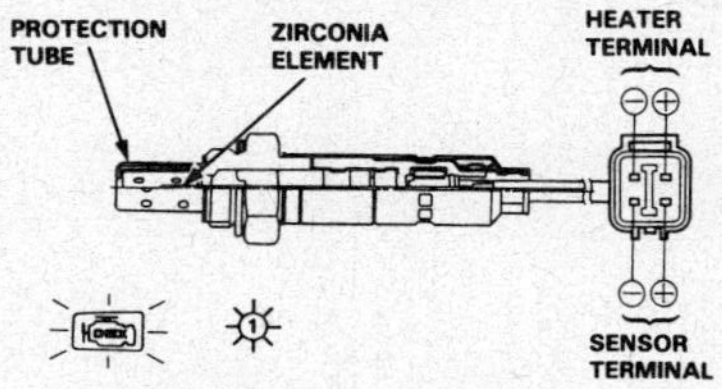

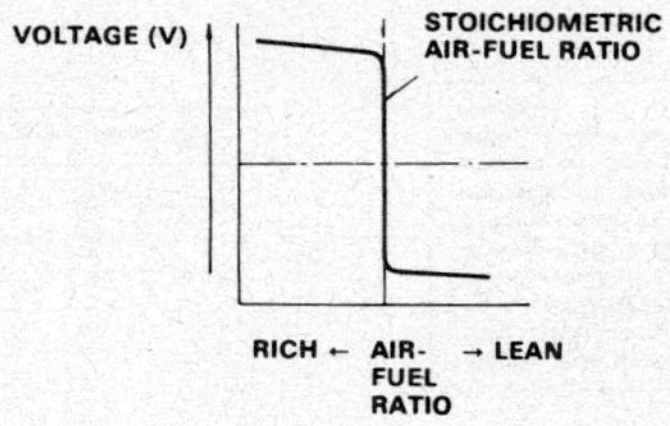

- – Check Engine warning light has been reported on. – LED indicates CODE 1.
- Turn the ignition switch OFF.
- Remove CLOCK fuse in the under-hood relay box for 10 seconds to reset ECU.
- Warm up engine to normal operating temperature (cooling fan comes on).
- Run engine for 60 seconds.
- Road test with the transmission in 2nd gear, accelerate using wide open throttle for at least 5 seconds. Then decelerate for at least 5 seconds with the throttle completely closed.
- Is Check Engine warning light on and does LED indicate CODE 1?
  - NO → Intermittent failure, system is OK at this time. Check for poor connections or loose wires at C236 (located at right shock tower) and at the O_2 sensor.
  - YES ↓
- test for CODE 43.

2.1L ENGINE—TROUBLESHOOTING—CODE 3—PRELUDE

Self-diagnosis LED indicates code 3: Most likely an electrical problem in the Manifold Absolute Pressure (MAP) Sensor system.

The MAP sensor converts manifold absolute pressure into electrical signals and inputs the ECU.

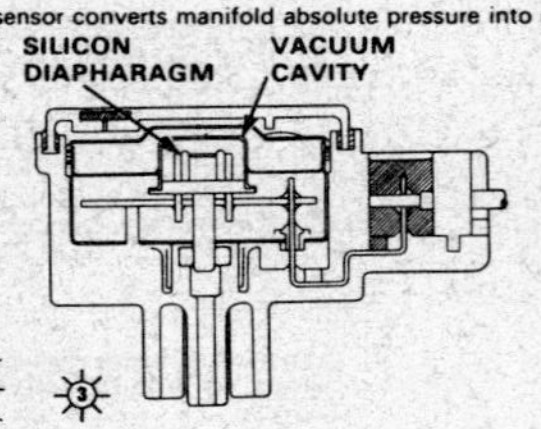

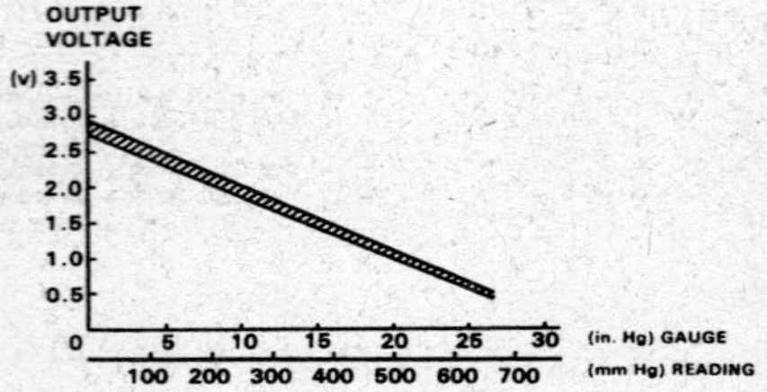

- Engine is warm and running.
- Check Engine warning light has been reported on.
- LED indicates CODE 3.

Turn the ignition switch OFF.

Remove CLOCK fuse in the under-hood relay box for 10 seconds to reset ECU.

Warm up engine to normal operating temperature (cooling fan comes on).

Is Check Engine warning light on and does LED indicate CODE 3?

NO → Intermittent failure, system is OK at this time (test drive may be necessary). Check for poor connection or loose wires at MAP sensor connector and ECU.

YES ↓

Turn the ignition switch OFF.

Disconnect the 3P connector from the MAP sensor.

Turn the ignition switch ON.

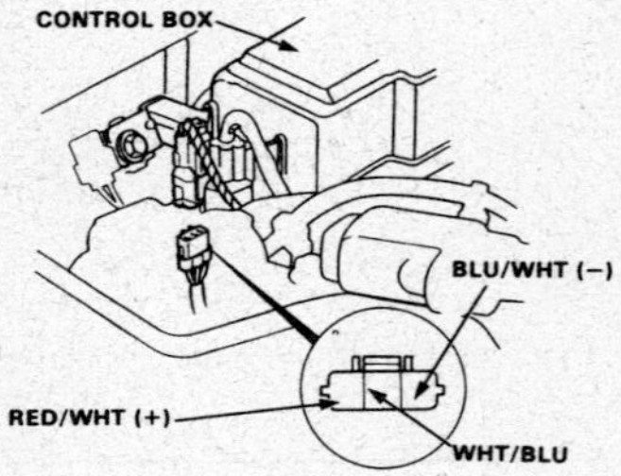

Measure voltage between RED/WHT (+) terminal and body ground.

Is there approx. 5V?

NO → Repair open in RED/WHT wire between ECU (D19) and MAP sensor. If wire is OK, substitute a known-good ECU and recheck. If prescribed voltage is now available, replace the original ECU.

YES ↓

Measure voltage between RED/WHT (+) terminal and BLU/WHT (−) terminal.

Is there approx. 5V?

NO → Repair open in BLU/WHT wire between ECU (D21) and MAP sensor. If wire is OK, substitute a known-good ECU and recheck. If prescribed voltage is now available, replace the original ECU.

YES ↓

Measure voltage between WHT/BLU (+) terminal and BLU/WHT (−) terminal.

Is there approx. 5V?

NO → Repair open or short in WHT/BLU wire between ECU (D17) and MAP sensor. If wire is OK, substitute a known-good ECU and recheck. If prescribed voltage is now available, replace the original ECU.

YES ↓

Turn the ignition switch OFF.

Reconnect the 3P connector to the MAP sensor.

Connect the ECU test harness between the ECU and connector

Turn the ignition switch ON.

2.1L ENGINE—TROUBLESHOOTING CODE 3—PRELUDE

Measure voltage between D17 (+) terminal and D21 (−) terminal.

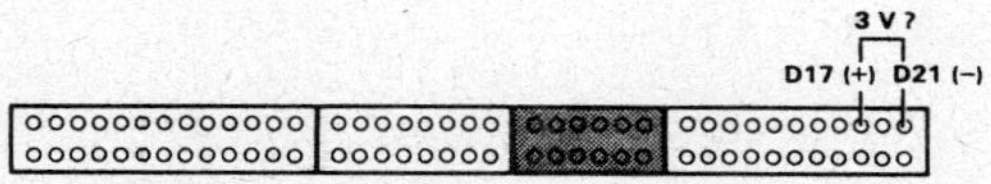

Is there approx. 3 V ?

NO → Replace MAP sensor.

YES → Substitute a known-good ECU and recheck. If symptom/indication goes away, replace the original ECU.

2.1L ENGINE—TROUBLESHOOTING CODE 4—PRELUDE

Self-diagnosis LED indicator blinks four times: A problem in the CRANK circuit of the TDC/CRANK Sensor.

The CRANK sensor determines timing for fuel injection and ignition of each cylinder and also detects engine RPM. The TDC sensor determines ignition timing at start-up (cranking) and when crank angle signal is abnormal.

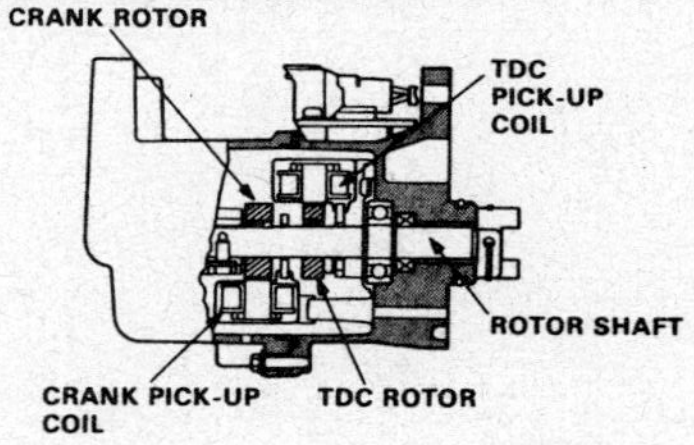

- Check Engine warning light has been reported on.
- LED indicates CODE 4.

Turn the ignition switch OFF.

Remove CLOCK fuse in the under-hood relay box for 10 seconds to reset ECU.

Start the engine.

Is Check Engine warning light on and does LED indicate CODE 4 ?

NO → Intermittent failure, system is OK at this time (test drive may be necessary). Check for poor connections or loose wires at distributor connector and C236 (located at right shock tower).

YES → Stop engine.

Disconnect the 4P connector from the TDC/CRANK sensor.

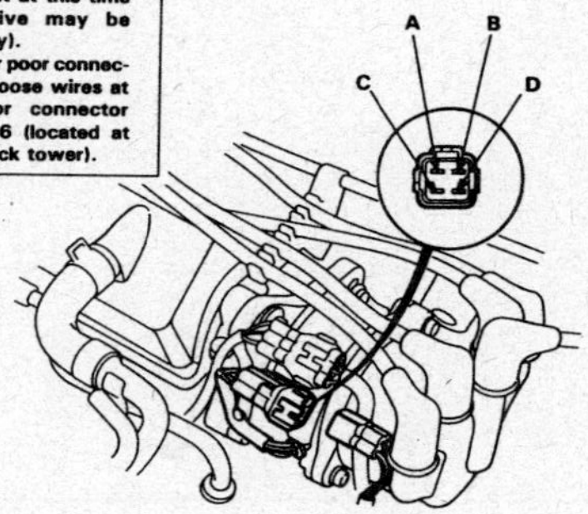

2.1L ENGINE—TROUBLESHOOTING CODE 4—PRELUDE

- Measure resistance between C terminal and D terminal.
- Is there 700–1,000 Ω ?
  - NO: Replace pulse generator assembly
  - YES: Check for continuity to body ground on C terminal and D terminal individually.
- Does continuity exist ?
  - YES: Replace pulse generator assembly
  - NO: Reconnect the connector.
- Connect the ECU test harness only to the main wire harness, but not to the ECU
- Measure resistance between B15 terminal and B16 terminal.

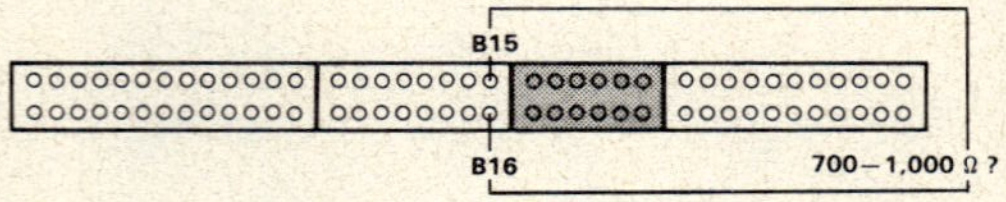

- Is there 700–1,000 Ω ?
  - NO: Repair open in BLU/YEL and/or BLU/GRN wires.
  - YES: Check for continuity to body ground on B15 terminal.
- Does continuity exist?
  - YES: Repair short in BLU/GRN wire between ECU (B15) and distributor connector.
  - NO: Substitute a known-good ECU and recheck. If symptom/indication goes away, replace the original ECU.

2.1L ENGINE—TROUBLESHOOTING CODE 5—PRELUDE

- – Check Engine warning light has been reported on.
  – LED indicates CODE 5.
- Turn the ignition switch OFF.
- Remove CLOCK fuse in the under-hood relay box for 10 seconds to reset ECU.
- Start the engine.
- Is Check Engine warning light on and does LED indicate CODE 5?
  - NO:
    - —Intermittent failure, system is OK at this time (test drive may be necessary).
    - —Check vacuum hoses, pipes and connections.
    - —Make sure all connectors are secure.
  - YES: Stop the engine.
- Disconnect #21 hose from the throttle body, connect vacuum pump to the hose and apply vacuum.

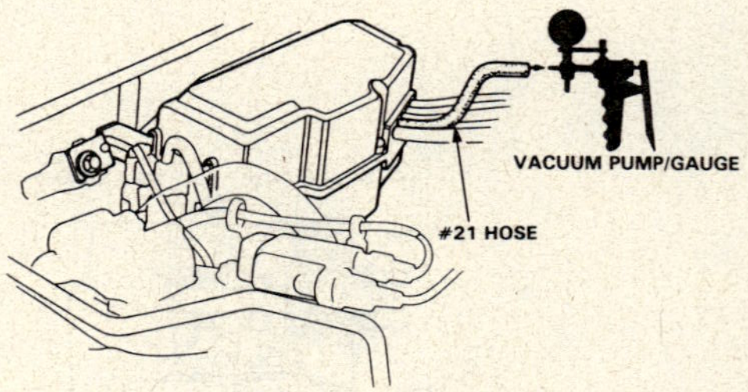

- Does it hold vacuum?
  - NO: Connect a vacuum pump to the MAP sensor and apply vacuum.
    - Does it hold vacuum?
      - NO: Replace MAP sensor.
      - YES: Replace #21 hose.
  - YES: Connect a T-fitting from a vacuum gauge between the throttle body and MAP sensor.

2.1L ENGINE—TROUBLESHOOTING CODE 5—PRELUDE

1. Start the engine.
2. Is there manifold vacuum?
   - NO: -Remove restriction from throttle body. -Replace throttle body.
   - YES: continue.
3. Stop the engine.
4. Connect the ECU test harness between the ECU and connector
5. Turn the ignition switch ON.
6. Measure voltage between D17 (+) terminal and D21 (−) terminal.

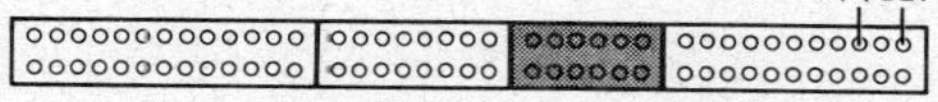

7. Is there approx. 3 V ?
   - NO: Replace the MAP sensor.
   - YES: continue.
8. Start the engine and allow it to idle.
9. Is there approx. 1 V?
   - NO: Replace MAP sensor.
   - YES: Substitute a known-good ECU and recheck. If symptom/indication goes away, replace the original ECU.

2.1L ENGINE—TROUBLESHOOTING CODE 6—PRELUDE

Self-diagnosis LED indicates code 6: Most likely a problem in the Coolant Temperature (TW) Sensor circuit.

The TW sensor is a temperature dependant resistor (thermistor). The resistance of the thermistor decreases as the coolant temperature increases as shown below.

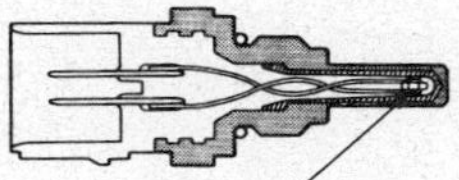

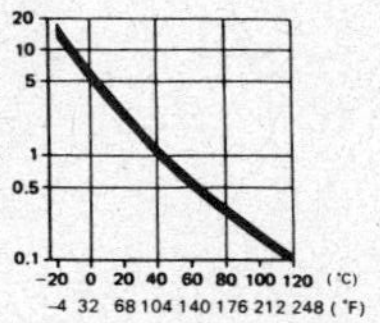

1. – Check Engine warning light has been reported on. – LED indicates CODE 6.
2. Turn the ignition switch OFF.
3. Remove CLOCK fuse in the under-hood relay box for 10 seconds to reset ECU.
4. Turn the ignition switch ON.
5. Is Check Engine warning light on and does LED indicate CODE 6?
   - NO: Intermittent failure, system is OK at this time (test drive may be necessary). Check for poor connections or loose wires at TW sensor and C235 (located at right shock tower).
   - YES: continue.
6. Warm up engine to normal operating temperature (cooling fan comes on).
7. Turn the ignition switch OFF.
8. Disconnect the 2P connector from the TW sensor.
9. Measure resistance between the 2 terminals on the TW sensor.
10. Is there 200—400 Ω ?
    - NO: Replace TW sensor.
    - YES

2.1L ENGINE—TROUBLESHOOTING CODE 6—PRELUDE

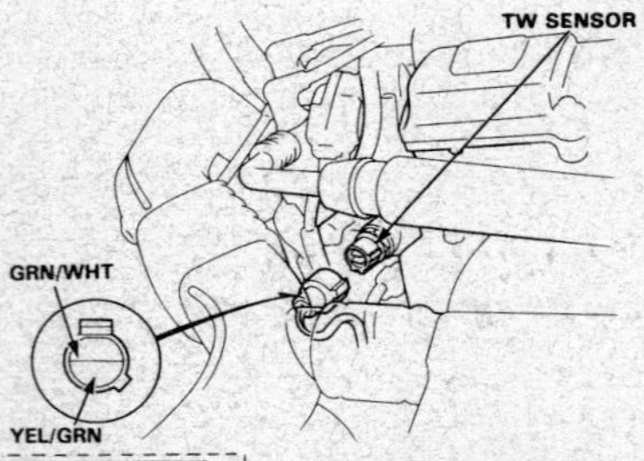

Turn the ignition switch ON.

Measure voltage between YEL/GRN and body ground.

Is there approx. 5V ?

- NO → A/T only:
  - Disconnect the 18P connector from the A/T control unit.
  - Is there approx. 5 V ?
    - YES → Replace the A/T control unit.
    - NO → Repair open or short in YEL/GRN wire between ECU (D13); A/T control unit and TW sensor. If wire is OK, substitute a known-good ECU and recheck. If prescribed voltage is now available, replace original ECU.
- YES ↓

Measure voltage between YEL/GRN (+) terminal and GRN/WHT (−) terminal.

Is there approx. 5V ?

- NO → Repair open in GRN/WHT wire between ECU (D22) and TW sensor.
- YES → Substitute a known-good ECU and recheck. If symptom/indication goes away, replace the original ECU.

2.1L ENGINE—TROUBLESHOOTING CODE 7—PRELUDE

Self-diagnosis LED indicates code 7: Most likely a problem in the Throttle Angle Sensor circuit.

The throttle angle sensor is a potentiometer. It is connected to the throttle valve shaft. As the throttle angle changes, the throttle angle sensor varies the voltage signal to the ECU.

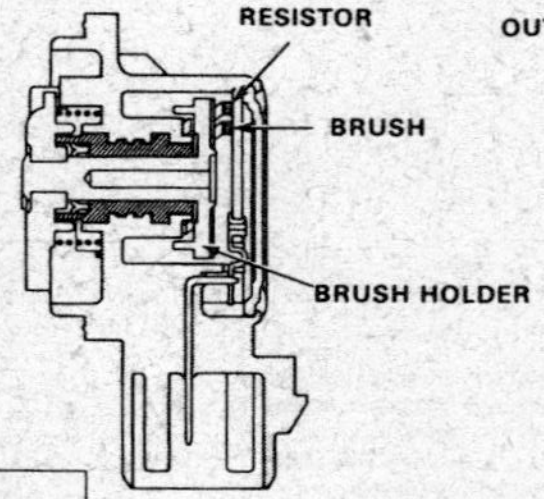

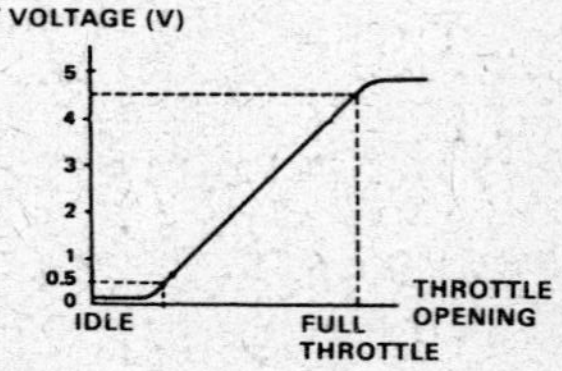

- Engine is running.
- Check Engine warning light has been reported on.
- LED indicates CODE 7.

Turn the ignition switch OFF.

Remove CLOCK fuse in the under-hood relay box for 10 seconds to reset ECU.

Start the engine.

Is Check Engine warning light on and does LED indicate CODE 7?

- NO → **Intermittent failure, system is OK at this time (test drive may be necessary). Check for poor connections or loose wires at throttle angle sensor and C235 (located at right shock tower).**
- YES ↓

Turn the ignition switch OFF.

Disconnect the 3P connector from the throttle angle sensor.

Turn the ignition switch ON.

Measure voltage between YEL/WHT (+) terminal and GRN/WHT (−) terminal.

Is there approx. 5V ?

- YES → Go to Path C
- NO → Measure voltage between YEL/WHT (+) terminal and body ground. → Go to Path D

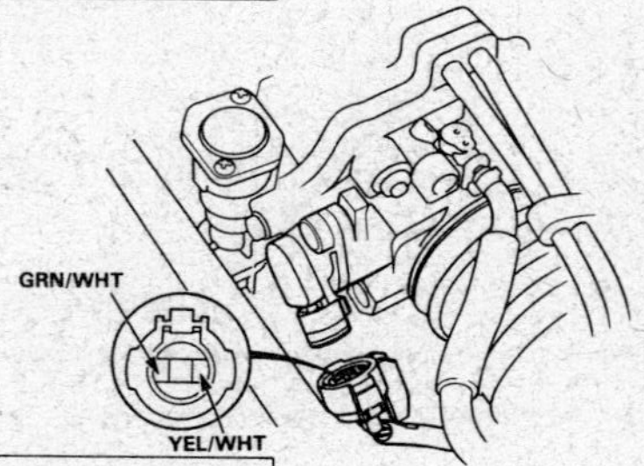

2.1L ENGINE—TROUBLESHOOTING CODE 7—PRELUDE

Path C

Turn the ignition switch OFF.

Reconnect the 3P connector.

Connect the ECU test harness between the ECU and connector

Turn the ignition switch ON.

Measure voltage between D11(+) terminal and D22 (−) terminal.

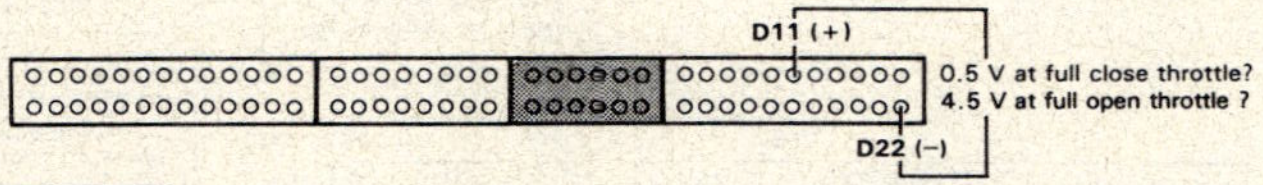

Is voltage approx. 0.5 V at full close throttle, and approx. 4.5 V at full open throttle?
NOTE: There should be a smooth transition from 0.5 V to 4.5 V as the throttle is depressed.

- YES → **Substitute a known-good ECU and recheck. If symptom/indication goes away, replace the original ECU.**
- NO → **A/T only**: Disconnect the 18P connector from the A/T control unit.

Is voltage approx. 0.5 V at full close throttle, and approx. 4.5 V at full open throttle?
NOTE: There should be a smooth transition from 0.5 V to 4.5 V as the throttle is depressed.

- YES → **Replace the A/T control unit.**
- NO → **– Replace throttle angle sensor. – Repair open or short in RED/YEL wire between ECU (D11), A/T control unit and throttle angle sensor.**

Path D

Is there approx. 5V ?

- YES → **Repair open in GRN/WHT wire between ECU (D22) and throttle angle sensor.**
- NO ↓

Turn the ignition switch OFF.

Connect the ECU test harness between the ECU and connector

Turn the ignition switch ON.

Measure voltage between D20 (+) terinal and D22 (−) terminal.

Is there approx. 5V ?

- YES → **Repair open in YEL/WHT wire between ECU (D20) and throttle angle sensor.**
- NO → **Substitute a known-good ECU and recheck. If prescribed voltage is now available, replace the original ECU.**

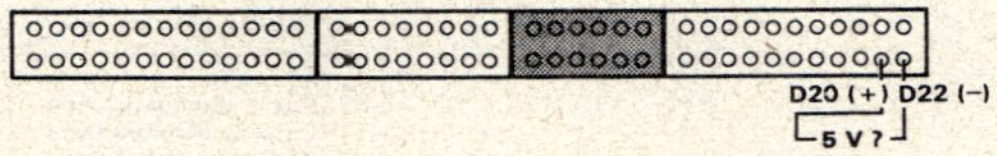

2.1L ENGINE—TROUBLESHOOTING CODE 8—PRELUDE

– Check Engine warning light has been reported on.
– LED indicates CODE 8.

Turn the ignition switch OFF.

Remove CLOCK fuse in the under-hood relay box for 10 seconds to reset ECU.

Start engine.

Is Check Engine warning light on and does LED indicate CODE 8 ?

- NO → **Intermittent failure, system is OK at this time (test drive may be necessary). Check for poor connections or loose wires at distributor connector and C236 (located at right shock tower).**
- YES ↓

Stop engine.

Disconnect the 4P connector from the TDC/CRANK sensor.

Measure resistance between A terminal and B terminal.

Is there 700—1,000 Ω ?

- NO → **Replace pulse generator assembly**
- YES ↓

Check for continuity to body ground on A terminal and B terminal individually.

Does continuity exist ?

- YES → **Replace pulse generator assembly**
- NO ↓

Reconnect the connector.

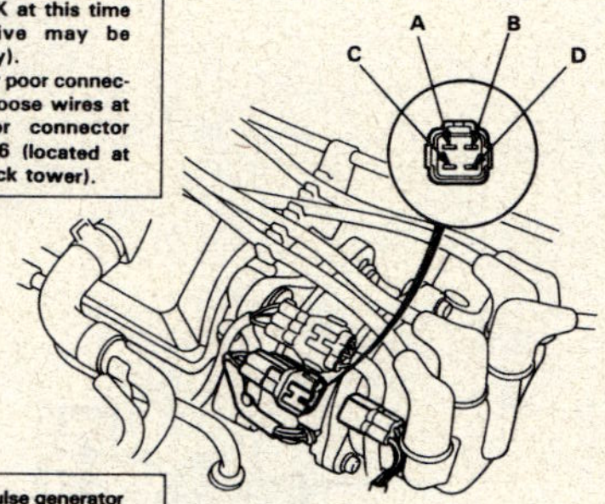

2.1L ENGINE—TROUBLESHOOTING CODE 8—PRELUDE

Connect the ECU test harness only to the main wire harness, but not to the ECU

Measure resistance between B13 terminal and B14 terminal.

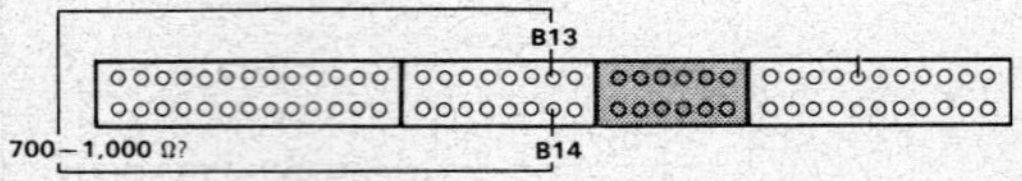

Is there 700 — 1,000 Ω ?

- NO: Repair open in ORN/BLU and/or WHT/BLU wires between the ECU and distributor connector.
- YES: Check for continuity to body ground on B13 terminal.

Does continuity exist?

- YES: Repair short in ORN/BLU wire between ECU (B13) and distributor connector.
- NO: Substitute a known-good ECU and recheck. If symptom/indication goes away, replace the original ECU.

2.1L ENGINE—TROUBLESHOOTING CODE 9—PRELUDE

Self-diagnosis LED indicates code 15: A problem in the CYL sensor.

The CYL sensor detects the position of NO.1 cylinder for sequential fuel injection to each cylinder.

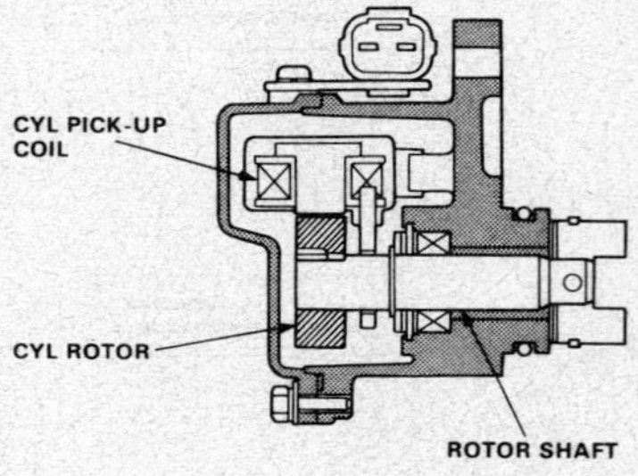

- Check Engine warning light has been reported on.
- LED indicates CODE 9.

Turn the ignition switch OFF.

Remove CLOCK fuse in the under-hood relay box for 10 seconds to reset ECU.

Start the engine.

Is Check Engine warning light on and does LED indicate CODE 9 ?

- NO: Intermittent failure, system is OK at this time (test drive may be necessary). Check for poor connections or loose wires at CYL Sensor connector and C236 (located at right shock tower).
- YES: Stop the engine.

Disconnect the 2P connector from the CYL sensor.

Measure resistance between 2 terminals on the CYL sensor.

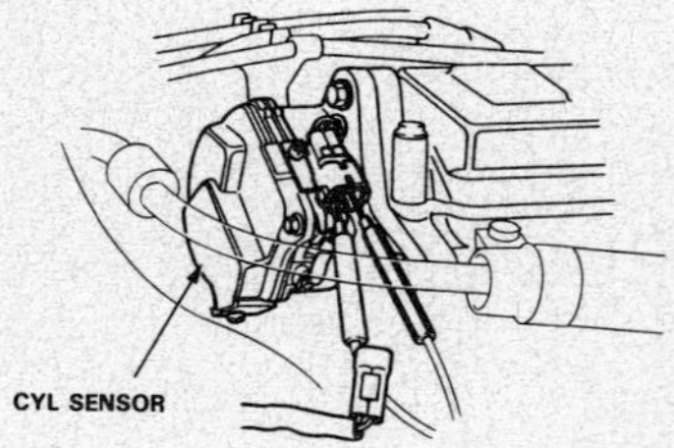

2.1L ENGINE—TROUBLESHOOTING CODE 9—PRELUDE

- Is there 700–1,100 Ω?
  - NO → **Replace CYL sensor assembly**
  - YES → Check for continuity to body ground on the 2 terminals individually.
- Does continuity exist ?
  - YES → **Replace CYL sensor assembly**
  - NO → Reconnect the connector.
- Connect the PGM-FI test harness only to the main wire harness, but not to the ECU
- Measure resistance between B11 terminal and B12 terminal.

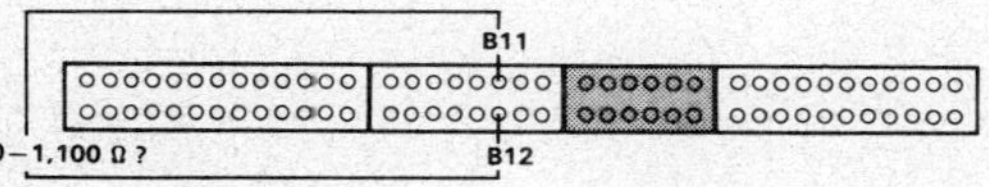

- Is there 700–1,100 Ω ?
  - NO → **Repair open in ORN and/or WHT wires.**
  - YES → Check for continuity to body ground on B11 terminal.
- Does continuity exist?
  - YES → **Repair short in ORN wire between ECU (B11) and CYL Sensor connector.**
  - NO → **Substitute a known-good ECU and recheck. If symptom/indication goes away, replace the original ECU.**

2.1L ENGINE—TROUBLESHOOTING CODE 10—PRELUDE

Self-diagnosis LED indicates code 10: Most likely a problem in the Intake Air Temperature (TA) Sensor circuit.

The TA sensor is a temperature dependant resistor (thermistor). The resistance of the thermistor decreases as the intake air temperature increases as shown below.

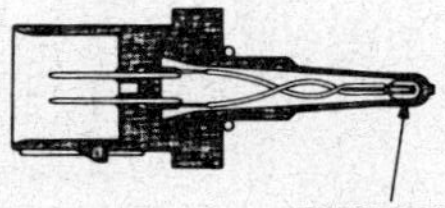

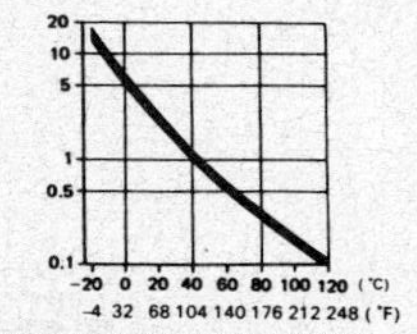

- **Check Engine warning light has beer reported on.**
- **LED indicates CODE 10.**

1. Turn the ignition switch OFF.
2. Remove CLOCK fuse in the under-hood relay box for 10 seconds to reset ECU.
3. Turn the ignition switch ON.
4. Is Check Engine warning light on and does LED indicate CODE 10?
   - NO → **Intermittent failure, system is OK at this time (test drive may be necessary). Check for poor connections or loose wires at TA sensor and C235 (located at right shock tower).**
   - YES → Turn the ignition switch OFF.
5. Disconnect the 2P connector from the TA sensor.
6. Measure resistance between the 2 terminals on the TA sensor.
7. Is there 1–4 kΩ ?
   - NO → **Replace TA sensor.**
   - YES

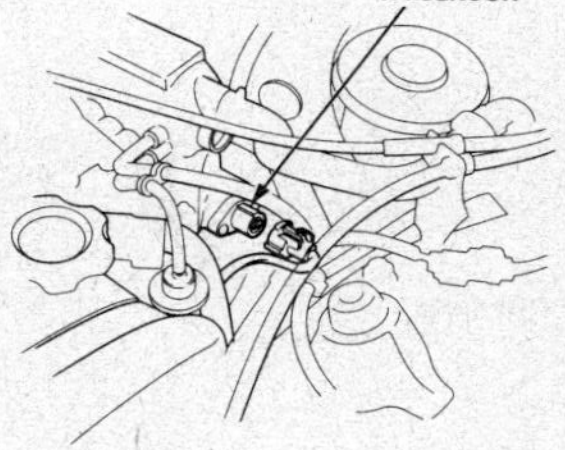

2.1L ENGINE—TROUBLESHOOTING CODE 10—PRELUDE

- Turn the ignition switch ON.
- Measure voltage between WHT/RED (+) terminal and body ground.
- Is there approx. 5 V ?
  - NO → Repair open or short in WHT/RED wire between ECU (D15) and TA sensor. If wire is OK, substitute a known-good ECU and recheck. If prescribed voltage is now available, replace original ECU.
  - YES → Measure voltage between WHT/RED (+) terminal and GRN/WHT (−)terminal.
- Is there approx. 5 V ?
  - NO → Repair open in GRN/WHT wire between ECU (D22) and TA sensor.
  - YES → Substitute a known-good ECU and recheck. If symptom/indication goes away, replace the original ECU.

2.1L ENGINE—TROUBLESHOOTING CODE 12—PRELUDE

- Check Engine warning light has been reported on.
- LED indicates CODE 12.

Turn the ignition switch OFF.

Remove CLOCK fuse in the under-hood relay box for 10 seconds to reset ECU.

Road test necessary: Warm up the engine to normal operating temperature (cooling fan comes on). Drive the car on the road for approx. 10 minutes. Try to keep the engine speed in the 1700— 2500 range.

Is Check Engine warning light on and does LED indicated CODE 12 ?

- NO → Intermittent failure, system is OK at this time. Check for poor connections or loose wires at EGR valve C235, C236 (located at right shock tower), control box.
- YES → With the engine at idle, disconnect the #16 hose from the EGR valve and connect a vacuum pump/gauge to the hose.

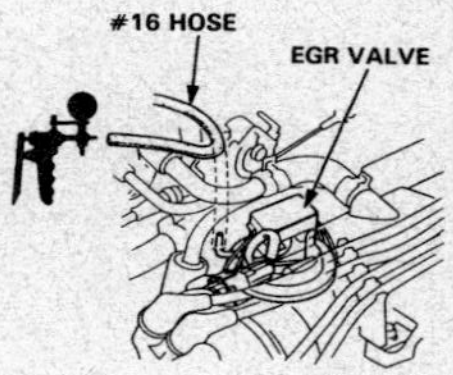

Is there any vacuum ?

- YES → Disconnect 4P connector from the control box and check the #16 hose for vacuum again. → Go to Path O
- NO → Move the vacuum pump/gauge to the EGR valve. → Go to Path N

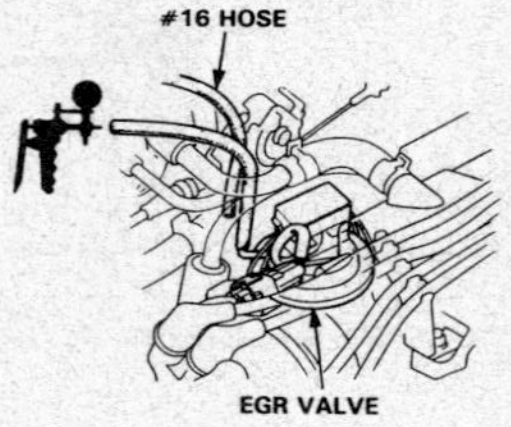

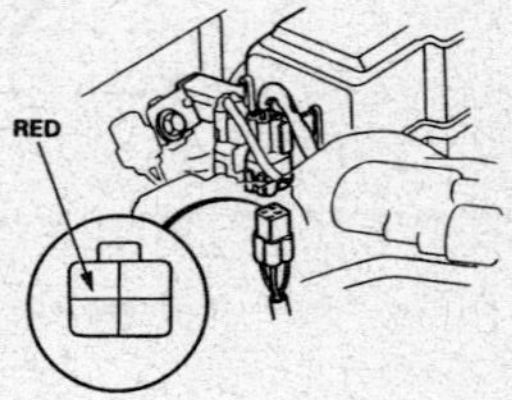

2.1L ENGINE—TROUBLESHOOTING—CODE 12—PRELUDE

Path N

- With the engine at idle, apply 8" of vacuum to the EGR valve.
- Does the engine stall and does the EGR valve hold vacuum ?
  - NO: **Replace EGR valve.**
  - YES:
- Disconnect the 4P connector from the control box.
- Measure voltage between BLK/YEL (+) terminal on the main wire harness and body ground.

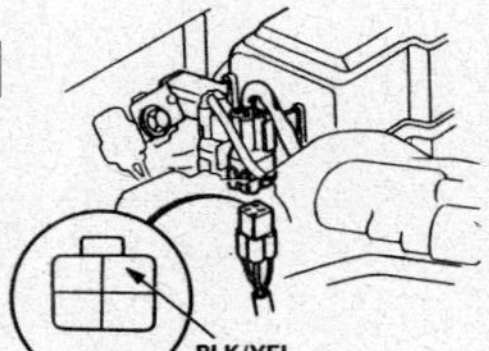

- Is there battery voltage ?
  - NO: **Repair open in BLK/YEL wire between the solenoid valve and No.12 fuse.**
  - YES:
- Reconnect the vacuum pump/gauge to the #16 hose.
- Start the engine and allow it to idle.

Path O

- Is there any vacuum ?
  - YES: **Check vacuum hose routing of the entire EGR system. If hose routing is OK, replace EGR control solenoid valve.**
  - NO: **Inspect for short to body ground in RED wire between EGR control solenoid valve and ECU (A11). If wire is OK, substitute a known-good ECU and retest. If symptom/indication goes away, replace the original ECU.**

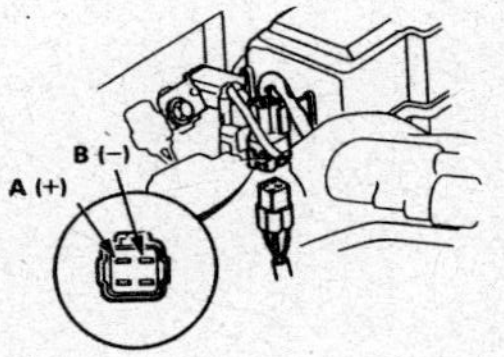

- Connect the battery positive terminal to the B terminal of the 4P connector. While watching the vacuum gauge, connect the battery negative terminal to the D terminal.
- Is there approx. 8" within 1 second?
  - NO: Turn the ignition switch OFF and inspect the #16 and #24 hoses for leaks, restrictions, or mis-routing.
    - Are the hoses OK?
      - NO: **Connect as necessary.**
      - YES:

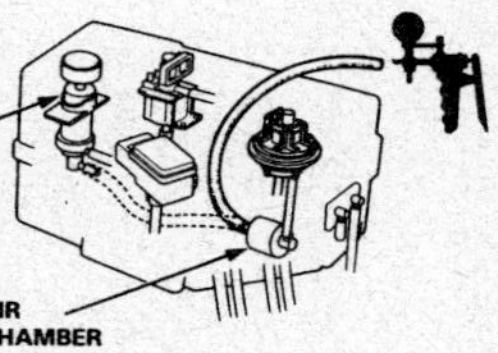

    - Disconnect the lower hose on EGR control solenoid valve and connect a vacuum gauge to the hose.
    - Start the engine and allow it to idle.
    - Is there 6"— 10" of vacuum ?
      - NO: **Replace CVC valve.**
      - YES: **Replace the EGR control solenoid valve.**
  - YES:
- Turn the ignition switch OFF and reconnect the 4P connector.

- Connect the test harness between the EGR valve lift sensor and engine wire harness.
- Turn the ignition switch ON.
- Measure voltage between RED (+) terminal and GRN (−) terminal.
- Is there approx. 5 V ?
  - NO:
    - **Repair open in YEL/WHT wire between EGR valve and ECU (D20).**
    - **Repair open in GRN/WHT wire between EGR valve and ECU (D22).**
  - YES

2.1L ENGINE—TROUBLESHOOTING—CODE 12—PRELUDE

Measure voltage between WHT (+) terminal and GRN (−) terminal.

Is there approx 1.2 V ?

- NO →
  - Replace EGR valve.
  - Repair short in WHT/BLK or YEL wire between ECU (D12) and EGR valve lift sensor.
  - Substitute a known-good ECU and recheck. If prescribed voltage is now available, replace the original ECU.
- YES ↓

While watching the voltmeter, slowly apply a continuous 8" of vacuum to the EGR valve. Repeat several times, completely releasing vacuum between.

Is the maximum voltage approx. 4 V ?

- NO → Replace EGR valve.
- YES ↓

Does the voltage consistently increase/decrease as the vacuum increases/decreases ?

- NO → Replace EGR valve.
- YES ↓

Reconnect the #16 hose to the EGR valve.

Turn the ignition switch OFF.

Connect the ECU test harness between the ECU and connector

Start the engine and allow it to idle.

Measure voltage between D12 (+) terminal and D22 (−) terminal.

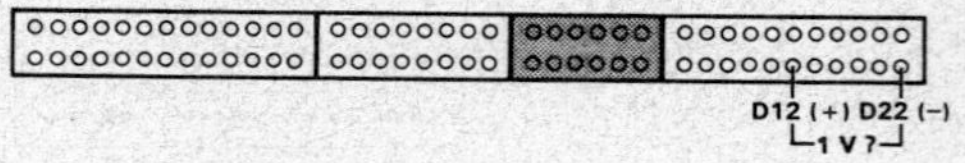

Is there approx. 1.2 V ?

- NO → Repair open in WHT/BLK or YEL wire between ECU (D12) and the sensor.
- YES ↓

Connect A11 terminal to A26 terminal with a jumper wire.

Did the engine stall ?

- NO → Repair open in RED wire between ECU (A11) and EGR control solenoid valve.
- YES ↓

Substitute a known-good ECU and recheck. If symptom/indication goes away, replace the original ECU.

2.1L ENGINE—TROUBLESHOOTING CODE 13—PRELUDE

Self-diagnosis LED indicates code 13: A problem in the Atmospheric Pressure (PA) Sensor.

The PA sensor is built into the ECU.

- Check Engine warning light has been reported on.
- LED indicates CODE 13.

Turn the ignition switch OFF.

Remove CLOCK fuse in the under-hood relay box for 10 seconds to reset ECU.

Turn the ignition switch ON.

Is Check Engine warning light on and does LED indicate CODE 13 ?

NO: Intermittent failure, system is OK at this time (test drive may be necessary).

YES: Substitute a known-good ECU and recheck. If symptom/indication goes away, replace the original ECU.

2.1L ENGINE—TROUBLESHOOTING CODE 14—PRELUDE

Self-diagnosis LED indicates code 14: A problem in the Electronic Air Control Valve (EACV) circuit.

The EACV changes the amount of air bypassing the throttle body in response to a current signal from the ECU in order to maintain the proper idle speed.

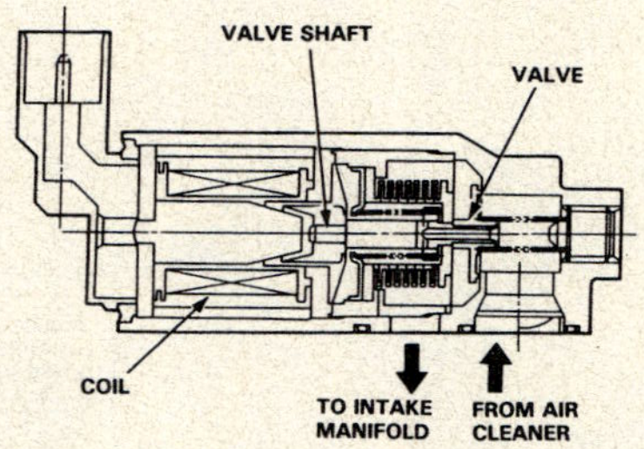

- Engine is running.
- Check Engine warning light has been reported on.
- LED indicates CODE 14.

Turn the ignition switch OFF.

Remove CLOCK fuse in the under-hood relay box for 10 seconds to reset ECU.

Start the engine.

Is Check Engine warning light on and does LED indicate CODE 14?

NO: Intermittent failure, system is OK at this time (test driving may be necessary). Check for poor connections or loose wires at EACV and at C235.

YES: Stop the engine.

Disconnect the 2P connector from the EACV.

Measure resistance between the 2 terminals on the EACV.

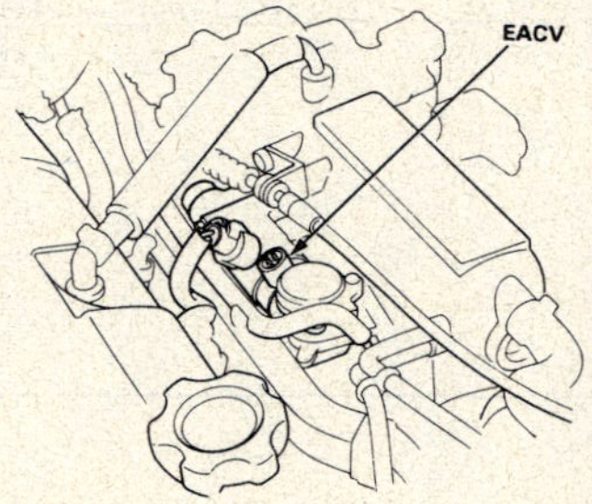

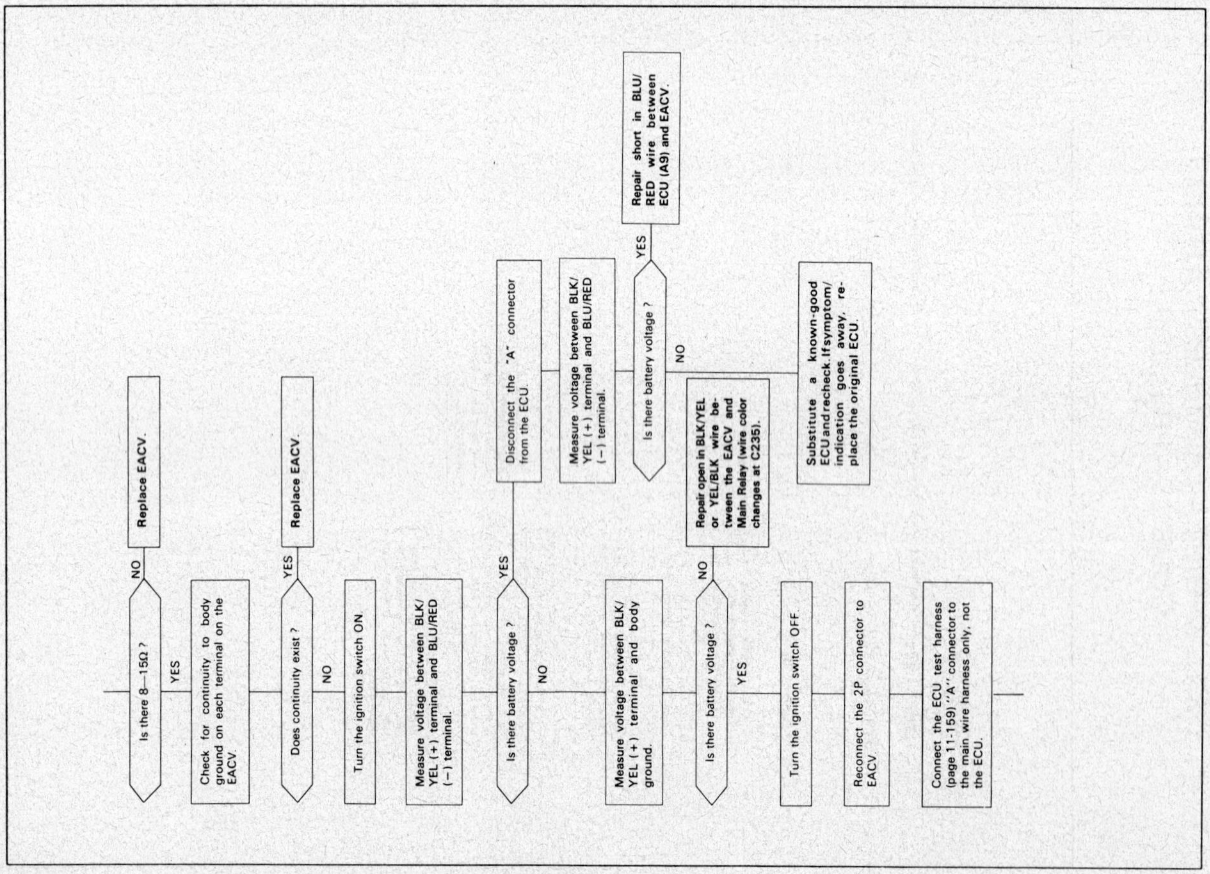
Is there 8–15Ω ?
NO
Replace EACV.
YES
Check for continuity to body ground on each terminal on the EACV.
Does continuity exist ?
YES
Replace EACV.
NO
Turn the ignition switch ON.
Measure voltage between BLK/YEL (+) terminal and BLU/RED (−) terminal.
Is there battery voltage ?
YES
Disconnect the "A" connector from the ECU.
Measure voltage between BLK/YEL (+) terminal and BLU/RED (−) terminal.
Is there battery voltage ?
YES
Repair short in BLU/RED wire between ECU (A9) and EACV.
NO
Substitute a known-good ECU and recheck. If symptom/indication goes away, replace the original ECU.
NO
Measure voltage between BLK/YEL (+) terminal and body ground.
Is there battery voltage ?
NO
Repair open in BLK/YEL or YEL/BLK wire between the EACV and Main Relay (wire color changes at C235).
YES
Turn the ignition switch OFF.
Reconnect the 2P connector to EACV.
Connect the ECU test harness (page 11-159) "A" connector to the main wire harness only, not the ECU.

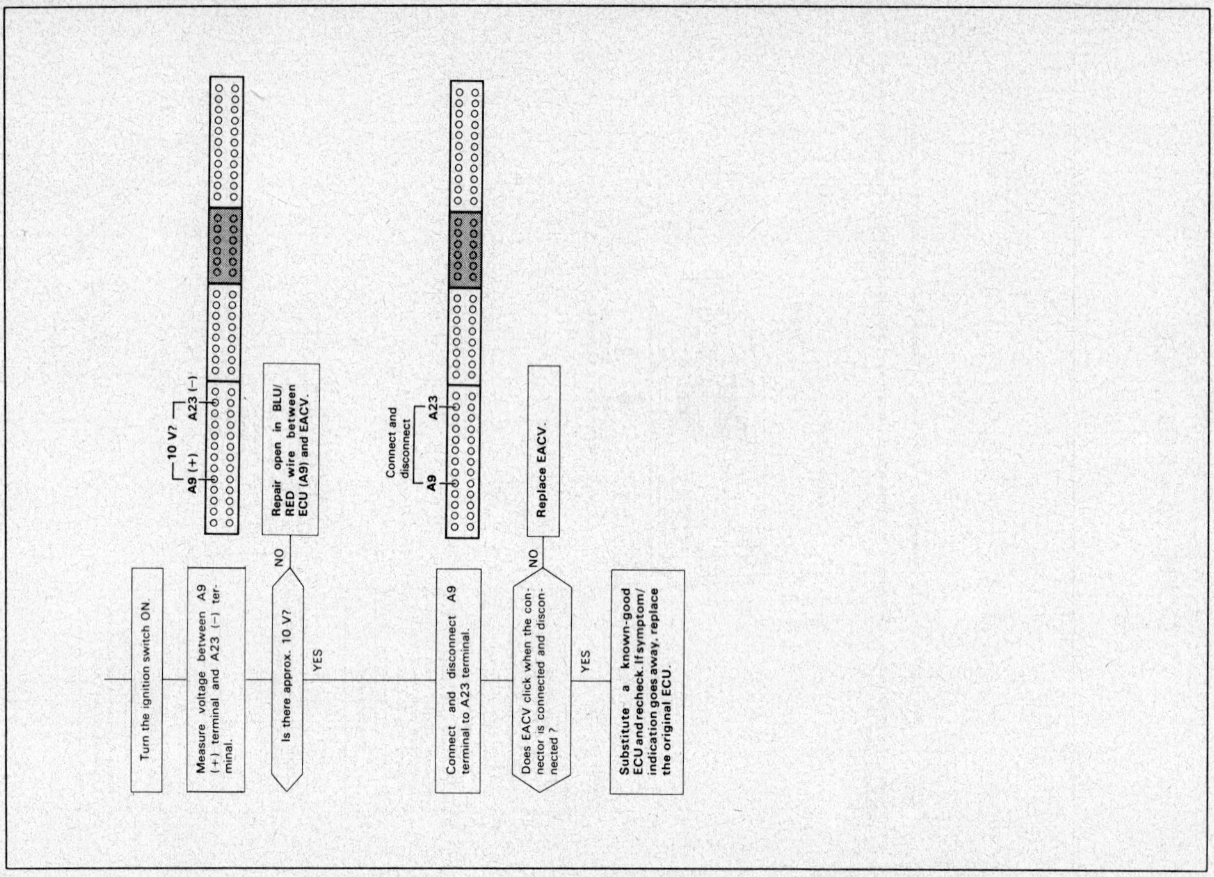
Turn the ignition switch ON.
Measure voltage between A9 (+) terminal and A23 (−) terminal.
10 V?
A9 (+)
A23 (−)
Is there approx. 10 V?
NO
Repair open in BLU/RED wire between ECU (A9) and EACV.
YES
Connect and disconnect A9 terminal to A23 terminal.
Connect and disconnect
A9
A23
Does EACV click when the connector is connected and disconnected ?
NO
Replace EACV.
YES
Substitute a known-good ECU and recheck. If symptom/indication goes away, replace the original ECU.

2.1L ENGINE—TROUBLESHOOTING—CODE 15—PRELUDE

Self-diagnosis LED indicator blinks fifteen times: A problem in the Ignition Output Signal circuit.

- **Check Engine warning light has been reported on.**
- **LED indicates CODE 15.**

Turn the ignition switch OFF.

Remove CLOCK fuse in the under-hood relay box for 10 seconds to reset ECU.

Turn the ignition switch to the Start Position.

Is Check Engine warning light on and does LED indicate CODE 15?

- NO → **Intermittent failure, system is OK at this time (test drive may be necessary). Check for poor connections or loose wires at the igniter unit connector**
- YES ↓

Turn the ignition switch OFF.

Disconnect the 6P connector on the igniter unit.

Turn the ignition switch ON.

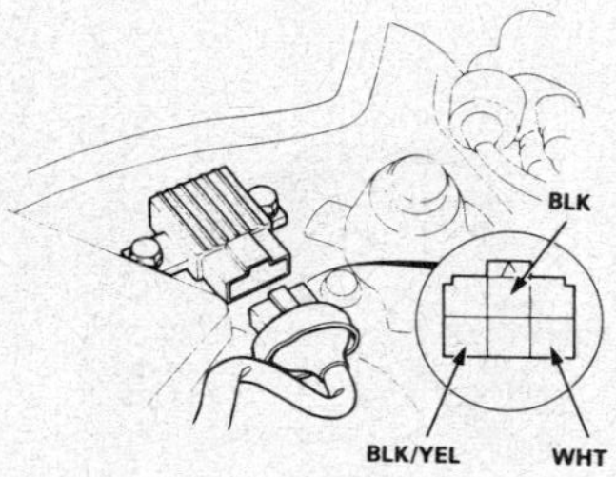

Measure voltage between the BLK/YEL (+) terminal and body ground.

Is there battery voltage ?

- NO → **Repair open in the BLK/YEL wire between the igniter unit and ignition switch.**
- YES ↓

Turn the ignition switch OFF.

Connect the ECU test harness between the ECU and connector

Reconnect the 6P connector on the igniter unit.

Turn the ignition switch ON.

Measure voltage individually between A21 (+), A22 (+) terminals and A26 (−) terminal.

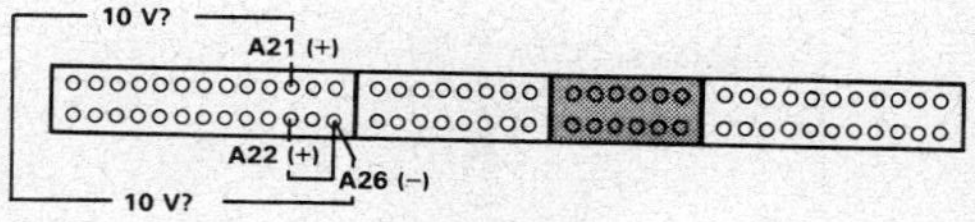

Is there approx. 10 V ?

- YES → **Substitute a known-good ECU and recheck. If symptom/indication goes away, replace the original ECU.**
- NO ↓

Turn the ignition switch OFF.

Disconnect the 6P connector on the igniter unit and the system checker harness from the ECU.

Check for continuity of WHT wires between the ECU (A21, A22) and the igniter unit.

Does continuity exist ?

- NO → **Repair open in WHT wires between the ECU (A21, A22) and the igniter unit.**
- YES ↓

Check for continuity between white terminal of 6P connector and body ground.

Does continuity exist ?

- YES → **Repair short in WHT wire.**
- NO → **Replace the igniter unit.**

NOTE: If the WHT wire was shorted, the igniter may be damaged.

Self-diagnosis LED indicates code 16: A problem in the fuel injector circuit.

The injectors are the solenoid-actuated constant-stroke pintle type consisting of a solenoid, plunger needle valve and housing. When current is applied to the solenoid coil, the valve lifts up and pressurized fuel is injected close to the intake valve. Because the needle valve lift and the fuel pressure are constant, the injection quantity is determined by the length of time that the valve is open (i.e., the duration the current is supplied to the solenoid coil). The injector is sealed by an O-ring and seal ring at the top and bottom. These seals also reduce operating noise.

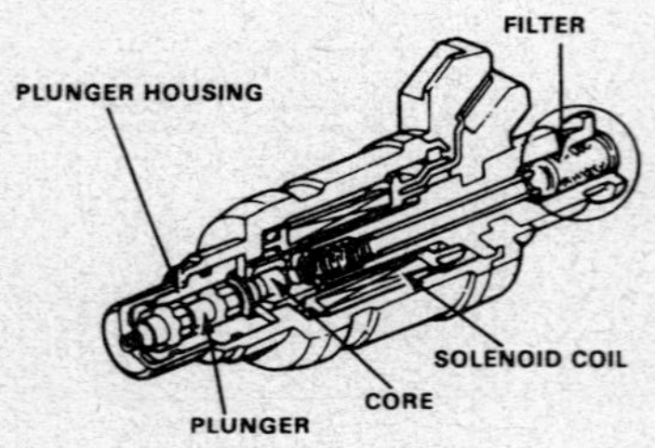

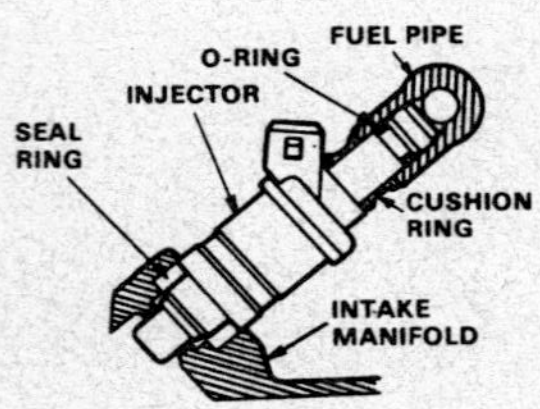

- Check Engine warning light has bee reported on.
- LED indicates CODE 16.

Turn the ignition switch OFF.

Remove CLOCK fuse in the under-hood relay box for 10 seconds to reset ECU.

Turn the ignition switch to START position.

Does the engine start?
- NO → Go to Path H
- YES → Is Check Engine warning light on and does LED indicate CODE 16?
  - NO → Intermittent failure, system is OK at this time (test drive may be necessary). Check for poor connections or loose wires at injectors, injector resistor and C235 (located at right shock tower).
  - YES → Go to Path G

Path G

Check the clicking sound of each injector by means of a stethoscope when the engine is idling.

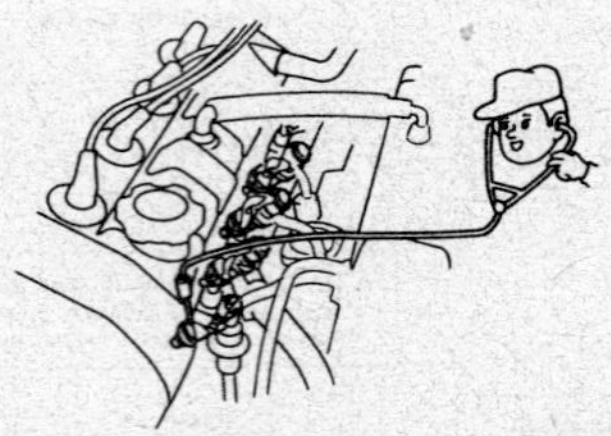

Do the injectors click?
- YES → Substitute a known-good ECU and recheck. If symptom/indication goes away, replace the original ECU.
- NO → Turn the ignition switch OFF.

Disconnect the 2P connector from the injector that does not click.

Measure resistance between the 2 terminals of the injector.

Path H

Turn the ignition switch OFF.

Disconnect the 2P connector from each injector.

Measure resistance between the 2 terminals of the injector.

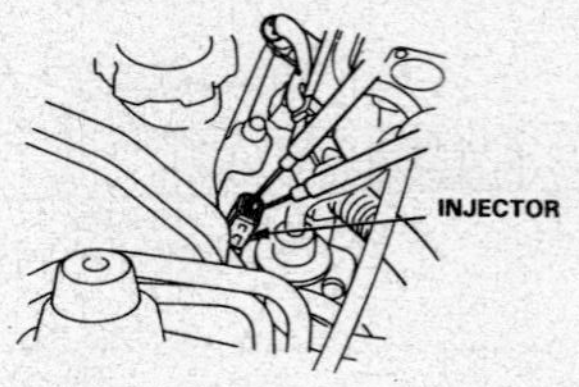

Is there 1.5–2.5Ω?
- NO → Replace the injector.
- YES → Turn the ignition switch ON.

2.1L ENGINE—TROUBLESHOOTING—CODE 16—PRELUDE

Measure voltage between RED/BLK (+) terminal on the 2P connector and body ground.

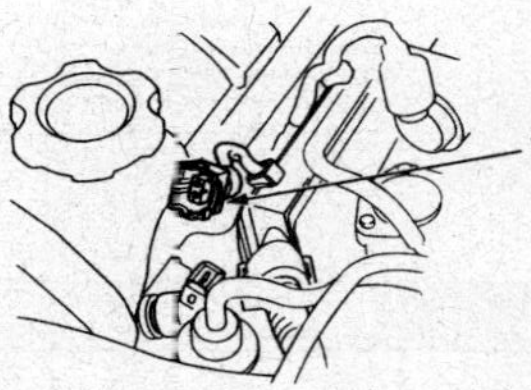

Is there battery voltage ?

- NO → Turn the ignition switch OFF.
  - Disconnect 6P connector on the injector resistor.
  - Turn the ignition switch ON.
  - Measure voltage between YEL/BLK (+) terminal and body ground.
  - Is there battery voltage ?
    - NO → Repair open in the YEL/BLK wire between the injector resistor and the main relay.
    - YES → —Replace the injector resistor.
      —Repair open RED/BLK wire between 2P connector and resistor.
- YES → Reconnect the 2P connector to the injector.
  - Turn the ignition switch OFF.

Connect the ECU test harness between the ECU and connector.

Turn the ignition switch ON.

Measure voltage between A23(−) terminal and following terminals.
- No.1 injector: A1 (+) terminal.
- No.2 injector: A3 (+) terminal.
- No.3 injector: A5 (+) terminal.
- No.4 injector: A2 (+) terminal.

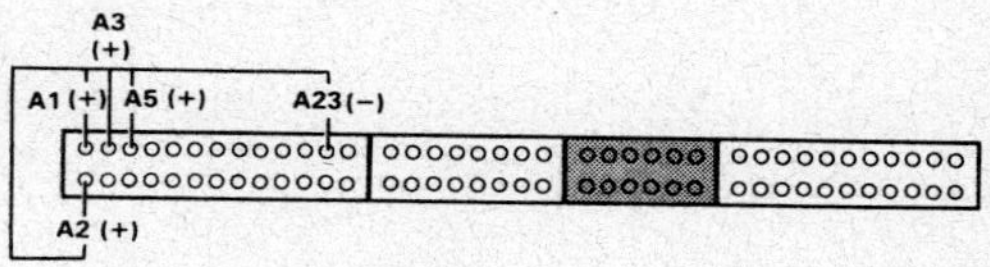

Is there battery voltage ?

- NO → Repair open in the wire between the ECU (A1, A3, A5 or A2) and the injector.
- YES → Substitute a known-good ECU and recheck. If symptom/indication goes away, replace the original ECU.

2.1L ENGINE—TROUBLESHOOTING CODE 17—PRELUDE

Self-diagnosis LED indicates code 17: A problem in the Vehicle Speed Sensor circuit.

The signal generated by the speed sensor, produces pulses when the front wheels turn.

- Check Engine warning light has been reported on.
- LED indicates CODE 17.

Turn the ignition switch OFF.

Remove CLOCK fuse in the under-hood relay box for 10 seconds to reset ECU.

Road test necessary.
In 2nd gear accelerate to 3,500 rpm, then decelerate to 1,500 rpm with throttle fully closed.

Is Check Engine warning light on and does LED indicate CODE 17?

NO → Intermittent failure, system is OK at this time. Check for poor connections or loose wires at C129 and C240.

YES ↓

Block rear wheels and set the parking brake. Jack up the front of the car and support with safety stands.

WARNING Block rear wheels before jacking up front of car.

Connect the ECU test harness between the ECU and connector

Turn the ignition switch ON.

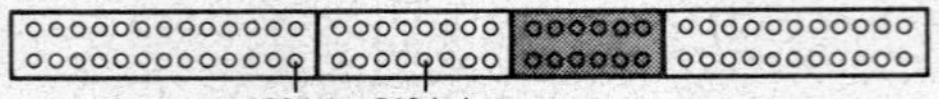

Slowly rotate left front wheel and measure voltage between B10 (+) terminal and A26 (−) terminal.

Does voltage pulse 0 V and 5 V?

NO →
- Repair open or short in WHT/BLU wire between ECU (B10) and the speed sensor.
- Faulty speed sensor.
- Substitute a known-good ECU and recheck. If symptom/indication goes away, replace the original ECU.

YES ↓

Substitute a known-good ECU and recheck. If symptom/indication goes away, replace the original ECU.

2.1L ENGINE—TROUBLESHOOTING CODE 41—PRELUDE

Self-diagnosis LED indicates code 41: A problem in the Oxygen (O_2) Sensor Heater circuit.

- Engine is running.
- Check Engine warning light has been reported on.
- LED indicates CODE 41.

Turn the ignition switch OFF.

Remove CLOCK fuse in the under-hood relay box for 10 seconds to reset ECU.

Start the engine

Is Check Engine warning light on and does LED indicate CODE 41 ?

NO → Intermittent failure, system is OK at this time (test drive may be necessary). Check for poor connections or loose wires at O_2 sensor, C236 (located at right shock tower), and ECU.

YES ↓

Stop the engine

Disconnect the 4P connector from the O_2 sensor.

Measure resistance between terminals A and B on the O_2 sensor.

Is there 15—40 Ω?

NO → Replace O_2 sensor.

YES ↓

Check for continuity to body ground on each terminal on the O_2 sensor.

Does continuity exist ?

YES → Replace O_2 sensor.

NO ↓

Check for continuity between terminal A and terminals C and D individually.

Does continuity exist ?

YES → Replace O_2 sensor.

NO

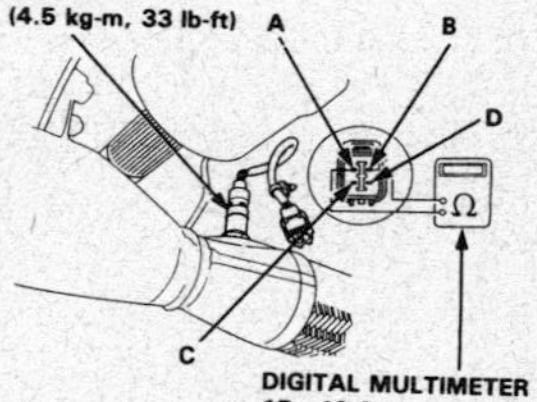

2.1L ENGINE—TROUBLESHOOTING CODE 41—PRELUDE

- Measure voltage between YEL/BLK (+) terminal and BLK (−) terminal.
- Is there battery voltage ?
  - YES → Disconnect the "A" connector from the ECU.
    - Measure voltage between YEL/BLK (+) terminal and BLK (−) terminal.
    - Is there battery voltage ?
      - YES → Repair short in BLK wire between ECU (A6) and O_2 sensor.
      - NO → Substitute a known-good ECU and recheck. If symptom/indication goes away, replace the original ECU.
  - NO → Measure voltage between YEL/BLK (+) terminal and body ground.
- Is there battery voltage ?
  - NO → Repair open in YEL/BLK wire between the O_2 sensor and Main Relay.
  - YES → Turn the ignition switch OFF.
- Reconnect the 4P connector to O_2 sensor.
- Connect the ECU test harness "A" connector to the main wire harness only, not the ECU
- Turn the ignition switch ON.
- Measure voltage between A6 (+) terminal and A23 (−) teminal.

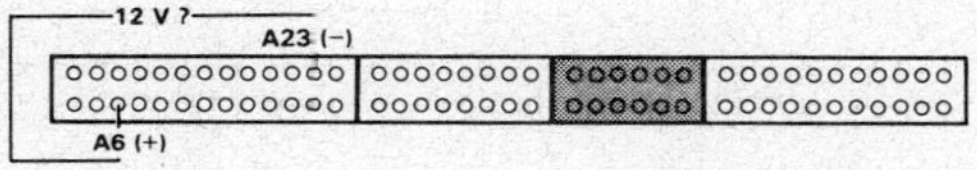

- Is there battery voltage ?
  - NO → Repair open in BLK wire between ECU (A6) and O_2 sensor.
  - YES → Substitute a known-good ECU and recheck. If symptom/indication goes away, replace the original ECU.

2.1L ENGINE—TROUBLESHOOTING CODE 43—PRELUDE

(43) Self-diagnosis LED indicates code 43: Most likely a problem in the Oxygen (O_2) Sensor circuit or a problem in the Fuel Supply System.

(43) NOTE: If 43 code is accompanied by the check engine warning light and poor driveability, go to Fuel Supply System.

- – Check Engine warning light has been reported on.
  – LED indicates CODE 43.
  – or continued from code 1.
- Turn the ignition switch OFF.
- Remove CLOCK fuse in the under-hood relay box for 10 seconds to reset ECU.
- Warm up engine to normal operating temperature (cooling fan comes on).
- Hold engine at 3,000 rpm for 2 minutes. (A/T: Transmission is [N] or [P].)
- Is the Check Engine warning light on and does LED indicate CODE 43?
  - NO → Intermittent failure, system is OK at this time (test drive may be necessary). Check for poor connections or loose wires at O_2 sensor C236 (located at right shock tower), and ECU.
  - YES → Turn the ignition switch OFF.
- Connect the ECU test harness between the ECU and connector
- With the ignition switch OFF, wait for at least two minutes.
- Turn the ignition switch ON.
- Measure voltage between D14 (+) terminal and A26 (−) terminal as soon as the ignition switch is turned on.

NOTE
- Use DIGITAL MULTIMETER
- Use 2 Volt range.

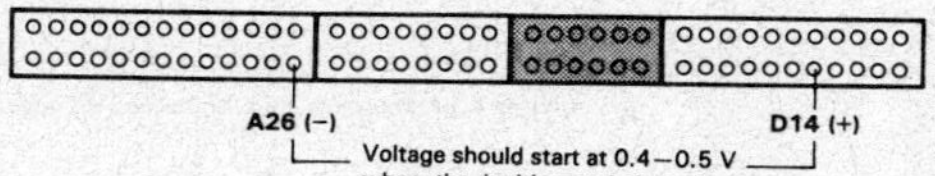

2.1L ENGINE—TROUBLESHOOTING CODE 43—PRELUDE

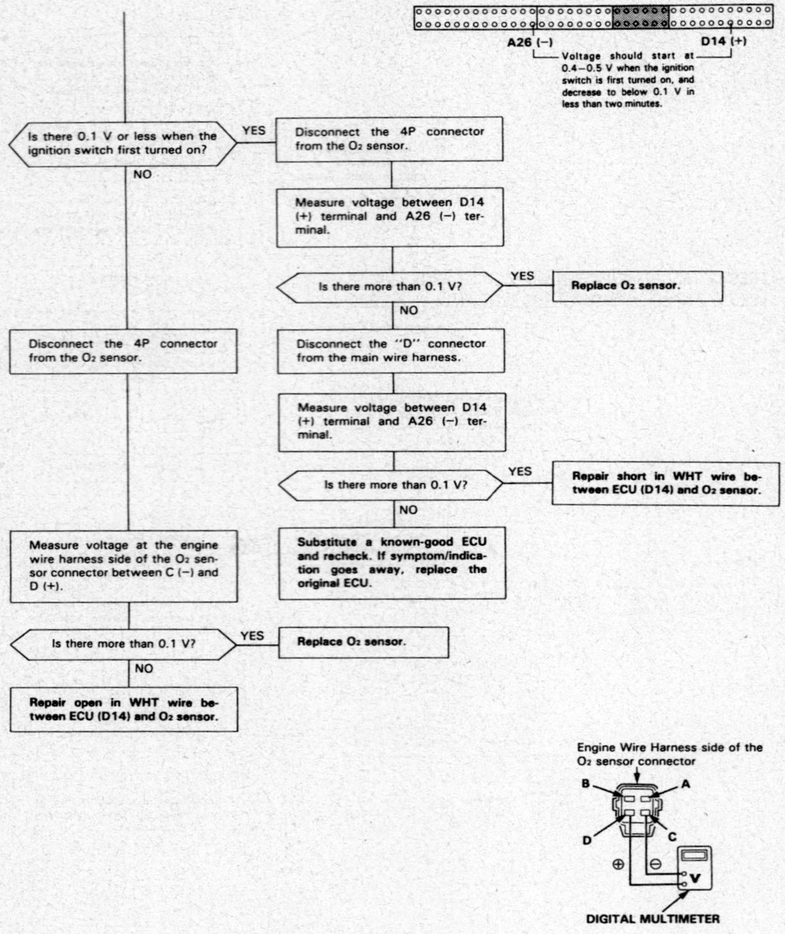

2.1L ENGINE—TROUBLESHOOTING AIR CONDITIONING SIGNAL—PRELUDE

This signals the PGM-FI ECU when there is a demand for cooling from the air conditioning system.

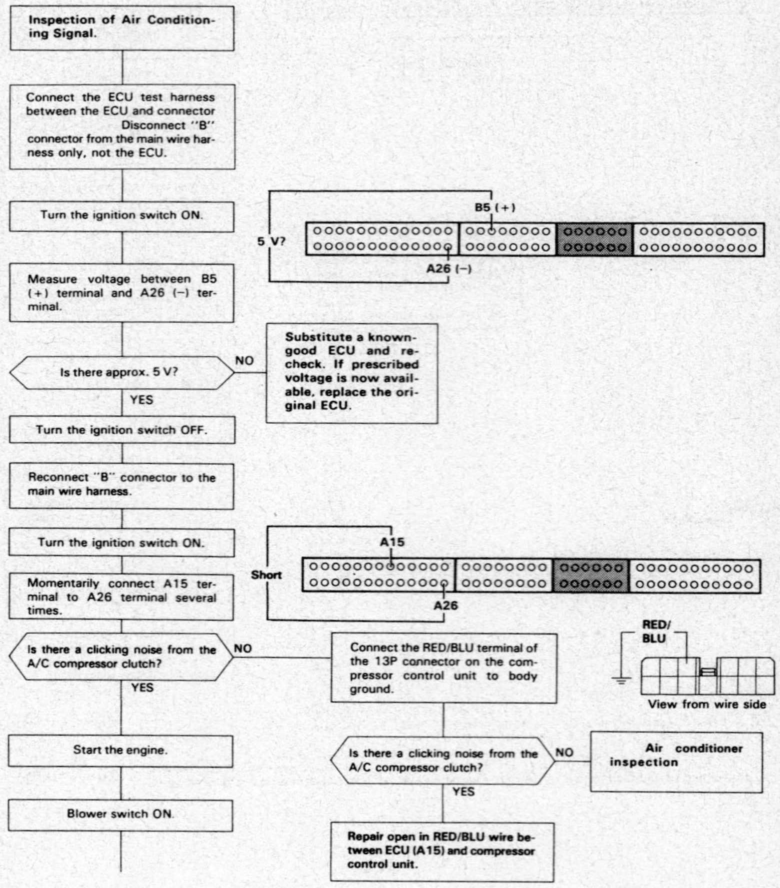

2.1L ENGINE—TROUBLESHOOTING
AIR CONDITIONING SIGNAL—PRELUDE

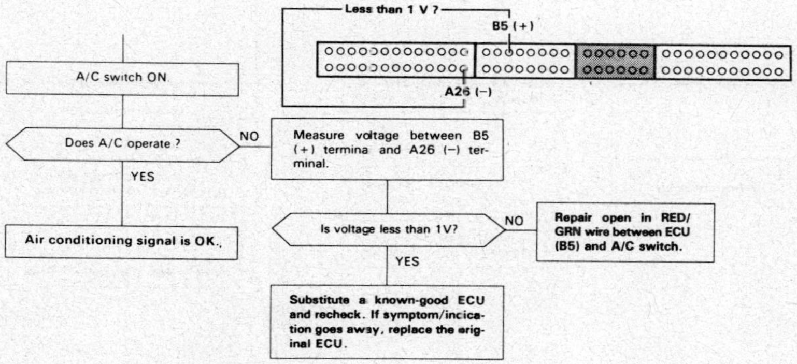

2.1L ENGINE—TROUBLESHOOTING
ALTERNATOR FR SIGNAL—PRELUDE

This signals the PGM-FI ECU when the alternator is charging.

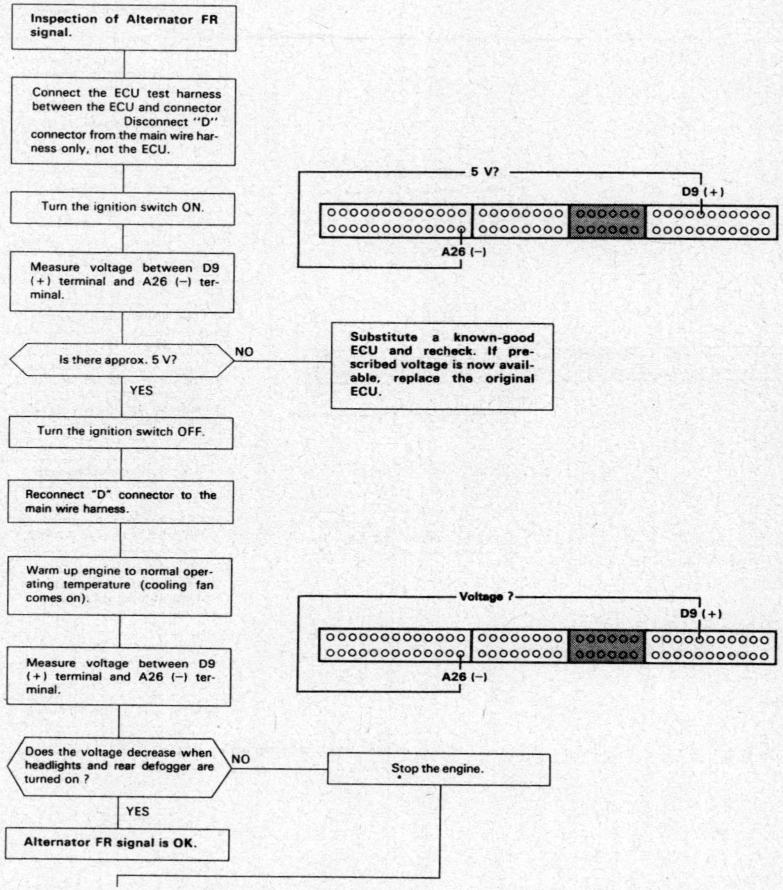

2.1L ENGINE—TROUBLESHOOTING ALTERNATOR FR SIGNAL—PRELUDE

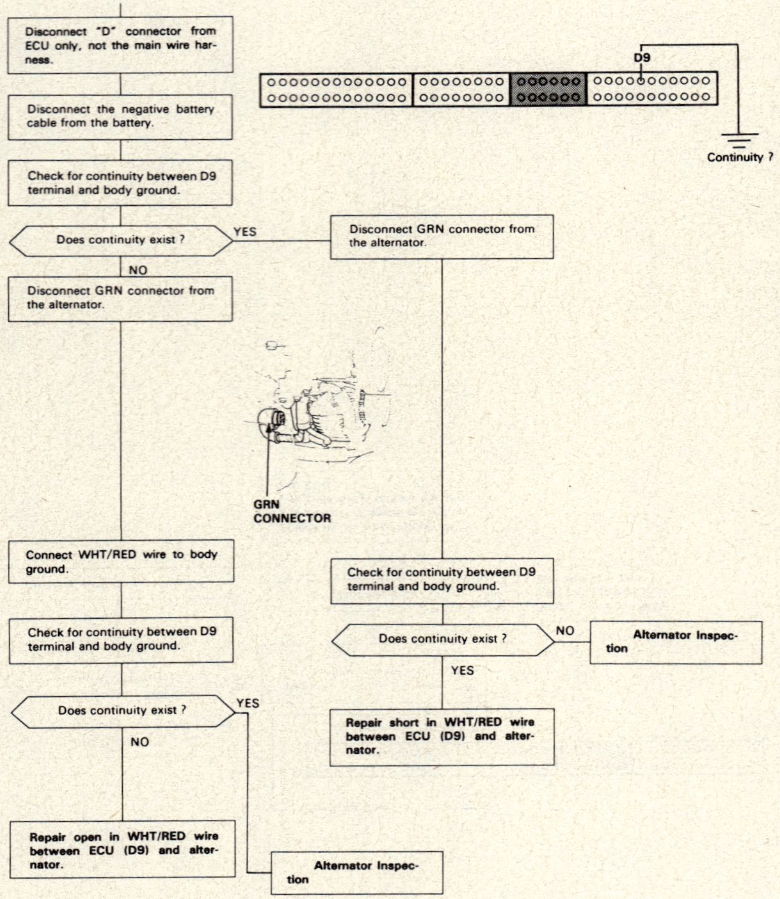

2.1L ENGINE—TROUBLESHOOTING AUTOMATIC TRANSAXLE SHIFT POSITION SIGNAL PRELUDE

This signals the PGM-FI ECU when the transmission is in Neutral or Park.

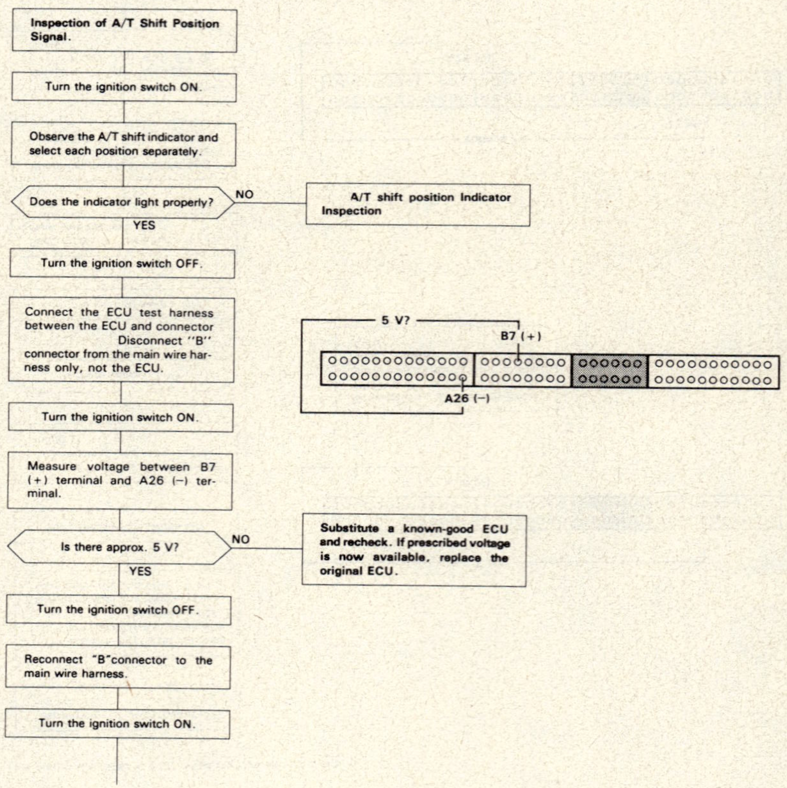

2.1L ENGINE—TROUBLESHOOTING
AUTOMATIC TRANSAXLE SHIFT POSITION SIGNAL PRELUDE

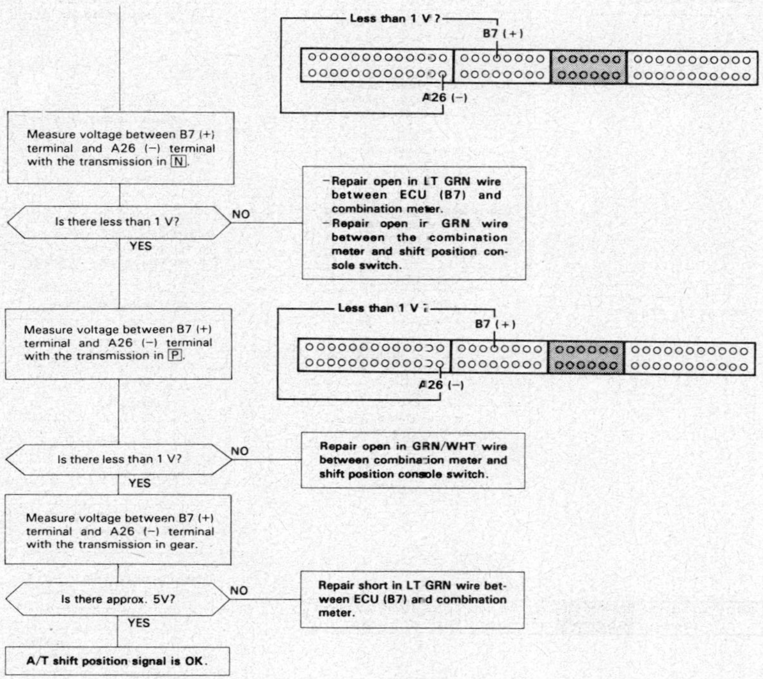

2.1L ENGINE—TROUBLESHOOTING
STARTER SWITCH SIGNAL—PRELUDE

This signals the PGM-FI ECU when the engine is cranking.

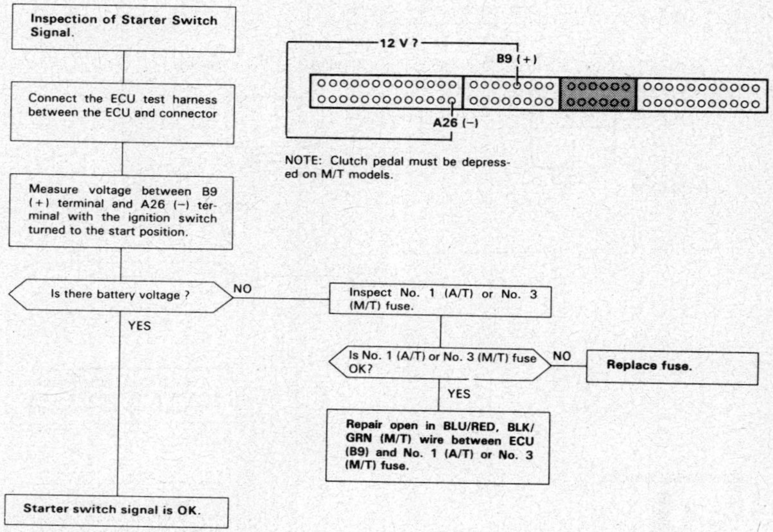

2.1L ENGINE—TROUBLESHOOTING POWER STEERING PRESSURE SIGNAL—PRELUDE

This signals the PGM-FI ECU when the power steering load is high.

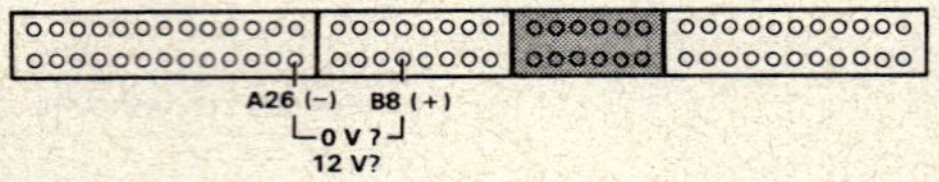

- **Inspection of P/S Oil Pressure Signal**
- Connect the ECU test harness between the ECU and connector
- Turn the ignition switch ON.
- Measure voltage between B8 (+) terminal and A26 (−) terminal.
- Is there voltage ?
  - YES → Disconnect the 2P connector on the P/S oil pressure switch.
    - Connect BLU/RED terminal to BLK terminal.
    - Is there voltage ?
      - NO → **Replace P/S oil pressure switch.**
      - YES → **Repair open in BLU/RED wire between ECU (B8) and P/S oil pressure switch or BLK wire between P/S oil pressure switch and G301.**
  - NO → Start the engine.
    - Turn steering wheel slowly.
    - Measure voltage between B8 (+) terminal and A26 (−) terminal while steering wheel is turning.
    - Is there battery voltage ?
      - YES → **P/S oil pressure signal is OK.**
      - NO → Turn the ignition switch OFF.
        - Disconnect "B" connector from main wire harness only, not the ECU.
        - Turn the ignition switch ON.
        - Is there battery voltage ?
          - NO → **Substitute a known-good ECU and recheck. If prescribed voltage is now available, replace the original ECU.**
          - YES → Reconnect "B" connector to main wire harness and disconnect 2P connector on the P/S oil pressure switch.
            - Is there battery voltage ?
              - YES → **Replace P/S oil pressure switch.**
              - NO → **Repair short in BLU/RED wire between ECU (B8) and the P/S oil pressure switch.**

2.1L ENGINE—TROUBLESHOOTING EVAPORATIVE EMISSION CONTROL SYSTEM PRELUDE

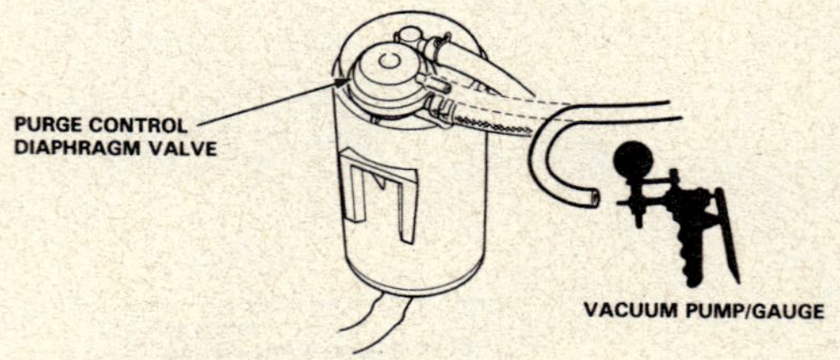

- **Inspection of Evaporative Emission Controls**
- Disconnect vacuum hose from the purge control diaphragm valve (on the charcoal canister) and connect a vacuum gauge to the hose.
- Start the engine and allow to idle. NOTE: Engine coolant temperature must be below 75°C (167°F).
- Is there vacuum ?
  - NO → Go to Path P
  - YES → Disconnect the 4P** or 2P* connector.
    - Measure voltage between BLK/YEL (+) terminal and YEL/BLK (−) terminal.
    - Is there battery voltage ?
      - YES → **Inspect vacuum hose routing. If OK, replace purge cut-off solenoid valve.**
      - NO → Measure voltage between BLK/YEL (+) terminal and body ground.
        - Go to Path Q

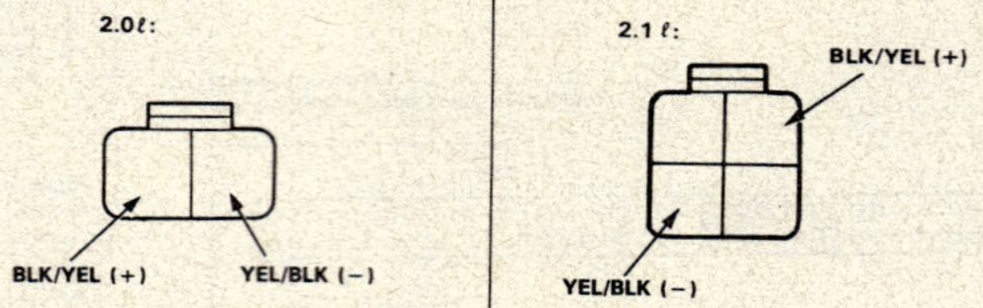

*: 2.0 ℓ
**: 2.1 ℓ

2.1L ENGINE – TROUBLESHOOTING – EVAPORATIVE EMISSION CONTROL SYSTEM – PRELUDE

Path P

Warm up the engine to normal operating temperature (cooling fan comes on).

Check for vacuum at #3 vacuum hose after starting the engine.

Is there manifold vacuum?
- NO → Disconnect the 4P** or 2P* connector.
  - Is there manifold vacuum?
    - NO → **Inspect vacuum hose running. If OK, replace purge cut-off solenoid valve.**
    - YES → **Inspect for a short to body ground in YEL/BLK wire between ECU (A20**, B2*) and the connector. If wire is OK, substitute a known-good ECU and recheck. If symptom/indication goes away, replace the original ECU.**
- YES → Reconnect the hose.

Remove fuel filler cap.

Path Q

Is there battery voltage ?
- NO → **Repair open in BLK/YEL wire between No.12 fuse and the 4P** or 2P* connector.**
- YES → **Inspect YEL/BLK wire for an open between ECU (A20**, B2*) and the connector. If wire is OK, substitute a known-good ECU and recheck. If symptom/indication goes away, replace the original ECU.**

*: 2.0 ℓ
**: 2.1 ℓ

Connect a vacuum gauge to canister purge air hose.

Start the engine and raise speed to 3,500 rpm.

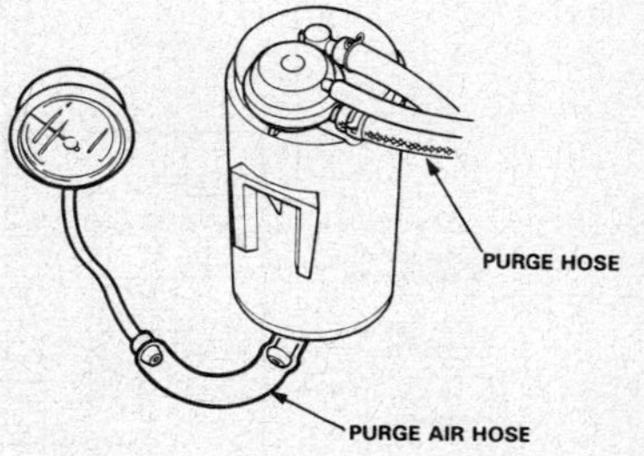

Does vacuum appear on gauge within 1 minute ?
- YES → **See two-way valve test to complete. Evaporative emission control are OK.**
- NO → Connect a vacuum gauge to the canister purge hose and raise the engine speed to 3,500 rpm.
  - Does vacuum appear on the gauge?
    - NO → **Inspect the purge hose and throttle body port for pinch or blockage.**
    - YES → **Replace the canister.**

2.1L ENGINE—VACUUM DIAGRAM—PRELUDE

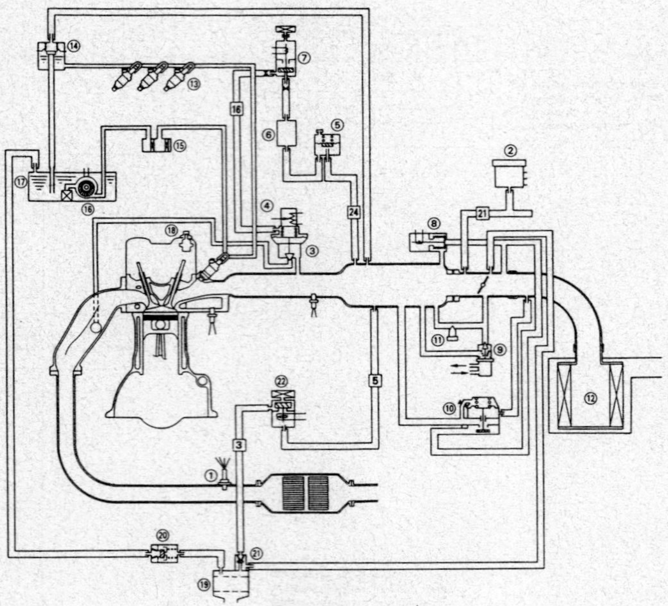

1. HEATED OXYGEN (O_2) SENSOR
2. MANIFOLD ABSOLUTE PRESSURE (MAP) SENSOR
3. EGR VALVE
4. EGR VALVE LIFT SENSOR
5. CONSTANT VACUUM CONTROL (CVC) VALVE
6. AIR CHAMBER
7. EGR CONTROL SOLENOID VALVE
8. ELECTRONIC AIR CONTROL VALVE(EACV)
9. FAST IDLE VALVE
10. AIR BOOST VALVE
11. IDLE ADJUSTING SCREW
12. AIR CLEANER
13. FUEL INJECTOR
14. PRESSURE REGULATOR
15. FUEL FILTER
16. FUEL PUMP
17. FUEL TANK
18. PCV VALVE
19. CHARCOAL CANISTER
20. TWO-WAY VALVE
21. PURGE CONTROL DIAPHRAGM VALVE
22. PURGE CUT-OFF SOLENOID VALVE

2.1L ENGINE—FUEL SYSTEM WIRING SCHEMATIC—PRELUDE

FUSES
1. EFI ECU (10A)*
2. No.12 (10A)
3. No. 1 (7.5A) (A/T)
   No. 3 (7.5A) (M/T)
4. No.13 (10A)
5. CLOCK (10A)*

*: In the under-hood relay fuse box

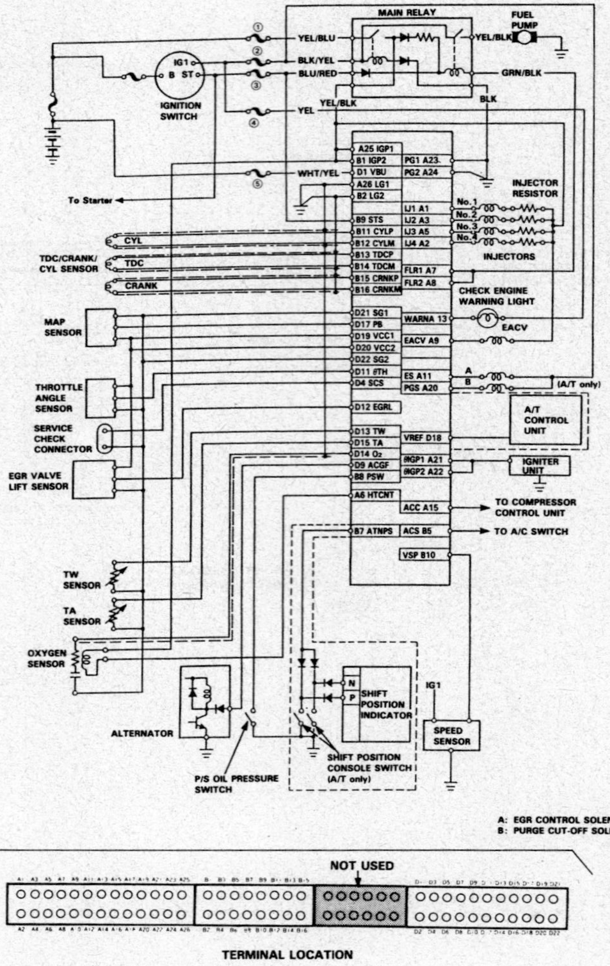

1.5L (D15B7, D15B8, D15Z1) AND 1.6L (D16Z6) ENGINES—TROUBLESHOOTING GUIDE—1992 CIVIC

NOTE: Across each row in the chart, the systems that could be sources of a symptom are ranked in the order they should be inspected starting with ①. Find the symptom in the left column, read across to the most likely source, then refer to the page listed at the top of that column. If inspection shows the system is OK, try the next most likely system ②, etc.

SYMPTOM \ SYSTEM		ECU	OXYGEN (D15Z1 engine: LAF)* SENSOR	MANIFOLD ABSOLUTE PRESSURE SENSOR	TDC/CRANK/CYL SENSOR	COOLANT TEMPERATURE SENSOR	THROTTLE ANGLE SENSOR	INTAKE AIR TEMPERATURE SENSOR	ATMOSPHERIC PRESSURE SENSOR	IGNITION OUTPUT SIGNAL	VEHICLE SPEED SENSOR
			PGM-FI								
CHECK ENGINE LIGHT** TURNS ON		(light off) or (light on)	(Check Engine light)	(Check Engine light)	(Check Engine light)	(Check Engine light)	(Check Engine light)	(Check Engine light)	(Check Engine light)	(Check Engine light)	(Check Engine light)
CHECK ENGINE LIGHT BLINKS		[illegible] or [illegible]	1 or 41 or 43* or 48	3 or 5	4 or 8 or 9	6	7	10	13	15	17
ENGINE WON'T START		①			③					③	
DIFFICULT TO START ENGINE WHEN COLD		(BU)		③	②	①					
IRREGULAR IDLING	WHEN COLD FAST IDLE OUT OF SPEC	(BU)				③					
	ROUGH IDLE	(BU)		③							
	WHEN WARM RPM TOO HIGH	(BU)									
	WHEN WARM RPM TOO LOW	(BU)									
FREQUENT STALLING	WHILE WARMING UP	(BU)				③					
	AFTER WARMING UP	(BU)									
POOR PERFORMANCE	MISFIRE OR ROUGH RUNNING	(BU)		②	③						
	FAILS EMISSION TEST	(BU)	③	②							
	LOSS OF POWER	(BU)		③			②				

SYMPTOM		LOCK-UP CONTROL SOLENOID VALVE	ELECTRIC LOAD DETECTOR	SPOOL SOLENOID VALVE	VALVE TIMING OIL PRESSURE SWITCH	ELECTRONIC AIR CONTROL VALVE	OTHER IDLE CONTROLS	FUEL INJECTOR	OTHER FUEL SUPPLY	AIR INTAKE	EGR CONTROL SYSTEM	OTHER EMISSION CONTROLS
		PGM-FI				IDLE CONTROL		FUEL SUPPLY			EMISSION CONTROL	
CHECK ENGINE LIGHT** TURNS ON		(Check Engine light)	(light off)	(Check Engine light)	(Check Engine light)	(Check Engine light)		(Check Engine light)				
CHECK ENGINE LIGHT BLINKS		19	20	21	22	14		16				
ENGINE WON'T START									②			
DIFFICULT TO START ENGINE WHEN COLD												
IRREGULAR IDLING	WHEN COLD FAST IDLE OUT OF SPEC					①	②					
	ROUGH IDLE					①		②			③	
	WHEN WARM RPM TOO HIGH					①	②					
	WHEN WARM RPM TOO LOW	③				①		②				
FREQUENT STALLING	WHILE WARMING UP					①	②		③			
	AFTER WARMING UP					③			①		②	
POOR PERFORMANCE	MISFIRE OR ROUGH RUNNING							①			③	
	FAILS EMISSION TEST											①
	LOSS OF POWER			③	③			③	①	③		

* If codes other than those listed above are indicated, count the number of blinks again. If the indicator is in fact blinking these codes, substitute a known-good ECU and recheck. If the indication goes away, replace the original ECU.

(BU) If the Check Engine light is on while the engine is running, jump the service check connector. If no code is displayed (Check Engine light stays on steady), the back-up system is in operation.
Substitute a known-good ECU and recheck. If the indication goes away, replace the original ECU.

** Check Engine light
USA: (CHECK engine symbol) CANADA: (engine symbol)

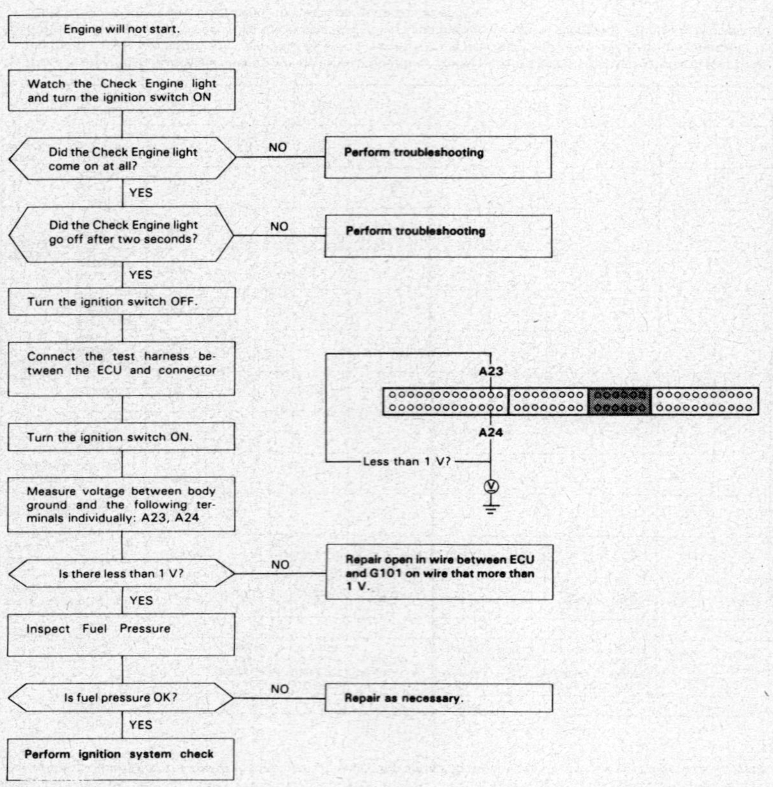
Engine will not start.
Watch the Check Engine light and turn the ignition switch ON
Did the Check Engine light come on at all?
NO
Perform troubleshooting
YES
Did the Check Engine light go off after two seconds?
NO
Perform troubleshooting
YES
Turn the ignition switch OFF.
Connect the test harness between the ECU and connector
A23
A24
Less than 1 V?
Turn the ignition switch ON.
Measure voltage between body ground and the following terminals individually: A23, A24
Is there less than 1 V?
NO
Repair open in wire between ECU and G101 on wire that more than 1 V.
YES
Inspect Fuel Pressure
Is fuel pressure OK?
NO
Repair as necessary.
YES
Perform ignition system check

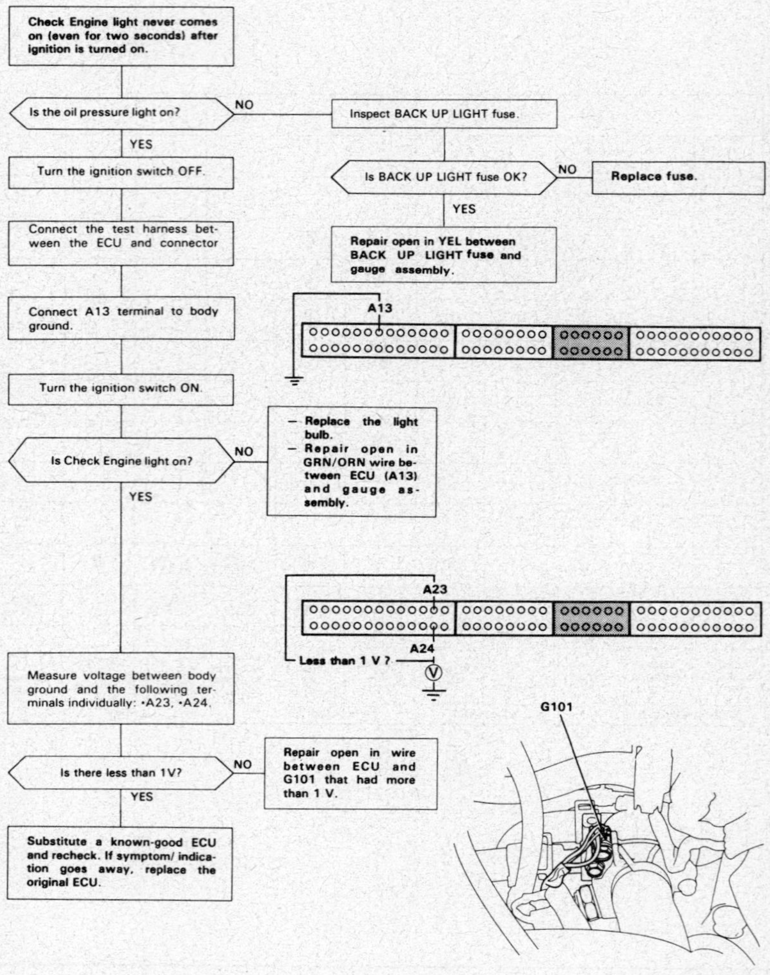
Check Engine light never comes on (even for two seconds) after ignition is turned on.
Is the oil pressure light on?
NO
Inspect BACK UP LIGHT fuse.
Is BACK UP LIGHT fuse OK?
NO
Replace fuse.
YES
Repair open in YEL between BACK UP LIGHT fuse and gauge assembly.
YES
Turn the ignition switch OFF.
Connect the test harness between the ECU and connector
Connect A13 terminal to body ground.
A13
Turn the ignition switch ON.
Is Check Engine light on?
NO
– Replace the light bulb.
– Repair open in GRN/ORN wire between ECU (A13) and gauge assembly.
YES
A23
A24
Less than 1 V ?
Measure voltage between body ground and the following terminals individually: •A23, •A24.
G101
Is there less than 1V?
NO
Repair open in wire between ECU and G101 that had more than 1 V.
YES
Substitute a known-good ECU and recheck. If symptom/indication goes away, replace the original ECU.

(cont'd)

1.5L (D15B7, D15B8, D15Z1) AND 1.6L (D16Z6) ENGINES—PGM-FI CONTROL SYSTEM—1992 CIVIC

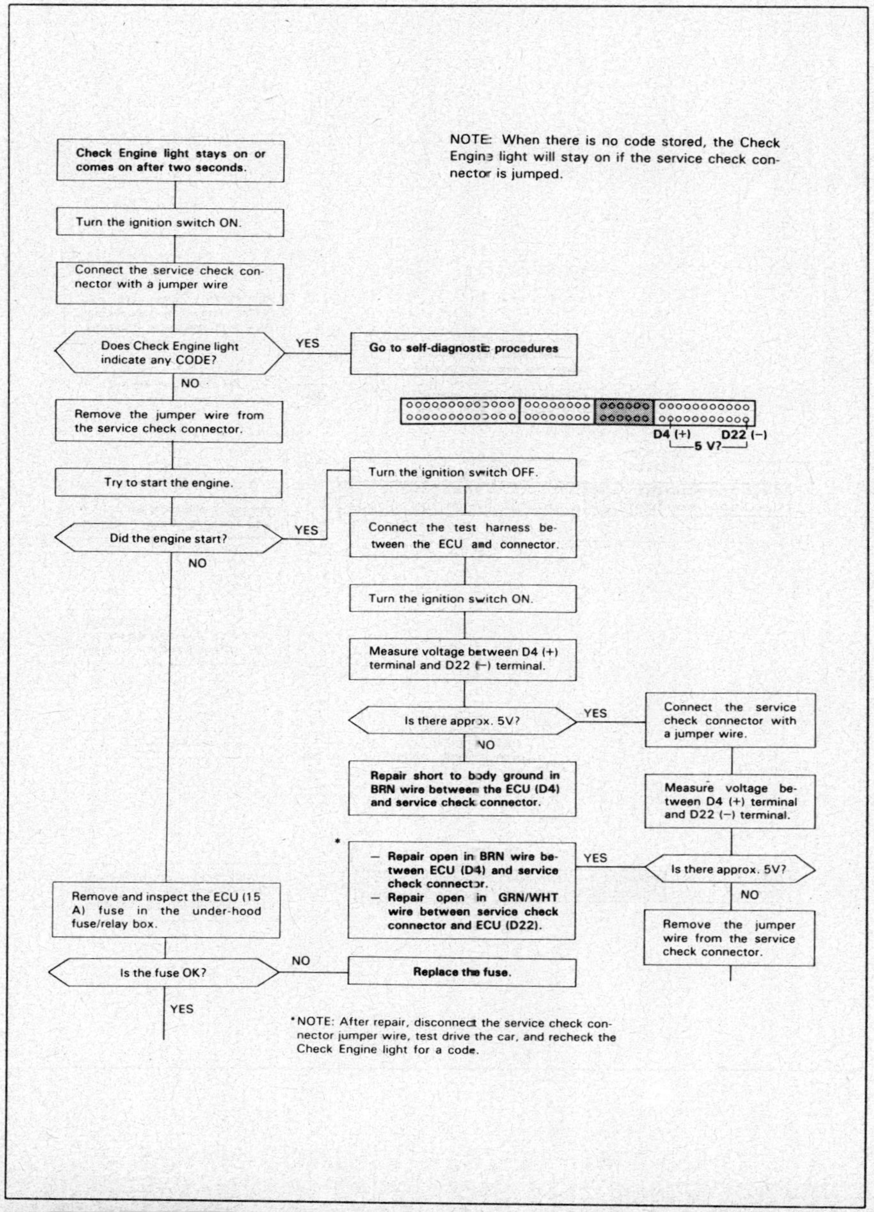

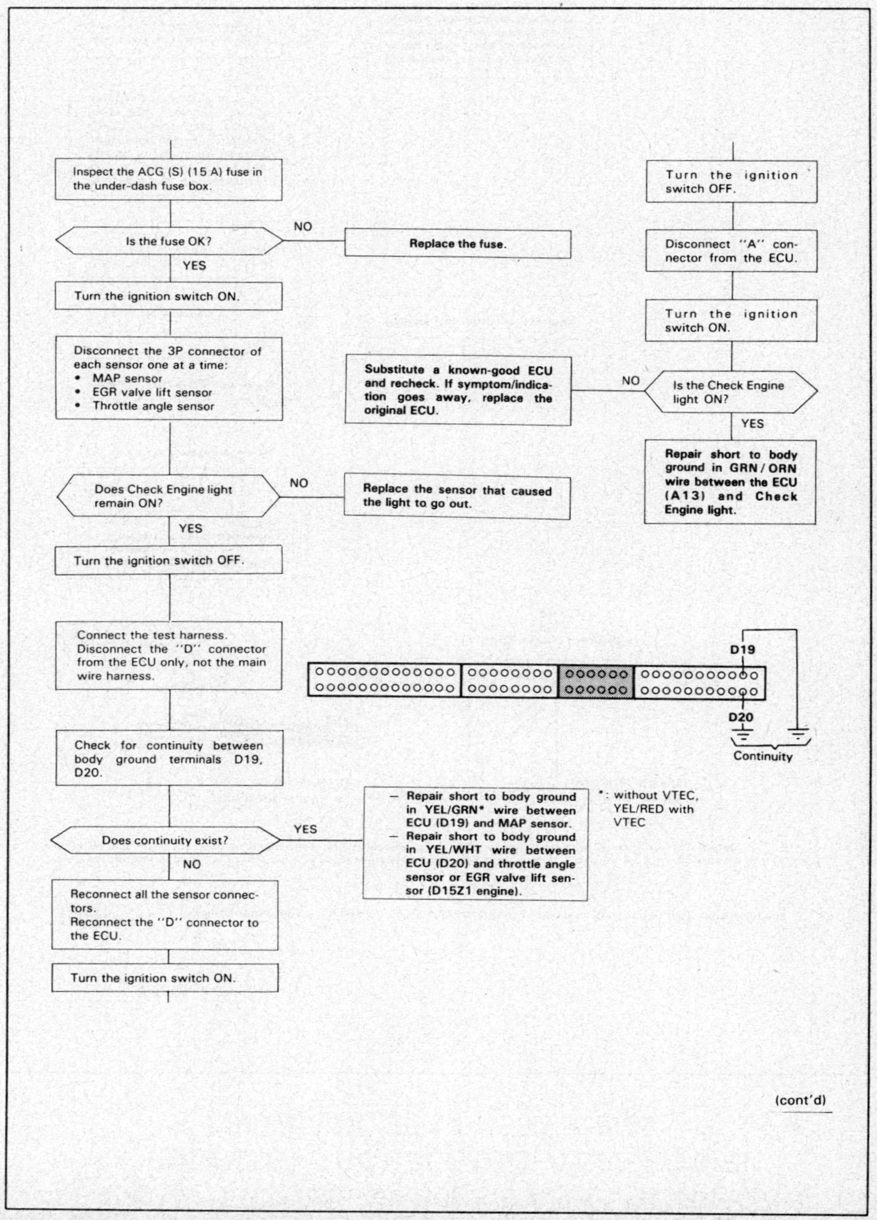

(cont'd)

Measure voltage between body ground and the following terminal individually: •A26, •B2.

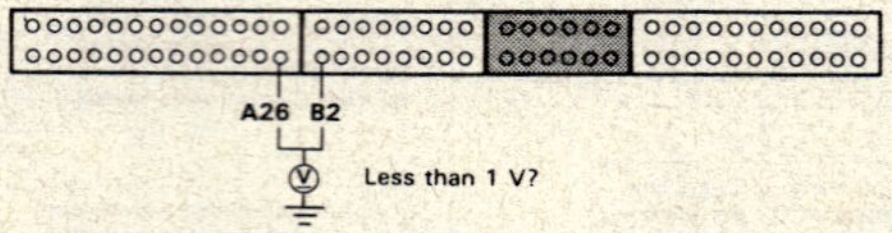

Is there less than 1 V?

NO: Repair open in BLK/RED (A26) or BRN/BLK (B2) and G101.

YES

Measure voltage between A26 (−) and the following: B1 (+) and A25 (+).

Measure voltage between A26 (−) and the following: B1 (+) and A25 (+).

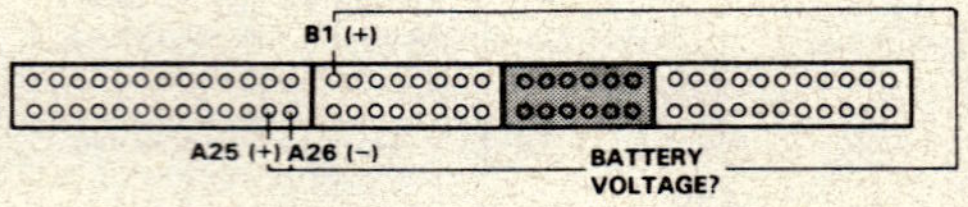

Is there battery voltage?

NO:
– Repair open in YEL/BLK wire between ECU (A25, B1) and main relay.
– Check main relay and wiring connectors at main relay.

YES

Substitute a known-good ECU and recheck. If symptom/indication goes away, replace the original ECU.

1.5L (D15B7, D15B8, D15Z1) AND 1.6L (D16Z6) ENGINES – TROUBLESHOOTING PGM-FI FAULT CODE 1 – 1992 CIVIC

CHECK 1 Self-diagnosis Check Engine light indicates code 1: A problem in the Oxygen (O_2) Sensor circuit.

The Oxygen sensor detects the oxygen content in the exhaust gas, and inputs the ECU. In operation, the ECU receives the signals from the sensor and varies the duration during which fuel is injected. The oxygen sensor is installed on the exhaust manifold.

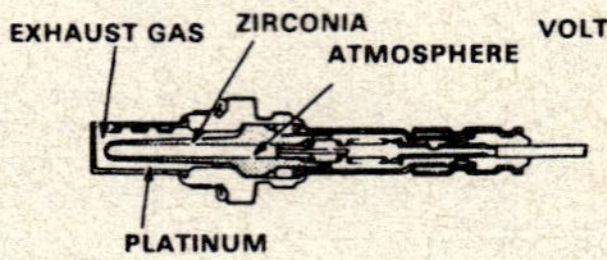

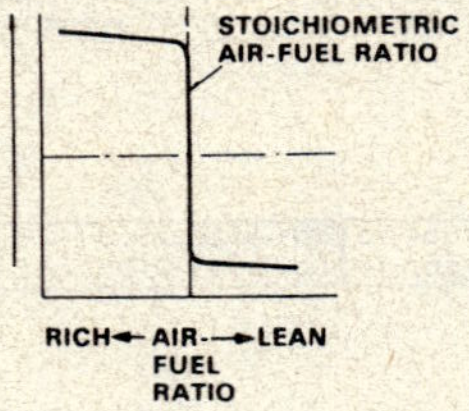

CHECK 1

– Check Engine light has been reported on.
– With service check connector jumped CODE 1 is indicated.

Do the ECU Reset Procedures

Inspect pressure regulator (page

Is it normal?

NO: Replace the pressure regulator

YES

Warm up engine to normal operating temperature (cooling fan comes on).

Road test with the transmission in 2nd gear (M/T: 4th gear). Starting at 1600 rpm, accelerate using wide open throttle for at least 5 seconds. Then decelerate for at least 5 seconds with the throttle completely closed.

Is Check Engine warning light on and does LED indicate CODE 1.

NO: Intermittent failure, system is OK at this time. Check for poor connections or loose wires at C213 (located at right shock tower), at the C111 (O_2 sensor) and ECU.

YES

(cont'd)

1.5L (D15B7, D15B8, D15Z1) AND 1.6L (D16Z6) ENGINES—TROUBLESHOOTING PGM-FI
FAULT CODE 1—1992 CIVIC

1 Self-diagnosis Check Engine light indicates code 1: A problem in the Oxygen (O_2) Sensor circuit.

The Oxygen sensor detects the oxygen content in the exhaust gas and signals the ECU. In operation, the ECU receives the signals from the sensor and varies the duration during which fuel is injected. The oxygen sensor has an internal heater. The heater stabilizes the sensor's output. The oxygen sensor is installed in the exhaust manifold.

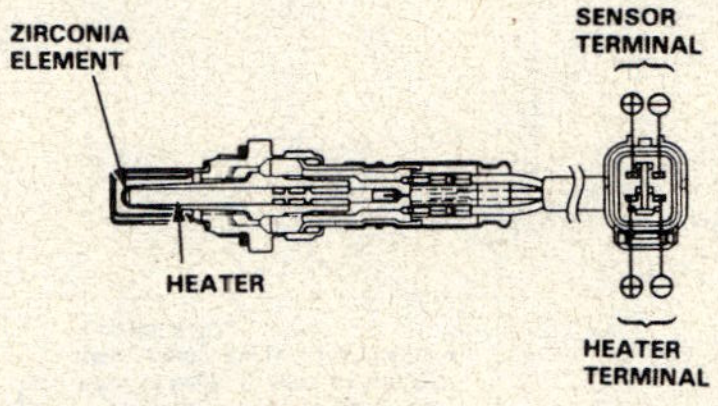

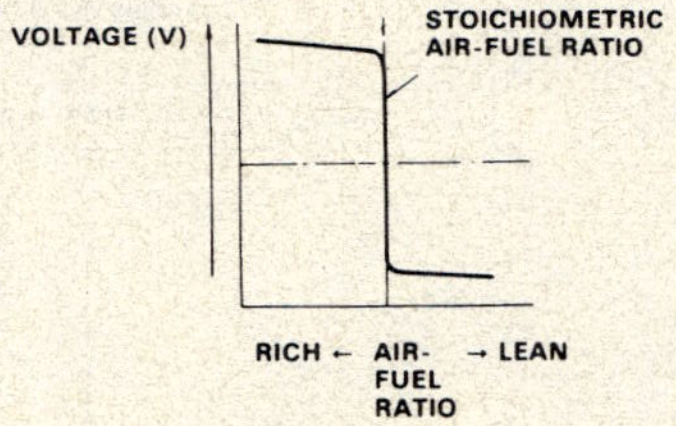

1

- Check Engine light has been reported on.
- With service check connector jumped CODE 1 is indicated.

↓

Do the ECU Reset Procedures

↓

Warm up engine to normal operating temperature (cooling fan comes on).

↓

Run engine for 60 seconds.

↓

Road test with the transmission in 2nd gear (M/T: 4th gear). Starting at 1600 rpm, accelerate using wide open throttle for at least 5 seconds. Then decelerate for at least 5 seconds with the throttle completely closed.

↓

Is Check Engine light on and does it indicate CODE 1?

- NO → Intermittent failure, system is OK at this time. Check for poor connections or loose wires at C213 (located at right shock tower, the C111 (O_2 sensor) and ECU.
- YES → Go to page and perform test for CODE 43

(cont'd)

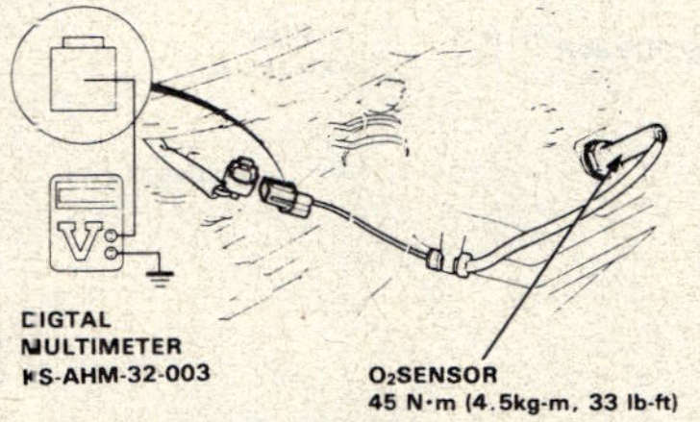

Disconnect engine wire harness from the O_2 sensor.

↓

Warm up engine to normal operating temperature again, then open the throttle wide open, then quickly release it.

↓

Measure voltage between the connector terminal and body ground.

↓

Is voltage above 0.6 V at wide open throttle and below 0.4 V when the throttle is quickly released ?

- NO → Replace O_2 sensor.
- YES ↓

Stop engine.

↓

Reconnect the oxygen sensor.

↓

Connect the test harness between the ECU and connector

↓

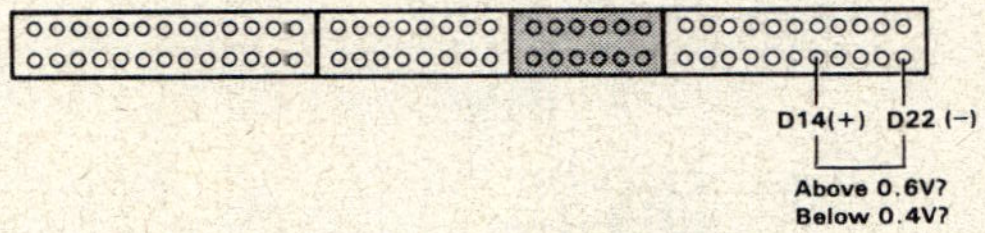

Restart and warm up engine to normal operating temperature, then open the throttle wide open, then quickly release it.

↓

Measure voltage between D14 (+) terminal and D22 (–) terminal.

↓

Is voltage above 0.6 V at wide open throttle and below 0.4 V when the throttle is quickly released ?

- NO → Repair short or open in WHT wire between ECU (D14) and O_2 sensor.
- YES ↓

Substitute a known-good ECU and recheck. If symptom/indication goes away, replace the original ECU.

1.5L (D15B7, D15B8, D15Z1) AND 1.6L (D16Z6) ENGINES—TROUBLESHOOTING PGM-FI FAULT CODES 3 AND 5—1992 CIVIC

3 — Self-diagnosis Check Engine light indicates code 3: An electrical problem in the Manifold Absolute Pressure (MAP) Sensor system.

5 — Self-diagnosis Check Engine light indicates code 45: Most likely a problem in the Fuel Metering System. Pressure (MAP) Sensor system.

The MAP sensor converts manifold absolute pressure into electrical signals and inputs the ECU.

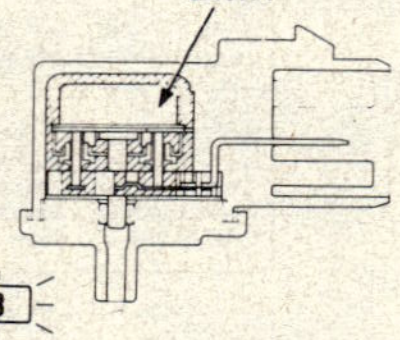

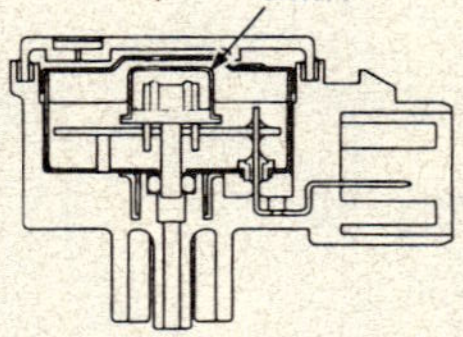

3

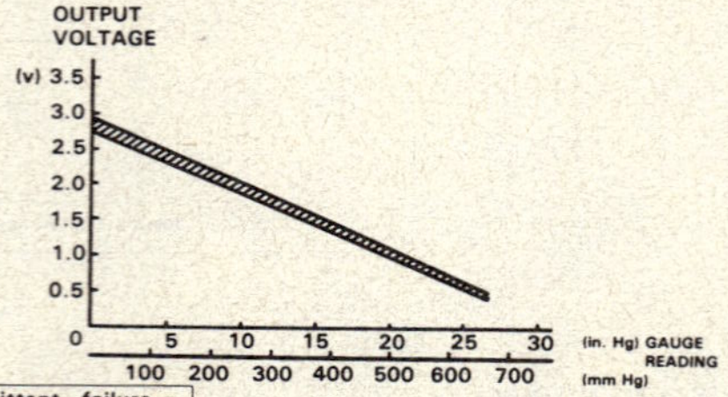

- Engine is warm and running.
- Check Engine light has been reported on.
- With service check connector jumped CODE 3 is indicated.

Do the ECU Reset Procedures

Start the engine and allow it to idle.

Is Check Engine light on and does it indicate CODE 3?

NO → Intermittent failure, system is OK at this time (test drive may be necessary). Check for poor connection or loose wires at C212 (located at right shock tower), C113, C201* (MAP sensor) and ECU.

YES → Turn the ignition switch OFF.

Disconnect the 3P connector from the MAP sensor.

Turn the ignition switch ON.

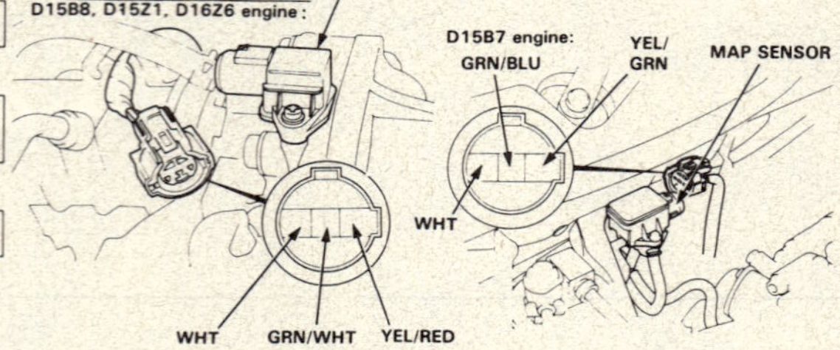

Measure voltage between YEL/RED (YEL/GRN)* (+) terminal and body ground.

Is there approx. 5V?

NO → Connect the test harness "D" connector to the ECU only, not to the main wire harness.

Measure voltage between D19 (+) terminal and D21 (−) terminal.

*: D15B7 engine

Is there approx. 5 V?

NO → Repair open in YEL/RED YEL/GRN (YEL/GRN)* wire between ECU (D19) and MAP sensor.

YES → Substitute a known-good ECU and recheck. If symptom/indication goes away, replace the original ECU.

YES → Measure voltage between YEL/RED (YEL/GRN)* (+) terminal and GRN/WHT (GRN/BLU)* (−) terminal.

*: D15B7 engine

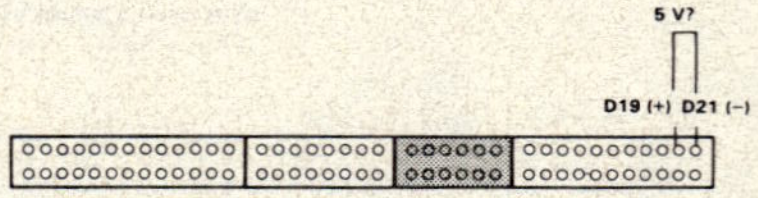

Is there approx. 5V?

NO → Repair open in GRN/WHT, GRN/BLU (GRN/BLU)* wire between ECU (D21) and MAP sensor.

YES → Measure voltage between WHT (+) terminal and GRN/WHT (GRN/BLU)* (−) terminal.

Is there approx. 5V?

NO

YES → Turn the ignition switch OFF.

Reconnect the 3P connector to the MAP sensor.

Connect the test harness between the ECU and connector (page 11-25).

Turn the ignition switch ON.

(cont'd)

1.5L (D15B7, D15B8, D15Z1) AND 1.6L (D16Z6) ENGINES—TROUBLESHOOTING PGM-FI FAULT CODES 3 AND 5—1992 CIVIC

Connect the test harness "D" connector to the ECU only, not to the main wire harness

Measure voltage between D17 (+) terminal and D21 (−) terminal.

Is there approx. 5 V?

- NO: Substitute a known-good ECU and recheck. If prescribed voltage is now available, replace the original ECU.
- YES:
  - Repair short in WHT wire between ECU (D17), and MAP sensor.
  - Repair open in WHT wire between ECU (D17) and MAP sensor.

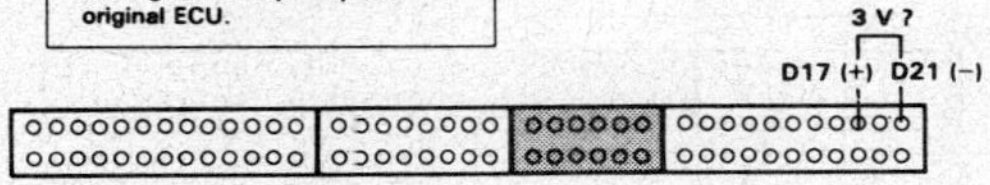

Measure voltage between D17 (+) terminal and D21 (−) terminal.

Is there approx. 3 V ?

- NO: Replace MAP sensor.
- YES: Substitute a known-good ECU and recheck. If symptom/indication goes away, replace the original ECU.

1.5L (D15B7, D15B8, D15Z1) AND 1.6L (D16Z6) ENGINES—TROUBLESHOOTING PGM-FI FAULT CODES 4, 8 AND 9—1992 CIVIC

CHECK – 4 – Self-diagnosis Check Engine light indicates code 4: A problem in the CRANK Sensor circuit.

CHECK – 8 – Self-diagnosis Check Engine light indicated code 8: A problem in the TDC Sensor circuit.

CHECK – 9 – Self-diagnosis Check Engine light indicates code 9: A problem in the CYL Sensor circuit.

The CRANK sensor determines timing for fuel injection and ignition of each cylinder and also detects engine RPM. The TDC sensor determines ignition timing at start-up (cranking) and when crank angle is abnormal. The CYL sensor detects the position of No. 1 cylinder for sequential fuel injection to each cylinder.

CHECK – 4 – 8 – or – 9

- Check Engine light has been reported on.
- With service check connector jumped CODE 4, 8 and/or 9 are indicated.

Do the ECU Reset Procedures

Start the engine.

Is Check Engine light on and does it indicate CODE 4, 8 or 9?

- NO: Intermittent failure, system is OK at this time (test drive may be necessary). Check for poor connections or loose wires at C213 (located at right shock tower), C107 (TDC/CRANK/CYL sensor) and ECU.
- YES: Stop the engine.

Disconnect the 8P connector from the TDC/CRANK/CYL sensor.

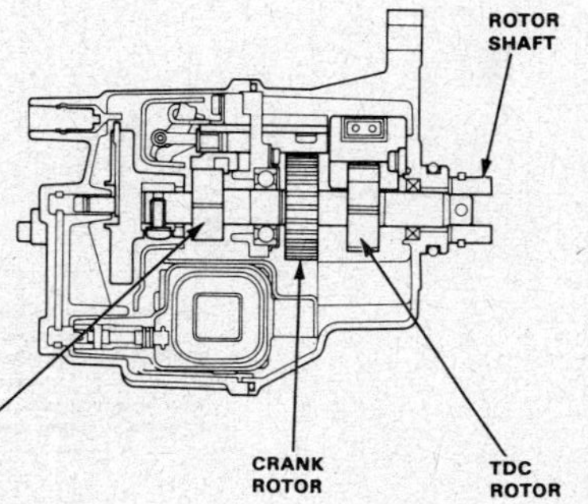

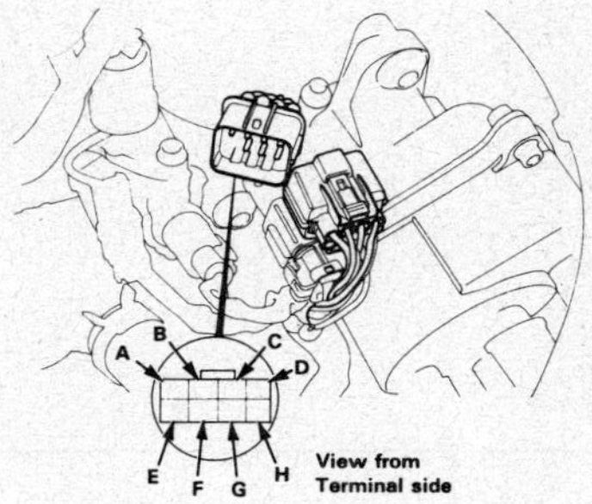

(cont'd)

1.5L (D15B7, D15B8, D15Z1) AND 1.6L (D16Z6) ENGINES—TROUBLESHOOTING PGM-FI FAULT CODES 4, 8 AND 9—1992 CIVIC

Measure resistance between terminals of the indicated sensor. *see table

Is there 350–700Ω?
- NO: Replace the distributor sub-assembly
- YES: Check for continuity to body ground on both terminals of the indicated sensor.

Does continuity exist?
- YES: Replace the distributor sub-assembly
- NO: Reconnect the connector.

Connect the test harness to the main wire harness only, not to the ECU

Measure resistance between terminals of the indicated sensor on test harness. *see table

Is there 350–700Ω?
- NO: Repair open in the indicated sensor wires. *see table
- YES: Check for continuity to body ground on B15, B13 and B11 terminals.

Does continuity exist?
- YES: Repair short to body ground in the indicated sensor wires. *see table
- NO: Substitute a known-good ECU and recheck. If symptom/indication goes away, replace the original ECU.

*:

SENSOR	CODE	SENSOR TERMINAL	ECU TERMINAL	WIRE COLOR
CRANK	4	B	B15	BLU/GRN
		F	B16	BLU/YEL
TDC	8	C	B13	ORN/BLU
		G	B14	WHT/BLU
CYL	9	D	B11	ORN
		H	B12	WHT

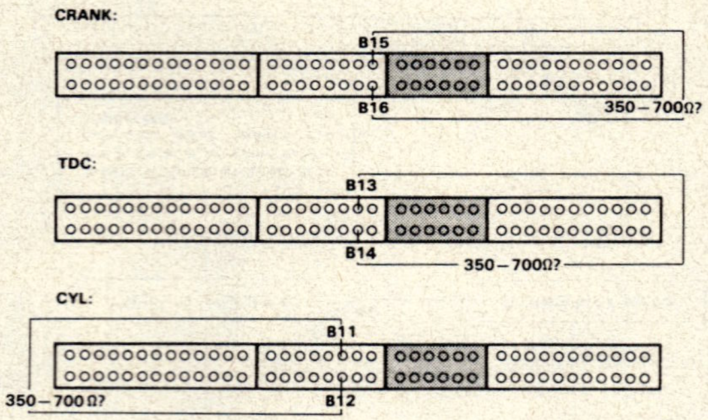

1.5L (D15B7, D15B8, D15Z1) AND 1.6L (D16Z6) ENGINES—TROUBLESHOOTING PGM-FI FAULT CODE 5—1992 CIVIC

cont'd)

5

- Check Engine light has been reported on.
- With service check connector jumped CODE 5 is indicated.

Do the ECU Reset Procedures

Start the engine and keep engine rpm at 2000 for one minute.

Is Check Engine light on and does it indicate CODE 5?
- NO: —Intermittent failure, system is OK at this time (test drive may be necessary). —Check vacuum hoses, pipes and connections. —Make sure all connectors are secure.
- YES: Stop the engine.

Remove the MAP sensor from throttle body.

Connect a vacuum pump to the MAP sensor and apply vacuum.

Does it hold vacuum?
- NO: Replace MAP sensor.
- YES: Start the engine.

Put your finger over the MAP port on throttle body.

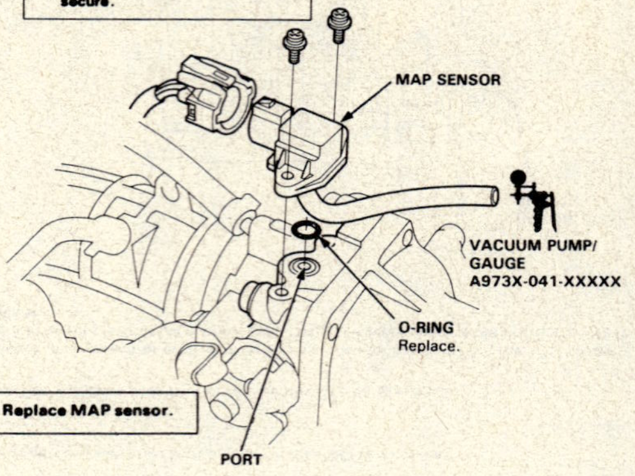

1.5L (D15B7, D15B8, D15Z1) AND 1.6L (D16Z6) ENGINES—TROUBLESHOOTING PGM-FI
FAULT CODE 5—1992 CIVIC

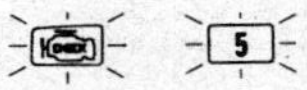

- Check Engine light has been reported on.
- With service check connector jumped CODE 5 is indicated.

Do the ECU Reset Procedures

Start the engine and keep engine rpm at 2000 for one minute.

Is Check Engine light on and does it indicate CODE 5?
- NO → —Intermittent failure, system is OK at this time (test drive may be necessary). —Check vacuum hoses, pipes and connections. —Make sure all connectors are secure.
- YES ↓

Stop the engine.

Disconnect #21 hose from the throttle body, connect vacuum pump to the hose and apply vacuum.

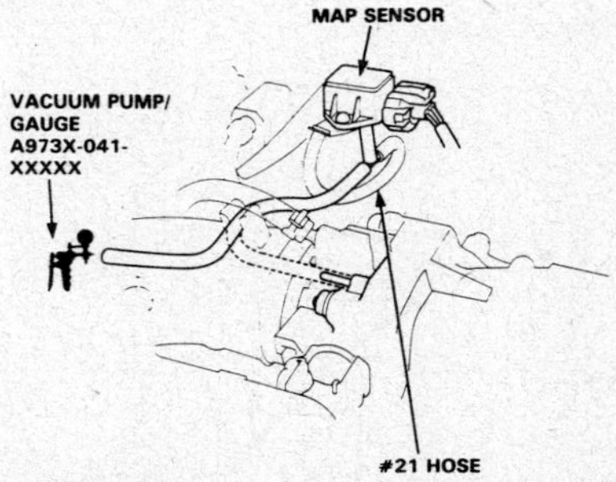

Does it hold vacuum?
- NO → Connect a vacuum pump to the MAP sensor and apply vacuum.
  - Does it hold vacuum?
    - NO → Replace MAP sensor.
    - YES → Repair vacuum leak in hose routing between MAP sensor and intake manifold.
- YES ↓

Connect a T-fitting from a vacuum gauge between the throttle body base and #21 hose.

Is there manifold vacuum?
- NO → Remove restriction from throttle body.
- YES ↓

Stop the engine.

Connect the test harness between the ECU and connector

Turn the ignition switch ON.

Measure voltage between D17 (+) terminal and D21 (−) terminal.

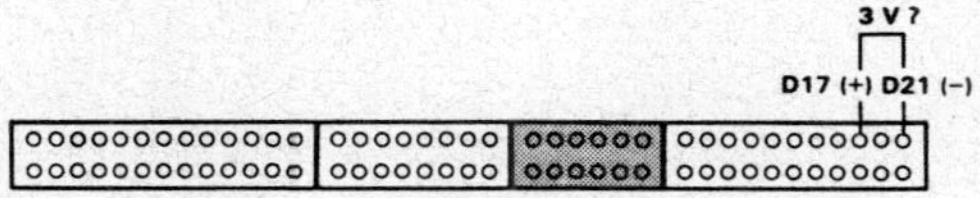

Is there approx. 3 V ?
- NO → Replace MAP sensor.
- YES ↓

Start the engine and allow it to idle.

Is there approx. 1 V?
- NO → Replace MAP sensor.
- YES ↓

Substitute a known-good ECU and recheck. If symptom/indication goes away, replace the original ECU.

1.5L (D15B7, D15B8, D15Z1) AND 1.6L (D16Z6) ENGINES—TROUBLESHOOTING PGM-FI FAULT CODE 5—1992 CIVIC

Start the engine.

Is there manifold vacuum?
- NO → Remove restriction from throttle body.
- YES ↓

Stop the engine.

Connect the test harness between the ECU and connector

Turn the ignition switch ON.

Measure voltage between D17 (+) terminal and D21 (−) terminal.

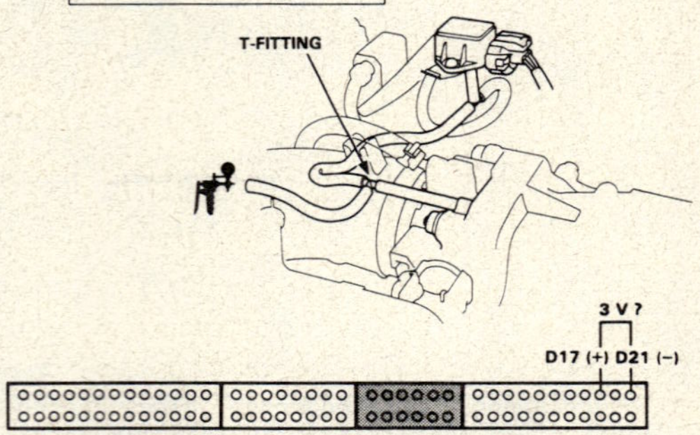

Is there approx. 3 V ?
- NO → Replace MAP sensor.
- YES ↓

Start the engine and allow it to idle.

Is there approx. 1 V?
- NO → Replace MAP sensor.
- YES ↓

Substitute a known-good ECU and recheck. If symptom/indication goes away, replace the original ECU.

1.5L (D15B7, D15B8, D15Z1) AND 1.6L (D16Z6) ENGINES—TROUBLESHOOTING PGM-FI FAULT CODE 6—1992 CIVIC

6 Self-diagnosis Check Engine light indicates code 6: A problem in the Coolant Temperature (TW) Sensor circuit.

The TW sensor is a temperature dependant resistor (thermistor). The resistance of the thermistor decreases as the coolant temperature increases as shown below.

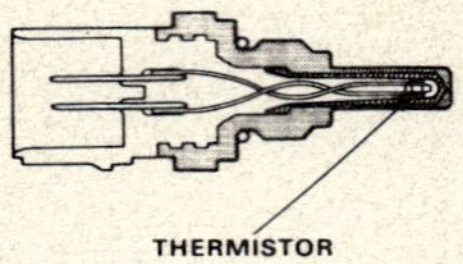

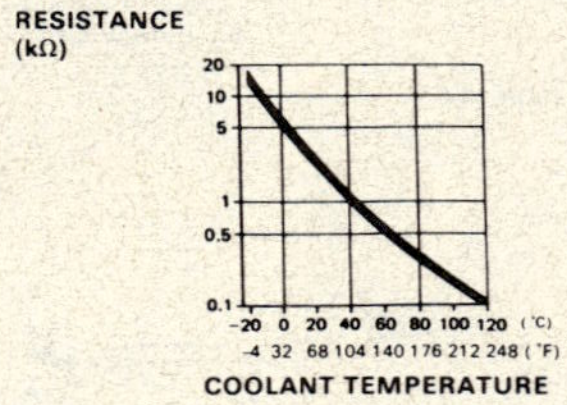

6

– Check Engine light has been reported on.
– With service check connector jumped CODE 6 is indicated.

Do the ECU Reset Procedures

Turn the ignition switch ON.

Is Check Engine light on and does it indicate CODE 6?
- NO → Intermittent failure, system is OK at this time (test drive may be necessary). Check for poor connections or loose wires at C212 (located at right shock tower), C109 (TW sensor) and ECU.
- YES ↓

Warm up engine to normal operating temperature (cooling fan comes on).

Turn the ignition switch OFF.

Disconnect the 2P connector from the TW sensor.

Measure resistance between the 2 terminals on the TW sensor.

Is there 200–400 Ω?
- NO → Replace TW sensor.
- YES

1.5L (D15B7, D15B8, D15Z1) AND 1.6L (D16Z6) ENGINES—TROUBLESHOOTING PGM-FI FAULT CODE 6—1992 CIVIC

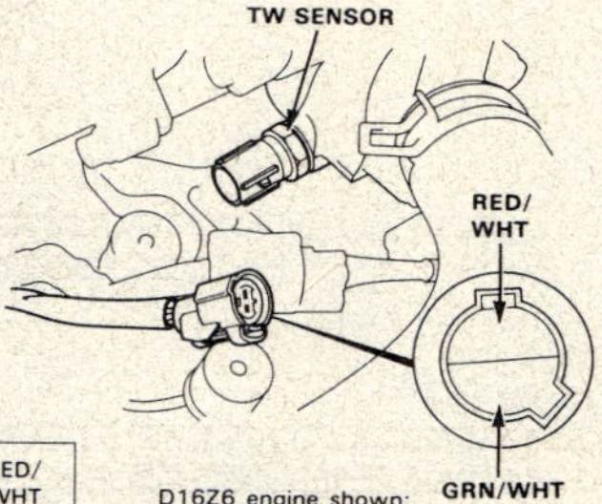

D16Z6 engine shown; D15B7, D15B8 and D15Z1 engine: TW sensor located under distributor.

- Turn the ignition switch ON.
- Measure voltage between RED/WHT and body ground.
- Is there approx. 5V ?
  - YES → Measure voltage between RED/WHT (+) terminal and GRN/WHT (−) terminal.
    - Is there approx. 5V ?
      - NO → Repair open in GRN/WHT wire between ECU (D22) and TW sensor.
      - YES → Substitute a known-good ECU and recheck. If symptom/indication goes away, replace the original ECU.
  - NO → Turn the ignition switch OFF.
- Connect the test harness "D" connector to the ECU only, not to the main wire harness
- Turn the ignition switch ON.
- Measure voltage between D13 (+) terminal and D22 (−) terminal.
- Is there approx. 5V?
  - YES → Repair open or short in RED/WHT wire between ECU (D13) and sensor.
  - NO → Substitute a known-good ECU and recheck. If symptom/indication goes away, replace the original ECU.

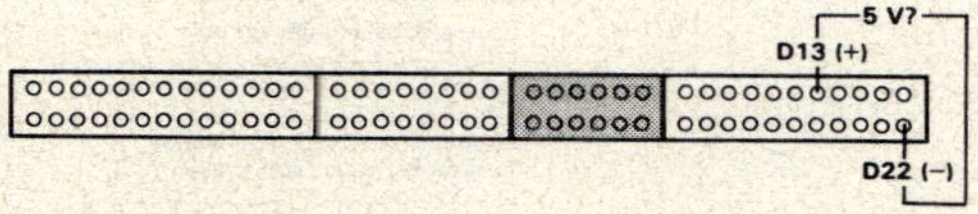

1.5L (D15B7, D15B8, D15Z1) AND 1.6L (D16Z6) ENGINES—TROUBLESHOOTING PGM-FI FAULT CODE 7—1992 CIVIC

7 Self-diagnosis Check Engine light indicates code 7: A problem in the Throttle Angle Sensor circuit.

The throttle angle sensor is a potentiometer. It is connected to the throttle valve shaft. As the throttle angle changes, the throttle angle sensor varies the voltage signal to the ECU.

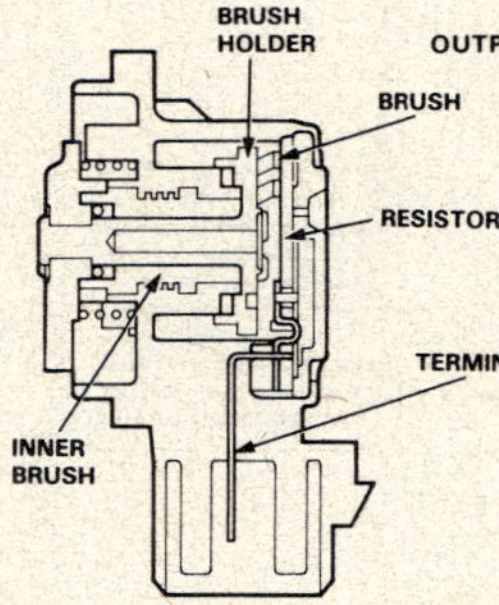

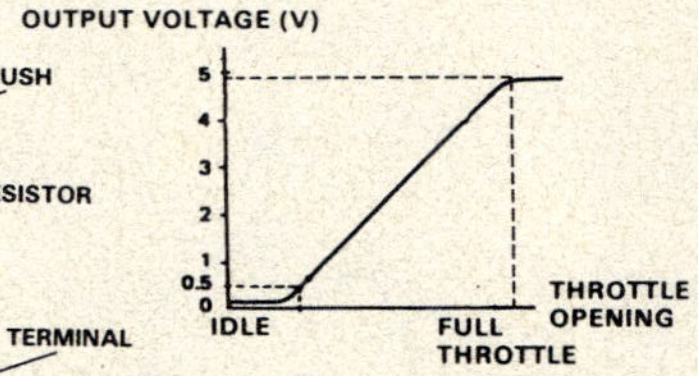

7

- Engine is running.
- Check Engine light has been reported on.
- With service check connector jumped CODE 7 is indicated.

- Do the ECU Reset Procedures
- Start the engine.
- Is Check Engine light on and does it indicate CODE 7?
  - NO → Intermittent failure, system is OK at this time (test drive may be necessary). Check for poor connections or loose wires at C212 (located at right shock tower), C114 (throttle angle sensor) and ECU.
  - YES → Turn the ignition switch OFF.
- Disconnect the 3P connector from the throttle angle sensor.
- Turn the ignition switch ON.
- Measure voltage between YEL/WHT (+) terminal and GRN/WHT (−) terminal.
- Is there approx. 5V ?
  - NO → Measure voltage between YEL/WHT (+) terminal and body ground.
  - YES

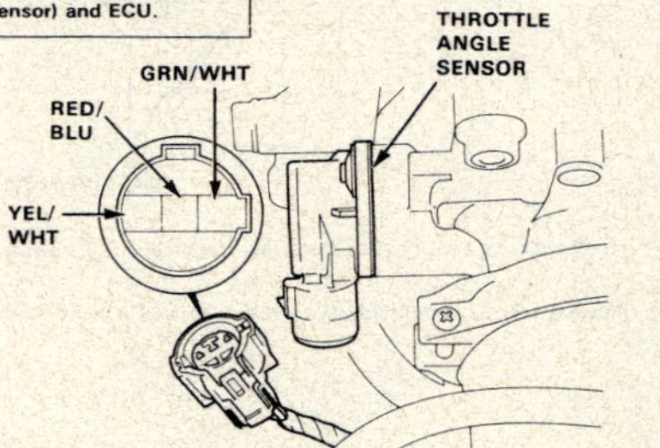

1.5L (D15B7, D15B8, D15Z1) AND 1.6L (D16Z6) ENGINES–TROUBLESHOOTING PGM-FI FAULT CODE 7–1992 CIVIC

Turn the ignition switch OFF.

Reconnect the 3P connector.

Connect the test harness between the ECU and connector

Turn the ignition switch ON.

Measure voltage between D11(+) terminal and D22 (−) terminal.

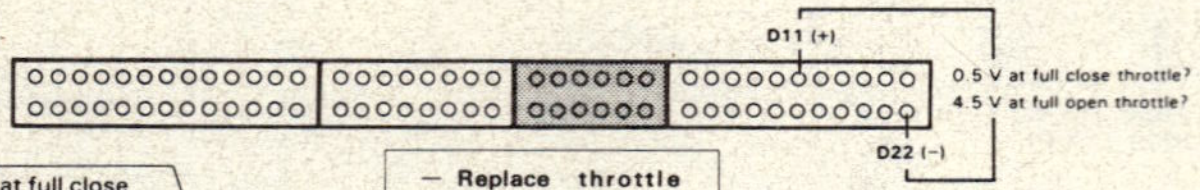

Is voltage 0.5 V at full close throttle, and approx. 4.5 V at full open throttle? NOTE: There should be a smooth transition from 0.5 V to approx. 4.5 V as the throttle is depressed.

- NO → – Replace throttle angle sensor. – Repair open or short in RED/BLU or LT GRN wire between ECU (D11), and throttle angle sensor.
- YES → Substitute a known-good ECU and recheck. If symptom/indication goes away, replace the original ECU.

Is there approx. 5V ?

- YES → Repair open in GRN/WHT wire between ECU (D22) and throttle angle sensor.
- NO ↓

Turn the ignition switch OFF.

Connect the test harness between the ECU and connector

Turn the ignition switch ON.

Measure voltage between D20 (+) terinal and D22 (−) terminal.

Is there approx. 5V ?

- YES → Repair open in YEL/WHT wire between ECU (D20) and throttle angle sensor.
- NO ↓

Substitute a known-good ECU and recheck. If prescribed voltage is now available, replace the original ECU.

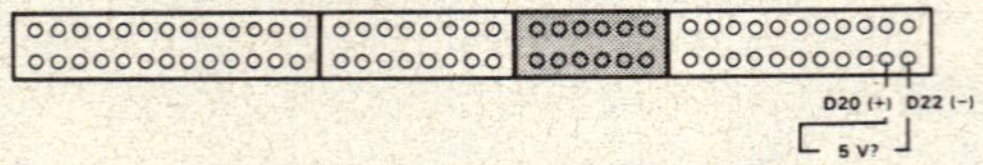

1.5L (D15B7, D15B8, D15Z1) AND 1.6L (D16Z6) ENGINES–TROUBLESHOOTING PGM-FI FAULT CODE 10–1992 CIVIC

(CHECK) – (10) Self-diagnosis Check Engine light indicates code 10: A problem in the Intake Air Temperature (TA) Sensor circuit.

The TA sensor is a temperature dependant resistor (thermistor). The resistance of the thermistor decreases as the intake air temperature increases as shown below.

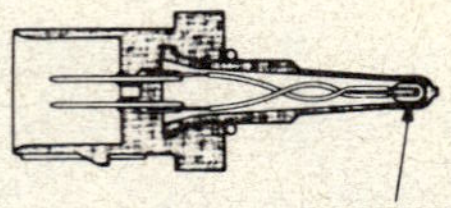

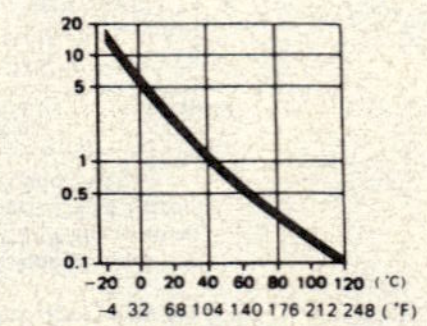

(CHECK) – (10)

- Check Engine light has been reported on.
- With service check connector jumped CODE 10 is indicated.

Do the ECU Reset Procedures

Turn the ignition switch ON.

Is Check Engine light on and does it indicate CODE 10?

- NO → Intermittent failure, system is OK at this time (test drive may be necessary). Check for poor connections or loose wires at C212 (located right shock tower), C116 (TA sensor) and ECU.
- YES ↓

Turn the ignition switch OFF.

Disconnect the 2P connector from the TA sensor.

Measure resistance between the 2 terminals on the TA sensor.

Is there 0.4–4.0 kΩ?

- NO → Replace TA sensor.
- YES

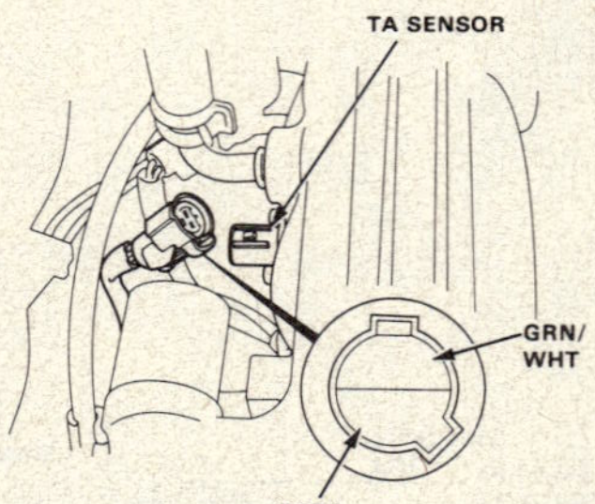

1.5L (D15B7, D15B8, D15Z1) AND 1.6L (D16Z6) ENGINES—TROUBLESHOOTING PGM-FI FAULT CODE 10—1992 CIVIC

Turn the ignition switch ON.

Measure voltage between RED/YEL (+) terminal and body ground.

Is there approx. 5 V ?

- YES → Measure voltage between RED/YEL (+) terminal and GRN/WHT (−) terminal.
  - Is there approx. 5 V ?
    - NO → Repair open in GRN/WHT wire between ECU (D22) and TA sensor.
    - YES → Substitute a known-good ECU and recheck. If symptom/indication goes away, replace the original ECU.
- NO → Turn the ignition switch OFF.

Connect the test harness "D" connector to the ECU only, not to the main wire harness

Turn the ignition switch ON.

Measure voltage between D15 (+) terminal and D22 (−) terminal.

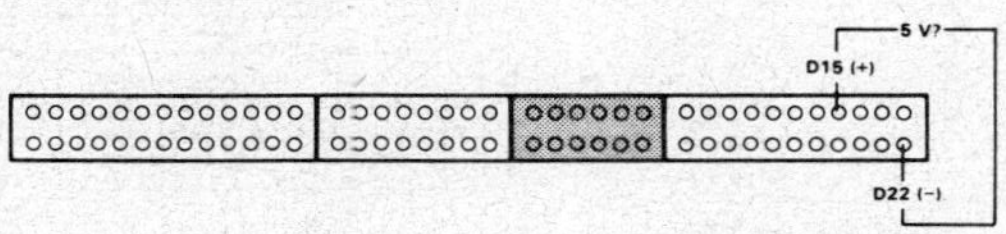

Is there approx. 5V?

- YES → Repair open or short in RED/YEL wire between ECU (D15) and TA sensor.
- NO → Substitute a known-good ECU and recheck. If prescribed voltage is now available, replace the original ECU.

1.5L (D15B7, D15B8, D15Z1) AND 1.6L (D16Z6) ENGINES—TROUBLESHOOTING PGM-FI FAULT CODE 12—1992 CIVIC

12

- Check Engine light has been reported on.
- With service check connector jumped CODE 12 is indicated.

Do the ECU Reset Procedures

Road test necessary: Warm up the engine to normal operating temperature (cooling fan comes on).
Drive the car on the road for approx. 10 minutes. With the transmission in low gear, keep the engine speed in the 1700–2500 range.

Is Check Engine light on and does it indicate CODE 12?

- NO → Intermittent failure, system is OK at this time. Check for poor connections or loose wires at C211 (located at right shock tower), C130 (EGR valve), C306 (control box), C420 (located at left shock tower) and ECU.

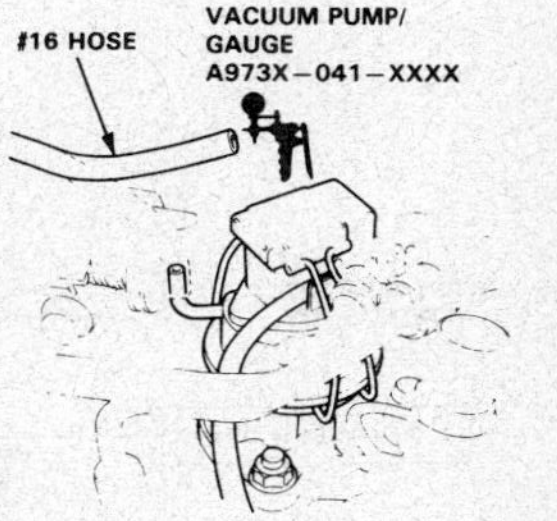

- YES → With the engine at idle, disconnect the #16 hose from the EGR valve and connect a vacuum pump/gauge to the hose.

Is there any vacuum?

- YES → Disconnect 2P connector from the control box and check the #16 hose for vacuum again.
- NO → Move the vacuum pump/gauge to the EGR valve.

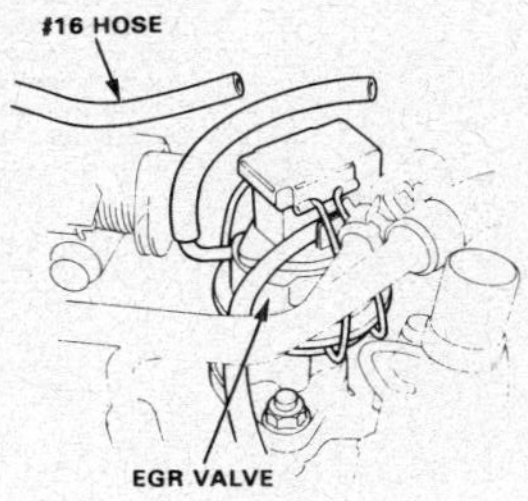

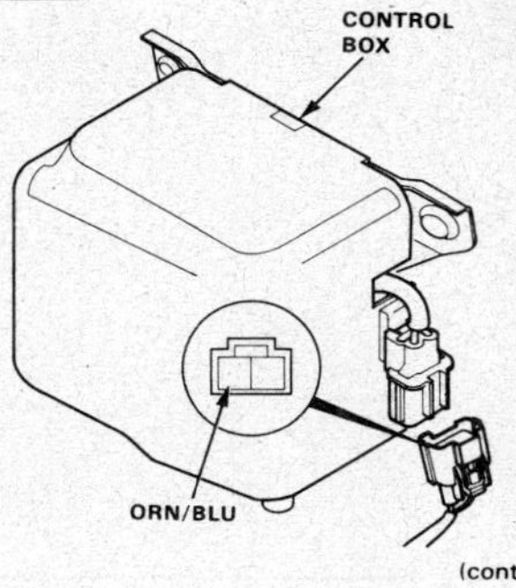

(cont'd)

1.5L (D15B7, D15B8, D15Z1) AND 1.6L (D16Z6) ENGINES—TROUBLESHOOTING PGM-FI
FAULT CODE 12—1992 CIVIC

With the engine at idle, apply 8" of vacuum to the EGR valve.

Does the engine stall or run rough and does the EGR valve hold vacuum?
- NO → Replace EGR valve.
- YES ↓

Disconnect the 2P connector from the control box.

Measure voltage between BLK/YEL (+) terminal on the main wire harness and body ground.

Is there battery voltage ?
- NO → Repair open in BLK/YEL wire between the solenoid valve and ACG (S) (15A) fuse.
- YES ↓

Reconnect the vacuum pump/gauge to the #16 hose.

Start the engine and allow it to idle.

Is there any vacuum ?
- YES → Check vacuum hose routing of the entire EGR system. If hose routing is OK, replace EGR control solenoid valve.
- NO ↓

Turn the ignition switch OFF and disconnect the "A" connector from the ECU.

Check for continuity to ground on ORN/BLU wire of 2P connector.

Does continuity exist?
- YES → Repair short in ORN/BLU wire between EGR control solenoid valve and ECU (A11).
- NO ↓

Substitute a known-good ECU and retest. If symptom/indication goes away, replace the original ECU.

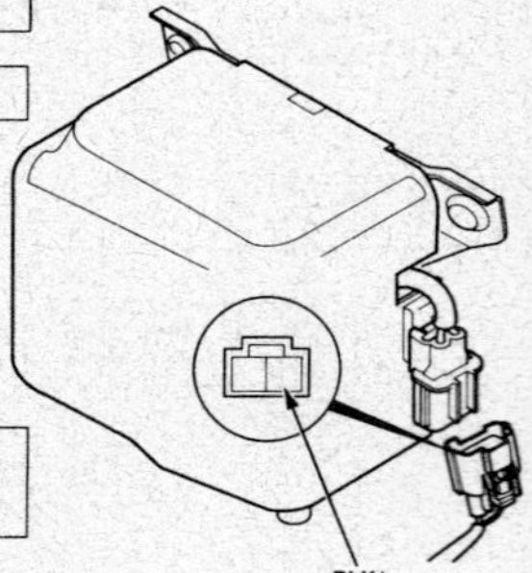

Connect the battery positive terminal to the A terminal of the 2P connector. While watching the vacuum gauge, connect the battery negative terminal to the B terminal.

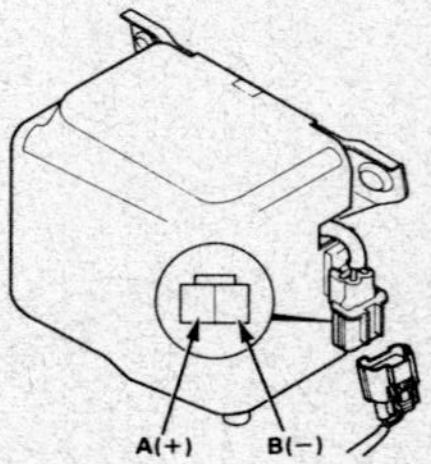

Is there approx. 8" within 1 second?
- NO → Turn the ignition switch OFF and inspect the #16 and #10 hoses for leaks, restrictions, or misrouting.
  - Are the hoses OK?
    - NO → Correct as necessary.
    - YES ↓
  - Disconnect the lower hose on EGR control solenoid valve and connect a vacuum gauge to the hose.
  - Start the engine and allow it to idle.
  - Is there 6"— 10" of vacuum ?
    - NO → Replace CVC valve.
    - YES → Replace the EGR control solenoid valve.
- YES ↓

Turn the ignition switch OFF and reconnect the 2P connector to the EGR control solenoid valve.

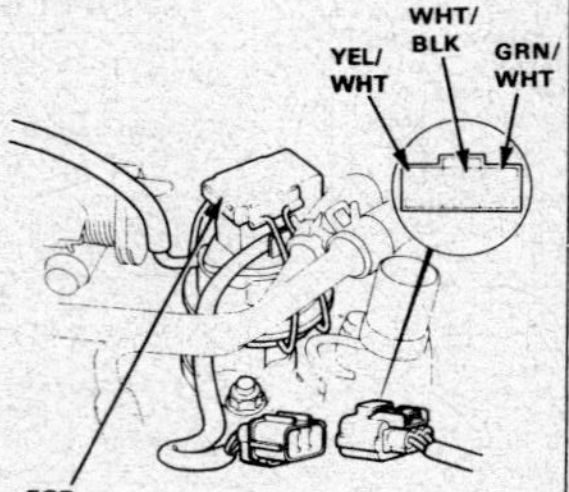

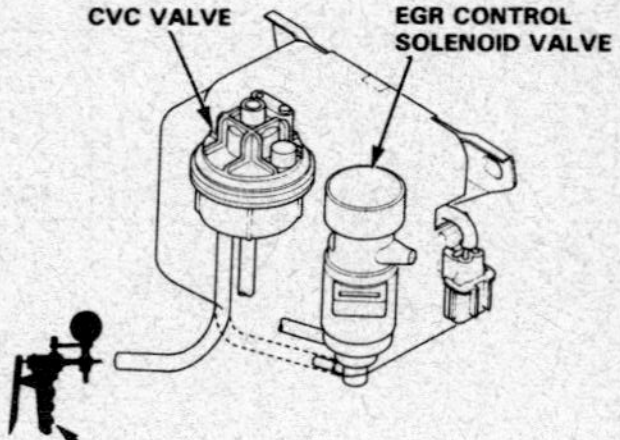

Disconnect 3P connector from the EGR valve.

Turn the ignition switch ON.

Measure voltage between YEL/WHT (+) terminal and GRN/WHT (−) terminal.

Is there approx. 5 V ?
- NO → Measure voltage between YEL/WHT (+) terminal and body ground.
- YES ↓

(cont'd)

1.5L (D15B7, D15B8, D15Z1) AND 1.6L (D16Z6) ENGINES—TROUBLESHOOTING PGM-FI
FAULT CODE 12—1992 CIVIC

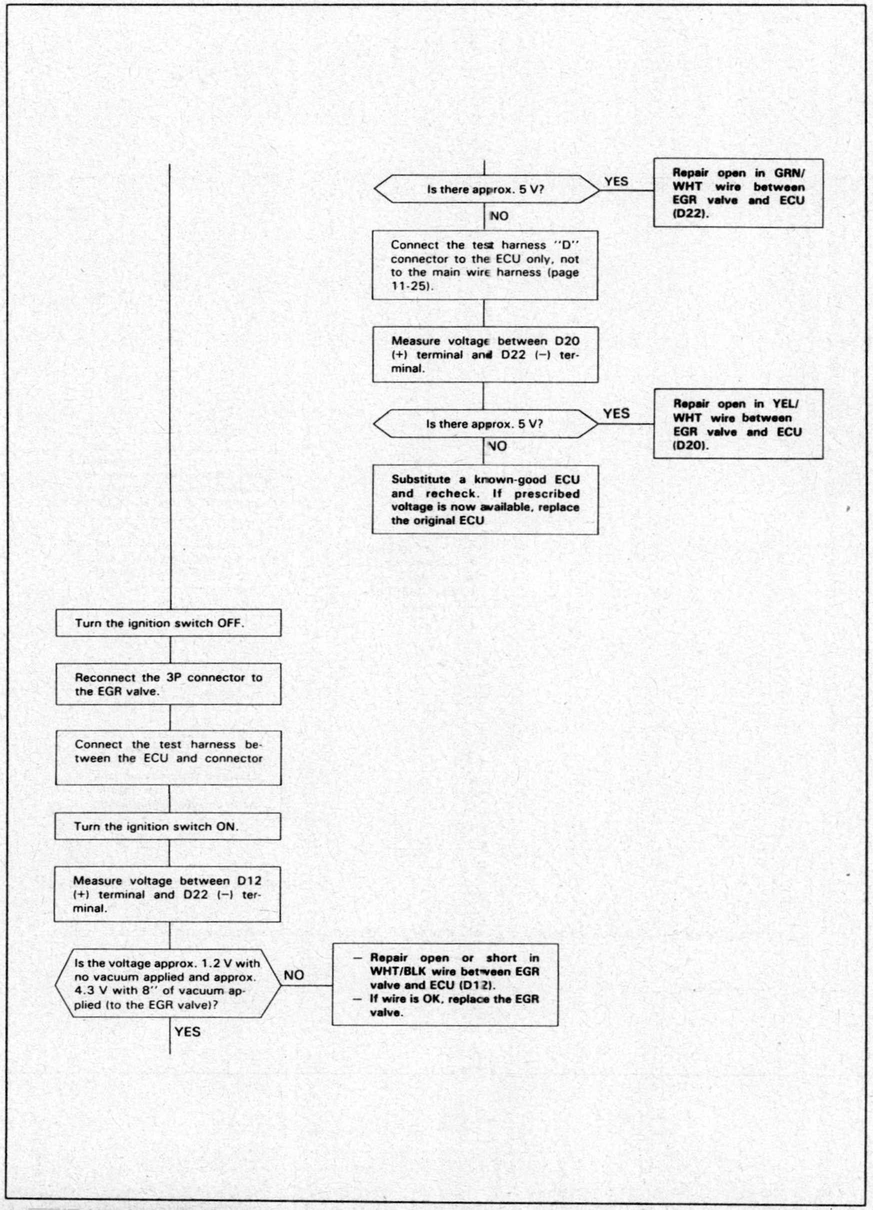

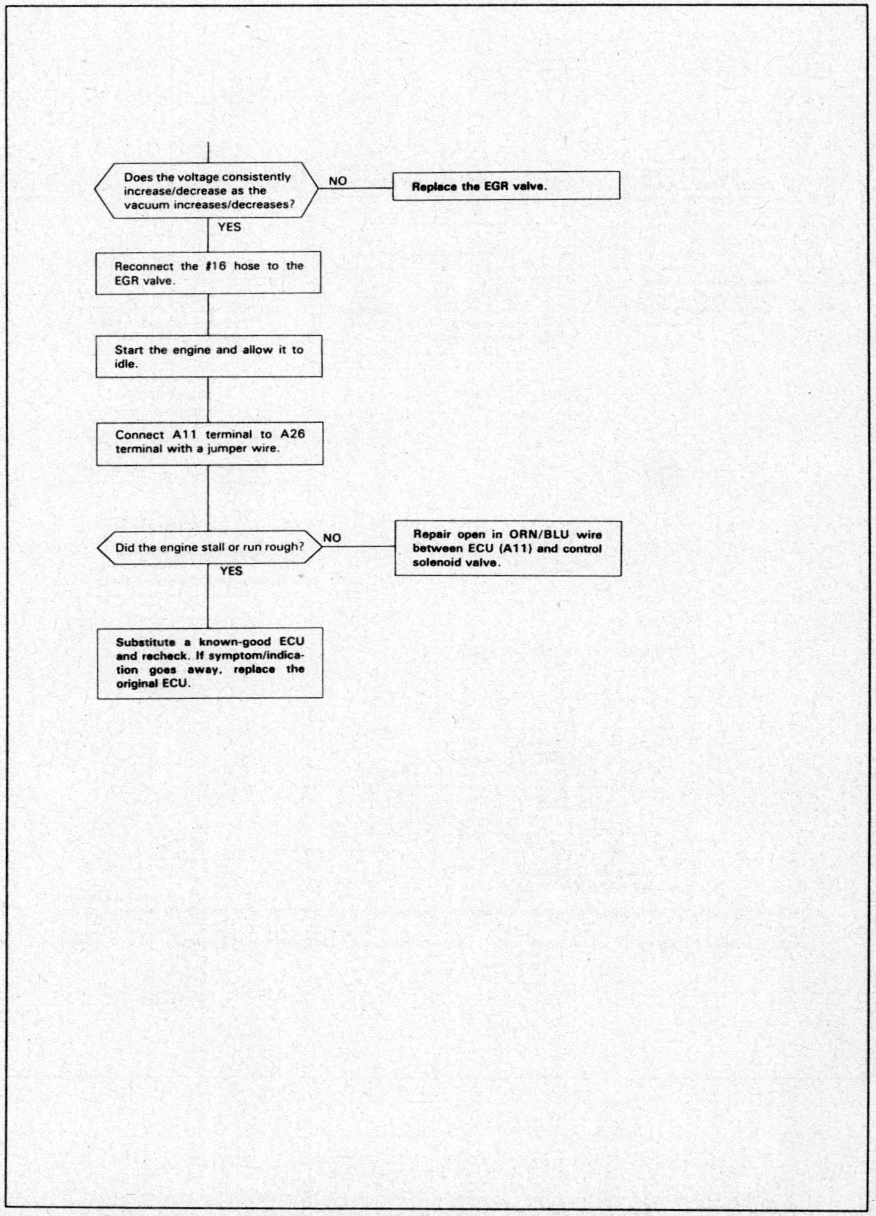

1.5L (D15B7, D15B8, D15Z1) AND 1.6L (D16Z6) ENGINES—TROUBLESHOOTING PGM-FI FAULT CODE 13—1992 CIVIC

CHECK 13 — Self-diagnosis Check Engine light indicates code 13: A problem in the Atmospheric Pressure (PA) Sensor.

The PA sensor is built into the ECU.

CHECK 13

- Check Engine light has been reported on.
- With service check connector jumped CODE 13 is indicated.

Do the ECU Reset Procedures

Turn the ignition switch ON.

Is Check Engine light on and does it indicate CODE 13?

- NO: Intermittent failure, system is OK at this time (test drive may be necessary).
- YES: Substitute a known-good ECU and recheck. If symptom/indication goes away, replace the original ECU.

1.5L (D15B7, D15B8, D15Z1) AND 1.6L (D16Z6) ENGINES—TROUBLESHOOTING PGM-FI FAULT CODE 14—1992 CIVIC

CHECK 14 — Self-diagnosis Check Engine light indicates code 14: A problem in the Electric Air Control Valve (EACV) circuit.

The EACV changes the amount of air bypassing the throttle body in response to a current signal from the ECU in order to maintain the proper idle speed.

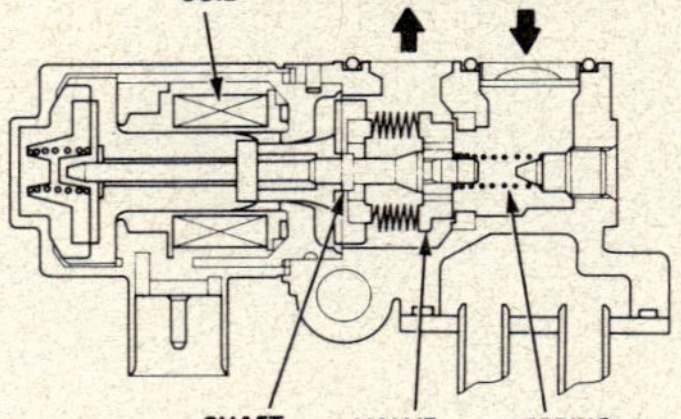

CHECK 14

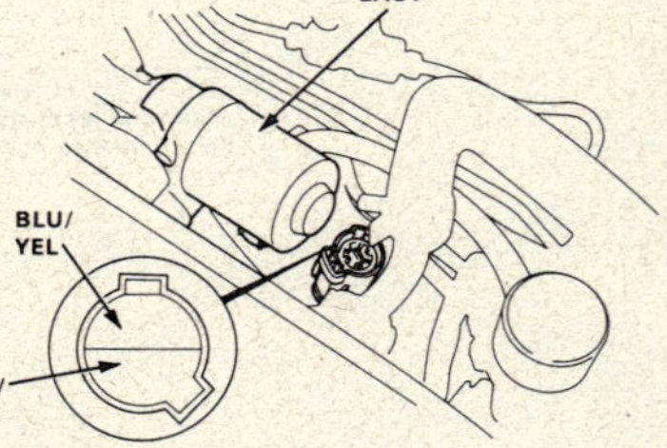

- Check Engine light has been reported on.
- With service check connector jumped CODE 14 is indicated.

Do the ECU Reset Procedures

Start the engine.

Is Check Engine light on and does it indicate CODE 14?

- YES: Remove the 2P connector from the EACV.
- NO: With the engine running and the accelerator pedal released, disconnect the 2P connector from the EACV.
  - Is there a reduction in engine rpm?
    - YES: Intermittent failure, system is OK at this time (test driving may be necessary). Check for poor connection or loose wires at C409 (located at left side under dash), C304 (located at left shock tower), C115 (EACV) and ECU.
    - NO: Substitute a known-good EACV and retest.

1.5L (D15B7, D15B8, D15Z1) AND 1.6L (D16Z6) ENGINES—TROUBLESHOOTING PGM-FI FAULT CODE 14—1992 CIVIC

Measure voltage between the YEL/BLK wire and body ground.

Is there battery voltage?
- NO → **Repair open in YEL/BLK wire between EACV and main relay.**
- YES ↓

Turn the ignition switch off and reconnect the 2P connector the EACV.

Connect the test harness "A" connector to the main wire harness only, not the ECU

Turn the ignition switch ON.

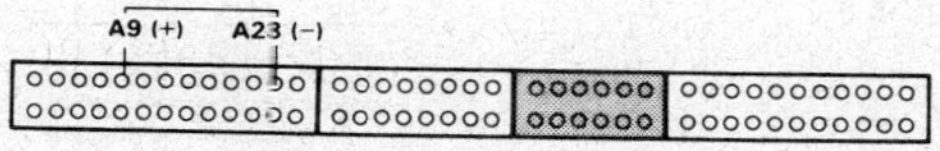

Momentarily connect A9 terminal to A23 terminal several times.

Does the EACV click?
- YES → **Substitute a known-good ECU and retest. If symptom/indication goes away, replace the original ECU.**
- NO → **Repair open or short in GRN/WHT or BLU/YEL wire between EACV and ECU (A9). If the wire is OK, replace the EACV.**

1.5L (D15B7, D15B8, D15Z1) AND 1.6L (D16Z6) ENGINES—TROUBLESHOOTING PGM-FI FAULT CODE 15—1992 CIVIC

15 Self-diagnosis Check Engine light indicates code 15: A problem in the Ignition Output Signal circuit.

— Check Engine light has been reported on.
— With service check connector jumped CODE 15 is indicated.

Do the ECU Reset Procedure

Start the engine.

NOTE: If the engine won't start, it may take 20 seconds of cranking to set the code.

Is Check Engine light on and does it indicate CODE 15?
- NO → **Intermittent failure, system is OK at this time (test drive may be necessary). Check for poor connections or loose wires at C212 (located at right shock tower), C107 (igniter unit) and ECU.**
- YES ↓

Turn the ignition switch OFF.

Disconnect the 2P connector from the distributor

Turn the ignition switch ON.

Measure voltage between BLK/YEL (+) terminal and body ground.

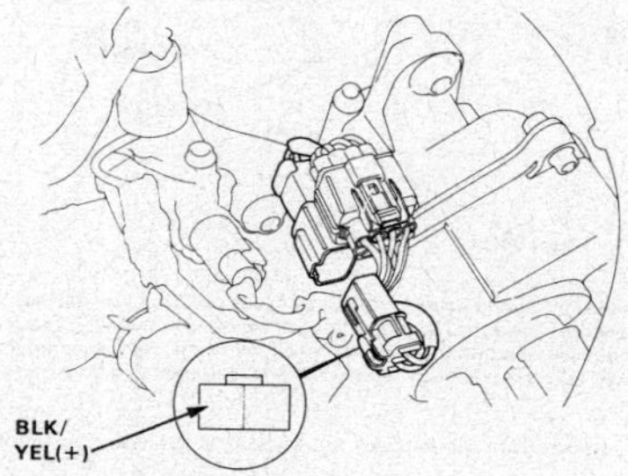

Is there battery voltage ?
- NO → **Repair open in BLK/YEL wire between the 2P connector and ignition switch.**
- YES

1.5L (D15B7, D15B8, D15Z1) AND 1.6L (D16Z6) ENGINES—TROUBLESHOOTING PGM-FI FAULT CODE 15—1992 CIVIC

Turn the ignition switch OFF.

Reconnect the 2P connector.

Connect the test harness between the ECU and connector

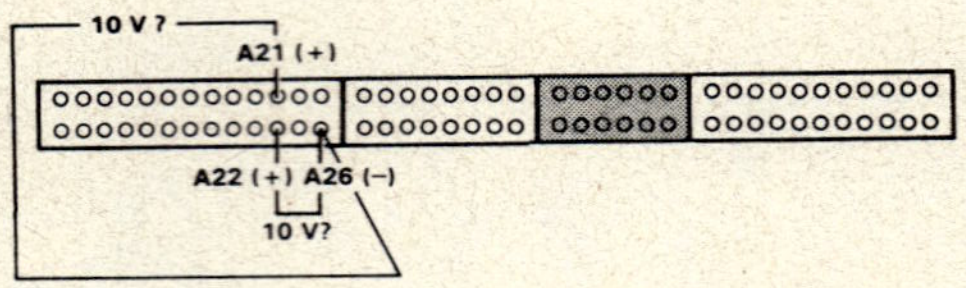

Turn the ignition switch ON.

Measure voltage individually between A21 (+), A22 (+) terminals and A26 (−) terminal.

Is there approx. 10 V?

NO:
- Replace the igniter unit.
- Repair open or short in YEL/GRN wire between igniter unit and ECU (A21 or A22).

NOTE: If the YEL/GRN wire was shorted, the igniter may be damaged.

YES:

Substitute a known-good ECU and recheck. If symptom/indication goes away, replace the original ECU.

1.5L (D15B7, D15B8, D15Z1) AND 1.6L (D16Z6) ENGINES—TROUBLESHOOTING PGM-FI FAULT CODE 16—1992 CIVIC

Troubleshooting Flowchart

16 Self-diagnosis Check Engine light indicates code 16: A problem in the fuel injector circuit.

The injectors are a solenoid-actuated constant-stroke pintle type consisting of a solenoid, plunger needle valve and housing. When current is applied to the solenoid coil, the valve lifts up and pressurized fuel is injected. Because the needle valve lift and the fuel pressure are constant, the injection quantity is determined by the length of time that the valve is open (i.e., the duration the current is supplied to the solenoid coil). The injector is sealed by an O-ring and seal ring at the top and bottom. These seals also reduce operating noise.

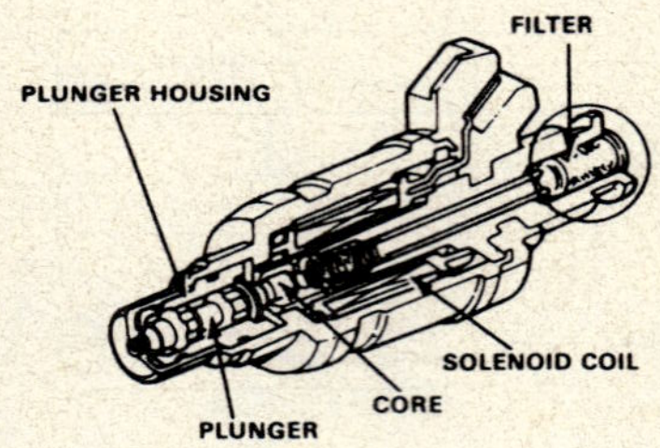

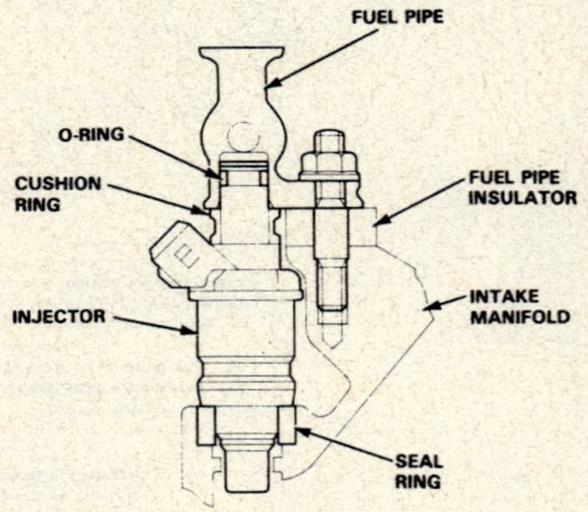

16

- Check Engine light has been reported on.
- With service check connector jumped CODE 16 is indicated.

Do the ECU Reset Procedures

Start the engine and allow it to idle.

NOTE: If engine will not start, it may take 10 seconds of cranking to set the code.

Is Check Engine light on and does it indicate CODE 16?

NO: Intermittent failure, system is OK at this time (test drive may be necessary). Check for poor connections or loose wires at C214 (located at right shock tower), C120, C121, C122, C123 (injector), and ECU.

YES

1.5L (D15B7, D15B8, D15Z1) AND 1.6L (D16Z6) ENGINES—TROUBLESHOOTING PGM-FI
FAULT CODE 16—1992 CIVIC

Turn the ignition switch OFF.

Disconnect the 2P connector from the injector that does not click.

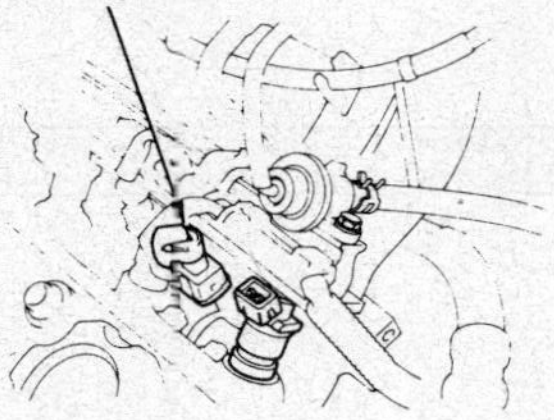

Measure resistance between the 2 terminals of the injector.

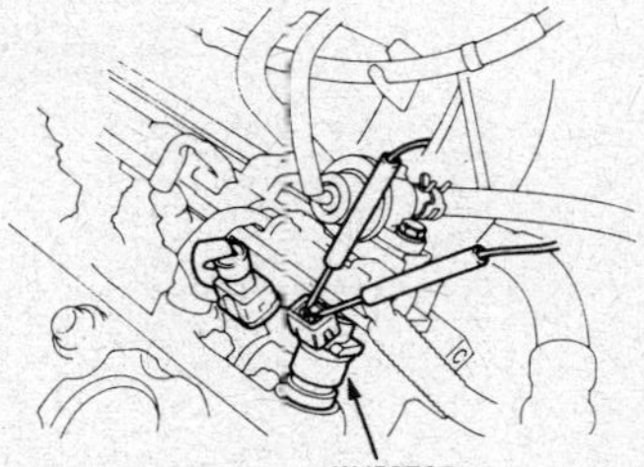

Is there 10—13 Ω?
- NO → Replace the injector/injectors that are not 10—13 Ω.
- YES ↓

Turn the ignition switch ON.

Measure voltage between YEL/BLK (+) terminal in the 2P connector and body ground.

Is there battery voltage ?
- NO → Repair open in the YEL/BLK wire between the injector and the main relay.
- YES ↓

(cont'd)

Reconnect the 2P connector to the injector.

Turn the ignition switch OFF.

Connect the test harness between the ECU and connector

Turn the ignition switch ON.

Measure voltage between A23 (−) terminal and following terminal:
- No.1 injector: A1 (+) terminal.
- No.2 injector: A3 (+) terminal.
- No.3 injector: A5 (+) terminal.
- No.4 injector: A2 (+) terminal.

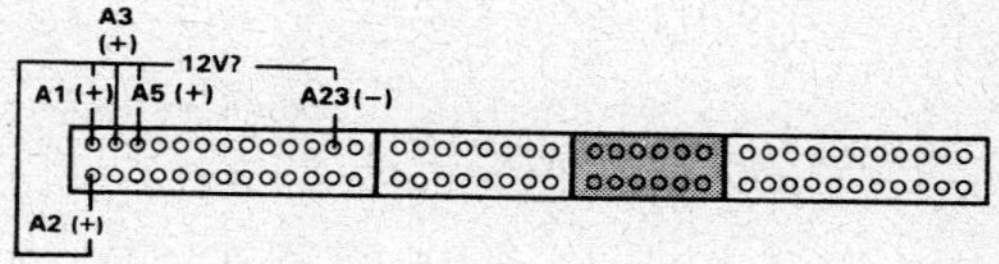

Is there battery voltage ?
- NO → Repair open in the wire between the ECU (A1, A3, A5 or A2) and the injector.
- YES ↓

Substitute a known-good ECU and recheck. If symptom/indication goes away, replace the original ECU.

1.5L (D15B7, D15B8, D15Z1) AND 1.6L (D16Z6) ENGINES—TROUBLESHOOTING PGM-FI
FAULT CODE 17—1992 CIVIC

17 Self-diagnosis Check Engine light indicates code 17: A problem in the Vehicle Speed Sensor circuit.

The signal generated by the speed sensor produces pulses when the front wheels turn.

- **Check Engine light has been reported on.**
- **With service check connector jumped CODE 17 is indicated.**

Do the ECU Reset Procedures

Road test necessary.
In 2nd gear accelerate to 4,000 rpm, then decelerate to 1,500 rpm with throttle fully closed.

Is Check Engine light on and does it indicate CODE 17?

NO → **Intermittent failure, system is OK at this time. Check for poor connections or loose wires at C409 (located at left side under dash), C304 (located at left shock tower), C112 (speed sensor) and ECU.**

YES ↓

Block rear wheels and set the parking brake. Jack up the front of the car and support with safety stands.

⚠WARNING Block rear wheels before jacking up front of car.

Connect the test harness between the ECU and connector

Turn the ignition switch ON.

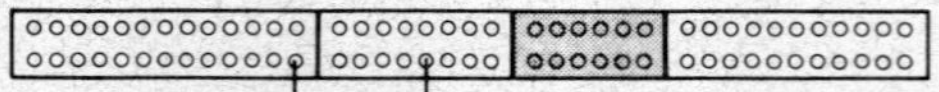

Block the right front wheel and slowly rotate left front wheel and measure voltage between B10 (+) terminal and A26 (−) terminal.

Does voltage pulse 0 V and 12 V?

NO → Turn the ignition switch OFF.

YES ↓

Substitute a known-good ECU and recheck. If symptom/indication goes away, replace the original ECU.

(From Turn the ignition switch OFF.)

Disconnect the B connector from the ECU only, not the main wire harness.

Turn the ignition switch ON.

Block the right front wheel and slowly rotate left front wheel and measure voltage between B10 (+) terminal and A26 (−) terminal.

Does voltage pulse 0 V and 12 V?

NO →
- **Repair short in YEL/BLU wire between ECU (B10) and the speed sensor, speedometer, or cruise control unit.**
- **Repair open in YEL/BLU wire between ECU (B10) and speed sensor.**
- **If wire is OK, test the speed sensor**

YES ↓

Substitute a known-good ECU and recheck. If symptom/indication goes away, replace the original ECU.

1.5L (D15B7, D15B8, D15Z1) AND 1.6L (D16Z6) ENGINES—TROUBLESHOOTING PGM-FI
FAULT CODE 19—1992 CIVIC

CHECK 19 Self-diagnosis Check Engine light indicates code 19: A problem in the Lock-up Control Solenoid Valve A (or B) circuit.

CHECK 19

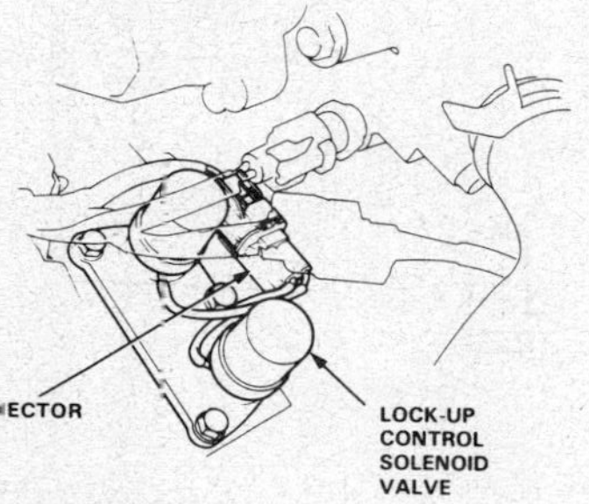

- **Check Engine light has been reported on.**
- **With service check connector jumped CODE 19 is indicated.**

Do the ECU Reset Procedures

Test drive necessary.
Drive the car for several miles so that the transmission upshfts and downshifts several times.

Does Check Engine light indicate CODE 19?

NO → **Intermittent failure, system is OK at this time. Check for poor connections or loose wires at C409 (located at left side under dash), C304 (located at left shock tower), C110 (Lock-up Control Solenoid Valves) and ECU.**

YES ↓

Turn the ignition switch OFF.

Connect the test harness to the main harness only, not to the ECU

Disconnect the 2P connector from the lock-up control solenoid valve.

Check for continuity between A19 or A17* and body ground.

Does continuity exist?

YES → **Repair short in YEL or GRN/BLK* wire between ECU (A19 or A17*) and the 2P connector.**

NO ↓

* : Lock-up Control Solenoid Valve B

Reconnect the 2P connector.

Measure resistance between A19 or A17* terminal and A26 terminal.

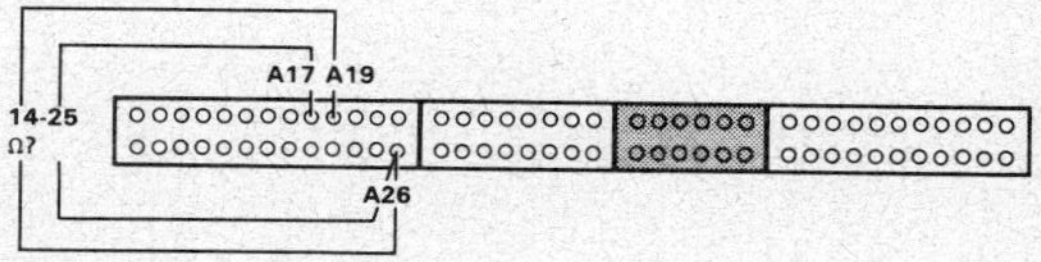

Is there 14-25 Ω ?

YES → **Substitute a known-good ECU and recheck. If symptom/indication goes away, replace the original ECU.**

NO → Check for continuity in YEL or GRN/BLK wire between ECU (A19 or A17*) and the 2P connector.

Does continuity exist?

YES → **Faulty lock-up control solenoid valve**

NO → **Repair open in YEL or GRN/BLK* wire between ECU (A19 or A17*) and the 2P connector.**

1.5L (D15B7, D15B8, D15Z1) AND 1.6L (D16Z6) ENGINES – TROUBLESHOOTING PGM-FI
FAULT CODE 20 – 1992 CIVIC

20 Self-diagnosis Check Engine light indicates code 20: A problem in the Electric Load Detector circuit.

With service check connector jumped CODE 20 is indicated.

Do the ECU Reset Procedure

Start engine and keep engine rpm at idle.
Turn on headlights.

Does Check Engine light indicate CODE 20?

- NO → **Intermittent failure, system is OK at this time (test drive may be necessary). Check for poor connections or loose wires at C202 (ELD) and ECU.**
- YES ↓

Turn the ignition switch OFF.

Remove the under-hood fuse/relay box and remove the fuse/relay box lower cover.

Disconnect the 3P connector from the electric load detector.

Turn the ignition switch ON.

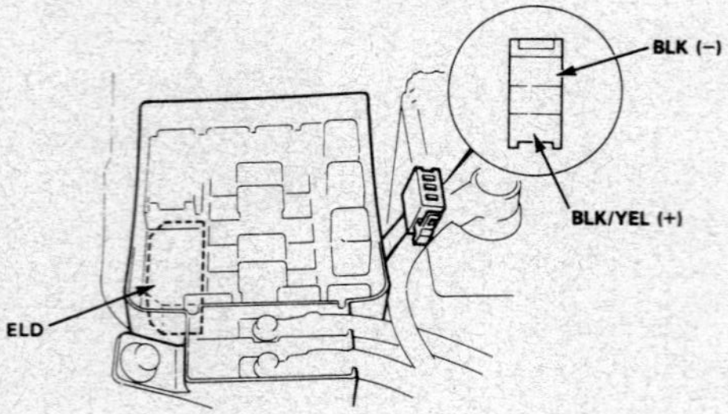

Measure voltage between BLK/YEL (+) terminal and BLK (–) terminal.

Is there battery voltage?

- NO → Measure voltage between BLK/YEL (+) terminal and body ground.
  - Is there battery voltage?
    - YES → **Repair open in BLK wire between the 3P connector and G201.**
    - NO → **Repair open in BLK/YEL wire between ACG (S) (15A) and the 3P connector.**
- YES ↓

Measure voltage between GRN/RED terminal and body ground.

Is there 4.5–5 V?

- NO → **Repair open or short in GRN/RED wire between ECU (D10) and the 3P connector. If wire is OK, substitute a known-good ECU and recheck.**
- YES ↓

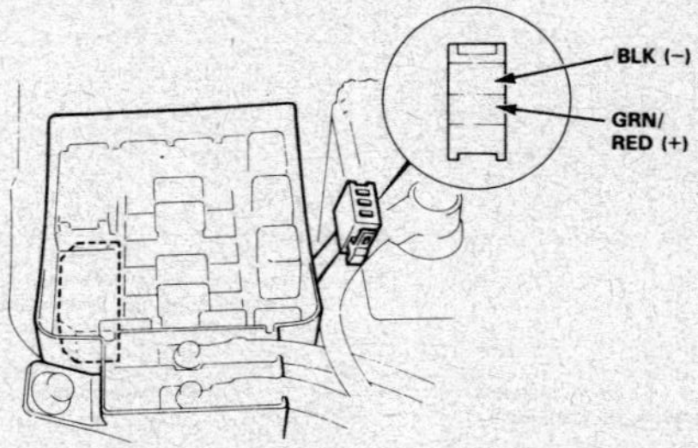

Turn the ignition switch OFF.

Connect the 3P connector to the electric load detector.

Connect the test harness between the ECU and connector (page 11-25).

(cont'd)

1.5L (D15B7, D15B8, D15Z1) AND 1.6L (D16Z6) ENGINES—TROUBLESHOOTING PGM-FI FAULT CODE 20—1992 CIVIC

Condition	Voltage
Headlight switch, first position (•)	2.5–3.5 V
Headlight switch, second position (●)	1.5–2.5 V

Start the engine and allow it to idle.

Under the conditions listed in the chart to the right, measure voltage between D10 (+) terminal and A26 (−) terminal.

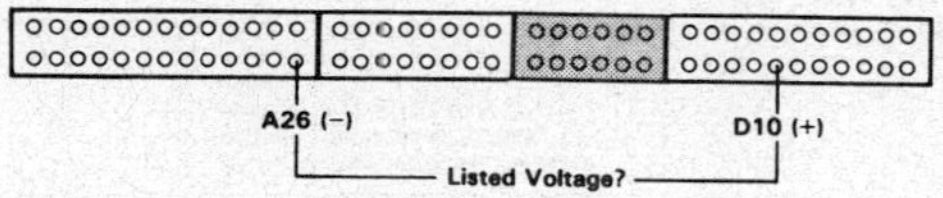

Is the voltage listed in the chart available?

NO — Faulty electric load detector.

YES — Substitute a known-good ECU and recheck. If symptom/indication goes away, replace the original ECU.

1.5L (D15B7, D15B8, D15Z1) AND 1.6L (D16Z6) ENGINES—TROUBLESHOOTING PGM-FI FAULT CODE 41—1992 CIVIC

41 — Self-diagnosis Check Engine light indicates code 41: A problem in the Oxygen (O_2) Sensor Heater circuit.

41

- Engine is running.
- Check Engine light has been reported on. With service check connector jumped CODE 41 is indicated.

Do the ECU Reset Procedures

Start the engine.

Is Check Engine light on and does it indicate CODE 41?

NO — Intermittent failure, system is OK at this time (test drive may be necessary). Check for poor connections or loose wires at C212 (located at right shock tower), C111 (O_2 sensor) and ECU.

YES — Stop the engine

Disconnect the 4P connector from the O_2 sensor.

Measure resistance between terminals C and D on the O_2 sensor.

Is there 10–40 Ω

NO — Replace O_2 sensor.

YES

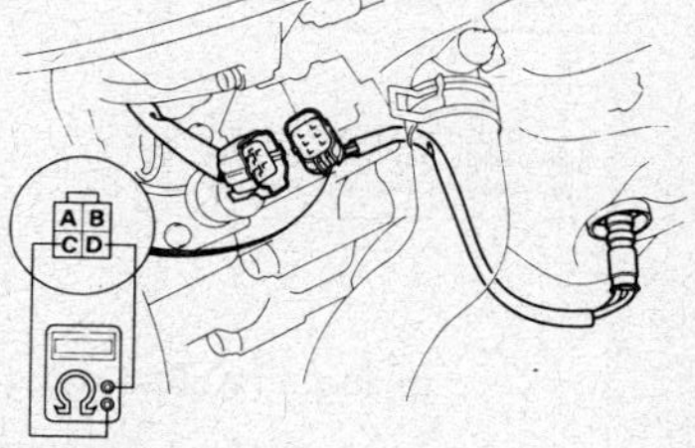

1.5L (D15B7, D15B8, D15Z1) AND 1.6L (D16Z6) ENGINES—TROUBLESHOOTING PGM-FI
FAULT CODE 41—1992 CIVIC

Check for continuity to body ground on terminals C and D of the O2 sensor.

Does continuity exist? — YES: Replace O2 sensor.

NO

Check for continuity between terminal D and terminals A and B individually.

Does continuity exist? — YES: Replace O2 sensor.

NO

Turn the ignition switch ON.

At O2 sensor harness, measure voltage between YEL/BLK (+) terminal and ORN/BLK (−) terminal.

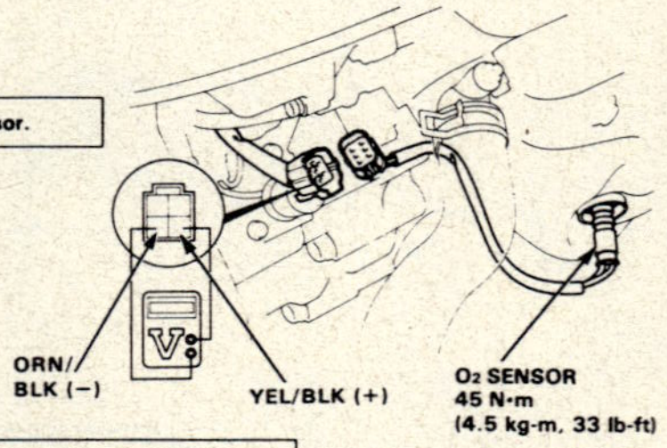

Is there battery voltage? — YES: Disconnect the "A" connector from the ECU.

- At O2 sensor harness, measure voltage between YEL/BLK (+) terminal and ORN/BLK (−) terminal.
- Is there battery voltage? — YES: Repair short in ORN/BLK wire between ECU (A6) and O2 sensor.
- NO: Reconnect the O2 sensor connector.

NO

Measure voltage between YEL/BLK (+) terminal and body ground.

Is there battery voltage? — NO: Repair open in YEL/BLK wire between O2 sensor and main relay.

YES

Turn the ignition switch OFF.

(cont'd)

(D15B7, D16Z6 engine) (cont'd)

Reconnect the 4P connector to O2 sensor.

Connect the test harness "A" connector to the main wire harness only, not the ECU

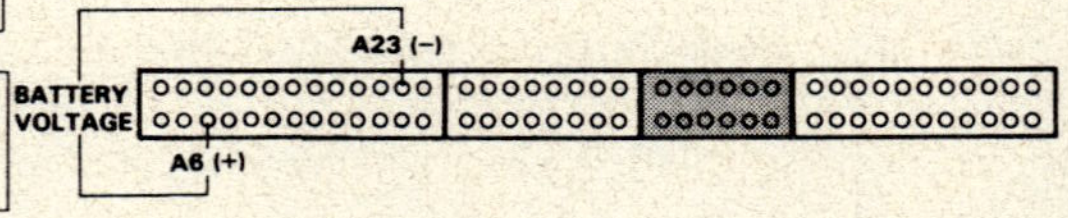

Turn the ignition switch ON.

Measure voltage between A6 (+) terminal and A23 (−) terminal.

Is there battery voltage? — NO: Repair open ORN/BLK wire between ECU (A6) and O2 sensor.

YES

Substitute a known-good ECU and recheck. If symptom/indication goes away, replace the original ECU.

Connect the test harness "A" connector to the main wire harness only, not the ECU

Connect an ammeter between terminals A6 (+) and A26 (−).

Is the current less than 0.1 A?* — YES: Replace O2 sensor.

NO

Substitute a known-good ECU and recheck. If symptom/indication goes away, replace the original ECU.

* Monitor over a 5 minutes period.

1.5L (D15B7, D15B8, D15Z1) AND 1.6L (D16Z6) ENGINES—TROUBLESHOOTING PGM-FI
FAULT CODE 41—1992 CIVIC

41 — Self-diagnosis Check Engine light indicates code 41: A problem in the Linear Air Fuel ratio (LAF) Sensor Heater circuit.

41

- Engine is running.
- Check Engine light has been reported on. With service check connector jumped CODE 41 is indicated.

Do the ECU Reset Procedures

Start the engine.

Is Check Engine light on and does it indicate CODE 41?

- NO → Intermittent failure, system is OK at this time (test drive may be necessary). Check for poor connections or loose wires at C212 (located at right shock tower), C111 (LAF sensor) and ECU.
- YES ↓

Stop the engine

Disconnect the 8P connector from the LAF sensor.

Measure resistance between terminals A and B on the LAF sensor.

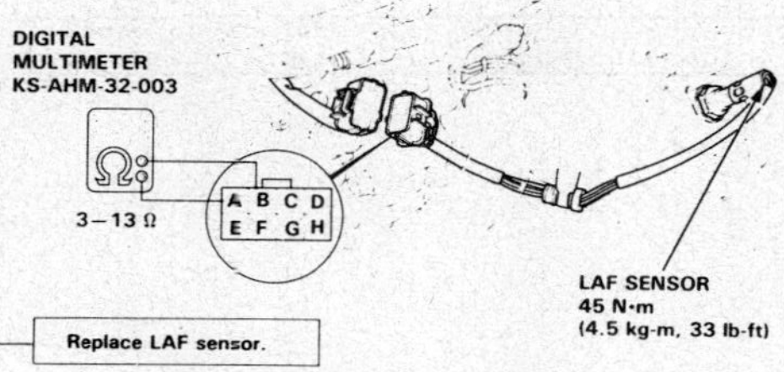

Is there 3–13 Ω?

- NO → Replace LAF sensor.
- YES ↓

Check for continuity to body ground on each terminal on the LAF sensor.

Does continuity exist?

- YES → Replace LAF sensor.
- NO ↓

Check for continuity between terminal A and terminals C–H individually.

Does continuity exist?

- YES → Replace LAF sensor.
- NO ↓

Start engine and keep engine rpm at idle.

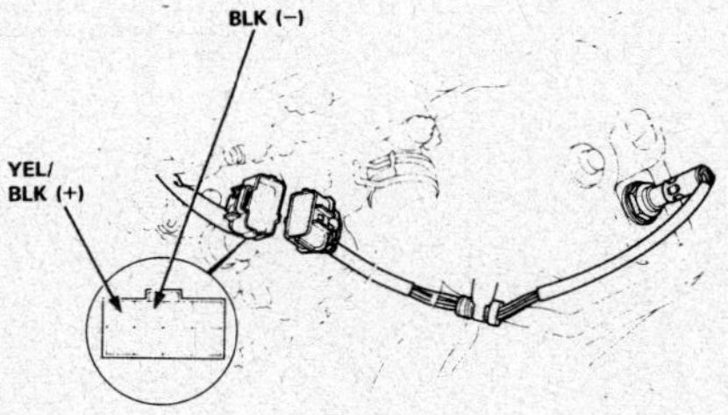

Measure voltage between YEL/BLK (+) terminal and body ground.

Is there approx. 7 V?

- YES → Measure voltage YEL/BLK (+) terminal and BLK (–) terminal.
  - Is there battery voltage?
    - NO → Repair open in BLK wire between LAF sensor and G101.
    - YES → Substitute a known-good ECU and recheck. If symptom/indication goes away, replace the original ECU.
- NO ↓

Turn the ignition switch OFF.

(cont'd)

1.5L (D15B7, D15B8, D15Z1) AND 1.6L (D16Z6) ENGINES – TROUBLESHOOTING PGM-FI FAULT CODE 41 – 1992 CIVIC

Connect the test harness between the ECU and connector. Disconnect "A" connector from the ECU only, not main wire harness

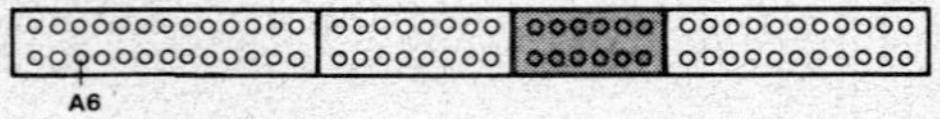

Check for continuity between A6 terminal and body ground.

Does continuity exist?
- YES → **Repair short in ORN/BLK, YEL/BLK wire between ECU (A6) and LAF sensor.**
- NO ↓

Disconnect the 8P connector from the LAF sensor.

Check for continuity ORN/BLK, YEL/BLK wire between ECU (A6) and LAF sensor.

Does continuity exist?
- YES → **Repair open in ORN/BLK, YEL/BLK wire between ECU (A6) and LAF sensor.**
- NO ↓

Substitute a known-good ECU and recheck. If symptom/indication goes away, replace the original ECU.

1.5L (D15B7, D15B8, D15Z1) AND 1.6L (D16Z6) ENGINES – TROUBLESHOOTING PGM-FI FAULT CODE 43 – 1992 CIVIC

43 — Self-diagnosis Check Engine light indicates code 43: Most likely a problem in the Oxygen (O_2) Sensor circuit or a problem in the Fuel Supply System.

43

- **Check Engine light has been reported on.**
- **With service check connector jumped CODE 43 is indicated.**
- **or continued from code 1.**

Is the 43 code accompanied by the Check Engine light and poor driveability?
- YES → **Go to Fuel Supply System**
- NO ↓

Do the ECU Reset Procedures

Warm up engine to normal operating temperature (cooling fan comes on).

Hold engine at 3,000 rpm for two minutes. (A/T: Transmission in N or P.)

Is the Check Engine light on and does it indicate CODE 43?
- NO → **Intermittent failure, system is OK at this time (test drive may be necessary). Check for poor connections or loose wires at C213 (located at right shock tower). C111 (O_2 sensor) and ECU.**
- YES ↓

Turn the ignition switch OFF.

Connect the test harness between the ECU and connector

With the ignition switch OFF, wait for at least two minutes.

Install a jumper wire on the test harness between A6 and A26.

Turn the ignition switch ON.

Measure voltage between D14 (+) terminal and A26 (−) terminal as soon as the ignition switch is turned on.

NOTE:
- Use DIGITAL MULTIMETER.
- Use 2 Volt range.

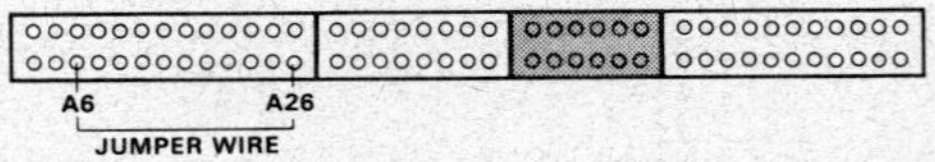

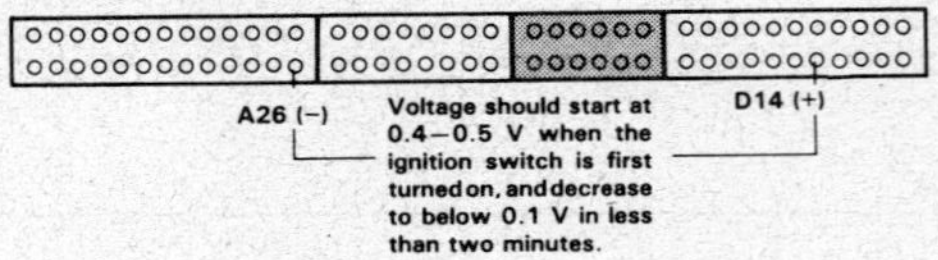

1.5L (D15B7, D15B8, D15Z1) AND 1.6L (D16Z6) ENGINES—TROUBLESHOOTING PGM-FI FAULT CODE 43—1992 CIVIC

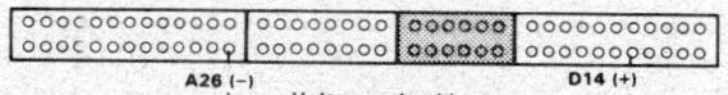

Is there 0.1 V or less when the ignition switch is first turned on?

- YES → Disconnect the 4P connector from the O_2 sensor.
  - Measure voltage between D14 (+) terminal and A26 (−) terminal.
  - Is there more than 0.1 V?
    - YES → Replace O_2 sensor.
    - NO → Disconnect the "D" connector from the main wire harness.
      - Measure voltage between D14 (+) terminal and A26 (−) terminal.
      - Is there more than 0.1 V?
        - YES → Repair short in WHT wire between ECU (D14) and O_2 sensor.
        - NO → Substitute a known-good ECU and recheck. If symptom/indication goes away, replace the original ECU.
- NO → Disconnect the 4P connector from the O_2 sensor.
  - Measure voltage at the engine wire harness side of the O_2 sensor connector between GRN/WHT (−) and WHT (+).
  - Is there more than 0.1 V?
    - YES → Replace O_2 sensor.
    - NO → Repair open in WHT wire between ECU (D14) and O_2 sensor. If wire is OK, check for open in GRN/WHT wire between ECU (D22) and O_2 sensor.

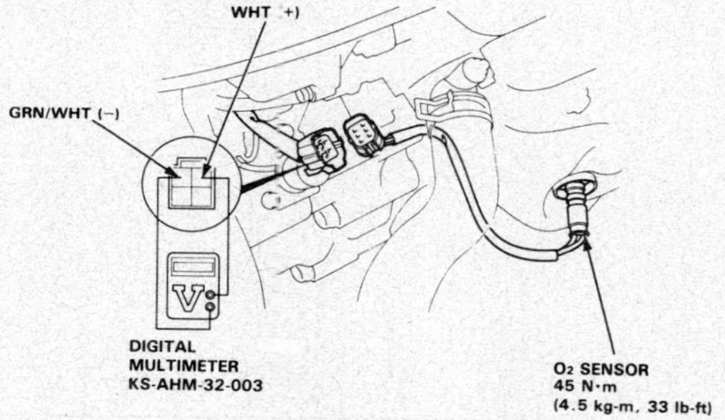

1.5L (D15B7, D15B8, D15Z1) AND 1.6L (D16Z6) ENGINES—TROUBLESHOOTING PGM-FI FAULT CODE 48—1992 CIVIC

CHECK — 48 — Self-diagnosis Check Engine light indicates code 48: A problem in the Linear Air Fuel ratio (LAF) Sensor circuit. The LAF sensor operates over a wide air/fuel range.

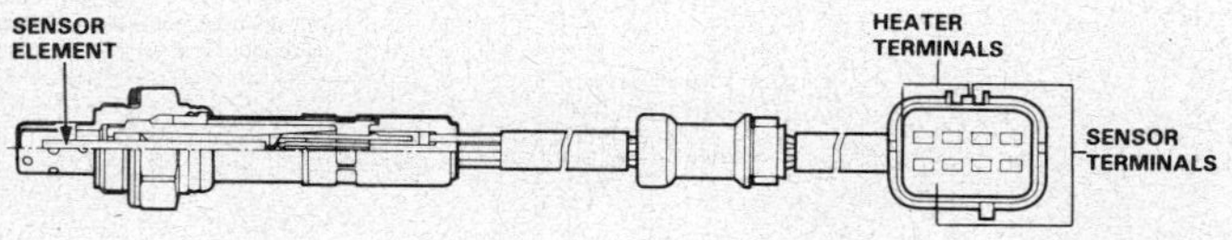

CHECK — 48

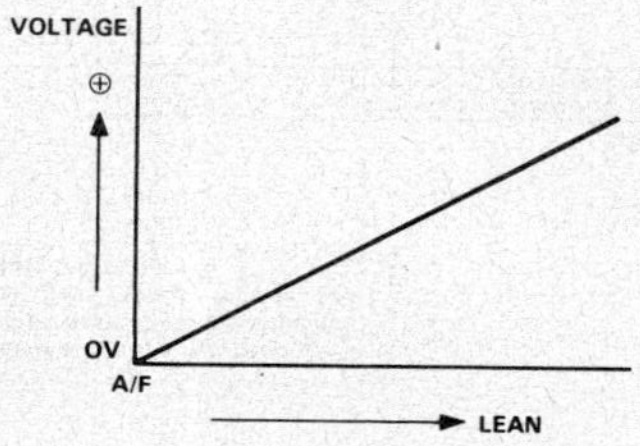

- Engine is running.
- Check Engine light has been reported on. With service check connector jumped CODE 48 is indicated.

Do the ECU Reset Procedures

Start the engine.

Is Check Engine light on and does it indicate CODE 48?

- YES → (continues)
- NO → Road test with the transmission in 3rd gear. Hold the engine speed 1500 rpm.
  - Is Check Engine light on and does it indicate CODE 48?
    - NO → Intermittent failure, system is OK at this time (test drive may be necessary). Check for poor connections or loose wires at C211 (located at right Shock Tower), C111 (LAF sensor) and ECU.
    - YES → Turn the ignition switch OFF.
      - Connect the test harness between the ECU and connector
      - (Go to wire harness open check: ECU (D14) terminal and LAF sensor)

NOTE: Use DIGITAL MULTIMETER

1.5L (D15B7, D15B8, D15Z1) AND 1.6L (D16Z6) ENGINES—TROUBLESHOOTING PGM-FI

FAULT CODE 48—1992 CIVIC

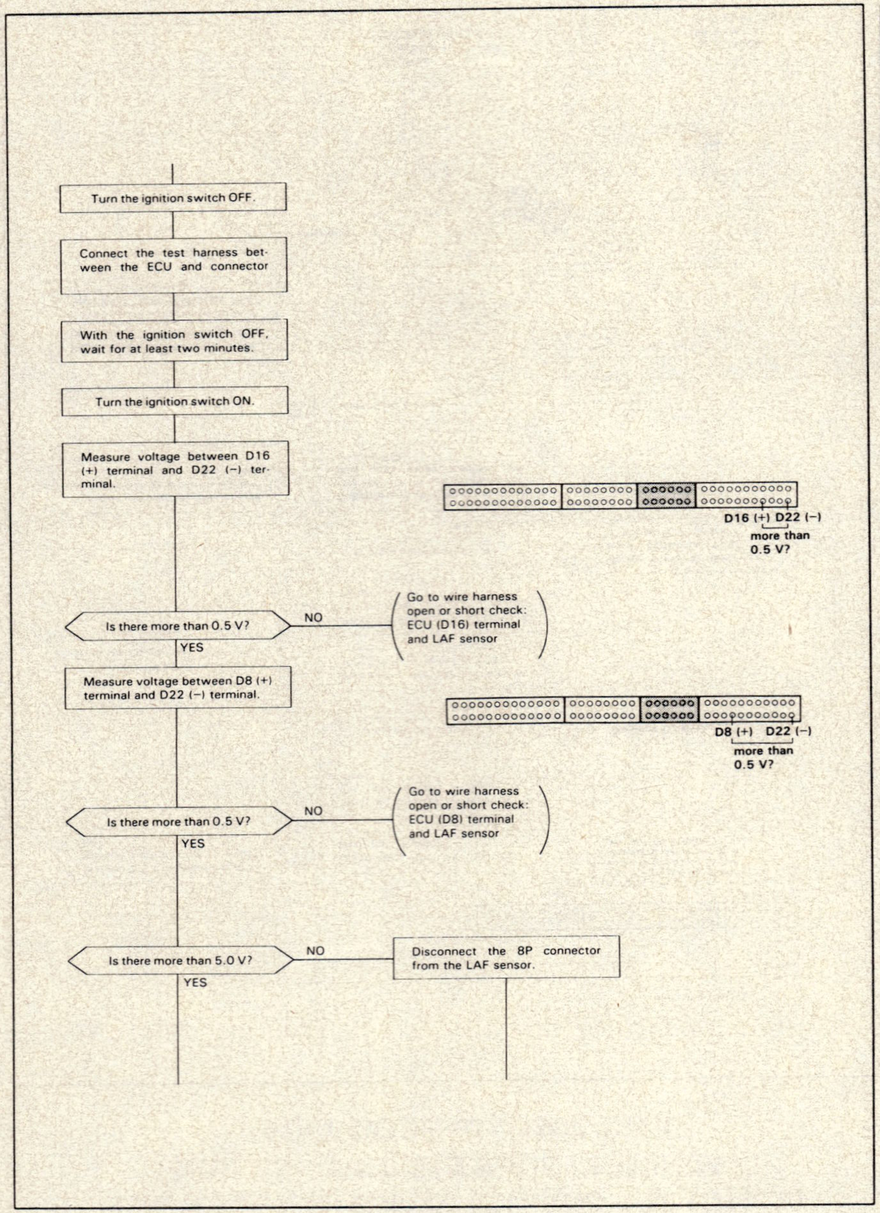

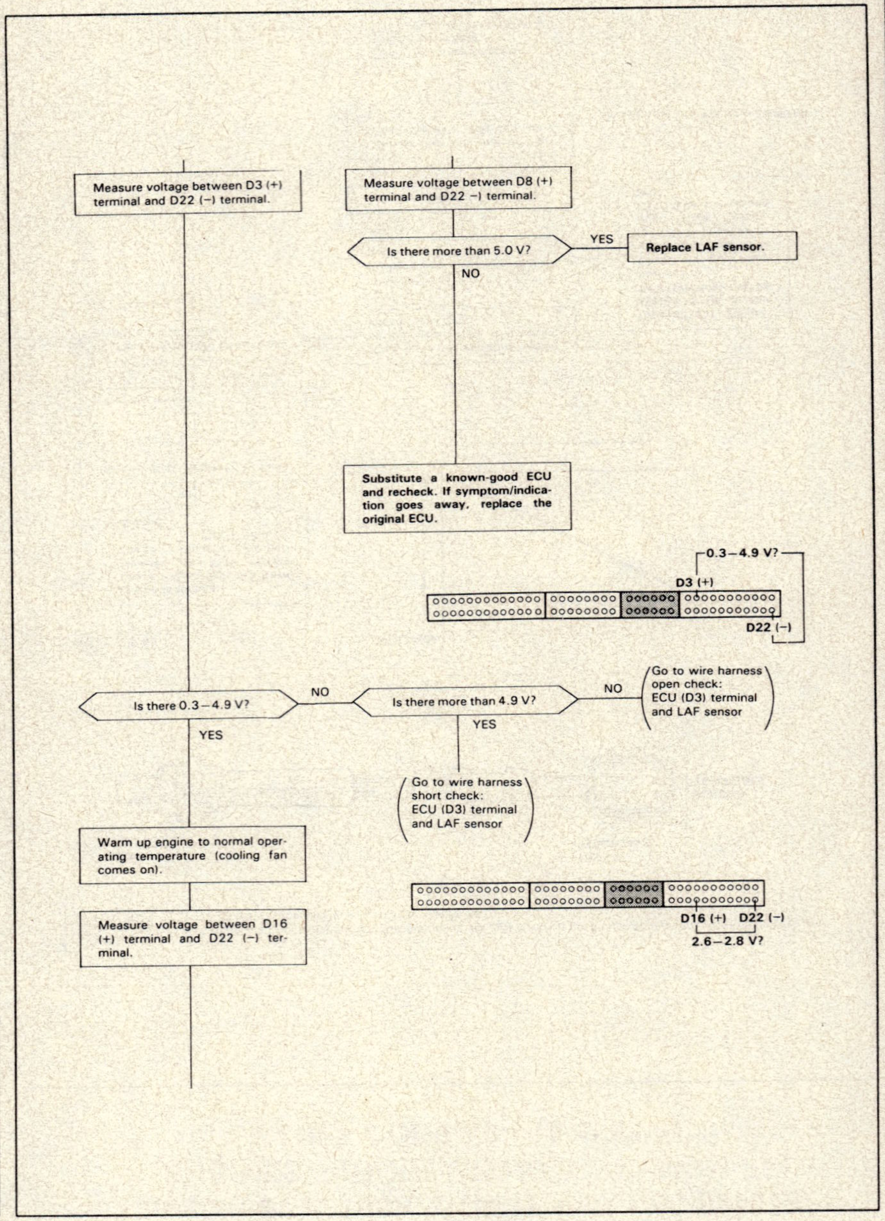

1.5L (D15B7, D15B8, D15Z1) AND 1.6L (D16Z6) ENGINES–TROUBLESHOOTING PGM-FI
FAULT CODE 48–1992 CIVIC

Is there 2.6–2.8 V?
- NO → Measure voltage between D8 (+) terminal and D22 (–) terminal.

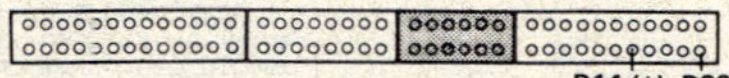

Is there less than 2.8 V?
- NO → **Replace LAF sensor.**
- YES → **Substitute a known-good ECU and recheck. If symptom/indicaton goes away, replace the original ECU.**

- YES → Measure voltage between D14 (+) terminal and D22 (–) terminal.

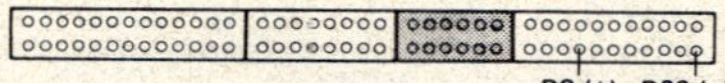

Is there more than 0.4 V?
- NO → (Go to wire harness short check: ECU (D14) terminal and LAF sensor)
- YES → **Substitute a known-good ECU and recheck. If symptom/indication goes away, replace the original ECU.**

Wire harness short check:

Turn the ignition switch OFF.

Disconnect "D" connector from the ECU only, not the main wire harness

Check for continuity to body ground on indicated terminals.
* see table

Does continuity exist?
- NO → **Substitute a known-good ECU and recheck. If symptom/indication goes away, replace the original ECU.**
- YES → Disconnect the 8p connector from the LAF sensor.

Check for continuity to body ground on indicated terminals.
* see table

Does continuity exist?
- YES → **Repair short in the indicated sensor wires.**
*** see table**
- NO → **Replace LAF sensor.**

* Table:

CODE	ECU TERMINAL	WIRE COLOR	
		MAIN WIRE HARNESS	ENGINE WIRE HARNESS
41	A6 (HTCNTL)	ORN/BLK	YEL/BLK
	A23 (PG1)	BLK	←
48	D22 (SG2)	GRN/WHT	←
	D3 (LABEL)	BLU/YEL	WHT
	D8 (VS +)	WHT/BLU	←
	D14 (IP +)	ORN/BLU	←
	D16 (IP –, VS –)	BLU/GRN	ORN

Wire harness open check:

Turn the ignition switch OFF.

Disconnect "D" connector from the ECU only, not the main wire harness

Disconnect the 8p connector from the LAF sensor.

Check for continuity between indicated terminals and wire harness.
* see table

Does continuity exist?
- NO → **Repair open in the indicated sensor wires.**
*** see table**
- YES → **Inspect for poor terminal to terminal contact at LAF sensor connector and ECU. If terminal contact is OK, replace LAF sensor.**

WIRE COLOR (ENGINE WIRE HARNESS SIDE):

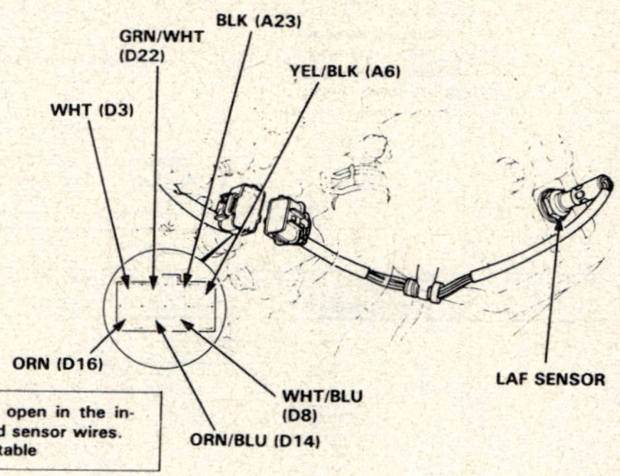

1.5L (D15B7, D15B8, D15Z1) AND 1.6L (D16Z6) ENGINES—TROUBLESHOOTING
AIR CONDITIONING SIGNAL— 1992 CIVIC

This signals the ECU when there is a demand for cooling from the air conditioning system.

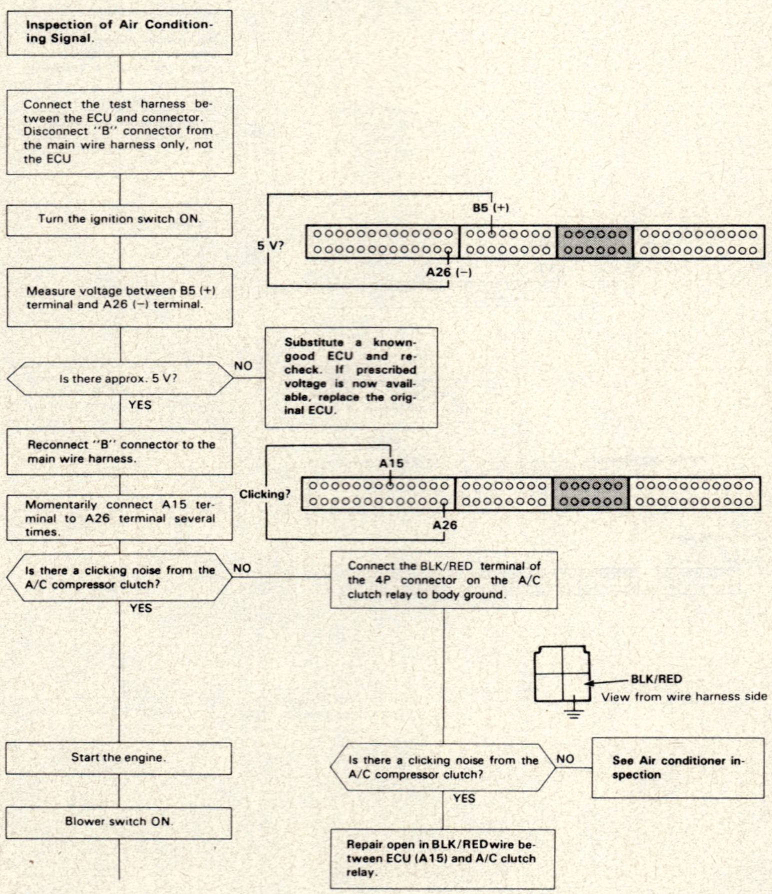

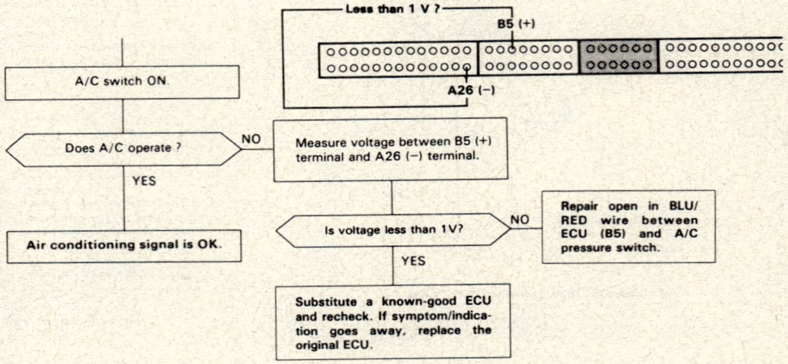

1.5L (D15B7, D15B8, D15Z1) AND 1.6L (D16Z6) ENGINES—TROUBLESHOOTING
ALTERNATOR FR SIGNAL—1992 CIVIC

This signals the ECU when the alternator is charging.

Inspection of Alternator FR signal.

Connect the test harness between the ECU and connector. Disconnect "D" connector from the main wire harness only, not the ECU

Turn the ignition switch ON.

Measure voltage between D9 (+) terminal and A26 (−) terminal.

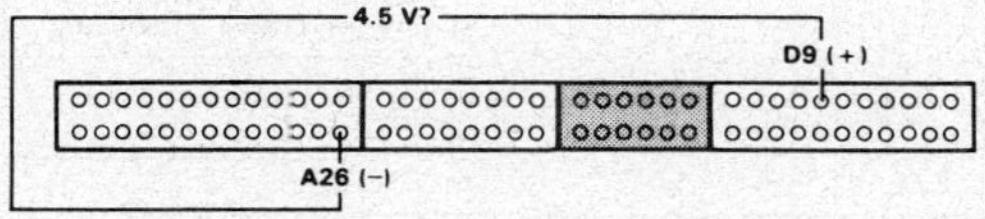

Is there approx. 4.5 V?

- NO → **Substitute a known-good ECU and recheck. If prescribed voltage is now available, replace the original ECU.**
- YES ↓

Turn the ignition switch OFF.

Reconnect "D" connector to the main wire harness.

Warm up engine to normal operating temperature (cooling fan comes on).

Measure voltage between D9 (+) terminal and A26 (−) terminal.

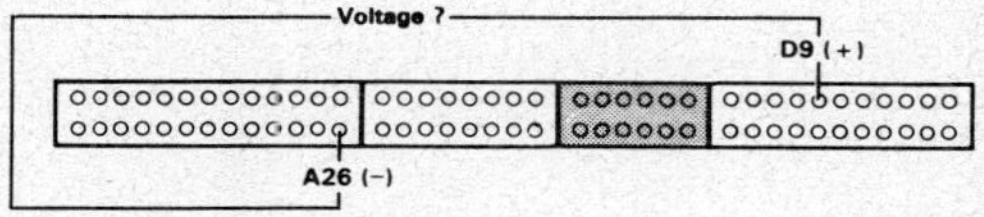

Does the voltage decrease when headlights and rear defogger are turned on ?

- NO → Stop the engine. ↓
- YES → Do the ECU Reset Procedure → **Alternator FR signal is OK.**

Disconnect "D" connector from ECU only, not the main wire harness.

Disconnect the negative battery cable from the battery.

Check for continuity between D9 terminal and body ground.

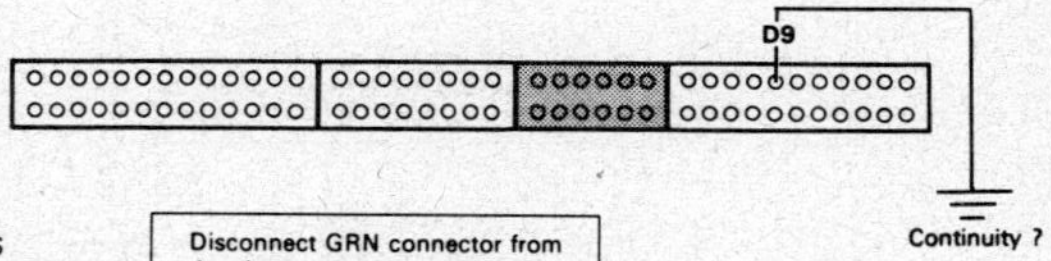

Does continuity exist ?

- YES → **Disconnect GRN connector from the alternator.** → **Check for continuity between D9 terminal and body ground.** → **Does continuity exist ?**
  - NO → **See Alternator Inspection**
  - YES → **Repair short in PNK or BLU wire between ECU (D9) and alternator.**
- NO → **Disconnect GRN connector from the alternator.**

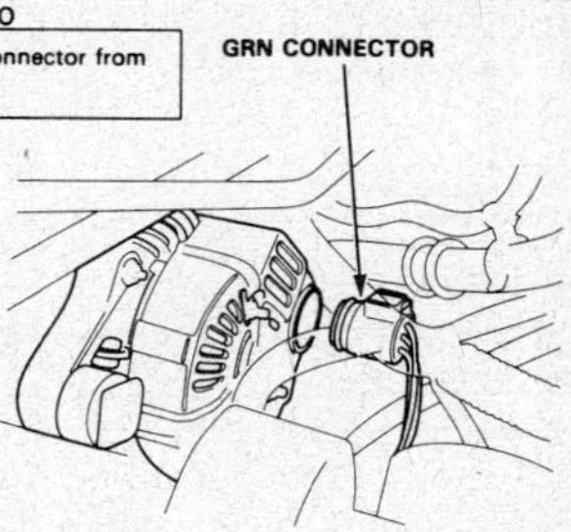

Connect BLU wire to body ground.

Check for continuity between D9 terminal and body ground.

Does continuity exist ?

- YES → **See Alternator Inspection**
- NO → **Repair open in PNK or BLU wire between ECU (D9) and alternator.**

1.5L (D15B7, D15B8, D15Z1) AND 1.6L (D16Z6) ENGINES—TROUBLESHOOTING

A/T SHIFT POSITION SIGNAL— 1992 CIVIC

This signals the ECU when the transmission is in Neutral or Park.

1. **Inspection of A/T Shift Position Signal.**
2. Turn the ignition switch ON.
3. Observe the A/T shift indicator and select each position separately.
4. Does the indicator light properly?
   - NO → See A/T Shift Position Indicator Inspection
   - YES ↓
5. Turn the ignition switch OFF.
6. Connect the test harness between the ECU and connector. Disconnect "B" connector from the main wire harness only, not the ECU
7. Turn the ignition switch ON.
8. Measure voltage between B7 (+) terminal and A26 (−) terminal.

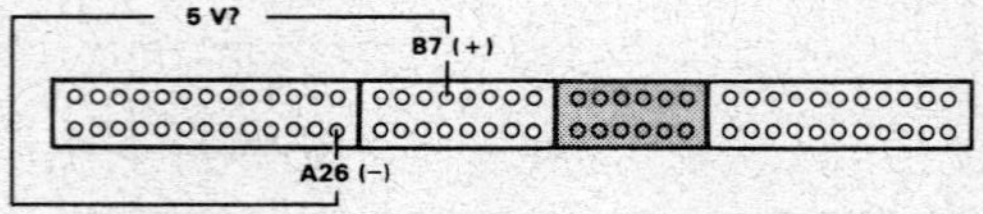

9. Is there approx. 5 V?
   - NO → Substitute a known-good ECU and recheck. If prescribed voltage is now available, replace the original ECU.
   - YES ↓
10. Turn the ignition switch OFF.
11. Reconnect "B" connector to the main wire harness.
12. Start the engine.

---

13. Measure voltage between B7 (+) terminal and A26 (−) terminal with the transmission in [N] and [P] individually.

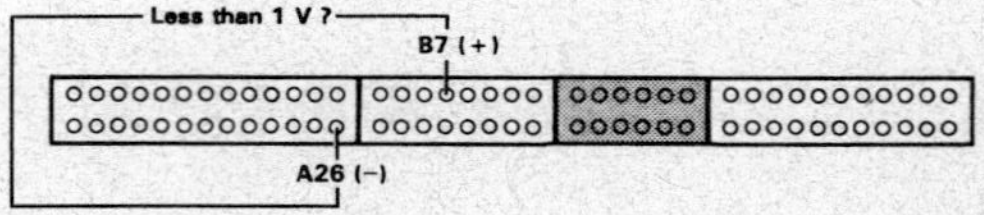

14. Is there less than 1 V?
    - NO → Repair open in GRN wire between ECU (B7) and gauge assembly.
    - YES ↓
15. Measure voltage between B7 (+) terminal and A26 (−) terminal with the transmission in gear.

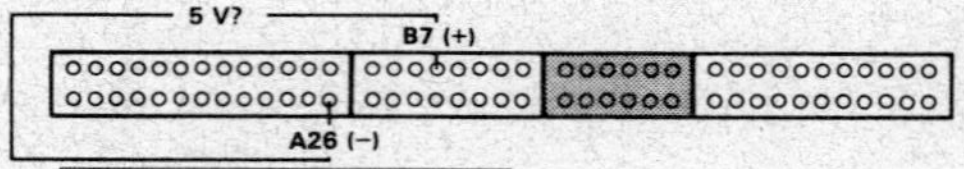

16. Is there approx. 5 V?
    - NO → Repair short in GRN wire between ECU (B7) and gauge assembly.
    - YES ↓
17. Measure voltage between B3 or B4* terminal and A26 (−) terminal with the transmission in [D3] or [D4] individually.

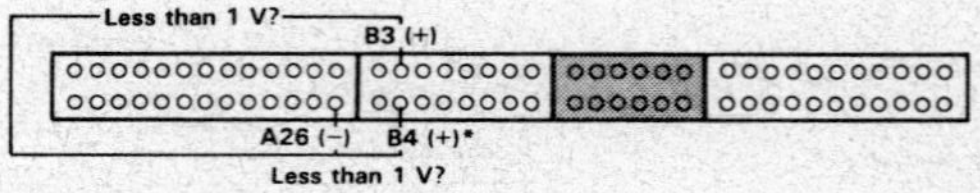

*: [D4]

18. Is there less than 1 V?
    - NO → Repair open in GRN/BLU or GRN/BLK wire between ECU (B3 or B4*) and Shift Position Console Switch.
    - YES ↓
19. Measure voltage between B3 or B4* (+) terminal and A26 (−) terminal with the transmission in [N] or [P].

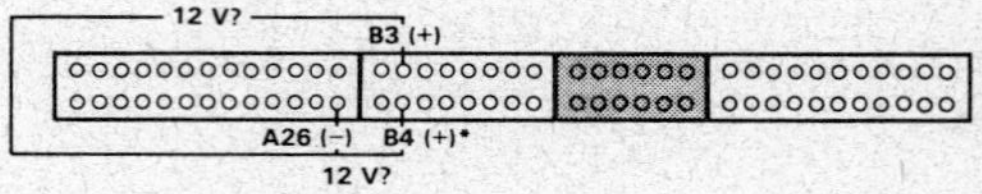

20. Is there approx. 12 V?
    - NO → Repair short in GRN/BLU or GRN/BLK* wire between ECU (B3 or B4*) and Shift Position Console Switch.
    - YES ↓
21. **A/T shift position signal is OK.**

1.5L (D15Z1) ENGINE – TROUBLESHOOTING M/T CLUTCH SWITCH SIGNAL – 1992 CIVIC

This signals the ECU when the clutch is engaged.

Inspection of clutch switch signal.

Connect the test harness between the ECU and connector

Turn the ignition switch ON.

Measure voltage between B7 (+) terminal and A26 (−) terminal.

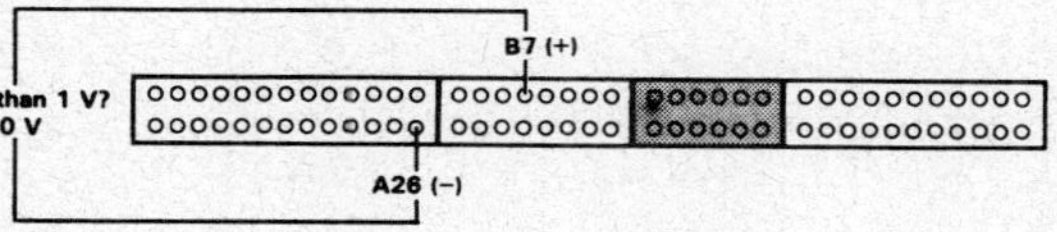

Is voltage less than 1 V?

- NO → Turn the ignition switch OFF.
  - Disconnect the 2P connector from the clutch switch.
  - Check for continuity between the A terminal and B terminal on the clutch switch.

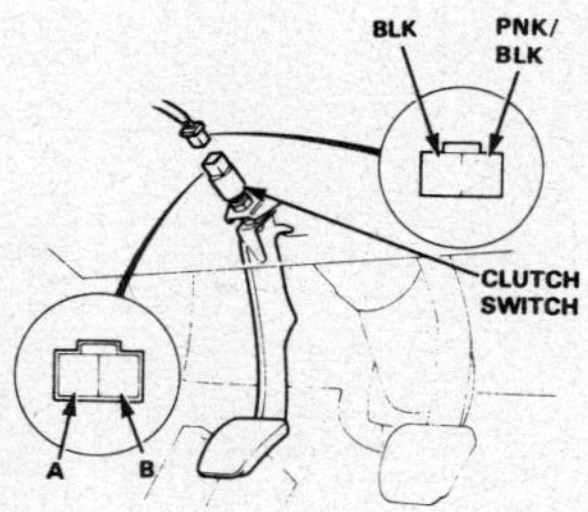

  - Does continuity exist?
    - NO → **Replace the clutch switch.**
    - YES → Turn the ignition switch ON.
      - Measure voltage between PNK/BLK (+) terminal and body ground.
      - Is there approx. 5 V?
        - NO → **Repair open in PNK/BLK wire between ECU (B7) and the clutch switch.**
        - YES → **Repair open in BLK wire between the clutch switch and G301.**
- YES → Depress the clutch pedal.
  - Measure voltage between B7 (+) terminal and A26 (−) terminal.
  - Is there approx. 5 V?
    - YES → **Clutch switch signal is OK.**
    - NO → Turn the ignition switch OFF.
      - Disconnect the 2P connector from the clutch switch.
      - Turn the ignition switch ON.
      - Measure voltage between B7 (+) terminal and A26 (−) terminal.
      - Is there approx. 5 V?
        - YES → **Replace the clutch switch.**
        - NO → Disconnect "B" connector from main wire harness only, not the ECU.
          - Is there approx. 5 V?
            - YES → **Repair short in PNK/BLK wire between ECU (B7) and the clutch switch.**
            - NO → **Substitute a known-good ECU and recheck. If prescribed voltage is now available, replace the original ECU.**

1.5L (D15B7, D15B8, D15Z1) AND 1.6L (D16Z6) ENGINES—TROUBLESHOOTING STARTER SWITCH SIGNAL— 1992 CIVIC

This signals the ECU when the engine is cranking.

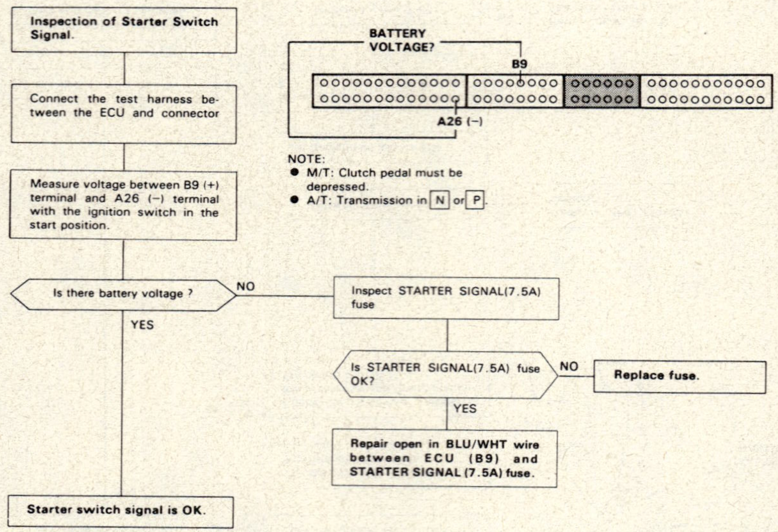

1.5L (D15B7, D15B8, D15Z1) AND 1.6L (D16Z6) ENGINES—TROUBLESHOOTING BRAKE SWITCH SIGNAL—1992 CIVIC

This signals the ECU when the brake pedal is depressed.

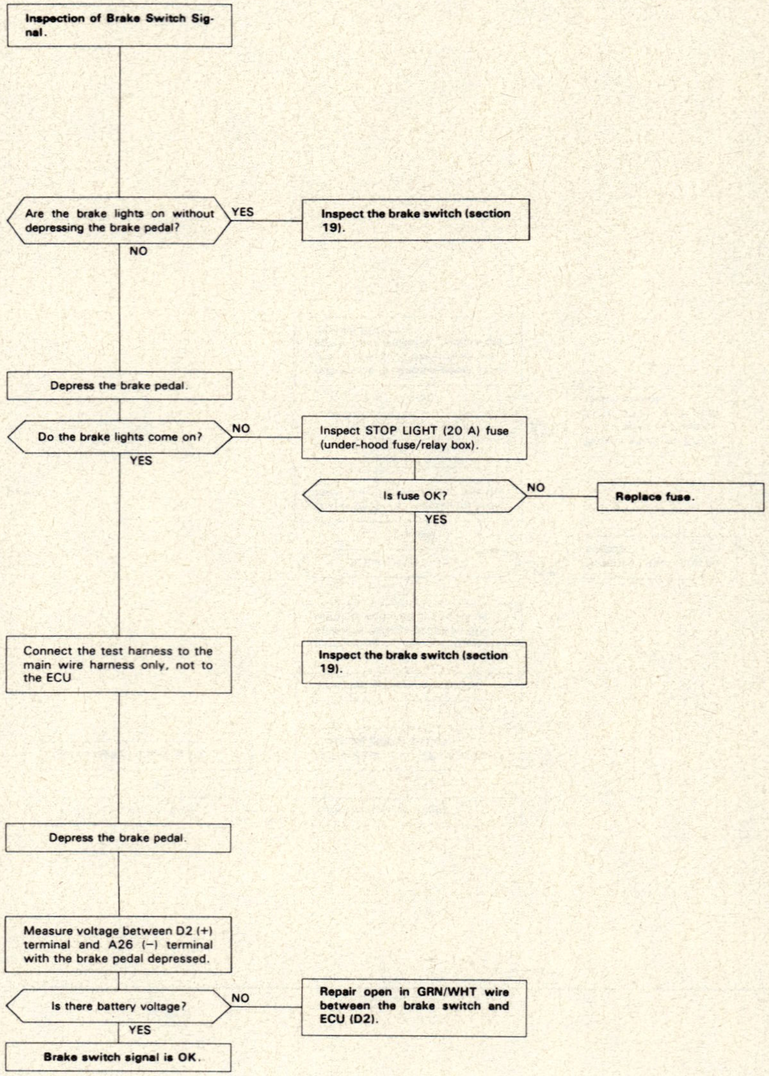

1.5L (D15B7, D15B8, D15Z1) AND 1.6L (D16Z6) ENGINES—TROUBLESHOOTING P/S OIL PRESSURE SIGNAL—1992 CIVIC

This signals the ECU when the power steering load is high.

Inspection of P/S Oil Pressure Signal

- Connect the test harness between the ECU and connector.
- Turn the ignition switch ON.
- Measure voltage between B8 (+) terminal and A26 (−) terminal.
- Is there more than 1V?
  - YES → Disconnect the 2P connector from the P/S oil pressure switch. → Connect BRN/RED terminal to BLK terminal. → Is there more than 1V?
    - NO → **Replace P/S oil pressure switch.**
    - YES → **Repair open in BRN/RED wire between ECU (B8) and P/S oil pressure switch or BLK wire between P/S oil pressure switch and G302.**
  - NO → Start the engine. → Turn steering wheel slowly. → Measure voltage between B8 (+) terminal and A26 (−) terminal while steering wheel is turning. → Is there battery voltage?
    - YES → **P/S oil pressure signal is OK.**
    - NO → Disconnect the 2P connector from the P/S oil pressure switch. → Is there battery voltage?
      - YES → **Replace P/S oil pressure switch.**
      - NO → **Repair short in BRN/RED wire between ECU (B8) and P/S oil pressure switch. If wire is OK, substitute a known-good ECU and recheck. If prescribed voltage is now available, replace the original ECU.**

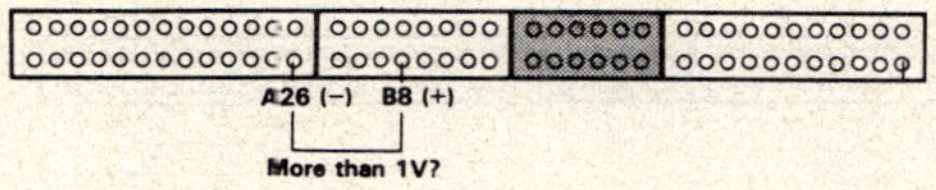

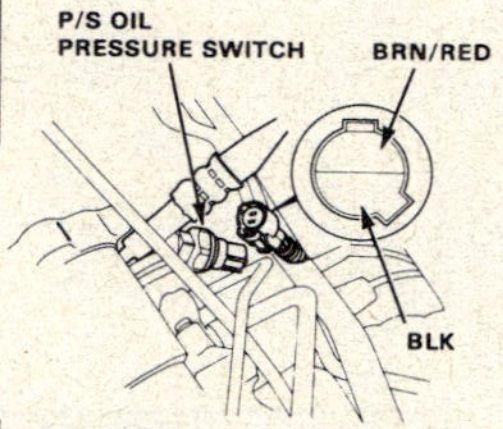

1.5L (D15B7, D15B8, D15Z1) AND 1.6L (D16Z6) ENGINES—TROUBLESHOOTING MAIN RELAY—1992 CIVIC

Troubleshooting Flowchart

- — Engine will not start.
- — Inspection of main relay and relay harness.

- Disconnect the main relay connectors.
- Check for continuity between BLK terminal ② and body ground.
- Does continuity exist?
  - NO → **Repair open in BLK wire between main relay and G101.**
  - YES → Measure the voltage between YEL/WHT terminal ① and body ground.
- Is there battery voltage?
  - NO → **— Replace ECU (15A) fuse. — Repair open in the YEL/WHT wire between the main relay and the ECU (15A) fuse.**
  - YES → Turn the ignition switch ON. → Measure the voltage between BLK/YEL terminal ⑤ and body ground.
- Is there battery voltage?
  - NO → **— Replace ACG (S) (15A) fuse. — Repair open in the BLK/YEL wire between the main relay and the ACG(S) (15A) fuse.**
  - YES → Turn the ignition switch to the START position. → Measure the voltage between BLU/WHT terminal ⑥ and body ground.
- Is there battery voltage?
  - NO → **Repair open in the BLU/WHT wire between the main relay and the ignition switch.**
  - YES

NOTE:
- M/T: Clutch pedal must be depressed.
- A/T: Transmission in [N] or [P].

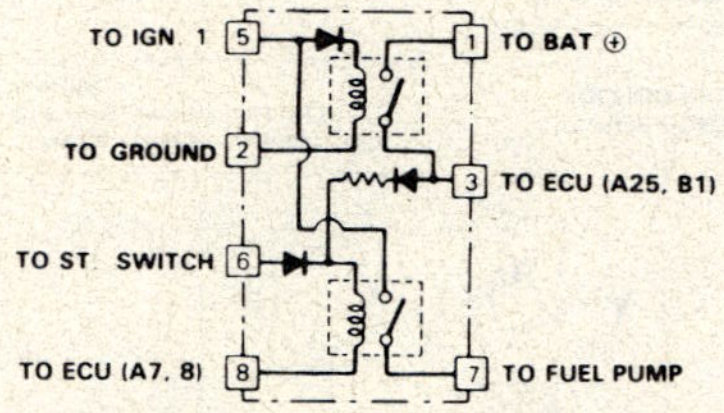

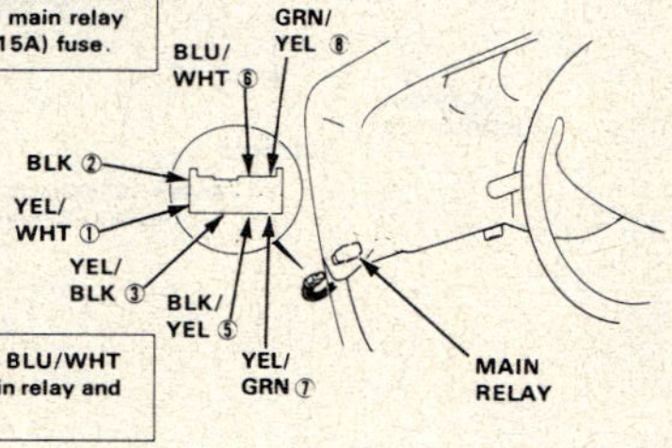

1.5L (D15B7, D15B8, D15Z1) AND 1.6L (D16Z6) ENGINES – TROUBLESHOOTING MAIN RELAY – 1992 CIVIC

Turn the ignition switch off.

Connect the test harness between the ECU and connector. Disconnect "A" connector from the ECU only, not the main wire harness

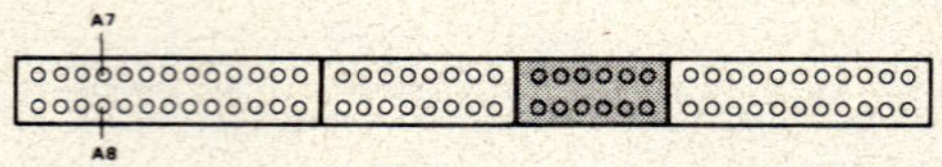

Check for continuity between GRN/YEL terminal ⑧ and the following terminals; A7, A8.

Does continuity exist?

- NO → Repair open in GRN/YEL wire between ECU (A7, A8) and main relay.

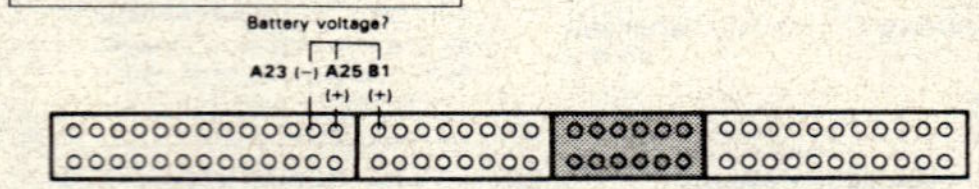

- YES → Reconnect "A" connector to the ECU.

Connect the main relay connector.

Turn the ignition switch ON.

Measure the voltage between A23 (−) terminal and the following terminals: A25 (+) B1 (+).

Is there battery voltage?

- NO →
  - Repair open in the YEL/BLK wire ① between the ECU (A25, B1) and main relay.
  - Replace main relay.

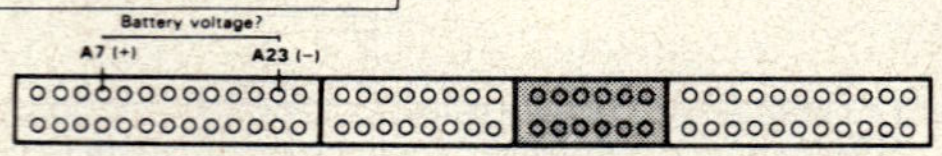

- YES → Turn the ignition switch OFF.

Connect a voltmeter between A7 (+) terminal and A23 (−) terminal.

Is there battery voltage for two seconds when the ignition switch is first turned on?

- YES → Substitute a known-good ECU and recheck. If symptom/indication goes away, replace the original ECU.
- NO → Check the main relay

1.5L (D15B7, D15B8, D15Z1) AND 1.6L (D16Z6) ENGINES – TROUBLESHOOTING EVAPORATIVE EMISSION CONTROL SYSTEM – 1992 CIVIC

Troubleshooting Flowchart

Inspection of Evaporative Emission Controls

Disconnect #7 vacuum hose from the purge control diaphragm valve (on the charcoal canister) and connect a vacuum gauge to the hose.

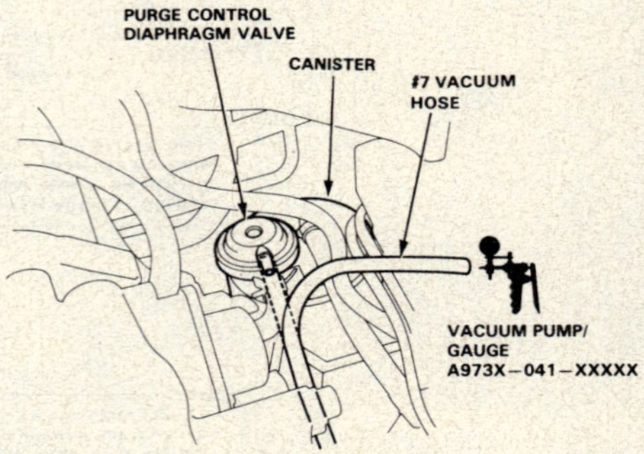

Start the engine and allow it to idle.
NOTE: Engine coolant temperature must be below D15Z1 engine: 75°C (167°F), Others: 70°C (158°F).

Is there vacuum ?

- YES → Disconnect the 2P connector from the purge control solenoid valve.

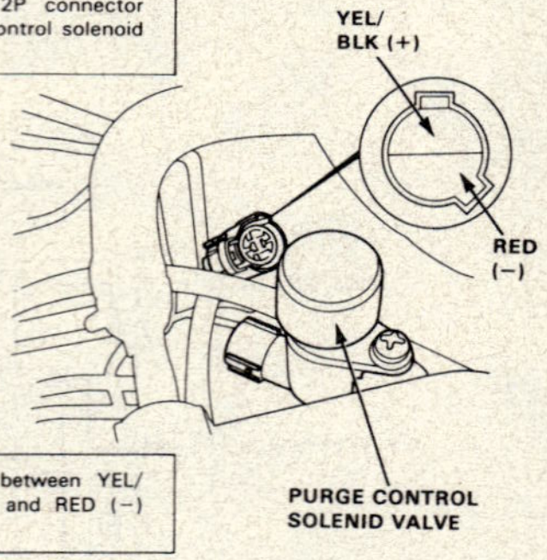

  Measure voltage between YEL/BLK (+) terminal and RED (−) terminal.

  Is there battery voltage ?

  - YES → Inspect vacuum hose routing. If OK, replace purge control solenoid valve.
  - NO → Measure voltage between YEL/BLK (+) terminal and body ground.
- NO

1.5L (D15B7, D15B8, D15Z1) AND 1.6L (D16Z6) ENGINES—TROUBLESHOOTING EVAPORATIVE EMISSION CONTROL SYSTEM—1992 CIVIC

Warm up the engine to normal operating temperature (cooling fan comes on).

Check for vacuum at #7 vacuum hose after starting the engine.

Is there manifold vacuum?

- NO → Disconnect the 2P connector.
  - Is there manifold vacuum?
    - NO → Inspect vacuum hose routing. If OK, replace purge control solenoid valve.
    - YES → Inspect for a short in RED wire between ECU (A20) and the connector. If wire is OK, substitute a known-good ECU and recheck. If symptom goes away, replace the original ECU.
- YES → Reconnect the hose.

Remove fuel filler cap.

Is there battery voltage ?

- NO → Repair open in YEL/BLK wire between main relay and the 2P connector.
- YES → Inspect RED wire for an open to body ground between ECU (A20) and the connector. If wire is OK, substitute a known-good ECU and recheck. If symptom goes away, replace the original ECU.

Connect a vacuum gauge to canister purge air hose.

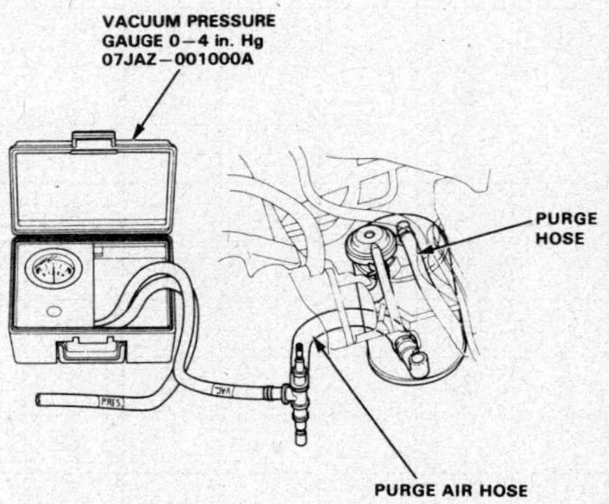

Start the engine and raise speed to 3,500 rpm.

Does vacuum appear on gauge within 1 minute ?

- NO → **Replace the canister.**
- YES → **See two-way valve test to complete. Evaporative emission controls are OK.**

Troubleshooting Guide

NOTE: Across each row in the chart, the systems that could be sources of a symptom are ranked in the order they should be inspected starting with ①. Find the symptom in the left column, read across to the most likely source, then refer to the page listed at the top of that column. If inspection shows the system is OK, try the next most likely system ②, etc.

SYMPTOM		PGM-FI: ECU	PGM-FI: OXYGEN SENSOR	PGM-FI: MANIFOLD ABSOLUTE PRESSURE SENSOR	PGM-FI: TDC/CRANK/CYL SENSOR	PGM-FI: COOLANT TEMPERATURE SENSOR	PGM-FI: THROTTLE ANGLE SENSOR	PGM-FI: INTAKE AIR TEMPERATURE SENSOR	PGM-FI: ATMOSPHERIC PRESSURE SENSOR	PGM-FI: IGNITION OUTPUT SIGNAL	PGM-FI: VEHICLE SPEED SENSOR
CHECK ENGINE LIGHT** TURNS ON		☐ or ☐	(lit)	(lit)	(lit)	(lit)	(lit)	(lit)	(lit)	(lit)	(lit)
CHECK ENGINE LIGHT BLINKS		[illegible] or [illegible]	1 or 41 or 43	3 or 5	4 or 8 or 9	6	7	10	13	15	17
ENGINE WON'T START		③			③					③	
DIFFICULT TO START ENGINE WHEN COLD		(BU)		③	②	①					
IRREGULAR IDLING	WHEN COLD FAST IDLE OUT OF SPEC	(BU)				③					
	ROUGH IDLE	(BU)		③							
	WHEN WARM RPM TOO HIGH	(BU)									
	WHEN WARM RPM TOO LOW	(BU)									
FREQUENT STALLING	WHILE WARMING UP	(BU)				③					
	AFTER WARMING UP	(BU)									
POOR PERFORMANCE	MISFIRE OR ROUGH RUNNING	(BU)		②	③						
	FAILS EMISSION TEST	(BU)	③	②							
	LOSS OF POWER	(BU)		③			②				

SYMPTOM		PGM-FI: ELECTRIC LOAD DETECTOR	PGM-FI: KNOCK SENSOR	PGM-FI: A/T FI SIGNAL A	PGM-FI: A/T FI SIGNAL B	IDLE CONTROL: ELECTRONIC AIR CONTROL VALVE	IDLE CONTROL: OTHER IDLE CONTROLS	FUEL SUPPLY: FUEL INJECTOR	FUEL SUPPLY: OTHER FUEL SUPPLY	AIR INTAKE	EMISSION CONTROL: EGR CONTROL SYSTEM	EMISSION CONTROL: OTHER EMISSION CONTROLS
CHECK ENGINE LIGHT** TURNS ON		☐	(lit)	☐	☐	(lit)		(lit)			(lit)	
CHECK ENGINE LIGHT BLINKS		20	23	30	31	14		16			12	
ENGINE WON'T START								②	①			
DIFFICULT TO START ENGINE WHEN COLD												
IRREGULAR IDLING	WHEN COLD FAST IDLE OUT OF SPEC					①	②					
	ROUGH IDLE					①		②			③	
	WHEN WARM RPM TOO HIGH					①	②					
	WHEN WARM RPM TOO LOW	③				①		②				
FREQUENT STALLING	WHILE WARMING UP					①	②		③			
	AFTER WARMING UP					③			①		②	
POOR PERFORMANCE	MISFIRE OR ROUGH RUNNING							①			③	
	FAILS EMISSION TEST											①
	LOSS OF POWER							③	①	③		

* If codes other than those listed above are indicated, count the number of blinks again. If the indicator is in fact blinking these codes, replace the original ECU.

(BU) If the Check Engine light is on while the engine is running, jump the service check connector. If no code is displayed (Check Engine light stays on steady), the back-up system may be in operation.

** Check Engine light

USA: (lit) CANADA: (lit)

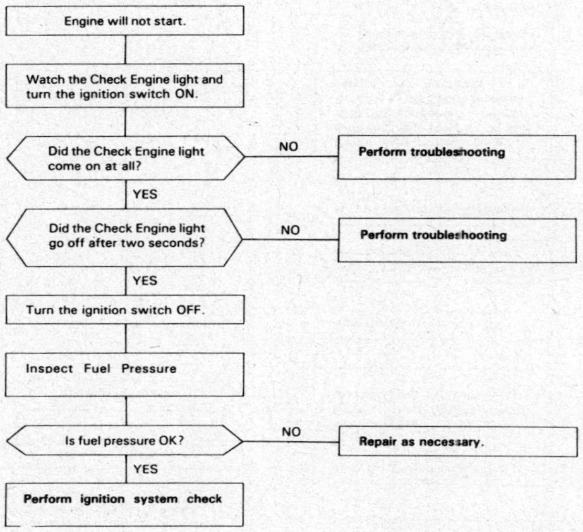
Engine will not start.
Watch the Check Engine light and turn the ignition switch ON.
Did the Check Engine light come on at all?
NO
Perform troubleshooting
YES
Did the Check Engine light go off after two seconds?
NO
Perform troubleshooting
YES
Turn the ignition switch OFF.
Inspect Fuel Pressure
Is fuel pressure OK?
NO
Repair as necessary.
YES
Perform ignition system check

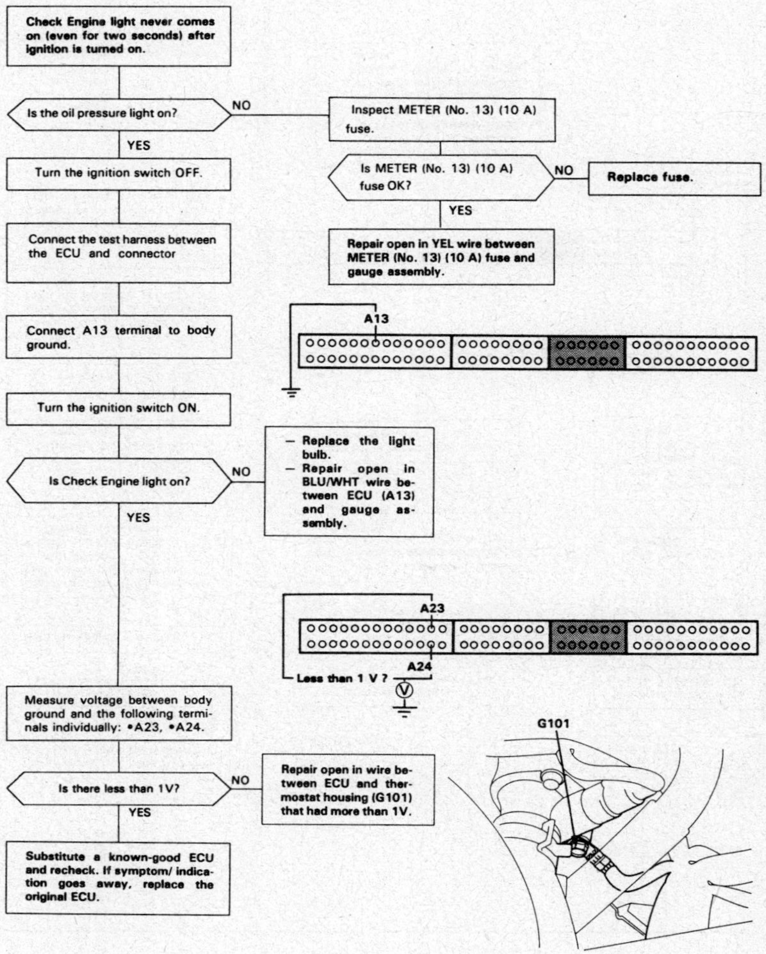
Check Engine light never comes on (even for two seconds) after ignition is turned on.
Is the oil pressure light on?
NO
Inspect METER (No. 13) (10 A) fuse.
Is METER (No. 13) (10 A) fuse OK?
NO
Replace fuse.
YES
Repair open in YEL wire between METER (No. 13) (10 A) fuse and gauge assembly.
YES
Turn the ignition switch OFF.
Connect the test harness between the ECU and connector
Connect A13 terminal to body ground.
A13
Turn the ignition switch ON.
Is Check Engine light on?
NO
– Replace the light bulb.
– Repair open in BLU/WHT wire between ECU (A13) and gauge assembly.
YES
A23
A24
Less than 1 V ?
Measure voltage between body ground and the following terminals individually: •A23, •A24.
Is there less than 1V?
NO
Repair open in wire between ECU and thermostat housing (G101) that had more than 1V.
G101
YES
Substitute a known-good ECU and recheck. If symptom/ indication goes away, replace the original ECU.

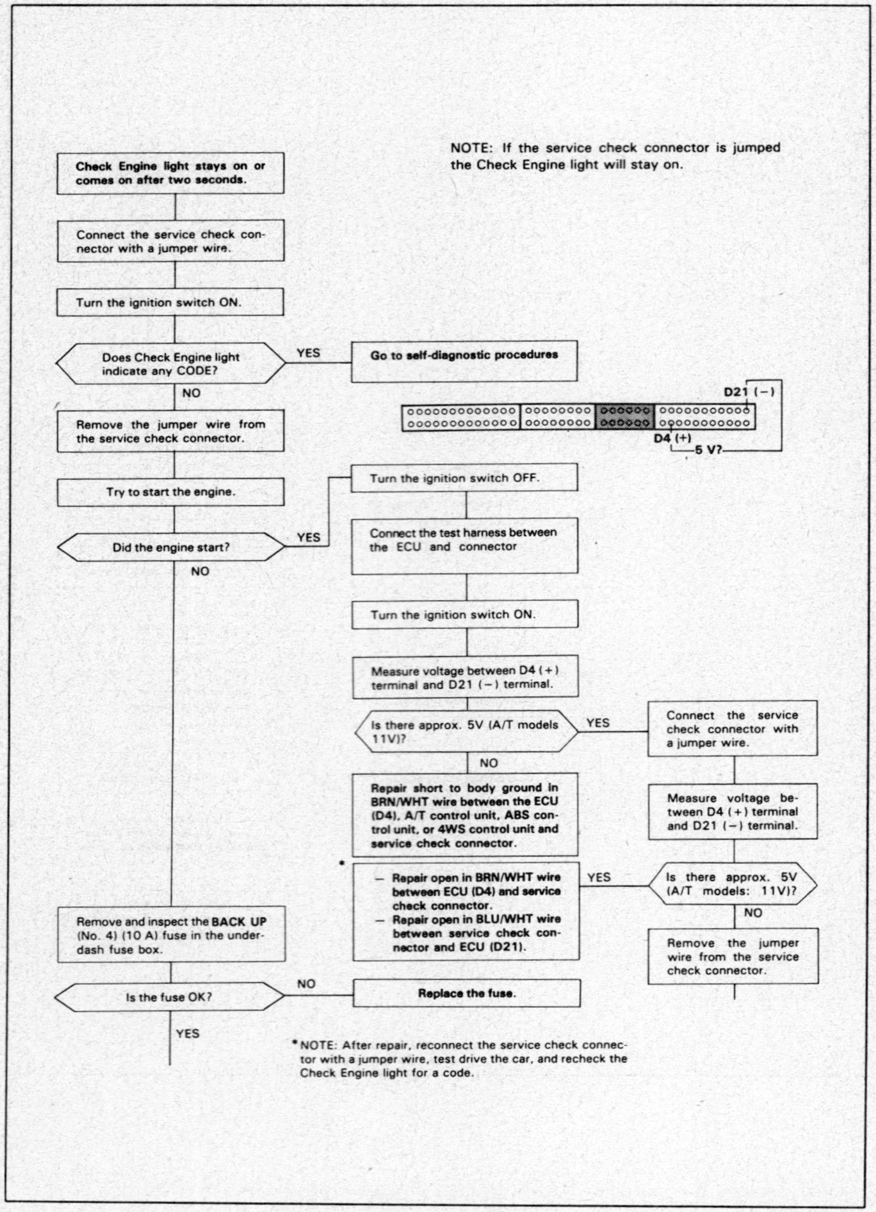
NOTE: If the service check connector is jumped the Check Engine light will stay on.
Check Engine light stays on or comes on after two seconds.
Connect the service check connector with a jumper wire.
Turn the ignition switch ON.
Does Check Engine light indicate any CODE?
YES
Go to self-diagnostic procedures
NO
Remove the jumper wire from the service check connector.
Try to start the engine.
Did the engine start?
YES
NO
Turn the ignition switch OFF.
Connect the test harness between the ECU and connector
Turn the ignition switch ON.
Measure voltage between D4 (+) terminal and D21 (−) terminal.
D21 (−)
D4 (+)
5 V?
Is there approx. 5V (A/T models 11V)?
YES
NO
Connect the service check connector with a jumper wire.
Measure voltage between D4 (+) terminal and D21 (−) terminal.
Is there approx. 5V (A/T models: 11V)?
YES
NO
Remove the jumper wire from the service check connector.
Repair short to body ground in BRN/WHT wire between the ECU (D4), A/T control unit, ABS control unit, or 4WS control unit and service check connector.
— Repair open in BRN/WHT wire between ECU (D4) and service check connector.
— Repair open in BLU/WHT wire between service check connector and ECU (D21).
Remove and inspect the BACK UP (No. 4) (10 A) fuse in the under-dash fuse box.
Is the fuse OK?
NO
Replace the fuse.
YES
*NOTE: After repair, reconnect the service check connector with a jumper wire, test drive the car, and recheck the Check Engine light for a code.

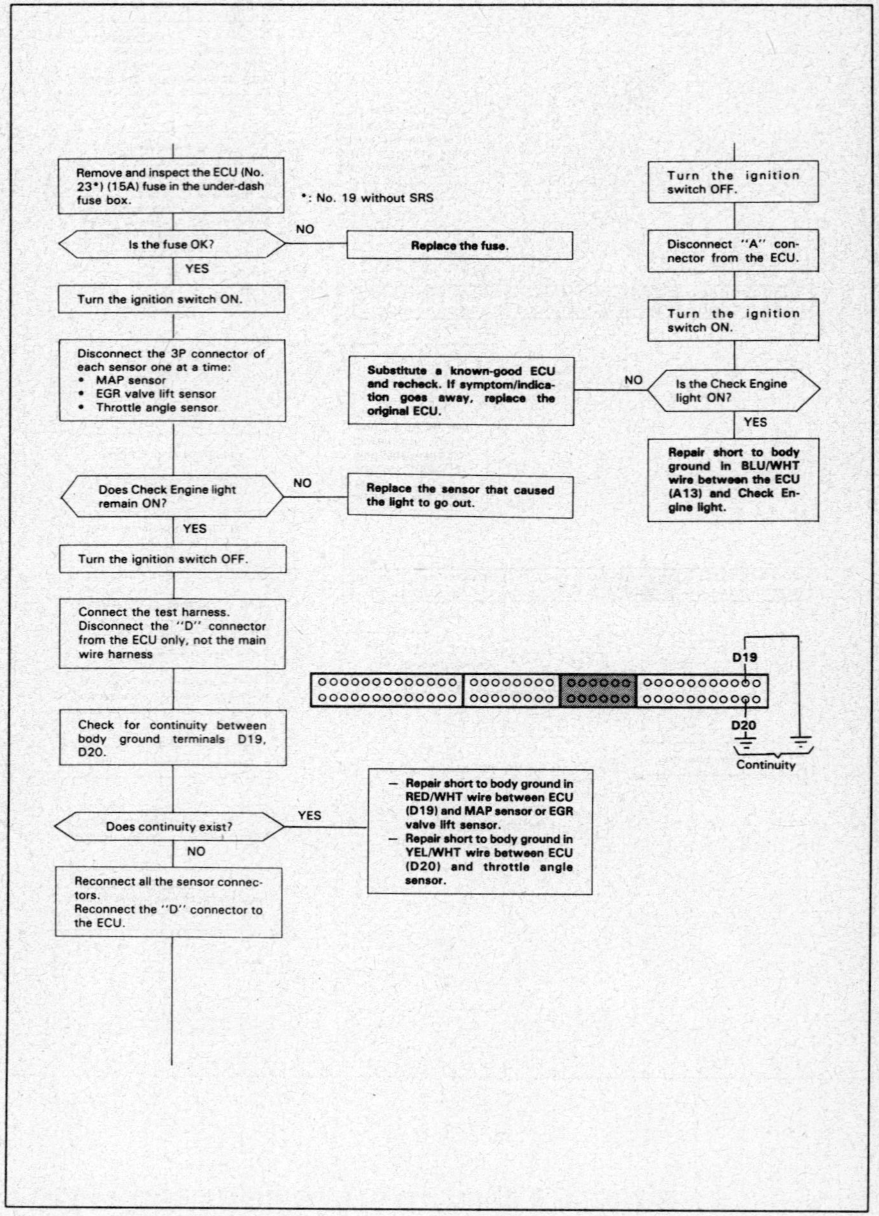
Remove and inspect the ECU (No. 23*) (15A) fuse in the under-dash fuse box.
*: No. 19 without SRS
Is the fuse OK?
NO
Replace the fuse.
YES
Turn the ignition switch ON.
Disconnect the 3P connector of each sensor one at a time:
• MAP sensor
• EGR valve lift sensor
• Throttle angle sensor
Does Check Engine light remain ON?
NO
Replace the sensor that caused the light to go out.
YES
Turn the ignition switch OFF.
Connect the test harness.
Disconnect the "D" connector from the ECU only, not the main wire harness
Check for continuity between body ground terminals D19, D20.
D19
D20
Continuity
Does continuity exist?
YES
— Repair short to body ground in RED/WHT wire between ECU (D19) and MAP sensor or EGR valve lift sensor.
— Repair short to body ground in YEL/WHT wire between ECU (D20) and throttle angle sensor.
NO
Reconnect all the sensor connectors.
Reconnect the "D" connector to the ECU.
Turn the ignition switch OFF.
Disconnect "A" connector from the ECU.
Turn the ignition switch ON.
Is the Check Engine light ON?
NO
Substitute a known-good ECU and recheck. If symptom/indication goes away, replace the original ECU.
YES
Repair short to body ground in BLU/WHT wire between the ECU (A13) and Check Engine light.

2.2L (F22A1) AND 2.3L (H23A1) ENGINES—PGM-FI CONTROL SYSTEM—1992 PRELUDE

Turn the ignition switch ON.

Measure voltage between body ground and the following terminal individually: •A26, •B2.

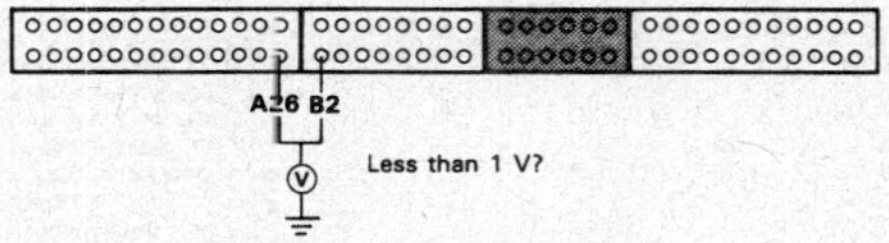

Is there less than 1 V?

NO → Repair open in BLK/RED (A26) or BRN/BLK (B2) and G101.

YES ↓

Measure voltage between A26 (−) and the following: B1 (+) and A25 (+).

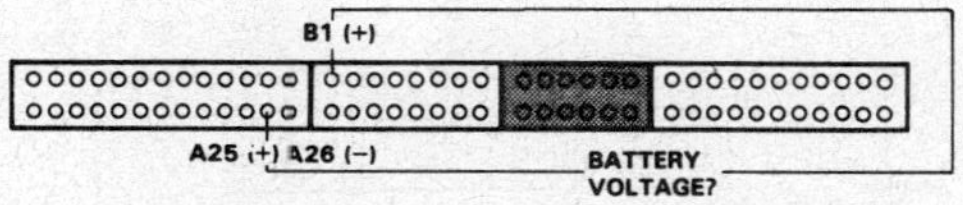

Is there battery voltage?

NO →
- Repair open in YEL/BLK wire between ECU (A25, B1) and main relay
- Check main relay and wiring connectors at main relay.

YES ↓

Substitute a known-good ECU and recheck. If symptom/indication goes away, replace the original ECU.

2.2L (F22A1) AND 2.3L (H23A1) ENGINES TROUBLESHOOTING PGM-FI FAULT CODE 1—1992 PRELUDE

Self-diagnosis Check Engine light indicates code 1: A problem in the Oxygen (O_2) Sensor circuit.

The Oxygen sensor detects the oxygen content in the exhaust gas and signals the ECU. In operation, the ECU receives the signals from the sensor and varies the duration during which fuel is injected. The oxygen sensor has an internal heater. The heater stabilizes the sensor's output. The oxygen sensor is installed in the exhaust manifold.

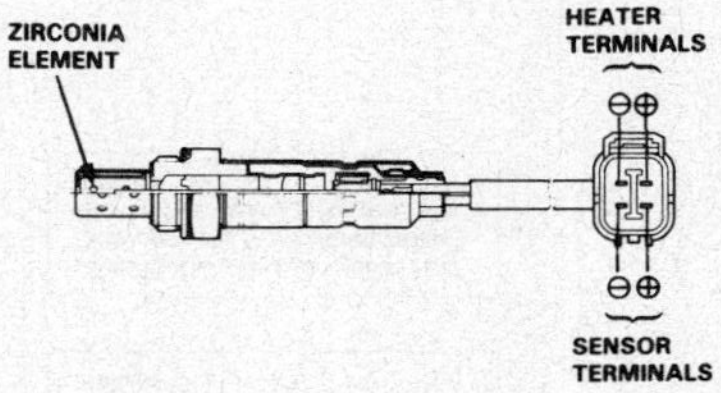

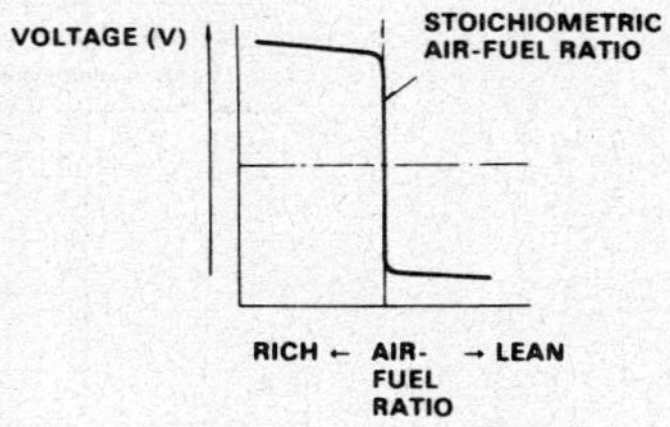

- Check Engine light has been reported on.
- With service check connector jumped CODE 1 is indicated.

Do the ECU Reset Procedure

Warm up engine to normal operating temperature (cooling fan comes on).

Run engine for 60 seconds.

Road test with the transmission in 2nd gear (M/T: 4th gear). Starting at 1600 rpm, accelerate using wide open throttle for at least 5 seconds. Then decelerate for at least 5 seconds with the throttle completely closed.

Is Check Engine light on and does it indicate CODE 1?

NO → Intermittent failure, system is OK at this time. Check for poor connections or loose wires at C425, C432, C433 (located at right shock tower), C114 (O_2 sensor) and ECU.

YES → Go to test for CODE 43.

2.2L (F22A1) AND 2.3L (H23A1) ENGINES—TROUBLESHOOTING PGM-FI FAULT CODES 3 AND 5—1992 PRELUDE

[CHECK] [3] Self-diagnosis Check Engine light indicates code 3: An electrical problem in the Manifold Absolute Pressure (MAP) Sensor system.

[CHECK] [5] Self-diagnosis Check Engine light indicates code 5: A mechanical problem (broken hose) in the Manifold Absolute Pressure (MAP) sensor System.

The MAP sensor converts manifold absolute pressure into electrical signals and inputs the ECU.

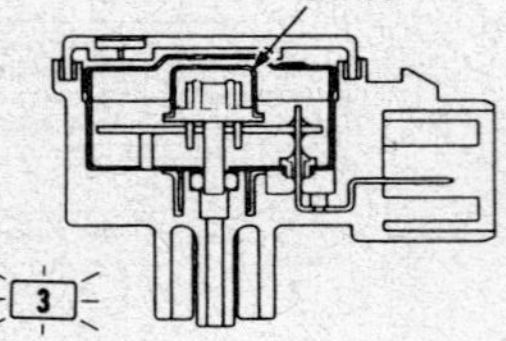

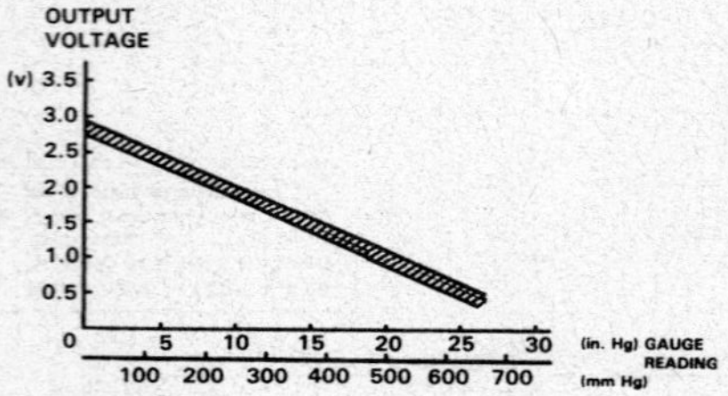

[CHECK] [3]

- Engine is warm and running.
- Check Engine light has been reported on.
- With service check connector jumped CODE 3 is indicated.

Do the ECU Reset Procedure

Start the engine and allow it to idle.

Is Check Engine light on and does it indicate CODE 3?

NO → Intermittent failure, system is OK at this time (test drive may be necessary). Check for poor connection or loose wires at C463 (located at left side under dash), C320 (MAP sensor) and ECU.

YES ↓

Turn the ignition switch OFF.

Disconnect the 3P connector from the MAP sensor.

Turn the ignition switch ON.

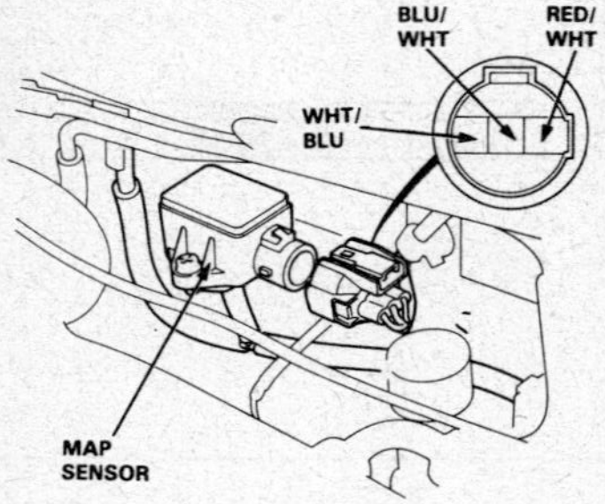

Measure voltage between RED/WHT (+) terminal and body ground.

Is there approx. 5V?

YES → Measure voltage between RED/WHT (+) terminal and BLU/WHT (−) terminal.

Is there approx. 5V?

- NO → Repair open in BLU/WHT wire between ECU (D21) and MAP sensor.
- YES → Measure voltage between WHT/BLU (+) terminal and BLU/WHT (−) terminal.

Is there approx. 5V?

- NO → (to Substitute a known-good ECU and recheck.)
- YES ↓

Turn the ignition switch OFF.

Reconnect the 3P connector to the MAP sensor.

Connect the test harness between the ECU and connector

Turn the ignition switch ON.

NO (from Is there approx. 5V? after RED/WHT to body ground) ↓

Connect the test harness "D" connector to the ECU only, not to the main wire harness

Turn the ignition switch ON.

Measure voltage between D19 (+) terminal and D21 (−) terminal.

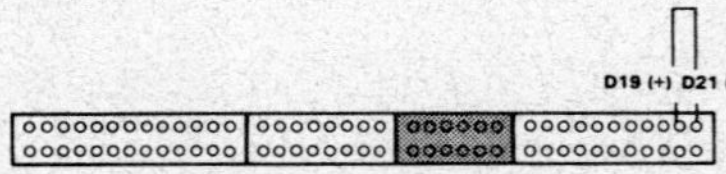

Is there approx. 5 V?

- YES → Repair open in RED/WHT wire between ECU (D19) and MAP sensor.
- NO → Substitute a known-good ECU and recheck. If symptom/indication goes away, replace the original ECU.

2.2L (F22A1) AND 2.3L (H23A1) ENGINES TROUBLESHOOTING PGM-FI FAULT CODES 3 AND 5—1992 PRELUDE

Connect the test harness "D" connector to the ECU only, not to the main wire harness

Measure voltage between D17 (+) terminal and D21 (−) terminal.

Is there approx. 5 V?

YES:
- Repair short in WHT/BLU wire between ECU (D17) and MAP sensor.
- Repair open in WHT/BLU wire between ECU (D17) and MAP sensor.

NO: Substitute a known-good ECU and recheck. If prescribed voltage is now available, replace the original ECU.

Measure voltage between D17 (+) terminal and D21 (−) terminal.

Is there approx. 3 V ?

NO: Replace MAP sensor.

YES: Substitute a known-good ECU and recheck. If symptom/indication goes away, replace the original ECU.

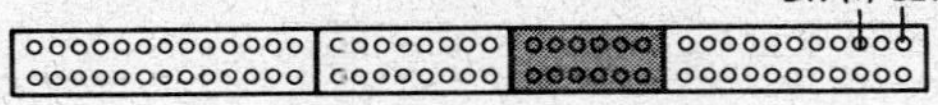

2.2L (F22A1) AND 2.3L (H23A1) ENGINES TROUBLESHOOTING PGM-FI FAULT CODES 4, 8 AND 9—1992 PRELUDE

4 — Self-diagnosis Check Engine light indicates code 4: A problem in the CRANK Sensor circuit.

8 — Self-diagnosis Check Engine light indicated code 8: A problem in the TDC Sensor circuit.

9 — Self-diagnosis Check Engine light indicates code 9: A problem in the CYL Sensor circuit.

The CRANK sensor determines timing for fuel injection and ignition of each cylinder and also detects engine RPM. The TDC sensor determines ignition timing at start-up (cranking) and when crank angle is abnormal. The CYL sensor detects the position of No. 1 cylinder for sequential fuel injection to each cylinder.

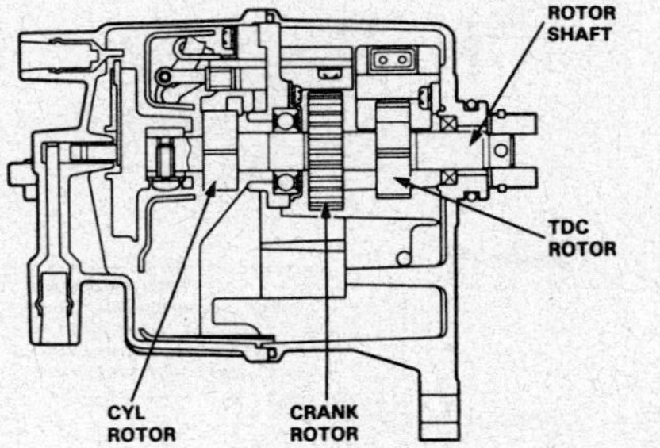

4 — 8 — or — 9

- Check Engine light has been reported on.
- With service check connector jumped CODE 4, 8 and/or 9 are indicated.

Do the ECU Reset Procedure

Start the engine.

Is Check Engine light on and does it indicate CODE 4, 8 or 9?

NO: Intermittent failure, system is OK at this time (test drive may be necessary). Check for poor connections or loose wires at C425 (located at right shock tower), C125 (TDC/CRANK/CYL sensor) and ECU.

YES: Stop the engine.

Disconnect the 8P connector from the TDC/CRANK/CYL sensor.

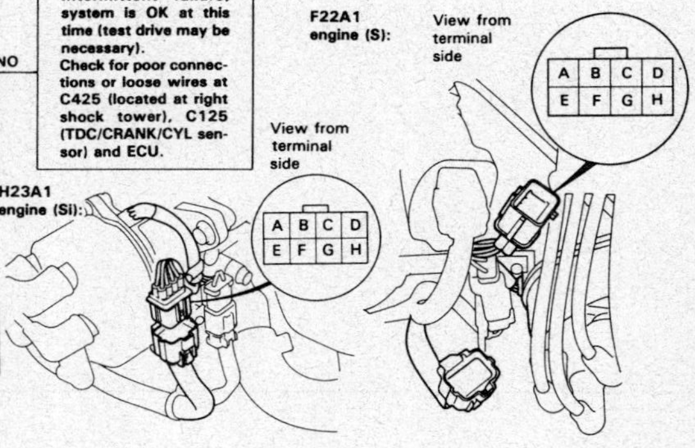

2.2L (F22A1) AND 2.3L (H23A1) ENGINES TROUBLESHOOTING PGM-FI FAULT CODES 4, 8 AND 9—1992 PRELUDE

- Measure resistance between terminals of the indicated sensor. *see table
- Is there 350 – 700 Ω?
  - NO: Replace the distributor sub-assembly
  - YES: Check for continuity to body ground on both terminals of the indicated sensor.
- Does continuity exist?
  - YES: Replace the distributor sub-assembly
  - NO: Reconnect the connector.
- Connect the test harness to the main wire harness only, not to the ECU
- Measure resistance between terminals of the indicated sensor on test harness. *see table
- Is there 350 – 700 Ω?
  - NO: Repair open in the indicated sensor wires. *see table
  - YES: Check for continuity to body ground on B15, B13 or B11 terminals.
- Does continuity exist?
  - YES: Repair short to body ground in the indicated sensor wires. *see table
  - NO: Substitute a known-good ECU and recheck. If symptom/indication goes away, replace the original ECU.

*:

SENSOR	CODE	SENSOR TERMINAL	ECU TERMINAL	WIRE COLOR
CRANK	4	B	B15	BLU/GRN
		F	B16	BLU/YEL
TDC	8	C	B13	ORN/BLU
		G	B14	WHT/BLU
CYL	9	D	B11	ORN
		H	B12	WHT

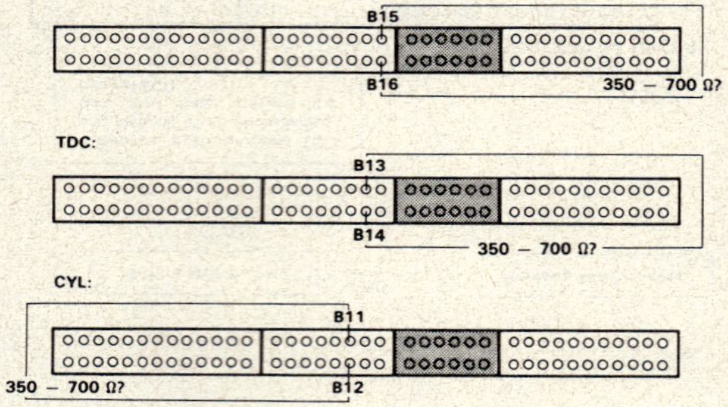

2.2L (F22A1) AND 2.3L (H23A1) ENGINES TROUBLESHOOTING PGM-FI FAULT CODE 5—1992 PRELUDE

- 5
- – Check Engine light has been reported on.
  – With service check connector jumped CODE 5 is indicated.
- Do the ECU Reset Procedure
- Start the engine and keep engine rpm at idle.
- Is Check Engine light on and does it indicate CODE 5?
  - NO: – Intermittent failure, system is OK at this time (test drive may be necessary). – Check vacuum hoses, pipes and connections.
  - YES: Stop the engine.
- Disconnect #21 hose from the throttle body, connect vacuum pump to the hose and apply vacuum.
- Does it hold vacuum?
  - NO: Connect a vacuum pump to the MAP sensor and apply vacuum.
    - Does it hold vacuum?
      - NO: Replace MAP sensor.
      - YES: Repair vacuum leak in hose routing between MAP sensor and intake manifold.
  - YES: Connect a T-fitting from a vacuum gauge between the throttle body base and #21 hose.

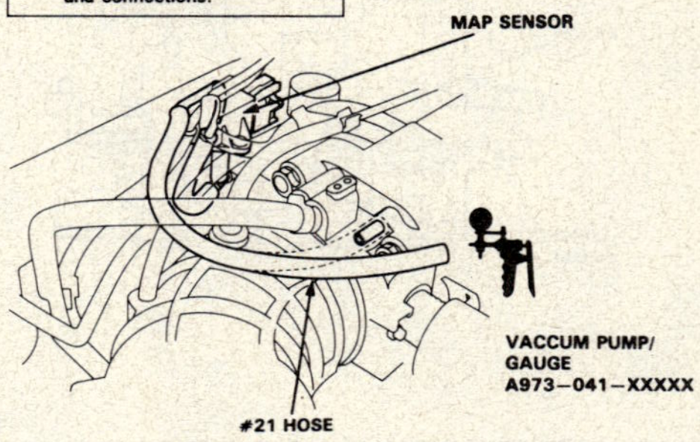

2.2L (F22A1) AND 2.3L (H23A1) ENGINES TROUBLESHOOTING PGM-FI FAULT CODE 5—1992 PRELUDE

Start the engine.

Is there manifold vacuum?
- NO → Remove restriction from throttle body.
- YES ↓

Stop the engine.

Connect the test harness between the ECU and connector

Turn the ignition switch ON.

Measure voltage between D17 (+) terminal and D21 (−) terminal.

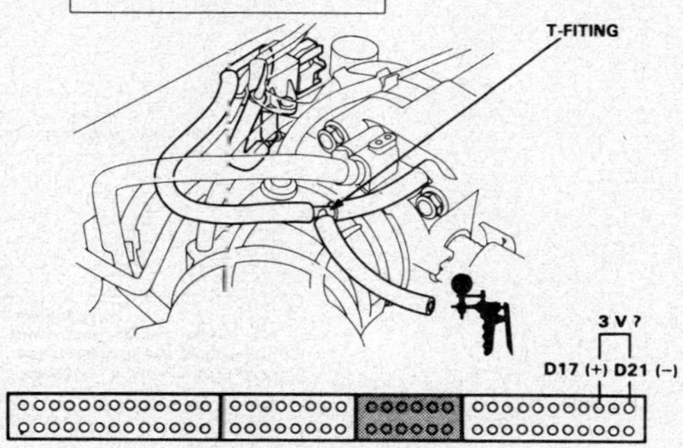

Is there approx. 3 V ?
- NO → Replace MAP sensor.
- YES ↓

Start the engine and allow it to idle.

Is there approx. 1 V?
- NO → Replace MAP sensor.
- YES ↓

Substitute a known-good ECU and recheck. If symptom/indication goes away, replace the original ECU.

2.2L (F22A1) AND 2.3L (H23A1) ENGINES TROUBLESHOOTING PGM-FI FAULT CODE 6—1992 PRELUDE

Self-diagnosis Check Engine light indicates code 6: A problem in the Coolant Temperature (TW) Sensor circuit.

The TW sensor is a temperature dependant resistor (thermistor). The resistance of the thermistor decreases as the coolant temperature increases as shown below.

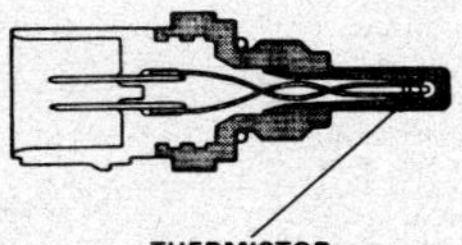

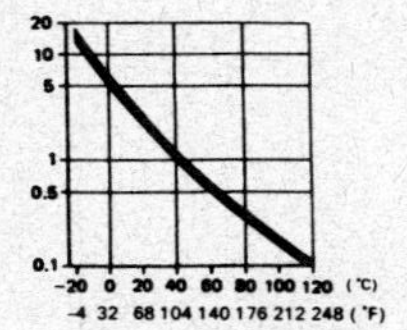

6

- Check Engine light has been reported on.
- With service check connector jumped CODE 6 is indicated.

Do the ECU Reset Procedure

Turn the ignition switch ON.

Is Check Engine light on and does it indicate CODE 6?
- NO → Intermittent failure, system is OK at this time (test drive may be necessary). Check for poor connections or loose wire at C425, C426 (located at right shock tower), C122 (TW sensor) and ECU.
- YES ↓

Warm up engine to normal operating temperature (cooling fan comes on).

Turn the ignition switch OFF.

Disconnect the 2P connector from the TW sensor.

Measure resistance between the 2 terminals on the TW sensor.

Is there 20—400 Ω?
- NO → Replace TW sensor.
- YES

2.2L (F22A1) AND 2.3L (H23A1) ENGINES TROUBLESHOOTING PGM-FI FAULT CODE 6—1992 PRELUDE

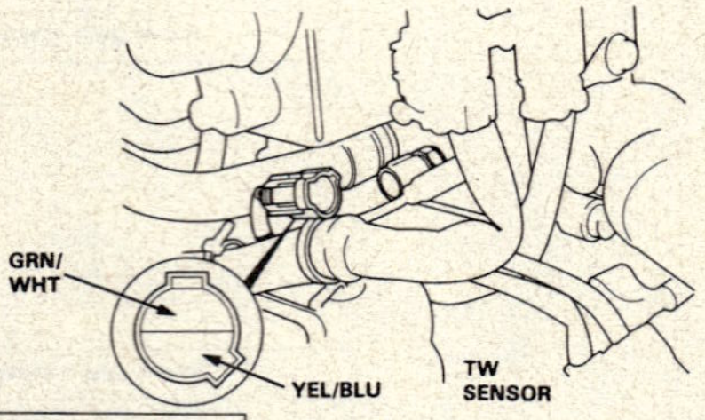

- Turn the ignition switch ON.
- Measure voltage between YEL/BLU and body ground.
- Is there approx. 5V ?
  - YES → Measure voltage between YEL/BLU (+) terminal and GRN/WHT (−) terminal.
    - Is there approx. 5V ?
      - NO → Repair open in GRN/WHT wire between ECU (D22) and TW sensor.
      - YES → Substitute a known-good ECU and recheck. If symptom/indication goes away, replace the original ECU.
  - NO ↓

A/T only

- Disconnect the 22P connector from the A/T control unit.
- Is there approx. 5 V?
  - YES → Replace the A/T control unit.
  - NO ↓
- Turn the ignition switch OFF.
- Connect the test harness "D" connector to the ECU only, not to the main wire harness
- Turn the ignition switch ON.
- Measure voltage between D13 (+) terminal and D22 (−) terminal.

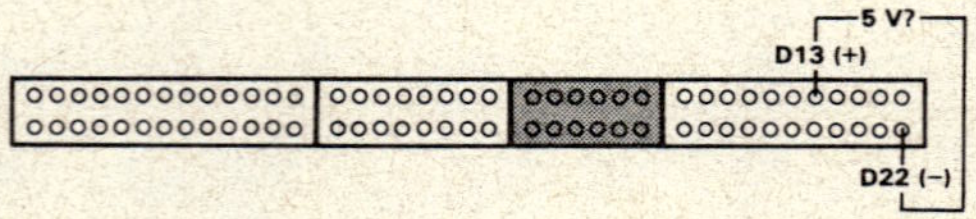

- Is there approx. 5V?
  - YES → Repair open or short in YEL/BLU wire between ECU (D13), A/T control unit and sensor.
  - NO → Substitute a known-good ECU and recheck. If symptom/indication goes away, replace the original ECU.

2.2L (F22A1) AND 2.3L (H23A1) ENGINES TROUBLESHOOTING PGM-FI FAULT CODE 7—1992 PRELUDE

[7] Self-diagnosis Check Engine light indicates code 7: A problem in the Throttle Angle Sensor circuit.

The throttle angle sensor is a potentiometer. It is connected to the throttle valve shaft. As the throttle angle changes, the throttle angle sensor varies the voltage signal to the ECU.

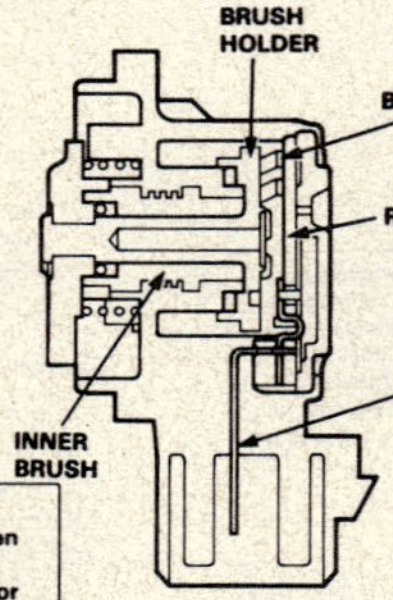

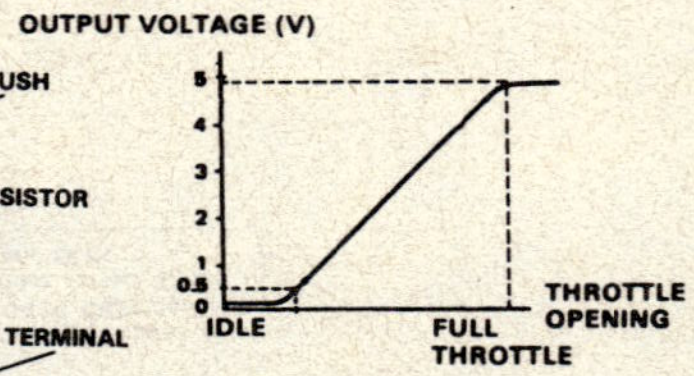

[7]

- — Engine is running.
  — Check Engine light has been reported on.
  — With service check connector jumped CODE 7 is indicated.
- Do the ECU Reset Procedure
- Start the engine.
- Is Check Engine light on and does it indicate CODE 7?
  - NO → Intermittent failure, system is OK at this time (test drive may be necessary). Check for poor connections or loose wires at C425, C433 (located at right shock tower), C113 (throttle angle sensor) and ECU.
  - YES ↓
- Turn the ignition switch OFF.
- Disconnect the 3P connector from the throttle angle sensor.
- Turn the ignition switch ON.
- Measure voltage between YEL/WHT (+) terminal and GRN/WHT (−) terminal.
- Is there approx. 5V ?
  - NO → Measure voltage between YEL/WHT (+) terminal and body ground.
  - YES

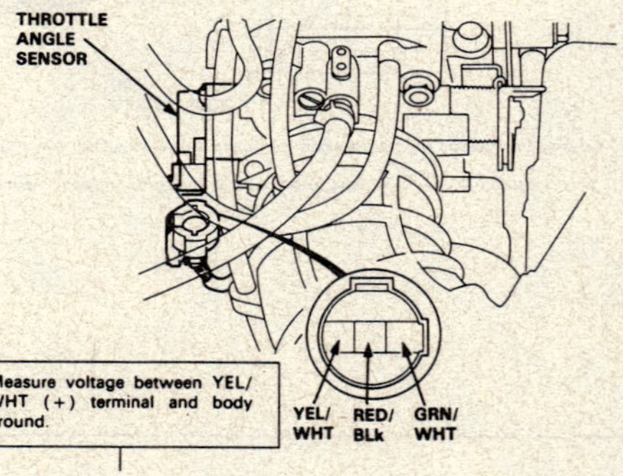

2.2L (F22A1) AND 2.3L (H23A1) ENGINES TROUBLESHOOTING PGM-FI FAULT CODE 7—1992 PRELUDE

Turn the ignition switch OFF.

Reconnect the 3P connector.

Connect the test harness between the ECU and connector

Turn the ignition switch ON.

Measure voltage between D11(+) terminal and D22 (−) terminal.

Is voltage 0.5 V at full close throttle, and approx. 4.5 V at full open throttle?
NOTE: There should be a smooth transition from 0.5 V to approx. 4.5 V as the throttle is depressed.

- YES → Substitute a known-good ECU and recheck. If symptom/indication goes away, replace the original ECU.
- NO → Is there approx. 5V ?
  - YES → Repair open in GRN/WHT wire between ECU (D22) and throttle angle sensor.
  - NO → Turn the ignition switch OFF.

Connect the test harness between the ECU and connector

Turn the ignition switch ON.

Measure voltage between D20 (+) terinal and D22 (−) terminal.

Is there approx. 5V ?

- YES → Repair open in YEL/WHT wire between ECU (D20) and throttle angle sensor.
- NO → Substitute a known-good ECU and recheck. If prescribed voltage is now available, replace the original ECU.

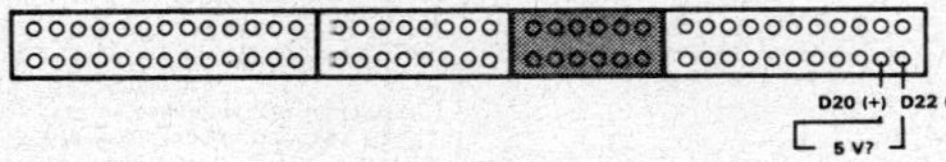

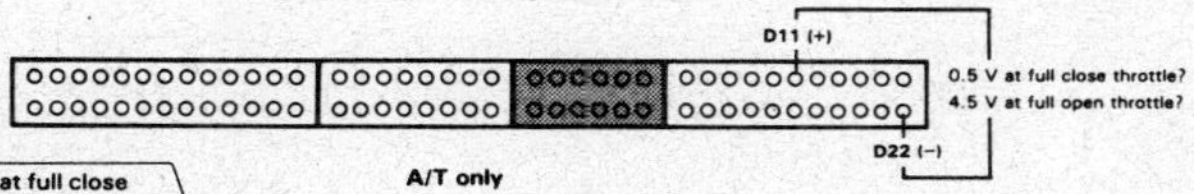

A/T only

Disconnect the 22P connector from the A/T control unit.

Is voltage 0.5 V at full close throttle, and 4.5 V at full open throttle?
NOTE: There should be a smooth transition from 0.5 V to 4.5 V as the throttle is depressed.

- YES → Replace the A/T control unit.
- NO →
  - Replace throttle angle sensor.
  - Repair open or short in RED/BLK wire between ECU (D11), A/T control unit and throttle angle sensor.

2.2L (F22A1) AND 2.3L (H23A1) ENGINES TROUBLESHOOTING PGM-FI FAULT CODE 10—1992 PRELUDE

10 — Self-diagnosis Check Engine light indicates code 10: A problem in the Intake Air Temperature (TA) Sensor circuit.

The TA sensor is a temperature dependant resistor (thermistor). The resistance of the thermistor decreases as the intake air temperature increases as shown below.

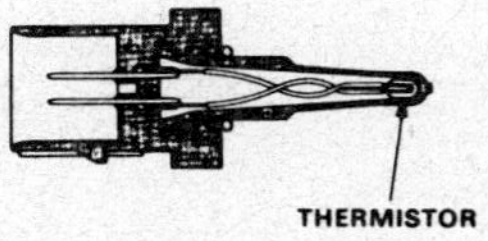

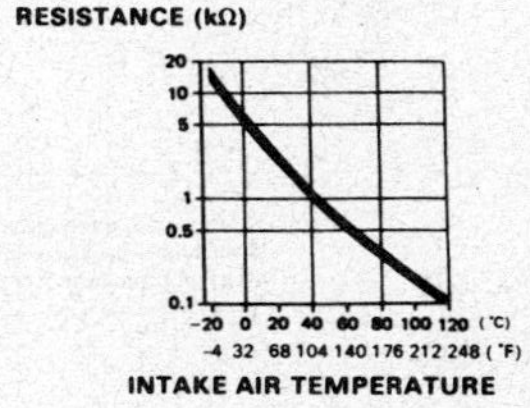

10

- Check Engine light has been reported on.
- With service check connector jumped CODE 10 is indicated.

Do the ECU Reset Procedure

Turn the ignition switch ON.

Is Check Engine light on and does it indicate CODE 10?

- NO → Intermittent failure, system is OK at this time (test drive may be necessary). Check for poor connections or loose wires at C425, C426 (located at right shock tower), C108 (TA sensor) and ECU.
- YES → Turn the ignition switch OFF.

Disconnect the 2P connector from the TA sensor.

Measure resistance between the 2 terminals on the TA sensor.

Is there 0.4—4.0 kΩ?

- NO → Replace TA sensor.
- YES

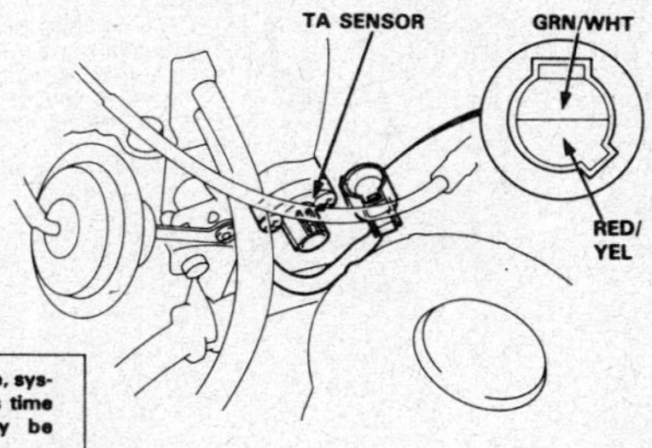

2.2L (F22A1) AND 2.3L (H23A1) ENGINES TROUBLESHOOTING PGM-FI FAULT CODE 10—1992 PRELUDE

Turn the ignition switch ON.

Measure voltage between RED/YEL (+) terminal and body ground.

Is there approx. 5 V ?

- YES → Measure voltage between RED/YEL (+) terminal and GRN/WHT (−) terminal.
  - Is there approx. 5 V ?
    - NO → Repair open in GRN/WHT wire between ECU (D22) and TA sensor.
    - YES → Substitute a known-good ECU and recheck. If symptom/indication goes away, replace the original ECU.
- NO → Turn the ignition switch OFF.

Connect the test harness "D" connector to the ECU only, not to the main wire harness

Turn the ignition switch ON.

Measure voltage between D15 (+) terminal and D22 (−) terminal.

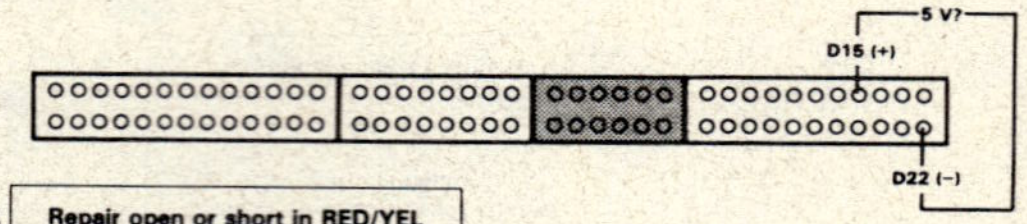

Is there approx. 5V?

- YES → Repair open or short in RED/YEL wire between ECU (D15) and TA sensor.
- NO → Substitute a known-good ECU and recheck. If symptom/indication goes away, replace the original ECU.

2.2L (F22A1) AND 2.3L (H23A1) ENGINES TROUBLESHOOTING PGM-FI FAULT CODE 12—1992 PRELUDE

12

- Check Engine light has been reported on.
- With service check connector jumped CODE 12 is indicated.

Do the ECU Reset Procedure

Road test necessary: Warm up the engine to normal operating temperature (cooling fan comes on). Drive the car on the road for approx. 10 minutes. Keep the engine speed in the 1700–2500 range.

Is Check Engine light on and does it indicate CODE 12?

- NO → Intermittent failure, system is OK at this time. Check for poor connections or loose wires at C426 (located at right shock tower), C117 (EGR valve), C310 (EGR control solenoid valve) and ECU.

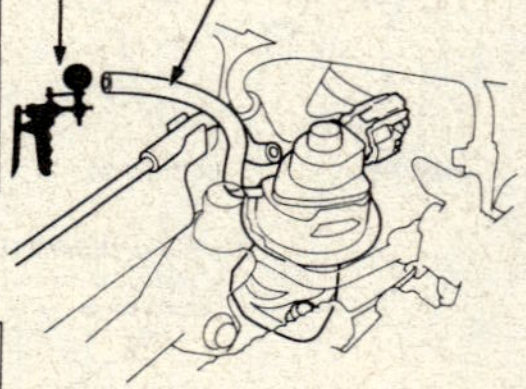

- YES → With the engine at idle, disconnect the # 16 hose from the EGR valve and connect a vacuum pump/gauge to the hose.

Is there any vacuum?

- YES → Disconnect 2P connector from the EGR control solenoid valve and check the # 16 hose for vacuum again.

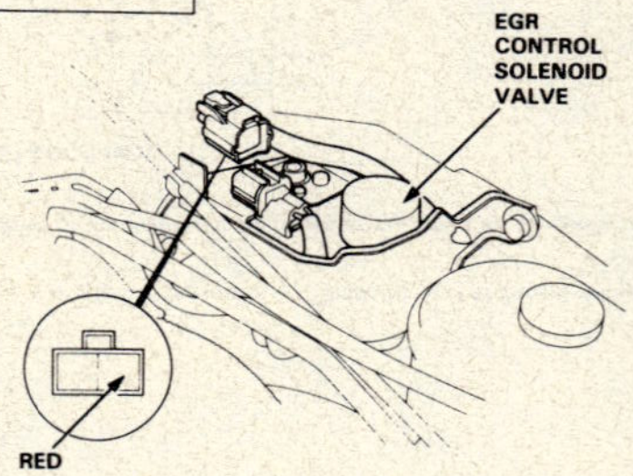

- NO → Move the vacuum pump/gauge to the EGR valve.

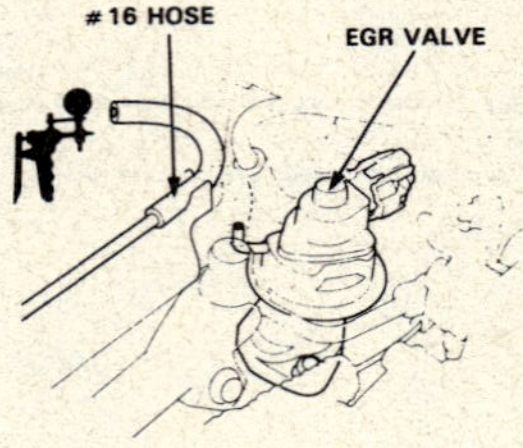

2.2L (F22A1) AND 2.3L (H23A1) ENGINES—TROUBLESHOOTING PGM-FI FAULT CODE 12—1992 PRELUDE

With the engine at idle, apply 8" of vacuum to the EGR valve.

Does the engine stall or run rought and does the EGR valve hold vacuum?

- NO → **Replace EGR valve.**
- YES ↓

Disconnect the 2P connector from the EGR control solenoid valve.

Measure voltage between BLK/YEL (+) terminal on the main wire harness and body ground.

Is there battery voltage ?

- NO → **Repair open in BLK/YEL wire between the solenoid valve and No. 23* ECU (15 A) fuse.**
- YES ↓

*: No. 19 without SRS

Reconnect the vacuum pump/gauge to the #16 hose.

Start the engine and allow it to idle.

Is there any vacuum ?

- YES → **Check vacuum hose routing of the entire EGR system. If hose routing is OK, replace EGR control solenoid valve.**
- NO ↓

Turn the ignition switch OFF and disconnect the "A" connector from the ECU.

Check for continuity to ground on RED wire of 2P connector.

Does continuity exist?

- YES → **Repair short in RED wire between EGR control solenoid valve and ECU (A11).**
- NO ↓

Substitute a known-good ECU and retest. If symptom/indication goes away, replace the original ECU.

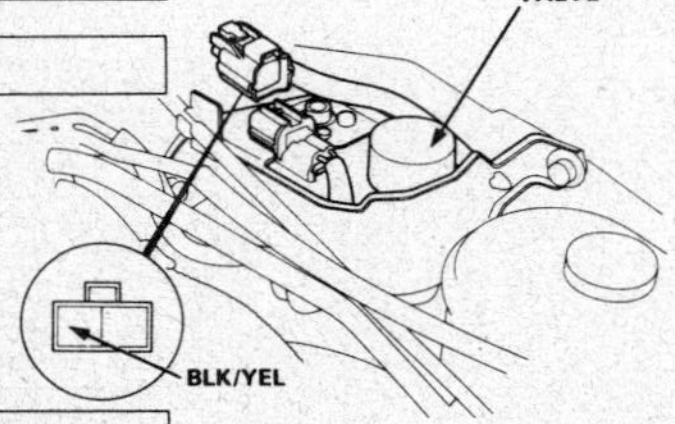

Connect the battery positive terminal to the B terminal of the 2P connector. While watching the vacuum gauge, connect the battery negative termainal to the A terminal.

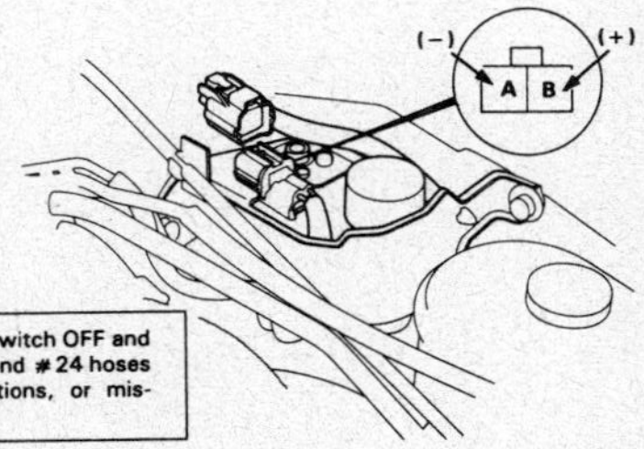

Is there approx. 8" within 1 second?

- NO → Turn the ignition switch OFF and inspect the #16 and #24 hoses for leaks, restrictions, or misrouting.
  - Are the hoses OK?
    - NO → **Correct as necessary.**
    - YES ↓
  - Disconnect the lower hose on EGR control solenoid valve and connect a vacuum gauge to the hose.
  - Start the engine and allow it to idle.
  - Is there 6"— 10" of vacuum ?
    - NO → **Replace CVC valve.**
    - YES → **Replace the EGR control solenoid valve.**
- YES ↓

Turn the ignition switch OFF and reconnect the 2P connector.

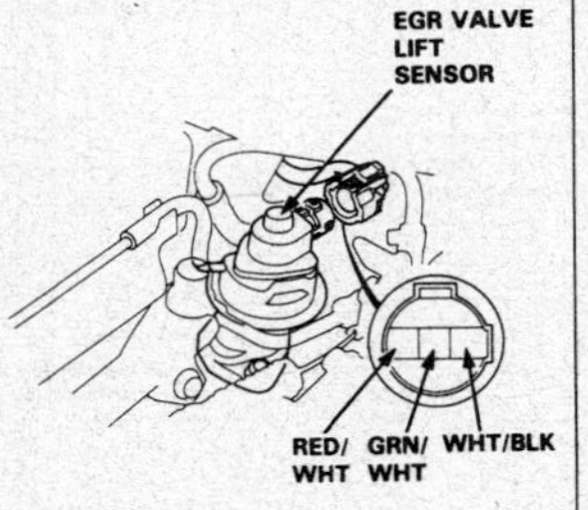

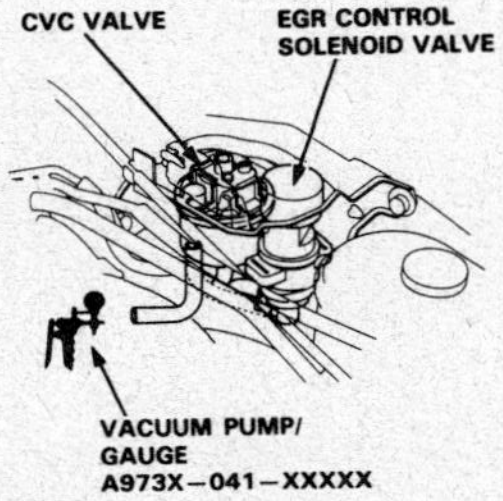

Disconnect 3P connector from the EGR valve.

Turn the ignition switch ON.

Measure voltage between RED/WHT (+) terminal and GRN/WHT (−) terminal.

Is there approx. 5 V ?

- NO → Measure voltage between RED/WHT (+) terminal and body ground.
- YES ↓

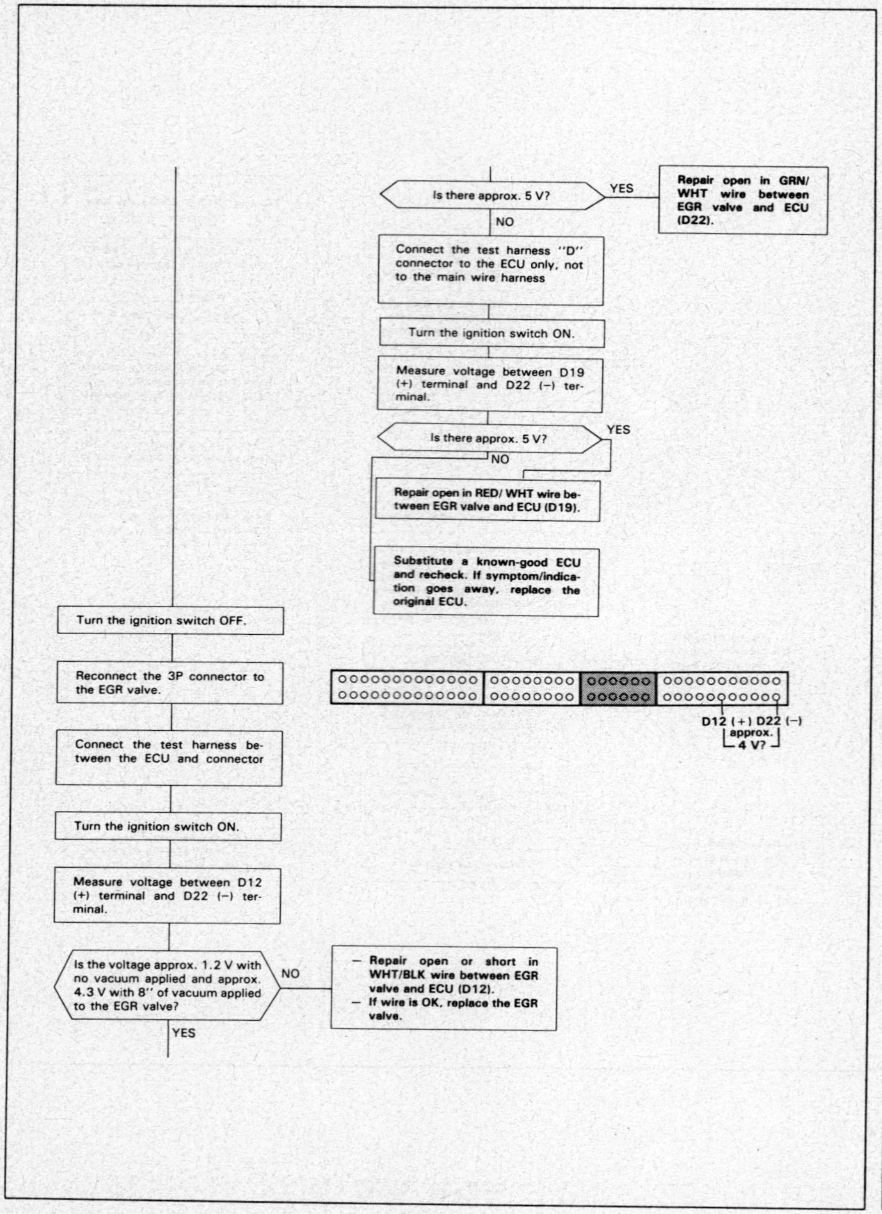
Is there approx. 5 V?
YES
Repair open in GRN/ WHT wire between EGR valve and ECU (D22).
NO
Connect the test harness "D" connector to the ECU only, not to the main wire harness
Turn the ignition switch ON.
Measure voltage between D19 (+) terminal and D22 (−) terminal.
Is there approx. 5 V?
YES
NO
Repair open in RED/ WHT wire between EGR valve and ECU (D19).
Substitute a known-good ECU and recheck. If symptom/indication goes away, replace the original ECU.
Turn the ignition switch OFF.
Reconnect the 3P connector to the EGR valve.
Connect the test harness between the ECU and connector
Turn the ignition switch ON.
Measure voltage between D12 (+) terminal and D22 (−) terminal.
D12 (+) D22 (−)
approx. 4 V?
Is the voltage approx. 1.2 V with no vacuum applied and approx. 4.3 V with 8" of vacuum applied to the EGR valve?
NO
– Repair open or short in WHT/BLK wire between EGR valve and ECU (D12).
– If wire is OK, replace the EGR valve.
YES

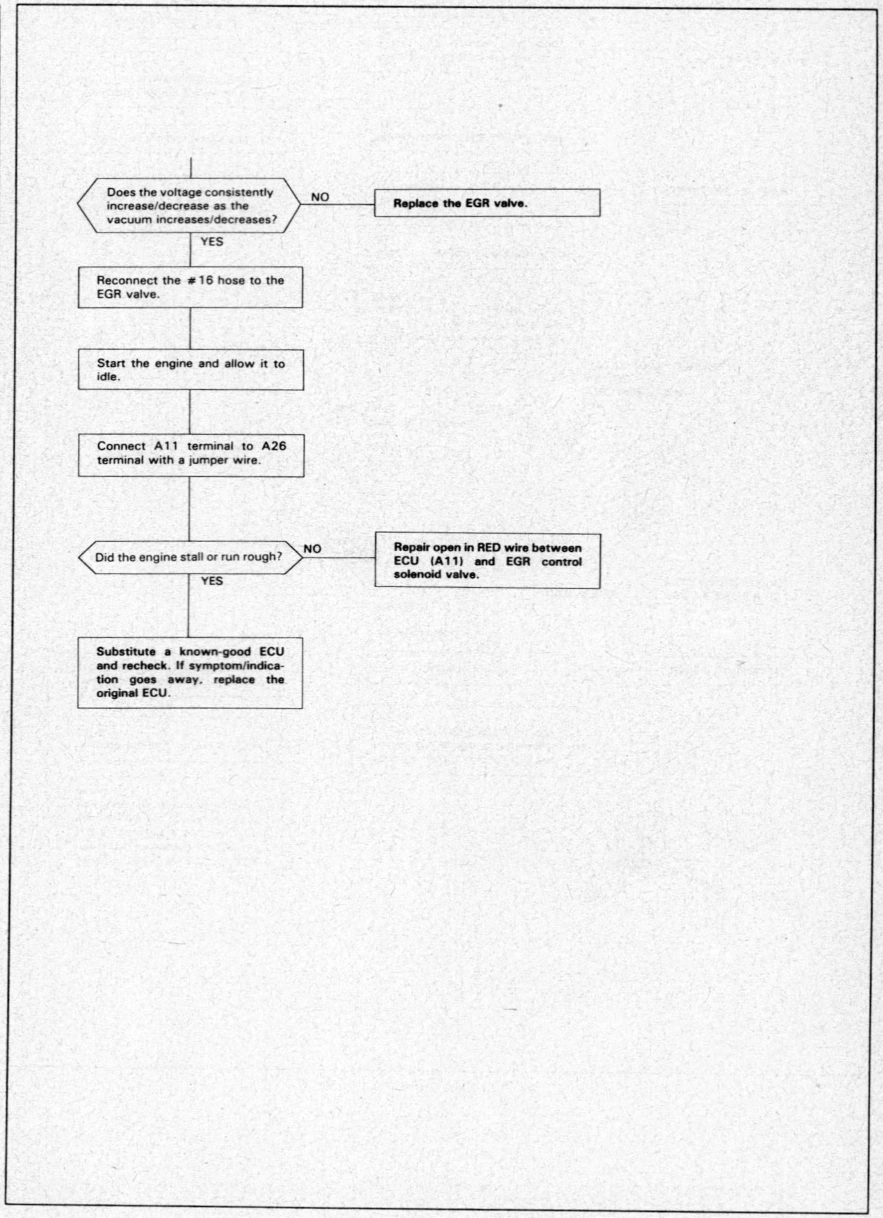
Does the voltage consistently increase/decrease as the vacuum increases/decreases?
NO
Replace the EGR valve.
YES
Reconnect the #16 hose to the EGR valve.
Start the engine and allow it to idle.
Connect A11 terminal to A26 terminal with a jumper wire.
Did the engine stall or run rough?
NO
Repair open in RED wire between ECU (A11) and EGR control solenoid valve.
YES
Substitute a known-good ECU and recheck. If symptom/indication goes away, replace the original ECU.

2.2L (F22A1) AND 2.3L (H23A1) ENGINES TROUBLESHOOTING PGM-FI FAULT CODE 13—1992 PRELUDE

Self-diagnosis Check Engine light indicates code 13: A problem in the Atmospheric Pressure (PA) Sensor.

The PA sensor is built into the ECU.

- Check Engine light has been reported on.
- With service check connector jumped CODE 13 is indicated.

↓

Do the ECU Reset Procedure

↓

Turn the ignition switch ON.

↓

Is Check Engine light on and does it indicate CODE 13?

- NO → Intermittent failure, system is OK at this time (test drive may be necessary).
- YES → Substitute a known-good ECU and recheck. If symptom/indication goes away, replace the original ECU.

2.2L (F22A1) AND 2.3L (H23A1) ENGINES TROUBLESHOOTING PGM-FI FAULT CODE 14—1992 PRELUDE

Self-diagnosis Check Engine light indicates code 14: A problem in the Electric Air Control Valve (EACV) circuit.

The EACV changes the amount of air bypassing the throttle body in response to a current signal from the ECU in order to maintain the proper idle speed.

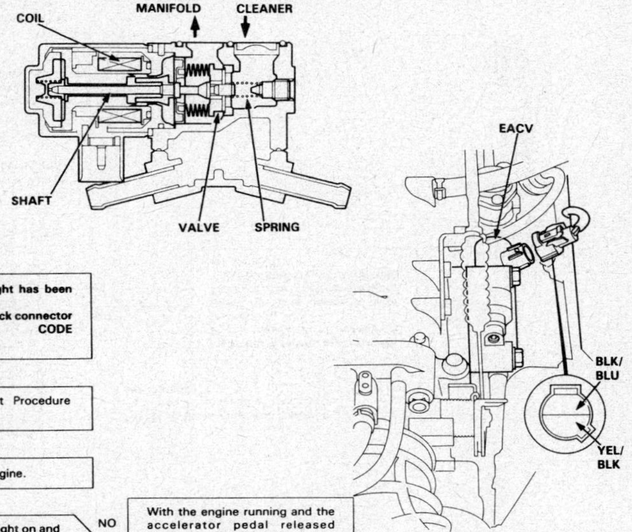

- Check Engine light has been reported on.
- With service check connector jumped CODE 14 is indicated.

↓

Do the ECU Reset Procedure

↓

Start the engine.

↓

Is Check Engine light on and does it indicate CODE 14?

- YES → Remove the 2P connector from the EACV.
- NO → With the engine running and the accelerator pedal released disconnect the 2P connector on the EACV.
  - Is there a reduction in engine rpm?
    - YES → Intermittent failure, system is OK at this time (test driving may be necessary). Check for poor connection or loose wires at C425, C433 (located at right shock tower), C106 (EACV) and ECU.
    - NO → Substitute a known-good EACV and retest.

2.2L (F22A1) AND 2.3L (H23A1) ENGINES TROUBLESHOOTING PGM-FI FAULT CODE 14—1992 PRELUDE

Measure voltage between the YEL/BLK wire and body ground.

Is there battery voltage?

- NO → Repair open in YEL/BLK wire between EACV and main relay.
- YES ↓

Turn the ignition switch off and reconnect the 2P connector the EACV.

Connect the test harness "A" connector to the main wire harness only, not the ECU

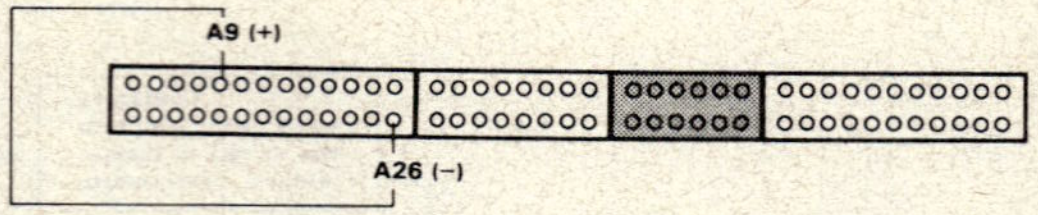

Turn the ignition switch ON.

Momentarily connect A9 terminal to A26 terminal several times.

Does the EACV click?

- YES → Substitute a known-good ECU and retest. If symptom/indication goes away, replace the original ECU.
- NO → Repair open or short in BLK/BLU wire between EACV and ECU (A9). If the wire is OK, replace the EACV.

2.2L (F22A1) AND 2.3L (H23A1) ENGINES TROUBLESHOOTING PGM-FI FAULT CODE 15—1992 PRELUDE

15 Self-diagnosis Check Engine light indicates code 15: A problem in the Ignition Output Signal circuit.

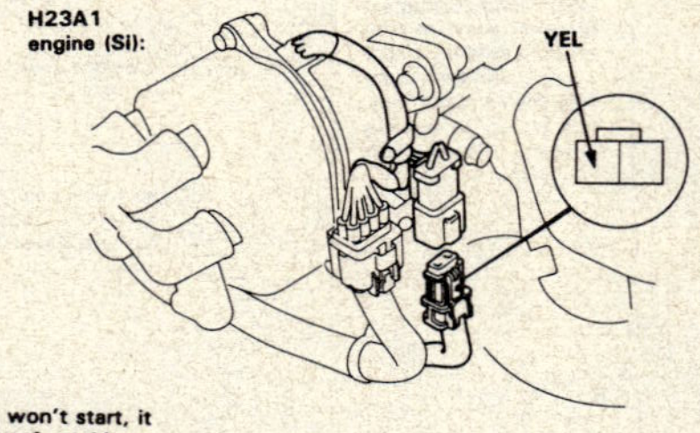

- Check Engine light has been reported on.
- With service check connector jumped CODE 15 is indicated.

Do the ECU Reset Procedure

Start the engine.

NOTE: If the engine won't start, it may take 20 seconds of cranking to set the code.

Is Check Engine light on and does it indicate CODE 15?

- NO → Intermittent failure, system is OK at this time (test drive may be necessary). Check for poor connections or loose wires at C125 (igniter unit), C433 (located at right shock tower) and ECU.
- YES ↓

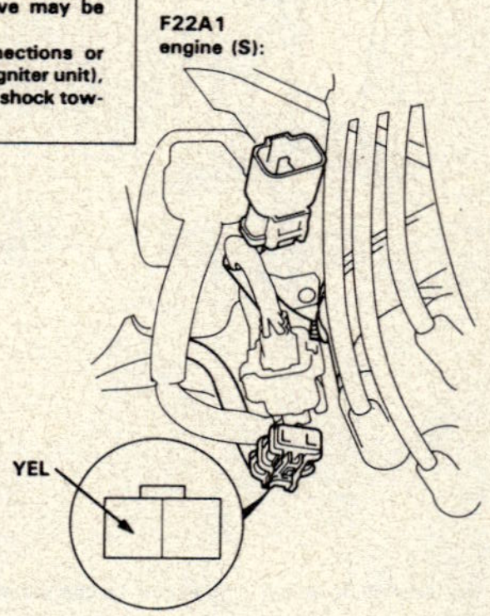

Turn the ignition switch OFF.

Disconnect the 2P connector at the distributor.

Turn the ignition switch ON.

Measure voltage between YEL (+) terminal and body ground.

Is there battery voltage ?

- NO
- YES

2.2L (F22A1) AND 2.3L (H23A1) ENGINES TROUBLESHOOTING PGM-FI FAULT CODE 15—1992 PRELUDE

Turn the ignition switch OFF.

Reconnect the 2P connector.

Connect the test harness between the ECU and connector

Turn the ignition switch ON.

Measure voltage individually between A21 (+), A22 (+) terminals and A26 (−) terminal.

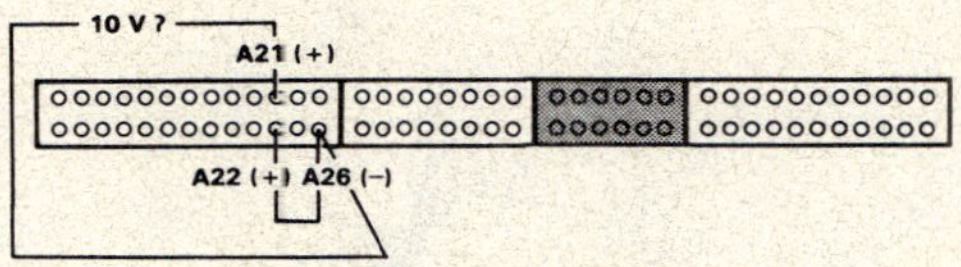

Is there approx. 10 V?

NO → **– Replace the igniter unit.**
– Repair open or short YEL/GRN wires between igniter unit and ECU (A21 or A22).

NOTE: If the YEL/GRN wire was shorted, the igniter may be damaged.

YES → **Substitute a known-good ECU and recheck. If symptom/indication goes away, replace the original ECU.**

Test the ignition coil

Is the ignition coil OK?

NO → **Replace the ignition coil.**

YES → **– Repair open in YEL wire between the 2P connector and ignition coil.**
– Repair open in BLK/YEL wire between the ignition coil and ignition switch.

2.2L (F22A1) AND 2.3L (H23A1) ENGINES TROUBLESHOOTING PGM-FI FAULT CODE 16—1992 PRELUDE

Troubleshooting Flowchart

CHECK 16 Self-diagnosis Check Engine light indicates code 16: A problem in the fuel injector circuit.

The injectors are a solenoid-actuated constant-stroke pintle type consisting of a solenoid, plunger needle valve and housing. When current is applied to the solenoid coil, the valve lifts up and pressurized fuel is injected. Because the needle valve lift and the fuel pressure are constant, the injection quantity is determined by the length of time that the valve is open (i.e., the duration the current is supplied to the solenoid coil). The injector is sealed by an O-ring and seal ring at the top and bottom. These seals also reduce operating noise.

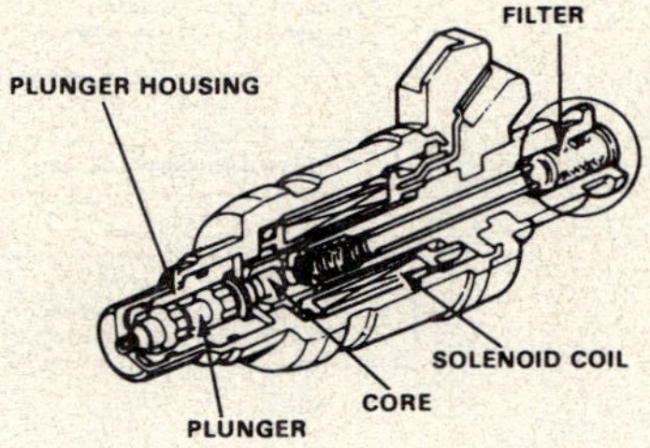

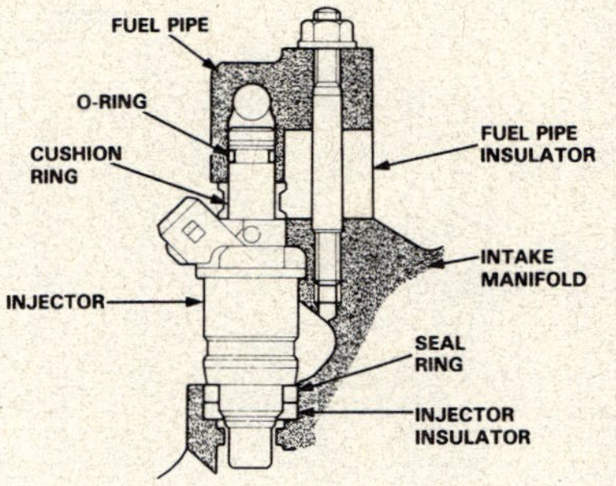

CHECK 16

– Check Engine light has been reported on.
– With service check connector jumped CODE 16 is indicated.

Do the ECU Reset Procedure

Start the engine and allow it to idle.

NOTE: If engine will not start, it may take 10 seconds of cranking to set the code.

Is Check Engine light on and does it indicate CODE 16?

YES

NO → **Intermittent failure, system is OK at this time (test drive may be necessary).**
Check for poor connections or loose wires at C425 (located at right shock tower), C102, C103, C104, C105 (injector), C104 (injector resistor) and ECU.

2.2L (F22A1) AND 2.3L (H23A1) ENGINES—TROUBLESHOOTING PGM-FI FAULT CODE 16—1992 PRELUDE

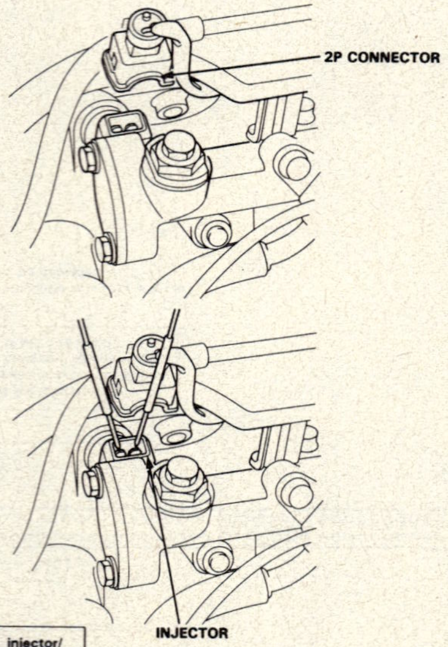

Start engine and listen at each injector for a clicking sound.

Turn the ignition switch OFF.

Disconnect the 2P connector from the injector that does not click.

Measure resistance between the 2 terminals of the injector.

Is there 1.5—2.5Ω ?
- NO → Replace the injector/injectors that are not 1.5–2.5 Ω.
- YES ↓

Turn the ignition switch ON.

Measure voltage between RED/BLK (+) terminal in the 2P connector and body ground.

Is there battery voltage ?
- NO → Turn the ignition switch OFF. → Disconnect 6P connector from the injector resistor. → Turn the ignition switch ON. → Measure voltage between YEL/BLK (+) terminal and body ground.
  - Is there battery voltage?
    - NO → Repair open in the YEL/BLK wire between the injector resistor and the main relay.
    - YES → Test the injector resistor
      - Does the injector resistor test OK?
        - NO → Replace the injector resistor.
        - YES → Repair open RED/BLK wire between 2P injector connector and resistor.
- YES ↓

Reconnect the 2P connector to the injector.

Turn the ignition switch OFF.

Connect the test harness between the ECU and connector

Turn the ignition switch ON.

Measure voltage between A23 (−) terminal and following terminal:
- No. 1 injector: A1 (+) terminal.
- No. 2 injector: A3 (+) terminal.
- No. 3 injector: A5 (+) terminal.
- No. 4 injector: A2 (+) terminal.

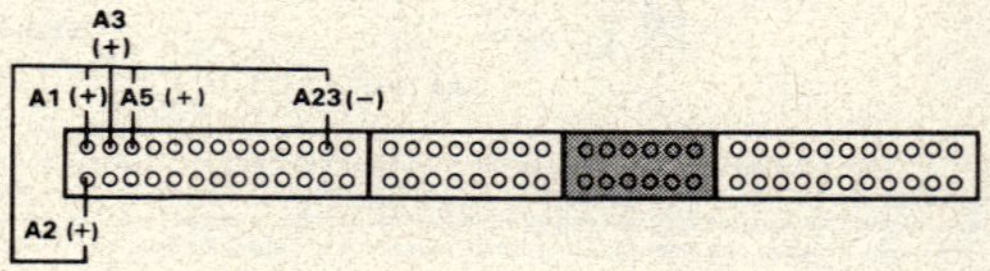

Is there battery voltage ?
- NO → Repair open in the wire between the ECU (A1, A3, A2 or A4) and the injector.
- YES → Substitute a known-good ECU and recheck. If symptom/indication goes away, replace the original ECU.

2.2L (F22A1) AND 2.3L (H23A1) ENGINES—TROUBLESHOOTING PGM-FI FAULT CODE 17—1992 PRELUDE

17 — Self-diagnosis Check Engine light indicates code 17: A problem in the Vehicle Speed Sensor circuit.

The signal generated by the speed sensor, produces pulses when the front wheels turn.

- **Check Engine light has been reported on.**
- **With service check connector jumped CODE 17 is indicated.**

Do the ECU Reset Precedure

Road test necessary. In 2nd gear accelerate to 4,000 rpm, then decelerate to 1,500 rpm with throttle fully closed.

Is Check Engine light on and does it indicate CODE 17?

- NO → **Intermittent failure, system is OK at this time. Check for poor connections or loose wires at C433 (located at right shock tower), C115 (speed sensor) and ECU.**
- YES ↓

Block rear wheels and set the parking brake. Jack up the front of the car and support with safety stands.

WARNING Block rear wheels before jacking up front of car.

Connect the test harness between the ECU and connector

Turn the ignition switch ON.

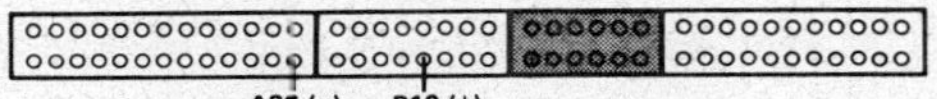

Block the right front wheel and slowly rotate left front wheel and measure voltage between B10 (+) terminal and A26 (−) terminal.

Does voltage pulse 0 V and 5 V?

- YES → **Substitute a known-good ECU and recheck. If symptom/indication goes away, replace the original ECU.**
- NO → Turn the ignition switch OFF.

Disconnect the B connector from the ECU only, not the main wire harness.

Turn the ignition switch ON.

Block the right front wheel and slowly rotate left front wheel and measure voltage between B10 (+) terminal and A26 (−) terminal.

Does voltage pulse 0 V and 5 V?

- YES → **Substitute a known-good ECU and recheck. If symptom/indication goes away, replace the original ECU.**
- NO →
  - **Repair short in ORN wire between ECU (B10) and the speed sensor, A/T control unit, speedometer, or cruise control unit.**
  - **Repair open in ORN wire between ECU (B10) and speed sensor.**
  - **If wire is OK, test the speed sensor**

2.2L (F22A1) AND 2.3L (H23A1) ENGINES—TROUBLESHOOTING PGM-FI FAULT CODE 20—1992 PRELUDE

[20] Self-diagnosis Check Engine light indicates code 20: A problem in the Electric Load Detector circuit.

- With service check connector jumped CODE 20 is indicated.
- Do the ECU Reset Procedure
- Start engine and keep engine rpm at idle. Turn on headlights.
- Does Check Engine light indicate CODE 20?
  - NO → Intermittent failure, system is OK at this time (test drive may be necessary). Check for poor connections or loose wires at C420 (ELD) and ECU.
  - YES ↓
- Turn the ignition switch OFF.
- Remove the under-hood fuse/relay box and remove the fuse/relay box lower cover.
- Disconnect the 3P connector from the electric load detector.
- Turn the ignition switch ON.
- Measure voltage between BLK/YEL (+) terminal and BLK (–) terminal.
- Is there battery voltage?
  - NO → Measure voltage between BLK/YEL (+) terminal and body ground.
    - Is there battery voltage?
      - YES → Repair open in BLK wire between the 3P connector and G403.
      - NO → Repair open in BLK/YEL wire between ECU (No. 23*) (15 A) fuse and the 3P connector.
  - YES ↓

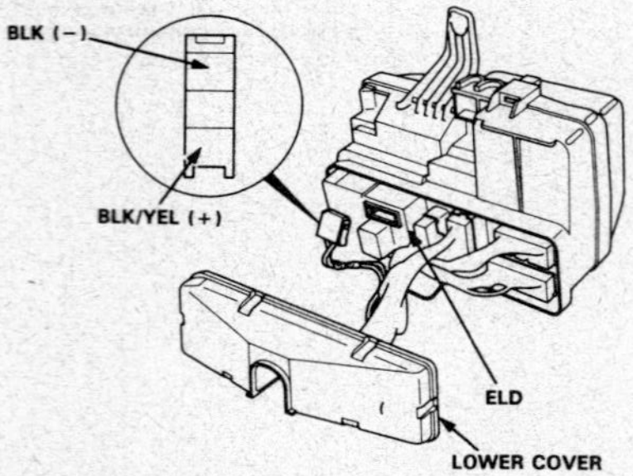

*: No. 19 without SRS

- Measure voltage between GRN/RED terminal and body ground.
- Is there 4.5–5 V?
  - NO → Repair open or short in GRN/BLK wire between ECU (D10) and the 3P connector. If wire is OK, substitute a known-good ECU and recheck.
  - YES ↓
- Turn the ignition switch OFF.
- Connect the 3P connector to the electric load detector.
- Connect the test harness between the ECU and connector

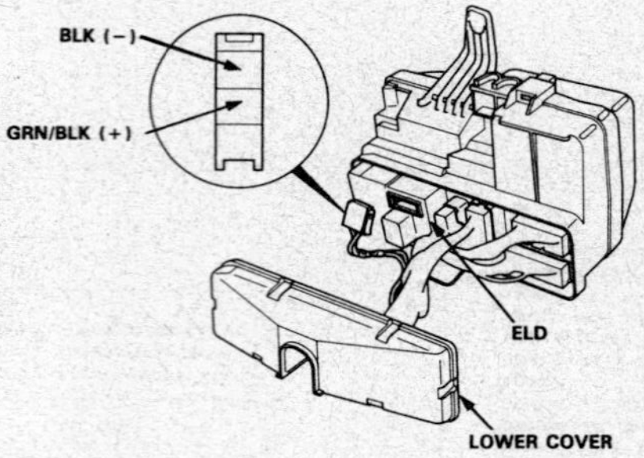

2.2L (F22A1) AND 2.3L (H23A1) ENGINES TROUBLESHOOTING PGM-FI FAULT CODE 20—1992 PRELUDE

Condition	Voltage
Headlight switch, first position (•)	1.5–3.0 V
Headlight switch, second position (●)	1.0–2.0 V

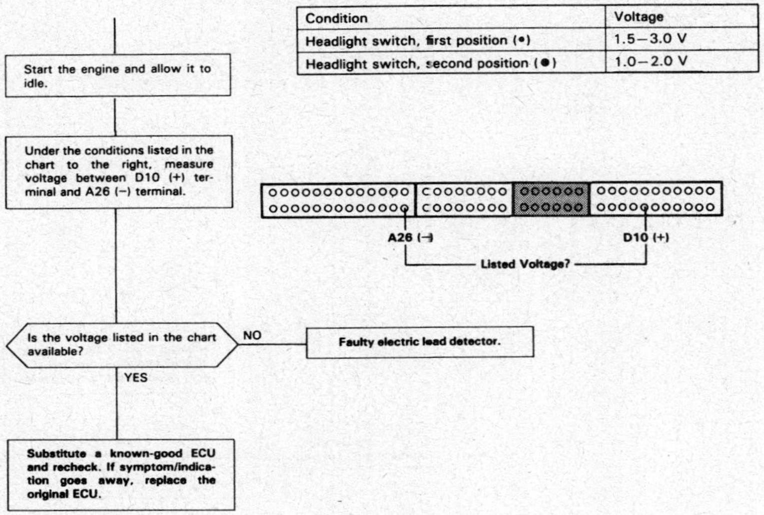

2.2L (F22A1) AND 2.3L (H23A1) ENGINES TROUBLESHOOTING PGM-FI FAULT CODE 23—1992 PRELUDE

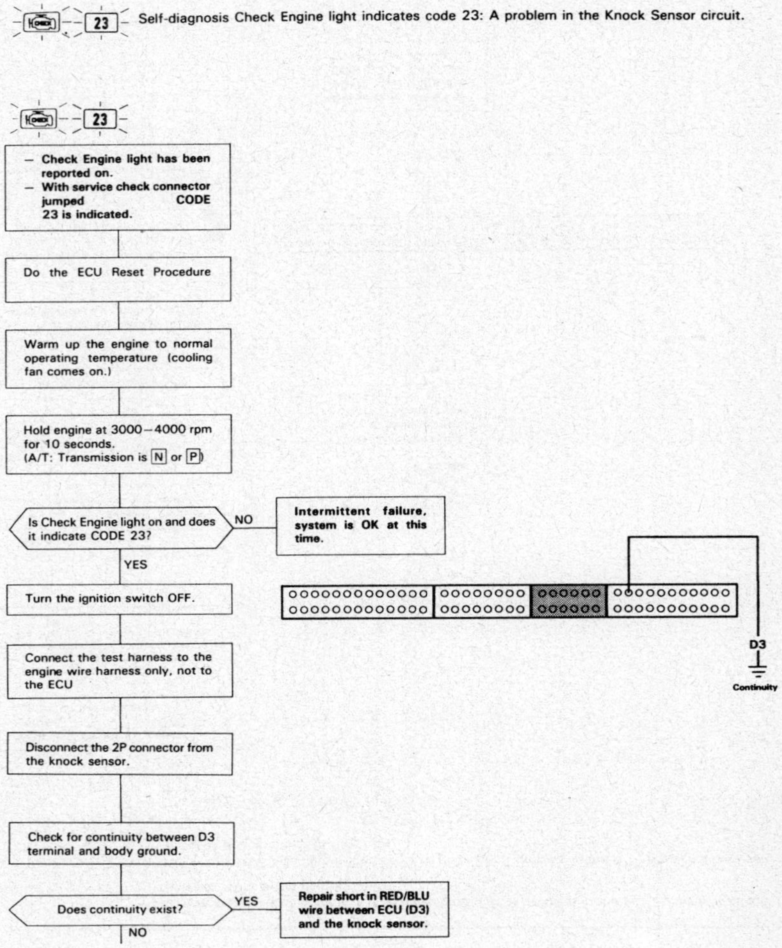

2.2L (F22A1) AND 2.3L (H23A1) ENGINES TROUBLESHOOTING PGM-FI FAULT CODES 30 AND 31—1992 PRELUDE

[30] Self-diagnosis Check Engine light indicates code 30: A problem in the A/T FI Signal A circuit between A/T control unit and ECU.

[31] Self-diagnosis Check Engine light indicates code 31: A problem in the A/T FI Signal B between A/T control unit and ECU.

[30] or [31]

With service check connector jumped CODE 30 and/or 31 are indicated.

Do the ECU Reset Procedure

Test drive necessary. Drive the car for several miles so that the transmission upshifts and downshifts several times.

Does Check Engine light indicate CODE 30 or 31?
- NO → Intermittent failure, system is OK at this time. Check for poor connections or loose wires at C411 (A/T control unit) and ECU.
- YES ↓

Turn the ignition switch OFF.

Connect the test harness to the main harness only, not to the ECU

Disconnect the 22P connector from the A/T control unit.

Check for continuity between B3 or B4* terminal and body ground.

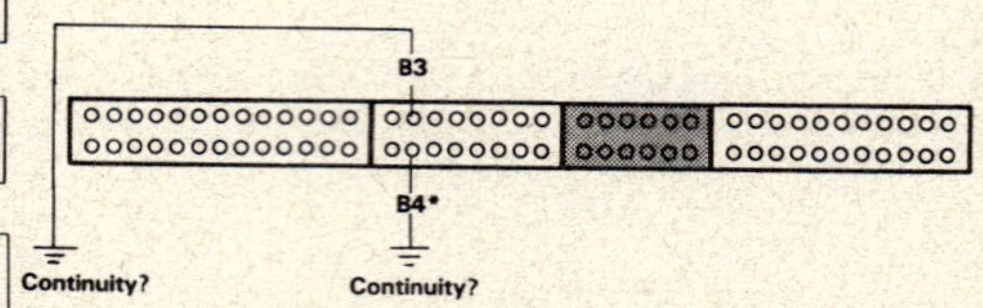

*: CODE 31 (A/T FI Signal B)

Does continuity exist?
- YES → Repair short in ORN or PNK* wire between ECU (B3 or B4*) and the A/T control unit.
- NO ↓

Check for continuity on ORN or PNK* wire between B3 or B4* and 22P connector of the A/T control unit.

Does continuity exist?
- NO → Repair open in ORN or PNK* wire between ECU (B3 or B4*) and the A/T control unit.
- YES ↓

Substitute a known-good ECU and recheck. If symptom/indication goes away, replace the original ECU.

2.2L (F22A1) AND 2.3L (H23A1) ENGINES TROUBLESHOOTING PGM-FI FAULT CODE 23—1992 PRELUDE

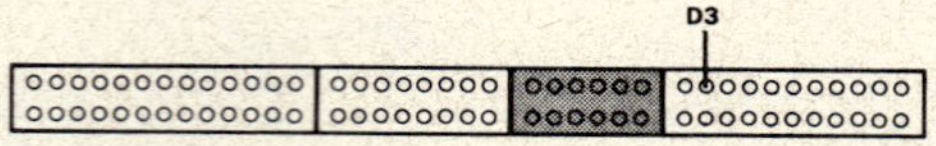

Check for continuity on RED/BLU wire between D3 terminal and 2P connector of engine wire harness.

Does continuity exist?
- NO → Repair open in RED/BLU wire between ECU (D3) and the knock sensor.

Substitute a known-good knock sensor and recheck.

Warm up the engine to normal operating temperature (cooling fan comes on).

Hold engine at 3000—4000 rpm for 10 seconds. (A/T: Transmission is [N] or [P])

Is Check Engine light on and does it indicate CODE 23?
- NO → Replace the original knock sensor.
- YES ↓

Substitute a known-good ECU and recheck. If symptom/indication goes away, replace the original ECU.

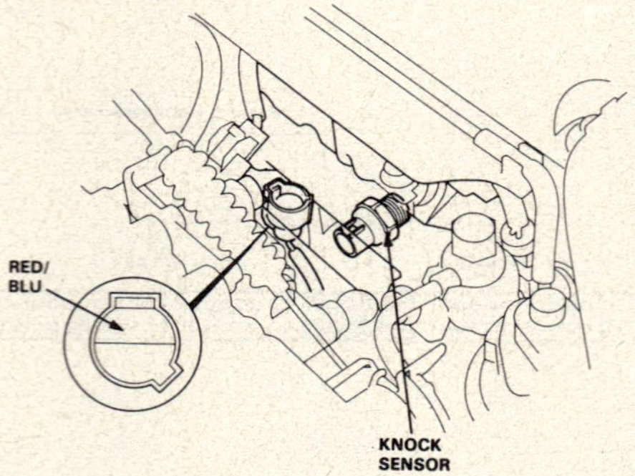

2.2L (F22A1) AND 2.3L (H23A1) ENGINES – TROUBLESHOOTING PGM-FI FAULT CODE 41 – 1992 PRELUDE

[CHECK] – [41] Self-diagnosis Check Engine light indicates code 41: A problem in the Oxygen (O_2) Sensor Heater circuit.

[CHECK] – [41]

- – Engine is running.
- – Check Engine light has been reported on. With service check connector jumped CODE 41 is indicated.

Do the ECU Reset Procedure

Start the engine.

Is Check Engine light on and does it indicate CODE 41?
- NO → Intermittent failure, system is OK at this time (test drive may be necessary). Check for poor connections or loose wires at C425, C432, C433 (located at right shock tower), C114 (O_2 sensor) and ECU.
- YES ↓

Stop the engine

Disconnect the 4P connector from the O_2 sensor.

Measure resistance between terminals C and D on the O_2 sensor.

Is there 10–40 Ω?
- NO → Replace O_2 sensor.
- YES ↓

F22A1 engine (S):

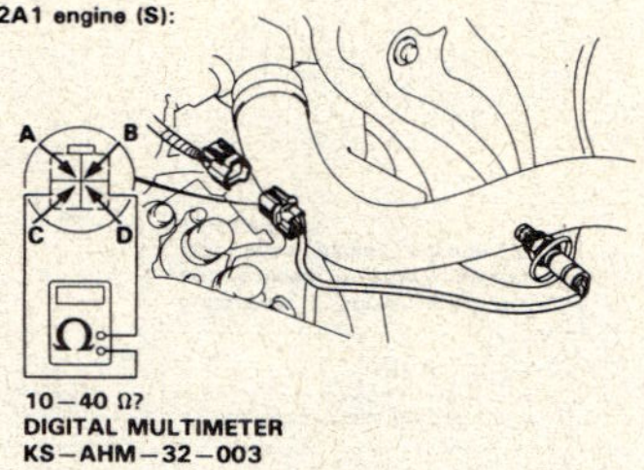

H23A1 engine (Si):

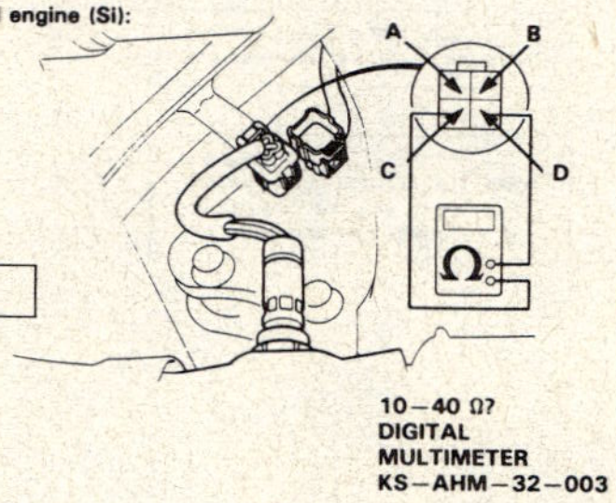

Check for continuity to body ground on each terminal on the O_2 sensor.

Does continuity exist?
- YES → Replace O_2 sensor.
- NO ↓

Check for continuity between terminal D and terminals A and B individually.

Does continuity exist?
- YES → Replace O_2 sensor.
- NO ↓

Turn the ignition switch ON.

At O_2 sensor harness, measure voltage between YEL/BLK (+) terminal and ORN/WHT (–) terminal.

Is there battery voltage?
- YES → Disconnect the "A" connector from the ECU.
  - At O_2 sensor harness, measure voltage between YEL/BLK (+) terminal and ORN/WHT (–) terminal.
  - Is there battery voltage?
    - YES → Repair short in ORN/WHT wire between ECU (A6) and O_2 sensor.
    - NO → Reconnect the O_2 sensor connector.
- NO ↓

Measure voltage between YEL/BLK (+) terminal and body ground.

Is there battery voltage?
- NO → Repair open in YEL/BLK wire between O_2 sensor and main relay.
- YES ↓

Turn the ignition switch OFF.

F22A1 engine (S):

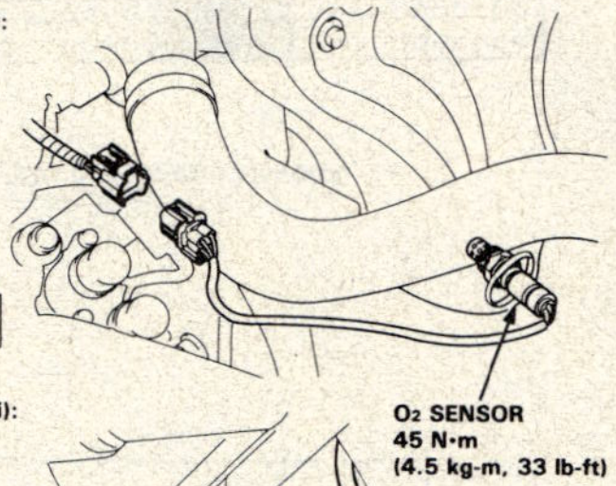

H23A1 engine (Si):

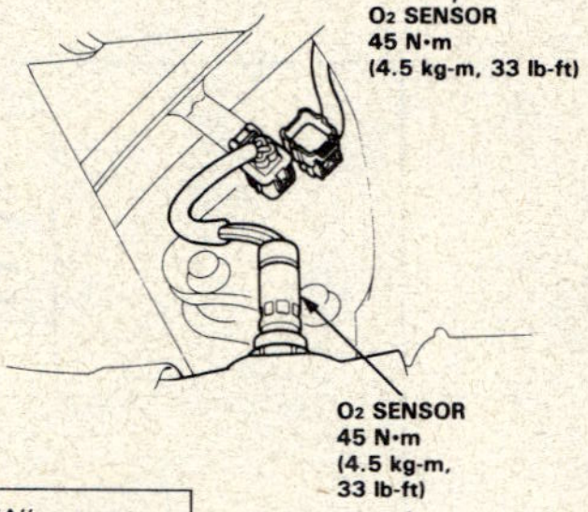

2.2L (F22A1) AND 2.3L (H23A1) ENGINES TROUBLESHOOTING PGM-FI FAULT CODE 41—1992 PRELUDE

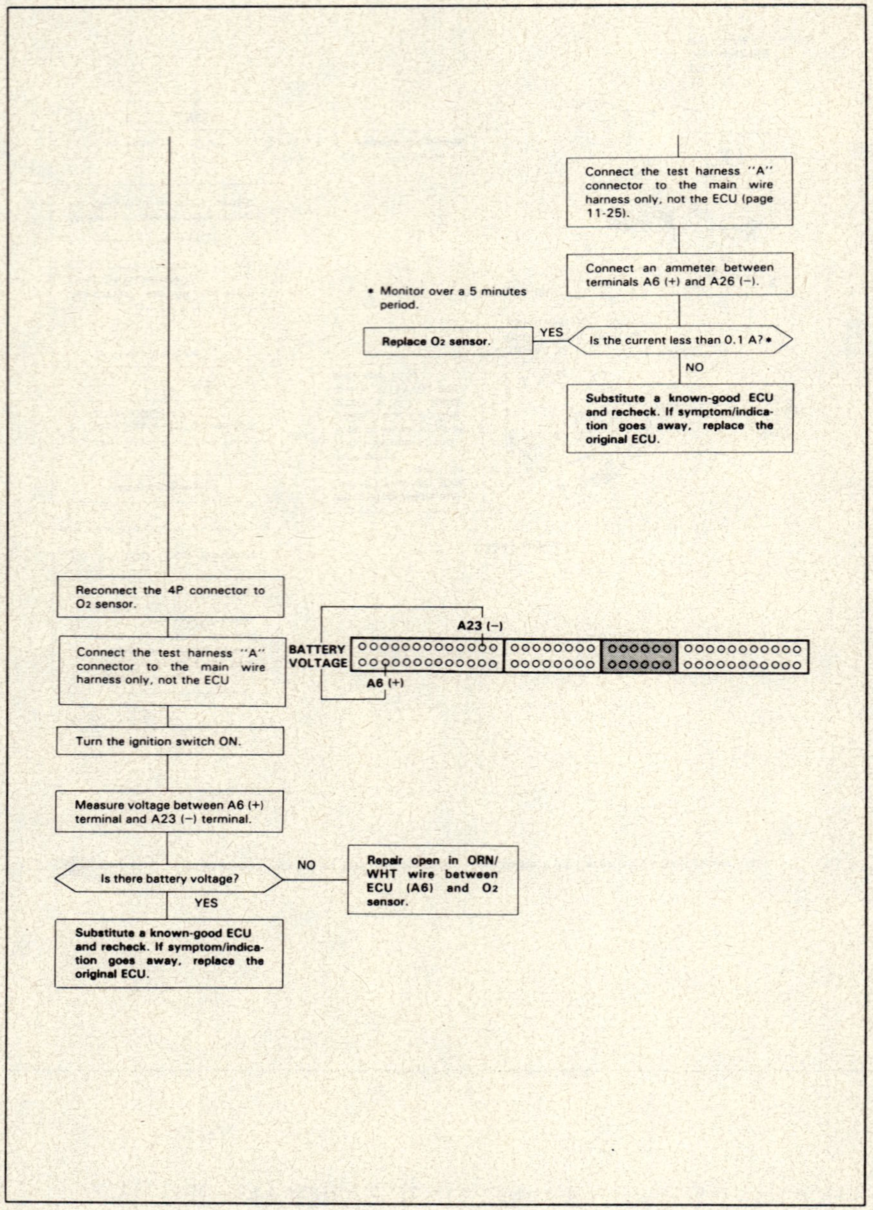

2.2L (F22A1) AND 2.3L (H23A1) ENGINES TROUBLESHOOTING PGM-FI FAULT CODE 43—1992 PRELUDE

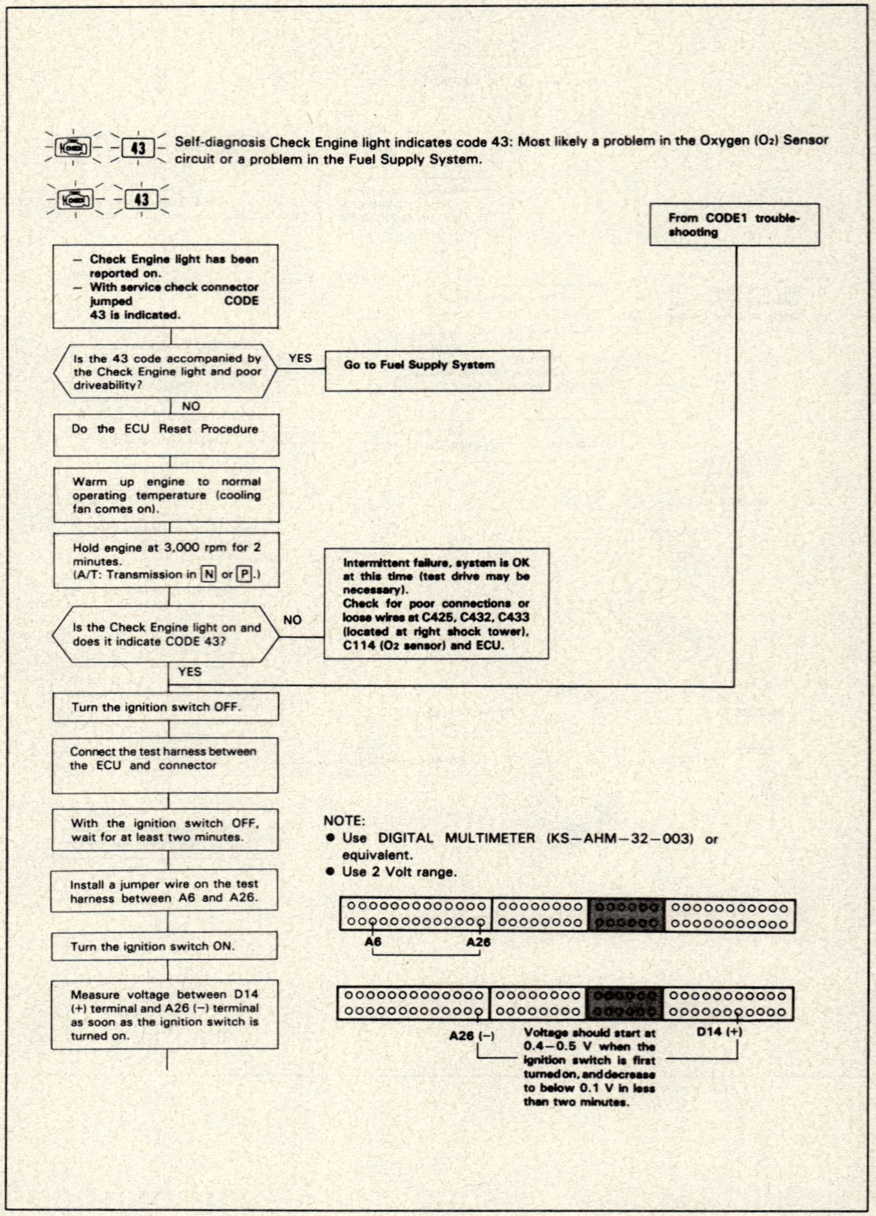

2.2L (F22A1) AND 2.3L (H23A1) ENGINES TROUBLESHOOTING PGM-FI FAULT CODE 43—1992 PRELUDE

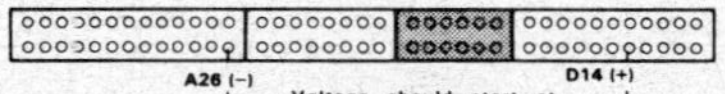

- Is there 0.1 V or less when the ignition switch is first turned on?
  - YES →
    - Disconnect the 4P connector from the O_2 sensor.
    - Measure voltage between D14 (+) terminal and A26 (−) terminal.
    - Is there more than 0.1 V?
      - YES → **Replace O_2 sensor.**
      - NO →
        - Disconnect the "D" connector from the main wire harness.
        - Measure voltage between D14 (+) terminal and A26 (−) terminal.
        - Is there more than 0.1 V?
          - YES → **Repair short in WHT wire between ECU (D14) and O_2 sensor.**
          - NO → **Substitute a known-good ECU and recheck. If symptom/indication goes away, replace the original ECU.**
  - NO →
    - Disconnect the 4P connector from the O_2 sensor.
    - Measure voltage at the engine wire harness side of the O_2 sensor connector between GRN/WHT (−) and WHT (+).
    - Is there more than 0.1 V?
      - YES → **Replace O_2 sensor.**
      - NO → **Check for open in WHT wire between ECU (D14) and O_2 sensor. If wire is OK, repair open in GRN/WHT wire between ECU (D22) and O_2 sensor.**

H23A1 engine (Si):

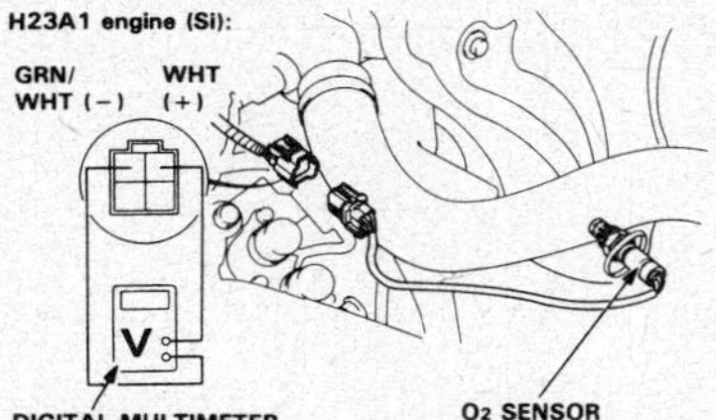

F22A1 engine (S):

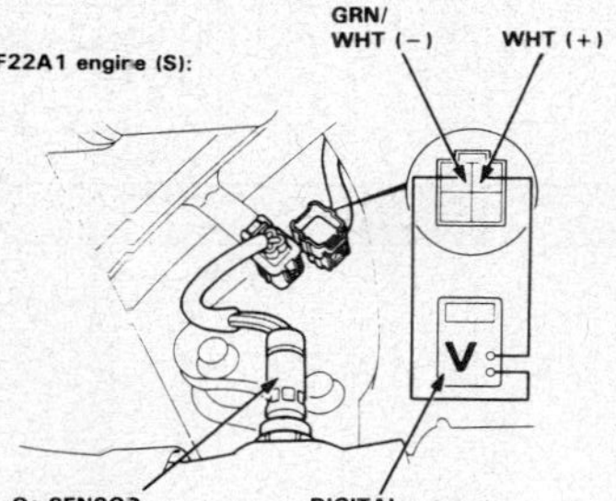

2.2L (F22A1) AND 2.3L (H23A1) ENGINES TROUBLESHOOTING P/S OIL PRESSURE SIGNAL—1992 PRELUDE

This signals the ECU when the power steering load is high.

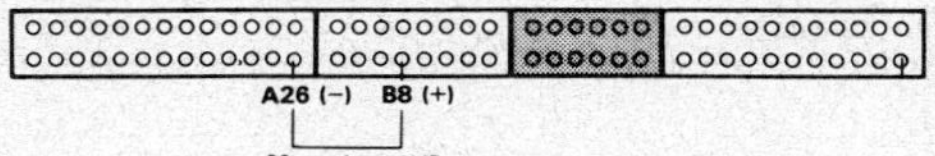

- **Inspection of P/S Oil Pressure Signal**
- Connect the test harness between the ECU and connector
- Turn the ignition switch ON.
- Measure voltage between B8 (+) terminal and A26 (−) terminal.
- Is there more than 1V?
  - YES →
    - Disconnect the 2P connector from the P/S oil pressure switch.
    - Connect RED/GRN terminal to BLK terminal.
    - Is there more than 1V?
      - NO → **Replace P/S oil pressure switch.**
      - YES → **Repair open in RED/GRN wire between ECU (B8) and P/S oil pressure switch or BLK wire between P/S oil pressure switch and G101.**
  - NO →
    - Start the engine.
    - Turn steering wheel slowly.
    - Measure voltage between B8 (+) terminal and A26 (−) terminal while steering wheel is turning.
    - Is there battery voltage?
      - YES → **P/S oil pressure signal is OK.**
      - NO →
        - Disconnect the 2P connector from the P/S oil pressure switch.
        - Is there battery voltage?
          - YES → **Replace P/S oil pressure switch.**
          - NO → **Repair short in RED/GRN wire between ECU (B8) and P/S oil pressure switch. If wire is OK, substitute a known-good ECU and recheck. If prescribed voltage is now available, replace the original ECU.**

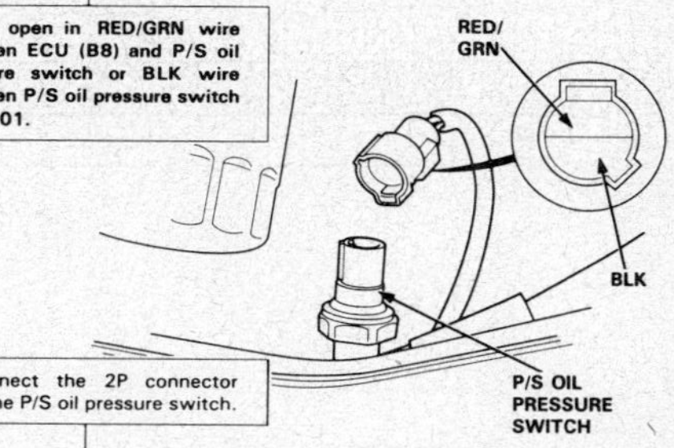

2.2L (F22A1) AND 2.3L (H23A1) ENGINES TROUBLESHOOTING AIR CONDITIONING SIGNAL—1992 PRELUDE

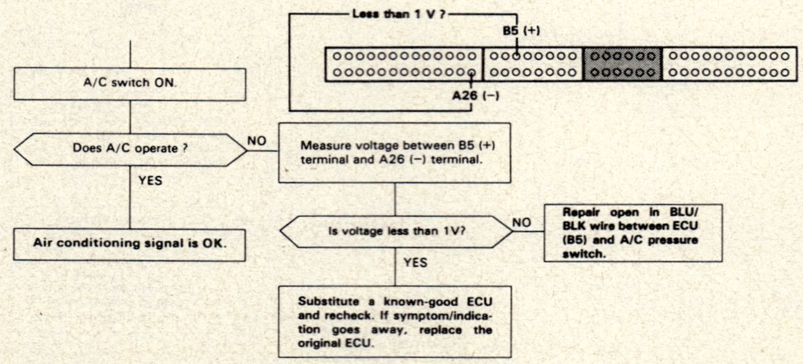

2.2L (F22A1) AND 2.3L (H23A1) ENGINES TROUBLESHOOTING ALTERNATOR FR SIGNAL—1992 PRELUDE

This signals the ECU when the alternator is charging.

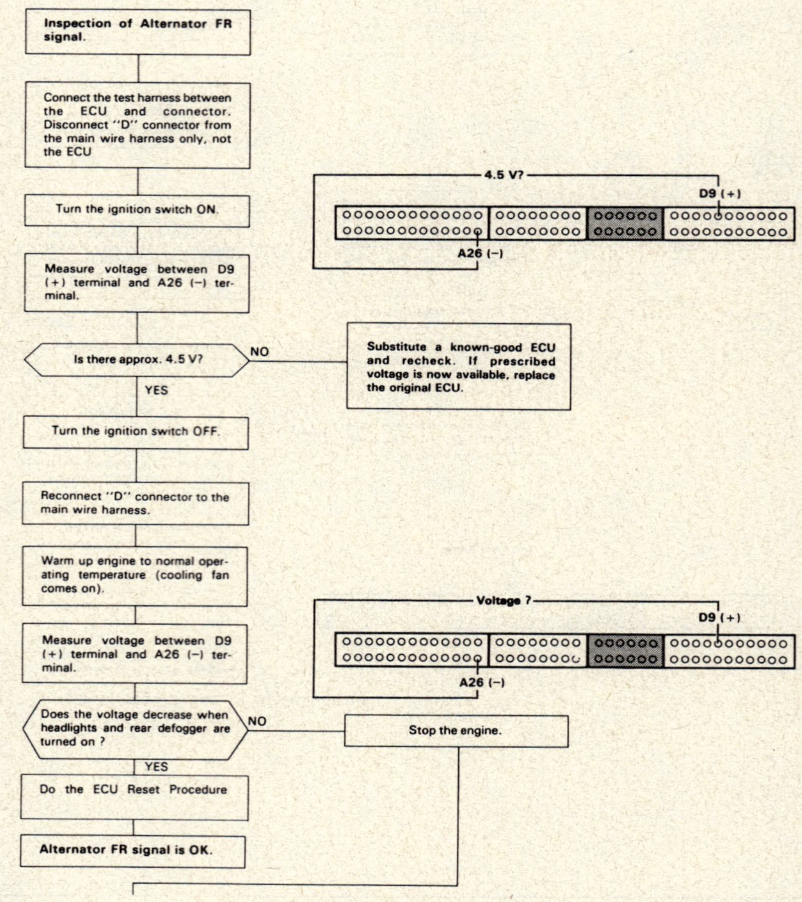

2.2L (F22A1) AND 2.3L (H23A1) ENGINES TROUBLESHOOTING ALTERNATOR FR SIGNAL—1992 PRELUDE

NOTE: The radio may have a coded theft protection circuit. Be sure you get the customer's code number before disconnecting the battery cable.

Disconnect "D" connector from ECU only, not the main wire harness.

Disconnect the negative battery cable from the battery.

Check for continuity between D9 terminal and body ground.

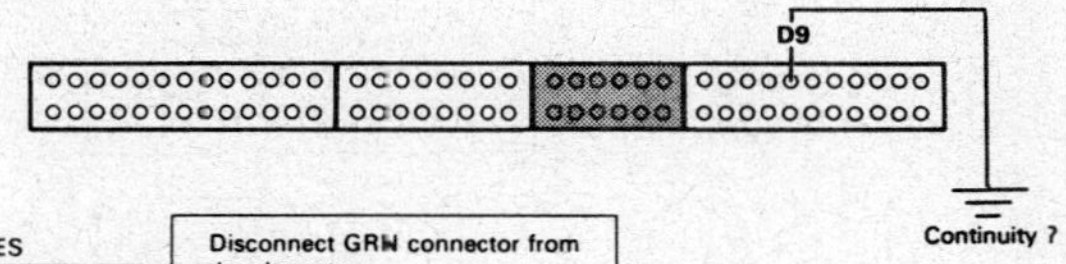

Does continuity exist ?

- YES → Disconnect GRN connector from the alternator. → Check for continuity between D9 terminal and body ground. → Does continuity exist ?
  - NO → See Alternator Inspection
  - YES → Repair short in WHT/RED wire between ECU (D9) and alternator.
- NO → Disconnect GRN connector from the alternator.

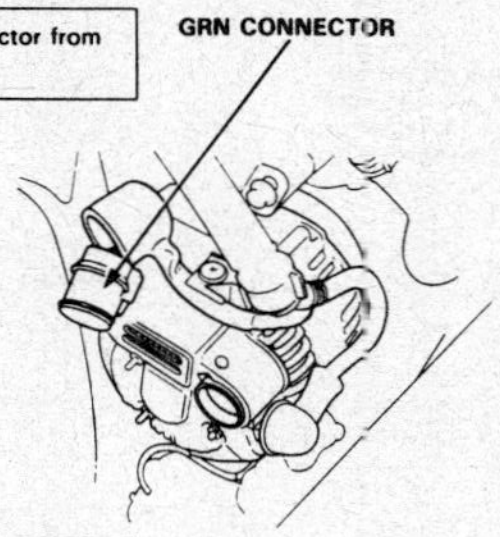

Connect WHT/RED wire to body ground.

Check for continuity between D9 terminal and body ground.

Does continuity exist ?

- YES → See Alternator Inspection
- NO → Repair open in WHT/RED wire between ECU (D9) and alternator.

2.2L (F22A1) AND 2.3L (H23A1) ENGINES TROUBLESHOOTING A/T SHIFT POSITION SIGNAL—1992 PRELUDE

This signals the ECU when the transmission is in Neutral or Park.

Inspection of A/T Shift Position Signal.

Turn the ignition switch ON.

Observe the A/T shift indicator and select each position separately.

Does the indicator light properly?

- NO → **See A/T Shift Position Indicator Inspection**
- YES ↓

Turn the ignition switch OFF.

Connect the test harness between the ECU and connector. Disconnect "B" connector from the main wire harness only, not the ECU

Turn the ignition switch ON.

Measure voltage between B7 (+) terminal and A26 (−) terminal.

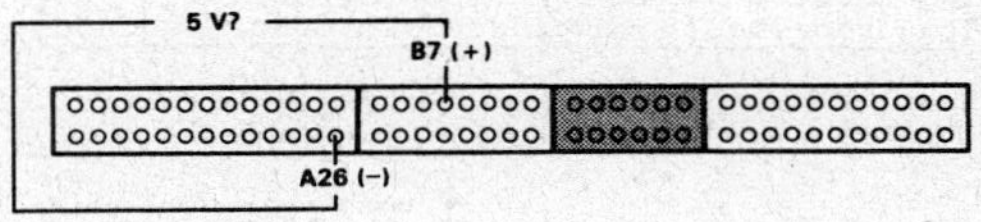

Is there approx. 5 V?

- NO → **Substitute a known-good ECU and recheck. If prescribed voltage is now available, replace the original ECU.**
- YES ↓

Turn the ignition switch OFF.

Reconnect "B" connector to the main wire harness.

Start the engine.

2.2L (F22A1) AND 2.3L (H23A1) ENGINES–TROUBLESHOOTING MAIN RELAY–1992 PRELUDE

Troubleshooting Flowchart

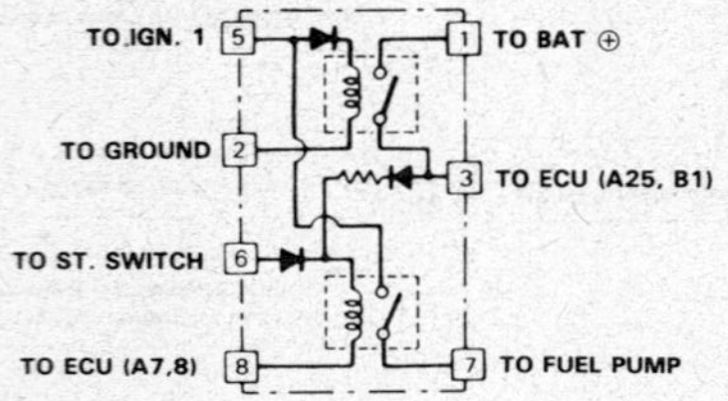

- **– Engine will not start.**
- **– Inspection of main relay.**

Disconnect the main relay connector.

Check for continuity between BLK terminal ② and body ground.

Does continuity exist?
- NO → **Repair open in BLK wire between main relay and G101.**
- YES ↓

Measure the voltage between YEL/BLU terminal ① and body ground.

Is there battery voltage?
- NO → **– Replace No. 4 BACK UP fuse. – Repair open in the YEL/BLU wire between the main relay and the No.4 BACK UP fuse.**
- YES ↓

Turn the ignition switch ON.

Measure the voltage between BLK/YEL terminal ⑤ and body ground.

Is there battery voltage?
- NO → **– Replace No. 23* ECU (15 A) fuse. – Repair open in the BLK/YEL wire between the main relay and the No. 23* ECU (15 A) fuse.**
- YES ↓

*: No. 19 without SRS

Turn the ignition switch to the START position.

NOTE:
- M/T: Clutch pedal must be depressed.
- A/T: Transmission in [N] or [P].

Measure the voltage between BLU/RED terminal ⑥ and body ground.

Is there battery voltage?
- NO → **Repair open in the BLU/RED wire between the main relay and the ignition switch.**
- YES ↓

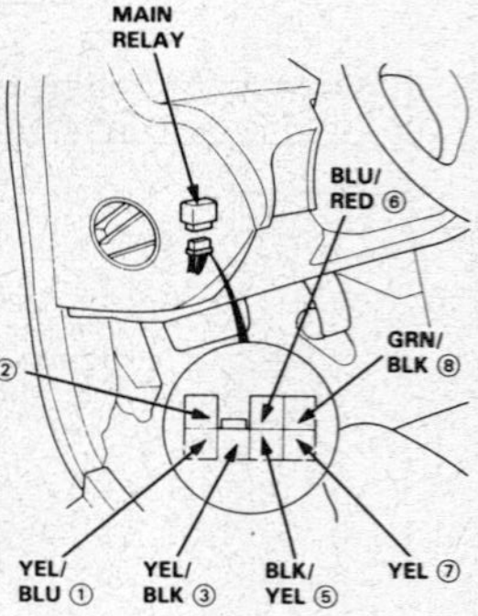

Turn the ignition switch off.

Connect the test harness between the ECU and connector. Disconnect "A" connector from the ECU only, not the main wire harness

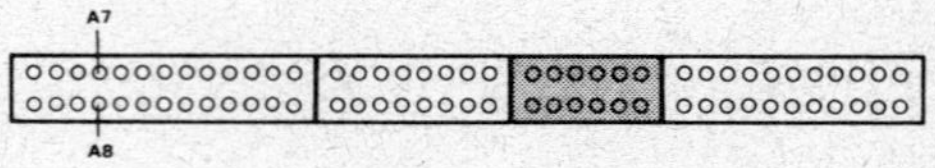

Check for continuity between GRN/BLK terminal ⑧ and the following terminals; A7, A8.

Does continuity exist?
- NO → **Repair open in GRN/BLK wire between ECU (A7, A8) and main relay.**
- YES ↓

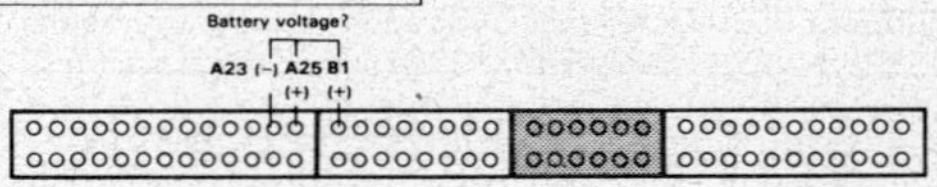

Reconnect "A" connector to the ECU.

Connect the main relay connector.

Turn the ignition switch ON.

Measure the voltage between A23 (–) terminal and the following terminals: A25 (+) B1 (+).

Is there battery voltage?
- NO → **– Repair open in the YEL/BLK ③ wire between the ECU (A25, B1) and main relay. – Replace main relay.**
- YES ↓

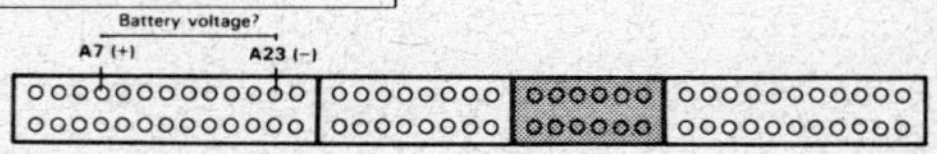

Turn the ignition switch OFF.

Connect a voltmeter between A7 (+) terminal and A23 (–) terminal.

Is there battery voltage for two seconds when the ignition switch is first turned on?
- YES → **Substitute a known-good ECU and recheck. If prescribed voltage is now available, replace the original ECU.**
- NO ↓

Check the main relay

2.3L (H23A1) ENGINE—TROUBLESHOOTING INTAKE CONTROL SYSTEM—1992 PRELUDE

Troubleshooting Flowchart

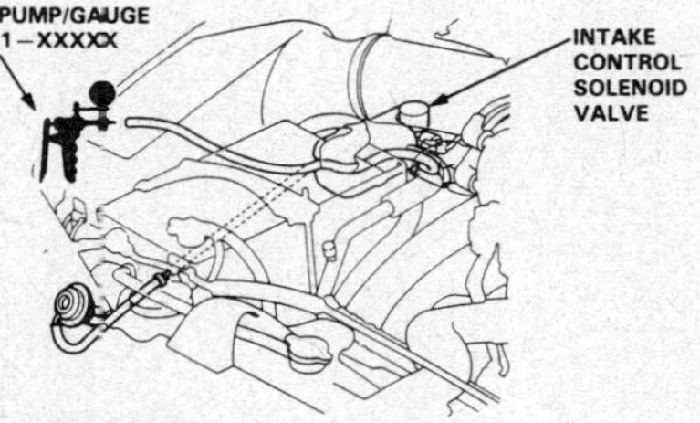

Inspection of Intake Control System

Start engine and allow to idle.

Remove the #8 vacuum hose from the hose joint and connect a vacuum gauge to the solenoid valve.

Is there vacuum?

- YES → Raise engine speed to 4,200 rpm. → Is there vacuum?
  - YES → Disconnect the 2P connector. → Is there vacuum?
    - YES → **Replace intake control solenoid valve.**
    - NO → Turn the ignition switch OFF. → Disconnect "A" connector from ECU. → Check for continuity to ground on WHT wire.
  - NO → (continue)
- NO → Disconnect the lower vacuum hose of the solenoid valve from the air chamber and connect a vacuum gauge to the hose. → Is there manifold vacuum?
  - NO → **– Check the vacuum line for proper connection, cracks, blockage or disconnected hose.**
    – Clean manifold port.
  - YES → Disconnect the 2P connector from the intake control solenoid valve. → Measure voltage between BLK/YEL (+) terminal and WHT (−) terminal. → Is there battery voltage*?
    - YES → **Replace the intake control solenoid valve.**
    - NO → Measure voltage between BLK/YEL (+) terminal and body ground. → Is there battery voltage?
      - NO → **Repair open in BLK/YEL wire between the 2P connector and No. 23* ECU (15 A) fuse.**
      - YES → Turn the ignition switch OFF. → Reconnect the 2P solenoid connector. → Connect the test harness between the ECU and connector → Turn the ignition switch ON. → Connect a jumper wire between ECU (A19) and ECU (A26). → Does solenoid valve click when the jumper is connected?
        - NO → **Repair open in WHT wire between ECU (A19) and the 2P connector.**
        - YES → **Substitute a known-good ECU and recheck. If symptom/indication goes away, replace the original ECU.**

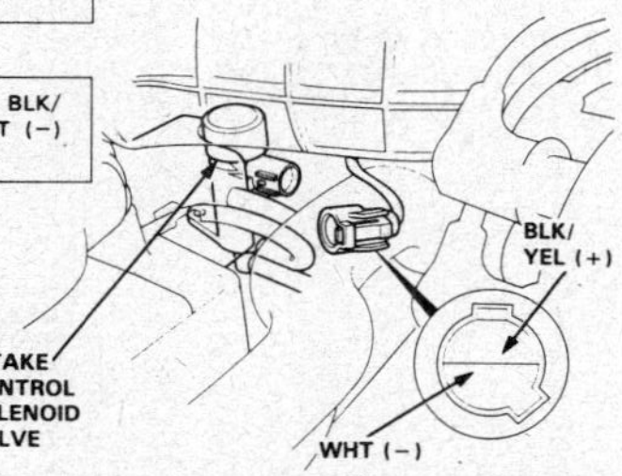

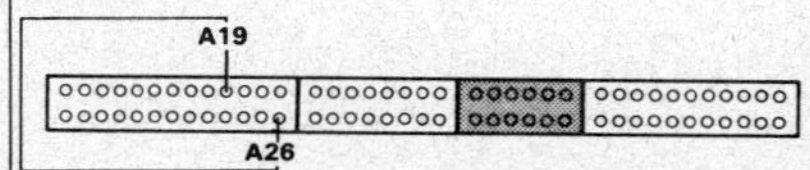

*: No. 19 without SRS

2.3L (H23A1) ENGINE—TROUBLESHOOTING INTAKE CONTROL SYSTEM—1992 PRELUDE

Is there continuity to ground?
- YES: Repair short to ground in WHT wire between ECU (A19) and the 2P connector.
- NO: Substitute a known-good ECU and recheck. If symptom goes away, replace the original ECU.

Connect a vacuum pump to the #8 vacuum hose.

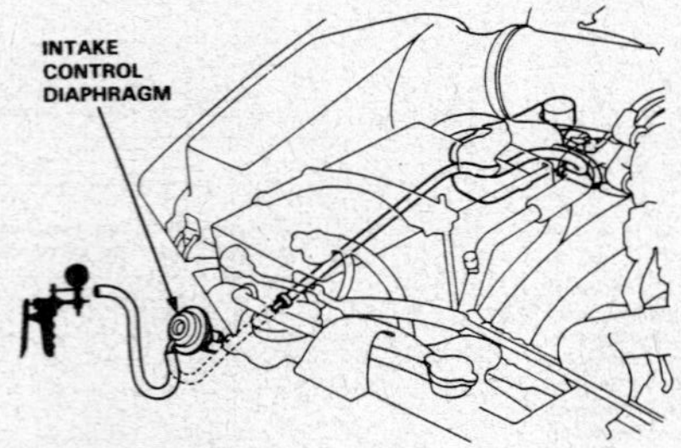

Apply vacuum.

Does it hold vacuum?
- NO: Check the vacuum line for proper connection or disconnected hose. If OK, replace the intake control diaphragm.
- YES: Intake control system is OK.

2.3L (H23A1) ENGINE—TROUBLESHOOTING BYPASS CONTROL SYSTEM—1992 PRELUDE

Troubleshooting Flowchart

Inspection of Bypass Control System

Start engine and allow to idle.

Remove #13 vacuum hose from the bypass control diaphragm and connect vacuum gauge to the hose.

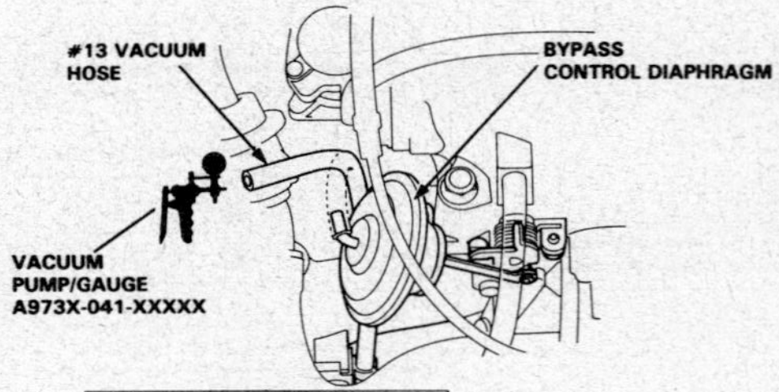

Is there vacuum ?
- YES
- NO: Disconnect the 2P connector from the bypass control solenoid valve.

Measure voltage between BLK/YEL (+) terminal and PNK (−) terminal.

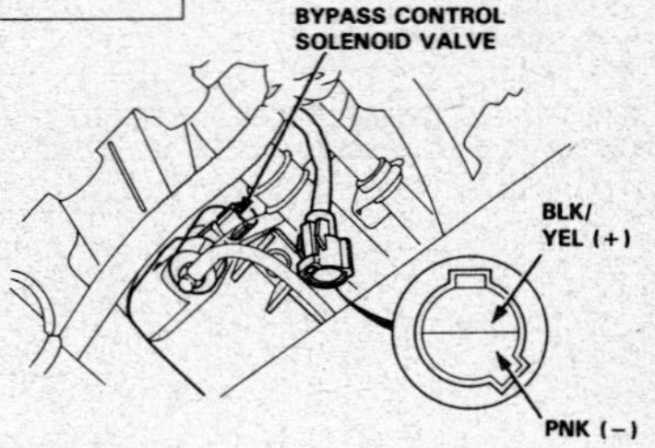

Is there battery voltage ?
- YES: Repair the blockage or vacuum leak between the intake manifold and diaphragm, if vacuum is OK, replace the bypass control solenoid valve.
- NO: Measure voltage between BLK/YEL (+) terminal and body ground.

Is there battery voltage ?
- NO: Repair open in BLK/YEL wire between the 2P connector and No. 23* ECU (15 A) fuse.
- YES: Turn the ignition switch OFF.

*: No. 19 without SRS

2.3L (H23A1) ENGINE—TROUBLESHOOTING BYPASS CONTROL SYSTEM—1992 PRELUDE

- Raise engine speed to 5,000 rpm.
- Check for vacuum at #13 vacuum hose.
- Is there vacuum ?
  - NO: Bypass control system is OK.
  - YES: Disconnect the 2P connector from the bypass control solenoid valve.
    - Is there vacuum ?
      - YES: **Replace bypass control solenoid valve.**
      - NO: Turn the ignition switch OFF.
        - Disconnect "A" connector from ECU.
        - Check for continuity to ground on the PNK wire.
        - Is there continuity to ground ?
          - YES: **Repair short to ground at PNK wire between ECU (A17) and the 2P connector.**
          - NO: **Substitute a known-good ECU and recheck. If symptom goes away, replace the original ECU.**

- Reconnect the 2P connector to the bypass control solenoid valve.
- Connect the test harness to the main wire harness only
- Turn the ignition switch ON and connect a jumper wire between ECU (A17) and ECU (A26).
- Does the solenoid valve click when the jumper is connected?
  - NO: **Repair open in PNK wire between ECU (A17) and the 2P connector. If the wire is OK, replace the bypass control solenoid valve.**
  - YES: **Substitute a known-good ECU and recheck. If prescribed voltage is now available replace the original ECU.**

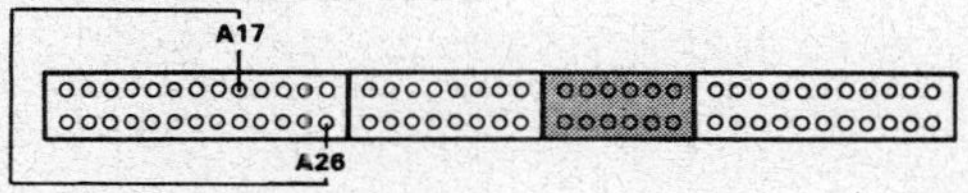

2.2L (F22A1) AND 2.3L (H23A1) ENGINES TROUBLESHOOTING EVAPORATIVE EMISSION CONTROL SYSTEM— 1992 PRELUDE

Troubleshooting Flowchart

- **Inspection of Evaporative Emission Controls**
- Disconnect #3 vacuum hose from the purge control diaphragm valve (on the charcoal canister) and connect a vacuum gauge to the hose.
- Start the engine and allow it to idle.
  NOTE: Engine coolant temperature must be below 75°C (167°F).
- Is there vacuum ?
  - NO
  - YES: Disconnect the 2P connector.
    - Measure voltage between BLK/YEL (+) terminal and RED/GRN (−) terminal.
    - Is there battery voltage ?
      - YES: **Inspect vacuum hose routing. If OK, replace purge control solenoid valve.**
      - NO: Measure voltage between BLK/YEL (+) terminal and body ground.

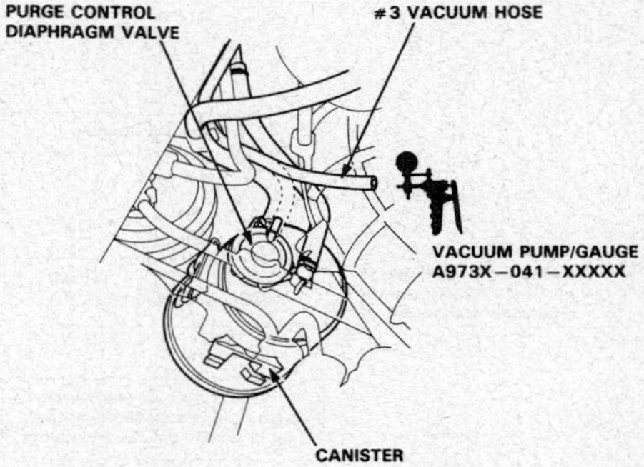

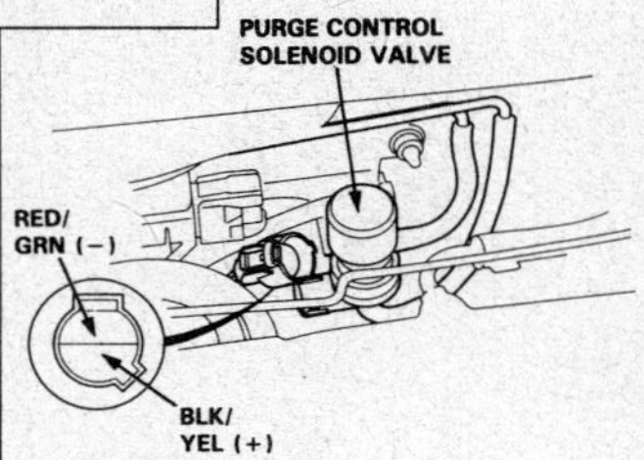

2.2L (F22A1) AND 2.3L (H23A1) ENGINES—TROUBLESHOOTING EVAPORATIVE EMISSION CONTROL SYSTEM— 1992 PRELUDE

Warm up the engine to normal operating temperature (cooling fan comes on).

Hold engine at 2,000 rpm.

Check for vacuum at # vacuum hose after starting th eengine.

Is there manifold vacuum?

- YES: Reconnect the hose.
- NO: Disconnect the 2P connector.
  - Is there manifold vacuum?
    - NO: Inspect vacuum hose routing. If OK, replace purge control solenoid valve.
    - YES: Inspect for a short in RED/GRN wire between ECU (A20) and the connector. If wire is OK, substitute a known-good ECU and recheck. If symptom goes away, replace the original ECU.

Remove fuel filler cap.

Is there battery voltage ?

- NO: Repair open in BLK/YEL wire between No. 23* ECU (15A) fuse and the 2P connector.
- YES: Inspect RED/GRN wire for an open between ECU (A20) and the connector. If wire is OK, substitute a known-good ECU and recheck. If symptom goes away, replace the original ECU.

*: No. 19 without SRS

Connect a vacuum gauge to canister purge air hose.

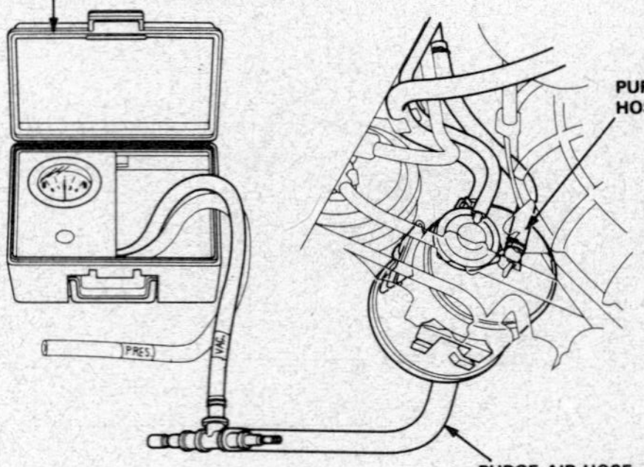

Start the engine and raise speed to 3,500 rpm.

Does vacuum appear on gauge within 1 minute ?

- YES: **See two-way valve test to complete. Evaporative emission controls are OK.**
- NO: Connect a vacuum gauge to the canister purge hose and raise the engine speed to 3,500 rpm.
  - Does vacuum appear on the gauge?
    - NO: **Inspect the purge hose and throttle body port for pinch or blockage.**
    - YES: **Replace the canister.**

Component Replacement

SERVICE PRECAUTIONS

- Disconnect the negative battery terminal.
- Whenever possible, use a flashlight instead of a drop light.
- Keep all open flame and smoking material out of the area.
- Use a shop cloth or similar to catch fuel when opening a fuel system.
- Always relieve fuel system pressure before loosening any line or component.
- Always use eye or full-face protection.
- Always keep a dry chemical (class B-C) fire extinguisher near the work area.

FUEL PRESSURE RELIEF

CAUTION

The fuel system is under constant pressure, even when the engine has not been run for a period of time. To avoid injury and/or property damage, always safely relieve the fuel system pressure before servicing any fuel or return line component.

1. Remove the fuel filler cap.
2. With the ignition **OFF**, disconnect the negative battery cable.
3. Locate the 6mm service bolt. On Accord and Civic, it is located at the fuel inlet line to the fuel filter. On Prelude, it is found on the fuel line connection to the fuel rail.
4. Use a wrench to hold the larger bolt below the service bolt. Use a box-end wrench on the service bolt. Place a rag or cloth over the wrenches and service bolt.
5. Slowly loosen the service bolt 1 full turn. The pressurized fuel will escape into the rag.
6. Remove the service bolt, replace the washer and reinstall the bolt. Tighten the bolt to 12 Nm (9 ft. lbs.). The washer must be replaced any time the service bolt is loosened or removed. Failure to replace the washer may result in high-pressure fuel leaks.
7. Dispose of the fuel-soaked rag in a manner correct for flammable items. Check the surrounding area for any spilled fuel.
8. Reconnect the battery cable. The pressure within the system will stay reduced until the ignition switch is turned **ON** or the fuel pump is operated by other means.

FUEL PUMP

Removal and Installation

ACCORD AND CIVIC

1. Relieve the fuel pressure.
2. Remove the fuel tank after disconnecting all fuel lines and wiring running to the tank.
3. Remove the fuel pump mounting bolts and remove the fuel pump assembly.
4. Install the fuel pump; tighten the retaining nuts to 6 Nm (4 ft. lbs.).
5. Reinstall the fuel tank

PRELUDE

1. Relieve the fuel pressure.
2. Lift the carpet in the cargo area. Remove the left maintenance access cover.
3. Disconnect the fuel lines and the electrical connector.
4. Remove the fuel pump mounting nuts and remove the pump.
5. Install the fuel pump, tightening the retaining nuts to 6 Nm (4 ft. lbs.).
6. Connect the fuel lines and wiring harnesses.
7. Install the maintenance cover and reposition the carpeting.

FUEL PRESSURE REGULATOR

Removal and Installation

1. Place a cloth or rag under the regulator. Be prepared to contain escaping fuel.
2. Relieve the fuel system pressure.
3. Disconnect the vacuum hose from the regulator.
4. Disconnect the fuel line from the regulator; plug the line to prevent spillage or the entry of dirt.
5. Remove the 2 retaining bolts and remove the regulator.
6. Remove the O-ring from the regulator and discard it.

To install:

7. Install a new O-ring on the regulator; coat the ring lightly with a thin coat of clean, fresh oil.
8. Install the regulator, making sure the O-ring is not distorted or damaged. Tighten the retaining bolts to 12 Nm (9 ft. lbs.).
9. Connect the fuel line.
10. Install the vacuum line.

FUEL RAIL AND INJECTORS

Removal and Installation

EXCEPT CIVIC WITH 1.5L TBI ENGINE

1. Relieve the fuel pressure.
2. Label and disconnect the electrical harnesses from the injectors.
3. Place a rag below the fuel pressure regulator. Disconnect the vacuum hose and disconnect the fuel line. Plug the fuel line to prevent spillage and the entry of dirt.
4. On Prelude, disconnect the fuel hose from the fuel rail.
5. On Prelude, remove the EACV from the intake manifold.
6. Remove the nuts holding the fuel pipe (rail) and the wiring harness.
7. Remove the fuel rail from the injectors, leaving the injectors in the manifold.
8. Remove each injector and remove the seal ring from each manifold port.
9. Remove the cushion ring and O-ring from each injector.

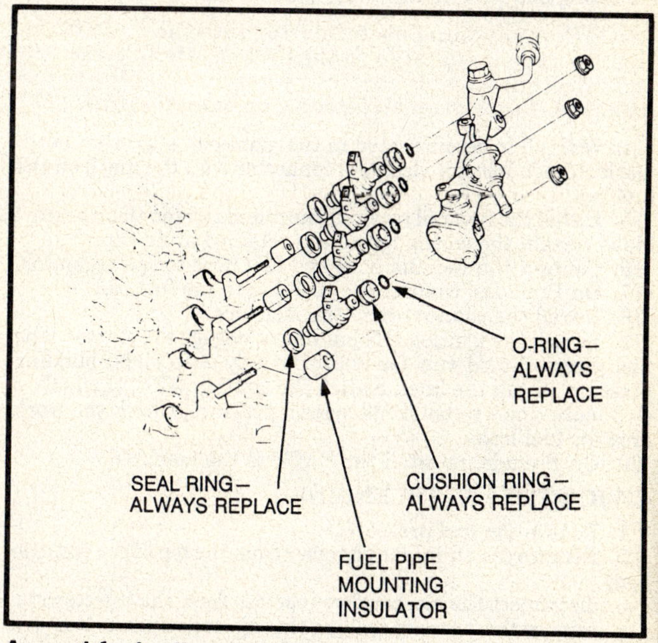

Accord fuel rail and injectors

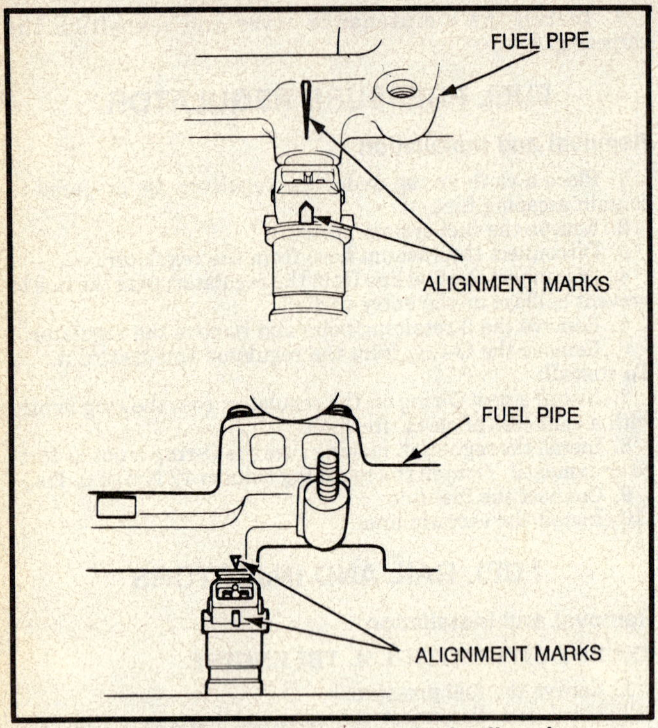

Examples of alignment marks on fuel rail and injectors

To install:

10. Install new cushion rings on each injector. Install new O-rings on each injector and coat the O-rings with a light coat of clean, thin oil.
11. Install the injectors into the fuel pipe. Make certain the O-rings seat properly and do not distort.
12. Coat new seal rings with a light coat of clean, thin oil and install the rings into the manifold.
13. Install the fuel rail and injectors to the intake manifold. Make certain the insulators are present on the Civic and Accord mounting bolts before installing the fuel rail.

NOTE: Assembling each injector into the rail off the engine prevents damage to the O-rings. Handle the rail and injector assembly carefully when reinstalling to the manifold. Don't drop an injector or bang the tips.

14. With all injectors seated in the manifold, align the center mark on each injector electrical connector with the mark on the fuel rail.
15. Install the fuel rail retaining nuts and tighten them evenly. Make certain the wiring harness is retained in its clips.
16. Connect the vacuum hose and fuel hose to the regulator.
17. On Preludes, connect the fuel line to the fuel rail.
18. Install the connectors to the injectors.
19. Switch the ignition **ON** but do not engage the starter. The fuel pump should run for approximately 2 seconds, building pressure within the lines. Switch the ignition **OFF**, then **ON** 2 or 3 more times to build full system pressure. Check the work area for fuel leaks.
20. On Prelude, reinstall the EACV to the manifold.

CIVIC WITH 1.5L TBI ENGINE

1. Relieve the fuel pressure.
2. Remove the air intake chamber from the top of the throttle body.
3. disconnect the electrical connectors from the injector(s).
4. remove the 2 screws holding the injector retainer. Remove the retainer. Lift the injector out of the bore.

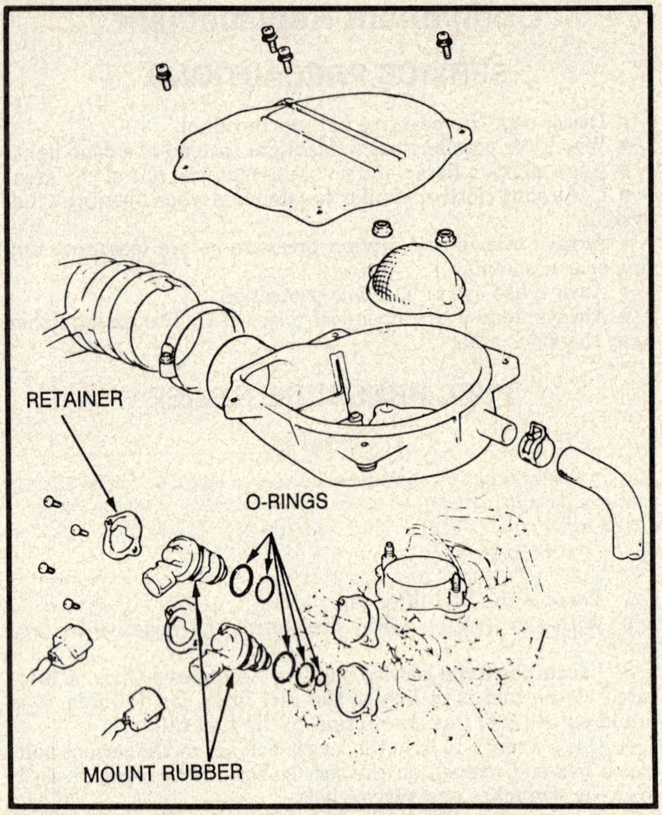

Injector installation—1.5L engine TBI Civic

5. Place a clean cloth over the throttle body and injector port to keep dirt and debris out of the system.

To install:

6. Replace the O-rings on the injector with new ones. Coat each O-ring lightly with clean, fresh oil.
7. Insert the injector into the bore. Make certain the O-rings are not deformed or out of place. Once installed, the injector should turn freely in the bore through approximately 30 degrees of rotation.
8. Install the retaining collar and tighten the screws.
9. Connect the wiring to the injector(s).
10. Switch the ignition **ON** but do not engage the starter. The fuel pump should run for approximately 2 seconds, building pressure within the lines. Switch the ignition **OFF**, then **ON** 2 or 3 more times to build full system pressure. Check the work area for fuel leaks.
11. Install the air intake chamber.

INJECTOR RESISTOR

Removal and Installation

1. The resistor is not used on 1.5L TBI Civic engines. On all other engines, switch the ignition **OFF**.
2. The resistor is located in the engine compartment ,at the extreme left of the Civic firewall or between the firewall and left shock tower for Accord or Prelude. Disconnect the wiring connector to the resistor.
3. Remove or disconnect any wiring harness clamps or retainers holding the resistor harness.
4. Remove the retaining bolts holding the resistor; remove the resistor.
5. Reassemble in reverse order.

TDC/CRANK ANGLE/CYLINDER SENSORS

Removal and Installation

CIVIC AND ACCORD

The sensors are contained within the distributor housing and are not individually serviceable. Should any of the sensors become faulty, the distributor housing must be replaced. All of the components external to the housing, such as the cap, rotor, coil and igniter may be transferred to the new unit.

PRELUDE

1. Remove the distributor.
2. Remove the distributor cap and seal.
3. Loosen the rotor retaining screw and remove the rotor.
4. Remove the toothed crank rotor; don't lose the retaining pin.
5. Remove the retaining screws and remove the crank angle sensor.
6. Remove the TDC rotor, then remove the sensor.

To install:

7. Fit the TDC sensor, rotor and crank angle sensor into place; install the retaining screws.
8. Install the crank rotor and pin
9. Install the rotor and tighten the retaining screw.
10. Install the seal and cap.

MAP SENSOR

Removal and Installation

The MAP sensor is located within the emission control box on the firewall for all vehicles except the Civic with 1.6L engine. For this Civic, the sensor is located on the firewall next to the purge cut-off solenoid valve.

Remove the control box from its mounts and open the box. Disconnect the wire harness connector to the sensor, then disconnect the vacuum hose(s). Remove the sensor from its mount. On 1.6L Civic, disconnect the wiring and vacuum hose, then remove the sensor.

When reassembling, handle the unit carefully. Install it securely in the mounting and attach the vacuum line(s) and wiring connector. Reassemble and install the emissions box.

INTAKE AIR TEMPERATURE (TA) SENSOR

Removal and Installation

The sensor is located on the intake manifold or air plenum. It is removed and installed in straightforward fashion; disconnect the wiring connector and remove the mounting screws. Remove the sensor.

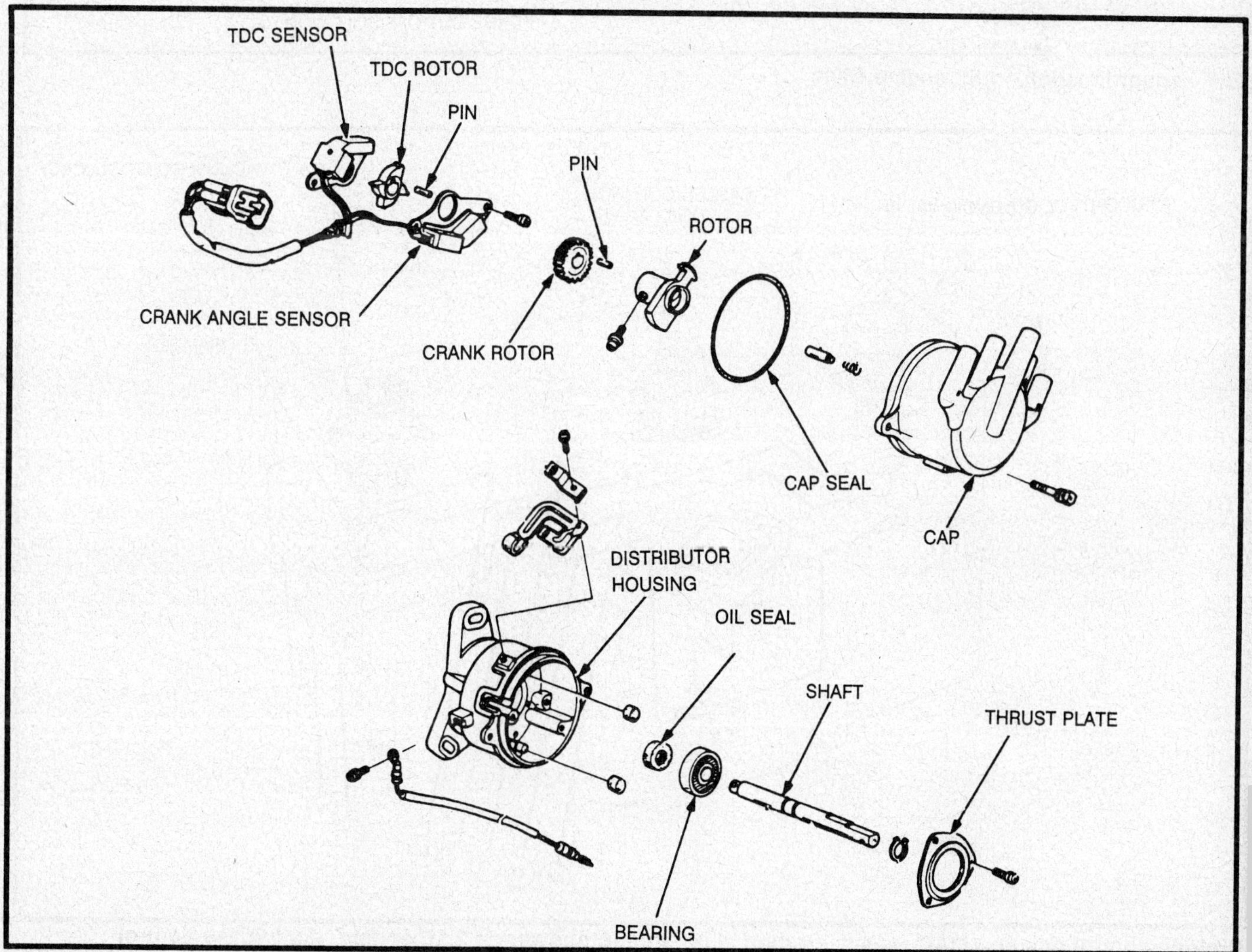

Prelude distributor with TDC and Crank Angle sensors

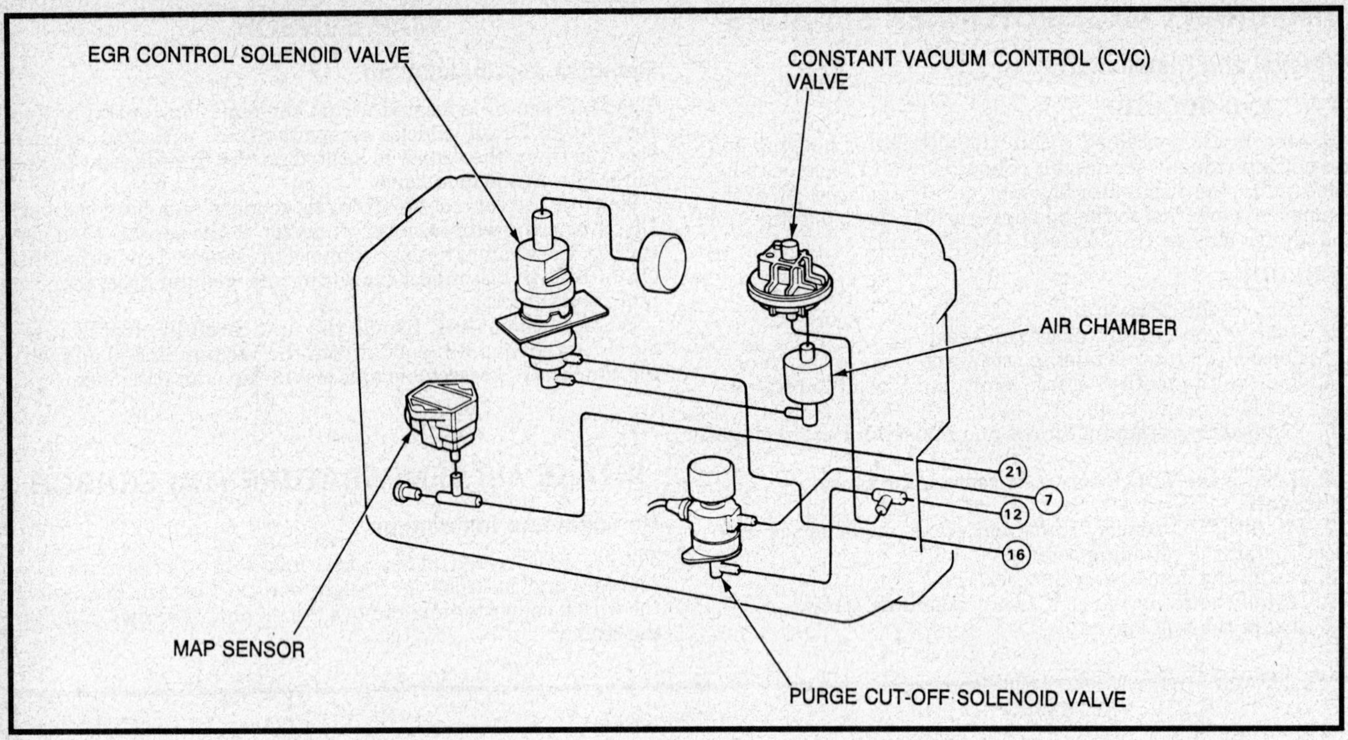

MAP sensor location—1.5L engine Civic

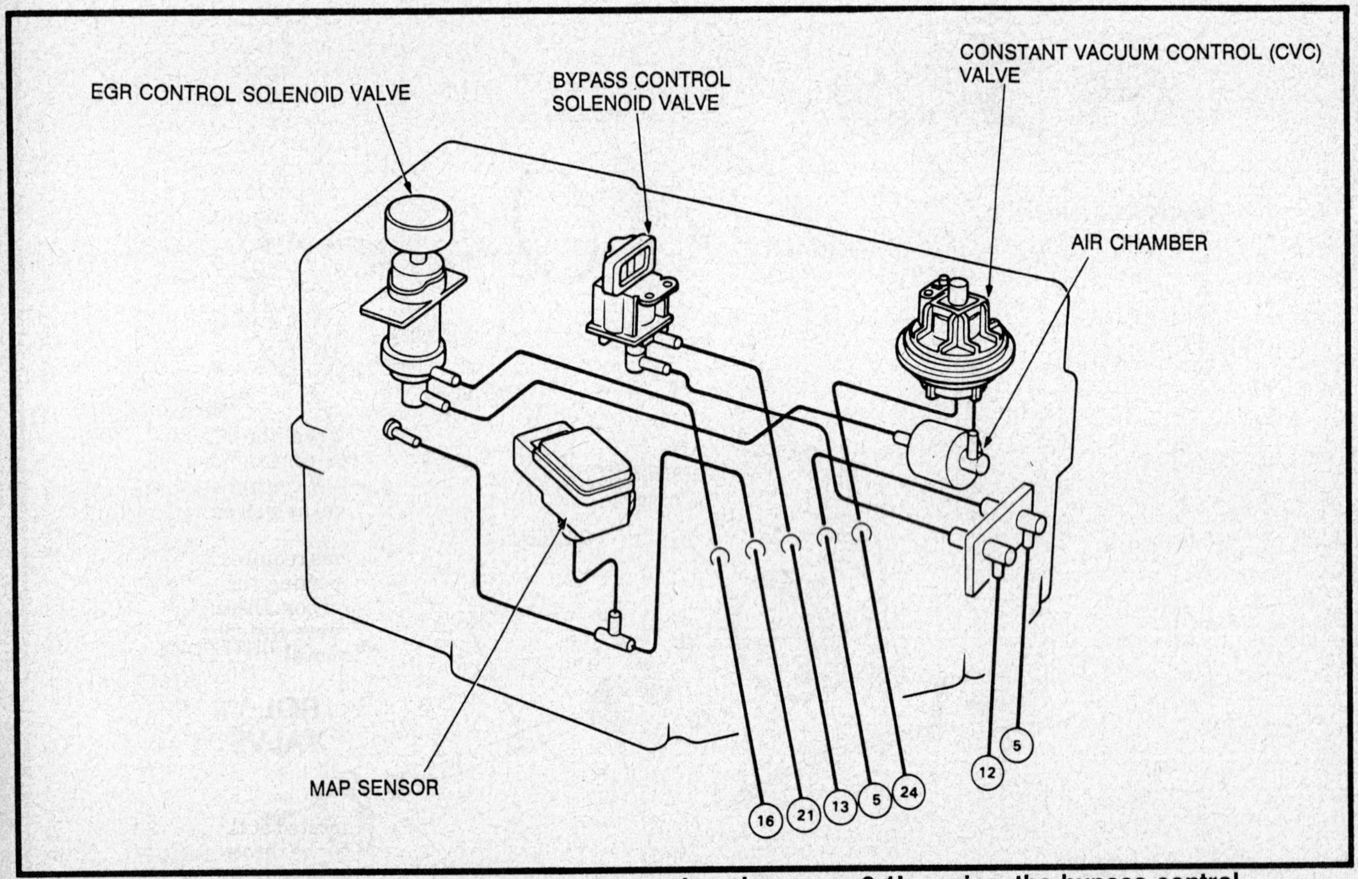

Prelude control box and MAP sensor location. 2.0L engine shown; on 2.1L engine, the bypass control valve is replaced by the purge cut-off solenoid valve in the same location

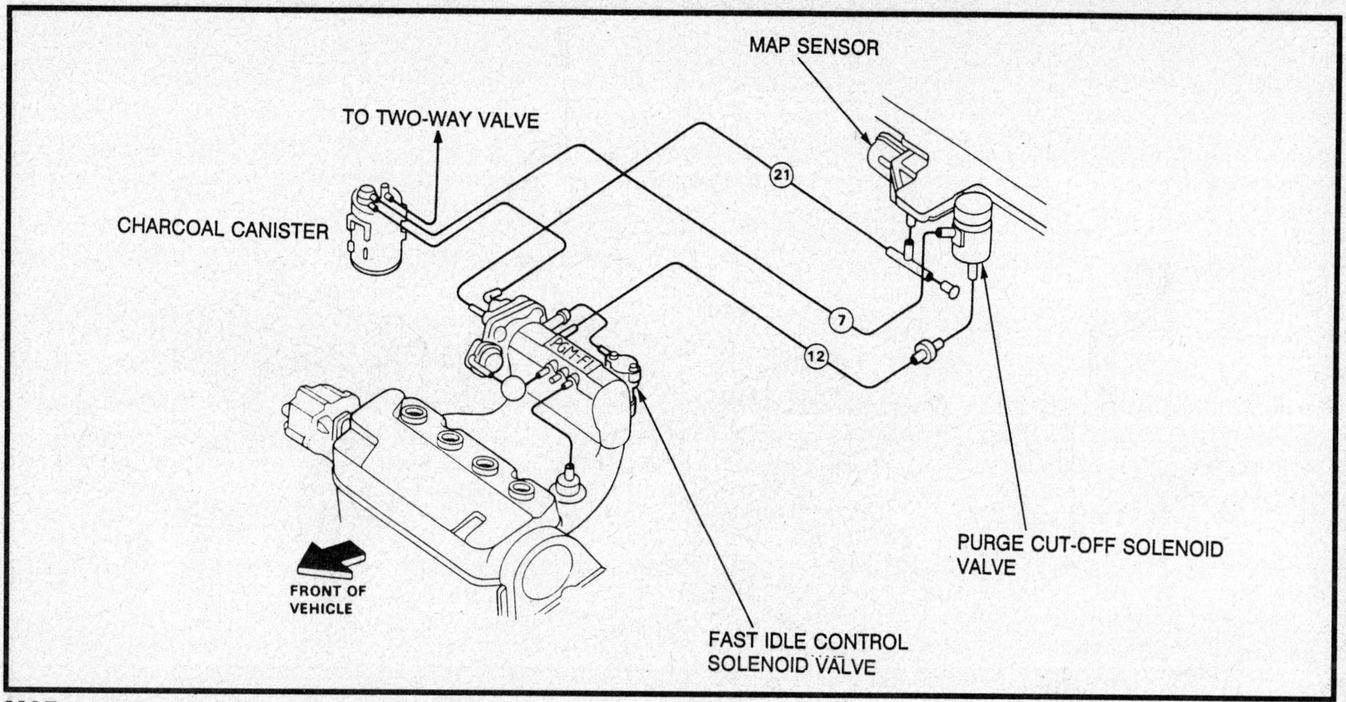

MAP sensor location—1.6L engine Civic

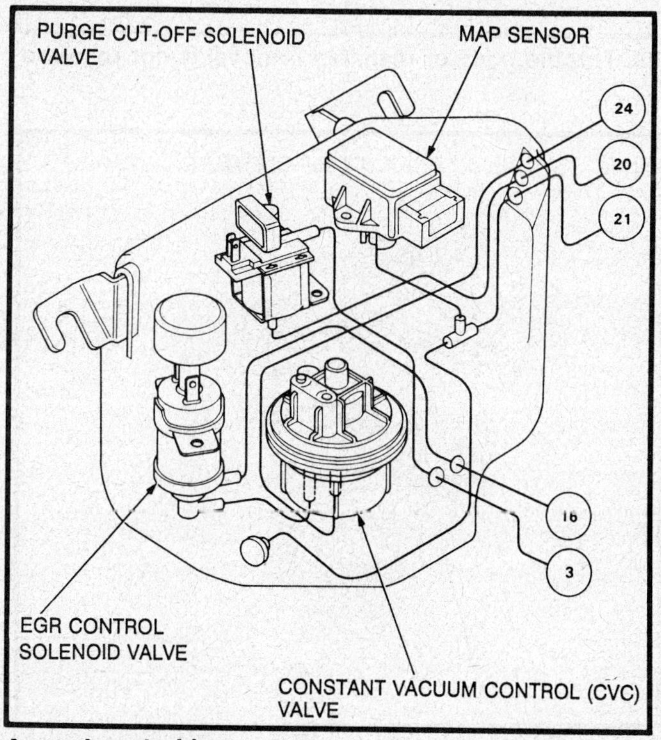

Accord control box

OXYGEN SENSOR

Removal and Installation

1. Allow the exhaust system to cool before beginning the work; the pipe and sensor may exceed 200 degrees when the engine has been operating.
2. Disconnect the wiring connector for the sensor(s). Free the wire from any retaining clips.
3. Using an oxygen sensor socket to unscrew the sensor. This tool prevents damage to the sensor or wiring harness.
4. If the sensor is to be reused, protect the tip from exposure to chemicals, oil, gasoline, etc.

To install:

5. Coat the threads of the sensor with an anti-seize compound. Take great care not to get any compound on the tip of the sensor. Install the sensor to the manifold or exhaust pipe.
7. Tighten the sensor to 33 ft. lbs. (45 Nm).
8. Install the wiring into the clips and connect the wiring connector.

EGR SYSTEM CONTROLS AND PURGE CONTROL

Removal and Installation

The controls associated with control of the EGR and purge control system are located within the control box on the firewall. These components include the EGR control solenoid valve, the CVC valve and air chamber if used, as well as the purge cut-off solenoid valve.

Replacement of any of these items requires removing the control box from its mounts, disconnecting the external electrical connectors and opening the control box. Disconnect the vacuum hose(s) to the component and remove the component. Installation is in the reverse order of removal.

ELECTRONIC AIR CONTROL VALVE (EACV) FAST IDLE VALVE

Removal and Installation

1. These air control valves are located at the side of the intake chamber or plenum downstream of the throttle body. Civic vehicles do not use the fast idle valve.
2. Disconnect the wiring connector at the valve.
3. Label and disconnect the vacuum hoses from the unit.

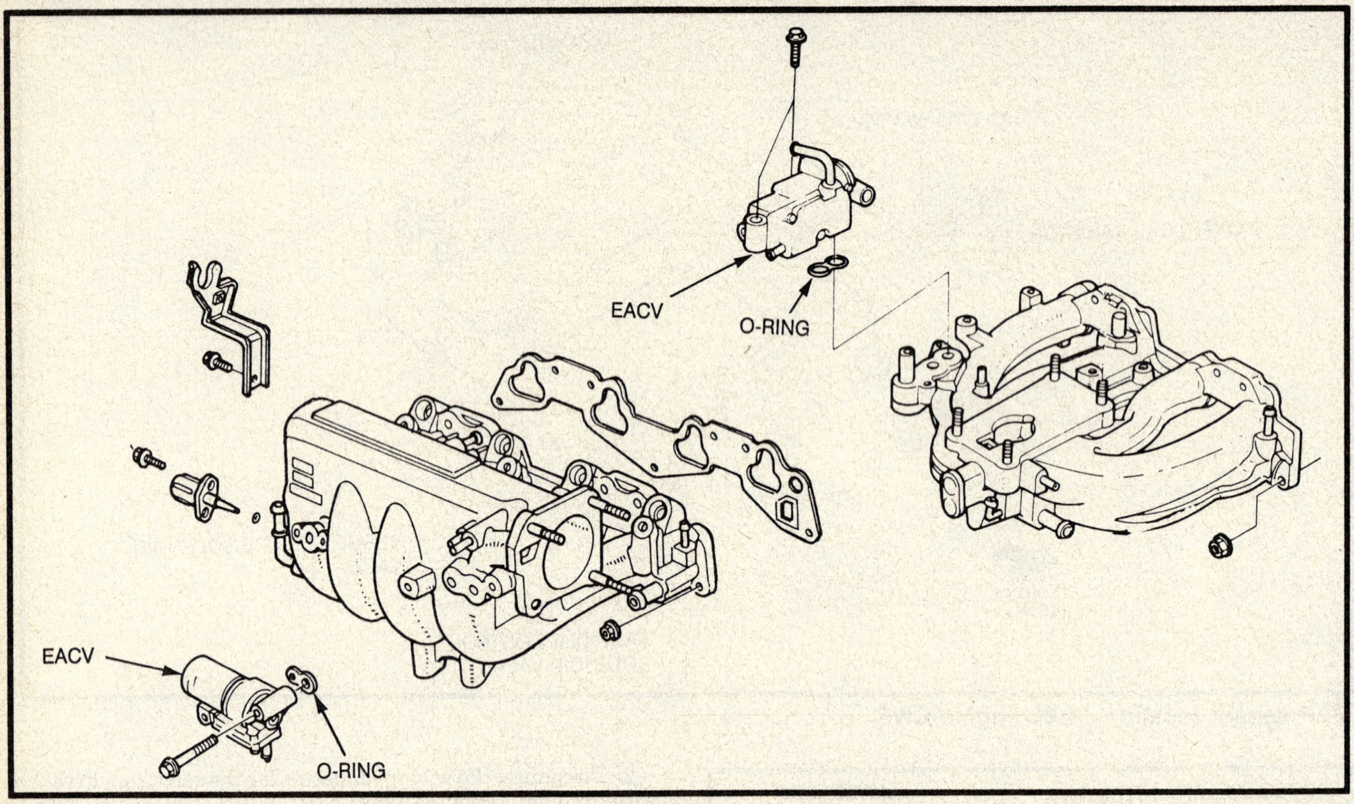

EACV valves for 1.5L engine (above) and 1.6L Civic engines. Throttle body or manifold removal is not required to remove the valves

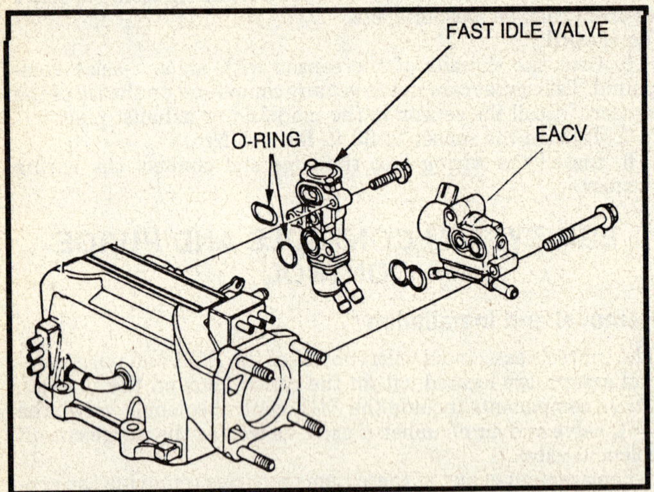

Accord EACV and fast idle valve. Removal of the throttle body or intake plenum is not required

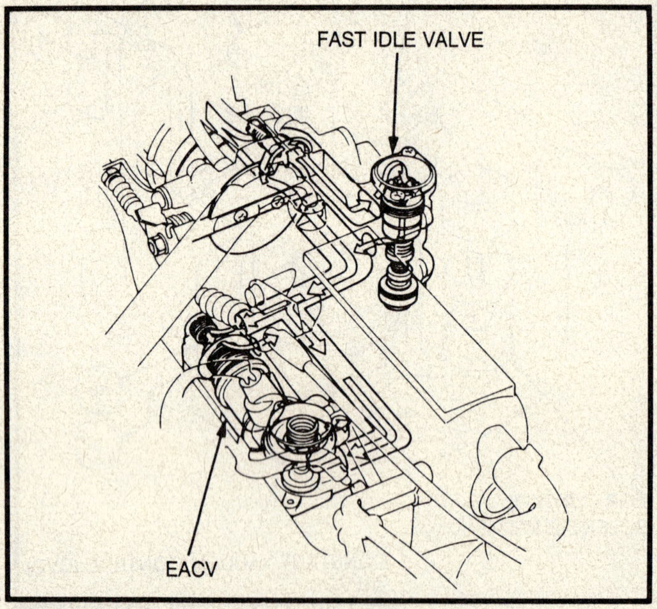

Prelude EACV and fast idle valve

4. Remove the mounting bolts and remove the valve. Recover and discard the O-rings from the vacuum ports.

To install:

5. Replace the O-rings and make certain they are properly seated.
6. Install the valve and the retaining bolts. The pintle or control valve for the EACV must seat correctly in its port. For Civic, tighten the EACV bolts to 9 ft. lbs. (12 Nm). For Accord and Prelude, tighten the EACV bolts to 16 ft. lbs. (22 Nm) and the fast idle valve bolts to 9 ft. lbs (12 Nm).
7. Connect the vacuum hoses to the valve; make certain each hose is connected to the proper port. Connect the electrical connector.

ELECTRIC LOAD DETECTOR

This sensor is used on Accord and Civic only and is located in the underhood fuse and relay box. The ELD unit is integral with the fuse box in which it is mounted; should the sensor fail, the fuse/relay box assembly must be replaced.

IGNITER

Removal and Installation

CIVIC AND ACCORD

1. Remove the distributor from the engine.
2. Remove the cap, rotor and seal.
3. Remove the protective cover under the cap
4. Remove the protective cover from the ignition coil and remove the coil.
5. Remove the retaining screws and remove the igniter from the distributor housing.

To install:

6. Install the igniter and tighten the retaining screws.
7. Install the coil and coil cover.
8. Reassemble the leak cover, rotor, seal and distributor cap.
9. Install the distributor.

PRELUDE

The Prelude igniter is mounted on the right front shock tower,

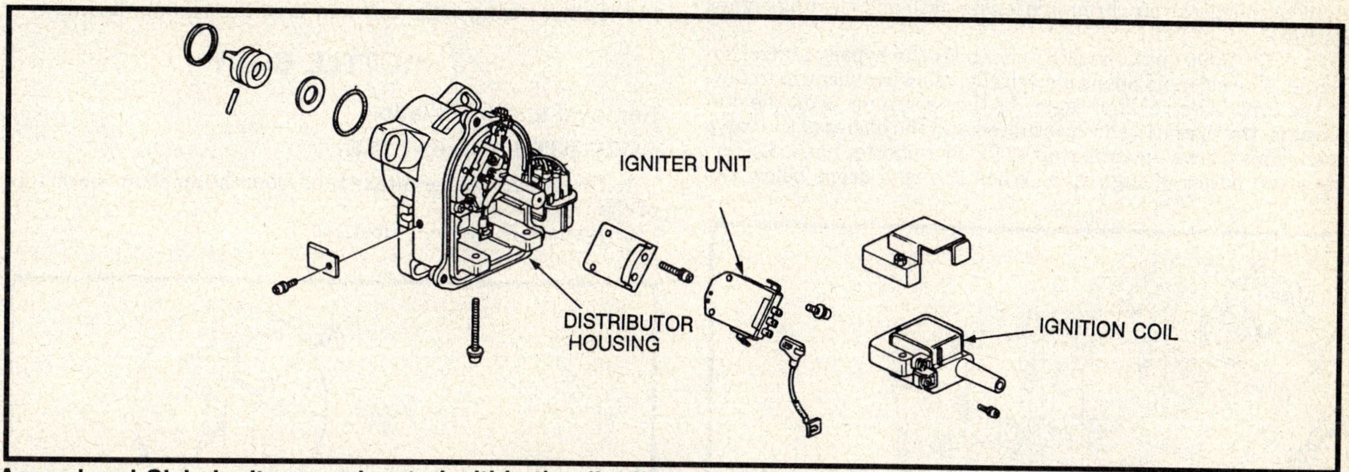

Accord and Civic igniters are located within the distributor

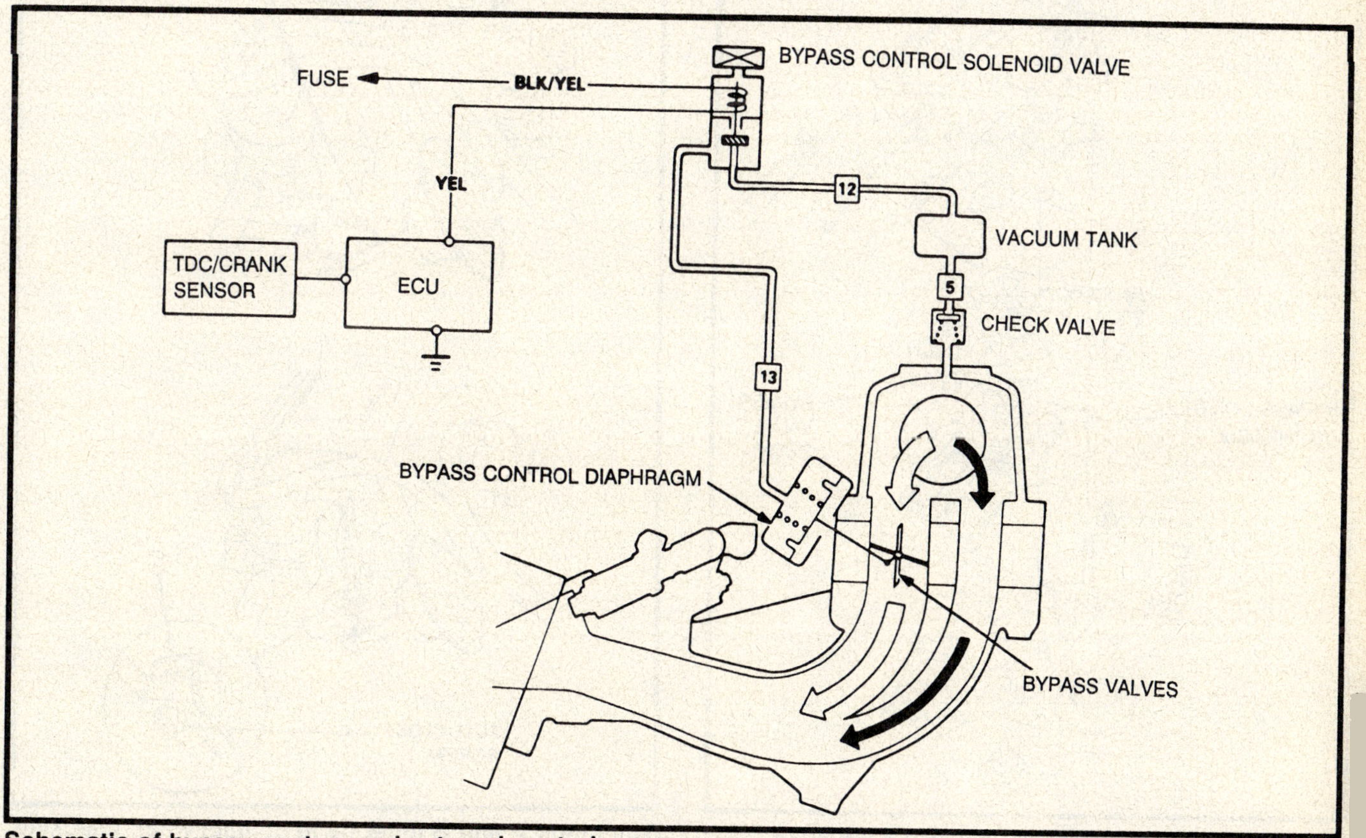

Schematic of bypass system and external controls. 2.0L engine Prelude shown; Accord similar except yellow wire is Pink/Blue

just behind the underhood fuse/relay box. To remove the igniter, turn the ignition switch **OFF**, disconnect the 6-pin connector and remove the retaining bolts. Reinstall in reverse order.

BYPASS VALVE SYSTEM DIAPHRAGM AND CONTROL SOLENOID

This system is found on Accord SE in the U.S. and Accord SE and Ex-R in Canada as well as Preludes with the 2.0L engine. The actual bypass valve assembly is a set of ports and control blades located under the air intake plenum or chamber. The valves remain closed during low and moderate rpm operation, routing the intake air through a longer path and yielding higher torque.

At 4700–5000 rpm, the ECU energizes the bypass control solenoid. The solenoid opens electrically, allowing vacuum to flow to the control valve diaphragm. As the vacuum acts on the diaphragm, the operating lever is moved and the blades of the valve open. This allows the intake air to follow a shorter path, delivering more power at high rpm. When the rpm drops below the trigger level, the signal from the ECU is discontinued, the solenoid shuts off the vacuum and the system reverts to its previous state.

The diaphragm and linkage may be inspected and cleaned with carburetor cleaner if it is not closing fully. The arm should rest on the limit screw with the vacuum hose connected, but the limit screw must not be adjusted; it is set at the factory and should not be changed. Additionally, the diaphragm is part of the bypass valve assembly. If the diaphragm must be replaced, the bypass valve must be removed from the intake manifold..

The control solenoid is located either in the Prelude control box on the firewall or at the left front of the Accord engine. The solenoid may be removed in straightforward fashion by disconnecting the wiring connector and then the vacuum hoses.

THROTTLE BODY

Removal and Installation

CIVIC WITH 1.5L ENGINE

1. Disconnect the air intake tube from the air chamber on top of the throttle body.
2. Remove the air chamber.

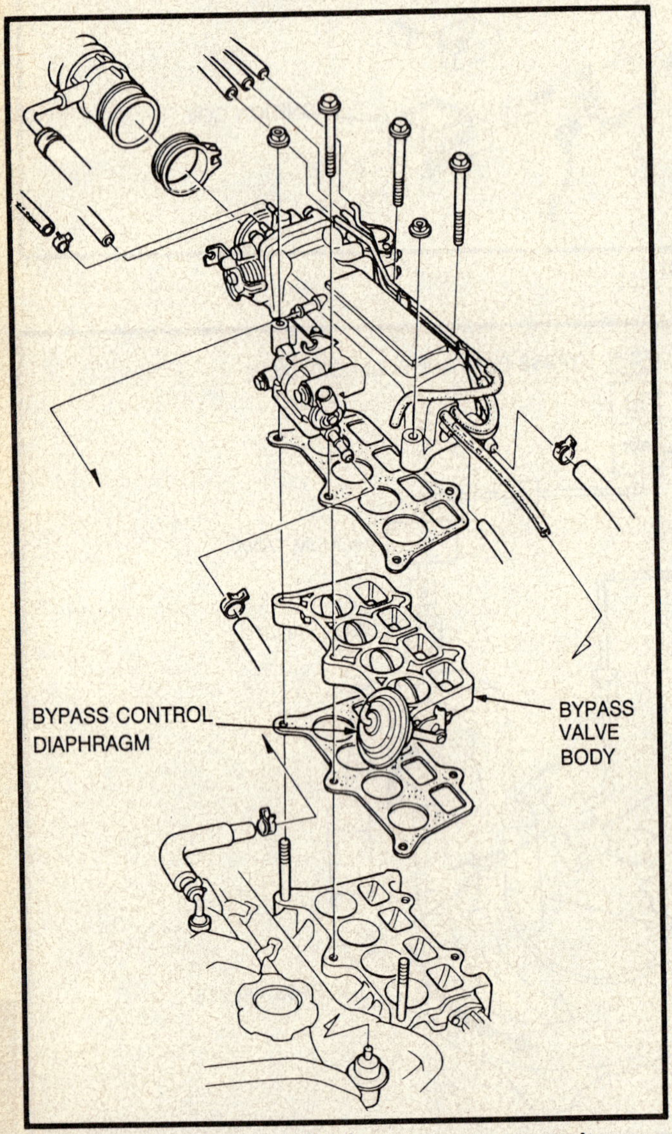

The bypass valve and diaphragm are a separate assembly between the intake manifold and air plenum

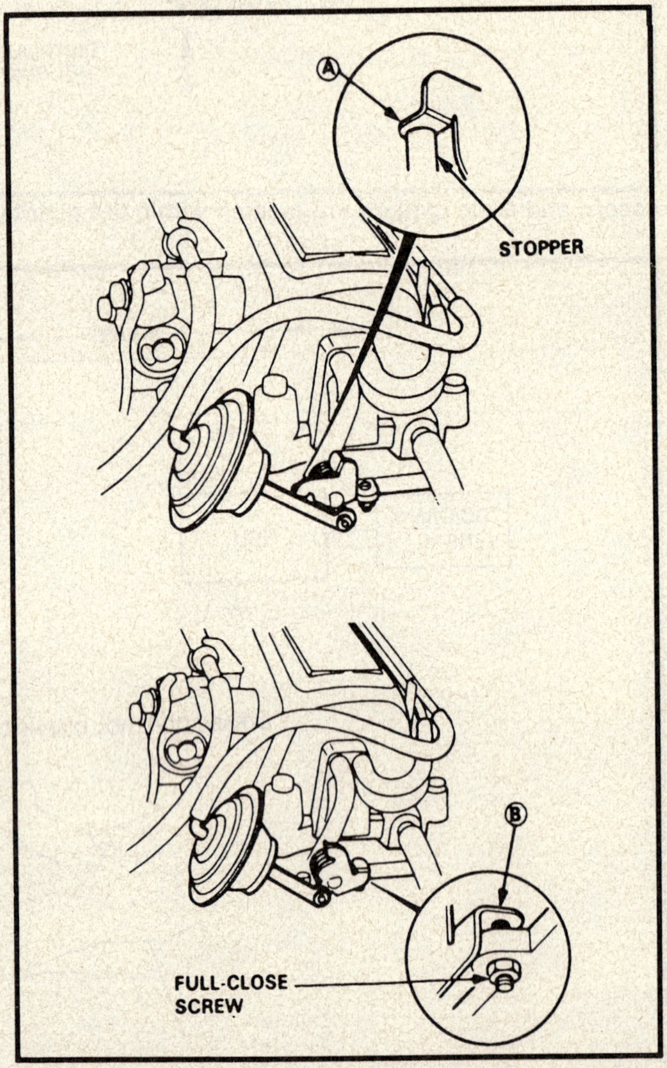

The stops A and B must be in contact when checking the diaphragm linkage, but limit screw B must never be changed

3. Drain the cooling system, at least to a level below the throttle body.
4. Disconnect the coolant hoses running to the throttle body.
5. Label and disconnect the vacuum hoses from the throttle body.
6. Relieve the fuel pressure using the correct procedure.
7. Disconnect the fuel supply line at the throttle body.
8. Label and disconnect the throttle control and/or accelerator cable. Do not kink the cable.
9. Remove the nuts holding the throttle body to the manifold. Loosen each nut 1–2 turns at a time to release the pressure evenly.
10. Remove the throttle body assembly. Recover the O-ring and discard it. Remove the insulator plate and lower gasket. Discard the gasket.

To install:

11. Install a new gasket on the manifold. Install the insulator.
12. Fit the O-ring to the throttle body and install the assembly onto the studs.
13. Install the retaining nuts. Tighten them evenly and alternately to 16 ft. lbs. (22 Nm).
14. Install the throttle control and/or accelerator cable.
15. Using new washers at the banjo bolt, connect the fuel supply line.
16. Connect the vacuum hoses.
17. Connect the coolant hoses.
18. Refill the cooling system.

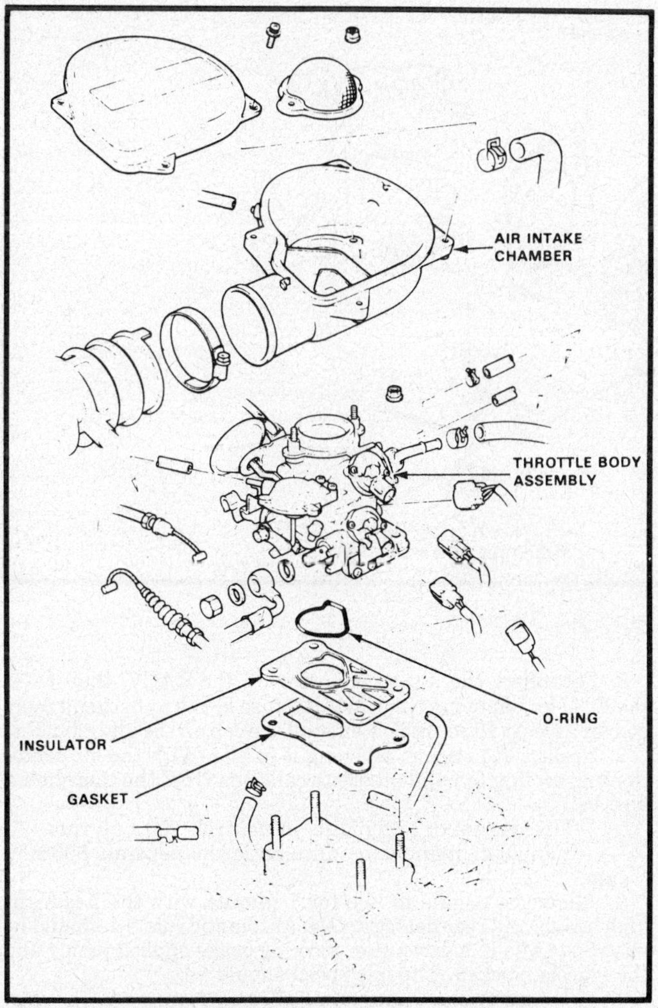

1.5L engine throttle body

19. Install the air chamber and connect the air intake ducts.
20. Adjust the throttle control cable and/or accelerator cable as necessary.

CIVIC WITH 1.6L ENGINE ACCORD AND PRELUDE

1. Disconnect the air intake tube from the throttle body.
2. Drain the cooling system, at least to a level below the throttle body.
3. Disconnect the coolant hoses running to the throttle body.
4. Label and disconnect the vacuum hoses from the throttle body.
5. Disconnect the wiring connector(s) from the throttle body.
6. Label and disconnect the throttle control and/or accelerator cable. Do not kink the cable.
7. Remove the nuts holding the throttle body to the intake chamber or plenum. Loosen each nut 1–2 turns at a time to release the pressure evenly.
8. Remove the throttle body assembly. Remove and discard the gasket.

To install:

9. Install a new gasket on the plenum; install the throttle body onto the studs.
10. Install the retaining nuts. Tighten them evenly and alternately to 16 ft. lbs. (22 Nm).
11. Install the throttle control and/or accelerator cables.
12. Connect the vacuum hoses.
13. Connect the coolant hoses.
14. Refill the cooling system.
15. Connect the air intake ducts.

THROTTLE ANGLE SENSOR

Removal and Installation

The throttle angle sensor is located on the side of the throttle body and connected to the throttle axle or pivot. Once the throttle body is removed, the sensor may be removed by removing the retaining bolts. When reinstalling, make certain the sensor is correctly mated with the throttle body shaft. Install the bolts.

IDLE SPEED ADJUSTMENT

NOTE: Except for the 1.5L Civic, idle is not adjusted at the throttle plate stop screw. That screw is set at the factory and is not adjustable. All idle adjustments are made by turning the idle adjusting screw on the top of the throttle body. All idle tests must be met and the idle must remain stable under all test conditions.

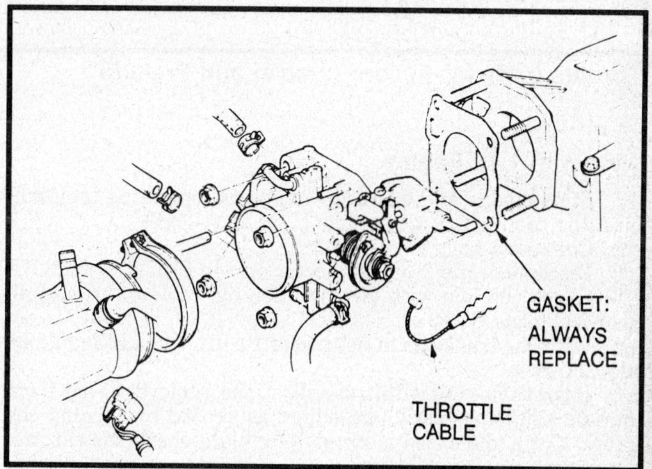

Throttle body—Civic with 1.6L engine

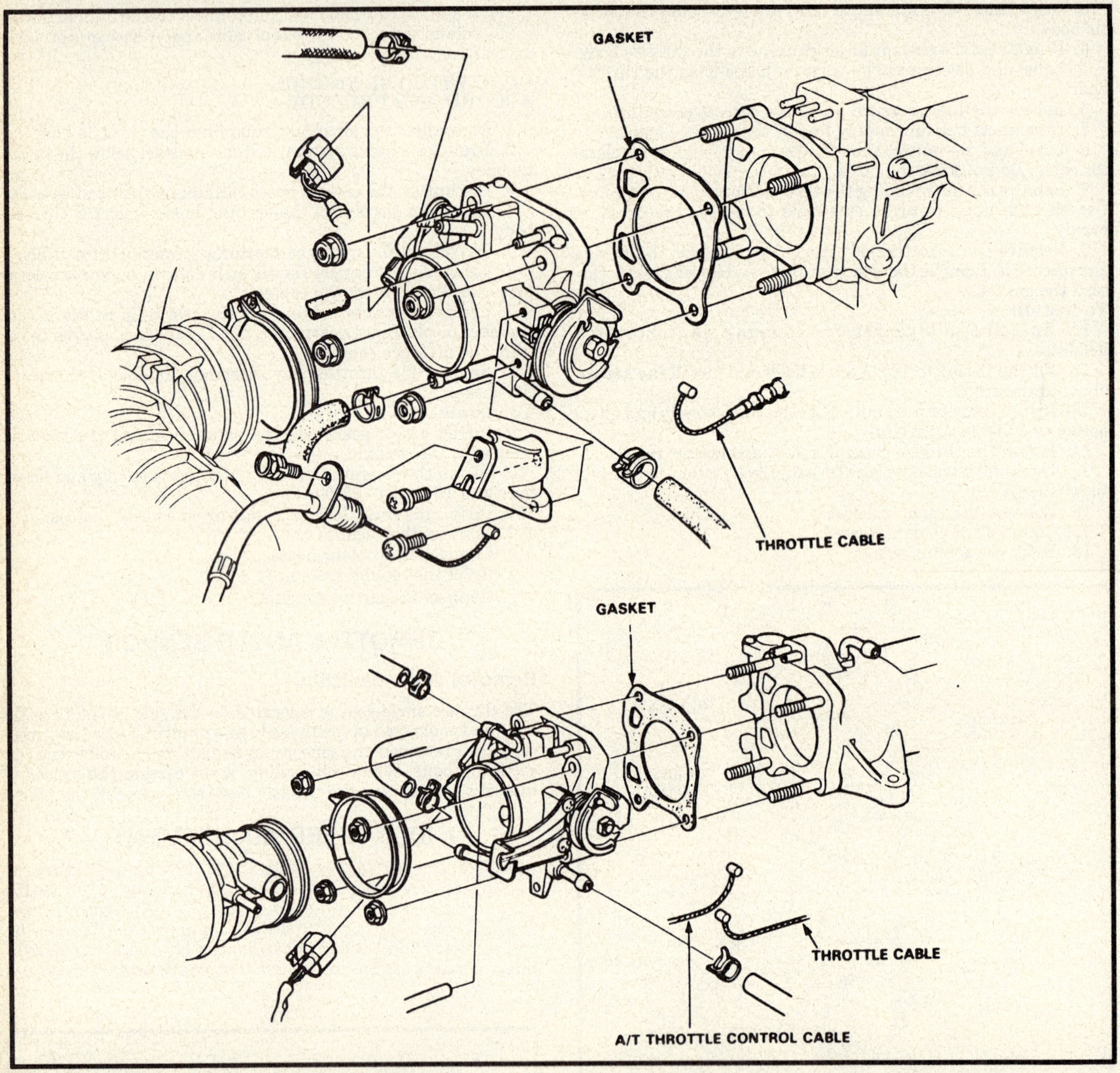

Throttle body for Accord (above) and Prelude

Civic with 1.5L Engine

1. Start engine and warm up to normal operating temperature; the cooling fan must come on at least once.
2. Connect a tachometer.
3. Disconnect the 2-pin connector on the EACV.
4. Check the idle with air conditioning, cooling fan and all electrical loads **OFF**.
5. With the transaxle in neutral or PARK, the idle should be 625±50 rpm.
6. If the rpm is not within specifications, remove the cap from the idle adjusting screw and adjust as needed by turning the screw. If the idle speed is excessively high, check the throttle control system.
7. Turn the ignition **OFF**.
8. Reconnect the 2-pin connector on the EACV, then reset the ECU by removing the HAZARD fuse in the underhood relay box for at least 10 seconds. Reinstall the cap on the idle adjuster.
9. Restart the engine, allowing it to idle. With the air conditioning, cooling fan and all electrical loads **OFF** the idle should now be:
   - US manual or automatic, in neutral: 750±50 rpm.
   - Canadian manual or automatic, in neutral: 800±50 rpm.
10. Allow the engine to idle for 1 minute with the headlight high beams and rear defogger **ON**. Automatic vehicles should be allowed to idle in a drive gear, parking brake applied firmly and the wheels blocked. The idle speed should be:
   - US manual or automatic, 810±50 rpm.
   - Canadian manual or automatic, 800±50 rpm.

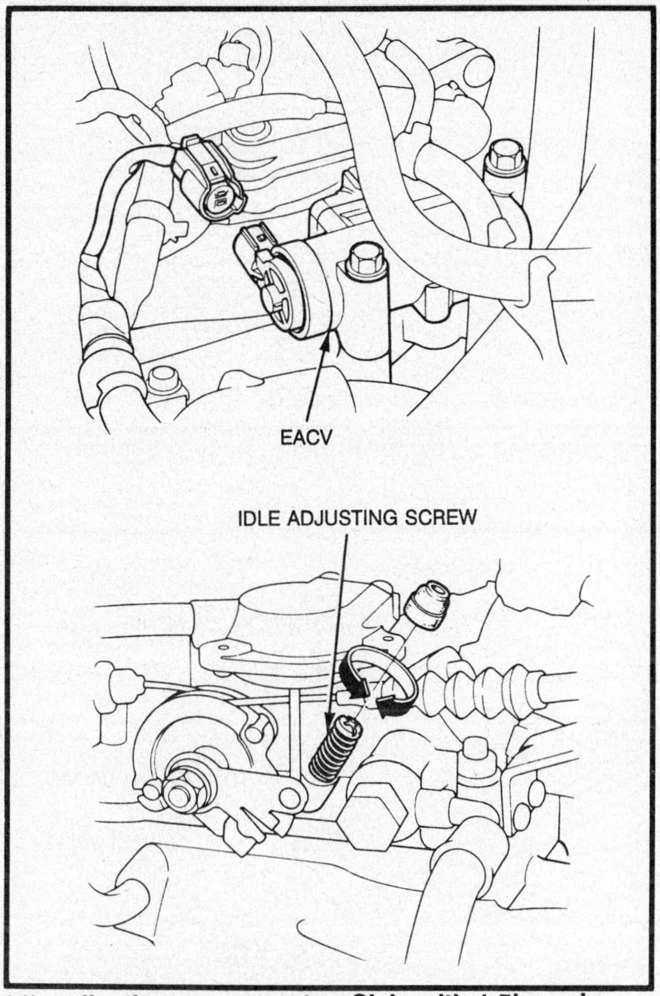

Idle adjusting components—Civic with 1.5L engine

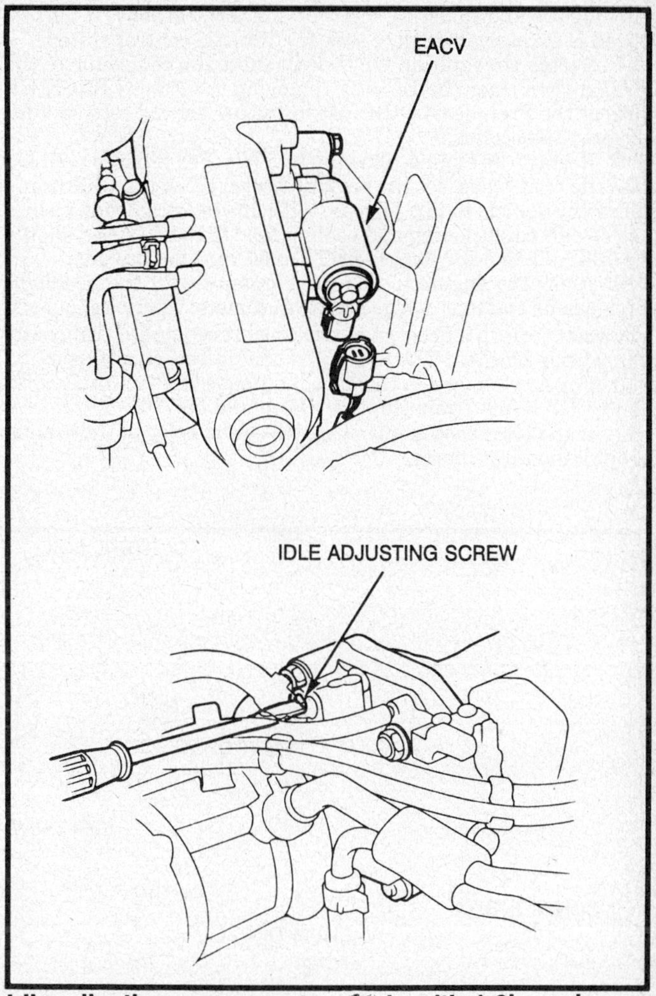

Idle adjusting components—Civic with 1.6L engine

11. Switch the air conditioner **ON** and the blower fan to **HI**. After 1 minute, the idle should be:
    - US manual or automatic, 810 ± 50 rpm.
    - Canadian manual or automatic, 800 ± 50 rpm.
12. If the idle speed is out of specification, refer to the system troubleshooting charts.

Civic with 1.6L Engine

1. Start engine and warm up to normal operating temperature; the cooling fan must come on at least once.
2. Connect a tachometer.
3. Disconnect the 2-pin connector on the EACV.
4. Check the idle with air conditioning, cooling fan and all electrical loads **OFF**.
5. The idle should be 550 ± 50 rpm for all vehicles.
6. If the rpm is not within specifications, adjust as needed by turning the idle adjusting screw on the throttle body. If the idle speed is excessively high, check the throttle control system.
7. Switch the ignition **OFF**. Reconnect the connector to the EACV, then clear the ECU by removing the HAZARD fuse in the underhood relay box for at least 10 seconds.
8. Restart the engine, allowing it to idle in neutral or PARK. With all electrical loads including the air conditioning and cooling fan **OFF**, idle speed should be:
    - All Civic except HF, Si and Canadian: 750 ± 50 rpm.
    - HF, 49 state: 600 ± 50 rpm.
    - HF, Calif. and Hi Altitude: 650 ± 50 rpm.
    - Si and Canadian: 780 ± 50 rpm.
9. Allow the engine to idle for 1 minute with the headlight high beams and rear defogger **ON**. Automatic vehicles should be allowed to idle in a drive gear, parking brake applied firmly and the wheels blocked. The idle speed should be:
    - All Civic except HF, Si and Canadian: 780 ± 50 rpm.
    - HF, 49 state: 650 ± 50 rpm.
    - HF, Calif. and Hi Altitude: 750 ± 50 rpm.
    - Si and Canadian: 780 ± 50 rpm.
10. Switch the air conditioner **ON** and the blower fan to **HI**. After 1 minute, the idle should be:
    - All Civic except HF, Si and Canadian: 810 ± 50 rpm.
    - HF, all: 750 ± 50 rpm.
    - Si and Canadian: 780 ± 50 rpm.
11. If the idle speed is out of specification, refer to the system troubleshooting charts.

Accord and Prelude

1. Apply the parking brake. Start engine and warm up to normal operating temperature; the cooling fan must come on at least once.
2. Connect a tachometer.
3. Disconnect the 2-pin connector on the EACV.
4. Check the idle with the transaxle in neutral or PARK and the air conditioning, cooling fan and all electrical loads **OFF**.
5. The idle should be 550 ± 50 rpm for Accord or 650 ± 50 rpm for Prelude.
6. If the rpm is not within specifications, adjust as needed by

turning the idle adjusting screw on the throttle body. If the idle speed is excessively high, check the throttle control system.

7. Switch the ignition **OFF**. Reconnect the connector to the EACV, then clear the ECU by removing the Accord BACK UP fuse or the Prelude CLOCK fuse in the underhood relay box for at least 10 seconds.

8. Restart the engine, allowing it to idle. For vehicles with automatic transaxles, set the parking brake, block the wheels and allow the vehicle to idle in gear. With all electrical loads including the air conditioning and cooling fan **OFF**, idle speed should be 700 ± 50 rpm for Accord or 770 ± 50 rpm for Prelude.

9. Allow the engine to idle for 1 minute with the headlight high beams and rear defogger **ON**. Automatic vehicles should be allowed to idle in a drive gear, parking brake applied firmly and the wheels blocked. The idle speed should be 770 ± 50 rpm.

10. For Accord, switch the air conditioner **ON** and the blower fan to **HI**. After 1 minute, the idle should be 770 ± 50 rpm.

11. If the idle speed is out of specification, refer to the system troubleshooting charts.

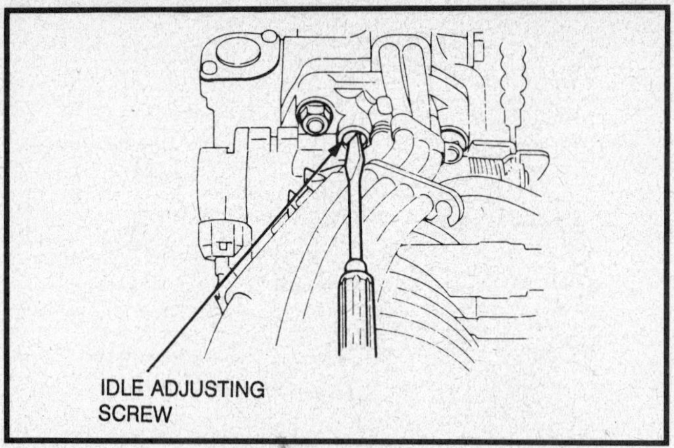

Idle adjusting screw for Prelude; Accord similar

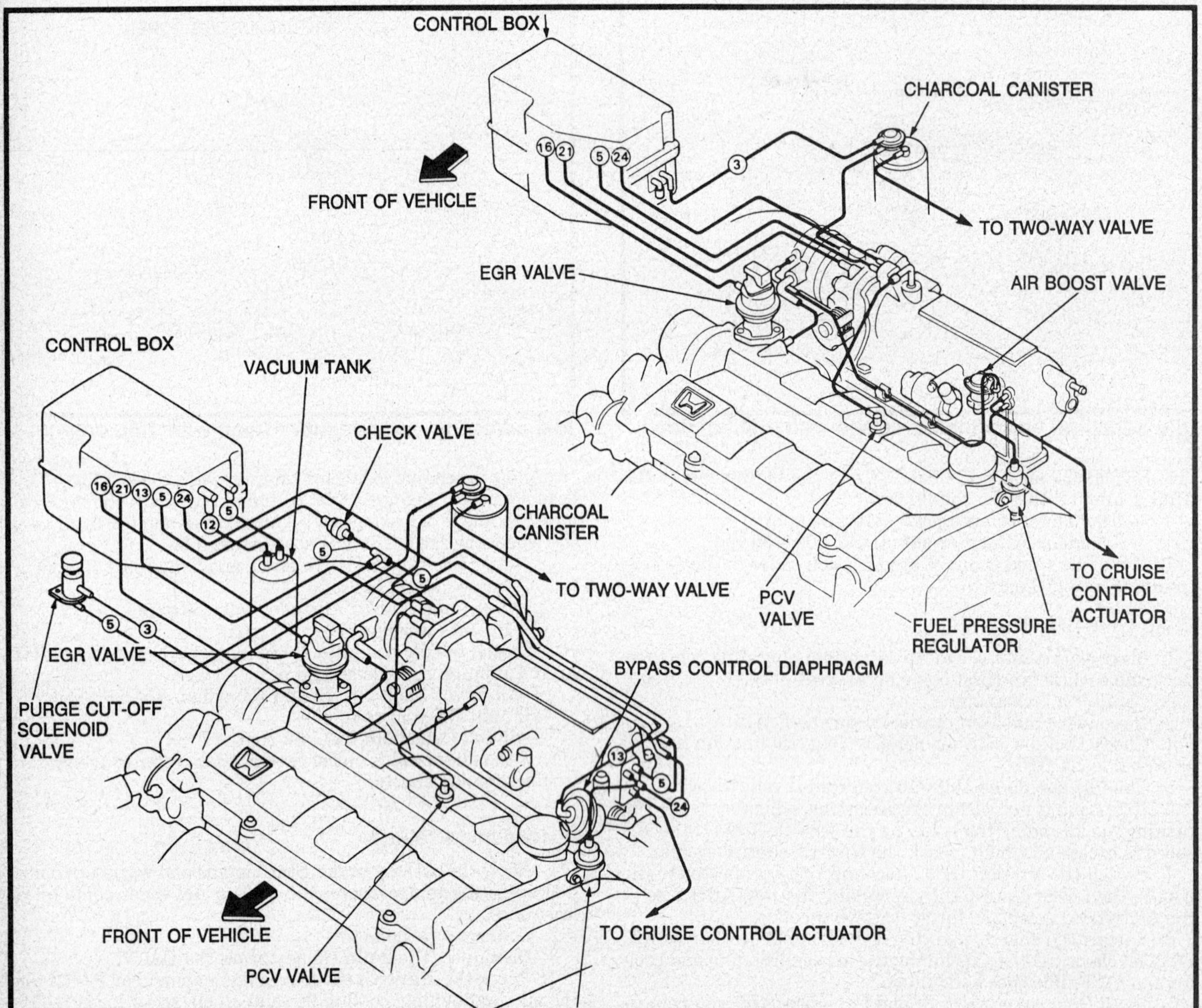

Vacuum hose routing, Prelude 2.1L (upper right) and 2.0L engines

Hyundai Fuel Injection

SECTION 9

HYUNDAI FUEL INJECTION SYSTEM

General Information

The fuel injection system used on Hyundai vehicles, is classified as a Multi-point Fuel Injection (MPI) system. The basic function of the system is to control the air/fuel ratio, based on input signals from various engine sensors. The air-fuel ratio is controlled by varying the injector driving time. The system is controlled by an Electronic Control Unit (ECU), which monitors the engine conditions, then calculates the injection timing and air/fuel ratio according to the signals from the sensors. The ECU consists of an 8-bit microprocessor, Random Access Memory (RAM), Read Only Memory (ROM) and input and output signal interface system.

The MPI system consists of 2 operating modes:

- Open Loop—air/fuel ratio is controlled by information programmed into the ECU by the manufacturer.
- Closed Loop—air/fuel ratio is varied by the ECU based on information supplied by the oxygen sensor.

An electric fuel pump supplies sufficient fuel to the injection system and the pressure regulator maintains a constant pressure to the injectors. These injectors inject a metered quantity of fuel into the intake manifold in accordance with signals from the Electronic Control Unit (ECU) or engine computer. After pressure regulation excess fuel is returned to the fuel tank.

The injectors have 2 modes (Injector Drive Timing) of operation:

- Non-synchronous Injection (Simultaneous Injection)
- Synchronous Injection (Sequential Injection)

Non-synchronous injection is activated during engine start-

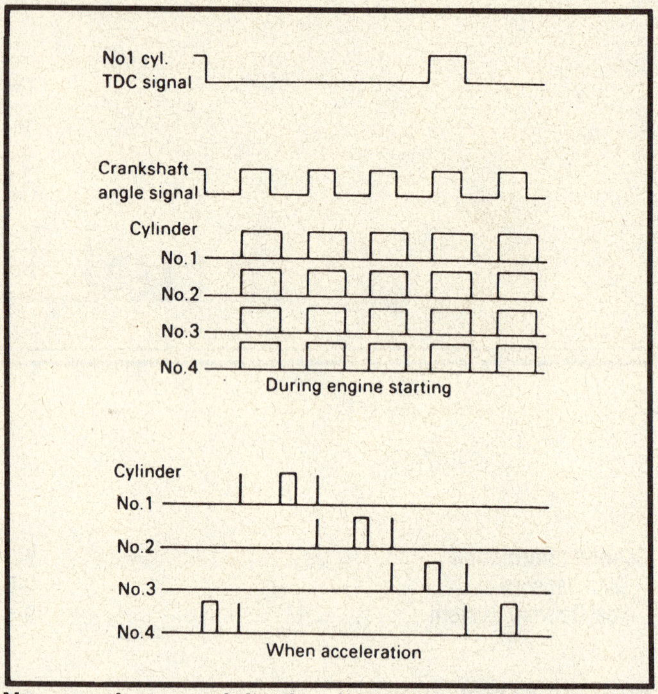

Non-synchronous injection (simultaneous injection)

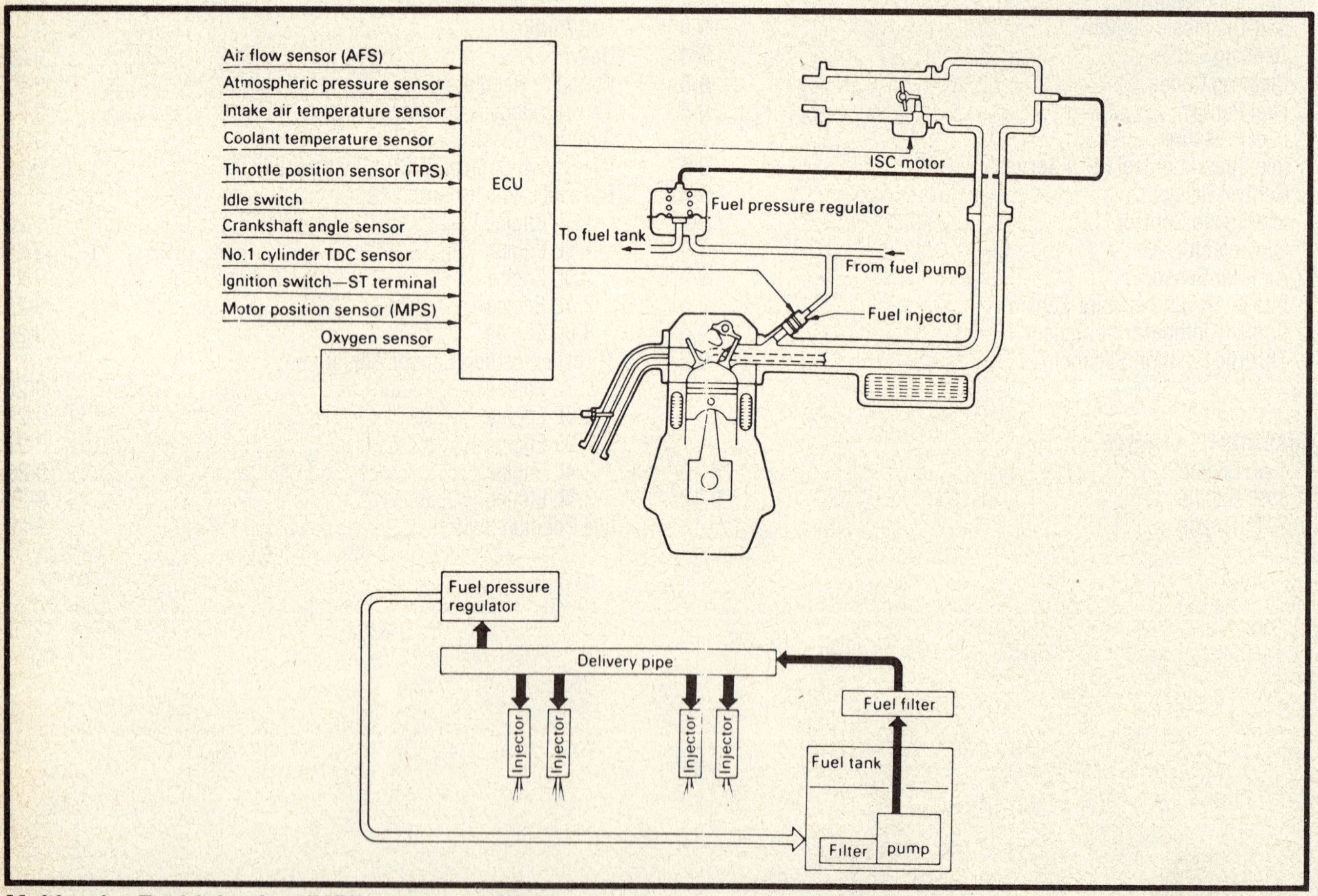

Multi-point Fuel Injection (MPI) system—typical

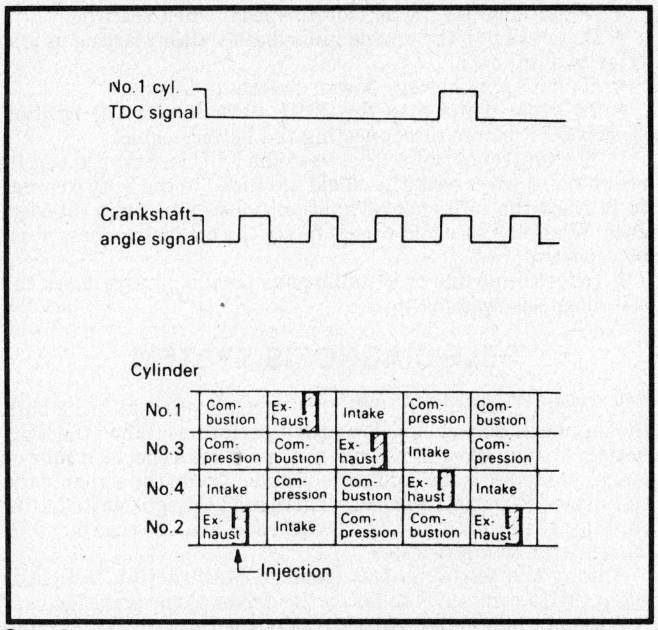

Synchronous injection (sequential injection)

ing (cranking). There are 2 fuel injections, for each engine rpm, to all 4 cylinders. Also, during acceleration, fuel proportionate to the magnitude of acceleration, is injected to 2 selected cylinders during the intake and exhaust strokes.

Synchronous Injection is activated after the engine have started. The injectors are activated at the exhaust stroke of each cylinder in a sequential manner, according to crankshaft angle sensor signal. There is 1 injection per cylinder for every 2 crankshaft revolution, according to firing order.

DATA SENSORS

Air Flow Sensor (AFS)

The AFS measures the intake air volume. The ECU uses this intake air volume signal to decide the basic fuel injection duration.

Atmosphere Pressure Sensor

The atmosphere pressure sensor signal is used by the ECU to compute the altitude of the vehicle and so the ECU can correct the ignition timing and air-fuel ratio. The atmosphere pressure sensor is contain in the AFS.

Intake Air Temperature Sensor

The air temperature sensor is a resistor based sensor for detecting the intake air temperature. The ECU provides fuel injection control based on this information. The air temperature sensor is in located on the AFS.

Engine Coolant Temperature Sensor

The coolant temperature sensor is located in the coolant passage of the intake manifold. The ECU uses this signal to determine the base fuel enrichment for cold and warm engine operation.

Throttle Position Sensor (TPS)

The TPS is a rotating type variable resistor that rotates with the throttle body shaft to sense the throttle valve opening. Based on TPS voltage signals, the ECU computes the throttle valve opening and accordingly corrects fuel for engine acceleration.

Idle Switch

The idle switch is a contact type switch. The switch is installed at the tip of the ISC. This switch provides the ECU with idle or off idle signal.

Motor Position Sensor (MPS)

The MPS is a variable resistor type sensor and is installed in the ISC servo. The MPS senses the ISC servo plunger position and sends the signal to the ECU. The ECU controls the valve opening, and consequently the idle speed by using the MPS signal, idle signal, engine coolant temperature signal, load signals and vehicle speed sensor.

Cylinder TDC and Crankshaft Angle Sensor

The No. 1 TDC sensor and the crankshaft angle sensor are composed of a disc and unit assembly in the distributor. The No. 1 cylinder TDC is detected by the signal obtained through the single inner slit of the disc. The ECU, based upon this signal, deter-

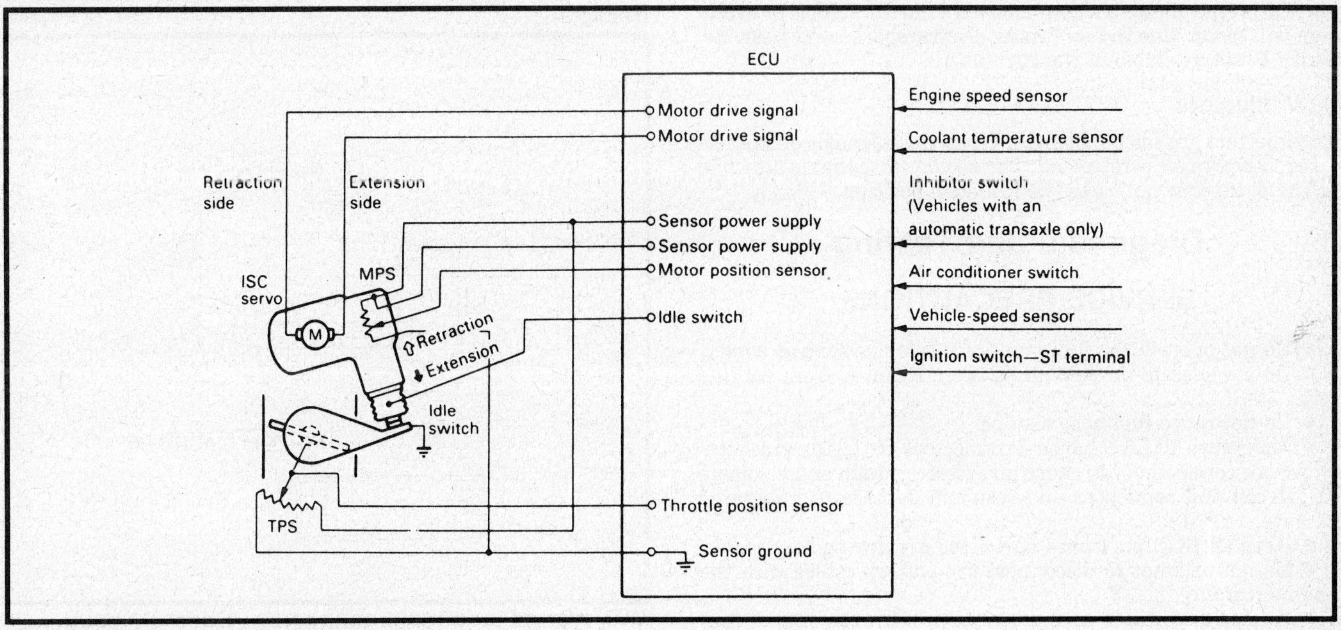

Idle speed control (ISC) system

mines the fuel injection cylinder. The crankshaft angle sensor signal comes from the the 4 slits at the outer circumference of the disc serve to detect the position of the crankshaft. The ECU, based on this signal, determines the fuel injection timing, and also calculates the amount of intake air, the ignition timing, etc. for each stroke of engine.

Oxygen Sensor

The oxygen sensor, by detecting the oxygen content in the exhaust gas, maintains the stoichiometric air/fuel ratio. In operation, the ECU receives the signals from the sensor and changes the duration during which fuel is injected. The oxygen sensor is located in the exhaust manifold.

Vehicle Speed Sensor

The vehicle speed sensor uses a reed type switch. The speed sensor is built into the speedometer and converts the speedometer gear revolution into pulse signals, which are sent to the ECU.

Neutral (Inhibitor) Switch

The ECU based on this signal senses when the automatic transaxle is in **NEUTRAL** or **PARK** and operates the ISC servo to keep the idle speed correct.

Idle Speed Control (ISC) Servo

The ISC servo consists of a motor, worm gear, worm wheel and plunger. The MPS is used to detect plunger position and an idle switch to detect closed throttle position.

As the motor rotates, according to signals from the ECU, the plunger extends or retracts depending on direction of the motor rotation. This actuates the throttle valve via the ISC lever. The idle speed is controlled by changing the throttle open by the ECU through this function.

FUEL DELIVERY SYSTEM

Fuel Pump

The fuel pump is a compact impeller design and is installed inside the fuel tank. The pump assembly consists of the impeller (driven by the motor), the pump casing (which forms the pumping chamber), and cover of the pump. This pump is called a wet-type pump, because the inside is also filled with fuel. Never operate this type of pump when removed from the vehicle or explosion will result, due to fuel fumes, electric sparks and fresh air, which is not available in the fuel tank.

Fuel Injectors

The injectors are solenoid valves. When the solenoid coil is energized, the plunger is retracted. The needle valve that is attached to the plunger is pulled to the full open position.

Diagnosis and Testing

SERVICE PRECAUTIONS

- Do not operate the fuel pump if the fuel system is empty.
- Do not operate the fuel pump when removed from the fuel tank.
- Do not reuse fuel hose clamps.
- Make sure all ECU harness connectors are fastened securely. A poor connection can cause an extremely high surge voltage in the coil and condenser and result in damage to integrated circuits.
- Keep all ECU parts and harnesses dry during service.
- Do not attempt to disconnect the battery cables with the engine running.
- Do not disconnect wiring connector with the engine running, unless instructed to do so.
- Do not depress the accelerator pedal when starting.
- Do not rev up the engine immediately after starting or just prior to shutdown.
- Do not apply battery power directly to injectors.
- To avoid damage to the ECU, always turn the ignition switch **OFF** before disconnecting the battery cable.
- The control harnesses between the ECU and oxygen sensor are shielded wires with the shield grounded to the body in order to prevent the influence of ignition noises and radio interference. When the shielded wire is faulty the control harness must be replaced.
- Before removing or installing any part(s), always check the self-diagnosis system.

SELF-DIAGNOSIS SYSTEM

The Engine Control Unit (ECU) is capable of monitoring both the input and output signals which it performed. When the ECU noticed that an irregularity has continued for a specified time or longer, it memorizes a trouble code and outputs the signal to the self-diagnosis output terminal. There are 14 diagnosis items, including the normal state. The trouble codes can be read out with a voltmeter or multi-tester.

Among the self-diagnosis items, a Malfunction Indicator Light (MIL) comes ON to notify the driver of the emission control items when an irregularity is detected.

Generally, the MIL will come ON for 5 seconds when the ignition switch is turned **ON**.

NOTE: Always check that the MIL illuminates for approximately 5 seconds, when the ignition switch is turned ON. If the light does not illuminate, check for an open circuit harness, blown fuse or blown bulb.

READING CODES

Using Voltmeter

1. Connect the voltmeter to the self-diagnostic connector, across the MPI diagnosis and ground terminals.
2. Turn the ignition switch **ON**. The ECU diagnostics memory will immediately start.
3. If the voltmeter displays a steady needle sweep, the system is normal and no codes are in the memory.
4. If the voltmeter displays a steady **HIGH**, signal the ECU is damaged.

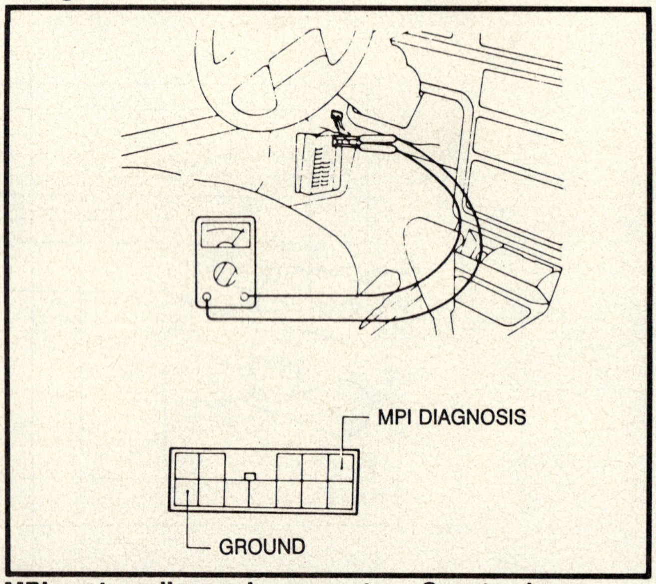

MPI system diagnosis connector—Sonata shown, Elantra similar

5. If the ECU has detected a malfunction, the voltmeter will deflect the indicating the diagnostic code.
6. Record the codes displayed by the voltmeter needle deflections.
7. The ECU will continue to send any memorized trouble codes to the self-diagnostic for as long as the ignition is **ON**, or until the codes have been cleared from memory.

Using Multi-Use Tester (MUT)

1. Turn the ignition switch **OFF**.
2. Connect the Multi-Use Tester (MUT), (Tool 09391–33200 or equivalent), to the diagnosis connector in the fuse box.
3. Connect the power-source terminal of the MUT to the cigarette lighter switch.
4. Turn the ignition switch **ON** and the ECU memory contents will immediately start.
5. Read the self-diagnosis output and check output of correct code.
6. After recording the fault codes, check and repair each part according to the check items in the Diagnosis Charts.

NOTE: If the battery voltage is too low, trouble codes cannot be read. If the battery or the ECU connector is disconnected, the diagnosis memory will be erased.

When the ignition key is in the **START** position, the cigarette lighter power is OFF. If it is necessary to perform tests, using the MUT in the crank mode, an auxiliary power supply will be necessary. Use the battery clamp harness provided.

CLEARING CODES

After completion of the repair or correction of the problem:

1. Turn the ignition switch **OFF**.
2. Disconnect the negative battery cable for approximately 15 seconds.
3. Reconnect the negative battery cable.
4. Check that no trouble codes are displayed by the MUT.
5. Remove the Multi-Use Tester (MUT).

FUEL PUMP

Operational Check

1. Turn the ignition switch **OFF**.
2. Apply battery voltage to the fuel pump drive connector and check that the fuel pump operates. The fuel pump is located inside the fuel tank and can be heard without removing the fuel tank cap.
3. Pinch the high pressure hose and note if fuel pressure is felt.

FUEL PRESSURE

Relieving

1. Disconnect the fuel pump harness at the rear of the tank.
2. Start engine and allow to idle.
3. After engine stalls, fuel pressure is reduced.
4. Disconnect the negative battery cable and reconnect the fuel pump wiring.
5. Remove the fuel tank cap.

NOTE: When disconnecting any fuel system hoses, always place a shop towel to prevent residual fuel from spaying.

Testing

1. Relieve fuel system pressure.
2. Turn ignition switch **OFF** and disconnect the negative battery cable.

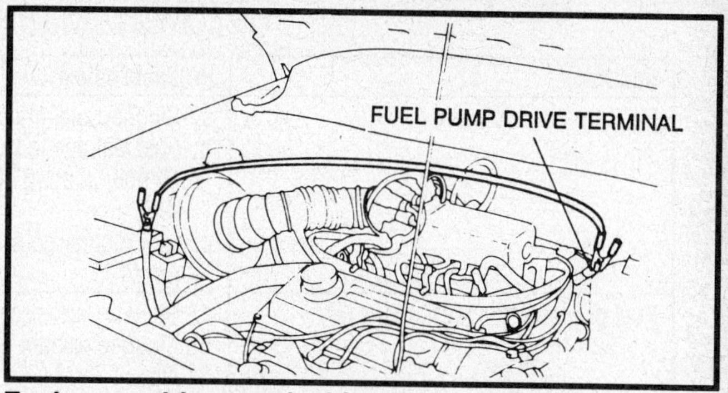

Fuel pump drive terminal location – Excel

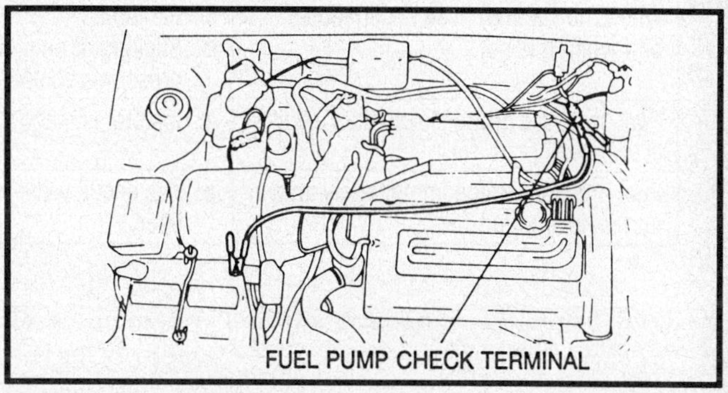

Fuel pump drive terminal location – Scoupe and Elantra

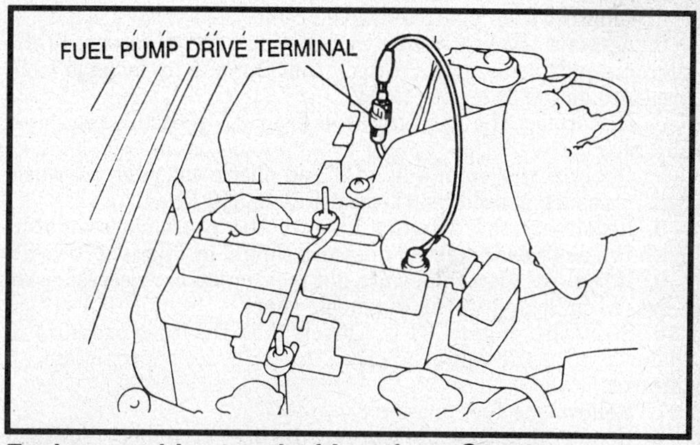

Fuel pump drive terminal location – Sonata

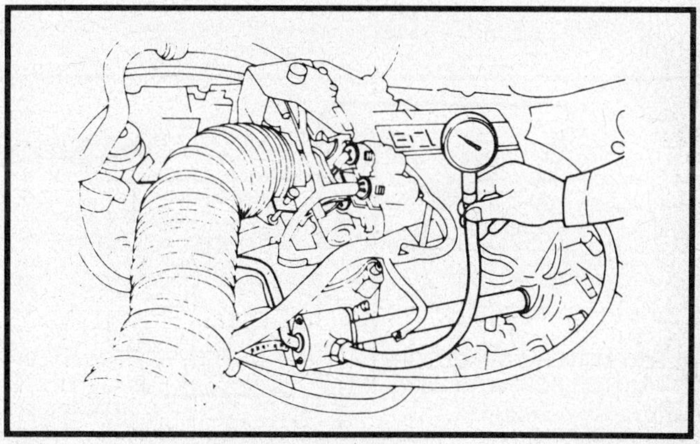
Checking fuel system pressure

FUEL SYSTEM PRESSURE—DIAGNOSIS

Condition	Probable cause	Remedy
Fuel pressure too low	a. Clogged fuel filter. b. Fuel leakage to the return side, caused by poor seating of the fuel-pressure regulator. c. Low discharge pressure of the fuel pump	a. Replace fuel filter b. Replace fuel pressure regulator. c. Check the in-tank fuel hose for leakage or replace the fuel pump.
Fuel pressure too high	a. Sticking fuel-pressure regulator b. Clogged or bent fuel return hose or pipe.	a. Replace fuel pressure regulator b. Repair or replace hose or pipe.
There is no difference in fuel pressure when the vacuum hose is connected and when it is not.	a. Clogging, or damaged vacuum hose or the nipple. b. Sticking or poor seating of the fuel-pressure regulator.	a. Repair or replace the vacuum hose or the nipple b. Repair or replace hose or pipe.
Fuel pressure drops slowly after engine is stopped	a. Injector leakage	a. Replace injector
Fuel pressure drops immediately after engine is stopped	a. The check valve within the fuel pump is open	a. Replace fuel pump

NOTE: When disconnecting any fuel system hoses, always place a shop towel to prevent residual fuel from spaying.

3. Disconnect the high pressure hose at the delivery side. Using a fuel pressure gauge and adapter (09353-24000 or equivalent), install the adapter to the delivery pipe.
4. Connect the negative battery cable.
5. Activate the fuel pump, by applying battery voltage to the fuel pump drive terminal. Check that there is no leakage from pressure gauge or connections.
6. Disconnect the vacuum hose from the pressure regulator and plug.
7. Operate the engine at idle and check the fuel pressure. Fuel pressure should be 47–50 psi (330–350 kPa).
8. Reconnect the vacuum hose to the pressure regulator. With the engine idling, fuel pressure should be 39 psi (270 kPa).
9. If the specified values are not met, make the necessary repairs, as indicated in the diagnosis chart.
10. Shut the engine OFF. Check that the fuel pressure is maintained at the value indicated in Step 8, for approximately 5 minutes.
11. Relieve the fuel pressure.
12. Remove the fuel pressure gauge and reconnect the high pressure hose, using a new O-ring.
13. Start the engine and check for leaks.

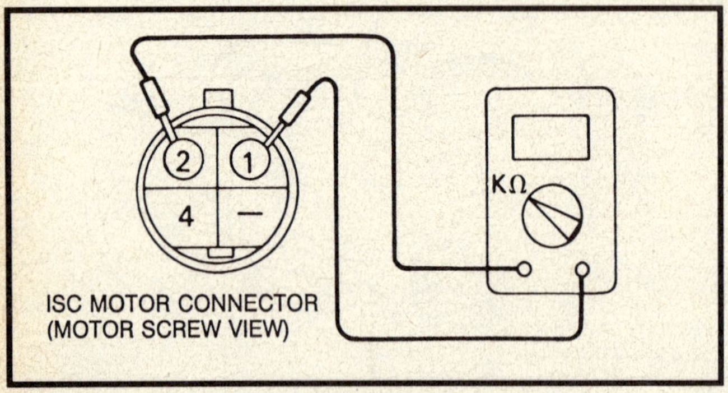

Check ISC motor for continuity

IDLE SPEED CONTROL (ISC) SERVO

Testing

SONATA

1. Disconnect the ISC motor connector.
2. Check the continuity of the ISC motor coil between terminal 1 and 2. Reading should be 5–35 ohms at 68°F (20°C).
3. Connect a 6 volt DC power supply (4 dry cells) between terminal 1 and 2. Check that the servo operates.

NOTE: Do not apply voltages higher than 6 volts DC or locking of the servo gears can result.

4. If the servo does not operate, replace it as an assembly.

CONTROL RELAY

Failure of the control relay will interrupt power supply to the fuel pump, injectors and the ECU, resulting in a No Start condition.

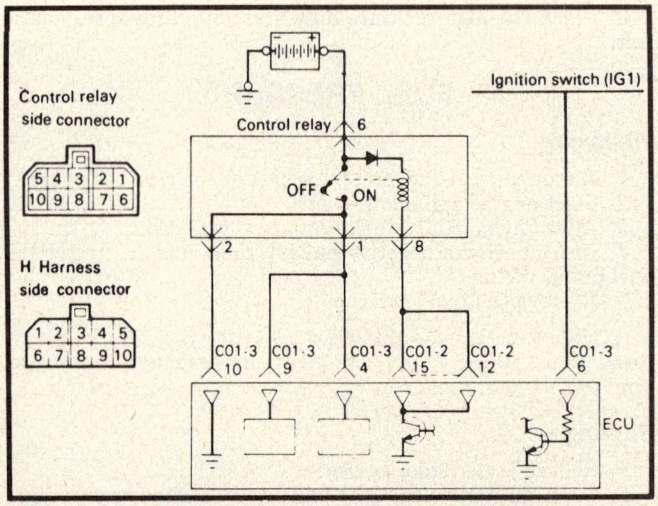

Control relay functional schematic—Elantra

Testing

ELANTRA

1. Apply 12 volts (+) power supply to terminal 6 of the control relay. Note the voltage at terminal 2 and at terminal 1 when the negative (−) lead is connected and disconnected from terminal 8. Should indicated 12 volts when connected and 0 volt when disconnected.
2. Apply 12 volts (+) power supply to terminal 3 of the control relay. Note the voltage at terminal 4 when the negative (−) lead is connected and disconnected from terminal 9. Should indicated 12 volts when connected and 0 volt when disconnected.
3. Connect the negative lead (−) of the power supply to terminal 10 of the control relay. Check for continuity between terminals 3 and 4 when the positive (+) terminal is connected and disconnected to terminal 7. Should indicated continuity when connected.
4. If the results is not satisfactory, replace the control relay.

EXCEL

1. Check for continuity between the terminals when the relay coil is not energized, as follow:
   a. Terminals 1 and 4. Should be infinity (non-conductive).
   b. Terminals 2 and 8. Should indicate a value of approximately 95 ohms.
   c. Terminals 3 and 8. Should indicate a value of approximately 95 ohms.
   c. Terminals 6 and 7. Should indicate a value of approximately 35 ohms.
   d. Terminals 1 and 2. Should be infinity (non-conductive).
2. Connect the ohmmeter across terminals 1 and 4. Jumper terminal 6 to ground and momentarily apply battery voltage to terminal 7 (Relay energized). Should indicate a value of 0 ohm.
3. Connect the ohmmeter across terminals 1 and 2. Jumper terminal 5 to ground and momentarily apply battery voltage to terminal 7 (Relay energized). Should indicate a value of 0 ohm.
4. If the results is not satisfactory, replace the control relay.

SCOUPE

1. Check for continuity between the terminals when the relay coil is not energized, as follow:
   a. Terminals 2 and 8. Should indicate a value of approximately 95 ohms.
   b. Terminals 3 and 8. Should indicate a value of approximately 35 ohms.
   c. Terminals 5 and 7. Should indicate continuity, but only in one direction.
2. Connect the ohmmeter across terminals 1 and 4. Jumper terminal 6 to ground and momentarily apply battery voltage to terminal 7 (Relay energized). Should indicate a value of 0 ohm. When the relay is de-energized it should indicated infinity.
3. Connect the ohmmeter across terminals 1 and 3. Jumper terminal 5 to ground and momentarily apply battery voltage to terminal 7 (Relay energized). Should indicate a value of 0 ohm. When the relay is de-energized it should indicated infinity.

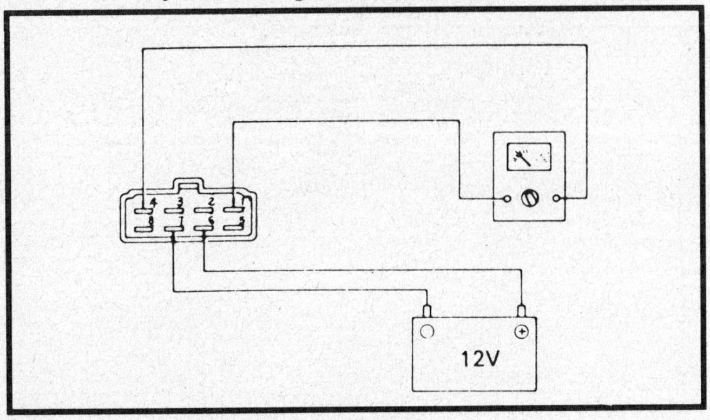

Checking control relay—Excel

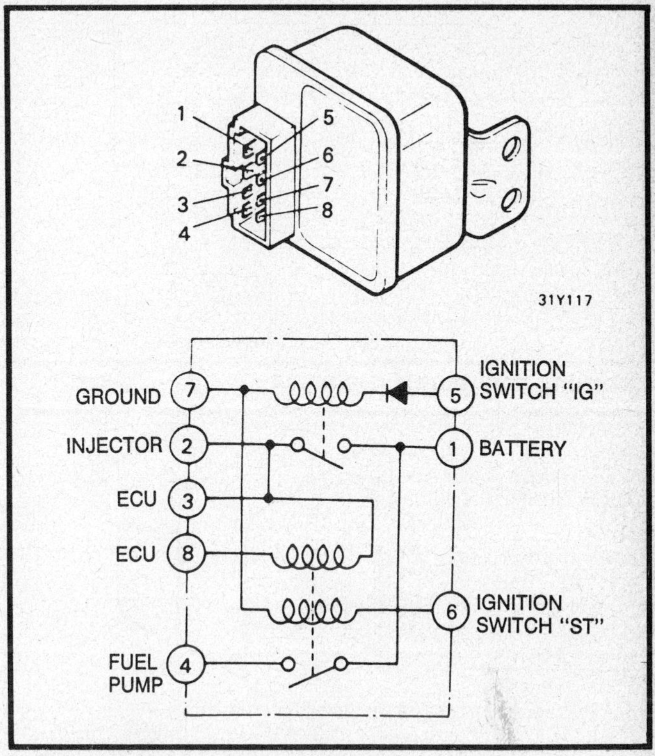

Control relay functional schematic—Excel

4. If the results is not satisfactory, replace the control relay.

SONATA

1. Check for continuity between the terminals when the relay coil is not energized, as follow:
   a. Terminals 1 and 4. Should be infinity (non-conductive).
   b. Terminals 2 and 8. Should indicate a value of approximately 95 ohms.
   c. Terminals 3 and 8. Should indicate a value of approximately 95 ohms.

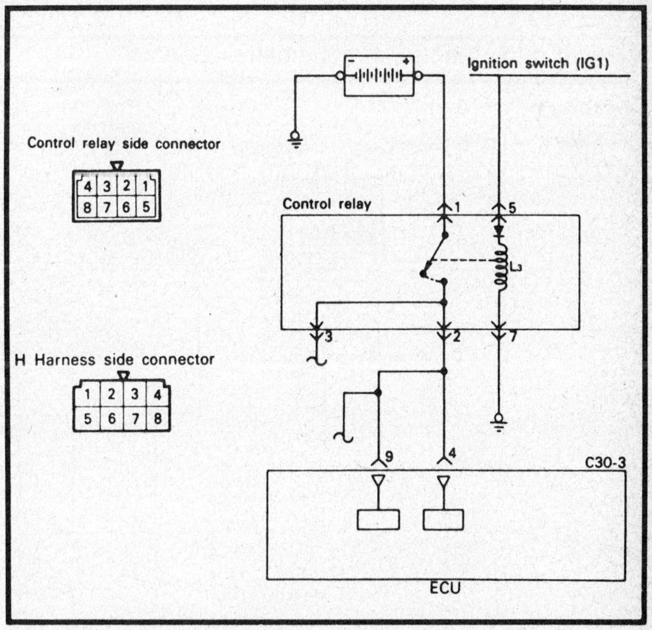

Control relay functional schematic—Scoupe

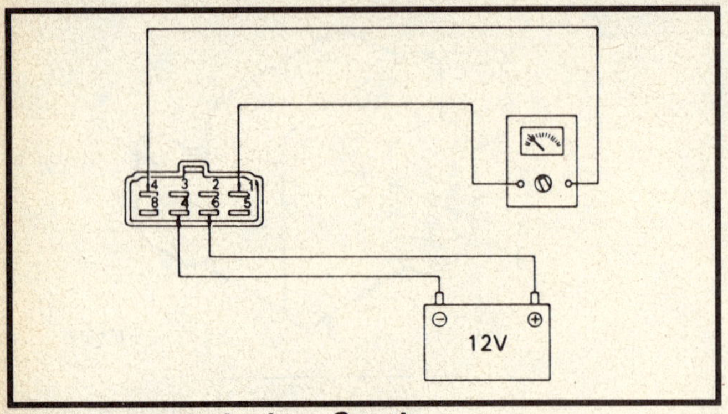

Checking control relay—Sonata

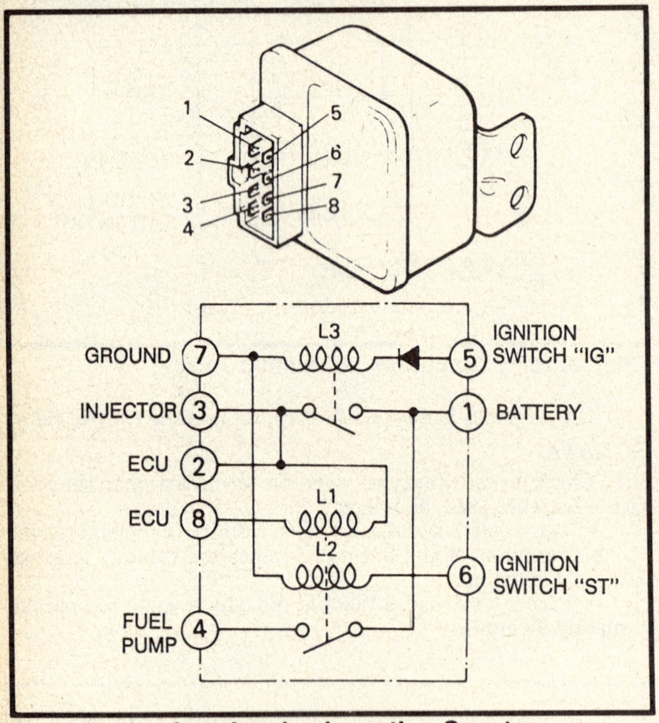

Control relay functional schematic—Sonata

c. Terminals 6 and 7. Should indicate a value of approximately 35 ohms.

d. Terminals 1 and 3. Should be infinity (non-conductive).

2. Connect the ohmmeter across terminals 1 and 4. Jumper terminal 6 to ground and momentarily apply battery voltage to terminal 7 (Relay energized). Should indicate a value of 0 ohm.

3. Connect the ohmmeter across terminals 1 and 3. Jumper terminal 5 to ground and momentarily apply battery voltage to terminal 7 (Relay energized). Should indicate a value of 0 ohm.

4. If the results is not satisfactory, replace the control relay.

IDLE SPEED CONTROL SERVO

Testing

ELANTRA

STEPPER MOTOR TYPE

1. Connect a 6 volt power supply source to the indicated terminals and check whether or not the stepper motor movement occurs, as follows:

a. Positive (+) lead to terminal 2 and 5 of the connector.
b. Negative (−) lead to terminal 1 and 4.
c. Negative (−) lead to terminal 3 and 4.
d. Negative (−) lead to terminal 3 and 6.
e. Negative (−) lead to terminal 1 and 6.

2. Measure the resistance between the respective terminals, as follows:

a. Terminals 2–3 and 1–28–33 ohms at 68°F (20°C)
b. Terminals 5–4 and 6–28–33 ohms at 68°F (20°C)

3. If the results are not as indicated in Step 1 and 2, replace the idle speed control servo.

SCOUPE

DC MOTOR TYPE

1. Check for an open-circuit or short-circuit to ground between the ECU terminals 1 and 2 to the ISC servo.

2. Disconnect the ISC servo connector.

3. Check for continuity across the terminals of the ISC. Should indicate a reading of 5–35 ohms at 68°F (20°C).

4. Connect a 6 volt DC power source across the ISC and check that the ISC servo operates.

NOTE: Do not apply more than 6 volts DC. A higher voltage may cause the servo gears to lock.

5. If the results are not as indicated, replace the ISC servo.

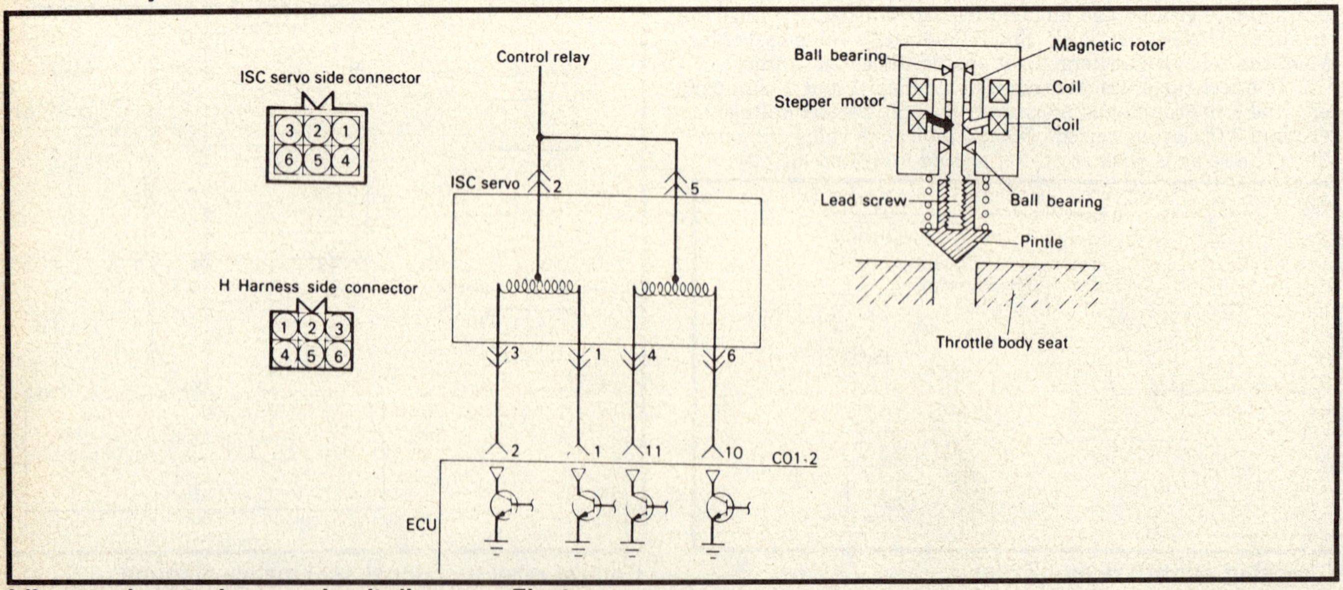

Idle speed control servo circuit diagram—Elantra

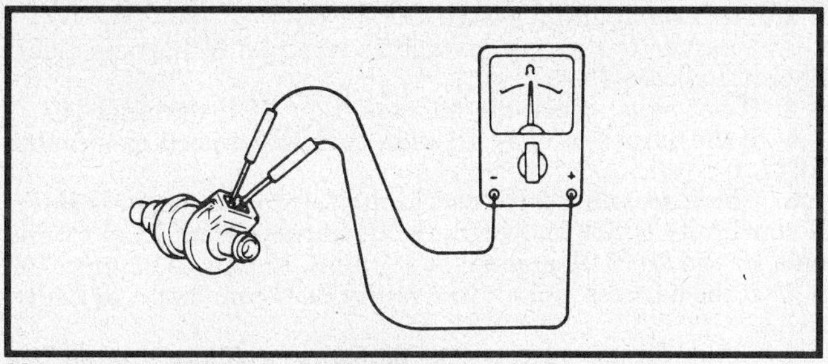

Measuring injector resistance

INJECTOR

Testing

1. Measure the resistance across the injector terminals.
2. Readings should be between 13–16 ohms at 68°F (20°C).
3. If the resistance is not within specifications, replace the injector.

AIR FLOW SENSOR (AFS)

Testing

1. Disconnect the harness connector to the AFS.
2. Turn the ignition switch **ON**.
3. Measure the supply voltage at terminal 2.
4. If source volt is not indicated, repair the harness from the control relay or the control relay is faulty.
5. Measure the voltage to terminal 3. Should indicate 4.8–5.2 volts.
6. If not within specification, repair harness and/or connector, as follow:
   1.5L engine – harness from the ECU terminal 3 (C30-1) or ECU is faulty.
   1.6L and 2.0L engines – harness from the ECU terminal 3 (C01-1) or ECU is faulty.
   3.0L engine – harness from the ECU terminal 15 (C81-1) or ECU is faulty.
7. If the harness is okay, its a faulty ECU connection or faulty ECU.
8. Check terminal 6 for continuity to ground. If continuity is not indicated, repair harness from ECU terminals 13 and 20 (except 3.0L engine) or terminal 13 only (3.0L engine).
9. If the harness is okay, substitute a known good ECU.
10. If problem still exist, replace the AFS.

INTAKE AIR TEMPERATURE SENSOR

Testing

1990 1.5L ENGINE

1. Disconnect the harness connector to the air flow sensor.
2. Measure the resistance between terminals 4 and 6 (AFS side connector). Resistance values should be as follow:
   32°F (0°C) – 5.4–6.6 ohms
   68°F (20°C) – 2.3–2.9.0 ohms
   176°F (80°C) – 0.3–0.4 ohms
3. Measure the resistance while heating the sensor with a hair drier. As the temperature increases, the resistance should decrease.
4. If the readings deviates from these values, replace the air flow sensor.

EXCEPT 1990 1.5L ENGINE

1. Disconnect the harness connector to the air flow sensor.
2. Turn the ignition switch **ON**.
3. Measure the supply voltage at terminal 4 (harness side). Should indicate 4.5–4.9 volts.
4. If not okay, check the harness from ECU terminal 3 (2.0L and 3.0L engines) or ECU terminal 5 (1.5L and 1.6L engines).
5. If the harness is okay, its a faulty ECU connection or faulty ECU.
6. Check terminal 6 for continuity to ground (harness side). If continuity is not indicated, repair harness from ECU terminals 13 and 20 (2.0L engine), terminal 13 (3.0L engine) or ECU is faulty.
7. Measure the resistance across the air flow sensor terminals. Readings should be 2.7 ohms at 68°F (20°C).
8. If the specified values are not obtained, replace the air flow sensor.

COOLANT TEMPERATURE SENSOR

Testing

1990 1.5L ENGINE

1. Disconnect the harness connector from the coolant sensor and remove the sensor from the intake manifold.
2. Measure the resistance across the sensor terminals, with the sensing portion of the sensor immersed in hot water. The sensor housing should be 0.12 inch (3mm) above the surface of the water.
3. Resistance values should be as follow:
   32°F (0°C) – 5.9 ohms
   68°F (20°C) – 2.5 ohms
   104°F (40°C) – 1.1 ohms
   176°F (80°C) – 0.3 ohms
4. If the readings deviates from these values, replace the coolant sensor.

EXCEPT 1990 1.5L ENGINE

1. Disconnect the harness connector to the sensor.
2. Turn the ignition switch **ON**.
3. Measure the supply voltage at terminal 2 (harness side). Should indicate 4.5–4.9 volts.
4. If not okay, check the harness from ECU terminal 17 or ECU is faulty.
5. If the harness is okay, its a faulty ECU connection or faulty ECU.
6. Check terminal 1 for continuity to ground (harness side). If continuity is not indicated, repair harness from ECU terminals 13 and 23 (1.5L engine), 13 and 20 (1.6L and 2.0L engine) or terminal 13 (3.0L engine).
7. If the harness is okay, its a faulty ECU connection or faulty ECU.
8. Measure the resistance across the air flow sensor terminals. Readings should be 2.5 ohms at 68°F (20°C).
9. If the specified values are not obtained, replace the air flow sensor.

THROTTLE POSITION SENSOR

Testing

1990 1.5L ENGINES

1. Turn the ignition switch **OFF** and disconnect the harness connector to the TPS sensor.
2. Measure the resistance between terminals 1 and 3 of the TPS. Should indicate a reading between 3.5–6.5 kilo-ohms.
3. Connect an analog-type ohmmeter between terminals 1 and 2.
4. Slowly open the throttle valve form the idle position to the fully open position and check that the resistance value change smoothly with the opening of the throttle valve. Should indicate a reading between 3.5–6.5 kilo-ohms.
5. If the specified values are not obtained, replace the TPS.

1.6L AND 1991–92 1.5L ENGINES

1. Disconnect the harness connector to the TPS sensor.

2. Turn the ignition switch **ON**.
3. Measure the supply voltage at terminal 4 (harness side). Should indicate 4.8–5.2 volts.
4. If not okay, check the harness from ECU terminal 14.
5. If the harness is okay, its a faulty ECU connection or faulty ECU.
6. Check terminal 1 for continuity to ground (harness side). If continuity is not indicated, repair harness from ECU terminals 13 and 23 (1.5L engine), 13 and 20 (1.6L engine).
7. If the harness is okay, its a faulty ECU connection or faulty ECU.
8. Check for an open-circuit or short-circuit to ground between the ECU terminal 18 and the TPS terminal 3.
9. If the supply voltage is obtained and the harness is okay, replace the TPS.

2.0L AND 3.0L ENGINES

1. Disconnect the harness connector to the TPS sensor.
2. Turn the ignition switch **ON**.
3. Measure the supply voltage at terminal 3 (harness side). Should indicate 4.8–5.2 volts.
4. If not okay, check the harness from ECU terminal 16.
5. If the harness is okay, its a faulty ECU connection or faulty ECU.
6. Check terminal 2 for continuity to ground (harness side). If continuity is not indicated, repair harness from ECU terminals 13 and 20 (2.0L engine) or terminal 13 only (3.0L engine).
7. If the harness is okay, its a faulty ECU connection or faulty ECU.
8. Check for an open-circuit or short-circuit to ground between the ECU terminal 18 and the TPS terminal 4.
9. If the supply voltage is obtained and the harness is okay, replace the TPS.

1.5L ENGINE—MPI DIAGNOSIS—EXCEL AND SCOUPE

Inspection Procedure (Using Voltmeter) J31DY3A

1. Connect the voltmeter to the self diagnosis terminal.
2. Turn the ignition switch, and the engine control unit memory contents will immediately start.
   After recording the abnormal item, check and repair each part according to the check items in "Diagnosis Chart".
3. When the defective parts have been repaired, disconnect the negative terminal of the battery cable for 15 seconds or more and connect it again to make sure that the trouble code has been erased.

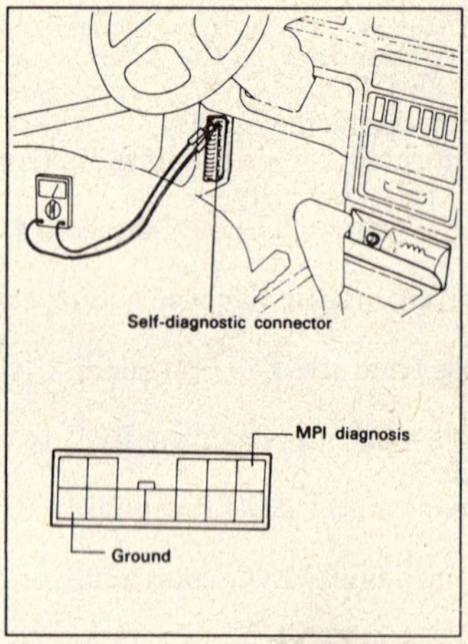

Diagnosis Chart

The 14 diagnosis items are provided as follows, and if plural items are activated, they are all indicated sequentially from the smallest code number.

Output preference order	Diagnosis item	Trouble code			Check item (Remedy)
		Output signal pattern	No.	Memory	
1	Engine control unit		—	—	(Replace engine control unit)
2	Oxygen sensor	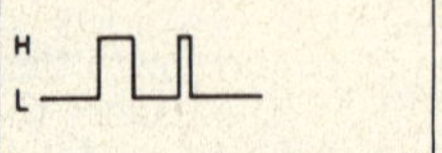	11	Retained	o Harness and connector o Fuel pressure o Injectors (Replace if defective.) o Intake air leaks o Oxygen sensor
3	Air flow sensor	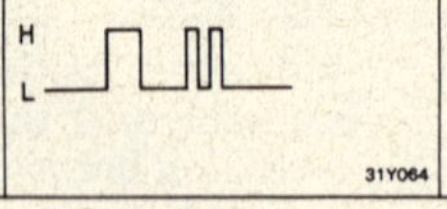	12	Retained	o Harness and connector (If harness and connector are normal, replace air flow sensor assembly.)

Output preference order	Diagnosis item	Malfunction code			Check item (Remedy)
		Output signal pattern	No.	Memory	
4	Air temperature sensor	 31Y065	13	Retained	o Harness and connector o Air temperature sensor
5	Throttle position sensor	 31Y066	14	Retained	o Harness and connector o Throttle position sensor o Idle position switch
6	Motor position sensor	 31Y067	15	Retained	o Harness and connector o Motor position sensor
7	Engine coolant temperature sensor	 31Y068	21	Retained	o Harness and connector o Engine coolant temperature sensor
8	Crank angle sensor	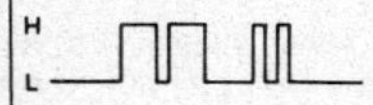 31Y069	22	Retained	o Harness and connector If harness and connector are normal, replace distributor assembly.)
9	No.1 cylinder top dead center sensor	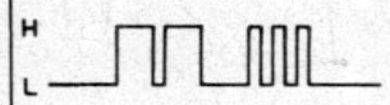 31Y070	23	Retained	o Harness and connector If harness and connector are normal, replace distributor assembly.)
10	Vehicle-speed sensor (reed switch)	 31Y071	24	Retained	o Harness and connector o Vehicle-speed sensor (reed switch)

Output preference order	Diagnosis item	Trouble code			Check item (Remedy)
		Output signal pattern	No.	Memory	
11	Barometric pressure sensor	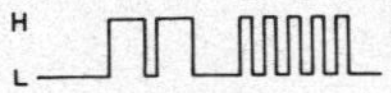 31Y072	25	Retained	o Harness and connector If harness and connector are normal, replace barometric pressure sensor assembly.)
12	Injector	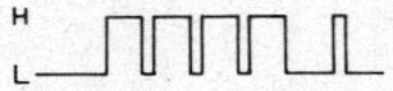 31Y073	41	Retained	o Harness and connector o Injector coil resistance
13	Fuel pump	 31Y074	42	Retained	o Harness and connector o Control relay
14	EGR*	 31Y075	43	Retained	o Harness and connector o EGR temperature sensor o EGR valve o EGR control solenoid valve o EGR valve control vacuum
15	Normal state	H L ⊓⊓⊓⊓⊓⊓⊓⊓⊓⊓ 31Y076	—	—	—

NOTE

1. Replace the engine control unit if a trouble code is read although the inspection reveals that there are no problems with the diagnosis item.
2. The diagnostic item marked* is applicable to the California vehicles only.

Inspection Procedure (Using Voltmeter) J31DY0A

1. Connect the voltmeter to the self diagnosis terminal.
2. Turn the ignition switch, and the engine control unit memory contents will immediately start.
   After recording the abnormal item, check and repair each part according to the check items in "Diagnosis Chart".
3. When the defective parts have been repaired, disconnect the negative terminal of the battery cable for 15 seconds or more and connect it again to make sure that the trouble code has been erased.

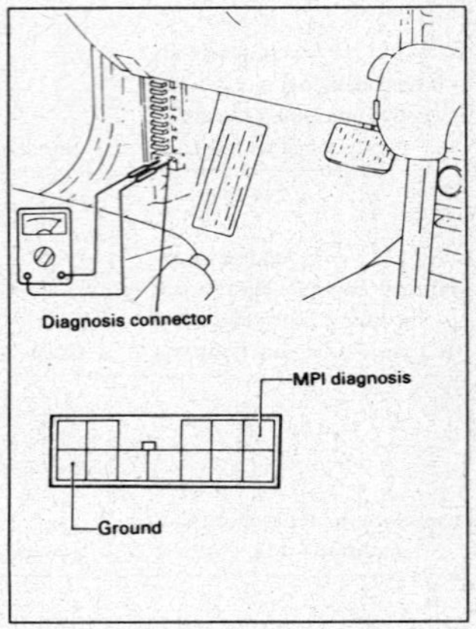

Diagnosis Chart

The 14 diagnosis items are provided as follows, and if plural items are activated, they are all indicated sequentially from the smallest code number.

Output preference order	Diagnosis item	Trouble code			Check item (Remedy)
		Output signal pattern	No.	Memory	
1	Engine control unit	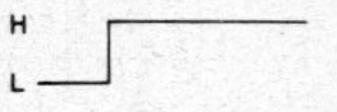 31J062	—	—	(Replace engine control unit)
2	Oxygen sensor	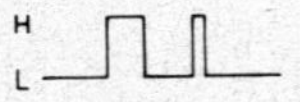 31J063	11	Retained	o Harness and connector o Fuel pressure o Injectors (Replace if defective.) o Intake air leaks o Oxygen sensor
3	Air flow sensor	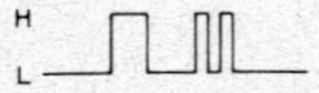 31Y064	12	Retained	o Harness and connector (If harness and connector are normal, replace air flow sensor assembly.)

Output preference order	Diagnosis item	Malfunction code			Check item (Remedy)
		Output signal pattern	No.	Memory	
4	Air temperature sensor	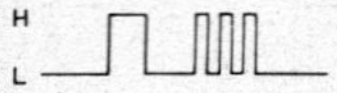 31Y065	13	Retained	o Harness and connector o Air temperature sensor
5	Throttle position sensor	 31Y066	14	Retained	o Harness and connector o Throttle position sensor o Idle position switch
6	Engine coolant temperature sensor	 31Y068	21	Retained	o Harness and connector o Engine coolant temperature sensor
7	Crank angle sensor	 31Y069	22	Retained	o Harness and connector If harness and connector are normal, replace crank angle sensor assembly.)
8	Top dead center sensor	 31Y070	23	Retained	o Harness and connector If harness and connector are normal, replace crank angle sensor assembly.)
9	Vehicle-speed sensor (reed switch)	 31Y071	24	Retained	o Harness and connector o Vehicle-speed sensor (reed switch)
10	Barometric pressure sensor	 31Y072	25	Retained	o Harness and connector If harness and connector are normal, replace barometric pressure sensor assembly.)

1.6L ENGINE—MPI DIAGNOSIS—ELANTRA

Output preference order	Diagnosis item	Trouble code: Output signal pattern	Trouble code: No.	Trouble code: Memory	Check item (Remedy)
11	Injector	H L 31Y073	41	Retained	o Harness and connector o Injector coil resistance
12	Fuel pump	H L 31Y074	42	Retained	o Harness and connector o Control relay
13	EGR*	H L 31Y075	43	Retained	o Harness and connector o EGR temperature sensor o EGR valve o EGR control solenoid valve o EGR valve control vacuum.
14	Ignition coil	H L 31J161	44	Retained	o Harness and connector o Ignition coil o Power transistor
15	Normal state	H L 31Y076	—	—	—

NOTE
1. Replace the engine control unit if a trouble code is read although the inspection reveals that there are no problems with the diagnosis item.
2. The diagnosis item marked* is applicable to California vehicles only.

2.0L ENGINE—MPI DIAGNOSIS—SONATA

Inspection Procedure (Using Voltmeter)

1. Connect the voltmeter to the self diagnosis terminal.
2. Turn the ignition switch, and the engine control unit memory contents will immediately start.
   After recording the abnormal item, check and repair each part according to the check items in "Diagnosis Chart".
3. When the defective parts have been repaired, disconnect the negative terminal of the battery cable for 15 seconds or more and connect it again to make sure that the trouble code has been erased.

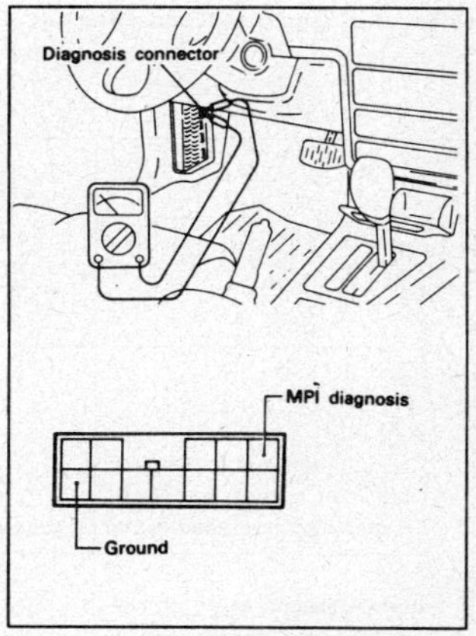

Diagnosis Chart

The 14 diagnosis items are provided as follows, and if plural items are activated, they are all indicated sequentially from the smallest code number.

Output preference order	Diagnosis item	Trouble code: Output signal pattern	Trouble code: No.	Trouble code: Memory	Check item (Remedy)
1	Engine control unit	H L 31J062	—	—	(Replace engine control unit)
2	Oxygen sensor	H L 31J063	11	Retained	o Harness and connector o Fuel pressure o Injectors (Replace if defective.) o Intake air leaks o Oxygen sensor
3	Air flow sensor	H L 31Y064	12	Retained	o Harness and connector (If harness and connector are normal, replace air flow sensor assembly.)

Output preference order	Diagnosis item	Malfunction code: Output signal pattern	Malfunction code: No.	Malfunction code: Memory	Check item (Remedy)
4	Air temperature sensor	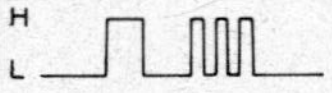 31Y065	13	Retained	o Harness and connector o Air temperature sensor
5	Throttle position sensor	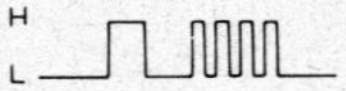 31Y066	14	Retained	o Harness and connector o Throttle position sensor o Idle position switch
6	Engine coolant temperature sensor	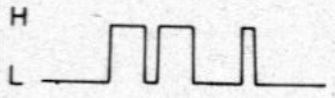 31Y068	21	Retained	o Harness and connector o Engine coolant temperature sensor
7	Crank angle sensor	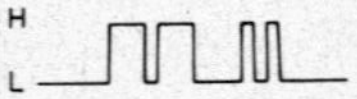 31Y069	22	Retained	o Harness and connector If harness and connector are normal, replace crank angle sensor assembly.
8	Top dead center sensor	 31Y070	23	Retained	o Harness and connector If harness and connector are normal, replace crank angle sensor assembly.
9	Vehicle-speed sensor (reed switch)	 31Y071	24	Retained	o Harness and connector o Vehicle-speed sensor (reed switch)
10	Barometric pressure sensor	 31Y072	25	Retained	o Harness and connector If harness and connector are normal, replace barometric pressure sensor assembly.

Output preference order	Diagnosis item	Trouble code: Output signal pattern	Trouble code: No.	Trouble code: Memory	Check item (Remedy)
11	Injector	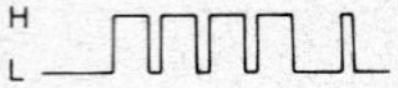 31Y073	41	Retained	o Harness and connector o Injector coil resistance
12	Fuel pump	 31Y074	42	Retained	o Harness and connector o Control relay
13	EGR*	 31Y075	43	Retained	o Harness and connector o EGR temperature sensor o EGR valve o EGR control solenoid valve o EGR valve control vacuum.
14	Ignition coil	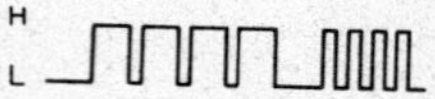 31J161	44	Retained	o Harness and connector o Ignition coil o Power transistor
15	Normal state	 31Y076	—	—	—

NOTE

1. **Replace the engine control unit if a trouble code is read although the inspection reveals that there are no problems with the diagnosis item.**
2. **The diagnosis item marked* is applicable to California vehicles only.**

2.4L ENGINE—MPI DIAGNOSIS—SONATA

Inspection Procedure (Using Voltmeter)

1. Connect the voltmeter to the self diagnosis connector.
2. Turn the ignition switch, and the engine control unit memory contents will immediately start.
   After recording the abnormal item, check and repair each part according to the check items in "Diagnosis Chart".
3. When the defective parts have been repaired, disconnect the negative terminal of the battery cable for 15 seconds or more and connect it again to make sure that the trouble code has been erased.

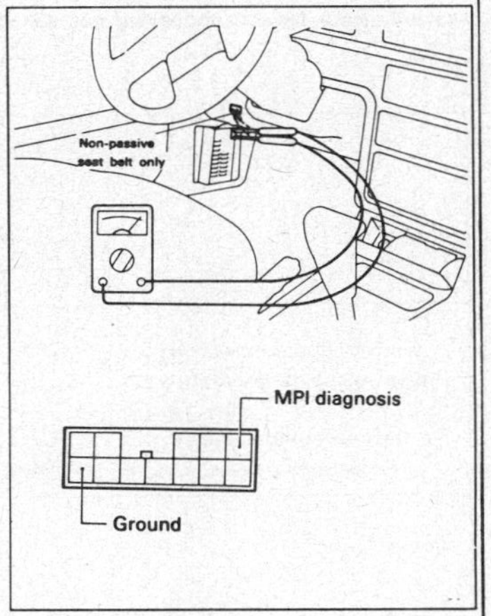

Diagnosis Chart

Output preference order	Diagnosis item	Trouble code: Output signal pattern	No.	Memory	Check item (Remedy)
1	Engine control unit	31Y062	—	—	(Replace engine control unit)
2	Oxygen sensor	31Y063	11	Retained	o Harness and connector o Fuel pressure o Injectors (Replace if defective.) o Intake air leaks o Oxygen sensor
3	Air flow sensor	31Y064	12	Retained	o Harness and connector (If harness and connector are normal, replace air flow sensor assembly.)

Output preference order	Diagnosis item	Malfunction code: Output signal pattern	No.	Memory	Check item (Remedy)
4	Air temperature sensor	31Y065	13	Retained	o Harness and connector o Air temperature sensor
5	Throttle position sensor	31Y066	14	Retained	o Harness and connector o Throttle position sensor o Idle position switch
6	Motor position sensor	31Y067	15	Retained	o Harness and connector o Motor position sensor
7	Engine coolant temperature sensor	31Y068	21	Retained	o Harness and connector o Engine coolant temperature sensor
8	Crank angle sensor	31Y069	22	Retained	o Harness and connector (If harness and connector are normal, replace distributor assembly.)
9	No.1 cylinder top dead center sensor	31Y070	23	Retained	o Harness and connector (If harness and connector are normal, replace distributor assembly.)
10	Vehicle-speed sensor (reed switch)	31Y071	24	Retained	o Harness and connector o Vehicle-speed sensor (reed switch)

Output preference order	Diagnosis item	Trouble code: Output signal pattern	Trouble code: No.	Trouble code: Memory	Check item (Remedy)
11	Barometric pressure sensor	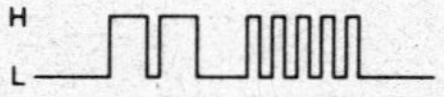 31Y072	25	Retained	o Harness and connector (If harness and connector are normal, replace barometric pressure sensor assembly.)
12	Injector	 31Y073	41	Retained	o Harness and connector o Injector coil resistance
13	Fuel pump	 31Y074	42	Retained	o Harness and connector o Control relay
14	EGR*	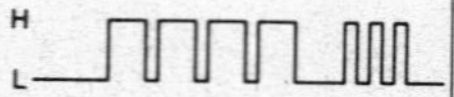 31Y075	43	Retained	o Harness and connector o EGR temperature sensor o EGR valve o EGR control solenoid valve o EGR valve control vacuum
15	Normal state	 31Y076	—	—	—

NOTE
1. **Replace the engine control unit if a trouble code is read although the inspection reveals that there are no problems with the diagnosis item.**
2. **The diagnosis item marked * is applicable to California vehicles only.**

CHECKING (USING THE MULTI-USE TESTER)

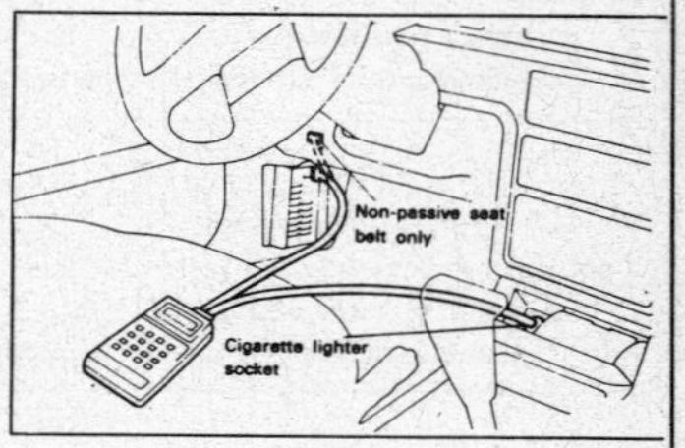

1. Turn OFF the ignition switch.
2. Connect the multi-use tester to the diagnosis connector in the fuse box.
3. Connect the power-source terminal of the multi-use tester to the cigarette lighter socket.
4. Turn ON the ignition switch.
5. Use the multi-use tester to make the various checks, and repair if a problem is found.

NOTE
If the malfunction indicator light (engine-check light) illuminates while the checks are being made, check the self-diagnosis output.

6. After repair, check once again to be sure the problem has been corrected.
7. Turn OFF the ignition switch.
8. Disconnect the multi-use tester.
9. Disconnect the battery terminals for 15 seconds or longer. This erases the self-diagnosis code.
10. Restart the engine. Check (by a driving test and other means) to be sure that the problem has been corrected.

Cranking Check

Item No.: Multi-use tester code number

Check points	Check conditions	Test specification	Probable cause of malfunction (or action)
Battery voltage o Service data o Item No.16	o Ignition switch: ON	11—13 V	o Low battery voltage o Power not supplied to the engine control unit 1) Check the power-supply circuit. 2) Check the ignition switch. 3) Check the control relay. o Malfunction of the engine control unit ground circuit

2.4L ENGINE—MPI DIAGNOSIS—SONATA

Check points	Check conditions		Test specification	Probable causes of malfunction (or action)
Malfunction code read out o Self diagnosis	o After holding for 15 seconds or longer with the ignition switch at "ON", move the ignition switch to "LOCK" and disconnect the ISC servo connector. o Crank for four seconds or longer. o Ignition switch: "ON" (Check for damage or disconnection of the injector, crank angle sensor circuit and other circuits.)		Normal	o Check in accordance with the diagnosis code. (Note that the diagnosis code will be erased if there is disconnection or damage of the engine control unit back-up power-supply circuit.) o If various diagnosis codes are output, the most frequent cause is damage or disconnection of the power-supply or ground circuit.
Fuel pump o Actuator test o Item No.7	Try under both conditions: o Engine cranking o Fuel pump forced activation	Pinch closed the return hose.	The pulsations of the fuel flow can be felt.	o Voltage is not supplied to the fuel pump. 1) Check the ignition switch (IG and ST) 2) Check the control relay. 3) Check the related circuits. o Malfunction of the fuel pump
		Listen close to the fuel tank.	The pump drive sound can be heard.	
Crank angle sensor o Service data o Item No.22	o Engine cranking o Tachometer connected (check, by using the tachometer for interruptions of the ignition coil primary current.)	Cranking speed (rpm)	Engine speed (rpm)	o If the tachometer reading is 0, there is no interruption of the ignition coil primary current. 1) Check the power transistor and the control circuit. 2) Check the ignition coil and the coil power supply circuit. o If the multi-use tester's indicated rpm is abnormal 1) Malfunction of the crank angle sensor circuit 2) Malfunction of the crank angle sensor 3) Malfunction of the timing belt
		Approx. 200	Approx. 200	

Check points	Check conditions		Test specification	Probable causes of malfunction (or action)
Injectors o Service data o Item No.41	o Engine cranking	Listen for operation sounds.	Operation sounds of the injectors can be heard.	o Injector malfunction o Improper contact of the connector and control relay contacts
		Engine coolant temperature [°C (°F)]	Injector activation time *2 (msec.)	o Malfunction of the engine coolant temperature sensor o Malfunction of the ignition switch—ST.
		0 (32)*1	Approx. 21	
		20 (68)	Approx. 43	
		80 (176)	Approx. 10	

NOTE

***1: When the engine coolant temperature is 0°C (32°F), injectors inject simultaneously at 4 cylinders.**

***2: Injector activation times are indicated at a battery voltage of 11V and a cranking speed of 250 rpm or lower.**

Sensor Check

Check points	Check conditions		Test specification	Probable cause of malfunction (or action)
Self-diagnosis output	o Engine: idling (2 minutes or more after engine start)		Normal	o Check in accordance with the diagnosis code. (Note that the diagnosis code will be erased if there is disconnection or damage of the engine control unit back-up power-supply circuit.) o If various diagnosis codes are output, the most frequent cause is damage or disconnection of the power-supply or earth circuit.
Oxygen sensor o Service data o Item No.11	o Engine warm-up (Make the mixture lean by engine speed reduction, and rich by racing.)	Engine condition	Voltage (mV)	o If the oxygen sensor output voltage is high during sudden deceleration 1) Check for injector leakage. 2) Check the oxygen sensor signal circuit. o If the oxygen sensor output voltage is low during engine racing 1) Check the oxygen sensor and signal circuit
		When sudden deceleration from 4,000 rpm	250 or lower	
		When engine is suddenly raced	600—1,000	
Air-flow sensor o Service data o Item No.12	o Engine warm-up	Engine condition	Frequency (Hz)	o If the air-flow sensor output frequency suddenly changes greatly, improper contact of the air-flow sensor or connector is probable. o If the output frequency of the air-flow sensor is unusually high or low, check the air cleaner element. o If the output frequency of the air-flow sensor is high, an increase of engine resistance or leakage of compression pressure is probable.
		750 rpm (Idling)	25—50	
		2,000 rpm	70—130	
		Racing	Increase caused by racing	
Air temperature sensor o Service data o Item No.13	o Ignition switch: ON, or engine running	Air temperature °C (°F)	Temperature °C (°F)	o Malfunction of air temperature or related circuitry
		-20 (-4)	-20 (-4)	
		0 (32)	0 (32)	
		20 (68)	20 (68)	
		40 (104)	40 (104)	
		80 (176)	80 (176)	
	o Ignition switch: ON	Warm by using hair dryer or other method.	Increases	
Throttle-position sensor o Service data o Item No.14	o Hold for 15 seconds or longer with the ignition switch at "ON".	Throttle valve	Voltage (mV)	o Throttle-position sensor maladjustment o Throttle-position sensor or related circuitry malfunction
		Idling position	480-520	
		Opens slowly.	Becomes higher in proportion to valve opening	
		Fully open	4,500—5,500	
Motor position sensor o Service data o Item No.15	o Engine: idling after warm-up (The idle position switch must be ON.)	Engine condition	Voltage (mV)	o If the voltage is low, check whether or not there is air intake. o If the voltage is high, the probable cause is: 1) Deposits adhered to the throttle valve 2) Increased engine resistance o If the voltage or idling speed is abnormal even though each part of the engine is normal, the probable cause is: 1) Improper adjustment of the idle speed control and/or the throttle position sensor 2) Malfunction of the motor position sensor or the related circuit
		Idle (750 rpm)	500—1,300	

Check points	Check conditions		Test specification	Probable cause of malfunction (or action)
Motor position sensor o Service data o Item No.15	(The compressor clutch must be activated when the air conditioner switch is switched ON.)	Air conditioner switch ON (900 rpm)	800—1,800	o If the engine speed does not increase when the air conditioner switch is switched from OFF to ON, check the DC motor drive circuit o Check the air conditioner system
		Air conditioner switch: ON Shift lever: "D" range (700 rpm)	900—1,900	o Check the inhibitor switch and the signal circuit
Crank signal o Service data o Item No.18	o Ignition switch: ON		OFF	o Ignition switch—ST signal circuit check o Ignition switch check
Engine coolant temperature sensor o Service data o Item No.21	o Ignition switch: ON, or engine running	Engine coolant temperature °C (°F)	Temperature °C (°F)	o Engine coolant temperature sensor or related circuitry malfunction
		-20 (-4)	-20 (-4)	
		0 (32)	0 (32)	
		20 (68)	20 (68)	
		40 (104)	40 (104)	
		80 (176)	80 (176)	
Crank angle sensor o Service data o Item No.22	o Engine: idling (Check with the ignition switch ON.)	Engine coolant temperature °C (°F)	Idling rpm	o If the idling speed suddenly becomes greater, a malfunction of the crank angle sensor or improper contact of the connector is probable
		-20 (-4)	1,500—1,700	
		0 (32)	1,350—1,550	
		20 (68)	1,200—1,400	
		40 (104)	900—1,100	
		80 (176)	650—850	

Check points	Check conditions		Test specification	Probable cause of malfunction (or action)
Barometric pressure sensor o Service data o Item No.25	o Ignition switch: ON	Altitude m (ft.)	Pressure mm Hg	o Barometric pressure sensor or related circuitry malfunction. (If the barometric pressure sensor pressure is low at high speed, restriction of the air cleaner element is probable.)
		0 (0)	760	
		600 (1,968)	710	
		1,200 (3,937)	660	
		1,800 (5,905)	610	
	o Engine: racing at 2,000 rpm	Gradually close the air-intake duct by using a hand.	Decreases.	
Idle switch o Service data o Item No.26	o Ignition switch: ON (Checking by using the accelerator pedal several times)	Throttle valve idling position	ON	o Idle position switch or related circuitry malfunction o Improper adjustment of the accelerator cable or cruise control cable.
		Open the throttle valve slightly.	OFF	o Malfunction of the setting of the fixed SAS.
Power-steering switch o Service data o Item No.27	o Engine: idling	Steering wheel neutral position (wheels straight-ahead direction)	OFF	o Power-steering oil-pressure switch or signal circuit malfunction
		Steering wheel half turn	ON	
Air-conditioner switch o Service data o Item No.28	o Engine: idling (The air conditioner compressor could be activated when the air-conditioner switch is ON.)	Air-conditioner switch "OFF"	OFF	o Check the air conditioner system.
		Air-conditioner switch "ON"	ON	
Inhibitor switch o Service data o Item No.29	o Ignition switch: ON (A/T models only)	Shift lever: "P" or "N"	"P", "N"	o Malfunction of the inhbititor switch or the signal circuit. o Improper adjustment of the control cable between the shift lever and the inhibitor switch.
		Shift lever: "D", "2", "L" or "R"	"D", "2", "L", "R"	

Check points	Check conditions		Test specification	Probable cause of malfunction (or action)
EGR temperature sensor (California) o Service data o Item No.43	o Engine: warmed up (Engine is maintained in a constant state for 2 minutes or more)	Engine condition	Temperature °C (°F)	o Check the EGR temperature sensor. o Check the EGR control system. o Check the EGR valve. o Check the EGR control solenoid valve. o Check the EGR control vacuum.
		750 rpm (idling)	100 (212) or less	
		o 3,500 rpm o Disconnect the vacuum hose (yellow stripe) from the A port nipple of the throttle body, and pinch the hose end with fingers.	150 (302) or more	
Injectors o Actuator test o Item No. 1—4	o Engine: idling after warm-up (Cut off the injectors in sequence during idling after engine warm-up; check the idling condition.)	Injector No.	Engine	o If the idling condition doesn't change, check the cylinder. 1) Check the injector operation sound. 2) Check the spark plug and high-tension cable. 3) Check the power transistor unit and control circuit.
		1	Idling condition changes more. (Becomes more unstable, or engine stalls.)	
		2		
		3		
		4		
Injector o Service data o Item No.41	o Engine: warmed up	Engine condition	Activation time (msec.)	o If the injector activation time is unusually long or short, there is a malfunction of the air-flow sensor, engine coolant temperature sensor, intake-air temperature sensor, or barometric pressure sensor. o If the injector activation time is long, increased engine resistance or leakage of compression pressure is probable.
		750 rpm (Idling)	2.5—4.0	
		2,000 rpm	2.5—4.0	
		Rapid racing	Increases.	
Ignition advance (power transistor) o Service data o Item No.44	o Engine: warmed up o Timing light: set (The timing light is set so as to check the actual ignition timing.)	Engine speed (rpm)	Ignition advance (°BTDC)	o If the ignition advance and actual ignition timing are different, adjust the ignition timing. [The ignition timing may fluctuate during idling, but this is not a problem. The advance is greater (approx. 5°) at high altitude.]
		750 (Idling)	5—15	
		2,000	30—40	

Check points	Check conditions		Test specifications	Probable cause of malfunction (or action)
Air conditioner relay o Service data o Item No.49	o Engine: idling after warm-up	Air conditioner switch	Air conditioner relay	o If the air conditioner relay output is abnormal, check the air conditioner signal input circuit and the air conditioner system. o If the activation of the air conditioner compressor clutch is not normal, check the compressor clutch and the relay circuit.
		OFF	OFF (compressor clutch non-activation)	
		ON	ON (compressor clutch activation)	
Purge control solenoid valve o Actuator test o Item No.8	o Ignition switch: ON (Engine stop)	Actuator forced actuation	Operation sound is audible	o Check the purge control solenoid valve. o Check the purge control solenoid valve drive circuit.
EGR control solenoid valve (California) o Actuator test o Item No.10	o Ignition switch: ON (Engine stop)	Actuator forced actuation	Operation sound is audible	o Check the EGR control solenoid valve. o Check the EGR control solenoid valve drive circuit.

3.0L ENGINE—MPI DIAGNOSIS—SONATA

Inspection Procedure (Using Voltmeter)

1. Connect the voltmeter to the self diagnosis terminal.
2. Turn the ignition switch, and the engine control unit memory contents will immediately start.
   After recording the abnormal item, check and repair each part according to the check items in "Diagnosis Chart".
3. When the defective parts have been repaired, disconnect the negative terminal of the battery cable for 15 seconds or more and connect it again to make sure that the trouble code has been erased.

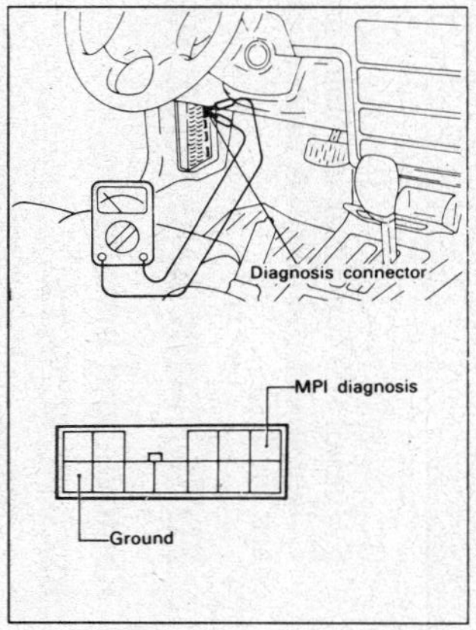

Diagnosis Chart

The 14 diagnosis items are provided as follows, and if plural items are activated, they are all indicated sequentially from the smallest code number.

Output preference order	Diagnosis item	Trouble code: Output signal pattern	Trouble code: No.	Trouble code: Memory	Check item (Remedy)
1	Engine control unit	H / L (31J062)	—	—	(Replace engine control unit)
2	Oxygen sensor	H / L (31J063)	11	Retained	o Harness and connector o Fuel pressure o Injectors (Replace if defective.) o Intake air leaks o Oxygen sensor
3	Air flow sensor	H / L (31Y064)	12	Retained	o Harness and connector (If harness and connector are normal, replace air flow sensor assembly.)

Output preference order	Diagnosis item	Malfunction code: Output signal pattern	Malfunction code: No.	Malfunction code: Memory	Check item (Remedy)
4	Air temperature sensor	H / L (31Y065)	13	Retained	o Harness and connector o Air temperature sensor
5	Throttle position sensor	H / L (31Y066)	14	Retained	o Harness and connector o Throttle position sensor o Idle position switch
6	Engine coolant temperature sensor	H / L (31Y068)	21	Retained	o Harness and connector o Engine coolant temperature sensor
7	Crank angle sensor	H / L (31Y069)	22	Retained	o Harness and connector If harness and connector are normal, replace crank angle sensor assembly.)
8	Top dead center sensor	H / L (31Y070)	23	Retained	o Harness and connector If harness and connector are normal, replace crank angle sensor assembly.)
9	Vehicle-speed sensor (reed switch)	H / L (31Y071)	24	Retained	o Harness and connector o Vehicle-speed sensor (reed switch)
10	Barometric pressure sensor	H / L (31Y072)	25	Retained	o Harness and connector If harness and connector are normal, replace barometric pressure sensor assembly.)

3.0L ENGINE—MPI DIAGNOSIS—SONATA

Output preference order	Diagnosis item	Trouble code			Check item (Remedy)
		Output signal pattern	No.	Memory	
11	Injector	 31Y073	41	Retained	o Harness and connector o Injector coil resistance
12	Fuel pump	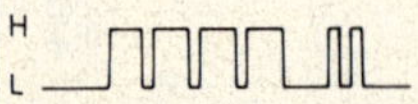 31Y074	42	Retained	o Harness and connector o Control relay
13	EGR*	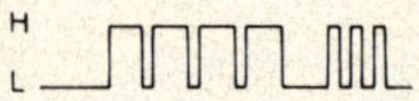 31Y075	43	Retained	o Harness and connector o EGR temperature sensor o EGR valve o EGR control solenoid valve o EGR valve control vacuum.
14	Normal state	 31Y076	—	—	—

NOTE
1. **Replace the engine control unit if a trouble code is read although the inspection reveals that there are no problems with the diagnosis item.**
2. **The diagnosis item marked* is applicable to California vehicles only.**

Components Replacement

RELIEVING SYSTEM PRESSURE

1. Disconnect the fuel pump harness at the rear of the tank.
2. Start engine and allow to idle.
3. After engine stalls, fuel pressure is reduced.
4. Disconnect the negative battery cable and reconnect the fuel pump wiring.
5. Remove the fuel tank cap.

NOTE: When disconnecting any fuel system hoses, always place a shop towel to prevent residual fuel from spaying.

FUEL FILTER

Removal and Installation

1. Release the fuel system pressure.
2. Turn the ignition switch **OFF**; then, disconnect the negative battery cable.

NOTE: Place a shop towel over fuel hose connections when removing, to prevent fuel spray.

3. Hold the fuel filter nuts tightly and remove the eye bolts.
4. Remove the fuel filter mounting bolts and remove the filter from the fuel clamp or bracket, as required.

To install:

5. Position the fuel filter into the retaining clamp or bracket and install the mounting bolts.
6. Install the fuel filter eye bolts and tighten.
7. Reconnect the negative battery cable, start the engine and check for leaks.

FUEL TANK AND FUEL PUMP ASSEMBLY

Removal and Installation

EXCEPT SONATA

1. Release the fuel system pressure.
2. Turn the ignition switch **OFF**; then, disconnect the negative battery cable.
3. Raise the vehicle and support it safely.
4. Remove the drain plug and drain the fuel.
5. Disconnect the fuel filler hose, leveling hose, vapor hose, fuel return hose and pressure hose from the tank.
6. Disconnect fuel sender and fuel pump electrical harness.
7. Loosen the self-locking nuts, which holds the tank in position and remove the tank retaining straps. Remove the fuel tank.
8. If the fuel pump is to be replaced, clean the area around the pump assembly. Remove the fuel pump retaining screws and remove the fuel pump assembly from the fuel tank.

To install:

9. If removed, install the fuel pump assembly into the fuel tank. Use a new gasket or O-ring, as required.
10. Place the fuel tank into position. Fit the retaining straps to the fuel tank and install the self-locking nuts.
11. Reconnect the electrical harness connector and fuel hoses. Tighten the high pressure hose flare nut to 23–30 ft. lbs. (32–42 Nm).

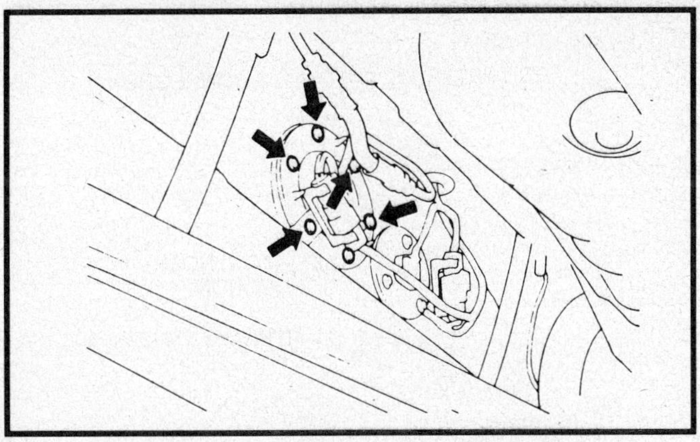

Fuel pump removal and installation—Sonata

12. Install the drain plug and tighten to 11–18 ft. lbs. (15–25 Nm).
13. Lower the vehicle and refill the fuel tank.
14. Reconnect the negative battery cable, start the engine and check for leaks.

FUEL PUMP

Removal and Installation

SONATA

1. Release the fuel system pressure.
2. Turn the ignition switch **OFF**; then, disconnect the negative battery cable.
3. Raise the vehicle and support it safely.
4. Disconnect the fuel pump connector.
5. Remove the fuel tank cap.
6. Drain the fuel.

NOTE: Place a shop towel over fuel hose connections when removing, to prevent fuel spray.

7. Disconnect the fuel high pressure hose.
8. Remove the fuel pump mounting screws and the bolt holding the fuel pump (at the lower side of the fuel tank).
9. Remove the fuel pump assembly from the fuel tank.

To install:

10. Align the 3 positioning projections of the packing with the holes in the fuel pump.
11. Place the fuel pump assembly into position and first install the holding bolt at the lower side of the fuel tank.
12. Install the fuel pump mounting screws.
13. Reconnect the fuel high pressure hose and fuel pump connector.
14. Lower the vehicle and refill the fuel tank.
15. Reconnect the negative battery cable, start the engine and check for leaks.

DELIVERY PIPE AND INJECTORS

Removal and Installation

1. Release the fuel system pressure.
2. Turn the ignition switch **OFF**; then, disconnect the negative battery cable.
3. Disconnect the wiring harness connectors from the injectors.

NOTE: Place a shop towel over fuel hose connections when removing, to prevent fuel spray.

4. Disconnect the high pressure fuel hose and return hose connections. Cover hoses immediately to prevent dirt from entering.
5. Disconnect the vacuum hose from the pressure regulator.
6. Remove the delivery pipe retaining bolts.

NOTE: When removing the delivery pipe, be careful not to drop any injectors. Also, be aware that fuel may flow out while removing.

7. Remove the delivery pipe with the fuel injectors and pressure regulator.
8. If removal of the injector(s) are necessary, separate the injector(s) from the delivery pipe.

To install:

9. If the injector(s) were removed:
   a. Install a new grommet(s) and O-ring(s) to the injector(s).
   b. Apply a coating of solvent (spindle oil or gasoline) to the O-ring(s) of the injector(s).
   c. Install the injector to the delivery pipe by twisting to the right and left. Make certain the injector turns smoothly.

NOTE: If the injector does not turn smoothly, the O-ring may be jammed; remove the injector and re-insert it into the delivery pipe.

10. Check that the insulators are properly installed in the delivery pipe assembly or on the intake manifold (2.4L engine). Install the delivery pipe to the engine. Install the retaining bolts and tighten to specifications.
11. Fit a new O-ring to the high pressure fuel hoses. Apply light oil or fuel to the O-ring, then insert. Be careful not to damage the O-ring during installation.
12. Reconnect the vacuum line to the pressure regulator.
13. Reconnect the wiring harness to the injectors.
14. Reconnect the negative battery cable, start the engine and check for leaks.

PRESSURE REGULATOR

Removal and Installation

1. Release the fuel system pressure.
2. Disconnect the negative battery cable.

NOTE: Place a shop towel over the pressure regulator when removing, to prevent fuel spray.

3. Disconnect the vacuum hose from the pressure regulator.
4. Remove the pressure regulator retaining bolts and remove the pressure regulator from the delivery pipe.

To install:

5. Fit a new O-ring to the pressure regulator. Apply light oil or fuel to the O-ring.
6. Install the pressure regulator to the delivery pipe. Be careful not to damage the O-ring during installation.
7. Install the pressure regulator retaining bolts. Tighten to specification.
8. Reconnect the vacuum line to the pressure regulator.9. Reconnect the negative battery cable, start the engine and check for leaks.

THROTTLE BODY ASSEMBLY

Removal and Installation

1. Disconnect the negative battery cable.
2. Drain the cooling system.
3. Disconnect the Throttle Position Sensor (TPS) electrical connector.

**NOTE: Do not remove the TPS, unless replacement is necessary.
Do not remove the throttle valve.
The ISC servo must not be removed. Do not disassemble the ISC servo.**

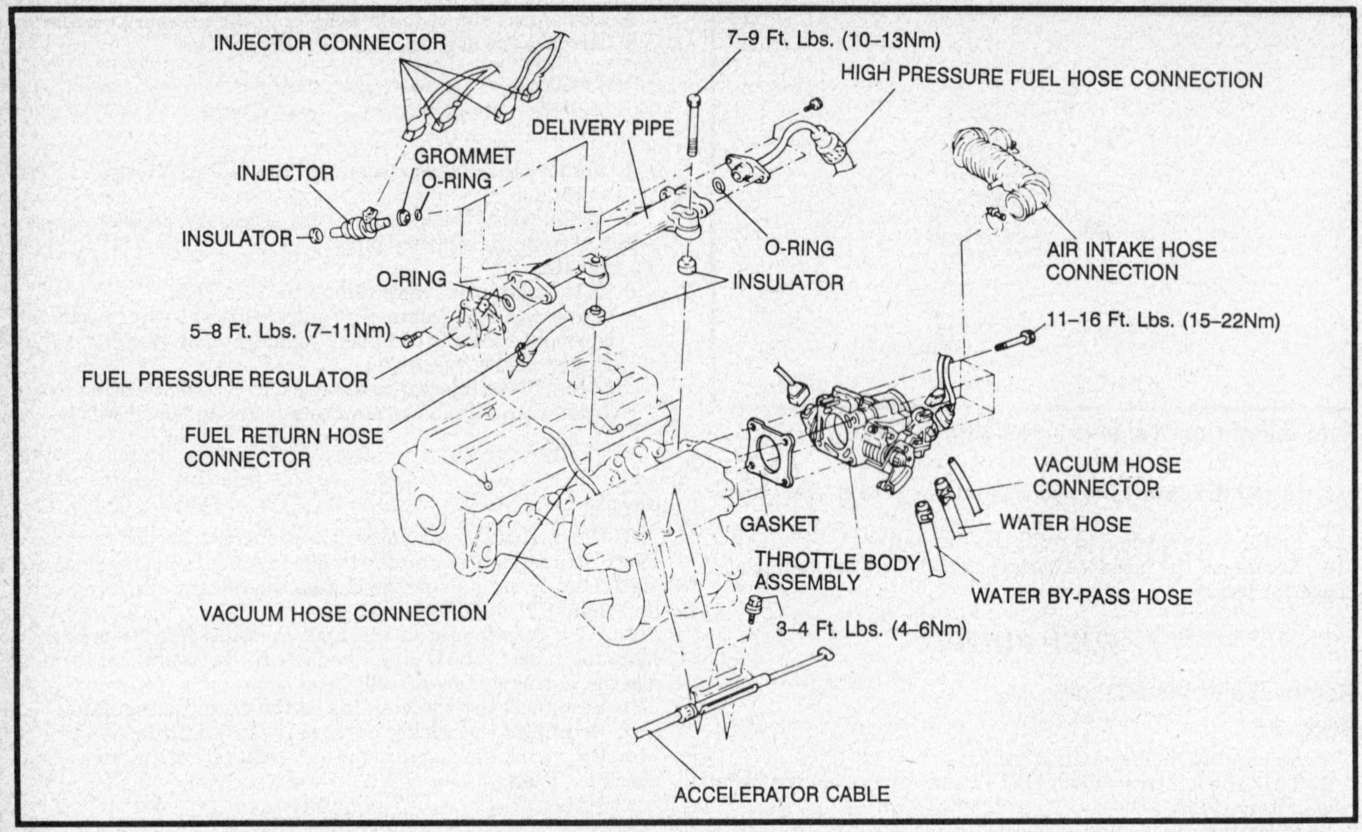

Injector and throttle body components – 1.5L engine

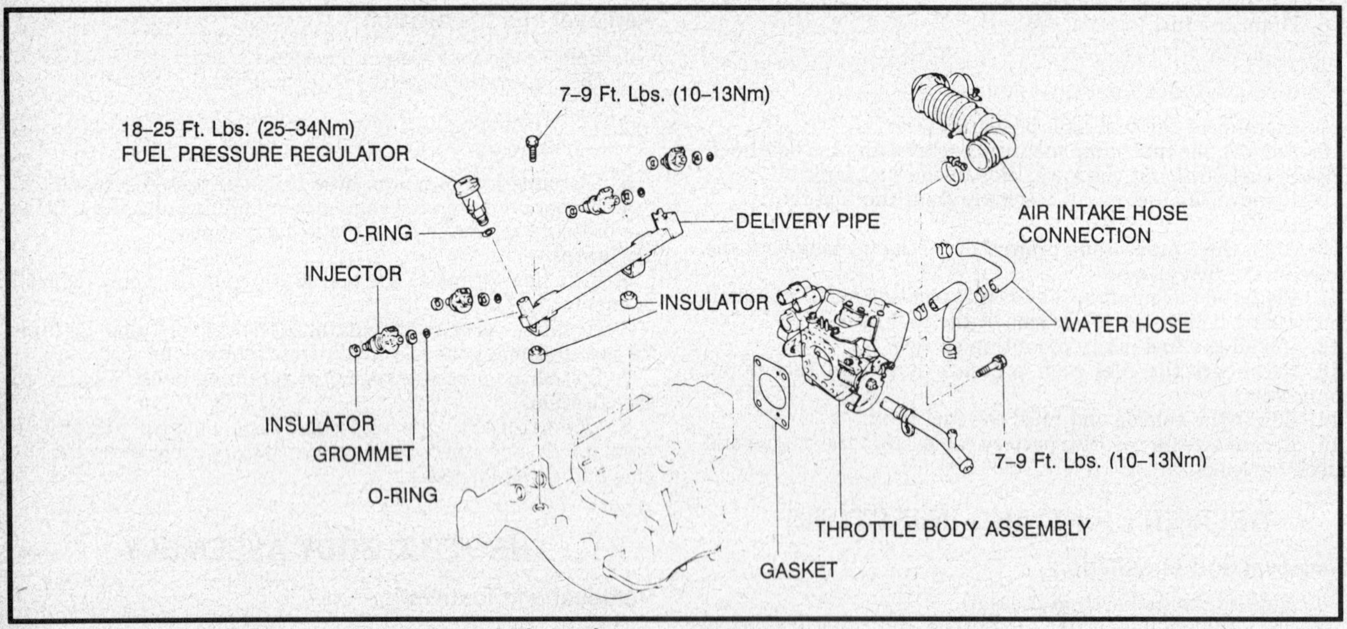

Injector and throttle body components – 2.4L engine

4. Remove the air intake hose and vacuum hose from the throttle body.
5. Remove the water hose(s) from the throttle body.
6. Remove the throttle body retaining bracket, is equipped.
7. Remove the throttle body retaining bolts and remove the throttle body assembly. Remove the throttle body gasket.

To install:

8. Clean the throttle body and housing mating surfaces.
9. Fit a new gasket to the throttle body and place the throttle body assembly into position. Install the mounting bolts and tighten to specifications.
10. Install the air intake hose and vacuum hose to the throttle body.

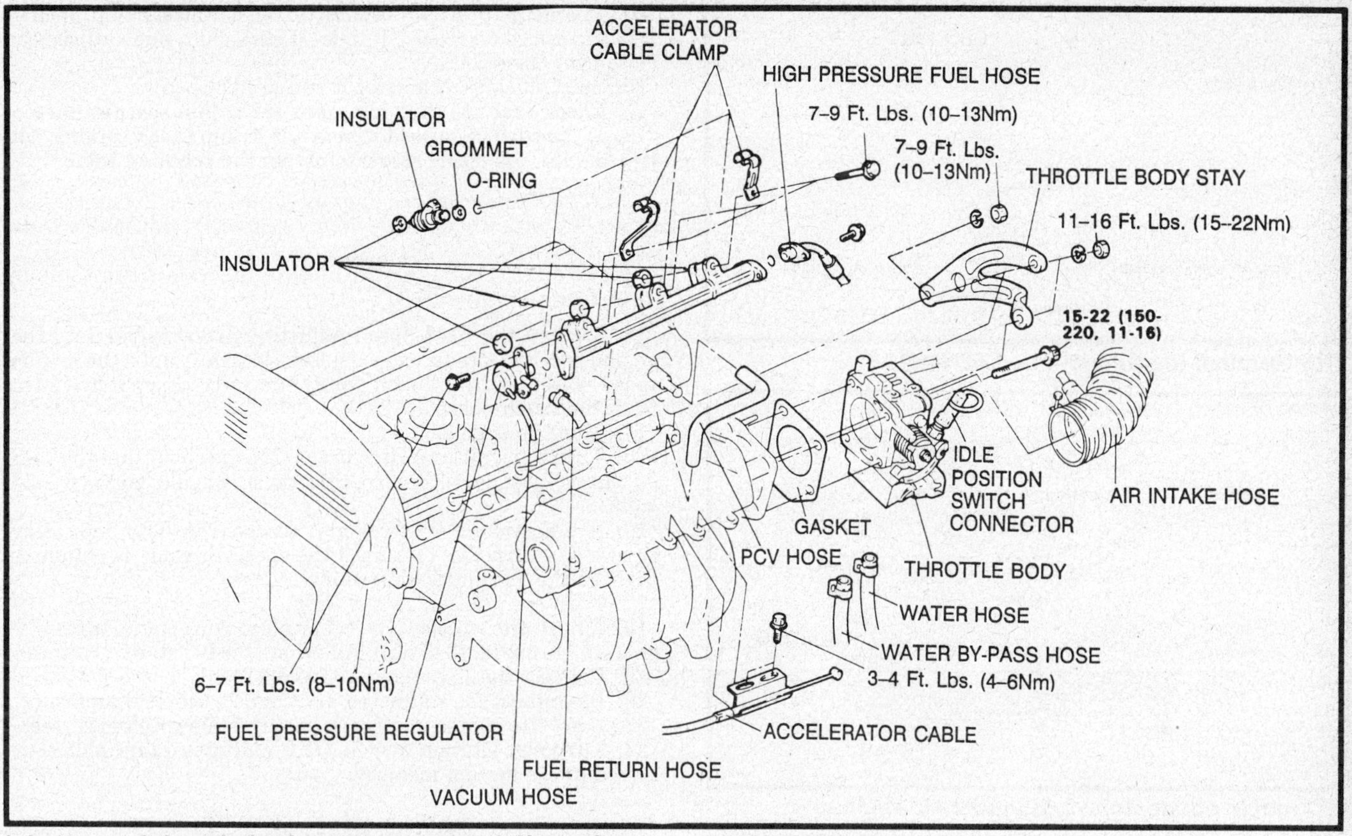

Injector and throttle body components – 1.6L engine

11. Install the water hose(s) to the throttle body.
12. Reconnect the TPS electrical connector.
13. Reconnect the negative battery cable.

Cleaning

1. Operate the engine until warm. Then, turn the engine **OFF**.
2. Disconnect the air intake hose from the throttle body and check the throttle valve surface for carbon deposits, rotating the valve.
3. On 2.0L engine, plug the bypass inlet of the throttle body.

NOTE: Make certain solvent does not enter the bypass passage.

4. Use a suitable carbon deposit spray and remove any carbon deposit, by spraying the solvent into the valve through the throttle body intake port. Let it sit for approximately 5 minutes.
5. Start the engine. Race it several times, then idle it for approximately 1 minute. Repeat this step, if necessary.
6. Attach the air intake hose.
7. On 2.0L and 3.0L engines, unplug the bypass inlet of the throttle body.
8. If necessary, perform the Idle Speed Control (ISC) and Throttle Position Sensor (TPS) adjustment on 1.5L and 2.4L engines.
9. If necessary, adjust the Basic Idle Speed on the 2.0L engine.

CURB IDLE RPM

Checking

1. Operate the engine until operating temperature is reached.
2. Turn OFF all accessories. Place the gear selector in **P** (automatic transaxle) or **N** (manual transaxle).
3. Set the steering wheel in straight ahead position. Apply the parking brake and block the drive wheels.
4. Check that the electric cooling fan is OFF.
5. Stop the engine and connect a tachometer and timing light. If available, the Multi-Use Tester (MUT) (Tool 09391-33200 or equivalent) may be used (Press Code 22 and read the idle rpm).
6. Operate the engine at curb idle.
7. Check basic ignition timing and adjust, if necessary. Timing should be 5–7 degrees BTDC.
8. Operate the engine for more than 5 seconds between 2000–3000 rpm.
9. Operate the engine at idle for 2 minutes; then check the curb idle rpm. Should be 600–700 rpm.

NOTE: Idle speed adjustment is usually unnecessary, since the ECU controls the idle speed.

10. If the curb idle rpm is not within specifications, use the following procedures:
    a. 1.5L and 2.4L Engines – go to Idle Speed Control and TPS system adjustment.
    b. 1.6L, 2.0L and 3.0L Engines – to to Basic Idle Speed Adjustment.

IDLE SPEED CONTROL (ISC) AND THROTTLE POSITION SENSOR (TPS)

Adjustment

1.5L AND 2.4L ENGINES

1. Operate the engine until operating temperature is reached.
2. Turn OFF all accessories. Place the gear selector in **P** (automatic transaxle) or **N** (manual transaxle).

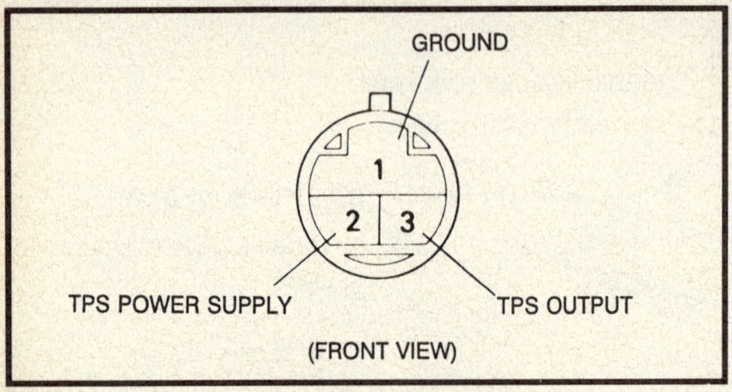

TPS terminal identification—1.5L engine

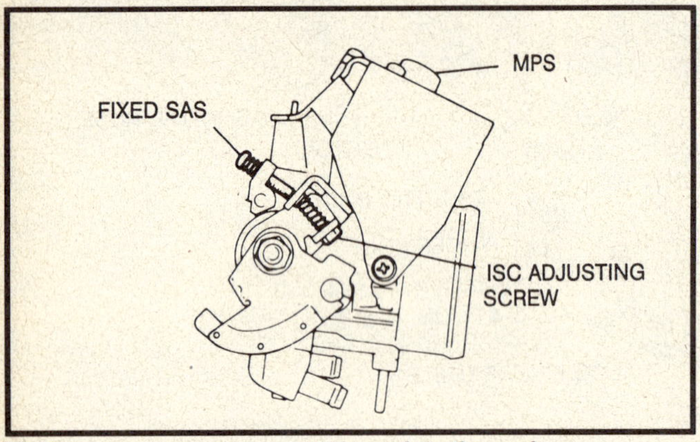

ISC motor connector—1.5L and 2.4L engines

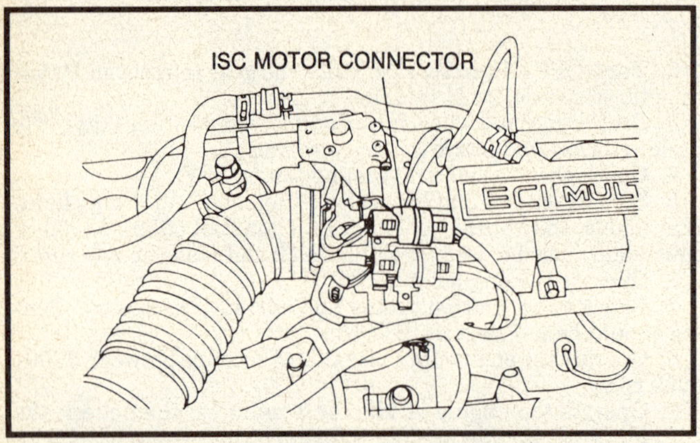

ISC adjusting screw location—1.5L engine

3. Set the steering wheel in straight ahead position. Apply the parking brake and block the drive wheels.
4. Loosen the accelerator cable.
5. Connect a digital type voltmeter between terminals 1 and 2 of the TPS. Insert the voltmeter probes at rear of the TPS connector. Do not disconnect the TPS connector from the throttle body.
6. Turn the ignition switch to the **ON** position for approximately 15 seconds; then check to be sure the idle speed control servo if fully retracted to the curb idle position. Do not start the engine.

NOTE: When the ignition switch is first turned ON, the ISC plunger extends to the fast idle position; after 15 seconds, it retracts to curb idle position.

7. Turn the ignition switch **OFF**.
8. Disconnect the ISC motor connector and secure the ISC motor at the fully retracted position.
9. In order to prevent the throttle valve from sticking, open it 2 or 3 times; then release it to let it click shut, and sufficiently loosen the fixed SAS.
10. Start the engine and let it run at idle speed.
11. Check that the curb idle rpms are within specifications.
12. If the curb idle speed rpm is not within specifications, adjust the ISC adjusting screw to obtain the specified value.

NOTE: The curb idle rpm on a new vehicle driven less than 300 miles (500 Km) may be 20–100 rpm lower than specifications, but no adjustment is necessary.
If the engine stalls or the engine speed is too low, clean the throttle valve area.

13. Tighten the fixed Speed Adjusting Screw (SAS) until the engine speed starts to increase; then loosen it until the engine speed ceases to drop (touch point) and then loosen a half turn from the touch point.
14. Stop the engine.
15. Turn the ignition switch to the **ON** and check that the TPS output voltage is within specifications. Should be 0.48–0.52 volts.
16. If adjustment is necessary, loosen the TPS mounting screws and turn the TPS until the specified value is obtained. Tighten the TPS mounting screws.
17. Turn the ignition switch **OFF**.
18. Adjust the accelerator cable play: Automatic transmission—0.12–0.2 in. (3–5mm). All others 0.04–0.08 in. (1–2mm).
19. Connect the ISC motor connector.
20. Disconnect the voltmeter, and connect the TPS connector.
21. Start the engine and check that the idle speed is correct.
22. Turn the ignition switch **OFF** and erase any fault code stored in the system memory.

BASIC IDLE SPEED

Adjustment

1.6L AND 2.0L ENGINES

1. Operate the engine until operating temperature is reached.
2. Turn OFF all accessories. Place the gear selector in **P** (automatic transaxle) or **N** (manual transaxle).
3. Set the steering wheel in straight ahead position. Apply the parking brake and block the drive wheels.
4. Loosen the accelerator cable.
5. Connect the Multi-Use Tester (MUT) to the diagnostic connector in the fuse box. If the MUT is not used, connect a tachometer to the engine and ground the self-diagnostic terminal.
6. Ground the ignition timing adjustment terminal.
7. Operate the engine for more than 5 seconds between 2000–3000 rpm.
8. Operate the engine at idle for 2 minutes; then check the curb idle rpm. Should be 700–800 rpm.
9. If the curb idle rpm is not within specifications, turn the Speed Adjusting Screw (SAS) until the specified value is obtained.

NOTE: If the idle speed is higher than specified, even with the SAS fully closed, check whether the idle switch (fixed SAS) moving mark exists or not. If it is found that the switch has moved, repair as required.

10. Turn the ignition switch **OFF**.
11. Disconnect the ground lead from the ignition timing adjusting terminal and remove the test equipments.
12. Operate the engine for approximately 10 minutes and check that the engine is in normal idling condition.

3.0L ENGINE

1. Operate the engine until operating temperature is reached.
2. Turn OFF all accessories. Place the gear selector in **P** (automatic transaxle) or **N** (manual transaxle).

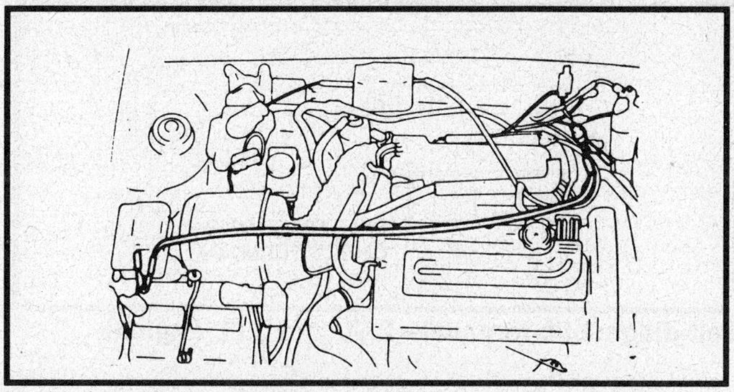

Ignition timing adjustment terminal – 1.6L engine

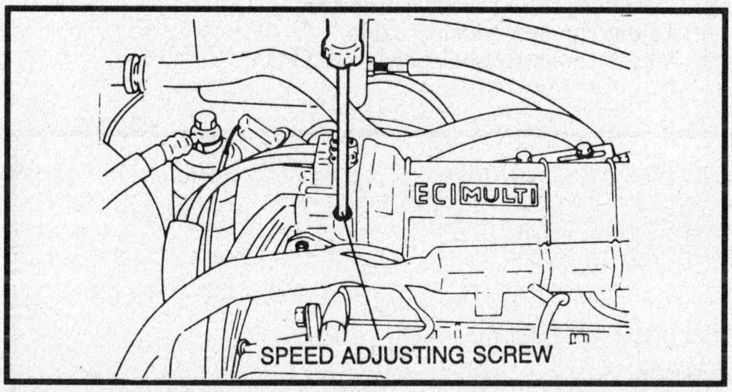

Speed Adjusting Screw (SAS) location – 1.6L and 2.0L engines

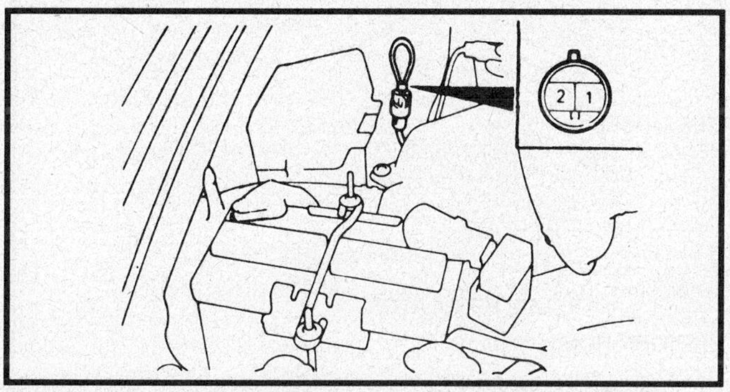

Ignition timing adjustment terminal – 2.0L and3.0L engines

Checking engine rpm, with tachometer – 3.0L engine

3. Set the steering wheel in straight ahead position. Apply the parking brake and block the drive wheels.
4. Loosen the accelerator cable.
5. Connect the Multi-Use Tester (MUT) to the diagnostic connector in the fuse box. If the MUT is not used, connect a tachometer to the engine and ground the ignition timing adjustment terminal.
6. Operate the engine at idle and check ignition timing. Timing should be 3–7 degrees BTDC.
7. Turn the ignition switch **OFF** and ground the self-diagnostic check terminal.
8. Operate the engine for more than 5 seconds between 2000–3000 rpm.
9. Operate the engine at idle for 2 minutes; then check the curb idle rpm. Should be 650–750 rpm.
10. If the curb idle rpm is not within specifications, turn the Speed Adjusting Screw (SAS) until the specified value is obtained.

NOTE: If the idle speed is higher than specified, even with the SAS fully closed, check whether the idle switch (fixed SAS) moving mark exists or not. If it is found that the switch has moved, repair as required.

11. Turn the ignition switch **OFF**.
12. Disconnect the ground lead from the ignition timing adjusting terminal and self-diagnosis check terminal. Remove the test equipments.
13. Clear the diagnostic memory.
14. Adjust the tension of the acceleration cable.
15. Operate the engine for approximately 10 minutes and check that the engine is in normal idling condition.

THROTTLE POSITION SENSOR (TPS)

Adjustment

2.0L ENGINE

1. Connect the MUT to the diagnostic connector in the fuse box. If the MUT is not used, connect a digital-type voltmeter, using special service tool (09351-33000 or equivalent).

NOTE: Do not disconnect the ECU connector from the ECU.

2. Turn the ignition switch **ON**. Do not start the engine.
3. Check the TPS output voltage. If the MUT is used, pressed code No. 14 and read the voltage. Voltage should be .480–.520.
4. If not within specifications, loosen the TPS mounting screws and adjust by turning the TPS. Tighten the TPS mounting screws.
5. Turn the ignition switch **OFF**.
6. Erase the ECU memory.

3.0L ENGINE

1. Loosen the acceleration cable tension.
2. Connect the MUT to the diagnostic connector in the fuse box. If the MUT is not used, connect a digital-type voltmeter between terminals 2 and 4 of the TPS.

NOTE: Do not disconnect the TPS.

3. Turn the ignition switch **ON**. Do not start the engine.
4. Check the TPS output voltage. Voltage should be .48–.52.
5. If not within specifications, loosen the TPS mounting screws and adjust by turning the TPS. Tighten the TPS mounting screws.
6. Turn the ignition switch **OFF**.
7. Start the engine and check the idle speed.
8. Erase the ECU memory.
9. Adjust the acceleration cable.

IDLE POSITION SWITCH (FIXED SAS)

Adjustment

2.0L AND 3.0L ENGINES

1. Loosen the acceleration cable tension.
2. Disconnect the idle switch (fixed SAS) harness connector.
3. Loosen the idle switch (fixed SAS) locknut.
4. Turn the idle switch counterclockwise until the throttle valve closes.
5. Connect an ohmmeter between the terminal of the switch and body.
6. Adjust the idle switch until continuity is indicated. Then turn the idle switch the amount of additional turns indicated:
   a. 2.0L engine—Then turn the idle switch and additional $^{15}/_{16}$ turns.
   b. 3.0L engine—Then turn the idle switch and additional 1¼ turns.

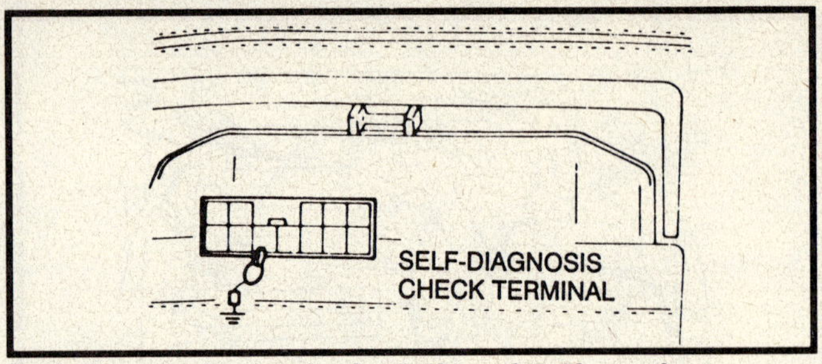

Self-diagnostic terminal—2.0L and 3.0L engines

7. Tighten the locknut, while keeping the switch from moving. Reconnect the harness connector.
8. Adjust the acceleration cable.
9. Adjust the curb idle speed and TPS.

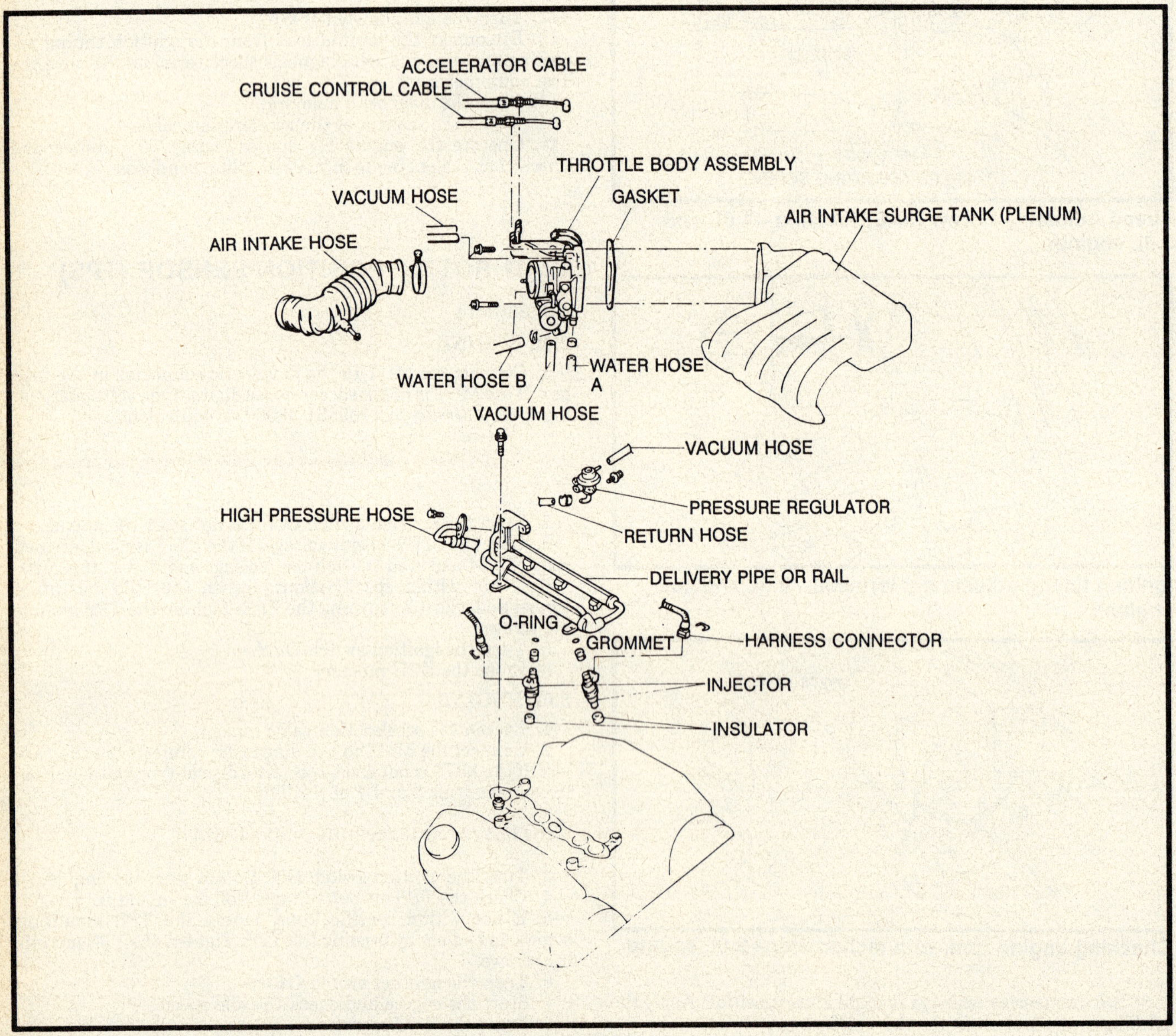

Sonata V-6 injector, throttle body and related components

Infiniti Fuel Injection

SECTION 10

INFINITI ELECTRONIC CONCENTRATED CONTROL (PORT FUEL INJECTION) SYSTEM (ECCS)

General Information

The Infiniti Electronic Concentrated Control System (ECCS) is an air flow controlled, sequential port fuel injection and engine control system. It is used on 1990–92 models equipped with 2.0L, 3.0L and 4.5L engines. The ECCS electronic control unit consists of a microcomputer, an inspection lamp, a diagnostic mode selector and connectors for signal input and output, powers and grounds.

The safety relay prevents electrical damage to the electronic control unit, or ECU, and the injectors in case the battery terminals are accidentally connected in reverse. The safety relay is built into the fuel pump control circuit.

Ignition timing is controlled in response to engine operating conditions. The optimum ignition timing in each driving condition is preprogrammed in the computer. The signal from the control unit is transmitted to the power transistor and this signal controls when the transistor turns the ignition coil primary circuit on and off (hence, the ignition timing). The idle speed is also controlled according to engine operating conditions, temperature and gear position. On manual transmission models, if battery voltage is less than 12 volts for a few seconds, a higher idle speed will be maintained by the control unit to improve charging function.

There is a fail-safe system built into the ECCS control unit. This system makes engine starting possible if a portion of the ECU's central processing unit circuit fails. Also, if a major component such as the crank angle sensor or the air flow meter were to malfunction, the ECU substitutes or borrows data to compensate for the fault. For example, if the output voltage of the air flow meter is extremely low, the ECU will substitute a preprogrammed value for the air flow meter signal and allow the vehicle to be driven as long as the engine speed is kept below 2000 rpm. Or, if the cylinder head temperature sensor circuit is open, the control unit clamps the warmup enrichment at a certain amount. This amount is almost the same as that when the cylinder head temperature is between 68–176°F (20–80°C).

If the fuel pump circuit malfunctions, the fuel pump relay comes on until the engine stops. This allows the fuel pump to receive power from the relay.

The electronic control unit controls the following functions:

- Injector pulse width
- Ignition timing
- Intake valve timing control (Q45)
- Air regulator control (G20)
- Exhaust gas recirculation (EGR) solenoid valve operation
- Exhaust gas sensor heater operation
- Idle speed
- FICD solenoid valve operation (G20 and M30)
- Fuel pump relay operation
- Fuel pump voltage (M30 and Q45)
- Fuel pressure regulator control (M30)
- AIV control (G20)
- Carbon canister control solenoid valve operation
- Air conditioner relay operation (During early wide-open throttle)
- Radiator fan operation (G20)
- Traction control system (TCS) operation (1991–92 Q45, if equipped)
- Self-diagnosis
- Fail-safe mode operation

DATA SENSORS

Crank Angle Sensor

The crank angle sensor is a basic component of the ECCS system. It monitors engine speed and piston position, as well as sending signals which the ECU uses to control fuel injection, ignition timing and other functions. The crank angle sensor has a rotor plate and a wave forming circuit. On all models, the rotor plate has 360 slits for 1 degree signals (crank angle), and 6 slits for 120 degree signal (engine speed).

The light emitting diodes (LED's) and photo diodes are built into the wave forming circuit. When the rotor plate passes the space between the LED and the photo diode, the slits of the rotor plate continually cut the light which is sent to the photo diode from the LED. This generates rough shaped pulses which are converted into ON/OFF pulses by the wave forming circuit and then sent to the ECU.

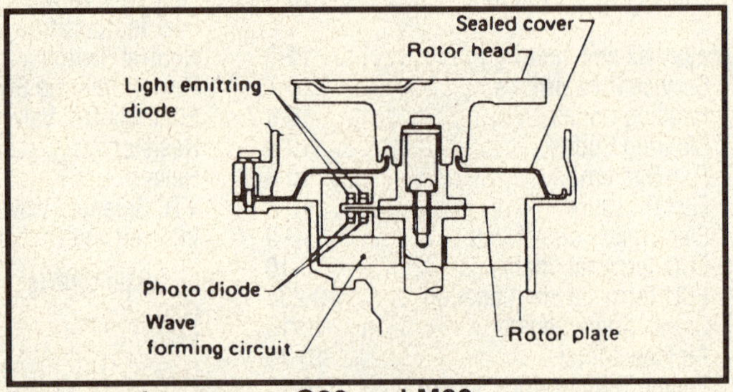

Crank angle sensor—G20 and M30

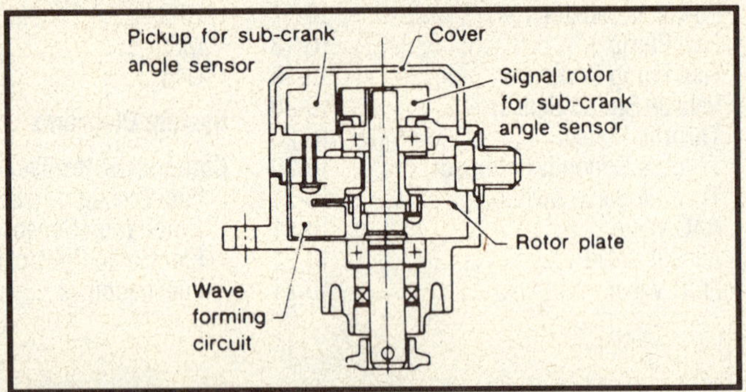

Crank angle sensor—Q45

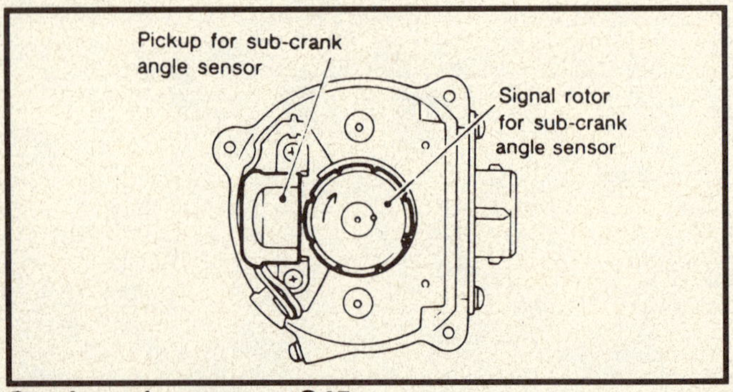

Crank angle sensor—Q45

Engine Temperature Sensor

The engine temperature sensor monitors changes in engine coolant temperature and transmits a signal to the ECU. The temperature sensing unit employs a thermistor which is sensitive to the change in temperature, with electrical resistance decreasing as temperature rises.

Air Flow Meter

The air flow meter measures the mass flow rate of intake air by taking a part of the entire flow entering the engine. Measurements are made in such a manner that the ECU receives electrical output signals that vary according to the amount of heat emitting from a hot wire (on G20 models) or a hot film (on M30 and Q45 models) placed in the stream of intake air.

When intake air flows through a route around the hot wire or film, some of the heat generated by the hot wire or film is taken away by the air. The amount of heat taken away depends on the amount of air flow. On the other hand , the temperature of the hot wire or film is automatically controlled to a certain number of degrees.

Thus, it is necessary to increase the amount of electric current to maintain the temperature of the wire or film. The ECU determines the air flow according to this current change.

Exhaust Gas Sensor

The exhaust gas sensor, which is placed in the exhaust tube, monitors the amount of oxygen in the exhaust gas. The sensor is made of ceramic titania which drastically changes electrical resistance at the ideal air/fuel ratio (14.7:1). The control unit supplies the sensor with approximately 1 volt and measures the output voltage of the sensor depending on its resistance. The oxygen sensor is equipped with a heater to bring it to operating temperature more quickly.

Throttle Sensor and Soft/Hard Idle Switch

The throttle sensor responds to the accelerator pedal movement. This sensor is a kind of potentiometer which transforms the throttle valve position into output voltage, and emits the voltage signal to the ECU In addition, the sensor detects the opening and closing speed of the throttle valve and feeds the voltage signal to the ECU.

Idle position of the throttle valve is determined by the ECU receiving the signal from the throttle sensor. This system is called "soft idle switch." This one controls engine operation such as fuel cut. The M30 and Q45 are also equipped with a system called "hard idle switch." It is built in the throttle sensor

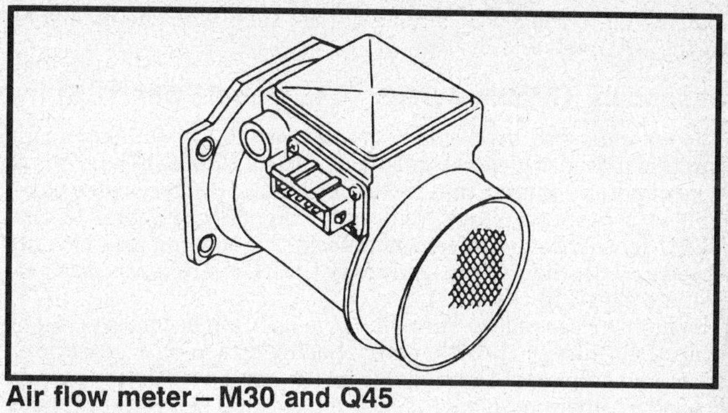

Air flow meter—M30 and Q45

unit and is used for engine control when the soft idle switch is malfunctioning on the M30, and for self-diagnosis on the Q45.

Detonation Sensor

The detonation sensor is attached to the cylinder block and senses engine knocking conditions. A knocking vibration from the cylinder block is applied as pressure to the piezoelectric element. This vibrational pressure is then converted into a voltage signal which is delivered as output. The Q45 is equipped with two of these sensors.

Vehicle Speed Sensor

The vehicle speed sensor provides a vehicle speed signal to the ECU. The speed sensor consists of a reed switch, which is installed in the speedometer unit and transforms vehicle speed into a pulse signal.

Exhaust Gas Temperature Sensor

The exhaust gas temperature sensor, which is found on California models, monitors exhaust gas temperature and transmits a signal to the ECU. The temperature sensing unit employs a thermistor which is sensitive to the change in temperature. Electric resistance of the thermistor decreases in response to the temperature rise.

Power Steering Oil Pressure Switch

The G20 and Q45 are equipped with a power steering oil pressure switch. This switch is attached to the power steering high-pressure tube and detects the power steering load, sending the

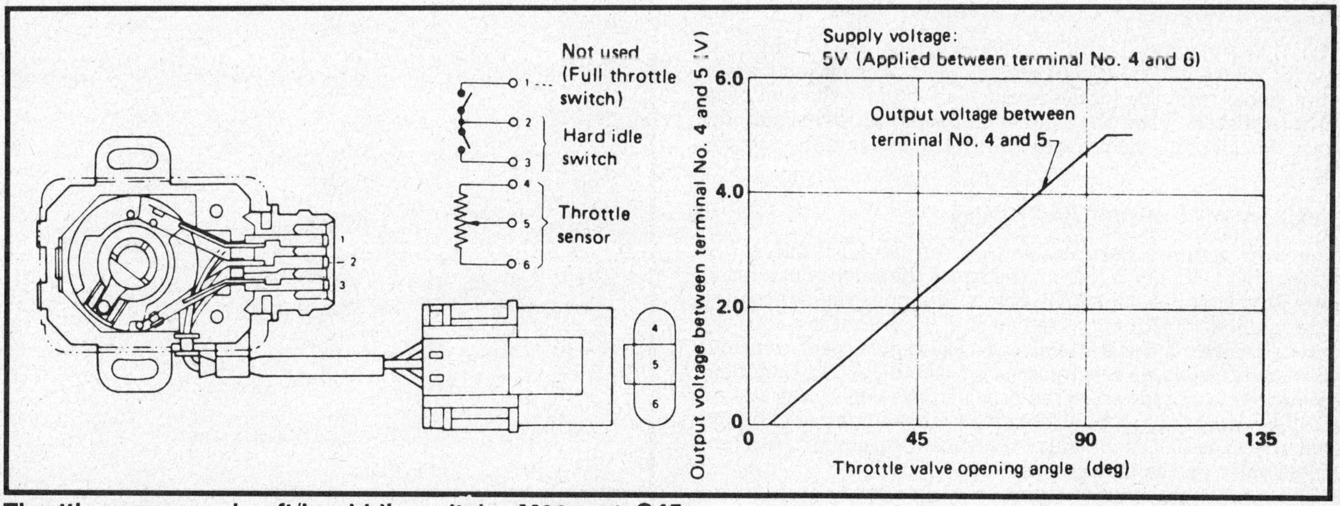

Throttle sensor and soft/hard idle switch—M30 and Q45

load signal to the ECU. The ECU then sends the idle-up signal to the AAC valve.

Secondary Throttle Sensor (Q45 Models with TCS)

The secondary throttle sensor responds to the movement of the throttle motor, which is controlled by the TCM. This sensor is a type of potentiometer that transforms the secondary valve position into output voltage, and emits the voltage signal to the TCM. In addition, the sensor detects the opening and closing speed and position of the secondary throttle valve and feeds the voltage signal to the TCM.

When the secondary throttle valve opening becomes smaller than the ordinary throttle valve opening due to TCS operation, then, and only then, the signal from the secondary throttle valve is used for engine control in place of the signal from the ordinary throttle sensor. The signal of the secondary throttle valve first enters the TCM, from where it is sent to the ECU.

CAUTION

Before touching the secondary throttle valve, be sure to disconnect the throttle motor connector; otherwise, personal injury may occur due to accidental actuation of the valve.

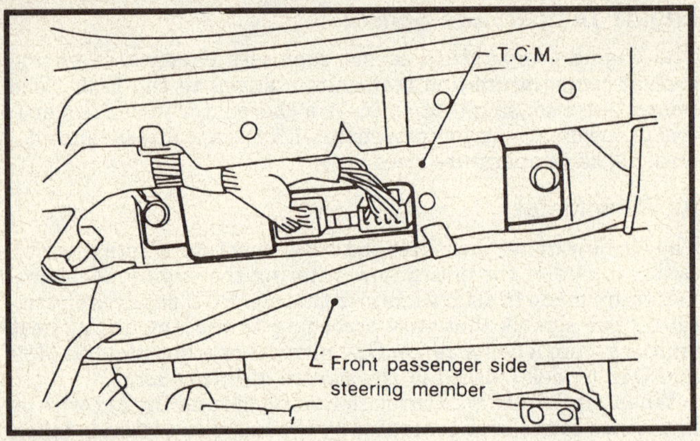

TCM location—Q45

CONTROLLED OUTPUTS

Power Transistor and Ignition Coil

The ignition signal from the ECU is amplified by a single power transistor on the G20 and M30, and by two transistors on the Q45, all of which turn the ignition coil primary circuit on and off, inducing the necessary high voltage in the secondary circuit to fire the spark plugs. This ignition signal is affected by a number of inputs:

1. Crank angle sensor (engine speed and piston position)
2. Air flow meter
3. Engine temperature sensor
4. Detonation sensor
5. Throttle sensor
6. Throttle valve switch (M30 and Q45)
7. Power steering oil pressure switch (Q45)
8. Neutral inhibitor switch (G20)
9. Inhibitor switch AT control unit (Q45)
10. Vehicle speed sensor
11. Ignition switch (start signal)
12. Air conditioner switch (Q45)
13. Battery (voltage)
14. Secondary throttle sensor (if the Q45 has traction control)

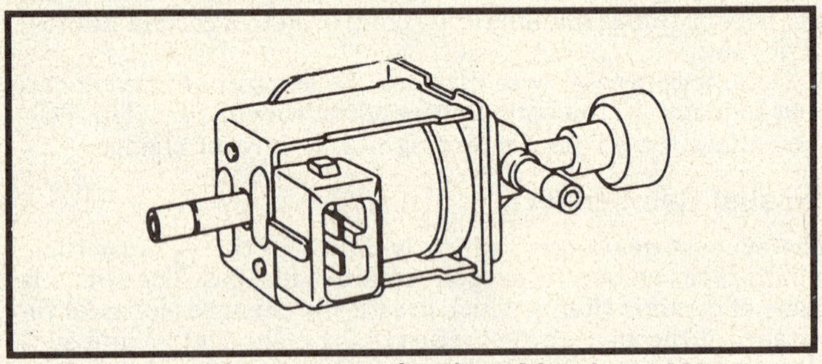
Pressure regulator control solenoid valve—M30

Pressure Regulator Control Solenoid Valve

The M30 is equipped with a solenoid valves, one of which responds to an ON/OFF signal from the ECU. When it is off, a vacuum signal from the intake manifold is fed into the fuel pressure regulator. When the control unit sends an ON signal, the solenoid coil pulls the plunger downward and cuts the vacuum signal.

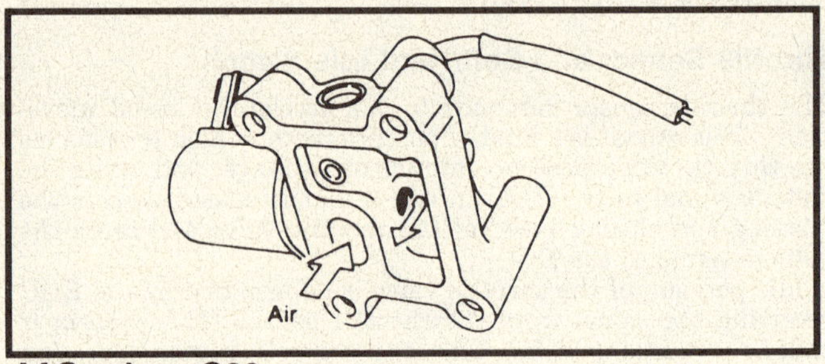

AAC valve—G20

Auxiliary Air Control (AAC) Valve

The ECU actuates the AAC valve by an ON/OFF pulse. The longer that ON duty is left on, the larger the amount of air that will flow through the AAC valve. Idle speed is controlled through fine adjustment of the amount of air that bypasses the throttle valve via the AAC valve so that engine speed coincides with the target value programmed into the ECU. The target engine speed is the lowest speed at which the engine can opeate steadily, depending on conditions. There are a number of inputs that affect the ECU's control of the AAC valve, and the FICD solenoid valve on the G20 and M30:

1. Crank angle sensor (engine speed and piston position)
2. Air flow meter (Q45)

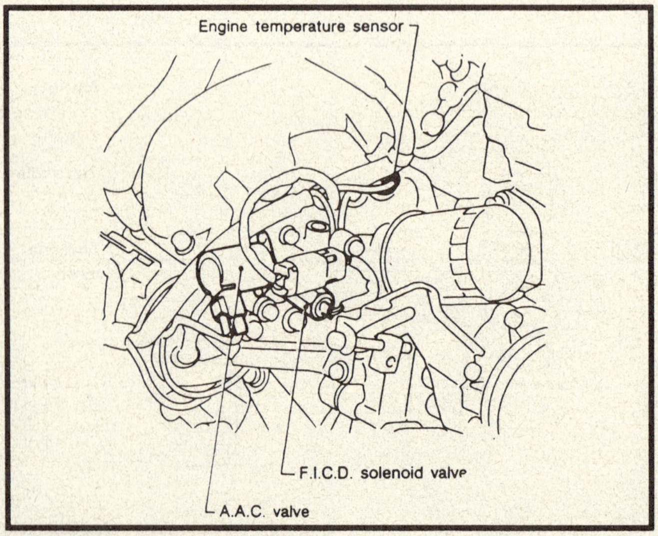

AAC valve location—G20

3. Engine temperature sensor
4. Power steering oil pressure switch (Q45)
5. Throttle sensor
6. Throttle valve switch (M30 and Q45)
7. Neutral inhibitor switch (G20)
8. AT control unit (M30; neutral position)
9. Inhibitor switch AT control unit (Q45)
10. Vehicle speed sensor
11. Ignition switch (start signal)
12. Air conditioner switch (Q45)
13. Battery (voltage)

Air Regulator

The G20 and M30 have an air regulator that provides an air bypass when the engine is cold for a fast idle during warm-up. The air regulator is controlled by the ECU at the same time as fuel pump **ON/OFF** control.

A bimetal heater and rotary shutter are built into the air regulator. When the bimetal temperature is low, the air bypass port opens. As the engine starts and electric current flows through a heater, the bimetal begins to turn the shutter to close the bypass port. The air passage remains closed until the engine stops and the bimetal temperature drops.

Fast Idle Cam (FIC)

The Q45 is equipped with an FIC that is installed on the throttle chamber to maintain adequate engine speed while the engine is cold. It is operated by a volumetric change in wax located inside the thermo-element. The thermo-element is controlled by engine coolant temperature.

Air Cut Valve

Q45 models made in 1991–92 that are equipped with traction control have an air cut valve. The function is the same as for the FIC. This valve is used to adjust the quantity of air that bypasses the secondary throttle valve. The air cut valve is operated by a volumetric change in wax located inside the thermo-element. The thermo-element is controlled by engine coolant temperature.

Accordingly, the quantity of bypass air is larger before engine warmup than after warmup.

Exhaust Gas Recirculation and Canister Control Solenoid Valve

The EGR systems and canister systems (G20 and Q45) are controlled by the ECU. Under certain conditions these systems are prevented from operating, since it would adversely affect the performance of the vehicle. These conditions are:

1. Low engine temperature.
2. Engine starting, idling and engine stopped.
3. High-speed engine operation.
4. Excessively high engine temperature.
5. Air flow meter malfunction (G20).
6. Vehicle speed below 12 MPH (20 km/h).

Backpressure Transducer (BPT) Valve

The G20 and Q45 are equipped with a BPT valve. This valve monitors exhaust pressure to activate the diaphragm, controlling throttle chamber vacuum applied to the EGR control valve. In other words, recirculated exhaust gas is controlled in response to positioning of the EGR control valve or to engine operation.

Idle Air Adjusting (IAA) Unit

The IAA unit, found on the G20 and M30, consists of the AAC valve, Fast Idle Control Device (FICD) solenoid valve and an idle adjusting screw. It receives signals from the ECU and controls the idle speed.

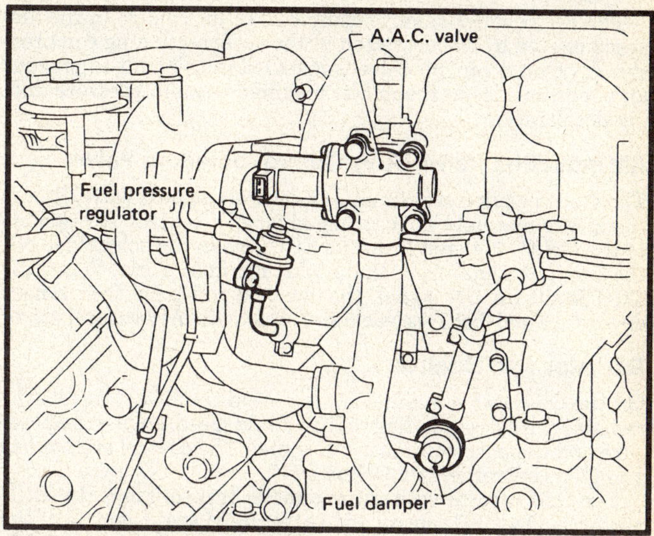

AAC valve location—Q45

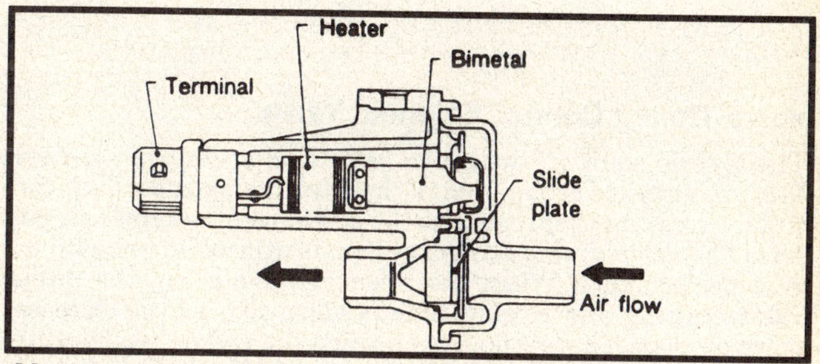

Air regulator—G20

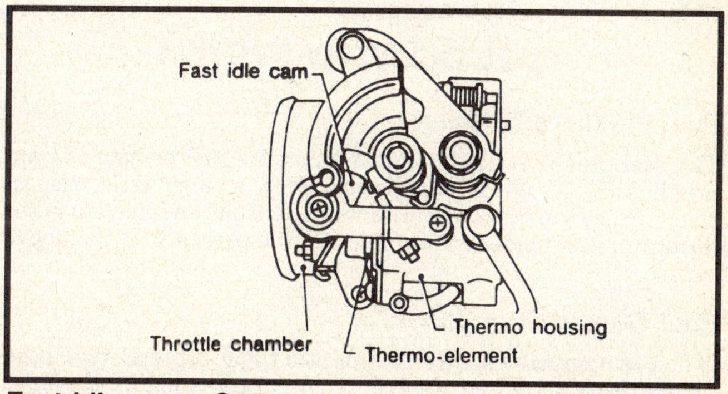

Fast idle cam—Q45

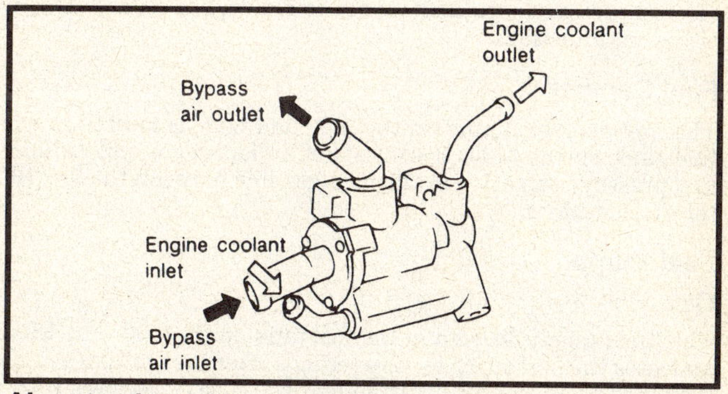

Air cut valve—Q45 with TCS

The FICD solenoid valve compensates for change in the idle speed caused by the operation of the air conditioning compressor. A vacuum control valve is installed in this unit to prevent an abnormal rise in the intake manifold vacuum pressure during deceleration.

Air Induction Valve (AIV) Control Solenoid Valve

The G20 is equipped with an AIV control solenoid valve. It cuts the intake manifold vacuum signal for AIV control. It responds to the ON/OFF signal from the ECU. When the solenoid is off, the vacuum signal from the intake manifold is cut. When the ECU sends an ON signal, the solenoid coil pulls the plunger downward and feeds the vacuum signal to the AIV control valve.

Radiator Fan Control

On the G20 with a manual transmission the ECU controls the radiator fan corresponding to the vehicle speed, engine temperature and the air conditioner ON signal. The control system has two-step control (either ON or OFF).

For G20 models with an automatic transmission the ECU performs ON/OFF control and LOW/HIGH speed control of the radiator fans. The factors involved are the same as for models that are equipped with manual transmissions.

NOTE: The radiator fan operates at HIGH if the ECU detects a fault in the engine temperature system.

Valve Timing Control Solenoid Valve

The Q45 is equipped with two valve timing control solenoids that are installed at the front of the intake camshafts. They control oil pressure that regulates the position of the camshaft pulleys. The valve timing control system is utilized to increase engine performance. When the solenoid valve is on, the intake valve opening and closing time is advanced, which increases valve overlap. The solenoid is on when the engine temperature is warm (not excessively hot), engine load is high, the inhibitor switch is off, the engine speed is below 4,600 rpm and is not at idle.

FUEL DELIVERY SYSTEM

Fuel Pressure Regulator

The fuel pressure regulator maintains the fuel pressure at approximately 34–43 psi, at idle and wide open throttle, respectively. Since the injected fuel amount depends on injection pulse duration, it is necessary to maintain the pressure at the above values.

Fuel Temperature Sensor

A fuel temperature sensor is built into the fuel pressure regulator on the M30. When the fuel temperature is higher than the preprogrammed level, the ECU will enrich the fuel injected to compensate for temperature expansion. The temperature sensor and pressure regulator should be replaced as an assembly if either malfunctions.

Fuel Damper

The fuel damper, found on the M30 and Q45, is located in the fuel tank on the M30. It consists of a diaphragm and reduces fuel pressure pulsation in the fuel feed line between the fuel filter and injectors.

Fuel Pump

G20

The fuel pump is located in the fuel tank on the G20. The ECU activates the fuel pump for five seconds after the ignition switch is turned on to improve startability. If the ECU receives a 180 degree signal from the crank angle sensor, it knows that the engine is rotating and energizes the ON/OFF fuel pump relay, which in turn controls the fuel pump. The ECU does not directly drive the fuel pump on the G20. If the engine stalls or is turned off, the ECU stops the fuel pump operation within one second, thereby improving safety, extending the life of the pump and preventing battery discharge.

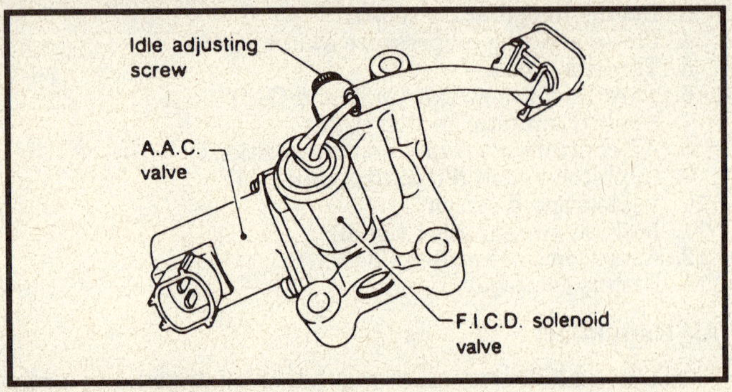

IAA unit—G20

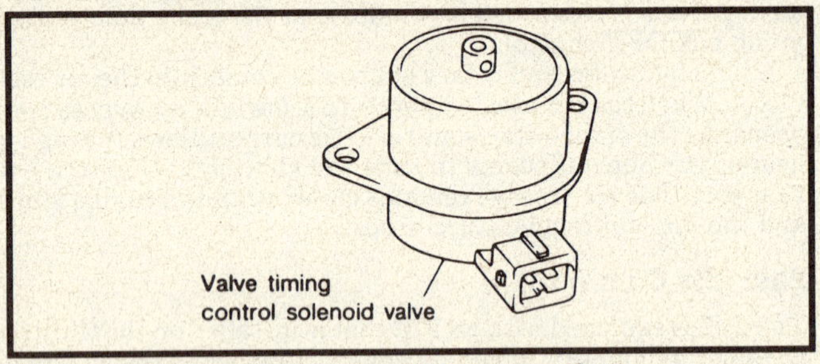

Valve timing control solenoid valve—Q45

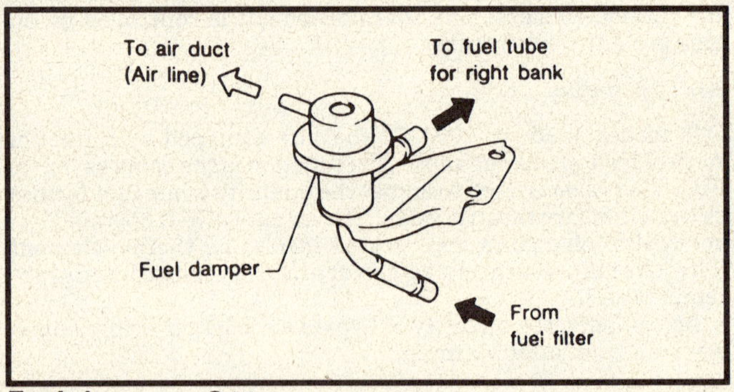

Fuel damper—Q45

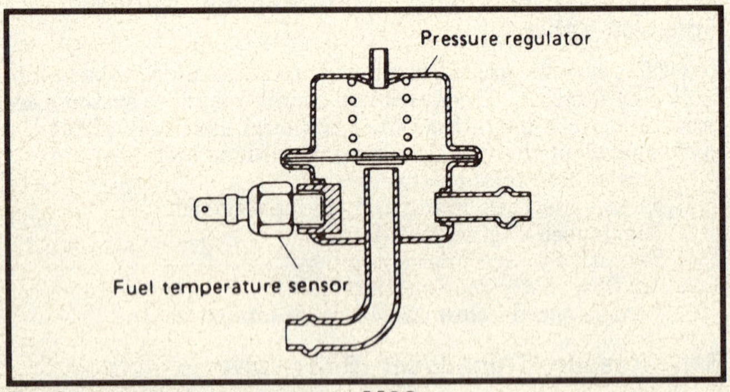

Fuel temperature sensor—M30

M30

The fuel pump is located in the fuel tank on the M30. The ground side of the fuel pump motor can be controlled three ways. It can be grounded through the relay when the ECU internally grounds pin 18, or it can be grounded directly by the ECU via pin 104, or the ECU can transmit a 20 Hz pulse signal to turn ON/OFF the ground at pin 104, resulting in a varying voltage that depends on operating conditions.

In the first case the ECU energizes the relay by grounding terminal 18 for five seconds when the ignition switch is first turned on. When the engine temperature is above 212°F (100°C) and the engine is being started, the relay will be energized as long as the engine is being cranked. After it is started with the engine temperature still above 212°F (100°C), the relay will stay energized for 30 seconds. When the engine stalls or is turned off, the relay is de-energized.

In the second case the ECU activates the fuel pump by applying a 100% ON time pulse, grounding terminal 104 for five seconds after the ignition switch is turned on when the engine is cold, when starting a cold engine, or when the engine is at idle and above 212°F (100°C). If the engine stalls or is stopped, the ON time becomes zero.

In the third case the resultant voltage at pin 104 is 0.1–0.5 volts under the following conditions:

1. Five seconds after the ignition switch is turned to **ON**.
2. Warm-engine cranking.
3. 30 seconds after starting (above 122°F or 50°C).
4. Engine temperature above 212°F (100°C), except at idle.
5. Engine temperature below 50°F (10°C).

In cases other than just listed, the resultant voltage will vary between 9–14 volts at pin 104.

Q45

The fuel pump is located in the fuel tank on the Q45. The ECU activates the fuel pump for five seconds after the ignition switch is turned on to improve startability. If the ECU receives a 1° signal from the crank angle sensor, it knows that the engine is rotating and energizes the ON/OFF fuel pump relay, which in turn controls the fuel pump. The ECU does not directly drive the fuel pump on the Q45. If the engine stalls or is turned off, the ECU stops the fuel pump operation within 1.5 seconds, thereby improving safety, extending the life of the pump and preventing battery discharge.

The fuel pump voltage can be controlled to three levels, however, and this is done by the fuel pump control unit. When the amount of fuel flow needed is high, the supplied voltage is 11–14 volts (battery voltage). It is needed during engine cranking, engine temperature below 32°F (0°C) or when the engine is running under a heavy load and high speed conditions. When the fuel need is moderate, the voltage supplied is approximately 7.8 volts. Under low fuel need conditions the voltage is approximately 5.6 volts.

Fuel Injector

The fuel injector is a small, precision solenoid valve. As the ECU sends an injection signal to each injector, the coil built into the injector pulls the needle valve back and fuel is injected through the nozzle and into the intake manifold. The amount of fuel injected is dependent on injection pulse duration and this duration depends on many input factors.

1. Crank angle sensor (engine speed and piston position)
2. Air flow meter
3. Engine temperature sensor
4. Exhaust gas sensor
5. Throttle sensor
6. Throttle valve switch (M30 and Q45)
7. Neutral inhibitor switch (G20)
8. AT control unit (M30)
9. Inhibitor switch AT control unit (Q45)
10. Vehicle speed sensor
11. Ignition switch (start signal)
12. Air conditioner switch (G20 and Q45)
13. Battery (voltage)
14. Secondary throttle sensor (if the Q45 has traction control)

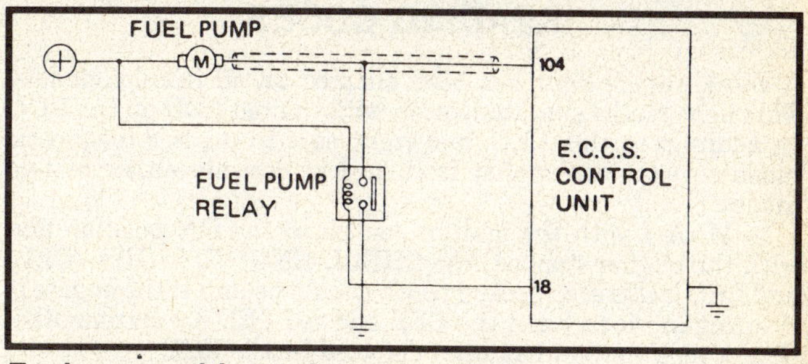

Fuel pump wiring schematic – M30

Diagnosis and Testing

SERVICE PRECAUTIONS

- Fuel system pressure must be relieved before disconnecting any fuel lines or attempting to remove any fuel system components.
- Do not operate the fuel pump when the fuel lines are empty.
- Do not reuse fuel hose clamps and tighten the new ones to the specified torque.
- Do not disconnect the injector harness connectors with the engine running.
- Do not apply battery power directly to the injectors.
- Always check for leaks after servicing the fuel system.
- Do not disconnect the ECCS harness connectors before the battery ground cable has been disconnected.
- Make sure all ECCS connectors are fastened securely. A poor connection can cause an extremely high surge voltage in the coil and condenser and result in damage to integrated circuits.
- When testing the ECU with a DVOM make sure that the probes of the tester never touch each other as this will result in damage to a transistor in the ECU.
- Keep the ECCS harness at least 4 in. away from adjacent harnesses to prevent an ECCS system malfunction due to external electronic "noise."
- Keep all parts and harnesses dry during service.
- Before attempting to remove any parts, turn **OFF** the ignition switch and disconnect the battery ground cable.
- Always use a 12 volt battery as a power source.
- Do not attempt to disconnect the battery cables with the engine running or the ignition key **ON**.
- If a battery terminal is disconnected (with the ignition key **OFF**), the memory will return to the ROM value. The ECCS will now self-control at its initial value. The engine operation can vary slightly when the terminal is disconnected. However, this is not an indication of a problem. Do not replace parts because of a slight variation.
- Do not clean the air flow meter with any type of detergent.
- Even a slight leak in the air intake system can cause serious problems.
- Do not depress the accelerator pedal when starting.
- Do not rev up the engine immediately after starting or just prior to shutdown.
- Do not attempt to disassemble the ECCS control unit under any circumstances.
- If installing a 2-way or CB radio, keep the antenna as far as possible away from the electronic control unit. Keep the antenna feeder line at least 8 in. away from the ECCS harness and do not let the 2 run parallel for a long distance. Be sure to ground the radio to the vehicle body.

READING CODES

A check engine light has been adopted on all Infiniti models. This light blinks simultaneously with the red LED on the ECU In addition to this LED, and right next to it , is a diagnostic mode selector screw that is turned to activate either of two modes.

In Mode I with the ignition switch in the **ON** position and with the engine stopped, the CHECK ENGINE LIGHT (CEL) and LED bulbs are checked (Both should be on.) If the engine is running in Mode I and the CEL and red LED are illuminated, then a malfunction has been detected by the ECU.

In Mode II with the ignition switch in the **ON** position and with the engine stopped, malfunction codes are indicated by the number of flashes from the red LED or the CEL When the engine is running in Mode II the exhaust gas sensor is monitored as long as the feedback system is in closed loop. The CEL and red LED are illuminated if the fuel mixture condition is lean, and off if the fuel mixture condition is rich. In open loop the bulbs remain off, regardless of the fuel mixture condition.

NOTE: If crank angle sensor and ignition signal-primary codes come out at the same time, check items causing a malfunction of the crank angle sensor circuit first.

The Nissan Consult scan-tool, or equivalent, can be used in several modes to access and erase codes, read sensor data, etc. The diagnostic connector for the Consult is located either behind the fuse box cover or to the rear of the hood release handle.

CLEARING CODES

The malfunction code is erased from the backup memory on the ECU when the diagnostic mode is changed from Mode II to Mode I. When the battery terminal is disconnected, the malfunction code will be lost from the backup memory within 24 hours, but before starting self-diagnosis, do not erase the stored memory before beginning self-diagnosis.

The Nissan Consult, or equivalent, can also be used to erase trouble codes quickly.

FUEL SYSTEM TESTING

Fuel Pressure Relief

CAUTION

Before disconnecting a fuel line, the fuel pressure must be released from the line to eliminate the danger of fire and personal injury.

The fuel pressure on the G20 and Q45 can be released by using the Consult, or equivalent, in the Work Support Mode. To release the fuel pressure without a scan tool see the procedure below.

G20 And M30

1. Remove the fuse for the fuel pump.
2. Start the engine.
3. After the engine stalls, crank it two or three times to release all the fuel pressure.
4. Turn the ignition switch off and reinstall the fuse.
5. On the M30 erase the memory, since a Code 22 will have been set.

Q45

1. Remove the fuse for the fuel pump.
2. Start the engine.
3. After the engine stalls, crank it two or three times to release all the fuel pressure.
4. Turn the ignition switch off and reinstall the relay and the fuse.

Fuel Pressure Check

1. After fuel pressure is released, disconnect fuel hose between fuel filter and fuel tube (engine side).
2. Install pressure guage between fuel filter and fuel tube.
3. Start engine and check for fuel leakage.
4. Read the fuel pressure guage.
   a. At idle with the fuel pressure regulator valve vacuum hose connected, the pressure should be approximately 34 psi.
   b. At idle with the vacuum hose disconnected, the pressure should be approximately 43 psi.
5. Stop the engine and disconnect the fuel pressure regulator vacuum hose from the intake manifold and plug the intake manifold with a rubber cap.
6. Connect a variable vacuum source to the fuel pressure regulator.
7. Start the engine and read the indication of the fuel pressure guage as the vacuum is changed. The fuel pressure should decrease as vacuum increases. If the results are unsatisfactory, replace the fuel pressure regulator.

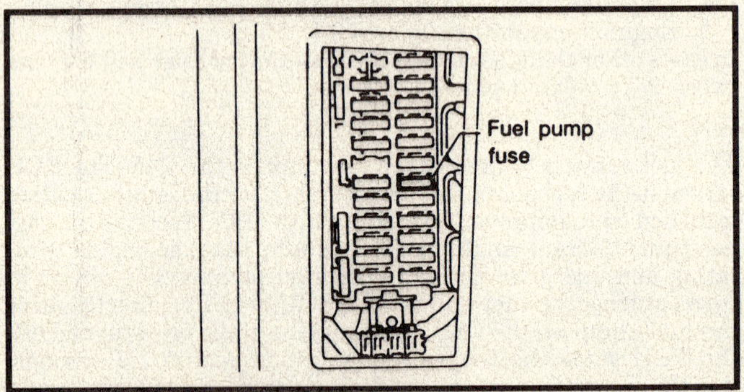

Fuel pump fuse location—G20

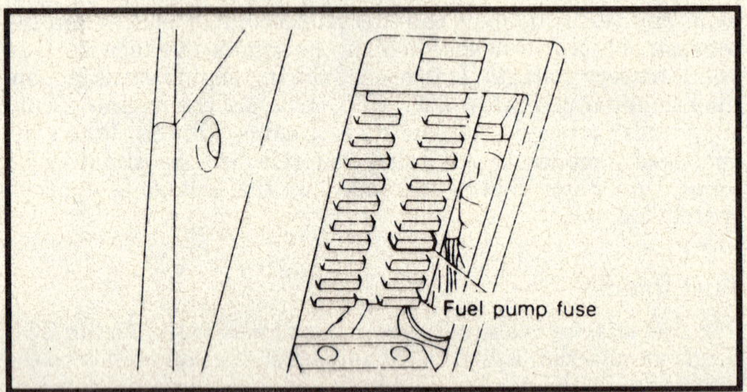

Fuel pump fuse location—M30

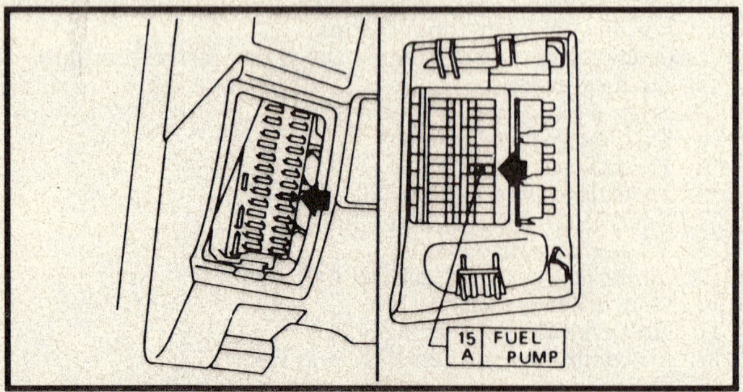

Fuel pump fuse location—Q45

ELECTRONIC CONTROL UNIT

Testing

Note: When testing the ECU with a DVOM make sure that the probes of the tester never touch each other as this will result in damage to a transistor in the ECU.

G20

1. The ECU is located behind the center console panel near the accelerator pedal. For this input/output signal inspection, remove the center console under cover.
2. Remove the plastic ECU harness protector to gain access to the ECU wiring.
3. Perform all the voltage measurements with the connectors connected. Modify the end of the tester probe with a thin piece of wire to perform the tests more easily and to avoid damaging the connectors.

M30

1. The ECU is located behind the right side dash panel. For this input/output signal inspection, remove the right side dash panel.
2. Remove the plastic ECU harness protector to gain access to the ECU wiring.
3. Perform all the voltage measurements with the connectors connected. Modify the end of the tester probe with a thin piece of wire to perform the tests more easily and to avoid damaging the connectors.

Q45

1. The ECU is located behind the front passenger side dash near the front right-side door. For this input/output signal inspection, remove the front passenger side dash.

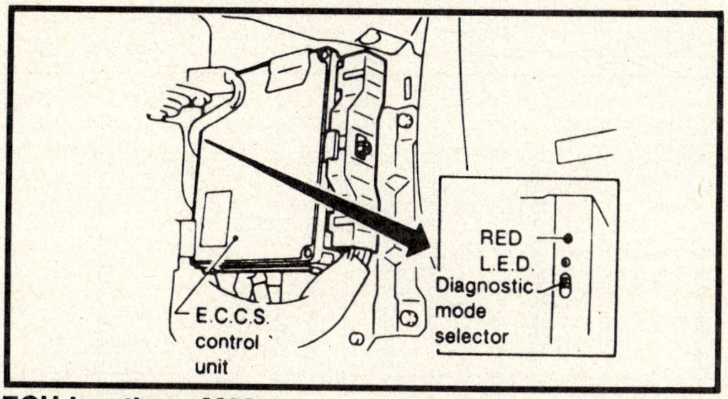

ECU location – M30

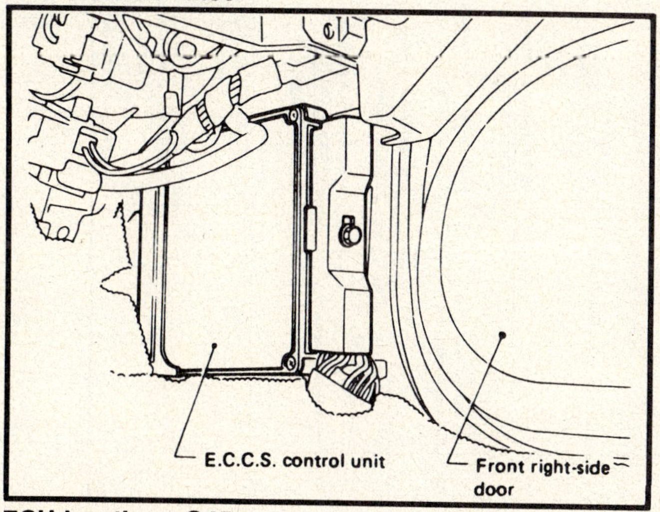

ECU location – Q45

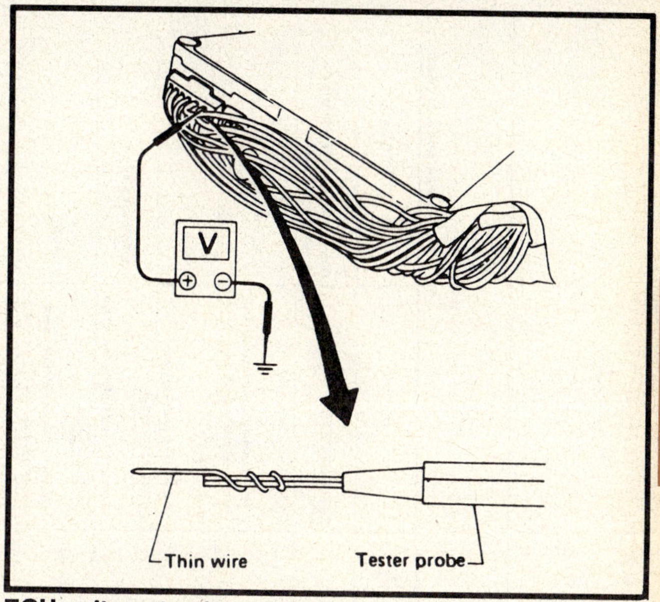

ECU voltage testing – Infiniti

2. Remove the plastic ECU harness protector to gain access to the ECU wiring.
3. Perform all the voltage measurements with the connectors connected. Modify the end of the tester probe with a thin piece of wire to perform the tests more easily and to avoid damaging the connectors.

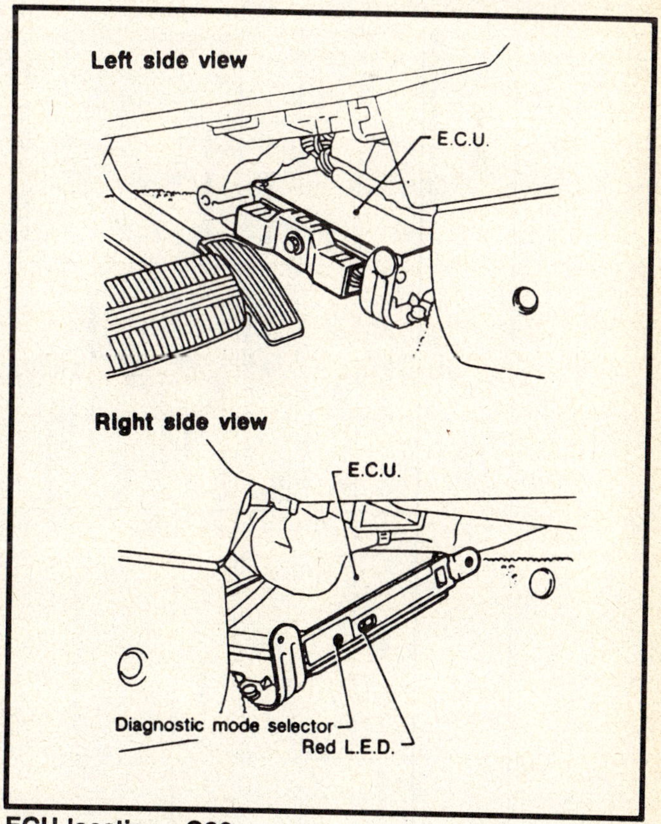

ECU location – G20

1990–92 2.0L ENGINE—ELECTRONIC CONTROL UNIT SPECIFICATIONS—G20

*Date are reference values.

TER-MINAL NO.	ITEM	CONDITION	*DATA
1	Ignition signal	Engine is running. — Idle speed	0.3 - 0.5V
		Engine is running. — Engine speed is 2,000 rpm	Approximately 0.7V
3	Ignition check	Engine is running. — Idle speed	Approximately 12V
4	E.C.C.S. relay (Self-shutoff)	Engine is running. — Ignition switch "OFF" — For a few seconds after turning ignition switch "OFF".	0 - 1V
		Ignition switch "OFF" — A few seconds after turning ignition switch "OFF"	BATTERY VOLTAGE (11 - 14V)
8	Exhaust gas temperature sensor	Engine is running. (Warm-up condition) — Idle speed	Less than 4.5V
		Engine is running. (Warm-up condition) — E.G.R. system is operating.	0 - 1.0V
9 10	Radiator fan relay (Low) Radiator fan relay (High)	Engine is running. — Radiator fan is not operating.	BATTERY VOLTAGE (11 - 14V)
		Engine is running. — Radiator fan is operating.	0.6 - 0.8V
11	Air conditioner relay	Engine is running. — Both A/C switch and blower switch are "ON".	0.6 - 0.8V
		Engine is running. — A/C switch is "OFF".	BATTERY VOLTAGE (11 - 14V)
16	Air flow meter	Engine is running. (Warm-up condition) — Idle speed	1.3 - 1.7V
		Engine is running. (Warm-up condition) — Engine speed is 2,000 rpm.	1.7 - 2.1V

*Date are reference values.

TER-MINAL NO.	ITEM	CONDITION	*DATA
18	Engine temperature sensor	Engine is running.	0 - 5.0V Output voltage varies with engine coolant temperature.
19	Exhaust gas sensor	Engine is running. — After warming up sufficiently and engine speed is 2,000 rpm.	0 - Approximately 1.0V
20	Throttle sensor	Ignition switch "ON" — Accelerator pedal released	0.45 - 0.55V
		Ignition switch "ON" — Accelerator pedal fully depressed	Approximately 4V
22 30	Crank angle sensor (Reference signal)	Engine is running. — **Do not run engine at high speed under no-load.**	0.2 - 0.5V
27	Detonation sensor	Engine is running. — Idle speed	2.0 - 3.0V
31 40	Crank angle sensor (Position signal)	Engine is running. — **Do not run engine at high speed under no-load.**	2.0 - 3.0V
34	Start signal	Ignition switch "ON"	Approximately 0V
		Ignition switch "START"	BATTERY VOLTAGE (11 - 14V)
35	Neutral switch (M/T models) Inhibitor switch (A/T models)	Ignition switch "ON" — Gear position is "Neutral" (M/T models) — Gear position is "N" or "P" (A/T models)	0V
		Ignition switch "ON" — Except the above gear position	BATTERY VOLTAGE (11 - 14V)
36	Ignition switch	Ignition switch "OFF"	0V
		Ignition switch "ON"	BATTERY VOLTAGE (11 - 14V)
37	Throttle sensor power supply	Ignition switch "ON"	Approximately 5V
38 47	Power supply for E.C.U.	Ignition switch "ON"	BATTERY VOLTAGE (11 - 14V)

1990–92 2.0L ENGINE – ELECTRONIC CONTROL UNIT SPECIFICATIONS – G20

*Date are reference values.

TER-MINAL NO.	ITEM	CONDITION	*DATA
41	Air conditioner switch	Engine is running. – Both air conditioner switch and blower switch are "ON".	Approximately 0V
		Engine is running. – Air conditioner switch is "OFF".	BATTERY VOLTAGE (11 - 14V)
43	Power steering oil pressure switch	Engine is running. – Steering wheel is being turned.	0V
		Engine is running. – Steering wheel is not being turned.	7 - 10V
46	Power supply (Back-up)	Ignition switch "OFF"	BATTERY VOLTAGE (11 - 14V)
101	Injector No. 1	Engine is running	BATTERY VOLTAGE (11 - 14V)
103	Injector No. 3		
110	Injector No. 2		
112	Injector No. 4		
102	A.I.V. control solenoid valve	Engine is running. (Warm-up condition) – Idle speed.	0.6 - 0.8V
		Engine is running. (Warm-up condition) – Accelerator pedal is depressed.	BATTERY VOLTAGE
104	Fuel pump relay	Ignition switch "ON" – For 5 seconds after turning ignition switch "ON" Engine is running.	0.7 - 0.9V
		Ignition switch "ON" – 5 seconds after turning ignition switch "ON"	BATTERY VOLTAGE (11 - 14V)
105	E.G.R. & canister control solenoid valve	Engine is running. (Warm-up condition) – Engine speed is 2,000 rpm.	BATTERY VOLTAGE (11 - 14V)
		Engine is running. (Warm-up condition) – Engine speed is 4,000 rpm.	0.6 - 0.8V

*Date are reference values.

TER-MINAL NO.	ITEM	CONDITION	*DATA
111	Exhaust gas sensor heater	Engine is running. – Engine speed is below 3,200 rpm.	0V
		Engine is running. – Engine speed is above 3,200 rpm.	BATTERY VOLTAGE (11 - 14V)
113	A.A.C. valve	Engine is running. – Idle speed	9 - 14V
		Engine is running. – Steering wheel is being turned. – Air conditioner is operating. – Rear defogger is "ON". – Headlamp are in high position.	5 - 9V

E.C.U. HARNESS CONNECTOR TERMINAL LAYOUT

1990–92 3.0L ENGINE—ELECTRONIC CONTROL UNIT SPECIFICATIONS—M30

*Data are reference values.

TERMINAL NO.	ITEM	CONDITION	*DATA
1	Ignition signal	Engine is running. — Idle speed	0.4 - 0.6V
		Engine is running. — Engine speed is 2,000 rpm	1.3 - 1.4V
2	Ignition check	Engine is running. — Idle speed	BATTERY VOLTAGE (11 - 14V)
4	A.A.C. valve	Engine is running. — Idle speed	9 - 12V
		Engine is running. — Air conditioner is operating. — Rear defogger is "ON". — Head lamps are in high position.	5 - 9V
9	Air conditioner relay	Engine is running. — Air conditioner switch "OFF"	BATTERY VOLTAGE (11 - 14V)
		Engine is running. — Air conditioner switch "ON"	0.7 - 0.9V
16	E.C.U. power source (Self-shutoff)	Engine is running. — Idle speed	0.8 - 1.0V
		Engine is not running. — For a few seconds after turning ignition switch "OFF"	BATTERY VOLTAGE (11 - 14V)
18	Fuel pump relay	Engine is running. — Normal condition	BATTERY VOLTAGE (11 - 14V)
		Engine is running. — Abnormal condition [Fuel pump voltage control circuit (E.C.U. terminal No. (104)) is inoperative.]	0.7 - 0.8V

*Data are reference values.

TERMINAL NO.	ITEM	CONDITION	*DATA
19	Air regulator	Ignition switch "ON" — For 5 seconds after turning ignition switch "ON" Engine is running.	0.7 - 0.9V
		Ignition switch "ON" — 5 seconds after turning ignition switch "ON"	BATTERY VOLTAGE (11 - 14V)
23	Detonation sensor	Engine is running. — Idle speed	3.5 - 4.0V
27	Air flow meter	Engine is running. (Warm-up condition) — Idle speed	1.0 - 1.5V
		Engine is running. (Warm-up condition) — Engine speed is 2,000 rpm.	1.4 - 1.9V
28	Engine temperature sensor	Engine is running.	0 - 5.0V Output voltage varies with engine temperature.
29	Exhaust gas sensor	Engine is running. — After warming up sufficiently and engine speed is 2,000 rpm.	0 ⟷ approximately 1.5V
36	Fuel temperature sensor	Engine is running.	0 - 5.0V Output voltage varies with fuel temperature.
38	Throttle sensor	Ignition switch "ON"	0.4 - 4.0V Output voltage varies with throttle valve opening angle.
39	Exhaust gas temperature sensor	Engine is running. (Warm-up condition) — Idle speed	Less than 4.5V
		Engine is running. (Warm-up condition) — E.G.R. system is operating.	0 - 1.0V
41 51	Crank angle sensor (Reference signal)	Engine is running. — **Do not run engine at high speed under no-load.**	0.3 - 0.5V Output voltage varies slightly with engine speed.
42 52	Crank angle sensor (Position signal)	Engine is running. — **Do not run engine at high speed under no-load.**	2.5 - 2.7V Output voltage varies slightly with engine speed.
43	Start signal	Ignition switch "ON"	Approx. 0V
		Ignition switch "START"	BATTERY VOLTAGE (11 - 14V)

1990–92 3.0L ENGINE—ELECTRONIC CONTROL UNIT SPECIFICATIONS—M30

*Data are reference values.

TER-MINAL NO.	ITEM	CONDITION	*DATA
44	A/T control unit (Neutral position)	Ignition switch "ON" — Gear position is "N" or "P"	Approx. 0V
		Ignition switch "ON" — Except the above conditions	7.0 - 9.0V
45	Ignition switch	Ignition switch "ON" — Engine stopped	BATTERY VOLTAGE (11 - 14V)
46	Air conditioner switch	Engine is running. — Air conditioner switch "OFF"	7.0 - 9.0V
		Engine is running. — Air conditioner switch "ON"	0.5 - 0.7V
48	Power source for sensors	Ignition switch "ON" — Engine stopped	Approximately 5.0V
49 59	Power supply	Ignition switch "ON" — Engine stopped	BATTERY VOLTAGE (11 - 14V)
54	Throttle valve switch (Idle position)	Ignition switch "ON" — Accelerator pedal is fully released (Engine running).	8.0 - 10.0V
		Ignition switch "ON" — Accelerator pedal is depressed (Engine running).	0V
57	Power source for idle switch	Ignition switch "ON" — Engine stopped	8.0 - 10.0V
101 103 105 110 112 114	Injectors	Ignition switch "OFF"	BATTERY VOLTAGE (11 - 14V)
102	E.G.R. control solenoid valve	Engine is running. (Warm-up condition) — Idle speed	0.7 - 0.8V
		Engine is running. (Warm-up condition) — Engine speed is 2,000 rpm.	BATTERY VOLTAGE (11 - 14V)
		Engine is running. (Warm-up condition) — Engine speed is above 3,100 rpm.	0.8 - 0.9V

*Data are reference values.

TER-MINAL NO.	ITEM	CONDITION	*DATA
104	Fuel pump voltage control	Ignition switch "ON" — For 5 seconds after turning ignition switch "ON"	Approx. 0.1V
		Engine is running. (Warm-up condition) — For 30 seconds after engine begins to run.	Approx. 0.5V
		Engine is running. — Racing (up to 4,000 rpm)	2.0 - 4.5V
111	P.R.V.R. control solenoid valve	Stop and restart engine after warming it up. — Fuel temperature is above 75°C (167°F)	0 - 1.0V (For 3 minutes after restarting engine)
			BATTERY VOLTAGE (11 - 14V) (3 minutes after restarting engine)
		Stop and restart engine after warming it up. — Fuel temperature is below 75°C (167°F)	BATTERY VOLTAGE(11 - 14V)

E.C.U. HARNESS CONNECTOR TERMINAL LAYOUT

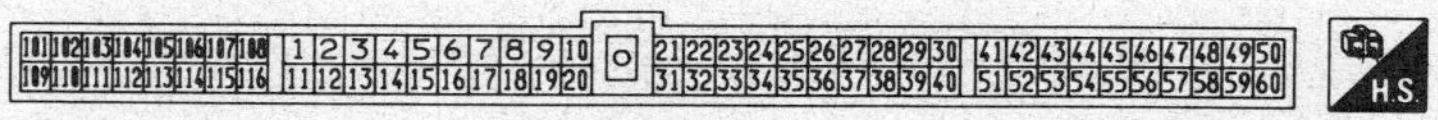

1990–92 4.5L ENGINE—ELECTRONIC CONTROL UNIT SPECIFICATIONS—Q45

*Data are reference values.

TERMINAL NO.	ITEM	CONDITION	*DATA
1, 2, 3, 11, 12, 13, 14, 15	Ignition signal	Engine is running. └ Idle speed	Approximately 70 mV
		Engine is running. └ Engine speed is 2,000 rpm.	Approximately 0.14V
6, 19	Fuel pump voltage control	Ignition switch "ON" └ For 5 seconds after turning ignition switch "ON"	Approximately 1.9V
		Ignition switch "ON" └ 5 seconds after turning ignition switch "ON"	Approximately 0V
		Engine is running. (Warm-up condition) └ Idle speed	1.9 - 2.0V
7	Tachometer	Engine is running. └ Idle speed	Approximately 1.5V
		Engine is running. └ Engine speed is 2,000 rpm.	Approximately 3.1V
8	Canister vacuum control solenoid valve	Ignition switch "ON"	Approximately 0.8V
		Engine is running. └ Idle speed	
		Engine is running. (Warm-up condition) ├ Engine speed is about 2,000 rpm. └ In "1" range	BATTERY VOLTAGE (11 - 14V)
9	Air conditioner relay	Engine is running. └ A/C switch is "ON".	Approximately 0V
		Engine is running. └ A/C switch is "OFF".	BATTERY VOLTAGE (11 - 14V)
16	E.C.C.S. relay (Self-shut off)	Engine is running. ↓ Ignition switch "OFF" └ For 6 seconds after turning ignition switch "OFF".	0 - 1V
		Ignition switch "OFF" └ In 6 seconds after turning ignition switch "OFF".	BATTERY VOLTAGE (11 - 14V)

*Data are reference values.

TERMINAL NO.	ITEM	CONDITION	*DATA
18	Fuel pump relay	Ignition switch "ON" └ For 5 seconds after turning ignition switch "ON"	Approximately 0.8V
		Ignition switch "ON" └ 5 seconds after turning ignition switch "ON".	BATTERY VOLTAGE (11 - 14V)
		Engine is running. └ Idle speed	Approximately 0.9V
23, 24	Detonation sensor	Engine is running. └ Idle speed	Approximately 2.5V
27	Air flow meter	Engine is running. (Warm-up condition) └ Idle speed	1.0 - 1.4V
		Engine is running. (Warm-up condition) └ Engine speed is 2,000 rpm	1.4 - 1.9V
28	Engine temperature sensor	Engine is running.	0 - 5.0V Output voltage varies with engine temperature.
29	Exhaust gas sensor R.H.	Engine is running. (Warm-up condition) └ Engine speed is 2,000 rpm	0 - Approximately 1.0V
55	Exhaust gas sensor L.H.		
36*	TCS signal	Ignition switch "ON"	Approximately 9V
		Ignition switch "ON" ├ Disconnect throttle motor harness connector. └ Fully close secondary throttle valve by hand.	Approximately 0V
37*	Secondary throttle sensor	Ignition switch "ON" └ Approximately 3 seconds after ignition switch "ON"	Approximately 3.4V
		Ignition switch "ON" ├ Disconnect throttle motor harness connector. └ Fully close secondary throttle valve by hand.	Approximately 0.4V
38	Throttle sensor	Engine is running. (Warm-up condition)	Approximately 0.4 - 4V Output voltage varies with the throttle valve opening angle.

***WARNING:**
Before touching the secondary throttle valve, be sure to disconnect the throttle motor connector; otherwise, injury may occur due to accidental actuation of the valve.

1990–92 4.5L ENGINE–ELECTRONIC CONTROL UNIT SPECIFICATIONS–Q45

*Data are reference values.

TERMINAL NO.	ITEM	CONDITION	*DATA
39	Exhaust gas temperature sensor	Engine is running. (Warm-up condition) └Idle speed	Less than 4.5V
		Engine is running. (Warm-up condition) └E.G.R. system is operating.	0 - 1.0V
40	Power steering oil pressure switch	Engine is running. └Steering wheel is in the "straight ahead" position.	Approximately 0V
		Engine is running. └Steering wheel is turned.	BATTERY VOLTAGE (11 - 14V)
41 51	Crank angle sensor (Reference signal)	Engine is running. Do not run engine at high speed under no-load.	0.8 - 1.0V
42	Crank angle sensor (Position signal)	Engine is running. Do not run engine at high speed under no-load.	2.4 - 2.7V Output voltage varies slightly with engine speed.
43	Start signal	Ignition switch "ON"	Approximately 0V
		Ignition switch "START"	BATTERY VOLTAGE (11 - 14V)
44	Inhibitor switch	Ignition switch "ON" └Gear position is "N" or "P".	Approximately 0V
		Ignition switch "ON" └Except the above gear position	8 - 10V
45	Ignition switch	Ignition switch "OFF"	Approximately 0V
		Ignition switch "ON"	BATTERY VOLTAGE (11 - 14V)
46	Air conditioner switch	Engine is running. └A/C switch is "ON".	0.5 - 0.7V
		Engine is running. └A/C switch is "OFF".	7 - 10V
48	Power source for sensors	Ignition switch "ON"	Approximately 5V
49 59	Power supply	Ignition switch "ON" └Engine is running.	BATTERY VOLTAGE (11 - 14V)
52	Sub-crank angle sensor	Engine is running. Do not run engine at high speed under no-load.	0 - Approximately 10 mV
54	Throttle valve switch	Engine is running. (Warm-up condition) └Accelerator pedal is fully released.	BATTERY VOLTAGE (11 - 14V)
		Engine is running. (Warm-up condition) └Accelerator pedal is depressed.	Approximately 0V

*Data are reference values.

TERMINAL NO.	ITEM	CONDITION	*DATA
56	Throttle opening signal	Ignition switch "ON"	0.3 - Approximately 3.3V Output voltage varies with the throttle valve opening angle.
58	Battery	Ignition switch "OFF"	BATTERY VOLTAGE (11 - 14V)
101 103 105 106 110 112 113 114	Injectors	Engine is running.	BATTERY VOLTAGE (11 - 14V)
102	E.G.R. vacuum control solenoid valve	Engine is running. (Warm-up condition) └Idle speed	Approximately 0V
		Engine is running. (Warm-up condition) └Engine speed is about 2,000 rpm.	BATTERY VOLTAGE (11 - 14V)
104	Valve timing control solenoid valve	Engine is running. └Idle speed	BATTERY VOLTAGE (11 - 14V)
		Engine is running. └Quickly depress accelerator pedal, then quickly release it.	Approximately 0V
111	A.A.C. valve	Engine is running. └Idle speed	10 - 11V
		Engine is running. ├Steering wheel is being turned. ├Air conditioner is operating. ├Rear defogger is "ON". └Head lamps are in high position.	6 - 9V
115	Exhaust gas sensor heater	Engine is running. └Engine speed is below 2,250 rpm.	Approximately 0V
		Engine is running. └Engine speed is above 2,250 rpm.	BATTERY VOLTAGE (11 - 14V)

E.C.U. HARNESS CONNECTOR TERMINAL LAYOUT

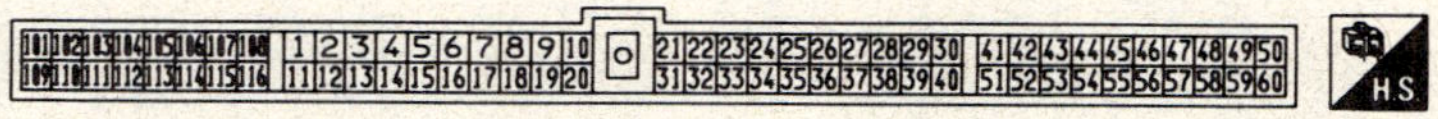

CRANK ANGLE SENSOR

Testing

NOTE: If crank angle sensor and ignition signal-primary codes come out at the same time, check items causing a malfunction of the crank angle sensor circuit first.

G20 AND M30

1. Visually check the signal plate for damage or dust.
2. Remove the distributor from the engine without disconnecting the crank angle sensor harness.
3. Disconnect the ignition wires.
4. Turn the ignition switch **ON**.
5. Rotate the distributor shaft slowly by hand and check the voltage between terminals "a", "d" and ground on the G20, and terminals 3 and ground, and 4 and ground on the M30.
6. If the tester's pointer does not fluctuate between 0–5 volts, replace the distributor assembly with crank angle sensor.

Q45

This procedure includes testing the crank angle sensor and the sub-crank angle sensor on the Q45.

1. Remove the crank angle sensor from the engine without disconnecting the sensor harness.
2. Turn the ignition switch **ON**.
3. Rotate the crank angle sensor shaft slowly by hand and check the voltage between terminals 1, 2 and ground. If the voltage does not fluctuate between 0–5 volts, replace the crank angle sensor.
4. Rotate the crank angle sensor shaft slowly by hand and check the voltage between terminals "a" and the "b" ground, using a circuit tester set in the 100 mV range, AC. If the tester pointer does not deflect, replace the crank angle sensor.
5. After this inspection, a malfunction code 11 might be displayed even though the crank angle sensor is functioning properly. In this case erase the stored memory.

AIR FLOW METER

Testing

G20

1. Fold back the air flow meter harness connector protector.
2. Start the engine and allow it to reach normal operating temperature.
3. Check the voltage between terminal "a" and ground. With the key on and engine off, there should be less than 1V. At idle there should be 1.3–1.7V.
4. If the readings are not within specifications, remove the air flow meter from the air duct and check the hot wire for damage or dust.

M30

1. Fold back the air flow meter harness connector protector.

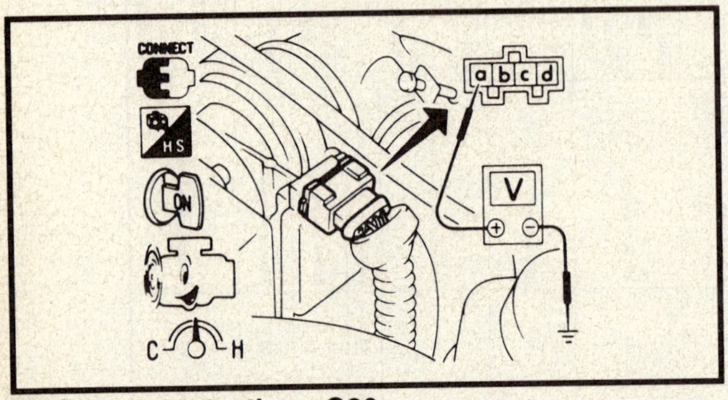

Air flow meter testing—G20

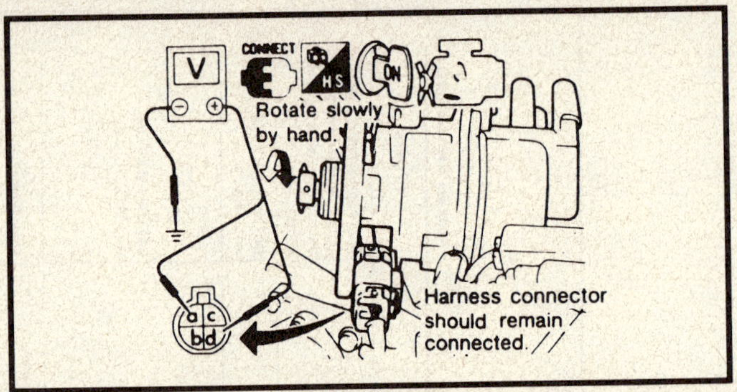

Crank angle sensor testing—G20

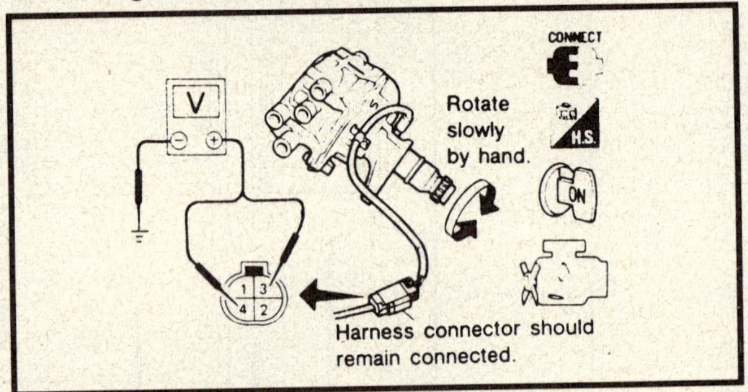

Crank angle sensor testing—M30

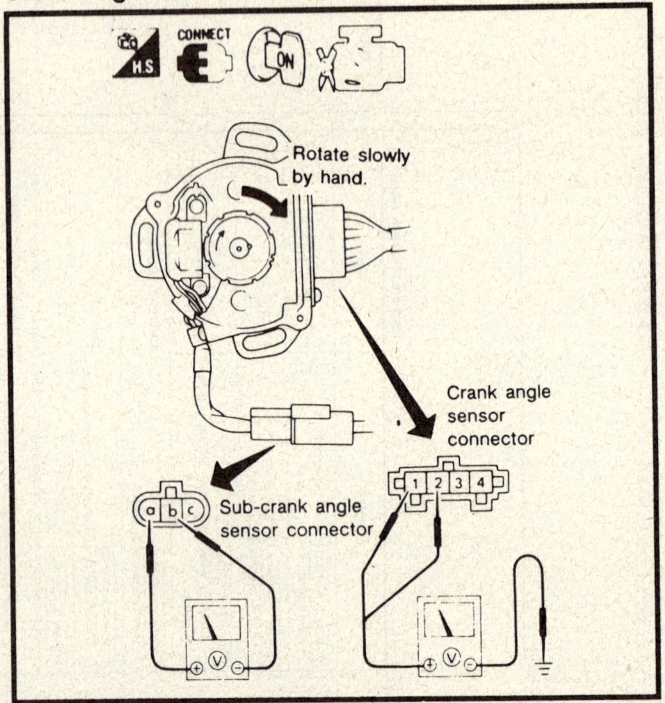

Crank angle sensor testing—Q45

2. Start the engine and allow it to reach normal operating temperature.
3. Check the voltage between terminal "a" and ground. With the key on and engine off, there should be less than 0.5V. At idle there should be approximately 1.0–1.3V.
4. If the readings are not within specifications, remove the air flow meter from the air duct and check the hot wire for damage or dust.

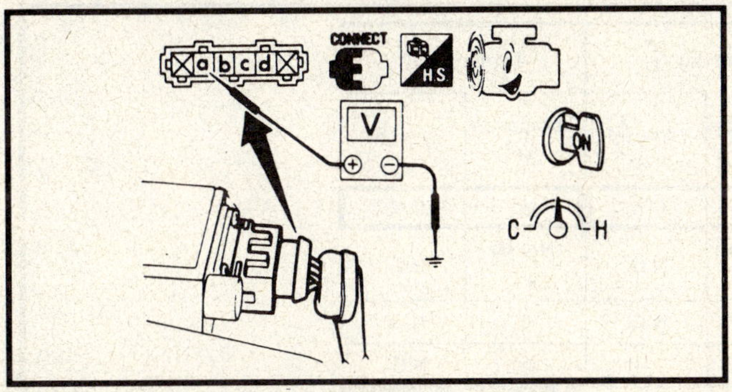

Air flow meter testing—M30

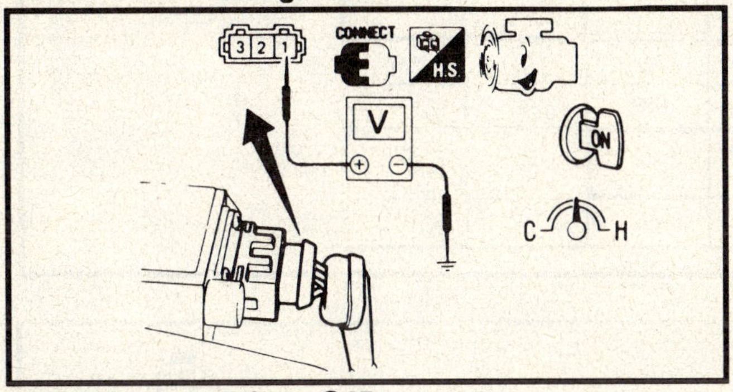

Air flow meter testing—Q45

Q45

1. Fold back the air flow meter harness connector protector.
2. Start the engine and allow it to reach normal operating temperature.
3. Check the voltage between terminal 1 and ground. With the key ON and engine OFF, there should be approximately 0.2V. At idle there should be approximately 1.0–1.4V.
4. If the readings are not within specifications, remove the air flow meter from the air duct and check the hot film for damage or dust.

ENGINE TEMPERATURE SENSOR

Testing

1. Disconnect the engine temperature sensor harness connector.
2. Check between the two terminals of the sensor for the resistance, which should be 2.1–2.9 kilo-ohms at 68°F, 0.68–1.00 at 122°F and 0.30–0.33 at 176°F.
3. If it is not within specifications, replace the engine temperature sensor.

IGNITION COIL

Testing

G20 AND M30

1. Disconnect the ignition coil harness connector.
2. Check the resistances between the following terminals. Between terminals 1 and 2 there should be approximately 1.0 ohm. Between terminals 1 and 3 there should be approximately 10 kilo-ohms.
3. If not within specifications, replace the ignition coil.

Q45

1. Disconnect the ignition coil harness connector.
2. Check the resistance between terminals 1 and 2, and between terminals 3 and 4. In each case there should be approximately 0.7 ohm.

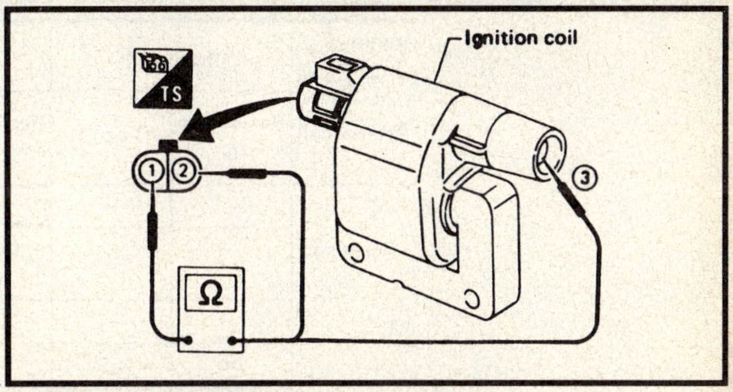

Ignition coil testing—G20 and M30

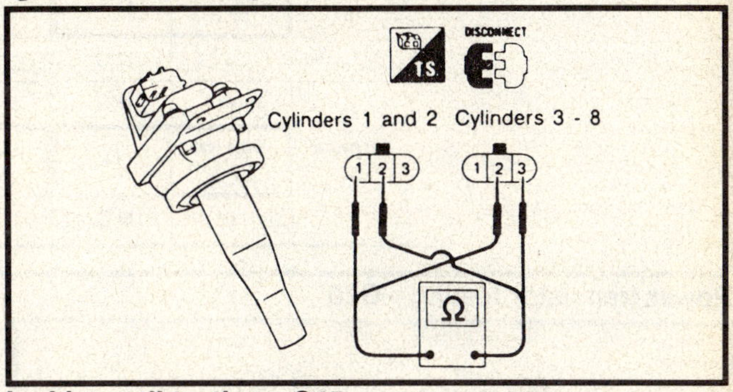

Ignition coil testing—Q45

3. If not within the specification, replace the ignition coil.

POWER TRANSISTOR

Testing

G20

1. Disconnect the power transistor harness connector.
2. Check the power transistor continuity between the appropriate terminals with a digital tester.
3. If it does not meet specifications, replace the power transistor.

M30

1. Disconnect the power transistor harness connector.
2. Check the power transistor continuity between the appropriate terminals with a digital tester.
3. If it does not meet specifications, replace the power transistor.

Q45

1. Disconnect the power transistor harness connector.
2. Check the power transistor continuity between the appropriate terminals with a digital tester.
3. If it does not meet specifications, replace the power transistor.

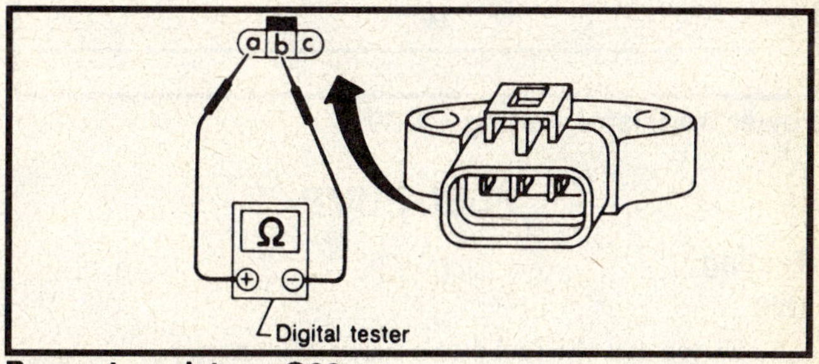

Power transistor—G20

⊕terminal side \ ⊖terminal side	Terminal ⓐ Resistance Ω	Terminal ⓐ Result	Terminal ⓑ Resistance Ω	Terminal ⓑ Result	Terminal ⓒ Resistance Ω	Terminal ⓒ Result
Terminal ⓐ	—	—	∞	O.K.	∞	O.K.
	—	—	Not ∞ or 0	N.G.	Not ∞ or 0	N.G.
	—	—	0	N.G.	0	N.G.
Terminal ⓑ	∞	N.G.	—	—	∞	N.G.
	Not ∞ or 0	O.K.	—	—	Not ∞ or 0	O.K.
	0	N.G.	—	—	0	N.G.
Terminal ⓒ	∞	N.G.	∞	N.G.	—	—
	Not ∞ or 0	O.K.	Not ∞ or 0	O.K.	—	—
	0	N.G.	0	N.G.	—	—

Power transistor testing—G20

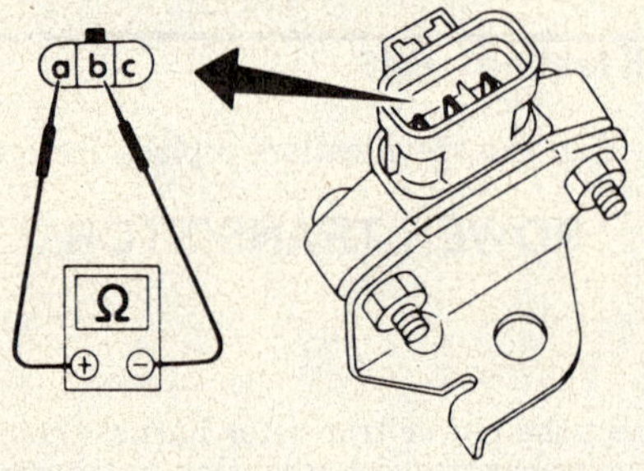

Power transistor—M30

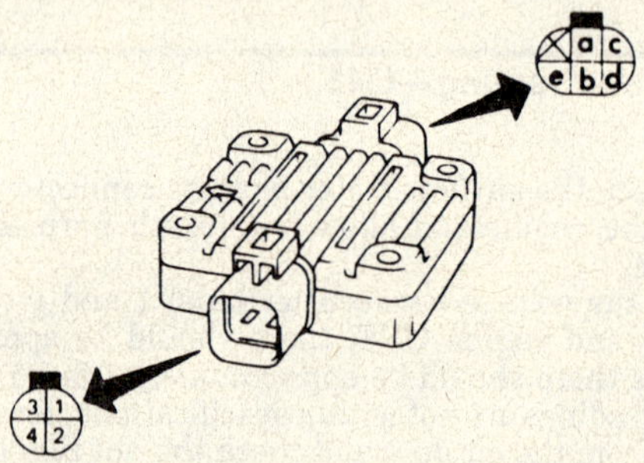

Power transistor—Q45

Terminal No.	Tester polarity	Continuity
ⓐ	⊕	Yes
ⓑ	⊖	
ⓐ	⊖	No
ⓑ	⊕	
ⓐ	⊕	Yes
ⓒ	⊖	
ⓐ	⊖	No
ⓒ	⊕	

Power transistor testing—M30

Terminal combination				Tester polarity	Continuity	Tester polarity	Continuity
e 1	e 2	e 3	e 4	⊕ ⊖	No	⊖ ⊕	Yes
e a	e b	e c	e d	⊕ ⊖	Yes	⊖ ⊕	Yes
1 a	2 b	3 c	4 d	⊕ ⊖	Yes	⊖ ⊕	No

Power transistor testing—Q45

FUEL PUMP

Testing

G20

1. Disconnect the fuel pump harness connector.
2. Check the resistance between terminals "a" and "b".
3. If the resistance is not approximately 0.7 ohm, replace the fuel pump.

M30

1. Disconnect the fuel pump harness connector.
2. Check the resistance between terminals "d" and "e".
3. If the resistance is not approximately 0.5 ohm, replace the fuel pump.

Q45

1. Disconnect the fuel pump harness connector.
2. Check the resistance between terminals "a" and "c".
3. If the resistance is not approximately 0.5 ohm, replace the fuel pump.

FUEL PUMP CONTROL UNIT

This component is used on the Q45 only.

Testing

1. With the ignition switch off, disconnect the fuel pump control unit harness connector.
2. Connect the proper wires between the fuel pump control unit harness terminals as shown in the chart.
3. Start the engine and check the voltage between terminals 7 and 4 on the fuel pump control unit in the three different configurations shown in the chart. If any voltage is out of range, replace the fuel pump control unit.

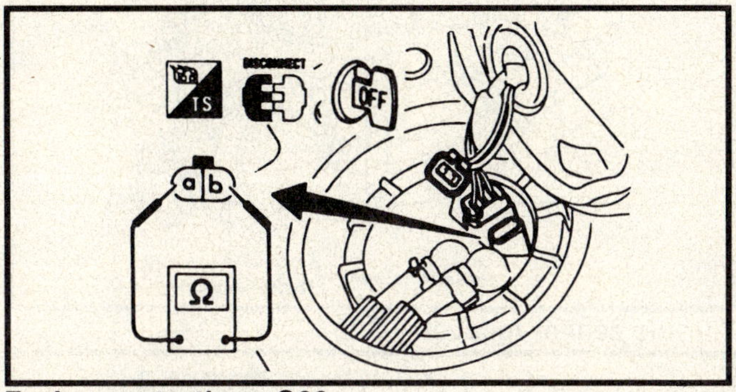

Fuel pump testing—G20

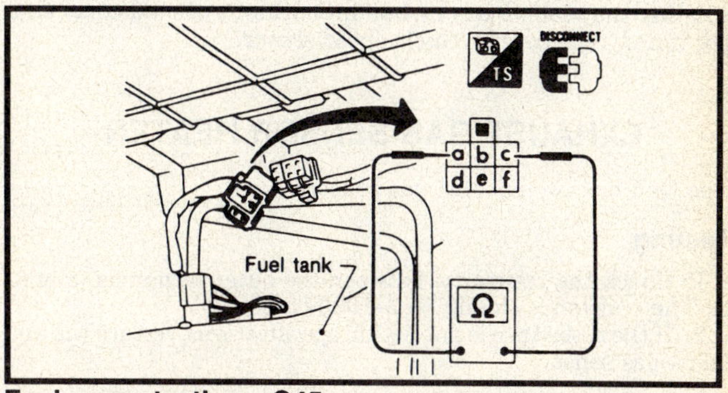

Fuel pump testing—Q45

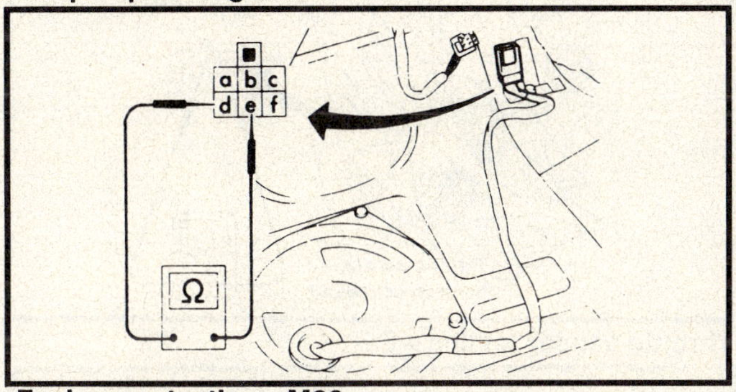

Fuel pump testing—M30

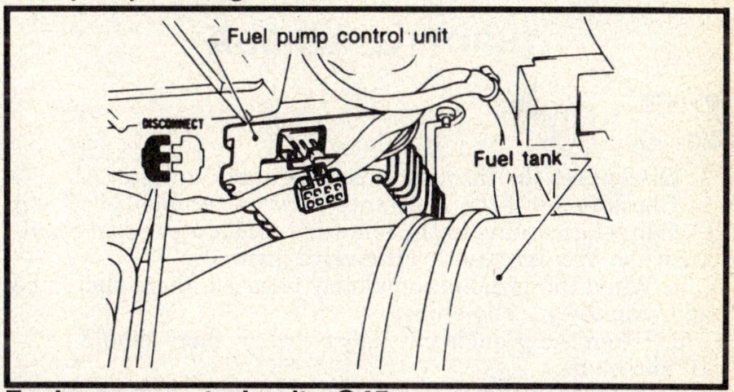

Fuel pump control unit—Q45

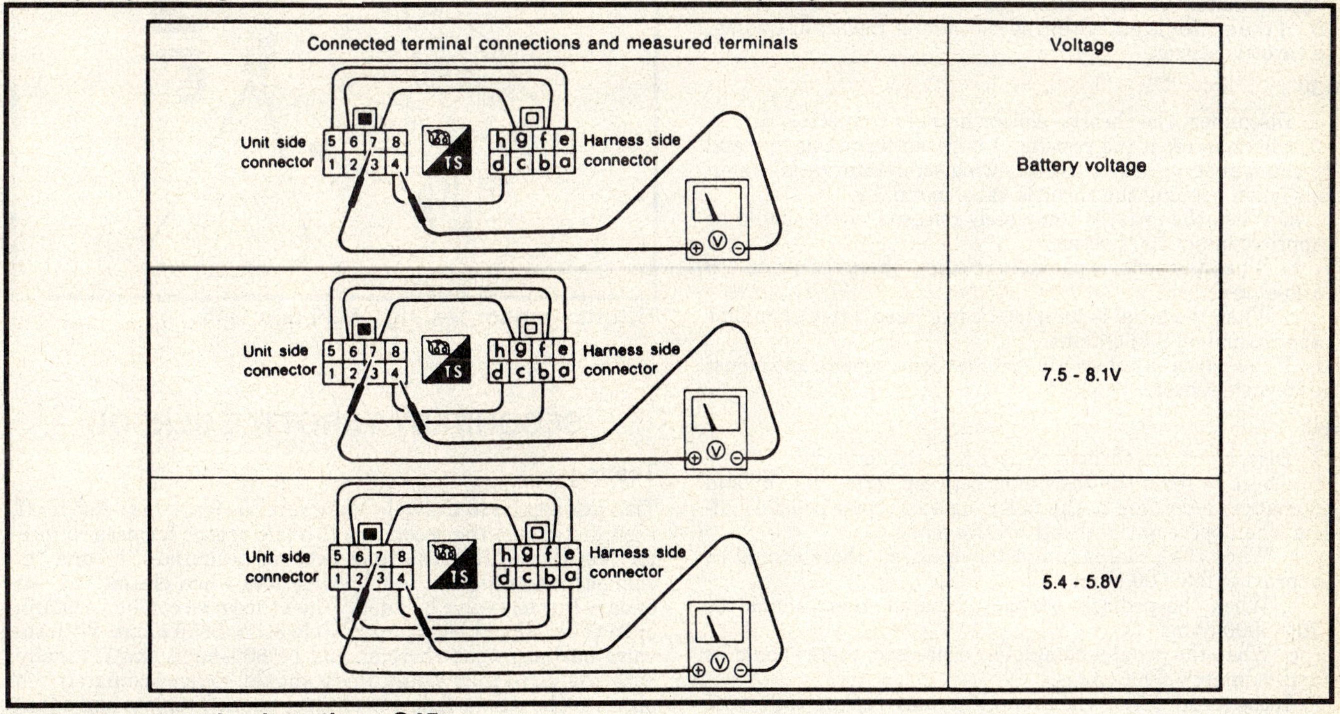

Fuel pump control unit testing—Q45

VEHICLE SPEED SENSOR

Testing

The following test is for the Q45. The G20 and M30 are similar.

1. Jack up the rear wheels and use safety stands to support the vehicle.
2. Disconnect the vehicle speed sensor harness connector.
3. Check the continuity between terminals "a" and "b" while rotating the rear wheels by hand. Continuity should come and go. If not, replace the vehicle speed sensor.

EXHAUST GAS SENSOR HEATER

The Q45 has two exhaust gas sensors.

Testing

1. Check the resistance between the outer terminals (a and c). The resistance should be 3–1000 ohms.
2. If the resistance is not within specifications, replace the exhaust gas sensor.

THROTTLE SENSOR

Testing

G20

1. Disconnect the throttle sensor harness connector.
2. Check to see if the resistance between terminals "a" and "b" changes according to the following accelerator pedal conditions when opening the throttle valve manually.
   a. When the pedal is completely released, there should be approximately 2 kilo-ohms.
   b. When the pedal is partially released, there should be 2–10 kilo-ohms.
   c. When the pedal is completely depressed, there should be approximately 10 kilo-ohms.
3. If the sensor is not within specifications, replace and adjust the throttle sensor.

M30

1. Disconnect the throttle sensor harness connector.
2. Check to see if the resistance between terminals "b" and "c" changes according to the following accelerator pedal conditions when opening the throttle valve manually.
   a. When the pedal is completely released, there should be approximately 1000 ohms.
   b. When the pedal is partially released, there should be 1–9 kilo-ohms.
   c. When the pedal is completely depressed, there should be approximately 9 kilo-ohms.
3. If the sensor is not within specifications, replace and adjust the throttle sensor.

Q45

1. Disconnect the throttle sensor harness connector.
2. Check to see if the resistance between terminals "b" and "c" changes according to the following accelerator pedal conditions when opening the throttle valve manually.
   a. When the pedal is completely released, there should be approximately 700 ohms.
   b. When the pedal is partially released, there should be 700–5000 ohms.
   c. When the pedal is completely depressed, there should be approximately 5000 ohms.
3. If the sensor is not within specifications, replace and adjust the throttle sensor.

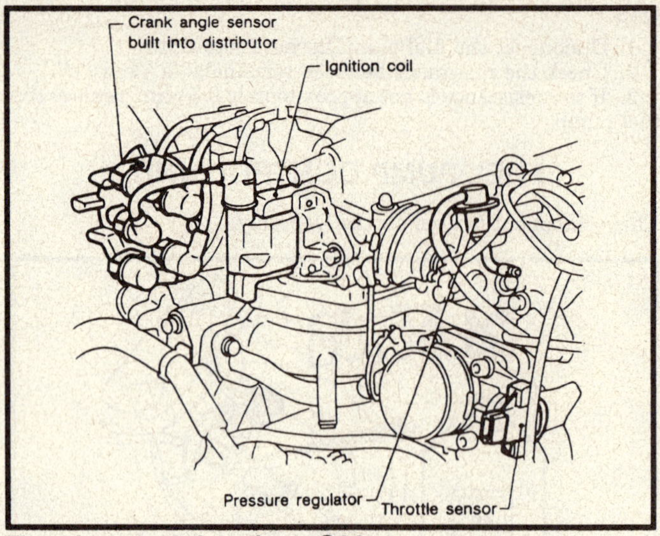

Throttle sensor location—G20

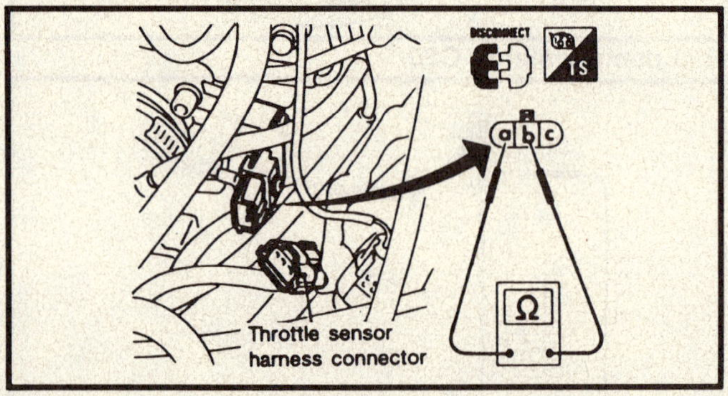

Throttle sensor testing—G20

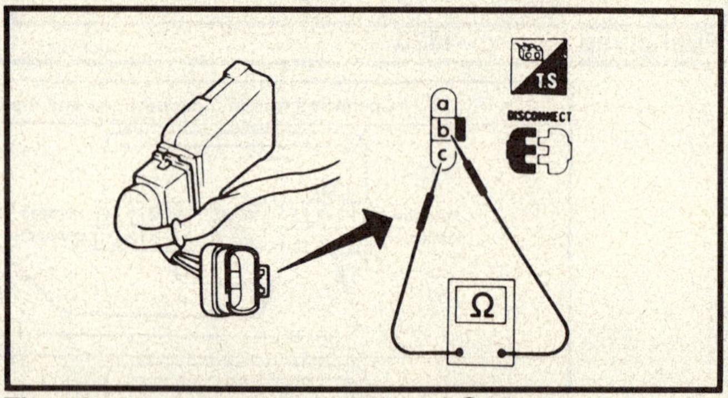

Throttle sensor testing—M30 and Q45

SECONDARY THROTTLE SENSOR

Testing

This test applies to the Q45. Make sure the ignition switch is off, then disconnect the secondary throttle sensor harness connector. Verify that the resistance between terminals "b" and "c" changes according to the following data when closing the secondary throttle valve by hand. When the valve condition is fully open there should be approximately 3000–5800 ohms. With the valve partially open there should be 800–5800 ohms. Finally, with the valve fully closed there should be approximately 800 ohms. If the sensor fails any of these checks, replace and adjust the secondary throttle sensor.

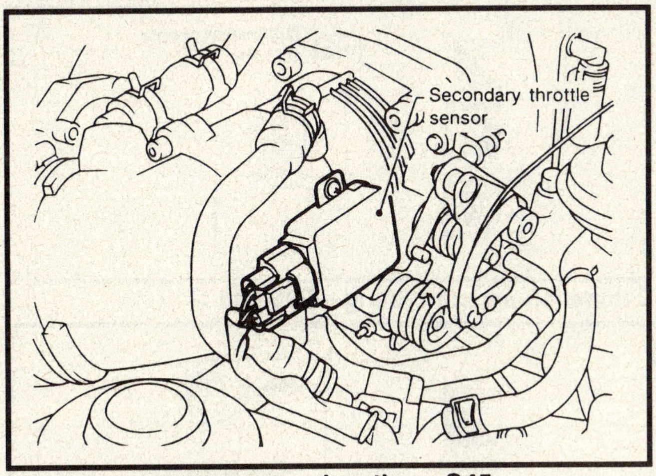

Secondary throttle sensor location—Q45

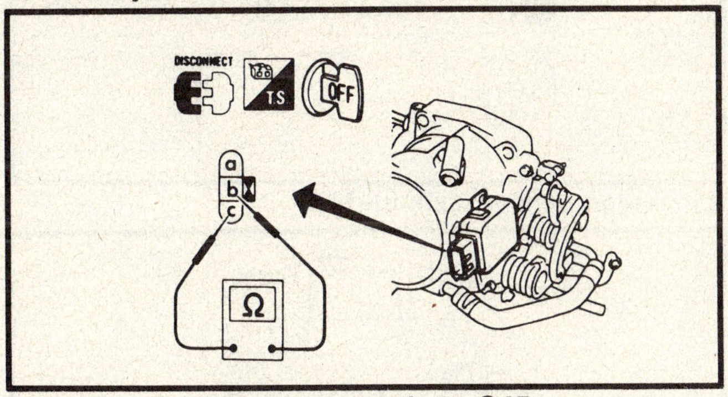

Secondary throttle sensor testing—Q45

THROTTLE VALVE SWITCH

Testing

This test applies to the Q45 and checks the status of the throttle valve switch in the idle position. With the engine warmed up sufficiently, make sure that the fast idle cam holds the cam follower in the released position. Then disconnect the idle switch harness connector. Check the continuity between terminals "a" and "b". With the accelerator pedal released there should be continuity. With the pedal depressed there should not be continuity. If the switch fails these checks, replace it.

AAC VALVE

Testing

Disconnect the AAC valve 2-wire harness connector and check the valve resistance for approximately 10 ohms. Also, check the plunger for seizing or sticking and check for a possible broken spring.

AIR CUT VALVE

Testing

This procedure applies to the 1991–92 Q45 equipped with traction control. Remove the air cut valve from the engine. Immerse the valve in hot, then cold water and check the air flow.

NOTE: Do not wet the gasket portion or the air chamber.

In ice-water, air should flow through the valve. Above 122°F almost no air should flow through the valve.

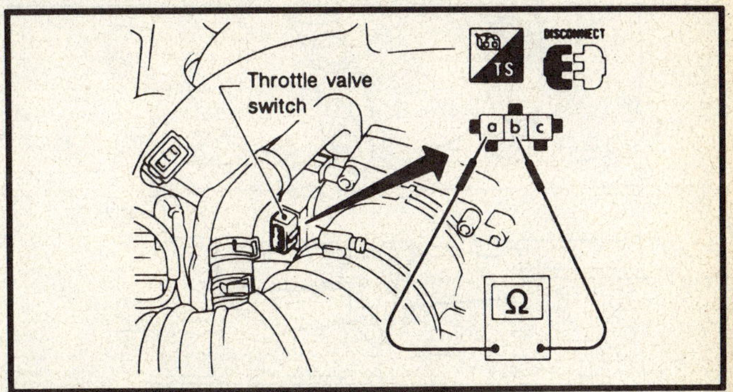

Throttle valve switch testing—Q45

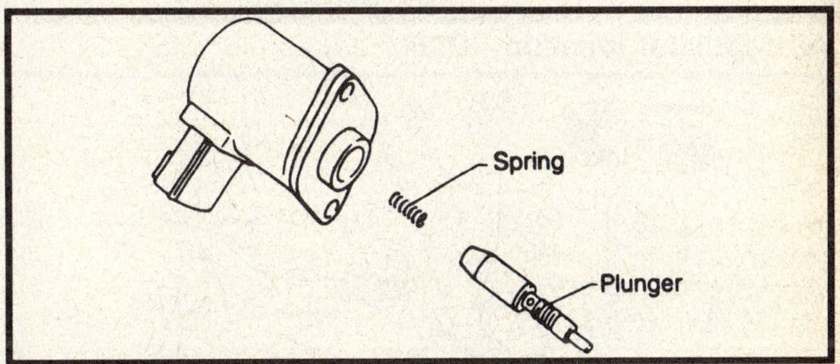

AAC valve—Infiniti

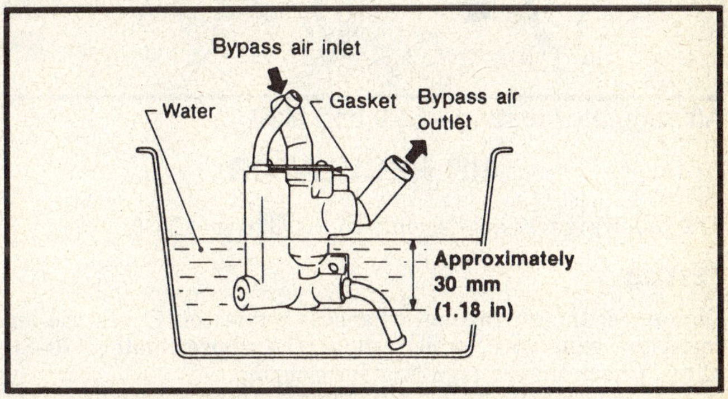

Air cut valve testing—Q45

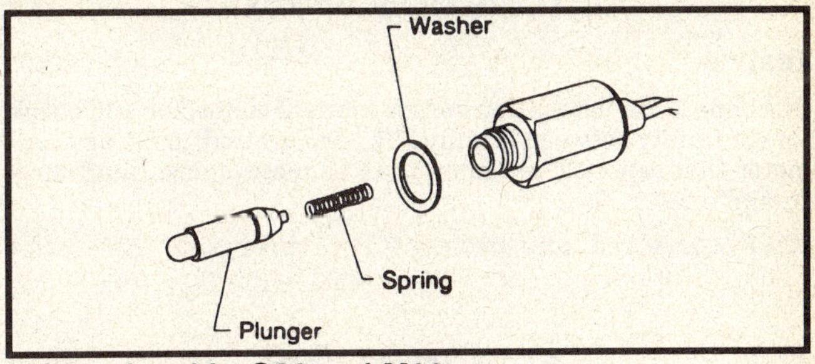

FICD solenoid—G20 and M30

FAST IDLE CONTROL DEVICE (FICD) SOLENOID VALVE

The following test applies only to the G20 and M30.

Testing

Disconnect the FICD solenoid valve harness connector. Check for a clicking sound when applying a 12V direct current to the terminals. Also, check the plunger for seizing or sticking and check for a broken spring.

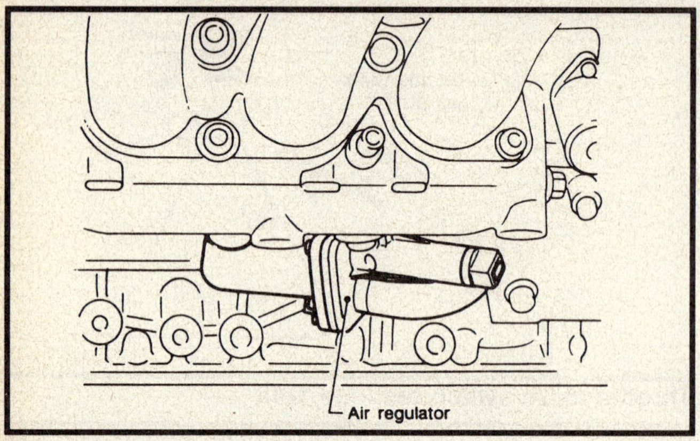

Air regulator location—G20

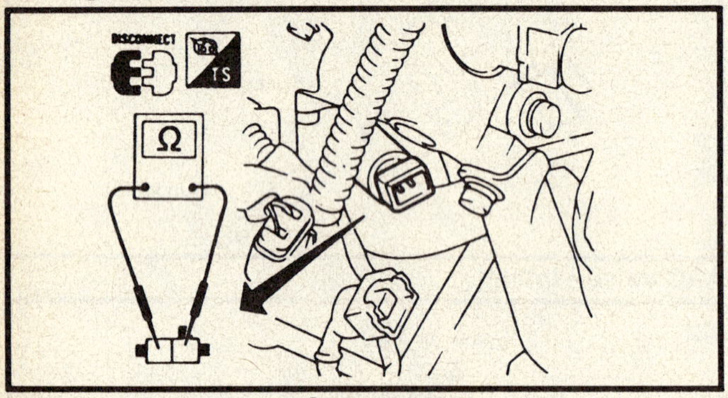

Air regulator testing—G20 and M30

AIR REGULATOR

The following test applies only to the G20 and M30.

Testing

Disconnect the air regulator harness connector. Check the air regulator resistance, which should be approximately 70–80 ohms. Check the air regulator for clogging.

DETONATION SENSOR

Testing

Disconnect the detonation sensor harness connector and check for continuity between terminal "a" and ground, using an ohmmeter that can measure more than 10 mega-ohms. Continuity should exist.

NOTE: Discard any detonation sensor that has been dropped or has undergone shocks; use a new one.

FUEL INJECTOR

Testing

Disconnect the injector harness connector and check the resistance between the two terminals. The resistance should be 10–14 ohms. If not, replace the injector.

FUEL TANK VACUUM RELIEF VALVE

Testing

After wiping clean the valve housing, suck air through the gas cap. A slight resistance accompanied by valve clicks indicates that the valve is in good mechanical condition. Note also that, by further sucking air, the resistance should disappear with valve clicks. If the valve is clogged or if no resistance is felt, replace the gas cap.

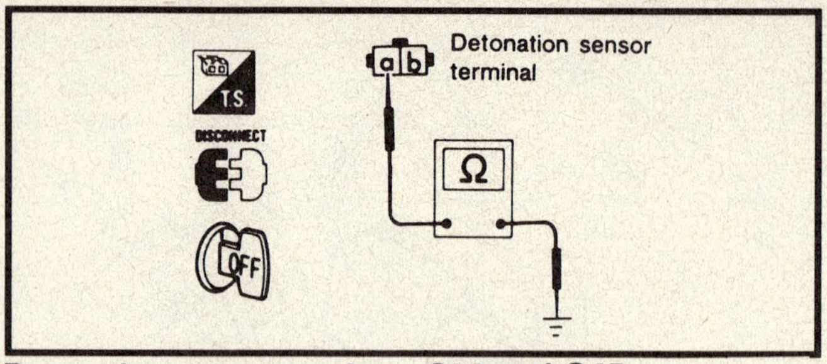

Detonation sensor testing—G20 and Q45

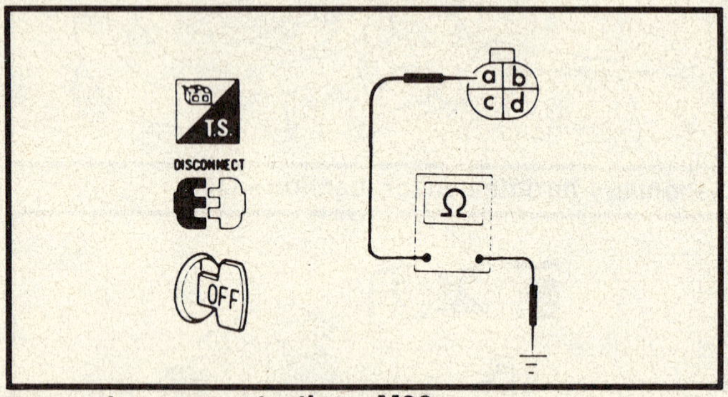

Detonation sensor testing—M30

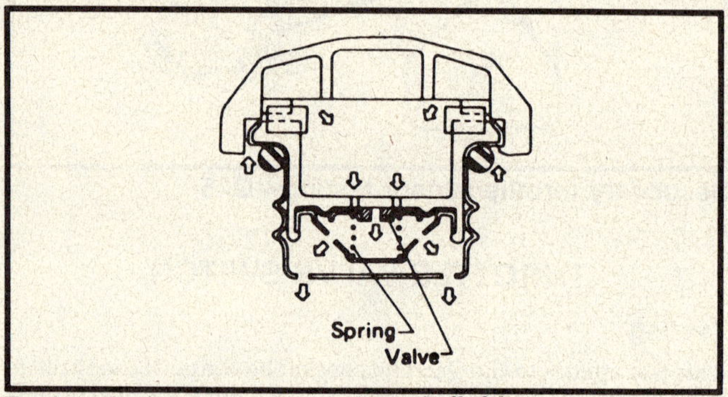

Fuel tank vacuum relief valve—Infiniti

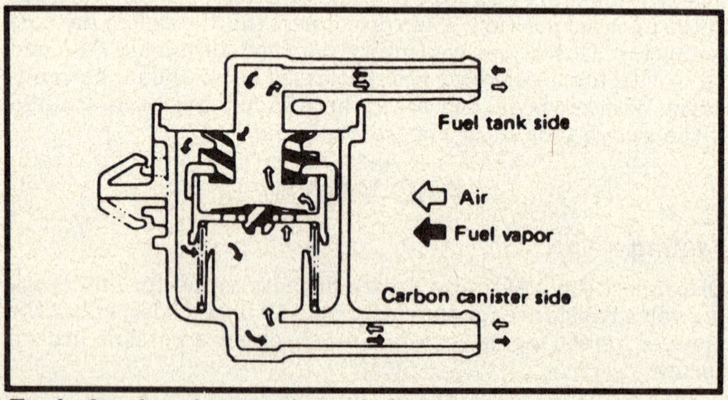

Fuel check valve testing—Infiniti

FUEL CHECK VALVE

Testing

Blow low-pressure air through the connector on the fuel tank side. A considerable resistance should be felt and a portion of air flow should be directed toward the canister. Now blow air through the connector on the cannister side. Air flow should be smoothly directed toward the fuel tank. If the check valve does not function as described, replace it.

EGR AND CANISTER CONTROL SOLENOID VALVE

Testing

The following checks apply to the G20. Apply 12 volts direct current to terminals 1 and 2. There should be air passage continuity between "a" and "b", and no air passage continuity between "b" and "c". With no power supplied to the solenoid there should be air passage continuity between "b" and "c", and no air passage continuity between "a" and "b".

AIV CONTROL SOLENOID VALVE

Testing

The following checks apply to the G20. Apply 12 volts direct current to terminals 1 and 2. There should be air passage continuity between "a" and "b", and no air passage continuity between "b" and "c". With no power supplied to the solenoid there should be air passage continuity between "b" and "c", and no air passage continuity between "a" and "b".

EGR CONTROL SOLENOID VALVE

Testing

The following checks apply to the M30. Disconnect the EGR control solenoid valve harness connector. While supplying 12 volts direct current between terminals 1 and 2, there should be air continuity between ports "a" and "b". With no voltage applied there should be no air continuity between ports "a" and "b".

PRESSURE REGULATOR CONTROL SOLENOID VALVE

Testing

The following checks apply to the M30. Disconnect the EGR control solenoid valve harness connector. While supplying 12 volts direct current between terminals 1 and 2, there should be air continuity between ports "a" and "b". With no voltage applied there should be no air continuity between ports "a" and "b".

EGR CONTROL SOLENOID VALVE

Testing

The following checks apply to the Q45. Apply 12 volts direct current to terminals 1 and 2. There should be air passage continuity between "a" and "b", and no air passage continuity between "a" and "c". With no power supplied to the solenoid there should be air passage continuity between "a" and "c", and no air passage continuity between "a" and "b".

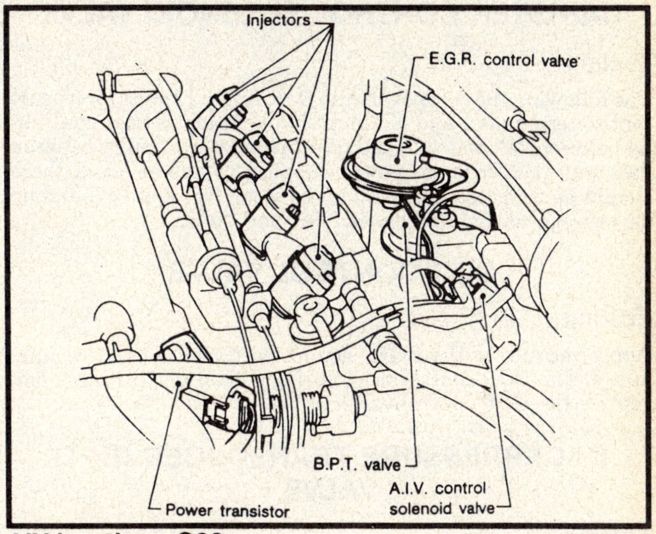

AIV location – G20

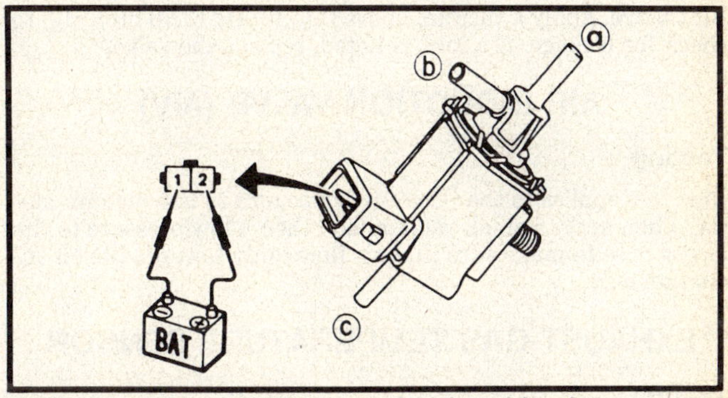

EGR, canister and AIV control solenoid valve testing – G20

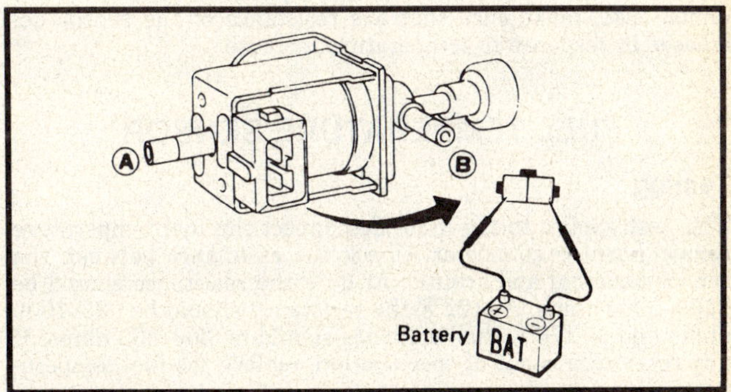

EGR and pressure regulator control solenoid valve testing – M30

CANISTER CONTROL SOLENOID VALVE

Testing

The following checks apply to the Q45. Apply 12 volts direct current to terminals 1 and 2. There should be air passage continuity between "a" and "b", and no air passage continuity between "a" and "c". With no power supplied to the solenoid there should be air passage continuity between "a" and "c", and no air passage continuity between "a" and "b".

EGR CONTROL VALVE

Testing

Apply vacuum to the EGR vacuum port with a hand vacuum pump. The EGR control valve spring should lift. If it does not, replace the EGR control valve.

BACKPRESSURE TRANSDUCER (BPT) VALVE

Testing

This test applies to the Q45. Plug one of the two ports of the BPT valve. Apply a vacuum above 0.15 in. Hg (3.70 mm. Hg) to check for leakage. If a leak is noted, replace the valve.

AIR INDUCTION VALVE (AIV)

Testing

This test applies to the G20. Apply vacuum to the vacuum motor. Then apply a small vacuum and then a low pressure to the larger hose to make sure that air flows only towards the air induction side.

EXHAUST GAS TEMPERATURE SENSOR

Testing

This test applies to G20 and Q45 vehicles with California emissions, and to the M30. Check the resistance value of the sensor when it is in boiling water (212°F). It should be 85,300 ± 8530 ohms. Also, make sure that the resistance of the sensor decreases in response to temperature increase.

FUEL TEMPERATURE SENSOR

Testing

This test applies to the M30. Disconnect the fuel temperature sensor harness connector. Check the resistance between the sensor terminal and ground. At 68°F the resistance should be 2100–2900 ohms. At 122°F the resistance should be 680–1000 ohms and at 176°F the resistance should be 300–330 ohms. If any resistance is out of specification, replace the fuel temperature sensor.

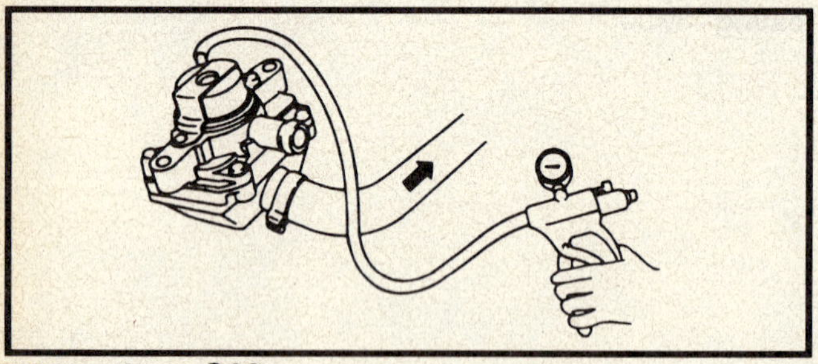

AIV testing – G20

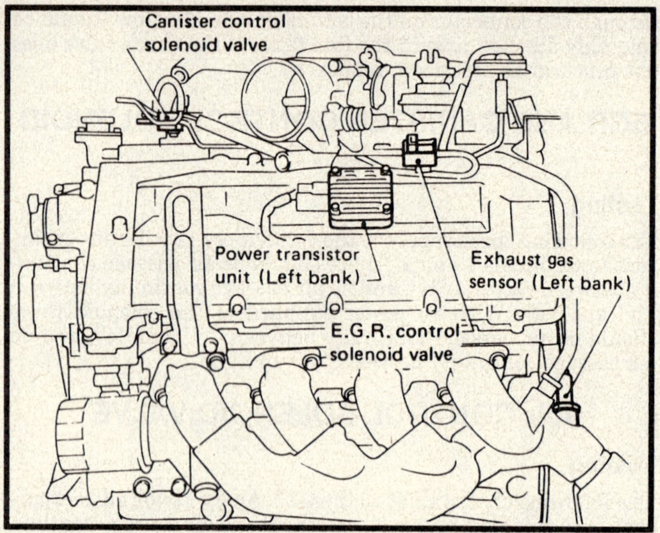

Canister control solenoid valve location – Q45

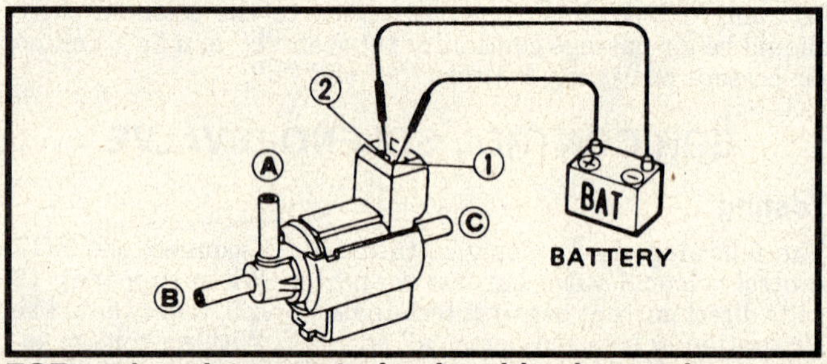

EGR and canister control solenoid valve testing – Q45

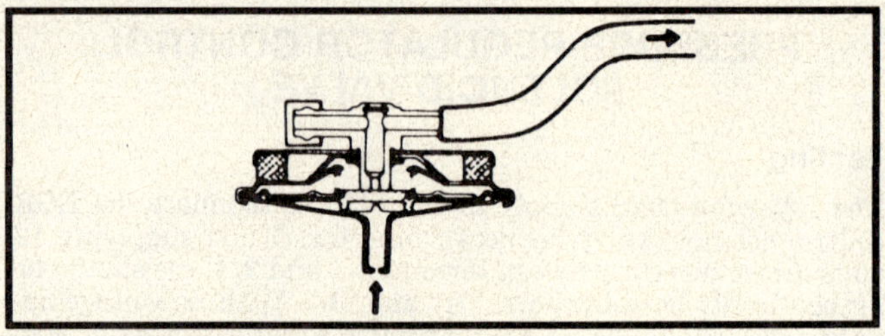

BPT valve testing – G20 and Q45

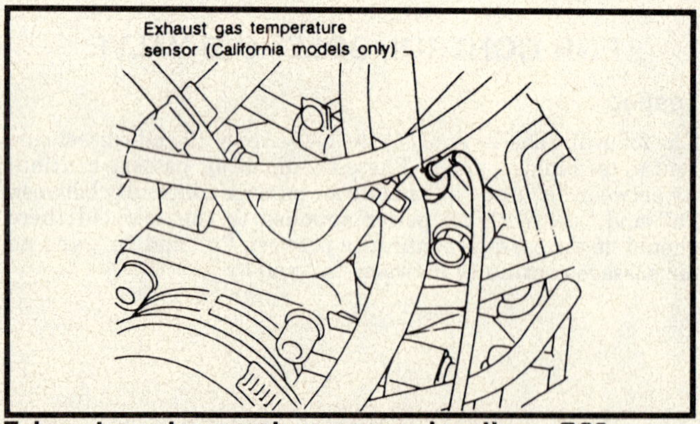

Exhaust gas temperature sensor location – G20

INHIBITOR SWITCH

Testing

This procedure applies to the G20 and the Q45.

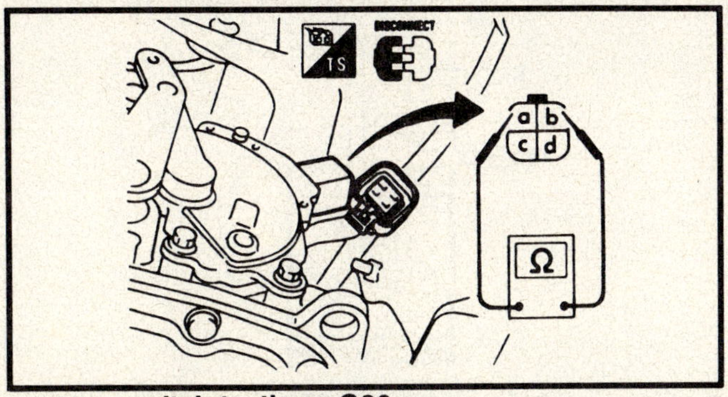

Inhibitor switch testing—G20

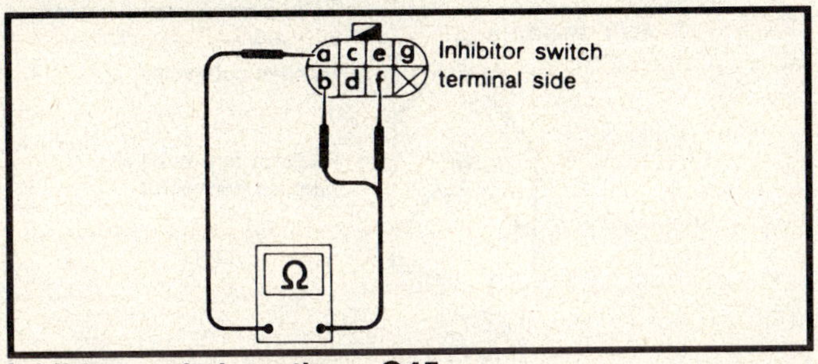

Inhibitor switch testing—Q45

G20

Check the continuity between terminals "a" and "b". With the gear selector in either the **P** or **N** positions there should be continuity. In any other position there should not be continuity.

Q45

Check the continuity between terminals "a" and "b", and between "a" and "f". With the gear selector in the **P** position there should be continuity between "a" and "b", but there should not be continuity between "a" and "f". With the selector in **N** there should not be continuity between "a" and "b", but there should be continuity between "a" and "f". With the selector in any other position there should neither be continuity between "a" and "b", nor between "a" and "f".

NEUTRAL SWITCH

Testing

This procedure applies to the G20. Check the continuity between terminals "a" and "b". With the gear selector in **N** there should be continuity. With the gear selector in any other position there should not be continuity. If the switch fails these checks, replace the neutral switch.

POWER STEERING OIL PRESSURE SWITCH

Testing

This procedure applies to the G20 and the Q45. Disconnect the power steering oil pressure switch harness connector, then start the engine. Check the continuity between the two terminals of the switch. While the steering wheel is being turned there should be continuity. When the steering wheel is not being turned there should not be continuity.

RESISTOR

Testing

This procedure applies to the G20 and M30. Disconnect the resistor harness connector and check the resistance between terminals "a" and "b". The resistance should be approximately 2200 ohms. If it is not, replace the resistor.

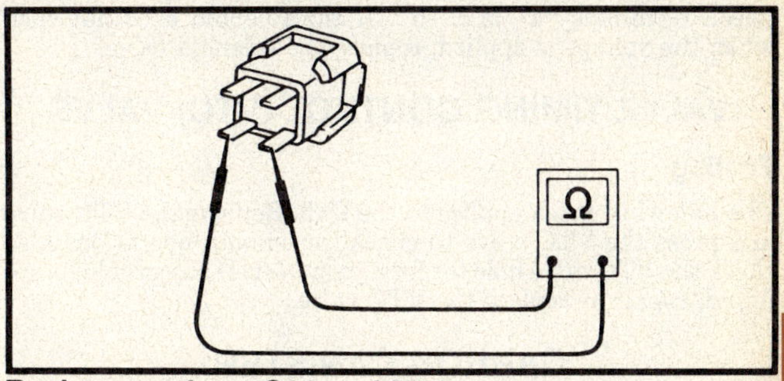

Resistor testing—G20 and M30

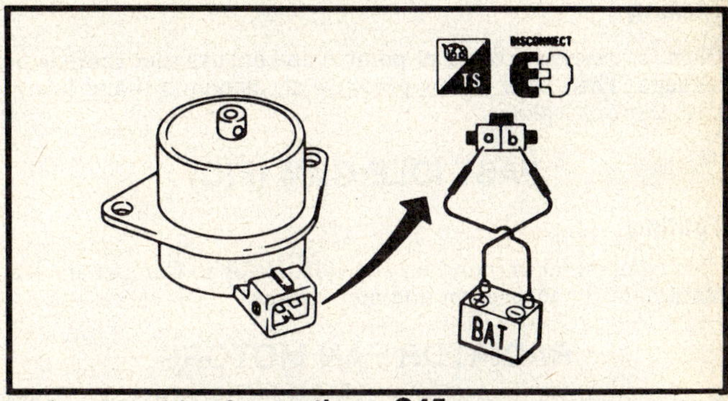

VTC solenoid valve testing—Q45

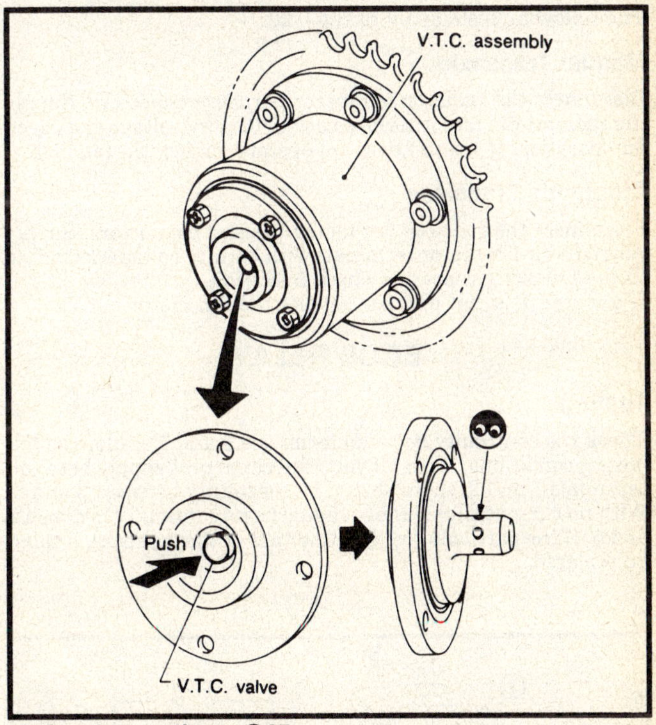

VTC valve testing—Q45

VALVE TIMING CONTROL (VTC) SOLENOID VALVE

Testing

The following check applies to the Q45. Disconnect the VTC solenoid valve harness connector and apply battery voltage between terminals "a" and "b". If the solenoid does not click when the voltage is applied, replace the solenoid valve.

VALVE TIMING CONTROL (VTC) VALVE

Testing

The following check applies to the Q45. Remove the VTC valve and press the VTC valve to check for smooth operation. Also check the VTC valve hole for foreign matter. If any problems are found, repair or replace the VTC valve.

CARBON CANISTER

Testing

Blow in low-pressure air at point A and ensure that there is no leakage. Then blow in low-pressure air at points B and insure that there is leakage.

FAST IDLE CAM (FIC)

Testing

This component is found on the Q45. Refer to Component Replacement for inspection and adjustment.

RADIATOR FAN MOTORS

Testing

The following tests apply to the G20.

Manual Transaxle

Disconnect the radiator fan motor harness connectors. Supply the radiator fan motor terminals with battery voltage and check for operation. If a motor does not operate, replace the fan motor.

Automatic Transaxle

Disconnect the radiator fan motor harness connectors. Supply the radiator fan motor terminals with battery voltage according to the following chart and check operation.
If a motor does not operate, replace the fan motor.

ECCS RELAY

Testing

Check the continuity between terminals 3 and 5, under the following conditions. With 12 volts direct current supply between terminals 1 and 2, there should be continuity between 3 and 5. With no current supply there should be no continuity between 3 and 5. If the particular relay fails either of these checks, it must be replaced.

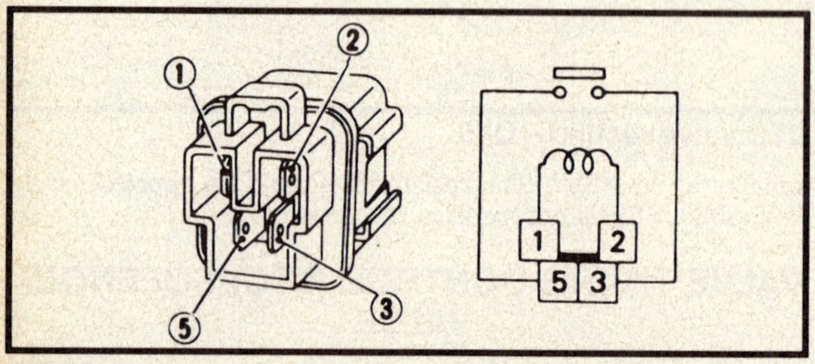

ECCS relay, radiator fan relay (G20 with manual transaxle), ignition coil relay (Q45) and fuel pump relay—Infiniti

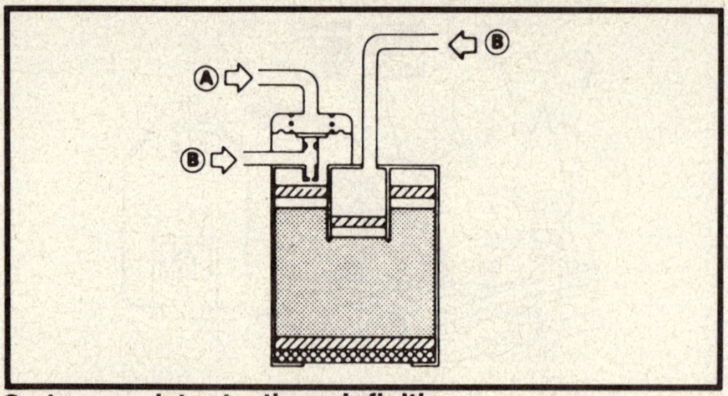

Carbon canister testing—Infiniti

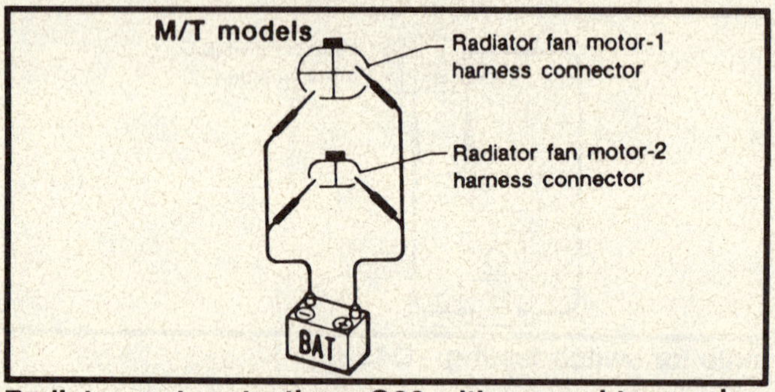

Radiator motors testing—G20 with manual transaxle

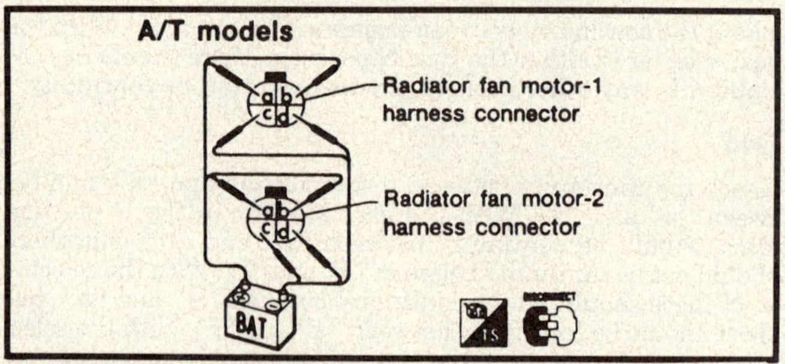

Radiator motors testing—G20 with automatic transaxle

	Speed	Terminals	
		(⊕)	(⊖)
Radiator fan motor-1	Low	ⓑ	ⓒ
	High	ⓐ, ⓑ	ⓒ, ⓓ
Radiator fan motor-2	Low	ⓑ	ⓒ
	High	ⓐ, ⓑ	ⓒ, ⓓ

Radiator motors testing—G20 with automatic transaxle

IGNITION COIL RELAY

Testing

This test applies to the Q45. Check the continuity between terminals 3 and 5, under the following conditions. With 12 volts direct current supply between terminals 1 and 2, there should be continuity between 3 and 5. With no current supply there should be no continuity between 3 and 5. If the particular relay fails either of these checks, it must be replaced.

FUEL PUMP RELAY

Testing

Check the continuity between terminals 3 and 5, under the following conditions. With 12 volts direct current supply between terminals 1 and 2, there should be continuity between 3 and 5. With no current supply there should be no continuity between 3 and 5. If the particular relay fails either of these checks, it must be replaced.

RADIATOR FAN RELAY

Testing

G20 WITH MANUAL TRANSAXLE

This test is for the radiator fan relay on the G20 with a manual transaxle. Check the continuity between terminals 3 and 5, under the following conditions. With 12 volts direct current supply between terminals 1 and 2, there should be continuity between 3 and 5. With no current supply there should be no continuity between 3 and 5. If the particular relay fails either of these checks, it must be replaced.

G20 WITH AUTOMATIC TRANSAXLE

This test applies to radiator fan relays 1, 2 and 3 on the G20 equipped with an automatic transaxle. Check the continuity between terminals 3 and 5, and between terminals 6 and 7 under the following conditions. With 12 volts direct current supplied between terminals 1 and 2, there should be continuity between the two pairs of terminals. With no current supplied there should be no continuity. The relay must be replaced if it fails either of these two checks.

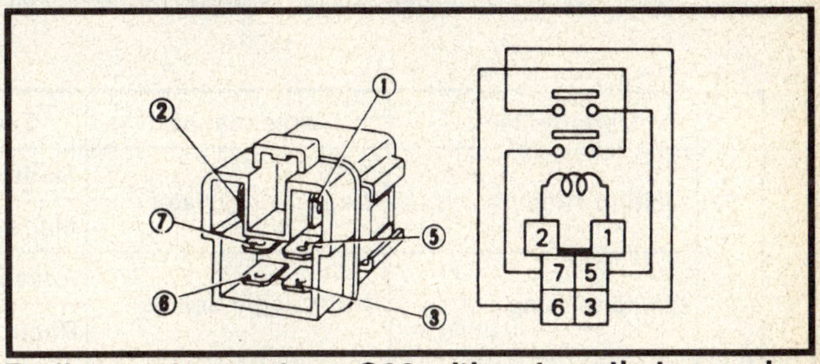

Radiator relay testing—G20 with automatic transaxle

SAFETY RELAY

Testing

This test applies to the M30. Check the continuity between terminals 2 and 3 in the following manner. Apply 12 volts direct current between terminals 1 and 2 with the negative lead going to terminal 1 and the positive lead going to terminal 2. There should be continuity between terminals 2 and 3. Now switch the polarity of the voltage supply so that the positive lead goes to terminal 1 and the negative lead to terminal 2. There should not be continuity between 2 and 3. Replace the relay if it is faulty.

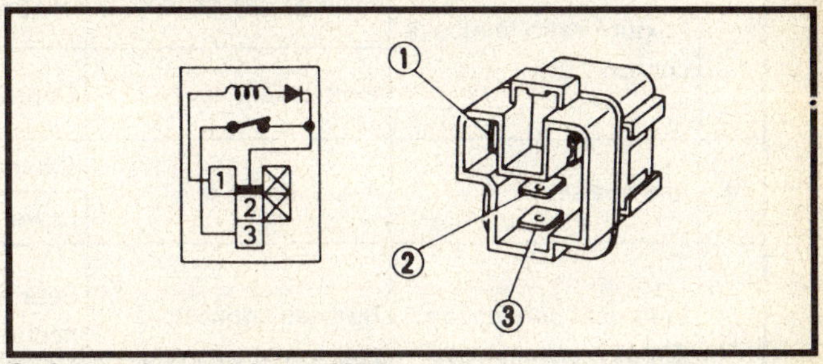

Safety relay testing—M30

NEUTRAL AND PARK RELAY

Testing

This test applies to the Q45. Check the continuity between terminals 3 and 4, using the following procedure. When applying 12 volts direct current supply between terminals 1 and 2, there should be continuity between terminals 3 and 4. With no current supply there should be no continuity between 3 and 4. Replace the relay if it fails either of these two checks.

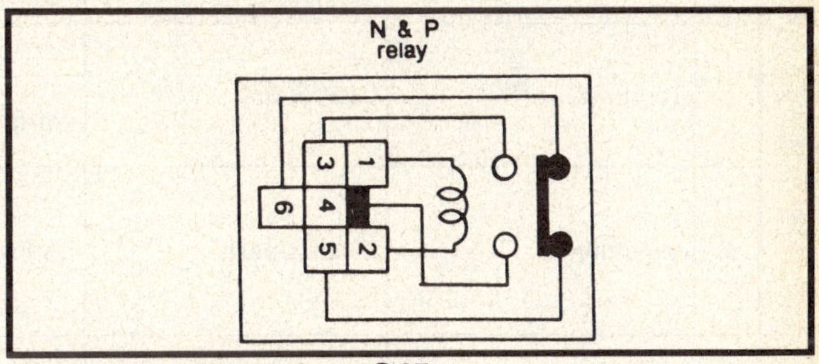

Neutral and park relay—Q45

INTERMITTENT PROBLEMS

There are techniques and procedures that can be used to better locate an intermittent problem. The goal of some of these techniques is to induce the problem to occur.

NOTE: Poor electrical connections and grounds are very frequently the cause of intermittent problems.

	Variable factor	Influential part	Target condition	Service procedure
1	Mixture ratio	Pressure regulator	Made lean	Remove vacuum hose and apply vacuum.
			Made rich	Remove vacuum hose and apply pressure.
2	Ignition timing	Crank angle sensor	Advanced	Rotate distributor clockwise.
			Retarded	Rotate distributor counterclockwise.
3	Mixture ratio feedback control	Exhaust gas sensor	Suspended	Disconnect exhaust gas sensor harness connector.
		Control unit	Operation check	Perform self-diagnosis (Mode II) at 2,000 rpm.
4	Idle speed	A.A.C. valve	Raised	Turn idle adjusting screw counterclockwise.
			Lowered	Turn idle adjusting screw clockwise.
5	Electrical connection (Electric continuity)	Harness connectors and wires	Poor electrical connection or improper wiring	Tap or wiggle.
				Race engine rapidly. See if the torque reaction of the engine unit causes electric breaks.
6	Temperature	Control unit	Cooled	Cool with an icing spray or similar device.
			Warmed	Heat with a hair drier. **[WARNING: Do not overheat the unit.]**
7	Moisture	Electric parts	Damp	Wet. **[WARNING: Do not directly pour water on components. Use a mist sprayer.]**
8	Electric loads	Load switches	Loaded	Turn on headlamps, air conditioner, rear defogger, etc.
9	Idle switch condition	Control unit	ON-OFF switching	Rotate throttle sensor body.
10	Ignition spark	Timing light	Spark power check	Try to flash timing light for each cylinder using ignition coil adapter (S.S.T.).

Intermittent trouble diagnosis – Infiniti

G20 ENGINE DIAGNOSTICS

The engine diagnostic procedures for each model are grouped into 3 areas: engine fault codes, engine symptoms and engine components. The proper diagnostic procedure to follow for testing is also given.

Engine Fault Codes

Code 11 – Crank Angle Sensor – Diagnostic Procedure 23.
Code 12 – Air Flow Meter – Diagnostic Procedure 24.
Code 13 – Engine Temperature Sensor – Diagnostic Procedure 25.
Code 14 – Vehicle Speed Sensor – Diagnostic Procedure 26.
Code 21 – Ignition Signal – Diagnostic Procedure 27.
Code 31 – ECCS Control Unit – Diagnostic Procedure 28.
Code 32 – EGR Function, EGR Control – Diagnostic Procedure 29.
Code 33 – Exhaust Gas Sensor – Diagnostic Procedure 30.
Code 34 – Detonation Sensor – Diagnostic Procedure 31.
Code 35 – Exhaust Gas Temperature Sensor – Diagnostic Procedure 32.
Code 43 – Throttle Sensor – Diagnostic Procedure 33.
Code 45 – Injector Leak – Diagnostic Procedure 34.

Engine Symptoms

High Idling after Warm-up – Diagnostic Procedure 1.
Hunting – Diagnostic Procedure 2.
Unstable Idle – Diagnostic Procedure 3.
Hard to Start or Impossible to Start when the Engine is Cold – Diagnostic Procedure 4.
Hard to Start or Impossible to Start when the Engine is Hot – Diagnostic Procedure 5.
Hard to Start or Impossible to Start under Normal Conditions – Diagnostic Procedure 6.
Hesitation when the Engine is Hot – Diagnostic Procedure 7.
Hesitation when the Engine is Cold – Diagnostic Procedure 8.
Hesitation under Normal Conditions – Diagnostic Procedure 9.

Engine Stalls when Turning—Diagnostic Procedure 10.
Engine Stalls when the Engine is Hot—Diagnostic Procedure 11.
Engine Stalls when the Engine is Cold—Diagnostic Procedure 12.
Engine Stalls when Momentarily Stepping on the Accelerator—Diagnostic Procedure 13.
Engine Stalls after Decelerating—Diagnostic Procedure 14.
Engine Stalls when Accelerating or when Driving at Constant Speed—Diagnostic Procedure 15.
Engine Stalls when the Electrical Load is Heavy—Diagnostic Procedure 16.
Lack of Power and Stumble—Diagnostic Procedure 17.
Detonation—Diagnostic Procedure 18.
Surge—Diagnostic Procedure 19.
Backfire through the Intake—Diagnostic Procedure 20.
Backfire through the Exhaust—Diagnostic Procedure 21.

Engine Components

Main Power Supply and Ground Circuit—Diagnostic Procedure 22.
Start Signal—Diagnostic Procedure 35.
AIV Control—Diagnostic Procedure 36.
Injector—Diagnostic Procedure 37.
Fuel Pump—Diagnostic Procedure 38.
Air Regulator—Diagnostic Procedure 39.
AAC Valve—Diagnostic Procedure 40.
Radiator Fan Control—Diagnostic Procedure 41.
Power Steering Oil Pressure Switch—Diagnostic Procedure 42.
Neutral Switch/Inhibitor Switch—Diagnostic Procedure 43.

M30 ENGINE DIAGNOSTICS

The engine diagnostic procedures for each model are grouped into 3 areas: engine fault codes, engine symptoms and engine components. The proper diagnostic procedure to follow for testing is also given.

Engine Fault Codes

Code 11—Crank Angle Sensor—Diagnostic Procedure 23.
Code 12—Air Flow Meter—Diagnostic Procedure 24.
Code 13—Engine Temperature Sensor—Diagnostic Procedure 25.
Code 14—Vehicle Speed Sensor—Diagnostic Procedure 26.
Code 21—Ignition Signal—Diagnostic Procedure 27.
Code 22—Fuel Pump Control—Diagnostic Procedure 28.
Code 31—ECCS Control Unit—Diagnostic Procedure 29.
Code 32—EGR Function and EGR Control—Diagnostic Procedure 30.
Code 33—Exhaust Gas Sensor—Diagnostic Procedure 31.
Code 34—Detonation Sensor—Diagnostic Procedure 32.
Code 35—Exhaust Gas Temperature Sensor—Diagnostic Procedure 33.
Code 42—Fuel Temperature Sensor—Diagnostic Procedure 34.
Code 43—Throttle Sensor—Diagnostic Procedure 35.
Code 45—Injector Leak—Diagnostic Procedure 36.
Code 51—Injector Circuit—Diagnostic Procedure 37.

Engine Symptoms

High Idling after Warm-up—Diagnostic Procedure 1.
Hunting—Diagnostic Procedure 2.
Unstable Idle—Diagnostic Procedure 3.
Hard to Start or Impossible to Start when the Engine is Cold—Diagnostic Procedure 4.
Hard to Start or Impossible to Start when the Engine is Hot—Diagnostic Procedure 5.
Hard to Start or Impossible to Start under Normal Conditions—Diagnostic Procedure 6.
Hesitation when the Engine is Hot—Diagnostic Procedure 7.
Hesitation when the Engine is Cold—Diagnostic Procedure 8.
Hesitation under Normal Conditions—Diagnostic Procedure 9.
Engine Stalls when Turning—Diagnostic Procedure 10.
Engine Stalls when the Engine is Hot—Diagnostic Procedure 11.
Engine Stalls when the Engine is Cold—Diagnostic Procedure 12.
Engine Stalls when Momentarily Stepping on the Accelerator—Diagnostic Procedure 13.
Engine Stalls after Decelerating—Diagnostic Procedure 14.
Engine Stalls when Accelerating or when Driving at Constant Speed—Diagnostic Procedure 15.
Engine Stalls when the Electrical Load is Heavy—Diagnostic Procedure 16.
Lack of Power and Stumble—Diagnostic Procedure 17.
Detonation—Diagnostic Procedure 18.
Surge—Diagnostic Procedure 19.
Backfire through the Intake—Diagnostic Procedure 20.
Backfire through the Exhaust—Diagnostic Procedure 21.

Engine Components

Main Power Supply and Ground Circuit—Diagnostic Procedure 22.
Throttle Valve Switch—Diagnostic Procedure 38.
Start Signal—Diagnostic Procedure 39.
PRVR Control—Diagnostic Procedure 40.
Air Regulator—Diagnostic Procedure 41.
AAC Valve—Diagnostic Procedure 42.
A/T Control Unit (Neutral Signal) Circuit—Diagnostic Procedure 43.
Fail-Safe System for Crank Angle Sensor and CPU of ECU—Diagnostic Procedure 44.

Q45 ENGINE DIAGNOSTICS

The engine diagnostic procedures for each model are grouped into 3 areas: engine fault codes, engine symptoms and engine components. The proper diagnostic procedure to follow for testing is also given.

Engine Fault Codes

Code 11—Crank Angle Sensor—Diagnostic Procedure 2.
Code 12—Air Flow Meter—Diagnostic Procedure 3.
Code 13—Engine Temperature Sensor—Diagnostic Procedure 4.
Code 14—Vehicle Speed Sensor—Diagnostic Procedure 5.
Code 16—TCS Signal—Diagnostic Procedure 6.
Code 21—Ignition Signal—Diagnostic Procedure 7.
Code 31—ECCS Control Unit—Diagnostic Procedure 8.
Code 32—EGR Function—Diagnostic Procedure 9.
Code 33 and 53—Exhaust Gas Sensors L.H. and R.H.—Diagnostic Procedure 10.
Code 34—Detonation Sensor—Diagnostic Procedure 11.
Code 35—Exhaust Gas Temperature Sensor—Diagnostic Procedure 12.
Code 43—Throttle Sensor—Diagnostic Procedure 13.
Code 45—Injector Leak—Diagnostic Procedure 14.
Code 46—Secondary Throttle Sensor—Diagnostic Procedure 15.
Code 51—Injector—Diagnostic Procedure 16.
Code 54—A/T Control—Diagnostic Procedure 17.

Engine Symptoms

High Idling after Warm-up—Diagnostic Procedure 32.
Hunting—Diagnostic Procedure 33.
Unstable Idle—Diagnostic Procedure 34.
Hard to Start or Impossible to Start when the Engine is Cold—Diagnostic Procedure 35.
Hard to Start or Impossible to Start when the Engine is Hot—Diagnostic Procedure 36.
Hard to Start or Impossible to Start under Normal Conditions—Diagnostic Procedure 37.
Hesitation when the Engine is Hot—Diagnostic Procedure 38.
Hesitation when the Engine is Cold—Diagnostic Procedure 39.
Hesitation under Normal Conditions—Diagnostic Procedure 40.
Engine Stalls when Turning—Diagnostic Procedure 41.
Engine Stalls when the Engine is Hot—Diagnostic Procedure 42.
Engine Stalls when the Engine is Cold—Diagnostic Procedure 43.
Engine Stalls when Momentarily Stepping on the Accelerator—Diagnostic Procedure 44.
Engine Stalls after Decelerating—Diagnostic Procedure 45.
Engine Stalls when Accelerating or when Driving at Constant Speed—Diagnostic Procedure 46.
Engine Stalls when the Electrical Load is Heavy—Diagnostic Procedure 47.
Lack of Power and Stumble—Diagnostic Procedure 48.
Detonation—Diagnostic Procedure 49.
Surge—Diagnostic Procedure 50.
Backfire through the Intake—Diagnostic Procedure 51.
Backfire through the Exhaust—Diagnostic Procedure 52.

Engine Components

Main Power Supply and Ground Circuit—Diagnostic Procedure 1.
Crank Angle Sensor—Diagnostic Procedure 18.
Sub-Crank Angle Sensor—Diagnostic Procedure 19.
Ignition Signal—Diagnostic Procedure 20.
EGR Control—Diagnostic Procedure 21.
Exhaust Gas Sensor Heater L.H and R.H.—Diagnostic Procedure 22.
Injector—Diagnostic Procedure 23.
Throttle Valve Switch—Diagnostic Procedure 24.
Start Signal—Diagnostic Procedure 25.
Fuel Pump Control—Diagnostic Procedure 26.
AAC Valve—Diagnostic Procedure 27.
Power Steering Oil Pressure Switch—Diagnostic Procedure 28.
Inhibitor Switch—Diagnostic Procedure 29.
Valve Timing Control—Diagnostic Procedure 30.
Canister Purge Control—Diagnostic Procedure 31.

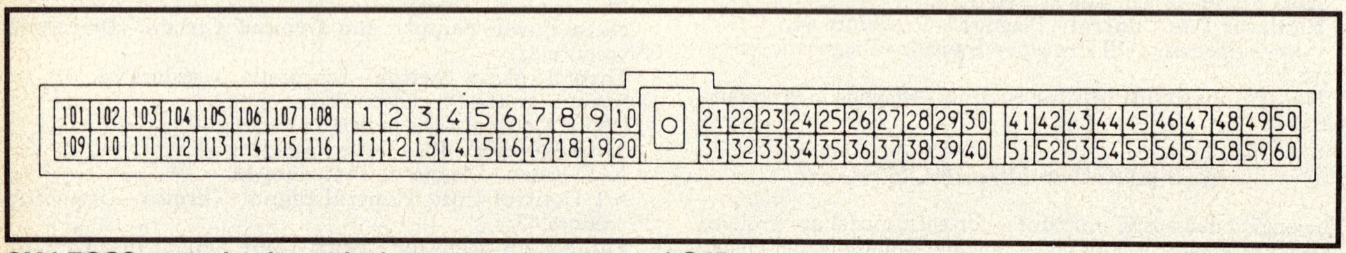

SMJ ECCS control unit terminal arrangement—M30 and Q45

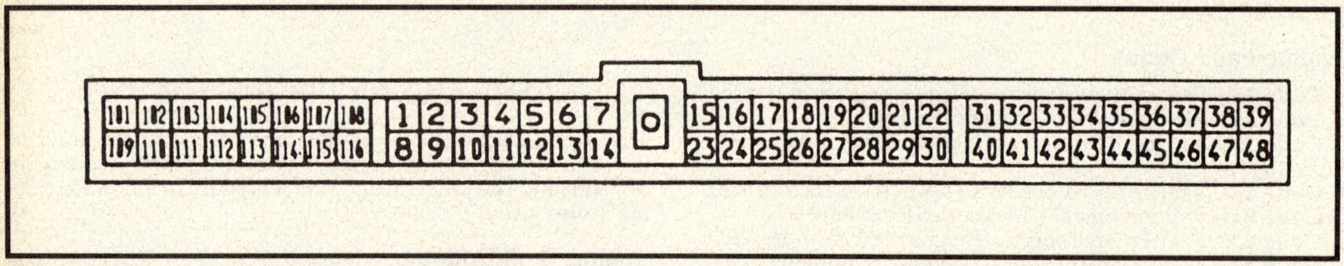

SMJ ECCS control unit terminal arrangement—G-20

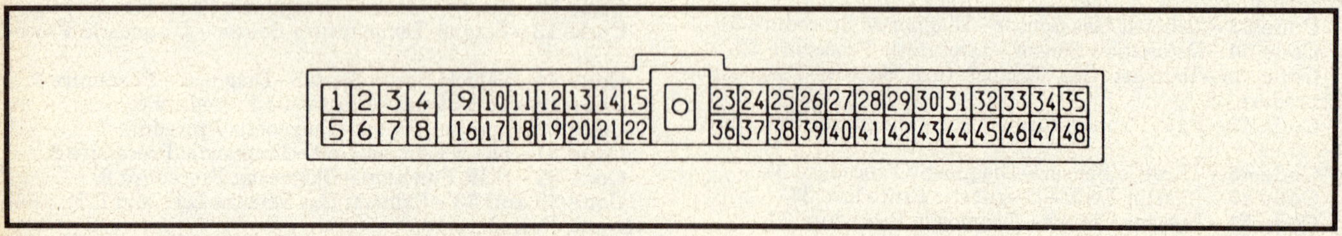

SMJ transmission control unit terminal arrangement

KEY TO SYMBOLS SIGNIFYING MEASUREMENTS OR PROCEDURES

Symbol	Symbol explanation	Symbol	Symbol explanation
	Check after disconnecting the connector to be measured.		Current should be measured with an ammeter.
	Check after connecting the connector to be measured.		Procedure with CONSULT
	Insert key into ignition switch.		Procedure without CONSULT
	Remove key from ignition switch.		A/C switch is "OFF".
	Turn ignition switch to "OFF" position.		A/C switch is "ON".
	Turn ignition switch to "ON" position.		Fan switch is "ON". (At any position except for "OFF" position)
	Turn ignition switch to "START" position.	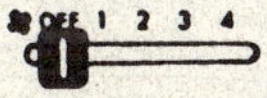	Fan switch is "OFF".
	Turn ignition switch from "OFF" to "ACC" position.		Apply battery voltage directly to components.
	Turn ignition switch from "ACC" to "OFF" position.		Drive vehicle.
	Turn ignition switch from "OFF" to "ON" position.		Disconnect battery negative cable.
	Turn ignition switch from "ON" to "OFF" position.		Depress brake pedal.
	Do not start engine, or check with engine stopped.		Release brake pedal.
	Start engine, or check with engine running.		Depress accelerator pedal.
	Apply parking brake.		Release accelerator pedal.
	Release parking brake.	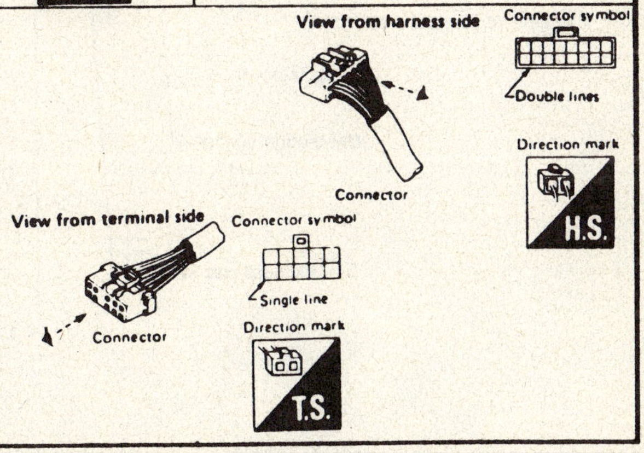	
	Check after engine is warmed up sufficiently.		
	Voltage should be measured with a voltmeter.		
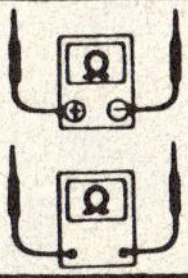	Circuit resistance should be measured with an ohmmeter.		

1990–92 2.0L ENGINE – ECCS COMPONENT LOCATIONS – G20

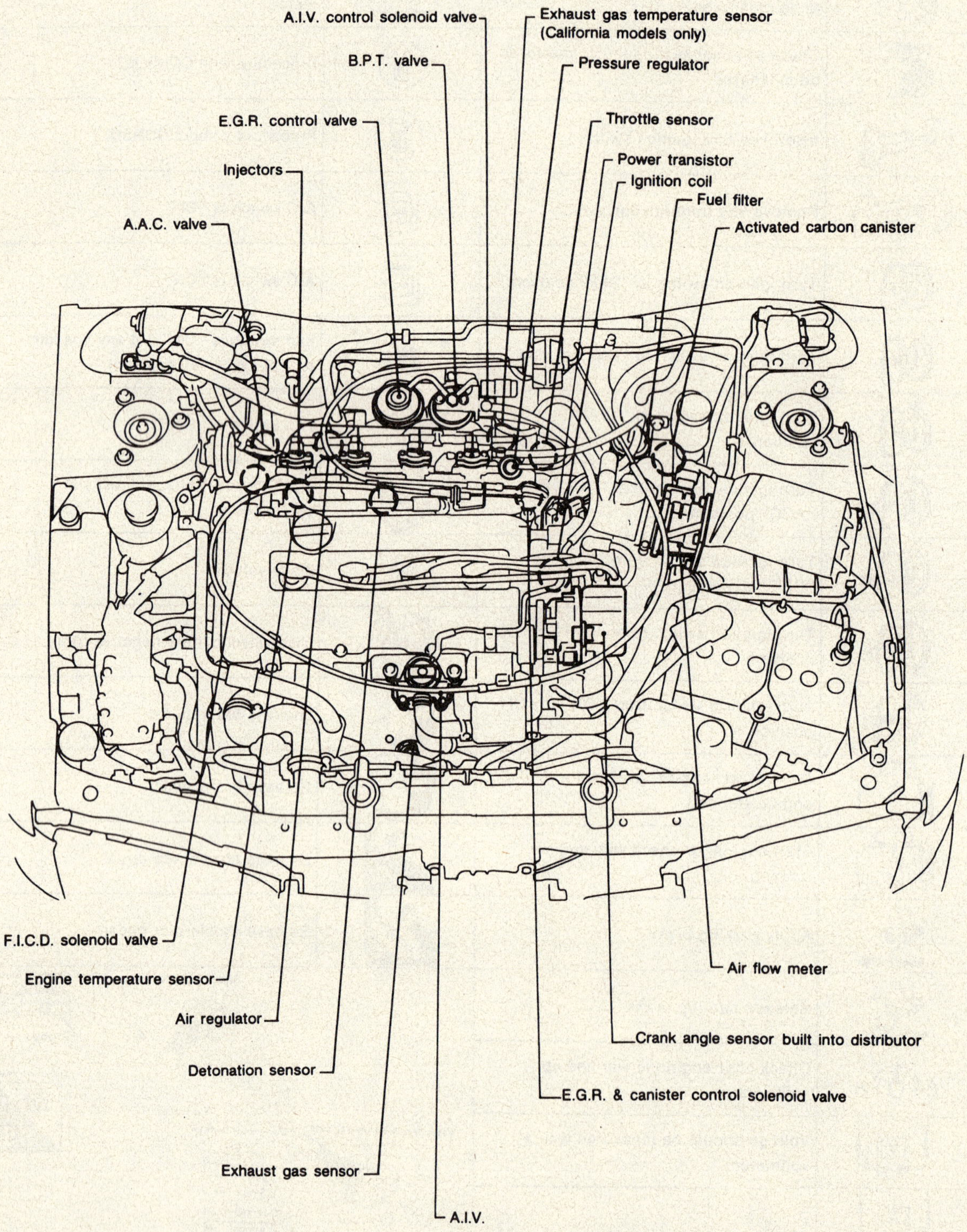

1990–92 2.0L ENGINE—ECCS SYSTEM SCHEMATIC—G20

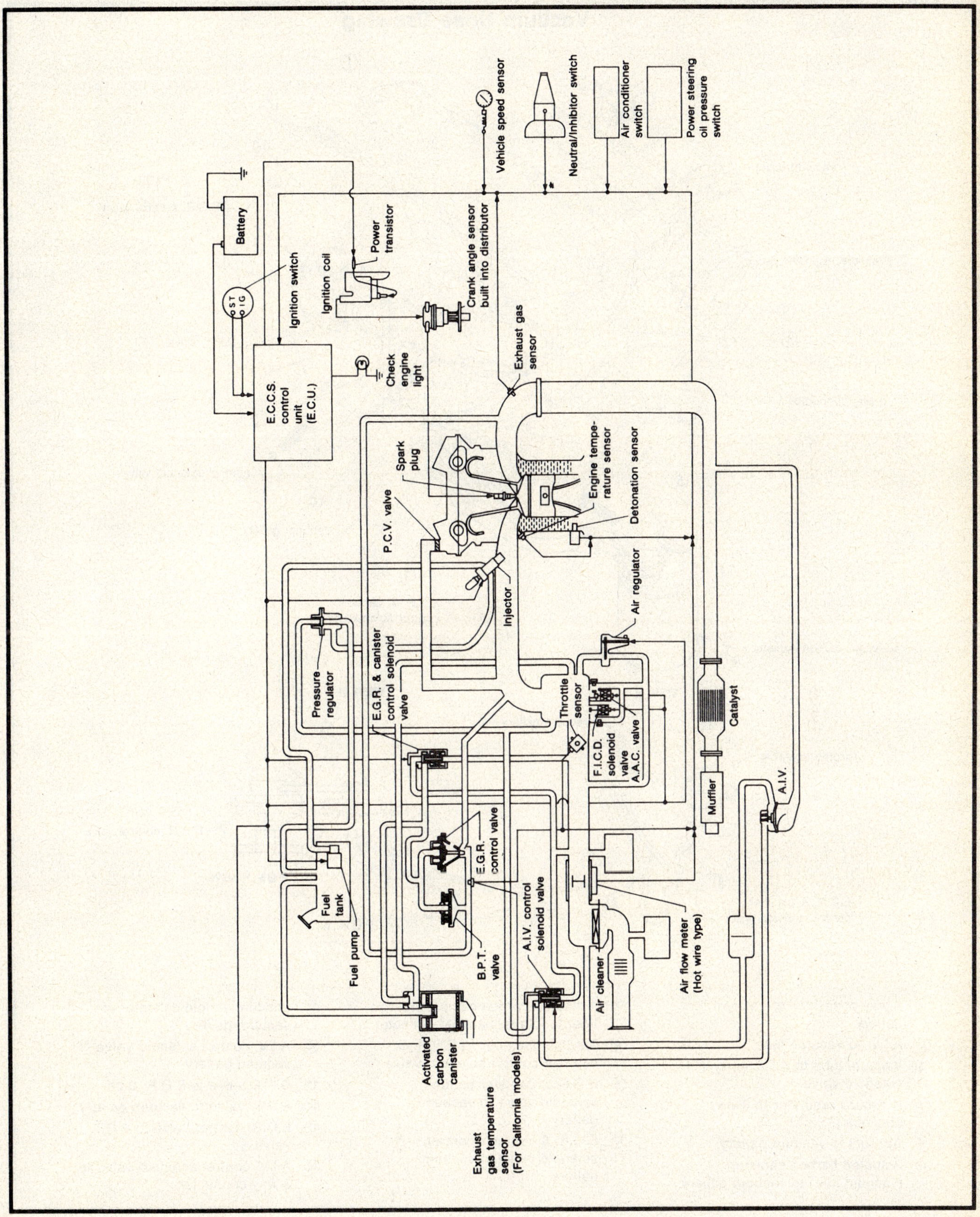

1990–92 2.0L ENGINE – ECCS VACUUM CONTROL SCHEMATIC – G20

Vacuum Hose Drawing

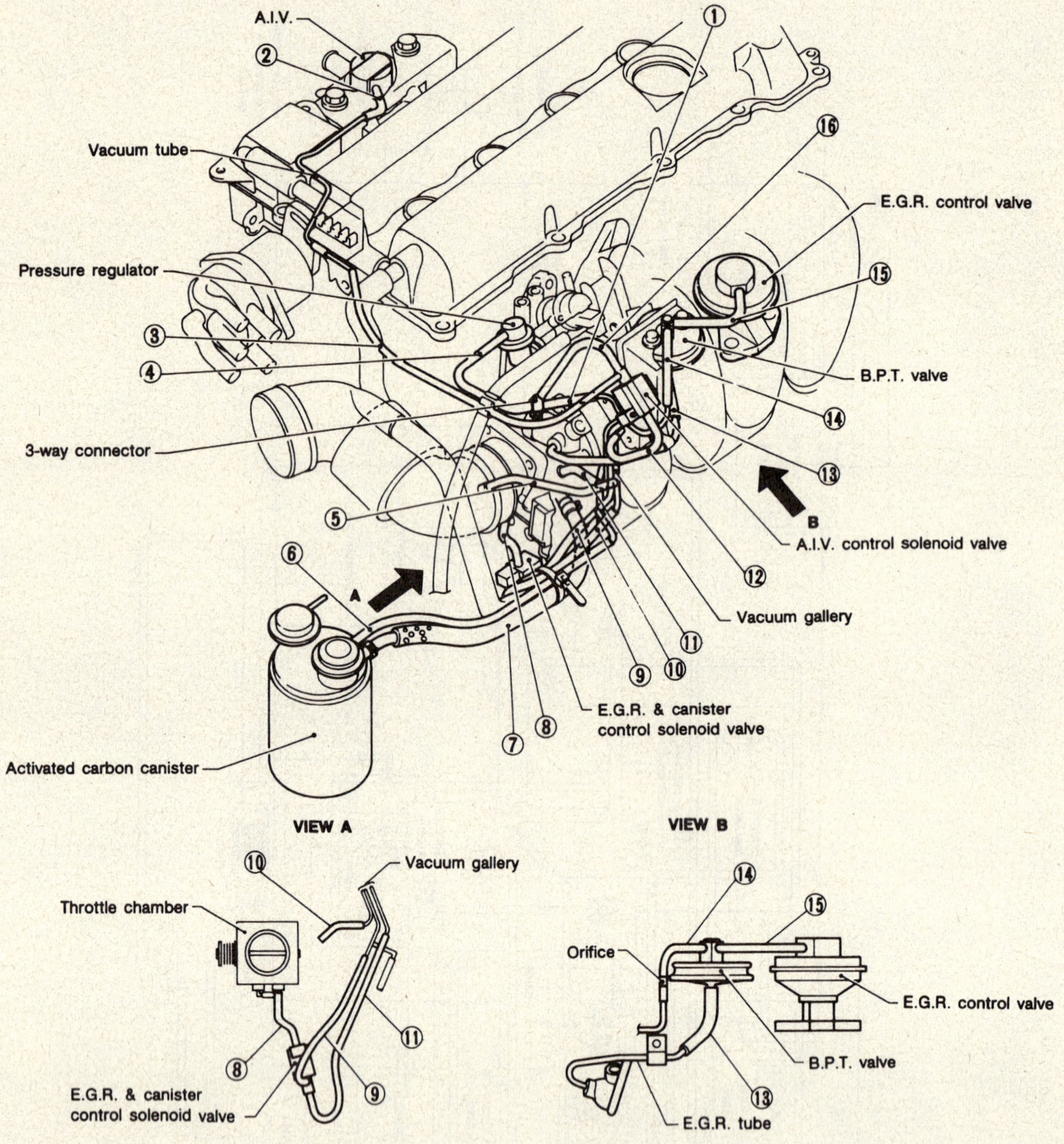

1. 3-way connector to vacuum gallery
2. A.I.V. to vacuum tube
3. Vacuum tube to A.I.V. control solenoid valve
4. Pressure regulator to 3-way connector
5. Air duct to vacuum gallery
6. Actuated carbon canister (vacuum line) to vacuum gallery
7. Activated carbon canister (purge line) to intake manifold collector
8. Throttle chamber to E.G.R. & canister control solenoid valve
9. E.G.R. & canister control solenoid valve to vacuum gallery
10. E.G.R. & canister control solenoid valve to vacuum gallery
11. Intake manifold collector to vacuum gallery
12. A.I.V. control solenoid valve to vacuum gallery
13. B.P.T. valve to E.G.R. tube
14. B.P.T. valve to vacuum gallery
15. E.G.R. control valve to B.P.T. valve
16. A.I.V. control solenoid valve to 3-way connector

1990–92 2.0L ENGINE—ECCS CIRCUIT DIAGRAM—G20

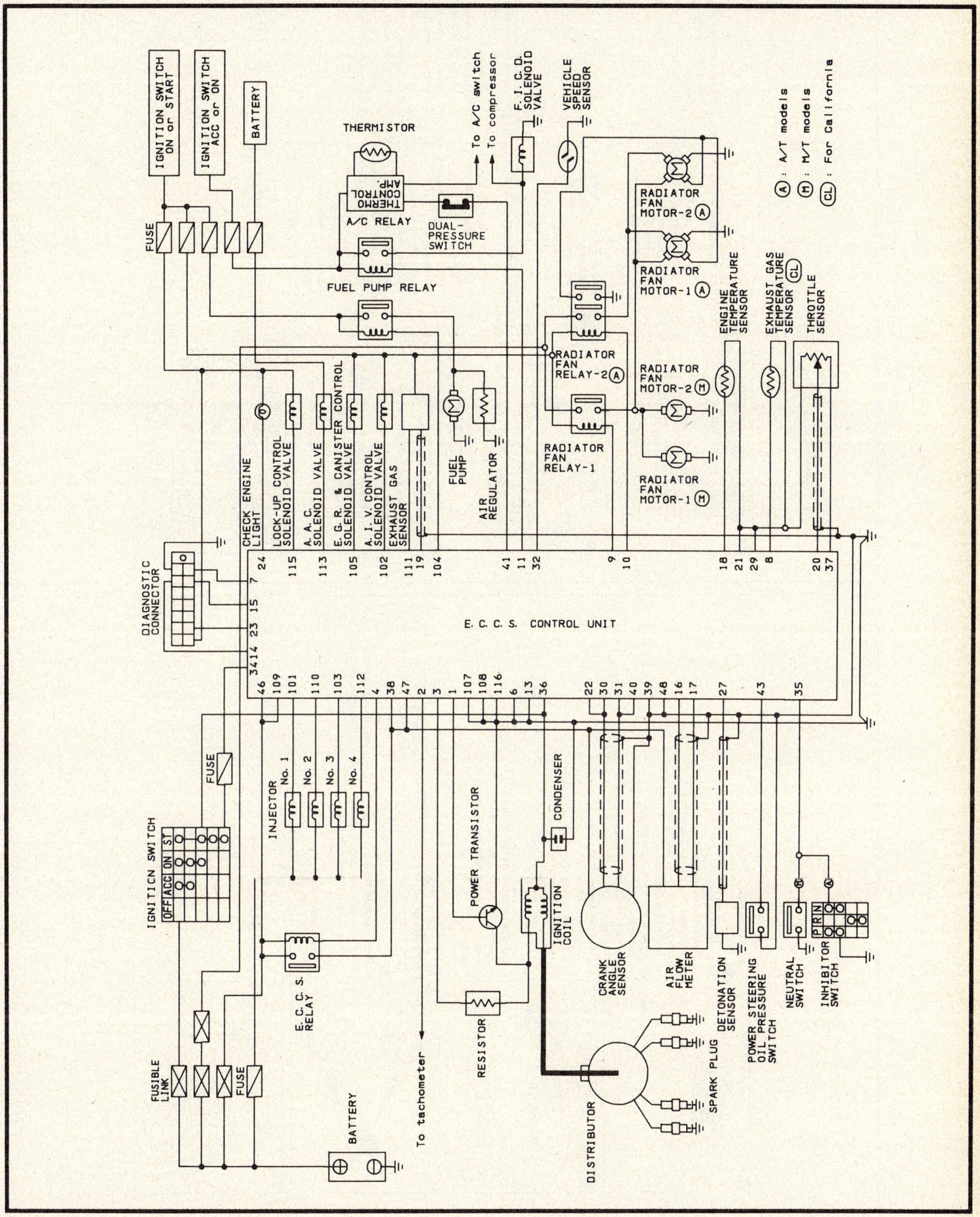

Basic Inspection

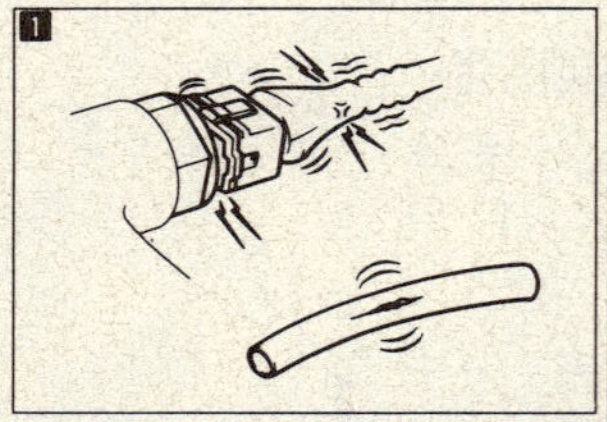

1 BEFORE STARTING

1. Check service records for any recent repairs that may indicate a related problem, or the current need for scheduled maintenance.
2. Open engine hood and check the following:
- Harness connectors for proper connections
- Vacuum hoses for splits, kinks, and proper connections
- Wiring for proper connections, pinches, and cuts

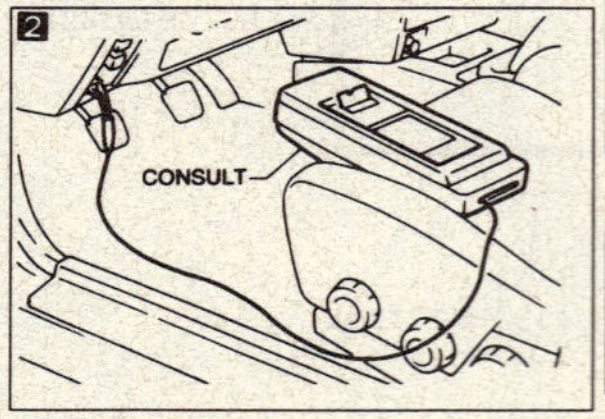

2 CONNECT CONSULT TO THE VEHICLE

Connect "CONSULT" to the diagnostic connector and select "ENGINE" from the menu.

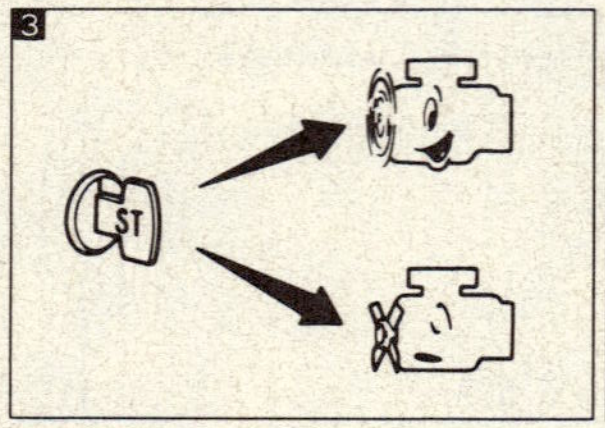

3 DOES ENGINE START?

No → Go to 6.

Yes ↓

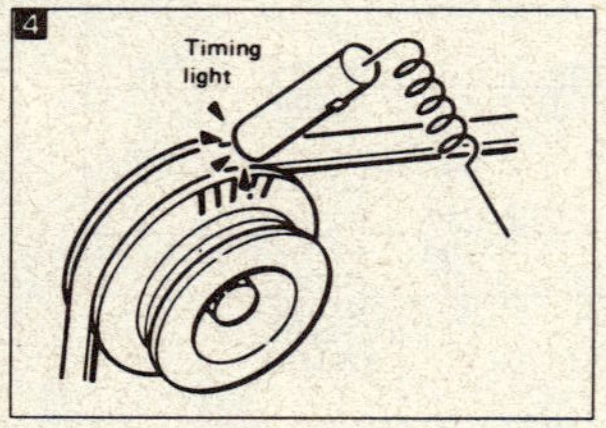

4 CHECK IGNITION TIMING.

Warm up engine sufficiently and check ignition timing at idle using timing light.

Ignition timing:
15° ± 2° B.T.D.C.

N.G. → Adjust ignition timing by turning crank angle sensor.

O.K. ↓

(Go to Ⓐ on next page.)

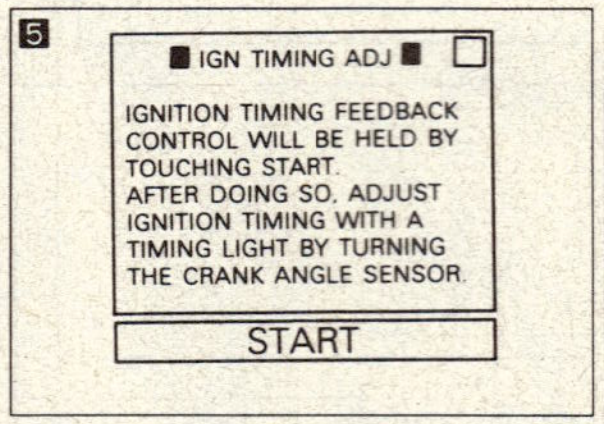

Ⓐ

5 CHECK IDLE ADJ. SCREW INITIAL SET RPM.

1. Select "IGN TIMING ADJ" in "WORK SUPPORT" mode.
2. When touching "START", does engine rpm fall to 750 ± 50 rpm (in "N" position)?

—OR—

When disconnecting throttle sensor harness connector, does engine rpm fall to 750 ± 50 rpm (in "N" position)?

No → Adjust engine rpm by turning idle adjusting screw.

Yes ↓

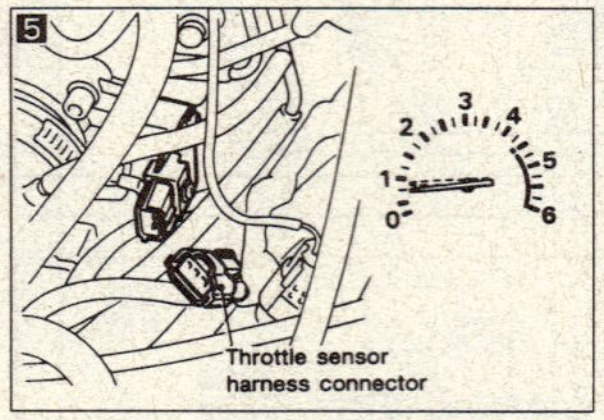

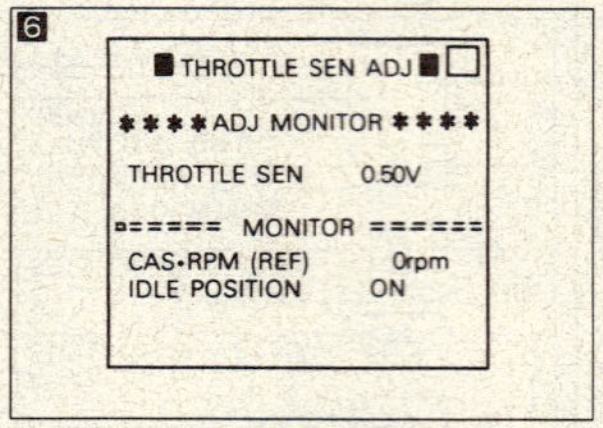

6 CHECK THROTTLE SENSOR IDLE POSITION.

1. Perform "THROTTLE SEN. ADJ" in "WORK SUPPORT" mode.
2. Check that output voltage of throttle sensor is 0.45 to 0.55V. (Throttle valve fully closes.) and "IDLE POSITION" stays "ON".

—OR—

Measure output voltage of throttle sensor using voltmeter, and check that it is 0.45 to 0.55V. (Throttle valve fully closed.)

N.G. →
1. Adjust output voltage by rotating throttle sensor body.
2. Disconnect throttle sensor harness connector for a few seconds and then reconnect it.
3. Confirm that "IDLE POSITION" stays "ON".

O.K. ↓

(Go to Ⓑ on next page.)

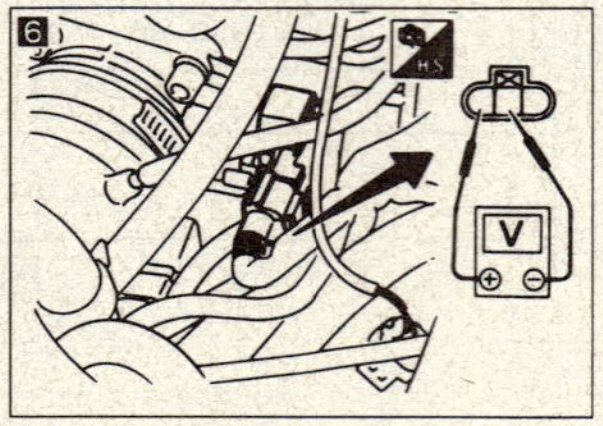

1990–92 2.0L ENGINE—BASIC INSPECTION—G20

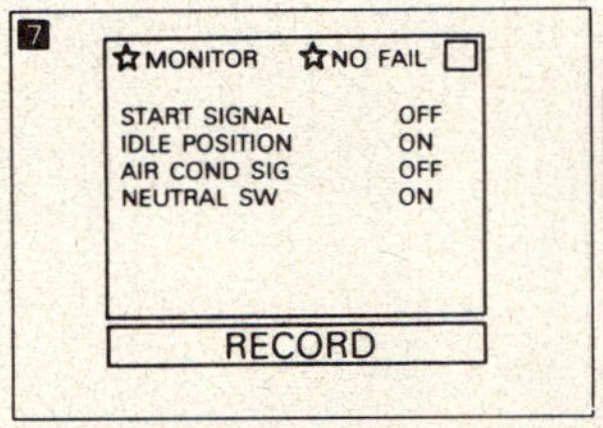

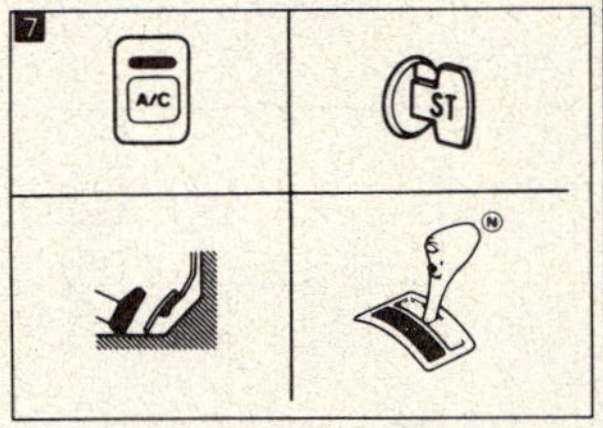

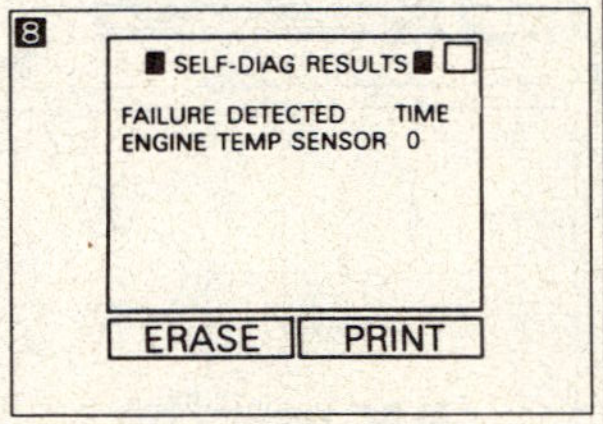

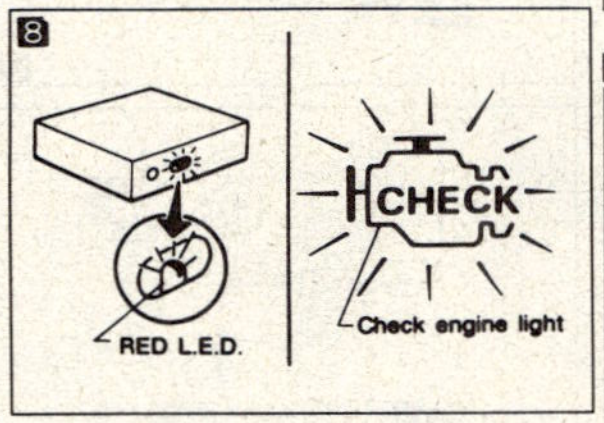

③

7 CHECK SWITCH INPUT SIGNAL.

Select the following switches in "DATA MONITOR" mode,

a) Start signal,
b) Idle position,
c) Air conditioner signal,
d) Neutral (Parking) switch,

and check the switches' ON-OFF operation.

—OR—

Remove E.C.U. from behind audio system panel and check the above switches' ON-OFF operation using voltmeter at each E.C.U. terminal.

Switch	Condition	Voltage (V)
Start signal	IGN ON → IGN START	0 → Battery voltage
Idle position	Accelerator pedal released → Accelerator pedal fully depressed	0.45 - 0.55 → Approx. 4.0
A/C signal	A/C OFF → A/C ON (Engine running)	Battery voltage → 0
Neutral (Parking) switch	Shift lever is Neutral position → Except Neutral position	0 → Battery voltage

N.G. → Repair or replace the malfunctioning switch or its circuit.

O.K. ↓

8 READ SELF-DIAGNOSTIC RESULTS.

1. Perform "SELF-DIAG RESULTS" mode.
2. Read out self-diagnostic results.
3. Is a failure detected?

—OR—

1. Set "Self-diagnostic results mode" in Mode II.
2. Count the number of RED L.E.D. or check engine light flashes and read out the codes.
3. Are the codes shown?

Yes → Go to the relevant inspection procedure.

No ↓

INSPECTION END

1990–92 2.0L ENGINE—DIAGNOSTIC PROCEDURE—G20

MAIN POWER SUPPLY AND GROUND CIRCUIT (Not self-diagnostic item)

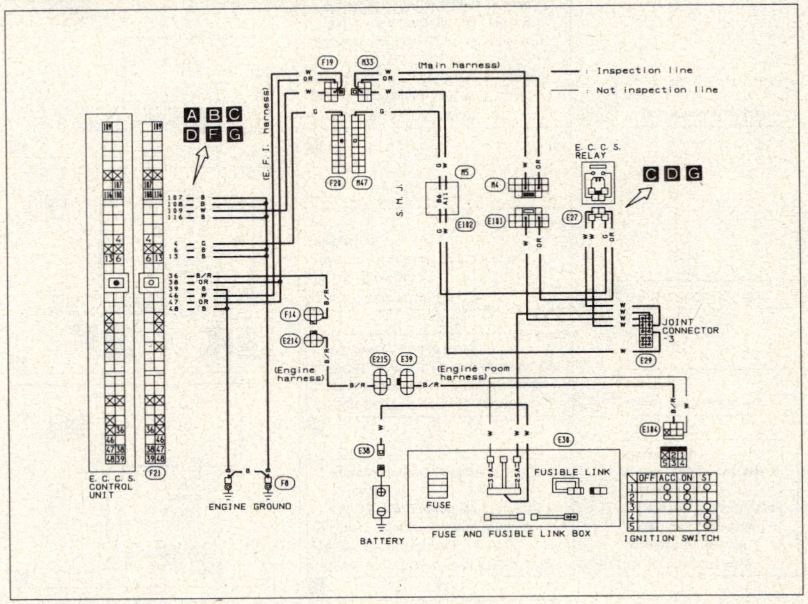

Harness layout

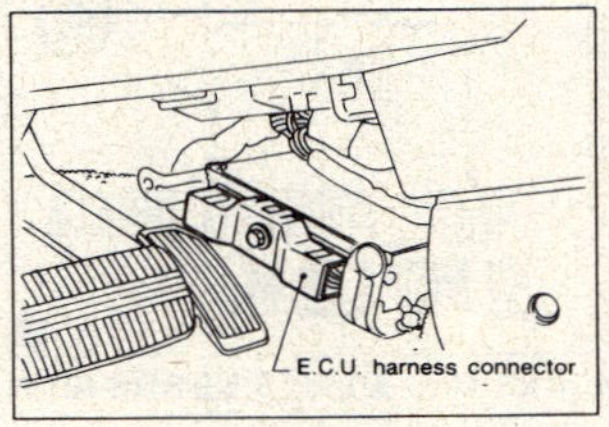

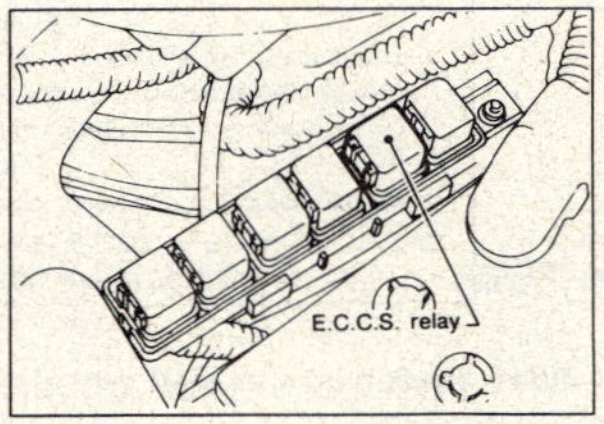

MAIN POWER SUPPLY AND GROUND CIRCUIT (Not self-diagnostic item)

A
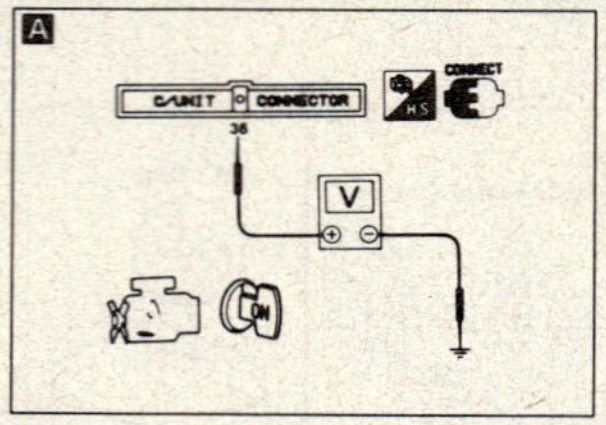

B
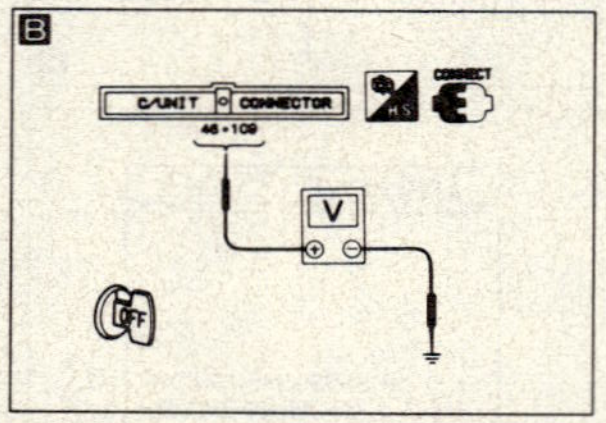

C
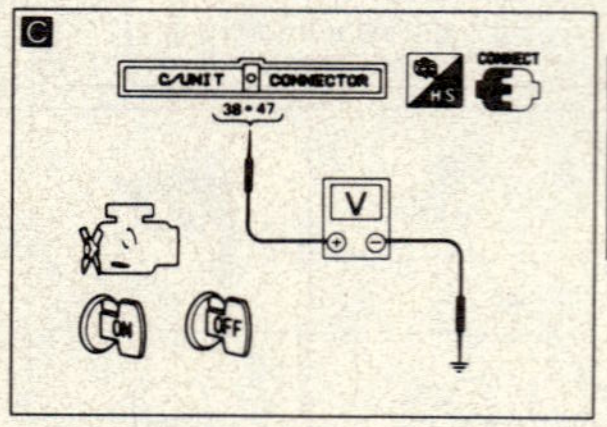

INSPECTION START

Start engine.
Is engine running?

No → A

A CHECK POWER SUPPLY-I.
1) Turn ignition switch "ON".
2) Check voltage between E.C.U. terminal (38) and ground with CONSULT or tester.
Voltage: Battery voltage
If N.G., check the following.
- Harness connectors (F14), (E214)
- Harness connectors (E215), (E39)
- Harness continuity between E.C.U. and ignition switch

If N.G., repair harness or connectors.

O.K. → Go to "CHECK GROUND CIRCUIT".

Yes ↓

B CHECK POWER SUPPLY-II.
1) Stop engine.
2) Check voltage between E.C.U. terminals (48), (109) and ground with CONSULT or tester.
Voltage: Battery voltage

N.G. → Check the following.
- Harness connectors (F19), (M33)
- Harness connectors (M4), (E181)
- Harness connectors (M5), (E182)
- Joint connector-3
- 25A fusible link
- Harness continuity between E.C.U. and battery

If N.G., repair harness or connectors.

O.K. ↓

C CHECK POWER SUPPLY-III.
1) Turn ignition switch "ON" and then "OFF".
2) Check voltage between E.C.U. terminals (38), (47) and ground with CONSULT or tester.
Voltage:
Ignition switch "ON" and for a few seconds after turning ignition switch "OFF"
Battery voltage
A few seconds after turning ignition switch "OFF"
Approximately 0V

O.K. → Go to "CHECK GROUND CIRCUIT".

N.G.:
Case-1: Battery voltage does not exist for a few seconds. → (A)
Case-2: Battery voltage exists for more than a few seconds. → Go to "CHECK COMPONENT (E.C.C.S. relay)".

MAIN POWER SUPPLY AND GROUND CIRCUIT (Not self-diagnostic item)

D
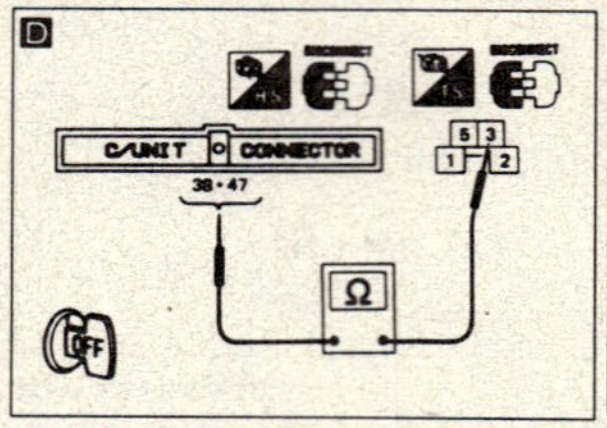

E
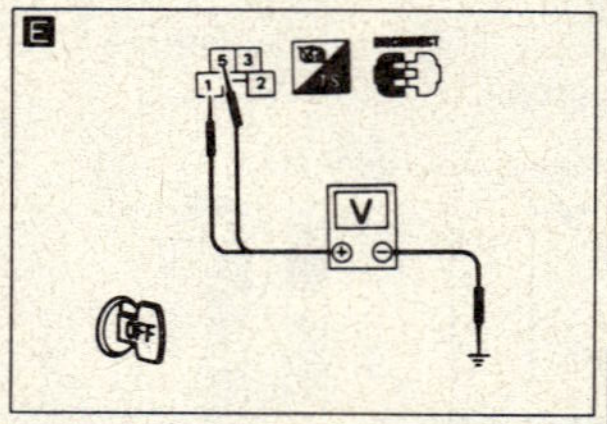

F
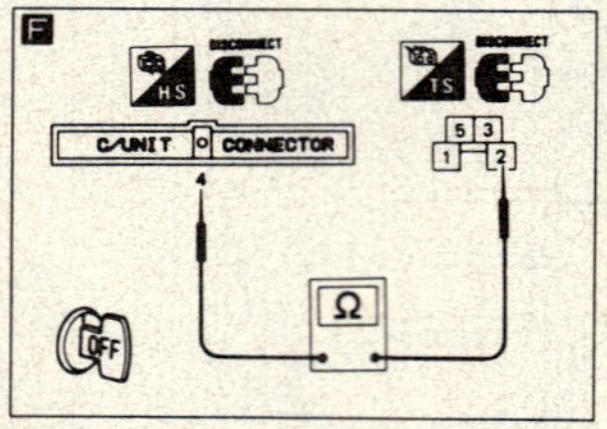

G
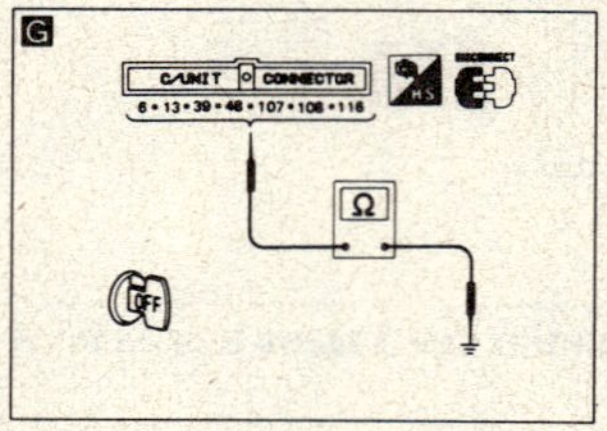

(A) ↓

D CHECK HARNESS CONTINUITY BETWEEN E.C.C.S. RELAY AND E.C.U.
1) Disconnect E.C.U. harness connector.
2) Disconnect E.C.C.S. relay.
3) Check harness continuity between E.C.U. terminals (38), (47) and terminal (3).
Continuity should exist.

N.G. → Repair harness or connectors.

O.K. ↓

E CHECK VOLTAGE BETWEEN E.C.C.S. RELAY AND GROUND.
1) Check voltage between terminals (1), (5) and ground with CONSULT or tester.
Voltage: Battery voltage

N.G. → Check the following.
- Joint connector-3
- Harness continuity between E.C.C.S. relay and battery

If N.G., repair harness or connectors.

O.K. ↓

F CHECK OUTPUT SIGNAL CIRCUIT.
1) Check harness continuity between E.C.U. terminal (4) and terminal (2).
Continuity should exist.

N.G. → Repair harness or connectors.

O.K. ↓

CHECK COMPONENT (E.C.C.S. relay).
Refer to "Electrical Components Inspection".

N.G. → Replace E.C.C.S. relay.

O.K. ↓

G CHECK GROUND CIRCUIT.
1) Turn ignition switch "OFF".
2) Disconnect E.C.U. harness connector.
3) Check harness continuity between E.C.U. terminals (6), (13), (39), (48), (107), (108), (116) and engine ground.
Continuity should exist.

N.G. → Repair harness or connectors.

O.K. ↓

Check E.C.U. pin terminals for damage or the connection of E.C.U. harness connector.

↓

INSPECTION END

Diagnostic Procedure 2

CRANK ANGLE SENSOR (Code No. 11)

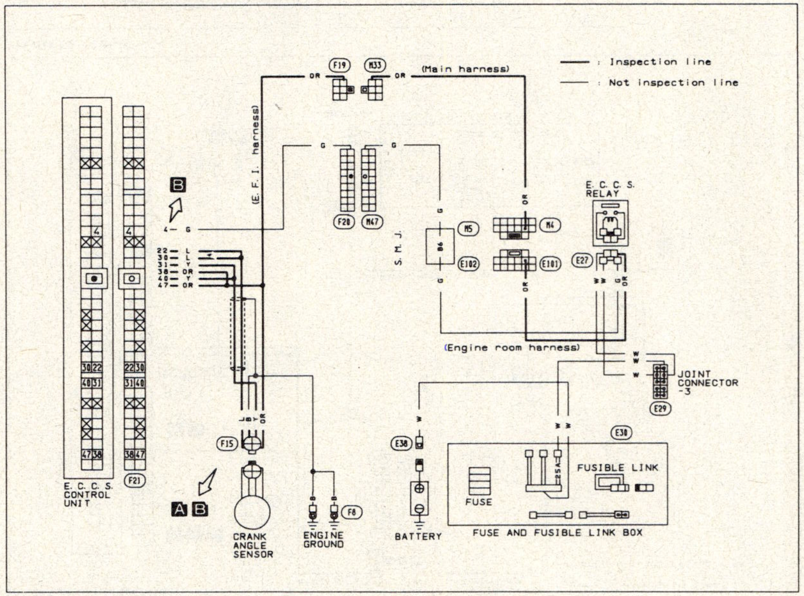

Harness layout

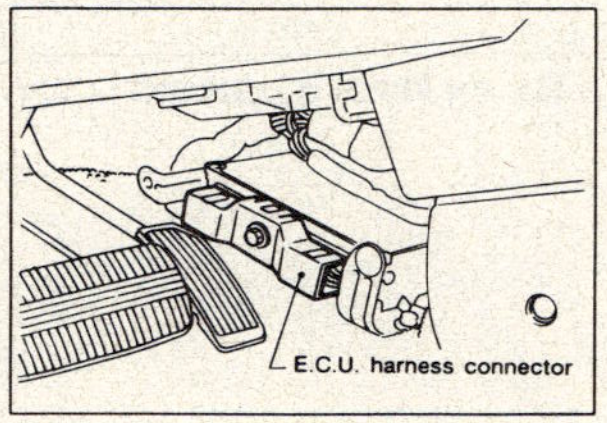

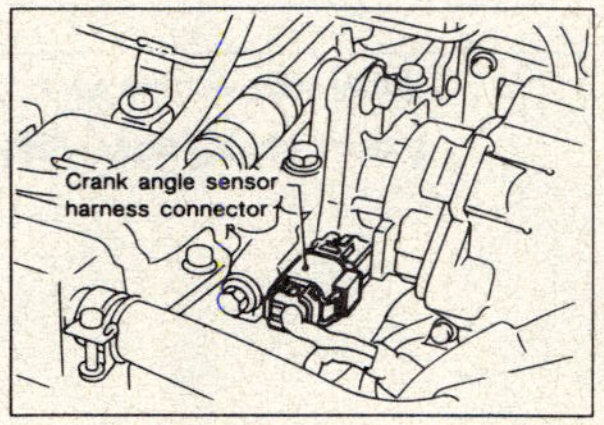

CRANK ANGLE SENSOR (Code No.11)

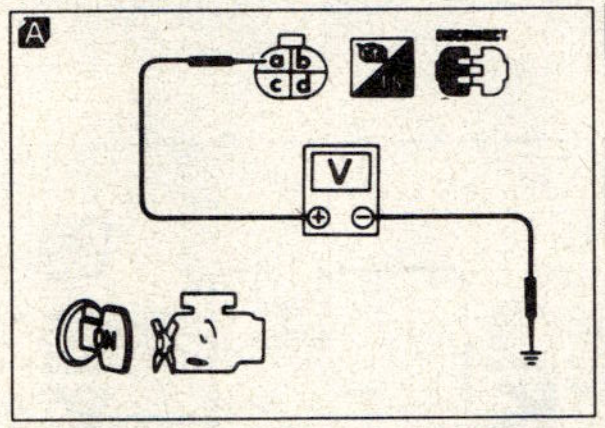

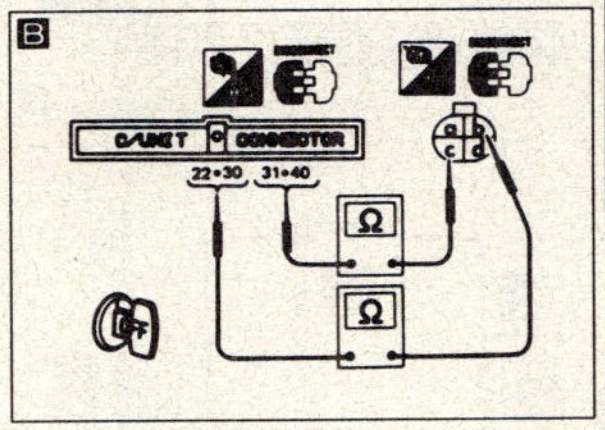

INSPECTION START

↓

PERFORM SELF-DIAGNOSIS.

- O.K. → Go to "CRANK ANGLE SENSOR (Not self-diagnostic item)".
- N.G. ↓

A CHECK POWER SUPPLY.
1) Turn ignition switch "OFF".
2) Disconnect crank angle sensor harness connector.
3) Turn ignition switch "ON".
4) Check voltage between terminal ⓐ and ground with CONSULT or tester.
Voltage: Battery voltage

- N.G. → Check the following.
  - Harness continuity between crank angle sensor and E.C.U.
  - Harness continuity between crank angle sensor and E.C.C.S. relay

  If N.G., repair harness or connectors.
- O.K. ↓

B CHECK INPUT SIGNAL CIRCUIT.
1) Turn ignition switch "OFF".
2) Disconnect E.C.U. harness connector.
3) Check harness continuity between terminal ⓑ and E.C.U. terminals ㉒, ㉚ (180° signal), terminal ⓒ and E.C.U. terminals ㉛, ㊵ (1° signal).
Continuity should exist.

- N.G. → Repair harness or connectors.
- O.K. ↓

CHECK COMPONENT
(Crank angle sensor).
Refer to "Electrical Components Inspection".

- N.G. → Replace crank angle sensor.
- O.K. ↓

Check E.C.U. pin terminals for damage or the connection of E.C.U. harness connector.

Diagnostic Procedure 3

AIR FLOW METER (Code No. 12) (Check engine light item)

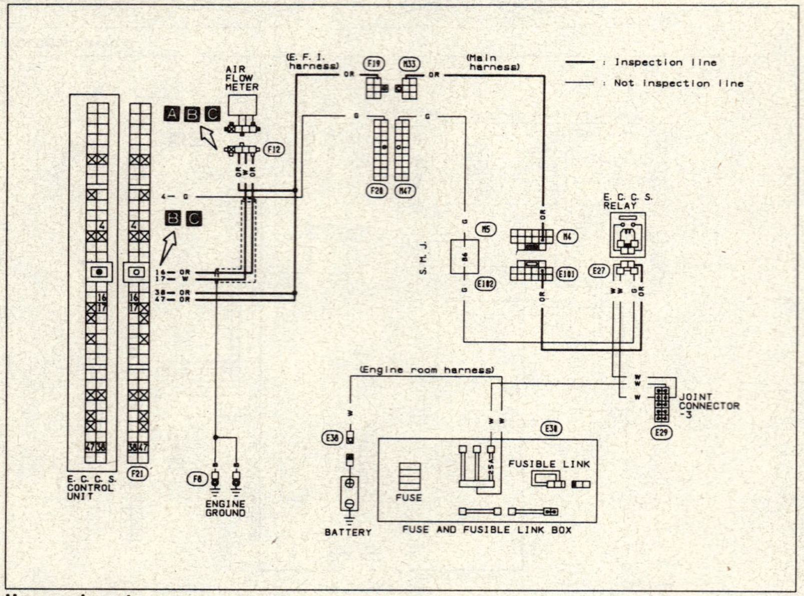

Harness layout

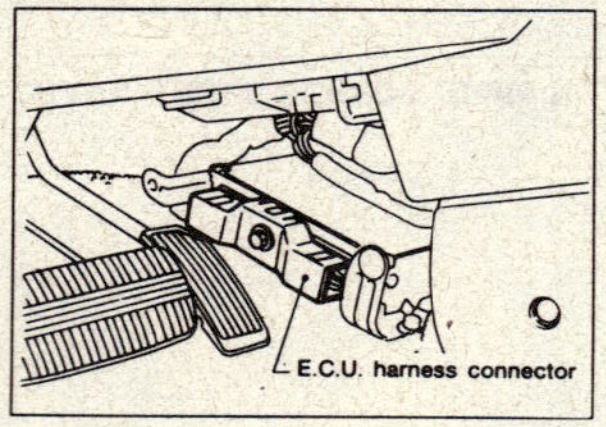

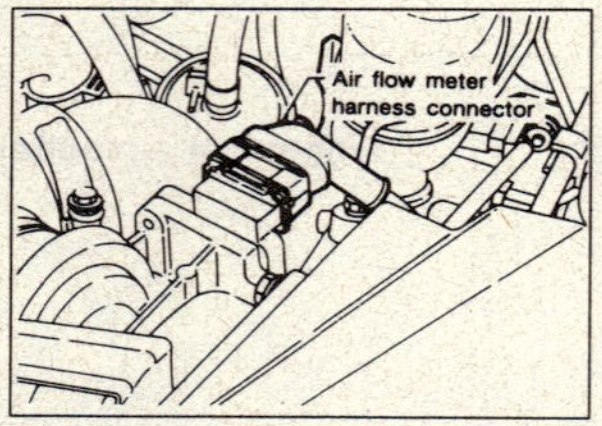

AIR FLOW METER (Code No.12) (Check engine light item)

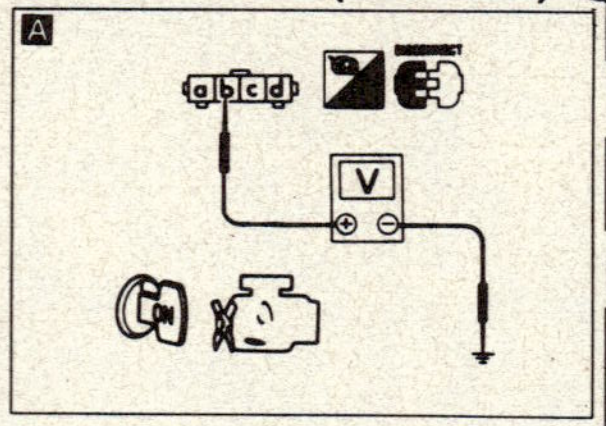

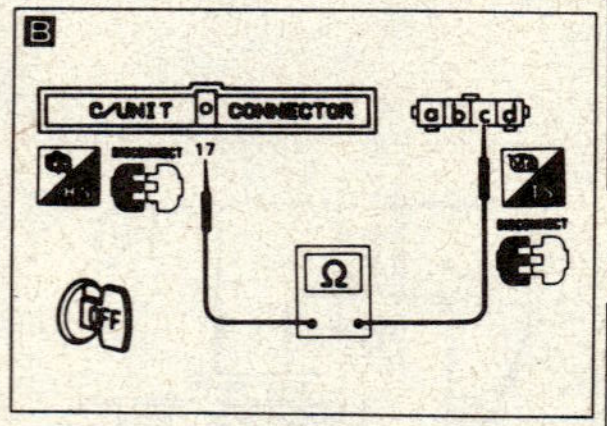

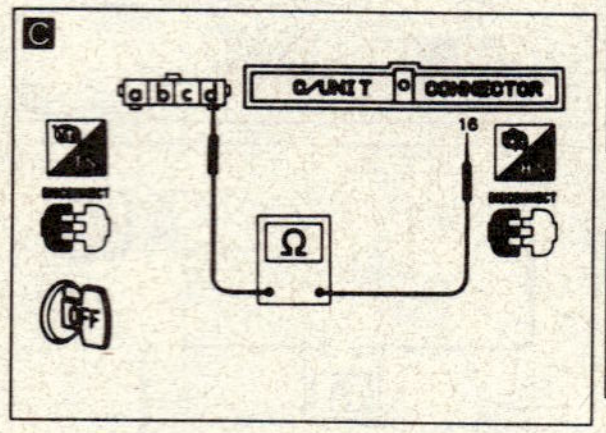

INSPECTION START
↓
PERFORM SELF-DIAGNOSIS. — O.K. → INSPECTION END
↓ N.G.

A CHECK POWER SUPPLY.
1) Turn ignition switch "OFF".
2) Disconnect air flow meter harness connector.
3) Turn ignition switch "ON".
4) Check voltage between terminal ⓑ and ground with CONSULT or tester.
Voltage: Battery voltage

N.G. → Check the following.
- Harness continuity between air flow meter and E.C.U.
- Harness continuity between air flow meter and E.C.C.S. relay

If N.G., repair harness or connectors.

↓ O.K.

B CHECK GROUND CIRCUIT.
1) Turn ignition switch "OFF".
2) Disconnect E.C.U. harness connector.
3) Check harness continuity between terminal ⓒ and E.C.U. terminal ⑰.
Continuity should exist.

N.G. → Repair harness or connectors.

↓ O.K.

C CHECK INPUT SIGNAL CIRCUIT.
1) Check harness continuity between terminal ⓓ and E.C.U. terminal ⑯.
Continuity should exist.

N.G. → Repair harness or connectors.

↓ O.K.

CHECK COMPONENT (Air flow meter).
Refer to "Electrical Components Inspection".

N.G. → Replace air flow meter.

↓ O.K.

Check E.C.U. pin terminals for damage or the connection of E.C.U. harness connector.

Diagnostic Procedure 4

ENGINE TEMPERATURE SENSOR (Code No. 13) (Check engine light item)

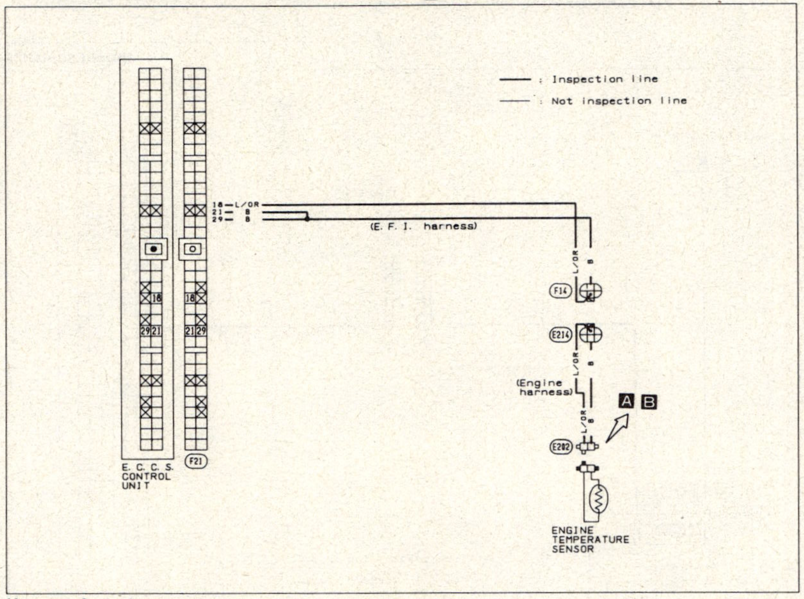

Harness layout

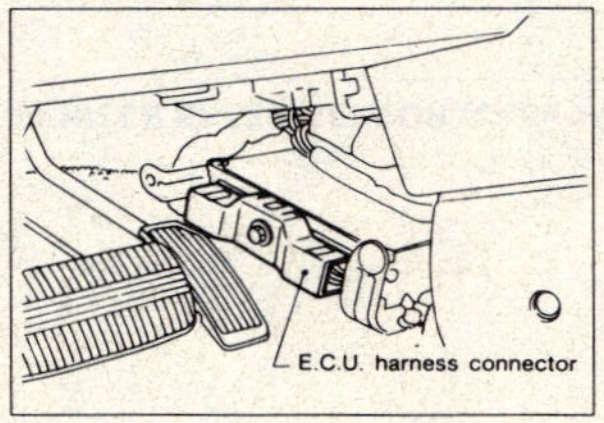

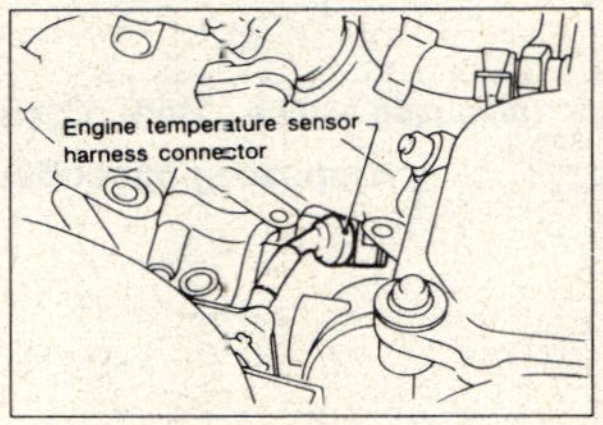

ENGINE TEMPERATURE SENSOR (Code No.13) (Check engine light item)

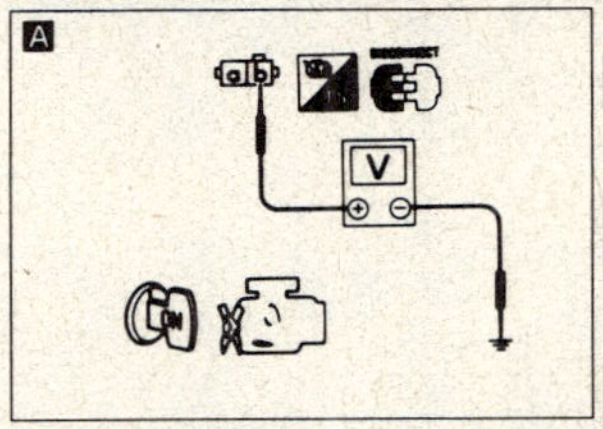

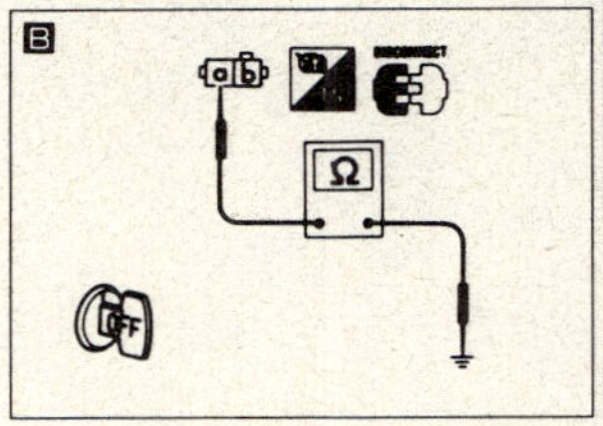

INSPECTION START

↓

PERFORM SELF-DIAGNOSIS. — O.K. → INSPECTION END

↓ N.G.

A CHECK POWER SUPPLY.
1) Turn ignition switch "OFF".
2) Disconnect engine temperature sensor harness connector.
3) Turn ignition switch "ON".
4) Check voltage between terminal ⓑ and ground with CONSULT or tester.
Voltage: Approximately 5V

— N.G. → Check the following.
- Harness connectors (F14), (E214)
- Harness continuity between E.C.U. and engine temperature sensor.

If N.G., repair harness or connectors.

↓ O.K.

B CHECK GROUND CIRCUIT.
1) Turn ignition switch "OFF".
2) Check harness continuity between terminal ⓐ and engine ground.
Continuity should exist.

— N.G. → Check the following.
- Harness connectors (F14), (E214)
- Harness continuity between E.C.U. and engine temperature sensor.

If N.G., repair harness or connectors.

↓ O.K.

CHECK COMPONENT
(Engine temperature sensor).
Refer to "Electrical Components Inspection".

— N.G. → Replace engine temperature sensor.

↓ O.K.

Check E.C.U. pin terminals for damage or the connection of E.C.U. harness connector.

Diagnostic Procedure 5

VEHICLE SPEED SENSOR (Code No. 14) (Check engine light item)

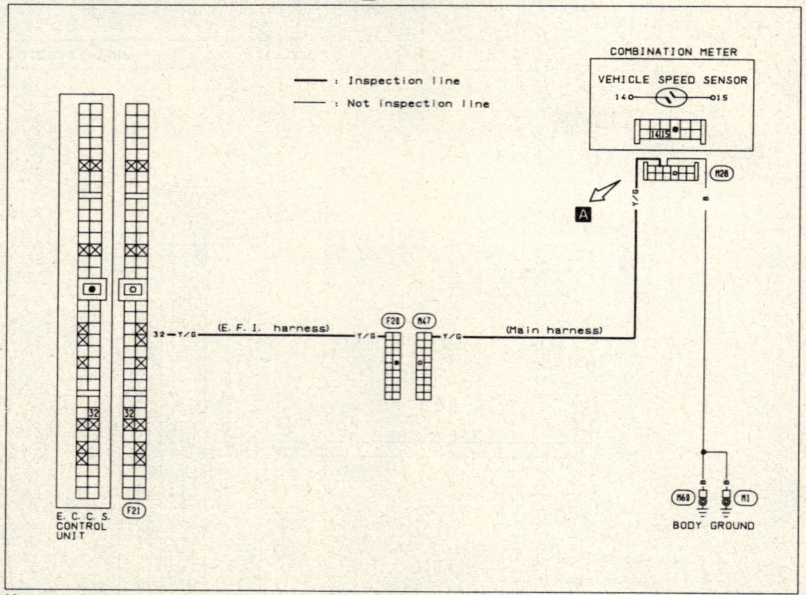

Harness layout

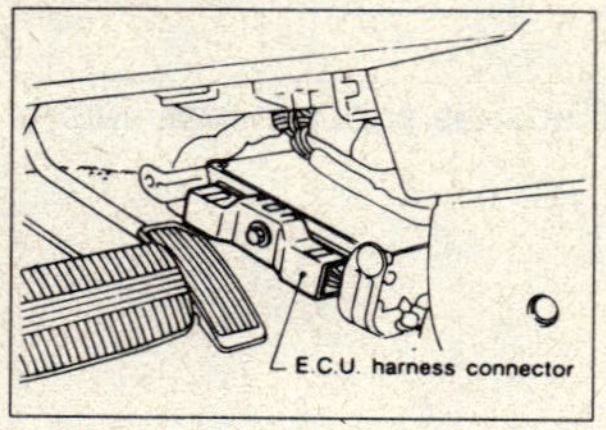

VEHICLE SPEED SENSOR (Code No.14) (Check engine light item)

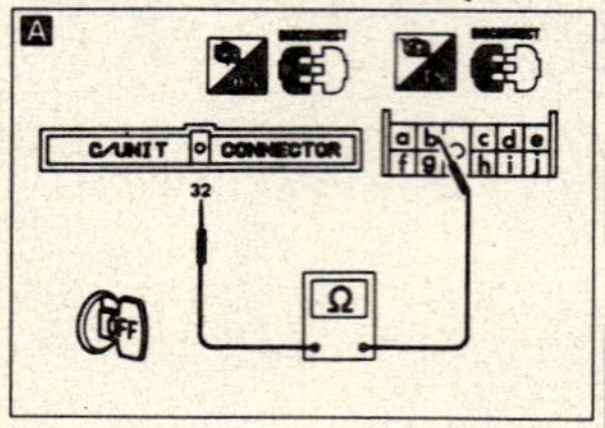

- INSPECTION START
- CHECK OVERALL FUNCTION.
  - O.K. → INSPECTION END
  - N.G. ↓
- CHECK SPEEDOMETER FUNCTION. Make sure that speedometer functions properly.
  - N.G. → Check vehicle speed sensor and its circuit.
  - O.K. ↓
- **A** CHECK INPUT SIGNAL CIRCUIT.
  1) Turn ignition switch "OFF".
  2) Disconnect E.C.U. harness connector and combination meter harness connector.
  3) Check harness continuity between E.C.U. terminal ㉜ and terminal ⓑ. **Continuity should exist.**
  - N.G. → Check the following.
    - Harness connectors F28, M47
    - Harness continuity between E.C.U. and combination meter

    If N.G., repair harness or connectors.
  - O.K. ↓
- Check E.C.U. pin terminals for damage or the connection of E.C.U. harness connector.

Perform FINAL CHECK by the following procedure after repair is completed.

- FINAL CHECK
- Erase the self-diagnosis memory.
- Perform test drive.
- Check overall function again.
  - N.G. → Recheck E.C.U. pin terminals for damage or the connection of E.C.U. harness connector.
  - O.K. ↓
- INSPECTION END

Diagnostic Procedure 6

IGNITION SIGNAL (Code No. 21)

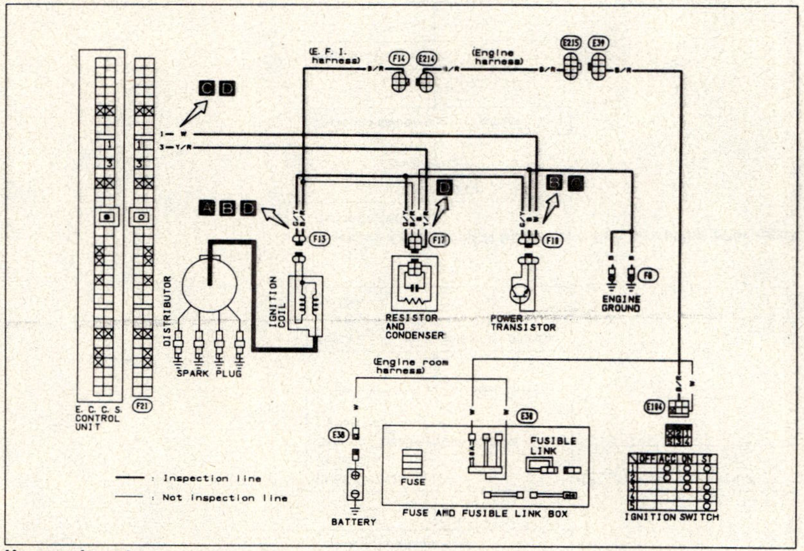

Harness layout

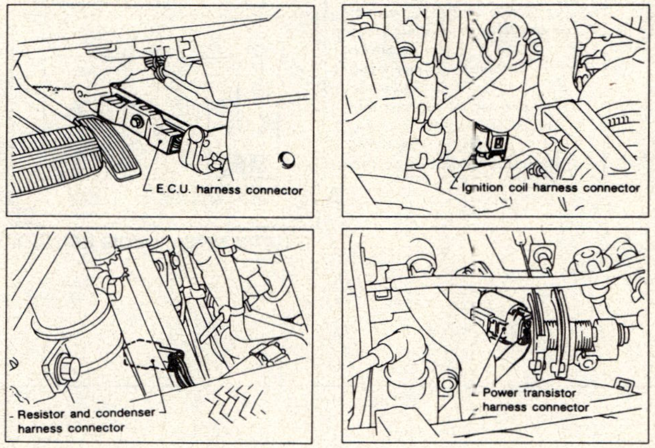

IGNITION SIGNAL (Code No.21)

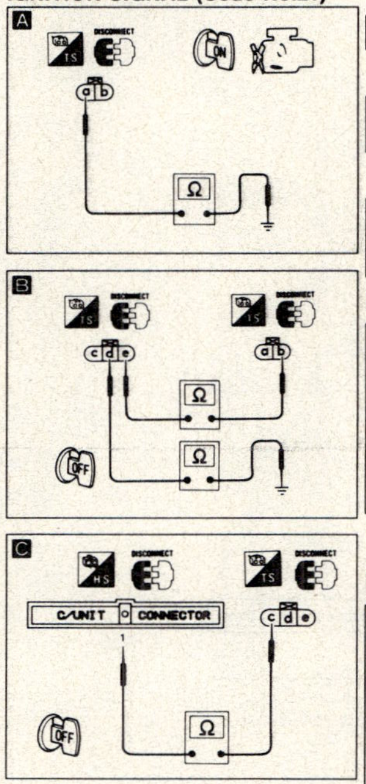

INSPECTION START

↓

PERFORM SELF-DIAGNOSIS.
- O.K. → INSPECTION END
- N.G. ↓

Turn ignition switch "OFF", and restart engine.
Is engine running?
- Yes → Ⓐ
- No ↓

A CHECK POWER SUPPLY.
1) Turn ignition switch "OFF".
2) Disconnect ignition coil harness connector.
3) Turn ignition switch "ON".
4) Check voltage between terminal ⓐ and ground with CONSULT or tester.
Voltage: Battery voltage
- N.G. → Check the following.
  - Harness connectors F14, E214
  - Harness connectors E215, E39
  - Harness continuity between ignition coil and ignition switch

  If N.G., repair harness or connectors.
- O.K. ↓

B CHECK GROUND CIRCUIT.
1) Turn ignition switch "OFF".
2) Disconnect power transistor harness connector.
3) Check harness continuity between terminal ⓑ and terminal ⓔ, terminal ⓓ and engine ground.
Continuity should exist.
- N.G. → Repair harness or connectors.
- O.K. ↓

C CHECK OUTPUT SIGNAL CIRCUIT.
1) Disconnect E.C.U. harness connector.
2) Check harness continuity between E.C.U. terminal ① and terminal ⓒ.
Continuity should exist.
- N.G. → Repair harness or connectors.
- O.K. ↓

CHECK COMPONENT
(Ignition coil, power transistor).
Refer to "Electrical Components Inspection".
- N.G. → Replace malfunctioning component(s).
- O.K. ↓

Check E.C.U. pin terminals for damage or the connection of E.C.U. harness connector.

IGNITION SIGNAL (Code No.21)

D

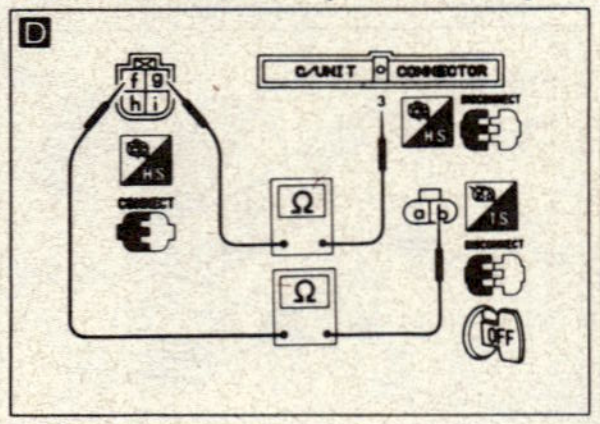

D

Ⓐ → CHECK INPUT SIGNAL CIRCUIT.
1) Stop engine.
2) Disconnect ignition coil harness connector and resistor and condenser.
3) Disconnect E.C.U. harness connector.
4) Check harness continuity between terminal ⓑ and terminal ⓕ, terminal ⓖ and E.C.U. terminal ③.
Continuity should exist.

- N.G. → Repair harness or connectors.
- O.K. ↓

CHECK COMPONENT (Resistor).
Refer to "Electrical Components Inspection".

- N.G. → Replace resistor and condenser.
- O.K. ↓

Check E.C.U. pin terminals for damage or the connection of E.C.U. harness connector.

Perform FINAL CHECK by the following procedure after repair is completed.

FINAL CHECK
↓
Erase the self-diagnosis memory.
↓
Perform test drive.
↓
Perform self-diagnosis again.

- N.G. → Recheck E.C.U. pin terminals for damage or the connection of E.C.U. harness connector.
- O.K. ↓

INSPECTION END

Diagnostic Procedure 7

E.C.C.S. CONTROL UNIT (Code No. 31)
(Check engine light item)

INSPECTION START
↓
1) Turn ignition switch "ON".
2) Erase the self-diagnosis memory.
↓
PERFORM SELF-DIAGNOSIS.
↓
Does E.C.U. display code No. 31 again?

- Yes → Replace E.C.U.
- No ↓

INSPECTION END

Diagnostic Procedure 8

E.G.R. FUNCTION (Code No. 32) (Check engine light item) : California models

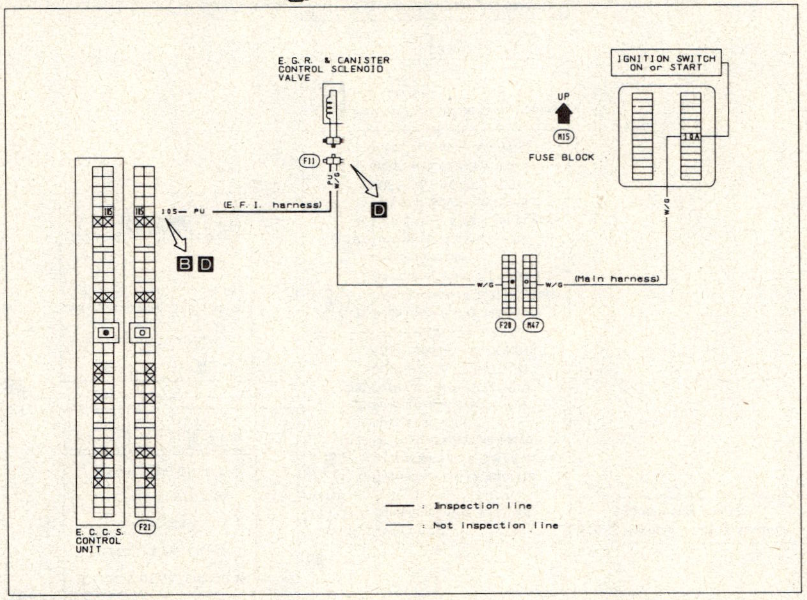

Harness layout

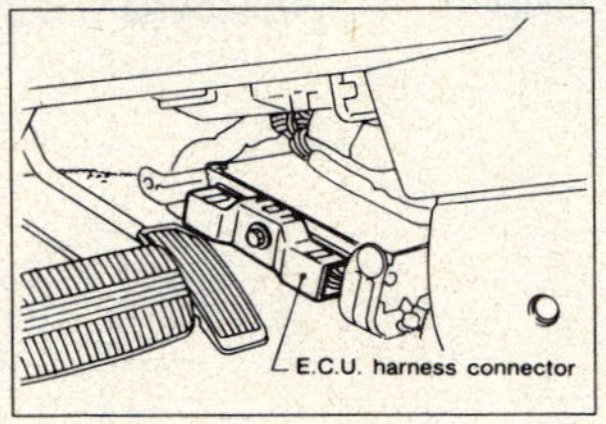

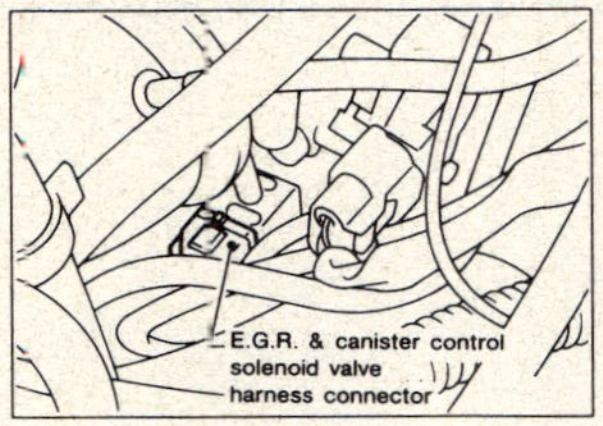

E.G.R. FUNCTION (Code No.32) (Check engine light item) : California models

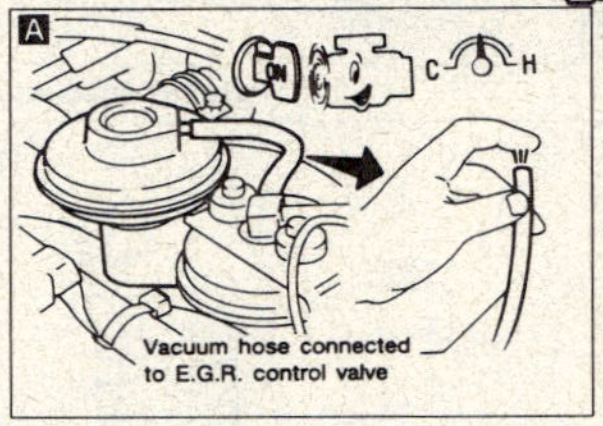

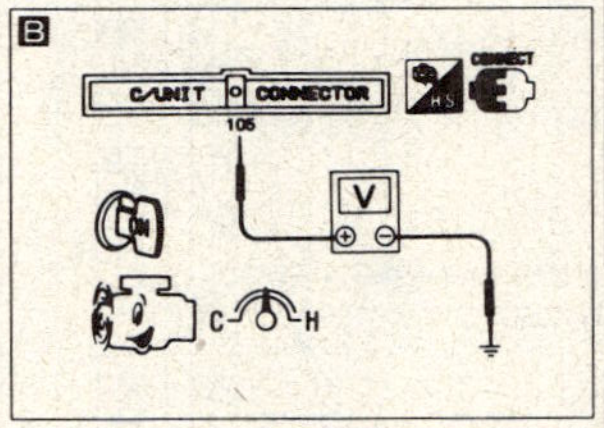

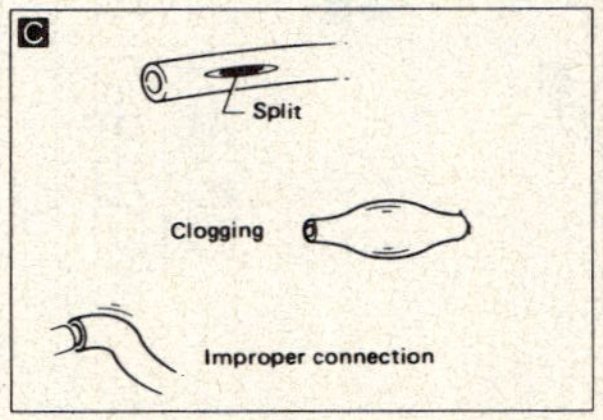

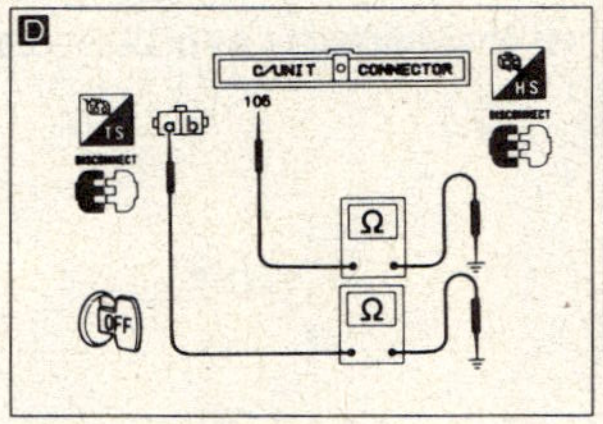

INSPECTION START

↓

PERFORM SELF-DIAGNOSIS. — O.K. → Go to E.G.R. CONTROL (Not self-diagnostic item).

↓ N.G.

A CHECK VACUUM SOURCE TO E.G.R. CONTROL VALVE.
1) Start engine and warm it up sufficiently.
2) Keep engine speed at about 2,000 rpm.
3) Disconnect vacuum hose to E.G.R. control valve.
4) Make sure that vacuum exists.
Vacuum should exist.

— O.K. → CHECK COMPONENTS (E.G.R. control valve, and B.P.T. valve). Refer to "Electrical Components Inspection". — N.G. → Replace malfunctioning component(s).

↓ N.G.

B CHECK CONTROL FUNCTION.
1) Check voltage between E.C.U. terminal (105) and ground under the following conditions with CONSULT or tester.
Voltage:
Engine speed is 4,000 rpm.
0.6 - 0.8V
Engine speed is 2,000 rpm.
Battery voltage

— O.K. → **C** CHECK VACUUM HOSE.
1) Check vacuum hose for clogging, cracks and proper connection.

↓ N.G.

D CHECK OUTPUT SIGNAL CIRCUIT.
1) Stop engine.
2) Disconnect E.C.U. harness connector.
3) Disconnect E.G.R. & canister control solenoid valve harness connector.
4) Check harness continuity between E.C.U. terminal (105) and ground, terminal (a) and ground.
Continuity should not exist.

— N.G. → Repair harness or connectors between E.C.U. and E.G.R. & canister control solenoid valve.

↓ O.K.

(A)

E.G.R. FUNCTION (Code No.32) (Check engine light item) : California models

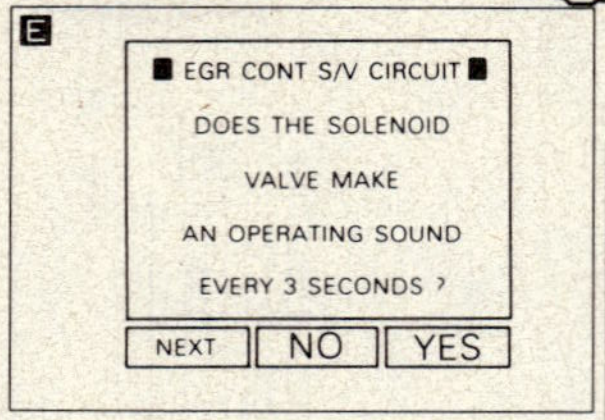

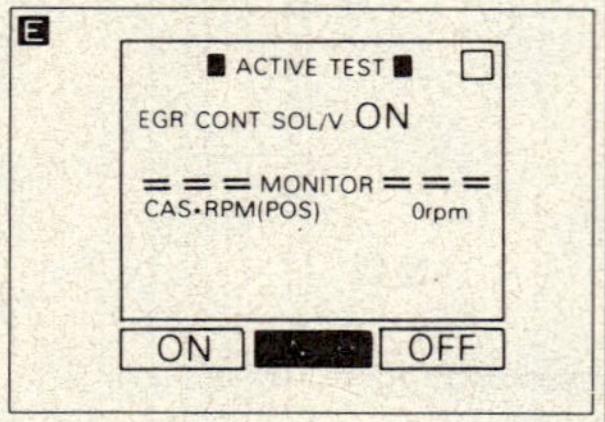

Ⓐ

E

CHECK COMPONENT
(E.G.R. & canister control solenoid valve).

1) Reconnect E.C.U. harness connector and E.G.R. & canister control solenoid valve harness connector.
2) Turn ignition switch "ON".
3) Perform "EGR CONT S/V CIRCUIT" in "FUNCTION TEST" mode with CONSULT.

OR

1) Reconnect E.C.U. harness connector and E.G.R. & canister control solenoid valve harness connector.
2) Turn ignition switch "ON".
3) Turn E.G.R. & canister control solenoid valve "ON" and "OFF" in "ACTIVE TEST" mode with CONSULT and check operating sound.

OR

Refer to "Electrical Components Inspection".

N.G. → Replace E.G.R. & canister control solenoid valve.

O.K. ↓

Check E.C.U. pin terminals for damage or the connection of E.C.U. harness connector.

↓

Check resistance of exhaust gas temperature sensor.

Diagnostic Procedure 9

EXHAUST GAS SENSOR (Code No. 33) (Check engine light item): California models only

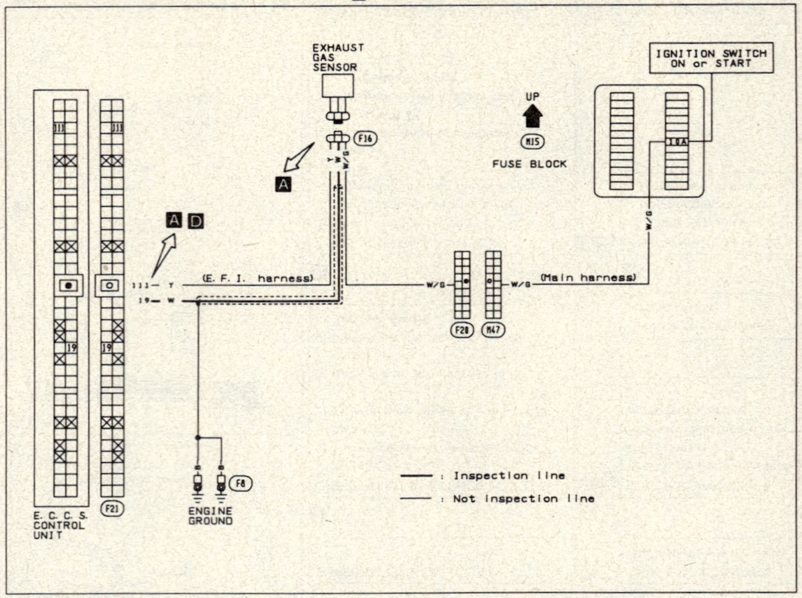

Harness layout

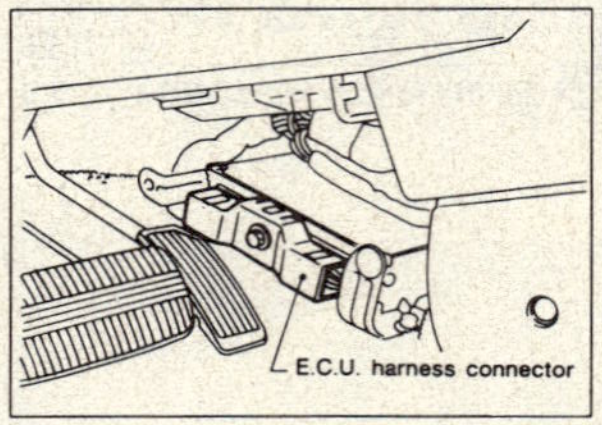

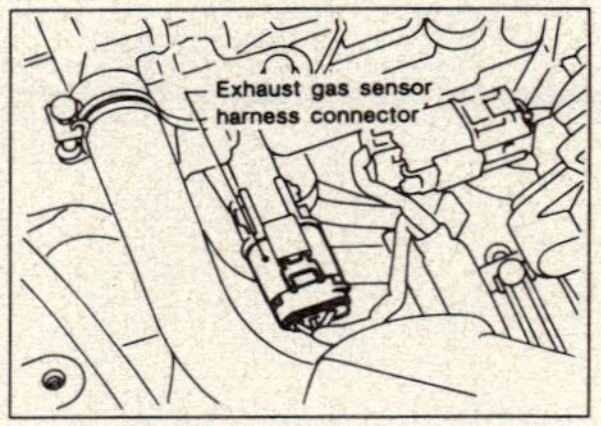

EXHAUST GAS SENSOR (Code No.33) (Check engine light item) : California models only

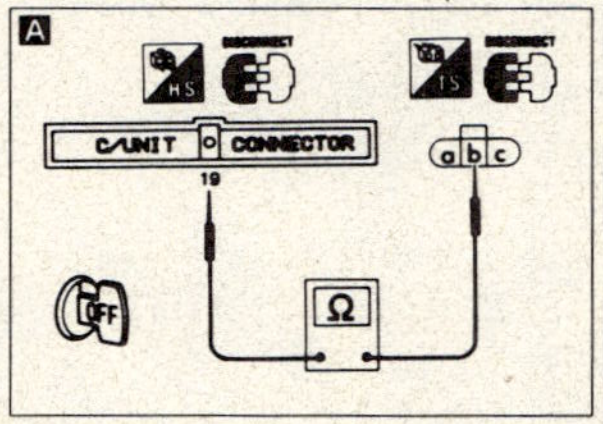

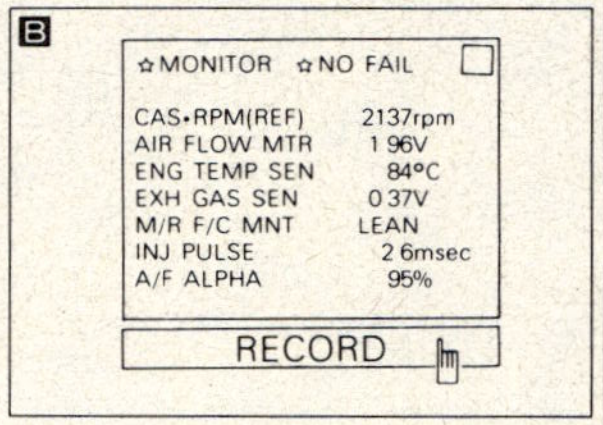

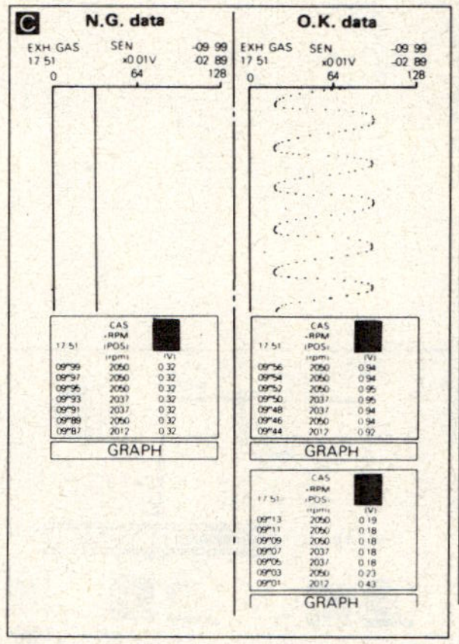

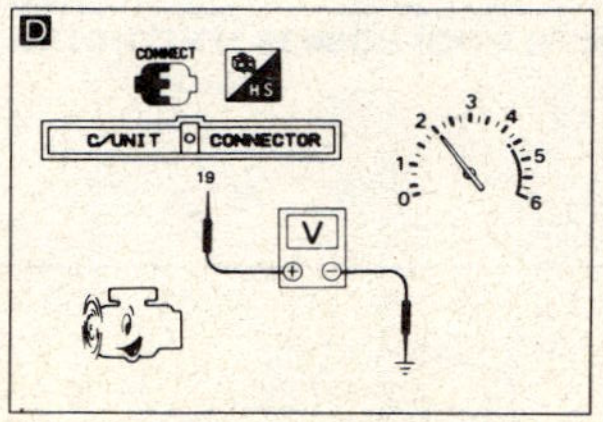

INSPECTION START

↓

CHECK OVERALL FUNCTION. — O.K. → INSPECTION END

↓ N.G.

A CHECK INPUT SIGNAL CIRCUIT.
1) Turn ignition switch "OFF".
2) Disconnect exhaust gas sensor harness connector and E.C.U. harness connector.
3) Check harness continuity between E.C.U. terminal ⑲ and terminal ⓑ.
Continuity should exist.

— N.G. → Repair harness or connectors.

↓ O.K.

CHECK VOLTAGE.
1) Reconnect exhaust gas sensor harness connector and E.C.U. harness connector.
2) Start engine and warm it up sufficiently.
B 3) Select "DATA MONITOR" mode with CONSULT.
4) Touch "RECORD" on CONSULT with engine speed held at approximately 2,000 rpm.
C 5) Check "EXH GAS SEN" voltage.
① **Approx. 0.3 volts (No voltage fluctuations) ... N.G.**
② **Other than ① above ... O.K.**

— OR —

D 3) Check voltage between E.C.U. terminal ⑲ and ground, with engine speed held at approximately 2,000 rpm.
① **Approx. 0.3 volts (No voltage fluctuations) ... N.G.**
② **Other than ① above ... O.K.**

— N.G. → Replace exhaust gas sensor.

↓ O.K.

Check E.C.U. pin terminals for damage or the connection of E.C.U. harness connector.

Diagnostic Procedure 10

DETONATION SENSOR (Code No. 34)

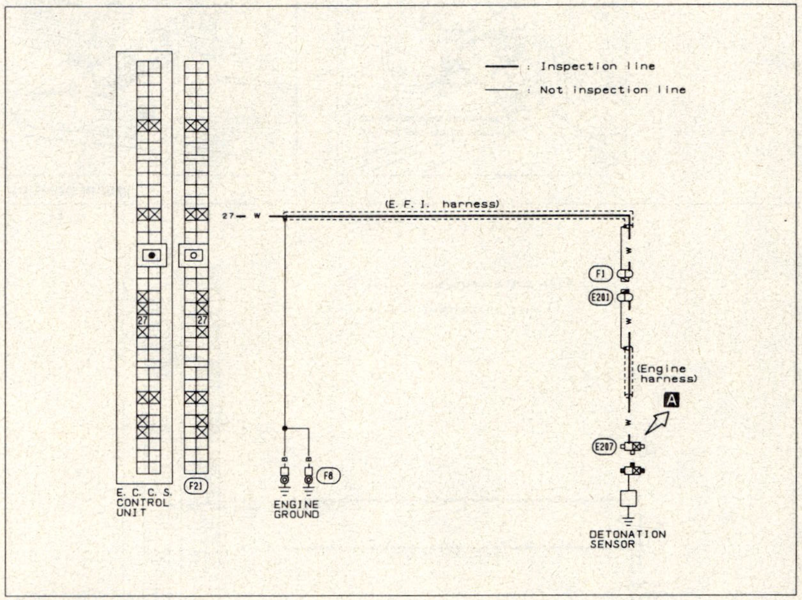

Harness layout

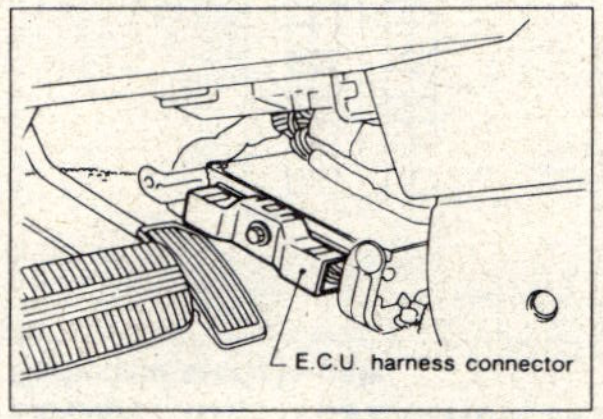

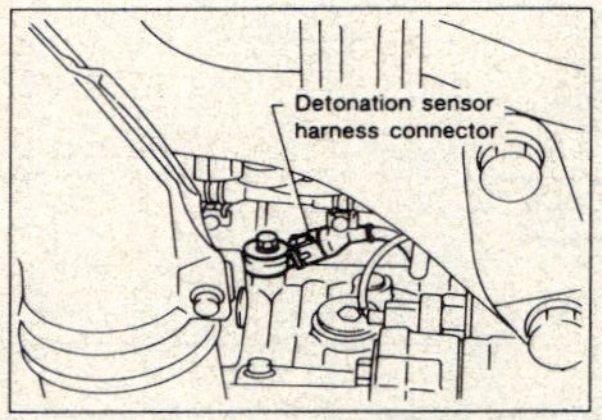

DETONATION SENSOR (Code No.34)

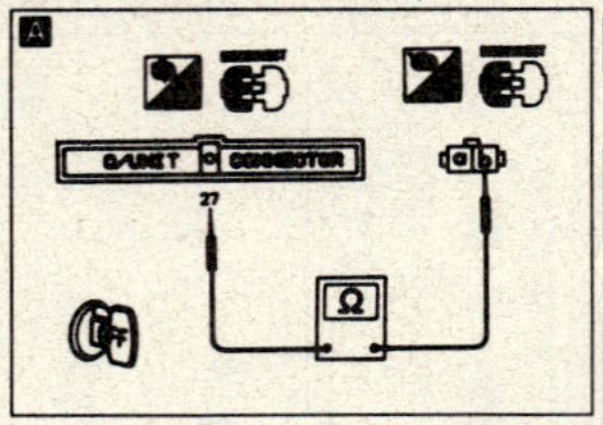

INSPECTION START
↓
PERFORM SELF-DIAGNOSIS. — O.K. → INSPECTION END
↓ N.G.

A

CHECK INPUT SIGNAL CIRCUIT.
1) Turn ignition switch "OFF".
2) Disconnect E.C.U. harness connector and detonation sensor harness connector.
3) Check harness continuity between terminal ⑧ and E.C.U. terminal ㉗.
Continuity should exist.

— N.G. → Check the following.
- Harness connectors (F1), (E281)
- Harness continuity between E.C.U. and detonation sensor

If N.G., repair harness or connectors.

↓ O.K.

CHECK COMPONENT (Detonation sensor).
Refer to "Electrical Components Inspection".

— N.G. → Replace detonation sensor.

↓ O.K.

Check E.C.U. pin terminals for damage or the connection of E.C.U. harness connector.

Perform FINAL CHECK by the following procedure after repair is completed.

FINAL CHECK
↓
Erase the self-diagnosis memory.
↓
Perform test drive.
↓
Perform self-diagnosis again. — N.G. → Recheck E.C.U. pin terminals for damage or the connection of E.C.U. harness connector.
↓ O.K.
INSPECTION END

Diagnostic Procedure 11

EXHAUST GAS TEMPERATURE SENSOR (Code No. 35) (Check engine light item): California models only

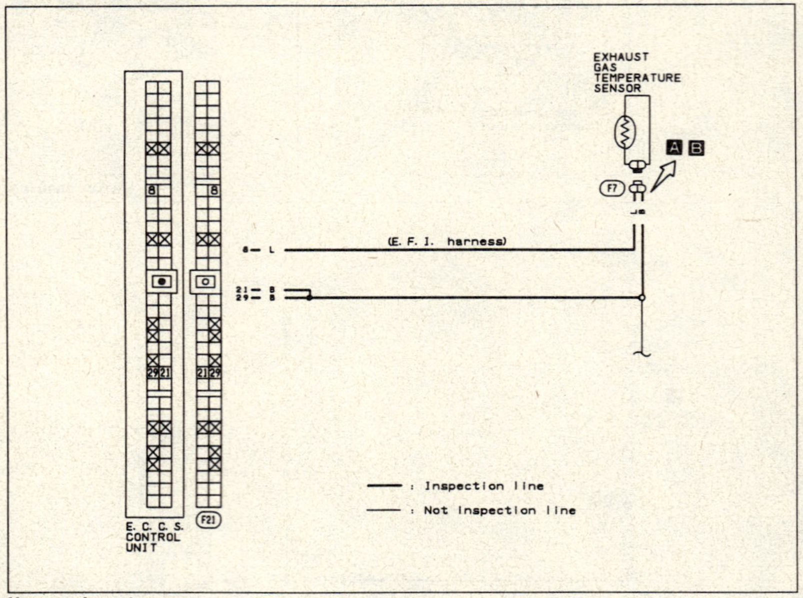

Harness layout

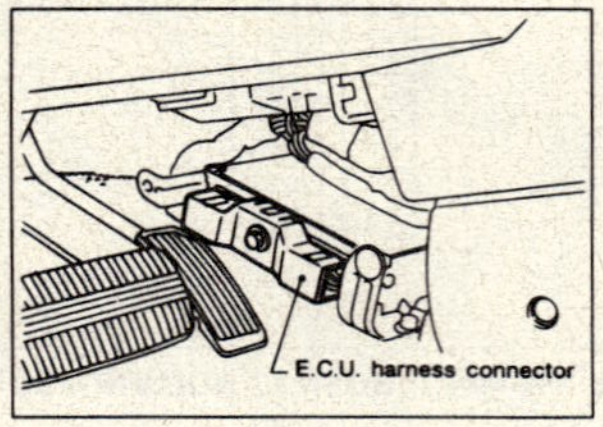

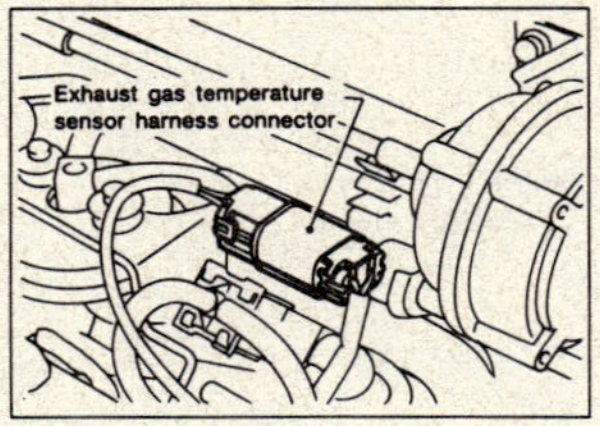

EXHAUST GAS TEMPERATURE SENSOR (Code No.35) (Check engine light item) : California models only

A
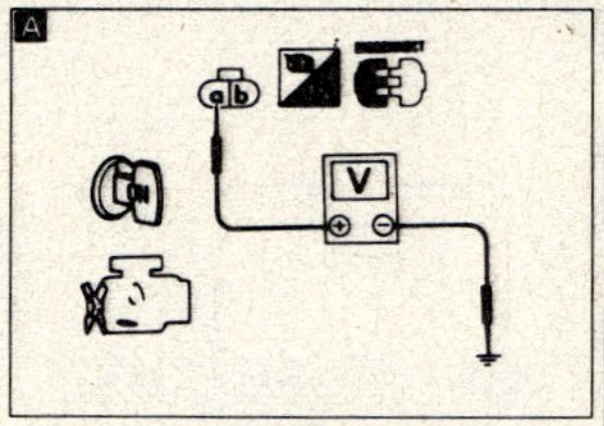

B
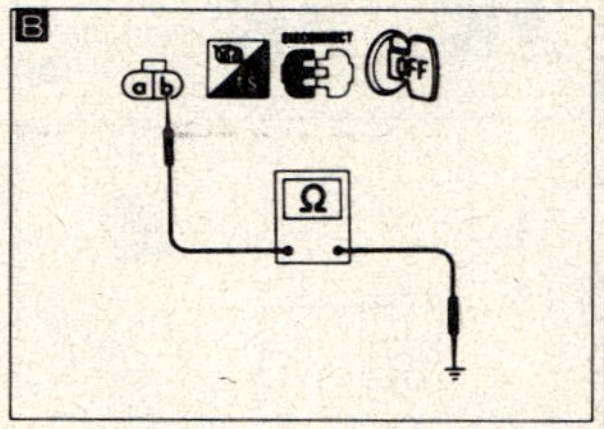

INSPECTION START
↓
PERFORM SELF-DIAGNOSIS. — O.K. → INSPECTION END
↓ N.G.

A CHECK POWER SUPPLY.
1) Turn ignition switch "OFF".
2) Disconnect exhaust gas temperature sensor harness connector.
3) Turn ignition switch "ON".
4) Check voltage between terminal ⓐ and ground with CONSULT or tester.
Voltage: Approximately 5V
— N.G. → Repair harness or connectors.
↓ O.K.

B CHECK GROUND CIRCUIT.
1) Turn ignition switch "OFF".
2) Check harness continuity between terminal ⓑ and engine ground.
Continuity should exist.
— N.G. → Repair harness or connectors.
↓ O.K.

CHECK COMPONENT
(Exhaust gas temperature sensor).
Refer to "Electrical Components Inspection".
— N.G. → Replace exhaust gas temperature sensor.
↓ O.K.

Check E.C.U. pin terminals for damage or the connection of E.C.U. harness connector.

Diagnostic Procedure 12

THROTTLE SENSOR (Code No. 43) (Check engine light item)

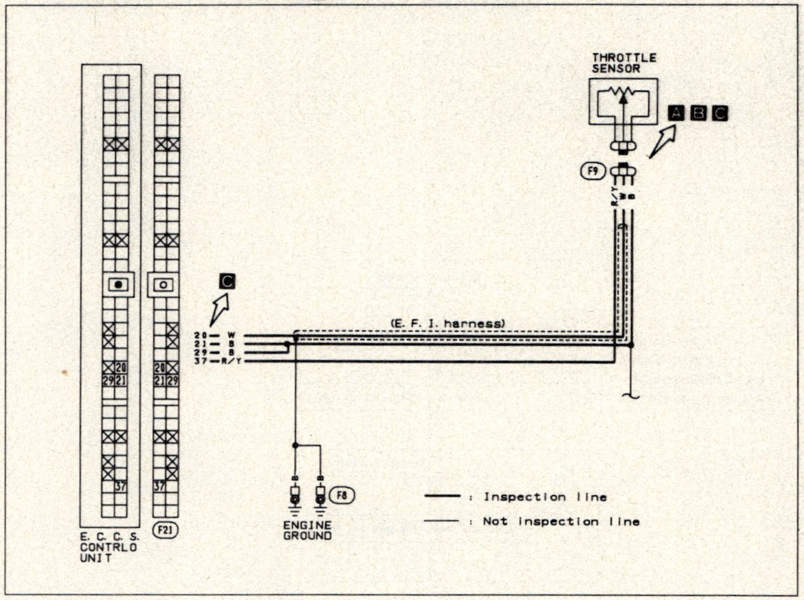

Harness layout

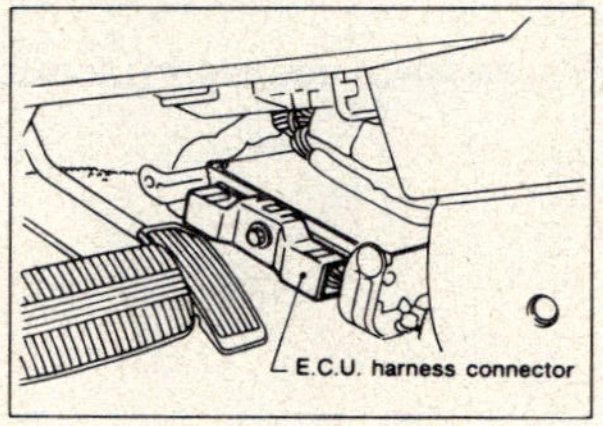

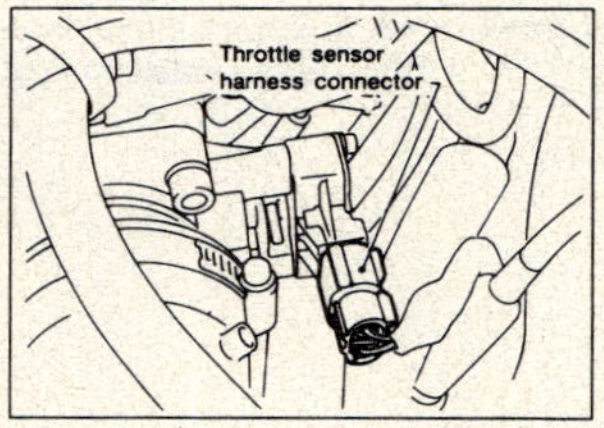

THROTTLE SENSOR (Code No.43) (Check engine light item)

A
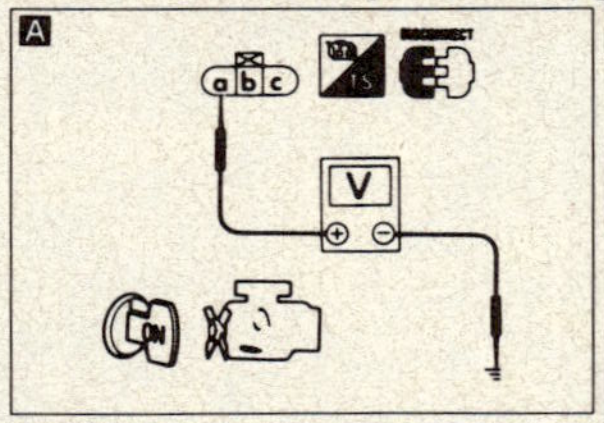

B
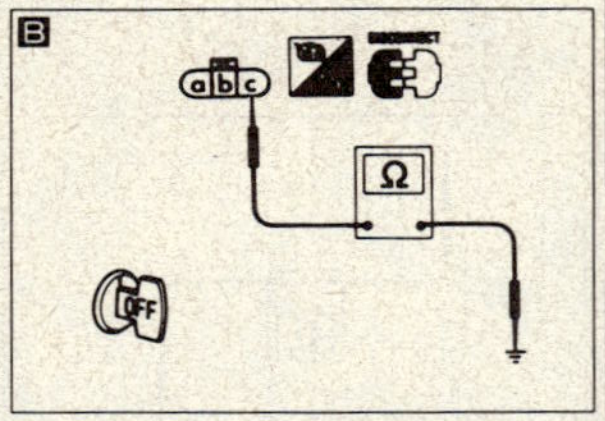

C
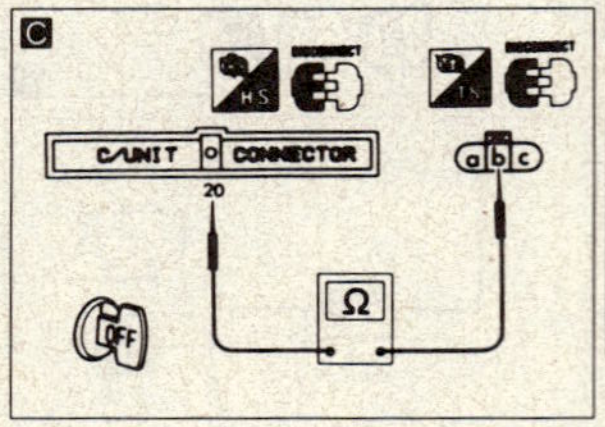

INSPECTION START
↓
PERFORM SELF-DIAGNOSIS. — O.K. → INSPECTION END
↓ N.G.

A CHECK POWER SUPPLY.
1) Turn ignition switch "OFF".
2) Disconnect throttle sensor harness connector.
3) Turn ignition switch "ON".
4) Check voltage between terminal ⓐ and ground with CONSULT or tester.
Voltage: Approximately 5V
— N.G. → Repair harness or connectors.
↓ O.K.

B CHECK GROUND CIRCUIT.
1) Turn ignition switch "OFF".
2) Check harness continuity between terminal ⓒ and engine ground.
Continuity should exist.
— N.G. → Repair harness or connectors.
↓ O.K.

C CHECK INPUT SIGNAL CIRCUIT.
1) Disconnect E.C.U. harness connector.
2) Check harness continuity between E.C.U. terminal ⑳ and terminal ⓑ.
Continuity should exist.
— N.G. → Repair harness or connectors.
↓ O.K.

CHECK COMPONENT
(Throttle sensor).
Refer to "Electrical Components Inspection".
— N.G. → Replace throttle sensor.
↓ O.K.

Check E.C.U. pin terminals for damage or the connection of E.C.U. harness connector.

THROTTLE SENSOR (Code No.43) (Check engine light item)
Perform FINAL CHECK by the following procedure after repair is completed.

FINAL CHECK
↓
Erase the self-diagnosis memory.
↓
Perform test drive.
↓
Perform self-diagnosis again. — N.G. → Recheck E.C.U. pin terminals for damage or the connection of E.C.U. harness connector.
↓ O.K.
INSPECTION END

Diagnostic Procedure 13

INJECTOR LEAK (Code No. 45) (Check engine light item): California models only

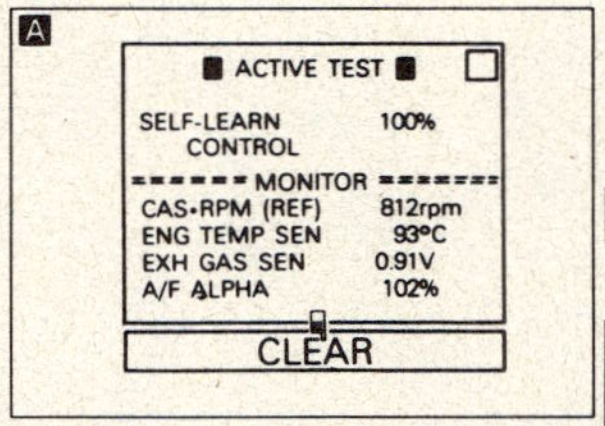

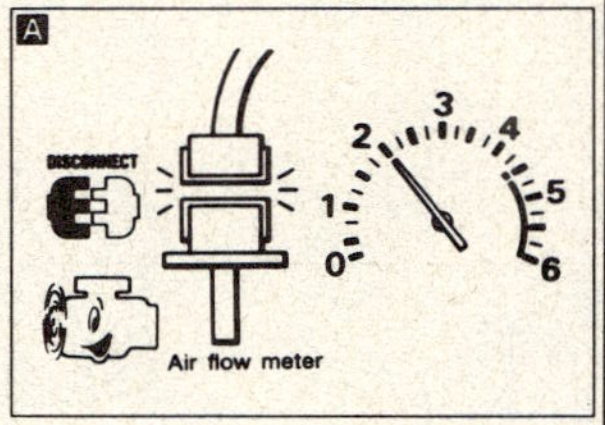

INSPECTION START

↓

A

Clear the self-learning data
1) Start engine and warm it up sufficiently.
2) Select "SELF- LEARNING CONT" in "ACTIVE TEST" mode with CONSULT.
3) Clear the self-learning control coefficient by touching "CLEAR".

—OR—

2) Disconnect air flow meter connector, and restart and run engine for at least 30 seconds at 2,000 rpm.
3) Stop engine and reconnect air flow meter connector.
4) Make sure code No. 12 is displayed in Mode II.
5) Erase the self- diagnosis memory. Make sure code No. 55 is displayed in Mode II.

↓

PERFORM SELF-DIAGNOSIS. — O.K. → INSPECTION END

↓ N.G.

Remove injector assembly.

Keep fuel hose and all injectors connected to injector gallery.

↓

Turn ignition switch "ON".
Make sure fuel does not drip from injector. — Drips → Replace the injectors from which fuel is dripping.

↓ Does not drip

Check E.C.U. pin terminals for damage or the connection of E.C.U. harness connector.

Diagnostic Procedure 14

CRANK ANGLE SENSOR (Not self-diagnostic item)

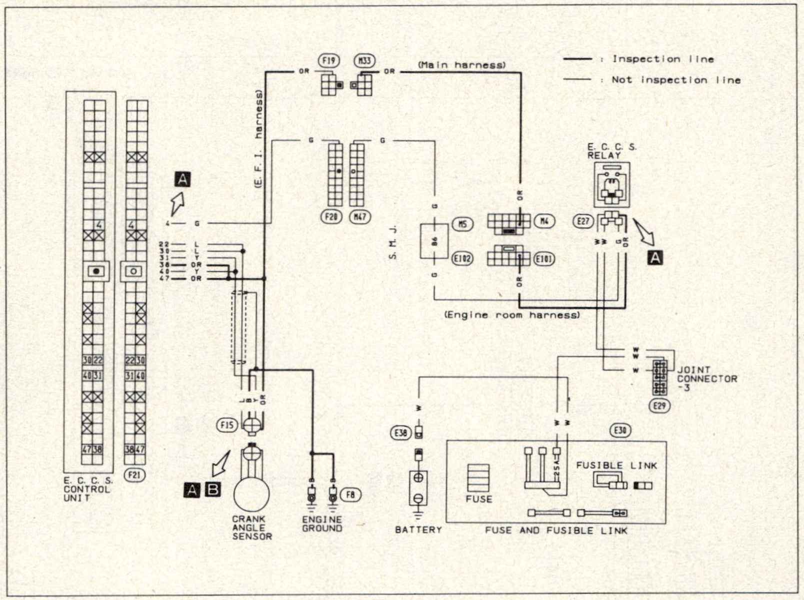

Harness layout

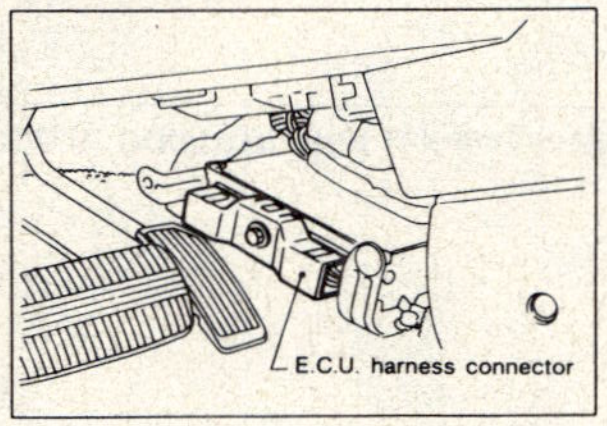

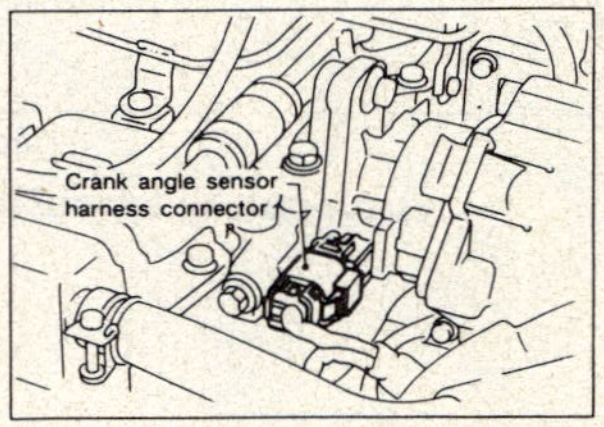

CRANK ANGLE SENSOR (Not self-diagnostic item)

A

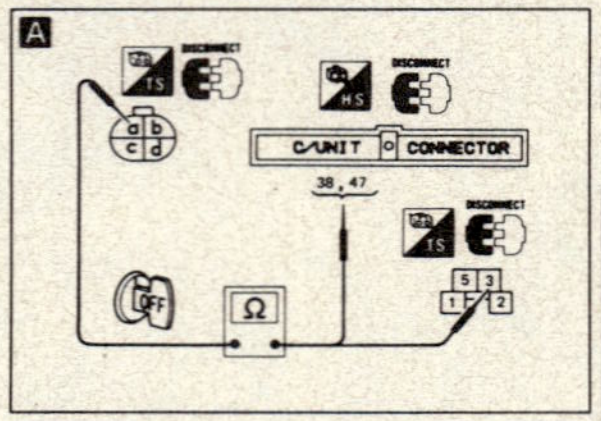

B

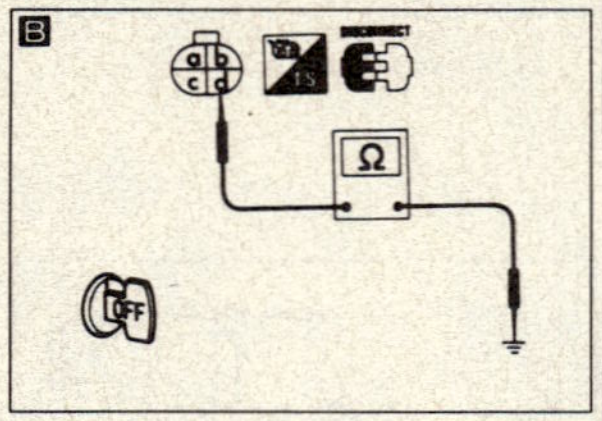

INSPECTION START

↓

A CHECK POWER SUPPLY CIRCUIT.
1) Turn ignition switch "OFF".
2) Disconnect crank angle sensor harness connector.
3) Disconnect E.C.U. harness connector and E.C.C.S. relay.
4) Check harness continuity between terminal ⓐ and terminal ③, E.C.U. terminals ㊳, ㊼.
Continuity should exist.

N.G. → Check the following.
- Harness connectors (F19), (M33)
- Harness connectors (M4), (E101)
- Harness continuity between crank angle sensor and E.C.C.S. relay
- Harness continuity between crank angle sensor and E.C.U.

If N.G., repair harness or connectors.

O.K. ↓

B CHECK GROUND CIRCUIT.
1) Check harness continuity between terminal ⓓ and engine ground.
Continuity should exist.

N.G. → Repair harness or connectors.

O.K. ↓

CHECK COMPONENT
(Crank angle sensor).
Refer to "Electrical Components Inspection".

N.G. → Replace crank angle sensor.

O.K. ↓

INSPECTION END

Diagnostic Procedure 15

E.G.R. CONTROL (Not self-diagnostic item)

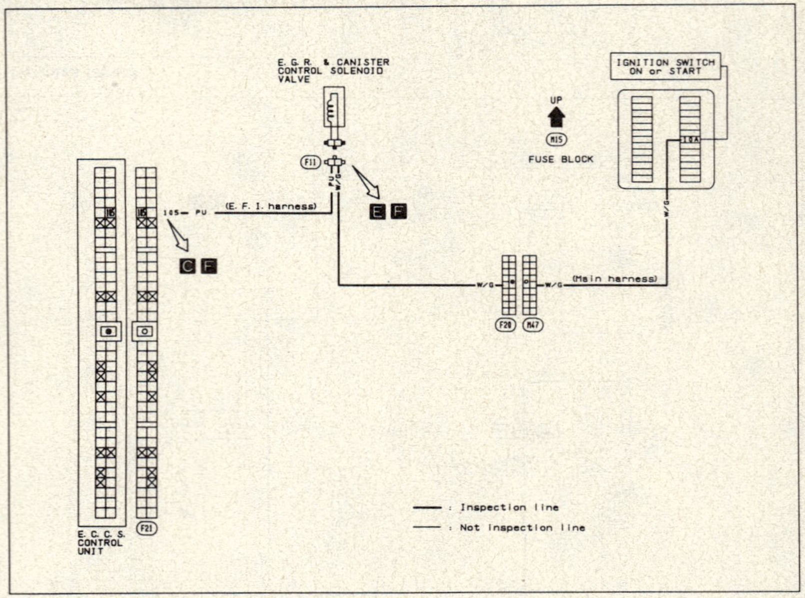

Harness layout

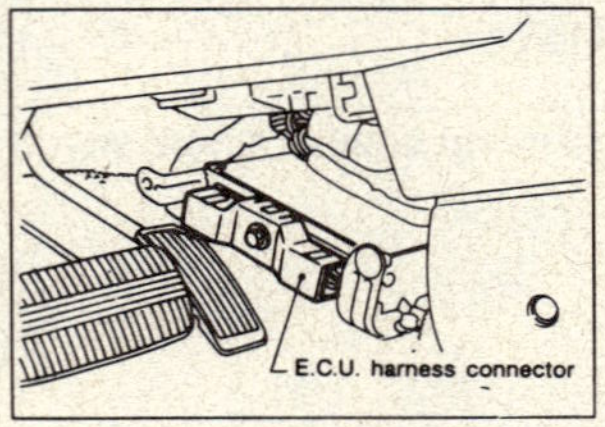

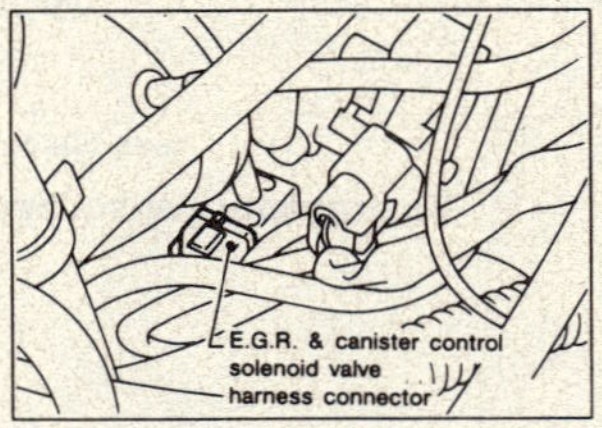

E.G.R. CONTROL (Not self-diagnostic item)

A
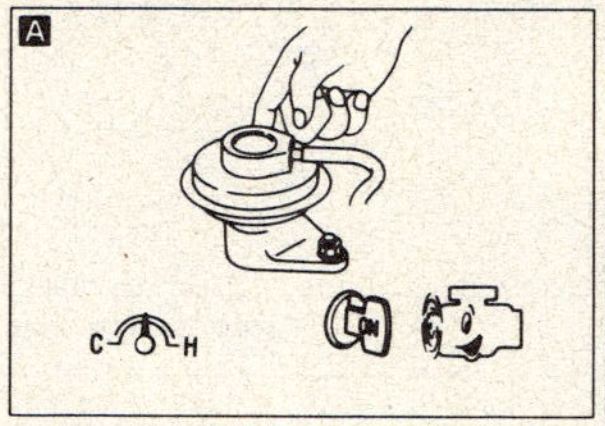

B
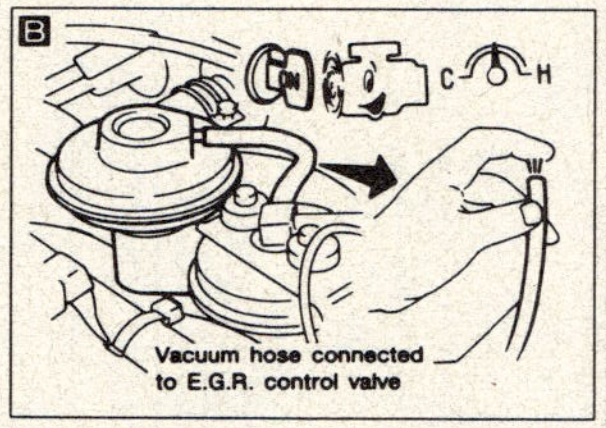

C
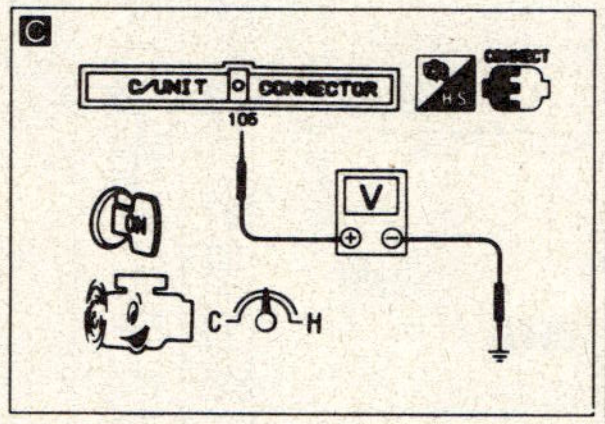

D
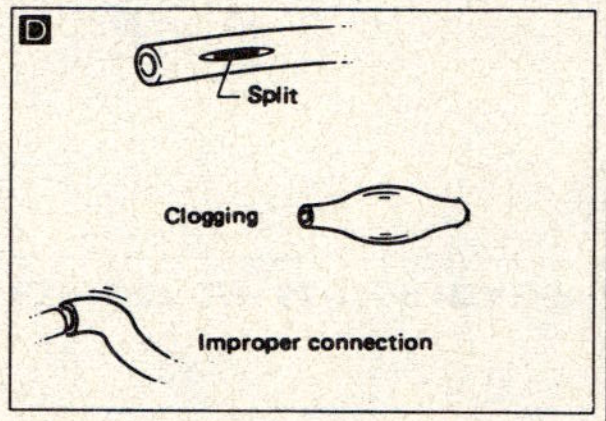

E
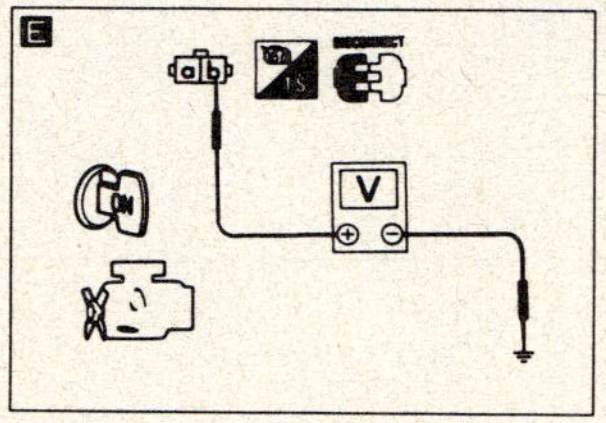

INSPECTION START

- Non-California models → A
- California models → Is lifted up and down. → INSPECTION END

A CHECK OVERALL FUNCTION.
1) Start engine and warm it up sufficiently.
2) Perform self-diagnosis (Mode II) with E.C.U.
Make sure that code No. 12 is not displayed.
3) Make sure that E.G.R. control valve spring is lifted up and down when racing engine (2,000 rpm).
(Use your finger.)

- Is lifted up and down. → INSPECTION END
- Is not lifted up and down. → B

B CHECK VACUUM SOURCE TO E.G.R. CONTROL VALVE.
1) Disconnect vacuum hose to E.G.R. control valve.
2) Make sure that vacuum exists under the following conditions.
Engine speed is 4,000 rpm:
Vacuum should not exist.
Engine speed is 2,000 rpm:
Vacuum should exist.

- O.K. → CHECK COMPONENTS (E.G.R. control valve and B.P.T. valve). Refer to "Electrical Components Inspection". → N.G. → Replace malfunctioning component(s).
- N.G. → C

C CHECK CONTROL FUNCTION.
1) Check voltage between E.C.U. terminal (105) and ground with CONSULT or tester under the following conditions.
Voltage:
Engine speed is 4,000 rpm
0.6 - 0.8V
Engine speed is 2,000 rpm
Battery voltage

- O.K. →
  - **D** (Non-California models) CHECK VACUUM HOSE. 1) Check vacuum hose for clogging, cracks and proper connection.
  - (California models) Go to "CHECK COMPONENT (E.G.R. & canister control solenoid valve)".
- N.G. → E

E CHECK POWER SUPPLY.
1) Stop engine.
2) Disconnect E.G.R. & canister control solenoid valve harness connector.
3) Turn ignition switch "ON".
4) Check voltage between terminal (b) and ground with CONSULT or tester.
Voltage: Battery voltage

- N.G. → Check the following.
  - Harness connectors (F28) (M47)
  - 10A fuse
  - Harness continuity between E.G.R. & canister control solenoid valve and fuse

  If N.G., repair harness or connectors.
- O.K. → Ⓐ

E.G.R. CONTROL (Not self-diagnostic item)

F
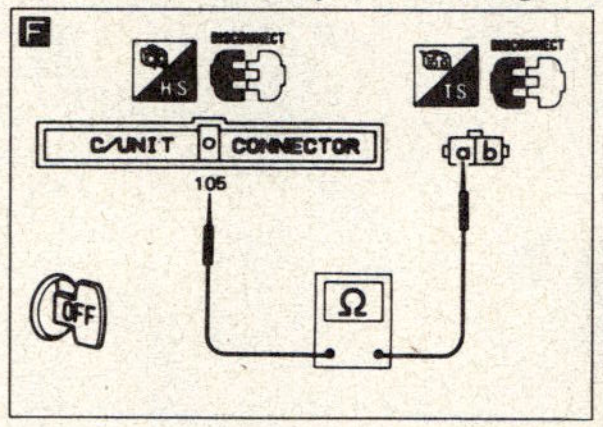

G
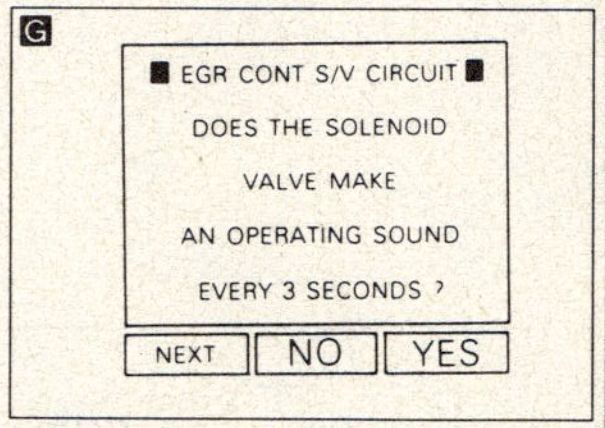

G
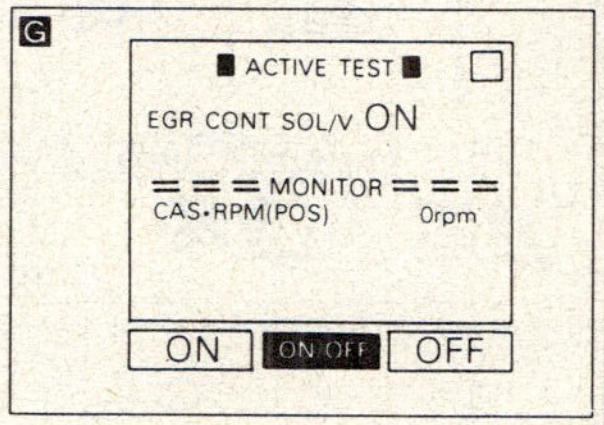

Ⓐ

F CHECK OUTPUT SIGNAL CIRCUIT.
1) Turn ignition switch "OFF".
2) Disconnect E.C.U. harness connector.
3) Check harness continuity between E.C.U. terminal (105) and terminal ⓐ.
Continuity should exist.

- N.G. → Repair harness or connectors.
- O.K. → G

G CHECK COMPONENT
(E.G.R., & canister control solenoid valve).
1) Reconnect E.C.U. harness connector and E.G.R. & canister control solenoid valve harness connector.
2) Turn ignition switch "ON".
3) Perform "EGR CONT S/V CIRCUIT" in "FUNCTION TEST" mode with CONSULT.

OR

1) Reconnect E.C.U. harness connector and E.G.R. & canister control solenoid valve harness connector.
2) Turn ignition switch "ON".
3) Turn E.G.R. & canister control solenoid valve "ON" and "OFF" in "ACTIVE TEST" mode with CONSULT and check operating sound.

OR

Refer to "Electrical Components Inspection".

- N.G. → Replace E.G.R. & canister control solenoid valve.
- O.K. → Check E.C.U. pin terminals for damage or the connection of E.C.U. harness connector.
  - California models → Check resistance of exhaust gas temperature sensor. → INSPECTION END
  - Non-California models → INSPECTION END

Diagnostic Procedure 16

EXHAUST GAS SENSOR HEATER (Not self-diagnostic item)

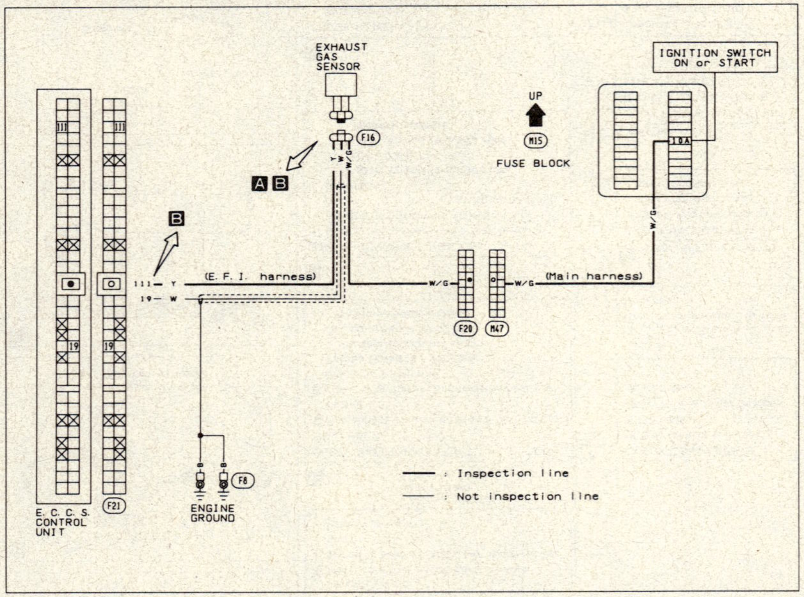

Harness layout

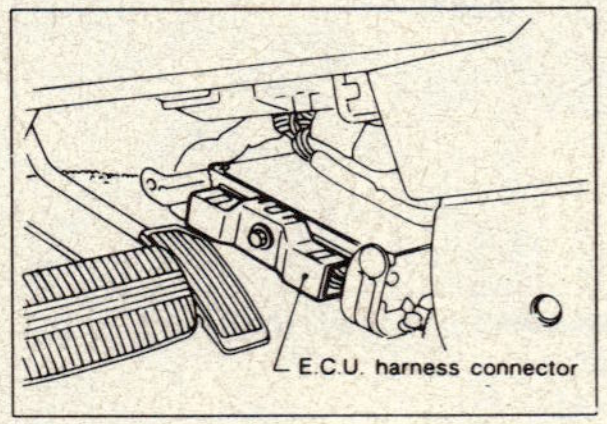

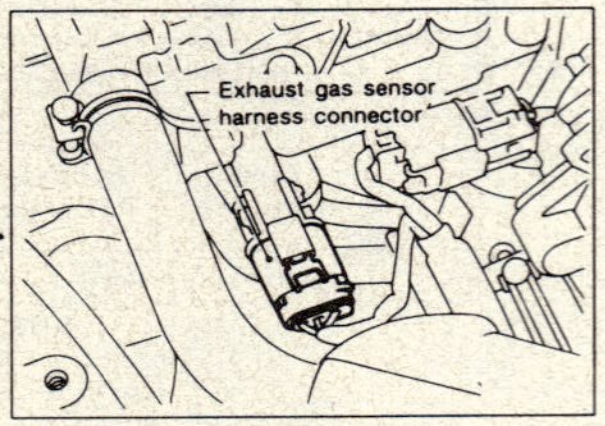

EXHAUST GAS SENSOR HEATER (Not self-diagnostic item)

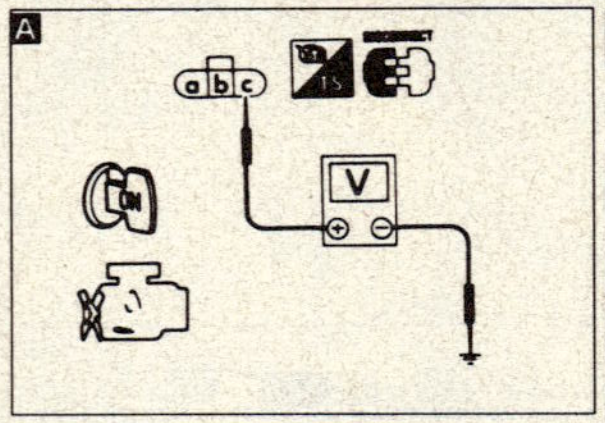

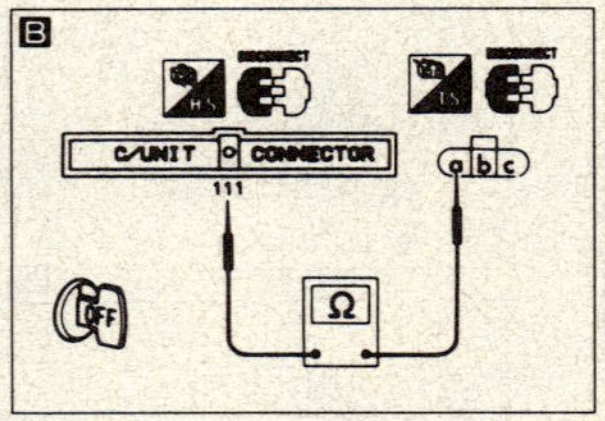

INSPECTION START

↓

A CHECK POWER SUPPLY.
1) Disconnect exhaust gas sensor harness connector.
2) Turn ignition switch "ON".
3) Check voltage between terminal ⓒ and ground.
Voltage: Battery voltage

N.G. → Check the following.
- Harness connectors (F28), (M47)
- 10A fuse
- Harness continuity between exhaust gas sensor and fuse

If N.G., repair harness or connectors.

O.K. ↓

B CHECK GROUND CIRCUIT.
1) Turn ignition switch "OFF".
2) Disconnect E.C.U. harness connector.
3) Check harness continuity between terminal ⓐ and E.C.U. terminal (111).
Continuity should exist.

N.G. → Repair harness or connectors.

O.K. ↓

CHECK COMPONENT
(Exhaust gas sensor heater).
Refer to "Electrical Components Inspection".

N.G. → Replace exhaust gas sensor.

O.K. ↓

Check E.C.U. pin terminals for damage or the connection of E.C.U. harness connector.

↓

INSPECTION END

Diagnostic Procedure 17

START SIGNAL (Not self-diagnostic item)

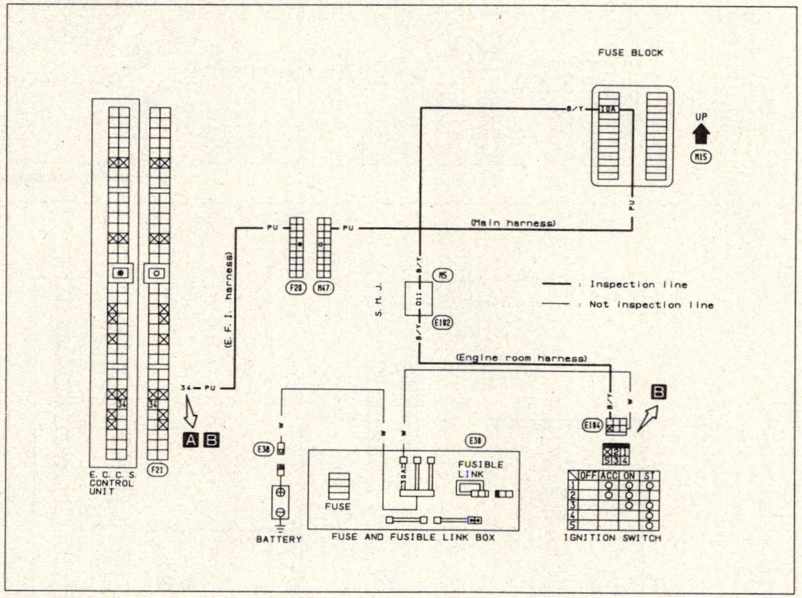

Harness layout

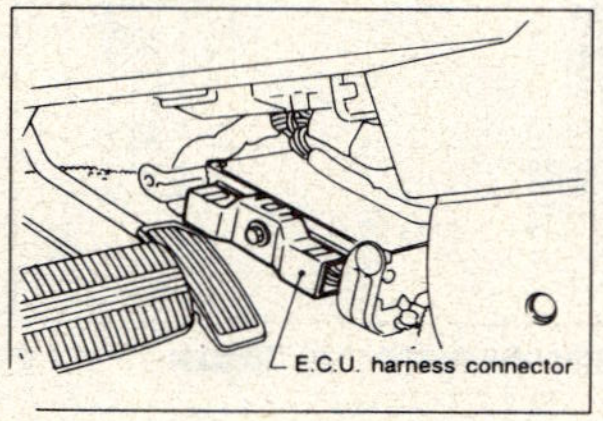

START SIGNAL (Not self-diagnostic item)

A

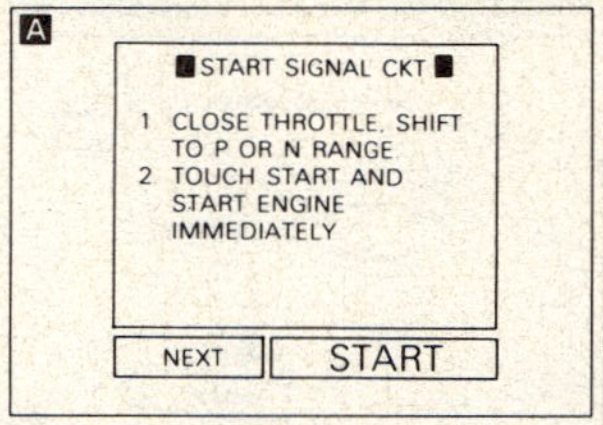

A

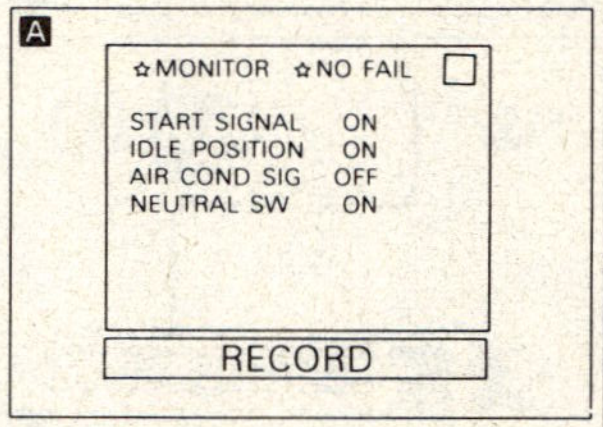

A

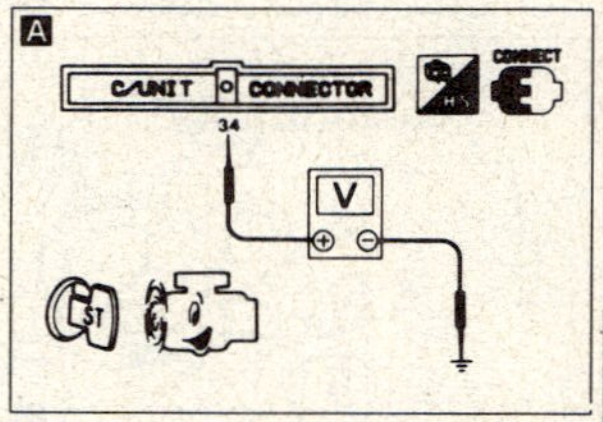

B

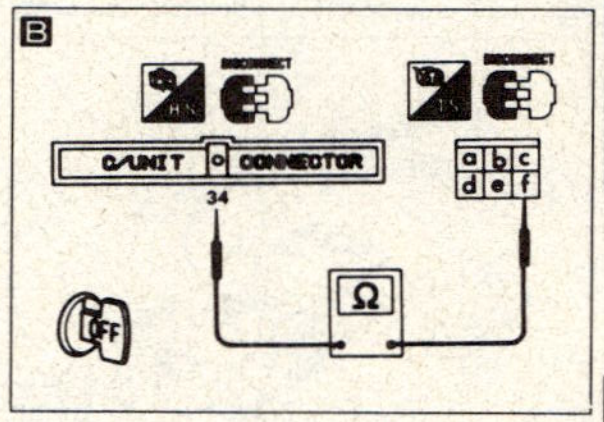

INSPECTION START

↓

A CHECK OVERALL FUNCTION.
1) Turn ignition switch "ON".
2) Perform "START SIGNAL CIRCUIT" in "FUNCTION TEST" mode with CONSULT.

— OR —

1) Turn ignition switch "ON".
2) Check start signal in "DATA MONITOR" mode with CONSULT.

IGN "ON"	OFF
IGN "START"	ON

— OR —

1) Turn ignition switch to "START".
2) Check voltage between E.C.U. terminal ㉞ and ground with CONSULT or tester.

Voltage:
Ignition switch "START"
Battery voltage
Except above
Approximately 0V

O.K. → INSPECTION END

N.G. ↓

B CHECK INPUT SIGNAL CIRCUIT.
1) Turn ignition switch "OFF".
2) Disconnect E.C.U. harness connector and ignition switch harness connector.
3) Check harness continuity between E.C.U. terminal ㉞ and terminal ①.
Continuity should exist.

N.G. → Check the following.
- Harness connectors (F28), (M47)
- Harness connectors (M5), (E102)
- 10A fuse
- Harness continuity between E.C.U. and ignition switch

If N.G., repair harness or connectors.

O.K. ↓

Check E.C.U. pin terminals for damage or the connection of E.C.U. harness connector.

↓

INSPECTION END

Diagnostic Procedure 18

A.I.V. CONTROL (Not self-diagnostic item)

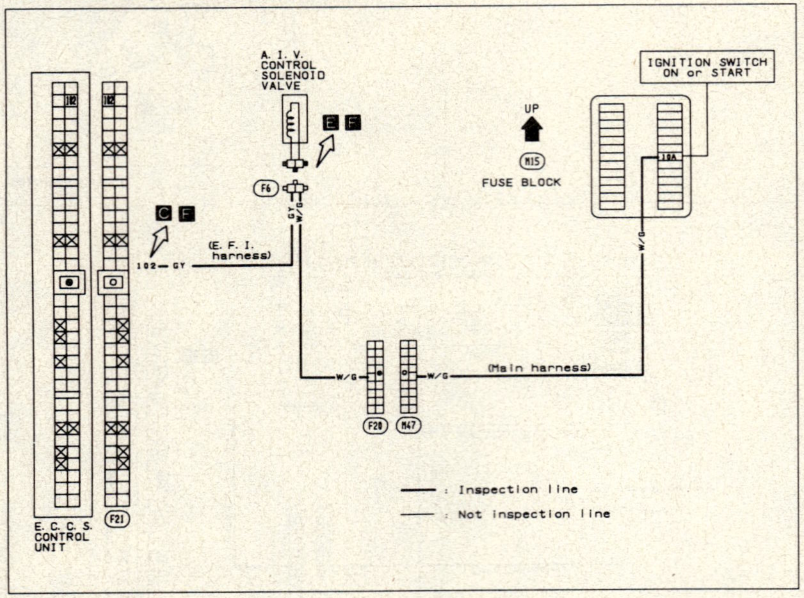

Harness layout

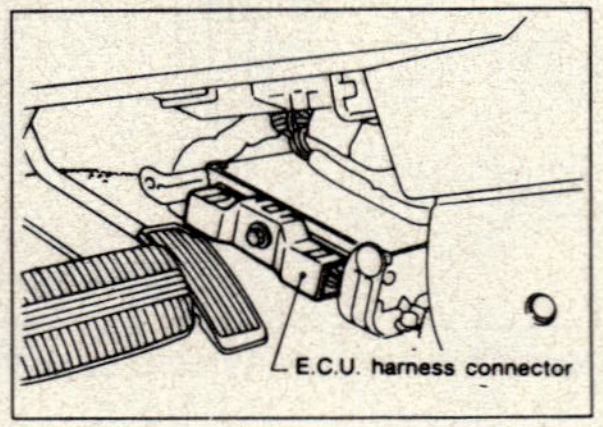

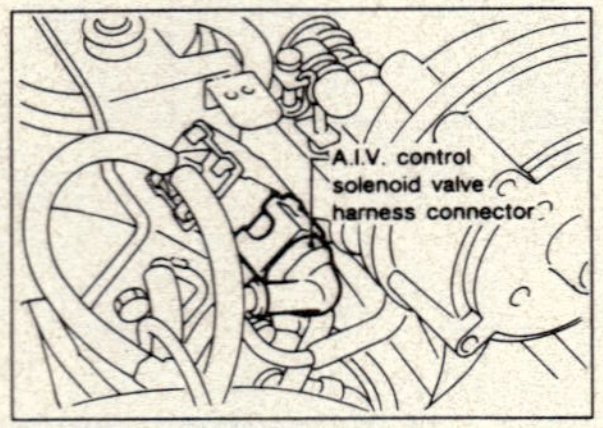

A.I.V. CONTROL (Not self-diagnostic item)

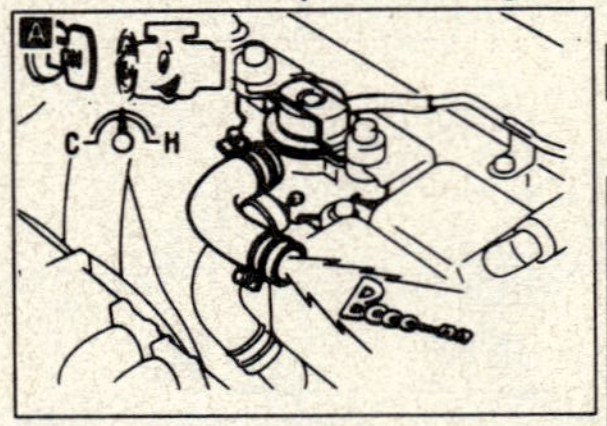

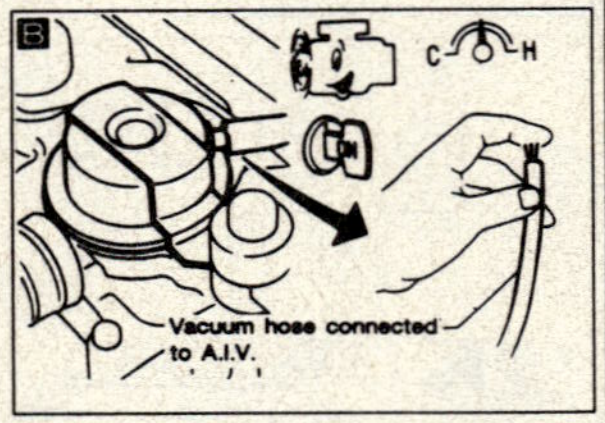

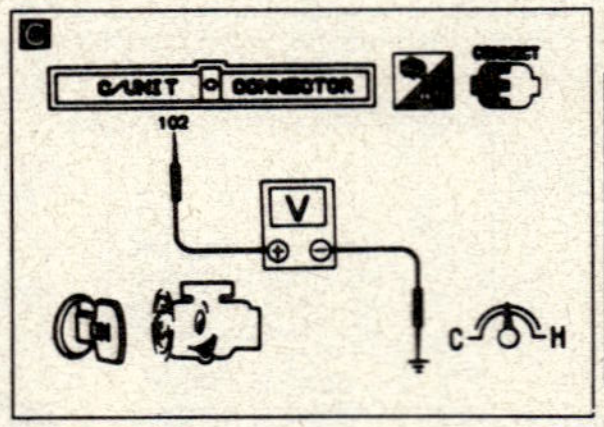

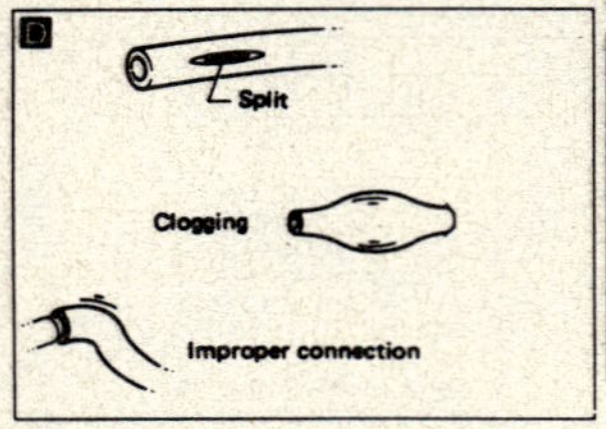

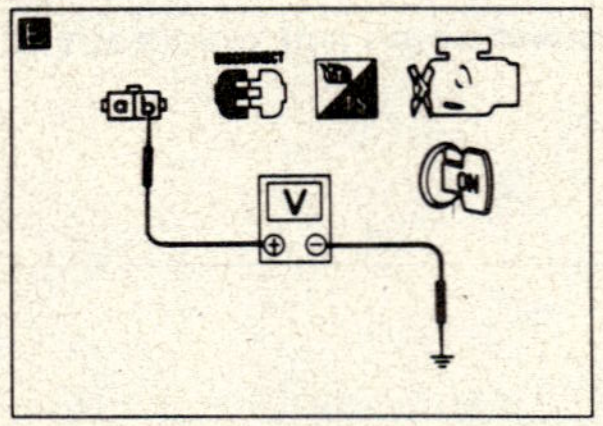

INSPECTION START

A CHECK OVERALL FUNCTION.
1) Start engine and warm it up sufficiently.
2) Run engine at about 2,000 rpm for about 2 minutes under no-load.
3) Make sure that the red L.E.D. on the E.C.U. or the check engine light on the instrument panel goes on and off periodically more than 5 times during 10 seconds at 2,000 rpm under no-load.
4) Release accelerator pedal fully, and run engine at idle.
5) Listen to A.I.V. operating sound.
At idle and during deceleration:
A.I.V. should operate.
Accelerator pedal is depressed:
A.I.V. should not operate.

O.K. → INSPECTION END

N.G. ↓

B CHECK VACUUM SOURCE TO A.I.V.
1) Disconnect vacuum hose to A.I.V.
2) Make sure that vacuum exists under the following conditions.
At idle and during deceleration:
Vacuum should exist.
Accelerator pedal is depressed:
Vacuum should not exist.

O.K. → CHECK COMPONENT (A.I.V.). Refer to "Electrical Components Inspection".

N.G. ↓

C CHECK CONTROL FUNCTION.
1) Check voltage between E.C.U. terminal (102) and ground with CONSULT or tester.
Voltage:
At idle and during deceleration
0.6 - 0.8V
Accelerator pedal is depressed:
Battery voltage

O.K. → **D** CHECK VACUUM HOSE. Check vacuum hose for clogging, cracks and proper connections.

N.G. ↓ (A)

A.I.V. CONTROL (Not self-diagnostic item)

F

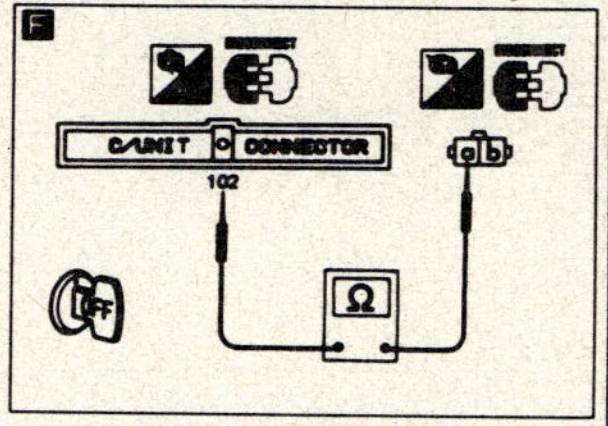

G

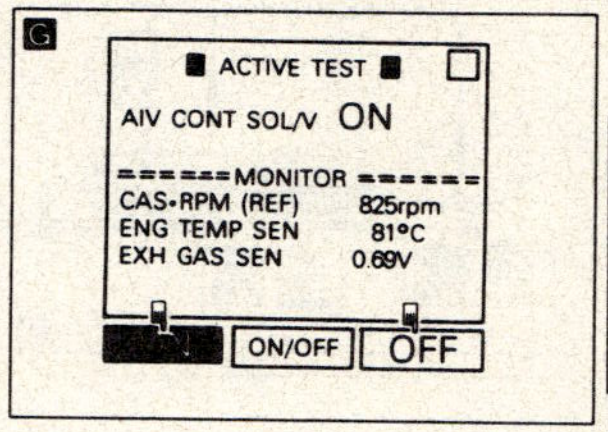

Ⓐ

E CHECK POWER SUPPLY.
1) Stop engine.
2) Disconnect A.I.V. control solenoid valve harness connector.
3) Turn ignition switch "ON".
4) Check voltage between terminal Ⓑ and ground with CONSULT or tester.
Voltage: Battery voltage

N.G. → Check the following:
- Harness connectors (F28), (M47)
- 10A fuse
- Harness continuity between fuse and A.I.V. control solenoid valve

If N.G., repair harness or connectors.

O.K. ↓

F CHECK OUTPUT SIGNAL CIRCUIT.
1) Turn ignition switch "OFF".
2) Disconnect E.C.U. harness connector.
3) Check harness continuity between E.C.U. terminal (102) and terminal Ⓐ.
Continuity should exist.

N.G. → Repair harness or connectors.

O.K. ↓

G CHECK COMPONENT
(A.I.V. control solenoid valve).
1) Reconnect E.C.U. harness connector and A.I.V. control solenoid valve harness connector.
2) Start engine.
3) Perform "AIV CONT SOL VALVE" in "ACTIVE TEST" mode with CONSULT.

— OR —

Refer to "Electrical Components Inspection".

N.G. → Replace A.I.V. control solenoid valve.

O.K. ↓

Check E.C.U. pin terminals for damage or the connection of E.C.U. harness connector.

↓

INSPECTION END

Diagnostic Procedure 19

INJECTOR (Not self-diagnostic item)

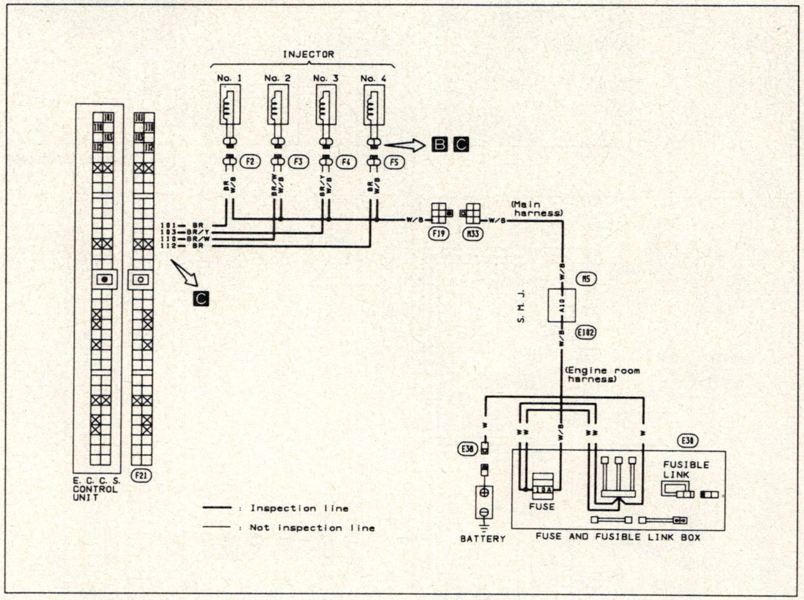

Harness layout

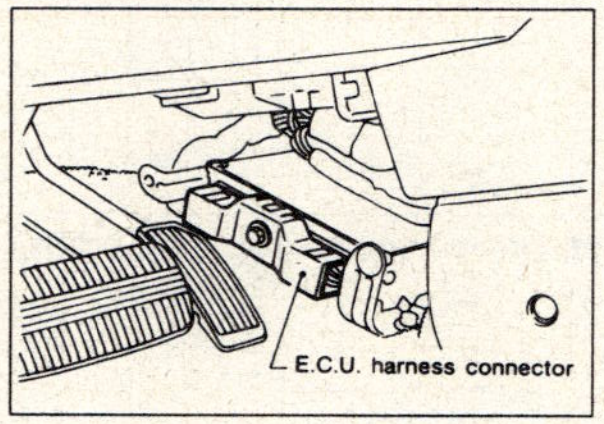

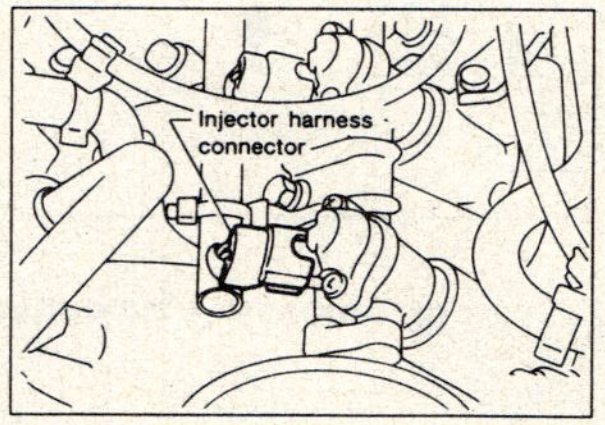

INJECTOR (Not self-diagnostic item)

A
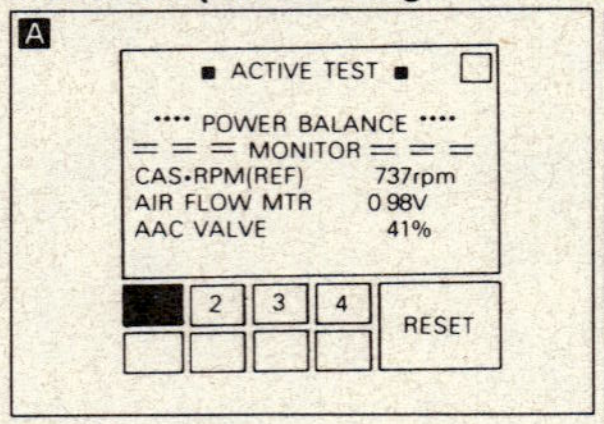

B
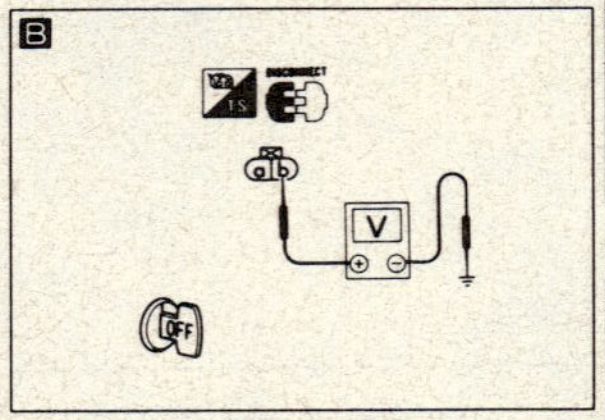

C
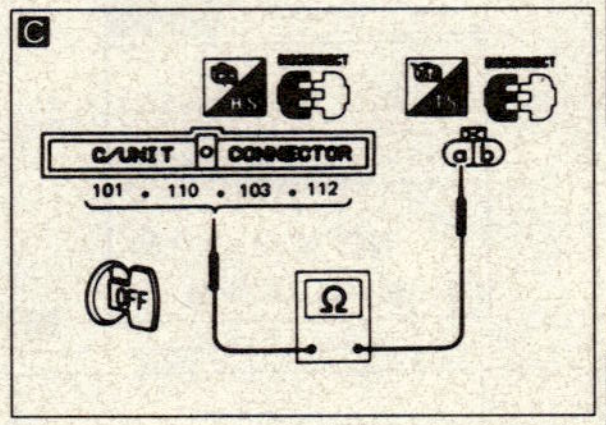

INSPECTION START

↓

CHECK OVERALL FUNCTION.
1) Start engine.
A 2) Perform "POWER BALANCE" in "ACTIVE TEST" mode with CONSULT.
3) Make sure that each circuit produces a momentary engine speed drop.
—OR—
2) Listen to each injector operating sound.
Clicking noise should be heard.

O.K. → INSPECTION END

N.G. ↓

B
CHECK POWER SUPPLY.
1) Stop engine.
2) Disconnect injector harness connector.
3) Check voltage between terminal ⓑ and ground with CONSULT or tester.
Voltage: Battery voltage

N.G. → Check the following:
- Harness connectors (F19), (M33)
- Harness connectors (M5), (E102)
- 25A fuse
- Harness continuity between injector and battery

If N.G., repair harness or connectors.

O.K. ↓

C
CHECK OUTPUT SIGNAL CIRCUIT.
1) Disconnect E.C.U. harness connector.
2) Check harness continuity between terminal ⓐ and E.C.U. terminals (101), (103), (110), (112).
Continuity should exist.

N.G. → Repair harness or connectors.

O.K. ↓

CHECK COMPONENT (Injector).
Refer to "Electrical Components Inspection".

N.G. → Replace injector.

O.K. ↓

Check E.C.U. pin terminals for damage or the connection of E.C.U. harness connector.

↓

INSPECTION END

Diagnostic Procedure 20

FUEL PUMP (Not self-diagnostic item)

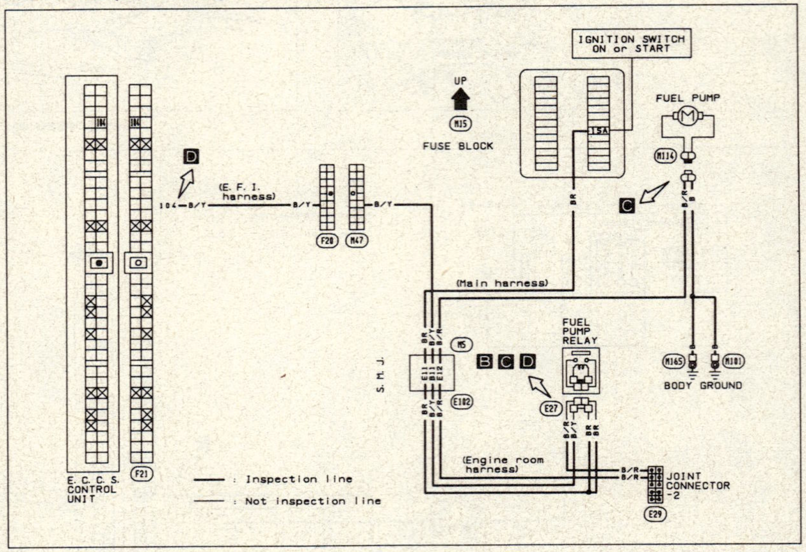

Harness layout

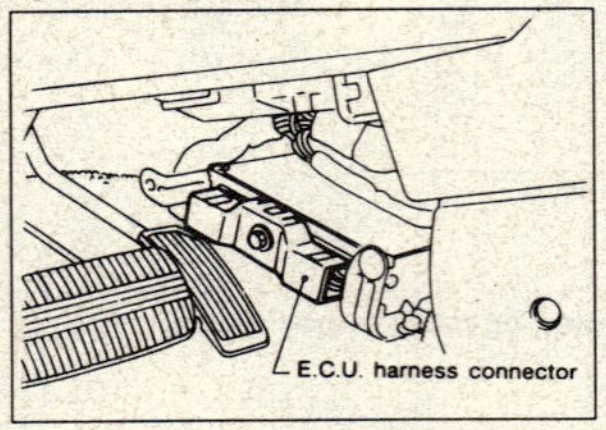

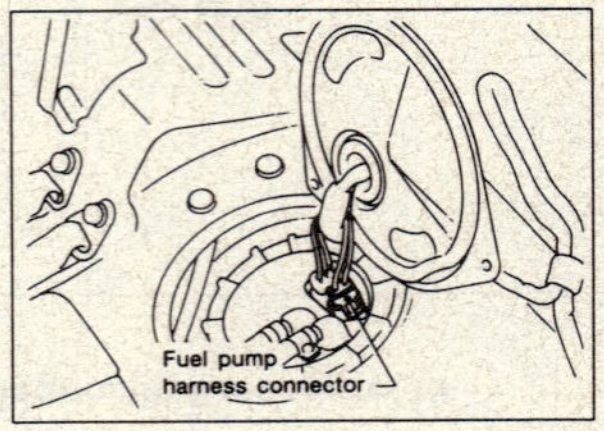

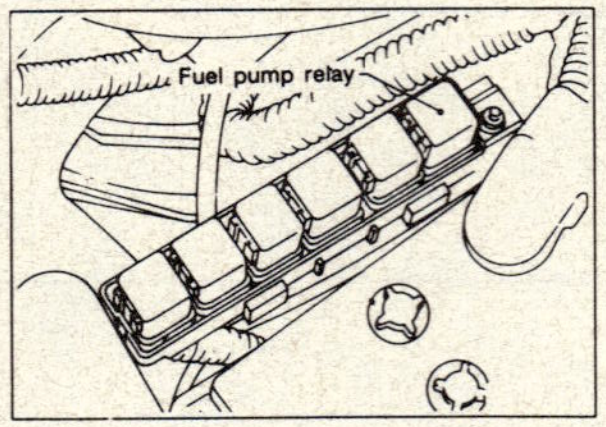

FUEL PUMP (Not self-diagnostic item)

A
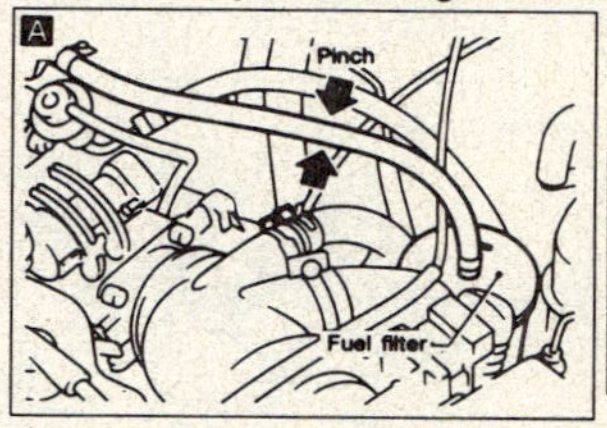

B
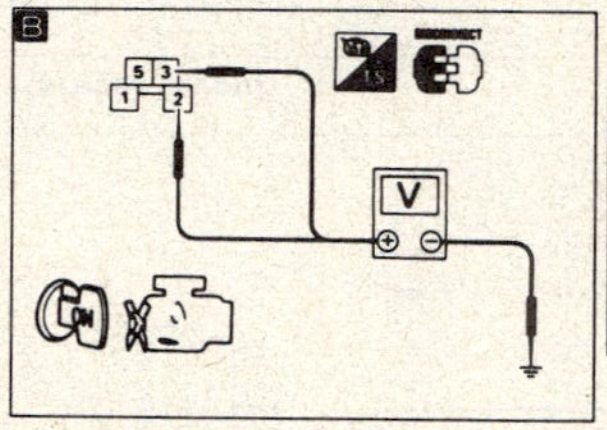

C
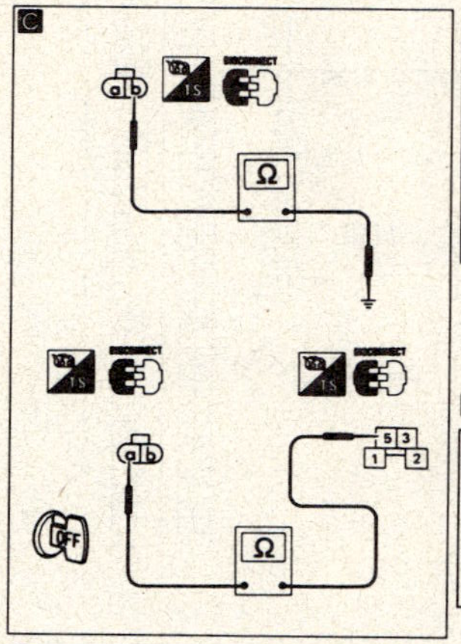

D
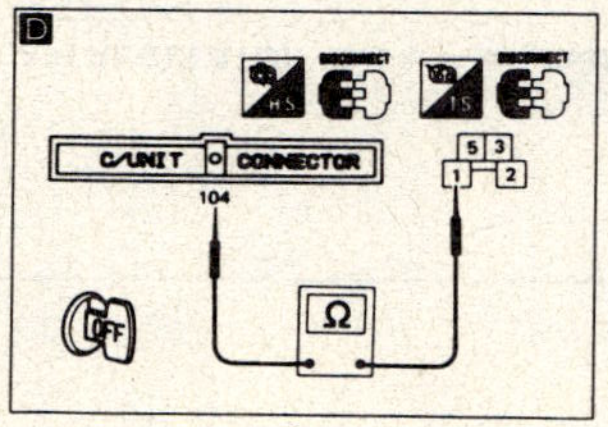

INSPECTION START

↓

A **CHECK OVERALL FUNCTION.**
1) Turn ignition switch "ON".
2) Pinch fuel feed hose with fingers.
Pressure pulsation should exist on the fuel feed hose for 5 seconds after ignition switch is turned "ON".

O.K. → INSPECTION END

N.G. ↓

B **CHECK POWER SUPPLY.**
1) Turn ignition switch "OFF".
2) Disconnect fuel pump relay.
3) Turn ignition switch "ON".
4) Check voltage between terminals ②, ③ and ground with CONSULT or tester.
Voltage: Battery voltage

N.G. → Check the following.
- Harness connectors (M5), (E182)
- 15A fuse
- Harness continuity between fuse and fuel pump relay

If N.G., repair harness or connectors.

O.K. ↓

C **CHECK GROUND CIRCUIT.**
1) Turn ignition switch "OFF".
2) Disconnect fuel pump harness connector.
3) Check harness continuity between terminal ⓑ and body ground, terminal ⓐ and terminal ⑤.
Continuity should exist.

N.G. → Check the following.
- Harness connectors (M5), (E182)
- Joint connector-2
- Harness continuity between fuel pump and body ground
- Harness continuity between fuel pump and fuel pump relay

If N.G., repair harness or connectors.

O.K. ↓

D **CHECK OUTPUT SIGNAL CIRCUIT.**
1) Disconnect E.C.U. harness connector.
2) Check harness continuity between E.C.U. terminal (104) and terminal ①.
Continuity should exist.

N.G. → Check the following.
- Harness connectors (F28), (M47)
- Harness connectors (M5), (E182)
- Harness continuity between E.C.U. and fuel pump relay

If N.G., repair harness or connectors.

O.K. ↓ Ⓐ

FUEL PUMP (Not self-diagnostic item)

E
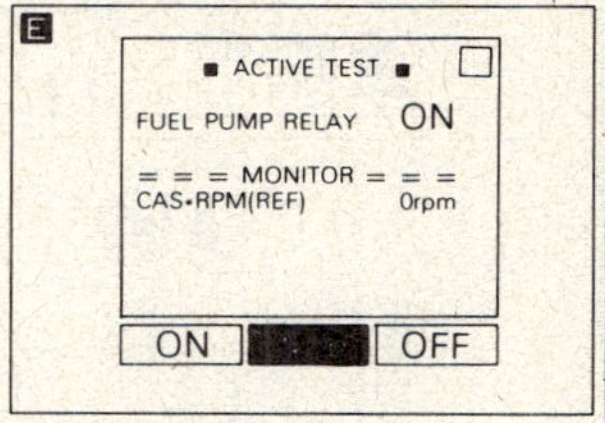

Ⓐ ↓

E **CHECK COMPONENT**
(Fuel pump relay).
1) Reconnect fuel pump relay, fuel pump harness connector and E.C.U. harness connector.
2) Ignition switch "ON".
3) Turn fuel pump relay "ON" and "OFF" in "ACTIVE TEST" mode with CONSULT and check operating sound.

— OR —

Refer to "Electrical Components Inspection".

N.G. → Replace fuel pump relay.

O.K. ↓

CHECK COMPONENT
(Fuel pump).
Refer to "Electrical Components Inspection".

N.G. → Replace fuel pump.

O.K. ↓

Check E.C.U. pin terminals for damage or the connection of E.C.U. harness connector.

↓

INSPECTION END

Diagnostic Procedure 21

AIR REGULATOR (Not self-diagnostic item)

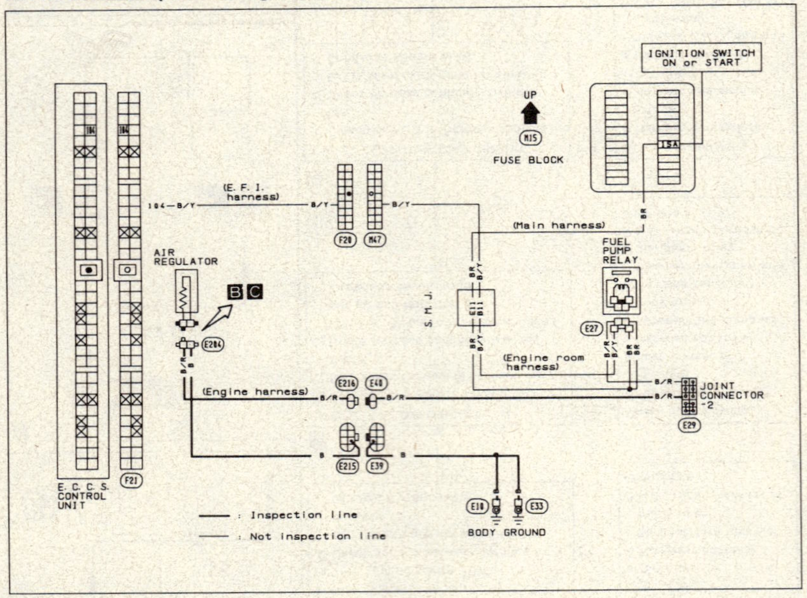

Harness layout

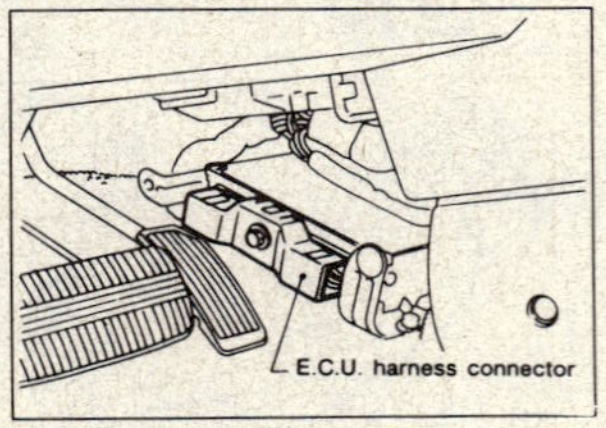

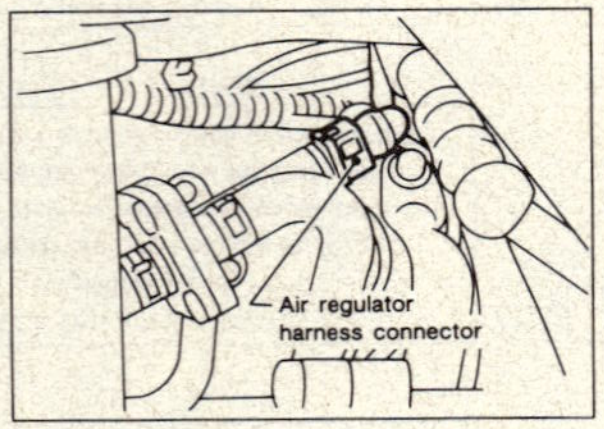

AIR REGULATOR (Not self-diagnostic item)

A

■ FUEL PUMP CIRCUIT ■
PINCH FUEL FEED HOSE WITH FINGERS. IS THERE ANY PRESSURE PULSATION ON THE FUEL FEED HOSE?
OR
DOES THE FUEL PUMP RELAY MAKE AN OPERATING SOUND EVERY 3 SECONDS?
NEXT | NO | YES

A

■ ACTIVE TEST ■
FUEL PUMP RELAY ON
= = = MONITOR = = =
CAS•RPM(REF) 0rpm
ON | OFF

A

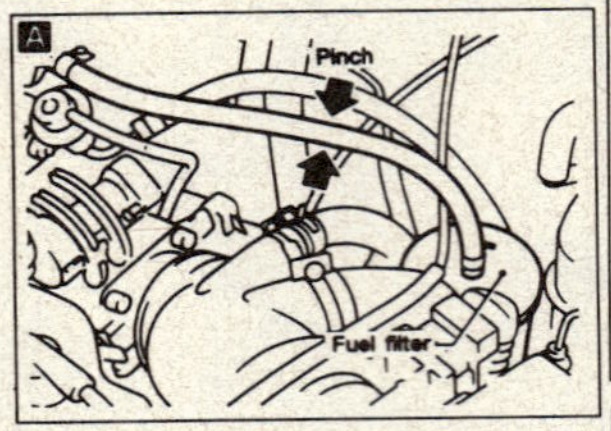

B

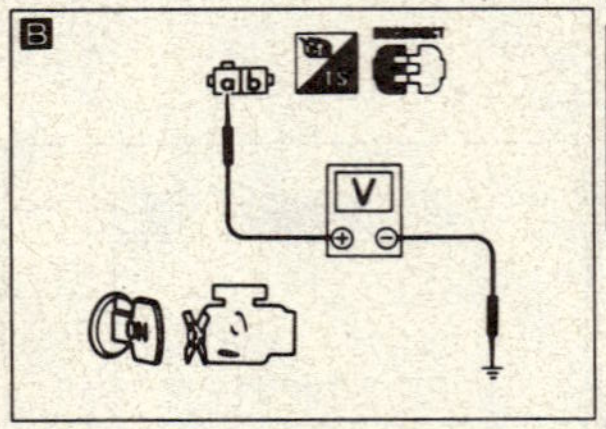

C

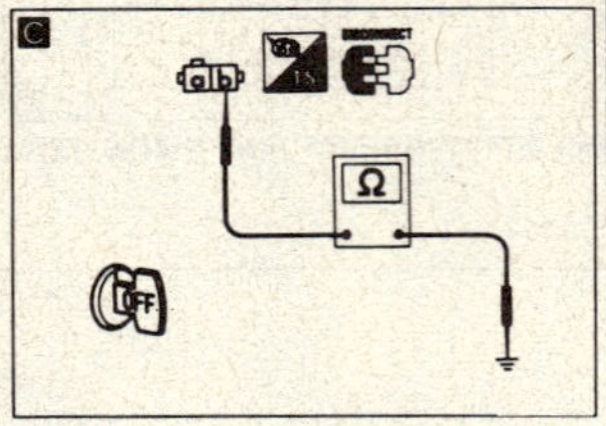

INSPECTION START
↓

A CHECK CONTROL FUNCTION.
1) Turn ignition switch "ON".
2) Perform "FUEL PUMP CIRCUIT" in "FUNCTION TEST" mode with CONSULT.
— OR —
2) Turn fuel pump relay "ON" and "OFF" in "ACTIVE TEST" mode with CONSULT and check operating sound.
— OR —
2) Pinch fuel feed hose with fingers.
Pressure pulsation should exist on the fuel feed hose for 5 seconds after ignition switch is turned "ON".

N.G. → Check fuel pump control circuit.

↓ O.K.

B CHECK POWER SUPPLY.
1) Turn ignition switch "OFF".
2) Disconnect air regulator harness connector.
3) Turn ignition switch "ON".
4) Check voltage between terminal ⓐ and ground with CONSULT or tester.
Battery voltage should exist for 5 seconds after ignition switch is turned "ON".

N.G. → Check the following.
- Harness connectors (E216), (E48)
- Joint connector-2
- Harness continuity between air regulator and joint connector-2

If N.G., repair harness or connectors.

↓ O.K.

C CHECK GROUND CIRCUIT.
1) Turn ignition switch "OFF".
2) Check harness continuity between terminal ⓑ and body ground.
Continuity should exist.

N.G. → Check the following.
- Harness connectors (E215), (E39)
- Harness continuity between air regulator and body ground

If N.G., repair harness or connectors.

↓ O.K.

CHECK COMPONENT
(Air regulator).
Refer to "Electrical Components Inspection".

N.G. → Replace air regulator.

↓ O.K.

INSPECTION END

Diagnostic Procedure 22

A.A.C. VALVE (Not self-diagnostic item)

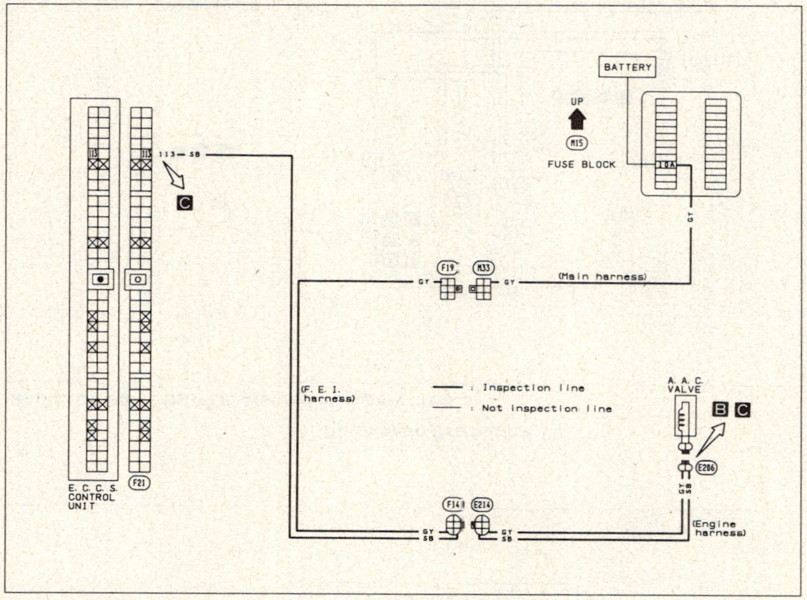

Harness layout

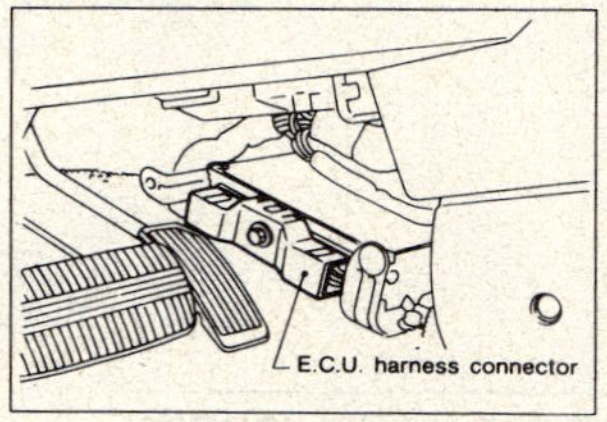

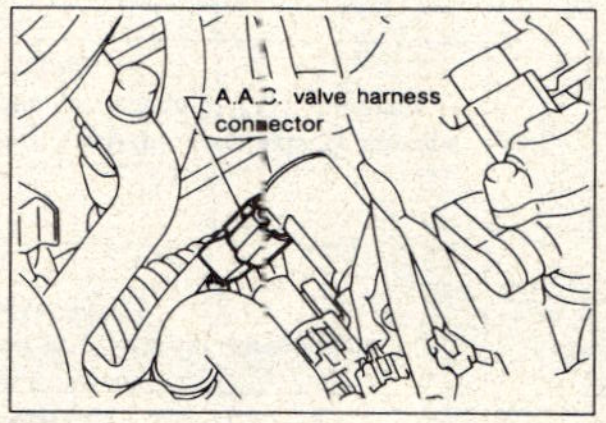

A.A.C. VALVE (Not self-diagnostic item)

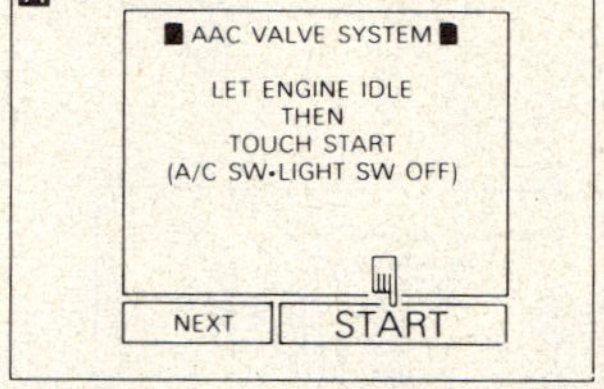

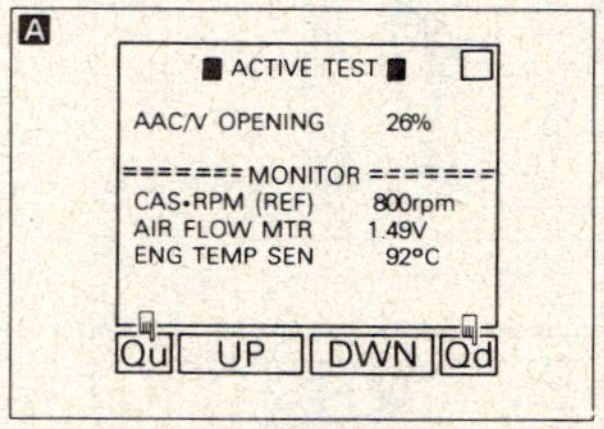

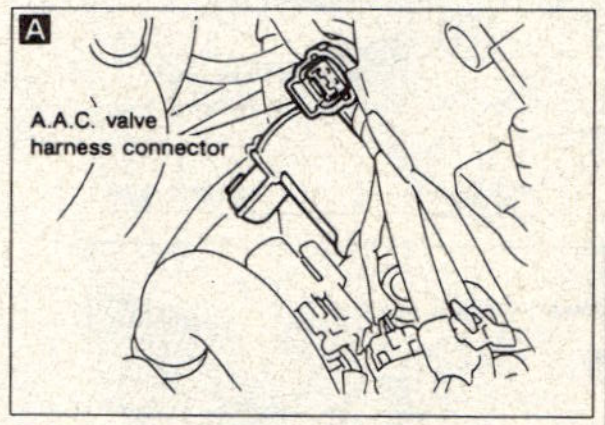

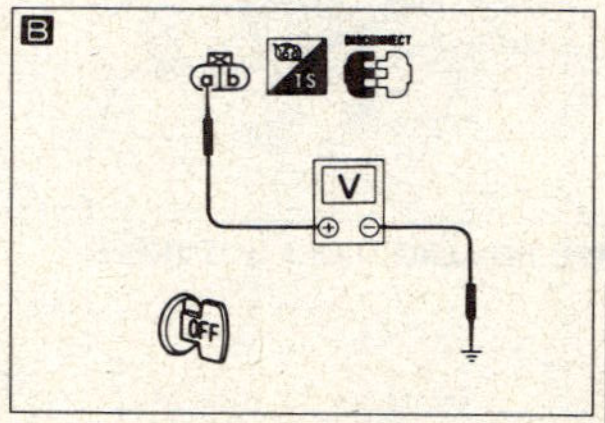

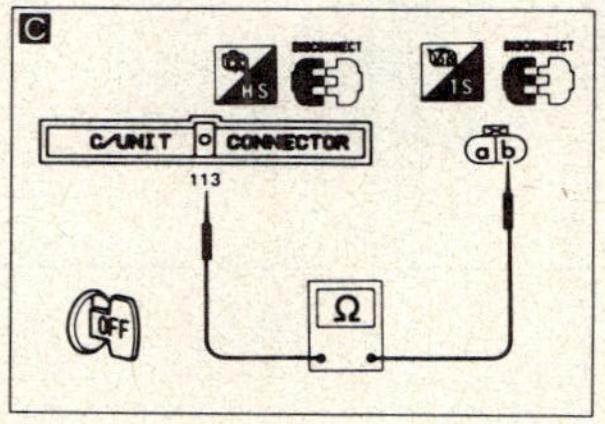

INSPECTION START

A CHECK OVERALL FUNCTION.
1) Start engine and warm it up sufficiently.
2) Check idle speed.
800 ± 50 rpm
If N.G., adjust idle speed.
3) Perform "AAC VALVE SYSTEM" in "FUNCTION TEST" mode with CONSULT.

—OR—

1) Select "AAC/V OPENING" in "ACTIVE TEST" mode with CONSULT.
2) When touching "Qu" and "Qd", check that engine speed changes according to the A.A.C. valve opening percentage.

—OR—

3) Disconnect A.A.C. valve harness connector.
4) Make sure that idle speed drops.

O.K. → INSPECTION END

N.G. ↓

B CHECK POWER SUPPLY.
1) Stop engine.
2) Disconnect A.A.C. valve harness connector.
3) Check voltage between terminal ⓐ and ground with CONSULT or tester.
Voltage: Battery voltage

N.G. → Check the following.
- Harness connectors F19, M33
- Harness connectors F14, E214
- 10A fuse
- Harness continuity between A.A.C. valve and fuse

If N.G., repair harness or connectors.

O.K. ↓

C CHECK OUTPUT SIGNAL CIRCUIT.
1) Disconnect E.C.U. harness connector.
2) Check harness continuity between E.C.U. terminal (113) and terminal ⓑ.
Continuity should exist.

N.G. → Check the following.
- Harness connectors F14, E214
- Harness continuity between E.C.U. and A.A.C. valve

If N.G., repair harness or connectors.

O.K. ↓

Ⓐ

A.A.C. VALVE (Not self-diagnostic item)

Ⓐ

↓

CHECK COMPONENT (A.A.C. valve). Refer to "Electrical Components Inspection". — N.G. → Replace A.A.C. valve.

↓ O.K.

Check E.C.U. pin terminals for damage or the connection of E.C.U. harness connector.

↓

INSPECTION END

Diagnostic Procedure 23

RADIATOR FAN CONTROL (Not self-diagnostic item)

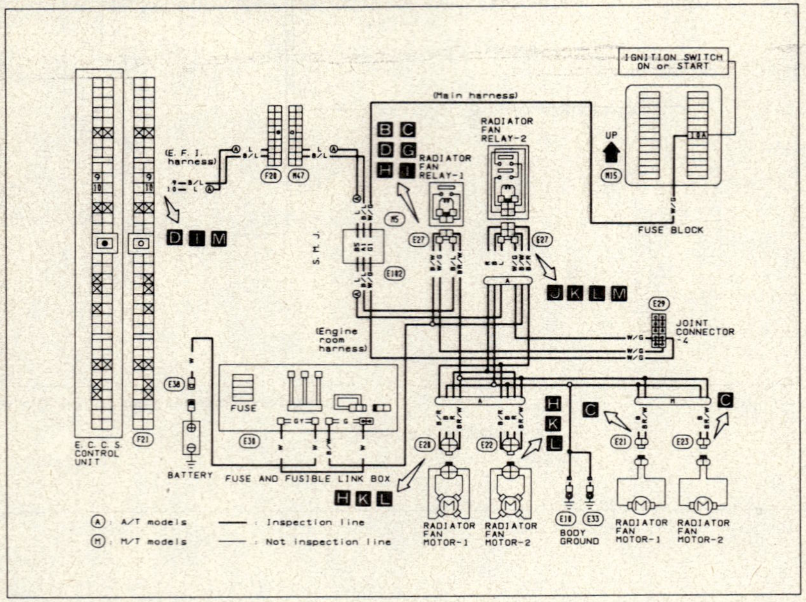

RADIATOR FAN CONTROL (Not self-diagnostic item)

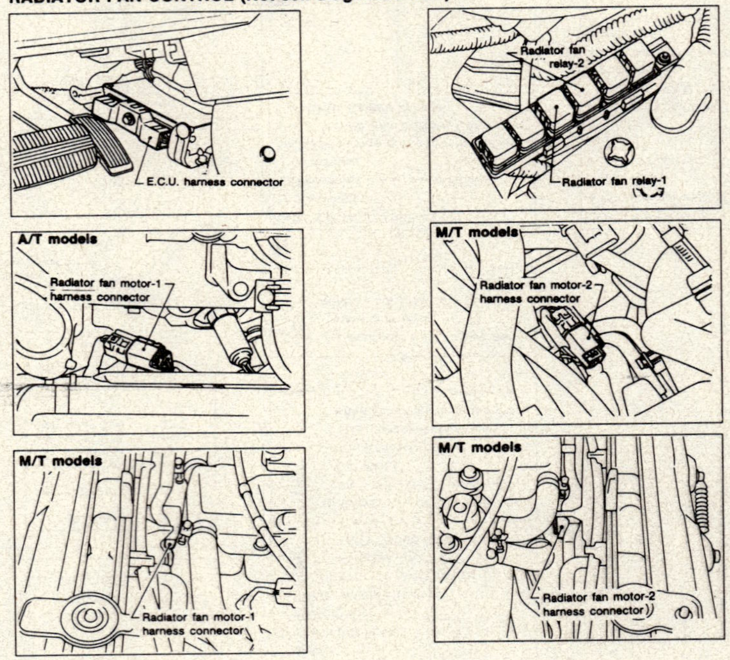

RADIATOR FAN CONTROL (Not self-diagnostic item)

A
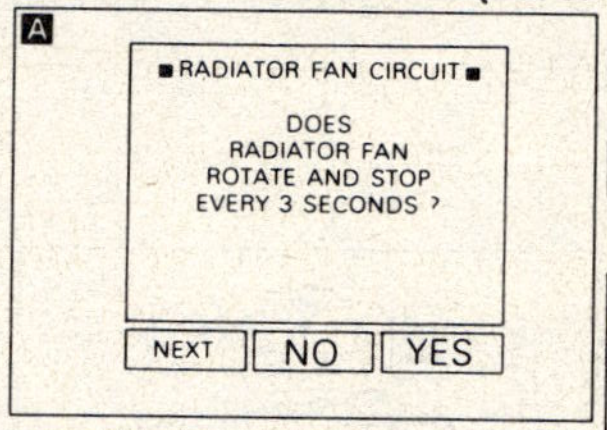

A
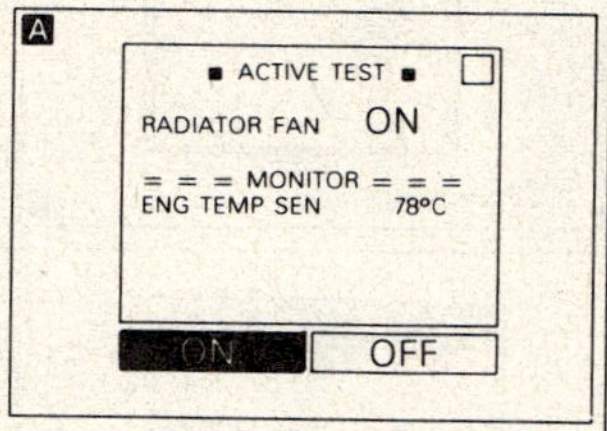

A
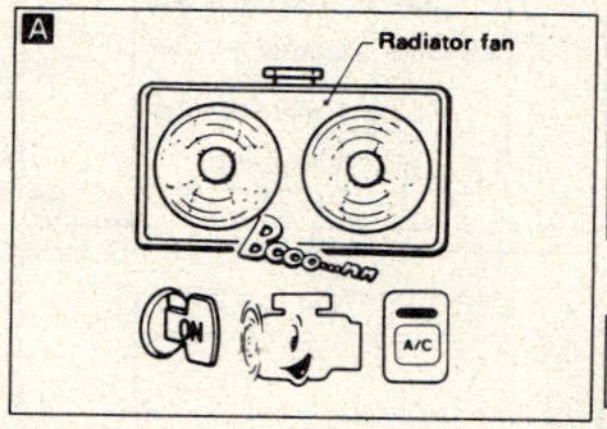

B
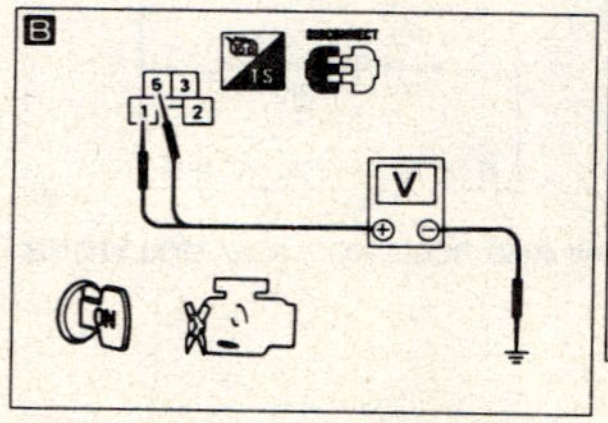

C
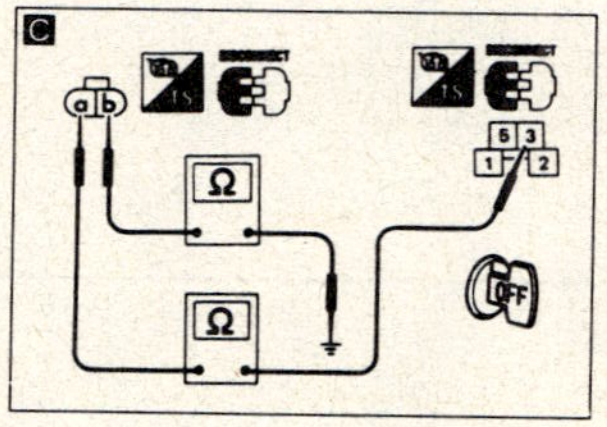

M/T models

INSPECTION START

↓

A **CHECK OVERALL FUNCTION.**
1) Turn ignition switch "ON".
2) Perform "RADIATOR FAN CIRCUIT" in "FUNCTION TEST" mode with CONSULT.
—OR—
1) Turn ignition switch "ON".
2) Perform "RADIATOR FAN" in "ACTIVE TEST" mode with CONSULT.
—OR—
1) Start engine.
2) Set temperature lever at full cold position.
3) Turn air conditioner switch "ON".
4) Turn blower fan switch "ON".
5) Run engine at idle for a few minutes with air conditioner operating.
6) Make sure that radiator fan operates.

O.K. → INSPECTION END

N.G. ↓

Turn air conditioner switch and blower fan switch "OFF".

↓

B **CHECK POWER SUPPLY.**
1) Turn ignition switch "OFF".
2) Disconnect radiator fan relay-1.
3) Turn ignition switch "ON".
4) Check voltage between terminals ①, ⑤ and ground with CONSULT or tester.
Voltage: Battery voltage

N.G. → Check the following.
- Harness connectors (R5), (E102)
- Joint connector-4
- "G" fusible link
- "GY" fusible link
- 10A fuse
- Harness continuity between battery and radiator fan relay-1
- Harness continuity between fuse and radiator fan relay-1

If N.G., repair harness or connectors.

O.K. ↓ Ⓐ

RADIATOR FAN CONTROL (Not self-diagnostic item)

D
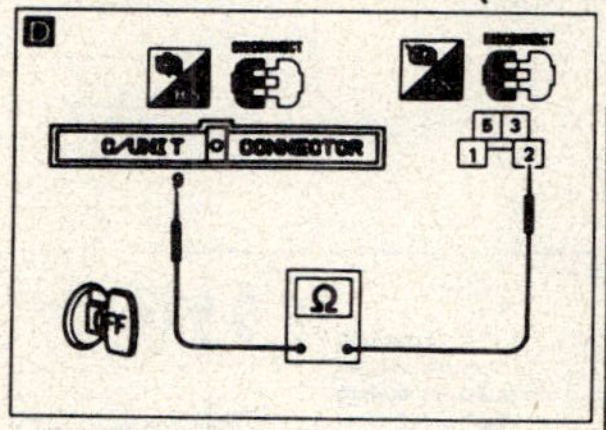

Ⓐ ↓

C **CHECK GROUND CIRCUIT.**
1) Turn ignition switch "OFF".
2) Disconnect radiator fan motor-1 harness connector and radiator fan motor-2 harness connector.
3) Check harness continuity between terminal ⓐ and terminal ③, terminal ⓑ and body ground.
Continuity should exist.

N.G. → Repair harness or connectors.

O.K. ↓

D **CHECK OUTPUT SIGNAL CIRCUIT.**
1) Disconnect E.C.U. harness connector.
2) Check harness continuity between E.C.U. terminal ⑨ and terminal ②.
Continuity should exist.

N.G. → Check the following.
- Harness connectors (F28), (M47)
- Harness connectors (R5), (E102)
- Harness continuity between E.C.U. and radiator fan relay-1

If N.G., repair harness or connectors.

O.K. ↓

CHECK COMPONENT
(Radiator fan relay-1).
Refer to "Electrical Components Inspection".

N.G. → Replace radiator fan relay.

O.K. ↓

CHECK COMPONENTS
(Radiator fan motors).
Refer to "Electrical Components Inspection".

N.G. → Replace radiator fan motors.

O.K. ↓

Check E.C.U. pin terminals for damage or the connection of E.C.U. harness connector.

↓

INSPECTION END

RADIATOR FAN CONTROL (Not self-diagnostic item)

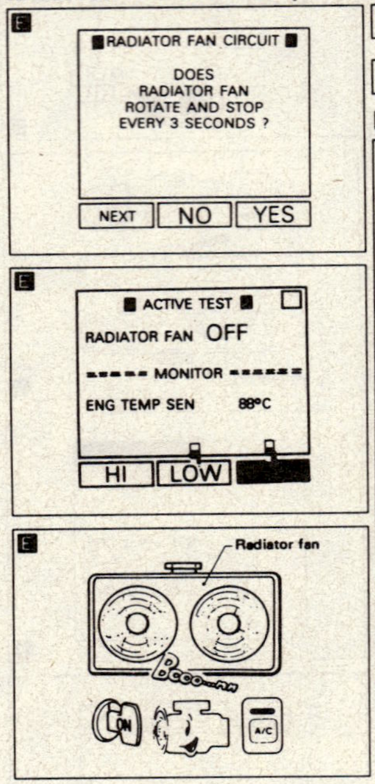

A/T models

INSPECTION START

E CHECK RADIATOR FAN LOW SPEED OPERATION.

1) Disconnect radiator fan relay-2.
2) Turn ignition switch "ON".
3) Perform "RADIATOR FAN CIRCUIT" in "FUNCTION TEST" mode with CONSULT.

— OR —

2) Turn ignition switch "ON".
3) Perform "RADIATOR FAN" in "ACTIVE TEST" mode with CONSULT.

— OR —

2) Start engine.
3) Set temperature lever of full cold position.
4) Turn air conditioner switch "ON".
5) Turn blower fan switch "ON".
6) Run engine at idle for a few minutes with air conditioner operating.
7) Make sure that radiator fan operates at low speed.

N.G. → Check radiator fan low speed control circuit. (Go to PROCEDURE A.)

O.K. → Ⓐ

RADIATOR FAN CONTROL (Not self-diagnostic item)

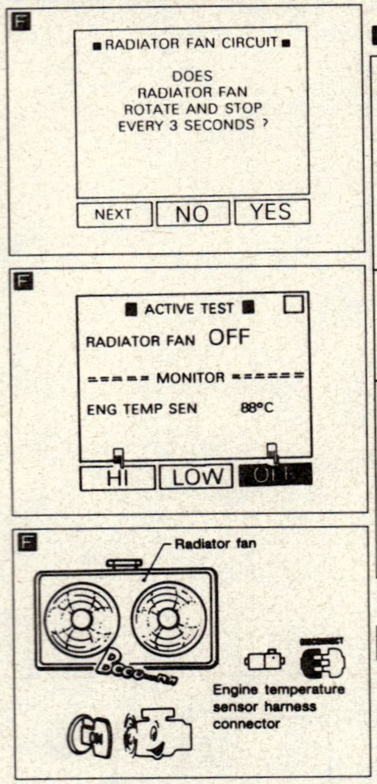

Ⓐ

F CHECK RADIATOR FAN HIGH SPEED OPERATION.

1) Turn ignition switch "OFF".
2) Reconnect radiator fan relay-2.
3) Disconnect radiator fan relay-1.
4) Turn ignition switch "ON".
5) Perform "RADIATOR FAN CIRCUIT" in "FUNCTION TEST" mode with CONSULT.

— OR —

4) Turn ignition switch "ON".
5) Perform "RADIATOR FAN" in "ACTIVE TEST" mode with CONSULT.

— OR —

4) Turn air conditioner switch and blower fan switch "OFF".
5) Disconnect engine temperature sensor harness connector.
6) Restart engine and make sure that radiator fan operates at higher speed than low speed.

N.G. → Check radiator fan high speed control circuit. (Go to PROCEDURE B.)

O.K. → INSPECTION END

RADIATOR FAN CONTROL (Not self-diagnostic item)

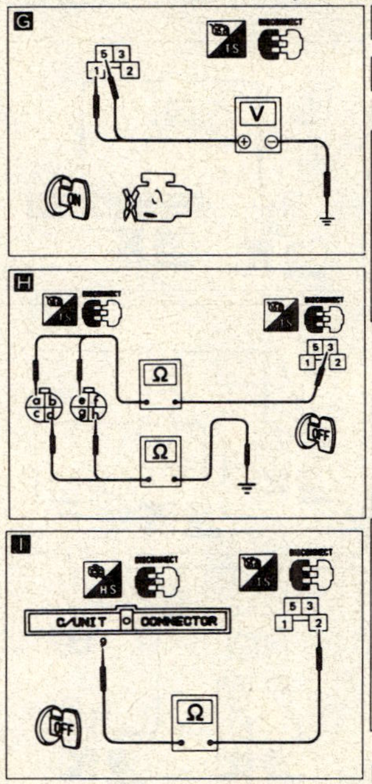

PROCEDURE A

INSPECTION START

G CHECK POWER SUPPLY.
1) Stop engine.
2) Disconnect radiator fan relay-1.
3) Turn ignition switch "ON".
4) Check voltage between terminals ①, ⑤ and ground with CONSULT or tester.
Voltage: Battery voltage

N.G. → Check the following.
- Harness connectors (M5), (E102)
- 10A fuse
- "G" fusible link
- "GY" fusible link
- Joint connector-4
- Harness continuity between radiator fan relay-1 and fuse
- Harness continuity between radiator fan relay-1 and battery

If N.G., repair harness or connectors.

O.K. ↓

H CHECK GROUND CIRCUIT.
1) Turn ignition switch "OFF".
2) Disconnect radiator fan motor-1 harness connector and radiator fan motor-2 harness connector.
3) Check harness continuity between terminals ⓐ, ⓔ and terminal ③, terminals ⓓ, ⓗ and body ground.
Continuity should exist.

N.G. → Repair harness or connectors.

O.K. ↓

I CHECK OUTPUT SIGNAL CIRCUIT.
1) Disconnect E.C.U. harness connector.
2) Check harness continuity between E.C.U. terminal ⑨ and terminal ②.
Continuity should exist.

N.G. → Check the following.
- Harness connectors (F20), (M47)
- Harness connectors (M5), (E102)
- Harness continuity between radiator fan relay-1 and E.C.U.

If N.G., repair harness or connectors.

O.K. ↓

CHECK COMPONENT
(Radiator fan relay-1).
Refer to "Electrical Components Inspection".

N.G. → Replace radiator fan relay.

O.K. ↓ Ⓐ

RADIATOR FAN CONTROL (Not self-diagnostic item)

Ⓐ ↓

CHECK COMPONENTS
(Radiator fan motors).
Refer to "Electrical Components Inspection".

N.G. → Replace radiator fan motors.

O.K. ↓

Check E.C.U. pin terminals for damage or the connection of E.C.U. harness connector.

↓

INSPECTION END

RADIATOR FAN CONTROL (Not self-diagnostic item)

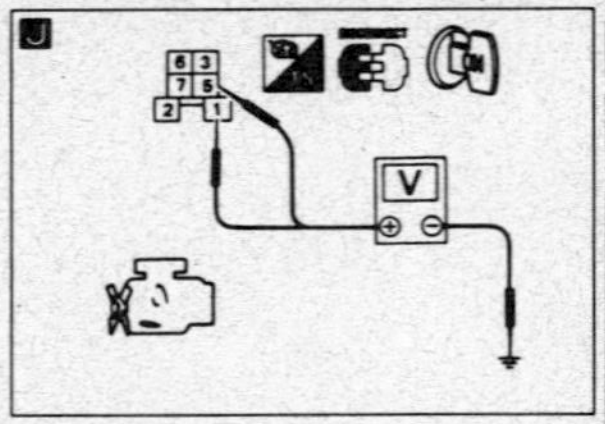

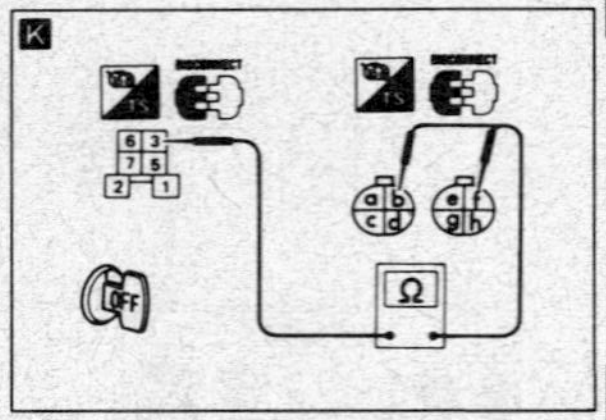

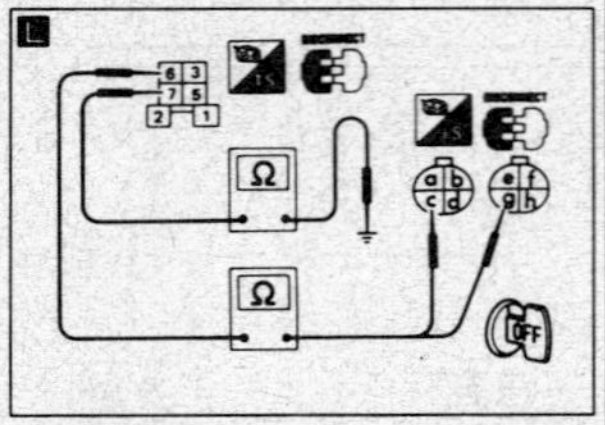

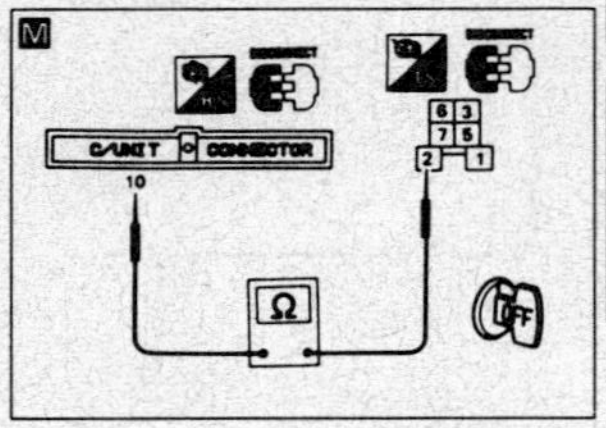

PROCEDURE B

INSPECTION START

↓

J CHECK POWER SUPPLY.
1) Stop engine.
2) Disconnect radiator fan relay-2.
3) Turn ignition switch "ON".
4) Check voltage between terminals ①, ⑤ and ground.
Voltage: Battery voltage

N.G. → Check the following.
- Harness connectors (M5), (E102)
- 10A fuse
- "G" fusible link
- "GY" fusible link
- Joint connector-4
- Harness continuity between radiator fan relay-2 and fuse
- Harness continuity between radiator fan relay-2 and battery

If N.G., repair harness or connectors.

O.K. ↓

CHECK GROUND CIRCUIT.
1) Turn ignition switch "OFF".
2) Disconnect radiator fan motor-1 harness connector and radiator fan motor-2 harness connector.
K 3) Check harness continuity between terminal ③ and terminals ⓑ, ⓕ.
Continuity should exist.
L 4) Check harness continuity between terminal ⑥ and terminals ⓒ, ⓖ, terminal ⑦ and body ground.
Continuity should exist.

N.G. → Repair harness or connectors.

O.K. ↓

M CHECK OUTPUT SIGNAL CIRCUIT.
1) Disconnect E.C.U. harness connector.
2) Check harness continuity between E.C.U. terminal ⑩ and terminal ②.
Continuity should exist.

N.G. → Check the following.
- Harness connectors (F28), (M7)
- Harness connectors (M5), (E102)
- Harness continuity between E.C.U. and radiator fan relay-2

If N.G., repair harness or connectors.

O.K. ↓

CHECK COMPONENT
(Radiator fan relay-2).
Refer to "Electrical Components Inspection".

N.G. → Replace radiator fan relay.

O.K. ↓

Ⓐ

RADIATOR FAN CONTROL (Not self-diagnostic item)

Ⓐ

↓

CHECK COMPONENTS
(Radiator fan motors).
Refer to "Electrical Components Inspection".

N.G. → Replace radiator fan motors.

O.K. ↓

Check E.C.U. pin terminals for damage or the connection of E.C.U. harness connector.

↓

INSPECTION END

Diagnostic Procedure 24

POWER STEERING OIL PRESSURE SWITCH (Not self-diagnostic item)

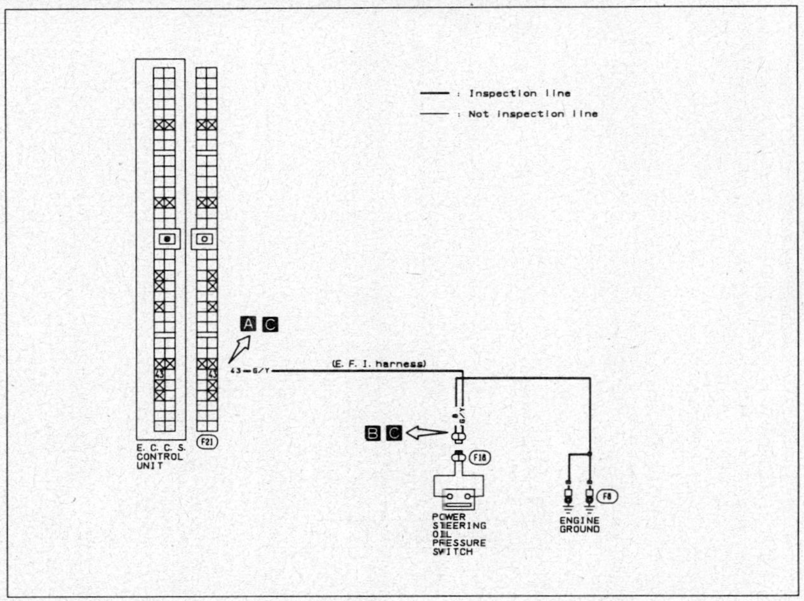

Harness layout

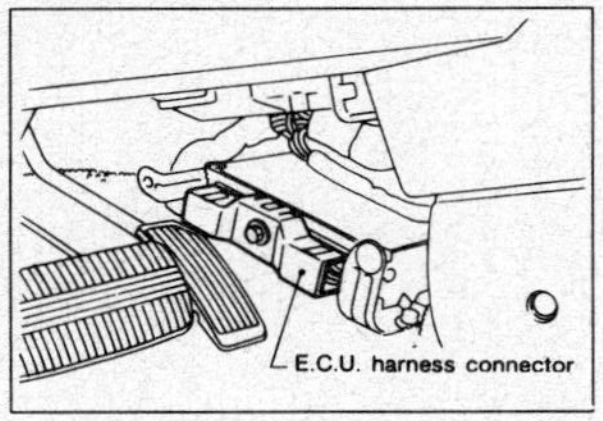

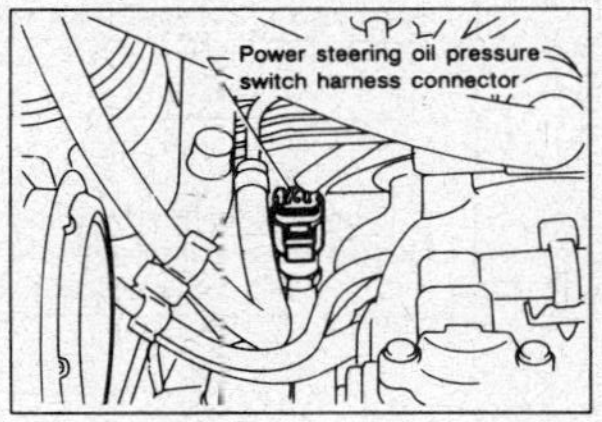

POWER STEERING OIL PRESSURE SWITCH (Not self-diagnostic item)

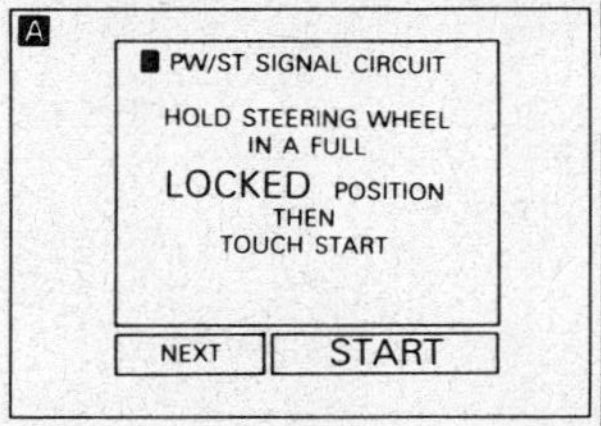

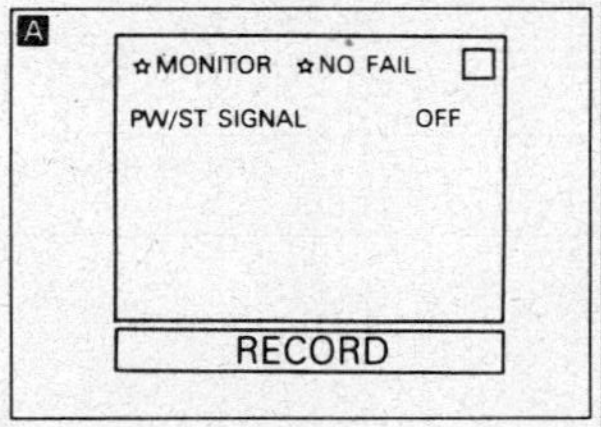

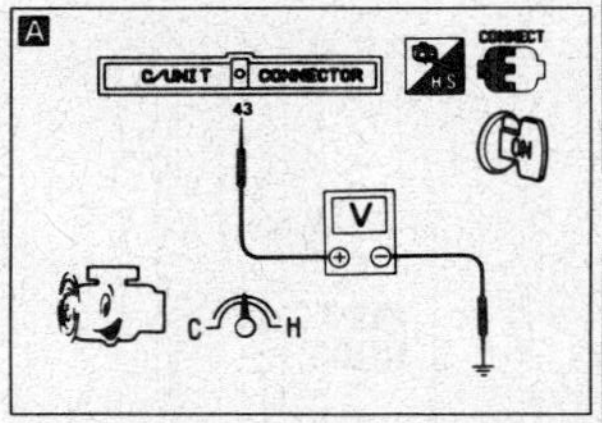

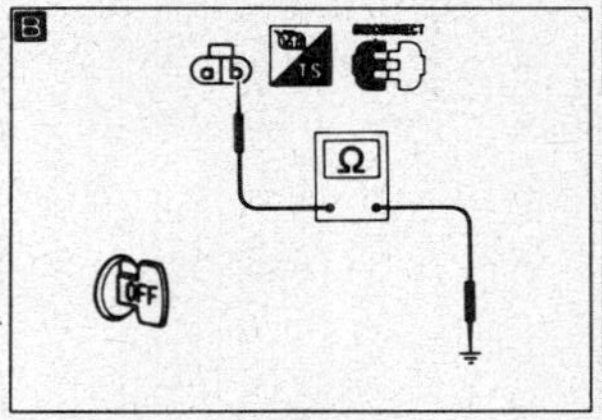

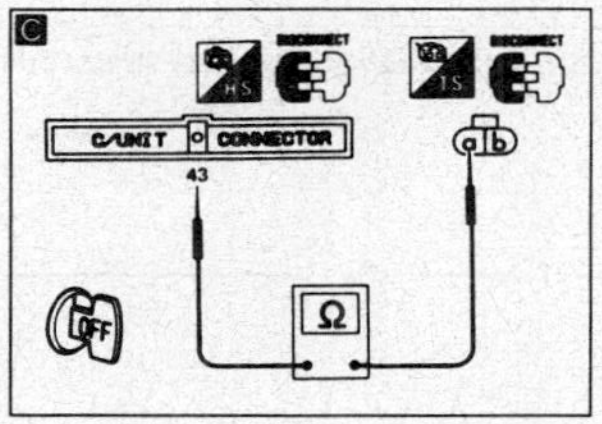

INSPECTION START

↓

A CHECK OVERALL FUNCTION.

1) Turn ignition switch "ON".
2) Perform "PW/ST SIGNAL CIRCUIT" in "FUNCTION TEST" mode with CONSULT.

— OR —

1) Start engine.
2) Check power steering oil pressure switch signal in "DATA MONITOR" mode with CONSULT.

Steering is neutral: OFF
Steering is turned: ON

— OR —

1) Start engine.
2) Check voltage between E.C.U. terminal ㊸ and ground with CONSULT or tester.

Voltage:
When steering wheel is turned quickly.
Approximately 0V
Except above
Approximately 7 - 9V

O.K. → INSPECTION END

N.G. ↓

B CHECK GROUND CIRCUIT.

1) Turn ignition switch "OFF".
2) Disconnect power steering oil pressure switch harness connector.
3) Check harness continuity between terminal ⓑ and engine ground.
Continuity should exist.

N.G. → Repair harness or connectors.

O.K. ↓

C CHECK INPUT SIGNAL CIRCUIT.

1) Disconnect E.C.U. harness connector.
2) Check harness continuity between E.C.U. terminal ㊸ and terminal ⓐ.
Continuity should exist.

N.G. → Repair harness or connectors.

O.K. ↓ Ⓐ

POWER STEERING OIL PRESSURE SWITCH (Not self-diagnostic item)

Ⓐ

CHECK COMPONENT
(Power steering oil pressure switch).
Refer to "Electrical Components Inspection".

N.G. → Replace power steering oil pressure switch.

O.K. ↓

Check E.C.U. pin terminals for damage or the connection of the E.C.U. harness connector.

↓

INSPECTION END

Diagnostic Procedure 25

NEUTRAL/INHIBITOR SWITCH (Not self-diagnostic item)

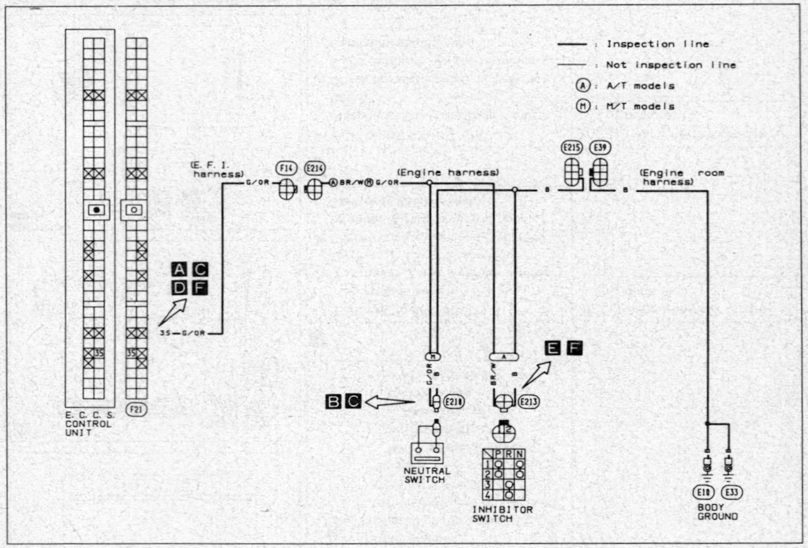

Harness layout

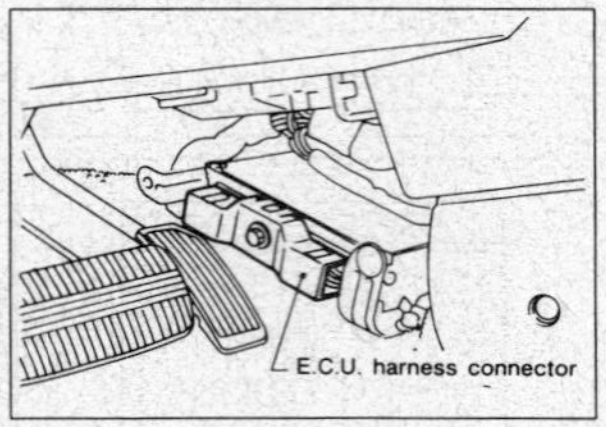

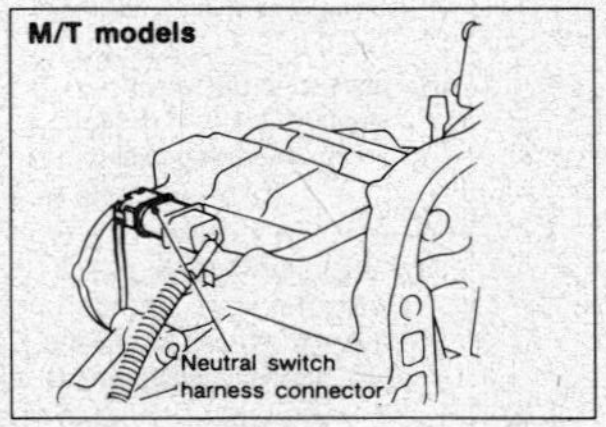

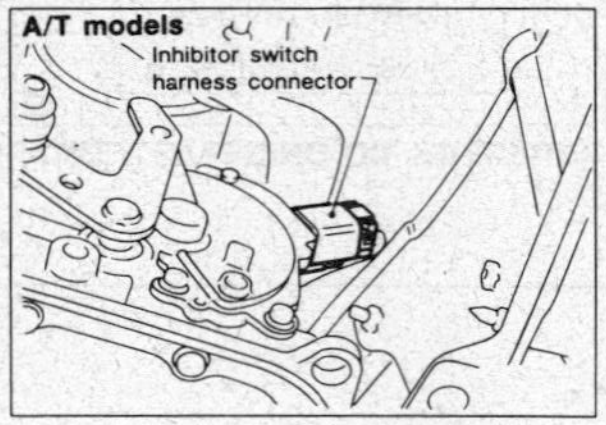

NEUTRAL/INHIBITOR SWITCH (Not self-diagnostic item)

A
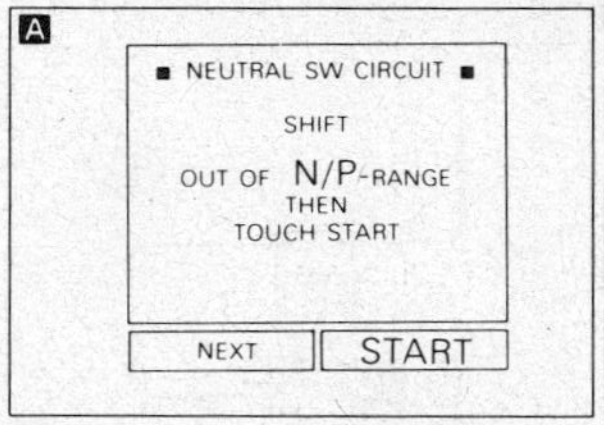

A
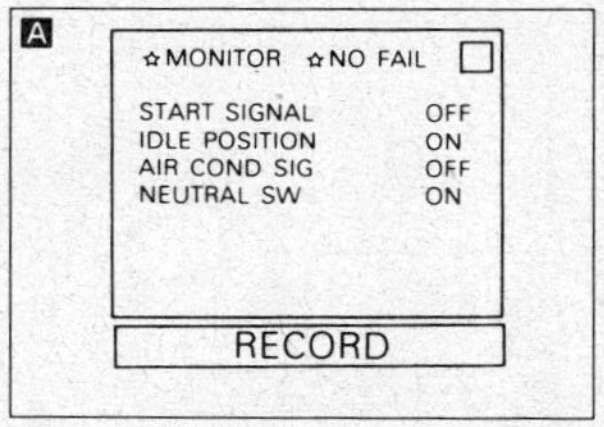

A
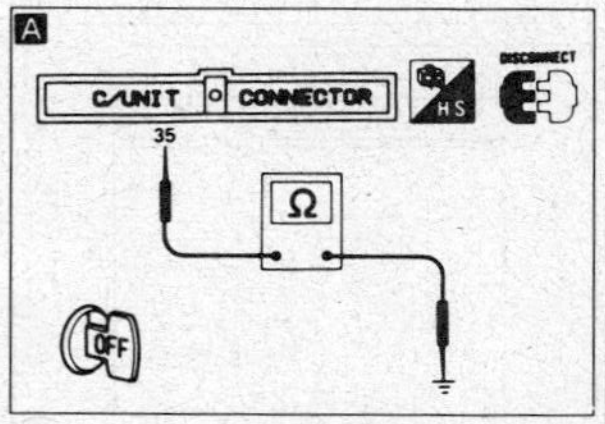

B
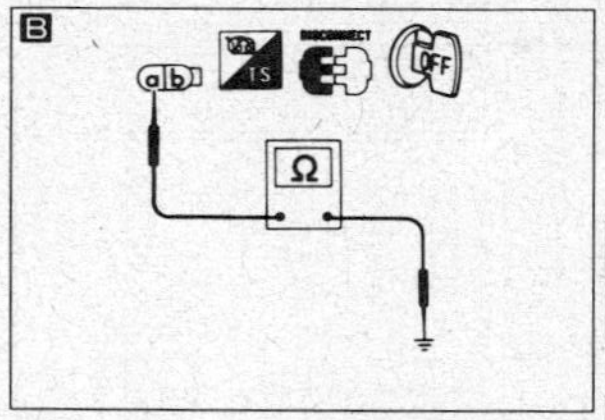

C
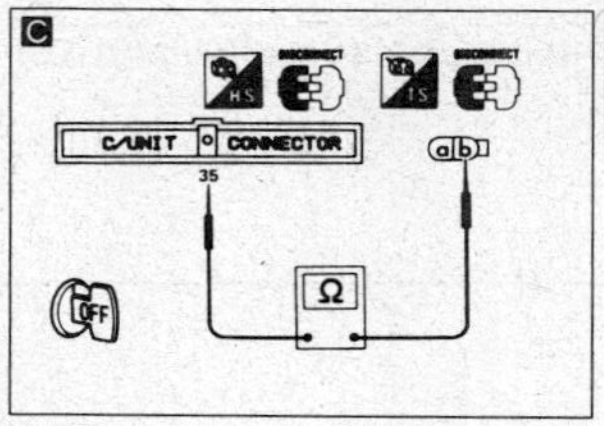

Neutral switch

INSPECTION START

↓

A **CHECK OVERALL FUNCTION.**
1) Turn ignition switch "ON".
2) Perform "NEUTRAL SW CIRCUIT" in "FUNCTION TEST" mode with CONSULT.

— OR —

1) Turn ignition switch "ON".
2) Check neutral switch signal in "DATA MONITOR" mode with CONSULT.
Neutral position: ON
Except above: OFF

— OR —

1) Set shift lever to the neutral position.
2) Disconnect E.C.U. harness connector.
3) Check harness continuity between E.C.U. terminal ㉟ and body ground.
Continuity should exist.

O.K. → INSPECTION END

N.G. ↓

Turn ignition switch "OFF".

↓

B **CHECK GROUND CIRCUIT.**
1) Disconnect neutral switch harness connector.
2) Check harness continuity between terminal ⓐ and body ground.
Continuity should exist.

N.G. → Check the following.
- Harness connectors (E215), (E39)
- Harness continuity between neutral switch and body ground

If N.G., repair harness or connectors.

O.K. ↓

Disconnect E.C.U. harness connector.

↓

C **CHECK INPUT SIGNAL CIRCUIT.**
1) Check harness continuity between E.C.U. terminal ㉟ and terminal ⓑ.
Continuity should exist.

N.G. → Check the following.
- Harness connectors (F14), (E214)
- Harness continuity between E.C.U. and neutral switch

If N.G., repair harness or connectors.

O.K. → Ⓐ

NEUTRAL/INHIBITOR SWITCH (Not self-diagnostic item)

Ⓐ

↓

CHECK COMPONENT (Neutral switch).
Refer to "Electrical Components Inspection".

N.G. → Replace neutral switch.

O.K. ↓

Check E.C.U. pin terminals for damage or the connection of E.C.U. harness connector.

↓

INSPECTION END

NEUTRAL/INHIBITOR SWITCH (Not self-diagnostic item)

D

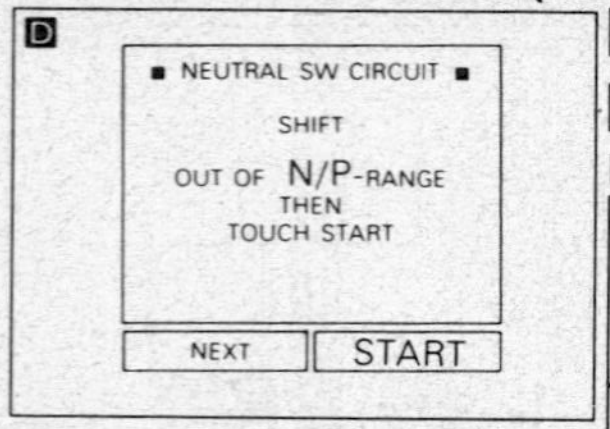

D

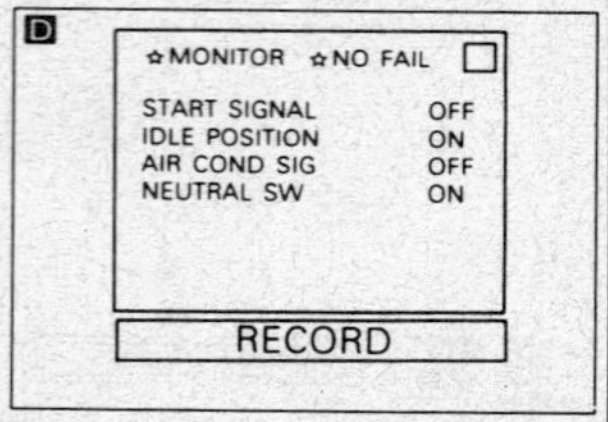

D

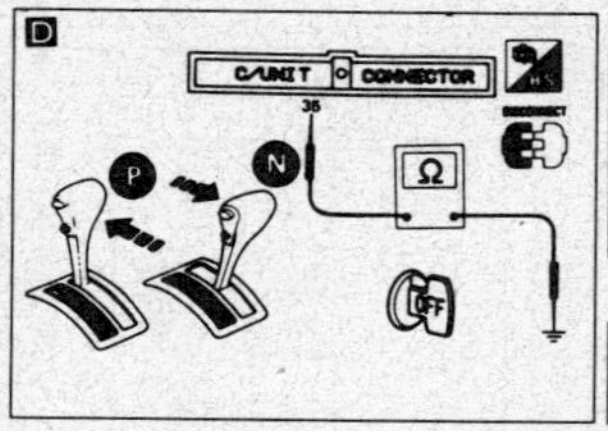

E

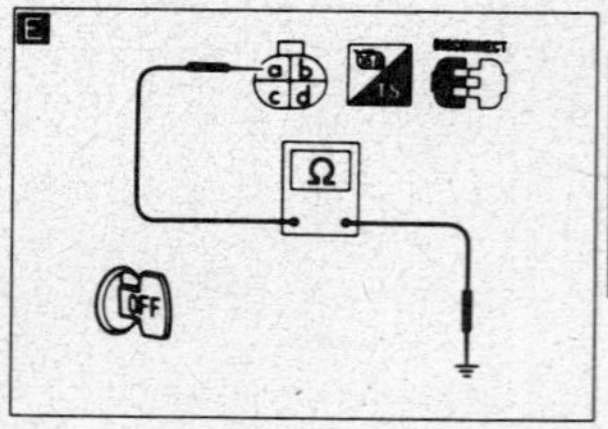

Inhibitor switch

INSPECTION START

D CHECK OVERALL FUNCTION.
1) Turn ignition switch "ON".
2) Perform "NEUTRAL SW CIRCUIT" in "FUNCTION TEST" mode with CONSULT.

— OR —

1) Turn ignition switch "ON".
2) Check neutral switch signal in "DATA MONITOR" mode with CONSULT.
"N" or "P": ON
Except above: OFF

— OR —

1) Shift selector lever to "P" range.
2) Disconnect E.C.U. harness connector.
3) Check harness continuity between E.C.U. terminal (35) and body ground.
Continuity should exist.
4) Shift selector lever to "N" range.
5) Check harness continuity between E.C.U. terminal (35) and body ground.
Continuity should exist.

O.K. → INSPECTION END

N.G. ↓

Turn ignition switch "OFF".

↓

E CHECK GROUND CIRCUIT.
1) Disconnect inhibitor switch harness connector.
2) Check harness continuity between terminal (c) and body ground.
Continuity should exist.

N.G. → Check the following.
- Harness connectors (E215), (E39)
- Harness continuity between inhibitor switch and body ground

If N.G., repair harness or connectors.

O.K. ↓

Disconnect E.C.U. harness connector.

↓ (A)

NEUTRAL/INHIBITOR SWITCH (Not self-diagnostic item)

F

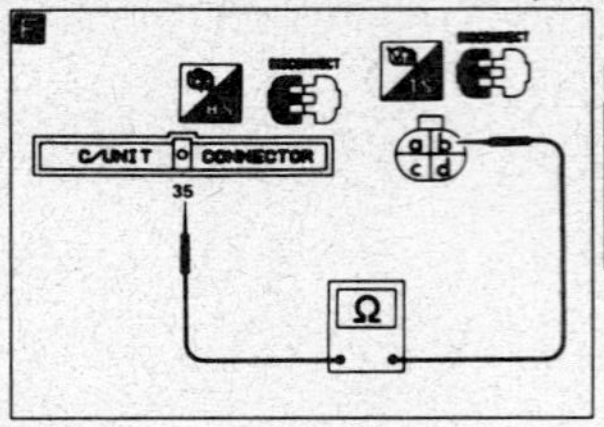

(A) ↓

F CHECK INPUT SIGNAL CIRCUIT.
1) Check harness continuity between E.C.U. terminal (35) and terminal (b).
Continuity should exist.

N.G. → Check the following.
- Harness connectors (F14), (E214)
- Harness continuity between E.C.U. and inhibitor switch

If N.G., repair harness or connectors.

O.K. ↓

CHECK COMPONENT (Inhibitor switch).
Refer to "Electrical Components Inspection".

N.G. → Replace inhibitor switch.

O.K. ↓

Check E.C.U. pin terminals for damage or the connection of E.C.U. harness connector.

↓

INSPECTION END

Diagnostic Procedure 26 — Symptom — High Idling after Warm-up

1

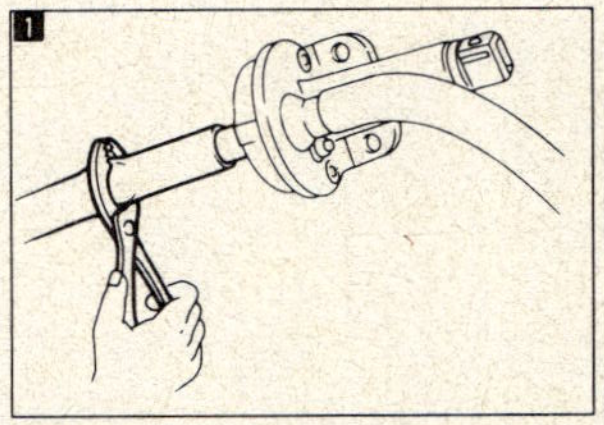

2

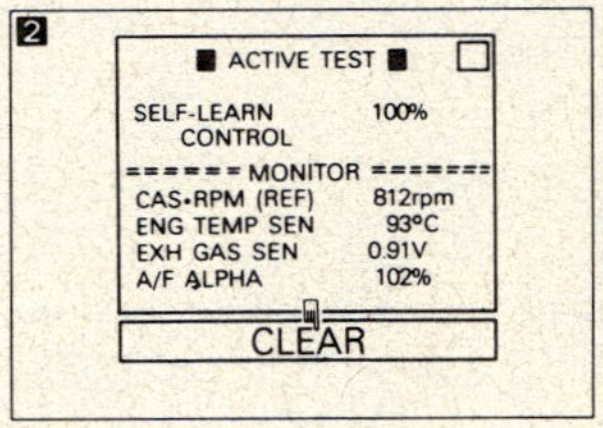

2

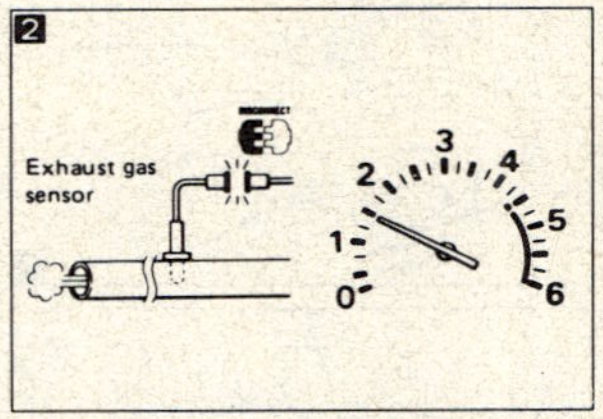

3

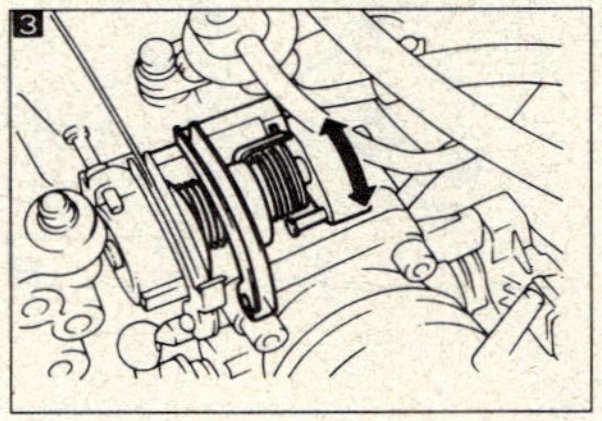

1 CHECK AIR REGULATOR.
When pinching the air regulator hose, does the engine speed drop?
- Yes → Check air regulator and circuit.
- No ↓

2 CHECK INTAKE AIR LEAK.
1. Select "SELF-LEARNING CONT" in "ACTIVE TEST" mode.
2. Clear the self-learning control coefficient by touching "CLEAR".
3. Does the engine speed drop?

— OR —

1. Disconnect exhaust gas sensor harness connector.
2. After racing engine at 2,000 rpm under no-load for about 30 seconds, does the engine speed drop?

- Yes → Discover air leak location and repair.
- No ↓

3 CHECK THROTTLE LINKAGE.
1. Check that throttle linkage moves smoothly.
2. Confirm that throttle valve both fully opens and fully closes.

- N.G. → Repair throttle linkage or sticking of throttle valve.
- O.K. ↓

INSPECTION END

Diagnostic Procedure 27 — Symptom — Hunting

1

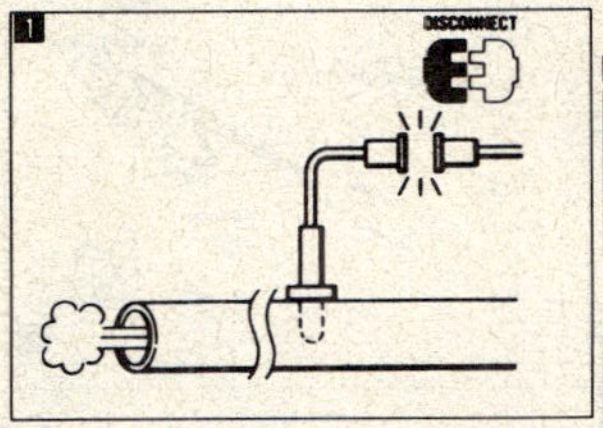

2

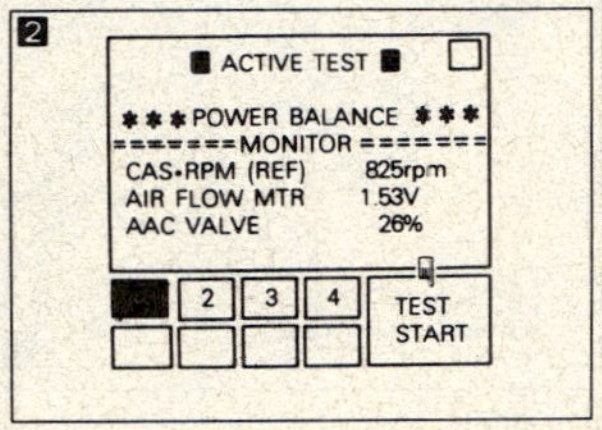

2

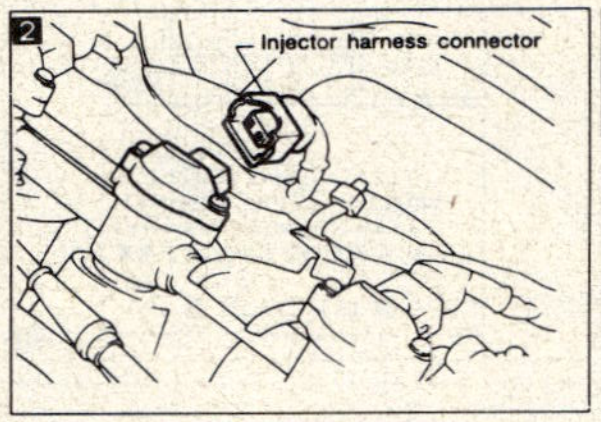

3

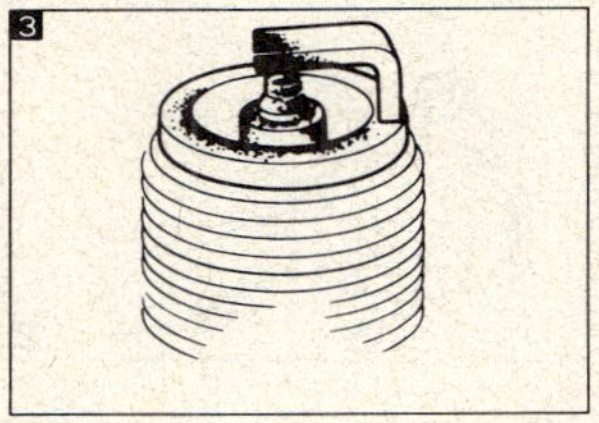

1 CHECK EXHAUST GAS SENSOR.
When disconnecting exhaust gas sensor harness connector, is the hunting fixed?
- Yes → Check exhaust gas sensor.
- No ↓

2 PERFORM POWER BALANCE TEST.
1. Perform "POWER BALANCE" in "ACTIVE TEST" mode.
2. Is there any cylinder which does not produce a momentary engine speed drop?

— OR —

When disconnecting each injector harness connector one at a time, is there any cylinder which does not produce a momentary engine speed drop?

- No → Go to 4.
- Yes ↓

3 CHECK SPARK PLUGS.
Remove the spark plugs and check for fouling, etc.
- N.G. → Repair or replace spark plug(s).
- O.K. ↓

(Go to Ⓐ on next page.)

Diagnostic Procedure 27 — Symptom — Hunting (Cont'd)

Ⓐ

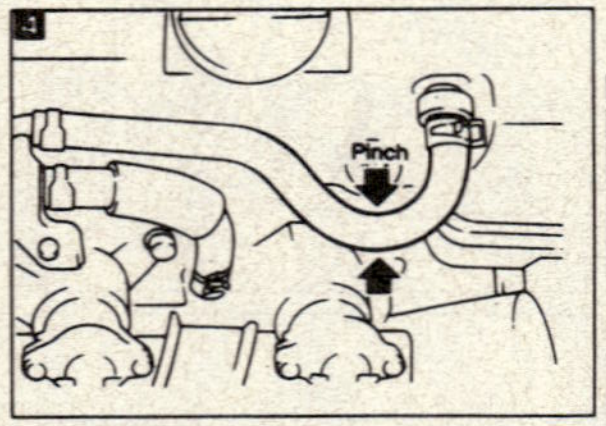

4 CHECK FOR INTAKE AIR LEAK.
When pinching blow-by hose (lowering the blow-by air supply), does the engine speed rise?

- Yes → Discover air leak location and repair.
- No ↓

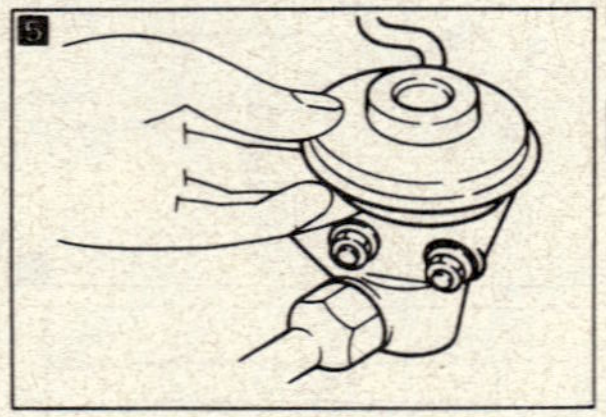

5 CHECK E.G.R. CONTROL VALVE
Check E.G.R. control valve for sticking

- N.G. → Repair or replace
- O.K. ↓

INSPECTION END

Diagnostic Procedure 28 — Symptom — Unstable Idle

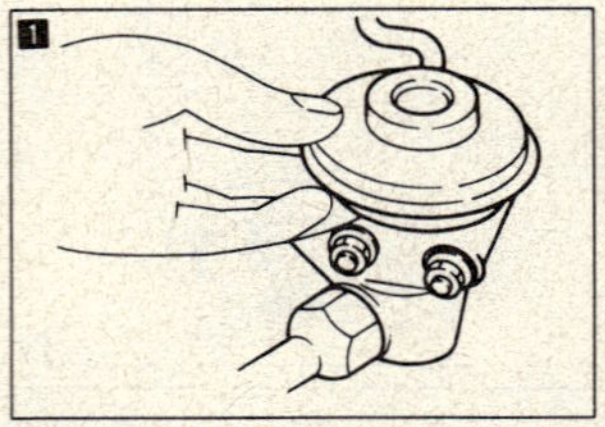

1 CHECK E.G.R. CONTROL VALVE.
Check E.G.R. control valve for sticking.

- N.G. → Repair or replace.
- O.K. ↓

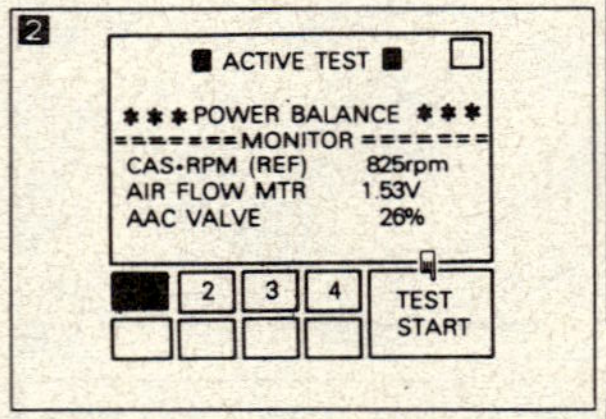

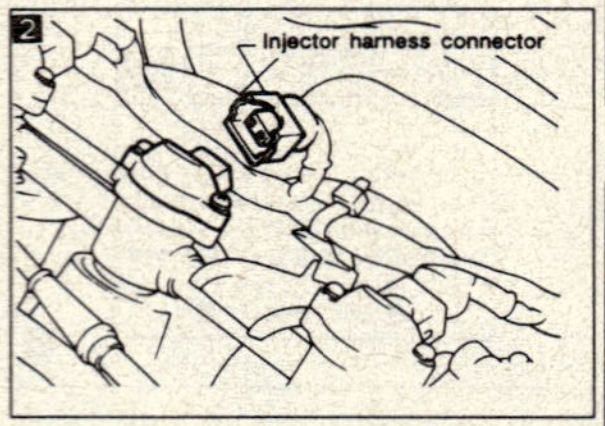

2 PERFORM POWER BALANCE TEST.
1. Perform "POWER BALANCE" in "ACTIVE TEST" mode.
2. Is there any cylinder which does not produce a momentary engine speed drop?

—OR—

When disconnecting each injector harness connector one at a time, is there any cylinder which does not produce a momentary engine speed drop?

- No → Go to 6.
- Yes ↓

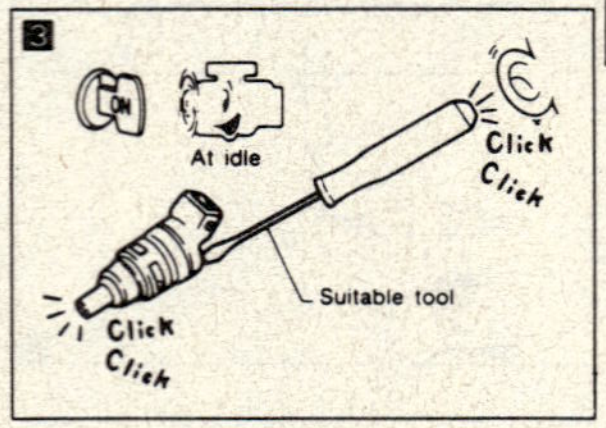

3 CHECK INJECTOR.
Does each injector make an operating sound at idle?

- No → Check injector(s) and circuit(s).
- Yes ↓

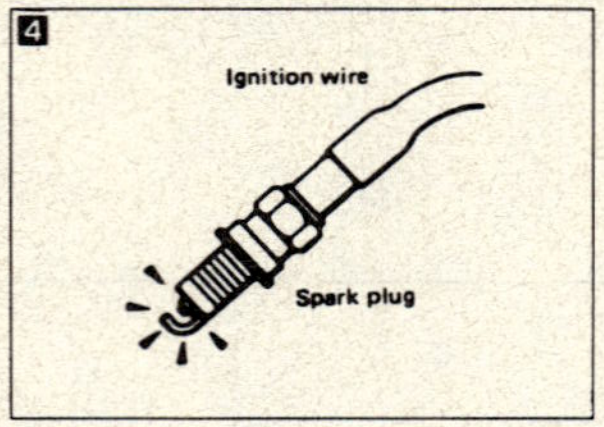

4 CHECK IGNITION SPARK.
1. Disconnect ignition wire from spark plug.
2. Connect a known good spark plug to the ignition wire.
3. Place end of spark plug against a suitable ground and crank engine.
4. Check for spark.

- N.G. → Check ignition coil, power transistor and their circuits.
- O.K. ↓

(Go to Ⓐ on next page.)

Diagnostic Procedure 28 — Symptom — Unstable Idle (Cont'd)

Ⓐ

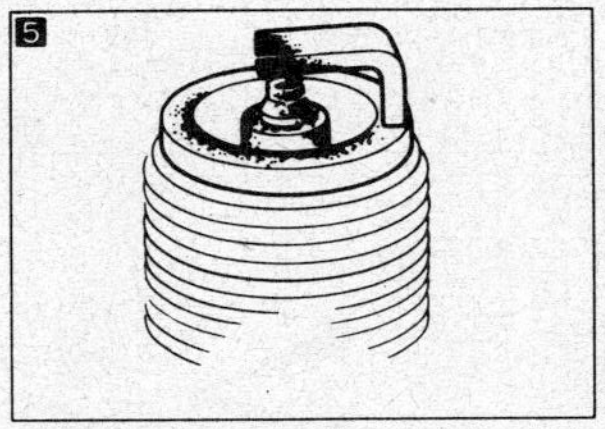

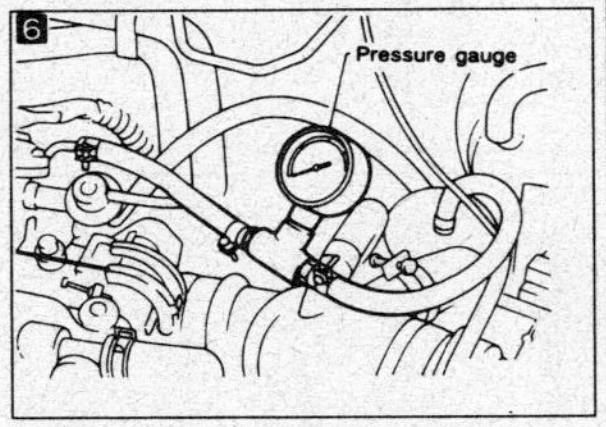

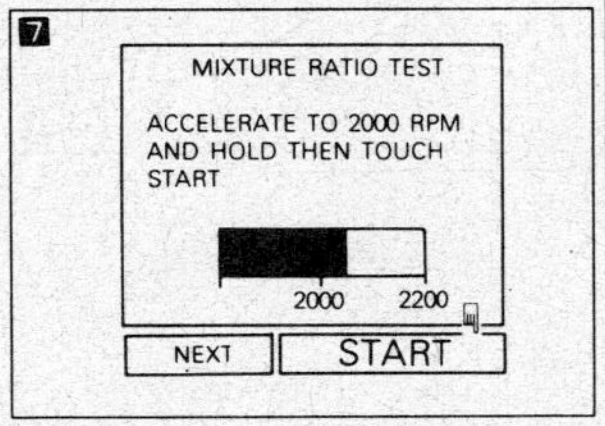

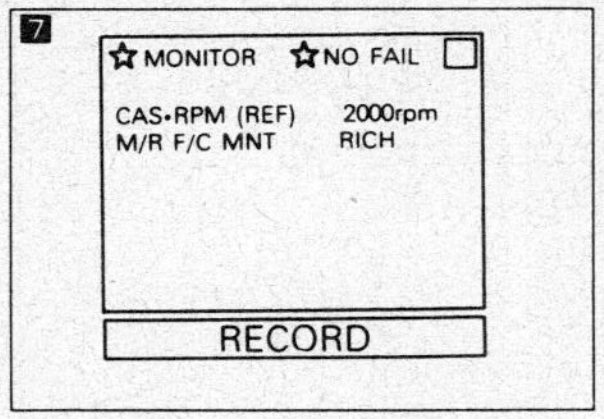

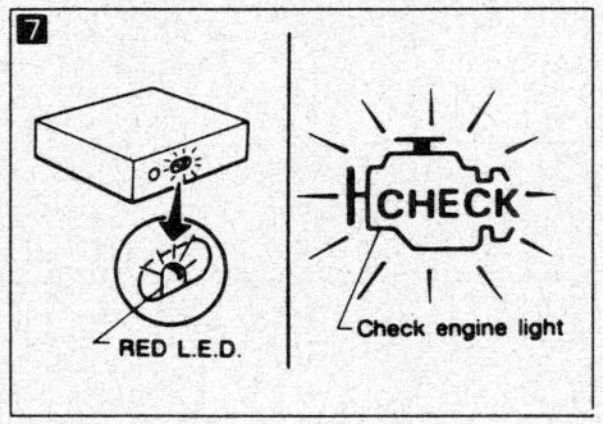

5 **CHECK SPARK PLUGS.**
Remove the spark plugs and check for fouling, etc.
- N.G. → Repair or replace spark plug(s).
- O.K. ↓

6 **CHECK FUEL PRESSURE.**
1. Release fuel pressure to zero.
2. Install fuel pressure gauge and check fuel pressure.
   At idle:
   Approx. 245 kPa (2.5 kg/cm², 36 psi)
- N.G. → Check fuel pump and circuit.
- O.K. ↓

7 **CHECK EXHAUST GAS SENSOR.**
1. Start engine and warm it up sufficiently.
2. Perform "MIXTURE RATIO TEST" in "FUNCTION TEST" mode.

— OR —

2. See "M/R F/C MNT" in "Data monitor" mode.
3. Maintaining engine at 2,000 rpm under no-load (engine is warmed up sufficiently.), check that the monitor fluctuates between "LEAN" and "RICH" more than 5 times during 10 seconds.
   RICH→LEAN→RICH→
   1 time 2 times
   LEAN→RICH.......

— OR —

2. Set "Exhaust gas sensor monitor" in the self-diagnostic Mode II.
3. Maintaining engine at 2,000 rpm under no-load, check to make sure that the RED L.E.D. on the E.C.U. or the check engine light on the instrument panel goes ON and OFF more than 5 times during 10 seconds.

- N.G. → Replace exhaust gas sensor.
- O.K. ↓

(Go to Ⓑ on next page.)

Diagnostic Procedure 28 — Symptom — Unstable Idle (Cont'd)

Ⓑ

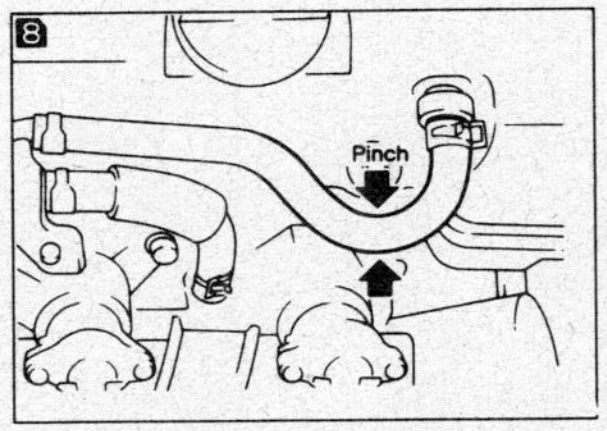

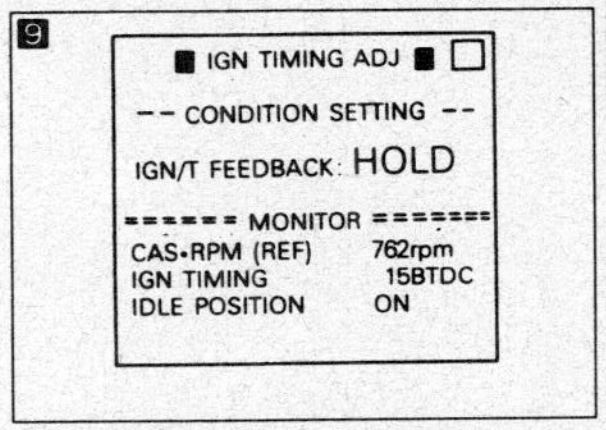

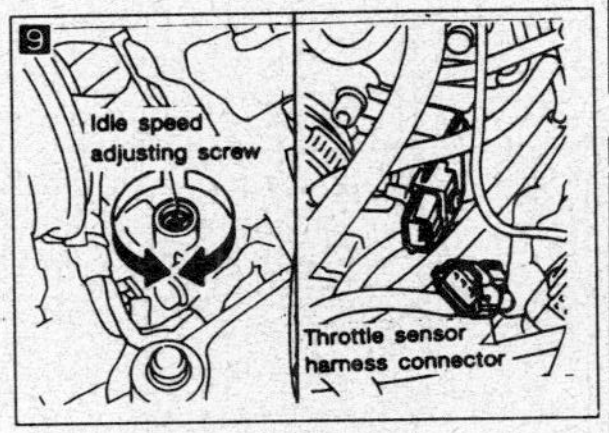

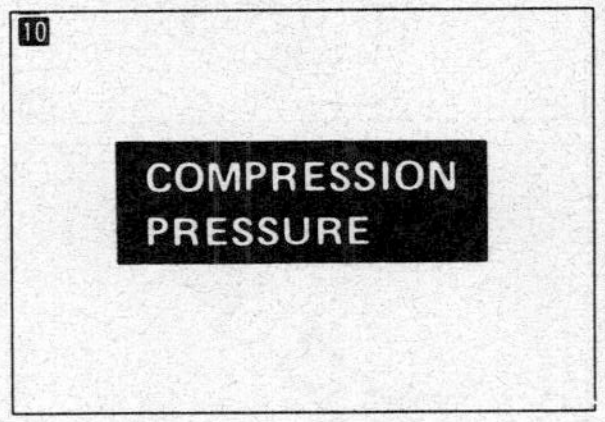

8 **CHECK FOR INTAKE AIR LEAK.**
When pinching blow-by hose (lowering the blow-by air supply), does the engine speed rise?
- Yes → Discover air leak location and repair.
- No ↓

9 **CHECK IDLE ADJ. SCREW CLOGGING.**
1. Perform "IGN TIMING ADJ" in "WORK SUPPORT" mode.
2. Can you set engine rpm at 750±50 rpm (in "N" position) by turning idle speed adjusting screw?

— OR —

1. Disconnect throttle sensor harness connector.
2. Can you set engine rpm at 750±50 rpm (in "N" position) by turning idle speed adjusting screw?

- No → Check for IAS clogging or throttle valve clogging.
- Yes ↓

10 **CHECK COMPRESSION PRESSURE.**
- Check compression pressure.
  Standard:
  kPa (kg/cm², psi)/300 rpm
  1,226 (12.5, 178)
  Minimum:
  kPa (kg/cm², psi)/300 rpm
  1,030 (10.5, 149)
  Difference between each cylinder:
  kPa (kg/cm², psi)/300 rpm
  98 (1.0, 14)
- N.G. → Check pistons, piston rings, valves, valve seats and cylinder head gaskets.
- O.K. ↓

11 **CHECK E.C.U. HARNESS CONNECTOR.**
Check the E.C.U. pin terminals for damage or poor connection of E.C.U. harness connector.
- N.G. → Repair or replace.
- O.K. ↓

12 **TRY A KNOWN GOOD E.C.U.***

↓

INSPECTION END

*: E.C.U. may be the cause of a problem, but this is rarely the case.

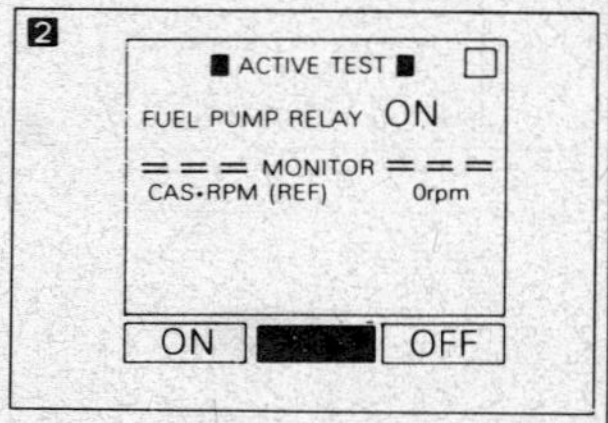

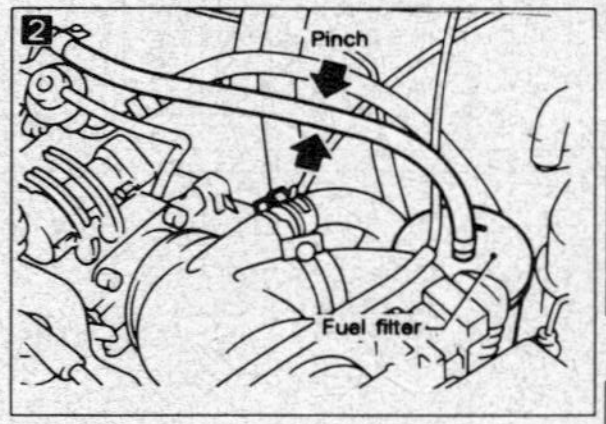

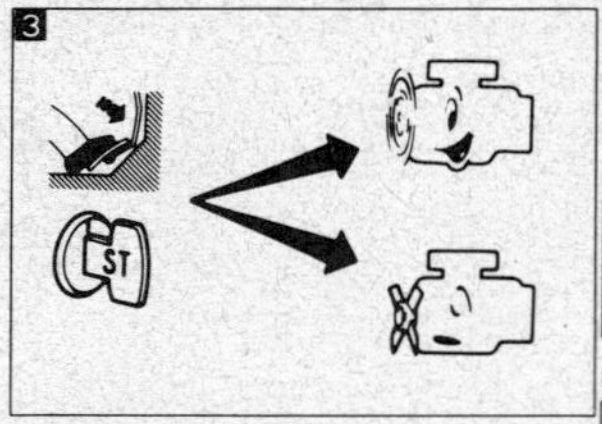

Diagnostic Procedure 29 — Symptom — Hard to Start or Impossible to Start when the Engine is Cold

1 CHECK BATTERY AND STARTER.
Check battery and starter condition.
- N.G. → Repair or replace.
- O.K. → go to 2.

2 CHECK FUEL PRESSURE.
1. Turn ignition switch "ON".
2. Perform "FUEL PUMP RELAY" in "ACTIVE TEST" mode.
3. Pinch fuel feed hose with fingers.

Is fuel pressure pulsation felt on the fuel feed hose?

— OR —

1. Pinch fuel feed hose with fingers.
2. When cranking the engine, is there any pressure on the fuel feed hose?

- No → Check fuel pump and circuit.
- Yes → go to 3.

3 CHECK AIR REGULATOR AND A.A.C. VALVE.
When pressing accelerator pedal fully, can you start the engine.
- Yes → Check A.A.C. valve, air regulator and circuits.
- No → go to 4.

4 CHECK INJECTOR.
1. Remove distributor from engine. (Crank angle sensor harness connector should remain connected.)
2. Disconnect ignition wires.
3. Turn ignition switch ON. (Do not start engine.)
4. When rotating distributor shaft slowly by hand, does each injector make an operating sound?

- No → Check injector(s) and circuit(s).
- Yes → go to 5.

5 CHECK IGNITION SPARK.
1. Disconnect ignition wire from spark plug.
2. Connect a known good spark plug to the ignition wire.
3. Place end of spark plug against a suitable ground and crank engine.
4. Check for spark.

- N.G. → Check ignition coil, power transistor and their circuits.
- O.K. → (Go to Ⓐ on next page.)

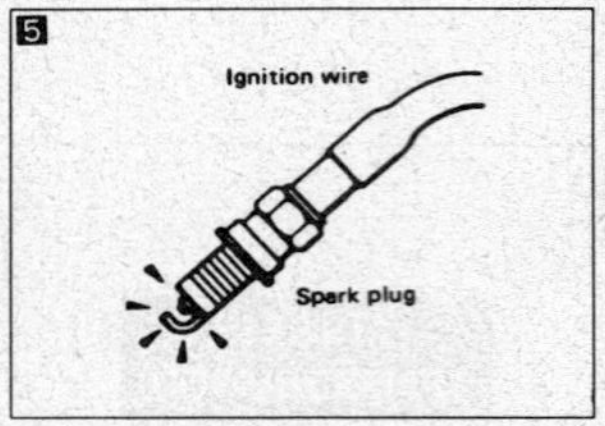

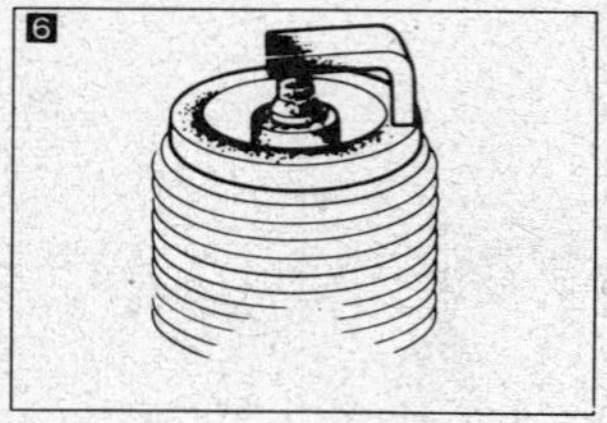

Diagnostic Procedure 29 — Symptom — Hard to Start or Impossible to Start when the Engine is Cold (Cont'd)

Ⓐ

6 CHECK SPARK PLUGS.
Remove the spark plugs and check for fouling, etc.
- N.G. → Repair or replace spark plug(s).
- O.K. → go to 7.

7 CHECK E.C.U. HARNESS CONNECTOR.
Check the E.C.U. pin terminals for damage or poor connection of E.C.U. harness connector.
- N.G. → Repair or replace.
- O.K. → go to 8.

8 CHECK E.C.U. POWER SUPPLY AND GROUND CIRCUIT.
- N.G. → Repair or replace.
- O.K. → go to 9.

9 TRY A KNOWN GOOD E.C.U.*

INSPECTION END

*: E.C.U. may be the cause of a problem, but this is rarely the case.

1990–92 2.0L ENGINE—DIAGNOSTIC PROCEDURE—G20

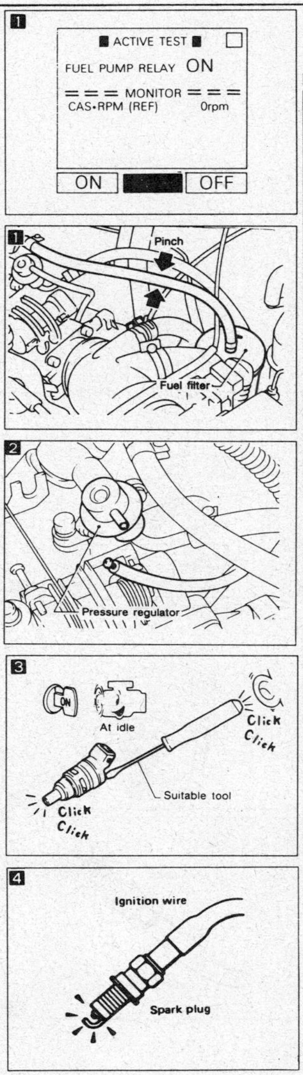

Diagnostic Procedure 30 — Symptom — Hard to Start or Impossible to Start when the Engine is Hot

1 CHECK FUEL PRESSURE.
1. Turn ignition switch "ON".
2. Perform "FUEL PUMP RELAY" in "ACTIVE TEST" mode.
3. Pinch fuel feed hose with fingers.

Is fuel pressure pulsation felt on the fuel feed hose?

—OR—

1. Pinch fuel feed hose with fingers.
2. When cranking the engine, is there any pressure on the fuel feed hose?

No → Check fuel pump and circuit.

Yes ↓

2 CHECK FUEL VAPOR.
1. Disconnect fuel pressure regulator vacuum hose and plug hose.
2. Can you start engine?

Yes → Check fuel properties.

No ↓

3 CHECK INJECTOR.
Does each injector make an operating sound at idle?

No → Check injector(s) and circuit(s).

Yes ↓

4 CHECK IGNITION SPARK.
1. Disconnect ignition wire from spark plug.
2. Connect a known good spark plug to the ignition wire
3. Place end of spark plug against a suitable ground and crank engine.
4. Check for spark.

N.G. → Check ignition coil, power transistor and circuits.

O.K. ↓

5 CHECK E.C.U. HARNESS CONNECTOR.
Check the E.C.U. pin terminals for damage or poor connection of E.C.U. harness connector.

N.G. → Repair or replace.

O.K. ↓

6 CHECK E.C.U. POWER SUPPLY AND GROUND CIRCUIT.

N.G. → Repair or replace.

O.K. ↓

7 TRY A KNOWN GOOD E.C.U.*

↓

INSPECTION END

*: E.C.U. may be the cause of a problem, but this is rarely the case.

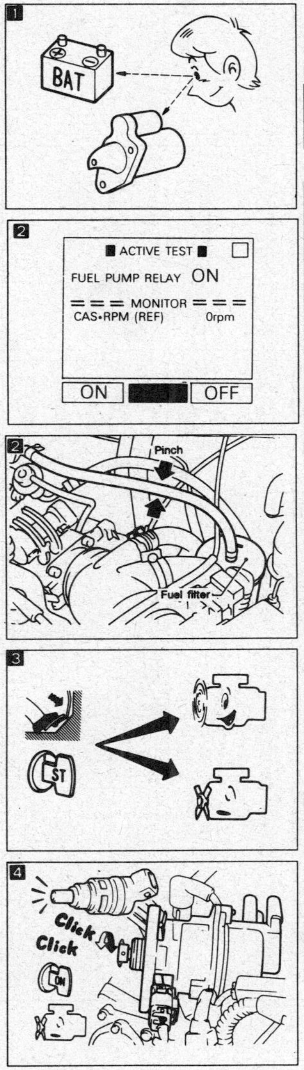

Diagnostic Procedure 31 — Symptom — Hard to Start or Impossible to Start under Normal Conditions

1 CHECK BATTERY AND STARTER.
Check battery and starter operation.

N.G. → Repair or replace.

O.K. ↓

2 CHECK FUEL PRESSURE.
1. Turn ignition switch "ON".
2. Perform "FUEL PUMP RELAY" in "ACTIVE TEST" mode.
3. Pinch fuel feed hose with fingers.

Is fuel pressure pulsation felt on the fuel feed hose?

—OR—

1. Pinch fuel feed hose with fingers.
2. When cranking the engine, is there any pressure on the fuel feed hose?

No → Check fuel pump and circuit.

Yes ↓

3 CHECK INJECTOR FOR LEAKAGE.
When pressing accelerator pedal fully, can you start the engine.

Yes → Check injector(s) for leakage.

No ↓

4 CHECK INJECTOR.
1. Remove distributor from engine. (Crank angle sensor harness connector should remain connected.)
2. Disconnect ignition wires.
3. Turn ignition switch ON. (Do not start engine.)
4. When rotating distributor shaft slowly by hand, does each injector make an operating sound?

No → Check injectors and circuits.

Yes ↓

5 CHECK IGNITION SPARK.
1. Disconnect ignition wire from spark plug.
2. Connect a known good spark plug to the ignition wire.
3. Place end of spark plug against a suitable ground and crank engine.
4. Check for spark.

N.G. → Check ignition coil, power transistor and circuits.

O.K. ↓

(Go to Ⓐ on next page.)

Diagnostic Procedure 31 — Symptom — Hard to Start or Impossible to Start under Normal Conditions (Cont'd)

5
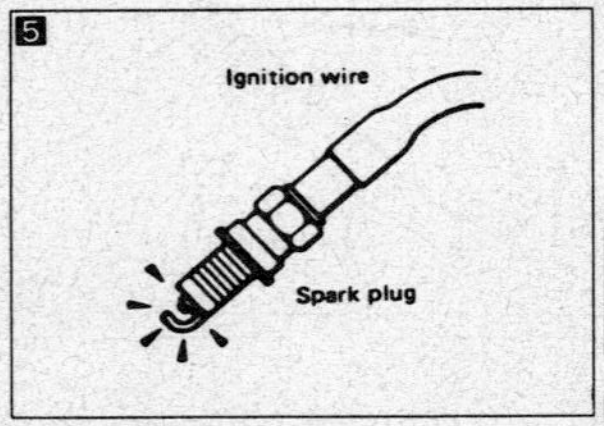

6
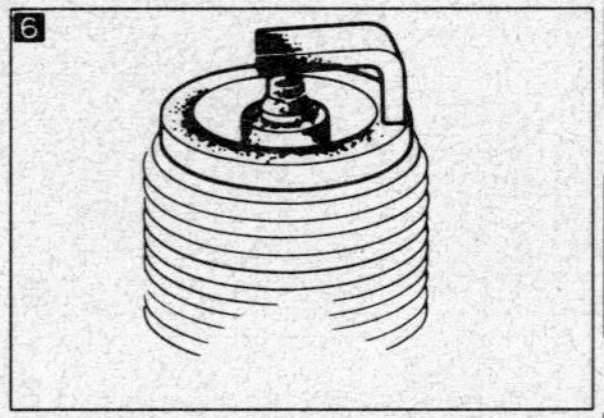

7
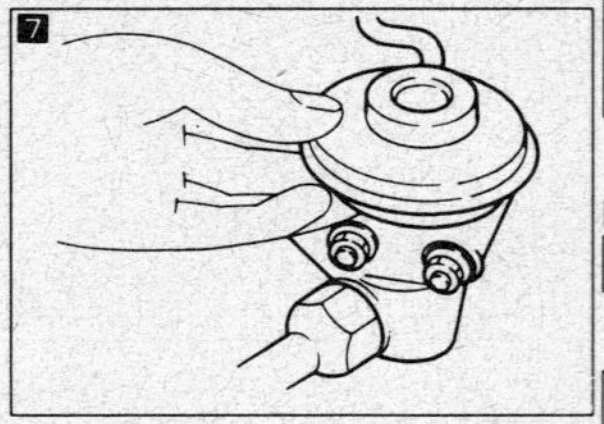

Ⓐ

6 **CHECK SPARK PLUGS.**
Remove the spark plugs and check for fouling, etc.
- N.G. → Repair or replace spark plug(s).
- O.K. ↓

7 **CHECK E.G.R. CONTROL VALVE.**
Check E.G.R. control valve for sticking.
- N.G. → Repair or replace.
- O.K. ↓

8 **CHECK E.C.U. HARNESS CONNECTOR.**
Check the E.C.U. pin terminals for damage or poor connection of E.C.U. harness connector.
- N.G. → Repair or replace.
- O.K. ↓

9 **CHECK E.C.U. POWER SUPPLY AND GROUND CIRCUIT.**
- N.G. → Repair or replace.
- O.K. ↓

10 **TRY A KNOWN GOOD E.C.U.***
- Trouble is fixed. → Replace E.C.U.
- Trouble is not fixed. ↓

INSPECTION END

*: E.C.U. may be the cause of a problem, but this is rarely the case.

1

2
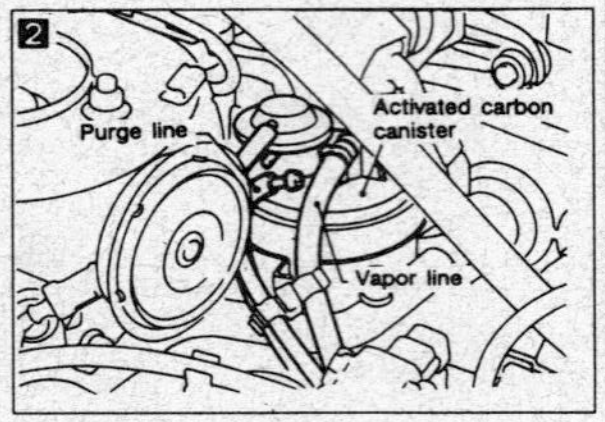

3
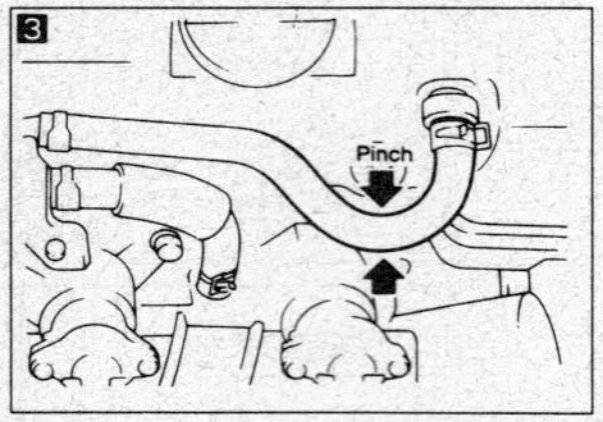

Diagnostic Procedure 32 — Symptom — Hesitation when the Engine is Hot

1 **CHECK FUEL VAPOR.**
1. Disconnect fuel pressure regulator vacuum hose and plug hose.
2. Perform cruise test.
3. Does the hesitation disappear?
- Yes → Check fuel properties.
- No ↓

2 **CHECK CANISTER PURGE.**
1. Disconnect canister purge line hose and plug hose.
2. Perform cruise test.
3. Does the hesitation disappear?
- Yes → Check purge and vacuum line.
- No ↓

3 **CHECK FOR INTAKE AIR LEAK.**
When pinching blow-by hose (lowering the blow-by air supply), does the engine speed rise?
- Yes → Discover air leak location and repair.
- No ↓

INSPECTION END

1
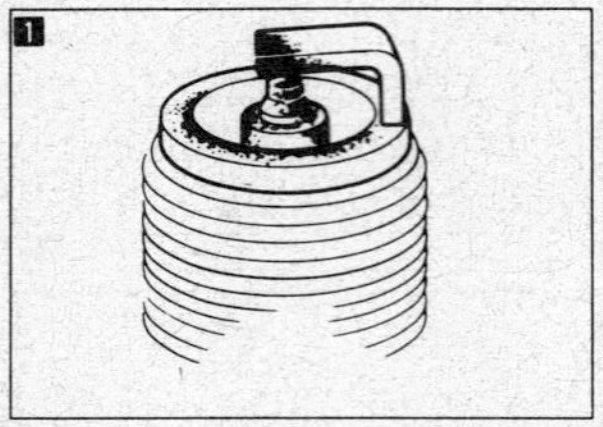

2
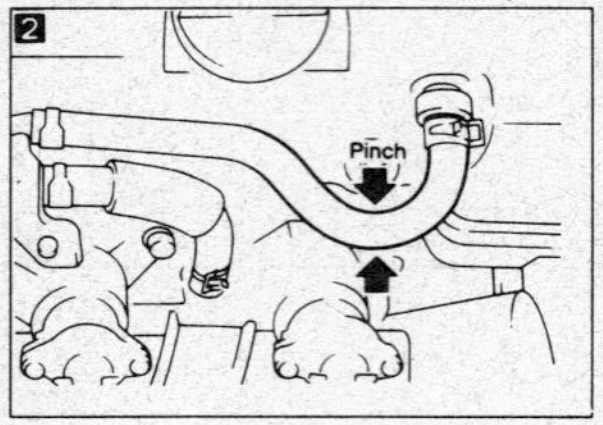

Diagnostic Procedure 33 — Symptom — Hesitation when the Engine is Cold

1 **CHECK SPARK PLUGS.**
Remove spark plugs and check for fouling, etc.
- N.G. → Repair or replace spark plug(s).
- O.K. ↓

2 **CHECK FOR INTAKE AIR LEAK.**
When pinching blow-by hose (lowering the blow-by air supply), does the engine speed rise?
- Yes → Discover air leak location and repair.
- No ↓

3 **TRY A KNOWN GOOD AIR FLOW METER.**
- Trouble is fixed. → Replace air flow meter.
- Trouble is not fixed. ↓

4 **CHECK FOR INTAKE VALVE DEPOSITS.**
If there are deposits on intake valves, clean them.
↓

INSPECTION END

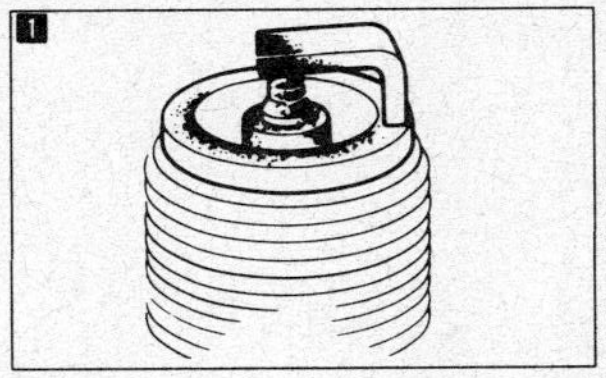

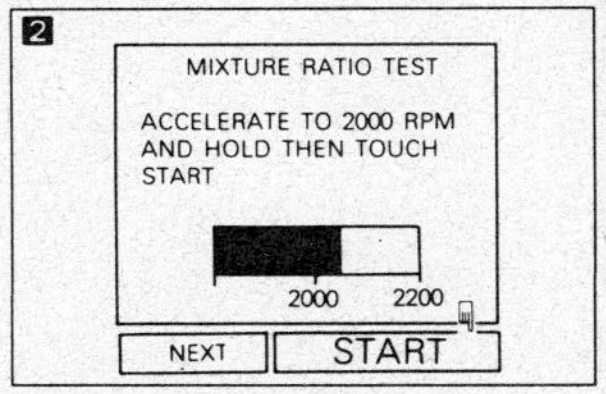

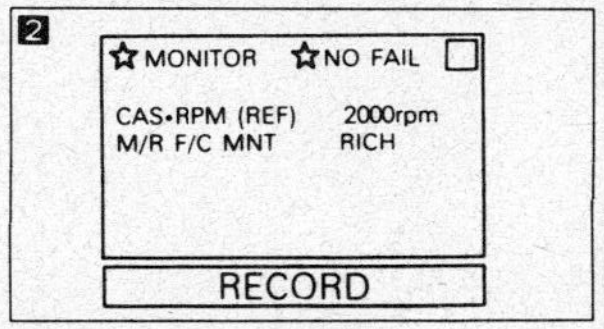

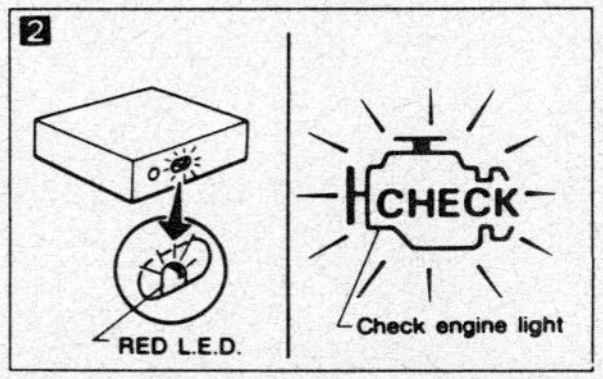

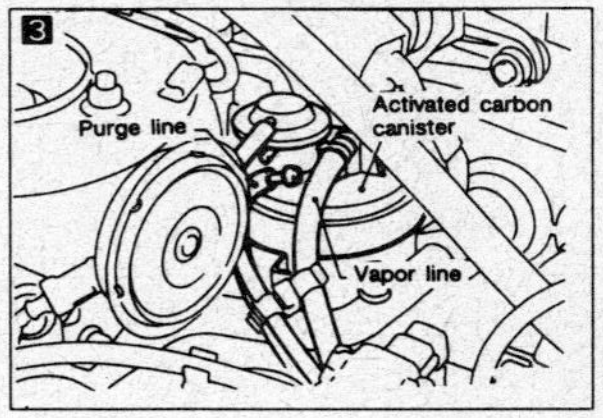

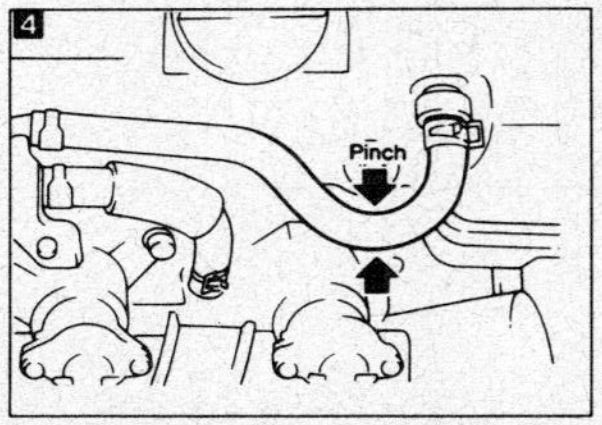

Diagnostic Procedure 34 — Symptom — Hesitation under Normal Conditions

1 CHECK SPARK PLUGS.
Remove spark plugs and check for fouling, etc.
- N.G. → Repair or replace spark plug(s).
- O.K. ↓

2 CHECK EXHAUST GAS SENSOR.
1. Start engine and warm it up sufficiently.
2. Perform "MIXTURE RATIO TEST" in "FUNCTION TEST" mode.

—OR—

2. See "M/R F/C MNT" in "DATA MONITOR" mode.
3. Maintaining engine at 2,000 rpm under no-load (with engine warmed up sufficiently.), check to make sure that the monitor fluctuates between "LEAN" and "RICH" more than 5 times during 10 seconds.
RICH→LEAN→RICH→
1 time 2 times
LEAN→RICH......

—OR—

2. Set "Exhaust gas sensor monitor" in the self-diagnostic Mode II.
3. Maintaining engine at 2,000 rpm under no-load, check that the RED L.E.D. on the E.C.U. or the check engine light on the instrument panel goes ON and OFF more than 5 times during 10 seconds.

- N.G. → Replace exhaust gas sensor.
- O.K. ↓

3 CHECK CANISTER PURGE.
1. Disconnect canister purge line hose and plug hose.
2. Perform cruise test.
3. Does the hesitation disappear?
- Yes → Check purge and vacuum line.
- No ↓

4 CHECK FOR INTAKE AIR LEAK.
When pinching blow-by hose (lowering the blow-by air supply) does the engine speed rise?
- Yes → Discover air leak location and repair.
- No ↓

INSPECTION END

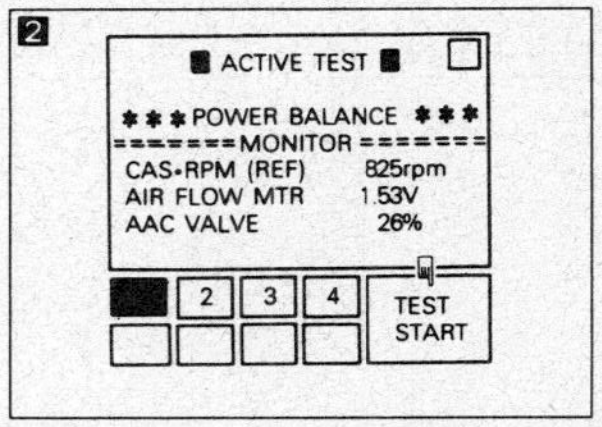

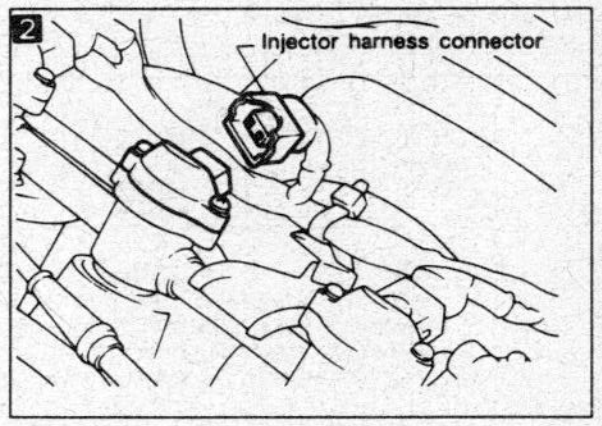

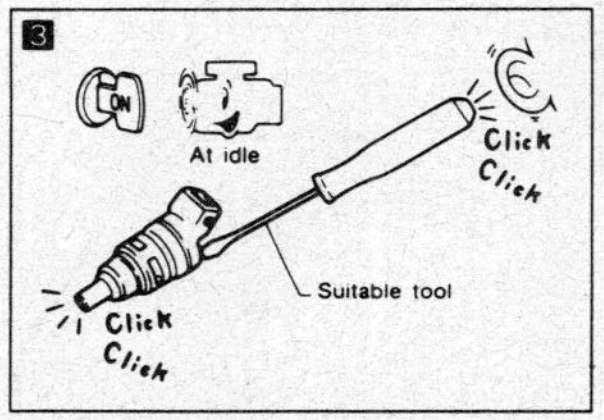

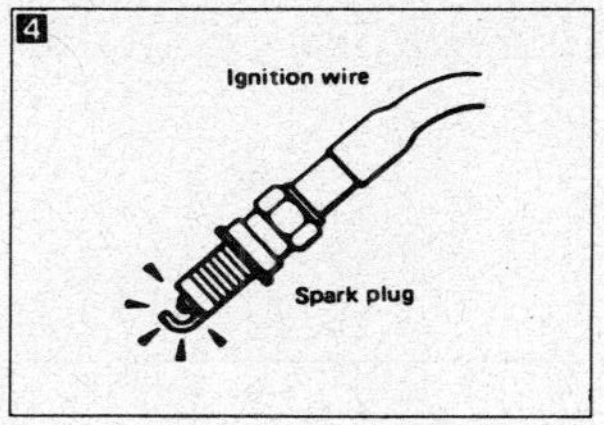

Diagnostic Procedure 35 — Symptom — Engine Stalls when Turning

1 CHECK FUEL LEVEL.
Check to see that there is enough fuel in tank.
- N.G. → Fill fuel tank with fuel.
- O.K. ↓

2 PERFORM POWER BALANCE TEST.
1. Perform "POWER BALANCE" in "ACTIVE TEST" mode.
2. Is there any cylinder which does not produce a momentary engine speed drop?

—OR—

When disconnecting each injector harness connector one at a time, is there any cylinder which does not produce a momentary engine speed drop?

- No → Go to 5.
- Yes ↓

3 CHECK INJECTOR.
Does each injector make an operating sound at idle?
- No → Check injector(s) and circuit(s).
- Yes ↓

4 CHECK IGNITION SPARK.
1. Disconnect ignition wire from spark plug.
2. Connect a known good spark plug to the ignition wire.
3. Place end of spark plug against a suitable ground and crank engine.
4. Check for spark.
- N.G. → Check ignition coil, power transistor and circuits.
- O.K. ↓

(Go to Ⓐ on next page.)

Diagnostic Procedure 35 — Symptom — Engine Stalls when Turning (Cont'd)

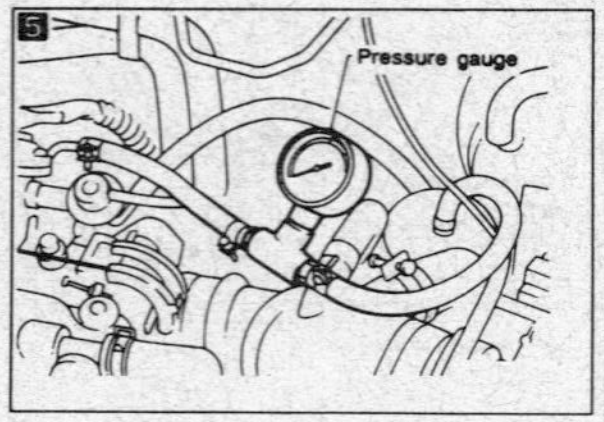

Ⓐ

5 CHECK FUEL PRESSURE.
1. Release fuel pressure to zero.
2. Install fuel pressure gauge and check fuel pressure.
   At idle:
   Approx. 245 kPa ($2.5\ kg/cm^2$, 36 psi)
   A few seconds after ignition switch is turned OFF to ON:
   Approx. 294 kPa ($3.0\ kg/cm^2$, 43 psi)

N.G. → Check fuel pressure regulator diaphragm.

O.K. ↓

6 CHECK E.C.U. HARNESS CONNECTOR.
Check the E.C.U. pin terminals for damage or poor connection of E.C.U. harness connector.

N.G. → Repair or replace.

O.K. ↓

7 CHECK E.C.U. POWER SUPPLY AND GROUND CIRCUIT.

N.G. → Repair or replace.

O.K. ↓

8 TRY A KNOWN GOOD E.C.U.*

↓

INSPECTION END

*: E.C.U. may be the cause of a problem, but this is rarely the case.

Diagnostic Procedure 36 — Symptom — Engine Stalls when the Engine is Hot

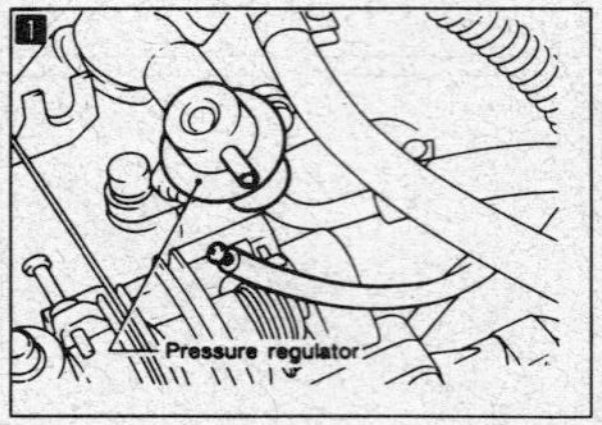

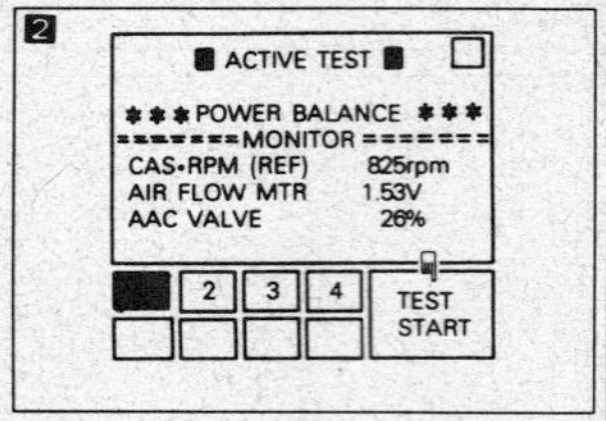

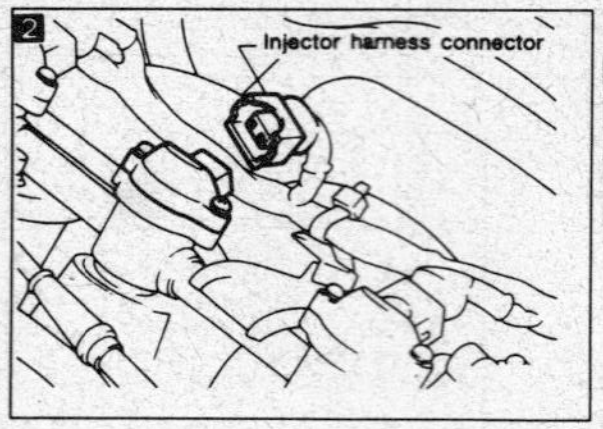

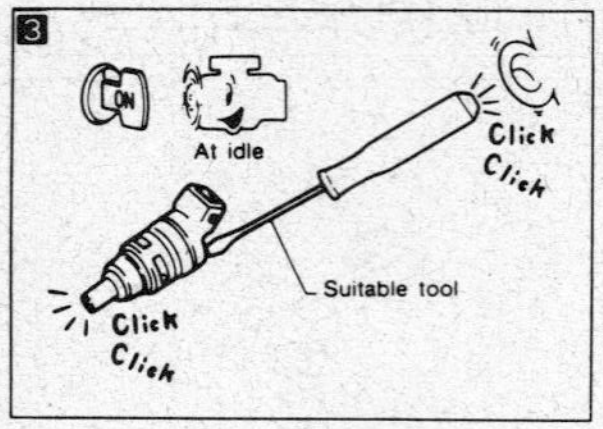

1 CHECK FUEL VAPOR.
1. Disconnect fuel pressure regulator vacuum hose and plug hose.
2. Perform cruise test.
3. Does the engine stall disappear?

Yes → Check fuel properties.

No ↓

2 PERFORM POWER BALANCE TEST.
1. Perform "POWER BALANCE" in "ACTIVE TEST" mode.
2. Is there any cylinder which does not produce a momentary engine speed drop?

— OR —

When disconnecting each injector harness connector one at a time, is there any cylinder which does not produce a momentary engine speed drop?

No → Go to 5.

Yes ↓

3 CHECK INJECTOR.
Does each injector make an operating sound at idle?

No → Check injector(s) and circuit(s).

Yes ↓

(Go to Ⓐ on next page.)

Diagnostic Procedure 36 — Symptom — Engine Stalls when the Engine is Hot (Cont'd)

4

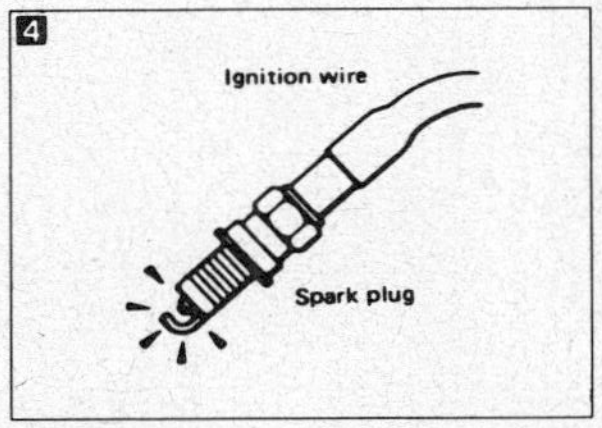

5

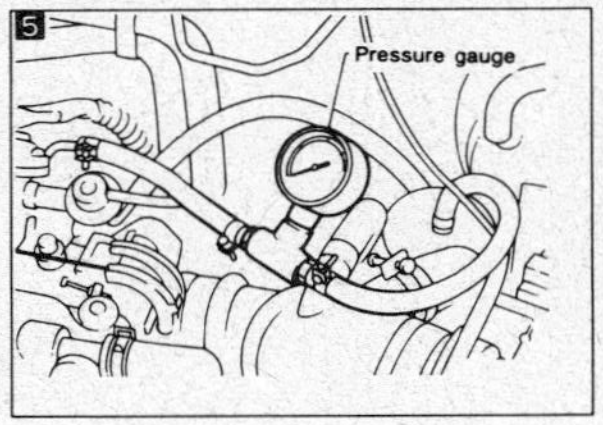

Ⓐ

4 CHECK IGNITION SPARK.
1. Disconnect ignition wire from spark plug.
2. Connect a known good spark plug to the ignition wire.
3. Place end of spark plug against a suitable ground and crank engine.
4. Check for spark.

N.G. → Check ignition coil, power transistor and their circuits.

O.K. ↓

5 CHECK FUEL PRESSURE.
1. Release fuel pressure to zero.
2. Install fuel pressure gauge and check fuel pressure
At idle:
Approx. 245 kPa (2.5 kg/cm², 36 psi)
A few seconds after ignition switch is turned OFF to ON:
Approx. 294 kPa (3.0 kg/cm², 43 psi)

N.G. → Check fuel pressure regulator diaphragm.

O.K. ↓

6 CHECK E.C.U. HARNESS CONNECTOR.
Check the E.C.U. pin terminals for damage or poor connection of E.C.U. harness connector.

N.G. → Repair or replace.

O.K. ↓

7 CHECK E.C.U. POWER SUPPLY AND GROUND CIRCUIT.

N.G. → Repair or replace.

O.K. ↓

8 TRY A KNOWN GOOD E.C.U.*

Trouble is fixed. → Replace E.C.U.

Trouble is not fixed. ↓

INSPECTION END

*: E.C.U. may be the cause of a problem, but this is rarely the case.

Diagnostic Procedure 37 — Symptom — Engine Stalls when the Engine is Cold

1

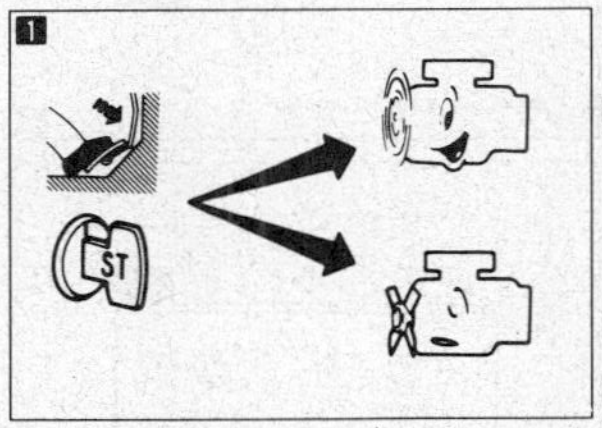

2

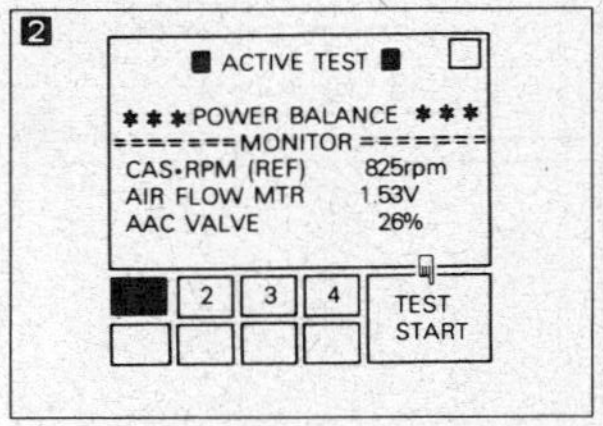

2

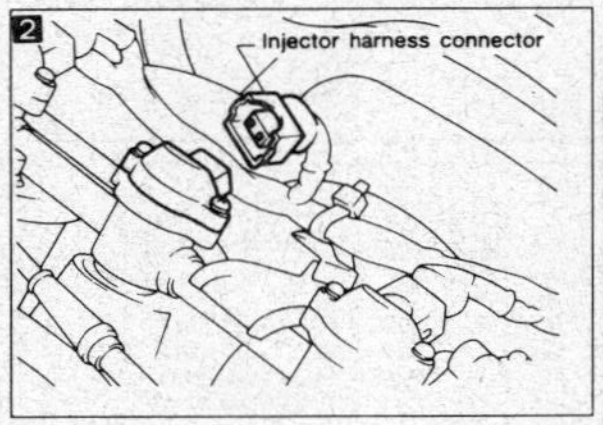

3

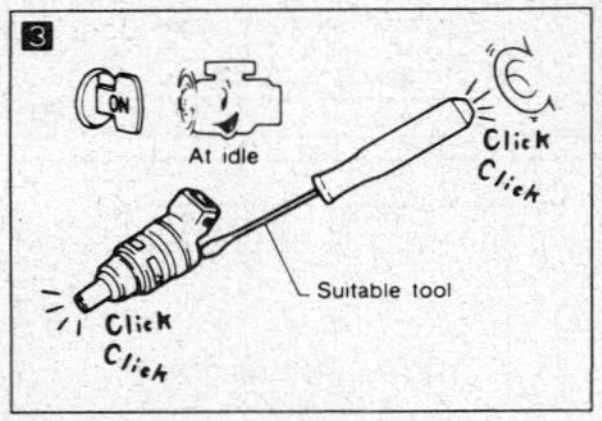

4

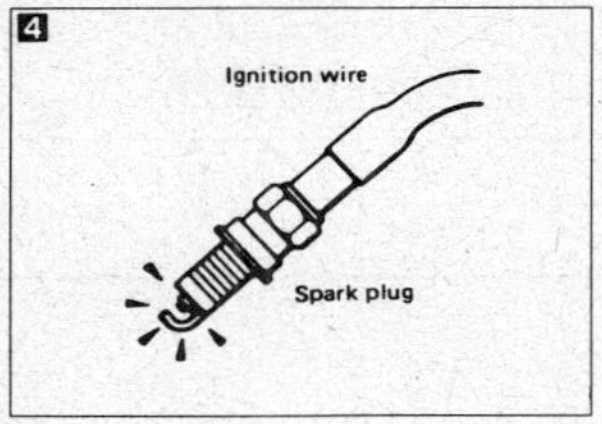

1 CHECK AIR REGULATOR AND A.A.C. VALVE.
When the engine is cold, can you start the engine when pressing accelerator pedal fully?

N.G. → Check A.A.C. valve, air regulator and circuits.

O.K. ↓

2 PERFORM POWER BALANCE TEST.
1. Perform "POWER BALANCE" in "ACTIVE TEST" mode.
2. Is there any cylinder which does not produce a momentary engine speed drop?

—OR—

When disconnecting each injector harness connector one at a time, is there any cylinder which does not produce a momentary engine speed drop?

N.G. → Go to 6.

O.K. ↓

3 CHECK INJECTOR.
Does each injector make an operating sound at idle?

N.G. → Check injector(s) and circuit(s).

O.K. ↓

4 CHECK IGNITION SPARK.
1. Disconnect ignition wire from spark plug.
2. Connect a known good spark plug to the ignition wire.
3. Place end of spark plug against a suitable ground and crank engine.
4. Check for spark.

N.G. → Check ignition coil, power transistor and circuits.

O.K. ↓

(Go to Ⓐ on next page.)

Diagnostic Procedure 37 — Symptom — Engine Stalls when the Engine is Cold (Cont'd)

5
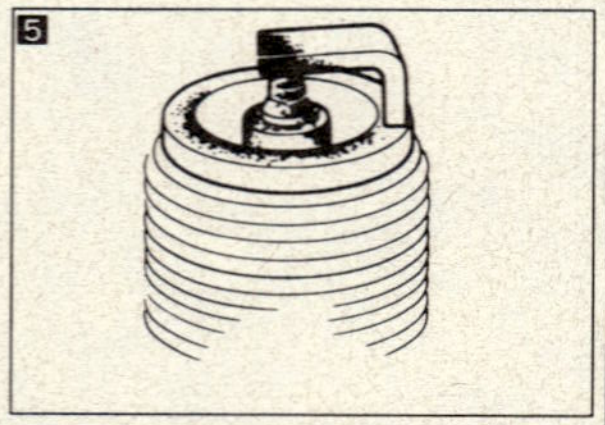

6
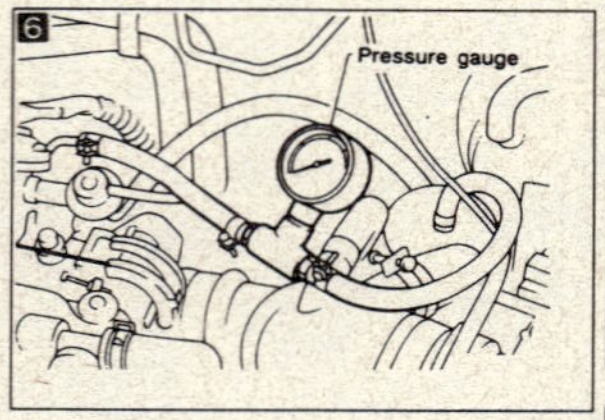

Ⓐ

5 CHECK SPARK PLUGS.
Remove the spark plugs and check for fouling, etc.
- N.G. → Repair or replace spark plug(s).
- O.K. ↓

6 CHECK FUEL PRESSURE.
1. Release fuel pressure to zero.
2. Install fuel pressure gauge and check fuel pressure.
   At idle:
   Approx. 245 kPa
   (2.5 kg/cm², 36 psi)
   A few seconds after ignition switch is turned OFF to ON:
   Approx. 294 kPa
   (3.0 kg/cm², 43 psi)
- N.G. → Check fuel pressure regulator diaphragm.
- O.K. ↓

7 CHECK E.C.U. HARNESS CONNECTOR.
Check the E.C.U. pin terminals for damage or poor connection of E.C.U. harness connector.
- N.G. → Repair or replace.
- O.K. ↓

8 CHECK E.C.U. POWER SUPPLY AND GROUND CIRCUIT.
- N.G. → Repair or replace.
- O.K. ↓

9 TRY A KNOWN GOOD E.C.U.*
↓
INSPECTION END

*: E.C.U. may be the cause of a problem, but this is rarely the case.

Diagnostic Procedure 38 — Symptom — Engine Stalls when Stepping on the Accelerator Momentarily

1
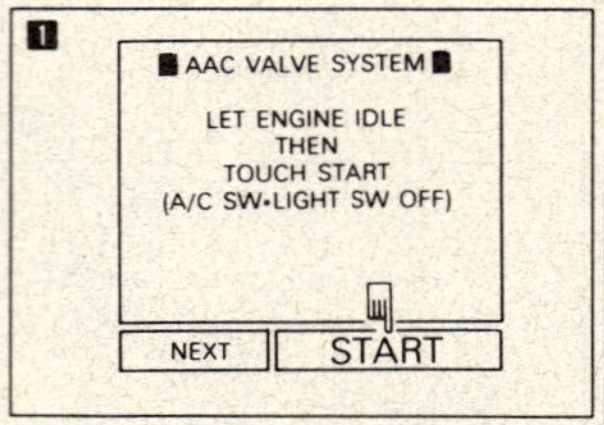

1
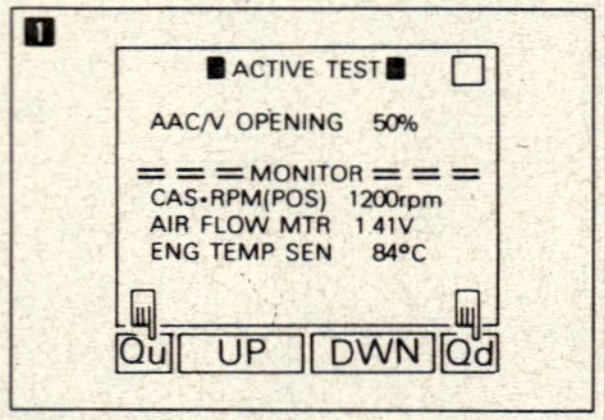

1
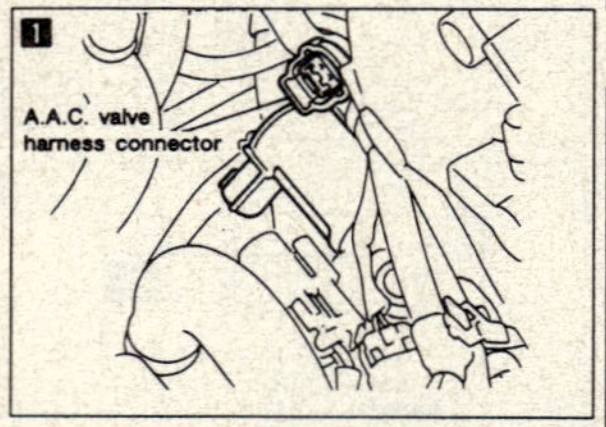

2
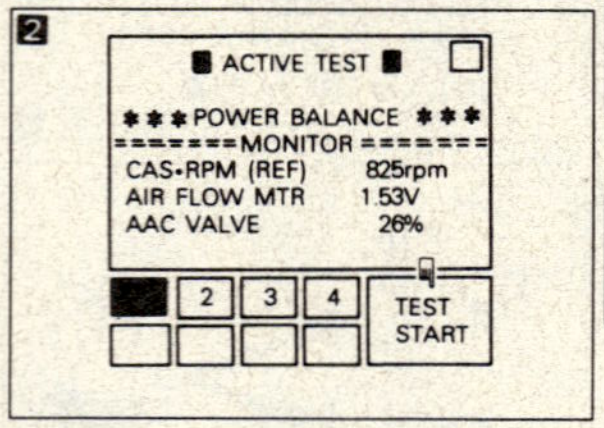

2
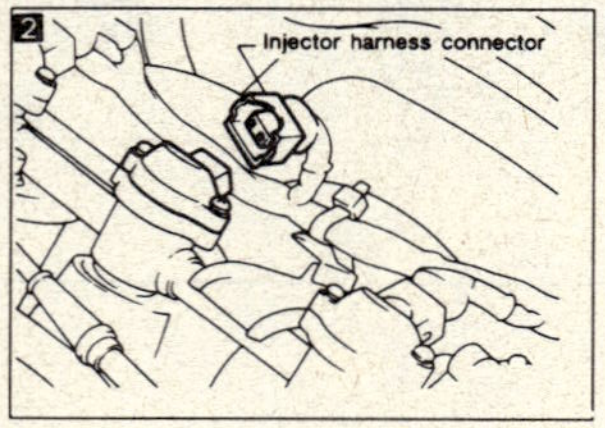

1 CHECK A.A.C. VALVE.
1. Start engine and warm it up sufficiently.
2. Perform "AAC VALVE SYSTEM" in "FUNCTION TEST" mode.

— OR —

1. Select "A.A.C. VALVE OPENING" in "ACTIVE TEST" mode.
2. When touching "Qu" and "Qd", does the engine speed change according to the percent of A.A.C. valve opening?

— OR —

When disconnecting A.A.C. valve harness connector, does the engine speed drop?
- No → Check A.A.C. valve and circuit.
- Yes ↓

2 PERFORM POWER BALANCE TEST.
1. Perform "POWER BALANCE" in "ACTIVE TEST" mode.
2. Is there any cylinder which does not produce a momentary engine speed drop?

— OR —

When disconnecting each injector harness connector one at a time, is there any cylinder which does not produce a momentary engine speed drop?
- No → Go to 5
- Yes ↓

3 CHECK INJECTOR.
Does each injector make an operating sound at idle?
- No → Check injector(s) and their circuit(s).
- Yes ↓

(Go to Ⓐ on next page.)

Diagnostic Procedure 38 — Symptom — Engine Stalls when Stepping on the Accelerator Momentarily (Cont'd)

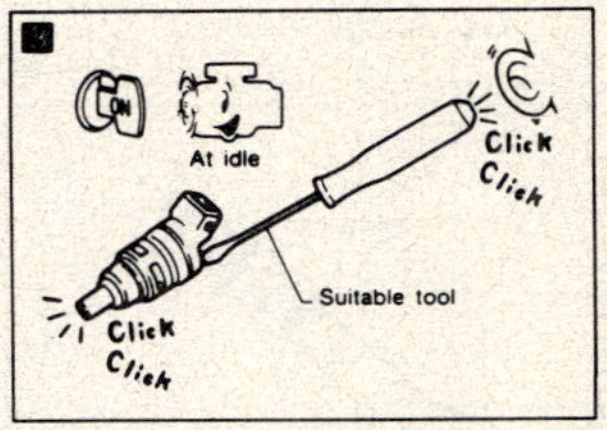

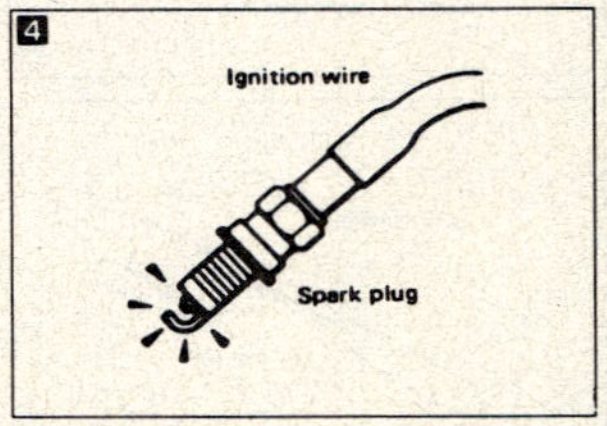

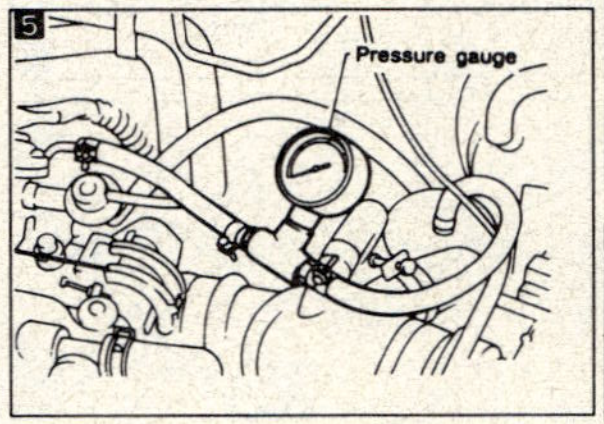

Ⓐ

4 CHECK IGNITION SPARK.
1. Disconnect ignition wire from spark plug.
2. Connect a known good spark plug to the ignition wire.
3. Place end of spark plug against an earth point with engine cranking.
4. Check for spark.

N.G. → Check ignition coil, power transistor and their circuits.

O.K. ↓

5 CHECK FUEL PRESSURE.
1. Release fuel pressure to zero.
2. Install fuel pressure gauge and check fuel pressure.
   At idle:
   Approx. 245 kPa (2.5 kg/cm², 36 psi)
   A few seconds after ignition switch is turned OFF to ON:
   Approx. 294 kPa (3.0 kg/cm², 43 psi)

N.G. → Check fuel pressure regulator diaphragm.

O.K. ↓

6 CHECK E.C.U. HARNESS CONNECTOR.
Check the E.C.U. pin terminals for damage or poor connection of E.C.U. harness connector.

N.G. → Repair or replace.

O.K. ↓

7 CHECK E.C.U. POWER SUPPLY AND GROUND CIRCUIT.

N.G. → Repair or replace.

O.K ↓

8 TRY A KNOWN GOOD E.C.U.*

↓

INSPECTION END

*: E.C.U. may be the cause of a problem, but this is rarely the case.

Diagnostic Procedure 39 — Symptom — Engine Stalls after Decelerating

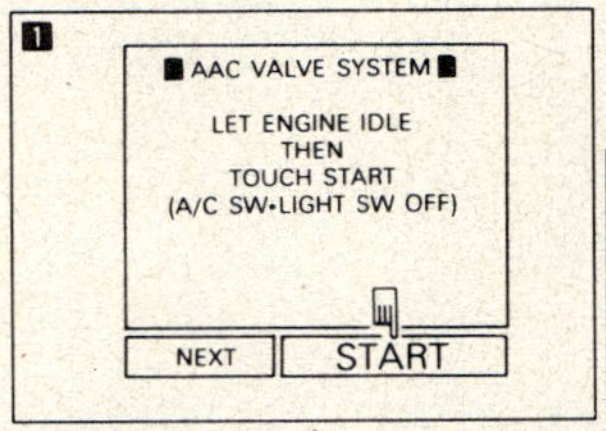

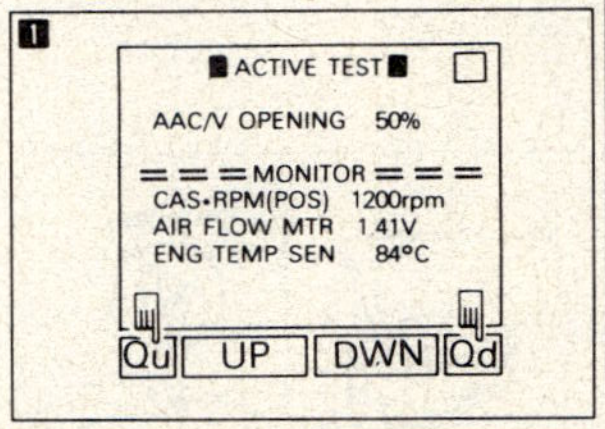

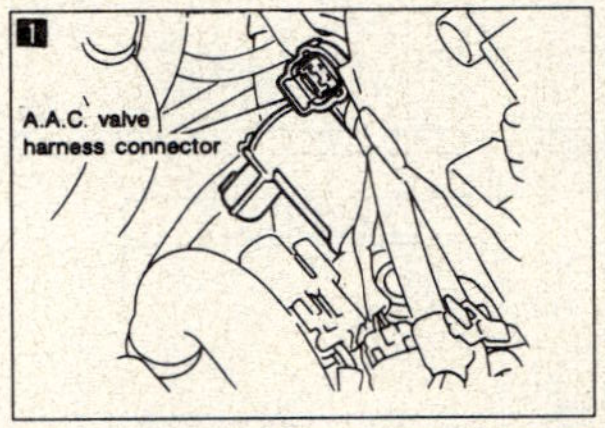

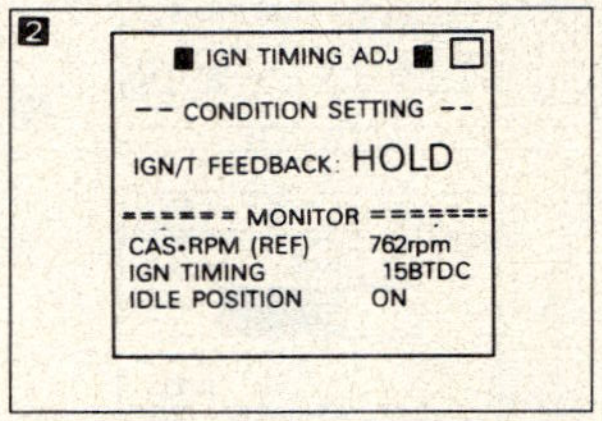

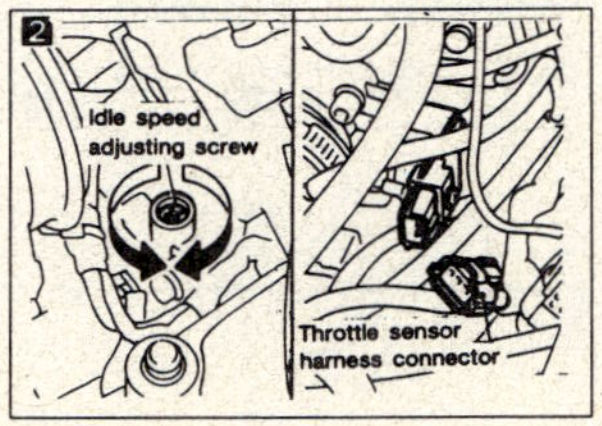

1 CHECK A.A.C. VALVE.
1. Start engine and warm it up sufficiently.
2. Perform "AAC VALVE SYSTEM" in "FUNCTION TEST" mode.

— OR —

1. Select "A.A.C. VALVE OPENING" in "ACTIVE TEST" mode.
2. When touching "Qu" and "Qd", does the engine speed change according to the percent of A.A.C. valve opening?

— OR —

When disconnecting A.A.C. valve harness connector, does the engine speed drop?

No → Check A.A.C. valve and circuit.

Yes ↓

2 CHECK IDLE ADJ. SCREW CLOGGING.
1. Perform "IGN TIMING ADJ" in "WORK SUPPORT" mode.
2. Can you set engine rpm at 750±50 rpm (in "N" position) by turning idle adjusting screw?

— OR —

1. Disconnect throttle sensor harness connector.
2. Can you set engine rpm at 750±50 rpm (in "N" position) by turning idle adjusting screw?

No → Check for IAS clogging or throttle chamber clogging.

Yes ↓

(Go to Ⓐ on next page.)

Diagnostic Procedure 39 — Symptom — Engine Stalls after Decelerating (Cont'd)

3
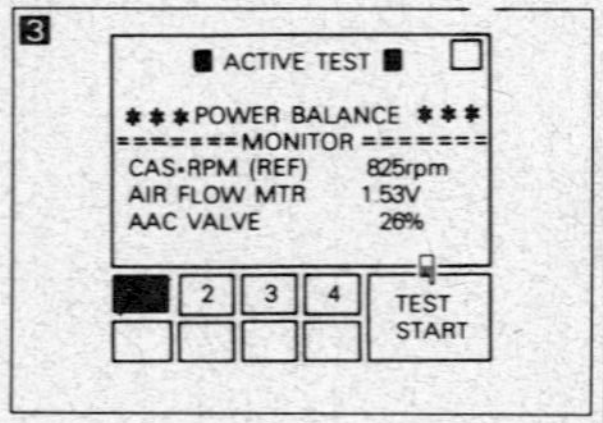

3
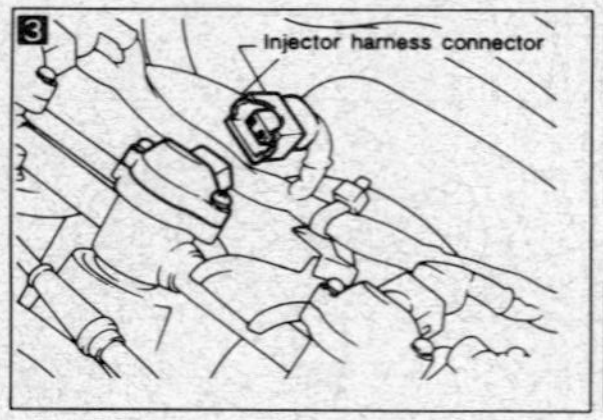

4
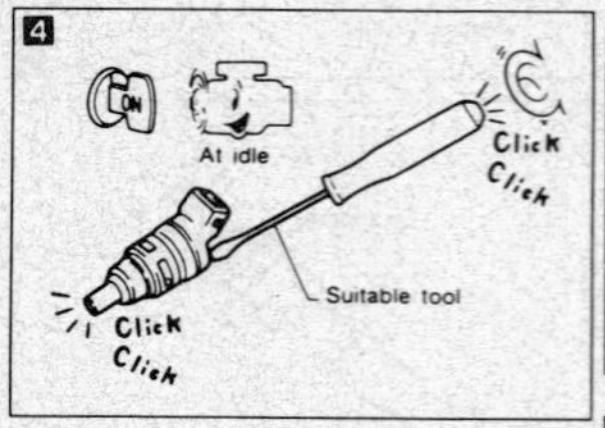

5
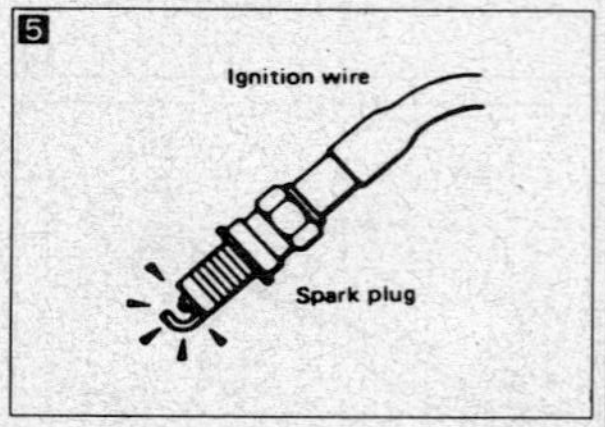

6
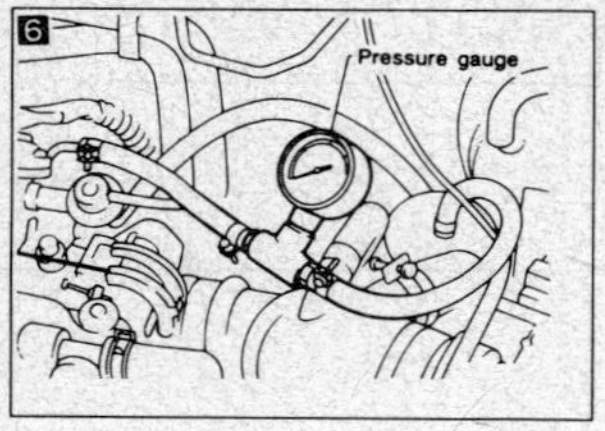

(A)

3 PERFORM POWER BALANCE TEST.
1. Perform "POWER BALANCE" in "ACTIVE TEST" mode.
2. Is there any cylinder which does not produce a momentary engine speed drop?

— OR —

When disconnecting each injector harness connector one at a time, is there any cylinder which does not produce a momentary engine speed drop?

No → Go to 6.

Yes ↓

4 CHECK INJECTOR.
Does each injector make an operating sound at idle?

No → Check injector(s) and circuit(s).

Yes ↓

5 CHECK IGNITION SPARK.
1. Disconnect ignition wire from spark plug.
2. Connect a known good spark plug to the ignition wire.
3. Place end of spark plug against a suitable ground and crank engine.
4. Check for spark.

No → Check ignition coil, power transistor and circuits.

Yes ↓

6 CHECK FUEL PRESSURE.
1. Release fuel pressure to zero.
2. Install fuel pressure gauge and check fuel pressure.
At idle:
Approx. 245 kPa (2.5 kg/cm², 36 psi)
A few seconds after ignition switch is turned OFF to ON:
Approx. 294 kPa (3.0 kg/cm², 43 psi)

N.G. → Check fuel pressure regulator diaphragm.

O.K. ↓

(Go to (B) on next page.)

Diagnostic Procedure 39 — Symptom — Engine Stalls after Decelerating (Cont'd)

7
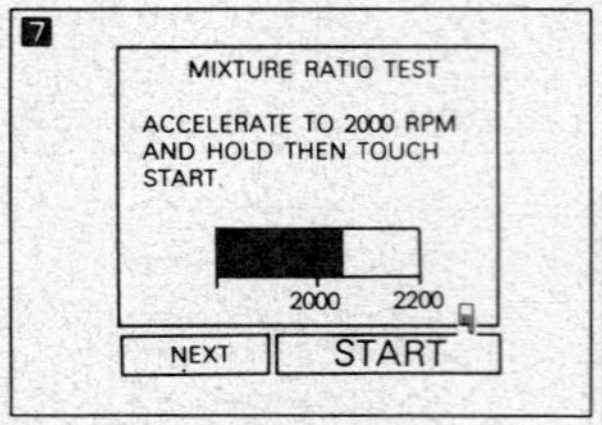

7
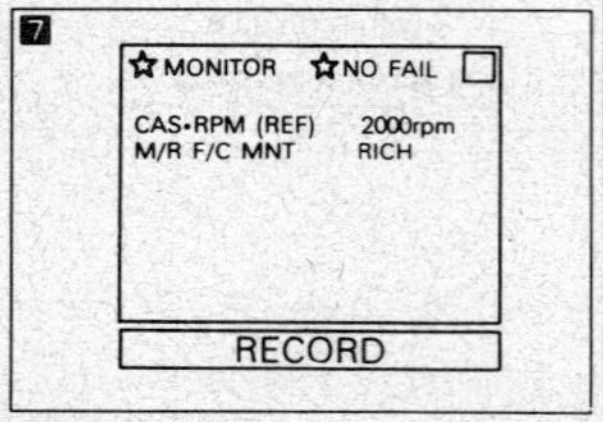

7
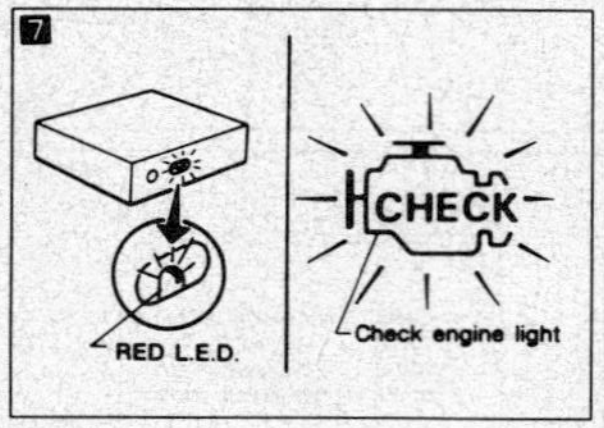

(B)

7 CHECK EXHAUST GAS SENSOR.
1. Start engine and warm it up sufficiently.
2. Perform "MIXTURE RATIO TEST" in "FUNCTION TEST" mode.

— OR —

2. See "M/R F/C MNT" in "DATA MONITOR" mode.
3. Maintaining engine at 2,000 rpm under no-load (with engine warmed up sufficiently.), check to make sure that the monitor fluctuates between "LEAN" and "RICH" more than 5 times during 10 seconds.
RICH→LEAN→RICH→ (1 time) (2 times)
LEAN→RICH.......

— OR —

2. Set "Exhaust gas sensor monitor" in the self- diagnostic Mode II.
3. Maintaining engine at 2,000 rpm under no-load, check that the RED L.E.D. on the E.C.U. or the check engine light on the instrument panel goes ON and OFF more than 5 times during 10 seconds.

N.G. → Replace exhaust gas sensor.

O.K. ↓

8 CHECK E.C.U. HARNESS CONNECTOR.
Check the E.C.U. pin terminals for damage or poor connection of E.C.U. harness connector.

N.G. → Repair or replace.

O.K. ↓

9 CHECK E.C.U. POWER SUPPLY AND GROUND CIRCUIT.

N.G. → Repair or replace.

O.K. ↓

10 TRY A KNOWN GOOD E.C.U.*

↓

INSPECTION END

*: E.C.U. may be the cause of a problem, but this is rarely the case.

1
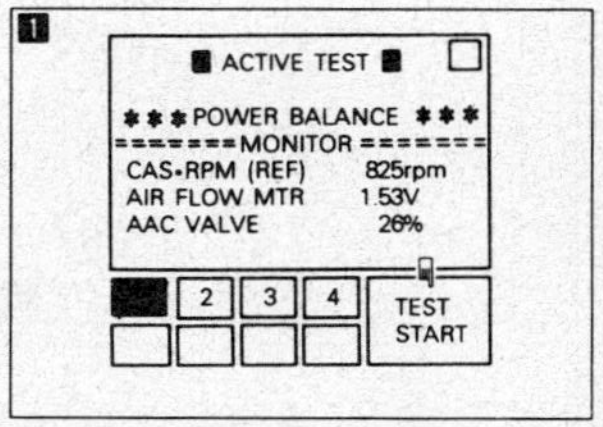

1
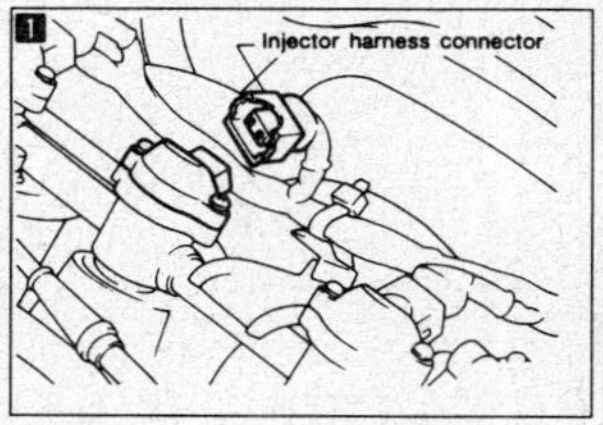

2
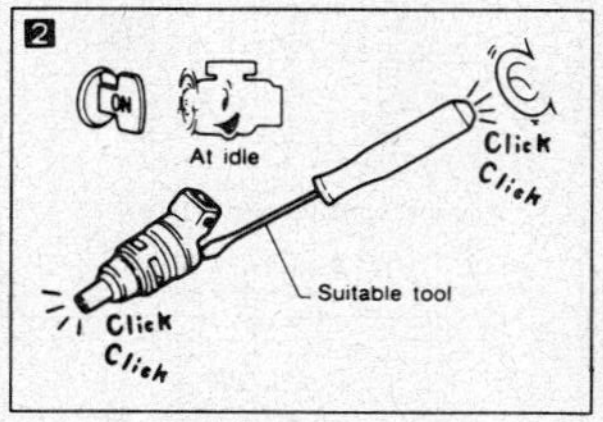

3
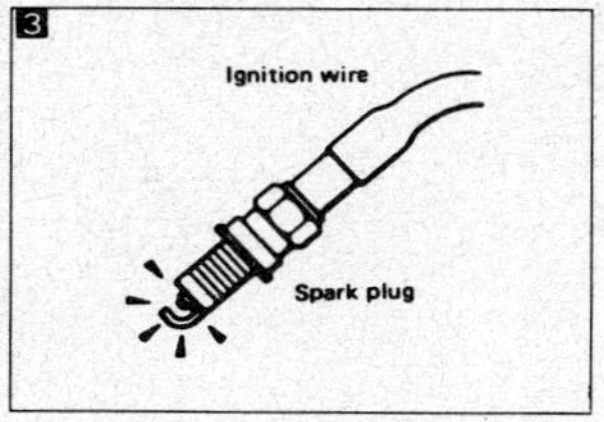

Diagnostic Procedure 40 — Symptom — Engine Stalls when Accelerating or when Driving at Constant Speed

1 **PERFORM POWER BALANCE TEST.**
1. Perform "POWER BALANCE" in "ACTIVE TEST" mode.
2. Is there any cylinder which does not produce a momentary engine speed drop?

— OR —

When disconnecting each injector harness connector one at a time, is there any cylinder which does not produce a momentary engine speed drop?

- No → Go to 4.
- Yes ↓

2 **CHECK INJECTOR.**
Does each injector make an operating sound at idle?

- No → Check injector(s) and circuit(s).
- Yes ↓

3 **CHECK IGNITION SPARK.**
1. Disconnect ignition wire from spark plug.
2. Connect a known good spark plug to the ignition wire.
3. Place end of spark plug against a suitable ground and crank engine.
4. Check for spark.

- No → Check ignition coil, power transistor and circuits.
- Yes ↓

(Go to Ⓐ on next page.)

4
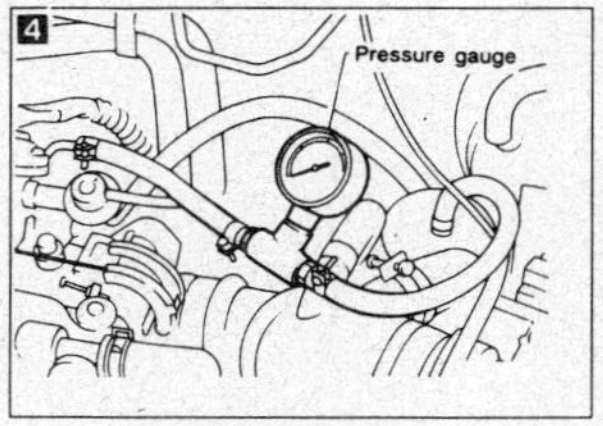

5
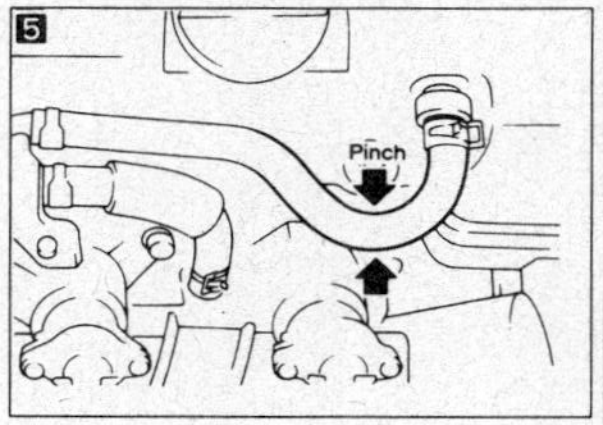

Diagnostic Procedure 40 — Symptom — Engine Stalls when Accelerating or when Driving at Constant Speed (Cont'd)

Ⓐ ↓

4 **CHECK FUEL PRESSURE.**
1. Release fuel pressure to zero.
2. Install fuel pressure gauge and check fuel pressure.

At idle:
Approx. 245 kPa (2.5 kg/cm², 36 psi)
A few seconds after ignition switch is turned OFF to ON:
Approx. 294 kPa (3.0 kg/cm², 43 psi)

- N.G. → Check fuel pump, circuit and fuel pressure regulator.
- O.K. ↓

5 **CHECK FOR INTAKE AIR LEAK.**
When pinching blow-by hose (lowering the blow-by air supply), does the engine speed rise?

- Yes → Discover air leak location and repair.
- No ↓

6 **CHECK E.C.U. HARNESS CONNECTOR.**
Check the E.C.U. pin terminals for damage or poor connection of E.C.U. harness connector.

- N.G. → Repair or replace.
- O.K. ↓

7 **CHECK E.C.U. POWER SUPPLY AND GROUND CIRCUIT.**

- Yes → Repair or replace.
- No ↓

8 **TRY A KNOWN GOOD E.C.U.***

↓

INSPECTION END

*: E.C.U. may be the cause of a problem, but this is rarely the case.

1
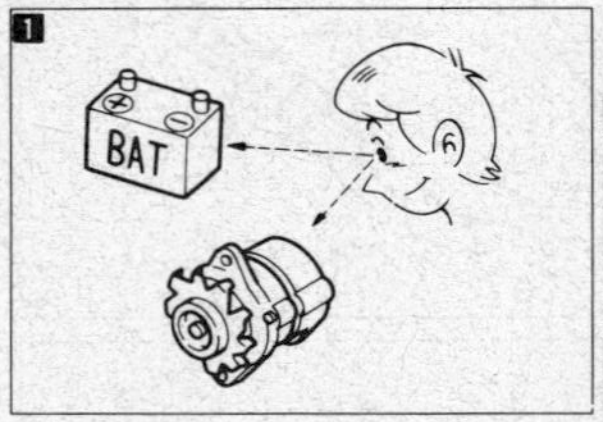

2
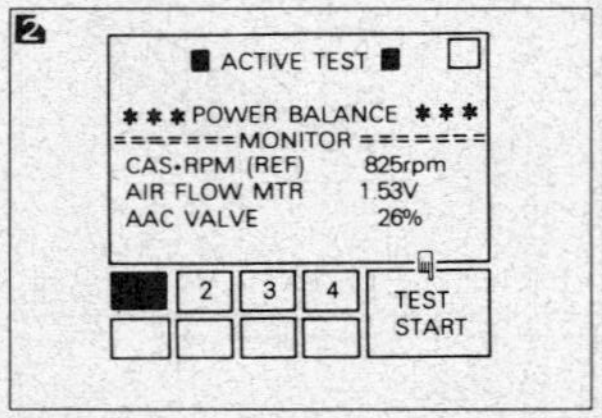

2
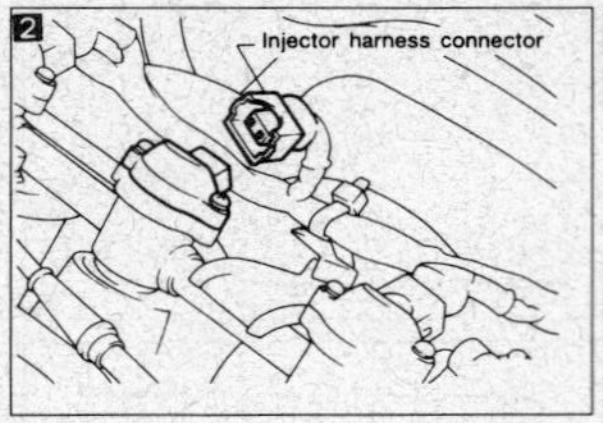

3
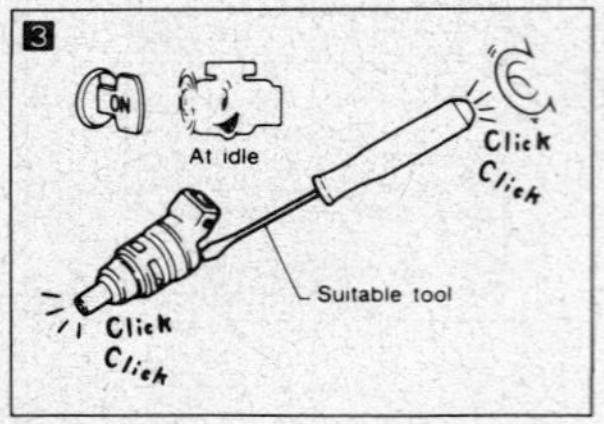

4
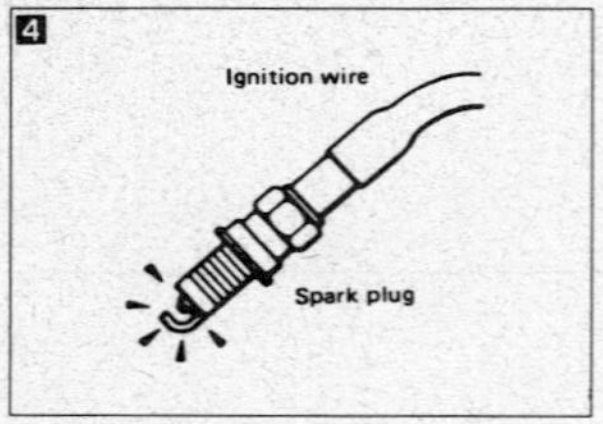

Diagnostic Procedure 41 — Symptom — Engine Stalls when the Electrical Load is Heavy

1 CHECK BATTERY AND ALTERNATOR.
Check battery and alternator condition.
- N.G. → Repair or replace.
- O.K. ↓

2 PERFORM POWER BALANCE TEST.
1. Perform "POWER BALANCE" in "ACTIVE TEST" mode.
2. Is there any cylinder which does not produce a momentary engine speed drop?

—OR—

When disconnecting each injector harness connector one at a time, is there any cylinder which does not produce a momentary engine speed drop?
- No → Go to 5.
- Yes ↓

3 CHECK INJECTOR.
Does each injector make an operating sound at idle?
- No → Check injector(s) and circuit(s).
- Yes ↓

4 CHECK IGNITION SPARK.
1. Disconnect ignition wire from spark plug.
2. Connect a known good spark plug to the ignition wire.
3. Place end of spark plug against a suitable ground and crank engine.
4. Check for spark.
- N.G. → Check ignition coil, power transistor and circuits.
- O.K. ↓

(Go to Ⓐ on next page.)

Diagnostic Procedure 41 — Symptom — Engine Stalls when the Electrical Load is Heavy (Cont'd)

5
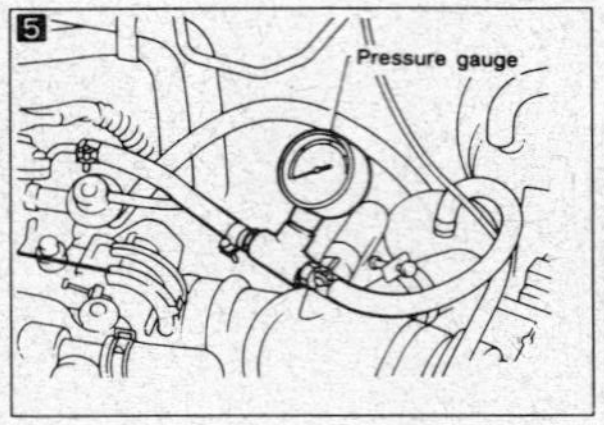

Ⓐ ↓

5 CHECK FUEL PRESSURE.
1. Release fuel pressure to zero.
2. Install fuel pressure gauge and check fuel pressure.
   At idle:
   Approx. 245 kPa
   (2.5 kg/cm², 36 psi)
   A few seconds after ignition switch is turned OFF to ON:
   Approx. 294 kPa
   (3.0 kg/cm², 43 psi)
- N.G. → Check fuel pressure regulator diaphragm.
- O.K. ↓

6 CHECK E.C.U. HARNESS CONNECTOR.
Check the E.C.U. pin terminals for damage or poor connection of E.C.U. harness connector.
- N.G. → Repair or replace.
- O.K. ↓

7 CHECK E.C.U. POWER SUPPLY AND GROUND CIRCUIT.
- N.G. → Repair or replace.
- O.K. ↓

8 TRY A KNOWN GOOD E.C.U.*

↓

INSPECTION END

*: E.C.U. may be the cause of a problem, but this is rarely the case.

1
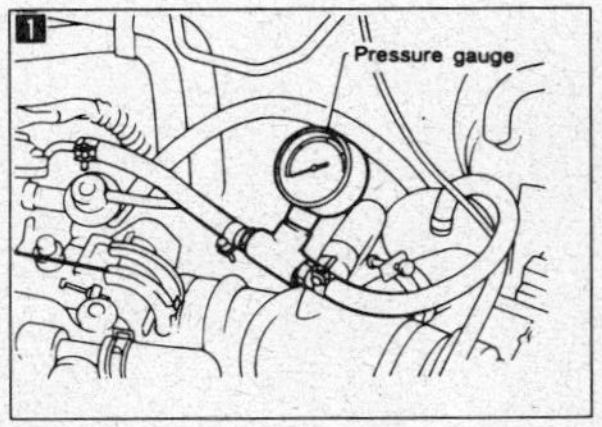

2
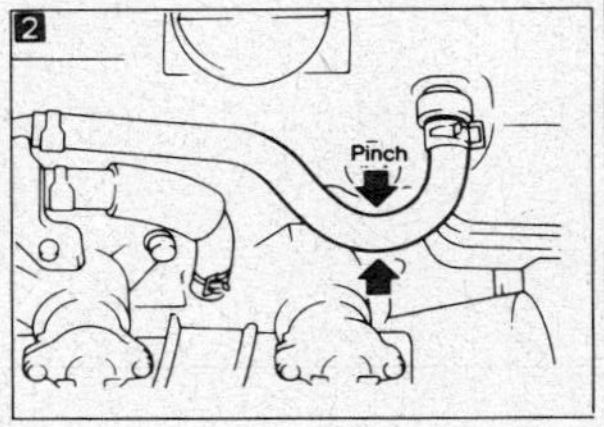

Diagnostic Procedure 42 — Symptom — Lack of Power and Stumble

1

CHECK FUEL PRESSURE.
1. Release fuel pressure to zero.
2. Install fuel pressure gauge and check fuel pressure.
   At idle:
   Approx. 245 kPa
   (2.5 kg/cm², 36 psi)
   A few seconds after ignition switch is turned OFF to ON:
   Approx. 294 kPa
   (3.0 kg/cm², 43 psi)

N.G. → Check fuel pressure regulator diaphragm.

O.K. ↓

2

CHECK FOR INTAKE AIR LEAK.
When pinching blow-by hose (lowering the blow-by air supply), does the engine speed rise?

Yes → Discover air leak location and repair.

No ↓

INSPECTION END

1
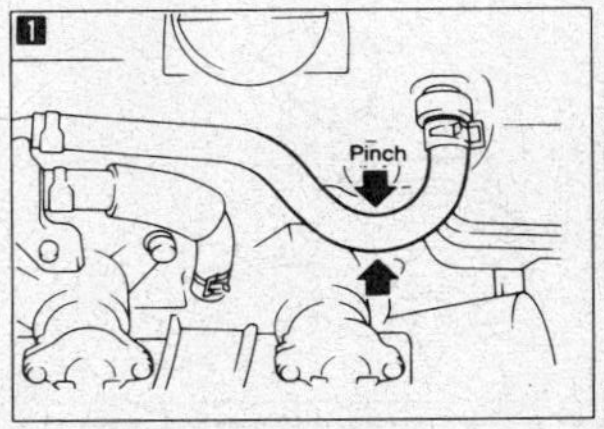

2
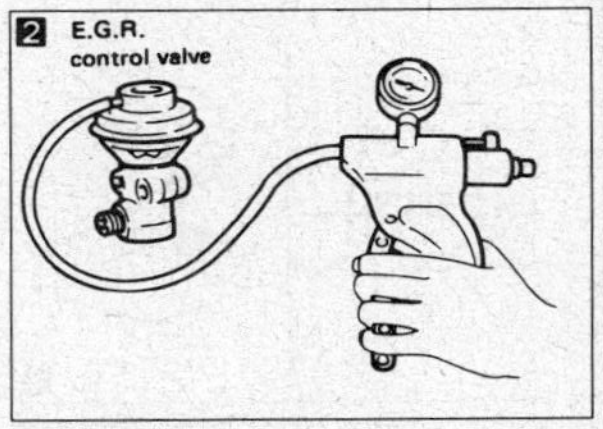

3
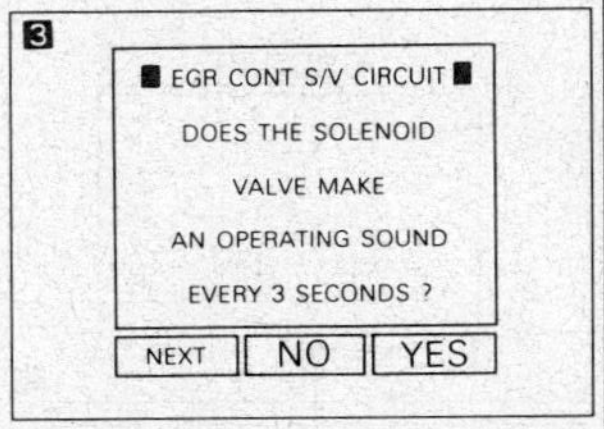

3
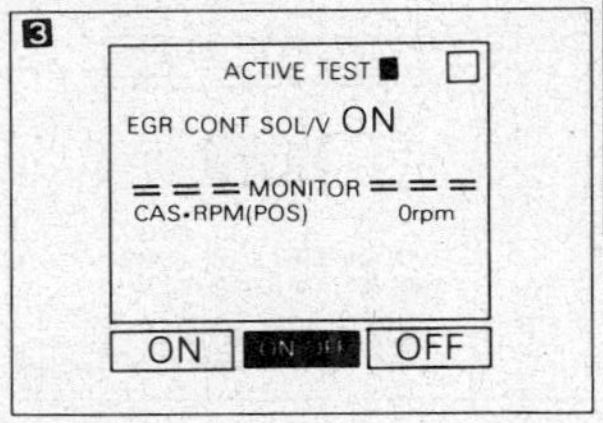

3
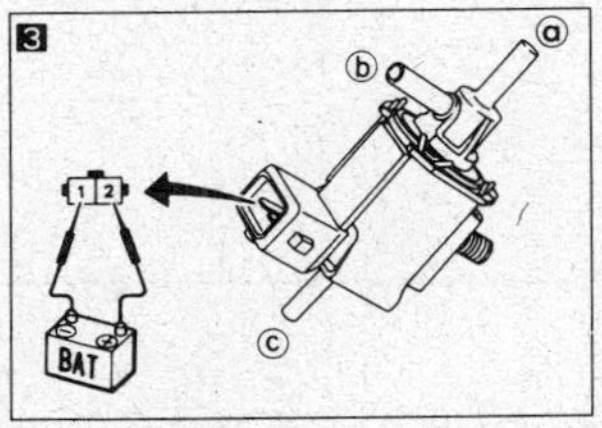

Diagnostic Procedure 43 — Symptom — Detonation

1

CHECK FOR INTAKE AIR LEAK.
When pinching blow-by hose (lowering the blow-by air supply), does the engine rpm rise?

Yes → Discover air leak location and repair.

No ↓

2

CHECK E.G.R. OPERATION.
1. Apply vacuum directly to the E.G.R. valve using a handy vacuum pump.
2. Check to see that the engine runs rough or dies.

No → Check E.G.R. valve for sticking.

Yes ↓

3

CHECK E.G.R. & CANISTER CONTROL SOLENOID VALVE.
1. Start engine and warm it up sufficiently.
2. Perform "EGR CONT S/V CIRCUIT" in "FUNCTION TEST" mode.

— OR —

1. Turn ignition switch "ON".
2. Select "E.G.R. CONT SOL VALVE" in "ACTIVE TEST" mode.
3. Turn E.G.R. & canister control solenoid valve ON and OFF.
4. Check operating sound.

— OR —

1. Disconnect E.G.R. & canister control solenoid valve harness connector.
2. Supply E.G.R. & canister control solenoid valve terminals with battery current and check operating sound.

N.G. → Check solenoid valve and circuit.

O.K. ↓

(Go to Ⓐ on next page.)

Diagnostic Procedure 43 — Symptom — Detonation (Cont'd)

4

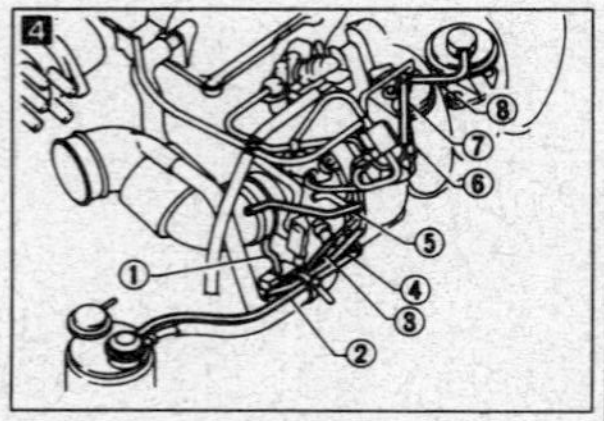

5

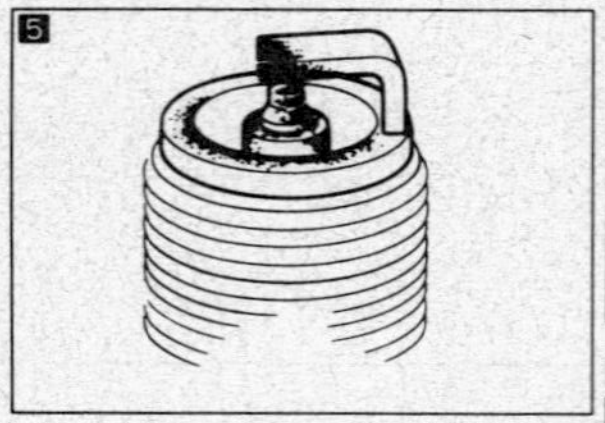

Ⓐ

4 CHECK VACUUM HOSES.
Check the following vacuum hoses for clogging, cracks and poor connection.
① Throttle chamber to E.G.R. & canister control solenoid valve
② Activated carbon canister to vacuum gallery
③ E.G.R. & canister control solenoid valve to vacuum gallery
④ E.G.R. & canister control solenoid valve to vacuum gallery
⑤ Air duct to vacuum gallery
⑥ B.P.T. valve to E.G.R. tube
⑦ B.P.T. valve to vacuum gallery
⑧ E.G.R. control valve to B.P.T. valve

N.G. → Repair or replace.

O.K. ↓

5 CHECK FOR OIL LEAK TO COMBUSTION CHAMBER.
Remove spark plugs and check for fouling with oil.

Yes → Check pistons, piston rings, valves, valve seats, valve oil seal, engine oil level, etc.

No ↓

INSPECTION END

Diagnostic Procedure 44 — Symptom — Surge

1

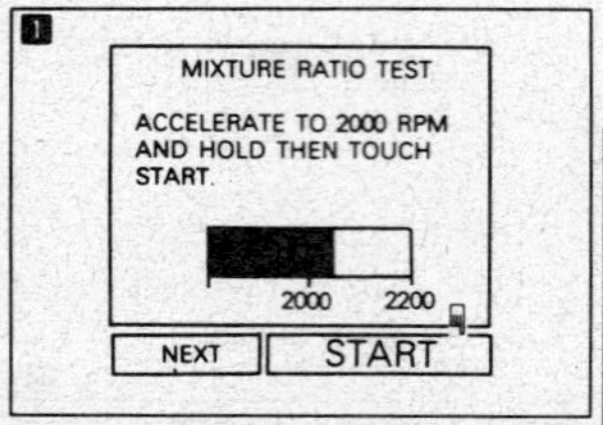

1

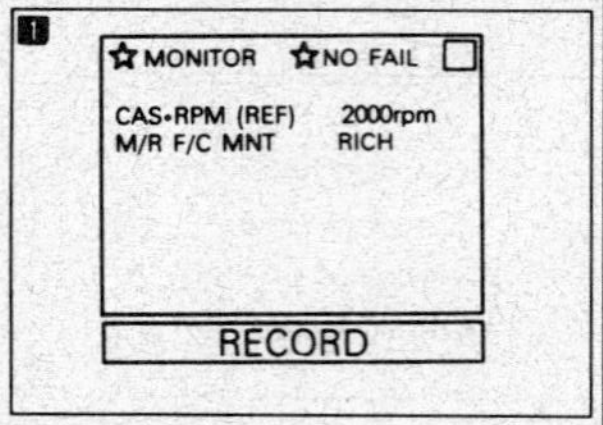

1

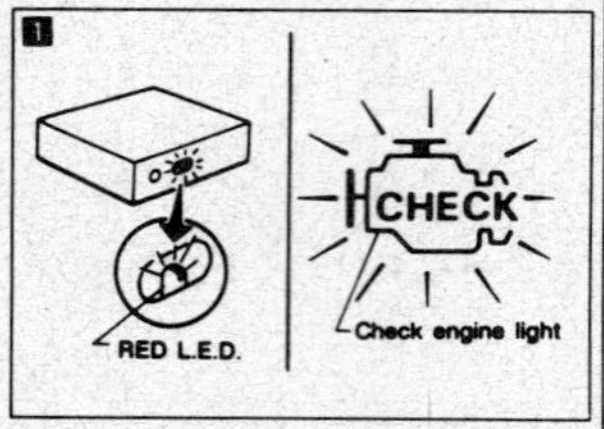

2

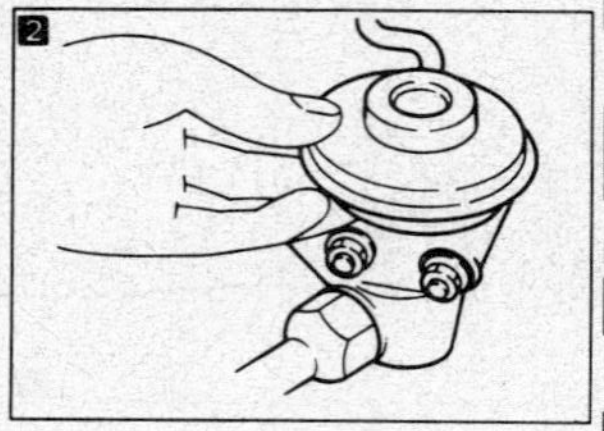

1 CHECK EXHAUST GAS SENSOR.
1. Start engine and warm it up sufficiently.
2. Perform "MIXTURE RATIO TEST" in "FUNCTION TEST" mode.

—OR—

2. See "M/R F/C MNT" in "DATA MONITOR" mode.
3. Maintaining engine at 2,000 rpm under no-load (with engine warmed up sufficiently.), check to make sure that the monitor fluctuates between "LEAN" and "RICH" more than 5 times during 10 seconds.
RICH→LEAN→RICH→ (1 time)
LEAN→RICH....... (2 times)

—OR—

2. Set "Exhaust gas sensor monitor" in the self-diagnostic Mode II.
3. Maintaining engine at 2,000 rpm under no-load, check that the RED L.E.D. on the E.C.U. or the check engine light on the instrument panel goes ON and OFF more than 5 times during 10 seconds.

N.G. → Replace exhaust gas sensor.

O.K. ↓

2 CHECK E.G.R. CONTROL VALVE.
Check E.G.R. control valve for sticking.

N.G. → Repair or replace.

O.K. ↓

3 TRY A KNOWN GOOD E.C.U.*

↓

INSPECTION END

*: E.C.U. may be the cause of a problem, but this is rarely the case.

1990–92 2.0L ENGINE – DIAGNOSTIC PROCEDURE – G20

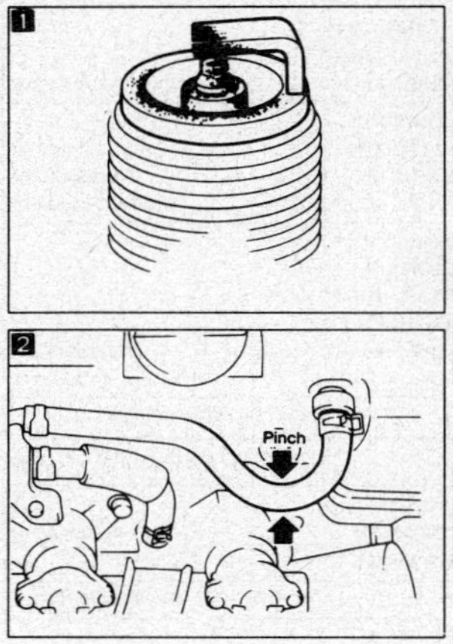

Diagnostic Procedure 45 — Symptom — Backfire through the Intake

1 CHECK SPARK PLUGS.
Remove the spark plugs and check for fouling, etc.
- N.G. → Repair or replace spark plug(s).
- O.K. ↓

2 CHECK INTAKE AIR LEAK.
When pinching blow-by hose (lowering the blow-by air supply), does the engine speed rise?
- Yes → Discover air leak location and repair.
- No ↓

3 CHECK FOR INTAKE VALVE DEPOSITS.
If there are deposits on intake valves, remove them.

↓

INSPECTION END

Diagnostic Procedure 46 — Symptom — Backfire through the Exhaust

CHECK ENGINE TEMPERATURE SENSOR.
Check engine temperature sensor and its circuit.
- N.G. → Replace or repair.
- O.K. ↓

INSPECTION END

1991–92 2.0L ENGINE – SCAN TOOL MODE APPLICATION – G20

E.C.C.S. COMPONENT PARTS APPLICATION

E.C.C.S. COMPONENT PARTS \ MODE		WORK SUPPORT	SELF-DIAGNOSTIC RESULTS	DATA MONITOR	ACTIVE TEST	FUNCTION TEST
INPUT	Crank angle sensor		X	X		
	Air flow meter		X	X		
	Engine temperature sensor		X	X	X	
	Exhaust gas sensors		X	X		X
	Vehicle speed sensors		X	X		X
	Throttle sensor	X	X	X		X
	Exhaust gas temperature sensor*		X	X		
	Detonation sensor		X			
	Ignition switch (start signal)			X		X
	Air conditioner switch			X		
	Neutral switch			X		X
	Power steering oil pump switch			X		X
	Battery			X		
OUTPUT	Injectors		X	X	X	X
	Power transistor (ignition timing)	X	X (Ignition signal)	X	X	X
	A.A.C. valve	X		X	X	X
	E.G.R. & canister control solenoid valve			X	X	X
	Air conditioner relay			X		
	Fuel pump relay	X		X	X	X
	Radiator fan			X	X	X
	A.I.V. control solenoid valve			X	X	

: The E.C.C.S. component part marked "" is applicable to vehicles for California only.
X: Applicable

1991–92 2.0L ENGINE—SCAN TOOL WORK SUPPORT MODE—G20

FUNCTION

Diagnostic mode	Function
Work support	This mode enables a technician to adjust some devices faster and more accurately by following the indications on the CONSULT unit.
Self-diagnostic results	Self-diagnostic results can be read and erased quickly.
Data monitor	Input/Output data in the control unit can be read.
Active test	Mode in which CONSULT drives some actuators apart from the control units and also shifts some parameters in a specified range.
E.C.U. part numbers	E.C.U. part numbers can be read.
Function test	Conducted by CONSULT instead of a technician to determine whether each system is "OK" or "NG".

WORK SUPPORT MODE

WORK ITEM	CONDITION	USAGE
THROTTLE SENSOR ADJUSTMENT	CHECK THE THROTTLE SENSOR SIGNAL. ADJUST IT TO THE SPECIFIED VALUE BY ROTATING THE SENSOR BODY UNDER THE FOLLOWING CONDITIONS. ● IGN SW "ON" ● ENG NOT RUNNING ● ACC PEDAL NOT PRESSED	When adjusting throttle sensor initial position.
IGNITION TIMING ADJUSTMENT	● IGNITION TIMING FEEDBACK CONTROL WILL BE HELD BY TOUCHING "START". AFTER DOING SO, ADJUST IGNITION TIMING WITH A TIMING LIGHT BY TURNING THE CRANK ANGLE SENSOR.	When adjusting initial ignition timing.
AAC VALVE ADJUSTMENT	SET ENGINE RPM AT THE SPECIFIED VALUE UNDER THE FOLLOWING CONDITIONS. ● ENGINE WARMED UP ● NO-LOAD	When adjusting idle speed.
FUEL PRESSURE RELEASE	● FUEL PUMP WILL STOP BY TOUCHING "START" DURING IDLING. CRANK A FEW TIMES AFTER ENGINE STALLS.	When releasing fuel pressure from fuel line.

1991–92 2.0L ENGINE—SCAN TOOL SELF-DIAGNOSTIC RESULTS MODE—G20

SELF-DIAGNOSTIC RESULTS MODE

DIAGNOSTIC ITEM	DIAGNOSTIC ITEM IS DETECTED WHEN ...	CHECK ITEM (REMEDY)
CRANK ANGLE SENSOR*	● Either 1° or 180° signal is not entered for the first few seconds during engine cranking. ● Either 1° or 180° signal is not input often enough while the engine speed is higher than the specified rpm.	● Harness and connector (If harness and connector are normal, replace crank angle sensor.)
AIR FLOW METER	● The air flow meter circuit is open or shorted. (An abnormally high or low voltage is entered.)	● Harness and connector (If harness and connector are normal, replace air flow meter.)
ENGINE TEMP SENSOR	● The engine temperature sensor circuit is open or shorted. (An abnormally high or low output voltage is entered.)	● Harness and connector ● Engine temperature sensor
CAR SPEED SENSOR	● The vehicle speed sensor circuit is open or shorted.	● Harness and connector ● Vehicle speed sensor (reed switch)
IGN SIGNAL-PRIMARY*	● The ignition signal in primary circuit is not entered during engine cranking or running.	● Harness and connector ● Power transistor unit
CONTROL UNIT	● E.C.U. calculation function is malfunctioning.	(Replace E.C.C.S. control unit.)
EGR SYSTEM**	● E.G.R. control valve does not operate. (E.G.R. control valve spring does not lift.)	● E.G.R. control valve ● E.G.R. & canister control solenoid valve
EXH GAS SENSOR	● The exhaust gas sensor circuit is open or shorted. (An abnormally high or low output voltage is entered.)	● Harness and connector ● Exhaust gas sensor ● Fuel pressure ● Injectors ● Intake air leaks
DETONATION SENSOR	● The detonation circuit is open or shorted. (An abnormally high or low voltage is entered.)	● Harness and connector ● Detonation sensor
EXH GAS TEMP SENSOR**	● The exhaust gas temperature sensor circuit is open or shorted. (An abnormally high or low voltage is entered.)	● Harness and connector ● Exhaust gas temperature sensor
THROTTLE SENSOR	● The throttle sensor circuit is open or shorted. (An abnormally high or low voltage is entered.)	● Harness and connector ● Throttle sensor
INJECTOR LEAK**	● Fuel leaks from injector.	● Injector

*: Check items causing a malfunction of crank angle sensor circuit first, if both "CRANK ANGLE SENSOR" and "IGN SIGNAL-PRIMARY" come out at the same time.

: The diagnostic item marked "" is applicable to vehicles for California only.

1991–92 2.0L ENGINE—DATA MONITOR MODE—G20

DATA MONITOR MODE

Remarks:

- The monitor item marked "**" is applicable to vehicles for California only.
- Specification data are reference values.
- Specification data are output/input values which are detected or supplied by the E.C.U. at the connector.
  - * Specification data may not be directly related to their components signals/values/operations.
  - i.e. Adjust ignition timing with a timing light before monitoring IGN TIMING, because the monitor may show the specification data in spite of the ignition timing not being adjusted to the specification data. This IGN TIMING monitors the data calculated by the E.C.U. according to the signals input from the crank angle sensor and other ignition timing related sensors.
- If the real-time diagnosis results are NG and the self-diagnosis results are OK when diagnosing the air flow meter, first check to see if the fuel pump control circuit is normal.

MONITOR ITEM	CONDITION		SPECIFICATION	CHECK ITEM WHEN OUTSIDE SPEC.
CAS. RPM (REF)	• Tachometer: Connect • Run engine and compare tachometer indication with the CONSULT value.		Almost the same speed as the CONSULT value.	• Harness and connector • Crank angle sensor
AIR FLOW MTR	• Engine: After warming up, idle the engine • A/C switch "OFF" • Shift lever "N"	Idle	1.3 - 1.7V	• Harness and connector • Air flow meter
		2,000 rpm	1.7 - 2.1V	
ENG TEMP SEN	• Engine: After warming up		More than 70°C (158°F)	• Harness and connector • Engine temperature sensor
EXH GAS SEN	• Engine: After warming up	Maintaining engine speed at 2,000 rpm	0 - 0.3V ↔ Approx. 0.6 - 1.0V	• Harness and connector • Exhaust gas sensor • Intake air leaks • Injectors
M/R F/C MNT			LEAN ↔ RICH Changes more than 5 times during 10 seconds.	
CAR SPEED SEN	• Turn drive wheels and compare speedometer indication with the CONSULT value		Almost the same speed as the CONSULT value	• Harness and connector • Vehicle speed sensor
BATTERY VOLT	• Ignition switch: ON (Engine stopped)		11 - 14V	• Battery • E.C.U. power supply circuit
THROTTLE SEN	• Ignition switch: ON (Engine stopped)	Throttle valve fully closed	0.45 - 0.55V	• Harness and connector • Throttle sensor • Throttle sensor adjustment
		Throttle valve fully opened	Approx. 4.0V	
EGR TEMP SEN**	• Engine: After warming up		Less than 4.5V	• Harness and connector • Exhaust gas temperature sensor
START SIGNAL	• Ignition switch: ON → START		OFF → ON	• Harness and connector • Starter switch
IDLE POSITION	• Ignition switch: ON (Engine stopped)	Throttle valve: Idle position	ON	• Harness and connector • Throttle sensor • Throttle sensor adjustment
		Throttle valve: Slightly open	OFF	
AIR COND SIG	• Engine: After warming up, idle the engine	A/C switch "OFF"	OFF	• Harness and connector • Air conditioner switch
		A/C switch "ON"	ON	
NEUTRAL SW	• Ignition switch: ON	Shift lever "P" or "N"	ON	• Harness and connector • Neutral switch
		Except above	OFF	

MONITOR ITEM	CONDITION			SPECIFICATION	CHECK ITEM WHEN OUTSIDE SPEC.
PW/ST SIGNAL	• Engine: After warming up, idle the engine		Steering wheel in neutral (forward direction)	OFF	• Harness and connector • Power steering oil pressure switch
			The steering wheel is turned	ON	
FUEL PUMP RLY	• Ignition switch is turned to ON (Operates for 5 seconds) • Engine running and cranking • When engine is stopped (stops in 1.0 seconds)			ON	• Harness and connector • Fuel pump relay
	Except as shown above			OFF	
RADIATOR FAN	• After warming up engine, idle the engine. • A/C switch "OFF"	M/T	Engine temperature is 99°C (210°F) or less	OFF	• Harness and connector • Radiator fan relay • Radiator fan
			Engine temperature is 100°C (212°F) or more	ON	
		A/T	Engine temperature is 94°C (201°F) or less	OFF	
			Engine temperature is between 95°C (203°F) and 104°C (219°F)	LOW	
			Engine temperature is 105°C (221°F) or more	HIGH	
INJ PULSE	• Engine: After warming up • A/C switch "OFF" • Shift lever "N" • No-load	Idle		2.4 - 3.2 msec.	• Harness and connector • Injector • Air flow meter • Intake air system
		2,000 rpm		1.9 - 2.8 msec.	
IGN TIMING	ditto	Idle		15° B.T.D.C.	• Harness and connector • Crank angle sensor
		2,000 rpm		More than 25° B.T.D.C.	
AAC VALVE	ditto	Idle		20 - 40%	• Harness and connector • A.A.C. valve
		2,000 rpm		—	
A/F ALPHA	• Engine: After warming up	Maintaining engine speed at 2,000 rpm		75 - 125%	• Harness and connector • Injectors • Air flow meter • Exhaust gas sensor • Canister purge line • Intake air system
AIR COND RLY	• Air conditioner switch OFF → ON			OFF → ON	• Harness and connector • Air conditioner switch • Air conditioner relay
EGR CONT S/V	• Engine: After warming up • A/C switch "OFF" • Shift lever "N" • No-load	2,000 rpm		OFF	• Harness and connector • E.G.R. & canister control solenoid valve
		4,000 rpm		ON	
AIV CONT S/V	• Engine: After warming up • Shift lever "N" • No-load	Idle		ON	• Harness and connector • A.I.V. control solenoid valve
		2,000 rpm		OFF	

1991–92 2.0L ENGINE – ACTIVE TEST MODE – G20

ACTIVE TEST MODE

TEST ITEM	CONDITION	JUDGMENT	CHECK ITEM (REMEDY)
FUEL INJECTION TEST	• Engine: Return to the original trouble condition • Change the amount of fuel injection using CONSULT.	If trouble symptom disappears, see CHECK ITEM.	• Harness and connector • Fuel injectors • Exhaust gas sensor
AAC/V OPENING TEST	• Engine: After warming up, idle the engine. • Change the AAC valve opening percent using CONSULT.	Engine speed changes according to the opening percent.	• Harness and connector • AAC valve
ENGINE TEMP TEST	• Engine: Return to the original trouble condition • Change the engine coolant temperature using CONSULT.	If trouble symptom disappears, see CHECK ITEM.	• Harness and connector • Engine temperature sensor • Fuel injectors
IGN TIMING TEST	• Engine: Return to the original trouble condition • Timing light: Set • Retard the ignition timing using CONSULT.	If trouble symptom disappears, see CHECK ITEM.	• Adjust initial ignition timing
POWER BALANCE TEST	• Engine: After warming up, idle the engine. • A/C switch "OFF" • Shift lever "N" • Cut off each injector signal one at a time using CONSULT.	Engine runs rough or dies.	• Harness and connector • Compression • Injectors • Power transistor • Spark plugs • Ignition coils
RADIATOR FAN TEST	• Ignition switch: ON • Turn the radiator fan "ON" and "OFF" using CONSULT.	Radiator fan moves and stops.	• Harness and connector • Radiator fan motor
FUEL PUMP RLY TEST	• Ignition switch: ON (Engine stopped) • Turn the fuel pump relay "ON" and "OFF" using CONSULT and listen to operating sound.	Fuel pump relay makes the operating sound.	• Harness and connector • Fuel pump relay
EGR CONT SOL/V TEST AIV CONT SOL/V TEST	• Ignition switch: ON • Turn solenoid valve "ON" and "OFF" with the CONSULT and listen to operating sound.	Each solenoid valve makes an operating sound.	• Harness and connector • Solenoid valve
SELF-LEARN CONT TEST	• In this test, the coefficient of self-learning control mixture ratio returns to the original coefficient by touching "CLEAR" on the screen.		

1992 2.0L ENGINE – FUNCTION TEST – G20

FUNCTION TEST MODE

FUNCTION TEST ITEM	CONDITION	JUDGEMENT		CHECK ITEM (REMEDY)
SELF-DIAG RESULTS	• Ignition switch: ON (Engine stopped) • Displays the results of self-diagnosis.	—		Objective system
IDLE POSITION (IDLE SWITCH CIRCUIT)	• Ignition switch: ON (Engine stopped) • Idle switch circuit is tested when throttle is opened and closed fully. ("IDLE POSITION" is the test item name for the vehicles in which idle is selected by throttle sensor.)	Throttle valve: opened	OFF	• Harness and connector • Throttle sensor (Idle switch) • Throttle sensor (Idle switch) adjustment • Throttle linkage • Verify operation in DATA MONITOR mode.
		Throttle valve: closed	ON	
THROTTLE SENSOR CKT	• Ignition switch: ON (Engine stopped) • Throttle sensor circuit is tested when throttle is opened and closed fully.	Range (Throttle valve fully opened — Throttle valve fully closed)	More than 3.0V	• Harness and connector • Throttle sensor • Throttle sensor adjustment • Throttle linkage • Verify operation in DATA MONITOR mode.
NEUTRAL SW CIRCUIT	• Ignition switch: ON (Engine stopped) • Neutral switch circuit is tested when shift lever is manipulated.	OUT OF N/P-RANGE	OFF	• Harness and connector • Neutral switch/Inhibitor switch • Linkage + Inhibitor switch adjustment
		IN N-RANGE	ON	
FUEL PUMP CIRCUIT	• Ignition switch: ON (Engine stopped) • Fuel pump circuit is tested by checking the pulsation in fuel pressure when fuel tube is pinched.	There is pressure pulsation on the fuel feed hose.		• Harness and connector • Fuel pump • Fuel pump relay • Fuel filter clogging • Fuel level
EGR CONT S/V CIRCUIT	• Ignition switch: ON (Engine stopped) • EGR control S/V circuit is tested by checking solenoid valve operating noise.	The solenoid valve makes an operating sound every 3 seconds.		• Harness and connector • EGR control solenoid valve
RADIATOR FAN CIRCUIT	• Ignition switch: ON (Engine stopped) • Radiator fan circuit is tested when radiator fan is rotated.	• The radiator fan rotates and stops every 3 seconds.		• Harness and connector • Radiator fan motor • Radiator fan relay

1992 2.0L ENGINE—FUNCTION TEST—G20

FUNCTION TEST ITEM	CONDITION	JUDGEMENT		CHECK ITEM (REMEDY)
START SIGNAL CIRCUIT	• Ignition switch: ON → START • Start signal circuit is tested when engine is started by operating the starter. Battery voltage and water temperature before cranking, and average battery voltage, air flow meter output voltage and cranking speed during cranking are displayed.	Start signal: OFF → ON		• Harness and connector • Ignition switch
PW/ST SIGNAL CIRCUIT	• Ignition switch: ON (Engine running) • Power steering circuit is tested when steering wheel is rotated fully and then set to a straight line running position.	Locked position	ON	• Harness and connector • Power steering oil pressure switch • Power steering oil pump
		Neutral	OFF	
CAR SPEED SEN CIRCUIT	• Vehicle speed sensor circuit is tested when vehicle is running at a speed of 10 km/h (6 MPH) or higher.	Vehicle speed sensor input signal is greater than 4 km/h (2 MPH)		• Harness and connector • Vehicle speed sensor • Electric speedometer
IGN TIMING ADJ	• After warming up, idle the engine. • Ignition timing adjustment is checked by reading ignition timing with a timing light and checking whether it agrees with specifications.	The timing light indicates the same value on the screen.		• Adjust ignition timing (by moving crank angle sensor or distributor) • Crank angle sensor drive mechanism
MIXTURE RATIO TEST	• Air-fuel ratio feedback circuit (injection system, ignition system, vacuum system, etc.) is tested by examining the exhaust gas sensor output at 2,000 rpm under non-loaded state.	• Exhaust gas sensor COUNT: More than 5 times during 10 seconds		• INJECTION SYS (Injector, fuel pressure regulator, harness or connector) • IGNITION SYS (Spark plug, power transistor, ignition coil, harness or connector) • VACUUM SYS (Intake air leaks) • Exhaust gas sensor circuit • Exhaust gas sensor operation • Fuel pressure high or low • Air flow meter

FUNCTION TEST ITEM	CONDITION	JUDGEMENT	CHECK ITEM (REMEDY)
POWER BALANCE	• After warming up, idle the engine. • Injector operation of each cylinder is stopped one after another, and resultant change in engine rotation is examined to evaluate combustion of each cylinder. (This is only displayed for models where a sequential injection system is used.)	Difference in engine rpm is greater than 25 rpm before and after cutting off the injector of each cylinder.	• Injector circuit (Injector, harness or connector) • Ignition circuit (Spark plug, power transistor, ignition coil, harness or connector) • Compression • Valve timing
AAC VALVE SYSTEM	• After warming up, idle the engine. • AAC valve system is tested by detecting change in engine rpm when AAC valve opening is changed to 0%, 20% and 80%.	Difference in engine rpm is greater than 150 rpm between when valve opening is at 80% (102 steps) and at 20% (25 steps).	• Harness and connector • AAC valve • Air passage restriction between air inlet and AAC valve • IAS (Idle adjusting screw) adjustment

1990–92 3.0L ENGINE—ECCS COMPONENT LOCATIONS—M30

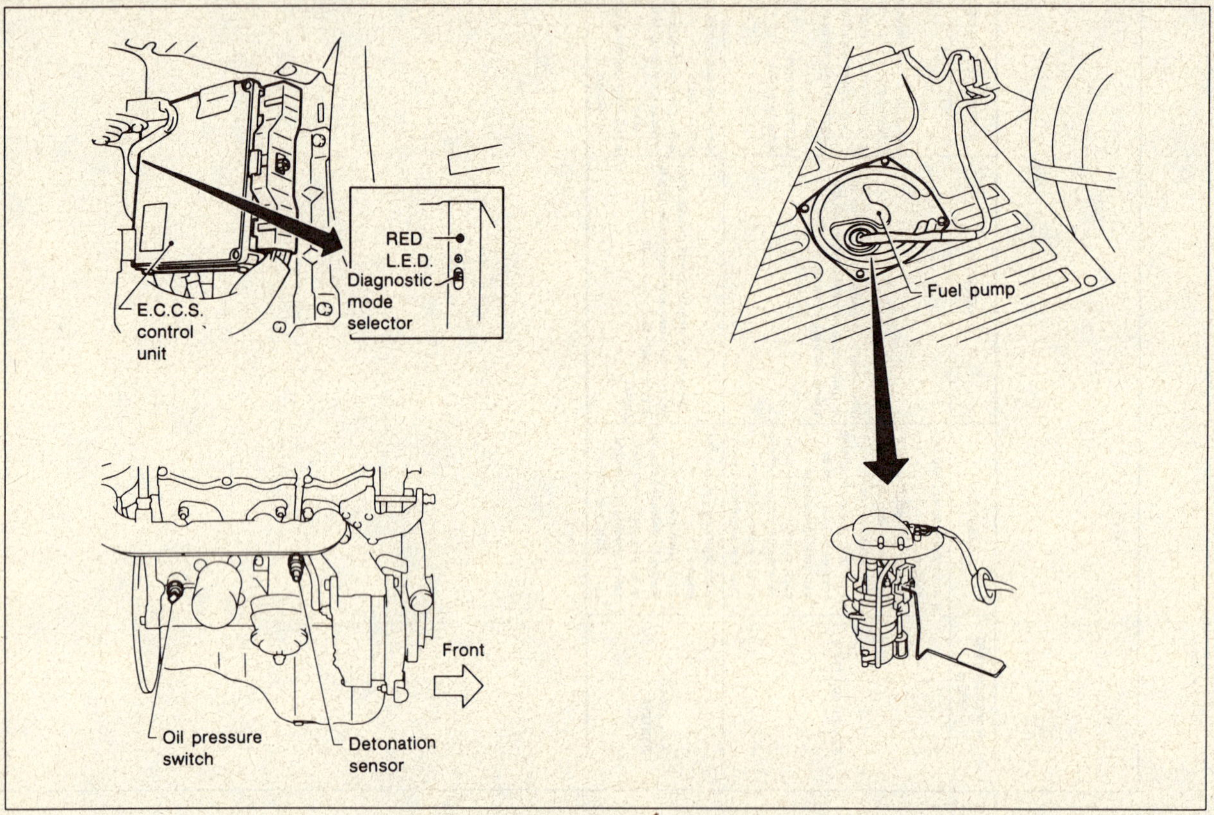

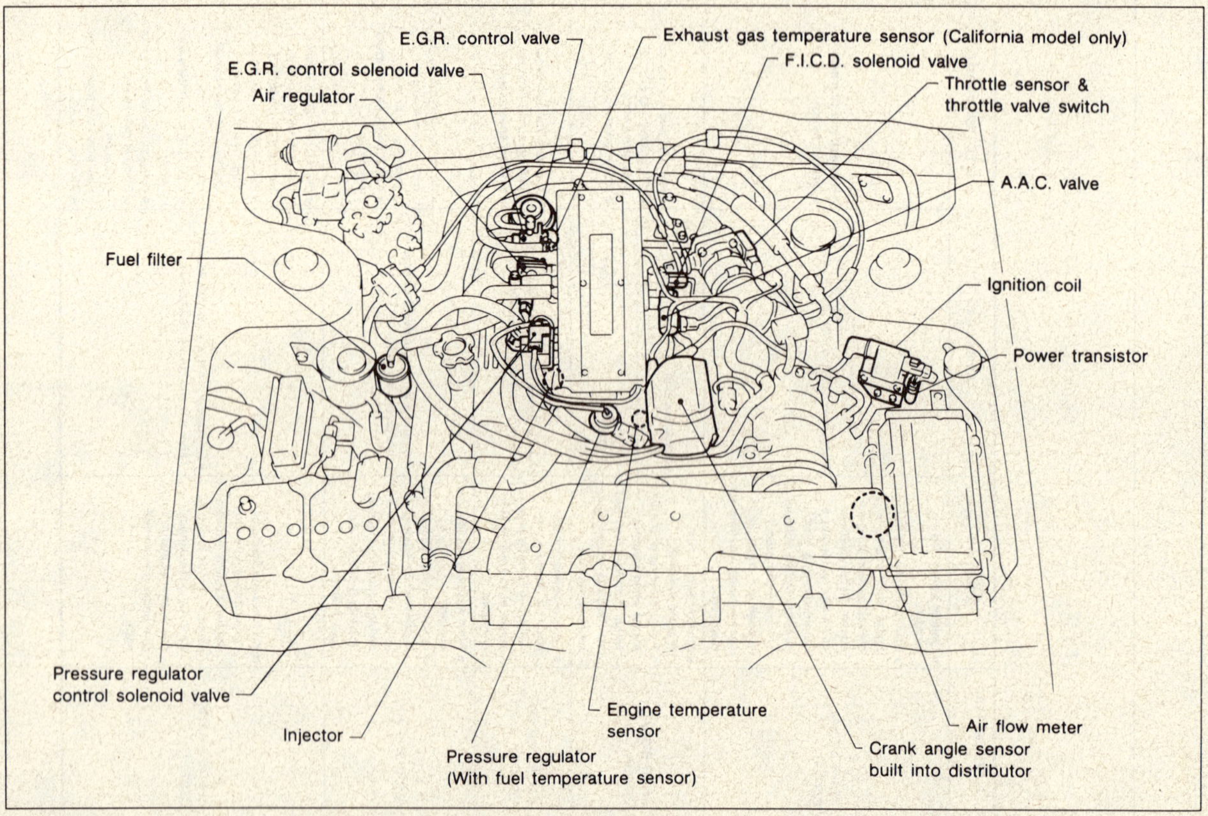

1990–92 3.0L ENGINE – ECCS SYSTEM SCHEMATIC – M30

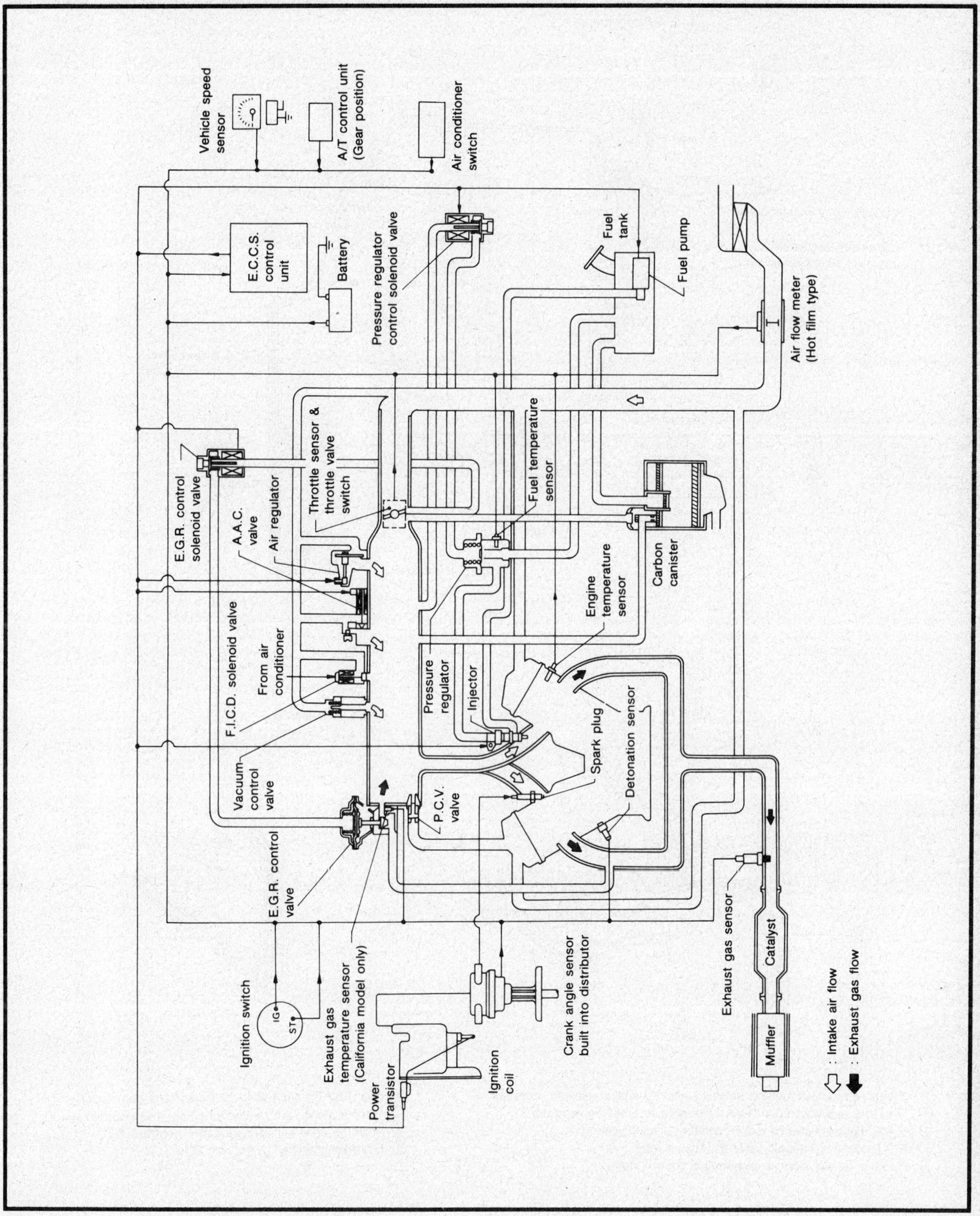

1990–92 3.0L ENGINE—ECCS VACUUM CONTROL SCHEMATIC—M30

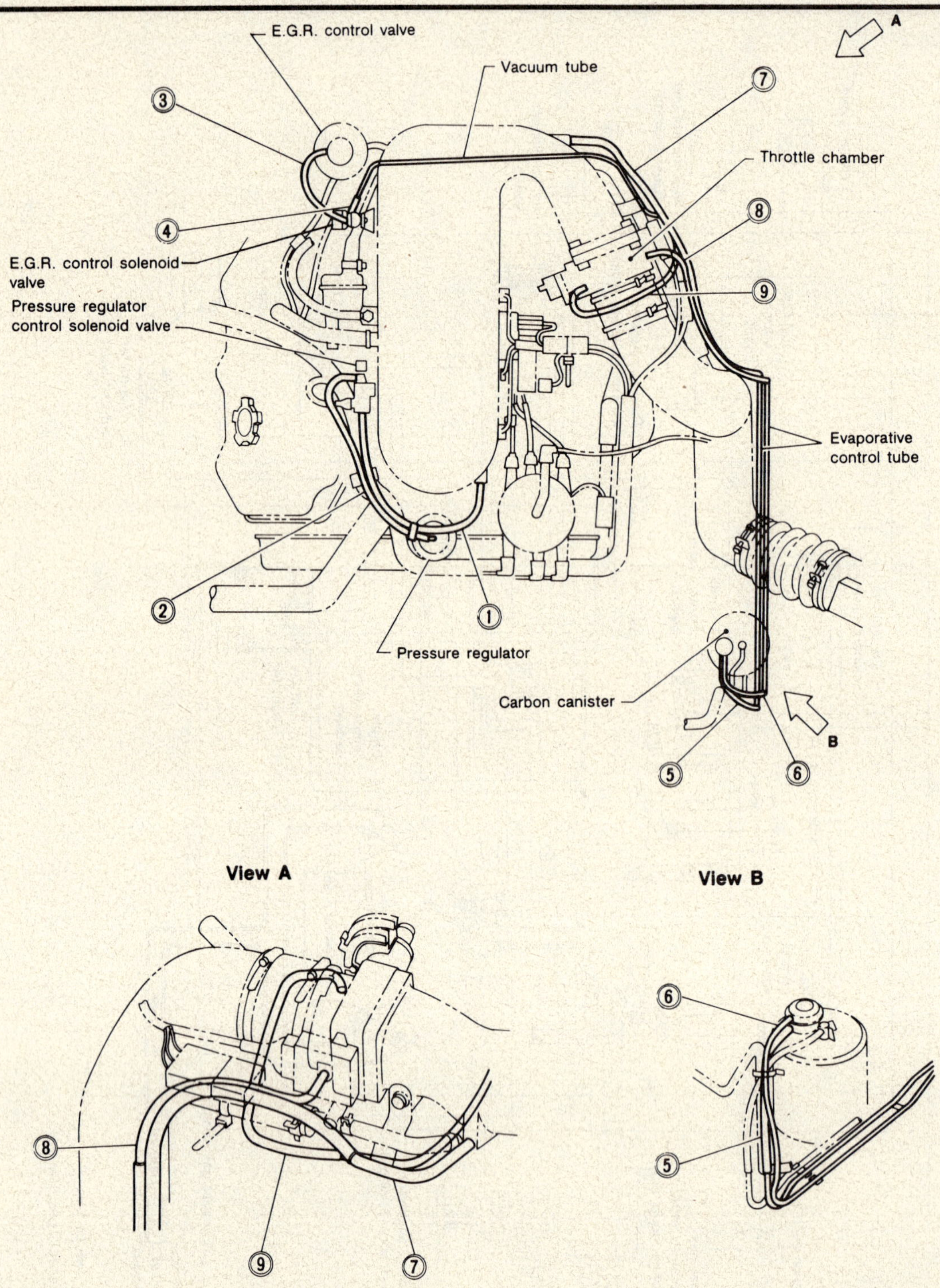

1. Pressure regulator control solenoid valve to intake manifold collector
2. Pressure regulator control solenoid valve to pressure regulator
3. E.G.R. control valve to E.G.R. control solenoid valve
4. E.G.R. control solenoid valve to vacuum tube
5. Canister (purge port) to evaporative control tube
6. Canister (vacuum port) to evaporative control tube
7. Intake manifold collector to evaporative control tube
8. Throttle chamber to evaporative control tube
9. Throttle chamber to vacuum tube

1990–92 3.0L ENGINE – ECCS CIRCUIT DIAGRAM – M30

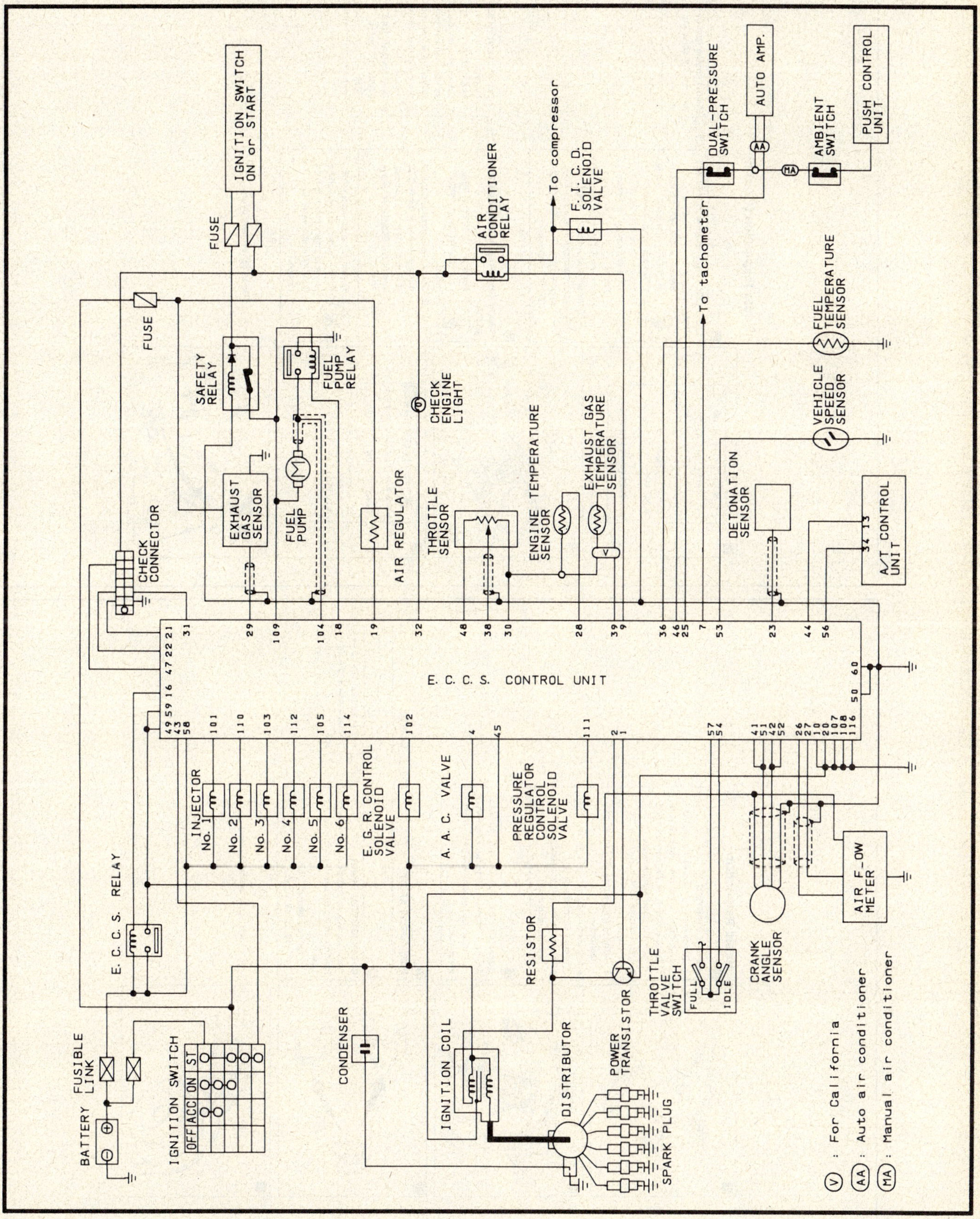

Basic Inspection

1
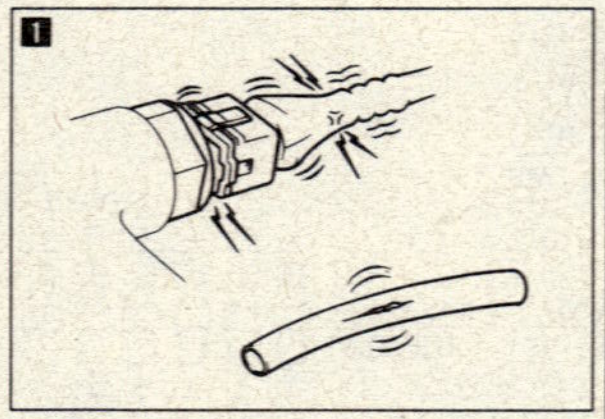

2
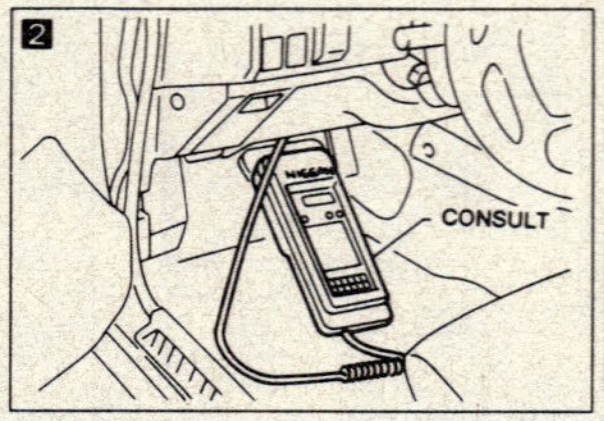

3
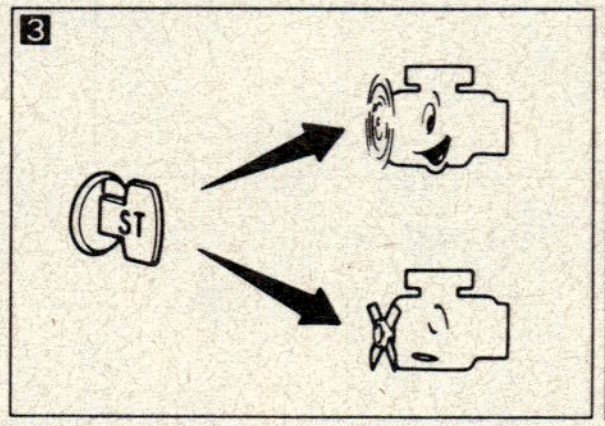

4
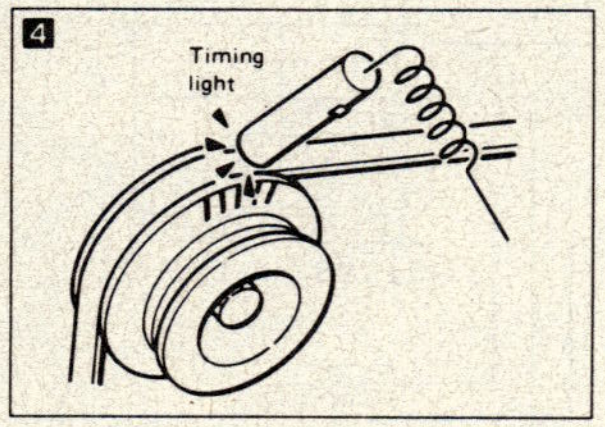

1 BEFORE STARTING

1. Check service records for any recent repairs that may indicate a related problem, or the current need for scheduled maintenance.
2. Open engine hood and check the following:
- Harness connectors for proper connections
- Vacuum hoses for splits, kinks, and proper connections
- Wiring for proper connections, pinches, and cuts

↓

2 CONNECT CONSULT TO THE VEHICLE

Connect "CONSULT" to the diagnostic connector and select "ENGINE" from the menu.

↓

3 DOES ENGINE START?

No → Go to 6.

Yes ↓

4 CHECK IGNITION TIMING.

Warm up engine sufficiently and check ignition timing at idle using timing light.

Ignition timing:
15° ± 2° B.T.D.C.

N.G. → Adjust ignition timing by turning crank angle sensor.

O.K. ↓

(Go to Ⓐ on next page.)

Basic Inspection (Cont'd)

Ⓐ ↓

5
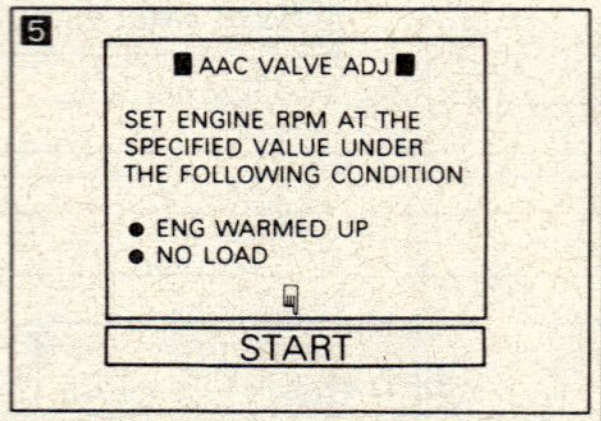

5
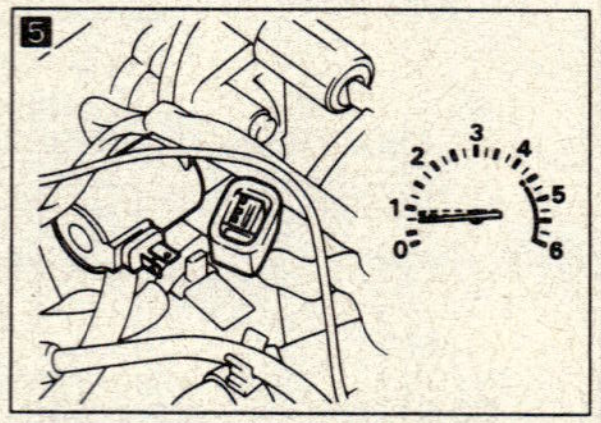

6
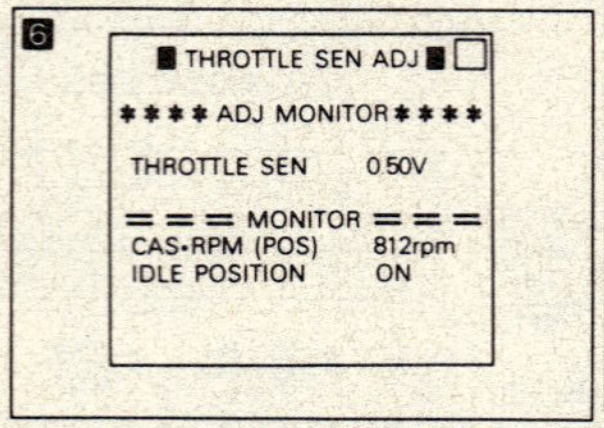

6
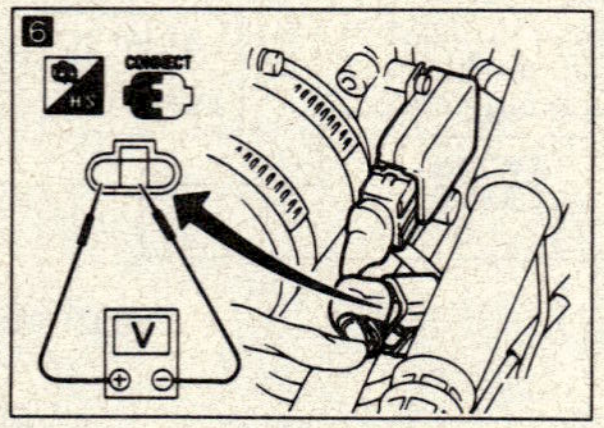

5 CHECK IDLE ADJ. SCREW INITIAL SET RPM.

1. Select "A.A.C. VALVE ADJ" in "WORK SUPPORT" mode.
2. When touching "START", does engine rpm fall to 700 ± 50 rpm (in "N" position)?

— OR —

When disconnecting A.A.C. valve harness connector, does engine rpm fall to 700 ± 50 rpm (in "N" position)?

No → Adjust engine rpm by turning idle adjusting screw.

Yes ↓

6 CHECK THROTTLE SENSOR IDLE POSITION.

1. Perform "THROTTLE SEN. ADJ" in "WORK SUPPORT" mode.
2. Check that output voltage of throttle sensor is 0.4 to 0.5V. (Throttle valve fully closes.) and "IDLE POSITION" stays "ON".

— OR —

Measure output voltage of throttle sensor using voltmeter, and check that it is 0.4 to 0.5V. (Throttle valve fully closed.)

N.G. →
1. Adjust output voltage by rotating throttle sensor body.
2. Disconnect throttle sensor harness connector for a few seconds and then reconnect it.
3. Confirm that "IDLE POSITION" stays "ON".

O.K. ↓

(Go to Ⓑ on next page)

Basic Inspection (Cont'd)

7

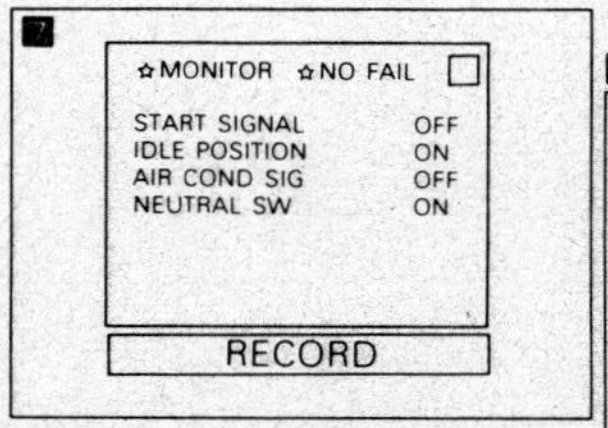

7

(B)

7 CHECK SWITCH INPUT SIGNAL.

Select the following switches in "DATA MONITOR" mode,
a) Start signal,
b) Idle position,
c) Air conditioner signal,
d) Neutral (Parking) switch,
and check the switches' ON-OFF operation.

— OR —

Remove E.C.U. from right dash side panel and check the above switches' ON-OFF operation using voltmeter at each E.C.U. terminal.

Switch	Condition	Voltage (V)
Start signal	IGN ON → IGN START	0 → Battery voltage
Idle position	Idle position → depress the accelerator pedal (Ignition switch "ON")	8.0 - 9.0 → 0
A/C signal	A/C OFF → A/C ON (Engine running)	Battery voltage → 0.5 - 0.7
Neutral (Parking) switch	Shift lever is "N" or "P" position → Except "N" and "P"	0 → 8.0 - 9.0

N.G. → Repair or replace the malfunctioning switch or its circuit.

O.K. ↓

8

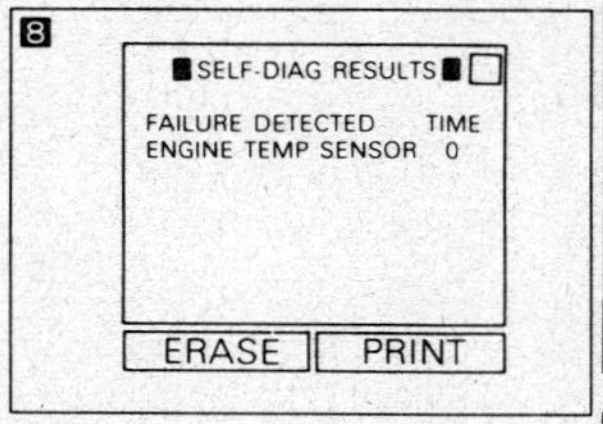

8

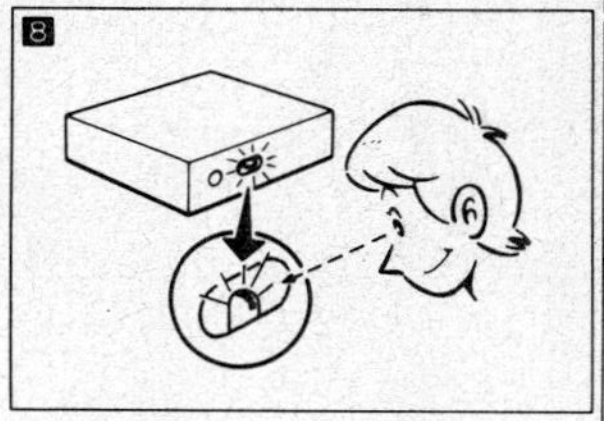

8 READ SELF-DIAGNOSTIC RESULTS.

1. Perform "SELF-DIAG RESULTS" mode.
2. Read out self-diagnostic results.
3. Is a failure detected?

— OR —

1. Set "Self-diagnostic results mode" in Mode II.
2. Count the number of RED L.E.D. flashes and read out the codes.
3. Are the codes being output?

Yes → Go to the relevant inspection procedure.

No ↓

INSPECTION END

Diagnostic Procedure 1 — High Idling after Warm-up

1

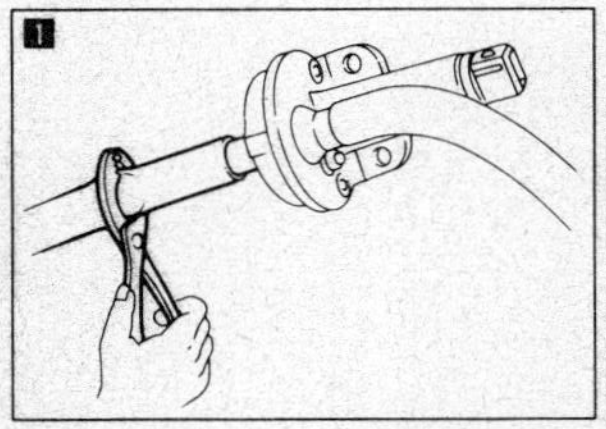

1 CHECK AIR REGULATOR.

When pinching the air regulator hose, does the engine speed drop?

Yes → Check air regulator and circuit.

No ↓

2

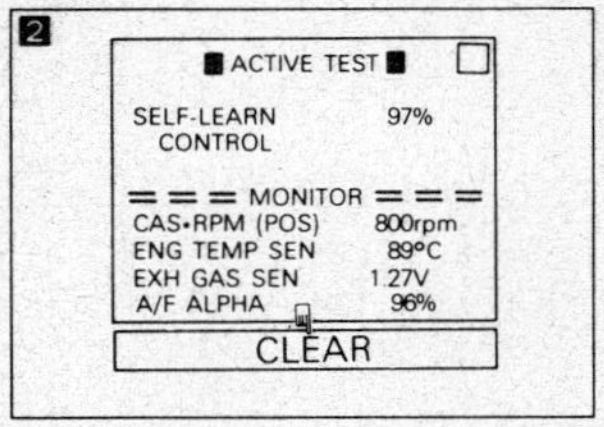

2

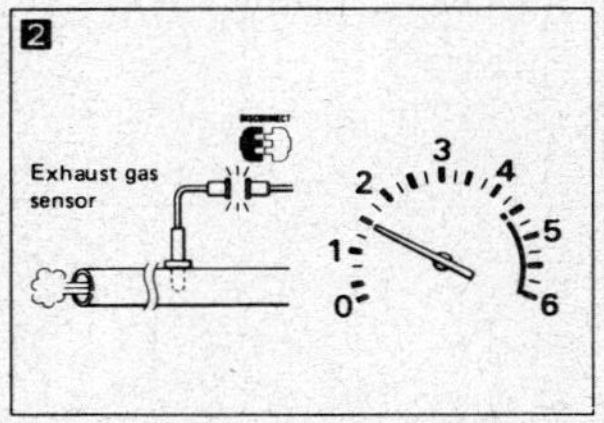

2 CHECK INTAKE AIR LEAK.

1. Select "SELF-LEARNING CONT" in "ACTIVE TEST" mode.
2. Clear the self-learning control coefficient by touching "CLEAR".
3. Does the engine speed drop?

— OR —

1. Disconnect exhaust gas sensor harness connector.
2. After racing engine at 1,500 rpm under no load for about 30 seconds, does the engine speed drop?

Yes → Discover air leak location and repair.

No ↓

3

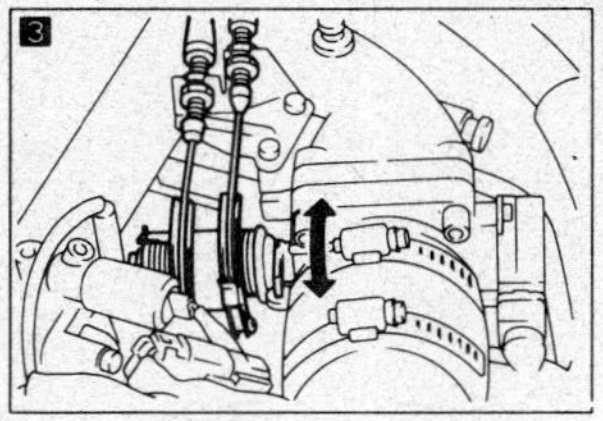

3 CHECK THROTTLE LINKAGE.

1. Check that throttle linkage moves smoothly.
2. Confirm that throttle valve both fully opens and fully closes.

N.G. → Repair throttle linkage or sticking of throttle valve.

O.K. ↓

INSPECTION END

1
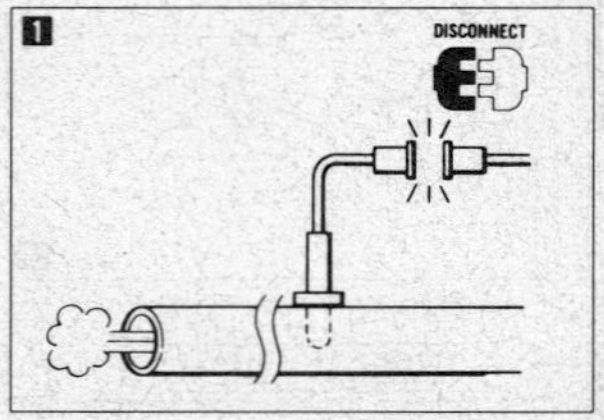

2
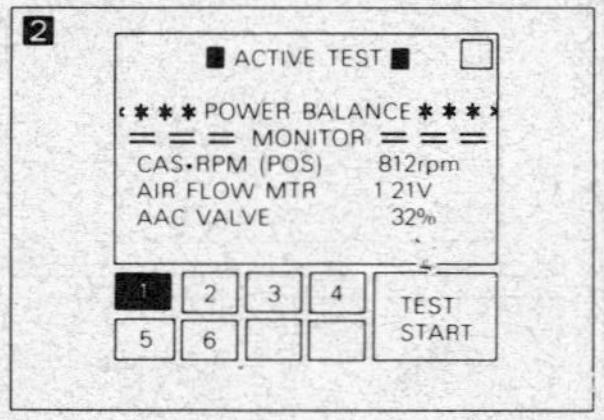

2
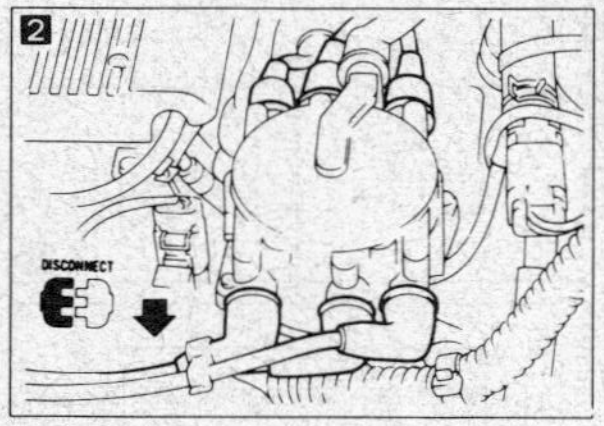

3
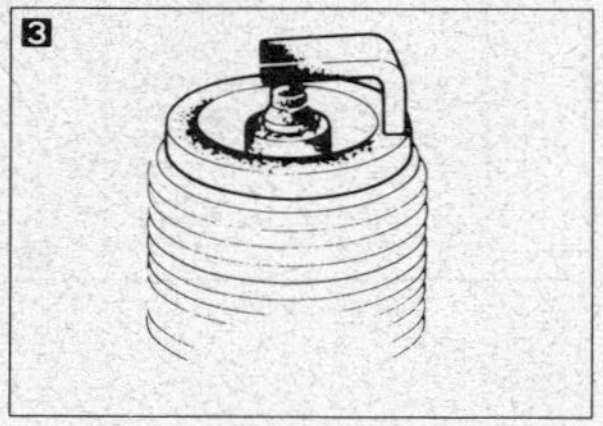

Diagnostic Procedure 2 — Hunting

1 **CHECK EXHAUST GAS SENSOR.**
When disconnecting exhaust gas sensor harness connector, is the hunting fixed?
- Yes → Check exhaust gas sensor.
- No ↓

2 **PERFORM POWER BALANCE TEST.**
1. Perform "POWER BALANCE" in "ACTIVE TEST" mode.
2. Is there any cylinder which does not produce a momentary engine speed drop?

—OR—

When disconnecting each ignition wire connector one at a time, is there any cylinder which does not produce a momentary engine speed drop?
- No → Go to 4.
- Yes ↓

3 **CHECK SPARK PLUGS.**
Remove the spark plugs and check for fouling, etc.
- N.G. → Repair or replace spark plug(s).
- O.K. ↓

(Go to (A) on next page.)

4

5
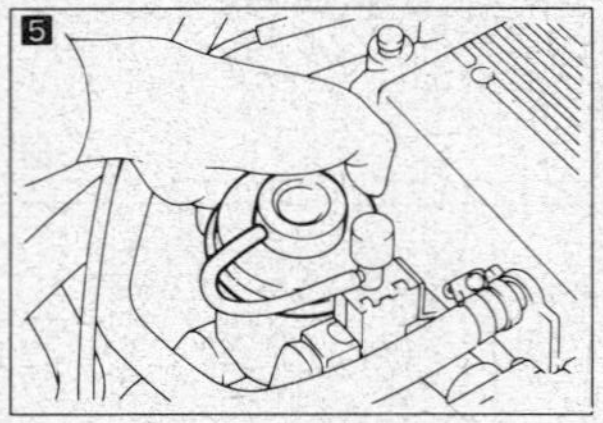

Diagnostic Procedure 2 — Hunting (Cont'd)

(A) ↓

4 **CHECK FOR INTAKE AIR LEAK.**
When pinching blow-by hose (lowering the blow-by air supply), does the engine speed rise?
- Yes → Discover air leak location and repair.
- No ↓

5 **CHECK E.G.R. CONTROL VALVE.**
Check E.G.R. control valve for sticking.
- N.G. → Repair or replace.
- O.K. ↓

INSPECTION END

1

2
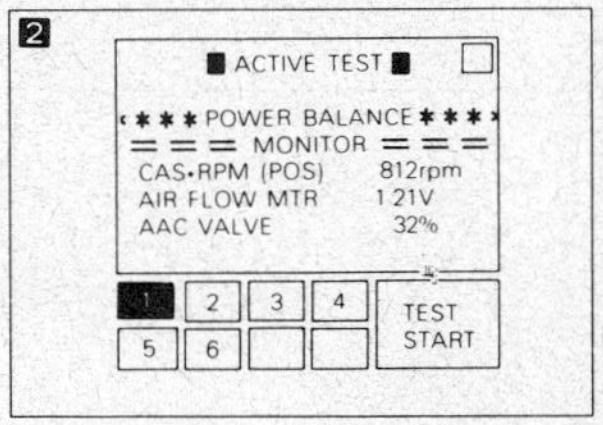

2
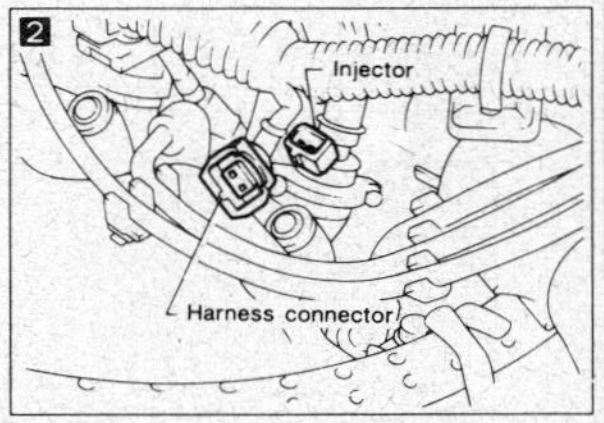

3

4
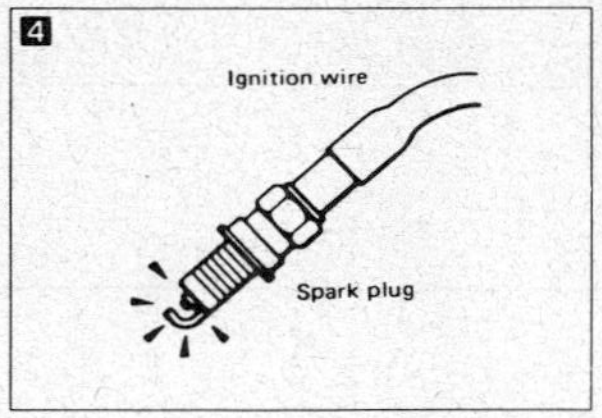

Diagnostic Procedure 3 — Unstable Idle

1

CHECK E.G.R. CONTROL VALVE.
Check E.G.R. control valve for sticking.

- N.G. → Repair or replace.
- O.K ↓

2

PERFORM POWER BALANCE TEST.
1. Perform "POWER BALANCE" in "ACTIVE TEST" mode.
2. Is there any cylinder which does not produce a momentary engine speed drop?

— OR —

When disconnecting each injector harness connector one at a time, is there any cylinder which does not produce a momentary engine speed drop?

- No → Go to 6.
- Yes ↓

3

CHECK INJECTOR.
1. Remove distributor from engine. (Crank angle sensor harness connector should remain connected.)
2. Turn ignition switch ON. (Do not start engine.)
3. When rotating distributor shaft slowly by hand, does each injector make an operating sound?

- No → Check injector(s) and circuit(s).
- Yes ↓

4

CHECK IGNITION SPARK.
1. Disconnect ignition wire from spark plug.
2. Connect a known good spark plug to the ignition wire.
3. Place end of spark plug against a suitable ground and crank engine.
4. Check for spark.

- N.G. → Check ignition coil, power transistor unit and their circuits.
- O.K. ↓

(Go to Ⓐ on next page.)

5
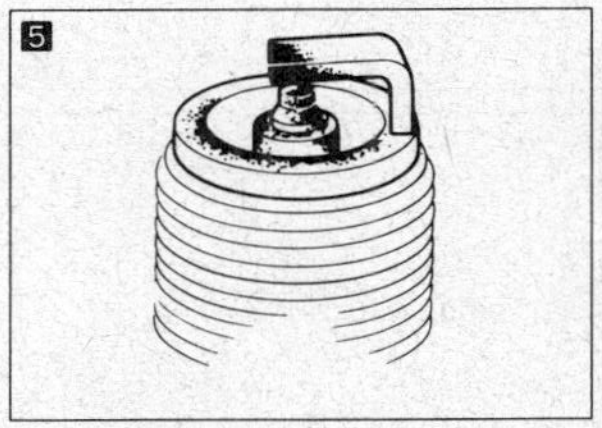

6
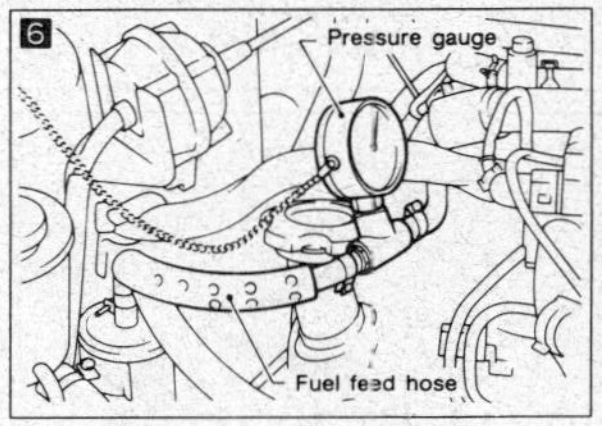

7
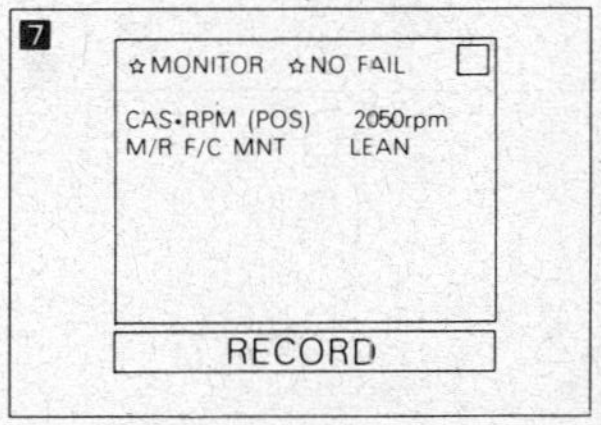

7
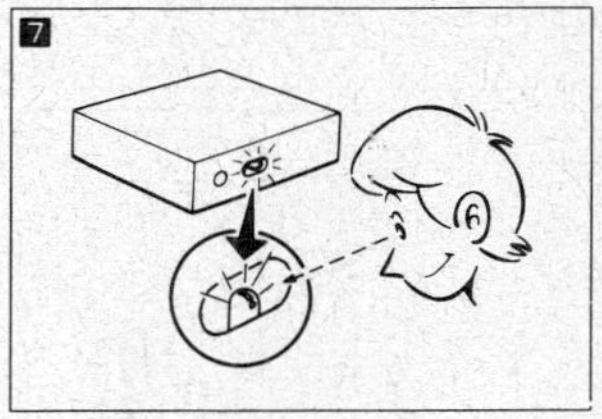

Diagnostic Procedure 3 — Unstable Idle (Cont'd)

Ⓐ ↓

5

CHECK SPARK PLUGS.
Remove the spark plugs and check for fouling, etc.

- N.G. → Repair or replace spark plug(s).
- O.K. ↓

6

CHECK FUEL PRESSURE.
1. Release fuel pressure to zero.
2. Install fuel pressure gauge and check fuel pressure.
   At idle:
   Approx. 235 kPa
   ($2.4\ kg/cm^2$, 34 psi)

- N.G. → Check fuel pump and circuit.
- O.K. ↓

7

CHECK EXHAUST GAS SENSOR.
1. See "M/R F/C MNT" in "Data monitor" mode.
2. Maintaining engine at 2,000 rpm under no-load (engine is warmed up sufficiently.), check that the monitor fluctuates between "LEAN" and "RICH" more than 5 times during 10 seconds.
   RICH→LEAN→RICH→
   1 time 2 times
   LEAN→RICH.......

— OR —

1. Set "Exhaust gas sensor monitor" in the self-diagnostic Mode II.
2. Maintaining engine at 2,000 rpm under no-load, check to make sure that RED L.E.D. on the E.C.U. goes ON and OFF more than 5 times during 10 seconds.

- N.G. → Replace exhaust gas sensor.
- O.K. ↓

(Go to Ⓑ on next page)

Diagnostic Procedure 3 — Unstable Idle (Cont'd)

8 

9 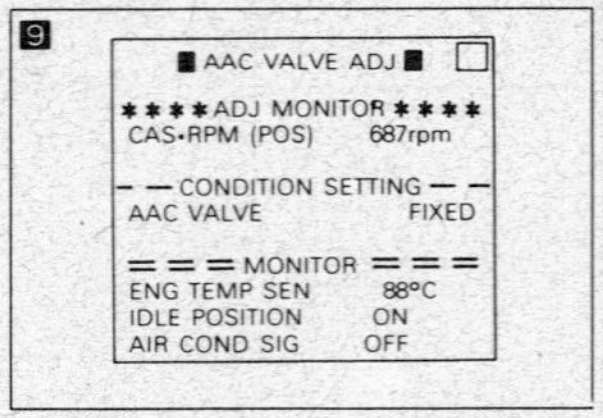

9 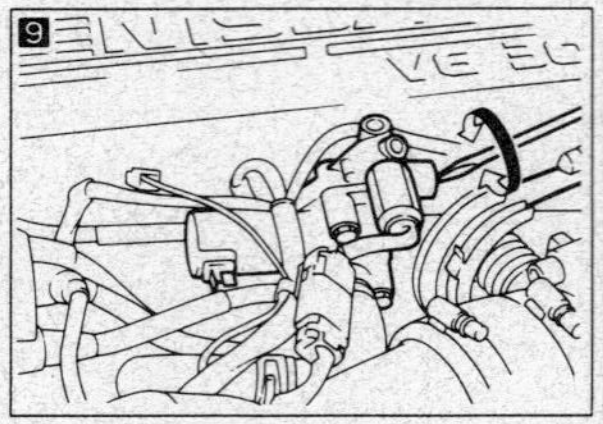

10 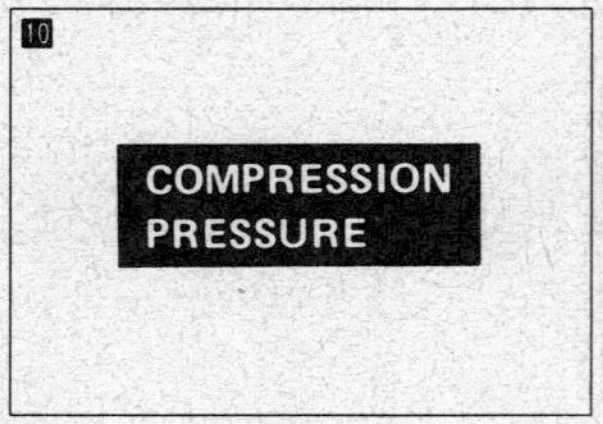

Ⓑ

8 CHECK FOR INTAKE AIR LEAK.
When pinching blow-by hose (lowering the blow-by air supply), does the engine speed rise?
- Yes → Discover air leak location and repair.
- No ↓

9 CHECK IDLE ADJ. SCREW CLOGGING.
1. Perform "AAC VALVE ADJ" in "WORK SUPPORT" mode.
2. Can you set engine rpm at 700±50 rpm (in "N" position) by turning idle adjusting screw?

— OR —

1. Disconnect A.A.C. valve harness connector.
2. Can you set engine rpm at 700±50 rpm (in "N" position) by turning idle adjusting screw?
- No → Check for IAS clogging or throttle valve clogging.
- Yes ↓

10 CHECK COMPRESSION PRESSURE.
- Check compression pressure.

Standard: kPa (kg/cm², psi)/300 rpm
1,196 (12.2, 173)
Minimum: kPa (kg/cm², psi)/300 rpm
883 (9.0, 128)
Difference between each cylinder:
kPa (kg/cm²,psi)/300 rpm
98 (1.0, 14)
- N.G. → Check pistons, piston rings, valves, valve seats and cylinder head gaskets.
- O.K. ↓

11 CHECK E.C.U. HARNESS CONNECTOR.
Check the E.C.U. pin terminals for damage or poor connection of E.C.U. harness connector.
- N.G. → Repair or replace.
- O.K. ↓

12 TRY A KNOWN GOOD E.C.U.
↓
INSPECTION END

Diagnostic Procedure 4 — Hard to Start or Impossible to Start when the Engine is Cold

1 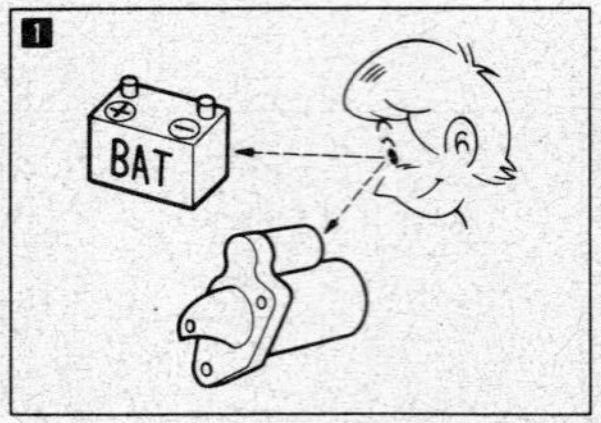

2 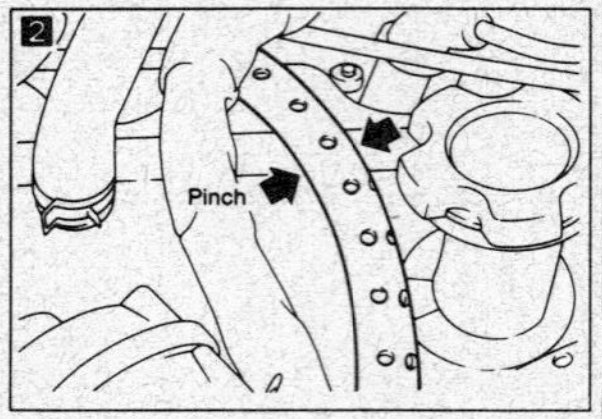

3 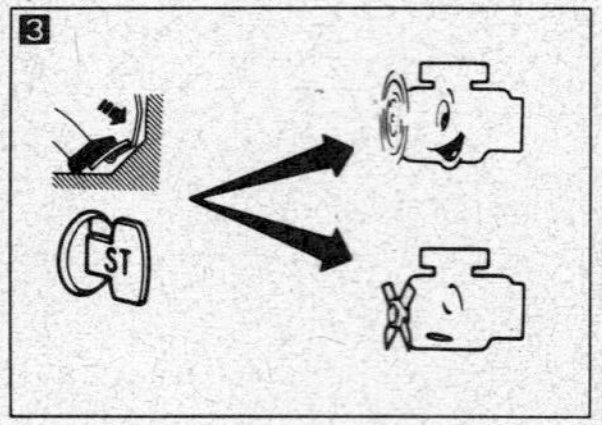

4 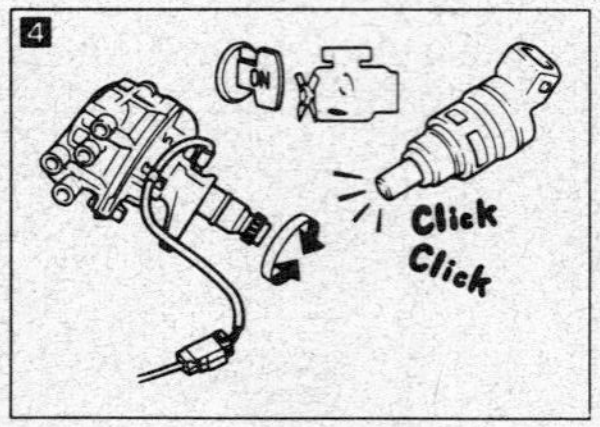

5 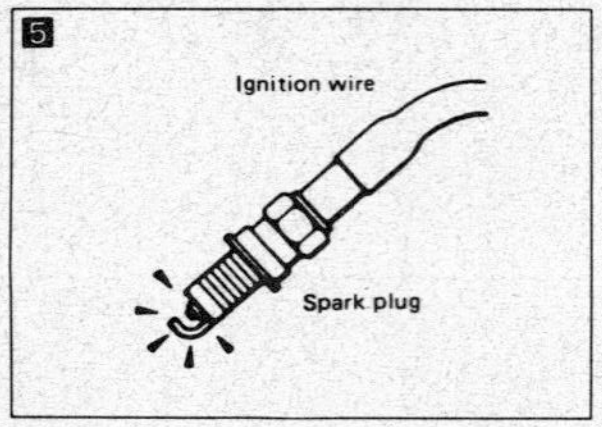

1 CHECK BATTERY AND STARTER.
Check battery and starter condition.
- N.G. → Repair or replace.
- O.K. ↓

2 CHECK FUEL PRESSURE.
1. Pinch fuel feed hose with fingers.
2. When cranking the engine, is there any pressure on the fuel feed hose?
- No → Check fuel pump and circuit.
- Yes ↓

3 CHECK AIR REGULATOR AND A.A.C. VALVE.
When pressing accelerator pedal fully, can you start the engine.
- Yes → Check A.A.C. valve, air regulator and circuits.
- No ↓

4 CHECK INJECTOR.
1. Remove distributor from engine. (Crank angle sensor harness connector should remain connected.)
2. Turn ignition switch ON. (Do not start engine.)
3. When rotating distributor shaft slowly by hand, does each injector make an operating sound?
- No → Check injector(s) and circuit(s).
- Yes ↓

5 CHECK IGNITION SPARK.
1. Disconnect ignition wire from spark plug.
2. Connect a known good spark plug to the ignition wire.
3. Place end of spark plug against a suitable ground and crank engine.
4. Check for spark.
- N.G. → Check ignition coil, power transistor unit and their circuits.
- O.K. ↓

(Go to Ⓐ on next page.)

Diagnostic Procedure 4 — Hard to Start or Impossible to Start when the Engine is Cold (Cont'd)

6

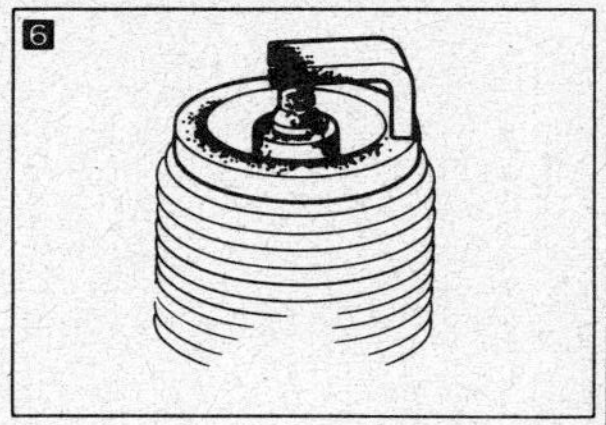

(A)

6 **CHECK SPARK PLUGS.**
Remove the spark plugs and check for fouling, etc.
- N.G. → Repair or replace spark plug(s).
- O.K. ↓

7 **CHECK E.C.U. HARNESS CONNECTOR.**
Check the E.C.U. pin terminals for damage or poor connection of E.C.U. harness connector.
- N.G. → Repair or replace.
- O.K. ↓

8 **CHECK E.C.U. POWER SUPPLY AND GROUND CIRCUIT.**
- N.G. → Repair or replace.
- O.K. ↓

9 **TRY A KNOWN GOOD E.C.U.**
↓
INSPECTION END

Diagnostic Procedure 5 — Hard to Start or Impossible to Start when the Engine is Hot

1

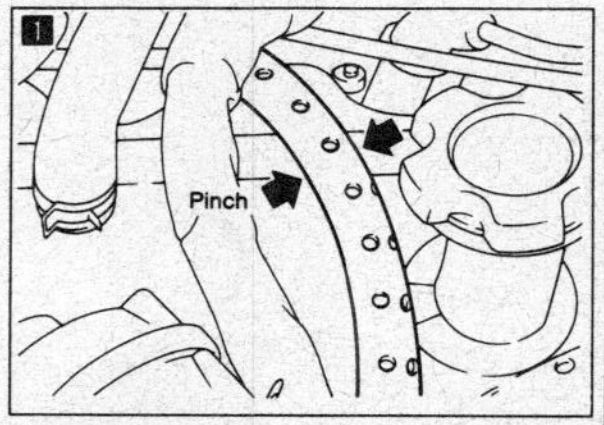

2

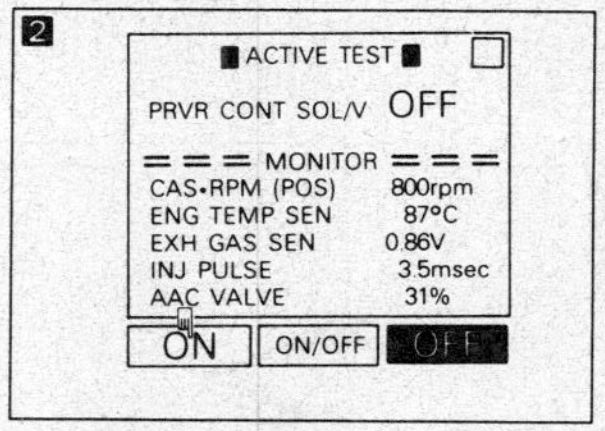

2

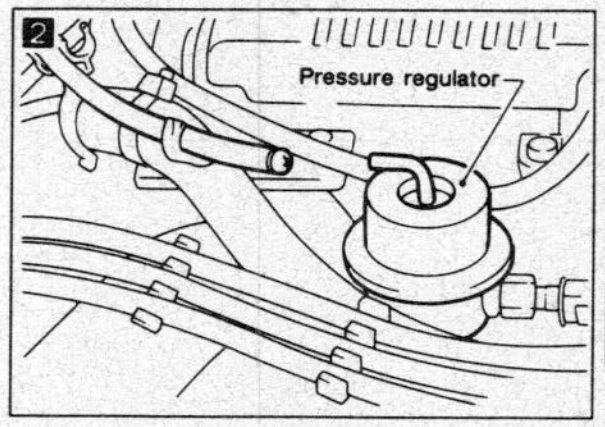

3

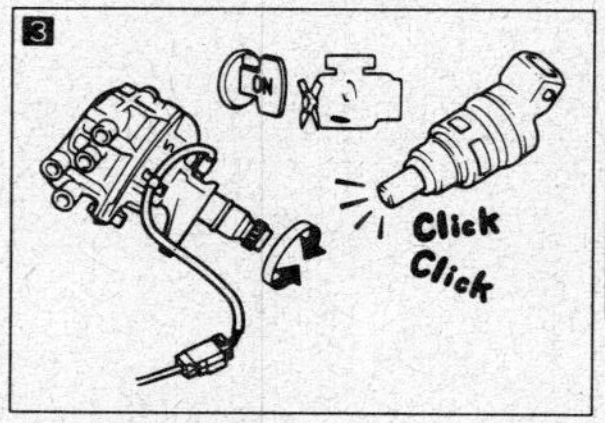

4

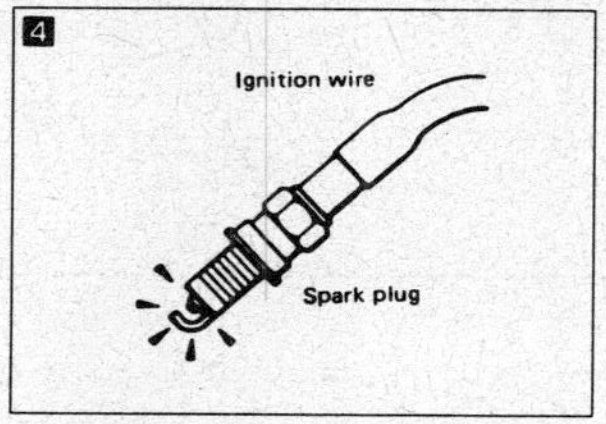

1 **CHECK FUEL PRESSURE.**
1. Pinch fuel feed hose with fingers.
2. When cranking the engine, is there any pressure on the fuel feed hose?
- No → Check fuel pump and circuit.
- Yes ↓

2 **CHECK FUEL VAPOR.**
1. Select "PRVR CONT SOL VALVE" in "ACTIVE TEST" mode.
2. After touching "ON", can you start the engine?

— OR —

1. Disconnect fuel pressure regulator vacuum hose and plug hose.
2. Can you start engine?
- Yes → Check fuel properties.
- No ↓

3 **CHECK INJECTOR.**
1. Remove distributor from engine. (Crank angle sensor harness connector should remain connected.)
2. Turn ignition switch ON. (Do not start engine.)
3. When rotating distributor shaft slowly by hand, does each injector make an operating sound?
- No → Check injector(s) and circuit(s).
- Yes ↓

4 **CHECK IGNITION SPARK.**
1. Disconnect ignition wire from spark plug.
2. Connect a known good spark plug to the ignition wire
3. Place end of spark plug against a suitable ground and crank engine.
4. Check for spark.
- N.G. → Check ignition coil, power transistor unit and circuits.
- O.K. ↓

(Go to (A) on next page.)

Diagnostic Procedure 5 — Hard to Start or Impossible to Start when the Engine is Hot (Cont'd)

(A)

5 CHECK E.C.U. HARNESS CONNECTOR.
Check the E.C.U. pin terminals for damage or poor connection of E.C.U. harness connector.
- N.G. → Repair or replace.
- O.K. ↓

6 CHECK E.C.U. POWER SUPPLY AND GROUND CIRCUIT.
- N.G. → Repair or replace.
- O.K. ↓

7 TRY A KNOWN GOOD E.C.U.
↓
INSPECTION END

Diagnostic Procedure 6 — Hard to Start or Impossible to Start under Normal Conditions

1
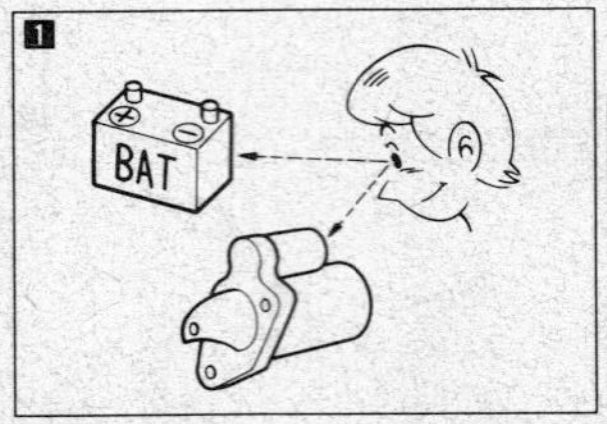

1 CHECK BATTERY AND STARTER.
Check battery and starter operation.
- N.G. → Repair or replace.
- O.K. ↓

2
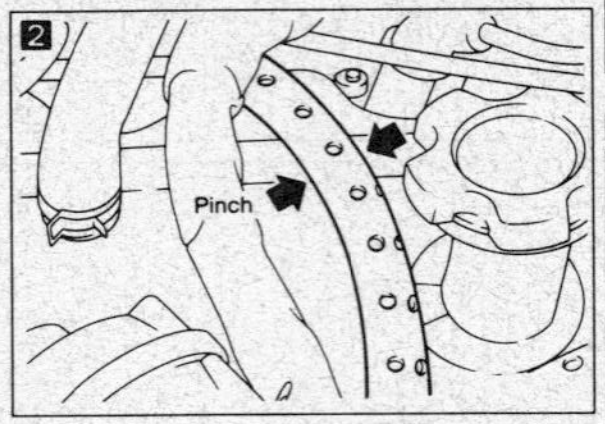

2 CHECK FUEL PRESSURE.
1. Pinch fuel feed hose with fingers.
2. When cranking the engine, is there any pressure on the fuel feed hose?
- No → Check fuel pump and circuit.
- Yes ↓

3
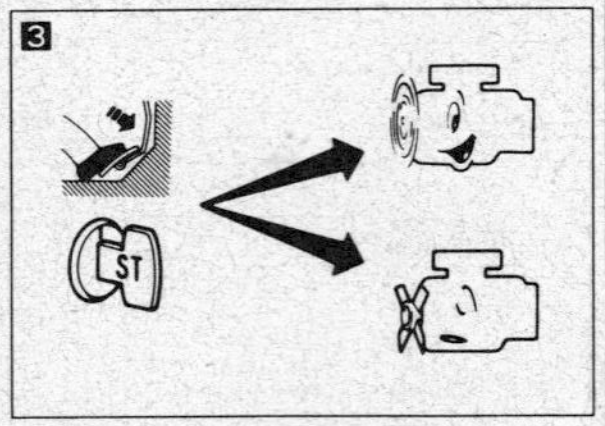

3 CHECK INJECTOR FOR LEAKAGE.
When pressing accelerator pedal fully, can you start the engine.
- Yes → Check injector(s) for leakage.
- No ↓

4
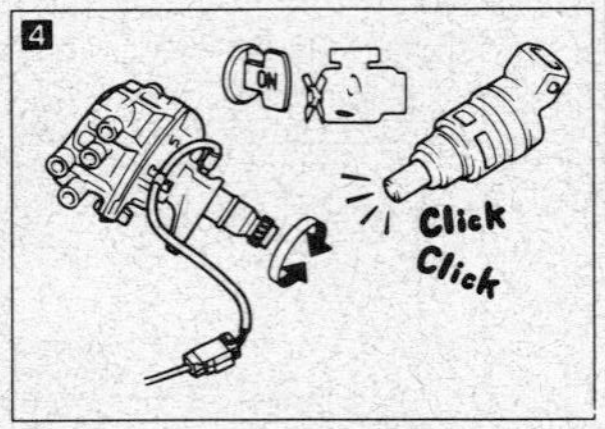

4 CHECK INJECTOR.
1. Remove distributor from engine. (Crank angle sensor harness connector should remain connected.)
2. Turn ignition switch ON. (Do not start engine.)
3. When rotating distributor shaft slowly by hand, does each injector make an operating sound?
- No → Check injector(s) for circuits.
- Yes ↓

5
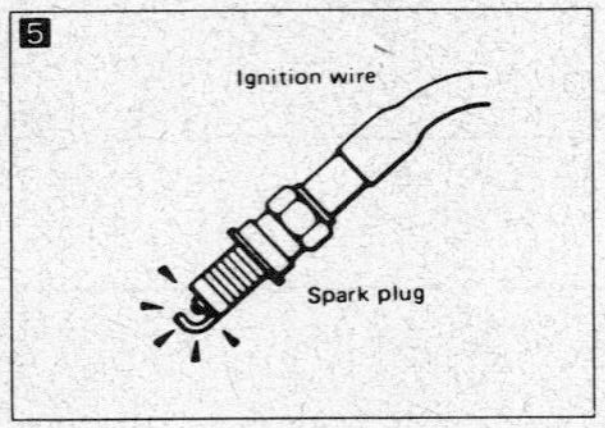

5 CHECK IGNITION SPARK.
1. Disconnect ignition wire from spark plug.
2. Connect a known good spark plug to the ignition wire.
3. Place end of spark plug against a suitable ground and crank engine.
4. Check for spark.
- N.G. → Check ignition coil, power transistor unit and circuits.
- O.K. ↓

(Go to (A) on next page.)

Diagnostic Procedure 6 — Hard to Start or Impossible to Start under Normal Conditions (Cont'd)

6
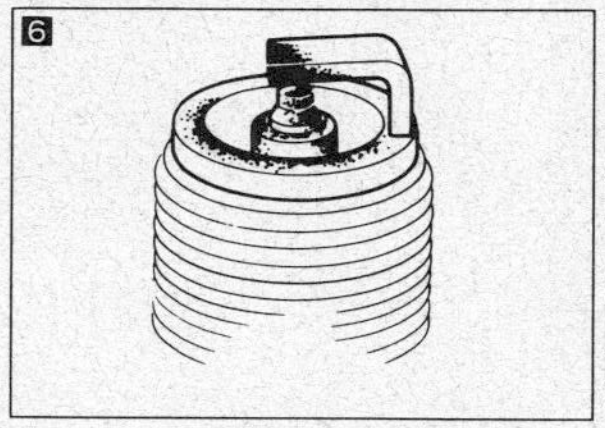

7
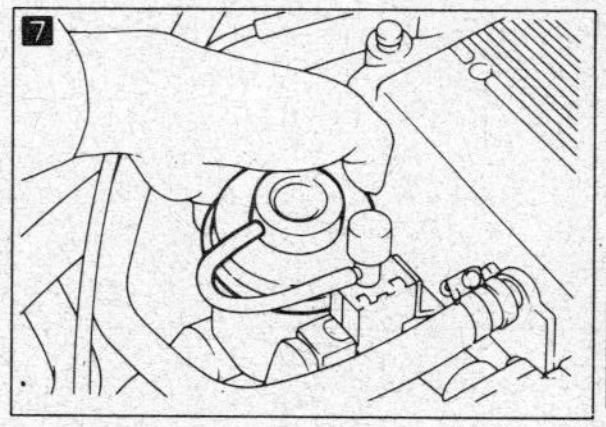

Ⓐ

6 CHECK SPARK PLUGS.
Remove the spark plugs and check for fouling, etc.
- N.G. → Repair or replace spark plug(s).
- O.K. ↓

7 CHECK E.G.R. CONTROL VALVE.
Check E.G.R. control valve for sticking.
- N.G. → Repair or replace.
- O.K. ↓

8 CHECK E.C.U. HARNESS CONNECTOR.
Check the E.C.U. pin terminals for damage or poor connection of E.C.U. harness connector.
- N.G. → Repair or replace.
- O.K. ↓

9 CHECK E.C.U. POWER SUPPLY AND GROUND CIRCUIT.
- N.G. → Repair or replace.
- O.K. ↓

10 TRY A KNOWN GOOD E.C.U.
- Trouble is fixed. → Replace E.C.U.
- Trouble is not fixed. ↓

11 CHECK TIMING BELT FOR PROPER INSTALLATION.
- N.G. → Replace timing belt.
- O.K. ↓

INSPECTION END

Diagnostic Procedure 7 — Hesitation when the Engine is Hot

1
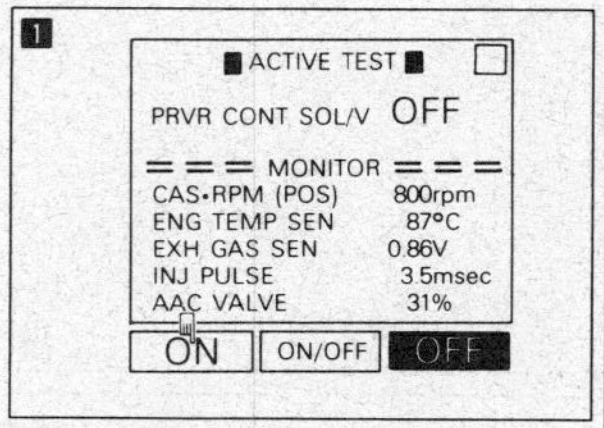

1
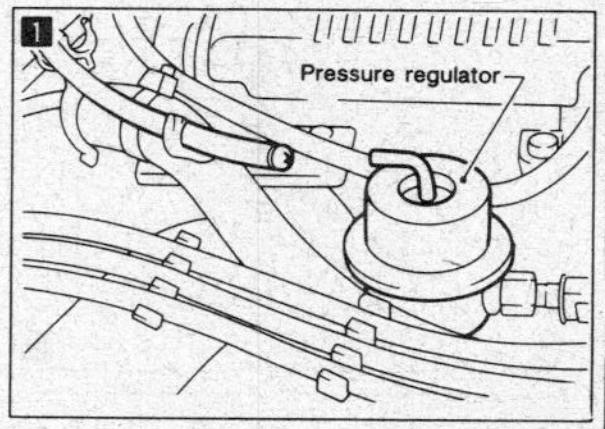

2
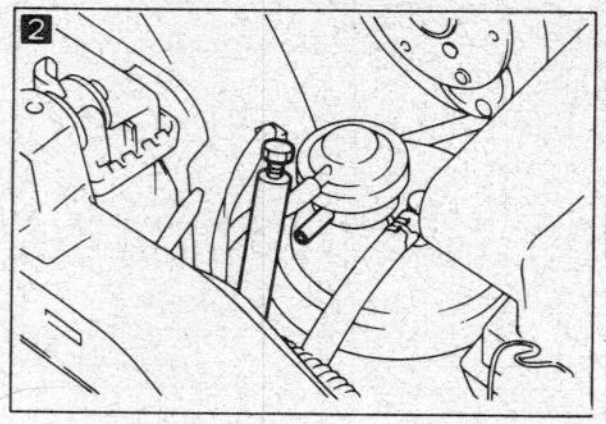

1 CHECK FUEL VAPOR.
1. Select "PRVR CONT SOL VALVE" in "ACTIVE TEST" mode.
2. After touching "ON", perform cruise test.
3. Does the hesitation disappear?

—OR—

1. Disconnect fuel pressure regulator vacuum hose and plug hose.
2. Perform cruise test.
3. Does the hesitation disappear?

- Yes → Check fuel properties.
- No ↓

2 CHECK CANISTER PURGE.
1. Disconnect canister purge line hose and plug hose.
2. Perform cruise test.
3. Does the hesitation disappear?

- Yes → Check purge and vacuum line.
- No ↓

INSPECTION END

Diagnostic Procedure 8 — Hesitation when the Engine is Cold

1
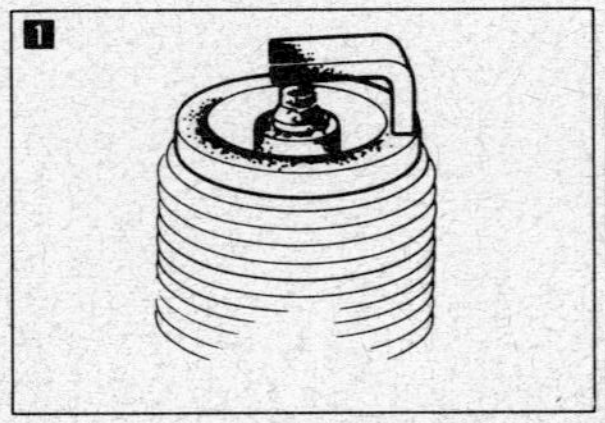

2

1 **CHECK SPARK PLUGS.**
Remove spark plugs and check for fouling, etc.
- N.G. → Repair or replace spark plug(s).
- O.K. ↓

2 **CHECK FOR INTAKE AIR LEAK.**
When pinching blow-by hose (lowering the blow-by air supply), does the engine speed rise?
- Yes → Discover air leak location and repair.
- No ↓

3 **TRY A KNOWN GOOD AIR FLOW METER.**
- Trouble is fixed. → Replace air flow meter.
- Trouble is not fixed. ↓

4 **CHECK FOR INTAKE VALVE DEPOSITS.**
If there are deposits on intake valves, remove them.
↓

INSPECTION END

Diagnostic Procedure 9 — Hesitation under Normal Conditions

1
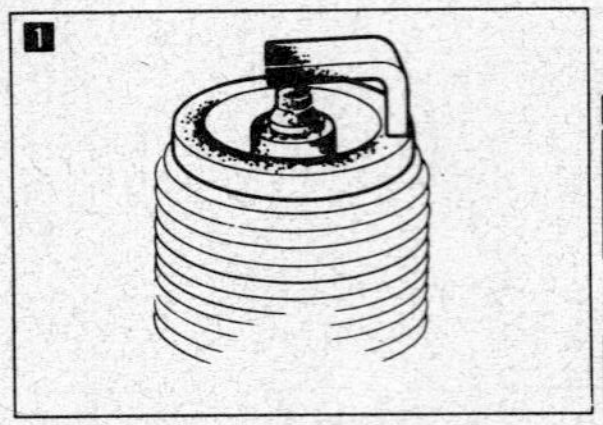

2
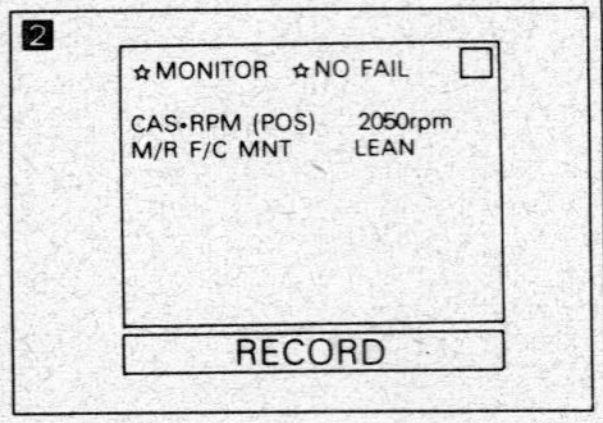

2
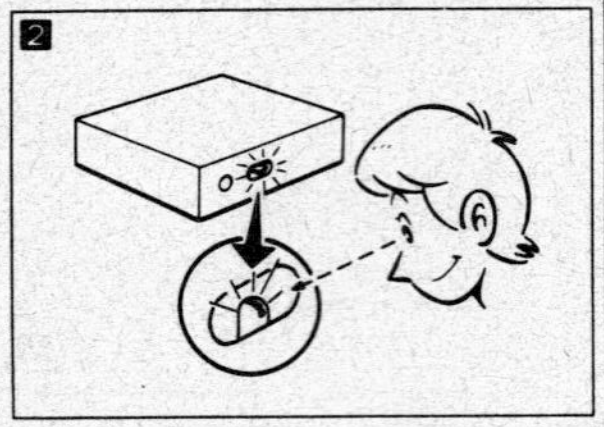

3
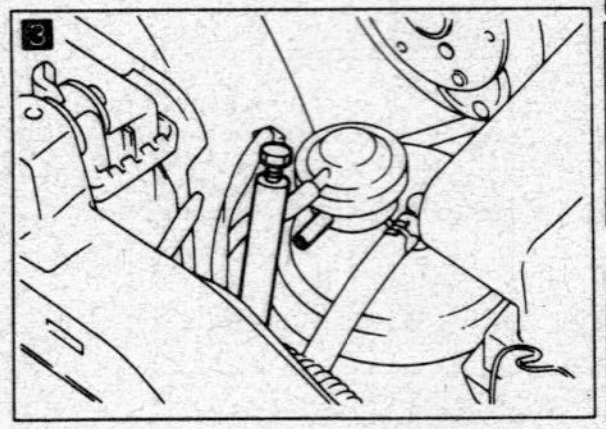

4

1 **CHECK SPARK PLUGS.**
Remove spark plugs and check for fouling, etc.
- N.G. → Repair or replace spark plug(s).
- O.K. ↓

2 **CHECK EXHAUST GAS SENSOR.**
1. See "M/R F/C MNT" in "DATA MONITOR" mode.
2. Maintaining engine at 2,000 rpm under no-load (with engine warmed up sufficiently.), check to make sure that the monitor fluctuates between "LEAN" and "RICH" more than 5 times during 10 seconds.

RICH→LEAN→RICH→ (1 time) LEAN→RICH....... (2 times)

— OR —

1. Set "Exhaust gas sensor monitor" in the self-diagnostic Mode II. (See page EF & EC-39.)
2. Maintaining engine at 2,000 rpm under no load, check that RED L.E.D. on the E.C.U. goes ON and OFF more than 5 times during 10 seconds.

- N.G. → Replace exhaust gas sensor.
- O.K. ↓

3 **CHECK CANISTER PURGE.**
1. Disconnect canister purge line hose and plug hose.
2. Perform cruise test.
3. Does the hesitation disappear?

- Yes → Check purge and vacuum line.
- No ↓

4 **CHECK FOR INTAKE AIR LEAK.**
When pinching blow-by hose (lowering the blow-by air supply), does the engine speed rise?
- Yes → Discover air leak location and repair.
- No ↓

INSPECTION END

Diagnostic Procedure 10 — Engine Stalls when Turning

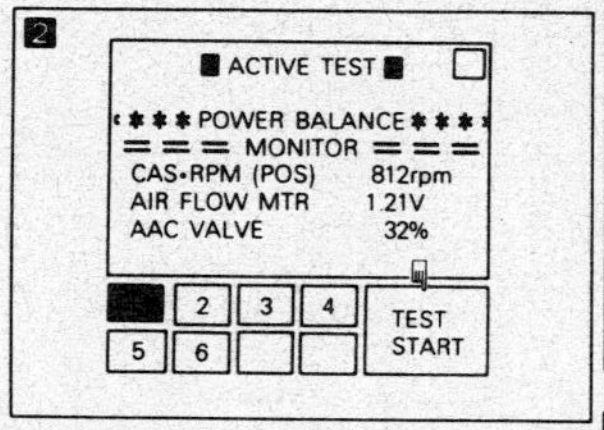

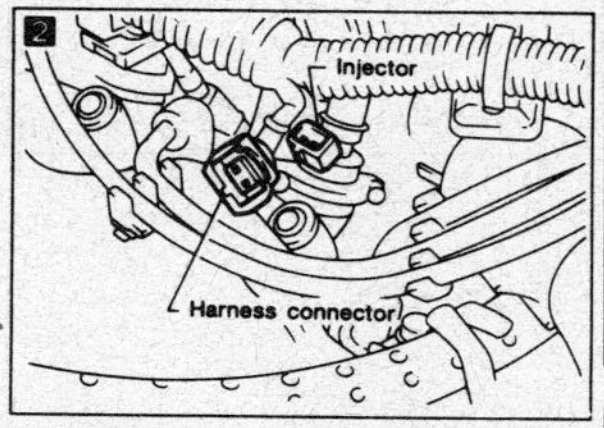

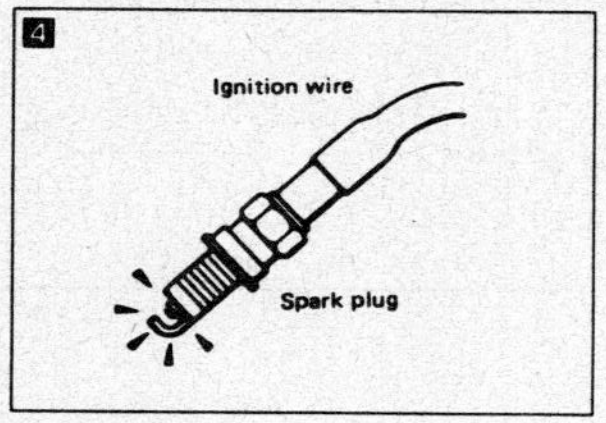

1 **CHECK FUEL LEVEL.**
Check to see that there is enough fuel in tank.

- N.G. → Fill fuel tank with fuel.
- O.K. ↓

2 **PERFORM POWER BALANCE TEST.**
1. Perform "POWER BALANCE" in "ACTIVE TEST" mode.
2. Is there any cylinder which does not produce a momentary engine speed drop?

— OR —

When disconnecting each injector harness connector one at a time, is there any cylinder which does not produce a momentary engine speed drop?

- No → Go to 5.
- Yes ↓

3 **CHECK INJECTOR.**
1. Remove distributor from engine. (Crank angle sensor harness connector should remain connected.)
2. Turn ignition switch ON. (Do not start engine.)
3. When rotating distributor shaft slowly by hand, does each injector make an operating sound?

- No → Check injector(s) and circuit(s).
- Yes ↓

4 **CHECK IGNITION SPARK.**
1. Disconnect ignition wire from spark plug.
2. Connect a known good spark plug to the ignition wire.
3. Place end of spark plug against a suitable ground and crank engine.
4. Check for spark.

- N.G. → Check ignition coil, power transistor unit and circuits.
- O.K. ↓

(Go to Ⓐ on next page.)

Diagnostic Procedure 10 — Engine Stalls when Turning (Cont'd)

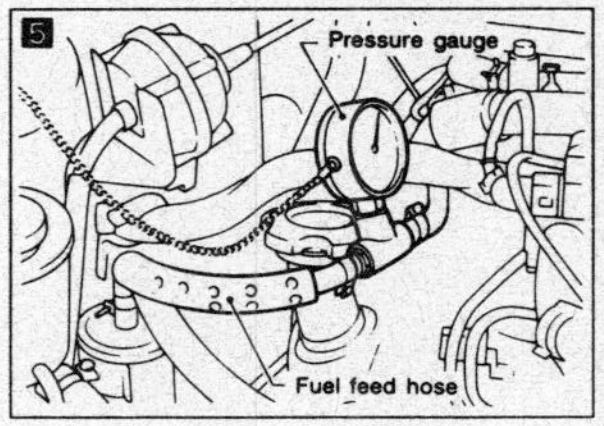

Ⓐ ↓

5 **CHECK FUEL PRESSURE.**
1. Release fuel pressure to zero.
2. Install fuel pressure gauge and check fuel pressure.
   At idle:
   Approx. 235 kPa (2.4 kg/cm², 34 psi)
   The moment throttle valve is fully open:
   Approx. 294 kPa (3.0 kg/cm², 43 psi)

- N.G. → Check fuel pressure regulator diaphragm.
- O.K. ↓

6 **CHECK E.C.U. HARNESS CONNECTOR.**
Check the E.C.U. pin terminals for damage or poor connection of E.C.U. harness connector.

- N.G. → Repair or replace.
- O.K. ↓

7 **CHECK E.C.U. POWER SUPPLY AND GROUND CIRCUIT.**

- N.G. → Repair or replace.
- O.K. ↓

8 **TRY A KNOWN GOOD E.C.U.**

↓

INSPECTION END

Diagnostic Procedure 11 — Engine Stalls when the Engine is Hot

1

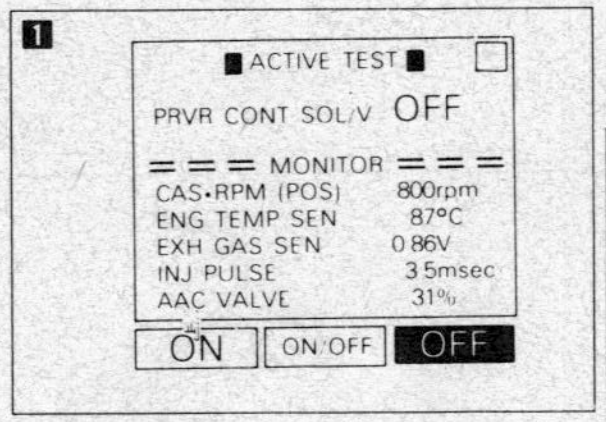

1

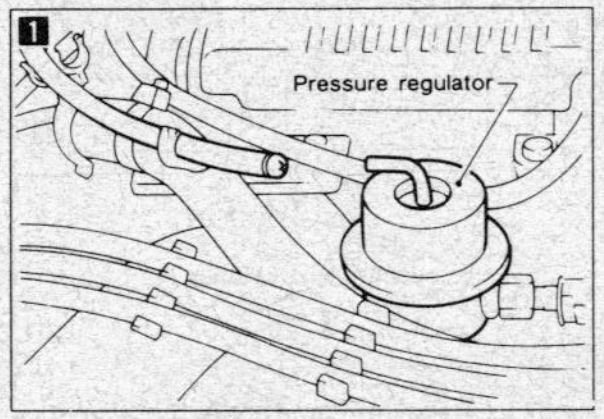

2

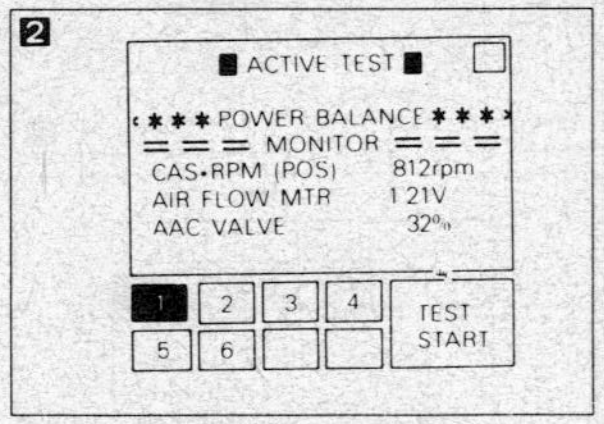

2

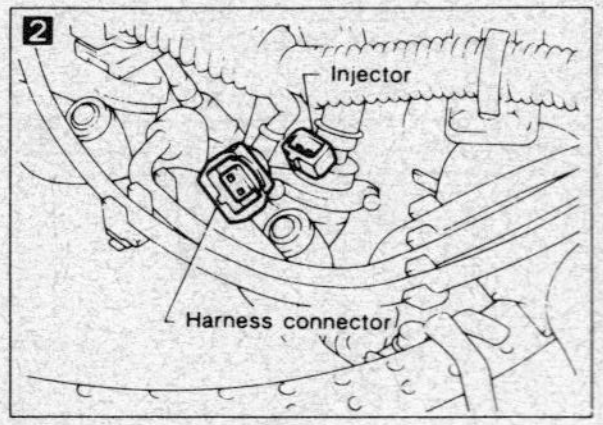

3

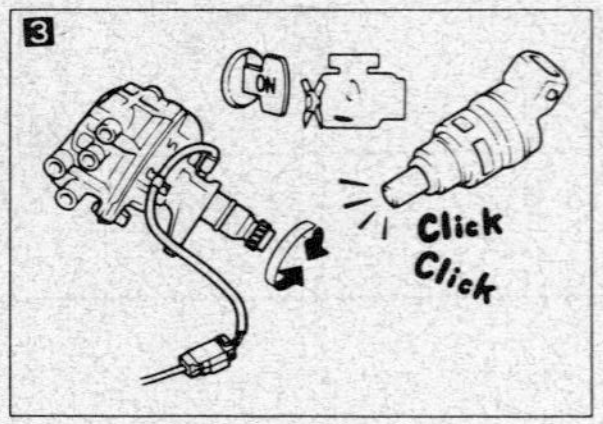

1 CHECK FUEL VAPOR.

1. Select "PRVR CONT SOL VALVE" in "ACTIVE TEST" mode.
2. After touching "ON", perform cruise test.
3. Does the engine stall disappear?

— OR —

1. Disconnect fuel pressure regulator vacuum hose and plug hose.
2. Perform cruise test.
3. Does the engine stall disappear?

Yes → Check fuel properties.

No ↓

2 PERFORM POWER BALANCE TEST.

1. Perform "POWER BALANCE" in "ACTIVE TEST" mode.
2. Is there any cylinder which does not produce a momentary engine speed drop?

— OR —

When disconnecting each injector harness connector one at a time, is there any cylinder which does not produce a momentary engine speed drop?

No → Go to 5.

Yes ↓

3 CHECK INJECTOR.

1. Remove distributor from engine. (Crank angle sensor harness connector should remain connected.)
2. Turn ignition switch ON. (Do not start engine.)
3. When rotating distributor shaft slowly by hand, does each injector make an operating sound?

No → Check injector(s) and circuit(s).

Yes ↓

(Go to A on next page.)

Diagnostic Procedure 11 — Engine Stalls when the Engine is Hot (Cont'd)

4

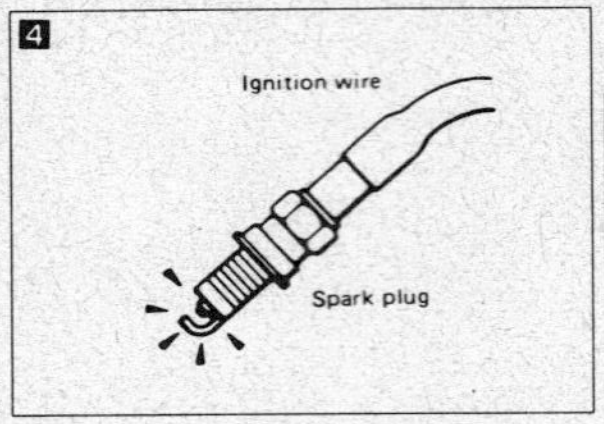

5

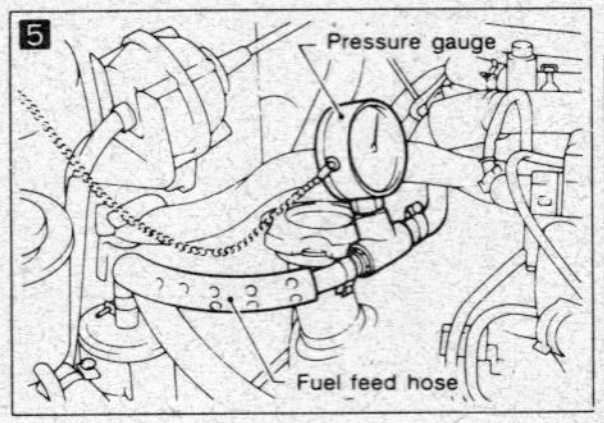

A ↓

4 CHECK IGNITION SPARK.

1. Disconnect ignition wire from spark plug.
2. Connect a known good spark plug to the ignition wire.
3. Place end of spark plug against a suitable ground and crank engine.
4. Check for spark.

N.G. → Check ignition coil, power transistor unit and their circuits.

O.K. ↓

5 CHECK FUEL PRESSURE.

1. Release fuel pressure to zero.
2. Install fuel pressure gauge and check fuel pressure.

At idle:
Approx. 235 kPa ($2.4 kg/cm^2$, 34 psi)
The moment throttle valve is fully open:
Approx. 294 kPa ($3.0 kg/cm^2$, 43 psi)

N.G. → Check fuel pressure regulator diaphragm.

O.K. ↓

6 CHECK E.C.U. HARNESS CONNECTOR.

Check the E.C.U. pin terminals for damage or poor connection of E.C.U. harness connector.

N.G. → Repair or replace.

O.K. ↓

7 CHECK E.C.U. POWER SUPPLY AND GROUND CIRCUIT.

N.G. → Repair or replace.

O.K. ↓

8 TRY A KNOWN GOOD E.C.U.

Trouble is fixed. → Replace E.C.U.

Trouble is not fixed. ↓

9 CHECK TIMING BELT FOR PROPER INSTALLATION.

N.G. → Replace timing belt and install new one.

O.K. ↓

INSPECTION END

Diagnostic Procedure 12 — Engine Stalls when the Engine is Cold

1

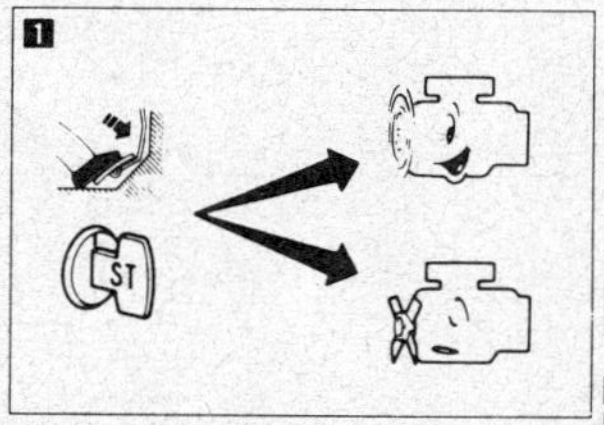

2

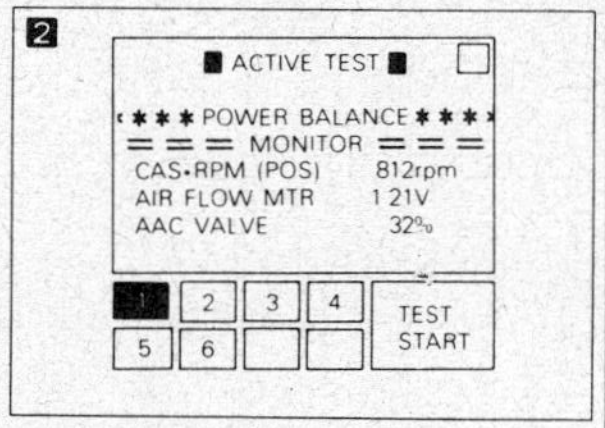

2

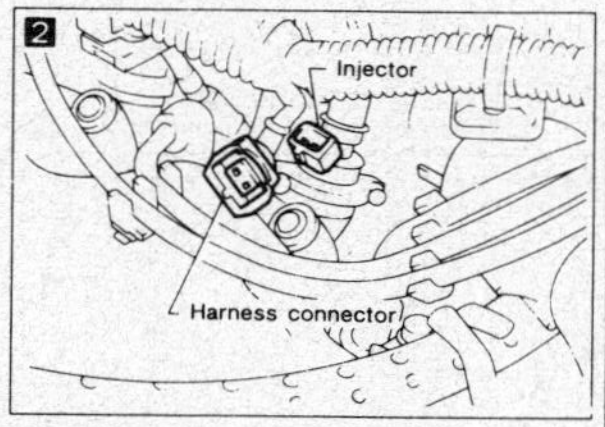

3

4

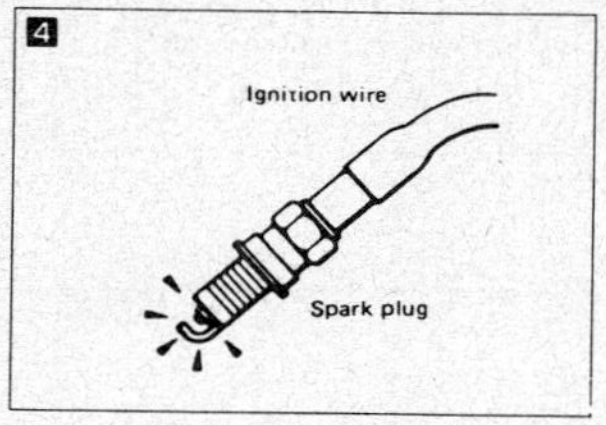

1 CHECK AIR REGULATOR AND A.A.C. VALVE.
When the engine is cold, can you start the engine when pressing accelerator pedal fully?

- N.G. → Check A.A.C. valve, air regulator and circuits.
- O.K. ↓

2 PERFORM POWER BALANCE TEST.
1. Perform "POWER BALANCE" in "ACTIVE TEST" mode.
2. Is there any cylinder which does not produce a momentary engine speed drop?

— OR —

When disconnecting each injector harness connector one at a time, is there any cylinder which does not produce a momentary engine speed drop?

- N.G. → Go to 6.
- O.K. ↓

3 CHECK INJECTOR.
1. Remove distributor from engine. (Crank angle sensor harness connector should remain connected.)
2. Turn ignition switch ON. (Do not start engine.)
3. When rotating distributor shaft slowly by hand, does each injector make an operating sound?

- N.G. → Check injector(s) and circuit(s).
- O.K. ↓

4 CHECK IGNITION SPARK.
1. Disconnect ignition wire from spark plug.
2. Connect a known good spark plug to the ignition wire.
3. Place end of spark plug against a suitable ground and crank engine.
4. Check for spark.

- N.G. → Check ignition coil, power transistor unit and circuits.
- O.K. ↓

(Go to Ⓐ on next page.)

Diagnostic Procedure 12 — Engine Stalls when the Engine is Cold (Cont'd)

5

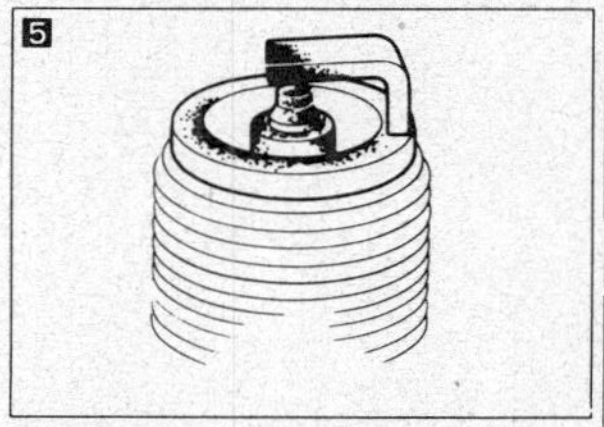

6

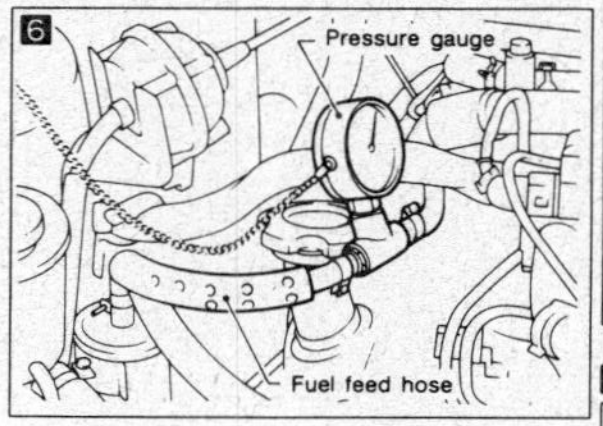

Ⓐ ↓

5 CHECK SPARK PLUGS.
Remove the spark plugs and check for fouling, etc.

- N.G. → Repair or replace spark plug(s).
- O.K. ↓

6 CHECK FUEL PRESSURE.
1. Release fuel pressure to zero.
2. Install fuel pressure gauge and check fuel pressure.
   At idle:
   Approx. 235 kPa (2.4 kg/cm², 34 psi)
   The moment throttle valve is fully open:
   Approx. 294 kPa (3.0 kg/cm², 43 psi)

- N.G. → Check fuel pressure regulator diaphragm.
- O.K. ↓

7 CHECK E.C.U. HARNESS CONNECTOR.
Check the E.C.U. pin terminals for damage or poor connection of E.C.U. harness connector.

- N.G. → Repair or replace.
- O.K. ↓

8 CHECK E.C.U. POWER SUPPLY AND GROUND CIRCUIT.

- N.G. → Repair or replace.
- O.K. ↓

9 TRY A KNOWN GOOD E.C.U.

↓

INSPECTION END

Diagnostic Procedure 13 — Engine Stalls when Stepping on the Accelerator Momentarily

1
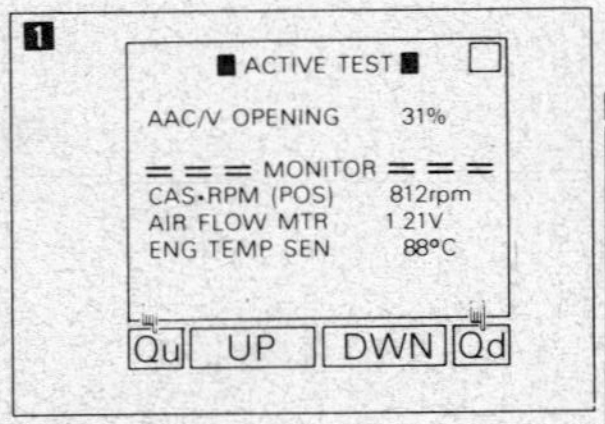

1
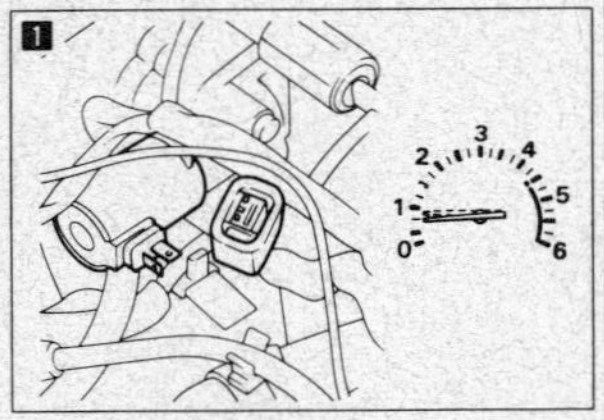

2
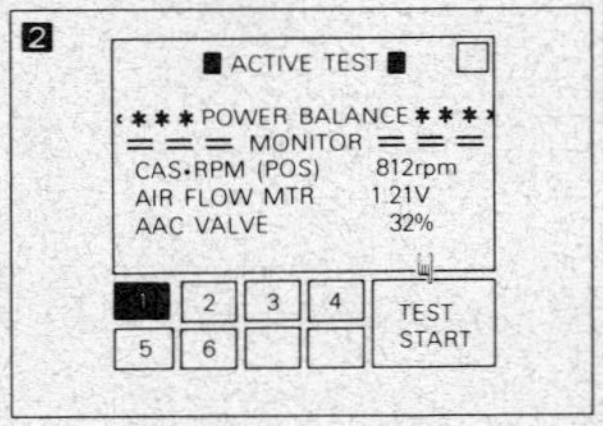

2
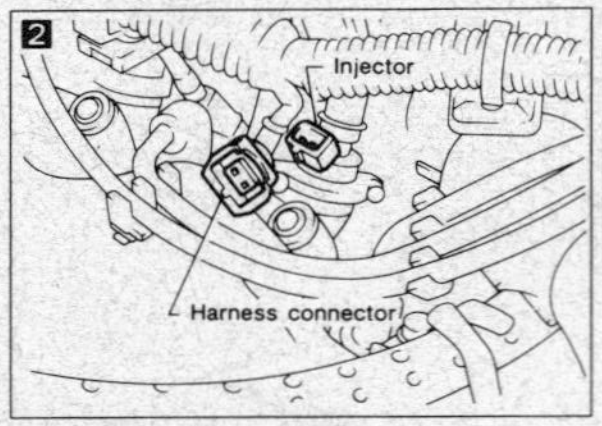

3

1 CHECK A.A.C. VALVE.

1. Select "A.A.C. VALVE OPENING" in "ACTIVE TEST" mode.
2. When touching "Qu" and "Qd", does the engine speed change according to the percent of A.A.C. valve opening?

— OR —

When disconnecting A.A.C. valve harness connector, does the engine speed drop?

- No → Check A.A.C. valve and circuit.
- Yes ↓

2 PERFORM POWER BALANCE TEST.

1. Perform "POWER BALANCE" in "ACTIVE TEST" mode.
2. Is there any cylinder which does not produce a momentary engine speed drop?

— OR —

When disconnecting each injector harness connector one at a time, is there any cylinder which does not produce a momentary engine speed drop?

- No → Go to 5.
- Yes ↓

3 CHECK INJECTOR.

1. Remove distributor from engine. (Crank angle sensor harness connector should remain connected.)
2. Turn ignition switch ON. (Do not start engine.)
3. When rotating distributor shaft slowly by hand, does each injector make an operating sound?

- No → Check injector(s) and their circuit(s).
- Yes ↓

(Go to Ⓐ on next page.)

Diagnostic Procedure 13 — Engine Stalls when Stepping on the Accelerator Momentarily (Cont'd)

4
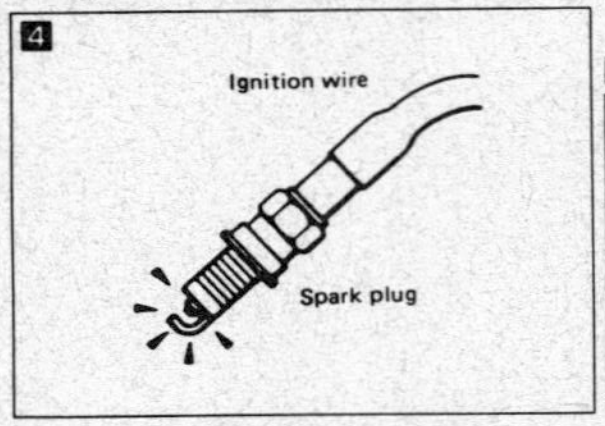

5
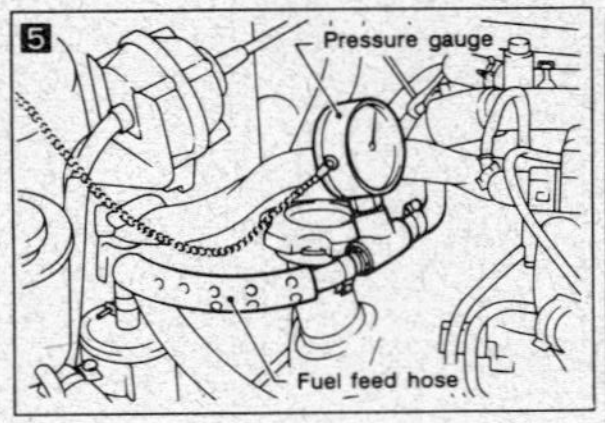

Ⓐ ↓

4 CHECK IGNITION SPARK.

1. Disconnect ignition wire from spark plug.
2. Connect a known good spark plug to the ignition wire.
3. Place end of spark plug against an earth point with engine cranking.
4. Check for spark.

- N.G. → Check ignition coil, power transistor unit and their circuits.
- O.K. ↓

5 CHECK FUEL PRESSURE.

1. Release fuel pressure to zero.
2. Install fuel pressure gauge and check fuel pressure.

At idle:
Approx. 235 kPa (2.4 kg/cm², 34 psi)
The moment throttle valve is fully open:
Approx. 294 kPa (3.0 kg/cm², 43 psi)

- N.G. → Check fuel pressure regulator diaphragm.
- O.K. ↓

6 CHECK E.C.U. HARNESS CONNECTOR.

Check the E.C.U. pin terminals for damage or poor connection of E.C.U. harness connector.

- N.G. → Repair or replace.
- O.K. ↓

7 CHECK E.C.U. POWER SUPPLY AND GROUND CIRCUIT.

- N.G. → Repair or replace.
- O.K. ↓

8 TRY A KNOWN GOOD E.C.U.

↓

INSPECTION END

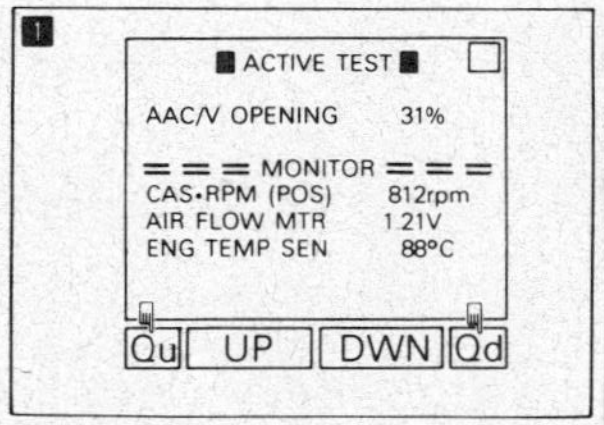

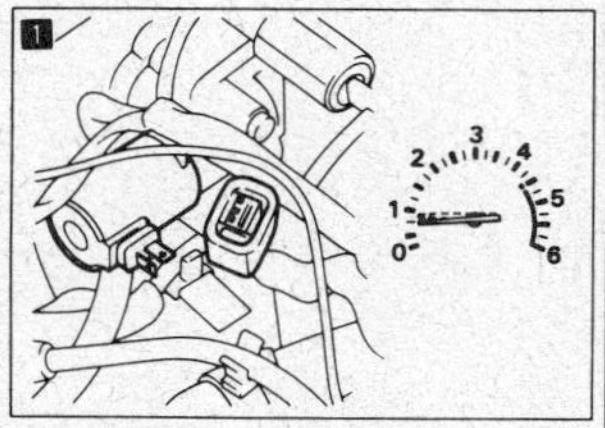

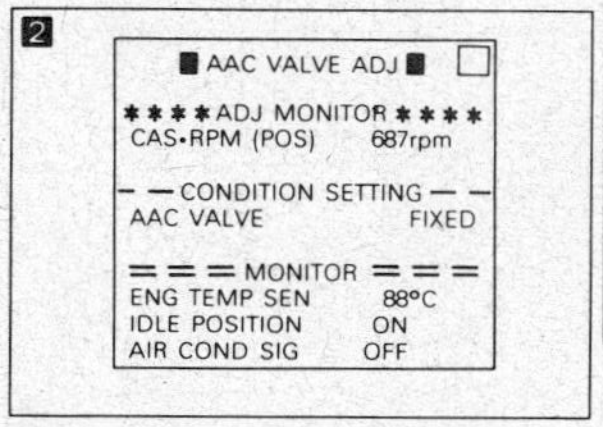

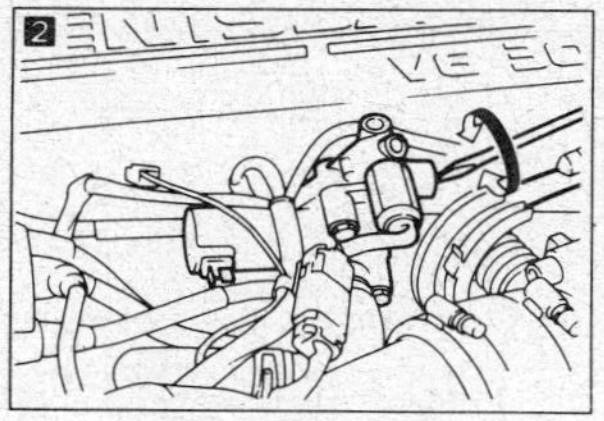

Diagnostic Procedure 14 — Engine Stalls after Decelerating

1

CHECK A.A.C. VALVE.

1. Select "A.A.C. VALVE OPENING" in "ACTIVE TEST" mode.
2. When touching "Qu" and "Qd", does the engine speed change according to the percent of A.A.C. valve opening?

— OR —

When disconnecting A.A.C. valve harness connector, does the engine speed drop?

No → Check A.A.C. valve and circuit.

Yes ↓

2

CHECK IDLE ADJ. SCREW CLOGGING.

1. Perform "A.A.C. VALVE ADJ" in "WORK SUPPORT" mode.
2. Can you set engine rpm at 700 ± 50 rpm (in "N" position) by turning idle adjusting screw?

— OR —

1. Disconnect A.A.C. valve harness connector.
2. Can you set engine rpm at 700 ± 50 rpm (in "N" position) by turning idle adjusting screw?

No → Check for IAS clogging or throttle chamber clogging.

Yes ↓

(Go to Ⓐ on next page.)

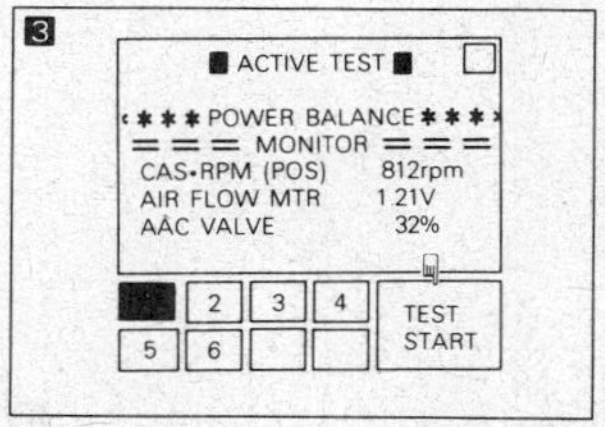

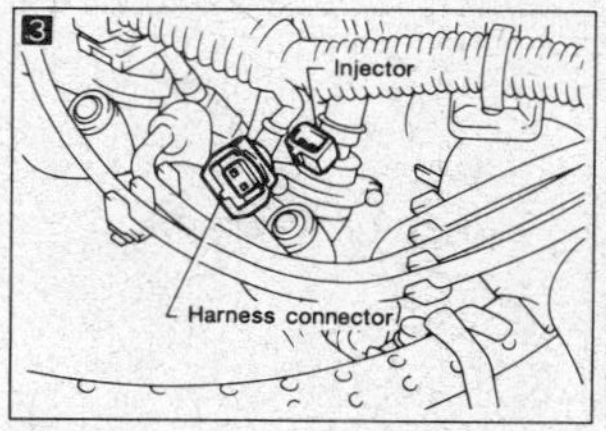

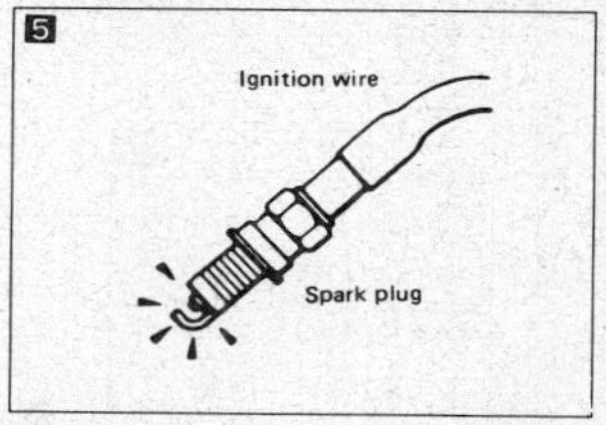

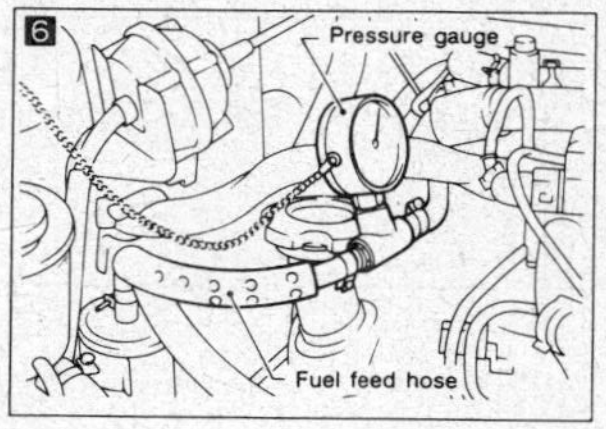

Diagnostic Procedure 14 — Engine Stalls after Decelerating (Cont'd)

Ⓐ ↓

3

PERFORM POWER BALANCE TEST.

1. Perform "POWER BALANCE" in "ACTIVE TEST" mode.
2. Is there any cylinder which does not produce a momentary engine speed drop?

— OR —

When disconnecting each injector harness connector one at a time, is there any cylinder which does not produce a momentary engine speed drop?

No → Go to **6**.

Yes ↓

4

CHECK INJECTOR.

1. Remove distributor from engine. (Crank angle sensor harness connector should remain connected.)
2. Turn ignition switch ON. (Do not start engine.)
3. When rotating distributor shaft slowly by hand, does each injector make an operating sound?

No → Check injector(s) and circuit(s).

Yes ↓

5

CHECK IGNITION SPARK.

1. Disconnect ignition wire from spark plug.
2. Connect a known good spark plug to the ignition wire.
3. Place end of spark plug against a suitable ground and crank engine.
4. Check for spark.

No → Check ignition coil, power transistor unit and circuits.

Yes ↓

6

CHECK FUEL PRESSURE.

1. Release fuel pressure to zero.
2. Install fuel pressure gauge and check fuel pressure.
   At idle:
   Approx. 235 kPa ($2.4\ kg/cm^2$, 34 psi)
   The moment throttle valve is fully open:
   Approx. 294 kPa ($3.0\ kg/cm^2$, 43 psi)

N.G. → Check fuel pressure regulator diaphragm.

O.K. ↓

(Go to Ⓑ on next page.)

Diagnostic Procedure 14 — Engine Stalls after Decelerating (Cont'd)

7

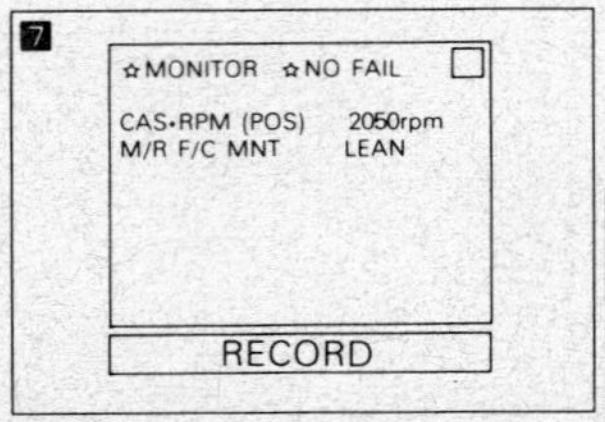

7

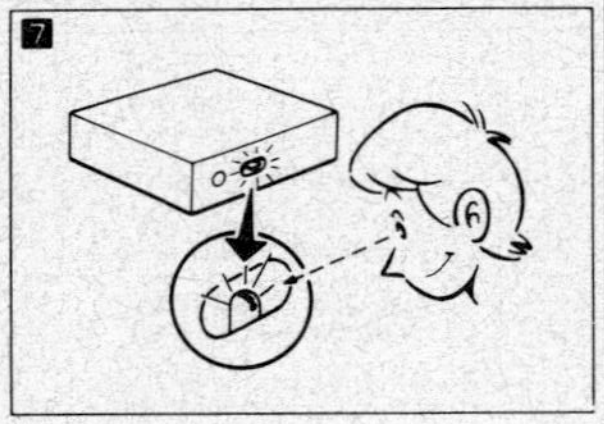

(B) ↓

7 **CHECK EXHAUST GAS SENSOR.**

1. See "M/R F/C MNT" in "DATA MONITOR" mode.
2. Maintaining engine at 2,000 rpm under no-load (with engine warmed up sufficiently.), check to make sure that the monitor fluctuates between "LEAN" and "RICH" more than 5 times during 10 seconds.
   RICH→LEAN→RICH→ (1 time) (2 times) LEAN→RICH.......

— OR —

1. Set "Exhaust gas sensor monitor" in the self- diagnostic Mode II.
2. Maintaining engine at 2,000 rpm under no load, check that RED LED on the E.C.U. goes ON and OFF more than 5 times during 10 seconds.

N.G. → Replace exhaust gas sensor.

O.K. ↓

8 **CHECK E.C.U. HARNESS CONNECTOR.** Check the E.C.U. pin terminals for damage or poor connection of E.C.U. harness connector.

N.G. → Repair or replace.

O.K. ↓

9 **CHECK E.C.U. POWER SUPPLY AND GROUND CIRCUIT.**

N.G. → Repair or replace.

O.K. ↓

10 **TRY A KNOWN GOOD E.C.U.**

↓

INSPECTION END

Diagnostic Procedure 15 — Engine Stalls when Accelerating or when Driving at Constant Speed

1

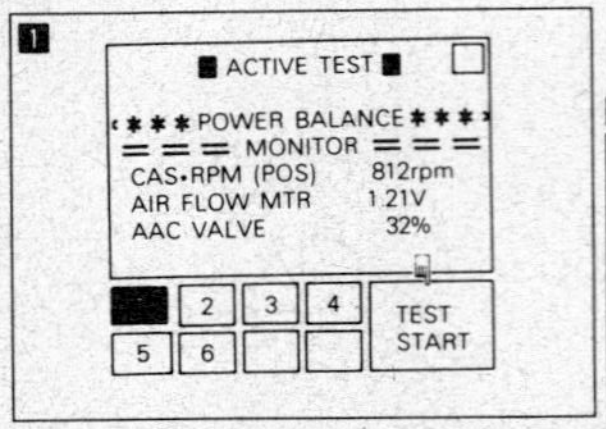

1

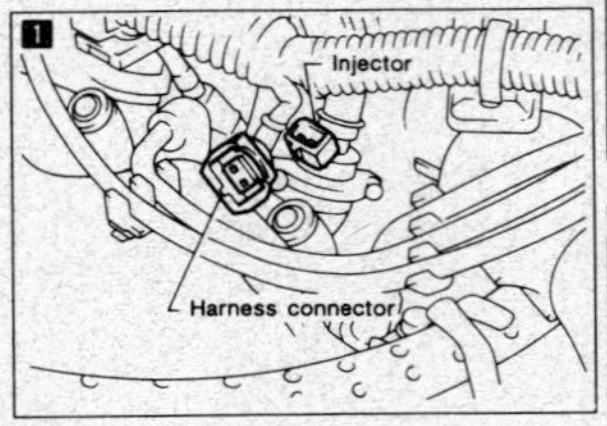

2

3

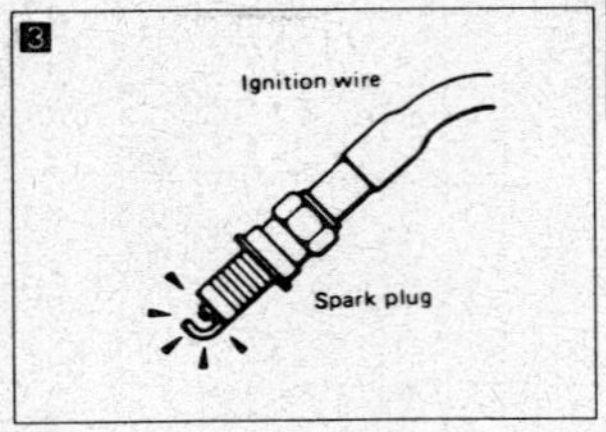

1 **PERFORM POWER BALANCE TEST.**

1. Perform "POWER BALANCE" in "ACTIVE TEST" mode.
2. Is there any cylinder which does not produce a momentary engine speed drop?

— OR —

When disconnecting each injector harness connector one at a time, is there any cylinder which does not produce a momentary engine speed drop?

No → Go to 4.

Yes ↓

2 **CHECK INJECTOR.**

1. Remove distributor from engine. (Crank angle sensor harness connector should remain connected.)
2. Turn ignition switch ON. (Do not start engine.)
3. When rotating distributor shaft slowly by hand, does each injector make an operating sound?

No → Check injector(s) and circuit(s).

Yes ↓

3 **CHECK IGNITION SPARK.**

1. Disconnect ignition wire from spark plug.
2. Connect a known good spark plug to the ignition wire.
3. Place end of spark plug against a suitable ground and crank engine.
4. Check for spark.

No → Check ignition coil, power transistor unit and circuits.

Yes ↓

(Go to (A) on next page.)

Diagnostic Procedure 15 — Engine Stalls when Accelerating or when Driving at Constant Speed (Cont'd)

4
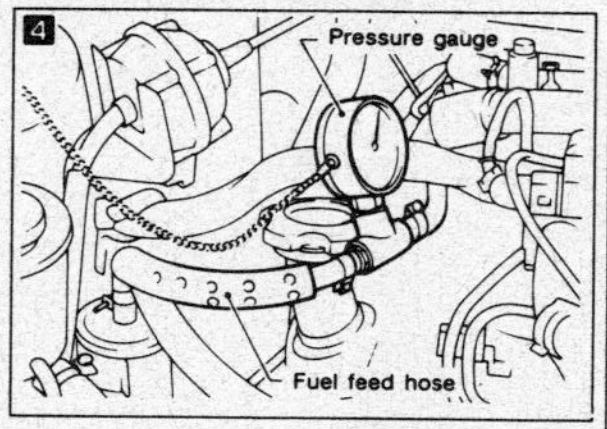

5

Ⓐ

4 CHECK FUEL PRESSURE.
1. Release fuel pressure to zero.
2. Install fuel pressure gauge and check fuel pressure.
   At idle:
   Approx. 235 kPa (2.4 kg/cm², 34 psi)
   The moment throttle valve is fully open:
   Approx. 294 kPa (3.0 kg/cm², 43 psi)

→ N.G.: Check fuel pump, circuit and fuel pressure regulator.

↓ O.K.

5 CHECK FOR INTAKE AIR LEAK.
When pinching blow-by hose (lowering the blow-by air supply), does the engine speed rise?

→ Yes: Discover air leak location and repair.

↓ No

6 CHECK E.C.U. HARNESS CONNECTOR.
Check the E.C.U. pin terminals for damage or poor connection of E.C.U. harness connector.

→ N.G.: Repair or replace.

↓ O.K.

7 CHECK E.C.U. POWER SUPPLY AND GROUND CIRCUIT.

→ Yes: Repair or replace.

↓ No

8 TRY A KNOWN GOOD E.C.U.

↓

INSPECTION END

Diagnostic Procedure 16 — Engine Stalls when the Electrical Load is Heavy

1
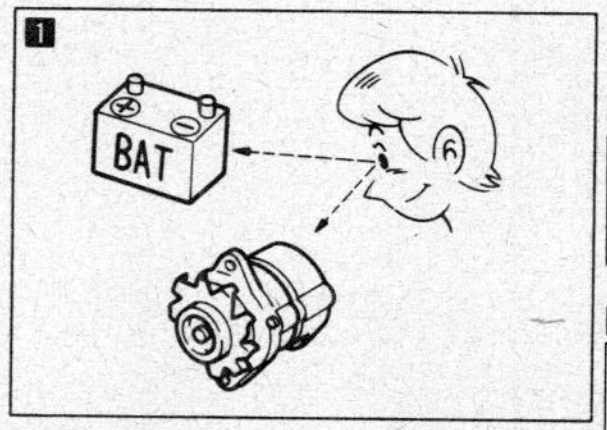

2
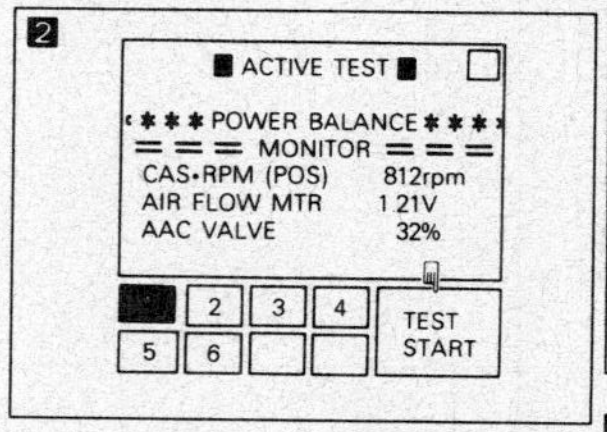

2
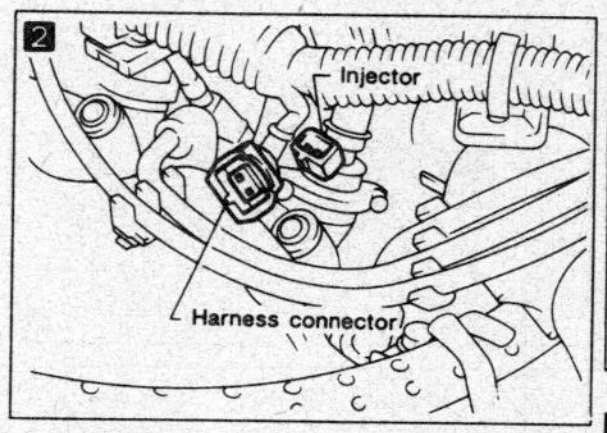

3

4
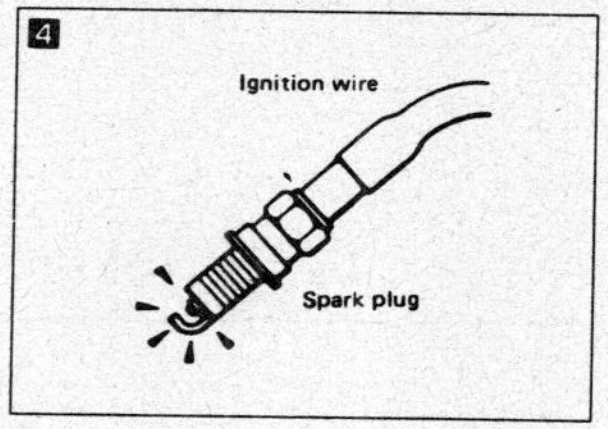

1 CHECK BATTERY AND ALTERNATOR.
Check battery and alternator condition.

→ N.G.: Repair or replace.

↓ O.K.

2 PERFORM POWER BALANCE TEST.
1. Perform "POWER BALANCE" in "ACTIVE TEST" mode.
2. Is there any cylinder which does not produce a momentary engine speed drop?

— OR —

When disconnecting each injector harness connector one at a time, is there any cylinder which does not produce a momentary engine speed drop?

→ No: Go to 5.

↓ Yes

3 CHECK INJECTOR.
1. Remove distributor from engine. (Crank angle sensor harness connector should remain connected.)
2. Turn ignition switch ON. (Do not start engine.)
3. When rotating distributor shaft slowly by hand, does each injector make an operating sound?

→ No: Check injector(s) and circuit(s).

↓ Yes

4 CHECK IGNITION SPARK.
1. Disconnect ignition wire from spark plug.
2. Connect a known good spark plug to the ignition wire.
3. Place end of spark plug against a suitable ground and crank engine.
4. Check for spark.

→ N.G.: Check ignition coil, power transistor unit and circuits.

↓ O.K.

(Go to Ⓐ on next page.)

Diagnostic Procedure 16 — Engine Stalls when the Electrical Load is Heavy (Cont'd)

5

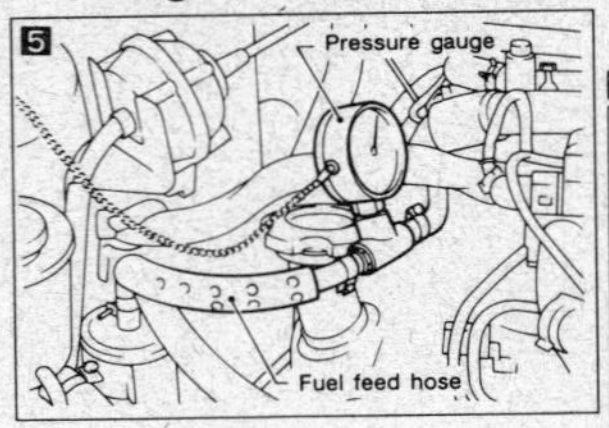

(A)

5 **CHECK FUEL PRESSURE.**
1. Release fuel pressure to zero.
2. Install fuel pressure gauge and check fuel pressure.
   At idle:
   Approx. 235 kPa (2.4 kg/cm², 34 psi)
   The moment throttle valve is fully open:
   Approx. 294 kPa (3.0 kg/cm², 43 psi)

N.G. → Check fuel pressure regulator diaphragm.

O.K. ↓

6 **CHECK E.C.U. HARNESS CONNECTOR.**
Check the E.C.U. pin terminals for damage or poor connection of E.C.U. harness connector.

N.G. → Repair or replace.

O.K. ↓

7 **CHECK E.C.U. POWER SUPPLY AND GROUND CIRCUIT.**

N.G. → Repair or replace.

O.K. ↓

8 **TRY A KNOWN GOOD E.C.U.**

↓

INSPECTION END

Diagnostic Procedure 17 — Lack of Power and Stumble

1

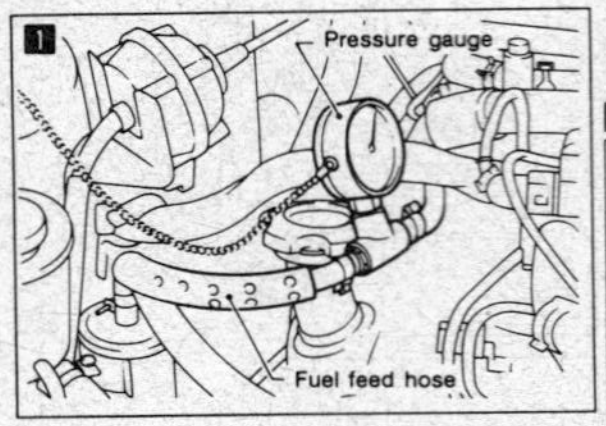

2

1 **CHECK FUEL PRESSURE.**
1. Release fuel pressure to zero.
2. Install fuel pressure gauge and check fuel pressure.
   At idle:
   Approx. 235 kPa (2.4 kg/cm², 34 psi)
   The moment throttle valve is fully open:
   Approx. 294 kPa (3.0 kg/cm², 43 psi)

N.G. → Check fuel pressure regulator diaphragm.

O.K. ↓

2 **CHECK FOR INTAKE AIR LEAK.**
When pinching blow-by hose (lowering the blow-by air supply), does the engine speed rise?

Yes → Discover air leak location and repair.

No ↓

CHECK TIMING BELT FOR PROPER INSTALLATION.

N.G. → Replace timing belt.

O.K. ↓

INSPECTION END

Diagnostic Procedure 18 — Detonation

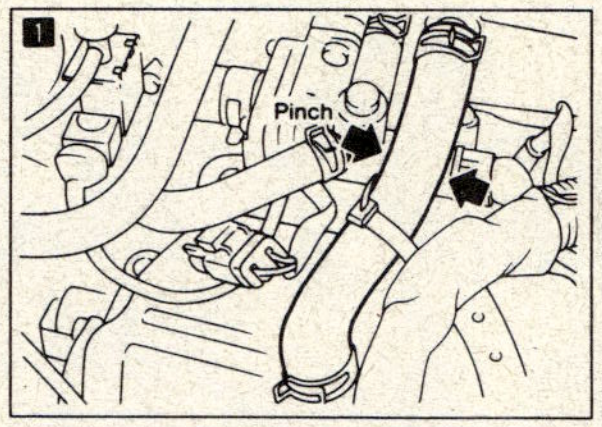

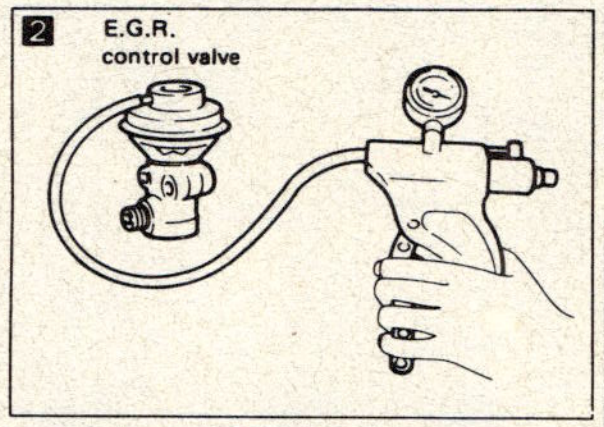

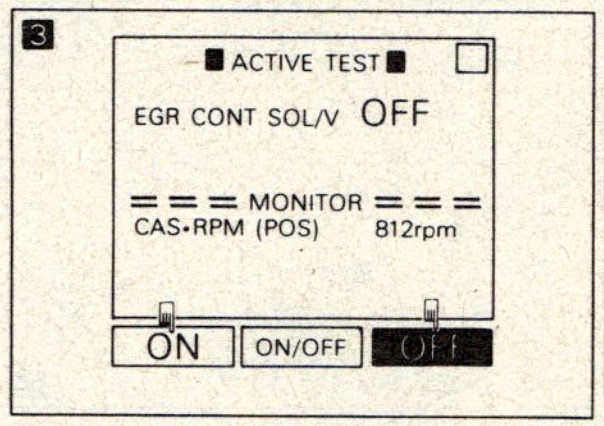

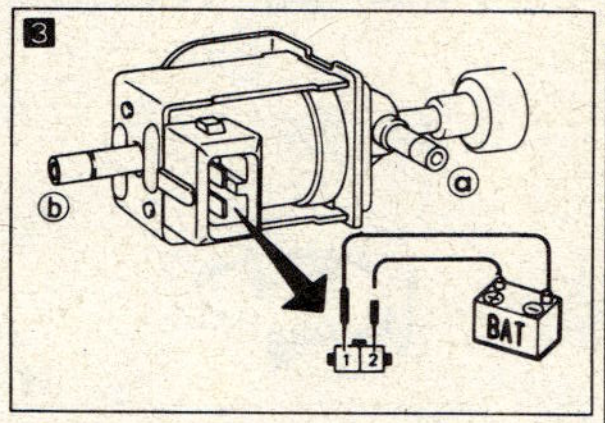

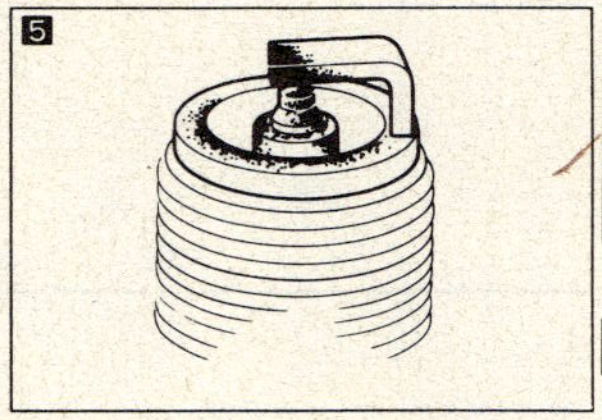

1 CHECK FOR INTAKE AIR LEAK.
When pinching blow-by hose (lowering the blow-by air supply), does the engine rpm rise?

- Yes → Discover air leak location and repair.
- No ↓

2 CHECK E.G.R. OPERATION.
1. Apply vacuum directly to the E.G.R. valve using a handy vacuum pump.
2. Check to see that the engine runs rough or dies.

- No → Check E.G.R. valve for sticking.
- Yes ↓

3 CHECK E.G.R. CONTROL SOLENOID VALVE.
1. Select "E.G.R. CONT SOL VALVE" in "ACTIVE TEST" mode.
2. Turn E.G.R. control solenoid valve ON and OFF.
3. Check operating sound.

— OR —

1. Disconnect E.G.R. control solenoid valve harness connector.
2. Supply E.G.R. control solenoid valve terminals with battery current and check operating sound.

- N.G. → Check solenoid valve and circuit.
- O.K. ↓

4 CHECK VACUUM HOSES.
Check the following vacuum hoses for clogging, cracks and poor connection.
a) Vacuum hose between E.G.R. control valve and E.G.R. control solenoid valve.
b) Vacuum hose between E.G.R. control solenoid valve and throttle chamber port.

- N.G. → Repair or replace.
- O.K. ↓

5 CHECK FOR OIL LEAK TO COMBUSTION CHAMBER.
Remove spark plugs and check for fouling with oil.

- Yes → Check pistons, piston rings, valves, valve seats, valve oil seal, engine oil level, etc.
- No ↓

INSPECTION END

Diagnostic Procedure 19 — Surge

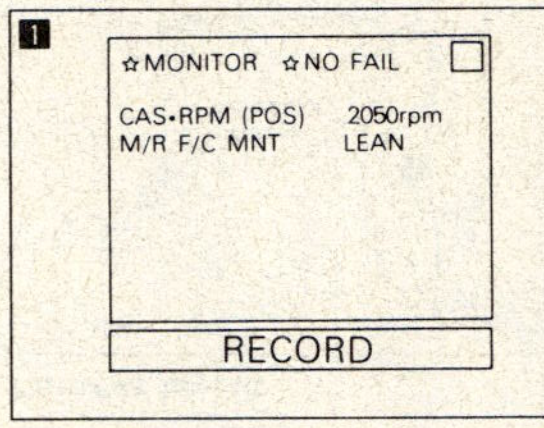

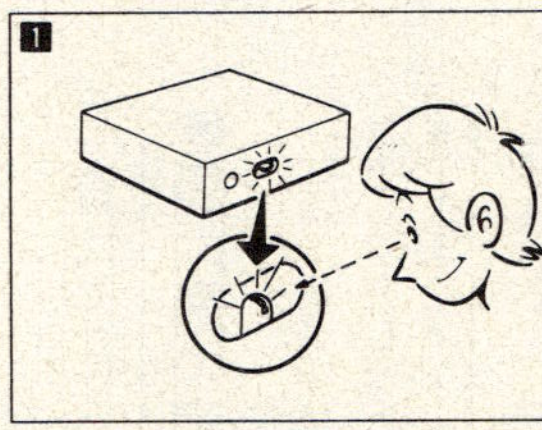

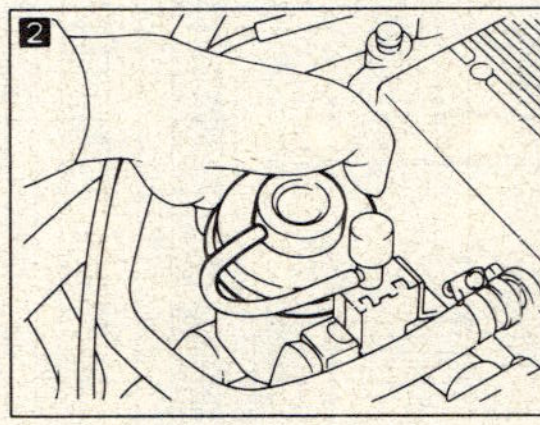

1 CHECK EXHAUST GAS SENSOR.
1. See "M/R F/C MNT" in "DATA MONITOR" mode.
2. Maintaining engine at 2,000 rpm under no-load (with engine warmed up sufficiently.), check to make sure that the monitor fluctuates between "LEAN" and "RICH" more than 5 times during 10 seconds.

RICH→LEAN→RICH→ (1 time)
LEAN→RICH....... (2 times)

— OR —

1. Set "Exhaust gas sensor monitor" in the self-diagnostic Mode II.
2. Maintaining engine at 2,000 rpm under no load, check that RED L.E.D. on the E.C.U. goes ON and OFF more than 5 times during 10 seconds.

- N.G. → Replace exhaust gas sensor.
- O.K. ↓

2 CHECK E.G.R. CONTROL VALVE.
Check E.G.R. control valve for sticking.

- N.G. → Repair or replace.
- O.K. ↓

3 TRY A KNOWN GOOD E.C.U.

↓

INSPECTION END

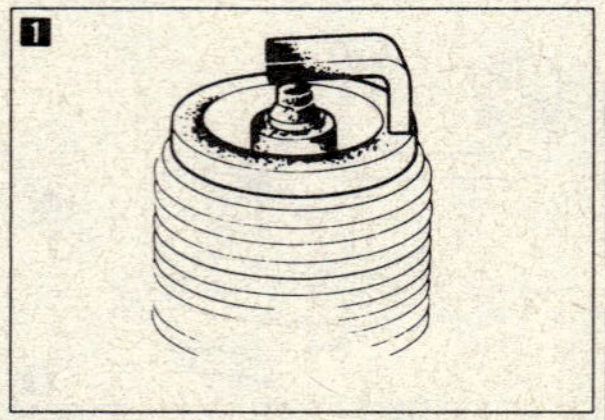

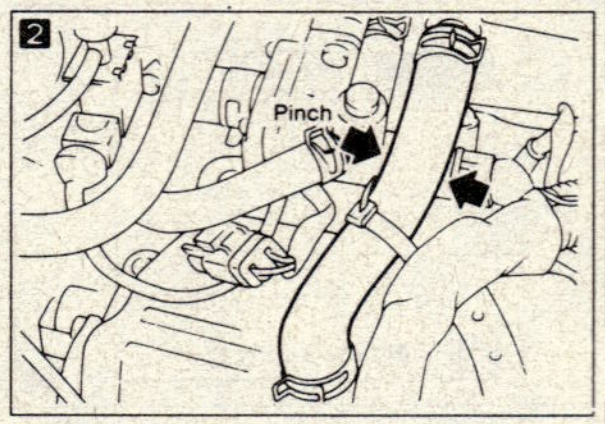

Diagnostic Procedure 20 — Backfire through the Intake

1 CHECK SPARK PLUGS.
Remove the spark plugs and check for fouling, etc.
- N.G. → Repair or replace spark plug(s).
- O.K. ↓

2 CHECK INTAKE AIR LEAK.
When pinching blow-by hose (lowering the blow-by air supply), does the engine speed rise?
- Yes → Discover air leak location and repair.
- No ↓

3 CHECK FOR INTAKE VALVE DEPOSITS.
If there are deposits on intake valves, remove them.
↓

INSPECTION END

Diagnostic Procedure 21 — Backfire through the Exhaust

1 CHECK ENGINE TEMPERATURE SENSOR.
Check engine temperature sensor and its circuit.
- N.G. → Replace or repair.
- O.K. ↓

INSPECTION END

Diagnostic Procedure 22

MAIN POWER SUPPLY AND GROUND CIRCUIT (Not self-diagnostic item)

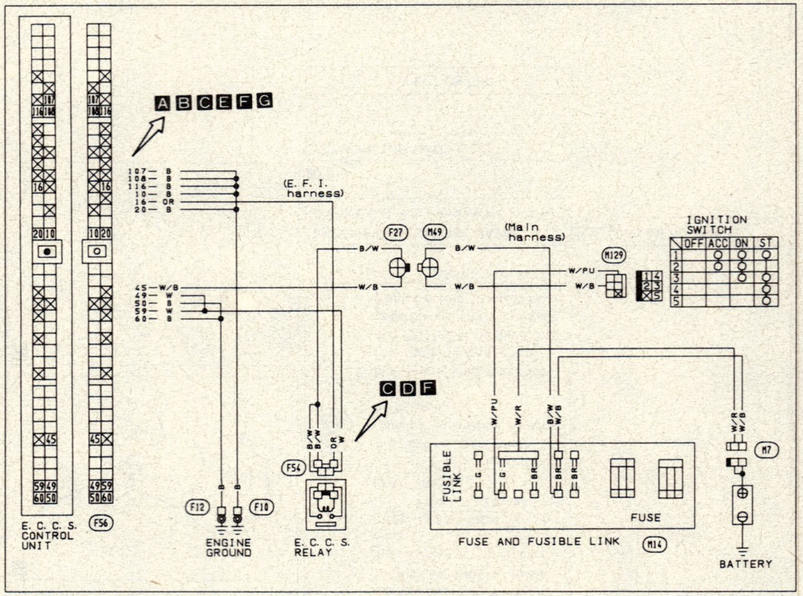

Harness layout

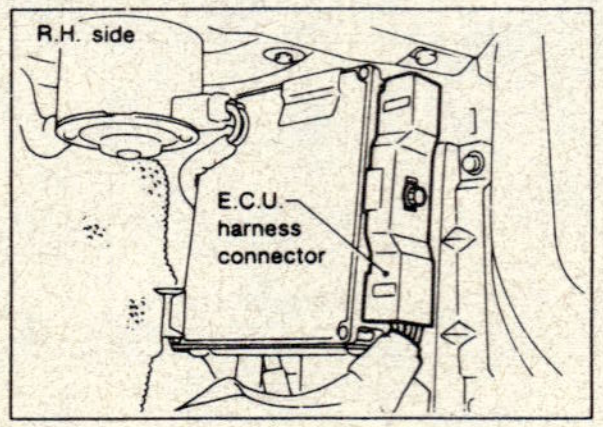

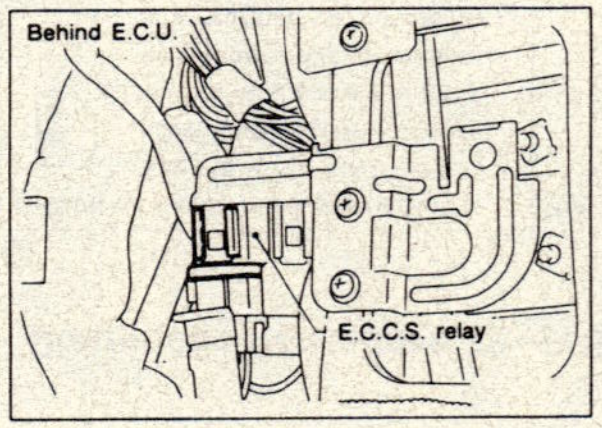

Diagnostic Procedure 22 (Cont'd)

A
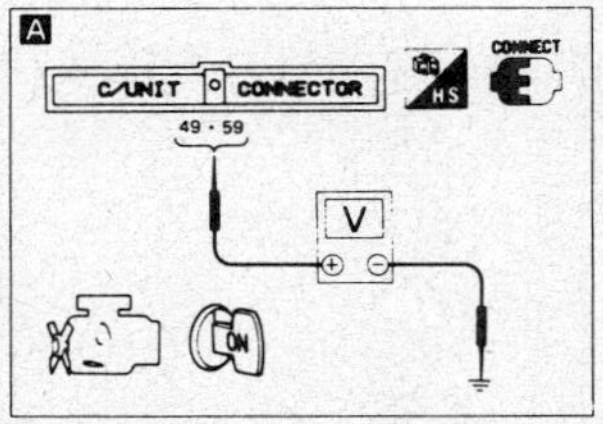

B
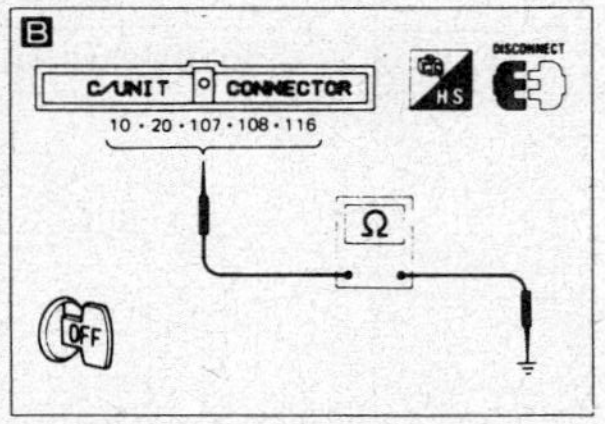

C
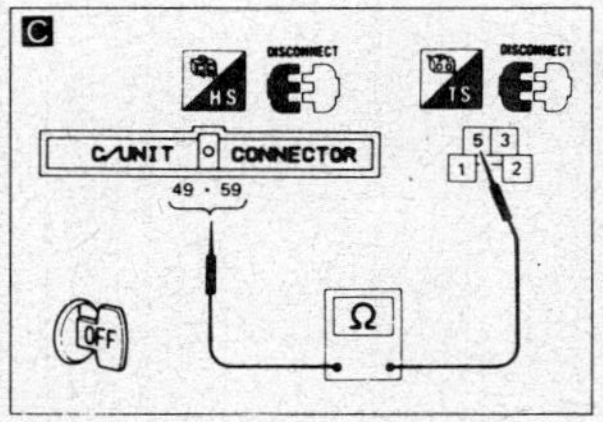

D
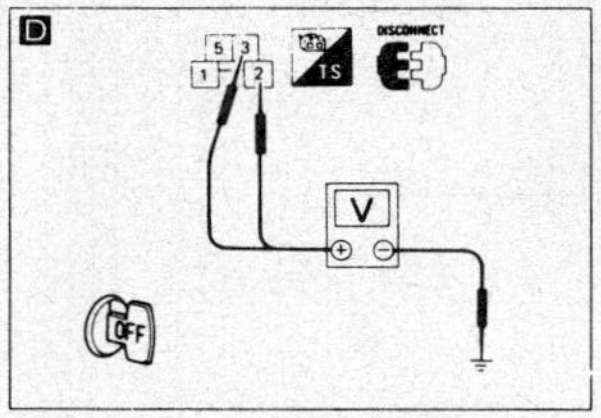

E
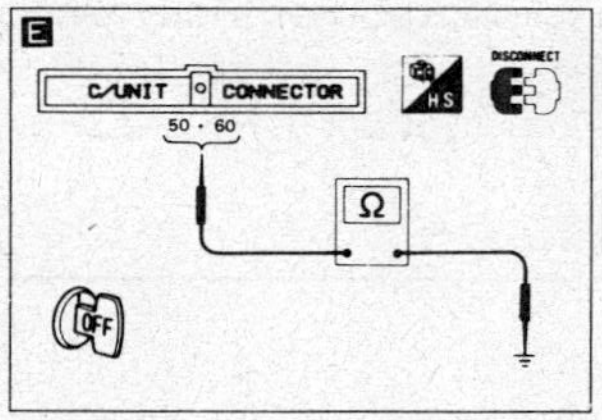

INSPECTION START

A CHECK POWER SUPPLY.
1) Turn ignition switch "ON".
2) Check voltage between E.C.U. terminals ㊾, ㊿ and ground.
Voltage: Battery voltage

O.K. → **B** CHECK GROUND CIRCUIT.
1) Turn ignition switch "OFF".
2) Disconnect E.C.U. harness connector.
3) Check harness continuity between E.C.U. terminals ⑩, ⑳, (107), (108), (116) and engine ground.
Continuity should exist.
If N.G., repair harness or connectors.

O.K. → Check E.C.U. pin terminals for damage or the connection of E.C.U. harness connector.

N.G. ↓

C CHECK HARNESS CONTINUITY BETWEEN E.C.C.S. RELAY AND E.C.U.
1) Turn ignition switch "OFF".
2) Disconnect E.C.U. harness connector.
3) Disconnect E.C.C.S. relay.
4) Check harness continuity between E.C.U. terminals ㊾, ㊿ and terminal ⑤.
Continuity should exist.

N.G. → Repair harness or connectors.

O.K. ↓

D CHECK VOLTAGE BETWEEN E.C.C.S. RELAY AND GROUND.
1) Check voltage between terminals ②, ③ and ground.
Voltage: Battery voltage

N.G. → Check the following.
- Harness connectors (F27), (M49)
- "BR" fusible link
- Harness continuity between E.C.C.S. relay and battery

If N.G., repair harness or connectors.

O.K. ↓

E CHECK GROUND CIRCUIT.
1) Check harness continuity between E.C.U. terminals ㊿, ⑥⓪ and engine ground.
Continuity should exist.

N.G. → Repair harness or connectors.

O.K. ↓ Ⓐ

Diagnostic Procedure 22 (Cont'd)

F
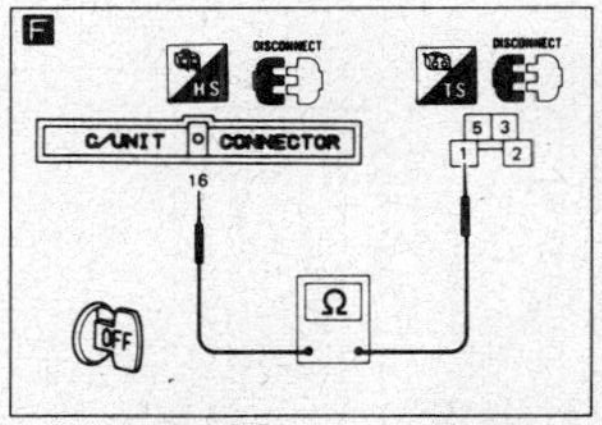

G
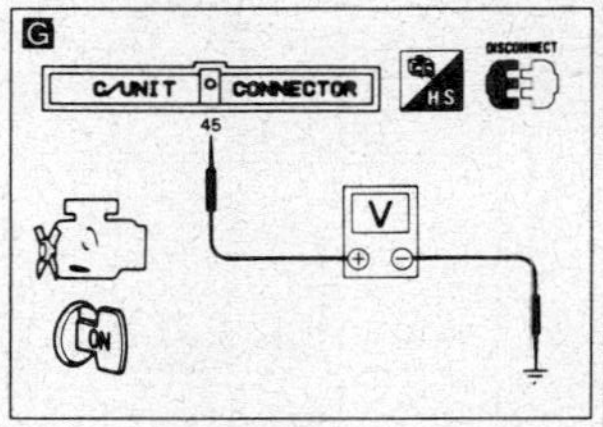

Ⓐ ↓

F CHECK OUTPUT SIGNAL CIRCUIT.
1) Check harness continuity between E.C.U. terminal ⑯ and terminal ①.
Continuity should exist.

N.G. → Repair harness or connectors.

O.K. ↓

G CHECK INPUT SIGNAL CIRCUIT.
1) Turn ignition switch "ON".
2) Check voltage between E.C.U. terminal ㊺ and ground.
Voltage: Battery voltage

N.G. → Check the following.
- Harness connectors (F27), (M49)
- Harness continuity between E.C.U. and ignition switch

If N.G., repair harness or connectors.

O.K. ↓

CHECK COMPONENT (E.C.C.S. relay).
Refer to "Electrical Components Inspection".

N.G. → Replace E.C.C.S. relay.

O.K. ↓

Check E.C.U. pin terminals for damage or the connection of E.C.U. harness connector.

Diagnostic Procedure 23

CRANK ANGLE SENSOR (Code No. 11)

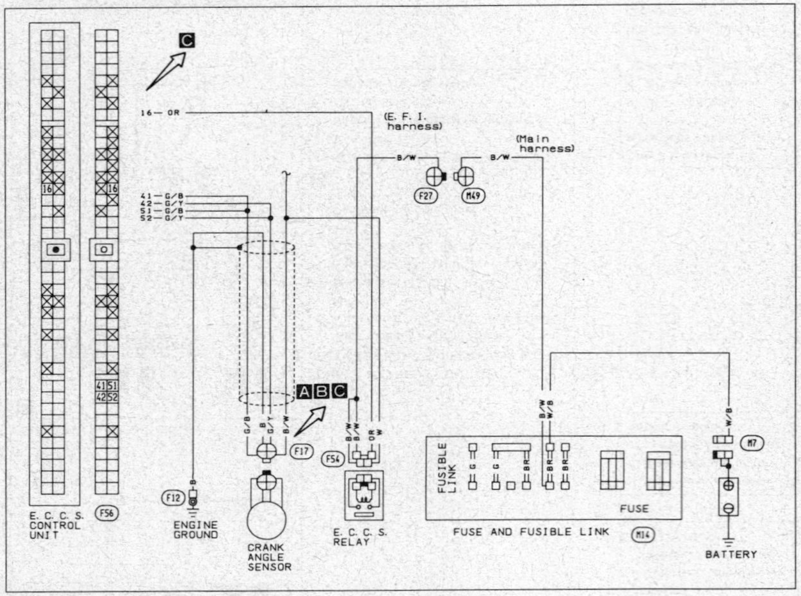

Harness layout

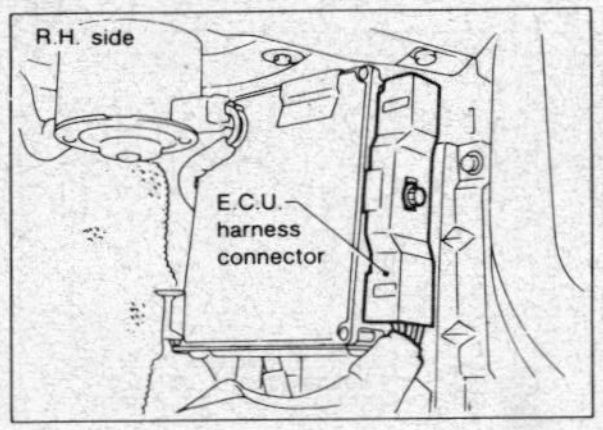

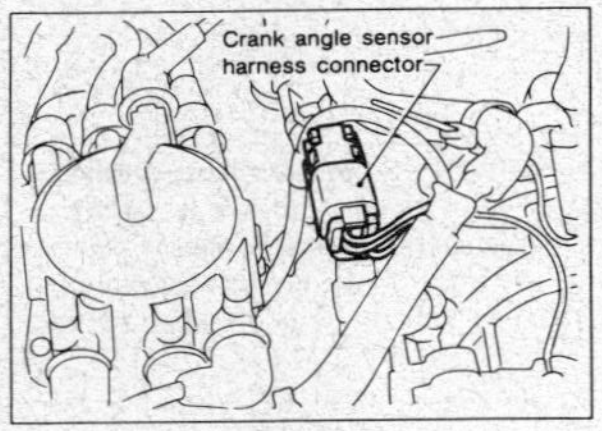

Diagnostic Procedure 23 (Cont'd)

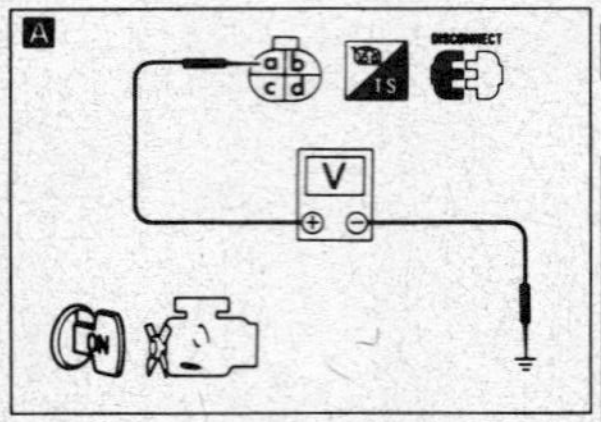

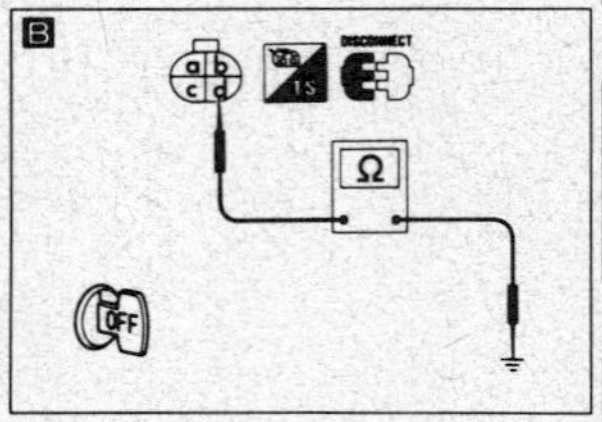

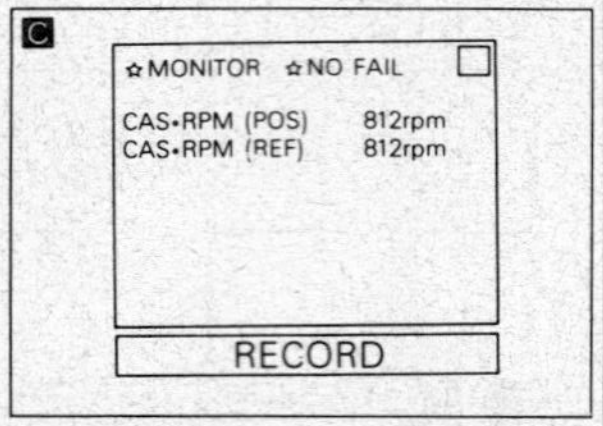

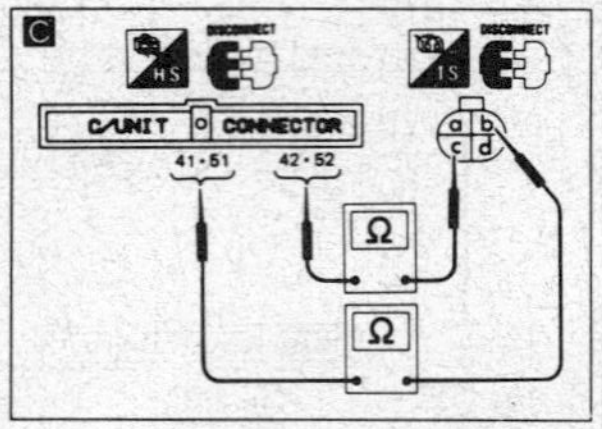

INSPECTION START

↓

A CHECK POWER SUPPLY.
1) Disconnect crank angle sensor harness connector.
2) Turn ignition switch "ON".
3) Check voltage between terminal ⓐ and ground.
Voltage: Battery voltage

N.G. → Check the following.
- Harness continuity between crank angle sensor and E.C.C.S. relay

If N.G., repair harness or connectors.

↓ O.K.

B CHECK GROUND CIRCUIT.
1) Turn ignition switch "OFF".
2) Check harness continuity between terminal ⓓ and engine ground.
Continuity should exist.

N.G. → Repair harness or connectors.

↓ O.K.

C CHECK INPUT SIGNAL CIRCUIT.
1) Reconnect crank angle sensor harness connector.
2) Start engine.
3) Read crank angle sensor signals in "DATA MONITOR" mode with CONSULT.
rpm: 800 ± 50

— OR —

1) Disconnect E.C.U. harness connector.
2) Check harness continuity between terminal ⓒ and E.C.U. terminals ㊷, ㊾ (1° signal), terminal ⓑ and E.C.U. terminals ㊶, ㊿ (120° signal).
Continuity should exist.

N.G. → Repair harness or connectors.

↓ O.K.

CHECK COMPONENT
(Crank angle sensor).
Refer to "Electrical Components Inspection".

N.G. → Replace crank angle sensor.

↓ O.K.

Check E.C.U. pin terminals for damage or the connection of E.C.U. harness connector.

Diagnostic Procedure 24

AIR FLOW METER (Code No. 12) (CHECK ENGINE LIGHT ITEM)

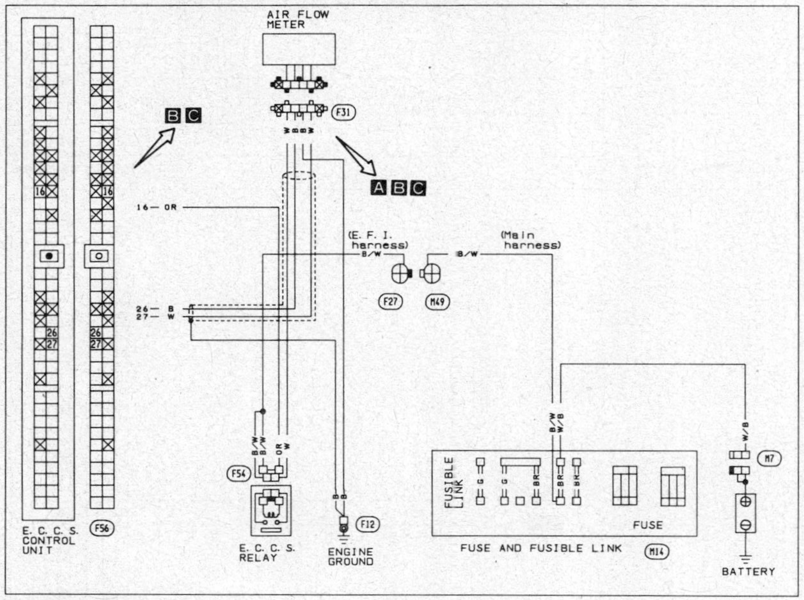

Harness layout

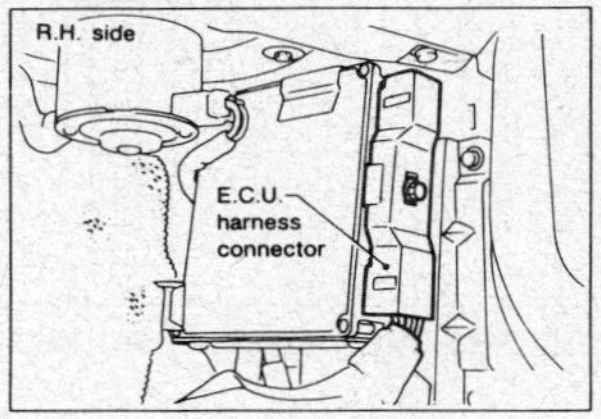

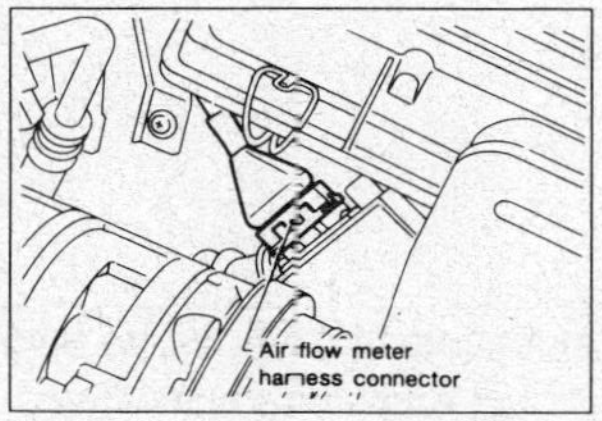

Diagnostic Procedure 24 (Cont'd)

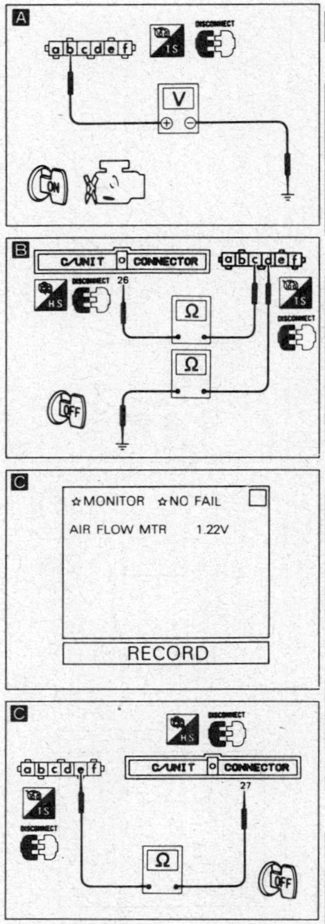

INSPECTION START

A CHECK POWER SUPPLY.
1) Disconnect air flow meter harness connector.
2) Turn ignition switch "ON".
3) Check voltage between terminal ⓑ and ground.
Voltage: Battery voltage

N.G. → Check the following.
- Harness continuity between air flow meter and E.C.C.S. relay

If N.G., repair harness or connectors.

O.K. ↓

B CHECK GROUND CIRCUIT.
1) Turn ignition switch "OFF".
2) Disconnect E.C.U. harness connector.
3) Check harness continuity between terminal ⓒ and E.C.U. terminal ㉖, terminal ⓓ and body ground.
Continuity should exist.

N.G. → Repair harness or connectors.

O.K. ↓

C CHECK INPUT SIGNAL CIRCUIT.
1) Reconnect air flow meter harness connector and E.C.U. harness connector.
2) Start engine and warm it up sufficiently.
3) Read air flow meter signal in "DATA MONITOR" mode with CONSULT.
Voltage: 1.0 - 1.5V

— OR —

1) Check harness continuity between terminal ⓔ and E.C.U. terminal ㉗.
Continuity should exist.

N.G. → Repair harness or connectors.

O.K. ↓

CHECK COMPONENT
(Air flow meter).
Refer to "Electrical Components Inspection".

N.G. → Replace air flow meter.

O.K. ↓

Check E.C.U. pin terminals for damage or the connection of E.C.U. harness connector.

Diagnostic Procedure 25

ENGINE TEMPERATURE SENSOR (Code No. 13) (CHECK ENGINE LIGHT ITEM)

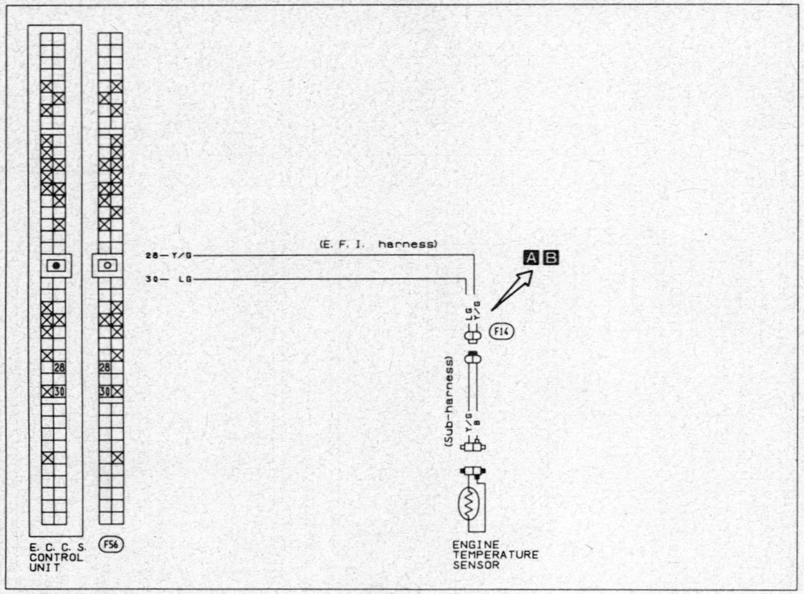

Harness layout

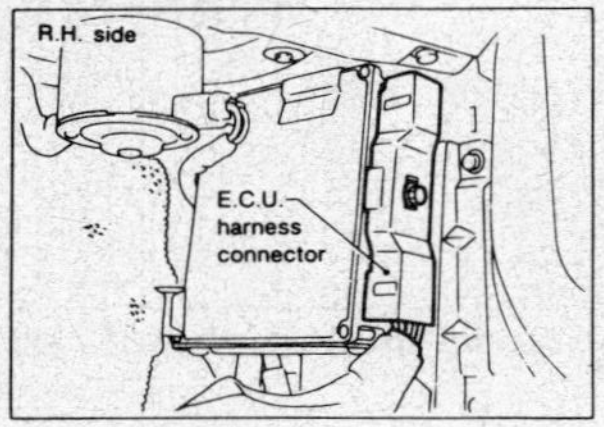

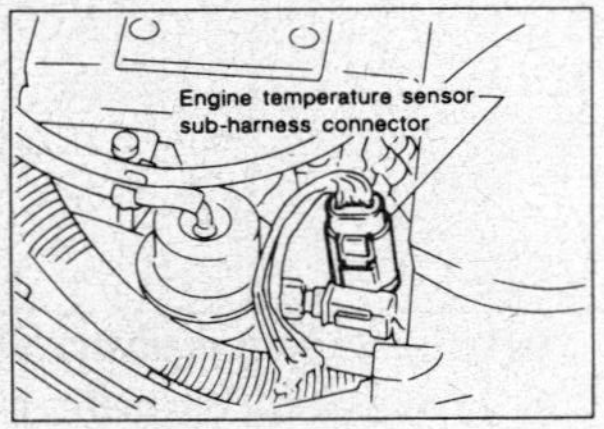

Diagnostic Procedure 25 (Cont'd)

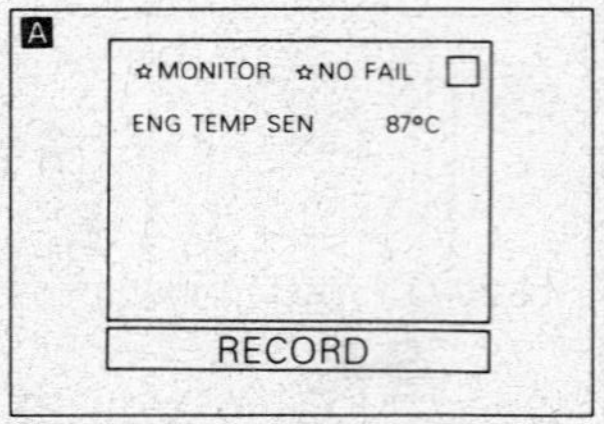

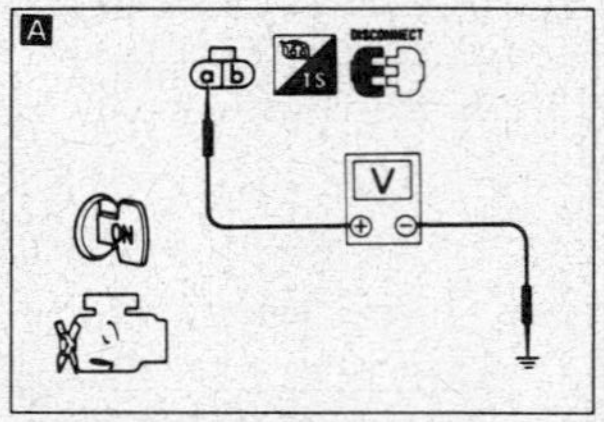

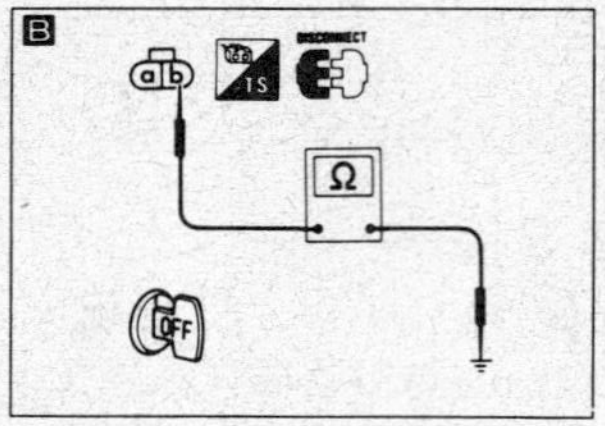

INSPECTION START

↓

A CHECK POWER SUPPLY.
1) Start engine.
2) Read engine temperature sensor signal in "DATA MONITOR" mode with CONSULT.
Engine temperature should gradually rise and reach more than 70°C (158°F) after engine warm-up.

— OR —

1) Disconnect engine temperature sensor sub-harness connector.
2) Turn ignition switch "ON".
3) Check voltage between terminal ⓐ and ground.
Voltage: Approximately 5V

N.G. → Repair harness or connectors.

↓ O.K.

B CHECK GROUND CIRCUIT.
1) Turn ignition switch "OFF".
2) Check harness continuity between terminal ⓑ and engine ground.
Continuity should exist.

N.G. → Repair harness or connectors.

↓ O.K.

CHECK COMPONENT
(Engine temperature sensor).
Refer to "Electrical Components Inspection".

N.G. → Replace engine temperature sensor.

↓ O.K.

Check E.C.U. pin terminals for damage or the connection of E.C.U. harness connector.

Perform FINAL CHECK by the following procedure after repair is completed.

FINAL CHECK

↓

Erase the self-diagnosis memory.

↓

Perform driving test.

↓

Perform self-diagnosis again.

N.G. → Recheck E.C.U. pin terminals for damage or the connection of E.C.U. harness connector.

↓ O.K.

INSPECTION END

Diagnostic Procedure 26

VEHICLE SPEED SENSOR (Code No. 14) (CHECK ENGINE LIGHT ITEM)

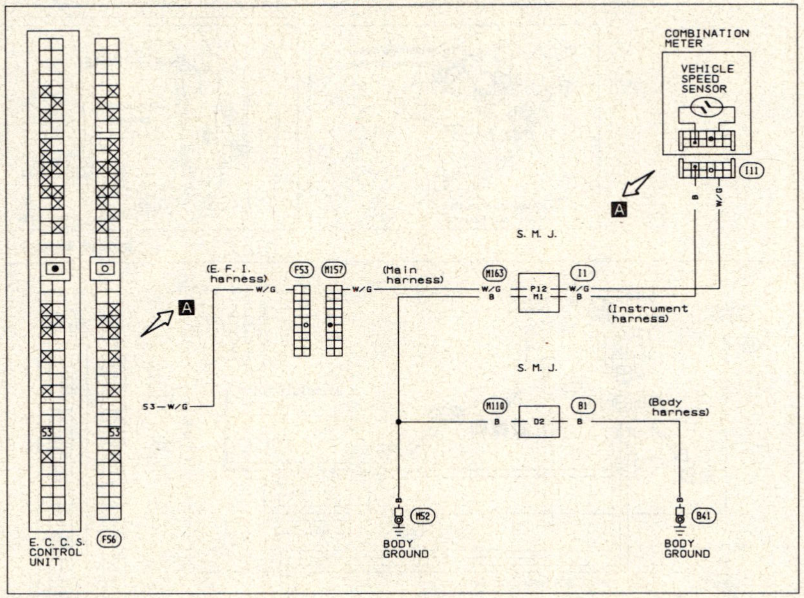

Harness layout

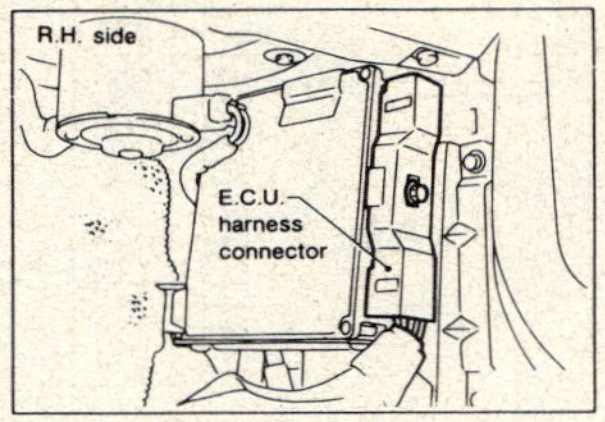

Diagnostic Procedure 26 (Cont'd)

INSPECTION START
↓
CHECK SPEEDOMETER FUNCTION.
Make sure that speedometer functions properly.
- N.G. → Check vehicle speed sensor and circuit.
- O.K. ↓

A CHECK INPUT SIGNAL CIRCUIT.
Perform "CAR SPEED SEN CIRCUIT" in "FUNCTION-TEST" mode with CONSULT.
—OR—
1) Read vehicle speed sensor signal in "DATA MONITOR" mode with CONSULT.
CONSULT value should be the same as the speedometer indication.
—OR—
1) Turn ignition switch "OFF".
2) Disconnect E.C.U. harness connector and combination meter harness connector.
3) Check harness continuity between E.C.U. terminal 53 and terminal e.
Continuity should exist.

- N.G. → Check the following.
  - Harness connectors F53, M157
  - Harness connectors M163, I1
  - Harness continuity between E.C.U. and combination meter

  If N.G., repair harness or connectors.
- O.K. ↓

Check E.C.U. pin terminals for damage or the connection of E.C.U. harness connector.

Perform FINAL CHECK by the following procedure after repair is completed.

FINAL CHECK
↓
Erase the self-diagnosis memory.
↓
Perform driving test.
↓
Perform self-diagnosis again.
- N.G. → Recheck E.C.U. pin terminals for damage or the connection of E.C.U. harness connector.
- O.K. ↓

INSPECTION END

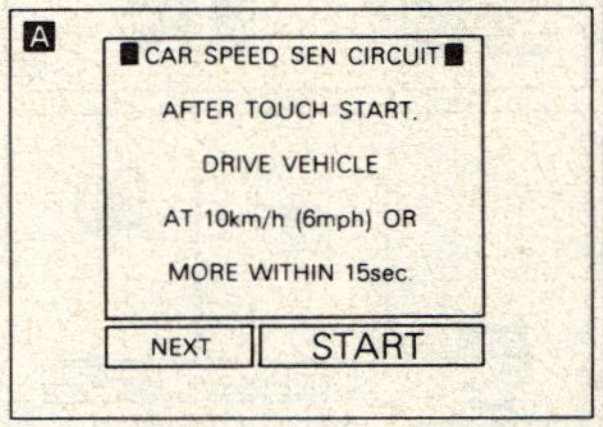

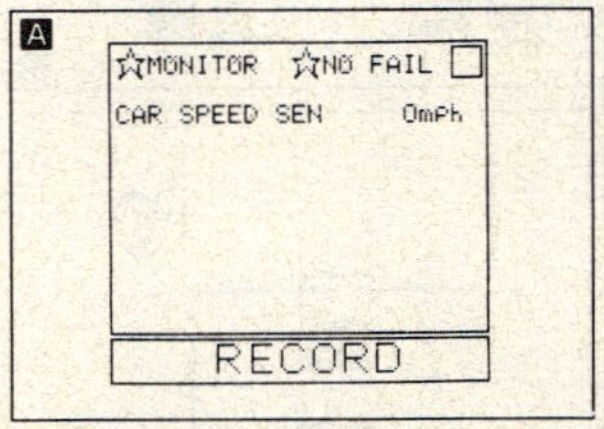

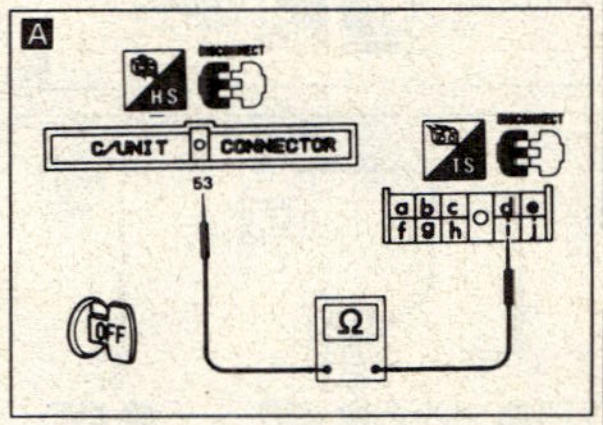

Diagnostic Procedure 27

IGNITION SIGNAL (Code No. 21)

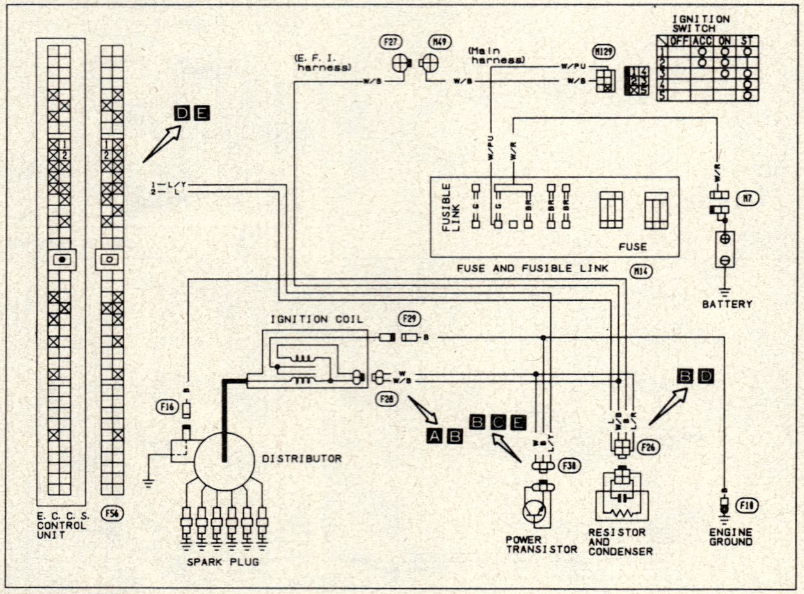

Harness layout

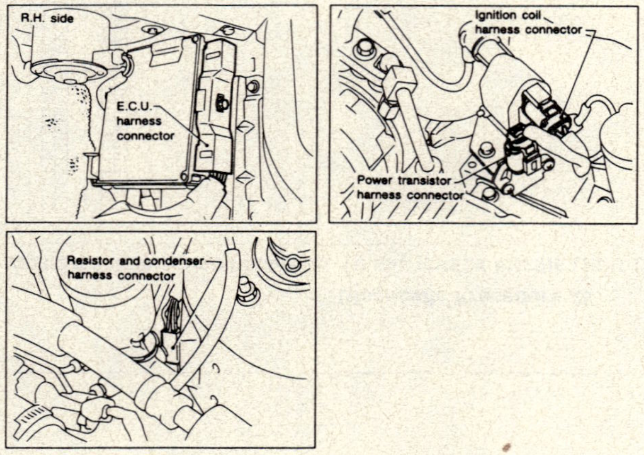

Diagnostic Procedure 27 (Cont'd)

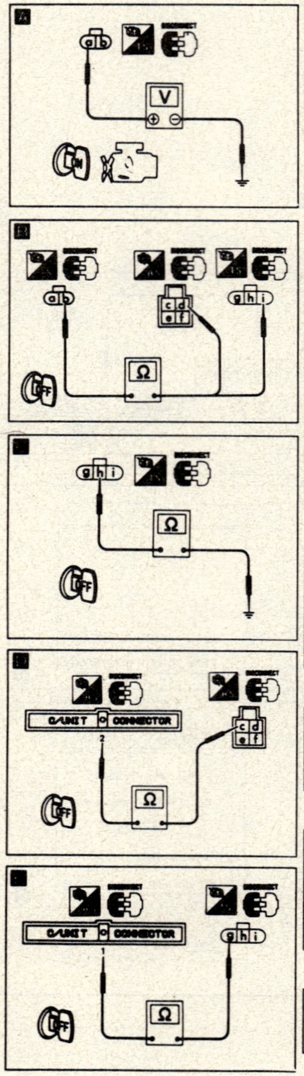

INSPECTION START

↓

A CHECK POWER SUPPLY.
1) Disconnect ignition coil harness connector.
2) Turn ignition switch "ON".
3) Check voltage between terminal ⓐ and ground.
Voltage: Battery voltage

N.G. → Check the following.
- Harness connectors (F27), (M49)
- Harness continuity between ignition coil and ignition switch

If N.G., repair harness or connectors.

O.K. ↓

B CHECK GROUND CIRCUIT.
1) Turn ignition switch "OFF".
2) Disconnect resistor and condenser harness connector.
3) Disconnect power transistor harness connector.
B 4) Check harness continuity between terminal ⓑ and ⓓ, ⓘ.
Continuity should exist.
C 5) Check harness continuity between terminal ⓑ and engine ground.
Continuity should exist.

N.G. → Repair harness or connectors.

O.K. ↓

D CHECK INPUT SIGNAL CIRCUIT.
1) Disconnect E.C.U. harness connector.
2) Check harness continuity between terminal ⓒ and E.C.U. terminal ②.
Continuity should exist.

N.G. → Repair harness or connectors.

O.K. ↓

E CHECK OUTPUT SIGNAL CIRCUIT.
1) Check harness continuity between terminal ⓖ and E.C.U. terminal ①.
Continuity should exist.

N.G. → Repair harness or connectors.

O.K. ↓

CHECK COMPONENTS
(Ignition coil, resistor and condenser, power transistor).
Refer to "Electrical Components Inspection".

N.G. → Replace malfunctioning component(s).

O.K. ↓

Check E.C.U. pin terminals for damage or the connection of E.C.U. harness connector.

Diagnostic Procedure 28

FUEL PUMP CONTROL (Code No. 22)

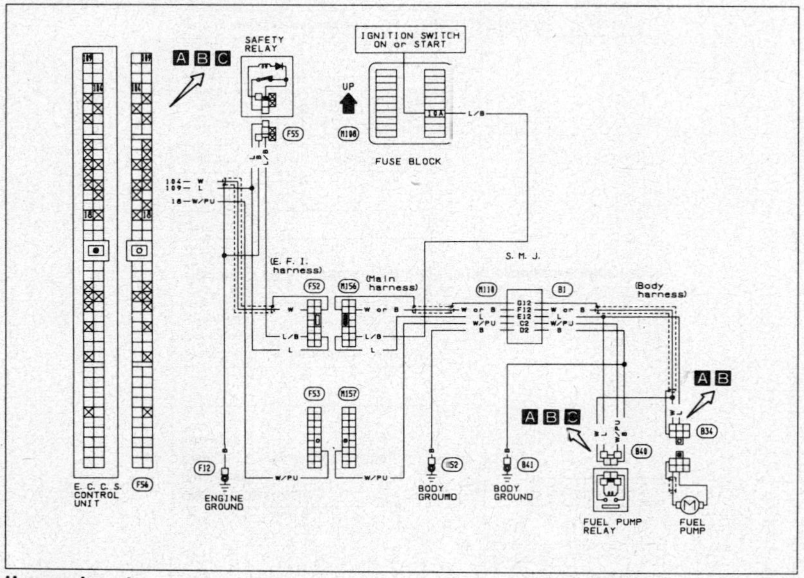

Harness layout

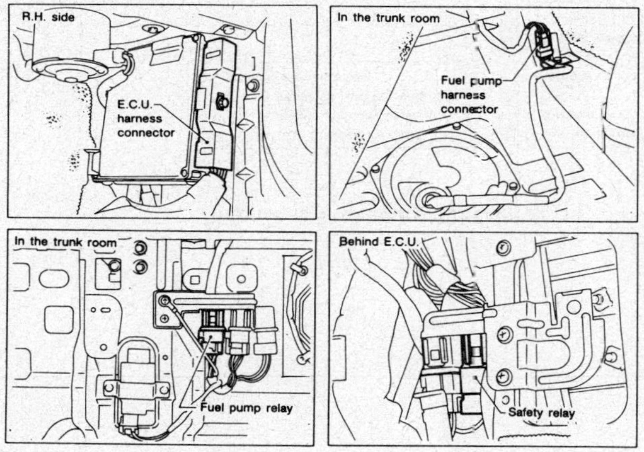

Diagnostic Procedure 28 (Cont'd)

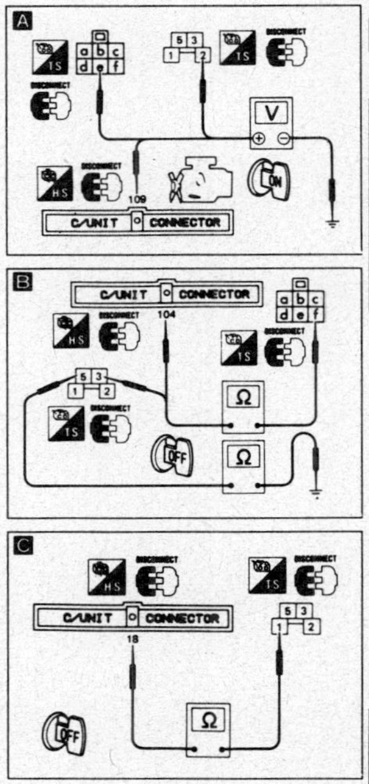

INSPECTION START

A CHECK POWER SUPPLY.
1) Disconnect E.C.U. harness connector, fuel pump harness connector and fuel pump relay.
2) Turn ignition switch "ON".
3) Check voltage between terminals ⑥, ②, E.C.U. terminal ⑩⑨ and ground.
Voltage: Battery voltage

N.G. → Check the following.
- Harness connectors M156, F52
- Harness connectors M118, B1
- Harness continuity between safety relay and fuse
- Harness continuity between safety relay and E.C.U., fuel pump, fuel pump relay
- 10A fuse

If N.G., repair harness or connectors.

O.K. → CHECK COMPONENT (Safety relay). Refer to "Electrical Components Inspection".

If N.G., replace safety relay.

O.K.

B CHECK GROUND CIRCUIT.
1) Turn ignition switch "OFF".
2) Check harness continuity between terminal ① and E.C.U. terminal ⑩④, terminal ① and ③, terminal ⑤ and body ground.
Continuity should exist.

N.G. → Check the following.
- Harness connectors B1, M118
- Harness connectors M156, F52
- Harness continuity between fuel pump and E.C.U., fuel pump and fuel pump relay, fuel pump relay and body ground

If N.G., repair harness or connectors.

O.K. → Ⓐ

Diagnostic Procedure 28 (Cont'd)

C (from Ⓐ)

CHECK OUTPUT SIGNAL CIRCUIT.
1) Check harness continuity between E.C.U. terminal ⑱ and terminal ①.
Continuity should exist.

- N.G. → Check the following.
  - Harness connectors (B1), (M110)
  - Harness connectors (M157), (F53)
  - Harness continuity between E.C.U. and fuel pump relay

  If N.G., repair harness or connectors.
- O.K. ↓

CHECK COMPONENT
(Fuel pump relay).
Refer to "Electrical Components Inspection".

- N.G. → Replace fuel pump relay.
- O.K. ↓

Check E.C.U. pin terminals for damage or the connection of E.C.U. harness connector.

Perform FINAL CHECK by the following procedure after repair is completed.

FINAL CHECK
↓
Erase the self-diagnosis memory.
↓
Perform driving test.
↓
Perform self-diagnosis again.

- N.G. → Recheck E.C.U. pin terminals for damage or the connection of E.C.U. harness connector.
- O.K. ↓

INSPECTION END

Diagnostic Procedure 29

E.C.C.S. CONTROL UNIT (Code No. 31) (CHECK) (CHECK ENGINE LIGHT ITEM)

INSPECTION START
↓
1) Turn ignition switch "ON".
2) Erase the self-diagnosis memory.
↓
Perform self-diagnosis again.
↓
Does E.C.U. display Code No. 31 again?

- Yes → Replace E.C.U.
- No ↓

INSPECTION END

Diagnostic Procedure 30

E.G.R. FUNCTION (Code No. 32) (CHECK ENGINE LIGHT ITEM): CALIFORNIA MODEL

E.G.R. CONTROL (Not self-diagnostic item): NON-CALIFORNIA MODEL

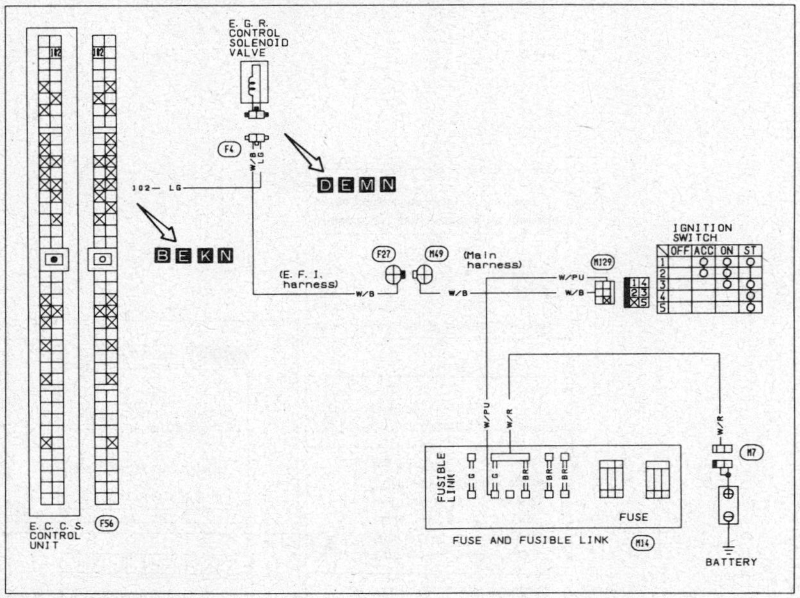

Harness layout

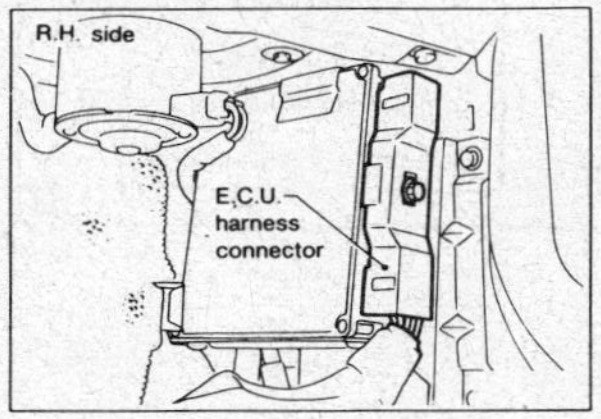

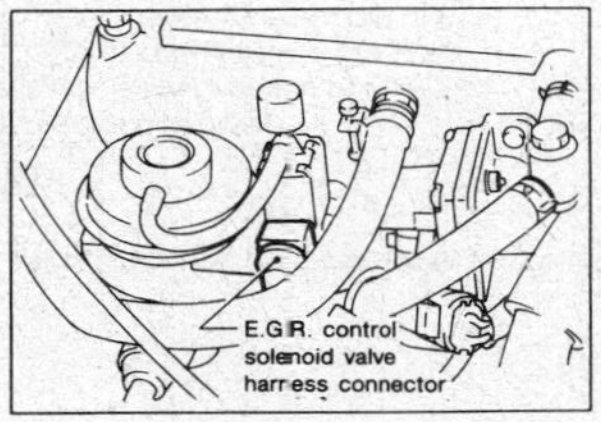

Diagnostic Procedure 30 (Cont'd)

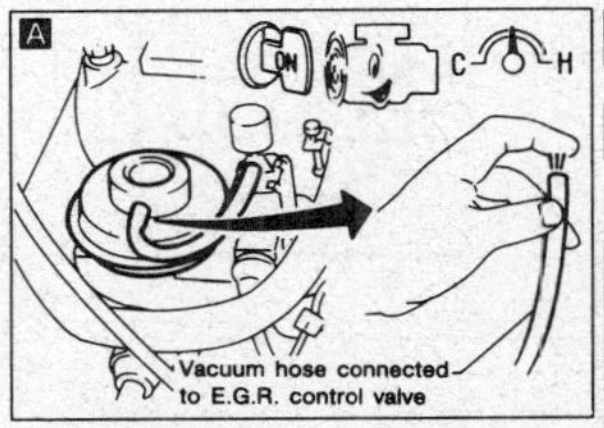

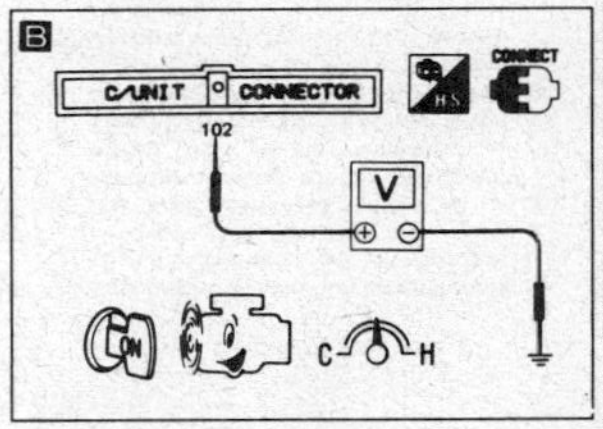

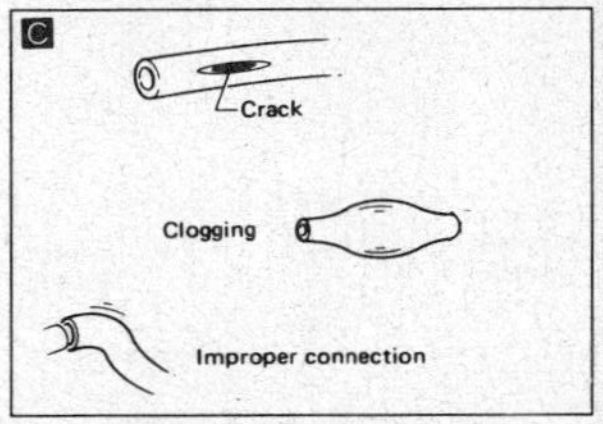

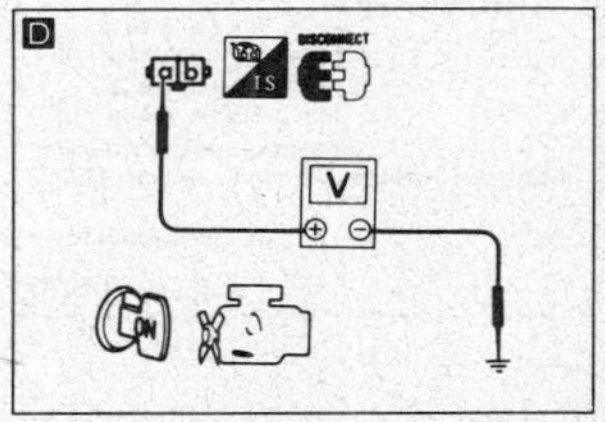

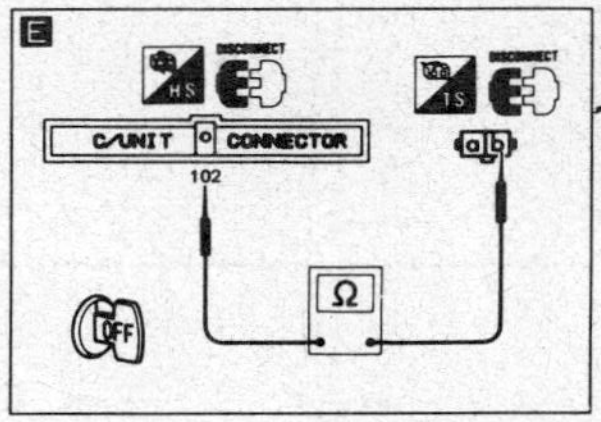

California model

INSPECTION START

A CHECK VACUUM SOURCE TO E.G.R. CONTROL VALVE.
1) Start engine and warm it up sufficiently.
2) Perform self-diagnosis.
Make sure that code No. 12 is not displayed.
Make sure that both crank angle sensor and E.C.U.'s C.P.U. are not in "fail-safe" state.
3) Keep engine speed at 2,000 rpm.
4) Disconnect vacuum hose to E.G.R. control valve.
5) Make sure that vacuum exists.
Vacuum should exist.

O.K. → CHECK COMPONENTS (E.G.R. control valve and exhaust gas temperature sensor). Refer to "Electrical Components Inspection".
N.G. → Replace malfunctioning component(s).

N.G. ↓

B CHECK CONTROL FUNCTION.
1) Check voltage between E.C.U. terminal (102) and ground under the following conditions.
Voltage:
At idle
0.7 - 0.8 V
Engine speed is about 2,000 rpm
Battery voltage

O.K. → **C** CHECK VACUUM HOSE.
1) Check vacuum hose for clogging, cracks and proper connection.

N.G. ↓

D CHECK POWER SUPPLY.
1) Stop engine.
2) Disconnect E.G.R. control solenoid valve harness connector.
3) Turn ignition switch "ON".
4) Check voltage between terminal ⓐ and ground.
Voltage: Battery voltage

N.G. → Check the following.
- Harness connectors (F27), (M49)
- Harness continuity between E.G.R. control solenoid valve and ignition switch

If N.G., repair harness or connectors.

O.K. ↓

E CHECK OUTPUT SIGNAL CIRCUIT.
1) Turn ignition switch "OFF".
2) Disconnect E.C.U. harness connector.
3) Check harness continuity between E.C.U. terminal (102) and terminal ⓑ.
Continuity should exist.

N.G. → Repair harness or connectors.

O.K. ↓ Ⓐ

Diagnostic Procedure 30 (Cont'd)

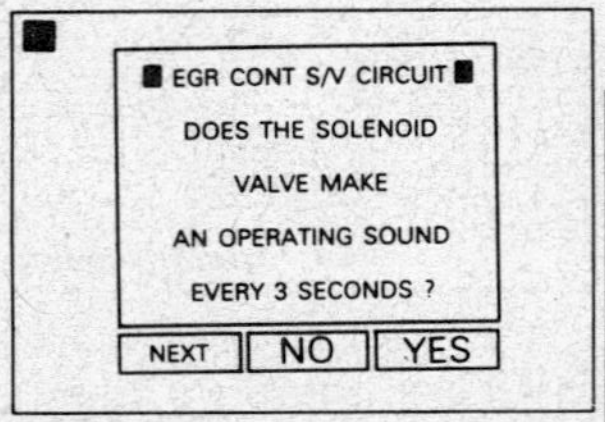

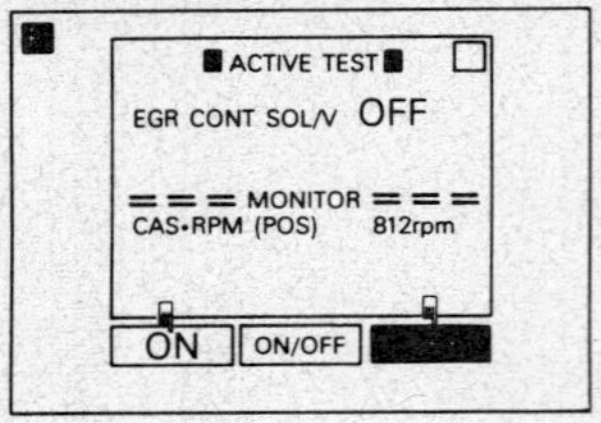

Ⓐ

CHECK COMPONENT
(E.G.R. control solenoid valve).

1) Turn ignition switch "ON".
2) Perform "EGR CONT S/V CIRCUIT" in "FUNCTION TEST" mode with CONSULT.

— OR —

1) Turn ignition switch "ON".
2) Turn E.G.R. control solenoid valve "ON" and "OFF" in "ACTIVE TEST" mode with CONSULT and check operating sound.

— OR —

Refer to "Electrical Components Inspection".

N.G. → Replace E.G.R. control solenoid valve.

O.K. ↓

Check resistance of exhaust gas temperature sensor.

O.K. ↓

Check E.C.U. pin terminals for damage or the connection of E.C.U. harness connector.

Diagnostic Procedure 30 (Cont'd)

Perform FINAL CHECK by the following procedure after repair is completed.

FINAL CHECK

↓

Erase the self-diagnosis memory.

↓

Perform driving test under the following conditions.
1) Warm up engine sufficiently.
2) Use test driving modes indicated in figure G

↓

Make sure that check engine light does not come on during driving test.

- Comes on → Recheck E.C.U. pin terminals for damage or the connection of E.C.U. harness connector.
- Does not come on → INSPECTION END

ROAD TEST

Test condition

Drive vehicle under the following conditions with a suitable shift position.

(1) Engine speed:
2,150±450 rpm

(2) Intake manifold vacuum:
−32.0±6.7 kPa
(−240±50 mmHg, −9.45±1.97 inHg)

Driving mode

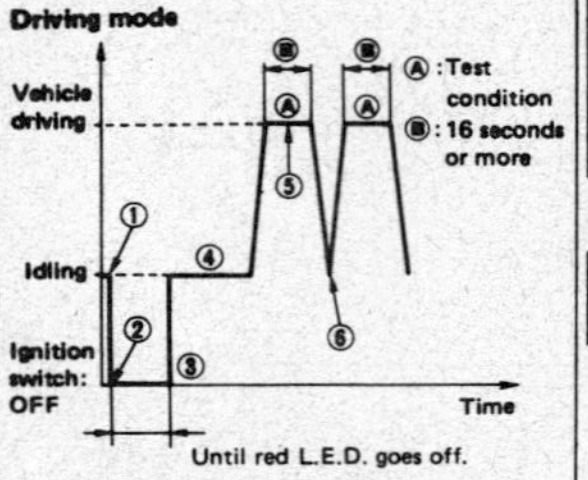

① Start engine and warm it up sufficiently.
② Turn off ignition switch and keep it off until red L.E.D. goes off.
③ Start engine and make sure that air conditioner switch and rear defogger are turned "OFF" during driving test.
④ Keep engine running for at least 4 minutes.
⑤ Shift to suitable gear position and drive in "Test condition" for at least 16 seconds.
⑥ Decrease engine revolution to less than 2,000 rpm.
⑦ Repeat steps ⑤ through ⑥ at least 1 time.

CHECK ENGINE LIGHT

Diagnostic Procedure 30 (Cont'd)

Non-California model

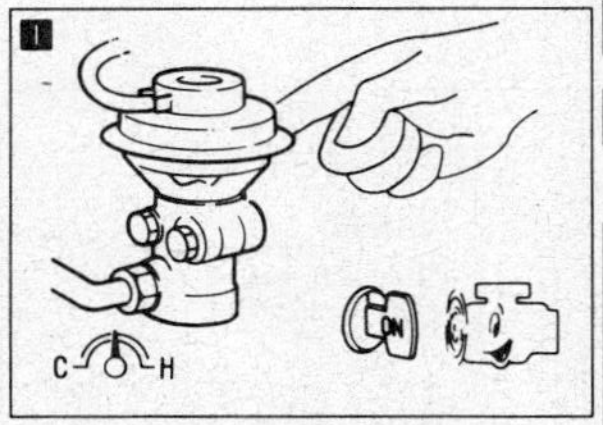

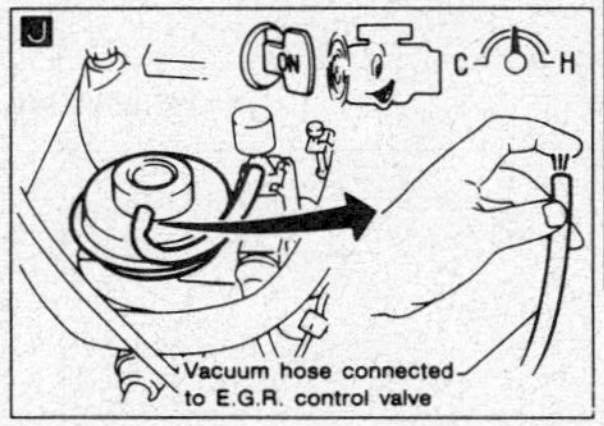

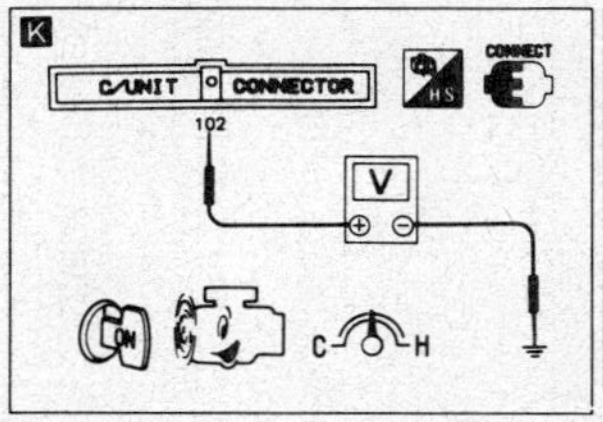

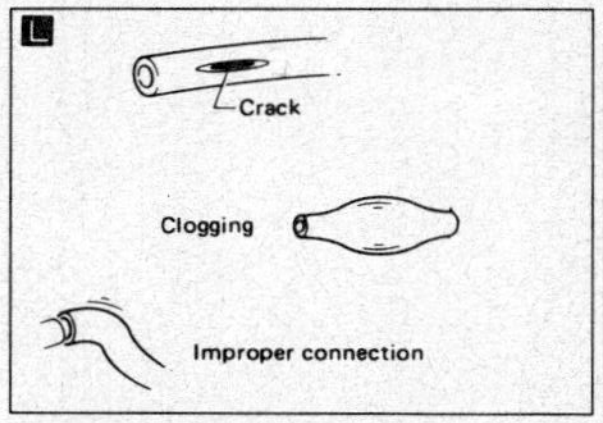

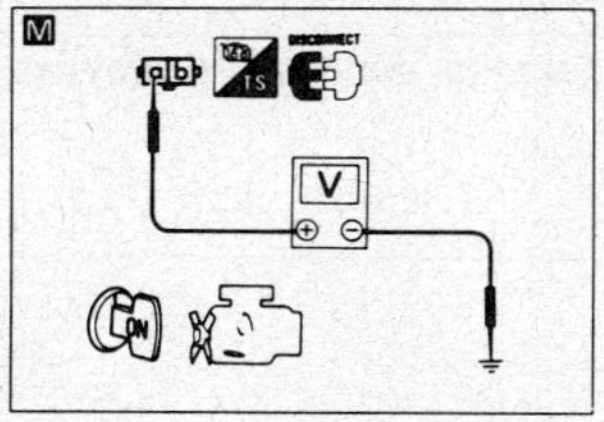

INSPECTION START

↓

I CHECK OVERALL FUNCTION.
1) Start engine and warm it up sufficiently.
2) Perform self-diagnosis.
Make sure that code No. 12 is not displayed.
Make sure that both crank angle sensor and E.C.U.'s C.P.U. are not in "fail-safe" state.
3) Make sure that E.G.R. control valve spring is lifted up and down when racing engine.
(Use your finger.)

→ Is lifted up and down → INSPECTION END

↓ Is not lifted up and down

J CHECK VACUUM SOURCE TO E.G.R. CONTROL VALVE.
1) Disconnect vacuum hose to E.G.R. control valve.
2) Make sure that vacuum exists under the following conditions.
At idle:
Vacuum should exist.
Engine speed is about 2,000 rpm:
Vacuum should not exist.

→ O.K. → CHECK COMPONENT (E.G.R. control valve). Refer to "Electrical Components Inspection".
→ N.G. → Replace malfunctioning component(s).

↓ N.G.

K CHECK CONTROL FUNCTION.
1) Check voltage between E.C.U. terminal (102) and ground under the following conditions.
Voltage:
At idle
0.7 - 0.8V
Engine speed is about 2,000 rpm
Battery voltage

→ O.K. → **L** CHECK VACUUM HOSE.
1) Check vacuum hose for clogging, cracks and proper connection.

↓ N.G.

Ⓐ

Diagnostic Procedure 30 (Cont'd)

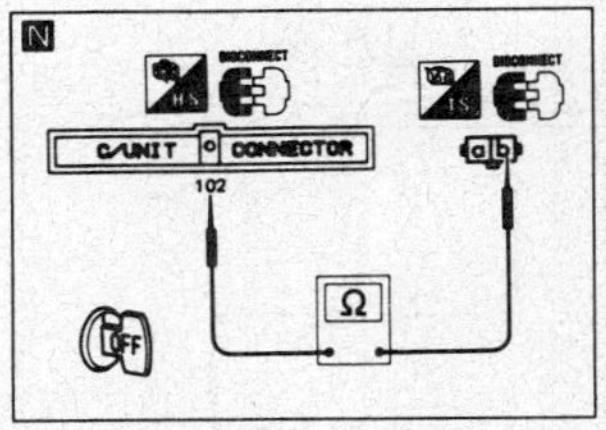

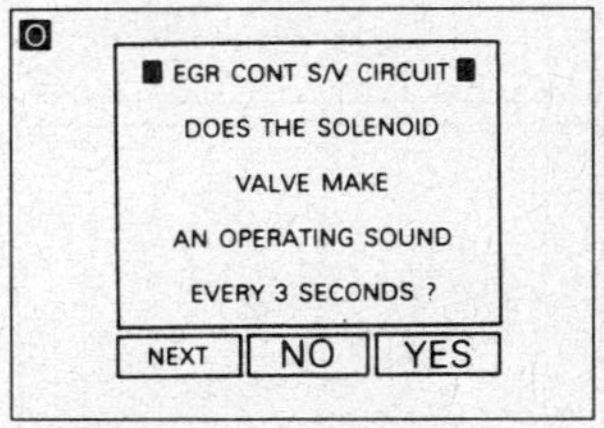

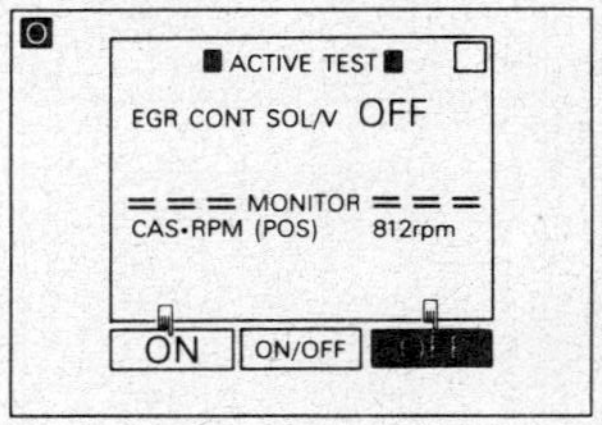

Ⓐ

↓

M CHECK POWER SUPPLY.
1) Stop engine.
2) Disconnect E.G.R. control solenoid valve harness connector.
3) Turn ignition switch "ON".
4) Check voltage between terminal ⓐ and ground.
Voltage: Battery voltage

→ N.G. → Check the following.
- Harness connectors (F27), (M49)
- Harness continuity between E.G.R. control solenoid valve and ignition switch

If N.G., repair harness or connectors.

↓ O.K.

N CHECK OUTPUT SIGNAL CIRCUIT.
1) Turn ignition switch "OFF".
2) Disconnect E.C.U. harness connector.
3) Check harness continuity between E.C.U. terminal (102) and terminal ⓑ.
Continuity should exist.

→ N.G. → Repair harness or connectors.

↓ O.K.

O CHECK COMPONENT (E.G.R. control solenoid valve).
1) Turn ignition switch "ON".
2) Perform "EGR CONT S/V CIRCUIT" in "FUNCTION TEST" mode with CONSULT.

— OR —

1) Turn ignition switch "ON".
2) Turn E.G.R. control solenoid valve "ON" and "OFF" in "ACTIVE TEST" mode with CONSULT and check operating sound.

— OR —

Refer to "Electrical Components Inspection".

→ N.G. → Replace E.G.R. control solenoid valve.

↓ O.K.

Check E.C.U. pin terminals for damage or the connection of E.C.U. harness connector.

Diagnostic Procedure 31

EXHAUST GAS SENSOR (Code No. 33)

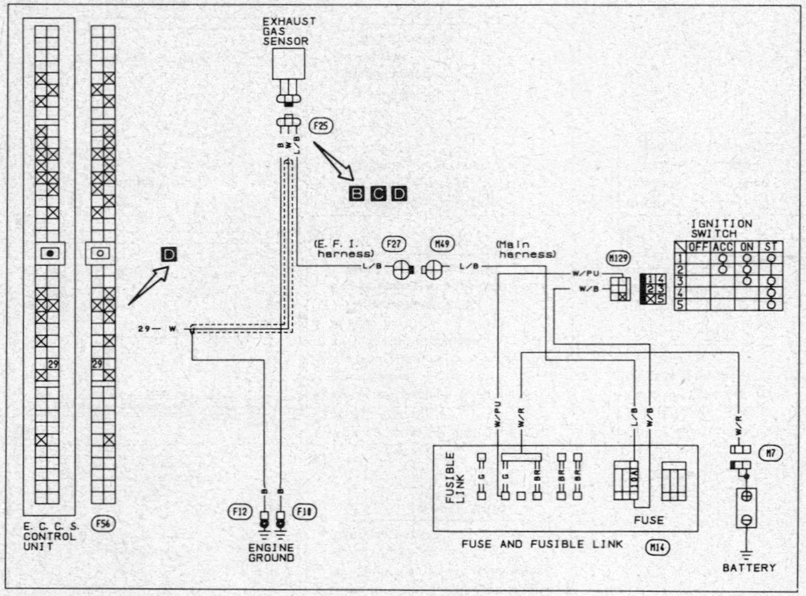

Harness layout

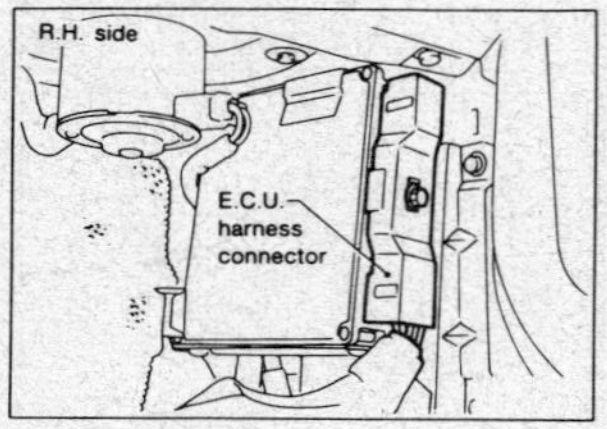

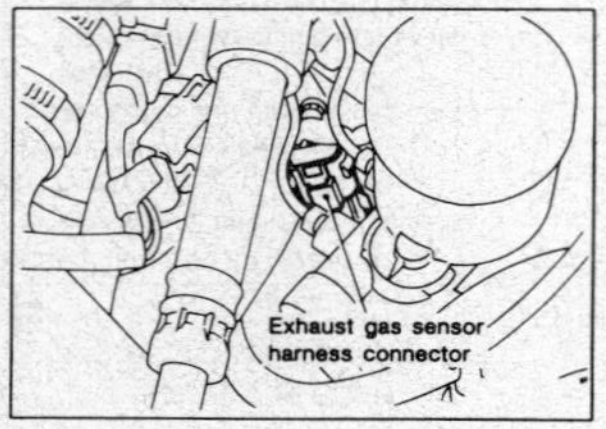

Diagnostic Procedure 31 (Cont'd)

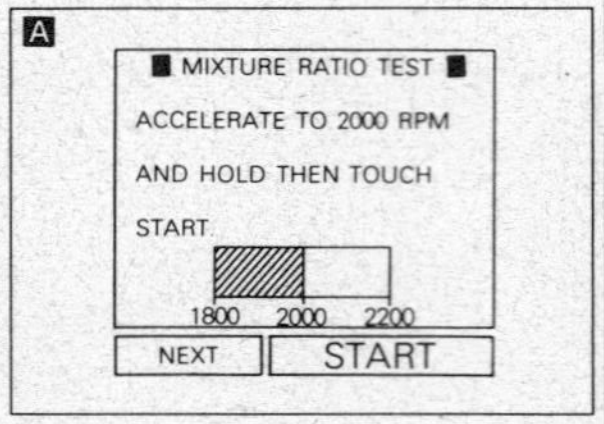

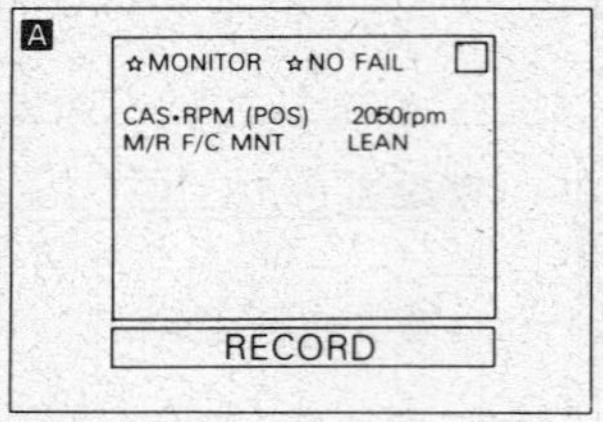

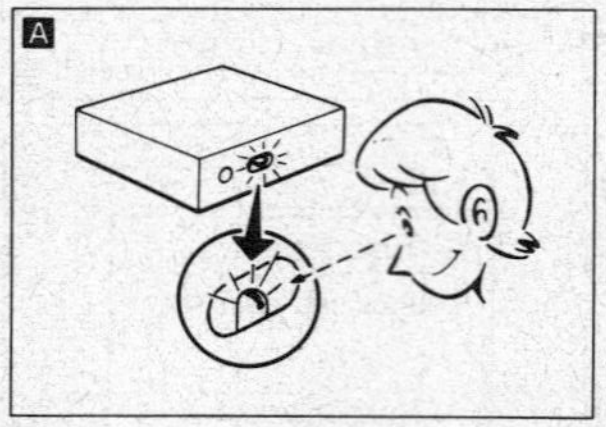

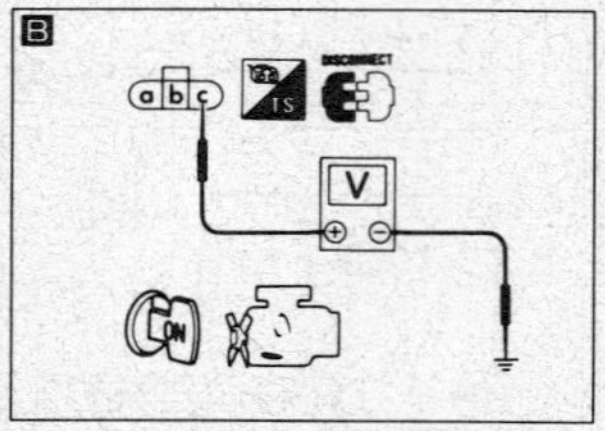

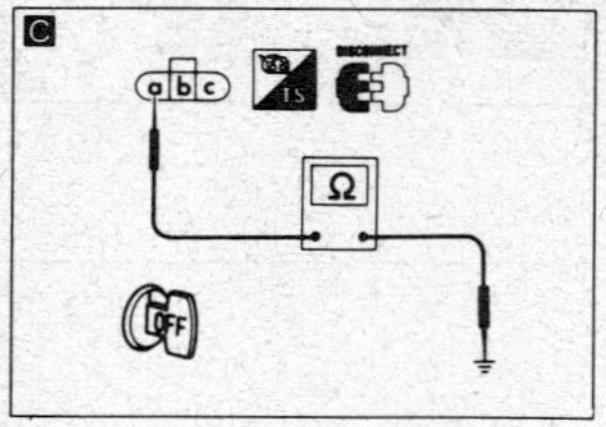

INSPECTION START

A CHECK EXHAUST GAS SENSOR CIRCUIT.

Perform "MIXTURE RATIO TEST" in "FUNCTION TEST" mode with CONSULT.

— OR —

1) Start engine and warm it up sufficiently.
2) Make sure that "M/R F/C MNT" in "DATA MONITOR" mode indicates "RICH" and "LEAN" periodically more than 5 times during 10 seconds at 2,000 rpm.

— OR —

1) Start engine and warm it up sufficiently.
2) Run engine at about 2,000 rpm for about 2 minutes under no- load.
3) Set E.C.U. diagnosis mode to Mode II.
4) Keep engine speed at 2,000 rpm and make sure that inspection lamp on E.C.U. goes on and off more than 5 times during 10 seconds.

O.K. → INSPECTION END

N.G. ↓

B CHECK POWER SUPPLY.
1) Disconnect exhaust gas sensor harness connector.
2) Turn ignition switch "ON".
3) Check voltage between terminal ⓒ and ground.

Voltage: Battery voltage

N.G. → Check the following.
- Harness connectors F27, M49
- 10A fuse
- Harness continuity between exhaust gas sensor and ignition switch

If N.G., repair harness or connectors.

O.K. ↓

C CHECK GROUND CIRCUIT.
1) Turn ignition switch "OFF".
2) Check harness continuity between terminal ⓐ and engine ground.

Continuity should exist.

N.G. → Repair harness or connectors.

O.K. → Ⓐ

Diagnostic Procedure 31 (Cont'd)

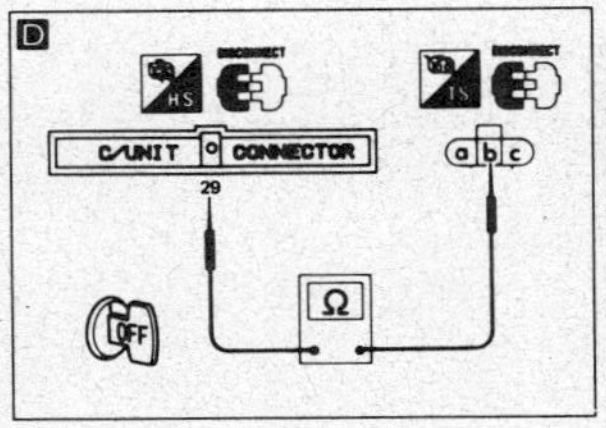

D

Ⓐ

CHECK INPUT SIGNAL CIRCUIT.
1) Disconnect E.C.U. harness connector.
2) Check harness continuity between E.C.U. terminal ㉙ and terminal ⓑ.
Continuity should exist.

N.G. → Repair harness or connectors.

O.K. ↓

Replace exhaust gas sensor.

Perform FINAL CHECK by the following procedure after repair is completed.

FINAL CHECK
↓
Erase the self-diagnosis memory.
↓
Perform driving test.
↓
Perform self-diagnosis again.

N.G. → Recheck E.C.U. pin terminals for damage or the connection of E.C.U. harness connector.

O.K. ↓

INSPECTION END

Diagnostic Procedure 32

DETONATION SENSOR (Code No. 34)

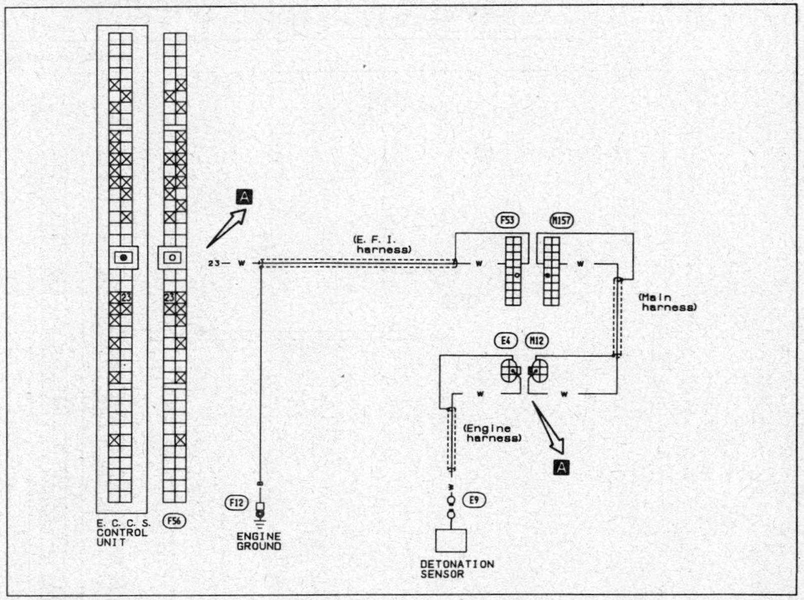

Harness layout

Diagnostic Procedure 32 (Cont'd)

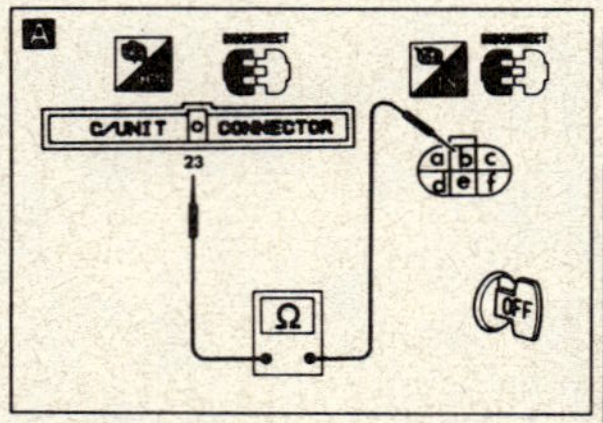

INSPECTION START

↓

A CHECK INPUT SIGNAL CIRCUIT.
1) Disconnect E.C.U. harness connector and detonation sensor sub-harness connector.
2) Check harness continuity between terminal ⓑ and E.C.U. terminal ㉓. **Continuity should exist.**

N.G. → Check the following.
- Harness connectors (F53), (M157)
- Harness continuity between E.C.U. and detonation sensor sub-harness connector

If N.G., repair harness or connectors.

O.K. ↓

CHECK COMPONENT (Detonation sensor). Refer to "Electrical Components Inspection".

N.G. → Replace detonation sensor.

O.K. ↓

Check E.C.U. pin terminals for damage or the connection of E.C.U. harness connector.

Perform FINAL CHECK by the following procedure after repair is completed.

FINAL CHECK

↓

Erase the self-diagnosis memory.

↓

Perform driving test.

↓

Perform self-diagnosis again.

N.G. → Recheck E.C.U. pin terminals for damage or the connection of E.C.U. harness connector.

O.K. ↓

INSPECTION END

Diagnostic Procedure 33

EXHAUST GAS TEMPERATURE SENSOR (Code No. 35) (CHECK ENGINE LIGHT ITEM): CALIFORNIA MODEL ONLY

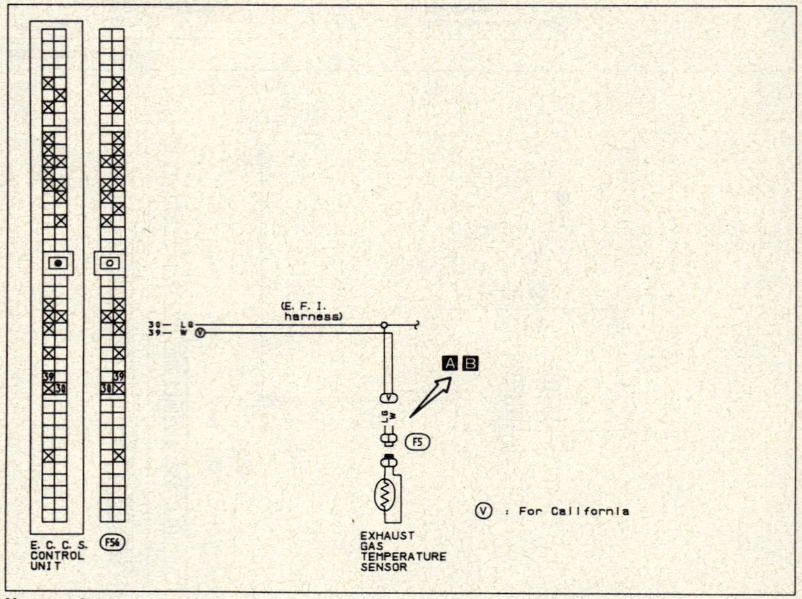

Harness layout

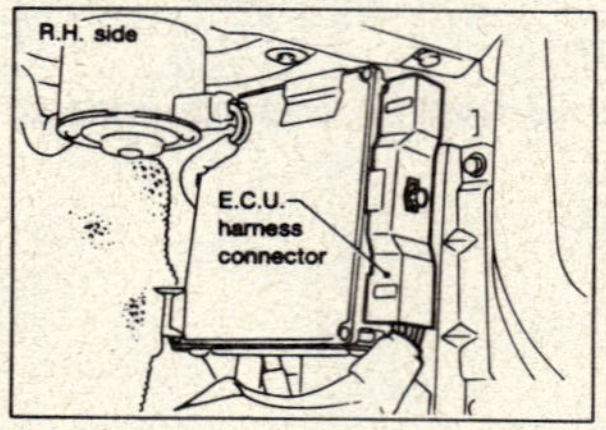

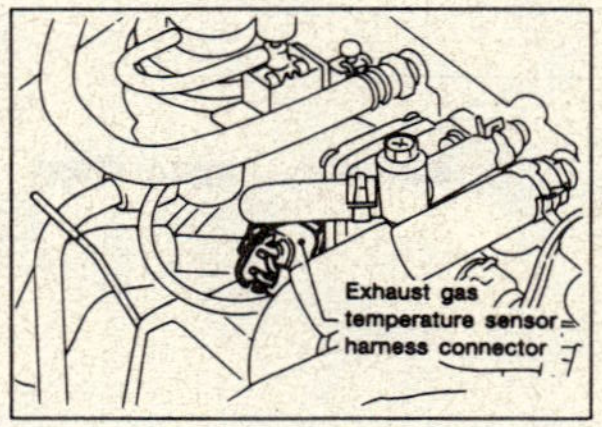

Diagnostic Procedure 33 (Cont'd)

A
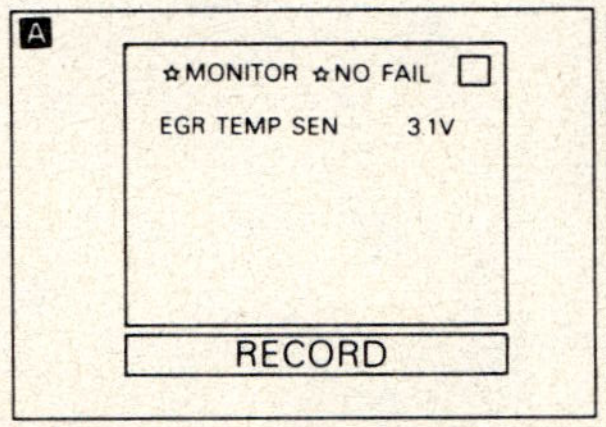

A
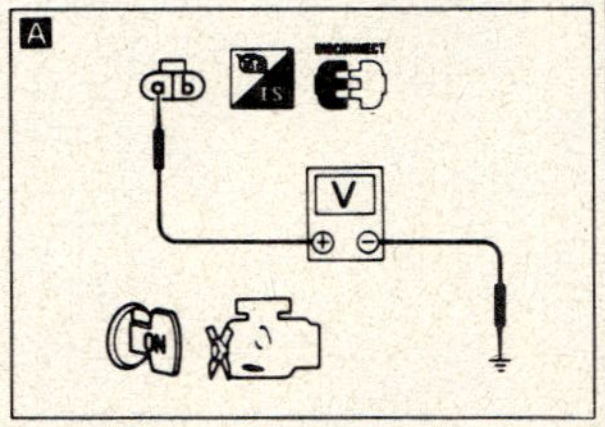

B
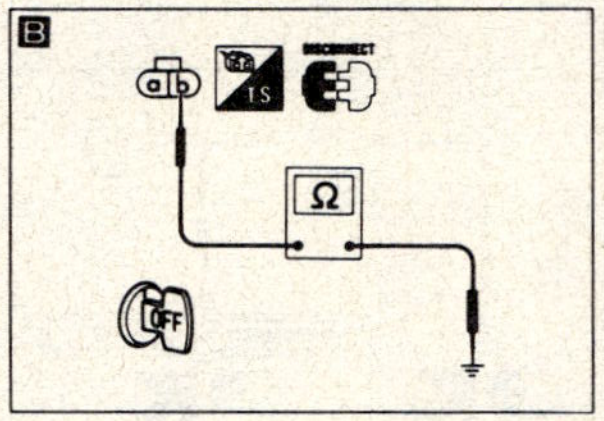

INSPECTION START

↓

A CHECK POWER SUPPLY.
1) Start engine and warm it up sufficiently.
2) Read exhaust gas temperature sensor signal in "DATA MONITOR" mode with CONSULT.
Voltage:
Less than 4.5V
— OR —
1) Disconnect exhaust gas temperature sensor harness connector.
2) Turn ignition switch "ON".
3) Check voltage between terminal ⓐ and ground.
Voltage:
Approximately 5V

N.G. → Repair harness or connectors.

↓ O.K.

B CHECK GROUND CIRCUIT.
1) Turn ignition switch "OFF".
2) Check harness continuity between terminal ⓑ and engine ground.
Continuity should exist.

N.G. → Repair harness or connectors.

↓ O.K.

CHECK COMPONENT
(Exhaust gas temperature sensor.)
Refer to "Electrical Components Inspection".

N.G. → Replace exhaust gas temperature sensor.

↓ O.K.

Check E.C.U. pin terminals for damage or the connection of E.C.U. harness connector.

Perform FINAL CHECK by the following procedure after repair is completed.

FINAL CHECK

↓

Erase the self-diagnosis memory.

↓

Perform driving test.

↓

Perform self-diagnosis again.

N.G. → Recheck E.C.U. pin terminals for damage or the connection of E.C.U. harness connector.

↓ O.K.

INSPECTION END

Diagnostic Procedure 34

FUEL TEMPERATURE SENSOR (Code No. 42) (CHECK ENGINE LIGHT ITEM)

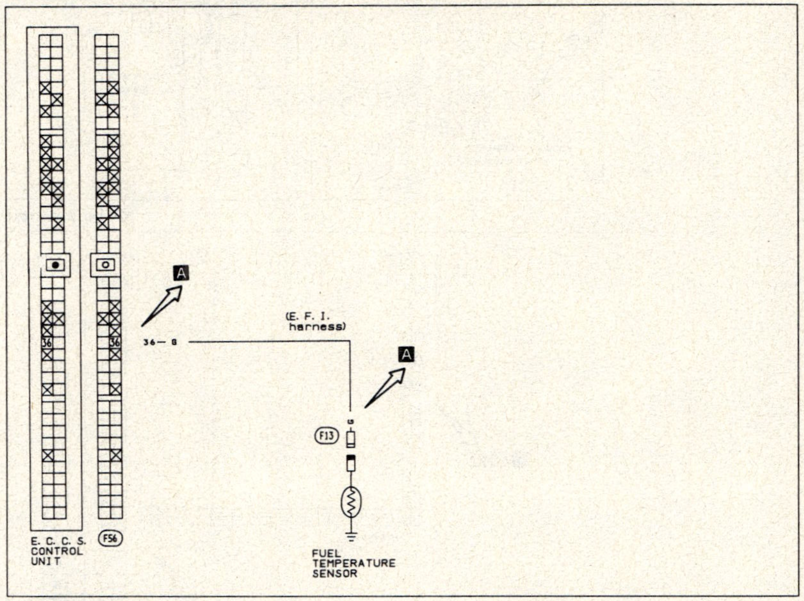

Harness layout

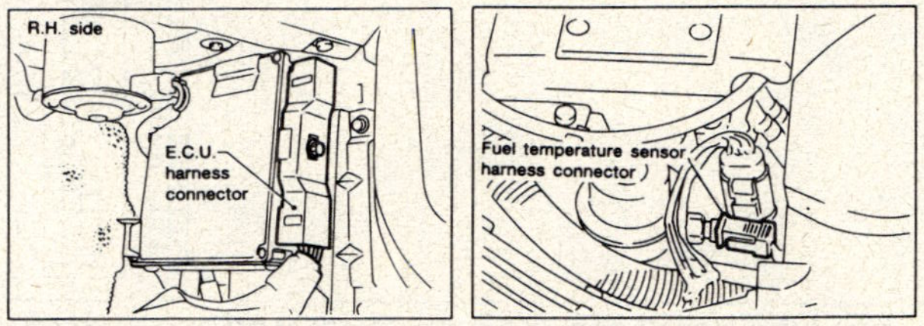

Diagnostic Procedure 34 (Cont'd)

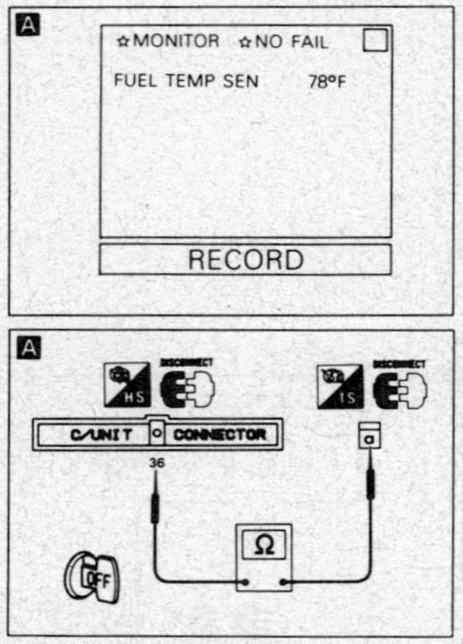

INSPECTION START

A — CHECK INPUT SIGNAL CIRCUIT.
Read fuel temperature sensor signal in "DATA MONITOR" mode with CONSULT after engine warm-up.
Fuel temperature:
20 - 60°C (68 - 140°F)
OR
1) Disconnect E.C.U. harness connector and fuel temperature sensor harness connector.
2) Check harness continuity between terminal ⓐ and E.C.U. terminal ㊱.
Continuity should exist.

N.G. → Repair harness or connectors.

O.K. ↓

CHECK COMPONENT
(Fuel temperature sensor).
Refer to "Electrical Components Inspection".

N.G. → Replace fuel temperature sensor.

O.K. ↓

Check E.C.U. pin terminals for damage or the connection of E.C.U. harness connector.

Perform FINAL CHECK by the following procedure after repair is completed.

FINAL CHECK
↓
Erase the self-diagnosis memory.
↓
Perform driving test.
↓
Perform self-diagnosis again.

N.G. → Recheck E.C.U. pin terminals for damage or the connection of E.C.U. harness connector.

O.K. ↓

INSPECTION END

Diagnostic Procedure 35

THROTTLE SENSOR (Code No. 43) (CHECK ENGINE LIGHT ITEM)

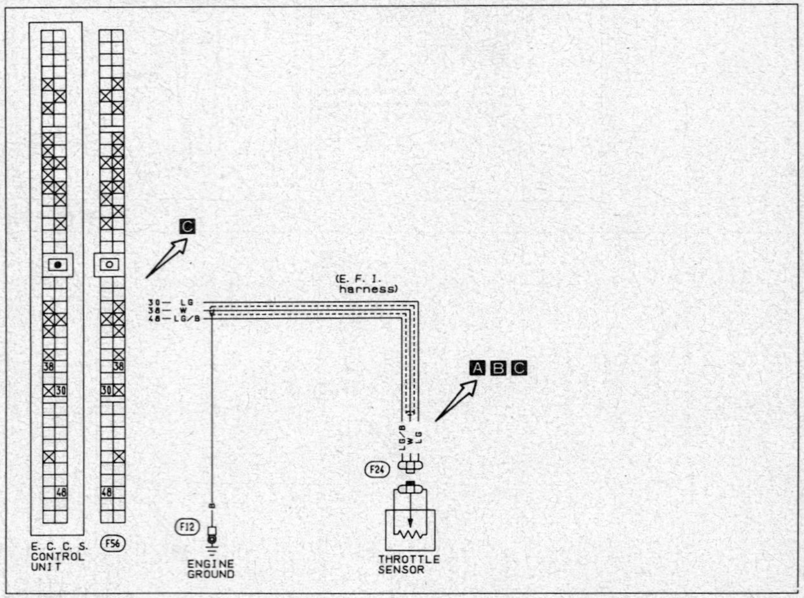

Harness layout

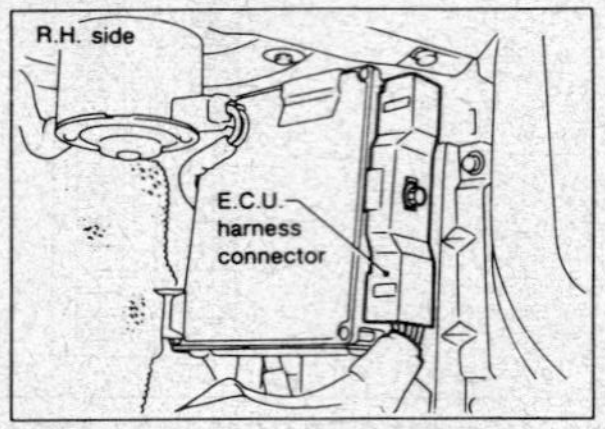

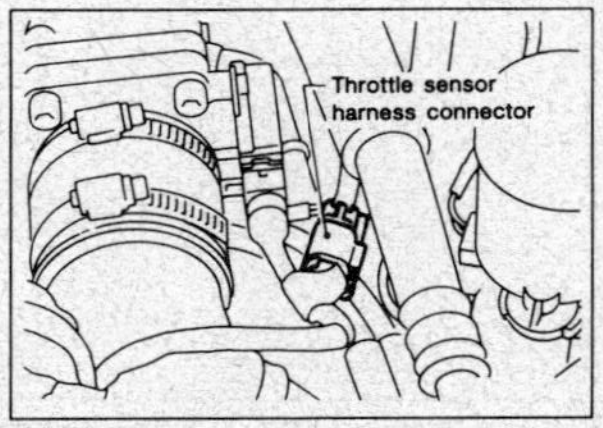

Diagnostic Procedure 35 (Cont'd)

A
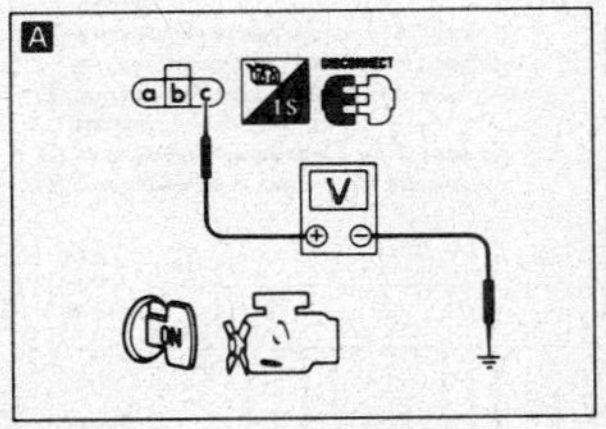

B
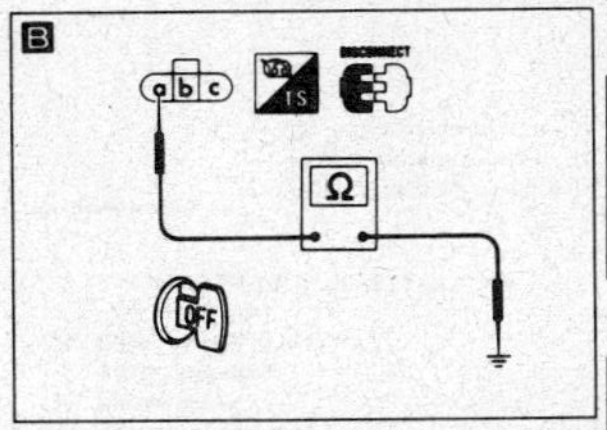

C
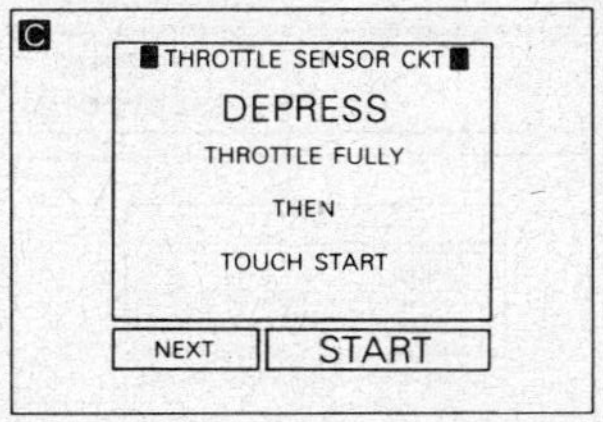

C
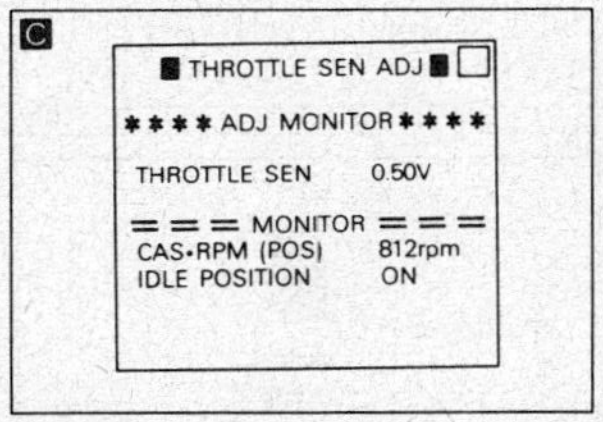

C
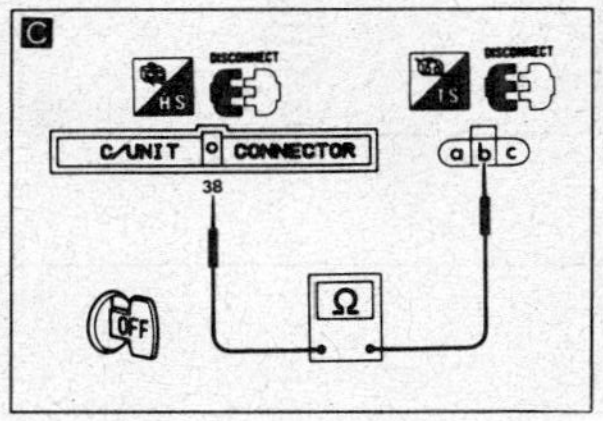

INSPECTION START

↓

A **CHECK POWER SUPPLY.**
1) Disconnect throttle sensor harness connector.
2) Turn ignition switch "ON".
3) Check voltage between terminal ⓒ and ground.
Voltage: Approximately 5V

N.G. → Repair harness or connectors.

O.K. ↓

B **CHECK GROUND CIRCUIT.**
1) Turn ignition switch "OFF".
2) Check harness continuity between terminal ⓐ and engine ground.
Continuity should exist.

N.G. → Repair harness or connectors.

O.K. ↓

C **CHECK INPUT SIGNAL CIRCUIT.**
Perform "THROTTLE SENSOR CKT" in "FUNCTION TEST" mode with CONSULT.

— OR —

1) Reconnect throttle sensor harness connector.
2) Turn ignition switch "ON".
3) Read throttle sensor output voltage in "WORK SUPPORT" mode with CONSULT.
Throttle valve fully closed: 0.4 - 0.5V
Throttle valve fully open: Approx. 4.0V

— OR —

1) Disconnect E.C.U. harness connector.
2) Check harness continuity between E.C.U. terminal ㊳ and terminal ⓑ.
Continuity should exist.

N.G. → Repair harness or connectors.

O.K. ↓

CHECK COMPONENT
(Throttle sensor).
Refer to "Electrical Components Inspection".

N.G. → Replace throttle sensor.

O.K. ↓

Check E.C.U. pin terminals for damage or the connection of E.C.U. harness connector.

Diagnostic Procedure 35 (Cont'd)

Ⓐ

↓

Check E.C.U. pin terminals for damage or the connection of E.C.U. harness connector.

Perform FINAL CHECK by the following procedure after repair is completed.

FINAL CHECK

↓

Erase the self-diagnosis memory.

↓

Perform driving test.

↓

Perform self-diagnosis again.

N.G. → Recheck E.C.U. pin terminals for damage or the connection of E.C.U. harness connector.

O.K. ↓

INSPECTION END

Diagnostic Procedure 36

INJECTOR LEAK (Code No. 45) (CHECK ENGINE LIGHT ITEM); CALIFORNIA MODEL ONLY

INSPECTION START

↓

Start engine and warm it up sufficiently.

↓

Make sure engine runs smoothly at idle after warming.
- Runs smoothly → Race engine two or three times under no-load, then run engine at idle speed. → Set diagnosis to Mode II and check that inspection lamps on E.C.U. blink at 2,000 rpm under no-load.
  - Does not blink → Check idle CO%.
  - Blinks → INSPECTION END
- Does not run smoothly → Set the diagnosis mode selector of E.C.U. to Mode II. → Check if inspection lamp on E.C.U. stays off for 10 seconds with engine at idle speed.
  - Does not stay off → Check mixture ratio feedback system.
  - Stays off → A

A Remove all spark plugs from cylinder head. Is any plug wet with fuel?
- Yes → Replace injectors which supply cylinders having wet spark plugs.
- No → Remove injector assembly.

Keep fuel hose and all injectors connected to injector gallery.

↓

Turn ignition switch "ON". Make sure fuel does not drip from injector.
- Drips → Replace the injectors from which fuel is dripping.
- Does not drip → Go to FINAL CHECK.

A

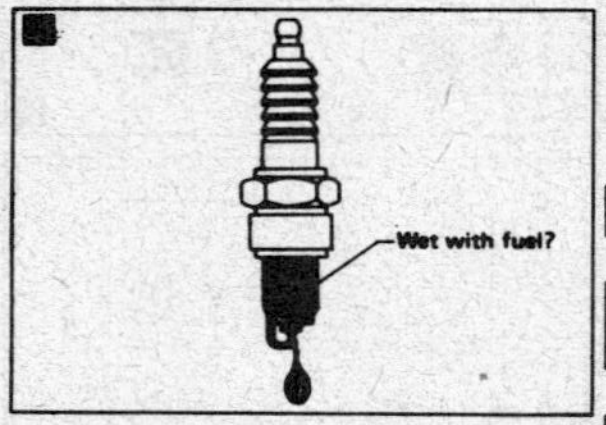

B

ACTIVE TEST

SELF-LEARN CONTROL	100%
=== MONITOR ===	
CAS-RPM (POS)	800rpm
ENG TEMP SEN	88°C
EXH GAS SEN	1.17V
A/F ALPHA	98%

CLEAR

C **ROAD TEST**

Test condition

Drive vehicle under the following conditions with a suitable shift position.

(1) Engine speed:
2,500±700 rpm

(2) Intake manifold vacuum:
−46.7±6.7 kPa
(−350±50 mmHg, −13.78±1.97 inHg)

Driving mode

Ⓐ : 60 seconds or more
Ⓑ : 5 seconds or more
Ⓒ : 10 seconds or more

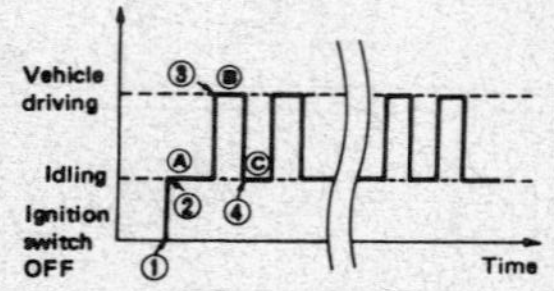

① Start engine and warm it up sufficiently.
② Keep engine at idle speed for at least 60 seconds.
③ Shift to a suitable gear position and drive in "Test condition" for at least 5 seconds.
④ Keep engine at idle speed for at least 10 seconds.
⑤ Repeat steps ② through ④ at least 10 times.

Diagnostic Procedure 36 (Cont'd)

Perform FINAL CHECK by the following procedure after repair is completed.

FINAL CHECK

↓

Erase the self-diagnosis memory. Make sure code No. 55 is displayed in Mode II.

↓

B Clear the self-learning data using the following procedure:
1) Start engine and warm it up sufficiently.
2) Select "SELF- LEARNING CONT" in "ACTIVE TEST" mode.
3) Clear the self-learning control coefficient by touching "CLEAR"

— OR —

2) Disconnect air flow meter connector, and restart and run engine for at least 30 seconds at 2,000 rpm.
3) Stop engine and reconnect air flow meter connector.
4) Make sure Code No. 12 is displayed in Mode II.
5) Erase the self- diagnosis memory.
Make sure Code No. 55 is displayed in Mode II.

↓

C Perform engine racing test as indicated in figure C.

↓

D Make sure that check engine light does not come on during engine test.
- Comes on → Recheck E.C.U. pin terminals for damage or the connection of E.C.U. harness connector.
- Does not come on → INSPECTION END

D

CHECK ENGINE LIGHT

Diagnostic Procedure 37

INJECTOR CIRCUIT (Code No. 51) (CHECK ENGINE LIGHT ITEM): CALIFORNIA MODEL
INJECTOR CIRCUIT (Not self-diagnostic item): NON-CALIFORNIA MODEL

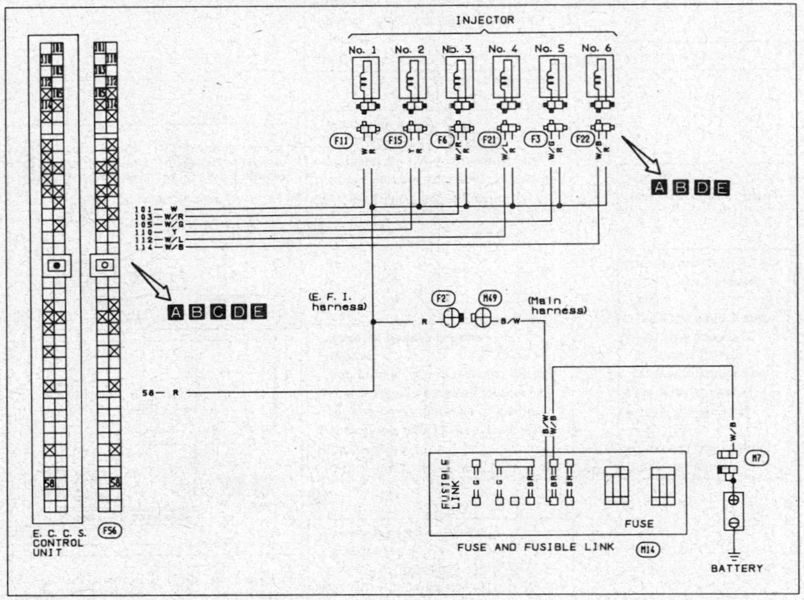

Harness layout

Diagnostic Procedure 37 (Cont'd)

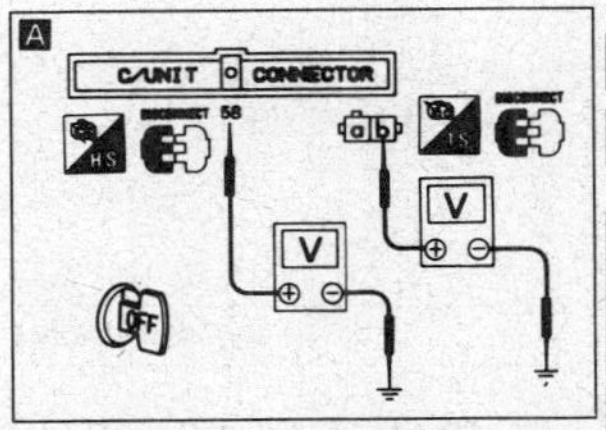

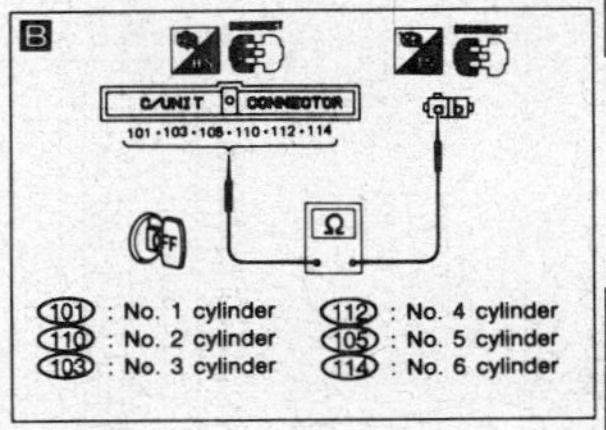

California model

INSPECTION START

↓

A CHECK POWER SUPPLY.
1) Disconnect injector harness connector and E.C.U. harness connector.
2) Check voltage between terminal ⓑ and ground, E.C.U. terminal (58) and ground.
Voltage: Battery voltage

N.G. → Check the following.
- Harness connectors (F27), (M49)
- "BR" fusible link
- Harness continuity between battery and injector
- Harness continuity between battery and E.C.U.

If N.G., repair harness or connectors.

O.K. ↓

B CHECK OUTPUT SIGNAL CIRCUIT.
1) Check harness continuity between terminal ⓐ and E.C.U. terminals (101), (103), (105), (110), (112), (114).
Continuity should exist.

N.G. → Repair harness or connectors.

O.K. ↓

CHECK COMPONENT (Injector).
Refer to "Electrical Components Inspection".

N.G. → Replace injector.

O.K. ↓

Check E.C.U. pin terminals for damage or the connection of E.C.U. harness connector.

Perform FINAL CHECK by the following procedure after repair is completed.

FINAL CHECK

↓

Erase the self-diagnosis memory.

↓

Perform driving test.

↓

Perform self-diagnosis again.

N.G. → Recheck E.C.U. pin terminals for damage or the connection of E.C.U. harness connector.

O.K. ↓

INSPECTION END

Diagnostic Procedure 37 (Cont'd)

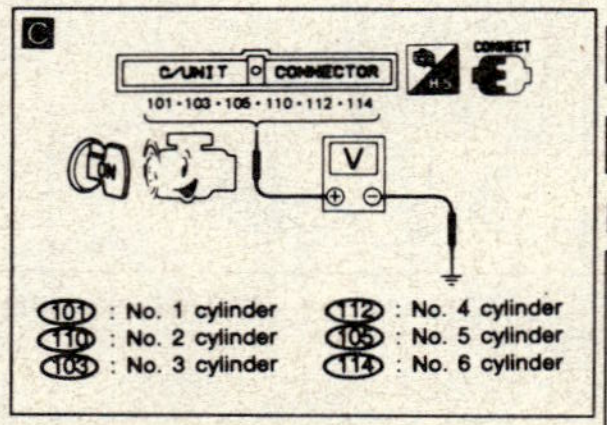

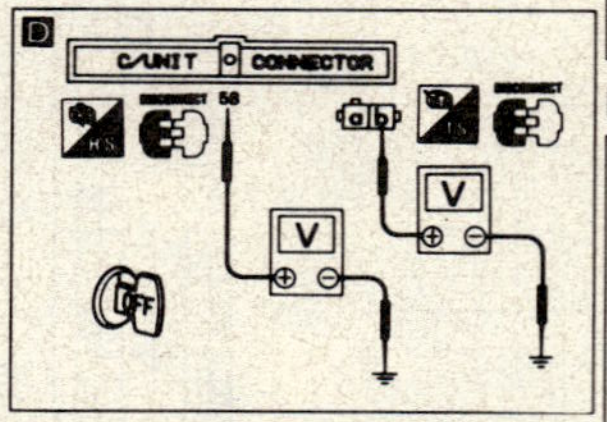

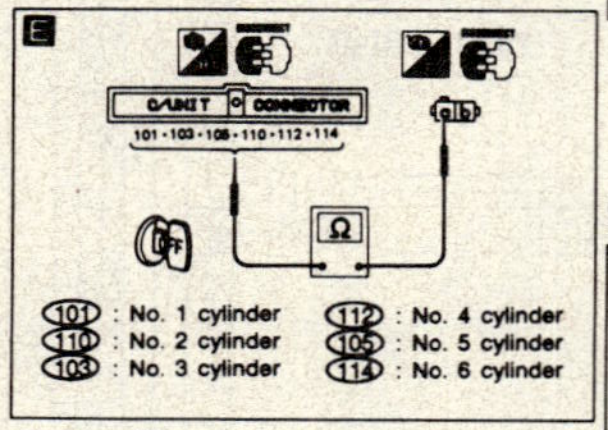

Non-California model

INSPECTION START
↓

C CHECK CONTROL FUNCTION.
1) Start engine.
2) Check voltage between E.C.U. terminals (101), (103), (105), (110), (112), (114) and ground.
Voltage: Battery voltage

O.K. → INSPECTION END

N.G. ↓

D CHECK POWER SUPPLY.
1) Stop engine.
2) Disconnect injector harness connector and E.C.U. harness connector.
3) Check voltage between terminal (b) and ground, E.C.U. terminal (58) and ground.
Voltage: Battery voltage

N.G. → Check the following.
- Harness connectors (F27), (M49)
- "BR" fusible link
- Harness continuity between battery and injector
- Harness continuity between battery and E.C.U.

If N.G., repair harness or connectors.

O.K. ↓

E CHECK OUTPUT SIGNAL CIRCUIT.
1) Check harness continuity between terminal (a) and E.C.U. terminals (101), (103), (105), (110), (112), (114)
Continuity should exist.

N.G. → Repair harness or connectors.

O.K. ↓

CHECK COMPONENT (Injector).
Refer to "Electrical Components Inspection".

N.G. → Replace injector.

O.K. ↓

Check E.C.U. pin terminals for damage or the connection of E.C.U. harness connector.

Diagnostic Procedure 38

THROTTLE VALVE SWITCH (Not self-diagnostic item)

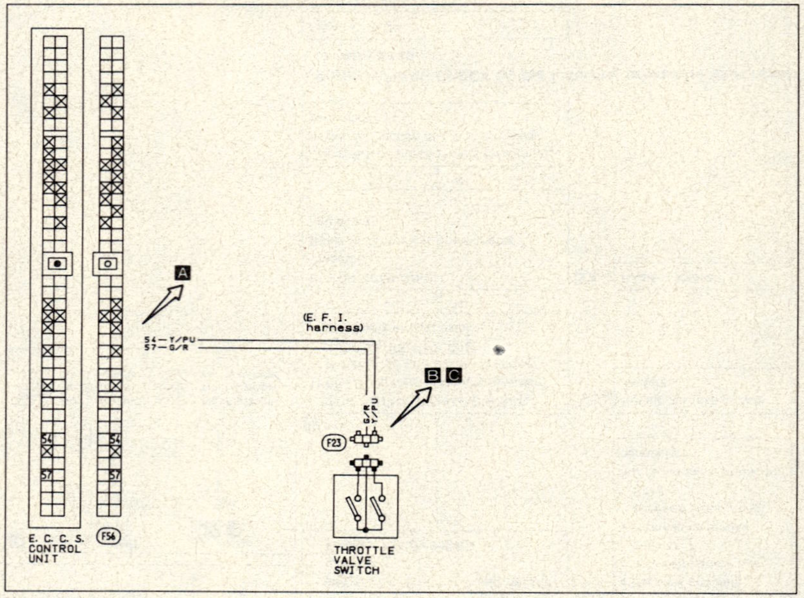

Harness layout

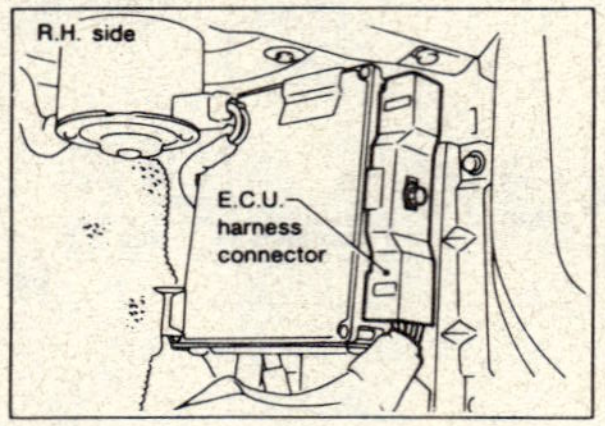

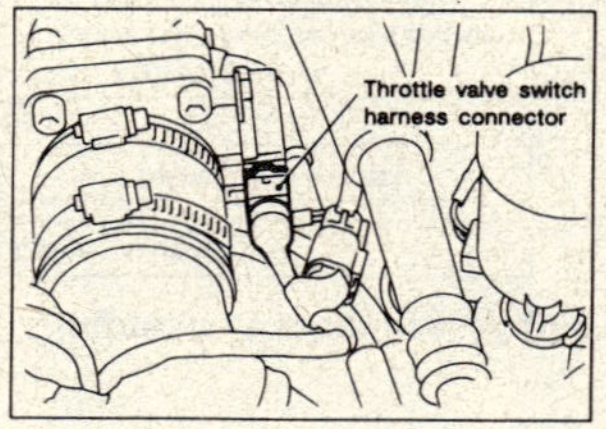

Diagnostic Procedure 38 (Cont'd)

A
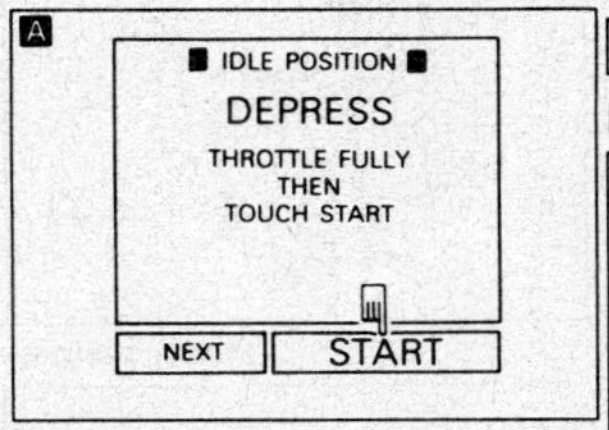

A
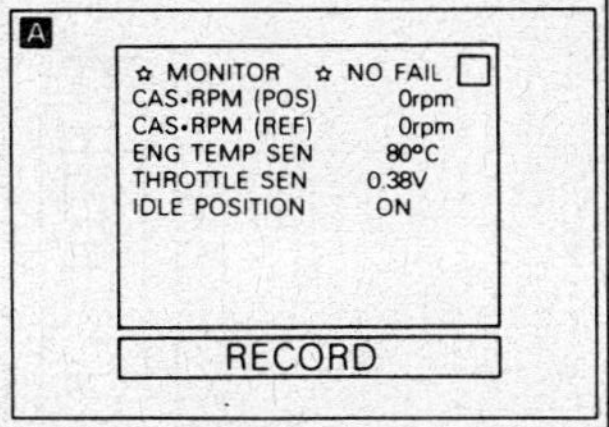

A
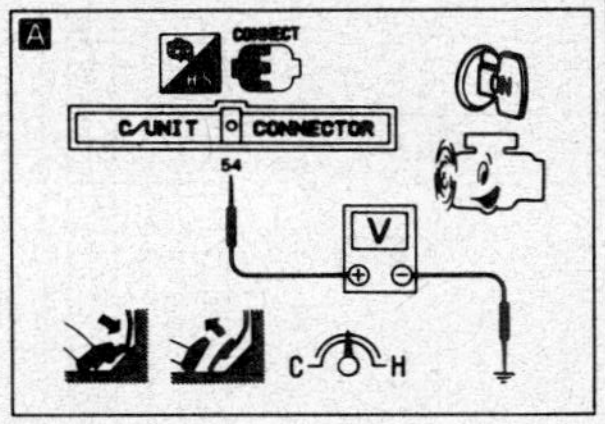

B
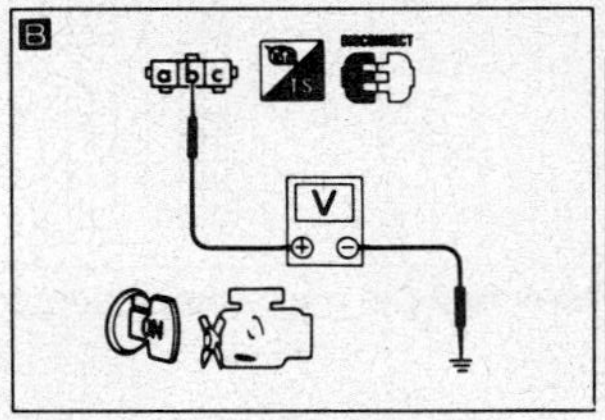

C
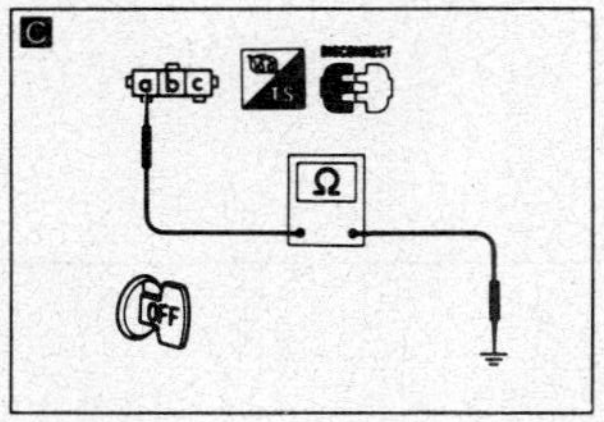

INSPECTION START

↓

A CHECK OVERALL FUNCTION.
1) Start engine and warm it up sufficiently.
2) Turn ignition switch "OFF".
3) Disconnect throttle sensor harness connector.
4) Turn ignition switch "ON".
5) Perform "IDLE POSITION" in "FUNCTION TEST" mode with CONSULT.

—OR—

5) Check "DATA MONITOR" mode with CONSULT.
Idle position:
Accelerator pedal is released ... ON
Accelerator pedal is depressed . OFF

—OR—

3) Turn ignition switch "ON".
4) Check voltage between E.C.U. terminal ㊹ and ground under the following conditions.
Voltage:
Accelerator pedal is released Approximately 8 - 10V
Accelerator pedal is depressed Approximately 0V

O.K. → INSPECTION END

N.G. ↓

B CHECK POWER SUPPLY.
1) Stop engine.
2) Disconnect throttle valve switch harness connector.
3) Turn ignition switch "ON".
4) Check voltage between terminal ⓑ and ground.
Voltage:
Approximately 8 - 10V

N.G. → Repair harness or connectors.

O.K. → Ⓐ

Diagnostic Procedure 38 (Cont'd)

Ⓐ

↓

■ CHECK GROUND CIRCUIT.
1) Turn ignition switch "OFF".
2) Check harness continuity between terminal ⓐ and engine ground.
Continuity should exist.

N.G. → Repair harness or connectors.

O.K. ↓

CHECK COMPONENT
(Throttle valve switch).
Refer to "Electrical Components Inspection".

N.G. → Replace throttle valve switch.

O.K. ↓

Check E.C.U. pin terminals for damage or the connection of E.C.U. harness connector.

Diagnostic Procedure 39

START SIGNAL (Not self-diagnostic item)

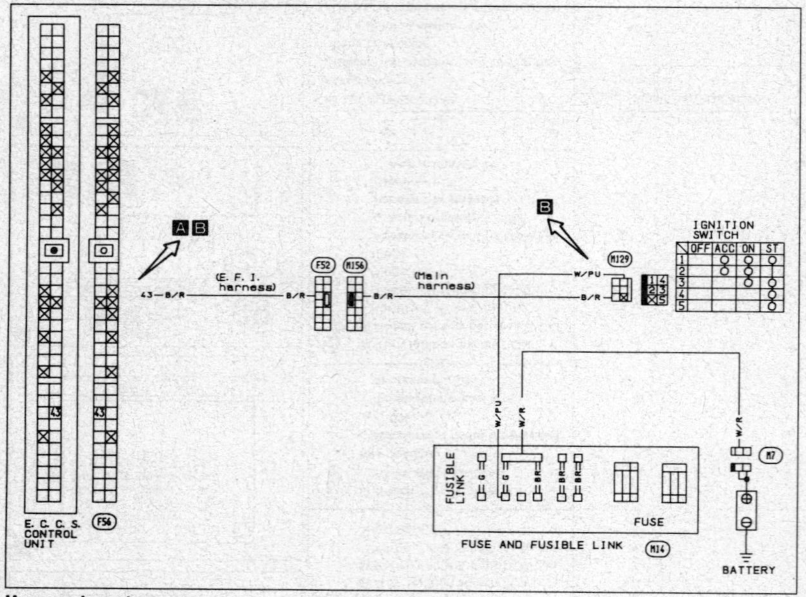

Harness layout

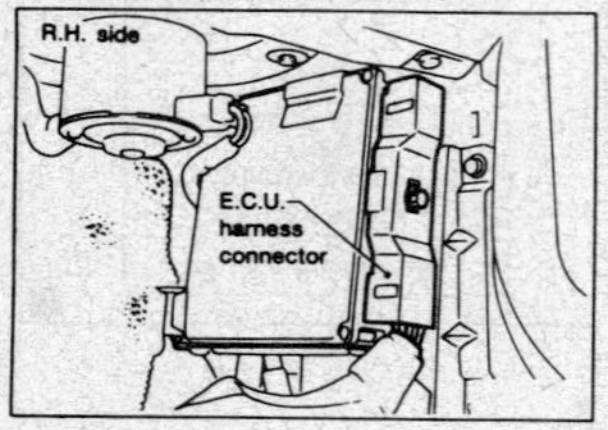

Diagnostic Procedure 39 (Cont'd)

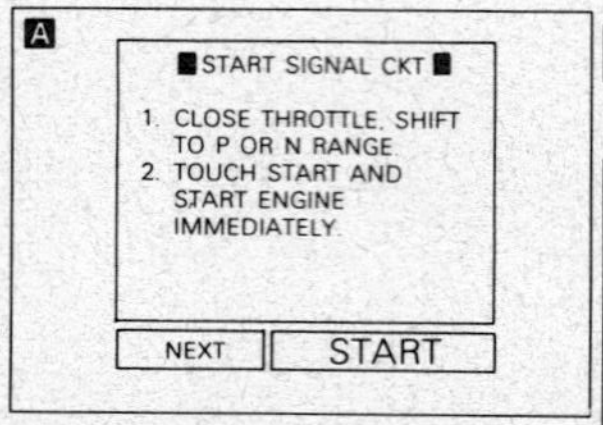

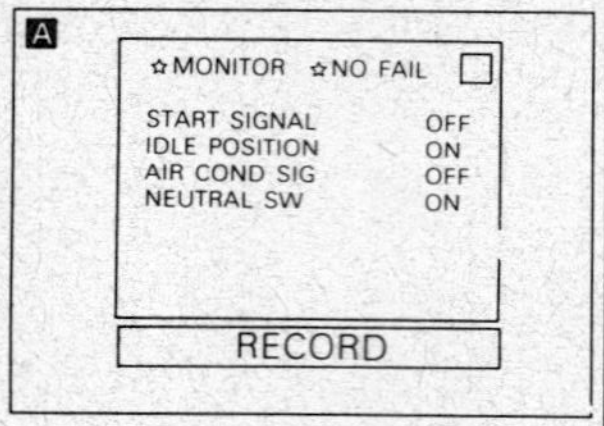

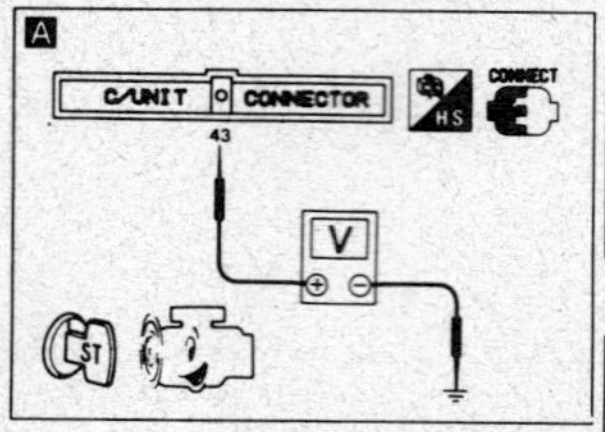

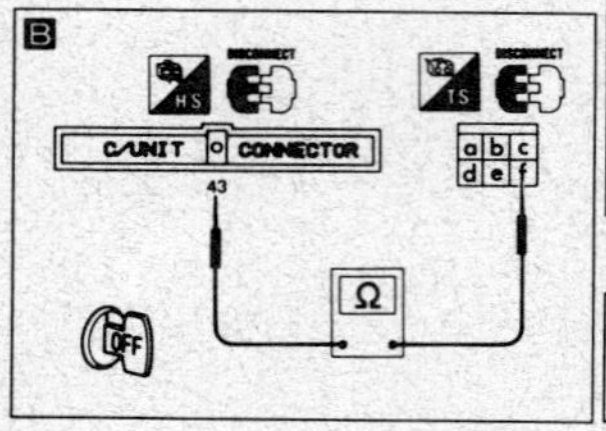

INSPECTION START

↓

A CHECK OVERALL FUNCTION.

1) Turn ignition switch "ON".
2) Perform "START SIGNAL CIRCUIT" in "FUNCTION TEST" mode with CONSULT.

—OR—

1) Turn ignition switch "ON".
2) Check start signal in "DATA MONITOR" mode with CONSULT.

IGN "ON"	OFF
IGN "START"	ON

—OR—

1) Turn ignition switch to "START".
2) Check voltage between E.C.U. terminal ㊸ and ground with CONSULT or tester.

Voltage:
Ignition switch "START"
Battery voltage
Except above
Approximately 0V

O.K. → INSPECTION END

↓ N.G.

B CHECK INPUT SIGNAL CIRCUIT.

1) Turn ignition switch "OFF".
2) Disconnect E.C.U. harness connector and ignition switch harness connector.
3) Check harness continuity between E.C.U. terminal ㊸ and terminal ⓕ.
Continuity should exist.

N.G. → Check the following.
- Harness connectors F52, M156
- Harness continuity between E.C.U. and ignition switch

If N.G., repair harness or connectors.

↓ O.K.

Check E.C.U. pin terminals for damage or the connection of E.C.U. harness connector.

Diagnostic Procedure 40 (Cont'd)

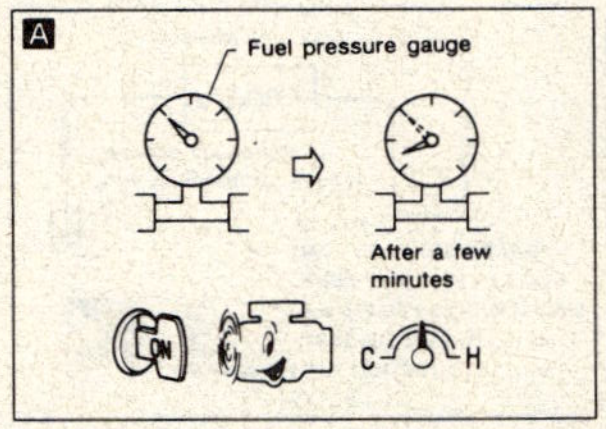

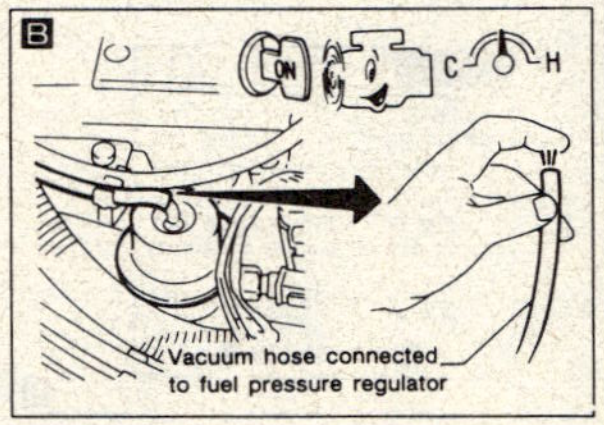

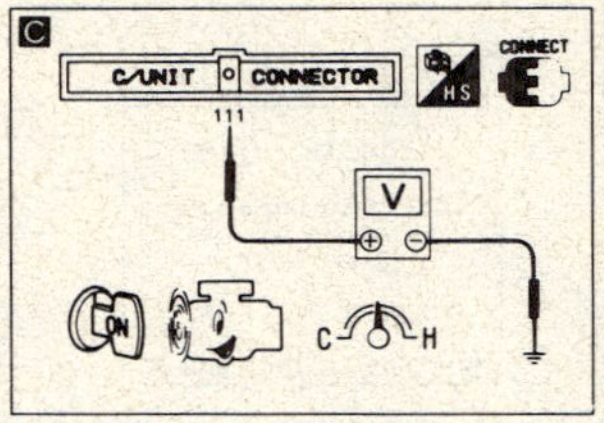

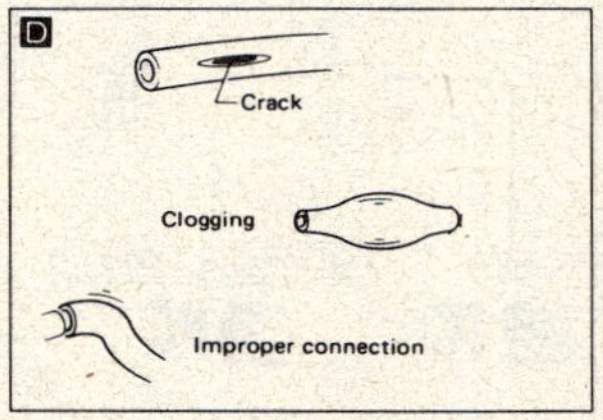

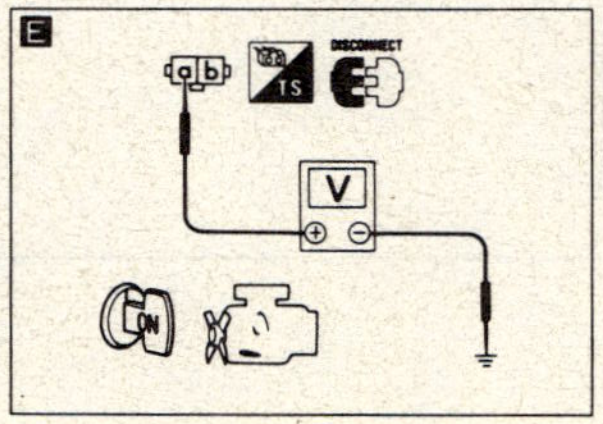

INSPECTION START

A CHECK OVERALL FUNCTION.
1) Install fuel pressure gauge in fuel feed hose.
2) Start engine and warm it up sufficiently.
3) Perform self-diagnosis. Make sure that code No. 42 is not displayed.
4) Stop engine.
5) Disconnect fuel temperature sensor harness connector and connect a resistor (0.1 kΩ) between fuel temperature sensor harness connector and ground. (*: See NOTE.)
6) Restart engine and check fuel pressure.
Fuel pressure:
Approximately 234 kPa (3.0 kg/cm², 43 psi)
7) Make sure that fuel pressure decreases after a few minutes.
Fuel pressure:
Approximately 235 kPa (2.4 kg/cm², 34 psi)

O.K. → INSPECTION END

N.G. ↓

B CHECK VACUUM SOURCE TO FUEL PRESSURE REGULATOR.
1) Stop engine.
2) Disconnect vacuum hose connected to fuel pressure regulator.
3) Restart engine.
4) Make sure that vacuum exists.

O.K. → CHECK COMPONENT (Fuel pressure regulator). Refer to "Fuel Pressure Check".

N.G. ↓

C CHECK CONTROL FUNCTION.
1) Stop engine and reconnect vacuum hose to fuel pressure regulator.
2) Restart engine.
3) Check voltage between E.C.U. terminal (111) and ground.
Voltage: 0 - 1.0V
4) After more than 3 minutes, recheck voltage between E.C.U. terminal (111) and ground.
Voltage: Battery voltage

O.K. → **D** CHECK VACUUM HOSE
1) Check vacuum hose for clogging, cracks and proper connection.

N.G. ↓ Ⓐ

Diagnostic Procedure 40

P.R.V.R. CONTROL (Not self-diagnostic item)

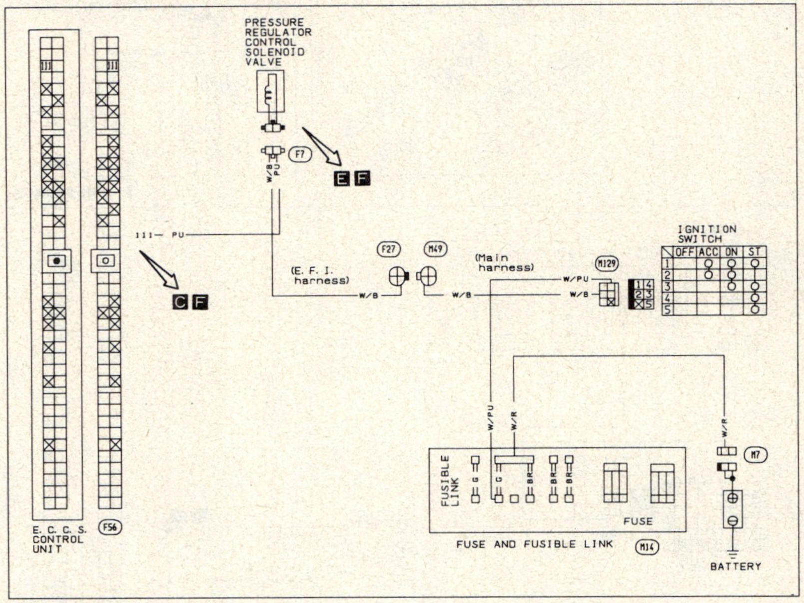

Harness layout

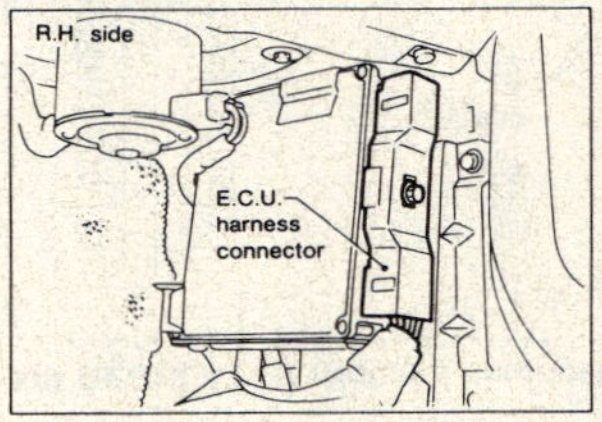

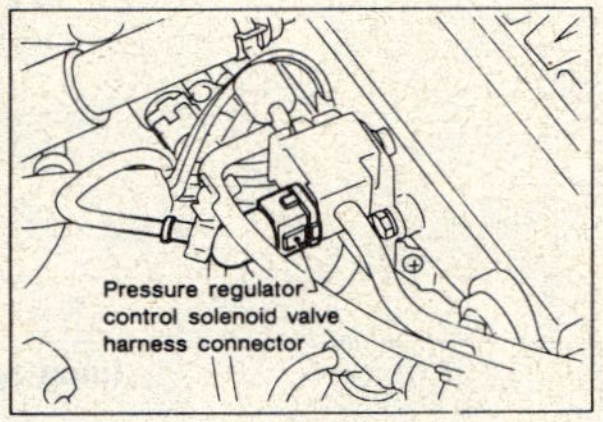

Diagnostic Procedure 40 (Cont'd)

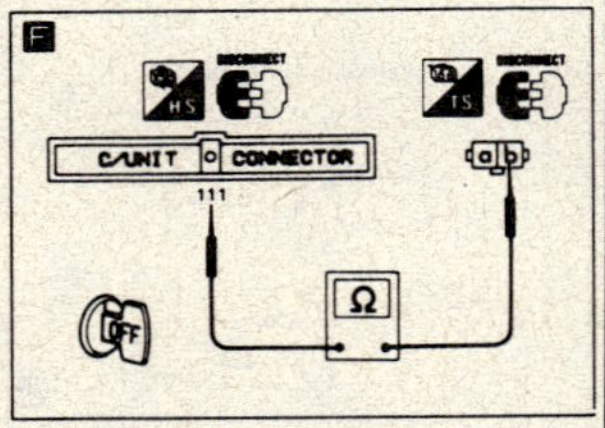

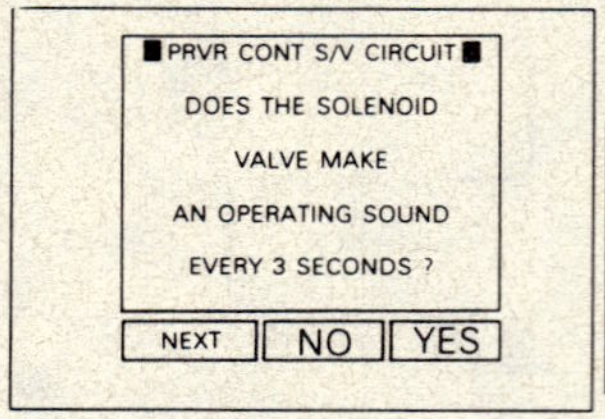

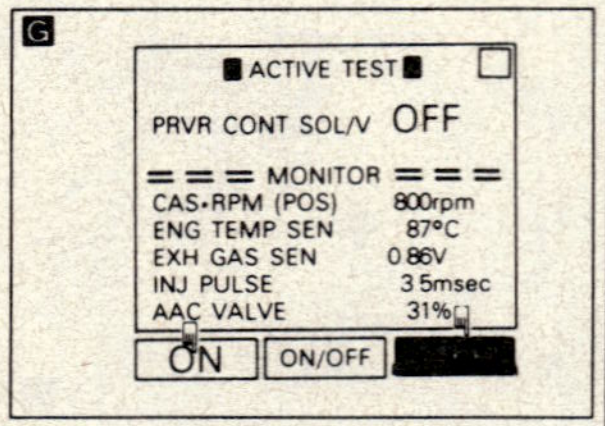

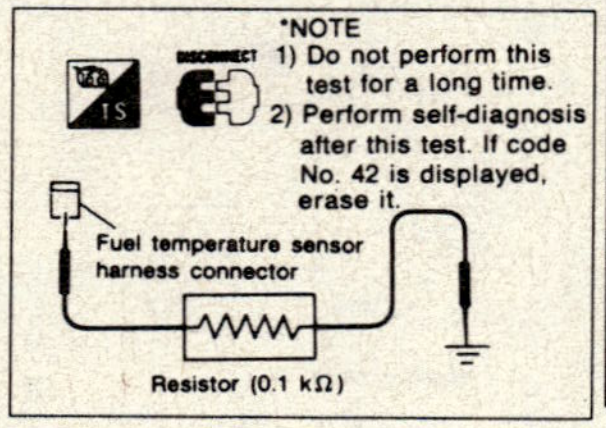

Ⓐ

E CHECK POWER SUPPLY.
1) Stop engine.
2) Disconnect pressure regulator control solenoid valve harness connector.
3) Turn ignition switch "ON".
4) Check voltage between terminal ⓐ and ground.
Voltage: Battery voltage

N.G. → Check the following.
- Harness connectors (F27), (M49)
- Harness continuity between pressure regulator control solenoid valve and ignition switch

If N.G., repair harness or connectors.

O.K. ↓

F CHECK OUTPUT SIGNAL CIRCUIT.
1) Turn ignition switch "OFF".
2) Disconnect E.C.U. harness connector.
3) Check harness continuity between E.C.U. terminal (111) and terminal ⓑ.
Continuity should exist.

N.G. → Repair harness or connectors.

O.K. ↓

G CHECK COMPONENT
(Pressure regulator control solenoid valve).

Perform "PRVR CONT S/V CIRCUIT" in "FUNCTION-TEST" mode with CONSULT.

—OR—

1) Reconnect pressure regulator control solenoid valve harness connector and E.C.U. harness connector.
2) Start engine.
3) Turn P.R.V.R. control solenoid valve "ON" and "OFF" in "ACTIVE TEST" mode with CONSULT and check operating sound.

—OR—

Refer to "Electrical Components Inspection".

N.G. → Replace pressure regulator control solenoid valve.

O.K. ↓

Check E.C.U. pin terminals for damage or the connection of E.C.U. harness connector.

Diagnostic Procedure 41

AIR REGULATOR (Not self-diagnostic item)

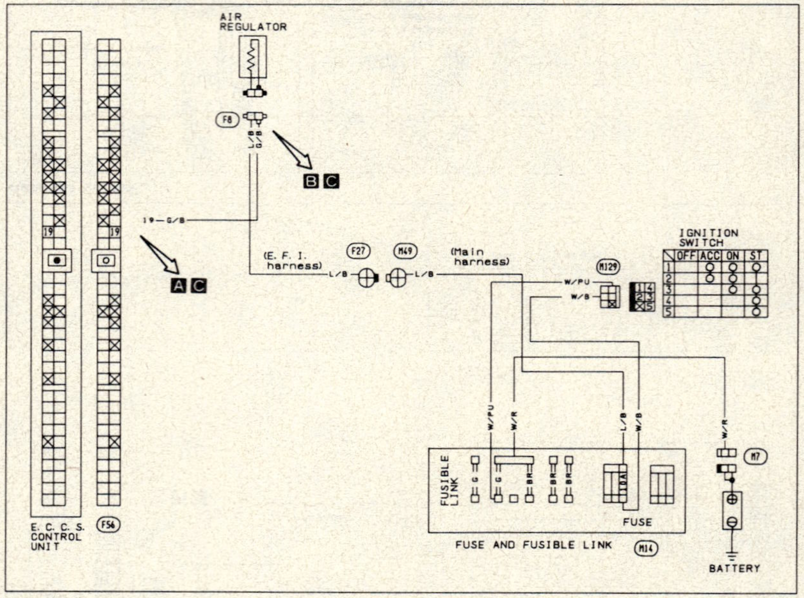

Harness layout

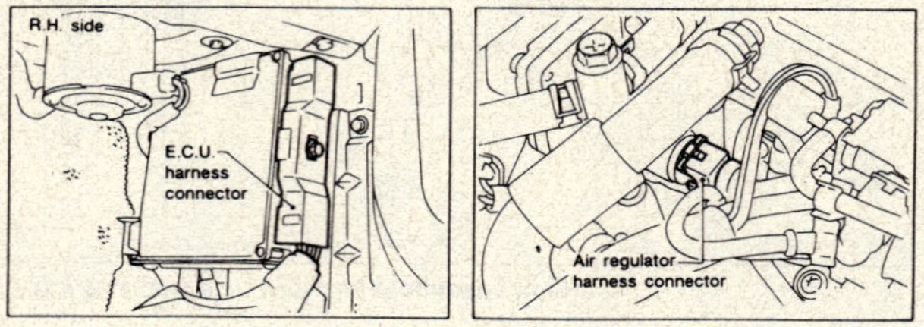

Diagnostic Procedure 41 (Cont'd)

A
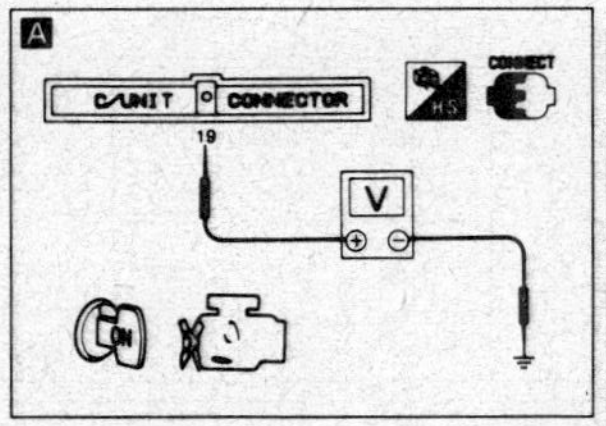

B
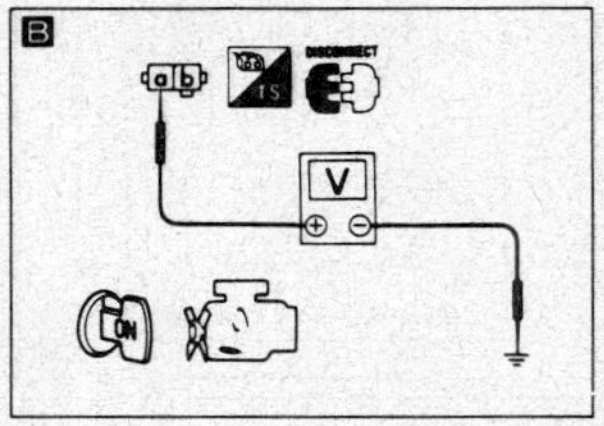

C
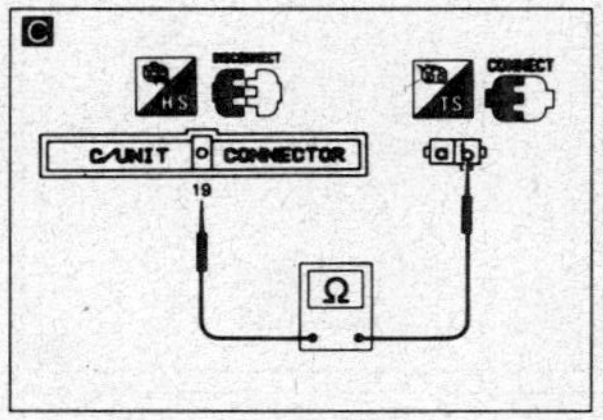

INSPECTION START

A CHECK CONTROL FUNCTION.
1) Turn ignition switch "ON".
2) Check voltage between E.C.U. terminal ⑲ and ground
Voltage:
0.7 - 0.9V
3) After 5 seconds, recheck voltage between E.C.U. terminal ⑲ and ground.
Voltage: Battery voltage

O.K. → INSPECTION END

N.G. ↓

B CHECK POWER SUPPLY.
1) Turn ignition switch "OFF".
2) Disconnect air regulator harness connector.
3) Turn ignition switch "ON".
4) Check voltage between terminal ⓐ and ground.
Voltage: Battery voltage

N.G. → Check the following.
- Harness connectors (F27), (M49)
- 10A fuse
- Harness continuity between air regulator and ignition switch

If N.G., repair harness or connectors.

O.K. ↓

C CHECK OUTPUT SIGNAL CIRCUIT.
1) Turn ignition switch "OFF".
2) Disconnect E.C.U. harness connector.
3) Check harness continuity between E.C.U. terminal ⑲ and terminal ⓑ.
Continuity should exist.

N.G. → Repair harness or connectors.

O.K. ↓

CHECK COMPONENT
(Air regulator).
Refer to "Electrical Components Inspection".

N.G. → Replace air regulator.

O.K. ↓

Check E.C.U. pin terminals for damage or the connection of E.C.U. harness connector.

Diagnostic Procedure 42

A.A.C. VALVE (Not self-diagnostic item)

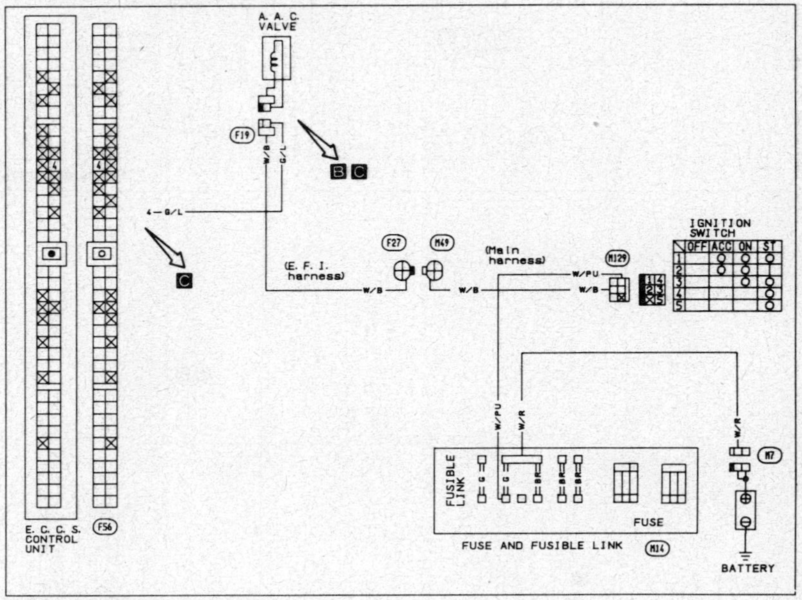

Harness layout

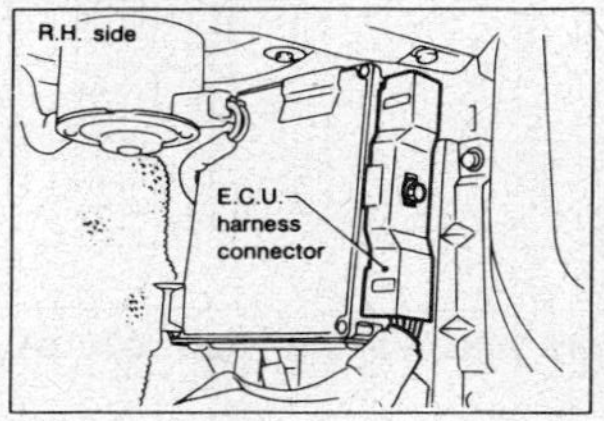

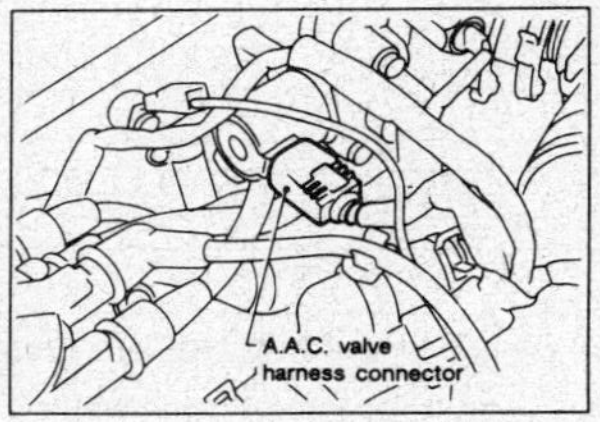

Diagnostic Procedure 42 (Cont'd)

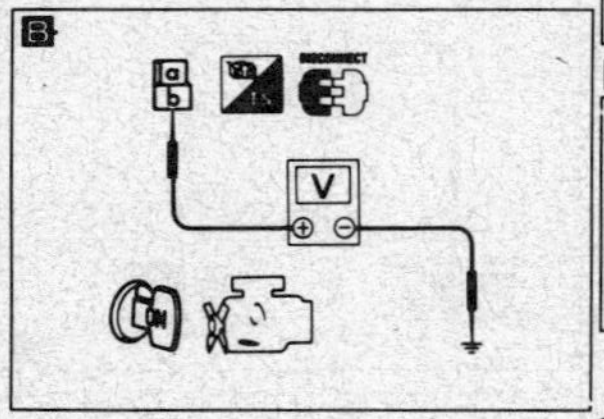

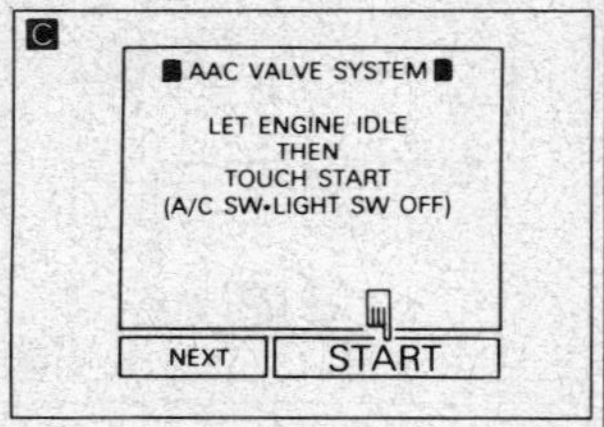

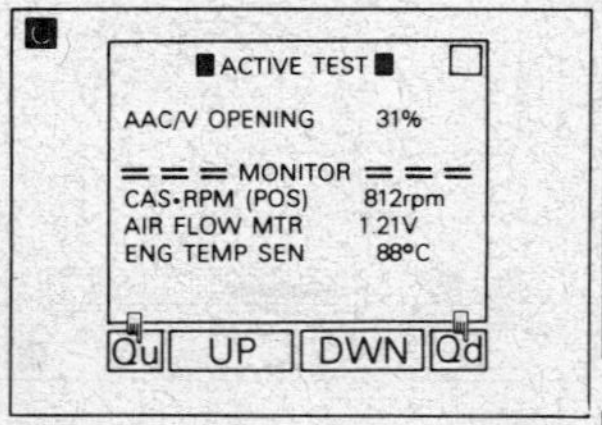

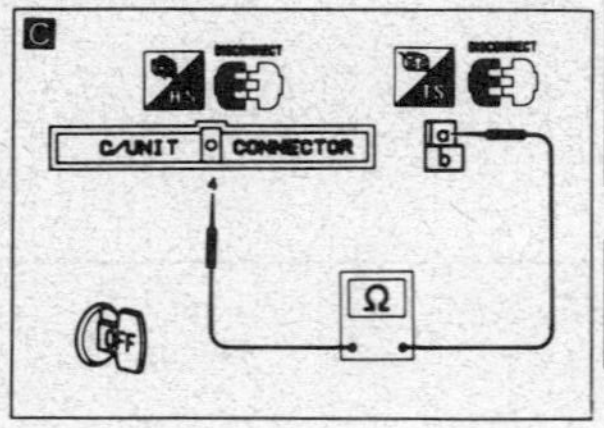

INSPECTION START

A CHECK OVERALL FUNCTION.
1) Start engine and warm it up sufficiently.
2) Check idle speed.
800 ± 50 rpm
If N.G., adjust idle speed.
3) Disconnect A.A.C. valve harness connector.
4) Make sure that idle speed drops.

- Drops → INSPECTION END
- Does not drop ↓

B CHECK POWER SUPPLY.
1) Stop engine.
2) Turn ignition switch "ON".
3) Check voltage between terminal ⓑ and ground.
Voltage: Battery voltage

- N.G. → Check the following.
  - Harness connectors (F27), (M49)
  - Harness continuity between A.A.C. valve and ignition switch

  If N.G., repair harness or connectors.
- O.K. ↓

C CHECK OUTPUT SIGNAL CIRCUIT.
1) Reconnect A.A.C. valve harness connector.
2) Perform "AAC VALVE SYSTEM" in "FUNCTION TEST" mode with CONSULT.

—OR—

1) Reconnect A.A.C. valve harness connector.
2) Perform "AAC VALVE OPENING TEST" in "ACTIVE TEST" mode with CONSULT.

—OR—

1) Turn ignition switch "OFF".
2) Disconnect E.C.U. harness connector.
3) Check harness continuity between E.C.U. terminal ④ and terminal ⓐ.
Continuity should exist.

- N.G. → Repair harness or connectors.
- O.K. ↓

CHECK COMPONENT
(A.A.C. valve).
Refer to "Electrical Components Inspection".

- N.G. → Replace A.A.C. valve.
- O.K. ↓

Check E.C.U. pin terminals for damage or the connection of E.C.U. harness connector.

Diagnostic Procedure 43

A/T CONTROL UNIT (NEUTRAL SIGNAL) CIRCUIT (Not self-diagnostic item)

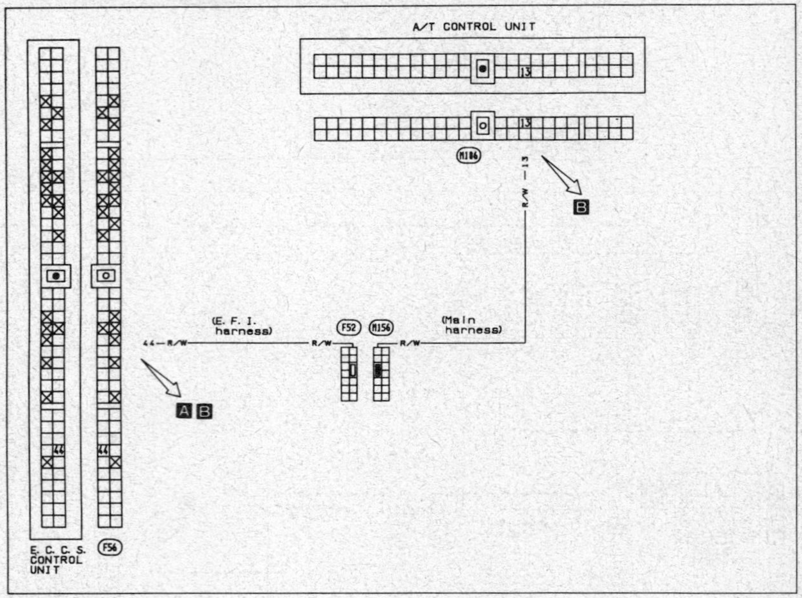

Harness layout

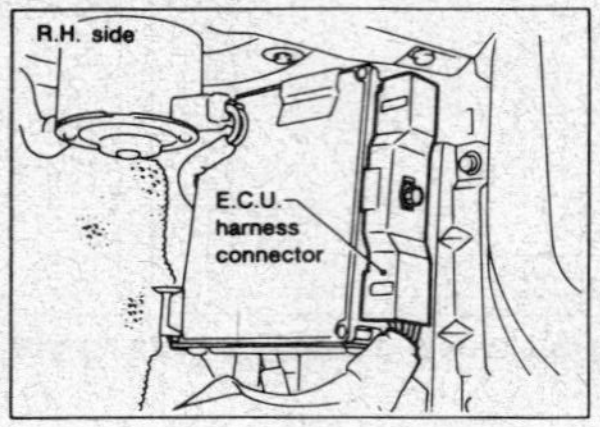

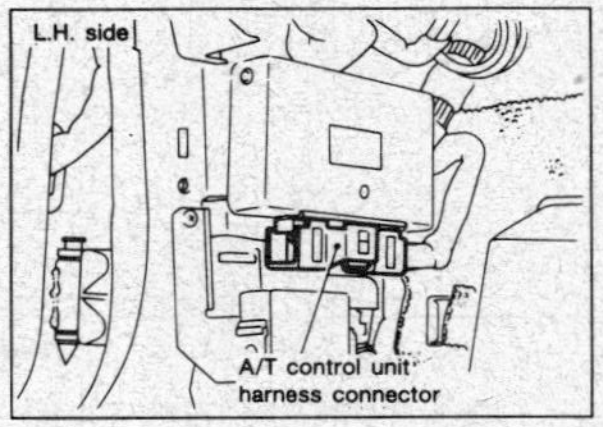

Diagnostic Procedure 43 (Cont'd)

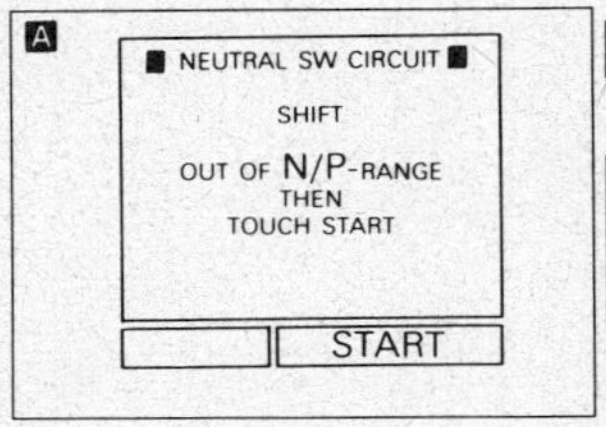

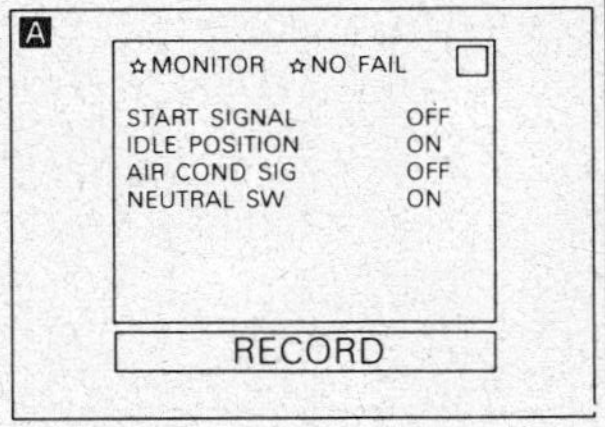

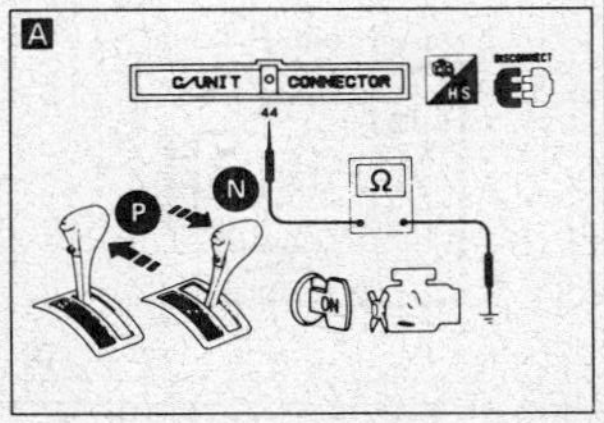

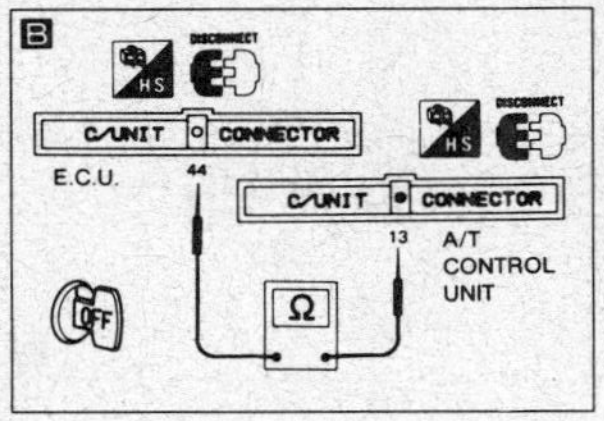

INSPECTION START

↓

A CHECK OVERALL FUNCTION.

1) Turn ignition switch "ON".
2) Perform "NEUTRAL SW CIRCUIT" in "FUNCTION TEST" mode with CONSULT.

— OR —

1) Turn ignition switch "ON".
2) Check neutral switch signal in "DATA MONITOR" mode with CONSULT.
"N" or "P": ON
Except above: OFF

— OR —

1) Make sure that inhibitor switch circuit functions properly. (Refer to AT section.)
2) Disconnect E.C.U. harness connector.
3) Shift selector lever to "P" range.
4) Turn ignition switch "ON".
5) Check harness continuity between E.C.U. terminal (44) and body ground.
Continuity should exist.
6) Shift selector lever to "N" range.
7) Check harness continuity between E.C.U. terminal (44) and body ground.
Continuity should exist.

O.K. → INSPECTION END

N.G. ↓

B CHECK INPUT SIGNAL CIRCUIT.

1) Turn ignition switch "OFF".
2) Disconnect A/T control unit harness connector.
3) Check harness continuity between E.C.U. terminal (44) and terminal (13).
Continuity should exist.

N.G. → Check the following.
- Harness connectors (F52), (M156)
- Harness continuity between E.C.U. and A/T control unit

If N.G., repair harness or connectors.

O.K. ↓

Check E.C.U. pin terminals for damage or the connection of E.C.U. harness connector.

Diagnostic Procedure 44

FAIL-SAFE SYSTEM FOR CRANK ANGLE SENSOR AND C.P.U. OF E.C.U.

INSPECTION START

↓

CHECK FAIL-SAFE SYSTEM FUNCTION.

1) Disconnect crank angle sensor harness connector.
2) Crank engine.

Engine begins to run in a few seconds → INSPECTION END

Engine does not runs after a few seconds ↓

Failure of engine to start may not be due to crank angle sensor circuit. Determine cause of problem using Diagnostic Procedure 4 through 6.

Diagnostic Procedure 44 (Cont'd)

TROUBLE DIAGNOSES FOR FAIL-SAFE SYSTEM

When fail-safe system activates, warning lamp (CHECK ENGINE LIGHT) in instrument panel blinks. When vehicle with such a problem is brought to dealer for checkup, conduct diagnostic procedures using the following chart as a guide.

A

B

CHECK ENGINE LIGHT

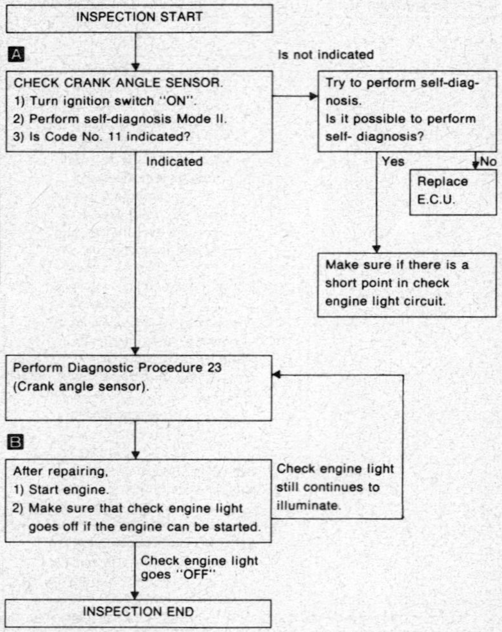

E.C.C.S. COMPONENT PARTS APPLICATION

E.C.C.S. COMPONENT PARTS \ MODE		WORK SUPPORT	SELF-DIAGNOSTIC RESULTS	DATA MONITOR	ACTIVE TEST	FUNCTION TEST
INPUT	Crank angle sensor		X	X		
	Air flow meter		X	X		
	Engine temperature sensor		X	X	X	
	Exhaust gas sensors		X	X		X
	Vehicle speed sensors		X	X		X
	Throttle sensor	X	X	X		X
	Fuel temperature sensor		X	X		
	Exhaust gas temperature sensor*		X	X		
	Detonation sensor		X			
	Ignition switch (start signal)			X		X
	Air conditioner switch			X		
	Neutral switch			X		X
	Battery			X		
OUT-PUT	Injectors		X	X	X	X
	Power transistor (ignition timing)		X (Ignition signal)	X	X	X
	A.A.C. valve	X		X	X	X
	Pressure regulator control solenoid valve			X	X	X
	E.G.R. control solenoid valve			X	X	X
	Air conditioner relay			X		
	Fuel pump		X			

*: The E.C.C.S. component part marked * is applicable to vehicles for California only.
X: Applicable

FUNCTION

Diagnostic mode	Function
Work support	This mode enables a technician to adjust some devices faster and more accurately by following the indications on the CONSULT unit.
Self-diagnostic results	Self-diagnostic results can be read and erased quickly.
Data monitor	Input/Output data in the control unit can be read.
Active test	Mode in which CONSULT drives some actuators apart from the control units and also shifts some parameters in a specified range.
E.C.U. part numbers	E.C.U. part numbers can be read.
Function test	Conducted by CONSULT instead of a technician to determine whether each system is "OK" or "NG".

1990–92 3.0L ENGINE–SCAN TOOL WORK SUPPORT MODE–M30

WORK SUPPORT MODE

WORK ITEM	CONDITION	USAGE
THROTTLE SENSOR ADJUSTMENT	CHECK THE THROTTLE SENSOR SIGNAL. ADJUST IT TO THE SPECIFIED VALUE BY ROTATING THE SENSOR BODY UNDER THE FOLLOWING CONDITIONS. • IGN SW "ON" • ENG NOT RUNNING • ACC PEDAL NOT PRESSED	When adjusting throttle sensor initial position.
IGNITION TIMING ADJUSTMENT*	• IGNITION TIMING FEEDBACK CONTROL WILL BE HELD BY TOUCHING "START". AFTER DOING SO, ADJUST IGNITION TIMING WITH A TIMING LIGHT BY TURNING THE CRANK ANGLE SENSOR.	When adjusting initial ignition timing.
AAC VALVE ADJUSTMENT	SET ENGINE RPM AT THE SPECIFIED VALUE UNDER THE FOLLOWING CONDITIONS. • ENGINE WARMED UP • NO-LOAD	When adjusting idle speed.

*: The ignition timing feedback control is not adopted on model INFINITI M30, so it is not necessary to perform IGNITION TIMING ADJUSTMENT.

1990–92 3.0L ENGINE–SCAN TOOL SELF-DIAGNOSTIC RESULTS MODE–M30

SELF-DIAGNOSTIC RESULTS MODE

DIAGNOSTIC ITEM	DIAGNOSTIC ITEM IS DETECTED WHEN ...	CHECK ITEM (REMEDY)
CRANK ANGLE SENSOR*	• Either 1° or 120° signal is not entered for the first few seconds during engine cranking. • Either 1° or 120° signal is not input often enough while the engine speed is higher than the specified rpm.	• Harness and connector (If harness and connector are normal, replace crank angle sensor.)
AIR FLOW METER	• The air flow meter circuit is open or shorted. (An abnormally high or low voltage is entered.)	• Harness and connector (If harness and connector are normal, replace air flow meter.)
ENGINE TEMP SENSOR	• The engine temperature sensor circuit is open or shorted. (An abnormally high or low output voltage is entered.)	• Harness and connector • Engine temperature sensor
CAR SPEED SENSOR	• The vehicle speed sensor circuit is open or shorted.	• Harness and connector • Vehicle speed sensor (reed switch)
IGN SIGNAL-PRIMARY*	• The ignition signal in primary circuit is not entered during engine cranking or running.	• Harness and connector • Power transistor unit
FUEL PUMP	• Abnormally high or low current supply to the fuel pump persists.	• Harness and connector • Fuel pump • Fuel pump relay
CONTROL UNIT	• E.C.U. calculation function is malfunctioning.	(Replace E.C.C.S. control unit.)
EGR SYSTEM**	• E.G.R. control valve does not operate. (E.G.R. control valve spring does not lift.)	• E.G.R. control valve • E.G.R. control solenoid valve
EXH GAS SENSOR	• The exhaust gas sensor circuit is open or shorted. (An abnormally high or low output voltage is entered.)	• Harness and connector • Exhaust gas sensor • Fuel pressure • Injectors • Intake air leaks
DETONATION SENSOR	• The detonation circuit is open or shorted. (An abnormally high or low voltage is entered.)	• Harness and connector • Detonation sensor
EXH GAS TEMP SENSOR**	• The exhaust gas temperature sensor circuit is open or shorted. (An abnormally high or low voltage is entered.)	• Harness and connector • Exhaust gas temperature sensor
FUEL TEMP SENSOR	• The fuel temperature sensor circuit is open or shorted. (An abnormally high or low voltage is entered.)	• Harness and connector • Fuel temperature sensor
THROTTLE SENSOR	• The throttle sensor circuit is open or shorted. (An abnormally high or low voltage is entered.)	• Harness and connector • Throttle sensor
INJECTOR LEAK**	• Fuel leaks from injector.	• Injector
INJECTOR OPEN**	• The injector circuit is open.	• Injector

*: Check items causing a malfunction of crank angle sensor circuit first, if both "CRANK ANGLE SENSOR" and "IGN SIGNAL-PRIMARY" come out at the same time.

**: The diagnostic item marked ** is applicable to vehicles for California only.

1990–92 3.0L ENGINE—DATA MONITOR MODE—M30

DATA MONITOR MODE

MONITOR ITEM	CONDITION		SPECIFICATION	CHECK ITEM WHEN OUTSIDE SPEC.
CAS, RPM (POS) / CAS, RPM (REF)	• Tachometer: Connect • Run engine and compare tachometer indication with the CONSULT value.		Almost the same speed as the CONSULT value.	• Harness and connector • Crank angle sensor
AIR FLOW MTR	• Engine: After warming up, idle the engine • A/C switch "OFF" • Shift lever "N"	Idle	1.0 - 1.5V	• Harness and connector • Air flow meter
		2,000 rpm	1.4 - 1.9V	
ENG TEMP SEN	• Engine: After warming up		More than 70°C (158°F)	• Harness and connector • Engine temperature sensor
EXH GAS SEN	• Engine: After warming up	Maintaining engine speed at 2,000 rpm	0 ↔ Approx. 1.5V	• Harness and connector • Exhaust gas sensor • Intake air leaks • Injectors
M/R F/C MNT			LEAN ↔ RICH Changes more than 5 times during 10 seconds.	
CAR SPEED SEN	• Turn drive wheels and compare speedometer indication with the CONSULT value		Almost the same speed as the CONSULT value	• Harness and connector • Vehicle speed sensor
BATTERY VOLT	• Ignition switch: ON (Engine stopped)		11 - 14V	• Battery • E.C.U. power supply circuit
THROTTLE SEN	• Ignition switch: ON (Engine stopped)	Throttle valve fully closed	0.4 - 0.5V	• Harness and connector • Throttle sensor • Throttle sensor adjustment
		Throttle valve fully opened	Approx. 4.0V	
FUEL TEMP SEN	• Engine: After warming up		20 - 60°C (68 - 140°F)	• Harness and connector • Fuel temp. sensor
EGR TEMP SEN*	• Engine: After warming up		Less than 4.5V	• Harness and connector • Exhaust gas temperature sensor
START SIGNAL	• Ignition switch: ON → START		OFF → ON	• Harness and connector • Starter switch
IDLE POSITION	• Ignition switch: ON (Engine stopped)	Throttle valve: Idle position	ON	• Harness and connector • Throttle sensor • Throttle sensor adjustment
		Throttle valve: Slightly open	OFF	
AIR COND SIG	• Engine: After warming up, idle the engine	A/C switch "OFF"	OFF	• Harness and connector • Air conditioner switch
		A/C switch "ON"	ON	
NEUTRAL SW	• Ignition switch: ON	Shift lever "P" or "N"	ON	• Harness and connector • Neutral switch
		Except above	OFF	
INJ PULSE	• Engine: After warming up • A/C switch "OFF" • Shift lever "N" • No-load	Idle	3.1 - 3.8 msec.	• Harness and connector • Injector • Air flow meter • Intake air system
		2,000 rpm	2.9 - 3.6 msec.	
IGN TIMING	ditto	Idle	15° B.T.D.C.	• Harness and connector • Crank angle sensor
		2,000 rpm	More than 25° B.T.D.C.	
AAC VALVE	ditto	Idle	15 - 40%	• Harness and connector • A.A.C. valve
		2,000 rpm	—	

Remarks: The monitor item marked * is applicable to vehicles for California only.
Specifications are reference values.

1990–92 3.0L ENGINE—ACTIVE TEST MODE—M30

DATA MONITOR MODE

MONITOR ITEM	CONDITION		SPECIFICATION	CHECK ITEM WHEN OUTSIDE SPEC.
AIR COND RLY	• Air conditioner switch OFF → ON		OFF → ON	• Harness and connector • Air conditioner switch • Air conditioner relay
PRVR CONT SOL VALVE	• Fuel temperature is above 75°C (167°F)	For 3 minutes after starting engine	ON	• Harness and connector • Pressure regulator control solenoid valve • Fuel temperature sensor
		3 minutes after starting engine	OFF	
EGR CONT S/V	• Engine: After warming up • A/C switch "OFF" • Shift lever "N" • No-load	Idle	ON	• Harness and connector • E.G.R. control solenoid valve
		2,000 rpm	OFF	

Remarks: The monitor item marked * is applicable to vehicles for California only.
Specifications are reference values.

ACTIVE TEST MODE

TEST ITEM	CONDITION	JUDGMENT	CHECK ITEM (REMEDY)
FUEL INJECTION TEST	• Engine: Return to the original trouble condition • Change the amount of fuel injection with the CONSULT.	If trouble symptom disappears, see CHECK ITEM.	• Harness and connector • Fuel injectors • Exhaust gas sensor
AAC/V OPENING TEST	• Engine: After warming up, idle the engine. • Change the AAC valve opening percent with the CONSULT.	Engine speed changes according to the opening percent.	• Harness and connector • AAC valve
ENGINE TEMP TEST	• Engine: Return to the original trouble condition • Change the engine coolant temperature with the CONSULT.	If trouble symptom disappears, see CHECK ITEM.	• Harness and connector • Engine temperature sensor • Fuel injectors
IGN TIMING TEST	• Engine: Return to the original trouble condition • Timing light: Set • Retard the ignition timing with the CONSULT.	If trouble symptom disappears, see CHECK ITEM.	• Adjust initial ignition timing
POWER BALANCE TEST	• Engine: After warming up, idle the engine. • A/C switch "OFF" • Shift lever "N" • Cut off each injector signal one at a time with the CONSULT.	Engine runs rough or dies.	• Harness and connector • Compression • Injectors • Power transistor • Spark plugs • Ignition coils
EGR CONT SOL/V TEST / PRVR CONT SOL/V TEST	• Ignition switch: ON • Turn solenoid valve "ON" and "OFF" with the CONSULT and listen to operating sound.	Each solenoid valve makes an operating sound.	• Harness and connector • Solenoid valve
SELF-LEARN CONT TEST	• In this test, the coefficient of self-learning control mixture ratio returns to the original coefficient by touching "CLEAR" on the screen.		

1992 3.0L ENGINE – FUNCTION TEST – M30

FUNCTION TEST MODE

FUNCTION TEST ITEM	CONDITION	JUDGEMENT		CHECK ITEM (REMEDY)
SELF-DIAG RESULTS	• Ignition switch: ON (Engine stopped) • Displays the results of self-diagnosis.	—		Objective system
IDLE POSITION (IDLE SWITCH CIRCUIT)	• Ignition switch: ON (Engine stopped) • Idle switch circuit is tested when throttle is opened and closed fully. ("IDLE POSITION" is the test item name for the vehicles in which idle is selected by throttle sensor.)	Throttle valve: opened	OFF	• Harness and connector • Throttle sensor (Idle switch) • Throttle sensor (Idle switch) adjustment • Throttle linkage • Verify operation in DATA MONITOR mode.
		Throttle valve: closed	ON	
THROTTLE SENSOR CKT	• Ignition switch: ON (Engine stopped) • Throttle sensor circuit is tested when throttle is opened and closed fully.	Range (Throttle valve fully opened — Throttle valve fully closed)	More than 3.0V	• Harness and connector • Throttle sensor • Throttle sensor adjustment • Throttle linkage • Verify operation in DATA MONITOR mode.
NEUTRAL SW CIRCUIT	• Ignition switch: ON (Engine stopped) • Neutral switch circuit is tested when shift lever is manipulated.	OUT OF N/P-RANGE	OFF	• Harness and connector • Neutral switch/Inhibitor switch • Linkage + Inhibitor switch adjustment
		IN N-RANGE	ON	
EGR CONT S/V CIRCUIT	• Ignition switch: ON (Engine stopped) • EGR control S/V circuit is tested by checking solenoid valve operating noise.	The solenoid valve makes an operating sound every 3 seconds.		• Harness and connector • EGR control solenoid valve
PRVR CONT S/V CIRCUIT	• Ignition switch : ON (Engine stopped) • PRVR CONT S/V circuit is tested by checking solenoid valve operating noise	The solenoid valve makes an operating sound every 3 seconds.		• Harness and connector • PRVR control solenoid valve
START SIGNAL CIRCUIT	• Ignition switch: ON → START • Start signal circuit is tested when engine is started by operating the starter. Battery voltage and water temperature before cranking, and average battery voltage, air flow meter output voltage and cranking speed during cranking are displayed.	Start signal: OFF → ON		• Harness and connector • Ignition switch

FUNCTION TEST ITEM	CONDITION	JUDGEMENT	CHECK ITEM (REMEDY)
CAR SPEED SEN CIRCUIT	• Vehicle speed sensor circuit is tested when vehicle is running at a speed of 10 km/h (6 mph) or higher.	Vehicle speed sensor input signal is greater than 4 km/h (2 MPH)	• Harness and connector • Vehicle speed sensor • Electric speedometer
IGN TIMING ADJ	• After warming up, idle the engine. • Ignition timing adjustment is checked by reading ignition timing with a timing light and checking whether it agrees with specifications.	The timing light indicates the same value on the screen.	• Adjust ignition timing (by moving crank angle sensor or distributor) • Crank angle sensor drive mechanism
MIXTURE RATIO TEST	• Air-fuel ratio feedback circuit (injection system, ignition system, vacuum system, etc.) is tested by examining the exhaust gas sensor output at 2,000 rpm under non-loaded state.	• O_2 SEN COUNT: More than 5 times during 10 seconds	• INJECTION SYS (Injector, fuel pressure regulator, harness or connector) • IGNITION SYS (Spark plug, power transistor, ignition coil, harness or connector) • VACUUM SYS (Intake air leaks) • Exhaust gas sensor circuit • Exhaust gas sensor operation • Fuel pressure high or low • Air flow meter
POWER BALANCE	• After warming up, idle the engine. • Injector operation of each cylinder is stopped one after another, and resultant change in engine rotation is examined to evaluate combustion of each cylinder. (This is only displayed for models where a sequential injection system is used.)	Difference in engine rpm is greater than 25 rpm before and after cutting off the injector of each cylinder.	• Injector circuit (Injector, harness or connector) • Ignition circuit (Spark plug, power transistor, ignition coil, harness or connector) • Compression • Valve timing
AAC VALVE SYSTEM	• After warming up, idle the engine. • AAC valve system is tested by detecting change in engine rpm when AAC valve opening is changed to 0%, 20% and 80%.	Difference in engine rpm is greater than 150 rpm between when valve opening is at 80% (102 steps) and at 20% (25 steps).	• Harness and connector • AAC valve • Air passage restriction between air inlet and AAC valve • IAS (Idle adjusting screw) adjustment

1990–92 4.5L ENGINE–ECCS COMPONENT LOCATIONS–Q45

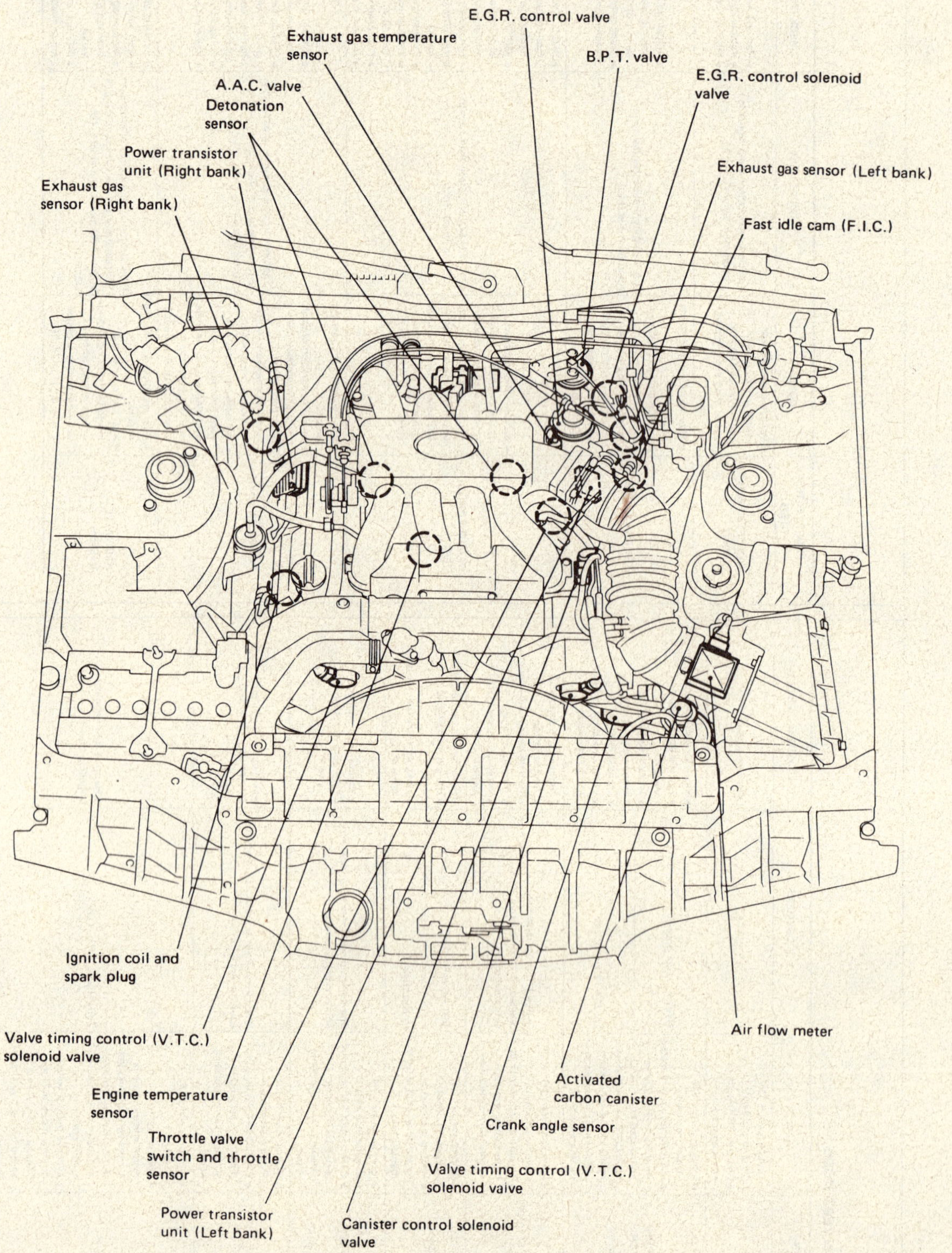

1990–92 4.5L ENGINE – ECCS SYSTEM SCHEMATIC – Q45

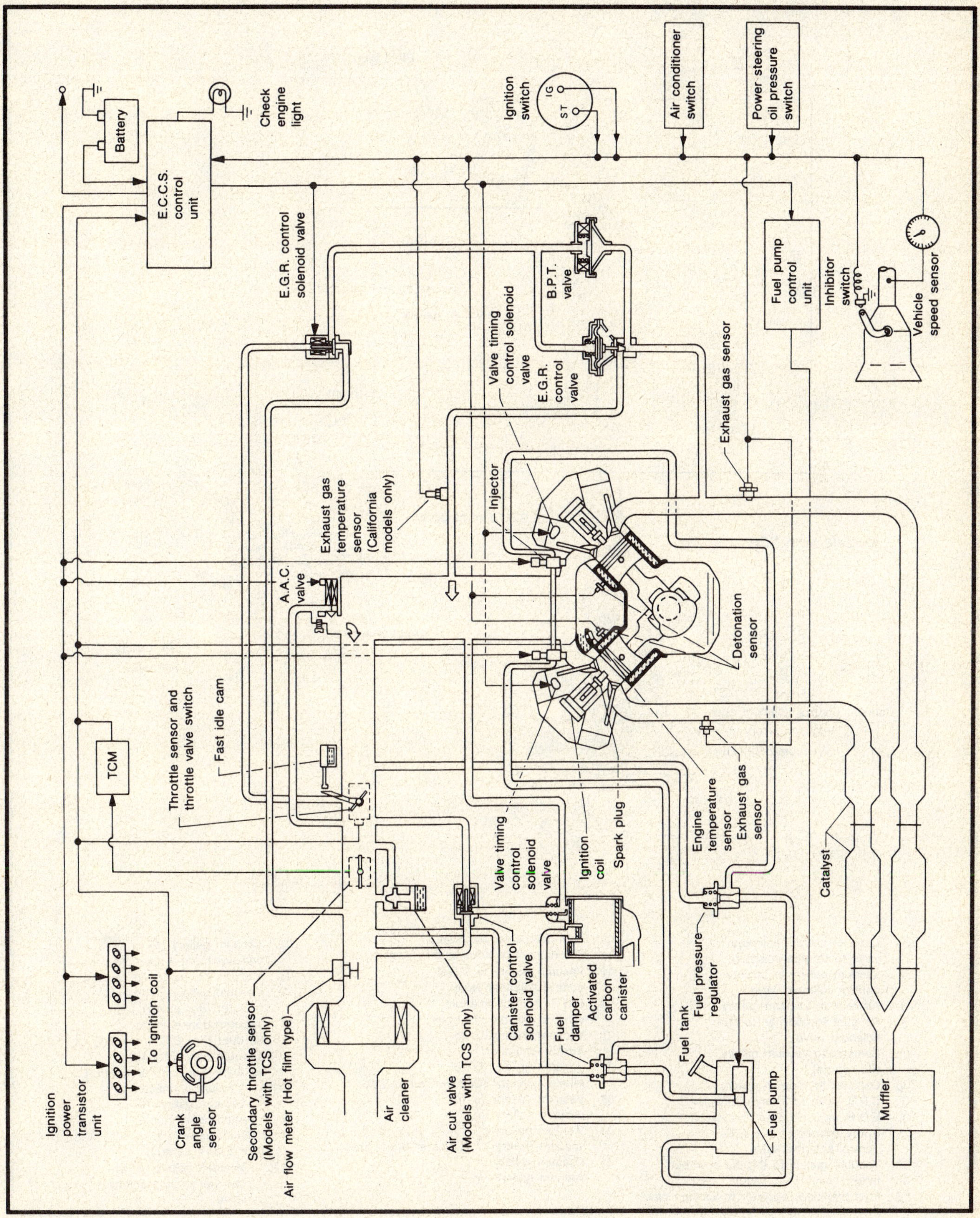

1990–92 4.5L ENGINE–ECCS VACUUM CONTROL SCHEMATIC–Q45

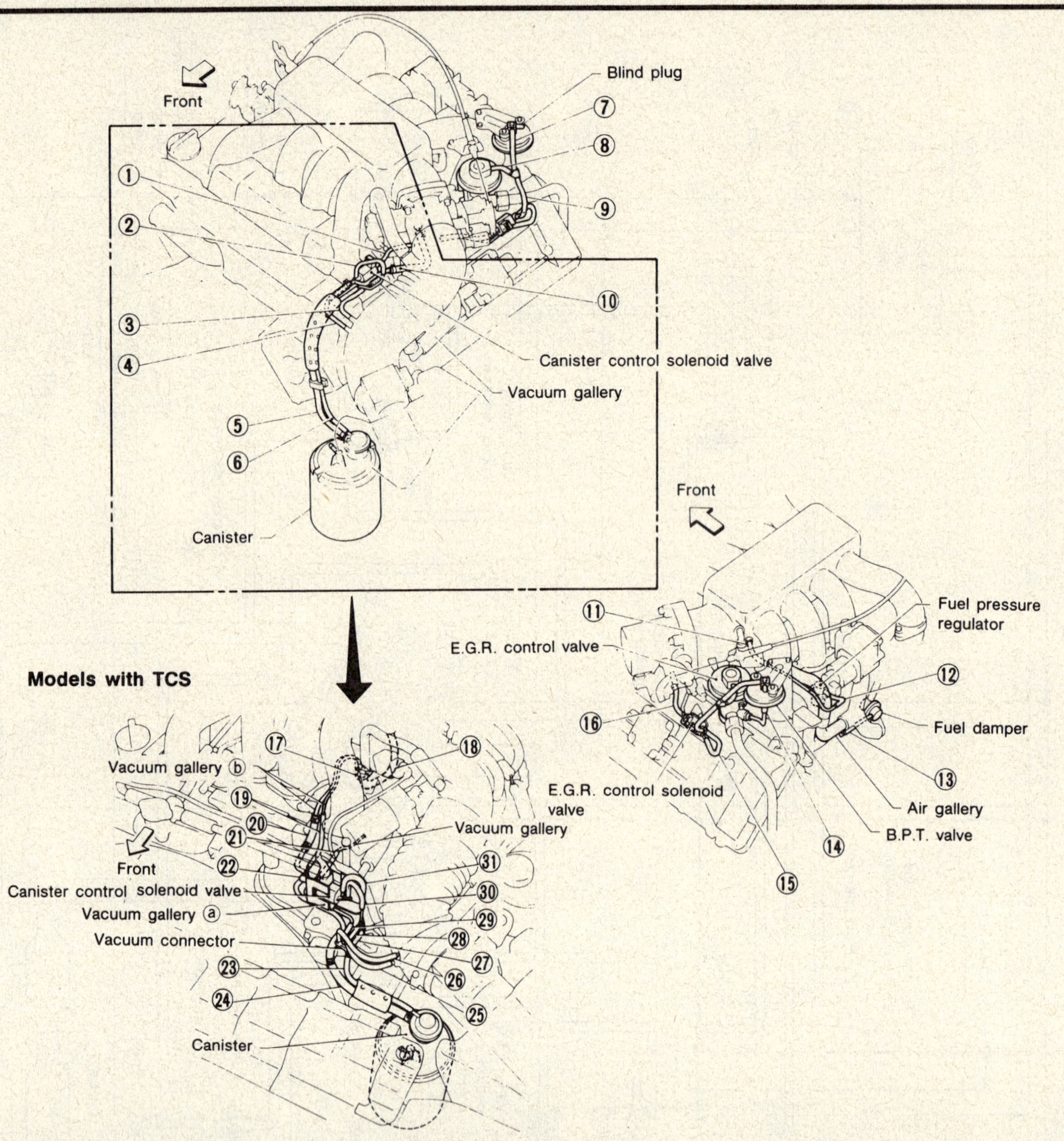

① Canister control solenoid valve to throttle chamber
② Vacuum gallery to canister control solenoid valve
③ Air duct to vacuum gallery
④ Air duct to canister control solenoid valve
⑤ Canister to vacuum gallery (For purge)
⑥ Canister to vacuum gallery
⑦ B.P.T. valve to 3-way connector
⑧ E.G.R. valve to 3-way connector
⑨ 3-way connector to E.G.R. control solenoid valve
⑩ Vacuum gallery to throttle chamber
⑪ Intake collector to vacuum pipe
⑫ Fuel pressure regulator to vacuum pipe
⑬ Fuel damper to air gallery
⑭ E.G.R. control valve to B.P.T. valve
⑮ Vacuum gallery to E.G.R. control solenoid valve
⑯ E.G.R. control solenoid valve to throttle chamber
⑰ Vacuum gallery ⓑ to throttle chamber
⑱ Vacuum gallery ⓑ to intake collector (Canister purge line)
⑲ Vacuum gallery ⓑ to vacuum gallery ⓐ
⑳ Vacuum gallery ⓑ to vacuum gallery ⓐ (Canister purge line)
㉑ Vacuum gallery to vacuum gallery ⓐ
㉒ Vacuum gallery ⓐ to canister control solenoid valve
㉓ Vacuum connector to canister
㉔ Vacuum gallery ⓐ to canister (For purge)
㉕ Vacuum connector to air duct (For E.G.R. control)
㉖ Vacuum connector to air duct (For canister control)
㉗㉘: Vacuum connector to vacuum gallery ⓐ
㉙ Vacuum connector to vacuum gallery ⓐ (For E.G.R. control)
㉚㉛: Vacuum gallery ⓐ to canister control solenoid valve

1990–92 4.5L ENGINE – ECCS CIRCUIT DIAGRAM – Q45

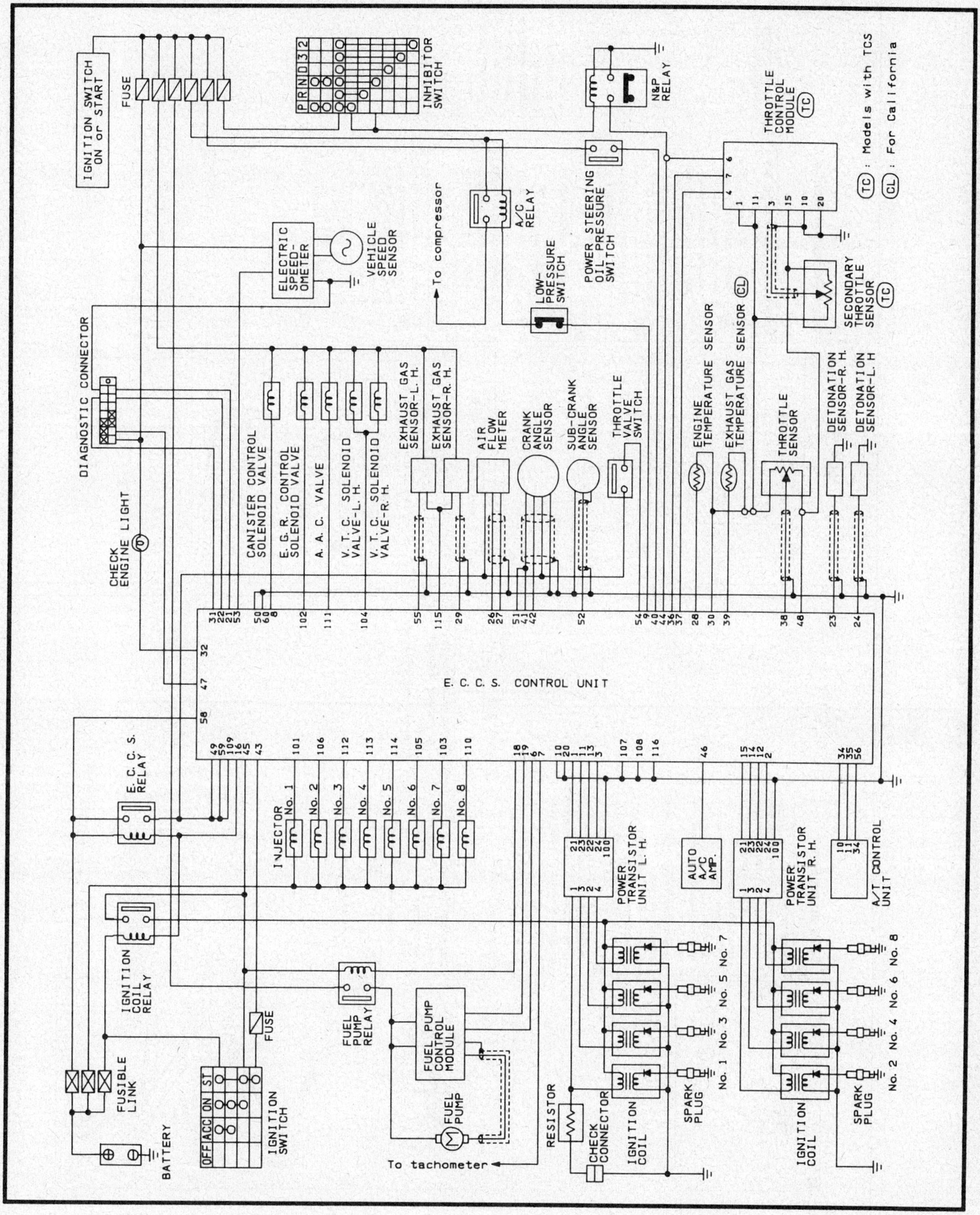

Basic Inspection

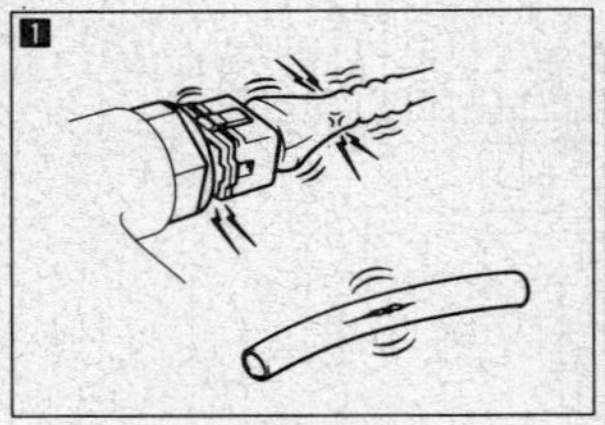

1 BEFORE STARTING

1. Check service records for any recent repairs that may indicate a related problem, or the current need for scheduled maintenance.
2. Open engine hood and check the following:

- Harness connectors for proper connections
- Vacuum hoses for splits, kinks, and proper connections
- Wiring for proper connections, pinches, and cuts

↓

2 CONNECT CONSULT TO THE VEHICLE.
Connect "CONSULT" to the diagnostic connector and select "ENGINE" from the menu.

↓

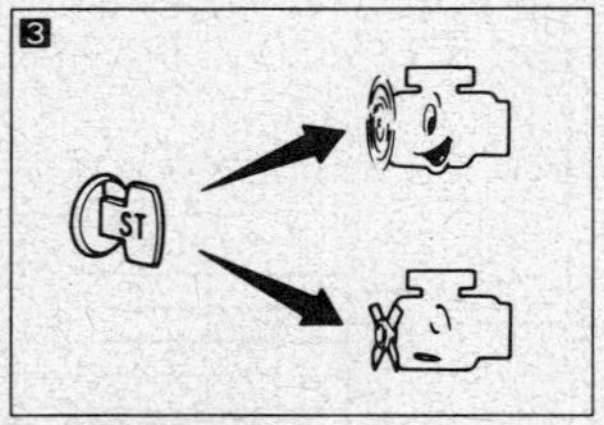

3 DOES ENGINE START?

- No → GO TO 6
- Yes ↓

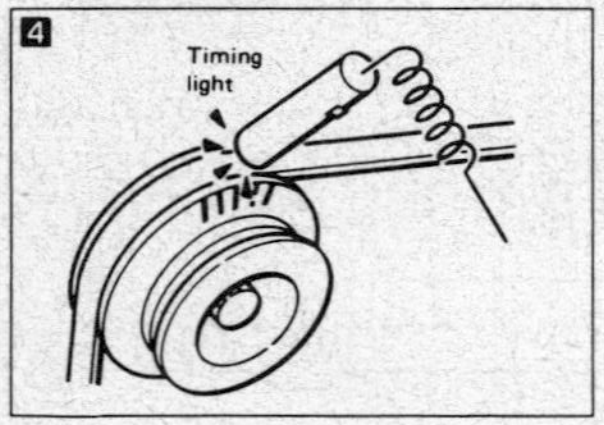

4 CHECK IGNITION TIMING.
Warm up engine sufficiently and check ignition timing at idle using timing light.

Ignition timing: 15° ± 2° B.T.D.C.

- N.G. → Adjust ignition timing by turning crank angle sensor.
- O.K. ↓

(Go to Ⓐ on next page.)

Basic Inspection (Cont'd)

Ⓐ ↓

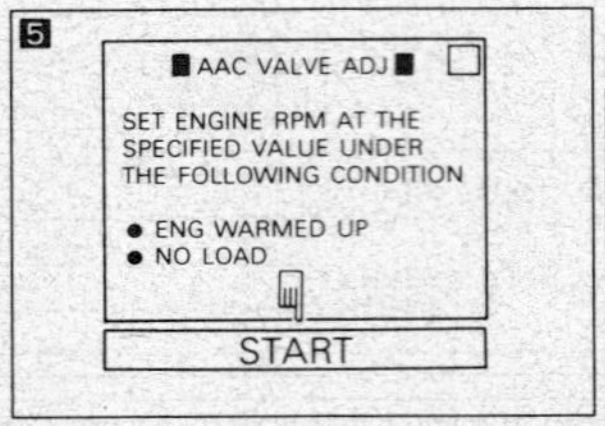

5 CHECK IDLE ADJ. SCREW INITIAL SET RPM.

1. Select "A.A.C. VALVE ADJ" in "WORK SUPPORT" mode.
2. When touching "START", does engine speed fall to 600 ± 25 rpm (A/T in "N" position)?

—OR—

When disconnecting A.A.C. valve harness connector, does engine speed fall to 600 ± 25 rpm (A/T in "N" position)?

- N.G. → Adjust engine speed by turning idle adjusting screw.
- O.K. ↓

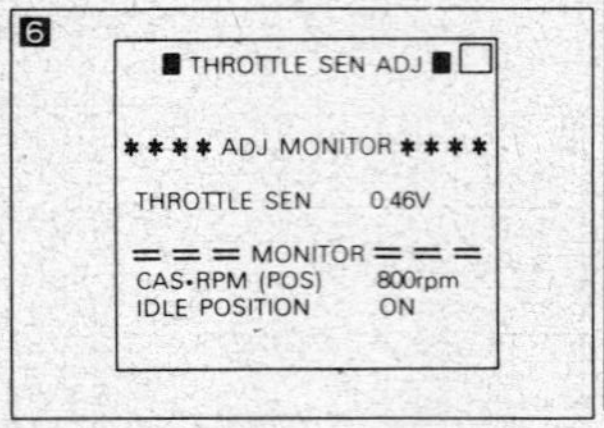

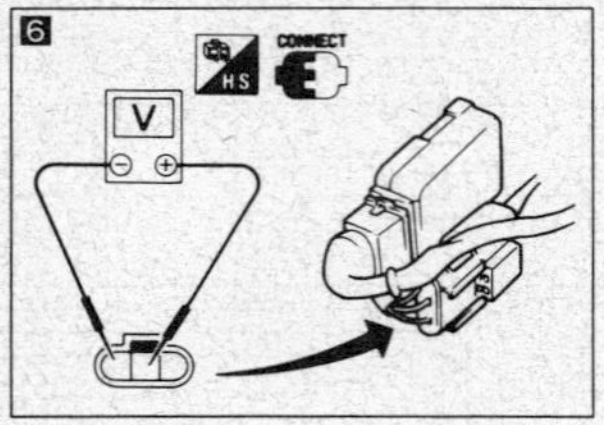

6 CHECK THROTTLE SENSOR IDLE POSITION.

1. Perform "THROTTLE SEN. ADJ." in "WORK SUPPORT" mode.
2. Check that output voltage of throttle sensor is approx. 0.4 to 0.5V. (Throttle valve fully closes.) and "IDLE POSITION" stays "ON".

—OR—

Measure output voltage of throttle sensor using voltmeter, and check that it is approx. 0.4 to 0.5V. (Throttle valve fully closed.)

- N.G. →
  1. Adjust output voltage by rotating throttle sensor body.
  2. Disconnect throttle sensor harness connector for a few seconds and then reconnect it.
  3. Confirm that "IDLE POSITION" stays "ON".
- O.K. ↓

(Go to Ⓑ on next page.)

Basic Inspection (Cont'd)

7

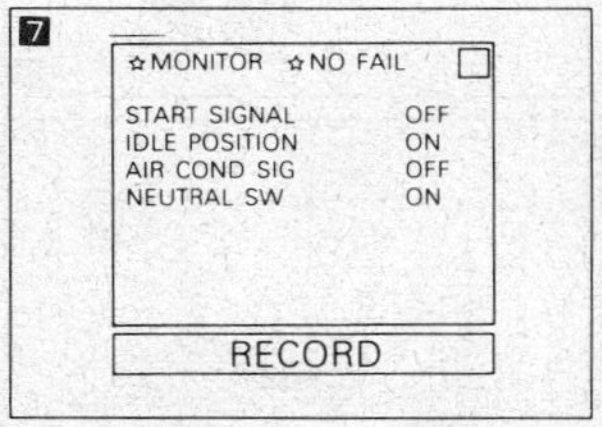

7

B

7 **CHECK SWITCH INPUT SIGNAL.**

Select the following switches in "DATA MONITOR" mode.

a) Start signal,
b) Idle position,
c) Air conditioner signal,
d) Neutral (Parking) switch, and check the switches' ON-OFF operation.

OR

Remove E.C.U. from front floor panel and check the above switches' ON-OFF operation using voltmeter at each E.C.U. terminal.

Switch	Condition	Voltage (V)
Start signal	IGN ON → IGN START	0 → Battery voltage
Idle position	Engine warmed up sufficiently Idle position → Depress the accelerator pedal.	Battery voltage → 0V
A/C signal	A/C OFF → A/C ON (Engine running)	7.0 - 10.0 → 0.5 - 0.7
Neutral (Parking) switch	Shift lever is "N" or "P" position → Except "N" and "P"	0 → 8.0 - 10.0

N.G. → Repair or replace the malfunctioning switch or its circuit.

O.K.

8

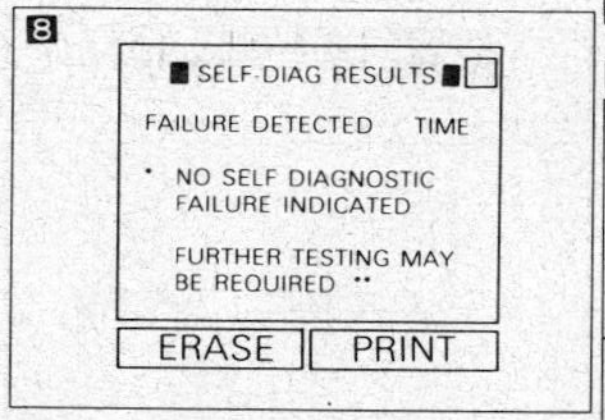

8

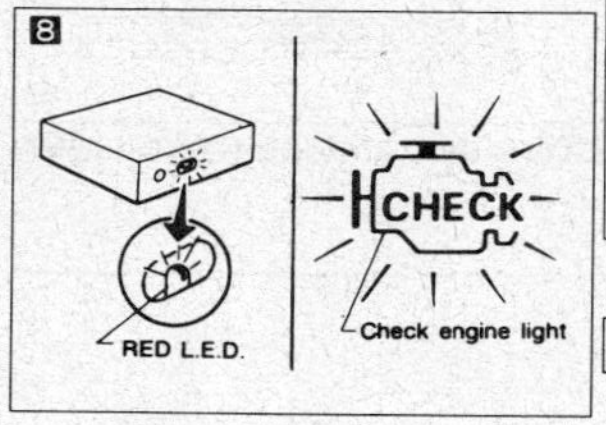

8 **READ SELF-DIAGNOSTIC RESULTS.**

1. Perform "SELF-DIAG RESULTS" mode.
2. Read out self-diagnostic results.
3. Is a failure detected?

OR

1. Set "Self-diagnostic results mode" in Mode II.
2. Count the number of check engine light and RED L.E.D. flashes and read out the codes.
3. Are the codes being output?

Yes → Go to the relevant inspection procedure.

No → INSPECTION END

Diagnostic Procedure 1

MAIN POWER SUPPLY AND GROUND CIRCUIT (Not self-diagnostic item)

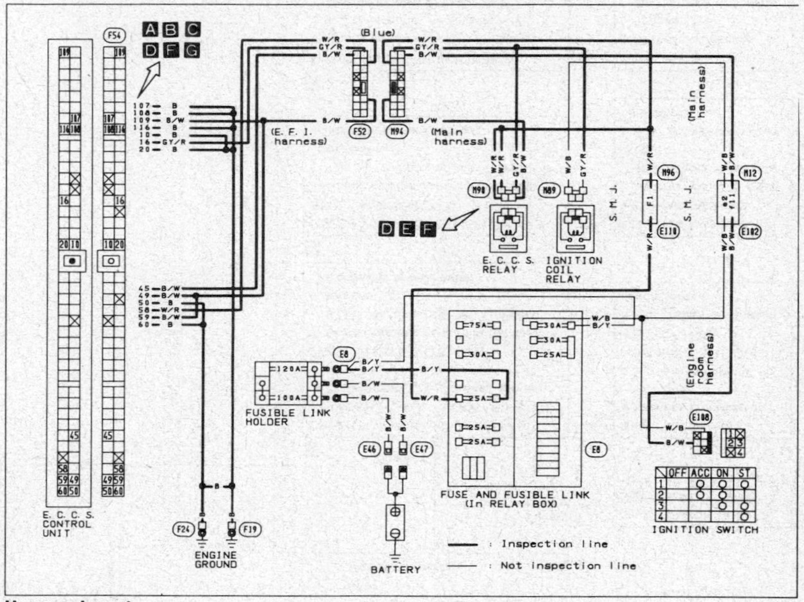

Harness layout

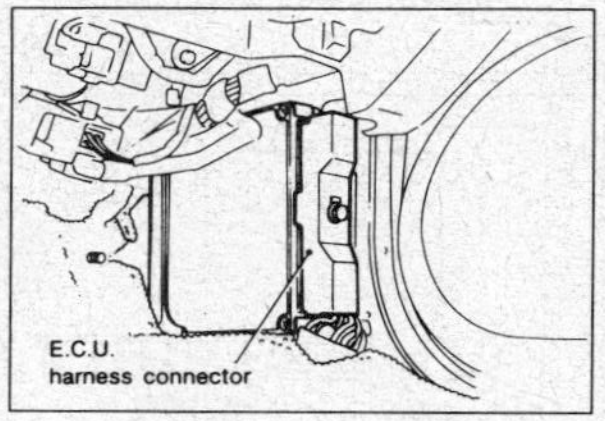

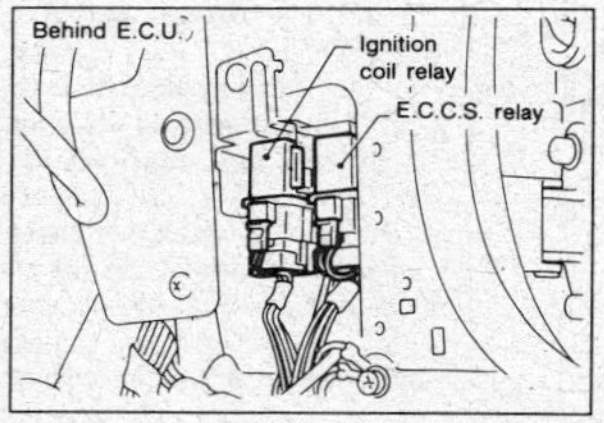

MAIN POWER SUPPLY AND GROUND CIRCUIT (Not self-diagnostic item)

A
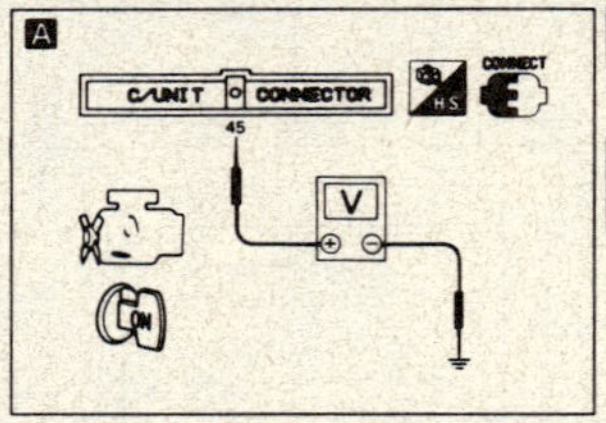

B
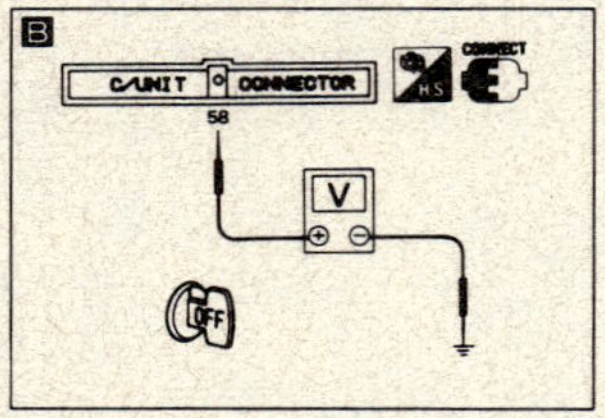

C
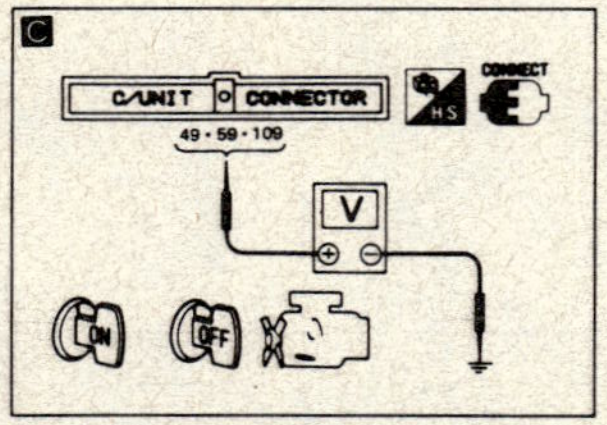

D
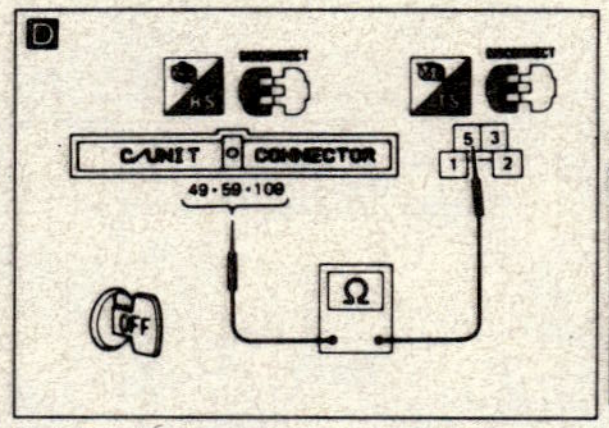

E
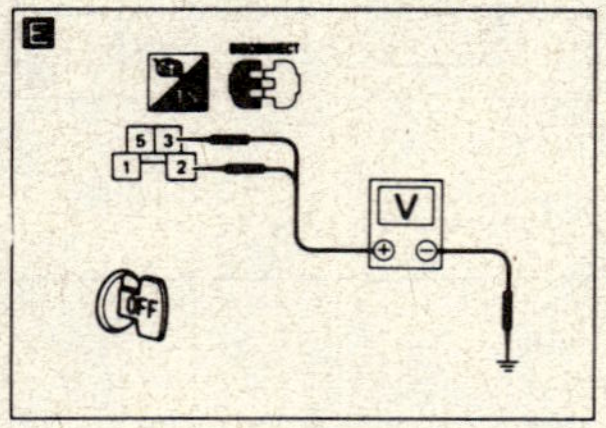

INSPECTION START

Start engine.
Is engine running?

No → Check the following.
- Harness continuity between E.C.U. and ignition coil relay

If N.G., repair harness or connectors.

O.K. ↓

A CHECK POWER SUPPLY-I.
1) Turn ignition switch "ON".
2) Check voltage between E.C.U. terminal 45 and ground with CONSULT or tester.
   Voltage: Battery voltage

If N.G., check the following.
- Harness connectors F52, M94
- Harness connectors M12, E102
- Harness continuity between E.C.U. and ignition switch

If N.G., repair harness or connectors.

O.K. → Go to "CHECK GROUND CIRCUIT".

Yes ↓

B CHECK POWER SUPPLY-II.
1) Stop engine.
2) Check voltage between E.C.U. terminals 58 and ground with CONSULT or tester.
   Voltage: Battery voltage

N.G. → Check the following.
- Harness connectors F52, M94
- Harness connectors M96, E110
- 25A fusible link
- Harness continuity between E.C.U. and fusible link holder

If N.G., repair harness or connectors.

O.K. → Ⓐ

MAIN POWER SUPPLY AND GROUND CIRCUIT (Not self-diagnostic item)

F
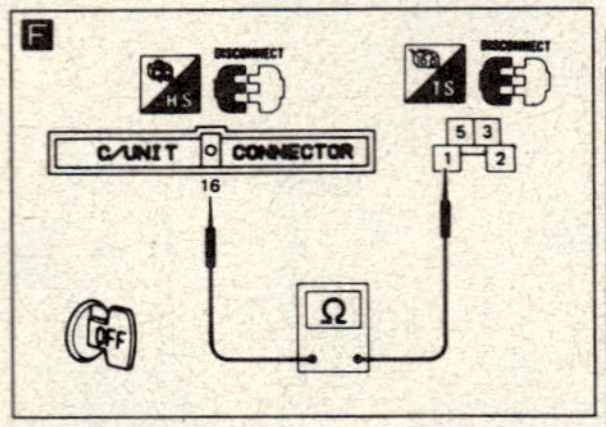

G
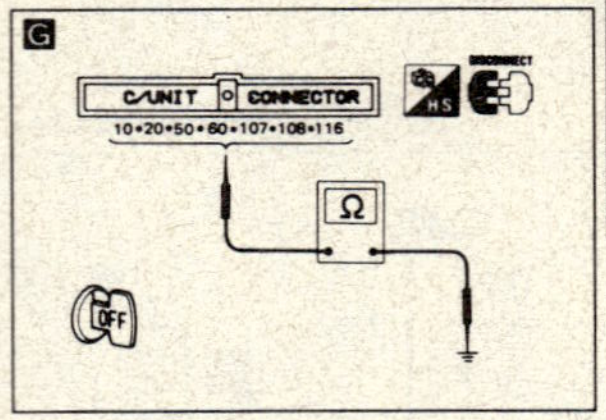

Ⓐ ↓

C CHECK POWER SUPPLY-III.
1) Turn ignition switch "ON" and then "OFF".
2) Check voltage between E.C.U. terminals 49, 58, 109 and ground with CONSULT or tester.
   Voltage:
   Ignition switch "ON" and for a few seconds after turning ignition switch "OFF"
   Battery voltage
   A few seconds after turning ignition switch "OFF"
   Approximately 0V

O.K. → Go to "CHECK GROUND CIRCUIT".

Case-1: Battery voltage does not exist for a few seconds
Case-2: Battery voltage exists for more than a few seconds

N.G. → Case-2 → Go to "CHECK COMPONENT (E.C.C.S. relay)".

N.G. → Case-1 ↓

D CHECK HARNESS CONTINUITY BETWEEN E.C.C.S. RELAY AND E.C.U.
1) Disconnect E.C.U. harness connector.
2) Disconnect E.C.C.S. relay.
3) Check harness continuity between E.C.U. terminals 49, 58, 109 and terminal 5.
   Continuity should exist.

N.G. → Check the following.
- Harness connectors F52, M94
- Harness continuity between E.C.U. and E.C.C.S. relay

If N.G., repair harness or connectors.

O.K. ↓

E CHECK VOLTAGE BETWEEN E.C.C.S. RELAY AND GROUND.
1) Check voltage between terminals 2, 3 and ground with CONSULT or tester.
   Voltage: Battery voltage

N.G. → Check the following.
- Harness continuity between E.C.C.S. relay and harness connector M96

If N.G., repair harness or connectors.

O.K. ↓

F CHECK OUTPUT SIGNAL CIRCUIT.
1) Check harness continuity between E.C.U. terminal 16 and terminal 1.
   Continuity should exist.

N.G. → Check the following.
- Harness continuity between E.C.C.S. relay and harness connector M94

If N.G., repair harness or connectors.

O.K. → Ⓑ

MAIN POWER SUPPLY AND GROUND CIRCUIT (Not self-diagnostic item)

Ⓑ

CHECK COMPONENT (E.C.C.S. relay). Refer to "Electrical Components Inspection".
- N.G. → Replace E.C.C.S. relay.
- O.K. ↓

Ⓖ **CHECK GROUND CIRCUIT.**
1) Turn ignition switch "OFF".
2) Disconnect E.C.U. harness connector.
3) Check harness continuity between E.C.U. terminals [illegible], [illegible], [illegible], [illegible], 107, 108, 116 and engine ground.
Continuity should exist.
- N.G. → Repair harness or connectors.
- O.K. ↓

Check E.C.U. pin terminals for damage or the connection of E.C.U. harness connector.
↓

INSPECTION END

Diagnostic Procedure 2

CRANK ANGLE SENSOR (Code No. 11)

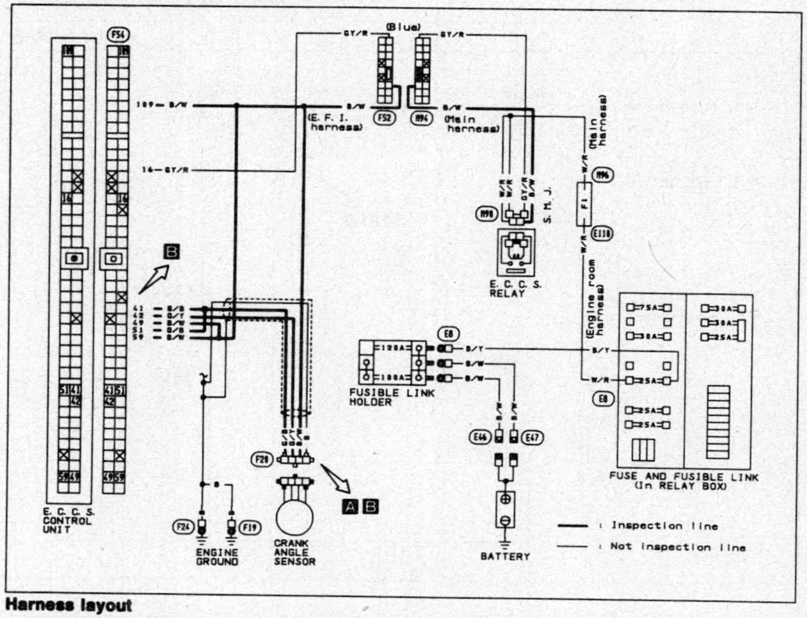

Harness layout

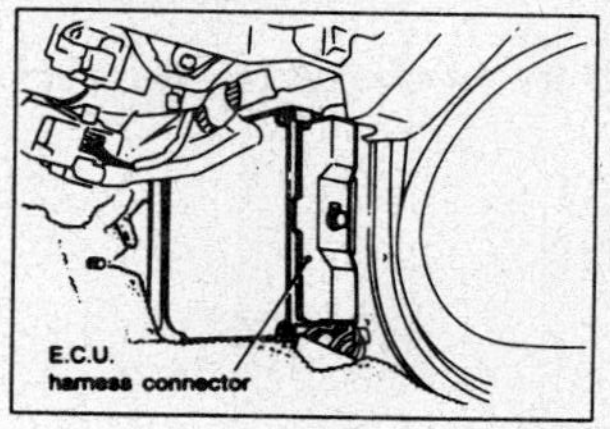

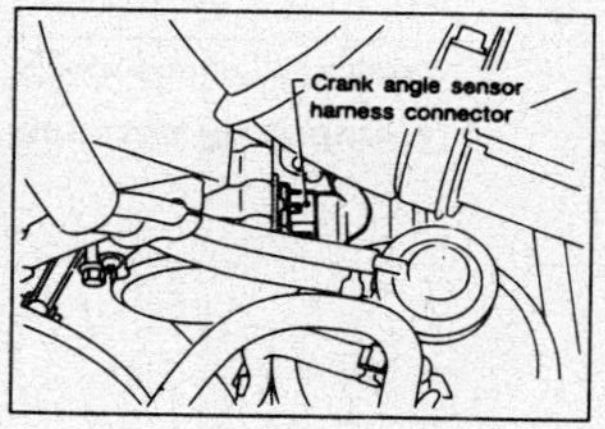

CRANK ANGLE SENSOR (Code No. 11)

A

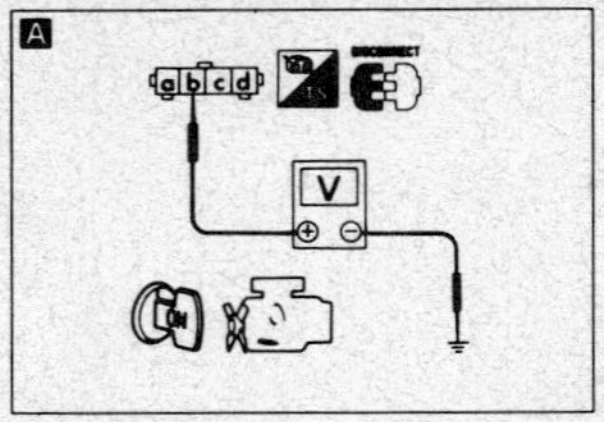

B

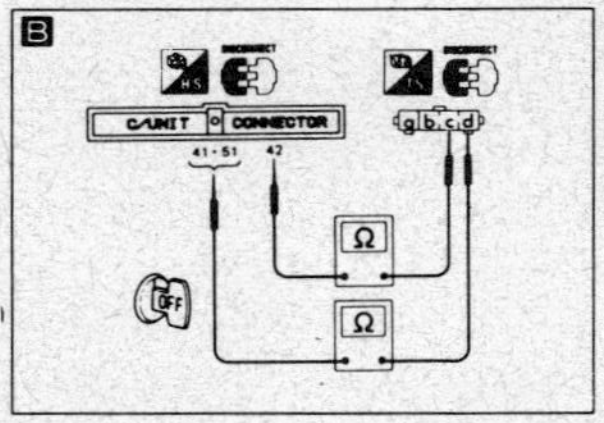

INSPECTION START

PERFORM SELF-DIAGNOSIS.

- O.K. → Go to "CRANK ANGLE SENSOR (Not self-diagnostic item)".
- N.G. ↓

A CHECK POWER SUPPLY.
1) Turn ignition switch "OFF".
2) Disconnect crank angle sensor harness connector.
3) Turn ignition switch "ON".
4) Check voltage between terminal ⓑ and ground with CONSULT or tester.
Voltage: Battery voltage

- N.G. → Check the following.
  - Harness connectors (F52), (M94)
  - Harness continuity between crank angle sensor and E.C.C.S. relay
  - Harness continuity between crank angle sensor and E.C.U.

  If N.G., repair harness or connectors.
- O.K. ↓

B CHECK INPUT SIGNAL CIRCUIT.
1) Turn ignition switch "OFF".
2) Disconnect E.C.U. harness connector.
3) Check harness continuity between terminal ⓓ and E.C.U. terminals (41), (51) (90° signal), terminal ⓒ and E.C.U. terminal (42) (1° signal).
Continuity should exist.

- N.G. → Repair harness or connectors.
- O.K. ↓

CHECK COMPONENT
(Crank angle sensor).
Refer to "Electrical Components Inspection".

- N.G. → Replace crank angle sensor.
- O.K. ↓

Check E.C.U. pin terminals for damage or the connection of E.C.U. harness connector.

Diagnostic Procedure 3

AIR FLOW METER (Code No. 12) (Check engine light item)

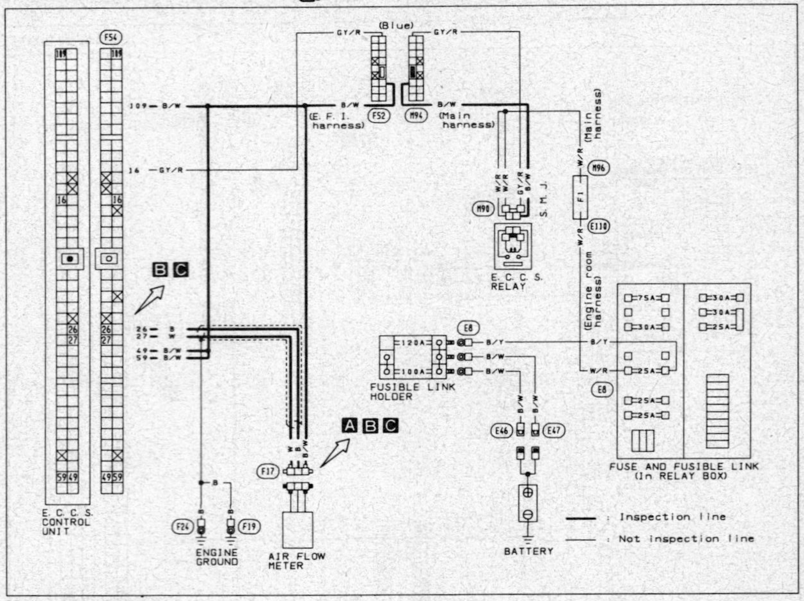

Harness layout

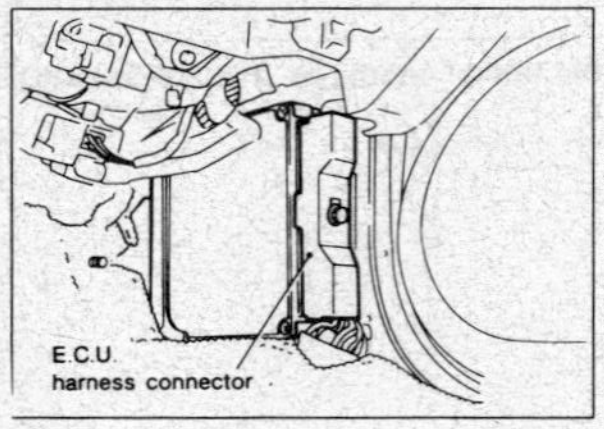

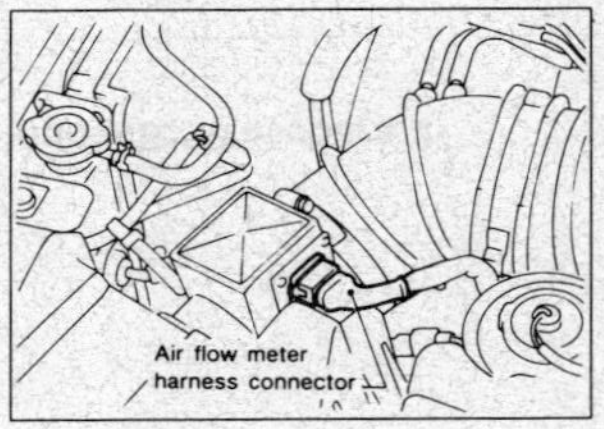

AIR FLOW METER (Code No. 12) (Check engine light item)

A

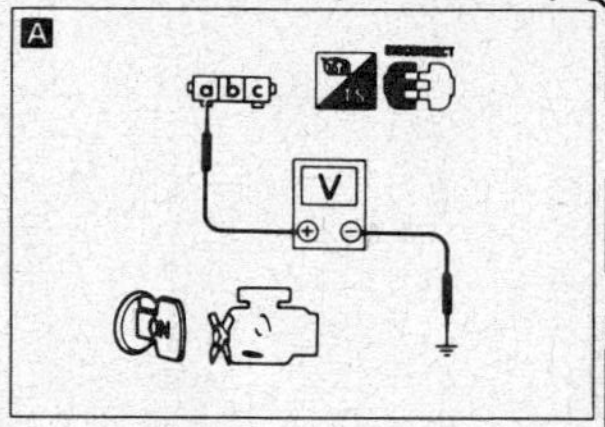

B

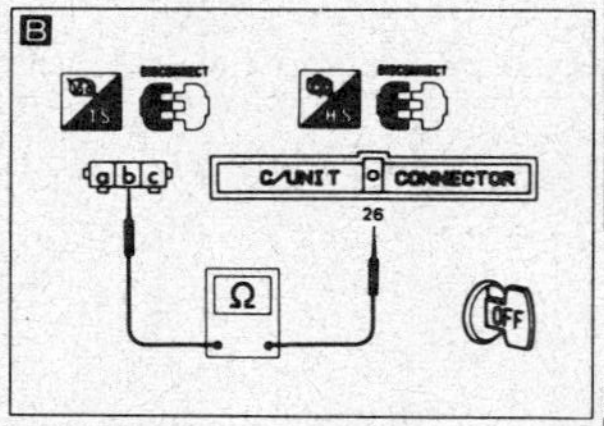

C

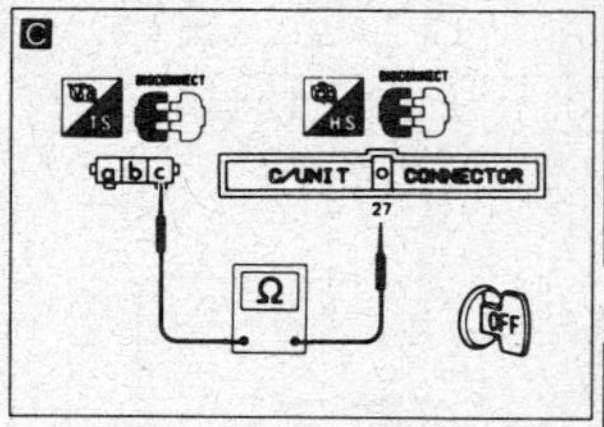

INSPECTION START

↓

PERFORM SELF-DIAGNOSIS. — O.K. → INSPECTION END

↓ N.G.

A CHECK POWER SUPPLY.
1) Turn ignition switch "OFF".
2) Disconnect air flow meter harness connector.
3) Turn ignition switch "ON".
4) Check voltage between terminal ⓐ and ground with CONSULT or tester.
Voltage: Battery voltage

— N.G. → Check the following.
- Harness connectors (F52), (M94)
- Harness continuity between air flow meter and E.C.C.S. relay
- Harness continuity between air flow meter and E.C.U.

If N.G., repair harness or connectors.

↓ O.K.

B CHECK GROUND CIRCUIT.
1) Turn ignition switch "OFF".
2) Disconnect E.C.U. harness connector.
3) Check harness continuity between terminal ⓑ and E.C.U. terminal ㉖.
Continuity should exist.

— N.G. → Repair harness or connectors.

↓ O.K.

C CHECK INPUT SIGNAL CIRCUIT.
1) Check harness continuity between terminal ⓒ and E.C.U. terminal ㉗.
Continuity should exist.

— N.G. → Repair harness or connectors.

↓ O.K.

CHECK COMPONENT
(Air flow meter).
Refer to "Electrical Components Inspection".

— N.G. → Replace air flow meter.

↓ O.K.

Check E.C.U. pin terminals for damage or the connection of E.C.U. harness connector.

Diagnostic Procedure 4

ENGINE TEMPERATURE SENSOR (Code No. 13) (Check engine light item)

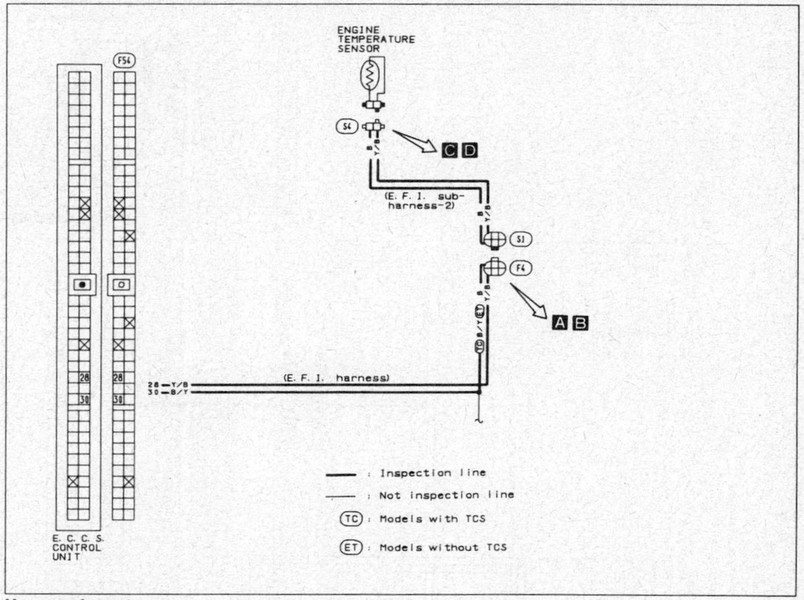

Harness layout

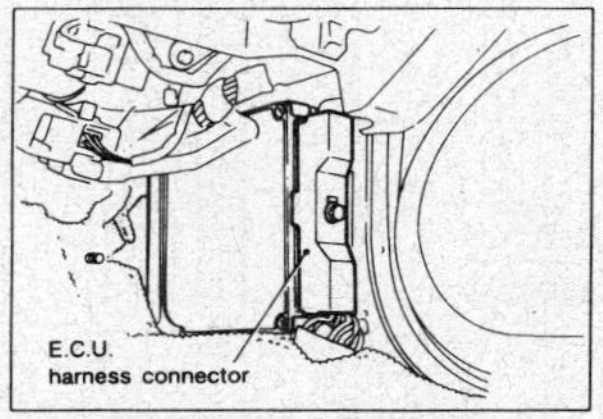

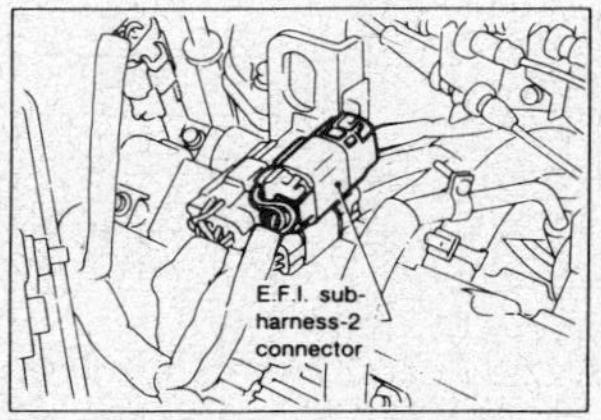

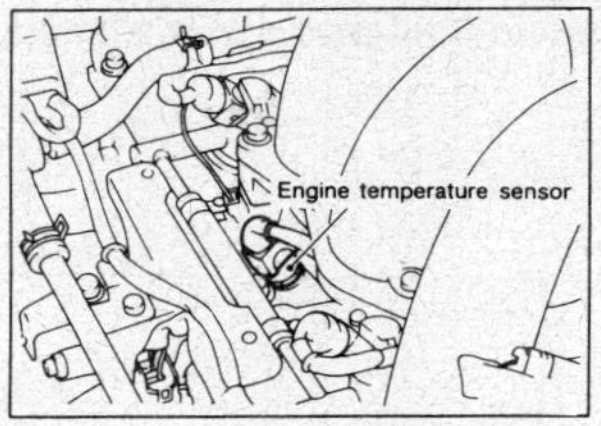

ENGINE TEMPERATURE SENSOR (Code No. 13) (Check engine light item)

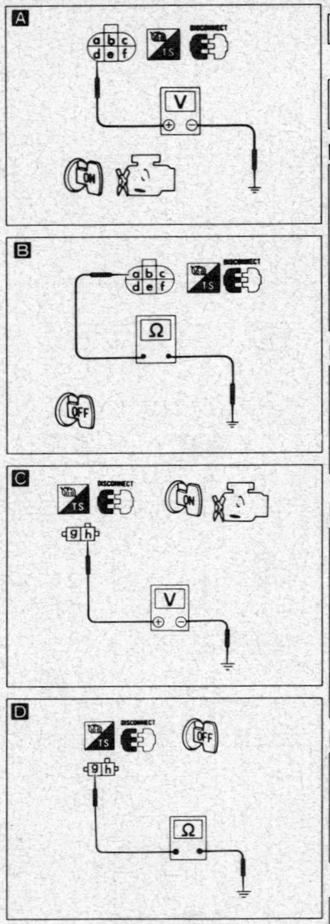

INSPECTION START

↓

PERFORM SELF-DIAGNOSIS. — O.K. → INSPECTION END

↓ N.G.

A CHECK POWER SUPPLY-I.
1) Turn ignition switch "OFF".
2) Disconnect E.F.I. sub-harness-2 connector.
3) Turn ignition switch "ON".
4) Check voltage between terminal ⓓ and ground with CONSULT or tester.
Voltage: Approximately 5V

N.G. → Repair harness or connectors.

↓ O.K.

B CHECK GROUND CIRCUIT-I.
1) Turn ignition switch "OFF".
2) Check harness continuity between terminal ⓐ and engine ground.
Continuity should exist.

N.G. → Repair harness or connectors.

↓ O.K.

C CHECK POWER SUPPLY-II.
1) Reconnect E.F.I. sub-harness-2 connector.
2) Disconnect engine temperature sensor harness connector.
3) Turn ignition switch "ON".
4) Check voltage between terminal ⓗ and ground with CONSULT or tester.
Voltage: Approximately 5V

N.G. → Repair harness or connectors.

↓ O.K.

D CHECK GROUND CIRCUIT-II.
1) Turn ignition switch "OFF".
2) Check harness continuity between terminal ⓖ and engine ground.
Continuity should exist.

N.G. → Repair harness or connectors.

↓ O.K.

CHECK COMPONENT
(Engine temperature sensor).
Refer to "Electrical Components Inspection".

N.G. → Replace engine temperature sensor.

↓ O.K.

Check E.C.U. pin terminals for damage or the connection of E.C.U. harness connector.

Diagnostic Procedure 5

VEHICLE SPEED SENSOR (Code No. 14) (Check engine light item)

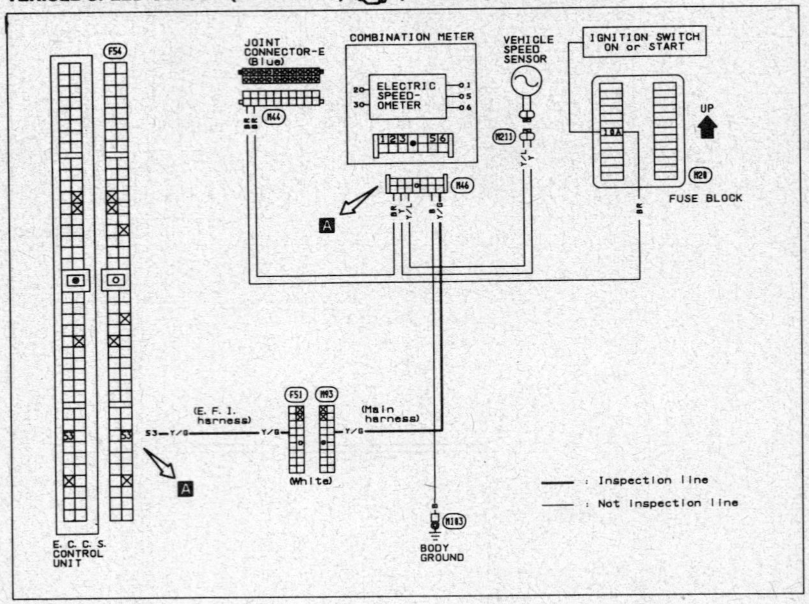

Harness layout

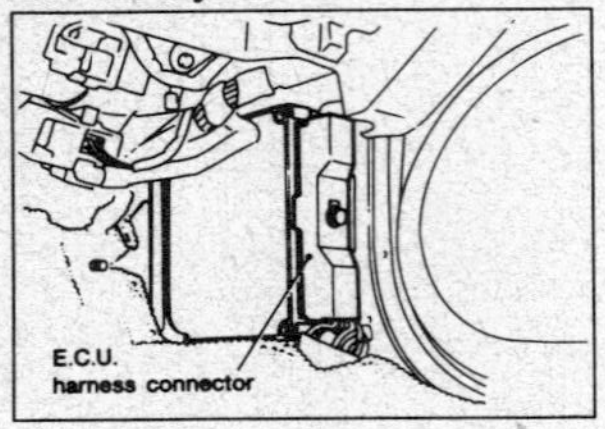

VEHICLE SPEED SENSOR (Code No. 14) (Check engine light item)

A

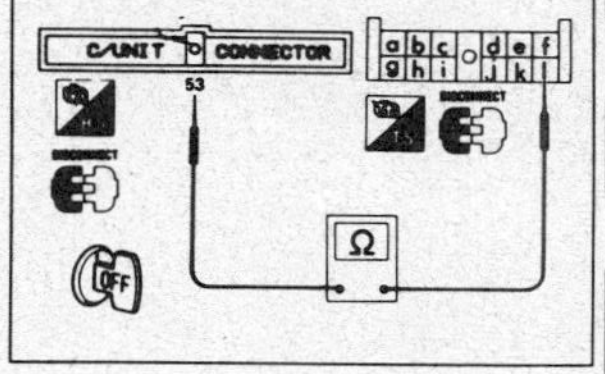

INSPECTION START
↓
CHECK OVERALL FUNCTION. — O.K. → INSPECTION END
↓ N.G.
CHECK SPEEDOMETER FUNCTION. Make sure that speedometer functions properly. — N.G. → Check vehicle speed sensor and its circuit.
↓ O.K.
A CHECK INPUT SIGNAL CIRCUIT.
1) Turn ignition switch "OFF".
2) Disconnect E.C.U. harness connector and combination meter harness connector.
3) Check harness continuity between E.C.U. terminal ⑬ and terminal ①.
Continuity should exist.
— N.G. → Check the following.
- Harness connectors (F51), (M93)
- Harness continuity between E.C.U. and combination meter

If N.G., repair harness or connectors.

↓ O.K.
Check E.C.U. pin terminals for damage or the connection of E.C.U. harness connector.

Perform FINAL CHECK by the following procedure after repair is completed.

FINAL CHECK
↓
Erase the self-diagnosis memory.
↓
Perform test drive.
↓
Check overall function again. — N.G. → Recheck E.C.U. pin terminals for damage or the connection of E.C.U. harness connector.
↓ O.K.
INSPECTION END

Diagnostic Procedure 6

TCS SIGNAL (Code No. 16): TCS models

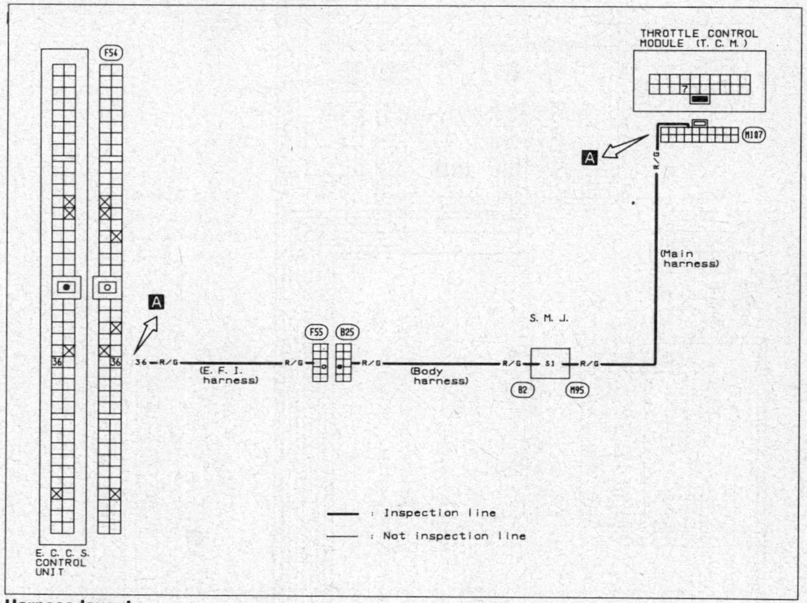

Harness layout

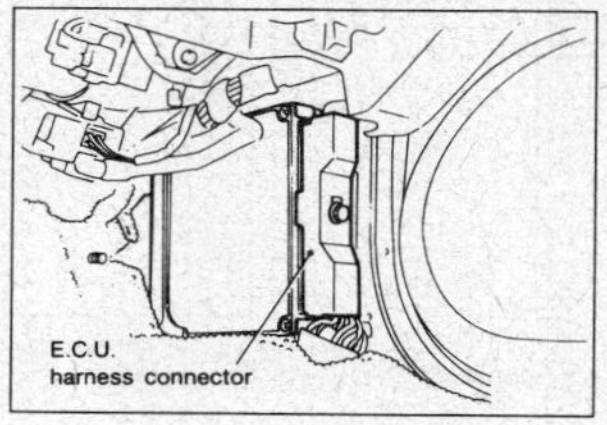

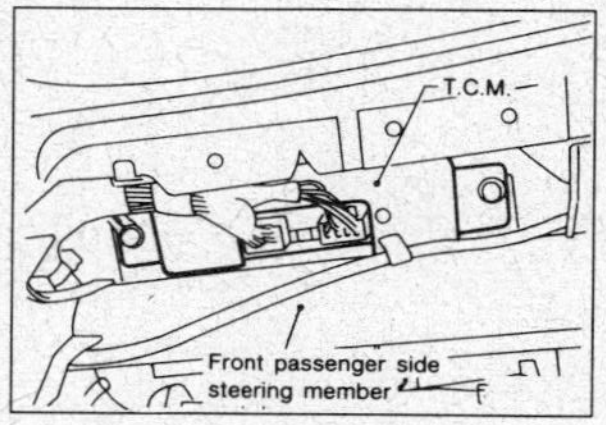

TCS SIGNAL (Code No. 16): TCS models

A

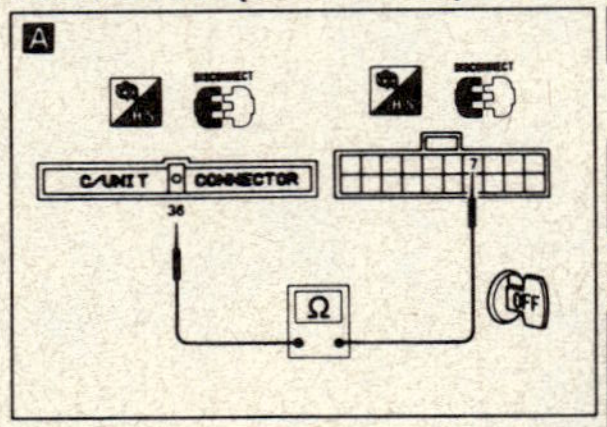

INSPECTION START

↓

PERFORM SELF-DIAGNOSIS. — O.K. → INSPECTION END

↓ N.G.

A CHECK INPUT SIGNAL CIRCUIT.
1) Turn ignition switch "OFF".
2) Disconnect E.C.U. harness connector and T.C.M. 20-pin harness connector.
3) Check harness continuity between E.C.U. terminal 38 and T.C.M. terminal 7.
Continuity should exist.

N.G. → Check the following.
- Harness connectors F55, B25
- Harness connectors B2, M95
- Harness continuity between E.C.U. and T.C.M.

If N.G., repair harness or connectors.

↓ O.K.

CHECK T.C.M. FUNCTION.
Make sure that T.C.M. and circuit function properly.

N.G. → Repair malfunctioning parts.

↓ O.K.

Check E.C.U. pin terminals for damage or the connection of E.C.U. harness connector.

Perform FINAL CHECK by the following procedure after repair is completed.

FINAL CHECK

↓

Erase the self-diagnosis memory.

↓

Perform test drive.

↓

Perform self-diagnosis again. — N.G. → Recheck E.C.U. pin terminals for damage or the connection of E.C.U. harness connector.

↓ O.K.

INSPECTION END

Diagnostic Procedure 7

IGNITION SIGNAL (Code No. 21)

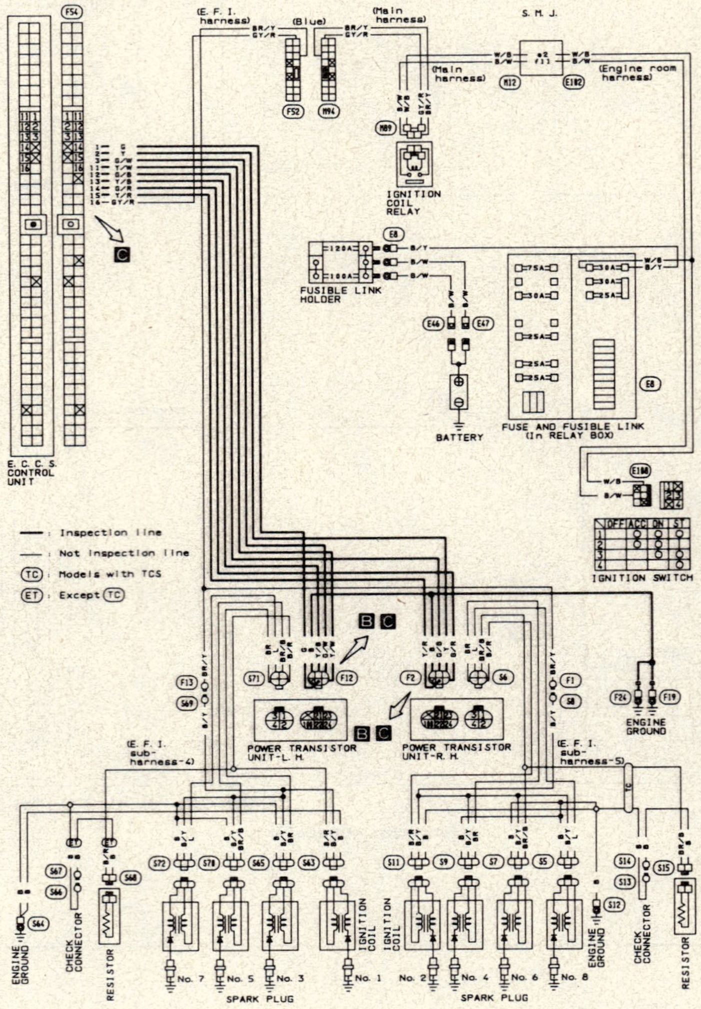

IGNITION SIGNAL (Code No. 21)

Harness layout

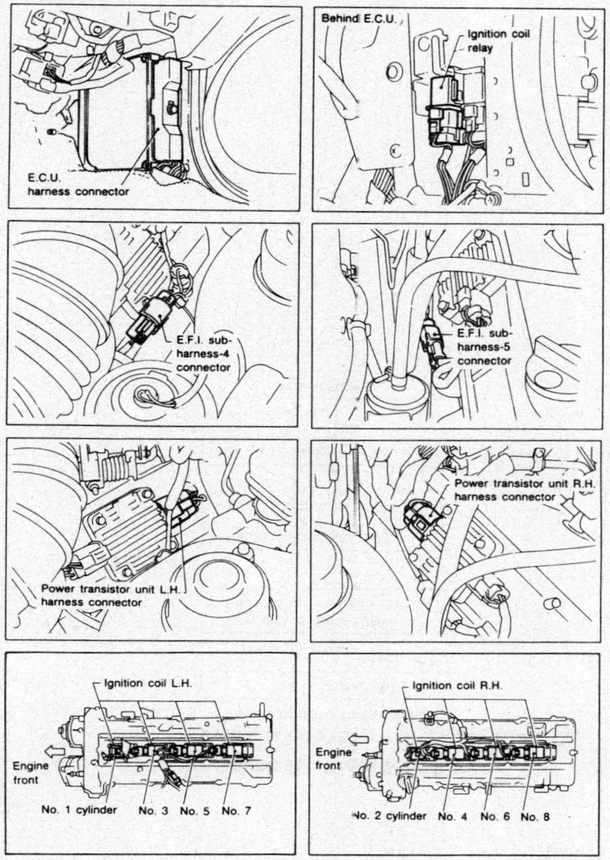

IGNITION SIGNAL (Code No. 21)

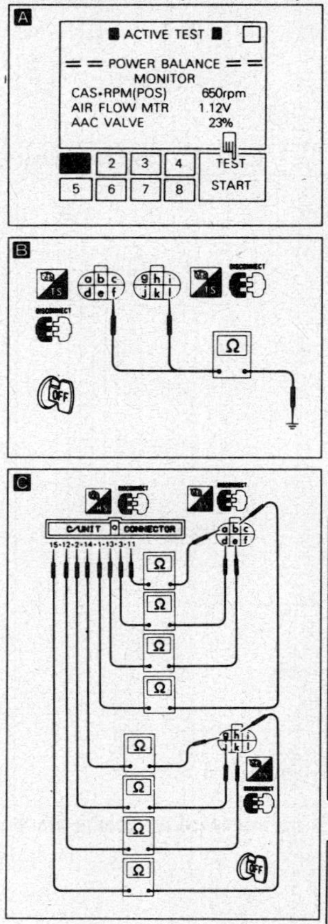

INSPECTION START

↓

PERFORM SELF-DIAGNOSIS.

- O.K. → Go to "IGNITION SIGNAL (Not self-diagnostic item-I)".
- N.G. ↓

A SEARCH FOR MALFUNCTIONING CIRCUIT.
1) Perform "POWER BALANCE" in "ACTIVE TEST" mode with CONSULT.
2) Search for circuit which does not produce a momentary engine speed drop.

- N.G. (from self-diagnosis, without CONSULT) → B
- Case 1: When either cylinders Nos. 1, 3, 5 and 7 or Nos. 2, 4, 6 and 8 are all N.G. → B
- Case 2: Except case 1 → C

B CHECK GROUND CIRCUIT.
1) Turn ignition switch "OFF".
2) Disconnect power transistor unit harness connector.
3) Check harness continuity between terminals ⓕ, ⓛ and engine ground.
Continuity should exist.

- N.G. → Repair harness or connectors.
- O.K. (without CONSULT) → C
- O.K. (with CONSULT) → CHECK COMPONENT

C CHECK OUTPUT SIGNAL CIRCUIT.
1) Turn ignition switch "OFF".
2) Disconnect E.C.U. harness connector.
3) Check harness continuity between
E.C.U. terminal ① and terminal ⓑ,
E.C.U. terminal ② and terminal ⓙ,
E.C.U. terminal ③ and terminal ⓓ,
E.C.U. terminal ⑪ and terminal ⓐ,
E.C.U. terminal ⑫ and terminal ⓚ,
E.C.U. terminal ⑬ and terminal ⓔ,
E.C.U. terminal ⑭ and terminal ⓖ,
E.C.U. terminal ⑮ and terminal ⓗ.
Continuity should exist.

- N.G. → Repair harness or connectors.
- O.K. ↓

CHECK COMPONENT
(Power transistor unit).
Refer to "Electrical Components Inspection".

- N.G. → Replace power transistor.
- O.K. ↓

Check E.C.U. pin terminals for damage or the connection of E.C.U. harness connector.

Diagnostic Procedure 8

E.C.C.S. CONTROL UNIT (Code No. 31)
(Check engine light item)

INSPECTION START
↓
1) Turn ignition switch "ON".
2) Erase the self-diagnosis memory.
↓
PERFORM SELF-DIAGNOSIS.
↓
Does E.C.U. display code No. 31 again?
- Yes → Replace E.C.U.
- No ↓

INSPECTION END

Diagnostic Procedure 9

E.G.R. FUNCTION (Code No. 32) (Check engine light item) : California models

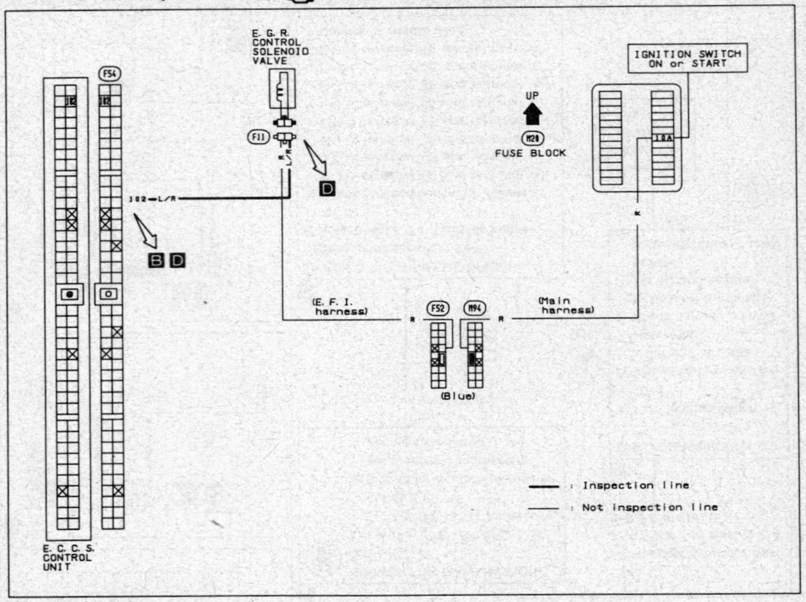

Harness layout

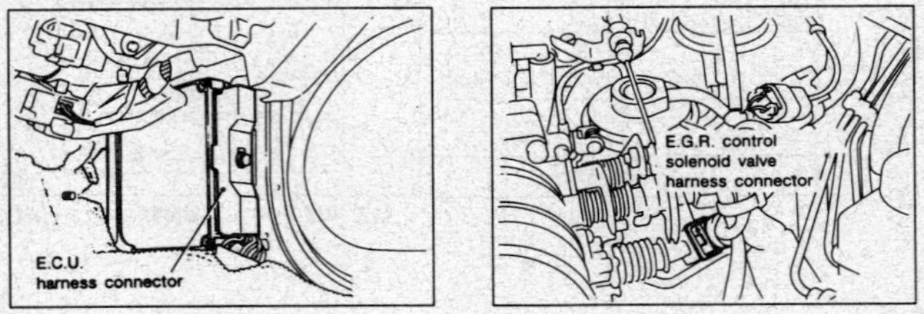

1990–92 4.5L ENGINE – DIAGNOSTIC PROCEDURE – Q45

E.G.R. FUNCTION (Code No. 32) (Check engine light item): California models

A
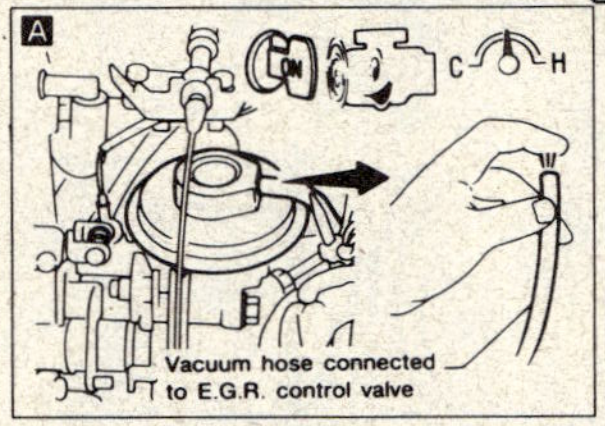

B
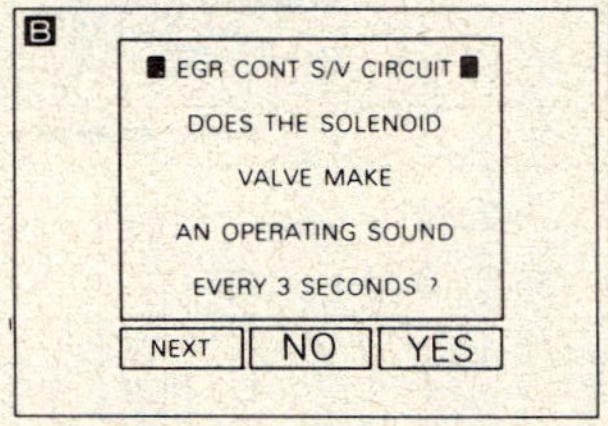

B
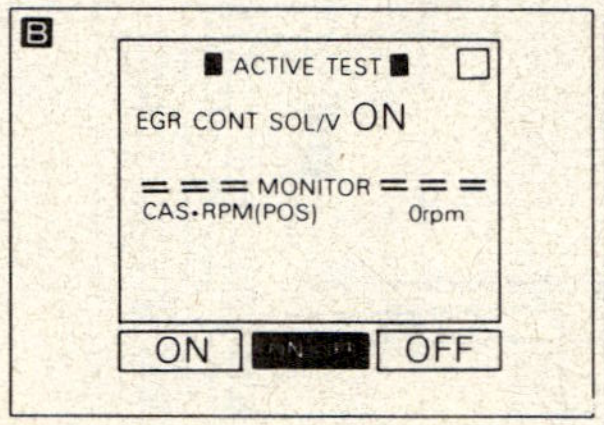

B
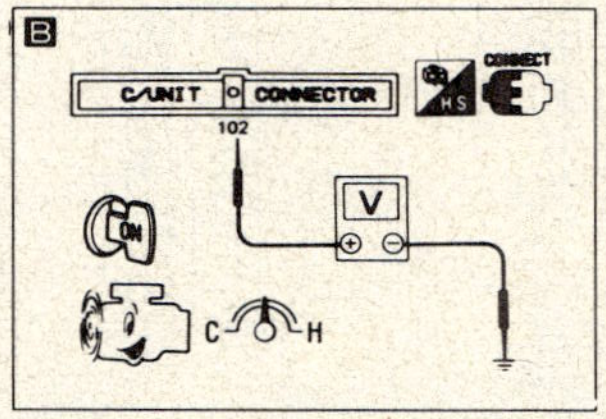

C
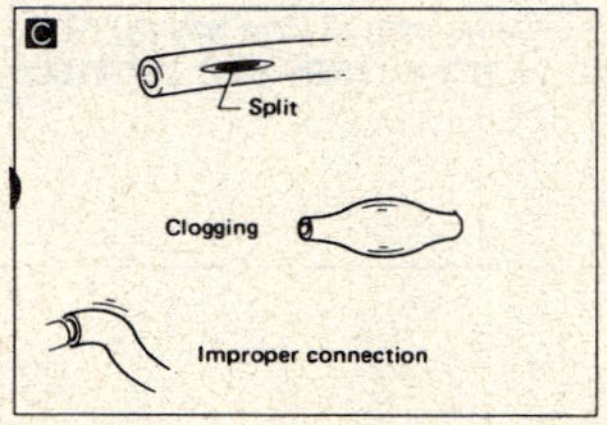

INSPECTION START
↓
PERFORM SELF-DIAGNOSIS. — O.K. → INSPECTION END
↓ N.G.

A CHECK VACUUM SOURCE TO E.G.R. CONTROL VALVE.
1) Start engine and warm it up sufficiently.
2) Perform self-diagnosis. Make sure that code No. 11 and 12 are not displayed.
3) Keep engine speed at about 2,000 rpm.
4) Disconnect vacuum hose to E.G.R. control valve.
5) Make sure that vacuum exists. **Vacuum should exist.**

O.K. → CHECK COMPONENTS (E.G.R. control valve, and B.P.T. valve). Refer to "Electrical Components Inspection". — N.G. → Replace malfunctioning component(s).

↓ N.G.

B CHECK CONTROL FUNCTION.
1) Stop engine.
2) Turn ignition switch "ON".
3) Perform "EGR CONT S/V CIRCUIT" in "FUNCTION TEST" mode with CONSULT.

— OR —

1) Stop engine.
2) Turn ignition switch "ON".
3) Turn E.G.R. control solenoid valve "ON" and "OFF" in "ACTIVE TEST" mode with CONSULT and check operating sound.

— OR —

1) Check voltage between E.C.U. terminal (102) and ground under the following conditions with CONSULT or tester.
Voltage:
At idle
Approximately 0V
Engine speed is about 2,000 rpm.
Battery voltage

O.K. → **C** CHECK VACUUM HOSE.
1) Check vacuum hose for clogging, cracks and proper connection.

↓ N.G.
Ⓐ

E.G.R. FUNCTION (Code No. 32) (Check engine light item): California models

D
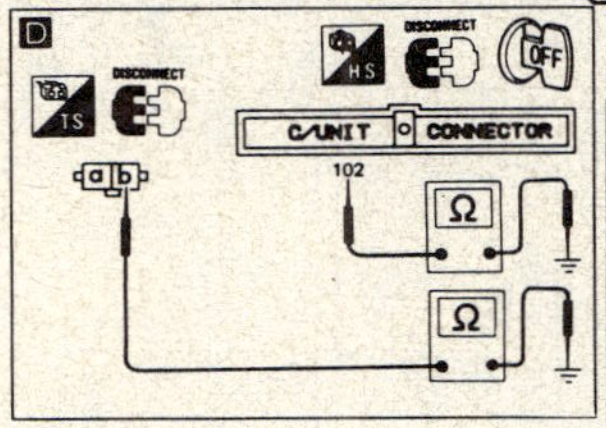

Ⓐ
↓

D CHECK OUTPUT SIGNAL CIRCUIT.
1) Turn ignition switch "OFF".
2) Disconnect E.C.U. harness connector.
3) Disconnect E.G.R. control solenoid valve harness connector.
4) Check harness continuity between E.C.U. terminal (102) and ground, terminal ⓑ and ground.
Continuity should not exist.

N.G. → Repair harness or connectors between E.C.U. and E.G.R. control solenoid valve.

↓ O.K.

CHECK COMPONENT (E.G.R. control solenoid valve). Refer to "Electrical Components Inspection"

N.G. → Replace E.G.R. control solenoid valve.

↓ O.K.

Check E.C.U. pin terminals for damage or the connection of E.C.U. harness connector.
↓
Check resistance of exhaust gas temperature sensor.

Perform FINAL CHECK by the following procedure after repair is completed.

FINAL CHECK
↓
Erase the self-diagnosis memory.
↓
Perform test drive.
↓
Perform self-diagnosis again. — N.G. → Recheck E.C.U. pin terminals for damage or the connection of E.C.U. harness connector.
↓ O.K.
INSPECTION END

Diagnostic Procedure 10

EXHAUST GAS SENSOR L.H. (Code No. 33) EXHAUST GAS SENSOR R.H. (Code No. 53)
(Check engine light item)

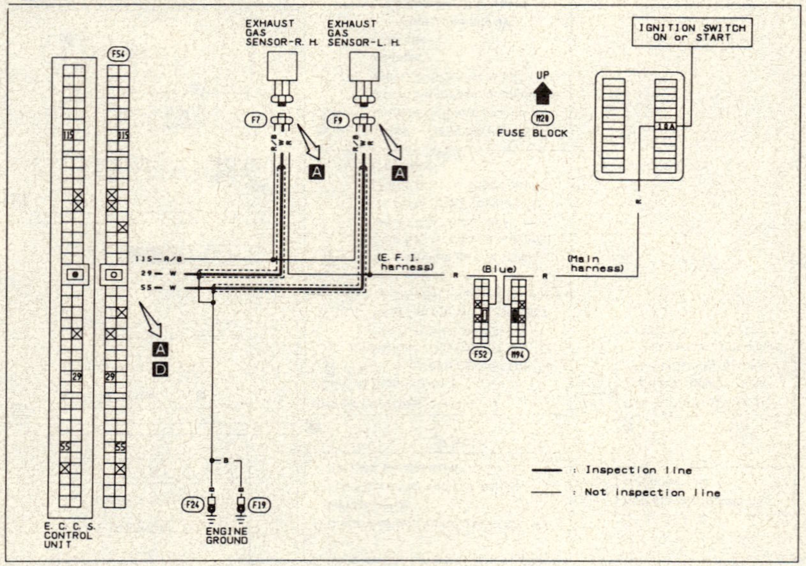

Harness layout

EXHAUST GAS SENSOR L.H. (Code No. 33) EXHAUST GAS SENSOR R.H. (Code No. 53) (Check engine light item)

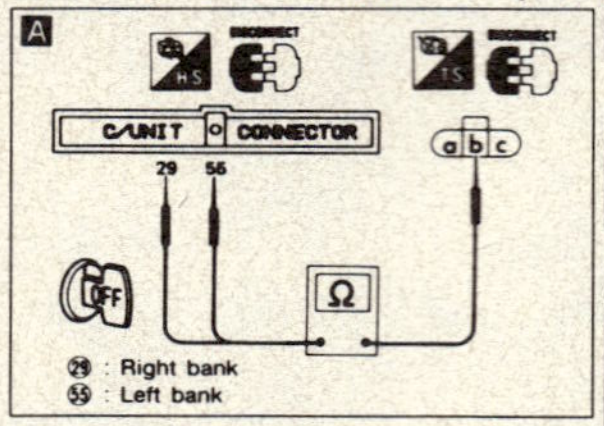

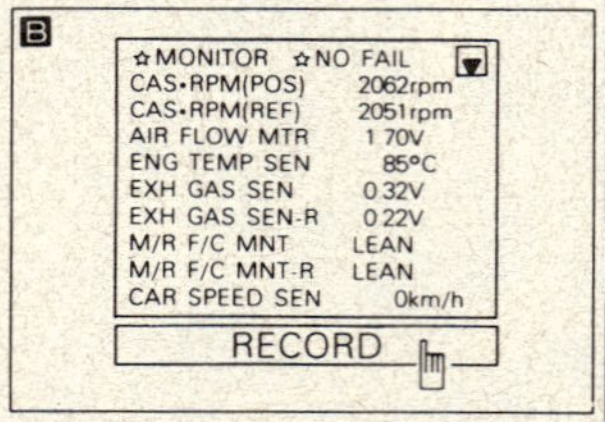

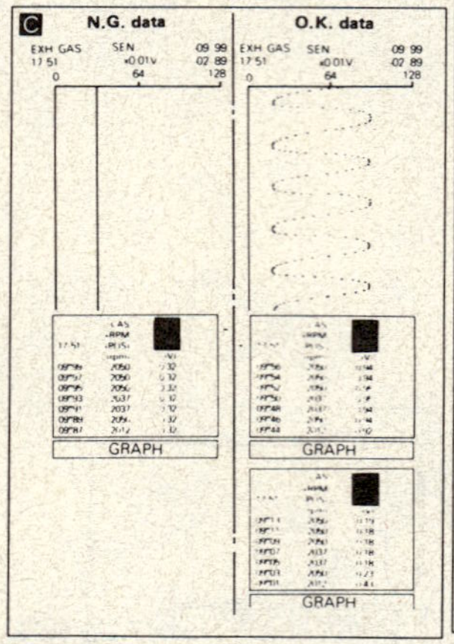

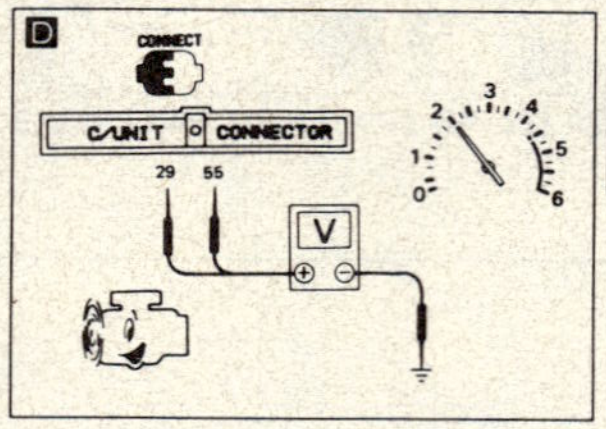

INSPECTION START

↓

CHECK OVERALL FUNCTION. — O.K. → INSPECTION END

↓ N.G.

A CHECK INPUT SIGNAL CIRCUIT.
1) Turn ignition switch "OFF".
2) Disconnect exhaust gas sensor harness connector and E.C.U. harness connector.
3) Check harness continuity between E.C.U. terminals 29, 55 and terminal b.
Continuity should exist.

— N.G. → Repair harness or connectors.

↓ O.K.

CHECK VOLTAGE.
1) Reconnect exhaust gas sensor harness connector and E.C.U. harness connector.
2) Start engine and warm it up sufficiently.
B 3) Select "DATA MONITOR" mode with CONSULT.
4) Touch "RECORD" on CONSULT with engine speed held at approximately 2,000 rpm.
C 5) Check "EXH GAS SEN (–R)" voltage.
① **Approx. 0.3 volts (No voltage fluctuations) ... N.G.**
② **Other than ① above ... O.K.**

—OR—

D 3) Check voltage between E.C.U. terminals 29, 55 and ground, with engine speed held at approximately 2,000 rpm.
① **Approx. 0.3 volts (No voltage fluctuations) ... N.G.**
② **Other than ① above ... O.K.**

— N.G. → Replace exhaust gas sensor.

↓ O.K.

Check E.C.U. pin terminals for damage or the connection of E.C.U. harness connector.

Diagnostic Procedure 11

DETONATION SENSOR (Code No. 34)

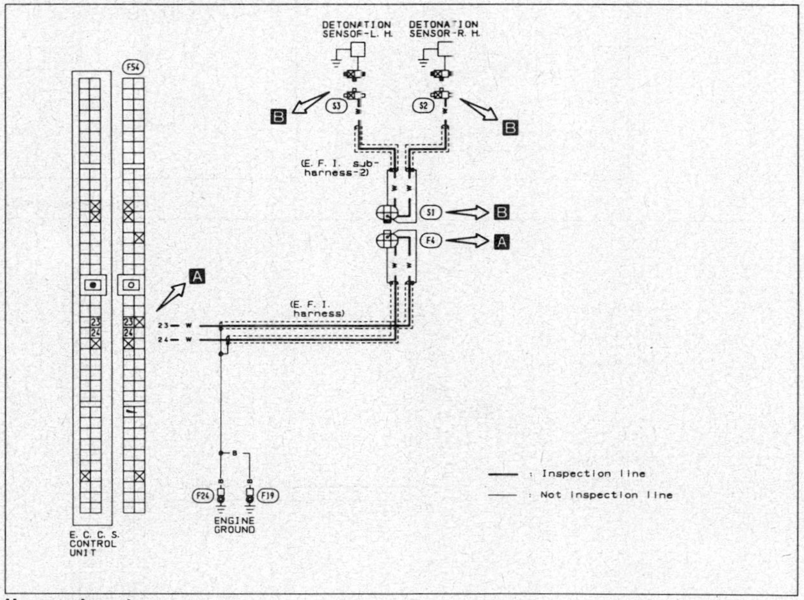

Harness layout

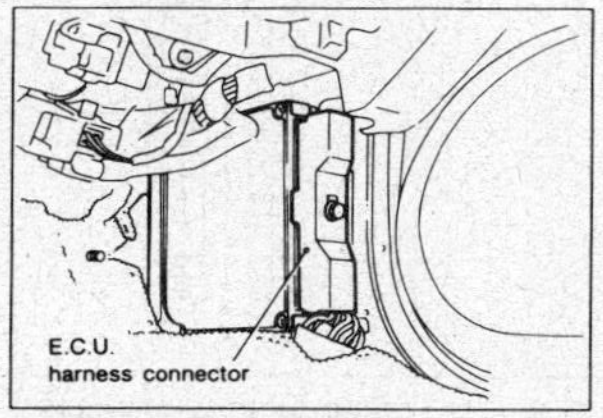

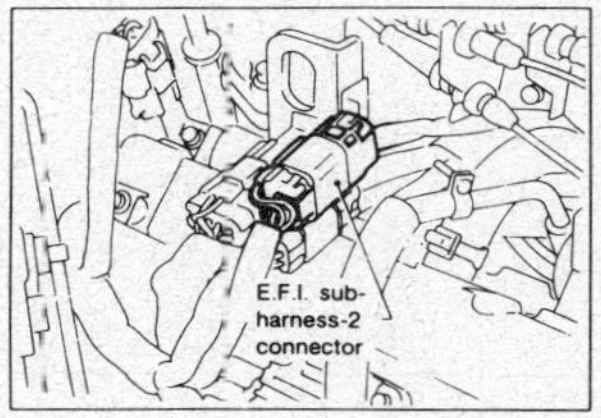

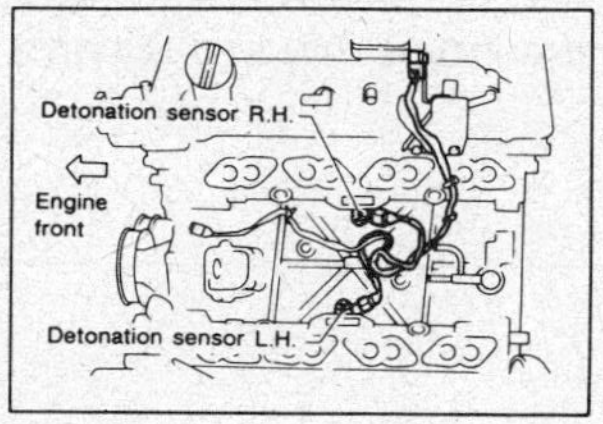

DETONATION SENSOR (Code No. 34)

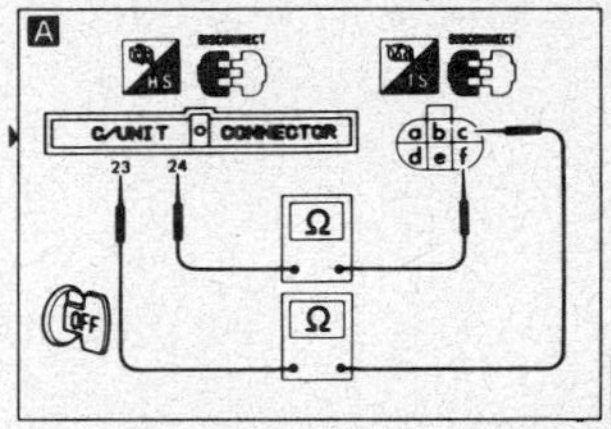

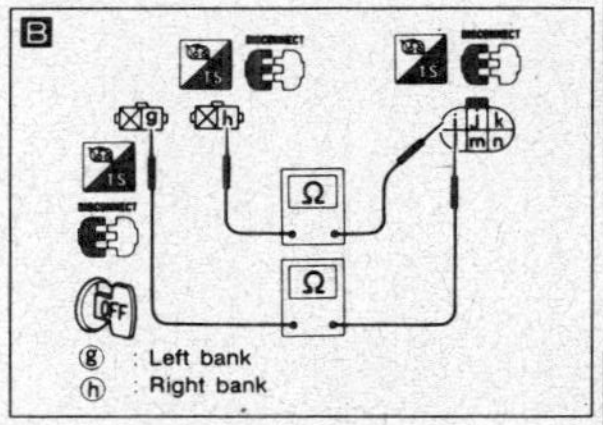

INSPECTION START

↓

PERFORM SELF-DIAGNOSIS. — O.K. → INSPECTION END

↓ N.G.

A CHECK INPUT SIGNAL CIRCUIT-I.
1) Turn ignition switch "OFF".
2) Disconnect E.C.U. harness connector and E.F.I. sub- harness-2 connector.
3) Check harness continuity between terminal ⓔ and E.C.U. terminal ㉓, terminal ⓕ and E.C.U. terminal ㉔.
Continuity should exist.

— N.G. → Repair harness or connectors.

↓ O.K.

B CHECK INPUT SIGNAL CIRCUIT-II.
1) Disconnect detonation sensor- L.H. & R.H. harness connectors.
2) Check harness continuity between the following terminals.
- ⓖ and ⓛ
- ⓗ and ⓛ

Continuity should exist.

— N.G. → Repair harness or connectors.

↓ O.K.

CHECK COMPONENT (Detonation sensor).
Refer to "Electrical Components Inspection".

— N.G. → Replace detonation sensor.

↓ O.K.

Check E.C.U. pin terminals for damage or the connection of E.C.U. harness connector.

Diagnostic Procedure 12

EXHAUST GAS TEMPERATURE SENSOR (Code No. 35) (Check engine light item):

California models only

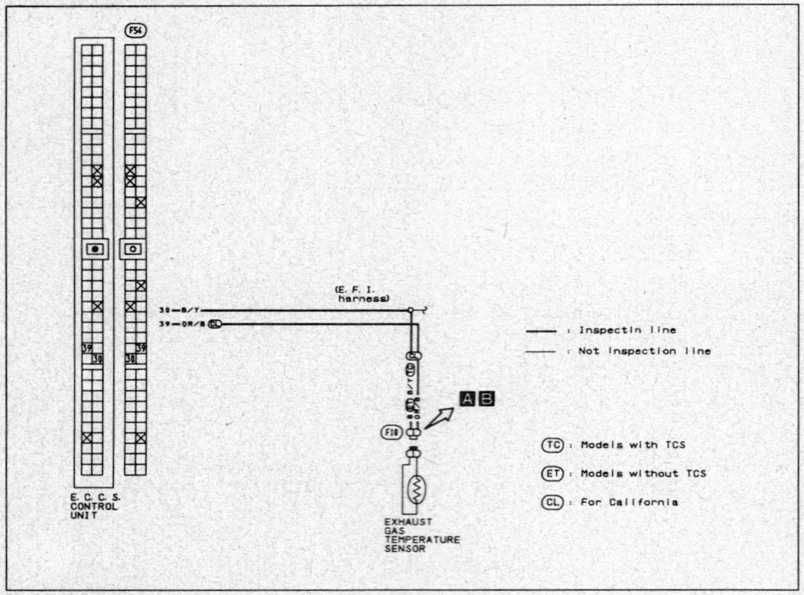

Harness layout

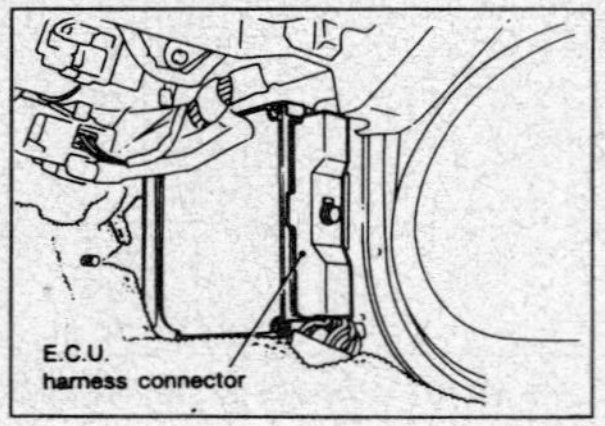

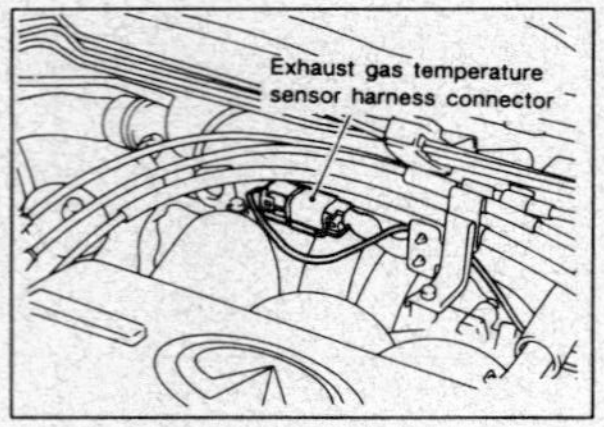

EXHAUST GAS TEMPERATURE SENSOR (Code No. 35) (Check engine light item):

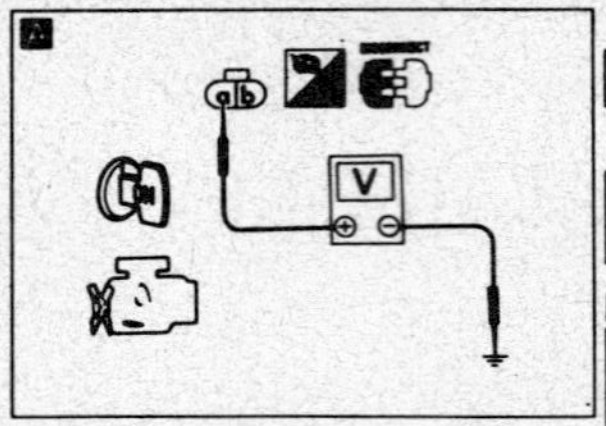

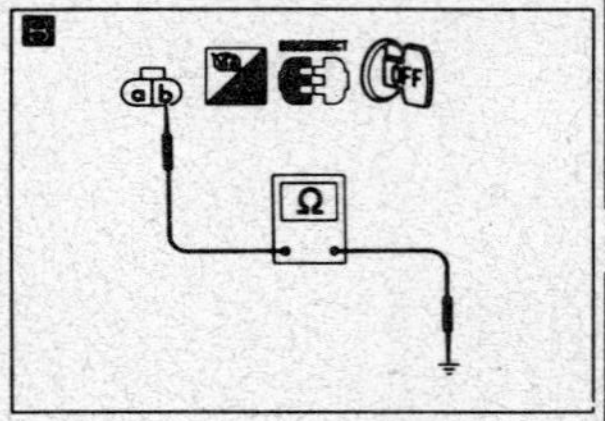

- INSPECTION START
- PERFORM SELF-DIAGNOSIS.
  - O.K. → INSPECTION END
  - N.G. ↓
- **A** CHECK POWER SUPPLY.
  1) Turn ignition switch "OFF".
  2) Disconnect exhaust gas temperature sensor harness connector.
  3) Turn ignition switch "ON".
  4) Check voltage between terminal ⓐ and ground with CONSULT or tester.
  Voltage: Approximately 5V
  - N.G. → Repair harness or connectors.
  - O.K. ↓
- **B** CHECK GROUND CIRCUIT.
  1) Turn ignition switch "OFF".
  2) Check harness continuity between terminal ⓑ and engine ground.
  Continuity should exist.
  - N.G. → Repair harness or connectors.
  - O.K. ↓
- CHECK COMPONENT (Exhaust gas temperature sensor). Refer to "Electrical Components Inspection".
  - N.G. → Replace exhaust gas temperature sensor.
  - O.K. ↓
- Check E.C.U. pin terminals for damage or the connection of E.C.U. harness connector.

Perform FINAL CHECK by the following procedure after repair is completed.

- FINAL CHECK
- Erase the self-diagnosis memory.
- Perform test drive.
- Perform self-diagnosis again.
  - N.G. → Recheck E.C.U. pin terminals for damage or the connection of E.C.U. harness connector.
  - O.K. ↓
- INSPECTION END

Diagnostic Procedure 13

THROTTLE SENSOR (Code No. 43) (Check engine light item)

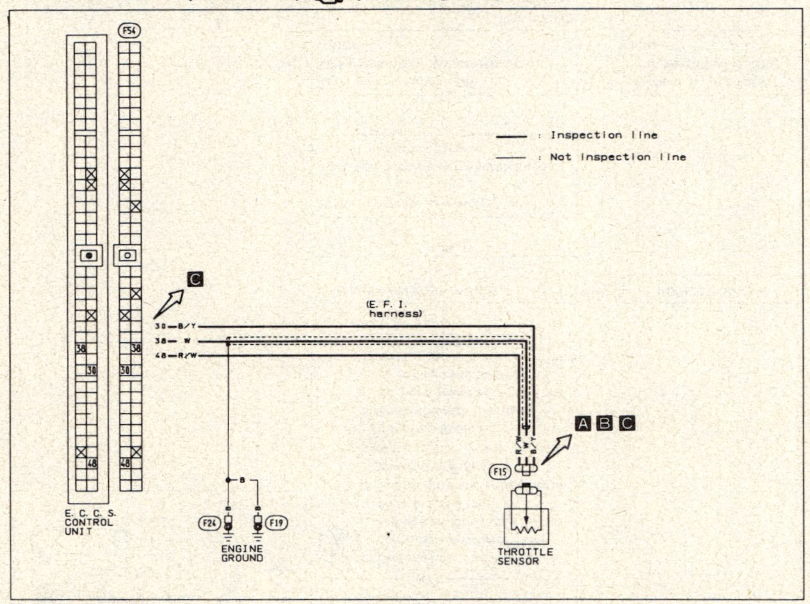

Harness layout

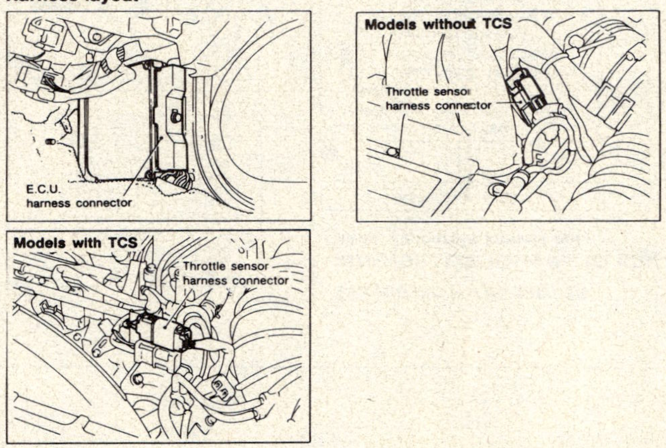

THROTTLE SENSOR (Code No. 43) (Check engine light item)

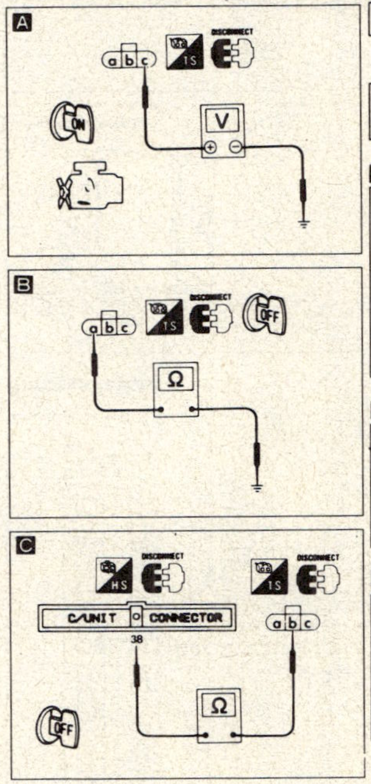

INSPECTION START

↓

PERFORM SELF-DIAGNOSIS. — O.K. → INSPECTION END

↓

A CHECK POWER SUPPLY.
1) Turn ignition switch "OFF".
2) Disconnect throttle sensor harness connector.
3) Turn ignition switch "ON".
4) Check voltage between terminal ⓒ and ground with CONSULT or tester.
Voltage: Approximately 5V

— N.G. → Repair harness or connectors.

↓ O.K.

B CHECK GROUND CIRCUIT.
1) Turn ignition switch "OFF".
2) Check harness continuity between terminal ⓐ and engine ground.
Continuity should exist.

— N.G. → Repair harness or connectors.

↓ O.K.

C CHECK INPUT SIGNAL CIRCUIT.
1) Disconnect E.C.U. harness connector.
2) Check harness continuity between E.C.U. terminal ㊳ and terminal ⓑ.
Continuity should exist.

— N.G. → Repair harness or connectors.

↓ O.K.

CHECK COMPONENT
(Throttle sensor).
Refer to "Electrical Components Inspection"

— N.G. → Replace throttle sensor.

↓ O.K.

Check E.C.U. pin terminals for damage or the connection of E.C.U. harness connector.

Diagnostic Procedure 14

INJECTOR LEAK (Code No. 45) (Check engine light item): California models only

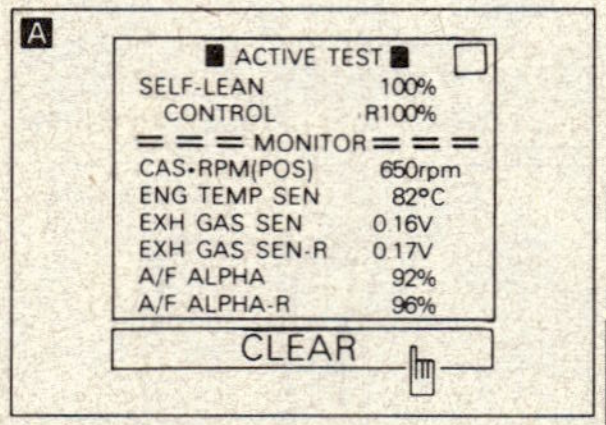

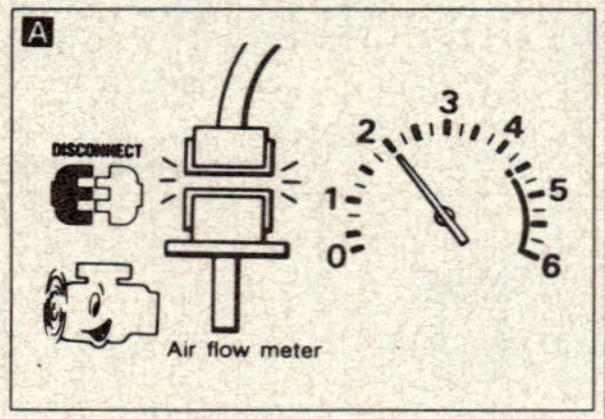

INSPECTION START

↓

A Clear the self-learning data
1) Start engine and warm it up sufficiently.
2) Select "SELF- LEARNING CONT" in "ACTIVE TEST" mode with CONSULT.
3) Clear the self-learning control coefficient by touching "CLEAR".

— OR —

2) Disconnect air flow meter connector, and restart and run engine for at least 30 seconds at 2,000 rpm.
3) Stop engine and reconnect air flow meter connector.
4) Make sure code No. 12 is displayed in Mode II.
5) Erase the self- diagnosis memory. Make sure code No. 55 is displayed in Mode II.

↓

PERFORM SELF-DIAGNOSIS. — O.K. → INSPECTION END

↓ N.G.

Remove injector assembly.

Keep fuel hose and all injectors connected to injector gallery.

↓

Turn ignition switch "ON". Make sure fuel does not drip from injector. — Drips → Replace the injectors from which fuel is dripping.

↓ Does not drip

Check E.C.U. pin terminals for damage or the connection of E.C.U. harness connector.

Diagnostic Procedure 15

SECONDARY THROTTLE SENSOR (Code No. 46): TCS models

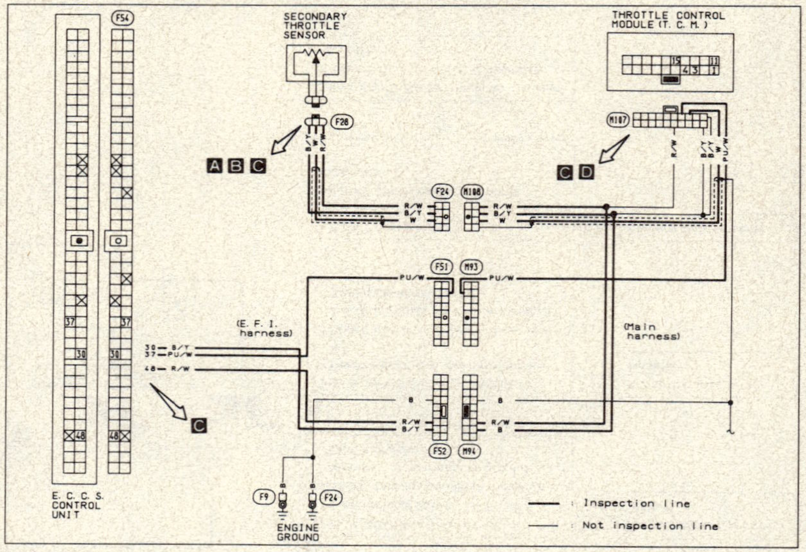

Harness layout

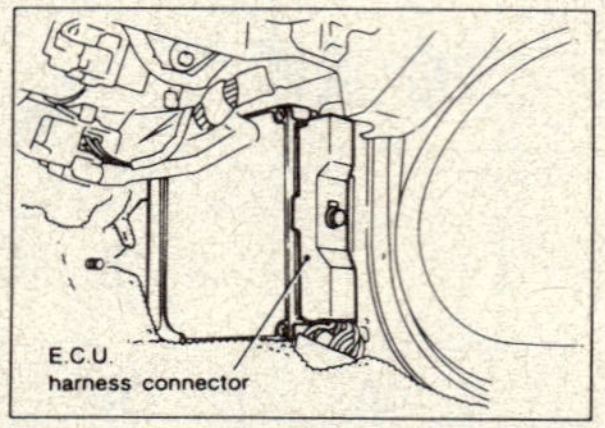

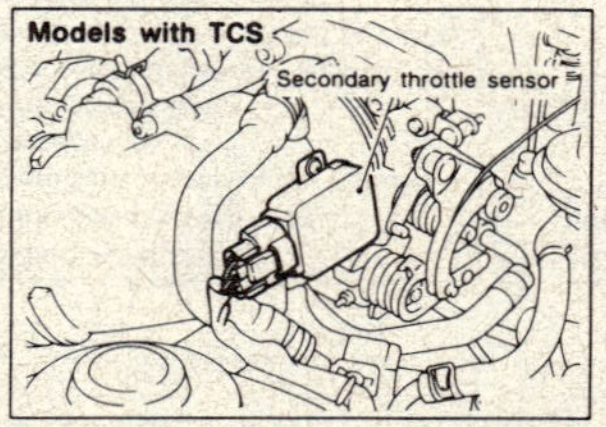

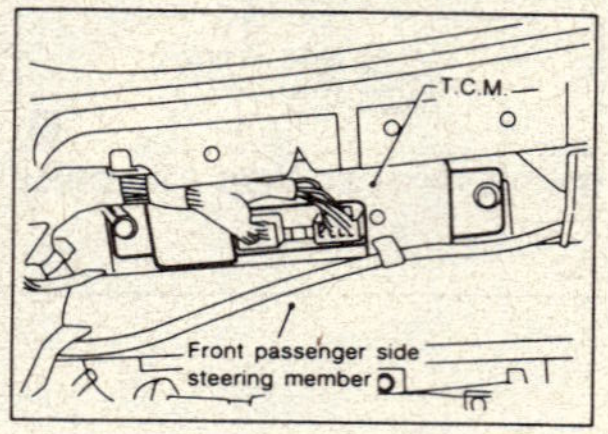

1990–92 4.5L ENGINE—DIAGNOSTIC PROCEDURE—Q45

SECONDARY THROTTLE SENSOR (Code No. 46): TCS models

A
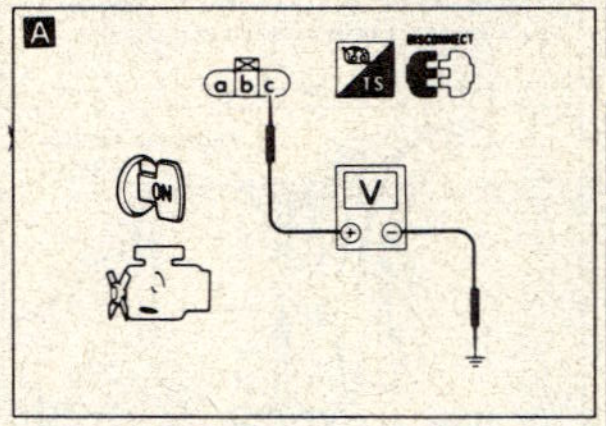

B
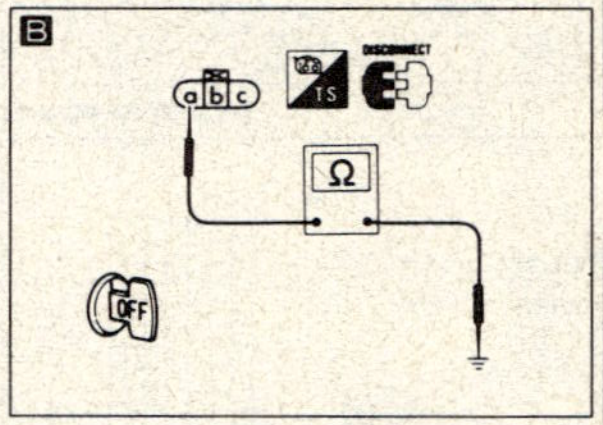

C
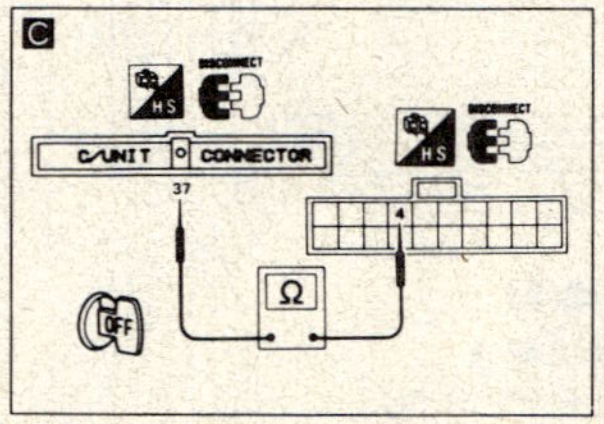

D
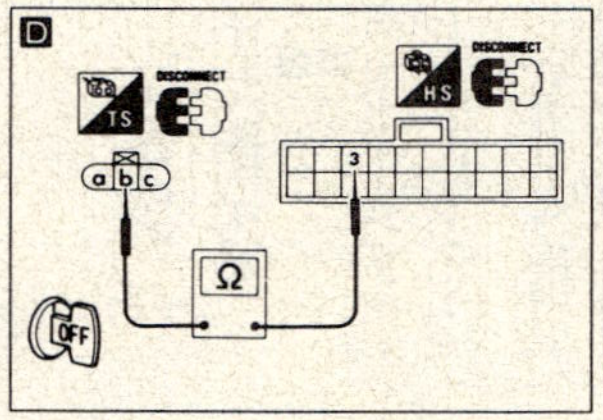

INSPECTION START
↓
PERFORM SELF-DIAGNOSIS. — O.K. → INSPECTION END
↓ N.G.

A CHECK POWER SUPPLY.
1) Turn ignition switch "OFF".
2) Disconnect secondary throttle sensor harness connector.
3) Turn ignition switch "ON".
4) Check voltage between terminal ⓒ and ground with CONSULT or tester.
Voltage: Approximately 5V

N.G. → Check the following.
- Harness connectors (F24), (M108)
- Harness connectors (M94), (F52)
- Harness continuity between secondary throttle sensor and E.C.U.

If N.G., repair harness or connectors.

↓ O.K.

B CHECK GROUND CIRCUIT.
1) Turn ignition switch "OFF".
2) Check harness continuity between terminal ⓐ and ground.
Continuity should exist.

N.G. → Check the following.
- Harness connectors (F24), (M108)
- Harness connectors (M94), (F52)
- Harness continuity between secondary throttle sensor and E.C.U.

If N.G., repair harness or connectors.

↓ O.K.

C CHECK INPUT SIGNAL CIRCUIT-I.
1) Disconnect E.C.U. harness connector and T.C.M. harness connector.
2) Check harness continuity between E.C.U. terminal (37) and T.C.M. terminal ④.
Continuity should exist.

N.G. → Check the following.
- Harness connectors (F51), (M93)
- Harness continuity between E.C.U. and T.C.M.

If N.G., repair harness or connectors.

↓ O.K.

D CHECK INPUT SIGNAL CIRCUIT-II.
1) Check harness continuity between T.C.M. terminal ③ and terminal ⓑ.
Continuity should exist.

N.G. → Check the following.
- Harness connectors (F24), (M108)
- Harness continuity between T.C.M. and secondary throttle sensor.

↓ O.K.

CHECK COMPONENT
(Secondary throttle sensor).
Refer to "Electrical Components Inspection".

N.G. → Replace secondary throttle sensor.

↓ O.K.
Ⓐ

SECONDARY THROTTLE SENSOR (Code No. 46): TCS models

Ⓐ
↓
CHECK T.C.M. FUNCTION.
Make sure that T.C.M. functions properly.

N.G. → Replace T.C.M.

↓ O.K.

Check E.C.U. pin terminals for damage or the connection of E.C.U. harness connector.

Perform FINAL CHECK by the following procedure after repair is completed.

FINAL CHECK
↓
Erase the self-diagnosis memory.
↓
Perform test drive.
↓
Perform self-diagnosis again. — N.G. → Recheck E.C.U. pin terminals for damage or the connection of E.C.U. harness connector.
↓ O.K.
INSPECTION END

Diagnostic Procedure 16

INJECTOR (Code No. 51) (Check engine light item): California models

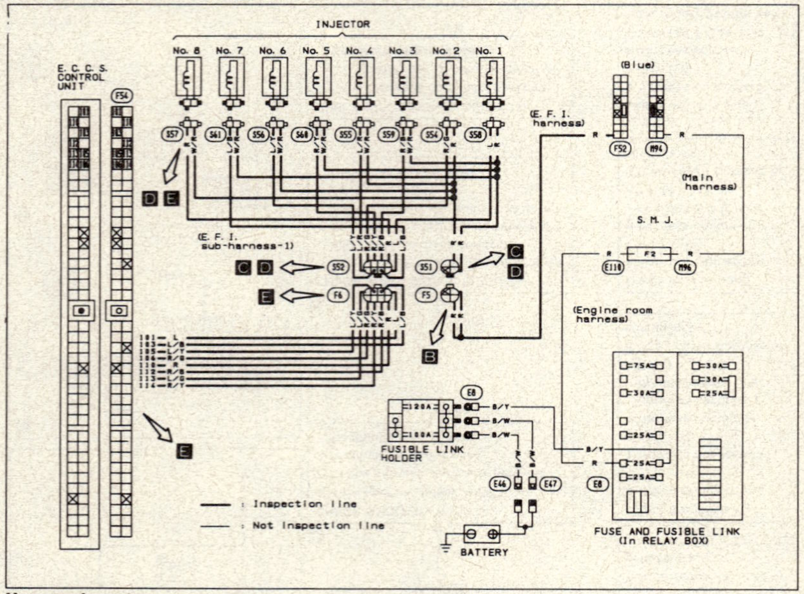

Harness layout

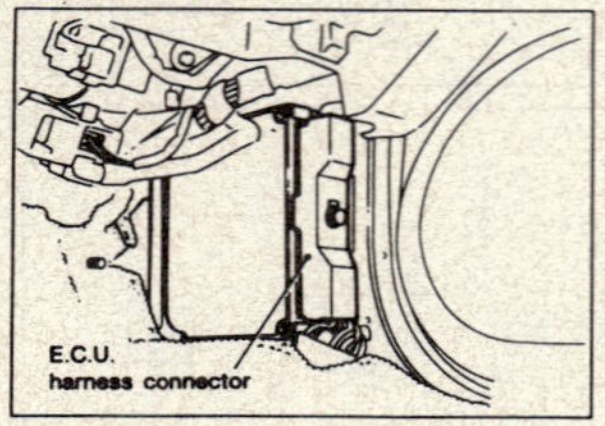

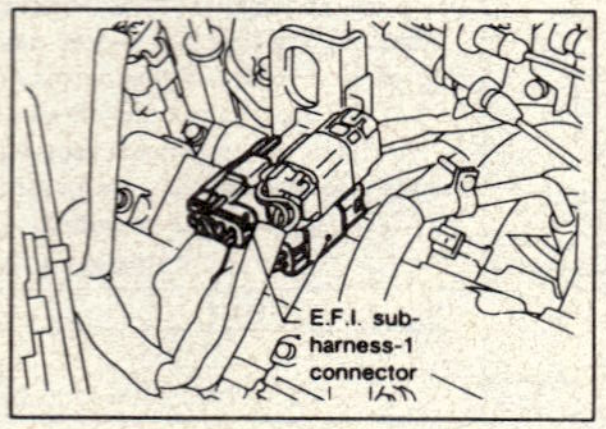

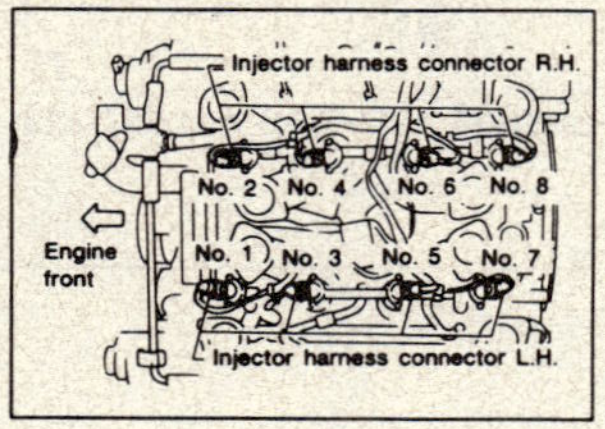

INJECTOR (Code No. 51) (Check engine light item): California models

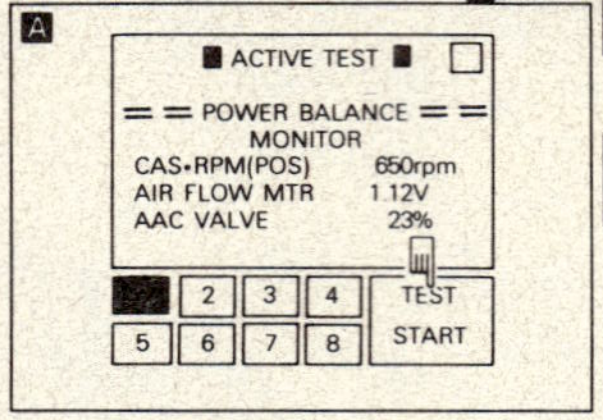

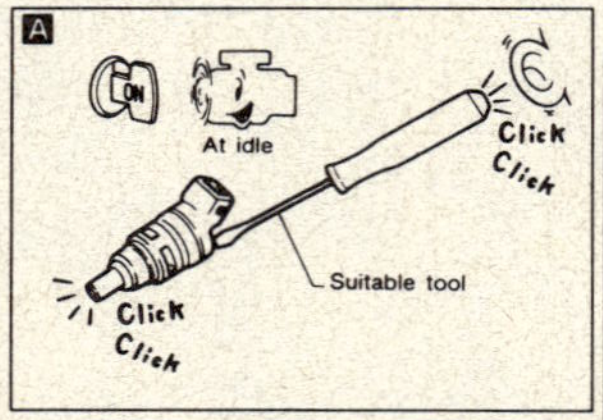

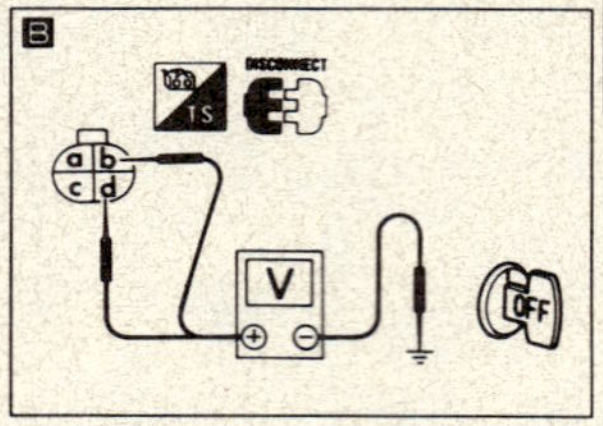

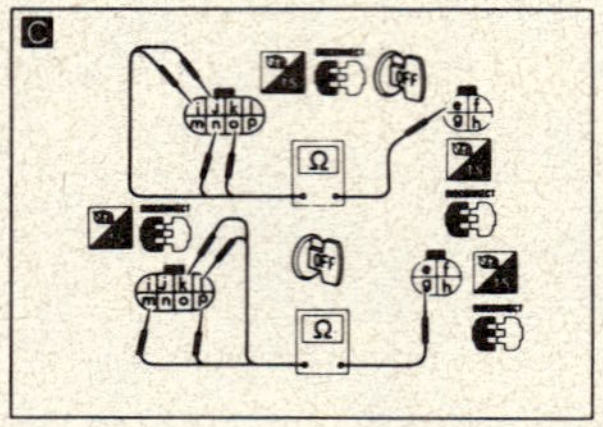

INSPECTION START

↓

PERFORM SELF-DIAGNOSIS.

- O.K. → INSPECTION END
- If engine speed does not exceed the idle speed or engine idle is unstable. → Go to "INJECTOR (Not self- diagnostic item)".
- N.G. ↓

A SEARCH FOR MALFUNCTIONING CIRCUIT.

1) Perform "POWER BALANCE" in "ACTIVE TEST" mode with CONSULT.
2) Search for circuit which does not produce a momentary engine speed drop.

— OR —

1) Search for circuit which does not produce injector operating sound at idle.

Case 1:
When either cylinders Nos. 1, 3, 5 and 7 or Nos. 2, 4, 6 and 8 are all N.G. →

Case 2:
Except case 1 → Ⓐ

B CHECK POWER SUPPLY-I.

1) Turn ignition switch "OFF".
2) Disconnect E.F.I. sub-harness- 1 connector (F5).
3) Check voltage between terminals ⓑ, ⓓ and ground with CONSULT or tester.

Voltage: Battery voltage

- O.K. → Ⓐ
- N.G. ↓

Check the following.

- Harness connector (F52), (M4)
- Harness continuity between E.F.I. sub-harness-1 connector (F5) and harness connector (F52).

If N.G., repair harness or connectors.

INJECTOR (Code No. 51) (Check engine light item): California models

D

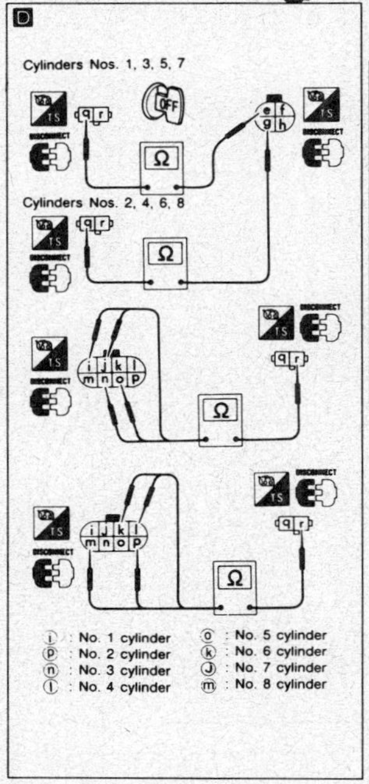

E

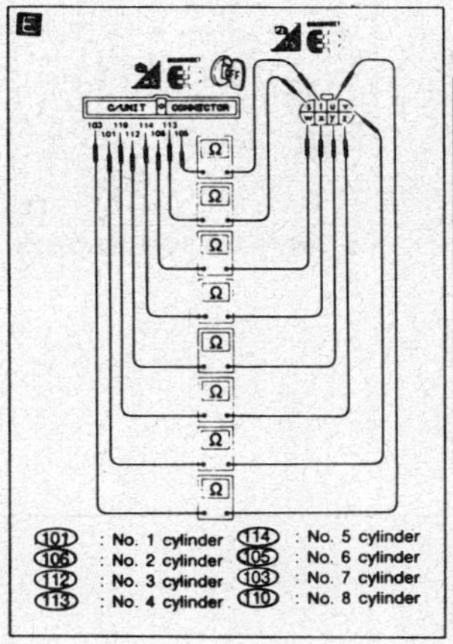

Ⓐ

C CHECK POWER SUPPLY-II.
1) Turn ignition switch "OFF".
2) Disconnect E.F.I. sub-harness- 1 connectors.
3) Check circuit continuity between the following terminals.
- ⓔ and ⓘ
- ⓔ and ⓙ
- ⓔ and ⓝ
- ⓔ and ⓞ
- ⓖ and ⓚ
- ⓖ and ⓛ
- ⓖ and ⓜ
- ⓖ and ⓟ

Continuity should exist.

N.G. → **D** Check the following.
1) Disconnect injector harness connectors.
2) Check harness continuity between the following terminals.
- ⓠ and ⓔ
- ⓠ and ⓖ
- ⓘ and ⓡ
- ⓙ and ⓡ
- ⓚ and ⓡ
- ⓛ and ⓡ
- ⓜ and ⓡ
- ⓝ and ⓡ
- ⓞ and ⓡ
- ⓟ and ⓡ

Continuity should exist.
If N.G., repair harness or connectors.

O.K. ↓

CHECK COMPONENT (Injector).
Refer to "Electrical Components Inspection".

If N.G. replace injector.

O.K. ↓

E CHECK OUTPUT SIGNAL CIRCUIT.
1) Disconnect E.C.U. harness connector.
2) Check continuity between the following terminals.
- ⓢ and (113)
- ⓣ and (105)
- ⓤ and (103)
- ⓥ and (101)
- ⓦ and (106)
- ⓧ and (114)
- ⓨ and (112)
- ⓩ and (110)

Continuity should exist.

N.G. → Repair harness or connectors.

O.K. ↓

Check E.C.U. pin terminals for damage or the connection of E.C.U. harness connector.

Diagnostic Procedure 17

A/T CONTROL (Code No. 54)

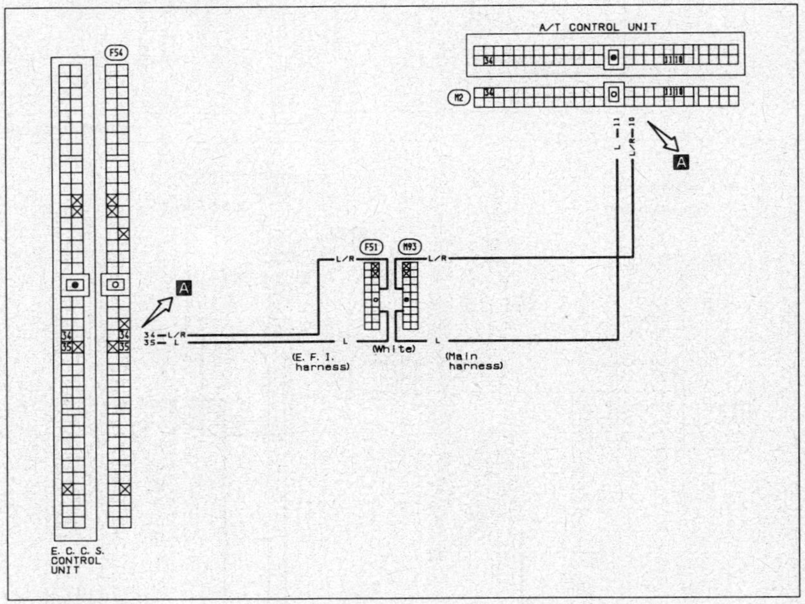

Harness layout

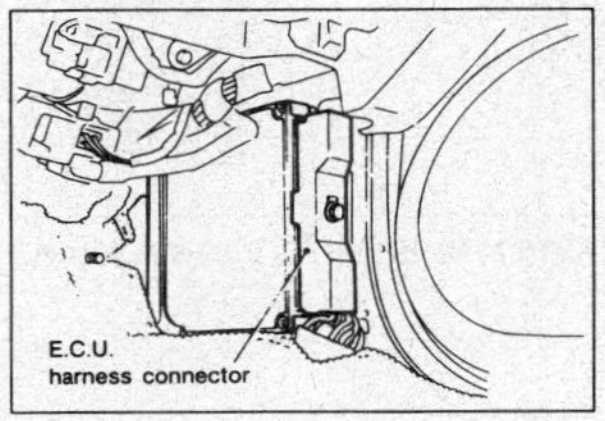

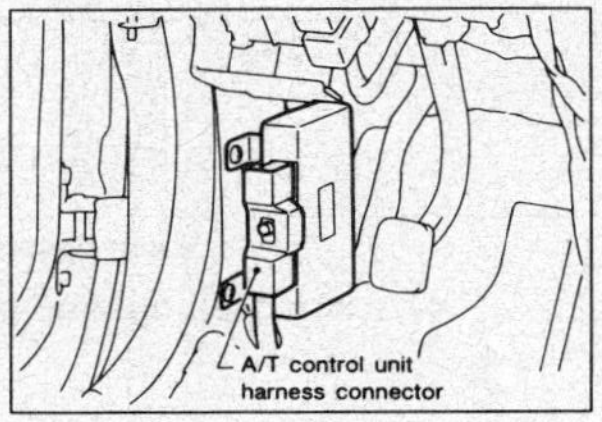

A/T CONTROL (Code No. 54)

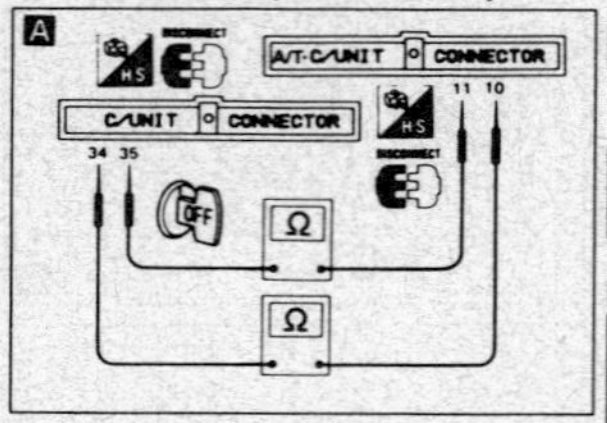

INSPECTION START

↓

PERFORM SELF-DIAGNOSIS. — O.K. → INSPECTION END

↓ N.G.

A CHECK INPUT SIGNAL CIRCUIT.
1) Turn ignition switch "OFF".
2) Disconnect E.C.U. harness connector and A/T control unit harness connector.
3) Check harness continuity between E.C.U. terminal ㉞ and terminal ⑩, E.C.U. terminal ㉟ and terminal ⑪.
Continuity should exist.

— N.G. → Check the following.
- Harness connectors (F51), (M93)
- Harness continuity between E.C.U. and A/T control unit

If N.G., repair harness or connectors.

↓ O.K.

Check E.C.U. pin terminals for damage or the connection of E.C.U. harness connector.

Perform FINAL CHECK by the following procedure after repair is completed.

FINAL CHECK

↓

Erase the self-diagnosis memory.

↓

Perform test drive.

↓

Perform self-diagnosis again. — N.G. → Recheck E.C.U. pin terminals for damage or the connection of E.C.U. harness connector.

↓ O.K.

INSPECTION END

Diagnostic Procedure 18

CRANK ANGLE SENSOR (Not self-diagnostic item)

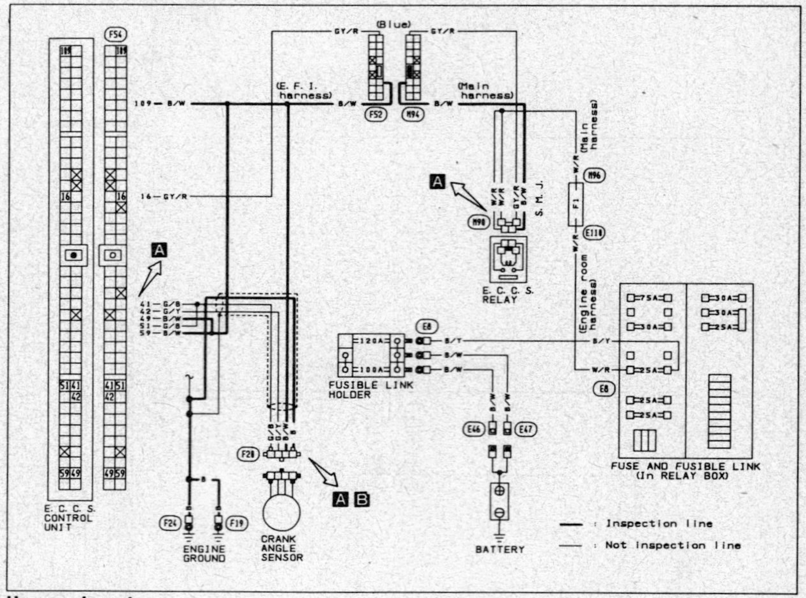

Harness layout

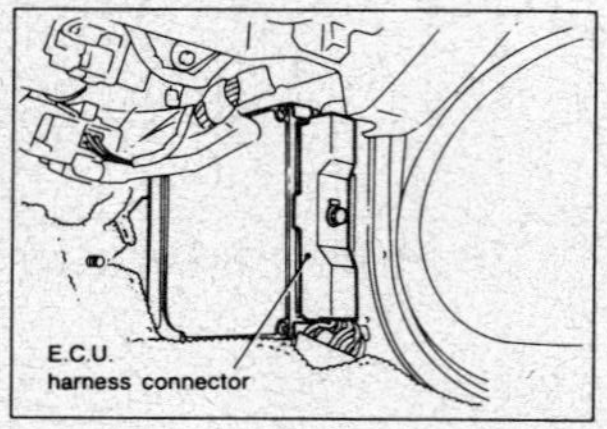

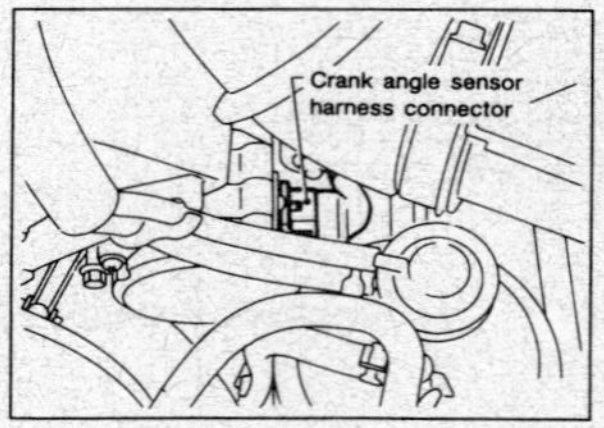

CRANK ANGLE SENSOR (Not self-diagnostic item)

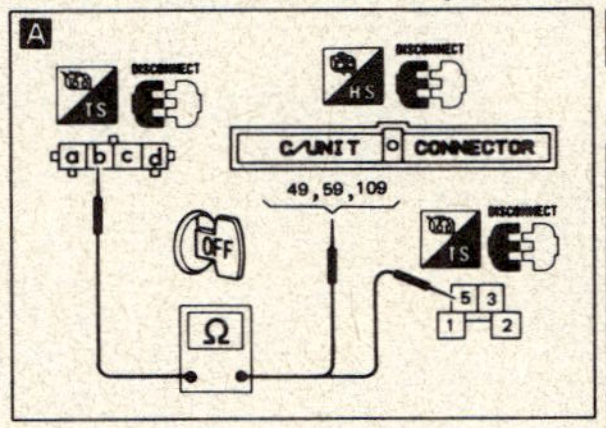

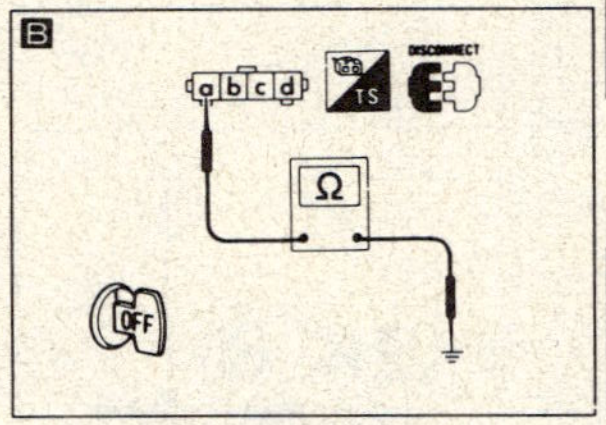

INSPECTION START

A CHECK POWER SUPPLY.
1) Turn ignition switch "OFF".
2) Disconnect crank angle sensor harness connector.
3) Disconnect E.C.U. harness connector and E.C.C.S. relay.
4) Check harness continuity between terminal ⓑ and terminal ⑤, E.C.U. terminals ㊾, ㊿, ⑩⑨.
Continuity should exist.

N.G. → Check the following.
- Harness connectors (F52), (M94)
- Harness continuity between crank angle sensor and E.C.C.S. relay
- Harness continuity between crank angle sensor and E.C.U.

If N.G., repair harness or connectors.

O.K. ↓

B CHECK GROUND CIRCUIT.
1) Check harness continuity between terminal ⓐ and engine ground.
Continuity should exist.

N.G. → Repair harness or connectors.

O.K. ↓

CHECK COMPONENT
(Crank angle sensor).
Refer to "Electrical Components Inspection".

N.G. → Replace crank angle sensor.

O.K. ↓

INSPECTION END

Diagnostic Procedure 19

SUB-CRANK ANGLE SENSOR (Not self-diagnostic item)

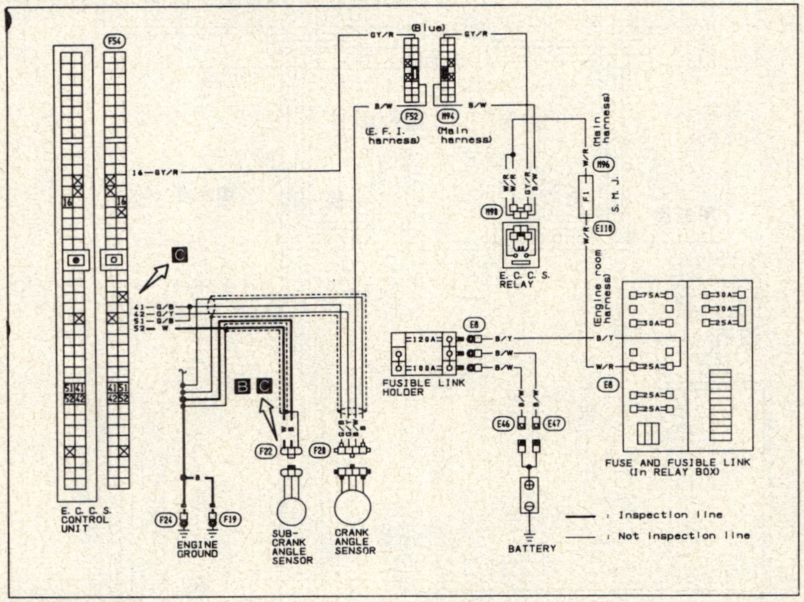

Harness layout

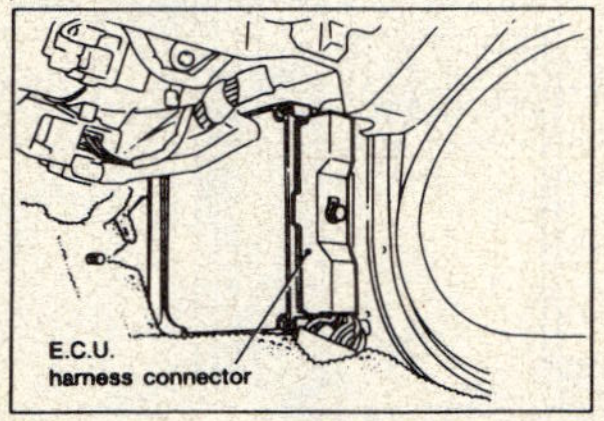

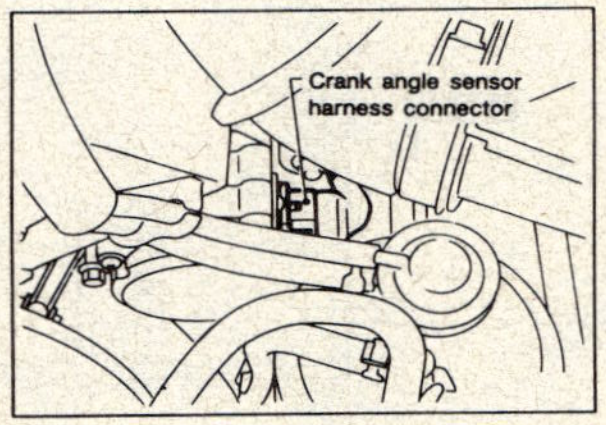

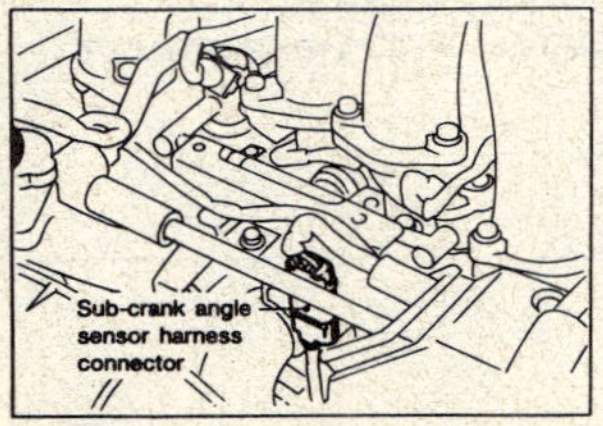

SUB-CRANK ANGLE SENSOR (Not self-diagnostic item)

A
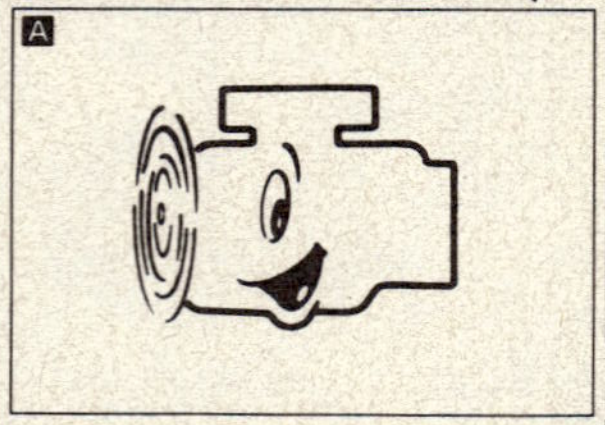

B
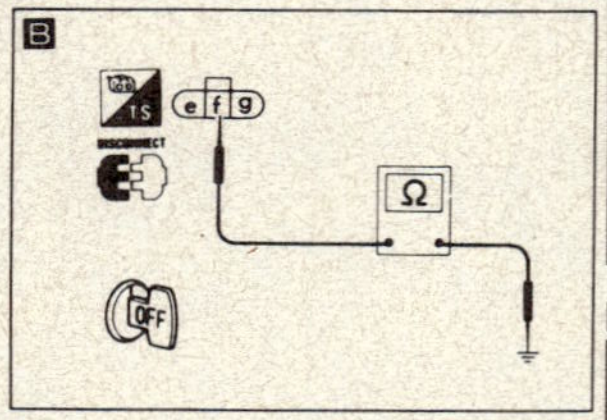

C
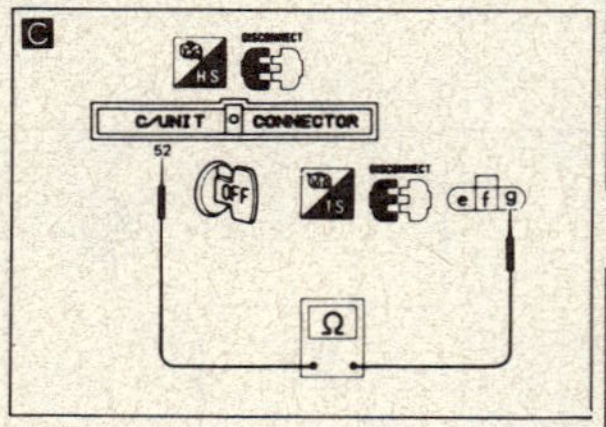

INSPECTION START

↓

A **CHECK OVERALL FUNCTION.**
1) Disconnect crank angle sensor harness connector.
2) Turn ignition switch "START".
Is engine running ?

Yes → INSPECTION END

No ↓

B **CHECK GROUND CIRCUIT.**
1) Turn ignition switch "OFF".
2) Disconnect sub-crank angle sensor harness connector.
3) Check harness continuity between terminal ⓕ and engine ground.
Continuity should exist.

N.G. → Repair harness or connectors.

O.K. ↓

C **CHECK INPUT SIGNAL CIRCUIT.**
1) Disconnect E.C.U. harness connector.
2) Check harness continuity between terminal ⓖ and E.C.U. terminal ㊷ (90° signal).
Continuity should exist.

N.G. → Repair harness or connectors.

O.K. ↓

CHECK COMPONENT
(Sub-crank angle sensor).
Refer to "Electrical Components Inspection".

N.G. → Replace sub-crank angle sensor.

O.K. ↓

Check E.C.U. pin terminals for damage or the connection of E.C.U. harness connector.

↓

Erase the self-diagnosis memory. Make sure Code No. 55 is displayed in Mode II.

↓

INSPECTION END

Diagnostic Procedure 20

IGNITION SIGNAL (Not self-diagnostic item)

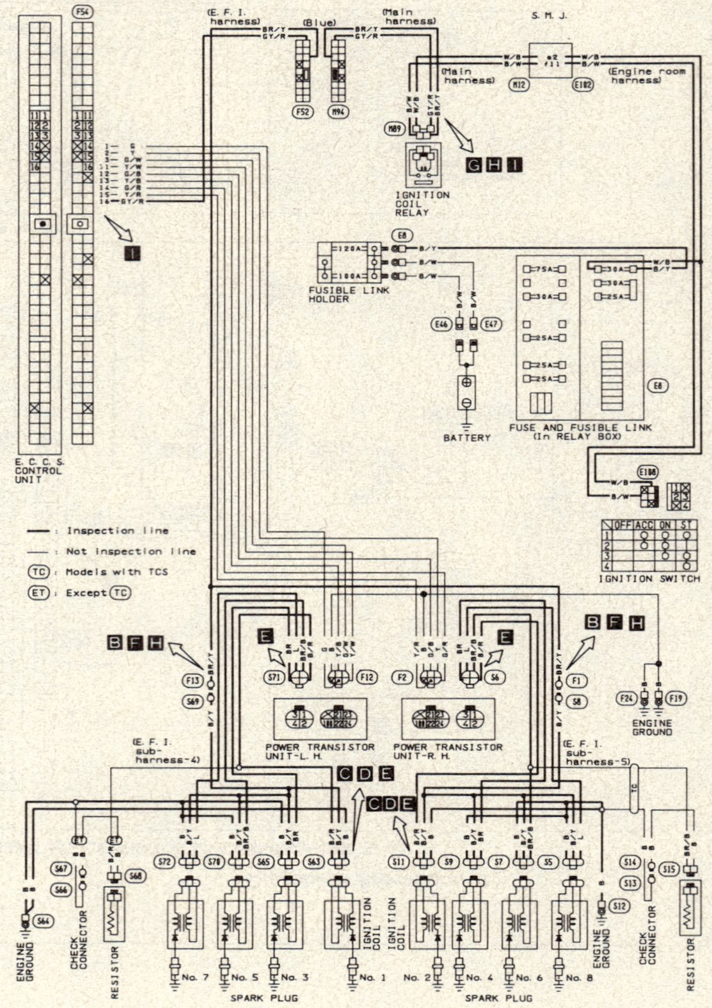

IGNITION SIGNAL (Not self-diagnostic item)

Harness layout

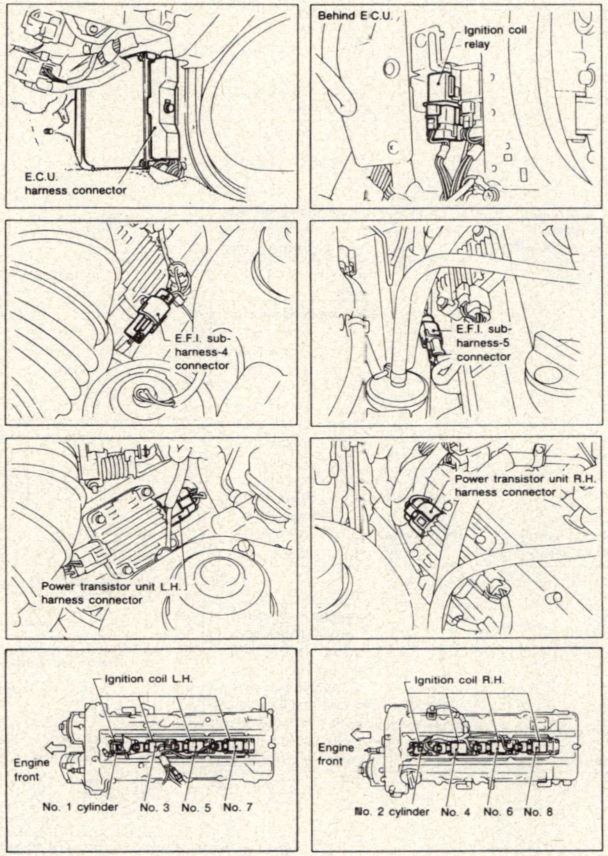

IGNITION SIGNAL (Not self-diagnostic item-I)

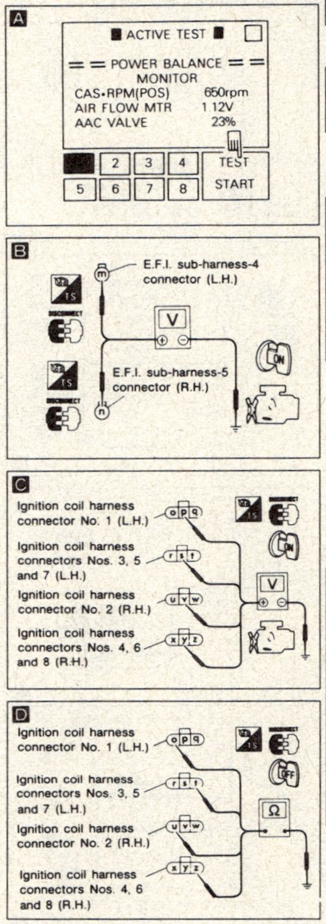

INSPECTION START

CHECK OVERALL FUNCTION.
1) Turn ignition switch "START".
2) Is any cylinder ignited?

- No → Go to "IGNITION SIGNAL (Not self-diagnostic item-II)".
- Yes → [A]

[A] **SEARCH FOR MALFUNCTIONING CIRCUIT.**
1) Perform "POWER BALANCE" in "ACTIVE TEST" mode with CONSULT.
2) Search for circuit which does not produce a momentary engine speed drop.

- Case 1: When either four cylinders Nos. 1, 3, 5 and 7 or Nos. 2, 4, 6 and 8 are ignited. → [B]
- Case 2: Except case 1 → [C]

[B] **CHECK POWER SUPPLY-I.**
1) Turn ignition switch "OFF".
2) Disconnect E.F.I. sub-harness- 4 or 5 connector.
3) Turn ignition switch "ON".
4) Check voltage between terminal (m) and ground or terminal (n) and ground with CONSULT or tester.
Voltage: Battery voltage
5) Turn ignition switch "OFF".
6) Reconnect E.F.I. sub-harness-4 or 5 connector.

- N.G. → Go to "IGNITION SIGNAL (Not self-diagnostic item-II)".

[C] **CHECK POWER SUPPLY-II.**
1) Turn ignition switch "OFF".
2) Disconnect ignition coil harness connectors.
3) Turn ignition switch "ON".
4) Check voltage between the following terminals and ground with CONSULT or tester.
- (p) and ground
- (s) and ground
- (v) and ground
- (y) and ground

Voltage: Battery voltage

- O.K. → [B] CHECK POWER SUPPLY-I.
- N.G. → Check the following.
  - Harness connectors (M69), (58)
  - Harness continuity between ignition coil and E.F.I. sub- harness connectors (M69), (58)

  If N.G., repair harness or connectors.
- O.K. → (A)

IGNITION SIGNAL (Not self-diagnostic item-I)

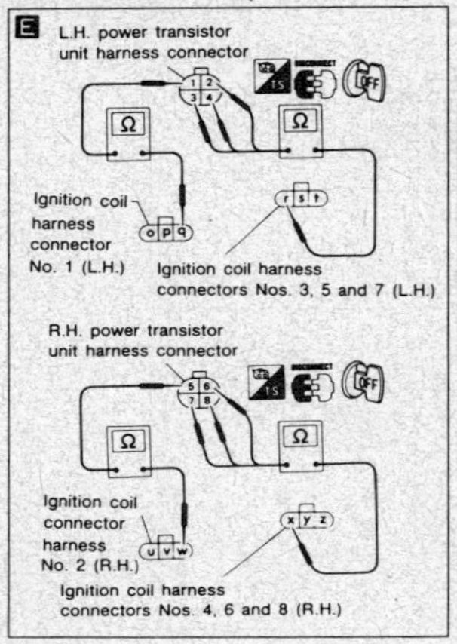

Ⓐ

D CHECK GROUND CIRCUIT.
1) Turn ignition switch "OFF".
2) Check harness continuity between the following terminals and ground.
- (o) and ground
- (t) and ground
- (u) and ground
- (z) and ground

Continuity should exist.

N.G. → Repair harness or connectors.

O.K. ↓

E CHECK OUTPUT SIGNAL CIRCUIT.
1) Disconnect power transistor unit harness connector.
2) Check harness continuity between the following terminals.
- (q) and ①
- (r) and ②
- (r) and ③
- (r) and ④
- (w) and ⑤
- (x) and ⑥
- (x) and ⑦
- (x) and ⑧

Continuity should exist.

N.G. → Repair harness or connectors.

O.K. ↓

CHECK COMPONENTS
(Power transistor unit and ignition coil).
Refer to "Electrical Components Inspection".

N.G. → Replace malfunctioning component(s).

O.K. ↓

INSPECTION END

IGNITION SIGNAL (Not self-diagnostic item-II)

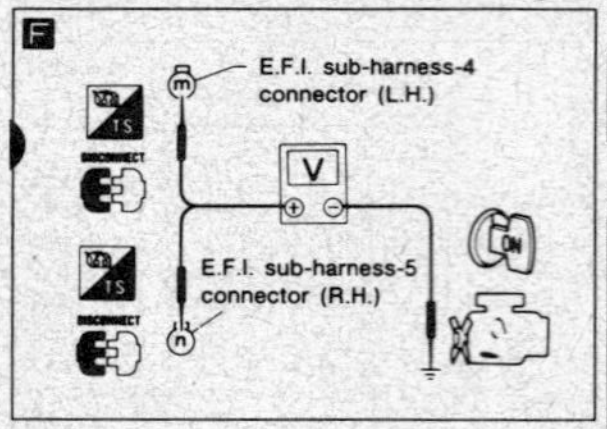

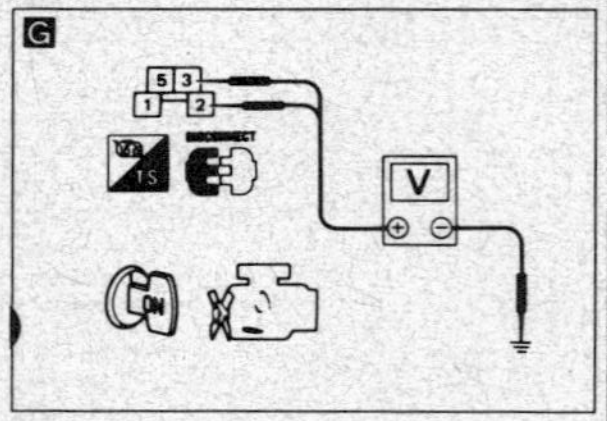

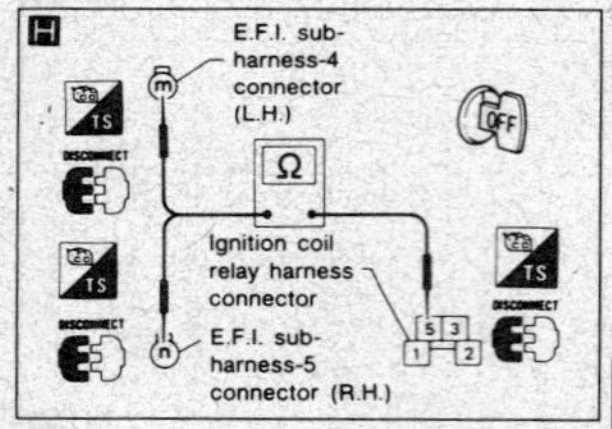

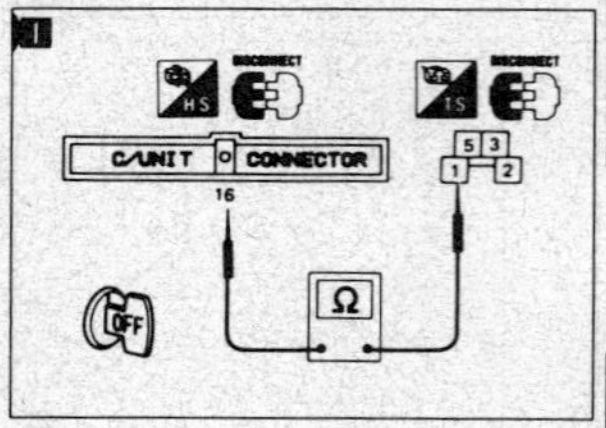

From "IGNITION SIGNAL (Not self-diagnostic item-I)".

↓

F CHECK OVERALL FUNCTION.
1) Turn ignition switch "OFF".
2) Disconnect E.F.I. sub-harness- 4 and 5 connectors.
3) Turn ignition switch "ON".
4) Check voltage between terminal (m) and ground, terminal (n) and ground with CONSULT or tester.

Voltage: Battery voltage

O.K. → Go to "CHECK POWER SUPPLY- II" in "IGNITION SIGNAL (Not self-diagnostic item-I)".

N.G. ↓

G CHECK POWER SUPPLY.
1) Turn ignition switch "OFF".
2) Disconnect ignition coil relay.
3) Turn ignition switch "ON".
4) Check voltage between terminals ②, ③ and ground with CONSULT or tester.

Voltage: Battery voltage

N.G. → Check the following.
- Harness connectors (M12), (E102)
- 30A fusible link
- Harness continuity between ignition coil relay and ignition switch
- Harness continuity between ignition coil relay and battery

If N.G., repair harness or connectors.

O.K. ↓

H CHECK GROUND CIRCUIT.
1) Turn ignition switch "OFF".
2) Check harness continuity between terminal (m) and terminal ⑤, or terminal (n) and terminal ⑤.

Continuity should exist.

N.G. → Check the following.
- Harness connectors (F52), (M94)
- Harness continuity between ignition coil relay and E.F.I. sub-harness-4 or 5

If N.G., repair harness or connectors.

O.K. ↓

I CHECK OUTPUT SIGNAL CIRCUIT.
1) Disconnect E.C.U. harness connector.
2) Check harness continuity between E.C.U. terminal ⑯ and terminal ①.

Continuity should exist.

N.G. → Check the following.
- Harness connectors (F52), (M94)
- Harness continuity between E.C.U. and ignition coil relay

If N.G., repair harness or connectors.

O.K. ↓

Ⓐ

IGNITION SIGNAL (Not self-diagnostic item-II)

Ⓐ

CHECK COMPONENTS (Ignition coil relay and ignition coil). Refer to "Electrical Components Inspection".

N.G.: Replace malfunctioning component(s).

O.K.

Check E.C.U. pin terminals for damage or the connection of E.C.U. harness connector.

INSPECTION END

Diagnostic Procedure 21

E.G.R. CONTROL (Not self-diagnostic item)

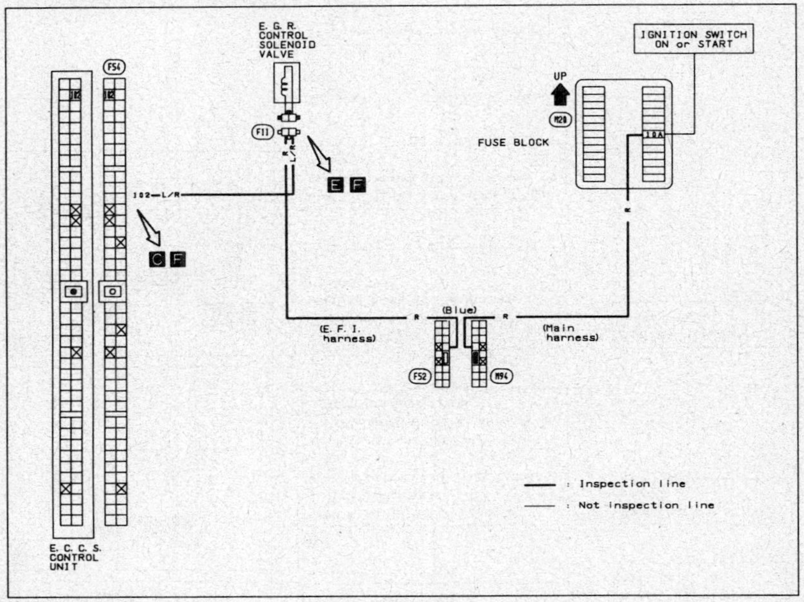

Harness layout

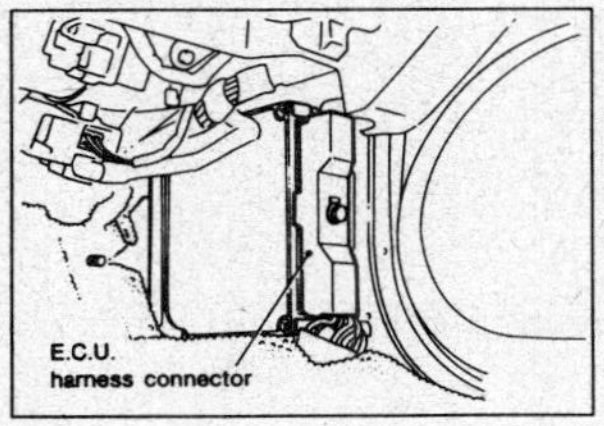

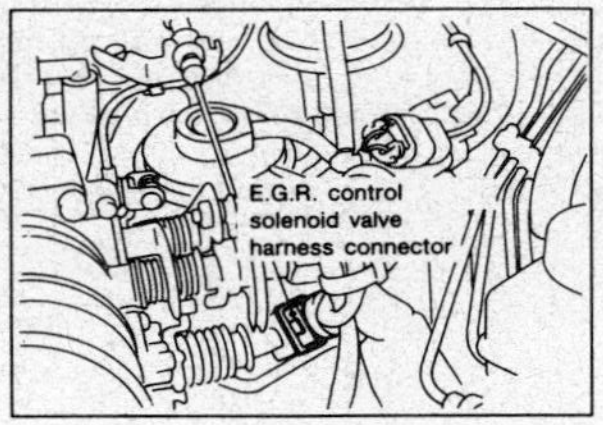

E.G.R. CONTROL (Not self-diagnostic item)

A
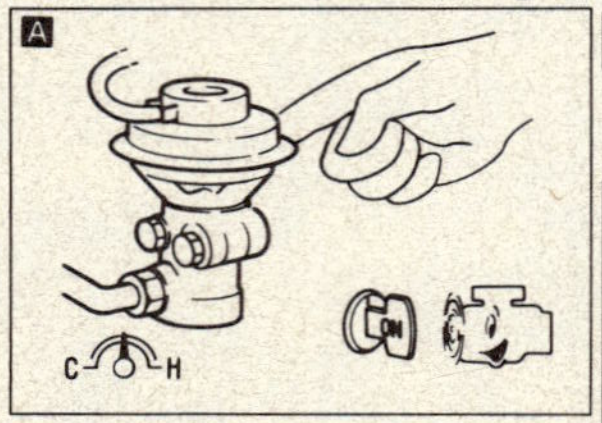

B
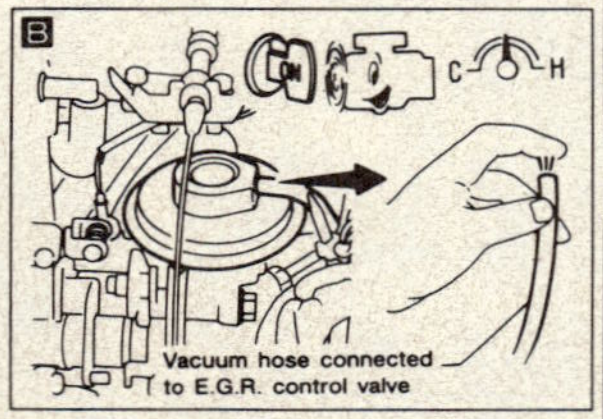

C
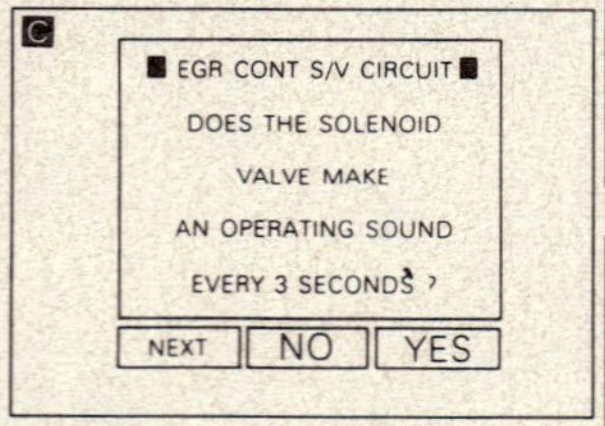

C
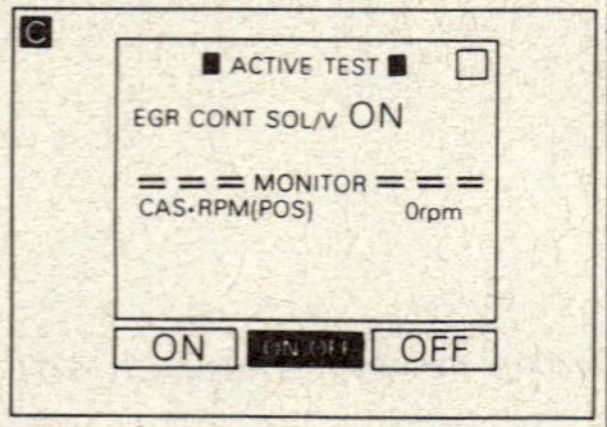

C
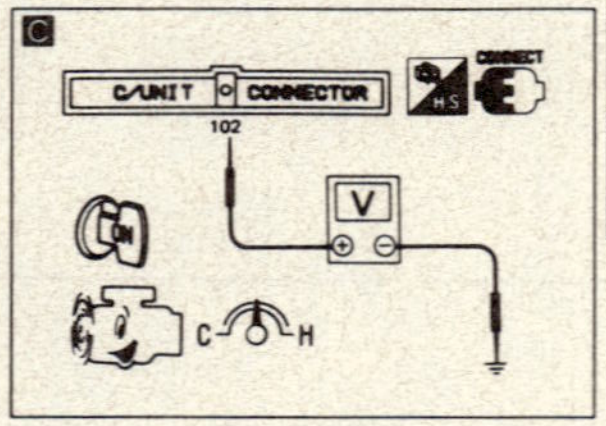

INSPECTION START

- Non-California models / California models

A CHECK OVERALL FUNCTION.
1) Start engine and warm it up sufficiently.
2) Perform self-diagnosis (Mode II) with E.C.U.
Make sure that code No. 11 and 12 are not displayed.
3) Make sure that E.G.R. control valve spring is lifted up and down when racing engine (2,000 rpm).
(Use your finger.)

- Is lifted up and down. → INSPECTION END
- Is not lifted up and down. ↓

B CHECK VACUUM SOURCE TO E.G.R. CONTROL VALVE.
1) Disconnect vacuum hose to E.G.R. control valve.
2) Make sure that vacuum exists under the following conditions with CONSULT or tester.
At idle:
Vacuum should not exist.
Engine speed is about 2,000 rpm:
Vacuum should exist.

- O.K. → CHECK COMPONENTS (E.G.R. control valve and B.P.T. valve). Refer to "Electrical Components Inspection".
  - N.G. → Replace malfunctioning component(s).
- N.G. ↓

C CHECK CONTROL FUNCTION.
1) Stop engine.
2) Turn ignition switch "ON".
3) Perform "EGR CONT S/V CIRCUIT" in "FUNCTION TEST" mode with CONSULT.

OR

1) Stop engine.
2) Turn ignition switch "ON".
3) Turn E.G.R. control solenoid valve "ON" and "OFF" in "ACTIVE TEST" mode with CONSULT and check operating sound.

OR

1) Check voltage between E.C.U. terminal (102) and ground under the following conditions with CONSULT or tester.
Voltage:
At idle
Approximately 0V
Engine speed is about 2,000 rpm.
Battery voltage

- O.K. → **D** (Non-California models) CHECK VACUUM HOSE.
1) Check vacuum hose for clogging, cracks and proper connection.
- O.K. → (California models) Go to "CHECK COMPONENT (E.G.R. control solenoid valve)".
- N.G. → Ⓐ

E.G.R. CONTROL (Not self-diagnostic item)

D
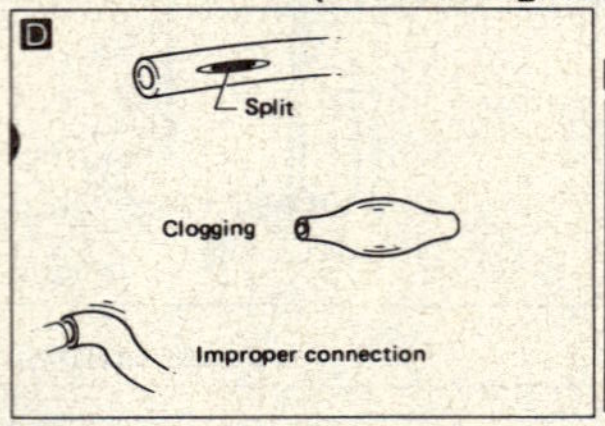

E
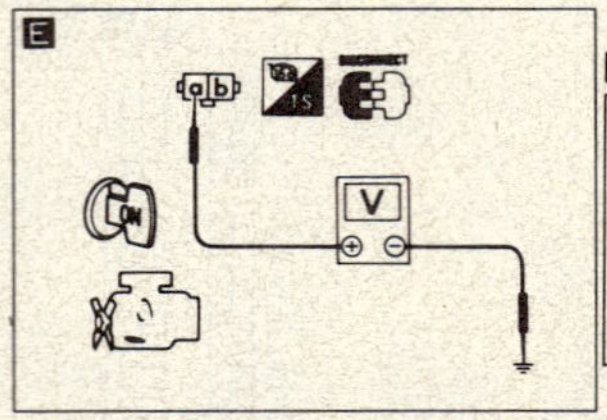

F
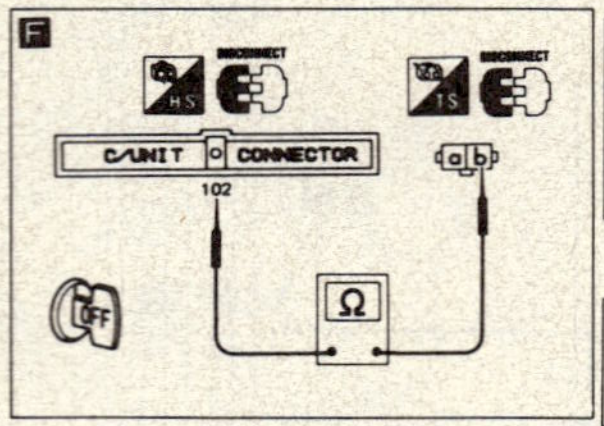

Ⓐ

E CHECK POWER SUPPLY.
1) Stop engine.
2) Disconnect E.G.R. control solenoid valve harness connector.
3) Turn ignition switch "ON".
4) Check voltage between terminal ⓐ and ground with CONSULT or tester.
Voltage: Battery voltage

- N.G. → Check the following.
  - Harness connectors (F52), (M4)
  - 10A fuse
  - Harness continuity between E.G.R. control solenoid valve and fuse
  - If N.G., repair harness or connectors.
- O.K. ↓

F CHECK OUTPUT SIGNAL CIRCUIT.
1) Turn ignition switch "OFF".
2) Disconnect E.C.U. harness connector.
3) Check harness continuity between E.C.U. terminal (102) and terminal ⓑ.
Continuity should exist.

- N.G. → Repair harness or connectors.
- O.K. ↓

CHECK COMPONENT (E.G.R. control solenoid valve).
Refer to "Electrical Components Inspection".

- N.G. → Replace E.G.R. control solenoid valve.
- O.K. ↓

Check E.C.U. pin terminals for damage or the connection of E.C.U. harness connector.

- California models → Check resistance of exhaust gas temperature sensor. → INSPECTION END
- Non-California models → INSPECTION END

Diagnostic Procedure 22

EXHAUST GAS SENSOR HEATER L.H. and R.H. (Not self-diagnostic item)

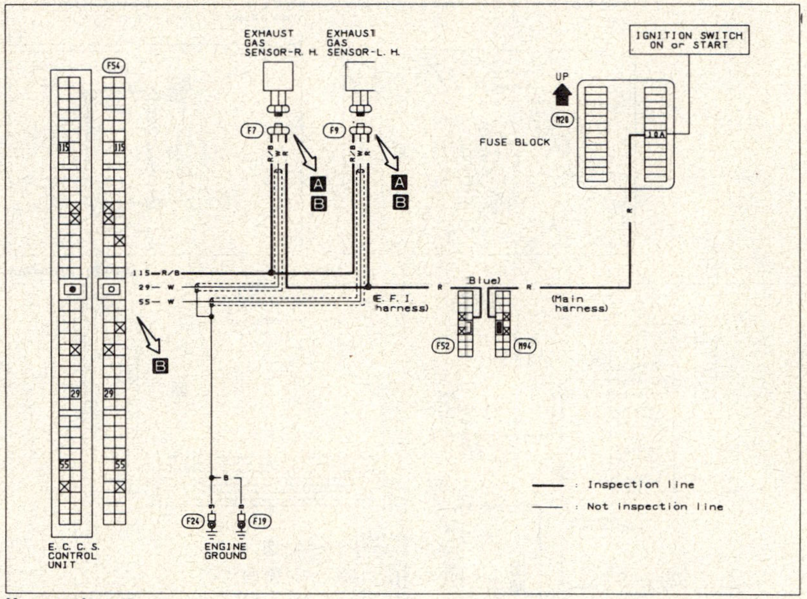

Harness layout

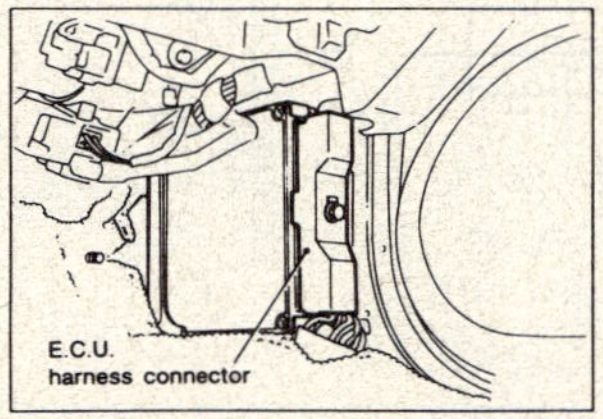

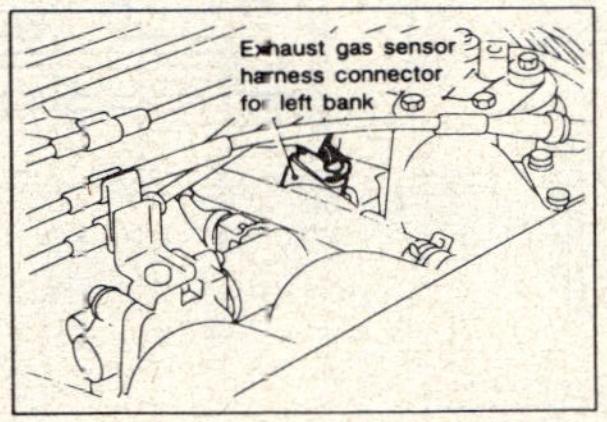

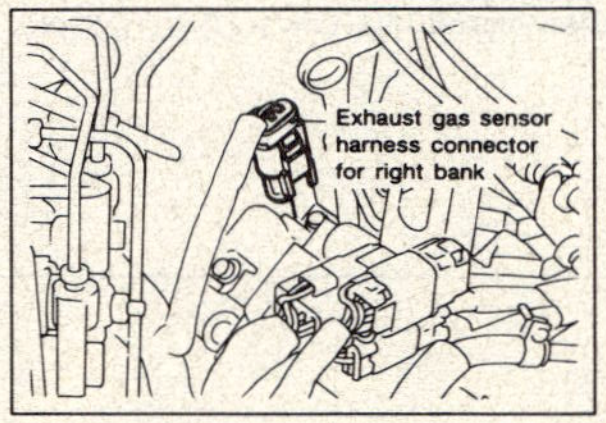

EXHAUST GAS SENSOR HEATER L.H. and R.H. (Not self-diagnostic item)

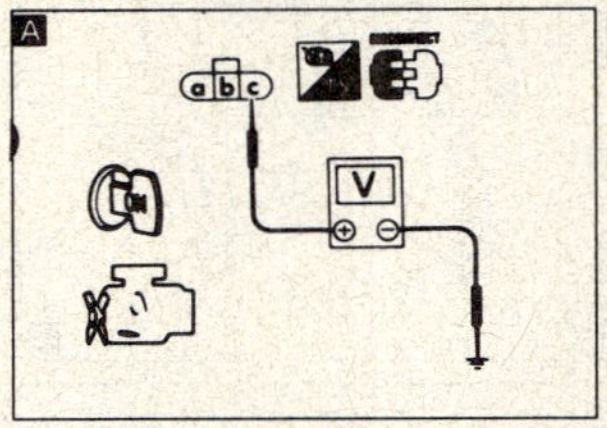

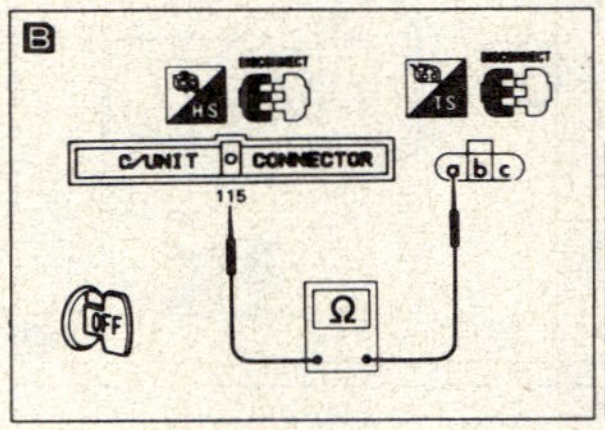

INSPECTION START

↓

A CHECK POWER SUPPLY.
1) Disconnect exhaust gas sensor harness connector.
2) Turn ignition switch "ON".
3) Check voltage between terminal ⓔ and ground.
Voltage: Battery voltage

N.G. → Check the following.
- Harness connectors (F52), (M94)
- 10A fuse
- Harness continuity between exhaust gas sensor and fuse

If N.G., repair harness or connectors.

O.K. ↓

B CHECK GROUND CIRCUIT.
1) Turn ignition switch "OFF".
2) Disconnect E.C.U. harness connector.
3) Check harness continuity between terminal ⓐ and E.C.U. terminal (115)
Continuity should exist.

N.G. → Repair harness or connectors.

O.K. ↓

CHECK COMPONENT
(Exhaust gas sensor heater).
Refer to "Electrical Components Inspection".

N.G. → Replace exhaust gas sensor.

O.K. ↓

Check E.C.U. pin terminals for damage or the connection of E.C.U. harness connector.

↓

INSPECTION END

Diagnostic Procedure 23

INJECTOR (Not self-diagnostic item-I)

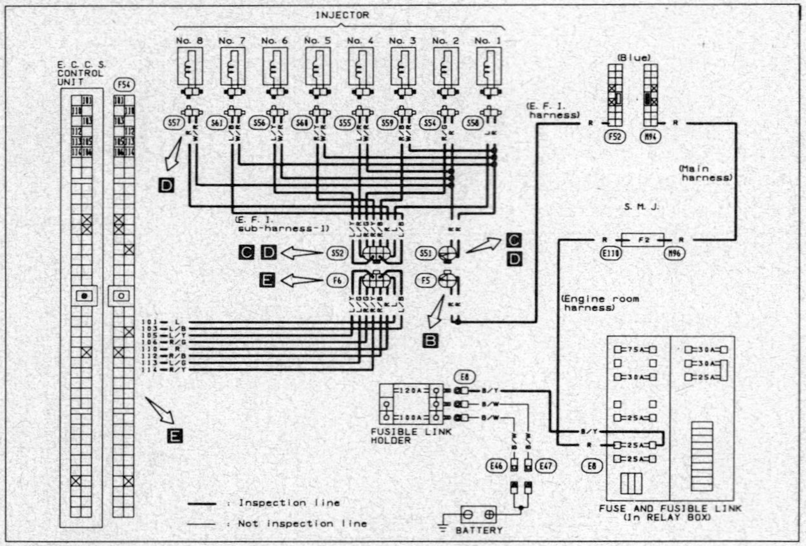

Harness layout

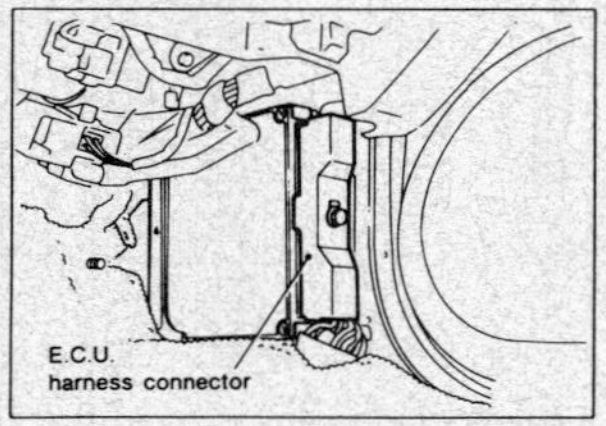

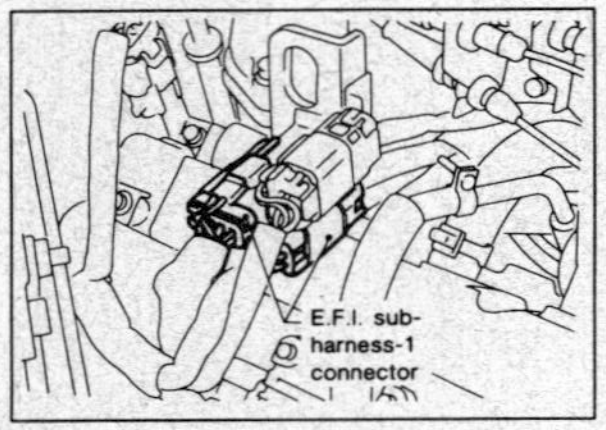

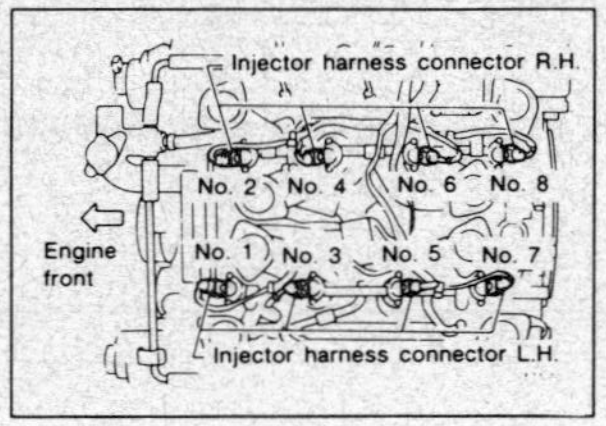

INJECTOR (Not self-diagnostic item-I)

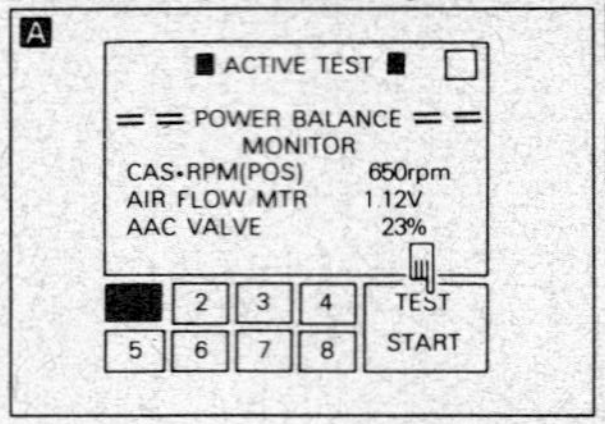

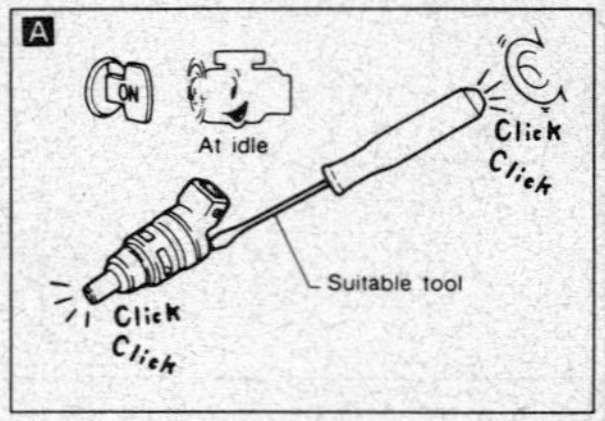

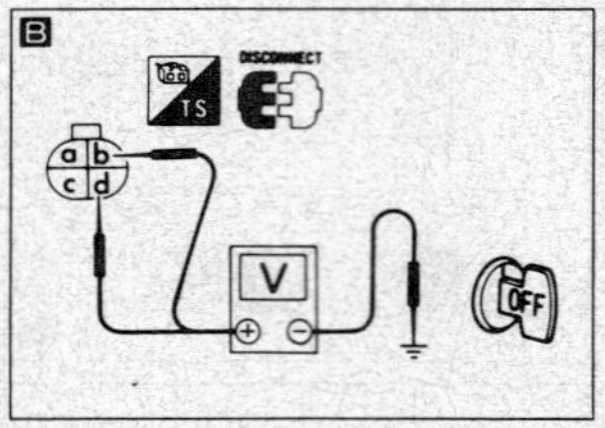

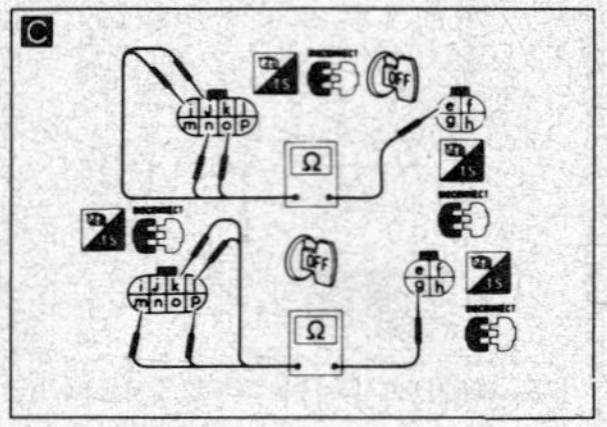

Non-California models

INSPECTION START

California models

INSPECTION START

Go to "CHECK POWER SUPPLY-I".

A CHECK OVERALL FUNCTION.
1) Start engine.
2) Perform "POWER BALANCE" in "ACTIVE TEST" mode with CONSULT.
3) Search for circuit which does not produce a momentary engine speed drop.
—OR—
2) Search for circuit which does not produce injector operating sound at idle.

All cylinders are O.K. → INSPECTION END

Case 1: When either cylinders Nos. 1, 3, 5 and 7 or Nos. 2, 4, 6 and 8 are all N.G., or no ignition occurs in any cylinder.

Case: 2
Except case 1 → Ⓐ

B CHECK POWER SUPPLY-I.
1) Turn ignition switch "OFF".
2) Disconnect E.F.I. sub-harness-1 connector (F5).
3) Check voltage between terminals ⓑ, ⓓ and ground with CONSULT or tester.
Voltage: Battery voltage

O.K. → Ⓐ

N.G.

Check the following.
- Harness connectors (F52), (M94)
- Harness connectors (M96), (E118)
- 25A fusible link
- Fusible link holder
- Harness continuity between E.F.I. sub-harness-1 connector and battery.

If N.G., repair harness or connectors.

INJECTOR (Not self-diagnostic item-I)

D

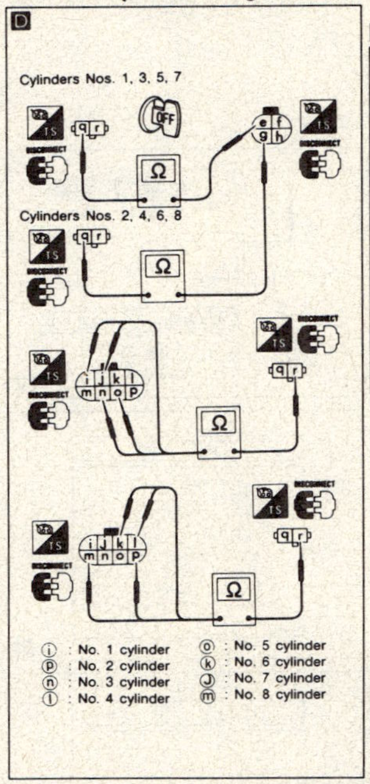

C

CHECK POWER SUPPLY-II.
1) Turn ignition switch "OFF".
2) Disconnect E.F.I. sub-harness- 1 connectors.
3) Check circuit continuity between the following terminals.
- (e) and (i)
- (e) and (j)
- (e) and (n)
- (e) and (o)
- (g) and (k)
- (g) and (l)
- (g) and (m)
- (g) and (p)

Continuity should exist.

N.G. → D

D

Check the following.
1) Disconnect injector harness connectors.
2) Check harness continuity between the following terminals.
- (q) and (e)
- (q) and (g)
- (i) and (r)
- (j) and (r)
- (k) and (r)
- (l) and (r)
- (m) and (r)
- (n) and (r)
- (o) and (r)
- (p) and (r)

Continuity should exist.

If N.G., repair harness or connectors.

O.K. ↓

CHECK COMPONENT (Injector).
Refer to "Electrical Components Inspection".

If N.G. replace injector.

O.K. (from C) ↓

E

CHECK OUTPUT SIGNAL CIRCUIT.
1) Disconnect E.C.U. harness connector.
2) Check continuity between the following terminals.
- (s) and (113)
- (t) and (105)
- (u) and (103)
- (v) and (101)
- (w) and (106)
- (x) and (114)
- (y) and (112)
- (z) and (110)

Continuity should exist.

N.G. → Repair harness or connectors.

O.K. ↓

Check E.C.U. pin terminals for damage or the connection of E.C.U. harness connector.

↓

INSPECTION END*

E

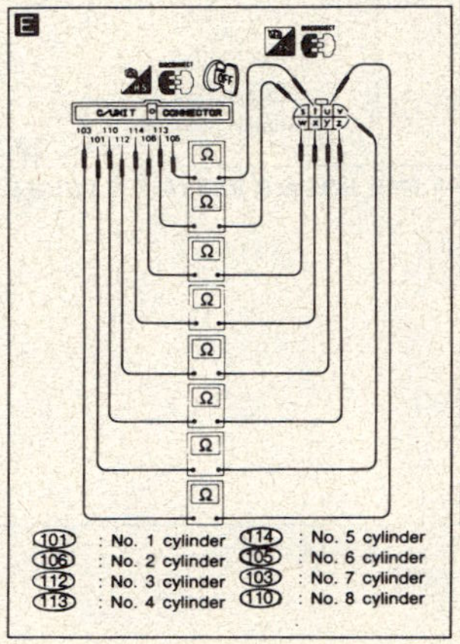

*: For California models, go to "CHECK OVERALL FUNCTION" if idle is still unstable. Then, perform "CHECK POWER SUPPLY-III" in "Diagnostic Procedure 1" after repair is completed.

Diagnostic Procedure 24

THROTTLE VALVE SWITCH (Not self-diagnostic item)

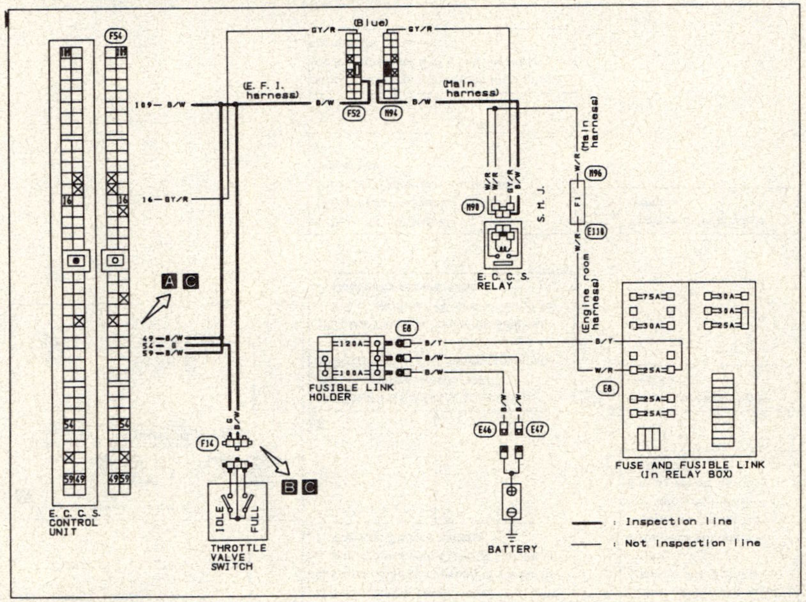

Harness layout

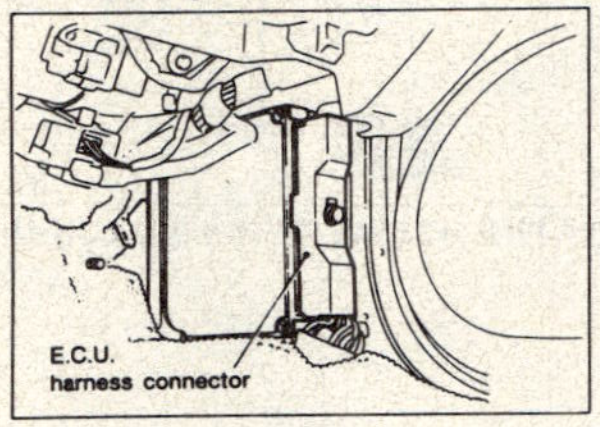

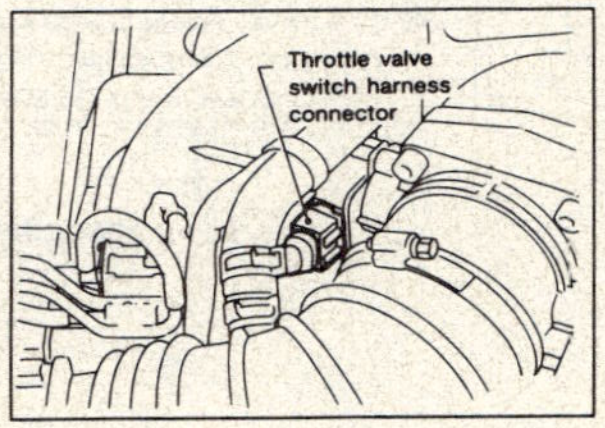

THROTTLE VALVE SWITCH (Not self-diagnostic item)

A

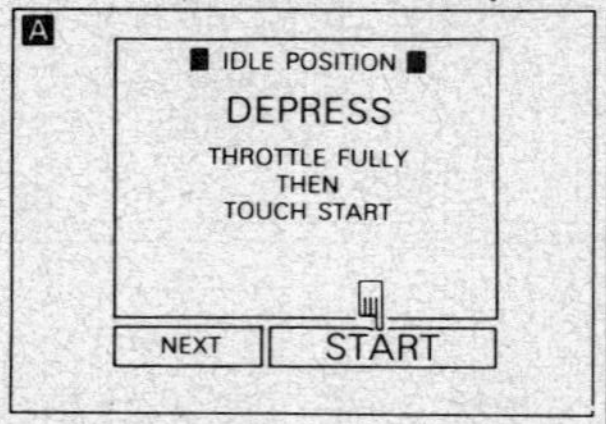

A

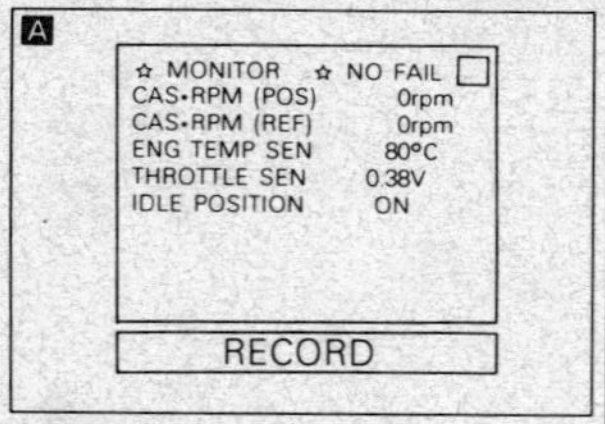

A

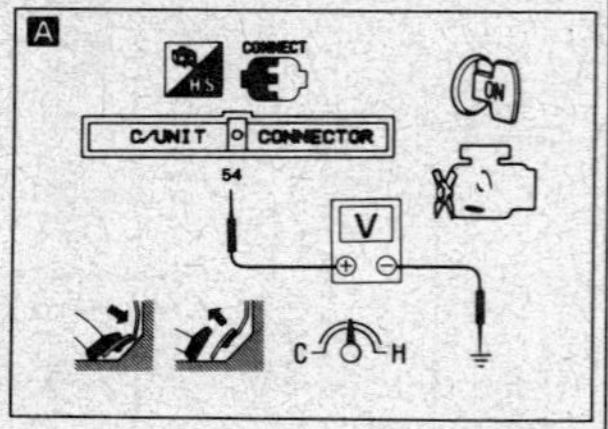

INSPECTION START

↓

A CHECK OVERALL FUNCTION.
1) Start engine and warm it up sufficiently.
2) Turn ignition switch "OFF".
3) Disconnect throttle sensor harness connector.
4) Turn ignition switch "ON".
5) Perform "IDLE POSITION" in "FUNCTION TEST" mode with CONSULT.

— OR —

3) Disconnect throttle sensor harness connector.
4) Turn ignition switch "ON".
5) Check "DATA MONITOR" mode with CONSULT.

Idle position:
Accelerator pedal is released ... ON
Accelerator pedal is depressed ... OFF

— OR —

3) Turn ignition switch "ON".
4) Check voltage between E.C.U. terminal 54 and ground with CONSULT or tester under the following conditions.

Voltage:
Accelerator pedal is released
Battery voltage
Accelerator pedal is depressed
Approximately 0V

O.K. → INSPECTION END

N.G. → Ⓐ

THROTTLE VALVE SWITCH (Not self-diagnostic item)

B

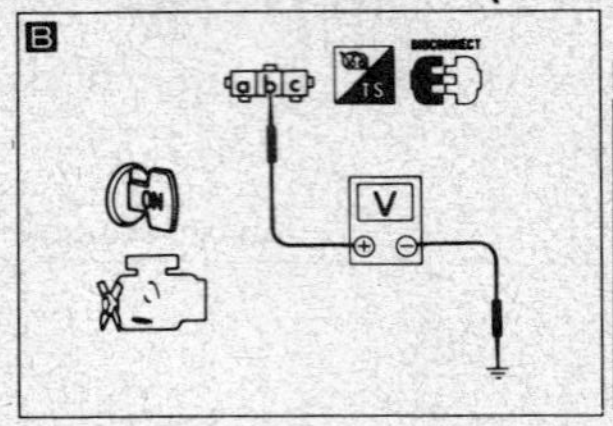

C

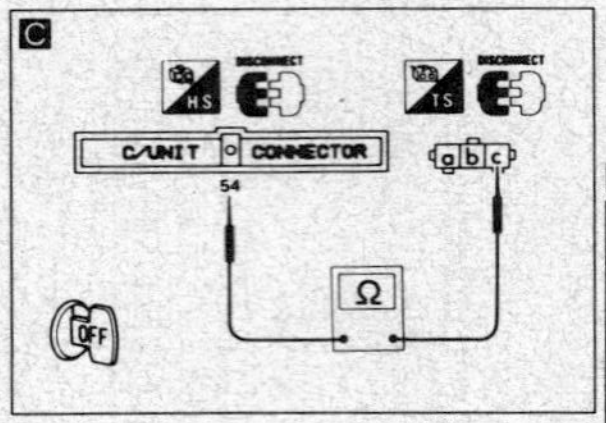

Ⓐ

↓

B CHECK POWER SUPPLY.
1) Turn ignition switch "OFF".
2) Disconnect throttle valve switch harness connector.
3) Turn ignition switch "ON".
4) Check voltage between terminal ⓑ and ground with CONSULT or tester.
Voltage: Battery voltage

N.G. → Check the following.
- Harness connectors (F52), (M94)
- Harness continuity between throttle valve switch and E.C.C.S. relay
- Harness continuity between throttle valve switch and E.C.U.

If N.G., repair harness or connectors.

O.K.

↓

C CHECK INPUT SIGNAL CIRCUIT.
1) Turn ignition switch "OFF".
2) Disconnect E.C.U. harness connector.
3) Check harness continuity between E.C.U. terminal 54 and terminal ⓒ.
Continuity should exist.

N.G. → Repair harness or connectors.

O.K.

↓

CHECK COMPONENT
(Throttle valve switch).
Refer to "Electrical Components Inspection".

N.G. → Replace throttle valve switch.

O.K.

↓

Check E.C.U. pin terminals for damage or the connection of E.C.U. harness connector.

↓

INSPECTION END

Diagnostic Procedure 25

START SIGNAL (Not self-diagnostic item)

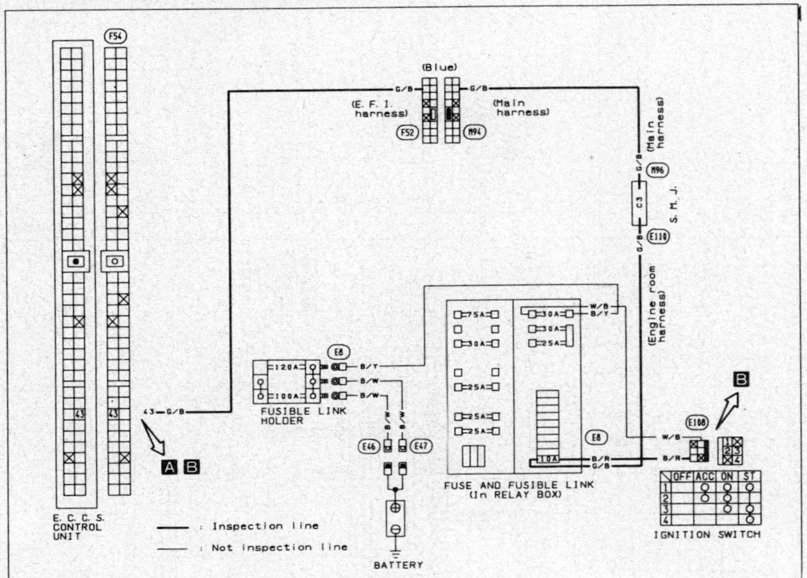

Harness layout

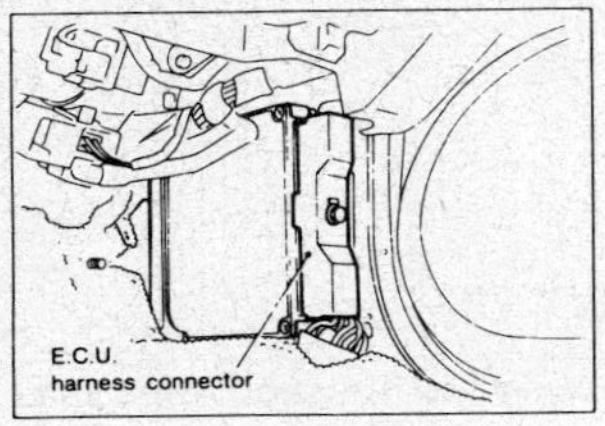

START SIGNAL (Not self-diagnostic item)

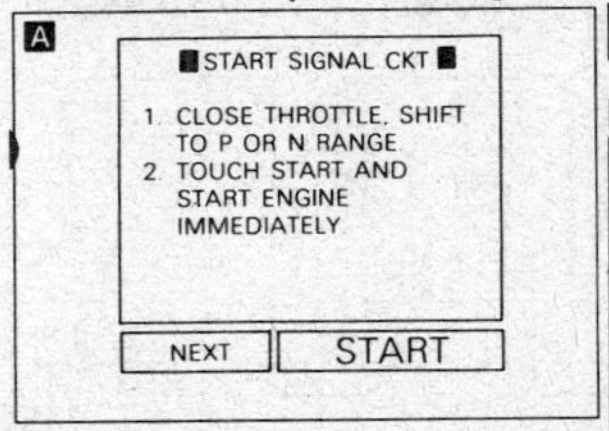

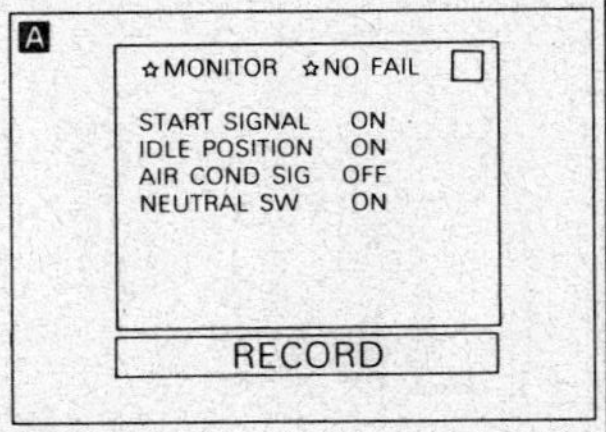

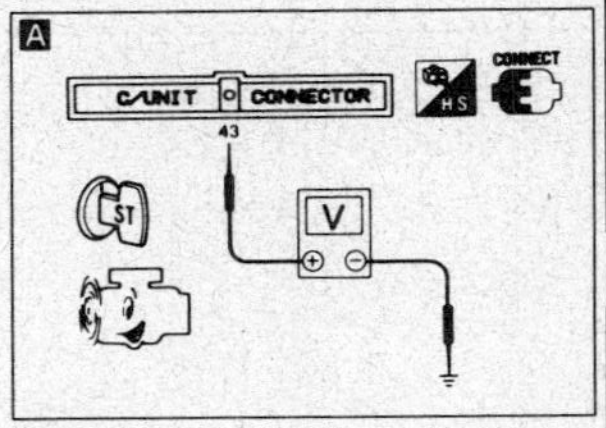

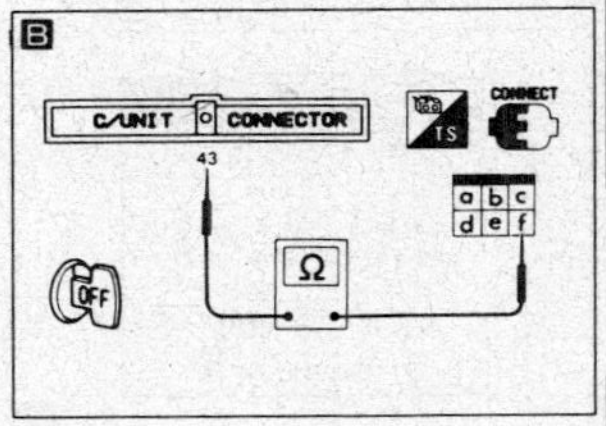

INSPECTION START

↓

A CHECK OVERALL FUNCTION.

1) Turn ignition switch "ON".
2) Perform "START SIGNAL CIRCUIT" in "FUNCTION TEST" mode with CONSULT.

— OR —

1) Turn ignition switch "ON".
2) Check start signal in "DATA MONITOR" mode with CONSULT.

IGN "ON"	OFF
IGN "START"	ON

— OR —

1) Turn ignition switch to "START".
2) Check voltage between E.C.U. terminal ㊸ and ground with CONSULT or tester.

Voltage:
Ignition switch "START"
Battery voltage
Except above
Approximately 0V

O.K. → INSPECTION END

N.G. ↓

B CHECK INPUT SIGNAL CIRCUIT.

1) Turn ignition switch "OFF".
2) Disconnect E.C.U. harness connector and ignition switch harness connector.
3) Check harness continuity between E.C.U. terminal ㊸ and terminal ①. **Continuity should exist.**

N.G. → Check the following.
- Harness connectors (F52), (M94)
- Harness connectors (M96), (E110)
- 10A fuse
- Harness continuity between E.C.U. and ignition switch

If N.G., repair harness or connectors.

O.K. ↓

Check E.C.U. pin terminals for damage or the connection of E.C.U. harness connector.

↓

INSPECTION END

Diagnostic Procedure 26

FUEL PUMP CONTROL (Not self-diagnostic item)

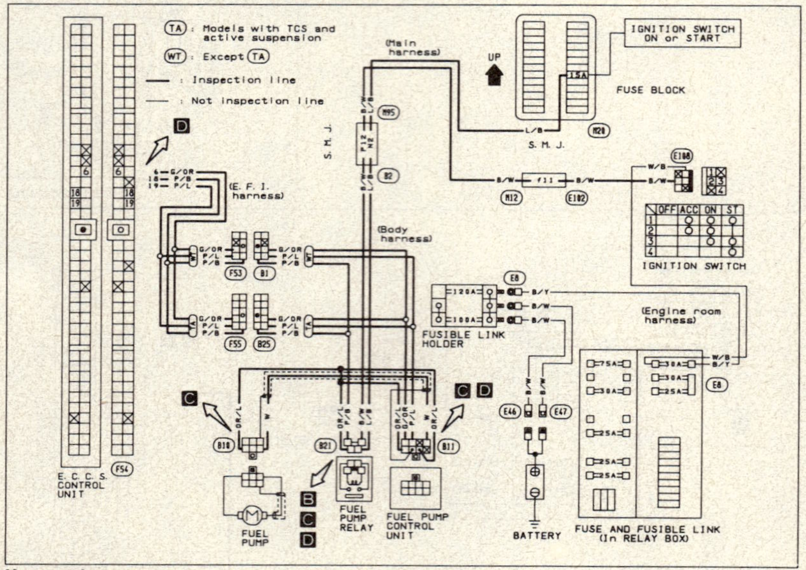

Harness layout

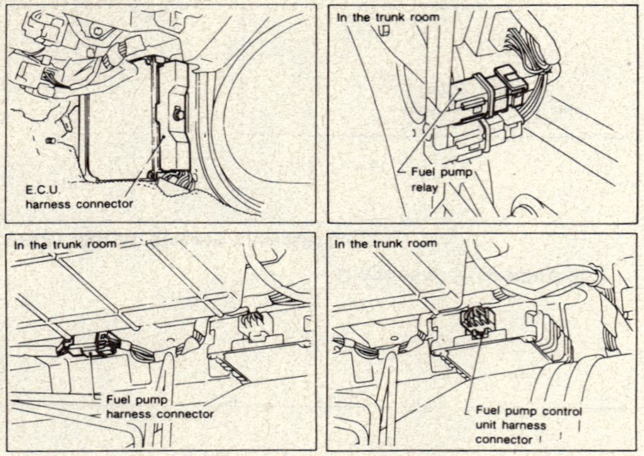

FUEL PUMP CONTROL (Not self-diagnostic item)

A

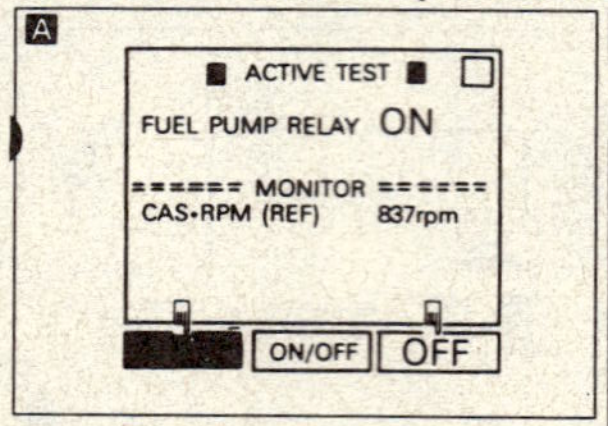

A

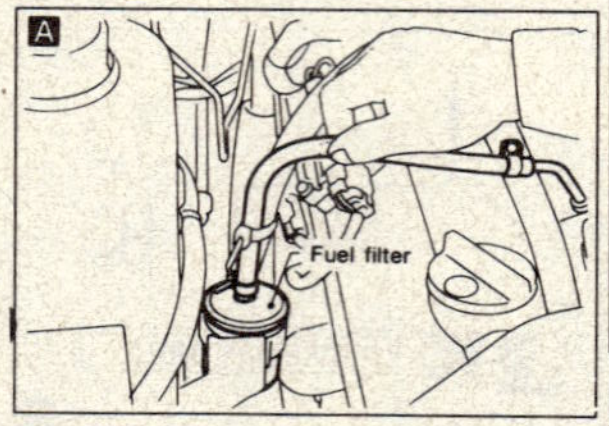

B

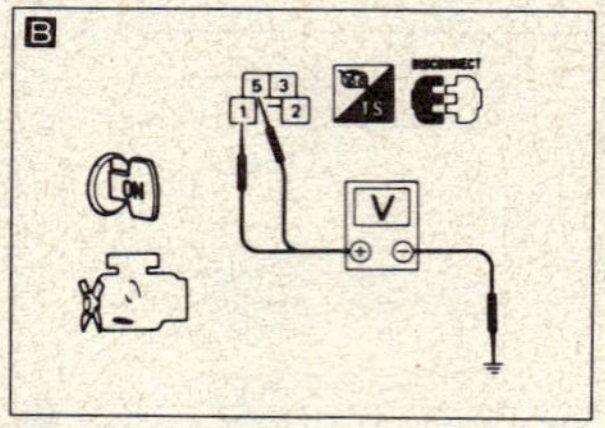

C

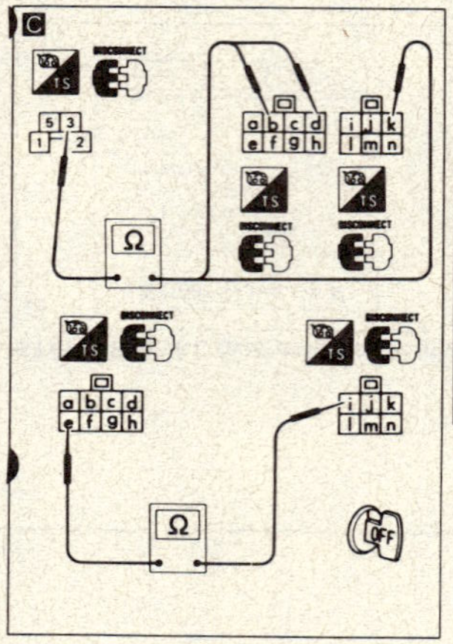

INSPECTION START

A CHECK OVERALL FUNCTION.
1) Turn ignition switch "ON".
2) Perform "FUEL PUMP RELAY" in "ACTIVE TEST" mode with CONSULT.
3) Pinch fuel feed hose with fingers.
Fuel pressure pulsation should be felt on the fuel feed hose.

—OR—

2) Pinch fuel feed hose with fingers.
Fuel pressure pulsation should be felt on the fuel feed hose for 5 seconds after ignition switch is turned "ON".

O.K. → *INSPECTION END

* If fuel pump noise is greater than normal during idling after completing warm-up, then go to "CHECK GROUND CIRCUIT".

N.G. ↓

B CHECK POWER SUPPLY.
1) Turn ignition switch "OFF".
2) Disconnect fuel pump relay.
3) Turn ignition switch "ON".
4) Check voltage between terminals ①, ⑤ and ground with CONSULT or tester.
Voltage: Battery voltage

N.G. → Check the following.
- Harness connectors (B2), (M95)
- Harness connectors (M12), (E102)
- 15A fuse
- Harness continuity between fuel pump relay and fuse
- Harness continuity between fuel pump relay and ignition switch

If N.G., repair harness or connectors.

O.K. ↓

C CHECK GROUND CIRCUIT.
1) Turn ignition switch "OFF".
2) Disconnect fuel pump harness connector and fuel pump control unit harness connector.
3) Check harness continuity between terminal ③ and ⓑ, ⓓ, ⓚ, terminal ⓔ and ⓘ.
Continuity should exist.

N.G. → Repair harness or connectors.

O.K. ↓ Ⓐ

FUEL PUMP CONTROL (Not self-diagnostic item)

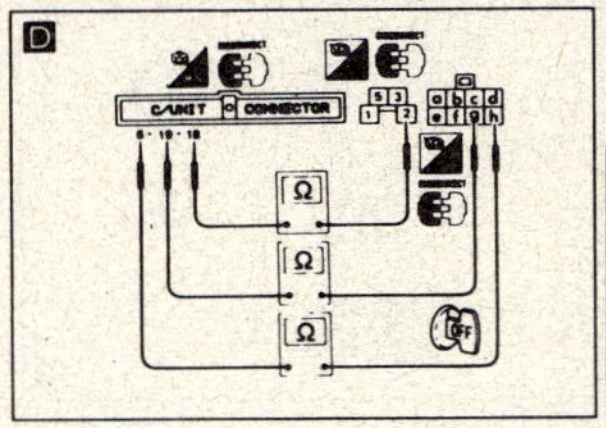

D

Ⓐ

CHECK OUTPUT SIGNAL CIRCUIT.
1) Disconnect E.C.U. harness connector.
2) Check harness continuity between E.C.U. terminal ⑱ and terminal ②, E.C.U. terminal ⑥ and terminal ⓗ, E.C.U. terminal ⑲ and terminal ⓖ.
Continuity should exist.

N.G. → Check the following.
- Harness connectors (F55), (B25)
- Harness connectors (F53), (B1)
- Harness continuity between E.C.U. and fuel pump relay
- Harness continuity between E.C.U. and fuel pump control unit

If N.G., repair harness or connectors.

O.K. ↓

CHECK COMPONENT
(Fuel pump relay).
Refer to "Electrical Components Inspection".

N.G. → Replace fuel pump relay.

O.K. ↓

CHECK COMPONENT
(Fuel pump control unit).
Refer to "Electrical Components Inspection".

N.G. → Replace fuel pump control unit.

O.K. ↓

Check E.C.U. pin terminals for damage or the connection of E.C.U. harness connector.

↓

INSPECTION END

Diagnostic Procedure 27

A.A.C. VALVE (Not self-diagnostic item)

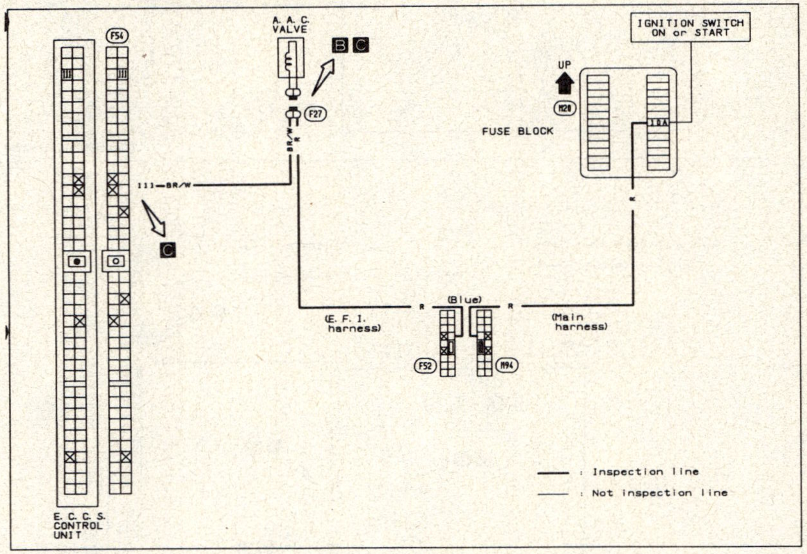

Harness layout

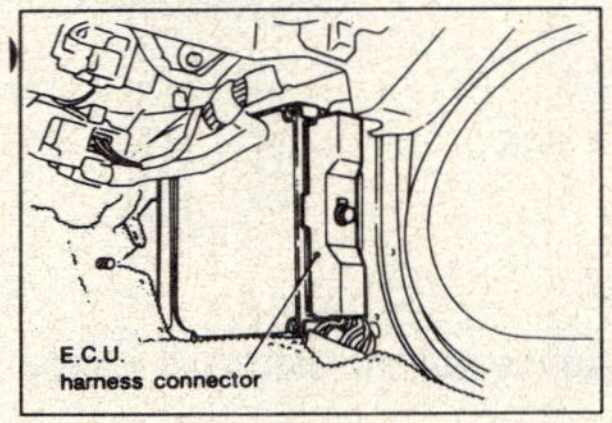

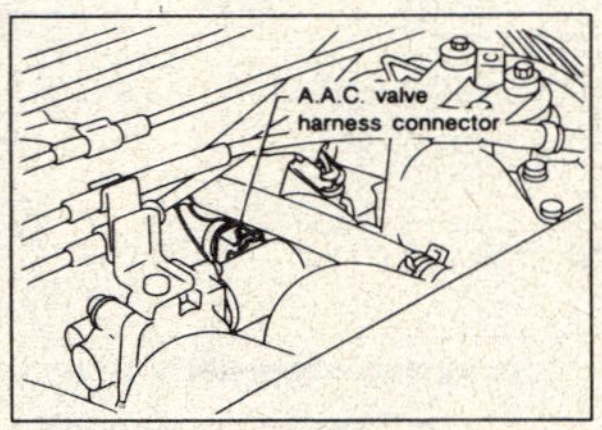

A.A.C. VALVE (Not self-diagnostic item)

A
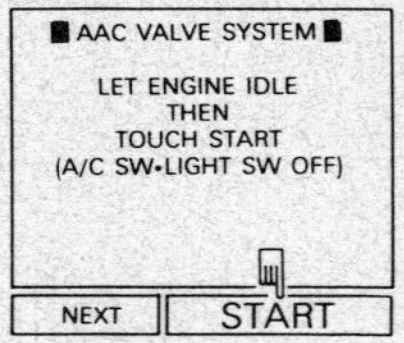

A
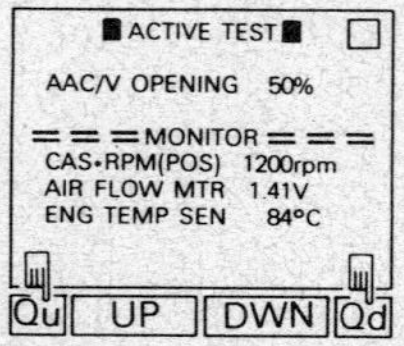

A
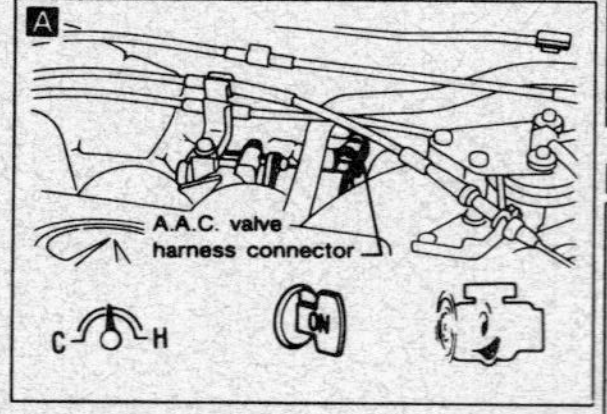

B
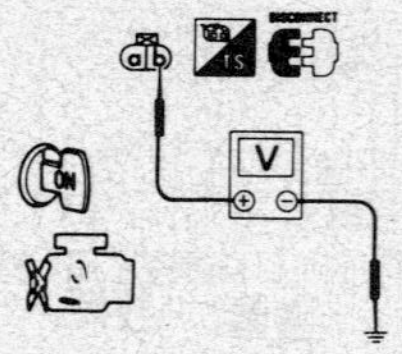

C
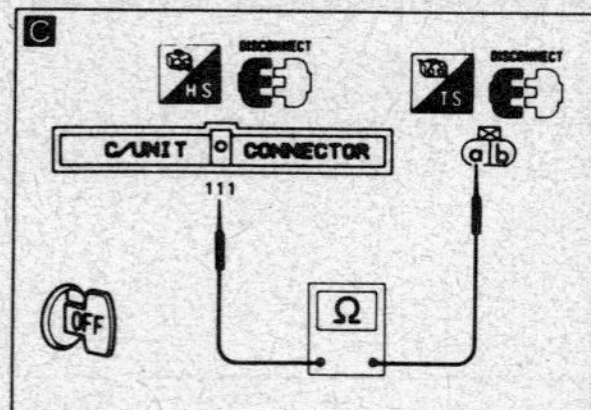

INSPECTION START

↓

A CHECK OVERALL FUNCTION.
1) Start engine and warm it up sufficiently.
2) Check idle speed.
650 ± 50 rpm
(A/T in "N" position)
3) Perform "AAC VALVE SYSTEM" in "FUNCTION TEST" mode with CONSULT.
— OR —
3) Select "AAC/V OPENING" in "ACTIVE TEST" mode with CONSULT.
4) When touching "Qu" and "Qd", check that engine speed changes according to the A.A.C. valve opening percentage.
— OR —
3) Disconnect A.A.C. valve harness connector.
4) Make sure that idle speed drops.

O.K. → INSPECTION END

N.G. ↓

B CHECK POWER SUPPLY.
1) Stop engine.
2) Disconnect A.A.C. valve harness connector.
3) Turn ignition switch "ON".
4) Check voltage between terminal ⓑ and ground with CONSULT or tester.
Voltage: Battery voltage

N.G. → Check the following.
- Harness connectors (F52), (M94)
- 10A fuse
- Harness continuity between A.A.C. valve and fuse

If N.G., repair harness or connectors.

O.K. ↓

C CHECK OUTPUT SIGNAL CIRCUIT.
1) Turn ignition switch "OFF".
2) Disconnect E.C.U. harness connector.
3) Check harness continuity between E.C.U. terminal (111) and terminal ⓐ.
Continuity should exist.

N.G. → Repair harness or connectors.

O.K. ↓

CHECK COMPONENT
(A.A.C. valve).
Refer to "Electrical Components Inspection".

N.G. → Replace A.A.C. valve.

O.K. ↓

Check E.C.U. pin terminals for damage or the connection of E.C.U. harness connector.

↓

INSPECTION END

Diagnostic Procedure 28

POWER STEERING OIL PRESSURE SWITCH (Not self-diagnostic item)

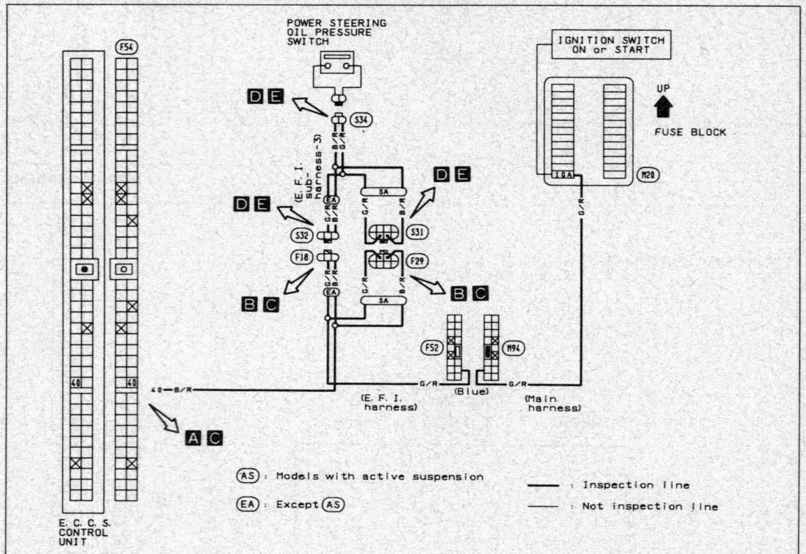

Harness layout

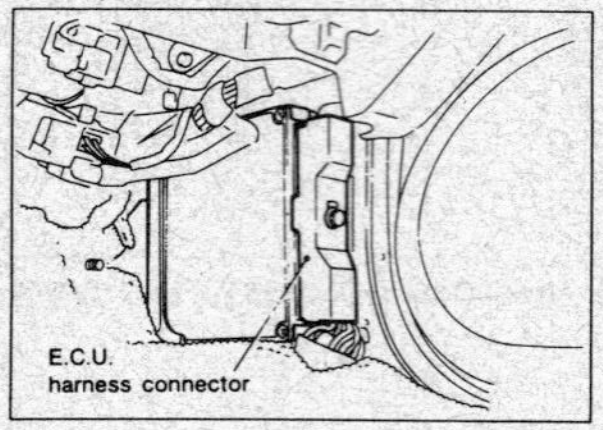

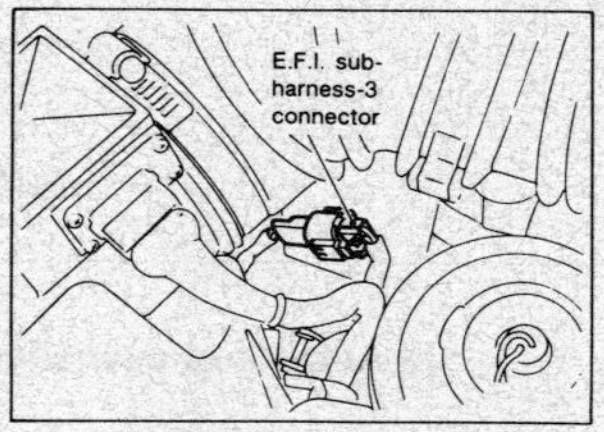

POWER STEERING OIL PRESSURE SWITCH (Not self-diagnostic item)

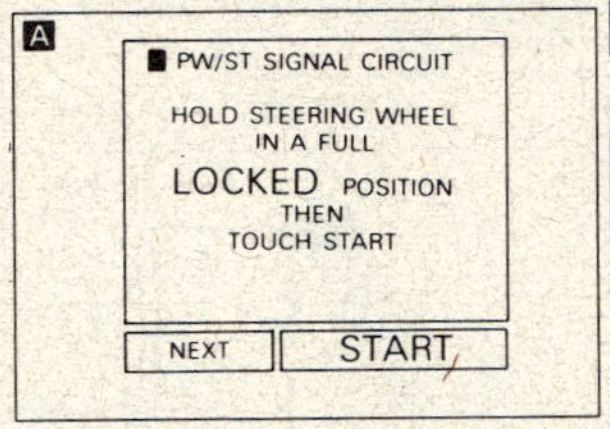

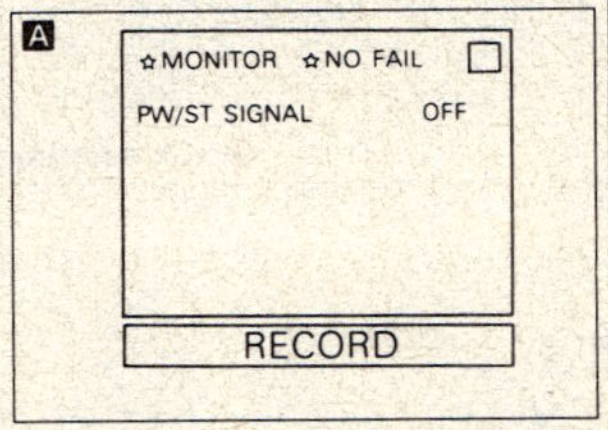

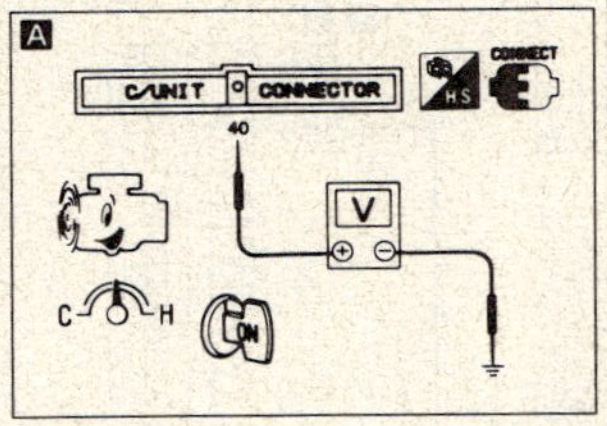

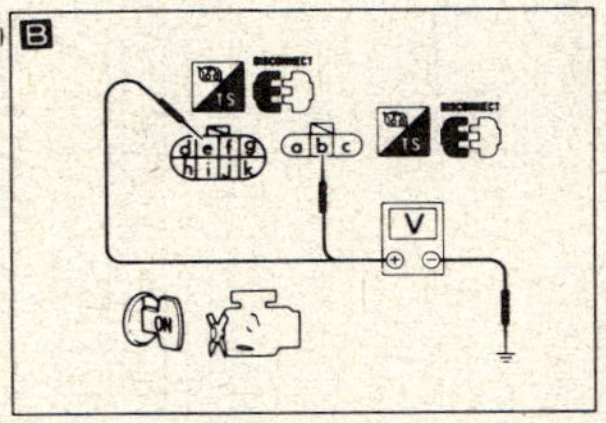

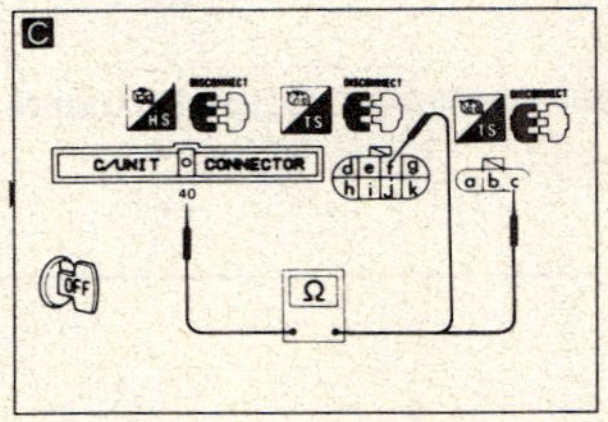

INSPECTION START

↓

A CHECK OVERALL FUNCTION.
1) Turn ignition switch "ON".
2) Perform "PW/ST SIGNAL CIRCUIT" in "FUNCTION TEST" mode with CONSULT.

— OR —

1) Start engine.
2) Check power steering oil pressure switch signal in "DATA MONITOR" mode with CONSULT.
Steering is neutral: OFF
Steering is turned: ON

— OR —

1) Start engine.
2) Check voltage between E.C.U. terminal ⑩ and ground with CONSULT or tester.
Voltage:
When steering wheel is turned quickly.
Battery voltage
Except above
Approximately 0V

→ O.K. → INSPECTION END

↓ N.G.

B CHECK POWER SUPPLY-I.
1) Stop engine.
2) Disconnect E.F.I. sub-harness- 3 connector.
3) Turn ignition switch "ON".
4) Check voltage between terminal ⓑ or ⓐ and ground with CONSULT or tester.
Voltage: Battery voltage

→ N.G. → Check the following.
- Harness connectors (F52), (M4)
- 10A fuse
- Harness continuity between E.F.I. sub-harness-3 connector and fuse

If N.G., repair harness or connectors.

↓ O.K.

C CHECK INPUT SIGNAL CIRCUIT-I.
1) Turn ignition switch "OFF".
2) Disconnect E.C.U. harness connector.
3) Check harness continuity between E.C.U. terminal ⑩ and terminal ⓒ or ⓕ.
Continuity should exist.

→ N.G. → Repair harness or connectors.

↓ O.K.

Ⓐ

POWER STEERING OIL PRESSURE SWITCH (Not self-diagnostic item)

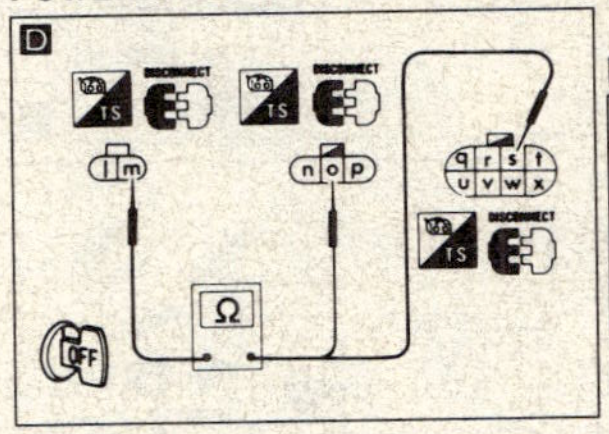

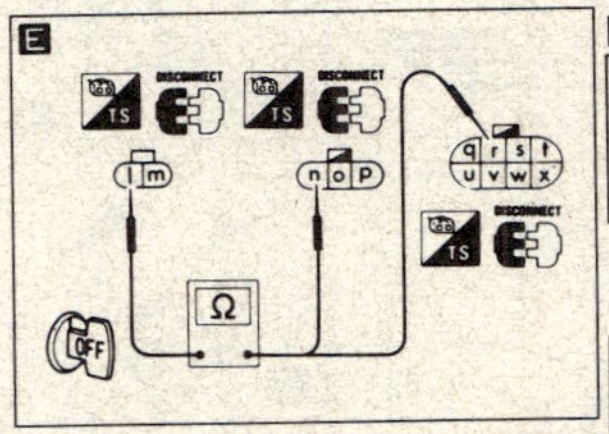

Ⓐ

↓

D CHECK POWER SUPPLY-II.
1) Disconnect power steering oil pressure switch harness connector.
2) Check harness continuity between terminal ⓜ and terminal ⓞ or ⓢ.
Continuity should exist.

→ N.G. → Check the following.
- Harness continuity between E.F.I. sub-harness-3 connector and power steering oil pressure switch.

If N.G., repair harness or connectors.

↓ O.K.

E CHECK INPUT SIGNAL CIRCUIT-II.
1) Check harness continuity between terminal ⓛ and terminal ⓟ or ⓣ.
Continuity should exist.

→ N.G. → (to Check the following, above)

↓ O.K.

CHECK COMPONENT
(Power steering oil pressure switch).
Refer to "Electrical Components Inspection".

→ N.G. → Replace power steering oil pressure switch.

↓ O.K.

Check E.C.U. pin terminals for damage or the connection of the E.C.U. harness connector.

↓

INSPECTION END

Diagnostic Procedure 29

INHIBITOR SWITCH (Not self-diagnostic item)

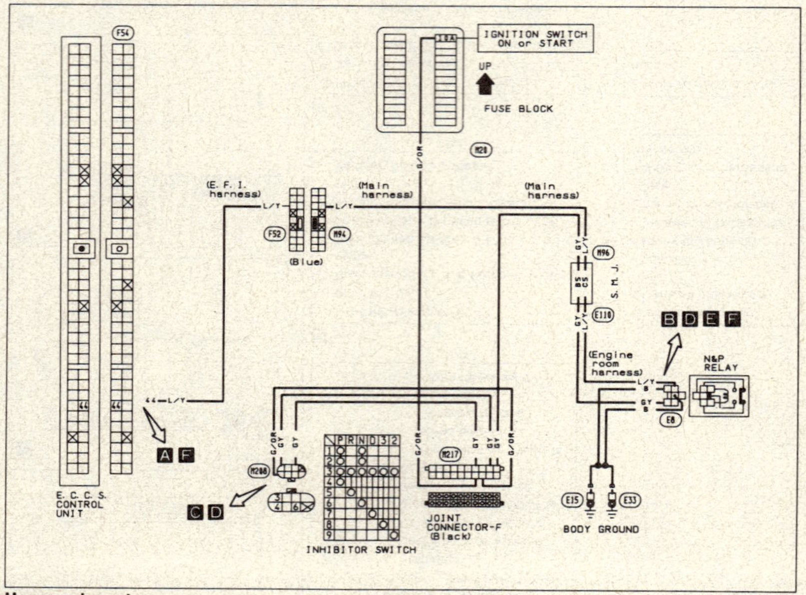

Harness layout

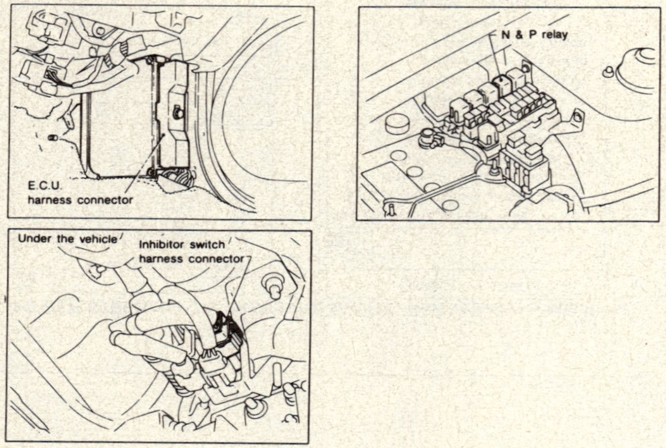

INHIBITOR SWITCH (Not self-diagnostic item)

A
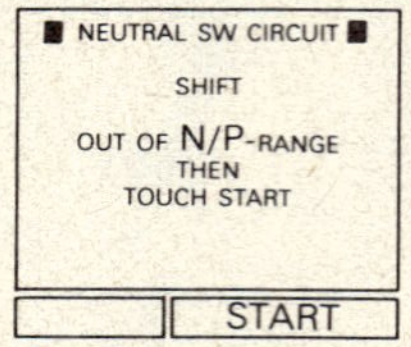

A
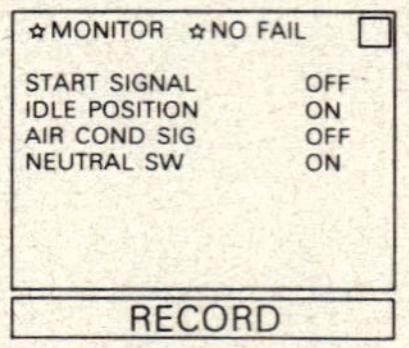

A

B
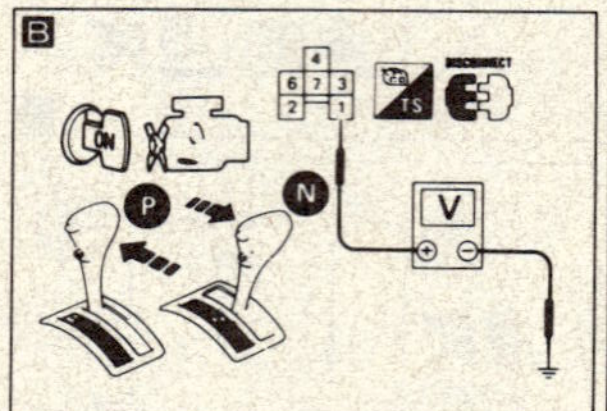

C
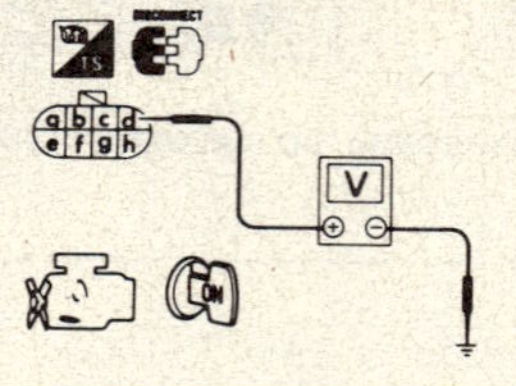

INSPECTION START

A CHECK OVERALL FUNCTION.
1) Turn ignition switch "ON".
2) Perform "NEUTRAL SW CIRCUIT" in "FUNCTION TEST" mode with CONSULT.

— OR —

1) Turn ignition switch "ON".
2) Perform "NEUTRAL SW" in "DATA MONITOR" mode with CONSULT.
"N" or "P": ON
Except above: OFF

— OR —

1) Shift selector lever to "P" range.
2) Disconnect E.C.U. harness connector.
3) Turn ignition switch "ON".
4) Check harness continuity between E.C.U. terminal ㊹ and body ground.
Continuity should exist.
5) Shift selector lever to "N" range.
6) Check harness continuity between E.C.U. terminal ㊹ and body ground.
Continuity should exist.

O.K. → INSPECTION END

N.G. ↓

B CHECK POWER SUPPLY.
1) Turn ignition switch "OFF".
2) Disconnect N & P relay.
3) Make sure that selector lever is in "N" range.
4) Turn ignition switch "ON".
5) Check voltage between terminal ① and ground with CONSULT or tester.
Voltage: Battery voltage
6) Shift selector lever into "P" range.
7) Check voltage between terminal ① and ground with CONSULT or tester.
Voltage: Battery voltage

O.K. → Ⓐ

N.G. →

Check the following.

C CHECK HARNESS CONTINUITY BETWEEN INHIBITOR SWITCH AND BATTERY.
1) Turn ignition switch "OFF".
2) Disconnect inhibitor switch harness connector.
3) Turn ignition switch "ON".
4) Check voltage between terminal ⓓ and ground with CONSULT or tester.
Voltage: Battery voltage
If N.G., check the following.
- Joint connector (M217)
- 10A fuse
- Harness continuity between fuse and inhibitor switch

If N.G., repair harness or connectors.

D CHECK HARNESS CONTINUITY BETWEEN INHIBITOR SWITCH AND N & P RELAY.
1) Turn ignition switch "OFF".
2) Check harness continuity between terminal ① and ⓕ ("N" range), terminal ① and ⓑ ("P" range).
Continuity should exist.
If N.G., check the following.
- Joint connector (M217)
- Harness connectors (M96), (E118)
- Harness continuity between inhibitor switch and N & P relay

If N.G., repair harness or connectors.

CHECK COMPONENT (Inhibitor switch).
Refer to "Electrical Components Inspection".

INHIBITOR SWITCH (Not self-diagnostic item)

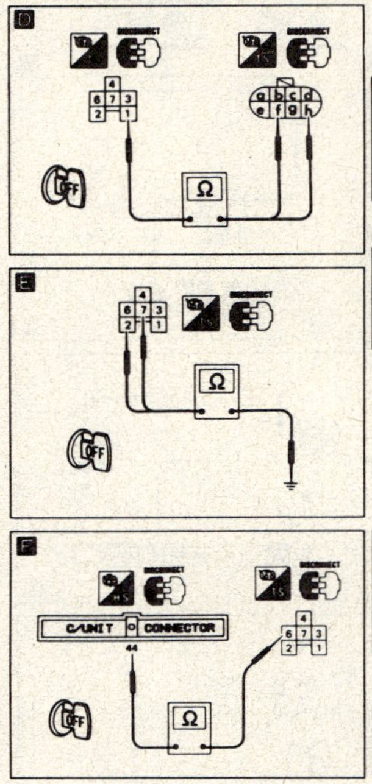

Ⓐ

E CHECK GROUND CIRCUIT.
1) Turn ignition switch "OFF".
2) Check harness continuity between terminals ②, ⑦ and body ground.
Continuity should exist.

N.G. → Repair harness or connectors.

O.K. ↓

F CHECK INPUT SIGNAL CIRCUIT.
1) Check harness continuity between E.C.U. terminal ㊹ and terminal ⑥.
Continuity should exist.

N.G. → Check the following.
- Harness connectors (F52), (M94)
- Harness connectors (M96), (E118)
- Harness continuity between E.C.U. and N & P relay

If N.G., repair harness or connectors.

O.K. ↓

CHECK COMPONENT (N & P relay).
Refer to "Electrical Components Inspection".

N.G. → Replace N&P relay.

O.K. ↓

Check E.C.U. pin terminals for damage or the connection of E.C.U. harness connector.

↓

INSPECTION END

Diagnostic Procedure 30

VALVE TIMING CONTROL (Not self-diagnostic item)

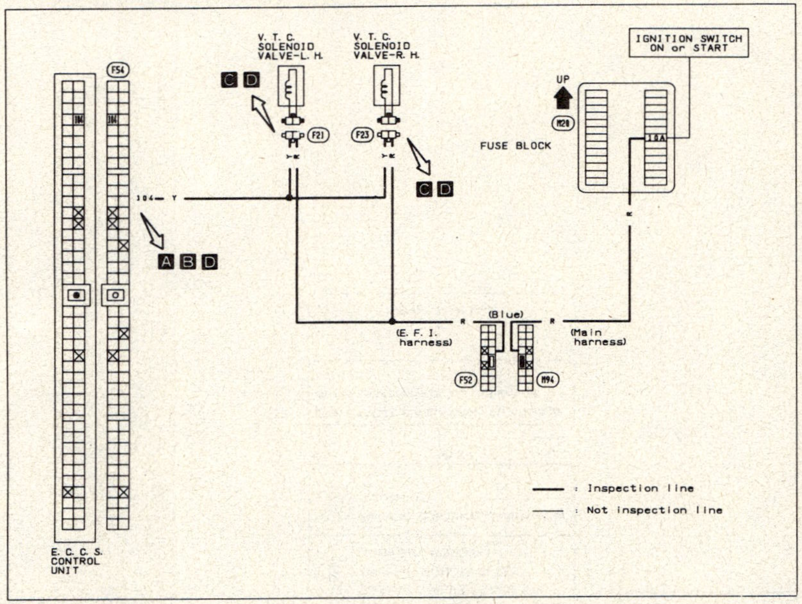

Harness layout

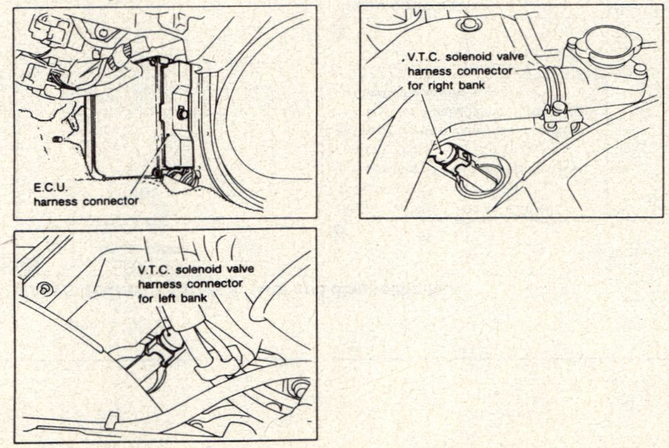

VALVE TIMING CONTROL (Not self-diagnostic item)

A
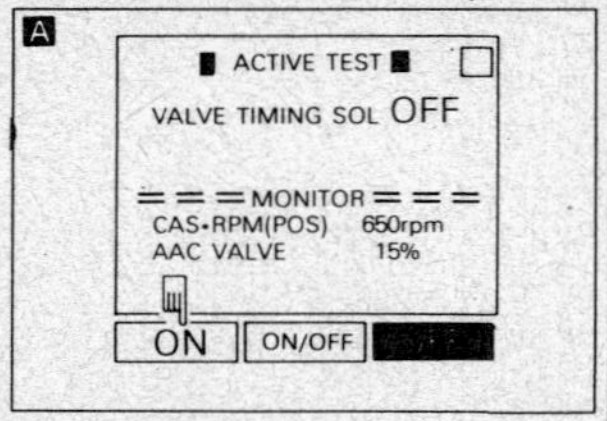

A
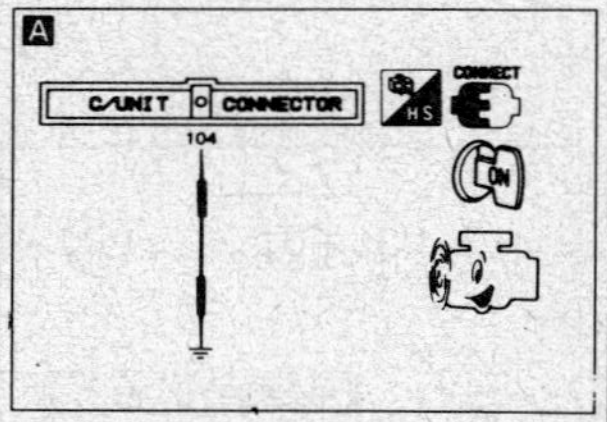

B
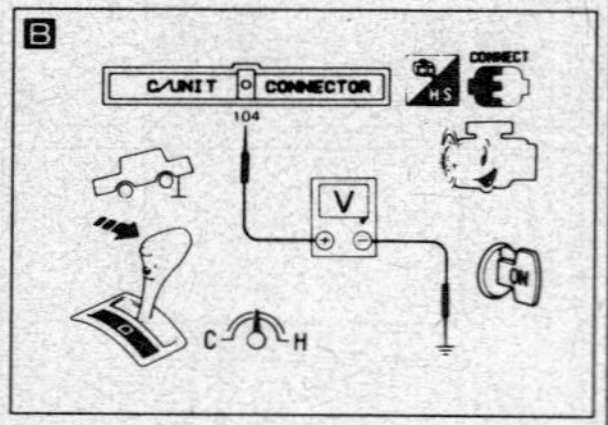

C
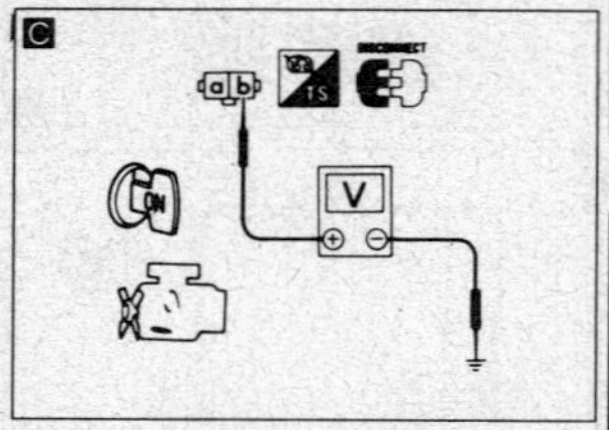

D
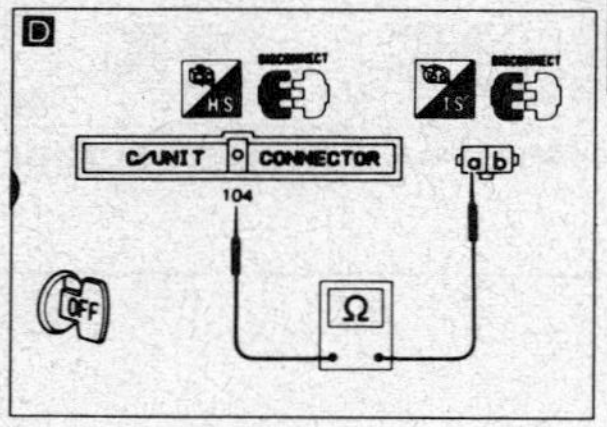

INSPECTION START

↓

A CHECK MECHANICAL FUNCTION.
1) Start engine and run it at idle.
2) Perform "VALVE TIMING SOL" in "ACTIVE TEST" mode with CONSULT and make sure that improper idle condition occurs.

—OR—

1) Connect a suitable jumper wire between E.C.U. terminal (104) and body ground.
2) Start engine and make sure that improper idle condition occurs.

Occurs. → INSPECTION END

Does not occur. ↓

B CHECK ELECTRICAL CONTROL FUNCTION.
1) Stop engine.
2) Jack up drive wheel.
3) Start engine and warm it up sufficiently.
4) Shift select lever to any range except "N" or "P" range.
5) Check voltage between E.C.U. terminal (104) and ground under the following conditions with CONSULT or tester.

Voltage:
Quickly depress accelerator pedal, then quickly release it.
Approximately 0V
At idle
Battery voltage

O.K. → CHECK COMPONENT (V.T.C. solenoid valve and V.T.C. valve). Refer to "Electrical Components Inspection".

N.G. → Replace malfunctioning component(s).

N.G. ↓

C CHECK POWER SUPPLY.
1) Stop engine.
2) Disconnect V.T.C. solenoid valve harness connector.
3) Turn ignition switch "ON".
4) Check voltage between terminal (b) and ground with CONSULT or tester.
Voltage: Battery voltage

N.G. → Check the following.
- Harness connectors (F52), (M94)
- 10A fuse
- Harness continuity between V.T.C. solenoid valve and fuse

If N.G., repair harness or connectors.

O.K. → (A)

VALVE TIMING CONTROL (Not self-diagnostic item)

E
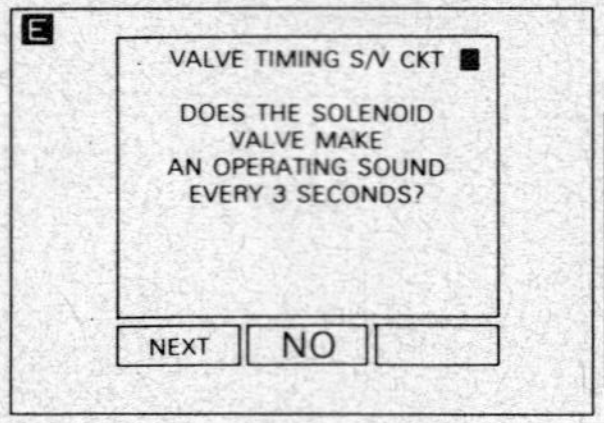

(A)

↓

D CHECK OUTPUT SIGNAL CIRCUIT.
1) Turn ignition switch "OFF".
2) Disconnect E.C.U. harness connector.
3) Check harness continuity between E.C.U. terminal (104) and terminal (a).
Continuity should exist.

N.G. → Repair harness or connectors.

O.K. ↓

E CHECK COMPONENT (V.T.C. solenoid valve).
Perform "VALVE TIMING S/V CKT" in "FUNCTION TEST" mode with CONSULT.

—OR—

Refer to "Electrical Components Inspection".

N.G. → Replace V.T.C. solenoid valve.

O.K. ↓

Check E.C.U. pin terminals for damage or the connection of E.C.U. harness connector.

↓

INSPECTION END

Diagnostic Procedure 31

CANISTER PURGE CONTROL (Not self-diagnostic item)

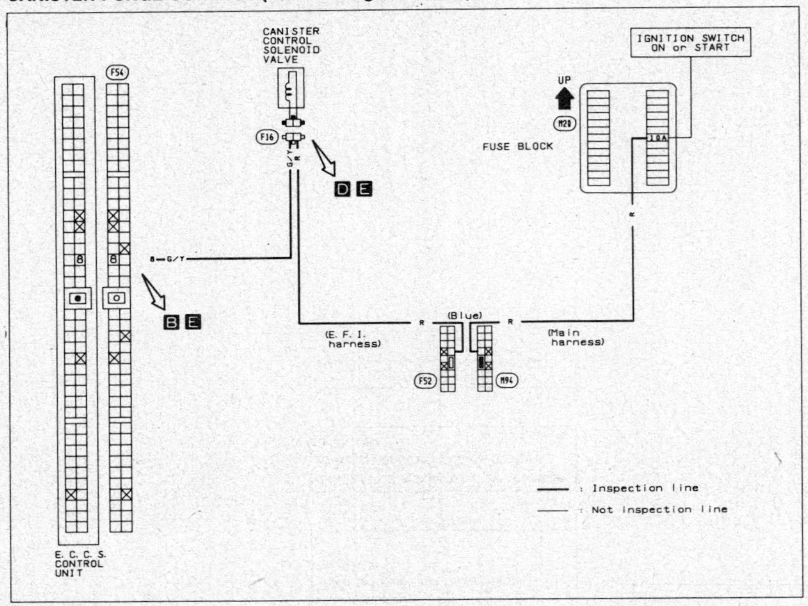

Harness layout

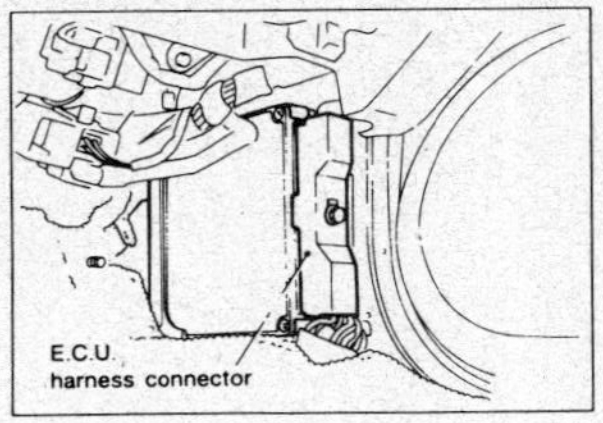

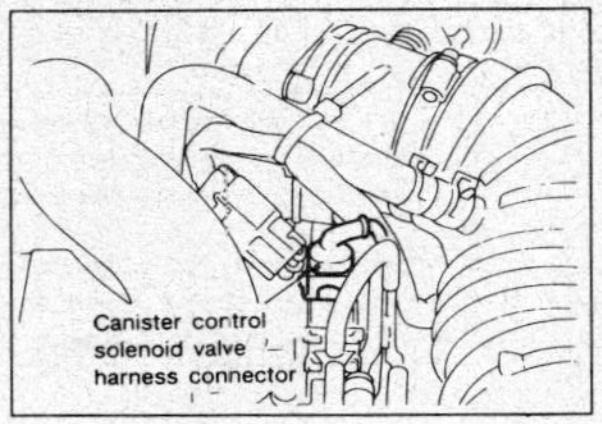

CANISTER PURGE CONTROL (Not self-diagnostic item)

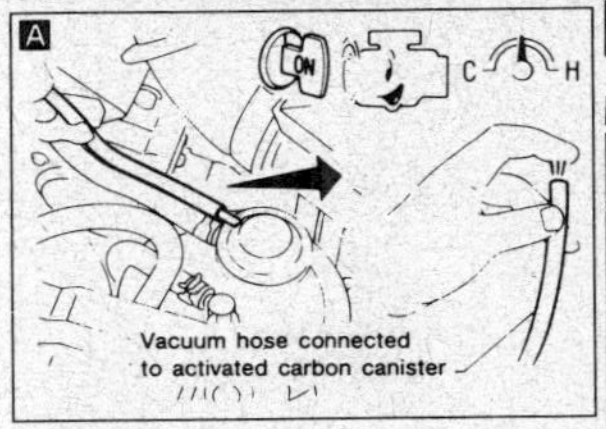

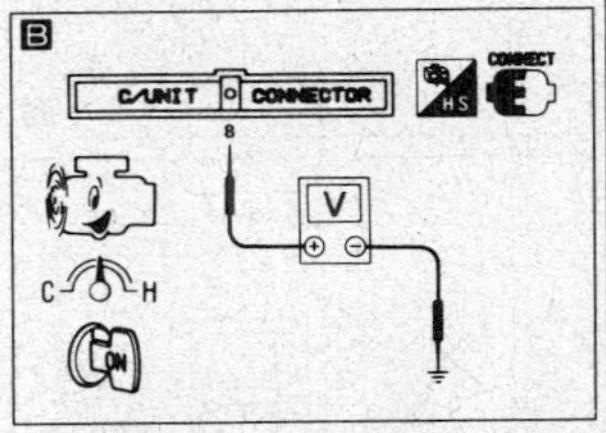

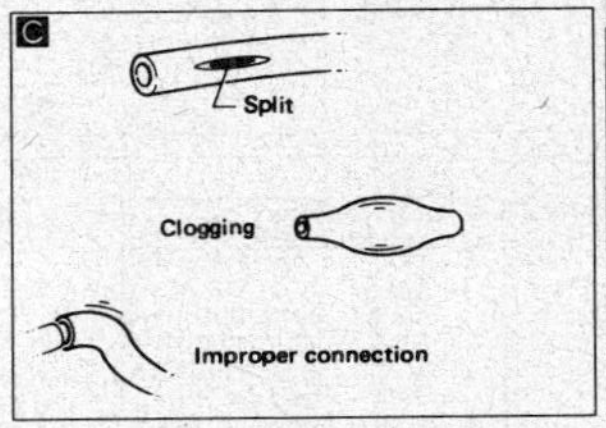

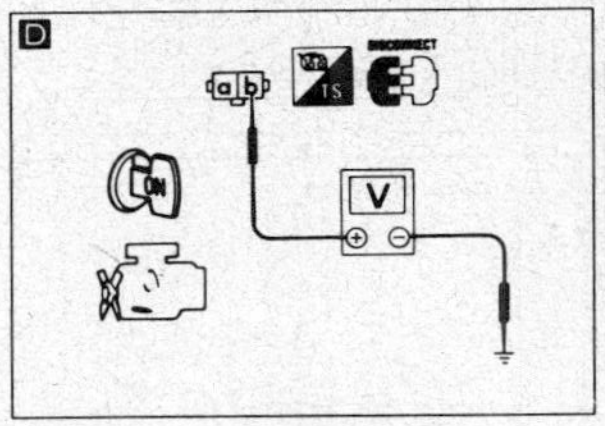

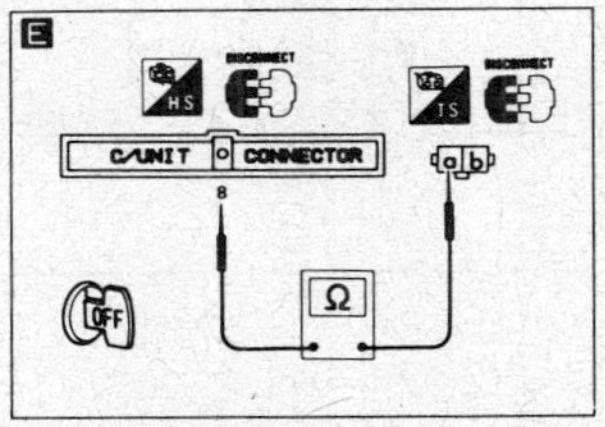

INSPECTION START

↓

A CHECK VACUUM SOURCE TO ACTIVATED CARBON CANISTER.
1) Start engine and warm it up sufficiently.
2) Perform self-diagnosis. Make sure that code No. 11 and 12 are not displayed.
3) Stop engine and jack up drive wheels.
4) Disconnect vacuum hose to activated carbon canister and restart engine.
5) Make sure that vacuum exists under the following conditions.
At idle:
Vacuum should not exist.
Engine speed is about 2,000 rpm (In "1" range)
Vacuum should exist.

O.K. → CHECK COMPONENT. (Activated carbon canister) Refer to "Electrical Components Inspection".
If N.G., replace activated carbon canister.

N.G. ↓

B CHECK CONTROL FUNCTION.
1) Check voltage between E.C.U. terminal ⑧ and ground with CONSULT or tester under the following conditions.
Voltage:
At idle
Approximately 0V
Engine speed is about 2,000 rpm (In "1" range)
Battery voltage

O.K. → **C** Check vacuum hose for clogging, cracks and proper connection.

N.G. ↓

D CHECK POWER SUPPLY.
1) Stop engine.
2) Disconnect canister control solenoid valve harness connector.
3) Turn ignition switch "ON".
4) Check voltage between terminal ⓑ and ground with CONSULT or tester.
Voltage: Battery voltage

N.G. → Check the following.
- Harness connectors F52, M94
- 10A fuse
- Harness continuity between canister control solenoid valve and fuse

If N.G., repair harness or connectors.

O.K. ↓

Ⓐ

CANISTER PURGE CONTROL (Not self-diagnostic item)

Ⓐ

E CHECK OUTPUT SIGNAL CIRCUIT.
1) Turn ignition switch "OFF".
2) Disconnect E.C.U. harness connector.
3) Check harness continuity between E.C.U. terminal ⑧ and terminal ⓐ.
Continuity should exist.

N.G. → Repair harness or connectors.

O.K. ↓

CHECK COMPONENTS
(Canister control solenoid valve).
Refer to "Electrical Components Inspection".

N.G. → Replace canister control solenoid valve.

O.K. ↓

Check E.C.U. pin terminals for damage or the connection of E.C.U. harness connector.

↓

INSPECTION END

Diagnostic Procedure 32 — Symptom — High Idling after Warm-up

1 **CHECK FAST IDLE CAM.**
Does fast idle cam keep cam follower lever released?

No → Check fast idle cam adjustment.

Yes ↓

2 **CHECK INTAKE AIR LEAK.**
1. Select "SELF-LEARNING CONT" in "ACTIVE TEST" mode.
2. Clear the self-learning control coefficient by touching "CLEAR".
3. Does the engine speed drop?

OR

1. Disconnect air flow meter harness connectors.
2. After starting and running engine for at least 30 seconds at 2,000 rpm, does the engine speed drop?

Yes → Discover air leak location and repair.

No ↓

3 **CHECK THROTTLE LINKAGE.**
1. Check that throttle linkage moves smoothly.
2. Confirm that throttle valve both fully opens and fully closes.

N.G. → Repair throttle linkage or sticking of throttle valve.

O.K. ↓

INSPECTION END

1
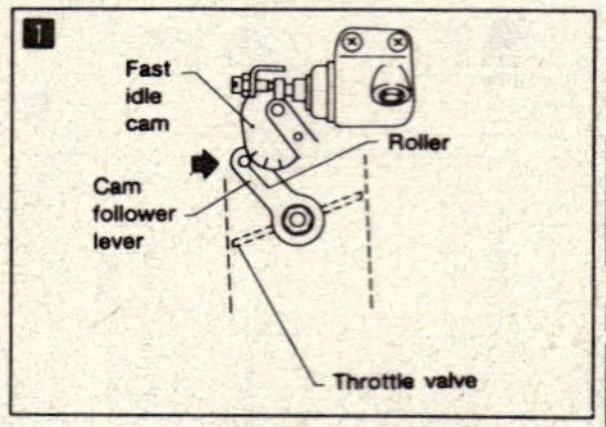

2
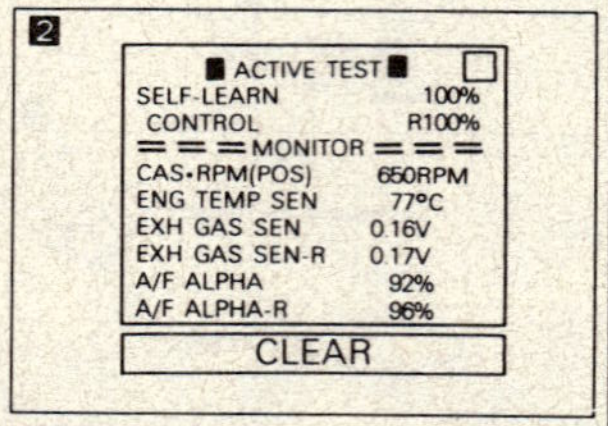

2
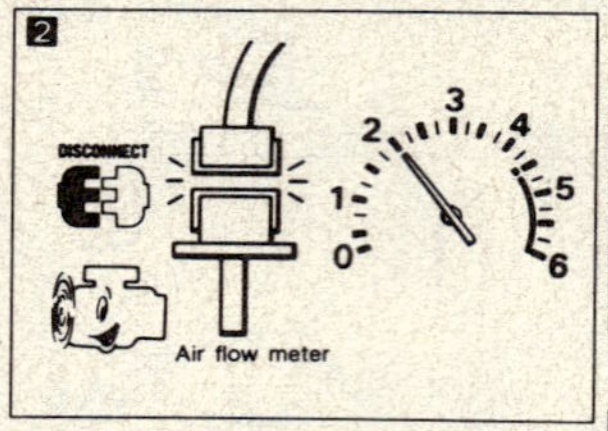

3
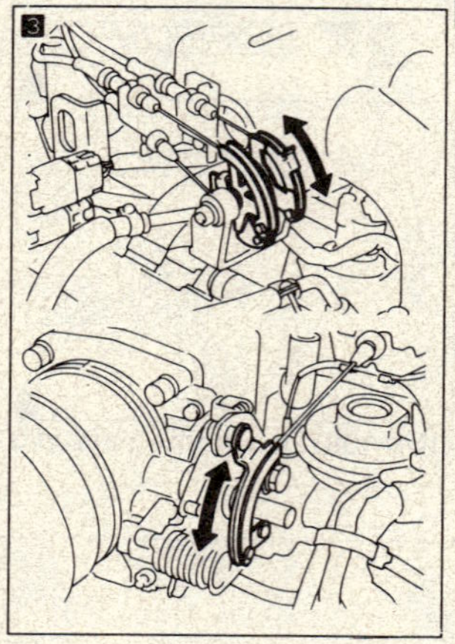

1990–92 4.5L ENGINE – DIAGNOSTIC PROCEDURE – Q45

Diagnostic Procedure 33 — Symptom — Hunting

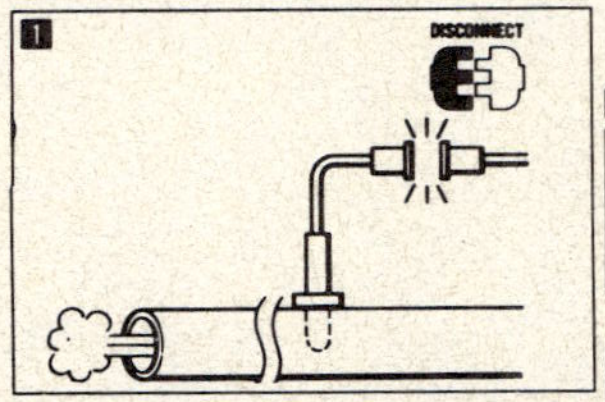

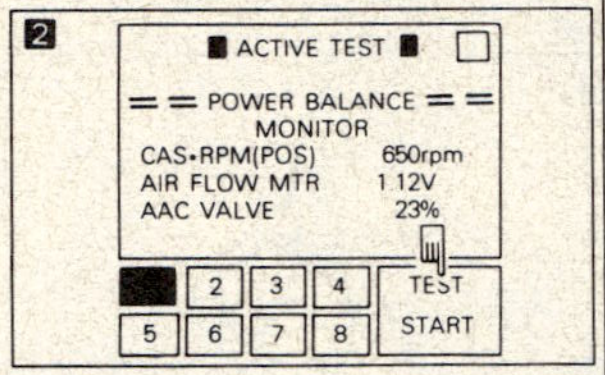

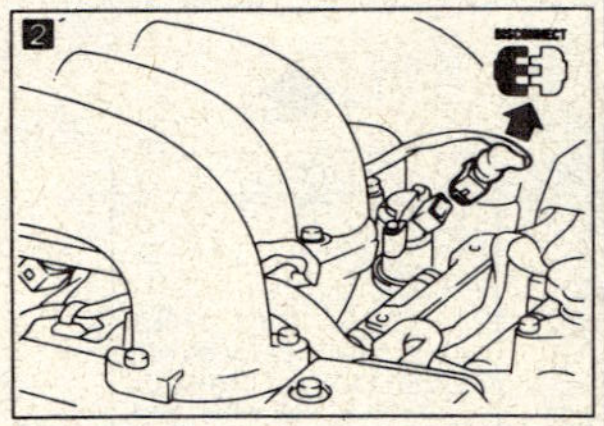

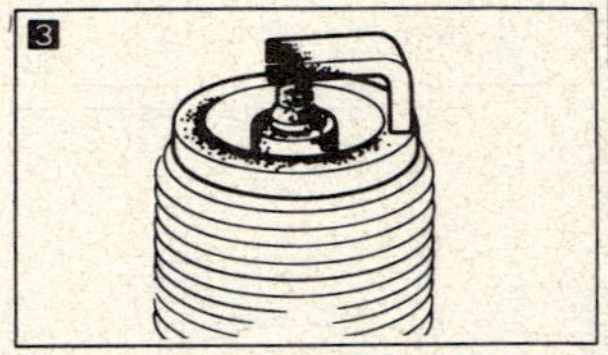

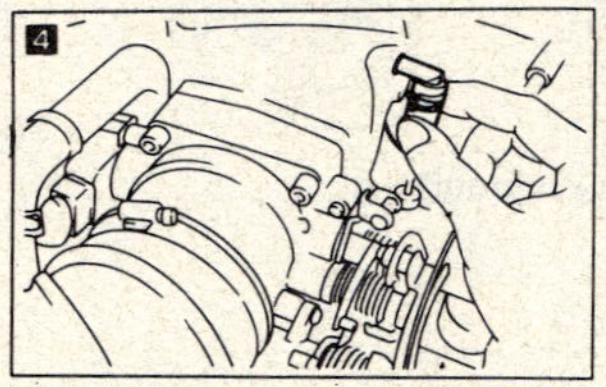

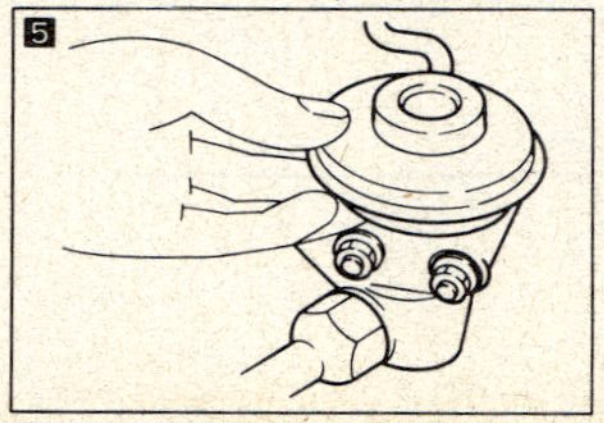

1 CHECK EXHAUST GAS SENSORS.
When disconnecting exhaust gas sensor harness connectors, is the hunting fixed?
- Yes → Check exhaust gas sensor(s).
- No ↓

2 PERFORM POWER BALANCE TEST.
1. Perform "POWER BALANCE" in "ACTIVE TEST" mode.
2. Is there any cylinder which does not produce a momentary engine speed drop?

— OR —

When disconnecting each injector harness connector one at a time, is there any cylinder which does not produce a momentary engine speed drop?
- No → Go to 4
- Yes ↓

3 CHECK SPARK PLUGS.
Remove the spark plugs and check for fouling, etc.
- N.G. → Repair or replace spark plug(s).
- O.K. ↓

4 CHECK FOR INTAKE AIR LEAK.
When pinching blow-by hose (lowering the blow-by air supply), does the engine speed rise?
- Yes → Discover air leak location and repair.
- No ↓

5 CHECK E.G.R. CONTROL VALVE.
Check E.G.R. control valve for sticking.
- N.G. → Repair or replace.
- O.K. ↓

INSPECTION END

Diagnostic Procedure 34 — Symptom — Unstable Idle

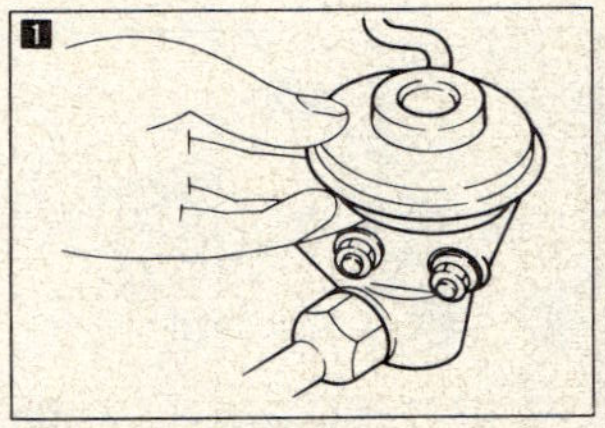

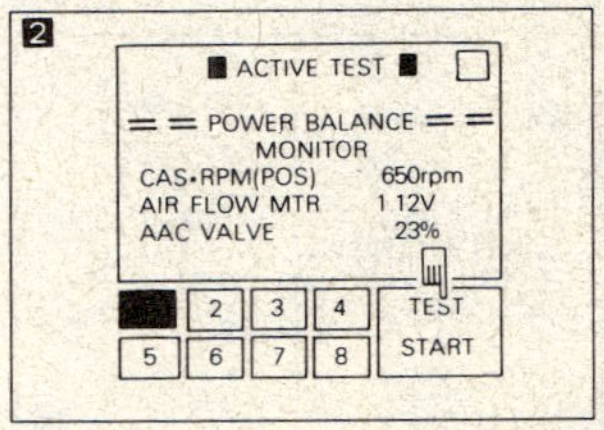

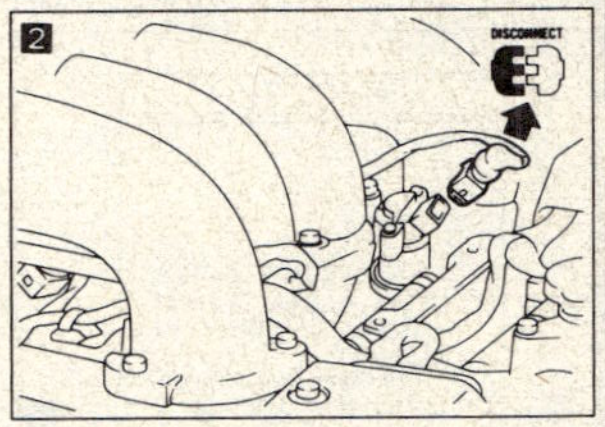

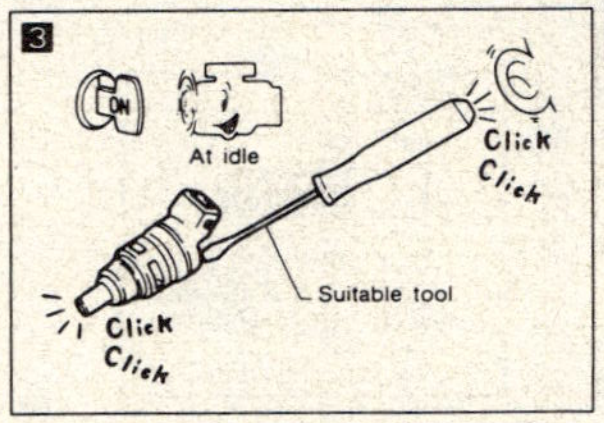

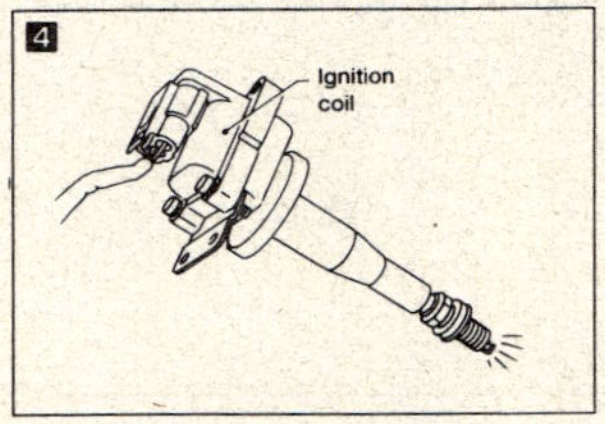

1 CHECK E.G.R. CONTROL VALVE.
Check E.G.R. control valve for sticking.
- N.G. → Repair or replace.
- O.K. ↓

2 PERFORM POWER BALANCE TEST.
1. Perform "POWER BALANCE" in "ACTIVE TEST" mode.
2. Is there any cylinder which does not produce a momentary engine speed drop?

— OR —

When disconnecting each injector harness connector one at a time, is there any cylinder which does not produce a momentary engine speed drop?
- No → Go to 6
- Yes ↓

3 CHECK INJECTOR.
Does each injector make an operating sound at idle?
- No → Check injector(s) and circuit(s).
- Yes ↓

4 CHECK IGNITION SPARK.
1. Disconnect ignition coil assembly from rocker cover.
2. Connect a known good spark plug to the ignition coil assembly.
3. Place end of spark plug against a suitable ground and crank engine.
4. Check for spark.
- N.G. → Check ignition coil, power transistor unit and their circuits.
- O.K. ↓

(Go to Ⓐ on next page.)

Diagnostic Procedure 34 — Symptom — Unstable Idle (Cont'd)

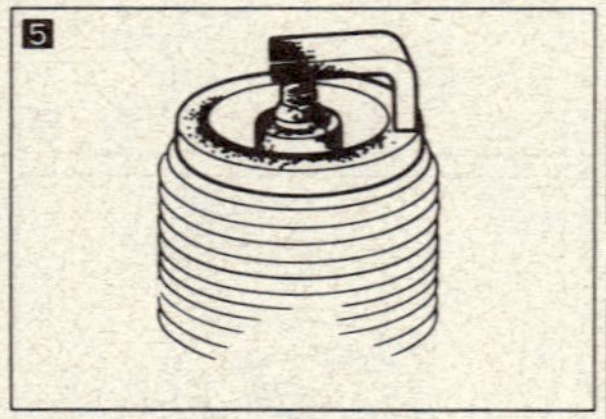

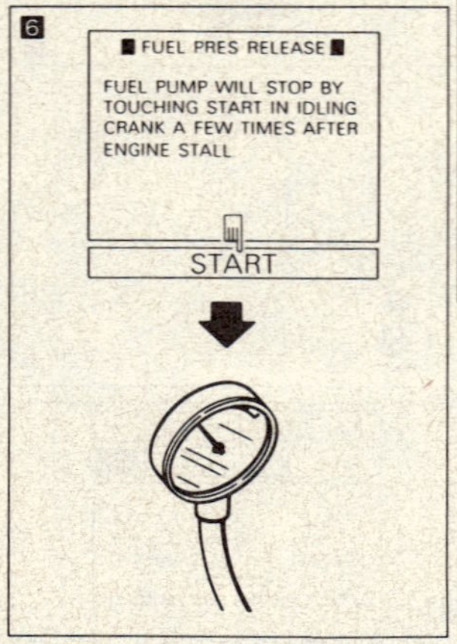

(A)

5 CHECK SPARK PLUGS.
Remove the spark plugs and check for fouling, etc.

N.G. → Repair or replace spark plug(s).

O.K. ↓

6 CHECK FUEL PRESSURE.
1. Perform "FUEL PRESSURE RELEASE" in "WORK SUPPORT" mode in order to release fuel pressure to zero.
2. Install fuel pressure gauge and check fuel pressure.
At idle approx. 235 kPa (2.4 kg/cm², 34 psi)

— OR —

1. Release fuel pressure to zero.
2. Install fuel pressure gauge and check fuel pressure.

N.G. → Check fuel pump and circuit.

O.K. ↓

(Go to (B) on next page.)

Diagnostic Procedure 34 — Symptom — Unstable Idle (Cont'd)

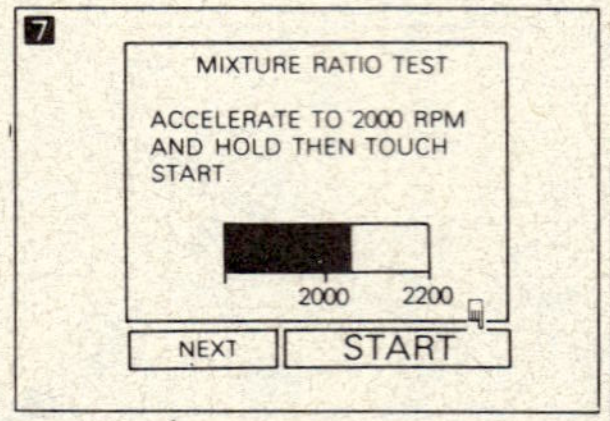

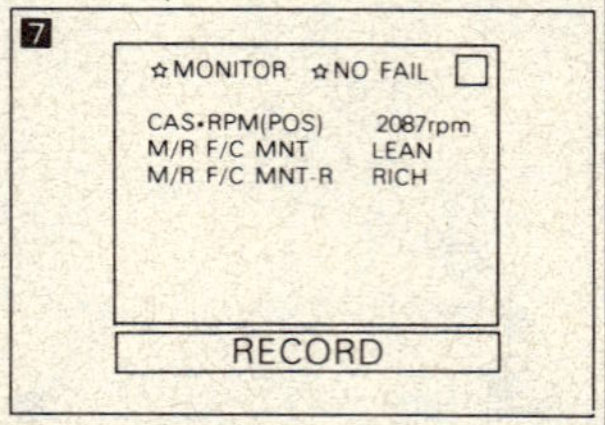

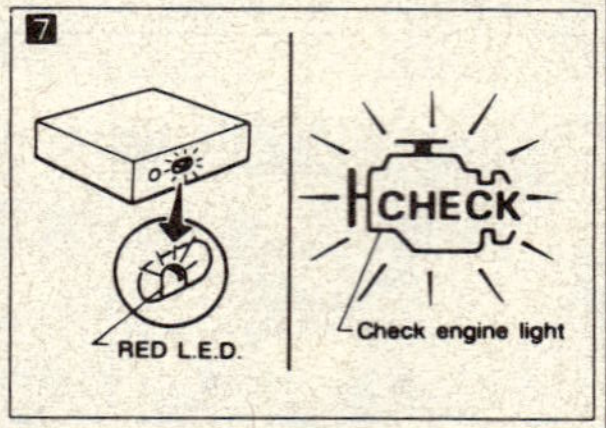

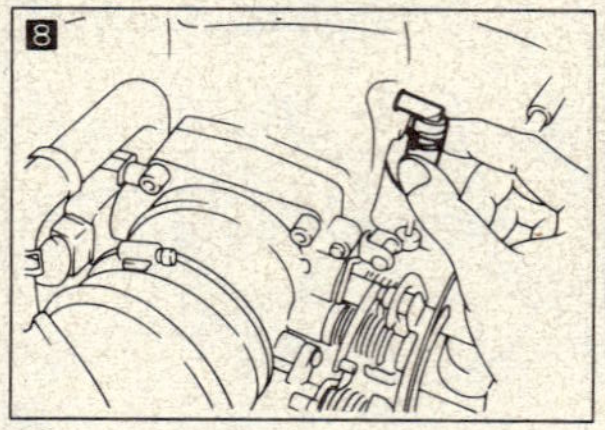

(B)

7 CHECK EXHAUST GAS SENSOR.
1. Start engine and warm it up sufficiently.
2. Perform "MIXTURE RATIO TEST" in "FUNCTION TEST" mode.

— OR —

2. See "M/R F/C MNT (right and left sides)" in "Data monitor" mode.
3. Maintaining engine at 2,000 rpm under no-load (engine is warmed up sufficiently.), check that the monitor fluctuates between "LEAN" and "RICH" more than 5 times during 10 seconds.
RICH→LEAN→RICH→
1 time 2 times
LEAN→RICH.......

— OR —

2. Set "Exhaust gas sensor monitor" in the self-diagnostic Mode II.
3. Maintaining engine at 2,000 rpm under no-load, check to make sure that check engine light and RED LED on the E.C.U. go ON and OFF more than 5 times during 10 seconds.

N.G. → Replace exhaust gas sensor(s).

O.K. ↓

8 CHECK FOR INTAKE AIR LEAK.
When pinching blow-by hose (lowering the blow-by air supply), does the engine speed rise?

Yes → Discover air leak location and repair.

No ↓

(Go to (C) on next page.)

Diagnostic Procedure 34 — Symptom — Unstable Idle (Cont'd)

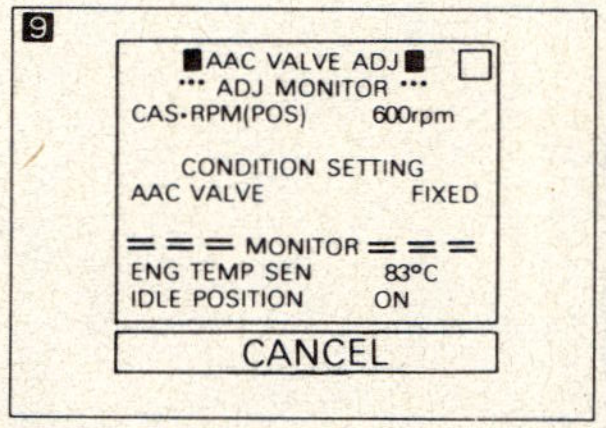

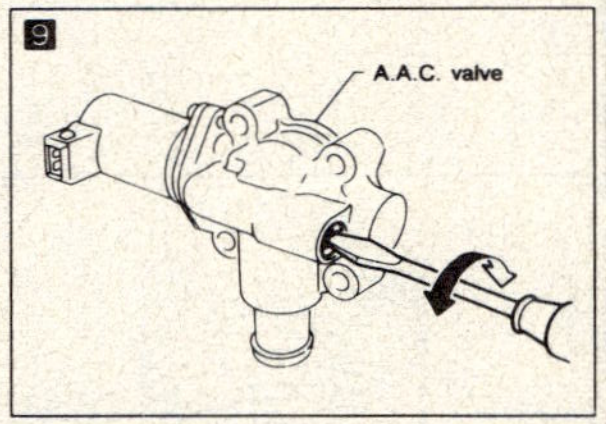

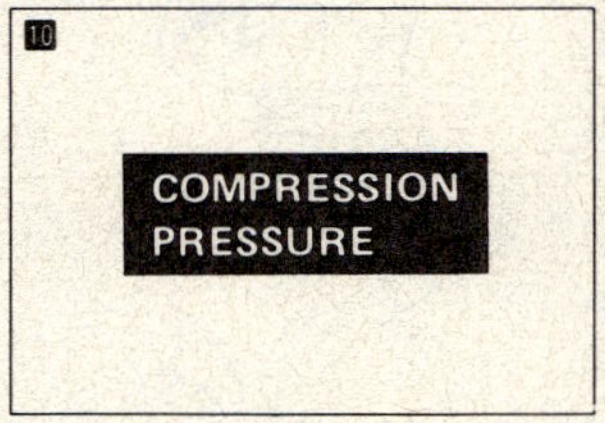

Ⓒ

9 CHECK IDLE ADJ. SCREW CLOGGING.
1. Perform "AAC VALVE ADJ" in "WORK SUPPORT" mode.
2. Can you set engine speed at 600 ± 25 rpm A/T in "N" position) by turning idle adjusting screw?

— OR —

1. Disconnect A.A.C. valve harness connector.
2. Can you set engine speed at 600 ± 25 rpm A/T in "N" position) by turning idle adjusting screw?

No → Check for A.A.C. valve clogging or throttle valve clogging.

Yes ↓

10 CHECK COMPRESSION PRESSURE.
- Check compression pressure.

Standard: kPa (kg/cm², psi)/300 rpm
1,275 (13.0, 185)
Minimum: kPa (kg/cm², psi)/300 rpm
981 (10.0, 142)
Difference between each cylinder: kPa (kg/cm², psi)/300 rpm
98 (1.0, 14)

N.G. → Check pistons, piston rings, valves, valve seats and cylinder head gaskets.

O.K. ↓

11 CHECK E.C.U. HARNESS CONNECTOR.
Check the E.C.U. pin terminals for damage or poor connection of E.C.U. harness connector.

N.G. → Repair or replace.

O.K. ↓

12 TRY A KNOWN GOOD E.C.U.*

↓

INSPECTION END

*: E.C.U. may be the cause of a problem, but this is rarely the case.

Diagnostic Procedure 35 — Symptom — Hard to Start or Impossible to Start when the Engine is Cold

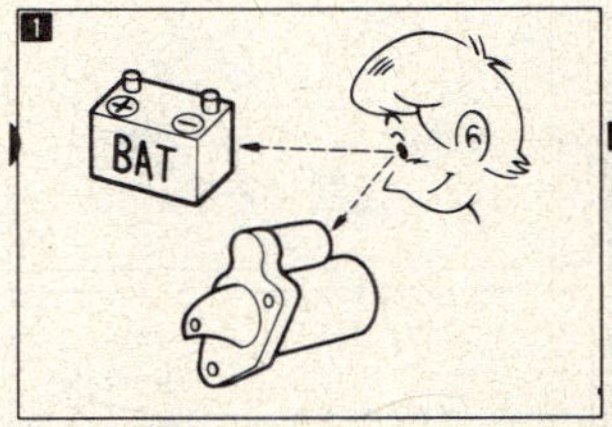

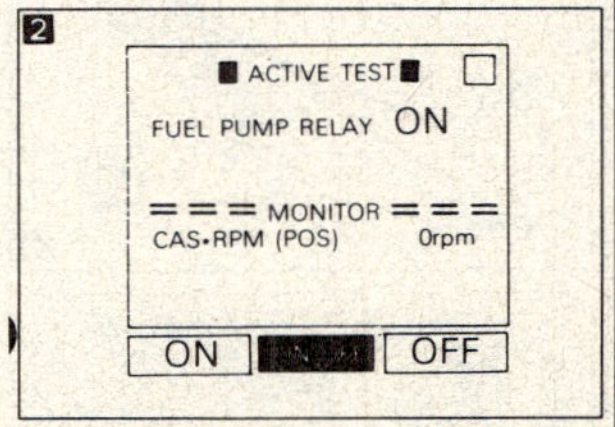

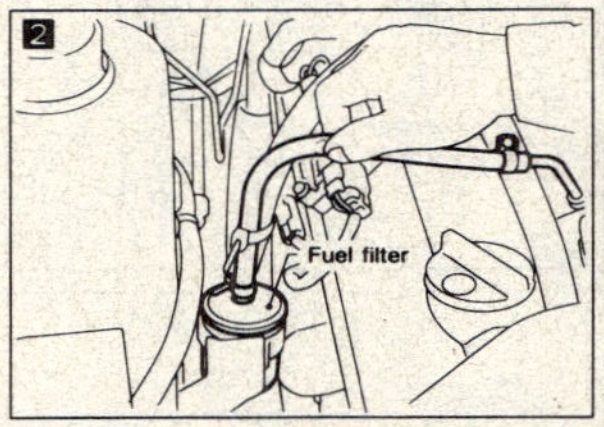

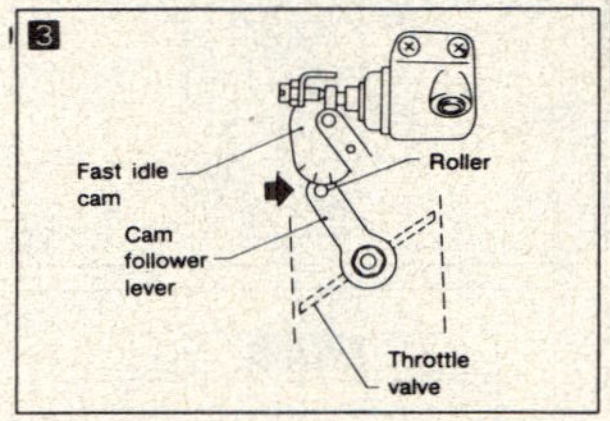

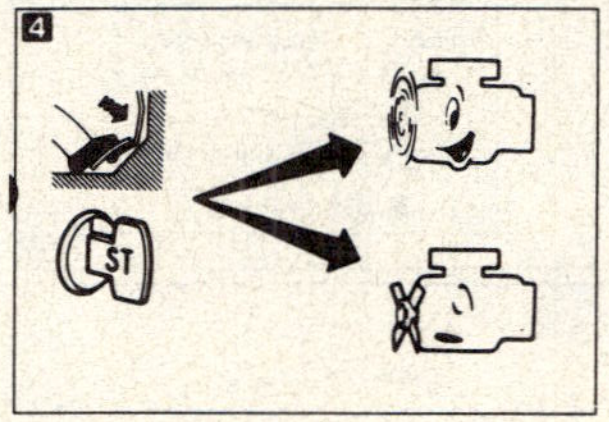

1 CHECK BATTERY AND STARTER.
Check battery and starter condition.

N.G. → Repair or replace.

O.K. ↓

2 CHECK FUEL PRESSURE.
1. Turn ignition switch "ON".
2. Perform "FUEL PUMP RELAY" in "ACTIVE TEST" mode.
3. Pinch fuel feed hose with fingers.

Is fuel pressure pulsation felt on the fuel feed hose?

— OR —

1. Pinch fuel feed hose with fingers.
2. When cranking the engine, is there any pressure on the fuel feed hose?

No → Check fuel pump and circuit.

Yes ↓

3 CHECK FAST IDLE CAM.
When the engine is cold, does fast idle cam keep cam follower lever in position?

No → Check fast idle cam adjustment.

Yes ↓

4 CHECK A.A.C. VALVE.
When pressing accelerator pedal fully, can you start the engine.

Yes → Check A.A.C. valve and circuit.

No ↓

(Go to Ⓐ on next page.)

Diagnostic Procedure 35 — Symptom — Hard to Start or Impossible to Start when the Engine is Cold (Cont'd)

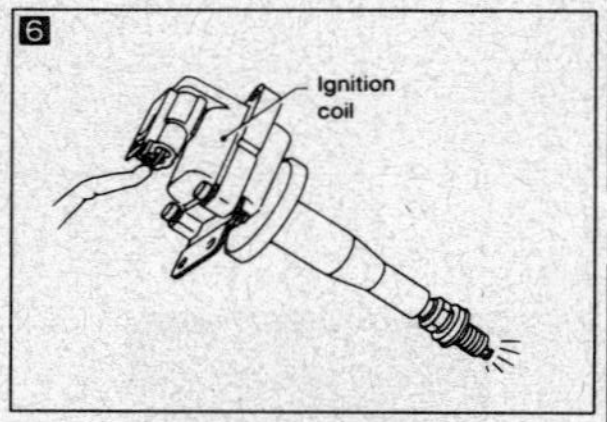

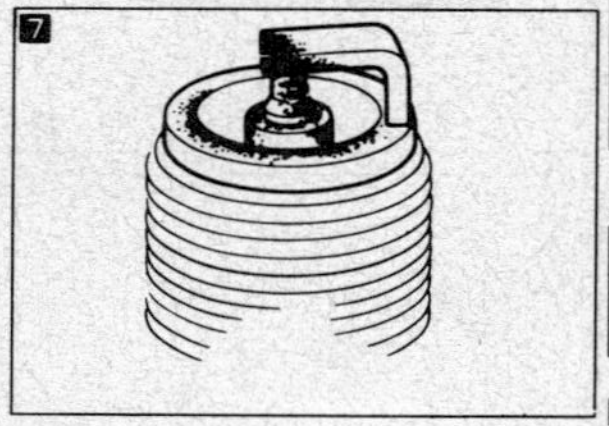

Ⓐ

5 CHECK INJECTOR.
1. Remove crank angle sensor from engine. (Harness connector should remain connected.)
2. Disconnect power transistor harness connector.
3. Turn ignition switch ON. (Do not start engine.)
4. When rotating crank angle sensor shaft, does each injector make an operating sound?

- No → Check injector(s) and circuit(s).
- Yes → 6

6 CHECK IGNITION SPARK.
1. Disconnect ignition coil assembly from rocker cover.
2. Connect a known good spark plug to the ignition coil assembly.
3. Place end of spark plug against a suitable ground and crank engine.
4. Check for spark.

- N.G. → Check ignition coil, power transistor unit and their circuits.
- O.K. → 7

7 CHECK SPARK PLUGS.
Remove the spark plugs and check for fouling, etc.

- N.G. → Repair or replace spark plug(s).
- O.K. → 8

8 CHECK E.C.U. HARNESS CONNECTOR.
Check the E.C.U. pin terminals for damage or poor connection of E.C.U. harness connector.

- N.G. → Repair or replace.
- O.K. → 9

9 CHECK E.C.U. POWER SUPPLY AND GROUND CIRCUIT.

- N.G. → Repair or replace.
- O.K. → 10

10 TRY A KNOWN GOOD E.C.U.*

↓

INSPECTION END

*: E.C.U. may be the cause of a problem, but this is rarely the case.

Diagnostic Procedure 36 — Symptom — Hard to Start or Impossible to Start when the Engine is Hot

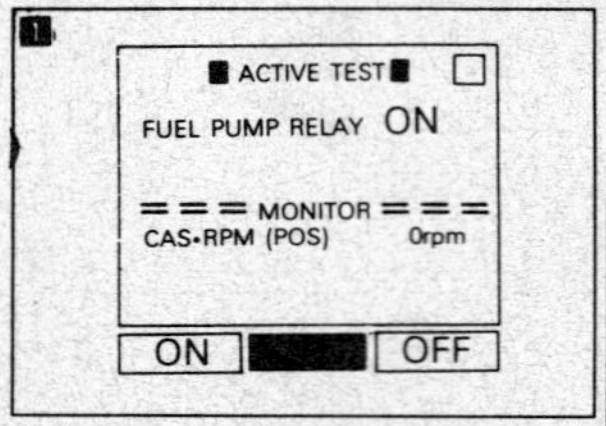

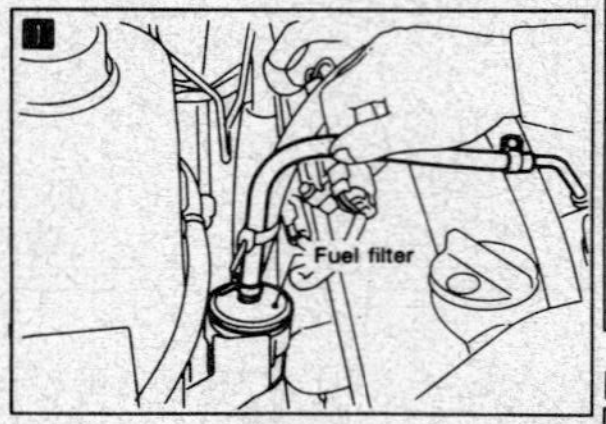

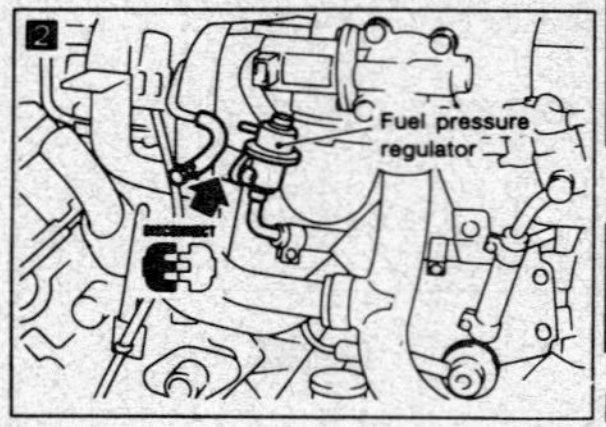

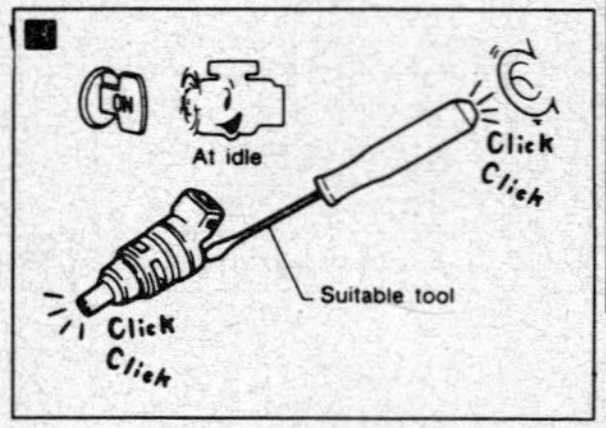

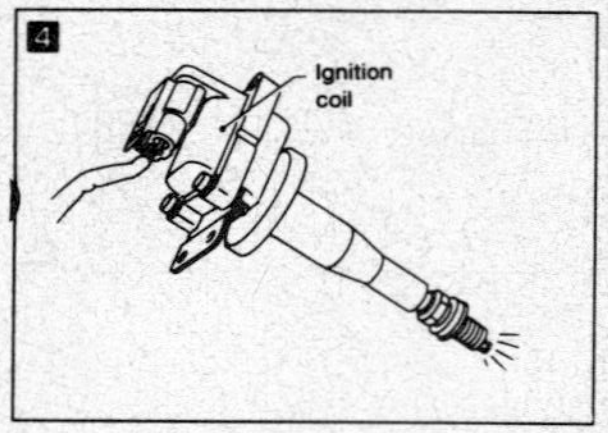

1 CHECK FUEL PRESSURE.
1. Turn ignition switch "ON".
2. Perform "FUEL PUMP RELAY" in "ACTIVE TEST" mode.
3. Pinch fuel feed hose with fingers.

Is fuel pressure pulsation felt on the fuel feed hose?

— OR —

1. Pinch fuel feed hose with fingers.
2. When cranking the engine, is there any pressure on the fuel feed hose?

- No → Check fuel pump and circuit.
- Yes → 2

2 CHECK FUEL VAPOR.
1. Disconnect fuel pressure regulator vacuum hose and plug hose.
2. Can you start engine?

- Yes → Check fuel properties.
- No → 3

3 CHECK INJECTOR.
Does each injector make an operating sound at idle?

- No → Check injector(s) and circuit(s).
- Yes → 4

4 CHECK IGNITION SPARK.
1. Disconnect ignition coil assembly from rocker cover.
2. Connect a known good spark plug to the ignition coil assembly.
3. Place end of spark plug against a suitable ground and crank engine.
4. Check for spark.

- N.G. → Check ignition coil, power transistor unit and circuits.
- O.K. → 5

5 CHECK E.C.U. HARNESS CONNECTOR.
Check the E.C.U. pin terminals for damage or poor connection of E.C.U. harness connector.

- N.G. → Repair or replace.
- O.K. → 6

6 CHECK E.C.U. POWER SUPPLY AND GROUND CIRCUIT.

- N.G. → Repair or replace.
- O.K. → 7

7 TRY A KNOWN GOOD E.C.U.*

↓

INSPECTION END

*: E.C.U. may be the cause of a problem, but this is rarely the case.

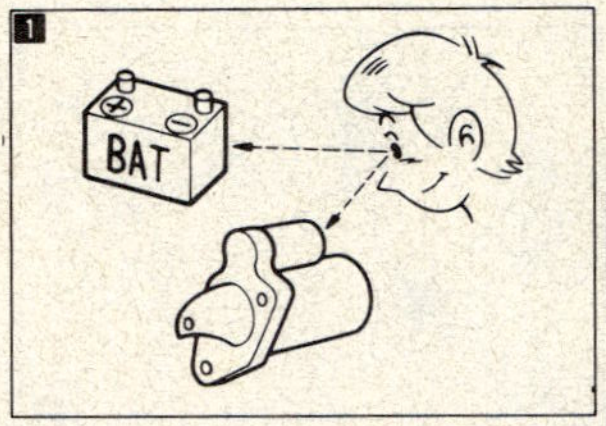

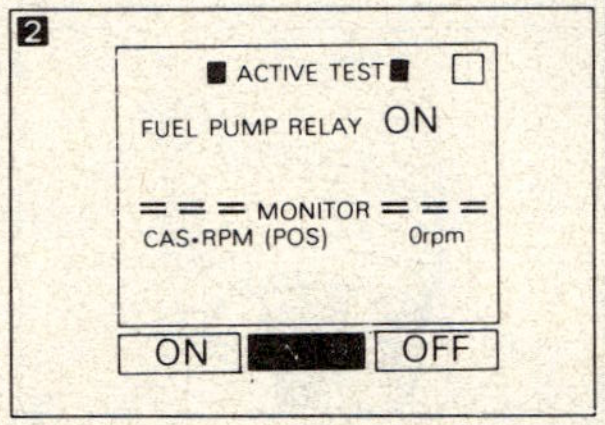

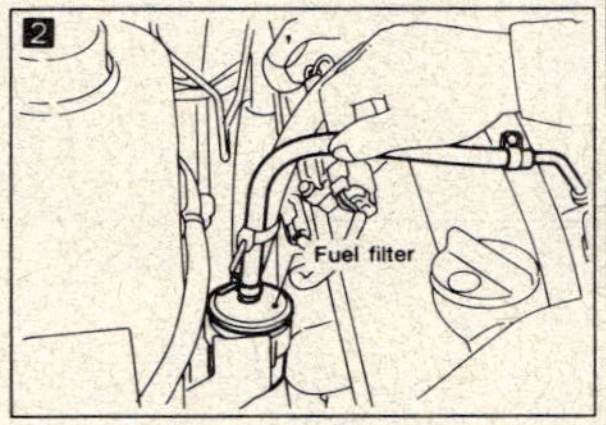

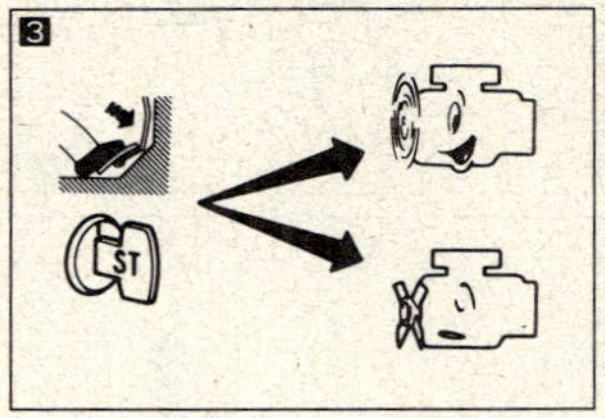

Diagnostic Procedure 37 — Symptom — Hard to Start or Impossible to Start under Normal Conditions

1 CHECK BATTERY AND STARTER.
Check battery and starter operation.
- N.G. → Repair or replace.
- O.K. ↓

2 CHECK FUEL PRESSURE.
1. Turn ignition switch "ON".
2. Perform "FUEL PUMP RELAY" in "ACTIVE TEST" mode.
3. Pinch fuel feed hose with fingers.

Is fuel pressure pulsation felt on the fuel feed hose?

— OR —

1. Pinch fuel feed hose with fingers.
2. When cranking the engine, is there any pressure on the fuel feed hose?

- No → Check fuel pump and circuit.
- Yes ↓

3 CHECK INJECTOR FOR LEAKAGE.
When pressing accelerator pedal fully, can you start the engine.
- Yes → Check injector(s) for leakage.
- No ↓

4 CHECK INJECTOR.
1. Remove crank angle sensor from engine. (Harness connector should remain connected.)
2. Disconnect power transistor harness connector.
3. Turn ignition switch ON. (Do not start engine.)
4. When rotating crank angle sensor shaft, does each injector make an operating sound?

- No → Check injectors and circuits.
- Yes ↓

(Go to Ⓐ on next page.)

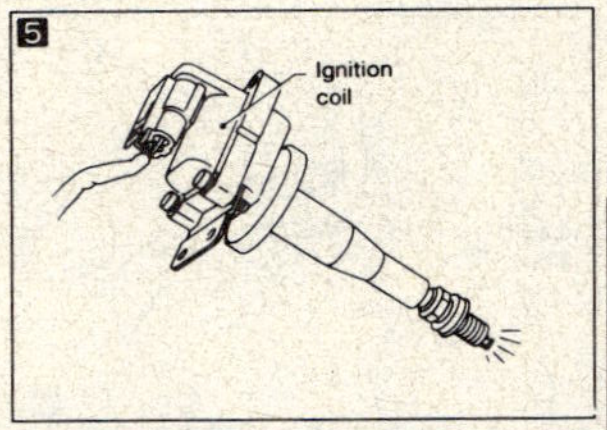

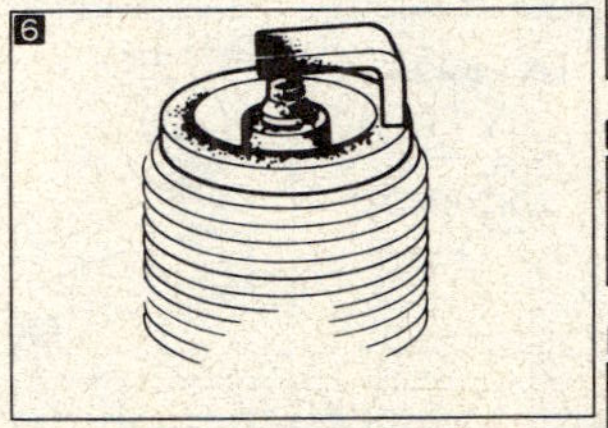

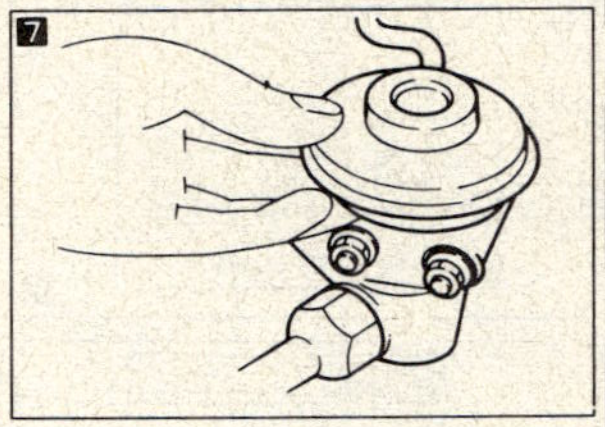

Diagnostic Procedure 37 — Symptom — Hard to Start or Impossible to Start under Normal Conditions (Cont'd)

Ⓐ ↓

5 CHECK IGNITION SPARK.
1. Disconnect ignition coil assembly from rocker cover.
2. Connect a known good spark plug to the ignition coil assembly.
3. Place end of spark plug against a suitable ground and crank engine.
4. Check for spark.

- N.G. → Check ignition coil, power transistor unit and circuits.
- O.K. ↓

6 CHECK SPARK PLUGS.
Remove the spark plugs and check for fouling, etc.
- N.G. → Repair or replace spark plug(s).
- O.K. ↓

7 CHECK E.G.R. CONTROL VALVE.
Check E.G.R. control valve for sticking.
- N.G. → Repair or replace.
- O.K. ↓

8 CHECK E.C.U. HARNESS CONNECTOR.
Check the E.C.U. pin terminals for damage or poor connection of E.C.U. harness connector.
- N.G. → Repair or replace.
- O.K. ↓

9 CHECK E.C.U. POWER SUPPLY AND GROUND CIRCUIT.
- N.G. → Repair or replace.
- O.K. ↓

10 TRY A KNOWN GOOD E.C.U.*

↓

INSPECTION END

*: E.C.U. may be the cause of a problem, but this is rarely the case.

Diagnostic Procedure 38 — Symptom — Hesitation when the Engine is Hot

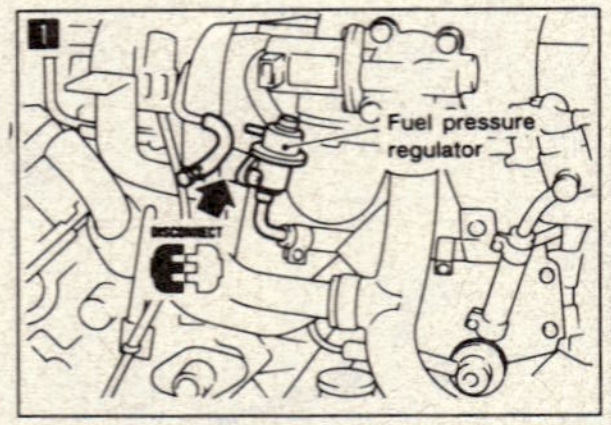

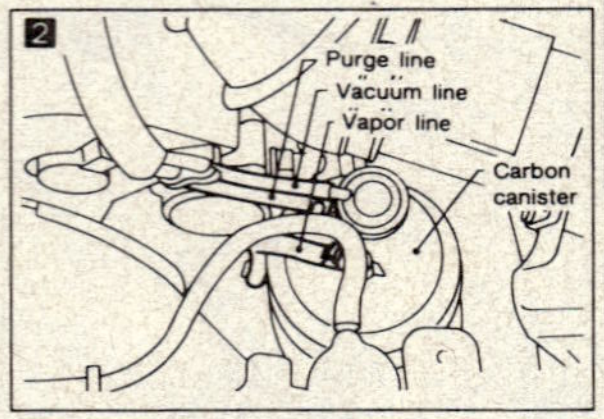

1 CHECK FUEL VAPOR.
1. Disconnect fuel pressure regulator vacuum hose and plug hose.
2. Perform cruise test.
3. Does the hesitation disappear?

Yes → Check fuel properties.

No ↓

2 CHECK CANISTER PURGE.
1. Disconnect canister purge line hose and plug hose.
2. Perform cruise test.
3. Does the hesitation disappear?

Yes → Check purge and vacuum lines.

No ↓

INSPECTION END

Diagnostic Procedure 39 — Symptom — Hesitation when the Engine is Cold

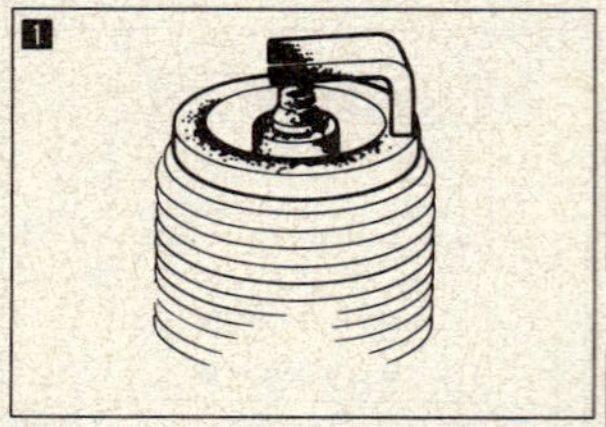

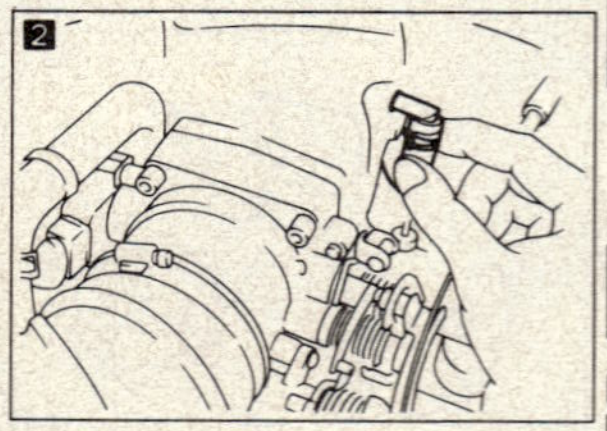

1 CHECK SPARK PLUGS.
Remove spark plugs and check for fouling, etc.

N.G. → Repair or replace spark plug(s).

O.K. ↓

2 CHECK FOR INTAKE AIR LEAK.
When pinching blow-by hose (lowering the blow-by air supply), does the engine speed rise?

Yes → Discover air leak location and repair.

No ↓

3 TRY A KNOWN GOOD AIR FLOW METER.

Trouble is fixed. → Replace air flow meter.

Trouble is not fixed. ↓

4 CHECK FOR INTAKE VALVE DEPOSITS.
If there are deposits on intake valves, remove them.

↓

INSPECTION END

Diagnostic Procedure 40 — Symptom — Hesitation under Normal Conditions

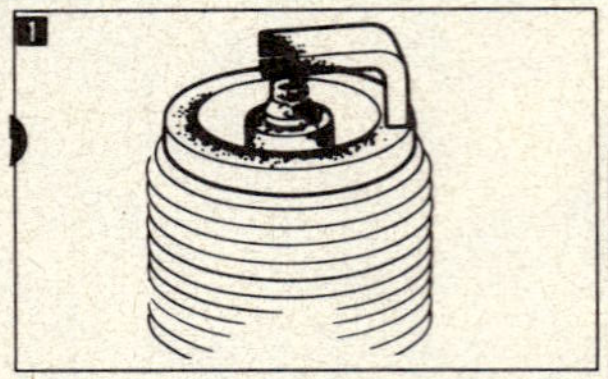

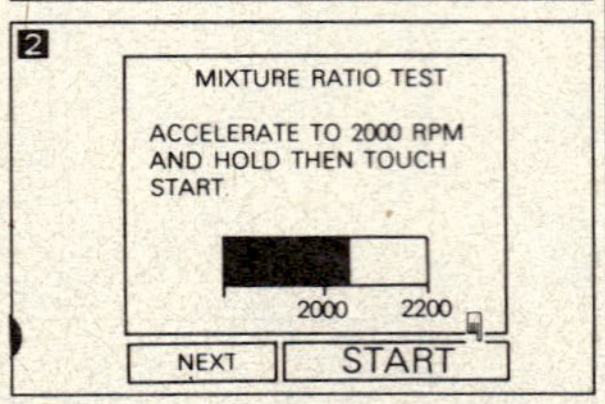

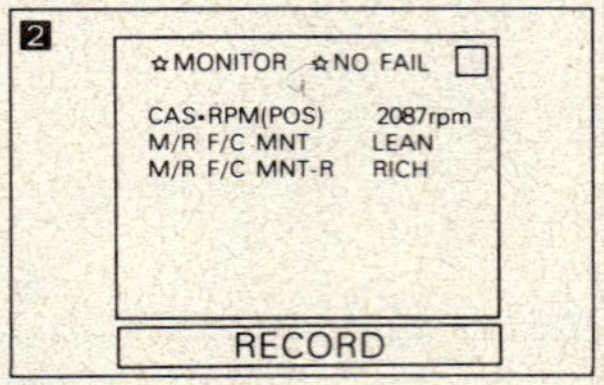

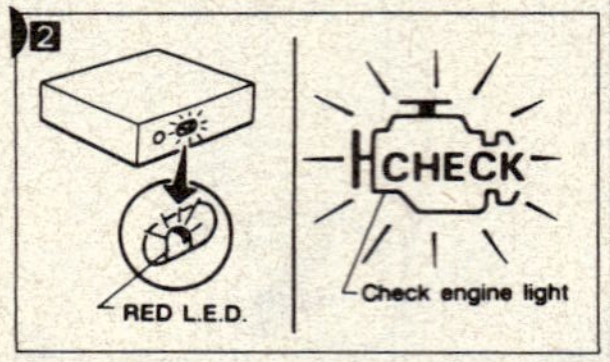

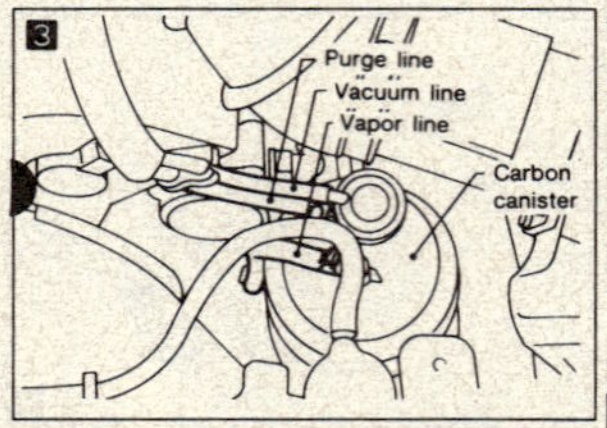

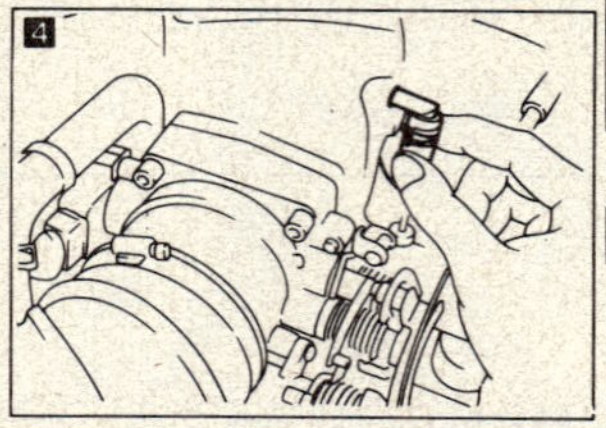

1 CHECK SPARK PLUGS.
Remove spark plugs and check for fouling, etc.

N.G. → Repair or replace spark plug(s).

O.K. ↓

2 CHECK EXHAUST GAS SENSOR.
1. Start engine and warm it up sufficiently.
2. Perform "MIXTURE RATIO TEST" in "FUNCTION TEST" mode.

— OR —

2. See "M/R F/C MNT (right and left sides)" in "DATA MONITOR" mode.
3. Maintaining engine at 2,000 rpm under no-load (with engine warmed up sufficiently.), check to make sure that the monitor fluctuates between "LEAN" and "RICH" more than 5 times during 10 seconds.
RICH→LEAN→RICH→ (1 time)
LEAN→RICH....... (2 times)

— OR —

2. Set "Exhaust gas sensor monitor" in the self-diagnostic Mode II.
3. Maintaining engine at 2,000 rpm under no load, check that check engine light and RED LED on the E.C.U. go ON and OFF more than 5 times during 10 seconds.

Yes → Replace exhaust gas sensor(s).

No ↓

3 CHECK CANISTER PURGE.
1. Disconnect canister purge line hose and plug hose.
2. Perform cruise test.
3. Does the hesitation disappear?

Yes → Check purge and vacuum lines.

No ↓

4 CHECK FOR INTAKE AIR LEAK.
When pinching blow-by hose (lowering the blow-by air supply), does the engine speed rise?

Yes → Discover air leak location and repair.

No ↓

INSPECTION END

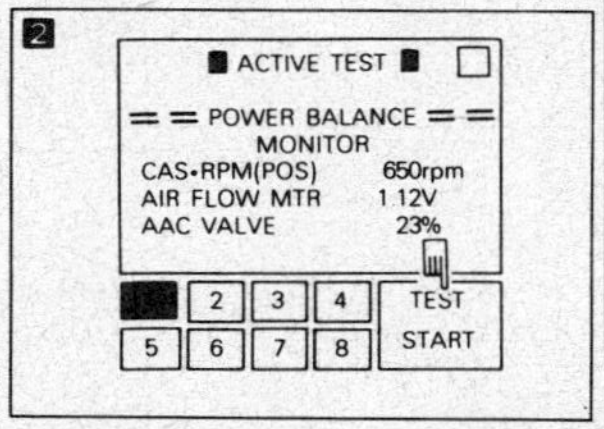

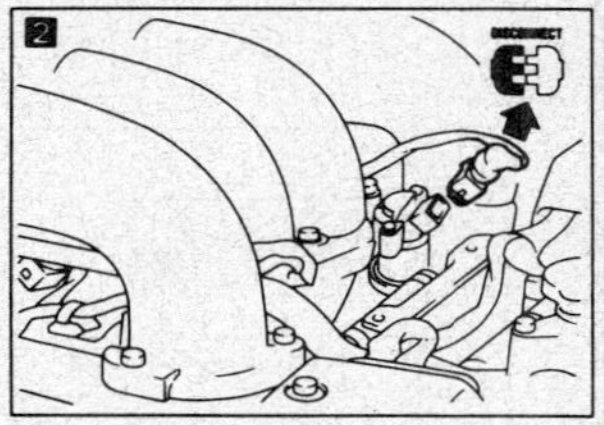

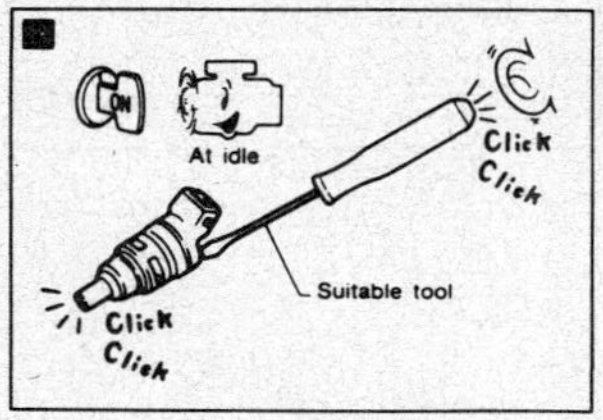

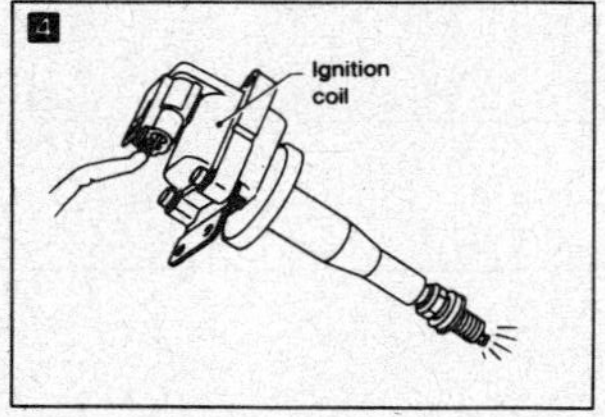

Diagnostic Procedure 41 — Symptom — Engine Stalls when Turning

1 **CHECK FUEL LEVEL.**
Check to see that there is enough fuel in tank.
- N.G. → Fill fuel tank with fuel.
- O.K. ↓

2 **PERFORM POWER BALANCE TEST.**
1. Perform "POWER BALANCE" in "ACTIVE TEST" mode.
2. Is there any cylinder which does not produce a momentary engine speed drop?

— OR —

When disconnecting each injector harness connector one at a time, is there any cylinder which does not produce a momentary engine speed drop?
- No → Go to 5
- Yes ↓

3 **CHECK INJECTOR.**
Does each injector make an operating sound at idle?
- No → Check injector(s) and circuit(s).
- Yes ↓

4 **CHECK IGNITION SPARK.**
1. Disconnect ignition coil assembly from rocker cover.
2. Connect a known good spark plug to the ignition coil assembly.
3. Place end of spark plug against a suitable ground and crank engine.
4. Check for spark.
- N.G. → Check ignition coil, power transistor unit and circuits.
- O.K ↓

(Go to Ⓐ on next page.)

Diagnostic Procedure 41 — Symptom — Engine Stalls when Turning (Cont'd)

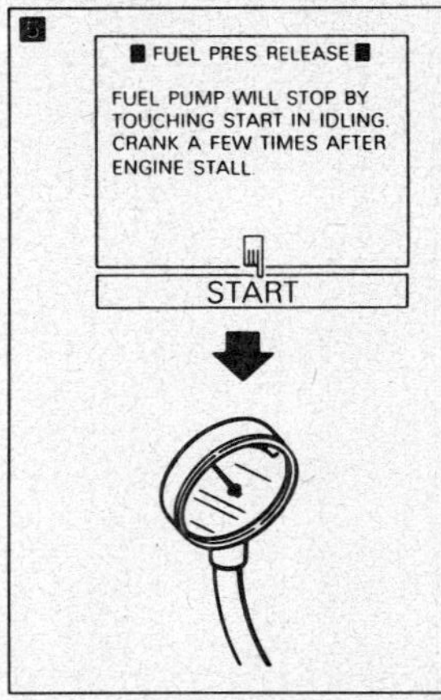

Ⓐ ↓

5 **CHECK FUEL PRESSURE.**
1. Perform "FUEL PRESSURE RELEASE" in "WORK SUPPORT" mode in order to release fuel pressure to zero.
2. Install fuel pressure gauge and check fuel pressure.
   At idle approx. 235 kPa (2.4 kg/cm², 34 psi)
   The moment throttle valve is fully open:
   approx. 294 kPa (3.0 kg/cm², 43 psi)

— OR —

1. Release fuel pressure to zero.
2. Install fuel pressure gauge and check fuel pressure.
- N.G. → Check fuel pressure regulator diaphragm.
- O.K. ↓

6 **CHECK E.C.U. HARNESS CONNECTOR.**
Check the E.C.U. pin terminals for damage or poor connection of E.C.U. harness connector.
- N.G. → Repair or replace.
- O.K. ↓

7 **CHECK E.C.U. POWER SUPPLY AND GROUND CIRCUIT.**
- N.G. → Repair or replace.
- O.K. ↓

8 **TRY A KNOWN GOOD E.C.U.***
↓

INSPECTION END

*: E.C.U. may be the cause of a problem, but this is rarely the case.

Diagnostic Procedure 42 — Symptom — Engine Stalls when the Engine is Hot

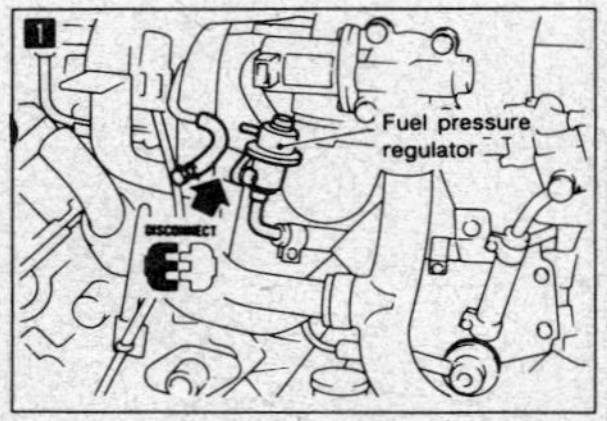

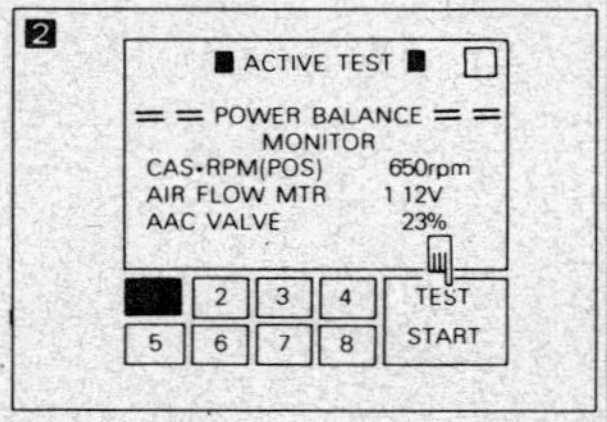

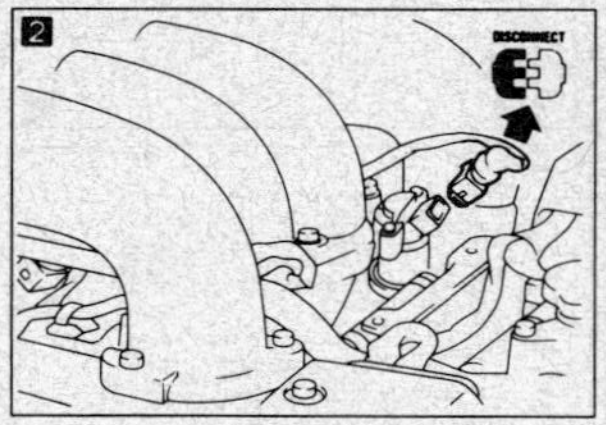

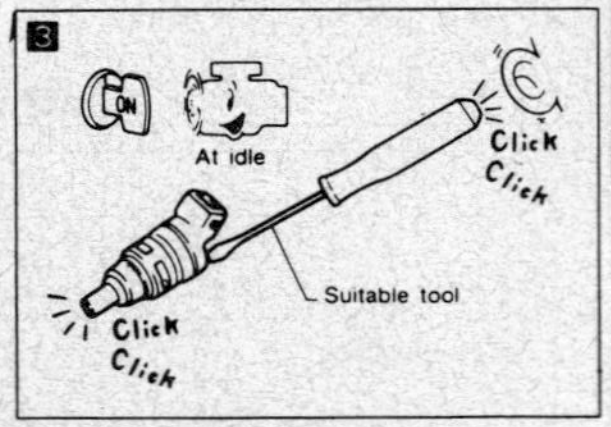

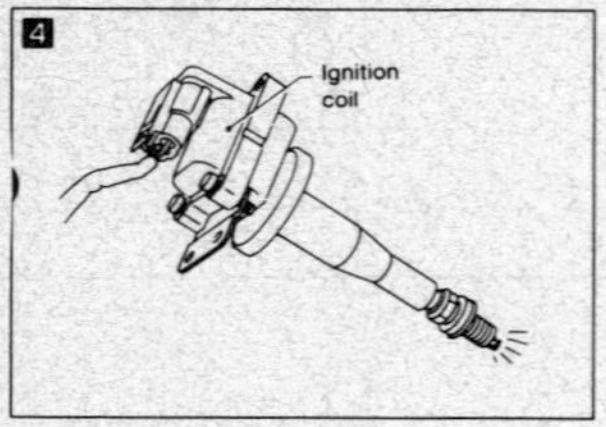

1 CHECK FUEL VAPOR.
1. Disconnect fuel pressure regulator vacuum hose and plug hose.
2. Perform cruise test.
3. Does the engine stall disappear?

- Yes → Check fuel properties.
- No ↓

2 PERFORM POWER BALANCE TEST.
1. Perform "POWER BALANCE" in "ACTIVE TEST" mode.
2. Is there any cylinder which does not produce a momentary engine speed drop?

— OR —

When disconnecting each injector harness connector one at a time, is there any cylinder which does not produce a momentary engine speed drop?

- No → Go to 5
- Yes ↓

3 CHECK INJECTOR.
Does each injector make an operating sound at idle?

- No → Check injector(s) and circuit(s).
- Yes ↓

4 CHECK IGNITION SPARK.
1. Disconnect ignition coil assembly from rocker cover.
2. Connect a known good spark plug to the ignition coil assembly.
3. Place end of spark plug against a suitable ground and crank engine.
4. Check for spark.

- N.G. → Check ignition coil, power transistor unit and their circuits.
- O.K. ↓

(Go to Ⓐ on next page.)

Diagnostic Procedure 42 — Symptom — Engine Stalls when the Engine is Hot (Cont'd)

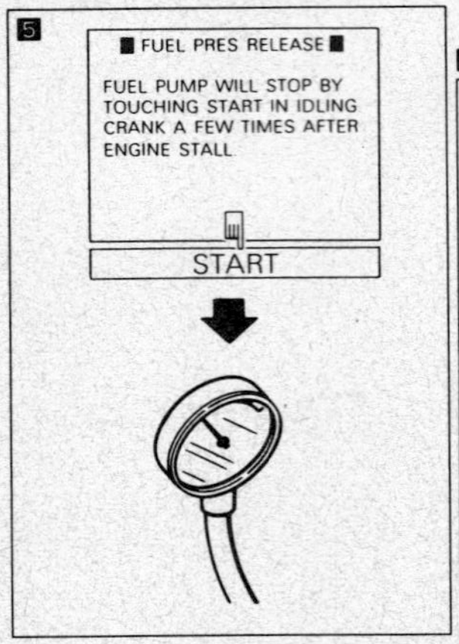

Ⓐ ↓

5 CHECK FUEL PRESSURE.
1. Perform "FUEL PRESSURE RELEASE" in "WORK SUPPORT" mode in order to release fuel pressure to zero.
2. Install fuel pressure gauge and check fuel pressure.
At idle approx. 235 kPa (2.4 kg/cm², 34 psi)
The moment throttle valve is fully open:
approx. 294 kPa (3.0 kg/cm², 43 psi)

— OR —

1. Release fuel pressure to zero.
2. Install fuel pressure gauge and check fuel pressure.

- N.G. → Check fuel pressure regulator diaphragm.
- O.K. ↓

6 CHECK E.C.U. HARNESS CONNECTOR.
Check the E.C.U. pin terminals for damage or poor connection of E.C.U. harness connector.

- N.G. → Repair or replace.
- O.K. ↓

7 CHECK E.C.U. POWER SUPPLY AND GROUND CIRCUIT.

- N.G. → Repair or replace.
- O.K. ↓

8 TRY A KNOWN GOOD E.C.U.*

↓

INSPECTION END

*: E.C.U. may be the cause of a problem, but this is rarely the case.

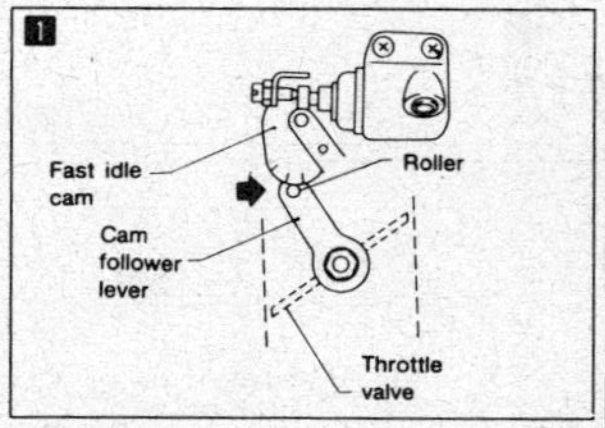

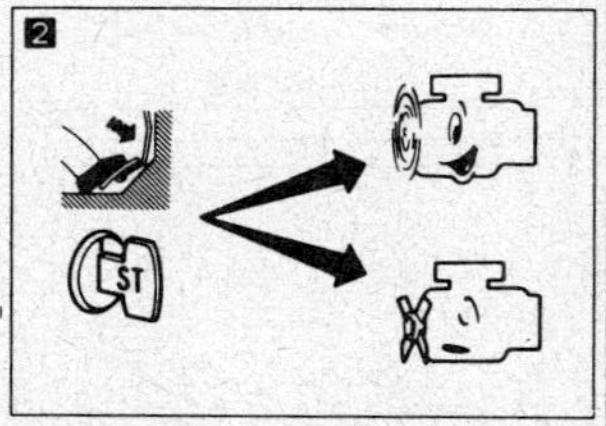

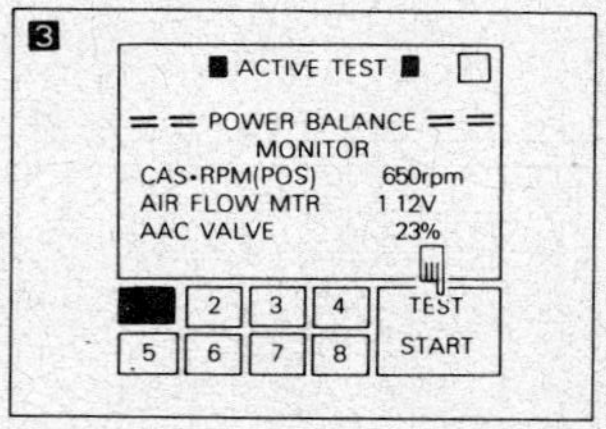

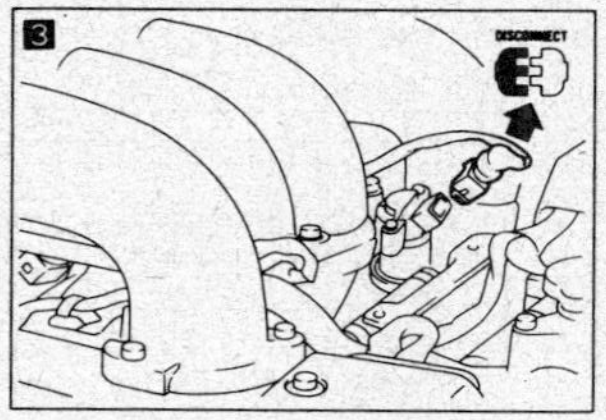

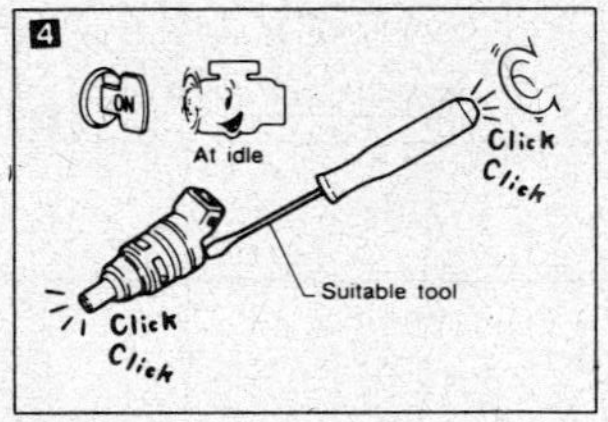

Diagnostic Procedure 43 — Symptom — Engine Stalls when the Engine is Cold

1 CHECK FAST IDLE CAM.
When the engine is cold, does fast idle cam keep cam follower lever in position?

- No → Check fast idle cam adjustment.
- Yes ↓

2 CHECK A.A.C. VALVE.
When the engine is cold, can you start the engine when pressing accelerator pedal fully?

- Yes → Check A.A.C. valve and circuits.
- No ↓

3 PERFORM POWER BALANCE TEST.
1. Perform "POWER BALANCE" in "ACTIVE TEST" mode.
2. Is there any cylinder which does not produce a momentary engine speed drop?

— OR —

When disconnecting each injector harness connector one at a time, is there any cylinder which does not produce a momentary engine speed drop?

- N.G. → Go to 7
- O.K. ↓

4 CHECK INJECTOR.
Does each injector make an operating sound at idle?

- N.G. → Check injector(s) and circuit(s).
- O.K. ↓

(Go to Ⓐ on next page.)

Diagnostic Procedure 43 — Symptom — Engine Stalls when the Engine is Cold (Cont'd)

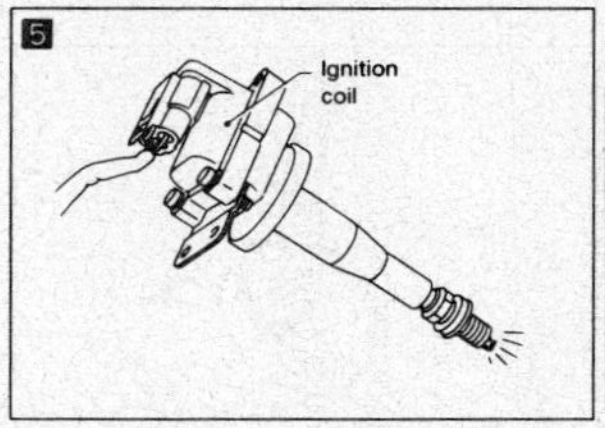

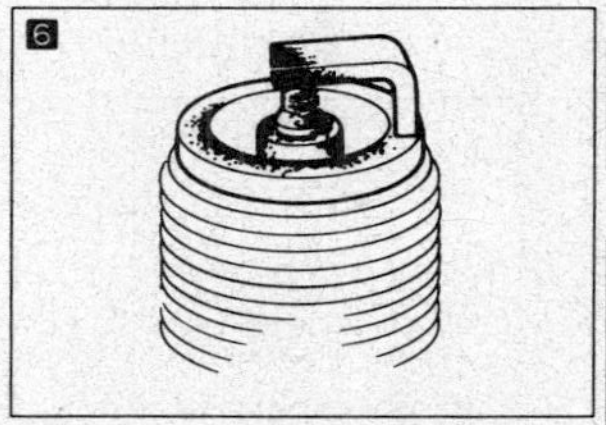

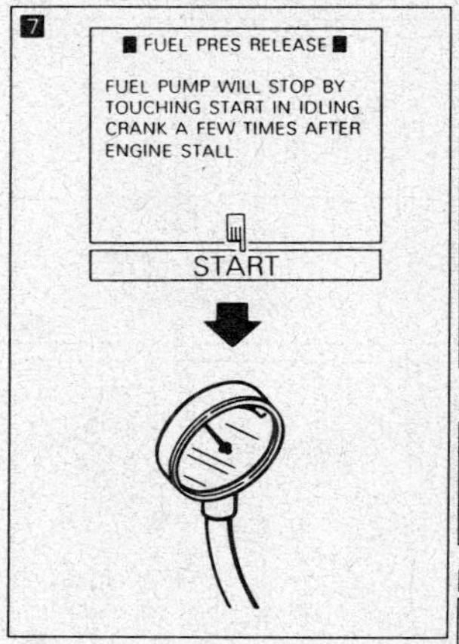

Ⓐ ↓

5 CHECK IGNITION SPARK.
1. Disconnect ignition coil assembly from rocker cover.
2. Connect a known good spark plug to the ignition coil assembly.
3. Place end of spark plug against a suitable ground and crank engine.
4. Check for spark.

- N.G. → Check ignition coil, power transistor unit and circuits.
- O.K. ↓

6 CHECK SPARK PLUGS.
Remove the spark plugs and check for fouling, etc.

- N.G. → Repair or replace spark plug(s).
- O.K. ↓

7 CHECK FUEL PRESSURE.
1. Perform "FUEL PRESSURE RELEASE" in "WORK SUPPORT" mode in order to release fuel pressure to zero.
2. Install fuel pressure gauge and check fuel pressure.
At idle approx. 235 kPa (2.4 kg/cm², 34 psi)
The moment throttle valve is fully open:
approx. 294 kPa (3.0 kg/cm², 43 psi)

— OR —

1. Release fuel pressure to zero.
2. Install fuel pressure gauge and check fuel pressure.

- N.G. → Check fuel pressure regulator diaphragm.
- O.K. ↓

8 CHECK E.C.U. HARNESS CONNECTOR.
Check the E.C.U. pin terminals for damage or poor connection of E.C.U. harness connector.

- N.G. → Repair or replace.
- O.K. ↓

9 CHECK E.C.U. POWER SUPPLY AND GROUND CIRCUIT.

- N.G. → Repair or replace.
- O.K. ↓

10 TRY A KNOWN GOOD E.C.U.*

↓

INSPECTION END

*: E.C.U. may be the cause of a problem, but this is rarely the case.

Diagnostic Procedure 44 — Symptom — Engine Stalls when Stepping on the Accelerator Momentarily

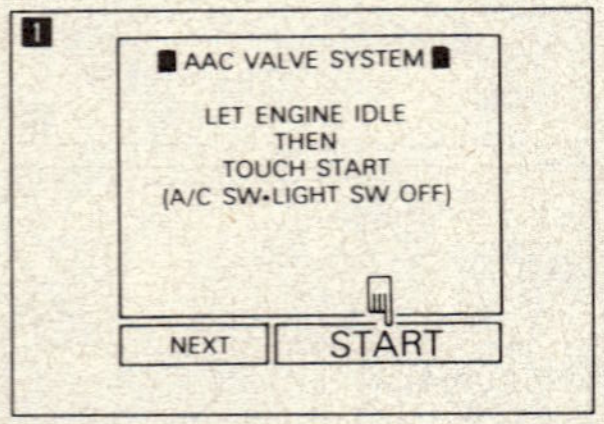

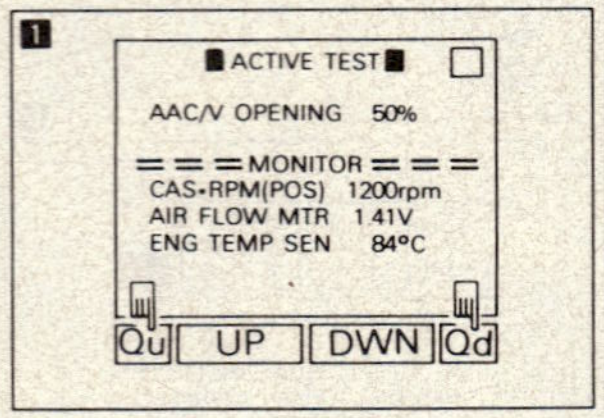

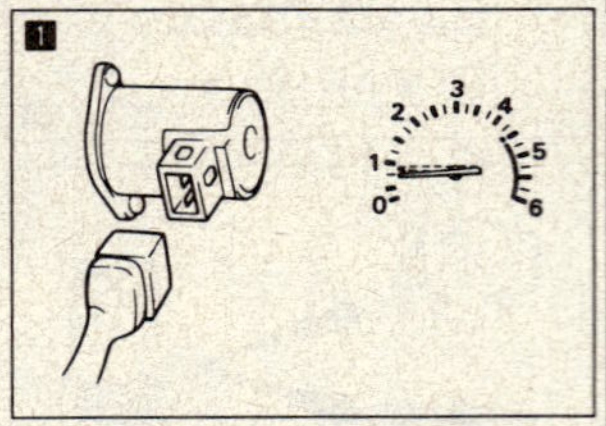

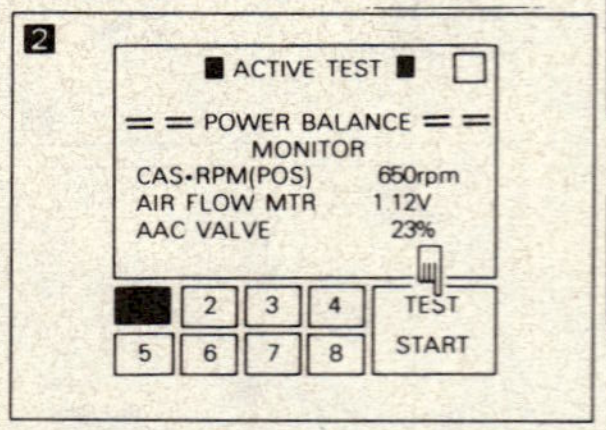

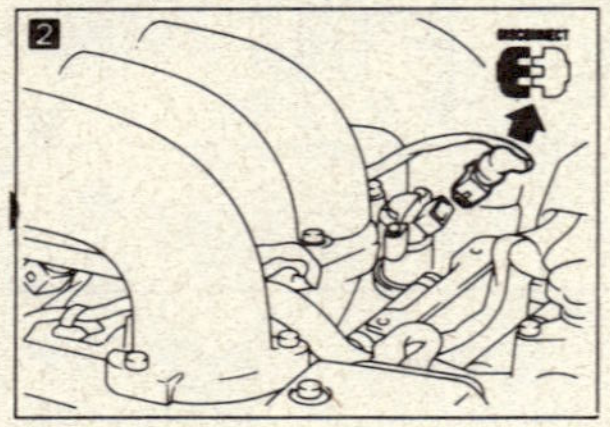

1 CHECK A.A.C. VALVE.

1. Start engine and warm it up sufficiently.
2. Perform "AAC VALVE SYSTEM" in "FUNCTION TEST" mode.

— OR —

1. Select "A.A.C. VALVE OPENING" in "ACTIVE TEST" mode.
2. When touching "Qu" and "Qd", does the engine speed change according to the percent of A.A.C. valve opening?

— OR —

When disconnecting A.A.C. valve harness connector, does the engine speed drop?

- No → Check A.A.C. valve and circuit.
- Yes ↓

2 PERFORM POWER BALANCE TEST.

1. Perform "POWER BALANCE" in "ACTIVE TEST" mode.
2. Is there any cylinder which does not produce a momentary engine speed drop?

— OR —

When disconnecting each injector harness connector one at a time, is there any cylinder which does not produce a momentary engine speed drop?

- No → Go to 5
- Yes ↓

3 CHECK INJECTOR.

Does each injector make an operating sound at idle?

- No → Check injector(s) and their circuit(s).
- Yes ↓

(Go to Ⓐ on next page.)

Diagnostic Procedure 44 — Symptom — Engine Stalls when Stepping on the Accelerator Momentarily (Cont'd)

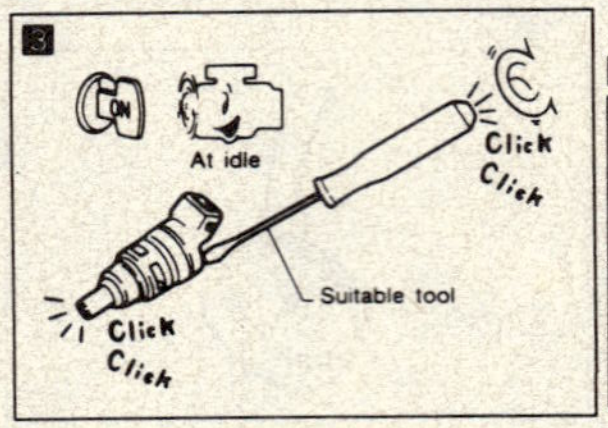

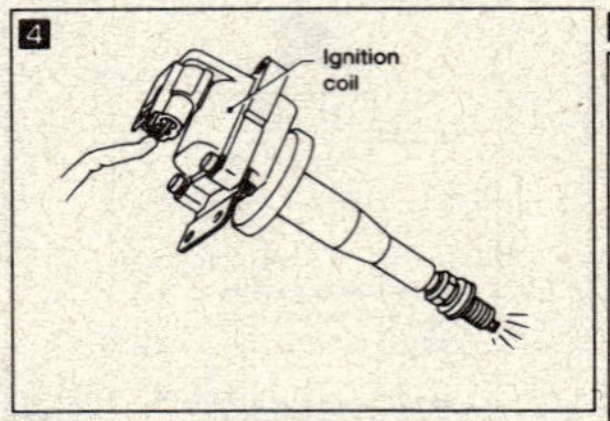

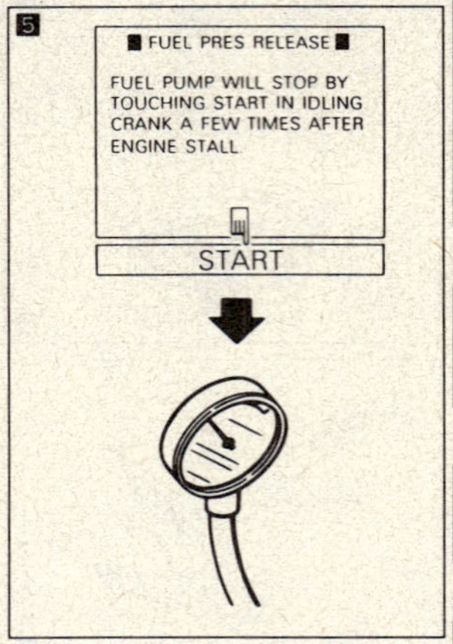

Ⓐ ↓

4 CHECK IGNITION SPARK.

1. Disconnect ignition coil assembly from rocker cover.
2. Connect a known good spark plug to the ignition coil assembly.
3. Place end of spark plug against an earth point with engine cranking.
4. Check for spark.

- N.G. → Check ignition coil, power transistor unit and their circuits.
- O.K. ↓

5 CHECK FUEL PRESSURE.

1. Perform "FUEL PRESSURE RELEASE" in "WORK SUPPORT" mode in order to release fuel pressure to zero.
2. Install fuel pressure gauge and check fuel pressure.
   At idle approx. 235 kPa (2.4 kg/cm², 34 psi)
   The moment throttle valve is fully open: approx. 294 kPa (3.0 kg/cm², 43 psi)

— OR —

1. Release fuel pressure to zero.
2. Install fuel pressure gauge and check fuel pressure.

- N.G. → Check fuel pressure regulator diaphragm.
- O.K. ↓

6 CHECK E.C.U. HARNESS CONNECTOR.

Check the E.C.U. pin terminals for damage or poor connection of E.C.U. harness connector.

- N.G. → Repair or replace.
- O.K. ↓

7 CHECK E.C.U. POWER SUPPLY AND GROUND CIRCUIT.

- N.G. → Repair or replace.
- O.K. ↓

8 TRY A KNOWN GOOD E.C.U.*

↓

INSPECTION END

*: E.C.U. may be the cause of a problem, but this is rarely the case.

Diagnostic Procedure 45 — Symptom — Engine Stalls after Decelerating

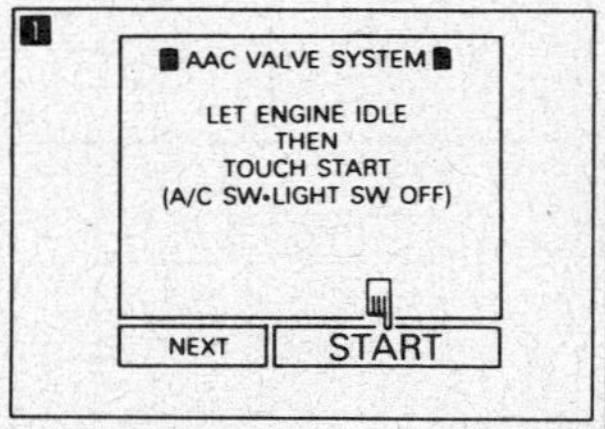

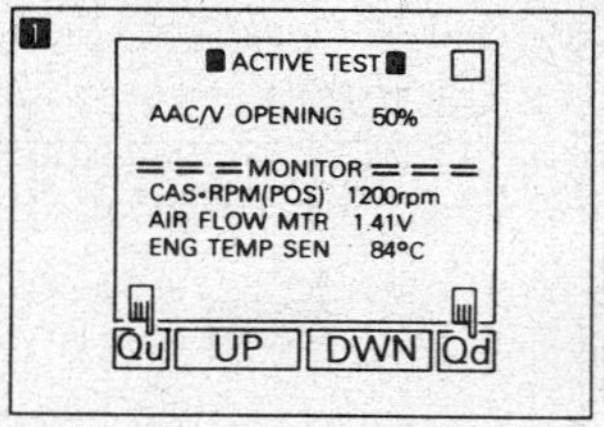

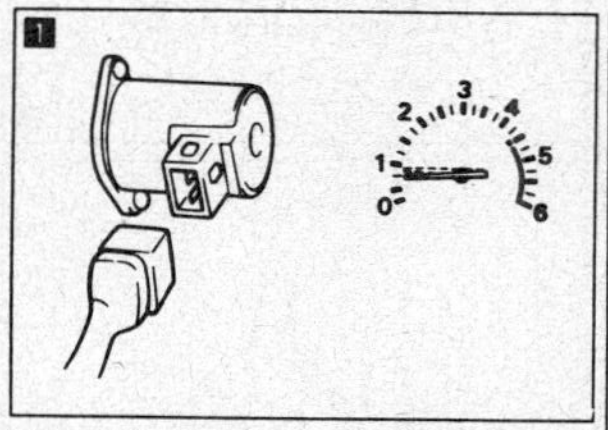

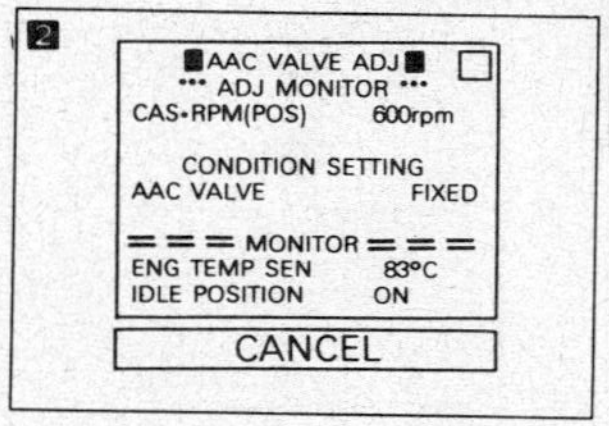

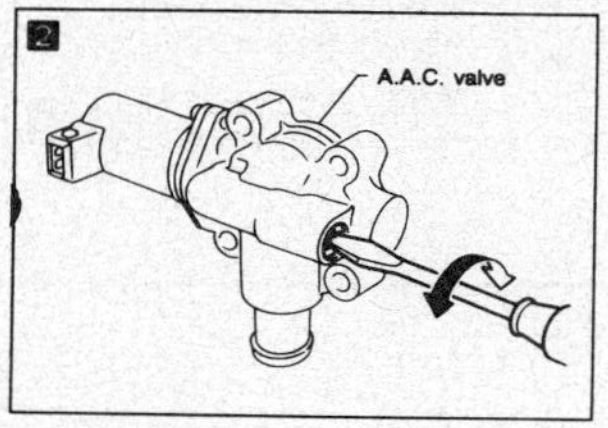

1 CHECK A.A.C. VALVE.

1. Start engine and warm it up sufficiently.
2. Perform "AAC VALVE SYSTEM" in "FUNCTION TEST" mode.

— OR —

1. Select "A.A.C. VALVE OPENING" in "ACTIVE TEST" mode.
2. When touching "Qu" and "Qd", does the engine speed change according to the percent of A.A.C. valve opening?

— OR —

When disconnecting A.A.C. valve harness connector, does the engine speed drop?

No → Check A.A.C. valve and circuit.

Yes ↓

2 CHECK IDLE ADJ. SCREW CLOGGING.

1. Perform "A.A.C. VALVE ADJ" in "WORK SUPPORT" mode.
2. Can you set engine speed at 600 ± 25 rpm (A/T in "N" position) by turning idle adjusting screw?

— OR —

1. Disconnect A.A.C. valve harness connector.
2. Can you set engine speed at 600 ± 25 rpm (A/T in "N" position) by turning idle adjusting screw?

No → Check for A.A.C. valve clogging or throttle chamber clogging.

Yes ↓

(Go to Ⓐ on next page.)

Diagnostic Procedure 45 — Symptom — Engine Stalls after Decelerating (Cont'd)

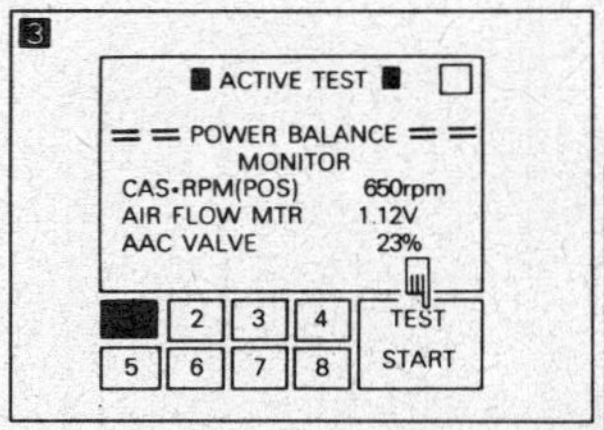

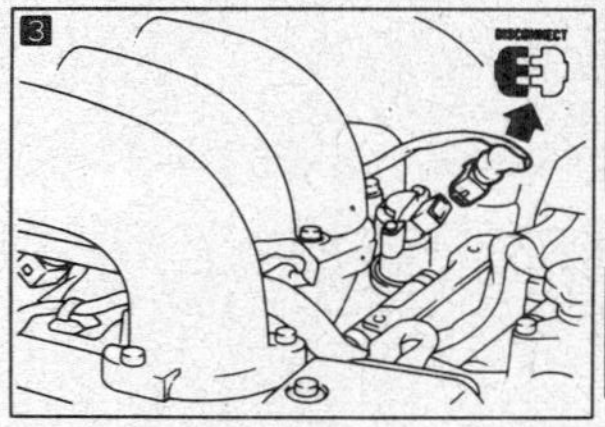

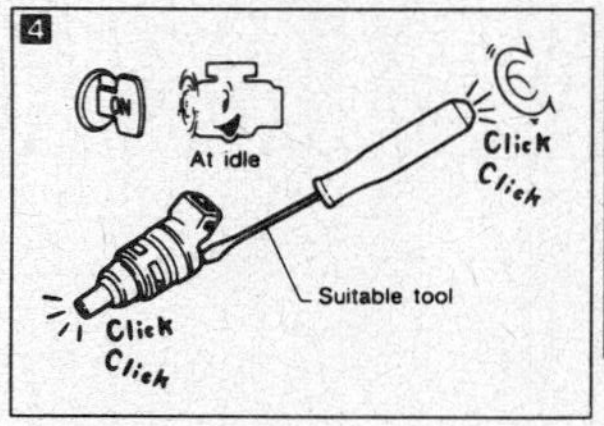

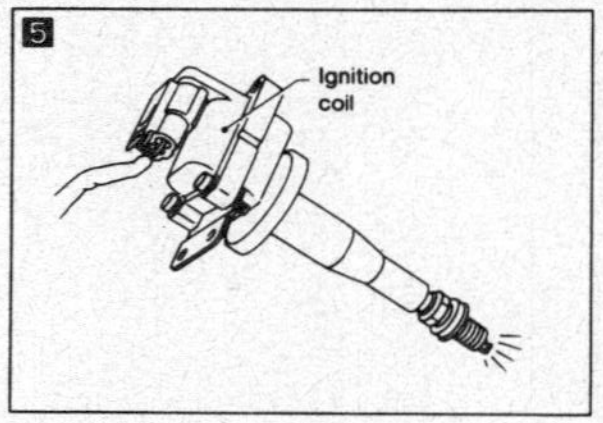

Ⓐ ↓

3 PERFORM POWER BALANCE TEST.

1. Perform "POWER BALANCE" in "ACTIVE TEST" mode.
2. Is there any cylinder which does not produce a momentary engine speed drop?

— OR —

When disconnecting each injector harness connector one at a time, is there any cylinder which does not produce a momentary engine speed drop?

No → Go to 6

Yes ↓

4 CHECK INJECTOR.

1. Does each injector make an operating sound at idle?

No → Check injector(s) and circuit(s).

Yes ↓

5 CHECK IGNITION SPARK.

1. Disconnect ignition coil assembly from rocker cover.
2. Connect a known good spark plug to the ignition coil assembly.
3. Place end of spark plug against a suitable ground and crank engine.
4. Check for spark.

N.G. → Check ignition coil, power transistor unit and circuits.

O.K. ↓

(Go to Ⓑ on next page.)

6
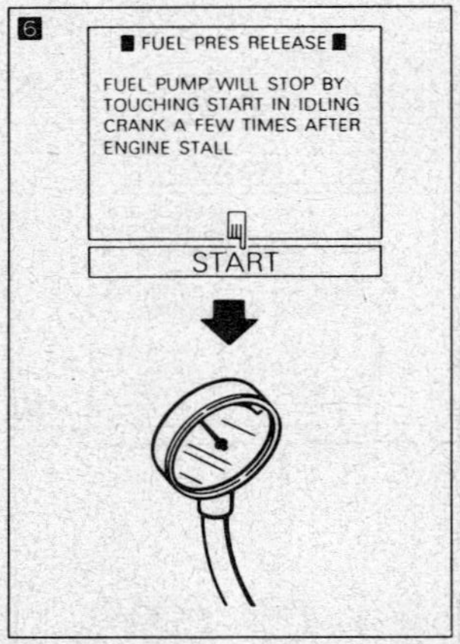

7
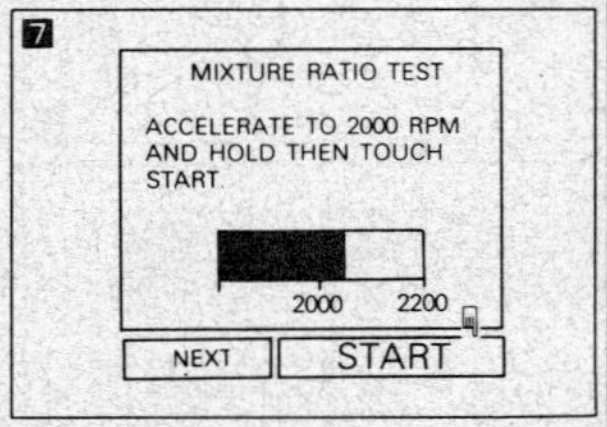

7
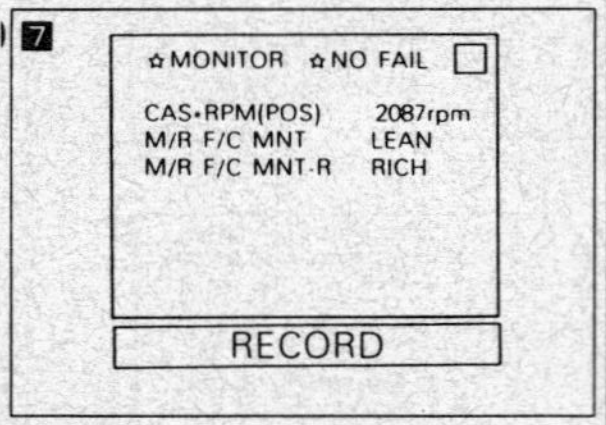

7
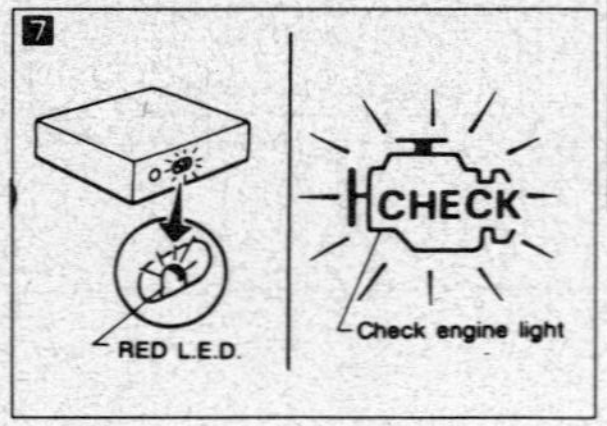

Diagnostic Procedure 45 — Symptom — Engine Stalls after Decelerating (Cont'd)

Ⓑ

6 CHECK FUEL PRESSURE.

1. Perform "FUEL PRESSURE RELEASE" in "WORK SUPPORT" mode in order to release fuel pressure to zero.
2. Install fuel pressure gauge and check fuel pressure.
   At idle approx. 235 kPa (2.4 kg/cm², 34 psi)
   The moment throttle valve is fully open:
   approx. 294 kPa (3.0 kg/cm², 43 psi)

— OR —

1. Release fuel pressure to zero.
2. Install fuel pressure gauge and check fuel pressure.

N.G. → Check fuel pressure regulator diaphragm.

O.K. ↓

7 CHECK EXHAUST GAS SENSOR.

1. Start engine and warm it up sufficiently.
2. Perform "MIXTURE RATIO TEST" in "FUNCTION TEST" mode.

— OR —

2. See "M/R F/C MNT (right and left sides)" in "DATA MONITOR" mode.
3. Maintaining engine at 2,000 rpm under no-load (with engine warmed up sufficiently.), check to make sure that the monitor fluctuates between "LEAN" and "RICH" more than 5 times during 10 seconds.
   RICH→LEAN→RICH→
   1 time **2 times**
   LEAN→RICH.......

— OR —

2. Set "Exhaust gas sensor monitor" in the self-diagnostic Mode II.
3. Maintaining engine at 2,000 rpm under no-load, check that check engine light and RED LED on the E.C.U. go ON and OFF more than 5 times during 10 seconds.

N.G. → Replace exhaust gas sensor(s).

O.K. ↓

(Go to Ⓒ on next page.)

Diagnostic Procedure 45 — Symptom — Engine Stalls after Decelerating (Cont'd)

Ⓒ

8 CHECK E.C.U. HARNESS CONNECTOR.
Check the E.C.U. pin terminals for damage or poor connection of E.C.U. harness connector.

N.G. → Repair or replace.

O.K. ↓

9 CHECK E.C.U. POWER SUPPLY AND GROUND CIRCUIT.

N.G. → Repair or replace.

O.K. ↓

10 TRY A KNOWN GOOD E.C.U.*

↓

INSPECTION END

*: E.C.U. may be the cause of a problem, but this is rarely the case.

Diagnostic Procedure 46 — Symptom — Engine Stalls when Accelerating or when Driving at Constant Speed

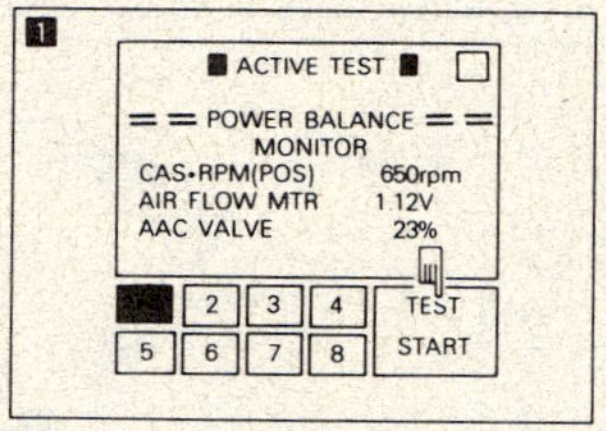

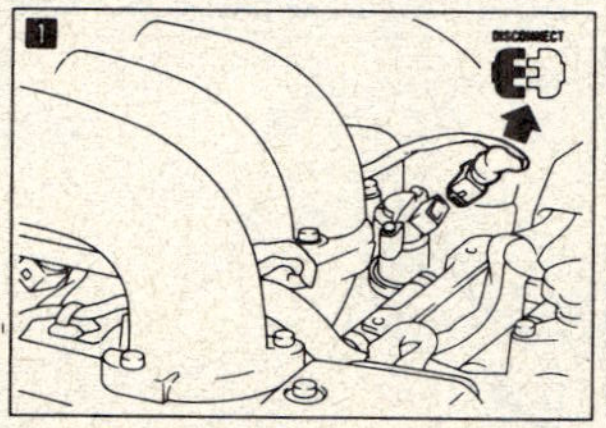

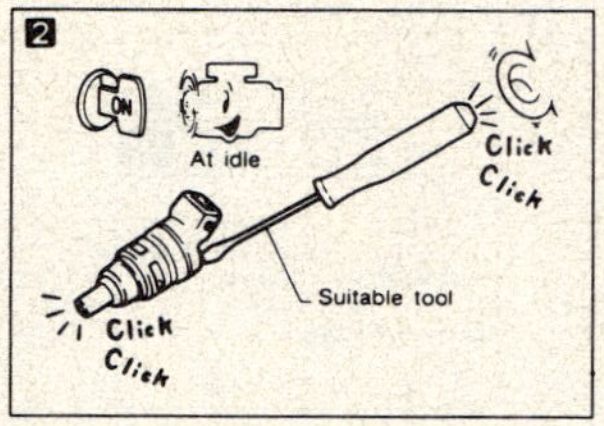

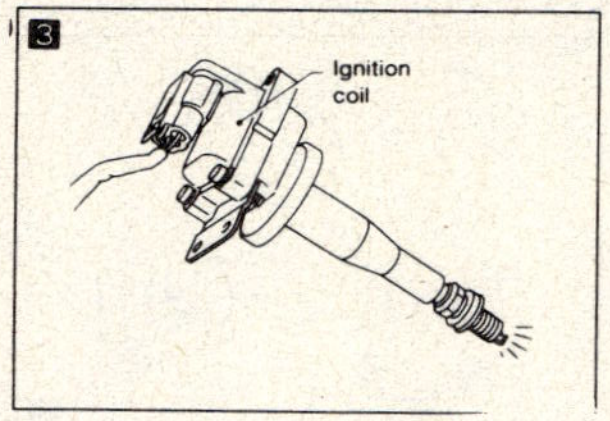

1 **PERFORM POWER BALANCE TEST.**
1. Perform "POWER BALANCE" in "ACTIVE TEST" mode.
2. Is there any cylinder which does not produce a momentary engine speed drop?

— OR —

When disconnecting each injector harness connector one at a time, is there any cylinder which does not produce a momentary engine speed drop?

- No → Go to **4**
- Yes ↓

2 **CHECK INJECTOR.**
Does each injector make an operating sound at idle?

- No → Check injector(s) and circuit(s).
- Yes ↓

3 **CHECK IGNITION SPARK.**
1. Disconnect ignition coil assembly from rocker cover.
2. Connect a known good spark plug to the ignition coil assembly.
3. Place end of spark plug against a suitable ground and crank engine.
4. Check for spark.

- N.G. → Check ignition coil, power transistor unit and circuits.
- O.K. ↓

(Go to Ⓐ on next page.)

Diagnostic Procedure 46 — Symptom — Engine Stalls when Accelerating or when Driving at Constant Speed (Cont'd)

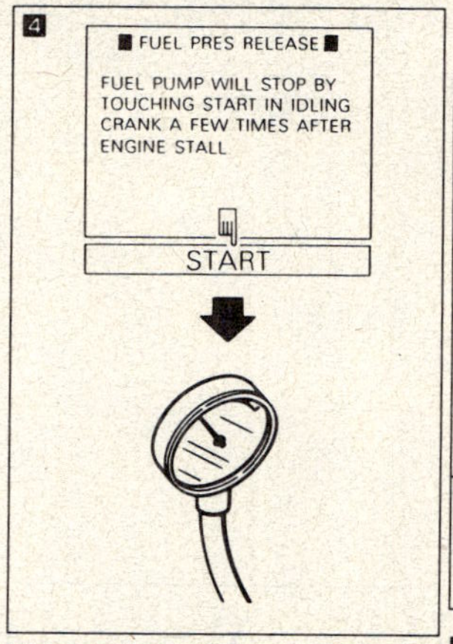

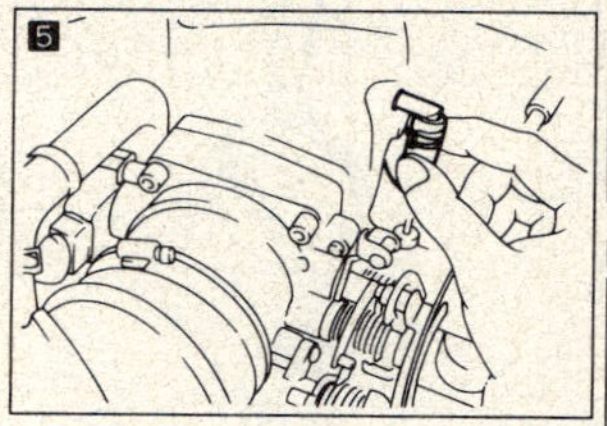

Ⓐ ↓

4 **CHECK FUEL PRESSURE.**
1. Perform "FUEL PRESSURE RELEASE" in "WORK SUPPORT" mode in order to release fuel pressure to zero.
2. Install fuel pressure gauge and check fuel pressure.
At idle approx. 235 kPa (2.4 kg/cm², 34 psi)
The moment throttle valve is fully open:
approx. 294 kPa (3.0 kg/cm², 43 psi)

— OR —

1. Release fuel pressure to zero.
2. Install fuel pressure gauge and check fuel pressure.

- N.G. → Check fuel pump, circuit and fuel pressure regulator.
- O.K. ↓

5 **CHECK FOR INTAKE AIR LEAK.**
When pinching blow-by hose (lowering the blow-by air supply), does the engine speed rise?

- Yes → Discover air leak location and repair.
- No ↓

6 **CHECK E.C.U. HARNESS CONNECTOR.**
Check the E.C.U. pin terminals for damage or poor connection of E.C.U. harness connector.

- N.G. → Repair or replace.
- O.K. ↓

7 **CHECK E.C.U. POWER SUPPLY AND GROUND CIRCUIT.**

- Yes → Repair or replace.
- No ↓

8 **TRY A KNOWN GOOD E.C.U.***

↓

INSPECTION END

*: E.C.U. may be the cause of a problem, but this is rarely the case.

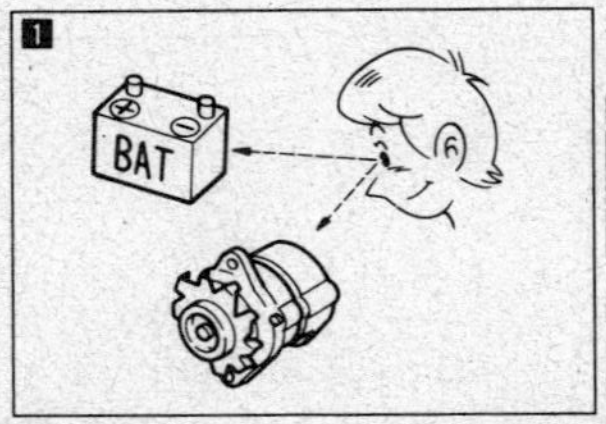

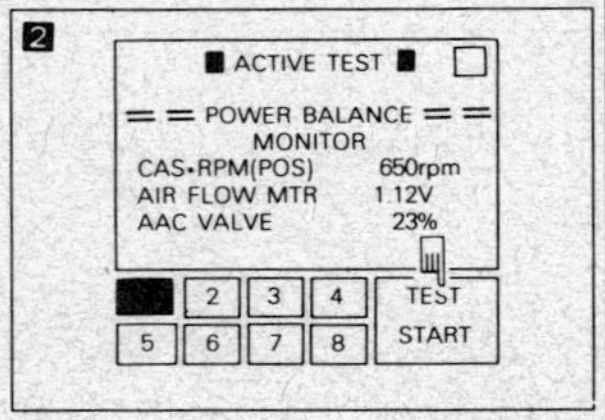

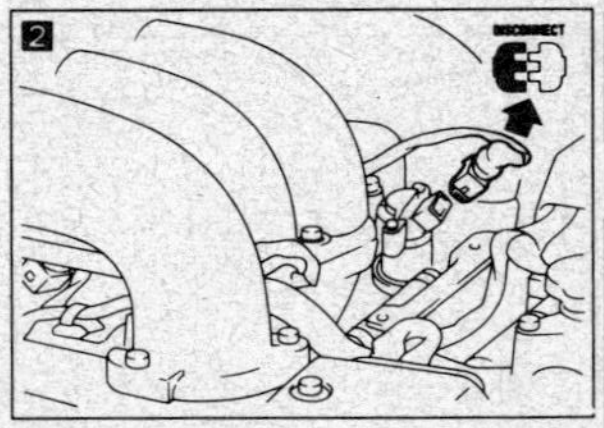

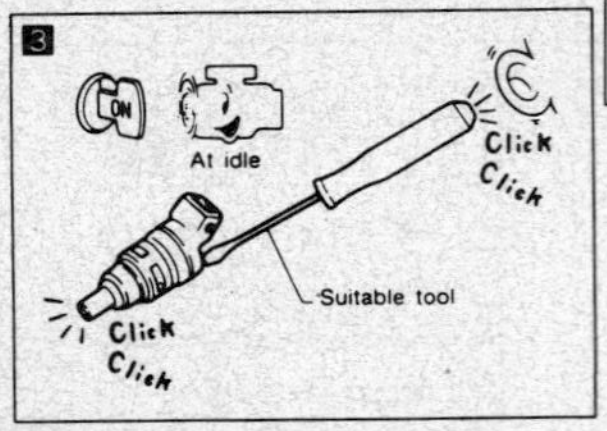

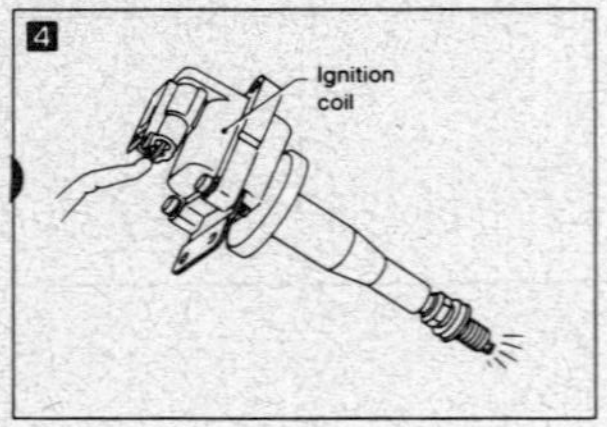

Diagnostic Procedure 47 — Symptom — Engine Stalls when the Electrical Load is Heavy

1 CHECK BATTERY AND ALTERNATOR.
Check battery and alternator condition.
- N.G. → Repair or replace.
- O.K. ↓

2 PERFORM POWER BALANCE TEST.
1. Perform "POWER BALANCE" in "ACTIVE TEST" mode.
2. Is there any cylinder which does not produce a momentary engine speed drop?

— OR —

When disconnecting each injector harness connector one at a time, is there any cylinder which does not produce a momentary engine speed drop?
- No → Go to 5
- Yes ↓

3 CHECK INJECTOR.
Does each injector make an operating sound at idle?
- No → Check injector(s) and circuit(s).
- Yes ↓

4 CHECK IGNITION SPARK.
1. Disconnect ignition coil assembly from rocker cover.
2. Connect a known good spark plug to the ignition coil assembly.
3. Place end of spark plug against a suitable ground and crank engine.
4. Check for spark.
- N.G. → Check ignition coil, power transistor unit and circuits.
- O.K. ↓

(Go to Ⓐ on next page.)

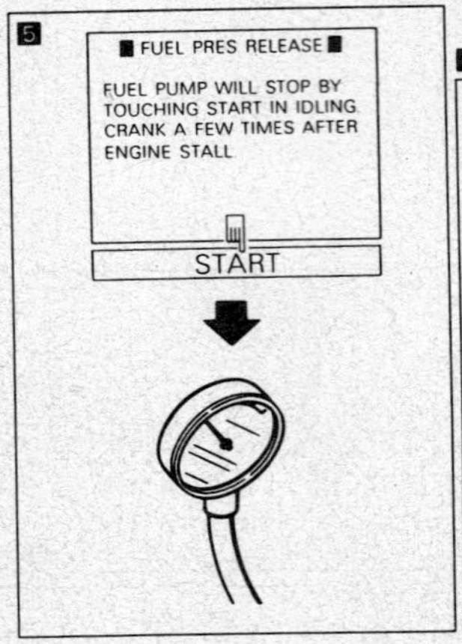

Diagnostic Procedure 47 — Symptom — Engine Stalls when the Electrical Load is Heavy (Cont'd)

Ⓐ

5 CHECK FUEL PRESSURE.
1. Perform "FUEL PRESSURE RELEASE" in "WORK SUPPORT" mode in order to release fuel pressure to zero.
2. Install fuel pressure gauge and check fuel pressure.
At idle approx. 235 kPa (2.4 kg/cm², 34 psi)
The moment throttle valve is fully open:
approx. 294 kPa (3.0 kg/cm², 43 psi)

— OR —

1. Release fuel pressure to zero.
2. Install fuel pressure gauge and check fuel pressure.
- N.G. → Check fuel pressure regulator diaphragm.
- O.K. ↓

6 CHECK E.C.U. HARNESS CONNECTOR.
Check the E.C.U. pin terminals for damage or poor connection of E.C.U. harness connector.
- N.G. → Repair or replace.
- O.K. ↓

7 CHECK E.C.U. POWER SUPPLY AND GROUND CIRCUIT.
- N.G. → Repair or replace.
- O.K. ↓

8 TRY A KNOWN GOOD E.C.U.*

↓

INSPECTION END

*: E.C.U. may be the cause of a problem, but this is rarely the case.

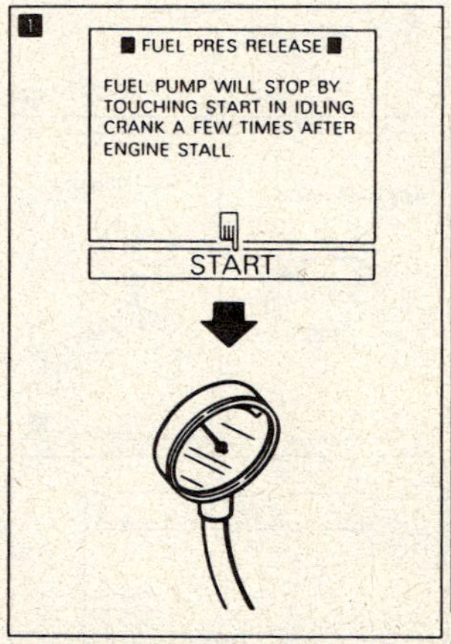

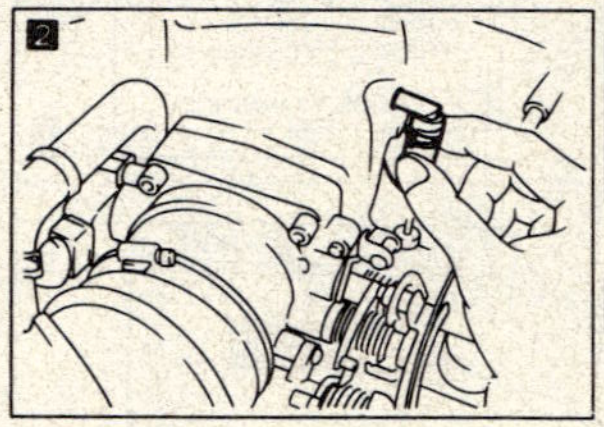

Diagnostic Procedure 48 — Symptom — Lack of Power and Stumble

1

CHECK FUEL PRESSURE.
1. Perform "FUEL PRESSURE RELEASE" in "WORK SUPPORT" mode in order to release fuel pressure to zero.
2. Install fuel pressure gauge and check fuel pressure.
At idle approx. 235 kPa (2.4 kg/cm², 34 psi)
The moment throttle valve is fully open:
approx. 294 kPa (3.0 kg/cm², 43 psi)

— OF —

1. Release fuel pressure to zero.
2. Install fuel pressure gauge and check fuel pressure.

N.G. → Check fuel pressure regulator diaphragm.

O.K. ↓

2

CHECK FOR INTAKE AIR LEAK.
When pinching blow-by hose (lowering the blow-by air supply), does the engine speed rise?

Yes → Discover air leak location and repair.

No ↓

INSPECTION END

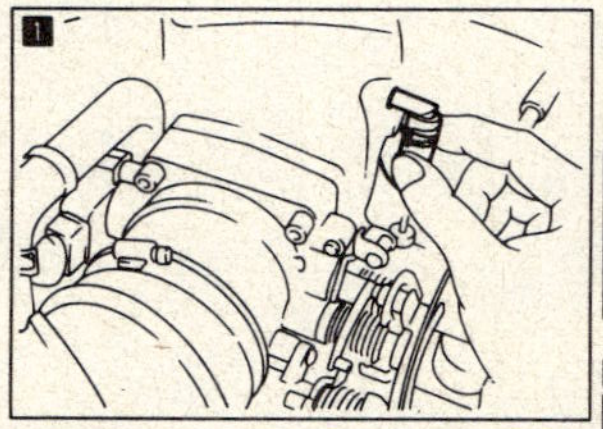

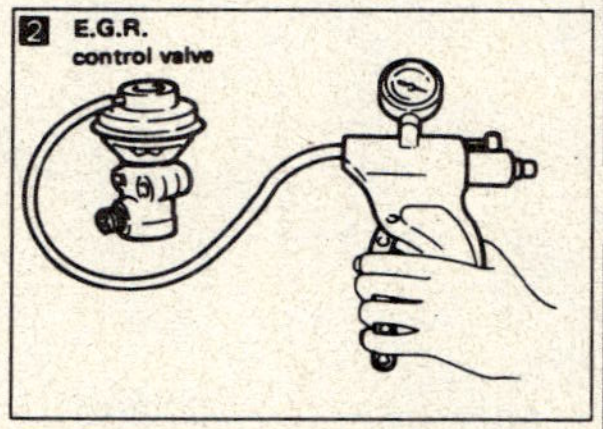

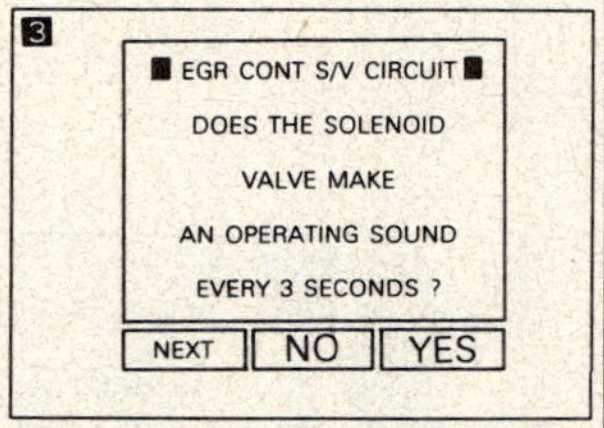

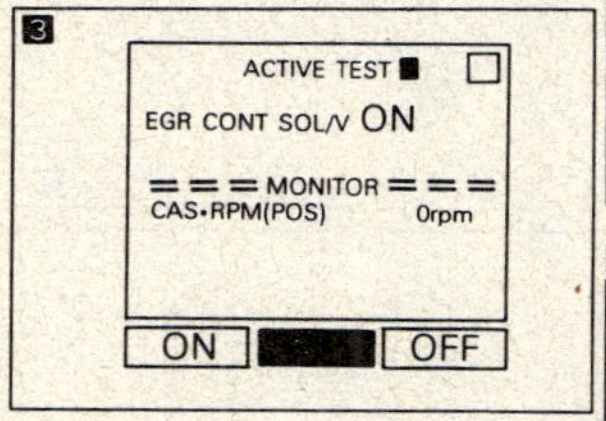

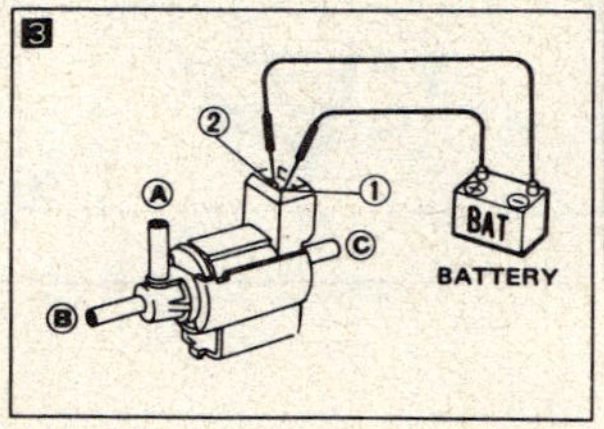

Diagnostic Procedure 49 — Symptom — Detonation

1

CHECK FOR INTAKE AIR LEAK.
When pinching blow-by hose (lowering the blow-by air supply), does the engine rpm rise?

Yes → Discover air leak location and repair.

No ↓

2

CHECK E.G.R. OPERATION.
1. Apply vacuum directly to the E.G.R. valve using a handy vacuum pump.
2. Check to see that the engine runs rough or dies.

No → Check E.G.R. valve for sticking.

Yes ↓

3

CHECK E.G.R. CONTROL SOLENOID VALVE.
1. Turn ignition switch "ON".
2. Perform "EGR CONT S/V CIRCUIT" in "FUNCTION TEST" mode.

— OR —

1. Turn ignition switch "ON".
2. Select "E.G.R. CONT SOL VALVE" in "ACTIVE TEST" mode.
3. Turn E.G.R. control solenoid valve ON and OFF.
4. Check operating sound.

— OR —

1. Disconnect E.G.R. control solenoid valve harness connector.
2. Supply E.G.R. control solenoid valve terminals with battery current and check operating sound.

N.G. → Check solenoid valve and circuit.

O.K. ↓

4

CHECK VACUUM HOSES.
Check the following vacuum hoses for clogging, cracks and poor connection.
a) Vacuum hose between E.G.R. control valve and E.G.R. control solenoid valve.
b) Vacuum hose between E.G.R. control solenoid valve and throttle chamber port.
c) Vacuum hose between E.G.R. control solenoid valve and air duct.

N.G. → Repair or replace.

O.K. ↓

(Go to Ⓐ on next page.)

Diagnostic Procedure 49 — Symptom — Detonation (Cont'd)

5

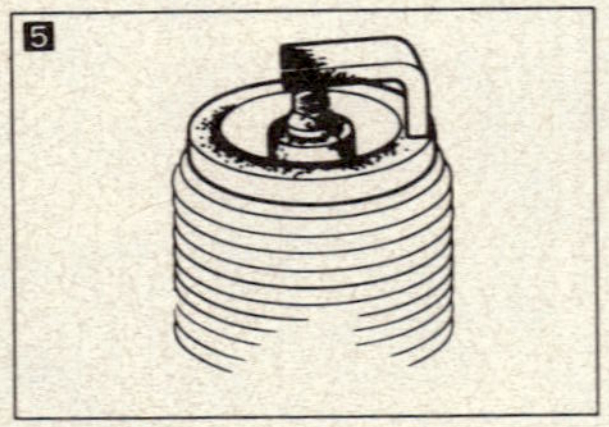

(A)

5 CHECK FOR OIL LEAK TO COMBUSTION CHAMBER.
Remove spark plugs and check for fouling with oil.

- Yes → Check pistons, piston rings, valves, valve seats, valve oil seal, engine oil level, etc.
- No → INSPECTION END

Diagnostic Procedure 50 — Symptom — Surge

1

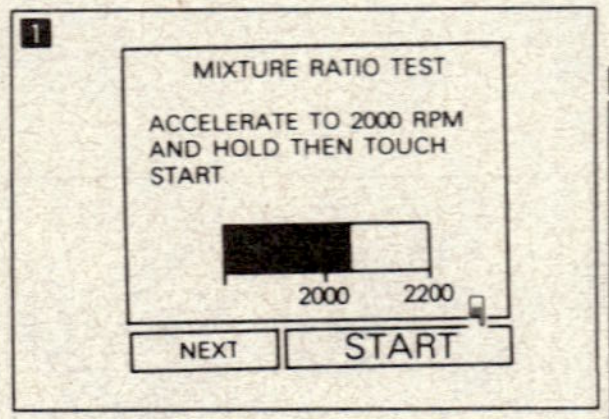

1

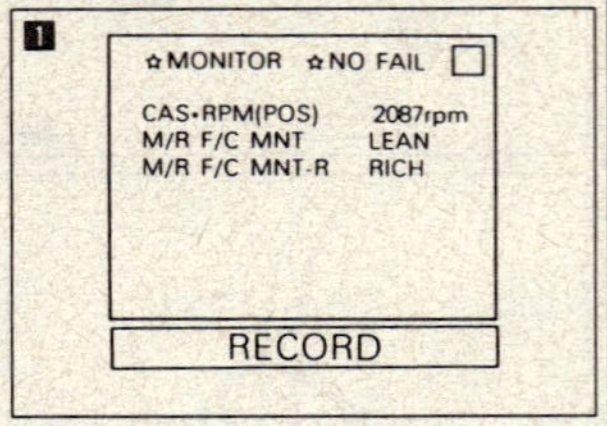

1

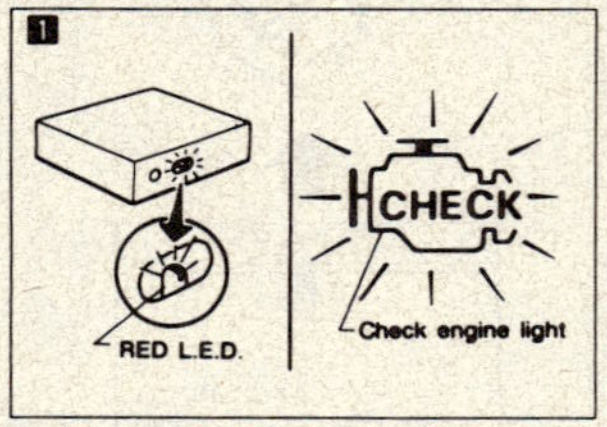

1 CHECK EXHAUST GAS SENSOR.
1. Start engine and warm it up sufficiently.
2. Perform "MIXTURE RATIO TEST" in "FUNCTION TEST" mode.

— OR —

2. See "M/R F/C MNT (right and left sides)" in "DATA MONITOR" mode.
3. Maintaining engine at 2,000 rpm under no-load (with engine warmed up sufficiently.), check to make sure that the monitor fluctuates between "LEAN" and "RICH" more than 5 times during 10 seconds.
RICH→LEAN→RICH→ (1 time)
LEAN→RICH....... (2 times)

— OR —

2. Set "Exhaust gas sensor monitor" in the self-diagnostic Mode II.
3. Maintaining engine at 2,000 rpm under no-load, check that check engine light and RED LED on the E.C.U. go ON and OFF more than 5 times during 10 seconds.

- N.G. → Replace exhaust gas sensor(s).
- O.K. ↓

2

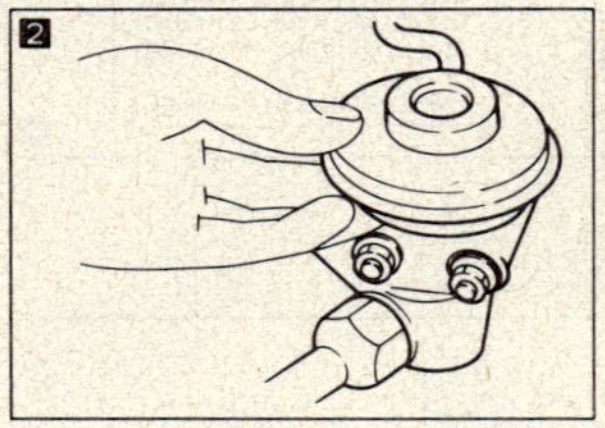

2 CHECK E.G.R. CONTROL VALVE.
Check E.G.R. control valve for sticking.

- N.G. → Repair or replace.
- O.K. ↓

3 TRY A KNOWN GOOD E.C.U.*

- O.K. → INSPECTION END

*: E.C.U. may be the cause of a problem, but this is rarely the case.

Diagnostic Procedure 51 — Symptom — Backfire through the Intake

1

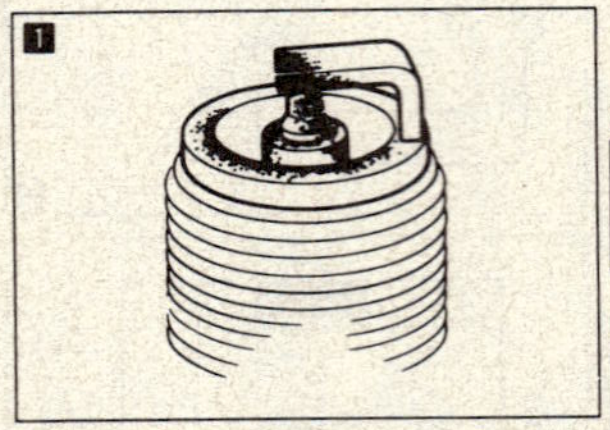

2

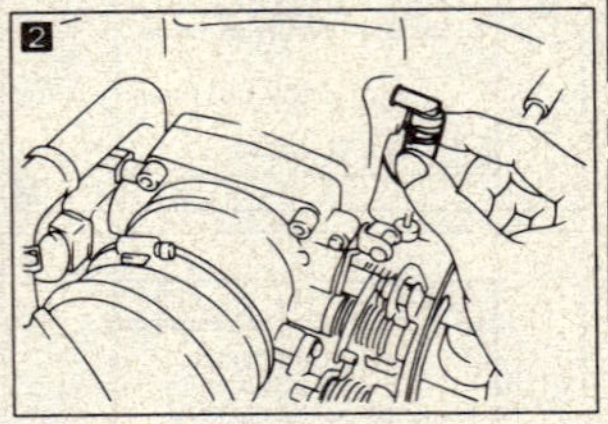

1 CHECK SPARK PLUGS.
Remove the spark plugs and check for fouling, etc.

- N.G. → Repair or replace spark plug(s).
- O.K. ↓

2 CHECK INTAKE AIR LEAK.
When pinching blow-by hose (lowering the blow-by air supply), does the engine speed rise?

- Yes → Discover air leak location and repair.
- No ↓

3 CHECK FOR INTAKE VALVE DEPOSITS.
If there are deposits on intake valves, remove them.

↓ INSPECTION END

Diagnostic Procedure 52 — Symptom — Backfire through the Exhaust

1 CHECK ENGINE TEMPERATURE SENSOR.
Check engine temperature sensor and its circuit.

- N.G. → Replace or repair.
- O.K. → INSPECTION END

1990–92 4.5L ENGINE–SCAN TOOL MODE APPLICATION–Q45

E.C.C.S. COMPONENT PARTS APPLICATION

	E.C.C.S. COMPONENT PARTS \ MODE	WORK SUPPORT	SELF-DIAGNOSTIC RESULTS	DATA MONITOR	ACTIVE TEST	FUNCTION TEST
INPUT	Crank angle sensor		X	X		
	Air flow meter		X	X		
	Engine temperature sensor		X	X	X	
	Exhaust gas sensors		X	X		X
	Vehicle speed sensors		X	X		X
	Throttle sensor	X	X	X		X
	Exhaust gas temperature sensor*		X	X		
	Detonation sensor		X			
	Ignition switch (start signal)			X		X
	Air conditioner switch			X		
	Neutral switch			X		X
	Power steering oil pressure switch			X		X
	Battery			X		
	A/T signal		X			
OUTPUT	Injectors		X	X	X	X
	Power transistor (ignition timing)		X (Ignition signal)	X	X	X
	A.A.C. valve	X		X	X	X
	Valve timing control solenoid valve			X	X	X
	E.G.R. control solenoid valve			X	X	X
	Air conditioner relay			X		
	Fuel pump relay	X		X	X	X

*: The E.C.C.S. component part marked * is applicable to vehicles for California only.
X: Applicable

FUNCTION

Diagnostic mode	Function
Work support	This mode enables a technician to adjust some devices faster and more accurately by following the indications on the CONSULT unit.
Self-diagnostic results	Self-diagnostic results can be read and erased quickly.
Data monitor	Input/Output data in the control unit can be read.
Active test	Mode in which CONSULT drives some actuators apart from the control units and also shifts some parameters in a specified range.
E.C.U. part numbers	E.C.U. part numbers can be read.
Function test	Conducted by CONSULT instead of a technician to determine whether each system is "OK" or "NG".

1990–92 4.5L ENGINE–SCAN TOOL WORK SUPPORT MODE–Q45

WORK SUPPORT MODE

WORK ITEM	CONDITION	USAGE
THROTTLE SENSOR ADJUSTMENT	CHECK THE THROTTLE SENSOR SIGNAL. ADJUST IT TO THE SPECIFIED VALUE BY ROTATING THE SENSOR BODY UNDER THE FOLLOWING CONDITIONS. • IGN SW "ON" • ENG NOT RUNNING • ACC PEDAL NOT PRESSED	When adjusting throttle sensor initial position
AAC VALVE ADJUSTMENT	SET ENGINE RPM AT THE SPECIFIED VALUE UNDER THE FOLLOWING CONDITIONS. • ENGINE WARMED UP • NO-LOAD	When adjusting idle speed
FUEL PRESSURE RELEASE	• FUEL PUMP WILL STOP BY TOUCHING "START" WHEN IDLING. CRANK A FEW TIMES AFTER ENGINE STALLS.	When releasing fuel pressure from fuel line

1990–92 4.5L ENGINE—SCAN TOOL SELF-DIAGNOSTIC RESULTS MODE—Q45

SELF-DIAGNOSTIC RESULTS MODE

DIAGNOSTIC ITEM	DIAGNOSTIC ITEM IS DETECTED WHEN	CHECK ITEM (REMEDY)
CRANK ANGLE SENSOR**	• Either 1° or 90° signal is not entered for the first few seconds during engine cranking. • Either 1° or 90° signal is not input often enough while the engine speed is higher than the specified rpm.	• Harness and connector (If harness and connector are normal, replace crank angle sensor.)
AIR FLOW METER	• The air flow meter circuit is open or shorted. (An abnormally high or low voltage is entered.)	• Harness and connector (If harness and connector are normal, replace air flow meter.)
ENGINE TEMP SENSOR	• The engine temperature sensor circuit is open or shorted. (An abnormally high or low output voltage is entered.)	• Harness and connector • Engine temperature sensor
CAR SPEED SENSOR	• The vehicle speed sensor circuit is open or shorted.	• Harness and connector • Vehicle speed sensor (reed switch)
TCS SIGNAL***	• The TCS signal circuit is open or shorted. (An abnormally high or low output voltage is entered.)	• Harness and connector • T.C.M.
IGN SIGNAL — PRIMARY**	• The ignition signal in primary circuit is not entered during engine cranking or running.	• Harness and connector • Power transistor unit
CONTROL UNIT	• E.C.U. calculation function is malfunctioning.	(Replace E.C.C.S. control unit.)
EGR SYSTEM*	• E.G.R. control valve does not operate. (E.G.R. control valve spring does not lift.)	• E.G.R. control valve • E.G.R. control solenoid valve
EXH GAS SENSOR EXH GAS SENSOR (R)	• The exhaust gas sensor circuit is open or shorted. (An abnormally high or low output voltage is entered.)	• Harness and connector • Exhaust gas sensor • Fuel pressure • Injectors • Intake air leaks
DETONATION SENSOR	• The detonation sensor circuit is open or shorted. (An abnormally high or low voltage is entered.)	• Harness and connector • Detonation sensor
EXH GAS TEMP SENSOR*	• The exhaust gas temperature sensor circuit is open or shorted. (An abnormally high or low voltage is entered.)	• Harness and connector • Exhaust gas temperature sensor
THROTTLE SENSOR	• The throttle sensor circuit is open or shorted. (An abnormally high or low voltage is entered.)	• Harness and connector • Throttle sensor
INJECTOR LEAK*	• Fuel leaks from injector.	• Injector
TCS THROTTLE SEN***	• The secondary throttle sensor circuit is open or shorted. (An abnormally high or low output voltage is entered.)	• Harness and connector • Secondary throttle sensor • T.C.M.
INJECTOR OPEN*	• The injector circuit is open or shorted.	• Injector
A/T COMM LINE	• The A/T communication line is open or shorted.	• Harness and connector

*: The diagnostic items marked * are applicable to vehicles for California only.
**: Check items causing a malfunction of crank angle sensor circuit first, if both "CRANK ANGLE SENSOR (No. 11)" and "IGN SIGNAL-PRIMARY (No. 21)" are displayed one after the other.
***: Models with TCS only

1990–92 4.5L ENGINE—DATA MONITOR MODE—Q45

DATA MONITOR MODE

Remarks :
- Specification data are reference values.
- Specification data are output/input values which are detected or supplied by E.C.U. at the connector.
  *Specification data may not be directly related to their components signals/values/operations.
  ie. Adjust ignition timing with a timing light before monitoring IGN TIMING, because the monitor may show the specification data in spite of the ignition timing being not adjusted to the specification data. This IGN TIMING monitors the calculated data by E.C.U. according to the input signals from crank angle sensor and other ignition timing related sensors.
- If the real-time diagnosis results are NG and the self-diagnosis results are OK when diagnosing the air flow meter, first check to see if the fuel pump control circuit is normal.
- The monitor item marked "★★" is applicable to vehicles for California only.

MONITOR ITEM	CONDITION		SPECIFICATION	CHECK ITEM WHEN OUTSIDE SPEC.
CAS·RPM (POS) CAS·RPM (REF)	• Tachometer: Connect • Run engine and compare tachometer indication with the CONSULT value.		Almost the same speed as the CONSULT value.	• Harness and connector • Crank angle sensor
AIR FLOW MTR	• Engine: After warming up, idle the engine • A/C switch "OFF" • Shift lever "N" • No-load	Idle	1.0 - 1.4V	• Harness and connector • Air flow meter
		2,000 rpm	1.4 - 1.9V	
ENG TEMP SEN	• Engine: After warming up		More than 70°C (158°F)	• Harness and connector • Engine temperature sensor
EXH GAS SEN EXH GAS SEN-R	• Engine: After warming up	Maintaining engine speed at 2,000 rpm	0 - 0.3V → 0.6 - 1.0V	• Harness and connector • Exhaust gas sensor • Intake air leaks • Injectors
M/R F/C MNT M/R F/C MNT-R			LEAN → RICH Changes more than 5 times during 10 seconds.	
CAR SPEED SEN	• Turn drive wheels and compare speedometer indication with the CONSULT value		Almost the same speed as the CONSULT value	• Harness and connector • Vehicle speed sensor
BATTERY VOLT	• Ignition switch: ON (Engine stopped)		11 - 14V	• Battery • E.C.U. power supply circuit
THROTTLE SEN	• Ignition switch: ON (Engine stopped)	Throttle valve fully closed (Engine: After warming up)	0.4 - 0.5V	• Harness and connector • Throttle sensor • Throttle sensor adjustment
		Throttle valve fully open	Approx. 4.0V	
THROTTLE SEN 2	• Ignition switch: ON (Engine stopped) • Disconnect throttle motor harness connector	Secondary throttle valve fully open	4.0 - 4.7V	• harness and connector • Secondary throttle sensor • Secondary throttle sensor adjustment • T.C.M.
		Secondary throttle valve fully closed	Approx. 0.5V (at least 0.3V)	
EGR TEMP SEN★★	• Engine: After warming up		Less than 4.5V	• Harness and connector • Exhaust gas temperature sensor
START SIGNAL	• Ignition switch: ON → START		OFF → ON	• Harness and connector • Starter switch

1990–92 4.5L ENGINE—DATA MONITOR MODE—Q45

MONITOR ITEM	CONDITION		SPECIFICATION	CHECK ITEM WHEN OUTSIDE SPEC.
IDLE POSITION	• Ignition switch: ON (Engine stopped)	Throttle valve: Idle position (Engine: After warming up)	ON	• Harness and connector • Throttle sensor • Throttle sensor adjustment • Throttle valve switch
		Throttle valve: Slightly open	OFF	
AIR COND SIG	• Engine: After warming up, idle the engine	A/C switch "OFF"	OFF	• Harness and connector • Air conditioner switch
		A/C switch "ON"	ON	
NEUTRAL SW	• Ignition switch: ON	Shift lever "P" or "N"	ON	• Harness and connector • Neutral switch
		Except above	OFF	
PW/ST SIGNAL	• Engine: After warming up, idle the engine	Steering wheel in neutral (forward direction)	OFF	• Harness and connector • Power steering oil pressure switch
		The steering wheel is turned	ON	
INJ PULSE	• Engine: After warming up • A/C switch "OFF" • Shift lever "N" • No-load	Idle	1.8 - 2.5 msec.	• Harness and connector • Injector • Air flow meter • Intake air system
INJ PULSE-R		2,000 rpm	1.7 - 2.4 msec.	
IGN TIMING	ditto	Idle	15° B.T.D.C.	• Harness and connector • Crank angle sensor
		2,000 rpm	More than 25° B.T.D.C.	
AAC VALVE	ditto	Idle	15 - 40%	• Harness and connector • A.A.C. valve
		2,000 rpm	—	
AIR COND RLY	Engine: After warming up, idle the engine Air conditioner switch OFF → ON		OFF → ON	• Harness and connector • Air conditioner switch • Air conditioner relay
FUEL PUMP RLY	• Ignition switch is turned to ON (Operates for 5 seconds) • Engine running and cranking • When engine is stopped (stops in 1.5 seconds)		ON	• Harness and connector • Fuel pump relay
	Except as shown above		OFF	
VALVE T/M SOL	• Jack up rear wheel • Engine: After warming up	• Idle	OFF	• Harness and connector • Valve timing solenoid valve
		• Shift select lever to any range except "N" or "P" range • Quickly depress accelerator pedal, then quickly release it	OFF → ON → OFF	
EGR CONT S/V	• Engine: After warming up • A/C switch "OFF" • Shift lever "N" • No-load	Idle	ON	• Harness and connector • EGR control solenoid valve
		2,000 rpm	OFF	

1990–92 4.5L ENGINE—ACTIVE TEST MODE—Q45

ACTIVE TEST MODE

TEST ITEM	CONDITION	JUDGEMENT	CHECK ITEM (REMEDY)
FUEL INJECTION TEST	• Engine: Return to the original trouble condition • Change the amount of fuel injection with the CONSULT.	If trouble symptom disappears, see CHECK ITEM.	• Harness and connector • Fuel injectors • Exhaust gas sensors
AAC/V OPENING TEST	• Engine: After warming up, idle the engine. • Change the AAC valve opening percent with the CONSULT.	Engine speed changes according to the opening percent.	• Harness and connector • AAC valve
ENGINE TEMP TEST	• Engine: Return to the original trouble condition • Change the engine coolant temperature with the CONSULT.	If trouble symptom disappears, see CHECK ITEM.	• Harness and connector • Engine temperature sensor • Fuel injectors
IGN TIMING TEST	• Engine: Return to the original trouble condition • Timing light: Set • Retard the ignition timing with the CONSULT.	If trouble symptom disappears, see CHECK ITEM.	• Adjust initial ignition timing
POWER BALANCE TEST	• Engine: After warming up, idle the engine. • A/C switch "OFF" • Shift lever "N" • Cut off each injector signal one at a time with the CONSULT.	Engine runs rough or dies.	• Harness and connector • Compression • Injectors • Power transistor • Spark plugs • Ignition coils
FUEL PUMP RLY TEST	• Ignition switch: ON (Engine stopped) • Turn the fuel pump relay "ON" and "OFF" with the CONSULT and listen to operating sound.	Fuel pump relay makes the operating sound.	• Harness and connector • Fuel pump relay
EGR CONT SOL/V TEST	• Ignition switch: ON • Turn solenoid valve "ON" and "OFF" with the CONSULT and listen to operating sound.	Each solenoid valve makes an operating sound.	• Harness and connector • Solenoid valve
VALVE TIM SOL TEST			
SELF-LEARN CONT TEST	• In this test, the coefficient of self-learning control mixture ratio returns to the original coefficient by touching "CLEAR" on the screen.		

1991–92 4.5L ENGINE—FUNCTION TEST—Q45

FUNCTION TEST MODE

FUNCTION TEST ITEM	CONDITION	JUDGEMENT		CHECK ITEM (REMEDY)
SELF-DIAG RESULTS	• Ignition switch: ON (Engine stopped) • Displays the results of self-diagnosis.	—		Objective system
IDLE POSITION (IDLE SWITCH CIRCUIT)	• Ignition switch: ON (Engine stopped) • Idle switch circuit is tested when throttle is opened and closed fully. ("IDLE POSITION" is the test item name for the vehicles in which idle is selected by throttle sensor.)	Throttle valve: opened	OFF	• Harness and connector • Throttle sensor (Idle switch) • Throttle sensor (Idle switch) adjustment • Throttle linkage • Verify operation in DATA MONITOR mode.
		Throttle valve: closed	ON	
THROTTLE SENSOR CKT	• Ignition switch: ON (Engine stopped) • Throttle sensor circuit is tested when throttle is opened and closed fully.	Range (Throttle valve fully opened — Throttle valve fully closed)	More than 3.0V	• Harness and connector • Throttle sensor • Throttle sensor adjustment • Throttle linkage • Verify operation in DATA MONITOR mode.
NEUTRAL SW CIRCUIT	• Ignition switch: ON (Engine stopped) • Neutral switch circuit is tested when shift lever is manipulated.	OUT OF N/P-RANGE	OFF	• Harness and connector • Neutral switch/Inhibitor switch • Linkage + Inhibitor switch adjustment
		IN N-RANGE	ON	
FUEL PUMP CIRCUIT	• Ignition switch: ON (Engine stopped) • Fuel pump circuit is tested by checking the pulsation in fuel pressure when fuel tube is pinched.	There is pressure pulsation on the fuel feed hose.		• Harness and connector • Fuel pump • Fuel pump relay • Fuel filter clogging • Fuel level
EGR CONT S/V CIRCUIT	• Ignition switch: ON (Engine stopped) • EGR control S/V circuit is tested by checking solenoid valve operating noise.	The solenoid valve makes an operating sound every 3 seconds.		• Harness and connector • EGR control solenoid valve
VALVE TIMING S/V CKT	• Ignition switch: ON (Engine stopped) • Valve timing S/V circuit is tested by checking solenoid valve operating noise.	The solenoid valve makes an operating sound every 3 seconds.		• Harness and connector • Valve timing solenoid valve
START SIGNAL CIRCUIT	• Ignition switch: ON → START • Start signal circuit is tested when engine is started by operating the starter. Battery voltage and water temperature before cranking, and average battery voltage, air flow meter output voltage and cranking speed during cranking are displayed.	Start signal: OFF → ON		• Harness and connector • Ignition switch

FUNCTION TEST ITEM	CONDITION	JUDGEMENT		CHECK ITEM (REMEDY)
PW/ST SIGNAL CIRCUIT	• Ignition switch: ON (Engine running) • Power steering circuit is tested when steering wheel is rotated fully and then set to a straight line running position.	Locked position	ON	• Harness and connector • Power steering oil pressure switch • Power steering oil pump
		Neutral	OFF	
CAR SPEED SEN CIRCUIT	• Car speed sensor circuit is tested when vehicle is running at a speed of 10 km/h (6 mph) or higher.	Car speed sensor input signal is greater than 4 km/h (2 MPH)		• Harness and connector • Car speed sensor • Electric speedometer
IGN TIMING ADJ	• After warming up, idle the engine. • Ignition timing adjustment is checked by reading ignition timing with a timing light and checking whether it agrees with specifications.	The timing light indicates the same value on the screen.		• Adjust ignition timing (by moving crank angle sensor or distributor) • Crank angle sensor drive mechanism
MIXTURE RATIO TEST	• Air-fuel ratio feedback circuit (injection system, ignition system, vacuum system, etc.) is tested by examining the O_2 sensor output at 2,000 rpm under non-loaded state.	• O_2 SEN COUNT: More than 5 times during 10 seconds (O_2 SEN-R COUNT: More than 5 times during 10 seconds)		• INJECTION SYS (Injector, fuel pressure regulator, harness or connector) • IGNITION SYS (Spark plug, power transistor, ignition coil, harness or connector) • VACUUM SYS (Intake air leaks) • O_2 sensor circuit • O_2 sensor operation • Fuel pressure high or low • Air flow meter
POWER BALANCE	• After warming up, idle the engine. • Injector operation of each cylinder is stopped one after another, and resultant change in engine rotation is examined to evaluate combustion of each cylinder. (This is only displayed for models where a sequential injection system is used.)	Difference in engine rpm is greater than 25 rpm before and after cutting off the injector of each cylinder.		• Injector circuit (Injector, harness or connector) • Ignition circuit (Spark plug, power transistor, ignition coil, harness or connector) • Compression • Valve timing
AAC VALVE SYSTEM	• After warming up, idle the engine. • AAC valve system is tested by detecting change in engine rpm when AAC valve opening is changed to 0%, 20% and 80%.	Difference in engine rpm is greater than 150 rpm between when valve opening is at 80% (102 steps) and at 20% (25 steps).		• Harness and connector • AAC valve • Air passage restriction between air inlet and AAC valve • IAS (Idle adjusting screw) adjustment

Component Replacement

SERVICE PRECAUTIONS

- Fuel system pressure must be relieved before disconnecting any fuel lines or attempting to remove any fuel system components.
- Do not operate the fuel pump when the fuel lines are empty.
- Do not reuse fuel hose clamps and tighten the new ones to the specified torque.
- Do not disconnect the injector harness connectors with the engine running.
- Always check for leaks after servicing the fuel system.
- Do not apply battery power directly to the injectors.
- Do not disconnect the ECCS harness connector before the ignition switch is **OFF** and the battery ground cable has been disconnected.
- When testing the ECU with a DVOM make sure that the probes of the tester never touch each other as this will result in damage to a transistor in the ECU.
- Make sure the ECCS connector is fastened securely. A poor connection can cause an extremely high surge voltage in the coil and condenser and result in damage to integrated circuits.
- When connecting the ECU harness connector, tighten the securing bolt until the red projection (right next to the bolt) is in line with the connector face.
- When connecting or disconnecting pin connectors into or from the ECU, take care not to bend or break the pin terminals.
- Before replacing an ECU, perform an ECU input/output signal inspection.
- When measuring the supply voltage of ECU controlled components, make sure that the tester probes do not make contact with each other during measurement. If this happens, the circuit will be shorted, resulting in damage to the control unit power transistor.
- Keep the ECCS harness at least 4 in. away from adjacent harnesses to prevent an ECCS system malfunction due to external electronic "noise."
- Keep all parts and harnesses dry during service.
- Before attempting to remove any parts, turn **OFF** the ignition switch and disconnect the battery ground cable.
- Always use a 12 volt battery as a power source.
- Do not attempt to disconnect the battery cables with the engine running or the ignition key **ON**.
- If a battery terminal is disconnected (with the ignition key **OFF**), the memory will return to the ROM value. The ECCS will now self-control at its initial value. The engine operation can vary slightly when the terminal is disconnected. However, this is not an indication of a problem. Do not replace parts because of a slight variation.
- Do not clean the air flow meter with any type of detergent.
- Even a slight leak in the air intake system can cause serious problems.
- Do not depress the accelerator pedal when starting.
- Do not rev up the engine immediately after starting or just prior to shutdown.
- Do not attempt to disassemble the ECCS control unit under any circumstances.
- If installing a 2-way or CB radio, keep the antenna as far as possible away from the electronic control unit. Keep the antenna feeder line at least 8 in. away from the ECCS harness and do not let the 2 run parallel for a long distance. Be sure to ground the radio to the vehicle body.
- If crank angle sensor and ignition signal-primary codes come out at the same time, check items causing a malfunction of the crank angle sensor circuit first.
- After a repair has been made, always perform a final check by erasing the self-diagnosis memory, test driving, and performing self-diagnosis again.

FUEL PRESSURE RELIEF

CAUTION

Before disconnecting a fuel line, the fuel pressure must be released from the line to eliminate the danger of fire and personal injury.

The fuel pressure on the G20 and Q45 can be released by using the Consult, or equivalent, in the Work Support Mode (See manufacturer's instructions). To release the fuel pressure without a scan tool see the list below.

G20 And M30

1. Remove the fuse for the fuel pump.
2. Start the engine.
3. After the engine stalls, crank it two or three times to release all the fuel pressure.
4. Turn the ignition switch off and reinstall the fuse.
5. On the M30 erase the memory, since a code 22 will have been set.

Q45

1. Remove the fuse for the fuel pump.
2. Start the engine.

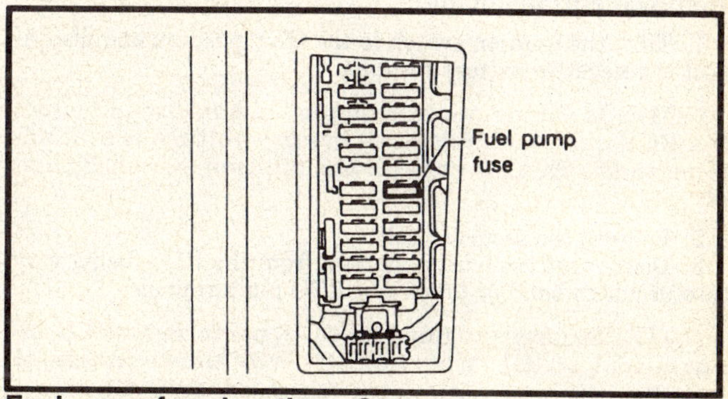

Fuel pump fuse location—G20

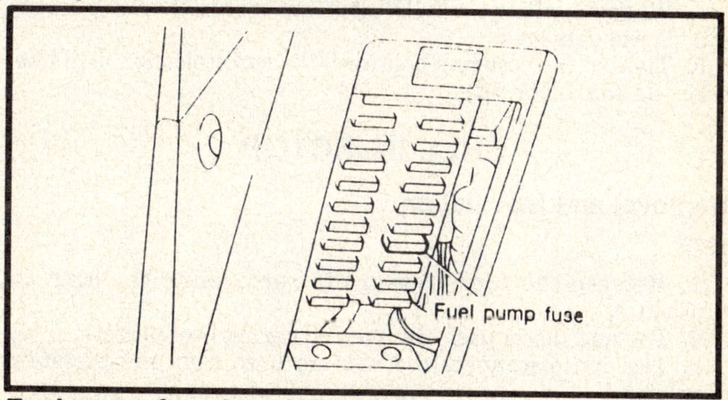

Fuel pump fuse location—M30

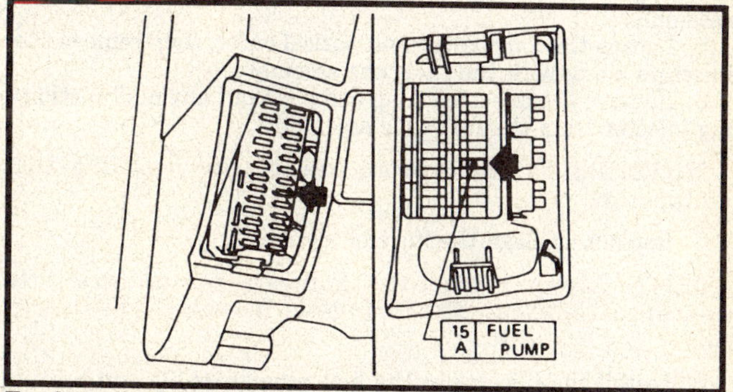

Fuel pump fuse location—Q45

3. After the engine stalls, crank it two or three times to release all the fuel pressure.
4. Turn the ignition switch off and reinstall the relay and the fuse.

DETONATION SENSOR

Removal and Assembly

1. Disconnect the electrical connector.
2. Unscrew the sensor or remove the attaching bolts, depending how the vehicle is equipped.

NOTE: Upon reassembly make sure that no foreign material is between the sensor and the mating surface. Also make sure that no other parts get caught on the sensor as it is being torqued.

3. Install the sensor and torque to specifications.
   G20, 25–33 ft. lbs. (34–44 Nm).
   M30, 15–20 ft. lbs. (21–26 Nm).
   Q45, 12–15 ft. lbs. (16–21 Nm).

ELECTRONIC CONTROL UNIT

Removal and Installation

1. Turn the ignition switch to the **OFF** position and disconnect the negative battery terminal.

NOTE: If the negative terminal is not disconnected, the ECU can be damaged as battery voltage is still applied to the ECU, even with the ignition switch turned off.

2. Remove the securing bolt.
3. Disconnect the wiring harness from the ECU, being very careful not to bend or break any ECU pin terminals.

NOTE: Before replacing an ECU, perform an ECU input/output signal inspection to verify the condition of the unit.

4. When installing an ECU, again make sure that no pin terminals are damaged.
5. Tighten the securing bolt until the red projection is in line with the connector face.

FUEL INJECTOR

Removal and Installation

G20

1. Release the fuel pressure to zero, using the correct procedure.
2. Tag and disconnect the injector harness connectors.
3. Tag and disconnect the vacuum hose from the pressure regulator.
4. Tag and disconnect the fuel hoses from the fuel tube assembly.
5. Loosen the four bolts in numerical order, then remove the injectors along with the fuel tube assembly.
6. Remove the injector clamp and push out any malfunctioning injector from the fuel tube assembly.

NOTE: Do not extract an injector by pinching the connector.

7. Replace or clean the injector as necessary.

NOTE: Always replace the O-rings and insulators with new ones. Lubricate the O-rings with a smear of silicone oil.

8. Install the injector in the fuel tube assembly and torque the clamp screws to 2.2–2.8 ft. lbs. (2.9–3.8 Nm).

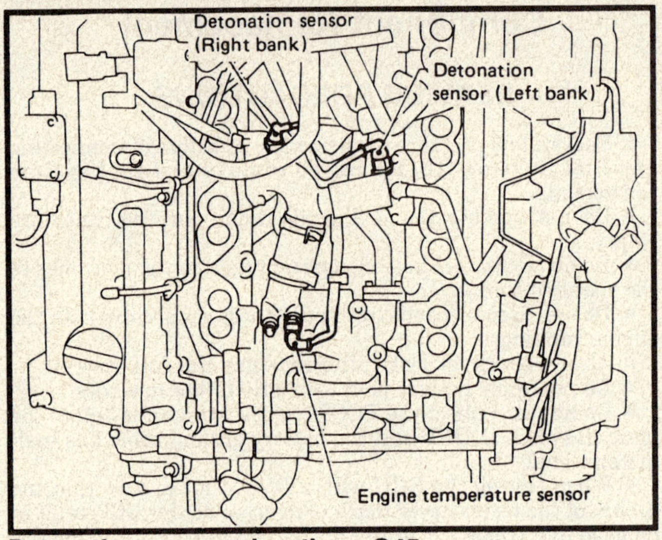

Detonation sensors location – Q45

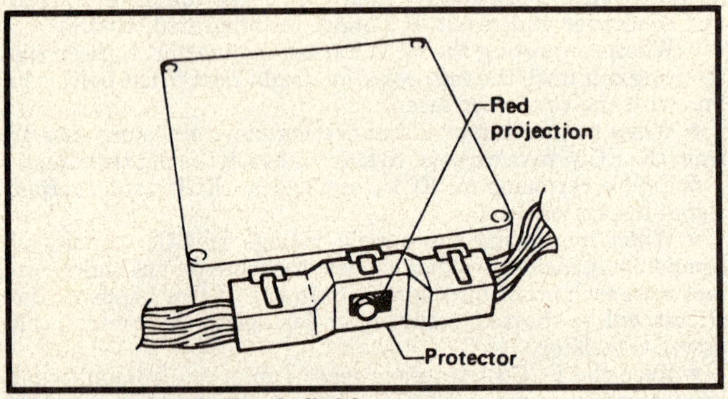

ECU securing bolt – Infiniti

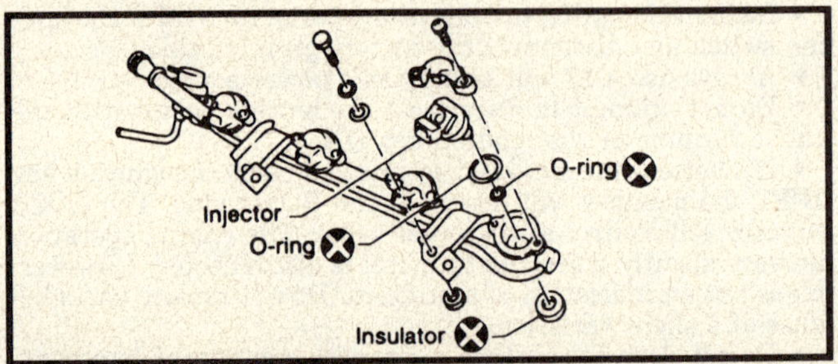

Injector removal – G20

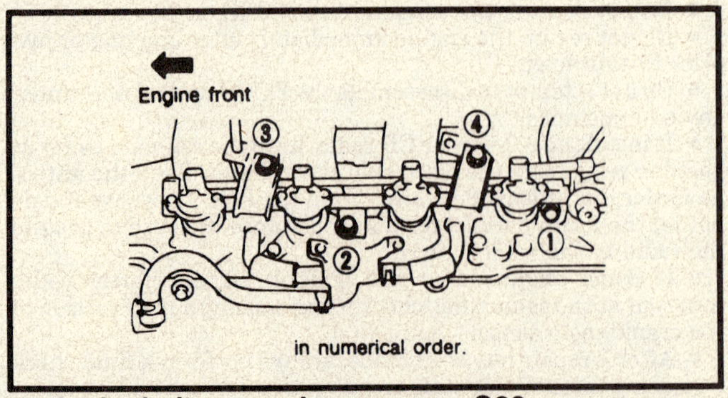

Fuel tube bolt removal sequence – G20

9. Install the injectors along with the fuel tube assembly to the intake manifold. Tighten the fuel tube bolts in sequence to the specified torques in two steps.
   a. Tighten all bolts to 6.9–8.0 ft. lbs. (9.3–10.8 Nm).
   b. Tighten all bolts to 15–20 ft. lbs. (21–26 Nm).
10. Lubricate the fuel hoses with a smear of silicone oil and install the hoses to the fuel tube assembly.
11. Reinstall any parts removed in reverse order of removal.

CAUTION

After properly connecting fuel hose to injector and fuel tube assembly, check the connections for fuel leakage to prevent fire and personal injury.

M30

1. Release the fuel pressure to zero using the proper procedure
2. Separate the ASCD and accelerator control wire from the intake manifold collector.
3. Remove the intake manifold collector from the engine by loosening the bolts in the correct sequence and discard the gasket.
4. Tag and disconnect the harness connectors for the AAC valve, FICD solenoid valve, throttle valve switch, throttle sensor, pressure regulator control solenoid valve, air regulator, EGR control solenoid valve, exhaust gas temperature sensor and earth harness.
5. Tag and disconnect the PCV hoses and the vacuum hoses for the brake master cylinder, EGR control solenoid valve, fuel pressure regulator and carbon canister.
6. Tag and remove the air hoses from the air duct, AAC valve and the air regulator.
7. Disconnect the water hoses for the throttle chamber and the air relief plug.
8. Disconnect the carbon canister purge hose and the EGR flair tube.
9. Remove the injector fuel tube assembly, loosening the bolts in sequence, after tagging and disconnecting the vacuum

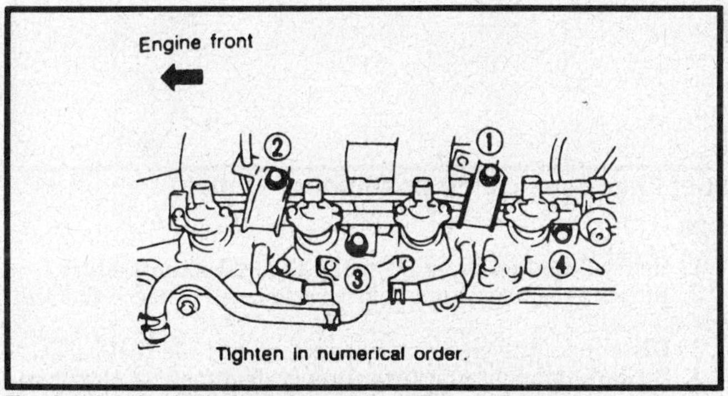

Fuel tube bolt tightening sequence – G20

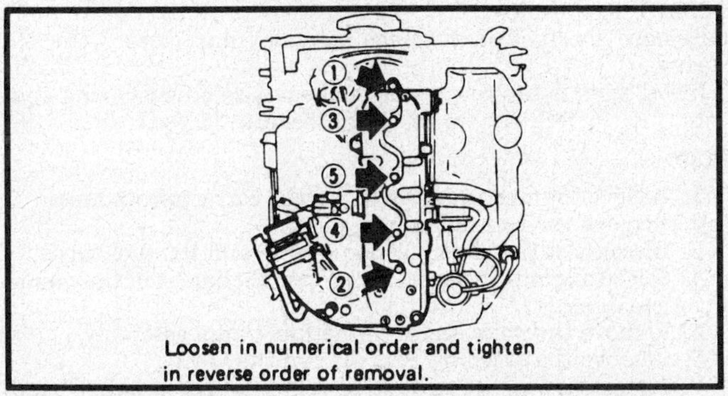

Intake manifold collector torque sequence – M30

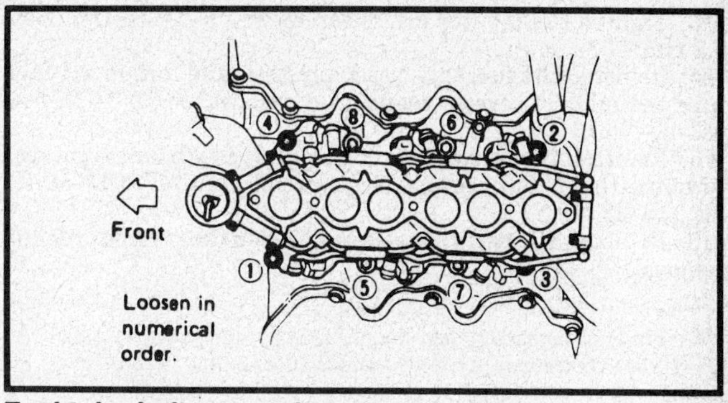

Fuel tube bolt removal sequence – M30

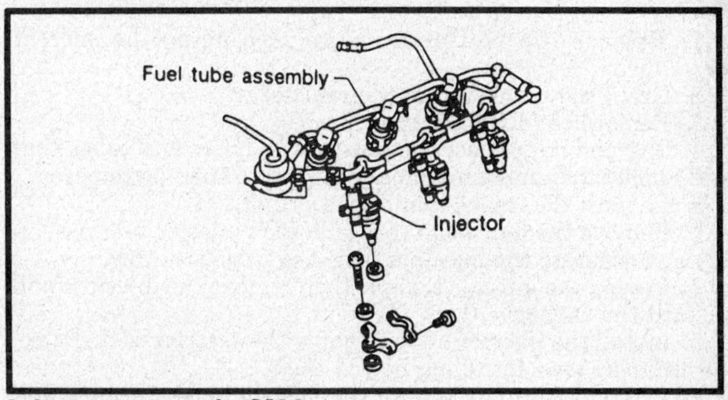

Injector removal – M30

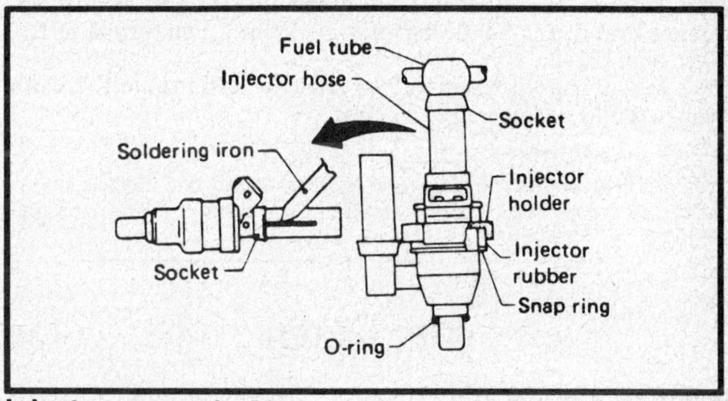

Injector removal – M30

hose for the fuel pressure regulator, fuel feed and return hoses, all injector harness connectors and the fuel temperature sensor harness connector.

10. Remove any malfunctioning injector from the injector fuel tube.
11. Remove the fuel hose by heating a soldering iron for 15 minutes, then cutting into the hose braided reinforcement from mark to socket end and fuel tube end. Then pull the rubber hose out by hand and discard the hose.

NOTE: Do not allow the soldering iron to cut all the way through the hose and touch the injector tail piece. Also, never place an injector in a vice when disconnecting the rubber hose.

12. Install a new fuel hose using the following procedure.
    a. Clean the exterior of the injector tail piece and the fuel tube end.
    b. Wet the inside of the new rubber hose with fuel.
    c. Push the end of the rubber hose with hose sockets onto the injector tail piece and fuel tube end by hand as far as they will go.

13. Tighten the injector clamp screws to 1.8–2.4 ft. lbs. (2.5–3.2 Nm).
14. Reinstall the fuel tube assembly first, and torque the fuel tube assembly in reverse sequence of removal to 12–14 ft. lbs. (16–20 Nm).
15. Reinstall the intake manifold collector with a new gasket, torquing the bolts in reverse sequence of removal to 13–16 ft. lbs. (18–22 Nm).
16. Reinstall any part that was removed in the reverse order of removal.

CAUTION

After properly connecting fuel hose to injector and fuel tube assembly, check the connections for fuel leakage to prevent fire and personal injury.

Q45

1. Release the fuel pressure to zero using the correct procedure.
2. Drain the coolant from the radiator.
3. Remove the EGR control valve.
4. Tag and disconnect the connectors, hoses and tubes that will enable the intake manifold collector to then be removed.
5. Remove the intake manifold collector.
6. Remove the four bolts that hold each injector tube assembly and separate the injectors from the tube assembly.
7. Loosen the injector clamps, then remove the injectors and discard the O-rings.
8. Install the injector after cleaning the exterior of the injector tail piece and installing new O-rings.
9. Assemble the injectors to the injector tube assembly and torque the injector clamps to 1.7–2.2 ft. lbs. (2.3–3.0 Nm).
10. Tighten the injector tube assemblies in the correct sequence first to 6.9–8.0 ft. lbs. (9.3–10.8 Nm), then torque to 15–20 ft. lbs. (21–26 Nm).
11. Install the intake manifold collector and torque the bolts evenly to 9–11 ft. lbs. (12–15 Nm).

CAUTION

After properly connecting fuel hose to injector and fuel tube assembly, check the connections for fuel leakage to prevent fire and personal injury.

FUEL FILTER

Removal and Installation

CAUTION

Release the fuel pressure using the appropriate procedure and use a shop towel to absorb spilled fuel to avoid the danger of fire and personal injury.

1. Loosen the fuel filter hose clamps and remove the fuel filter.

NOTE: Do not use a synthetic resinous type fuel filter. Use only a high-pressure type filter. Tighten the high-pressure rubber hose clamp so that the clamp end is 3 mm (0.12 in.) from the hose end. Ensure that the screw does not contact adjacent parts.

2. Install the filter so that it is pointing in the right direction and then tighten the clamps.
3. Start the engine and check for leaks.

FUEL PUMP

Removal and Installation

NOTE: The fuel pump can be removed without removing the fuel tank on the G20 and the M30.

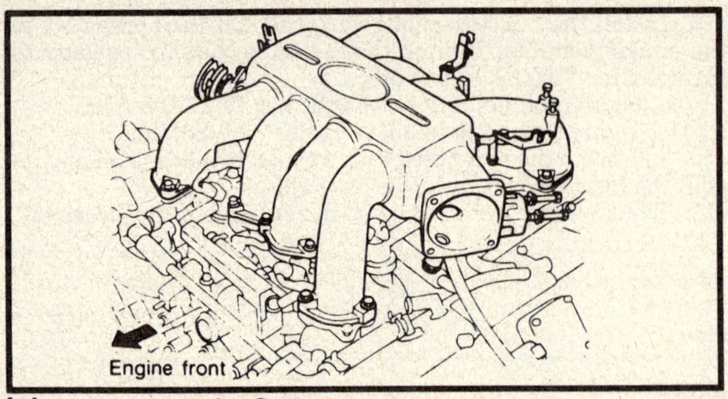

Injector removal—Q45

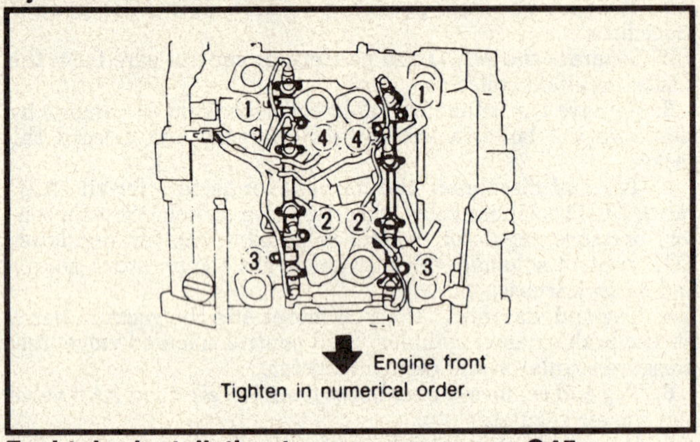

Fuel tube installation torque sequence—Q45

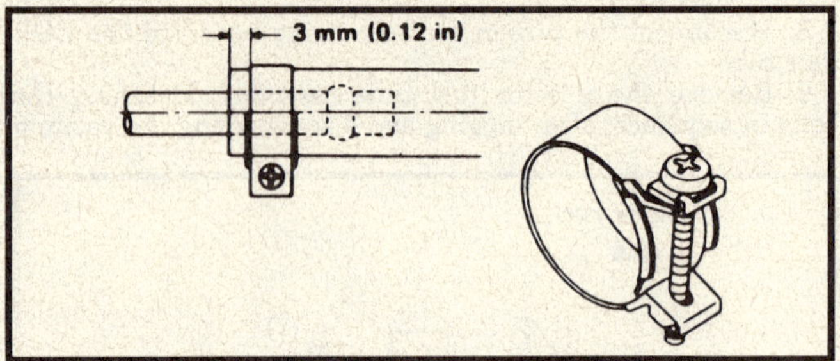

Fuel filter hose clamp positioning—Infiniti

G20

1. Relieve the fuel pressure using the correct procedure.
2. Remove the inspection hole cover located beneath the rear seat.
3. Disconnect the electrical connectors and the fuel tubes.
4. Scribe alignment marks for proper alignment of the pump upon reassembly.
5. Remove the special lock ring, using special service tool X38879 or equivalent.
6. Remove the fuel pump/guage assembly and discard the O-ring.
7. Installation is the reverse of removal. Use a new O-ring and torque the lock ring to 22–26 ft. lbs. (30–35 Nm).

M30

1. Relieve the fuel pressure using the correct procedure.
2. Remove the inspection hole cover.
3. Disconnect the electrical connectors and the fuel tubes.
4. Scribe alignment marks for proper alignment of the pump upon reassembly.
5. Remove the bolts securing the fuel pump assembly.
6. Remove the assembly and discard the O-ring.
7. Installation is the reverse of removal. Use a new O-ring and torque the bolts to 1.4–1.9 ft. lbs. (2.0–2.5 Nm).

Q45

1. Relieve the fuel pressure using the correct procedure.
2. Drain the fuel tank into an approved container.
3. Remove the fuel tank.
4. Scribe alignment marks for proper alignment of the pump upon reassembly.
5. Remove the bolts securing the fuel pump assembly, then remove the assembly.
6. Discard all 6 of the O-rings encountered.
7. Installation is the reverse of removal. Use a new O-rings and torque the fuel pump assembly bolts to 1.4–1.9 ft. lbs. (2.0–2.5 Nm).
8. Torque the filler tube flange bolts to 2.4–3.1 ft. lbs. (3.2–4.2 Nm).

THROTTLE SENSOR

Removal and Installation

Disconnect the throttle sensor harness connector. Remove the sensor mounting bolts and remove the sensor. Installation is the reverse of removal. Adjustment is necessary after installation.

Adjustment

G20 and M30

1. Install the throttle sensor body in the throttle chamber, but leave the bolts loose.
2. Connect the throttle sensor harness connector, then start the engine and warm it up to normal operating temperature.
3. Either perform THROTTLE SEN ADJ in the WORK SUPPORT MODE using a scanner, or measure the output voltage of the throttle sensor using a digital voltmeter.
4. Make the adjustment by rotating the throttle sensor body so that the output voltage is 0.45–0.55 volts for the G20, and 0.4–0.5 volts for the M30.
5. Tighten the mounting bolts.
6. Disconnect the throttle sensor harness connector for a few seconds, then reconnect it.

Q45

This procedure includes adjustment for the throttle sensor and the throttle valve switch.

1. Install the throttle sensor body in the throttle chamber, but leave the bolts loose.
2. Connect the throttle sensor and idle switch harness connector, then start the engine and warm it up to normal operating temperature.
3. Disconnect the idle switch harness connector.
4. With a circuit tester check the idle switch **OFF⇨ON** speed while closing the throttle valve manually. The idle switch **OFF⇨ON** speed should be 810 ± 150 rpm with the gear selector in the **N** position.

SECONDARY THROTTLE SENSOR

Adjustment

This procedure applies to the Q45. If the secondary throttle sensor is removed or replaced, it must be adjusted.

CAUTION

Before touching the secondary throttle valve, be sure to disconnect the throttle motor connector; otherwise, personal injury may occur due to accidental actuation of the valve.

1. Disconnect the rubber air duct.
2. Install the secondary throttle sensor body in the throttle chamber, but do not tighten the bolts yet.
3. Disconnect the throttle motor harness connector.
4. Connect the secondary throttle sensor harness connector.
5. Turn the ignition switch **ON**.

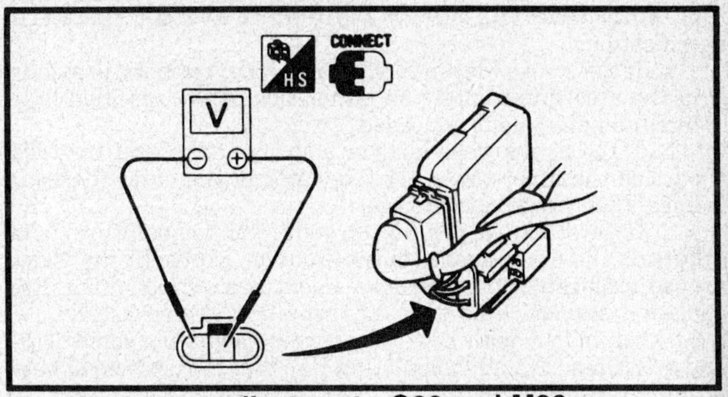

Throttle sensor adjustment – G20 and M30

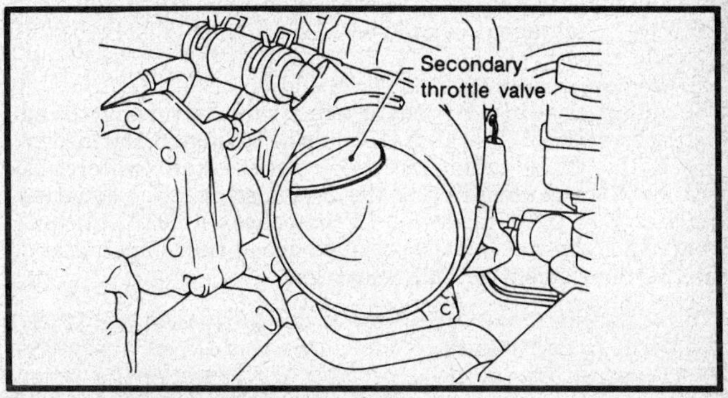

Secondary throttle valve – Q45

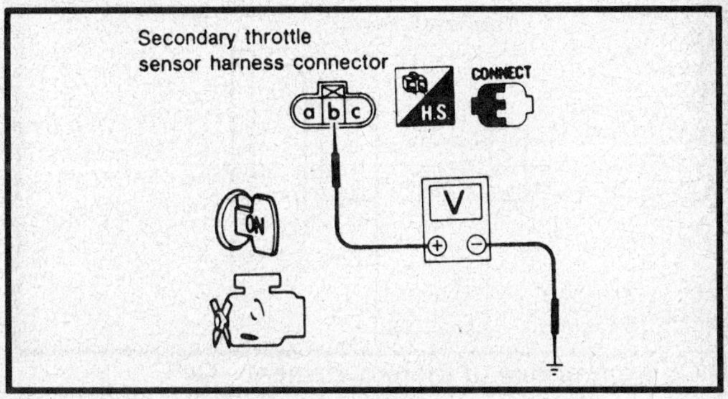

Secondary throttle sensor adjustment – Q45

6. Make sure that the voltage between terminal "b" and ground changes according to the following conditions when closing the secondary throttle valve by hand.
   a. Secondary throttle valve fully open: 4.4–4.6 volts.
   b. Secondary throttle valve partially open: 0.4–4.6 volts.
   c. Secondary throttle valve fully closed: 0.4 volts or more.

FAST IDLE CAM (FIC)

Removal and Installation

These procedures apply to the Q45.

1. Remove the four bolts that secure the throttle chamber to the engine.
2. Unplug the electrical connectors and remove the throttle chamber, discarding the gasket.
3. Wait for at least three hours to stabilize the temperature of the thermo-element at room temperature.
4. Measure the thermo-element stroke (L) and the room temperature.

5. Compare the thermo-element stroke (L) with specifications.

a. If the stroke (depending on your room temperature) falls in the area immediately on either side of the specified line, then no adjustment is needed.

b. If the stroke (depending on your room temperature) falls outside the inner three lines, but within the outer (dashed) lines, then procede to step 6.

c. If the stroke (depending on your room temperature) falls outside the outer dashed lines (in other words, in the wide-open area), then the thermo-element is malfunctioning. Replace it and continue with the inspection at step 3 again.

6. Align hole A with hole B by turning adjusting screw (S_1), and insert pin through holes A and B so that the position of hole A will not be changed.

7. Adjust clearance C between the throttle stopper and the throttle adjusting screw to specification by turning adjusting screw (S_2). Clearance C should be 0.62–0.74 mm (0.0244–0.0291 inches).

8. Remove the pin from holes A and B.

9. Adjust clearance D to the specified value by turning the adjusting screw (S_1). Clearance D should be 3.8 mm (0.150 inches).

10. Rotate the adjusting screw (S_1) clockwise or counterclockwise by **Z** turns according to the S_1 adjusting screw equation, which is: **Z** = (L (in inches) − Ls (in inches)) ÷ 0.0197 (inches), where Ls is the value of the specified line at the temperature of the thermo-element actually measured.

NOTE: When the direction of the adjusting screw (S_1) rotation is counterclockwise, this results in a positive (+) Z value. When the direction of the adjusting screw (S_1) rotation is clockwise, this results in a negative (−) Z value.

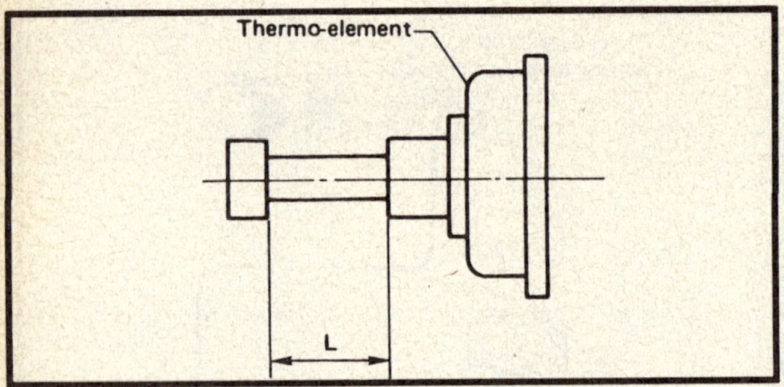

Measuring stroke of thermo-element—Q45

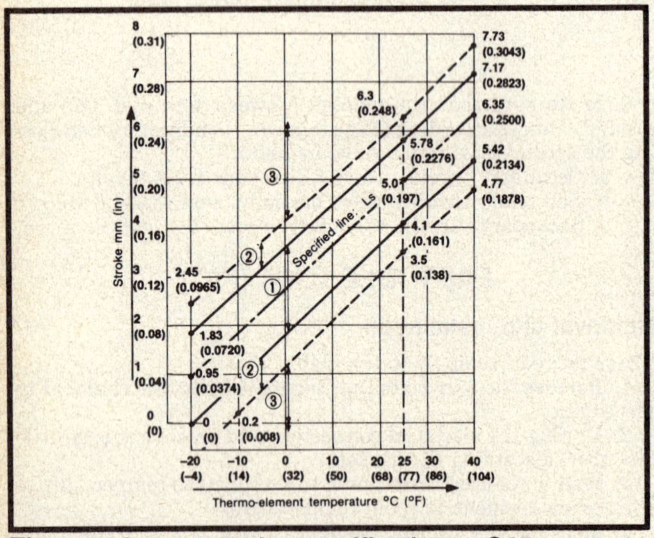

Thermo-element stroke specifications—Q45

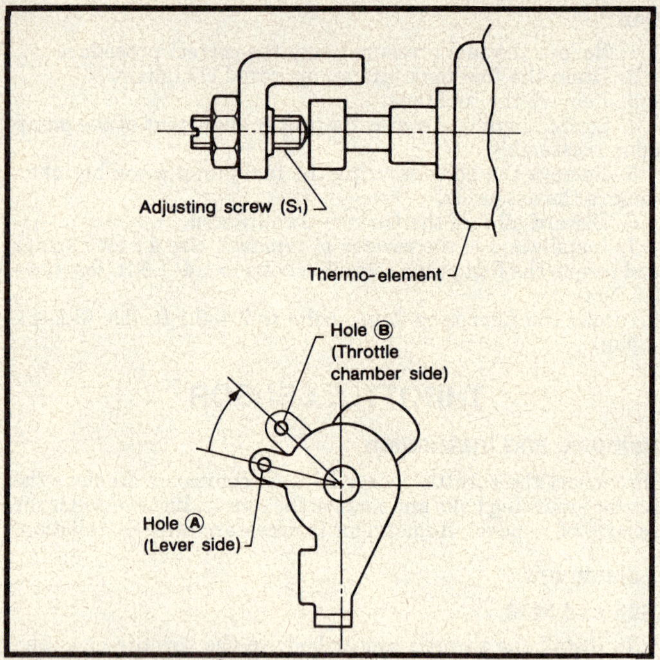

FIC adjustment—Q45

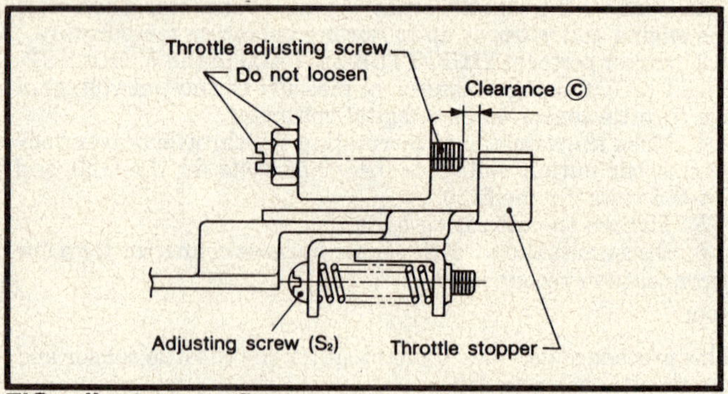

FIC adjustment—Q45

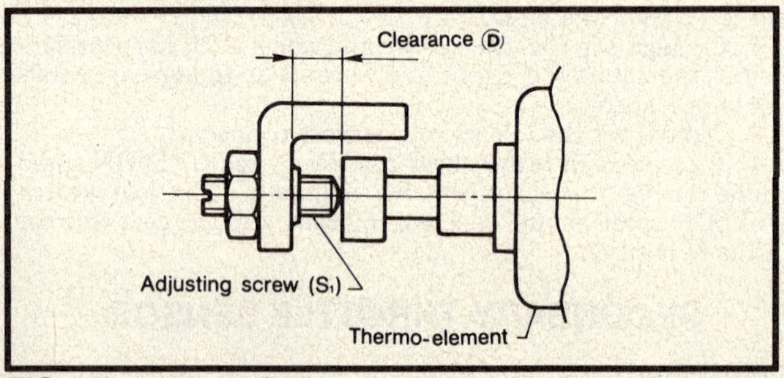

FIC adjustment—Q45

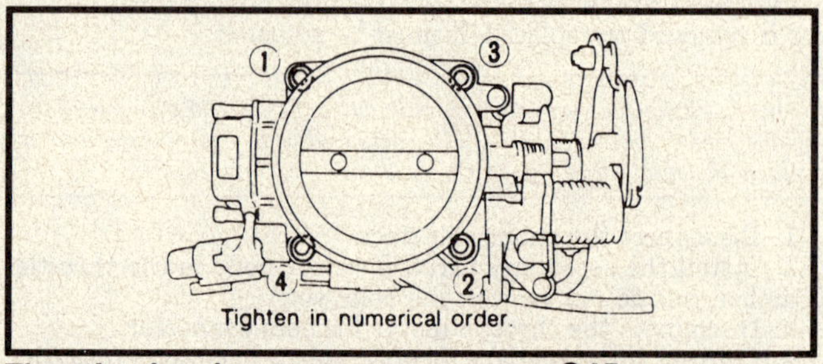

Throttle chamber torque sequence—Q45

	Case I	Case II
Thermo-element temperature °C (°F)	25 (77)	0 (32)
Thermo-element specified stroke (Ls) mm (in)	5.0 (0.197)	2.75 (0.1083)
Thermo-element stroke (L) mm (in)	5.5 (0.217)	2.00 (0.0787)
Revolutions of adjusting screw (Z) mm /in	$Z = \frac{5.5 - 5.0}{0.50} = 1.0$/ $\frac{0.217 - 0.197}{0.0197} = 1.0$	$Z = \frac{2.00 - 2.75}{0.50} = -1.5$/ $\frac{0.0787 - 0.1083}{0.0197} = -1.5$
Direction of revolution	Counterclockwise	Clockwise

Example of S_1 adjusting screw equation – Q45

11. Tighten the adjusting screw lock nut.
11. Using a new gasket, install the throttle chamber, tightening the bolts in sequence and to the specified torques in two steps. First, tighten all the bolts to 6.5–8.0 ft. lbs., then tighten all the bolts to 13–16 ft. lbs.

IDLE SPEED, IGNITION TIMING AND IDLE MIXTURE RATIO INSPECTION

Adjustment

This idle speed adjustment is actually the minimum idle speed adjustment. Under normal conditions the idle speed is feedback controlled and needs no adjustments.

1990–92 2.0L ENGINE – IDLE SPEED, IGNITION TIMING AND IDLE MIXTURE RATIO INSPECTION – G20

PREPARATION

1. **Make sure that the following parts are in good order.**
   - **Battery**
   - **Ignition system**
   - **Engine oil and coolant levels**
   - **Fuses**
   - **E.C.U. harness connector**
   - **Vacuum hoses**
   - **Air intake system (Oil filler cap, oil level gauge, etc.)**
   - **Fuel pressure**
   - **Engine compression**
   - **Throttle valve**
2. **On air conditioner equipped models, checks should be carried out while the air conditioner is "OFF".**
3. **On automatic transaxle equipped models, when checking idle rpm, ignition timing and mixture ratio, checks should be carried out while shift lever is in "N" position.**
4. **When measuring "CO" percentage, insert probe more than 40 cm (15.7 in) into tail pipe.**
5. **Turn off headlamps, heater blower, rear defogger.**
6. **Keep front wheels pointed straight ahead.**
7. **Make the check after the radiator fan has stopped.**

Overall inspection sequence

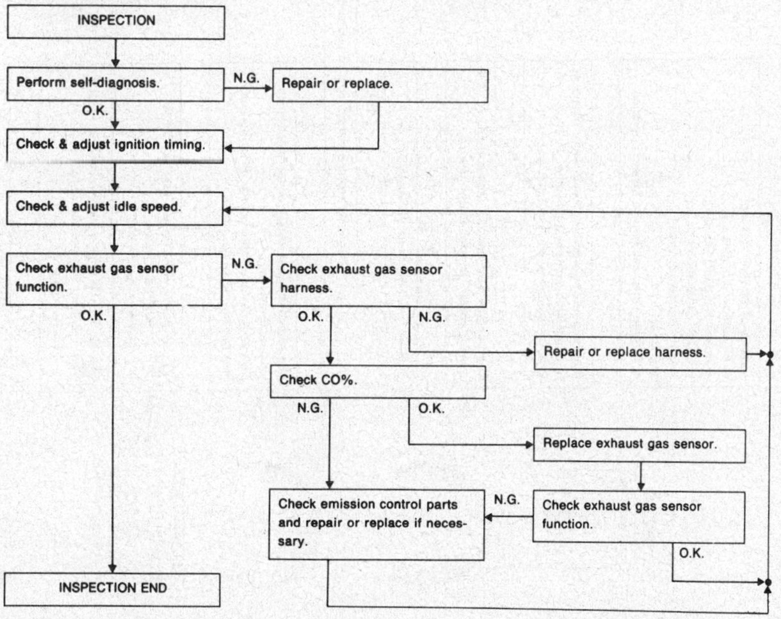

1990–92 2.0L ENGINE—IDLE SPEED, IGNITION TIMING AND IDLE MIXTURE RATIO INSPECTION—G20

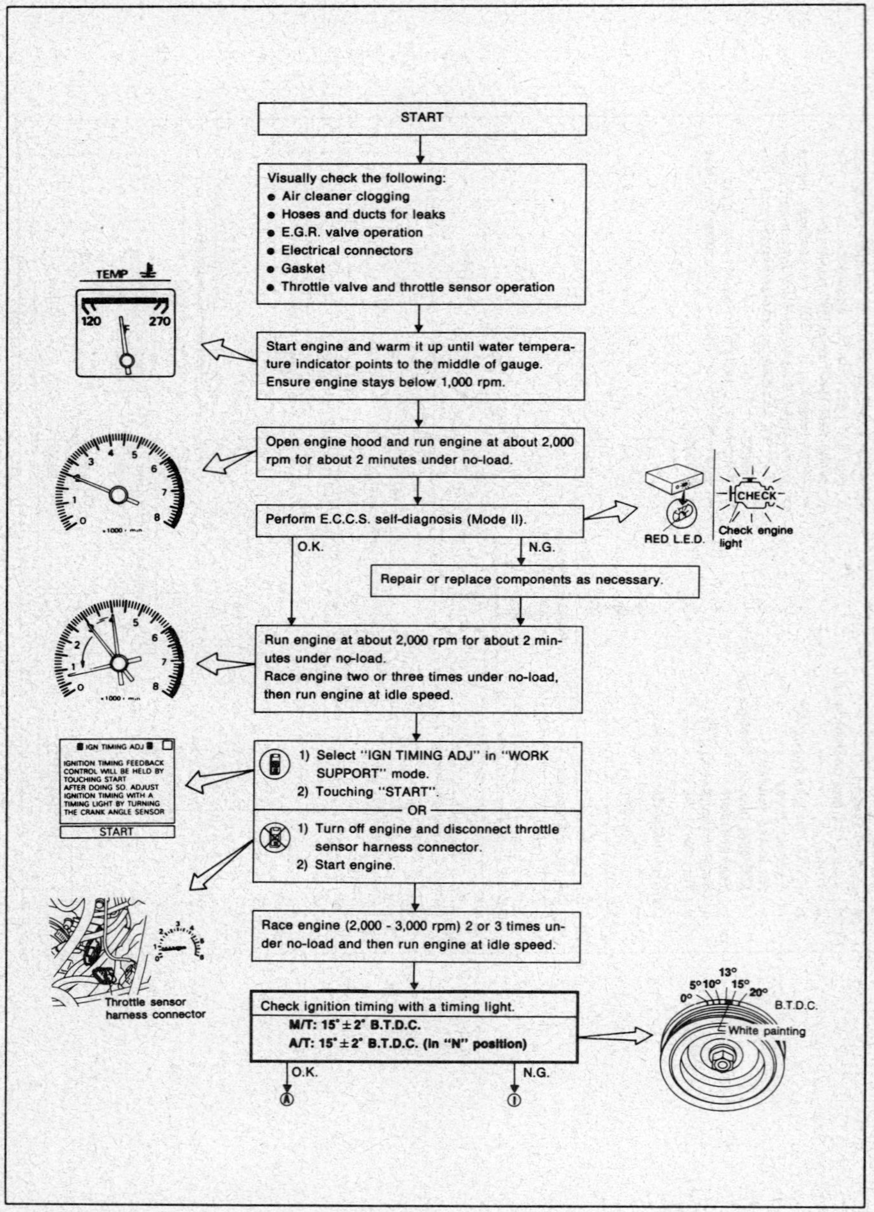

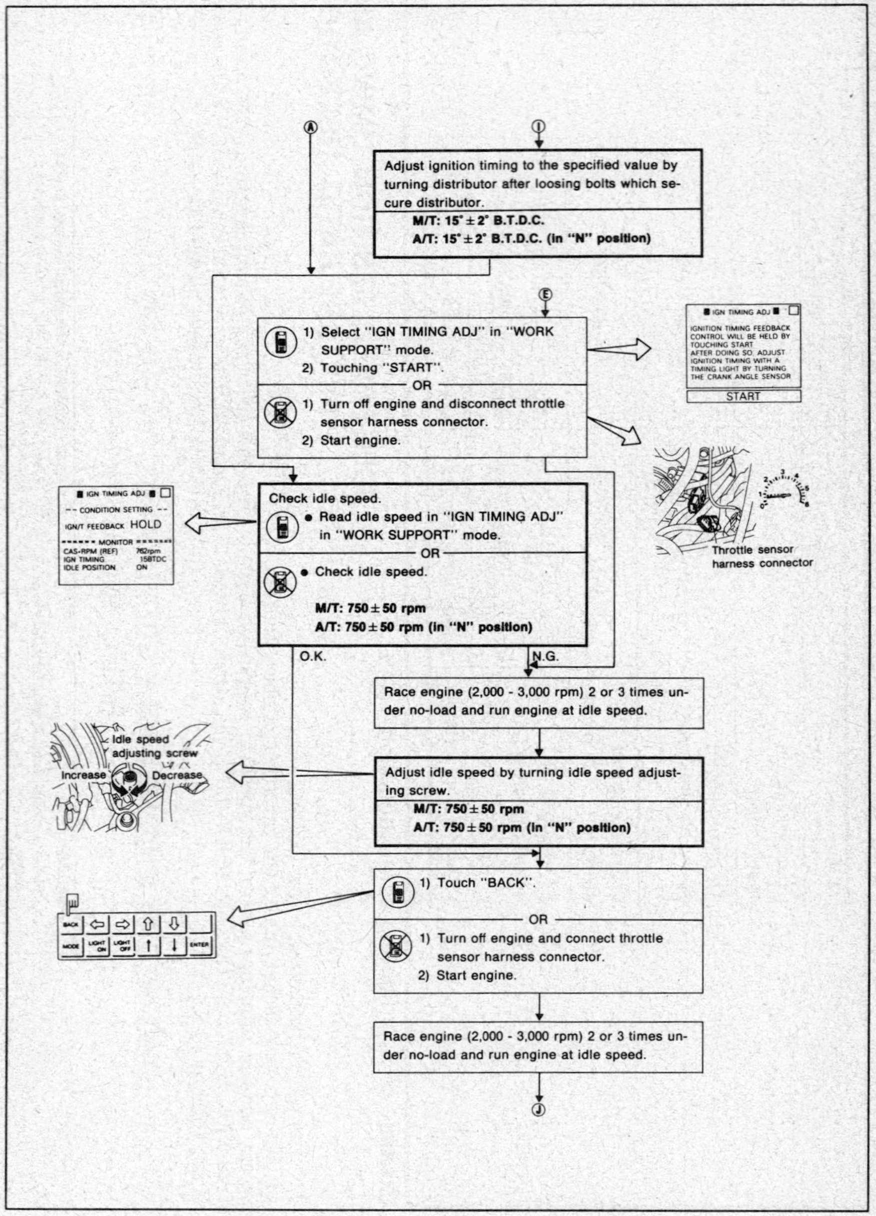

1990–92 2.0L ENGINE – IDLE SPEED, IGNITION TIMING AND IDLE MIXTURE RATIO INSPECTION – G20

Ⓙ

Check idle speed.
- Read idle speed in "DATA MONITOR" mode with CONSULT.

— OR —

- Check idle speed.

M/T: 800 ± 50 rpm
A/T: 800 ± 50 rpm (in "N" position)

O.K. → (to box below *)
N.G. ↓

Check A.A.C. valve and replace if necessary.

↓

Check A.A.C. valve harness and repair if necessary.

↓

Check E.C.U. function* by substituting another known good E.C.U.

↓

* E.C.U. may be the cause of a problem, but this is rarely the case.

↓

1. See "M/R F/C MNT" in "Data monitor" mode.
2. Run engine at about 2,000 rpm for about 2 minutes under no-load.
3. Maintaining engine at 2,000 rpm under no-load (engine is warmed up sufficiently.), check that the monitor fluctuates between "LEAN" and "RICH" more than 5 times during 10 seconds.
   RICH → LEAN → RICH → (1 time, 2 times)
   LEAN → RICH .....

— OR —

1. Set "Exhaust gas sensor monitor" in the self-diagnostic mode II.
2. Run engine at about 2,000 rpm for about 2 minutes under no-load.
3. Maintaining engine at 2,000 rpm under no-load, check to make sure that the RED L.E.D. on the E.C.U. or the check engine light on the instrument panel goes ON and OFF more than 5 times during 10 seconds.

N.G. → Ⓒ

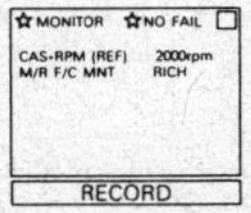

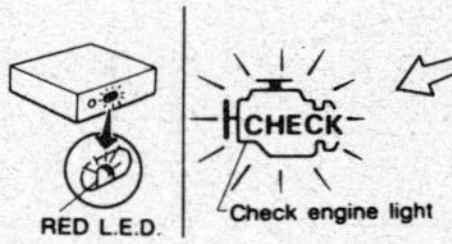

↓ O.K.

END

Ⓒ

Check exhaust gas sensor harness:
1) Turn off engine and disconnect battery ground cable.
2) Disconnect E.C.U. S.M.J. harness connector from E.C.U.
3) Disconnect exhaust gas sensor harness connector and connect main harness side terminal for exhaust gas sensor to ground with a jumper wire.
4) Check for continuity between terminal No. 19 of E.C.U. S.M.J. harness connector and body ground.

Continuity exists .... **O.K.**
Continuity does not exist .... **N.G.**

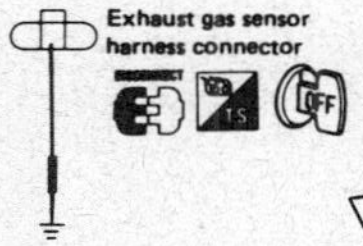

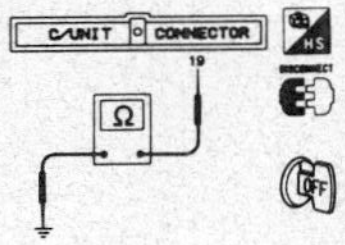

N.G. → Repair or replace harness. → Ⓔ

O.K. ↓

Connect E.C.U. S.M.J. harness connector to control unit.

↓

Disconnect A.I.V. control solenoid valve harness connector.

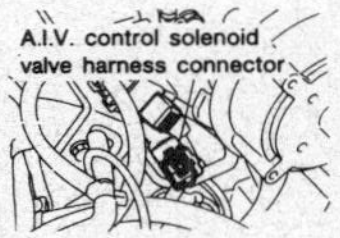

↓

1) Connect battery ground cable.
2) Select "ENG TEMPERATURE" in "ACTIVE TEST" mode.
3) Set "ENGINE TEMP" to 20°C (68°F) by touching "Qu" and "Qd" and "UP", "DOWN".

— OR —

1) Disconnect engine temperature sensor harness connector.
2) Connect a resistor (2.5 kΩ) between terminals of engine temperature sensor harness connector.
3) Connect battery ground cable.

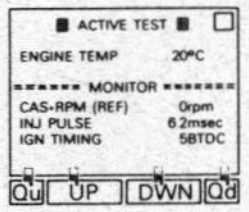

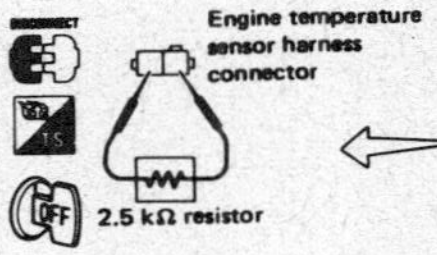

↓

Start engine and warm it up until water temperature indicator points to middle of gauge.

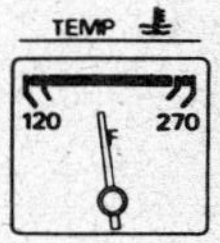

↓

Ⓓ

Ⓓ

Race engine two or three times under no-load, then run engine at idle speed.

Check "CO"%.

Idle CO: Less than 10%

After checking CO%,

1) Touch "BACK".
2) Connect A.I.V. control solenoid valve harness connector.

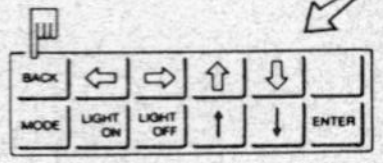

1) Disconnect the resistor from terminals of engine temperature sensor harness connector.
2) Connect engine temperature sensor harness connector to engine temperature sensor.
3) Connect A.I.V. control solenoid valve harness connector.

N.G. → Ⓕ

O.K.

Replace exhaust gas sensor.

1. See "M/R F/C MNT" in "Data monitor" mode.
2. Maintaining engine at 2,000 rpm under no-load (engine is warmed up sufficiently.), check that the monitor fluctuates between "LEAN" and "RICH" more than 5 times during 10 seconds.

RICH → LEAN → RICH →
1 time **2 times**
LEAN → RICH .....

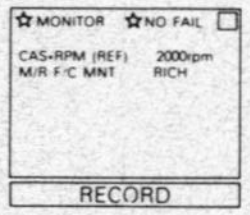

— OR —

1. Set "Exhaust gas sensor monitor" in the self-diagnostic mode II.
2. Maintaining engine at 2,000 rpm under no-load, check to make sure that the RED L.E.D. on the E.C.U. or the check engine light on the instrument panel goes ON and OFF more than 5 times during 10 seconds.

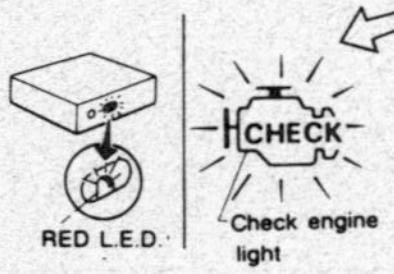

N.G. → Ⓕ

O.K. → Ⓔ

Ⓕ

Connect exhaust gas sensor harness connector to exhaust gas sensor.

↓

Check fuel pressure regulator.

↓

Check air flow meter and its circuit.

↓

Check injector and its circuit.

Clean or replace if necessary.

↓

Check engine temperature sensor and its circuit.

↓

Check E.C.U. function* by substituting another known good E.C.U.

*: E.C.U. may be the cause of a problem, but this is rarely the case.

↓

Ⓔ

1990–92 3.0L ENGINE—IDLE SPEED, IGNITION TIMING AND IDLE MIXTURE RATIO INSPECTION—M30

PREPARATION

1. **Make sure that the following parts are in good order.**
   - **Battery**
   - **Ignition system**
   - **Engine oil and coolant levels**
   - **Fuses**
   - **E.C.U. harness connector**
   - **Vacuum hoses**
   - **Air intake system (Oil filler cap, oil level gauge, etc.)**
   - **Fuel pressure**
   - **Engine compression**
   - **E.G.R. control valve operation**
   - **Throttle valve**
2. **On air conditioner equipped models, checks should be carried out while the air conditioner is "OFF".**
3. **When checking idle rpm, ignition timing and mixture ratio, checks should be carried out while shift lever is in "N" position.**
4. **When measuring "CO" percentage, insert probe more than 40 cm (15.7 in) into tail pipe.**
5. **Turn off headlamps, heater blower, rear defogger.**
6. **Keep front wheels pointed straight ahead.**
7. **Make the check after the radiator fan has stopped.**

WARNING:

a. **When selector lever is shifted to "D" position, apply parking brake and block both front and rear wheels with chocks.**

b. **Depress brake pedal while racing the engine to prevent forward surge of vehicle.**

c. **After the adjustment has been made, shift the lever to the "N" or "P" position and remove wheel chocks.**

Overall inspection sequence

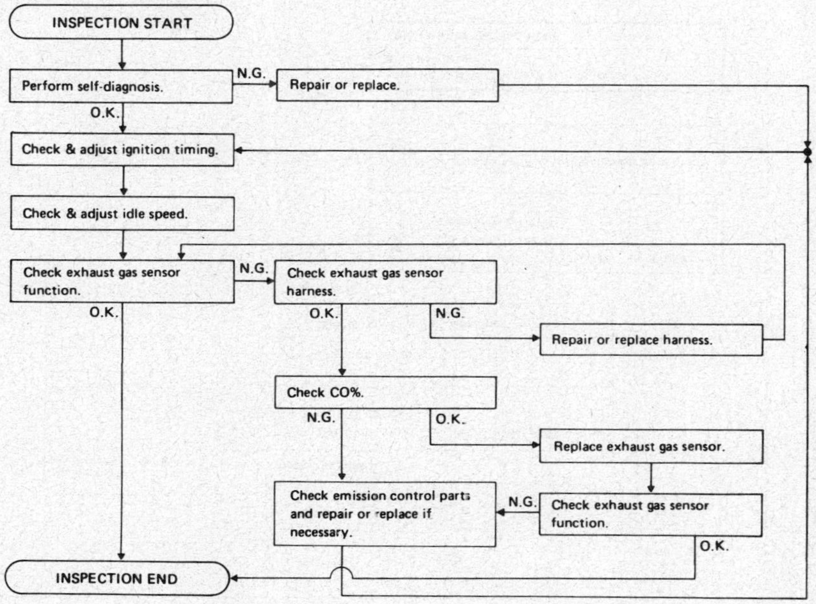

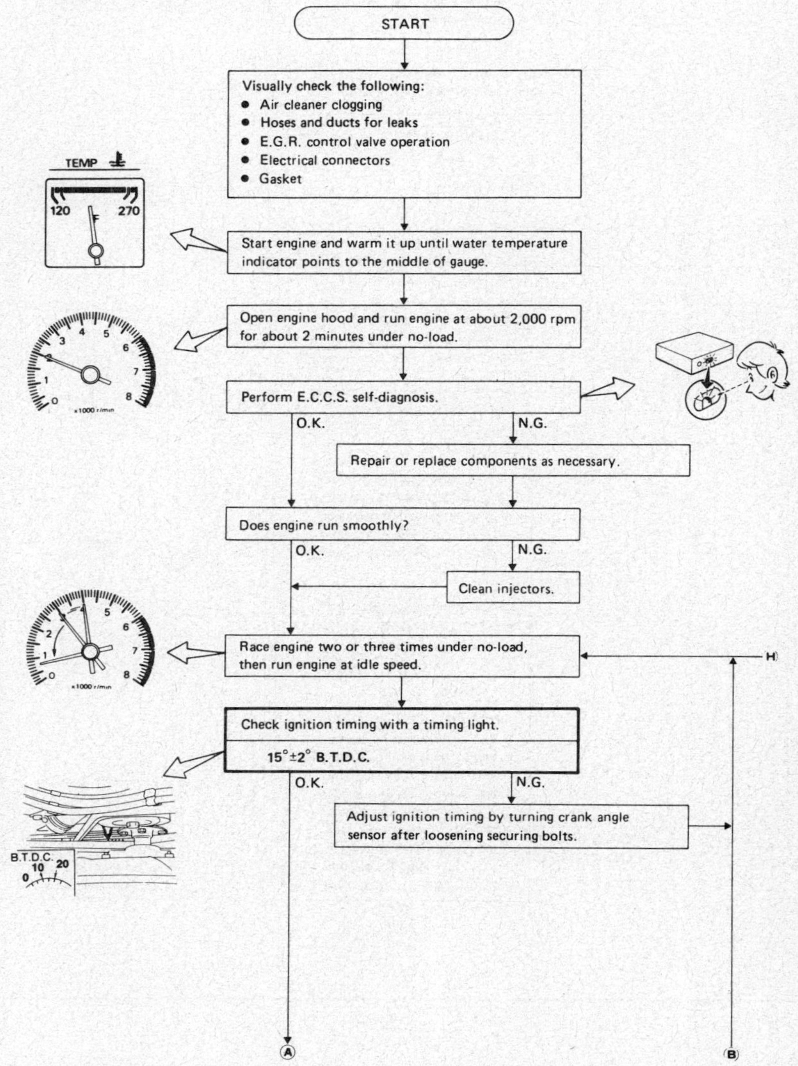

1990–92 3.0L ENGINE—IDLE SPEED, IGNITION TIMING AND IDLE MIXTURE RATIO INSPECTION—M30

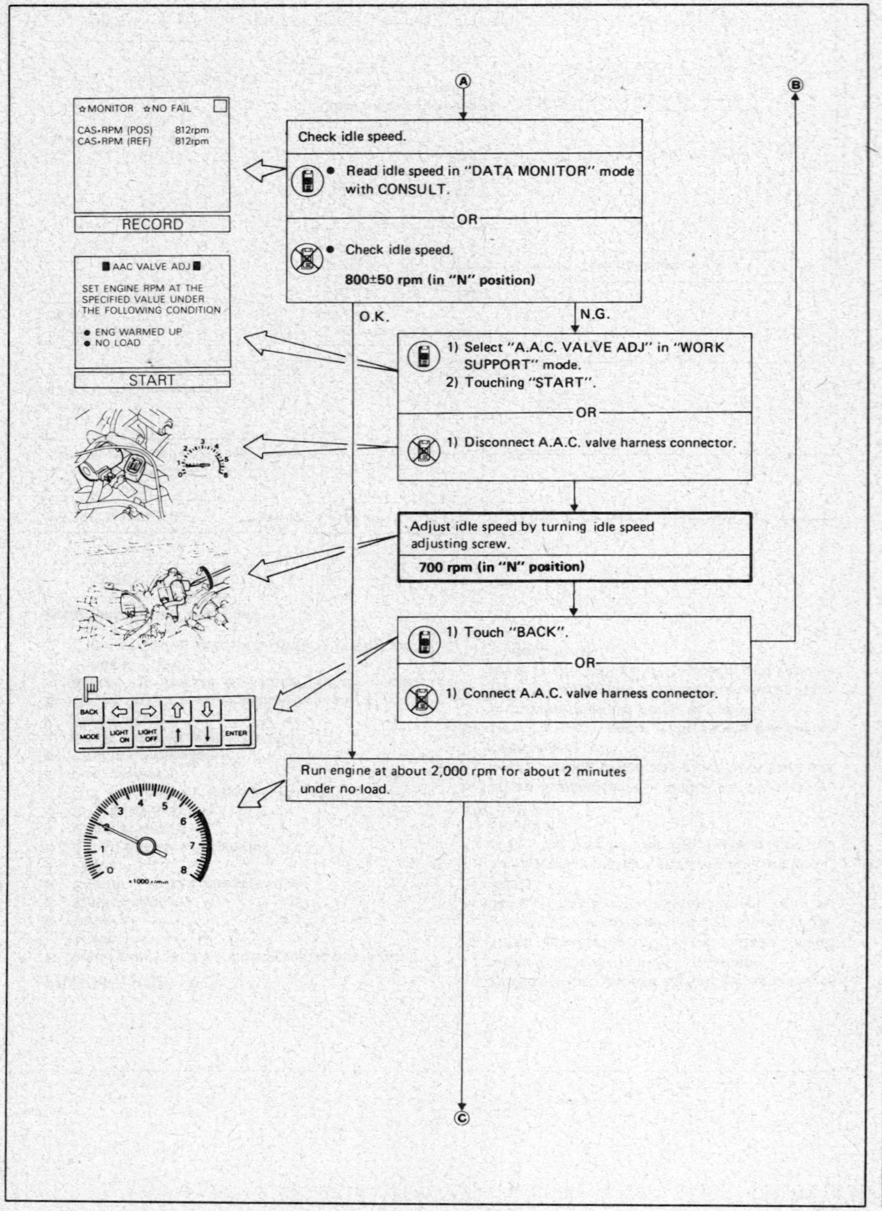

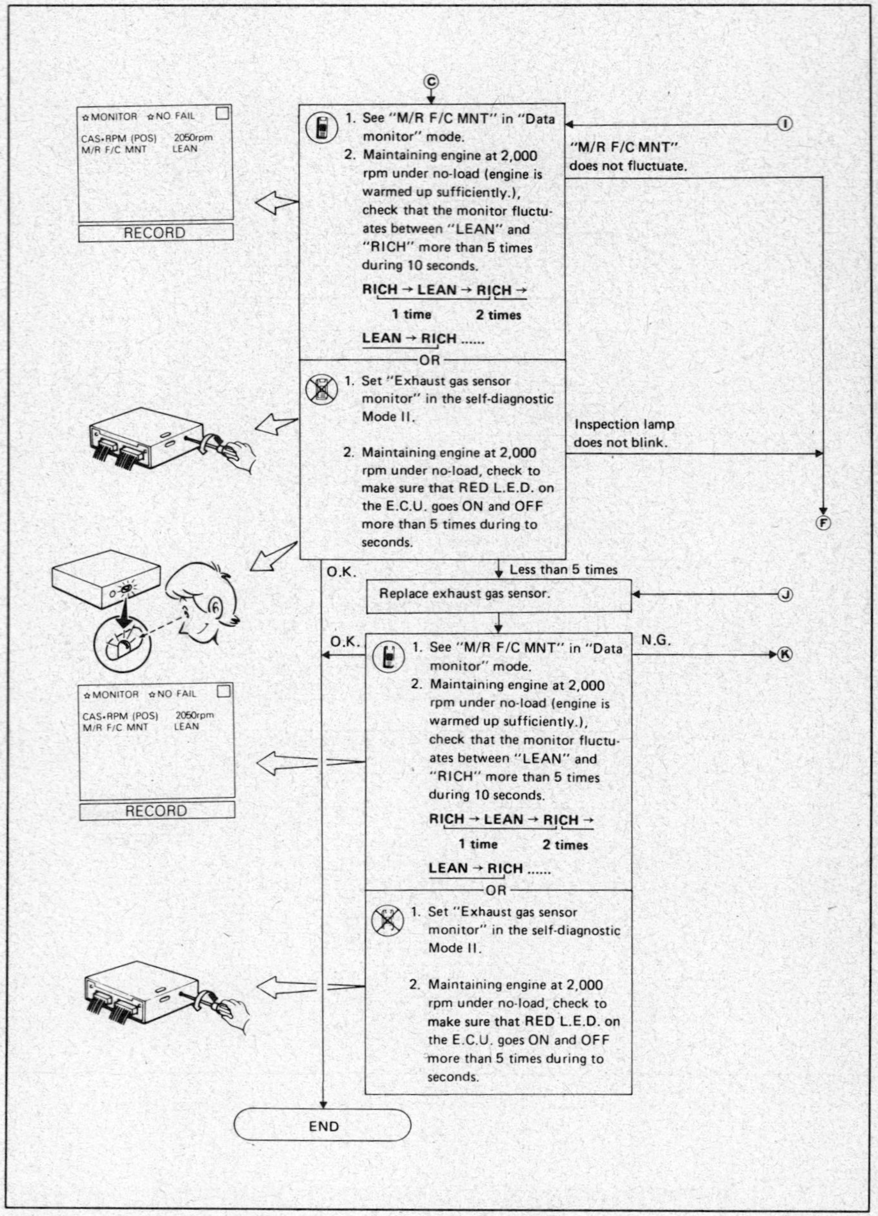

1990–92 3.0L ENGINE—IDLE SPEED, IGNITION TIMING AND IDLE MIXTURE RATIO INSPECTION—M30

(F)

Check exhaust gas sensor harness:
1) Turn off engine and disconnect battery ground cable.
2) Disconnect E.C.U. S.M.J. harness connector from E.C.U.
3) Disconnect exhaust gas sensor harness connector and connect main harness side terminal for exhaust gas sensor to ground with a jumper wire.
4) Check for continuity between terminal No. 29 of E.C.U. S.M.J. harness connector and body ground

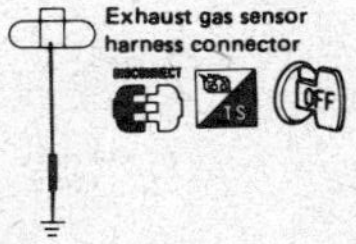

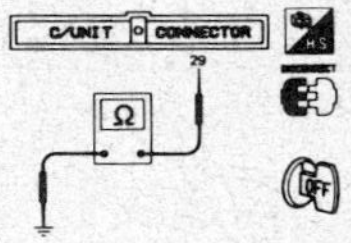

Continuity exists .... O.K.
Continuity does not exist .... N.G.

N.G. → Repair or replace harness. → (I)

O.K. → Connect 76-pin connector to control unit.

↓

Connect battery ground cable

↓

1) Select "ENG TEMPERATURE" in "ACTIVE TEST" mode.
2) Set "ENGINE TEMP" to 20°C (68°F) by touching "Qu" and "Qd" and "UP", "DOWN".

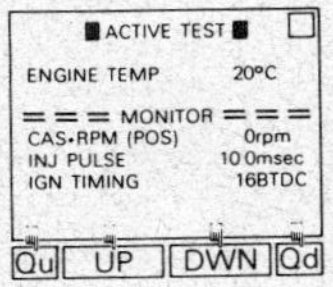

—OR—

1) Disconnect engine temperature sensor harness connector.
2) Connect a resistor (2.5 kΩ) between terminals of engine temperature sensor harness connector.

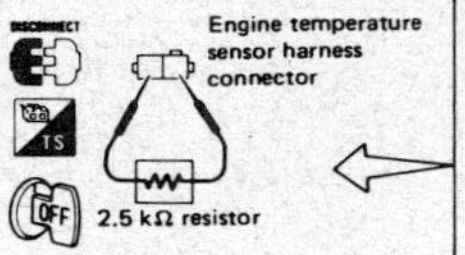

↓

Disconnect jumper wire connected to exhaust gas sensor harness connector (main harness side).

↓

Start engine and warm it up until water temperature indicator points to middle of gauge.

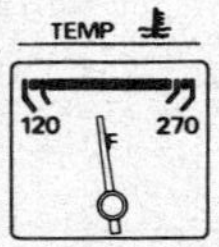

↓

(G)

(G)

Race engine two or three times under no-load, then run engine at idle speed.

↓

Check "CO" %.

Idle CO: 0.2 - 8.0%

After checking CO%,
1) Touch "BACK".

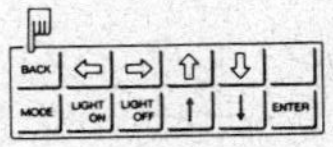

1) Disconnect the resistor from terminals of engine temperature sensor harness connector.
2) Connect engine temperature sensor harness connector to engine temperature sensor.

O.K. → (J)

N.G. → Connect exhaust gas sensor harness connector to exhaust gas sensor.

↓

Check fuel pressure regulator.

↓

Check air flow meter and its circuit.

↓

Check injector and its circuit.

Clean or replace if necessary.

↓

Check engine temperature sensor and its circuit.

↓

Check E.C.U. function* by substituting another known good E.C.U.

↓

(H)

*: E.C.U. may be the cause of a problem, but this is rarely the case.

Direct Ignition System

CHECKING IDLE SPEED AND IGNITION TIMING

Idle speed

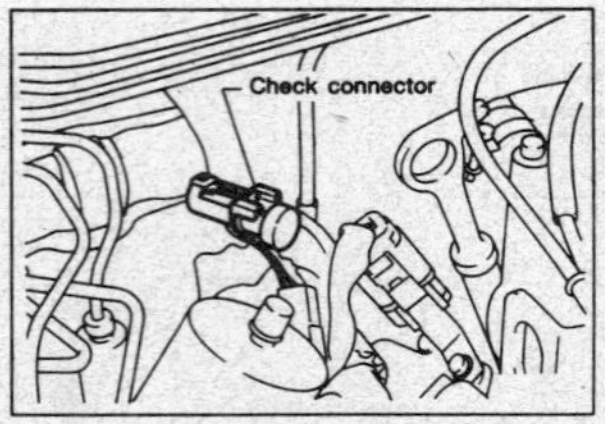

1. Disconnect check connector for voltage type tachometer.

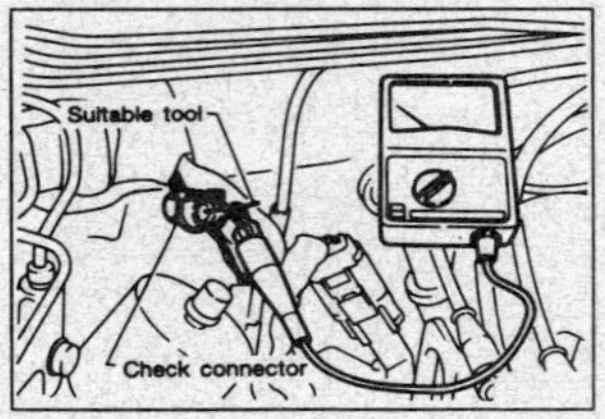

2. Connect tachometer using a suitable tool.

Ignition timing

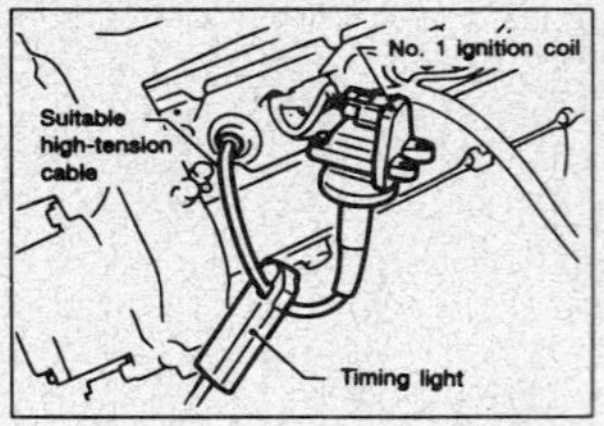

- **Method A (Without S.S.T.)**

1. Remove No. 1 or No. 6 ignition coil.
2. Connect No. 1 or No. 6 ignition coil and No. 1 or No. 6 spark plug with a suitable high-tension wire as shown, and attach timing light the above procedures, enlarge the end of a suitable high-tension wire with insulating tape as shown.
3. Check ignition timing.

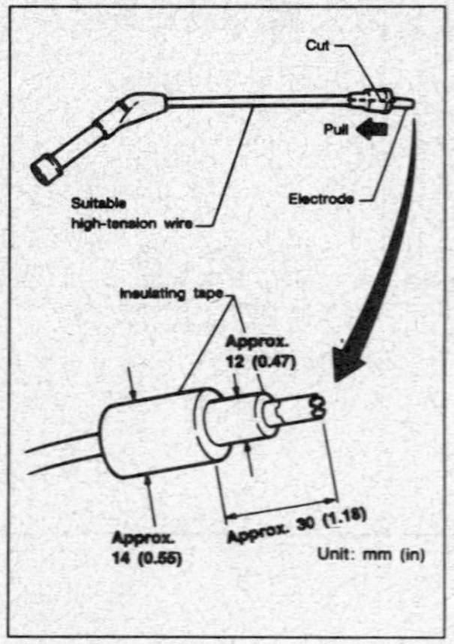

4. For the above procedures, enlarge the end of a suitable high-tension wire with insulating tape as shown.

Direct Ignition System (Cont'd)

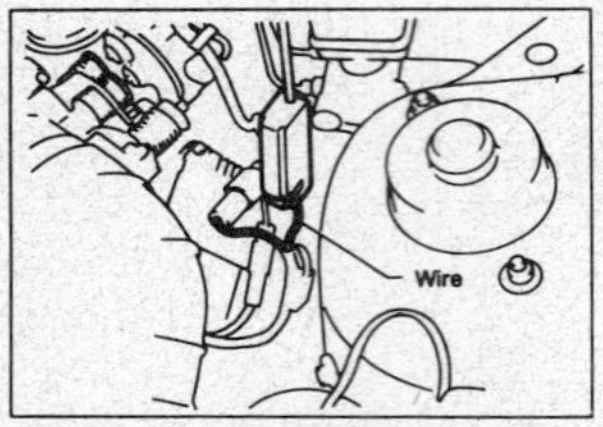

- **Method B (Without S.S.T.)**

Clamp wire as shown.
This connector is installed at the lower end of the left bank power transistor on some models, and the right bank power transistor on other models.

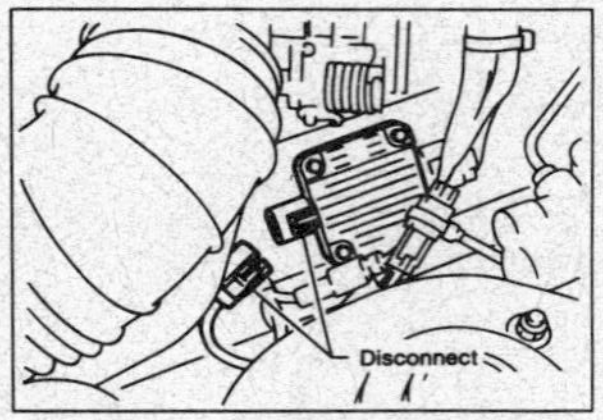

- **Method C (With S.S.T.)**

1. Disconnect No. 1 ignition coil connector.

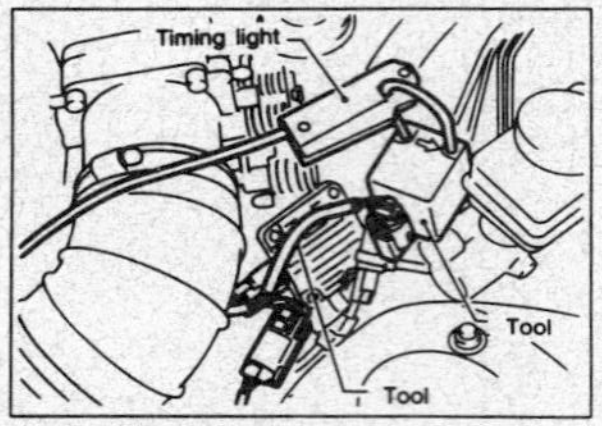

2. Connect S.S.T. and clamp wire with timing light as shown.
3. Check ignition timing.

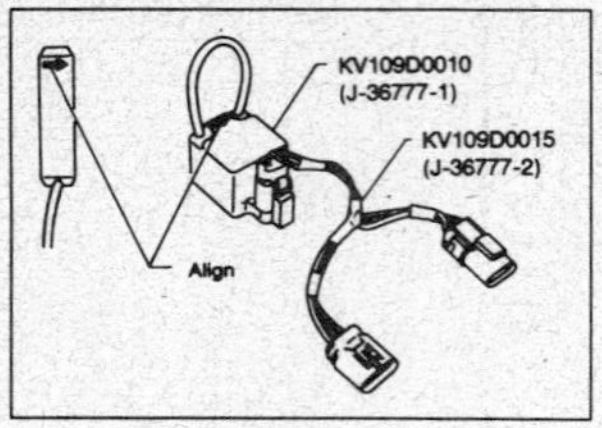

Align direction marks on S.S.T. and timing light clamp if aligning mark is punched.

1990–92 4.5L ENGINE—IDLE SPEED, IGNITION TIMING AND IDLE MIXTURE RATIO INSPECTION—Q45

PREPARATION

1. Make sure that the following parts are in good order.
   - Battery
   - Ignition system
   - Engine oil and coolant levels
   - Fuses
   - E.C.U. harness connector
   - Vacuum hoses
   - Air intake system (Oil filler cap, oil level gauge, etc.)
   - Fuel pressure
   - Engine compression
   - E.G.R. control valve operation
   - Throttle valve
2. On air conditioner equipped models, checks should be carried out while the air conditioner is "OFF".
3. On automatic transmission equipped models, when checking idle rpm, ignition timing and mixture ratio, checks should be carried out while shift lever is in "N" position.
4. When measuring "CO" percentage, insert probe more than 40 cm (15.7 in) into tail pipe.
5. Turn off headlamps, heater blower, rear defogger.
6. Keep front wheels pointed straight ahead.
7. Make the check after the radiator fan has stopped.

Overall inspection sequence

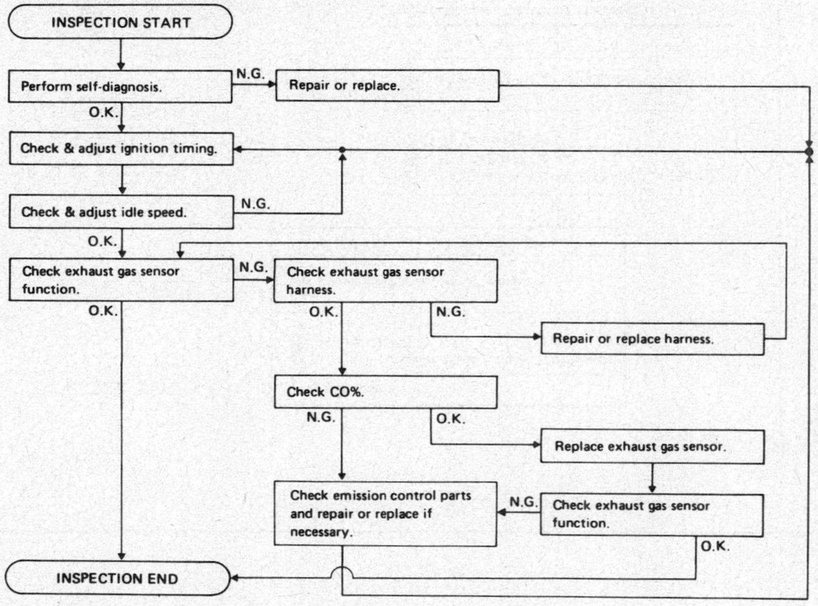

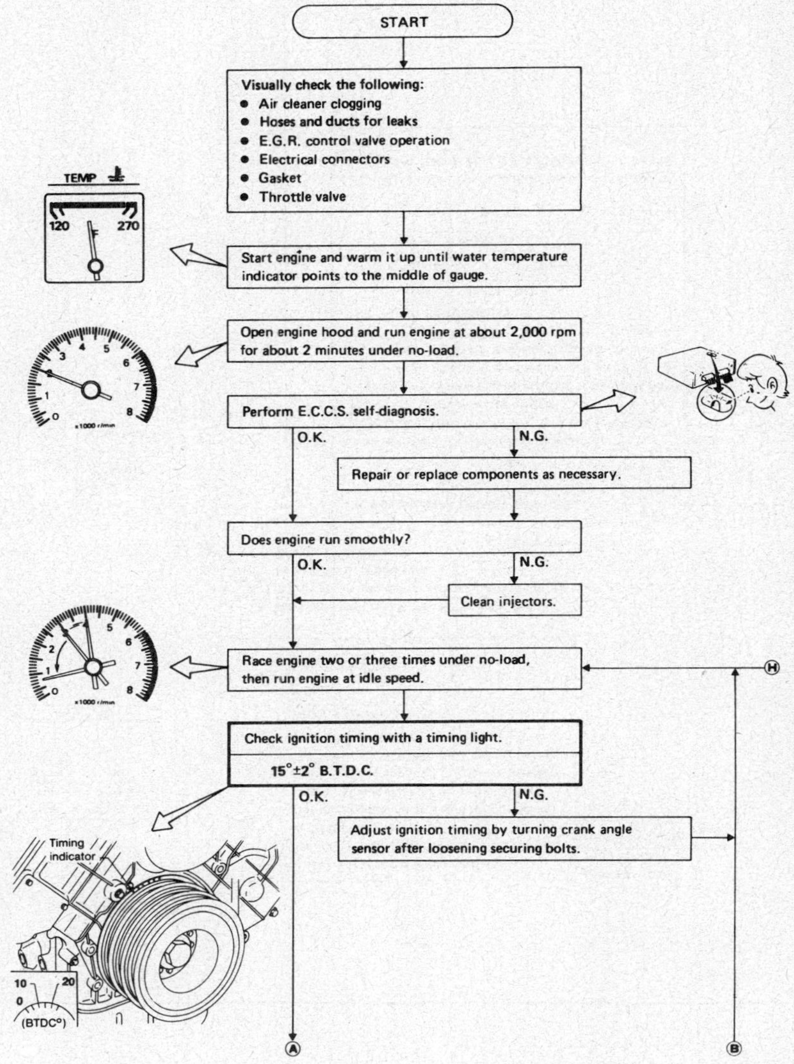

1990–92 4.5L ENGINE—IDLE SPEED, IGNITION TIMING AND IDLE MIXTURE RATIO INSPECTION—Q45

Ⓐ

Check idle speed.

Read idle speed in "DATA MONITOR" mode with CONSULT.

—OR—

Check idle speed in "N" position

1990: 750 ± 50 rpm; 1991–92: 650 ± 50 rpm

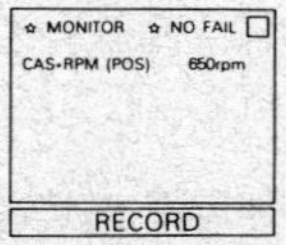

O.K. → Run engine at about 2,000 rpm for about 2 minutes under no-load.

N.G. →

1) Select "A.A.C. VALVE ADJ" in "WORK SUPPORT" mode.
2) Touch "START".

—OR—

Disconnect A.A.C. valve harness connector.

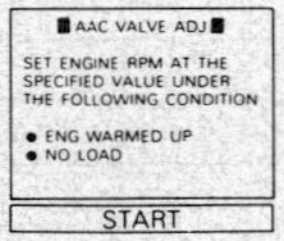

Adjust idle speed by turning idle speed adjusting screw.

1990: 670 ± 25 rpm; 1991–92: 600 ± 25 rpm

1) Touch "BACK". → Ⓑ

1) Connect A.A.C. valve harness connector.

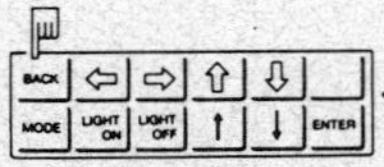

Run engine at about 2,000 rpm for about 2 minutes under no-load.

Set the diagnosis mode of E.C.U. to Mode II. ← Ⓘ

→ Ⓓ

Ⓓ

Check left bank exhaust gas sensor signal.
1) See "M/R F/C MNT (right and left sides)" in "Data monitor" mode.
2) Maintaining engine at 2,000 rpm under no-load (engine is warmed up sufficiently.), check that the monitor fluctuates between "LEAN" and "RICH" more than 5 times during 10 seconds.

RICH → LEAN → RICH →
1 time 2 times
LEAN → RICH .....

—OR—

Make sure that check engine light and red L.E.D. go on and off more than 5 times during 10 seconds at 2,000 rpm.

Monitor does not fluctuate. / Inspection lamp does not blink. → Ⓕ

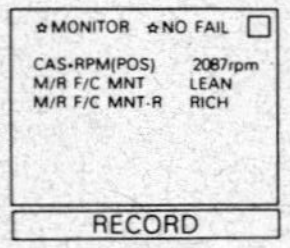

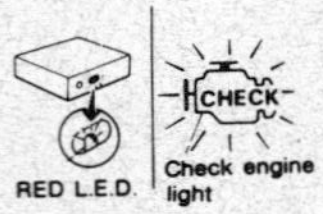

O.K. → Ⓛ

Less than 5 times →

Replace left bank exhaust gas sensor. ← Ⓙ

Check left bank exhaust gas sensor signal.
1) See "M/R F/C MNT (right and left sides)" in "Data monitor" mode.
2) Maintaining engine at 2,000 rpm under no-load (engine is warmed up sufficiently.), check that the monitor fluctuates between "LEAN" and "RICH" more than 5 times during 10 seconds.

RICH → LEAN → RICH →
1 time 2 times
LEAN → RICH .....

—OR—

Make sure that inspection lamp goes on and off more than 5 times during 10 seconds at 2,000 rpm.

O.K. → Ⓛ

N.G. → Ⓚ, Ⓔ

1990–92 4.5L ENGINE—IDLE SPEED, IGNITION TIMING AND IDLE MIXTURE RATIO INSPECTION—Q45

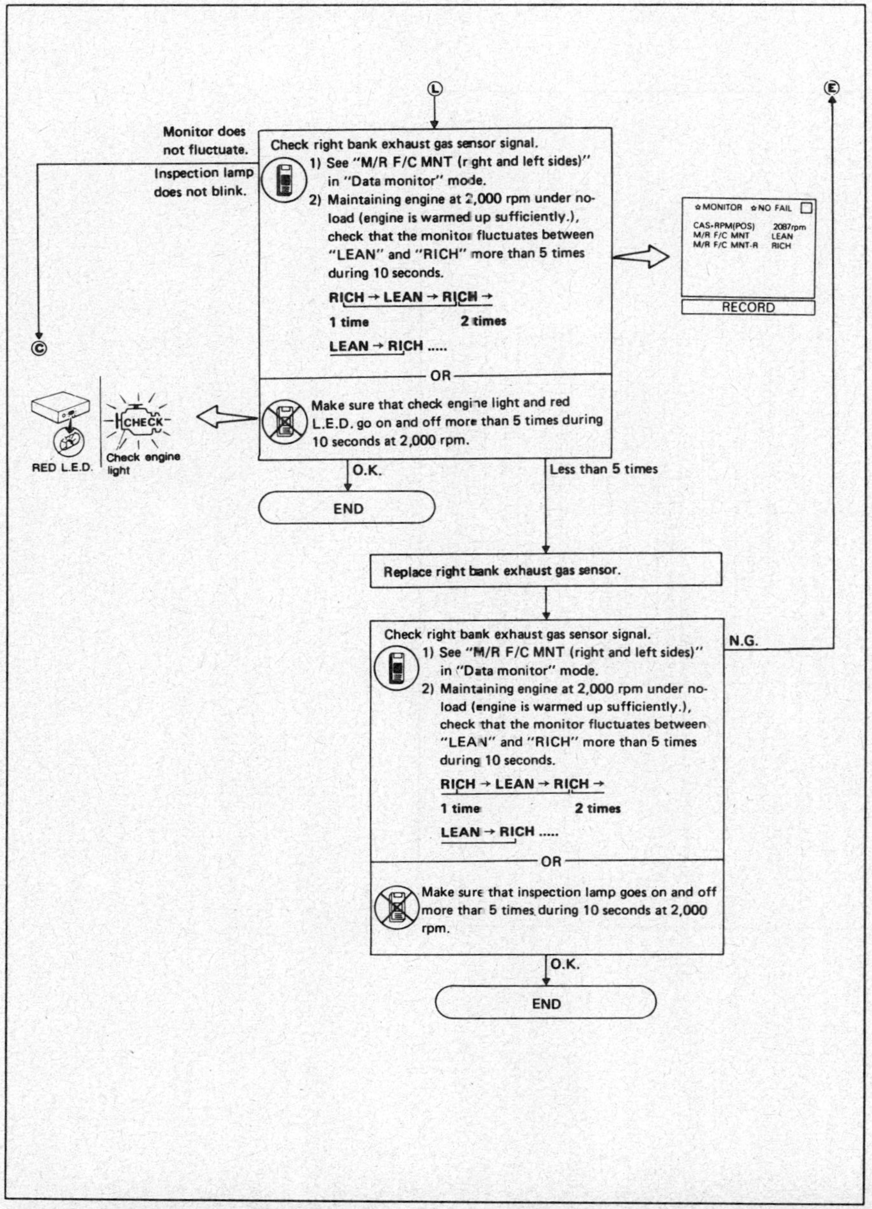

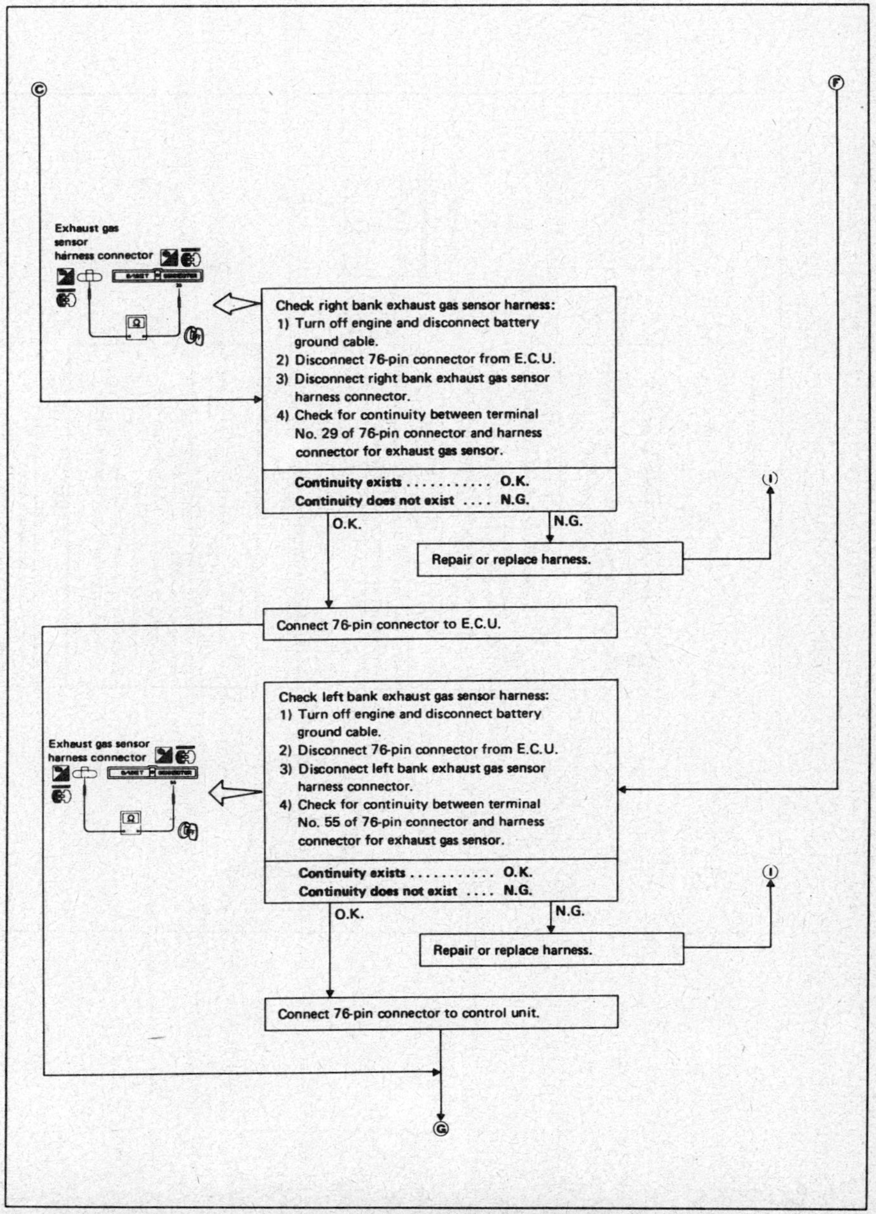

1990–92 4.5L ENGINE – IDLE SPEED, IGNITION TIMING AND IDLE MIXTURE RATIO INSPECTION – Q45

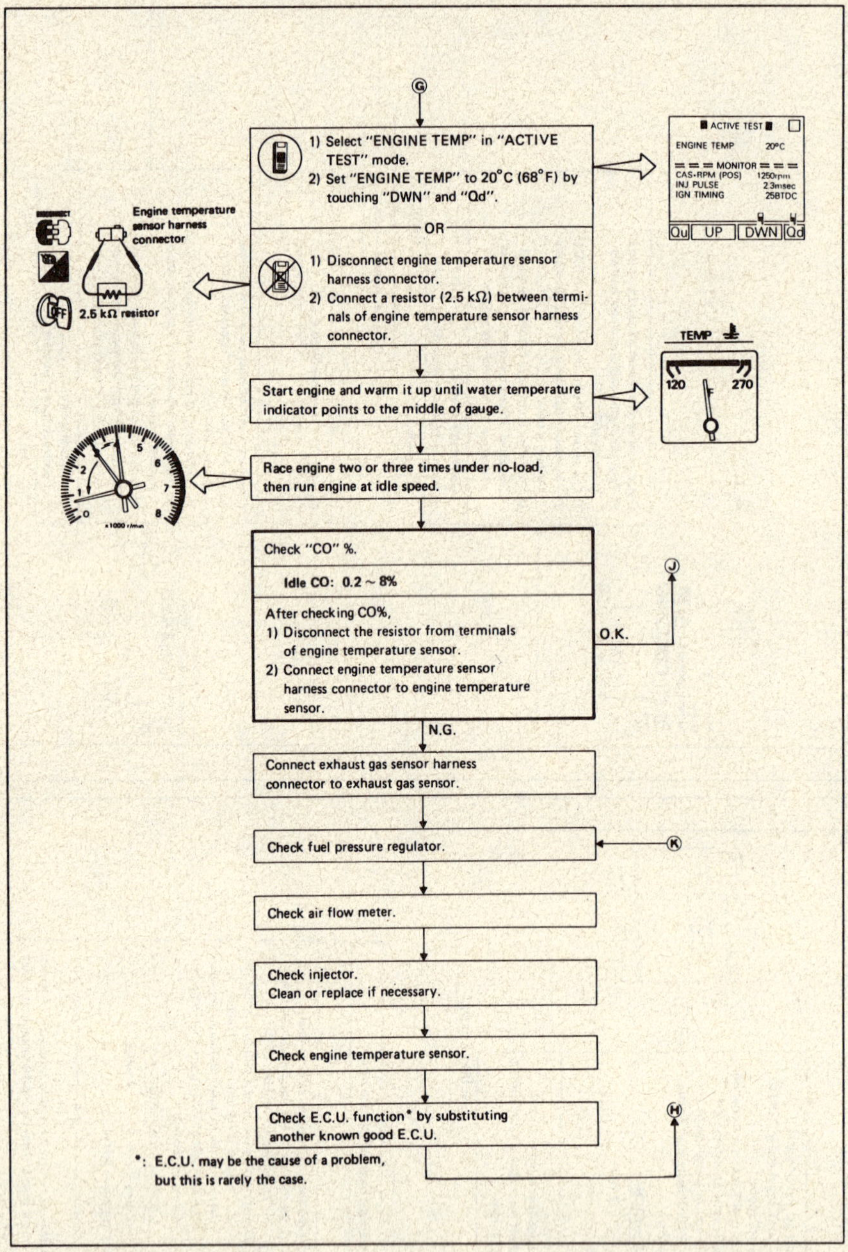

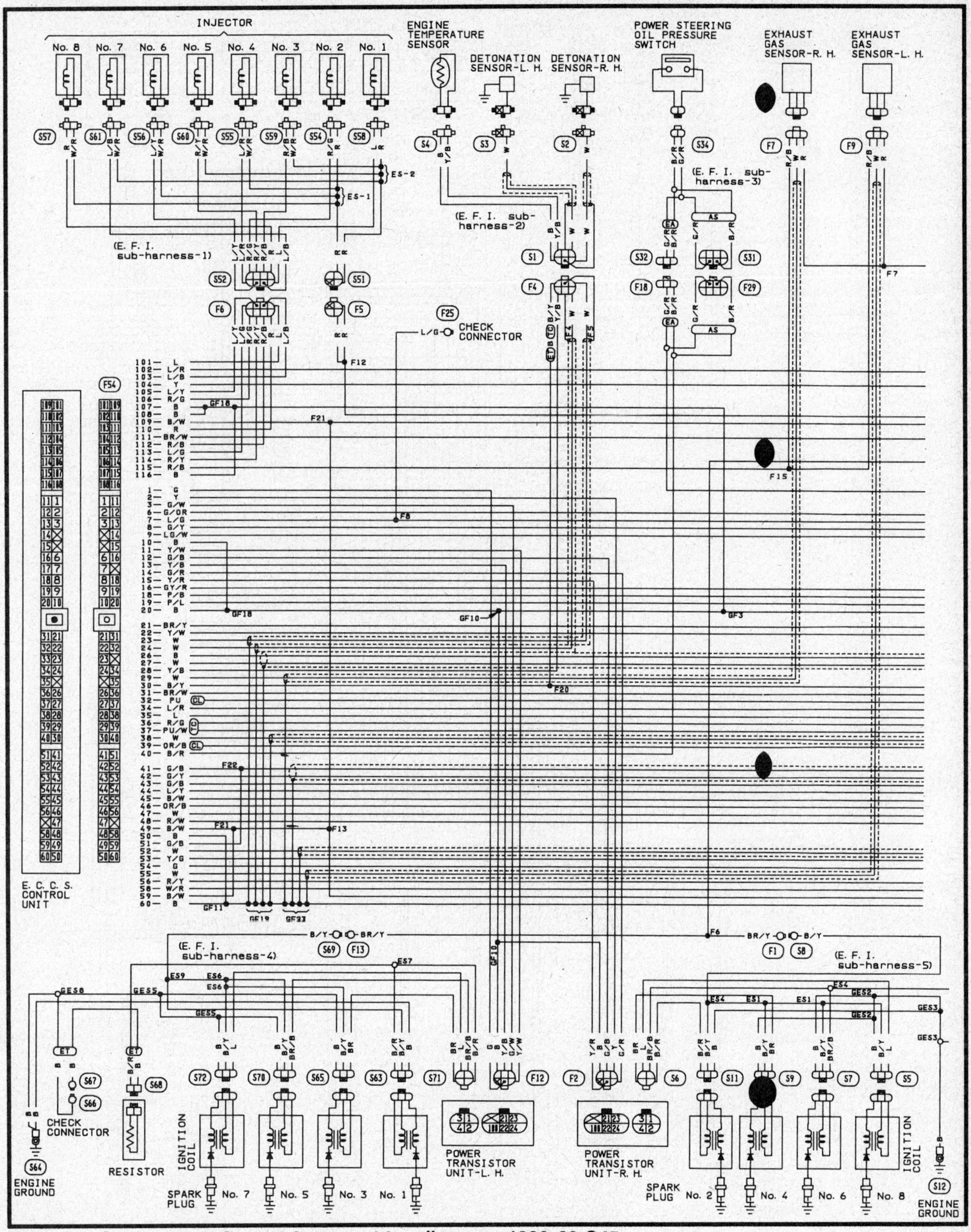

Electronic Concentrated Control System wiring diagram—1990–92 Q45

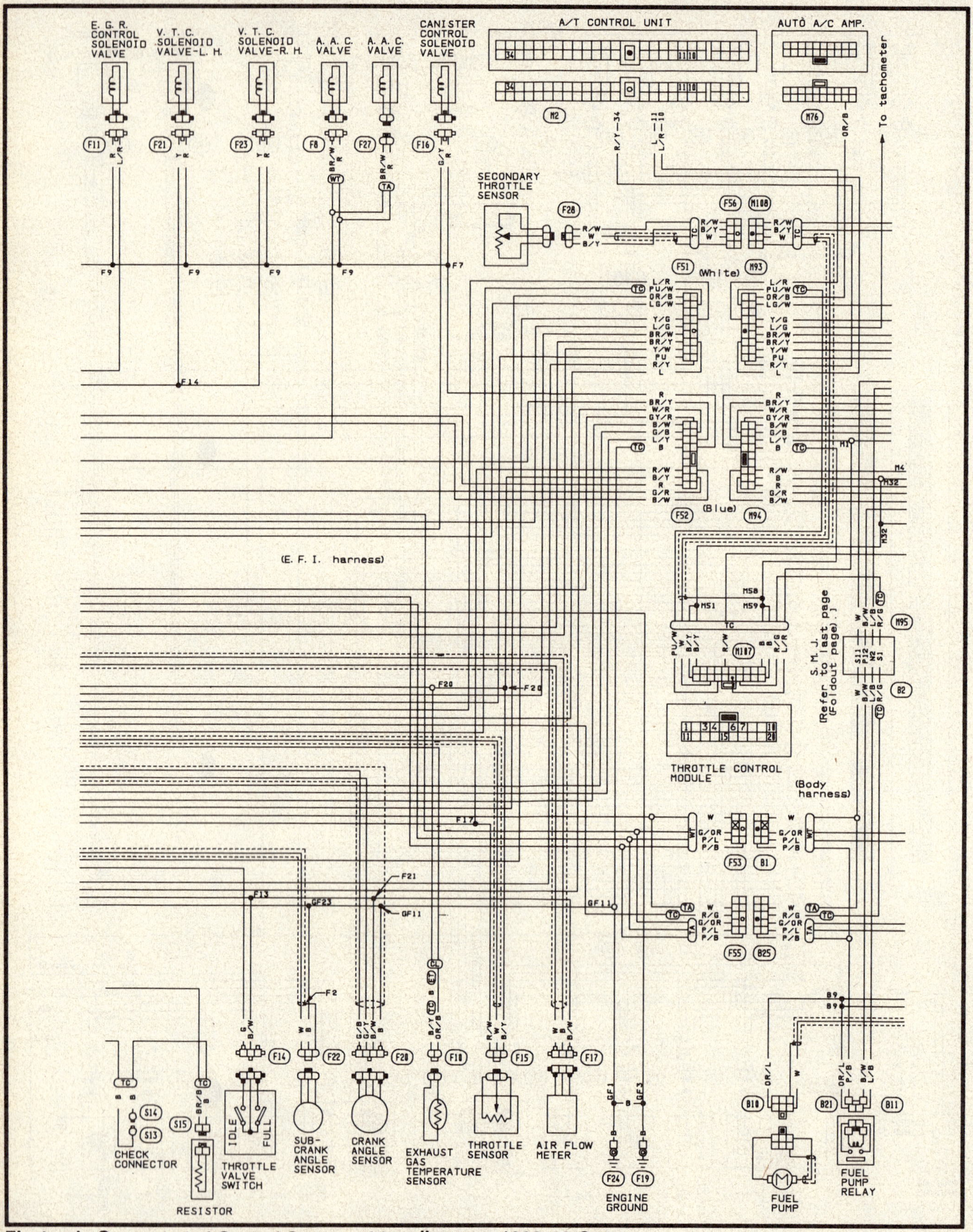

Electronic Concentrated Control System wiring diagram—1990–92 Q45

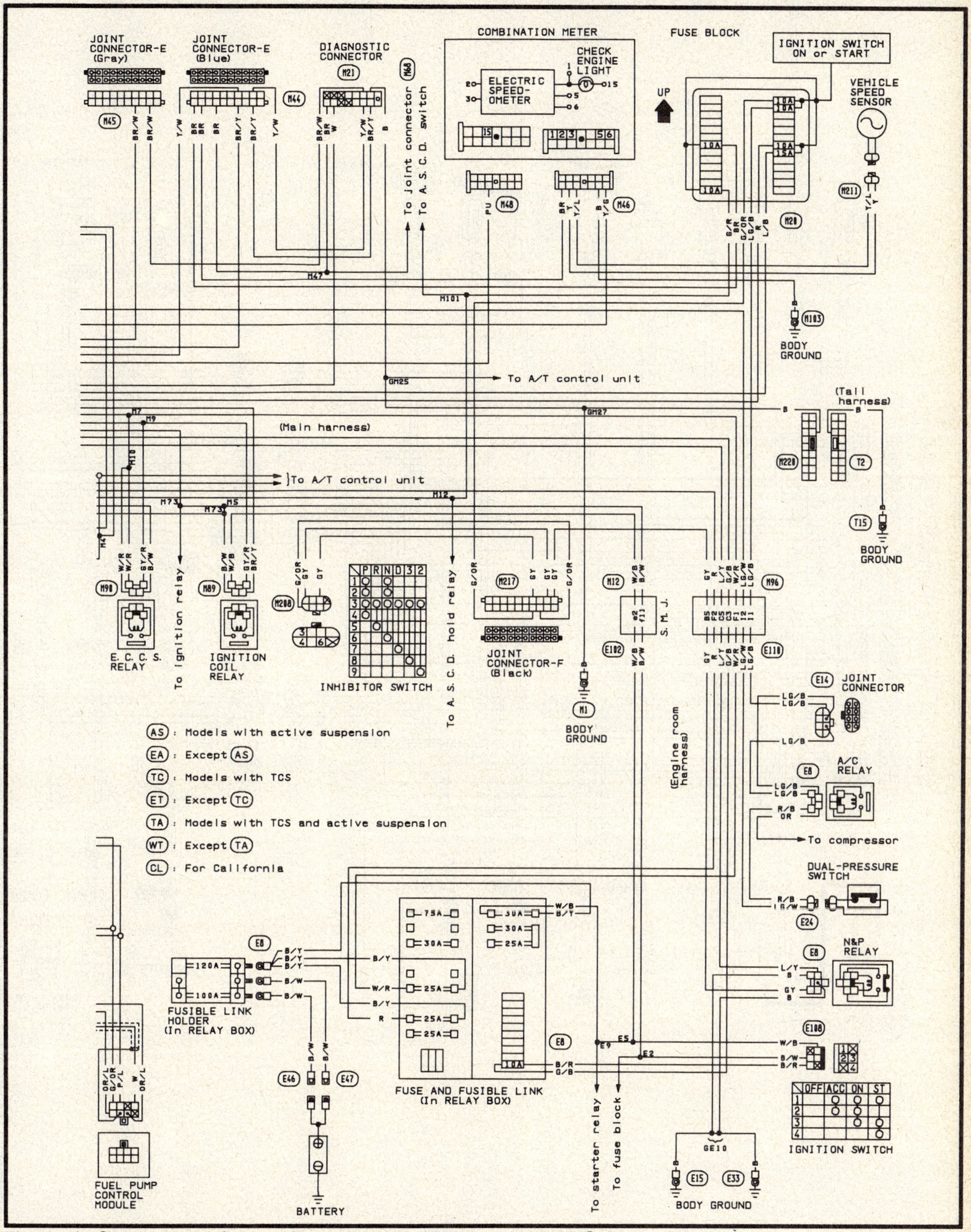

Electronic Concentrated Control System wiring diagram—1990–92 Q45

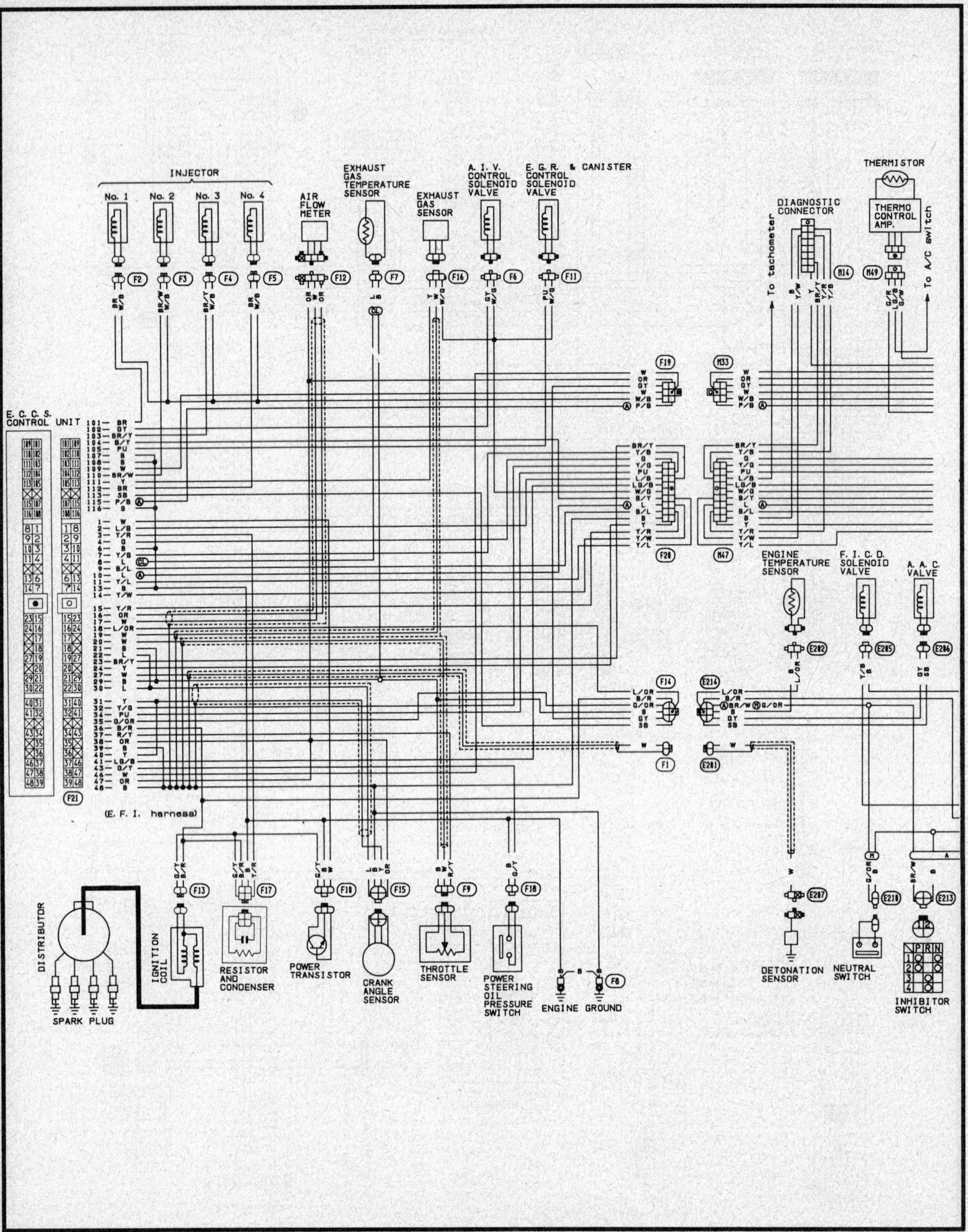

Electronic Concentrated Control System wiring diagram—1990–92 G20

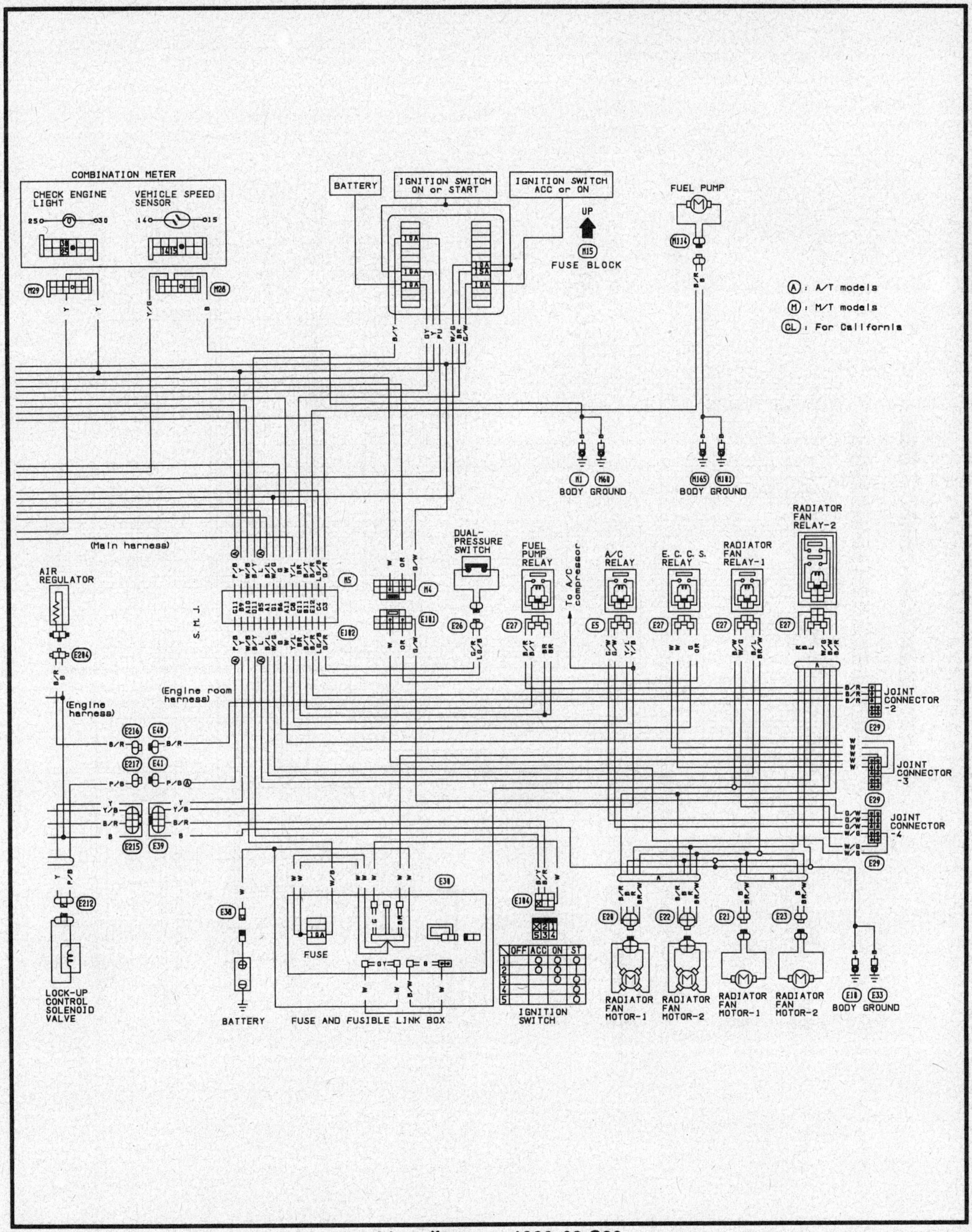

Electronic Concentrated Control System wiring diagram—1990–92 G20

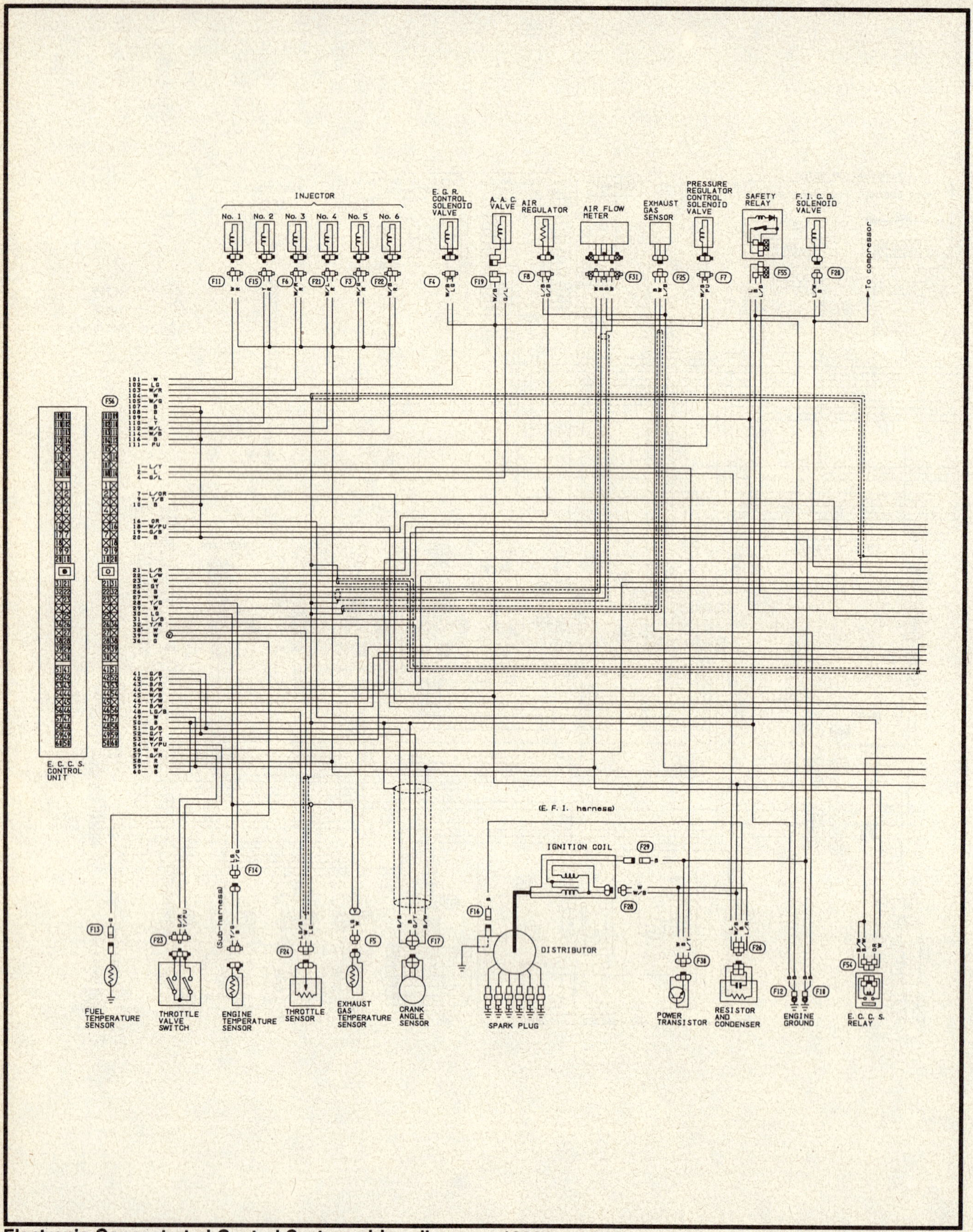

Electronic Concentrated Control System wiring diagram—1990–92 M30

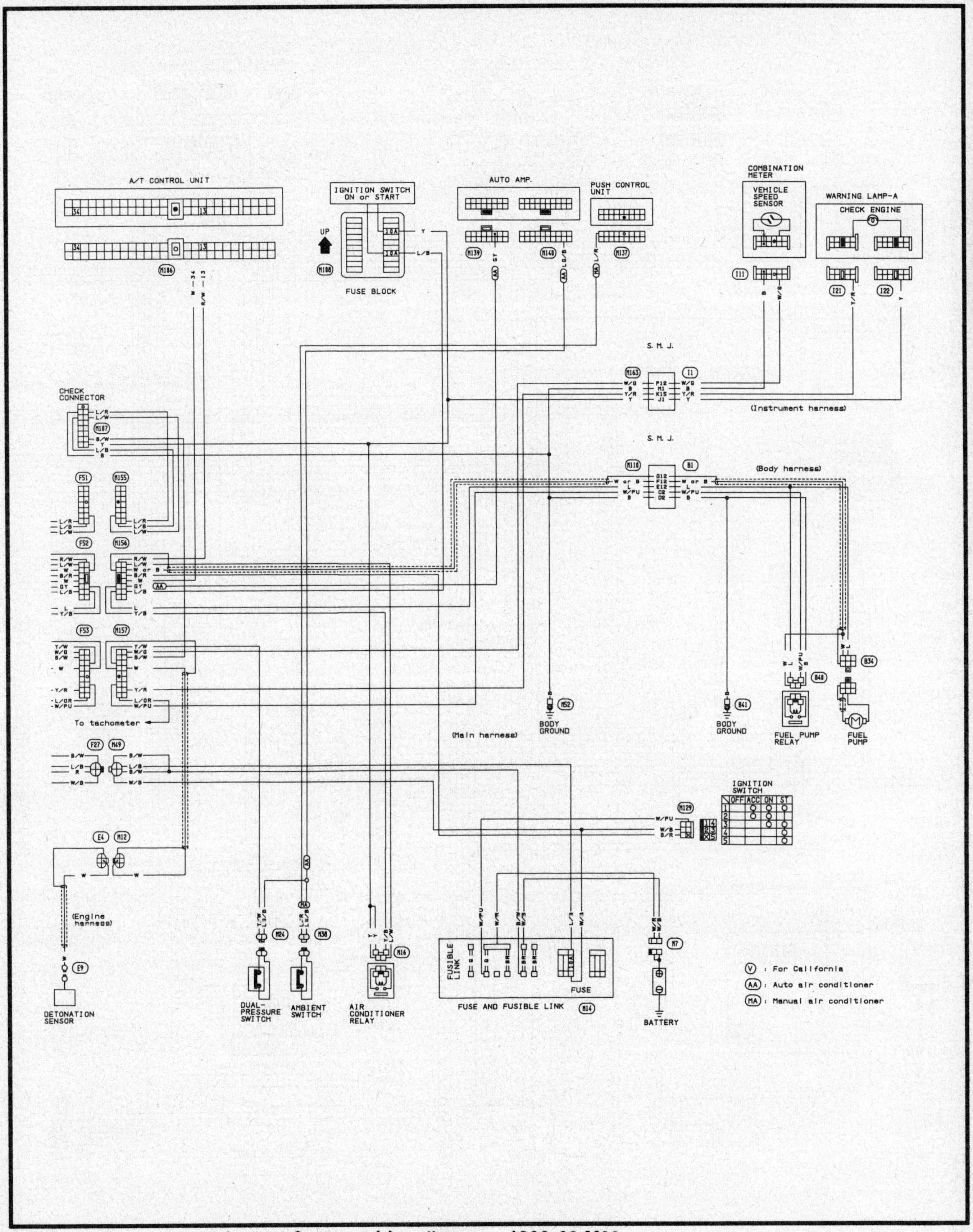

Electronic Concentrated Control System wiring diagram—1990–92 M30

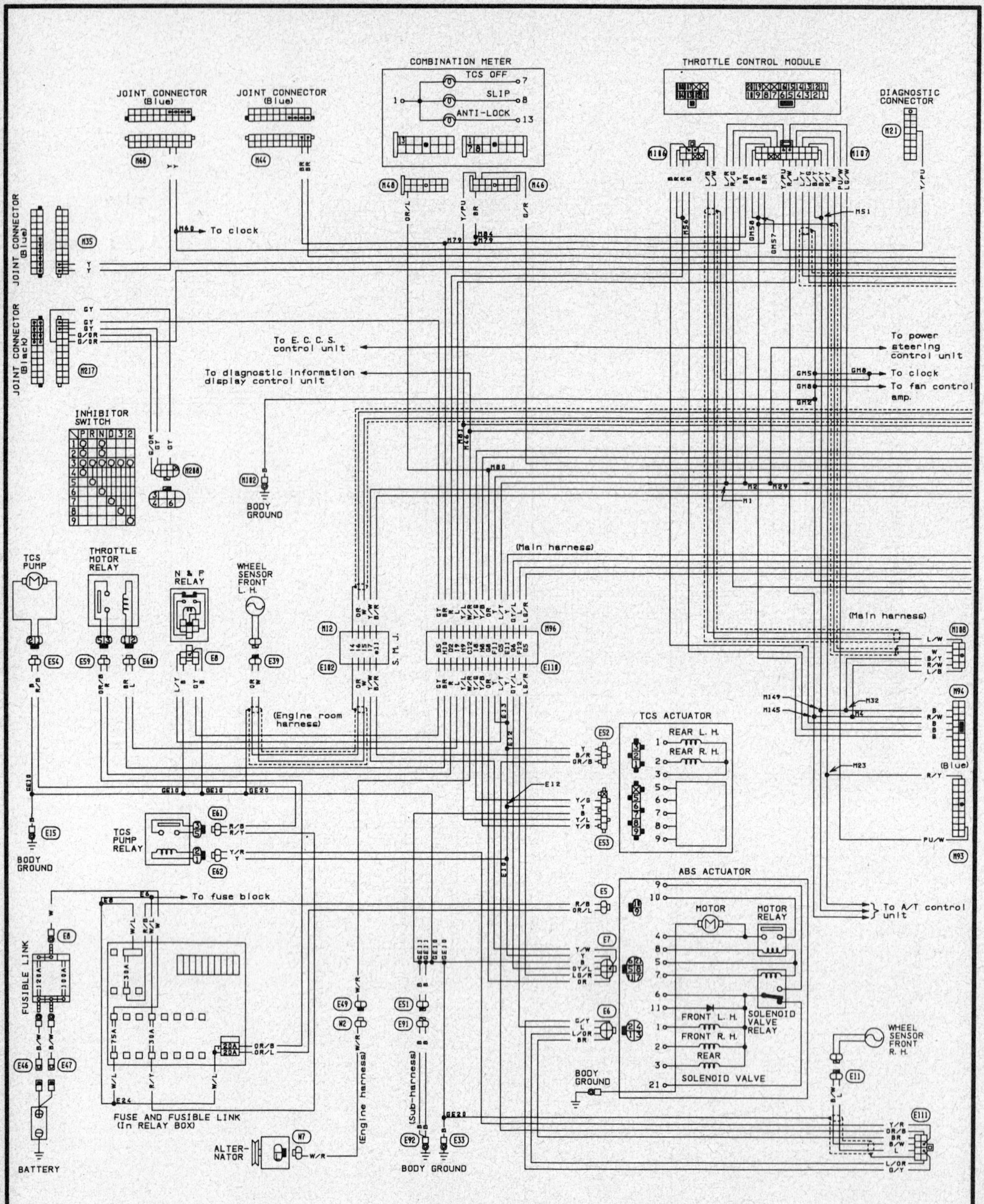

Traction Control System wiring diagram—1990–92 Q45

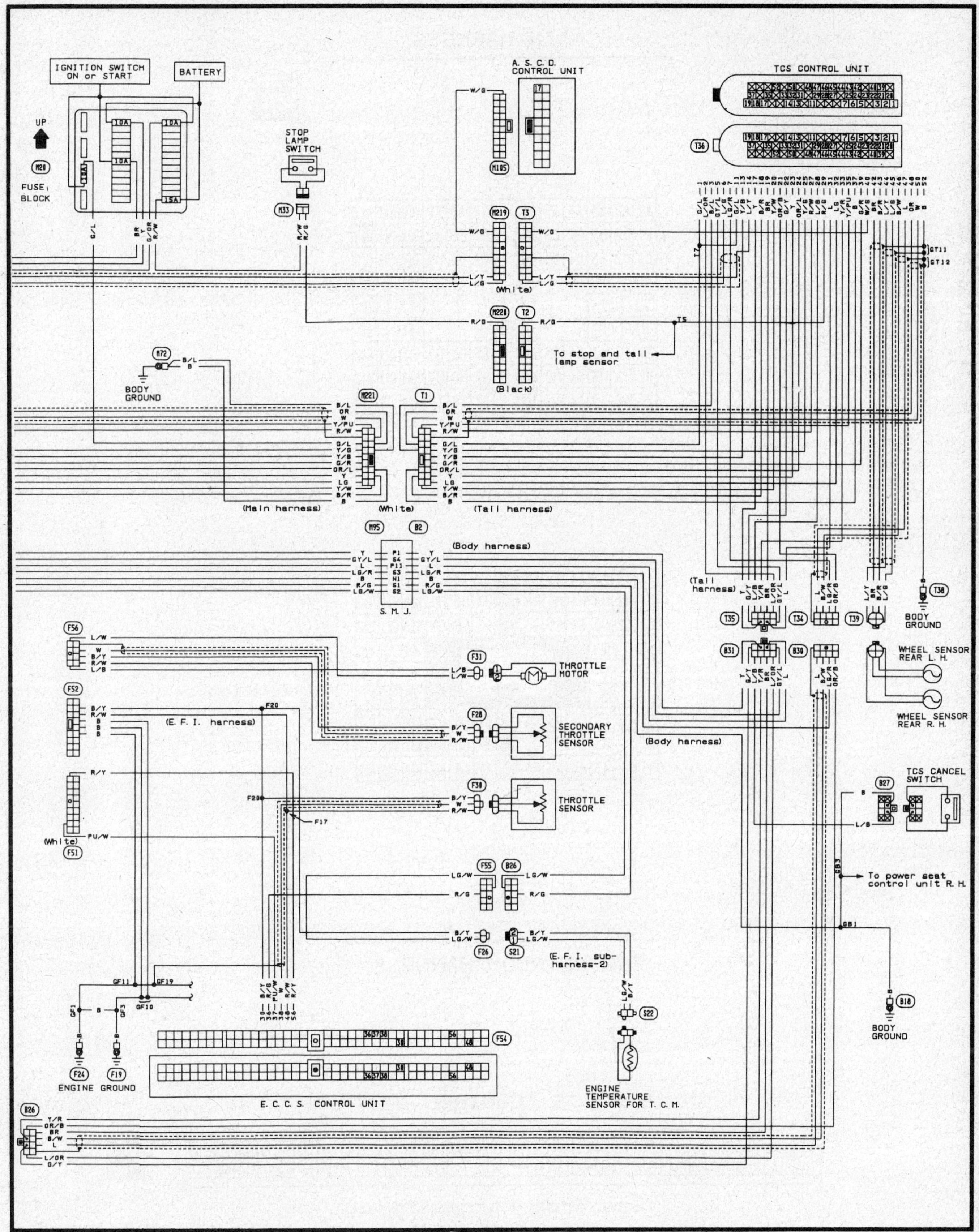

Traction Control System wiring diagram—1990–92 Q45

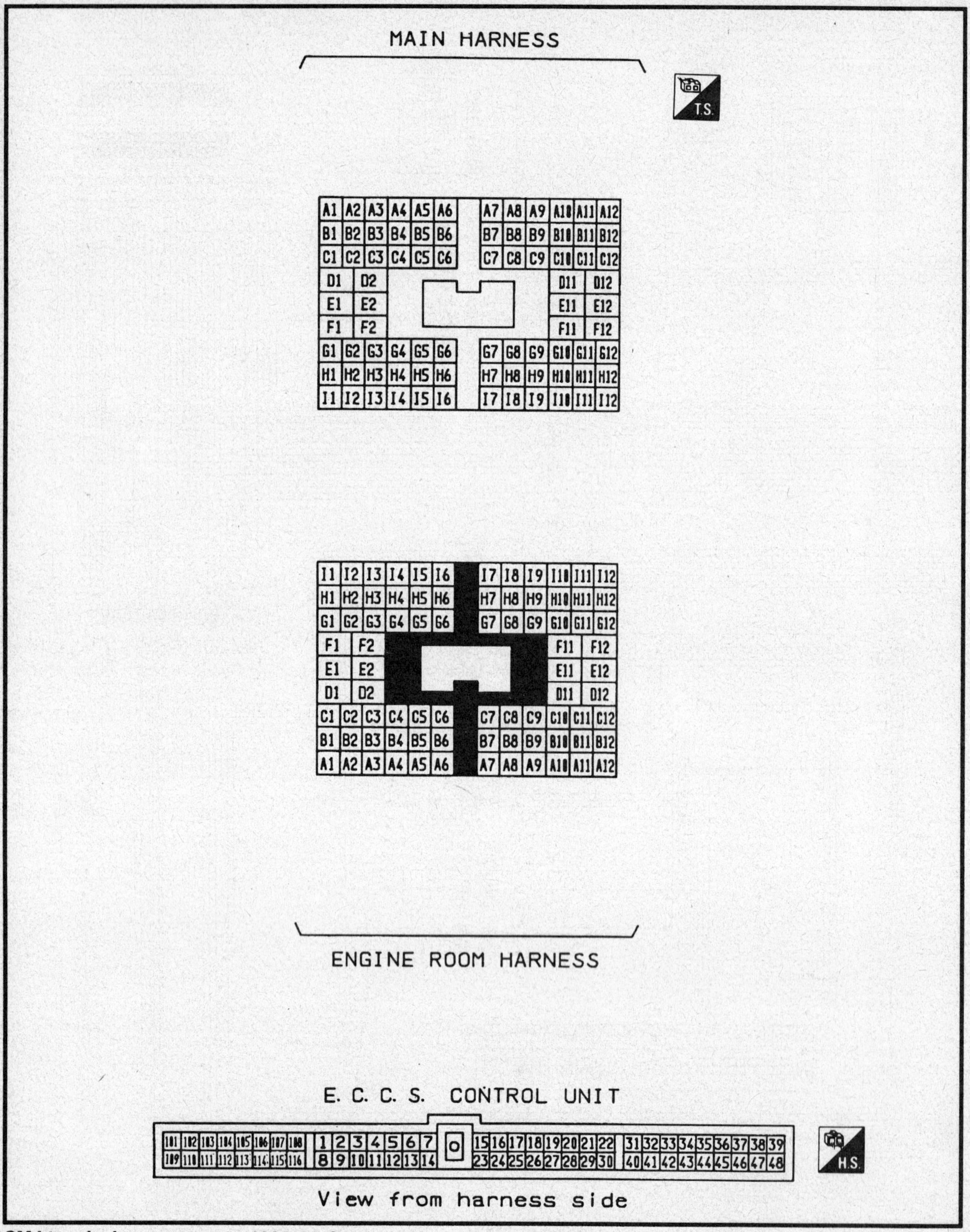

SMJ terminal arrangement – 1990–92 G20

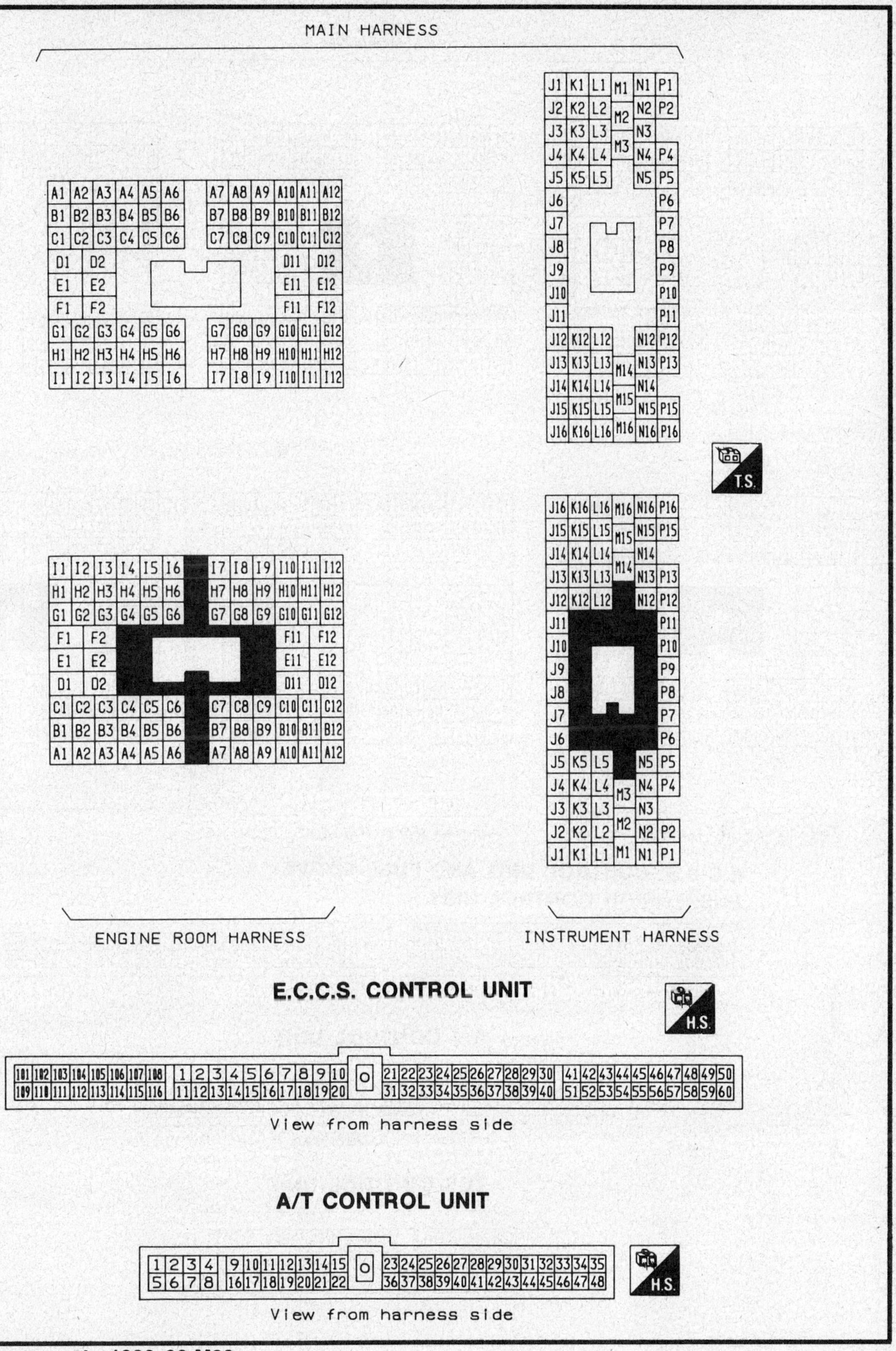

SMJ terminal arrangement—1990–92 M30

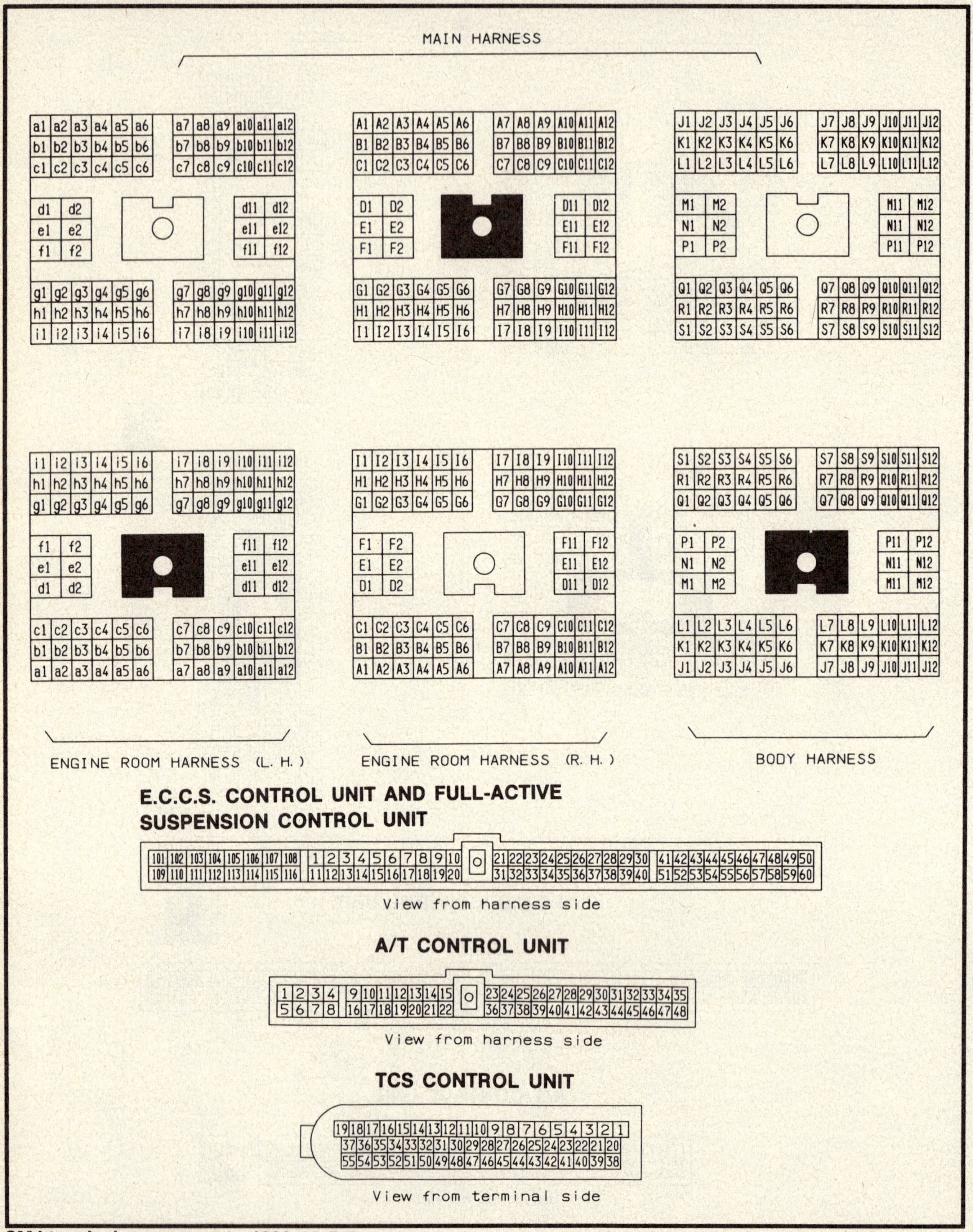

SMJ terminal arrangement—1990–92 Q45

Isuzu Fuel Injection

11 SECTION

ISUZU FUEL SYSTEM

General Information

ELECTRONIC CONTROL MODULE (ECM)

The Electronic Control Module constantly monitors and controls engine operation, which in turn helps lower emissions while maintaining optimum fuel economy and driveability. The ECM controls the fuel injection system, the ignition system and the turbocharger control system, if equipped. The ECM has built in back-up, diagnostic and fail safe control systems.

ECM Locations:

Imuplse and Stylus—under the instrument panel, to the left of the steering column

Amigo/Pickup and Rodeo—behind the left front kick panel and below the A pillar.

Trooper—Under the storage compartment in the center console

The ignition system consists of the following components; the crank angle sensor, the throttle position switch, the vehicle speed sensor, the coolant temperature sensor, air flow sensor, knock sensor, transistorized ignition coil and the ECM.

High Energy Ignition (HEI) System

EXCEPT 1.6L DOHC TURBOCHARGED ENGINE

The High Energy Ignition (HEI) system controls fuel combustion by providing a spark to ignite the air/fuel mixture at the appropriate time. To provide improved engine performance, fuel economy and control of exhaust emissions, the ECM controls distributor spark advance (timing) with the Electronic Spark Timing (EST) system.

The HEI system includes a distributor with pick-up coil assembly and secondary circuit wiring, an HEI distributor module that works in conjunction with the EST system and the ECM.

Direct Ignition System (DIS)

1.6L DOHC TURBOCHARGED ENGINE

The Direct Ignition System (DIS) consists of the following components: a coil pack, DIS module, cam angle sensor and the ECM. The coil pack consists of 2 separate, interchangeable, ignition coils. These coils operate in the same manner as previous coils. Two coils are needed because each coil fires 2 cylinders. The ignition module is located under the coil pack and is connected to the ECM by a 6 pin connector. The ignition module controls the primary circuits to the coils, turning them on and off and controls spark timing below 400 rpm and if the ECM bypass circuit becomes open or grounded.

The cam angle sensor is mounted at the front of the cylinder head and is driven by the cam shaft. The cam angle sensor supplies information to the DIS module concerning crankshaft position and engine speed.

This system uses EST and control wires from the ECM, as with the distributor systems. The ECM controls the timing using crankshaft position, engine rpm, engine temperature and manifold absolute pressure sensing.

Electronic Spark Timing (EST)

2.8L AND 3.1L ENGINES

The EST distributor contains no vacuum or centrifugal advance and uses a 7-terminal distributor module. It also has 4 wires going to a 4-terminal connector in addition to the connectors normally found on HEI distributors. A reference pulse, indicating both engine rpm and crankshaft position, is sent to the ECM. The ECM determines the proper spark advance for the engine operating conditions and sends an EST pulse to the distributor.

The EST system is designed to optimize spark timing for better control of exhaust emissions and for fuel economy improvements. The ECM monitors information from various engine sensors, computes the desired spark timing and changes the timing accordingly. A backup spark advance system is incorporated in the module in case of EST failure.

Electronic Spark Control (ESC)

2.8L AND 3.1L ENGINES

When engines are equipped with ESC in conjunction with EST, ESC is used to reduce spark advance under conditions of detonation. A knock sensor signals a separate ESC controller to retard the timing when it senses engine knock. The ESC controller signals the ECM which reduces spark advance until no more signals are received from the knock sensor.

DATA SENSORS

Air Flow Sensor

EXCEPT 2.8L AND 3.1L ENGINES

The air flow sensor is usually located in the air cleaner housing assembly. The purpose of the air flow sensor is to measure the volume (rate) of air that is coming into the engine.

Coolant Temperature Sensor

EXCEPT 2.8L AND 3.1L ENGINES

This sensor is usually located on the engine block, under the intake manifold. It sends the coolant temperature information back to the ECU. The ECU then uses this information to determine the engine temperature for calculating the required air/fuel mixture.

2.8L AND 3.1L ENGINES

The coolant sensor is a thermistor (a resistor which changes value based on temperature) mounted in the engine coolant stream. As the temperature of the engine coolant changes, the resistance of the coolant sensor changes. Low coolant temperature produces a high resistance (100,000 ohms at −40°F/−40°C), while high temperature causes low resistance (70 ohms at 266°F/130°C).

The ECM supplies a 5 volt signal to the coolant sensor and measures the voltage that returns. By measuring the voltage change, the ECM determines the engine coolant temperature. The voltage will be high when the engine is cold and low when the engine is hot. This information is used to control fuel management, IAC, spark timing, EGR, canister purge and other engine operating conditions.

A failure in the coolant sensor circuit should either set a Code 14 or 15. These codes indicate a failure in the coolant temperature sensor circuit.

Cam Angle Sensor

1.6L DOHC TURBOCHARGED ENGINE

Based on the pulses received from the inductive pick-up sensor in the cam angle sensor, the DIS module sends rpm reference signals to the ECM which are used to indicate crankshaft position and engine speed. The magnetic pickup sensor provides a cam signal to identify correct firing sequence and crank signals to trigger each coil at the proper time.

Crank Angle Sensor

2.6L ENGINE

The crank angle sensor is located inside the distributor housing.

The sensor is used to detect engine speed and the relative position of each piston. Using these parameters, the ECU calculates the proper ignition timing and dwell angle. The ECU then sends a signal to the transistorized ignition coil to create a spark.

Crankshaft Position/Reference Signal

1.6L SOHC AND DOHC NON-TURBOCHARGED, 1.8L DOHC, 2.8L AND 3.1L ENGINES

The HEI ignition system supplies a crankshaft position/reference signal to the ECM via the distributor HEI module. The signal is supplied by the ignition pick-up coil in the distributor to terminals **P and** N of the HEI module.

Knock Sensor

This sensor is usually located in the engine block near the cylinders or in the intake manifold at the rear of the engine. The purpose of this sensor is to send electrical impulses to the ECU when ever engine "knock" occurs. The ECU uses these impulses to retard the ignition timing.

Manifold Absolute Pressure (MAP) Sensor

The Manifold Absolute Pressure (MAP) sensor measures the changes in the intake manifold pressure which result from engine load and speed changes. The pressure measured by the MAP sensor is the difference between barometric pressure (outside air) and manifold pressure (vacuum). A closed throttle engine coastdown would produce a relatively low MAP value (approximately 20–35 kPa), while wide-open throttle would produce a high value (100 kPa). This high value is produced when the pressure inside the manifold is the same as outside the manifold, and 100% of outside air (or 100 kPa) is being measured. This MAP output is the opposite of what you would measure on a vacuum gauge. The use of this sensor also allows the ECM to adjust automatically for different altitude.

The ECM sends a 5 volt reference signal to the MAP sensor. As the MAP changes, the electrical resistance of the sensor also changes. By monitoring the sensor output voltage the ECM can determine the manifold pressure. A higher pressure, lower vacuum (high voltage) requires more fuel, while a lower pressure, higher vacuum (low voltage) requires less fuel. The ECM uses the MAP sensor to control fuel delivery and ignition timing. A failure in the MAP sensor circuit should set a Code 33 or Code 34.

Manifold Air Temperature (MAT) Sensor

1.6L AND 1.8L ENGINES

The MAT sensor is a thermistor mounted in the intake manifold. A thermistor is a resistor which changes resistance based on temperature. Low manifold air temperature produces a high resistance (100,000 ohms at −40°F/−40°C), while high temperature causes low resistance (70 ohms at 266°F/130°C).

The ECM supplies a 5 volt signal to the MAT sensor through a resistor in the ECM and monitors the voltage. The voltage will be high when the manifold air is cold and low when the air is hot. By monitoring the voltage, the ECM calculates the air temperature and uses this data to help determine the fuel delivery and spark advance. A failure in the MAT circuit should set either a Code 23 or Code 25.

Oil Pressure Switch

The oil pressure switch is usually mounted on the back of the engine, just below the intake manifold. Some vehicles use the oil pressure switch as a parallel power supply (with the fuel pump relay) and will provide voltage to the fuel pump after approximately 4 psi (28 kPa) of oil pressure is reached. This switch will also help prevent engine seizure by shutting off the power to the fuel pump and causing the engine to stop when the oil pressure is lower than 4 psi.

Oxygen Sensor

EXCEPT 2.8L AND 3.1L ENGINES

This sensor is usually threaded into the exhaust manifold. The oxygen sensor measures and produces an electrical signal proportional to the amount of the oxygen present in the exhaust gases.

2.8L AND 3.1L ENGINES

The exhaust oxygen sensor is mounted in the exhaust system where it can monitor the oxygen content of the exhaust gas stream. The oxygen content in the exhaust reacts with the oxygen sensor to produce a voltage output. This voltage ranges from approximately 100 millivolts (high oxygen—lean mixture) to 900 millivolts (low oxygen—rich mixture).

By monitoring the voltage output of the oxygen sensor, the ECM will determine what fuel mixture command to give to the injector (lean mixture—low voltage—rich command, rich mixture—high voltage—lean command).

Remember that the oxygen sensor indicates to the ECM what is happening in the exhaust. It does not cause things to happen. It is a type of gauge: high oxygen content = lean mixture; low oxygen content = rich mixture. The ECM adjusts fuel to keep the system working.

The oxygen sensor, if open, should set a Code 13. A constant low voltage in the sensor circuit should set a Code 44 while a constant high voltage in the circuit should set a Code 45. Codes 44 and 45 could also be set as a result of fuel system problems.

Park/Neutral Switch

2.8L AND 3.1L ENGINES

NOTE: Vehicle should not be driven with the park/neutral switch disconnected as idle quality may be affected in PARK or NEUTRAL and a Code 24 (VSS) may be set.

This switch indicates to the ECM when the transmission is in P or N. The information is used by the ECM for control on the torque converter clutch, EGR, and the idle air control valve operation.

Power Steering Pressure Switch

IMPULSE AND STYLUS

The power steering pressure switch is used so that the power steering pump load will not effect the engine idle. Turning the steering wheel increases the power steering oil pressure and pump load on the engine. The power steering pressure switch will close before the load can cause an idle problem.

Throttle Position Sensor

EXCEPT 2.8L AND 3.1L ENGINES

The Throttle Position Sensor (TPS) is used on turbocharged models only. The sensor controls the fuel cut system.

2.8L AND 3.1L ENGINES

The Throttle Position Sensor (TPS) is connected to the throttle shaft and is controlled by the throttle mechanism. A 5 volt reference signal is sent to the TPS from the ECM. As the throttle valve angle is changed (accelerator pedal moved), the resistance of the TPS also changes. At a closed throttle position, the resistance of the TPS is high, so the output voltage to the ECM will be low (approximately 0.5 volts). As the throttle plate opens, the resistance decreases so that, at wide open throttle, the output voltage should be approximately 5 volts. At closed throttle position, the voltage at the TPS should be less than 1.25 volts.

By monitoring the output voltage from the TPS, the ECM can determine fuel delivery based on throttle valve angle (driver demand). The TPS can either be misadjusted, shorted, open or loose. Misadjustment might result in poor idle or poor wide-open throttle performance. An open TPS signals the ECM that the throttle is always closed, resulting in poor performance. This usually sets a Code 22. A shorted TPS gives the ECM a constant wide-open throttle signal and should set a Code 21. A loose TPS indicates to the ECM that the throttle is moving. This causes intermittent bursts of fuel from the injector and an unstable idle. On some vehicles, the TPS is adjustable and therefore can be adjusted to correct any complications caused by too high or too low a voltage signal.

Throttle Valve Position Switch

EXCEPT 2.8L AND 3.1L ENGINES

The throttle valve position switch is usually located on the throttle body. It is used to detect the throttle valve position at engine idle, part throttle and wide open throttle.

Turbocharger Control Unit

EXCEPT 2.8L AND 3.1L ENGINES

This control unit is usually located in the rear of the vehicle, behind the driver's side tail light cover in the luggage compartment. The turbocharger control unit controls the operation of the turbocharger system.

Vacuum Switching Valve

EXCEPT 2.8L AND 3.1L ENGINES

The vacuum switching valve is controlled by a signal from the ECU. This valve controls the fuel pressure according to the vacuum developed in the intake manifold.

Vehicle Speed Sensor

EXCEPT 2.8L AND 3.1L ENGINES

The vehicle speed sensor is incorporated into the speedometer. The ECU receives electrical impulses from the vehicle speed sensor.

2.8L AND 3.1L ENGINES

NOTE: A vehicle equipped with a speed sensor, should not be driven without the speed sensor connected, as idle quality may be affected.

The Vehicle Speed Sensor (VSS) is mounted behind the speedometer in the instrument cluster or on the transmission/speedometer drive gear. It provides electrical pulses to the ECM from the speedometer head. The pulses indicate the road speed. The ECM uses this information to operate the IAC, canister purge, and TCC.

Some vehicles equipped with digital instrument clusters use a Permanent Magnet (PM) generator to provide the VSS signal. The PM generator is located in the transmission and replaces the speedometer cable. The signal from the PM generator drives a stepper motor which drives the odometer. A failure in the VSS circuit should set a Code 24.

CONTROLLED OUTPUTS

Air Management System

2.8L AND 3.1L ENGINES

The air management system is used to reduce the levels of carbon monoxide and hydrocarbon emissions. The Air Injection Reaction (AIR) system adds oxygen to the exhaust manifold to continue combustion after the exhaust gases leave the combustion chamber. This added air also brings the catalytic converter up to operating temperature more quickly when the engine is cold.

When the engine is cold or in the wide open throttle condition, the ECM engergizes the solenoid on the valve and air is directed to the exhaust manifold ports. When the coolant temperature increases, the solenoid is deenergized and air goes into the air cleaner which also act as as a silencer. At higher engine speeds, air is directed to the air cleaner through the pressure relief valve even though the solenoid may be energized. There should be no air going to the exhaust ports while operating in the "Closed Loop" mode.

Air is supplied to the AIR system by a belt-driven air pump.

Exhaust Gas Recirculation (EGR)

The EGR system recirculates a small amount of exhaust gas back into the intake manifold to help reduce oxides of nitrogen emission levels by lowering combustion chamber temperatures.

The ECM operates a solenoid to control the EGR valve. This solenoid is normally closed. By providing a ground path, the

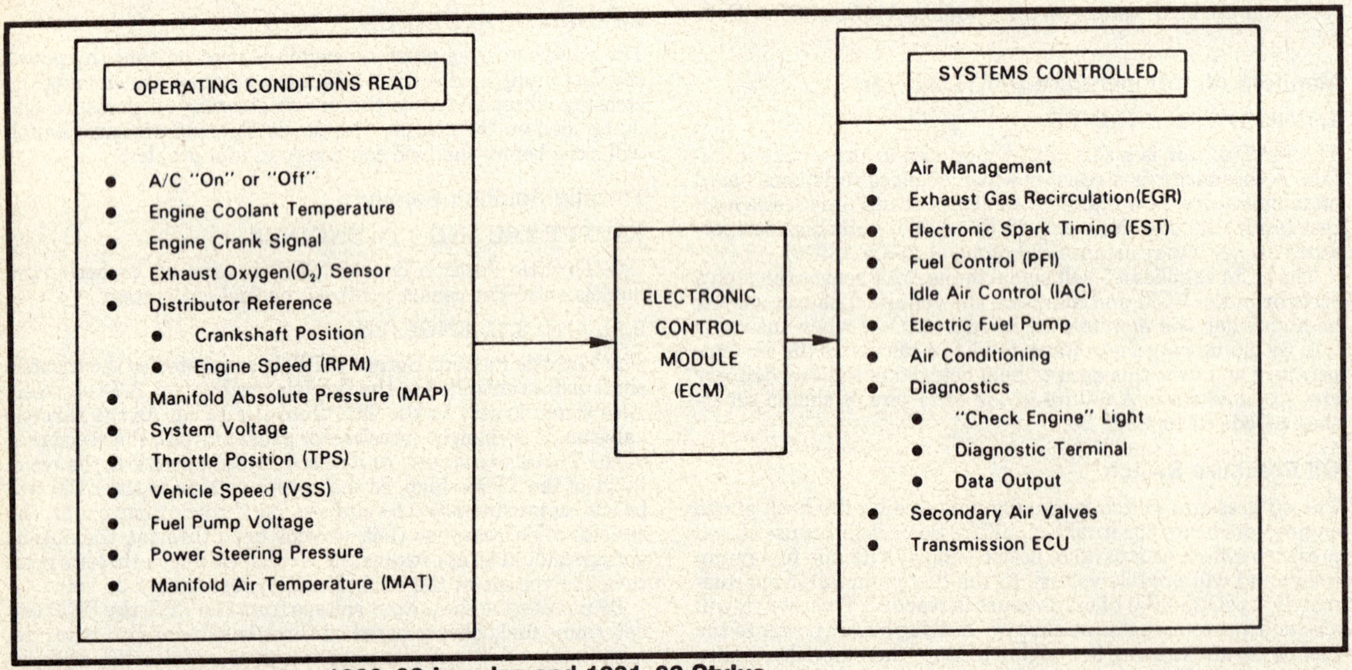

ECM inputs and outputs—1990–92 Impulse and 1991–92 Stylus

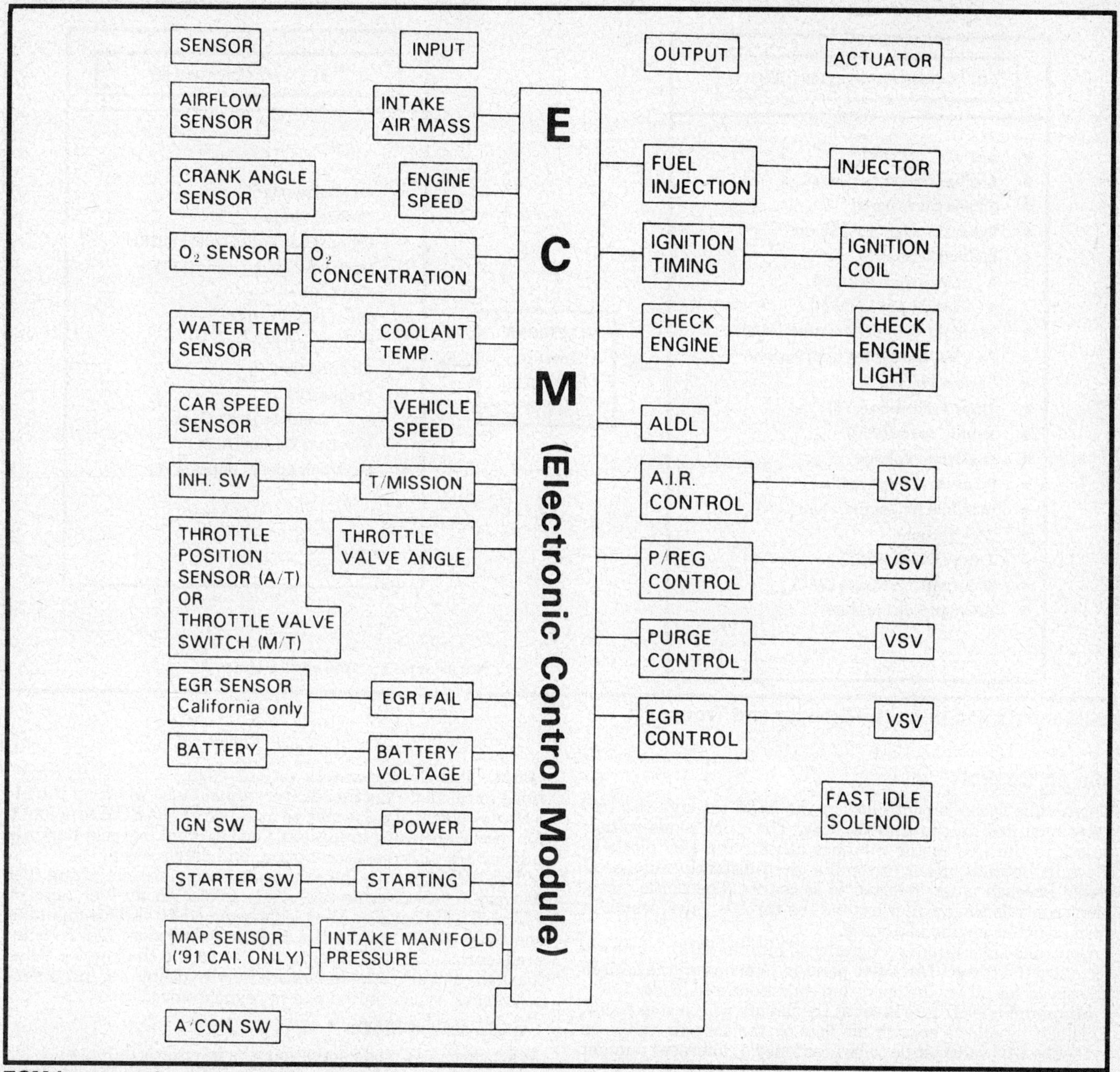

ECM inputs and outputs—2.6L engine

ECM energizes the solenoid which then allows vacuum to pass to the EGR valve.

Electronic Spark Timing (EST)

The EST system is controlled by the ECM. The ECM receives data pertaining to crankshaft position, engine speed, engine load (as determined through manifold pressure), atmospheric pressure and engine temperature to control ignition timing.

The ECM uses information from the MAP and coolant sensors, in addition to rpm, to calculate spark advance as follows:

Low MAP output voltage, more spark advance
Cold engine, more spark advance
High MAP output voltage, less spark advance
Hot engine, less spark advance.

Electronic Spark Control (ESC)

2.8 AND 3.1L ENGINES

The ESC system operates above 900 rpm with the engine at normal operating temperature to retard engine timing when the knock sensor detects detonation. The ESC system includes a knock sensor, ESC module, ESC portion of the ECM and associated wiring.

Electronic spark control is accomplished through the ESC module that sends a voltage signal to the ECM. As the knock sensor detects engine knock, the voltage from the ESC module to the ECM drops and this signals the ECM to retard timing.

When the ECM reads low voltage at ECM terminal B7 for longer than 5 seconds with the engine running or the system has failed the functional check, a code 43 will be set.

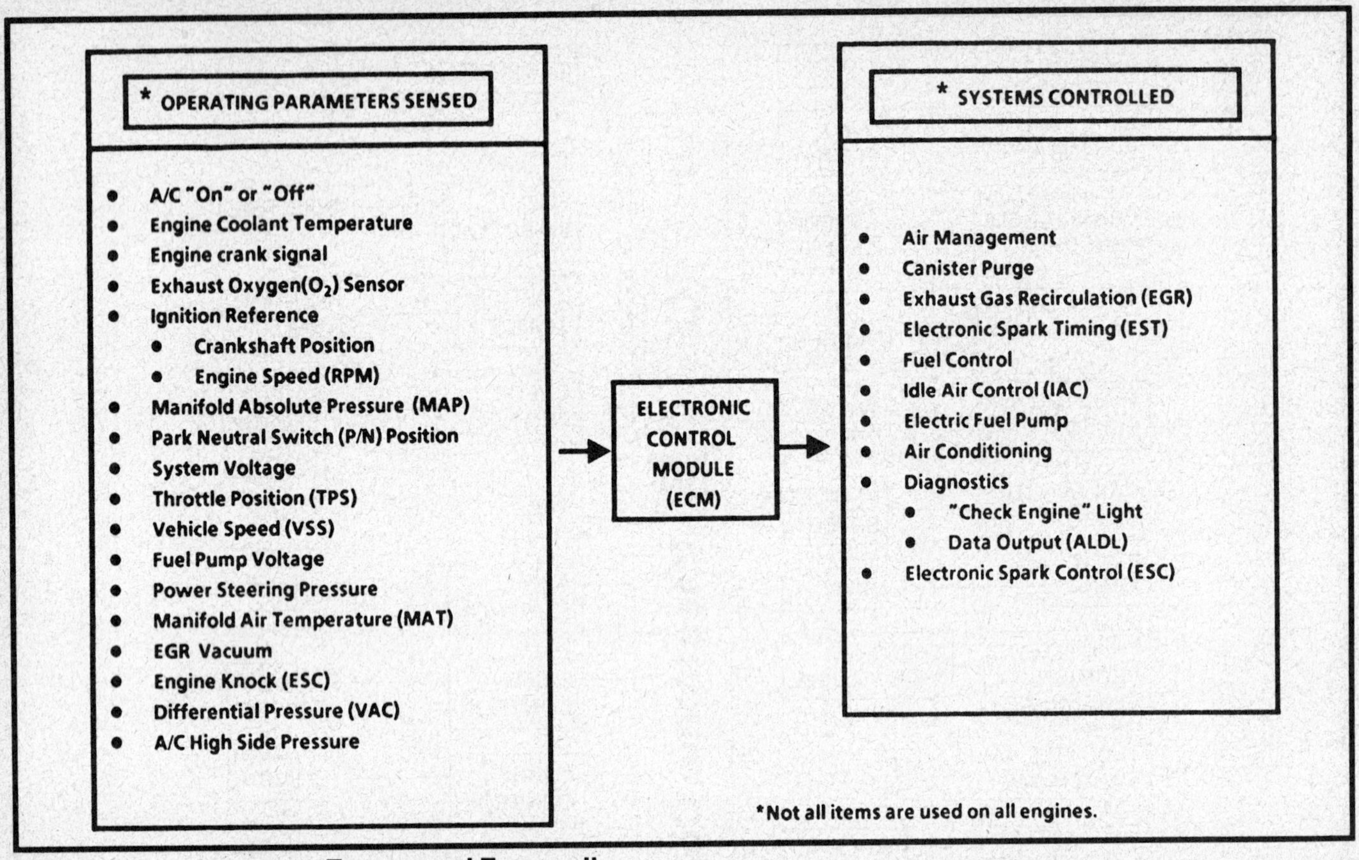

ECM inputs and outputs—Trooper and Trooper II

Idle Air Control

Engine idle speed is controlled by the ECM through the IAC valve mounted on the throttle body. The ECM sends voltage pulses to the IAC motor windings causing the IAC pintle to move in (retract) or out (extend) a given distance (number of steps) for each pulse, referred to as counts. The pintle movement controls the air flow around the throttle valve, which in turn, controls engine idle speed.

Controlled idle speed is programmed into the ECM, which determines the correct IAC valve position to maintain the desired idle speed for all engine operating conditions and loads.

Minimum idle air rate is set at the factory with a stop screw. This setting allows enough air flow by the throttle valves to cause the IAC valve pintle to be positioned a calibrated number of steps from the seat during controlled idle operation.

FUEL DELIVERY SYSTEM

The fuel injection system consists of the following components; the crank angle sensor, the throttle valve switch, the vehicle speed sensor, the coolant temperature sensor, air flow sensor, oxygen sensor, fuel injectors and the ECU.

If equipped, the turbocharger system consists of the following components; the throttle valve switch, the throttle position sensor, the knock sensor, the coolant temperature sensor, air flow sensor, oxygen sensor, the stepping motor, the turbocharger controller and the ECU.

2.8L AND 3.1L ENGINES

The electronic throttle body fuel injection system used on 2.8L and 3.1L engines is a fuel metering system with the amount of fuel delivered by the throttle body injector(s) determined by an electronic signal supplied by the Electronic Control Module (ECM). The ECM monitors various engine and vehicle conditions to calculate the fuel delivery time (pulse width) of the injector(s). The fuel pulse may be modified by the ECM to account for special operating conditions, such as cranking, cold starting, high or low altitude, acceleration, and deceleration.

The ECM controls the exhaust emissions by modifying fuel delivery to achieve, as near as possible, an air/fuel ratio of 14.7:1. The injector on-time is determined by various inputs to the ECM. By increasing the injector pulse, more fuel is delivered, enriching the air/fuel ratio. Decreasing the injector pulse leans the air/fuel ratio. Pulses are sent to the injector in 2 different modes: synchronized and nonsynchronized.

Synchronized Mode

In synchronized mode operation, the injector is pulsed once for each distributor reference pulse. In dual injector throttle body systems, the injectors are pulse alternately.

Non-synchronized Mode

In non-synchronized mode operation, the injector is pulsed once every 12.5 milliseconds or 6.25 milliseconds depending on calibration. This pulse time is totally independent of distributor reference pulses.

Non-synchronized mode results only under the following conditions:

1. The fuel pulse width is too small to be delivered accurately by the injector (approximately 1.5 milliseconds).
2. During the delivery of prime pulses (prime pulses charge the intake manifold with fuel during or just prior to engine starting).
3. During acceleration enrichment.
4. During deceleration leanout.

The basic TBI unit is made up of 2 major casting assemblies: (1) a throttle body with a valve to control airflow and (2) a fuel body assembly with an integral pressure regulator and fuel injector to supply the required fuel. An electronically operated device to control the idle speed and a device to provide information regarding throttle valve position are included as part of the TBI unit.

Each fuel injector is a solenoid-operated device controlled by the ECM. The incoming fuel is directed to the lower end of the injector assembly which has a fine screen filter surrounding the injector inlet. The ECM actuates the solenoid, which lifts a normally closed ball valve off a seat. The fuel under pressure is injected in a conical spray pattern at the walls of the throttle body bore above the throttle valve. The excess fuel passes through a pressure regulator before being returned to the vehicle's fuel tank.

The pressure regulator is a diaphragm-operated relief valve with injector pressure on one side and air cleaner pressure on the other. The function of the regulator is to maintain a constant pressure drop across the injector throughout the operating load and speed range of the engine.

The throttle body portion of the TBI may contain ports located at, above or below the throttle valve. These ports generate the vacuum signals for the EGR valve, MAP sensor, and the canister purge system.

The Throttle Position Sensor (TPS) is a variable resistor used to convert the degree of throttle plate opening to an electrical signal to the ECM. The ECM uses this signal as a reference point of throttle valve position. In addition, an Idle Air Control (IAC) assembly, mounted in the throttle body is used to control idle speeds. A cone-shaped valve in the IAC assembly is located in an air passage in the throttle body that leads from the point beneath the air cleaner to below the throttle valve. The ECM monitors idle speeds and, depending on engine load, moves the IAC cone in the air passage to increase or decrease air bypassing the throttle valve to the intake manifold for control of idle speeds.

Diagnosis and Testing

SERVICE PRECAUTIONS

- Be careful not to get water on any fuel injection system component. Pay close attention to the relay box and throttle valve switch connector. The connector is not water proofed and will be damaged by water.
- When charging the battery be sure to remove it from the vehicle first. Never disconnect the battery cable from the battery when the engine is running. The generation of surge voltage may damage the control unit and other electrical parts such as the multi-drive monitor.
- When replacing parts or checking the system, make sure to turn the ignition switch to the **OFF** position. When measuring voltage at the control unit harness connector, disconnect all the control unit harnesses first, then turn the ignition switch to the **ON** position.
- When checking the electrical terminals of the control unit with a tester, do not apply the probe to terminal directly but insert a pin into the terminal from the harness side and perform the measurement through the pin. If the tester probe is held against the terminal directly, the terminal will be deformed, causing poor contact. Connect each harness correctly and firmly to insure a good contact.
- The wiring connectors for the fuel injector(s), throttle valve switch, air regulator and water temperature sensor are provided with locked wires. To unlock the connector, pull and shake it gently.
- System cables must be placed at least 4 in. away from the tension cables. Be careful not to apply any shock to the system components such as the air flow sensor, crank angle sensor, and control unit. Component parts of the fuel injection system are precisely set. Even a slight distortion or dent will seriously affect performance.

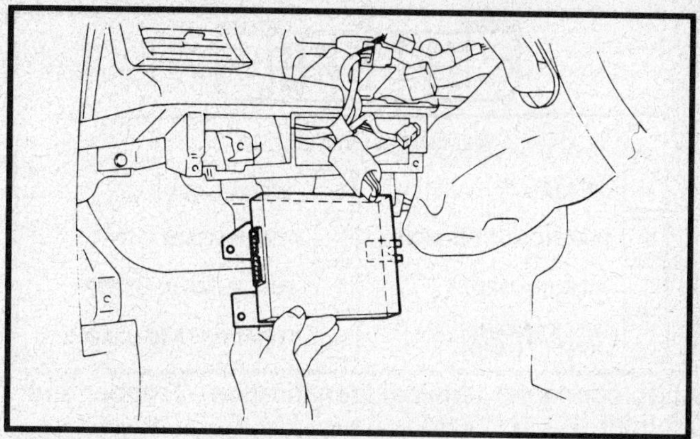

ECM location—1990–92 Impulse and 1991–92 Stylus

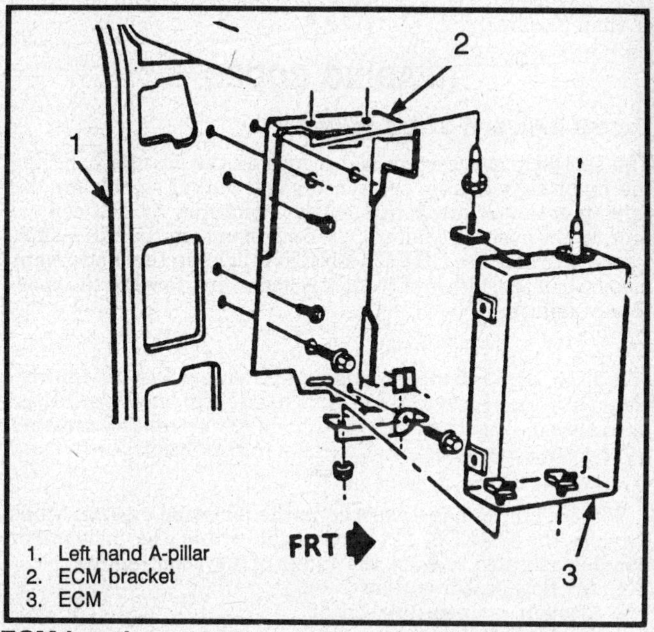

ECM location—1991–92 Amigo, Pick-Up and Rodeo

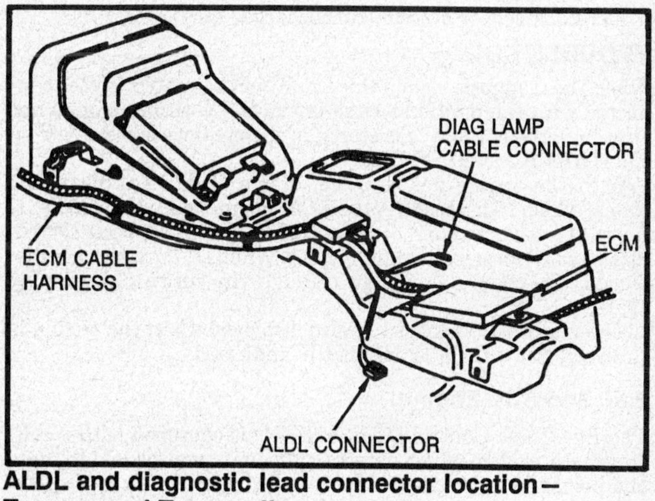

ALDL and diagnostic lead connector location—Trooper and Trooper II

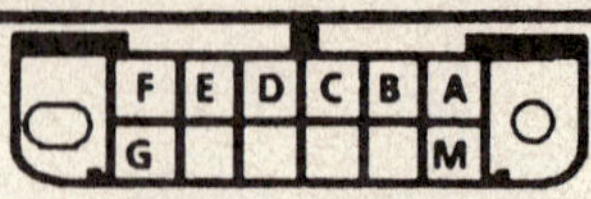

TERMINAL IDENTIFICATION

A	GROUND	E	SERIAL DATA
B	DIAGNOSTIC TERMINAL	F	TCC (IF USED)
C	A.I.R. (IF USED)	G	FUEL PUMP (IF USED)
D	CHECK ENGINE LIGHT (IF USED)	M	SERIAL DATA (IF USED)

ALDL connector terminal identification—Trooper and Trooper II

- The fuel pump must not be operated without fuel. Since fuel lubricates the pump, noise or other serious problems such as parts seizure will result. It is also prohibited to use any fuel other than gasoline.

READING CODES

Except 2.8L and 3.1L Engines

The self-diagnosis system is designed so the circuits handling the input signals from the sensors and output signals for the driving actuator are continuously monitored by the control unit. In the event of a failure, the control unit stores it in memory and operates the **CHECK ENGINE** light on the instrument panel when the nature of trouble is important, to warn the operator of failure.

NOTE: The self-diagnosis system is capable of troubleshooting the electrical circuits in the Closed Loop Emission Control system only and does not cover the trouble in the sensors, actuators or the engine itself.

When a failure has developed in the following systems while driving, the **CHECK ENGINE** light within the instrument panel is operated to warn the driver of a system failure:

1. Air flow sensor system
2. Coolant temperature sensor
3. Fuel injector system
4. Micro computer
5. Oxygen sensor
6. Vehicle speed sensor
7. Knock sensor system (turbocharged only).

TROUBLE CODES

When the diagnosis lead in the vicinity of the control unit is connected with the ignition switch in the **ON** position (engine not running), the trouble code stored in the memory is displayed by the **CHECK ENGINE** light.

With the engine not running, first a Code 12 is flashed 3 times, which indicates that the ECM is functioning normally. If there are other trouble codes stored in the ECM, each code is displayed 3 times in ascending order. When all the codes are displayed, the computer will go through the entire code display again.

The malfunction codes are also displayed when the engine is running, but a Code 12 will not be indicated.

2.8L and 3.1L Engine

The Electronic Control Module (ECM) is equipped with a self-diagnostic system which detects component and system failures and helps the technician to identify the faulty circuit by displaying a trouble code.

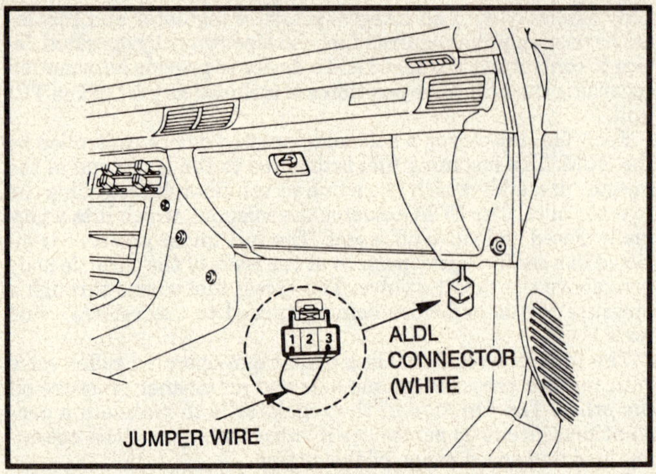

ALDL connector location 1990–92 Impulse and 1991–92 Stylus

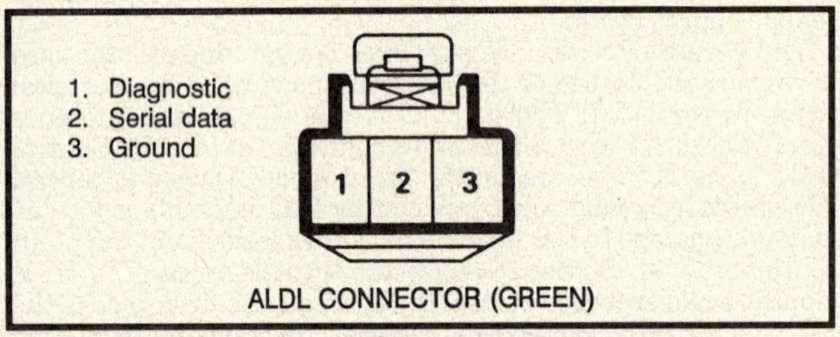

ALDL connector pin identification (except 2.8L and 3.1L engine)

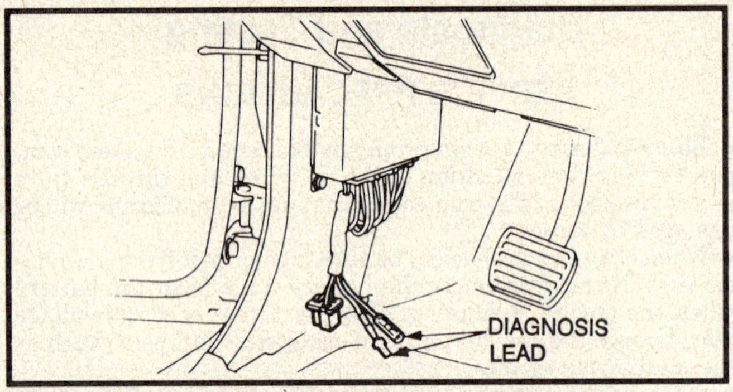

Diagnostic lead location—Amigo and Pick-up

CHECK ENGINE LIGHT

The **CHECK ENGINE** light is located on the instrument panel and has 2 functions:

1. It alerts the driver that a problem has occurred in the system and that the car should be taken to the service station for evaluation as soon as possible.
2. Once in the shop, the light allows the technician to read Trouble Codes to help diagnose system problems. Normally, the light will go on with the key **ON** and the engine not running. If the ECM has detected a problem within the system, the light will remain on. If the problem goes away, the light will go out in most cases but a trouble code will remain stored in the ECM.

INTERMITTENT CHECK ENGINE LIGHT

Fault codes are generally divided into 2 categories: intermittent and hard.

An intermittent code is defined as one that does not reset itself, but the fault condition is not active when the technician is working on the car. The most probable cause of intermittent faults are loose wiring connections.

A hard code is defined as one that is present when the vehicle is brought in for service and one that the ECM recognizes and displays to the technician.

TROUBLE CODES

The ECM uses a network of sensors throughout the engine to measure a number of engine operating parameters. It compares these readings to pre-programmed reference values stored in the ECM's memory. If a sensor reading is not what the ECM thinks it should be, the ECM will illuminate the **CHECK ENGINE** light on the instrument panel, assign a trouble code to the fault and store the code in the ECM's memory. The trouble code does not always identify the faulty component, but alerts the technician to what circuit the fault is in. An engine circuit consists of the sensor, attendant wiring, connectors and the ECM.

ALDL CONNECTOR

The Assembly Line Diagnostic Link (ALDL) connector is a diagnostic connector located in the passenger compartment. The connector has terminals used at the point of assembly to make sure that the engine is operating to design specifications before leaving the plant. Terminal **1** of the connector is the diagnostic terminal and can be connected to terminal **3** to enter the DIAGNOSTIC mode or ground to enter the FIELD SERVICE mode.

DIAGNOSTIC MODE

NOTE: In the Diagnostic mode, codes can be read only with the engine stopped. Grounding the diagnostic terminal with the engine running sends the ECM into the FIELD SERVICE MODE described below.

If the diagnostic terminal of the ALDL connector is grounded with the ignition switch in the **ON** position and the engine not running, the ECM will enter the Diagnostic mode. In this mode the ECM will perform the following functions:

1. Display a Code 12 by flashing the **CHECK ENGINE** light which indicates that the system is functioning properly. A Code 12 consists of 1 flash, followed by a short pause, then 2 quick flashes. The code will be repeated 3 times. If no other codes are stored in the system, code will continue to flash until the diagnostic terminal is ungrounded.
2. Display any stored trouble codes by flashing the **CHECK ENGINE** light. Each code will be code will be flashed 3 times, then Code 12 will be flashed again. If a trouble code is displayed, use the diagnostic trouble code charts to troubleshoot the system.
3. Energize the all ECM relays and solenoids except the fuel pump relay.
4. Move the IAC valve to the fully extended position.

FIELD SERVICE MODE

The Field Service Mode is entered by grounding the diagnostic terminal with the engine running and will indicate to the technician whether the ECM is in "Open" or "Closed" loop.

In Open Loop, the **CHECK ENGINE** light flashes 2½ times a second.

In Closed Loop, the **CHECK ENGINE** light flashes once per second.

In Closed Loop, the light will stay out most of the time if the system is too lean. The light will stay lit most of the time if the system is too rich.

When the system is in the Field Service Mode:

1. New trouble codes cannot be stored in the ECM.
2. The closed loop timer is bypassed.

CLEARING TROUBLE CODES

Except 2.8L and 3.1L Engines

After completion of the service operation, clear the trouble codes stored in memory by disconnecting the 60A main fuse in the fuse junction block, then check that only Code 12 is displayed.

NOTE: All the codes stored in memory will be cleared automatically when the 13-pole connector in the control unit is disconnected. Since all the memory will be cleared when number 4 fuse is disconnected, it will be necessary to reset the clock and other electrical equipment.

2.8L and 3.1L Engines

When the ECM sets a trouble code, the **CHECK ENGINE** light will illuminate and the code will be stored in the computer's memory. If the problem is intermittent, the light will go out in 10 seconds when the fault goes away. However, the code will remain in the ECM's memory until battery voltage to the ECM is removed. Removing battery voltage for 30 seconds will clear all stored trouble codes. Trouble codes will be cleared from the ECM after the fault is repaired.

NOTE: To prevent damage to the ECM, the ignition key must be turned to the OFF position when disconnecting voltage to the ECM (battery cable, ECM pigtail, ECM fuse, jumper cables).

ECM LEARNING ABILITY

The ECM has an inherent "learning" ability which allows the computer to make corrections for minor variations in the fuel system to improve driveability. If the battery cable was disconnected to clear a diagnostic code, or if a repair has been performed, the "learning" process has to be reinitiated. After a repair, a change may be noticed in the vehicle's normal performance. To "teach" the vehicle, warm up the engine to normal operating temperature and drive at part throttle with moderate accelerator and idle conditions until normal performance resumes.

SCAN TOOLS USED WITH INTERMITTENTS

In some scan tool applications, such as when trying to detect an intermittent problem which lasts for a very short time, the data update rate makes the tool less effective than a voltmeter. However, the scan tool does allow manipulation of the wiring harness or components under the hood with the engine not running while observing the scan tool's readout.

The scan tool can be plugged in and observed while driving the vehicle under the condition when the **CHECK ENGINE** light turns on momentarily or when the engine driveability is momentarily poor. If the problem seems to be related to certain parameters that can be checked on the scan tool, they should be checked while driving the vehicle. If there does not seem to be any correlation between the problem and any specific circuit, the scan tool can be checked on each position. Watching for a period of time to see if there is any change in the reading that indicates intermittent operation.

The scan tool is also an easy way to compare the operating parameters of a poorly operating engine with those of a known good one. For example, a sensor may shift in value but not set a trouble code. Comparing the sensor's reading with those of a known good vehicle may uncover the problem.

The scan tool has the ability to save time in diagnosis and prevent the replacement of good parts. The key to using the scan tool successfully for diagnosis lies in the technician's ability to understand the system he is trying to diagnose as well as an understanding of the scan tool's operation and limitations. The technician should read the tool manufacturer's operating manual to become familiar with the tool's operation.

FUEL CUT-OFF SYSTEM

Inspection

1. Start the engine and let it run to reach normal operating temperature. Turn the ignition switch to the **OFF** position.
2. Disconnect the throttle valve switch harness at the connector, insert a fine wire into the idle terminal (upper), then make the connections at the connector with the wire pulled part way out.
3. Disconnect the throttle position sensor at the connector. Use a suitable jumper wire to connect the harness side connector terminals.

NOTE: Do not connect the harness side connector and throttle position sensor with the piece of wire in place. Shorting will result.

4. Start the engine and keep the engine running at 2500–3000 rpm. Check to make sure the engine speed drops to approximately 1250 rpm when battery voltage is applied to the wire extending from the idle terminal.

NOTE: Engine hunting may occur at approximately 1250 rpm, but does not indicate abnormal conditions.

AIR REGULATOR FAST IDLE

Inspection

NOTE: The engine must be cold to perform this inspection.

1. The engine idling speed must be slightly higher than normal immediately after starting the engine.
2. The engine idle speed should normalize as the engine temperature increases.
3. When fast idle speed is not obtained, use the following procedure:
   a. Disconnect the air regulator harness at the connector.
   b. Using a suitable ohmmeter, measure the resistance across the air regulator terminals.
   c. The standard resistance should be 38–42 ohms.
   d. If the measured resistance deviates from the specified range, replace the air regulator assembly.
4. Turn the ignition switch to the **ON** position and, using a suitable voltmeter, check for voltage at the air regulator connectors.
   a. When voltage is present, check the wiring leading to the air regulator and correct as necessary. The fuel pump may be defective; replace it as necessary.
   b. When voltage is present, the air regulator is defective and should be replaced.

AIR REGULATOR

Leakage Inspection

1. Disconnect the hose between the air regulator and throttle valve body at the throttle valve body side.
2. Close the hole in the throttle valve body. Start the engine.
3. Check for variation in the engine speed by closing and opening the end of the air regulator hose (this test is made to determine the function of the air regulator).
4. The variation of the engine speed should be less than 50 rpm. If the variation in engine speed is more than specified, replace the air regulator assembly.

THROTTLE VALVE SWITCH

Inspection and Testing

1. Check that the operation of the accelerator pedal and control cable is smooth.
2. Check that the valve stopper returns and makes a firm contact with the throttle valve stopper screw when the throttle valve is released.

NOTE: The throttle valve stopper screw is factory set and sealed with paint. Setting of the screw should not be disturbed unless absolutely necessary.

3. Test the throttle switch as follows:
   a. Disconnect the throttle valve switch harness connector.
   b. Using a suitable ohmmeter, make a continuity test across the terminals with the accelerator pedal depressed in steps.
   c. With the accelerator pedal NOT depressed, the resistance should be 0 ohms at idle contact and infinity at full contact when testing terminals **I** and **P**.
   d. With the accelerator pedal slightly depressed, the resistance should be infinity ohms at idle contact and infinity at full contact when testing terminals **I and P** and **P and F**.
   e. With the accelerator pedal fully depressed, the resistance should be infinity at idle contact and 0 ohms at full contact when testing terminals **I and P**.

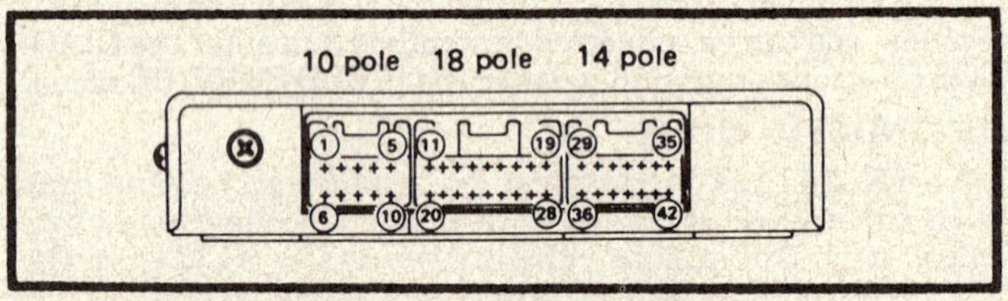

Control unit harness connector terminal identification—Amigo, Pick-up, Trooper and Trooper II (except 2.8L engine)

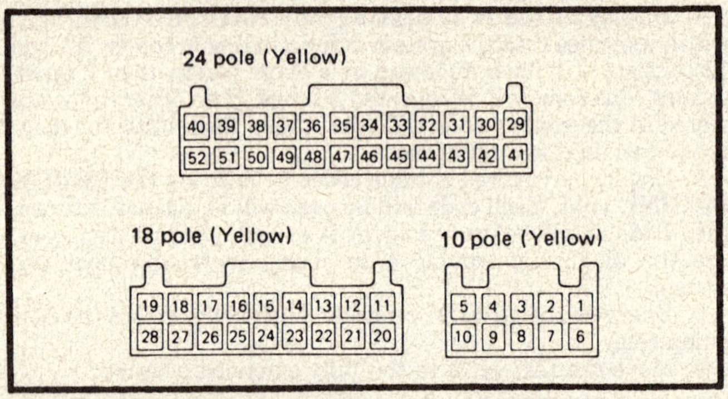

Control unit harness connector terminal identification—Impulse

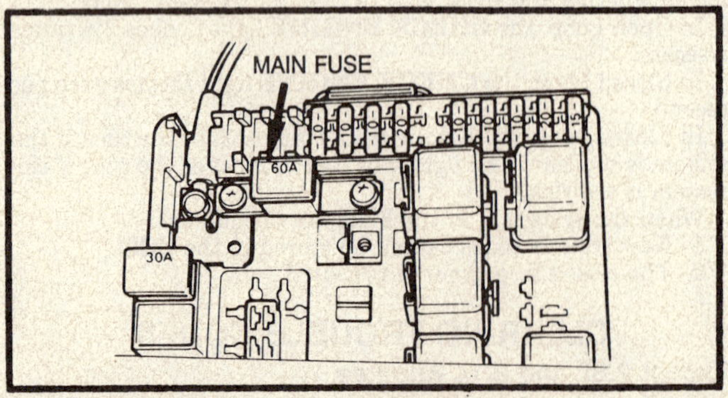

Main fuse location—Amigo and Pick-Up

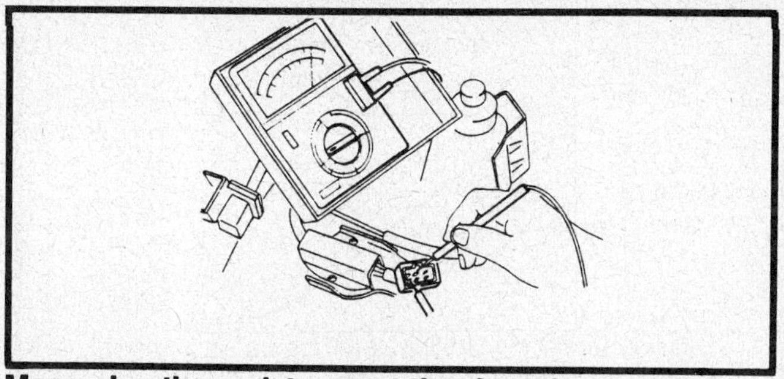

Measuring the resistance at the dropping resistor

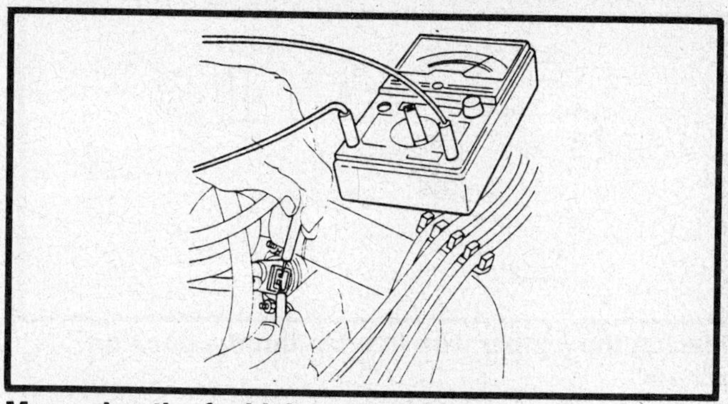

Measuring the fuel injector resistance

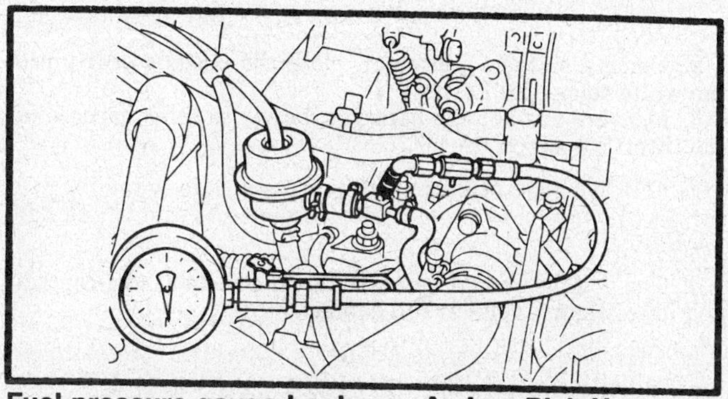

Fuel pressure gauge hook-up—Amigo, Pick-Up, Trooper and Trooper II (except 2.8L engine)

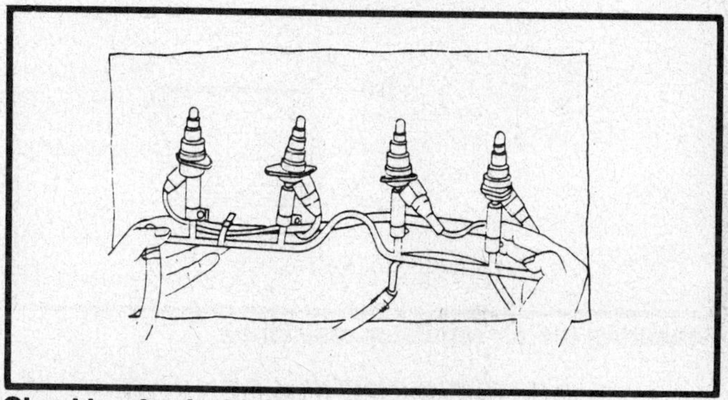

Checking for fuel injector leak down

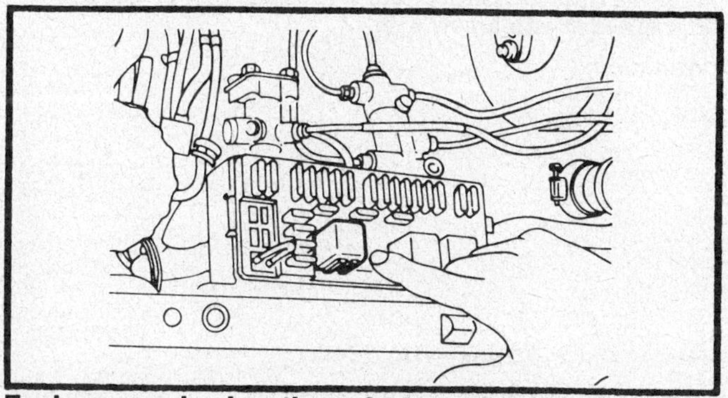

Fuel pump relay location—Amigo, Pick-Up, Trooper and Trooper II (except 2.8L engine)

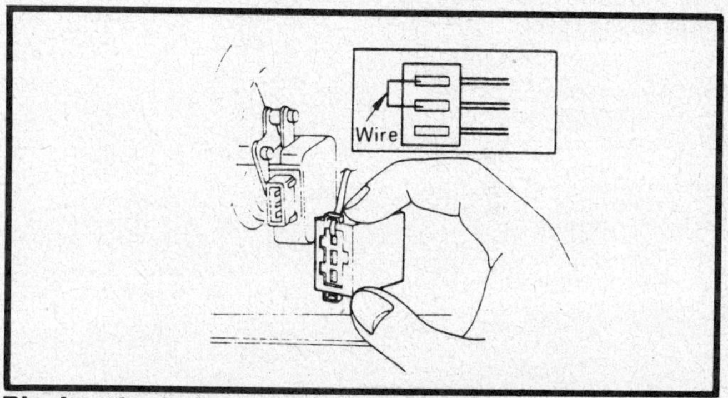

Placing the jumper wire into the throttle valve switch harness

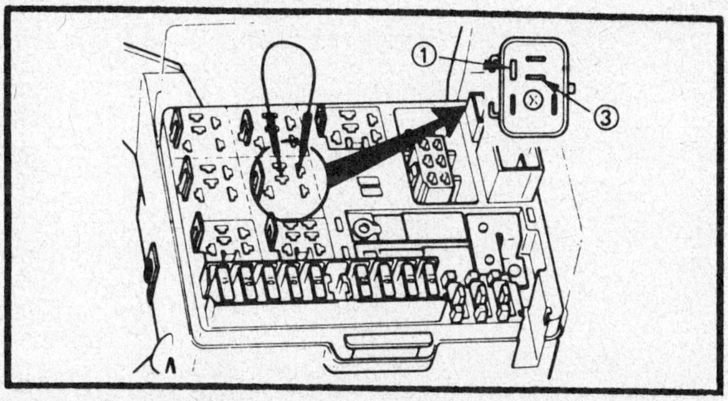

Short the fuel pump relay terminals—Amigo, Pick-Up and Trooper and Trooper II (except 2.8L engine)

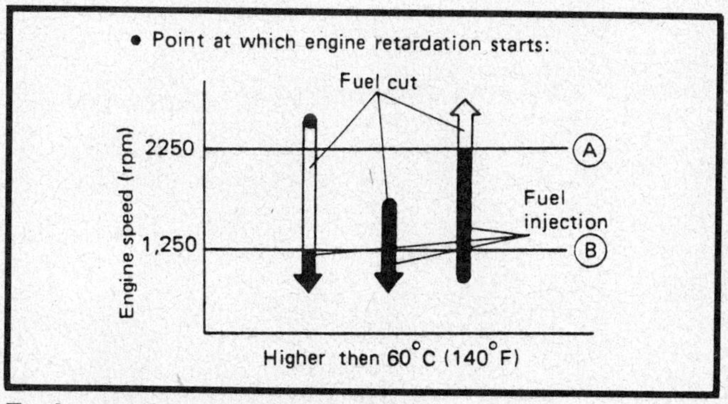

Fuel cut off system graph chart

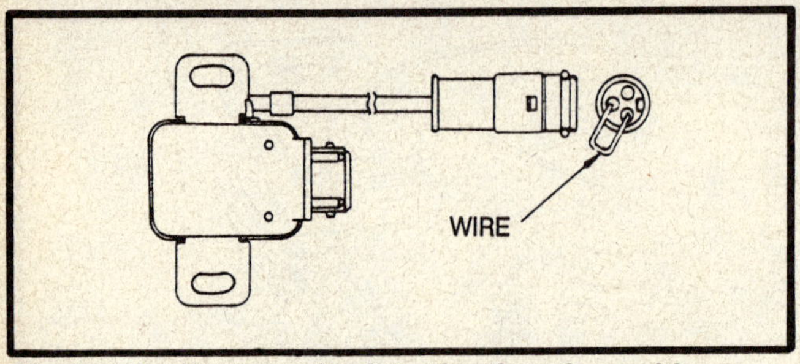

Placing the jumper wire into the throttle position sensor

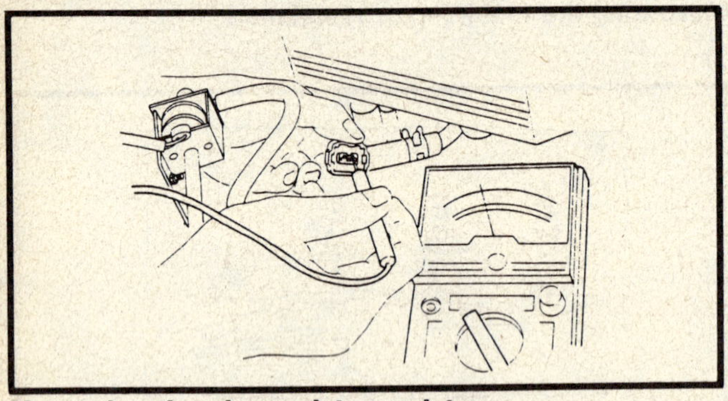

Measuring the air regulator resistance

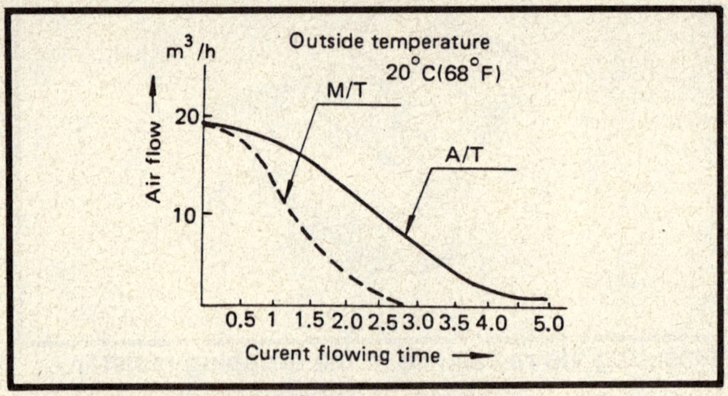

Air regulator fast idle flow graph chart

THROTTLE POSITION SENSOR

Inspection

1. Turn the ignition switch to the **ON** position. Remove the water shield cover.
2. Using a suitable voltmeter, place the positive probe into the white color wire harness.
3. Measure the voltage between the white color harness of the throttle position sensor.

NOTE: The throttle position sensor has 3 leads, red for 5 volt power source, white for output and black for ground.

4. The voltage should be higher then 4 volts at idle contact and lower than 2 volts at full contact.

NOTE: Make sure that 5 volts (±0.5 volts) is measured at the red colored harness before making the voltage check in Step 4.

5. Confirm the difference in the voltage of the idle contact and full contact is 3.6 ±1 volt.
6. If the throttle position sensor fails any part of this inspection, replace it with a new one.

1.6L SOHC ENGINE—ENGINE COMPONENT LOCATION STYLUS

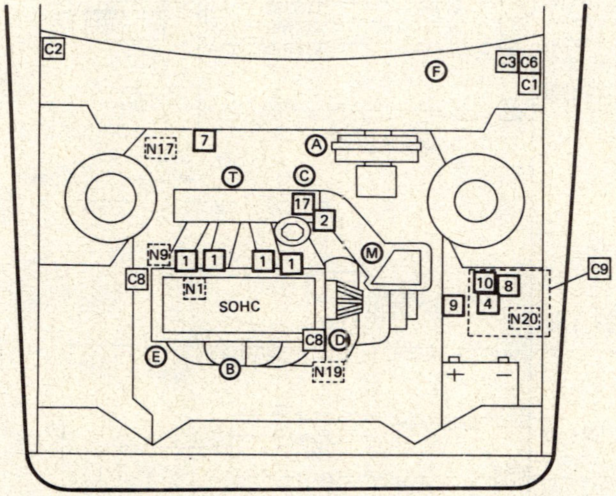

COMPUTER HARNESS

- C1 Electronic Control Module (ECM)
- C2 ALDL Connector
- C3 "Check Engine" Light
- C6 Fuse Panel
- C8 ECM Grounds
- C9 Underhood Relay/Fuse Center

NOT ECM CONNECTED

- N1 Crankcase Vent Valve (PCV) Diaphragm Type
- N9 Fuel Pressure Regulator Valve
- N17 Fuel Vapor Canister
- N19 Fan Switch
- N20 A/C Compressor Relay

CONTROLLED DEVICES

- 1 Fuel Injectors
- 2 IAC Valve
- 4 EFI Main Relay
- 7 EGR Vacuum Switching Valve (VSV)
- 8 Cooling Fan Relay
- 9 Ignition Coil
- 10 Fuel Pump Relay
- 17 Throttle Body

INFORMATION SENSORS

- A Manifold Absolute Pressure (MAP) Sensor
- B Oxygen (O_2) Sensor
- C Throttle Position Sensor (TPS)
- D Coolant Temperature Sensor (CTS)
- E Power Steering Pressure Switch*
- F Vehicle Speed Sensor (VSS)
- T Manifold Air Temperature (MAT) Sensor
- Inhibitor Switch

Exhaust Gas Recirculation (EGR) Valve

NOTE: *Power Steering Pressure Switch is part of the Power Steering Pump.

1.6L DOHC ENGINE—ENGINE COMPONENT LOCATION IMPULSE AND STYLUS

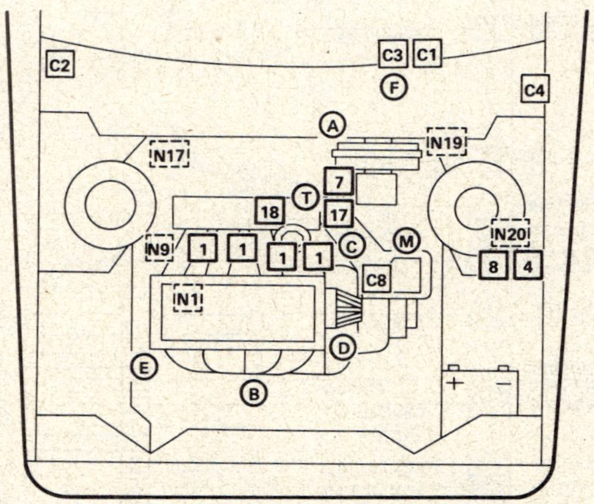

COMPUTER HARNESS

- C1 Electronic Control Module (ECM)
- C2 ALDL White Connector
- C3 "Check Engine" Light
- C6 Fuse Panel
- C8 ECM Grounds

NOT ECM CONNECTED

- N1 Crankcase Vent Valve (PCV)
- N9 Fuel Pressure Regulator Valve
- N17 Fuel Vapor Canister
- N19 Fan Switch
- N20 A/C Compressor Relay

CONTROLLED DEVICES

- 1 Fuel Injectors
- 2 Ignition Coil
- 3 IAC Valve
- 4 Main Relay
- 7 EGR Vacuum Switching Valve (VSV)
- 8 Cooling Fan Relay
- 17 Throttle Body
- 18 Secondary Air Vacuum Switching Valve (VSV) (Located Under Intake Plenum)

INFORMATION SENSORS

- A Manifold Absolute Pressure (MAP) Sensor
- B Oxygen (O_2) Sensor
- C Throttle Sensor (TS)
- D Coolant Temperature Sensor (CTS)
- E Power Steering Pressure Switch
- F Vehicle Speed Sensor (VSS)
- M Park/Neutral (P/N) Switch
- T Manifold Air Temperature (MAT) Sensor

Exhaust Gas Recirculation (EGR) Valve

NOTE: *Power Steering Pressure Switch is part of the Power Steering Pump.

1.6L ENGINE—ECM WIRING DIAGRAM IMPUSLE AND STYLUS

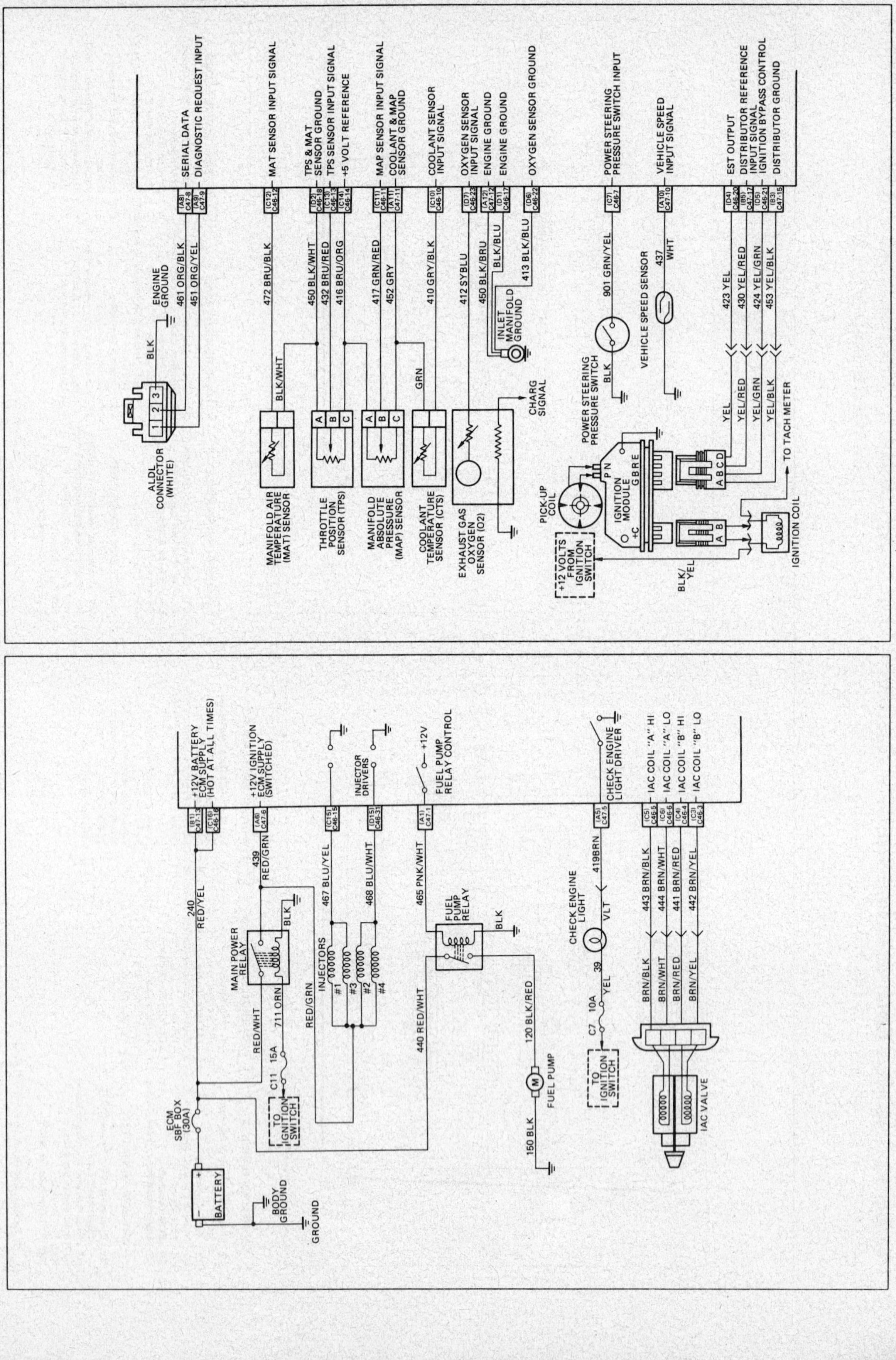

1.6L ENGINE—ECM WIRING DIAGRAM IMPULSE AND STYLUS

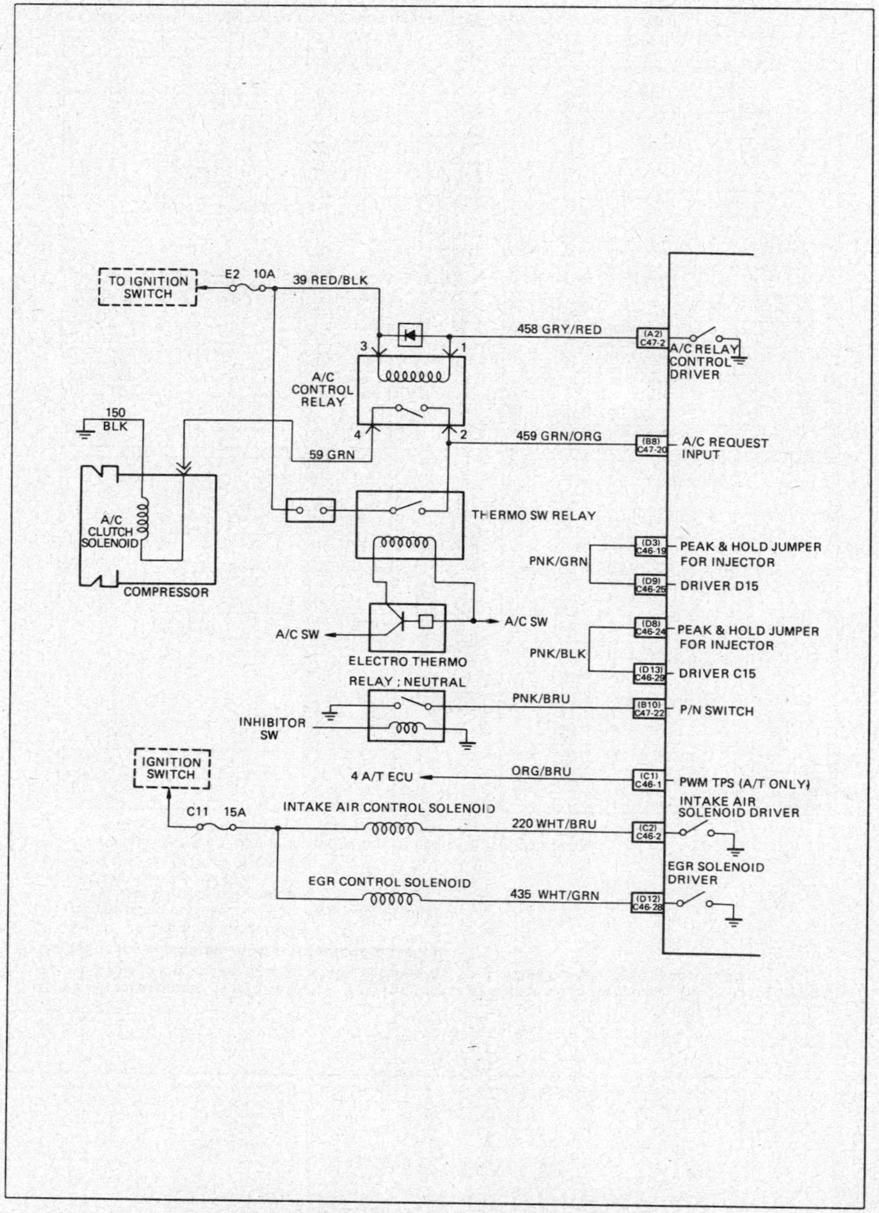

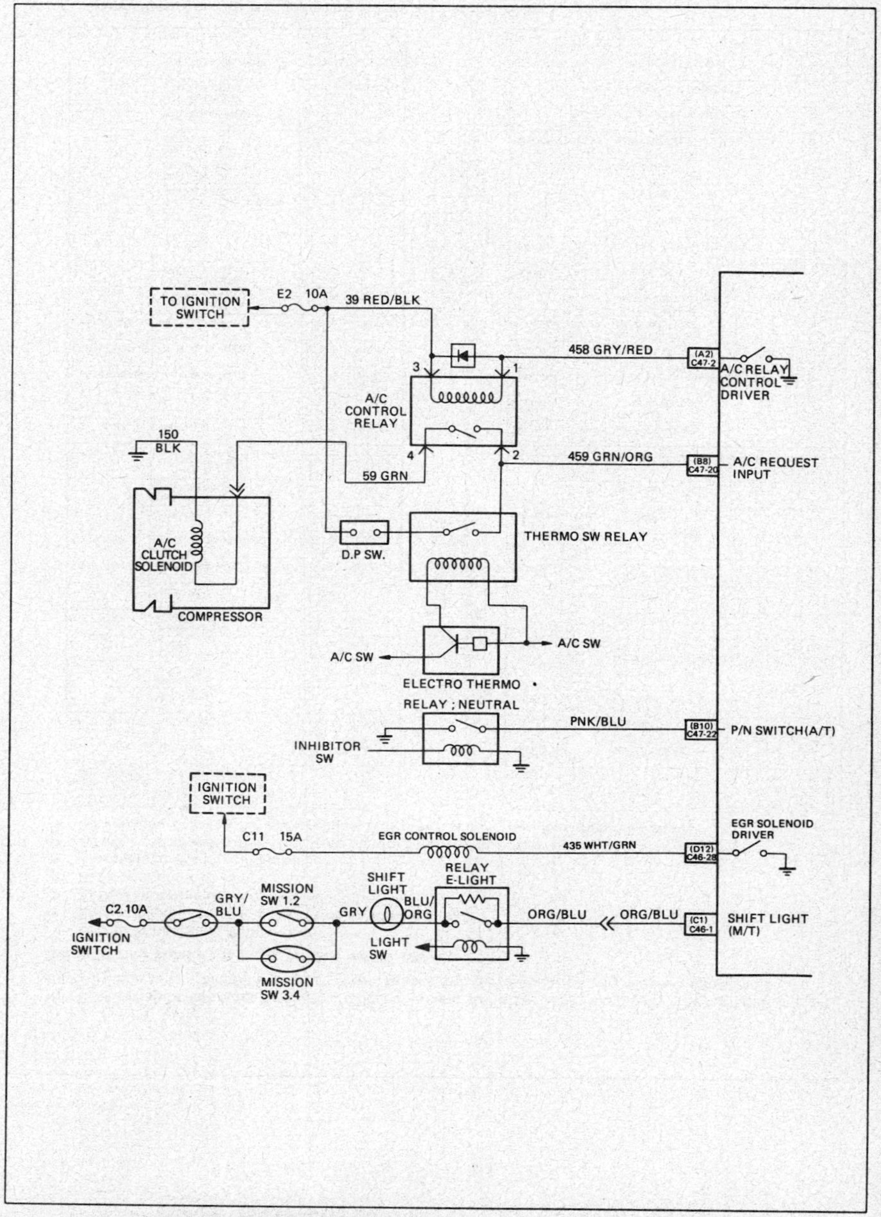

ECM CONNECTOR IDENTIFICATION 1991–92 STYLUS

This ECM voltage chart is for use with a digital voltmeter to further aid in diagnosis. The voltages you get may vary due to low battery charge or other reasons, but they should be very close to specification.

The following conditions must be met before testing:

- Engine at operating temperature
- Engine idling in close loop (for "Engine Run" column)
- ALDL Test terminal not grounded
- "SCAN" tool not installed
- Transaxle in "Park" or "Neutral"
- All voltages showing "B+" indicates battery or charging system voltage
- *All voltages showing "0" are ground circuits, and should read less than 0.5 volts

VOLTAGE DC KEY "ON"	VOLTAGE DC ENG. RUN	CIRCUIT	PIN	WIRE COLOR
① B+	B+	FUEL PUMP RELAY	(A1) C47-1	PNK/WHT
0	0-ON B+-OFF	A/C CUT-OUT	(A2) C47-2	GRY/RED
		NOT USED	(A3) C47-3	
		NOT USED	(A4) C47-4	
0*	B+	CHECK ENGINE LIGHT	(A5) C47-5	VLT
B+	B+	IGNITION FEED MAIN RELAY	(A6) C47-6	RED/GRN
		NOT USED	(A7) C47-7	
5.0	5.0	SERIAL DATA OUTPUT	(A8) C47-8	ORG/BLK
B+	B+	DIAGNOSTIC REQUEST	(A9) C47-9	ORG/YEL
0 or B+	B+	VSS SIGNAL INPUT	(A10) C47-10	WHT
0*	0*	COOLANT, MAP SENSOR GROUND	(A11) C47-11	GRY
0	0	ECM GROUND	(A12) C47-12	BLK/BLU

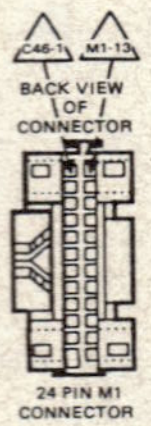

WIRE COLOR	PIN	CIRCUIT	VOLTAGE DC KEY "ON"	VOLTAGE DC ENG. RUN
RED/WHT	(B1) C47-13	SBF BOX (30A)	B+	B+
	(B2) C47-14	NOT USED		
YEL/BLK	(B3) C47-15	DISTRIBUTOR REFERENCE GROUND	0*	0*
	(B4) C47-16	NOT USED		
YEL/RED	(B5) C47-17	DISTRIBUTOR REFERENCE INPUT SIGNAL	0*	1.0
	(B6) C47-18	NOT USED		
	(B7) C47-19	NOT USED		
GRN/ORG	(B8) C47-20	AIR CONDITIONING REQUEST INPUT	0	0-OFF B+-ON
	(B9) C47-21	NOT USED		
PNK/BLU	(B10) C47-22	P/N SWITCH (A/T)	0	0
	(B11) C47-23	NOT USED		
GRY/BLW	(B12) C47-24	EGR GAS TEMPERATURE SENSOR INPUT		

VOLTAGE DC KEY "ON"	VOLTAGE DC ENG. RUN	CIRCUIT	PIN	WIRE COLOR
B+	B+	SHIFT INDI. RELAY (M/T)	(C1) C46-1	ORG/BLU
		NOT USED	(C2) C46-2	
NOT USABLE	NOT USABLE	IAC (COIL B LOW)	(C3) C46-3	BRN/YEL
NOT USABLE	NOT USABLE	IAC (COIL B HIGH)	(C4) C46-4	BRN/RED
NOT USABLE	NOT USABLE	IAC (COIL A HIGH)	(C5) C46-5	BRN/BLK
NOT USABLE	NOT USABLE	IAC (COIL A LOW)	(C6) C46-6	BRN/WHT
B+	B+	POWER STEERING SWITCH INPUT	(C7) C46-7	GRN/YEL
		NOT USED	(C8) C46-8	
		NOT USED	(C9) C46-9	
②	②	COOLANT TEMP SENSOR SIGNAL INPUT	(C10) C46-10	GRY/BLK
4.79V	⑤	MAP SENSOR SIGNAL INPUT	(C11) C46-11	GRY/RED
②	②	MAT SENSOR SIGNAL INPUT	(C12) C46-12	BLU/BLK
0.64	0.64	TPS SIGNAL INPUT	(C13) C46-13	YEL/BLU
5.0	5.0	+5V REFERENCE VOLTAGE OUTPUT	(C14) C46-14	BLU/ORG
B+	B+	INJECTION 3&4 DRIVE	(C15) C46-15	BLU/YEL
B+	B+	SBF BOX (30A)	(C16) C46-16	RED/WHL

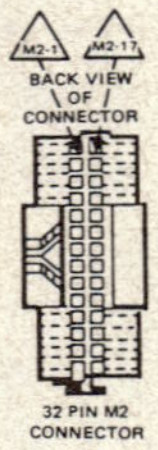

WIRE COLOR	PIN	CIRCUIT	VOLTAGE DC KEY "ON"	VOLTAGE DC ENG. RUN
BLK/BLU	(D1) C46-17	ECM GROUND	0	0
BLK	(D2) C46-18	TPS, MAT SENSOR GROUND	0	0
BLK/BLU	(D3) C46-19	ENGINE GROUND	0	0
YEL	(D4) C46-20	EST OUTPUT	0	1.15
YEL/GRN	(D5) C46-21	IGNITION MODULE BYPASS CIRCUIT	0	5
BLK/BLU	(D6) C46-22	OXYGEN SENSOR ENGINE GROUND CIRCUIT	0	0
SKYBLU	(D7) C46-23	OXYGEN SENSOR INPUT SIGNAL	0.45	0.1-0.9
BLK/BLU	(D8) C46-24	ENGINE GROUND	0	0
	(D9) C46-25	NOT USED		
	(D10) C46-26	NOT USED		
	(D11) C46-27	NOT USED		
WHT/GRN	(D12) C46-28	EGR CONTROL	B+	B+
	(D13) C46-29	NOT USED		
	(D14) C46-30	NOT USED		
BLU/WHT	(D15) C46-31	INJECTION 1&2 DRIVE	B+	B+
	(D16) C46-32	NOT USED		

① Reads battery voltage for 2 seconds after ignition "ON" then should read 0 volts.
② Varies depending on temperature.
③ B+ in R, D or L.
④ Varies with vehicle speed.
⑤ Varies depending on engine load.

ECM CONNECTOR IDENTIFICATION 1990 IMPULSE

This ECM voltage chart is for use with a digital voltmeter to further aid in diagnosis. The voltages you get may vary due to low battery charge or other reasons, but they should be very close to specification.

The following conditions must be met before testing:

- Engine at operating temperature
- Engine idling in close loop (for "Engine Run" column)
- ALDL Test terminal not grounded
- ALDL tool not installed
- Transaxle in "Park" or "Neutral"
- All voltages showing "B+" indicates battery or charging system voltage
- All voltages showing "O" are ground circuits, and should read less than 0.5 volts

VOLTAGE DC KEY "ON"	VOLTAGE DC ENG. RUN	CIRCUIT	PIN	WIRE COLOR
B+	B+	FUEL PUMP RELAY	(A1) C47-1	PNK/WHT
B+	0-ON B+-OFF	A/C CUT-OUT	(A2) C47-2	GRY/RED
		NOT USED	(A3) C47-3	
		INTAKE AIR SOLENOID	(A4) C47-4	
0	B2	CHECK ENGINE LAMP	(A5) C47-5	VGLT
B+	B+	IGNITION FEED MAIN RELAY	(A6) C47-6	RED/GRN
		NOT USED	(A7) C47-7	
5.0V	5.0V	SERIAL DATA OUTPUT	(A8) C47-8	VLT
B+	B+	DIAGNOSTIC REQUEST	(A9) C47-9	ORG/YEL
0	B+	VSS SIGNAL INPUT	(A10) C47-10	WHT
0	0	COLANT, MAP SENSOR GROUND	(A11) C47-11	GRN
0	0	ENGINE GROUND	(A12) C47-12	BLK/BLU

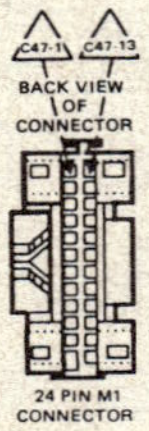

WIRE COLOR	PIN	CIRCUIT	VOLTAGE DC KEY "ON"	VOLTAGE DC ENG. RUN
RED/WHT	(B1) C47-13	SBF BOX (30A)	B+	B+
	(B2) C47-14	NOT USED		
VLT/BLK	(B3) C47-15	DISTRIBUTOR REFERENCE GROUND	0	0
	(B4) C47-16	NOT USED		
YEL/RED	(B5) C47-17	DISTRIBUTOR REFERENCE INPUT SIGNAL	0	1.0V
	(B6) C47-18	NOT USED		
	(B7) C47-19	NOT USED		
GRN/ORG	(B8) C47-20	AIR CONDITIONING REQUEST INPUT	0	B+
	(B9) C47-21	NOT USED		
PNK/BLU	(B10) C47-22	P/N SWITCH	0	0
	(B11) C47-23	NOT USED		
	(B12) C47-24	NOT USED		

VOLTAGE DC KEY "ON"	VOLTAGE DC ENG. RUN	CIRCUIT	PIN	WIRE COLOR
	3.3	PWM TPS	(C1) C46-1	ORG/BLU
		NOT USED	(C2) C46-2	
NOT USABLE	NOT USABLE	IAC (COIL B LOW)	(C3) C46-3	BRN/YEL
NOT USABLE	NOT USABLE	IAC (COIL B HIGH)	(C4) C46-4	BRN/RED
NOT USABLE	NOT USABLE	IAC (COIL A HIGH)	(C5) C46-5	BRN/BLK
		IAC (COIL A LOW)	(C6) C46-6	BRN/WHT
B+	B+	POWER STEERING SWITCH INPUT	(C7) C46-7	GRN/YEL
		NOT USED	(C8) C46-8	
		NOT USED	(C9) C46-9	
2.59	2.3	COOLANT TEMP SENSOR SIGNAL INPUT	(C10) C46-10	GRY/BLK
4.79	1.9	MAP SENSOR SIGNAL INPUT	(C11) C46-11	GRY/RED
1-2	3-4	MAT SENSOR SIGNAL INPUT	(C12) C46-12	BLU/BLK
.9	.56	TPS SIGNAL INPUT	(C13) C46-13	BLU/RED
5.0	5.0	+5V REFERENCE VOLTAGE OUTPUT	(C14) C46-14	BLU/ORG
B+	B+	INJECTION 1&3 DRIVE	(C15) C46-15	BLU/YEL
B+	B+	SBF BOX (30A)	(C16) C46-16	RED/WHL

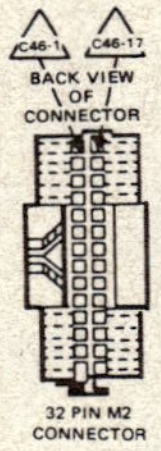

WIRE COLOR	PIN	CIRCUIT	VOLTAGE DC KEY "ON"	VOLTAGE DC ENG. RUN
BLK/BLU	(D1) C46-17	ENGINE GROUND	0	0
BLK/WHT	(D2) C46-18	TPS, MAT SENSOR GROUND	0	0
PNK/GRN	(D3) C46-19	PEAK & HOLD JUMPER D9	0	0
YEL	(D4) C46-20	EST OUTPUT	0	1.15
YEL/GRN	(D5) C46-21	IGNITION MODULE BYPASS CIRCUIT	0	4.6
BLK	(D6) C46-22	OXYGEN SENSOR ENGINE GROUND CIRCUIT	0.43V	0.1-0.9
SKY/BLU	(D7) C46-23	OXYGEN SENSOR INPUT SIGNAL		
PNK/BLK	(D8) C46-24	PEAK & HOLD JUMPER D13		
PNK/GRN	(D9) C46-25	PEAK & HOLD JUMPER D3		
	(D10) C46-26	NOT USED		
	(D11) C46-27	NOT USED		
WHT/GRN	(D12) C46-28	EGR CONTROL	B+	B+
	(D13) C46-29	PEAK & HOLD JUMPER D8		
	(D14) C46-30	PEAK & HOLD JUMPER		
BLU/WHT	(D15) C46-31	INJECTION 2&4 DRIVE	B+	B+
	(D16) C46-32	NOT USED		

- All voltages showing "B+" indicates battery or charging system voltage
- All voltages showing "0" are ground circuits, and should read less than 0.10 volts

ECM CONNECTOR IDENTIFICATION
1991–92 IMPULSE

This ECM voltage chart is for use with a digital voltmeter to further aid in diagnosis. The voltages you get may vary due to low battery charge or other reasons, but they should be very close to specification.

The following conditions must be met before testing:

- Engine at operating temperature
- Engine idling in close loop (for "Engine Run" column)
- ALDL Test terminal not grounded
- ALDL tool not installed
- Transaxle in "Park" or "Neutral"

VOLTAGE DC KEY "ON"	VOLTAGE DC ENG. RUN	CIRCUIT	PIN	WIRE COLOR
① B+	B+	FUEL PUMP RELAY	(A1) C47-1	PNK/WHT
0*	0-ON B+-OFF	A/C CUT-OUT RELAY	(A2) C47-2	GRY/RED
		NOT USED	(A3) C47-3	
B+	0 ③	INTAKE AIR SOLENOID	(A4) C47-4	WHT/BLU
0	B+	CHECK ENGINE LIGHT	(A5) C47-5	VLT
B+	B+	IGNITION FEED MAIN RELAY	(A6) C47-6	RED/GRN
		NOT USED	(A7) C47-7	
4.7V	4.7V	SERIAL DATA OUTPUT	(A8) C47-8	ORG/BLK
B+	B+	DIAGNOSTIC REQUEST	(A9) C47-9	ORG/YEL
⑤ 0*	0*	VSS SIGNAL INPUT	(A10) C47-10	WHT
0*	0*	COOLANT, MAP SENSOR GROUND	(A11) C47-11	GRY
0*	0*	ENGINE GROUND	(A12) C47-12	BLK/BLU

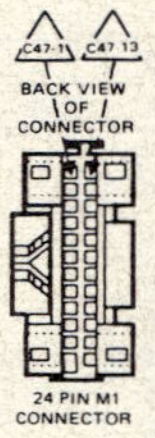

WIRE COLOR	PIN	CIRCUIT	VOLTAGE DC KEY "ON"	VOLTAGE DC ENG. RUN
RED/WHT	(B1) C47-13	SBF BOX (30A)	B+	B+
	(B2) C47-14	NOT USED		
YEL/BLK	(B3) C47-15	DISTRIBUTOR REFERENCE GROUND	0*	0*
	(B4) C47-16	NOT USED		
YEL/RED	(B5) C47-17	DISTRIBUTOR REFERENCE INPUT SIGNAL	0*	1.1V
	(B6) C47-18	NOT USED		
	(B7) C47-19	NOT USED		
GRN/ORG	(B8) C47-20	AIR CONDITIONING REQUEST INPUT	0	B+
	(B9) C47-21	NOT USED		
PNK/BLU	(B10) C47-22	P/N SWITCH	0*	0* ④
	(B11) C47-23	NOT USED		
	(B12) C47-24	NOT USED		

VOLTAGE DC KEY "ON"	VOLTAGE DC ENG. RUN	CIRCUIT	PIN	WIRE COLOR
⑥ 4.0V	3.3 IDLE	PWM TPS	(C1) C46-1	ORG/BLU
		NOT USED	(C2) C46-2	
NOT USABLE	NOT USABLE	IAC (COIL B LOW)	(C3) C46-3	BRN/YEL
NOT USABLE	NOT USABLE	IAC (COIL B HIGH)	(C4) C46-4	BRN/RED
NOT USABLE	NOT USABLE	IAC (COIL A HIGH)	(C5) C46-5	BRN/BLK
NOT USABLE	NOT USABLE	IAC (COIL A LOW)	(C6) C46-6	BRN/WHT
B+	B+	POWER STEERING SWITCH INPUT	(C7) C46-7	GRN/YEL
		NOT USED	(C8) C46-8	
		NOT USED	(C9) C46-9	
②	②	COOLANT TEMP SENSOR SIGNAL INPUT	(C10) C46-10	GRY/BLK
4.7	1.9	MAP SENSOR SIGNAL INPUT	(C11) C46-11	GRY/RED
②	②	MAT SENSOR SIGNAL INPUT	(C12) C46-12	BLU/BLK
.56V	.56V	TPS SIGNAL INPUT	(C13) C46-13	BLU/RED
5.0V	5.0V	+5V REFERENCE VOLTAGE OUTPUT	(C14) C46-14	BLU/ORG
B+	B+	INJECTION 1&3 DRIVE	(C15) C46-15	BLU/YEL
B+	B+	SBF BOX (30A)	(C16) C46-16	RED/WHL

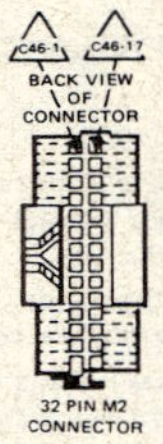

WIRE COLOR	PIN	CIRCUIT	VOLTAGE DC KEY "ON"	VOLTAGE DC ENG. RUN
BLK/BLU	(D1) C46-17	ENGINE GROUND	0*	0*
BLK/WHT	(D2) C46-18	TPS, MAT SENSOR GROUND	0*	0*
PNK/GRN	(D3) C46-19	PEAK & HOLD JUMPER D9	0*	0*
YEL	(D4) C46-20	EST OUTPUT	0*	1.1V
YEL/GRN	(D5) C46-21	IGNITION MODULE BYPASS CIRCUIT	0*	4.6V
BLK/BLU	(D6) C46-22	OXYGEN SENSOR ENGINE GROUND CIRCUIT	0	0
SKYBLU	(D7) C46-23	OXYGEN SENSOR INPUT SIGNAL	.38	VARIES
PNK/BLK	(D8) C46-24	PEAK & HOLD JUMPER D13	0*	0*
PNK/GRN	(D9) C46-25	PEAK & HOLD JUMPER D3	0*	0*
	(D10) C46-26	NOT USED		
	(D11) C46-27	NOT USED		
WHT/GRN	(D12) C46-28	EGR CONTROL	0*	B+
PNK/BLK	(D13) C46-29	PEAK & HOLD JUMPER D8	0*	0*
	(D14) C46-30	NOT USE		
BLU/WHT	(D15) C46-31	INJECTION 1&3 DRIVE	B+	B+
	(D16) C46-32	NOT USED		

- All voltages showing "B+" indicates battery or charging system voltage
- *All voltages showing "0" are ground circuits, and should read less than 0.10 volts

① Reads battery voltage for 2 seconds after ignition "ON" then should read 0 volts
② Varies depending on temperature
③ B+ above 5000 rpm
④ B+ in R, D or L
⑤ Varies with vehicle speed
⑥ For M/T transaxle B+

1.6L SOHC ENGINE – DIAGNOSTIC CIRCUIT CHECK
STYLUS

The "Scan" Diagnostic Circuit Check is an organized approach to identifying a problem created by an Electronic Engine Control System malfunction. It must be the starting point for any driveability complaint diagnosis, because it directs the Service Technician to the next logical step in diagnosing the complaint.
The "Scan Data" listed in the table may be used for comparison, after completing the Diagnostic Circuit Check and finding the on-board diagnostics functioning properly and no trouble codes displayed. The "Typical Values" are an average of display values recorded from normally operating vehicles and are intended to represent what a normally functioning system would typically display.

A "SCAN" tool that displays faulty data should not be used, and the problem should be reported to the manufacturer. The use of a faulty "SCAN" tool can result in misdiagnosis and unnecessary parts replacement.

Only the parameters listed below are used in this manual for diagnosis. If a "Scan" reads other parameters, the values are not recommended by ISUZU Motors for use in diagnosis.

"SCAN" DATA

Test under following condition.
Idle / Upper radiator hose hot / Closed throttle / Park or neutral / "Closed Loop" / All Accessories "OFF"

"SCAN" position	Units displayed	Typical data value
Desired RPM	RPM	ECM idle command (Varies with temperature)
RPM	RPM	± 50 RPM from desired RPM in drive
Coolant Temp.	C°	85 - 105
MAT Temp.	C°	10 - 90 (depends on underhood temperature)
MAP	Kpa/Volts	25-50/0.5 - 2.0 (depends on Vac. & Baro pressure)
BPW (base pulse width)	m/Sec	.8 - 3.0
O_2	Milli Volts	100-900 (varies continuously)
Idle Throttle	Yes/No	Yes
Block Learn Emable	Yes/No	Yes
TPS	Volts	.4 - 0.9
Throttle Angle	0 - 100%	0 - 1
IAC	Counts (steps)	1 - 50
P/N Switch	P-N and R-D-L	Park/Neutral (P/N) (AUTO)
INT (Integrator)	Counts	100 - 150
BLM (Block Learn Memory)	Counts	100 - 150
Open/Closed Loop	Open/Closed	Closed Loop (Open with extended idle)
Block Learn Cell	O/I	1
Vehicle Speed	MPH	0
Spark Advance	Degrees	Varies
Battery	Volts	12.5 - 14.5
EGR Solenoid	On/Off	Off
Time From Start	00:00	
A/C Request	Yes/No	No
A/C Clutch	On/Off	Off
Power Steering Switch	Normal/Hi pressure	Normal
Shift Light (M/T)	On/Off	"Off"

If all values are within the range illustrated, refer to symptoms

1.6L DOHC ENGINE—DIAGNOSTIC CIRCUIT CHECK IMPULSE AND STYLUS

The "Scan" Diagnostic Circuit Check is an organized approach to identifying a problem created by an Electronic Engine Control System malfunction. It must be the starting point for any driveability complaint diagnosis, because it directs the Service Technician to the next logical step in diagnosing the complaint. The "Scan Data" listed in the table may be used for comparison, after completing the Diagnostic Circuit Check and finding the on-board diagnostics functioning properly and no trouble codes displayed. The "Typical Values" are an average of display values recorded from normally operating vehicles and are intended to represent what a normally functioning system would typically display.

A "SCAN" tool that displays faulty data should not be used, and the problem should be reported to the manufacturer. The use of a faulty "SCAN" tool can result in misdiagnosis and unnecessary parts replacement.

Only the parameters listed below are used in this manual for diagnosis. If a "Scan" reads other parameters, the values are not recommended by ISUZU Motors for use in diagnosis.

"SCAN" DATA

Test under following condition.
Idle / Upper radiator hose hot / Closed throttle / Park or neutral / "Closed Loop" / All Accessories "OFF"

"SCAN" position	Units displayed	Typical data value
Desired RPM	RPM	ECM idle command (Varies with temperature)
RPM	RPM	± 50 RPM from desired RPM in drive
Coolant Temp.	C°	85 - 105
MAT Temp.	C°	10 - 90 (depends on underhood temperature)
MAP	Kpa/Volts	25-50/0.5 - 2.0 (depends on Vac. & Baro pressure)
BPW (base pulse width)	m/Sec	.8 - 3.0
O_2	Milli Volts	100-900 (varies continuously)
Idle Throttle	Yes/No	No
Block Learn Enable	Yes/No	No
TPS	Volts	.4 - 1.25
Throttle Angle	0 - 100%	0
IAC	Counts (steps)	1 - 50
P/N Switch	P-N and R-D-L	Park/Neutral (P/N) (AUTO)
INT (Integrator)	Counts	110 - 145
BLM (Block Learn Memory)	Counts	118 - 138
Open/Closed Loop	Open/Closed	Closed Loop (Open with extended idle)
Block Learn Cell	O/I	0
Vehicle Speed	MPH	0
Spark Advance	Degrees	Varies
Battery	Volts	13.5 - 14.5
EGR Solenoid	On/Off	Off
Time From Start	00:00	
A/C Request	Yes/No	No
A/C Clutch	On/Off	"Off"
Power Steering Switch	Normal/Hi pressure	Normal
Shift Light (M/T)	On/Off	"Off"
EGR Temp. Input	Volts	0 - 5

If all values are within the range illustrated, refer to symptoms

1.6L ENGINE—DIAGNOSTIC CIRCUIT CHECK CHART IMPULSE AND STYLUS

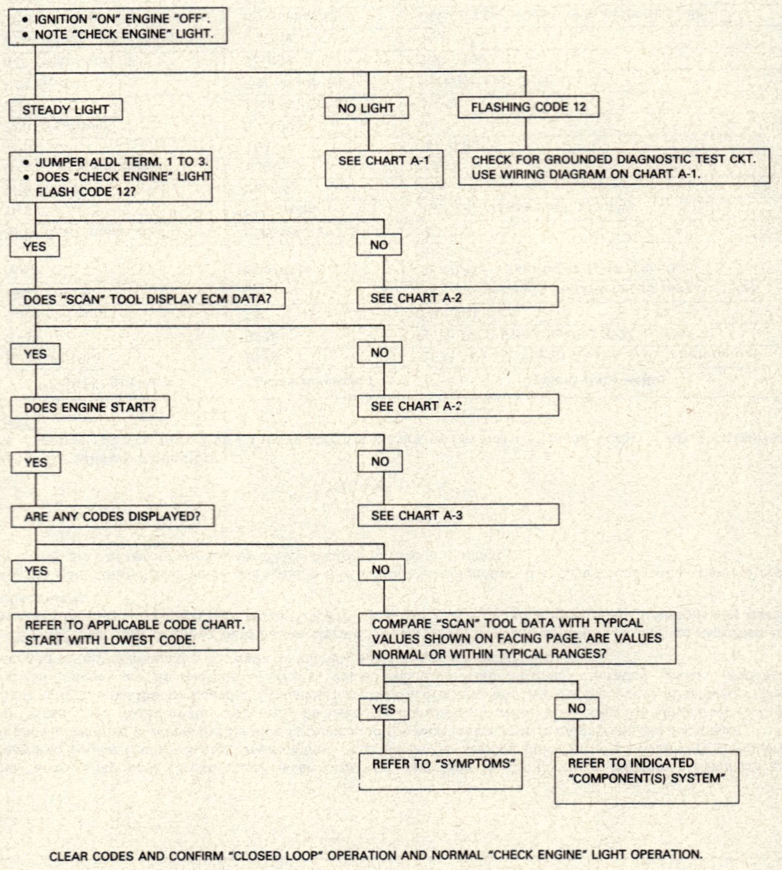

1.6L ENGINE—PRELIMINARY CHECK CHARTS—IMPULSE AND STYLUS

CHART A-1

NO "CHECK ENGINE" LIGHT

Circuit Description:

There should always be a steady "Check Engine" light, when the ignition is "ON" and engine stopped. Battery voltage is supplied directly to the light bulb from fuse C-7. The Electronic Control Module (ECM) will control the light and turn it "ON" by providing a ground path through CKT 419 to the ECM.

Test Description:

Numbers below refer to circled numbers on the diagnostic chart.

① Battery feed CKT 240 is protected by a fusible link in relay/fuse box under the hood near the battery.
② If CKT 240 has voltage, the ECM grounds or the ECM are faulty.
③ Using a test light connected to 12 volts, probe each of the system ground circuits to be sure a good ground is present. Refer to the ECM terminal end view in front of this section for ECM pin locations of ground circuits.

Diagnostic Aids;

Engine runs OK, check;
- Faulty light bulb.
- CKT 419 open.
- Fuse C-7 10A blown. This will result in no oil or generator lights, seat belt reminder, etc.

Engine cranks, but will not run.
- Continuous battery-fuse or fusible link open.
- ECM ignition fuse open.
- Battery CKT 240 to ECM open.
- Ignition CKT 439 to ECM open.
- Poor connection to ECM.

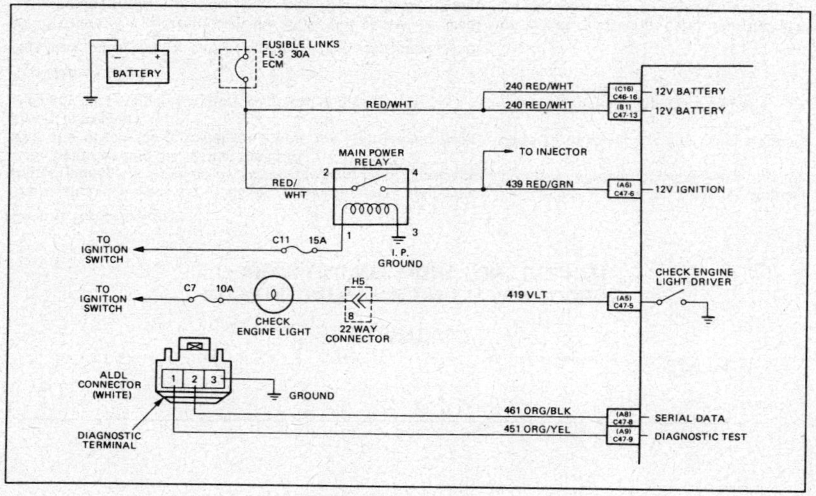

CHART A-1

NO "CHECK ENGINE" LIGHT

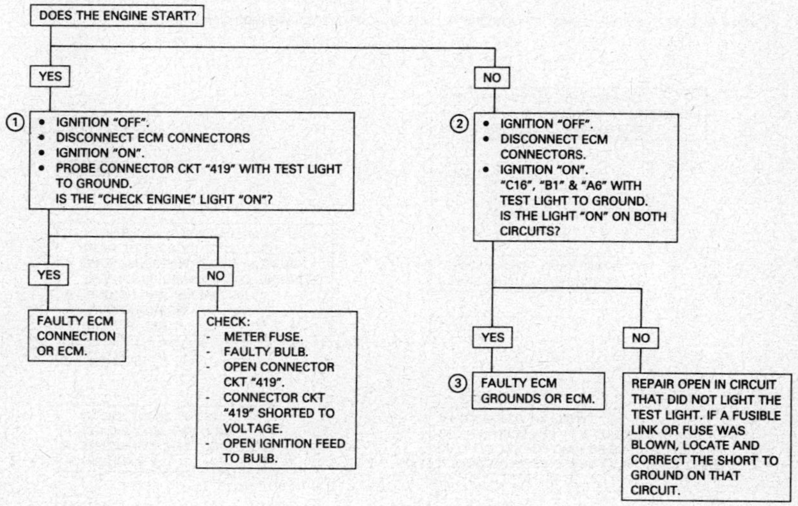

CLEAR CODES AND CONFIRM "CLOSED LOOP" OPERATION AND NORMAL "CHECK ENGINE" LIGHT OPERATION.

CHART A-2

NO ALDL DATA OR WON'T FLASH CODE 12 ("CHECK ENGINE" LIGHT "ON" STEADY)

Circuit Description:

There should always be a steady "Check Engine" light when the ignition is "ON" and engine stopped. Battery voltage is supplied directly to the light bulb. The Electronic Control Module (ECM) will turn the light "ON" by grounding CKT 419 at the ECM.
With the diagnostic terminal grounded, the light should flash a Code 12, followed by any trouble code(s) stored in memory.
A steady light suggests a short to ground in the light control CKT 419, or an open in diagnostic CKT 451.

Test Description:

Numbers below refer to circled numbers on the diagnostic chart.

① If there is a problem with the ECM that causes a "scan" tool to not read serial data, then the ECM should not flash a code 12. If code 12 does flash, be sure that the "scan" tool is working properly on another vehicle. If the "scan" tool is functioning properly and CKT 461 is OK, the ECM may be at fault for the No ALDL Symptom.

② If the light goes "OFF" when the ECM connector is disconnected, the CKT 419 is not shorted to ground.

③ This step will check for an open diagnostic CKT 451.

④ At this point, the "Check Engine" light wiring is OK. The problem is a faulty ECM. If code 12 does not flash, the ECM should be replaced.

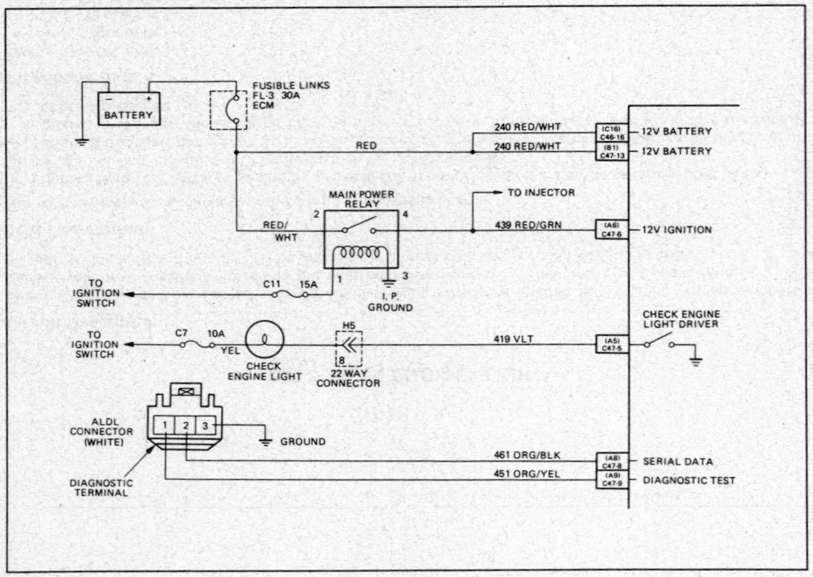

CHART A-2

NO ALDL DATA OR WON'T FLASH CODE 12 ("CHECK ENGINE" LIGHT "ON" STEADY)

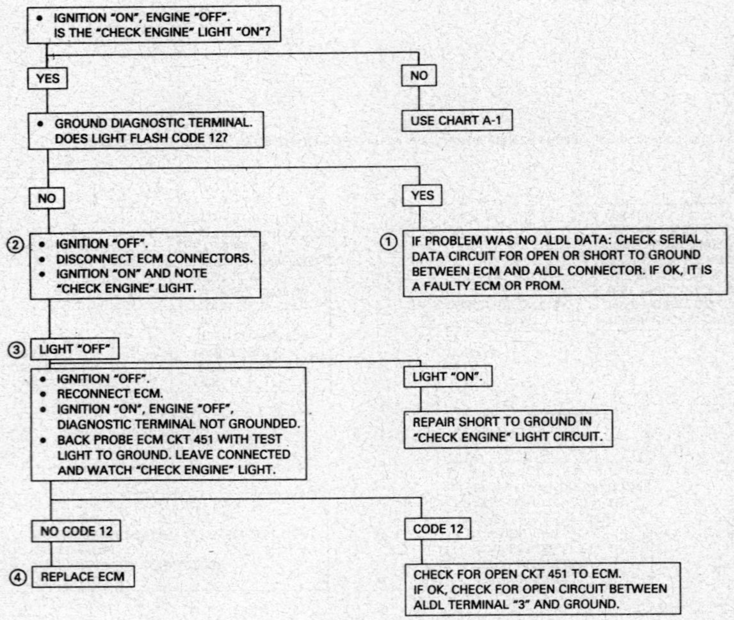

CLEAR CODES AND CONFIRM "CLOSED LOOP" OPERATION AND NORMAL "CHECK ENGINE" LIGHT OPERATION.

1.6L ENGINE – PRELIMINARY CHECK CHARTS – IMPULSE AND STYLUS

CHART A-3

ENGINE CRANKS BUT WON'T RUN

Circuit Description:

Before using this chart, battery condition, engine cranking speed, and fuel quantity should be checked and verified as being OK.

Test Description:

Numbers below refer to circled numbers on the diagnostic chart.

① A "Check Engine" light "ON" is a basic test to determine if there is battery voltage and ignition voltage supplied to the ECM. No ALDL data may be due to an ECM problem. CHART A-2 will diagnose the ECM. If TPS voltage is over 2.5 volts, the engine may be in the "Clear Flood" mode, which will cause starting problems. The engine will not start without reference pulses and, therefore, the "Scan" tool should indicate engine speed during cranking.

② If engine speed was indicated during crank, the ignition module is receiving a crank signal, but "No Spark" at this test indicates the ignition module is not triggering the coil or there is a secondary ignition problem.

③ The test light should flash, indicating the ECM is controlling the injectors. The brightness of the light is not important. However, the test light should be a J-34730 or equivalent.

④ This test will determine if the ignition module is not generating the reference pulse, or if the wiring or ECM are at fault. By touching and removing a test light to battery voltage on CKT 430, a reference pulse should be generated. If engine speed is indicated, the ECM and wiring are OK.

Diagnostic Aids:

- Water or foreign material can cause a no start condition during freezing weather. The engine may start after 5 or 6 minutes in a heated shop. The problem may recur after parking overnight in freezing temperatures.
- An EGR sticking open can cause a low air/fuel ratio during cranking. Unless the system enters "Clear Flood" at the first indication of a flooding condition it can result in a no start.
- Fuel pressure: Low fuel pressure can result in a very lean air/fuel ratio. See CHART A-7.

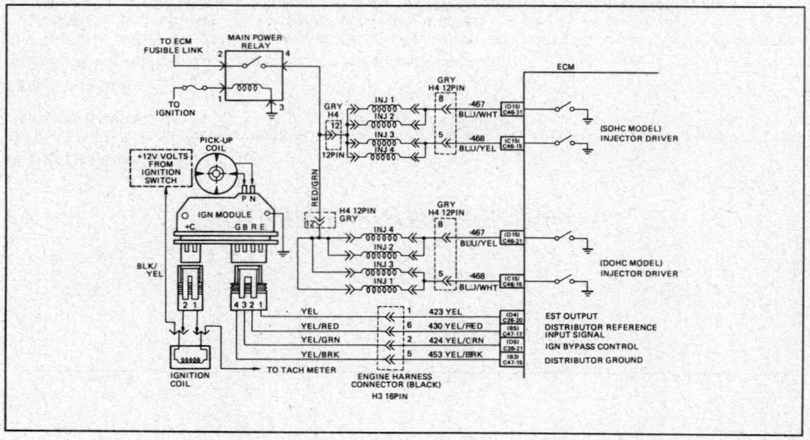

CHART A-3

1 OF 2
ENGINE CRANKS BUT WON'T RUN

①
- TURN IGNITION "ON".
- IF "CHECK ENGINE" LIGHT IS "OFF", USE CHART A-1.
- INSTALL "SCAN" TOOL-IF "NO ALDL" DATA, USE CHART A-2.
- CHECK THE FOLLOWING:
  - TPS - IF OVER 2.5V AT CLOSED THROTTLE, SEE CODE 21.
  - COOLANT - IF BELOW −30°C, SEE CODE 14.
  - IS RPM INDICATED DURING CRANKING?

YES →

②
- USING A ST-125 (SPARK CHECKER), J26792 OR EQUIVALENT, CHECK FOR SPARK WHILE CRANKING (CHECK TWO WIRES). IS SPARK PRESENT?

  YES →
  - PROBE FUEL PUMP RELAY CONNECTOR CKT 465 WITH A TEST LIGHT TO GROUND.
  - IGNITION "OFF" FOR 10 SECONDS.
  - IGNITION "ON".
  - TEST LIGHT SHOULD LIGHT FOR 2 SECONDS AFTER IGNITION "ON". DOES IT?

    YES →

    ③
    - CONNECT INJECTOR TEST LIGHT J34730 OR EQUIVALENT TO INJECTOR HARNESS CONNECTOR.
    - CHECK CYLINDERS 1 & 4.
    - CRANK ENGINE. DOES TEST LIGHT FLASH ON BOTH?

      YES →
      - IGNITION "OFF".
      - INSTALL FUEL PRESSURE GAUGE, REFER TO CHART A-7 (2 OF 2) FOR INSTALLATION PROCEDURE
      - IGNITION "ON".
      - FUEL PRESSURE SHOULD BE 207-276 kPa (35-42 psi). IS IT?

        YES → CHECK FOR FOULED SPARK PLUGS OR EGR VALVE STUCK OPEN. SEE "DIAGNOSTIC CIRCUIT CHECK"

        NO → USE FUEL SYSTEM DIAGNOSIS CHART A-7.

      NO → USE CHART A-3

    NO → USE FUEL PUMP RELAY CIRCUIT CHART A-5.

  NO → BASIC HEI PROBLEM.

NO →
- USING A ST-125 (SPARK CHECKER), J26792 OR EQUIVALENT, CHECK FOR SPARK WHILE CRANKING (CHECK TWO WIRES). IS SPARK PRESENT?

  YES →

  ④
  - IGNITION "OFF"
  - DISCONNECT DISTRIBUTOR 4-WAY CONNECTOR.
  - IGNITION "ON".
  - MOMENTARILY TOUCH HARNESS CONN TERMINAL "B5" WITH A TEST LIGHT TO B+.
  - "SCAN" TOOL SHOULD INDICATE ENGINE SPEED WHEN TEST IS PERFORMED. DOES IT?

    YES → FAULTY CONNECTION OR IGNITION MODULE.

    NO → CKT 430 OPEN, SHORTED TO GROUND OR FAULTY ECM.

  NO → CHECK FOR BATTERY VOLTAGE TO IGNITION SYSTEM. IF OK, THERE IS A BASIC HEI PROBLEM.

NOTICE: PFI SYSTEM UNDER PRESSURE. TO AVOID FUEL SPILLAGE, REFER TO FIELD SERVICE PROCEDURES FOR TESTING OR MAKING REPAIRS REQUIRING DISASSEMBLY OF FUEL LINES OR FITTINGS

CLEAR CODES AND CONFIRM "CLOSED LOOP" OPERATION AND NORMAL "CHECK ENGINE" LIGHT OPERATION.

CHART A-3

ENGINE CRANKS BUT WON'T RUN

Circuit Description

Before using this chart, battery condition, engine cranking speed, and fuel quantity should be checked and verified as being OK.

Test Description:

Numbers below refer to circled numbers on the diagnostic chart.

1. This step checks for ignition voltage at the injector harness connector. Disconnect H4-GRAY connector before probing terminal "12". Reconnect connector after test.
2. Checks for open CKT 467 or 468 from connector to ECM. Be sure injector harness is connected.
3. Disconnect ECM C-D connector and check injector jumper circuits with a DVOM.

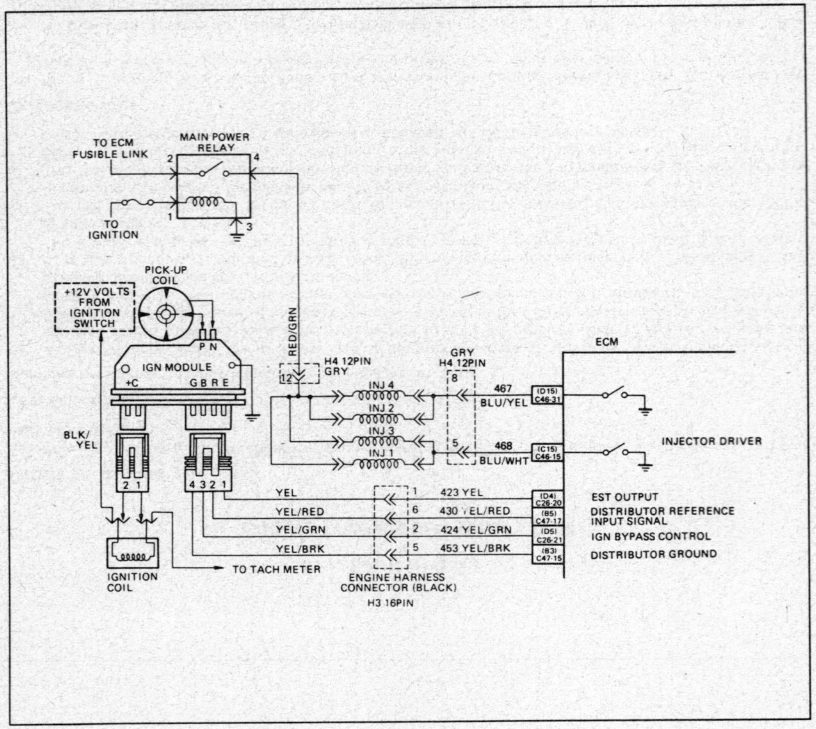

CHART A-3

2 of 2
ENGINE CRANKS BUT WON'T RUN

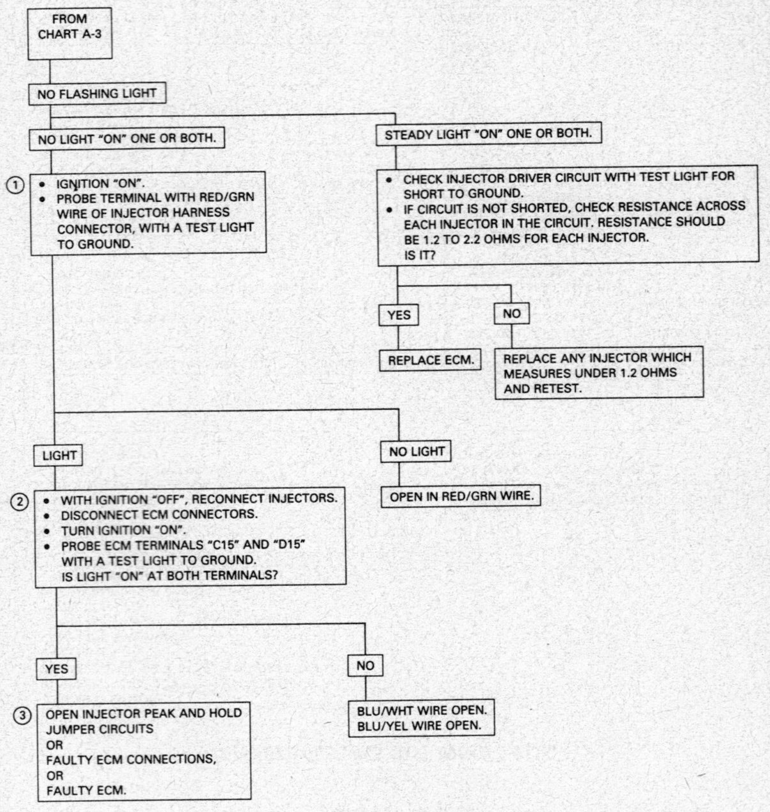

CLEAR CODES AND CONFIRM "CLOSED LOOP" OPERATION AND NORMAL "CHECK ENGINE" LIGHT OPERATION.

CHART A-5

FUEL PUMP RELAY CIRCUIT

Circuit Description:

When the ignition switch is turned "ON", the Electronic Control Module (ECM) will activate the fuel pump relay and run the in-take fuel pump. The fuel pump will operate as long as the engine is cranking or running and the ECM is receiving ignition reference pulses. If there are no reference pulses, the ECM will shut "Off" the fuel pump within 2 seconds after the key is turned to "ON".

Test Description:

Numbers below refer to circled numbers on the diagnostic chart.

① Checks for the presence of B+ at switching portion of Fuel Pump Relay.
② Checks circuit between Fuel Pump Relay and ground (including the fuel pump).
③ This test will confirm if voltage is available to the relay coil.

Diagnostic Aids:

A visual inspection of wiring and connectors should be made if an intermittent problem exists. For more information, see "Intermittents" in "Symptoms"

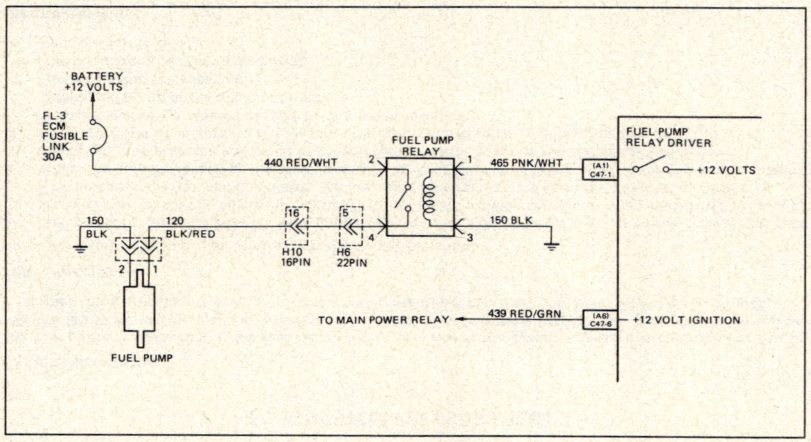

CHART A-5

FUEL PUMP RELAY CIRCUIT

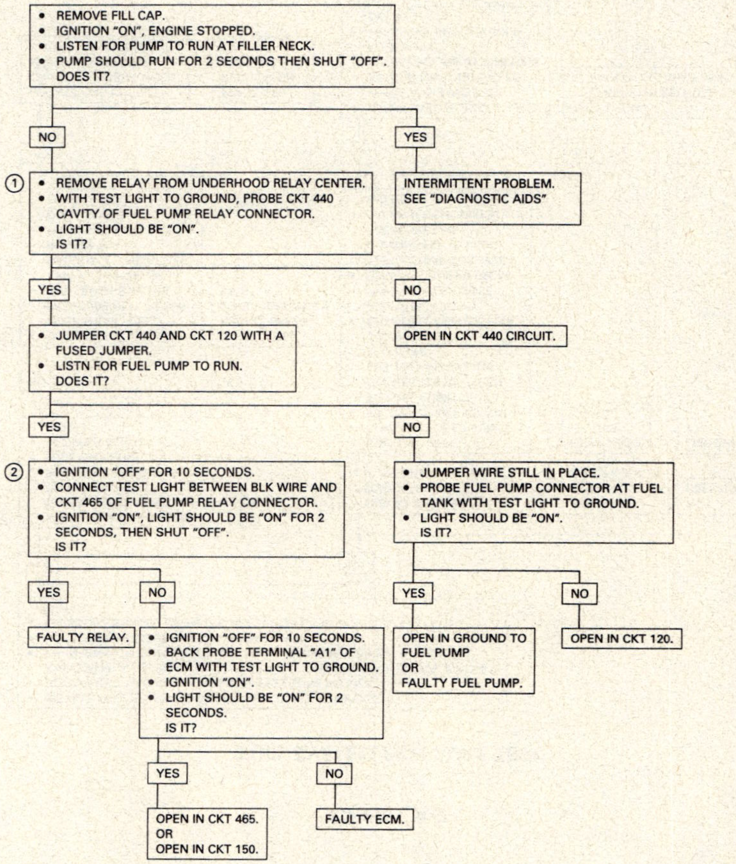

CLEAR CODES AND CONFIRM "CLOSED LOOP" OPERATION AND NORMAL "CHECK ENGINE" LIGHT OPERATION.

CHART A-7

FUEL SYSTEM PRESSURE TEST

Circuit Description:

The fuel pump delivers fuel to the fuel rail and injectors, where the system pressure is controlled to 245 to 286 kPa (35 to 42 psi) by the pressure regulator. Excess fuel is returned to the fuel tank. When the engine is stopped, the pump can be energized by jumpering the fuel pump relay connector with a fused jumper.

Test Description:

Numbers below refer to circled numbers on the diagnostic chart.

① Use pressure gauge J-34730-1 and gauge adapter J-35957-10. Wrap a shop towel around the fuel pressure tap to absorb any small amount of fuel leakage that may occur when installing the gauge. (The pressure will not leak down after the fuel pump is stopped on a correctly functioning system.)

② While the engine is idling, manifold pressure is low (vacuum). When applied to the fuel regulator diaphragm, the pressure will result in a lower fuel pressure at about 190-200 kPa (25-30 psi).

③ The application of vacuum to the pressure regulator should result in a fuel pressure drop.

④ Pressure leak-down may be caused by one of the following:

- In-tank fuel pump check valve not holding
- Pump coupling hose leaking
- Fuel pressure regulator valve leaking
- Injector sticking open

Diagnostic Aids:

Improper fuel system pressure may contribute to one or all of the following symptoms:

- Cranks but won't run
- Code 44 or 45
- Cutting out (May feel like ignition problem)
- Hesitation, loss of power or poor fuel economy

Refer to "Symptoms"

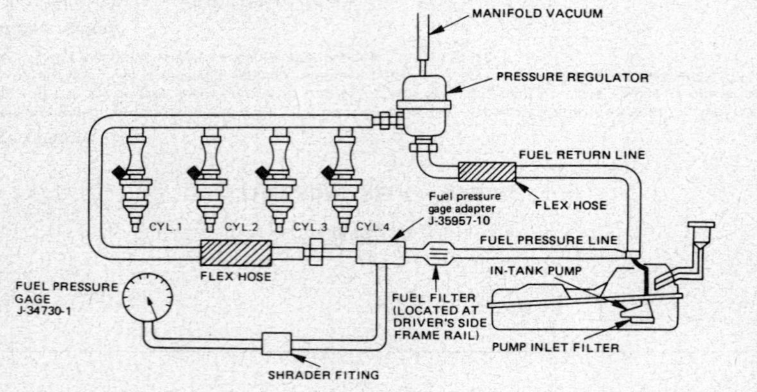

CHART A-7

FUEL SYSTEM PRESSURE TEST

①
- INSTALL FUEL PRESSURE GAUGE, J34730-1 OR EQUIVALENT.
- DISCONNECT VACUUM HOSE FROM PRESSURE REGULATOR.
- IGNITION MUST BE "OFF" FOR 10 SECONDS AND ENSURE THAT A/C IS "OFF".
- TURN IGNITION "ON". FUEL PUMP SHOULD RUN FOR ABOUT 2 SECONDS.
- NOTE FUEL PRESSURE AFTER PUMP STOPS.
- PRESSURE SHOULD BE 245-256 kPa (35-38 PSI) AND HOLDING STEADY.

IS IT?

YES →

②
- START AND IDLE ENGINE AT NORMAL OPERATING TEMPERATURE.
- RECONNECT VACUUM LINE.
- FUEL PRESSURE SHOULD HAVE DROPPED SLIGHTLY.

DID IT?

- **YES** → NO TROUBLE FOUND. REFER TO "DIAGNOSTIC AIDS"
- **NO** →

③
- DISCONNECT VACUUM HOSE FROM FUEL PRESSURE REGULATOR.
- WITH ENGINE IDLING. APPLY 12-14" Hg OF VACUUM TO PRESSURE REGULATOR
- FUEL PRESSURE SHOULD DROP SLIGHTLY.

DID IT?

- **YES** → LOCATE AND CORRECT CAUSE OF INTERRUPTED MANIFOLD PRESSURE SUPPLY TO REGULATOR.
- **NO** → PRESSURE REGULATOR ASSEMBLY IS FAULTY.

NO →

- ④ PRESSURE WITHIN SPECIFICATION, BUT NOT HOLDING.
  - IGNITION "OFF".
  - CONNECT FUSED JUMPER FROM BATTERY VOLTAGE TO FUEL PUMP RELAY CONNECTOR BETWEEN CKT 120 AND CKT 440 IN ENGINE COMPARTMENT.
  - GRADUALLY BLOCK FUEL PRESSURE LINE BY PINCHING FLEX HOSE.
  - REMOVE JUMPER FROM TEST CONNECTOR WHILE CONTINUING TO BLOCK PRESSURE LINE.
  - PRESSURE SHOULD HOLD.

  DID IT?
  - **YES** → CHECK FOR
    - LEAKING PUMP HOSE
    - FAULTY IN-TANK PUMP.
  - **NO** →
    - REPEAT ABOVE PROCEDURE WHILE BLOCKING FUEL RETURN LINE INSTEAD OF PRESSURE LINE BY PINCHING FLEX HOSE.
    - PRESSURE SHOULD HOLD.

    DID IT?
    - **YES** → PRESSURE REGULATOR ASSEMBLY IS FAULTY.
    - **NO** → LOCATE AND REPLACE LEAKING INJECTOR AND DRY SPARK PLUGS FROM FLOODED CYLINDERS.
- PRESSURE OUT OF SPECIFICATION. → USE PAGE 2 OF THIS CHART.
- NO PRESSURE → USE CHART A-5

CLEAR CODES AND CONFIRM "CLOSED LOOP" OPERATION AND NORMAL "CHECK ENGINE" LIGHT OPERATION.

CHART A-7

FUEL SYSTEM PRESSURE TEST

Circuit Description:

The fuel pump delivers fuel to the fuel rail and injectors, where the system pressure is controlled to 245 to 286 kPa (35 to 42 psi) by the pressure regulator. Excess fuel is returned to the fuel tank. When the engine is stopped, the pump can be energized by jumpering the fuel pump relay connect with a fused jumper.

Test Description:

Numbers below refer to circled numbers on the diagnostic chart.

⑤ Pressure less than 245 kPa (35 psi) may be caused by one of two problems.
- The regulated fuel pressure is too low. The system will be running lean and may set Code 44. Also, hard cold starting and overall poor performance is possible.
- Restricted flow is causing a pressure drop. Normally, a vehicle with a fuel pressure loss at idle will not be drivable. However, if the pressure drop occurs only while driving, the engine will surge and then stop as pressure begins to drop rapidly.

⑥ Restricting the fuel return line allows the fuel pump to build above regulated pressure. When battery voltage is applied to the pump test terminal, pressure should be above 450 kPa (65 psi).

⑦ This test determines if the high fuel pressure is due to a restricted fuel return line or a pressure regulator problem.

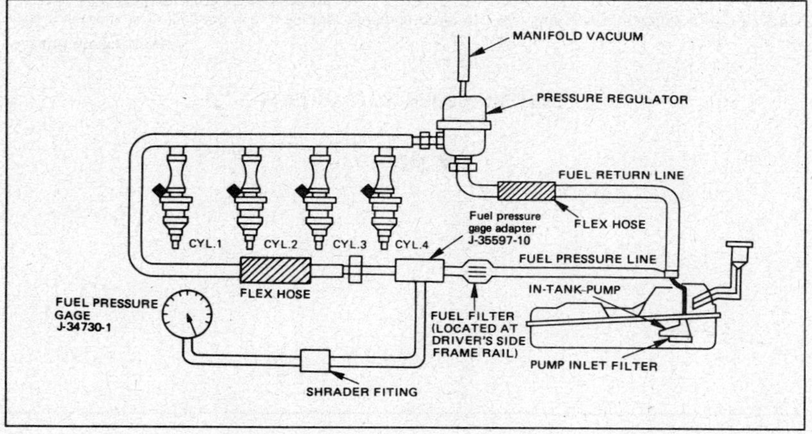

CHART A-7

FUEL SYSTEM PRESSURE TEST

FROM CHART A-7
PRESSURE OUT OF SPEC.
207-276 kPa (30-39 PSI)

LOW PRESSURE

⑤
- CHECK FOR RESTRICTED IN-LINE FILTER.
IS FILTER OK?

- NO → REPLACE FILTER AND RE-TEST.
- YES →

⑥
- IGNITION "OFF".
- APPLY BATTERY VOLTAGE TO FUEL PUMP RELAY CONNECTOR USING A FUSED JUMPER WIRE WITH FUEL PUMP RELAY REMOVED.
- GRADUALLY BLOCK FUEL RETURN LINE BY PINCHING FLEXIBLE HOSE AND NOTE PRESSURE.
- PRESSURE SHOULD BE ABOVE 450 kPa (65 psi).
IS IT?

- YES → PRESSURE REGULATOR ASSEMBLY IS FAULTY.
- NO → CHECK IN-TANK PUMP ASSEMBLY FOR
  - FAULTY FUEL PUMP
  - LEAKING COUPLING HOSE
  - INCORRECT PUMP.

HIGH PRESSURE

⑦
- DISCONNECT VACUUM LINE TO PRESSURE REGULATOR.
- DISCONNECT FUEL RETURN LINE FLEXIBLE HOSE.
- ATTACH A 5/16 IN. FLEX HOSE TO PRESSURE REGULATOR SIDE OF RETURN LINE AND INSERT THE OTHER END INTO AN APPROVED GASOLINE CONTAINER.
- NOTE FUEL PRESSURE WITHIN 2 SECONDS AFTER IGNITION IS TURNED "ON"
- PRESSURE SHOULD BE 245-286 kPa (35-42 PSI).
IS IT?

- NO → PRESSURE REGULATOR ASSEMBLY IS FAULTY.
- YES → LOCATE AND REPAIR RESTRICTED FUEL RETURN LINE.

CLEAR CODES AND CONFIRM "CLOSED LOOP" OPERATION AND NORMAL "CHECK ENGINE" LIGHT OPERATION.

CHART A-7

FUEL SYSTEM PRESSURE TEST

Circuit Description

The fuel rail is not equipped with a shrader fitting. In order to install fuel pressure gauge J-34730-1, gauge adapter J-23597-10 must be installed as shown.

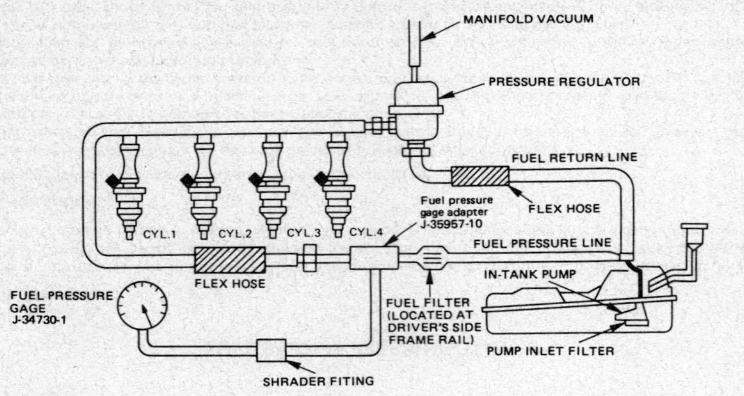

CHART A-7

FUEL SYSTEM PRESSURE TEST

Tools Required: J34730-1 FUEL PRESSURE GAUGE
J35957-10 FUEL PRESSURE GAUGE ADAPTER

CAUTION: TO REDUCE THE RISK OF FIRE AND PERSONAL INJURY:
- IT IS NECESSARY TO RELIEVE FUEL SYSTEM PRESSURE BEFORE CONNECTING A FUEL PRESSURE GAUGE.
- A SMALL AMOUNT OF FUEL MAY BE RELEASED WHEN DISCONNECTING THE FUEL LINES. COVER FUEL LINE FITTINGS WITH A SHOP TOWEL BEFORE DISCONNECTING, TO CATCH ANY FUEL THAT MAY LEAK OUT. PLACE TOWEL IN APPROVED CONTAINER WHEN DISCONNECT IS COMPLETED.

FUEL PRESSURE RELIEF PROCEDURE
1. REMOVE FUEL CAP.
2. REMOVE FUEL PUMP RELAY FROM UNDERHOOD RELAY CENTER.
3. START ENGINE AND ALLOW TO STALL.
4. CRANK ENGINE FOR ADDITIONAL 30 SECONDS.
5. REMOVE NEGATIVE BATTERY CABLE.

FUEL GAUGE ADAPTER INSTALLATION
1. DISCONNECT FUEL PRESSURE LINE AT THE FUEL FILTER.
   USE A SHOP TOWEL TO CATCH ANY ADDITIONAL FUEL THAT WAS LEFT IN THE LINE.
2. INSTALL FUEL GAUGE ADAPTER J35957-10 BETWEEN FUEL PRESSURE LINE AND FUEL FILTER AT FRAME RAIL.
3. INSTALL FUEL GAUGE J34730-1 TO ADAPTER.
4. REINSTALL FUEL PUMP RELAY.
5. CONNECT NEGATIVE BATTERY CABLE.
6. INSTALL FUEL CAP.
7. START ENGINE AND CHECK FOR LEAKS.

NOTE: REVERSE THE ABOVE PROCEDURE ABOVE TO REMOVE FUEL GAUGE ADAPTER.

CLEAR CODES AND CONFIRM "CLOSED LOOP" OPERATION AND NORMAL "CHECK ENGINE" LIGHT OPERATION.

CODE 13

OXYGEN SENSOR CIRCUIT
(OPEN/SHORTED CIRCUIT)

Circuit Description:

The ECM supplies a voltage of about .45 volt between terminals "D7" and "D6". (If measured with a 10 megohm digital voltmeter, this may read as low as .32 volts).
When the O_2 sensor reaches operating temperature, it varies this voltage from about .1 volt (exhaust is lean) to about .9 (exhaust is rich). The O_2 sensor has a heater incorporated as part of the sensor. The heater is used to heat the sensor up, thus causing the O_2 sensor to produce voltage sooner so that the ECM can monitor the exhaust gauges. This will allow the ECM to take control of the fuel sooner. The heater portion of the sensor will not affect how Code 13 sets.
The sensor is like an open circuit and produces no voltage when it is below 360°C (600°F). An open sensor circuit cold causes "Open Loop" operation.

Test Description:

Numbers below refer to circled numbers on the diagnostic chart.

① Code 13 will set under the following conditions:

- Engine operating temperature above 69.5°C (157.1°F).
- At least 2 minutes have elapsed since engine start-up.
- O_2 signal voltage is steady between 0 and .347 volts or steady above .547 volts.
- Throttle angle above 5%.
- All above conditions are met for about 40 seconds.

If the conditions for a Code 13 exist, the system will not go "Closed Loop".

② This test determines if the O_2 sensor is the problem or if the ECM and wiring are at fault.

③ Use only a 10 megohm digital voltmeter. This test checks the continuity of "D7" and "D6". If "D6" is open, the ECM voltage on "D7" will be over .6 volts (600mV).

Diagnostic Aids:

Normal "Scan" tool O_2 sensor voltage varies between 100mV to 999mV (.1 and 1.0 volt) while in "Close Loop". Code 13 sets in one minute if sensor signal voltage remains between 0.347mV and 0.547mV, but the system will go to "Open Loop" in about 15 seconds. Verify a clean, tight ground connection for "D6". Open "D7" or "D6" will result in a Code 13.

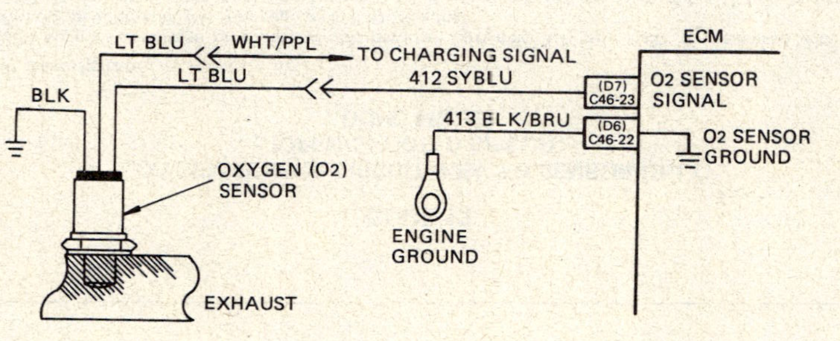

CODE 13

OXYGEN SENSOR CIRCUIT (NON-HEATED SENSOR ONLY)
(OPEN/GROUNDED CIRCUIT)
SOHC MODEL

①
- ENGINE AT NORMAL OPERATING TEMPERATURE (ABOVE 69.5°C/157°F).
- RUN ENGINE ABOVE 1200 RPM FOR TWO MINUTES.
- DOES "SCAN" TOOL INDICATE "CLOSED LOOP"?

YES → CODE 13 IS INTERMITTENT. IF NO ADDITIONAL CODES WERE STORED, REFER TO "DIAGNOSTIC AIDS"

NO →

②
- DISCONNECT O_2 SENSOR.
- JUMPER HARNESS TERMINAL "D7" (ECM SIDE) TO GROUND.
- "SCAN" TOOL SHOULD DISPLAY O_2 VOLTAGE BELOW .2 VOLTS (200mV) WITH ENGINE RUNNING. DOES IT?

YES → FAULTY O_2 SENSOR CONNECTION OR FAULTY O_2 SENSOR

NO →

③
- REMOVE JUMPER.
- IGNITION "ON", ENGINE "OFF".
- CHECK VOLTAGE OF O_2 SENSOR SIGNAL CKT (ECM SIDE) AT O_2 SENSOR HARNESS CONNECTOR USING A DVOM.

.3-.6 VOLTS (300-600mV) → FAULTY ECM

OVER .6 VOLT (600mV) → OPEN O_2 SENSOR GROUND CKT OR FAULTY CONNECTION OR FAULTY ECM.

LESS THAN .3 VOLT (300mV) → OPEN O_2 SENSOR SIGNAL CKT OR FAULTY ECM CONNECTION OR FAULTY ECM.

CLEAR CODES AND CONFIRM "CLOSED LOOP" OPERATION AND NORMAL "CHECK ENGINE" LIGHT OPERATION.

CODE 13

OXYGEN SENSOR CIRCUIT (HEATED SENSOR ONLY) (OPEN/SHORTED CIRCUIT) DOHC MODEL

Circuit Description:

The ECM supplies a voltage of about .45 volt between terminals "D7" and "D6". (If measured with a 10 megohm digital voltmeter, this may read as low as .32 volts).
When the O_2 sensor reaches operating temperature, it varies this voltage from about .1 volt (exhaust is lean) to about .9 (exhaust is rich). The O_2 sensor has a heater incorporated as part of the sensor. The heater is used to heat the sensor up, thus causing the O_2 sensor to produce voltage sooner so that the ECM can monitor the exhaust gauges. This will allow the ECM to take control of the fuel sooner. The heater portion of the sensor will not affect how Code 13 sets.
The sensor is like an open circuit and produces no voltage when it is below 360°C (600°F). An open sensor circuit cold causes "Open Loop" operation.

Test Description:

Numbers below refer to circled numbers on the diagnostic chart.

① Code 13 will set under the following conditions:

- Engine operating temperature above 69.5°C (157.1°F).
- At least 2 minutes have elapsed since engine start-up.
- O_2 signal voltage is steady between 0 and .347 volts or steady above .547 volts.
- Throttle angle above 5%.
- All above conditions are met for about 40 seconds.

If the conditions for a Code 13 exist, the system will not go "Closed Loop".

② This test determines if the O_2 sensor is the problem or if the ECM and wiring are at fault.

③ Use only a 10 megohm digital voltmeter. This test checks the continuity of "D7" and "D6". If "D6" is open, the ECM voltage on "D7" will be over .6 volts (600mV).

Diagnostic Aids:

Normal "Scan" tool O_2 sensor voltage varies between 100mV to 999mV (.1 and 1.0 volt) while in "Close Loop". Code 13 sets in one minute if sensor signal voltage remains between 0.347mV and 0.547mV, but the system will go to "Open Loop" in about 15 seconds. Verify a clean, tight ground connection for "D6". Open "D7" or "D6" will result in a Code 13.

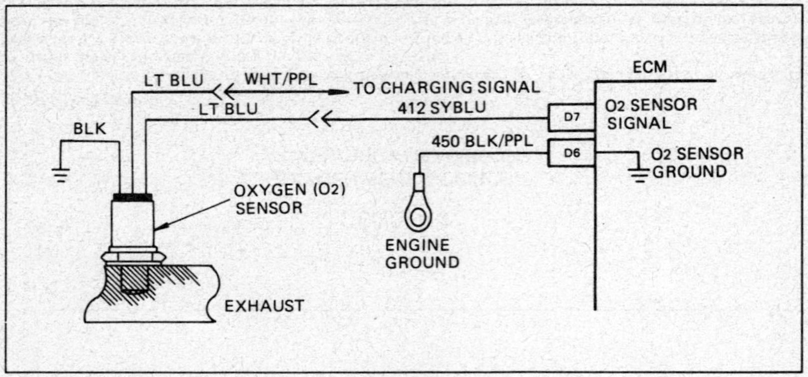

CODE 13

OXYGEN SENSOR CIRCUIT (HEATED SENSOR ONLY) (OPEN/SHORTED CIRCUIT) DOHC MODEL

① ENGINE AT NORMAL OPERATING TEMPERATURE (ABOVE 69.5°C/157°F).
- RUN ENGINE ABOVE 1200 RPM FOR TWO MINUTES.
- DOES "SCAN" TOOL INDICATE "CLOSED LOOP"?

YES → CODE 13 IS INTERMITTENT. IF NO ADDITIONAL CODES WERE STORED, REFER TO "DIAGNOSTIC AIDS"

NO →

②
- DISCONNECT O_2 SENSOR.
- JUMPER HARNESS TERMINAL "D7" (ECM SIDE) TO GROUND.
- "SCAN" TOOL SHOULD DISPLAY O_2 VOLTAGE BELOW .2 VOLTS (200mV) WITH ENGINE RUNNING. DOES IT?

YES → FAULTY O_2 SENSOR CONNECTION OR FAULTY O_2 SENSOR

NO →

③
- REMOVE JUMPER.
- IGNITION "ON", ENGINE "OFF".
- CHECK VOLTAGE OF O_2 SENSOR SIGNAL CKT (ECM SIDE) AT O_2 SENSOR HARNESS CONNECTOR USING A DVOM.

.3-6 VOLTS (300-600mV) → FAULTY ECM

OVER .6 VOLT (600mV) → OPEN O_2 SENSOR GROUND CKT OR FAULTY CONNECTION OR FAULTY ECM.

LESS THAN .3 VOLT (300mV) → OPEN O_2 SENSOR SIGNAL CKT OR FAULTY ECM CONNECTION OR FAULTY ECM.

CLEAR CODES AND CONFIRM "CLOSED LOOP" OPERATION AND NORMAL "CHECK ENGINE" LIGHT OPERATION.

CODE 14A

COOLANT TEMPERATURE SENSOR CIRCUIT (HIGH TEMPERATURE INDICATED)

Circuit Description:

The Coolant Temperature Sensor uses a thermistor to control the signal voltage to the ECM. The ECM applies a reference voltage on CKT 410 to the sensor. When the engine is cold, the sensor (thermistor) resistance is high. The ECM will then sense a high signal voltage.

As the engine warms, the sensor resistance decreases, and the voltage drops. At normal engine operating temperature, the voltage will measure about 1.5 to 2.0 volts at the ECM terminal "A11".
Coolant temperature is one of the inputs used to control the following:

- Fuel delivery
- Electronic Spark Timing (EST)
- Idle Air Control (IAC)

Test Description:

Numbers below refer to circled numbers on the diagnostic chart.

① Checks to see if code was set as result of hard failure or intermittent condition.
Code 14A will set if:
- Signal Voltage indicates a coolant temperature above 135°C (275°F) for 3 seconds.

② This test simulates conditions for Code 14B. If the ECM recognizes the open circuit (high voltage), and displays a low temperature, the ECM and wiring are OK.

Diagnostic Aids:

A "Scan" tool reads engine temperature in degrees Celsius.
After the engine is started, the temperature should rise steadily to about 90°, then stabilize, when the thermostat opens.
If the engine has been allowed to cool to an ambient temperature (overnight), coolant temperature and manifold air temperature (MAT) may be checked with a "Scan" tool and should read close to each other.
When a code 14A is set, the ECM will turn "ON" the engine cooling fan.
A Code 14A will result if CKT 410 is shorted to ground.

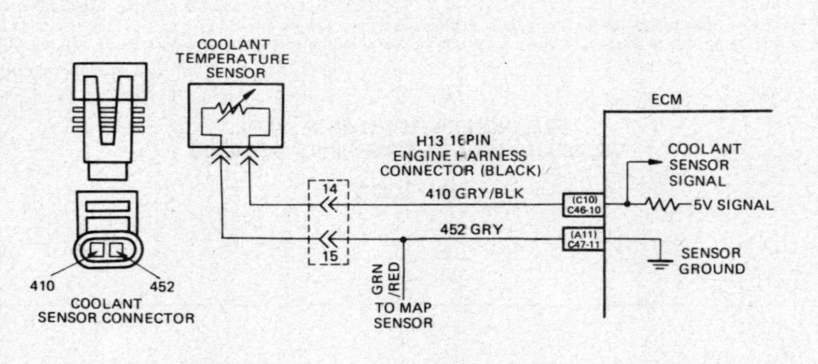

CODE 14A

COOLANT TEMPERATURE SENSOR CIRCUIT (HIGH TEMPERATURE INDICATED)

① DOES "SCAN" TOOL DISPLAY COOLANT TEMPERATURE OF 130°C OR HIGHER?

- YES → ② • DISCONNECT SENSOR. "SCAN" TOOL SHOULD DISPLAY TEMPERATURE BELOW −30°C. DOES IT?
  - YES → REPLACE SENSOR.
  - NO → CTS SIGNAL SHORTED TO GROUND, OR CTS SIGNAL SHORTED TO SENSOR GROUND CIRCUIT OR FAULTY ECM.
- NO → DOES "SCAN TOOL" DISPLAY COOLANT TEMPERATURE −30°C OR LESS?
  - NO → CODE 14 IS INTERMITTENT. IF NO ADDITIONAL CODES WERE STORED, REFER TO "DIAGNOSTIC AIDS"
  - YES → REFER TO CHART 14B

DIAGNOSTIC AID

COOLANT SENSOR		
TEMPERATURE TO RESISTANCE VALUES (APPROXIMATE)		
°F	°C	OHMS
210	100	185
160	70	450
100	38	1,800
70	20	3,400
40	4	7,500
20	−7	13,500
0	−18	25,000
−40	−40	100,700

CLEAR CODES AND CONFIRM "CLOSED LOOP" OPERATION AND NORMAL "CHECK ENGINE" LIGHT OPERATION.

CODE 14B

COOLANT TEMPERATURE SENSOR CIRCUIT (LOW TEMPERATURE INDICATED)

Circuit Description:

The Coolant Temperature Sensor uses a thermistor to control the signal voltage at the ECM. The ECM applies a voltage on CKT 410 to the sensor. When the engine is cold, the sensor (thermistor) resistance is high, therefore, the ECM will see high signal voltage.

As the engine warms, the sensor resistance becomes less, and the voltage drops. At normal engine operating temperature, the voltage will measure about 1.5 to 2.0 volts at the ECM terminal "A-11".
Coolant temperature is one of the inputs used to control:

- Fuel delivery
- Engine Spark Timing (EST)
- Idle (IAC)

Test Description:

Numbers below refer to circled numbers on the diagnostic charts.

① Checks to see if code was set as result of hard failure or intermittent condition.
Code 14B will set if:
- The engine has been running for 2 minute.
- Signal Voltage indicates a coolant temperature below −30°C (−22°F).

② This test simulates conditions for Code 14A. If the ECM recognizes the grounded circuit (low voltage), and displays a high temperature, the ECM and wiring are OK.

③ This test will determine if there is a wiring problem or a faulty ECM.

Diagnostic Aids:

A "Scan" tool reads engine temperature in degrees centigrade.
After the engine is started, the temperature should rise steadily to about 90°C, then stabilize, when the thermostat opens.
If the engine has been allowed to cool to an ambient temperature (overnight), coolant and MAT temperatures may be checked with a "Scan" tool and should read close to each other.
A Code 14B will result if CKT 410 or CKT 452 are open.

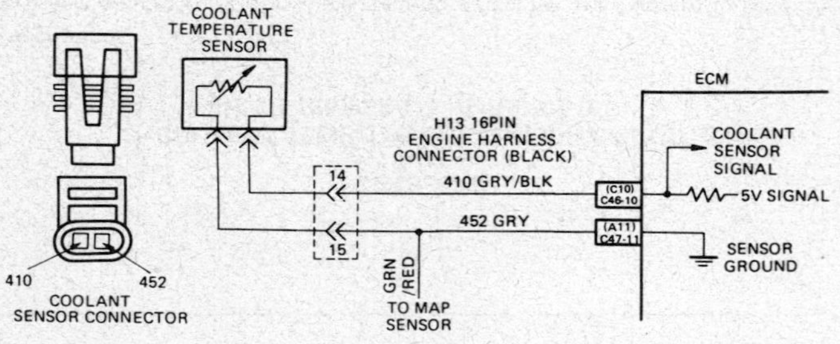

CODE 14B

COOLANT TEMPERATURE SENSOR CIRCUIT (LOW TEMPERATURE INDICATED)

① • DOES "SCAN" TOOL DISPLAY COOLANT TEMPERATURE OF −30°C OR LESS?

- NO → CODE 14 IS INTERMITTENT. IF NO ADDITIONAL CODES WERE STORED, REFER TO "DIAGNOSTIC AIDS"
- YES →

② • DISCONNECT SENSOR.
JUMPER HARNESS TERMINALS TOGETHER.
• "SCAN" TOOL SHOULD DISPLAY 130°C OR MORE.
DOES IT?

- YES → FAULTY CONNECTION OR SENSOR.
- NO →

③ • JUMPER GREY WIRE TO GROUND.
• "SCAN" TOOL SHOULD DISPLAY OVER 130°C.
DOES IT?

- YES → OPEN SENSOR GROUND CIRCUIT, FAULTY CONNECTION OR FAULTY ECM.
- NO → OPEN GREY WIRE, FAULTY CONNECTION AT ECM, OR FAULTY ECM.

DIAGNOSTIC AID

COOLANT SENSOR		
TEMPERATURE TO RESISTANCE VALUES (APPROXIMATE)		
°F	°C	OHMS
210	100	185
160	70	450
100	38	1,800
70	20	3,400
40	4	7,500
20	−7	13,500
0	−18	25,000
−40	−40	100,700

CLEAR CODES AND CONFIRM "CLOSED LOOP" OPERATION AND NORMAL "CHECK ENGINE" LIGHT OPERATION.

CODE 21A

THROTTLE POSITION SENSOR(TPS) CIRCUIT (SIGNAL VOLTAGE HIGH)

Circuit Description:

The Throttle Position Sensor (TPS) provides a voltage signal that changes relative to the throttle valve. Signal voltage will vary from less than 0.9 volts at idle to about 4.5 volts at wide open throttle(WOT).
The TPS signal is one of the most important inputs used by the ECM for fuel control and for many of the ECM controlled outputs.

Test Description:

Numbers below refer to circled numbers on the diagnostic chart.

① This step checks to see if Code 21A is the result of a hard failure or an intermittent condition.
A Code 21A will set if.
- TPS reading above 2.5 volts.
- Engine speed less than 1800 rpm.
- MAP reading below 60 kPa.
- All of the above conditions present for 2 seconds.

② This step simulates conditions for a Code 21B. If the ECM recognizes the change, the ECM and CKT's 416 and 417 are OK.

③ This step isolates a faulty sensor, ECM, or an open CKT 452. If the CKT 452 is open, there may also be a Code 23A.

Diagnostic Aids:

A "Scan" tool displays throttle position in volts. Closed throttle voltage should be less than 1.25 volts. TPS voltage should increase at a steady rate as throttle is moved to WOT.
A Code 21A will result if CKT 452 is open or CKT 417 is shorted to voltage.

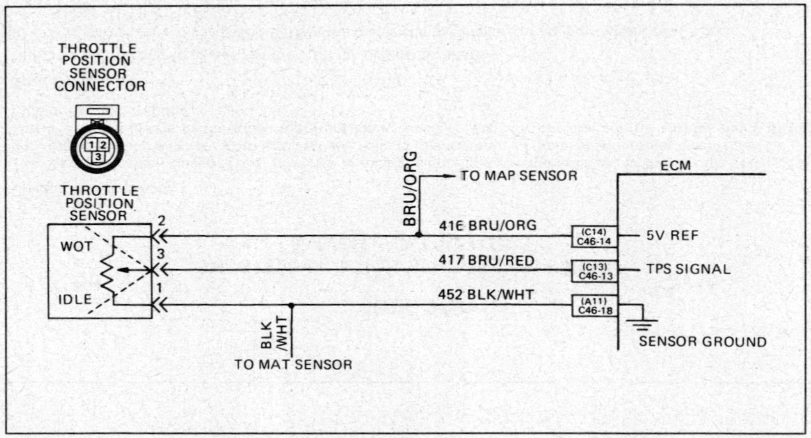

CODE 21A

THROTTLE POSITION SENSOR(TPS) CIRCUIT (SIGNAL VOLTAGE HIGH)

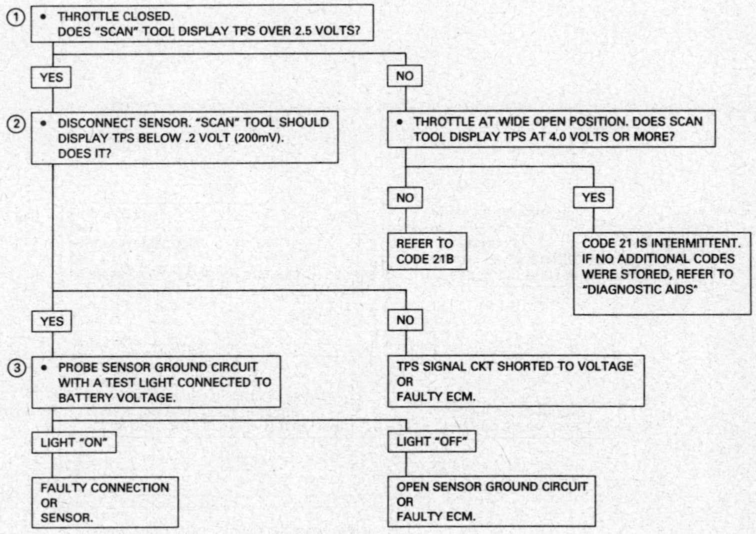

CLEAR CODES AND CONFIRM "CLOSED LOOP" OPERATION AND NORMAL "CHECK ENGINE" LIGHT OPERATION.

CODE 21B

THROTTLE POSITION SENSOR(TPS) CIRCUIT (SIGNAL VOLTAGE LOW)

Circuit Description:

The Throttle Position Sensor (TPS) provides a voltage signal that changes relative to the throttle valve position. Signal voltage will vary from less than 0.9 volts at idle to about 4.5 volts at wide open throttle(WOT). The TPS signal is one of the most important inputs used by the ECM for fuel control and for many of the ECM controlled outputs.

Test Description:

Numbers below refer to circled numbers on the diagnostic chart.

① This step checks to see if Code 21B is the result of a hard failure or an intermittent condition.
A Code 21B will set if:
- The engine is running
- TPS voltage is below .20 volts(200mv).

② This step simulates conditions for a Code 21A. If a Code 21A is set, or the "Scan" tool displays over 4 volts, the ECM and wiring are OK.

③ The "Scan" tool may not display 12 volts. The important thing is that the ECM recognizes the voltage as over 4 volts, indicating that CKT 417 and the ECM are OK.

④ If CKT 416 is open or shorted to ground, there may also be a stored Code 33B.

Diagnostic Aids:

A "Scan" tool displays throttle position in volts. Closed throttle voltage should be less than 1.25 volts. TPS voltage should increase at a steady rate as throttle is moved to WOT.

An open or grounded CKT 416 or CKT 417 will result in a Code 21B.

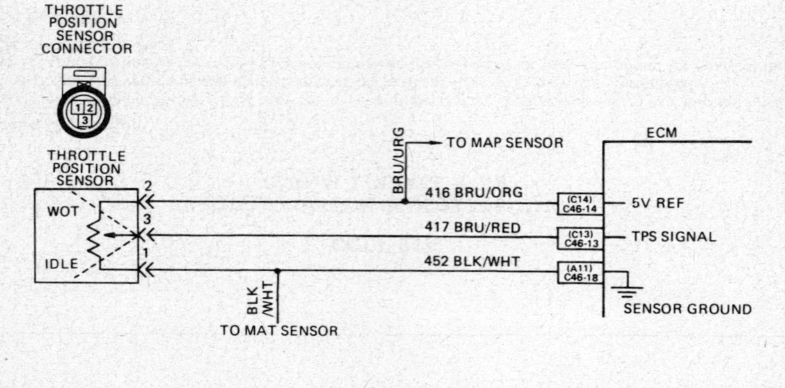

CODE 21B

THROTTLE POSITION SENSOR(TPS) CIRCUIT (SIGNAL VOLTAGE LOW)

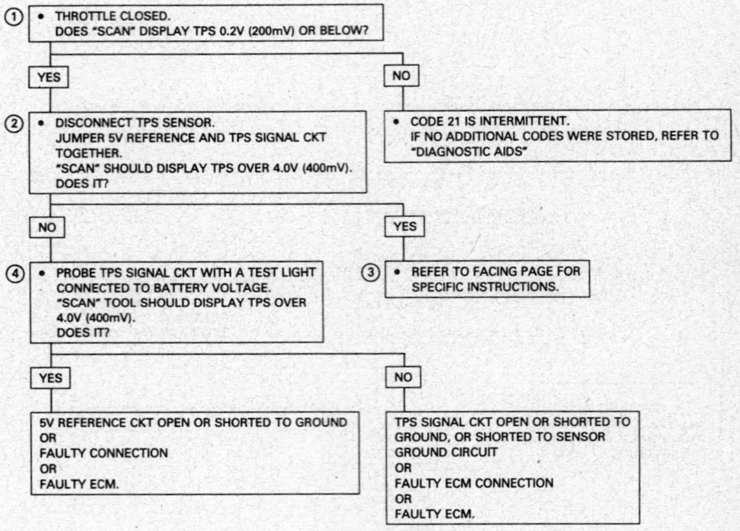

CLEAR CODES AND CONFIRM "CLOSED LOOP" OPERATION AND NORMAL "CHECK ENGINE" LIGHT OPERATION.

CODE 23A

MANIFOLD AIR TEMPERATURE(MAT) SENSOR CIRCUIT (LOW TEMPERATURE INDICATED)

Circuit Description:

The Manifold Air Temperature Sensor uses a thermistor to control the signal voltage to the ECM. The ECM applies a voltage (4-6 volts) to CKT 472 of the sensor. When manifold air is cold, the sensor (thermistor) resistance is high, therefore, the ECM will see a high signal voltage. As the air warms, the sensor resistance becomes less and the voltage drops.

MAT LOW Test Description:

Numbers below refer to circled numbers on the diagnostic chart.

① This step checks to see if Code 23A is the result of a hard failure or an intermittent condition.
A Code 23A will set if:
- Signal voltage indicates a MAT temperature less than -30°C
- Engine has been running for longer than 2 minutes.

② This test simulates conditions for a Code 23B. If the "Scan" tool displays a high temperature, the ECM and wiring are OK.

③ This step checks continuity of CKT's 472 and 452. If CKT 452 is open there may also be a Code 21A.

Diagnostic Aids:

If the engine has been allowed to cool to an ambient temperature (overnight), coolant and MAT temperatures may be checked with a "Scan" tool and should read nearly the same.

A Code 23A will result if CKT's 472 or 452 become open.

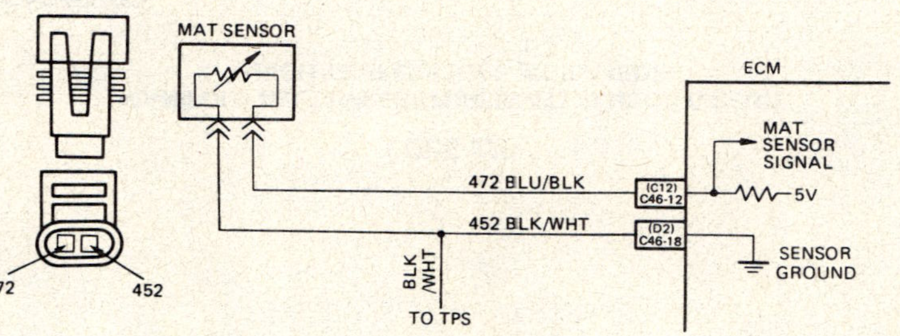

CODE 23A

MANIFOLD AIR TEMPERATURE(MAT) SENSOR CIRCUIT (LOW TEMPERATURE INDICATED)

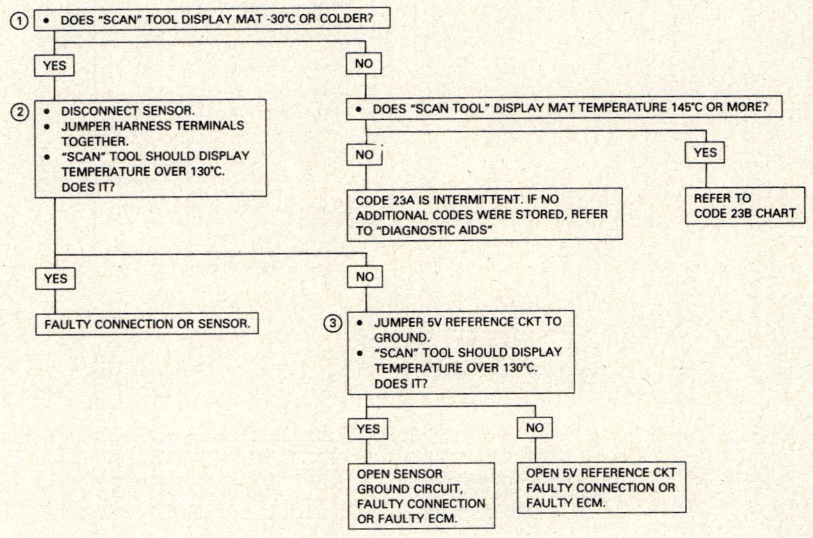

DIAGNOSTIC AID

COOLANT SENSOR		
TEMPERATURE TO RESISTANCE VALUES (APPROXIMATE)		
°F	°C	OHMS
210	100	185
160	70	450
100	38	1,800
70	20	3,400
40	4	7,500
20	−7	13,500
0	−18	25,000
−40	−40	100,700

CLEAR CODES AND CONFIRM "CLOSED LOOP" OPERATION AND NORMAL "CHECK ENGINE" LIGHT OPERATION.

CODE 23B

MANIFOLD AIR TEMPERATURE(MAT) SENSOR CIRCUIT (HIGH TEMPERATURE INDICATED)

Circuit Description:

The Manifold Air Temperature Sensor uses a thermistor to control the signal voltage to the ECM. The ECM applies a voltage (4-6 volts) on CKT 472 to the sensor. When manifold air is cold, the sensor (thermistor) resistance is high, therefore, the ECM will see a high signal voltage. As the air warms, the sensor resistance becomes less and the voltage drops.

MAT HIGH Test Description:

Numbers below refer to circled numbers on the diagnostic chart.

① This check determines if the Code 23B is the result of a hard failure or an intermittent condition.
A Code 23B will set if:
- A MAT temperature greater than 14°C is detected for a time longer than 2 seconds.

Diagnostic Aids:

If the engine has been allowed to cool to an ambient temperature (overnight), coolant and MAT temperatures may be checked with a "Scan" tool and should read nearly the same.

A Code 23B will result if CKT's 472 or 42 become open, or shorted to ground.

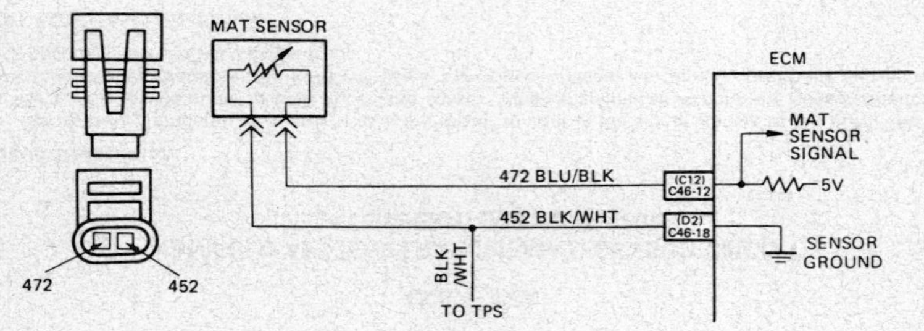

CODE 23B

MANIFOLD AIR TEMPERATURE(MAT) SENSOR CIRCUIT (HIGH TEMPERATURE INDICATED)

① • DOES "SCAN" TOOL DISPLAY MAT OF 145°C (293°F) OR HOTTER?

- YES → • DISCONNECT SENSOR. "SCAN" TOOL SHOULD DISPLAY TEMPERATURE BELOW −30°C (−22°F). DOES IT?
  - YES → REPLACE SENSOR.
  - NO → CKT 472 SHORTED TO GROUND, OR TO SENSOR GROUND, OR ECM IS FAULTY.
- NO → CODE 23B INTERMITTENT. IF NO ADDITIONAL CODES WERE STORED, REFER TO "DIAGNOSTIC AIDS"

DIAGNOSTIC AID

COOLANT SENSOR		
TEMPERATURE TO RESISTANCE VALUES (APPROXIMATE)		
°F	°C	OHMS
210	100	185
160	70	450
100	38	1,800
70	20	3,400
40	4	7,500
20	−7	13,500
0	−18	25,000
−40	−40	100,700

CLEAR CODES AND CONFIRM "CLOSED LOOP" OPERATION AND NORMAL "CHECK ENGINE" LIGHT OPERATION.

CODE 24

VEHICLE SPEED SENSOR(VSS) CIRCUIT

Circuit Description:

Vehicle speed information is provided to the ECM by the Vehicle Speed Sensor (VSS), which alternately grounds CKT 347 located in the meter panel when drive wheels are turning. This pulsing action takes place 4096 times per mile, and the ECM will calculate vehicle speed based on the time between "pulses".

Test Description:

Numbers below refer to circled numbers on the diagnostic chart.

① Code 24 will set if vehicle speed equals 0 mph when;
- Engine speed is between 1500–4400 (1990); 2400–4400 (1991–92)
- TPS is at 0% (closed throttle)
- Low speed condition (low MAP voltage, high manifold vacuum)
- Not in park or neutral.
- All conditions met for 3 seconds.

These conditions are met during a road load deceleration. Disregard Code 24 that sets when drive wheels are not turning.

Diagnostic Aids:

"Scan" should indicate a vehicle speed whenever the drive wheels are turning greater than 3 mph.
Check the WHT wire for proper connections to be sure they are clean and tight and the harness is routed correctly. Refer to "Intermittents" in "Symptoms"
(A/T) — A faulty or misadjusted Park/Neutral (P/N) switch can result in a false Code 24. Use a "Scan" tool and check for proper signal while in drive. Refer to CHART C-1A for P/N switch check.

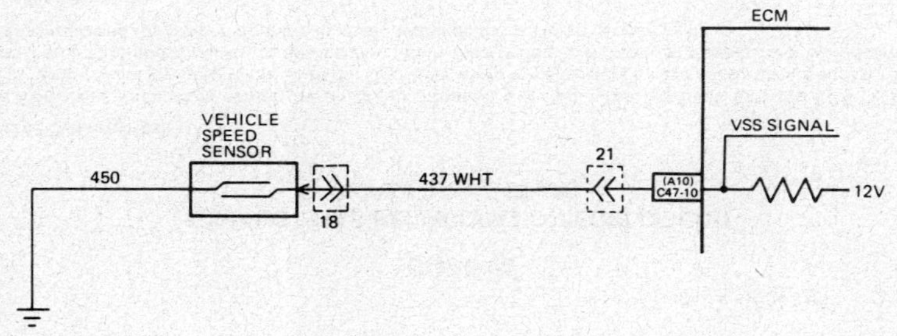

CODE 24

VEHICLE SPEED SENSOR(VSS) CIRCUIT

DISREGARD CODE 24 IF SET WHILE DRIVE WHEELS ARE NOT TURNING.

- VERIFY SPEEDOMETER IS WORKING, IF NOT REFER TO APPROPRIATE SECTION OF THE SERVICE MANUAL FOR NECESSARY REPAIR PROCEDURES.
- CLEAR CODE FROM ECM AND RETEST. IF CODE RESETS PROCEED WITH CODE 24 CHART.

①
- RAISE DRIVE WHEELS
- "NOTICE": DO NOT PERFORM THIS TEST WITHOUT SUPPORTING THE LOWER CONTROL ARMS SO THAT THE DRIVE AXLES ARE IN A NORMAL HORIZONTAL POSITION. RUNNING THE VEHICLE IN GEAR WITH THE WHEELS HANGING DOWN AT FULL TRAVEL MAY DAMAGE THE DRIVE AXLES.
- WITH ENGINE IDLING IN GEAR, "SCAN" TOOL SHOULD DISPLAY VEHICLE SPEED ABOVE 0.
DOES IT?

NO → DOES SPEEDOMETER WORK?

YES → CODE 24 IS INTERMITTENT. IF NO ADDITIONAL CODES WERE STORED, REFER TO "DIAGNOSTIC AIDS"

DOES SPEEDOMETER WORK?

NO → USING A DVOM, BACKPROBE ECM TERMINAL "A10" WITH DRIVE WHEELS ROTATING. VOLTAGE SHOULD BE VARYING. IS IT?

YES → CHECK FOR OPEN IN WHITE WIRE AT ECM TERMINAL "A10". IF OK REPLACE ECM.

USING A DVOM, BACKPROBE ECM TERMINAL "A10" WITH DRIVE WHEELS ROTATING. VOLTAGE SHOULD BE VARYING. IS IT?

NO → OPEN IN WHITE WIRE OR FAULTY ECM

YES → FAULTY SPEEDOMETER

CLEAR CODES AND CONFIRM "CLOSED LOOP" OPERATION AND NORMAL "CHECK ENGINE" LIGHT OPERATION.

CODE 32

EXHAUST GAS RECIRCULATION(EGR) CIRCUIT (CALIFORNIA ONLY)

Circuit Description:

The EGR valve vacuum is control by an ECM controlled solenoid. The ECM will turn the EGR "ON" and "OFF" (Duty cycle) by grounding CKT435. The duty cycle is calculated by the ECM, based on information from the coolant and throttle position sensor. With the ignition "ON", engine stopped, the EGR solenoid is de-energized and, by grounding the diagnostic terminal, the solenoid is energized.

Test Description:

① Code 32 will set:
- Engine at normal operating temperature.
- Difference of EGR gas temperature between EGR off and EGR on is less than 3°C.
- Or signal voltage indicates a EGR gas temperature above 350°C or below −36°C.

② This test determines if the EGR gas temperature sensor or the EGR valve is the problem, or, if the ECM and wiring are at fault.

Diagnostic Aids:

A "Scan" tool reads EGR gas temperature in degrees celsius.
After the engine is started, the temperature should nearly reach manifold air temperature, when EGR is off. But when the EGR is on, the EGR gas temperature should be about between 100°C and 250°C.

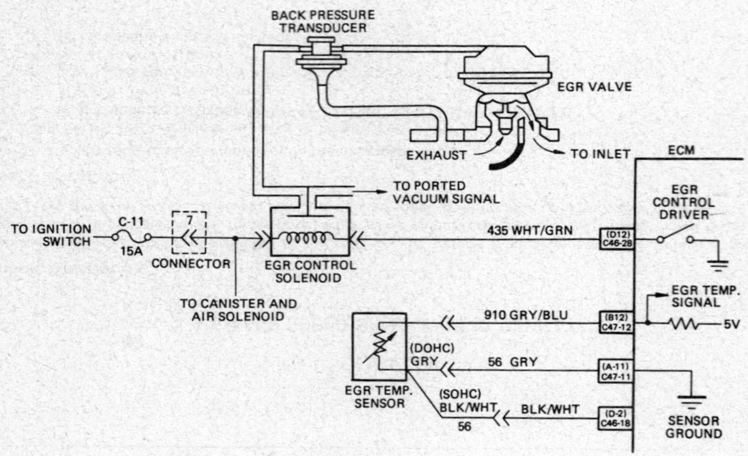

CODE 32

EXHAUST GAS RECIRCULATION(EGR) CIRCUIT (CALIFORNIA ONLY)

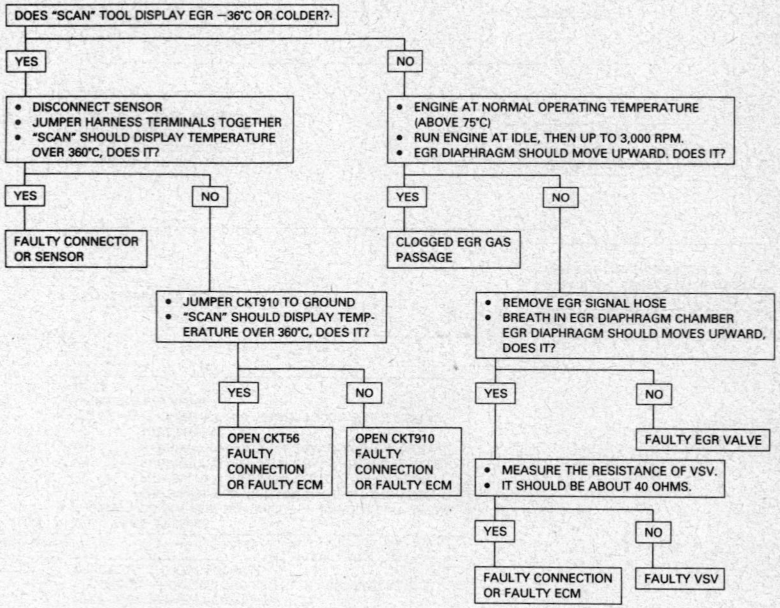

EGR GAS TEMP. SENSOR

TEMPERATURE TO RESISTANCE VALUES (APPROXIMATE)		
°F	°C	OHMS
−40	−40	200,000
32	0	28,000
104	40	5,800
212	100	1,000
392	200	135

CLEAR CODES AND CONFIRM "CLOSED LOOP" OPERATION AND NORMAL "CHECK ENGINE" LIGHT OPERATION.

CODE 33A

MANIFOLD ABSOLUTE PRESSURE(MAP) SENSOR CIRCUIT (SIGNAL VOLTAGE HIGH-LOW VACUUM)

Circuit Description:

The Manifold Absolute Pressure (MAP) Sensor responds to changes in manifold pressure (vacuum). The ECM receives this information as a signal voltage that will vary from about 1 to 1.5 volts, at closed throttle idle, to 4.5 volts at wide open throttle (low vacuum), 0.5–2.0V for 1990 vehicles.
If the MAP sensor fails, the ECM will substitute a fixed MAP value based on Throttle Position Sensor (TPS) to control fuel delivery.

Test Description:

Numbers below refer to circled numbers on the diagnostic chart.

① This step will determine if Code 33 is the result of a hard failure or an intermittent condition.
A Code 33 will set if:
- MAP signal indicates greater than 84 kPa (low vacuum), 95 kPa for 1990 vehicles.
- TPS less than 5%.
- No Code 21A or Code 21B is set.
- These conditions are present for a time longer than 5 seconds.

② This step simulates conditions for a Code 33B. If the ECM recognizes the change, the ECM, and CKT's 416 and 432, are OK. If CKT 452 is open, there may also be a Code 23 stored, code 21 for 1990.

Diagnostic Aids:

With the ignition "ON" and the engine stopped, the manifold pressure is equal to atmospheric pressure and the signal voltage will be high. This information is used by the ECM as an indication of vehicle altitude and is referred to as BARO. Comparison of this BARO reading with a known good vehicle with the same sensor is a good way to check accuracy of a "suspect" sensor. Readings should be the same ±0.4 volt.

A Code 33A will result if CKT 452 is open, or if CKT 432 is shorted to voltage or to CKT 416.

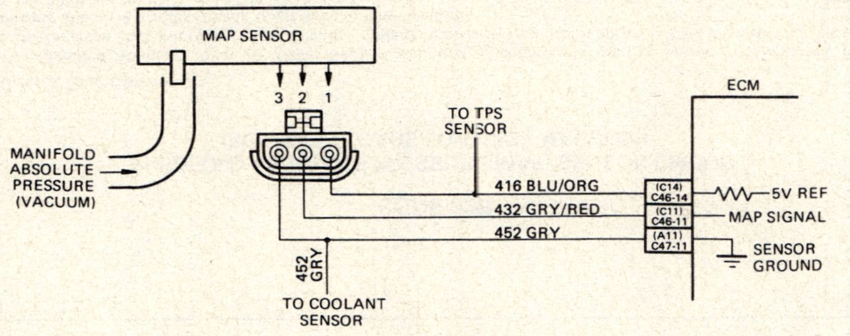

CODE 33A

MANIFOLD ABSOLUTE PRESSURE(MAP) SENSOR CIRCUIT (SIGNAL VOLTAGE HIGH-LOW VACUUM)

①
- IF ENGINE IDLE IS ROUGH, UNSTABLE, OR INCORRECT, CORRECT CONDITION BEFORE USING CHART. SEE "SYMPTOMS"
- ENGINE IDLING.
- DOES "SCAN" TOOL DISPLAY A MAP VOLTAGE OF 4.0 VOLTS OR OVER?

YES →

②
- IGNITION "OFF".
- DISCONNECT MAP SENSOR ELECTRICAL CONNECTOR.
- IGNITION "ON".
- "SCAN" TOOL SHOULD READ A VOLTAGE OF 1 VOLTS OR LESS. DOES IT?

  YES →
  - PROBE SENSOR GROUND CIRCUIT WITH A TEST LIGHT TO BATTERY VOLTAGE.
  - TEST LIGHT SHOULD LIGHT. DOES IT?

    YES → PLUGGED OR LEAKING SENSOR VACUUM HOSE OR FAULTY MAP SENSOR.

    NO → OPEN SENSOR GROUND CIRCUIT.

  NO → MAP SIGNAL CKT SHORTED TO VOLTAGE, SHORTED TO 5 VOLT REFERENCE CKT, OR FAULTY ECM.

NO →
- DOES "SCAN TOOL" DISPLAY MAP SENSOR VOLTAGE LESS THAN 1 VOLT?

  NO → CODE 33A IS INTERMITTENT. IF NO ADDITIONAL CODES WERE STORED, REFER TO "DIAGNOSTIC AIDS"

  YES → REFER TO CODE 33B CHART.

IGNITION "ON" ENGINE STOPPED VOLTAGES

ALTITUDE		VOLTAGE RANGE
Meters	Feet	
Below 305	Below 1,000	3.8–5.5V
305– 610	1,000– 2,000	3.6–5.3V
610– 914	2,000– 3,000	3.5–5.1V
914–1219	3,000– 4,000	3.3–5.0V
1219–1524	4,000– 5,000	3.2–4.8V
1524–1829	5,000– 6,000	3.0–4.6V
1829–2133	6,000– 7,000	2.9–4.5V
2133–2438	7,000– 8,000	2.8–4.3V
2438–2743	8,000– 9,000	2.6–4.2V
2743–3048	9,000–10,000	2.5–4.0V
LOW ALTITUDE — HIGH PRESSURE — HIGH VOLTAGE		

CLEAR CODES AND CONFIRM "CLOSED LOOP" OPERATION AND NORMAL "CHECK ENGINE" LIGHT OPERATION.

CODE 33B

MANIFOLD ABSOLUTE PRESSURE(MAP) SENSOR CIRCUIT (SIGNAL VOLTAGE LOW-HIGH VACUUM)

Circuit Description:

The Manifold Absolute Pressure (MAP) Sensor responds to changes in manifold pressure (vacuum). The ECM receives this information as a signal voltage that will vary from about 0.5 to 2.0 volts, at closed throttle idle, to 4-5 volts at wide open throttle (low vacuum).

If the MAP sensor fails, the ECM will substitute a fixed MAP value based on Throttle Position Sensor (TPS) to control fuel delivery.

Test Description:

Numbers below refer to circled numbers on the diagnostic chart.

① This step determine if Code 33B is the result of a hard failure or an intermittent condition. A Code 33B will set when.
- Engine rpm is greater than 1200 rpm.
- TPS is greater than 15%
- MAP signal voltage is too low.
- No Code 21A or Code 21B is set.

② Jumpering harness terminals "2" to "1", 5 volt to signal, will determine if the sensor is at fault, or if there is a problem with the ECM or wiring.

③ The "Scan" tool may not display 12 volts. The important thing is that the ECM recognizes the voltage as more than 4 volts, indicating that the ECM and CKT 432 are OK.

Diagnostic Aids:

With the ignition "ON" and the engine stopped, the manifold pressure is equal to atmospheric pressure and the signal voltage will be high. This information is used by the ECM as an indication of vehicle altitude and is referred to as BARO. Comparison of this BARO reading with a known good vehicle with the same sensor is a good way to check accuracy of a "suspect" sensor. Readings should be the same ±0.4 volt.

A Code 33B will result if CKT 452 is open, or if CKT 432 is shorted to voltage or to CKT 416.

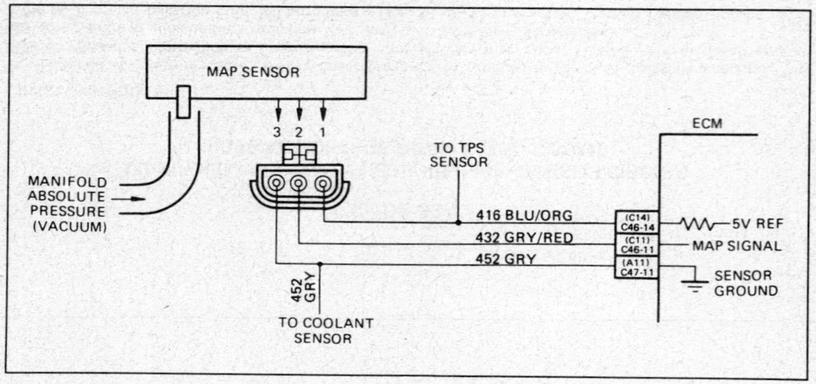

CODE 33B

MANIFOLD ABSOLUTE PRESSURE(MAP) SENSOR CIRCUIT (SIGNAL VOLTAGE LOW-HIGH VACUUM)

①
- ENGINE IDLING
- DOES "SCAN" TOOL DISPLAY MAP VOLTAGE BELOW .25 VOLTS?

NO → CODE 33B IS INTERMITTENT. IF NO ADDITIONAL CODES WERE STORED, REFER TO "DIAGNOSTIC AIDS"

YES ↓

②
- IGNITION "OFF".
- DISCONNECT SENSOR ELECTRICAL CONNECTOR.
- JUMPER HARNESS TERMINALS "2" TO "1".
- IGNITION "ON".
- MAP VOLTAGE SHOULD READ OVER 4 VOLTS. DOES IT?

YES → FAULTY CONNECTION OR SENSOR.

NO ↓

③
- IGNITION "OFF".
- REMOVE JUMPER WIRE.
- PROBE TERMINAL "B" GREY/RED WIRE WITH A TEST LIGHT TO BATTERY VOLTAGE.
- IGNITION "ON".
- "SCAN" TOOL SHOULD READ OVER 4 VOLTS. DOES IT?

YES → 5 VOLT REFERENCE CIRCUIT OPEN OR SHORTED TO GROUND OR FAULTY ECM.

NO → CKT 432 OPN OR CKT 432 SHORTED TO GROUND OR CKT 432 SHORTED TO SENSOR GROUND OR FAULTY ECM.

IGNITION "ON" ENGINE STOPPED VOLTAGES

ALTITUDE		VOLTAGE RANGE
Meters	Feet	
Below 305	Below 1,000	3.8~5.5V
305~ 610	1,000~ 2,000	3.6~5.3V
610~ 914	2,000~ 3,000	3.5~5.1V
914~1219	3,000~ 4,000	3.3~5.0V
1219~1524	4,000~ 5,000	3.2~4.8V
1524~1829	5,000~ 6,000	3.0~4.6V
1829~2133	6,000~ 7,000	2.9~4.5V
2133~2438	7,000~ 8,000	2.8~4.3V
2438~2743	8,000~ 9,000	2.6~4.2V
2743~3048	9,000~10,000	2.5~4.0V
LOW ALTITUDE = HIGH PRESSURE = HIGH VOLTAGE		

CLEAR CODES AND CONFIRM "CLOSED LOOP" OPERATION AND NORMAL "CHECK ENGINE" LIGHT OPERATION.

CODE 42

ELECTRONIC SPARK TIMING (EST) CIRCUIT

Circuit Description:

The ignition module sends a reference signal (CKT 430) to the ECM, when the engine is cranking. While the engine speed is under 600 rpm, the ECM applies 5 volts to the "bypass" line (CKT 424) to switch the timing to ECM control (EST CKT 423).

When the system is running on the ignition module, that is, nc voltage on the bypass line, the ignition module grounds the EST signal. The ECM expects to see no voltage on the EST line during this condition. If it sees a voltage, it sets Code 42 and will not go into the EST mode.

When the rpm for EST is reached (about 400 rpm), voltage will be applied to the bypass line, the EST should no longer be grounded in the ignition module, so the EST voltage should be varying.

If the bypass line is open or grounded, the ignition module will not switch to EST mode, so, the EST voltage will be low and Code 42 will be set.

If the EST line is grounded, the ignition module will switch to EST but, because the line is grounded, there will be no EST signal. A Code 42 will be set.

Test Description:

Numbers below refer to circled numbers on the diagnostic chart.

① Code 42 means the ECM has seen an open or short to ground in the EST or bypass circuits. This test confirms Code 42 and that the fault causing the code is present.

② Checks for a normal EST ground path through the ignition module. An EST CKT 423, shorted to ground, will also read less than 500 ohms, however, this will be checked later.

③ As the test light voltage touches CKT 424, the module should switch, causing the ohmmeter to "over-range", if the meter is in the 1000-20000 ohms position. Selecting the 10-20,000 ohms position will indicate above 5000 ohms. The important thing is that the module "switched".

④ The module did not switch and this step checks for:
- EST CKT 423 shorted to ground.
- Bypass CKT 424 open.
- Faulty ignition module connection or module.

⑤ Confirms that Code 42 is a faulty ECM and not an intermittent in CKTs 423 or 424.

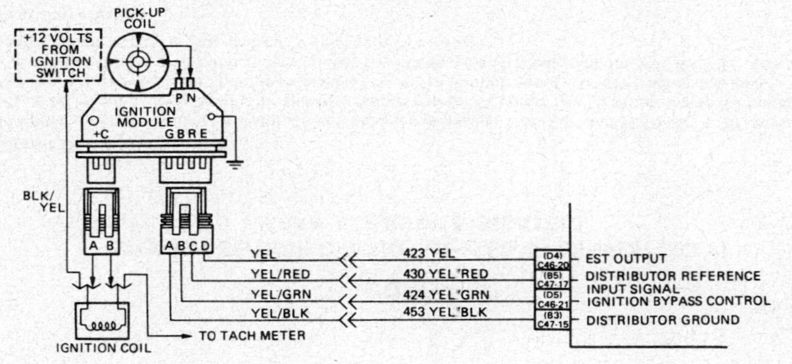

CODE 42

ELECTRONIC SPARK TIMING (EST) CIRCUIT

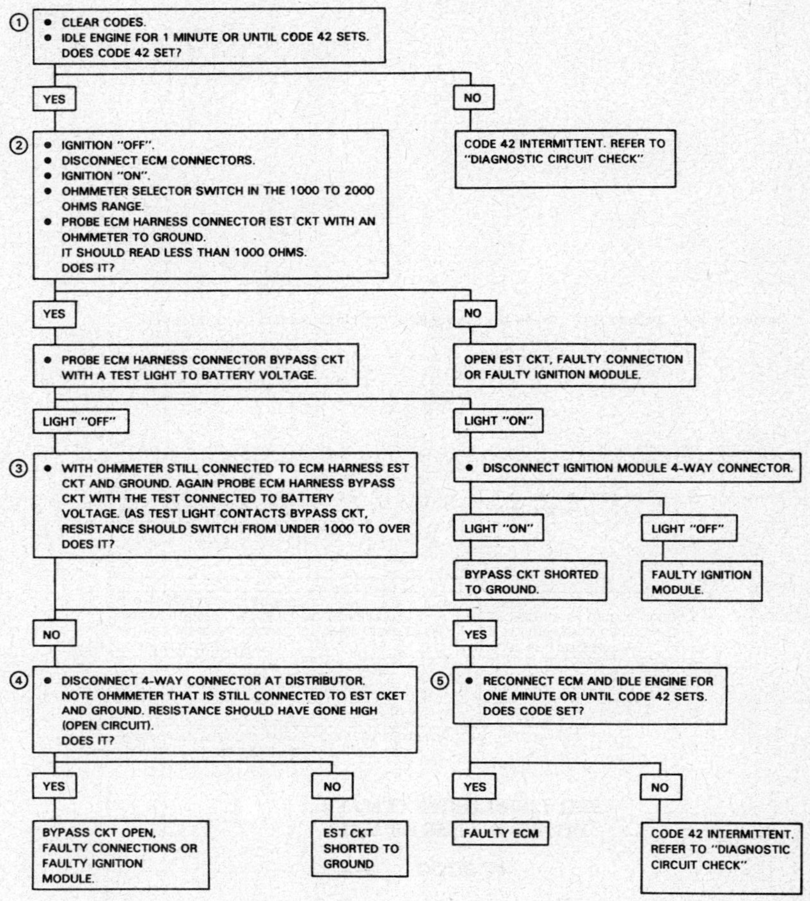

CLEAR CODES AND CONFIRM "CLOSED LOOP" OPERATION AND NORMAL "CHECK ENGINE" LIGHT OPERATION.

CODE 44

OXYGEN SENSOR CIRCUIT (HEATED O_2 SENSOR ONLY) (LEAN EXHAUST INDICATED)

Circuit Description:

The ECM supplies a voltage of about .45 volt between terminals "C46-23" and "C46-22". (If measured with a 10 megohm digital voltmeter, this may read as low as .32 volts.) The O_2 sensor varies the voltage within a range of about 1 volt, if the exhaust is rich, down through about .10 volt, if exhaust is lean. The sensor is like an open circuit and produces no voltage, when it is below about 360°(600°F). An open sensor circuit, or cold sensor, causes "Open Loop" operation.

Test Description:

Numbers below refer to circled numbers on the diagnostic chart.

① Code 44 will set under the following conditions:

- No code 21A/21B or Code 33A/33B
- Engine Running for 50 seconds or more
- O_2 voltage below .278 mV for 99 seconds
- System is in Closed Loop
- Coolant Temperature above 69.5°C
- TPS is above 39°C
- Integrator not at 128

Diagnostic Aids:

Using the "Scan" tool, observe the block learn value at different rpms. The "Scan" tool, also displays the block cells, so the block learn values can be checked in each of the cells, to determine when the Code 44 may have been set. If the conditions for Code 44 exist, the block learn values will be around 150.

- O_2 Sensor Wire - Sensor pigtail may be mispositioned and contacting the exhaust manifold.
- Check for ground in wire between connector and sensor.
- **Fuel Contamination** — Water, even in small amounts, near the in-tank fuel pump inlet can be delivered to the injector. The water causes a lean exhaust and can set a Code 44.
- **Fuel Pressure** — System will be lean if pressure is too low. It may be necessary to monitor fuel pressure, while driving the car at various road speeds and/or loads to confirm. See Fuel System diagnosis CHART A-7
- **Exhaust Leaks** — If there is an exhaust leak, the engine can cause outside air to be pulled into the exhaust and past the sensor. Vacuum or crankcase leaks can cause a lean condition.
- If the Heater part of the O_2 sensor fails, this may cause a Code 44 or Code 45 to set

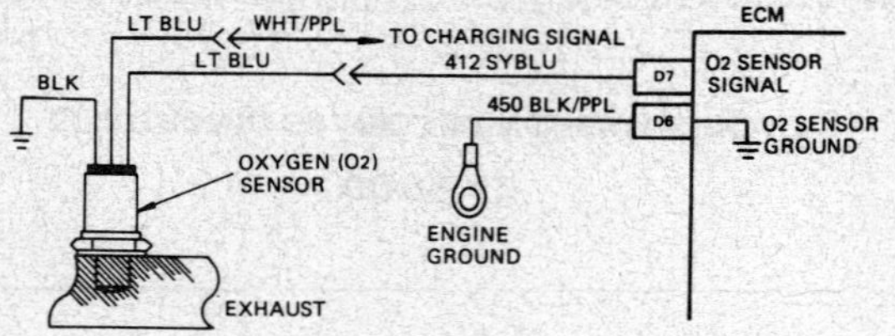

CODE 44

OXYGEN SENSOR CIRCUIT (LEAN EXHAUST INDICATED)

①
- RUN WARM ENGINE (69.5°C/155°F) AT 1200 RPM
- DOES "SCAN" TOOL INDICATE O_2 VOLTAGE FIXED BELOW .28 VOLTS(280 mV)?

YES:
- DISCONNECT O_2 SENSOR
- WITH ENGINE IDLING "SCAN" TOOL SHOULD DISPLAY O_2 SENSOR BETWEEN .35 VOLTS AND .55 VOLTS (350 mV AND 550 mV). DOES IT?

  YES: REFER TO "DIAGNOSTIC AIDS"

  NO: O_2 SENSOR SIGNAL CKT SHORTED TO GROUND OR FAULTY ECM

NO: CODE 44 IS INTERMITTENT. IF NO ADDITIONAL CODES WERE STORED, REFER TO "DIAGNOSTIC AIDS"

CLEAR CODES AND CONFIRM "CLOSED LOOP" OPERATION AND NORMAL "CHECK ENGINE" LIGHT OPERATION.

CODE 44

OXYGEN SENSOR CIRCUIT (NON-HEATED O_2 SENSOR ONLY) (LEAN EXHAUST INDICATED) SOHC MODEL

Circuit Description:

The ECM supplies a voltage of about .45 volt between terminals "D7" and "D6". (If measured with a 10 megohm digital voltmeter, this may read as low as .32 volts.) The O_2 sensor varies the voltage within a range of about 1 volt, if the exhaust is rich, down through about .10 volt, if exhaust is lean.
The sensor is like an open circuit and produces no voltage, when it is below about 360° (600°F). An open sensor circuit, or cold sensor, causes "Open Loop" operation.

Test Description:

Numbers below refer to circled numbers on the diagnostic chart.

① Code 44 will set when the following occurs:
- No code 21A/21B or Code 33A/33B
- Engine Running for 50 seconds or more
- O_2 voltage below .278 mV for 99 seconds
- System is in "Closed Loop"
- Coolant Temperature above 69.5°C
- TPS is above 3%
- Integrator not at 128

Diagnostic Aids:

Using the "Scan" tool, observe the block learn value at different rpms. The "Scan" tool also displays the block cells, so the block learn values can be checked in each of the cells, to determine when the Code 44 may have been set. If the conditions for Code 44 exist, the block learn values will be around 150.

- O_2 Sensor Wire - Sensor pigtail may be mispositioned and contacting the exhaust manifold.
- Check for ground in wire between connector and sensor.
- **Fuel Contamination** — Water, even in small amounts, near the in-tank fuel pump inlet can be delivered to the injector. The water causes a lean exhaust and can set a Code 44.
- **Fuel Pressure** — System will be lean if pressure is too low. It may be necessary to monitor fuel pressure, while driving the car at various road speeds and/or loads to confirm. See Fuel System diagnosis CHART A-7.
- **Exhaust Leaks** — If there is an exhaust leak, the engine can cause outside air to be pulled into the exhaust and past the sensor. Vacuum or crankcase leaks can cause a lean condition.

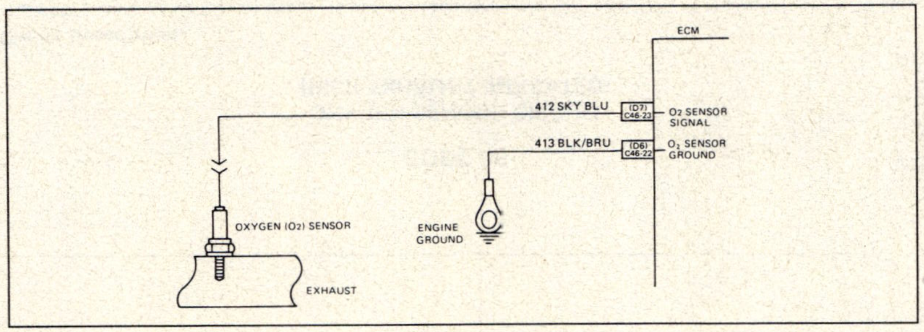

CODE 44

OXYGEN SENSOR CIRCUIT (NON-HEATED O_2 SENSOR ONLY) (LEAN EXHAUST INDICATED) SOHC MODEL

①
- RUN WARM ENGINE (69.5°C/155°F) AT 1200 RPM
- DOES "SCAN" TOOL INDICATE O_2 SENSOR VOLTAGE FIXED BELOW .28 VOLTS(280 mV)?

YES:
- DISCONNECT O_2 SENSOR
- WITH ENGINE IDLING "SCAN" TOOL SHOULD DISPLAY O_2 SENSOR BETWEEN .35 VOLTS AND .55 VOLTS (350 mV AND 550 mV). DOES IT?

  YES: REFER TO "DIAGNOSTIC AIDS"

  NO: O_2 SENSOR SIGNAL CKT SHORTED TO GROUND OR FAULTY ECM

NO: CODE 44 IS INTERMITTENT. IF NO ADDITIONAL CODES WERE STORED, REFER TO "DIAGNOSTIC AIDS"

CLEAR CODES AND CONFIRM "CLOSED LOOP" OPERATION AND NORMAL "CHECK ENGINE" LIGHT OPERATION.

CODE 45

OXYGEN SENSOR CIRCUIT (RICH EXHAUST INDICATED)

Circuit Description:

The ECM supplies a voltage of about .45 volt between terminals "D7" and "D6". (If measured with a 10 megohm digital voltmeter, this may read as low as .32 volts.) The O_2 sensor varies the voltage within a range of about 1 volt, if the exhaust is rich, down through about .10 volt, if exhaust is lean.

The sensor is like an open circuit and produces no voltage, when it is below about 360°C (600°F). An open sensor circuit, or cold sensor, causes "Open Loop" operation.

Test Description:

Numbers below refer to circled numbers on the diagnostic chart.

① Code 45 is set, when O_2 sensor signal voltage is fixed above .75 volts (750 mV) under the conditions:

- TPS greater than 3%
- No Code 21A/21B or 33A/33B
- System is operating in "Closed Loop"
- Temperature above 69.5°C (155°F)
- Integrator Not at 128

Diagnostic Aids:

The Code 45, or rich exhaust, is most likely caused by one of the flowing:

- **Fuel Pressure** — System will go rich, if pressure is too high. The ECM can compensate for some increase. However, if it gets too high, a Code 45 will be set. See Fuel System diagnosis CHART A-5.
- **HEI Shielding** — An open ground CKT 450 may result in EMI, or induced electrical "noise". The ECM looks at this "noise" as reference pulses. The additional pulses result in a higher than actual engine speed signal. The ECM then delivers too much fuel, causing system to go rich. Engine tachometer will, also, show higher than actual engine speed, which can help in diagnosing this problem.
- **Canister Purge** — Check for fuel saturation. If full of fuel, check canister control and hoses.
- **MAP Sensor** — An output that causes the ECM to sense a higher than normal manifold pressure (low vacuum) can cause the system to go rich. Disconnecting the MAP sensor will allow the ECM to set a fixed value for the MAP sensor. Substitute a different MAP sensor, if the rich condition is gone, while the sensor is disconnected.
- **TPS** — An intermittent TPS output will cause the system to go rich, due to a false indication of the engine accelerating.
- **O_2 Sensor Contamination** — Inspect Oxygen Sensor for silicone contamination from fuel, or use of improper RTV sealant. The sensor may have a white, powdery coating and result in a high, but false signal voltage (rich exhaust indication). The ECM will then reduce the amount of fuel delivered to the engine, causing a severe surge driveability problem.
- **EGR** — Valve sticking open at idle, usually accompanied by a rough idle, or stall complaint, especially at idle.

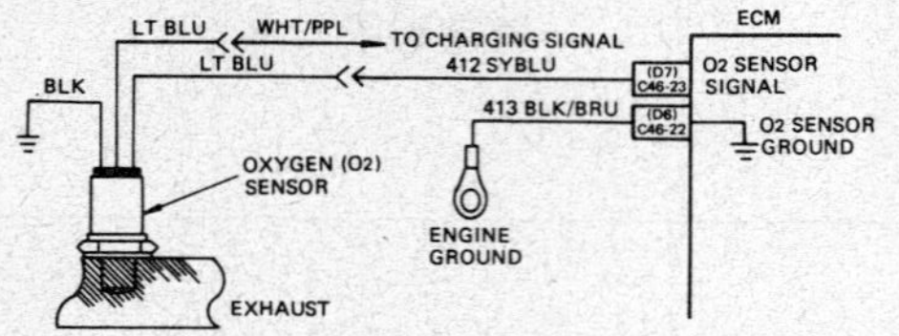

CODE 45

OXYGEN SENSOR CIRCUIT (RICH EXHAUST INDICATED)

①
- RUN WARM ENGINE (69.5°C/155°F) AT 1200 RPM
- DOES "SCAN" TOOL DISPLAY O_2 SENSOR VOLTAGE FIXED ABOVE .75 VOLTS(750 mV)?

YES:
- DISCONNECT O_2 SENSOR AND JUMPER HARNESS CKT 412 TO GROUND.
- "SCAN" TOOL SHOULD DISPLAY O_2 BELOW .35 VOLTS(350 mV). DOES IT?

  YES: REFER TO DIAGNOSTIC AIDS

  NO: FAULTY ECM

NO: CODE 45 IS INTERMITTENT. IF NO ADDITIONAL CODES WERE STORED, REFER TO "DIAGNOSTIC AIDS"

CLEAR CODES AND CONFIRM "CLOSED LOOP" OPERATION AND NORMAL "CHECK ENGINE" LIGHT OPERATION.

1.6L ENGINE—TROUBLE CODE CHARTS IMPULSE AND STYLUS

CODE 51

ECM FAILURE (ECM FAILED OR EEPROM FAILURE)

CHECK THAT ALL ECM CONNECTIONS ARE GOOD
IF OK, CLEAR MEMORY AND RECHECK ECM.
IF CODE 51 REAPPEARS, REPLACE ECM AND REPROGRAM
FOR ECM REPLACEMENT AND REPROGRAMMING
REFER TO SECTION C1 "ECM AND SENSORS".

CLEAR CODES AND CONFIRM "CLOSED LOOP" OPERATION AND NORMAL "CHECK ENGINE" LIGHT OPERATION.

1.6L ENGINE—DRIVEABILITY SYMPTOMS IMPULSE AND STYLUS

BEFORE STARTING

Before using this section you should have performed a careful visual/physical check. This check should include:

- ECM and engine grounds for being clean and tight.
- ECM B+ wires for being clean and tight at starter and/or junction block.
- Wiring for proper connections, pinches, and cuts.
- Vacuum hoses for splits, kinks, and proper connections, as shown on emission control information label.
- Air leaks at throttle body mounting and intake manifold.
- Air leaks between MAP sensor and throttle body.
- Ignition wires for cracking, hardness, proper routing, and carbon tracking.

The importance of this step cannot be stressed too strongly — it can lead to correcting a problem without further checks and can save valuable time.

You should also perform the DIAGNOSTIC CIRCUIT CHECK to determine that:

1. The ECM and "Check Engine" light are operating properly.
2. There are no trouble codes stored, or there is a trouble code but no "Check Engine" light.
3. The fuel control system is operating OK (by performing field service mode check).

Verify the customer complaint, and locate the correct SYMPTOM. Check the items indicated under that symptom.

If the engine starts but immediately dies, repeatedly, use CHART A-3.

If the ENGINE CRANKS BUT WILL NOT RUN, see CHART A-3.

If the complaint is extended crank times, see CHART A-7.

INTERMITTENTS

Definition: Problem may or may not activate the "Check engine" light or store a trouble code.

DO NOT use the trouble code charts in Section "A" for intermittent problems. The fault must be present to locate the problem. If a fault is intermittent, the use of trouble code charts may result in the replacement of good parts.

- Most intermittent problems are caused by faulty electrical connections or wiring. Perform careful checks of suspected circuits for
  - Poor mating of the connector halves and terminals not fully seated in the connector body (backed out).
  - Improperly formed or damaged terminals. All connector terminals in the problem circuit should be carefully reformed to increase contact tension.
  - Poor terminal to wire connection. Perform this connector body check
- If a visual and physical check does not locate the cause of the problem, the car can be driven with a voltmeter connected to a suspected circuit, or a "Scan" tool may be used. An abnormal voltage reading while the problem occurs indicates that the problem may be in that circuit.
- Check for loss of trouble code memory. To check, disconnect the TPS and allow the engine to idle until the "Check engine" light turns "ON". Code 21 should be stored and kept in memory when the ignition is turned "OFF" for at least 10 seconds. If not, the ECM is faulty.
- An intermittent Check engine light and no trouble codes may be caused by
  - Electrical system interference caused by a defective relay, ECM driven solenoid, or switch. They can cause a sharp electrical surge. Normally, the problem will occur when the faulty component is operated.
  - Improper installation of electrical options, such as lights, 2-way radios, etc.
  - EST wires which should be routed away from spark plug wires, ignition system components, and generator. Ground wire from ECM to ignition system may be faulty.
  - Ignition secondary wire shorted to ground.
  - "Check engine" light and diagnostic "test" terminal circuits intermittently shorted to ground.
  - Faulty ECM grounds.

HARD START

Definition: Engine cranks well but does not start for a long time. Engine does eventually start, but may or may not continue to run.

Perform careful visual and physical check

Perform "Diagnostic Circuit Check".

- **CHECK**
  - Fuel for poor quality, "stale" fuel, and water contamination
  - Ignition wires for shorts or faulty insulation
  - Ignition coil connections
  - Fuel pump relay. Connect test light between pump "test" terminal and BAT +. Light should be "OFF" for 2 seconds following ignition "ON". If not, use CHART A-5.
  - Secondary ignition voltage output with ST-125 tester
  - Spark plugs. Look for wetness, cracks, improper gap, burned electrodes, and heavy deposits. Visually inspect ignition system for moisture, dust, cracks, burns, etc.
  - For faulty ECM and ignition grounds
  - EE PROM for correct application. Spray plug wires with fine water mist to check for shorts.
  - For possibility of misfiring, crossfiring, or cutting out under load or at idle.
  - For improper pick-up coil resistance or faulty connections
  - EGR operation.
  - Fuel system for restricted filter or improper pressure. Use CHART A-7.
  - Injector assembly for leakage. Pressurize system by energizing fuel pump through the underhood fuel pump relay connector.
  - Coolant sensor for a shift in calibration. Use Code 14A or Code 14B chart.
- Check spark plugs following:
  - Cracks
  - Wear
  - Improper gap
  - Burned electrode
  - Heavy deposits
- **CHECK**
  - Pick-up coil resistance and connections
  - Worn dist. shaft
  - Moisture in dist. cap
  - TPS for sticking or binding. TPS voltage should read less than 1.25 volts on a "Scan" tool.
  - In-tank fuel pump check valve. A faulty valve would allow the fuel in the lines to drain back to the tank after the engine is stopped. To check for this condition, use CHART A-7.
  - For the possibility of an exhaust restriction or improper valve timing by performing the following test.
    1. With engine at normal operating temperature, connect a vacuum gauge to any convenient vacuum port on intake manifold.
    2. Run engine at 1000 rpm and record vacuum reading.
    3. Increase engine speed slowly to 2500 rpm. Note vacuum reading at steady 2500 rpm.
    4. If vacuum at 2500 rpm decreases more than 3" Hg from reading at 1000 rpm, the exhaust system should be inspected for restrictions.
    5. Disconnect exhaust pipe from engine and repeat Steps 3 & 4. If vacuum still drops more than 3" Hg with exhaust disconnected, check valve timing.
  - Engine valve timing and compression.

HESITATION, SAG, OR STUMBLE

Definition: Momentary lack of response as the accelerator is pushed down. Can occur at all vehicle speeds. Usually most severe when first trying to make the car move, as from a stop sign. May cause the engine to stall if severe enough.

Perform careful visual and physical check

Perform "Diagnostic Circuit Check".

- **CHECK**
  - Ignition wires for shorts or faulty insulation
  - Ignition system for moisture, dust, cracks, burns, etc. Spray plug wires with fine water mist to check for shorts.
  - For possibility of misfiring, crossfiring, or cutting out under load or at idle.
  - Secondary ignition voltage output with ST-125 tester
  - Ignition coil connections
  - ECM and ignition system for faulty grounds
  - Proper operation of EST.
  - Spark plugs. Lock for wetness, cracks, improper gap, burned electrodes, and heavy deposits.
  - Spark plugs for correct heat range
  - Fuel system for restricted filter or improper pressure.
  - EGR operation.
  - For vacuum leaks at intake manifold gasket
  - Air cleaner element (filter) for dirt or plugging
  - Idle Air Control system.
  - Check electrical system voltage. IAC valve will not move if voltage is below 9 volts or greater than 17.8 volts. Also check battery cables and ground straps for poor contact. Erratic voltage will cause the IAC valve to change its position, resulting in poor idle quality.
  - PCV valve for proper operation by placing finger over inlet hole in valve end several times. Valve should snap back. If not, replace valve. Ensure that valve is correct part. Also check PCV hose.
  - Canister purge system for proper operation.
  - Throttle shaft or TPS for sticking or binding TPS voltage should read less than 1.25 volts on a "Scan" tool with the throttle closed.
  - MAP sensor output.
  - Oxygen sensor for silicone contamination from contaminated fuel or use of improper RTV sealant. The sensor will have a white, powdery coating and will cause a high but false signal voltage (rich exhaust indication). The ECM will reduce the amount of fuel delivered to the engine, causing a severe driveability problem.
  - Coolant sensor for a shift in calibration. Use Code 14A or Code 14B charts.
  - A/C relay operation. A/C should cut out at wide open throttle.
  - A/C refrigerant pressure for high pressure. Check for overcharging or faulty pressure switch.
  - P/N switch circuit on vehicle with automatic transmission.
  - Alternator output voltage. Repair if less than 9 volts or more than 16 volts.
  - Transmission torque converter operation.
  - Transmission for proper shift points.
  - Transmission torque converter clutch operation.
  - Vehicle brakes for dragging
  - Engine valve timing and compression.
  - For the possibility of an exhaust restriction or improper valve timing by performing the following test.
    1. With engine at normal operating temperature, connect a vacuum gauge to any convenient vacuum port on intake manifold.
    2. Run engine at 1000 rpm and record vacuum reading.
    3. Increase engine speed slowly to 2500 rpm. Note vacuum reading at steady 200 rpm.
    4. If vacuum at 2500 rpm decreases more than 3" Hg from reading at 1000 rpm, the exhaust system should be inspected for restrictions.
    5. Disconnect exhaust pipe from engine and repeat Steps 3 & 4. If vacuum still drops more than 3" Hg with exhaust disconnected, check valve timing.
  - For worn or incorrect basic engine parts such as cam, heads, pistons, etc. Also check for bent pushrods, worn rocker arms, and broken or weak valve springs.
  - For overheating and possible causes. Look for the following.
    - Low or incorrect coolant solution. It should be a 50/50 mix of anti-freeze coolant and water.
    - Restricted air flow to radiator, or restricted water flow through radiator
    - Faulty or incorrect thermostat
    - Inoperative electric cooling fan circuit.
- If the system is running RICH (block learn less than 118), refer to "Diagnostic Aids" on facing page of Code 45.
- If the system is running LEAN (block learn greater than 138), refer to "Diagnostic Aids" on facing page of Code 44.

1.6L ENGINE—DRIVEABILITY SYMPTOMS—IMPULSE AND STYLUS

SURGES AND/OR CHUGGLE

Definition: Engine power variation under steady throttle or cruise.
Feels like the car speeds up and slows down with no change in the accelerator pedal.

Perform careful visual and physical check

Perform "Diagnostic Circuit Check".

- **CHECK**
  - Ignition wires for shorts or faulty insulation
  - Ignition system for moisture, dust, cracks, burns, etc. Spray plug wires with fine water mist to check for shorts.
  - For possibility of misfiring, crossfiring, or cutting out under load or at idle.
  - Secondary ignition voltage output with ST-125 tester.
  - Ignition coil connections.
  - ECM and ignition system for faulty grounds.
  - Proper operation of EST.
  - Spark plugs. Look for wetness, cracks, improper gap, burned electrodes, and heavy deposits.
  - Spark plugs for correct heat range.
  - Fuel for poor quality, "stale" fuel, and water contamination.
  - Fuel system for restricted filter or improper pressure. Use CHART A-7.
  - Injector assembly for leakage. Pressurize the fuel system by jumpering the fuel pump relay located under the hood connector.
  - EGR operation.
  - For vacuum leaks at intake manifold gasket.
  - Electrical system voltage. IAC valve will not move if voltage is below 9 volts or greater than 17.8 volts. Also check battery cables and ground straps for poor contact. Erratic voltage will cause the IAC valve to change its position, resulting in poor idle quality.
  - Ignition Timing Check
  - PCV valve for proper operation by placing finger over inlet hole in valve end several times. Valve should snap back. If not replace valve. Ensure that valve is correct part. Also check PCV hose.
  - Canister purge system for proper operation.
  - Throttle shaft or TPS for sticking or binding. TPS voltage should read less than 1.25 volts on a "Scan" tool with the throttle closed.
  - MAP sensor output.
  - Oxygen sensor for silicone contamination from contaminated fuel or use of improper RTV sealant. The sensor will have a white, powdery coating and will cause a high but false signal voltage (rich exhaust indication). The ECM will reduce the amount of fuel delivered to the engine, causing a severe driveability problem.
  - Coolant sensor for a shift in calibration. Use Code 14A or Code 14B charts.
  - Vehicle speed sensor (VSS) input with a "Scan" tool to make sure reading of VSS matches that of vehicle speedometer.
  - A/C relay operation. A/C should cut out at wide open throttle.
  - P/N switch circuit on vehicle with automatic tranmission.
  - Transmission torque coverter clutch operation.

LACK OF POWER, SLUGGISH, OR SPONGY

Definition: Engine delivers less than expected power. There is little or no increased when the accelerator pedal is depressed partially.

Perform careful visual and physical check as

Perform "Diagnostic Circuit Check".

- **CHECK**
  - Ignition wires for shorts or faulty insulation
  - Ignition system for moisture, dust, cracks, burns, etc. Spray plug wires with fine water mist to check for shorts.
  - For possibility of misfiring, crossfiring, or cutting out under load or at idle.
  - Secondary ignition voltage output with ST-125 tester.
  - Ignition coil connections.
  - ECM and ignition system for faulty grounds.
  - Proper operation of EST.
  - Spark plugs. Look for wetness, cracks, improper gap, burned electrodes, and heavy deposits.
  - Spark plugs for correct heat range.
  - Fuel for poor quality, "stale" fuel, and water contamination.
  - Fuel system for restricted filter or improper pressure. Use CHART A-7.
  - EGR operation.
  - For vacuum leaks at intake manifold gasket
  - Air cleaner element (filter) for dirt or plugging
  - Throttle shaft or TPS for sticking or binding. TPS voltage should read less than 0.9 volts on a "Scan" tool with the throttle closed.
  - MAP sensor output.
  - Oxygen sensor for silicone contamination from contaminated fuel or use of improper RTV sealant. The sensor will have a white, powdery coaitng and will cause a high but false signal voltage (rich exhaust indication). The ECM will reduce the amount of fuel delivered to the engine, causing a severe driveability problem.
  - Coolant sensor for a shift in calibration. Use Code 14A or Code 14B charts.
  - Vehicle Speed Sensor (VSS) input with a "Scan" tool to make sure reading of VSS matches that of vehicle speedometer.
  - Engine for improper or worn camshaft. See General Engine Mechanical
  - A/C relay operation. A/C should cut out at wide open throttle.
  - A/C refrigerant pressure for high pressure. Check for oevercharging or faulty pressure switch.
  - Alternator output voltage. Repair if less than 9 volts or more than 16 volts.
  - Cooling fan operation.
  - Power steering. The ECM should compensate for power steering loads. Loss of this signal would be most noticeable when steering loads are high such as during parking.
  - Transmission torque converter operation.
  - Transmission for proper shift points.
  - Transmission torque converter clutch operation.
  - Vehicle brakes for dragging
  - Engine valve timing and compression. See Engine Diagnosis
  - For worn or incorrect basic engine parts such as cam, heads, pistons, etc. Also check for, worn rocker arms, and broken or weak valve springs. See General engine Mechanical
  - For the possibility of an exhaust restriction or improper valve timing by performing the following test.
    1. With engine at normal operating temperature, connect a vacuum gauge to any convenient vacuum port on intake manifold.
    2. Run engine at 1000 rpm and record vacuum reading.
    3. Increase engine speed slowly to 2500 rpm. Note vacuum reading at steady 2500 rpm.
    4. IF vacuum at 2500 rpm decreases more than 3" Hg from reading at 1000 rpm, the exhaust system should be inspected for restrictions.
    5. Disconnect exhaust pipe from engine and repeat Steps 3 & 4. If vacuum still drops more than 3" Hg with exhaust disconnected, check valve timing.
  - For overheating and possible causing. Look for the following:
  - Low or incorrect coolant solution. It should be a 50/50 mix of anti-freeze coolant and water.
    - Restricted air flow to radiator, or restricted water flow through radiator.
    - Faulty or incorrect thermostat.
    - Inoperative electric cooling fan circuit.
- If the system is running RICH (block learn less than 118), refer to "Diagnostic Aids" on facing page of Code 45.
- If the system is running LEAN (block learn greater than 138), refer to "Diagnostic Aids" on facing page of Code 44.

DETONATION/SPARK KNOCK

Definition: A mild to severe ping, usually worse under acceleration. The engine makes sharp metallic makes sharp metallic knocks that change with throttle opening.

Perform careful visual and physical check

Perform "Diagnostic Circuit Check".

- **CHECK**
  - Ignition wires for shorts or faulty insulation
  - For possibility of misfiring, crossfiring, or cutting out under load or at idle.
  - Spark plugs for correct heat range
  - Ignition timing refer to underhood emissions label for correct procedure.
  - Fuel for poor quality, "stale" fuel, and water contamination
  - Fuel system for restricted filter or improper pressure. Use CHART A-7.
  - For excessive oil entering combustion chamber. Oil will reduce the effective octane of fuel.
  - EGR operation.
  - For vacuum leaks at intake manifold gasket
  - PCV valve for proper operation by placing finger over inlet hole in valve end several times. Valve should snap back. If not, replace valve. Ensure that valve is correct part. Also check PCV hose.
  - MAP sensor output.
  - Coolant sensor for a shift in calibration
  - Oxygen sensor for silicone contamination from contaminated fuel or use of improper RTV sealant. The sensor will have a white, powdery coating and will cause a high but false signal voltage (rich exhaust indication). The ECM will reduce the amount of fuel delivered to the engine, causing a severe driveability problem.
  - Vehicle Speed Sensor (VSS) input with a "Scan" tool to make sure reading of VSS matches that of vehicle speedometer.
  - Transmission torque converter operation.
  - Transmission for proper shift points.
  - Transmission torque converter clutch operation.
  - Vehicle brakes for dragging
  - For overheating and possible causes. Look for the following.
    - Low or incorrect coolant solution. It should be a 50/50 mix of anti-freeze coolant and water.
    - Restricted air flow to radiator or restricted water flow through radiator
    - Faulty or incorrect thermostat
    - Inoperative electric cooling fan circuit
  - Engine valve timing and compresson.
  - For worn or incorrect basic engine parts such as cam, hads, pistons, etc. Also check for bent pushrods, worn rocker arms, and broken or weak valve springs.
  - For the possibility of an exhaust restriction or improper valve timing by performing the following test.
    1. With engine at normal operating temperature, connect a vacuum port on intake manifold.
    2. Run engine at 1000 rpm and record vacuum reading.
    3. Increase engine speed slowly to 2500 rpm. Note vacuum reading at steady 2500 rpm.
    4. If vacuum at 2500 rpm decreases more than 3" Hg from reading at 1000 rpm, the exhaust system should be inspected for restrictions.
    5. Disconect exhaust pipe from engine and repeat Steps 3 & 4. If vacuum still disconnected, check valve timing.
- Remove internal engine carbon with top engine cleaner.
- If the system is running LEAN, (block learn greater than 138), refer to "Diagnostic Aids" on facing page of Code 44.

CUTS OUT OR MISSES

Definition: Steady pulsation of jerking that follows engine speed, usually more pronounced as engine load increases. The exhaust has a steady spitting sound at idle or low speed.

Perform careful visual and physical check

Perform "Diagnostic Circuit Check".

- **CHECK**
  - Ignition wires for shorts or faulty insulation
  - Ignition system for moisture, dust, cracks, burns, etc. Spray plug wires with fine water mist to check for shorts.
  - For possibility of misfiring, crossfiring, or cutting out under load or at idle.
  - Secondary ignition voltage output with ST-125 tester
  - Ignition coil connections
  - ECM and ignition system for faulty grounds
  - Proper operation of EST.
  - Spark plugs. Look for wetness, cracks, improper gap, burned electrodes, and heavy deposits.
  - Spark plugs for correct heat range
  - For improper crank sensor resistance or faulty connections
  - Fuel for poor quality, "stale" fuel, and water contamination
  - Fuel system for restricted filter or improper pressure. Use CHART A-7.
  - Throttle shaft or TPS for sticking or binding. TPS voltage should read less than 0.9 volts on a "Scan" tool with the throttle closed.

For possibility of misfiring, crossfiring, or cutting under load or at idle. Locate misfiring cylinder(s) by performing the following test.

1. Start engine. Disconnect idle air control valve. Remove one spark plug wire from a spark plug and ground it against the engine.
2. Note drop in engine speed.
3. Repeat for all four cylinders.
4. Stop engine and reconnect idle air control valve.

If the engine speed dropped equally (within 50 rpm) on all cylinders, refer to "Rough, Unstable, or Incorrect Idle, Stalling" symptom. If there was no drop or excessive variation in engine speed on one or more cylinders, check for spark on the respective cylinder(s) with J 26792 (ST-125) spark tester or equivalent. If there is no spark is present, remove the spark plugs from the cylinder(s) and check for the following.

Cracks
Wear
Improper gap
Burned electrode
Heavy deposit

POOR FUEL ECONOMY

Definition: Gas mileage, as measured by an actual road test, is noticeably lower than expected. Gas mileage is noticeably lower than it was during a previous actual road test.

Perform careful visual and physical check

Perform "Diagnostic Circuit Check".

- **CHECK**
  - Proper operation of EST.
  - For possibility of misfiring, crossfiring, or cutting out under load or at idle.
  - Spark plugs. Look for wetness, cracks, improper gap, burned electrodes, and heavy deposits.
  - Spark plugs for correct heat range.
  - Fuel for poor quality, "stale" fuel, and water contamination.
  - Fuel system for restricted filter or improper pressure.
  - Injector assembly for leakage. Pressurize system by energizing fuel pump relay. Jumper the relay connector located in the underhood relay center.
  - EGR operation.
  - Distributor for worn shaft.
  - For vacuum leaks at intake manifold gasket.
  - Air cleaner element (filter) for dirt or plugging.
  - Idle Air Control system.
  - Canister purge system for proper operation.
  - Throttle shaft or TPS for sticking or binding. TPS voltage should read less than 0.9 volts on a "Scan" tool with the throttle closed.

1.6L ENGINE—DRIVEABILITY SYMPTOMS—IMPULSE AND STYLUS

- MAP sensor output.

- Oxygen sensor for silicone contamination from contaminated fuel or use of improper RTV sealant. The sensor will have a white, powdery coating and will cause a high but false signal voltage (rich exhaust indication). The ECM will reduce the amount of fuel delivered to the engine, causing a severe driveability problem.
- Coolant sensor for a shift in calibration. Use Code 14A or Code 14B charts.
- Vehicle Speed Sensor (VSS) input with a "Scan" tool to make sure reading of VSS matches that of vehicle speedometer.
- A/C relay operation. A/C should cut out at wideopen throttle.
- A/C refrigerant pressure for high pressure. Check for overcharging or faulty pressure switch.
- Alternator output voltage. Repair if less than 9 volts or more than 16 volts.
- Cooling fan operation.

- Power steering. The ECM should compensate for power steering loads. Loss of this signal would be most noticeable when steering loads are high such as during parking.
- Transmission torque converter operation.

- Tranmission torque converter clutch operation.
- Engine valve timing and compression.

- For worn or incorrect basic engine parts such as cam, heads, pistons, etc. Also check for, worn rocker arms, and broken or weak valve springs.

- For the possibility of an exhaust restriction or improper valve timing by performing the following test.
  1. With engine at normal operating temperature, connect a vacuum gauge to any convenient vacuum port on intake manifold.
  2. Run engine at 1000 rpm and record vacuum reading.
  3. Increase engine speed slowly to 2500 rpm. Note vacuum reading at steady 2500 rpm.
  4. If vacuum at 2500 rpm decreases more than 3" Hg from reading at 1000 rpm, the exhaust system should be inspected for restrictions.
  5. Disconnect exhaust pipe from engine and repeat Steps 3 & 4. If vacuum still drops more than 3" Hg with exhaust disconnected, check valve timing.
- Thermostat for incorrect heat range or being inoperative
- Check driver's driving habits and vehicle conditions which affect gas mileage.
  - Suggest driver read "Important Facts on Fuel Economy" in Owner's Manual.
  - Is A/C "ON" full time (Defroster mode "ON")?
  - Are tires at correct pressure?
  - Is acceleration often heavy?
  - Are the wheels aligned correctly?
  - Is the speedometer calibrated correctly?
  - Are the vehicle brakes dragging?
  - Is the brake switch applying excessive force on the brake pedal?
- If the system is running RICH, (block learn less than 118), refer to "Diagnostic Aids" on facing page of Code 45.

ROUGH, UNSTABLE, OR INCORRECT IDLE, STALLING

Definition: The engine runs unevenly at idle. If severe, the car may shake. Also, the idle speed may vary (called "hunting"). Either condition may be severe enough to cause stalling. Engine idles at incorrect speed.

Perform careful visual and physical check as

Perform "Diagnostic Circuit Check".

- **CHECK**
  - MAP sensor. Use CHART C1-D.
  - Throttle for sticking shaft or binding linkage. This will cause a high TPS voltage (open throttle indication) and the ECM will not control idle. TPS voltage should be less than 0.9 volts with throttle closed.
  - Battery cables and ground straps for poor contact. Erratic voltage will cause the IAC valve to change its position, resulting in poor idle quality.
  - Ignition wires for shorts or faulty insulation.
  - Ignition system for moisture, dust, cracks, burns, etc. Spray plug wires with fine water mist to check for shorts.
  - For possibility of misfiring, crossfiring, or cutting out under load or at idle.

- Secondary ignition voltage output with ST-1 tester.
- For worn distributor shaft.
- For moisture in distributor cap.
- Ignition pick-up coil resistance and connection.
- Ignition pick-up coil resistance and connection.
- Ignition coil connections.
- ECM and ignition system for faulty grounds.
- Proper operation of EST.
- Spark plugs. Look for wetness, cracks, improper gap, burned electrodes, and heavy deposits.
- Fuel system for restricted filter or improper pressure. Use CHART A-7.
- Injector assembly for leakage. Pressurize system by energizing fuel pump relay. Jumper the relay connector in the underhood relay center.
- EGR operation.
- For vacuum leaks at intake manifold gasket.
- Idle Air Control system.

- Electrical system voltage. IAC valve will not move if voltage is below 9 volts or greater than 17.8 volts.
- A/C system for being overcharged. Refer to A/C Section in this service manual.
- PCV valve for proper operation by placing finger over inlet hole in valve end several times. Valve should snap back. If not, replace valve. Ensure that valve is correct part. Also check PCV hose.
- Canister purge system for proper operation.
- Throttle shaft or TPS for sticking or binding. TPS voltage should read less than 1.25 volts on a "Scan" tool with the throttle closed.
- MAP sensor output.

- Oxygen sensor for silicone contamination from contaminated fuel or use of improper RTV sealant. The sensor will have a white, powdery coating and will cause a high but false signal voltage (rich exhaust indication). The ECM will reduce the amount of fuel delivered to the engine, causing a severe driveability problem.
- Coolant sensor for a shift in calibration. Use Code 14A or Code 14B charts.
- P/N switch circuit on vehicle with automatic tranmission.

- Alternator output voltage. Repair if less than 9 volts or more than 16 volts.
- Power steering. The ECM should compensate for power steering loads. Loss of this signal would be most noticeable when steering loads are high such as during parking.
- Engine valve timing and compression.

- For worn or incorrect basic engine parts such as cam, heads, pistons, etc. Also check for, worn rocker arms, and broken or weak valve springs.

- Check ignition timing. Refer to underhood emission label for proper procedure.
- For the possibility of an exhaust restriction or improper valve timing, perform the following test.
  1. With engine at normal operating temperature, connect a vacuum gauge to any convenient vacuum port on intake manifold.
  2. Run engine at 1000 rpm and record vacuum reading.
  3. Increase engine speed slowly to 2500 rpm. Note vacuum reading at steady 2500 rpm.
  4. If vacuum at 2500 rpm decreases more than 3" Hg from reading at 1000 rpm, the exhaust system should be inspected for restrictions.
  5. Disconnect exhaust pipe from engine and repeat Steps 3 & 4. If vacuum still drops more than 3" Hg with exhaust disconnected, check valve timing.
- For overheating and possible causes. Look for the following.
  - Low or incorrect coolant solution. It should be a 50/50 mix of anti-freeze coolant and water.
  - Loose water pump belt.
  - Restricted air flow to radiator, or restricted water flow through radiator.
  - Faulty or incorrect thermostat.
  - Inoperative electric cooling fan circuit.
- If the system is running RICH, (block learn less than 118), refer to "Diagnostic Aids" on facing page of Code 45.
- If the system is running LEAN, (block learn greater than 138), refer to "Diagnostic Aids" on facing page of Code 44.

CHECK

- For possibility of misfiring, crossfiring, or cutting under load or at idle. Locate misfiring cylinder(s) by performing the following test.

1.6L ENGINE – DRIVEABILITY SYMPTOMS IMPULSE AND STYLUS

1. Start engine. Disconnect idle air control valve. Remove one spark plug wire from a spark plug and ground it against the engine.
2. Note drop in engine speed.
3. Repeat for all four cylinders.
4. Stop engine and reconnect idle air control valve.

If the engine speed dropped equally (within 50 rpm) on all cylinders, proceed through the causes listed. If there was no drop or excessive variation in engine speed on one or more cylinders, check for spark on the respective cylinder(s) with J 26792 (ST-125) spark tester or equivalent.

EXCESSIVE EXHAUST EMISSIONS OR ODORS

Definition: Vehicle fails an emission test or vehicle has excessive "rotten egg" smell. (Excessive odors do not necessarily indicate excessive emissons).

Perform careful visual and physical check

Perform "Diagnostic Circuit Check".

- **CHECK**
  - EGR valve not opening.
  - Vacuum leaks
  - Faulty coolant system and/or coolant fan operation.
  - Remove carbon with top engine cleaner. Follow instructions on can.
- If the system is running RICH (block learn less than 118), refer to "Diagnostic Aids" on facing page of Code 45.
- If the system is running LEAN (block learn greater than 138), refer to "Diagnostic Aids" on facing page of Code 44.
- If emission test indicates excessive NOx, check for items which cause car to run lean or too hot.
- If emission test indicates excessive HC and CO or exhaust has excessive odors, check for items which cause car to run RICH.
  - Incorrect fuel pressure. Use CHART A-7.
  - Fuel loading of evaporative vapor canister.
  - PCV valve plugging, sticking, or blocked PCV hose. Check for fuel in crankcase.
  - Catalytic converter lead contamination (Look for removal of fuel filler neck restrictor).
  - Improper fuel cap installation
  - Faulty spark plugs, plug wires, or ignition components.

DIESELING, RUN-ON

Definition: Engine continues to run after key is turned "OFF", but runs very roughly. (If engine runs smoothly, check ignition switch).

Perform careful visual and physical check

Perform "Diagnostic Circuit Check".

- **CHECK**
  Injector assembly for leakage. Pressurize system by energizing fuel pump through the fuel pump test connector.

BACKFIRE

Definition: Fuel ignites in intake manifold or in exhaust system, making a loud popping sound.

Perform careful visual and physical check

Perform "Diagnostic Circuit Check".

- **CHECK**
  - EGR operation for valve being open all the time.
  - Intake manifold gasket for leaks
  - For possibility of misfiring, crossfiring, or cutting out under load or at idle.
  - Spark plugs. Look for wetness, cracks, improper gap, burned electrodes, and heavy deposits.
  - Ignition coil connections
  - Ignition system for moisture, dust, cracks, burns, etc. Spray plug wires with fine water mist to check for shorts.
  - ECM and ignition system for faulty grounds
  - Secondary ignition voltage output with ST-125 tester
  - For vacuum leaks at intake manifold gasket
  - Engine valve timing and compression.
  - For worn or incorrect basic engine parts such as cam, heads, pistons, etc. Also check for, worn rocker arms, and broken or weak valve springs.

2.6L ENGINE – FUEL INJECTION SYSTEM DIAGRAM (FEDERAL) AMIGO, PICK-UP, TROOPER AND TROOPER II

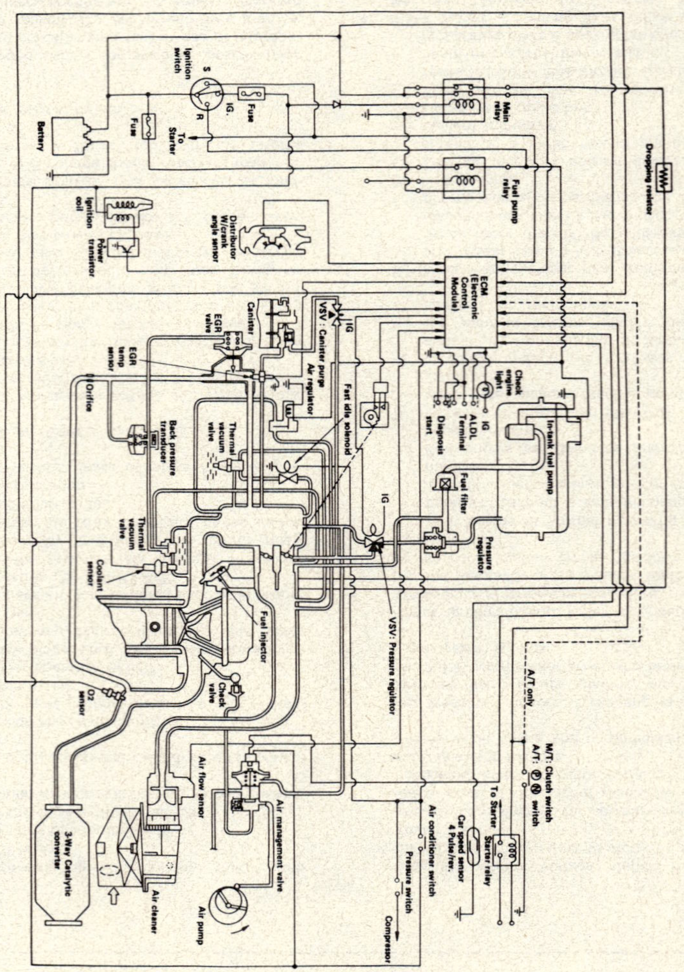

2.6L ENGINE—FUEL INJECTION SYSTEM DIAGRAM (CALIFORNIA) AMIGO, PICK-UP, TROOPER AND TROOPER II

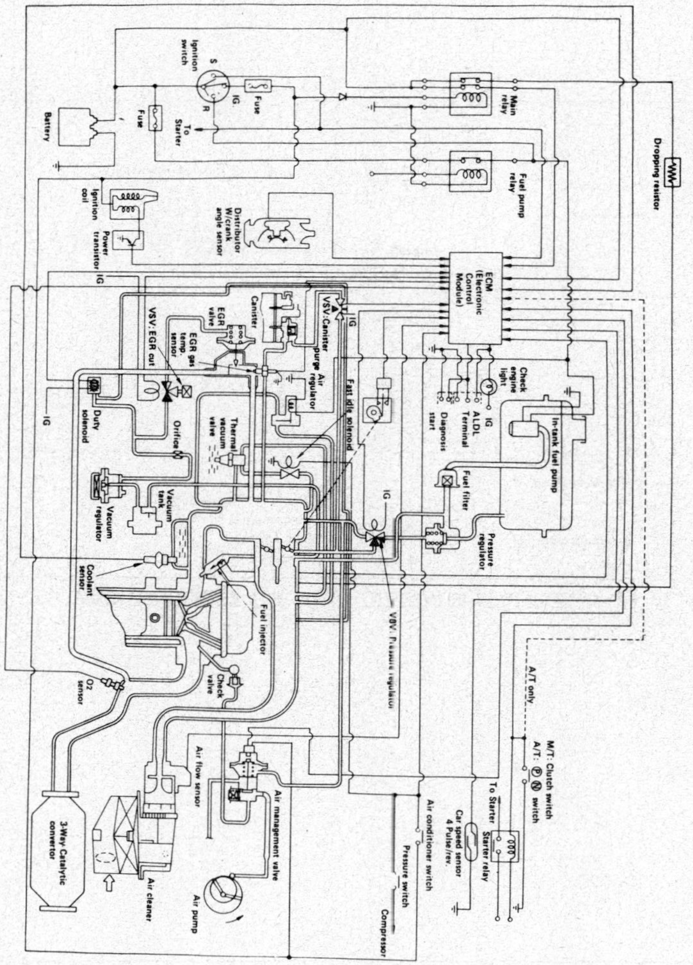

2.6L ENGINE—VACUUM DIAGRAM (FEDERAL AND 1990 CALIFORNIA) AMIGO, PICK-UP, TROOPER AND TROOPER II

VACUUM HOSE ROUTING DIAGRAM (FEDERAL)

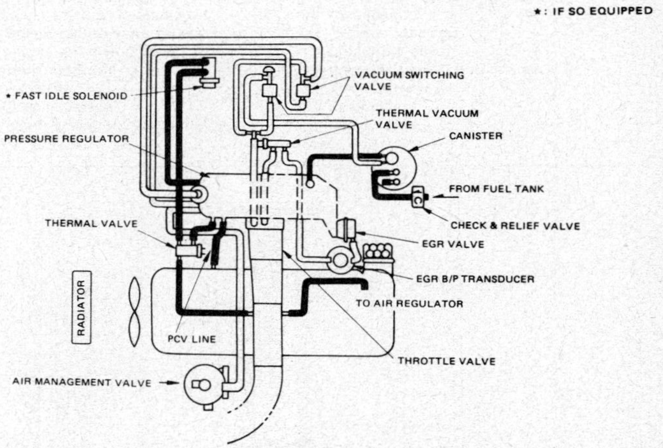

VACUUM HOSE ROUTING DIAGRAM (CALIFORNIA)

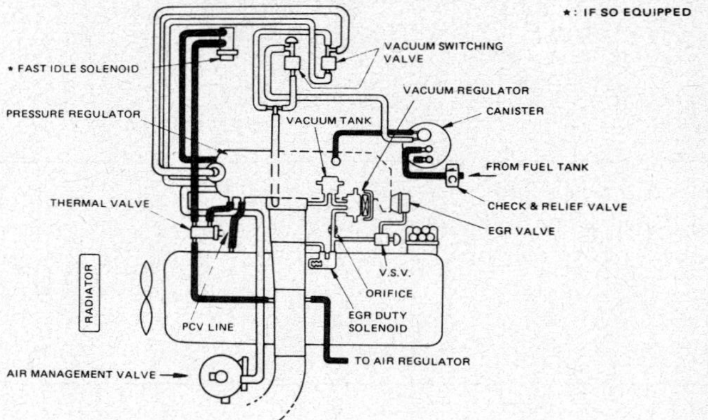

2.6L ENGINE—VACUUM DIAGRAM (1991–92 CALIFORNIA) AMIGO, PICK-UP, TROOPER AND TROOPER II

VACUUM HOSE ROUTING DIAGRAM (CALIFORNIA)

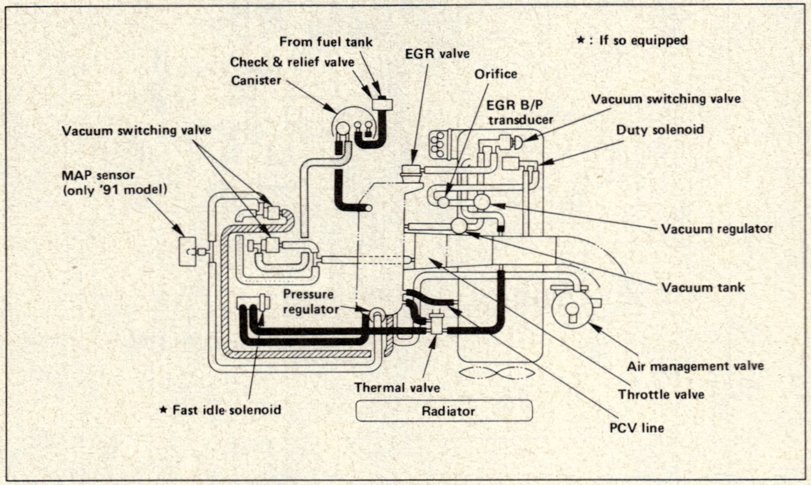

2.6L ENGINE—ENGINE COMPONENT LOCATION AMIGO, PICK-UP, TROOPER AND TROOPER II

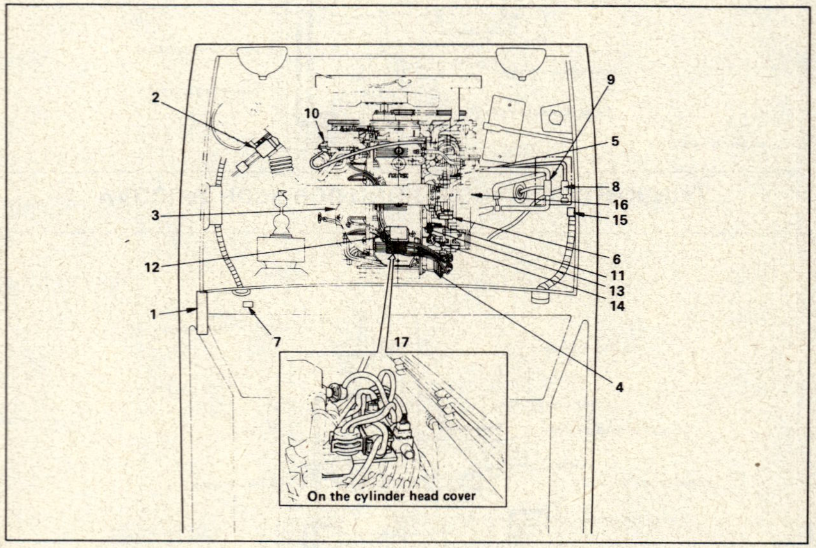

1. Electronic control module (ECM)
2. Air flow sensor
3. O_2 sensor
4. Crank angle sensor (Built in the distributor)
5. Coolant temperature sensor (Under the intake manifold)
6. Throttle valve switch (M/T or A/T)
7. Car speed sensor (Built in to speedometer)
8. Vacuum switching valve (Purge control)
9. Vacuum switching valve (Pressure regulator)
10. Air management valve (AIR control)
11. Fuel injector
12. Ignition coil
13. Power switch
14. EGR gas temperature sensor (California only)
15. Dropping resistor
16. Air regulator (Under the common chamber)
17. EGR control system (California only)
    - Duty solenoid
    - Vacuum tank
    - Vacuum regulator
    - VSV ; EGR cut
    - ORIFICE

2.6L ENGINE—FUEL INJECTION SYSTEM WIRING DIAGRAM 1990 AMIGO, PICK-UP, TROOPER AND TROOPER II

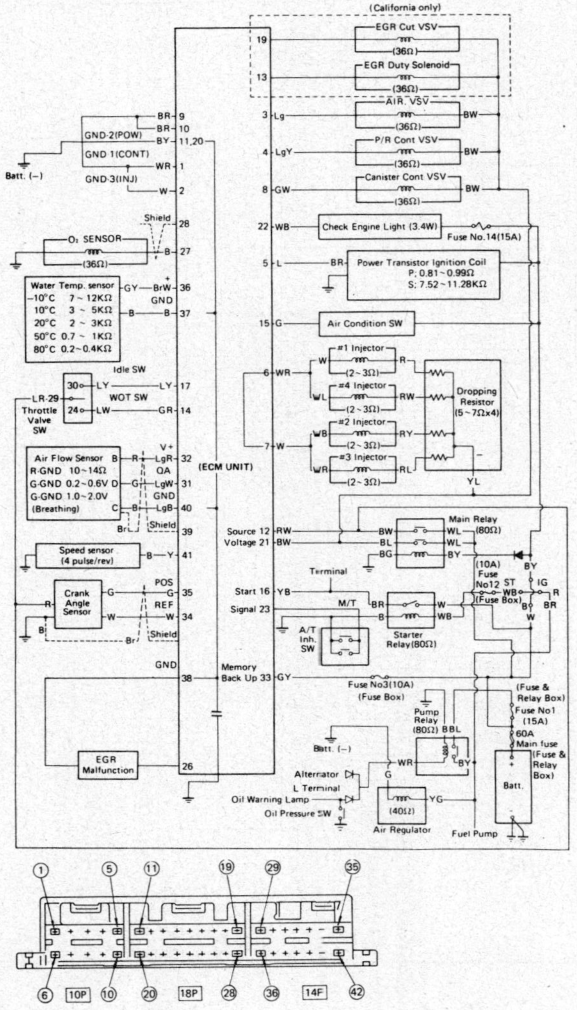

2.6L ENGINE—FUEL INJECTION SYSTEM WIRING DIAGRAM 1991 AMIGO, PICK-UP, TROOPER AND TROOPER II

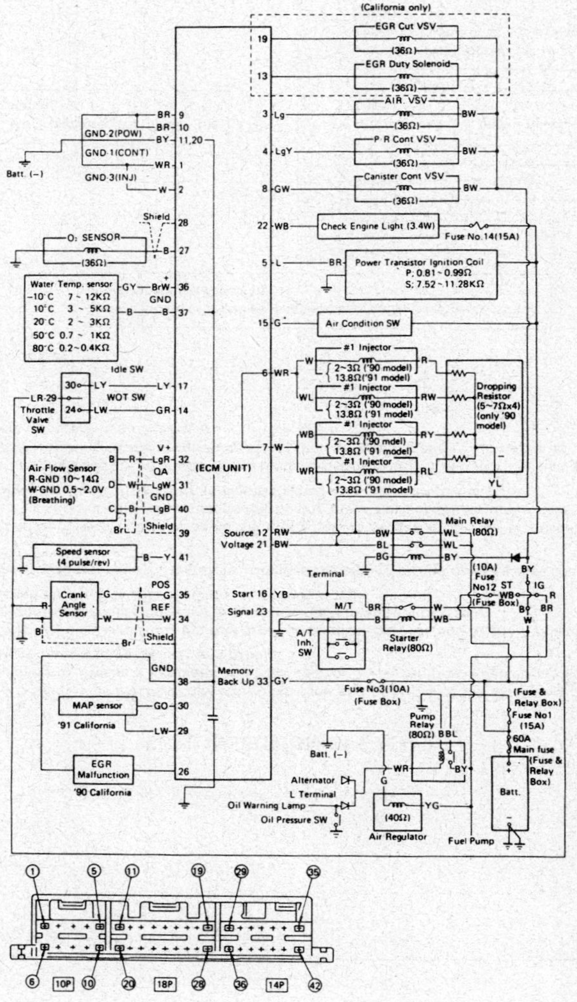

2.6L ENGINE—FUEL INJECTION SYSTEM WIRING DIAGRAM 1992 AMIGO, PICK-UP, TROOPER AND TROOPER II

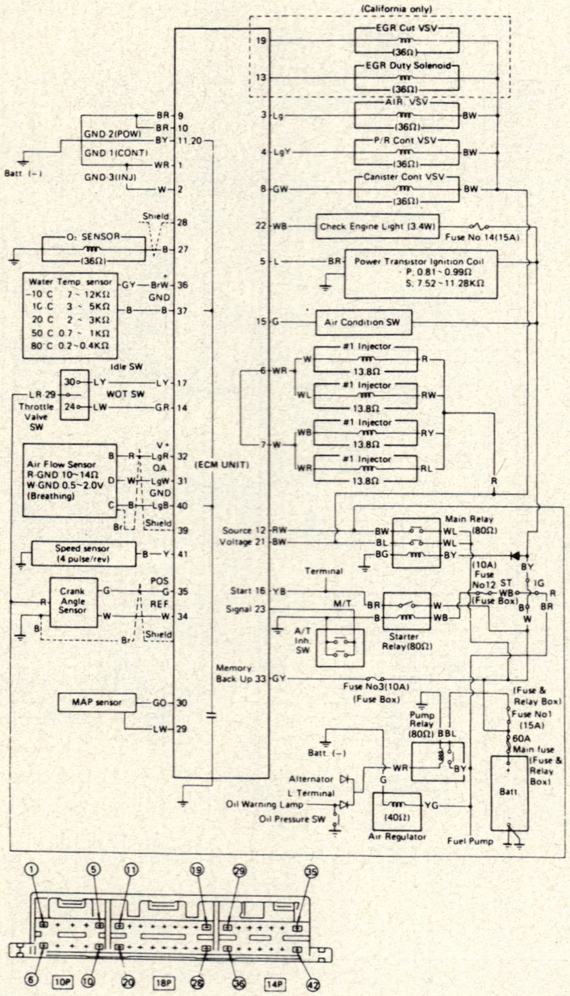

FUEL INJECTION SYSTEM

The fuel injection system is of electronically controlled type and consists of the Electronic Control Module (ECM), Electric Fuel Pump, Pressure Regulator, Air Flow Sensor, Fuel Injectors, Crank Angle Sensor (built into Distributor), Coolant Temperature Sensor and others.

The pressure of the fuel delivered from the fuel pump is controlled at the end of the fuel line, so that it can keep a certain amount of pressure differential over the intake manifold pressure.
This is accomplished by the pressure regulator of overflow type.

The volume of the fuel to be injected, therefore, changes depending on the opening duration of the injector valves.

Regarding the air to be drawn into the engine, its mass flow is sensed by the air flow sensor of hot wire type. This sensor generates voltage corresponding to the mass flow of the air drawn into the engine. The changing voltage is transformed into an electric signal and provided to the ECM.

With receipt of the signals sent from the air flow sensor and others, the ECM determines an appropriate fuel injection pulse width referring to the programmed data thereof, feeding such information to the fuel injector valves to effect an appropriate air/fuel ratio.

Fuel Injection Control

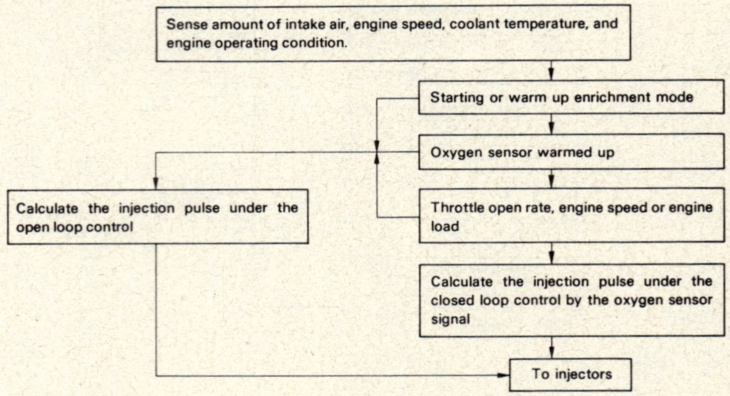

2.6L ENGINE – FUEL INJECTION SYSTEM DESCRIPTION AMIGO, PICK-UP, TROOPER AND TROOPER II

Fuel Shut-Off System:

The fuel shut-off system functions to cut off an extra fuel supply to the engine, when the vehicle is decelerated, to ensure an improved fuel economy and to prevent the occurrence of the after-fire in the exhaust. This system starts functioning when the vehicle is held standing or when the vehicle is running at speeds over the specified speed level, both with the throttle valve closed and with the engine speed still remaining high beyond the specified level. This system ceases to function when the engine speed comes down low to a point where the operation of the system is released or when any other vehicle operation condition comes to act enough against the continued operation of the system. The engine speed to activate or disactivate this system changes correspondingly as the engine coolant temperature changes.

Fuel Enrichment for Engine Start and/or Warming Up:

When the engine is started and/or is warmed up, the ECM starts working to enrich the air/fuel ratio of the air/fuel mixture. This is accomplished by the work of the injector valves according to the signal fed from the ECM. In such a particular case where the engine is started and/or is warmed up in cold weather, extra air is provided by the air regulator in addition to the air/fuel enrichment mentioned above. The air flow rate first set by the air regulator to match its surrounding temperature is changed best to suit to the specific condition of the engine operation. This is accomplished by optimum changing the air valve opening within the air regulator by means of the bi-metal to work with the electric heater.

Transient Enrichment of the Air/Fuel Ratio:

While the engine is warming up, the acceleration enrichment is provided for the short period when the throttle valve switch is moved toward IDLE OFF position from IDLE, or when the intake air flow becomes high. Both the quantity of the enrichment and its working time are decreased gradually with increasing engine temperature and are not provided in normal operating engine temperature.

Full-Load Time Enrichment of the Air/Fuel Ratio:

The ECM also works to ensure the enrichment of the air/fuel ratio at the time of full-load running. With the throttle valve opening wider, the contact point of the throttle valve switch moves toward FULL ON position, where the vehicle is running with fully loaded. This is signaled to the ECM. The ECM determines an appropriate injection pulse width referring to the programmed data stored in its memory unit, sending the signal to the injector valves to effect the most appropriate injector valve opening.

2.6L ENGINE – IGNITION SYSTEM SCHEMATIC AMIGO, PICK-UP, TROOPER AND TROOPER II

This is an electronically controlled ignition system. The ignition timing is controlled by the electronic control module according to the engine operating conditions.

For determination of ignition timing, the signals from sensors (coolant temperature sensor, air flow sensor, throttle valve switch, crank angle sensor and car speed sensor) are used. The signal from the electronic control module is transmitted to the ignitor, and the signal controls the ignition timing.

ELECTRONIC IGNITION TIMING CONTROL

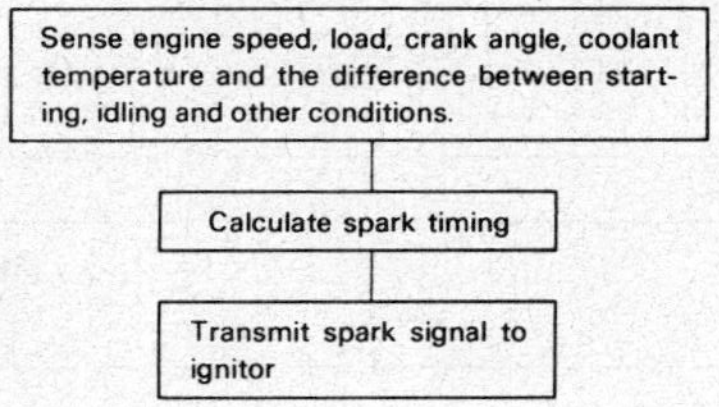

DIAGRAM OF IGNITION SYSTEM

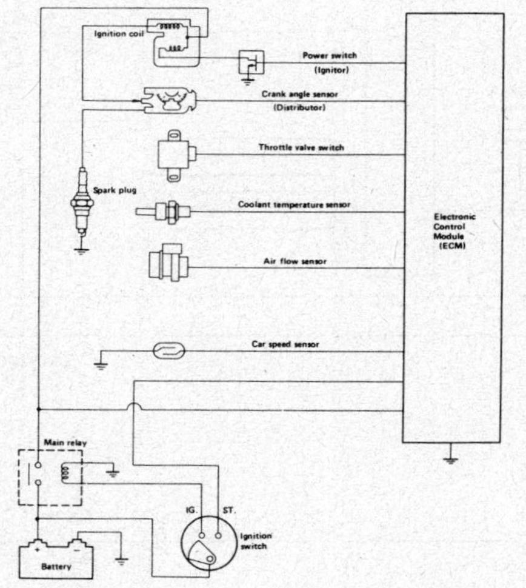

2.6L ENGINE—VACUUM SWITCHING VALVE CONTROL SYSTEM AMIGO, PICK-UP, TROOPER AND TROOPER II

Canister purge control

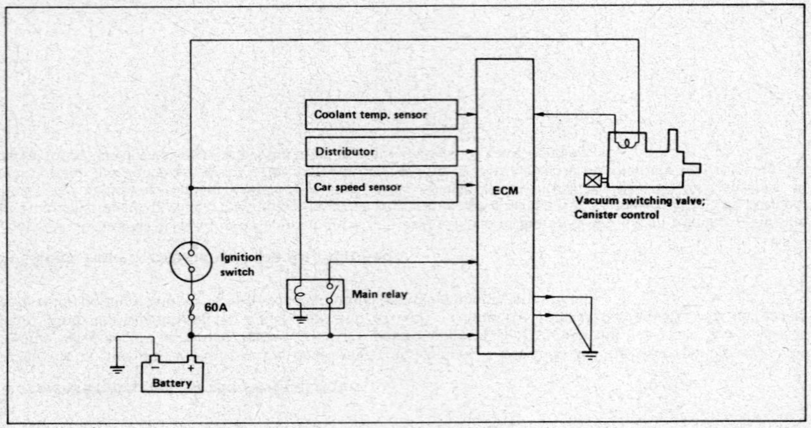

Pressure regulator control

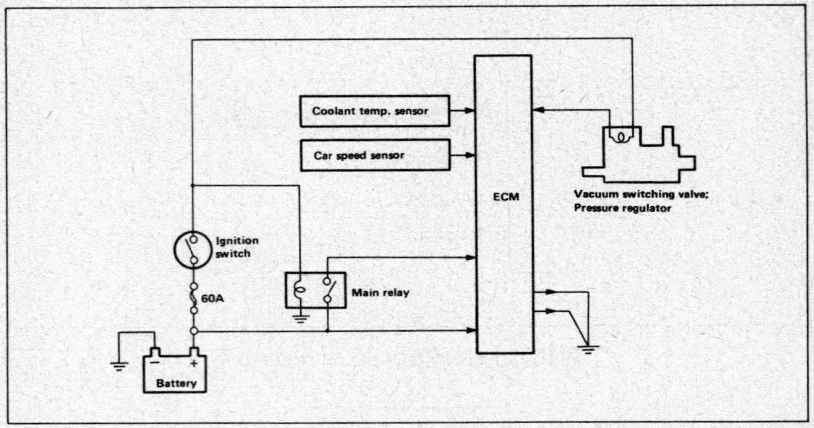

Air injection reactor control

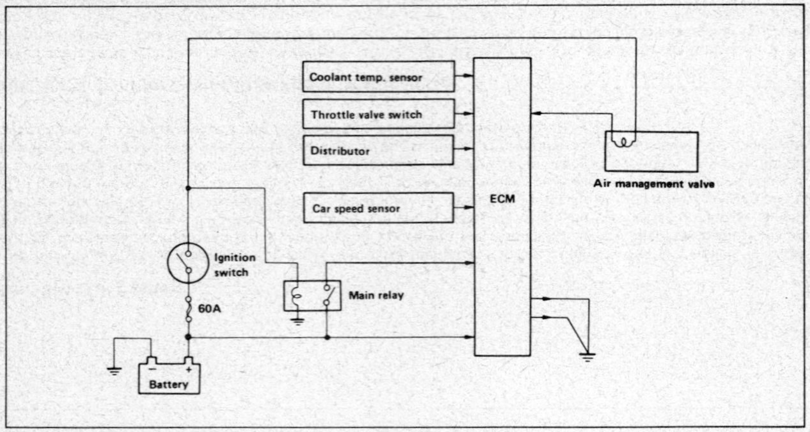

EGR control (California only)

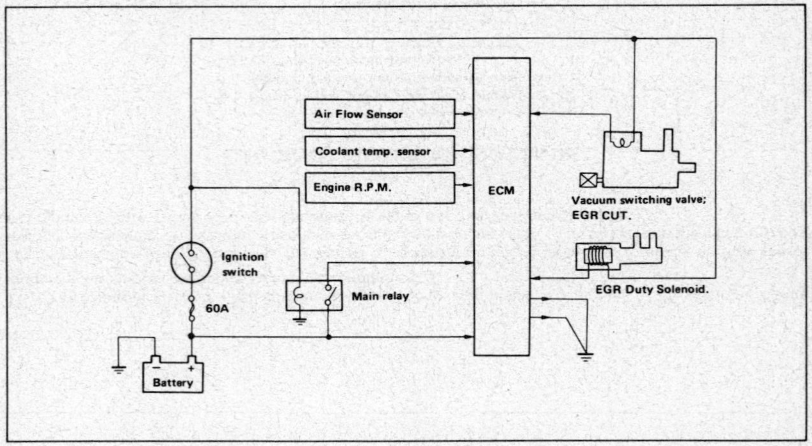

2.6L ENGINE—FAIL-SAFE OPERATION AMIGO, PICK-UP, TROOPER AND TROOPER II

If a failure has developed in one of the following sensor systems, driveability or safety may be adversely affected. To give an extra margin of safety in the event of system failure, the output of the ECM will be switched automatically to the fail-safe mode.

- O_2 sensor system
- Coolant temperature sensor system
- Throttle valve switch system
- Air flow sensor system

BACK-UP CONTROL

In the event a malfunction has developed in the micro computer within the ECM, the back-up control system works to maintain the minimum necessary functions of the ECM to permit continuous operation of the vehicle.

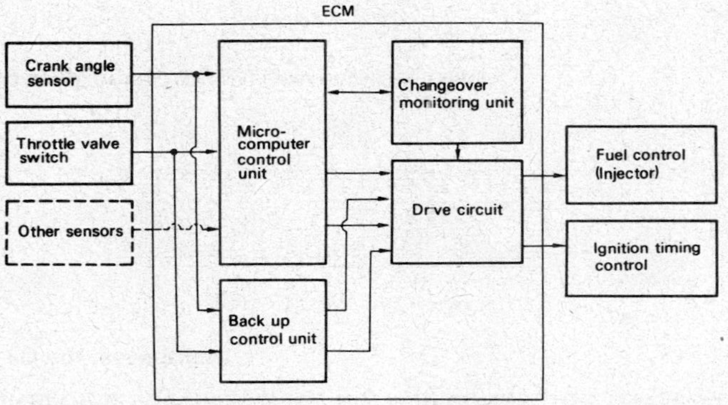

2.6L ENGINE—CONTROL UNIT CONNECTOR PIN IDENTIFICATION 1990 AMIGO, PICK-UP, TROOPER AND TROOPER II

[FACE VIEW]

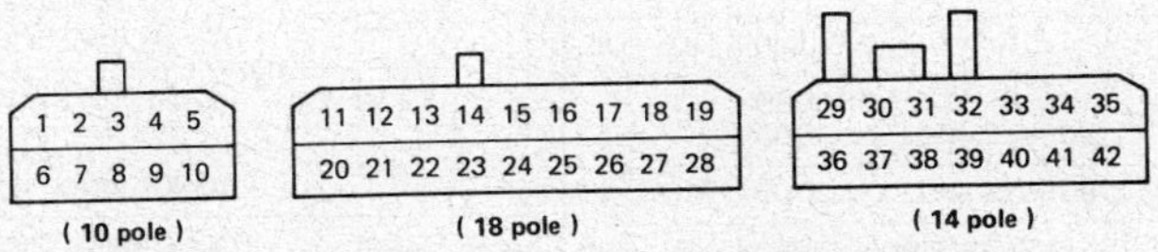

Note: Written numbers in connector terminal sections correspond to ECM pin NO.

No.	COLOR	MAIN CONNECTION	MARKING
1	BG	GROUND; Injector	GND-1
2	BG	GROUND; Injector	GND-1
3	Lg	AIR CONTROL	V.S.V. (AIR)
4	LgY	PRESSURE REGULATOR CONTROL	V.S.V. (P/REG)
5	L	IGNITION	IGN.
6	WR	INJECTOR	INJ #1/#2
7	W	INJECTOR	INJ #3/#4
8	GW	CANISTER PURGE	V.S.V. (Purge cont)
9	BR	GROUND (POWER) (IGN.)	GND (PW)
10	BR	GROUND (POWER) (IGN.)	GND (PW)
11	B	GROUND; Power	GND (control unit)
12	RW	POWER	(VB1)
13	Lg	EGR DUTY SOLENOID	EGR control
14	GR	FULL; Throttle valve	WOT
15	G	AIR CONDITIONER SWITCH	A/C
16	YB	STARTER SWITCH	ST
17	LY	IDLE SWITCH; Throttle valve	IDL
18	–	–	–
19	WG	EGR CUT	V.S.V. (EGR cut)
20	B	GROUND; Power	GND (control unit) Joint to 11

No.	COLOR	MAIN CONNECTION	MARKING
21	BW	POWER	VB1
22	WB	CHECK ENGINE LIGHT	DISP
23	WL	INH SWITCH	A/T Pos.
24	BR	ALDL	ALDL
25	–	–	–
26	YG	EGR GAS TEMP SENSOR	EGR sensor
27	B	O_2 SENSOR	O_2
28	SHIELD	O_2 SENSOR-SHIELD	GND-S
29	–	–	–
30	–	–	–
31	W	AIR FLOW SENSOR	Vout
32	R	AIR FLOW SENSOR (POWER)	V+
33	GY	BACK UP (POWER)	VB2
34	BW	CRANK ANGLE SENSOR	REF
35	GY	CRANK ANGLE SENSOR	POS
36	BrW	COOLANT TEMP SENSOR	TW
37	B	COOLANT TEMP SENSOR GROUND	TWGND
38	BG	GND (Joint to 37)	GND
39	SHIELD	AIR FLOW SENSOR SHIELD	GND-S
40	B	AIR FLOW SENSOR GND	AFSGND
41	Y	CAR SPEED SENSOR	4P
42	–	–	–

2.6L ENGINE—CONTROL UNIT CONNECTOR PIN IDENTIFICATION 1990 AMIGO, PICK-UP, TROOPER AND TROOPER II

CORRESPONDENCE OF ECM PIN NO. AND CONNECTOR TERMINAL NO.

1) ECM PIN NO. ARRANGEMENT

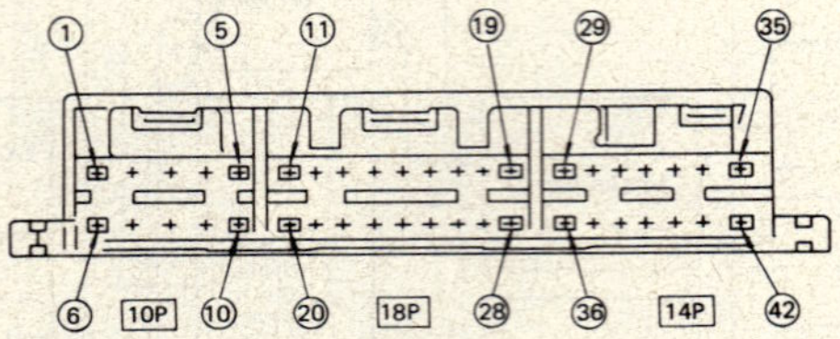

2) ECM 3 CONNECTORS TERMINAL ARRANGEMENT (Female)

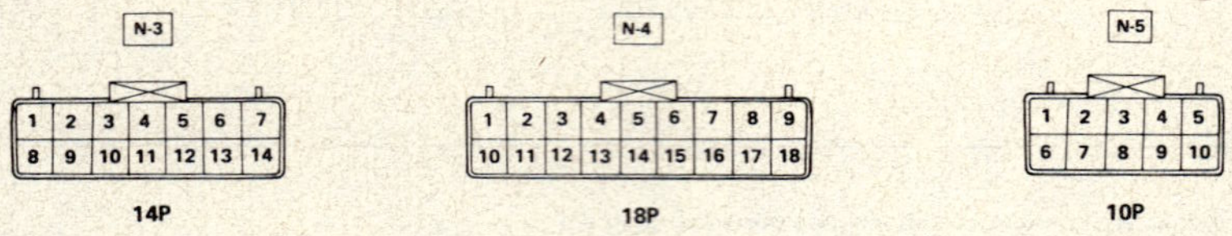

3) CORRESPONDENCE TABLE

ECM CONNECTOR TERMINAL NO. / ECM PIN NO.

	1	2	3	4	5	6	7	8	9	10	11	12	13	14	15	16	17	18	19	20	21	22	23	24
N5 (10P)	5	4	3	2	1	10	9	8	7	6														
N4 (18P)											9	8	7	6	5	4	3	2	1	18	17	16	15	14
N3 (14P)																								

ECM CONNECTOR TERMINAL NO. / ECM PIN NO.

	25	26	27	28	29	30	31	32	33	34	35	36	37	38	39	40	41	42
N5 (10P)																		
N4 (18P)	13	12	11	10														
N3 (14P)					7	6	5	4	3	2	1	14	13	12	11	10	9	8

2.6L ENGINE—CONTROL UNIT CONNECTOR PIN IDENTIFICATION 1991–92 AMIGO, PICK-UP, TROOPER AND TROOPER II

[FACE VIEW]

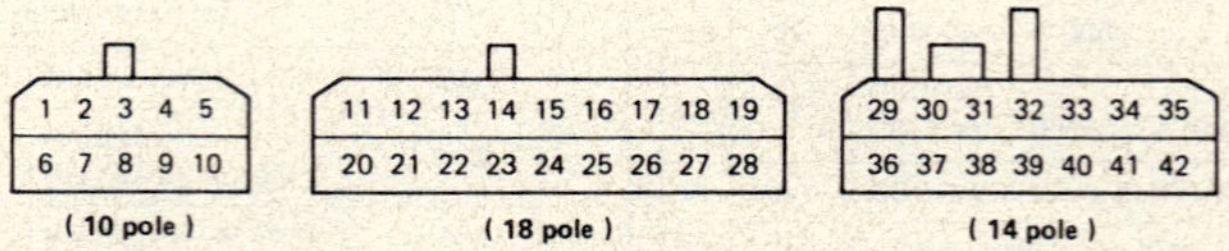

Note: Written numbers in connector terminal sections correspond to ECM pin NO.

No.	COLOR	MAIN CONNECTION	MARKING
1	BG	GROUND; Injector	GND-1
2	BG	GROUND; Injector	GND-1
3	Lg	AIR CONTROL	V.S.V. (AIR)
4	LgY	PRESSURE REGULATOR CONTROL	V.S.V. (P/REG)
5	L	IGNITION	IGN.
6	WR	INJECTOR	INJ #1/#2
7	W	INJECTOR	INJ #3/#4
8	GW	CANISTER PURGE	V.S.V. (Purge cont)
9	BR	GROUND (POWER) (IGN.)	GND (PW)
10	BR	GROUND (POWER) (IGN.)	GND (PW)
11	B	GROUND; Power	GND (control unit)
12	LgR	POWER	(VB1)
13	Lg	EGR DUTY SOLENOID	EGR control
14	GR	FULL; Throttle valve	WOT
15	G	AIR CONDITIONER SWITCH	A/C
16	YB	STARTER SWITCH	ST
17	LY	IDLE SWITCH; Throttle valve	IDL
18	—	—	—
19	WG	EGR CUT	V.S.V. (EGR cut)
20	B	GROUND; Power	GND (control unit) Joint to 11

No.	COLOR	MAIN CONNECTION	MARKING
21	LgB	POWER	VB1
22	WB	CHECK ENGINE LIGHT	DISP
23	WL	INH SWITCH	A/T Pos.
24	BR	ALDL	ALDL
25	—	—	—
26	—	—	—
27	B	O_2 SENSOR	O_2
28	SHIELD	O_2 SENSOR-SHIELD	GND-S
29	LW	MAP SENSOR	VOUT
30	GO	MAP SENOSR	5.1V
31	W	AIR FLOW SENSOR	Vout
32	R	AIR FLOW SENSOR (POWER)	V+
33	GY	BACK UP (POWER)	VB2
34	W	CRANK ANGLE SENSOR	REF
35	G	CRANK ANGLE SENSOR	POS
36	BrW	COOLANT TEMP SENSOR	TW
37	B	COOLANT TEMP SENSOR GROUND	TWGND
38	LR	MAP GND	GND
39	SHIELD	AIR FLOW SENSOR SHIELD	GND-S
40	B	AIR FLOW SENSOR GND	AFSGND
41	Y	CAR SPEED SENSOR	4P
42	—	—	—

2.6L ENGINE—CONTROL UNIT CONNECTOR PIN IDENTIFICATION 1991–92 AMIGO, PICK-UP, TROOPER AND TROOPER II

Code No.	Item	Measure Terminal		Key SW Posit.	Condition to Check	Standard Value
		Tester + Side	Tester – Side			
	Main Relay	Terminal (WR)	Terminal (B)	OFF		Approx. 80 Ω
	Charge Relay	Terminal (WG, B)	Terminal (B)	OFF		Approx. 80 Ω
	Fuel Pump Relay	Terminal (WR)	Terminal (B)	OFF		Approx. 80 Ω

2.6L ENGINE—INSPECTION AND MAINTENANCE CHART 1990–91 AMIGO, PICK-UP, TROOPER AND TROOPER II

NORMAL CONDITION;
- SIGNAL CODE NO. 12 WILL DISPLAY
- THE MEMORY CLEARED BY MAIN FUSE

I. DISCONNECT THE SENSOR OR ACTUATOR WIRING CONNECTOR (WHEN CHECKING THE PART/COMPONENT INDEPENDENTLY)

Code No.	Item		Measure Terminal		Key SW Posit.	Condition to Check		Standard Value
			Tester + Side	Tester – Side				
21 43 65	Throttle valve SW.		Center Terminal (Terminal P)	Upper Terminal (Terminal I)	OFF	Ac. Pedal Not Depressed		0 Ω
						Ac. Pedal Slightly Depressed		∞ Ω
						Ac. Pedal Fully Depressed		∞ Ω
			Center Terminal (Terminal P)	Lower Terminal (Terminal F)	OFF	Ac. Pedal Not Depressed		∞ Ω
						Ac. Pedal Slightly Depressed		∞ Ω
						Ac. Pedal Fully Depressed		0 Ω
33 64	Fuel Injector	#1	Terminal	Terminal	OFF			2~3 Ω
		#2	Terminal	Terminal	OFF			2~3 Ω
		#3	Terminal	Terminal	OFF			2~3 Ω
		#4	Terminal	Terminal	OFF			2~3 Ω
23 35 54	Ignition Coil		+ Terminal	– Terminal	OFF	Primary		0.81~0.99 Ω
			+ Terminal	Center Terminal	OFF	Secondary		7.52~11.28K Ω
25 53	Vacuum Switching Valve		Terminal	Terminal	OFF			Approx. 36 Ω
14 15	Water Temperature Sensor		Terminal	Terminal	OFF	Water Temp. °C(°F)	–10 (14) 10 (50) 20 (68) 50 (122) 80 (176)	7~12K Ω 3~5K Ω 2~3K Ω 0.7~1K Ω 0.2~0.4K Ω
33	Dropping Resistor		Center Terminal	Another Terminal	OFF			5~7 Ω/Pc.
63	Car Speed Sensor		41	Ground	OFF	Turn the Inner Cable Slowly		4 Pulse/rev.
	Air Regulator		Terminal	Terminal	OFF			Approx. 40 Ω

2.6L ENGINE—INSPECTION AND MAINTENANCE CHART
1992 AMIGO, PICK-UP, TROOPER AND TROOPER II

NORMAL CONDITION; * SIGNAL CODE NO. 12 WILL DISPLAY
* THE MEMORY CLEARED BY MAIN FUSE

I. DISCONNECT THE SENSOR OR ACTUATOR WIRING CONNECTOR (WHEN CHECKING THE PART/COMPONENT INDEPENDENTLY)

Code No.	Item		Measure Terminal		Key SW Posit.	Condition to Check	Standard Value
			Tester + Side	Tester − Side			
21 23 65	Throttle valve SW.		Center Terminal (Terminal P)	Upper Terminal Terminal I)	OFF	Ac. Pedal Not Depressed	0Ω
						Ac. Pedal Slightly Depressed	∞Ω
						Ac. Pedal Fully Depressed	∞Ω
			Center Terminal (Terminal P)	Lower Terminal Terminal F)	OFF	Ac. Pedal Not Depressed	∞Ω
						Ac. Pedal Slightly Depressed	∞Ω
						Ac. Pedal Fully Depressed	0Ω
33 4	Fuel Injector	#1	Terminal	Terminal	OFF		1.38Ω
		#2	Terminal	Terminal	OFF		1.38Ω
		#3	Terminal	Terminal	OFF		1.38Ω
		#4	Terminal	Terminal	OFF		1.38Ω
23 5 54	Ignition Coil		+ Terminal	− Terminal	OFF	Primary	0.81~0.99Ω
			+ Terminal	Center Terminal	OFF	Secondary	7.52~11.28KΩ
25 53	Vacuum Switching Valve		Terminal	Terminal	OFF		Approx. 36Ω
14 15	Water Temperature Sensor		Terminal	Terminal	OFF	Water Temp. °C(°F) −10 (14) 10 (50) 20 (68) 50 (122) 80 (176)	7~12KΩ 3~ 5KΩ 2~ 3KΩ 0.7~1KΩ 0.2~0.4KΩ
63	Car Speed Sensor		41	Ground	OFF	Turn the Inner Cable Slowly	4 Pulse/rev.
	Air Regulator		Terminal	Terminal	OFF		Approx. 40Ω

INSPECTION AND MAINTENANCE

II. DISCONNECT THE ECM UNIT WIRING CONNECTORS

Code No.	Item		Measure Terminal		Key SW Posit.	Condition to Check	Standard Value
			Tester + Side	Tester − Side			
	Power Supply Circuit		33	Ground (11 or 20)	ON		10~14V
			12 (21)	Ground (11 or 20)	ON		10~14V
			9 & 10	Ground (11 or 20)	OFF		0Ω
			1 & 2	Ground (11 or 20)	OFF		0Ω
			11 or 20	Ground (Body)	OFF		0Ω
33 64	Fuel Injector	#1, #4	6	Ground (1 or 2)	ON		10~14V
		#2, #3	7	Ground (1 or 2)	ON		10~14V
22	Starter Signal		16	Ground (11 or 20)	ST	Disconnect Starter Relay	10~14V
25 53	P/R V.S.V. (Vacuum Switching Valve)		4	Ground (9 or 10)	ON		10~14V
21 43 65	Throttle Valve SW.		17	Ground (11 or 20)	ON	Ac. Pedal Fully Depressed	10~14V
						Ac. Pedal Slightly Depressed	0V
						Ac. Pedal Fully Depressed	0V
			14	Ground (11 or 20)	ON	Ac. Pedal Not Depressed	0V
						Ac. Pedal Slightly Depressed	0V
						Ac. Pedal Fully Depressed	10~14V
26 27	Canister V.S.V.		8	Ground (9 or 10)	ON		10~14V
23 35 54	Ignition-Terminal (Power Transistor for Ignition)		Ground (Dry Batt. + Side)	P-Transistor Yellow Harness (Dry Batt. − Side)	ON	Connected ECM Wiring Connector. Keep 5~6 mm Gap in Between the High Tension Cable and Ground. and Then, Disconnect either the Plus or Minus Probe and See if There are Sparks. If Sparks occur, This System is Normal.	
13 44 45	O_2 sensor		27	Ground (11 or 20)	ON	Engine Speed is About 2000 RPM For Approx. 10 Min.	0.3~0.8V

2.6L ENGINE—MALFUNCTION CODES VS FAULTY COMPONENTS AMIGO, PICK-UP, TROOPER AND TROOPER II

	Code No.	12	13	14	15	21	22	23	25	26	29	32	33	34	35	41	43	44	45	51	52	53	54	61	62	63	64	65
Input Sensors	Air Flow Sensor		X														X	X	X					(X)	(X)	X		X
	Coolant Temp. Sensor		X	(X)	(X)							X		X				X	X									
	Oxygen Sensor		(X)															(X)	(X)									
	Crank Angle Sensor	X					X					X				(X)	X	X	X					X	X	X		X
	Idle/WOT Position		X			(X)						X		X			(X)	X	X									(X)
	Starter Switch						(X)									X												
	EGR Gas Temp. Sensor											(X)		(X)														
	Ignition Coil	X					(X)					X				X		X	X									
	Computer																			(X)	(X)			X	X			
	Car Speed Sensor		X				X					X		X					X							(X)		
Output Actuaters	Fuel Injector												(X)															
	Power Switch (Ignition Coil Driver)							X							X								(X)				X	
	V.S.V.; Pressure regulator								(X)													X						

Note:

Code 12 does not indicate malfunction, but checks whether ECM is functioning normally. Listed malfunction codes are only those which show malfunction of sensed parameters.

ECM determines whether a malfunction of that code exists by evaluating the parameters marked "(X)". Parameters marked "X" are used to set the zones in which existence or nonexistence of a malfunction is to be determined.

Parameters marked "X" cannot be main factors, but are used to judge the existance of these relative codes, together with other parameters.

2.6L ENGINE—DIAGNOSTIC CODE DESCRIPTION CHART AMIGO, PICK-UP, TROOPER AND TROOPER II

DIAGNOSTIC CODES

DIAGNOSED ITEM		DIAGNOSED CONTENT: FAULT MODE	DIAGNOSED CONTENT: MICRO-COMPUTER INPUT	CODE NO.		CODE MEMORY	FAIL-SAFE FUNCTION OF MICRO-COMPUTER	ENGINE CONDITION	VEHICLE CONDITION
Engine is not started		—	Engine speed less than 200 rpm	12		No	None	Normal	Normal
O_2 sensor system		Harness open. Sensor deterior-ation	Intermediate voltage	13		Yes	Fuel is not compensated by O_2 signal	Exhaust emission is worsened	No noticeable abnormal operation
Fuel metering system		Incorrect signal (Lean)	Lean signal (Low voltage)	44		Yes			
		Incorrect signal (Rich)	Rich signal (High voltage)	45		Yes			
Water temperature sensor system		Shorted with ground	Insufficient signal	14		Yes	Coolant temperature is assumed to be 85°C (185°F)	· The engine does not operate normally when temper-ature is low and signal is insufficient, or when temperature is high and signal is excessive. Or the engine cannot be started. · The engine does not operate smoothly when temperature is low after the fail-safe function is actuated.	
		Harness open	Excessive signal	15		Yes			
Throttle valve switch system	Both idle contact and full contact	Both idle contact and full make contact simultaneously	Both contacts make simultaneously	21		Yes	Both signals are assumed to be OFF.	Coasting Fuel-cut is not made, after the fail-safe function is actuated, and lean fuel is resulted when the throttle valve is fully opened.	Fuel consumption rate is worsened after the fail-safe function is actuated. Fuel tends to be lean when the throttle valve is fully opened.
	Idle contact	Continuously make contact	Signal is sent continuously Not diagnosed when the air flow sensor system is defective	43		Yes	Assumed to be OFF.	· Fuel cut range appears during running before fail-safe function is actuated. · Coasting Fuel-cut is not made, after the fail-safe function is actuated.	· The vehicle does not run smoothly. · Fuel consumption rate is worsened after the fail-safe function is actuated.
	Full contact	Continuously make contact	Signal is sent continuously	65		Yes	Assumed to be OFF.	· Air-fuel ratio is high when partially loaded. · Fuel tends to be lean when the throttle valve is fully opened after the fail-safe function is actuated.	· Fuel consumption rate and exhaust are worsen-ed. The spark plugs are carboned or the engine stalls depending on the condition. · Fuel tends to be lean when the throttle valve is fully opened after the fail-safe function is actuated.

2.6L ENGINE—DIAGNOSTIC CODE DESCRIPTION CHART AMIGO, PICK-UP, TROOPER AND TROOPER II

DIAGNOSED ITEM	DIAGNOSED CONTENT		CODE NO.		CODE MEMORY	FAIL-SAFE FUNCTION OF MICRO-COMPUTER	ENGINE CONDITION	VEHICLE CONDITION
	FAULT MODE	MICRO-COMPUTER INPUT						
Starter signal system	Harness open or ground shorted	Signal is not input	22		Yes	None	Normal	No abnormality is felt.
Crank angle sensor system	· Harness open · Faulty signal	· No angle sensor signal is input. · Idling speed is lower than the actual speed.	41		Yes	None	· The engine stalls. The engine cannot be started. · Ignition timing may be changed.	· Stalls or cannot be started. · The driveability is affected.
Air flow sensor system	Harness open, shorting with ground, or broken hot wire.	Insufficient signal	61		Yes	Injection pulse is changed over at throttle valve position.	· The engine stalls depending on the condition. · The engine may be slightly unstable after the fail-safe function is actuated.	· The engine stalls depending on the condition. · The accelerator response is worsened after the fail-safe function is actuated. The engine may not operate smoothly.
	Broken cold wire.	Excessive signal	62		Yes			
Car speed sensor system	· Harness open or ground shorted. · Car speed sensor broken.	No signal is input (Not diagnosed with the air flow sensor system is defective)	63		Yes	None	The fuel cut operates even when running under a low speed.	Does not run smoothly and shakes.
Micro-computer unit	Abnormal LSI (1)	–	51		Yes	· Injection pulse is changed over at throttle valve position · Fixed ignition timing.	· In the worst case, the engine stalls, or the engine does not operate smoothly at a certain time. · The engine does not stall but it does not satisfy the specification.	· In the worst case, the engine stalls, or the engine does not operate smoothly at a certain time. · The exhaust and running performance deviate from the specification.
	Abnormal LSI (2)	–	52		Yes			
Power transistor system for ignition	Output terminal shorted with ground.	–	23		Yes	None	· The engine stalls. · Cannot be started.	· The engine stalls. · Cannot be started.
	Harness open.	–	35		Yes			
	Defective transistor or grounding system.	–	54		Yes			

DIAGNOSED ITEM	DIAGNOSED CONTENT		CODE NO.		CODE MEMORY	FAIL-SAFE FUNCTION OF MICRO-COMPUTER	ENGINE CONDITION	VEHICLE CONDITION
	FAULT MODE	MICRO-COMPUTER INPUT						
Vacuum switching valve system for canister purge	· Open circuit harness connector VSV or inside of ECM	–	26		Yes	None	Air/Fuel ratio tends to be rich under the idle or partial conditions before warmed up	Fuel consumption rate is worsened when engine is cold.
	· Ground shorted harness connector VSV or inside of ECM	–					Air/Fuel ratio tends to be lean at hot restart	The engine can not be restarted smoothly when coolant temp. is very high.
	· Driver transister in ECM open	–	27		Yes			
Vacuum switching valve system for pressure regulator	· Open circuit harness connector VSV or inside of ECM		25		Yes	None	Air/Fuel ratio tends to be rich under the idle or partial conditions before warmed up	Fuel consumption rate is worsened when engine is cold.
	· Ground shorted harness connector VSV or inside of ECM						Air/Fuel ratio tends to be lean at hot restart	The engine can not be restarted smoothly when coolant temp. is very high.
	· Driver transister in ECM open	–	53		Yes			
Fuel injector system.	· Open circuit harness connector injector, dropping resister, or inside of ECM	–	33		Yes	None	Air/Fuel ratio tends to be rich under the idle or partial conditions before warmed up	Fuel consumption rate is worsened when engine is cold.
	· Ground shorted harness connector injector, dropping resister, or inside of ECM	–					Air/Fuel ratio tends to be lean at hot restart	The engine can not be restarted smoothly when coolant temp. is very high.
	· Driver transister in ECM open	–	64		Yes			
EGR temperature sensor system	Mulfunction	–	32		Yes	None	Exhaust is worsened	No noticeable abnormal operation
	EGR temperature sensor or harness NG	–	34		Yes	None	Normal	No abnormality is felt.

2.6L ENGINE – PRELIMINARY CHECKS
AMIGO, PICK-UP, TROOPER AND TROOPER II

BEFORE STARTING

1. Verify the customer complaint and locate the correct symptoms.

2. Several of the symptom procedures below call for a careful visual check. This check should include:
   - Vacuum hoses for splits, kinks, and proper connections, as shown on Emission Control Information label.
   - Air leaks at throttle valve mounting, common chamber mounting and intake manifold.
   - Ignition wires for cracking, harness, proper routing and carbon tracking.
   - Wiring for proper connections, pinches, and cuts.
     The importance of this step cannot be stressed too strongly — it can lead to correcting a problem without further checks and can save valuable time.

3. Be careful to check the system components (Closed Loop Emission Control System), especially handling the electrical parts, ECM etc..

   - When checking the ECM harness connector terminals from the connector wire side with the tester probe, be careful not to touch the other terminal by mistake.
     Do not insert the circuit tester test probes into the connector open side to test the continuity.
   - Before starting with a voltmeter connected to a suspected circuit, check to see if the battery is in full charged condition.
   - Before disconnecting the ECM connectors, check the ignition switch "OFF" condition, and check and record the stored memory in ECM.
     When disconnecting the ECM connectors, all three connectors (10P, 18P, 14P) should be done, and note that all stored codes also be cleared at the same time.
   - To avoid the connector terminals being damaged or touching the wrong terminal, short jumper cables which are staked terminals are useful.
   - When connecting a suspected circuit to the Battery + terminal or Engine ground, it will be advisable to install a test light (3.2W, 12V) in between, to protect electrical parts overcharged or damaged.

INTERMITTENTS CHECK PROCEDURES

Problem may or may not turn "on" the "CHECK ENGINE" light, or store a code.

"INTERMITTENTS" means that "CHECK ENGINE" light comes on at times, but does not stay on.

If the trouble is found "INTERMITTENTS" when using the "Trouble Code Chart", refer to this column to locate the problem.

If a fault is intermittent, use of Trouble Code Charts may result in replacement of good parts.

- Most intermittent problems are caused by faulty electrical connections or wiring. Perform careful check as described below.
- Poor mating of the connector halves, or terminals not fully seated in the connector body (backed out).
- Improperly formed or damaged terminals. All connector terminals in problem circuit should be carefully reformed to increase contact tension.
- Poor terminal to wire connection. This requires removing the terminal from the connector body to check.
- Poor ground connections.
- If a visual check does not find the cause of the problem, the car can be driven with a voltmeter connected to a suspected circuit. An abnormal voltage reading when the problem occurs indicates the problem may be in that circuit.
  An intermittent "CHECK ENGINE" light with no stored code may be caused by;
- Ignition coil to ground and arcing at spark plug wires or plugs.
- Main relay which is in power supply circuit is faulty.
- Poor matings of power supply circuits connections.
- "CHECK ENGINE" light wire to ECM shorted to ground (circuit (22)).
- Diagnostic "Test" Terminal wire to ECM, shorted to ground (circuit (24)).
- ECM power grounds are poor connected to Intake maniflold. See ECM wiring diagrams.
- Check for an electrical system interference caused by a defective relay, ECM driven solenoid, or switch. They can cause a sharp electrical surge. Normally, the problem will occur when the faulty component is operated.
- Check for improper installation of electrical options, such as lights, etc.

DIAGNOSTIC CIRCUIT CHECK

The Diagnostic Circuit Check is an organized approach for identifying a problem caused by the Fuel Injection System.

Driver comments normally fall into one of the following areas:

- Steady "CHECK ENGINE" light with engine running
- Driveability Problem
- Engine "Cranks But Will Not Run"

Understanding the chart and using it correctly will reduce diagnosis time and prevent the unneccessary replacement of parts.

1. A steady "CHECK ENGINE" light with the ignition "ON" and engine stopped confirms battery and ignition voltage to the Electronic Control Module (ECM).
2. Connect "Trouble Code Test" terminals with each other, near ECM connector.
   The ECM will cause the "CHECK ENGINE" light to flash Code 12, indicating that the ECM diagnostics are working. Code 12 will flash three (3) times, followed by any other trouble codes stored in the memory. Each additional code will flash three (3) times, starting with the lowest code, and then start over again with Code 12. If there are no other codes, Code 12 will flash until the "Trouble Code Test" terminals are disconnected or the engines is started.
3. Record all stored codes except for Code 12.
4. If no additional codes were recorded, see driveability symptoms and recommended service procedures.
5. Road test of the system should be done, under the safety conditions and traffic rules.
   If available, it will be better to use a chassis dynamometer.
6. Clearing codes. Ignition off; remove fuse No. 3 (10A) located in the lower-left part of the Instrument panel or Main fuse (60A) in the fuse and relay box of Engine room or remove Battery – terminal, for at least 10 seconds.

DIAGNOSTIC CIRCUIT CHECK
(for 4ZE1)

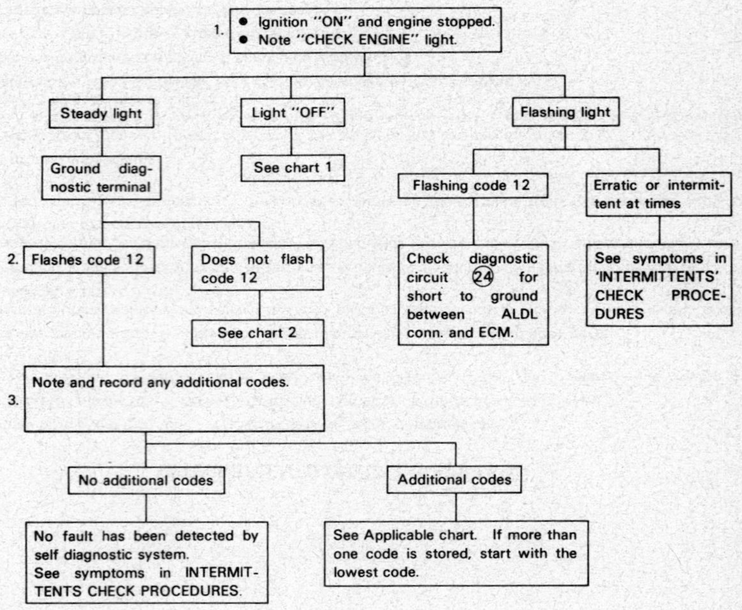

CHART 1

NO "CHECK ENGINE" LIGHT

4ZE1 Fuel Injection

There should always be a steady "CHECK ENGINE" Light when the ignition is "on" and engine stopped. Battery voltage is supplied directly to the light bulb. The Electronic Control Module (ECM) will control the light and turn it on by providing a ground path through circuit (22) to the ECM.

Engine runs ok, check:

- Faulty light bulb.
- Circuit (22) open.
- Gage fuse blown. This will result in no meter lights, turn signal lights or reminder buzzer etc.

Engine cranks but will not run.

- Continuous battery - fuse or main fuse open.
- ECM ignition fuse open.
- Battery circuit (12), (21) to ECM open.
- Ignition circuit (5) to ECM open.
- Poor connection to ECM.

1. Solenoids and relays are turned "ON" and "OFF" by the ECM, using internal electronic switches called "drivers". Solenoid and relay coil resistance must measure more than 20 ohms. Less resistance will cause early failure of the ECM "driver".

   Failure of outer load such as shorted mode can damage the ECM driver.

   Before replacing ECM, be sure to check the coil resistance of all solenoids and relays controlled by the ECM. See ECM wiring diagram for the solenoid(s) and relay(s) and the coil terminal identification.

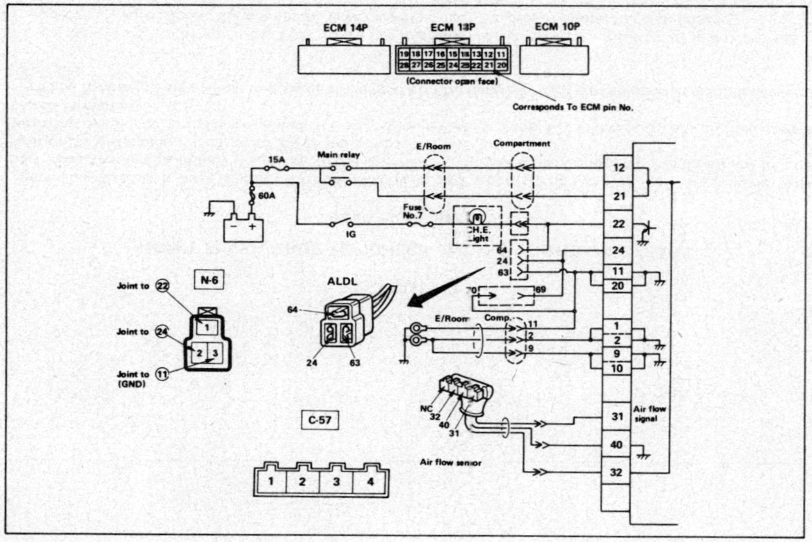

CHART 1

NO "CHECK ENGINE" LIGHT

4ZE1 Fuel Injection

Disconnect Air flow sensor connector.
Uncover waterproof rubber boot of the connector.
Key "ON", engine stopped.
Check the voltage of the terminal (32) on Air flow sensor connector.
It should be 10 ~ 14V (Battery voltage).

OK

- Key "OFF"
- Disconnect all ECM connectors
- Key "ON"
- ALDL Terminal of circuit (64) with a test light to ground and note "CHECK ENGINE" light

"CHECK ENGINE" light on

1. • Check ECM ground circuits. If ECM ground circuits are OK, it is poor connection at ECM or faulty ECM. See note.**

"CHECK ENGINE" light off

Check;
- Faulty bulb
- Gage fuse
- Open circuit (22)
- Open ignition feed

Not OK

Check power supply circuit (12), (21) (from Battery to ECM) including 15A fuse.

Note: **

Before replacing ECM, use an ohmmeter and check resistance of each ECM controlled relay and solenoid coil.
See ECM wiring diagram for coil terminal identification for solenoid(s) and relay(s) to be checked.
Replace any relay or solenoid if the coil resistance measures less than 20 ohms.

CHART 2

WON'T FLASH CODE 12/"CHECK ENGINE" LIGHT ON STEADY

4ZE1 Fuel Injection

There should always be a steady "CHECK ENGINE" Light when the ignition is "on" and engine stopped. Battery ignition voltage is supplied to the light bulb. The Electronic Control Module (ECM) will turn the light on by grounding circuit (64) or (22) at the ECM.
With the diagnostic terminal grounded, the light should flash a Code 12, followed by any trouble code(s) stored in memory.
A steady light suggests a short to ground in the light control circuit (64) or (22) or an open in diagnostic circuit (24) or (69).

1. If the light goes off when the ECM connector is disconnected, then circuit (64) or (22) is not shorted to ground. Also, check the connector terminals physically for proper contact at this time.
2. This step will check for an open diagnostic circuit (24) or (69).
3. At this point the "CHECK ENGINE" light wiring is okay. The problem is a faulty ECM connector or ECM.

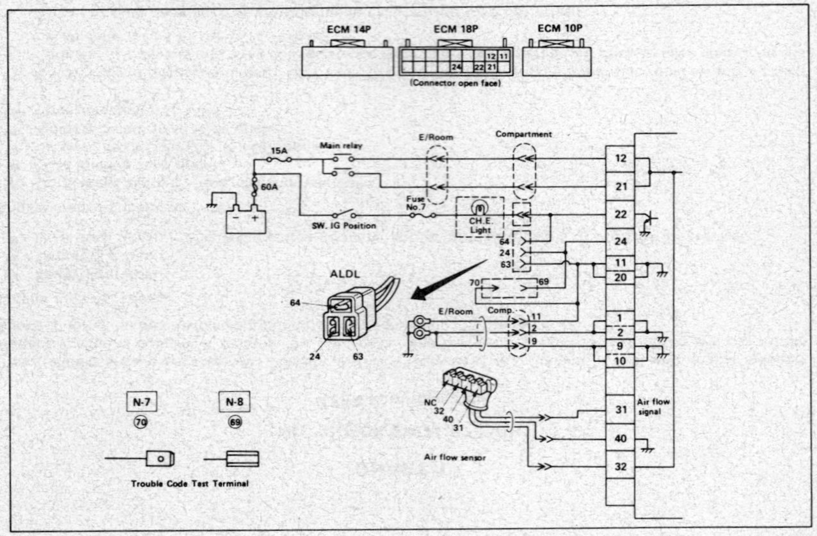

CHART 2

WON'T FLASH CODE 12 "CHECK ENIGNE" LIGHT ON STEADY

4ZE1 Fuel Injection

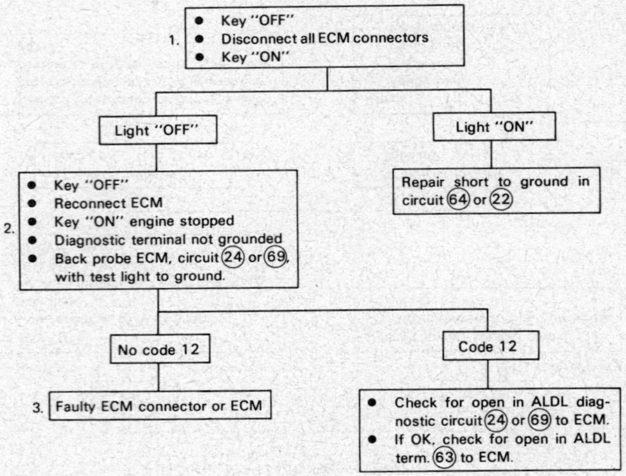

TROUBLE SHOOTING PROCEDURE

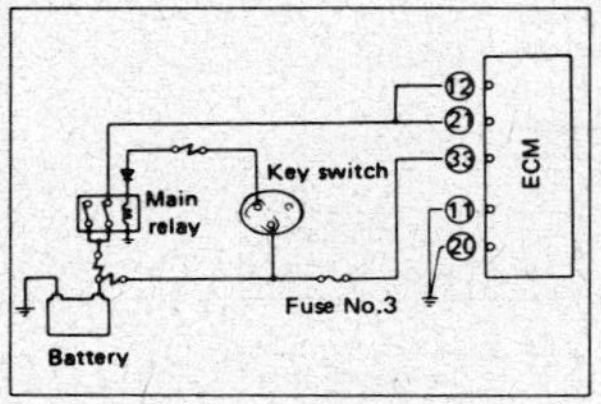

Checking ECM Unit Power Supply Circuit

Confirmation

1. Disconnect ECM unit harnesses at connectors.
2. Turn key switch to ON position.
3. Check that power voltage is present at connector terminals (33), (12) or (21).
4. Check that continuity exists between the connector terminals (11), (20) and ground.

Inspection and Correction

1. When power voltage is not present at terminals (33) and (12) or (21).
   Check the main relay, starter switch, fuses, fusible link wires and related circuits and correct or replace the parts as necessary.
2. If the circuit tester does not indicate continuity across the terminals (11), (20) and ground, correct or replace grounding cable.
3. Connect the ECM unit harnesses at connector properly.

CODE NO. 12; ECM DIAGNOSTICS WORKING

Signal Code No. 12

This code No. 12 is always displayed with the key "ON" and the engine not running and it only means that the engine is not running. This code No. 12 is not stored in memory.

Note: If the signal code No. 12 only is displayed when the function key is operated after clearing the memory, it indicates that the ECM electrical circuits and self diagnosis system are functioning normally.

CODE NO. 13, 44 and 45; NO SIGNAL OR ABNORMAL SIGNAL FROM OXYGEN SENSOR

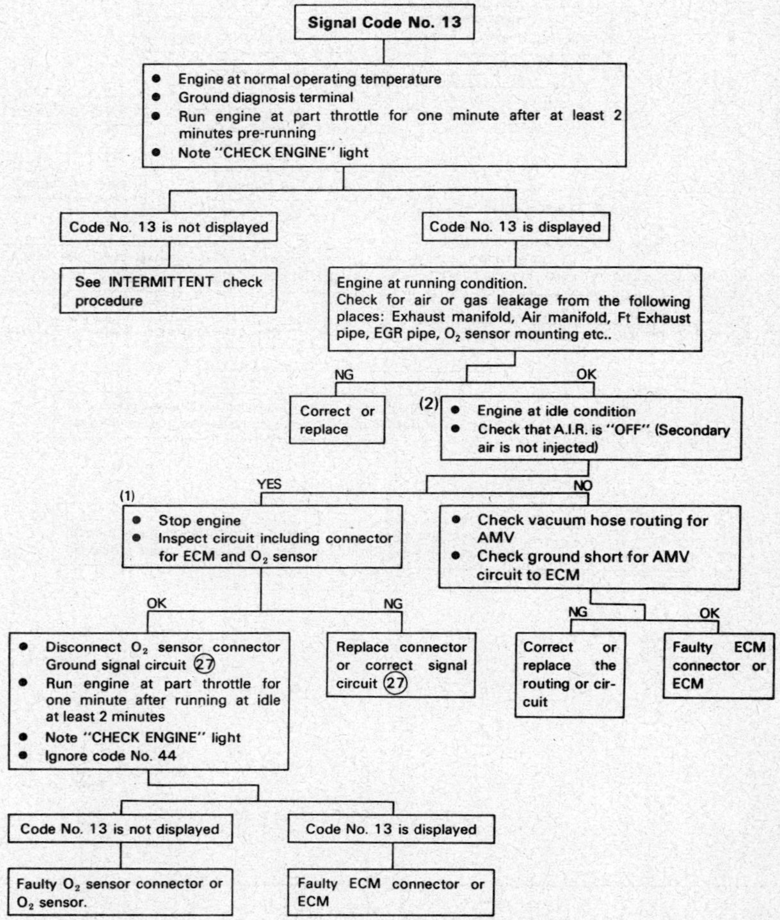

CODE NO. 44, and 45; FUEL METERING FAULTY (LEAN OR RICH)

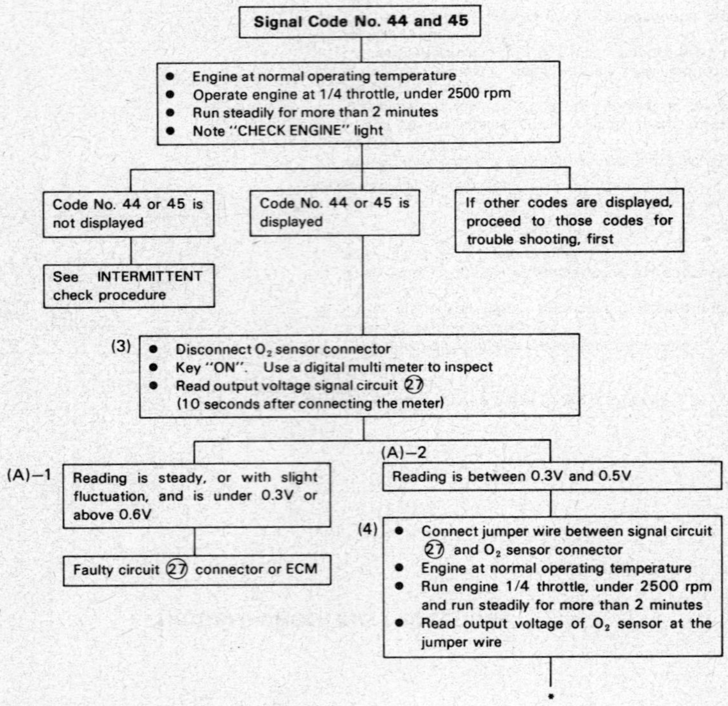

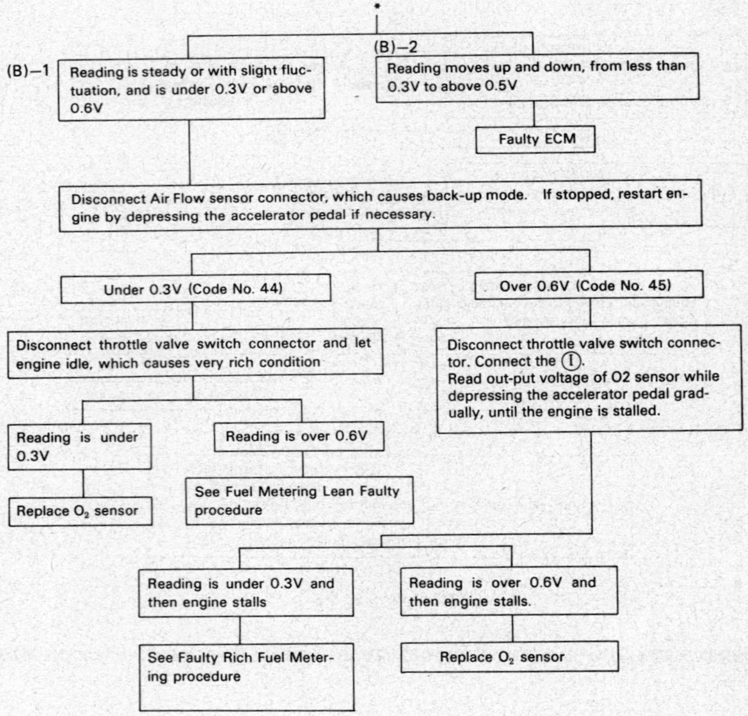

FAULTY RICH FUEL METERING INSPECTION PROCEDURE

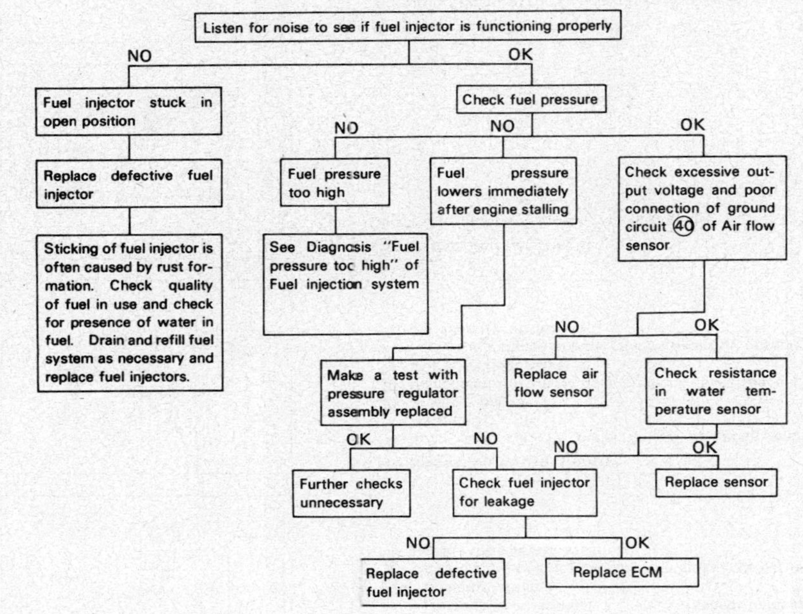

FAULTY LEAN FUEL METERING INSPECTION PROCEDURE

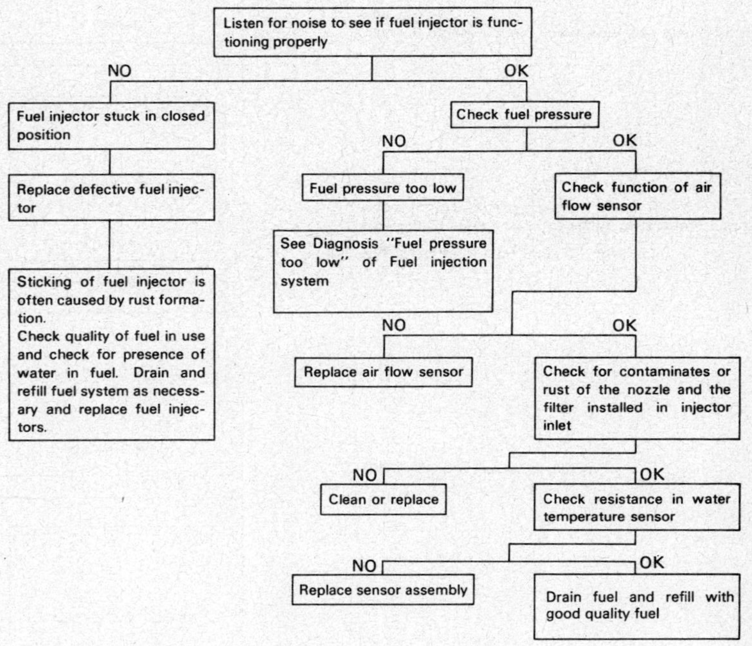

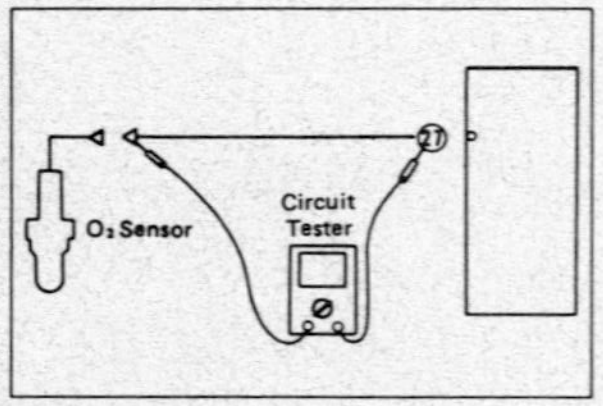

Inspection for Signal Code No. 13, 44 and 45

1. Circuit Continuity Check

1) Check that continuity exists between the ECM harness connector terminal 27 and O_2 sensor harness connector terminal.
2) If the tester does not indicate continuity, correct or replace the harness assembly.

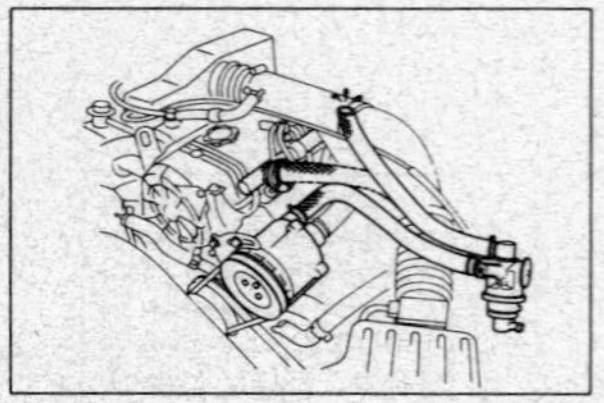

2. Inspection for AIR Control

1) Keep engine at idle, under normal operating temperature.
2) Stop the engine.
3) Disconnect the rubber hose at check valve side, and plug the check valve.
4) Restart the engine and let idle. Immediately observe the direction of the air.

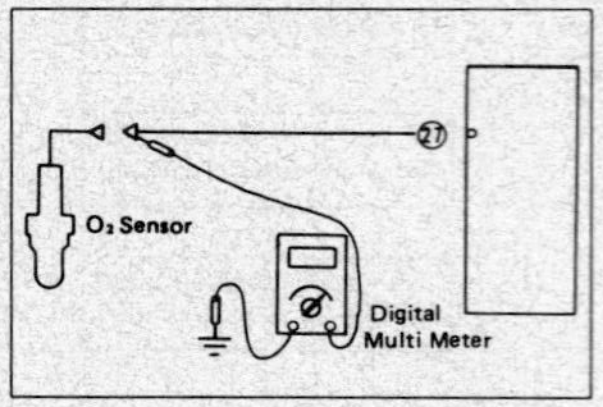

3. Output Voltage Check

Check the output voltage of the signal circuit 27 .

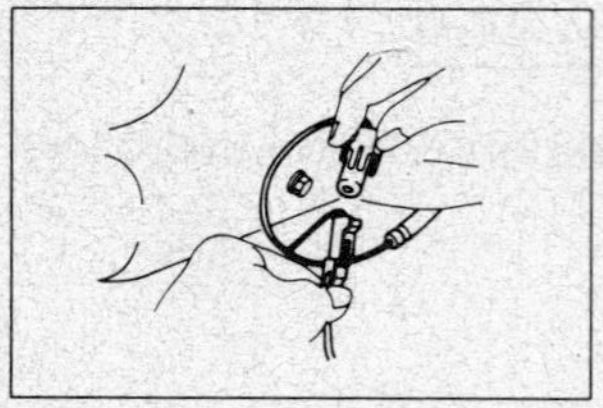

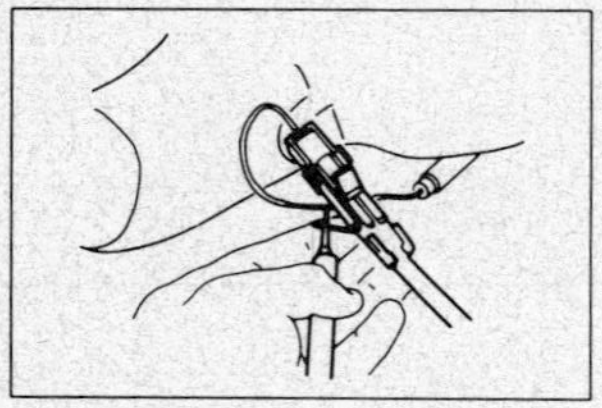

4. Inspection of O_2 Sensor

The output voltage of the O_2 sensor can be checked with a voltmeter having a minimum input resistance of one megohm.

1) Disconnect the O_2 sensor from the vehicle harness.
2) Insert a jumper with exposed wire between the sensor and vehicle harness connectors.
3) Set the voltmeter to the approximately two volt range.
4) Connect the positive (+) lead of the voltmeter to the jumper and the negative (–) lead to vehicle ground.
5) Start the engine and run at elevated speed until the emission system has gone "closed loop". The meter should move between approximately 0.50 and 0.80 volts.

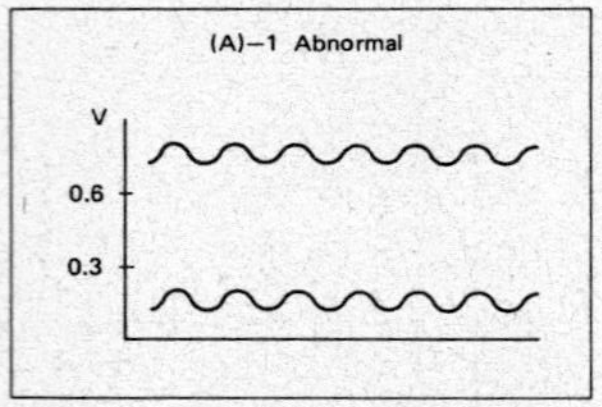

(A) Reading output voltage of circuit (27) and ECM.

Reading is steady or with slight fluctuation, and is under 0.3V or above 0.6V.

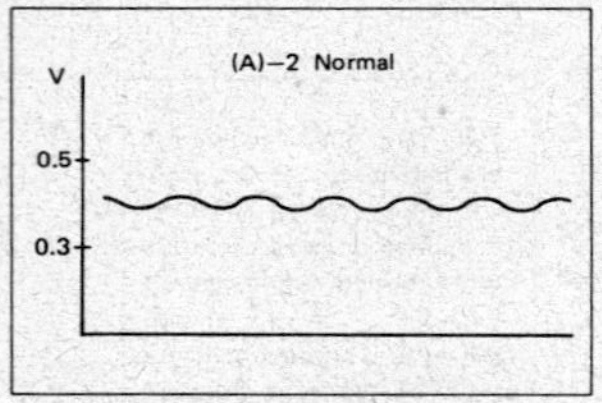

Reading is between 0.3V to 0.5V.

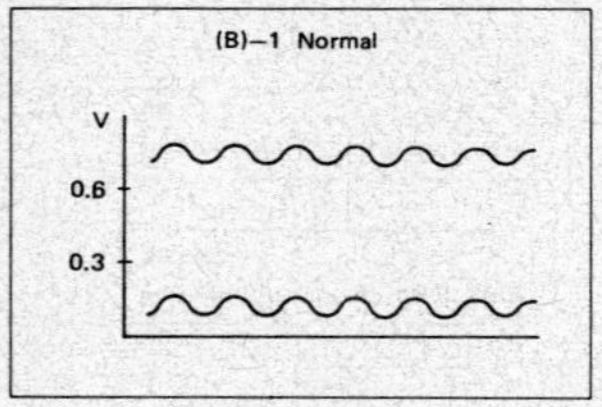

(B) Reading output voltage of O_2 sensor,

Reading is steady or with slight fluctuation, and is under 0.3V or above 0.6V.

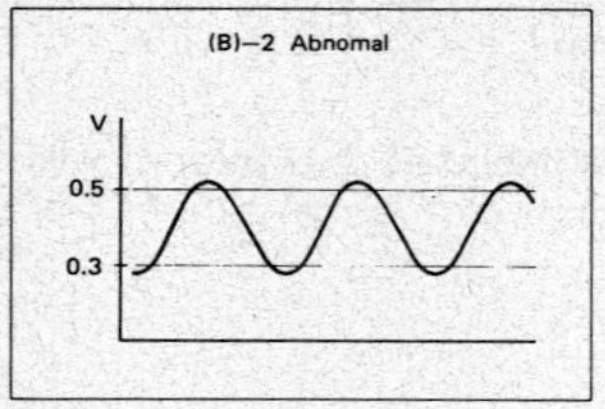

Reading moves up and down from less than 0.3V to above 0.5V.

CODE NO. 14 and 15; ABNORMAL SIGNAL FROM COOLANT TEMPERATURE SENSOR

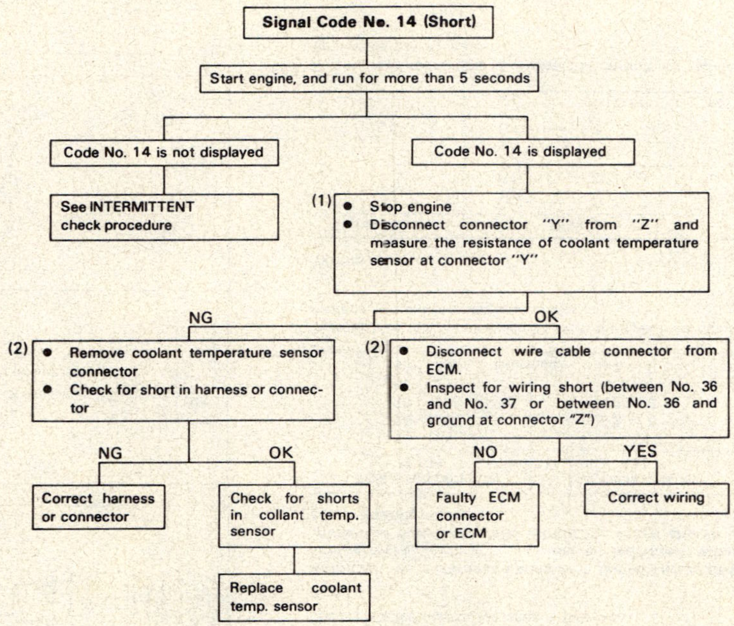

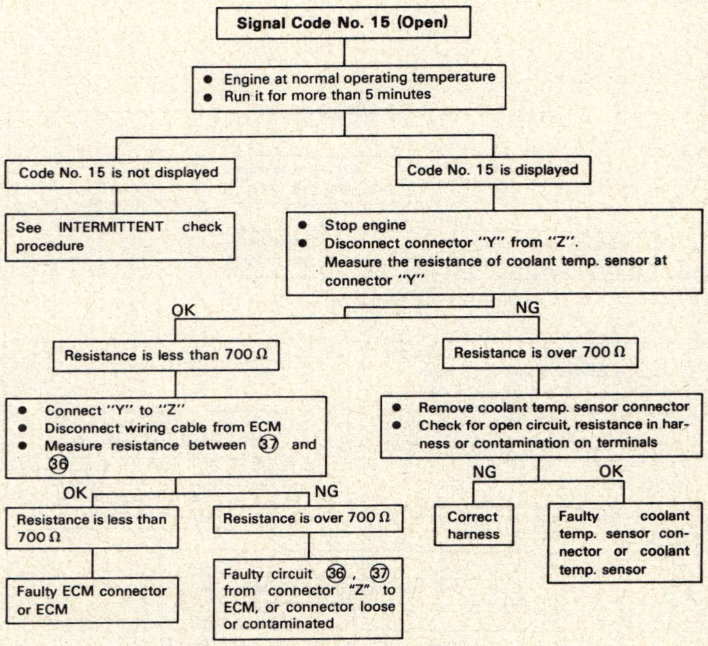

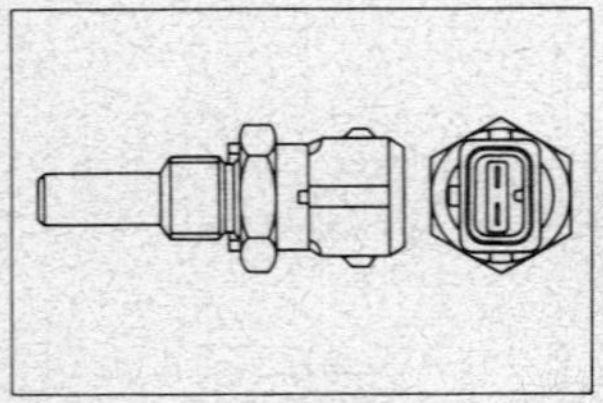

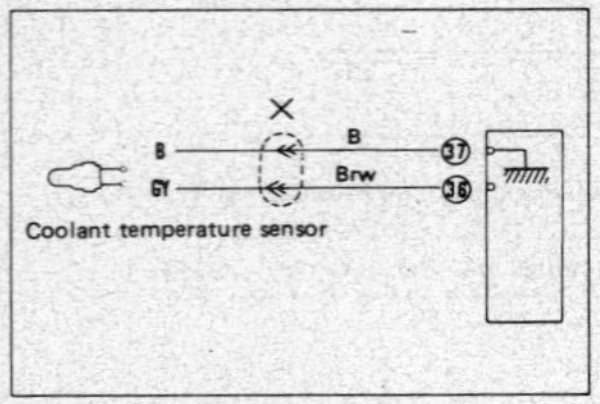

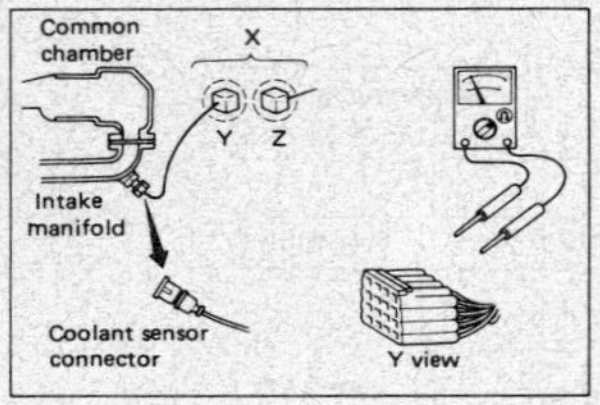

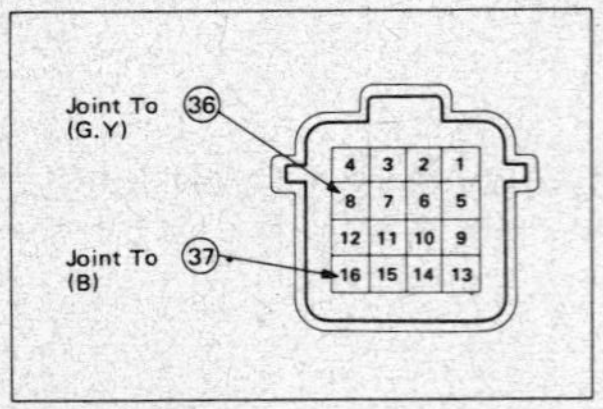

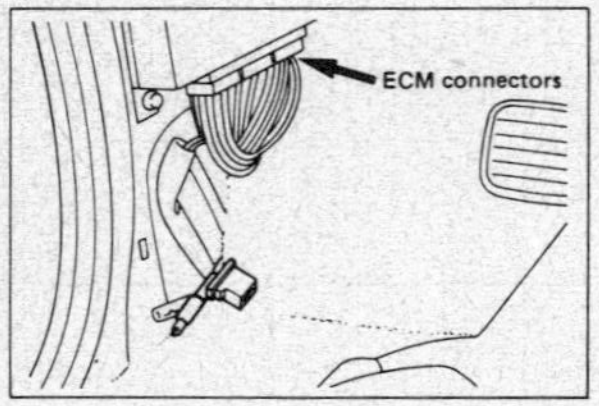

Inspection for Signal Code No. 14, 15

Measurement of Resistance in Sensor

Disconnect sensor harnesses at connector and measure the resistance across the sensor terminals.

Note:

Although the resistance varies with the engine coolant temperature, normal condition is indicated when measured resistance falls within the range shown in the following table:

Coolant temperature °C(°F)	Normal resistance KΩ
−10 (14)	7 — 12
10 (50)	3 — 5
20 (68)	2 — 3
50 (122)	0.7 — 1
80 (176) (warmed)	0.2 — 0.4

If measured resistance deviates from the specified range, replace coolant temperature sensor.

Wiring

Signal code No.	Wiring	Ω
14	(36) —— B	∞
	(36) —— Ground	∞
15	(36) —— GY	0
	(37) —— B	0
	(37) —— Ground	∞

If abnormal condition is indicated, correct or replace applicable harnesses.

CODE NO. 21; THROTTLE VALVE SWITCH FAULTY

Signal Code No. 21 (Idle "ON" and/or Full "ON")

- Run engine at idle for more than 5 sec.

Code No. 21 is displayed (Faulty Full sw circuit)
→ See THROTTLE VALVE SW. inspection procedure

Code No. 21 is not displayed
→
- Drive on highway appropriate for the following test.
- Keep accelerator pedal fully depressed at about 40 MPH for 5 sec.
  Be careful not to exceed the speed limit of the road or over-rev the engine.

→ Code No. 21 is not displayed
→ See INTERMITTENT check procedure

→ Code No. 21 is displayed (Faulty Idle sw circuit)
→ See THROTTLE VALVE SW. inspection procedure

2.6L ENGINE—TROUBLE CODE CHARTS—AMIGO, PICK-UP, TROOPER AND TROOPER II

CODE NO. 43 and 65; THROTTLE VALVE SWITCH SYSTEM FAULTY

Signal Code No. 43 and 65

- Check code No. 61, 62 are displayed simultaneously

NO:
- Code No. 43 was displayed
  - Drive on highway appropriate for the following test.
  - Keep accelerator pedal fully depressed at about 40 MPH for 5 sec.
  - Be careful not to exceed the speed limit of the road or over rev the engine.
- Code No. 65 was displayed
  - Run at Idle more than 5 sec.

YES:
- Before reconfirming that code No. 43 and 65 are displayed, repair the system for diagnosis codes No. 61 and 62.

Recheck if code No. 43 or 65 is displayed

NO: See INTERMITTENT check procedure

YES: See THROTTLE VALVE SW inspection procedure

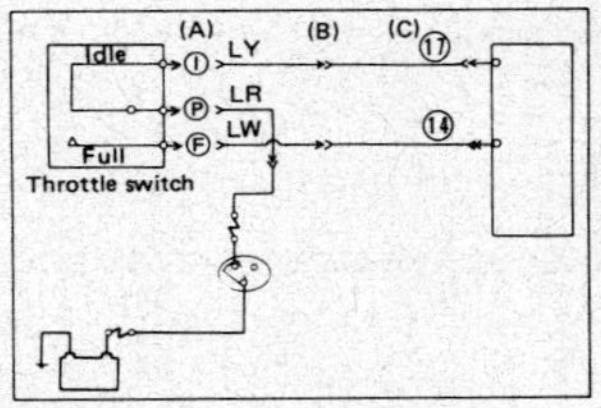

Circuits

When throttle switch is functioning properly.

Signal code No.	Wiring	Ω
21, 43	(14) —— LW	∞
21, 65	(17) —— LY	∞

If circuit is found to be at fault, correct or replace applicable harness assembly.

THROTTLE VALVE SWITCH CIRCUIT INSPECTION PROCEDURE

Key "OFF"

Disconnect (A) at throttle valve switch connector. Check switch is working. See Fig. 17, 18

NG: Replace throttle valve switch

OK: Key "ON"
Measure voltage of (I) or (F) against the ground at (A) of harness side. See Fig. 17, 18

- Voltage of (I) or (F) is less than 1V
  - Run engine more than 5 sec at idle or partial
  - Note "CHECK ENGINE" light
    - Code No. 21 is displayed → Faulty ECM
    - Code No. 21 is not displayed →
      - Poor connection at connector (A)
      - Clean both connector of throttle valve sw and the harness
- Voltage of (I) or (F) is over 1V
  - Disconnect (C)
    Check insulation between (P) and (I) or (F) in the harness from connector (A) to (C)
    - NG: Correct circuit (I) or (F)
    - OK: Faulty ECM connector or ECM

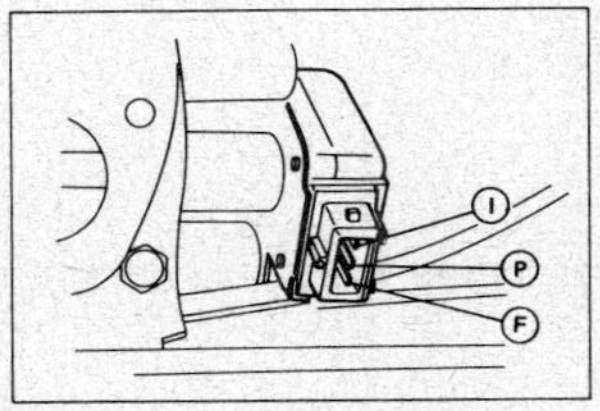

Throttle Switch

Refer to "Throttle switch" for throttle valve inspection procedure.

Accelerator pedal position	Continuity I - P	Continuity P - F
Not depressed	0 Ω	∞ Ω
Slightly depressed	∞ Ω	∞ Ω
Fully depressed	∞ Ω	0 Ω

Note:

Trouble-shooting will not be performed for troubles indicated by the signal code No. 43 and 65 when air flow sensor circuit is malfunctioning.

CODE NO. 22; NO SIGNAL FROM THE STARTER

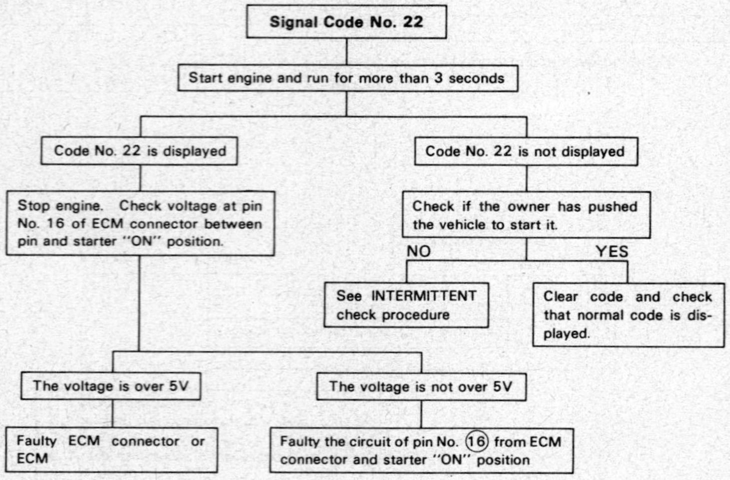

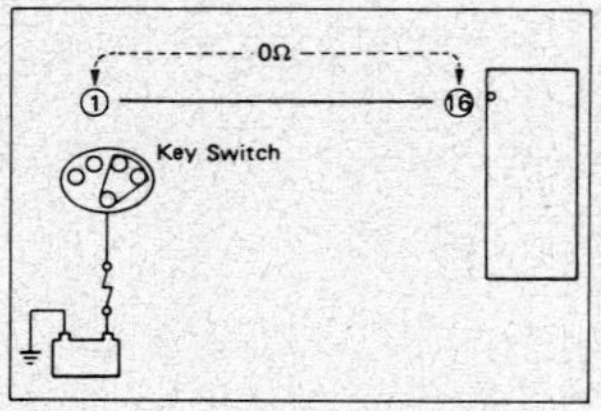

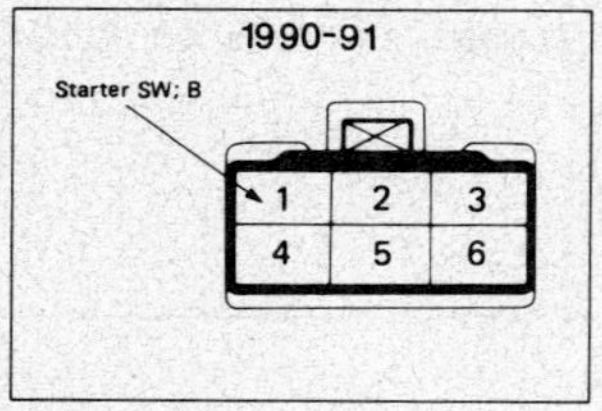

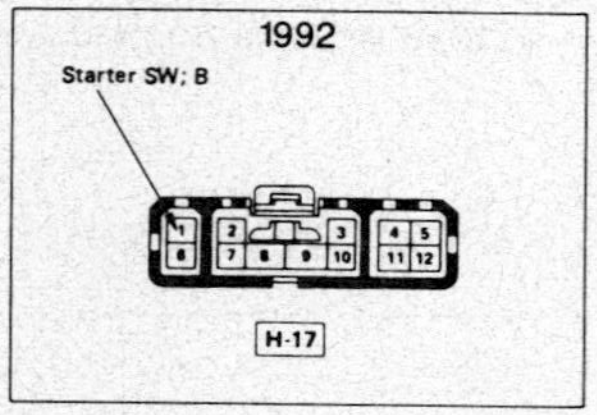

Signal Code No. 22 Inspection

Circuit

When the circuit is found to be at fault, correct or replace applicable harness assembly.

CODE NO. 41; CRANK ANGLE SENSOR SYSTEM HARNESS OPEN

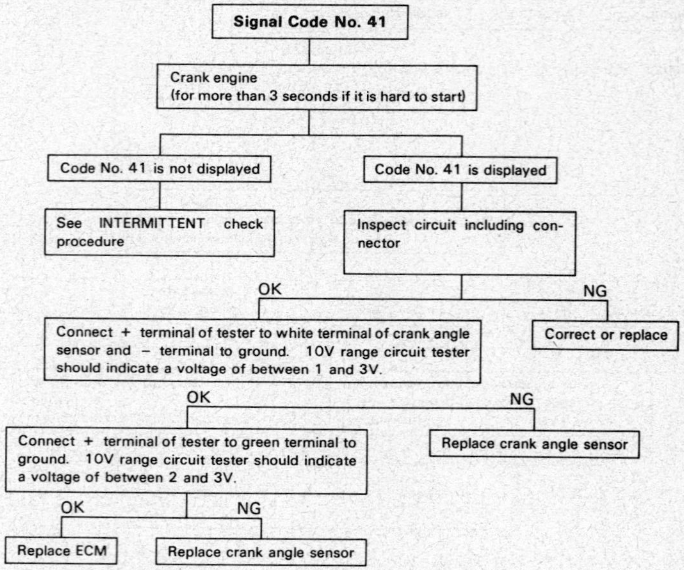

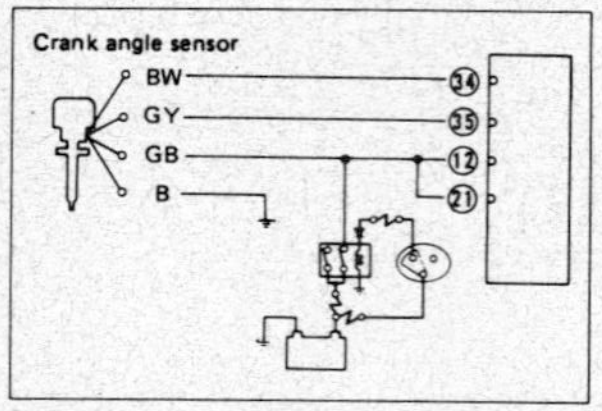

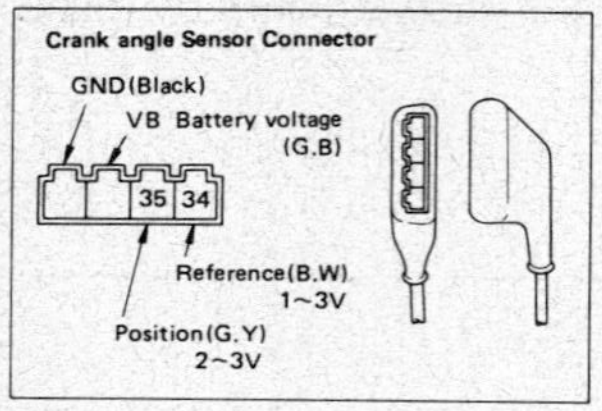

Inspection for Signal Code No. 22 and 41

Circuit

1. Disconnect crank angle sensor harnesses at crank angle sensor connector.
2. Disconnect all connectors at ECM.
3. Turn ignition on.

Wiring	Ω	V
(34) —— BW	0	—
(35) —— GY	0	—
GB —— Ground	—	10 — 14
B —— Ground	0	—

4. Measure resistances and voltage shown in chart above.
   When circuit is found to be at fault, correct or replace applicable harness assembly.

When circuit is found to be in normal condition, replace distributor assembly.

2.6L ENGINE—TROUBLE CODE CHARTS—AMIGO, PICK-UP, TROOPER AND TROOPER II

CODE NO. 61, and 62; AIR FLOW SENSOR SYSTEM FAULTY

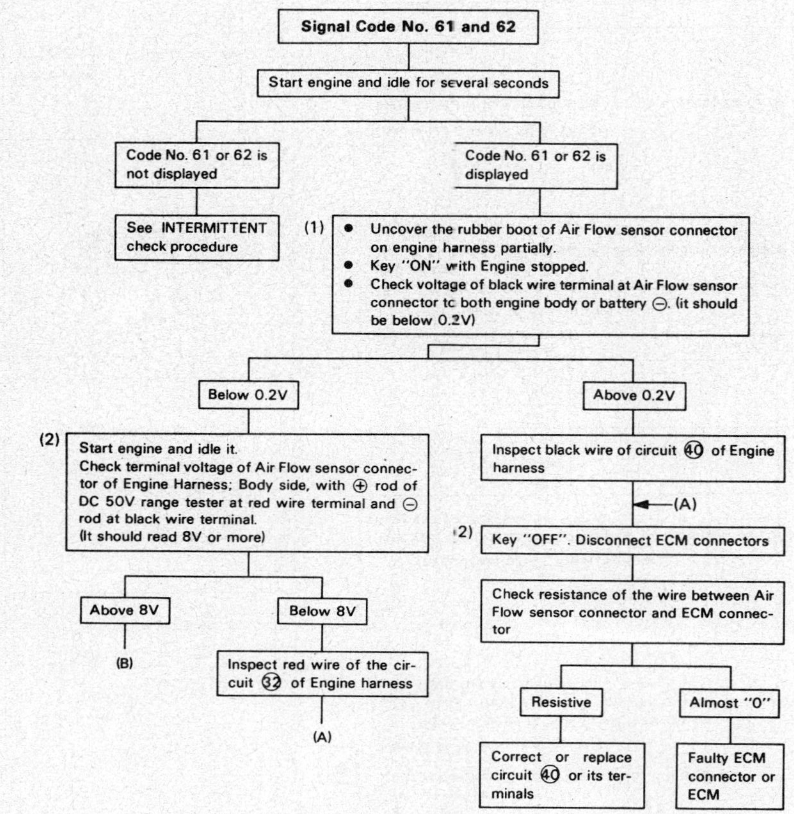

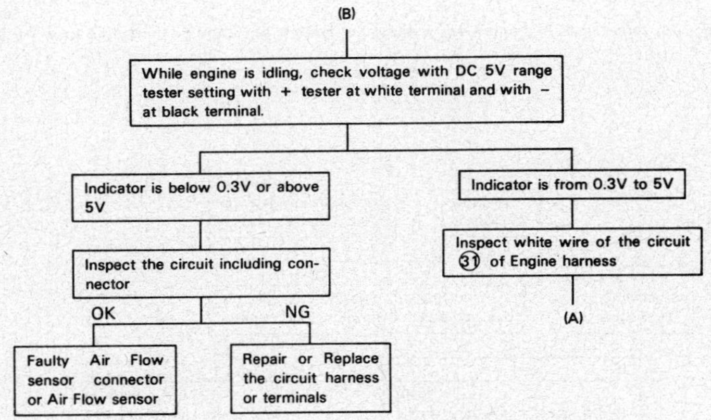

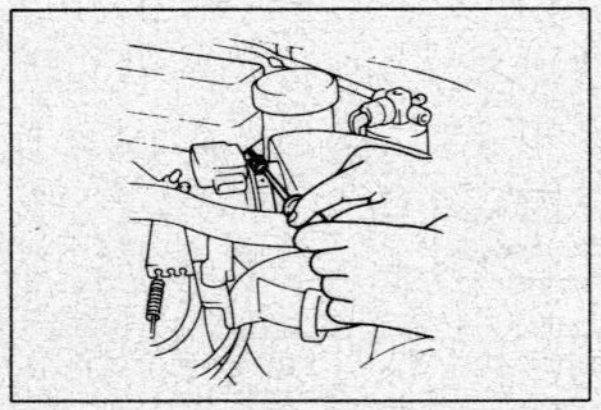

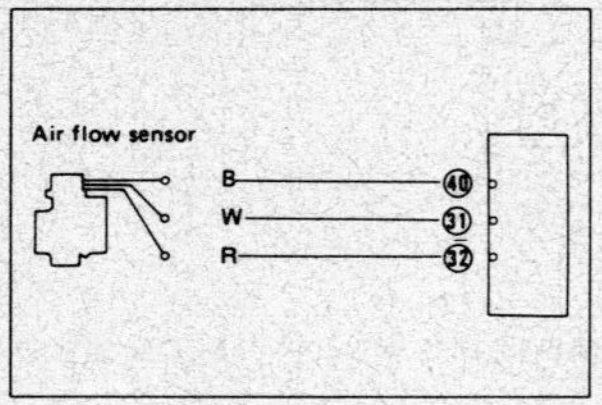

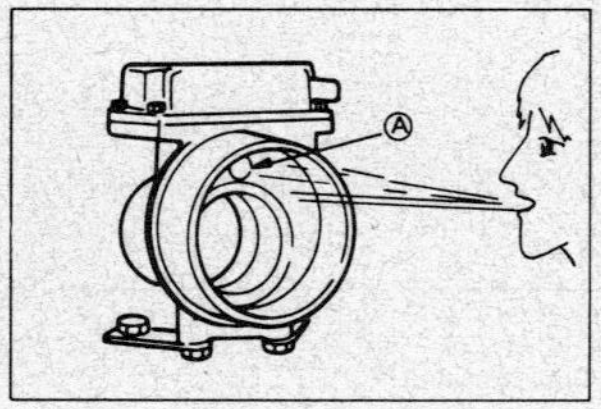

Inspection for Signal Code No. 61 and 62

Removal of Air Flow Sensor

1. Remove bolts attaching air flow sensor.
2. Loosen the clamp bolt and remove air flow sensor from the air cleaner.

1. Inspection of Air Flow Sensor

Uncover the rubber boot on the air flow sensor side of the harness connector.

Wiring	Ignition switch	Condition	V	Ω
A. Red-Ground	ON	—	10—14	—
	ON	—	Less than 0.2	—
B. White-Ground	ON	When breathing	0.5—1.5	—
C. Black-Ground*	ON/OFF	—	Less than 0.2	0

* Ground means engine ground.

When A. and C. are abnormal, check circuit and correct as necessary.
When A. and C. are normal but B. is abnormal, replace air flow sensor.

Inspection of Air Flow Sensor

Install the air flow sensor assembly in the reverse order of removal.

2. Circuits

Wiring	Ω
W —— (31)	0
W —— Ground	∞
R —— (32)	0
R —— Ground	∞
B —— (40)	0

When circuit is found to be at fault, correct or replace applicable harness assembly.

Note:

If the air flow sensor has been removed, check by-pass circuit (A) for contamination and clean with carburetor cleaner as necessary.

CODE NO. 63; CAR SPEED SENSOR FALUTY

Signal Code No. 63

- Code No. 22 or 65 is displayed at the same time
  - Do trouble shooting for code No. 22 or 65, first
- Code No. 22 or 65 is not displayed at the same time
  - Drive on highway appropriate for the following test.
  - Keep accelerator pedal fully depressed at about 40 MPH, for 5 seconds.
  - Be careful not to exceed the speed limit or over-rev the engine.
    - Code No. 63 is not displayed
      - See INTERMITTENT check procedure
    - Code No. 63 is displayed
      - (1) Inspect Car speed sensor equipment
        - NG: (2) Inspect circuit including connector and Auto Cruise Controller if equipped
          - OK: (3) Inspect car speed sensor. Replace sensor if necessary
          - NG: Correct or Replace
        - OK: Faulty ECM connector or ECM

Note:

This code No. may be displayed even the system is normal if the vehicle is used under the following conditions.

A/T; **In "D" range, while depressing the brake pedal, the accel pedal is depressed over 1/2 throttle for more than 3 sec.**

M/T; **Under severe climbing conditions, the vehicle is kept stopped while the accel pedal is depressed over 1/2 throttle, with the clutch engaged partially, for more than 3 sec.**

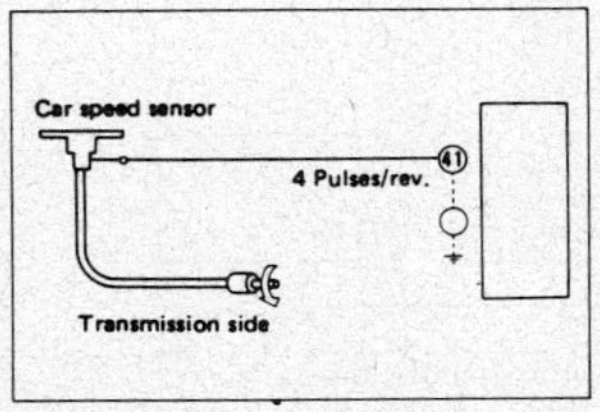

Inspection for Signal Code No. 63

1. Sensor and Equipment

1) Connect a circuit tester between the control unit harness connector terminal (41) and ground.
2) Disconnect speedometer cable at transmission end and turn the inner cable slowly.
3) If the tester indicates continuity and open circuit condition alternately as the cable is turned, the car speed sensor and circuit are operating normally.

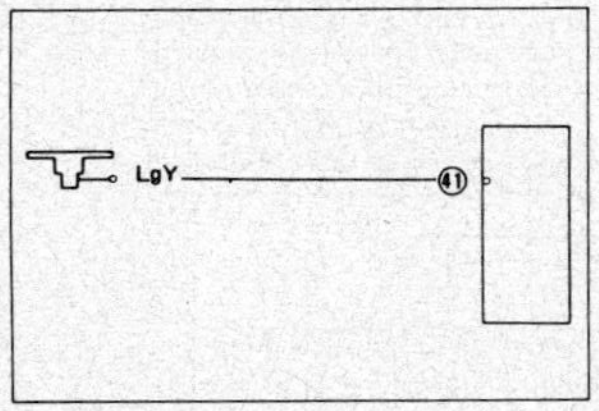

2. Circuits

Wiring	Ω
LgY ——— (41)	0
B ——— Ground	∞

When circuit is found to be at fault, correct or replace applicable harness assembly.

Install the speedometer assembly.

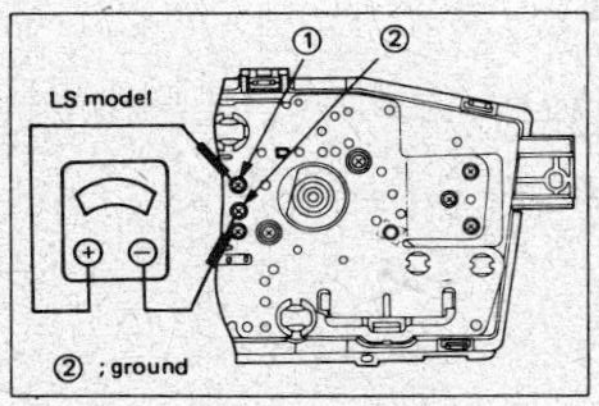

3. Car Speed Sensor (Incorporated in the Speedometer)

1) Remove the meter assembly.
2) Connect the leads of a circuit tester across the terminals (①, ②) at rear face of the meter assembly as illustrated.

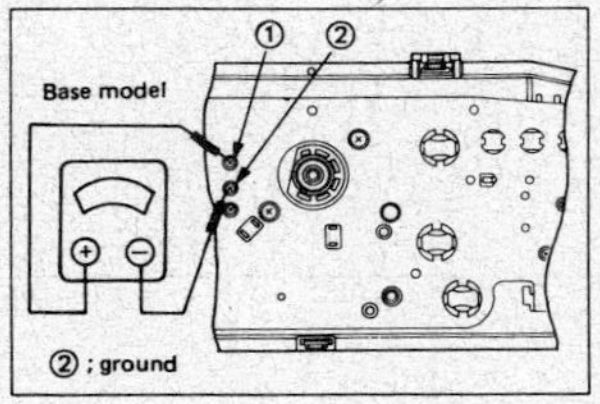

3) Turn the inner shaft slowly and check that the tester indicates a continuity and open circuit condition alternately.
If car speed sensor is inoperative, replace speedometer assembly.

CODE NO. 51, 52; ROM (CODE NO. 51) NG, RAM (CODE NO. 52) NG

Signal Code No. 51, 52
→ Start engine
→ Is code 51, 52 displayed?
- NO → END
- YES → Replace ECM

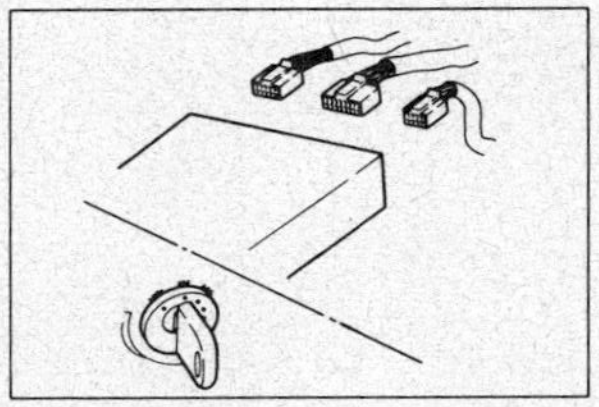

Inspection for Signal Code No. 51 and 52

Circuit Connections

1. Disconnect the control unit harnesses at connector and check for proper connections at terminals.
2. Regardless of terminal conditions, road test the vehicle with all the harnesses reconnected at the connector.
3. Obtain display of trouble codes and see if the trouble has been corrected.
If the trouble persists, replace the ECM.

CODE NO. 23, 35 and 54; POWER TRANSISTOR SYSTEM FAULTY

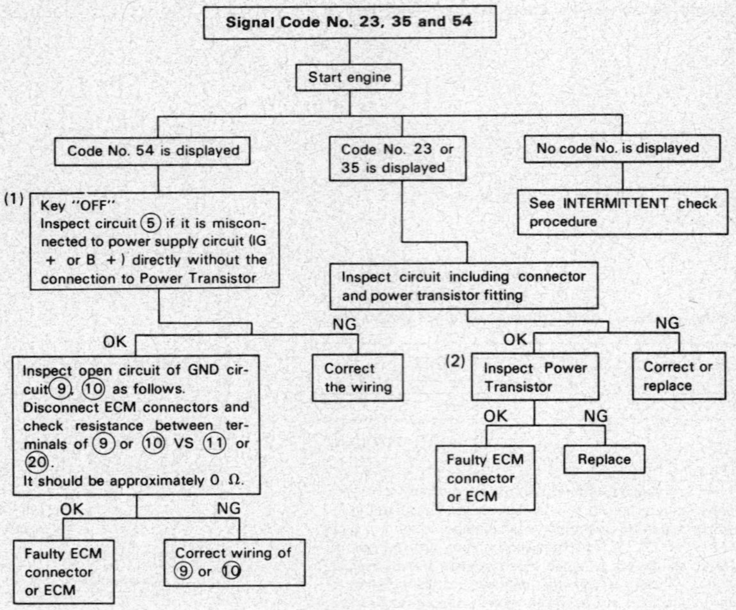

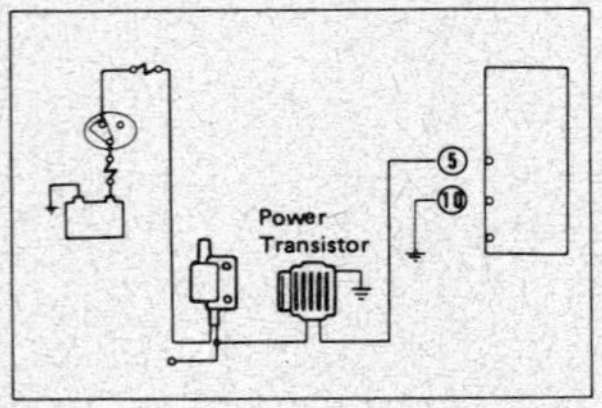

Inspection for Signal Code No. 23, 35 and 54

1. Circuits

Check wiring by referring to the illustration and wiring diagram.

Signal code No.	Wiring	Ω
23	⑤ ——— Ground	∞
35	STAY of power Transister — Intake manifold	0
54	⑩ ——— Ground	0

If any abnormal condition was found, correct or replace faulty parts.

2. Power Transistor

1. Disconnect the high-tension cable between the ignition coil and the distributor at the distributor side.
2. Attach a spark test tool (Kent-Moore J-125 or equivalent) to the end of the high-tension cable from the coil. If a spark test tool is not available, hold the end of the disconnected high tension cable approximately 5-6 mm (1/4 in.) away from a metal surface with a pair of insulated pliers.
3. Using a jack or lift, elevate the right front side of the vehicle. Support the vehicle with a jack stand.
4. Remove the right front tire.
5. Unsnap the rubber cover on the inner fender.
6. Remove the power transistor from the bottom of the intake manifold.
7. Disconnect the power transistor harness at the connector.
8. Connect a jumper lead from the harness side connector to the power transistor connector.
9. Connect a second jumper lead from the poer transistor connector to the negative (−) side of a 1.5 V dry cell battery.
10. Connect a third jumper lead from the harness side connector to the positive (+) side of the dry cell battery.
11. Ground the power transistor to the vehicle frame.
12. Turn the ignition switch to the "ON" position.
13. Disconnect and reconnect the jumper lead at the power transistor connector. Verify that a spark occurs across the gap in the spark tool or the air gap between the high tension cable and ground. Repeat this step several times.

If spark does not occur, replace the power transistor together with the bracket.

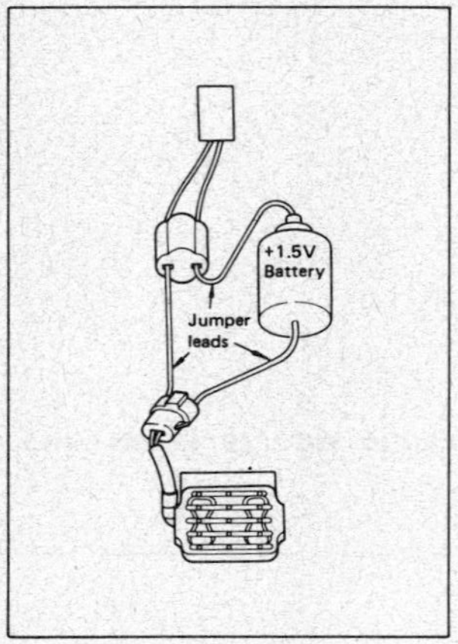

CODE NO. 25 and 53; VSV FAULTY PRESSURE REGULATOR

Signal Code No. 25 and 53

- Engine start, warmed up, and vehicle stopped (Coolant temperature should be more than 80°C)
- Keep the vehicle about 10 sec. under the above condition
- And then drive faster than 24 km/h (15 MPH), for more than 10 seconds, and note "CHECK ENGINE" light.

Code No. 25, 53 is not displayed → See INTERMITTENT check procedure

Code No. 25 or 53 is displayed →

(1) Key "OFF".
Disconnect VSV connector from VSV. Check the resistance of VSV.
It should be approximately 35 Ω

In case of Code No. 25:

- > 40 Ω → Replace VSV
- 30 Ω ≤ ≤ 40 Ω →

(2) Inspect circuit ④ including connector for open circuit or poor insulation with GND.
Key "OFF". Disconnect ECM connectors from ECM. Measure the resistance of circuit ④ between both end terminals of ECM connector and VSV connector.
It should be less than 1 Ω.
Also measure the resistance between circuit ④ and the intake manifold or other grounded portion.
It should be ∞.

- NG → Correct or replace circuit ④
- OK → Faulty ECM connector or ECM

In case of Code No. 53:

- 30 Ω > → Replace VSV
- 30 Ω ≤ ≤ 40 Ω →

(2) Check for poor insulation of circuit ④ with power supply circuit.
Key "ON", engine stopped.
Measure the voltage of circuit ④ terminal at VSV connector against the intake manifold of engine.
It should be almost "0"V.

- Under 1V → Replace ECM
- Over 1V →

Key "OFF"
Disconnect ECM connector from ECM.
Key "ON"
Measure the voltage of circuit ④ against the intake manifold.

- Over 1V → Correct or replace circuit ④
- Under 1V → Replace ECM

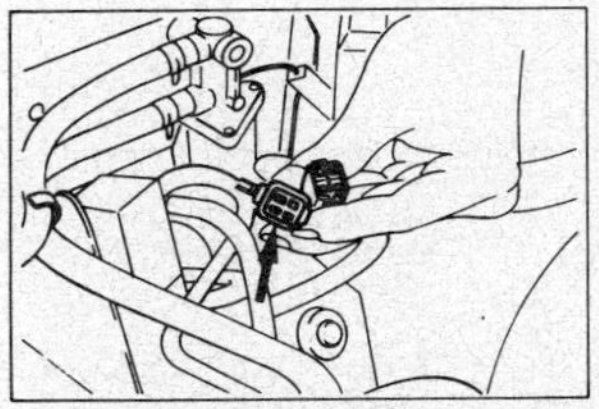

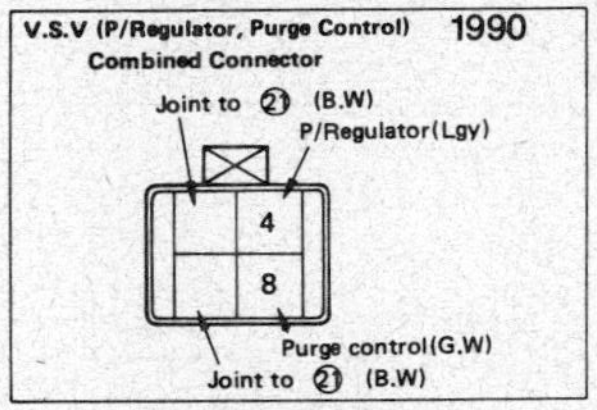

*

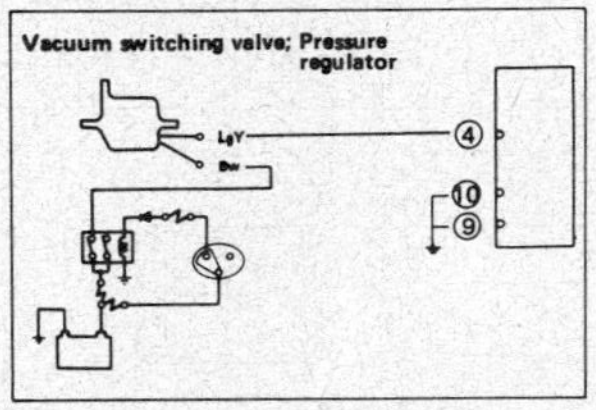

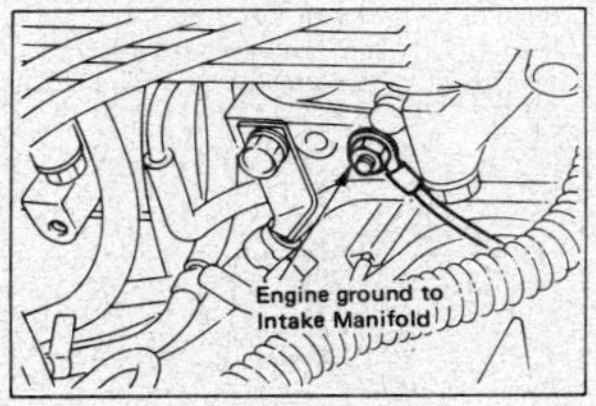

Inspection for Signal Code No. 25 and 53

1. **Check the Resistance of Vacuum Switching Valve**

(Ω)

Standard resistance in vacuum switching valve	Approx. 35

Note:
Checking operation of vacuum switching valve will also be useful.
Check operation of vacuum switching valve by applying battery voltage directly to the vacuum switching valve.
Replace vacuum switching valve if found to be defective.

2. **Circuits**

Signal code No.	Wiring	Ω	V
25	Bw —— Ground	—	12
	LgY —— ④	0	—
53	⑨, ⑩ —— Ground	0	—

If any abnormal condition is found, correct or replace applicable harness assembly.

Note:
Normal operation of the entire vacuum switching valve system including control unit is indicated if the vacuum switching valve operates properly when the starter switch is turned to ON position with the engine stationary and with the diagnosis lead connected.

*

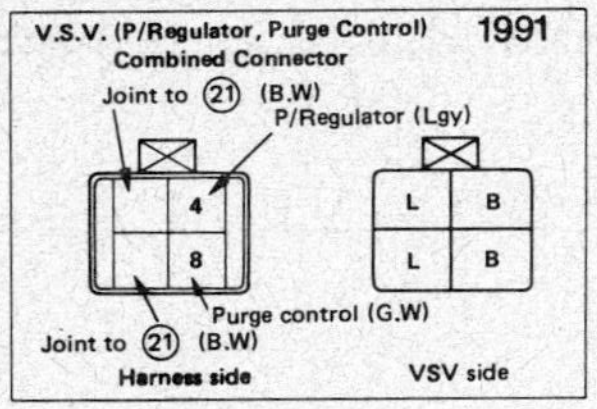

*

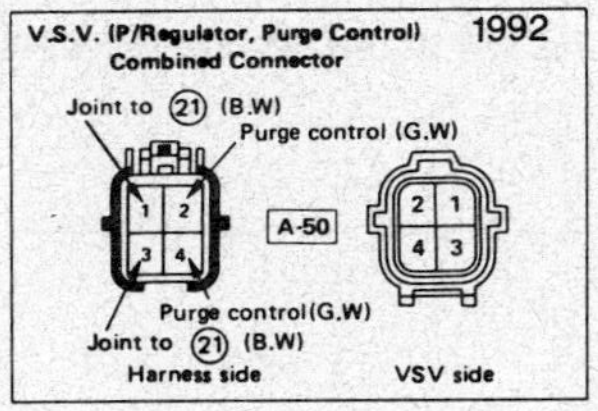

CODE NO. 26, 27; CANISTER VSV DOES NOT OPERATE OR OPERATES ABNORMALLY

Signal Code No. 26, 27

- Start engine
- Run engine at normal temperature, for 10 sec. at idle, then for 10 sec. at faster than 2500 rpm
- Note "CHECK ENGINE" light

Code No. 26, 27 is not displayed → See INTERMITTENT check procedure

Code No. 26 or 27 is displayed →

(1) Key "OFF". Disconnect VSV connector from VSV. Check the resistance of VSV. It should be approximately 35 Ω

In case of Code No. 26

- Over 40 Ω → Replace VSV
- Between 30 Ω and 40 Ω →

(2) Inspect circuit ⑧ including connector for open circuit or insulation with GND. Key "OFF". Disconnect ECM connectors from ECM. Measure the resistance of circuit ⑧ between both end terminals of ECM connector and VSV connector. It should be less than 1 Ω. Also measure the resistance between circuit ⑧ and the intake manifold or other grounded spot. It should be ∞.

- NG → Correct or replace circuit ⑧
- OK → Faulty ECM connector or ECM

In case of Code No. 27

- Under 30 Ω → Replace VSV
- Between 30 Ω and 40 Ω →

(2) Check for poor insulation of circuit ⑧ with power supply circuit. Key "ON", engine stopped. Measure the voltage of circuit ⑧ terminal at VSV connector against the intake manifold of engine. It should be almost "0"V.

- Under 1V → Replace ECM
- Over 1V → Key "OFF" Disconnect ECM connector from ECM. Key "ON" Measure the voltage of circuit ⑧ against the intake manifold.
  - Over 1V → Correct or replace circuit ⑧
  - Under 1V → Replace ECM

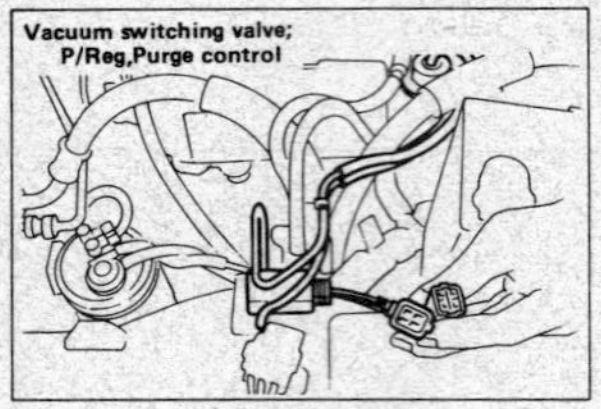

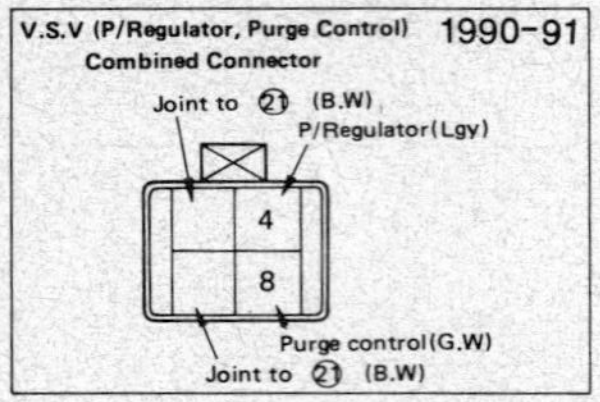

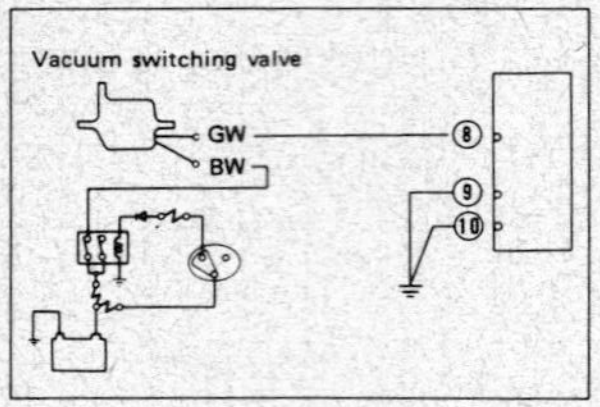

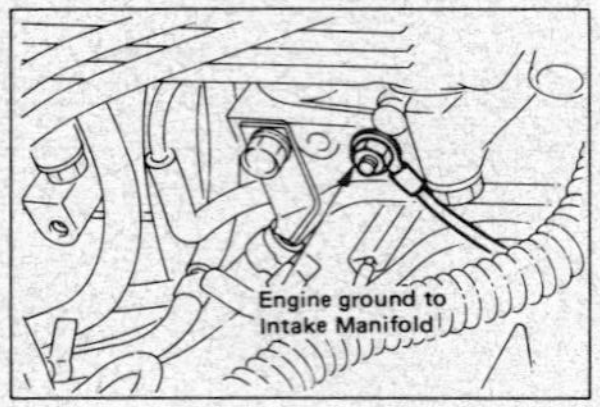

Inspection for Signal Code No. 26 and 27

1. Check the Resistance of Vacuum Switching Valve

Standard resistance in vacuum switching valve (Ω)	Approx. 35

Note:

Checking operation of vacuum switching valve will also be useful.
Check operation of vacuum switching valve by applying battery voltage directly to the vacuum switching valve.
Replace vacuum switching valve if found to be defective.

2. Circuits

Signal code No.	Wiring	Ω	V
26	BW —— Ground	–	12
	GW —— ⑧	0	–
27	⑨, ⑩ —— Ground	0	–

If any abnormal condition is found, correct or replace applicable harness assembly.

Note:

Normal operation of the entire vacuum switching valve system including control unit is indicated if the vacuum switching valve operates properly when the starter switch is turned to ON position with the engine stationary and with the diagnosis lead connected.

*

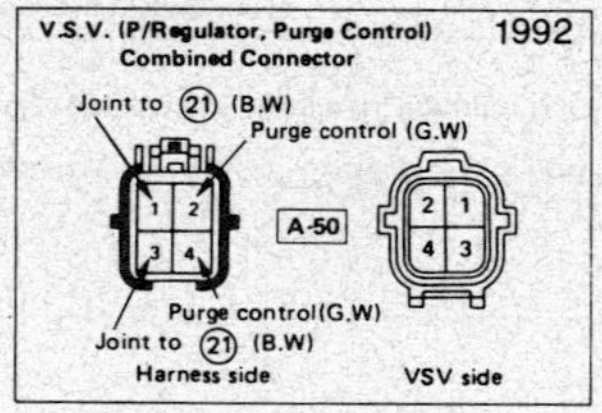

2.6L ENGINE–TROUBLE CODE CHART 33
1990 AMIGO, PICK-UP, TROOPER AND TROOPER II

CODE NO. 33; FUEL INJECTOR ERROR

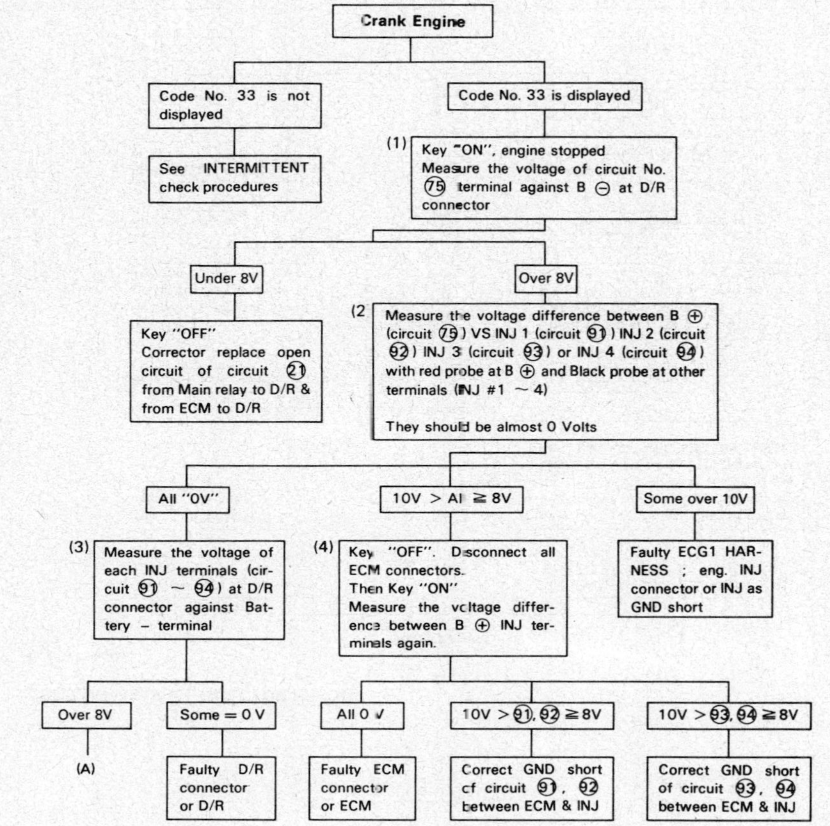

(D/R; Dropping Resistor
INJ; Injector)

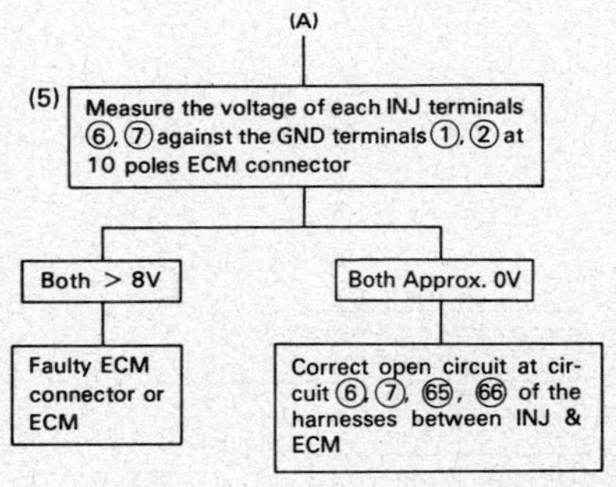

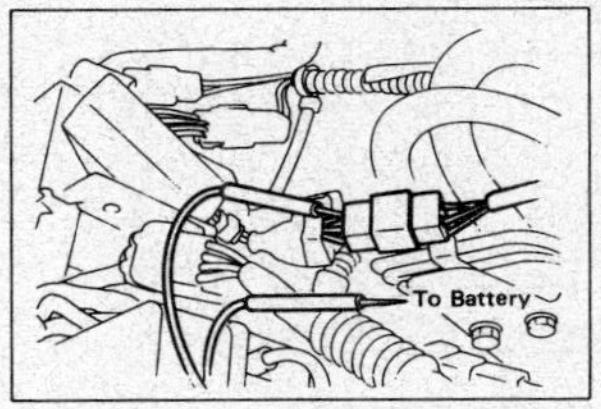

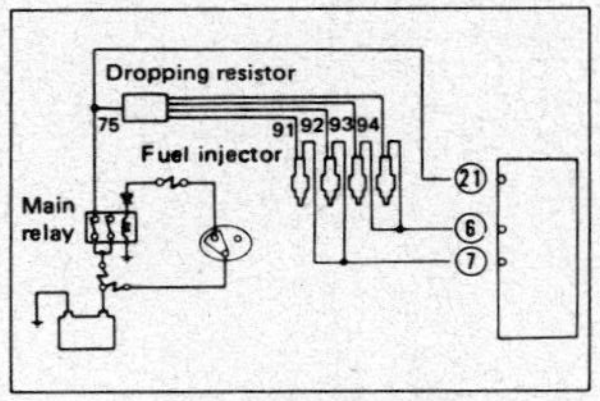

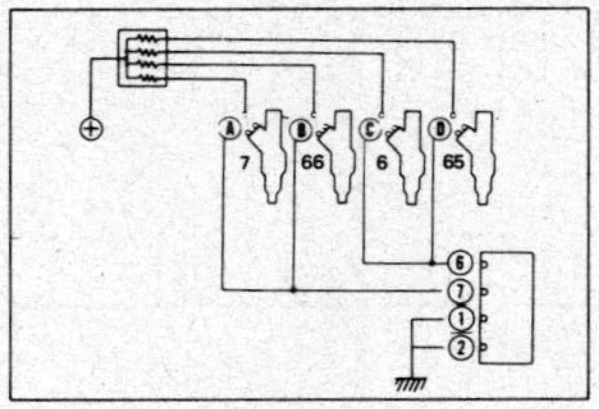

Inspection for Signal Code No. 33

(1) Check the circuit 21
Key "ON", engine stopped.
Measure the voltage of the terminal 75, against Battery ⊖ at Dropping Resistor connector.
Over 8V is normal

(2) Measure the voltage difference between Battery ⊕ and each injector at D/Resistor connector.
Almost OV

(3) Measure the voltage difference between Battery ⊖ and each injector at D/Resistor connector.
Over 8V is normal

(4) Key "OFF". Disconnect all ECM connectors. Then Key "ON".
Measure the voltage between Battery ⊕ against injector terminals.

(5) Measure the voltage of each injector terminls No. 6, 7 against the GND terminals No. 1, 2 at ECM connector (10 poles).

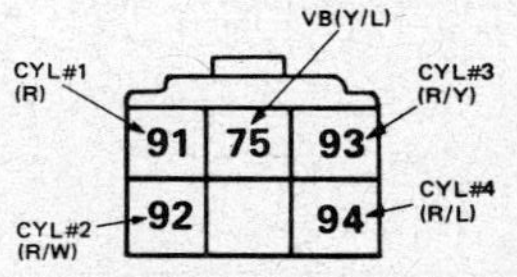

Dropping resistor connector

CODE NO. 33; FUEL INJECTOR ERROR

For '91 model

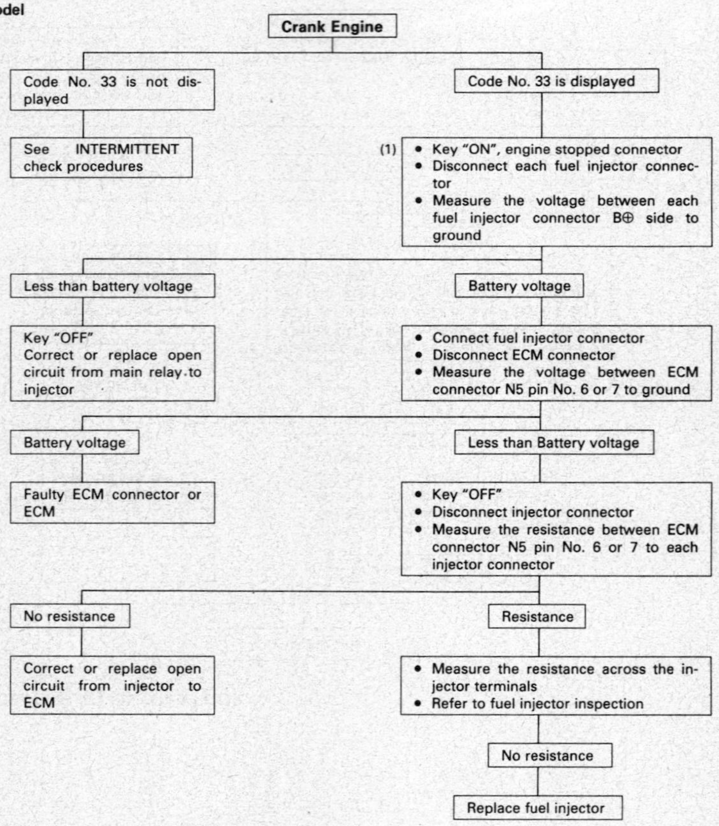

CODE NO. 64; FUEL INJECTOR ERROR

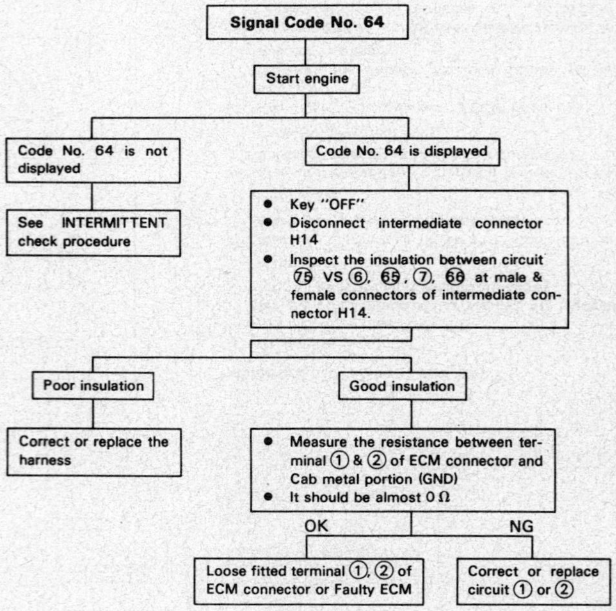

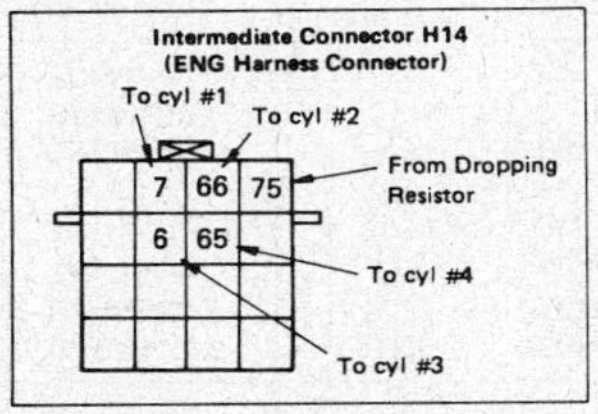

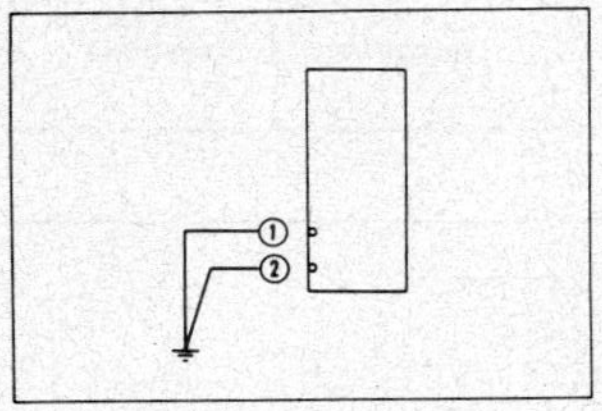

Inspection for Signal Code No. 64

1. Check the insulation for Injectors and Dropping resistor circuit.

- Key "OFF"
- Disconnect the connector H14
- Check the insulation between circuit ⑦⑤ VS ⑥, ⑥⑤, ⑦, ⑥⑥ at both male and female terminals.

2. Check for open at GND circuit ①, ②.

Circuit

Wiring	Ω
① ——— Ground	0
② ——— Ground	0

If any abnormal condition was found, correct or replace faulty parts.

Note:

This diagnostic procedure does not disclose troubles in switches, injectors such as poor contact, restrictions, etc.

In such an instance, the control unit may analyze the case and give abnormal air flow sensor or O_2 sensor signal code.

Inspection for Signal Code No. 33 and 64

ENG. HARNESS ECM

CABLE HARNESS; Eng room

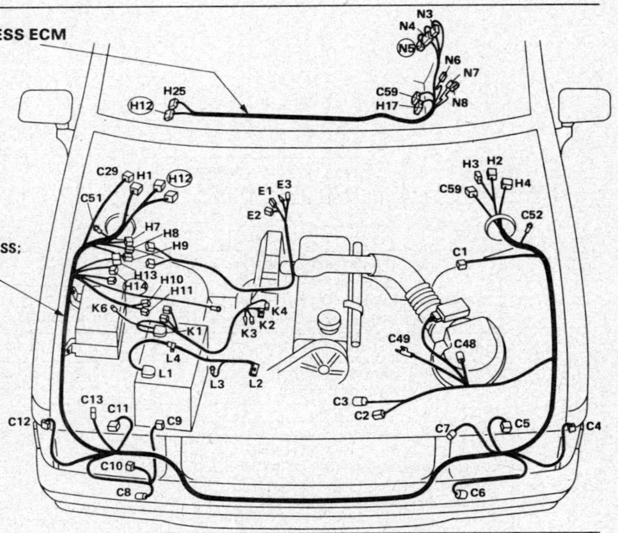

ECGI HARNESS ; ENGINE

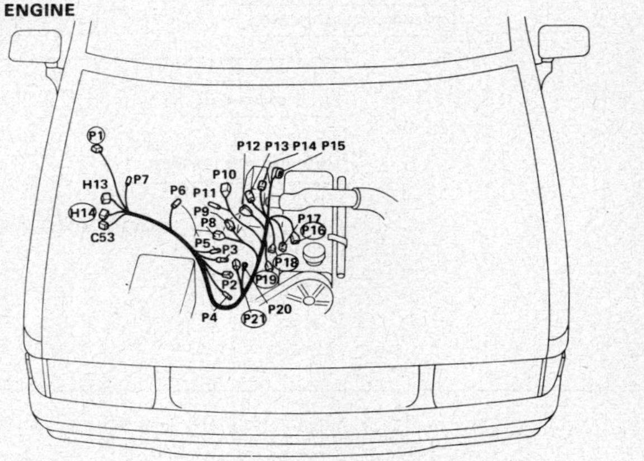

FUEL INJECTION CIRCUITS AND CONNECTORS

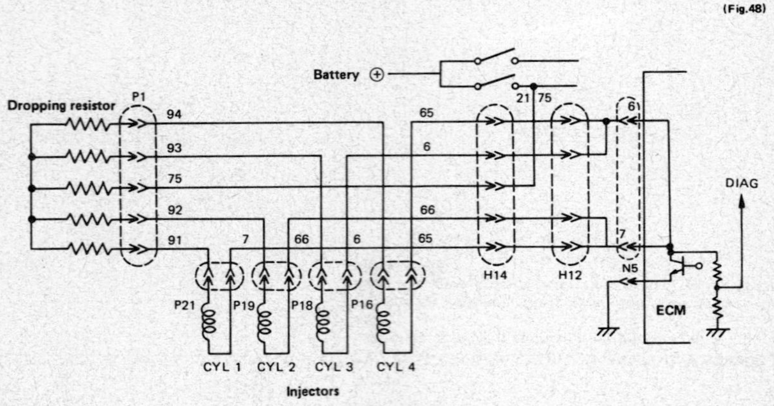

Harness Name	Connector No.	Parts Name	Refer to Figure	
			Wire Routing Diagram	System Circuit Diagram
CABLE HARNESS; Eng. room	H12 H14	Engine room harness—Engine harness ECM (RH) Engine room harness—closed loop harness	Fig.-46	Fig.-48
ENGINE HARNESS ECM	N5	ECM controller 10 pin	Fig.-46	Fig.-48
ECGI HARNESS; Engine	P1 P16 P18 P19 P21	Dropping resistor Resistor ; fuel injector (No. 4) Resistor ; fuel injector (No. 3) Resistor ; fuel injector (No. 2) Resistor ; fuel injector (No. 1)	Fig.-47	Fig.-48

CODE NO. 64; FUEL INJECTOR ERROR

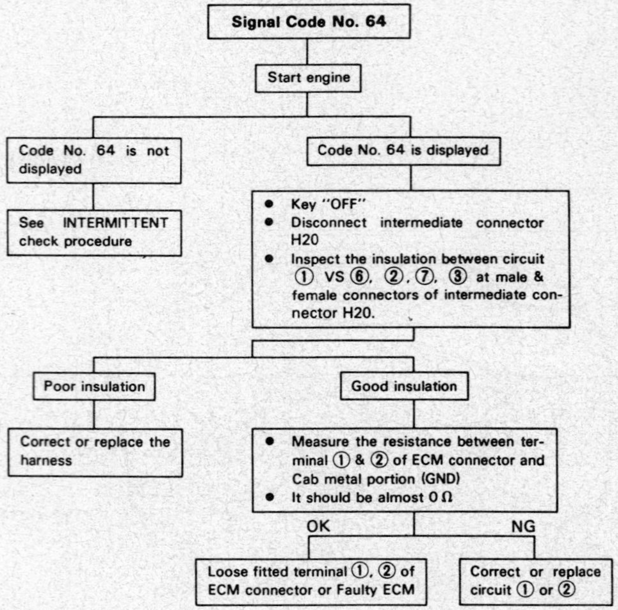

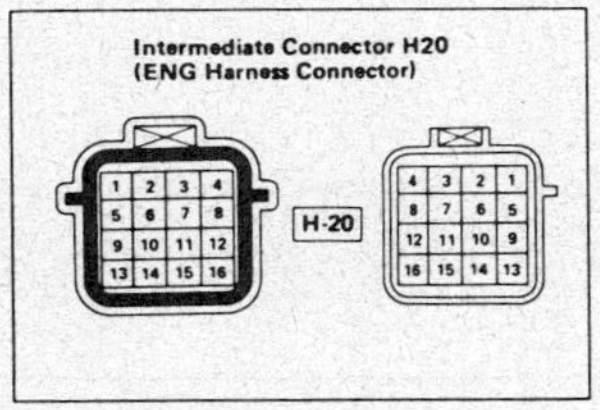

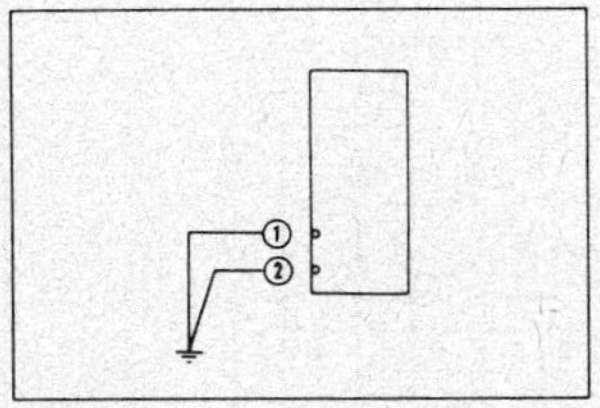

Inspection for Signal Code No. 64

1. Check the insulation for Injectors circuit.
- Key "OFF"
- Disconnect the connector H-20
- Check the insulation between circuit ① VS ⑥, ②, ⑦, ③ at both male and female terminals.
2. Check for open at GND circuit ①, ②.

Circuit

Wiring		Ω
①	Ground	0
②	Ground	0

If any abnormal condition was found, correct or replace faulty parts.

Note:

This diagnostic procedure does not disclose troubles in switches, injectors such as poor contact, restrictions, etc.
In such an instance, the control unit may analyze the case and give abnormal air flow sensor or O_2 sensor signal code.

Inspection for Signal Code No. 33 and 64

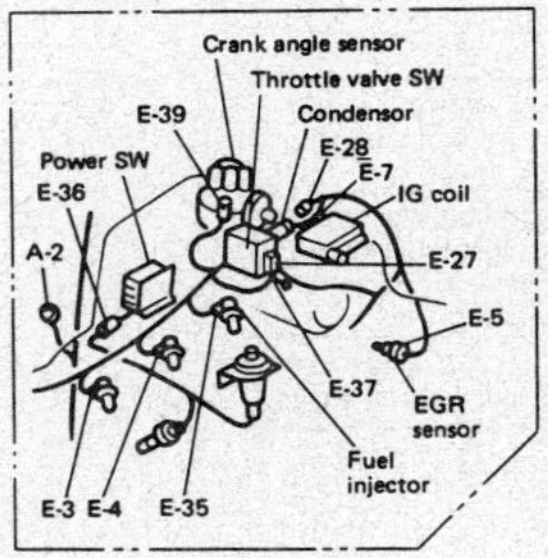

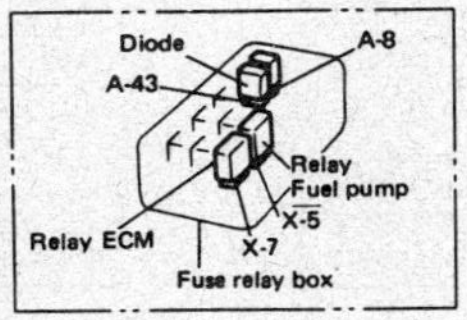

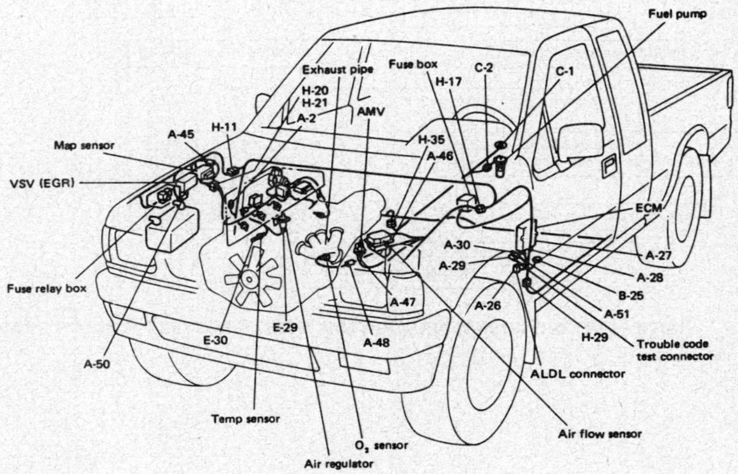

FUEL INJECTION CIRCUITS AND CONNECTORS

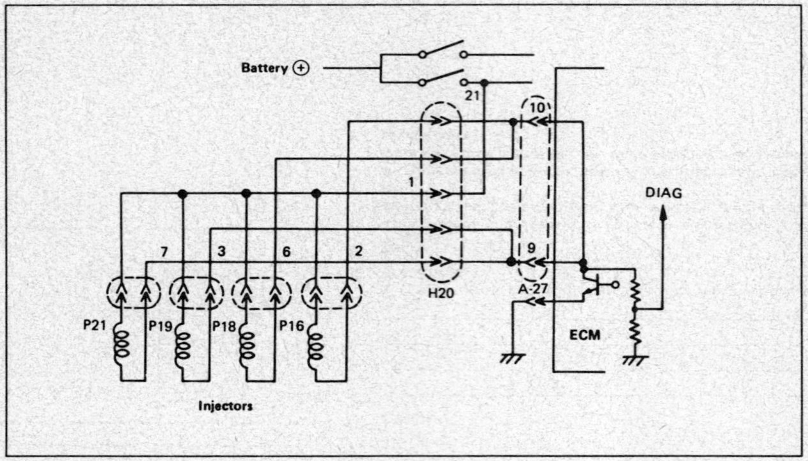

CODE NO. 32 and 34; EGR SYSTEM FAILURE AND SENSOR CIRCUIT FAILURE

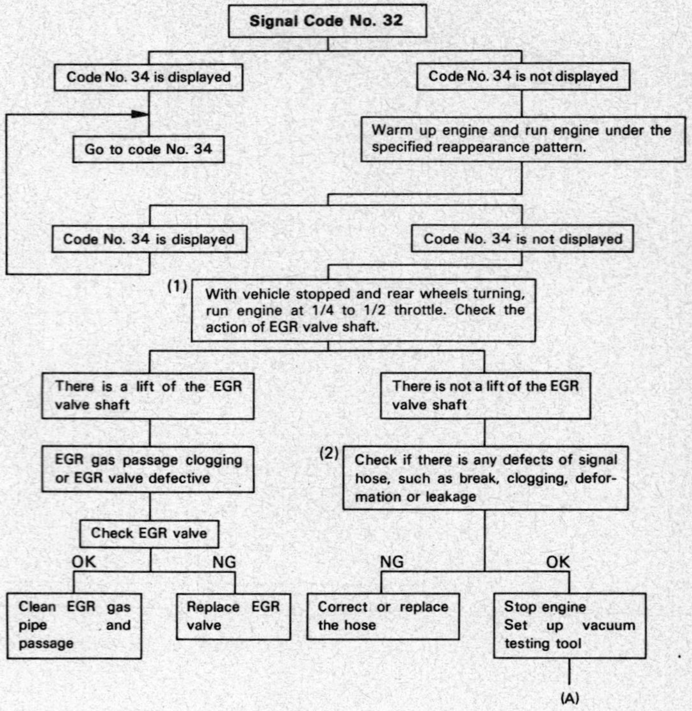

2.6L ENGINE – TROUBLE CODE CHART 32 and 34
1990 AMIGO, PICK-UP, TROOPER AND TROOPER II

2.6L ENGINE – TROUBLE CODE CHART 32 and 34
1991–92 AMIGO, PICK-UP, TROOPER AND TROOPER II

CODE NO. 32 and 34; EGR SYSTEM FAILURE AND SENSOR CIRCUIT FAILURE

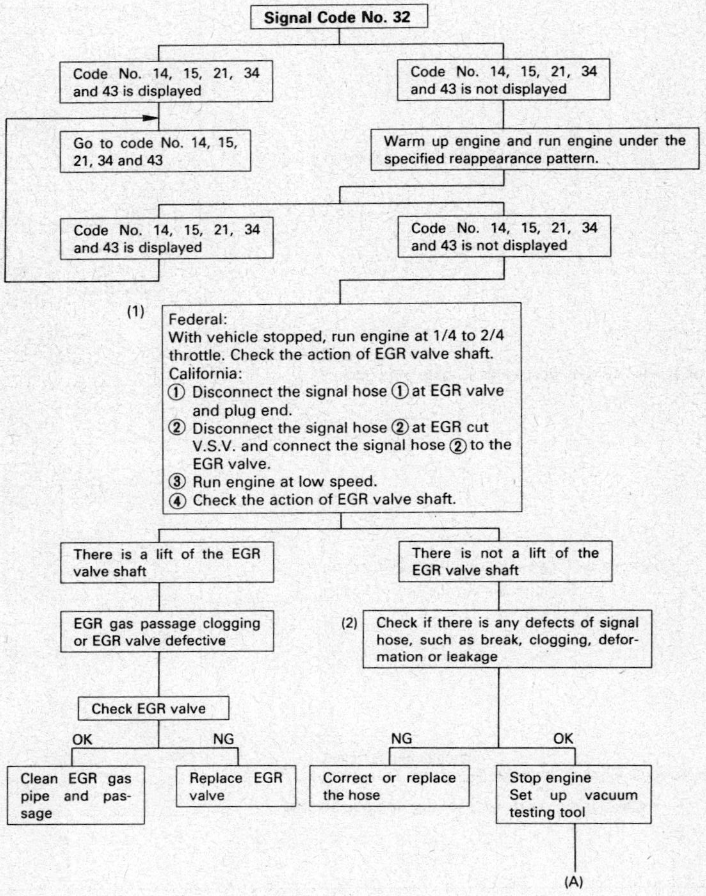

Federal

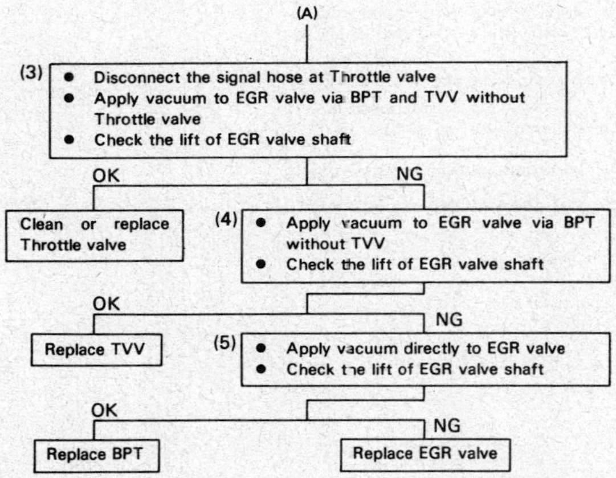

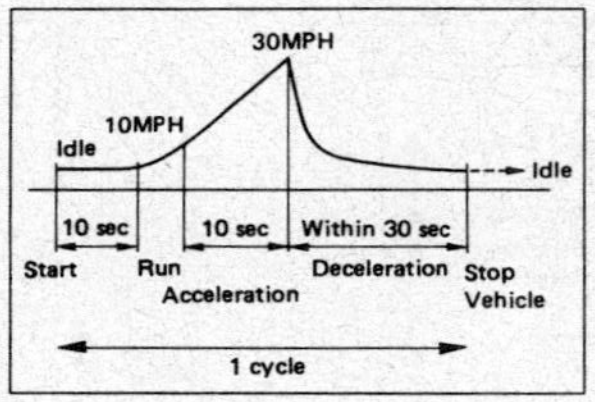

Inspection for Signal Code No. 32 and 34

Specified reappearance pattern

① Run engine at idle condition, at normal operating temperature.

② Keep at idle for more than 10 seconds, with vehicle stopped.

③ Drive vehicle at just over 10 MPH. Accelerate quickly to more than 30 MPH (must reach 30 MPH within 10 seconds of starting acceleration).

④ Within 30 seconds, decelerate safely to a complete stop.

⑤ Repeat above procedure ② through ④ several times.

2.6L ENGINE – TROUBLE CODE CHART 32 and 34
1991–92 AMIGO, PICK-UP, TROOPER AND TROOPER II

(A)

(3)
- Disconnect the signal hose at Throttle valve
- Apply vacuum to EGR valve via BPT and TVV without Throttle valve
- Check the lift of EGR valve shaft

OK → Clean or replace Throttle valve

NG → (4)
- Apply vacuum to EGR valve via BPT without TVV
- Check the lift of EGR valve shaft

OK → Replace TVV

NG → (5)
- Apply vacuum directly to EGR valve
- Check the lift of EGR valve shaft

OK → Replace BPT

NG → Replace EGR valve

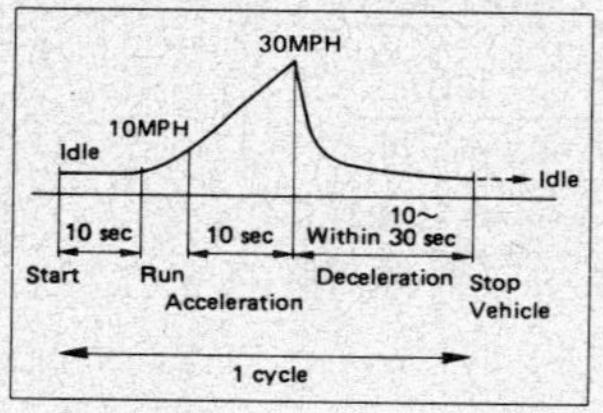

Inspection for Signal Code No. 32 and 34

Specified reappearance pattern

① Run engine at idle condition, at normal operating temperature.

② Keep at idle for more than 10 seconds, with vehicle stopped.

③ Drive vehicle at just over 10 MPH. Accelerate quickly to more than 30 MPH (must reach 30 MPH within 10 seconds of starting acceleration).

④ Within 10~30 seconds, decelerate safely to a complete stop.

⑤ Repeat above procedure ② through ④ several times.

2.6L ENGINE – TROUBLE CODE CHARTS
AMIGO, PICK-UP, TROOPER AND TROOPER II

Inspection for Signal Code No. 32 (Federal)

1. Inspect for EGR valve operation of valve shaft by applying signal vacuum.

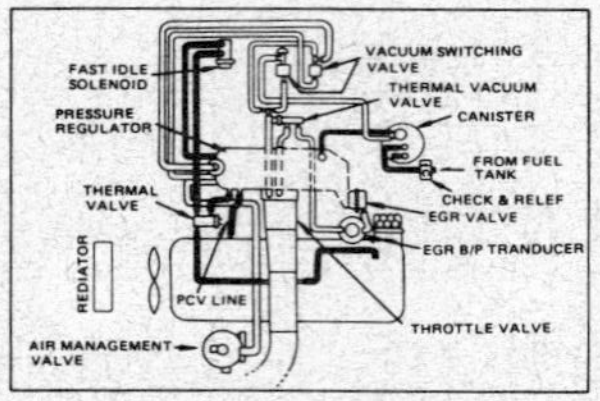

2. Check for defects of signal hoses.
   Intake manifold → TVV → BPT → EGR valve

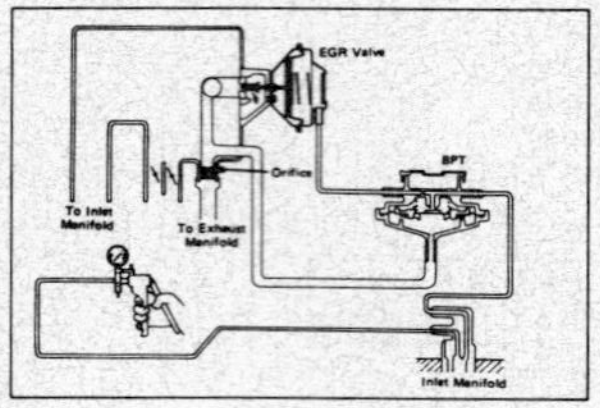

3. Check if the vacuum signal from throttle valve is correct.

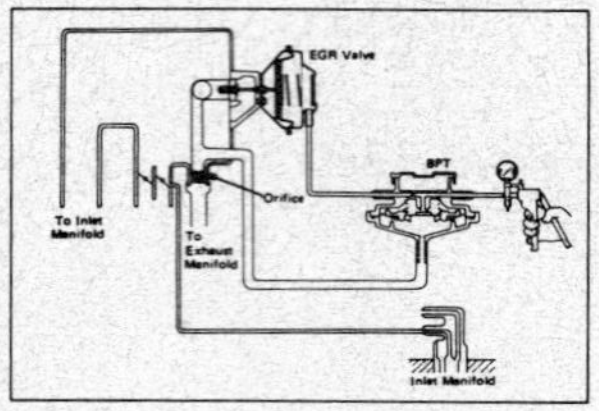

4. Check if TVV is working properly.

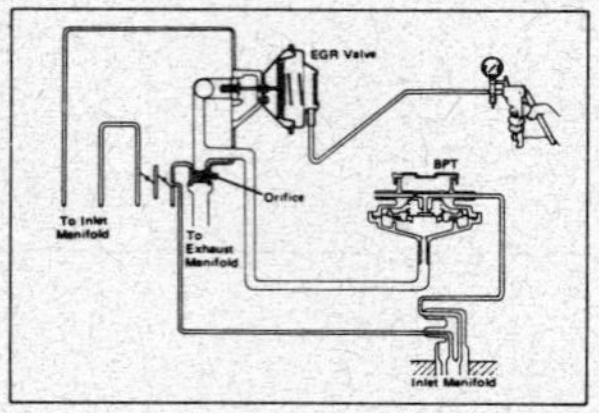

5. Check if B.P.T. is working properly.

2.6L ENGINE—TROUBLE CODE CHARTS—AMIGO, PICK-UP, TROOPER AND TROOPER II

California only

(A)

(3)
- Disconnect the signal hose at Throttle valve and plug.
- Apply vacuum to EGR valve via Duty solenoid assembly without via Throttle valve
- With the wheels turning and the engine running at 1/4 to 1/2 throttle, check the lift of EGR valve shaft

OK → Clean or replace Throttle valve

(4)
- Apply vacuum directly to EGR valve
- Check the lift of EGR valve shaft

OK → Replace Duty solenoid

NG → Replace EGR valve

NG →

(4)
- Apply vacuum directly to EGR valve
- Check the lift of EGR valve shaft

OK → Replace Duty Solenoid

NG → Replace EGR valve

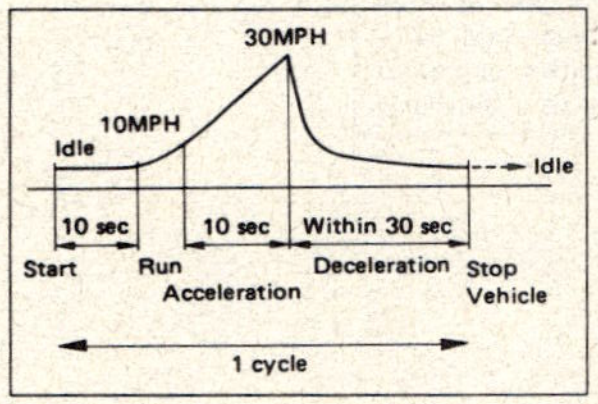

Inspection for Signal Code No. 32 and 34

Specified reappearance pattern

① Run engine at idle condition, at normal operating temperature.

② Keep at idle for more than 10 seconds, with vehicle stopped.

③ Drive vehicle at just over 10 MPH. Accelerate quickly to more than 30 MPH (must reach 30 MPH within 10 seconds of starting acceleration).

④ Within 30 seconds, decelerate safely to a complete stop.

⑤ Repeat above procedure ② through ④ several times.

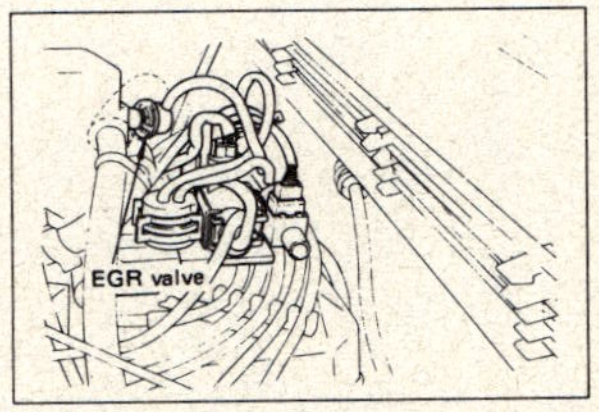

Inspection for Signal Code No. 32 (California only)

1. Check for operation of EGR valve shaft by applying signal vacuum.

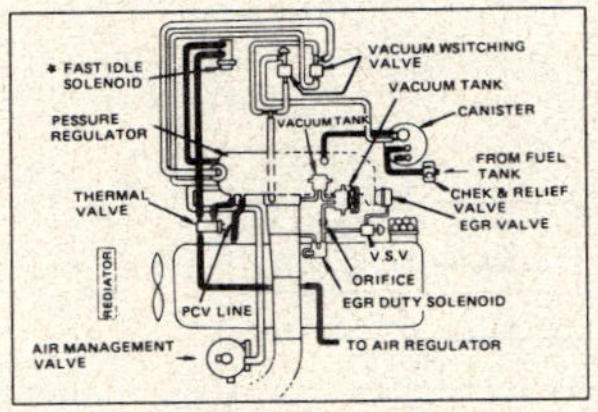

2. Check for defects of signal hoses.
   Throttle valve → Vacuum tank → Vacuum regulator → Duty solenoid → VSV ; EGR cut → EGR valve

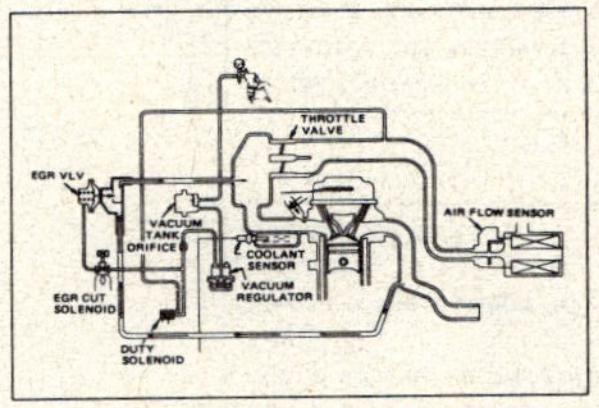

3. Check if the vacuum signal from throttle valve is correct.

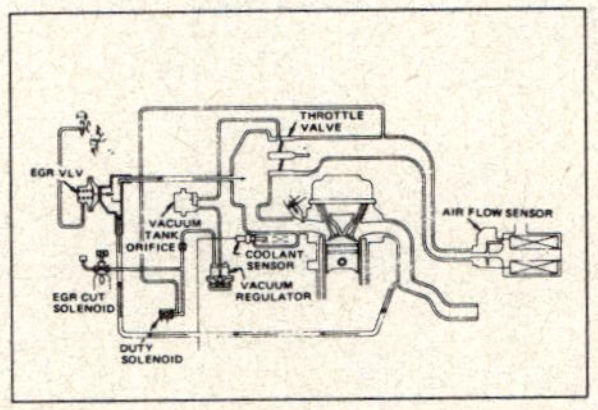

5. Check if Duty solenoid is working properly.

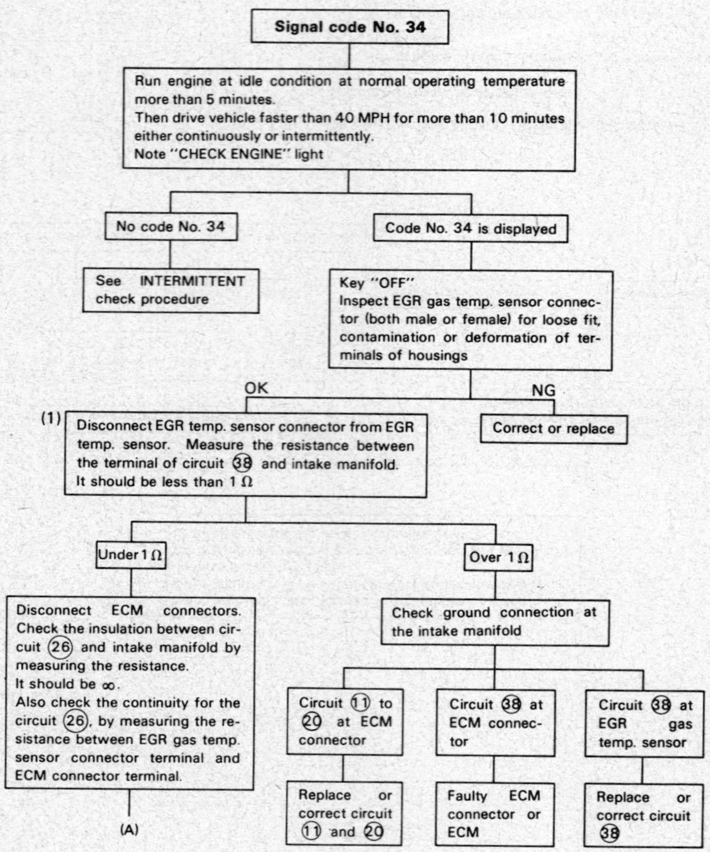
Signal code No. 34
Run engine at idle condition at normal operating temperature more than 5 minutes.
Then drive vehicle faster than 40 MPH for more than 10 minutes either continuously or intermittently.
Note "CHECK ENGINE" light
No code No. 34
See INTERMITTENT check procedure
Code No. 34 is displayed
Key "OFF"
Inspect EGR gas temp. sensor connector (both male or female) for loose fit, contamination or deformation of terminals of housings
OK
NG
Correct or replace
(1)
Disconnect EGR temp. sensor connector from EGR temp. sensor. Measure the resistance between the terminal of circuit 38 and intake manifold.
It should be less than 1 Ω
Under 1 Ω
Disconnect ECM connectors.
Check the insulation between circuit 26 and intake manifold by measuring the resistance.
It should be ∞.
Also check the continuity for the circuit 26, by measuring the resistance between EGR gas temp. sensor connector terminal and ECM connector terminal.
(A)
Over 1 Ω
Check ground connection at the intake manifold
Circuit 11 to 20 at ECM connector
Replace or correct circuit 11 and 20
Circuit 38 at ECM connector
Faulty ECM connector or ECM
Circuit 38 at EGR gas temp. sensor
Replace or correct circuit 38

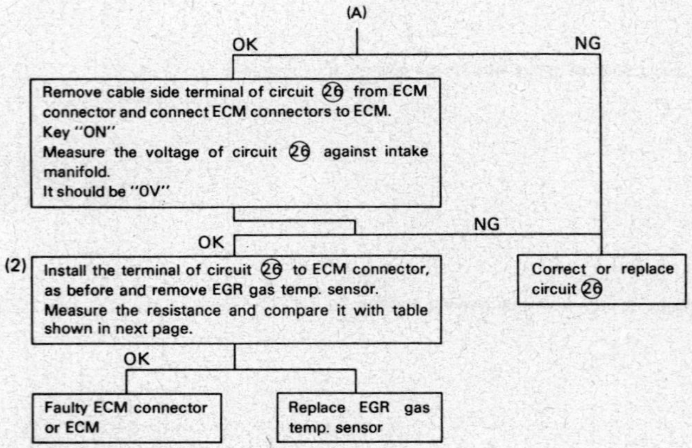
(A)
OK
NG
Remove cable side terminal of circuit 26 from ECM connector and connect ECM connectors to ECM.
Key "ON"
Measure the voltage of circuit 26 against intake manifold.
It should be "0V"
OK
NG
(2)
Install the terminal of circuit 26 to ECM connector, as before and remove EGR gas temp. sensor.
Measure the resistance and compare it with table shown in next page.
Correct or replace circuit 26
OK
Faulty ECM connector or ECM
Replace EGR gas temp. sensor

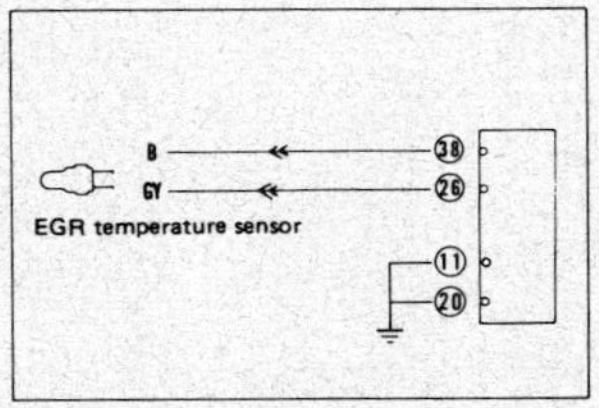

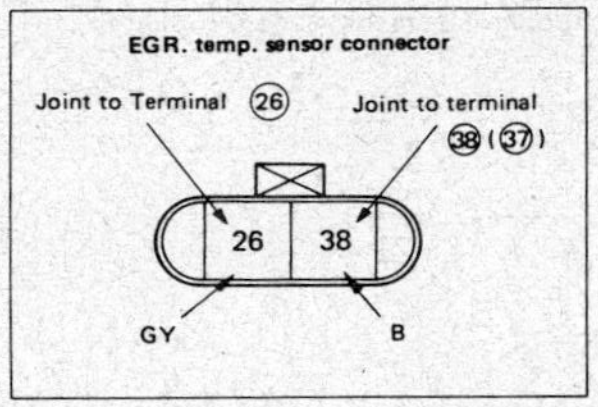

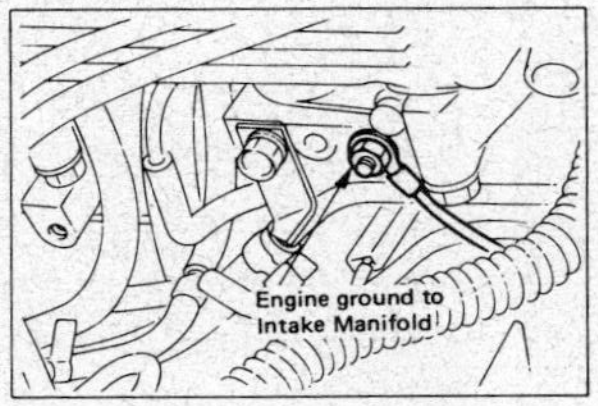

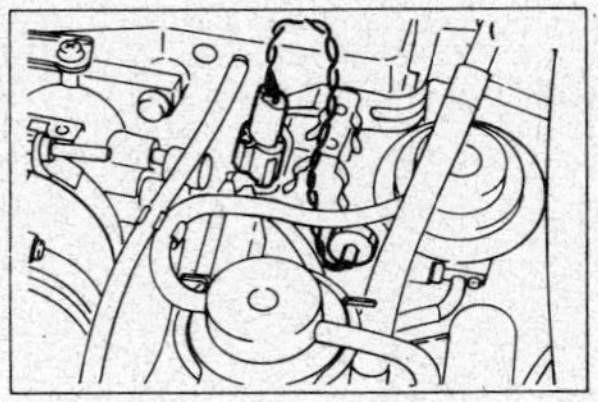

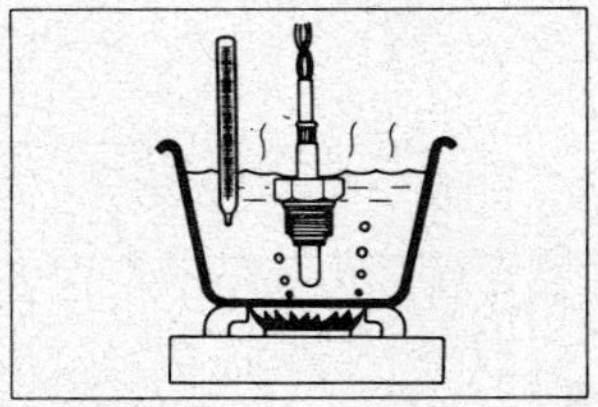

Inspection for Signal Code No. 34

Wiring

Signal code No.	Wiring	Ω
34	(26)——— GY	0
	(38)——— B	0
	(38)——— Ground	0
	(11) (20)——— Ground	0

When abnormal condition is indicated, correct or replace applicable harnesses.

Measurement of Resistance in EGR Temperature Sensor

Remove the sensor and put in the oil. Measure the resistance across the sensor terminals.

Note:

Although the resistance varies with the oil, normal condition is indicated when measured resistance falls within the range shown in the following table:

Temperature °C(°F)	Resistance KΩ (approx)
0 (32)	27.0
20 (68)	12.1
40 (109)	5.9
80 (176)	1.7
100 (212)	1.0

If measured resistance deviates from the specified range, replace EGR temperature sensor.

'91 model

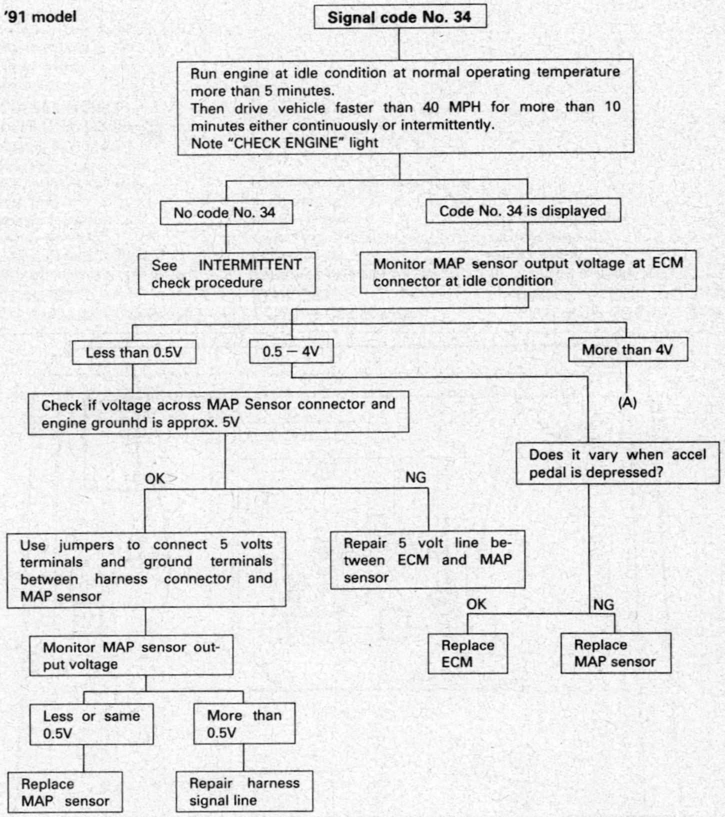

2.6L ENGINE – TROUBLE CODE CHARTS AMIGO, PICK-UP, TROOPER AND TROOPER II

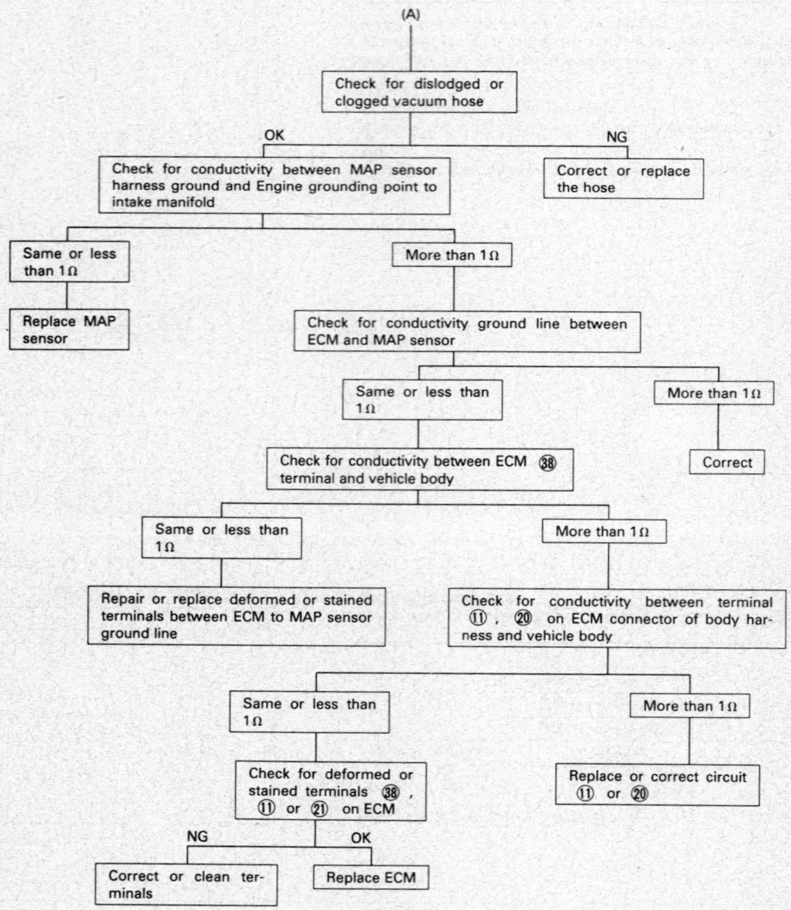

2.8L ENGINE – ENGINE COMPONENT LOCATIONS TROOPER AND TROOPER II

"KT" SERIES RPO: LL2 ENGINE CODE: R 2.8L V6

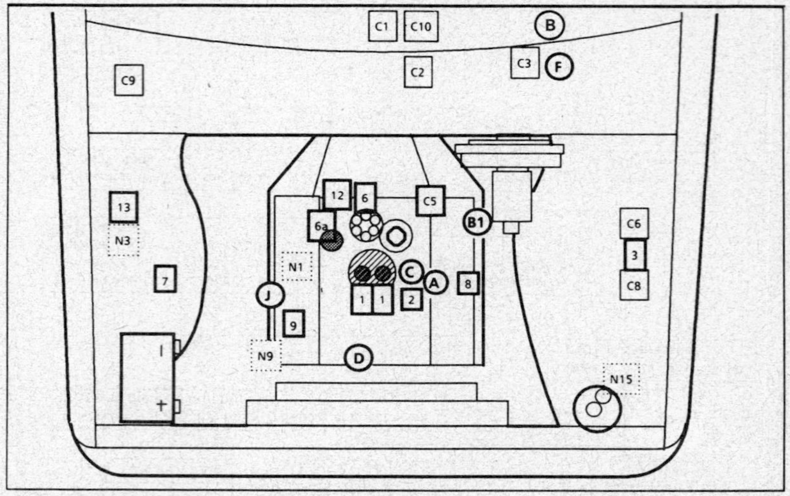

COMPUTER COMMAND CONTROL

- C1 Electronic Control Module (E.C.M.)
- C2 ALDL diagnostic connector
- C3 Check engine light
- C5 ECM harness ground
- C6 Fuse panel
- C8 Fuel pump test connector
- C9 Elapsed Timer Module
- C10 Set Timing Connector

ECM CONTROLLED COMPONENTS

- 1 Fuel injector
- 2 Idle air control
- 3 Fuel pump relay
- 6 Electronic Spark Timing Distributor (E.S.T.)
- 6a Remote ignition coil
- 7 Electronic Spark Control module (E.S.C.)
- 8 Oil pressure switch
- 9 Electric Air Control solenoid (E.A.C.)
- 12 Exhaust Gas Recirculation Vacuum Solenoid
- 13 A/C Relay

ECM INFORMATION SENSORS

- A Manifold Absolute Pressure (M.A.P.)
- B Exhaust oxygen
- C Throttle position (T.P.S.)
- D Coolant temperature
- F Vehicle speed (V.S.S.)
- J Electronic Spark Control Knock (E.S.C.)

EMISSION COMPONENTS (NOT ECM CONTROLLED)

- N1 Crankcase vent valve (PCV)
- N3 Main Relay
- N9 Air Pump
- N15 Fuel Vapor Canister

2.8L ENGINE—ECM WIRING DIAGRAM—TROOPER AND TROOPER II

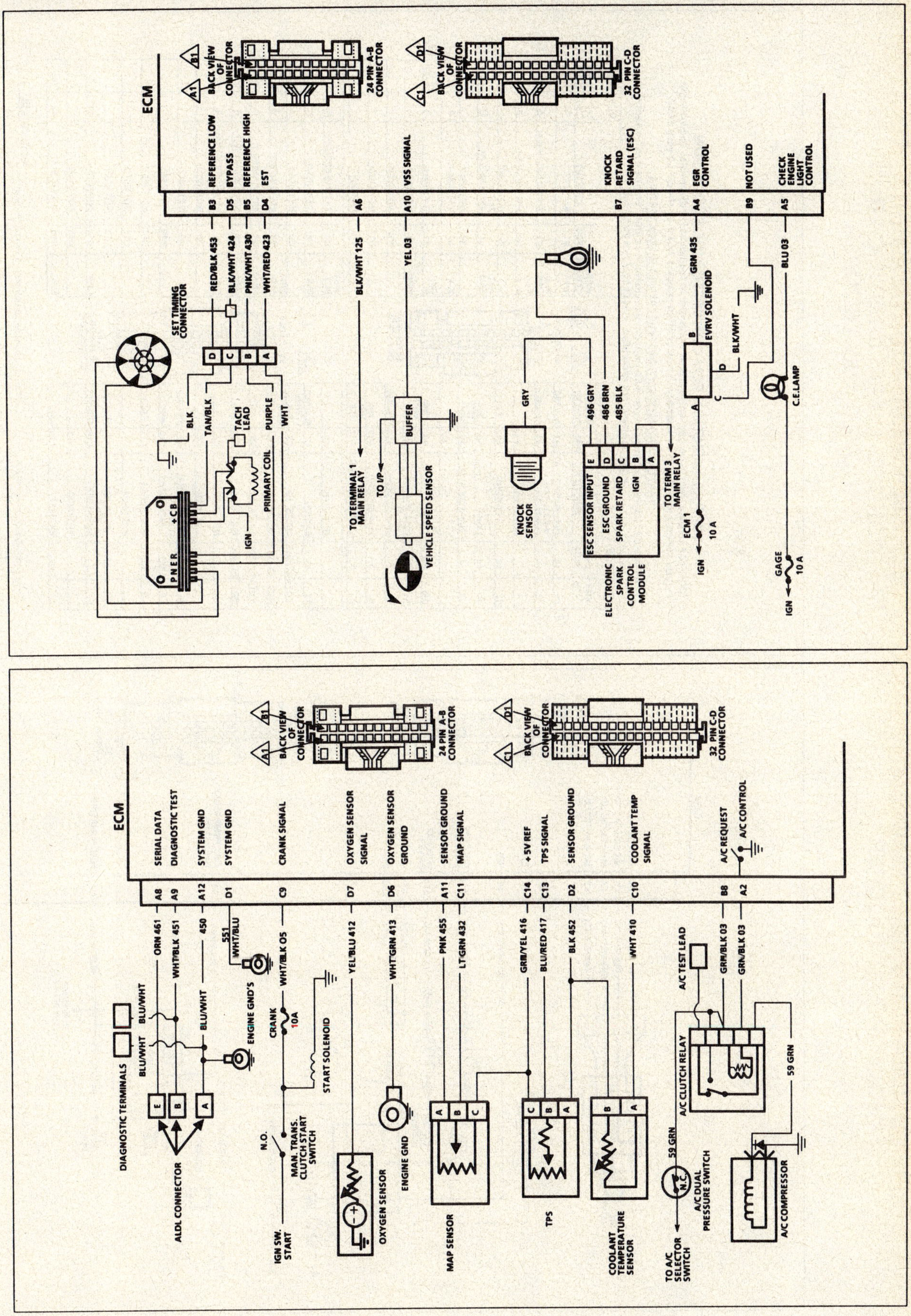

2.8L ENGINE—ECM WIRING DIAGRAM TROOPER AND TROOPER II

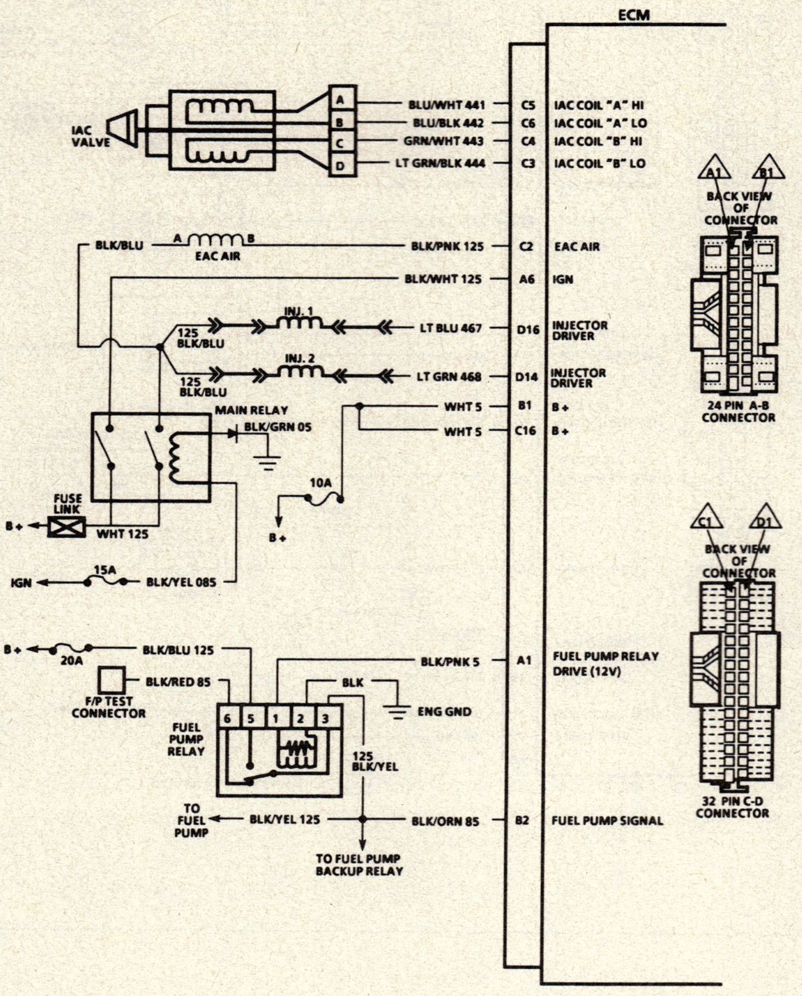

2.8L ENGINE—CONTROL UNIT CONNECTOR PIN IDENTIFICATION TROOPER AND TROOPER II

TBI FUEL INJECTION ECM CONNECTOR IDENTIFICATION

This ECM voltage chart is for use with a digital voltmeter to further aid in diagnosis. The voltages you get may vary due to low battery charge or other reasons, but they should be very close.

THE FOLLOWING CONDITIONS MUST BE MET BEFORE TESTING:

- Engine at operating temperature
- Engine idling in closed loop (for "Engine Run" column)
- Test terminal not grounded
- ALCL tool not installed

Note	VOLTAGE KEY "ON"	VOLTAGE ENG. RUN	CIRCUIT	PIN	WIRE COLOR
4	0	14	FUEL PUMP RELAY	A1	BLK/PNK
7	12	14	A/C CLUTCH CONTROL	A2	GRN/BLK
			NOT USED	A3	
	12	14	EGR CONTROL	A4	GRN
	0	14	"CHECK ENGINE" LIGHT	A5	BLU
	12	14	IGN - ECM FUSE	A6	BLK/WHT
				A7	
	2-5	2.5	SERIAL DATA	A8	ORN
	5	5	DIAG. TERM.	A9	WHT/BLU
	①	①	SPEED SENSOR SIGNAL	A10	YELLOW
	0	0	MAP GROUND	A11	PNK
	0	0	SYSTEM GROUND	A12	WHT/BLK
			NOT USED	C1	
9	12	14	EAC SOL.	C2	BLK/PNK
	NOT USEABLE		IAC "B" LO	C3	LT GRN/BLK
	NOT USEABLE		IAC "B" HI	C4	GRN/WHT
	NOT USEABLE		IAC "A" HI	C5	BLU/WHT
	NOT USEABLE		IAC "A" LO	C6	BLU/BLK
			NOT USED	C7	
			NOT USED	C8	
	0	0	CRANK SIGNAL	C9	WHT/BLK
2	1.9	1.7	COOLANT TEMP. SIGNAL	C10	WHT
	4.9	2.0	MAP SIGNAL	C11	LT GRN
			NOT USED	C12	
5	.48	.48	TPS SIGNAL	C13	BLU/RED
	5	5	5 VOLT REFERENCE	C14	GRN/YEL
			NOT USED	C15	
	12	14	BATTERY	C16	WHT

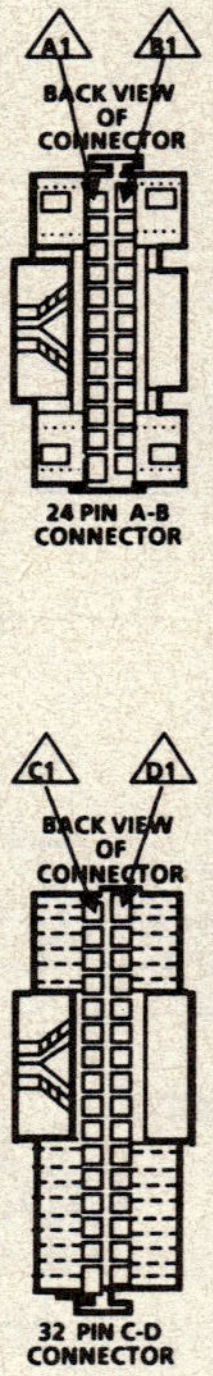

WIRE COLOR	PIN	CIRCUIT	VOLTAGE KEY "ON"	VOLTAGE ENG. RUN	Note
WHT	B1	BATTERY	12	14	
BLK/ORN	B2	FUEL PUMP SIGNAL	0	14	4
RED/BLK	B3	EST REF LOW	0	0	
	B4	NOT USED			
PNK/WHT	B5	EST REFERENCE HI	0	1.6	
	B6	NOT USED			
BLK	B7	ESC SIGNAL	9	9	
GRN/BLU	B8	A/C SIGNAL	0	0/12	8
BRN	B9	NOT USED			
	B10	NOT USED			
	B11	NOT USED			
	B12	NOT USED			
WHT/BLU	D1	SYSTEM GROUND	0	0	
BLK	D2	5V RETURN	0	0	
	D3	NOT USED			
WHT/RED	D4	EST CONTROL	0	1.0	
BLK/WHT	D5	EST BYPASS	0	4.75	
WHT/GRN	D6	SENSOR GRD. (0_2)	0	0	
YEL/BLU	D7	0^2 SENSOR SIGNAL	0	③	
	D8	NOT USED			
	D9	NOT USED			
	D10	NOT USED			
	D11	NOT USED			
	D12	NOT USED			
	D13	NOT USED			
GRN	D14	INJECTOR B	12	14	
	D15	NOT USED			
BLU	D16	INJECTOR A	12	14	

1 Varies from .60 to battery voltage depending on position of drive wheels.
2 Varies with temperature.
3 Varies. (toggles)
4 12V First two seconds.
5 Measured between terminals C13 and D2 (± .06V).
6 12 V in gear
7 0 V with A/C "on"
8 12 V with A/C "on"
9 12 V when cold

ENGINE 2.8L LL2

2.8L ENGINE—DIAGNOSTIC CIRCUIT CHECK—TROOPER AND TROOPER II

The Diagnostic Circuit Check is an organized approach to identifying a problem created by an electronic engine control system malfunction. It must be the starting point for any driveability complaint diagnosis, because it directs the service technician to the next logical step in diagnosing the complaint.

The "Scan Data" listed in the table may be used for comparison, after completing the diagnostic circuit check and finding the on-board diagnostics functioning properly and no trouble codes displayed. The "Typical Values" are an average of display values recorded from normally operating vehicles and are intended to represent what a normally functioning system would typically display.

A "SCAN" TOOL THAT DISPLAYS FAULTY DATA SHOULD NOT BE USED, AND THE PROBLEM SHOULD BE REPORTED TO THE MANUFACTURER. THE USE OF A FAULTY "SCAN" CAN RESULT IN MISDIAGNOSIS AND UNNECESSARY PARTS REPLACEMENT.

Only the parameters listed below are used in this manual for diagnosis. If a "Scan" tool reads other parameters, the values are not recommended by Isuzu for use in diagnosis.

"SCAN" DATA

Idle / Upper Radiator Hose Hot / Closed Throttle / Park or Neutral / Closed Loop / Acc. off

"SCAN" Position	Units Displayed	Typical Data Value
Coolant Temp.	C°	85°-105°
TPS	Volts	.48 +-.06
MAP	Volts	1 - 2 (depends on Vac. & Baro pressure)
INT (Integrator)	Counts	Varies
BLM (Block Learn)	Counts	118-138
IAC	Counts (steps)	1 - 50
RPM	RPM	650 ± 100 RPM
O_2	Volts	001 - 999 and varies
Open/Closed Loop	Open/Closed	Closed Loop (will go open at idle)
A/C Request	Yes/No	No (yes, with A/C requested)
VSS	MPH	0
Battery	Volts	13.5 - 14.5
Air Divert	Normal/Divert	Normal
Air Switch	Converter/Port	Converter
Knock Signal	Yes/No	No

DIAGNOSTIC CIRCUIT CHECK

2.8L (VIN R) "KT" (TBI)

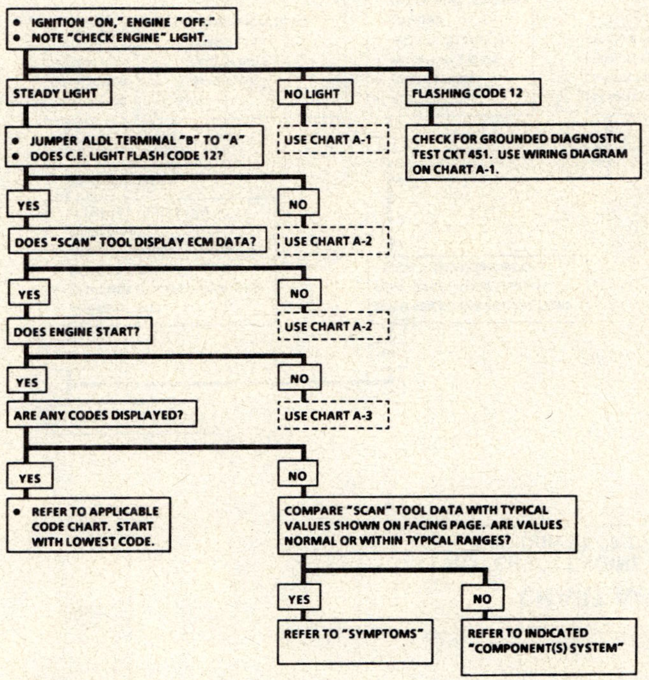

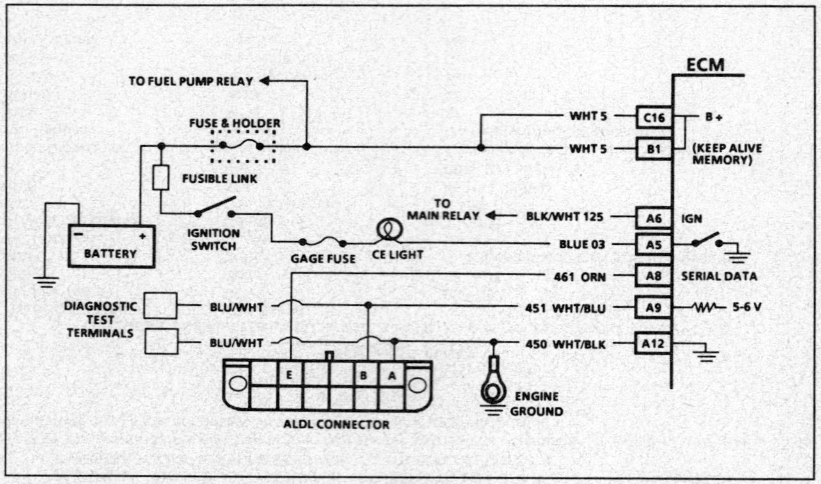

CHART A-1

NO "CHECK ENGINE" LIGHT
2.8L (VIN R) "KT" (TBI)

Circuit Description:

There should always be a steady "Check Engine" light, when the ignition is "ON" and engine stopped. Ignition voltage is supplied directly to the light bulb. The electronic control module (ECM) will control the light and turn it "ON" by providing a ground path through CKT 03 to the ECM.

Test Description: Numbers below refer to circled numbers on the diagnostic chart.

1. Battery feed CKT 5 is protected by a 10A in-line fuse. If this fuse was blown, refer to wiring diagram on the facing page of Code 54.
2. Using a test light connected to 12 volts, probe each of the system ground circuits to be sure a good ground is present. See ECM terminal end view in front of this section for ECM pin locations of ground circuits.

Diagnostic Aids:

Engine runs OK, check:
- Faulty light bulb
- CKT 03 open
- Gage fuse blown. This will result in no oil, or generator lights, seat belt reminder, etc.

Engine cranks, but will not run.
- Continuous battery - fuse or fusible link open
- ECM ignition fuse open
- Battery CKT 5 to ECM open
- Ignition CKT 125 to ECM open
- Poor connection to ECM

CHART A-1

NO "CHECK ENGINE " LIGHT
2.8L (VIN R) "KT" (TBI)

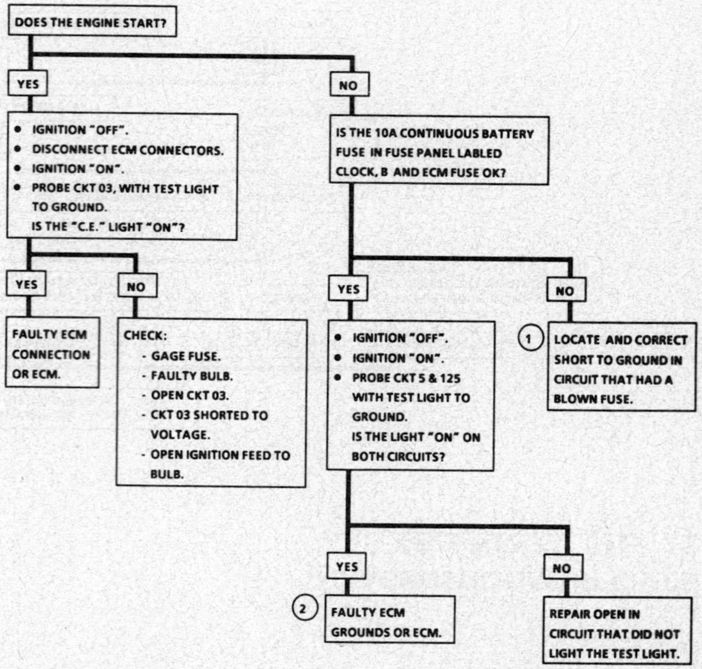

CLEAR CODES AND CONFIRM "CLOSED LOOP" OPERATION AND NO "CHECK ENGINE" LIGHT.

2.8L ENGINE – PRELIMINARY CHECK CHARTS – TROOPER AND TROOPER II

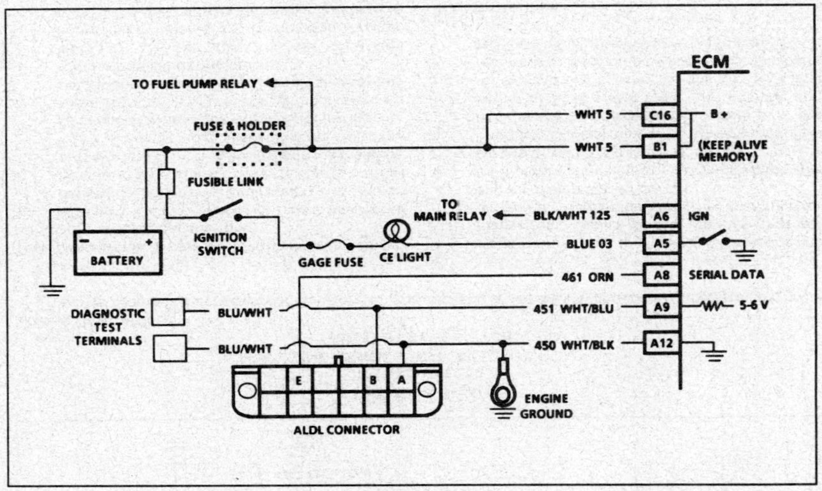

CHART A-2

NO ALDL DATA OR WON'T FLASH CODE 12 "CHECK ENGINE" LIGHT "ON" STEADY 2.8L (VIN R) "KT" (TBI)

Circuit Description:

There should always be a steady "Check Engine" light, when the ignition is "ON" and engine stopped. Ignition voltage is supplied directly to the light bulb. The electronic control module (ECM) will turn the light "ON" by grounding CKT 03 at the ECM.

With the diagnostic terminal grounded, the light should flash a Code 12, followed by any trouble code(s) stored in memory.

A steady light suggests a short to ground in the light control CKT 03, or an open in diagnostic CKT 451.

Test Description: Numbers below refer to circled numbers on the diagnostic chart.

1. If there is a problem with the ECM that causes a "Scan" tool to not read serial data, then the ECM should not flash a Code 12. If Code 12 does flash, be sure that the "Scan" tool is working properly on another vehicle. If the "Scan" is functioning properly and CKT 461 is OK, the Mem-Cal or ECM may be at fault for the NO ALDL symptom.
2. If the light goes "OFF," when the ECM connector is disconnected, then CKT 03 is not shorted to ground.
3. This step will check for an open diagnostic CKT451.
4. At this point, the "Check Engine" light wiring is OK. The problem is a faulty ECM or PROM. If Code 12 does not flash, the ECM should be replaced using the original PROM. Replace the PROM only after trying an ECM, as a defective PROM is an unlikely cause of the problem.

CHART A-2

NO ALDL DATA OR WON'T FLASH CODE 12 "CHECK ENGINE" LIGHT "ON" STEADY 2.8L (VIN R) "KT" (TBI)

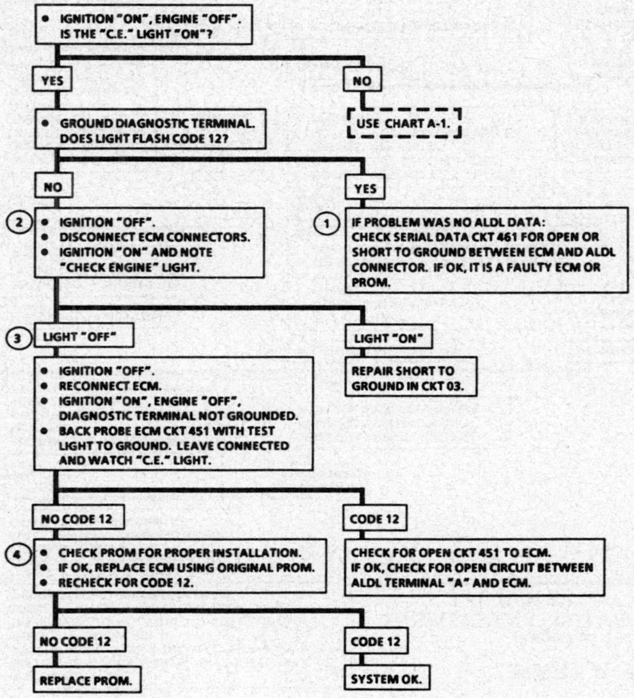

CLEAR CODES AND CONFIRM "CLOSED LOOP" OPERATION AND NO "CHECK ENGINE" LIGHT.

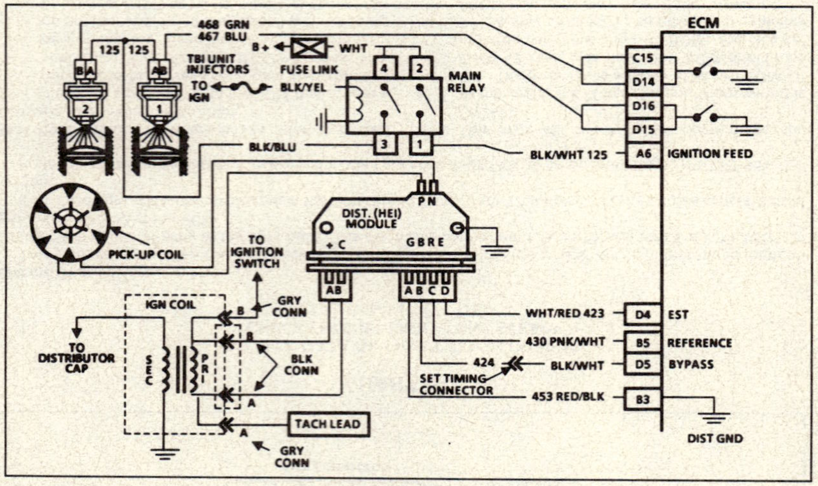

CHART A-3
(Page 1 of 2)
ENGINE CRANKS BUT WILL NOT RUN
2.8L (VIN R) "KT" (TBI)

Circuit Description:

This chart assumes that battery condition and engine cranking speed are OK, and there is adequate fuel in the tank.

Test Description: Numbers below refer to circled numbers on the diagnostic chart.

1. A "Check Engine" light "ON" is a basic test to determine if there is a 12 volt supply and ignition 12 volts to ECM. No ALDL may be due to an ECM problem and CHART A-2 will diagnose the ECM. If TPS is over 2.5 volts, the engine may be in the clear flood mode which will cause starting problems. The engine will not start without reference pulses and therefore, the "Scan" should read rpm (reference) during crank.
2. No spark may be caused by one of several components related to the ignition system. CHART C-4 will address all problems related to the causes of a no spark condition.
3. Fuel spray from the injector(s) indicates that fuel is available, however, the engine could be severely flooded due to too much fuel.
4. While cranking engine, there should be no fuel spray with injector disconnected. Replace an injector if it sprays fuel or drips like a leaking water faucet.
5. The fuel pressure will drop after the fuel pump stops running due to a controlled bleed in the fuel system. Use of the fuel pressure gage will determine if fuel system pressure is enough for engine to start and run.
6. No fuel spray from injector indicates a faulty fuel system or no ECM control of injector.
7. This test will determine if the ignition module is not generating the reference pulse, or if the wiring or ECM are at fault. By touching and removing a test light to 12 volts on CKT 430, a reference pulse should be generated. If injector test light blinks, the ECM and wiring are OK.

Diagnostic Aids:

- Water or foreign material can cause a no start during freezing weather.
- An EGR sticking open can cause a low air/fuel ratio during cranking.
- Fuel pressure: Low fuel pressure can result in a very lean air/fuel ratio. See CHART A-7.
- A grounded CKT 423 (EST) may cause a "No-Start" or a "Start then Stall" condition.

CHART A-3
(Page 1 of 2)
ENGINE CRANKS BUT WILL NOT RUN
2.8L (VIN R) "KT" (TBI)

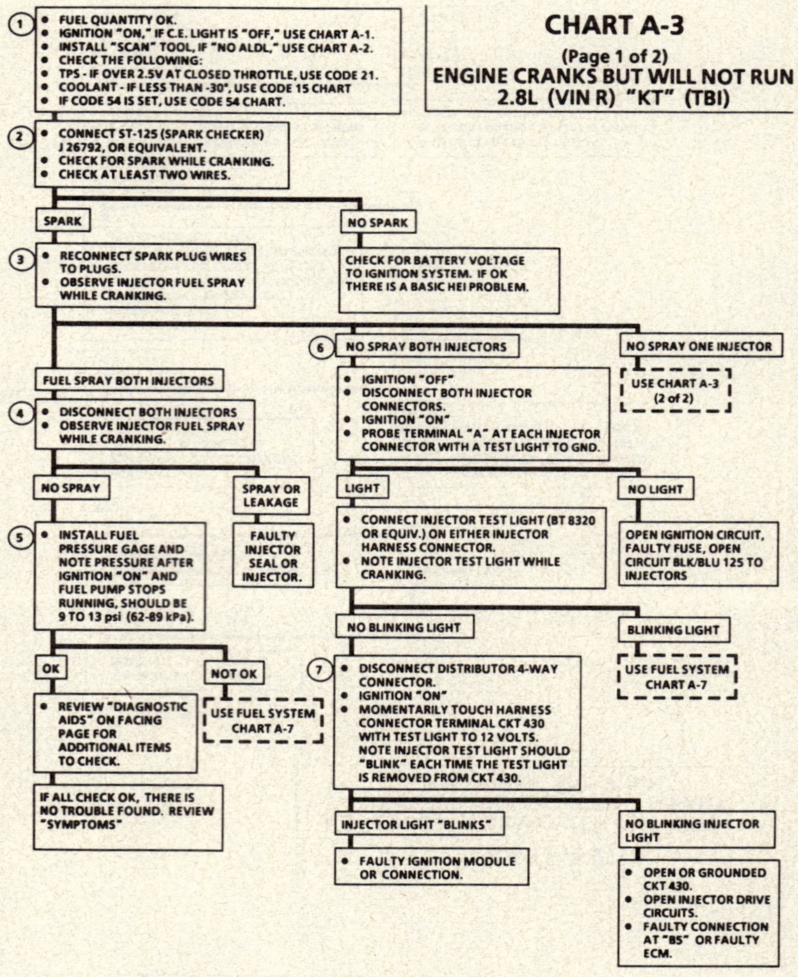

CLEAR CODES AND CONFIRM "CLOSED LOOP" OPERATION AND NO "CHECK ENGINE" LIGHT.

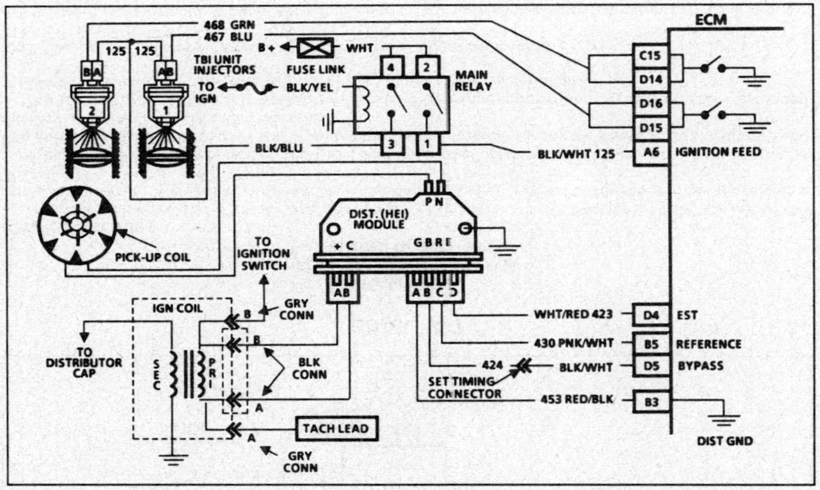

CHART A-3

(Page 2 of 2)
ENGINE CRANKS BUT WILL NOT RUN
2.8L (VIN R) "KT" (TBI)

Circuit Description:
This chart assumes that battery condition and engine cranking speed are OK, and there is adequate fuel in the tank.

Test Description: Numbers below refer to circled numbers on the diagnostic chart.

1. No fuel spray from one injector indicates a faulty fuel injector or no ECM control of injector. If the test light "blinks" while cranking, then ECM control should be considered OK. Be sure test light makes good contact between connector terminals during test. The light may be a little dim when "blinking." This is due to current draw of the test light. How bright it "blinks" is not important. The test light bulb should be a J36400-1 or equivalent.
2. CKTs BLK/BLU 125 supply ignition voltage to the injectors. Probe each connector terminal with a test light to ground. There should be a light "ON" at one terminal. If the test light confirms ignition voltage at the connector, the ECM injector control CKT 467 or 468 may be open. Reconnect the injector and using a test light connected to ground, check for a light at the applicable ECM connector terminal "D14" or "D16". A light at this point indicates that the injector drive circuit involved is OK
If an ECM repeat failure has occurred, the injector is shorted. Replace the injector and ECM.

CHART A-3

(Page 2 of 2)
ENGINE CRANKS BUT WILL NOT RUN
2.8L (VIN R) "KT" (TBI)

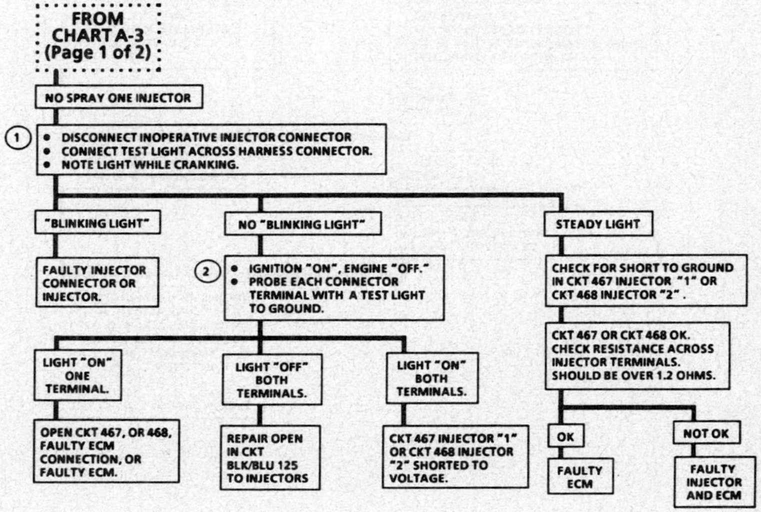

CLEAR CODES AND CONFIRM "CLOSED LOOP" OPERATION AND NO "CHECK ENGINE" LIGHT.

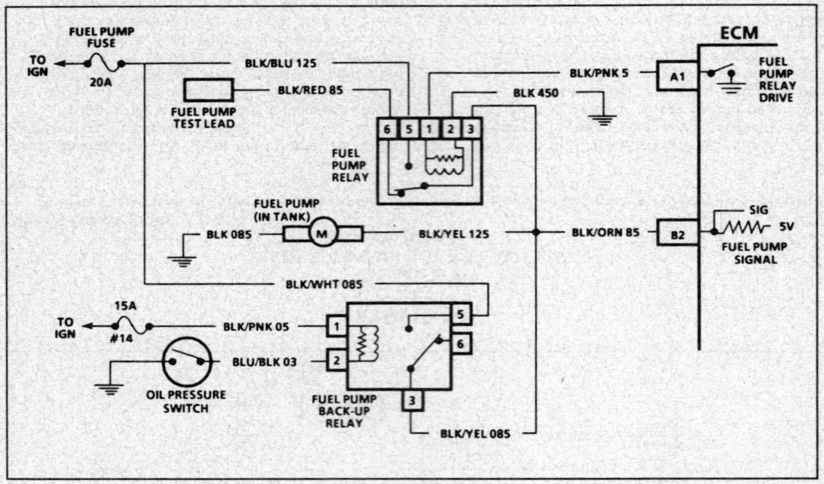

CHART A-7

(Page 1 of 2)
FUEL SYSTEM DIAGNOSIS
2.8L (VIN R) "KT" (TBI)

Circuit Description:

When the ignition switch is turned "ON," the electronic control module (ECM) will turn "ON" the in-tank fuel pump. It will remain "ON" as long as the engine is cranking or running, and the ECM is receiving ignition reference pulses.

If there are no reference pulses, the ECM will shut "OFF" the fuel pump within 2 seconds after key "ON." The pump will deliver fuel to the TBI unit, where the system pressure is controlled to 62 to 90 kPa (9 to 13 psi). Excess fuel is then returned to the fuel tank.

The fuel pump test terminal is located in the left side of the engine compartment. When the engine is stopped, the pump can be turned "ON" by applying battery voltage to the test terminal.

Test Description: Numbers below refer to circled numbers on the diagnostic chart.

1. Fuel pressure should be noted while fuel pump is running. Fuel pressure will drop immediately after fuel pump stops running due to a controlled bleed in the fuel system.

Diagnostic Aids:

Improper fuel system pressure can result in one of the following symptoms:

- Cranks, but won't run.
- Code 44.
- Code 45.
- Cuts out, may feel like ignition problem.
- Poor fuel economy, loss of power.
- Hesitation.

CHART A-7

(Page 1 of 2)
FUEL SYSTEM DIAGNOSIS
2.8L (VIN R) "KT" (TBI)

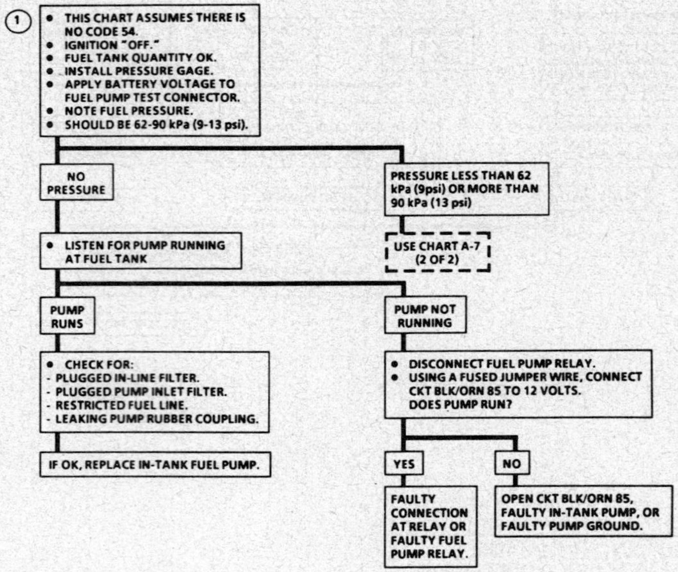

CLEAR CODES AND CONFIRM "CLOSED LOOP" OPERATION AND NO "CHECK ENGINE" LIGHT.

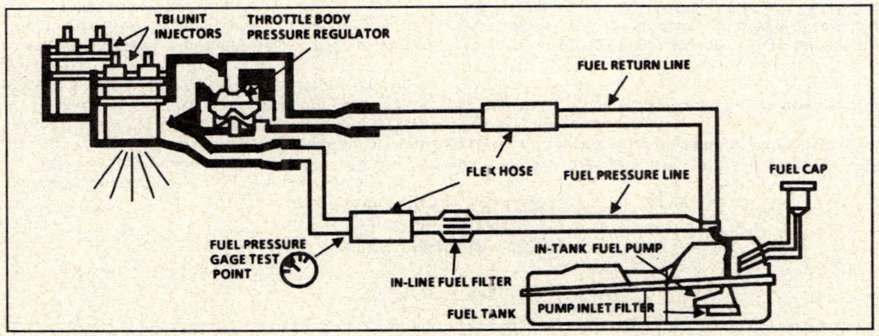

CHART A-7

(Page 2 of 2)
FUEL SYSTEM DIAGNOSIS
2.8L (VIN R) "KT" (TBI)

Test Description: Numbers below refer to circled numbers on the diagnostic chart.

1. Pressure than 62 kPa (9 psi) falls into two areas:
   - Regulated pressure than 62 kPa (9 psi). Amount of fuel to injectors OK but pressure is too low. System will be lean and may set Code 44. Also, hard starting cold and poor overall performance.
   - Restricted flow causing pressure drop. Normally, a vehicle with a fuel pressure of less than 62 kPa (9 psi) at idle will not be driveable. However, if the pressure drop occurs only while driving, the engine will normally surge then stop as pressure begins to drop rapidly.
2. Restricting the fuel return line allows the fuel pump to develop its maximum pressure (dead head pressure). When battery voltage is applied to the pump test terminal, pressure should be from 90 to 124 kPa (13 to 18 psi).
3. This test determines if the high fuel pressure is due to a restricted fuel return line or a throttle body pressure regulator problem.

CHART A-7

(Page 2 of 2)
FUEL SYSTEM DIAGNOSIS
2.8L (VIN R) "KT" (TBI)

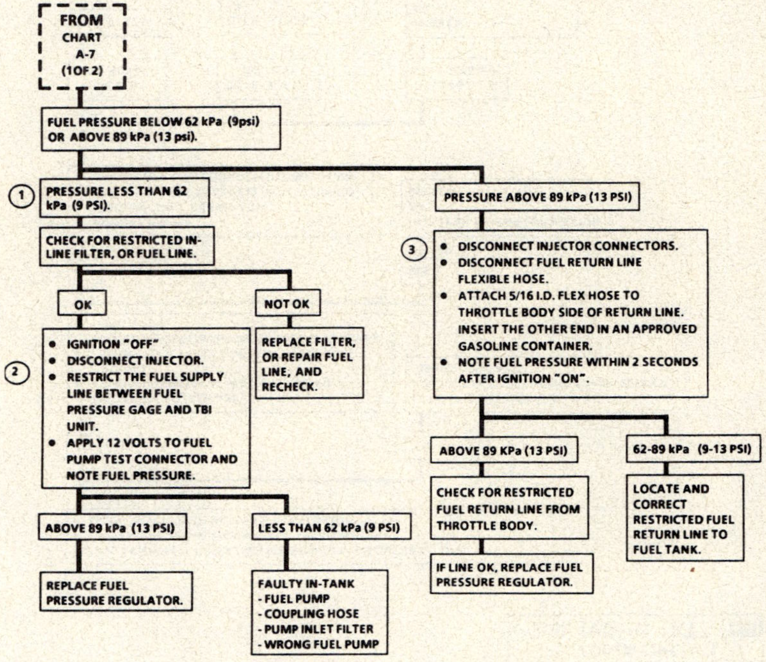

CLEAR CODES AND CONFIRM "CLOSED LOOP" OPERATION AND NO "CHECK ENGINE" LIGHT.

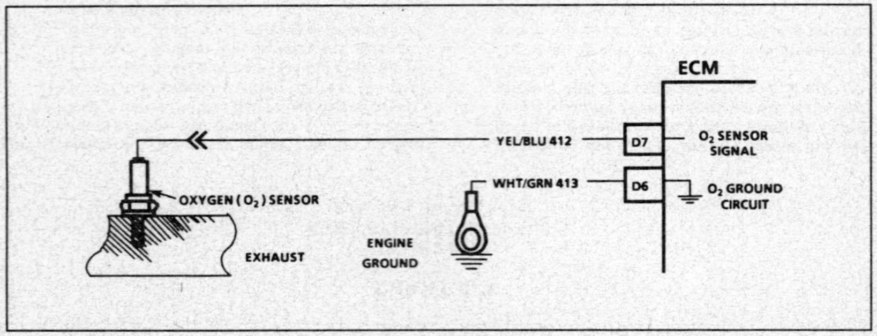

CODE 1 3

OXYGEN SENSOR CIRCUIT
(OPEN CIRCUIT)
2.8L (VIN R) "KT" (TBI)

Circuit Description:

The ECM supplies a voltage of about .45 volt between terminals "D7" and "D6". (If measured with a 10 megohm digital voltmeter, this may read as low as .32 volt). The O_2 sensor varies the voltage within a range of about 1 volt if the exhaust is rich, down through about .10 volt if exhaust is lean.

The sensor is like an open circuit and produces no voltage when it is below 315°C (600°F). An open sensor circuit or cold sensor causes "Open Loop" operation.

Test Description: Numbers below refer to circled numbers on the diagnostic chart.

1. Code 13 will set:
   - Engine at normal operating temperature.
   - At least 2 minutes engine time after start.
   - O_2 signal voltage steady between .35 and .55 volt.
   - rpm above 1600.
   - Throttle position sensor signal above 5% (about .3 volt above closed throttle voltage).
   - All conditions must be met for about 60 seconds.

   If the conditions for a Code 13 exist, the system will not go "Closed Loop."
2. This will determine if the sensor is at fault or the wiring or ECM is the cause of the Code 13.
3. In doing this test, use only a high impedance digital volt ohmmeter. This test checks the continuity of CKT(s) 412 and 413. If CKT 413 is open, the ECM voltage on CKT 412 will be over .6 volt (600 mV).

Diagnostic Aids:

Normal "Scan" voltage varies between 100 mv to 999 mV (.1 and 1.0 volt), while in "Closed Loop." Code 13 sets in one minute, if voltage remains between .35 and .55 volt, but the system will go "Open Loop" in about 15 seconds.

Refer to "Intermittents"

CODE 13

OXYGEN SENSOR CIRCUIT
(OPEN CIRCUIT)
2.8L (VIN R) "KT" (TBI)

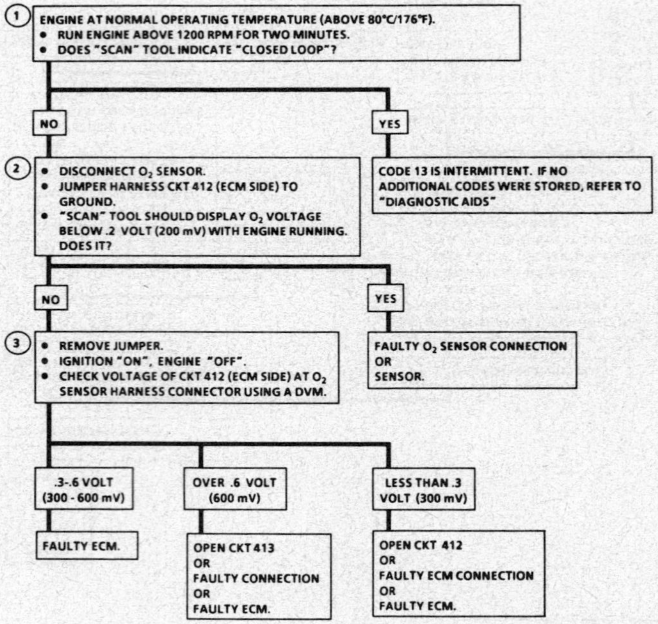

CLEAR CODES AND CONFIRM "CLOSED LOOP" OPERATION AND NO "CHECK ENGINE" LIGHT.

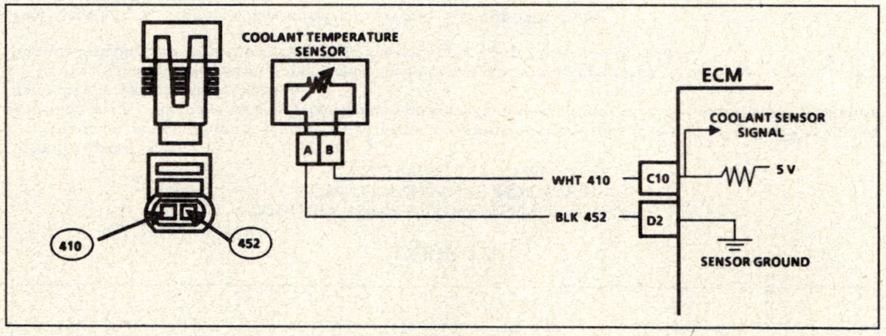

CODE 14

COOLANT TEMPERATURE SENSOR CIRCUIT (HIGH TEMPERATURE INDICATED) 2.8L (VIN R) "KT" (TBI)

Circuit Description:

The coolant temperature sengor uses a thermistor to control the signal voltage to the ECM. The ECM applies a voltage on CKT 410 to the sensor. When the engine is cold the sensor (thermistor) resistance is high, therefore the ECM will see high signal voltage.

As the engine warms, the sensor resistance becomes less, and the voltage drops. At normal engine operating temperature (85°C to 95°C) the voltage will measure about 1.5 to 2.0 volts.

Test Description: Numbers below refer to circled numbers on the diagnostic chart.

1. Code 14 will set if:
   - Signal voltage indicates a coolant temperature above 135°C (275°F) for 2 seconds.
2. This test will determine if CKT 410 is shorted to ground which will cause the conditions for Code 14.

Diagnostic Aids:

Check harness routing for a potential short to ground in CKT 410.

"Scan" tool displays engine temp. in degrees centigrade. After engine is started, the temperature should rise steadily to about 90°C then stabilize when thermostat opens.

Refer to "Intermittents"

CODE 14

COOLANT TEMPERATURE SENSOR CIRCUIT (HIGH TEMPERATURE INDICATED) 2.8L (VIN R) "KT" (TBI)

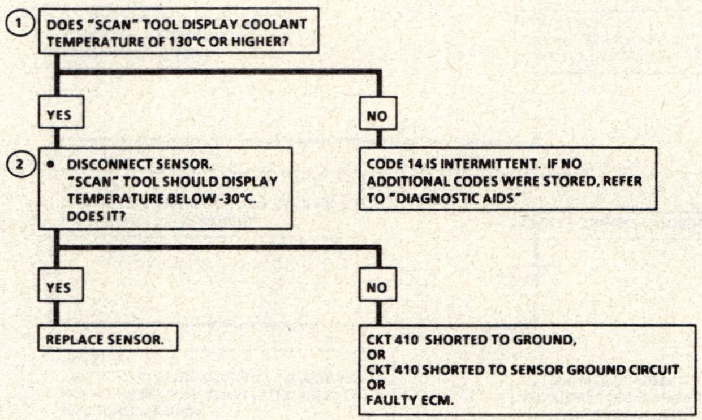

DIAGNOSTIC AID

COOLANT SENSOR		
TEMPERATURE VS. RESISTANCE VALUES (APPROXIMATE)		
°F	°C	OHMS
210	100	185
160	70	450
100	38	1,800
70	20	3,400
40	4	7,500
20	-7	13,500
0	-18	25,000
-40	-40	100,700

CLEAR CODES AND CONFIRM "CLOSED LOOP" OPERATION AND NO "CHECK ENGINE" LIGHT.

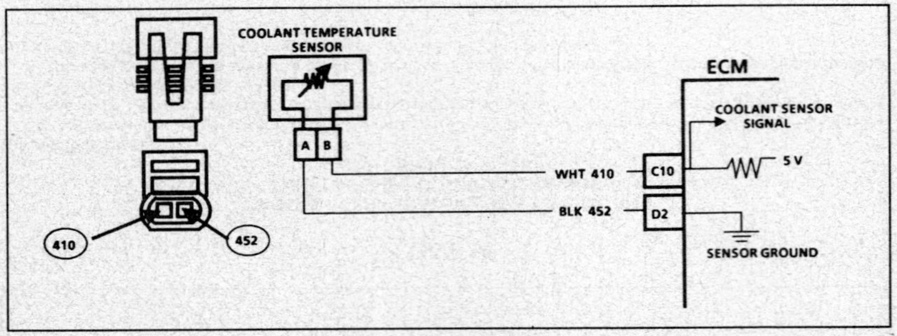

CODE 15

COOLANT TEMPERATURE SENSOR CIRCUIT
(LOW TEMPERATURE INDICATED)
2.8L (VIN R) "KT" (TBI)

Circuit Description:

The coolant temperature sensor uses a thermistor to control the signal voltage to the ECM. The ECM applies a voltage on CKT 410 to the sensor. When the engine is cold the sensor (thermistor) resistance is high, therefore the ECM will see high signal voltage.

As the engine warms, the sensor resistance becomes less, and the voltage drops. At normal engine operating temperature (85°C to 95°C) the voltage will measure about 1.5 to 2.0 volts at the ECM.

Test Description: Numbers below refer to circled numbers on the diagnostic chart.

1. Code 15 will set if:
   - Engine running longer than 30 seconds.
   - Coolant temp. less than -30°C (-22°F), for 3 seconds.
2. This test simulates a Code 14. If the ECM recognizes the low signal voltage, (high temperature) and "Scan" reads 130°C or above, the ECM and wiring are OK.
3. This test will determine if CKT 410 is open. There should be 5 volts present at sensor connector if measured with a DVM.

Diagnostic Aids

A "Scan" tool reads engine temperature in degrees centigrade. After engine is started the temperature should rise steadily to about 90°C then stabilize when thermostat opens.

If Code 21 is also set, check CKT 452 for faulty wiring or connections. Check terminals at sensor for good contact.

Refer to "Intermittents"

CODE 15

COOLANT TEMPERATURE SENSOR CIRCUIT
(LOW TEMPERATURE INDICATED)
2.8L (VIN R) "KT" (TBI)

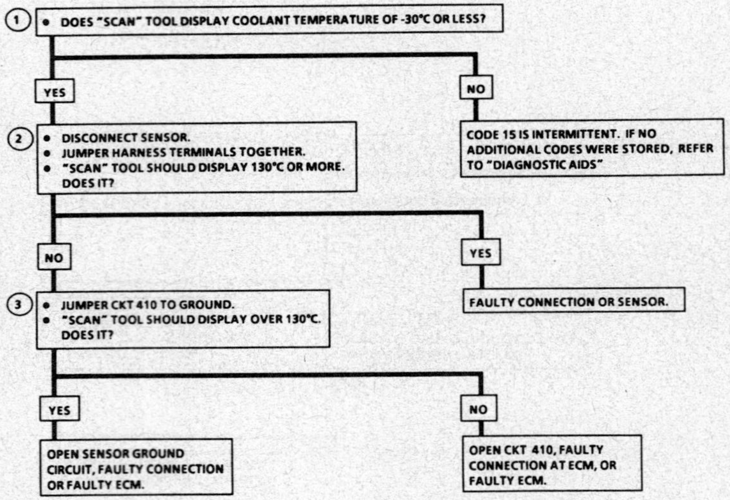

DIAGNOSTIC AID

COOLANT SENSOR
TEMPERATURE TO RESISTANCE VALUES (APPROXIMATE)

°F	°C	OHMS
210	100	185
160	70	450
100	38	1,800
70	20	3,400
40	4	7,500
20	-7	13,500
0	-18	25,000
-40	-40	100,700

CLEAR CODES AND CONFIRM "CLOSED LOOP" OPERATION AND NO "CHECK ENGINE" LIGHT.

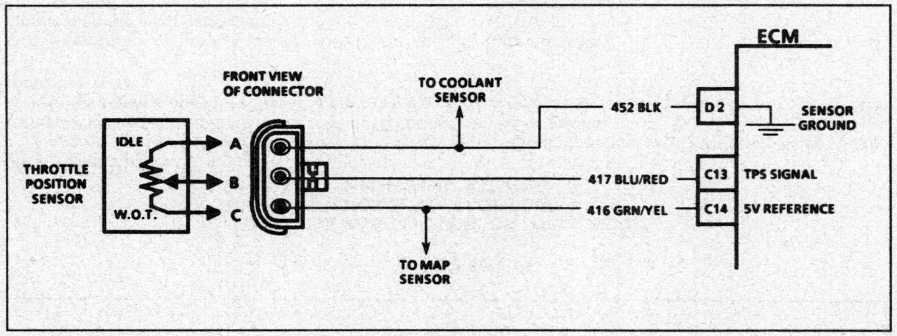

CODE 21

THROTTLE POSITION SENSOR (TPS) CIRCUIT
(SIGNAL VOLTAGE HIGH)
2.8L (VIN R) "KT" (TBI)

Circuit Description:

The throttle position sensor (TPS) provides a voltage signal that changes relative to the throttle blade. Signal voltage will vary from about .5 at idle to about 5 volts at wide open throttle.

The TPS signal is one of the most important inputs used by the ECM for fuel control and for most of the ECM control outputs.

Test Description: Numbers below refer to circled numbers on the diagnostic chart.

1. Code 21 will set if:
   - TPS signal voltage is greater than 2.5 volts.
   - All conditions met for 8 seconds.
   - MAP less than 52 kPa.
2. With the TPS sensor disconnected, the TPS voltage should go low if the ECM and wiring are OK.
3. Probing CKT 452 with a test light to 12 volts checks the sensor ground CKT. A faulty sensor ground will cause a Code 21.

Diagnostic Aids:

A "Scan" tool reads throttle position in volts. Should read less than 1.25 volts with throttle closed and ignition "ON" or at idle. Voltage should increase at a steady rate as throttle is moved toward WOT.

An open in CKT 452 will result in a Code 21. Refer to "Intermittents"

CODE 21

THROTTLE POSITION SENSOR (TPS) CIRCUIT
(SIGNAL VOLTAGE HIGH)
2.8L (VIN R) "KT" (TBI)

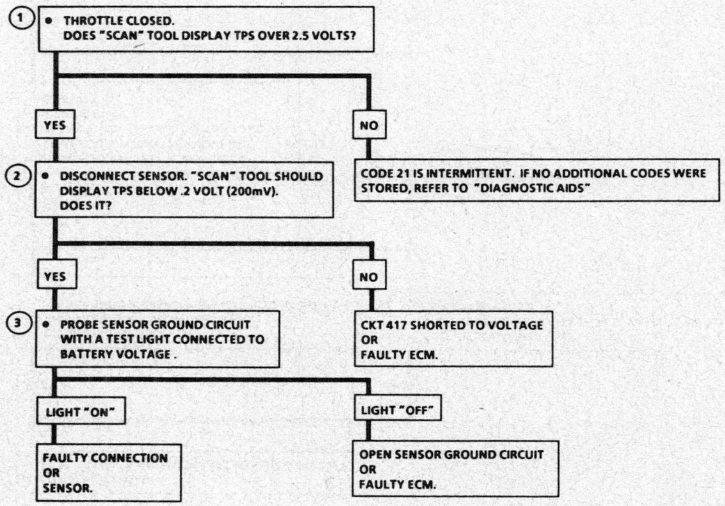

CLEAR CODES AND CONFIRM "CLOSED LOOP" OPERATION AND NO "CHECK ENGINE" LIGHT.

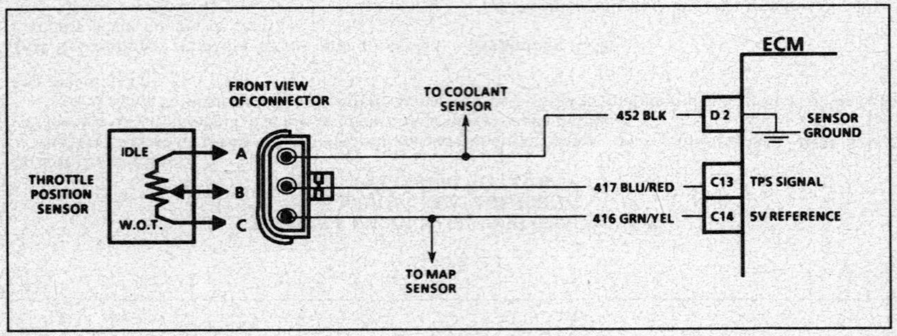

CODE 22

THROTTLE POSITION SENSOR (TPS) CIRCUIT (SIGNAL VOLTAGE LOW) 2.8L (VIN R) "KT" (TBI)

Circuit Description:

The throttle position sensor (TPS) provides a voltage signal that changes relative to the throttle blade. Signal voltage will vary from about .5 at idle to about 5 volts at wide open throttle.

The TPS signal is one of the most important inputs used by the ECM for fuel control and for most of the ECM control outputs.

Test Description: Numbers below refer to circled numbers on the diagnostic chart.

1. Code 22 will set if:
   - Engine is running
   - TPS signal voltage is less than about .2 volt for 2 seconds.
2. Simulates Code 21: (high voltage) If the ECM recognizes the high signal voltage then the ECM and wiring are OK.
3. This simulates a high signal voltage to check for an open in CKT 417. The "Scan" tool will not read up to 12 volts, but what is important is that the ECM recognizes the signal on CKT 417.
4. There should be 5 volts at terminal "C" if measured with a DVM when ignition is "ON."

Diagnostic Aids:

A "Scan" tool reads throttle position in volts. Should read less than 1.25 volts with throttle closed and ignition "ON" or at idle. Voltage should increase at a steady rate as throttle is moved toward WOT.

An open or short to ground in CKTs 416 or 417 will result in a Code 22.

If a Code 22 is also set check CKT 416 carefully for open or short to ground.

Refer to "Intermittents"

CODE 22

THROTTLE POSITION SENSOR (TPS) CIRCUIT (SIGNAL VOLTAGE LOW) 2.8L (VIN R) "KT" (TBI)

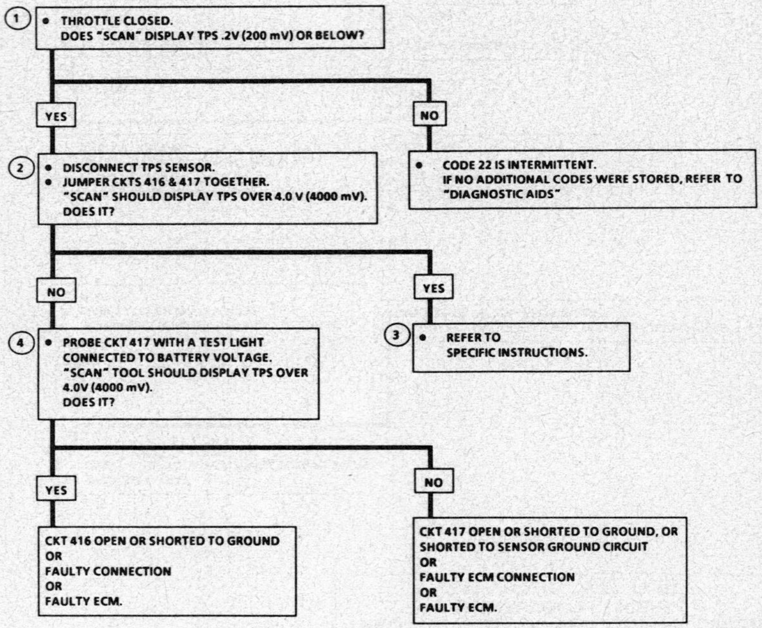

CLEAR CODES AND CONFIRM "CLOSED LOOP" OPERATION AND NO "CHECK ENGINE" LIGHT.

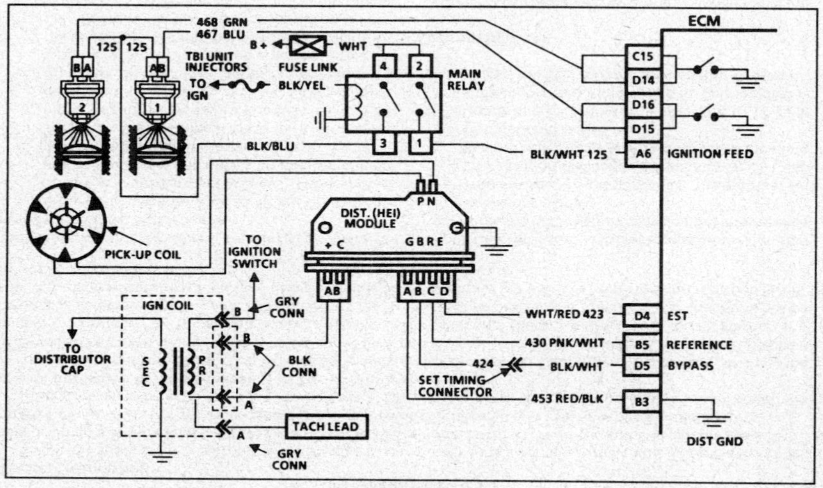

CODE 42

ELECTRONIC SPARK TIMING (EST) CIRCUIT
2.8L (VIN R) "KT" (TBI)

Circuit Description:

When the system is running on the ignition module, that is, no voltage on the bypass line, the ignition module grounds the EST signal. The ECM expects to see no voltage on the EST line during this condition. If it sees a voltage, it sets Code 42 and will not go into the EST mode.

When the rpm for EST is reached (about 400 rpm), and bypass voltage applied, the EST should no longer be grounded in the ignition module so the EST voltage should be varying.

If the bypass line is open or grounded, the ignition module will not switch to EST mode so the EST voltage will be low and Code 42 will be set.

If the EST line is grounded, the ignition module will switch to EST, but because the line is grounded there will be no EST signal. A Code 42 will be set.

Test Description: Numbers below refer to circled numbers on the diagnostic chart.

1. Code 42 means the ECM has seen an open or short to ground in the EST or bypass circuits. This test confirms Code 42 and that the fault causing the code is present.
2. Checks for a normal EST ground path through the ignition module. An EST CKT 423 shorted to ground will also read less than 500 ohms; however, this will be checked later.
3. As the test light voltage touches CKT 424, the module should switch causing the ohmmeter to "overrange" if the meter is in the 1000-2000 ohms position. Selecting the 10-20,000 ohms position will indicate above 5000 ohms. The important thing is that the module "switched".
4. The module did not switch and this step checks for:
   - EST CKT 423 shorted to ground.
   - Bypass CKT 424 open.
   - Faulty ignition module connection or module.
5. Confirms that Code 42 is a faulty ECM and not an intermittent in CKTs 423 or 424.

Diagnostic Aids:

If a Code 42 was stored and the customer complains of a "hard start," the problem is most likely a grounded EST line (CKT 423).

The "Scan" tool does not have any ability to help diagnose a Code 42 problem.

A PROM not fully seated in the ECM can result in a Code 42.

Refer to "Intermittents"

CODE 42

ELECTRONIC SPARK TIMING (EST) CIRCUIT
2.8L (VIN R) "KT" (TBI)

(1)
- CLEAR CODES.
- IDLE ENGINE FOR 1 MINUTE OR UNTIL CODE 42 SETS. DOES CODE 42 SET?

YES → (2)
- IGNITION "OFF".
- DISCONNECT ECM CONNECTORS.
- IGNITION "ON".
- OHMMETER SELECTOR SWITCH IN THE 1000 TO 2000 OHMS RANGE.
- PROBE ECM HARNESS CONNECTOR CKT 423 WITH AN OHMMETER TO GROUND. IT SHOULD READ LESS THAN 1000 OHMS. DOES IT?

NO → CODE 42 INTERMITTENT. REFER TO "DIAGNOSTIC AIDS"

(2) YES →
- PROBE ECM HARNESS CONNECTOR CKT 424 WITH A TEST LIGHT TO BATTERY VOLTAGE.

(2) NO → OPEN CKT 423, FAULTY CONNECTION OR FAULTY IGNITION MODULE.

LIGHT "OFF" → (3)
- WITH OHMMETER STILL CONNECTED TO ECM HARNESS CKT 423 AND GROUND. AGAIN PROBE ECM HARNESS CKT 424 WITH THE TEST LIGHT CONNECTED TO BATTERY VOLTAGE. (AS TEST LIGHT CONTACTS CKT 424, RESISTANCE SHOULD SWITCH FROM UNDER 1000 TO OVER 2000 OHMS.) DOES IT?

LIGHT "ON" →
- DISCONNECT IGNITION MODULE 4-WAY CONNECTOR.

LIGHT "ON" → CKT 424 SHORTED TO GROUND.

LIGHT "OFF" → FAULTY IGNITION MODULE.

(3) NO → (4)
- DISCONNECT DIST. 4-WAY CONNECTOR. NOTE OHMMETER THAT IS STILL CONNECTED TO CKT 423 AND GROUND. RESISTANCE SHOULD HAVE GONE HIGH (OPEN CIRCUIT). DOES IT?

(3) YES → (5)
- RECONNECT ECM AND IDLE ENGINE FOR ONE MINUTE OR UNTIL CODE 42 SETS. DOES CODE SET?

(4) YES → CKT 424 OPEN, FAULTY CONNECTIONS OR FAULTY IGNITION MODULE.

(4) NO → CKT 423 SHORTED TO GROUND.

(5) YES → FAULTY ECM

(5) NO → CODE 42 INTERMITTENT. REFER TO "DIAGNOSTIC AIDS"

CLEAR CODES AND CONFIRM "CLOSED LOOP" OPERATION AND NO "CHECK ENGINE" LIGHT.

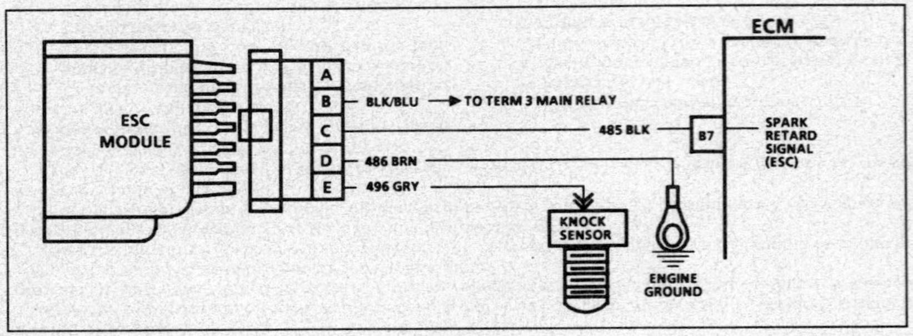

CODE 43

ELECTRONIC SPARK CONTROL (ESC) CIRCUIT
2.8L (VIN R) "KT" (TBI)

Circuit Description:

Electronic spark control is accomplished with a module that sends a voltage signal to the ECM. As the knock sensor detects engine knock, the voltage from the ESC module to the ECM drops, and this signals the ECM to retard timing. The ECM will retard the timing when knock is detected and rpm is above about 900 rpm.

Code 43 means the ECM has been low voltage at CKT 485 terminal "B7" for longer than 5 seconds with the engine running or the system has failed the functional check.

This system performs a functional check once per start up to check the ESC system. To perform this test the ECM will advance the spark when coolant is above 95°C and at a high load condition (near W.O.T.). The ECM then checks the signal at "B7" to see if a knock is detected. The functional check is performed once per start up and if knock is detected when coolant is below 95°C (194°F) the test has passed and the functional check will not be run. If the functional check fails, the "Service Engine Soon" light will remain "ON" until ignition is turned "OFF" or until a knock signal is detected.

Test Description: Numbers below refer to circled numbers on the diagnostic chart.

1. If the conditions for a Code 43 are present the "Scan" will always display "yes." There should not be a knock at idle unless an internal engine problem, or a system problem exists.
2. This test will determine if the system is functioning at this time. Usually a knock signal can be generated by tapping on the right exhaust manifold. If no knock signal is generated try tapping on block close to the area of the sensor.
3. Because Code 43 sets when the signal voltage on CKT 485 remains low this test should cause the signal on CKT 485 to go high. The 12 volts signal should be seen by the ECM as "no knock" if the ECM and wiring are OK.
4. This test will determine if the knock signal is being detected on CKT 496 or if the ESC module is at fault.
5. If CKT 496 is routed to close to secondary ignition wires the ESC module may see the interference as a knock signal.
6. This checks the ground circuit to the module. An open ground will cause the voltage on CKT 485 to be about 12 volts which would cause the Code 43 functional test to fail.
7. Contacting CKT 496 with a test light to 12 volts should generate a knock signal. This will determine if the ESC module is operating correctly.

Diagnostic Aids:

Code 43 can be caused by a faulty connection at the knock sensor at the ESC module or at the ECM. Also check CKT 485 for possible open or short to ground.

Refer to "Intermittents"

CODE 43

ELECTRONIC SPARK CONTROL (ESC) CIRCUIT
2.8L (VIN R) "KT" (TBI)

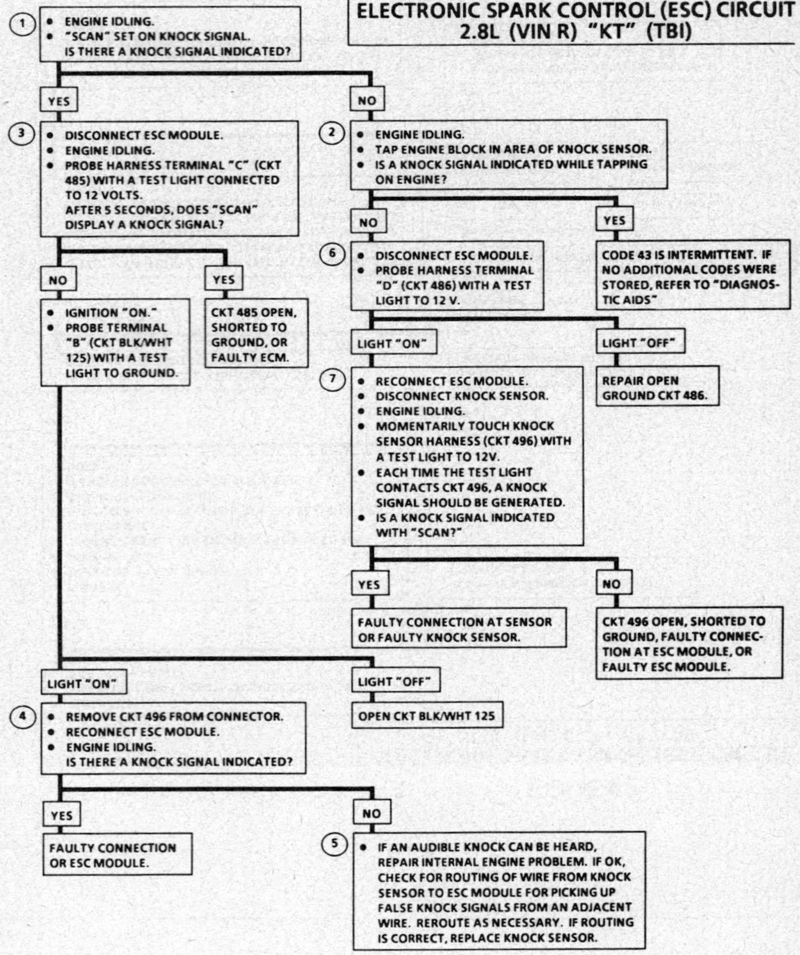

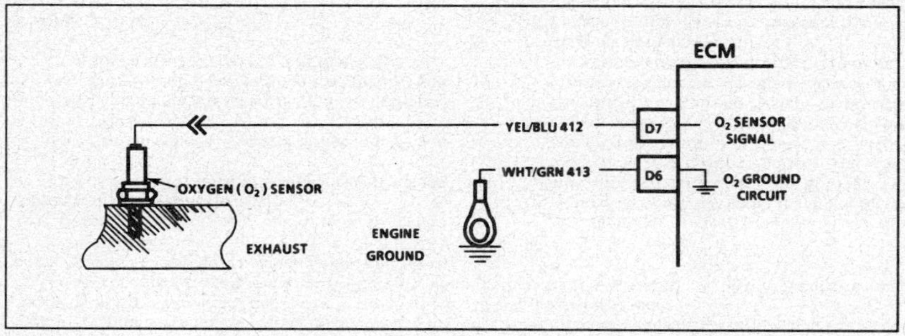

CODE 44

OXYGEN SENSOR CIRCUIT (LEAN EXHAUST INDICATED) 2.8L (VIN R) "KT" (TBI)

Circuit Description:

The ECM supplies a voltage of about .45 volt between terminals "D6" and "D7". (If measured with a 10 megohm digital voltmeter, this may read as low as .32 volt.) The O_2 sensor varies the voltage within a range of about 1 volt if the exhaust is rich, down through about .10 volt if exhaust is lean.

The sensor is like an open circuit and produces no voltage when it is below about 360°C (600°F) An open sensor circuit or cold sensor causes "Open Loop" operation.

Test Description: Numbers below refer to circled numbers on the diagnostic chart.

1. Code 44 is set when the O_2 sensor signal voltage on CKT 412.
   - Remains below .2 volt for 50 seconds.
   - And the system is operating in "Closed Loop."

Diagnostic Aids:

Using the "Scan," observe the block learn values at different rpm and air flow conditions to determine when the Code 44 may have been set. If the conditions for Code 44 exists the block learn values will be around 150.

- O_2 Sensor Wire. Sensor pigtail may be mispositioned and contacting the exhaust manifold.
- Check for intermittent ground in wire between connector and sensor.
- MAP Sensor. A (MAP) sensor output that causes the ECM to sense a higher than normal vacuum will cause the system to go lean. Disconnect the MAP sensor and if the lean condition is gone, replace the sensor.
- Lean Injector(s).
- Fuel Contamination. Water, even in small amounts, near the in-tank fuel pump inlet can be delivered to the injectors. The water causes a lean exhaust and can set a Code 44.
- Fuel Pressure. System will be lean if pressure is too low. It may be necessary to monitor fuel pressure while driving the car at various road speeds and/or loads to confirm. See "Fuel System Diagnosis" CHART A-7.
- Exhaust Leaks. If there is an exhaust leak, the engine can pull outside air into the exhaust and past the sensor. Vacuum or crankcase leaks can cause a lean condition.
- If the above are OK, it is a faulty oxygen sensor.

CODE 44

OXYGEN SENSOR CIRCUIT (LEAN EXHAUST INDICATED) 2.8L (VIN R) "KT" (TBI)

(1)
- RUN WARM ENGINE (75°C/167°F TO 95°C/203°F) AT 1200 RPM.
- DOES "SCAN" TOOL INDICATE O_2 SENSOR VOLTAGE FIXED BELOW .35 VOLT (350 mV)?

YES:
- DISCONNECT O_2 SENSOR.
- WITH ENGINE IDLING, "SCAN" TOOL SHOULD DISPLAY O_2 SENSOR VOLTAGE BETWEEN .35 VOLT AND .55 VOLT (350 mV AND 550 mV). DOES IT?

  YES: REFER TO "DIAGNOSTIC AIDS"

  NO: CKT 412 SHORTED TO GROUND OR FAULTY ECM.

NO: CODE 44 IS INTERMITTENT. IF NO ADDITIONAL CODES WERE STORED, REFER TO "DIAGNOSTIC AIDS"

CLEAR CODES AND CONFIRM "CLOSED LOOP" OPERATION AND NO "CHECK ENGINE" LIGHT.

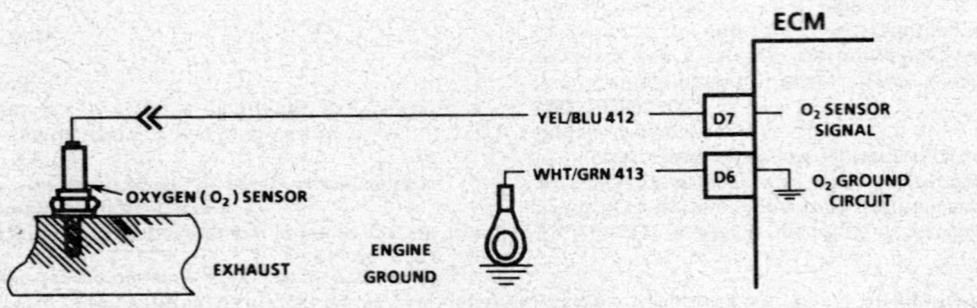

CODE 45

OXYGEN SENSOR CIRCUIT
(RICH EXHAUST INDICATED)
2.8L (VIN R) "KT" (TBI)

Circuit Description:
The ECM supplies a voltage of about .45 volt between terminals "D6" and "D7". (If measured with a 10 megohm digital voltmeter, this may read as low as .32 volt.) The O_2 sensor varies the voltage within a range of about 1 volt if the exhaust is rich, down through about . 10 volt if exhaust is lean.

The sensor is like an open circuit and produces no voltage when it is below about 360°C (600°F) An open sensor circuit or cold sensor causes "Open Loop" operation.

Test Description: Numbers below refer to circled numbers on the diagnostic chart.

1. Code 45 is set when the O_2 sensor signal voltage or CKT 412.
   - Remains above .7 volt for 50 seconds; and in "Closed Loop."
   - Engine time after start is 1 minute or more.
   - Throttle angle greater than 2% (about .2 volt above idle voltage) but less than 25%.

Diagnostic Aids:

Using the "Scan," observe the block learn values at different rpm conditions to determine when the Code 45 may have been set. If the conditions for Code 45 exists, The block learn values will be around 115.

- Fuel Pressure. System will go rich if pressure is too high. The ECM can compensate for some increase. However, if it gets too high, a Code 45 may be set. See "Fuel System Diagnosis" CHART A-7.
- Leaking injector. See CHART A-7.
- Check for fuel contaminated oil.
- HEI Shielding. An open ground CKT 453 (ignition system ref. low) may result in EMI, or induced electrical "noise." The ECM looks at this "noise" as reference pulses. The additional pulses result in a higher than actual engine speed signal. The ECM then delivers too much fuel, causing system to go rich. Engine tachometer will also show higher than actual engine speed, which can help in diagnosing this problem.
- Canister purge. Check for fuel saturation. If full of fuel, check canister control and hoses.
- MAP sensor. An output that causes the ECM to sense a lower than normal vacuum can cause the system to go rich. Disconnecting the MAP sensor will allow the ECM to set a fixed value for the sensor. Substitute a different MAP sensor if the rich condition is gone while the sensor is disconnected.
- TPS. An intermittent TPS output will cause the system to go rich, due to a false indication of the engine accelerating.

CODE 45

OXYGEN SENSOR CIRCUIT
(RICH EXHAUST INDICATED)
2.8L (VIN R) "KT" (TBI)

(1)
- RUN WARM ENGINE (75°C/167°F TO 95°C/203°F) AT 1200 RPM.
- DOES "SCAN" TOOL DISPLAY O_2 SENSOR VOLTAGE FIXED ABOVE .75 VOLT (750 mV)?

YES:
- DISCONNECT O_2 SENSOR AND JUMPER HARNESS CKT 412 TO GROUND.
- "SCAN" TOOL SHOULD DISPLAY O_2 BELOW .35 VOLT (350 mV). DOES IT?

  YES: REFER TO "DIAGNOSTIC AIDS"

  NO: REPLACE ECM.

NO: CODE 45 IS INTERMITTENT. IF NO ADDITIONAL CODES WERE STORED, REFER TO "DIAGNOSTIC AIDS"

CLEAR CODES AND CONFIRM "CLOSED LOOP" OPERATION AND NO "CHECK ENGINE" LIGHT.

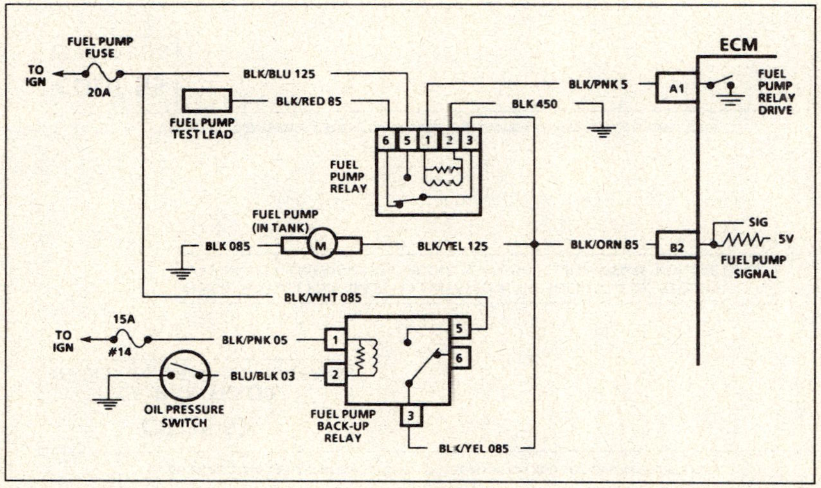

CODE 54

FUEL PUMP CIRCUIT
(LOW VOLTAGE)
2.8L (VIN R) "KT" (TBI)

Circuit Description:

When the ignition switch is turned "ON," the electronic control module (ECM) will activate the fuel pump relay and run the in-tank fuel pump. The fuel pump will operate as long as the engine is cranking or running, and the ECM is receiving ignition reference pulses.

If there are no reference pulses, the ECM will shut "OFF" the fuel pump within 2 seconds after key "ON." Should the fuel pump relay, or the 12 volt relay drive from the ECM fail, the fuel pump will be run through an oil pressure switch which activates the back up relay circuit.

Diagnostic Aids:

An inoperative fuel pump relay can result in long cranking times, particularly if the engine is cold or engine oil pressure is low. The extended crank period is caused by the time necessary for oil pressure to build enough to close the oil pressure switch and turn "ON" the fuel pump.

Intermittent problems may be caused by:
- Poor wiring connections
- Corrosion
- All connections should be checked visually

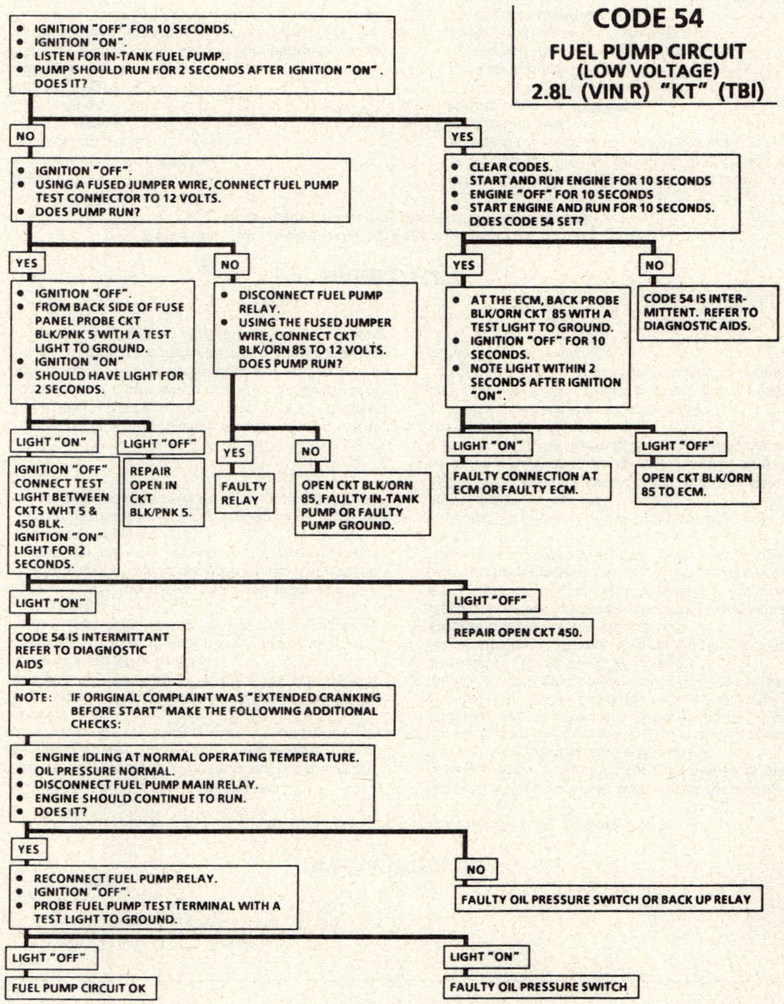

2.8L ENGINE – TROUBLE CODE CHARTS TROOPER AND TROOPER II

CODE 51

PROM ERROR (FAULTY OR INCORRECT PROM)

CHECK THAT ALL PINS ARE FULLY INSERTED IN THE SOCKET. IF OK, REPLACE PROM, CLEAR MEMORY AND RECHECK. IF CODE 51 REAPPEARS, REPLACE ECM.

CLEAR ALL CODES AND CONFIRM "CLOSED LOOP" OPERATION AND NO "CHECK ENGINE" LIGHT

CODE 52

CALPAK ERROR (FAULTY OR INCORRECT CALPAK)

CHECK THAT ALL PINS ARE FULLY INSERTED IN THE SOCKET. IF OK, REPLACE CALPAK, CLEAR MEMORY AND RECHECK. IF CODE 52 REAPPEARS, REPLACE ECM.

CLEAR ALL CODES AND CONFIRM "CLOSED LOOP" OPERATION AND NO "CHECK ENGINE" LIGHT

CODE 55

ECM ERROR

REPLACE ELECTRONIC CONTROL MODULE (ECM).

CLEAR ALL CODES AND CONFIRM "CLOSED LOOP" OPERATION AND NO "CHECK ENGINE" LIGHT

2.8L ENGINE – DRIVEABILITY SYMPTOMS

INTERMITTENTS

Problem may or may not turn "ON" the "Check Engine" light, or store a code.

DO NOT use the Trouble Code Charts for intermittent problems. The fault must be present to locate the problem. If a fault is intermittent, use of Trouble Code Charts may result in replacement of good parts.

- Most intermittent problems are caused by faulty electrical connections or wiring. Perform careful check of suspect circuits for:
  - Poor mating of the connector halves, or terminals, not fully seated in the connector body (backed out).
  - Improperly formed or damaged terminals. All connector terminals in problem circuit should be carefully reformed or replaced to insure proper contact tension.
  - Poor terminal to wire connection. This requires removing the terminal from the connector body
- If a visual (physical) check does not find the cause of the problem, the car can be driven with a voltmeter connected to a suspected circuit or a "Scan" tool may be used. An abnormal voltage reading, when the problem occurs, indicates the problem may be in that circuit. If the wiring and connectors check OK, and a Trouble Code was stored for a circuit having a sensor, except for Codes 44 and 45, substitute a known good sensor and recheck.
- Loss of trouble code memory. To check, disconnect TPS and idle engine until "Check Engine" light comes "ON." Code 22 should be stored, and kept in memory, when ignition is turned "OFF" for at least 10 seconds. If not, the ECM is faulty.
- An intermittent "Check Engine" light, and No Trouble Codes, may be caused by:
  - Electrical system interference caused by a defective relay, ECM driven solenoid, or switch. They can cause a sharp electrical surge. Normally, the problem will occur when the faulty component is operated.
  - Improper installation of electrical options, such as lights, 2-way radios, etc.
  - EST wires should be routed away from spark plug wires, ignition system components, and generator. Wire for CKT 453 from ECM to ignition system should be a good ground.
  - Ignition secondary shorted to ground.
  - CKTs BLU/PNK 03 ("Check Engine" light) or 451 (Diagnostic Test) intermittently shorted to ground.
  - ECM power grounds.

HARD START

Definition: Engine cranks OK, but does not start for a long time. Does eventually run, or may start but immediately dies.

- **CHECK:**
  - For water contaminated fuel.
  - Fuel system pressure CHART A-7.
  - TPS for sticking or binding should read less than 1.25 volts on a "Scan" tool.
  - No crank signal
  - EGR operation
  - Fuel pump relay - Connect test light between pump test terminal and ground. Light should be "ON" for 2 seconds following ignition "ON." If not, refer to CODE 54 CHART.
  - For a faulty in-tank fuel pump check valve, which would allow the fuel in the lines to drain back to the tank after the engine is stopped. To check for this condition:
    1. Ignition "OFF."
    2. Disconnect fuel line at the filter.
    3. Remove the tank filler cap .
    4. Connect a radiator test pump to the line and apply 103 kPa (15 psi) pressure. If the pressure will hold for 60 seconds, the check valve is OK.
- Check ignition system for:
  - Proper output with ST-125.
  - Worn shaft.
  - Bare and shorted wires.
  - Pickup coil resistance and connections.
  - Loose ignition coil connections.
  - Moisture in distributor cap.
  - Spark plugs, wet plugs, cracks, wear, improper gap, burned electrodes, or heavy deposits.
- If engine starts but then, immediately stalls, open distributor bypass line. If engine then starts, and runs OK, replace distributor pickup coil.
- Check CKT 423 (EST) for short to ground.

2.8L ENGINE – DRIVEABILITY SYMPTOMS

SURGES AND/OR CHUGGLE

Definition: Engine power variation, under steady throttle or cruise. Feels like the car speeds up and slows down, with no change in the accelerator pedal.

- Use a "Scan" tool to make sure reading of VSS matches vehicle speedometer.
- **CHECK:**
  - For intermittent EGR at idle.
  - Ignition timing. See Important Vehicle Information label.
  - Inline fuel filter for dirt or restriction.
  - Fuel pressure. See CHART A-7.
  - Generator output voltage. Repair, if less than 9, or more than 16 volts.
- Inspect oxygen sensor for silicon contamination from fuel, or use of improper RTV sealant. The sensor may have a white, powdery coating and result in a high but false signal voltage (rich exhaust indication). The ECM will then reduce the amount of fuel delivered to the engine, causing a severe driveability problem.
- Remove spark plugs. Check for cracks, wear, improper gap, burned electrodes, or heavy deposits. Also, check condition of the rest of the Ignition System.

LACK OF POWER, SLUGGISH, OR SPONGY

Definition: Engine delivers less than expected power. Little or no increase in speed, when accelerator pedal is pushed down part way.

- Compare customer's car to similar unit. Make sure the customer's car has an actual problem.
- Remove air cleaner and check air filter for dirt, or for being plugged. Replace as necessary.
- If there is spray from only one injector, then there is a malfunction in the injector assembly, or in the signal to the injector assembly. The malfunction can be isolated, by switching the injector connectors. If the problem remains with the original injector, after switching the connector, then, the injector is defective. Replace the injector. If the problem moves with the injector connector, then, the problem is an improper signal in the injector circuits, see CHART A-3.
- **CHECK:**
  - Ignition timing. See the Important Vehicle Information label.
  - For restricted fuel filter, contaminated fuel or improper fuel pressure. See CHART A-7.
  - ECM grounds.
  - EGR operation for being open, or partly open, all the time
  - Generator output voltage. Repair, if less than 9, or more than 16 volts.
  - Engine valve timing and compression.
  - Engine, for proper or worn camshaft.
  - Secondary ignition voltage, using a scope or ST-125.
  - Proper operation of EST.
- Check exhaust system for restriction
  1. With engine at normal operating temperature, connect a vacuum gage to any convenient vacuum port on intake manifold.
  2. Run engine at 1000 rpm and record vacuum reading.
  3. Increase rpm slowly to 2500 rpm. Note vacuum reading at steady 2500 rpm.
  4. If vacuum at 2500 rpm decreases more than 3" from reading at 1000 rpm, the exhaust system should be inspected for restrictions.
  5. Disconnect exhaust pipe from engine and repeat steps 3 & 4. If vacuum still drops more than 3" with exhaust disconnected, check valve timing.

DETONATION / SPARK KNOCK

Definition: A mild to severe ping, usually worse under acceleration. The engine makes sharp metallic knocks that change with throttle opening.

- CHECK for obvious overheating problems.
  - Low coolant.
  - Loose water pump belt.
  - Restricted air flow to radiator, or restricted
  - water flow through radiator.
  - Faulty or incorrect thermostat.
  - Coolant sensor, which has shifted in value.
  - Correct coolant solution - should be a 50/50 mix of anti-freeze coolant and water.
- **CHECK:**
  - For poor fuel quality, proper octane rating.
  - For correct PROM.
  - Spark plugs for correct heat range.
  - Ignition timing. See Important Vehicle Information label.
  - Proper operation of Thermac.
  - Fuel system for low pressure. See CHART A-7.
  - Check EGR system for not opening
  - For incorrect basic engine parts such as cam, heads, pistons, etc.
- Excessive oil entering combustion chamber.
- Remove carbon with top engine cleaner. Follow instructions on can.
- If there is spray from only one injector, then, there is a malfunction in the injector assembly, or in the signal to the injector assembly. The malfunction can be isolated, by switching the injector connectors. If the problem remains with the original injector, after switching the connector, then, the injector is defective. Replace the injector. If the problem moves with the injector connector, then, the problem is an improper signal in the injector circuits, see CHART A-3.

HESITATION, SAG, STUMBLE

Definition: Momentary lack of response as the accelerator is pushed down. Can occur at all car speeds. Usually most severe when first trying to make the car move, as from a stop sign. May cause the engine to stall if severe enough.

- Perform careful visual (physical) check
- **CHECK:**
  - Fuel pressure. See CHART A-7.
  - Water contaminated fuel.
  - TPS for binding or sticking.
  - Ignition timing. See Important Vehicle Information label.
  - Generator output voltage. Repair if less than 9 or more than 16 volts.
  - For open Ignition System ground, CKT 453.
  - Canister purge system for proper operation.
  - EGR valve operation

CUTS OUT, MISSES

Definition: Steady pulsation or jerking that follows engine speed, usually more pronounced as engine load increases. The exhaust has a steady spitting sound at idle or low speed.

- Perform careful visual (physical) check
- If ignition system is suspected of causing a miss at idle or cutting, out under load:
- Check for missing cylinder by:
  1. Disconnect IAC motor. Start engine. Remove one spark plug wire at a time, using insulated pliers.
  2. If there is an rpm drop, on all cylinders, (equal to within 50 rpm), go to ROUGH, UNSTABLE, OR INCORRECT IDLE, OR STALLING symptom. Reconnect IAC motor.
  3. If there is no rpm drop on one or more cylinders, or excessive variation in drop, check for spark, on the suspected cylinder(s) with J 26792 (ST-125) spark tester or equivalent. If no spark, see "Engine Electrical" for Intermittent Operation or Miss. If there is spark, remove spark plug(s) in these cylinders and check for:
     - Cracks
     - Wear
     - Improper gap
     - Burned electrodes
     - Heavy deposits
     - Perform compression check on questionable cylinder.

2.8L ENGINE – DRIVEABILITY SYMPTOMS

- Check wire resistance (should not exceed 30,000 ohms), also, check rotor and distributor cap.
- If the previous checks did not find the problem:
- Visually inspect ignition system for moisture, dust, cracks, burns, etc. Spray plug wires with fine water mist to check for shorts.
- Fuel System - Plugged fuel filter, water, low pressure. See CHART A-7.
- Perform compression check.
- Valve Timing.
- Remove rocker covers. Check for bent pushrods, worn rocker arms, broken or weak valve springs, worn camshaft lobes. Repair as necessary. See "Engine Mechanical".
- If there is spray from only one injector, then, there is a malfunction in the injector assembly, or in the signal to the injector assembly. The malfunction can be isolated, by switching the injector connectors. If the problem remains with the original injector, after switching the connector, then, the injector is defective. Replace the injector. If the problem moves with the injector connector, then, the problem is an improper signal in the injector circuits, see CHART A-3.

POOR FUEL ECONOMY

Definition: Fuel economy, as measured by an actual road test, is noticeably lower than expected. Also, economy is noticeably lower than it was on this car at one time, as previously shown by an actual road test.

- **CHECK:**
- Engine thermostat for faulty part (always open) or for wrong heat range.
- Fuel Pressure. See CHART A-7.
- Check owner's driving habits.
- Is A/C "ON" full time (Defroster mode "ON")?
- Are tires at correct pressure?
- Are excessively heavy loads being carried?
- Is acceleration too much, too often?
- Suggest driver read "Driving for Best Fuel Economy" in Owner's Manual.
- Perform "Diagnostic Circuit Check."
- Check air cleaner element (filter) for dirt or being plugged.
- Check for proper calibration of speedometer.
- Visually (physically) Check:
- Vacuum hoses for splits, kinks, and proper connections, as shown on Important Vehicle Information label.
- Ignition wires for cracking, hardness, and proper connections.
- Check Ignition timing. See Important Vehicle Information label.
- Remove spark plugs. Check for cracks, wear, improper gap, burned electrodes or heavy deposits. Repair or replace, as necessary.
- Check compression. See "Engine Mechanical".
- Check for dragging brakes.
- Suggest owner fill fuel tank and recheck fuel economy.
- Check for exhaust system restriction.

ROUGH, UNSTABLE, OR INCORRECT IDLE, STALLING

Definition: The engine runs unevenly at idle. If bad enough, the car may shake.
Also, the idle may vary in rpm (called "hunting"). Either condition may be severe enough to cause stalling. Engine idles at incorrect speed.

- **CHECK:**
- Ignition timing. See Important Vehicle Information label.
- For injector(s) leaking. Check Fuel Pressure CHART A-7.
- IAC
- If a sticking throttle shaft or binding linkage causes a high TPS voltage (open throttle indication), the ECM will not control idle.
  Monitor TPS voltage. "Scan" and/or voltmeter should read less than 1.25 volts with throttle closed.
- Vacuum leaks can cause higher than normal idle.
- EGR "ON," while idling, will cause roughness, stalling, and hard starting. CHART C-7.
- Battery cables and ground straps should be clean and secure. Erratic voltage will cause IAC to change its position, resulting in poor idle quality.
- IAC valve will not move, if system voltage is below 9, or greater than 17.8 volts.
- Use "Scan" tool to determine if ECM is receiving A/C request signal.
- MAP Sensor - Ignition "ON," engine stopped. Compare MAP voltage with known good vehicle. Voltage should be the same kk 400 mV (.4 volt).
  or
  Start and idle engine. Disconnect sensor electrical connector. If idle improves, substitute a known good sensor and recheck.
- A/C Refrigerant Pressure too high. Check for overcharge or faulty pressure switch.
- PCV valve for proper operation by placing finger over inlet hole in valve end several times. Valve should snap back. If not, replace valve.
- Run a cylinder compression check.
- Inspect oxygen sensor for silicon contamination from fuel, or use of improper RTV sealant. The sensor will have a white, powdery coating, and will result in a high but false signal voltage (rich exhaust indication). The ECM will then reduce the amount of fuel delivered to the engine, causing a severe driveability problem.

EXCESSIVE EXHAUST EMISSIONS OR ODORS

Definition: Vehicle fails an emission test. Vehicle has excessive "rotten egg" smell.
Excessive odors do not necessarily indicate excessive emissions.

- Perform "Diagnostic Circuit Check."
- IF TEST SHOWS EXCESSIVE CO AND HC, (or also has excessive odors):
  - Check items which cause car to run RICH.
    - Make sure engine is at normal operating temperature.
- **CHECK:**
  - Fuel pressure. See CHART A-7.
  - Incorrect timing. See Important Vehicle Information label.
  - Canister for fuel loading.
  - PCV valve for being plugged, stuck, or blocked PCV hose, or fuel in the crankcase.
  - Spark plugs, plug wires, and ignition components.
  - Check for lead contamination of catalytic converter (look for removal of fuel filler neck restrictor).
  - Check for properly installed fuel cap.
- If the system is running rich, (block learn less than 118), refer to "Diagnostic Aids" on facing page of Code 45.
- IF TEST SHOWS EXCESSIVE NOx:
  - Check items which cause car to run LEAN, or to run too hot.
    - EGR valve for not opening.
    - Vacuum leaks.
    - Coolant system and coolant fan for proper operation.
    - Remove carbon with top engine cleaner. Follow instructions on can.
    - Check ignition timing for excessive base advance. See Important Vehicle Information label.
- If the system is running lean, (block learn greater than 138), refer to "Diagnostic Aids" on facing page of Code 44.

DIESELING, RUN-ON

Definition: Engine continues to run after key is turned "OFF," but runs very roughly. If engine runs smoothly, check ignition switch and adjustment.

- Check injector for leaking. Apply 12 volts to fuel pump test terminal to turn "ON" fuel pump and pressurize fuel system.
  Visually check injector and TBI assembly for fuel leakage.

BACKFIRE

Definition: Fuel ignites in intake manifold, or in exhaust system, making a loud popping noise.

- **CHECK:**
- EGR operation for being open all the time.
- Output voltage of ignition coil(s).
- For crossfire between spark plugs (distributor cap, spark plug wires, and proper routing of plug wires).
- Engine timing - See Important Vehicle Information label.
- For faulty spark plugs and/or plug wires or boots.
- Perform a compression check - look for sticking or leaking valves.
- For proper valve timing.
- Broken or worn valve train parts.

3.1L ENGINE—ENGINE COMPONENT LOCATIONS RODEO AND 1991–92 PICK-UP

"UT" CARLINE RPO: LG6 ENGINE CODE: D 3.1L V6

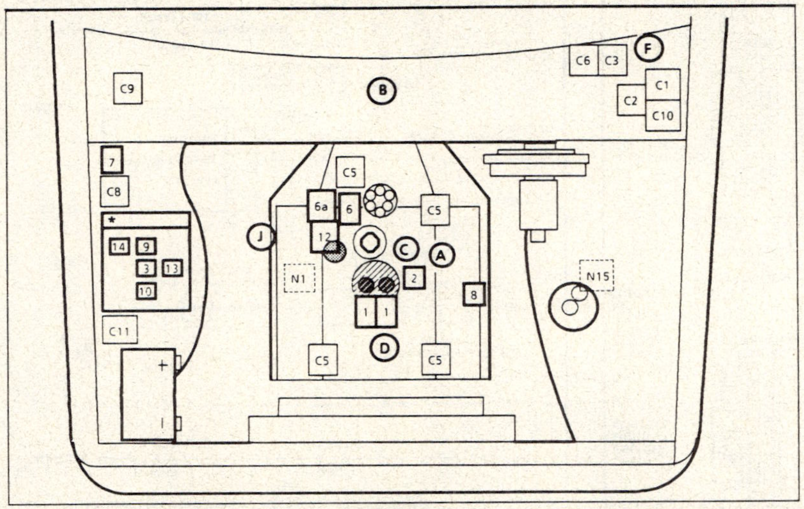

COMPUTER COMMAND CONTROL

- C1 Electronic Control Module (ECM)
- C2 ALDL diagnostic connector
- C3 Check engine light
- C5 ECM harness ground
- C6 Fuse panel
- C8 Fuel pump test connector
- C9 Elapsed timer module
- C10 Set timing connector
- C11 A/C test connector

ECM CONTROLLED COMPONENTS

- 1 Fuel injector
- 2 Idle air control
- 3 Fuel pump relay
- 6 Electronic Spark Timing Distributor (EST)
- 6a Remote ignition coil
- 7 Electronic Spark Control Module (ESC)
- 8 Oil pressure switch
- 9 ECM relay
- 10 Starter relay
- 12 Exhaust gas recirculation vacuum solenoid
- 13 A/C relay
- 14 60 amp Main-1 fuse

* FUSE AND RELAY CONVENIENCE CENTER

ECM INFORMATION SENSORS

- A Manifold Absolute Pressure (MAP)
- B Exhaust oxygen (Federal)
- C Throttle Position Sensor (TPS)
- D Coolant temperature
- F Vehicle Speed Sensor (VSS)
- J Electronic Spark Control Knock (ESC)

EGR VALVE

EMISSION COMPONENTS (NOT ECM CONTROLLED)

- N1 Crankcase Vent Valve (PCV)
- N15 Fuel vapor canister

3.1L ENGINE—ECM WIRING DIAGRAMS RODEO AND 1991–92 PICK-UP

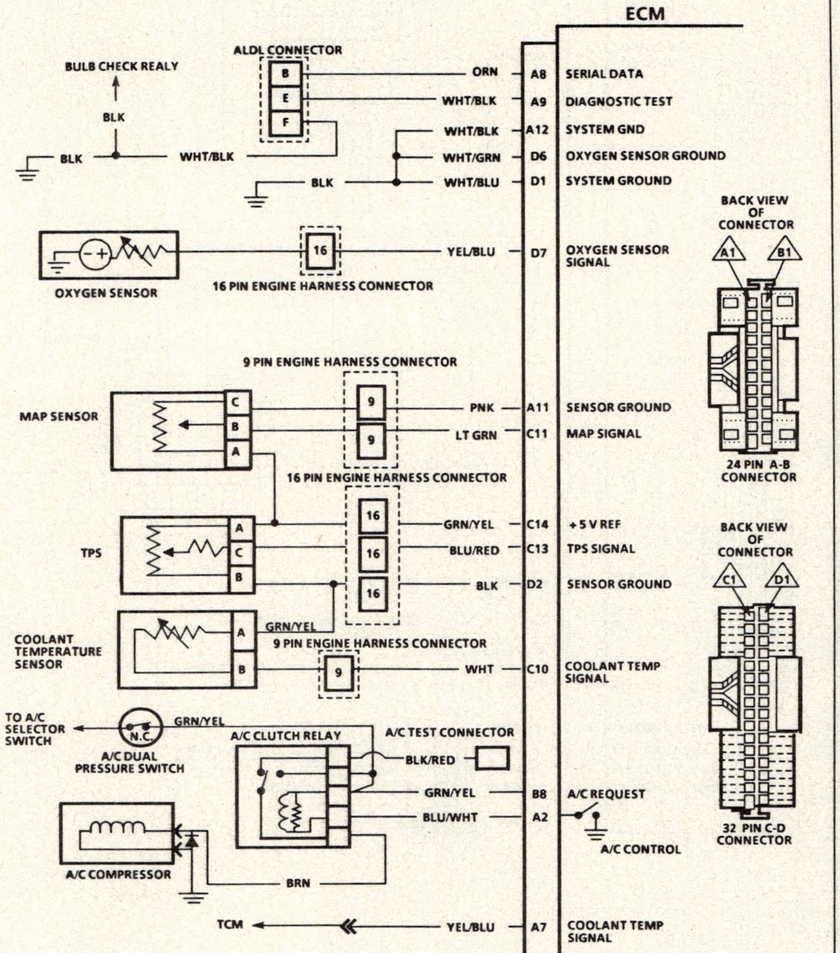

3.1L ENGINE—ECM WIRING DIAGRAMS—RODEO AND 1991–92 PICK-UP

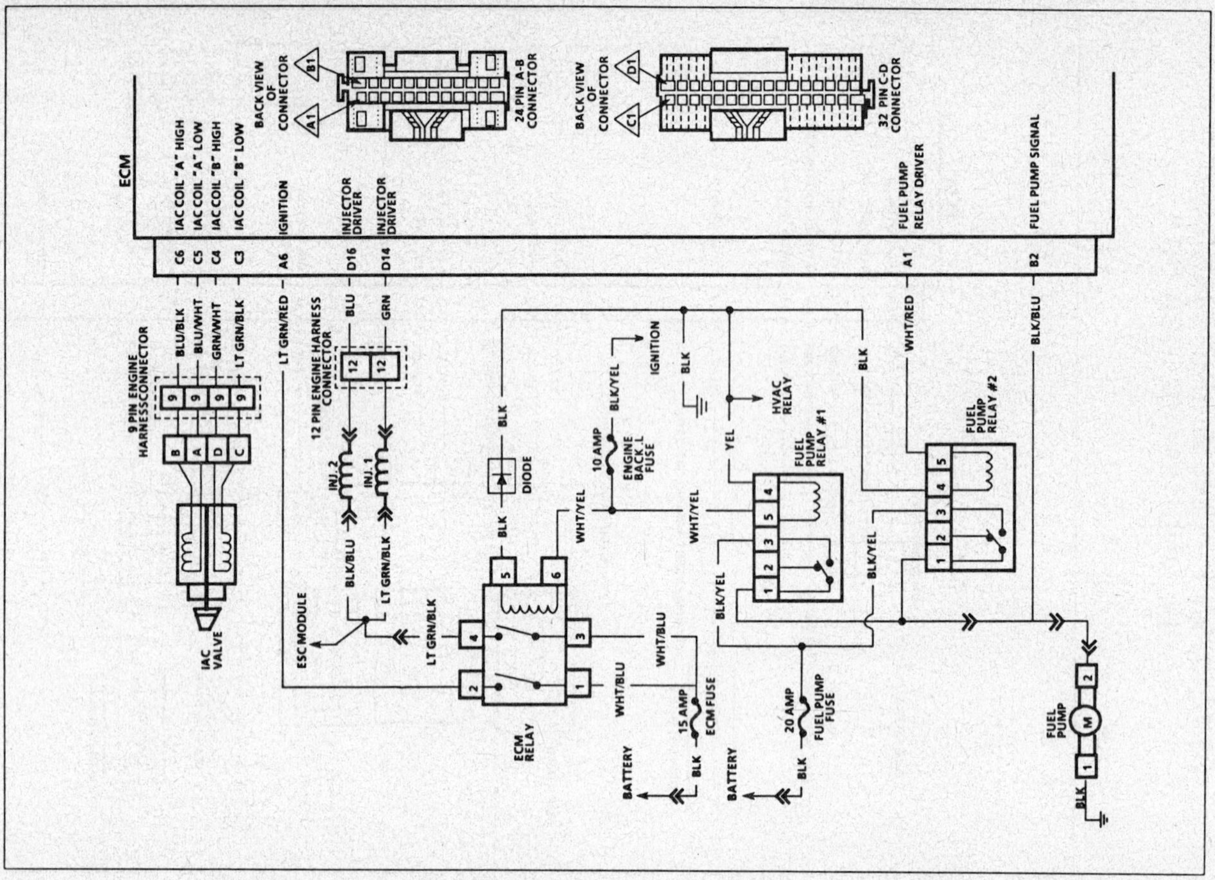

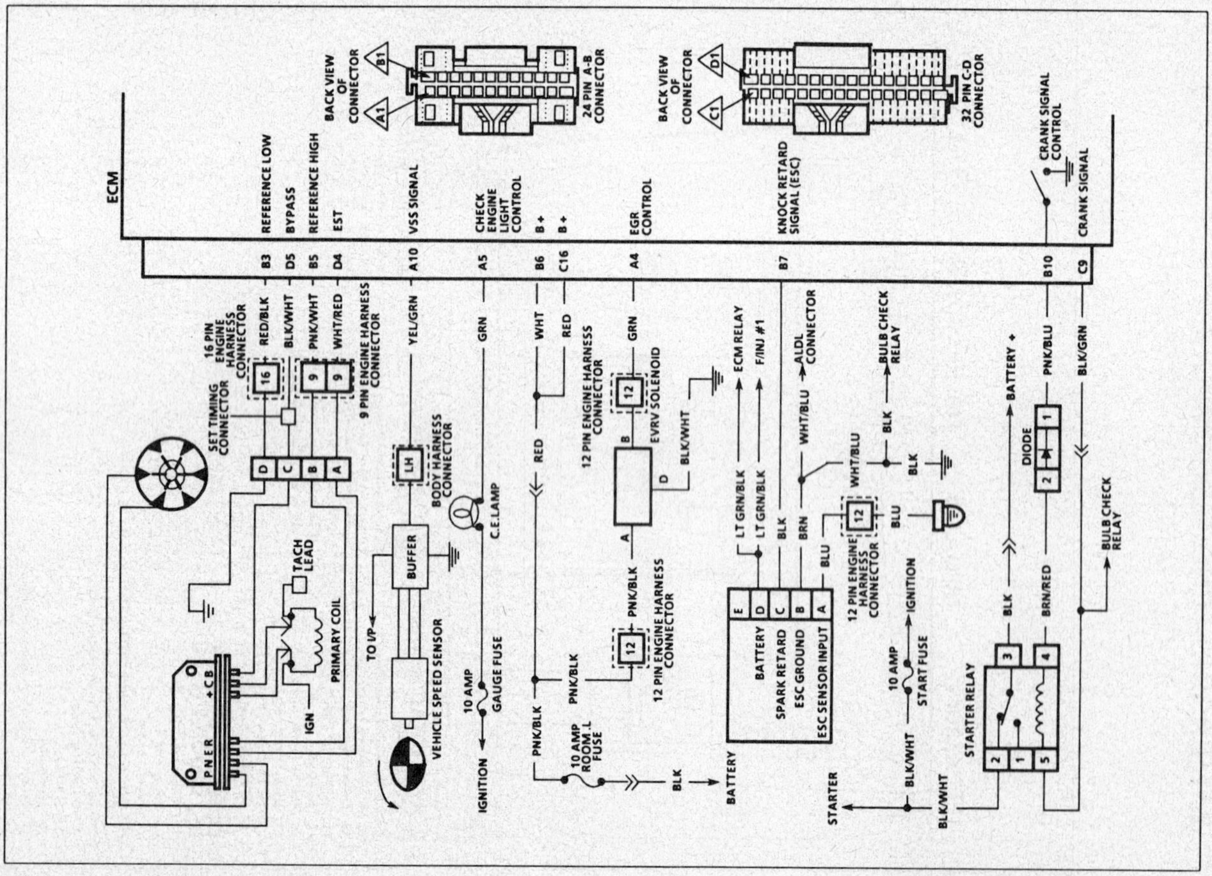

3.1L ENGINE—CONTROL UNIT CONNECTOR PIN IDENTIFICATION RODEO AND 1991–92 PICK-UP

TBI FUEL INJECTION ECM CONNECTOR IDENTIFICATION

This ECM voltage chart is for use with a digital voltmeter to further aid in diagnosis. The voltages you get may vary due to low battery charge or other reasons, but they should be very close.

THE FOLLOWING CONDITIONS MUST BE MET BEFORE TESTING:

- Engine at operating temperature
- Engine idling in "Closed Loop" (for "Engine Run" column)
- Test terminal not grounded
- ALCL tool not installed

	KEY "ON"	ENG. RUN	CIRCUIT	PIN	WIRE COLOR
(4)	0	14	FUEL PUMP RELAY	A1	WHT/RED
(7)	12	14	A/C CLUTCH CONTROL	A2	BLU/WHT
			NOT USED	A3	
	12	14	EGR CONTROL	A4	GRN
	0	14	"CHECK ENGINE" LIGHT	A5	GRN
	12	14	ECM RELAY	A6	LT GRN/RED
			TCM	A7	YEL/BLU
	2-5	2-5	SERIAL DATA	A8	ORN
	5	5	DIAGNOSTIC TERMINAL	A9	WHT/BLU
	①	①	SPEED SENSOR SIGNAL	A10	YEL/GRN
	0	0	SENSOR GROUND	A11	PNK
	0	0	SYSTEM GROUND	A12	WHT/BLK
			NOT USED	C1	
			NOT USED	C2	
	NOT	USE-ABLE	IAC "B" LOW	C3	LT GRN/BLK
	NOT	USE-ABLE	IAC "B" HIGH	C4	GRN/WHT
	NOT	USE-ABLE	IAC "A" LOW	C5	BLU/WHT
	NOT	USE-ABLE	IAC "A" HIGH	C6	BLU/BLK
			NOT USED	C7	
			NOT USED	C8	
	0	0	CRANK SIGNAL	C9	BLK/GRN
(2)	1.9	1.7	COOLANT TEMP. SIGNAL	C10	WHT
	4.9	2.0	MAP SIGNAL	C11	LT GRN
			NOT USED	C12	
(5)	.48	.48	TPS SIGNAL	C13	BLU/RED
	5	5	5 VOLT REFERENCE	C14	GRN/YEL
			NOT USED	C15	
			BATTERY	C16	RED

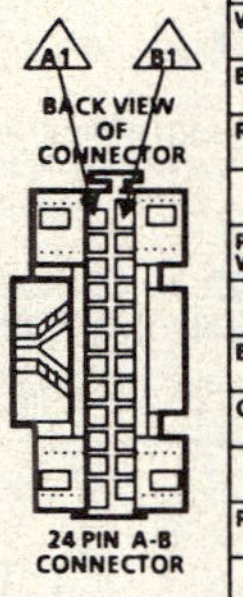

24 PIN A-B CONNECTOR

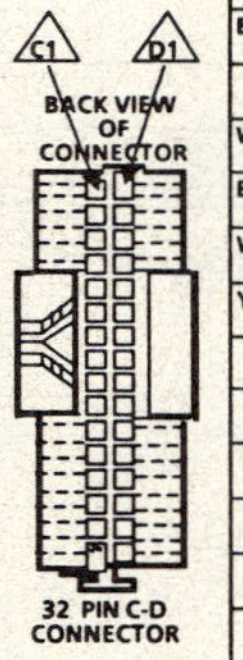

32 PIN C-D CONNECTOR

WIRE COLOR	PIN	CIRCUIT	KEY "ON"	ENG. RUN	
WHT	B1	BATTERY	12	14	
BLK/BLU	B2	FUEL PUMP SIGNAL	0	14	(4)
RED/BLK	B3	EST REF LOW	0	0	
	B4	NOT USED			
PNK/WHT	B5	EST REFERENCE HI	0	1.6	
	B6	NOT USED			
BLK	B7	ESC SIGNAL	9	9	
GRN/YEL	B8	A/C SIGNAL	0	0/12	(8)
	B9	NOT USED			
PNK/BLU	B10	STARTER RELAY			
	B11	NOT USED			
	B12	NOT USED			
WHT/BLU	D1	SYSTEM GROUND	0	0	
BLK	D2	SENSOR GROUND	0	0	
	D3	NOT USED			
WHT/RED	D4	EST CONTROL	0	1.0	
BLK/WHT	D5	EST BYPASS	0	4.75	
WHT/GRN	D6	SENSOR GROUND	0	0	
YEL/BLU	D7	O_2 SENSOR SIGNAL	0	③	
	D8	NOT USED			
	D9	NOT USED			
	D10	NOT USED			
	D11	NOT USED			
	D12	NOT USED			
	D13	NOT USED			
GRN	D14	INJECTOR #1	12	14	
	D15	NOT USED			
BLU	D16	INJECTOR #2	12	14	

(1) Varies from .60 to battery voltage depending on position of drive wheels.
(2) Varies with temperature.
(3) Varies. (toggles)
(4) 12V First two seconds.
(5) Measured between terminals C13 and D2 (± .06V).
(6) 12 V in gear
(7) 0 V with A/C "on"
(8) 12 V with A/C "on"
(9) 12 V when cold

ENGINE 3.1L LG6

3.1L ENGINE—DIAGNOSTIC CIRCUIT CHECK RODEO AND 1991–92 PICK-UP

The Diagnostic Circuit Check is an organized approach to identifying a problem created by an electronic engine control system malfunction. It must be the starting point for any driveability complaint diagnosis, because it directs the service technician to the next logical step in diagnosing the complaint.

The "Scan Data" listed in the table may be used for comparison, after completing the diagnostic circuit check and finding the on-board diagnostics functioning properly and no trouble codes displayed. The "Typical Values" are an average of display values recorded from normally operating vehicles and are intended to represent what a normally functioning system would typically display.

A "SCAN" TOOL THAT DISPLAYS FAULTY DATA SHOULD NOT BE USED, AND THE PROBLEM SHOULD BE REPORTED TO THE MANUFACTURER. THE USE OF A FAULTY "SCAN" CAN RESULT IN MISDIAGNOSIS AND UNNECESSARY PARTS REPLACEMENT.

Only the parameters listed below are used in this manual for diagnosis. If a "Scan" tool reads other parameters, the values are not recommended by General Motors for use in diagnosis.

Idle / Upper Radiator Hose Hot / Closed Throttle / Park or Neutral / Closed Loop / Acc. off

"SCAN" Position	Units Displayed	Typical Data Value
Coolant Temp.	°C/°F	85°C–105°C (185°F–221°F)
Battery Voltage	Volts	13.5–14.5
MAP pressure	kPa/Volts	29–48 kPa (1–2 volts) depends on Vac. & Baro
Throttle Position	Volts	.48 ± .06 volts
Oxygen Sensor	M/Volts	.001–999 and varying
Open/Closed Loop	Open/Closed	Closed Loop (will go open at idle)
Rich/Lean Flag	Rich /Lean	Varies
O_2 Cross Counts	Counts	Greater than 1
Engine Speed	rpm	650 ± 100 rpm
Knock Signal	No/Yes	No
Fuel Integrator	Counts	Varies
Block Learn	Counts	118–138
Idle Air Control	Counts (steps)	1–50
Park/Neutral	P/N and RDL	Park/Neutral (P/N)
MPH/KPH	mph/kph	0
Torque Converter Clutch	Off/On	Off (On, with TCC commanded)
A/C Clutch	Off/On	Off (On, with A/C commanded on)
Idle Air Control	Counts (steps)	1–50
Prom ID	Varies	Varies

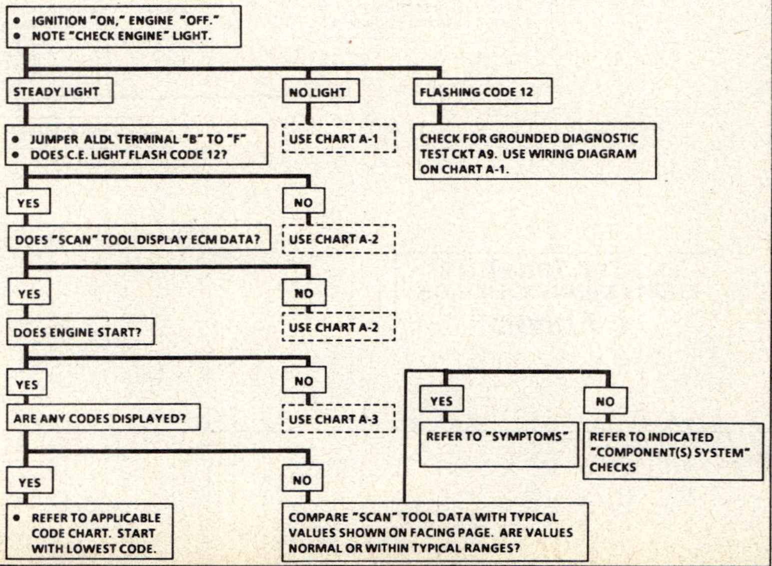

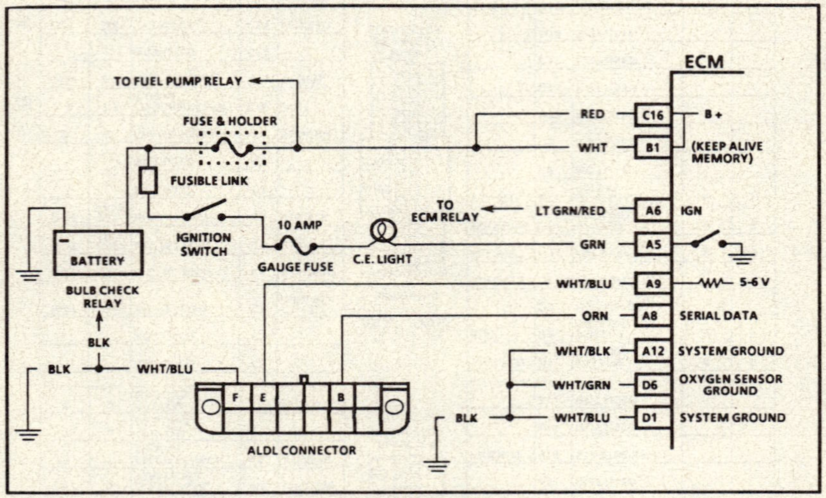

CHART A-1

NO "CHECK ENGINE" LIGHT 3.1L (VIN D) "UT" (TBI)

Circuit Description:

There should always be a steady "Check Engine" light, when the ignition is "ON" and engine stopped. Ignition voltage is supplied directly to the light bulb. The Electronic Control Module (ECM) will control the light and turn it "ON" by providing a ground path through CKT 03 to the ECM.

Test Description: Number(s) below refer to circled number(s) on the diagnostic chart.

1. Battery feed CKTs C16 and B1 are protected by a 60 amp main-1 fuse. Located at fuse and relay convenience center. (Engine compartment R. H. side). If this fuse was open, refer to CHASSIS ELECTRICAL
2. Using a test light connected to 12 volts, probe each of the system ground circuits to be sure a good ground is present. Refer to "ECM Terminal End View" in front of this section for ECM pin locations of ground circuits.

Engine cranks, but will not start.

- Fusible link, open
- 15 amp ignition fuse (fuse box under dash L. H. side)
- 15 amp ECM fuse (fuse and relay convenience center, engine compartment R. H. side)
- 10 amp room L fuse (fuse box under dash L. H. side)
- Battery CKTs C16 and B1 to ECM open
- Ignition CKT A6 to ECM open
- Poor connection to ECM

Diagnostic Aids:

Engine runs OK, check:

- Faulty light bulb
- CKT A5 open
- Gauge fuse blown. This will result in no oil, or generator lights, seat belt reminder, etc.

CHART A-1

NO "CHECK ENGINE" LIGHT 3.1L (VIN D) "UT" (TBI)

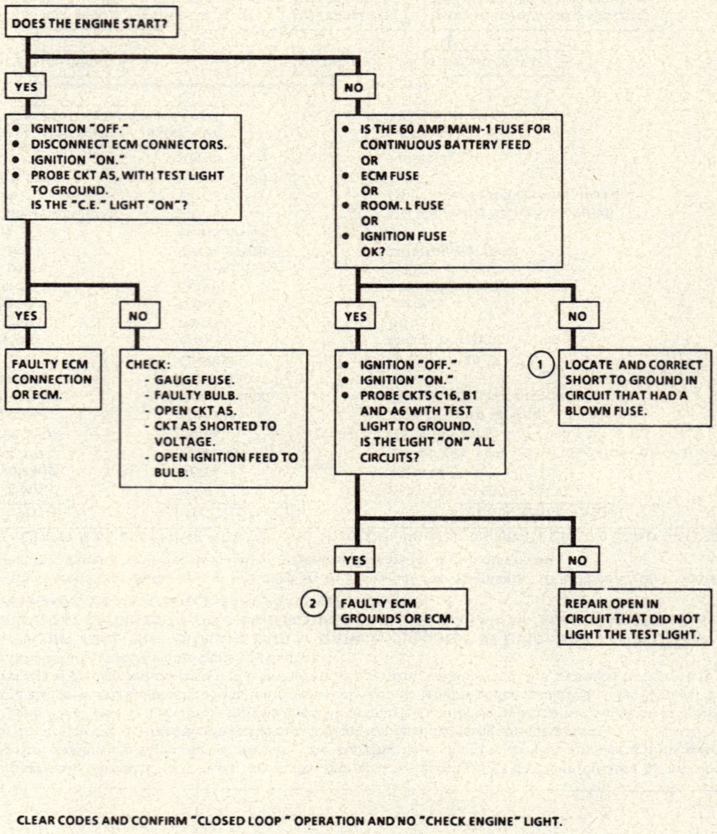

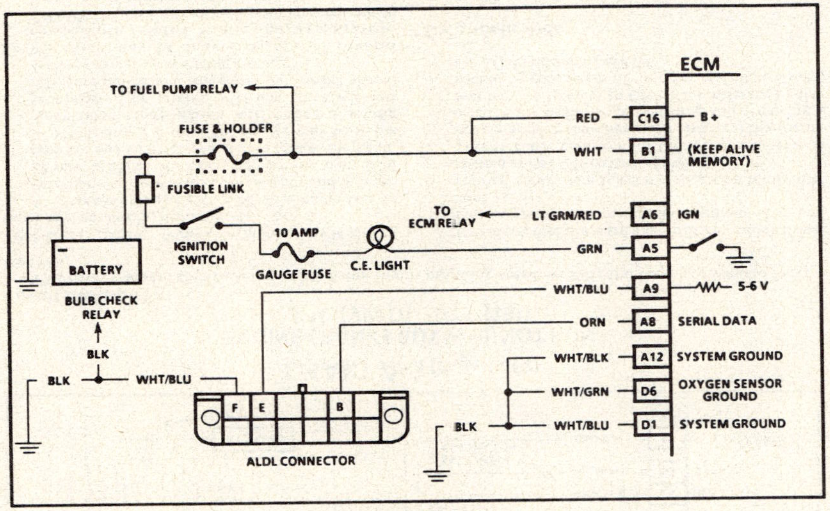

CHART A-2

NO ALDL DATA OR WILL NOT FLASH CODE 12 "CHECK ENGINE" LIGHT "ON" STEADY 3.1L (VIN D) "UT" (TBI)

Circuit Description:

There should always be a steady "Check Engine" light, when the ignition is "ON" and engine stopped. Ignition voltage is supplied directly to the light bulb. The Electronic Control Module (ECM) will turn the light "ON" by grounding CKT A5 at the ECM.

With the diagnostic terminal grounded, the light should flash a Code 12, followed by any trouble code(s) stored in memory.

A steady light suggests a short to ground in the light control CKT A5, or an open in diagnostic CKT A9.

Test Description: Number(s) below refer to circled number(s) on the diagnostic chart.

1. If there is a problem with the ECM that causes a "Scan" tool to not read serial data, then the ECM should not flash a Code 12. If Code 12 does flash, be sure that the "Scan" tool is working properly on another vehicle. If the "Scan" is functioning properly and CKT A8 is OK, the MEM-CAL or ECM may be at fault for the NO ALDL symptom.
2. If the light goes "OFF," when the ECM connector is disconnected, then CKT A5 is not shorted to ground.
3. This step will check for an open diagnostic CKT A9.
4. At this point, the "Check Engine" light wiring is OK. The problem is a faulty ECM or PROM. If Code 12 does not flash, the ECM should be replaced using the original PROM. Replace the PROM only after trying an ECM, as a defective PROM is an unlikely cause of the problem.

CHART A-2

NO ALDL DATA OR WILL NOT FLASH CODE 12 "CHECK ENGINE" LIGHT "ON" STEADY 3.1L (VIN D) "UT" (TBI)

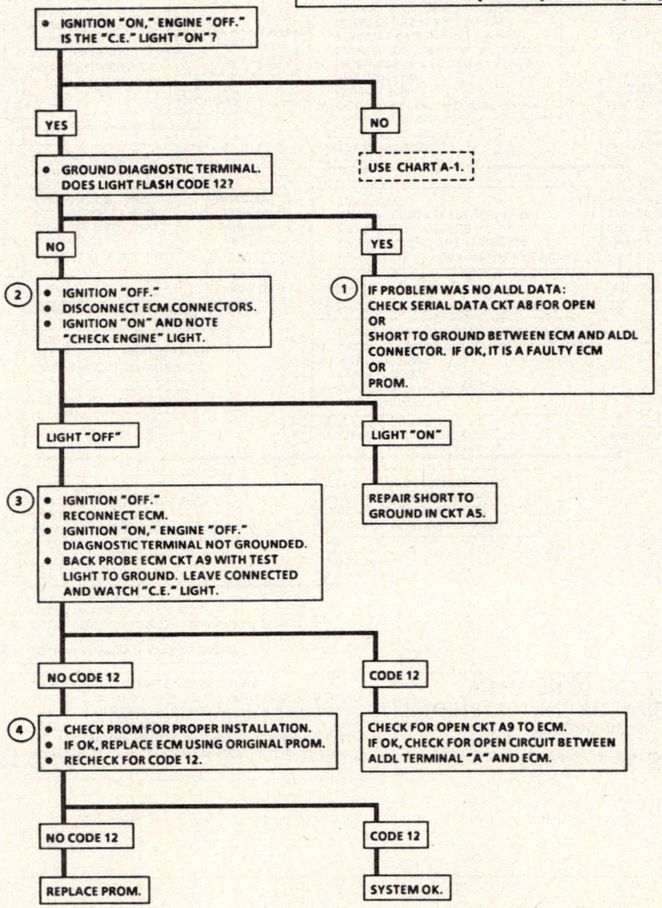

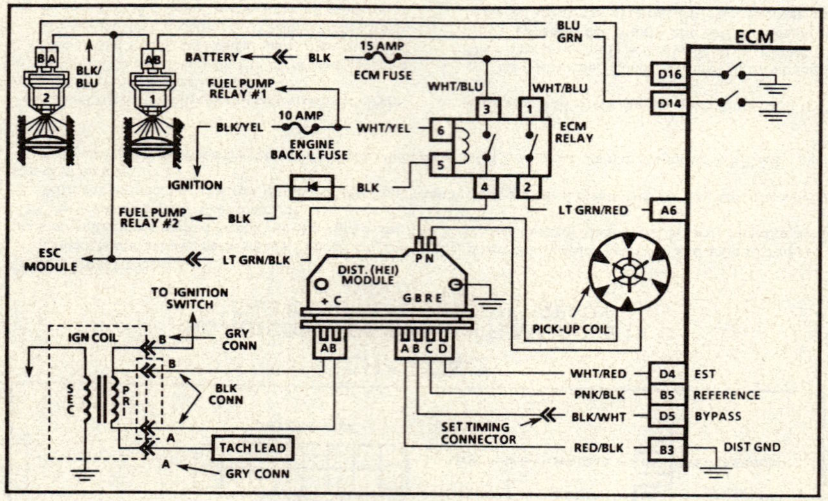

CHART A-3 (Page 1 of 2)

ENGINE CRANKS BUT WILL NOT RUN
3.1L (VIN D) "UT" (TBI)

Circuit Description:
This chart assumes that battery condition and engine cranking speed are OK, and there is adequate fuel in the tank.

Test Description: Number(s) below refer to circled number(s) on the diagnostic chart.

1. A "Check Engine" light "ON" is a basic test to determine if there is a 12 volt supply and ignition 12 volts to ECM. No ALDL may be due to an ECM problem and CHART A-2 will diagnose the ECM. If TPS is over 2.5 volts, the engine may be in the clear flood mode which will cause starting problems. The engine will not start without reference pulses and therefore, the "Scan" should read rpm (reference) during crank.
2. No spark may be caused by one of several components related to the ignition system. CHART C-4 will address all problems related to the causes of a no spark condition.
3. Fuel spray from the injector(s) indicates that fuel is available, however, the engine could be severely flooded due to too much fuel.
4. While cranking engine, there should be no fuel spray with injector disconnected. Replace an injector if it sprays fuel or drips like a leaking water faucet.
5. The fuel pressure will drop after the fuel pump stops running due to a controlled bleed in the fuel system. Use of the fuel pressure gage will determine if fuel system pressure is enough for engine to start and run.
6. No fuel spray from injector indicates a faulty fuel system or no ECM control of injector.
7. This test will determine if the ignition module is not generating the reference pulse, or if the wiring or ECM are at fault. By touching and removing a test light to 12 volts on CKT B5, a reference pulse should be generated. If injector test light blinks, the ECM and wiring are OK.

Diagnostic Aids:

- Water or foreign material can cause a no start during freezing weather.
- An EGR sticking open can cause a low air/fuel ratio during cranking.
- Fuel pressure: Low fuel pressure can result in a very lean air/fuel ratio. See CHART A-7.
- A grounded CKT D4 (EST) may cause a "No-Start" or a "Start then Stall" condition.

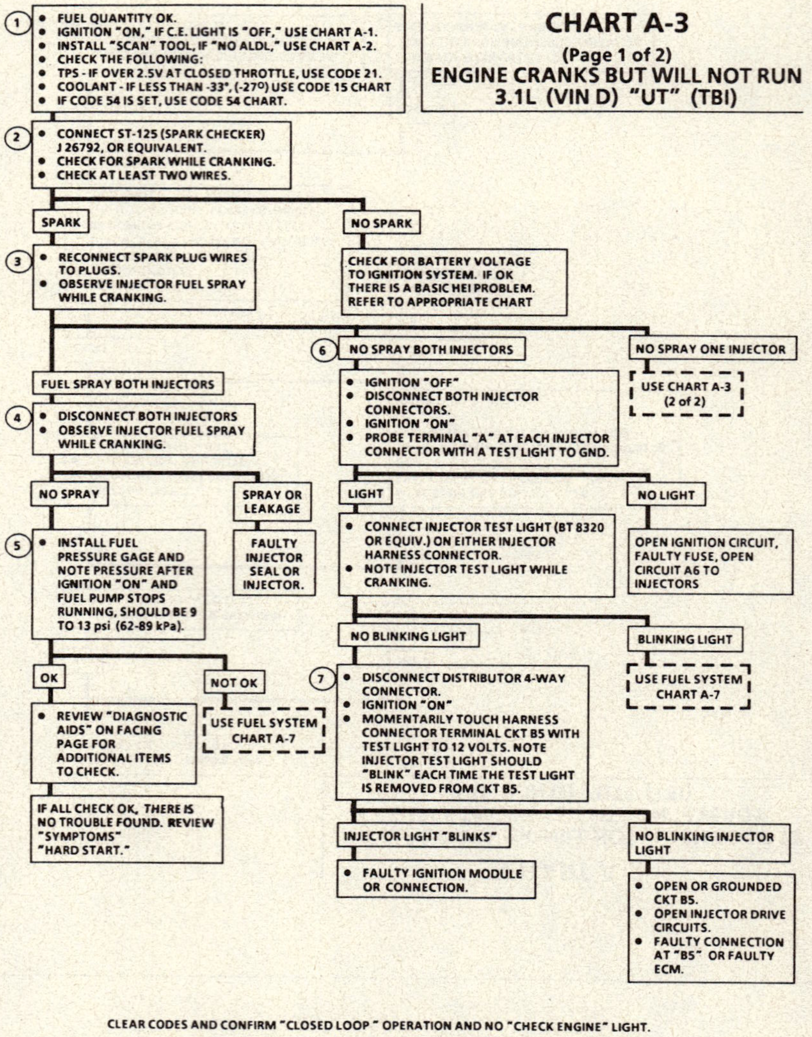

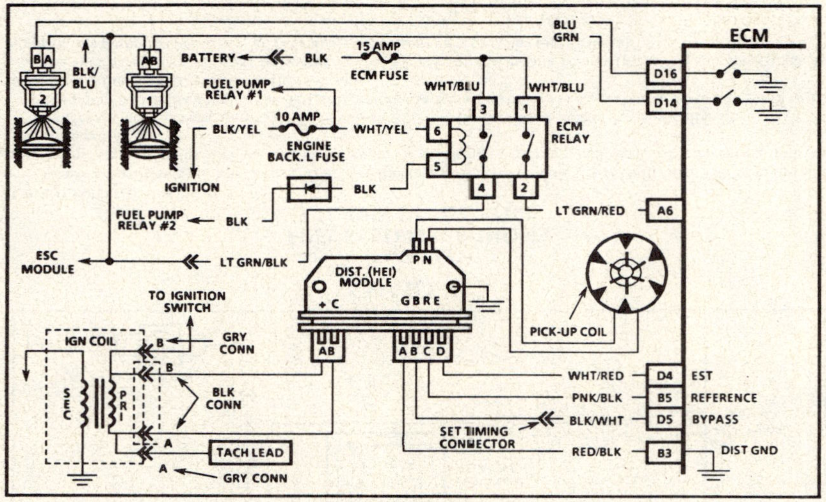

CHART A-3

(Page 2 of 2)

ENGINE CRANKS BUT WILL NOT RUN
3.1L (VIN D) "UT" (TBI)

Circuit Description:

This chart assumes that battery condition and engine cranking speed are OK, and there is adequate fuel in the tank.

Test Description: Number(s) below refer to circled number(s) on the diagnostic chart.

1. No fuel spray from one injector indicates a faulty fuel injector or no ECM control of injector. If the test light "blinks" while cranking, then ECM control should be considered OK. Be sure test light makes good contact between connector terminals during test. The light may be a little dim when "blinking." This is due to current draw of the test light. How bright it "blinks" is not important. The test light bulb should be a J 36400-1 or equivalent.
2. CKT A6 supply ignition voltage to the injectors. Probe each connector terminal with a test light to ground. There should be a light "ON" at one terminal. If the test light confirms ignition voltage at the connector, the ECM injector control CKT D16 or D14 may be open. Reconnect the injector and using a test light connected to ground, check for a light at the applicable ECM connector terminal "D16" or "D14". A light at this point indicates that the injector drive circuit involved is OK.
   If an ECM repeat failure has occurred, the injector is shorted. Replace the injector and ECM.

CHART A-3

(Page 2 of 2)

ENGINE CRANKS BUT WILL NOT RUN
3.1L (VIN D) "UT" (TBI)

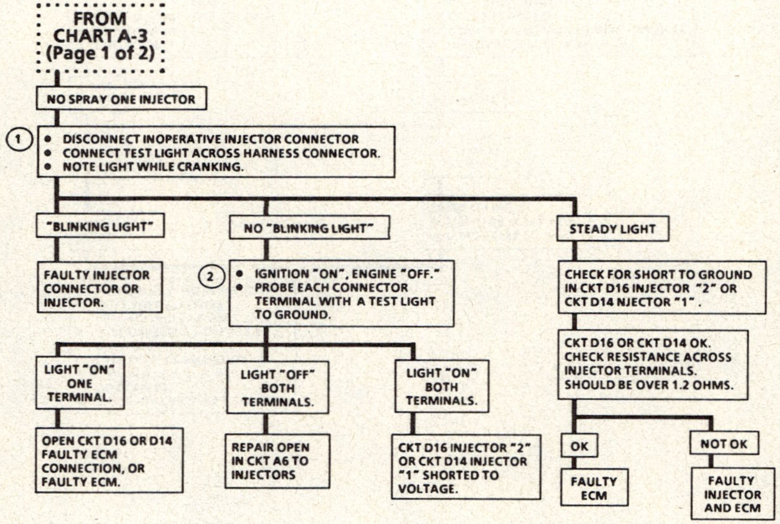

CLEAR CODES AND CONFIRM "CLOSED LOOP" OPERATION AND NO "CHECK ENGINE" LIGHT.

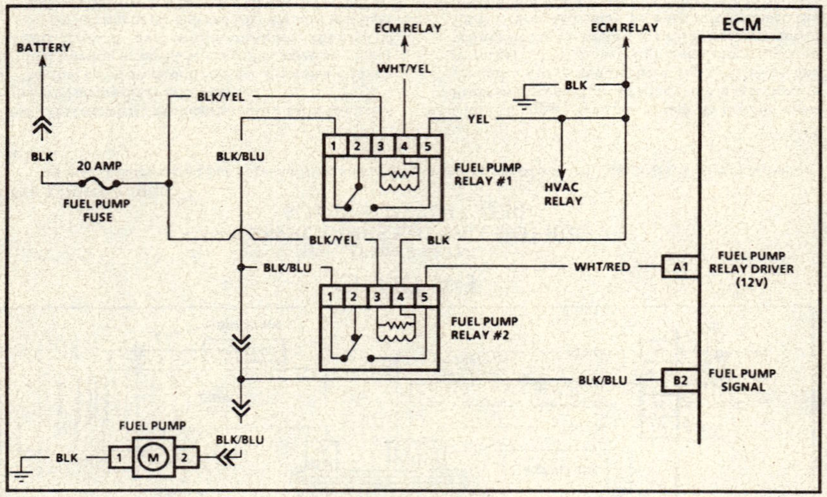

CHART A-7
(Page 1 of 2)
FUEL SYSTEM DIAGNOSIS
3.1L (VIN D) "UT" (TBI)

Circuit Description:

When the ignition switch is turned "ON," the Electronic Control Module (ECM) will turn "ON" the in-tank fuel pump. It will remain "ON" as long as the engine is cranking or running, and the ECM is receiving ignition reference pulses.

If there are no reference pulses, the ECM will shut "OFF" the fuel pump within 2 seconds after key "ON."

The pump will deliver fuel to the TBI unit, where the system pressure is controlled to 62 to 90 kPa (9 to 13 psi). Excess fuel is then returned to the fuel tank.

The fuel pump test terminal is located in the left side of the engine compartment. When the engine is stopped, the pump can be turned "ON" by applying battery voltage to the test terminal.

Test Description: Number(s) below refers to circled number(s) on the diagnostic chart.

1. Fuel pressure should be noted while fuel pump is running. Fuel pressure will drop immediately after fuel pump stops running due to a controlled bleed in the fuel system.

Diagnostic Aids:

Improper fuel system pressure can result in one of the following symptoms:

- Cranks, but will not run.
- Code 44.
- Code 45.
- Cuts out, may feel like ignition problem.
- Poor fuel economy, loss of power.
- Hesitation.

CHART A-7
(Page 1 of 2)
FUEL SYSTEM DIAGNOSIS
3.1L (VIN D) "UT" (TBI)

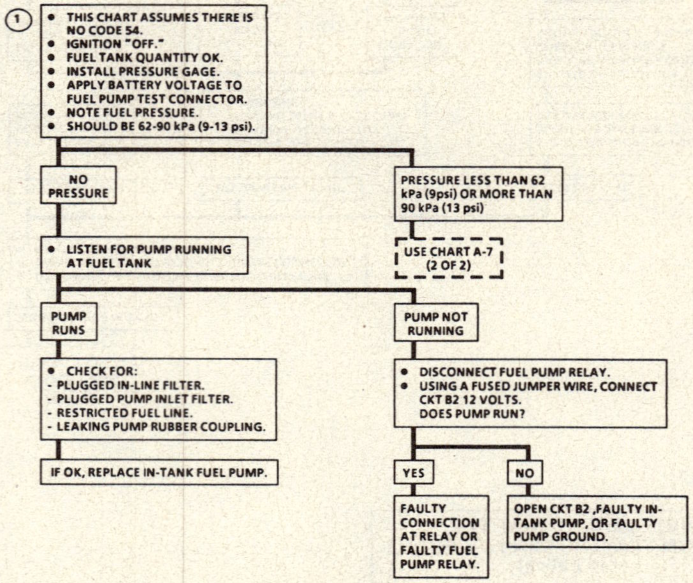

CLEAR CODES AND CONFIRM "CLOSED LOOP" OPERATION AND NO "CHECK ENGINE" LIGHT.

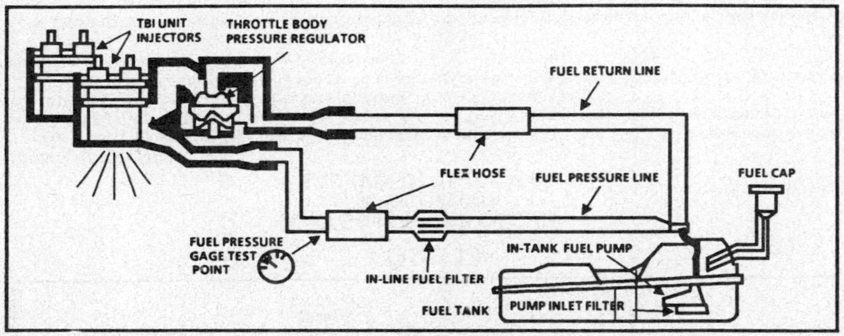

CHART A-7

(Page 2 of 2)
FUEL SYSTEM DIAGNOSIS
3.1L (VIN D) "UT" (TBI)

Test Description: Number(s) below refer to circled number(s) on the diagnostic chart.

1. Fuel pressure less than 62 kPa (9 psi) falls into two areas:
   - Regulated pressure less than 62 kPa (9 psi). Amount of fuel to injectors OK but pressure is too low. System will be lean and may set Code 44. Also, hard starting cold and poor overall performance.
   - Restricted flow causing pressure drop. Normally, a vehicle with a fuel pressure of less than 62 kPa (9 psi) at idle will not be drivable. However, if the pressure drop occurs only while driving, the engine will normally surge then stop as pressure begins to drop rapidly.
2. Restricting the fuel return line allows the fuel pump to develop its maximum pressure (dead head pressure). When battery voltage is applied to the pump test terminal, pressure should be from 90 to 124 kPa (13 to 18 psi).
3. This test determines if the high fuel pressure is due to a restricted fuel return line or a throttle body pressure regulator problem.

CHART A-7

(Page 2 of 2)
FUEL SYSTEM DIAGNOSIS
3.1L (VIN D) "UT" (TBI)

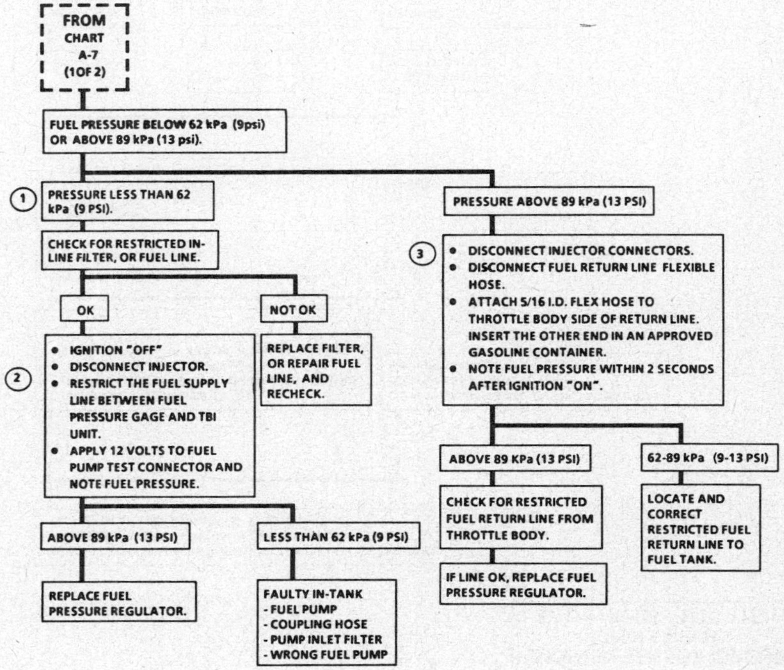

"AFTER REPAIRS," CONFIRM "CLOSED LOOP" OPERATION AND NO "SERVICE ENGINE SOON" LIGHT.

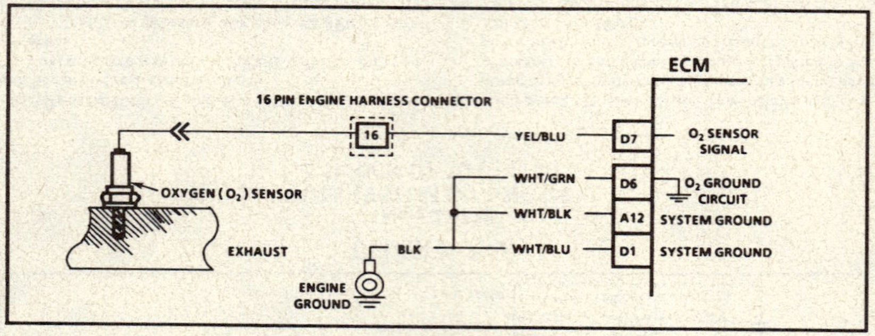

CODE 13
OXYGEN SENSOR CIRCUIT
(OPEN CIRCUIT)
3.1L (VIN D) "UT" (TBI)

Circuit Description:

The ECM supplies a voltage of about .45 volt between terminals "D7" and "D6". (If measured with a 10 megohm digital voltmeter, this may read as low as .32 volt). The O_2 sensor varies the voltage within a range of about 1 volt if the exhaust is rich, down through about .10 volt if exhaust is lean.

The sensor is like an open circuit and produces no voltage when it is below 315°C (600°F). An open sensor circuit or cold sensor causes "Open Loop" operation.

Test Description: Number(s) below refer to circled number(s) on the diagnostic chart.

1. Code 13 will set:
   - At least 2 minutes engine time after start.
   - O_2 signal voltage steady between .35 and .55 volt.
   - Coolant temperature above 70°C (158°F).
   - Throttle position sensor signal above 5% (about .3 volt above closed throttle voltage).
   - All conditions must be met for about 60 seconds.

   If the conditions for a Code 13 exist, the system will not go "Closed Loop."
2. This will determine if the sensor is at fault or the wiring or ECM is the cause of the Code 13.
3. In doing this test, use only a high impedance digital volt ohmmeter. This test checks the continuity of CKT(s) D7 and D6. If CKT D6 is open, the ECM voltage on CKT D7 will be over .6 volt (600 mV).

Diagnostic Aids:

Normal "Scan" voltage varies between 100 mV to 999 mV (.1 and 1.0 volt), while in "Closed Loop." Code 13 sets in one minute, if voltage remains between .35 and .55 volt, but the system will go "Open Loop" in about 15 seconds.

Refer to "Intermittents" in "Symptoms"

CODE 13
OXYGEN SENSOR CIRCUIT
(OPEN CIRCUIT)
3.1L (VIN D) "UT" (TBI)

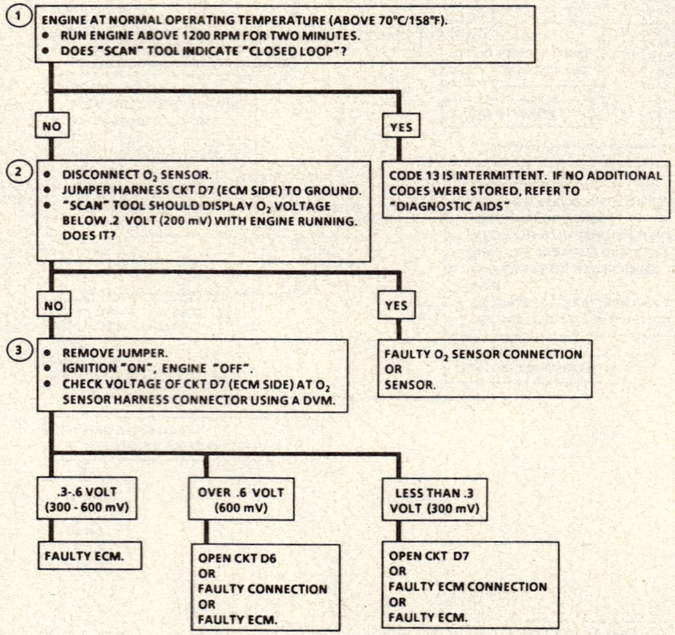

CLEAR CODES AND CONFIRM "CLOSED LOOP" OPERATION AND NO "CHECK ENGINE" LIGHT.

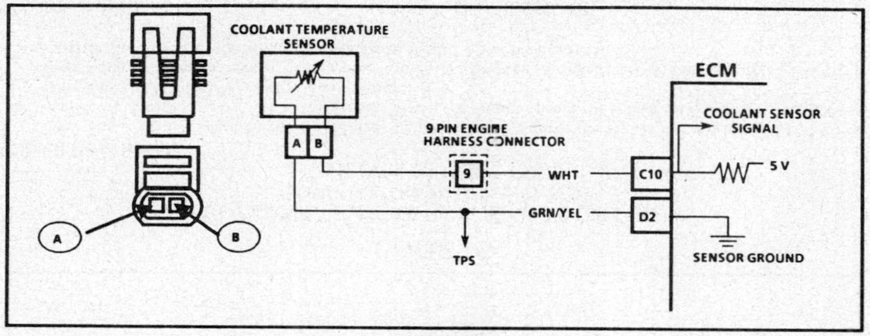

CODE 14

COOLANT TEMPERATURE SENSOR CIRCUIT
(HIGH TEMPERATURE INDICATED)
3.1L (VIN D) "UT" (TBI)

Circuit Description:

The Coolant Temperature Sensor uses a thermistor to control the signal voltage to the ECM. The ECM applies a voltage on CKT C10 to the sensor. When the engine is cold the sensor (thermistor) resistance is high, therefore the ECM will see high signal voltage.

As the engine warms, the sensor resistance becomes less, and the voltage drops. At normal engine operating temperature (85°C to 95°C) the voltage will measure about 1.5 to 2.0 volts.

Test Description: Number(s) below refer to circled number(s) on the diagnostic chart.

1. Code 14 will set if:
   - Signal voltage indicates a coolant temperature above 135°C (275°F) for 6 seconds.
2. This test will determine if CKT C10 is shorted to ground which will cause the conditions for Code 14.

Diagnostic Aids:

Check harness routing for a potential short to ground in CKT C10.

"Scan" tool displays engine temperature in degrees centigrade. After engine is started, the temperature should rise steadily to about 90°C then stabilize when thermostat opens.

Refer to "Intermittents" in "Symptoms"

CODE 14

COOLANT TEMPERATURE SENSOR CIRCUIT
(HIGH TEMPERATURE INDICATED)
3.1L (VIN D) "UT" (TBI)

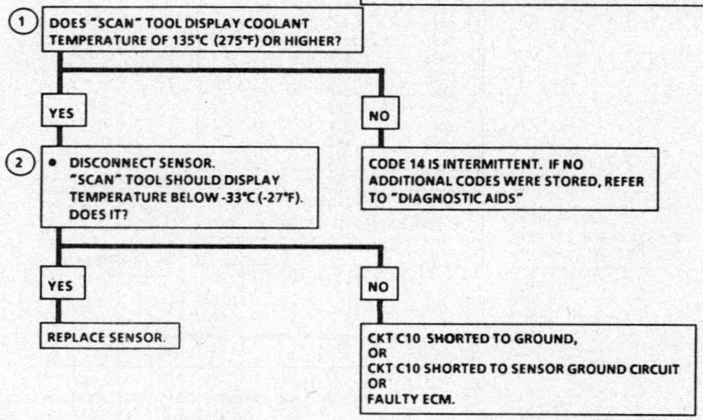

DIAGNOSTIC AID

COOLANT SENSOR
TEMPERATURE VS. RESISTANCE VALUES
(APPROXIMATE)

°C	°F	OHMS
100	212	177
90	194	241
80	176	332
70	158	467
60	140	667
50	122	973
45	113	1188
40	104	1459
35	95	1802
30	86	2238
25	77	2796
20	68	3520
15	59	4450
10	50	5670
5	41	7280
0	32	9420
-5	23	12300
-10	14	16180
-15	5	21450
-20	-4	28680
-30	-22	52700
-40	-40	100700

CLEAR CODES AND CONFIRM "CLOSED LOOP" OPERATION AND NO "CHECK ENGINE" LIGHT.

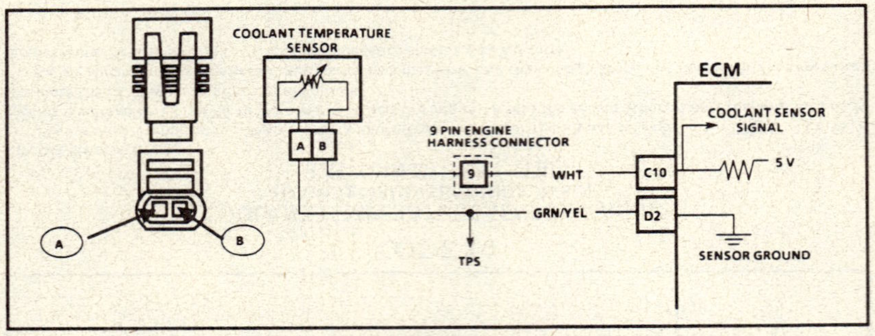

CODE 15

COOLANT TEMPERATURE SENSOR CIRCUIT
(LOW TEMPERATURE INDICATED)
3.1L (VIN D) "UT" (TBI)

Circuit Description:

The Coolant Temperature Sensor (CTS) uses a thermistor to control the signal voltage to the ECM. The ECM applies a voltage on CKT C10 to the sensor. When the engine is cold the sensor (thermistor) resistance is high, therefore the ECM will see high signal voltage.

As the engine warms, the sensor resistance becomes less, and the voltage drops. At normal engine operating temperature (85°C to 95°C) the voltage will measure about 1.5 to 2.0 volts at the ECM.

Test Description: Number(s) below refer to circled number(s) on the diagnostic chart.

1. Code 15 will set if:
   - Engine running longer than 30 seconds.
   - Coolant temperature less than -33°C (-27°F), for 3 seconds.
2. This test simulates a Code 14. If the ECM recognizes the low signal voltage, (high temperature) and "Scan" reads 135°C (275°F) or above, the ECM and wiring are OK.
3. This test will determine if CKT C10 is open. There should be 5 volts present at sensor connector if measured with a DVM.

Diagnostic Aids:

A "Scan" tool reads engine temperature in degrees centigrade. After engine is started the temperature should rise steadily to about 90°C then stabilize when thermostat opens.

If Code 21 is also set, check CKT D2 for faulty wiring or connections. Check terminals at sensor for good contact.

Refer to "Intermittents" in "Symptoms."

CODE 15

COOLANT TEMPERATURE SENSOR CIRCUIT
(LOW TEMPERATURE INDICATED)
3.1L (VIN D) "UT" (TBI)

(1) • DOES "SCAN" TOOL DISPLAY COOLANT TEMPERATURE OF -33°C (-27°F) OR LESS?

- NO → CODE 15 IS INTERMITTENT. IF NO ADDITIONAL CODES WERE STORED, REFER TO "DIAGNOSTIC AIDS"
- YES →

(2) • DISCONNECT SENSOR.
• JUMPER HARNESS TERMINALS TOGETHER.
• "SCAN" TOOL SHOULD DISPLAY 135°C (275°F) OR MORE. DOES IT?

- YES → FAULTY CONNECTION OR SENSOR.
- NO →

(3) • JUMPER CKT C10 TO GROUND.
• "SCAN" TOOL SHOULD DISPLAY OVER 135°C (275°F). DOES IT?

- YES → OPEN SENSOR GROUND CIRCUIT, FAULTY CONNECTION OR FAULTY ECM.
- NO → OPEN CKT C10, FAULTY CONNECTION AT ECM, OR FAULTY ECM.

DIAGNOSTIC AID

COOLANT SENSOR
TEMPERATURE VS. RESISTANCE VALUES
(APPROXIMATE)

°C	°F	OHMS
100	212	177
90	194	241
80	176	332
70	158	467
60	140	667
50	122	973
45	113	1188
40	104	1459
35	95	1802
30	86	2238
25	77	2796
20	68	3520
15	59	4450
10	50	5670
5	41	7280
0	32	9420
-5	23	12300
-10	14	16180
-15	5	21450
-20	-4	28680
-30	-22	52700
-40	-40	100700

CLEAR CODES AND CONFIRM "CLOSED LOOP" OPERATION AND NO "CHECK ENGINE" LIGHT.

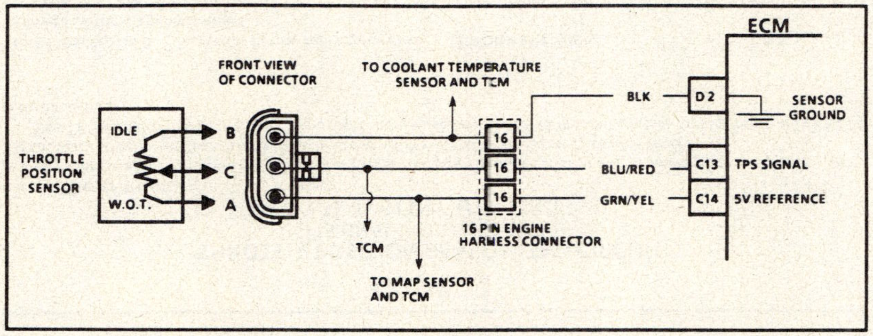

CODE 21

THROTTLE POSITION SENSOR (TPS) CIRCUIT
(SIGNAL VOLTAGE HIGH)
3.1L (VIN D) "UT" (TBI)

Circuit Description:

The Throttle Position Sensor (TPS) provides a voltage signal that changes relative to the throttle blade. Signal voltage will vary from about .5 at idle to about 5 volts at Wide Open Throttle (WOT).

The TPS signal is one of the most important inputs used by the ECM for fuel control and for most of the ECM control outputs.

Test Description: Number(s) below refer to circled number(s) on the diagnostic chart.

1. Code 21 will set if:
   - TPS signal voltage is greater than 2.5 volts.
   - All conditions met for 8 seconds.
   - MAP less than 52 kPa.
2. With the TPS sensor disconnected, the TPS voltage should go low if the ECM and wiring are OK.
3. Probing CKT D2 with a test light to 12 volts checks the sensor ground CKT. A faulty sensor ground will cause a Code 21.

Diagnostic Aids:

A "Scan" tool reads throttle position in volts. Should read less than 1.25 volts with throttle closed and ignition "ON" or at idle. Voltage should increase at a steady rate as throttle is moved toward WOT.

An open in CKT D2 will result in a Code 21.

Refer to "Intermittents" in "Symptoms"

CODE 21

THROTTLE POSITION SENSOR (TPS) CIRCUIT
(SIGNAL VOLTAGE HIGH)
3.1L (VIN D) "UT" (TBI)

(1) • THROTTLE CLOSED.
DOES "SCAN" TOOL DISPLAY TPS OVER 2.5 VOLTS?

- YES → (2)
- NO → CODE 21 IS INTERMITTENT. IF NO ADDITIONAL CODES WERE STORED, REFER TO "DIAGNOSTIC AIDS"

(2) • DISCONNECT SENSOR. "SCAN" TOOL SHOULD DISPLAY TPS BELOW .2 VOLT (200mV).
DOES IT?

- YES → (3)
- NO → CKT C13 SHORTED TO VOLTAGE OR FAULTY ECM.

(3) • PROBE SENSOR GROUND CIRCUIT WITH A TEST LIGHT CONNECTED TO BATTERY VOLTAGE.

- LIGHT "ON" → FAULTY CONNECTION OR SENSOR.
- LIGHT "OFF" → OPEN SENSOR GROUND CIRCUIT OR FAULTY ECM.

CLEAR CODES AND CONFIRM "CLOSED LOOP" OPERATION AND NO "CHECK ENGINE" LIGHT.

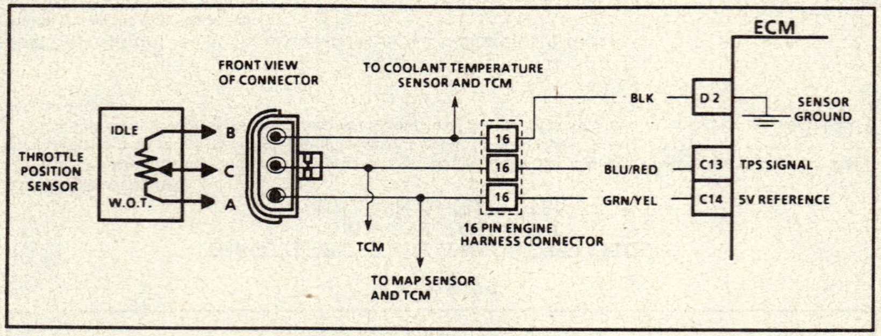

CODE 22

THROTTLE POSITION SENSOR (TPS) CIRCUIT (SIGNAL VOLTAGE LOW) 3.1L (VIN D) "UT" (TBI)

Circuit Description:

The Throttle Position Sensor (TPS) provides a voltage signal that changes relative to the throttle blade. Signal voltage will vary from about .5 at idle to about 5 volts at Wide Open Throttle (WOT).

The TPS signal is one of the most important inputs used by the ECM for fuel control and for most of the ECM control outputs.

Test Description: Numbers below refer to circled numbers on the diagnostic chart.

1. Code 22 will set if:
   - Engine is running
   - TPS signal voltage is less than about .2 volt for 2 seconds.
2. Simulates Code 21: (high voltage) If the ECM recognizes the high signal voltage then the ECM and wiring are OK.
3. This simulates a high signal voltage to check for an open in CKT C13. The "Scan" tool will not read up to 12 volts, but what is important is that the ECM recognizes the signal on CKT C13.
4. There should be 5 volts at terminal "C" if measured with a DVM when ignition is "ON."

Diagnostic Aids:

A "Scan" tool reads throttle position in volts. Should read less than 1.25 volts with throttle closed and ignition "ON" or at idle. Voltage should increase at a steady rate as throttle is moved toward WOT.

An open or short to ground in CKTs C14 or C13 will result in a Code 22.

If a Code 22 is also set check CKT C14 carefully for open or short to ground.

Refer to "Intermittents" in "Symptoms."

CODE 22

THROTTLE POSITION SENSOR (TPS) CIRCUIT (SIGNAL VOLTAGE LOW) 3.1L (VIN D) "UT" (TBI)

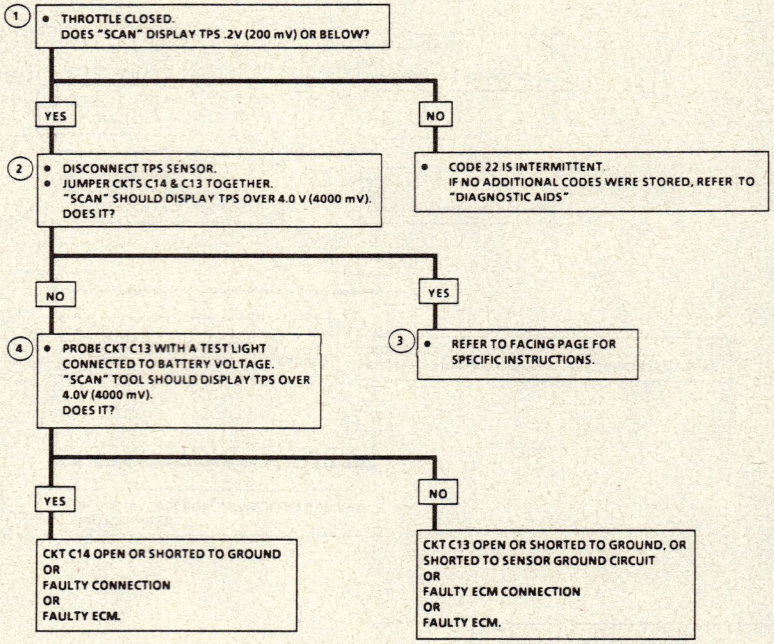

CLEAR CODES AND CONFIRM "CLOSED LOOP" OPERATION AND NO "CHECK ENGINE" LIGHT.

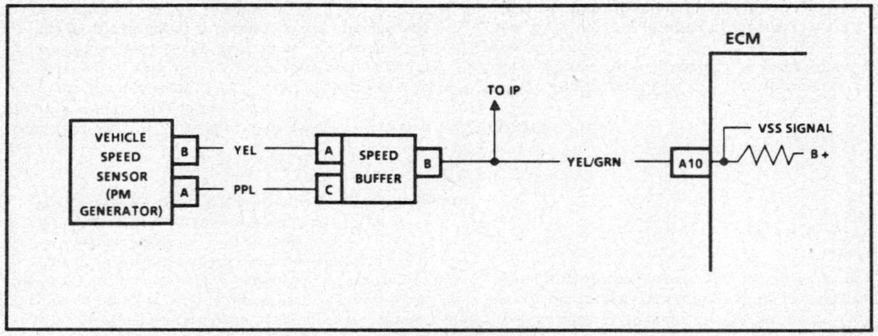

CODE 24

VEHICLE SPEED SENSOR (VSS) CIRCUIT 3.1L (VIN D) "UT" (TBI)

Circuit Description:

The ECM applies and monitors 12 volts on CKT A10. CKT A10 connects to the vehicle speed sensor which alternately grounds CKT A10 when drive wheels are turning. This pulsing action takes place about 2000 times per mile and the ECM will calculate vehicle speed based on the time between "pulses."

"Scan" reading should closely match with speedometer reading with drive wheels turning. Disregard a Code 24 set when drive wheels are not turning.

Test Description: Number(s) below refer to circled number(s) on the diagnostic chart.

1. Code 24 will set if vehicle speed equals 2 mph when:
   - Engine speed is greater than 1400 rpm
   - TPS is less than 5% (closed throttle)
   - Low load condition (high vacuum). Less than 25 kPa.
   - All conditions met for 5 seconds.

   These conditions are met during a road load deceleration. Disregard Code 24 that sets when drive wheels are not turning.
2. 8-12 volts, at the I/P connector, indicates CKT A10 is open between the I/P connector and the VSS, or there is a faulty vehicle speed sensor. A voltage of less than 1 volt, at the I/P connector, indicates that CKT A10 is shorted to ground. If, after disconnecting CKT A10 at the vehicle speed sensor, the voltage reads above 10 volts, the vehicle speed sensor is faulty. If voltage remains less than 8 volts, then CKT A10 is grounded. If CKT A10 is not grounded, there is a faulty connection at the ECM, or a faulty ECM.

Diagnostic Aids:

"Scan" should indicate a vehicle speed whenever the drive wheels are turning.

If all OK, refer to "Intermittents" in "Symptoms."

CODE 24

VEHICLE SPEED SENSOR (VSS) CIRCUIT 3.1L (VIN D) "UT" (TBI)

NOTE: DISREGARD CODE 24 IF SET WHEN DRIVE WHEELS ARE NOT TURNING.

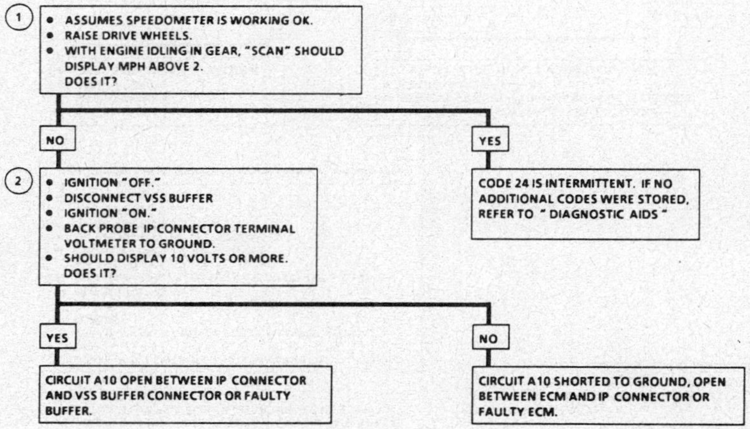

CLEAR CODES AND CONFIRM "CLOSED LOOP" OPERATION AND NO "CHECK ENGINE" LIGHT.

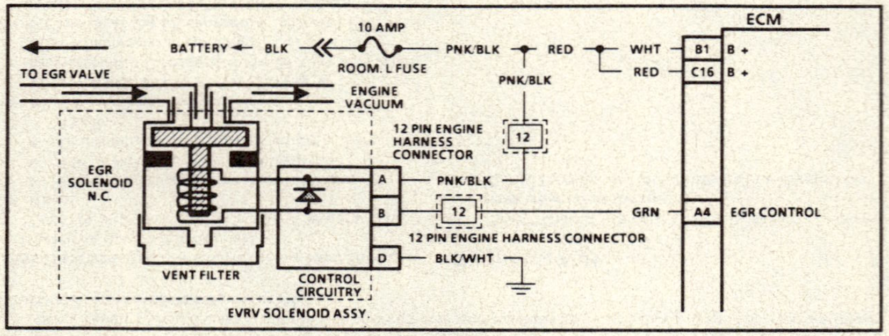

CODE 32

EGR SYSTEM FAILURE
3.1L (VIN D) "UT" (TBI)

Circuit Description:

The ECM operates a solenoid to control the Exhaust Gas Recirculation (EGR) valve. This solenoid is normally closed. By providing a ground path, the ECM energizes the solenoid which then allows vacuum to pass to the EGR valve.

The ECM monitors EGR effectiveness by de-energizing the EGR control solenoid thereby shutting "OFF" vacuum to the EGR valve diaphragm. With the EGR valve closed, fuel integrator counts will be greater than they were during normal EGR operation. If the change is not within the calibrated window, a Code 32 will be set.

The ECM will check EGR operation when:

- Vehicle speed is above 50 mph.
- Engine vacuum is between 40 and 51 kPa.
- No change in throttle position while test is being run.

Test Description: Number(s) below refer to circled number(s) on the diagnostic chart.

1. With the ignition "ON," engine stopped, the solenoid should not be energized and vacuum should not pass to the EGR valve. Grounding the diagnostic terminal will energize the solenoid and allow vacuum to pass to valve.
2. Checks for plugged EGR passages. If passages are plugged, the engine may have severe detonation on acceleration.
3. The vehicle must be driven during this test in order to produce sufficient engine load to operate the EGR. Lightly accelerating (approximately 1/4 throttle) will produce a large and stable enough reading to determine if the ECM is commanding the system "ON."

Diagnostic Aids:

- Before replacing ECM, use an ohmmeter and check the resistance of each ECM controlled relay and solenoid coil.
  See ECM wiring diagram for coil terminal identification of solenoid(s) and relay(s) to be checked.
  Replace any solenoid where resistance measures less than 20 ohms.

CODE 32

EGR SYSTEM FAILURE
3.1L (VIN D) "UT" (TBI)

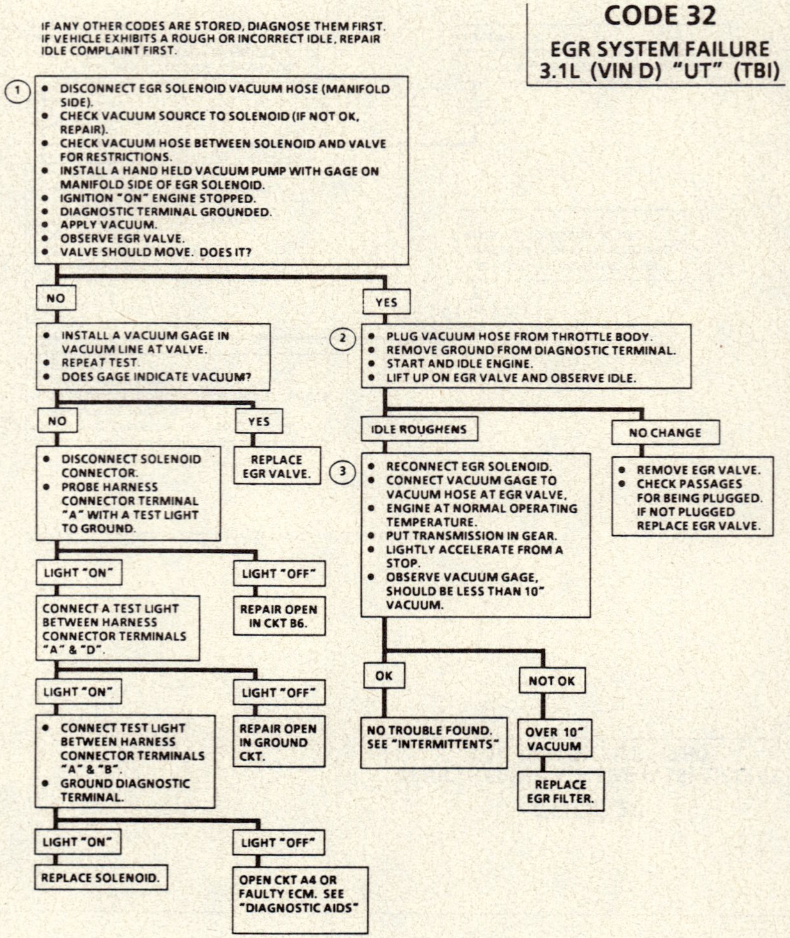

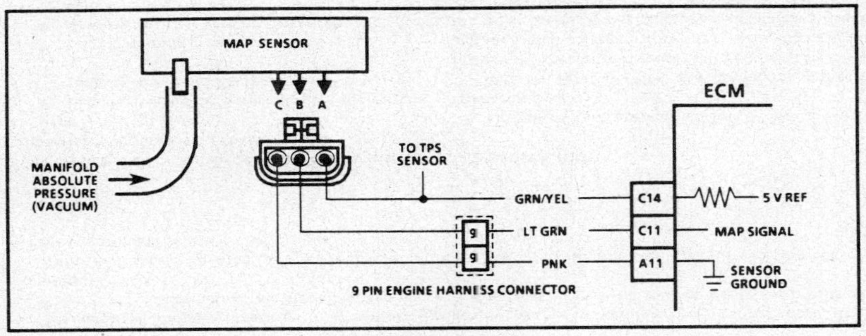

CODE 33

MANIFOLD ABSOLUTE PRESSURE (MAP) SENSOR CIRCUIT
(SIGNAL VOLTAGE HIGH - LOW VACUUM)
3.1L (VIN D) "UT" (TBI)

Circuit Description:

The Manifold Absolute Pressure (MAP) sensor responds to changes in manifold pressure (vacuum). The ECM receives this information as a signal voltage that will vary from about 1-1.5 volts at idle to 4.5-4.8 volts at Wide Open Throttle (WOT).

A "Scan" displays manifold pressure in volts. Low pressure (high vacuum) reads a low voltage while a high pressure (low vacuum) reads a high voltage.

If the MAP sensor fails the ECM will substitute a fixed MAP value and use the Throttle Position Sensor (TPS) to control fuel delivery.

Test Description: Number(s) below refer to circled number(s) on the diagnostic chart.

1. Code 33 will set when:
   - Signal is too high, (kPa greater than 82 kPa), for a time greater than 5 seconds.
   - TPS less than 4%.
   - Vehicle speed is above 2 mph.

   Engine misfire or a low unstable idle may set Code 33. Disconnect MAP sensor and system will go into backup mode. If the misfire or idle condition remains
2. If the ECM recognizes the low MAP signal, the ECM and wiring are OK.

Diagnostic Aids:

With the ignition "ON" and the engine stopped, the manifold pressure is equal to atmospheric pressure and the signal voltage will be high.

This information is used by the ECM as an indication of vehicle altitude. Comparison of this reading with a known good vehicle with the same sensor is a good way to check accuracy of a "suspect" sensor. Readings should be the same ± .4 volt.

If idle is rough or unstable refer to "Symptoms," for items which can cause an unstable idle.

An open in CKT A11 or the connection will result in a Code 33.

- Check all connections.

NOTE: Make sure electrical connector remains securely fastened.

- Disconnect sensor from bracket and twist sensor by hand (only) to check for intermittent connections. Output changes greater than .1 volt indicates a bad connector or connection. If OK, replace sensor.
- Refer to "Intermittents" in "Symptoms "

CODE 33

MANIFOLD ABSOLUTE PRESSURE (MAP) SENSOR CIRCUIT
(SIGNAL VOLTAGE HIGH - LOW VACUUM)
3.1L (VIN D) "UT" (TBI)

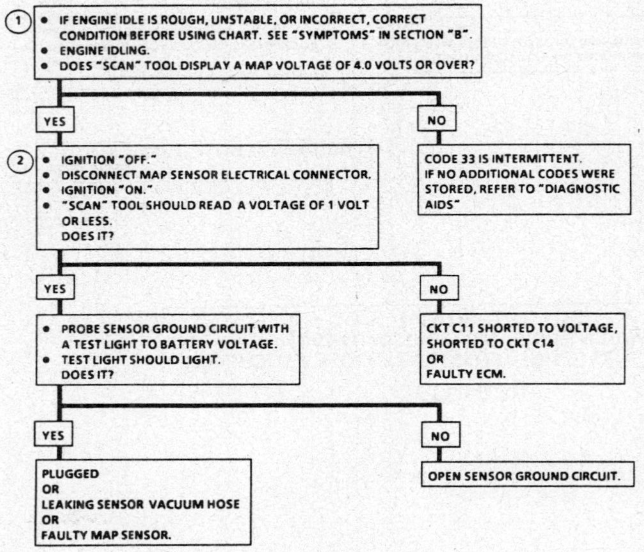

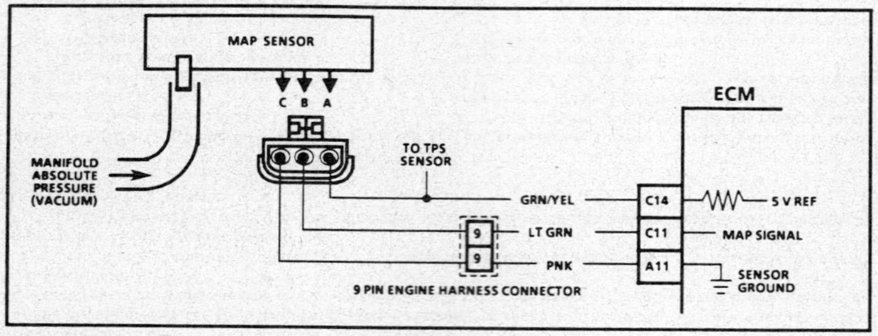

CODE 34

MANIFOLD ABSOLUTE PRESSURE (MAP) SENSOR CIRCUIT
(SIGNAL VOLTAGE LOW - HIGH VACUUM)
3.1L (VIN D) "UT" (TBI)

Circuit Description:

The Manifold Absolute Pressure (MAP) sensor responds to changes in manifold pressure (vacuum). The ECM receives this information as a signal voltage that will vary from about 1–1.5 volts at idle to 4–4.5 volts at Wide Open Throttle (WOT).

If the MAP sensor fails the ECM will substitute a fixed MAP value and use the Throttle Position Sensor (TPS) to control fuel delivery.

Test Description: Number(s) below refer to circled number(s) on the diagnostic chart.

1. Code 34 will set when:
   - Signal is too low (kPa less than 14) and engine running less than 1200 rpm.

     OR
   - Engine running greater than 1200 rpm.
   - Throttle position greater than 15% (over 1.5 volts).
2. If the ECM recognizes the high MAP signal, the ECM and wiring are OK.
3. The "Scan" tool may not display 12 volts. The important thing is that the ECM recognizes the voltage as more than 4 volts, indicating that the ECM and CKT C11 are OK.

Diagnostic Aids:

An intermittent open in CKTs C11 or C14 will result in a Code 34.

With the ignition "ON" and the engine stopped, the manifold pressure is equal to atmospheric pressure and the signal voltage will be high. This information is used by the ECM as an indication of vehicle altitude and is referred to as BARO. Comparison of this BARO reading with a known good vehicle with the same sensor is a good way to check accuracy of a "suspect" sensor. Reading should be the same ± .4 volt.

Refer to "Intermittents" in "Symptoms"

CODE 34

MANIFOLD ABSOLUTE PRESSURE (MAP) SENSOR CIRCUIT
(SIGNAL VOLTAGE LOW - HIGH VACUUM)
3.1L (VIN D) "UT" (TBI)

(1)
- ENGINE IDLING.
- DOES "SCAN" TOOL DISPLAY MAP VOLTAGE BELOW .25 VOLT?

NO: CODE 34 IS INTERMITTENT. IF NO ADDITIONAL CODES WERE STORED, REFER TO "DIAGNOSTIC AIDS"

YES:

(2)
- IGNITION "OFF."
- DISCONNECT SENSOR ELECTRICAL CONNECTOR.
- JUMPER HARNESS TERMINALS "B" TO "C".
- IGNITION "ON."
- MAP VOLTAGE SHOULD READ OVER 4.7 VOLTS. DOES IT?

YES: FAULTY CONNECTION OR SENSOR. REFER TO "DIAGNOSTIC AIDS"

NO:

(3)
- IGNITION "OFF."
- REMOVE JUMPER WIRE.
- PROBE TERMINAL "B" (MAP SIGNAL CIRCUIT) WITH A TEST LIGHT TO BATTERY VOLTAGE.
- IGNITION "ON."
- "SCAN" TOOL SHOULD READ OVER 4 VOLTS. DOES IT?

YES: 5 VOLT REFERENCE CIRCUIT OPEN OR SHORTED TO GROUND OR FAULTY ECM.

NO: CKT C11 OPEN OR CKT C11 SHORTED TO GROUND OR CKT C11 SHORTED TO SENSOR GROUND OR FAULTY ECM.

3.1L ENGINE—TROUBLE CODE CHARTS—RODEO AND 1991–92 PICK-UP

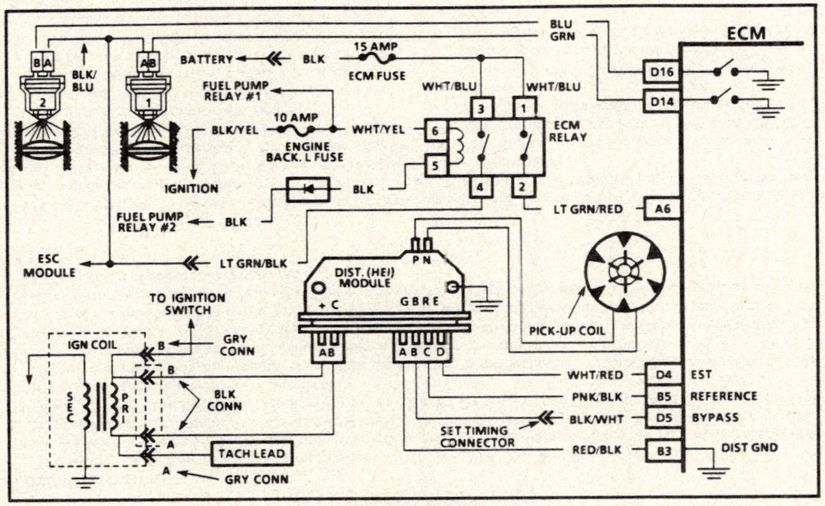

CODE 42

ELECTRONIC SPARK TIMING (EST) CIRCUIT
3.1L (VIN D) "UT" (TBI)

Circuit Description:

When the system is running on the ignition module, that is, no voltage on the bypass line, the ignition module grounds the Electronic Spark Timing (EST) signal. The ECM expects to see no voltage on the EST line during this condition. If it sees a voltage, it sets Code 42 and will not go into the EST mode.

When the rpm for EST is reached (about 400 rpm), and bypass voltage applied, the EST should no longer be grounded in the ignition module so the EST voltage should be varying.

If the bypass line is open or grounded, the ignition module will not switch to EST mode so the EST voltage will be low and Code 42 will be set.

If the EST line is grounded, the ignition module will switch to EST, but because the line is grounded, there will be no EST signal. A Code 42 will be set.

Test Description: Number(s) below refer to circled number(s) on the diagnostic chart.

1. Code 42 means the ECM has seen an open or short to ground in the EST or bypass circuits. This test confirms Code 42 and that the fault causing the code is present.
2. Checks for a normal EST ground path through the ignition module. An EST CKT D4 shorted to ground will also read less than 500 ohms; however, this will be checked later.
3. As the test light voltage touches CKT D5, the module should switch causing the ohmmeter to "overrange" if the meter is in the 1000-2000 ohms position. Selecting the 10-20,000 ohms position will indicate above 5000 ohms. The important thing is that the module "switched."
4. The module did not switch and this step checks for:
   - EST CKT D4 shorted to ground.
   - Bypass CKT D5 open.
   - Faulty ignition module connection or module.
5. Confirms that Code 42 is a faulty ECM and not an intermittent in CKTs D4 or D5.

Diagnostic Aids:

If a Code 42 was stored and the customer complains of a "hard start," the problem is most likely a grounded EST line (CKT D4).

The "Scan" tool does not have any ability to help diagnose a Code 42 problem.

A PROM not fully seated in the ECM can result in a Code 42.

Refer to "Intermittents" in "Symptoms."

CODE 42

ELECTRONIC SPARK TIMING (EST) CIRCUIT
3.1L (VIN D) "UT" (TBI)

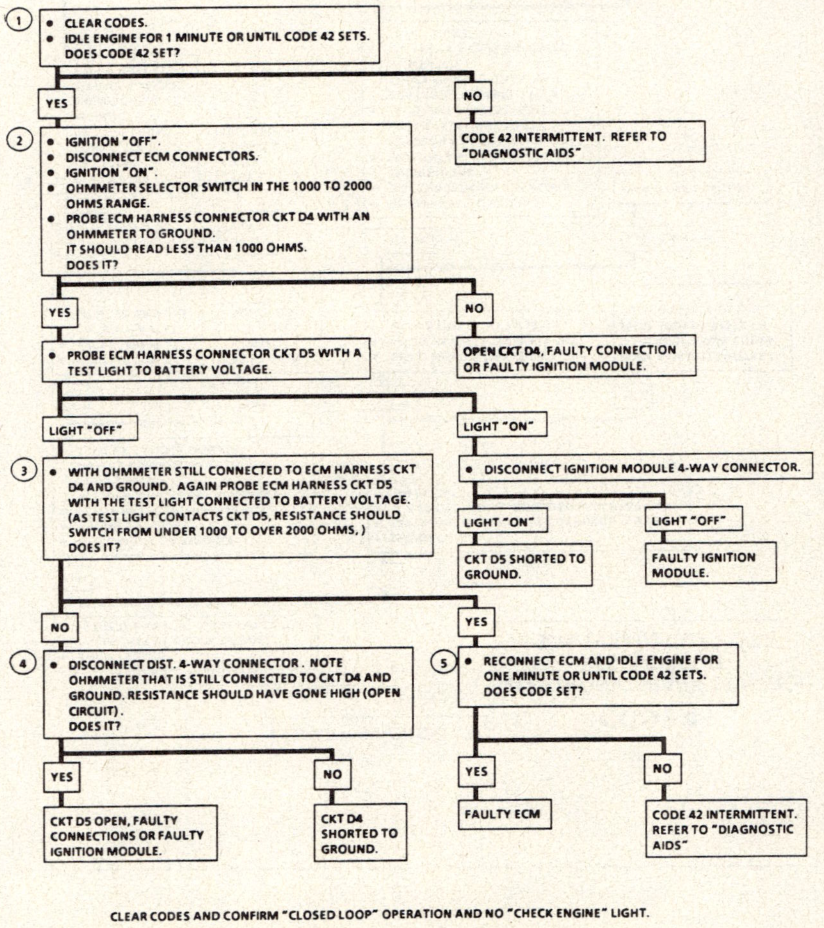

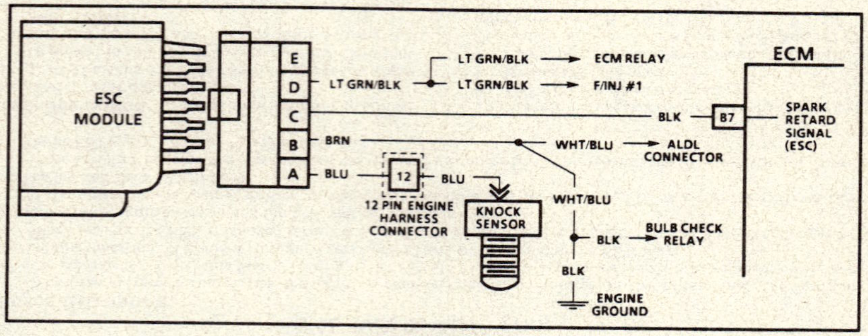

CODE 43

ELECTRONIC SPARK CONTROL (ESC) CIRCUIT
3.1L (VIN D) "UT" (TBI)

Circuit Description:

Electronic Spark Control (ESC) is accomplished with a module that sends a voltage signal to the ECM. As the knock sensor detects engine knock, the voltage from the ESC module to the ECM drops, and this signals the ECM to retard timing. The ECM will retard the timing when knock is detected and rpm is above about 900 rpm.

Code 43 means the ECM has been low voltage at CKT B7 for longer than 5 seconds with the engine running or the system has failed the functional check.

This system performs a functional check once per start up to check the ESC system. To perform this test the ECM will advance the spark when coolant is above 95°C and at a high load condition (near WOT). The ECM then checks the signal at "B7" to see if a knock is detected. The functional check is performed once per start up and if knock is detected when coolant is below 95°C (194°F) the test has passed and the functional check will not be run. If the functional check fails, the "Service Engine Soon" light will remain "ON" until ignition is turned "OFF" or until a knock signal is detected.

Test Description: Number(s) below refer to circled number(s) on the diagnostic chart.

1. If the conditions for a Code 43 are present the "Scan" will always display "YES." There should not be a knock at idle unless an internal engine problem, or a system problem exists.
2. This test will determine if the system is functioning at this time. Usually a knock signal can be generated by tapping on the right exhaust manifold. If no knock signal is generated try tapping on block close to the area of the sensor.
3. Because Code 43 sets when the signal voltage on CKT B7 remains low this test should cause the signal on CKT B7 to go high. The 12 volts signal should be seen by the ECM as "no knock" if the ECM and wiring are OK.
4. This test will determine if the knock signal is being detected on CKT E or if the ESC module is at fault.
5. If CKT E is routed to close to secondary ignition wires the ESC module may see the interference as a knock signal.
6. This checks the ground circuit to the module. An open ground will cause the voltage on CKT B7 to be about 12 volts which would cause the Code 43 functional test to fail.
7. Contacting CKT E with a test light to 12 volts should generate a knock signal. This will determine if the ESC module is operating correctly.

Diagnostic Aids:

Code 43 can be caused by a faulty connection at the knock sensor at the ESC module or at the ECM. Also check CKT B7 for possible open or short to ground.

Refer to "Intermittents" in "Symptoms"

"SCAN" DIAGNOSTICS

CODE 43

ELECTRONIC SPARK CONTROL (ESC) CIRCUIT
3.1L (VIN D) "UT" (TBI)

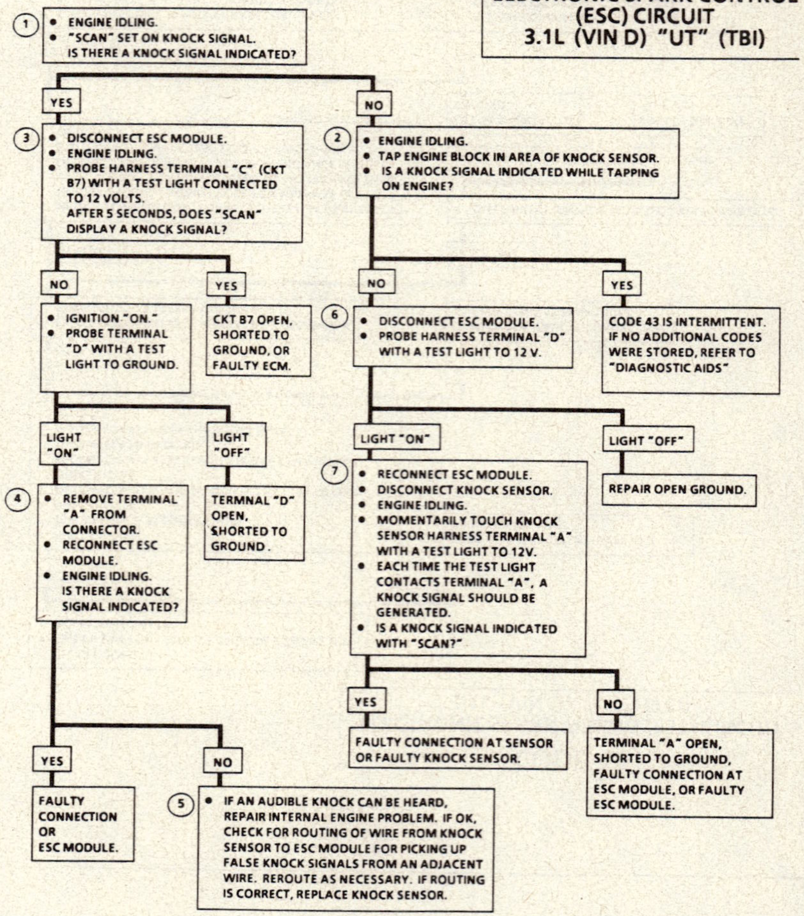

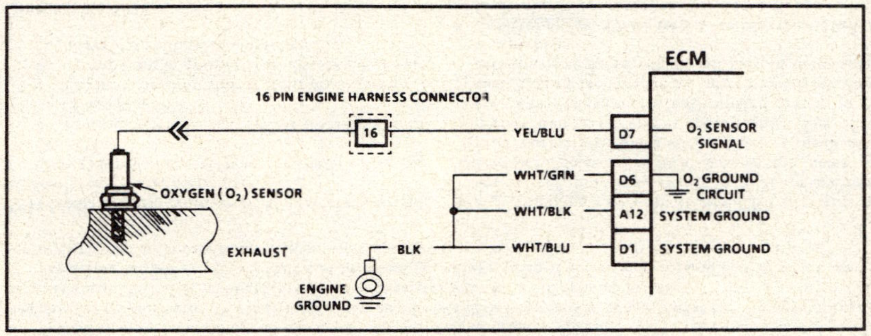

CODE 44

OXYGEN (O_2) SENSOR CIRCUIT
(LEAN EXHAUST INDICATED)
3.1L (VIN D) "UT" (TBI)

Circuit Description:

The ECM supplies a voltage of about .45 volt between terminals "D6" and "D7". (If measured with a 10 megohm digital voltmeter, this may read as low as .32 volt). The Oxygen (O_2) sensor varies the voltage within a range of about 1 volt if the exhaust is rich, down through about .10 volt if exhaust is lean.

The sensor is like an open circuit and produces no voltage when it is below about 360°C (600°F). An open sensor circuit or cold sensor causes "Open Loop" operation.

Test Description: Number(s) below refers to circled number(s) on the diagnostic chart.

1. Code 44 is set when the O_2 sensor signal voltage on CKT D7.
   - Remains below .2 volt for 50 seconds.
   - And the system is operating in "Closed Loop."

Diagnostic Aids:

Using the "Scan," observe the block learn values at different rpm and air flow conditions to determine when the Code 44 may have been set. If the conditions for Code 44 exists the block learn values will be around 150.

- O_2 Sensor Wire. Sensor pigtail may be mispositioned and contacting the exhaust manifold.
- Check for intermittent ground in wire between connector and sensor.
- MAP Sensor. A (MAP) sensor output that causes the ECM to sense a higher than normal vacuum will cause the system to go lean. Disconnect the MAP sensor and if the lean condition is gone, replace the sensor.
- Lean Injector(s).
- Fuel Contamination. Water, even in small amounts, near the in-tank fuel pump inlet can be delivered to the injectors. The water causes a lean exhaust and can set a Code 44.
- Fuel Pressure. System will be lean if pressure is too low. It may be necessary to monitor fuel pressure while driving the car at various road speeds and/or loads to confirm. See "Fuel System Diagnosis" CHART A-7.
- Exhaust Leaks. If there is an exhaust leak, the engine can pull outside air into the exhaust and past the sensor. Vacuum or crankcase leaks can cause a lean condition.
- If the above are OK, it is a faulty oxygen sensor.

CODE 44

OXYGEN (O_2) SENSOR CIRCUIT
(LEAN EXHAUST INDICATED)
3.1L (VIN D) "UT" (TBI)

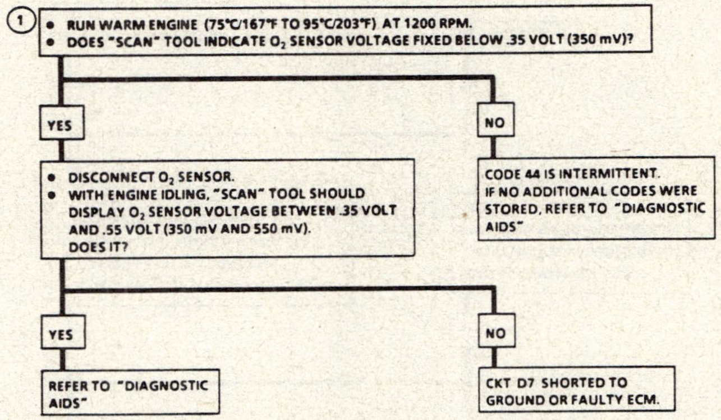

CLEAR CODES AND CONFIRM "CLOSED LOOP" OPERATION AND NO "CHECK ENGINE" LIGHT.

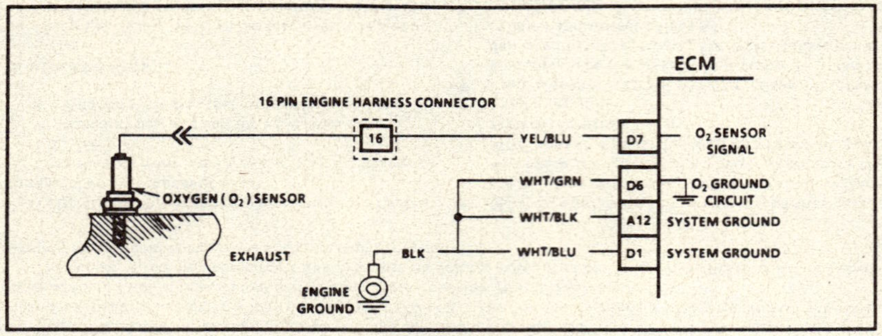

CODE 45

OXYGEN (O_2) SENSOR CIRCUIT
(RICH EXHAUST INDICATED)
3.1L (VIN D) "UT" (TBI)

Circuit Description:

The ECM supplies a voltage of about .45 volt between terminals "D6" and "D7". (If measured with a 10 megohm digital voltmeter, this may read as low as .32 volt). The Oxygen (O_2) sensor varies the voltage within a range of about 1 volt if the exhaust is rich, down through about .10 volt if exhaust is lean.

The sensor is like an open circuit and produces no voltage when it is below about 360°C (600°F) An open sensor circuit or cold sensor causes "Open Loop" operation.

Test Description: Number(s) below refers to circled number(s) on the diagnostic chart.

1. Code 45 is set when the O_2 sensor signal voltage or CKT D7.
   - Remains above .7 volt for 50 seconds; and in "Closed Loop."
   - Engine time after start is 1 minute or more.
   - Throttle angle greater than 2% (about .2 volt above idle voltage) but less than 25%.

Diagnostic Aids:

Using the "Scan," observe the block learn values at different rpm conditions to determine when the Code 45 may have been set. If the conditions for Code 45 exists, The block learn values will be around 115.

- Fuel Pressure. System will go rich if pressure is too high. The ECM can compensate for some increase. However, if it gets too high, a Code 45 may be set. See "Fuel System Diagnosis" CHART A-7.
- Leaking injector. See CHART A-7.
- Check for fuel contaminated oil.
- HEI Shielding. An open ground CKT B3 (ignition system ref. low) may result in EMI, or induced electrical "noise." The ECM looks at this "noise" as reference pulses. The additional pulses result in a higher than actual engine speed signal. The ECM then delivers too much fuel, causing system to go rich. Engine tachometer will also show higher than actual engine speed, which can help in diagnosing this problem.
- Canister purge. Check for fuel saturation. If full of fuel, check canister control and hoses.
- MAP sensor. An output that causes the ECM to sense a lower than normal vacuum can cause the system to go rich. Disconnecting the MAP sensor will allow the ECM to set a fixed value for the sensor. Substitute a different MAP sensor if the rich condition is gone while the sensor is disconnected.
- TPS. An intermittent TPS output will cause the system to go rich, due to a false indication of the engine accelerating.

CODE 45

OXYGEN (O_2) SENSOR CIRCUIT
(RICH EXHAUST INDICATED)
3.1L (VIN D) "UT" (TBI)

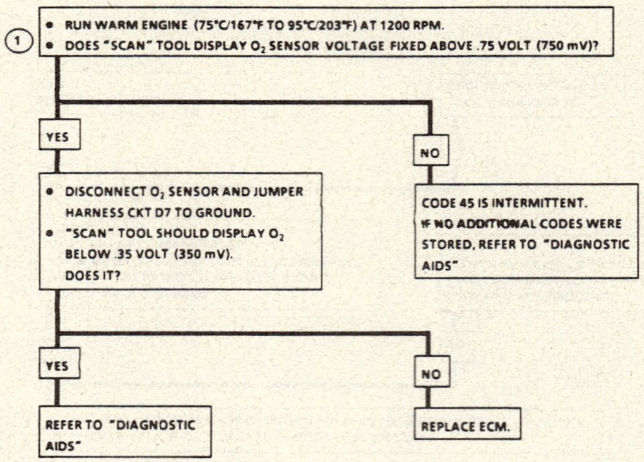

CLEAR CODES AND CONFIRM "CLOSED LOOP" OPERATION AND NO "CHECK ENGINE" LIGHT.

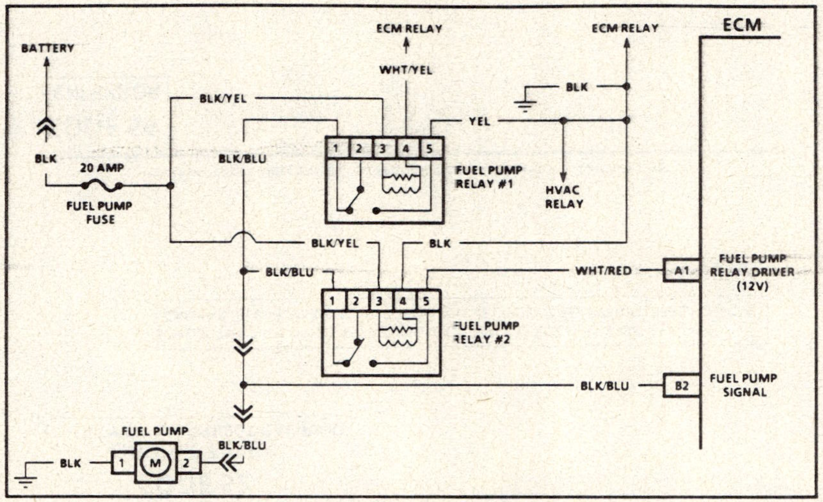

CODE 54

FUEL PUMP CIRCUIT
(LOW VOLTAGE)
3.1L (VIN D) "UT" (TBI)

Circuit Description:

When the ignition switch is turned "ON," the Electronic Control Module (ECM) will activate the fuel pump relay and run the in-tank fuel pump. The fuel pump will operate as long as the engine is cranking or running, and the ECM is receiving ignition reference pulses.

If there are no reference pulses, the ECM will shut "OFF" the fuel pump within 2 seconds after key "ON."

Should the fuel pump relay, or the 12 volt relay drive from the ECM fail, the fuel pump will be run through an oil pressure switch which activates the back up relay circuit.

Diagnostic Aids:

An inoperative fuel pump relay can result in long cranking times, particularly if the engine is cold or engine oil pressure is low. The extended crank period is caused by the time necessary for oil pressure to build enough to close the oil pressure switch and turn "ON" the fuel pump.

Intermittent problems may be caused by:

- Poor wiring connections
- Corrosion
- All connections should be checked visually

CODE 54

FUEL PUMP CIRCUIT
(LOW VOLTAGE)
3.1L (VIN D) "UT" (TBI)

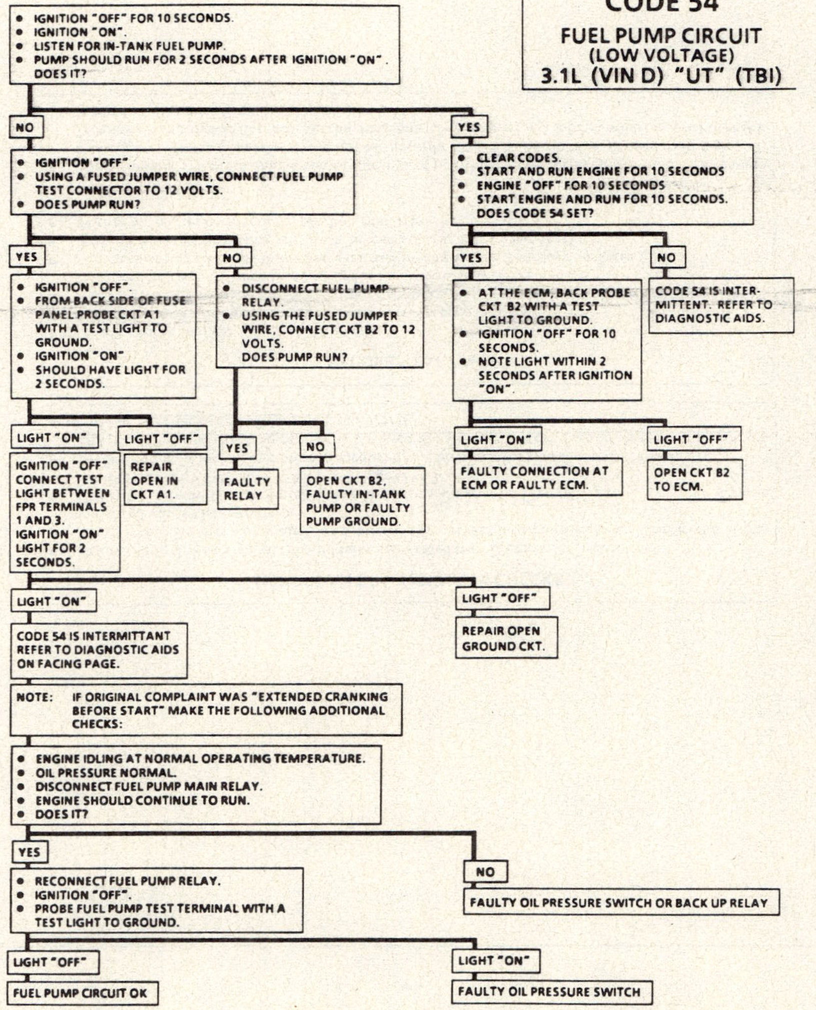

3.1L ENGINE—TROUBLE CODE CHARTS RODEO AND 1991–92 PICK-UP

CODE 51
PROM ERROR
(FAULTY OR INCORRECT CAL-PAK)

CHECK THAT ALL PINS ARE FULLY INSERTED IN THE SOCKET. IF OK, REPLACE PROM, CLEAR MEMORY AND RECHECK. IF CODE 51 REAPPEARS, REPLACE ECM.

CLEAR ALL CODES AND CONFIRM "CLOSED LOOP" OPERATION AND NO "CHECK ENGINE" LIGHT

CODE 52
CAL-PAK ERROR
(FAULTY OR INCORRECT CAL-PAK)

CHECK THAT ALL PINS ARE FULLY INSERTED IN THE SOCKET. IF OK, REPLACE CAL-PAK, CLEAR MEMORY AND RECHECK. IF CODE 52 REAPPEARS, REPLACE ECM.

CLEAR ALL CODES AND CONFIRM "CLOSED LOOP" OPERATION AND NO "CHECK ENGINE" LIGHT

CODE 55
ECM ERROR

REPLACE ELECTRONIC CONTROL MODULE (ECM).

"AFTER REPAIRS," CONFIRM "CLOSED LOOP" OPERATION AND NO "SERVICE ENGINE SOON" LIGHT.

3.1L ENGINE—DRIVEABILITY CHARTS

IMPORTANT PRELIMINARY CHECKS

- Before using this section you should have performed the "Diagnostic Circuit Check."
- Verify the customer complaint, and locate the correct symptom below. Check the items indicated under that symptom.
- Refer to "Engine Components/Wiring Diagrams/Diagnostic Charts," Engine Cranks But Will-Not Run,CHART A-3 Page (1 of 2).
- Several of the following symptom procedures call for a careful visual/physical check. The importance of this step cannot be stressed too strongly - it can lead to correcting a problem without further checks and can save valuable time.

BEFORE STARTING

This check should include:

- ECM grounds for being clean, tight, and in their proper location.
- Vacuum hoses for splits, kinks and proper connections, as shown on "Vehicle Emission Control Information" label. Check thoroughly for any type of leak or restriction.
- Air leaks at throttle body mounting area and intake manifold sealing surfaces.
- Ignition wires for cracking, hardness, proper routing and carbon tracking.
- Wiring for proper connections, pinches, and cuts.

- The following symptom charts contain groups of possible causes for each symptom that covers this engine. **These procedures are not necessarily meant to be done in consecutive order.** If "Scanner" readings do not indicate the problems, then proceed in a logical order, easiest to check or most likely cause first.

3.1L ENGINE – DRIVEABILITY CHARTS

INTERMITTENTS

(Page 1 of 2)

Definition: Problem may or may not turn "ON" the "Service Engine Soon" light, or store a code.

PRELIMINARY CHECKS

- Perform the careful visual/physical checks as described at start of this section.

DIAGNOSTIC CODE CHARTS IN "Driveability and Emissions," Section "6E2".

- DO NOT use the Diagnostic Code Charts in "Driveability and Emission," for intermittent problems. The fault must be present to locate the problem. If a fault is intermittent, use of diagnostic code charts may result in replacement of good parts.

FAULTY ELECTRICAL CONNECTIONS OR WIRING

- Most intermittent problems are caused by faulty electrical connections or wiring. Perform careful check of suspect circuits for:
  - Poor mating of the connector halves, or terminals, not fully seated in the connector body (backed out).
  - Improperly formed or damaged terminals. All connector terminals in problem circuit should be carefully reformed or replaced to insure proper contact tension.
  - Poor terminal to wire connection. This requires removing the terminal from the connector body to check.

ROAD TEST

- If a visual/physical check does not find the cause of the problem, the vehicle can be driven with a voltmeter connected to a suspected circuit or a "Scan" tool may be used. An abnormal voltage or "Scan" reading, when the problem occurs, indicates the problem may be in that circuit. If the wiring and connectors check OK, and a diagnostic code was stored for a circuit having a sensor, except for Codes 44 and 45, replace sensor.

INTERMITTENTS

(Page 2 of 2)

Definition: Problem may or may not turn "ON" the "Service Engine Soon" light, or store a code.

INTERMITTENT "SERVICE ENGINE SOON LIGHT"

- An intermittent "Service Engine Soon" light, and No Diagnostic Codes, may be caused by:
  - Electrical system interference caused by a defective relay, ECM driven solenoid, or switch. They can cause a sharp electrical surge. Normally, the problem will occur when the faulty component is operated.
  - Improper installation of electrical devices, such as lights, 2-way radios, electric motors, etc.
  - EST wires should be routed away from spark plug wires, ignition system components and generator. Wire for CKT B3 from ECM to ignition system should be a good ground.
  - Ignition secondary shorted to ground.
  - CKT 419 ("Service Engine Soon" light) or CKT A9 (diagnostic "test" terminal) intermittently shorted to ground.
  - ECM grounds.

LOSS OF DIAGNOSTIC CODE MEMORY

- To check, disconnect TPS and idle engine until "Service Engine Soon" light comes "ON." Code 22 should be stored, and kept in memory when the ignition is turned "OFF" for at least 10 seconds. If not, the ECM is faulty.

3.1L ENGINE – DRIVEABILITY CHARTS

HARD START
(Page 1 of 2)

Definition: Engine cranks OK, but does not start for a long time. Does eventually run, or may start but immediately dies.

PRELIMINARY CHECKS

- Perform the careful visual/physical checks as described at start of this section.
- Make sure the driver is using the correct starting procedure.

SENSORS

- **CHECK:** Coolant Temperature Sensor (CTS) - Using a "Scan" tool, compare coolant temperature with ambient temperature on a cold engine.
  If coolant temperature reading is 5 degrees greater than or less than ambient air temperature on a cold engine, check for high resistance in coolant sensor circuit or sensor itself. Compare resistance value to the "Diagnostic Aids" chart in Code 15 chart.
- **CHECK:** Throttle Position Sensor (TPS) - If a sticking throttle shaft or binding linkage causes a high TPS voltage (open throttle indication), the ECM will not control idle. Monitoring TPS voltage. With a "Scan" tool and/or voltmeter should read less than 1.25 volts with throttle closed.

FUEL SYSTEM

- **NOTE:** Fuel pump relay operation - pump should turn "ON" for 2 seconds when ignition is turned "ON."
- **CHECK:** Fuel Pressure. Refer to Fuel System Diagnosis CHART A-7, Page (1 of 2).
- **CHECK:** For water contaminated fuel.
- **CHECK:** For a faulty in-tank fuel pump check valve, which would allow the fuel in the lines to drain back to the tank after the engine is stopped. To check for this condition:
  1. Ignition "OFF."
  2. Disconnect fuel line at the filter.
  3. Remove the tank filler cap.
  4. Connect a radiator test pump to the fuel line and apply 103 kPa (15 psi) pressure. If the pressure will hold for 60 seconds, the check valve is OK.
- **CHECK:** Fuel pump relay - Connect test light between fuel pump "test" terminal and ground.

IGNITION SYSTEM

- **CHECK:** Ignition system for:
  - Proper ignition voltage output with spark tester J 26792 or equivalent (ST-125).
  - Spark plugs, wet plugs, cracks, wear, improper gap, burned electrodes or heavy deposits.
  - Bare and shorted wires
  - Moisture in distributor cap
  - Worn distributor shaft
  - Pickup coil resistance and connections
  - Loose ignition coil connections
- **NOTE:** If engine starts, but then immediately stalls, disconnect the set timing connector. If engine then starts, and runs OK, replace distributor pickup coil.
- **CHECK:** CKT D4 (EST) for short to ground.

ADDITIONAL CHECKS

- **CHECK:** IAC operation
- **CHECK:** No crank signal
- **CHECK:** EGR operation
- **CHECK:** Service Bulletins for PROM updates

SURGES AND/OR CHUGGLES

Definition: Engine power variation, under steady throttle or cruise. Feels like the vehicle speeds up and slows down, with no change in the accelerator pedal.

PRELIMINARY CHECKS

- Perform the careful visual/physical checks as described at start of this section.
- Be sure driver understands Torque Converter Clutch (TCC) and A/C compressor operation in owner's manual.
- Use a "Scan" tool to make sure reading of VSS matches vehicle speedometer.

SENSORS

- **CHECK:** Oxygen (O_2) sensor for silicon contamination from fuel, or use of improper RTV sealant. The sensor may have a white, powdery coating and result in a high but false signal voltage (rich exhaust indication). The ECM will then reduce the amount of fuel delivered to the engine, causing a severe driveability problem.

FUEL SYSTEM

- **NOTE:** To determine if the condition is caused by a rich or lean system, the vehicle should be driven at the speed of the complaint. Monitoring block learn will help identify a problem.
  Lean - Block learn greater than 150. Refer to "Diagnostic Aids" on facing page of Code 44.
  Rich - Block learn less than 115. Refer to "Diagnostic Aids" on facing page of Code 45.
- **CHECK:** Fuel pressure while condition exists. Refer to "Engine Components/Wiring Diagrams/Diagnostic Charts." Fuel System Diagnosis CHART A-7 Page (1 of 2).
- **CHECK:** In-line fuel filter. Replace if dirty or plugged.

IGNITION SYSTEM

- **CHECK:** For proper ignition voltage output voltage using spark tester J 26792 or equivalent (ST-125).
- **CHECK:** Spark plugs. Remove spark plugs, check for wet plugs, cracks, wear, improper gap, burned electrodes, or heavy deposits. Repair or replace as necessary.
- **CHECK:** Ignition timing. See "Vehicle Emission Control Information" label.

ADDITIONAL CHECKS

- **CHECK:** ECM grounds for being clean, tight, and in their proper locations.
- **CHECK:** Generator output voltage. Repair if less than 9 or more than 16 volts.
- **CHECK:** Vacuum lines for kinks or leaks.
- **CHECK:** For intermittent EGR.
- **CHECK:** TCC operation.

3.1L ENGINE—DRIVEABILITY CHARTS

LACK OF POWER, SLUGGISH, OR SPONGY

(Page 1 of 2)

Definition: Engine delivers less than expected power. Little or no increase in speed, when accelerator pedal is pushed down part way.

PRELIMINARY CHECKS

- Perform the careful visual/physical checks as described at start of this section.
- Compare customer's vehicle to similar unit. Make sure the customer has an actual problem.
- Remove air filter and check air filter for dirt, or for being plugged, replace as necessary.
- Transmission shift pattern and down shift operation.
- If there is spray from only one injector, then, there is a malfunction in the injector assembly, or in the signal to the injector assembly. The malfunction can be isolated, by switching the injector connectors. If the problem remains with the original injector, after switching the connector, the injector is defective. Replace the injector. If the problem moves with the injector connector, the problem is an improper signal in the injector circuits. Refer to "Engine Components/Wiring Diagrams/Diagnostic Charts," Engine Cranks But Will Not Run, CHART A-3, Page (1 of 2).

FUEL SYSTEM

- **CHECK:** For contaminated fuel.
- **CHECK:** For restricted fuel filter, contaminated fuel or improper fuel pressure, refer to "Engine Components/Wiring Diagrams/Diagnostic Charts," Fuel System Diagnosis, CHART A-7, Page (1 of 2).

IGNITION SYSTEM

- **CHECK:** Proper ignition voltage output with spark tester J 26792 or equivalent (ST-125).
- **CHECK:** Ignition timing. See "Vehicle Emission Control Information" label.
- **CHECK:** Proper operation of EST.

EXHAUST SYSTEM

- **CHECK:** Exhaust System for possible restriction: refer to CHART B-1.
  Inspect exhaust system for damaged or collapsed pipes. Inspect muffler for heat distress or possible internal failure.
  1. With engine at normal operating temperature, connect a vacuum gage to any convenient vacuum port on intake manifold.
  2. Run engine at 1000 rpm and record vacuum reading.
  3. Increase rpm slowly to 2500 rpm. Note vacuum reading at steady 2500 rpm.
  4. If vacuum at 2500 rpm decreases more than 3", from reading at 1000 rpm, the exhaust system should be inspected for restrictions.
  5. Disconnect exhaust pipe from engine and repeat Steps 3 & 4. If vacuum still drops more than 3", with exhaust disconnected, check for exhaust manifold restriction and valve timing.

LACK OF POWER, SLUGGISH, OR SPONGY

(Page 2 of 2)

Definition: Engine delivers less than expected power. Little or no increase in speed, when accelerator pedal is pushed down part way.

ADDITIONAL CHECKS

- **CHECK:** ECM Grounds for being clean, tight, and in their proper location.
- **CHECK:** EGR operation for being open or partly open all the time.
- **CHECK:** Torque Converter Clutch (TCC) operation.
- **CHECK:** A/C operation.
- **CHECK:** Generator output voltage. Repair if less than 9 or more than 16 volts.

ENGINE MECHANICAL

- **CHECK:** Engine compression.
- **CHECK:** Engine valve timing.
- **CHECK:** Engine for proper or worn camshaft.

3.1L ENGINE—DRIVEABILITY CHARTS

DETONATION/SPARK KNOCK

(Page 1 of 2)

Definition: A mild to severe ping, usually worse under acceleration. The engine makes sharp metallic knocks that change with throttle opening.

PRELIMINARY CHECKS

- Perform the careful/physical visual checks as described at start of this section.
- Make sure the customer has an actual problem.
- If there is spray from only one injector, then, there is a malfunction in the injector assembly, or in the signal to the injector assembly. The malfunction can be isolated, by switching the injector connectors. If the problem remains with the original injector, after switching the connector, the injector is defective. Replace the injector. If the problem moves with the injector connector, the problem is an improper signal in the injector circuits, Refer to Engine Cranks But Will Not Run, CHART A-3, Page (1 of 2).
- Park/Neutral (P/N) switch. Be sure "Scan" indicates drive with gear selector in drive.

COOLING SYSTEM

- **CHECK:** For obvious over heating problems.
- **CHECK:** Low engine coolant.
- **CHECK:** Loose water pump belt
- **CHECK:** Restricted air flow to radiator, or restricted coolant flow.
- **CHECK:** Faulty or incorrect thermostat.
- **CHECK:** Correct coolant solution - should be a 50/50 mix of GM #1052753 anti-freeze coolant (or equivalent) and water.

SENSOR

- **CHECK:** Coolant Temperature Sensor (CTS), which has shifted in value. Compare CTS resistance to the "Diagnostic Aids" on Code 15 chart.

FUEL SYSTEM

- **NOTE:** To determine if the condition is caused by a lean system, the vehicle should be driven at the speed of the complaint. Monitoring block learn will help identify the problem. Lean - Block learn greater than 150. Refer to "Diagnostic Aids" on facing page of Code 44.
- **CHECK:** Fuel pressure, refer to "Engien Component/Wiring Diagrams/Diagnostic Charts," Fuel System Diagnosis, CHART A-7, Page (1 of 2).
- **CHECK:** For poor fuel quality, proper octane rating.
- **NOTE:** If "Scan" tool readings are normal and there are no engine mechanical faults, fill fuel tank with a premium gasoline that has a minimum octane rating of 92 and re-evaluate vehicle performance.

DETONATION/SPARK KNOCK

(Page 2 of 2)

Definition: A mild to severe ping, usually worse under acceleration. The engine makes sharp metallic knocks that change with throttle opening.

IGNITION SYSTEM

- **CHECK:** Spark plugs for proper heat range.
- **CHECK:** ESC system operation
- **CHECK:** Ignition timing. See "Vehicle Emission Control Information" label.

ENGINE MECHANICAL

- **CHECK:** For carbon buildup. Remove carbon with top engine cleaner. Follow instructions on can.
- **CHECK:** For incorrect basic engine parts such as cam, heads, pistons, etc.
- **CHECK:** For excessive oil entering combustion chamber.

ADDITIONAL CHECKS

- **CHECK:** Proper operation of EGR valve.
- **CHECK:** Proper operation of Thermac.
- **CHECK:** For proper transmission shift points.
- **CHECK:** Torque Converter Clutch (TCC) operation.
- **CHECK:** For correct PROM (see Service Bulletins).

3.1L ENGINE – DRIVEABILITY CHARTS

HESITATION, SAG, STUMBLE

Definition: Momentary lack of response as the accelerator is pushed down. Can occur at any vehicle. Usually most severe when first trying to make the vehicle move, as from a stop sign. May cause the engine to stall if severe enough.

PRELIMINARY CHECKS

- Perform the careful visual/physical checks as described at start of this section.

FUEL SYSTEM

- **CHECK:** Fuel pressure.
- **CHECK:** Throttle Position Sensor (TPS) - Check TPS for binding or sticking. Voltage should increase at a steady rate as throttle is moved toward Wide Open Throttle (WOT).
- **CHECK:** MAP sensor response and accuracy.
- **CHECK:** Water/Contaminated fuel.
- **CHECK:** Canister purge system for proper operation.

IGNITION SYSTEM

- **CHECK:** Spark plug wires for being faulty.
- **CHECK:** Spark plugs for being fouled.
- **CHECK:** Open ignition system ground, CKT B3.
- **CHECK:** Ignition timing. See "Vehicle Emission Control Information" label.

ADDITIONAL CHECKS

- **CHECK:** For correct PROM (see Service Bulletins).
- **CHECK:** Generator output voltage. Repair, if less than 9 or more than 16 volts.
- **CHECK:** EGR valve operation.

CUTS OUT, MISSES

Definition: Steady pulsation or jerking that follows engine speed, usually more pronounced as engine load increases. The exhaust has a steady spitting sound at idle or low speed.

PRELIMINARY CHECKS

- Perform the careful visual checks as described at start of this section.
- If there is spray from only one injector, then, there is a malfunction in the injector assembly, or in the signal to the injector assembly. The malfunction can be isolated, by switching the injector connectors. If the problem remains with the original injector, after switching the connector, the injector is defective. Replace the injector. If the problem moves with the injector connector, the problem is an improper signal in the injector circuits, Refer to Engine Cranks But Will Not Run, CHART A-3, Page (1 of 2).

IGNITION SYSTEM

- **CHECK:** For cylinder miss by:
  1. Start engine, allow engine to stabilize then disconnect IAC motor. Remove one spark plug wire at a time, using insulated pliers.

CAUTION: Do not perform this test for more than 2 minutes, as this may cause damage to the catalytic converter.

  2. If there is an rpm drop, on all cylinders, (equal to within 50 rpm), go to "Rough, Unstable Or Incorrect Idle, Stalling" symptom. Reconnect IAC motor with ignition "OFF."
  3. If there is no rpm drop on one or more cylinders, or excessive variation in drop, check for spark, on the suspected cylinder(s) with J 26792 (ST-125) Spark Tester or equivalent. If no spark, refer to "Ignition System/EST" If there is spark, remove spark plug(s) in these cylinders and check for:
     - Insulation Cracks
     - Wear
     - Improper Gap
     - Burned Electrodes
     - Heavy Deposits
- **CHECK:** Spark plug wire resistance (should not exceed 30,000 ohms), also, check rotor and distributor cap.
- **NOTE:** If the previous checks did not find the problem:
  - Visually inspect ignition system for moisture, dust, cracks, burns, etc. With engine running spray plug wires with fine water mist to check for shorts.

FUEL SYSTEM

- **CHECK:** Fuel Pressure, refer to "Engine Components/Wiring Diagrams/Diagnostic Charts," Fuel System Diagnosis, CHART A-7, Page (1 of 2)

ENGINE MECHANICAL

- **CHECK:** For proper valve timing. Remove rocker covers. Check for bent pushrods, worn rocker arms, broken or weak valve springs, worn camshaft lobes. Repair as necessary.
- **CHECK:** Low compression. Perform compression check.
- **CHECK:** Intake and exhaust manifold passages for casting flash.

POOR FUEL ECONOMY

Definition: Fuel economy, as measured by an actual road test, is noticeably lower than expected. Also, economy is noticeably lower than it was on this vehicle at one time, as previously shown by an actual road test.

PRELIMINARY CHECKS

- Perform the careful visual/physical checks as described at start of this section.
- Check air cleaner element (filter) for dirt or being plugged.
- Visually (physically) check: Vacuum hoses for splits, kinks, and proper connections as shown on "Vehicle Emission Control Information" label.
- Perform "Diagnostic Circuit Check."
- Check owner's driving habits.
  - Is A/C "ON" full time (Defroster mode "ON")?
  - Are tires at correct pressure?
  - Are excessively heavy loads being carried?
  - Is acceleration too much, too often?
- **NOTE:** Suggest owner fill fuel tank and recheck fuel economy.
  Suggest driver read "Important Facts on Fuel Economy" in Owner's Manual.

FUEL SYSTEM

- **CHECK:** Fuel type, quality and alcohol content.
- **CHECK:** Fuel pressure. Refer to "Engine Components/Wiring Diagrams/Diagnostic Charts." Fuel System Diagnosis, CHART A-7, Page (1 of 2)

IGNITION SYSTEM

- **CHECK:** Spark plugs. Remove spark plugs, check for wet plugs, cracks, wear, improper gap, burned electrodes, or heavy deposits. Repair or replace as necessary.
- **CHECK:** Ignition wires for cracking, hardness, and proper connections.
- **CHECK:** ESC operation.
- **CHECK:** Ignition timing. See "Vehicle Emission Control Information" label.

COOLING SYSTEM

- **CHECK:** Engine coolant level.
- **CHECK:** Engine thermostat for faulty part (always open) or for wrong heat range.

ADDITIONAL CHECKS

- **CHECK:** Transmission shift pattern.
- **CHECK:** TCC Operation
- **CHECK:** For proper calibration of speedometer.
- **CHECK:** For dragging brakes.

ROUGH, UNSTABLE OR INCORRECT IDLE STALLING

(Page 1 of 2)

Definition: The engine runs unevenly at idle. If bad enough, the vehicle may shake. Also, the idle may vary in rpm (called "hunting"). Either condition may be severe enough to cause stalling. Engine idles at incorrect speed.

PRELIMINARY CHECKS

- Perform the careful visual/physical checks as described at start of this section.

SENSORS

- **CHECK:** Oxygen (O2) sensor - Inspect sensor for silicon contamination from fuel, or use of improper RTV sealant. The sensor will have a white, powdery coating, and will result in a high but false signal voltage (rich exhaust indication). The ECM will then reduce the amount of fuel delivered to the engine, causing a severe driveability problem.
- **CHECK:** Throttle Position Sensor (TPS) - If a sticking throttle shaft or binding linkage causes a high TPS voltage (open throttle indication, the ECM will not control idle. Monitor TPS voltage. A "Scan" tool and/or voltmeter should read less than 1.25 volts with throttle closed. Refer to "Fuel Metering System."
- **CHECK:** Coolant Temperature Sensor (CTS) - Using a "Scan" tool compare coolant temperature with ambient temperature on a cold engine.
  If coolant temperature reads 5 degrees greater than or less than ambient air temperature. Check for high resistance in coolant sensor circuit or sensor itself. Compare resistance value to "Diagnostic Aids" on facing page of Code 15.
- **CHECK:** MAP sensor response and accuracy

FUEL SYSTEM

- **NOTE:** To determine if the condition is caused by a rich or lean system, the vehicle should be driven at the speed of the complaint. Monitoring block learn will help identify problem.
  Lean - Block learn greater than 150. Refer to "Diagnostic Aids" on facing page of Code 44.
  Rich - Block learn less than 115. Refer to "Diagnostic Aids" on facing page of Code 45.
- **CHECK:** Evaporative Emission Control System
- **CHECK:** Perform a cylinder compression check.
- **CHECK:** For injector(s) leaking. Check fuel pressure, refer to "Engine Components/Wiring Diagrams/Diagnostic Charts," Fuel System Diagnosis, CHART A-7, Page (1 of 2).

IGNITION SYSTEM

- **CHECK:** Ignition System.
- **CHECK:** Ignition Timing. See "Vehicle Emission Control Information" label.

3.1L ENGINE – DRIVEABILITY CHARTS

STALLING, ROUGH, UNSTABLE OR INCORRECT IDLE

(Page 2 of 2)

Definition: The engine runs unevenly at idle. If bad enough, the vehicle may shake. Also, the idle may vary in rpm (called "hunting"). Either condition may be severe enough to cause stalling. Engine idles at incorrect speed.

ADDITIONAL CHECKS

- **CHECK:** Vacuum leaks can cause higher than normal idle and low IAC counts.
- **CHECK:** IAC operation - See Code 35.
- **CHECK:** ECM grounds for clean, tight, and proper routing.
- **CHECK:** P/N switch circuit. "Scan" tool. indicates vehicle is in drive with gear selector in drive or overdrive
- **NOTE:** Use "Scan" tool to determine if ECM is receiving A/C request signal.

  If problem exists with A/C "ON," check A/C system opelation.
- **CHECK:** EGR "ON," while idling, will cause roughness, stalling and hard starting.
- **CHECK:** Battery cables and ground straps should be clean and secure. Erratic voltage will cause IAC to change its position. resulting in poor idle quality.
- **CHECK:** IAC valve will not move, if system voltage is below 9 or greater than 16 volts.
- **CHECK:** A/C refrigerant pressure too high.
- **CHECK:** For overcharge or faulty high pressure switch.
- **CHECK:** PCV valve for proper operation by placing finger over inlet hole in valve end several times. Valve should snap back.

ENGINE MECHANICAL

- **CHECK:** For broken motor mounts.
- **CHECK:** For low compression

EXCESSIVE EXHAUST EMISSIONS OR ODORS

Definition: Vehicle fails an emission test. Vehicle has excessive "rotten egg" smell. Excessive odors do not necessarily indicate excessive emissions.

PRELIMINARY CHECKS

- Perform "Diagnostic Circuit Check."
- If EMISSION TEST shows excessive CO and HC check items which cause car to run RICH. (Block Learn Memory (BLM) less than 118) refer to "Diagnostic Aids" on facing page of Code 45. Make sure engine is at normal operating temperature.
- If EMISSION TEST shows excessive NOx check items which cause car to run LEAN or too hot.

SENSORS

- **NOTE:** If the "Scan" tool indicates a very high coolant temperature and the system is running LEAN: Check the Cooling System and Cooling Fan for proper operation.

FUEL SYSTEM

- **NOTE:** If the system is running rich, (block learn less than 118), refer to "Diagnostic Aids" on facing page of Code 45. If the system is running lean, (block learn greater than 138), refer to "Diagnostic Aids" on facing page of Code 44
- **CHECK:** For properly installed fuel cap.
- **CHECK:** Fuel pressure, Refer to "Engine Components/Wiring "Diagrams/Diagnostic charts." Fuel System Diagnosis, CHART A-7, Page (1 of 2).
- **NOTE:** If test shows excessive NOx, check items which cause car to run LEAN or too hot.
- **CHECK:** Canister for fuel loading

IGNITION SYSTEM

- **CHECK:** Ignition system.
- **CHECK:** For incorrect timing or excessive advance. See "Vehicle Emission Control Information" label.
- **CHECK:** Spark plugs, plug wires, and igniton components.

ADDITIONAL CHECKS

- **CHECK:** For vacuum leaks.
- **CHECK:** For lead contamination of catalytic converter (look for the removal of fuel filler neck restrictor).
- **CHECK:** Carbon build-up. Remove carbon with top engine cleaner. Follow instructions on can.
- **CHECK:** EGR valve for not opening.
- **CHECK:** PCV valve for being plugged, stuck, or blocked PCV hose, or fuel in the crankcase.
- **CHECK:** For presence of fuel in crankcase.

DIESELING, RUN-ON

Definition: Engine continues to run after key is turned "OFF," but runs very roughly. If engine runs smoothly, check ignition switch and adjustment.

PRELIMINARY CHECKS

- Perform the careful visual/physical checks as described at start of this section.

FUEL SYSTEM

- **CHECK:** Evaporative system and fuel tank venting.
- **CHECK:** Injector(s) for leaking. Apply 12 volts to fuel pump "test" terminal to turn "ON" fuel pump and pressurize fuel system. Visually check injector(s) and TBI assembly for fuel leakage.

BACKFIRE

Definition: Fuel ignites in intake manifold, or in exhaust system, making a loud popping noise.

PRELIMINARY CHECKS

- Perform the careful visual/physical checks as described at start of this section.

IGNITION SYSTEM

- **CHECK:** Proper ignition coil output voltage with spark tester J 26792 or equivalent (ST-125).
- **CHECK:** Spark plugs. Remove spark plugs, check for wet plugs, cracks, wear, improper gap, burned electrodes, or heavy deposits. Repair or replace as necessary.
- **CHECK:** Ignition system.
- **CHECK:** For crossfire between spark plugs (distributor cap, spark plug wires, and proper routing of plug wires.)
- **CHECK:** Ignition timing. See "Vehicle Emission Control Information" label.

ENGINE MECHANICAL

- **CHECK:** Compression. Perform a compression check - look for sticking or leaking valves.
- **CHECK:** Valve timing.
- **CHECK:** Intake manifold gasket for vacuum leaks.
- **CHECK:** EGR operation for being open all the time.
- **CHECK:** Intake and exhaust system for casting flash or other restrictions.

Component Replacement

SERVICE PRECAUTIONS

- Be careful not to get water on any fuel injection system component. Pay close attention to the relay box and throttle valve switch connector. The connector is not water proofed and will be damaged by water.
- When charging the battery be sure to remove it from the vehicle first. Never disconnect the battery cable from the battery when the engine is running. The generation of surge voltage may damage the control unit and other electrical parts such as the multi-drive monitor.
- When replacing parts or checking the system, make sure to turn the ignition switch to the **OFF** position. When measuring voltage at the control unit harness connector, disconnect all the control unit harnesses first, then turn the ignition switch to the **ON** position.
- When checking the electrical terminals of the control unit with a tester, do not apply the probe to terminal directly but insert a pin into the terminal from the harness side and perform the measurement through the pin. If the tester probe is held against the terminal directly, the terminal will be deformed, causing poor contact. Connect each harness correctly and firmly to insure a good contact.
- The wiring connectors for the fuel injector(s), throttle valve switch, air regulator and water temperature sensor are provided with locked wires. To unlock the connector, pull and shake it gently.
- System cables must be placed at least 4 in. away from the tension cables. Be careful not to apply any shock to the system components such as the air flow sensor, crank angle sensor, and control unit. Component parts of the fuel injection system are precisely set. Even a slight distortion or dent will seriously affect performance.
- The fuel pump must not be operated without fuel. Since fuel lubricates the pump, noise or other serious problems such as parts seizure will result. Do not any fuel other than gasoline.

FUEL PRESSURE RELIEF

1.6L, 1.8L AND 2.6L ENGINES

1. Remove the fuel filler cap.
2. Remove the fuel pump relay from the underhood relay center.
3. Start the engine and allow to stall.
4. Crank the engine for an additional 30 seconds.
5. Disconnect the negative battery cable.

2.8L AND 3.1L ENGINES

1. Disconnect the negative battery cable.
2. Loosen the fuel filler cap.
3. The TBI unit is designed with an internal constant bleed feature to relieve fuel system pressure. Therefore, no further steps need to be taken to relieve the pressure.

FUEL METER COVER

Removal and Installation

2.8L AND 3.1L ENGINES

1. Disconnect the negative battery cable.
2. Relieve the fuel system pressure.
3. Remove the air cleaner assembly.
4. Disconnect the fuel injector electrical connectors by squeezing the tabs and pulling straight up.
5. Remove the fuel meter cover attaching screws.
6. Remove the fuel meter cover assembly.
7. Remove the fuel meter outlet gasket, cover gasket and pressure regulator dust seal. Discard the gaskets and seal.

To install:

8. Install a new pressure regulator dust seal, fuel meter outlet passage gasket and cover gasket.
9. Install the fuel meter cover assembly.

NOTE: If the fuel meter cover is removed for service, a suitable thread locking compound must be applied to the screw threads of the attaching screws before reinstalling.

10. Install the fuel meter cover attaching screws, coated with a suitable thread locking compound.
11. Connect the fuel injector electrical connectors.
12. Tighten the fuel filler cap.
13. Connect the negative battery cable.
14. Turn the ignition switch to the **ON** position for 2 seconds, then turn to the **OFF** position for 10 seconds. Turn the ignition switch to the **ON** position and check for leaks.
15. Install the air cleaner assembly.

FUEL INJECTOR

Removal and Installation

1.6L, 1.8L AND 2.6L ENGINES

1. Relieve the fuel pressure.
2. Disconnect the negative battery cable.
3. Label and disconnect the electrical connectors from the fuel injectors.
4. Disconnect the fuel rail from the fuel system.
5. Remove the fuel rail from the intake manifold; pull the fuel rail with the injectors connected from the intake manifold.
6. Separate the fuel injectors from the fuel rail.

To install:

7. Replace the fuel injector O-rings.
8. Install the fuel injectors to the fuel rail.
9. Lubricate the fuel injector O-rings with automatic transmission fluid and press them, with the fuel rail, into the intake manifold.
10. Install the fuel rail-to-intake manifold bolts.
11. Connect the fuel rail to the fuel system.
12. Connect the electrical connectors to the fuel injectors.
13. Connect the negative battery cable.
14. Turn the ignition switch **ON** and check for fuel leaks at the fuel rail.

2.8L AND 3.1L ENGINES

1. Relieve the fuel pressure.
2. Disconnect the negative battery cable.
3. Remove the air cleaner.
4. At the injector electrical connectors, squeeze the 2 tabs together and pull them straight upward.
5. Remove the fuel meter cover and leave the cover gasket in place.
6. Using a small prybar, carefully pry the injectors upward until they are free of the throttle body.
7. Remove the small O-ring from the nozzle end of the injector. Carefully rotate the injector's fuel filter back and forth to remove it from the base of the injector.
8. Discard the fuel meter cover gasket.
9. Remove the large O-ring and back-up washer from the top of the counterbore of the fuel meter body injector cavity.

To install:

10. Lubricate the O-rings with automatic transmission fluid and push them into the fuel injector cavities.
11. Install the new fuel meter cover gasket and cover.
12. Install the electrical connectors to the injectors.
13. Install the air cleaner and connect the negative battery cable.

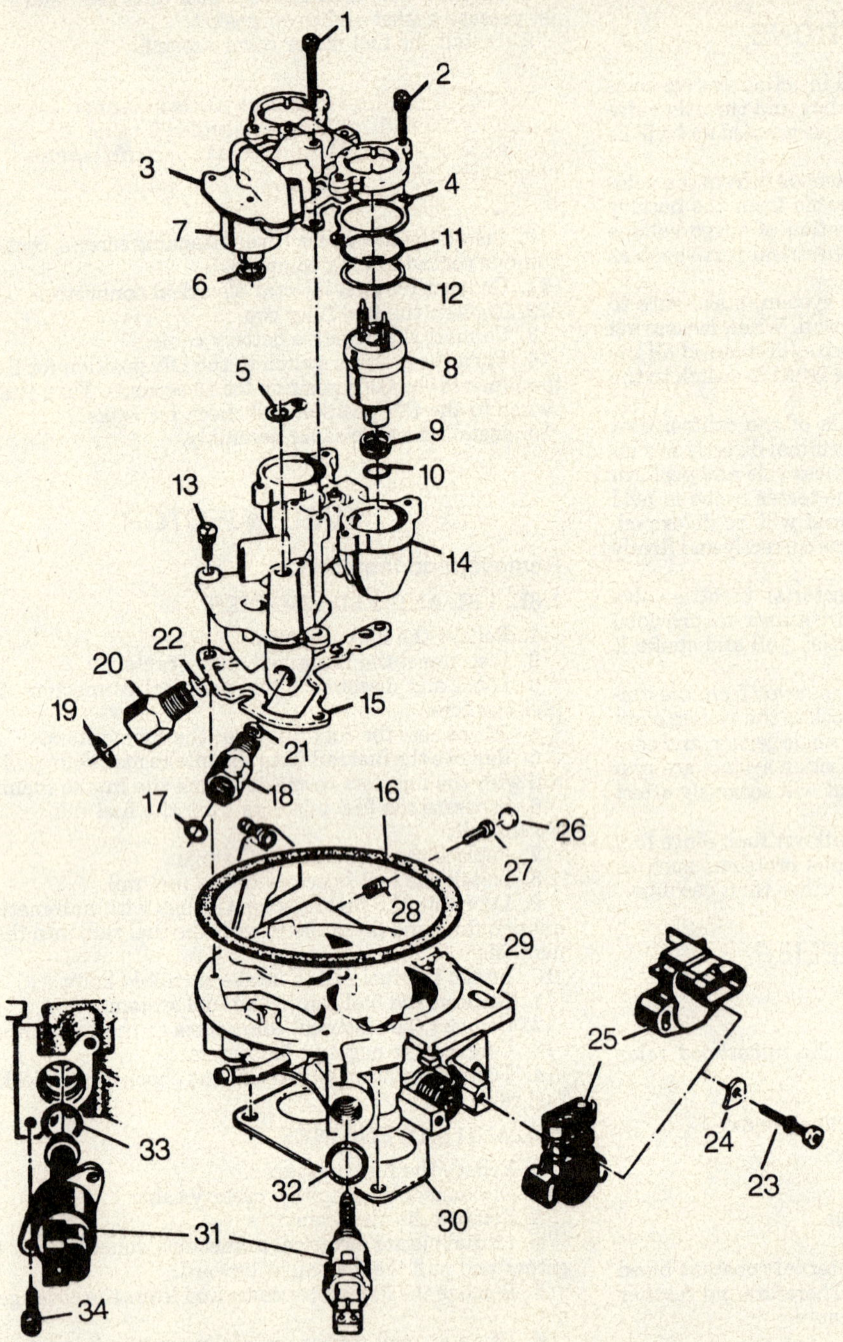

1. Long fuel meter screw assembly
2. Short fuel meter screw assembly
3. Fuel meter cover assembly
4. Fuel meter cover gasket
5. Fuel meter outlet gasket
6. Pressure regulator seal
7. Pressure regulator
8. Fuel injector
9. Fuel injector inlet filter
10. Lower fuel injector O-ring
11. Upper fuel injector O-ring
12. Fuel injector washer
13. Fuel meter body-to-throttle body attaching screw assembly
14. Fuel meter body assembly
15. Throttle body-to-fuel meter body gasket
16. Air cleaner gasket
17. Fuel return line O-ring
18. Fuel outlet nut
19. Fuel inlet line O-ring
20. Fuel inlet nut
21. Fuel outlet nut gasket
22. Fuel inlet nut gasket
23. TPS attaching screw assembly
24. TPS attaching retainer screw
25. Throttle Position Sensor
26. Idle stop screw plug
27. Idle stop screw assembly
28. Idle stop screw spring
29. Throttle body assembly
30. Flange gasket
31. Idle air control valve assembly
32. Idle air control valve assembly gasket
33. IAC valve O-ring
34. IAC valve attaching screw assembly

Exploded diagram of throttle body assembly—2.8L and 3.1L engine

FUEL METER BODY

Removal and Installation

2.8L AND 3.1L ENGINES

1. Disconnect the negative battery cable.
2. Relieve the fuel system pressure.
3. Remove the air cleaner assembly.
4. Disconnect the fuel injector electrical connectors by squeezing the plastic tabs and pulling straight up.
5. Remove the fuel meter cover and fuel injectors.
6. Disconnect the fuel feed and return lines.
7. Remove the fuel line O-rings and discard.

NOTE: Observe the location of nuts for proper reassembly. Inlet nut has a larger passage than the outlet nut.

8. Remove the fuel inlet and outlet nuts and gaskets from the fuel meter body assembly. Discard the gaskets.

To install:

9. Install new throttle body-to-fuel meter body gasket.

NOTE: Be sure to match the cut-out portions in the gasket with the openings in the throttle body.

10. Install the fuel meter body assembly on the throttle body assembly.

NOTE: If reinstalling the original fuel meter body assembly, check to see that the attaching screws have enough thread locking compound remaining to insure proper locking of the screws. If not, apply additional compound.

11. Install the fuel meter body-to-throttle body ataching screws.
12. Install the fuel inlet and outlet nuts with new gaskets. Tighten the inlet nut to 30 ft. lbs. (40 Nm) and the outlet nut to 21 ft. lbs. (29 Nm).
13. Install new O-rings on the fuel feed and return lines. Connect the fuel inlet and outlet lines. Using a back-up wrench on the TBI unit nuts, tighten the fuel line nuts to 20 ft. lbs. (27 Nm).
14. Install the fuel injectors and the fuel meter cover.
15. Connect the fuel injector electrical connectors.
16. Tighten the fuel filler cap.
17. Connect the negative battery cable.
18. Turn the ignition switch to the **ON** position for 2 seconds, then turn to the **OFF** position for 10 seconds. Turn the switch to the **ON** position and check for fuel leaks.
19. Install the air cleaner assembly.

FUEL PRESSURE REGULATOR

Removal and Installation

1.6L, 1.8L AND 2.6L ENGINES

1. Relieve the fuel system pressure.
2. Disconnect the vacuum line(s).
3. If mounted to the end of the fuel rail, remove the attaching bolt(s).
4. If connected by flexible fuel lines, slide the clamps back and disconnect the fuel lines, noting their positions.
5. Remove the pressure regulator.

To install:

6. If equipped, replace the O-rings.
7. If connected by flexible fuel lines, connect the fuel lines and reposition the clamps.
8. If mounted to the end of the fuel rail, install the pressure regulator and attaching bolt(s).
9. Connect the vacuum lines.
10. Connect the negative battery cable.
11. Replace the fuel pump relay. Turn the ignition switch to the **ON** position and check for leaks.
12. Start the engine and recheck for leaks.

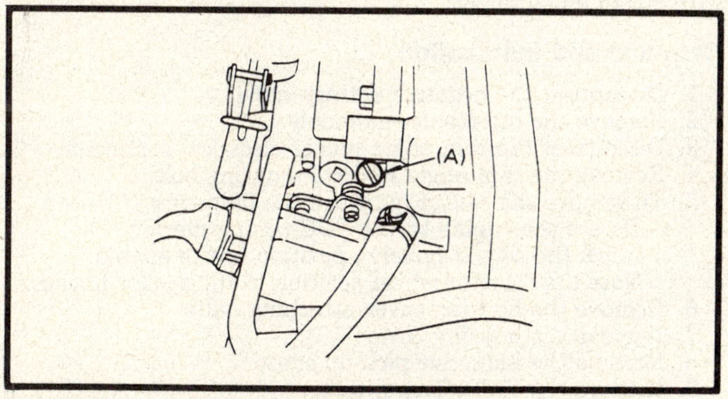

Location of the idle speed screw

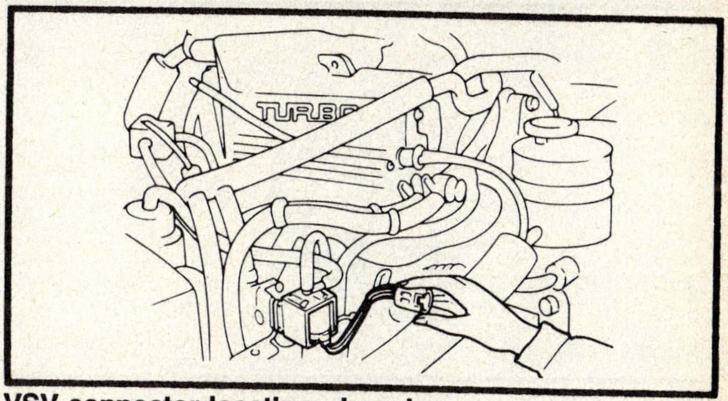

VSV connector location—Impulse

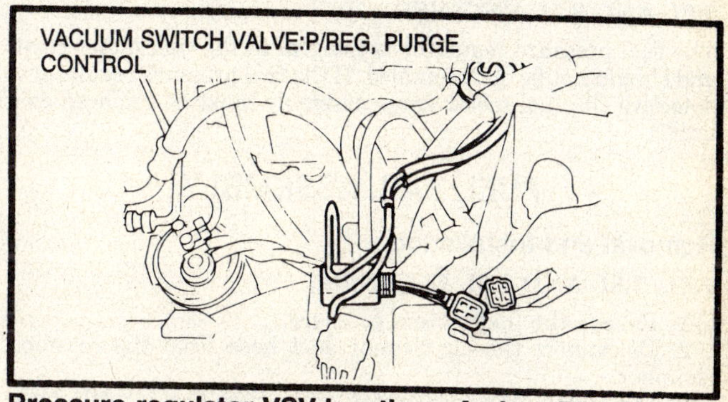

Pressure regulator VSV location—Amigo, Pick-Up, Trooper and Trooper II

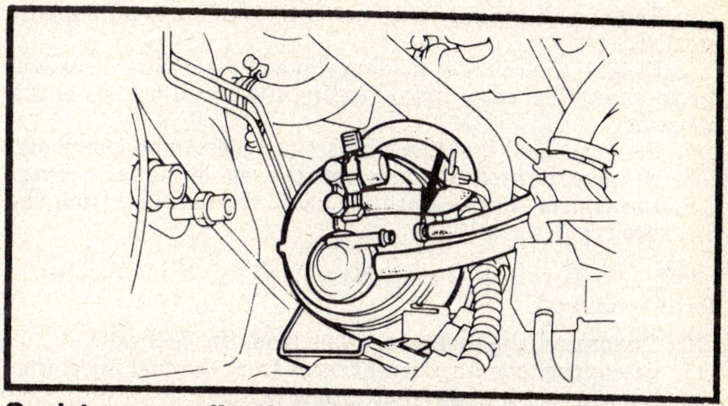

Canister purge line location—Amigo, Pick-Up, Trooper and Trooper II

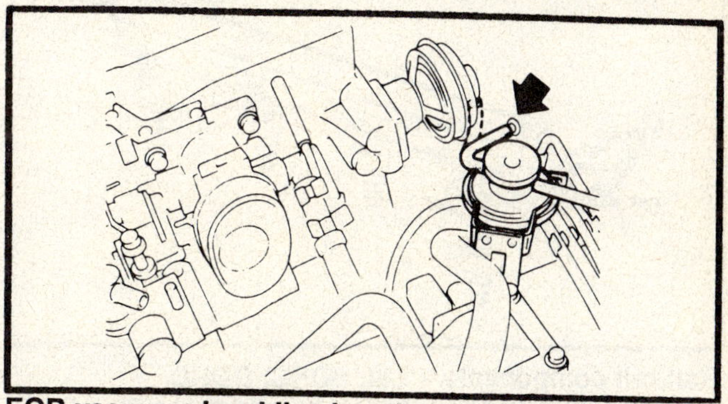

EGR vacuum signal line location—Amigo, Pick-Up, Trooper and Trooper II

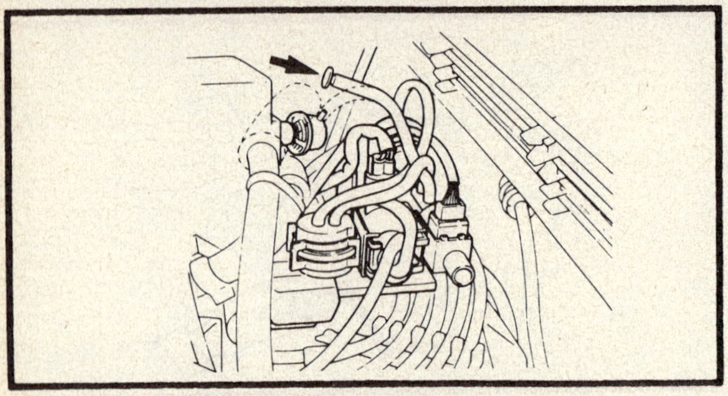

EGR cut VSV line location—California Amigo and Pick-Up

2.8L AND 3.1L ENGINES

The fuel pressure regulator is integral to the fuel meter cover and should not be disassembled. If the fuel pressure regulator is defective the fuel meter cover assembly must be replaced as a unit.

FUEL RAIL ASSEMBLY

Removal and Installation

1.6L, 1.8L AND 2.6L ENGINES

1. Relieve the fuel system pressure.
2. Disconnect the air cleaner duct hose from the common chamber.
3. Disconnect the accelerator cable from the throttle body and at the common chamber.
4. Remove the throttle valve assembly from the common chamber.
5. Remove the common chamber-to-cylinder head support bracket.
6. Release the canister holding clip and reposition the canister to allow clearance. Disconnecting the vacuum lines is not necessary.
7. Remove the 3 fuel line clips from the common chamber.
8. Remove the breather hose from the rear of the valve cover.
9. Disconnect the pressure regulator vacuum line from the pressure regulator.

NOTE: Ensure the entire fuel system has been depressurized.

10. Disconnect the fuel return line from the fuel rail.
11. Disconnect the injection harness from the fuel injectors.
12. Remove the 2 fuel rail attaching bolts.
13. Remove the fuel rail assembly.

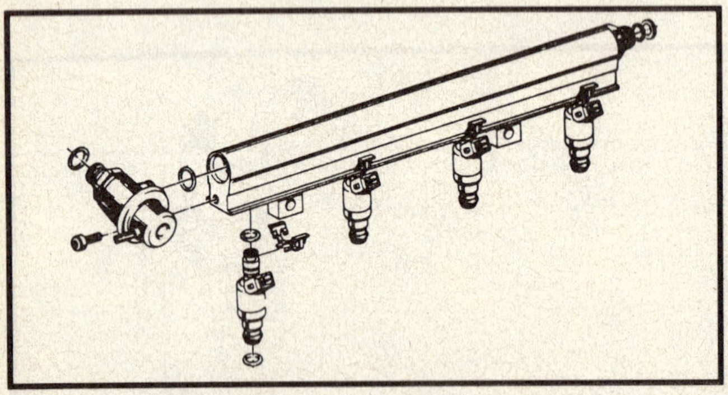

Fuel rail components—1.6L SOHC, DOHC Turbocharged and Non-turbocharged and 1.8L DOHC engine—Impulse and Sytlus

To install:

14. Replace the O-ring seals.
15. Place the fuel rail into position and install the attaching bolts.
16. Connect the injection harness to the fuel injectors.
17. Connect the fuel return line to the fuel rail.
18. Connect the pressure regulator vacuum line to the pressure regulator.
19. Install the breather hose to the rear of the valve cover.
20. Install the 3 fuel line clips to the common chamber.
21. Place the canister into the holder. Ensure the vacuum lines are routed properly and free of kinks.
22. Install the common chamber-to-cylinder head support bracket.
23. Install the throttle valve assembly to the common chamber.
24. Connect the accelerator cable to the throttle body and at the common chamber.
25. Connect the air cleaner duct hose to the common chamber.
26. Connect the negative battery cable.

IDLE AIR CONTROL (IAC) VALVE

Removal and Installation

1. Disconnect the negative battery cable.
2. Remove the air cleaner assembly.
3. Disconnect the IAC valve electrical connector.
4. Remove the IAC valve assembly and gasket.

To install:

5. Clean the IAC valve gasket sealing surface, pintle valve seat and air passage with carburetor cleaner. Do not methyl ethyl ketone solvent.
6. If installing a new IAC valve, measure the distance between the tip of the IAC valve pintle and mounting surface. If greater than 1.102 in. (28mm), use finger pressure to slowly retract the pintle. The force required to retract the pintle of a new IAC valve will not damage the valve.
7. Install a new gasket and install the IAC valve assembly. Using a socket or box wrench, tighten to 13 ft. lbs. (18 Nm).
8. Connect the IAC valve electrical connector.
9. Install the air cleaner assembly.
10. Connect the negative battery cable.
11. Reset the IAC valve pintle position by performing the following:
    a. Depress the accelerator pedal slightly.
    b. Start the engine and run for 5 seconds.
    c. Turn the ignition to the **OFF** position for 10 seconds.
    d. Restart the engine and check for proper idle.

CAM ANGLE SENSOR

Only the 1991-92 1.6L DOHC turbocharged engine is equipped with a cam angle sensor.

Removal and Installation

1. Disconnect the negative battery cable.
2. Remove the intercooler assembly.
3. Disconnect the cam angle sensor electrical connector.
4. Remove the cam angle sensor mounting bolt.
5. To remove the sensor, perform the following:
   a. Rotate the engine to the TDC timing mark.
   b. Mark the sensor housing position to the engine.
   c. Note the reluctor wheel position to the sensor housing.
6. Remove the housing cover attaching bolts.
7. Remove the housing cover.
8. Remove the inductive pick-up sensor.
9. Remove the Hall-effect switch.
10. Remove the dual cam-angle sensor housing.

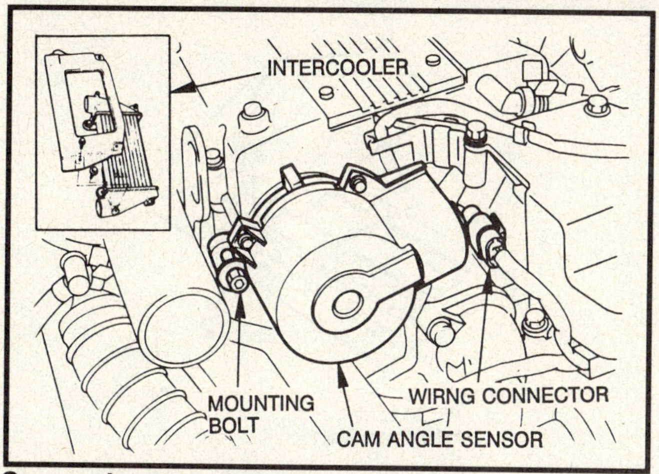

Cam angle sensor location—1.6L DOCH Turbocharged engine—Impulse

To install:

11. Inspect the sensor O-ring for wear, cracks or leakage. Replace as required. Lubricate the new O-ring with engine oil prior to installation.
12. Install the dual cam-angle sensor housing.
13. Install the Hall-effect switch.
14. Install the inductive pick-up sensor.
15. Install the housing cover.
16. Install the attaching bolts.
17. Install the cam angle sensor.
18. Install the mounting bolt.
19. Connect the cam angle sensor electrical connector.
20. Install the intercooler assembly.
21. Connect the negative battery cable.

COOLANT TEMPERATURE SENSOR

Removal and Installation

1. Disconnect the negative battery cable.
2. Disconnect the coolant sensor electrical connector.
3. Carefully, back out the coolant sensor.

To install:

4. Install the coolant sensor in the engine.
5. Connect the coolant sensor electrical connector.
6. Connect the negative battery cable.

OXYGEN SENSOR

Removal and Installation

1. Disconnect the negative battery cable.
2. Raise and safely support the vehicle, as required.
3. Disconnect the oxygen sensor electrical connector.
4. Using a box wrench, remove the oxygen sensor from the exhaust pipe.

NOTE: The oxygen sensor may be difficult to remove if the engine temperature is below 120°F (45°C). The use of excessive force may damage the threads in the exhaust pipe.

To install:

NOTE: The threads on new sensors are coated with a special anti-seize compound consisting of liquid graphite and glass beads which combine to make the oxygen sensor easier to remove. If the original oxygen sensor is being reinstalled, the threads must be coated with a similar compound to make future removal easier.

5. Coat the threads of the oxygen sensor with anti-seize compound or equivalent.
6. Install the oxygen sensor. Tighten to 30 ft. lbs. (41 Nm).
7. Connect the oxygen sensor electrical connector.
8. If raised, lower the vehicle.
9. Connect the negative battery cable.

MANIFOLD ABSOLUTE PRESSURE (MAP) SENSOR

Removal and Installation

1. If the MAP sensor is determined to be faulty, remove the top of the air cleaner.
2. Slide the retaining clip off.
3. Disconnect the vacuum line and electrical connector.
4. Remove the MAP sensor.

To install:

5. Install the new MAP sensor in the air cleaner.
6. Install the retaining clip.
7. Connect the vacuum line and electrical connector.
8. Install the top of the air cleaner.

MANIFOLD AIR TEMPERATURE (MAT) SENSOR

Removal and Installation

1.6L AND 1.8L ENGINES

1. Disconnect the negative battery cable.
2. Disconnect the MAT sensor electrical connector.
3. Carefully, back out the MAT sensor.

To install:

4. Apply a suitable sealer to the new MAT sensor threads.
5. Install the MAT sensor. Tighten to 22 ft. lbs. (30 Nm).
6. Connect the negative battery cable.

POWER STEERING PRESSURE SWITCH

Removal and Installation

IMPULSE AND STYLUS

1. Disconnect the negative battery cable.
2. Disconnect the power steering pressure switch electrical connector.
3. Carefully, remove the power steering pressure switch.

To install:

4. Install the power steering pressure switch.
5. Connect the electrical connector.
6. Connect the negative battery cable.
7. Bleed the power steering hydraulic system.

ELECTRONIC CONTROL MODULE

Removal and Installation

1. Disconnect the negative battery cable.
2. Remove the trim cover to gain access to the ECM.
3. Remove the ECM mounting hardware.
4. Disconnect the ECM connectors.
5. Remove the ECM.

To install:

6. Install the ECM.
7. Connect the ECM connectors.
8. Install the ECM mounting hardware.
9. Install the trim cover.
10. Connect the negative battery cable.

MEM-CAL

Removal and Installation

1. Disconnect the negative battery cable.
2. Remove the ECM access cover.
3. Remove the MEM-CAL, using a suitable removal tool. Grasp the MEM-CAL carrier at the narrow end only. Rock the carrier from end to end while applying a firm upward force.

To install:

4. Install the MEM-CAL. If a service MEM-CAL is being installed, ensure it has the same part number as the MEM-CAL that was removed. Do not press on the MEM-CAL, apply pressure only to the carrier.

NOTE: The small notch of the carrier must be aligned with the small notch in the socket. Press on the MEM-CAL carrier until it is firmly seated in the socket.

5. Install the ECM access cover.
6. Install the ECM in its proper location.
7. Connect the negative battery cable.
8. Perform a "Diagnostic Circuit Check" to confirm proper installation.

PROM

Removal and Installation

1. Disconnect the negative battery cable.
2. Remove the ECM.
3. Remove the PROM access cover.
4. Remove the PROM assembly.
5. Using a rocker-type PROM removal tool, engage 1 end of the PROM carrier with the hook end of the tool. Press on the vertical bar end of the tool and rock the engine end of the PROM carrier up as far as possible. Engage the opposite end of the PROM carrier in the same manner and rock this end up as far as possible. Repeat this process until the PROM carrier and PROM are free of the PROM socket.

To install:

6. Inspect the new PROM for same part number as old. Do not remove the PROM from the carrier to check the PROM number.
7. Place the new PROM carrier in PROM socket.

NOTE: The small notch of the carrier should be aligned with the small notch in the socket. Press on the PROM carrier until it is firmly seated in the socket. Do not press only on the PROM carrier.

8. Install the access cover on the ECM.

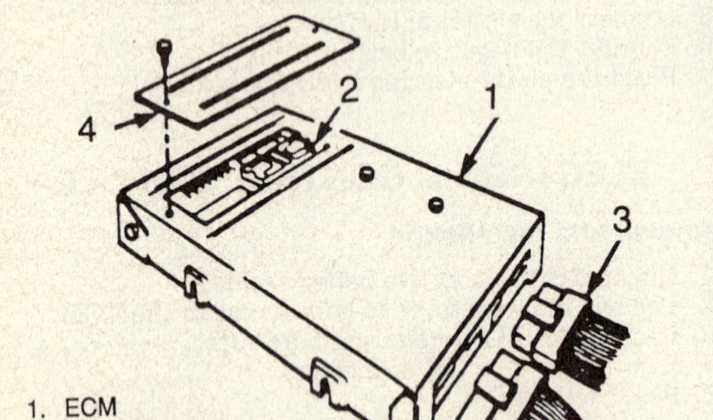

1. ECM
2. PROM carrier
3. ECM harness connectors to ECM
4. PROM access cover

PROM access cover location

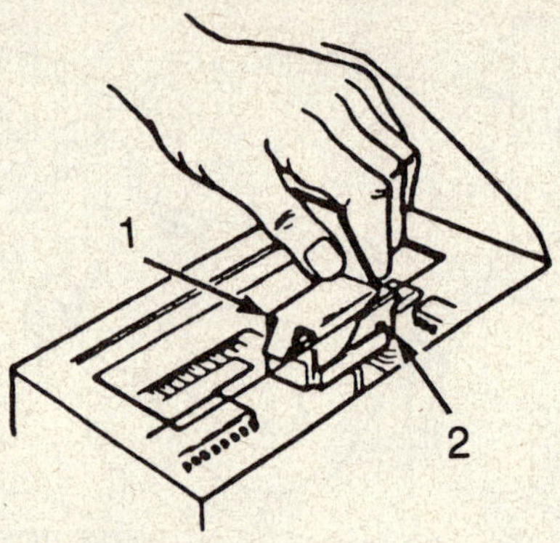

1. PROM carrier
2. PROM removal tool

Removing the PROM using removal tool

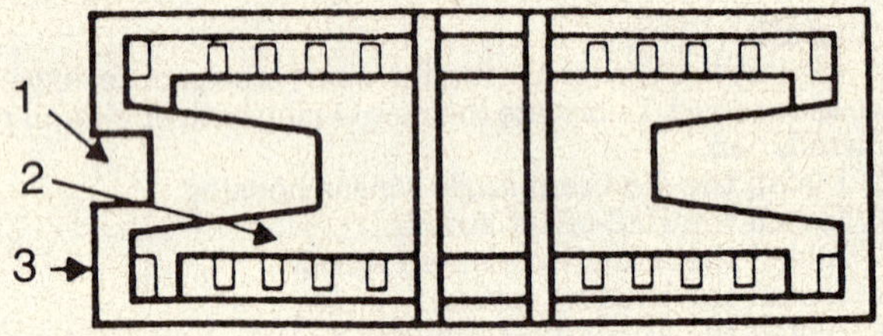

1. Reference end
2. PROM
3. PROM carrier

Proper positioning of PROM in carrier

9. Install the ECM.
10. Connect the negative battery cable.

CAL-PAK

Removal and Installation

1. Disconnect the negative battery cable.
2. Remove the ECM access cover.
3. Remove the CAL-PAK, using a suitable removal tool. Grasp the CAL-PAK carrier at the narrow end only. Rock the carrier from end to end while applying a firm upward force.

To install:

4. Install the CAL-PAK. If a service CAL-PAK is being installed, ensure it has the same part number as the CAL-PAK that was removed. Do not press on the CAL-PAK, apply pressure only to the carrier.

NOTE: The small notch of the carrier must be aligned with the small notch in the socket. Press on the CAL-PAK carrier until it is firmly seated in the socket.

5. Install the ECM access cover.
6. Install the ECM in its proper location.
7. Connect the negative battery cable.
8. Perform a "Diagnostic Circuit Check" to confirm proper installation.

THROTTLE BODY

Removal and Installation

2.8L AND 3.1L ENGINES

1. Disconnect the negative battery cable.

2. Relieve the fuel system pressure.
3. Remove the air cleaner assembly.
4. Remove the electrical connectors to the IAC valve, TPS and fuel injectors.
5. Reposition injector wiring harness aside.
6. Disconnect the throttle cable and, if equipped, the transmission control and cruise control cables.
7. Remove the cable support bracket attaching screws and bracket.
8. Label and disconnect the vacuum hoses. Remove the support bracket, as required.
9. Disconnect the fuel feed and return lines. Use a back-up wrench on the TBI fuel nuts to prevent them from turning.
10. Remove the fuel line O-rings and discard.
11. Remove the TBI unit attaching bolts.
12. Remove the TBI unit and flange gasket. Discard the gasket.

NOTE: Stuff a shop towel into the manifold opening to prevent dirt and other contaminants from entering the engine.

To install:

13. Clean the gasket surface, taking care not to use a gasket scraper that is too sharp or use excessive pressure in removing the gasket which could damage machined aluminum surfaces.
14. Install a new TBI flange gasket.
15. Install the TBI unit.
16. Install the TBI unit attaching bolts. Tighten to 18 ft. lbs. (25 Nm).
17. Install new O-rings on the fuel lines.
18. Connect the fuel feed and return lines. Using a back-up wrench on the TBI fuel nuts, tighten the fuel lines nuts to 20 ft. lbs. (27 Nm).
19. If removed, install the vacuum hose brackets. Connect the vacuum hoses.
20. Install the support bracket and attaching screws.
21. Connect the throttle cable and, if equipped, connect the transmission control and cruise control cables.

NOTE: Ensure that the throttle and cruise control cables do not hold the throttle open.

22. Position the injector harness and connect the injector electrical connectors, making sure the connectors are fully seated and latched.
23. Tighten the fuel filler cap.
24. Connect the negative battery cable.

THROTTLE POSITION SWITCH

Removal and Installation

1. Disconnect the negative battery cable. Remove the air cleaner.
2. Disconnect the TPS electrical connector.
3. Remove the 2 mounting screws.
4. Remove the TPS and, if equipped, TPS seal from the throttle body.

To install:

5. Place the TPS in position. Align the TPS lever with the TPS drive lever on the throttle body.
6. Install the 2 TPS mounting screws.

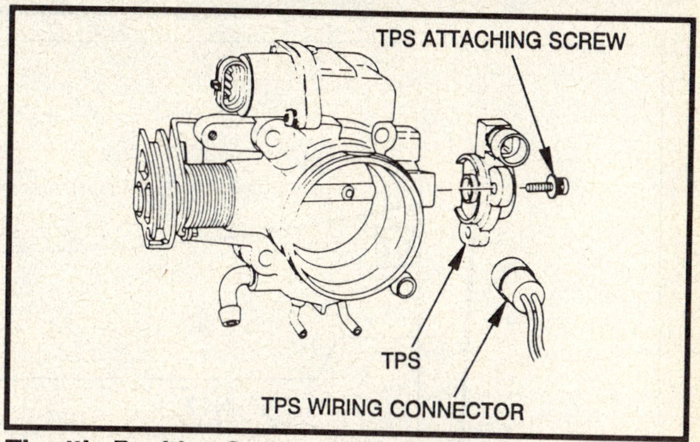

Throttle Position Switch (TPS) installation – 1.6L SOHC, DOHC Turbocharged and Non-turbocharged and 1.8L DOHC engine – Impulse and Sytlus

7. Connect the electrical connector.
8. Connect the negative battery cable. Install the air cleaner.

IDLE SPEED ADJUSTMENT

1.6L AND 1.8L ENGINES

Idle speed is controlled by the ECM. No adjustment is necessary.

2.6L ENGINE

1. Prior to adjusting engine idle speed, confirm that the following conditions are met:
   a. Engine is at normal operating temperature.
   b. Throttle valve is completely closed and idle contact is on.
   c. If equipped with power steering, front wheels straight ahead.
   d. If equipped, air conditioning system is off.
   e. Transmission selector in **N** or **P**.
   f. Harness connector of Vacuum Switching Valve (VSV) for pressure regulator disconnected.
   g. Canister purge line disconnected and plugged.
2. Connect a tachometer.
3. Adjust the idle speed to 850–950 rpm.

NOTE: Check and clean the idle port as necessary. Restrictions in the port cause fluctuations in idle speed.

4. Connect the VSV harness at the connector.
5. Connect the canister purge line.

2.8L AND 3.1L ENGINES

Idle speed is controlled by the ECM. No adjustment is necessary.

IDLE MIXTURE ADJUSTMENT

The idle mixture is controlled by the ECM. No adjustment is necessary.

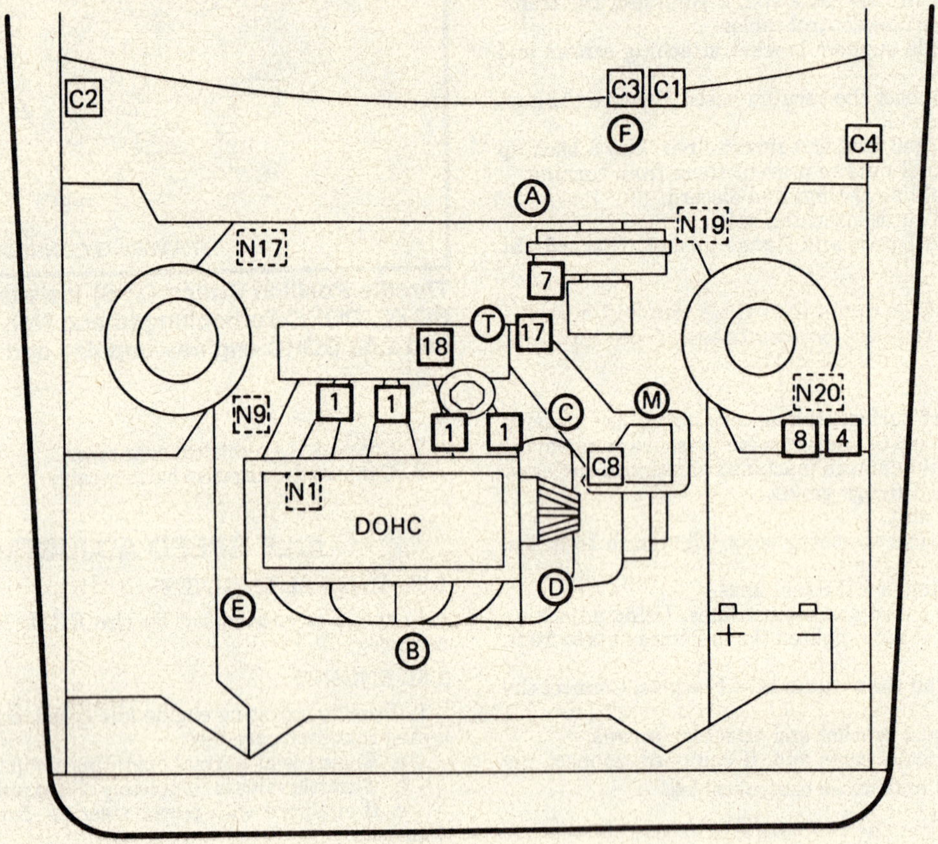

COMPUTER HARNESS

- C1 Electronic Control Module (ECM)
- C2 ALDL White Connector
- C3 "Check Engine" Light
- C6 Fuse Panel
- C8 ECM Grounds

NOT ECM CONNECTED

- N1 Crankcase Vent Valve (PCV)
- N9 Fuel Pressure Regulator Valve
- N17 Fuel Vapor Canister
- N19 Fan Switch
- N20 A/C Compressor Relay

CONTROLLED DEVICES

- 1 Fuel Injectors
- 2 Ignition Coil
- 3 IAC Valve
- 4 Main Relay
- 7 EGR Vacuum Switching Valve (VSV)
- 8 Cooling Fan Relay
- 17 Throttle Body
- 18 Secondary Air Vacuum Switching Valve (VSV) (Located Under Intake Plenum)

INFORMATION SENSORS

- A Manifold Absolute Pressure (MAP) Sensor
- B Oxygen (O_2) Sensor
- C Throttle Sensor (TS)
- D Coolant Temperature Sensor (CTS)
- E Power Steering Pressure Switch
- F Vehicle Speed Sensor
- M Park/Neutral (P/N) Switch
- T Manifold Air Temperature (MAT) Sensor

Exhaust Gas Recirculation (EGR) Valve

NOTE: *Power Steering Pressure Switch is part of the Power Steering Pump.

Engine component location—1.6L DOHC Turbocharged and Non-turbocharged and 1.8L DOHC Engines—Impulse and Stylus

Lexus Fuel Injection System

12 SECTION

LEXUS ELECTRONIC FUEL INJECTION (EFI) SYSTEM

General Information

This system is broken down into 3 major systems: the Fuel System, Air Induction System and the Electronic Control System.

AIR INDUCTION SYSTEM

The air induction system provides sufficient, clean air for the engine operation. This includes the throttle body, air intake ducting and cleaner and idle control components. Lexus equips the ES300, SC300 and SC400 with a system of induction tuning that changes the induction path length. The Intake Air Control Valve (IACV) changes the length of the induction path to broaden the power curve by matching the resonance characteristics of the intake charge with the engine speed. When the IACV opens the effective length of the intake tract is shortened boosting top end power. With the IACV closed the intake tract is long, boosting the low end torque.

ELECTRONIC CONTROL SYSTEM

Lexus engines are equipped with a computer which centrally controls the electronic fuel injection, electronic spark advance and the exhaust gas recirculation valve. The systems can be diagnosed by means of an Electronic Control Unit (ECU).

Electronic Fuel Injection (EFI)

The ECU receives signals from the various sensors indicating changing engine operations conditions such as:

1. Intake air flow
2. Intake air temperature
3. Coolant temperature sensor
4. Engine rpm
5. Acceleration/deceleration
6. Exhaust oxygen content

These signals are utilized by the ECU to determine the injection duration necessary for an optimum air-fuel ratio.

The Electronic Spark Advance (ESA)

The ECU is programmed with data for optimum ignition timing during any and all operating conditions. Using the data provided by sensors which monitor various engine functions (rpm, intake air volume, coolant temperature, etc.), the microcomputer (ECU) triggers the spark at precisely the right moment.

Idle Speed Control (ISC)

The ECU is programmed with specific engine speed values to respond to different engine conditions (coolant temperature, air conditioner on/off, etc.). Sensors transmit signals to the ECU which controls the flow of air through the by-pass of the throttle valve and adjusts the idle speed to the specified value.

Fail-Safe Function

In the event of a computer malfunction, a backup circuit will take over to provide minimal driveability. Simultaneously, the "Check Engine" warning light is activated.

DATA SENSORS

The data sensors provide information to the ECU for processing. The information is used to adjust the controlled outputs for maximum engine control under changing situations.

Airflow Meter

The airflow meter provides information on the amount of air being drawn into the engine. There are 2 methods of determining the airflow. The most basic method uses a deflecting vane; the amount the vane is moved by the incoming air represents the airflow into the engine. The Karman–Vortex type meter relies on a purpose made airflow disruption that form a vortex in the airstream. The frequency of the vortex turbulence acts on a flexible reflecting surface. A light emitting diode and phototransistor combination measures the vibration of the reflecting surface. The output from the phototransistor is fed to the ECU as a measure of the airflow into the engine.

Throttle Position Sensor

The throttle position sensor feeds the ECU information on the position of the throttle at any given time. The ECU uses this information to adjust the fuel flow at the injectors to keep pace with the engines demand.

Subthrottle Position Sensor

Provides the same information as the throttle position sensor but used on vehicles equipped with traction control systems.

Water Temperature Sensor

Provides temperature information to the ECU to adjust the operation parameters for proper warm-up. The water temperature sensor plays a large role in the proper operation of the fuel injection system and should be checked during the troubleshooting routine as matter of course.

EGR Gas Temperature Sensor

Provides temperature information to the ECU to adjust the operation parameters for proper EGR gas mixture.

Knock Sensors

Informs the ECU when detonation occurs. This allows the ECU to adjust for fuel quality and provide optimum performance. Also helps prevent engine damage from continued detonation.

Oxygen Sensors

The oxygen sensors provide information on the oxygen content of the exhaust gases. The catalytic converters need to be run at a fixed oxygen content level for efficient exhaust gas treatment to occur. The engine must be run at stoichimetric ratio and the oxygen sensor provides the information required by the ECU to adjust the air to fuel ratio to achieve the proper ratio.

Engine Position Sensors

Provides the ECU with information about the position of the engine and keep the proper timing of the fuel injection function.

CONTROLLED OUTPUTS

These are systems controlled by the ECU providing the adjustments to the operating parameters. The ECU adjusts these systems in response from inputs from the data sensors.

Cold Start Injector

The cold start injector provides the extra fuel needed for a cold start. The length of time of injection is regulated by the cold start time switch. The operation of the cold start injector is provided by the ECU.

Fuel Injectors

The fuel injectors introduce the fuel into the airstream. The ECU controls the pulsing of the injectors as a method of controlling the amount of fuel injected per given unit of time.

EGR Valve

The Exhaust Gas Recirculation valve is controlled by the ECU to allow an amount of exhaust gas into the combustion chamber. The ECU monitors the parameters that determine how much EGR can be used for efficient pollution control, economy and performance.

Idle Stability Control (ISC) Valve

The idle stability control valve opens to allow additional air into the engine during idle. As the idle speed requirements decrease as the engine warms, the ISC decreases the amount of air into the engine, thus reducing the idle speed.

FUEL SYSTEM

An electric fuel pump supplies sufficient fuel, under a constant pressure, to the EFI injectors. These injectors pass a metered quantity of fuel into the intake manifold in accordance with signals from the EFI computer.

Diagnosis and Testing

SERVICE PRECAUTIONS

- Do not operate the fuel pump when the fuel lines are empty.
- Do not reuse fuel hose clamps.
- Make sure all EFI harness connectors are fastened securely. A poor connection can cause an extremely high surge voltage and result in damage to integrated circuits.
- Keep the EFI harness at least 4 in. away from adjacent harnesses to prevent an EFI system malfunction due to external electronic "noise."
- Keep EFI all parts and harnesses dry during service.
- Before attempting to remove any parts, turn **OFF** the ignition switch and disconnect the battery ground cable.
- Always use a 12 volt battery as a power source.
- Do not attempt to disconnect the battery cables with the engine running.
- Do not depress the accelerator pedal when starting.
- Do not rev up the engine immediately after starting or just prior to shutdown.
- Do not attempt to disassemble the EFI control unit under any circumstances.
- If installing a 2-way or CB radio, keep the antenna as far as possible away from the electronic control unit. Keep the antenna feeder line at least 8 in. away from the EFI harness and do not let them run parallel for a long distance. Be sure to ground the radio to the vehicle body.
- Handle air flow meter carefully to avoid damage.
- Do not disassemble air flow meter or clean meter with any type of detergent.

TESTING PRECAUTIONS

- Before connecting or disconnecting control unit ECU harness connectors, make sure the ignition switch is **OFF** or the negative battery cable is disconnected to avoid the possibility of damage to the control unit.
- When performing ECU input/output signal diagnosis, remove the water-proofing rubber plug, if equipped, from the connectors to make it easier to insert tester probes into the connector.
- When connecting or disconnecting pin connectors from the ECU, take care not to bend or break any pin terminals. Check that there are no bends or breaks on ECU pin terminals before attempting any connections.
- Before replacing any ECU, perform the ECU input/output signal diagnosis to make sure the ECU is functioning properly or not.
- When measuring supply voltage of ECU controlled components with a circuit tester, separate a tester probe from another. If the 2 tester probes accidentally make contact with each other during measurement, a short circuit will result and damage the ECU.

SELF-DIAGNOSIS

The ECU contains a built-in self diagnosis system by which troubles with the engine signal the engine signal network are detected and a "Check Engine" warning light on the instrument panel flashes Code No. 12–71 (these code numbers vary from model to model). The "Check Engine" light on the instrument panel informs the driver that a malfunction has been detected. The light goes out automatically when the malfunction has been cleared.

The diagnostic code can be read by the number of blinks of the "Check Engine" warning light when the proper terminals of the check connector are connected.

Check Engine Warning Light

1. The "Check Engine" warning light will come on when the ignition switch is placed **ON** and the engine is not running.
2. When the engine is started, the "Check Engine" warning light should go out.
3. If the light remains on, the diagnosis system has detected a malfunction in the system.

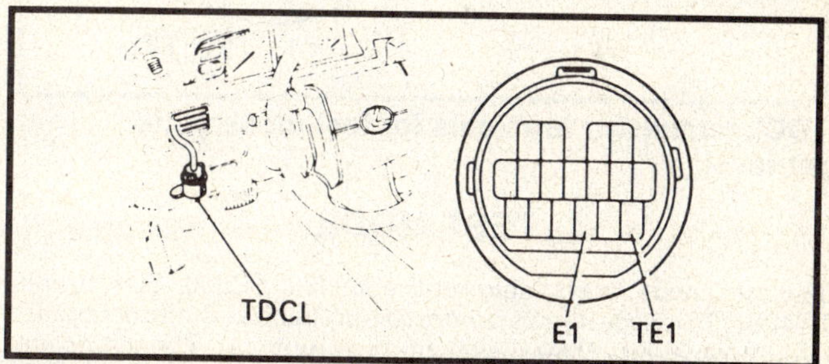

TDCL connector for retrieving codes

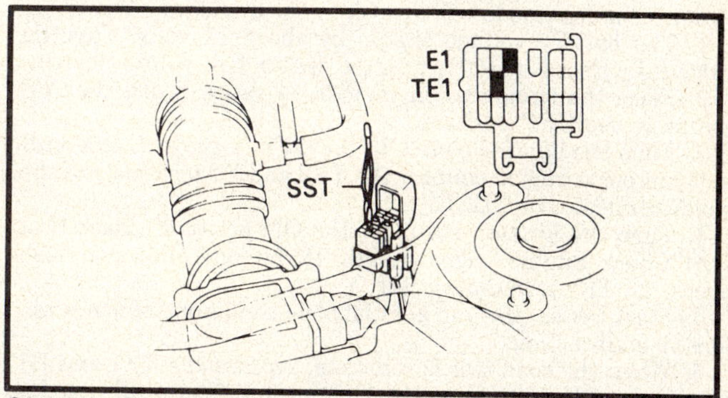

Check connector for retrieving codes

READING CODES

1. The battery voltage should be above 11 volts. Throttle valve fully closed (throttle position sensor IDL points closed).
2. Place the transmission in **P** or **N** range. Turn the A/C switch **OFF**.
3. Turn the ignition switch to the **ON** position. Do not start the engine. Use a suitable jumper wire and connect the TE1 and E1 terminals of the check connector or TDCL.
4. Read the diagnostic code as indicated by the number of flashes of the "Check Engine" warning light.
5. If the system is operating normally (no malfunction), the light will blink once every 0.25 seconds.
6. In the event of a malfunction, the light will blink once every 0.5 seconds. The 1st number of blinks will equal the 1st digit of a 2 digit diagnostic code. After a 1.5 second pause, the 2nd number of blinks will equal the 2nd number of a 2 digit diagnostic code. If there are 2 or more codes, there will be a 2.5 second pause between each.
7. After all the codes have been output, there will be a 4.5 second pause and they will be repeated as long as the terminals of the check connector are connected.

NOTE: In event of multiple trouble codes, indication will begin from the smaller value and continue to the larger in order.

8. After the diagnosis check, remove the jumper wire from the check connector.

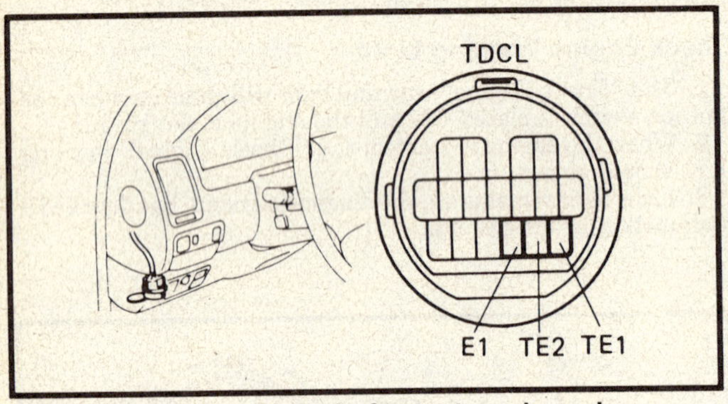

TDCL connector terminals for test mode code retrieval

TEST MODE

The test mode is available on the ES300, SC300, SC400 and LS400. This mode allows a more comprehensive malfunction diagnosis capability than the standard output test. The test mode can detect malfunctions in the cam position sensors, the IDL contact signal from the throttle position sensor, the air conditioner signal, neutral start switch signal and the starter signal circuit. The test mode will also show the standard outputs.

1. The battery voltage should be above 11 volts. Throttle valve fully closed (throttle position sensor IDL points closed).
2. Place the transmission in **P** or **N** range. Turn the A/C switch to the **OFF** position.
3. Turn the ignition switch to the **OFF** position. Use a suitable jumper wire and connect the TE2 and E1 terminals of the check connector or TDCL.
4. Turn the ignition switch to the **ON** position. Check that the "Check Engine" light flashes. If the light does not flash check the TE2 terminal circuit.
5. Start the engine and simulate the conditions of the problem or malfunction.
6. When the road test is complete, connect the TE1 and E1 terminals of the TDCL or check connector. The trouble codes will be output.
7. After the codes are output disconnect the jumpers from the connector. The test mode cannot be initiated if the connection between TE2 and E1 are made with the key in the **ON** position. If the engine is not cranked a code for starter signal and cam position sensors will be set, but in this case these are not abnormal codes. If any of the sensed switches are used, the transmission shift lever, throttle or air conditioner, the switch condition code will be set, but this is not abnormal either.

CLEARING CODES

1. After repairing the trouble area, the diagnostic code that is retained in the ECU memory must be canceled out by removing the EFI (15A, 20A or 30A depending on model and year) fuse for 30 seconds or more, depending on the ambient temperature (the lower temperature, the longer the fuse must be left out with the ignition switch **OFF**).

NOTE: Cancellation can also be done by removing the battery negative terminal, but keep in mind, when removing the negative battery cable, the other memory systems (radio, ETR, clock, etc.) will also be canceled out.

2. If the diagnostic code is not canceled out, it will be retained by the ECU and appear along with a new code in event of future trouble. If it is necessary to work on engine components requiring removal of the battery terminal, a check must first be made to see if a diagnostic code is detected.
3. After cancellation, perform a road test, if necessary, confirm that a normal code is now read on the "Check Engine" warning light or super monitor display.
4. If the same diagnostic code is still indicated, it indicates that the trouble area has not been repaired thoroughly.

Diagnosis Indication

1. Including "Normal", the ECU is preprogrammed with diagnostic codes.
2. When more than a single code is indicated, the lowest number code will appear first. However, no other code will appear along with Code 11 which indicates normal operation.
3. All detected diagnostic codes, except Code 51 and 53, will be retained in memory by the ECU from the time of detection until canceled out.
4. Once the malfunction is cleared, the "Check Engine" warning light on the instrument panel will go out but the diagnostic code(s) remain stored in the ECU memory (except for Code 51).

FUEL SYSTEM

Fuel Pump Operation

1. Turn the ignition switch **ON**, but do not start the engine.
2. Jumper the +B and FP terminals of the check connector. Check that there is pressure in the fuel inlet hose.

NOTE: At this point, listen for fuel return noise from the pressure regulator.

3. Remove the jumper wire and turn **OFF** the ignition switch.
4. If there is no pressure, check the following components:
   a. Fusible link
   b. EFI and Ignition fuses
   c. EFI main relay
   d. Circuit opening relay
   e. Fuel pump
   f. Wiring connector

Checking the inlet hose for pressure – LS400

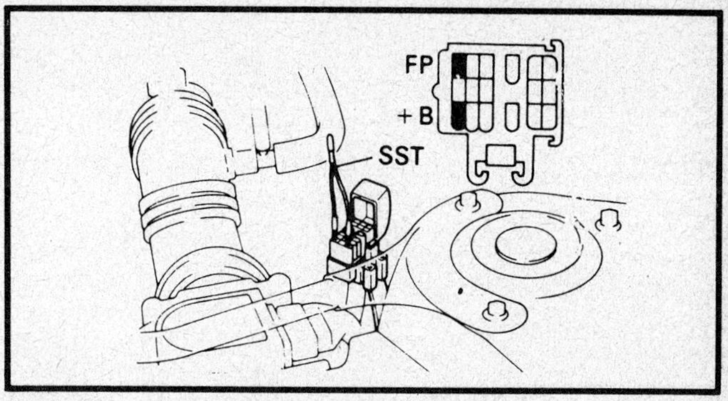

Activating the fuel pump from the check connector – ES250

Fuel Pressure Check

1. Disconnect the negative battery cable. Disconnect the wiring connector from the cold start injector on the ES250 and ES300. Remove the Vacuum Switching Valve (VSV) for the EGR on the LS400. Remove the VSV filter on the SC400.
2. Place a suitable container or shop towel under the fuel connection.
3. On the ES250 slowly loosen the union bolt of the cold start injector hose and remove the bolt and 2 gaskets from the delivery pipe. On the LS400 and SC400 slowly loosen the union bolt from the left side of the rear fuel pipe and remove the bolt and 2 gaskets from the delivery pipe. On the SC300 Remove the union bolt attaching the fuel delivery pipe to the fuel rail.
4. Drain the fuel from the delivery pipe. Install a gasket, a pressure gauge (tool SST-09268-45012 or equivalent), another gasket and union bolt to the delivery pipe. This forms a stacked connection of the 2 unions. Torque the connection to:
   ES250 – 14 ft. lbs. (20 Nm)
   ES300 – 11 ft. lbs. (15 Nm⁰
   SC300 – 30 ft. lbs. (42 Nm)
   SC400 – 29 ft. lbs. (39 Nm)
   LS400 – 22 ft. lbs. (29 Nm)
5. Wipe up any excess gasoline, reconnect the negative battery cable.
6. Jumper the terminals on the check connector (Fp and +B) and turn the ignition switch to the **ON** position, then take a fuel pressure reading.
7. The fuel pressure should read 38–44 psi (265–304 kPa).
8. Remove the service wire from the check connector. Start the engine. Disconnect the vacuum sensing hose from the pressure regulator and pinch it off. Measure the fuel pressure at idle. The fuel pressure should be the same as the above readings.
9. If the fuel pressure is too high, replace the pressure regulator.
10. If the fuel pressure is low, check the following components:
    a. Fuel hoses and fuel connections
    b. Fuel pump
    c. Fuel filter
    d. Pressure regulator
11. Reconnect the vacuum sensing hose to the pressure regulator. Measure the fuel pressure at idling:
    ES250 – 33–37 psi (225–265 kPa)
    ES300 – 33–37 psi (225–265 kPa)
    SC300 – 28–34 psi (200–240 kPa)
    SC400 – 28–34 psi (200–240 kPa)
    LS400 – 28–34 psi (200–240 kPa)
12. If the fuel pressure is not as specified, check the vacuum sensing hose and pressure regulator. Stop the engine. Check that the fuel pressure remains above 21 psi for 5 minutes after the engine has been shut off.
13. If not within specifications, check the fuel pump, pressure regulator and or the injectors.

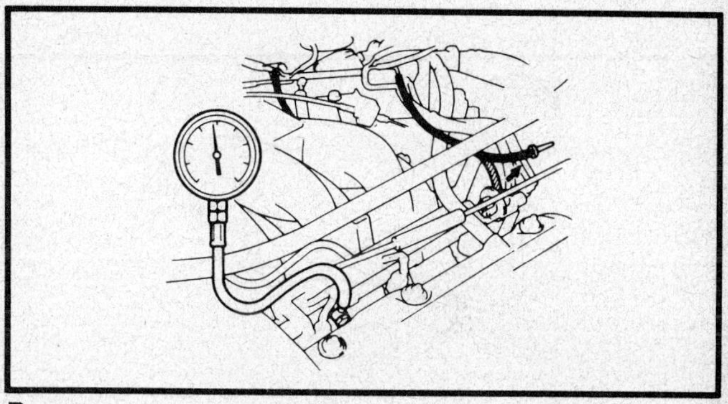

Pressure gauge connection and vacuum hose – ES250

Activating the fuel pump from the check connector – LS400

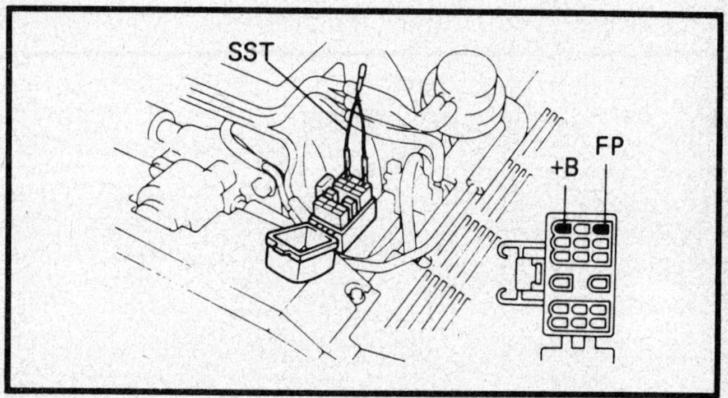

Pressure gauge connection and vacuum hose – LS400

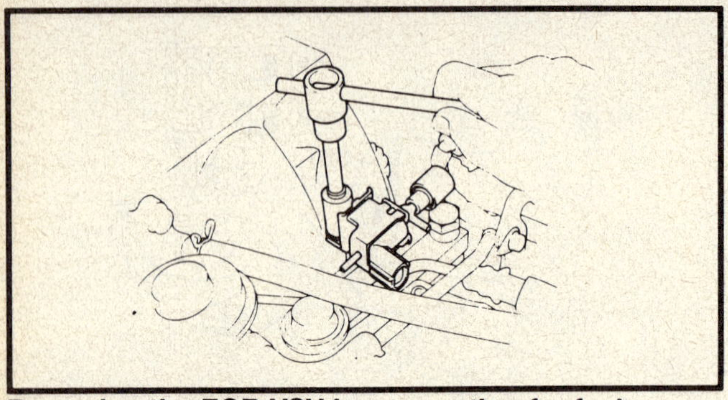

Removing the EGR VSV in preparation for fuel pressure test–LS400

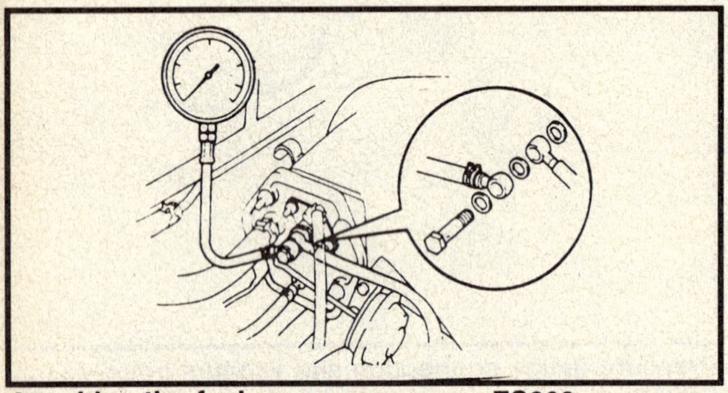

Attaching the fuel pressure gauge–ES300

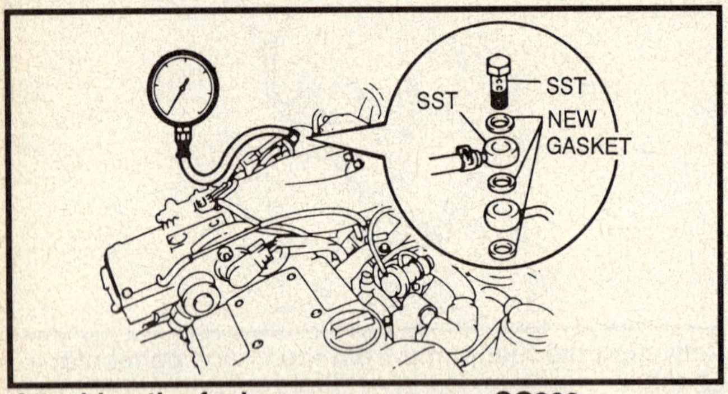

Attaching the fuel pressure gauge–SC300

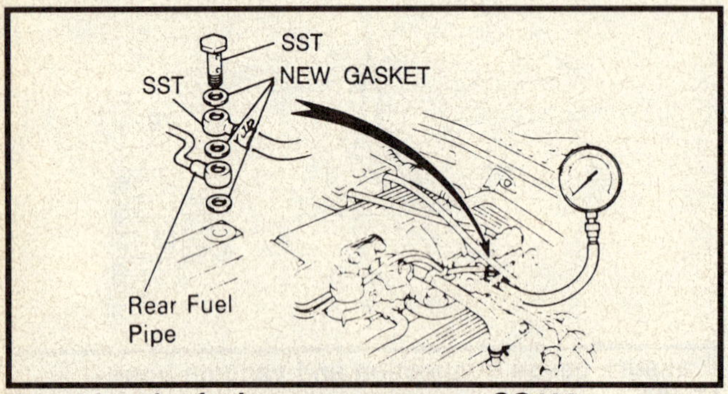

Attaching the fuel pressure gauge–SC400

14. After checking the fuel pressure, disconnect the battery ground cable and carefully remove the pressure gauge to prevent gasoline from splashing.

15. Using new gaskets, reconnect the cold start injector hose or fuel pipe to the delivery pipe (install the cold start injector valve if it was removed). Torque to:

ES250–14 ft. lbs. (20 Nm)
ES300–11 ft. lbs. (15 Nm0
SC300–30 ft. lbs. (42 Nm)
SC400–29 ft. lbs. (39 Nm)
LS400–22 ft. lbs. (29 Nm)

16. Connect the wiring connector to the cold start injector. Connect the VSV for the EGR. Torque to 13 ft. lbs. (185 Nm). Attach the VSV filter on the SC400. Check for fuel leakage.

IDLE SPEED CONTROL (ISC) VALVE

Testing

1. Check for operating sound from the ISC valve.
2. Disconnect the ISC valve connector.
3. Using an ohmmeter, measure the resistance between the B1 terminal (upper middle terminal) and the S1 or S3 (upper outer terminals), and between B2 terminal (lower middle terminal) and the S2 or S4 (lower outer terminals).
4. If the resistance is not 10–30 ohms, replace the ISC valve.

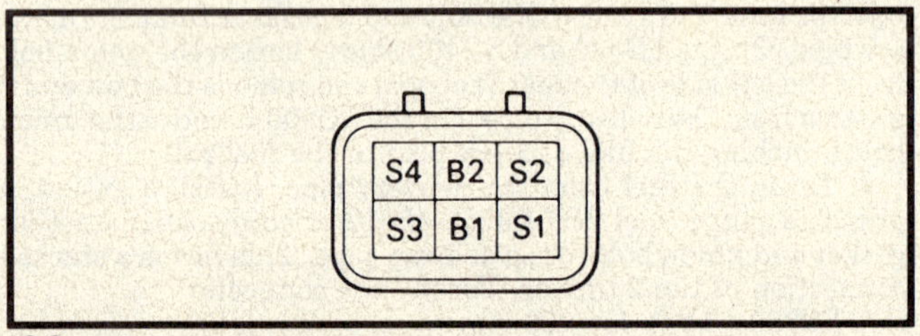

ISC valve terminals

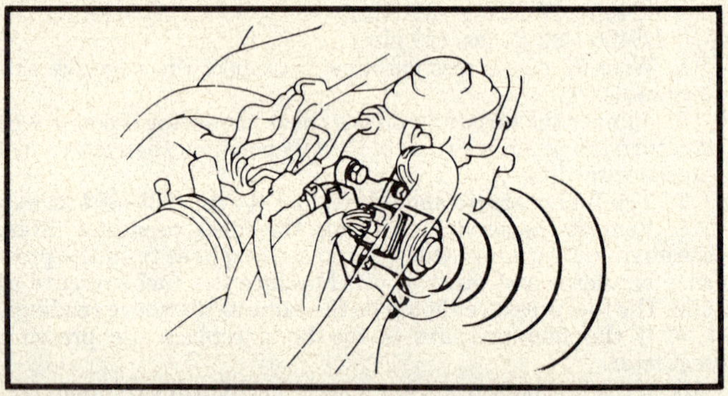

Listening for ISC valve operation–ES250

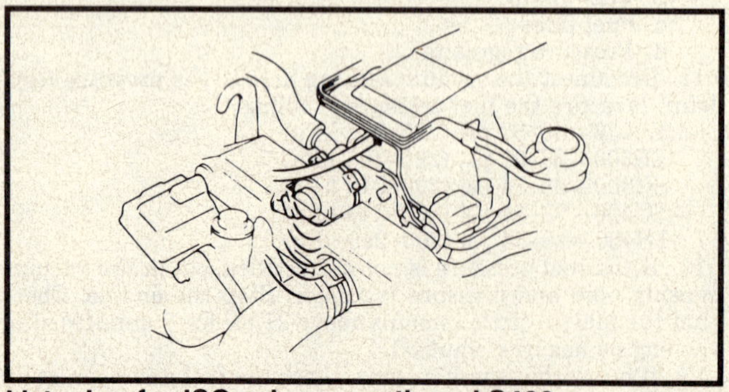

Listening for ISC valve operation–LS400

5. Apply battery voltage to terminals B1 and B2 while repeatedly grounding S1, S2, S3, S4, S1 in sequence. Check that valve moves toward the closed position.

6. Apply battery voltage to terminals B1 and B2 while repeatedly grounding S4, S3, S2, S1, S4 in sequence. Check that valve moves toward the opened position.

7. If valve does not function as specified, replace the ISC valve.

THROTTLE POSITION SENSOR

Testing

1. Disconnect the sensor connector.
2. Insert a feeler gauge between the throttle stop screw and the stop lever.
3. Using a ohmmeter measure the resistance between the terminals. Check that the resistance falls within specification for the gap set.

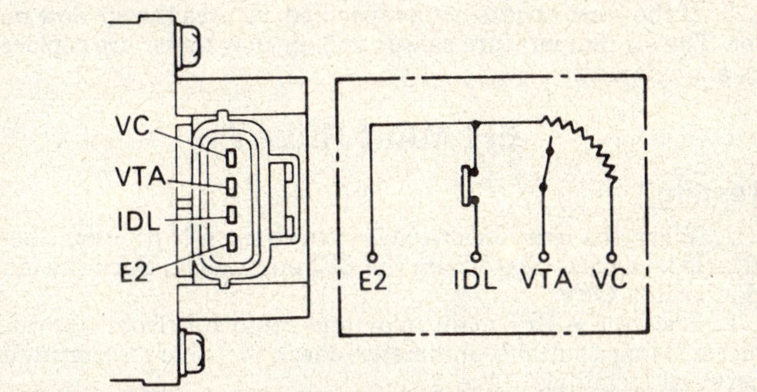

Throttle position sensor terminals

Clearance between lever and stop screw	Between terminals	Resistance
0 mm (0 in).	VTA - E2	0.3 - 6.3 kΩ
0.30 mm (0.012 in.)	IDL - E2	2.3 kΩ or less
0.70 mm (0.028 in.)	IDL - E2	Infinity
Throttle valve fully open	VTA - E2	3.5 - 10.3 kΩ
–	VC - E2	4.25 - 8.25 kΩ

Throttle position sensor specifications–ES250

Clearance between lever and stop screw	Between terminals	Resistance
0 mm (0 in.)	VTA – E2	0.2 – 1.2 kΩ
0.40 mm (0.0157 in.)	IDL – E2	Less than 2.3 kΩ
0.85 mm (0.0335 in.)	IDL – E2	Infinity
Throttle valve fully opened position	VTA – E2	3.3 – 10 kΩ
–	VC – E2	4 – 9 kΩ

Throttle position sensor specifications–LS400

Clearance between lever and stop screw	Between terminals	Resistance
0 mm (0 in.)	VTA – E2	0.28 – 6.4 kΩ
0.35 mm (0.014 in.)	IDL – E2	0.5 kΩ or less
0.70 mm (0.028 in.)	IDL – E2	Infinity
Trottle valve fully open	VTA – E2	2.0 – 11.6 kΩ
–	VC – E2	2.7 – 7.7 kΩ

Throttle position sensor specifications–ES300

Clearance between lever and stop screw	Between terminals	Resistance
0 mm (0 in.)	VTA1 - E2	0.2 - 0.8 Ω
0.45 mm (0.018 in.)	IDL1 - E2	2.3 kΩ or less
0.55 mm (0.022 in.)	IDL1 - E2	Infinity
Throttle valve fully open	VTA1 - E2	3.3 - 10.0 kΩ
-	VC - E2	4.0 - 9.0 kΩ

Throttle position sensor specifications–SC300

Clearance between lever and stop screw	Between terminals	Resistance
0 mm (0 in.)	VTA1 - E2	0.2 - 0.8 Ω
0.40 mm (0.016 in.)	IDL1 - E2	2.3 kΩ or less
0.65 mm (0.026 in.)	IDL1 - E2	Infinity
Throttle valve fully open	VTA1 - E2	3.3 - 10.0 kΩ
-	VC - E2	4.0 - 9.0 kΩ

Throttle position sensor specifications–SC400

AIR FLOW METER AND TEMPERATURE SENSOR

The air flow meter and temperature sensor are tested and serviced as a single unit. If only testing the air temperature sensor, perform the resistance tests on terminals THA and E2.

On the LS400, SC400 and SC300 only the temperature sensor can be tested due to the method of air flow measurement which uses no moving parts. On these air flow meters, called Karman Vortex types, photo-optics are used to determine airflow. Only physical integrity can be inspected on the Karman Vortex type meters. If damaged they must be replaced.

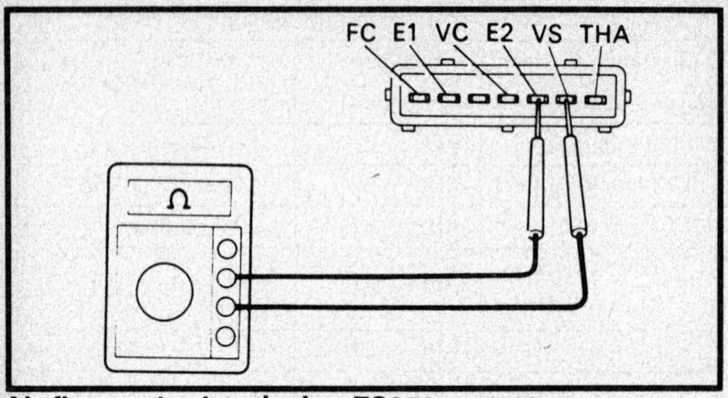

Air flow meter terminals—ES250 and ES300

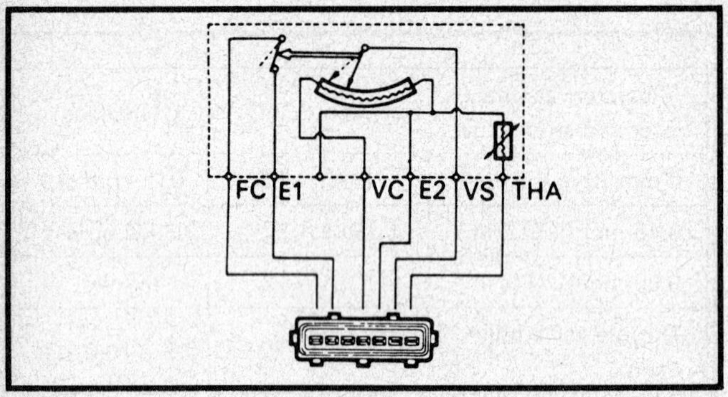

Air flow meter internal circuit—ES250 and ES300

Between terminals	Resistance (Ω)	Temp. °C(°F)
VS – E2	200 – 600	–
VC – E2	200 – 400	–
THA – E2	10,000 – 20,000 4,000 – 7,000 2,000 – 3,000 900 – 1,300 400 – 700	-20 (-4) 0 (32) 20 (68) 40 (104) 60 (140)
FC – E1	Infinity	–

Air flow meter specifications—ES250 and ES300

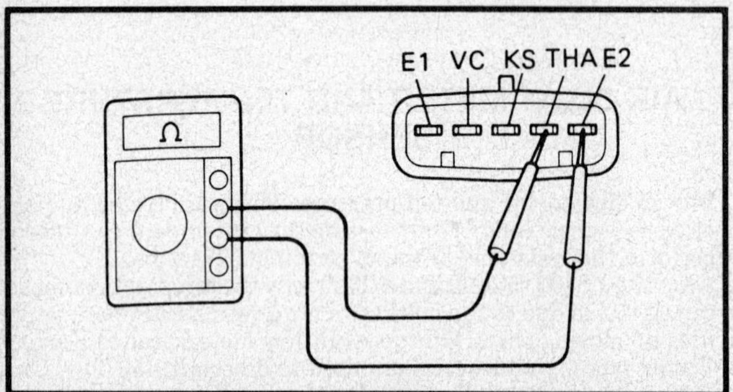

Air flow meter terminals—LS400, SC400 and SC300

Terminals	Resistance (kΩ)	Temp. (°C/°F)
THA–E2	10 – 20 4 – 7 2 – 3 0.9 – 1.3 0.4 – 0.7	–20 (–4) 0 (32) 20 (68) 40 (104) 60 (140)

Air flow meter temperature sensor specifications—LS400, SC400 and SC300

Testing

1. Disconnect the air flow meter connector.
2. Using a ohmmeter measure the resistance between the terminals.
3. If the resistance is not as specified, replace the air flow meter. The air temperature sensor and air flow meter are replaced as a single unit.

EFI MAIN RELAY

Testing

1. When the ignition switch is turn to the **ON** position, there should be a noise heard from the EFI main relay. Turn the ignition switch **OFF**.
2. Locate the EFI main relay and remove it from the relay block. Using a suitable ohmmeter, check that there is continuity between terminals 1 and 3.
3. Check that there is no continuity between terminals 2 and 4 or 3 and 4.
4. If the continuity is not a specified, replace the relay.
5. Apply battery voltage across terminals 1 and 3. Check that there is continuity between terminals 2 and 4.
6. If the continuity is not a specified, replace the relay.

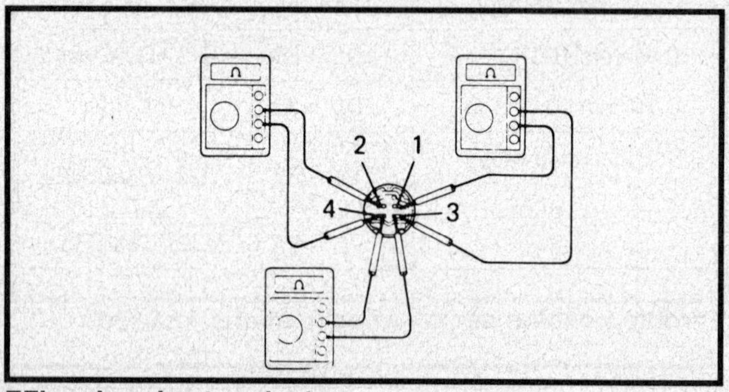

EFI main relay terminals

CIRCUIT OPENING RELAY

Testing

1. Check that there is continuity between terminals ST(A) and E1.
2. Check that there is continuity between terminals IG (B on ES250) and FC.
3. Check that there is no continuity between terminals B and FP.
4. Apply battery voltage across terminals ST(A) and E1. There now should be continuity between terminals B and FP.

5. Apply battery voltage across terminals IG (B on ES250) and FC. There now should be continuity between terminals B and FP.

6. If the continuity is not a specified, replace the relay.

COLD START INJECTOR TIME SWITCH

Testing

1. Disconnect the start injector time switch electrical connector.
2. Using a suitable ohmmeter, measure the resistance between the terminals.
3. If the resistance is not correct, replace the timer.

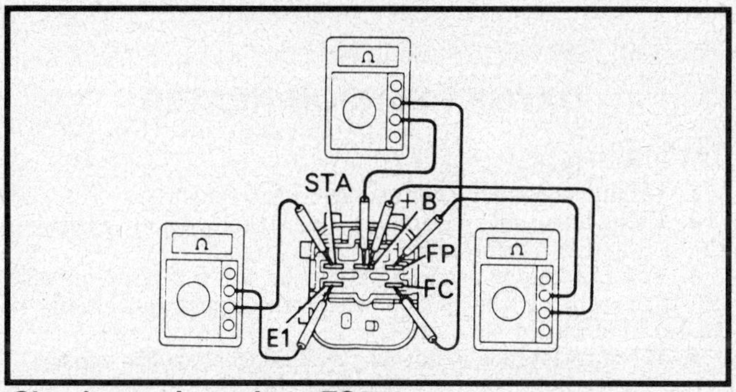

Circuit opening relay—ES250

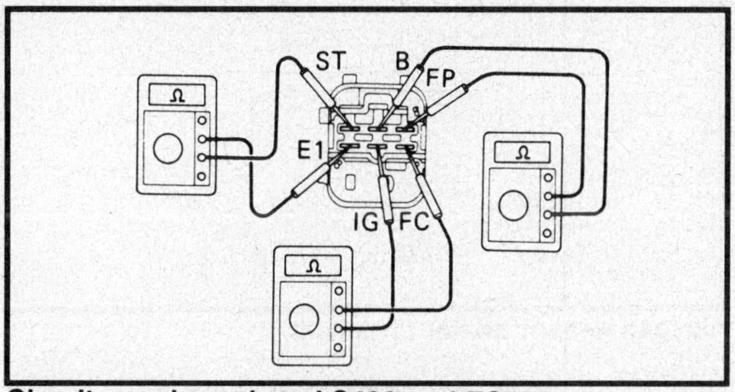

Circuit opening relay—LS400 and ES300

Between terminals	Resistance (Ω)	Coolant temperature
STA – STJ	25 – 45	below 15°C (59°F)
	65 – 85	above 30°C (86°F)
STA – Ground	25 – 85	—

Cold start injector time switch specifications

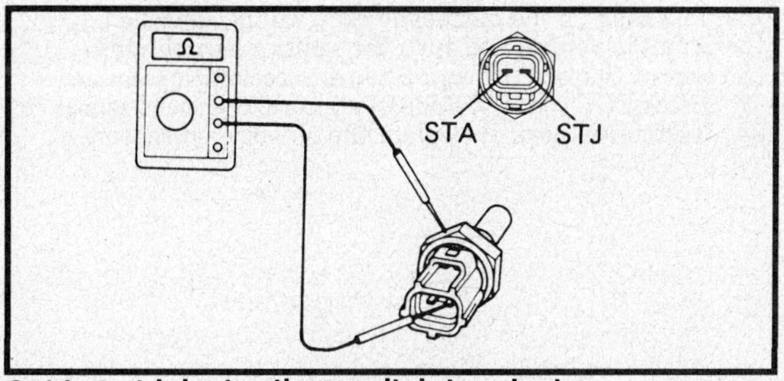

Cold start injector time switch terminals

WATER THERMO SENSOR

Testing

1. Disconnect the electrical connector to the water thermo sensor.
2. Using a suitable ohmmeter, measure the resistance between both terminals. Refer to the water thermo sensor resistance chart.
3. Replace sensor if not within specification.

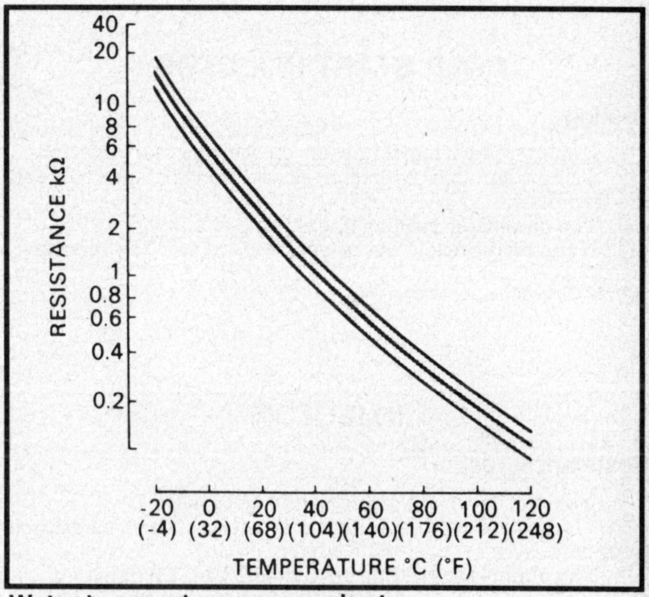

Water temperature sensor chart

FUEL CUT RPM

Testing

1. Connect a tachometer to the IG terminal of the check connector. Remove the V-bank cover on the ES300 to expose the injectors. Start the engine and let it run until it reaches normal operating temperature. Make sure the AC is off.

NOTE: The vehicles wheels should be blocked and the parking brake applied.

2. On the LS400 and ES250 disconnect the throttle position sensor connector from the throttle position sensor.
3. On the LS400 and ES250 short circuit terminals E_2 and IDL on the wire connector side. Gradually raise the engine rpm and check that there is fluctuation between the fuel cut and fuel return points. The fuel cut rpm should be 1800 rpm. The fuel return point should be 1400 rpm on the LS400 and 1200 rpm on the ES250.

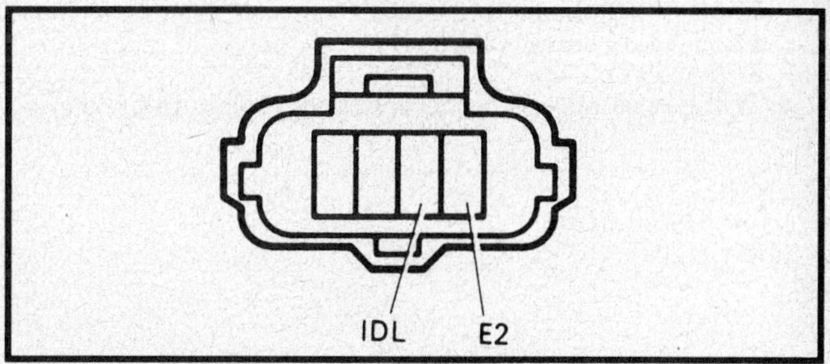

Throttle position sensor terminals

4. On the SC400, SC300 and ES300 raise the rpm to above 2500 rpm and release the throttle. Injector sounds should stop momentarily and resume at 1200 rpm on the ES300, 1400 rpm on the SC400 and 1700 rpm on the SC300.
5. Stop the engine and remove the tachometer. Install the V-bank cover on the ES300.

COLD START INJECTOR

Testing

1. Unplug the wiring connector on the cold start injector.
2. Using a suitable ohmmeter, check the resistance between the terminals.
3. The ohmmeter reading should be 2–4 ohms.
4. If the resistance is not as specified, replace the injector.

INJECTOR

Resistance Test

1. Unplug the wiring connector on the injector.
2. Using a suitable ohmmeter, check the resistance terminals.
3. The ohmmeter reading should be 13.4–14.2 ohms.
4. If the resistance is not as specified, replace the injector.

Operation Test

1. Remove engine covers to expose injectors, if equipped.
2. Start the engine.
3. Using a sound scope or equivalent check the operating sound of each injector.
4. The sound should be even and proportional to the speed of the engine. If there is no sound check the connection at the injector and the wiring harness.

EGR TEMPERATURE SENSOR

Testing

1. Unplug the wiring connector on the sensor.
2. Using a suitable ohmmeter, check the resistance between the terminals.
3. The ohmmeter reading should be as follows:
   69.40–88.50 ohms@112°F (50°C)
   11.89–14.37 ohms@212°F (100°C)
   2.79–3.59 ohms@302°F (150°C)
4. If the resistance is not as specified, replace the sensor.

OXYGEN SENSOR

Testing

1. Start the engine and allow to warm to operation temperature.
2. Connect the negative lead of a voltmeter to terminal E1 of the check connector. Connect the positive lead to VF1 for the left cylinder bank or VF2 for the right cylinder bank. On the SC300 with the inline engine VF1 is for oxygen sensor No.1 and VF2 is for oxygen sensor No.2.
3. Follow the flow chart to troubleshoot the oxygen sensor system.

OXYGEN SENSOR HEATER

Testing

1. Unplug the wiring connector on the sensor.
2. Using a suitable ohmmeter, check the continuity between the +B and HT terminals.
3. The ohmmeter reading should be 5.1–6.3 ohms at 68°F (20°C), except on the ES300 where the resistance should be 11.7–14.3 ohms at 68°F (20°C).
4. If the resistance is not as specified, replace the sensor.

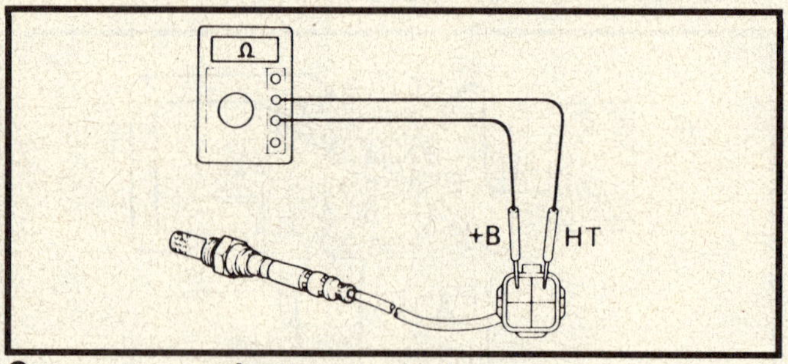

Oxygen sensor heater terminals

SUB-OXYGEN SENSOR

Testing

NOTE: The sub-oxygen sensor is only used on California vehicles. Only follow these procedures if a Code 27 is displayed.

1. Clear all codes.
2. Bring engine up to normal operating temperature.
3. For the LS400 drive at least 5 minutes at less than 50 mph in **D** range. For the ES250 drive at least 5 minutes between 50 mph and 62 mph in **D** range or 4th or 5th gear.
4. Fully depress the accelerator for 2 seconds or more. Do not exceed 62 mph or the diagnostic code will be cancelled.
5. Stop the vehicle and turn the ignition switch **OFF**.
6. Repeat Steps 2 through 5 to test acceleration action.
7. Check ECU codes. If Code 27 appears, check the sensor circuit. If circuit is normal, replace the sub-oxygen sensor.

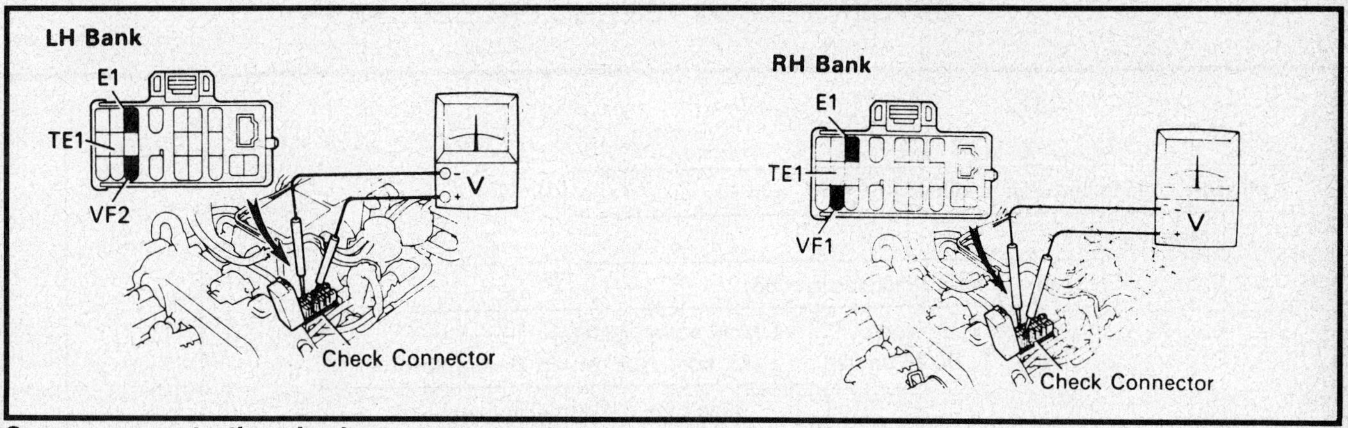

Oxygen sensor testing check connector terminals

OXYGEN SENSOR TESTING PROCEDURE

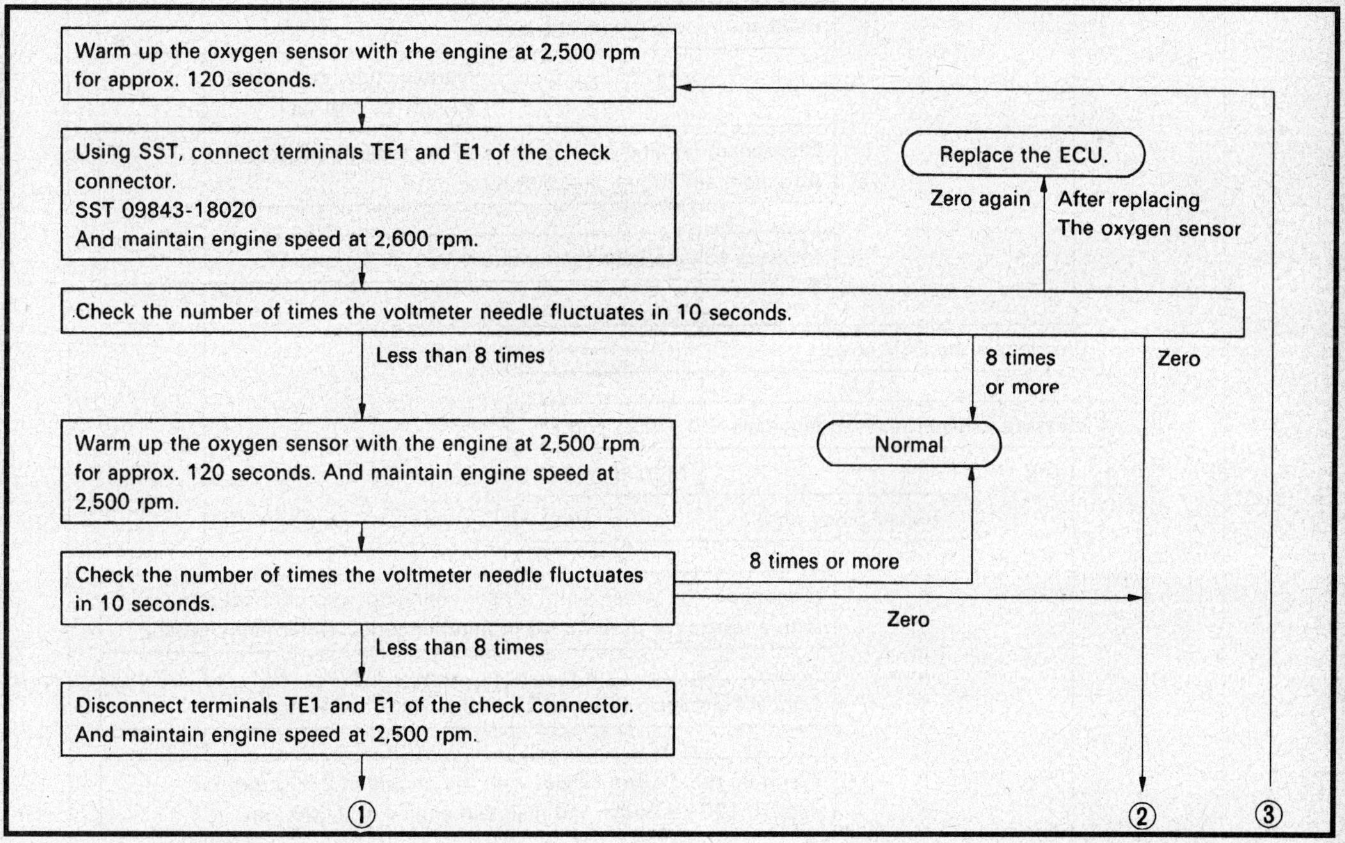

OXYGEN SENSOR TESTING PROCEDURE

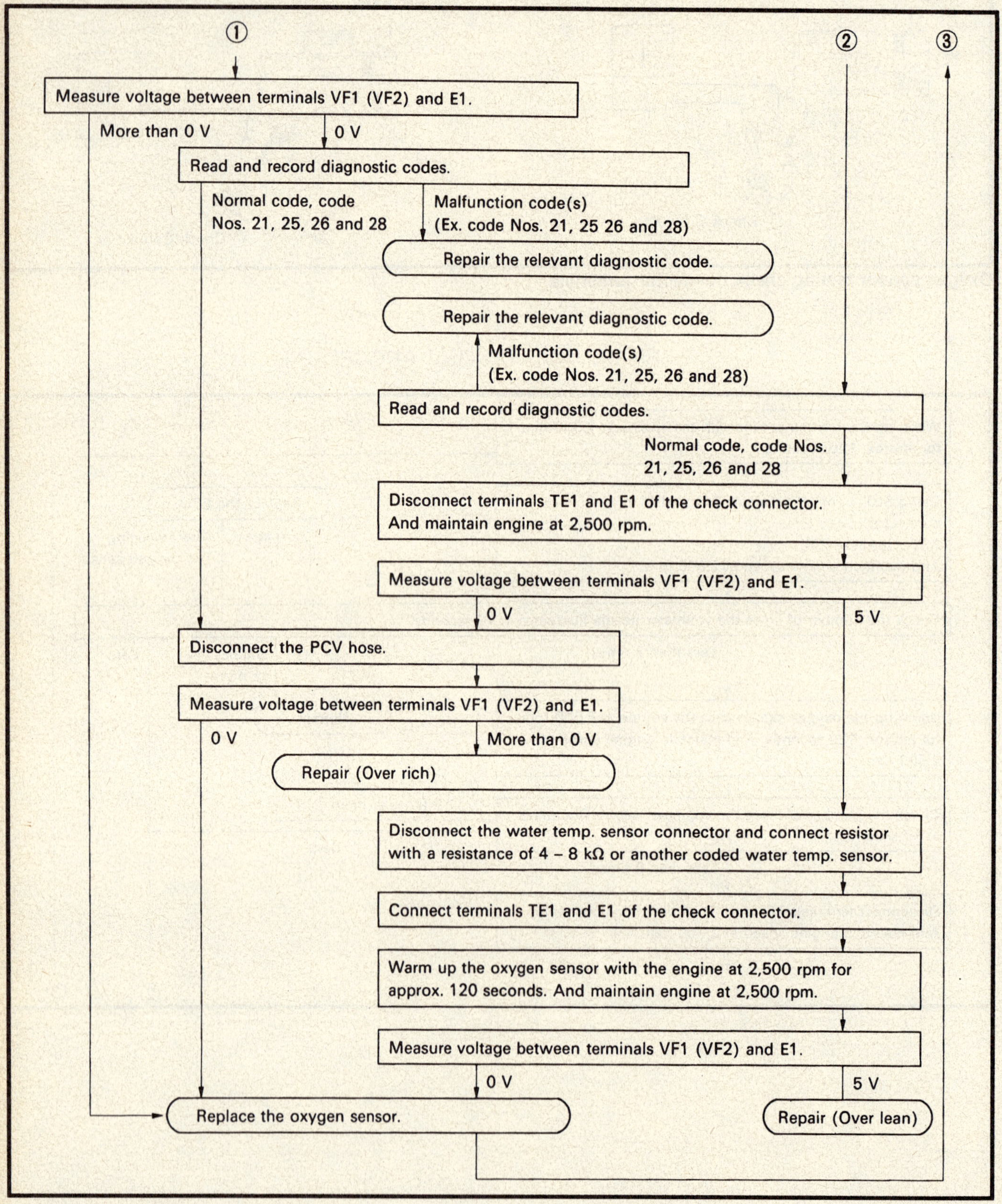

2.5L ENGINE–DIAGNOSTIC CODES–ES250

Code No.	Number to blinks "CHECK" Engine	System	*1 "CHECK" Engine Warning Light	Diagnosis	Trouble Area	*2 Memory
–		Normal	–	Output when no other code is recorded.	–	–
12		RPM Signal	ON	No G or NE signal is input to the ECU for 2 sec. or more after STA turns ON.	• Open or short in NE, G circuit • Distributor • Open or short in STA circuit • ECU	○
13		RPM Signal	ON	NE signal not input to ECU for 50 msec. or more when engine speed is 1,000 rpm or more.	• Open or short in NE circuit • Distributor • ECU	○
14		Ignition Signal	ON	IGF signal from igniter is not input to EC'' for 6 consecutive ignition.	• Open or short in IGF or IGT circuit from igniter to ECU • Igniter • ECU	○
16		ECT Control Signal	ON	Normal signal is not output from ECU of ECT.	• ECU	X
21		Main Oxygen Sensor Signal	ON	(1) Open or short in heater circuit of main oxygen sensor for 0.5 sec. or more. (HT) (2) At normal driving speed (below 60 mph and engine speed is above 1,500 rpm), amplitude of oxygen sensor signal (OX1) is reduced to between 0.35 – 0.70 V continuously for 60 sec. or more. (2 trip detection logic)*6	• Open or short in heater circuit of main oxygen sensor • Main oxygen sensor heater • ECU • Open or short in main oxygen sensor circuit • Main oxygen sensor • ECU	○
22		Water Temp. Sensor Signal	ON	Open or short in water temp. sensor circuit for 0.5 sec. or more. (THW)	• Open or short in water temp. sensor circuit • Water temp. sensor • ECU	○
24		Intake Air Temp. Sensor Signal	ON*3	Open or short in intake air temp. sensor circuit for 0.5 sec. or more (THA)	• Open or short in intake air temp. circuit • Intake air temp. sensor • ECU	○
25		Air-Fuel Ratio Lean Malfunction	ON	(1) Oxygen sensor output is less than 0.45 V for at least 90 sec. when oxygen sensor is warmed up (racing at 2,000 rpm). (2)*4 When air-fuel ratio feedback correction value or adaptive control value continues at the upper (lean) or lower (rich) limit for a certain period of time or adaptive control value is not renewed for a certain period of time.	• Engine ground bolt loose • Open in E1 circuit • Open in injector circuit • Fuel line pressure (Injector blockage, etc.) • Open or short in oxygen sensor circuit • Oxygen sensor • Ignition system • Water temp. sensor • Air flow meter (air intake) • Foreign object caught in valve • ECU	○
26		Air-Fuel Ratio Rich Malfunction	ON	(3)*4 When marked variation is detected in engine revolutions for each cylinder during idle switch on and feedback condition. (2 trip detection logic) *6	• Engine ground bolt loose • Open in E1 circuit • Short in injector circuit • Fuel line pressure (Injector leakage, etc.) • Open or short in cold start injector circuit • Cold start injector • Open or short in oxygen sensor circuit • Oxygen sensor • Water temp. sensor • Air flow meter • Compression pressure • Foreign object caught in valve • ECU	○

Code No.	Number to blinks "CHECK" Engine	System	*1 "CHECK" Engine Warning Light	Diagnosis	Trouble Area	*2 Memory
27*5		Sub-Oxygen Sensor Signal	ON	When sub-oxygen sensor is warmed up apply acceleration with the throttle fully open for 2 seconds, output of main oxygen sensor is 0.45 V or more (rich) and output of sub-oxygen sensor is 0.45 V or less (lean). (OX2) (2 trip detection logic)*6	• Short or open in sub-oxygen sensor circuit • Sub-oxygen sensor • ECU	○
31		Air Flow Meter Signal	ON	At idling, open or short detected continuously for 0.5 sec. or more in air flow meter circuit.	• Open (VC) or short (VC-E2) in air flow meter circuit • Air flow meter • ECU	○
32		Air Flow Meter Signal	ON	Open or short detected continuously for 0.5 sec. or more in air flow meter circuit.	• Open (E2) or short (VS-VC) in air flow meter circuit • Air flow meter • ECU	○
41		Throttle Position Sensor Signal	ON*3	Open or short detected in throttle position sensor signal (VTA) for 0.5 sec. or more. IDL contact is ON and VTA output exceeds 1.45 V.	• Open or short in throttle position sensor circuit • Throttle position sensor • ECU	○
42		Vehicle Speed Sensor Signal	OFF	SPD signal is not input to the ECU for at least 8 seconds during high load driving with engine speed between 2,500 rpm and 4,500 rpm.	• Open or short in vehicle speed sensor circuit • Vehicle speed sensor • ECU	○
43		Starter Signal	OFF	Starter signal (STA) is not input to ECU even once until engine reaches 800 rpm or more when cranking.	• Open or short in starter signal circuit • Open or short in IG SW circuit • ECU	○
52		Knock Sensor Signal	ON	With engine speed between 1,600 rpm – 5,200 rpm, signal from knock sensor is not input to ECU for 6 revolution. (KNK)	• Open or short in knock sensor circuit • Knock sensor (looseness etc.) • ECU	○
53		Knock Control Signal	ON	Engine speed is between 650 rpm and 5,200 rpm and engine control computer (for knock control) malfunction is detected.	• ECU	X
71*5		EGR System Malfunction	ON	With engine warmed up and EGR operation possible, code is displayed when EGR gas temp. sensor is 55°C (131°F) for M/T, 60°C (140°F) for ECT or less when vehicle is driven for 90 seconds. (THG) (2 trip detection logic)*6	• Open in EGR gas temp. sensor circuit • EGR or BVSV vacuum hose disconnected, valve stuck • Clogged EGR gas passage • ECU	○
51		Switch Condition Signal	OFF	Displayed when A/C is ON, IDL contact OFF or shift position in "R", "D", "2", or "1" renges with the check terminals E1 and TE1 connected.	• A/C switch circuit • Throttle position sensor IDL circuit • Neutral start switch circuit • Accelerator pedal, cable • ECU	X

REMARKS:
*1: "ON" displayed in the diagnosis mode column indicates that the "CHECK" Engine Warning Light is lighted up when a malfunction is detected.
"OFF" indicates that the "CHECK" does not light up during malfunction diagnosis, even if a malfunction is detected.
*2: "○" in the memory column indicates that a diagnostic code is recorded in the ECU memory when a malfunction occurs. "X" indicates that a diagnostic code is not recorded in the ECU memory even if a malfunction occurs. Accordingly, output of diagnostic results is performed with the IG switch ON.
*3: The "CHECK" Engine Warning Light light up only for California specifications if malfunction occurs.
*4: No. (2) and (3) in the diagnostic contents of codes No. 25 and 26 apply to California specification vehicles only, while (1) applies to all models.
*5: Codes 27 and 71 are used only for California specifications.

*6: This indicates items for which "2 trip detection logic" is used. With this logic, when a logic malfunction is first detected, the malfunction is temporarily stored in the ECU memory. If the same case is detected again during the second drive test, this second detection causes the "CHECK" Engine Warning Light to light up.
The 2 trip repeats the same mode a 2nd time. (However, the IG switch must be turned OFF between the 1st trip and 2nd trip).

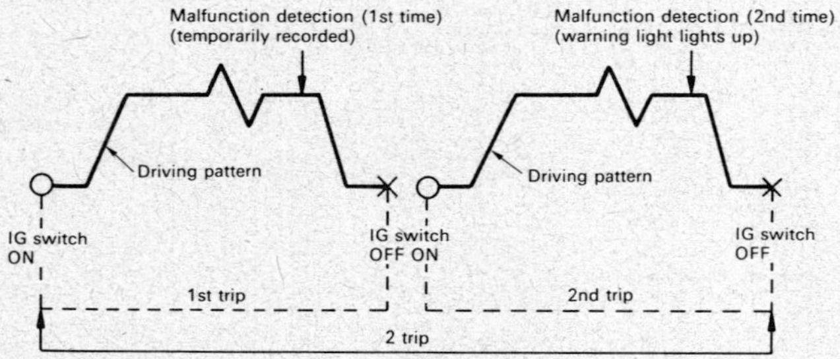

HINT:
- When 2 or more codes are indicated, the lowest number (code) will appear first.
- When there is no code display (the light does not blink), the ECU is faulty or open circuit in terminal TE1 of TDCL or check connector.
- If the light stays on, the wire harness is shorted.
- If a code not listed on the chart is displayed, then the ECU is faulty.

DIAGNOSTIC CODE CLEARANCE

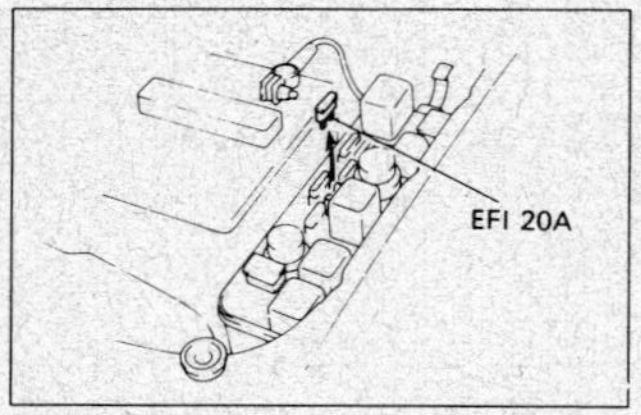

1. After repair of the trouble areas, the diagnostic code retained in the ECU memory must be cancelled out by removing the EFI fuse (20A) for 10 seconds or more, depending on ambient temperature (the lower the temperature, the longer the fuse must be left out) with the ignition switch OFF.
   HINT:
   - Cancellation can also be done by removing the battery negative (–) terminal, but in this case, other memory systems (clock, radio etc.) will also be cancelled out.
   - If the diagnostic code is not cancelled out, it will be retained in memory and appear along with a new code in the event of future trouble.
   - If it is necessary to work on engine components requiring removal of the battery terminal, a check must first be made to see if a diagnostic code has been recorded.
2. After cancellation, road test the vehicle to check that a normal code is now read on the "CHECK" engine warning light.
   If the same diagnostic code appears, it indicates that the trouble area has not been repaired thoroughly.

SYMPTOM SIMULATION

"Intermittent troubles or problems" are the malfunctions about which the customer has a complaint, but which do not occur and can not be confirmed in the workshop. The intermittent problems also include complaints about the "CHECK" engine warning light going on and off erratically.
The self-diagnostic system stores the circuit of the intermittent problem in memory even if the ignition switch is turned off.

The problem simulation methods described below are the effective ways for this nature of problem to produce the problem conditions by applying vibration, heat, and humidity.

1 VIBRATION METHOD: When vibration seems to be the major cause.

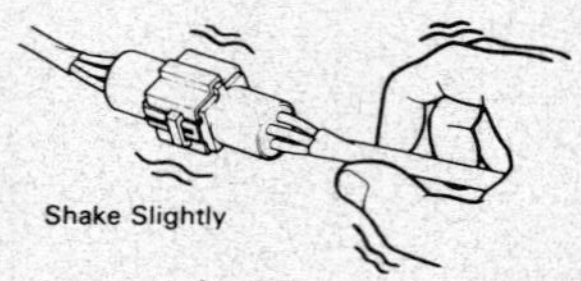

CONNECTORS
Slightly shake the connector vertically and horizontally.

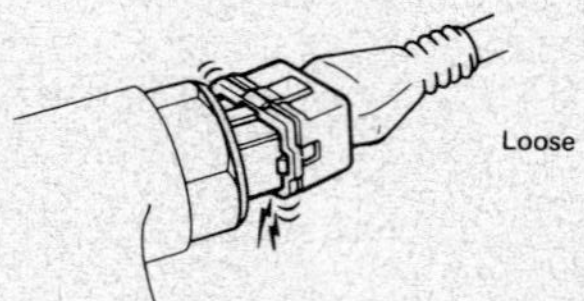

(Inspection of connectors)
(a) Is it loosely connected?

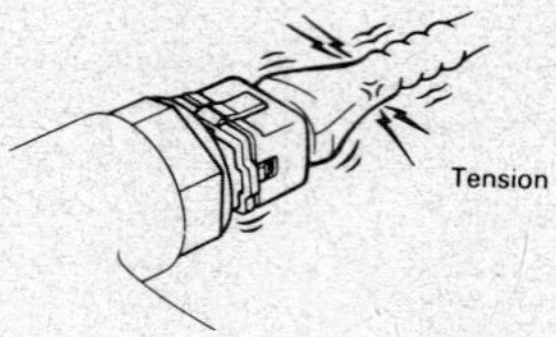

(b) Does the wire harness connecting with its corresponding part have insufficient slack?

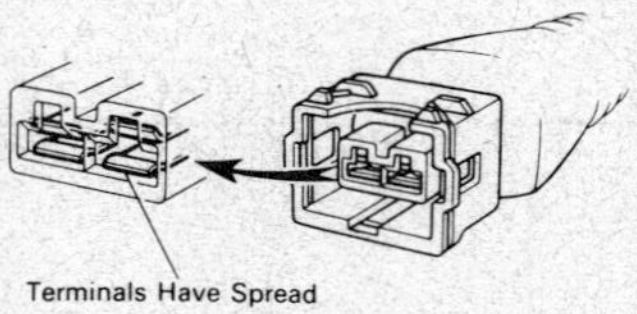

(c) Are the terminals dirty?
(d) Are the terminals making loose contact due to terminals spread?

1 VIBRATION METHOD: When vibration seems to be the major cause. (cont'd)

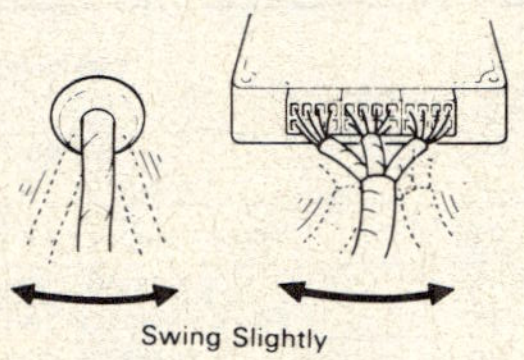

WIRE HARNESS

Slightly shake the wire harness vartically and horizontally. The connector joint, fulcrum of the vibration, and body through portion are the major areas to be checked thoroughly.

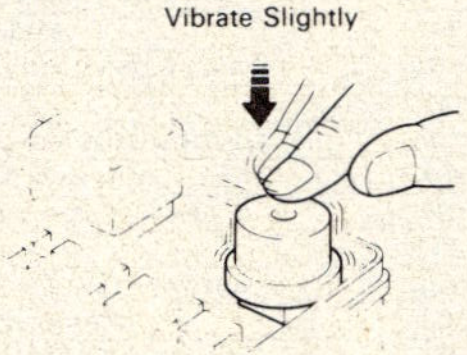

PARTS AND SENSORS

Apply vibration slightly by a finger to the part or sensor considered to be the problem cause and check if the malfunction will occur.

HINT: Strong vibration to the relays may result in open relays.

2 HEAT METHOD: When the problem seems to occur when the suspect area is heated.

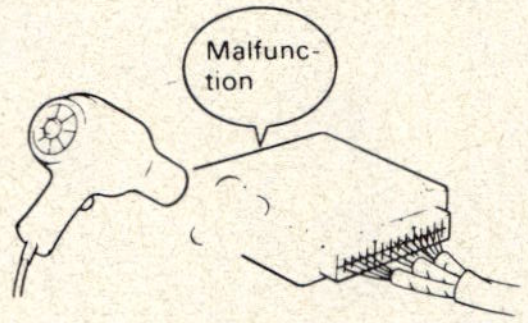

Heat the component that is likely the cause of the malfunction with a hair dryer or similar object. Check to see if the malfunction will occur.

NOTICE:

- **Do not heat to more than 60°C (140°F) (Temperature limit that the component can be touched with a hand.).**
- **Do not apply heat directly to part in the ECU.**

3 WATER SPRINKLING METHOD: When the malfunction seems to occur on a rainy day or in a high-humidity condition.

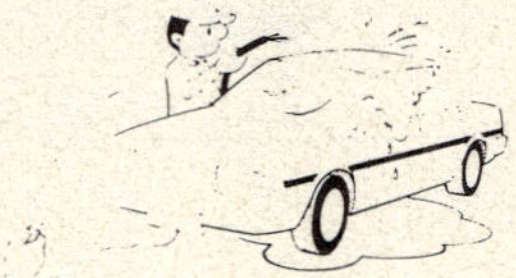

Sprinkle water onto the vehicle and check to see if the malfunction will occur.

NOTICE:

- **Never sprinkle water directly into the engine compartment, but indirectly change the temperature and humidity by applying water spray onto the radiator front surface.**
- **Never apply water directly onto the electronic components.**

HINT:

- If a vehicle is subject to water leakage, the leaked water may contaminate the ECU. When testing a vehicle with a water leakage problem, special caution must be paid.

4 OTHER: When a malfunction seems to occur when electrical load is excessive.

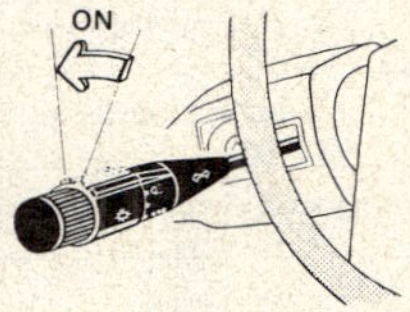

Turn on all electrical loads including the heater blower, head lamps, rear window defogger, etc. and check to see if the malfunction will occur.

2.5L ENGINE—ECU TERMINALS—ES250

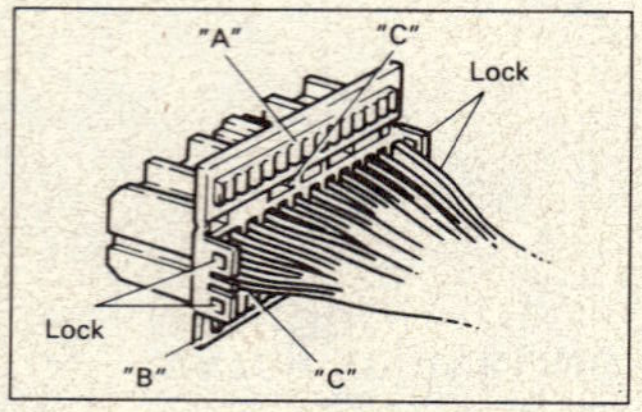

TERMINALS OF ECU

When measuring the voltage or resistance of the connector part of the ECU, remove the locks as required in the following procedure and always insert the test probe into the connector from the wire harness side.

PREPARATION

(a) Turn the ignition switch to LOCK position.

(b) Disconnect the connectors form the ECU.

(c) To remove section "A" and "B", carefully pull outward the lock on both sides as shown in the illustration.

NOTICE: Pay attention to section "C" in the illustration which can be easily broken.

(d) Reconnect the connectors to the ECU.

TERMINALS OF ECU

Symbol	Terminal Name	Symbol	Terminal Name	Symbol	Terminal Name
E01	POWER GROUND	G2	DISTRIBUTOR	A/C	A/C COMPRESSOR
E02	POWER GROUND	G1	DISTRIBUTOR	*1OD1	CRUISE CONTROL ECU
No. 10	INJECTOR (No. 1 and 6)	NE	DISTRIBUTOR	SP1	No. 1 SPEED SENSOR (Meter side)
No. 30	INJECTOR (No. 4 and 5)	G⊖	DISTRIBUTOR	*1OD2	OD MAIN SWITCH
No. 20	INJECTOR (No. 2 and 3)	VF	CHECK CONNECTOR	*1SP2	No. 2 SPEED SENSOR (A/T side)
E1	ECU GROUND	T2	CHECK CONNECTOR		–
STJ	COLD START INJECTOR	*1PWR	PATTERN SELECT SWITCH	*1DG	CHECK CONNECTOR
	–	T1	CHECK CONNECTOR		–
FPU	FUEL PRESSURE VSV	OX1	OXYGEN SENSOR	*2CHK	SUB-OXYGEN SENSOR
ACT	A/C AMPLIFIER	KNK	KNOCK CONTROL SENSOR		–
HT	OXYGEN SENSOR HEATER	*2OX2	SUB-OXYGEN SENSOR	W	CHECK ENGINE WARNING LIGHT
	–	*1BK	BRAKE SWITCH	*1N	SHIFT POSITION SWITCH
ISC1	ISC MOTOR NO. 1 COIL	THW	WATER TEMP. SENSOR	M-REL	EFI MAIN RELAY (COIL)
IGT	IGNITER	IDL	THROTTLE POSITION SENSOR	*12	SHIFT POSITION SWITCH
ISC2	ISC MOTOR NO. 2 COIL	THA	AIR TEMP. SENSOR		–
*1S1	ECT SOLENOID	VTA	THROTTLE POSITION SENSOR	*1L	SHIFT POSITION SWITCH
ISC3	ISC MOTOR NO. 3 COIL	VS	AIR FLOW METER	IG SW	IGNITION SWITCH
*1S2	ECT SOLENOID	*2THG	EGR GAS TEMP. SENSOR	B1	EFI MAIN RELAY
ISC4	ISC MOTOR NO. 4 COIL	VC	AIR FLOW METER THROTTLE POSITION SENSOR	BATT	BATTERY
*1SL	ECT SOLENOID	E2	SENSOR GROUND	B	EFI MAIN RELAY
IGF	IGNITER	STA	STARTER SWITCH		
	–	*1NSW	NEUTRAL START SWITCH		

*1 w/ECT
*2 CALIF. only

ECU Terminals

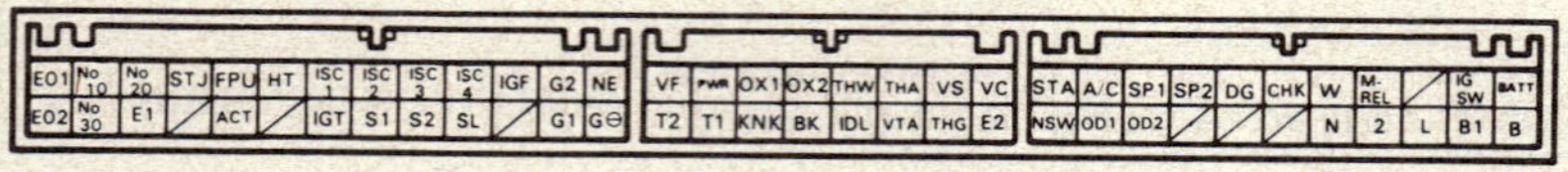

2.5L ENGINE—DIAGNOSTIC CODE CHART—ES250

DIAGNOSTIC CODE CHART

After confirming the symptoms, check the diagnosis codes. If a malfunction code is output, check the circuit corresponding to that circuit as found in the diagnostic code chart below.

Code No.	Inspection Circuit	
12	RPM Signal	
13	RPM Signal	
14	Ignition Signal	
16	ECT Control Signal	
21	Main Oxygen Sensor Signal	
	Main Oxygen Sensor Heater Signal	
22	Water Temp. Sensor Signal	
24	Intake Air Temp. Sensor Signal	
25	Air-fuel Ratio Lean Malfunction	
26	Air-fuel Ratio Rich Malfunction	
27	Sub-oxygen Sensor Signal (California Only)	
31	Air Flow Meter Signal	
32	Air Flow Meter Signal	
41	Throttle Position Sensor Signal	
42	Vehicle Speed Sensor Signal	
43	Starter Signal	
52	Knock Sensor Signal	
53	Knock Control Signal	
71	EGR System Malfunction (California Only)	
51	Switch Condition Signal	

2.5L ENGINE – BASIC INSPECTION – ES250

BASIC INSPECTION

When the normal code is displayed in the diagnostic code check, troubleshooting should be performed in the order for all possible circuits to be considered as the causes of the problems.
In many cases, by carrying out the basic engine check shown in the following flow chart, the location causing the problem can be found quickly and efficiently. Therefore, use of this check is essential in engine troubleshooting.

1 Is battery voltage 11 V or more when engine is stopped?

YES — NO: **Charge or replace battery.**

2 Does engine crank?

YES — NO: **Go to Matrix Chart of Problem Symptoms**

3 Does engine start?

YES — NO: **Go to step 8.**

4 Check air filter.

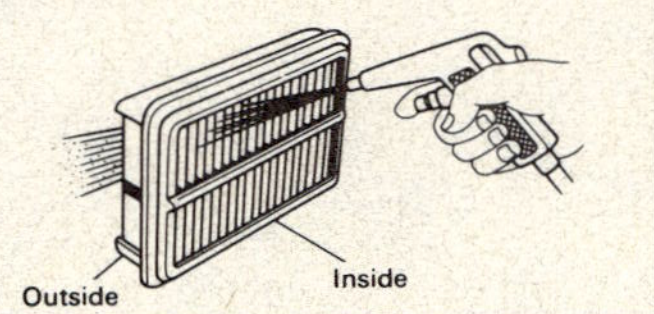

P: Remove air filter.

C: Visually check if the air cleaner element is not excessively damaged or oily.

Hint: If necessary clean the element with compressed air. First blow from inside thoroughly, then blow off outside of element.

OK — NG: **Replace air cleaner element.**

5 Is air leaking into air intake system?

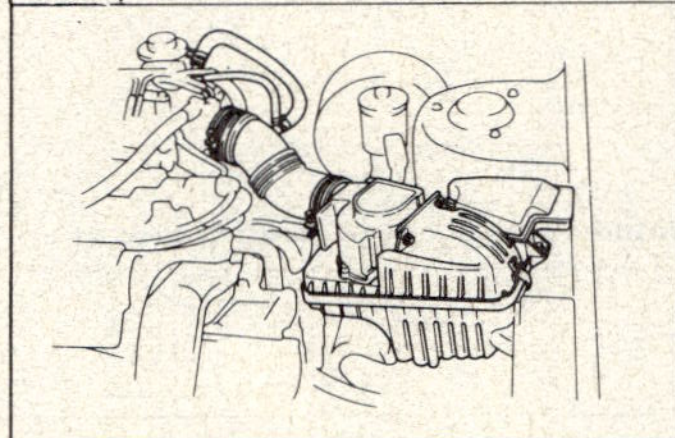

C: Is air leaking into air intake system?

Hint: Loose or missing engine oil dipstick, oil filler cap, PCV hose, etc., can allow air to leak into air intake system, causing mixture to become too lean.

OK: No air leaking into air intake system between air flow meter and cylinder head.

YES — NO: **Repair air leakage.**

6 Check idle speed.

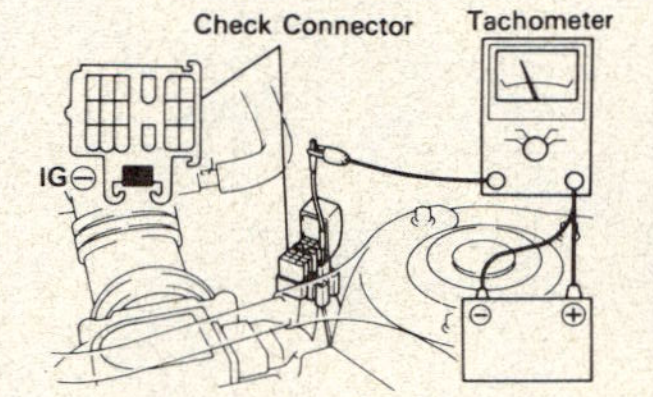

P:
(1) Shift transmission into "N" range.
(2) Warm up engine at normal operating temperature.
(3) Switch off all accessories.
(4) Switch off air conditioner.
(5) Connect tachometer test probe to terminal IG⊖ of check connector.

C: Check idle speed.

OK: Idle speed: 700 ± 50 rpm

Caution:
- **NEVER allow tachometer test probe to touch ground as it could result in damage to igniter and/or ignition coil.**
- **As some tachometers are not compatible with this ignition system, we recommend you to confirm the compatibility of your unit before use.**

OK — NG: **Go to Matrix Chart of Problem Symptoms**

7 Check ignition timing.

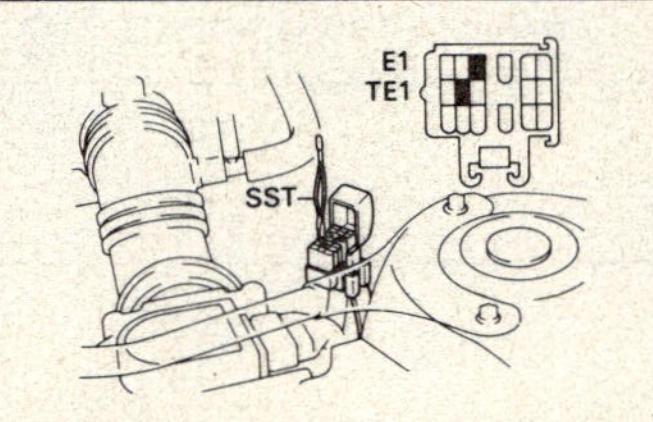

P:
(1) Connect timing light to engine.
(2) Keep engine at idle.
(3) Using SST, connect terminals TE1 and E1 of check connector.
SST 09843-18020

7 Check ignition timing (Cont'd).

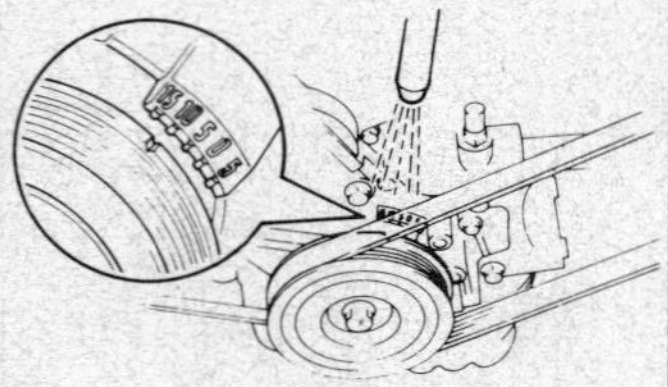

C Check ignition timing using a timing light.

OK Ignition timing: 10° BTDC at idle

Hint Transmission in "N" range.

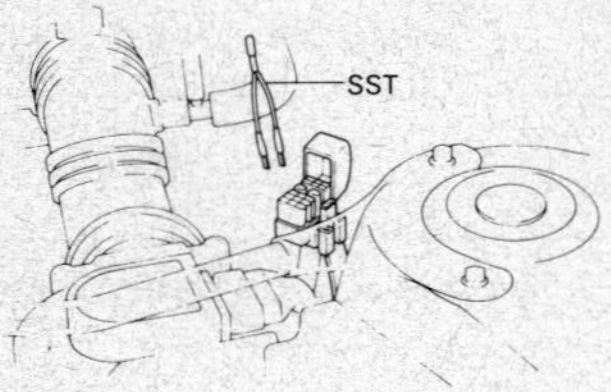

FURTHER CHECK

P Disconnect SST.

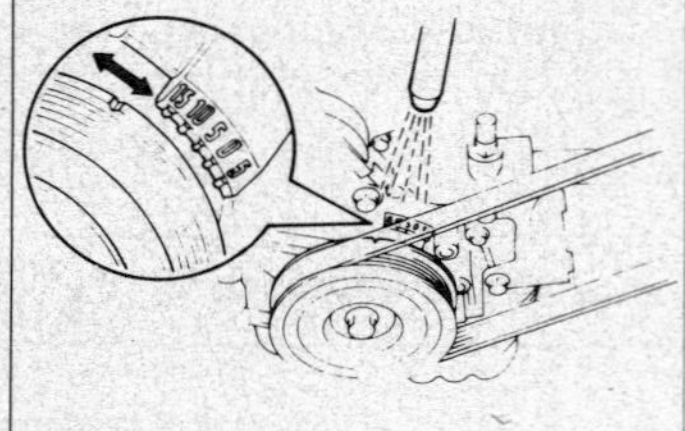

C Check if the ignition timing advances using a timing light.

OK Ignition timing:The timing mark moves in a range between 13° and 27°.

Hint Transmission in "N" range.

OK → Go to Matrix Chart of Problem Symptoms

NG → Adjust ignition timing

8 Check fuel pressure.

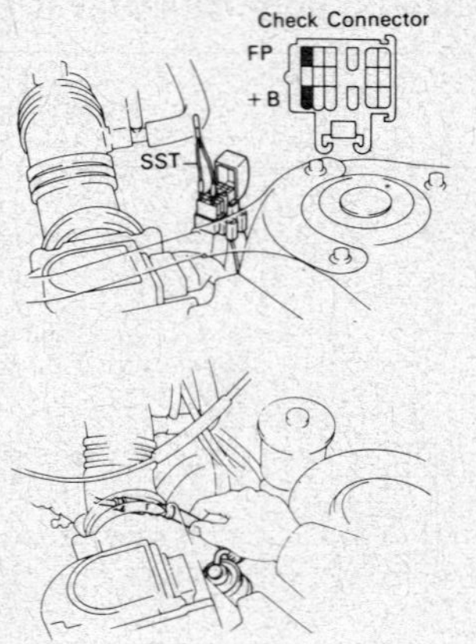

P (1) Be sure that fuel is enough in tank.
(2) Turn ignition switch ON.
(3) Using SST, connect terminals +B and FP of check connector.
SST 09843-18020

C Check if there is a pressure in hose from fuel filter by pinching hose by fingers.

Hint At this time, you will hear fuel return noise.

OK Fuel pressure can be felt.

OK

NG → Go to EFI SYSTEM

9 Check for spark.

OK → Go to Matrix Chart of Problems Symptoms

NG → Go to IGNITION SYSTEM

2.5L ENGINE—MATRIX CHART OF PROBLEM SYMPTOMS—ES250

MATRIX CHART OF PROBLEM SYMPTOMS

When the malfunction code is not confirmed in the diagnostic code check and the problem still can not be confirmed in the basic inspection, then proceed to this step and perform troubleshooting according to the number in order given in the table below.

The circuits indicated by ※ on the matrix chart can be inspected using the TCCS checker.

HINT: Before proceeding with troubleshooting using this matrix chart of problem symptoms, basic inspection should be performed first.

	Symptom \ Suspect area	RPM Signal Circuit ※	Ignition Circuit	Main Oxygen Sensor Circuit ※	Water Temp. Sensor Circuit ※	Intake Air Temp. Sensor Circuit	Sub-oxygen Sensor Circuit	Air Flow Meter Circuit ※	Throttle Position Sensor Circuit ※	STA Signal Circuit ※	Knock Sensor Circuit	NSW Signal Circuit ※	A/C Signal Circuit	Fuel Pump	Fuel Pressure Regulator	Fuel Lines	Injectors	Cold Start System
Does not start	Engine does not crank																	
	Starter runs – engine does not crank																	
	No initial combustion	12	2					5						6				13
	No complete combustion				4			1							3		9	10
	Engine cranks slowly												2					
Difficult to start	Difficult to start ordinaly	12	13		4	14								7	6	8	16	17
	Difficult to start in cold				1	6				2				8	7	9	10	5
	Difficult to start in hot				1	5								8	7	9	10	6
Pour idling	Incorrect first idle				3													
	High engine idle speed				3	5			6			8	7				9	10
	Low engine idle speed				1			4									5	
	Rough idling		18		2			12						7	6	8	16	17
	Misfire		4		6			8									9	10
Poor drivability	Hesitation Poor acceleration			12	10	11		9	8					14	13	15	18	19
	Back fire			6	3	7		5	4					9	8	10	11	
	Muffler explosion (after fire)			8	3	7		5	6						4		9	10
	Serging														1		4	
	Knocking										1							
Engine stall	Engine stall soon after starting				8			7						3	2	4	9	10
	After acceleration pedal depressed							1	3						5	6	7	
	After acceleration pedal released							3										
	During A/C operation												1					
	When N to D shift											1						
Others	Poor fuel economy			21	16	22	23	18	17			19	20				14	15
	Engine overheat										9							
	Engine overcool																	
	Excessive oil consumption																	
	Low oil pressure																	
	High oil pressure																	
	Starter keeps running																	
	Battery often discharge																	

	Symptom \ Suspect area	Idle Speed Control Valve	EFI Main Relay	Circuit Opening Relay	Fuel Cut System	Fuel Pressure Control System	Engine and ECT ECU	Theft Deterrent ECU	Fuel Quality	Fuel Leakage	Coolant Leakage	Oil Leakage	Starter Relay	Neutral Start SW or Clutch Start SW	Starter	Spark Plug	Distributor	EGR System
Does not start	Engine does not crank							4					1	3	2			
	Starter runs – engine does not crank														1			
	No initial combustion	8	3	4			14		7								1	
	No complete combustion	2					11											
	Engine cranks slowly														1			
Difficult to start	Difficult to start ordinaly	3					18		1							2	15	5
	Difficult to start in cold	4					11		3									
	Difficult to start in hot	3				4	11		2									
Poor idling	Incorrect first idle	4					5											
	High engine idle speed	4					11											
	Low engine idle speed	2					6											3
	Rough idling	9					19		1							4	5	3
	Misfire						11		1							3	5	2
Poor drivability	Hesitation Poor acceleration						20		3							5	6	4
	Back fire						12											1
	Muffler explosion (after fire)				1		11											
	Serging						5									2	3	
	Knocking						10		2							3		4
Engine stall	Engine stall soon after starting	6					11		1									5
	After acceleration pedal depressed						8									4		2
	After acceleration pedal released	1					4											2
	During A/C operation	2					3											
	When N to D shift	2					3											
Others	Poor fuel economy				8		24		2	1						9	10	7
	Engine overheat										1					8		
	Engine overcool																	
	Excessive oil consumption											1						
	Low oil pressure											1						
	High oil pressure																	
	Starter keeps running												1		2			
	Battery often discharge																	

2.5L ENGINE—MATRIX CHART OF PROBLEM SYMPTOMS
ES250

Accelerator Pedal Link	Brakes drag even when released	Cooling Fan System	Dash Pot	Clutch	Compression	Valve Clearance	Valve Timing	Timing Belt	Water Pump	Valve Stem Guide Bushing	Oil Pump	Connecting Rod Bearing	Crankshaft Bearing	Cylinder Head	Piston Ring	Flywheel or Drive Plate	Radiator and Radiator Cap	Thermostat	Drive Belt	Water Temp. Sender Gauge	Oil Plessure Switch	Alternator	Cylinder Block
																2							
					9		10	11															
					5		7	8							6								
												3	4										
					9		11								10								
1			2																				
1			2																				
					10	13	14	15						20	11								
					7																		
	2			1	7	16	17																
							2																
							2																
		7					5		9								6	8					
3	6		4	5	11		13								12								
		2					7	5	6		10			11			3	4		13			12
		1																2		3			
					3					2				5	4								6
											2	3	4								5		
											1										2		
																			1			2	

2.5L ENGINE
CIRCUIT INSPECTION AND TROUBLESHOOTING
ES250

Diag. Code	12	RPM Signal

CIRCUIT DESCRIPTION

The distributor in the Engine Control System contains three pick-up coils (G1, G2 and NE).
The G1, G2 signals inform the ECU of the standard crankshaft angle.
The NE signals inform the ECU of the crankshaft angle and the engine speed.

Code No.	Diagnosis	Trouble Area
12	No G or NE signal is input to the ECU for 2 sec. or more after STA turns ON.	• Open or short in NE, G circuit • Distributor • Open or short in STA circuit • ECU

DIAGNOSTIC CHART

Step	Check	Result	Action
1	Does engine crank?	NO	Go to Matrix Chart of Problem Symptom
		YES	Go to step 2
2	Check air gap.	NG	Replace distributor.
		OK	Go to step 3
3	Check pickup coils in distributor.	NG	Replace distributor.
		OK	Go to step 4
4	Check resistance of pickup coils at ECU connector.	NG	Check and repair harness and connector between ECU and distributor.
		OK	Replace ECU.*

* When diag code 12 is not displayed, proceed to next circuit inspection shown on matrix chart

WIRING DIAGRAM AND ECU CONNECTORS

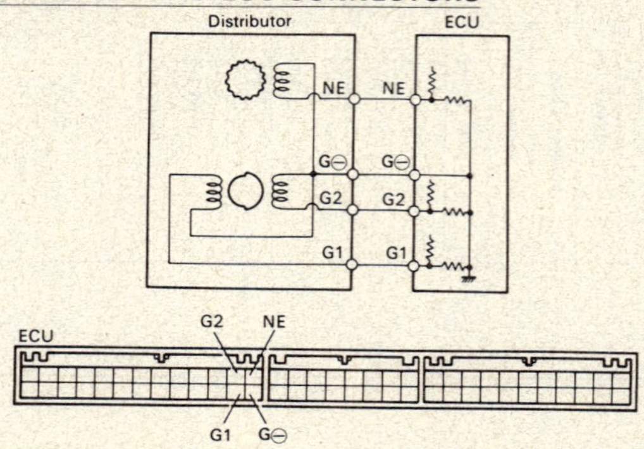

INSPECTION PROCEDURE

1 Does engine crank?

YES

NO → Go to Matrix Chart of Problem Symptom

2 Check air gap.

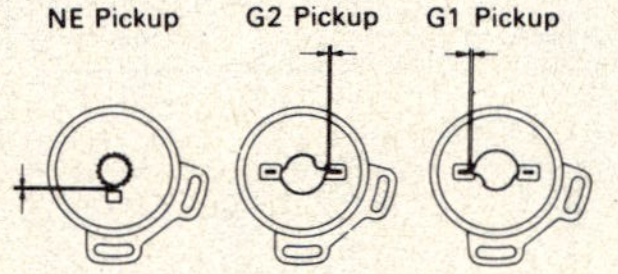

P: Remove distributor cap.

C: Using a thickness gauge, measure the gap between signal rotor projection and pickup coil.

OK: Air gap: 0.2 mm (0.008 in) or more

OK

NG → Replace distributor.

3 Check pickup coils in distributor.

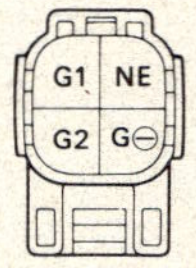

P: Disconnect distributor connector.

C: Using an ohmmeter, measure resistance between each terminal shown in table below:

OK:

Coil	Terminals	Resistance
G1 pickup coil	G1 – G⊖	140 – 180Ω
G2 pickup coil	G2 – G⊖	140 – 180Ω
NE pickup coil	NE – G⊖	140 – 180Ω

OK

NG → Replace distributor.

4 Check resistance of pickup coils at ECU connector.

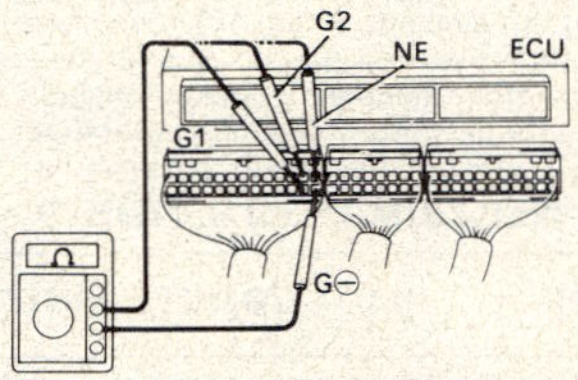

P: (1) Connect distributor connector.
(2) Turn ignition switch LOCK.
(3) Disconnect ECU wiring connectors from ECU.

C: Using an ohmmeter, measure resistance between each terminal shown in table below:

OK:

Coil	Terminals	Resistance
G1 pickup coil	G1 – G⊖	140 – 180Ω
G2 pickup coil	G2 – G⊖	140 – 180Ω
NE pickup coil	NE – G⊖	140 – 180Ω

OK

NG → Check and repair harness and connector between ECU and distributor.

Replace ECU. *

* However, when diag code 12 ia not displayed, proceed to next circuit inspection shown on matrix chart

Diag. Code	13	RPM Signal

CIRCUIT DESCRIPTION

The distributor in the Engine Control System contains three pick-up coils (G1, G2 and NE).
The G1, G2 signals inform the ECU of the standard crankshaft angle.
The NE signals inform the ECU of the crankshaft angle and the engine speed.

Code No.	Diagnosis	Trouble Area
13	NE signal not input to ECU for 50 msec. or more when engine speed is 1,000 rpm or more.	• Open or short in NE circuit • Distributor • ECU

DIAGNOSTIC CHART

This code indicates that a momentary interruption of the RPM signal from the distributor to the ECU has occurred, but that it is returned to normal. Note that although this problem may not necessarily appear at the time of inspection, it cannot be ignored because this diagnostic code is output, indicating that there is or was a malfunction in the RPM signal circuit; this "malfunction" is usually a loose connector.
The distributor connector and the NE terminal of the ECU connector must therefore be checked for the following:

1. Loose connectors
2. Dirty connector terminals
3. Loose connector terminals

WIRING DIAGRAM AND ECU CONNECTORS

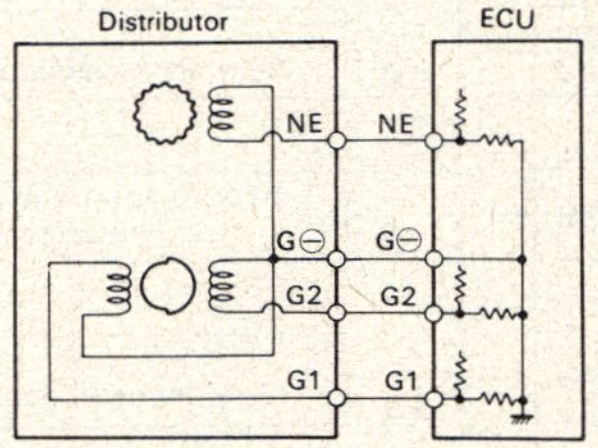

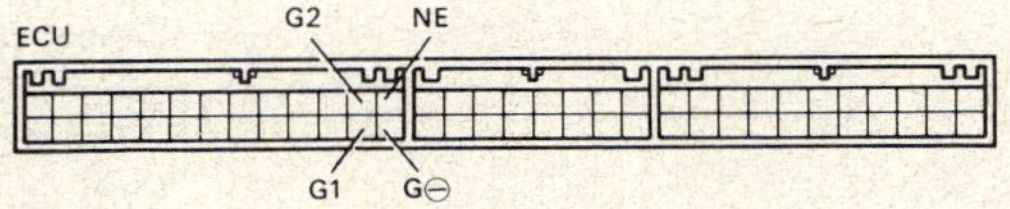

Diag. Code	14	Ignition Signal

CIRCUIT DESCRIPTION

The ECU determines the ignition timing, turns on TR1 at a predetermined angle before the desired ignition timing, and outputs an ignition signal (IGT) "1" to the ignitor. Since the length of the IGT signal is constant, the dwell angle control circuit in the ignitor determines the time the control circuit starts primary current flow to the ignition coil based on the engine rpm and ignition timing in one revolution ago, that is, the time the TR2 turns on. When it reaches the ignition timing, the ECU turns TR1 off and outputs the IGT signal "0". This turns TR2 off, interrupting the primary current flow and generating a high voltage in the secondary coil which causes the spark plug to spark.
Also, by the counter electromotive force generated when the primary current is interrupted, the ignitor sends an ignition confirmation signal (IGF) to the ECU.
The ECU stops fuel injection as a fail safe function when the IGF signal is not input to the ECU.

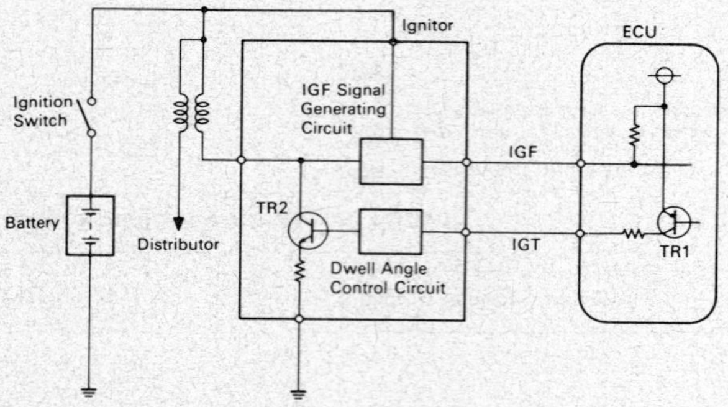

Code No.	Diagnosis	Trouble Area
14	IGF signal from igniter is not input to ECU for 6 consecutive ignition.	● Open or short in IGF or IGT circuit from igniter to ECU. ● Igniter ● ECU

DIAGNOSTIC CHART

1 Check for spark.
- NG → go to 5
- OK ↓

2 Check voltage between terminal IGF of ECU connector and body ground.
- OK → Replace ECU.*
- NG ↓

3 Check voltage between terminol IGF of ignitor connector and body ground.
- OK → Replace ignitor.
- NG ↓

4 Check harness and connector between ECU and ignitor.
- NG → Repair or replace harness or connector.
- OK ↓

Replace ECU.

5 Check power source line voltage of ignitor.
- NG → Repair or replace harness or connector between ignitor and ignition switch.
- OK ↓

6 Check power source line voltage of ignition coil.
- NG → Repair harness or connector between ignitor and ignition coil or replace ignitor.
- OK ↓

7 Check ignition coil.
- NG → Replace ignition coil.
- OK ↓

8 Check voltage between terminal IGT of ignitor connector and body ground.
- OK → Replace ignitor.
- NG ↓

9 Check voltage between terminal IGT of ECU connector and body ground.
- OK → Check and repair harness or connector between ECU and ignitor.
- NG ↓

10 Check harness and connector between ECU and ignitor.
- NG → Repair or replace harness or connector.
- OK ↓

Replace ECU .

* When diag. code 14 is not displayed, proceed to next circuit inspection shown on matrix chart

DIAGNOSTIC CHART

WIRING DIAGRAM AND ECU CONNECTORS

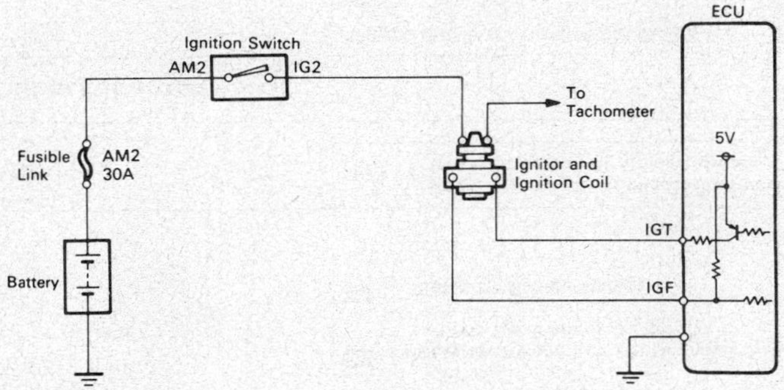

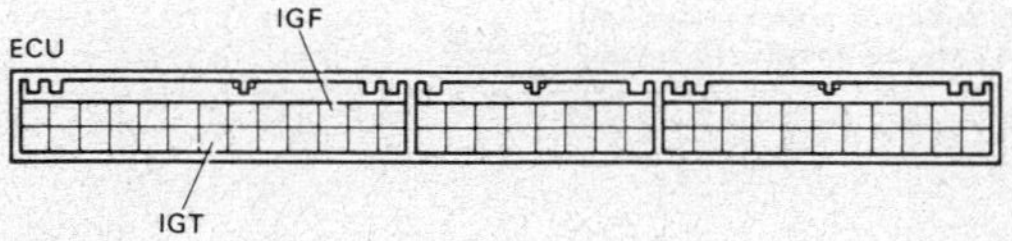

INSPECTION PROCEDURE

1 Check for spark.

OK ↓ | **NG** → Go to step **5**.

2 Check voltage between terminal IGF of ECU connector and body ground.

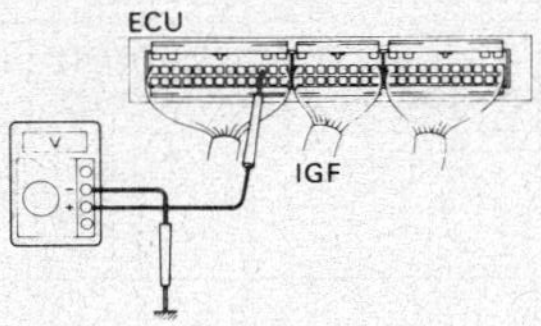

C Measure voltage between terminal IGF of ECU connector and body ground, both with ignition switch ON but engine not cranking, and with engine cranking.

OK Voltage: Approx. 1.0V for both conditions

NG ↓ | **OK** → Replace ECU.

3 Check voltage between terminal IGF of ignitor connector and body ground.

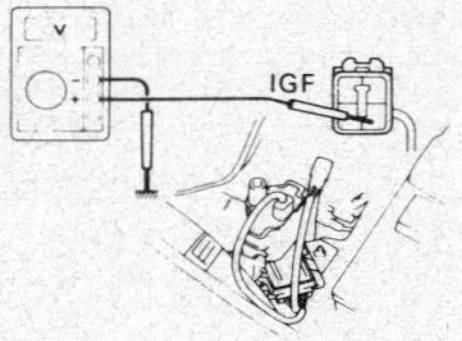

P (1) Disconnect ignitor connector.
(2) Turn ignition switch ON.

C Measure voltage between terminal IGF of ignitor connector and body ground.

OK Voltage: 5V

NG ↓ | **OK** → Replace ignitor.

4 Check harness and connector between ECU and ignitor.

OK ↓ | **NG** → Repair or replace harness or connector.

Replace ECU.

5 Check power source line voltage of ignitor.

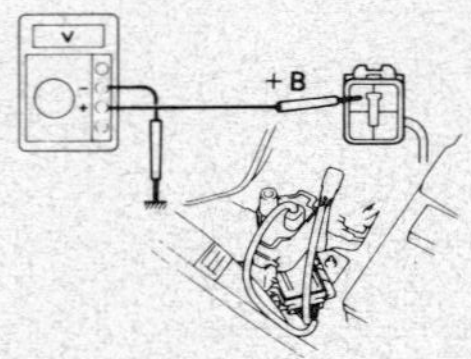

P (1) Disconnect ignitor connector.
(2) Turn ignition switch ON.

C Measure voltage between terminal +B of ignitor connector and body ground.

OK Voltage: Battery voltage

NG → Repair or replace harness or connector between ignitor and ignition switch.

OK ↓

6 Check power source line voltage of ignition coil.

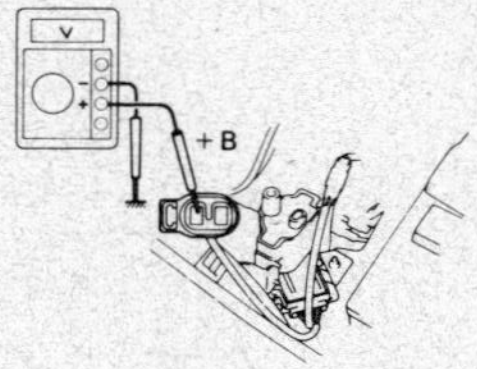

P (1) Connect ignitor connector.
(2) Disconnect ignition coil connector.
(3) Turn ignition switch ON.

C Measure voltage between terminal +B of ignition coil connector and body ground.

OK Voltage: Battery voltage

NG → Repair harness or connector between ignitor and ignition coil or replace ignitor.

OK ↓

7 Check ignition coil.

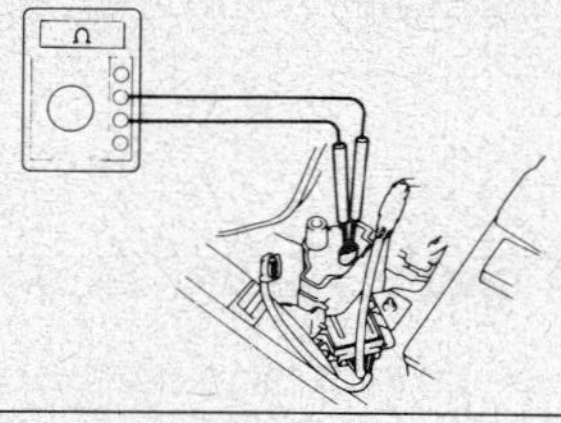

CHECK PRIMARY COIL RESISTANCE

C Measure resistance between terminal +B and E of ignition coil connector.

OK Primary coil resistance (cold): 0.41 – 0.50 Ω

* However, when diag. code 14 is not displayed, proceed to next circuit inspection shown on matrix chart

7 Check ignition coil. (cont'd)

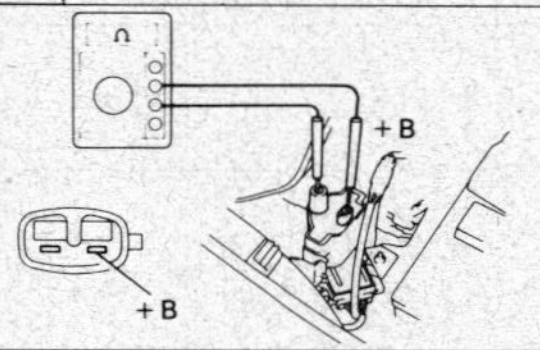

CHECK SECONDARY COIL RESISTANCE

C Measure resistance between terminals +B and high tension terminal of ignition coil.

OK Secondary coil resistance (cold): 10.2 – 13.8 kΩ

NG → Replace ignition coil.

OK ↓

8 Check voltage between terminal IGT of ignitor connector and body ground.

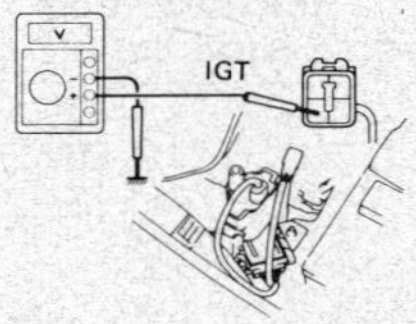

P (1) Connect ignition coil connector.
(2) Disconnect ignitor connector.

C Measure voltage between terminal IGT of ignitor connector and body ground when engine is cranked.

OK Voltage : Approx. 1.0V
(Neither 0V nor 5V)

OK → Replace ignitor.

NG ↓

9 Check voltage between terminal IGT of ECU connector and body ground.

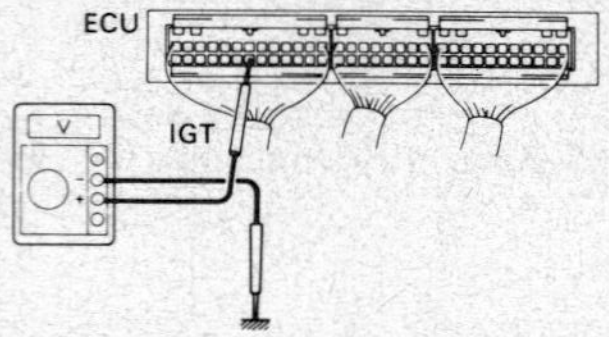

C Measure voltage between terminal IGT of ECU connector and body ground when engine is cranked.

OK Voltage: Approx. 1.0V
(Neither 0V nor 5V)

OK → Check and repair harness or connector between ECU and ignitor.

NG ↓

10 Check harness and connector between ECU and ignitor.

NG → Repair or replace harness or connector.

OK ↓

Replace ECU.

Diag. Code 16 ECT Control Signal

CIRCUIT DESCRIPTION

Code No.	Diagnosis	Trouble Area
16	Normal signal is not output from ECU of ECT.	• ECU

When the ECU stores diagnostic code 16 in memory, torque control which performs smooth gear shifting is prohibited.

DIAGNOSTIC CHART

1 Are there any other codes (besides code 16) being output?

YES → Go to relevant Diag. Code Chart.

NO → Replace ECU.

Diag. Code 21 Main Oxygen Sensor and Heater Signal

CIRCUIT DESCRIPTION

The main oxygen sensor is located in the exhaust manifold. It indirectly determines whether the fuel mixture is rich or lean by detecting the concentration of oxygen present in the exhaust gas. The main oxygen sensor is provided with a heater. The heater is controlled by the ECU. When the intake air volume is low (the temperature of the exhaust gas is low) current flows to the heater to heat the sensor for accurate detection.

Code No.	Diagnosis	Trouble Area
21	(1) Open or short in heater circuit of main oxygen sensor for 0.5 sec. or more (HT).	• Open or short in heater circuit of main oxygen sensor • Main oxygen sensor heater • ECU
	(2) At normal driving speed (below 60 mph and engine speed is above 1,500 rpm), amplitude of oxygen sensor signal (OX1) is reduced to between 0.35 – 0.70 V continuously for 60 sec. or more. (2 trip detection logic)	• Open or short in main oxygen sensor circuit • Main oxygen sensor • ECU

CIRCUIT DESCRIPTION (Cont'd)

DIAGNOSIS CODE DETECTION DRIVING PATTERN

Purpose of the driving patten.

(a) After the diag. code is erased, the malfunction is reproduced and the diag. code is recorded again.

(b) To check that the malfunction is cleared once the repair is completed (confirm that diag. code is no longer detected).

Malfunction: Deterioration of Main Oxygen Sensor

For A/T Vehicle

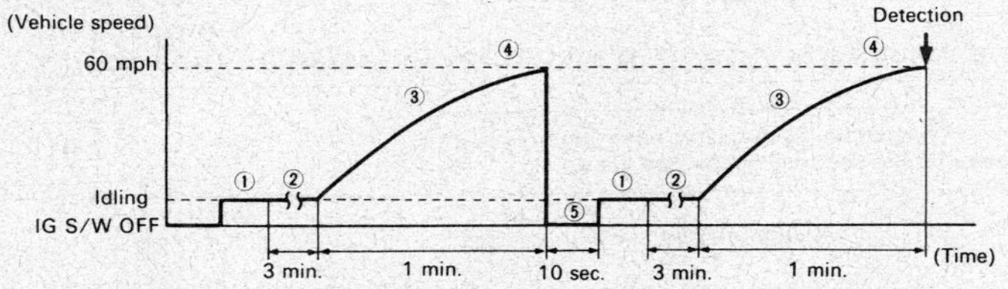

① Start engine and warm engine up.
② After engine is warmed up, let it idle for 3 min.
③ With the A/C ON and the transmission in D range (O/D ON), perform gradual acceleration with in the range 1,800-2,200 rpm (Centered around 2,000 rpm). (Take care that the engine speed does not fall below 1,200 rpm when shifting. Gradually depress the accelerator pedal and kept it. Steady so that engine braking does not occur.)
④ Keep vehicle speed below 60 mph.
⑤ Stop the vehicle, then 10 sec. or more after turning IG switch to OFF (not Acc), perform ①, ②, ③, ④ again.

For M/T Vehicle

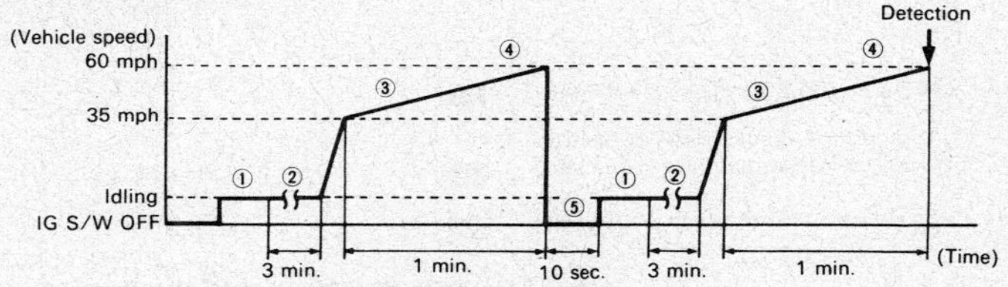

① Start engine and warm engine up.
② After engine is warmed up, let it idle for 3 min.
③ With the A/C ON and vehicle speed approx. 35 mph, shift into 4th gear and gradually accelerate so that engine speed does not drop below, 1,500 rpm (Keep the accelerator pedal position constant to avoid engine braking).
④ Keep vehicle speed below 60 mph.
⑤ Stop the vehicle, then 10 sec. or more after turning IG switch to OFF (not Acc), perform ①, ②, ③, ④ again.

NOTICE: **If the conditions in this test are not strictly observed, detection of the malfunction will not be possible.**

DIAGNOSTIC CHART

	Step	Result	Action
1	Check main oxygen sensor heater.	NG	Replace main oxygen sensor.
	↓ OK		
2	Disconnect ECU connector and check voltage between terminal HT of ECU connector and body ground.	NG	Check and repair harness and connector between ECU and EFI main relay.
	↓ OK		
3	Check voltage between terminal HT of ECU connector and body ground.	NG	Replace ECU.
	↓ OK		
	Replace main oxygen sensor.		

WIRING DIAGRAM AND ECU CONNECTORS

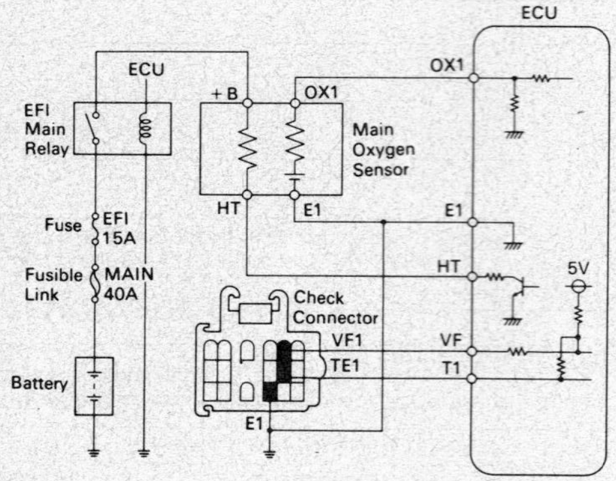

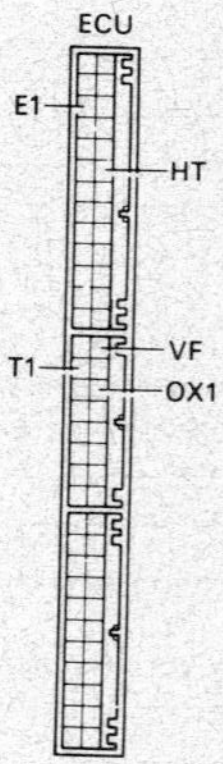

INSPECTION PROCEDURE

1 Check main oxygen sensor heater.

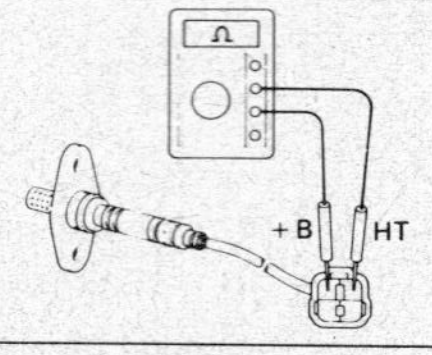

P Disconnect main oxygen sensor connector.

C Measure resistance between terminals +B and HT of main oxygen sensor connector.

OK Resistance: 5.1 – 6.3 Ω at 20°C (68°F)

NG 〉 **Replace main oxygen sensor.**

OK

2 Disconnect ECU connector and check voltage between terminal HT of ECU connector and body ground.

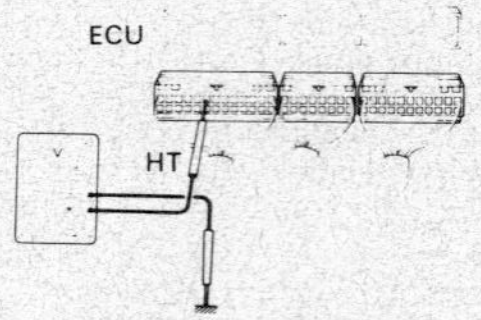

P (1) Connect main oxygen sensor connector.
(2) Disconnect ECU connector.
(3) Turn ignition switch ON.

C Measure voltage between terminal HT of ECU connector and body ground.

OK Voltage: Battery voltage

NG 〉 **Check and repair harness and connector between ECU and EFI main relay.**

OK

3 Check voltage between terminal HT of ECU connector and body ground.

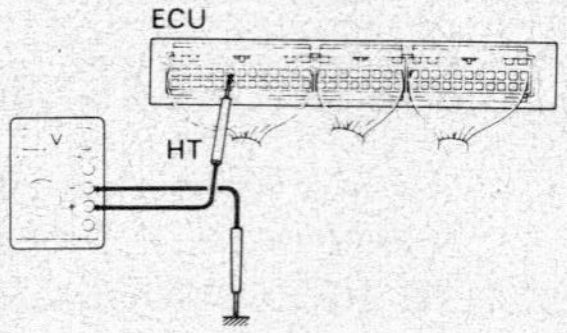

P (1) Turn ignition switch LOCK.
(2) Connect ECU connector.
(3) Warm up engine at normal operating temperature.

C Measure voltage between terminal HT of ECU connector and body ground, when engine is idling and racing at 4000 rpm.

OK

Engine rpm	Voltage
Idling	Approx. 0 V
4000 rpm	Battery voltage

Hint In the above check, continue racing engine at 4000 rpm for approx. 20 seconds or more.

NG 〉 **Replace ECU.**

OK

Replace main oxygen sensor.

Diag. Code	22	Water Temp. Sensor Signal

CIRCUIT DESCRIPTION

The water temperature sensor senses the coolant temperature. The resistance value of a thermistor built in the sensor changes according to the coolant temperature. The lower the coolant temperature, the larger the resistance value becomes.
If diagnostic code 22 is recorded in memory, the fail safe function operates and fixes the coolant temperature at a constant 80°C (176°F).

Code No.	Diagnosis	Trouble Area
22	Open or short in water temp. sensor circuit for 0.5 sec. or more (THW).	• Open or short in water temp. sensor circuit. • Water temp. sensor • ECU

DIAGNOSTIC CHART

1 Check water temp. sensor — NG → Replace water temp. sensor.
↓ OK
2 Check voltage between terminals THW and E2 of ECU connector. — OK → Replace ECU. *
↓ NG
3 Check harness and connector between ECU and water temp. sensor. — NG → Repair or replace harness or connector.
↓ OK
Replace ECU.

* When diag. code 22 is not displayed, proceed to next circuit inspection shown on matrix chart

WIRING DIAGRAM AND ECU CONNECTORS

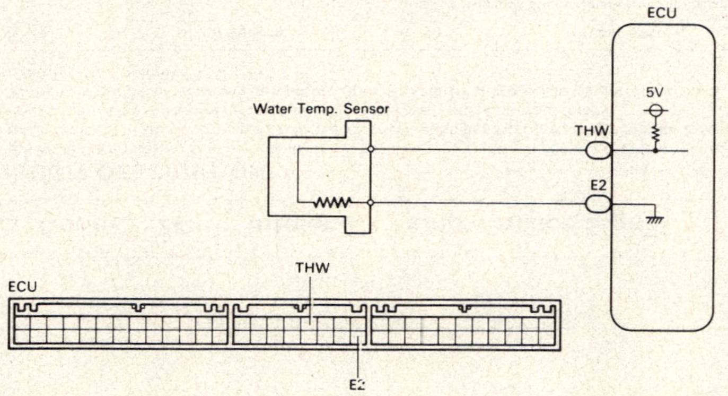

INSPECTION PROCEDURE

1 Check water temp. sensor.

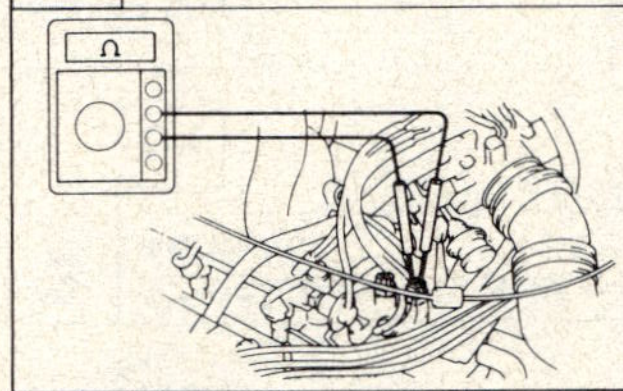

P: Disconnect water temp. sensor connector.

C: Using an ohmmeter, measure resistance between both terminals.

OK:

Water temp. °C (°F)	Resistance
20 (68)	2 – 3 kΩ
80 (176)	0.2 – 0.4 kΩ

NG → **Replace water temp. sensor.**

OK ↓

2 Check voltage between terminals THW and E2 of ECU connector.

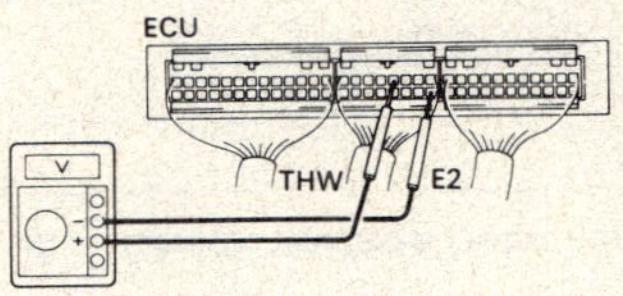

P: (1) Connect water temp. sensor connector.
(2) Turn ignition switch ON.

C: Measure voltage between terminals THW and E2 of ECU connector.

OK: Voltage: Neither 0V nor 5V

OK → **Replace ECU.** *

NG ↓

3 Check harness and connector between ECU and water temp. sensor.

NG → **Repair or replace harness or connector.**

OK ↓

Replace ECU.

* When diag. code 22 is not displayed, proceed to next circuit inspection shown on matrix chart

Diag. Code	24	Intake Air Temp. Sensor Signal

CIRCUIT DESCRIPTION

The intake air temperature sensor is built in the air flow meter and senses the intake air temperature. The resistance value of a thermistor built in sensor changes according to the intake air temperature. The lower the intake air temperature, the larger the resistance value becomes.
If diagnostic code 24 is recorded in memory, the fair safe function operates and fixes the intake air temperature at a constant 20°C (68°F).

Code No.	Diagnosis	Trouble Area
24	Open or short in intake air temp. sensor circuit for 0.5 sec. or more (THA).	● Open or short in intake air temp. circuit. ● Intake air temp. sensor ● ECU

DIAGNOSTIC CHART

1 Check intake air temp. sensor. — NG → Replace air flow meter.
↓ OK
2 Check voltage between terminals THA and E2 of ECU connector. — OK → Replace ECU. *
↓ NG
3 Check harness and connector between ECU and intake air temp. sensor. — NG → Repair or replace harness or connector.
↓ OK
Replace ECU.

* When diag. code 24 is not displayed, proceed to next circuit inspection shown on matrix chart

WIRING DIAGRAM AND ECU CONNECTORS

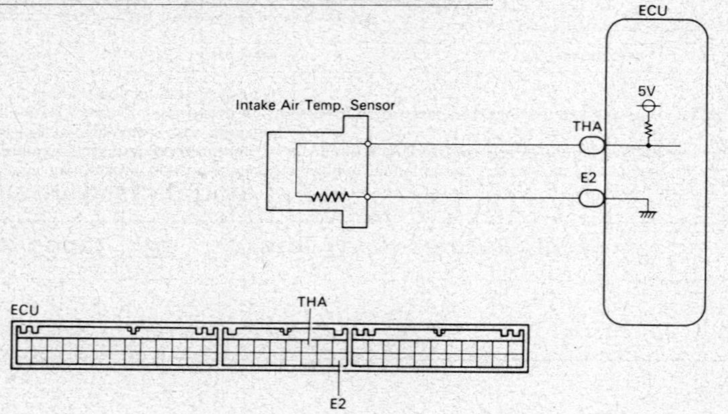

INSPECTION PROCEDURE

1 Check intake air temp. sensor.

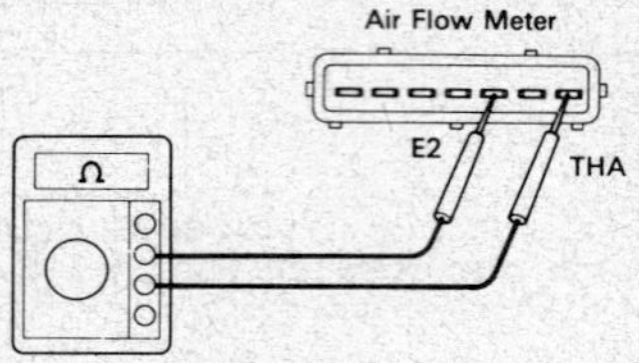

P Disconnect air flow meter connector.

C Using an ohmmeter, measure resistance between terminals THA and E2.

OK

Air temp. °C (°F)	Resistance
0 (32)	4 – 7 kΩ
20 (68)	2 – 3 kΩ

NG → **Replace air flow meter.**

OK ↓

2 Check voltage between terminals THA and E2 of ECU connector.

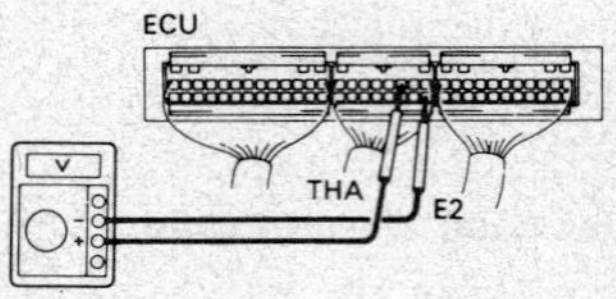

P (1) Connect air flow meter connector.
(2) Turn ignition switch ON.

C Measure voltage between terminals THA and E2 of ECU connector.

OK Voltage: Neither 0V nor 5V

OK → **Replace ECU.** *

NG ↓

3 Check harness and connector between ECU and intake air temp. sensor.

NG → **Repair or replace harness or connector.**

OK ↓

Replace ECU.

* When diagnostic code 24 is not displayed, proceed to next circuit inspection shown on matrix chart

Diag. Code	25	Air-fuel Ratio Lean Malfunction
	26	Air-fuel Ratio Rich Malfunction

CIRCUIT DESCRIPTION

The main oxygen sensor is located in the exhaust manifold. It indirectly determines whether the fuel mixture is rich or lean by detecting the concentration of oxygen present in the exhaust gas.

Code No.	Diagnosis	Trouble Area
25	(1) Oxygen sensor output is less than 0.45 V for at least 90 sec. when oxygen sensor is warmed up (racing at 2,000 rpm). *1 (2) When air-fuel ratio feed-back correction value or adaptive control value continues at the upper (lean) or lower (rich) limit for a certain period of time or adaptive control value is not renewed for a certain period of time.	● Engine ground bolt loose ● Open in E1 circuit ● Open in injector circuit ● Fuel line pressure (Injector blockage, etc.) ● Open or short in oxygen sensor circuit ● Oxygen sensor ● Ignition system ● Water temp. sensor ● Air flow meter (air intake) ● Foreign object caught in valve ● ECU
26	*1 (3) When marked variation is detected in engine revolutions for each cylinder during idle switch on and feedback condition. (2 trip detection logic)	● Engine ground bolt loose ● Open in E1 circuit ● Short in injector circuit ● Fuel line pressure (Injector leakage, etc.) ● Open or short in cold start injector circuit ● Cold start injector ● Open or short in oxygen sensor circuit ● Oxygen sensor ● Water temp. sensor ● Air flow meter ● Compression pressure ● Foreign object caught in valve ● ECU

*1: No. (2) and (3) in the diagnostic contents of codes No. 25 and 26 apply to California specification vehicles only, while (1) applies to all models.

CIRCUIT DESCRIPTION (Cont'd)

DIAGNOSIS CODE DETECTION DRIVING PATTERN

Purpose of the driving patten.

(a) After the diag. code is erased, the malfunction is reproduced and the diag. code is recorded again.

(b) To check that the malfunction is cleared once the repair is completed (confirm that diag. code is no longer detected).

Malfunction: Open or Short in Main Oxygen Sensor

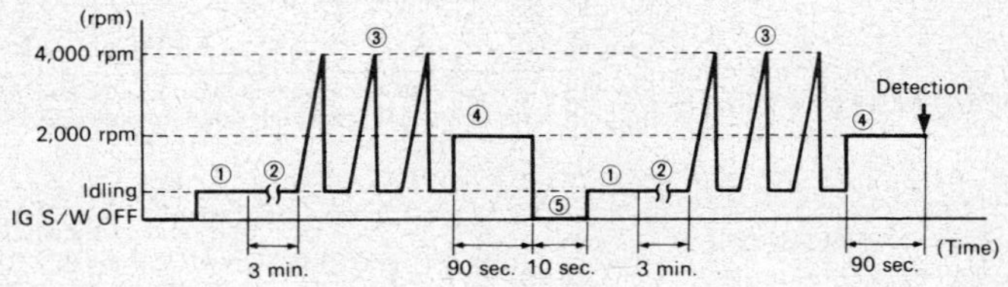

① Start engine and warm engine up.
② After engine is warmed up, let it idle for 3 min.
③ Perform sudden racing to 4,000 rpm three times. (Rapidly depress the accelerator pedal, then remove your foot from the pedal at 4,000 rpm.)
④ After performing the racing in ③, perform racing at 2,000 rpm for 90 sec.
⑤ Keep the IG switch at OFF (not Acc) for at least 10 sec., then perform ①, ②, ③, ④ again.

NOTICE: **If the conditions in this test are not strictly observed, detection of the malfunction will not be possible.**

DIAGNOSTIC CHART

(For vehicles other than for California)

1 Check voltage between terminals VF1 and E1 of check connector.
- OK → go to step 3
- NG → go to step 2

2 Check harness or connector between ECU and main oxygen sensor, ECU and check connector.
- NG → Repair or replace harness or connector.
- OK → go to step 3

3 Check each item found to be a possible cause of problem.
- NG → Repair or replace.
- OK → go to step 4

4 Does malfunction disappear when a known good main oxygen sensor is installed?
- YES → Replace main oxygen sensor.
- NO → Replace ECU.

DIAGNOSTIC CHART

(California specification vehicles)

1 Check voltage between terminals VF1 and E1 of check connector.
- OK →
  - When code 25 is displayed, go to step 3.
  - When code 26 is displayed, go to step 5.
  - When codes 25 and 26 are displayed, go to step 3.
- NG →
  - When voltage stays at 5 V, go to step 5.
  - When needle fluctuates less than 8 times for every 10 seconds, replace main oxygen sensor.
- NG (Voltage stays at 0 V.) → go to step 2

2 Check harness or connector between ECU and main oxygen sensor, ECU and check connector.
- NG → Repair or replace harness or connector.
- OK → go to step 3

3 Check each item found to be a possible cause of problem.
- NG → Repair or replace.
- OK → go to step 4

4 Does malfunction disappear when a known good main oxygen sensor is installed?
- YES → Replace main oxygen sensor.
- NO → Replace ECU.

5 Check each item found to be a possible cause of problem.
- NG → Repair or replace.
- OK → Replace ECU.

DIAGNOSTIC CHART
WIRING DIAGRAM AND ECU CONNECTORS

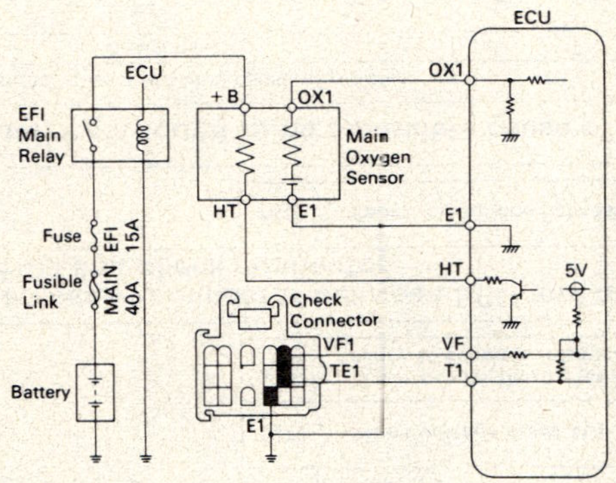

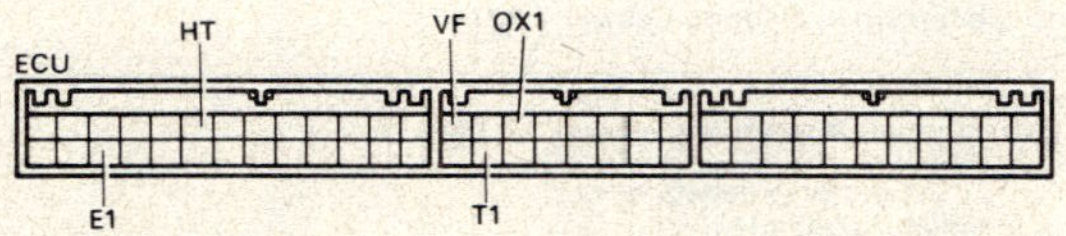

INSPECTION PROCEDURE
(Vehicles other than for California)

1 Check voltage between terminals VF1 and E1 of check connector.

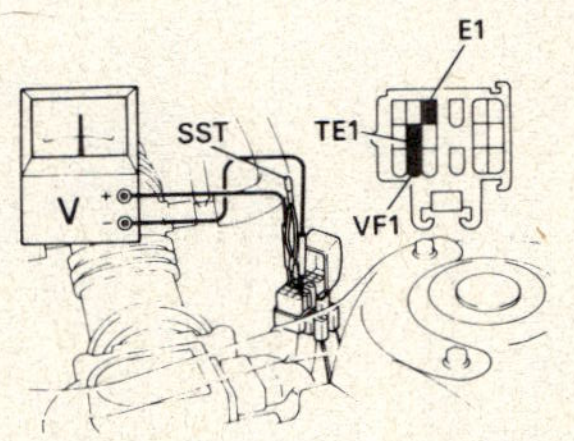

P Warm up engine at normal operating temperature.

C (1) Using SST, connect terminals TE1 and E1 of check connector.
SST 09843-18020

(2) Using an analog voltmeter, connect positive prove to VF1 terminal and negative prove to E1 terminal of check connector, and set it to 5 – 20 V range.

(3) Warm up oxygen sensor by running engine at 2,500 rpm for about 2 minutes.

(4) Then, maintaining engine at 2,500 rpm, count how many times needle of voltmeter fluctuates between 0 and 5 V.

OK Needle fluctuates a minimum of 8 times for every 10 seconds.

NG ↓ **OK** Go to step **3**.

2 Check harness or connector between ECU and main oxygen sensor, ECU and check connector.

OK ↓ **NG** Repair or replace harness or connector.

3 Check each item found to be a possible cause of problem.

Check each item found to be a possible cause of problem.

Check item
Check air leakage.
Check installation of main oxygen sensor.
Check ignition system (misfire).
Check fuel system.
Check water temp. sensor (characteristics deviation).
Check air flow meter (characteristics deviation).
Check intake air temp. sensor (characteristics deviation).
Check valve timing.

OK ↓ **NG** Repair or replace.

4 Does malfunction disappear when a known good main oxygen sensor installed?

NO ↓ **YES** Replace main oxygen sensor.

Replace ECU.

(California specification vehicles)

1 Check voltage between terminals VF1 and E1 of check connector.

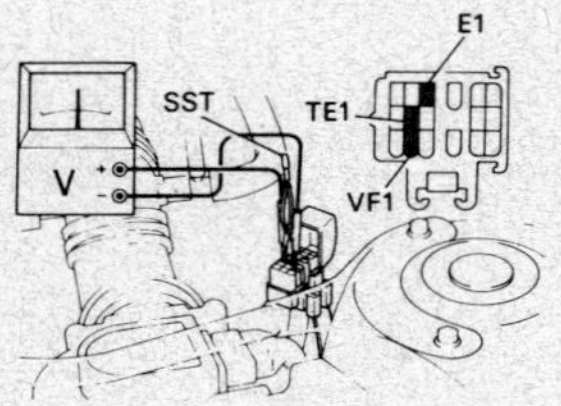

P Warm up engine at normal operating temperature.

C
(1) Using SST, connect terminals TE1 and E1 of check connector.
SST 09843-18020
(2) Using an analog voltmeter, connect positive prove to VF1 terminal and negative prove to E1 terminal of check connector, and set it to 5 – 20 V range.
(3) Warm up oxygen sensor by running engine at 2,500 rpm for about 2 minutes.
(4) Then, maintaining engine at 2,500 rpm, count how many times needle of voltmeter fluctuates between 0 and 5 V.

OK Needle fluctuates a minimum of 8 times for every 10 seconds.

OK When code 25 is displayed, go to step **3**.

OK When code 26 is displayed, go to step **5**.

OK When codes 25 and 26 are displayed, go to step **3**.

NG When voltage stays at 5 V, go to step **5**.

NG When needle fluctuates less than 8 times for every 10 seconds, replace main oxygen sensor.

NG (Voltage stays at 0 V.)

2 Check harness or connector between ECU and main oxygen sensor, ECU and check connector.

OK

NG Repair or replace harness or connector.

3 Check each item found to be a possible cause of problem.

Check each item found to be a possible cause of problem.

Check item
Check air leakage.
Check installation of main oxygen sensor.
Check ignition system (misfire).
Check fuel system.
Check water temp. sensor (characteristics deviation).
Check air flow meter (characteristics deviation).
Check intake air temp. sensor (characteristics deviation).
Check valve timing.

OK

NG Repair or replace.

4 Does malfunction disappear when a known good main oxygen sensor is installed?

YES Replace main oxygen sensor.

NO

Replace ECU.

5 Check each item found to be a possible cause of problem.

Check each item found to be a possible cause of problem.

Check item	
Check fuel system.	
Check water temp. sensor (characteristics deviation).	
Check air flow meter (characteristics deviation).	
Check intake air temp. sensor (characteristics deviation).	

NG Repair or replace.

OK

Replace ECU.

Diag. Code 27 Sub-oxygen Sensor Signal (California Only)

CIRCUIT DESCRIPTION

The sub-oxygen sensor is located in the exhaust pipe behind the TWC. It rechecks the exhaust emission level after it is purified by the TWC and feeds the result back to the ECU.

Code No.	Diagnosis	Trouble Area
27	When sub-oxygen sensor is warmed up and apply acceleration with the throttle fully open for 2 seconds, output of main oxygen sensor is 0.45 V or more (rich) and output of sub-oxygen sensor is 0.45 V or less (lean). (OX2) (2 trip detection logic)	• Short or open in sub-oxygen sensor circuit. • Sub-oxygen sensor • ECU

DIAGNOSIS CODE DETECTION DRIVING PATTERN

Purpose of the driving patten.

(a) After the diag. code is erased, the malfunction is reproduced and the diag. code is recorded again.

(b) To check that the malfunction is cleared once the repair is completed (confirm that diag. code is no longer detected).

Malfunction: Open or Short in Sub-Oxygen Sensor

For A/T Vehicle

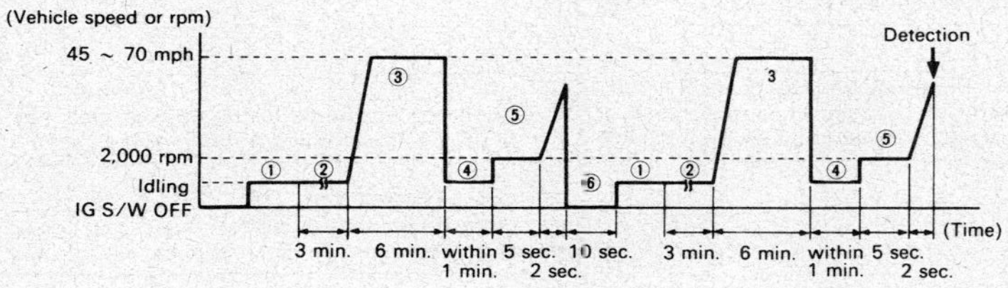

① Start engine and warm engine up.
② After engine is warmed up, let it idle for 3 min.
③ With the A/C ON and the transmission in D range (O/D OFF), drive at 45-70 mph for 6 min.
④ After the driving in ③, stop the vehicle.
⑤ Within 1 min. of stopping the vehicle, shift into 2 range and gradually accelerate for 5 sec. while gradually increasing the engine speed approx. 2,000 rpm. Then, without releasing the accelerator pedal at all accelerate with throttle wide open for 2 sec.
⑥ Stop the vehicle, then 10 sec. or more after turning IG switch to OFF (not Acc), perform ①, ②, ③, ④, ⑤ again.

CIRCUIT DESCRIPTION (Cont'd)

DIAGNOSIS CODE DETECTION DRIVING PATTERN

For M/T Vehicle

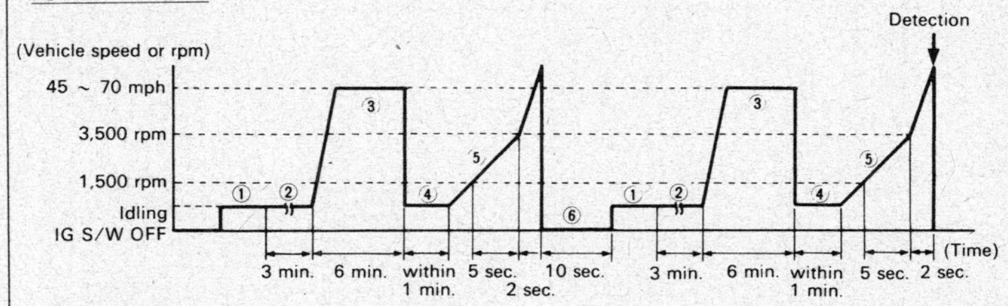

① Start engine and warm engine up.
② After engine is warmed up, let it idle for 3 min.
③ With the A/C ON and the transmission in 4th gear, drive at 45-70 mph for 6 min.
④ After the driving in ③, stop the vehicle.
⑤ Within 1 min. of stopping the vehicle, shift into 2nd gear and gradually accelerate for 5 sec. while gradually increasing the engine speed from 1,500 rpm up to 3,500 rpm. Then, without releasing the accelerator pedal at all, accelerate with throttle wide open for 2 sec.

NOTICE: **If the conditions in this test are not strictly observed, detection of the malfunction will not be possible.**

DIAGNOSTIC CHART

Step	Check	Result	Action
1	Check connection of sub-oxygen sensor connector.	NG	Repair.
		OK	Go to 2
2	Check continuity between terminals CHK and GND of sub-oxygen sensor connector.	NG	Replace sub-oxygen sensor.
		OK	Go to 3
3	Check voltage between terminals CHK and GND of sub-oxygen sensor connector.	OK	Go to 6
		NG	Go to 4
4	Check voltage between terminal CHK of sub-oxygen sensor connector and body ground.	OK	Check and repair harness and connector between sub-oxygen sensor and body ground.
		NG	Go to 5
5	Check voltage between terminal CHK of ECU connector and body ground.	OK	Check and repair harness and connector between ECU and sub-oxygen sensor.
		NG	Replace ECU.
6	Check air leakage.	NG	Repair air leakage.
		OK	Go to 7
7	Check voltage between terminals OX2 and E2 of ECU connector.	OK	Replace ECU.
		NG	Go to 8
8	Check harness or connector between ECU, sub-oxygen sensor and body ground.	NG	Repair or replace harness or connector.
		OK	Go to 9
9	Check fuel system	NG	Repair fuel system.
		OK	Replace sub-oxygen sensor.

DIAGNOSTIC CHART
WIRING DIAGRAM AND ECU CONNECTORS

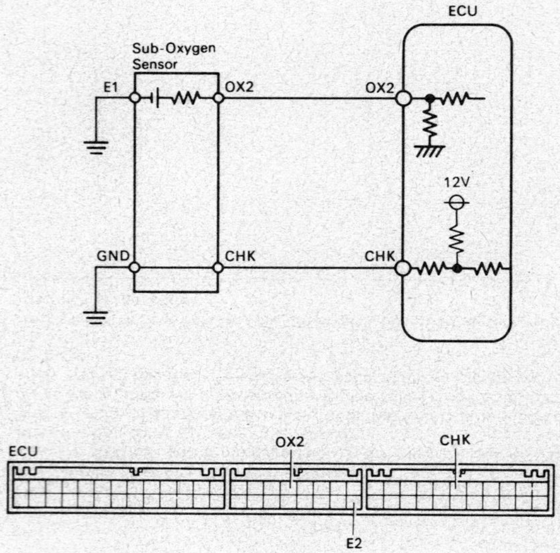

INSPECTION PROCEDURE

HINT: If other codes are output at the same time code No. 27 is output, check the other codes first.

1 Check connection of sub-oxygen sensor connector.

OK

NG ⟩ **Repair.**

2 Check continuity between terminals CHK and GND of sub-oxygen sensor connector.

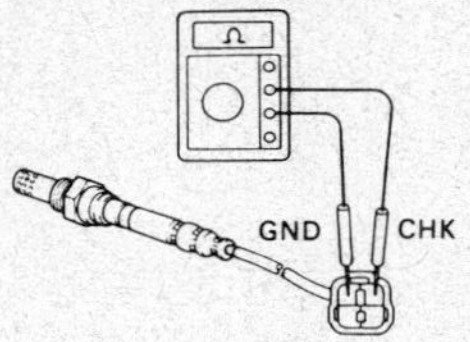

P Disconnect sub-oxygen sensor connector.

C Check continuity between terminals CHK and GND of sub-oxygen sensor connector.

OK Continuity

OK

NG ⟩ **Replace sub-oxygen sensor.**

3 Check voltage between terminals CHK and GND of sub-oxygen sensor connector.

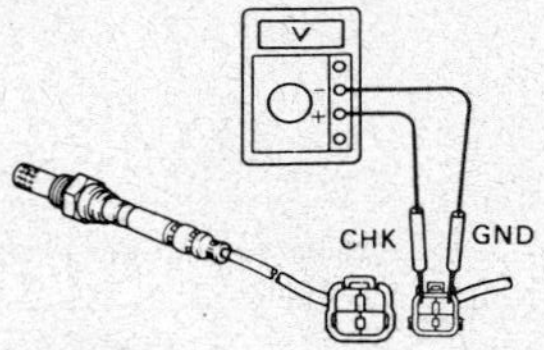

P Turn ignition switch ON.

C Measure voltage between terminals CHK and GND of sub-oxygen sensor connector.

OK Voltage: 12V

NG

OK ⟩ **Go to step 6.**

4 Check voltage between terminal CHK of sub-oxygen sensor connector and body ground.

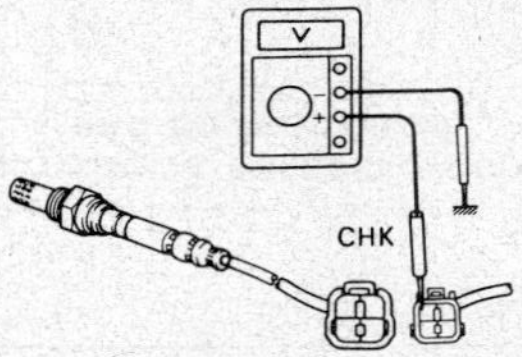

C Measure voltage between terminal CHK of sub-oxygen sensor connector and body ground.

OK Voltage: 12V

NG

OK ⟩ **Check and repair harness and connector between sub-oxygen sensor and body ground.**

5 Check voltage between terminal CHK of ECU connector and body ground.

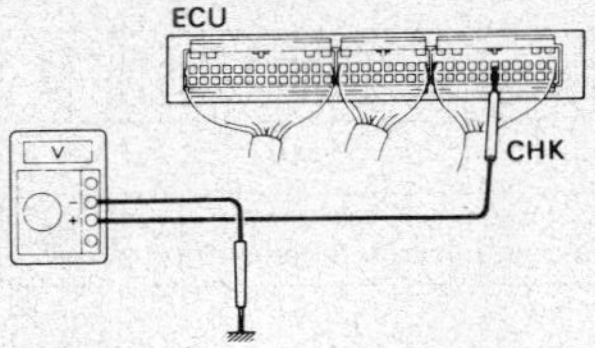

P (1) Disconnect sub-oxygen sensor connector.
(2) Turn ignition switch ON.

C Measure voltage between terminal CHK of ECU connector and body ground.

OK Voltage: 12V

OK ⟩ **Check and repair harness and connector between ECU and sub-oxygen sensor.**

NG

Replace ECU.

6 Check air leakage.

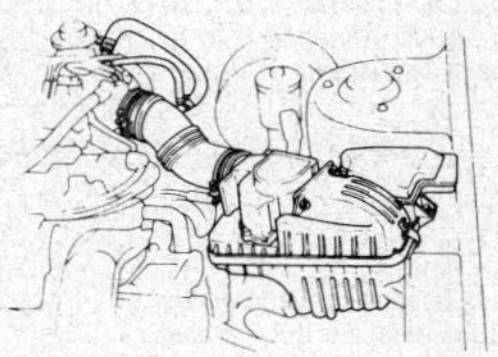

C Check air leakage.

Hint Loose or missing engine oil dipstick, oil filler cap, PCV hose, etc., can allow air to leak into air intake system, causing mixture to become too lean.

OK No air leaking into air intake system between air flow meter and cylinder head.

OK

NG ⟩ **Repair air leakage.**

7 Check voltage between terminals OX2 and E2 of ECU connector.

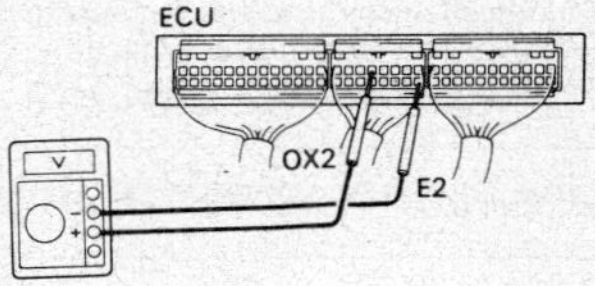

P (1) Warm up engine at normal operating temperature.
(2) Engine continue racing at 4,000 rpm to approx. 3 minutes.

C Measure voltage between terminals OX2 and E2 of ECU connector when engine is suddenly raced to full throttle.

OK Voltage should be 0.5 V or higher at least once.

NG

OK ⟩ **Replace ECU.**

8 Check harness or connector between ECU, sub-oxygen sensor and body ground.

- OK
- NG → Repair or replace harness or connector.

9 Check fuel system

- OK → Replace sub-oxygen sensor.
- NG → Repair fuel system.

Diag. Code 31 or 32 Air Flow Meter Signal

CIRCUIT DESCRIPTION

The air flow meter is located directly behind the air cleaner, and detects the intake air volume. It converts the intake air readings into a voltage signal by means of a potentiometer. Intake air volume signal is used to calculate the basic injection duration and basic ignition advance angle.

Code No.	Diagnosis	Trouble Area
31	At Idling, open or short detected continuously for 0.5 sec. or more in air flow meter circuit.	● Open (VC) or short (VC-E2) in air flow meter circuit ● Air flow meter ● ECU
32	Open or short detected continuously for 0.5 sec. or more in air flow meter circuit.	● Open (E2) or short (VS-VC) in air flow meter circuit ● Air flow meter ● ECU

DIAGNOSTIC CHART

1 Check voltage between terminals VC and E2 of air flow meter connector.
- OK → go to 2
- NG → go to 4

2 Check voltage between terminals VS and E2 of air flow meter connector.
- OK → go to 3
- NG → Go to step 7

3 Check voltage between terminals VS and E2 of ECU connector.
- OK → Replace ECU. *
- NG → Check for open circuit between ECU and air flow meter.

* When diag code. 31 or 32 is not displayed, proceed to next circuit inspection shown on matrix chart

4 Check voltage between terminals VC and E2 of air flow meter connector.
- OK → Replace air flow meter.
- NG → go to 5

5 Check voltage between terminals VC and E2 of ECU connector.
- OK → Check for open circuit between ECU and air flow meter.
- NG → go to 6

6 Check VC circuit for short circuit between ECU and air flow meter.
- OK → Replace ECU.
- NG → Repair or replace harness or connector.

Go to step 7

DIAGNOSTIC CHART (Cont'd)

From step **2**

7 Check VS circuit for short circuit between ECU and air flow meter.

- OK → Replace ECU.
- NG → Repair or replace harness or connector.

WIRING DIAGRAM AND ECU CONNECTORS

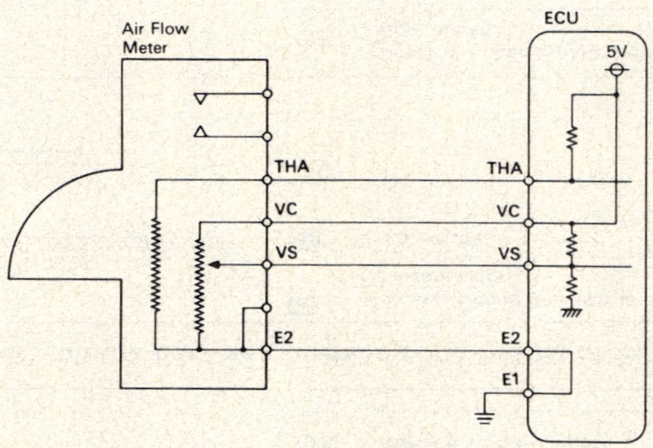

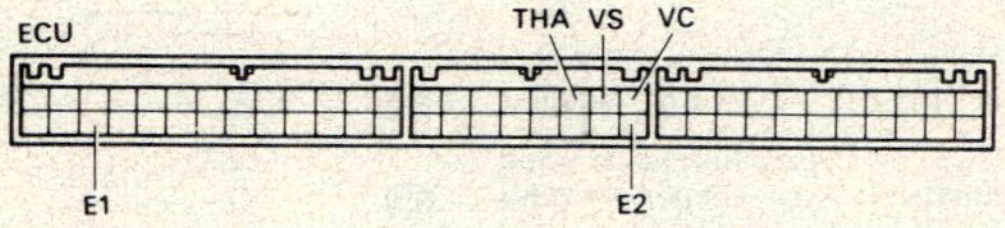

INSPECTION PROCEDURE

1 Check voltage between terminals VC and E2 of air flow meter connector.

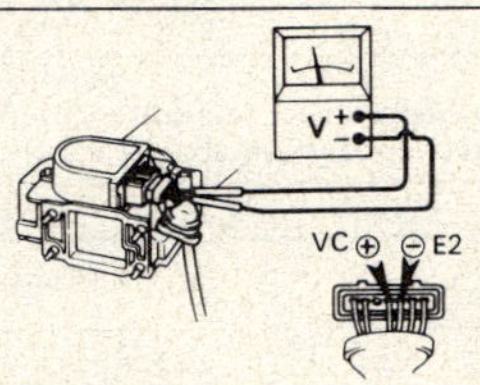

P (1) Carefully remove water-proofing rubber cover of air flow meter connector.
(2) Turn ignition switch ON.

C Measure voltage between terminals VC and E2 of air flow meter connector.

OK Voltage: 5V

OK ↓

NG Go to step **4**.

2 Check voltage between terminals VS and E2 of air flow meter connector.

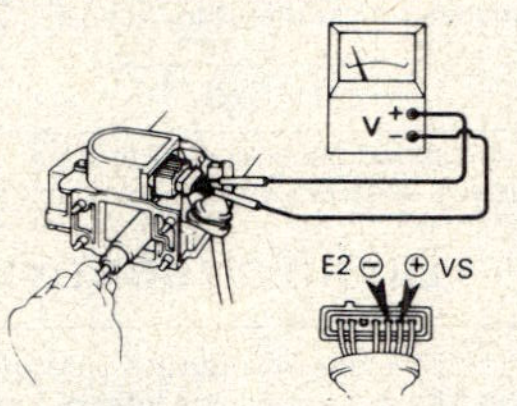

P (1) Turn ignition switch LOCK.
(2) Remove the air cleaner case.
(3) Turn ignition switch ON.

C Check change in voltage between terminals VS and E2 of air flow meter connector while slowly pushing air flow meter measuring plate from closed to open position with a screwdriver or similar object.

OK Voltage changes smoothly from about 5V to about 0V; i.e., does not suddenly jump up to 5V or down to 0V.

OK ↓

NG Go to step **7**.

3 Check voltage between terminals VS and E2 of ECU connector.

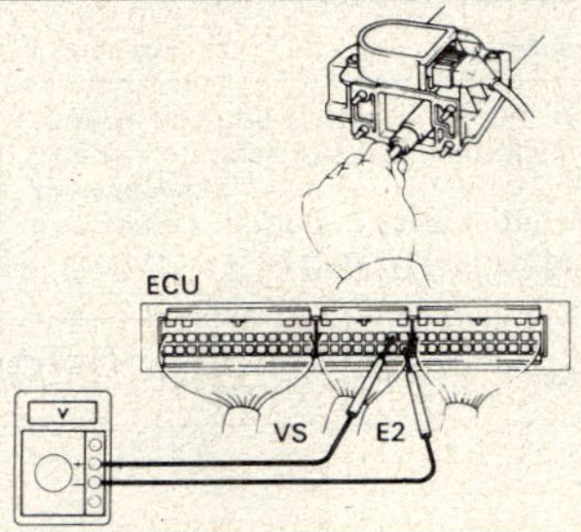

C Check change in voltage between terminals VS and E2 of ECU connector while slowly pushing air flow meter measuring plate from closed to open position with a screwdriver or similar object.

OK Voltage changes smoothly from about 5V (when closed) to about 0V (when open); i.e., does not suddenly jump up to 5V or down to 0V.

OK → Replace ECU.*

NG Check for open circuit between ECU and air flow meter.

* When diag. code 31 or 32 is not displayed, proceed to next circuit inspection shown on matrix chart

4 Check voltage between terminals VC and E2 of air flow meter connector.

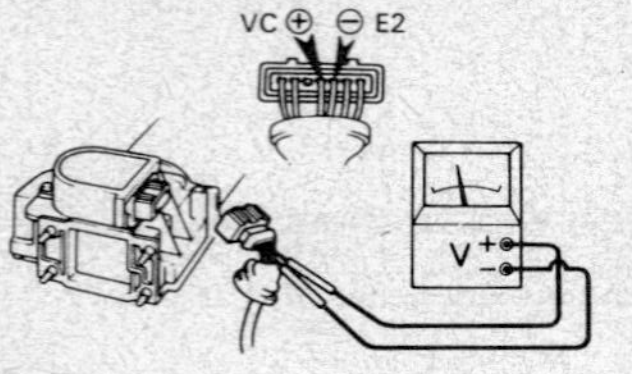

P (1) Turn ignition switch LOCK.
(2) Disconnect air flow meter connector.
(3) Turn ignition switch ON.

C Measure voltage between terminals VC and E2 of air flow meter connector.

OK Voltage: 5V

OK → **Replace air flow meter.**

NG ↓

5 Check voltage between terminals VC and E2 of ECU connector.

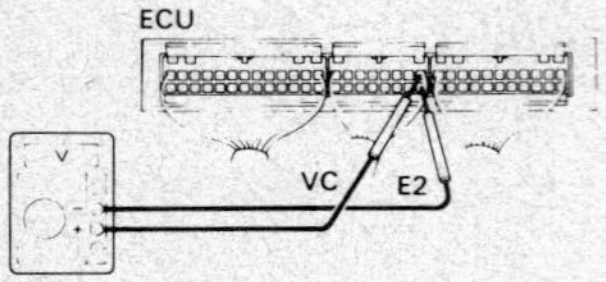

C Measure voltage between terminals VC and E2 of ECU connector.

OK Voltage: 5V

OK → **Check for open circuit between ECU and air flow meter.**

NG ↓

6 Check VC circuit for short circuit between ECU and air flow meter.

OK → **Replace ECU.**

NG ↓

Repair or replace harness or connector.

7 Check VS circuit for short circuit between ECU and air flow meter.

OK → **Replace ECU.**

NG ↓

Repair or replace harness or connector.

Diag. Code 41 Throttle Position Sensor Signal

CIRCUIT DESCRIPTION

The throttle position sensor is provided to the throttle body and detects the opening angle of the throttle valve.
The ECU judges the vehicle driving conditions from the signals input from the terminal VTA of the throttle position sensor.

Code No.	Diagnosis	Trouble Area
41	Open or short detected in throttle position sensor signal (VTA) for 0.5 sec. or more.	• Open or short in throttle position sensor circuit.
	IDL contact is ON and VTA output exceeds 1.45 V.	• Throttle position sensor • ECU

HINT: When the connector for the throttle position sensor is disconnected, diagnostic code 41 is not output. Diagnostic code 41 is output only when there is an open or short in the VTA signal circuit of the throttle position sensor.

DIAGNOSTIC CHART

1 Check voltage between terminals VC and E2 of throttle position sensor connector.
- NG → go to 4
- OK ↓

2 Check throttle position sensor.
- NG → Adjust or replace throttle position sensor.
- OK ↓

3 Check voltage between terminals VTA and E2 of ECU connector.
- NG → Check harness and connector between ECU and throttle position sensor.
- OK ↓

Replace ECU. *

* When diag. code 41 is not displayed, proceed to next circuit inspection shown on matrix chart

4 Check voltage between terminal VC of throttle position sensor connector and body ground.
- OK → Check harness and connector between ECU and throttle position sensor.
- NG ↓

5 Check voltage between terminals VC and E2 of ECU connector.
- OK → Check for open circuit between ECU and throttle position sensor.
- NG ↓

6 Check for short circuit between ECU and throttle position sensor.
- OK → Replace ECU.
- NG ↓

Repair or replace harness or connector.

DIAGNOSTIC CHART

WIRING DIAGRAM AND ECU CONNECTORS

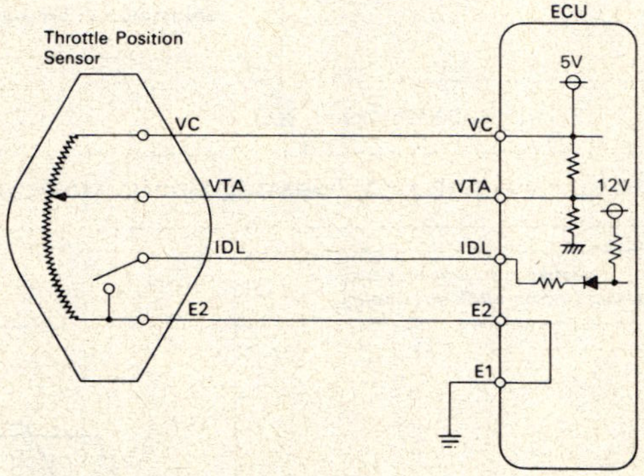

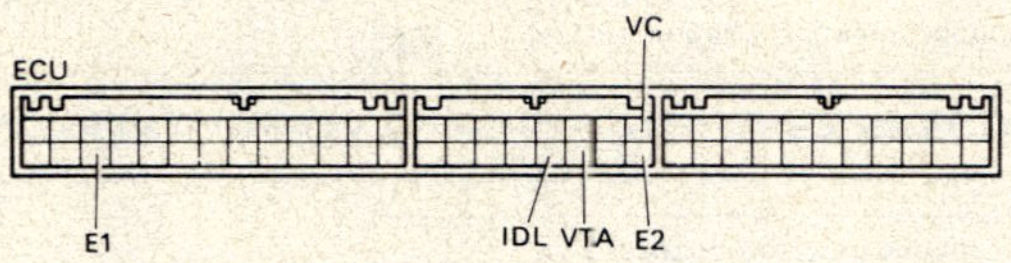

INSPECTION PROCEDURE

1 Check voltage between terminals VC and E2 of throttle position sensor connector.

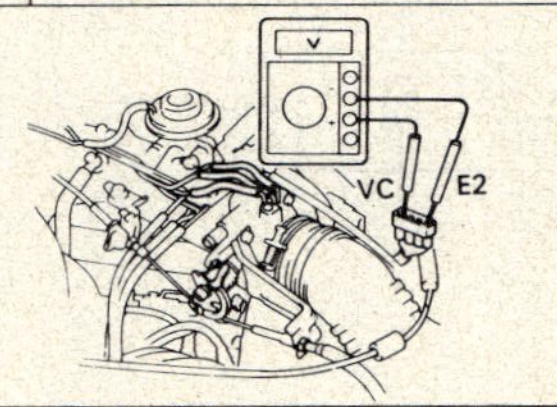

P (1) Disconnect throttle position sensor connector.
(2) Turn ignition switch ON.

C Measure voltage between terminals VC and E2 of throttle position sensor connector.

OK Voltage: 5V

OK ↓

NG → **Go to step 4.**

2 Check throttle position sensor.

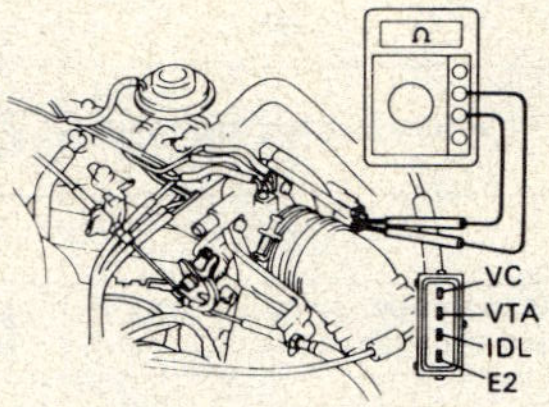

P Turn ignition switch LOCK.

C Measure resistance between each terminal shown below:

OK

Terminals		Resistance
VC – E2		4.25 – 8.25 kΩ
VTA – E2	Throttle valve fully closed	0.3 – 6.3 kΩ
	Throttle valve full open	3.5 – 10.3 kΩ

- Resistance changes smoothly from 0.3 – 6.3 kΩ (when closed) to 3.5 – 10.3 kΩ (when open).

OK ↓

NG → **Adjust or replace throttle position sensor.**

3 Check voltage between terminals VTA and E2 of ECU connector.

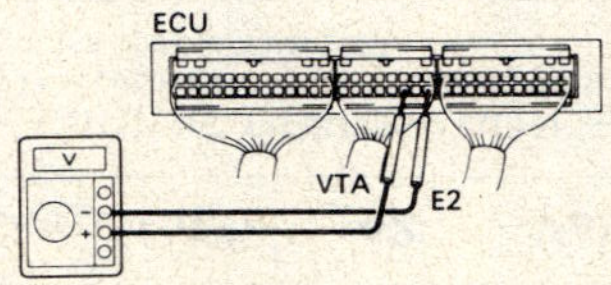

P (1) Connect throttle position sensor connector.
(2) Turn ignition switch ON.

C Check change in voltage between terminals VTA and E2 of ECU connector while slowly opening throttle valve from closed to open position.

OK Voltage changes smoothly from 0.1V (when closed) to about 5V (when open); i.e., it does not suddenly jump up to 5V or down to 0.1V.

OK ↓

NG → **Check harness and connector between ECU and throttle position sensor.**

Replace ECU. *

* When diag. code 41 is not displayed, proceed to next circuit inspection shown on matrix chart

4 Check voltage between terminal VC of throttle position sensor connector and body ground.

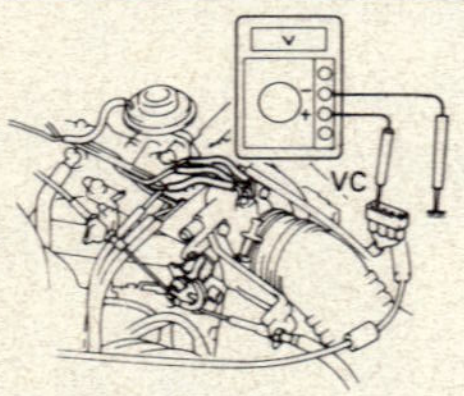

C: Measure voltage between terminal VC of throttle position sensor connector and body ground.

OK: Voltage: 5V

OK → Check harness and connector between ECU and throttle position sensor.

NG ↓

5 Check voltage between terminals VC and E2 of ECU connector.

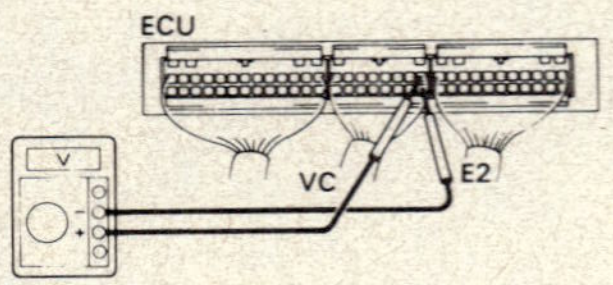

C: Measure voltage between terminals VC and E2 of ECU connector.

OK: Voltage: 5V

OK → Check for open circuit between ECU and throttle position sensor.

NG ↓

6 Check for short circuit between ECU and throttle position sensor.

OK → Replace ECU.

NG ↓

Repair or replace harness or connector.

Diag. Code	42	Vehicle Speed Sensor Signal

CIRCUIT DESCRIPTION

The speed sensor is located inside the speedometer, and outputs four-pulse signals for each revolution of the speedometer cable.

Code No.	Diagnosis	Trouble Area
42	SPD signal is not input to the ECU for at least 8 seconds during high load driving with engine speed between 2,500 rpm and 4,500 rpm.	• Open or short in vehicle speed sensor circuit • Vehicle speed sensor • ECU

DIAGNOSTIC CHART

1 Check continuity between terminal SP1 of ECU connector and body ground. — OK → Replace ECU.

NG ↓

2 Check No. 1 speed sensor. — OK → Check harness and connector between ECU and speed sensor in speedometer.

NG ↓

Repair or replace No. 1 speed sensor.

WIRING DIAGRAM AND ECU CONNECTORS

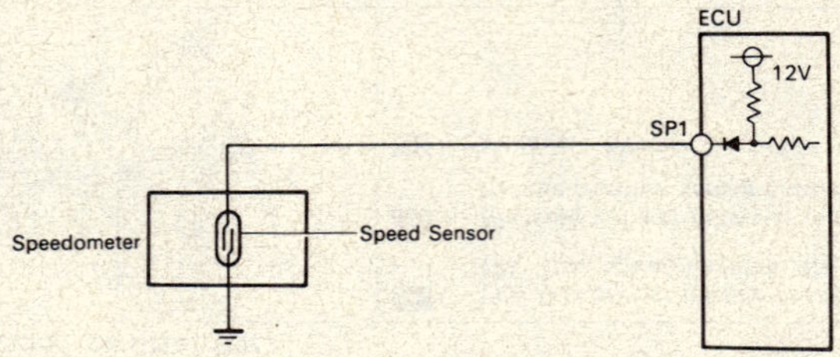

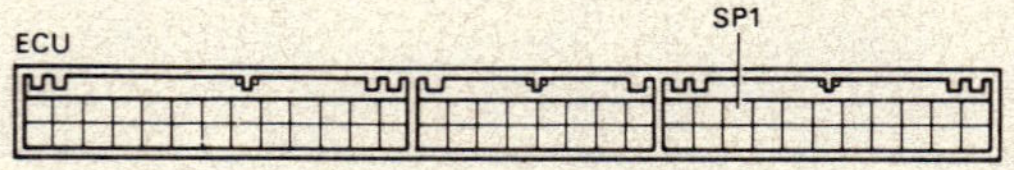

Diag. Code 43 Starter Signal

CIRCUIT DESCRIPTION

When the engine is cranked, the intake air flow is slow, so fuel vaporization is poor. A rich-mixture is therefore necessary in order to achieve good startability. While the engine is being cranked, the battery voltage is applied to terminal STA of the ECU. The STA signal is mainly used to increase the fuel injection volume for the starting injection control and after-start injection control.

Code No.	Diagnosis	Trouble Area
43	Starter signal (STA) is not input to ECU even once until engine reaches 800 rpm or more when cranking.	● Open or short in starter signal circuit. ● Open or short in IG switch circuit ● ECU

DIAGNOSTIC CHART

1	Check starter operation.	NG →	Check starter circuit.
	↓ OK		
2	Check voltage between terminals STA and E1 of ECU connector.	NG →	Check harness and connector between ECU and terminal STA (50) of starter.
	↓ OK		
	Replace ECU. *		

* When diag. code 43 is not displayed, proceed to next circuit inspection shown on matrix chart

WIRING DIAGRAM AND ECU CONNECTORS

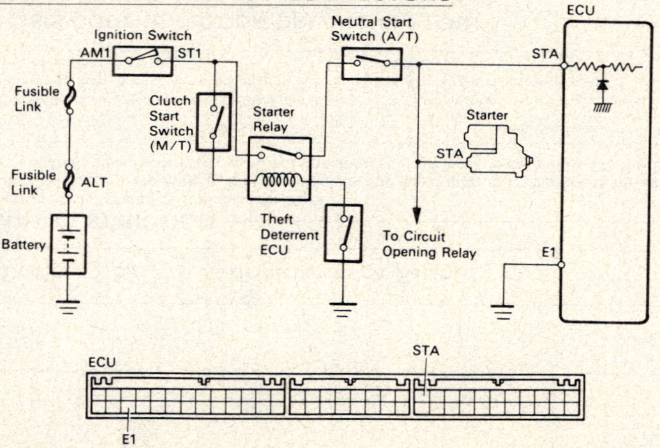

INSPECTION PROCEDURE

1 **Check continuity between terminal SP1 of ECU connector and body ground.**

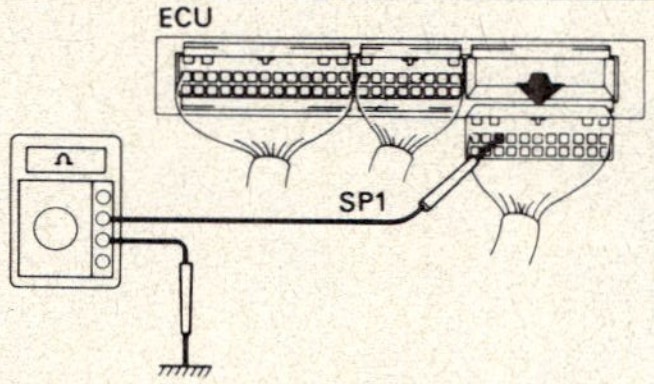

P (1) Disconnect ECU connector.
(2) Jack up a front wheel on one side.

C Check continuity between terminal SP1 of ECU connector and body ground while turning wheel slowly by hand.

OK Resistance changes from ∞Ω to 0Ω.

NG ↓ OK → **Replace ECU.**

2 **Check No. 1 speed sensor.**

NG ↓ OK → **Check harness and connector between ECU and speed sensor in speedometer.**

Repair or replace No. 1 speed sensor.

INSPECTION PROCEDURE

1 Check starter operation.

OK ↓ | NG → **Check starter circuit.**

2 Check voltage between terminals STA and E1 of ECU circuit.

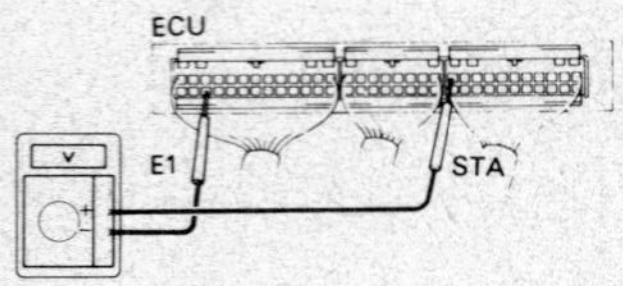

P Turn ignition switch STA.

C Measure voltage between terminals STA and E1 of ECU connector.

OK Voltage: 6 V or more

OK ↓ | NG → **Check harness and connector between ECU and terminal STA (50) of starter.**

Replace ECU. *

* When diag. code 43 is not displayed, proceed to next circuit inspection shown on matrix chart

Diag. Code	52	Knock Sensor Signal

CIRCUIT DESCRIPTION

The knock sensor is located on the cylinder block.
If knocking develops in the engine, the knock sensor, by resonating with the knocking vibration, generates a voltage which corresponds to the knocking strength and sends a signal to the ECU.

Code No.	Diagnosis	Trouble Area
52	With engine speed between 1,600 rpm-5,200 rpm, signal from knock sensor is not input to ECU for revolution. (KNK)	● Open or short in knock sensor circuit. ● Knock sensor (looseness etc.) ● ECU

DIAGNOSIS CODE DETECTION DRIVING PATTERN

Purpose of the driving patten.

(a) After the diag. code is erased, the malfunction is reproduced and the diag. code is recorded again.

(b) To check that the malfunction is cleared once the repair is completed (confirm that diag. code is no longer detected).

Malfunction: Open or Short in Knock Sensor

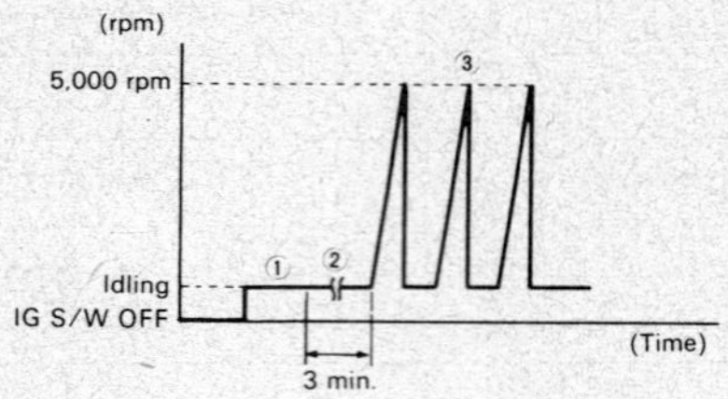

① Start engine and warm engine up.
② After engine is warmed up, let it idle for 3 min.
③ With the A/C ON, perform sudden racing (5,000 rpm) three times.
(Rapidly depress the accelerator pedal and suddenly release it.)

HINT: If a malfunction exists, the "CHECK" engine warning light will light up when sudden racing is performd.

NOTICE: If the conditions in this test are not strictly absorved, detection of the malfunction will not be possible.

DIAGNOSTIC CHART

1 Check continuity between terminal KNK of ECU connector and body ground.
- OK → **3**
- NG → **2**

2 Check knock sensor.
- NG → Replace knock sensor.
- OK → Check for short circuit between EUC and knock sensor.

3 Check for open circuit between ECU and knock sensor.
- NG → Repair or replace harness and connector.
- OK → **4**

4 Dose malfunction disappear when a known good knock sensor is installed?
- YES → Replace knock sensor.
- NO → Replace ECU. *

* When diag. code 52 is not displayed, proceed to next circuit inspection shown on matrix chart

WIRING DIAGRAM AND ECU CONNECTORS

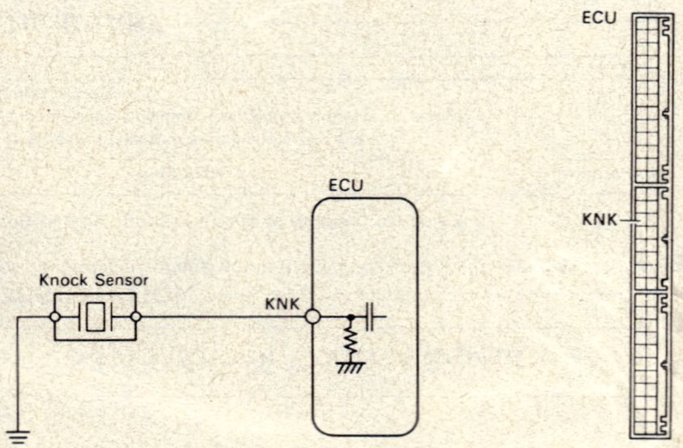

INSPECTION PROCEDURE

1 Check continuity between terminal KNK of ECU connector and body ground.

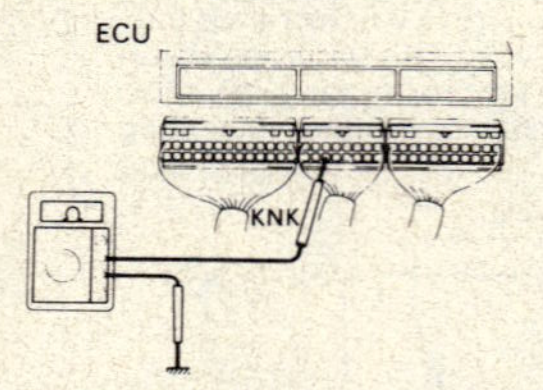

P Disconnect ECU connector.

C Check continuity between terminal KNK of ECU connector and body ground.

OK No continuity

NG | OK Go to step 3.

2 Check knock sensor.

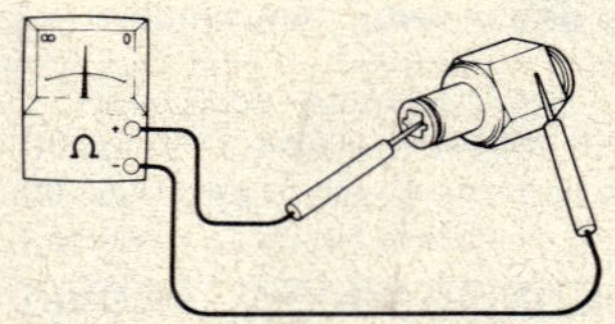

P Disconnect knock sensor connector.

C Check continuity between knock sensor terminal and body.

OK No continuity

OK | NG Replace knock sensor.

Check for short circuit between ECU and knock sensor.

3 Check for open circuit between ECU and knock sensor.

OK | NG Repair or replace harness or connector.

4 Does malfunction disappear when a known good knock sensor is installed?

NO | YES Replace knock sensor.

Replace ECU. *

* When diag. code 52 is not displayed, proceed to next circuit inspection shown on matrix chart

Diag. Code 53 Knock Control Signal

CIRCUIT DESCRIPTION

The knock sensor generates a voltage when the cylinder block vibrates due to knocking and sends a signal to the ECU.
If engine knocking occurs, ignition timing is retarded to reduce it.

Code No.	Diagnosis	Trouble Area
53	Engine speed is between 650 rpm and 5,200 rpm and engine control computer (for knock control) malfunction is detected.	● ECU

DIAGNOSTIC CHART

1 Are there any other codes (besides code 53) being output?

- YES → Go to relevant Diag. Code Chart.
- NO → Replace ECU.

Diag. Code 71 EGR System Malfunction (California Only)

CIRCUIT DESCRIPTION

To reduce NOx emission, part of the exhaust gases are recirculated through the EGR valve to the intake manifold to lower the maximum combustion temperature.

Code No.	Diagnosis	Trouble Area
71	With engine warmed up and EGR operation possible, code is displayed when EGR gas temp. sensor is 55°C (131°F) for M/T, 60 °C (140°F) for ECT or less when vehicle is driven for 90 seconds (THG). (2 trip detection logic)	● Open in EGR gas temp. sensor circuit ● EGR or BVSV vacuum hose disconnected, valve stuck ● Clogged EGR gas passage ● ECU

DIAGNOSIS CODE DETECTION DRIVING PATTERN

Purpose of the driving patten.

(a) After the diag. code is erased, the malfunction is reproduced and the diag. code is recorded again.

(b) To check that the malfunction is cleared once the repair is completed (confirm that diag. code is no longer detected).

Malfunction: Open in EGR Gas Temp. Sensor, Loose EGR or BVSV Vacuum Hose, Valve Stuck Clogged EGR Gas Passage

For A/T Vehicle

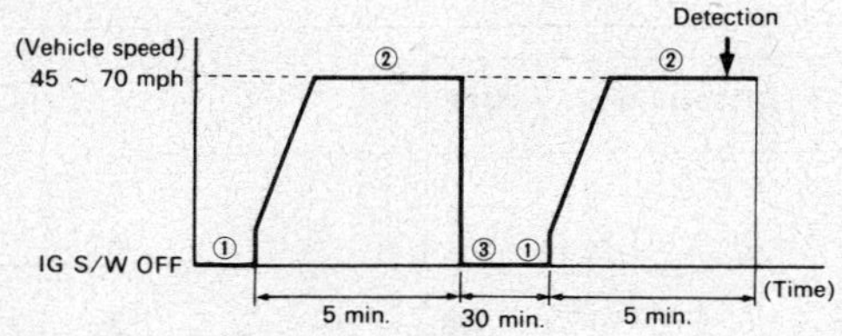

① Start engine.
② Soon after starting the engine (no warming up), drive for 5 min. at 45-70 mph in D range (O/D OFF) with the air conditioner ON.
(Keep the accelerator pedal position constant).
③ Turn the IG switch to OFF (not Acc) when you finish driving, then 30 min. later (to cool the engine) perform ①, ② again.

CIRCUIT DESCRIPTION (Cont'd)

DIAGNOSIS CODE DETECTION DRIVING PATTERN

For M/T Vehicle

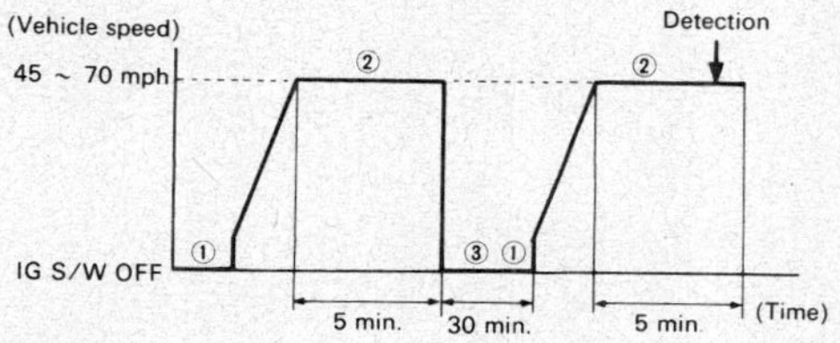

① Start engine.

② Soon after starting the engine (no warming up), drive for 5 min. at 45-70 mph in 4th gear with air conditioner ON. (Keep the accelerator pedal position constant).

③ Turn the IG switch to OFF (not Acc) when you finish driving, then 30 min. later (to cool the engine) perform ①, ② again.

NOTICE: **If the conditions in this test are not strictly observed, detection of the malfunction will not be possible.**

DIAGNOSTIC CHART

1 Check EGR system
- NG → Repair EGR system.
- OK ↓

2 Check voltage between terminals THG and E2 of EGR gas temp. sensor connector.
- NG → go to step 5
- OK ↓

3 Check EGR gas temp. sensor.
- NG → Replace EGR gas temp. sensor.
- OK ↓

4 Check EGR gas temp. sensor connector.
- NG → Repair or replace EGR gas temp. sensor connector.
- OK ↓

Replace ECU.

5 Check voltage between terminal THG of EGR gas temp. sensor connector and body ground.
- OK → Check harness and connector between terminal E2 of ECU connector and EGR gas temp. sensor.
- NG ↓

6 Check voltage between terminals THG and E2 of ECU connector.
- OK → Check for open circuit between ECU and EGR gas temp. sensor.
- NG ↓

7 Check for short circuit between ECU and EGR gas temp. sensor.
- OK → Replace ECU.
- NG ↓

Repair or replace harness or connector.

DIAGNOSTIC CHART

WIRING DIAGRAM AND ECU CONNECTORS

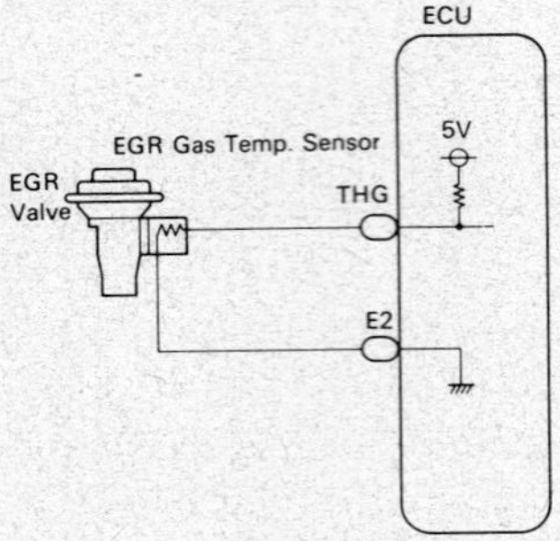

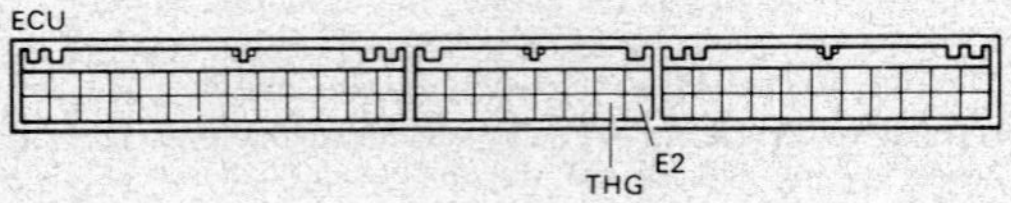

INSPECTION PROCEDURE

1 **Check EGR system**

OK

NG 〉 **Repair EGR system.**

2 **Check voltage between terminals THG and E2 of EGR gas temp. sensor connector.**

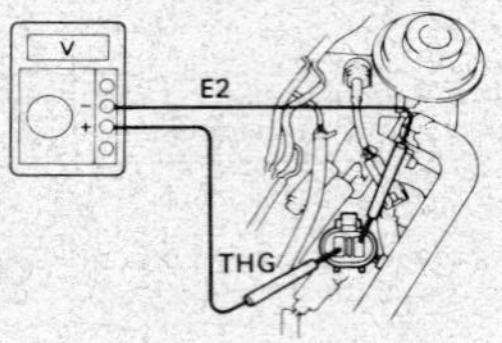

P (1) Disconnect EGR gas temp. sensor connector.
(2) Turn ignition switch ON.

C Measure voltage between terminals THG and E2 of EGR gas temp. sensor connector on harness side.

OK Voltage: 5V

OK

NG 〉 **Go to step 5.**

3 **Check EGR gas temp. sensor.**

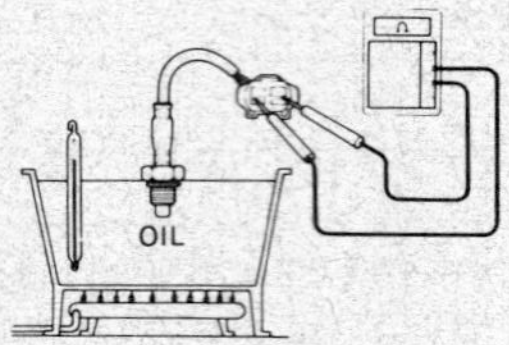

P Remove EGR gas temp. sensor.

C Measure resistance between terminals of EGR gas temp. sensor connector.

OK

Oil temp.	Resistance
50°C (122°F)	69.40 – 88.50 kΩ
100°C (212°F)	11.89 – 14.37 kΩ
150°C (302°F)	2.79 – 3.59 kΩ

OK

NG 〉 **Replace EGR gas temp. sensor.**

4 **Check EGR gas temp. sensor connector.**

OK

NG 〉 **Repair or replace EGR gas temp. sensor connector.**

Replace ECU.

5 Check voltage between terminal THG of EGR gas temp. sensor connector and body ground.

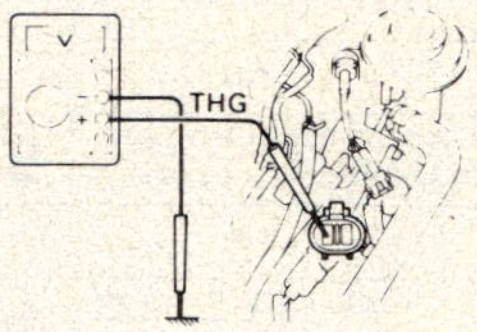

C Measure voltage between terminal THG of EGR gas temp. sensor connector and body ground.

OK Voltage: 5V

NG → (go to 6)

OK → Check harness and connector between terminal E2 of ECU connector and EGR gas temp. sensor.

6 Check voltage between terminals THG and E2 of ECU connector.

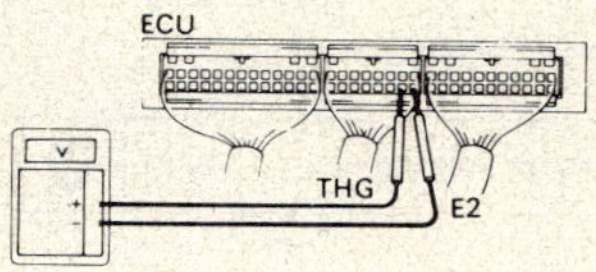

P (1) Disconnect EGR gas temp. sensor connector.
(2) Turn ignition switch ON.

C Measure voltage between terminals THG and E2 of ECU connector.

OK Voltage: 5V

NG → (go to 7)

OK → Check for open circuit between ECU and EGR gas temp. sensor.

7 Check for short circuit between ECU and EGR gas temp. sensor.

NG → Repair or replace harness or connector.

OK → Replace ECU.

Diag. Code	51	Switch Condition Signal

CIRCUIT DESCRIPTION

Neutral Start Switch Signal
The ECU uses the signals from the neutral start switch to determine whether the transmission is in park or neutral, or in some other gear.

Air Conditioner Switch Signal
The ECU uses the signal from the air conditioner switch to determine whether or not the air conditioner is operating, and increases the idling speed of the engine.

Throttle Position Sensor IDL Signal
The IDL contacts are provided in the throttle position sensor, and detects the idle condition.

Code No.	Diagnosis	Trouble Area
51	Displayed when A/C is ON, IDL contact OFF or shift position in "R", "D", "2", or "1" renges with the check terminals E1 and TE1 connected.	• A/C switch circuit • Throttle position sensor IDL circuit • Neutral start switch circuit • Accelerator pedal, cable • ECU

DIAGNOSTIC CHART

1 Is code 51 being output?
- NO → System is normal.
- Yes → Go to relevant switch signal circuit inspection.

• **Neutral Start Switch Signal**

1 Check operation of starter.
- NG → (go to 3)
- OK → (go to 2)

2 Check voltage between terminal NSW of ECU connector and body ground.
- NG → Check harness or connector between ECU and neutral start switch.
- OK → Replace ECU.

3 Check neutral start switch.
- NG → Replace neutral start switch.
- OK → Check starter circuit.

DIAGNOSTIC CHART

• Air Conditioner Switch Signal

1 Check operation of A/C system.
- NG → Repair A/C system.
- OK ↓

2 Check voltage between terminal A/C of ECU connector and body ground.
- NG → Check harness or connector between ECU and A/C amplifier.
- OK ↓

Replace ECU. *

• Throttle Position Sensor IDL Signal

1 Check voltage between terminals IDL and E2 of throttle position sensor connector.
- NG → go to 3
- OK ↓

2 Check throttle position sensor.
- NG → Replace throttle position sensor.
- OK ↓

Replace ECU. *

3 Check voltage between terminal IDL of throttle position sensor connector and body ground.
- OK → Check harness and connector between ECU and throttle position sensor.
- NG ↓

4 Check voltage between terminals IDL and E2 of ECU connector.
- OK → Check for open circuit between ECU and throttle position sensor.
- NG ↓

5 Check for short circuit between ECU and throttle position sensor.
- OK → Replace ECU.
- NG ↓

Repair or replace harness or connector.

* When diag. code 51 is not displayed, proceed to next circuit inspection shown on matrix chart

DIAGNOSTIC CHART
WIRING DIAGRAM AND ECU CONNECTORS

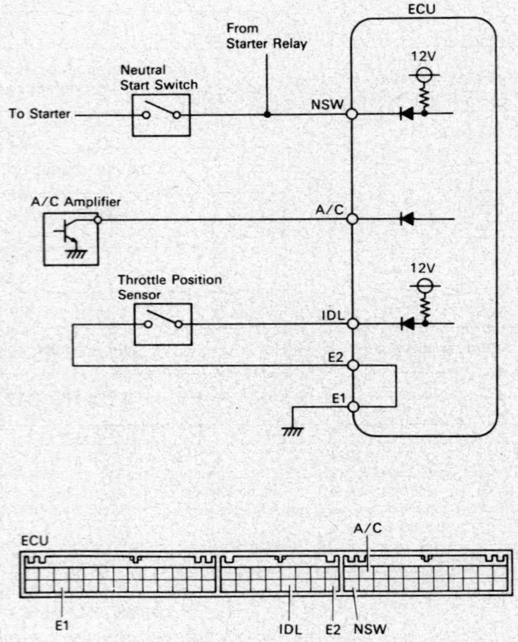

INSPECTION PROCEDURE

1 Is code 51 being output?

P (1) Shift automatic transmission into "P" or "N" range.
(2) Turn air conditioner switch OFF.
(3) Set throttle valve to fully closed (throttle position sensor IDL contacts closed).

C Is code 51 being output?

NO → **System is normal.**

YES ↓

Go to relevant switch signal circuit inspection.

Neutral Start Switch Signal

1 Check operation of starter.

NG → **Go to step 3.**

OK ↓

2 Check voltage between terminal NSW of ECU connector and body ground.

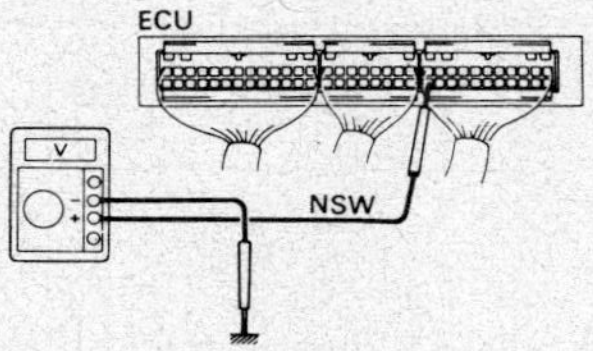

P Turn ignition switch ON.

C Measure voltage between terminal NSW of ECU connector and body ground.

OK

Switch condition	Voltage
"P" or "N" range	0V
"D", "2", "1" or "R" range	12V

NG → **Check harness or connector between ECU and neutral start switch.**

OK ↓

Replace ECU. *

* When diag. code 51 is not displayed, proceed to next circuit inspection shown on matrix chart

3 Check neutral start switch

NG → **Replace neutral start switch.**

OK ↓

Check starter circuit.

Air Conditioner Switch Signal

1 Check operation of A/C system.

NG → **Repair A/C system.**

OK ↓

2 Check voltage between terminal A/C of ECU connector and body ground.

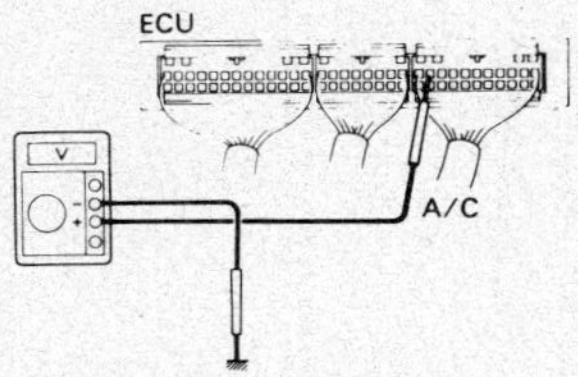

P Turn ignition switch ON.

C Measure voltage between terminal A/C of ECU connector and body ground.

OK

Switch condition	Voltage
A/C switch ON	0V
A/C switch OFF	12V

NG → **Check harness or connector between ECU and A/C amplifier.**

OK ↓

Replace ECU. *

* When diag. code 51 is not displayed, proceed to next circuit inspection shown on matrix chart

Throttle Position Sensor IDL Signal

1 Check voltage between terminals IDL and E2 of throttle position sensor connector.

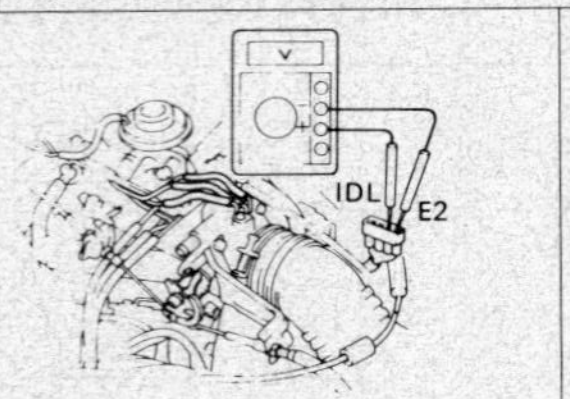

P (1) Disconnect throttle position sensor connector.
(2) Turn ignition switch ON.

C Measure voltage between terminals IDL and E2 of throttle position sensor connector.

OK Voltage: 5V

NG 〉 Go to step **3**.

OK

2 Check throttle position sensor.

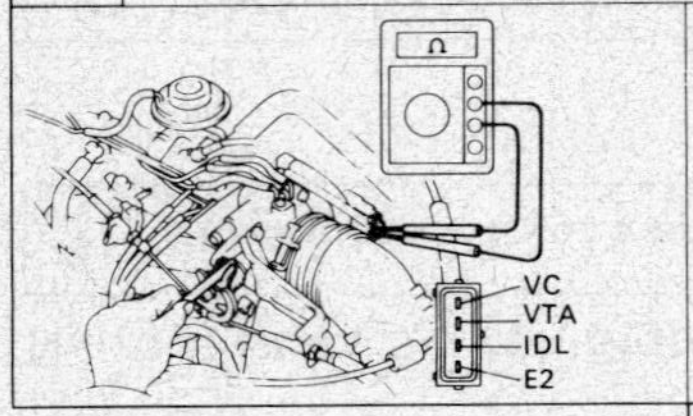

P (1) Turn ignition switch LOCK.
(2) Insert a feeler gauge between the throttle stop screw and stop lever.

C Measure resistance between terminals IDL and E2 of throttle position sensor connector.

OK

Clearance between stop lever and stop screw	Resistance
0.30 mm (0.012 in.)	2.3 kΩ or less
0.70 mm (0.028 in.)	∞ (Infinity)

NG 〉 Adjust or replace throttle position sensor.

OK

Replace ECU. *

* When diag. code 51 is not displayed, proceed to next circuit inspection shown on matrix chart

3 Check voltage between terminal IDL of throttle position sensor connector and body ground.

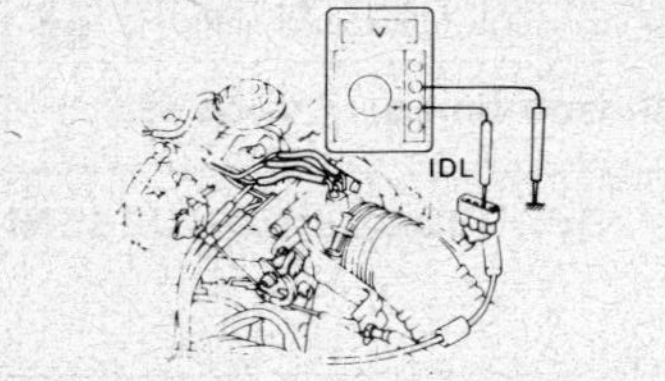

C Measure voltage between terminal IDL of throttle position sensor connector and body ground.

OK Voltage: 5V

OK 〉 Check harness and connector between ECU and throttle position sensor.

NG

4 Check voltage between terminals IDL and E2 of ECU connector.

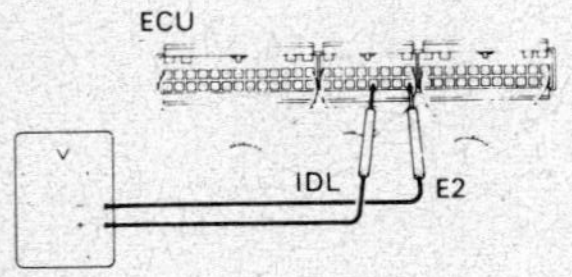

P Connect throttle position sensor connector.

C Measure voltage between terminals IDL and E2 of ECU connector.

OK

Condition	Voltage
Acceleration pedal released	0V
Acceleration pedal depressed	5V

OK 〉 Check for open circuit between ECU and throttle position sensor.

NG

5 Check for short circuit between ECU and throttle position sensor.

OK 〉 Replace ECU.

NG

Repair or replace harness or connector.

DIAGNOSTIC CODE CHART

HINT: Parameters listed in the chart may not be exactly same as your reading due to type of the instruments or other factors.

If a malfunction code is displayed during the diagnostic code check in test mode, check the circuit for that code listed in the table below

Code No.	Number of Check Engine Blinks	Circuit	Diagnostic Code Detecting Condition	Trouble Area	"CHECK" Engine Warning Light*1 Normal Mode	"CHECK" Engine Warning Light*1 Test Mode	Memory*2	
—		Normal	No code is recorded.	————	—	—	———	
12		RPM Signal	No "NE" or "G1" and "G2" signal to ECU within 2 sec. after cranking. Open in "G⊖" circuit.	• Open or short in NE, G circuit. • Distributor • Open or short in STA circuit. • ECU	ON	N.A.	○	
13		RPM Signal	No NE signal to ECU for 0.1 sec. or more at 1,000 rpm or more.	• Open or short in NE circuit. • Distributor • ECU	ON	N.A.	○	
			No 12 pulses of NE to ECU during the interval between G1 and G2 pulses.	• Open or short in NE circuit. • Distributor • ECU	N.A.	ON		
14		Ignition Signal	No IGF signal to ECU for 6 consecutive IGT signal.	• Open or short in IGF or IGT circuit from igniter to ECU. • Igniter • ECU	ON	N.A.	○	
16		ECT Control Signal	Fault in communications between the engine CPU and ECT CPU in the ECU.	• ECU	ON	N.A.	×	
21		Main Oxygen Sensor Signal (on left bank)	Main oxygen sensor signal voltage is reduced to between 0.35 V and 0.70 V for 60 sec. under conditions (a) ~ (d). (2 trip detection logic)*4 (a) Coolant temp.: Between 70°C (158°F) and 90°C (194°F). (b) Engine speed: 1,500 rpm or more. (c) Load driving (EX. ECT in 4th speed, A/C ON, Flat road, 50 mph (80 km/h)) (d) Main oxygen sensor signal voltage: Alternating above and below 0.45 V.	• Main oxygen sensor circuit • Main oxygen sensor	ON	ON	○	

BE3931

DIAGNOSTIC CODE CHART (Cont'd)

Code No.	Number of Check Engine Blinks	Circuit	Diagnostic Code Detecting Condition	Trouble Area	"CHECK" Engine Warning Light*1 Normal Mode	"CHECK" Engine Warning Light*1 Test Mode	Memory*2	
22	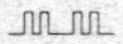	Water Temp. Sensor Signal	Open or short in water temp. sensor circuit for 0.5 sec. or more.	• Open or short in water temp. sensor circuit. • Water temp. sensor • ECU	ON	ON	○	
24		Intake Air Temp. Sensor Signal	Open or short in intake air temp. sensor circuit for 0.5 sec. or more.	• Open or short in intake air temp. sensor circuit. • Intake air temp. sensor • ECU	OFF ON*3	ON	○	
25		Air-Fuel Ratio Lean Malfunction	(1) Main oxygen sensor voltage is 0.45 V or less (lean) for 90 sec. under conditions (a) and (b). (2 trip detection logic)*4 (a) Coolant temp.: 50°C (122°F) or more. (b) Engine speed: 1,500 rpm or more.	• Open or short in main oxygen sensor circuit. • Main oxygen sensor • Ignition system • ECU	ON	ON	○	
			(2)*3 Difference of air-fuel ratio feedback compensation value between right and left banks is more than 15 percentage for 20 sec. or more under conditions (a) and (b). (2 trip detection logic)*4 (a) Engine speed: Idling (b) Coolant temp.: Between 70°C (158°F) and 95°C (203°F).	• Fuel line pressure (injector leak, blockage) • Mechanical system malfunction (skipping teeth of timing belt) • Ignition system • Compression pressure • Air flow meter • ECU				
26*3		Air-Fuel Ratio Rich Malfunction	Difference of air-fuel ratio feedback compensation value between right and left banks is more than 15 percentage for 20 sec. or more under conditions (a) and (b). (2 trip detection logic)*4 (a) Engine speed: Idling (b) Coolant temp.: Between 70°C (158°F) and 95°C (203°F).	• Fuel line pressure (injector leak, blockage) • Mechanical system malfunction (skipping teeth of timing belt) • Ignition system • Compression pressure • Air flow meter • ECU	ON	ON	○	
27*3		Sub-Oxygen Sensor Signal	(1) Open or short in heater circuit of sub-oxygen sensor for 0.5 sec. or more.	• Open or short in heater circuit of sub-oxygen sensor. • Sub-oxygen sensor heater. • ECU	ON	N.A.	○	
			(2) Main oxygen sensor signal is 0.45 V or more and sub-oxygen sensor signal is 0.45 V or less under conditions (a)~(c). (2 trip detection logic)*4 (a) Coolant temp.: 80°C (176°F) or more. (b) Engine speed: 1,500 rpm or more. (c) Accel-pedal: Full depressed for 2 sec. or more.	• Open or short in sub-oxygen sensor circuit. • Sub-oxygen sensor • ECU	ON	ON		
28	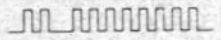	Main Oxygen Sensor Signal (on right bank)	Same as Code No. 21	• Same as Code No. 21	ON	ON	○	

3.0L ENGINE—DIAGNOSTIC CODES—ES300

DIAGNOSTIC CODE CHART (Cont'd)

Code No.	Number of Check Engine Blinks	Circuit	Diagnostic Code Detecting Condition	Trouble Area	"CHECK" Engine Warning Light*1 Normal Mode	"CHECK" Engine Warning Light*1 Test Mode	Memory*2	
31		Air Flow Meter Signal	All conditions below are detected continuously for 0.5 sec. or more. (a) Idle switch: ON (IDL) (b) Voltage cf terminal VS: 0.1 V or less (c) Engine speed: 1,500 rpm or less	• Open (VC) or short (VC-E2) in air flow meter circuit. • Air flow meter • ECU	ON	ON	○	
32		Air Flow Meter Signal	VS exceeds 4.9 V for 0.5 sec. or more.	• Open (E2) or short (VC-VS) in air flow meter circuit. • Air flow meter • ECU	ON	ON	○	
41		Throttle Position Sensor Signal	Open or short in throttle position sensor circuit (VTA) for 0.5 sec. cr more.	• Open or short in throttle position sensor circuit. • Throttle position sensor • ECU	OFF ON*3	ON	○	
42		Vehicle Speed Sensor Signal (for ECT)	All conditions below are detected continuously for 8 sec. or more. (a) Vehicle speed signal: 0 km/h (mph) (b) Engine speed: 3,000 rpm or more (c) Neutral start switch (NSW): OFF	• No. 1 speed sensor • Combination meter • Open or short in No. 1 speed sensor circuit. • ECU	OFF	OFF	○	
		Vehicle Speed Sensor Signal (for M/T)	All conditions below are detected continuously for 8 sec. or more. (a) Vehicle speed signal: 0 km/h (mph) (b) Engine speed: Between 3,000 rpm and 4,500 rpm (c) Coolant temp.: 80°C (176°F) or more (d) Load driving					
43		Starter Signal	No starter signal to ECU.	• Open or short in starter signal circuit. • Open or short in ignition switch or starter relay circuit. • ECU	N.A.	OFF	×	
52		Knock Sensor Signal (on left bank)	No No. 2 knock sensor signal to ECU for 2 crank revolutions with engine speed between 1,600 rpm and 5,200 rpm.	• Open or short in knock sensor (on left bank) circuit. • Knock sensor of left bank (looseness) • ECU	ON	N.A.	○	
53		Knock Control Signal	Engine control computer (for knock control) malfunction at engine speed between 650 rpm and 5,200 rpm.	• ECU	ON	N.A.	×	
55		Knock Sensor Signal (on right bank)	No No. 1 knock sensor signal to ECU for 2 crank revolutions with engine speed between 1,600 rpm and 5,200 rpm.	• Open or short in knock sensor (on right bank) circuit. • Knock sensor of right bank (looseness) • ECU	ON	N.A.	○	

DIAGNOSTIC CODE CHART (Cont'd)

Code No.	Number of Check Engine Blinks	Circuit	Diagnostic Code Detecting Condition	Trouble Area	"CHECK" Engine Warning Light*1 Normal Mode	"CHECK" Engine Warning Light*1 Test Mode	Memory*2	
71*3		EGR System Malfunction	EGR gas temp. is 70°C (158°F) or below for 1～4 min. under conditions (a) and (b). (2 trip detection logic)*4 (a) Coolant temp.: 60°C (140°F) or more. (b) EGR operation possible (EX. ECT in D range, A/C ON, 60 mph (96 km/h), Flat road.).	• Open in EGR gas temp. sensor circuit. • EGR or BVSV vacuum hose disconnected, valve stuck. • Clogged EGR gas passage. • ECU	ON	ON	○	
51		Switch Condition Signal	(1) 3 sec. or more after engine starts, idle switch OFF (IDL). (2) Neutral start switch OFF (NSW). (Shift position in "R", "D", "2" or "1" ranges). (3) A/C switch ON.	• A/C switch circuit • Throttle position sensor IDL circuit • Neutral start switch circuit • Accelerator pedal and cable • ECU	N.A.	OFF	×	

*1: "ON" displayed in the diagnosis mode column indicates that the "CHECK" Engine Warning Light is lighted up when a malfunction is detected. "OFF" indicates that the "CHECK" does not light up during malfunction diagnosis, even if a malfunction is detected. "N.A." indicates that the item is not included in malfunction diagnosis.

*2: "○" in the memory column indicates that a diagnostic code is recorded in the ECU memory when a malfunction occurs. "×" indicates that a diagnostic code is not recorded in the ECU memory even if a malfunction occurs. Accordingly, output of diagnostic results in normal or test mode is performed with the IG switch ON.

*3: Only for California specification vehicles.

*4: This indicates items for which "2 trip detection logic" is used. With this logic, when a logic malfunction is first detected, the malfunction is temporarily stored in the ECU memory. If the same case is detected again during the second drive test, this second detection causes the "CHECK" Engine Warning Light to light up.
The 2 trip repeats the same mode a 2nd time. (However, the IG switch must be turned OFF between the 1st trip and 2nd trip).
In the Test Mode, the "CHECK" Engine Warning Light lights up the 1st trip a malfunction is detected.

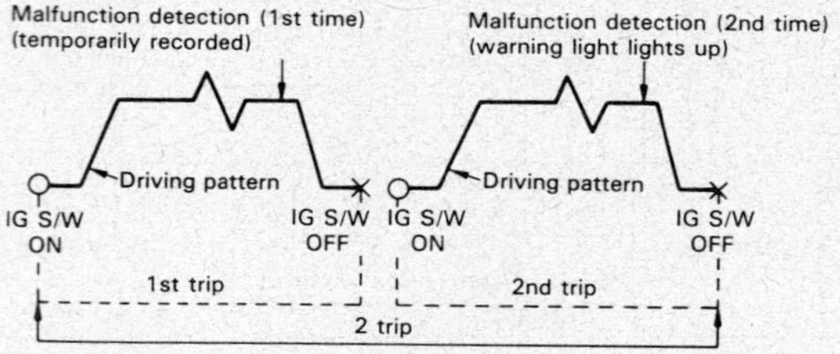

FAIL-SAFE CHART

If any of the following codes is recorded, the ECU enters fail-safe mode.

Code No.	Fail-Safe Operation	Fail-Safe Deactivation Conditions
14	Fuel cut	1 IGF1 detected in consecutive 3 ignitions.
16	Torque control prohibited	Returned to normal condition.
22	THW is fixed at 80°C (176°F)	Returned to normal condition.
24	THA is fixed at 20°C (68°F)	Returned to normal condition.
31 · 32	• Ilgnition timing fixed at 10° BTDC. • Injection time fixed (IDL ON....3.0 m sec. IDL OFF...5.6 m sec.	Returned to normal condition
41	VTA is fixed at 0°	The following 1 or 2 must be repeated at least 2 times consecutively. ① 0.1 V ≦ VTA ≦ 0.95 V ② IDL ON
52	Max. timing retardation.	IG switch OFF.
53	Max. timing retardation.	Returned to normal condition.
55	Max. timing retardation.	IG switch OFF.

Back-Up Function

If there is trouble with the program in the ECU and the ignition signals (IGT) are not output, the ECU controls fuel injection and ignition timing at predetermined levels as a back-up function to make it possible to continue to operate the vehicle.
Furthermore, the injection duration is calculated from the starting signal (STA) and the throttle position signal (IDL). Also, the ignition timing is fixed at the initial ignition timing, 5° BTDC, without relation to the engine speed.

HINT: If the engine is controlled by the back-up function, the CHECK engine warning light lights up to warn the driver of the malfunction but the diagnostic code is not output.

BASIC INSPECTION

When the normal code is displayed in the diagnostic code check, troubleshooting should be performed in the order for all possible circuits to be considered as the causes of the problems.
In many cases, by carrying out the basic engine check shown in the following flow chart, the location causing the problem can be found quickly and efficiently. Therefore, use of this check is essential in engine troubleshooting.

1 **Is battery voltage 11 V or more when engine is stopped?**

NO — **Charge or replace battery.**

YES

2 **Is engine cranked?**

NO — **Proceed to matrix chart of problem symptoms**

YES

3 **Does engine start?**

NO — **Go to step 7.**

YES

4 **Check air filter.**

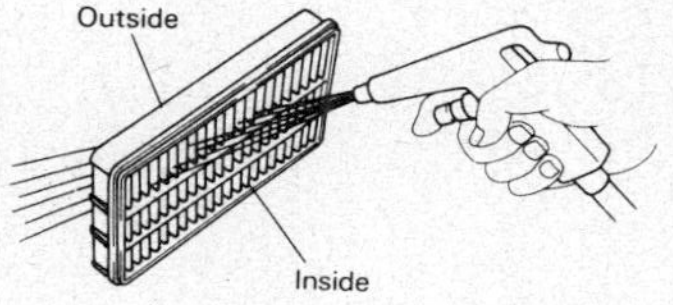

P Remove air filter.

C Visually check that the air filter is not excessively damaged or oily.

Hint If necessary, clean the air filter with compressed air. First blow from inside thoroughly, then blow off outside of the air filter.

NG — **Repair or replace.**

OK

Go to step 5.

5 Check idle speed.

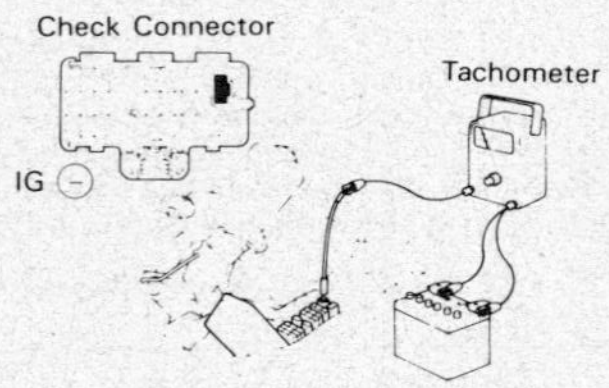

P
(1) Shift transmission into "N" range or neutral.
(2) Warm up engine at normal operating temperature.
(3) Switched off all accessories.
(4) Switched off air conditioner.
(5) Connect tachometer test probe to terminal IG ⊖ of check connector.

C Check idle speed.

OK **Idle speed: 650 ~ 750 rpm**

Caution
- **NEVER allow tachometer test probe to touch ground as it could result in damage to igniter and/or ignition coil.**
- **As some tachometers are not compatible with this ignition system, we recommended that you confirm compatibility of your unit before use.**

OK ↓

NG → **Proceed to matrix chart of problem symptoms**

6 Check ignition timing.

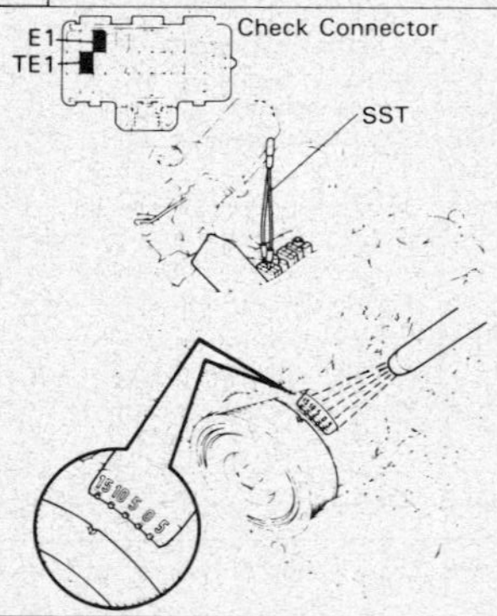

P
(1) Shift transmission into "N" range or neutral.
(2) Warm up engine at normal operating temperature.
(3) Keep the engine speed at idle.
(4) Using SST, connect terminals TE1 and E1 of check connector.
SST 09843-18020
(5) Using a timing light, connect the tester to No. 4 high-tension cord.

C Check ignition timing.

OK **Ignition timing: 10° BTDC at idle**

NG → **Proceed to troubleshoot.**

OK ↓

Proceed to matrix chart of problem symptoms

7 Check fuel pressure.

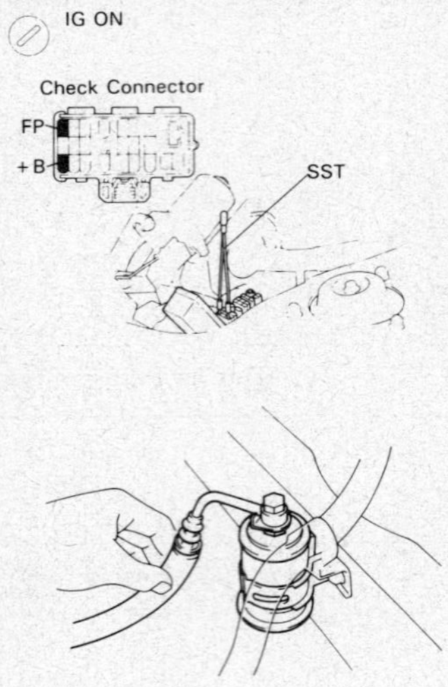

P
(1) Be sure that fuel is enough in tank.
(2) Turn ignition switch on.
(3) Using SST, connect terminals FP and +B of check connector.
SST 09843-18020

C Check that there is pressure in the hose from the fuel filter.

Hint At this time, you will hear fuel return noise.

Caution
Never make a mistake with the terminal connection position as this will cause a malfunction.

OK ↓

NG → **Proceed to troubleshoot.**

8 Check for spark.

C Disconnect the high-tension cord from the distributor and, hold the end about 12.5 mm (1/2") from the ground, see if spark occurs while the engine is being cranked.

Hint To prevent excessive fuel injected from the injectors during this test, don't crank the engine for more than 1–2 seconds at a time.

NG → **Proceed to troubleshoot.**

OK ↓

Proceed to matrix chart of problem symptoms

3.0L ENGINE—PARTS LOCATION—ES300

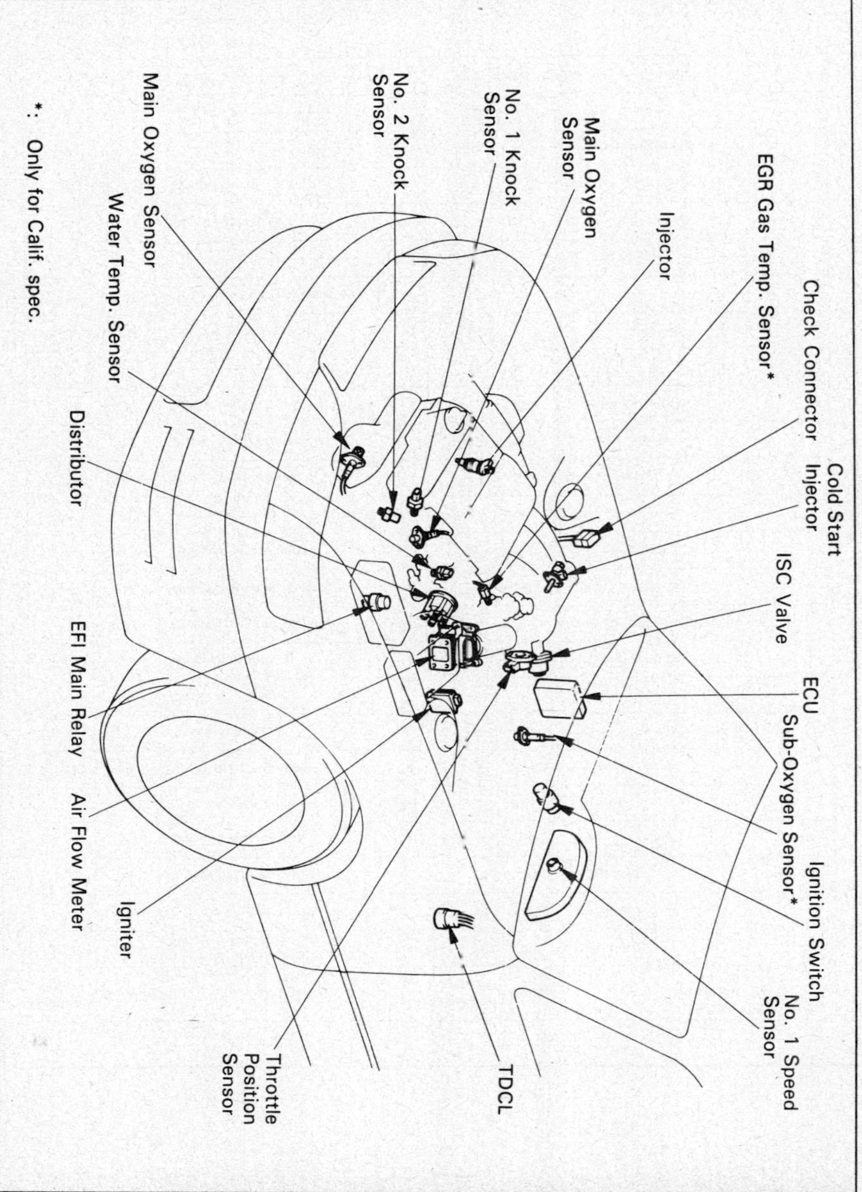

*: Only for Calif. spec.

3.0L ENGINE—ECU TERMINALS—ES300

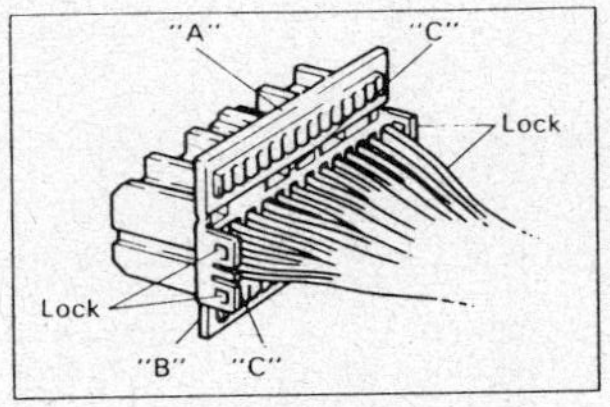

TERMINALS OF ECU

When measuring the voltage or resistance of the connector part of the ECU, remove the locks as required in the following procedure and always insert the test probe into the connector from the wire harness side.

PREPARATION

(a) Turn the ignition switch to LOCK position.

(b) Disconnect the connectors from the ECU.

(c) To remove section "A" and "B", carefully pull outward the lock on both sides as shown in the illustration.

NOTICE: Pay attention to section "C" in the illustration which can be easily broken.

(d) Reconnect the connectors to the ECU.

ECU Terminals

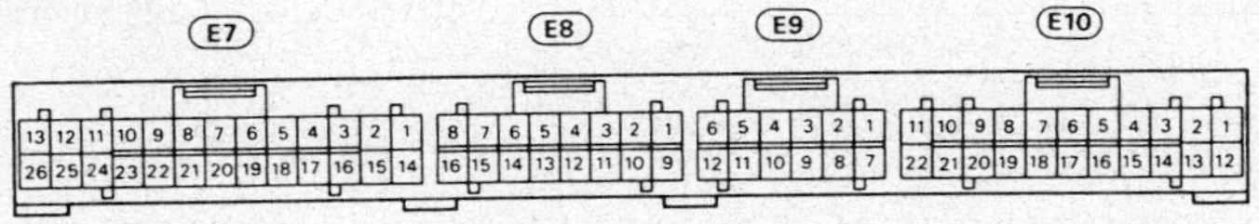

Terminal No.	Symbol	Connection	Terminal No.	Symbol	Connection
E7-1	HT*2	Sub-oxygen sensor heater	E7-14	VF2	Check connector
-2	2*1	Neutral start switch	-15	VF1	Check connector
-3	L*1	Neutral start switch	-16	–	———
-4	ISC4	ISC valve	-17	IGF	Igniter
-5	ISC3	ISC valve	-18	IGT	Igniter
-6	ISC2	ISC valve	-19	–	———
-7	ISC1	ISC valve	-20	#60	No. 6 injector
-8	SL*1	ECT solenoid	-21	#50	No. 5 injector
-9	S2*1	ECT solenoid	-22	#40	No. 4 injector
-10	S1*1	ECT solenoid	-23	#30	No. 3 injector
-11	#20	No.2 injector	-24	E1	ECU ground
-12	#10	No.1 injector	-25	STJ	Cold start injector
-13	E01	Power ground	-26	E02	Power ground

3.0L ENGINE—ECU TERMINALS—ES300

Terminal No.	Symbol	Connection	Terminal No.	Symbol	Connection
E8-1	VC	Throttle position sensor Air flow meter	E9-10	G2	Distributor
-2	VS	Air flow meter	-11	G1	Distributor
-3	THA	Intake air temp. sensor	-12	NE	Distributor
-4	THW	Water temp. sensor	E10-1	+B1	EFI main relay
-5	OX1	Main oxygen sensor (on right bank)	-2	BATT	Battery
-6	KNK1	No. 1 knock sensor	-3	M-REL	EFI main relay
-7	TE1	Check connector	-4	W	"CHECK" engine warning light
-8	DG	Check connector	-5	—	———
-9	E2	Sensor ground	-6	ACT	A/C control assembly
-10	VTA	Throttle position sensor	-7	A/C	A/C control assembly
-11	IDL	Throttle position sensor	-8	SP1	Speed sensor No.1
-12	THG*2	EGR gas temp. sensor	-9	OD2*1	O/D main switch
-13	OX2	Main oxygen sensor (on left bank)	-10	—	———
-14	KNK2	No. 2 knock sensor	-11	STA	Starter relay
-15	TE2	Check connector	-12	+B	EFI main relay
-16	SP2*1	Speed sensor No. 2 (for ECT)	-13	IGSW	Ignition switch
E9-1	—	———	-14	BK*1	Stop light switch Stop light
-2	FPU	Fuel pressure up VSV	-15	—	———
-3	IACV	VSV (intake air control valve)	-16	—	———
-4	—	———	-17	—	———
-5	OX3*2	Sub-oxygen sensor	-18	OD1*1	Cruise control ECU
-6	G⊖	Distributor	-19	R*1	Neutral start switch
-7	—	———	-20	P*1	ECT pattern select switch
-8	—	———	-21	—	———
-9	—	———	-22	NSW*1	Neutral start switch

*1 : Only vehicles with ECT.
*2 : Only for California specification vehicles.

3.0L ENGINE
ECU TERMINALS STANDARD VALUES—ES300

STANDARD VALUE OF ECU TERMINALS

ECU Terminals

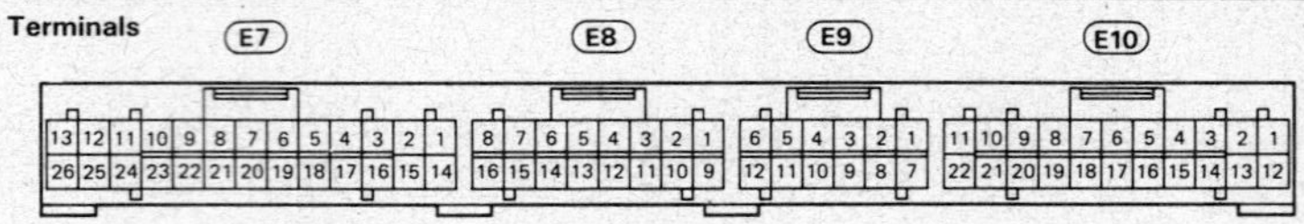

Symbols (Terminals No.)	Wiring Color	STD Voltage (V)	Condition
BATT (E10-2)–E1 (E7-24)	W-L ↔ BR	10~14	Always
IGSW (E10-13) +B (E10-12)–E1 (E7-24) +B1 (E10-1)	BO (IGSW) B-O (+B) ↔ BR B-O (+B1)	10~14	IG switch ON.
VC (E8-1)–E2 (E8-9)	L-R ↔ BR	4.0~6.0	IG switch ON.
IDL (E8-11)–E2 (E8-9)	L ↔ BR	Below 1.0	IG switch ON. Throttle valve fully closed
		10~14	IG switch ON. Throttle valve fully open
VTA (E8-10)–E2 (E8-9)	B ↔ BR	0.3~0.8	IG switch ON. Throttle valve fully closed
		2.7~5.2	IG switch ON. Throttle valve fully open
VS (E8-2)–E2 (E8-9)	Y-L ↔ BR	3.7~4.3	IG switch ON.
		2.0~4.0	Idling
THA (E8-3)–E2 (E8-9)	L-B ↔ BR	1.0~3.0	Idling, Intake air temp. 20°C (68°F)
THW (E8-4)–E2 (E8-9)	G ↔ BR	0.1~1.0	Idling, Water temp. 80°C (176°F)
#10 (E7-12) #20 (E7-11) #30 (E7-23) – E01 (E7-13) #40 (E7-22) #50 (E7-21) #60 (E7-20)	W (#10) Y (#20) GR (#30) ↔ W-B L (#40) R (#50) G (#60)	10~14	IG switch ON.
		Pulse generation	Idling
IGT (E7-18)–E1 (E7-24)	W ↔ BR	Pulse generation (0 and 4~6)	Idling
IGF (E7-17)–E1 (E7-24)	W-R ↔ BR	Below 1.0	IG switch ON.
		Pulse generation (1.0 and 4~6)	Idling
G1 (E9-11) G2 (E9-10) – G⊖ (E9-6)	R (G1), L(G2) ↔ Y	Pulse generation (−1.3~ +1.3 or higher)	Idling
NE (E9-12)–G⊖ (E9-6)	B ↔ Y	Pulse generation (−0.6~ +0.6 or higher)	Idling
M-REL (E10-3)–E1 (E7-24)	RL ↔ BR	10~14	IG switch ON.

3.0L ENGINE
ECU TERMINALS STANDARD VALUES—ES300

Symbols (Terminals No.)	Wiring Color	STD Voltage (V)	Condition
FPU (E9-2)–E1 (E7-24)	B-R ↔ BR	10~14	IG switch ON.
		Below 2.0	Restarting high temp. (Coolant temp.: 100°C (212°F) or more / Intake air temp.: 65°C (149°F) or more)
ISC1 (E7-7) ISC2 (E7-6) ISC3 (E7-5) ISC4 (E7-4) – E1 (E7-24)	W-L (ISC1) G-W (ISC2) L-R (ISC3) R-B (ISC4) ↔ BR	10~14	IG switch ON.
VF1 (E7-15) VF2 (E7-14) –E1 (E7-24)	R-W (VF1) B (VF2) ↔ BR	1.0~4.0	Maintain engine speed at 2,500 rpm for two minutes after warming up then return to idling.
OX1 (E8-5) OX2 (E8-13) –E1 (E7-24)	W (OX1) R-L (OX2) ↔ BR	Pulse generation (0~1.0)	Maintain engine speed at 2,500 rpm for two minutes after warming up.
HT (E7-1)–E01 (E7-13)	P-B ↔ W-B	10~14	IG switch ON.
		Below 2.0	Idling
KNK1 (E8-6) KNK2 (E8-14) –E1 (E7-24)	W (KNK1) R-L (KNK2) ↔ BR	Pluse generation (Frequency: 7.6 kHz)	Idling
NSW (E10-22)–E1 (E7-24)	B-W ↔ BR	10~14	IG switch ON. Other shift position in "P", "N" ranges.
		Below 2.0	IG switch ON. Shift position in "P", "N" ranges.
SP1 (E10-8)–E1 (E7-24)	V-Y ↔ BR	Pulse generation (0 and 4~6)	IG switch ON. Rotate driving wheel slowly.
TE1 (E8-7)–E1 (E7-24)	GR ↔ BR	10~14	IG switch ON.
TE2 (E8-15)–E1 (E7-24)	G-W ↔ BR	10~14	IG switch ON.
W (E10-4)–E1 (E7-24)	G-R ↔ BR	Below 2.0	IG switch ON.
		10~14	Idling
OD1 (E10-18)–E1 (E7-24)	Y-B ↔ BR	10~14	IG switch ON.
A/C (E10-7)–E1 (E7-24)	B-Y ↔ BR	Below 2.0	A/C switch ON. (At idling)
		10~14	A/C switch OFF.
ACT (E10-6)–E1 (E7-24)	LG-R ↔ BR	10~14	A/C switch ON. (At idling)
		Below 2.0	A/C switch OFF.
IACV (E9-3)–E01 (E7-13)	LG ↔ W-B	10~14	IG switch ON.
STA (E10-11)–E1 (E7-24)	B-W ↔ BR	6.0 or more	Cranking
STJ (E7-25)–E01 (E7-13)	G ↔ W-B	10~14	Starting

3.0L ENGINE
MATRIX CHART OF PROBLEM SYMPTOMS—ES300

MATRIX CHART OF PROBLEM SYMPTOMS

When the malfunction code is not confirmed in the diagnostic code check and the problem still can not be confirmed in the basic inspection, then proceed to this step and perform troubleshooting according to the numbered order given in the table below.

The circuits indicated by ※ on the matrix chart can be inspected using the TCCS checker.

Symptom \ Suspect area		Air flow meter circuit ※	Starter signal circuit ※	EGR System	Switch condition signal circuit	Neutral start switch circuit ※	ECU power source circuit	Back up power source circuit	Injector circuit ※	Cold start injector circuit	ISC valve circuit	Fuel system circuit	Fuel pressure control VSV circuit	A/C cut control circuit	A/C signal circuit (Compressor circuit) ※	Starter and starter relay	Ignition signal circuit (Spark test)	Spark plug	Ignition coil	Distributor	Compression	ECT faulty	Theft deterrent ECU	Engine (& ECT) ECU
Does not start	Engine does not crank					2										1							3	
	No initial combustion						1		5	6		3					2	4						
	No complete combustion								6	7		1					5	4	2	3				
Difficult to start	Difficult to start normally		1						8	9	2	3						6	4	5	7			
	Difficult to start cold		1						5	3	2	4						8	6	7				
	Difficult to start hot		1						5	6	2	4	3					9	7	8				
Poor Idling	Incorrect first idle				1						2													
	High engine idle speed				1	5	4	6			2				3									
	Low engine idle speed	7			3	4		8	6		1	5			2									
	Rough idling	3		5	1			12	4	13	2	8					6	11	9	10	7			
	Hunting	3			1		4				2	5												
Poor Driveability	Hesitation/Poor acceleration	2			1				3			4		9			5	8	6	7		10		
	Muffler explosion (after fire)				1				5									4	2	3				
	Surging				1				5			2						4		3				
Engine Stall	Engine stalls soon after starting	2								4	3	1												
	After acceleration pedal depressed	2			1					3														
	After acceleration pedal released								1	3	2													4
	During A/C operation										1				2									3
	When N to D shift					1				3	2													

LOCATION OF CONNECTORS

Location of Connectors in Engine Compartment

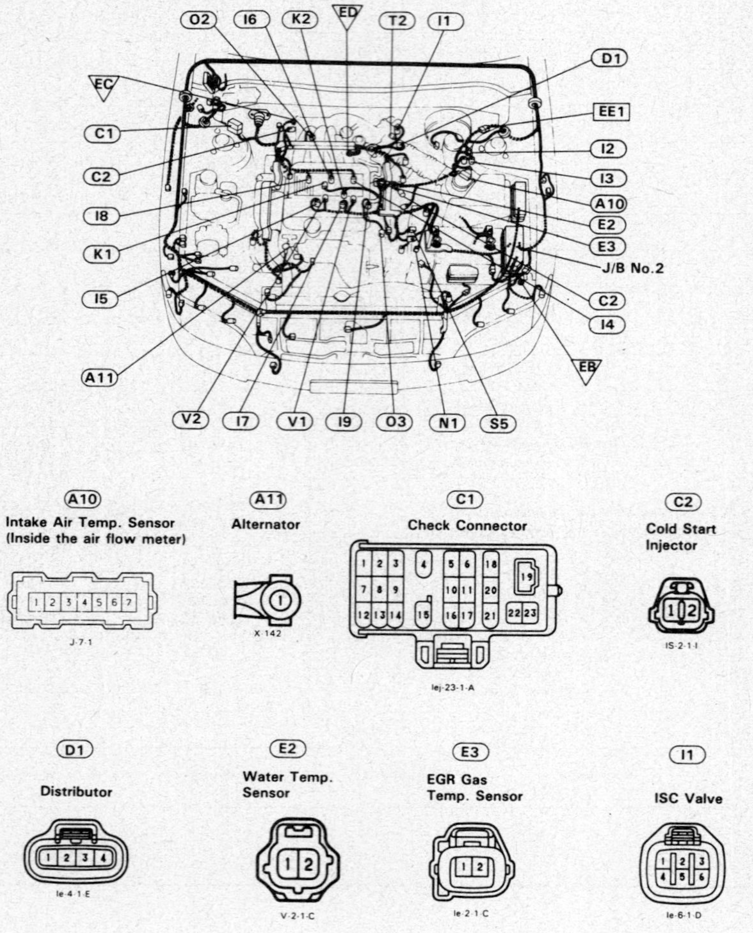

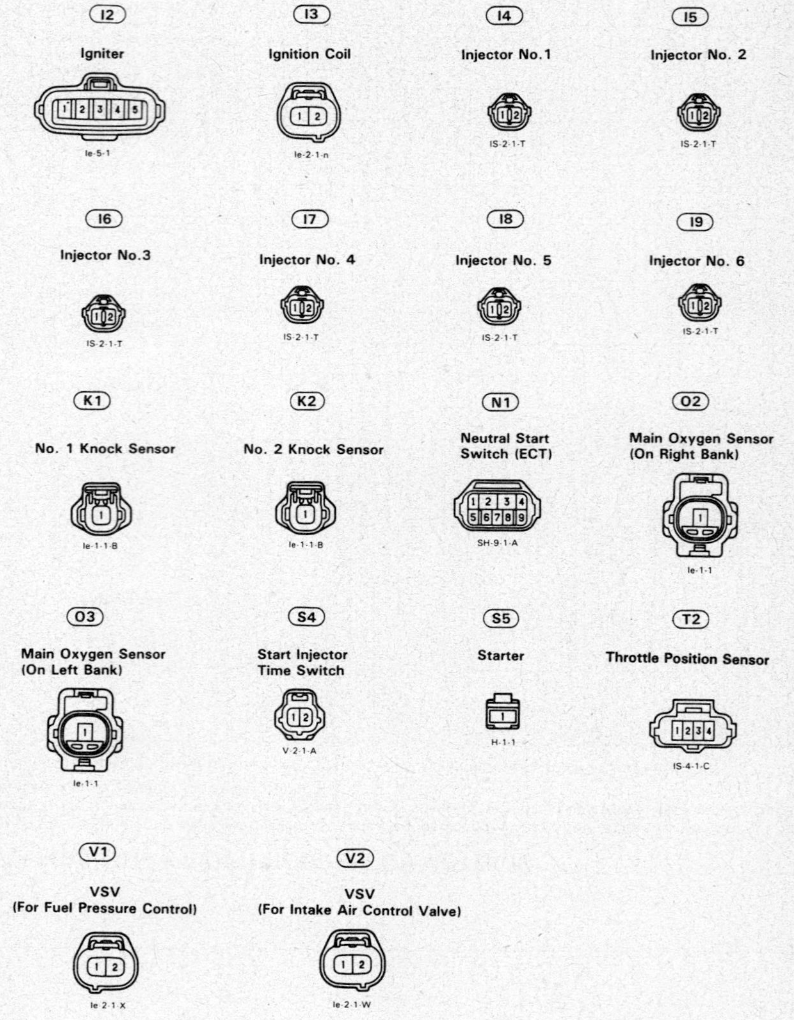

3.0L ENGINE—CONNECTOR LOCATIONS—ES300

Location of Connectors in Engine Compartment (Cont'd)

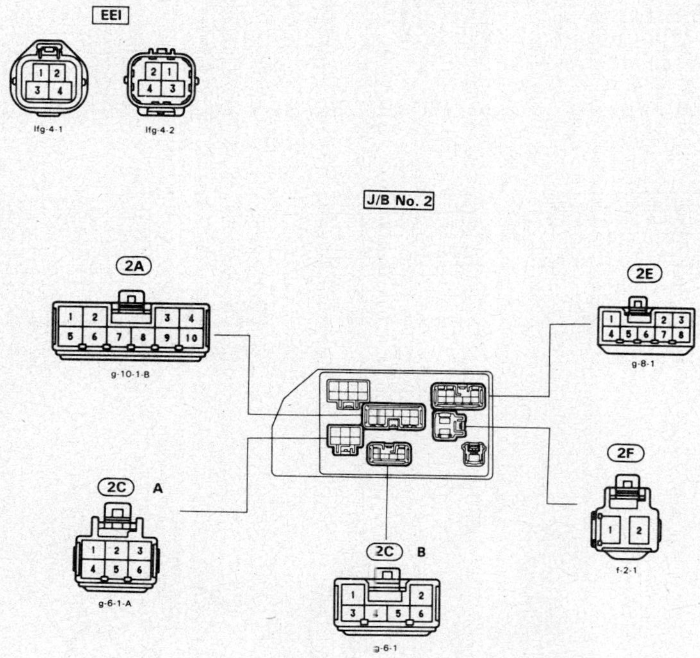

Location of Connectors in Body

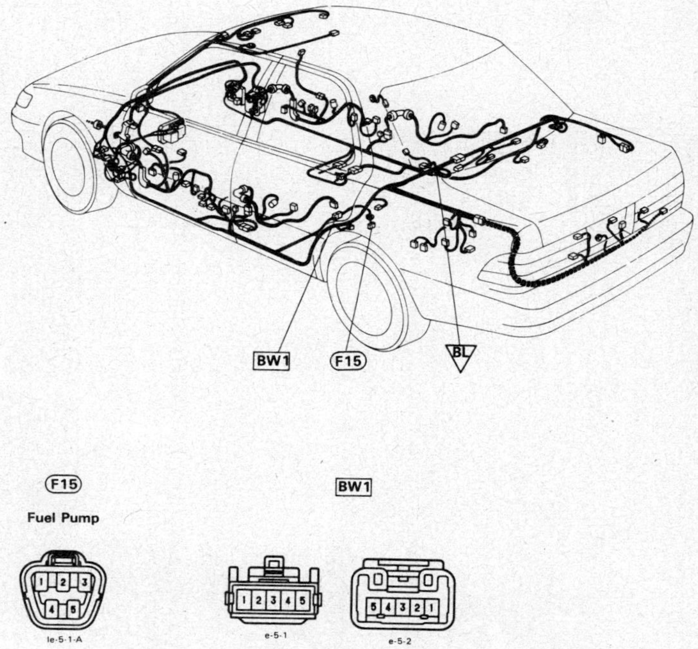

Location of Connectors in Instrument Panel

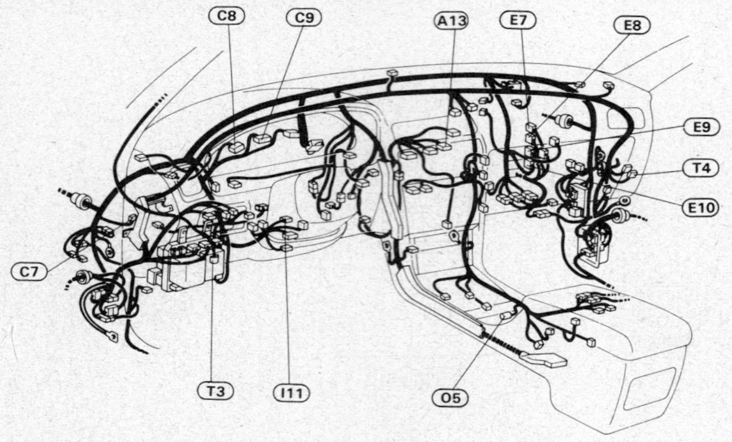

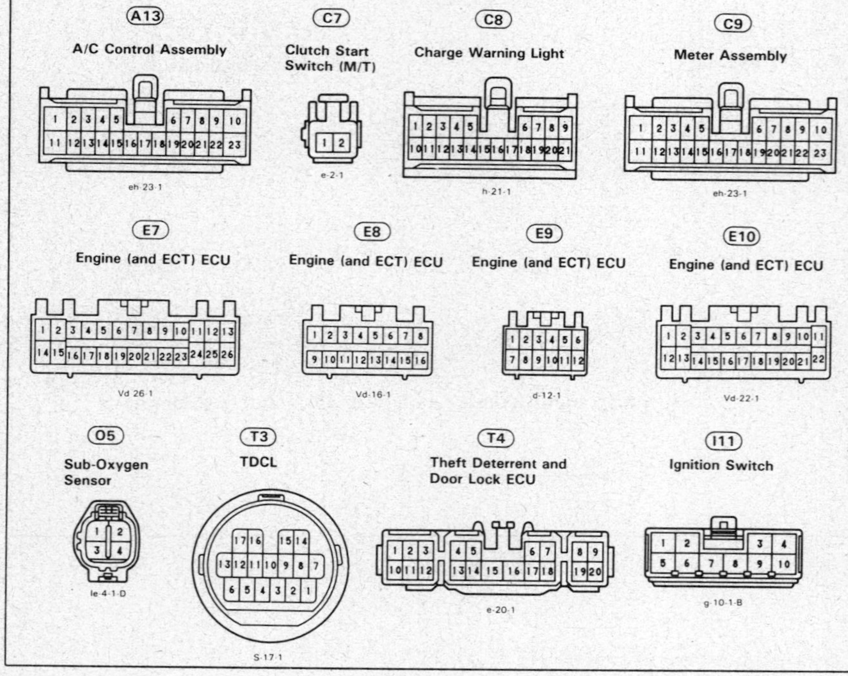

Location of Connectors in Instrument Panel (Cont'd)

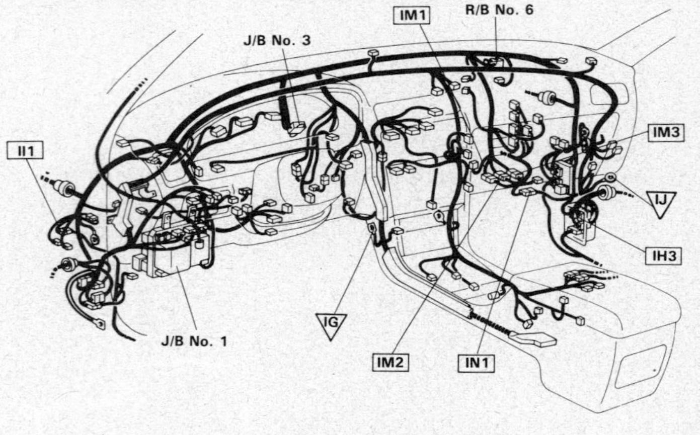

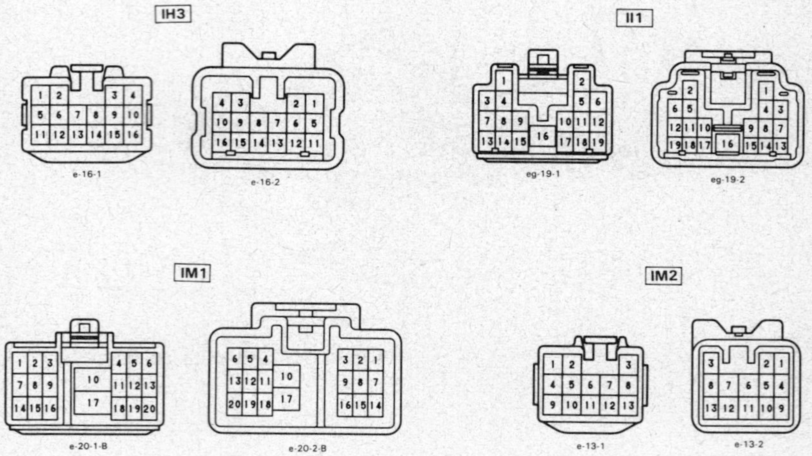

3.0L ENGINE – CONNECTOR LOCATIONS – ES300

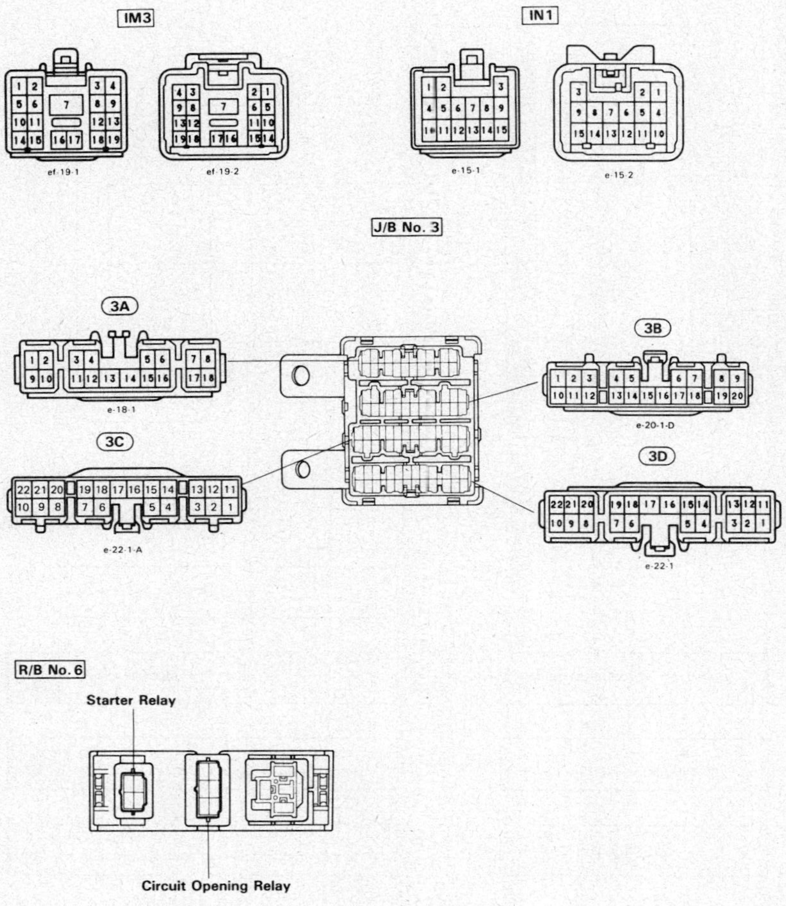

3.0L ENGINE CIRCUIT INSPECTION AND TROUBLESHOOTING – ES300

CIRCUIT INSPECTION

Diag. Code	12	RPM Signal Circuit (No. 1)

CIRCUIT DESCRIPTION

The distributor in the Engine Control System contains three pick-up coils (G1, G2 and NE).
The G1, G2 signals inform the ECU of the standard crankshaft angle.
The NE signals inform the ECU of the crankshaft angle and the engine speed.

Code No.	Diagnostic Code Detecting Condition	Trouble Area
12	No "NE" or "G1" and "G2" signal to ECU within 2 sec. after cranking.	• Open or short in NE, G circuit. • Distributor • Open or short in STA circuit. • ECU
	Open in "G⊖" circuit.	

Reference — INSPECTION USING OSCILLOSCOPE

G, NE signal waveforms

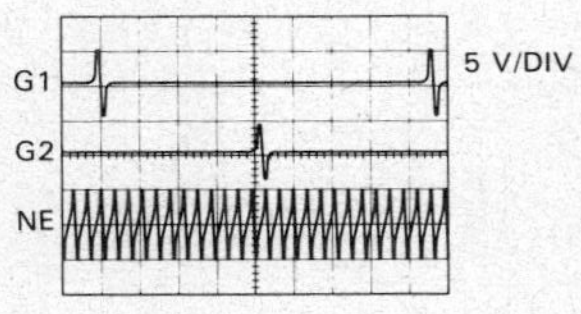

20 m sec/Division (Idling)

- During cranking or idling, check between terminals G1, G2, NE and G⊖ of engine (& ECT) ECU.

HINT: The correct waveforms appear as shown in the illustration on the left.

DIAGNOSTIC CHART

1 Check resistance of each pickup coils in distributor. — NG → Replace distributor.

OK ↓

2 Check for open and short in harness and connector between ECU and distributor. — NG → Repair on replace harness or connector.

OK ↓

3 Check air gap. — NG → Replace distributor.

OK ↓

Check and replace ECU.

WIRING DIAGRAM

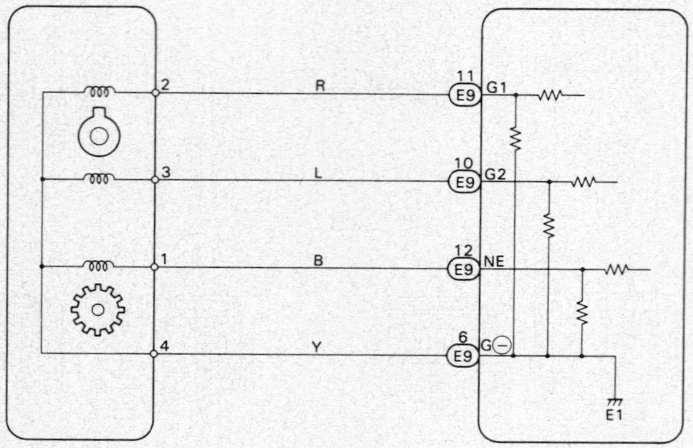

INSPECTION PROCEDURE

1 Check resistance of each pickup coils in distributor.

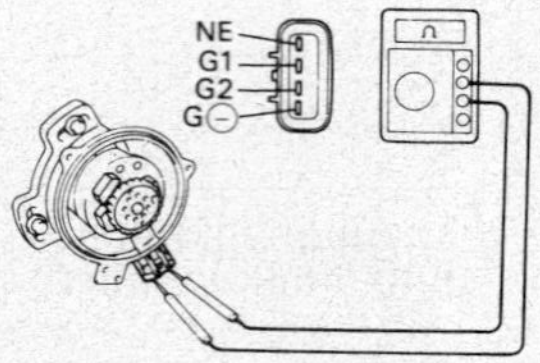

P Disconnect distributor connector.

C Measure resistance between each terminal shown in table below.

OK

	Terminals	Resistance
G1 pickup coil (Cold)	G1–G ⊖	125–190 Ω
G2 pickup coil (Cold)	G2–G ⊖	125–190 Ω
NE pickup coil (Cold)	NE–G ⊖	155–240 Ω

OK

NG Replace distributor.

2 Check for open and short in harness and connector between engine (& ECT) ECU and distributor

OK

NG Repair or replace harness or connector.

3 Check air gap.

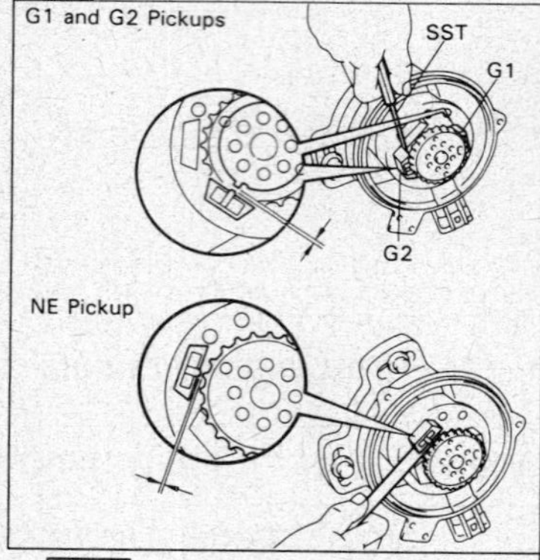

P Remove distributor cap and rotor.

C Using SST (G1 and G2 pickups) and a thickness gauge (NE pickup), measure the air gap between the signal rotor projection and pickup coil.
SST 09240-00020 for G1 and G2 pickups

OK Air gap: 0.2 – 0.4 mm (0.008 – 0.016 in.)

OK

NG Replace distributor.

Check and replace engine (& ECT) ECU.

Diag. Code 13 RPM Signal Circuit (No. 2)

CIRCUIT DESCRIPTION

Refer to RPM signal circuit (No. 1)

Code No.	Diagnostic Code Detecting Condition	Trouble Area
13	No NE signal to ECU for 0.1 sec. or more at 1,000 rpm or more. No 12 pulses of NE to ECU during the interval between G1 and G2 pulses.	• Open or short in NE circuit. • Distributor • ECU

DIAGNOSTIC CHART

This code indicates that a momentary interruption of the RPM signal from the distributor to the ECU has occurred, but that it is returned to normal. Note that although this problem may not necessarily appear at the time of inspection, it cannot be ignored because this diagnostic code is output, indicating that there is or was a malfunction in the RPM signal circuit; this "malfunction" is usually a loose connector.
The distributor connector and the NE terminal of the ECU connector must therefore be checked for the following:

1. Loose connectors
2. Dirty connector terminals
3. Loose connector terminals

Diag. Code 14 Ignition Signal Circuit

The ECU determines the ignition timing, turns on Tr_1 at a predetermined angle (°CA) before the desired ignition timing and outputs an ignition signal (IGT) "1" to the igniter.
Since the width of the IGT signal is constant, the dwell angle control circuit in the igniter determines the time the control circuit starts primary current flow to the ignition coil based on the engine rpm and ignition timing one revolution ago, that is, the time the Tr_2 turns on.
When it reaches the ignition timing, the ECU turns Tr_1 off and outputs the IGT signal "0".
This turns Tr_2 off, interrupting the primary current flow and generating a high voltage in the secondary coil which causes the spark plug to spark. Also, by the counter electromotive force generated when the primary current is interrupted, the igniter sends an ignition confirmation signal (IGF) to the ECU. The ECU stops fuel injection as a fail safe function when the IGF signal is not input to the ECU.

Code No.	Diagnostic Code Detecting Condition	Trouble Area
14	No IGF signal to ECU for 6 consecutive IGT signal.	• Open or short in IGF or IGT circuit from igniter to ECU. • Igniter • ECU

WIRING DIAGRAM

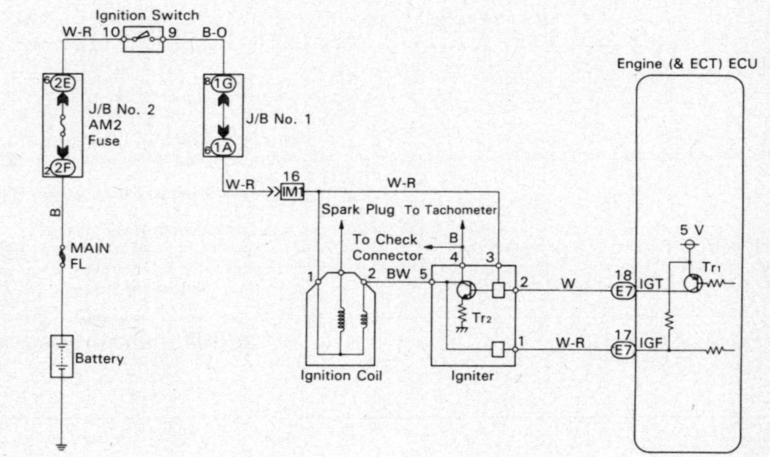

DIAGNOSTIC CHART

1. Check for spark.
   - NG → go to 4 (Check voltage of terminal IGT).
   - OK ↓
2. Check for open and short in IGF signal circuit between igniter and ECU.
   - NG → Repair or replace harness or connector.
   - OK ↓
3. Check voltage of terminal IGF.
   - OK → Replace igniter.
   - NG ↓ Check and replace ECU.
4. Check voltage of terminal IGT.
   - NG → go to 8 (Check voltage of terminal IGT (Disconnect igniter connector)).
   - OK ↓
5. Check voltage of igniter power source.
   - NG → Check and repair igniter power source circuit.
   - OK ↓
6. Check for open and short in primary coil circuit.
   - NG → Repair or replace harness or connector.
   - OK ↓
7. Check ignition coil.
   - NG → Replace ignition coil.
   - OK ↓ Replace igniter.
8. Check voltage of terminal IGT (Disconnect igniter connector).
   - OK → Replace igniter.
   - NG ↓
9. Check for open and short in IGT signal circuit between igniter and ECU.
   - NG → Repair or replace harness or connector.
   - OK ↓ Check and replace ECU.

INSPECTION PROCEDURE

1 Check for spark.

C Disconnect the high-tension cord from the distributor, hold its end about 12.5 mm (1/2") from the ground, see if spark occurs while the engine is being cranked.

OK **Spark should be generated.**

Hint To prevent excessive fuel injected from the injectors during this check, don't crank the engine for more than 1 – 2 seconds at a time.

OK ↓ | **NG** → Go to step **4**.

2 Check for open and short in harness and connector in IGF signal circuit between engine (& ECT) ECU and igniter

OK ↓ | **NG** → **Repair or replace harness or connector.**

3 Disconnect igniter connector and check voltage between terminal IGF of engine (& ECT) ECU connector and body ground.

ON
IG ON

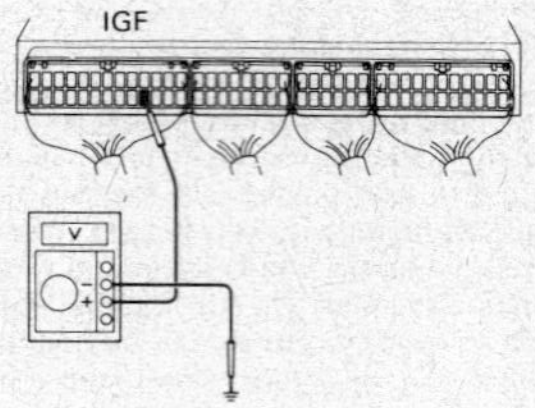

P (1) Disconnect igniter connector.
(2) Remove glove compartment.
(3) Turn ignition switch on.

C Measure voltage between terminal IGF of engine (& ECT) ECU connector and body ground.

OK **Voltage: 4 – 6 V**

OK → **Replace igniter.**

NG ↓

Check and replace engine (& ECT) ECU.

4 Check voltage between terminal IGT of engine (& ECT) ECU connector and body ground.

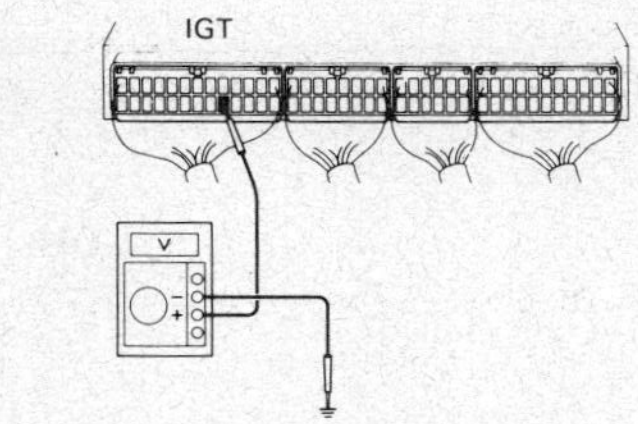

P Remove glove compartment.

C Measure voltage between terminal IGT of engine (& ECT) ECU connector and body ground when engine is cranked.

OK Voltage: 0.5 – 1.0 V
(Neither 0 V nor 5 V)

Reference **INSPECTION USING OSCILLOSCOPE**

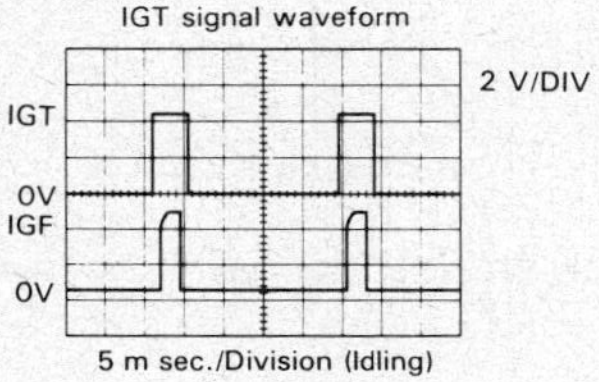

- During cranking or idling, check waveform between terminal IGT and E1 of engine (& ECT) ECU.

HINT: The correct waveform appears as shown in the illustration on the left, with rectangle waves.

OK | **NG** Go to step **8**.

5 Check voltage between terminal 3 of igniter connector and body ground.

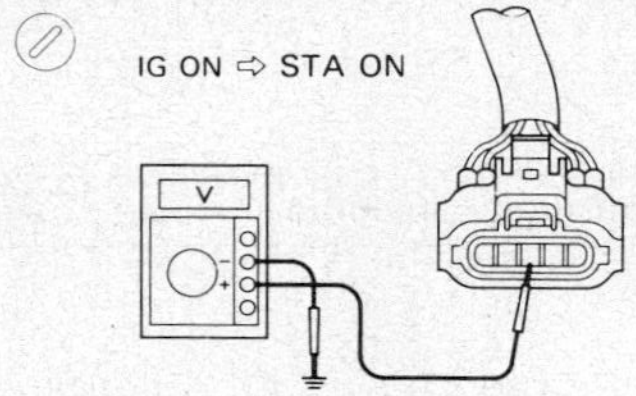

P Disconnect igniter connector.

C Measure voltage between terminal 3 of igniter connector and body ground, when ignition switch is turned to "ON" and "STA" position.

OK Voltage: 10 – 14 V

OK | **NG** **Check and repair igniter power source circuit.**

6 Check for open and short in harness and connector between ignition switch and ignition coil, ignition coil and igniter

OK | **NG** **Repair or replace harness or connector.**

7 Check ignition coil.

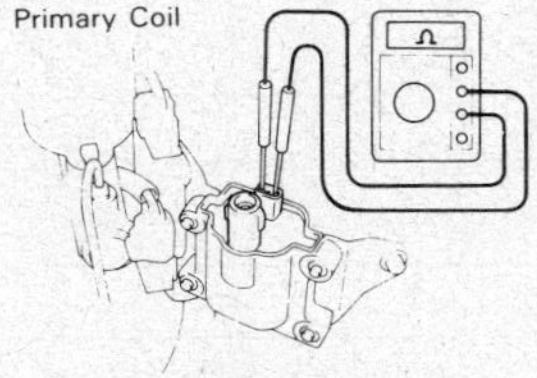

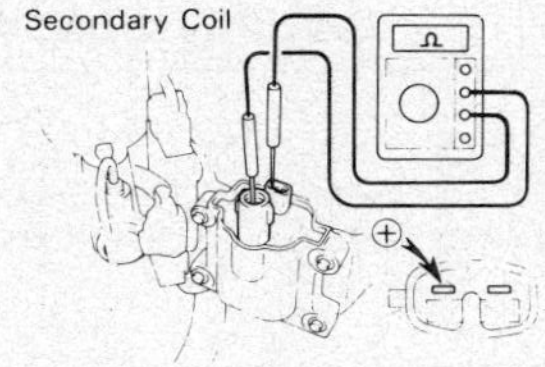

P Disconnect ignition coil connector.

C (1) Check primary coil.
Measure resistance between terminals of ignition coil connector.

(2) Check secondary coil.
Measure resistance between terminal ⊕ of ignition coil connector and high-tension terminal.

OK

	Resistance
Primary Coil (Cold)	0.2 – 0.3 Ω
Secondary Coil (Cold)	6 – 11 kΩ

OK | **NG** **Replace ignition coil.**

Replace igniter.

8 Disconnect igniter connector and check voltage between terminal IGT of engine (& ECT) ECU connector and body ground.

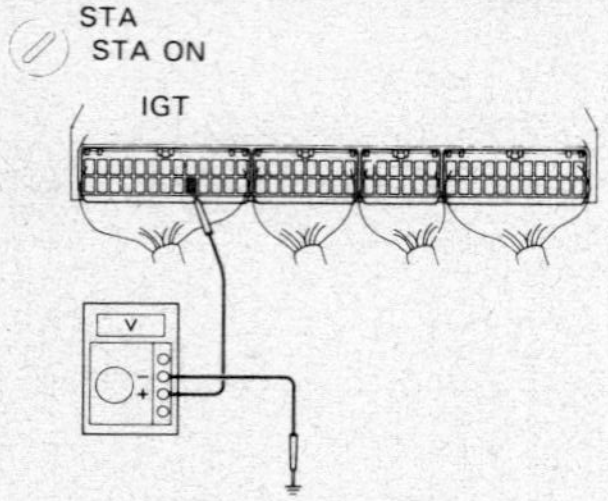

P Disconnect igniter connector.

C Measure voltage between terminal IGT of engine (& ECT) ECU connector and body ground when engine is cranked.

OK Voltage: 0.5 – 1.0 V
(Neither 0 V nor 5 V)

Reference — INSPECTION USING OSCILLOSCOPE

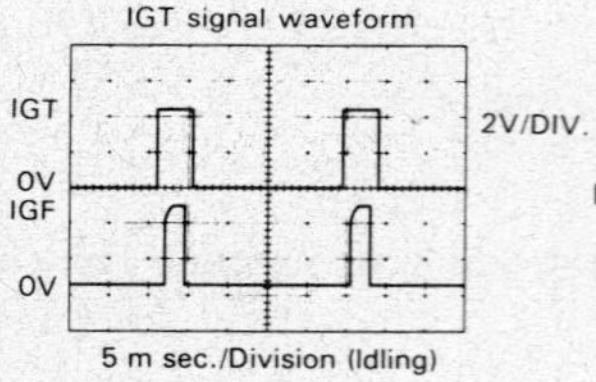

- During cranking or idling, measure between terminal IGT and E1 of engine (& ECT) ECU.

HINT: The correct waveform appears as shown in the illustration on the left, with rectangle waves.

NG ↓

OK → **Replace igniter.**

9 Check for open and short in harness and connector in IGT signal circuit between engine (& ECT) ECU and igniter

NG → **Repair or replace harness or connector.**

OK ↓

Check and replace engine (& ECT) ECU.

Diag. Code 16 — ECT Control Signal Malfunction

CIRCUIT DESCRIPTION

The signal from the ECT CPU retards the ignition timing of the engine during ECT gear shifting, thus momentarily reducing torque output of the engine for smooth clutch operation inside the transmission and reduced shift shock.

Code No.	Diagnostic Code Detecting Condition	Trouble Area
16	Fault in communications between the engine CPU and ECT CPU in the ECU.	• ECU

If the ECU records the diagnostic code "16" in memory, it prohibits the torque control of the ECT which performs smooth gear shifting.

DIAGNOSTIC CHART

1 Are there any other codes (besides Code 16) being output?

YES → Go to relevant diag. code chart.

NO ↓

Replace engine (& ECT) ECU.

Diag. Code	21, 28	Main Oxygen Sensor Circuit

CIRCUIT DESCRIPTION

To obtain a high purification rate for the CO, HC and NOx components of the exhaust gas, a three-way catalyst is used, but for most efficient use of the three-way catalyst, the air-fuel ratio must be precisely controlled so that it is always close to the stoichiometric air-fuel ratio.

The oxygen sensor has the characteristic whereby its output voltage changes suddenly in the vicinity of the stoichiometric air-fuel ratio. This characteristic is used to detect the oxygen concentration in the exhaust gas and provide feedback to the computer for control of the air-fuel ratio.

When the air-fuel ratio becomes LEAN, the oxygen concentration in the exhaust increases and the oxygen sensor informs the ECU of the LEAN condition (small electromotive force: 0 V).

When the air-fuel ratio is RICHER than the stoichiometric air-fuel ratio the oxygen concentration in the exhaust gas is reduced and the oxygen sensor informs the ECU of the RICH condition (large electromotive force: 1 V).

The ECU judges by the electromotive force from the oxygen sensor whether the air-fuel ratio is RICH or LEAN and controls the injection time accordingly. However, if malfunction of the oxygen sensor causes output of abnormal electromotive force, the ECU is unable to perform accurate air-fuel ratio control.

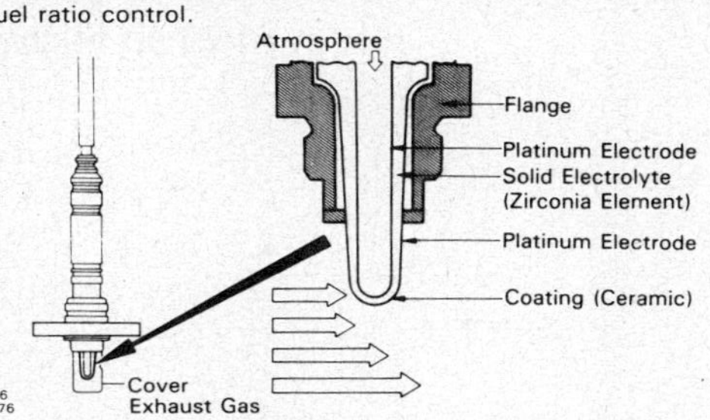

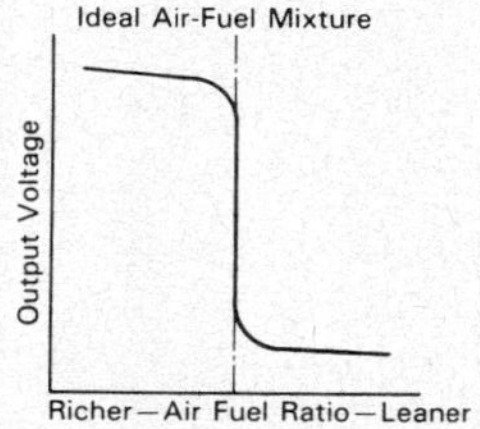

Code No.	Diagnostic Code Detecting Condition	Trouble Area
21 · 28	Main oxygen sensor signal voltage is reduced to between 0.35 V and 0.70 V for 60 sec. under conditions (a) ~ (d). (2 trip detection logic) (a) Coolant temp.: Between 70°C (158°F) and 90°C (194°F). (b) Engine speed: 1,500 rpm or more. (c) Load driving (Ex. ECT in 4th speed, A/C ON, Flat road, 50 mph (80 km/h)). (d) Main oxygen sensor signal voltage: Alternating above and below 0.45 V.	• Main oxygen sensor cirucit • Main oxygen sensor

HINT: Diag. code 21 is for the left bank main oxygen sensor circuit. Diag. code 28 is for the right bank main oxygen sensor circuit.

CIRCUIT DESCRIPTION (Cont'd)

DIAGNOSIS CODE DETECTION DRIVING PATTERN

Purpose of the driving pattern.
(a) To simulate diag. code detecting condition after diag. code is recorded.
(b) To check that the malfunction is corrected when the repair is completed confirming that diag. code is no longer detected.

Malfunction: Main Oxygen Sensor Deterioration

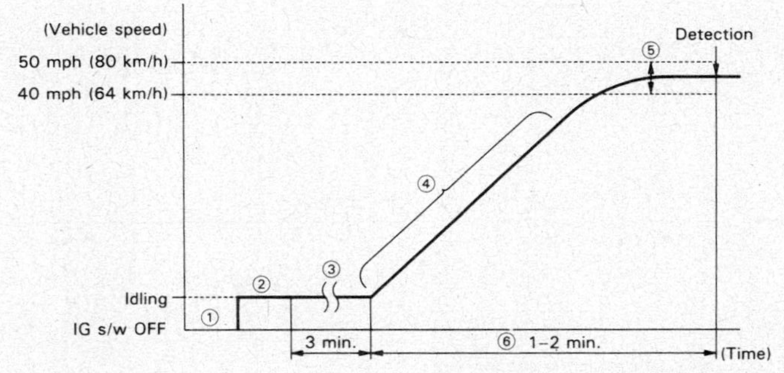

① Disconnect the EFI fuse (15 A) for 10 sec. or more, with IG switch OFF.
Initiate test mode (Connect terminal TE2 and E1 of check connector or TDCL with IG switch OFF).
② Start the engine and warm up with all ACC switch OFF.
③ After the engine is warmed up, let it idle for 3 min.
④ After performing the idling in ③, perform gradual acceleration with in the range 1,300 ~ 1,700 rpm (centered around 1,500 rpm) with the A/C switch ON and D range for A/T (5th for M/T).
(Take care that the engine speed does not fall below 1,200 rpm when shifting. Gradually depress the accelerator pedal and kept it.
Steady so that engine braking does not occur).
⑤ Maintain the vehicle speed at 40 – 50 mph (64 – 80 km/h).
⑥ Keep the vehicle running for 1 – 2 min. after starting acceleration.

HINT: If a malfunction exists, the "CHECK" engine warning light will light up after approx. 60 sec. from the start of acceleration.

NOTICE: If the conditions in this test are not strictly followed, detection of the malfunction will not be possible.

DIAGNOSTIC CHART

1 Are there any other codes (besides code 21 or 28) being output?

- YES → Go to relevant diag. code chart.
- NO → Replace oxygen sensor.

HINT: If diag. code 21 is output, replace the left bank main oxygen sensor.
If diag. code 28 is output, replace the right bank main oxygen sensor.

WIRING DIAGRAM

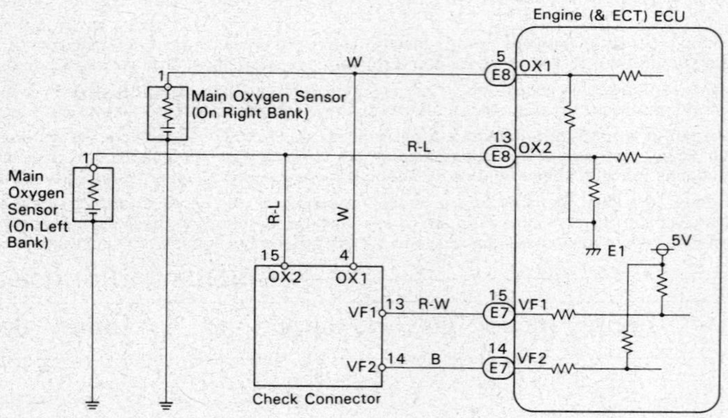

Reference | **INSPECTION USING OSCILLOSCOPE**

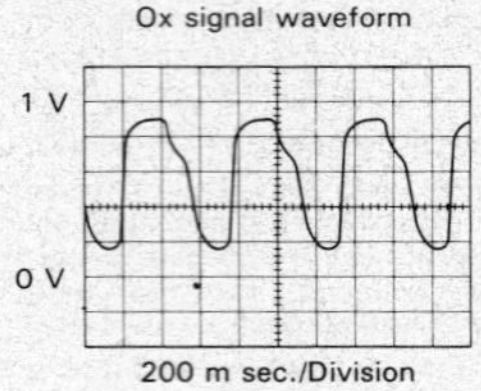

- With the engine racing (4,000 rpm) measure between terminals OX1, OX2 and E1 of engine (& ECT) ECU.

HINT: The correct waveform appears as shown in the illustration on the left, oscillating between approx. 0.1 V and 0.9 V.

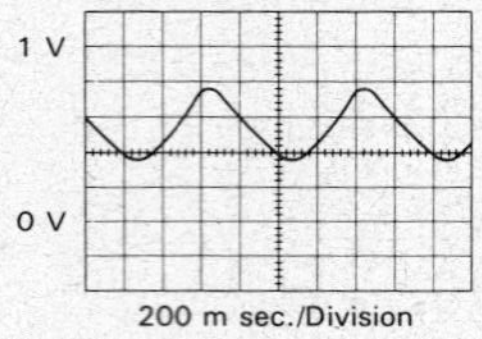

If the oxygen sensor is deteriorated, the amplitude of the voltage is reduced as shown on the left.

Diag. Code	22	Water Temp. Sensor Circuit

CIRCUIT DESCRIPTION

The water temperature sensor senses the coolant temperature. A thermistor built in the sensor changes the resistance value according to the coolant temperature. The lower the coolant temperature, the greater the thermistor resistance value, and the higher the coolant temperature, the lower the thermistor resistance value (See Fig. 1).
The water temperature sensor is connected to the ECU (See next page). The 5 V power source voltage in the ECU is applied to the water temperature sensor from the terminal THW via a resistor R. That is, the resistor R and the water temperature sensor are connected in series. When the resistance value of the water temperature sensor changes in accordance with changes in the coolant temperature, the potential at the terminal THW also changes. Based on this signal, the ECU increases the fuel injection volume to improve driveability during cold engine operation.
If the ECU records the diagnostic code 22, it operates the fail safe function, keeping the coolant temperature at a constant 80°C (176°F).

(fig-1)

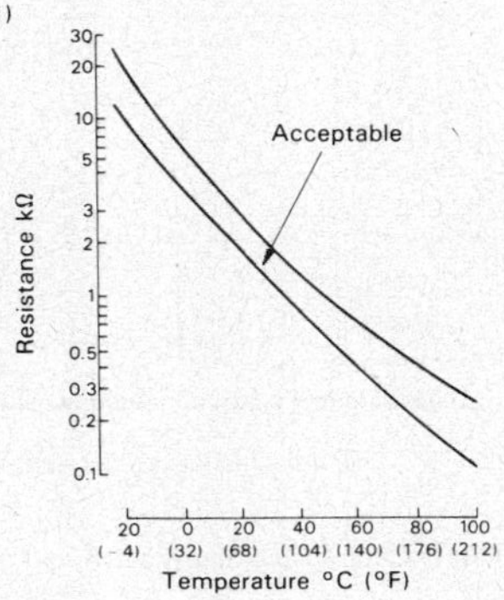

< Reference >

Water Temp. °C (°F)	Resistance (kΩ)	Voltage (V)
−20 (−4)	16.0	4.3
0 (32)	5.9	3.4
20 (68)	2.5	2.4
40 (104)	1.2	1.5
60 (140)	0.6	0.9
80 (176)	0.3	0.5
100 (212)	0.2	0.3

Code No.	Diagnostic Code Detecting Condition	Trouble Area
22	Open or short in water temp. sensor circuit for 0.5 sec. or more.	• Open or short in water temp. sensor circuit • Water temp. sensor • ECU

DIAGNOSTIC CHART

HINT: If diagnostic codes 22, 24, 32 and 41 are output simultaneously, E2 (Sensor ground) may be open.

1 Check voltage of sensor. — OK → Check for momentary interruption

NG ↓

2 Check resistance of sensor. — NG → Replace water temp. sensor.

OK ↓

3 Check for open and short in harness and connector between ECU and water temp. sensor. — NG → Repair or replace harness or connector.

OK ↓

Check and replace ECU.

WIRING DIAGRAM

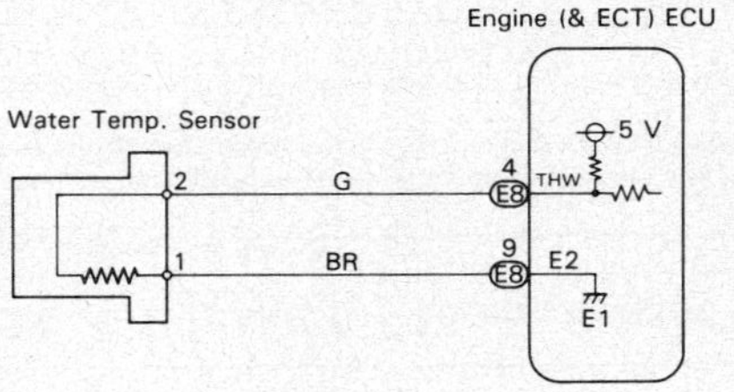

INSPECTION PROCEDURE

HINT: If diagnostic codes 22, 24, 32 and 41 are output simultaneously, E2 (sensor ground) may be open.

1 Check voltage between terminals THW and E2 of engine (& ECT) ECU connector.

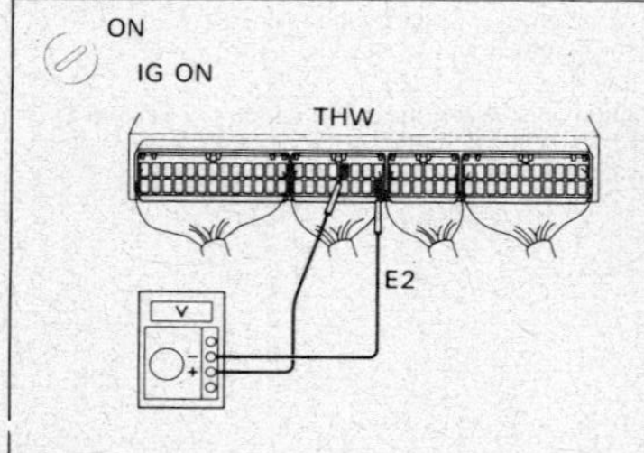

P (1) Remove glove compartment.

(2) Turn ignition switch on.

C Measure voltage between terminals THW and E2 of engine (& ECT) ECU connector.

OK

Water Temp. °C (°F)	Voltage
20 (68) (Engine is cool)	1 – 3 V
80 (176) (Engine is hot)	0.1 – 1.0 V

OK → **Check for momentary interruption**

NG ↓

2 Check water temp. sensor.

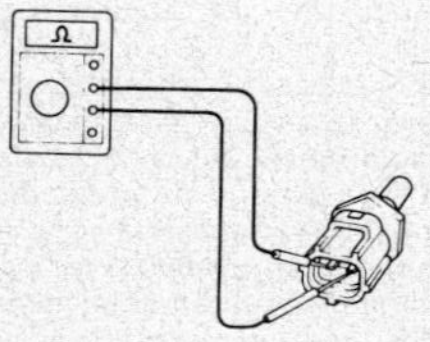

P Disconnect the water temp. sensor connector.

C Measure resistance between terminals.

OK **Resistance is within Acceptable Zone on chart.**

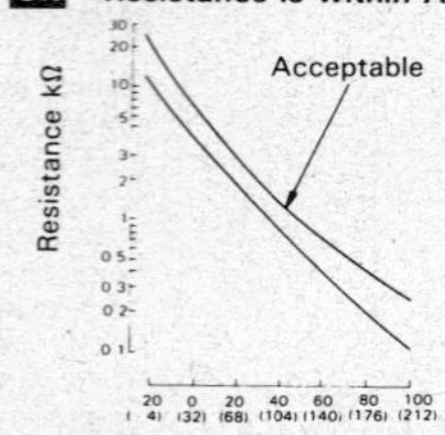

Water temp. °C (°F)	Resistance
20 (68)	2 – 3 kΩ
80 (176)	0.2 – 0.4 kΩ

NG → **Replace water temp. sensor.**

OK ↓

3 Check for open and short in harness and connector between engine (& ECT) ECU and water temp. sensor

NG → **Repair or replace harness or connector.**

OK ↓

Check and replace engine (& ECT) ECU.

Diag. Code	24	Intake Air Temp. Sensor Circuit

CIRCUIT DESCRIPTION

The intake air temp. sensor is built into the air flow meter and senses the intake air temperature. The structure of the sensor and connection to the ECU is the same as in the water temp. sensor

If the ECU records the diagnostic code "24", it operates the fail safe function, keeping the intake air temperature at a constant 20°C (68°F).

Code No.	Diagnostic Code Detecting Condition	Trouble Area
24	Open or short in intake air temp. sensor circuit for 0.5 sec. or more.	• Open or short in intake air temp. sensor circuit. • Intake air temp. sensor • ECU

DIAGNOSTIC CHART

HINT: If diagnostic codes 22, 24, 32 and 41 are output simultaneously, E2 (sensor ground) may be open.

1. Check voltage of sensor. — OK → Check for momentary interruption
   - NG ↓
2. Check resistance of sensor. — NG → Replace intake air temp. sensor.
   - OK ↓
3. Check for open and short in harness and connector between ECU and intake air temp. sensor. — NG → Repair or replace harness or connector.
   - OK ↓

Check and replace ECU.

WIRING DIAGRAM

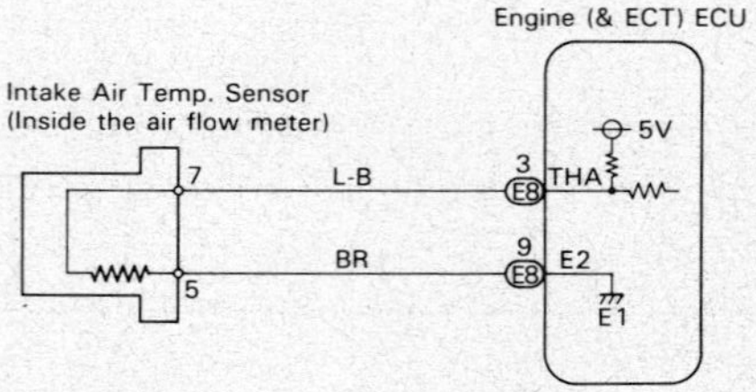

INSPECTION PROCEDURE

1 Check voltage between terminals THA and E2 of engine (& ECT) ECU connector.

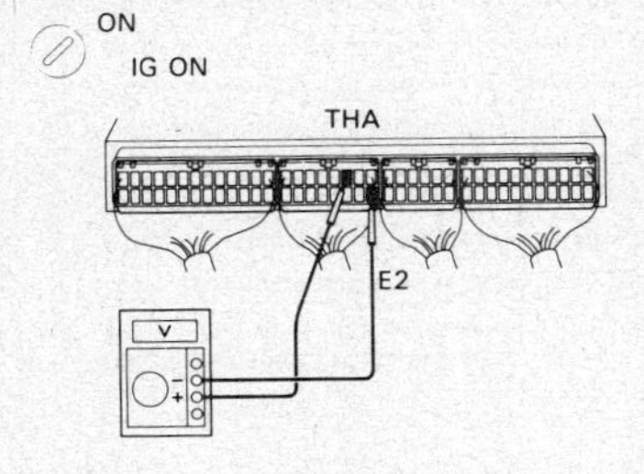

P (1) Remove glove compartment.

(2) Turn ignition switch on.

C Measure voltage between terminals THA and E2 of engine (& ECT) ECU connector.

OK

Intake air temp. °C (°F)	Voltage
20 (68)	1 – 3 V
60 (140)	0.5 – 1.0 V

NG

OK → **Check for momentary interruption**

2 Check intake air temp. sensor.

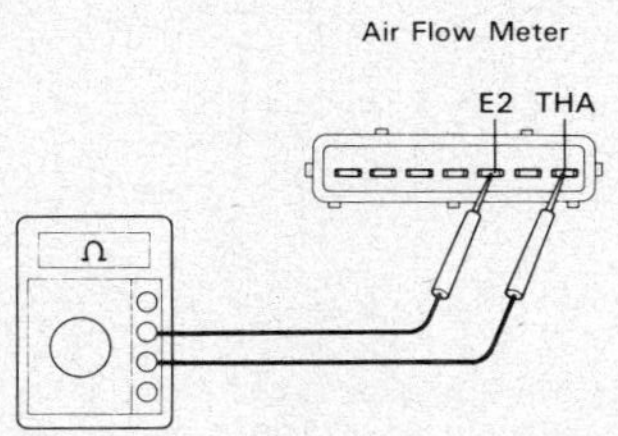

P Disconnect the air flow meter connector.

C Measure resistance between terminals THA and E2 of air flow meter connector

OK **Resistance is within Acceptable Zone on chart.**

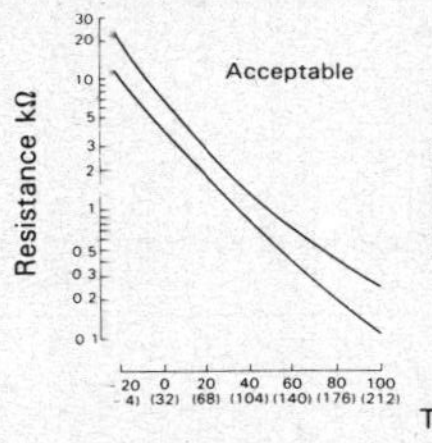

Intake air temp. °C (°F)	Resistance
20 (68)	2 – 3 kΩ
60 (140)	0.4 – 0.7 kΩ

OK

NG → **Replace intake air temp. sensor (Replace air flow meter).**

3 Check for open and short in harness and connector between engine (& ECT) ECU and intake air temp. sensor

OK

NG → **Repair or replace harness or connector.**

Check and replace engine (& ECT) ECU.

Diag. Code		
	25	Air-Fuel Ratio Lean Malfunction
	26	Air-Fuel Ratio Rich Malfunction

CIRCUIT DESCRIPTION

The main oxygen sensor is located in the exhaust manifold.
It indirectly determines whether the fuel mixture is rich or lean by detecting the concentration of oxygen present in the exhaust gas.

Code No.	Diagnostic Code Detecting Condition	Trouble Area
25	Main oxygen sensor voltage is 0.45 V or less (lean) for 90 sec. under conditions (a) and (b). (2 trip detection logic) (a) Coolant temp.: 50°C (122°F) or more. (b) Engine speed: 1,500 rpm or more.	• Open or short in main oxygen sensor circuit • Main oxygen sensor • Ignition system • ECU
	*1 Difference of air-fuel ratio feedback compensation value between right and left banks is more than 15 percentage for 20 sec. or more under conditions (a) and (b). (2 trip detection logic) (a) Coolant temp.: Between 70°C (158°F) and 95°C (203°F). (b) Engine speed: Idling	• Fuel line pressure (injector leak, blockage) • Mechanical system malfunction (skipping teeth of timing belt) • Ignition system • Compression pressure • Air flow meter • ECU
26*1	Difference of air-fuel ratio feedback compensation value between right and left banks is more than 15 percentage for 20 sec. or more under conditions (a) and (b). (2 trip detection logic) (a) Coolant temp.: Between 70°C (158°F) and 95°C (203°F). (b) Engine speed: Idling	• Fuel line pressure (injector leak, blockage) • Mechanical system malfunction (skipping teeth of timing belt) • Ignition system • Compression pressure • Air flow meter • ECU

*1: Only for California specification vehicles.

CIRCUIT DESCRIPTION (Cont'd)

DIAGNOSIS CODE DETECTION DRIVING PATTERN

Purpose of the driving pattern.

(a) To simulate diag. code detecting condition after diag. code is recorded.

(b) To check that the malfunction is corrected when the repair is completed confirming that diag. code is no longer detected.

Malfunction: Open or Short in Main Oxygen Sensor
Open or Short in Injector circuit, Injector Leak or Blockage

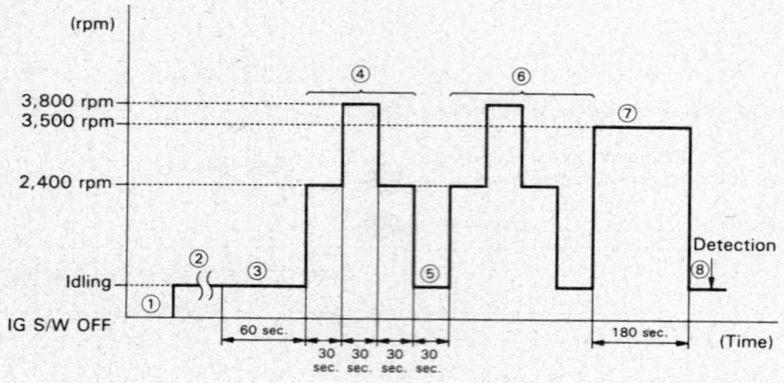

① Disconnect the EFI fuse (15 A) for 10 sec. or more, with IG switch OFF.
Initiate test mode (Connect terminal TE2 and E1 of check connector or TDCL with IG S/W OFF.)

② Start the engine and warm up, with all ACC switch OFF.

③ After the engine is warmed up, let it idle for 60 sec.

④ After performing the idling in ③, perform racing at 2,400 rpm, 3,800 rpm then 2,400 rpm again for 30 sec. each.

⑤ After performing the racing in ④, perform idling for 30 sec.

⑥ After performing the idling in ⑤, perform steps ④ and ⑤ again.

⑦ After performing the idling in ⑥, perform idling at 3,500 rpm for 180 sec.

⑧ After performing the racing in ⑦, perform idling 30 sec.

HINT: If a malfunction exists, the "CHECK" engine warning light will light up during step ⑧.

NOTICE: If the conditions in this test are not strictly followed, detection of the malfunction will not be possible.

DIAGNOSTIC CHART

1. Check voltage of terminal VF. — NG Type I → 2; OK → 7; NG Type II → 7
2. Check voltage of terminal OX. — NG → 3; OK → 7
3. Check for open and short in harness and connector. — NG → Repair or replace harness or connector. — OK → 4
4. Check each item found to be a possible cause of problem. — NG → Repair or replace. — OK → 5
5. Check compression. — NG → Repair or replace. — OK → 6
6. Does malfunction disappear when a good main oxygen sensor is installed? — YES → Replace main oxygen sensor. — NO → Check and replace ECU.
7. Check each item found to be a possible cause of problem. — NG → Repair or replace. — OK → 8
8. Check compression. — NG → Repair or replace. — OK → 9
9. Does malfunction disappear when a good main oxygen sensor is installed? — YES → Replace main oxygen sensor. — NO → Check and replace ECU.

INSPECTION PROCEDURE

1 Check voltage between terminals VF1, VF2 and E1 of check connector.

Left Bank

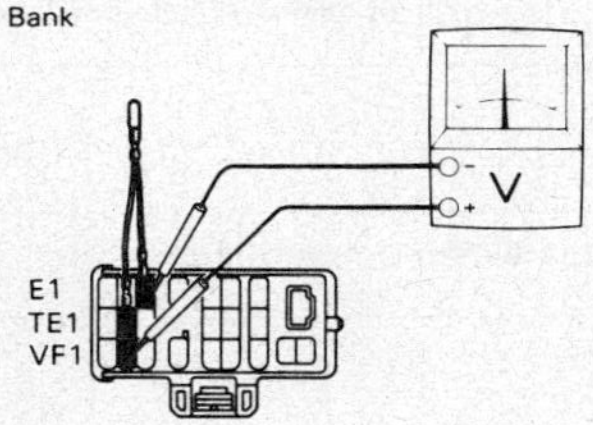

Right Bank

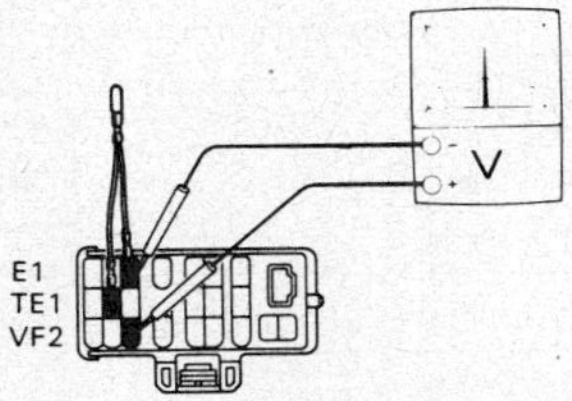

P (1) Warm up engine at normal operating temperature.
(2) Connect terminals TE1 and E1 of check connector.
(3) Connect positive prove to terminal VF1, VF2 and negative prove to terminal E1 of check connector.

C (1) Warm up the oxygen sensor by running engine at 2,500 rpm for about 2 minutes.
(2) Then, maintaining engine at 2,500 rpm, count how many times needle of voltmeter fluctuates between 0 and 5 V.

Result

Result	
Needle fluctuates of 8 times for every ten seconds	OK
Continue at 0 V	NG Type I
Continue at 5 V	NG Type II

NG Type I

NG Type II → Go to step 7

2 Check voltage between terminals OX1, OX2 and E1 of check connector.

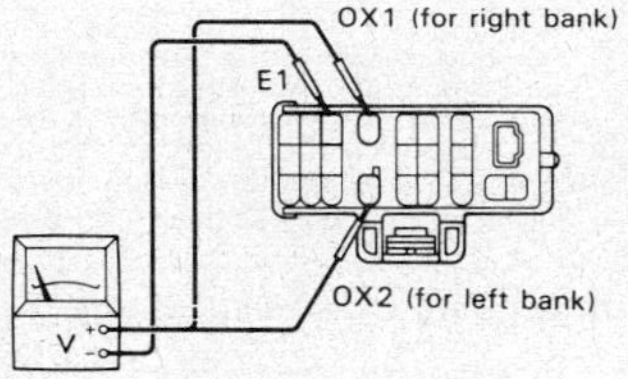

P Warm up engine at normal operating temperature.

C Measure voltage between terminals OX1, OX2 and E1 of check connector when engine is suddenly raced to full throttle.

OK **The voltage should be 0.5 V or higher at least once.**

Hint Perform inspection within 1 second.

NG

OK → Go to step 7

3 Check for open and short in harness and connector between engine (& ECT) ECU and main oxygen sensor, engine (& ECT) ECU and check connector

OK

NG → Repair or replace harness or connector.

4 Check each item found to be a possible cause of problem.

Check each circuit found to be a possible cause of trouble according to the results of the check in 1 or 2. The numbers in the table below show the order in which the checks should be performed.

Main oxygen sensor signal from either side continues at 0 V.	Main oxygen sensor signals from both sides continue at 0 V.	Possible Cause	
1		Faulty sensor installation.	
3		Injector circuit	
2	3	Misfire	
4		Valve timing	
	1	Air leakage	
	2	Fuel system	
	6	Characteristics deviation in air flow meter.	
	4	Characteristics deviation in water temp. sensor.	
	5	Characteristics deviation in intake air temp. sensor.	

OK

NG → Repair or replace.

5 Check compression

OK

NG → Repair or replace.

6 Does malfunction disappear when a good main oxygen sensor is installed?

NO

YES → Replace main oxygen sensor.

Check and replace engine (& ECT) ECU.

7 Check each item found to be a possible cause of problem.

Check each circuit found to be a possible cause of trouble according to the results of the check in 1. The numbers in the table below show the order in which the checks should be performed.

Main oxygen sensor signal from either side continues at 5.0 V.	Main oxygen sensor signals from both sides continue at 5.0 V.	Main oxygen sensor signals from both sides are normal.	Possible Cause	
1		7	Injector circuit	
		3	Misfire	
2		4	Valve timing	
		1	Air leakage	
	1	2	Fuel system	
3	4		Cold start injector circuit	
	5	8	Characteristics deviation in air flow meter.	
	2	5	Characteristics deviation in water temp. sensor.	
	3	6	Characteristics deviation in intake air temp. sensor.	

OK ↓ | **NG** → Repair or replace.

8 Check compression

OK ↓ | **NG** → Repair or replace.

9 Does malfunction disappear when a good main oxygen sensor is installed?

NO ↓ | **YES** → Replace main oxygen sensor.

Check and replace engine (& ECT) ECU.

Diag. Code 27 — Sub-Oxygen Sensor Circuit (Only for California spec.)

CIRCUIT DESCRIPTION

The sub-oxygen sensor is installed on the exhaust pipe. Its construction and operation is the same as the main oxygen sensor

Code No.	Diagnostic Code Detecting Condition	Trouble Area
27	(1) Open or short in heater circuit of sub-oxygen sensor for 0.5 sec. or more.	• Open or short in heater circuit of sub-oxygen sensor. • Sub-oxygen sensor heater. • ECU
	(2) Main oxygen sensor signal is 0.45 V or more and sub-oxygen sensor signal is 0.45 V or less under conditions (a) ~ (c). (2 trip detection logic) (a) Coolant temp.: 80°C (176°F) or more. (b) Engine speed: 1,500 rpm or more. (c) Accel. pedal: Fully depressed for 2 sec. or more.	• Open or short in sub-oxygen sensor cirucit. • Sub-oxygen sensor • ECU

CIRCUIT DESCRIPTION (Cont'd)

DIAGNOSIS CODE DETECTION DRIVING PATTERN

Purpose of the driving pattern.

(a) To simulate diag. code detecting condition after diag. code is recorded.

(b) To check that the malfunction is corrected when the repair is completed confirming that diag. code is no longer detected.

Malfunction: Open or Short in Sub-Oxygen Sensor

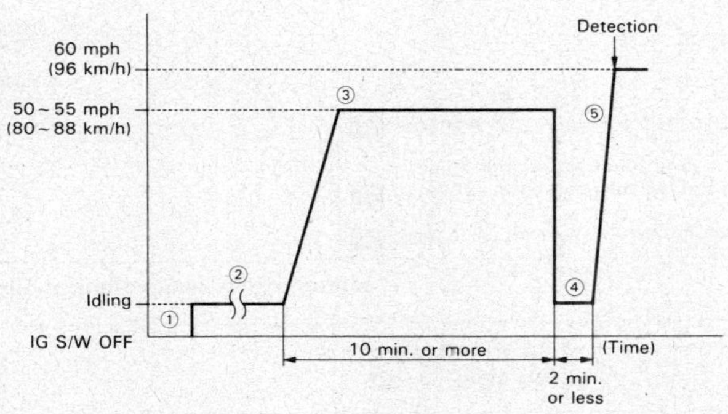

① Disconnect the EFI fuse (15 A) for 10 sec. or more, with IG switch OFF.
Initiate test mode (Connect terminal TE2 and E1 of check connector with IG switch OFF).

② Start the engine and warm up, with all ACC switch OFF.

③ After the engine is warmed up, let it drive at 50 ~ 55 mph (80 ~ 88 km/h) for 10 min. or more.

④ After driving, stop at a safe place and perform idling for 2 min. or less.

⑤ After performing the idling in ④, perform acceleration to 60 mph (96 km/h) with the throttle valve fully open.

HINT: If a malfunction exists, the "CHECK" engine warning light will light up during step ⑤.

NOTICE: If the conditions in this test are not strictly followed, detection of the malfunction will not be possible.

DIAGNOSTIC CHART

HINT:

- When other codes are output in addition to 27 at the same time, check the circuits for other codes first.

1 Check voltage of terminal HT.
- OK → go to 4
- NG ↓

2 Check resistance of sub-oxygen sensor heater.
- NG → Replace sub-oxygen sensor.
- OK ↓

3 Check for open and short in harness and connector between EFI main relay and ECU.
- NG → Repair or replace harness or connector.
- OK ↓

Check and replace ECU.

4 Check operation of sub-oxygen sensor heater.
- OK → Replace sub-oxygen sensor.*
- NG ↓

Check and replace ECU.

*: In this case, oxygen sensor can be deteriorated.

WIRING DIAGRAM

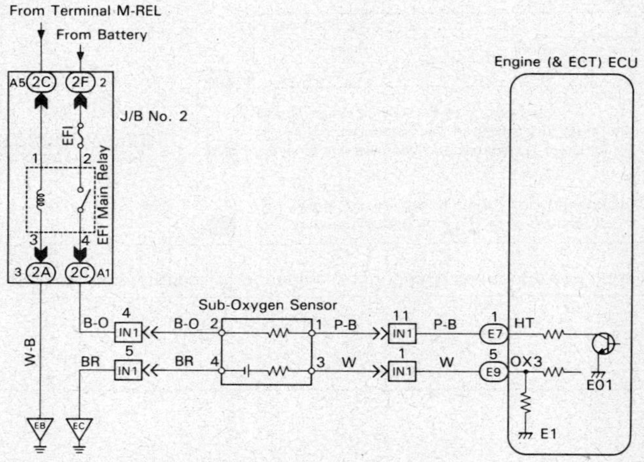

INSPECTION PROCEDURE

1 Check voltage between terminal HT of engine (& ECT) ECU connector and body ground.

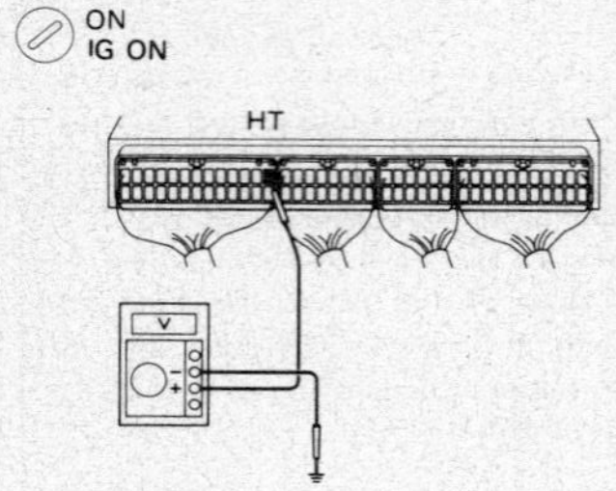

P (1) Remove glove compartment.
(2) Turn ignition switch on.

C Measure voltage between terminal HT of engine (& ECT) ECU connector and body ground.

OK **Voltage: 10 – 14 V**

OK → **Go to step 4.**

NG ↓

2 Check sub-oxygen sensor heater.

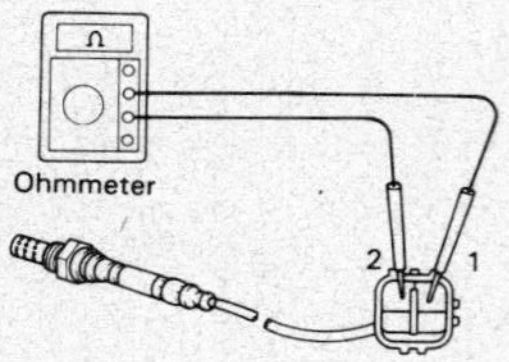

P Disconnect sub-oxygen sensor connector.

C Measure resistance between terminals 1 and 2 of sub-oxygen sensor connector.

OK **Resistance: 11.7 – 14.3 Ω 20°C (68°F)**

NG → **Replace sub-oxygen sensor.**

OK ↓

3 Check for open and short in harness and connector between EFI main relay and engine (& ECT) ECU

NG → **Repair or replace harness or connector.**

OK ↓

Check and replace engine (& ECT) ECU.

4 Check voltage between terminal HT of engine (& ECT) ECU connector and body ground.

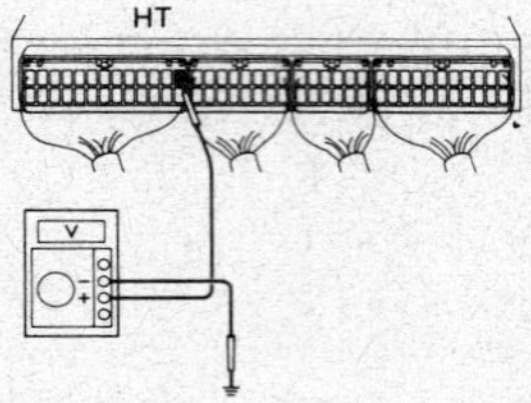

P (1) Connect engine (& ECT) ECU connector.
(2) Warm up engine to normal operating temperature.

C Measure voltage between terminal HT of engine (& ECT) ECU connector and body ground, when engine is idling and racing at 2,500 rpm.

OK

	Voltage
Idling	0 V
Racing at 2,500 rpm	10 – 14 V

Hint In the 2,500 rpm racing check, continue engine racing at 2,500 rpm for approx. 20 seconds or more.

OK → **Replace sub-oxygen sensor.***

*: In this case, oxygen sensor can be deteriorated.

NG ↓

Check and replace engine (& ECT) ECU.

Diag. Code 31 — Air Flow Meter Circuit (No.1)

CIRCUIT DESCRIPTION

The air flow meter is located directly behind the air cleaner, and detects the intake air volume. It converts the intake air readings into a voltage signal by means of a potentiometer. Intake air volume signal is used to calculate the basic injection duration and basic ignition advance angle.

Code No.	Diagnostic Code Detecting Condition	Trouble Area
31	All conditions below are detected continuously for 0.5 sec. or more. (a) Idle switch: ON (IDL) (b) Voltage of terminal VS: 0.1 V or less (c) Engine speed: 1,500 rpm or less	• Open (VC) or short (VC-E2) in air flow meter circuit. • Air flow meter • ECU

DIAGNOSTIC CHART

1 Check voltage of terminal VC. — OK → Check for momentary interruption.
NG ↓

2 Check air flow meter. — NG → Replace air flow meter.
OK ↓

3 Check for open and short in harness and connector between ECU and air flow meter. — NG → Repair or replace harness or connector.
OK ↓

Check and replace ECU.

WIRING DIAGRAM

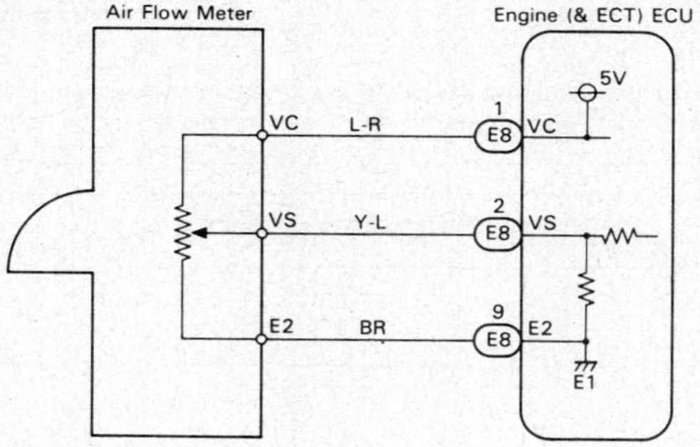

INSPECTION PROCEDURE

1 **Check voltage between terminals VC and E2 of engine (& ECT) ECU.**

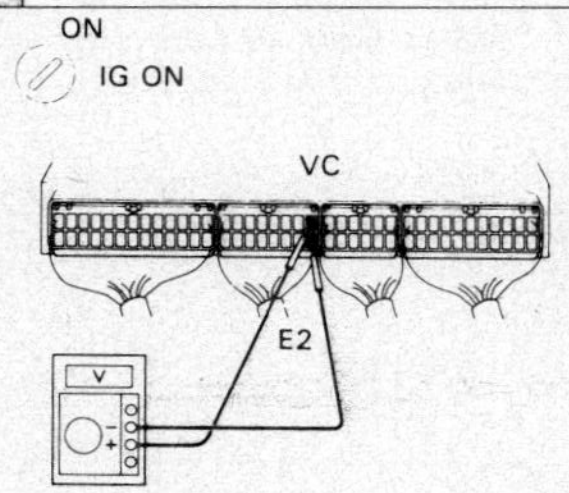

P (1) Remove glove compartment.

(2) Turn ignition switch ON.

C Measure voltage between terminals VC and E2 of engine (& ECT) ECU.

OK **Voltage: 4 – 6 V**

OK → **Check for momentary interruption**

NG ↓

2 **Check resistance between terminals VC and E2 of air flow meter.**

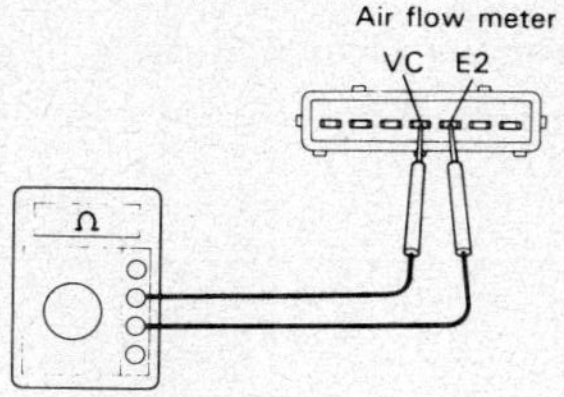

P Disconnect the air flow meter connector.

C Measure resistance between terminals VC and E2 of air flow meter.

OK **Resistance: 200 – 400 Ω**

NG → **Replace air flow meter.**

OK ↓

3 **Check for open and short in harness and connector between engine (& ECT) ECU and air flow meter**

NG → **Repair or replace harness or connector.**

OK ↓

Check and replace engine (& ECT) ECU.

Diag. Code	32	Air Flow Meter Circuit (No. 2)

CIRCUIT DESCRIPTION

The air flow meter is located directly behind the air cleaner, and detects the intake air volume. It converts the intake air readings into a voltage signal by means of a potentiometer. Intake air volume signal is used to calculate the basic injection duration and basic ignition advance angle.

Code No.	Diagnostic Code Detecting Condition	Trouble Area
32	VS exceeds 4.9 V for 0.5 sec. or more.	• Open (E2) or short (VC-VS) in air flow meter circuit. • Air flow meter • ECU

DIAGNOSTIC CHART

HINT: If diagnostic codes 22, 24, 32 and 41 are output simultaneously, E2 (sensor ground) may be open.

1. Check voltage of terminal VS. — OK → Check for momentary interruption.
   - NG ↓
2. Check air flow meter. — NG → Replace air flow meter.
   - OK ↓
3. Check for open and short in harness and connector between ECU and air flow meter. — NG → Repair or replace harness or connector.
   - OK ↓
   - Check and replace ECU.

WIRING DIAGRAM

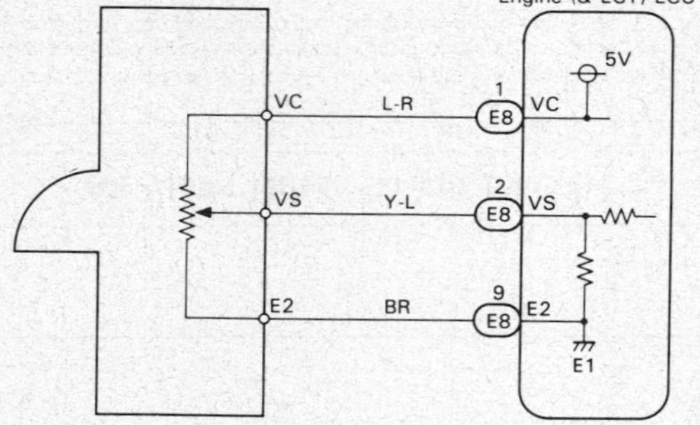

INSPECTION PROCEDURE

1 Check voltage between terminals VS and E2 of engine (& ECT) ECU.

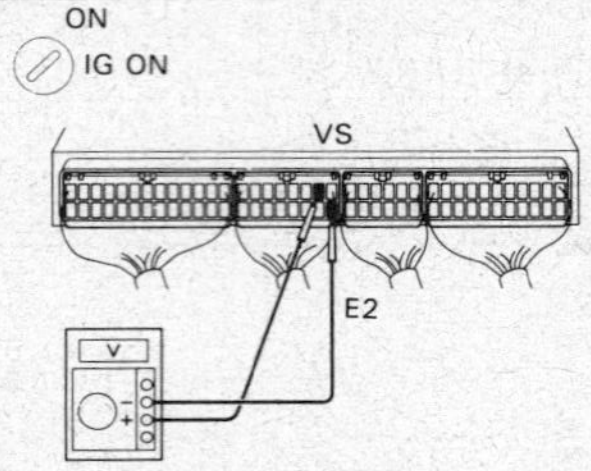

P (1) Remove air cleaner cap.
(2) Remove glove compartment.
(3) Turn ignition switch ON.

C Measure voltage between terminals VS and E2 of engine (& ECT) ECU.

OK

Measuring Plate	Voltage
Fully closed	3.7 – 4.3 V
Fully open	0.2 – 0.5 V

OK → Check for momentary interruption

NG ↓

2 Check resistance between terminals VS and E2 of air flow meter.

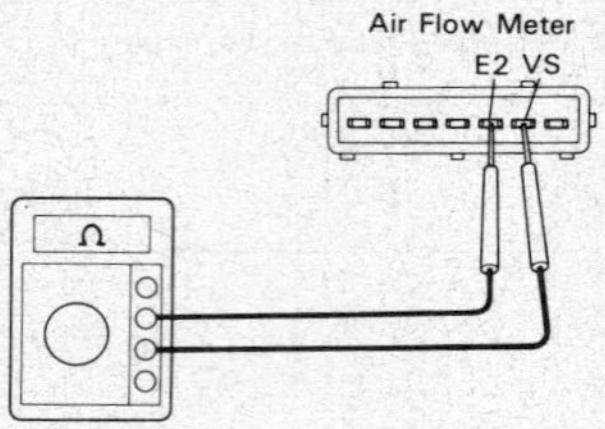

P Disconnect the air flow meter connector.

C Measure resistance between terminals VS and E2 of air flow meter.

OK

Measuring Plate	Resistance
Fully closed	200 – 600 Ω
Fully open	20 – 1,200 Ω

NG → Replace air flow meter.

OK ↓

3 Check for open and short in harness and connector between engine (& ECT) ECU and air flow meter

NG → Repair or replace harness or connector.

OK ↓

Check and replace engine (& ECT) ECU.

Diag. Code 41 | Throttle Position Sensor Circuit

CIRCUIT DESCRIPTION

The throttle position sensor is mounted in the throttle body and detects the throttle valve opening angle. When the throttle valve is fully closed, the IDL contacts in the throttle position sensor are on, so the voltage at the terminal IDL of the ECU become 0 V. At this time, a voltage of approximately 0.7 V is applied to the terminal VTA of the ECU. When the throttle valve is opened, the IDL contacts go off and thus the power source voltage of approximately 12 V in the ECU is applied to the terminal IDL of the ECU. The voltage applied to the terminal VTA of the ECU increases in proportion to the opening angle of the throttle valve and becomes approximately 3.5 – 5.0 V when the throttle valve is fully opened. The ECU judges the vehicle driving conditions from these signals input from the terminals VTA and IDL, and uses them as one of the conditions for deciding the air-fuel ratio correction, power increase correction and fuel-cut control etc.

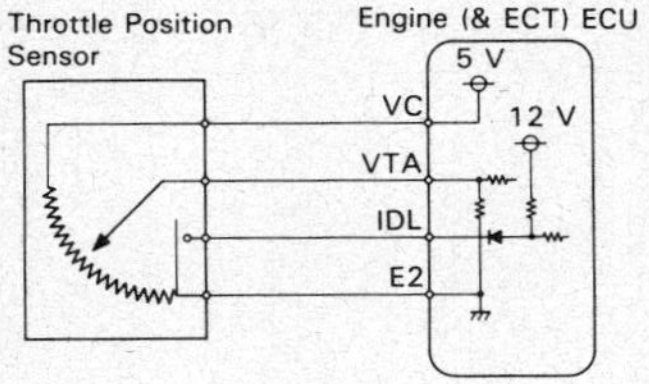

Code No.	Diagnostic Code Detecting Condition	Trouble Area
41	Open or short in throttle position sensor circuit (VTA) for 0.5 sec. or more.	• Open or short in throttle position sensor circuit. • Throttle position sensor. • ECU

HINT: When the connector for the throttle position sensor is disconnected, diagnostic code 41 is not displayed. Diagnostic code 41 is displayed only when there is an open or short in the VTA signal circuit of the throttle position sensor.

DIAGNOSTIC CHART

HINT: If diagnostic codes 22, 24, 32 and 41 are output simultaneously, E2 (sensor ground) may be open.

1. Check voltage of terminal VC. — OK → go to 4; NG → 2
2. Check throttle position sensor. — NG → Replace throttle position sensor.; OK → 3
3. Check for open and short in harness and connector between ECU and throttle position sensor. — NG → Repair or replace harness or connector.; OK → Check and replace ECU.
4. Check voltage of terminal VTA. — OK → Check for momentary interruption.; NG → 5
5. Check throttle position sensor. — NG → Replace throttle position sensor.; OK → 6
6. Check for open and short in harness and connector between ECU and throttle position sensor — NG → Repair or replace harness or connector.; OK → Check and replace ECU.

WIRING DIAGRAM

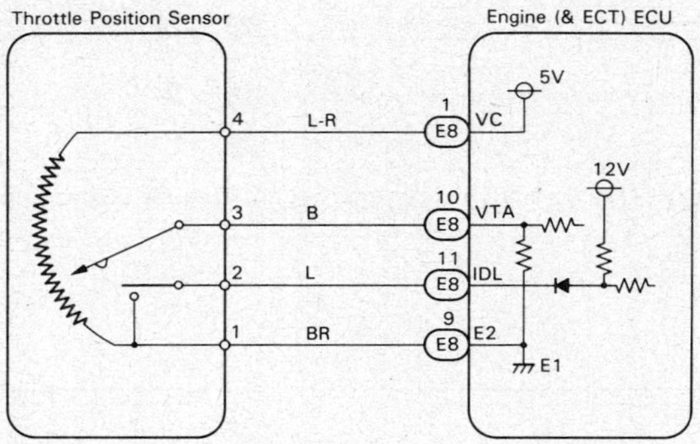

INSPECTION PROCEDURE

HINT: If diagnostic codes 22, 24, 32 and 41 are output simultaneously, E2 (sensor ground) may be open.

1 Check voltage between terminals VC and E2 of engine (& ECT) ECU.

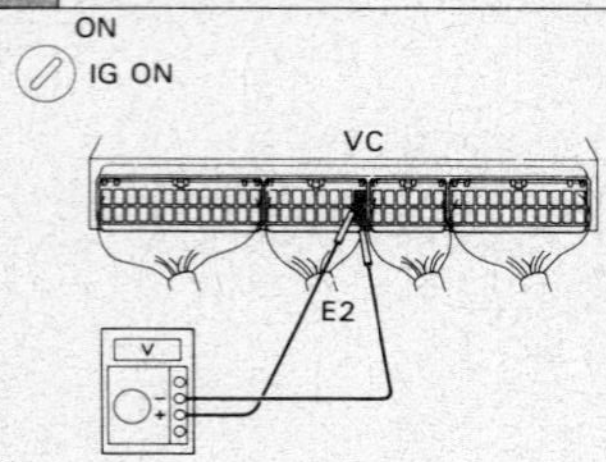

P (1) Remove glove compartment.

(2) Turn ignition switch ON.

C Measure voltage between terminals VC and E2 of engine (& ECT) ECU.

OK **Voltage: 4 – 6 V**

OK → Go to step **4**.

NG ↓

2 Check resistance between terminals 4 and 1 of throttle position sensor.

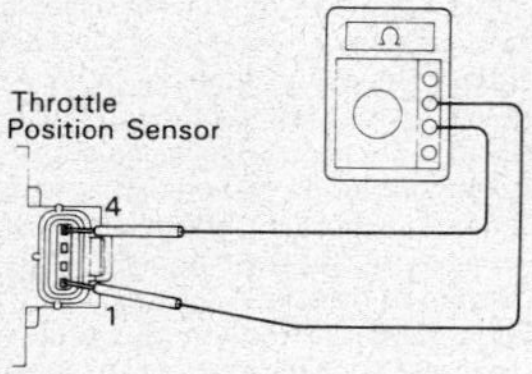

P Disconnect the throttle position sensor connector.

C Measure resistance between terminals 4 and 1 of throttle position sensor.

OK **Resistance: 4.25 – 8.25 kΩ**

NG → Replace throttle position sensor.

OK ↓

3 Check for open and short in harness and connector between engine (& ECT) ECU and throttle position sensor

NG → Repair or replace harness or connector.

OK ↓

Check and replace engine (& ECT) ECU.

4 Check voltage between terminals VTA and E2 of engine (& ECT) ECU.

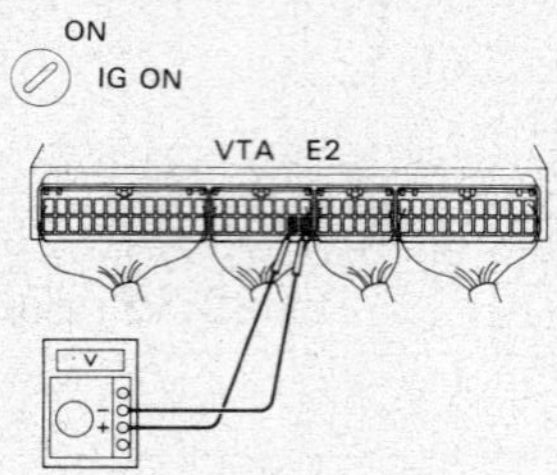

P Turn ignition switch ON.

C Measure voltage between terminals VTA and E2 of engine (& ECT) ECU.

OK

Throttle Valve	Voltage
Fully closed	0.3 – 0.8 V
Fully open	2.7 – 5.2 V

OK → Check for momentary interruption

NG ↓

5 Check resistance between terminals 3 and 1 of throttle position sensor.

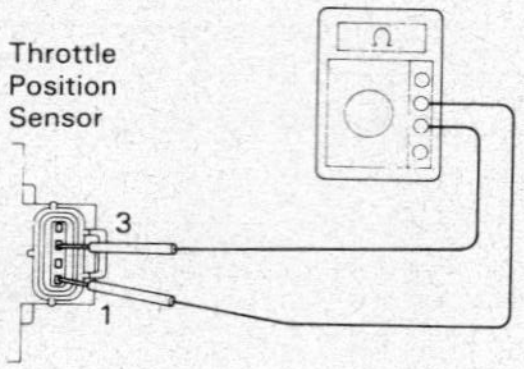

P Disconnect the throttle position sensor connector.

C Measure resistance between terminals 3 and 1 of throttle position sensor.

OK

Throttle Valve	Resistance
Fully closed	0.3 – 6.3 kΩ
Fully open	3.5 – 10.3 kΩ

NG → Replace throttle position sensor.

OK ↓

6 Check for open and short in harness and connector between engine (& ECT) ECU and throttle position sensor

NG → Repair or replace harness or connector.

OK ↓

Check and replace engine (& ECT) ECU.

Diag. Code	42	Vehicle Speed Sensor Signal Circuit

CIRCUIT DESCRIPTION

The No. 1 speed sensor outputs a 4-pulse signal for every revolution of the rotor shaft, which is rotated by the transmission output shaft via the driven gear. After this signal is converted into a more precise rectangular waveform by the waveform shaping circuit inside the combination meter, it is then transmitted to the engine (& ECT) ECU. The ECU determines the vehicle speed based on the frequency of these pulse signals.

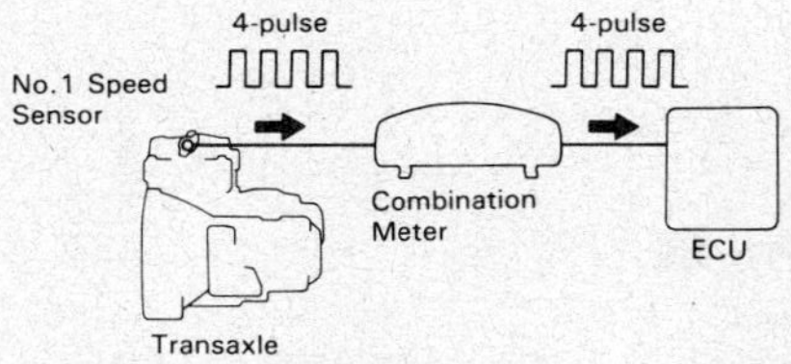

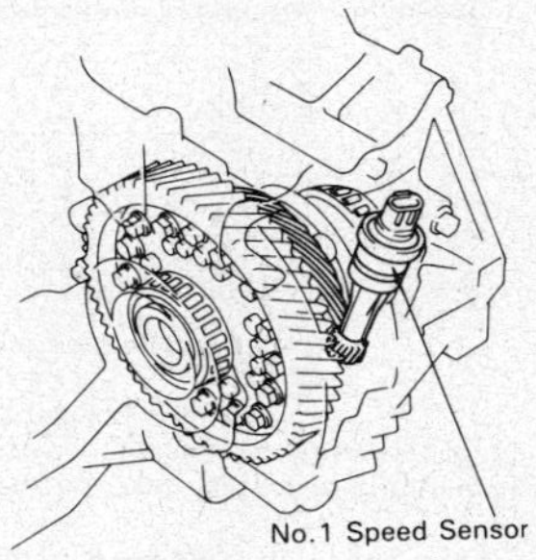

Code No.	Diagnostic Code Detecting Condition	Trouble Area
42	For ECT All conditions below are detected continuously for 8 sec. or more. (a) Vehicle speed signal: 0 km/h (mph) (b) Engine speed: 3,000 rpm or more (c) Neutral start switch (NSW): OFF	• No. 1 speed sensor • Combination meter • Open or short in No. 1 speed sensor circuit. • ECU
	For M/T All conditions below are detected continuously for 8 sec. or more. (a) Vehicle speed signal: 0 km/h (mph) (b) Engine speed: Between 3,000 ~ 4,500 rpm. (c) Coolant temp.: 80°C (176°F) or more (d) Load driving	

HINT: In test mode, Diag. code 42 is output when vehicle speed is 5 km/h (3 mph) or below.

DIAGNOSTIC CHART

1	Check operation of speedometer.	NG	Check speedometer circuit.
	OK ↓		
2	Check voltage of terminal SP1.	NG	Check speedometer circuit.
	OK ↓		
	Check and replace ECU.		

WIRING DIAGRAM

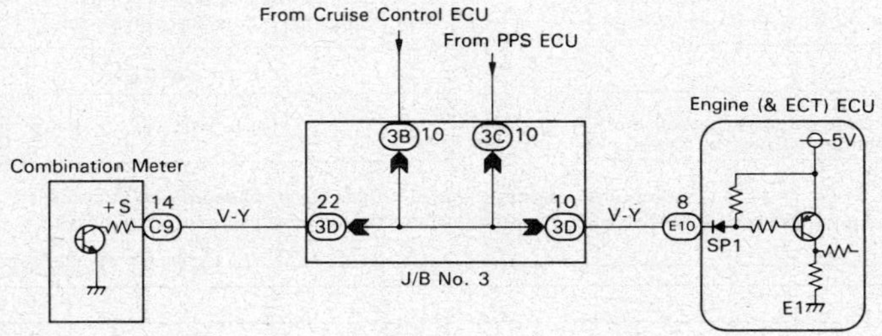

INSPECTION PROCEDURE

1 Check operation of speedometer.

C Drive the vehicle and check if the operation of the speedometer in the combination meter is normal.

Hint The No. 1 speed sensor is operating normally if the speedometer display is normal.

OK ↓

NG 〉 **Check speedometer circuit. See combination meter troubleshooting**

2 Check voltage between terminal SP1 of engine (& ECT) ECU connector and body ground.

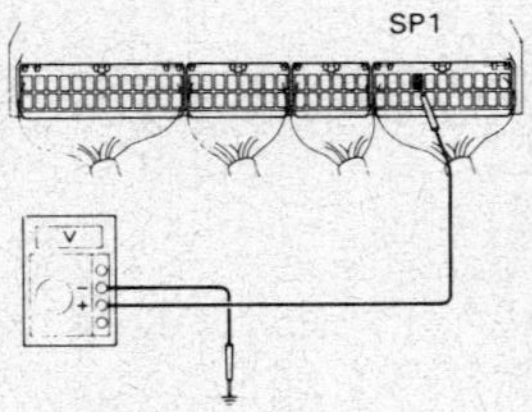

P (1) Remove glove compartment.
(2) Disconnect power sterring ECU connector and cruise control ECU connector.
(3) Shift the shift lever to N range or neutral.
(4) Jack up a front wheel on one side.
(5) Turn ignition switch on.

C Measure voltage between terminal SP1 of engine (& ECT) ECU connector and body ground when the wheel is turned slowly.

OK **Voltage is generated intermittently.**

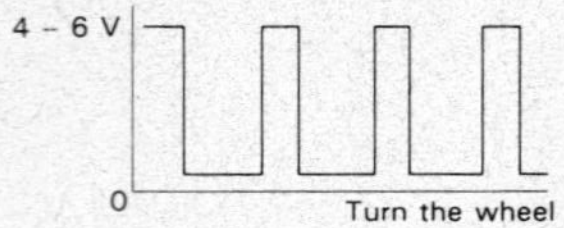

OK ↓

NG 〉 **Check speedometer circuit. See combination meter troubleshooting**

Check and replace engine (& ECT) ECU.

Diag. Code	43	Starter Signal Circuit

CIRCUIT DESCRIPTION

When the engine is cranked, the intake air flow is slow, so fuel vaporization is poor. A rich mixture is therefore necessary in order to achieve good startability. While the engine is being cranked, the battery voltage is applied to terminal STA of the ECU. The starter signal is mainly used to increase the fuel injection volume for the starting injection control and after-start injection control.

Code No.	Diagnostic Code Detecting Condition	Trouble Area
43	No starter signal to ECU.	• Open or short in starter signal circuit. • Open or short in ignition switch or starter relay circuit. • ECU

DIAGNOSTIC CHART

HINT: This diagnostic chart is based on the premise that the engine is cranked normally. If the engine is not cranked, proceed to the matrix chart of problem symptoms

1 Check for the test mode. — OK → Proceed to next circuit inspection shown on matrix chart

↓ NG

2 Check for open in harness and connectors between ECU and starter relay. — NG → Repair or replace harness or connector.

↓ OK

Check and replace ECU.

WIRING DIAGRAM

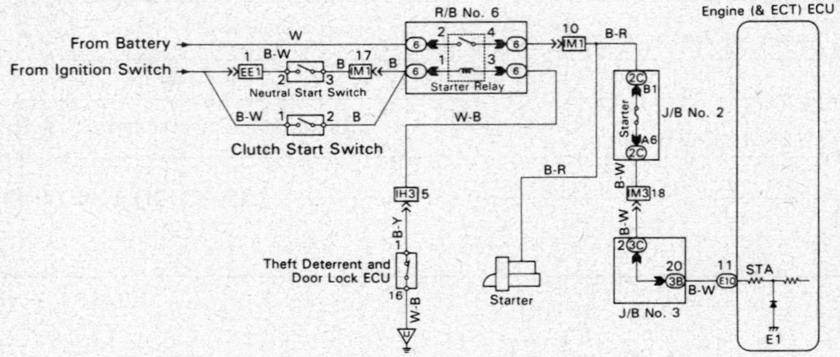

INSPECTION PROCEDURE

1 Check output condition of diag. code 43.

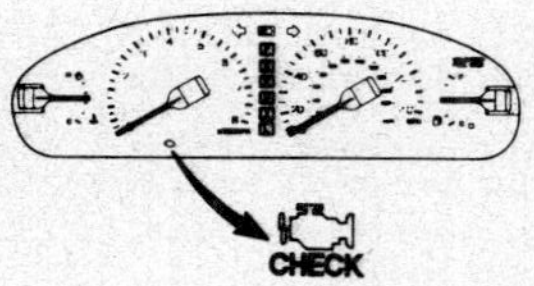

P Setting the test mode.
(1) Turn ignition switch OFF.
(2) Connect terminals TE2 and E1 of TDCL.
(3) Turn ignition switch ON.
(Don't start the engine)
(4) Connect terminals TE1 and E1 of TDCL.

C Check if code "43" is output by the "CHECK" engine warning light.

OK **Code "43" is output.**

C Start the engine.
Check if the code "43" disappear.

OK **Code "43" is not output.**

OK → **Proceed to next circuit inspection shown on matrix chart**

NG ↓

2 Check for open in harness and connectors between engine (& ECT) ECU and starter relay

NG → **Repair or replace harness or connector.**

OK ↓

Check and replace engine (& ECT) ECU.

Diag. Code 52,53,55 Knock Sensor Circuit

CIRCUIT DESCRIPTION

Knock sensors are fitted one each to the right bank and left bank of the cylinder block to detect engine knocking. This sensor contains a piezoelectric element which generates a voltage when it becomes deformed, which occurs when the cylinder block vibrates due to knocking. If engine knocking occurs, ignition timing is retarded to suppress it.

Code No.	Diagnostic Code Detecting Condition	Trouble Area
52	No No. 2 knock sensor signal to ECU for 4 crank revolutions with engine speed between 1,600 rpm ~ 5,200 rpm.	• Open or short in No. 2 knock sensor circuit. • No. 2 knock sensor (looseness) • ECU
53	Engine control computer (for knock control) malfunction at engine speed between 650 rpm and 5,200 rpm.	• ECU
55	No No. 1 knock sensor signal to ECU for 4 crank revolutions with engine speed between 1,600 rpm ~ 5,200 rpm.	• Open or short in No. 1 knock sensor circuit. • No. 1 knock sensor (looseness) • ECU

If the ECU detects the above diagnosis conditions, it operates the fail safe function in which the corrective retard angle value is set to the maximum value.

CIRCUIT DESCRIPTION (Cont'd)

DIAGNOSIS CODE DETECTION DRIVING PATTERN

Purpose of the driving pattern.

(a) To simulate diag. code detecting condition after diag. code is recorded.

(b) To check that the malfunction is corrected when the repair is completed confirming that diag. code is no longer detected.

Malfunction: Open or Short in Knock Sensor

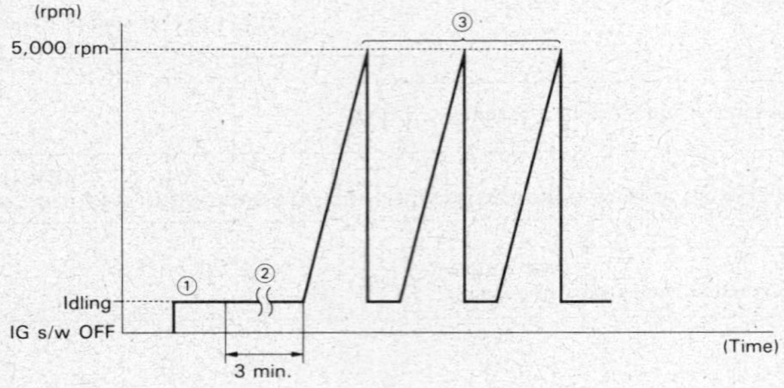

① Start the engine and warm up.
② After the engine is warmed up, let it idle for 3 min.
③ With the A/C ON, perform quick racing (5,000 rpm) three times.
(Rapidly depress the accelerator pedal and suddenly release it.)

HINT: If a malfunction exists, the "CHECK" engine warning light will light up when sudden racing is performed.

NOTICE: If the conditions in this test are not strictly followed, detection of the malfunction will not be possible.

DIAGNOSTIC CHART

HINT: If diag. code 52 is displayed, check No. 2 knock sensor (for left bank) circuit.
If diag. code 55 is displayed, check No. 1 knock sensor (for right bank) circuit.
If diag. code 53 is displayed, replace engine (& ECT) ECU.

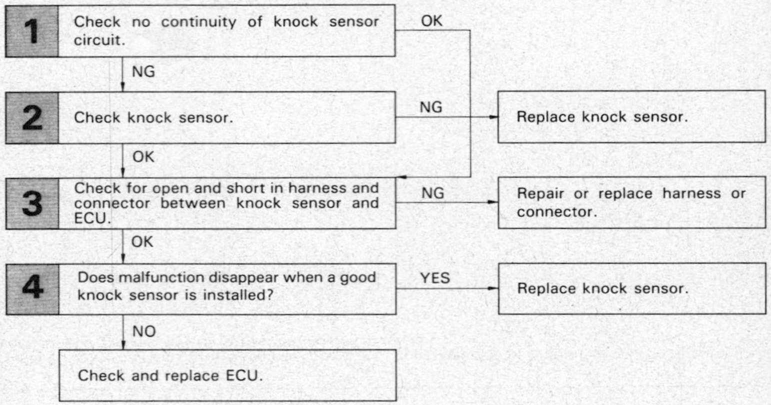

WIRING DIAGRAM

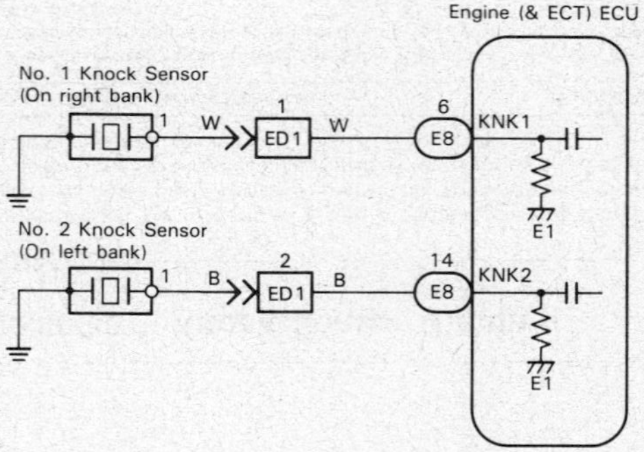

INSPECTION PROCEDURE

1 Check continuity between terminals KNK1, KNK2 of engine (& ECT) ECU connector and body ground.

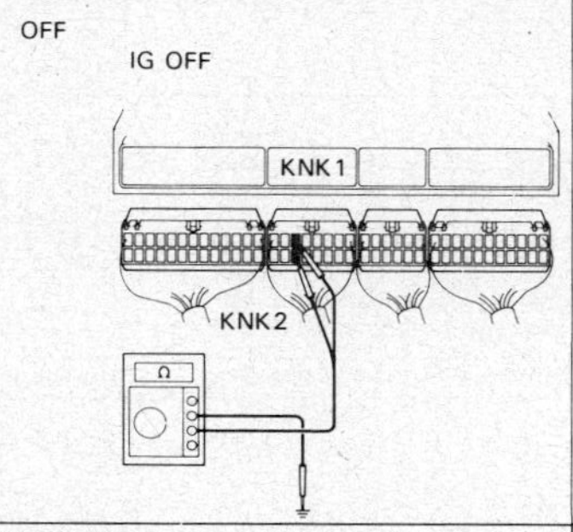

P (1) Remove glove compartment.

(2) Disconnect the engine (& ECT) ECU connector.

C Measure resistance between terminals KNK1, KNK2 of engine (& ECT) ECU connector and body ground.

OK **Resistance: 1 MΩ or higher**

NG ↓

OK 〉 **Go to step 3.**

2 Check knock sensor.

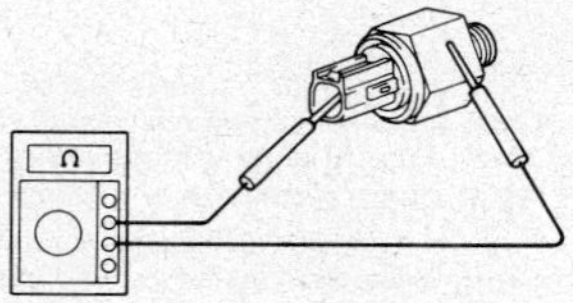

P Disconnect knock sensor connector.

C Measure resistance between the knock sensor terminal and body.

OK **Resistance: 1 MΩ or higher**

OK ↓

NG 〉 **Replace knock sensor**

3 Check for open and short in harness and connectors between engine (& ECT) ECU and knock sensor

NG 〉 **Repair or replace harness or connector.**

OK ↓

4 Does malfunction disappear when a good knock sensor is installed?

YES 〉 **Replace knock sensor**

NO ↓

Check and replace engine (& ECT) ECU.

Reference — INSPECTION USING OSCILLOSCOPE

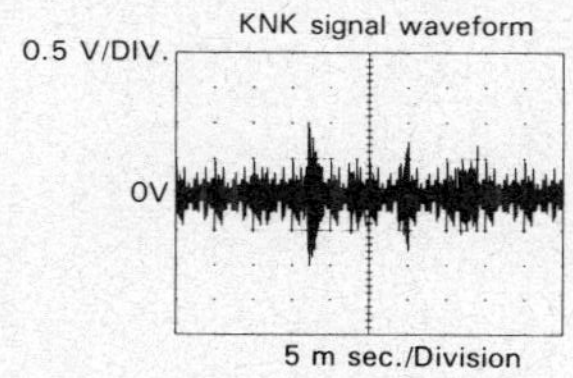

- With the engine racing (4,000 rpm) measure between terminals KNK1, KNK2 of engine (& ECT) ECU and body ground.

HINT: The correct waveform appears as shown in the illustration on the left.

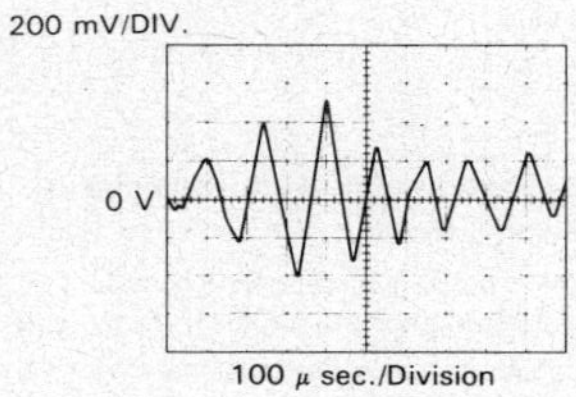

- Spread the time on the horizontal axis, and confirm that period of the wave is 132 μ sec.
(Normal mode vibration frequency of knock sensor: 7.6 KHz).

HINT: If normal mode vibration frequency is not 7.6 KHz, the sensor is malfunctioning.

Diag. Code	71	EGR System Malfunction (Only for California spec.)

CIRCUIT DESCRIPTION

The EGR system recirculates exhaust gas, which is controlled to the proper quantity to suit the driving conditions, into the intake air mixture to slow down combustion, reduce the combustion temperature and reduce NOx emissions. The amount of EGR is regulated by the EGR vacuum modulator according to the engine load.

Under the following conditions, EGR is cut to maintain driveability.

- Coolant temp. below 35°C (95°F)
- During deceleration (throttle valve closed)
- Light engine load (amount of intake air very small)
- Engine idling

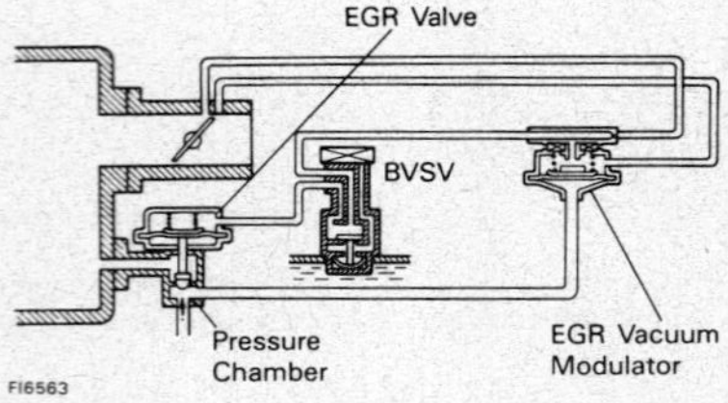

Code No.	Diagnostic Code Detecting Condition	Trouble Area
71	EGR gas temp. is 70°C (158°F) or below for 1 ~ 4 min. under conditions (a) and (b). (2 trip detection logic) (a) Coolant temp.: 60°C (140°F) or more. (b) EGR operation possible (EX. ECT in D range, A/C ON, 60 mph (96 km/h), Flat road).	• Open in EGR gas temp. sensor circuit. • EGR or BVSV vacuum hose disconnected valve stuck. • Clogged EGR gas passage. • ECU

CIRCUIT DESCRIPTION (Cont'd)

DIAGNOSIS CODE DETECTION DRIVING PATTERN

Purpose of the driving pattern.

(a) To simulate diag. code detecting condition after diag. code is recorded.

(b) To check that the malfunction is corrected when the repair is completed confirming that diag. code is no longer detected.

Malfunction: Open in EGR gas temp. sensor circuit

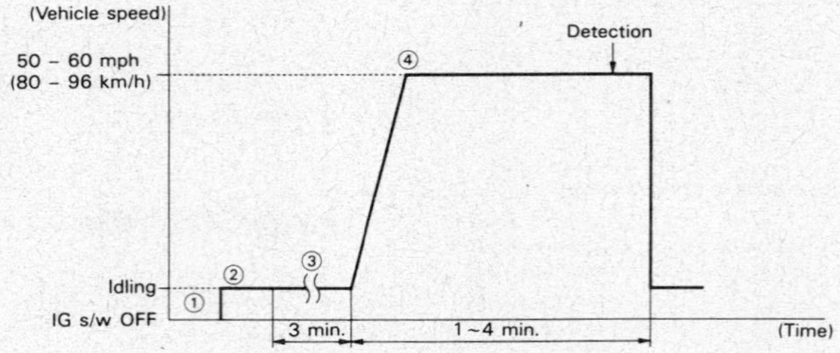

① Disconnect the EFI fuse (15 A) for 10 sec. or more, with IG switch OFF.
Initiate test mode (Connect terminal TE2 and E1 of check connector or TDCL with IG switch OFF).

② Start the engine and warm up.

③ After the engine is warmed up, let it idle for 3 min.

④ With the A/C ON and transmission in 5th gear (D range for A/T), drive at 50 ~ 60 mph (80 ~ 96 km/h) for 4 min. or less.

HINT: If a malfunction exists, the "CHECK" engine warning light will light up during step ④.

NOTICE: If the conditions in this test are not strictly observed, detection of the malfunction will not be possible.

DIAGNOSTIC CHART

Step	Check	If NG
1	Check EGR system.	Repair EGR system.
OK ↓		
2	Check resistance of EGR gas temp. sensor.	Replace EGR gas temp. sensor.
OK ↓		
3	Check for open and short in harness and connector between ECU and EGR gas temp. sensor.	Repair or replace harness or connector.
OK ↓		
	Check and replace ECU.	

WIRING DIAGRAM

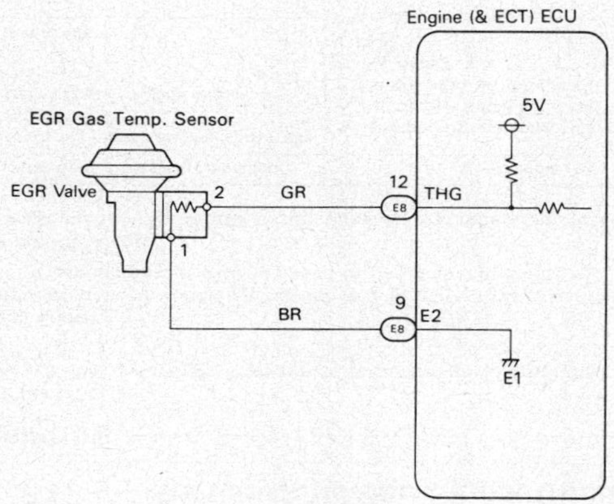

INSPECTION PROCEDURE

1 Check EGR system

OK → go to step 2

NG → Repair EGR system.

2 Check resistance of EGR gas temp. sensor.

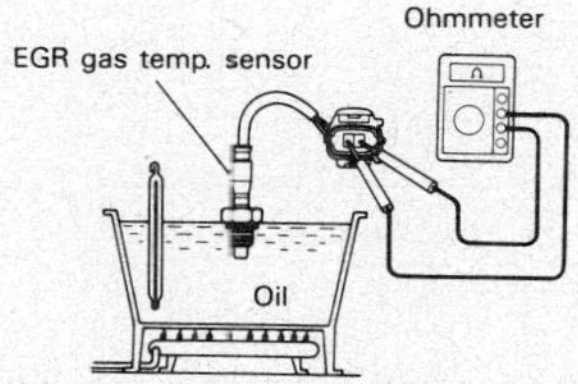

P Remove EGR gas temp. sensor.

C Measure resistance between terminals of EGR gas temp. sensor connector.

OK Resistance: 69 – 89 kΩ at 50°C (122°F)
12 – 15 kΩ at 100°C (212°F)
2 – 4 kΩ at 150°C (302°F)

OK → go to step 3

NG → Replace EGR gas temp. sensor.

3 Check for open in harness and connector between EGR gas temp. sensor and engine (& ECT) ECU

OK → Check and replace engine (& ECT) ECU.

NG → Repair or replace harness or connector.

Diag. Code	51	Switch Condition Signal Circuit

CIRCUIT DESCRIPTION

Neutral Start Switch Signal*
The ECU uses the signals from the neutral start switch to determine whether the transmission is in park or neutral, or in some other gear.

Air Conditioner Switch Signal
The ECU uses the output from the air conditioner switch to determine whether or not the air conditioner is operating so that it can increase the idling speed of the engine if necessary.

Throttle Position Sensor IDL Signal
The IDL contacts are mounted in the throttle position sensor, and detects the idle condition.

Code No.	Diagnostic Code Detecting Condition	Trouble Area
51	(1) 3 sec. or more after engine starts idle switch OFF (IDL1). (2) *Neutral start switch OFF (NSW). (Shift position in "R", "D", "2" or "1" ranges). (3) A/C switch ON.	• Throttle position sensor IDL circuit. • Accelerator pedal and cable • Neutral start switch circuit • A/C switch circuit • ECU

*: Only vehicles with ECT.
HINT: In this circuit, diagnosis can only be made in the test mode.

DIAGNOSTIC CHART

WIRING DIAGRAM

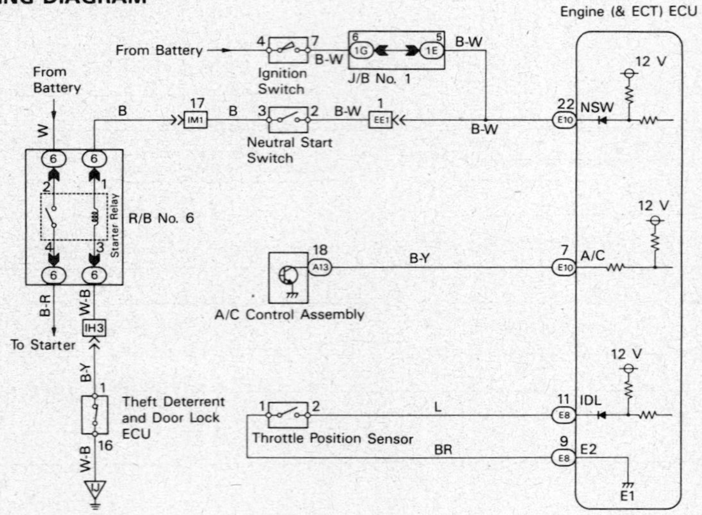

DIAGNOSTIC CHART

1 Check for input signal.
- NG (NSW) → See Neutral start switch circuit
- NG (A/C) → go to 3
- NG (IDL) → go to 2
- OK → Proceed to next circuit inspection shown on matrix chart

2 Check resistance of throttle position sensor.
- NG → Adjust or replace throttle position sensor
- OK → Check and repair harness or connector between sensor and ECU.

3 Check input circuit of A/C.
- OK → Check and replace A/C control assembly.
- NG → go to 4

4 Check for open and short in harness and connector between A/C control assy. and ECU.
- NG → Repair or replace harness or connector.
- OK → Check and replace ECU.

INSPECTION PROCEDURE

1 Check output condition of diag. code 51.

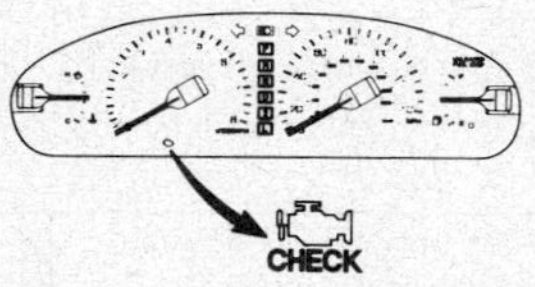

P Setting the test mode.
(1) Turn ignition switch OFF.
(2) Connect terminals TE2 and E1 of TDCL.
(3) Turn ignition switch ON.
(For checking terminal A/C, start the engine.)
(4) Connect terminals TE1 and E1 of TDCL.

C Check if code "51" is output by the "CHECK" engine warning light.

OK

	Condition	Code
Neutral Start Switch (NSW)	P or N range	Normal*
	R,D,2 or L range	51*
Throttle Position Sensor (IDL)	Accelerator pedal released	Normal*
	Accelerator pedal depressed	51*
A/C Switch (A/C)	A/C SW ON	51
	A/C SW OFF	Normal

*: Before the STA signal is input (ST is not ON), diagnostic code 43 is also output.

Hint Diag. code 42 is output with vehicle speed 5 km/h (3 mph) or below.

OK → Proceed to next circuit inspection shown on matrix chart

NG → IDL....Go to step 2 NSW....
A/C....Go to step 3

2 Check throttle position sensor.

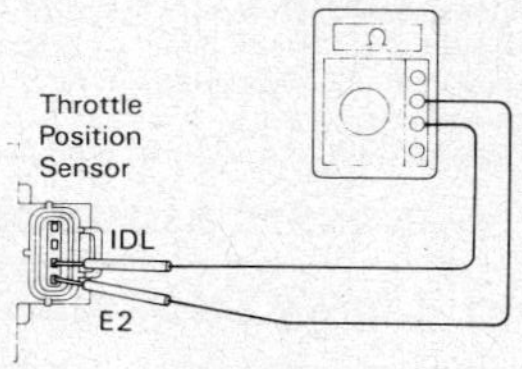

P Disconnect throttle position sensor connector.

C Measure resistance between terminals IDL and E2 of throttle position sensor connector.

OK

Throttle Valve	Resistance
Fully closed	Less than 0.5 kΩ
Opened	1 MΩ or higher

NG → Adjust or replace throttle position sensor

OK → Check and repair harness or connector between engine (& ECT) ECU and throttle position sensor.

3 Disconnect A/C control assembly connector, check voltage between terminal AC1 of A/C control assembly connector and body ground.

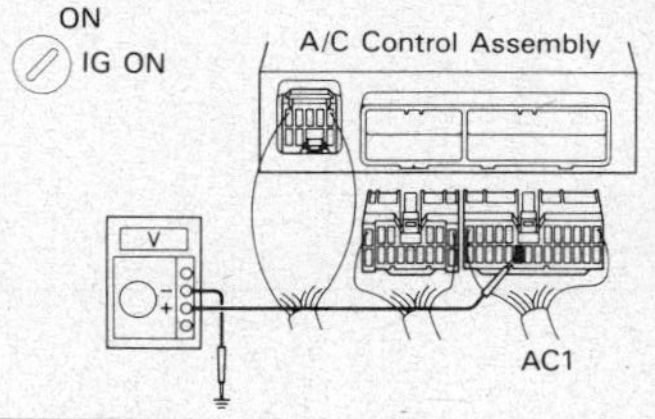

P (1) Remove air conditioner control assembly.
(2) Disconnect air conditioner control assembly connector.
(3) Turn ignition switch on.

C Measure voltage between terminal AC1 of air conditioner control assembly connector and body ground.

OK Voltage: 10 – 14 V

OK → Check and replace A/C control assembly.

NG ↓

4 Check for open and short in harness and connector between engine (& ECT) ECU and A/C control assembly

NG → Repair or replace harness or connector.

OK → Check and replace engine (& ECT) ECU.

Neutral Start Switch Circuit (Only vehicles with ECT)

CIRCUIT DESCRIPTION

The neutral start switch goes on when the shift lever is in the N or P shift position. When it goes on the terminal NSW of the ECU is grounded to body ground via the starter relay and theft deterrent ECU, thus the terminal NSW voltage becomes 0V. When the shift lever is in the D, 2, L or R position, the neutral start switch goes off, so the voltage of ECU terminal NSW becomes battery voltage, the voltage of the ECU internal power source.
If the shift lever is moved from the N range to the D range, this signal is used for air-fuel ratio correction and for idle speed control (estimated control), etc.
When the neutral start switch is off, code "51" is output in the test mode diagnosis. (This is not abnormal.)

DIAGNOSTIC CHART

HINT: This diagnosis chart is based on premise that the engine is cranked normally. If the engine is not cranked, proceed to the matrix chart of problem symptoms

1 Check output condition of diag. code 51.
- OK → Proceed to next circuit inspection shown on matrix chart
- NG Type I → go to 2
- NG Type II → go to 3

2 Check for open in harness and connector between neutral start switch and ECU.
- OK → Check and replace ECU.
- NG → Repair or replace harness or connector.

3 Check neutral start switch.
- NG → Replace neutral start switch.
- OK → Check and replace ECU.

WIRING DIAGRAM

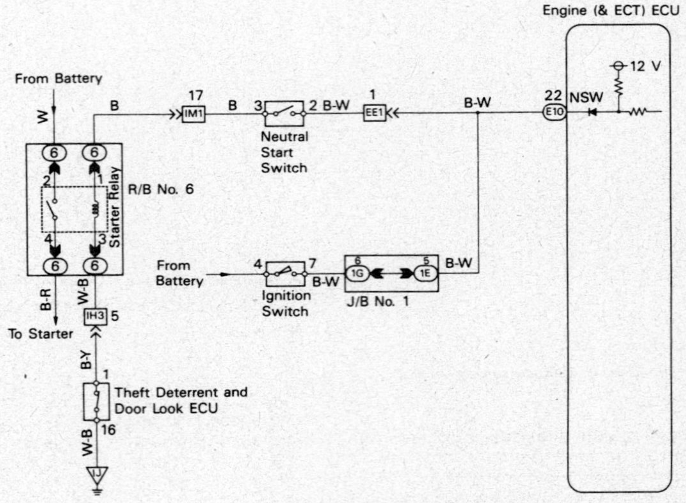

INSPECTION PROCEDURE

1 Check output condition of diag. code 51.

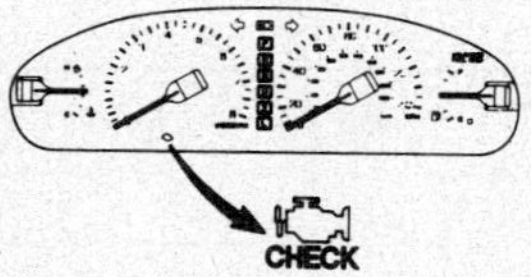

P
(1) Connect terminals TE2 and E1 of TDCL.
(2) Turn ignition switch on.
(3) Crank the engine.
(4) Connect terminals TE1 and E1 of TDCL.

C Check if diagnostic code "51" is output when the shift lever is in the P and D shift positions.

Result

Shift Position	Result		
	OK	NG Type I	NG Type II
"P"	Normal Code	Code 51	Normal Code
"D"	Code 51	Code 51	Normal Code

NG Type I → Go to step **2**.

NG Type II → Go to step **3**.

OK

Proceed to next circuit inspection shown on matrix chart

2 Check for open in harness and connector between engine (& ECT) ECU and neutral start switch

OK → Check and replace engine (& ECT) ECU.

NG

Repair or replace harness or connector.

3 Check neutral start switch

NG → Replace neutral start switch.

OK

Check and replace engine (& ECT) ECU.

ECU Power Source Circuit

CIRCUIT DESCRIPTION

When the ignition switch is turned on, battery voltage is applied to the terminal IGSW of the ECU, and the main relay control circuit in the ECU sends a signal to the terminal M-REL of the ECU, switching on the main relay. This signal causes current to flow to the coil, closing the contacts of the main relay and supplying power to the terminals +B and +B1 of the ECU.
If the ignition switch is turned off, the ECU continues to switch on the main relay for a maximum of 2 seconds for the initial setting of the ISC valve.

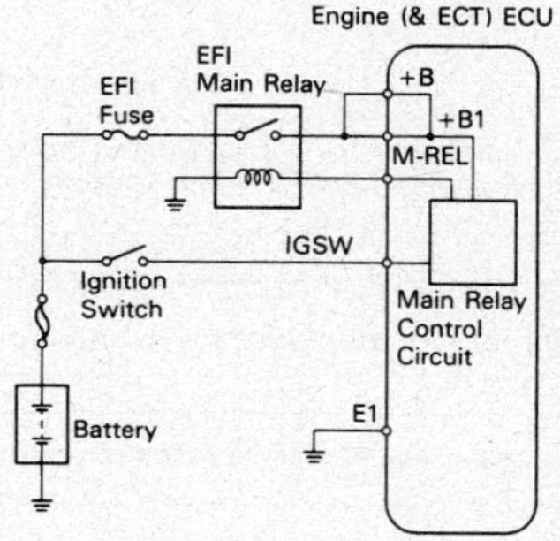

DIAGNOSTIC CHART

See next page for the DIAGNOSTIC CHART.

WIRING DIAGRAM

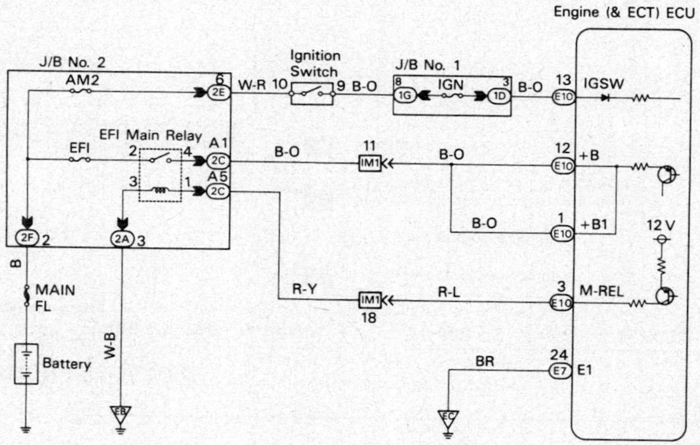

DIAGNOSTIC CHART

1 Check voltage of ECU power source.
- OK → Proceed to next circuit inspection shown on matrix chart
- NG ↓

2 Check continuity between terminals E1 and body ground.
- NG → Repair or replace harness or connector.
- OK ↓

3 Check voltage of terminal IGSW.
- OK → (to step 6)
- NG ↓

4 Check IGN fuse.
- NG → Check for short in all the harness and components connected to IGN fuse.
- OK ↓

5 Check ignition switch.
- NG → Replace ignition switch.
- OK ↓

Check for open in harness and connector between battery and ECU.

6 Check voltage of terminal M-REL.
- NG → Check and replace ECU.
- OK ↓

7 Check EFI fuse.
- NG → Check for short in all the harness and components connected to EFI fuse.
- OK ↓

8 Check EFI main relay.
- NG → Replace EFI main relay.
- OK ↓

9 Check for open and short in harness and connector between terminal M-REL and body ground.
- NG → Repair or replace harness or connector.
- OK ↓

Check and repair harness or connector between EFI fuse and battery.

INSPECTION PROCEDURE

1 Check voltage between termianls +B, +B1 and E1 of engine (& ECT) ECU connector.

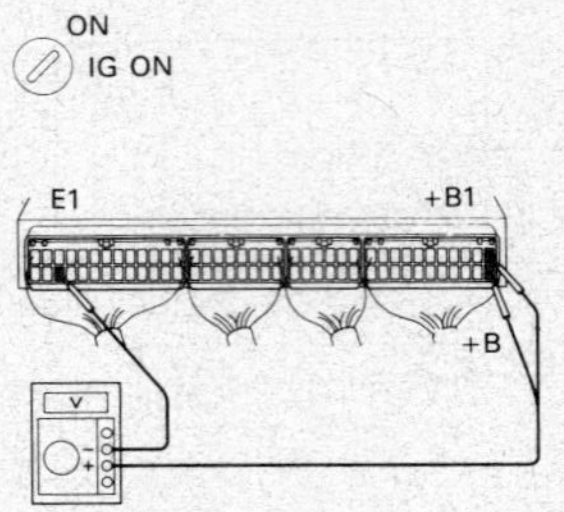

P (1) Remove glove compartment.

(2) Turn ignition switch on.

C Measure voltage between terminals +B, +B1 and E1 of engine (& ECT) ECU connector.

OK Voltage: 10 – 14 V

OK → Proceed to next circuit inspection shown on matrix chart

NG ↓

2 Check for open in harness and connector between terminal E1 of engine (& ECT) ECU and body ground

NG → Repair or replace harness or connector.

OK ↓

3 Check voltage between terminal IGSW of engine (& ECT) ECU connector and body gound.

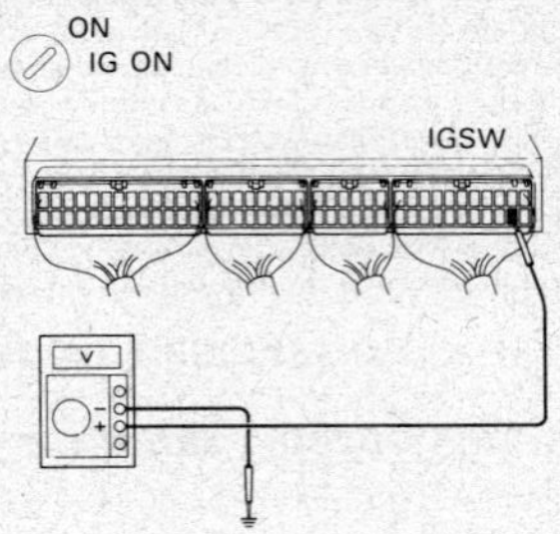

P Turn ignition switch on.

C Measure voltage between terminal IGSW of engine (& ECT) ECU connector and body ground.

OK Voltage: 10 – 14 V

OK → Go to step 6.

NG ↓

4 Check IGN fuse.

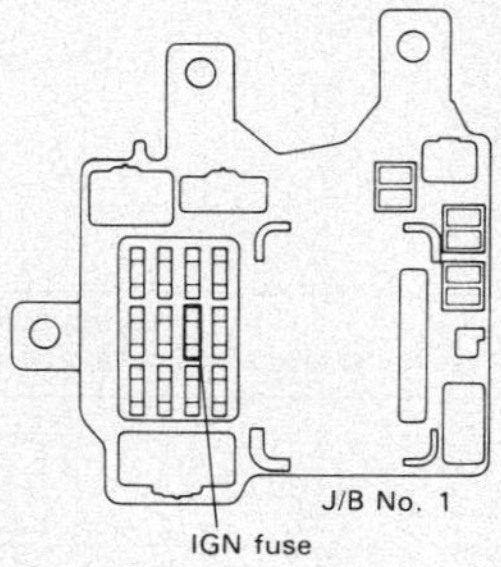

P Remove IGN fuse from J/B No. 1.

C Check continuity of IGN fuse.

OK **Continuity**

OK

NG → **Check for short in all the harness and components connected to IGN fuse**

5 Check ignition switch.

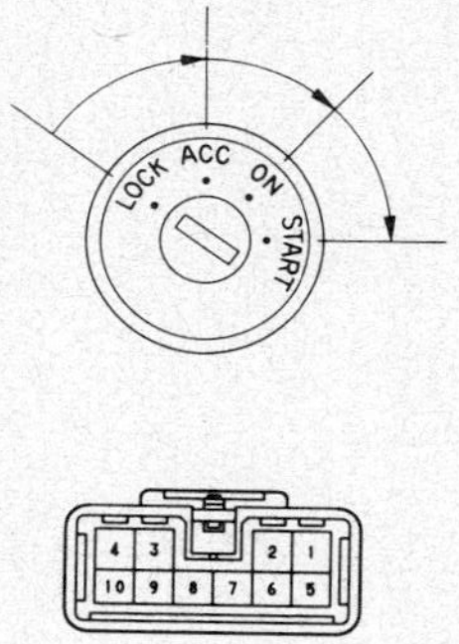

P Remove under cover and finish panel.

C Check continuity between terminals.

OK

○—○ continuity

Terminal / Switch position	2 IG1	3 ACC	4 AM1	6 ST2	7 ST1	9 IG2	10 AM2
LOCK							
ACC		○	○				
ON	○	○	○			○	○
START	○		○	○	○	○	○

OK

NG → **Replace ignition switch.**

Check and repair harness and connector between battery and ignition switch, ignition switch and engine (& ECT) ECU.

6 Check voltage between terminal M-REL of engine (& ECT) ECU connector and body ground.

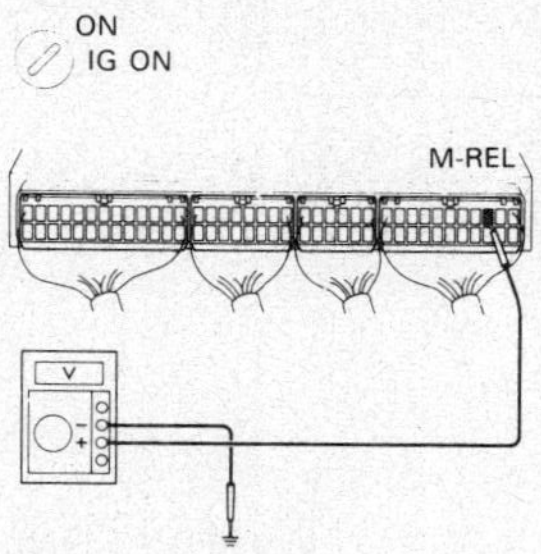

P Turn ignition switch on.

C Measure voltage between terminal M-REL of engine (& ECT) ECU connector and body ground.

OK **Voltage: 10 – 14 V**

OK

NG → **Check and replace engine (& ECT) ECU.**

7 Check EFI fuse.

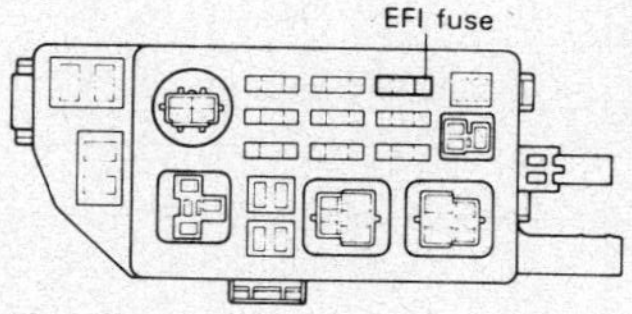

P Remove EFI fuse from J/B No. 2.

C Check continuity of EFI fuse.

OK **Continuity**

OK

NG → **Check for short in all the harness and components connected to EFI fuse**

8 Check EFI main relay.

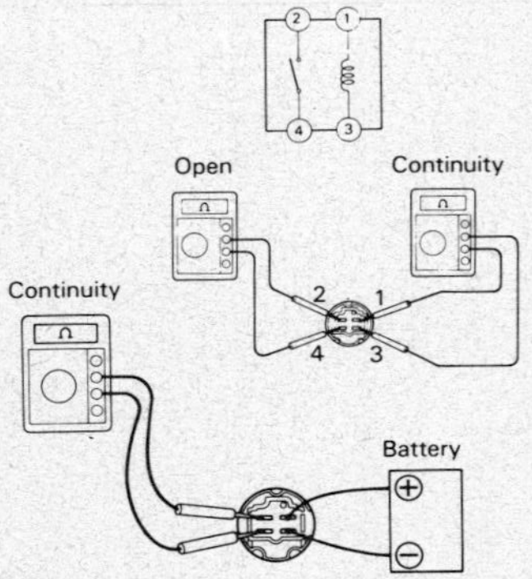

P Remove EFI main relay from J/B No.2.

C Check continuity between terminals of EFI main relay shown below.

OK

Terminals 2 and 4	Open
Terminals 1 and 3	Continuity (Reference value 72 Ω)

C (1) Apply battery voltage between terminals 1 and 3.
(2) Check continuity between terminals 2 and 4.

OK

Terminals 2 and 4	Continuity

OK

NG — Replace EFI main relay.

9 Check for open and short in harness and connector between terminal M-REL of engine (& ECT) ECU and body ground

NG — Repair or replace harness or connector.

OK

Check and repair harness or connector between EFI fuse and battery.

Back Up Power Source Circuit

CIRCUIT DESCRIPTION

Battery voltage is supplied to terminal BATT of the ECU even when the ignition switch is off for use by the diagnostic code memory and air-fuel ratio adaptive control value memory, etc.

DIAGNOSTIC CHART

1 Check EFI fuse. — NG → Check for short in all the harness and components connected to EFI fuse.

OK

2 Check voltage of terminal BATT. — NG → Check and repair harness or connector between battery, EFI fuse and ECU.

OK

3 Check operation for the back up. — NG → Check and replace ECU.

OK

Proceed to next circuit inspection shown on matrix chart

WIRING DIAGRAM

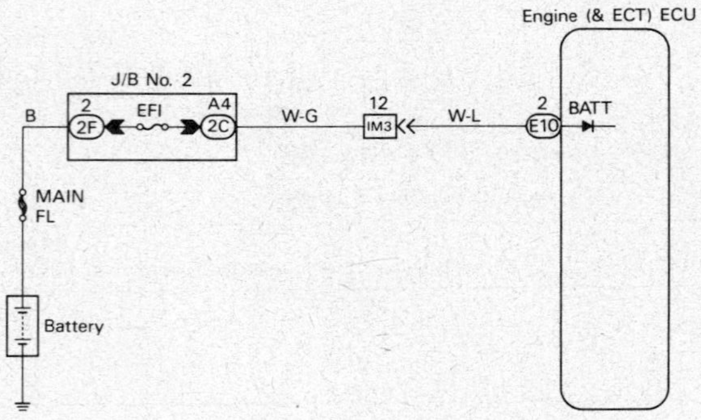

INSPECTION PROCEDURE

1 Check EFI fuse.

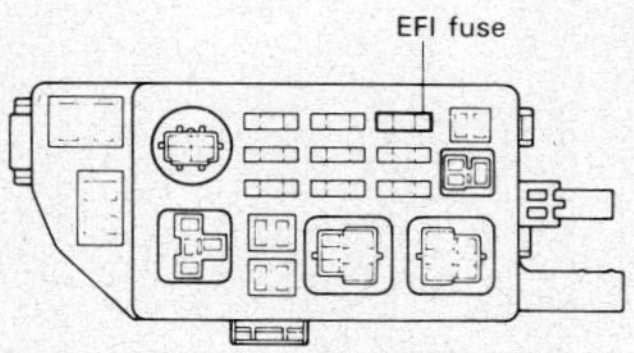

P Remove EFI fuse from J/B No. 2.

C Check continuity of EFI fuse.

OK **Continuity**

OK ↓

NG 〉 **Check for short in all the harness and components connected to EFI fuse**

2 Check voltage between terminal BATT of engine (& ECT) ECU connector and body ground.

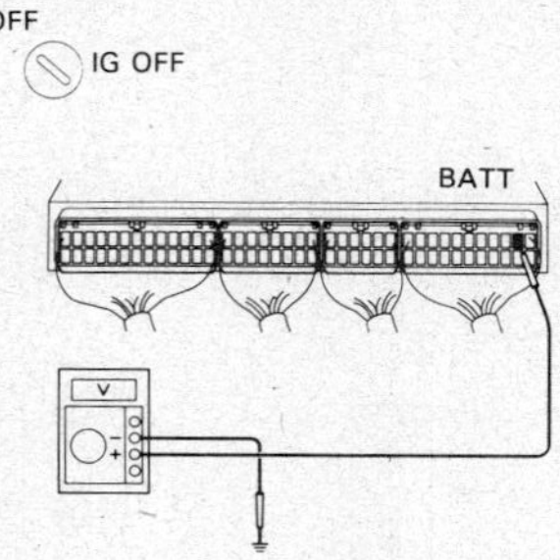

P Remove glove compartment.

C Measure voltage between terminal BATT of engine (& ECT) ECU connector and body ground.

OK **Voltage: 10 – 14 V**

OK ↓

NG 〉 **Check and repair harness or connector between engine (& ECT) ECU and EFI fuse, EFI fuse and battery.**

3 Are the diagnostic codes still in the memory when the ignition switch is turned OFF?

YES ↓

NO 〉 **Check and replace engine (& ECT) ECU.**

Proceed to next circuit inspection shown on matrix chart

Injector Circuit

CIRCUIT DESCRIPTION

The injectors are provided to the intake manifold. They inject fuel into the cylinders based on the signals from the engine (& ECT) ECU.

Reference **INSPECTION USING OSCILLOSCOPE**

INJECTOR SIGNAL WAVEFORM

- With the engine idling measure between terminals # 10 ~ # 60 and E01 of engine (& ECT) ECU.

HINT: The correct waveform appears as shown in the illustration on the below.

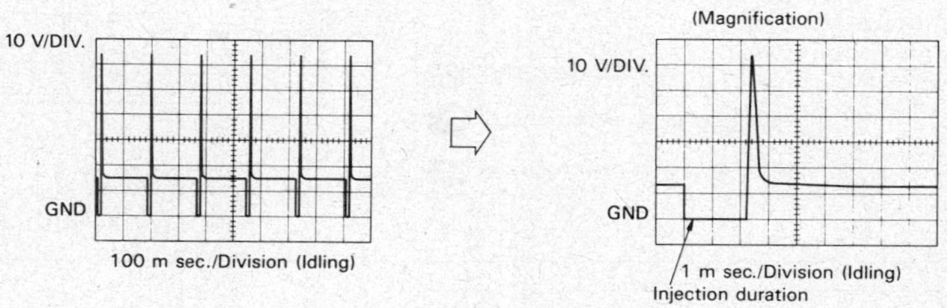

DIAGNOSTIC CHART

1 Check voltage of terminals # 10 ~ # 60. — OK → go to 3; NG ↓

2 Check AM2 fuse. — NG → Check for short in all the harness and components connected to AM2 fuse.; OK ↓

Check for open in harness and connector between ECU and battery.

3 Check continuity between terminals E01, E02 and body ground. — NG → Repair or replace harness or connector.; OK ↓

4 Check operation for injectors. — NG → Replace injector.; OK ↓

Check and replace ECU.

WIRING DIAGRAM

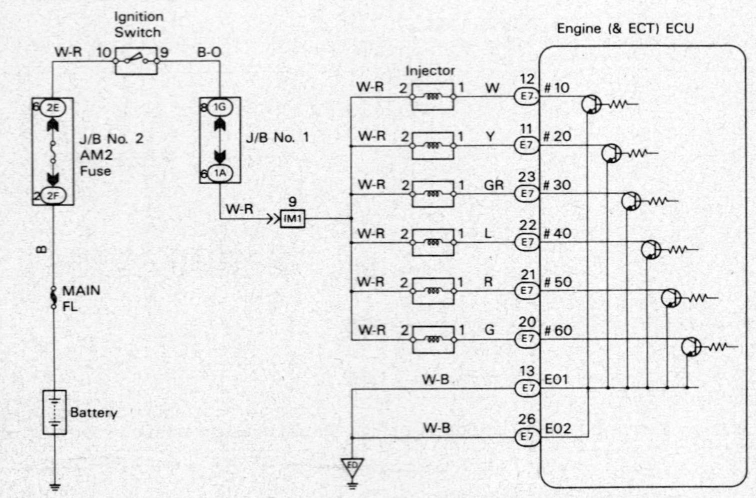

INSPECTION PROCEDURE

1 Check voltage between terminals #10~60 of engine (& ECT) ECU and body ground.

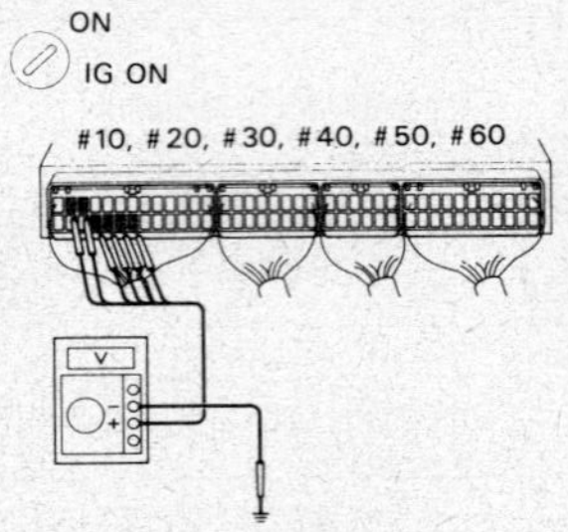

P (1) Remove glove compartment.

(2) Turn ignition switch on.

C Measure voltage between terminals #10~60 of engine (&ECT) ECU and body ground.

OK Voltage: 10 – 14 V

OK Go to step **3**.

NG

2 Check AM2 fuse.

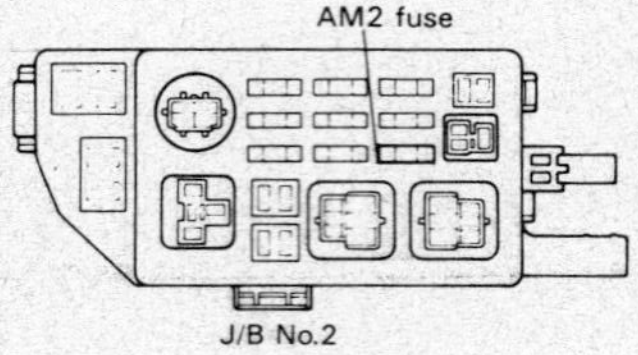

P Remove AM2 fuse from J/B No. 2.

C Check continuity of AM2 fuse.

OK Continuity

NG Check for short in all the harness and components connected to AM2 fuse.

OK

Check for open in harness and connector between engine (& ECT) ECU and battery.

3 Check for open in harness and connector between terminal E01, E02 of engine (& ECT) ECU connector and body ground

OK

NG Repair or replace harness or connector.

4 Check injectors.

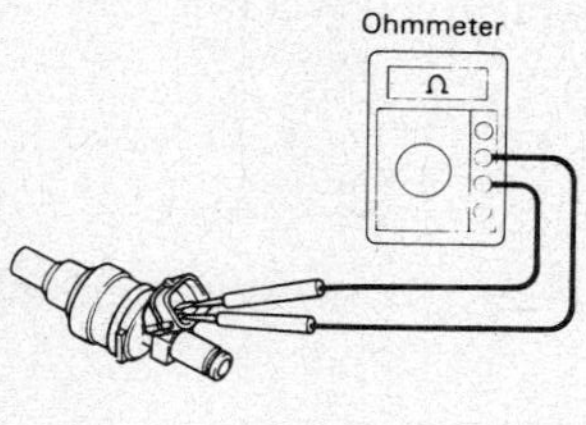

P Disconnect injector connector

C Measure resistance of injector.

OK **Resistance: 13.4 – 14.2 Ω at 20°C (68°F)**

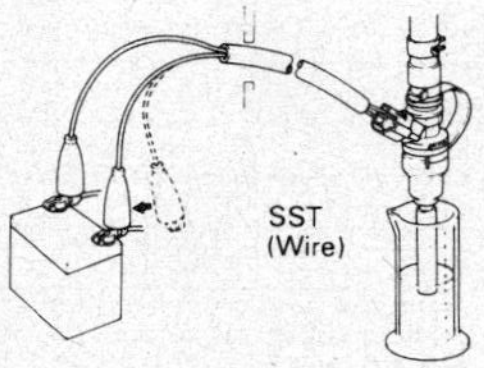

C Check injection volume of injector

OK
- **Injection volume**
  58 – 68 cc/15 sec. (3.5 – 4.1 cu in.)
  Difference between each injector:
  Less than 5 cc (0.3 cu in.)
- **Leakage**
  Fuel drop: One drop or less per minute.

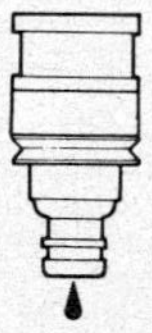

OK

NG Replace injector.

Check and replace engine (& ECT) ECU.

Cold Start Injector Circuit

CIRCUIT DESCRIPTION

The cold start injector is used to maintain the engine startability when it is cold. The injection volume, i,e, the length of time the injector is energized, is controlled by the ECU and the cold start injector time switch.
During a cold start, when the starter turns the contacts in the cold start injector time switch close. Thus current flows to the cold start injector coil, injecting fuel. At the same time, a bimetal in the heat coil is energized and heats up. This soon causes the contacts to open, cutting off the current flow to the injector coil and stopping fuel injection.

The injection duration of the cold start injector is determined by the coolant temperature and the length of time current flows to the heat coil. When the engine is warm, the contacts are opened by the bimetal and the cold start injector does not operate.
When the engine is hard to start and the starter is operated continuously, heat coil (2) heats up the bimetal keeping the contacts open to prevent spark plugs from becoming fouled, which is caused by the cold start injector operation when the open contacts close again.

When the engine is started at a coolant temperature of 22°C (72°F) or lower, the cold start injector operation time is controlled by the cold start injector time switch.
When the coolant temperature is in the normal temperature range 22°C (72°F) or higher, the contacts of the cold start injector time switch are open and the time switch is off, instead, the ECU controls the operating time of the cold start injector.
In this way, the CO and HC levels can be reduced while the engine is being started and the engine startability is maintained. Control by the ECU ends when the coolant temperature reaches 60°C (140°F).

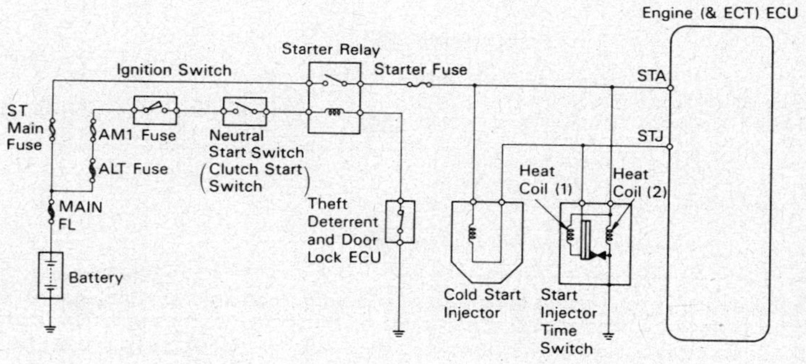

DIAGNOSTIC CHART

1 Check voltage of cold start injector power source.
- NG → **2**
- OK → **3**

2 Check cold start injector.
- NG → Replace cold start injector.
- OK → Check for open in harness and connector between battery and ECU.

3 Check start injector time switch.
- NG → Replace start injector time switch.
- OK → **4**

4 Check for open and short in harness and connector between cold start injector and start injector time switch.
- NG → Repair or replace harness or connector.
- OK → **5**

5 Check fuel injection of cold start injector.
- NG → Replace cold start injector.
- OK → Proceed to next circuit inspection shown on matrix chart

WIRING DIAGRAM

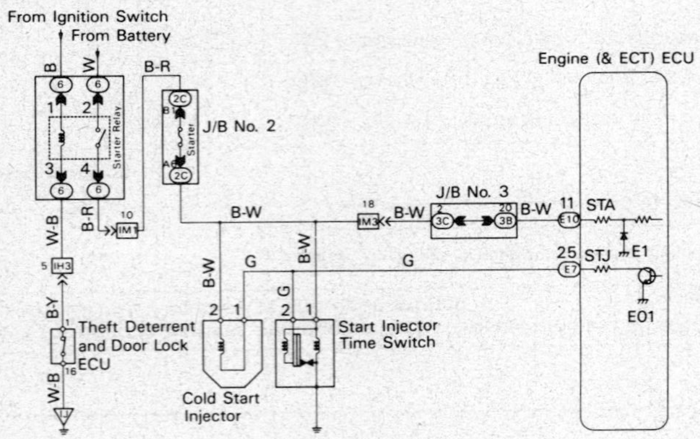

INSPECTION PROCEDURE

1 Check voltage between terminal STJ of engine (& ECT) ECU connector and body ground.

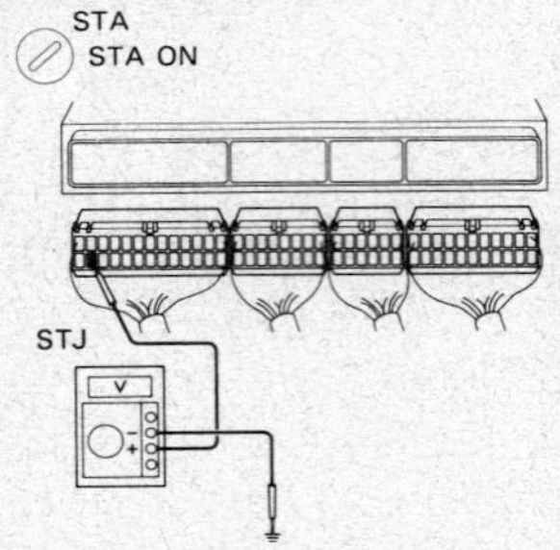

P (1) Remove glove compartment.
(2) Disconnect start injector time switch connector.
(3) Disconnect engine (& ECT) ECU connector.

C Measure voltage between terminal STJ of engine (& ECT) ECU connector and body ground when ignition switch is turned to STA.

OK **Voltage: 10 – 14 V**

NG → **2**

OK → Go to step **3**.

2 Check resistance of cold start injector.

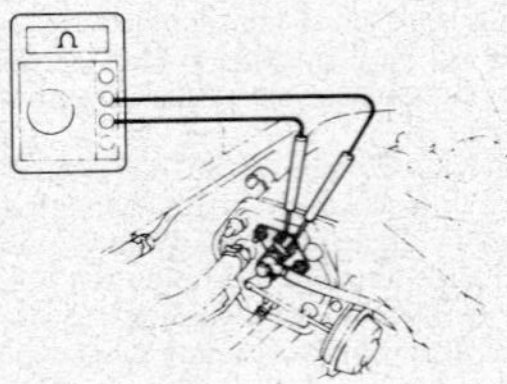

P Disconnect cold start injector connector.

C Measure resistance between terminals of cold start injector.

OK **Resistance: 2 – 4 Ω at 20°C (68°F)**

NG → **Replace cold start injector.**

OK → **Check and repair harness and connector between engine (& ECT) ECU and cold start injector, cold start injector and battery.**

3 Check start injector time switch.

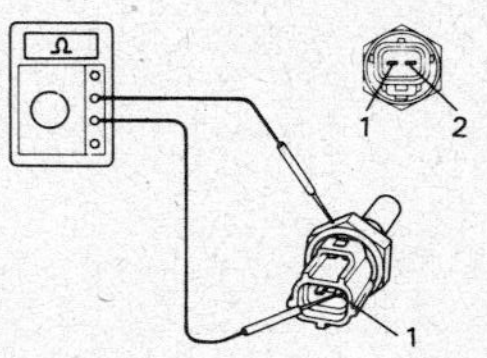

P Remove start injector time switch (See page EG-219).

C Measure resistance between terminals shown below.

OK

Terminals	Resistance	Coolant temperature
1 – 2	25 – 45 Ω	below 15°C (59°F)
	65 – 85 Ω	above 30°C (86°F)
1-Ground	25 – 85 Ω	–

OK

NG → **Replace start injector time switch.**

4 Check for open and short in harness and connector between cold start injector and start injector time switch

OK

NG → **Repair or replace harness or connector.**

5 Check fuel injection of cold start injector.

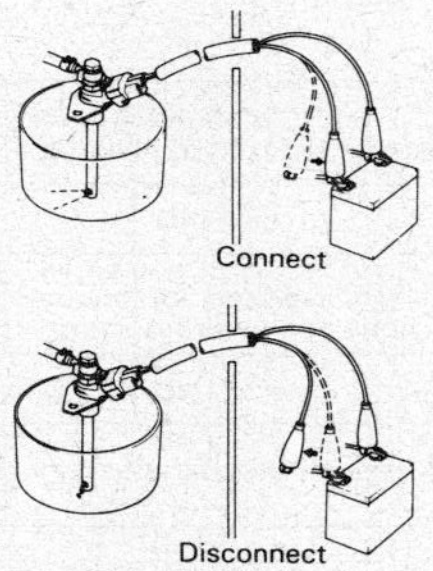

P Remove cold start injector

C (1) Check fuel injection of cold start injector

(2) Check fuel leakage of cold start injector

OK **(1) Fuel is injected normally.**
(2) Fuel leakage is less than one drop per minute.

Notice **Perform this check within the shortest possible time.**

OK

NG → **Repair or replace cold start injector.**

Proceed to next circuit inspection shown on matrix chart

ISC Valve Circuit

CIRCUIT DESCRIPTION

The ISC valve is provided on the intake air chamber and intake air bypassing the throttle valve is directed to the ISC valve through a hose.
A step motor is built into the ISC valve. It consists of four coils, the magnetic rotor, valve shaft and valve.
When current flows to the coils due to signals from the ECU, the rotor turns and moves the valve shaft forward or backward, changing the clearance between the valve and the valve seat.
In this way the intake air volume bypassing the throttle valve is regulated, controlling the engine speed.
There are 125 possible positions to which the valve can be opened.

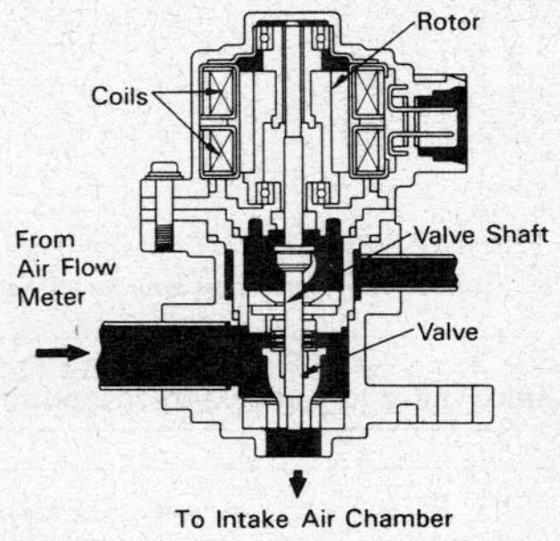

Reference **INSPECTION USING OSCILLOSCOPE**

ISC SIGNAL WAVEFORM

A/C switch OFF → ON
(ISC valve open)

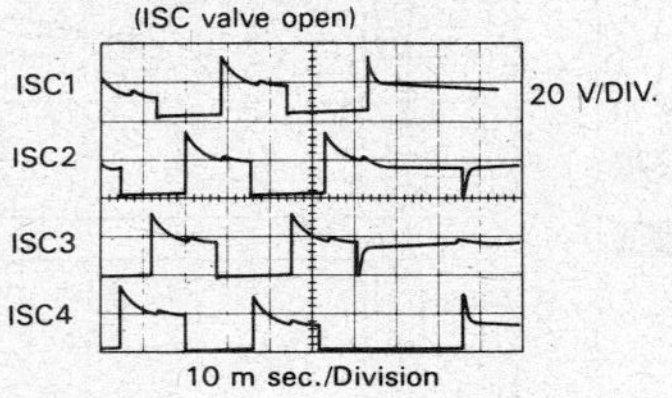

A/C switch ON → OFF
(ISC valve close)

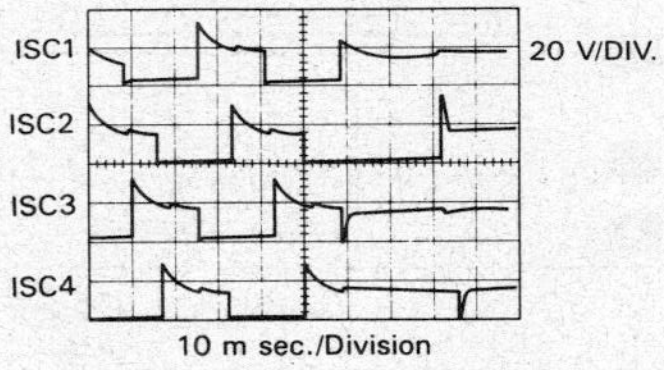

- With the engine idling measure between terminals ISC1, ISC2, ISC3, ISC4 and E01 of engine (& ECT) ECU when A/C switch ON or OFF.

HINT:
The correct waveform appears as shown in the illustration on the left.

DIAGNOSTIC CHART

1 Check operation for ISC valve. — NG → Replace ISC valve.

OK ↓

2 Check for open and short in harness and connector between EFI main relay, ISC valve and ECU. — NG → Repair or replace harness or connector.

OK ↓

Proceed to next circuit inspection shown on matrix chart

WIRING DIAGRAM

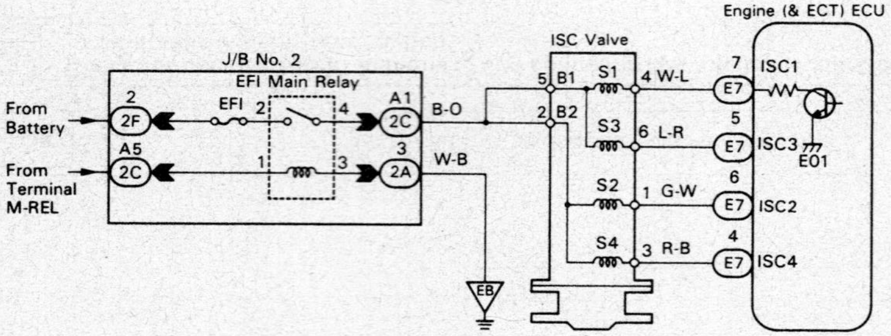

INSPECTION PROCEDURE

1 Check ISC valve.

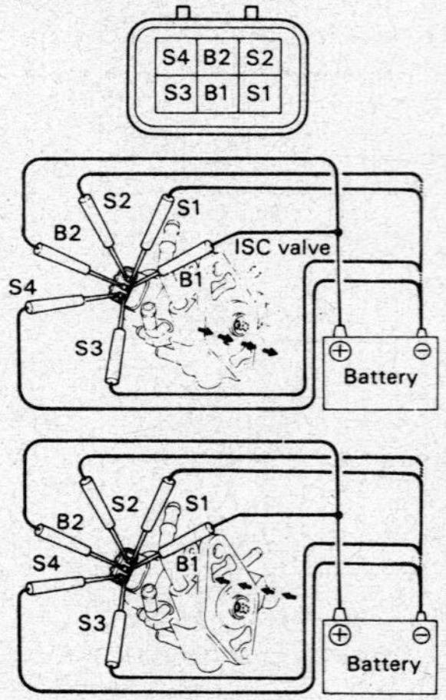

P Disconnect ISC valve connector.

C Measure resistance between terminals shown below.

OK

Terminal	Resistance
B1 – S1	10 Ω – 30 Ω
B1 – S3	10 Ω – 30 Ω
B2 – S2	10 Ω – 30 Ω
B2 – S4	10 Ω – 30 Ω

P Remove ISC valve.

C (1) Connect the battery positive lead to terminals B1 and B2, and the negative lead to terminals S1-S2-S3-S4 in that order.
(2) Connect the battery positive lead to terminals B1 and B2, and the negative lead to terminals S4-S3-S2-S1 in that order.

OK **(1) The valve moves in the closing direction.**
(2) The valve moves in the opening direction.

NG → **Replace ISC valve.**

OK ↓

2 Check for open and short in harness and connector between EFI main relay and ISC valve, ISC valve and engine (& ECT) ECU

NG → **Repair or replace harness or connector.**

OK ↓

Proceed to next circuit inspection shown on matrix chart

Fuel System Circuit

CIRCUIT DESCRIPTION

In the diagram below, when the engine is cranked, current flows from terminal ST of the ignition switch to the starter relay coil, the starter relay switches on and current flows to coil L1 of the circuit opening relay. Thus the circuit opening relay switches on, power is supplied to the fuel pump and the fuel pump operates. After the engine cranks, the cylinders begining in air, causing the measuring plate inside the air flow meter to open. This turns on the fuel pump switch, which is connected to the measuring plate, and current flows to the L2 coil of the circuit opening relay.

After the engine starts and the ignition switch is turned from START back to ON, current flowing to the L1 coil of the circuit opening relay is cut off. However current continues to flow to the L2 coil even after the engine starts due to the fuel pump switch inside the air flow meter. As a result, the circuit opening relay stays on, allowing the fuel pump to continue operating.

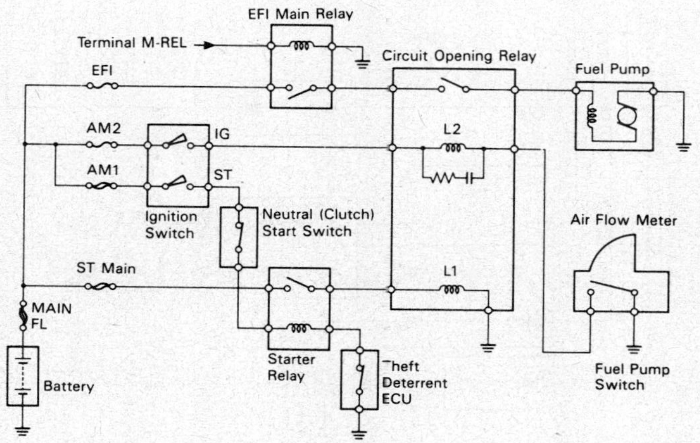

DIAGNOSTIC CHART

Step	Check	Result	Action
1	Check starter fuse.	NG	Check for short in all the harness and components connected to starter fuse.
	↓ OK		
2	Check for circuit opening relay.	NG	Replace circuit opening relay.
	↓ OK		
3	Check for open in harness and connector between starter fuse and relay, relay and body ground.	NG	Repair or replace harness or connector.
	↓ OK		
4	Check IGN fuse.	NG	Check for short in all the harness and components connected to IGN fuse.
	↓ OK		
5	Check fuel pump switch.	NG	Replace air flow meter.
	↓ OK		
6	Check for open in harness and connector between IGN fuse and relay, relay and body ground.	NG	Repair or replace harness or connector.
	↓ OK		
7	Check ECU power source circuit.	NG	Repair or replace.
	↓ OK		
8	Check fuel pump.	NG	Replace fuel pump.
	↓ OK		
	Check for open in harness and connector between J/B No. 2 and relay, relay and body ground.		

WIRING DIAGRAM

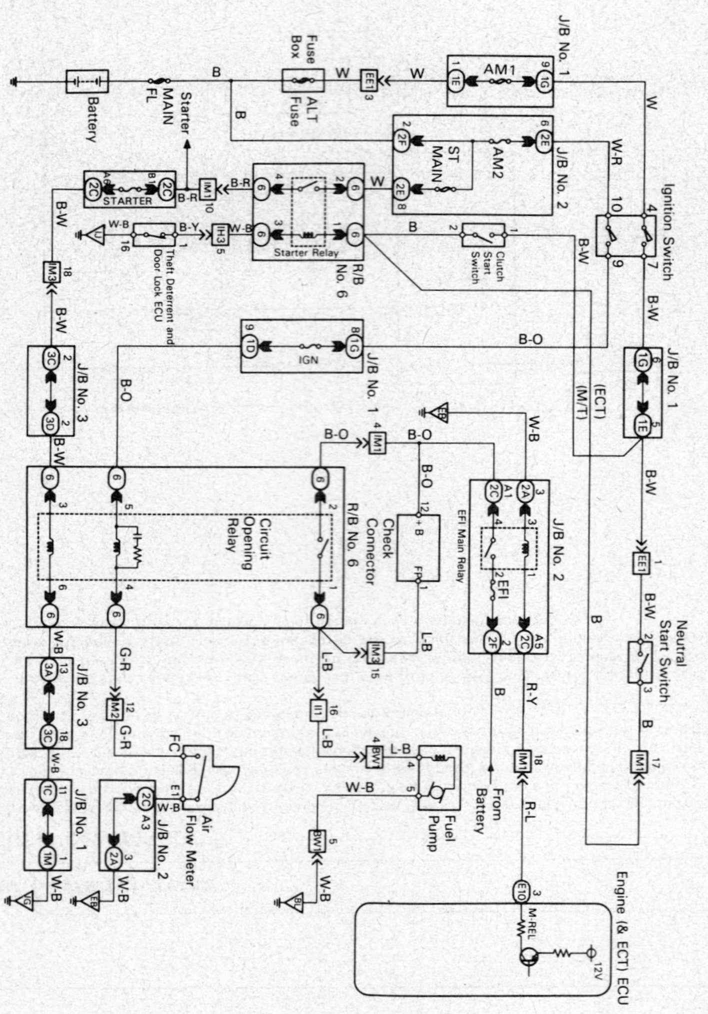

INSPECTION PROCEDURE

1 Check STARTER fuse.

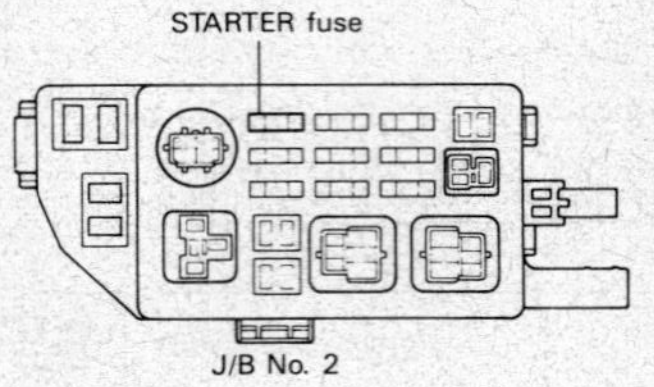

P Remove STARTER fuse from J/B No. 2.

C Check continuity of STARTER fuse.

OK **Continuity**

OK

NG **Check for short in all the harness and components connected to STARTER fuse.**

2 Check circuit opening relay.

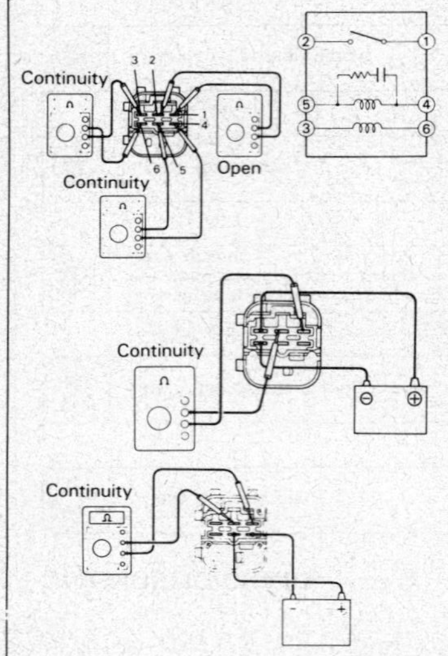

P (1) Remove glove compartment

(2) Remove circuit opening relay from R/B No. 6.

C Check continuity between terminals of circuit opening relay shown below.

OK

Terminals 1 and 2	Open
Terminals 3 and 6	Continuity (Reference value 30 Ω)
Terminals 4 and 5	Continuity (Reference value 150 Ω)

C (1) Apply battery voltage between terminals 3 and 6, 4 and 5.

(2) Check continuity between terminals 1 and 2.

OK

Terminals 1 and 2	Continuity

OK

NG **Replace circuit opening relay.**

3 Check for open in harness and connector between STARTER fuse and circuit opening relay, relay and body ground

NG ⟩ Repair or replace harness or connector.

OK

4 Check IGN fuse.

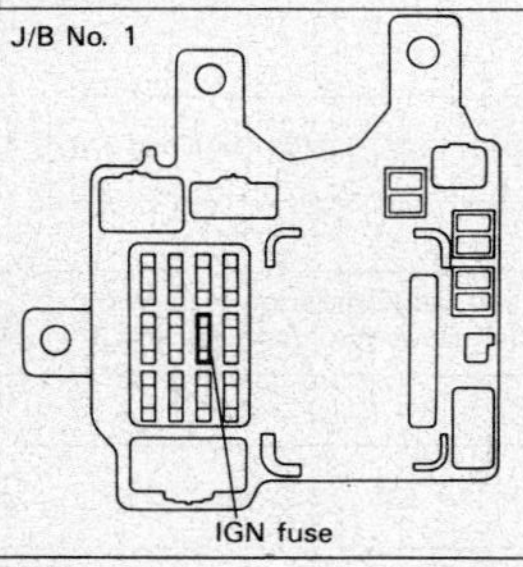

P Remove IGN fuse from J/B No. 1.

C Check continuity of IGN fuse.

OK **Continuity**

NG ⟩ Check for short in all the harness and components connected to IGN fuse.

OK

5 Check fuel pump switch (inside the air flow meter).

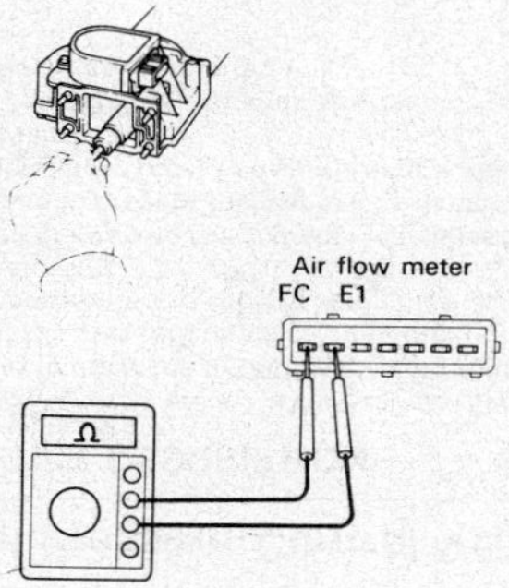

P (1) Remove air cleaner cap.
(2) Disconnect air flow meter connector.

C Measure resistance between terminals FC and E1 of air flow meter connector when measuring plate is open and close with a screwdriver.

OK

Measuring plate	Resistance
Open	Continuity
Closed	1 MΩ or higher

NG ⟩ Replace air flow meter.

OK

6 Check for open in harness and connector between IGN fuse and circuit opening relay, relay and body ground

NG ⟩ Repair or replace harness or connector.

OK

7 Check ECU power source circuit

NG ⟩ Repair or replace.

OK

8 Check fuel pump

NG ⟩ Replace fuel pump.

OK

Check for open in harness and connector between J/B No. 2 and circuit opening relay, relay and body ground

Fuel Pressure Control VSV Circuit

CIRCUIT DESCRIPTION

The ECU turns on a VSV (Vacuum Switching Valve) to draw the air into the diaphragm chamber of the pressure regulator if it detects that the temperature of the coolant is too high during engine starting.
The air drawn into the chamber increases the fuel pressure to prevent fuel vapor lock at high engine temperature in order to help the engine start when it is warm.
Fuel pressure control ends approx. 120 sec. after the engine is started.

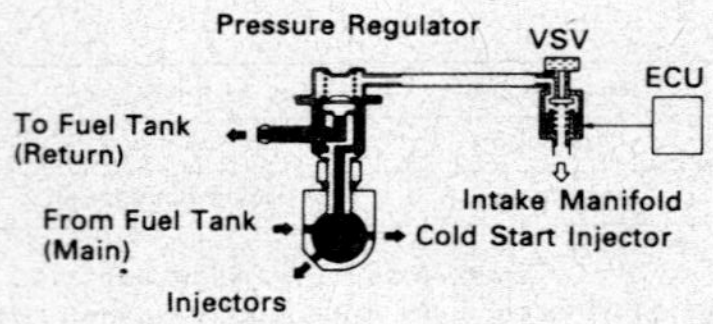

DIAGNOSTIC CHART

1 Check operation of the fuel pressure control VSV. — NG → Replace fuel pressure control VSV.

↓ OK

2 Check voltage of VSV power source. — OK → Proceed to next circuit inspection shown on matrix chart

↓ NG

3 Check for open and short in harness and connector between main relay and ECU. — NG → Repair or replace harness or connector.

↓ OK

Check and replace ECU.

WIRING DIAGRAM

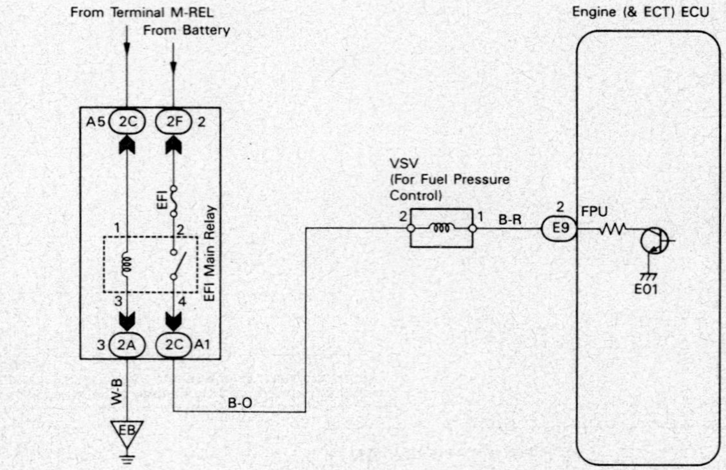

INSPECTION PROCEDURE

1 Check fuel pressure control VSV.

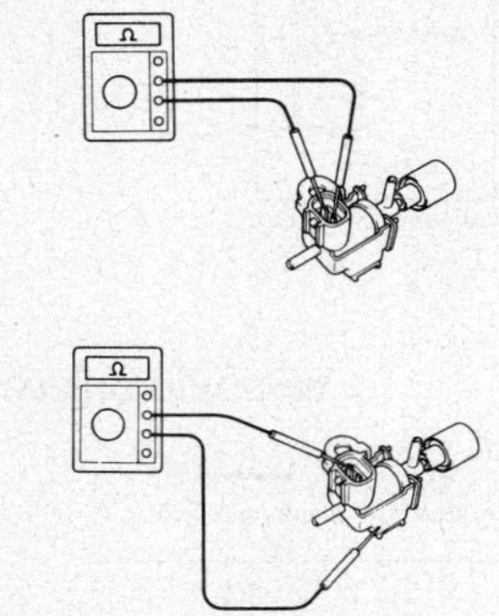

P (1) Remove fuel pressure control VSV.
(2) Disconnect fuel pressure control VSV connector.

C (1) Measure resistance between terminals.
(2) Measure resistance between each terminal and the body.

OK (1) **Resistance: 30 – 50 Ω at 20°C (68°F)**
(2) **Resistance: 1 MΩ or higher**

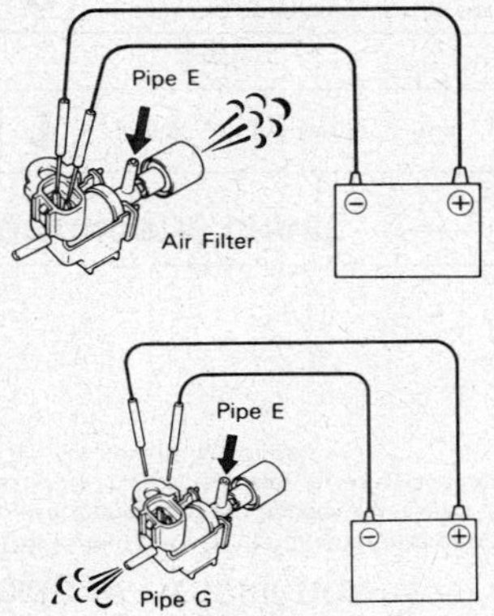

C Check operation of fuel pressure control VSV when battery voltage is applied to the terminals of fuel pressure control VSV connector or not.

OK **Battery voltage is applied:**
The air from pipe E is flowing out through the air filter.

Battery voltage is not applied:
The air from pipe E is flowing out through pipe G.

OK → Go to step 2.

NG → **Replace fuel pressure control VSV.**

2 Check voltage between terminal FPU of engine (& ECT) ECU connector and body ground.

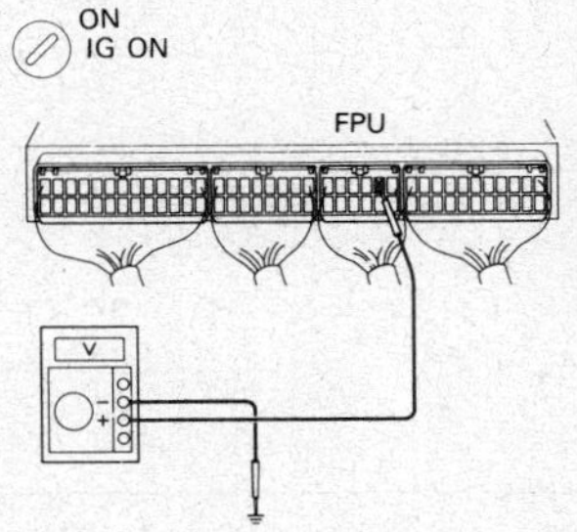

P (1) Remove glove compartment.
(2) Turn ignition switch on.

C Measure voltage between terminal FPU of engine (& ECT) ECU connector and body ground.

OK **Voltage: 10 – 14 V**

OK → **Proceed to next circuit inspection shown on matrix chart**

NG ↓

3 Check for open and short in harness and connector between engine (& ECT) ECU and VSV, VSV and EFI main relay

NG → **Repair or replace harness or connector.**

OK ↓

Check and replace engine (& ECT) ECU.

Air Conditioner Cut Control Circuit

CIRCUIT DESCRIPTION

This circuit cuts air conditioner operation during vehicle acceleration in order to increase acceleration performance. During acceleration with the vehicle speed at 25 km/h (16 mph) or less, engine speed at 1,600 rpm or less and throttle valve opening angle at 60° or more, the A/C magnet switch is turned OFF for several seconds.

DIAGNOSTIC CHART

1 Check voltage terminal ACT. — OK → Check and replace engine (& ECT) ECU.

NG ↓

2 Check for open and short in harness and connector between engine (& ECT) ECU and A/C control assembly. — NG → Repair or replace harness or connector.

OK ↓

Check and replace A/C control assembly.

WIRING DIAGRAM

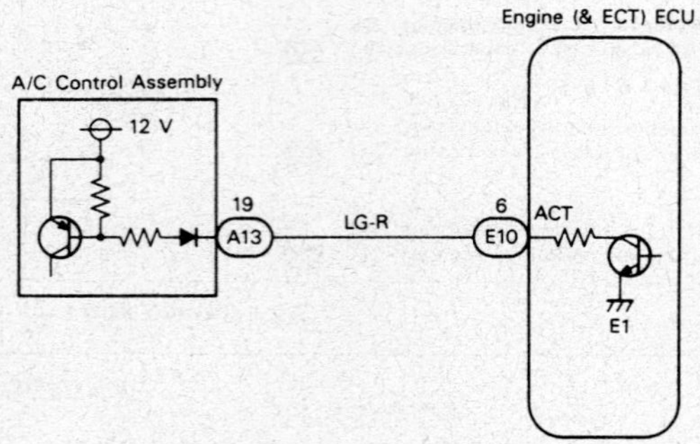

INSPECTION PROCEDURE

1 **Check voltage between terminal ACT of engine (& ECT) ECU and body ground.**

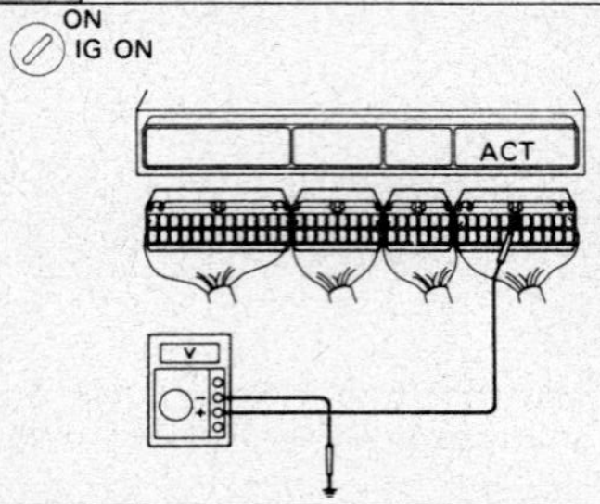

P (1) Remove glove compartment.

(2) Disconnect engine (& ECT) ECU connector.
(3) Turn ignition switch ON.

C Measure voltage between terminal ACT of engine (& ECT) ECU and body ground when A/C switch is turned to ON.

OK **Voltage: 10 – 14 V**

OK → **Check and replace engine (& ECT) ECU.**

NG ↓

2 **Check for open and short in harness and connector between engine (& ECT) ECU and A/C control assembly.**

NG → **Repair or replace harness or connector.**

OK ↓

Check and replace A/C control assembly.

IACV Control VSV Circuit

CIRCUIT DESCRIPTION

This circuit opens and closes the IACV (Intake Air Control Valve) in response to the engine load in order to increase the intake efficiency (ACIS: Acoustic Control Induction System).
When the engine speed is 3,900 rpm or less and the throttle valve opening angle is 60° or more, the ECU turns the VSV ON and closes the IACV. At all other times, the VSV is OFF, so the IACV is open.

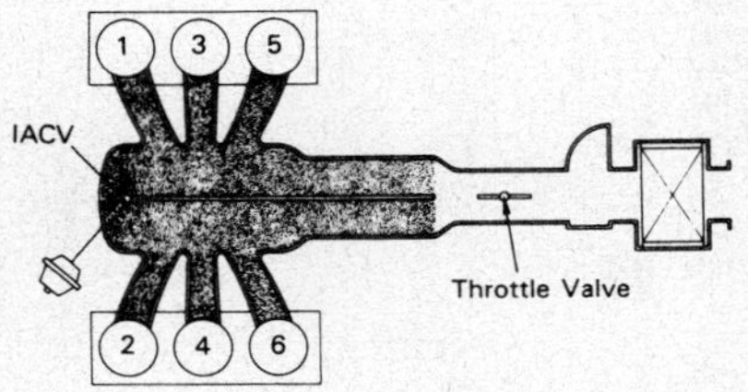

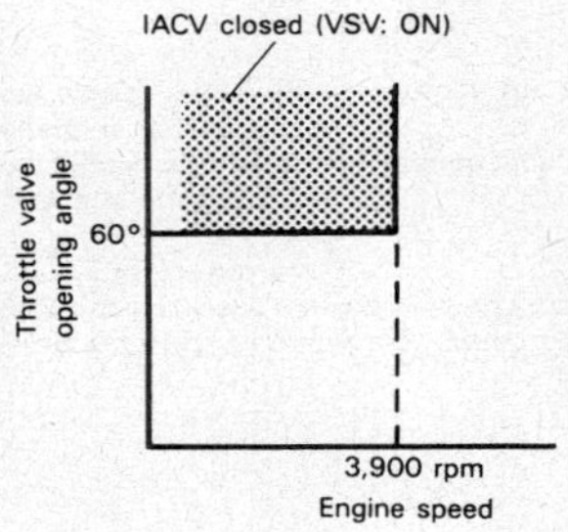

DIAGNOSTIC CHART

Step	Check	Result	Action
1	Check operation of the IACV control VSV.	NG	Replace IAVC control VSV.
		OK	Go to 2
2	Check voltage of VSV power source.	OK	Check for vacuum tank.
		NG	Go to 3
3	Check for open and short in harness and connector between main relay and ECU.	NG	Repair or replace harness or connector.
		OK	Check and replace ECU.

WIRING DIAGRAM

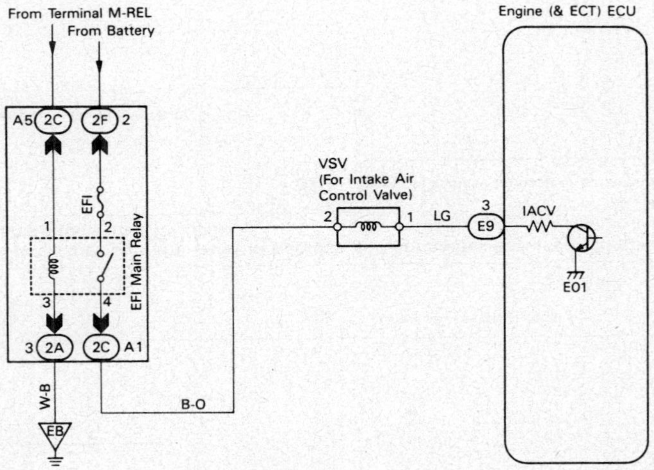

INSPECTION PROCEDURE

1 Check IACV control VSV.

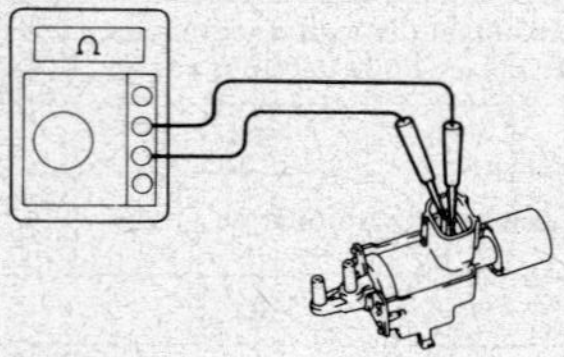

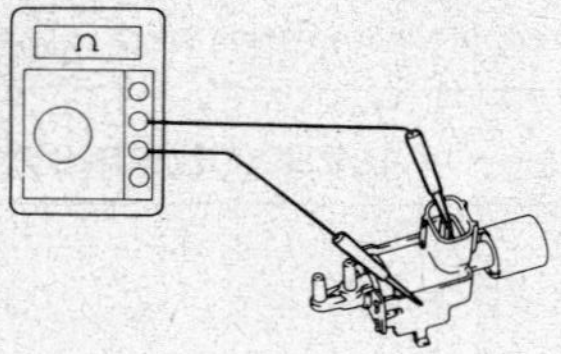

P (1) Remove IACV control VSV.
(2) Disconnect IACV control VSV connector.

C (1) Measure resistance between terminals.
(2) Measure resistance between each terminal and the body.

OK **(1) Resistance: 33 – 39 Ω at 20°C (68°F)**
(2) Resistance: 1 MΩ or higher.

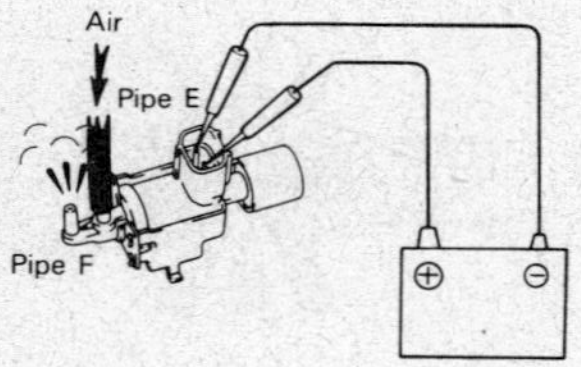

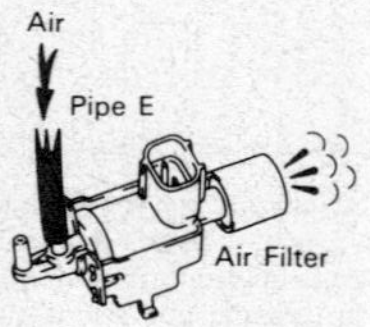

C Check operation of IACV control VSV when battery voltage is applied to the terminals of IACV control VSV connector or not.

OK **Battery voltage is applied:**
The air from pipe E is flowing out through pipe F.
Battery voltage is not applied:
The air from pipe E is flowing out through the air filter.

NG Replace IACV control VSV.

OK Go to step **2**.

2 Check voltage between terminal IACV of engine (& ECT) ECU connector and body ground.

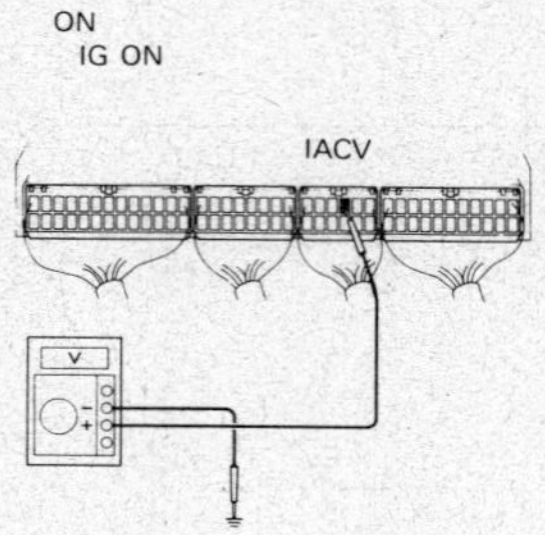

P (1) Remove glove compartment.
(2) Turn ignition switch on.

C Measure voltage between terminal IACV of engine (& ECT) ECU connector and body ground.

OK **Voltage: 10 – 14 V**

OK Check for vacuum tank.

NG

3 Check for open and short in harness and connector between EFI main relay and engine (& ECT) ECU

NG Repair or replace harness or connector.

OK Check and replace engine (& ECT) ECU.

TE1, TE2 Terminal Circuit

CIRCUIT DESCRIPTION

Terminals TE1 and TE2 are located in the check connector and TDCL. The check connector located in the engine compartment and the TDCL located in the cabin. When these terminals are connected with the E1 terminal, diagnostic codes in normal mode or test mode can be read from the "CHECK" engine warning light on the combination meter.

DIAGNOSTIC CHART

HINT: If terminals TE1 and TE2 are connected with terminal E1, diagnostic code is not output or test mode is not activated.
Even though terminal TE1 is not connected with terminal E1, the "CHECK" engine warning light blinks.
For the above phenomenon, the likely cause is an open or short in the wire harness, or malfunction inside the ECU.

1 Check voltage of terminals TE1, TE2. — OK → Check and replace ECU.

NG ↓

2 Check continuity between terminal E1 and body ground. — NG → Repair or replace harness or connector.

OK ↓

3 Check for open and short in harness and connector between check connector, TDCL and ECU. — NG → Repair or replace harness or connector.

OK ↓

Check and replace ECU.

WIRING DIAGRAM

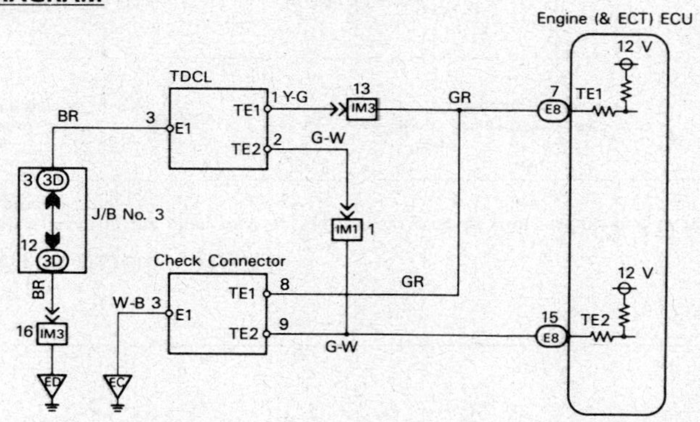

INSPECTION PROCEDURE

1 Check voltage between terminals TE1, TE2 and E1 of check connector, TDCL.

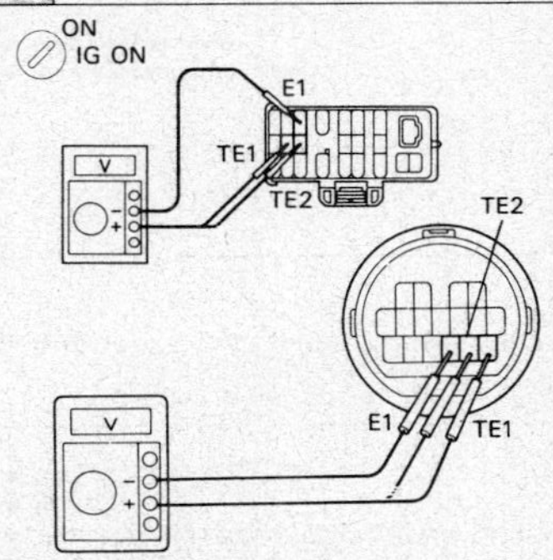

P Turn ignition switch on.

C Measure voltage between terminals TE1, TE2 and E1 of check connector, TDCL.

OK **Voltage: 10 – 14 V**

OK → **Check and replace engine (& ECT) ECU.**

NG ↓

2 Check continuity between terminal E1 of check connector, TDCL and body ground.

NG → **Repair or replace harness or connector.**

OK ↓

3 Check for open and short in harness and connector between engine (& ECT) ECU and check connector, TDCL

NG → **Repair or replace harness or connector.**

OK ↓

Check and replace engine (& ECT) ECU.

DIAGNOSTIC CODE CHART

HINT: Parameters listed in the chart may not be exactly same as your reading due to type of the instruments or other factors.

If a malfunction code is displayed during the diagnostic code check in test mode, check the circuit for that code listed in the table below

Code No.	Number of Check Engine Blinks	Circuit	Diagnostic Code Detecting Condition	Trouble Area	"CHECK" Engine Warning Light[*1] Normal Mode	"CHECK" Engine Warning Light[*1] Test Mode	Memory[*2]	
—		Normal	No code is recorded.	————	—	—	——	——
12		RPM Signal	No NE or G1 and G2 signal to ECU within 2 sec. after cranking. Open in "G ⊖" circuit.	• Open or short in NE, G circuit • Distributor • Open or short in STA circuit • ECU	ON	N.A.	○	
13		RPM Signal	No NE signal to ECU for 0.1 sec. or more at 1,000 rpm or more.	• Open or short in NE circuit • Distributor • ECU	ON	N.A.	○	
			No 12 pulses of NE to ECU during the interval between G1 and G2 pulses.	• Open or short in NE circuit • Distributor • ECU	N.A.	ON		
14		Ignition Signal	No IGF signal to ECU for 6 consecutive IGT signal.	• Open or short in IGF or IGT circuit from igniter to ECU. • Igniter • ECU	ON	N.A.	○	
16		ECT Control Signal	Fault in communications between the engine CPU and ECT CPU in the ECU.	• ECU	ON	N.A.	X	
21		Main Oxygen Sensor Signal (front side)	(1) Main Oxygen sensor signal voltage is reduced to between 0.35 V and 0.70 V for 60 sec. under conditions (a) ~ (d). (2 trip detection logic)[*5] (a) Coolant temp.: Between 80°C (176°F) and 95°C (203°F). (b) Engine speed: 1,500 rpm or more. (c) Load driving (Ex. ECT in 4th speed (5th for M/T), A/C ON, Flat road, 50 mph (80 km/h)). (d) Main oxygen sensor signal voltage: Alternating above and below 0.45 V	• Main oxygen sensor circuit • Main oxygen sensor	ON	ON	○	
			(2) Main oxygen sensor signal voltage exceeds 0.70 V for 3 sec. or more during fuel cut.	• Main oxygen sensor circuit • Main oxygen sensor.				

3.0L ENGINE—DIAGNOSTIC CODES—SC300

Code No.	Number of Check Engine Blinks	Circuit	Diagnostic Code Detecting Condition	Trouble Area	"CHECK" Engine Warning Light*1 Normal Mode	"CHECK" Engine Warning Light*1 Test Mode	Memory*2	
22		Water Temp. Sensor Signal	Open or short in water temp. sensor circuit for 0.5 sec. or more.	• Open or short in water temp. sensor circuit. • Water temp. sensor. • ECU	ON	ON	○	
24		Intake Air Temp. Sensor Signal	Open or short in intake air temp. sensor circuit for 0.5 sec. or more.	• Open or short in intake air temp. sensor circuit. • Intake air temp. sensor • ECU	OFF ON*3	ON	○	
25		Air-Fuel Ratio Lean Malfunction	(1). Main oxygen sensor voltage is 0.45 V or less (lean) for 90 sec. under conditions (a) and (b). (2 trip detection logic)*5 (a) Coolant temp.: 70°C (158°F) or more. (b) Engine speed: 1,500 rpm or more.	• Open or short in main oxygen sensor circuit. • Main oxygen sensor • Ignition system. • ECU	ON	ON	○	
			(2)*3 Difference of air-fuel ratio feedback compensation value between front (No.1 ~ 3 cylinders) and rear (No.4 ~ 6 cylinders) is more than 15 percentage for 20 sec. or more under conditions (a) and (b). (2 trip detection logic)*5 (a) Engine speed: 2,000 rpm or more. (b) Coolant temp.: Between 60°C (140°F) and 95°C (203°F).	• Open or short in injector circuit. • Fuel line pressure (injector leak, blockage) • Mechanical system malfunction (skipping teeth of timing belt) • Ignition system • Compression pressure (foreign object caught in valve) • Air flow meter (air intake) • ECU				
			(3)*3 Engine speed varies by more than 15 rpm over the preceding crank angle period during a period of 20 sec. or more under conditions (a) and (b). (2 trip detection logic)*5 (a) Engine speed: Idling (b) Coolant temp.: Between 60°C (140°F) and 95°C (203°F)	• Open or short in injector circuit • Fuel line pressure (injector leak, blockage) • Mechanical system malfunction (skipping teeth of timing belt) • Ignition system • Compression pressure (foreign object caught in valve) • Air flow meter (air intake) • ECU				

Code No.	Number of Check Engine Blinks	Circuit	Diagnostic Code Detecting Condition	Trouble Area	"CHECK" Engine Warning Light*1 Normal Mode	"CHECK" Engine Warning Light*1 Test Mode	Memory*2	
26*3		Air-Fuel Ratio Rich Malfunction	(1) Difference of air-fuel ratio feedback compensation value between front (No.1 ~ 3 cylinders) and rear (No.4 ~ 6 cylinders) is more than 15 percentage for 20 sec. or more under conditions (a) and (b). (2 trip detection logic)*5 (a) Engine speed: 2,000 rpm or more. (b) Coolant temp.: Between 60°C (140°F) and 95°C (203°F).	● Open or short in injector circuit ● Fuel line pressure (injector leak, blockage) ● Mechanical system malfunction (skipping teeth of timing belt) ● Ignition system ● Compression pressure (foreign object caught in valve) ● Air flow meter (air intake) ● ECU	ON	ON	○	
			(2) Engine speed varies by more than 15 rpm over the preceding crank angle period during a period of 20 sec. or more under conditions (a) and (b). (2 trip detection logic)*5 (a) Engine speed: 2,000 rpm or more. (b) Coolant temp.: Between 60°C (140°F) and 95°C (203°F).	● Open or short in injector circuit ● Fuel line pressure (injector leak, blockage) ● Mechanical system malfunction (skipping teeth of timing belt) ● Ignition system ● Compression pressure (foreign object caught in valve) ● Air flow meter (air intake) ● ECU				
27*3		Sub-Oxygen Sensor Signal	(1) Open or short in heater circuit of sub-oxygen sensor for 0.5 sec. or more.	● Open or short in heater circuit of sub-oxygen sensor. ● Sub-oxygen sensor heater ● ECU	ON	N.A.	○	
			(2) Main oxygen sensor signal is 0.45 V or more and sub-oxygen sensor signal is 0.45 V or less under conditions (a) ~ (c). (2 trip detection logic)*5 (a) Coolant temp.: 80°C (176°F) or more. (b) Engine speed: 1,500 rpm or more. (c) Accel. pedal: Fully depressed for 2 sec. or more.	● Open or short in sub-oxygen sensor circuit. ● Sub-oxygen sensor ● ECU	ON	ON		
28		Main Oxygen Sensor Signal (rear side)	Same as code No.21.	Same as code No.21.	ON	ON	○	

Code No.	Number of Check Engine Blinks	Circuit	Diagnostic Code Detecting Condition	Trouble Area	"CHECK" Engine Warning Light*1 Normal Mode	"CHECK" Engine Warning Light*1 Test Mode	Memory*2	
31		Air Flow Mater Signal	All conditions below are detected. (a) No air-flow meter signal to ECU for 2 sec. when engine speed is above 300 rpm. (b) Engine stall.	● Open or short in air flow meter circuit ● Air flow meter ● ECU	ON	N.A.	○	
35		HAC Sensor Signal	Open or short in HAC sensor circuit for 0.5 sec. or more.	● ECU	ON	ON	○	
41		Throttle Position Sensor Signal	(1) Open or short in throttle position sensor circuit for 0.5 sec. or more.	● Open or short in throttle position sensor circuit.	OFF	ON	○	
			(2)*4IDL1 contact is ON and VTA1 output exceeds 1.5 V for 0.5 sec. or more.	● Throttle position sensor. ● ECU	(ON*3)			
42		Vehicle Speed Sensor Signal (for ECT)	All conditions below are detected continuously for 8 sec. or more. (a) Vehicle speed signal: 0 km/h (mph). (b) Engine speed: 2,800 rpm or more. (c) Neutral start switch (NSW): OFF (d) Stop light switch: OFF	● No.1 speed sensor. ● Combination meter. ● Open or short in No.1 speed sensor circuit ● ECU	OFF	OFF	○	
		Vehicle Speed Sensor Signal (for M/T)	All conditions below are detected continuously for 8 sec. or more. (a) Vehicle speed signal: 0 km/h (mph). (b) Engine speed: Between 2,350 rpm and 5,000 rpm. (c) Coolant temp.: 80°C (176°F) or more. (d) Load driving					
43		Starter Signal	No starter signal to ECU.	● Open or short in starter signal circuit. ● Open or short in ignition switch or starter relay circuit. ● ECU	N.A.	OFF	×	

Code No.	Number of Check Engine Blinks	Circuit	Diagnostic Code Detecting Condition	Trouble Area	"CHECK" Engine Warning Light[*1] Normal Mode	"CHECK" Engine Warning Light[*1] Test Mode	Memory[*2]	
47[*4]		Sub-Throttle Position Sensor Signal	(1) Open or short in sub-throttle position sensor circuit for 0.5 sec. or more. (2) IDL 2 contact is ON and VTA2 output exceeds 1.5 V for 0.5 sec. or more.	● Open or short in sub-throttle position sensor circuit. ● Sub-throttle position sensor. ● ECU	OFF	ON	○	
52		Knock Sensor Signal(front side)	No No.1 knock sensor signal to ECU for 4 crank revolutions with engine speed between 1,600 rpm and 5,200 rpm.	● Open or short in No.1 knock sensor circuit. ● No.1 knock sensor (looseness). ● ECU	ON	N.A.	○	
53	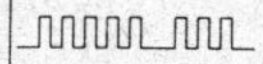	Knock Control Signal	Engine control computer (for knock control) malfunction at engine speed between 650 rpm and 5,200 rpm.	● ECU	ON	N.A.	X	
55	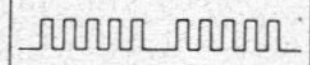	Knock Sensor Signal (rear side)	No No.2 knock sensor signal to ECU for 4 crank revolutions with engine speed between 1,600 rpm and 5,200 rpm.	● Open or short in No.2 knock sensor circuit ● No.2 knock sensor (looseness). ● ECU	ON	N.A.	○	
71[*3]		EGR System Malfunction	EGR gas temp. is 70°C (158°F) or below for 1 ~ 4 min. under conditions (a) and (b). (2 trip detection logic)[*5] (a) Coolant temp.: 63°C (145°F) or more. (b) EGR operation possible (Ex. ECT in 3rd speed (5th for M/T), A/C ON, 60 mph (96 km/h), Flat road).	● Open in EGR gas temp. sensor circuit. ● Short in VSV circuit for EGR. ● EGR hose disconnected, valve stuck. ● Clogged EGR gas passage. ● ECU	ON	ON	○	

Code No.	Number of Check Engine Blinks	Circuit	Diagnostic Code Detecting Condition	Trouble Area	"CHECK" Engine Warning Light*1 Normal Mode	"CHECK" Engine Warning Light*1 Test Mode	Memory*2	
78		Fuel Pump Control Signal	(1) Open or short in fuel pump circuit for 1 sec. or more with engine speed 1,000 rpm or less. (2 trip detection logic)*5 (2) Open in input circuit of fuel pump ECU (FPC) with engine speed 1,000 rpm or less. (2 trip detection logic)*5 (3) Open or short in diagnostic signal line (DI) of fuel pump ECU with engine speed 1,000 rpm or less. (2 trip detection logic)*5	• Open or short in fuel pump ECU circuit. • Fuel pump ECU • Fuel pump • Engine (& ECT) ECU power source circuit. • Engine (& ECT) ECU	N.A.	ON	○	
51		Switch Condition Signal	(1) 3 sec. or more after engine starts, idle switch OFF (IDL1). (2) Neutral start switch OFF (NSW). (Shift position in "R", "D", "2", or "1" ranges). (3) A/C switch ON.	• A/C switch circuit. • Throttle position sensor IDL circuit. • Neutral start switch circuit. • Accelerator pedal and cable. • ECU	N.A.	OFF	X	

*1: "ON" displayed in the diagnosis mode column indicates that the "CHECK" Engine Warning Light is lighted up when a malfunction is detected. "OFF" indicates that the "CHECK" does not light up during malfunction diagnosis, even if a malfunction is detected. "N.A." indicates that the item is not included in malfunction diagnosis.

*2: "○" in the memory column indicates that a diagnostic code is recorded in the ECU memory when a malfunction occurs. "X" indicates that a diagnostic code is not recorded in the ECU memory even if a malfunction occurs. Accordingly, output of diagnostic results in normal or test mode is performed with the IG switch ON.

*3: Only for USA specification vehicles.

*4: Only vehicles with TRAC.

*5: This indicates items for which "2 trip detection logic" is used. With this logic, when a logic malfunction is first detected, the malfunction is temporarily stored in the ECU memory. If the same case is detected again during the second drive test, this second detection causes the "CHECK" Engine Warning Light to light up.
The 2 trip repeats the same mode a 2nd time. (However, the IG switch must be turned OFF between the 1st trip and 2nd trip).
In the Test Mode, the "CHECK" Engine Warning Light lights up the 1st trip a malfunction is detected.

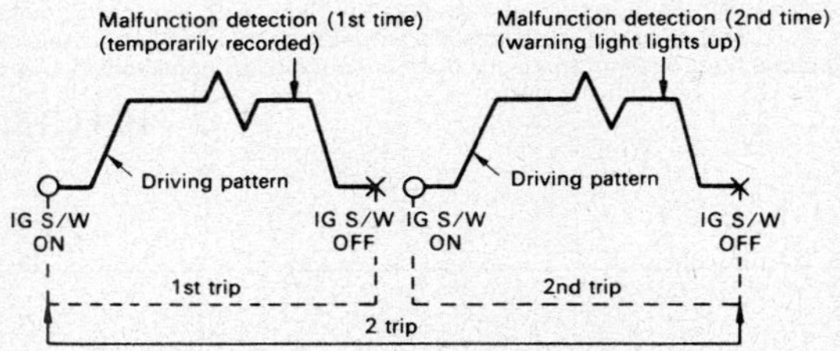

FAIL-SAFE CHART

If any of the following codes is recorded, the ECU enters fail-safe mode.

Code No.	Fail-Safe Operation	Fail-Safe Deactivation Conditions
14	Fuel cut	1 IGF detected in consecutive 3 ignitions.
16	Torque control prohibited.	Returned to normal condition.
22	THW is fixed at 80°C (176°F).	Returned to normal condition.
24	THA is fixed at 20°C (68°F).	Returned to normal condition.
31	• Ignition timing fixed at 10° BTDC. • Injection time fixed Starting ..... 9 m sec. IDL ON ..... 3.6 m sec. IDL OFF .... 6.7 m sec.	KS input 15 times/sec. or more.
35	Atmospheric pressure is fixed at 101.3 kPa (760 mmHg, 29.92 in.Hg).	Returned to normal condition.
41	VTA1 is fixed at 0°.	The following must each be repeated at least 2 time consecutively. (w/o TRAC) • 0.1 V ≦ VTA ≦ 0.95 V • IDL : ON (w/ TRAC) • 0.25 V ≦ VTA ≦ 0.95 V • IDL : ON
52	Max. timing retardation.	IG switch OFF.
53	Max. timing retardation.	Returned to normal condition.
55	Max. timing retardation.	IG switch OFF.

Back-Up Function

If there is trouble with the program in the ECU and the ignition signals (IGT) are not output, the ECU controls fuel injection and ignition timing at predetermined levels as a back-up function to make it possible to continue to operate the vehicle.

Furthermore, the injection duration is calculated from the starting signal (STA) and the throttle position signal (IDL). Also, the ignition timing is fixed at the initial ignition timing, 10° BTDC, without relation to the engine speed.

HINT: If the engine is controlled by the back-up function, the "CHECK" engine warning light lights up to warn the driver of the malfunction but the diagnostic code is not output.

BASIC INSPECTION

When the normal code is displayed in the diagnostic code check, troubleshooting should be performed in the order for all possible circuits to be considered as the causes of the problems.

In many cases, by carrying out the basic engine check shown in the following flow chart, the location causing the problem can be found quickly and efficiently. Therefore, use of this check is essential in engine troubleshooting.

1 Is battery voltage 11 V or more when engine is stopped ?

NO → Charge or replace battery.

YES ↓

2 Is engine cranked ?

NO → Proceed to matrix chart of problem symptoms

YES ↓

3 Does engine start ?

NO → Go to step 7.

YES ↓

4 Check air filter.

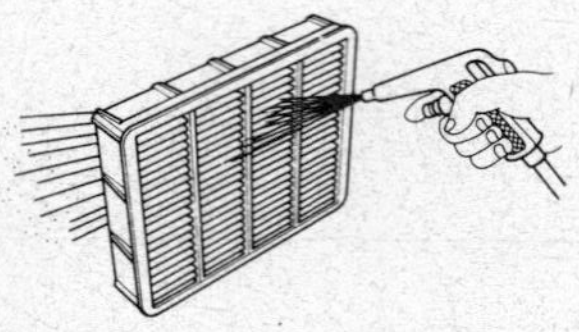

P Remove air filter.

C Visually check that the air filter is not excessively damaged or oily.

Hint If necessary, clean the air filter with compressed air. First blow from inside thoroughly, then blow off outside of the air filter.

NG → Repair or replace.

OK ↓

Go to step 5.

5 Check idle speed.

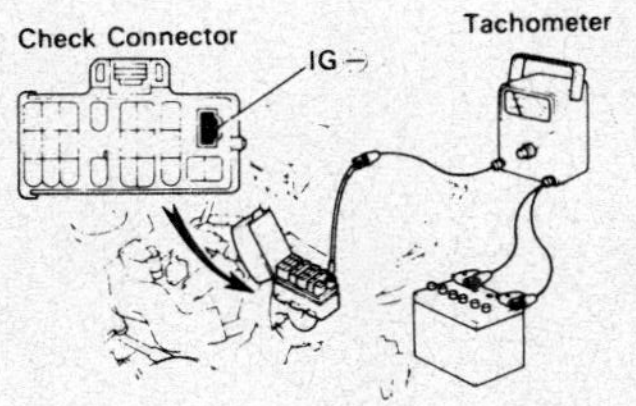

P
(1) Shift transmission into "N" range or neutral.
(2) Warm up engine at normal operating temperature.
(3) Switched off all accessories.
(4) Switched off air conditioner.
(5) Connect tachometer test probe to terminal IG – of check connector.

C Check idle speed.

OK **Idle speed: Reference**
M/T 700 rpm
A/T 650 rpm (N range)

Caution
- **NEVER allow tachometer test probe to touch ground as it could result in damage to igniter and/or ignition coil.**
- **As some tachometers are not compatible with this ignition system, we recommended that you confirm compatibility of your unit before use.**

NG → **Proceed to matrix chart of problem symptoms**

OK ↓

6 Check ignition timing.

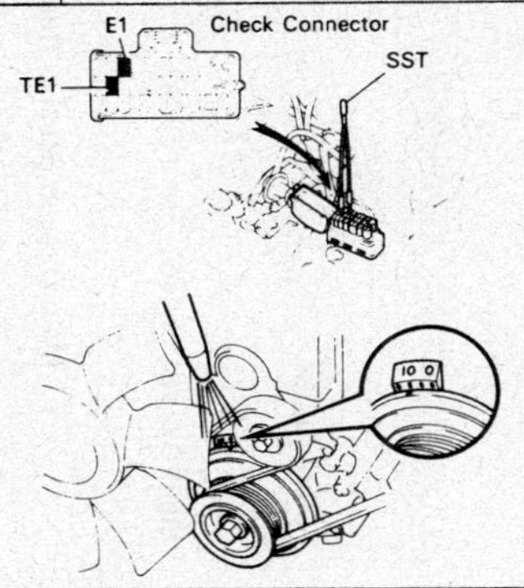

P
(1) Shift transmission into "N" range or neutral.
(2) Warm up engine at normal operating temperature.
(3) Keep the engine speed at idle.
(4) Using SST, connect terminals TE1 and E1 of check connector.

SST 09843-18020

(5) Using a timing light, connect the tester to No. 1 high-tension cord.

C Check ignition timing.

OK **Ignition timing: 10° BTDC at idle**

NG → **Proceed to troubleshoot.**

OK ↓

Proceed to matrix chart of problem symptoms

7 Check fuel pressure.

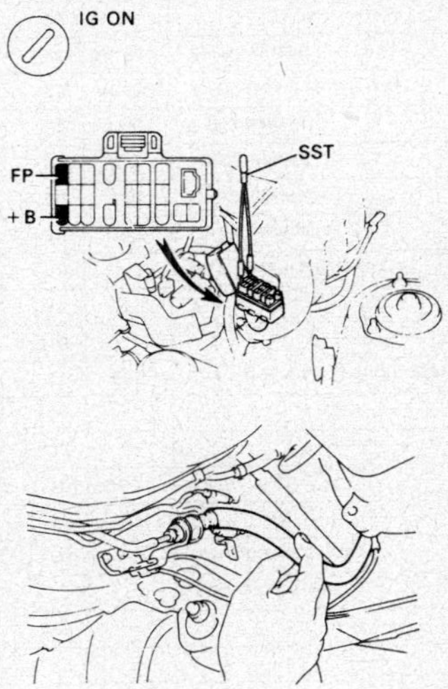

P
(1) Be sure that fuel is enough in tank.
(2) Turn ignition switch on.
(3) Using SST, connect terminals FP and +B of check connector.

SST 09843-18020

C Check that there is pressure in the hose from the fuel filter.

Hint At this time, you will hear fuel return noise.

Caution
Never make a mistake with the terminal connection position as this will cause a malfunction.

NG → **Proceed to troubleshoot.**

OK ↓

8 Check for spark.

C Disconnect the high-tension cord from the distributor and, hold the end about 12.5 mm (1/2") from the ground, see if spark occurs while the engine is being cranked.

Hint To prevent excessive fuel injected from the injectors during this test, don't crank the engine for more than 1 – 2 seconds at a time.

NG → **Proceed to troubleshoot.**

OK ↓

Proceed to matrix chart of problem symptoms

TERMINALS OF ECU

ECU Terminals

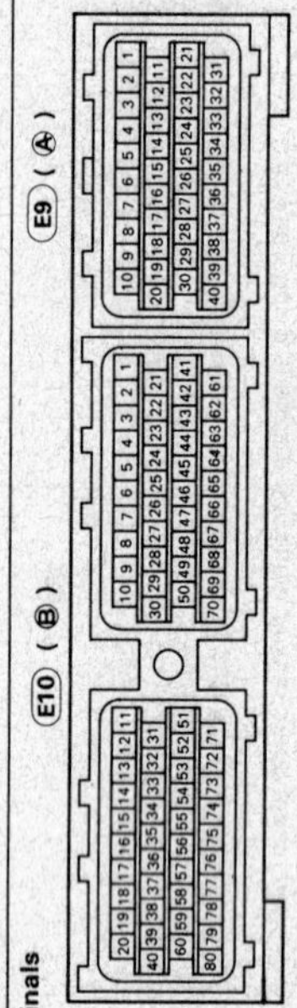

Terminal No.	Symbol	Connection	Terminal No.	Symbol	Connection
E9 – 1	IGSW	Ignition switch	E9 – 31	+B	EFI main relay
2	SP1	Speed sensor No. 1	32	+B1	EFI main relay
3	KD*1	Kick-down switch	33	BATT	Battery
4	STP	Stop light switch	34	A/C	A/C control assembly
5	—	—	35	PS*1	PS pressure switch
6	W	"CHECK" engine warning light	36	—	—
7	—	—	37	TR	TRAC ECU
8	—	—	38	NEO	TRAC ECU
9	2*1	Neutral start switch	39	VTO2	TRAC ECU
10	L*1	Neutral start switch	40	VTO1	TRAC ECU
11	—	—	E10 – 1	—	—
12	OD1*1	Cruise control ECU	2	—	—
13	—	—	3	SP2⊖*1	No. 2 speed sensor
14	—	—	4	E11	ECU ground
15	ELS	Defogger relay, Taillight relay	5	—	—
16	—	—	6	—	—
17	TT*1	TDCL	7	G⊖	Distributor
18	P*1	ECT pattern select switch	8	S3*1	ECT No. 3 solenoid
19	TE2	Check connector	9	S2*1	ECT No. 2 solenoid
20	TE1	Check connector	10	S1*1	ECT No. 1 solenoid
21	DI	Fuel pump ECU	11	—	—
22	FPC	Fuel pump ECU	12	—	—
23	ACMG	A/C magnet clutch relay	13	—	—
24	M-REL	EFI main relay	14	—	—
25	—	—	15	#60	No. 6 injector
26	—	—	16	#50	No. 5 injector
27	—	—	17	#40	No. 4 injector
28	OD2*1	O/D main switch	18	#30	No. 3 injector
29	—	—	19	#20	No. 2 injector
30	—	—	20	#10	No. 1 injector

*1: Only vehicles with ECT.

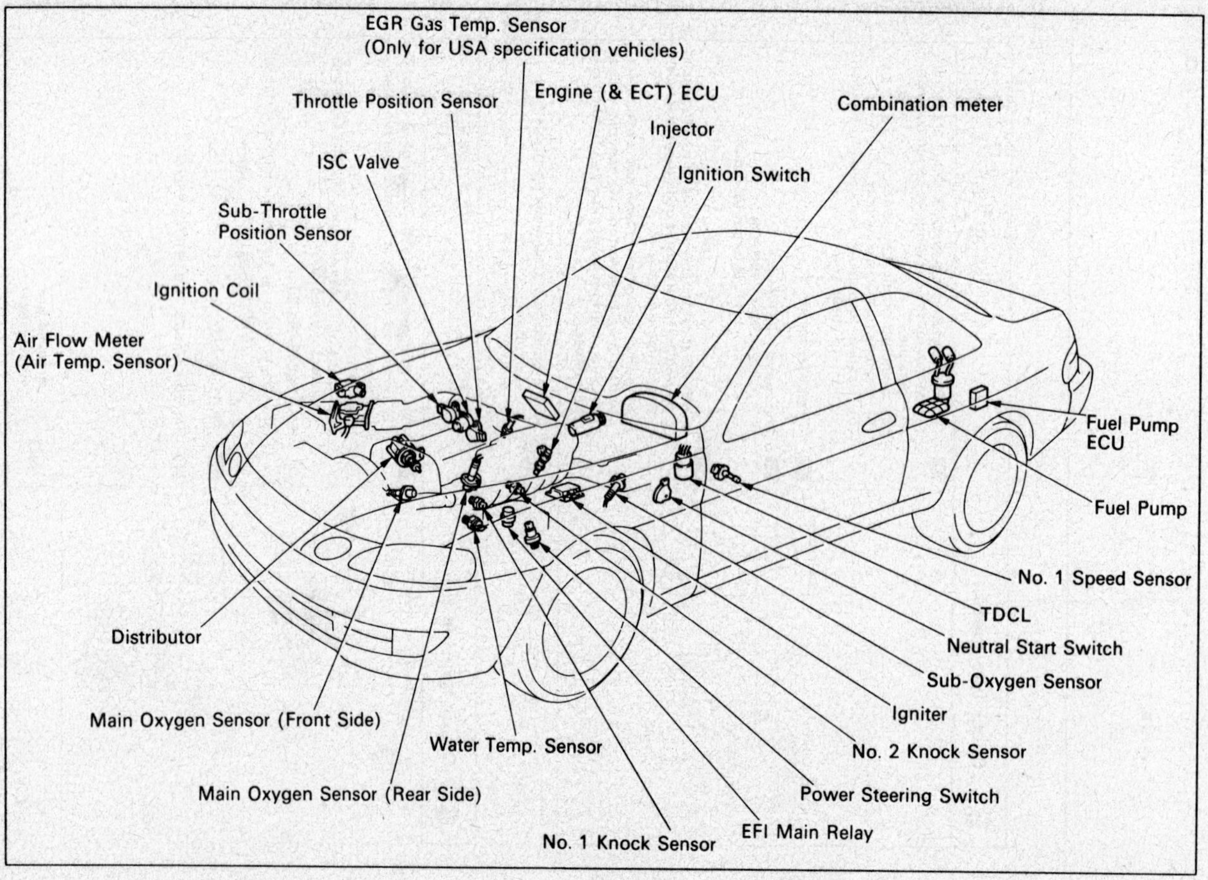

3.0L ENGINE – ECU TERMINALS – SC300

TERMINALS OF ECU (Cont'd)

ECU Terminals E10 (B) E9 (A)

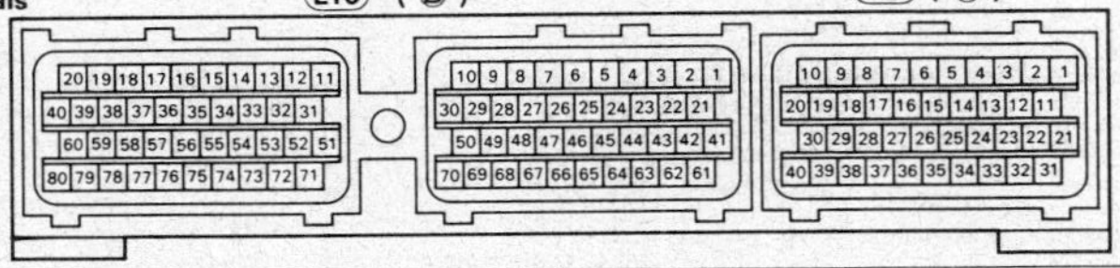

Terminal No.	Symbol	Connection	Terminal No.	Symbol	Connection
E10 – 21	—	———	E10 – 51	—	———
22	—	———	52	—	———
23	SP2*1	No. 2 speed sensor	53	—	———
24	—	———	54	—	———
25	G2	Distributor	55	—	———
26	G1	Distributor	56	—	———
27	NE	Distributor	57	IGT	igniter
28	VF2	Check connector	58	IGF	igniter
29	VF1	Check connector	59	—	———
30	—	———	60	—	———
31	—	———	61	OX3*2	Sub-oxygen sensor
32	ISC4	ISC valve	62	—	———
33	ISC3	ISC valve	63	IDL2	Sub-throttle position sensor
34	ISC2	ISC valve	64	IDL1	Throttle position sensor
35	ISC1	ISC valve	65	E2	Sensor ground
36	—	———	66	KS	Air flow meter
37	—	———	67	—	———
38	—	———	68	—	———
39	ACIS	VSV (intake air control valve)	69	E1	ECU ground
40	—	———	70	—	———
41	VC	Throttle position sensor. Air flow meter.	71	HT*2	Sub-oxygen sensor heater
42	VTA2	Sub-throttle position sensor	72	—	———
43	VTA1	Throttle position sensor	73	—	———
44	THW	Water temp. sensor	74	EVAP	EVAP purge VSV
45	THA	Intake air temp. sensor	75	EGR	EGR VSV
46	THG*2	EGR gas temp. sensor	76	NSW	Neutral start switch
47	OX2	Main oxygen sensor (rear side)	77	STA	Starter relay
48	OX1	Main oxygen sensor (front side)	78	—	———
49	KNK2	No. 2 knock sensor (rear side)	79	EO2	Power ground
50	KNK1	No. 1 knock sensor (front side)	80	EO1	Power ground

*1: Only vehicles with ECT.
*2: Only for USA specification vehicles.

3.0L ENGINE – ECU TERMINALS STANDARD VALUES SC300

STANDARD VALUE OF ECU TERMINALS

(※: A = E9, B = E10)

Symbols (Terminals No.) ※	STD Voltage (V)	Condition
BATT (A33) – E1 (B69)	10 ~ 14	Always
IGSW (A1), +B (A31), +B1 (A32) – E1 (B69)	10 ~ 14	IG switch ON
VC (B41) – E2 (B65)	4.0 ~ 6.0	IG switch ON
IDL1 (B64), IDL2 (B63) – E2 (B65)	Below 1.0	IG switch ON Main or sub throttle valve fully closed.
	10 ~ 14	IG switch ON Main or sub throttle valve fully open.
VTA1 (B43), VTA2 (B42) – E2 (B65)	0.1 ~ 1.0	IG switch ON Throttle valve fully closed.
	3.0 ~ 6.0	IG switch ON Throttle valve fully open.
KS (B66) – E1 (B69)	Pulse generation (0 and 4 ~ 6)	Idling
THA (B45) – E2 (B65)	1.0 ~ 3.0	Idling, Intake air temp. 20°C (68°F).
THW (B44) – E2 (B65)	0.1 ~ 1.0	Idling, Water temp. 80°C (176°F).
STA (B77) – E1 (B69)	6.0 or more	Cranking
#10 (B20), #20 (B19), #30 (B18), #40 (B17), #50 (B16), #60 (B15) – E01 (B80)	10 ~ 14	IG switch ON
	Pulse generation	Idling
IGT (B57) – E1 (B69)	Pulse generation (0 and 4 ~ 6)	Idling
IGF (B58) – E1 (B69)	Below 1.0	IG switch ON
	Pulse generation (0 and 4 ~ 6)	Idling
G1 (B26), G2 (B25) – G ⊖ (B6)	Pulse generation (Positive voltage peak 1.3 or higher)	Idling
NE (B27) – G ⊖ (B6)	Pulse generation (Positive voltage peak 0.6 or higher)	Idling
M-REL (A24) – E1 (B69)	10 ~ 14	IG switch ON
FPC (A22) – E1 (B69)	4.2 ~ 6.0	Cranking, Sudden racing (6,000 rpm)
	Pulse generation (0 and 4 ~ 6)	Idling
DI (A21) – E1 (B69)	7.5 or more	Idling
ACIS (B39) – E01 (B80)	10 ~ 14	IG switch ON
EVAP (B74) – E01 (B80)	10 ~ 14	IG switch ON
EGR (B75) – E01 (B80)	Below 2.0	Idling
	10 ~ 14	Engine speed at 3,500 rpm

STANDARD VALUE OF ECU TERMINALS

(※ : A=(E9), B=(E10))

※ Symbols (Terminals No.)	STD Voltage (V)	Condition
ISC1 (B35), ISC2 (B34) ISC3 (B33), ISC4 (B32) – E01 (B80)	10 ∼ 14	IG switch ON
VF1 (B29), VF2 (B28) – E1 (B69)	1.0 ∼ 4.0	Maintain engine speed at 2,500 rpm for two minutes after warming up then return to idling.
OX1 (B48), OX2 (B47) – E1 (B69)	Pulse generation (0 ∼ 1.0)	Maintain engine speed at 2,500 rpm for two minutes after warming up.
HT (B71) – E01 (B80)*	Below 2.0	Idling
	10 ∼ 14	IG switch ON
KNK1 (B50), KNK2 (B49) – E1 (B69)	Pulse generation (Frequency : 8.1 KHz)	Idling
NSW (B76) – E1 (B69)	10 ∼ 14	IG switch ON Other shift position in "P", "N" ranges.
	Below 1.0	IG switch ON Shift position in "P", "N" ranges.
SP1 (A2) – E1 (B69)	Pulse generation (0 and 4 ∼ 6)	IG switch ON Rotate driving wheel slowly.
TE1 (A20) – E1 (B69)	10 ∼ 14	IG switch ON
TE2 (A19) – E1 (B69)	10 ∼ 14	IG switch ON
W (A6) – E1 (B69)	10 ∼ 14	Idling
	Below 2.0	IG switch ON
ODI (A12) – E1 (B69)	10 ∼ 14	IG switch ON
A/C (A34) – E1 (B69)	Bellow 2.0	A/C switch ON (At idling)
	10 ∼ 14	A/C switch OFF
ACMG (A23) – E1 (B69)	Below 2.0	A/C switch ON (At idling)
	10 ∼ 14	A/C switch OFF
TR (A37) – E1 (B69)	10 ∼ 14	IG switch ON
VTO1 (A40) VTO2 (A39) – E2 (B65)	Below 1.0	IG switch ON Main or sub throttle valve fully closed.
	3.0 ∼ 5.5	IG switch ON Main or sub throttle valve fully open.
NEO (A38) – E1 (B69)	4.0 ∼ 6.0	IG switch ON

* Only for USA specification vehicles.

MATRIX CHART OF PROBLEM SYMPTOMS

When the malfunction code is not confirmed in the diagnostic code check and the problem still can not be confirmed in the basic inspection, then proceed to this step and perform troubleshooting according to the numbered order given in the table below.

The circuits indicated by ※ on the matrix chart can be inspected using the TCCS checker.

Symptom \ Suspect area		Air flow meter circuit ※	Starter signal circuit ※	EGR System	Switch condition signal circuit	Neutral start switch circuit ※	ECU power source circuit	Back up power source circuit	Injector circuit ※	ISC valve circuit ※	Fuel pump control circuit	A/C signal circuit (Compressor circuit)	Starter and Starter relay	Ignition signal circuit (Spark test)	Spark plug	Ignition coil	Distributor	Compression	ECT faulty	Theft deterrent ECU	Engine (& ECT) ECU
Does not start	Engine does not crank					2							1							3	
	No initial combustion						1		5		3			2	4						
	No complete combustion								6		1			5	4	2	3				
Difficult to start	Difficult to start normally		1						8	2	3				6	4	5	7			
	Difficult to start cold		1						4	2	3				7	5	6				
	Difficult to start hot		1						4	2	3				7	5	6				
Poor Idling	Incorrect first idle				1					2											
	High engine idle speed				1	5	4	6		2		3									
	Low engine idle speed	7			3	4		8	6	1	5	2									
	Rough idling	3		5	1			12	4	2	8			6	11	9	10	7			
	Hunting	3			1		4			2	5										
Poor Driveability	Hesitation/Poor acceleration	2			1				3		4			5	8	6	7		9		
	Muffler explosion (after fire)				1				5						4	2	3				
	Surging				1				5		2				4		3				
Engine Stall	Engine stalls soon after starting	2								3	1										
	After acceleration pedal depressed	2			1																
	After acceleration pedal released								1	2											3
	During A/C operation									1		2									3
	When N to D shift					1				2											

3.0L ENGINE—CONNECTOR LOCATIONS—SC300

LOCATION OF CONNECTORS

Location of Connectors in Engine Compartment

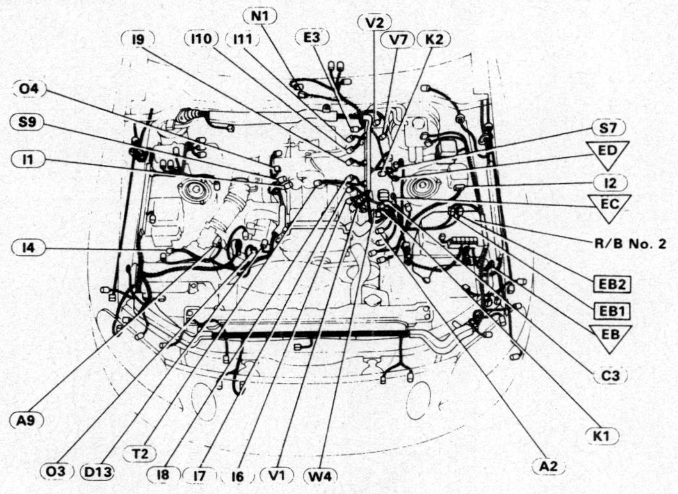

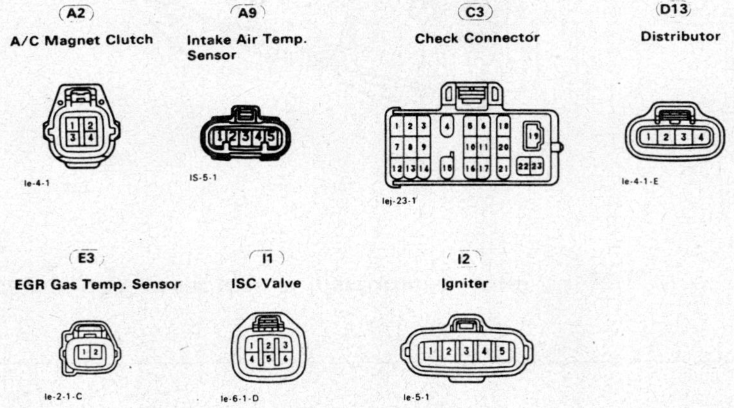

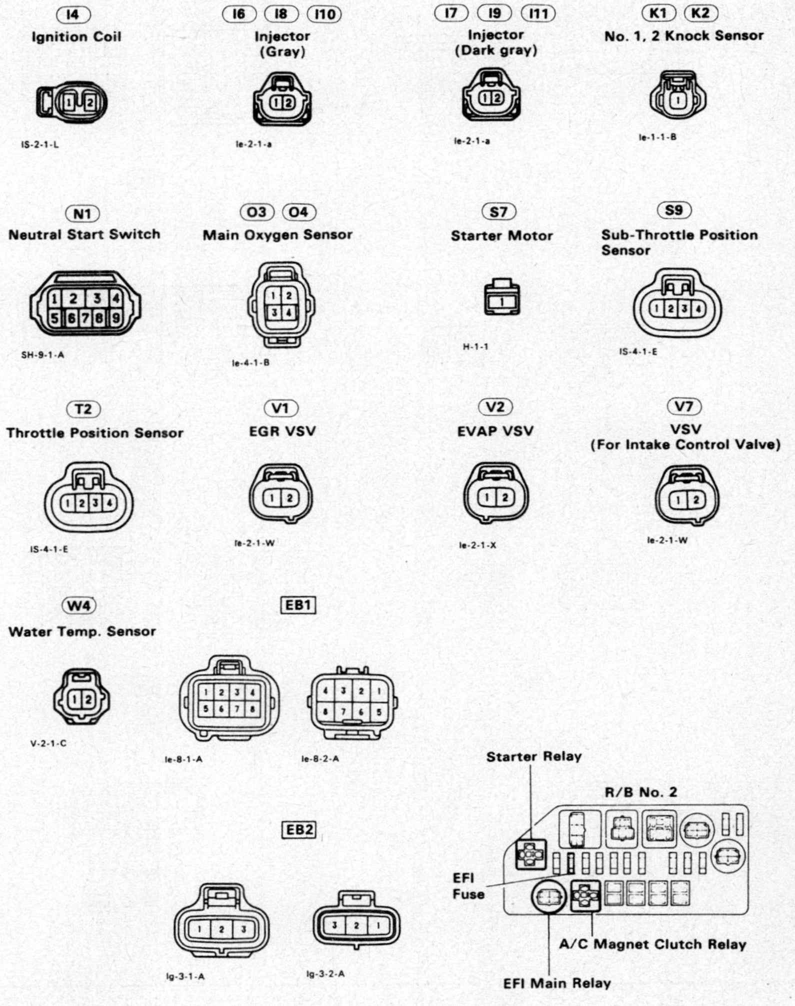

Location of Connectors in Instrument Panel

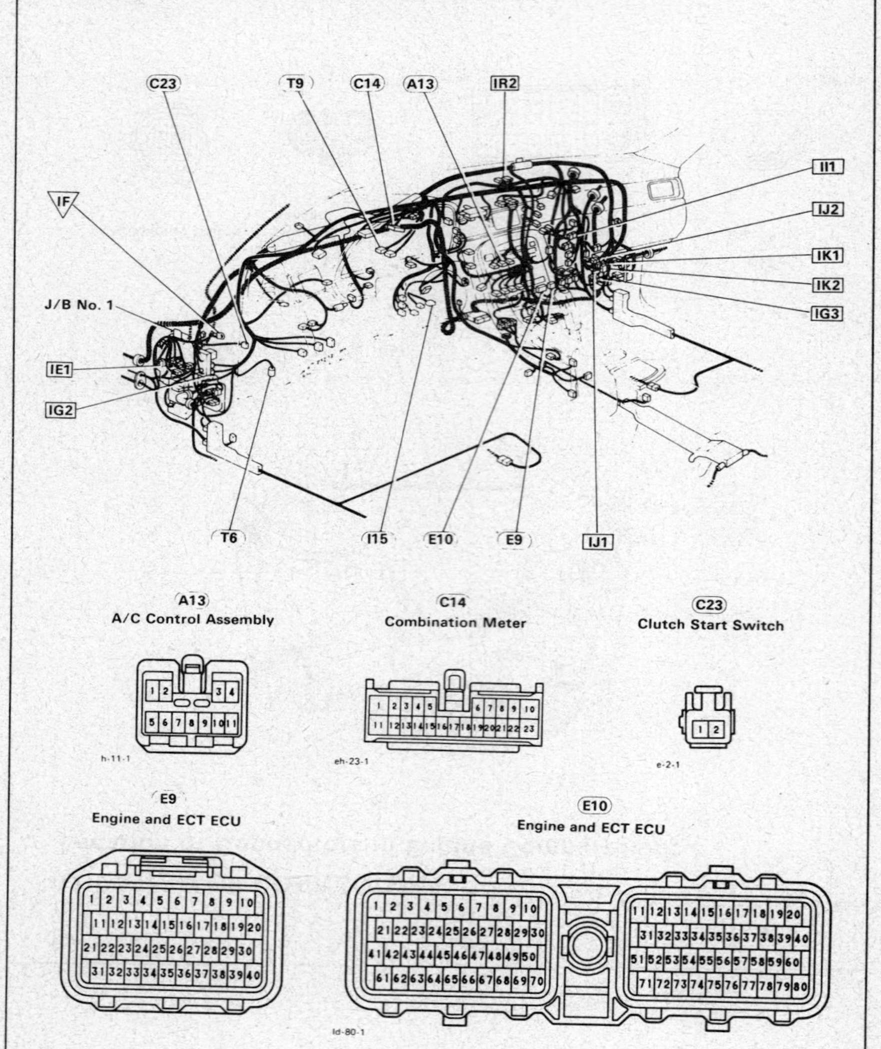

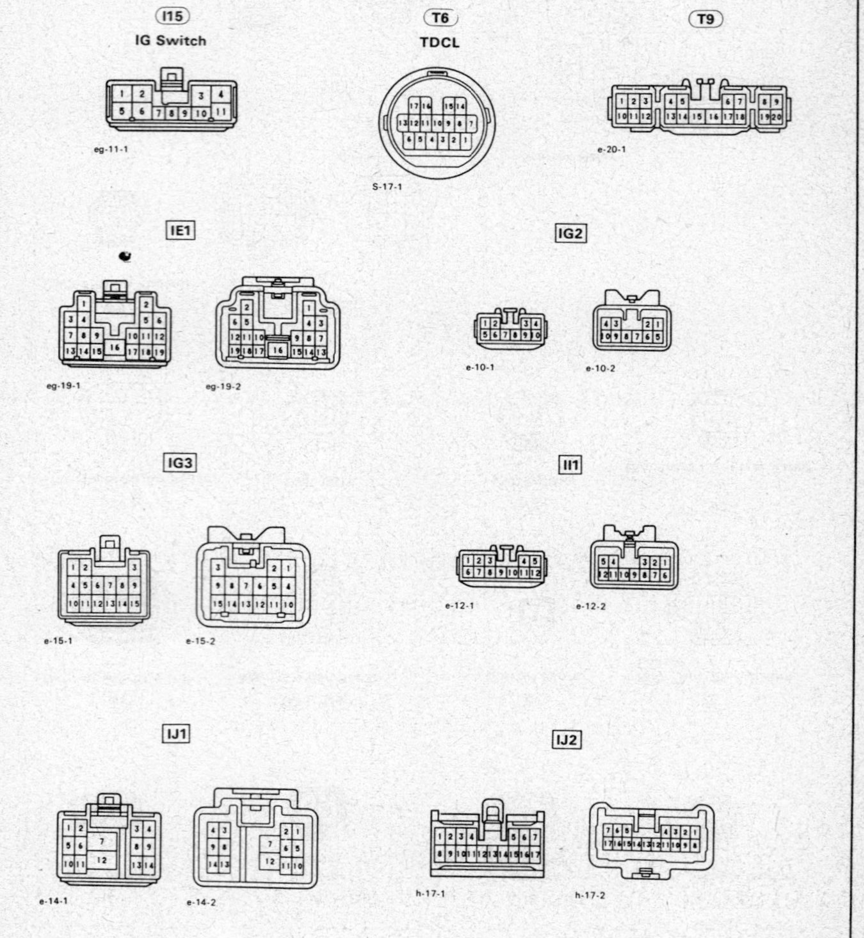

Location of Connectors in Instrument Panel (Cont'd)

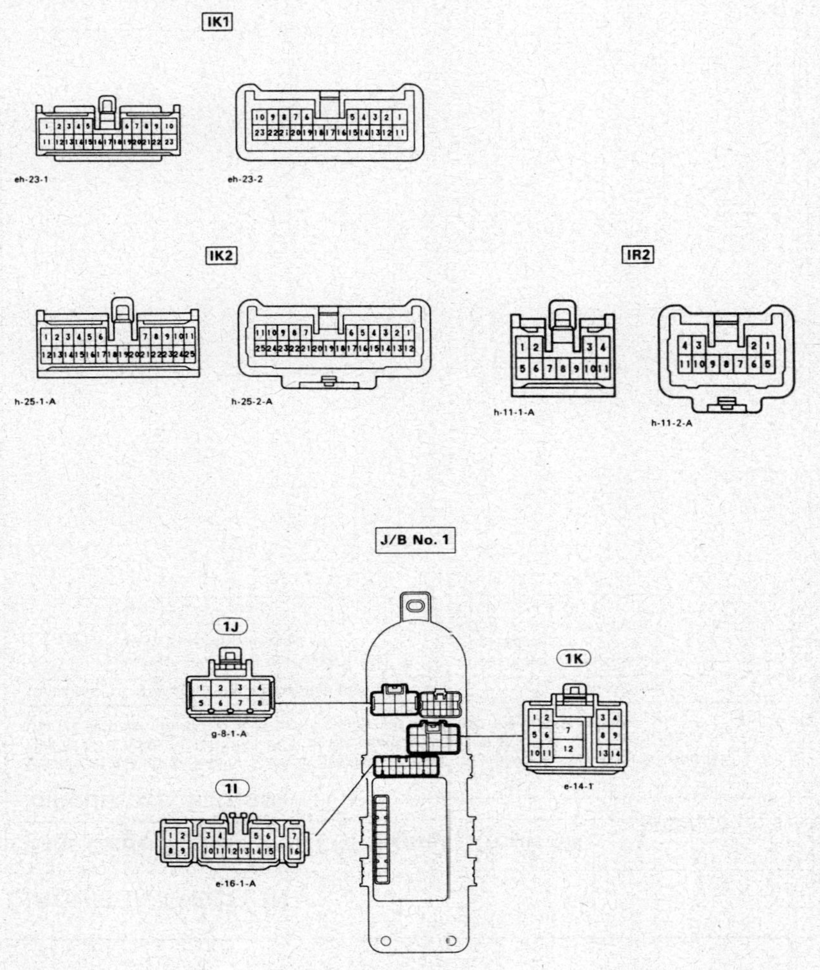

Location of Connectors in Body

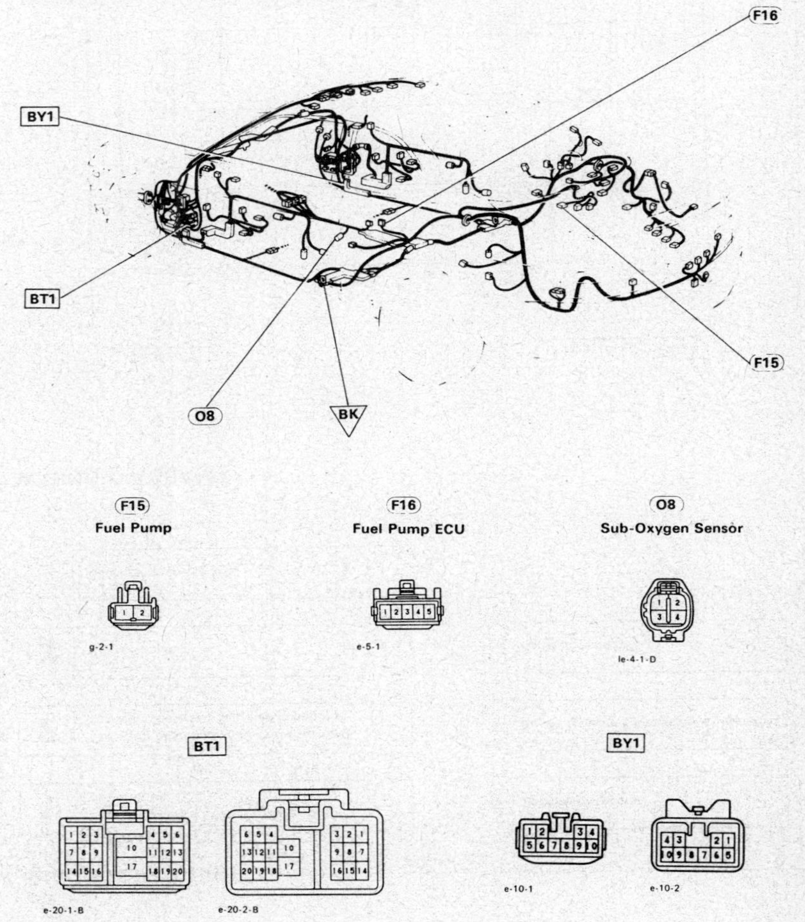

CIRCUIT INSPECTION

Diag. Code	12	RPM Signal Circuit (No. 1)

CIRCUIT DESCRIPTION

The distributor in the Engine Control System contains three pick-up coils (G1, G2 and NE).
The G1, G2 signals inform the ECU of the standard crankshaft angle.
The NE signals inform the ECU of the crankshaft angle and the engine speed.

Code No.	Diagnostic Code Detecting Condition	Trouble Area
12	No "NE" or "G1" and "G2" signal to ECU within 2 sec. after cranking.	• Open or short in NE, G circuit. • Distributor • Open or short in STA circuit. • ECU
	Open in "G⊖" circuit.	

DIAGNOSTIC CHART

Step	Check	Result	Action
1	Check resistance of each pickup coils in distributor.	NG	Replace distributor.
	↓ OK		
2	Check for open and short in harness and connector between ECU and distributor.	NG	Repair or replace harness or connector.
	↓ OK		
3	Check air gap.	NG	Replace distributor.
	↓ OK		
	Check and replace ECU.		

WIRING DIAGRAM

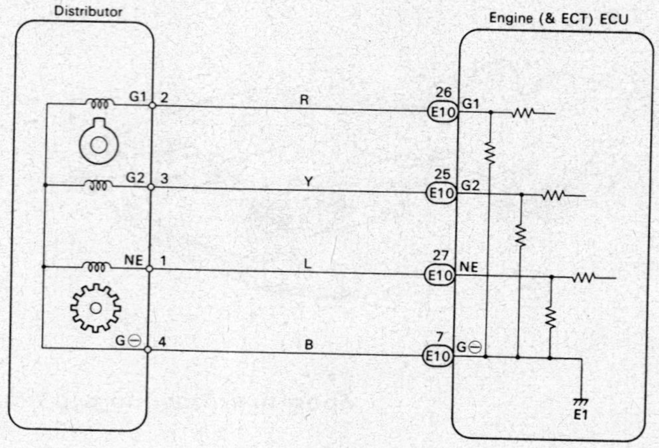

INSPECTION PROCEDURE

1 Check resistance of each pickup coils in distributor.

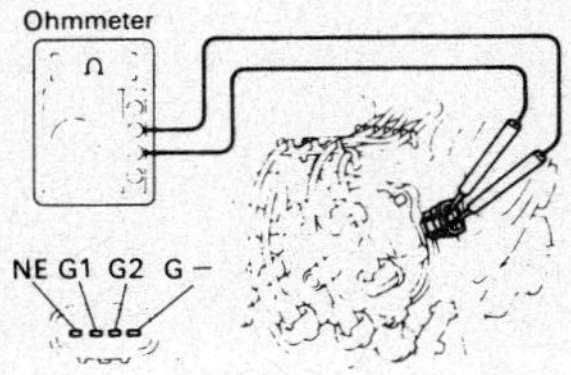

P Disconnect distributor connector.

C Measure resistance between each terminal shown in table below.

OK

	Terminals	Resistance
G1 pickup coil (Cold)	G1 – G⊖	125 – 190 Ω
G2 pickup coil (Cold)	G2 – G⊖	125 – 190 Ω
NE pickup coil (Cold)	NE – G⊖	155 – 240 Ω

Reference **INSPECTION USING OSCILLOSCOPE**

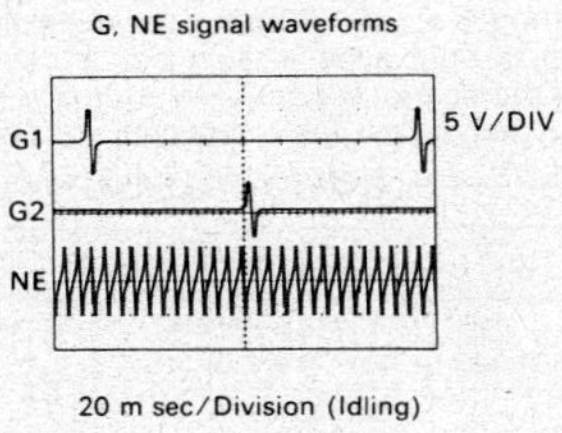

- During cranking or idling, check between terminals G1, G2, NE and G⊖ of engine (& ECT) ECU.

HINT: The correct waveforms appear as shown in the illustration on the left.

OK | **NG** **Replace distributor.**

2 Check for open and short in harness and connector between engine (& ECT) ECU and distributor

OK | **NG** **Repair or replace harness or connector.**

3 Check air gap.

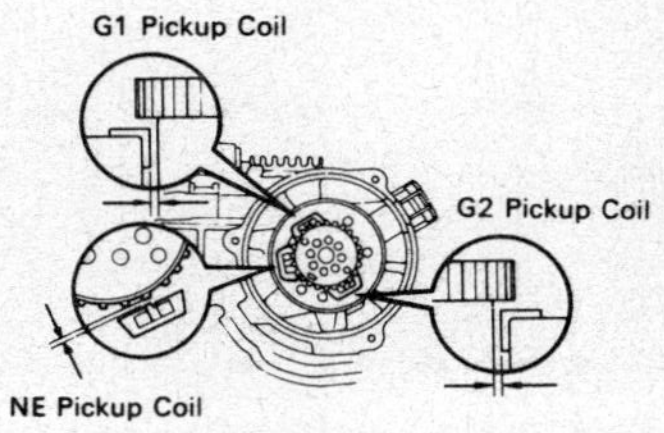

P Remove distributor cap & rotor.

C Using SST (G1 and G2 pickups) and a thickness gauge (NE pickup), measure the air gap between the signal rotor projection and pickup coil.
SST 09240-00020 for G1 and G2 pickups

OK **Air gap: 0.2 – 0.4 mm (0.008 – 0.016 in.)**

OK | **NG** **Replace distributor.**

Check and replace engine (& ECT) ECU.

Diag. Code 13 RPM Signal Circuit (No. 2)

CIRCUIT DESCRIPTION

Refer to RPM signal circuit (No. 1) on page TR-46.

Code No.	Diagnostic Code Detecting Condition	Trouble Area
13	No NE signal to ECU for 0.1 sec. or more at 1,000 rpm or more.	• Open or short in NE circuit • Distributor • ECU
	No 12 pulses of NE to ECU during the interval between G1 and G2 pulses.	

DIAGNOSTIC CHART

This code indicates that a momentary interruption of the RPM signal from the distributor to the ECU has occurred, but that it is returned to normal. Note that although this problem may not necessarily appear at the time of inspection, it cannot be ignored because this diagnostic code is output, indicating that there is or was a malfunction in the RPM signal circuit; this "malfunction" is usually a loose connector.

The distributor connector and the NE terminal of the ECU connector must therefore be checked for the following:

1. Loose connectors
2. Dirty connector terminals
3. Loose connector terminals

Diag. Code 14 Ignition Signal Circuit

CIRCUIT DESCRIPTION

The ECU determines the ignition timing, turns on Tr_1 at a predetermined angle (°CA) before the desired ignition timing and outputs an ignition signal (IGT) "1" to the igniter.
Since the width of the IGT signal is constant, the dwell angle control circuit in the igniter determines the time the control circuit starts primary current flow to the ignition coil based on the engine rpm and ignition timing one revolution ago, that is, the time the Tr_2 turns on.
When it reaches the ignition timing, the ECU turns Tr_1 off and outputs the IGT signal "O".
This turns Tr_2 off, interrupting the primary current flow and generating a high voltage in the secondary coil which causes the spark plug to spark. Also, by the counter electromotive force generated when the primary current is interrupted, the igniter sends an ignition confirmation signal (IGF) to the ECU.
The ECU stops fuel injection as a fail safe function when the IGF signal is not input to the ECU.

Code No.	Diagnostic Code Detecting Condition	Trouble Area
14	No IGF signal to ECU for 6 consecutive IGT signal.	• Open or short in IGF or IGT circuit from igniter to ECU. • Igniter • ECU

WIRING DIAGRAM

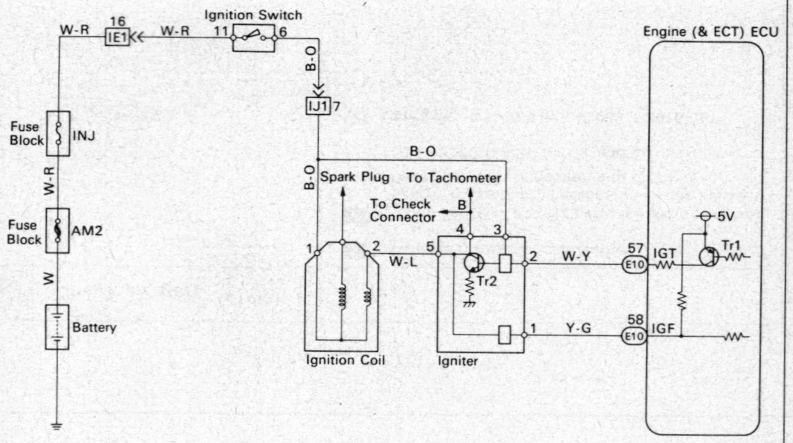

DIAGNOSTIC CHART

1 Check for spark. — NG → (go to 4)
↓ OK
2 Check for open and short in IGF signal circuit between igniter and ECU. — NG → Repair or replace harness or connector.
↓ OK
3 Check voltage of terminal IGF. — OK → Replace igniter.
↓ NG
Check and replace ECU.

4 Check voltage of terminal IGT. — NG → (go to 8)
↓ OK
5 Check voltage of igniter power source. — NG → Check and repair igniter power source circuit.
↓ OK
6 Check for open and short in primary coil circuit. — NG → Repair or replace harness or connector.
↓ OK
7 Check ignition coil. — NG → Replace ignition coil.
↓ OK
Replace igniter.

8 Check voltage of terminal IGT (Disconnect igniter connector). — OK → Replace igniter.
↓ NG
9 Check for open and short in IGT signal circuit between igniter and ECU. — NG → Repair or replace harness or connector.
↓ OK
Check and replace ECU.

INSPECTION PROCEDURE

1 Check for spark.

C Disconnect the high-tension cord from the distributor, hold its end about 12.5 mm (1/2") from the ground, see if spark occurs while the engine is being cranked.

OK **Spark should be generated.**

Hint To prevent excessive fuel injected from the injectors during this check. Don't crank the engine for more than 1 – 2 seconds at a time.

OK | **NG** **Go to step 4.**

2 Check for open and short in harness and connector in IGF signal circuit between engine (& ECT) ECU and igniter

OK | **NG** **Repair or replace harness or connector.**

3 Disconnect igniter connector and check voltage between terminal IGF of engine (& ECT) ECU connector and body ground.

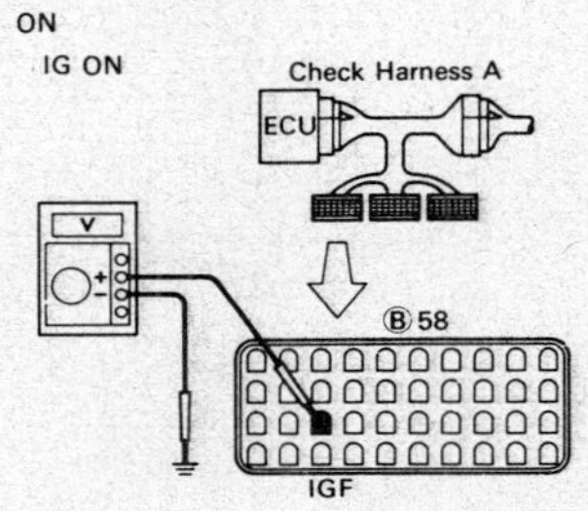

P (1) Disconnect igniter connector.
(2) Connect the Check Harness A.
(3) Turn ignition switch on.

C Measure voltage between terminal IGF of engine (& ECT) ECU connector and body ground.

OK **Voltage: 4 – 6 V**

OK **Replace igniter.**

NG

Check and replace engine (& ECT) ECU.

4 Check voltage between terminal IGT of engine (& ECT) ECU connector and body ground.

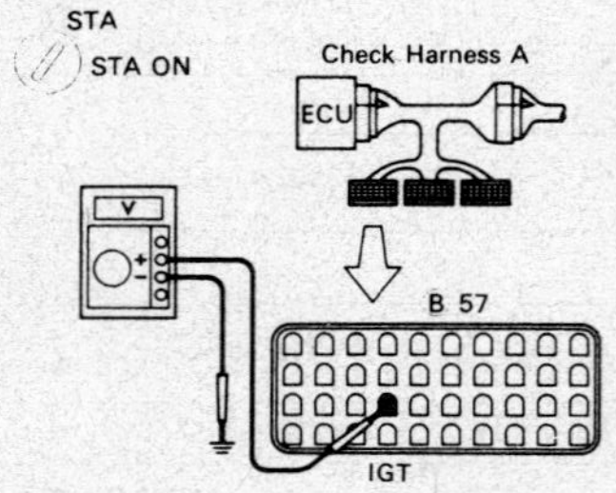

P Connect the Check Harness A.

C Measure voltage between terminal IGT of engine (& ECT) ECU connector and body ground when engine is cranked.

OK Voltage: 0.5 – 1.0 V
(Neither 0 V nor 5 V)

Reference **INSPECTION USING OSCILLOSCOPE**

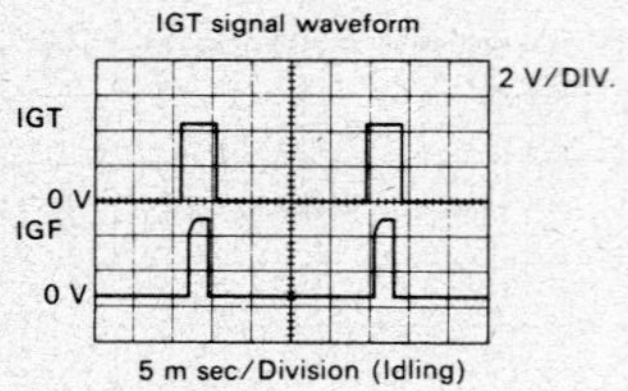

- During cranking or idling, check waveform between terminal IGTand E1 of engine (& ECT) ECU.

HINT: The correct waveform appears as shown in the illustration on the left, with rectangle waves.

OK

NG Go to step **8**.

5 Check voltage between terminal 3 of igniter connector and body ground.

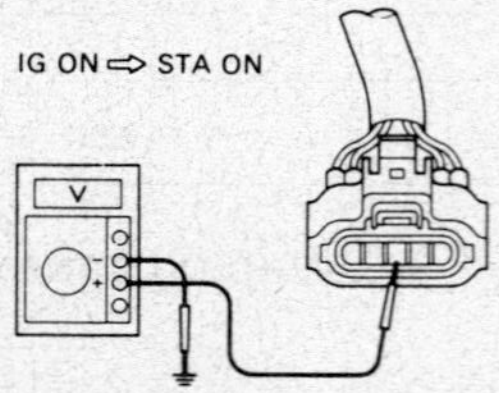

P Disconnect igniter connector.

C Measure voltage between terminal 3 of igniter connector and body ground, when ignition switch is turned to "ON" and "STA" position.

OK Voltage: 10 – 14 V

OK

NG Check and repair igniter power source circuit.

6 Check for open and short in harness and connector between ignition switch and ignition coil, ignition coil and igniter

OK

NG Repair or replace harness or connector.

7 Check ignition coil.

Primary Coil

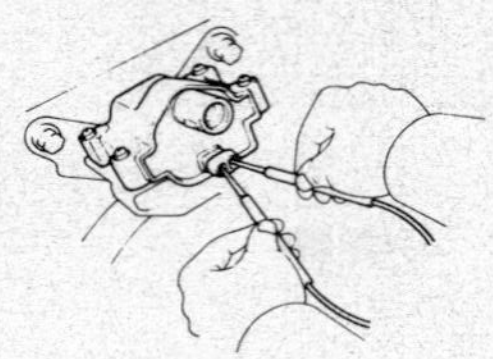

Secondary Coil

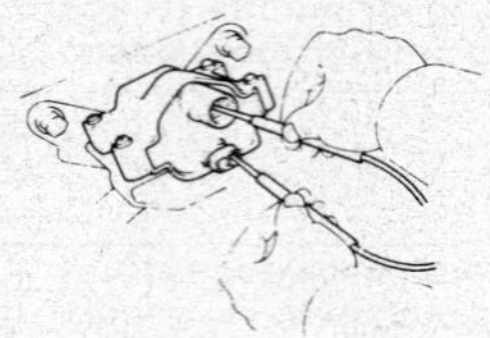

P Disconnect ignition coil connector.

C (1) Check primary coil.
Measure resistance between terminals of ignition coil connector.
(2) Check secondary coil.
Measure resistance between terminal ⊕ of ignition coil connector and high-tension terminal.

OK

	Resistance
Primary Coil (Cold)	0.2 – 0.3 Ω
Secondary Coil (Cold)	6 – 11 kΩ

OK

NG Replace ignition coil.

Replace igniter.

8 Disconnect igniter connector and check voltage between terminal IGT of engine (& ECT) ECU connector and body ground.

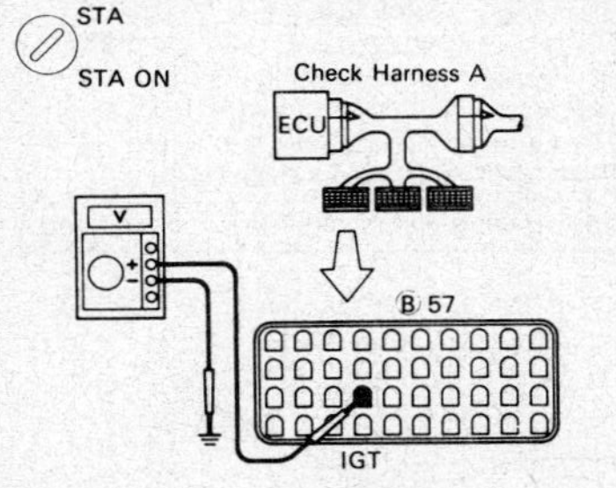

P Disconnect igniter connector.

C Measure voltage between terminal IGT of engine (& ECT) ECU connector and body ground when engine is cranked.

OK Voltage: 0.5 – 1.0 V
(Neither 0 V nor 5 V)

Reference INSPECTION USING OSCILLOSCOPE

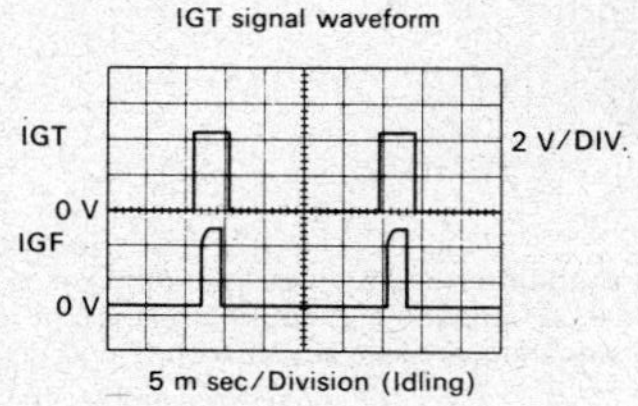

- During cranking or idling, measure between terminal IGT and E1 of engine (& ECT) ECU.

HINT: The correct waveform appears as shown in the illustration on the left, with rectangle waves.

NG

OK Replace igniter.

9 Check for open and short in harness and connector in IGT signal circuit between engine (& ECT) ECU and igniter

OK

NG Repair or replace harness or connector.

Check and replace engine (& ECT) ECU.

Diag. Code 16 ECT Control Signal Malfunction

CIRCUIT DESCRIPTION

The signal from the ECT CPU retards the ignition timing of the engine during ECT gear shifting, thus momentarily reducing torque output of the engine for smooth clutch operation inside the transmission and reduced shift shock.

Code No.	Diagnostic Code Detecting Condition	Trouble Area
16	Fault in communications between the engine CPU and ECT CPU in the ECU.	• ECU

If the ECU records the diagnostic code "16" in memory, it prohibits the torque control of the ECT which performs smooth gear shifting.

DIAGNOSTIC CHART

1 Are there any other codes (besides Code 16) being output?

YES → Go to relevant diag. code chart.

NO → Replace engine (& ECT) ECU.

Diag. Code 21, 28 — Main Oxygen Sensor Circuit

CIRCUIT DESCRIPTION

To obtain a high purification rate for the CO, HC and NOx components of the exhaust gas, a three-way catalyst is used, but for most efficient use of the three-way catalyst, the air-fuel ratio must be precisely controlled so that it is always close to the stoichiometric air-fuel ratio.

The oxygen sensor has the characteristic whereby its output voltage changes suddenly in the vicinity of the stoichiometric air-fuel ratio. This characteristic is used to detect the oxygen concentration in the exhaust gas and provide feedback to the computer for control of the air-fuel ratio.

When the air-fuel ratio becomes LEAN, the oxygen concentration in the exhaust increases and the oxygen sensor informs the ECU of the LEAN condition (small electromotive force: 0 V).

When the air-fuel ratio is RICHER than the stoichiometric air-fuel ratio the oxygen concentration in the exhaust gas is reduced and the oxygen sensor informs the ECU of the RICH condition (large electromotive force: 1V).

The ECU judges by the electromotive force from the oxygen sensor whether the air-fuel ratio is RICH or LEAN and controls the injection time accordingly. However, if malfunction of the oxygen sensor causes output of abnormal electromotive force, the ECU is unable to perform accurate air-fuel ratio control.

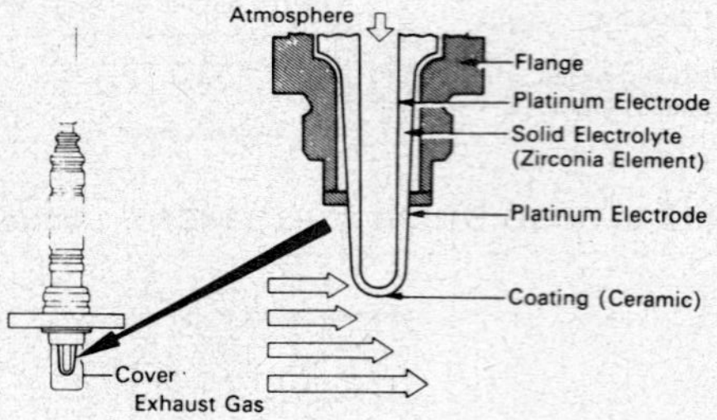

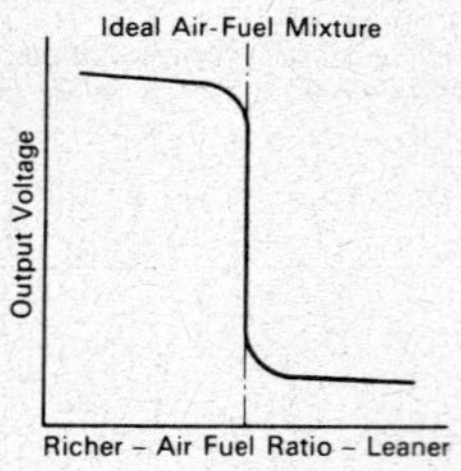

Code No.	Diagnostic Code Detecting Condition	Trouble Area
21 · 28	(1) Main oxygen sensor signal voltage is reduced to between 0.35 V and 0.70 V for 60 sec. under conditions (a) ~ (d). (2 trip detection logic) (a) Coolant temp. : Between 80°C (176°F) and 95°C (203°F). (b) Engine speed : 1,500 rpm or more. (c) Load driving (EX. ECT in 4th (5th for M/T) speed, A/C ON, Flat road, 50 mph (80km/h)). (d) Main oxygen sensor signal voltage : Alternating above and below 0.45 V. (2) Main oxygen sensor signal voltage exceeds 0.70 V for 3 sec. or more during fuel cut.	• Main oxygen sensor circuit • Main oxygen sensor

HINT: Diag. code 21 is for the front side main oxygen sensor circuit. Diag. code 28 is for the rear side main oxygen sensor circuit.

CIRCUIT DESCRIPTION (Cont'd)

DIAGNOSIS CODE DETECTION DRIVING PATTERN

Purpose of the driving pattern.
(a) To simulate diag. code detecting condition after diag. code is recorded.
(b) To check that the malfunction is corrected when the repair is completed confirming that diag. code is no longer detected.

Malfunction: Main Oxygen Sensor Deterioration

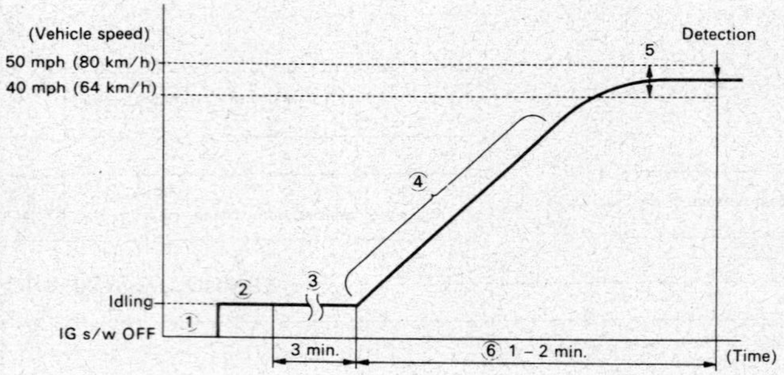

① Disconnect the EFI fuse (15 A) for 10 sec. or more, with IG switch OFF. Initiate test mode (Connect terminal TE2 and E1 of check connector or TDCL with IG switch OFF).
② Start the engine and warm up with all ACC switch OFF.
③ After the engine is warmed up, let it idle for 3 min.
④ After performing the idling in ③, perform gradual acceleration with in the range 1,300 ~ 1,700 rpm (centered around 1,500 rpm) with the A/C switch ON and D range for A/T (5th for M/T).
(Take care that the engine speed does not fall below 1,200 rpm when shifting. Gradually depress the accelerator pedal and kept it. Steady so that engine braking does not occur).
⑤ Maintain the vehicle speed at 40 – 50 mph (64 – 80 km/h).
⑥ Keep the vehicle running for 1 – 2 min. after starting acceleration.

HINT: If a malfunction exists, the "CHECK" engine warning light will light up after approx. 60 sec. from the start of acceleration.

NOTICE: **If the conditions in this test are not strictly followed, detection of the malfunction will not be possible.**

DIAGNOSTIC CHART

1 Are there any other codes (besides code 21 or 28) being output ?

- YES → Go to relevant diag. code chart.
- NO → Replace oxygen sensor.

HINT: If diag. code 21 is output, replace the front side main oxygen sensor.
If diag. code 28 is output, replace the rear side main oxygen sensor.

WIRING DIAGRAM

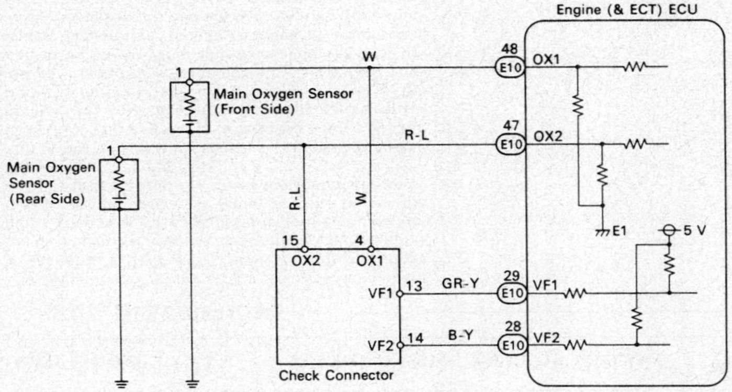

Reference INSPECTION USING OSCILLOSCOPE

Ox signal waveform

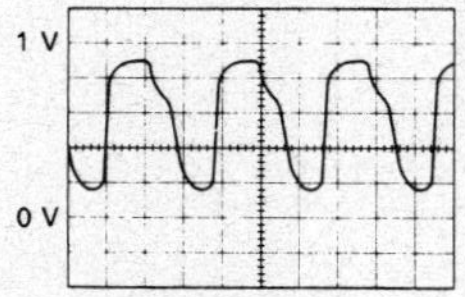

200 m sec./Division

- With the engine racing (4,000 rpm) measure between terminals OX1, OX2 and E1 of engine (& ECT) ECU.

HINT: The correct waveform appears as shown in the illustration on the left, oscillating between approx. 0.1 V and 0.9 V

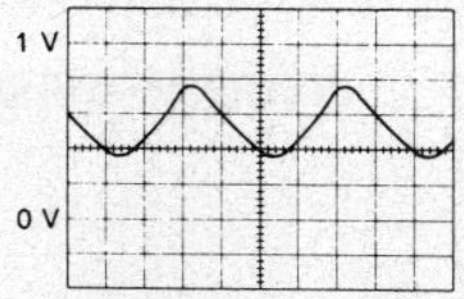

200 m sec./Division

If the oxygen sensor is deteriorated, the amplitude of the voltage is reduced as shown on the left.

Diag. Code 22 Water Temp. Sensor Circuit

CIRCUIT DESCRIPTION

The water temperature sensor senses the coolant temperature. A thermistor built in the sensor changes the resistance value according to the coolant temperature. The lower the coolant temperature, the greater the thermistor resistance value, and the higher the coolant temperature, the lower the thermistor resistance value (See Fig. 1.).
The water temperature sensor is connected to the ECU (See next page). The 5 V power source voltage in the ECU is applied to the water temperature sensor from the terminal THW via a resistor R. That is, the resistor R and the water temperature sensor are connected in series. When the resistance value of the water temperature sensor changes in accordance with changes in the coolant temperature, the potential at the terminal THW also changes. Based on this signal, the ECU increases the fuel injection volume to improve driveability during cold engine operation. If the ECU records the diagnostic code 22, it operates the fail safe function, keeping the coolant temperature at a constant 80°C (176°F).

(fig-1)

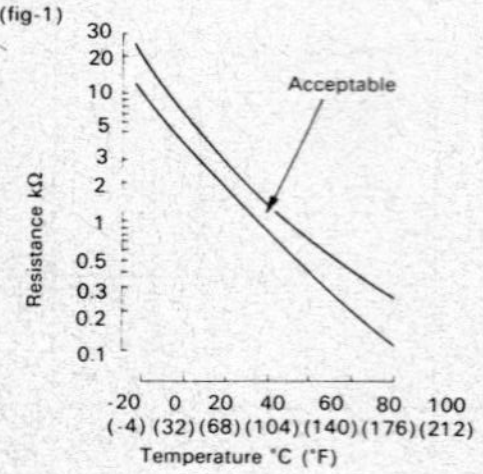

< Reference >

Water Temp. °C (°F)	Resistance (kΩ)	Voltage (V)
-20 (-4)	16.0	4.3
0 (32)	5.9	3.4
20 (68)	2.5	2.4
40 (104)	1.2	1.5
60 (140)	0.6	0.9
80 (176)	0.3	0.5
100 (212)	0.2	0.3

Code No.	Diagnostic Code Detecting Condition	Trouble Area
22	Open or short in water temp. sensor circuit for 0.5 sec. or more.	• Open or short in water temp. sensor circuit • Water temp. sensor • ECU

DIAGNOSTIC CHART

HINT: If diagnostic codes "22" (water temperature sensor circuit), "24" (intake air temperature sensor circuit) and "41" (throttle position sensor circuit) are output simultaneously, E2 (sensor ground) may be open.

1 Check voltage of sensor. — OK → Check for momentary interruption
NG ↓

2 Check resistance of sensor. — NG → Replace water temp. sensor.
OK ↓

3 Check for open and short in harness and connector between ECU and water temp. sensor. — NG → Repair or replace harness or connector.
OK ↓

Check and replace ECU.

WIRING DIAGRAM

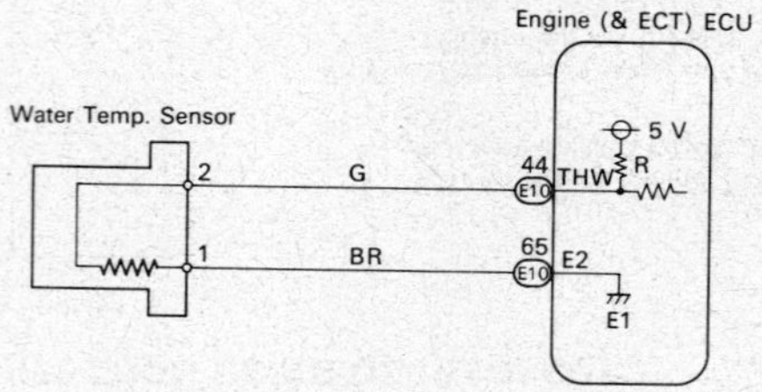

Diag. Code	24	Intake Air Temp. Sensor Circuit

CIRCUIT DESCRIPTION

The intake air temp. sensor is built into the air flow meter and senses the intake air temperature. The structure of the sensor and connection to the ECU is the same as in the water temp. sensor shown on page TR-64.
If the ECU records the diagnostic code "24", it operates the fail safe function, keeping the intake air temperature at a constant 20°C (68°F).

Code No.	Diagnostic Code Detecting Condition	Trouble Area
24	Open or short in intake air temp. sensor circuit for 0.5 sec. or more.	• Open or short in intake air temp. sensor circuit. • Intake air temp. sensor • ECU

DIAGNOSTIC CHART

HINT: If diagnostic codes "22" (water temperature sensor circuit), "24" (intake air temperature sensor circuit) and "41" (throttle position sensor circuit) are output simultaneously, E2 (sensor ground) may be open.

1 Check voltage of sensor. — OK → Check for momentary interruption
NG ↓

2 Check resistance of sensor. — NG → Replace intake air temp. sensor.
OK ↓

3 Check for open and short in harness and connector between ECU and intake air temp. sensor. — NG → Repair or replace harness or connector.
OK ↓

Check and replace ECU.

WIRING DIAGRAM

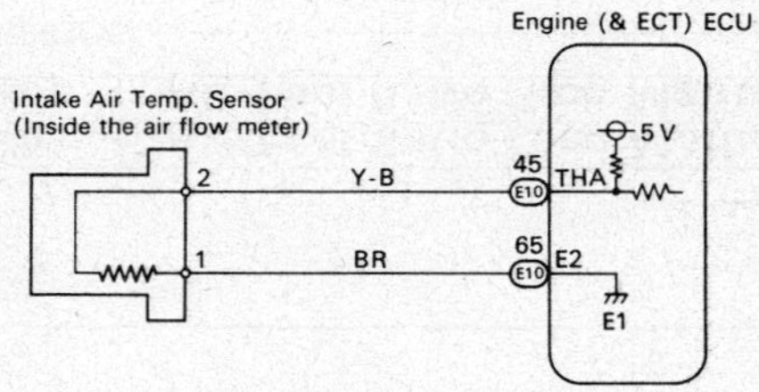

INSPECTION PROCEDURE

HINT: If diagnostic codes "22" (water temperature sensor circuit), "24" (intake air temperature sensor circuit) and "41" (throttle position sensor circuit) are output simultaneously, E2 (sensor ground) may be open.

1 Check voltage between terminals THW and E2 of engine (& ECT) ECU connector.

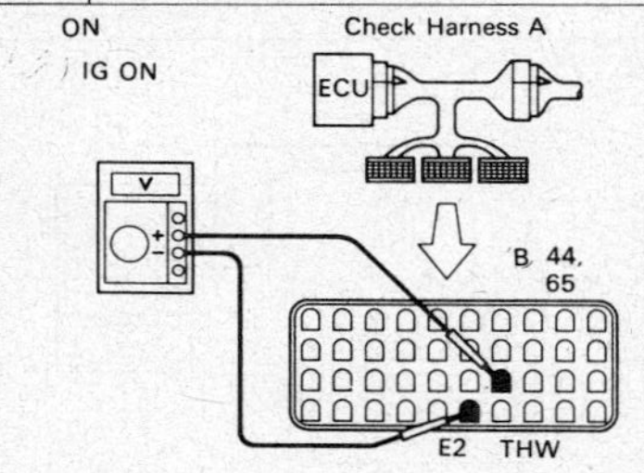

P (1) Connect the Check Harness A.

(2) Turn ignition switch on.

C Measure voltage between terminals THW and E2 of engine (& ECT) ECU connector.

OK

Water Temp. °C (°F)	Voltage
20 (68) (Engine is cool)	1 – 3 V
80 (176) (Engine is hot)	0.1 – 1.0 V

OK → **Check for momentary imterruption**

NG ↓

2 Check water temp. sensor.

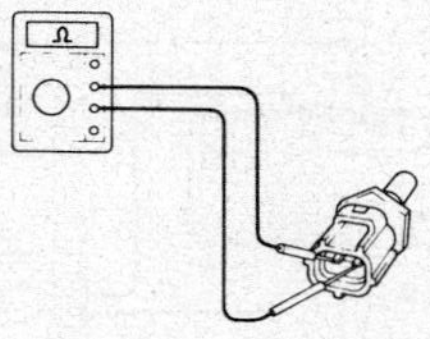

P Disconnect the water temp. sensor connector.
C Measure resistance between terminals.
OK **Resistance is within Acceptable Zone on chart.**

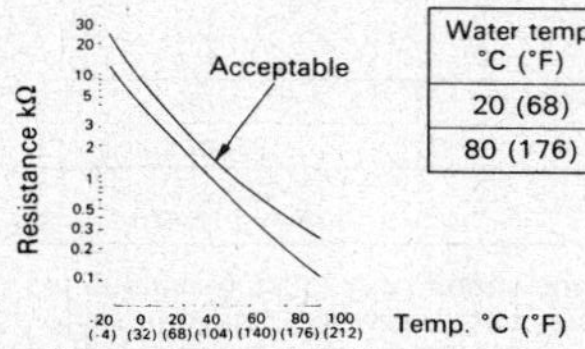

Water temp. °C (°F)	Resistance
20 (68)	2 – 3 kΩ
80 (176)	0.2 – 0.4 kΩ

NG → **Replace water temp. sensor.**

OK ↓

3 Check for open and short in harness and connector between engine (& ECT) ECU and water temp. sensor

NG → **Repair or replace harness or connector.**

OK ↓

Check and replace engine (& ECT) ECU.

Diag. Code	25	Air-Fuel Ratio Lean Malfunction
	26	Air-Fuel Ratio Rich Malfunction

CIRCUIT DESCRIPTION

Code No.	Diagnostic Code Detecting Condition	Trouble Area
25	(1) Main oxygen sensor voltage is 0.45 V or less (lean) for 90 sec. under conditions (a) and (b). (2 trip detection logic) (a) Coolant temp. : 70°C (158°F) or more. (b) Engine speed : 1,500 rpm or more.	• Open or short in main oxygen sensor circuit • Main oxygen sensor • Ignition system • ECU
	(2)*1 Difference of air-fuel ratio feedback compensation value between front (No. 1 ~ 3 cylinders) and rear (No. 4 ~ 6 cylinders) is more than 15 percentage for 20 sec. or more under conditions (a) and (b). (2 trip detection logic) (a) Engine speed: 2,000 rpm or more. (b) Coolant temp.: Between 60°C (140°F) and 95°C (203°F).	• Open and short in injector circuit. • Fuel line pressure (injector leak, blockage) • Mechanical system malfunction (skipping teeth of timing belt) • Ignition system • Compression pressure (foreign object caught in valve) • Air flow meter (air intake) • ECU
	(3)*1 Engine speed varies by more than 15 rpm over the preceding crank angle period during a period of 20 sec. or more under conditions (a) and (b). (2 trip detection logic) (a) Engine speed: Idling (b) Coolant temp.: Between 60°C (140°F) and 95°C (203°F).	• Open and short in injector circuit. • Fuel line pressure (injector leak, blockage) • Mechanical system malfunction (skipping teeth of timing belt) • Ignition system • Compression pressure (foreign object caught in valve) • Air flow meter (air intake) • ECU
26*1	(1) Difference of air-fuel ratio feedback compensation value between front (No.1 ~ 3 cylinders) and rear (No.4 ~ 6 cylinders) is more than 15 percentage for 20 sec. or more under conditions (a) and (b). (2 trip detection logic) (a) Engine speed: 2,000 rpm or more. (b) Coolant temp.: Between 60°C (140°F) and 95°C (203°F).	• Open and short in injector circuit. • Fuel line pressure (injector leak, blockage) • Mechanical system malfunction (skipping teeth of timing belt) • Ignition system • Compression pressure (foreign object caught in valve) • Air flow meter (air intake) • ECU
	(2) Engine speed varies by more than 15 rpm over the preceding crank angle period during a period of 20 sec. or more under conditions (a) and (b). (2 trip detection logic) (a) Engine speed: Idling (b) Coolant temp.: Between 60°C (140°F) and 95°C (203°F).	• Open and short in injector circuit. • Fuel line pressure (injector leak, blockage) • Mechanical system malfunction (skipping teeth of timing belt) • Ignition system • Compression pressure (foreign object caught in valve • Air flow meter (air intake) • ECU

*1: Only for USA specification vehicles.

INSPECTION PROCEDURE

1 Check voltage between terminals THA and E2 of engine (& ECT) ECU connector.

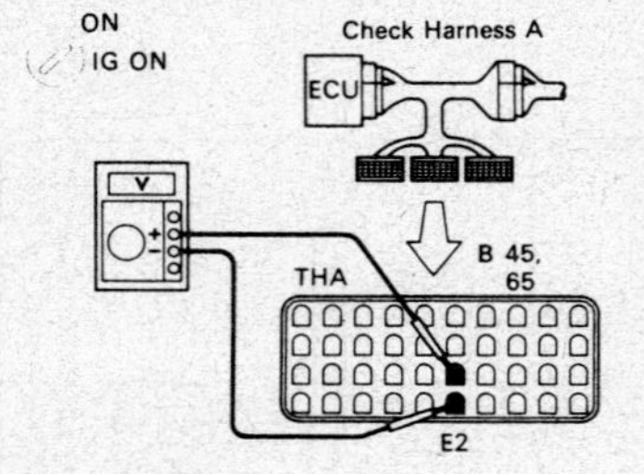

P (1) Connect the Check Harness A.

(2) Turn ignition switch on.

C Measure voltage between terminals THA and E2 of engine (& ECT) ECU connector.

OK

Intake air temp. °C (°F)	Voltage
20 (68)	1 – 3 V
60 (140)	0.5 – 1.0 V

OK → **Check for momentary interruption**

NG ↓

2 Check intake air temp. sensor.

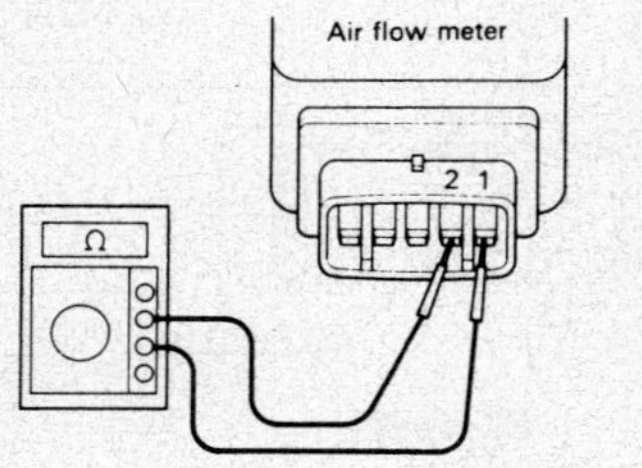

P Disconnect the air flow meter connector.

C Measure resistance between terminals 1 and 2 of air flow meter connector.

OK **Resistance is within Acceptable Zone on chart.**

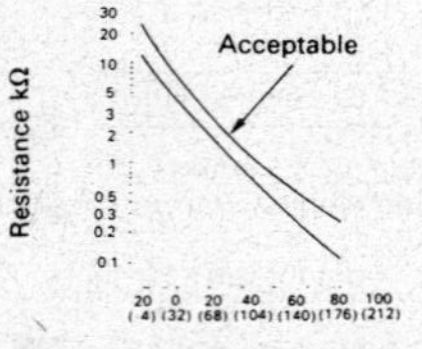

Intake air temp. °C (°F)	Resistance
20 (68)	2 – 3 kΩ
60 (140)	0.4 – 0.7 kΩ

NG → **Replace intake air temp. sensor (Replace air flow meter).**

OK ↓

3 Check for open and short in harness and connector between engine (& ECT) ECU and intake air temp. sensor

NG → **Repair or replace harness or connector.**

OK ↓

Check and replace engine (& ECT) ECU.

CIRCUIT DESCRIPTION (Cont'd)

DIAGNOSIS CODE DETECTION DRIVING PATTERN

Purpose of the driving pattern.

(a) To simulate diag. code detecting condition after diag. code is recorded.

(b) To check that the malfunction is corrected when the repair is completed confirming that diag. code is no longer detected.

Malfunction: Open or Short in Injector circuit, Injector Leak or Blockage

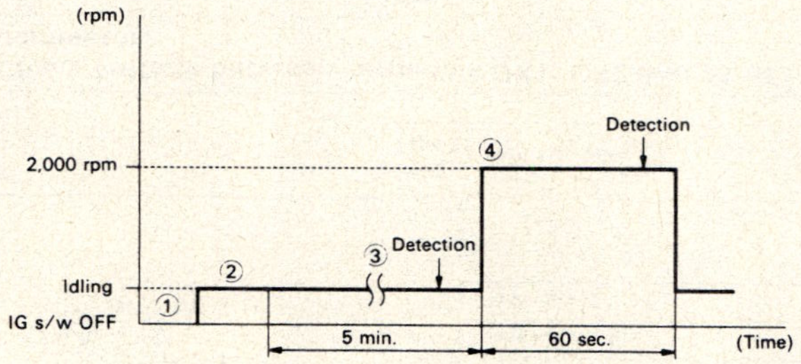

① Disconnect the EFI fuse (15 A) for 10 sec. or more. with IG switch OFF. Initiate test mode (Connect terminal TE2 and E1 of check connector or TDCL with IG switch OFF).

② Start engine and warm up with all ACC switch OFF.

③ After engine is warmed up, let it idle for 5 min. (After the engine is started, do not depress the accelerator pedal.)

④ If the malfunction is not detected during idling, perform racing without any load at approx. 2,000 rpm for 60 sec.

HINT: If a malfunction exists, the "CHECK" engine warning light will light up during the 5 min. idling period or within 60 sec. of starting racing.

NOTICE: If the conditions in this test are not strictly followed, detection of the malfunction will not be possible.

DIAGNOSTIC CHART

Step	Check	Result → Next	Result → Action
1	Check voltage of terminal VF.	NG Type I → 2	NG Type II → 7
2	Check voltage of terminal OX.	NG → 3	OK → 7
3	Check for open and short in harness and connector.	OK → 4	NG → Repair or replace harness or connector.
4	Check each item found to be a possible cause of problem.	OK → 5	NG → Repair or replace.
5	Check compression.	OK → 6	NG → Repair or replace.
6	Does malfunction disappear when a good main oxygen sensor is installed?	NO → Check and replace ECU.	YES → Replace main oxygen sensor.
7	Check each item found to be a possible cause of problem.	OK → 8	NG → Repair or replace.
8	Check compression.	OK → 9	NG → Repair or replace.
9	Does malfunction disappear when a good main oxygen sensor is installed?	NO → Check and replace ECU.	YES → Replace main oxygen sensor.

INSPECTION PROCEDURE

1 Check voltage between terminals VF1, VF2 and E1 of check connector.

Front Side

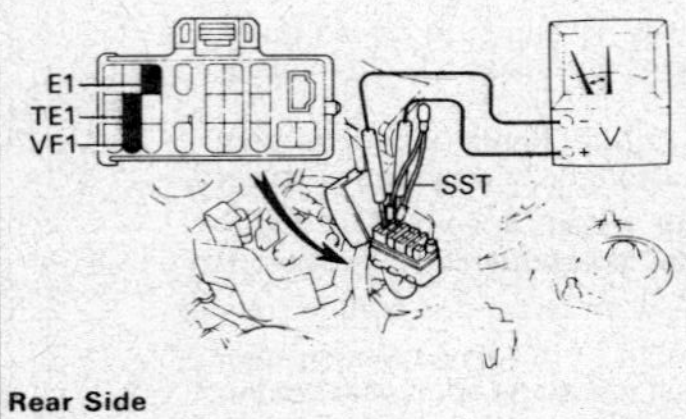

Rear Side

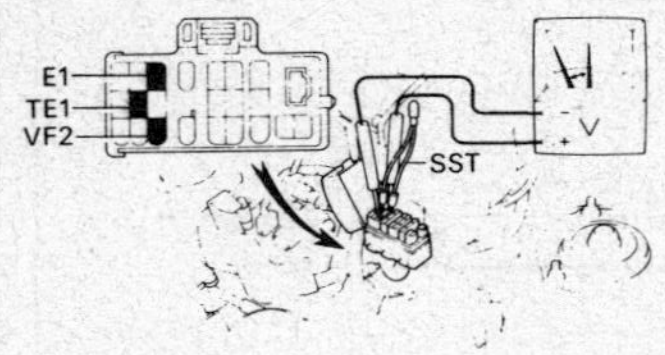

P
(1) Warm up engine at normal operating temperature.
(2) Connect terminals TE1 and E1 of check connector.
(3) Connect positive prove to terminal VF1, VF2 and negative prove to terminal E1 of check connector.

C
(1) Warm up the oxygen sensor by running engine at 2,500 rpm for about 2 minutes.
(2) Then, maintaining engine at 2,500 rpm, count how many times needle of voltmeter fluctuates between 0 and 5 V.

Result

Result	
Needle fluctuates of 8 times for every ten seconds	OK
Continue at 0 V	NG Type I
Continue at 5 V	NG Type II

NG Type II ⟩ Go to step 7.

NG Type I

2 Check voltage between terminals OX1, OX2 and E1 of check connector.

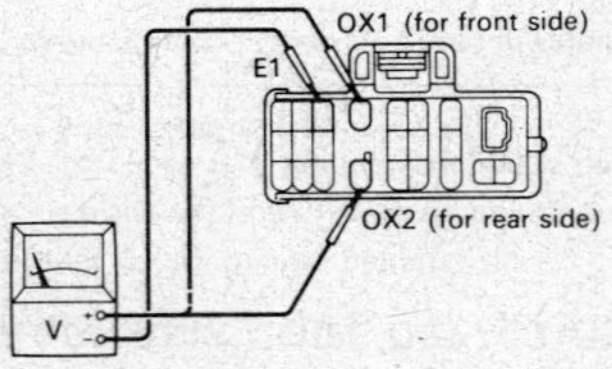

P Warm up engine at normal operating temperature.

C Measure voltage between terminals OX1, OX2 and E1 of check connector when engine is suddenly raced to full throttle.

OK **The voltage should be 0.5 V or higher at least once.**

Hint Perform inspection within 1 second.

OK ⟩ Go to step 7.

NG

3 Check for open and short in harness and connector between engine (& ECT) ECU and main oxygen sensor, engine (& ECT) ECU and check connector

NG ⟩ **Repair or replace harness or connector.**

OK

4 Check each item found to be a possible cause of problem.

Check each circuit found to be a possible cause of trouble according to the results of the check in 1 or 2. The numbers in the table below show the order in which the checks should be performed.

Main oxygen sensor signal from either side continues at 0 V.	Main oxygen sensor signals from both sides continue at 0 V.	Possible Cause	
1		Faulty sensor installation.	
3		Injector circuit	
2	3	Misfire	
4		Valve timing	
	1	Air leakage	
	2	Fuel system	
	6	Characteristics deviation in air flow meter.	
	4	Characteristics deviation in water temp. sensor.	
	5	Characteristics deviation in intake air temp. sensor.	

NG ⟩ **Repair or replace.**

OK

5 Check compression

NG ⟩ **Repair or replace.**

OK

6 Does malfunction disappear when a good main oxygen sensor is installed?

YES ⟩ **Replace main oxygen sensor.**

NG

Check and replace engine (& ECT) ECU.

7 Check each item found to be a possible cause of problem.

Check each circuit found to be a possible cause of trouble according to the results of the check in [1]. The numbers in the table below show the order in which the checks should be performed.

Main oxygen sensor signal from either side continues at 5.0 V.	Main oxygen sensor signals from both sides continue at 5.0 V.	Main oxygen sensor signals from both sides are normal.	Possible Cause	
1		7	Injector circuit	
		3	Misfire	
2		4	Valve timing	
		1	Air leakage	
	1	2	Fuel system	
	4	8	Characteristics deviation in air flow meter.	
	2	5	Characteristics deviation in water temp. sensor.	
	3	6	Characteristics deviation in intake air temp. sensor.	

OK

NG — Repair or replace.

8 Check compression

OK

NG — Repair or replace.

9 Does malfunction disappear when a good main oxygen sensor is installed?

YES — Replace main oxygen sensor.

NG

Check and replace engine (& ECT) ECU.

Diag. Code 27 — Sub-Oxygen Sensor Circuit (Only for USA spec.)

CIRCUIT DESCRIPTION

The sub-oxygen sensor is installed on the exhaust pipe. Its construction and operation is the same as the main oxygen sensor

Code No.	Diagnostic Code Detecting Condition	Trouble Area
27	(1) Open or short in heater circuit of sub-oxygen sensor for 0.5 sec. or more.	• Open or short in heater circuit of sub-oxygen sensor. • Sub-oxygen sensor heater. • ECU
	(2) Main oxygen sensor signal is 0.45 V or more and sub-oxygen sensor signal is 0.45 V or less under conditions (a) ~ (c). (2 trip detection logic) (a) Coolant temp. : 80°C (176°F) or more. (b) Engine speed : 1,500 rpm or more. (c) Accel. pedal : Fully depressed for 2 sec. or more.	• Open or short in sub-oxygen sensor circuit. • Sub-oxygen sensor • ECU

CIRCUIT DESCRIPTION (Cont'd)

DIAGNOSIS CODE DETECTION DRIVING PATTERN

Purpose of the driving pattern.

(a) To simulate diag. code detecting condition after diag. code is recorded.

(b) To check that the malfunction is corrected when the repair is completed confirming that diag. code is no longer detected.

Malfunction: Open or Short in Sub-Oxygen Sensor

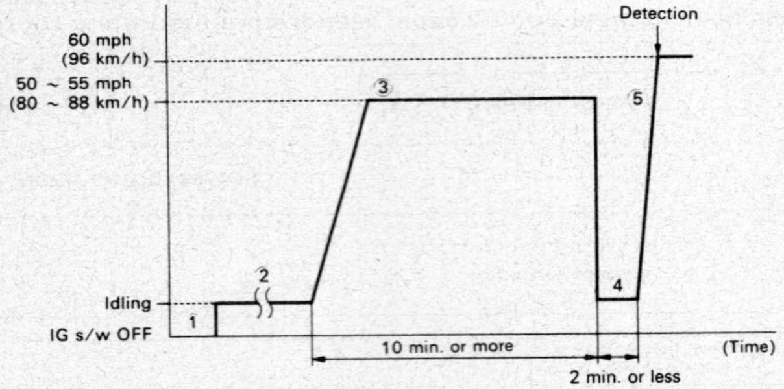

① Disconnect the EFI fuse (15 A) for 10 sec. or more, with IG switch OFF. Initiate test mode (Connect terminal TE2 and E1 of check connector with IG switch OFF).

② Start the engine and warm up with all ACC switch OFF.

③ After the engine is warmed up, let it drive at 50 ~ 55 mph (80 ~ 88 km/h) for 10 min. or more.

④ After driving, stop at a safe place and perform idling for 2 min. or less.

⑤ After performing the idling in ④, perform acceleration to 60 mph (96 km/h) with the throttle valve fully open.

HINT: If a malfunction exists, the "CHECK" engine warning light will light up during step ⑤.

NOTICE: **If the conditions in this test are not strictly followed, detection of the malfunction will not be possible.**

DIAGNOSTIC CHART

HINT: When other codes are output in addition to 27 at the same time, check the circuits for other codes first.

1 Check voltage of terminal HT.
- OK → go to 4
- NG ↓

2 Check resistance of sub-oxygen sensor heater.
- NG → Replace sub-oxygen sensor.
- OK ↓

3 Check for open and short in harness and connector between EFI main relay and ECU.
- NG → Repair or replace harness or connector.
- OK ↓

Check and replace ECU.

4 Check operation of sub-oxygen sensor heater.
- OK → Replace sub-oxygen sensor.*
- NG ↓

Check and replace ECU.

* In this case, oxygen sensor can be deteriorated.

WIRING DIAGRAM

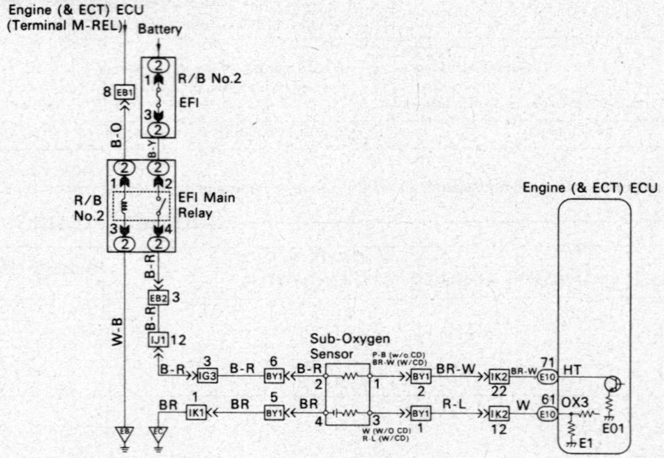

INSPECTION PROCEDURE

1 Check voltage between terminal HT of engine (& ECT) ECU connector and body ground.

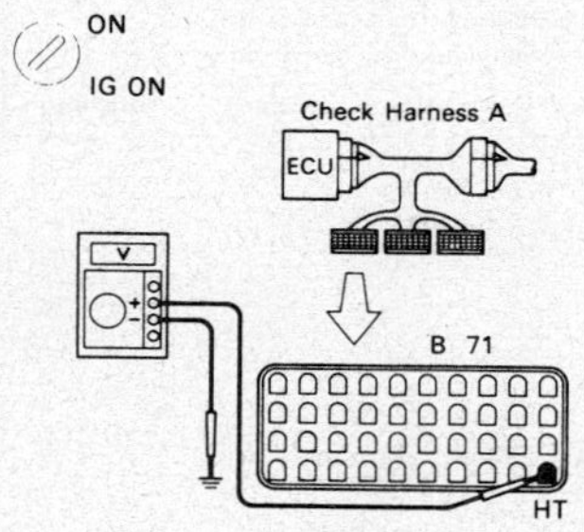

P (1) Connect the Check Harness A.

(2) Turn ignition switch on.

C Measure voltage between terminal HT of engine (& ECT) ECU connector and body ground.

OK **Voltage: 10 – 14 V**

OK → Go to step 4.

NG ↓

2 Check sub-oxygen sensor heater.

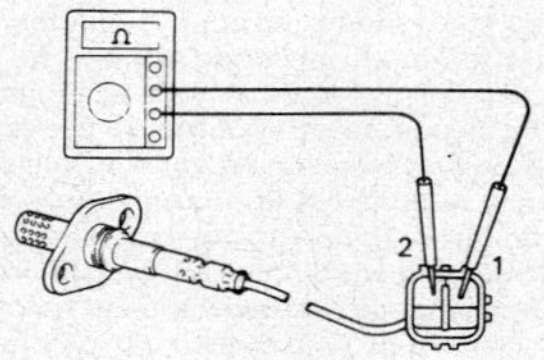

P Disconnect sub-oxygen sensor connector.

C Measure resistance between terminals 1 and 2 of sub-oxygen sensor connector.

OK **Resistance: 5.1 – 6.3 Ω 20°C (68°F)**

NG → **Replace sub-oxygen sensor.**

OK ↓

3 Check for open and short in harness and connector between EFI main relay and engine (& ECT) ECU

NG → **Repair or replace harness or connector.**

OK ↓

Check and replace engine (& ECT) ECU.

4 Check voltage between terminal HT of engine (& ECT) ECU connector and body ground.

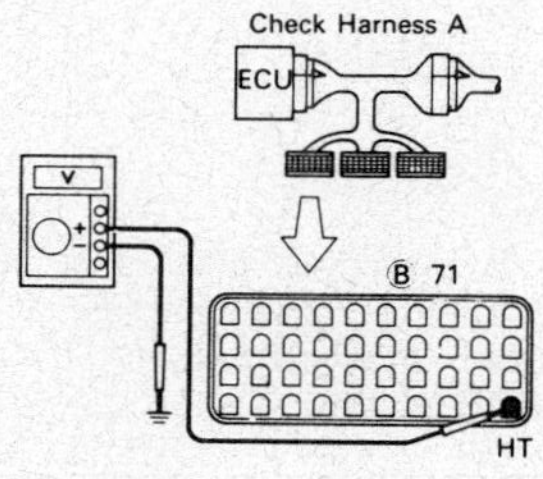

FI6615

P Warm up engine to normal operating temperature.

C Measure voltage between terminal HT of engine (& ECT) ECU connector and body ground, when engine is idling and racing at 3,500 rpm.

OK

	Voltage
Idling	0 V
Racing at 3,500 rpm	10 – 14 V

OK → **Replace sub-oxygen sensor.***

NG ↓

Check and replace engine (& ECT) ECU.

*: **In this case, oxygen sensor can be deteriorated.**

Diag. Code	31	Air Flow Meter Circuit

CIRCUIT DESCRIPTION

As shown in the figure at right, when a pillar (Vortex generating body) is placed in the path of a uniform flow, vortices called Karman-Vortex are generated downstream of the object. Using this principle, a vortex generator is placed inside the air flow meter. By measuring the frequency of the vortices generated, the ECU can determine the volume of air flowing through the air flow meter. The vortices are detected by their exerting pressure on thin metal foil (mirror) surfaces and a light emitting element and light receptor (LED and photo transistor) positioned opposite the mirror which senses the vibrations in the mirror optically. The ECU uses these signals mainly for calculation of the basic injection volume and the basic ignition advance angle.

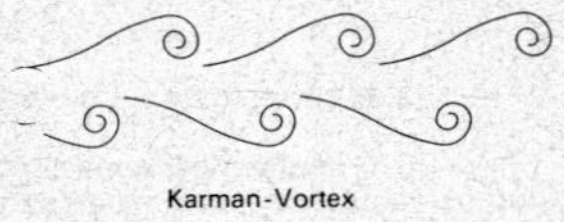

FI4504

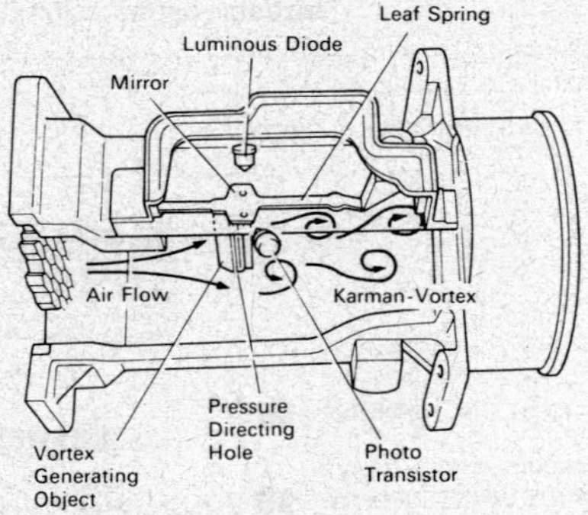

Code No.	Diagnostic Code Detecting Condition	Trouble Area
31	All conditions below are detected. (a) No air-flow meter signal to ECU for 2 sec. when engine speed is above 300 rpm. (b) Engine stall.	• Open or short in air flow meter circuit • Air flow meter • ECU

If the ECU records diagnostic code "31", it operates the fail safe function, keeping the ignition timing and fuel injection volume constant and making it possible to drive the vehicle.

DIAGNOSTIC CHART

1 Check voltage of terminal KS.
- OK → When diag. code 31 is displayed, check and replace ECU.
- NG ↓

2 Check for open and short in harness and connector between air flow meter and ECU.
- NG → Repair or replace harness or connector.
- OK ↓

3 Check voltage terminal KS of ECU.
- NG → Check and replace ECU.
- OK ↓

4 Check voltage of air flow meter power source.
- OK → Replace air flow meter.
- NG ↓

Check and replace ECU.

WIRING DIAGRAM

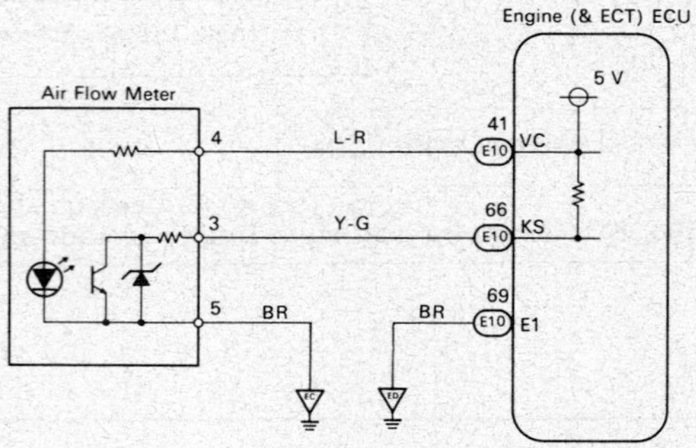

INSPECTION PROCEDURE

1 Check voltage between terminals KS and E1 of engine (& ECT) ECU connector.

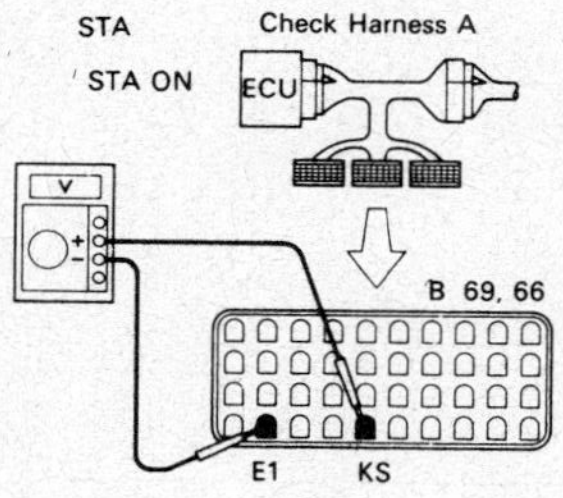

P Connect the Check Harness A.

C Measure voltage between terminals KS and E1 of engine (& ECT) ECU connector while engine is cranked.

OK **Voltage: 2.0 – 4.0 V**
(Neither 0 V nor 5 V)

Reference **INSPECTION USING OSCILLOSCOPE**

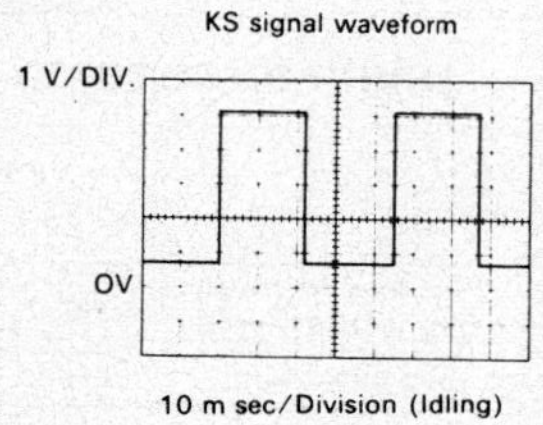

- During cranking or idling, measure between terminals KS and E1 of engine (& ECT) ECU.

HINT: The correct waveform appears as shown in the illustration on the left, with rectangle waves.

NG / **OK** → **When diag. code 31 is displayed, check and replace engine (& ECT) ECU.**

2 Check for open and short in harness and connector between engine (& ECT) ECU and air flow meter

OK / **NG** → **Repair or replace harness or connector.**

3 Disconnect air flow meter connector and check voltage between terminals KS and E1 of engine (& ECT) ECU connector.

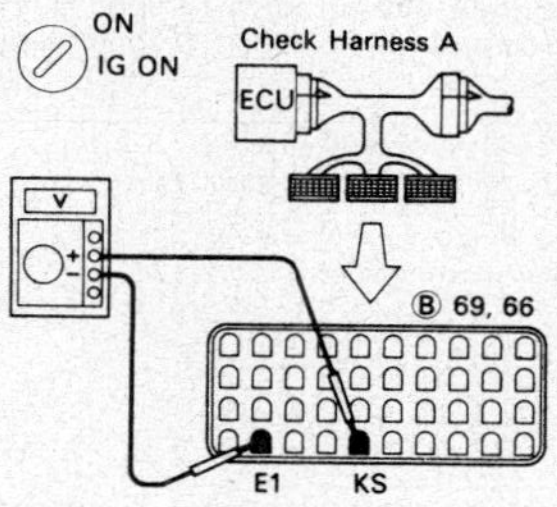

P (1) Disconnect the air flow meter connector.
(2) Turn ignition switch on.

C Measure voltage between terminals KS and E1 of engine (& ECT) ECU connector.

OK **Voltage: 4 – 6 V**

OK / **NG** → **Check and replace engine (& ECT) ECU.**

4 Disconnect air flow meter connector and check voltage between terminals VC and E1 of engine (& ECT) ECU connector.

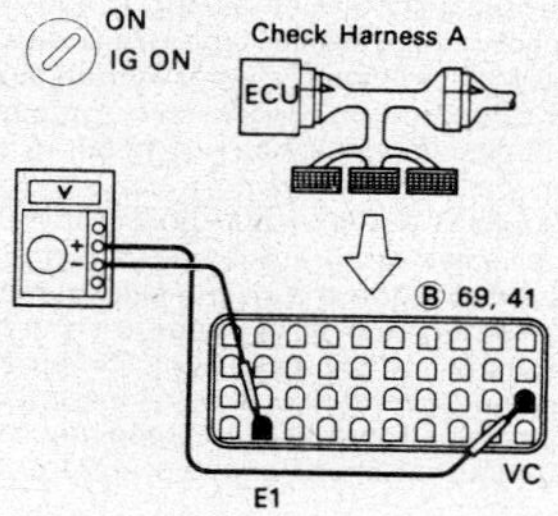

P (1) Disconnect the air flow meter connector.
(2) Turn ignition switch on.

C Measure voltage between terminals VC and E1 of engine (& ECT) ECU connector.

OK **Voltage: 4 – 6 V**

NG / **OK** → **Replace air flow meter.**

Check and replace engine (& ECT) ECU.

Diag. Code 35 High Altitude Compensator Sensor (HAC Sensor) Circuit

CIRCUIT DESCRIPTION

The HAC sensor is built into the ECU. This is a semiconductor pressure sensor with properties which cause its electrical resistance to change when stress is applied to the sensor's crystal (silicon) (piezoelectric effect). This sensor is used to detect the atmospheric (absolute) pressure and outputs corresponding electrical signals. Fluctuations in the air pressure cause changes in the intake air density, which can cause deviations in the air-fuel ratio. The signals from HAC sensor are used to make corrections for these fluctuations. If the ECU records diagnostic code "35", the fail safe function operates and the atmospheric pressure is set at a constant 101.3 kPa (760 mmHg, 29.92 in.Hg).

Code No.	Diagnostic Code Detecting Condition	Trouble Area
35	Open or short in HAC sensor circuit for 0.5 sec. or more.	• ECU

DIAGNOSTIC CHART

1 Are there any other codes (besides Code 16) being output?

- YES → Go to relevant diag. code chart.
- NO → Replace engine (& ECT) ECU.

Diag. Code 41, 47 (Sub-) Throttle Position Sensor Circuit

CIRCUIT DESCRIPTION

The throttle position sensor is mounted in the throttle body and detects the throttle valve opening angle. When the throttle valve is fully closed, the IDL contacts in the throttle position sensor are on, so the voltage at the terminal IDL of the ECU become 0 V. At this time, a voltage of approximately 0.7 V is applied to the terminal VTA of the ECU. When the throttle valve is opened, the IDL contacts go off and thus the power source voltage of approximately 12 V in the ECU is applied to the terminal IDL of the ECU. The voltage applied to the terminal VTA of the ECU increases in proportion to the opening angle of the throttle valve and becomes approximately 3.5 – 5.0 V when the throttle valve is fully opened. The ECU judges the vehicle driving conditions from these signals input from the terminals VTA and IDL, and uses them as one of the conditions for deciding the air-fuel ratio correction, power increase correction and fuel-cut control etc. The sub-throttle position sensor is built and operates in the same way as the main throttle position sensor. This sensor is used for traction control. The sub-throttle valve is opened and closed by the sub-throttle actuator according to signals from the TRAC ECU to control the engine output.

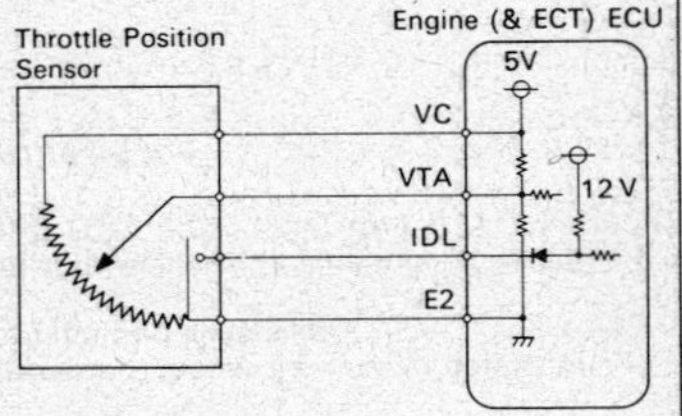

Code No.	Diagnostic Code Detecting Condition	Trouble Area
41	(1) Open or short in throttle position sensor circuit (VTA1) for 0.5 sec. or more.	• Open or short in throttle position sensor circuit.
	(2) IDL1 contact is ON and VTA1 output exceeds 1.5 V for 0.5 sec. or more.	• Throttle position sensor • ECU
47	(1) Open or short in sub-throttle position sensor circuit (VTA2) for 0.5 sec. or more.	• Open or short in sub-throttle position sensor circuit.
	(2) IDL2 contact is ON and VTA2 output exceeds 1.5 V for 0.5 sec. or more.	• Sub-throttle position sensor • ECU

HINT:
Diag. code 41 is for the throttle position sensor circuit.
Diag. code 47 is for the sub-throttle position sensor circuit.

- When the connector for the (sub-) throttle position sensor is disconnected, diagnostic code 41 or 47 is not displayed. Diagnostic code 41 or 47 is displayed only when there is an open or short in the VTA signal circuit of the (sub-) throttle position sensor.
- Signals from the (sub-) throttle position sensor are also input to the TRAC ECU, So when a malfunction occurs on the TRAC side, code 41 or 47 may be displayed.

DIAGNOSTIC CHART

HINT:
- If diag. code 41 is displayed, check throttle position sensor circuit; if diag. code 47 is displayed, check sub-throttle position sensor circuit.
- If diag. codes "22" (water temperature sensor circuit), "24" (intake air temperature sensor circuit) and "41" (throttle position sensor circuit) are output simultaneously, E2 (sensor ground) may be open.

Step	Check	Result	Action
1	Check voltage of throttle position sensor.	OK	Check for momentary interruption
		NG	Go to 2
2	Check operation for throttle position sensor.	NG	Adjust or replace throttle position sensor.
		OK	Go to 3
3	Check for open and short in harness and connector between throttle position sensor and ECU.	NG	Repair or replace harness or connector.
		OK	Check and replace ECU.

WIRING DIAGRAM

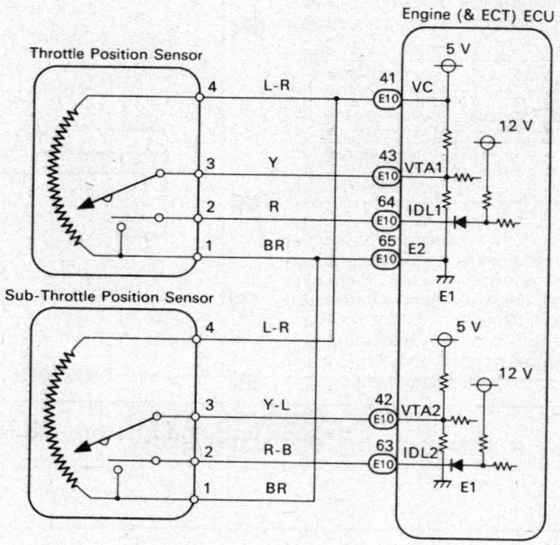

INSPECTION PROCEDURE

HINT:
- If diag. code 41 is displayed, check throttle position sensor circuit. If diag. code 47 is displayed, check sub-throttle position sensor circuit.
- If diag. codes "22" (water temperature sensor circuit), "24" (intake air temperature sensor circuit) and "41" (throttle position sensor circuit) are output simultaneously, E2 (sensor ground) may be open.

1 Check voltage between terminals VTA1, 2, IDL1, 2 and E2 of engine (& ECT) ECU connector.

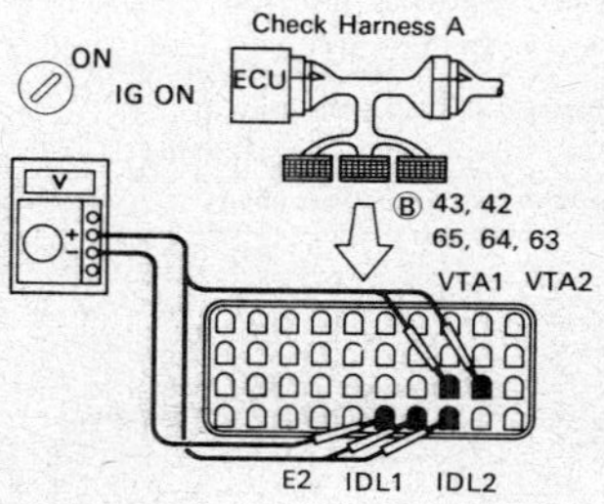

P (1) Connect the Check Harness A.

(2) Turn ignition switch on.
(3) For sub-throttle position sensor, remove intake air duct and disconnect sub-throttle valve step motor connector.

C Measure voltage between terminals VTA1, 2, IDL1, 2 and E2 of engine (& ECT) ECU connector when the (sub-) throttle valve is opened gradually from the closed condition.

OK

Terminal / Throttle Valve	VTA1, 2	IDL1, 2
Fully Closed	0.1 – 1.0 V	Below 1.0 V
Fully Open	3 – 5 V	10 – 14 V

Hint: The voltage should increase steadily in proportion to the throttle valve opening angle.

OK → Check for momentary interruption

NG → Go to step 2.

2 Check (sub-) throttle position sensor.

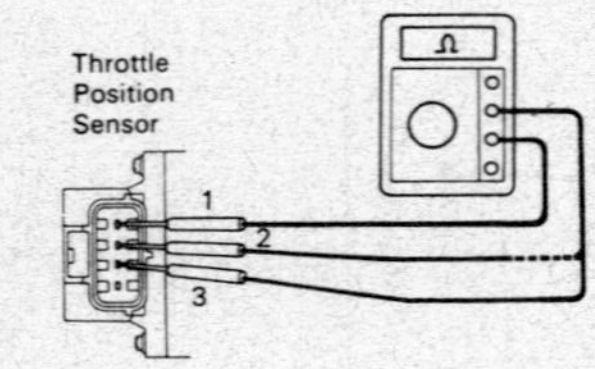

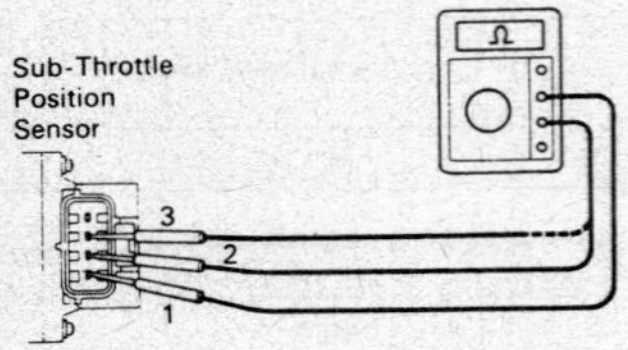

P (1) Remove intake air duct.
(2) Disconnect (sub-) throttle position sensor connector.

C Measure resistance between terminals 3, 2 and 1 of (sub-) throttle position sensor connector when the throttle valve is opened gradually from the closed condition.

OK

Terminal / Throttle Valve	3 – 1	2 – 1
Fully Closed	0.2 – 0.8 kΩ	Less than 0.5 kΩ
Fully Open	2.8 – 8.0 kΩ	1MΩ or higher

Hint Resistance between terminals 3 and 1 should increase gradually in accordance with the throttle valve opening angle.

OK ↓ | NG → Adjust or replace (sub-) throttle position sensor

3 Check for open and short in harness and connector between engine (& ECT) ECU and (sub-) throttle position sensor

OK ↓ | NG → Repair or replace harness or connector.

Check and replace engine (& ECT) ECU.

Diag. Code 42 Vehicle Speed Sensor Signal Circuit

CIRCUIT DESCRIPTION

The No.1 speed sensor outputs a 4-pulse signal for every revolution of the rotor shaft, which is rotated by the transmission output shaft via the driven gear. After this signal is converted into a more precise rectangular waveform by the waveform shaping, circuit inside the combination meter, it is then transmitted to the engine (& ECT) ECU. The ECU determines the vehicle speed based on the frequency of these pulse signals.

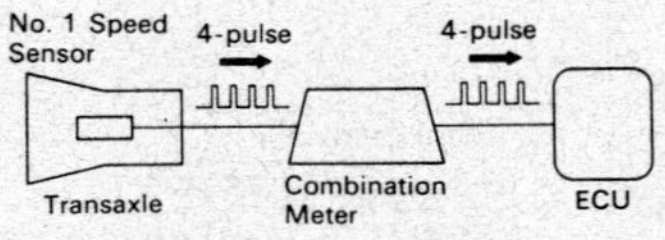

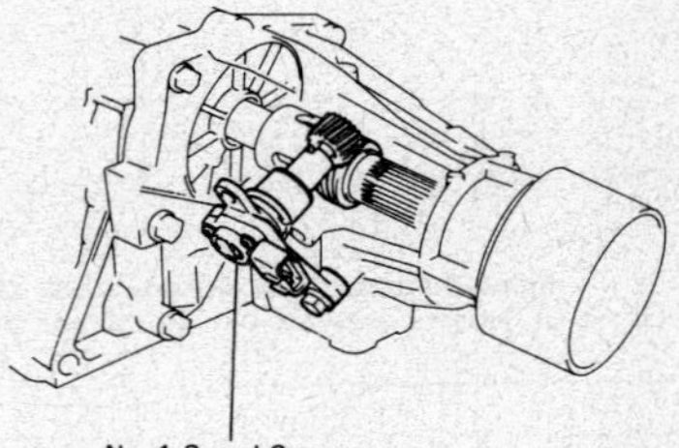

Code No.	Diagnostic Code Detecting Condition	Trouble Area
42	For ECT All conditions below are detected continuously for 8 sec. or more. (a) Vehicle speed signal: 0 km/h (mph) (b) Engine speed: 2,800 rpm or more (c) Neutral start switch (NSW): OFF (d) Stop light switch: OFF For M/T All conditions below are detected continuously for 8 sec. or more. (a) Vehicle speed signal: 0 km/h (mph) (b) Engine speed: Between 2,350 rpm and 5,000 rpm (c) Coolant temp.: 80°C (176°F) or more (d) Load driving	• No. 1 speed sensor • Combination meter • Open or short in No. 1 speed sensor circuit. • ECU

HINT: In test mode, diag. code 42 is output when vehicle speed is 5 km/h (3 mph) or below.

DIAGNOSTIC CHART

1	Check operation of speedometer.	NG →	Check speedometer circuit.
	↓ OK		
2	Check voltage of terminal SP1.	NG →	Check speedometer circuit.
	↓ OK		
	Check and replace ECU.		

WIRING DIAGRAM

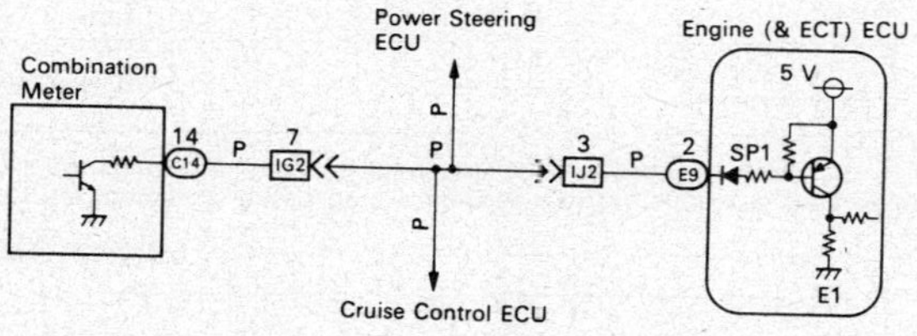

INSPECTION PROCEDURE

1 Check operation of speedometer.

C Drive the vehicle and check if the operation of the speedometer in the combination meter is normal.

Hint The No. 1 speed sensor is operating normally if the speedometer display is normal.

OK ↓ | **NG** > **Check speedometer circuit.**

2 Check voltage between terminal SP1 of engine (& ECT) ECU connector and body ground.

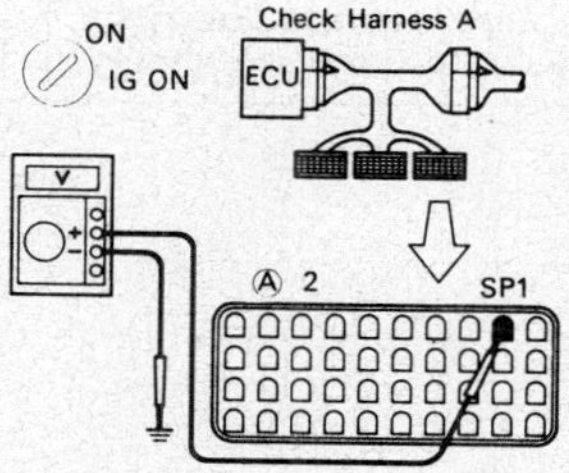

P (1) Shift the shift lever to N range or neutral.
(2) Jack up a rear wheel on one side.
(3) Connect the Check Harness A.
(4) Disconnect power sterring ECU connector and cruise control ECU connector.
(5) Turn ignition switch on.

C Measure voltage between terminal SP1 of engine (& ECT) ECU connector and body ground when the wheel is turned slowly.

OK **Voltage is generated intermittently.**

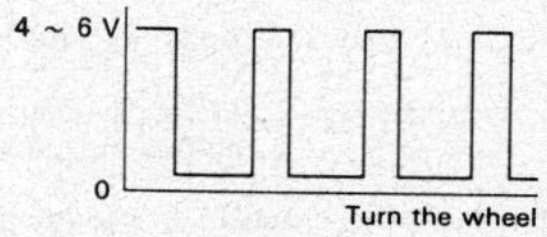

AT7809

OK ↓ | **NG** > **Check speedometer circuit.**

Check and replace engine (& ECT) ECU.

Diag. Code 43 Starter Signal Circuit

CIRCUIT DESCRIPTION

When the engine is cranked, the intake air flow is slow, so fuel vaporization is poor. A rich mixture is therefore necessary in order to achieve good startability. While the engine is being cranked, the battery voltage is applied to terminal STA of the ECU. The starter signal is mainly used to increase the fuel injection volume for the starting injection control and after-start injection control.

Code No.	Diagnostic Code Detecting Condition	Trouble Area
43	No starter signal to ECU.	• Open or short in starter signal circuit. • Open or short in ignition switch or starter relay circuit. • ECU

DIAGNOSTIC CHART

HINT: This diagnostic chart is based on the premise that the engine is cranked normally. If the engine is not cranked, proceed to the matrix chart of problem symptoms

1 Check for the test mode.
- OK → Proceed to next circuit inspection shown on matrix chart
- NG ↓

2 Check for open in harness and connector between ECU and relay.
- NG → Repair or replace harness or connector.
- OK ↓

Check and replace ECU.

WIRING DIAGRAM

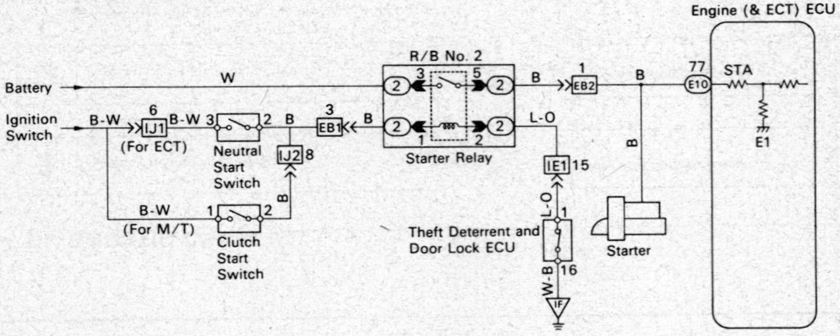

INSPECTION PROCEDURE

1 **Check output condition of diag. code 43.**

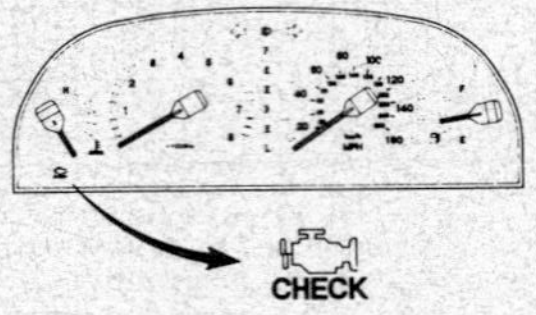

P Setting the test mode.
(1) Turn ignition switch OFF.
(2) Connect terminals TE2 and E1 of TDCL.
(3) Turn ignition switch ON.
(Don't start the engine)
(4) Connect terminals TE1 and E1 of TDCL.

C Check if code "43" is output by the "CHECK" engine warning light.

OK **Code "43" is output.**

C Start the engine.
Check if the code "43" disappear.

OK **Code "43" is not output.**

OK → **Proceed to next circuit inspection shown on matrix chart**

NG ↓

2 **Check for open in harness and connector between engine (& ECT) ECU and starter relay**

NG → **Repair or replace harness or connector.**

OK ↓

Check and replace engine (& ECT) ECU.

Diag. Code 52,53,55 Knock Sensor Circuit

CIRCUIT DESCRIPTION

Knock sensors are fitted one each to the front and rear of the left side of the cylinder block to detect engine knocking. This sensor contains a piezoelectric element which generates a voltage when it becomes deformed, which occurs when the cylinder block vibrates due to knocking. If engine knocking occurs, ignition timing is retarded to suppress it.

Code No.	Diagnostic Code Detecting Condition	Trouble Area
52	No No. 1 knock sensor signal to ECU for 4 crank revolutions with engine speed between 1,600 rpm ~ 5,200 rpm.	• Open or short in No. 1 knock sensor circuit. • No. 1 knock sensor (looseness) • ECU
53	Engine control computer (for knock control) malfunction at engine speed between 650 rpm and 5,200 rpm.	• ECU
55	No No. 2 knock sensor signal to ECU for 4 crank revolutions with engine speed between 1,600 rpm ~ 5,200 rpm.	• Open or short in No. 2 knock sensor circuit. • No. 2 knock sensor (looseness) • ECU

If the ECU detects the above diagnosis conditions, it operates the fail safe function in which the corrective retard angle value is set to the maximum value.

CIRCUIT DESCRIPTION (Cont'd)

DIAGNOSIS CODE DETECTION DRIVING PATTERN

Purpose of the driving pattern.

(a) To simulate diag. code detecting condition after diag. code is recorded.

(b) To check that the malfunction is corrected when the repair is completed confirming that diag. code is no longer detected.

Malfunction: Open or Short in Knock Sensor

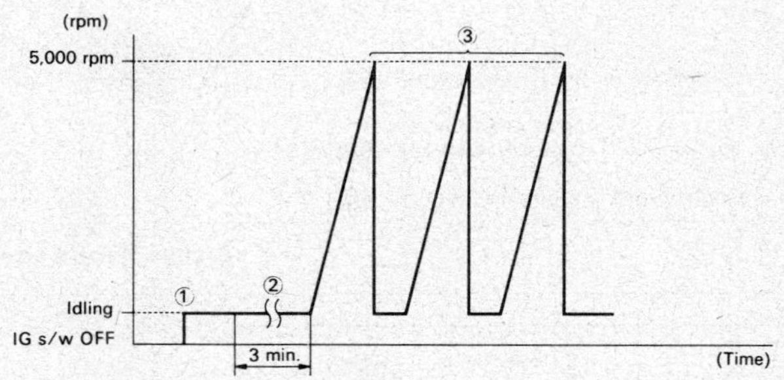

① Start the engine and warm up.
② After the engine is warmed up, let it idle for 3 min.
③ With the A/C ON, perform quick racing (5,000 rpm) three times. (Rapidly depress the accelerator pedal and suddenly release it.)

HINT: If a malfunction exists, the "CHECK" engine warning light will light up when sudden racing is performed.

NOTICE: If the conditions in this test are not strictly followed, detection of the malfunction will not be possible.

DIAGNOSTIC CHART

HINT: If diag. code 52 is displayed, check No. 1 knock sensor (for front side) circuit.
If diag. code 55 is displayed, check No. 2 knock sensor (for rear side) circuit.
If diag. code 53 is displayed, replace engine (& ECT) ECU.

1. Check no continuity of knock sensor circuit. — OK → go to 3; NG → go to 2
2. Check knock sensor. — NG → Replace knock sensor.; OK → go to 3
3. Check for open and short in harness and connector between knock sensor and ECU. — NG → Repair or replace harness or connector.; OK → go to 4
4. Does malfunction disappear when a good knock sensor is installed? — YES → Replace knock sensor.; NO → Check and replace ECU.

WIRING DIAGRAM

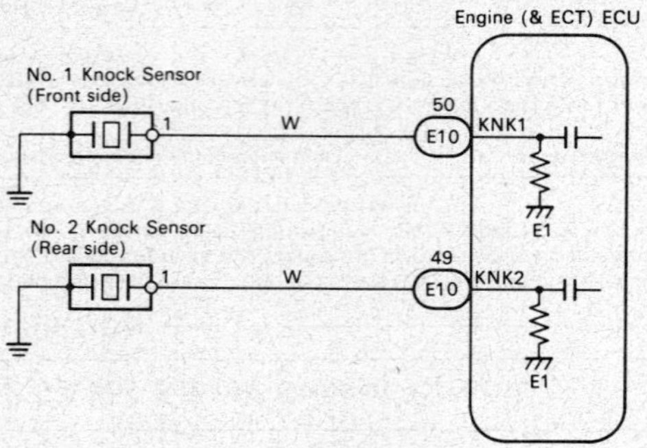

INSPECTION PROCEDURE

1 Check continuity between terminals KNK1, KNK2 of engine (& ECT) ECU connector and body ground.

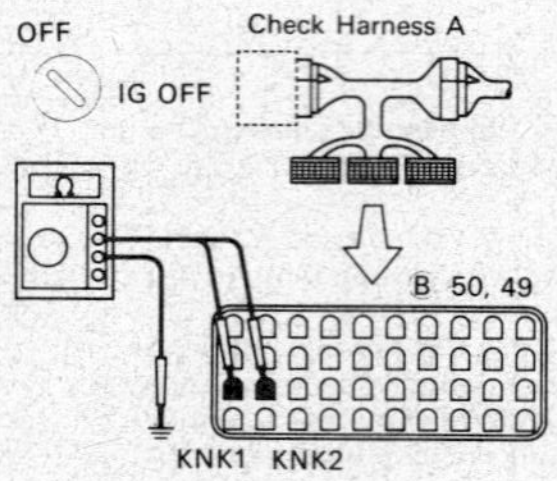

P (1) Connect the Check Harness A.

(2) Disconnect the engine (& ECT) ECU.

C Measure resistance between terminals KNK1, KNK2 of engine (& ECT) ECU connector and body ground.

OK **Resistance: 1 MΩ or higher**

OK → **Go to step 3.**

NG ↓

2 Check knock sensor.

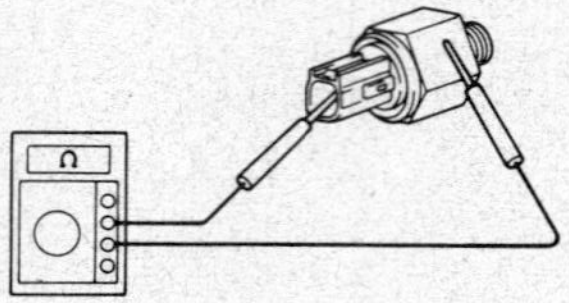

P Disconnect knock sensor connector.

C Measure resistance between the knock sensor terminal and body.

OK **Resistance: 1 MΩ or higher**

NG → **Replace knock sensor**

OK ↓

3 Check for open and short in harness and connector between engine (& ECT) ECU and knock sensor

OK / **NG** ⟩ Repair or replace harness or connector.

4 Does malfunction disappear when a good knock sensor is installed?

NO / **YES** ⟩ Replace knock sensor

Check and replace engine (& ECT) ECU.

Reference — **INSPECTION USING OSCILLOSCOPE**

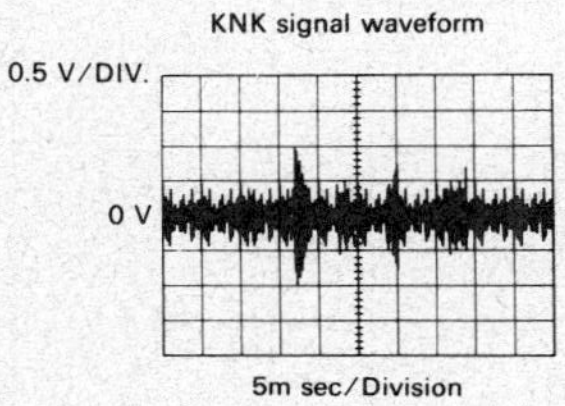

- With the engine racing (4,000 rpm) measure between terminals KNK1, KNK2 of engine (& ECT) ECU and body ground.

HINT: The correct waveform appears as shown in the illustration on the left.

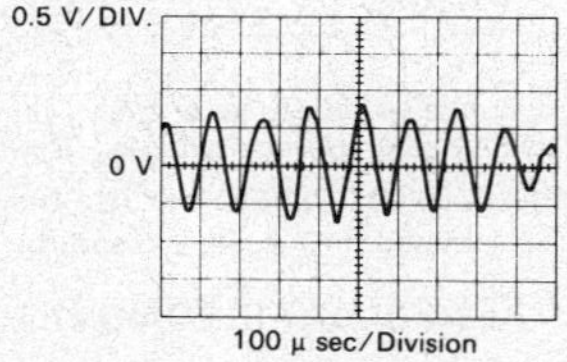

- Spread the time on the horizontal axis, and confirm that period of the wave is 123 μ sec. (Normal mode vibration frequency of knock sensor: 8.1 KHz).

HINT: If normal mode vibration frequency is not 8.1 KHz, the sensor is malfunctioning.

Diag. Code	71	EGR System Malfunction (Only for USA spec.)

CIRCUIT DESCRIPTION

The EGR system is designed to recirculate the exhaust gas properly controlled according to the driving condition back into the intake air-fuel mixture. It helps to slow down combustion in the cylinder and thus lower the combustion temperature which, in turn, reduces the amount of NOx emission. The amount of EGR is regulated by the EGR vacuum modulator according to the engine load.

If even one of the following conditions is fulfilled, the VSV is turned ON by a signal from the ECU. This results in atmospheric air acting on the EGR valve, closing the EGR valve and shutting off the exhaust gas (EGR cut-OFF).

- Coolant temp. below 50°C (122°F)
- During deceleration (throttle valve closed)
- Light engine load (amount of intake air very small).
- Engine speed over 5,200 rpm.
- Traction control is operating.

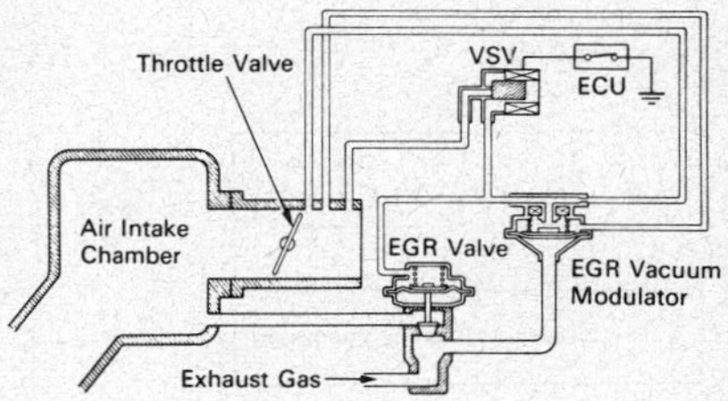

Code No.	Diagnostic Code Detecting Condition	Trouble Area
71	EGR gas temp. is 70°C (158°F) or below for 1~4 min. under conditions (a) and (b). (2 trip detection logic) (a) Coolant temp.: 63°C (145°F) or more. (b) EGR operation possible (EX. ECT in 3rd speed (5th for M/T), A/C ON, 60 mph (96 km/h), Flat road).	• Open in EGR gas temp. sensor circuit. • Short in VSV circuit for EGR. • EGR hose disconnected, valve stuck. • Clogged EGR gas passage. • ECU

CIRCUIT DESCRIPTION (Cont'd)

DIAGNOSIS CODE DETECTION DRIVING PATTERN

Purpose of the driving pattern.

(a) To simulate diag. code detecting condition after diag. code is recorded.

(b) To check that the malfunction is corrected when the repair is completed confirming that diag. code is no longer detected.

Malfunction: Open in EGR gas temp. sensor circuit

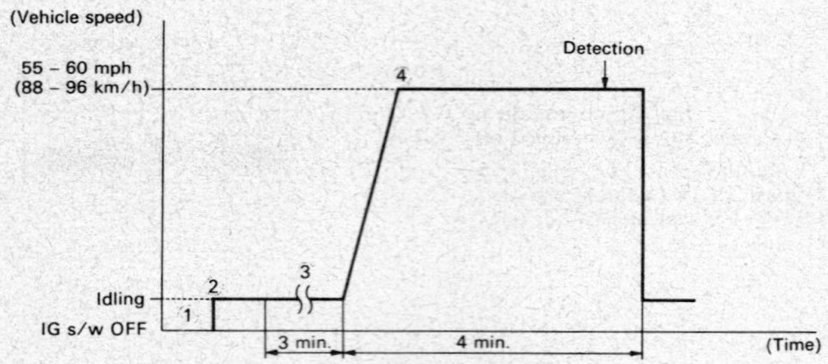

① Disconnect the EFI fuse (15 A) for 10 sec. or more, with IG switch OFF. Initiate test mode (Connect terminal TE2 and E1 of check connector or TDCL with IG switch OFF).

② Start the engine and warm up.

③ After the engine is warmed up, let it idle for 3 min.

④ With the A/C ON and transmission in 5th gear (ECT in 3rd speed) drive at 55 ~ 60 mph (88 ~ 96 km/h) for 4 min or less.

HINT: If a malfunction exists, the "CHECK" engine warning light will light up during step ④.

NOTICE: If the conditions in this test are not strictly followed, detection of the malfunction will not be possible.

DIAGNOSTIC CHART

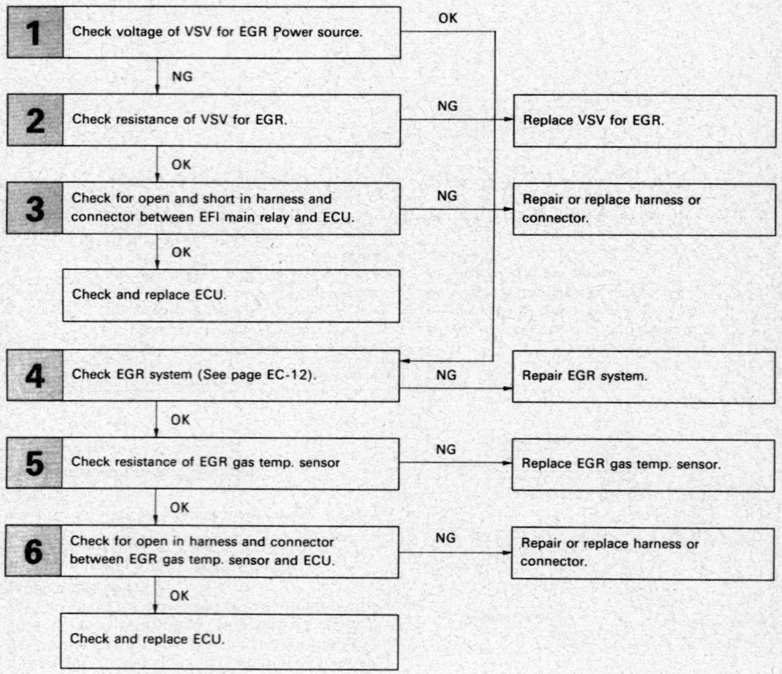

WIRING DIAGRAM

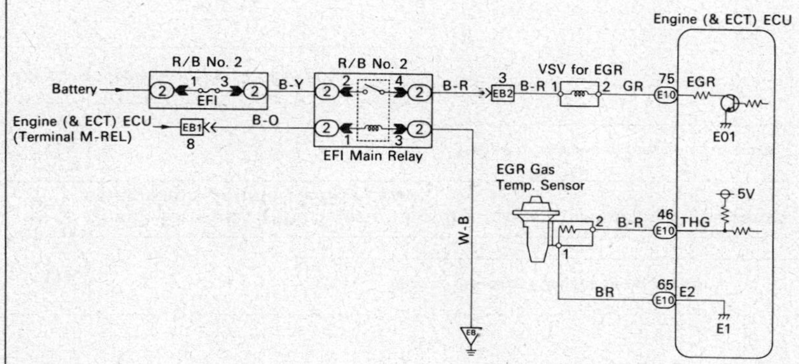

INSPECTION PROCEDURE

1 Check voltage between terminal EGR of engine (& ECT) ECU connector and body ground.

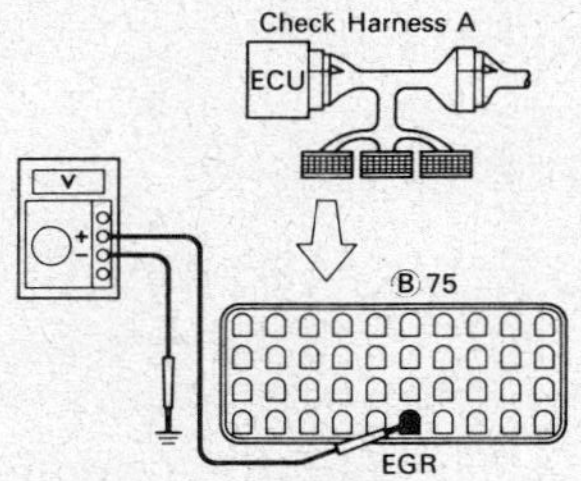

P (1) Connect the Check Harness A.

(2) Warm up engine to normal operating temperature.

C Measure voltage between terminal EGR of engine (& ECT) ECU connector and body ground.

OK **Voltage: 10 – 14 V**

OK → Go to step **4**.

NG ↓

2 Check resistance between terminals of VSV for EGR.

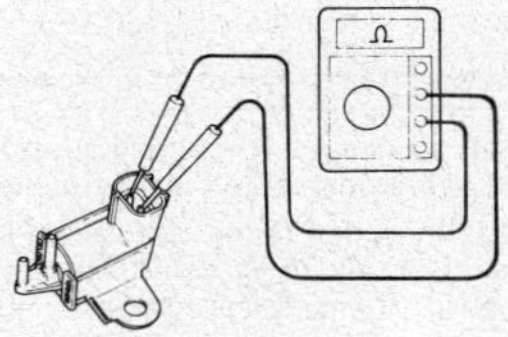

P Remove VSV for EGR.

C Measure resistance between terminals of VSV for EGR.

OK **Resistance: 38.5 – 44.5 Ω at 20°C (68°F)**

NG → Replace VSV for EGR.

OK ↓

3 Check for open and short in harness and connector between EFI main relay and VSV for EGR, VSV for EGR and engine (& ECT) ECU.

NG → Repair or replace harness or connector.

OK ↓

Check and replace engine (& ECT) ECU.

4 Check EGR system

OK

NG → Repair EGR system.

5 Check resistance of EGR gas temp. sensor.

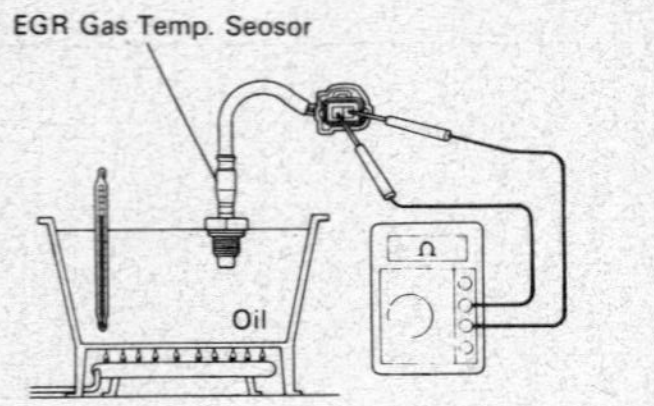

P Remove EGR gas temp. sensor.

C Measure resistance between terminals of EGR gas temp. sensor connector.

OK Resistance: 69 – 89 kΩ at 50°C (122°F)
12 – 15 kΩ at 100°C (212°F)
2 – 4 kΩ at 150°C (302°F)

OK

NG → Replace EGR gas temp. sensor.

6 Check for open in harness and connector between EGR gas temp. sensor and engine (& ECT) ECU.

OK

NG → Repair or replace harness or connector.

Check and replace engine (& ECT) ECU.

Diag. Code	78	Fuel Pump Control Circuit

CIRCUIT DESCRIPTION

The fuel pump speed is controlled at two steps (high speed,low speed) by the condition of the engine (starting, light load, heavy load), when the engine starts (STA ON), the engine (& ECT) ECU sends a Hi signal to the fuel pump ECU (FPC terminal).

The fuel pump ECU then outputs Hi voltage (battery voltage) to the fuel pump so that the fuel pump operates at high speed.

After the engine starts, during idling or light loads, the engine (& ECT) ECU outputs a High Low signal to the fuel pump ECU, the fuel pump ECU outputs Lo battery voltage (about 9V) to the fuel pump and causes the fuel pump to operate at low speed.

If the intake air volume increases (high engine load), the engine (& ECT) ECU sends a Hi signal to the fuel pump ECU and causes the fuel pump to operate at high speed.

Code No.	Diagnostic Code Detecting Condition	Trouble Area
78	(1) Open or short in fuel pump circuit for 1 sec. or more with engine speed 1,000 rpm or less. (2 trip detection logic) (2) Open in input circuit of fuel pump ECU (FPC) with engine speed 1,000 rpm or less. (2 trip detection logic) (3) Open or short in diagnostic signal line (DI) of fuel pump ECU with engine speed 1,000 rpm or less. (2 trip detection logic)	• Open or short in fuel pump ECU circuit. • Fuel pump ECU • Engine (& ECT) ECU power source circuit. • Fuel pump • Engine (& ECT) ECU

WIRING DIAGRAM

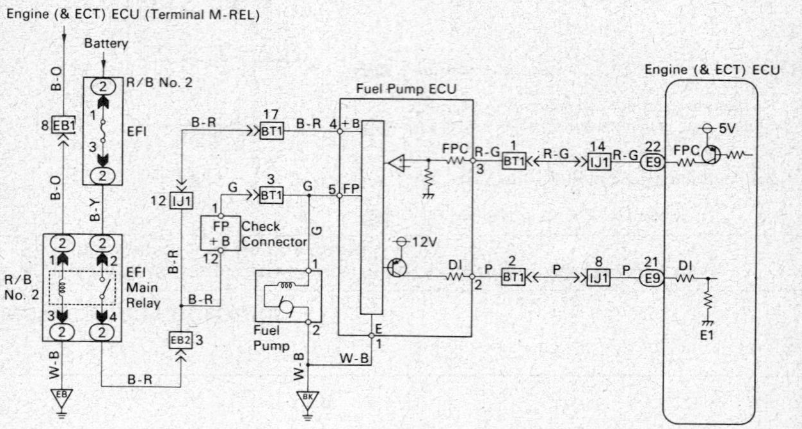

DIAGNOSTIC CHART

Step	Check	Result	Action
1	Check fuel pump operation.	OK	Go to 2
		NG	Go to 3
2	Check for open and short in harness and connector between check connector and fuel pump ECU.	OK	Go to 5
		NG	Repair or replace harness or connector.
3	Check voltage of terminal +B of check connector.	OK	Go to 4
		NG	Check for ECU power source circuit.
4	Check for open and short in harness and connector between check connector and fuel pump.	OK	Repair or replace fuel pump.
		NG	Repair or replace harness or connector.
5	Check voltage of terminal FPC.	OK	Replace fuel pump ECU.
		NG	Go to 6
6	Check for open and short in harness and connector between ECU and fuel pump ECU. (FPC)	OK	Go to 7
		NG	Repair or replace harness or connector.
7	Check for open and short in harness and connector between ECU and fuel pump ECU. (DI)	OK	Check and replace ECU.
		NG	Repair or replace harness or connector.

INSPECTION PROCEDURE

1 Check fuel pump operation.

ON
IG ON

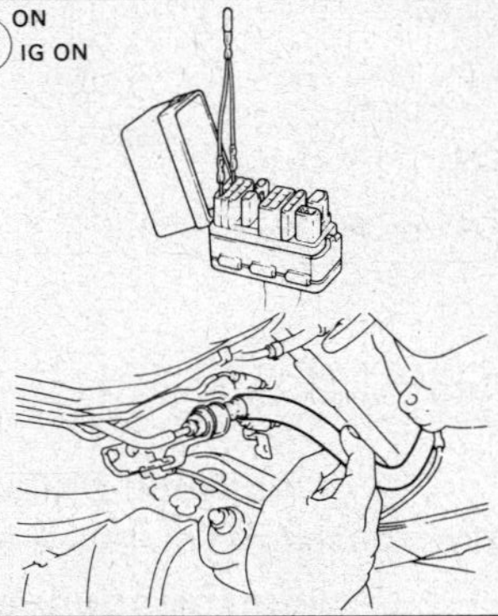

P (1) Turn ignition switch ON.
(2) Using SST, connect terminals +B and FP of check connector.
SST 09843-18020

C Check that there is pressure in the hose from the fuel filter.

OK **Fuel pressure can be felt.**

OK → (go to 2)

NG → **Go to step 3.**

2 Check for open and short in harness and connector between terminals +B ↔ +B, FP ↔ FP of the check connector and fuel pump ECU

NG → **Repair or replace harness or connector.**

OK → **Go to step 5.**

3 Check voltage of terminal +B of check connector.

ON
IG ON

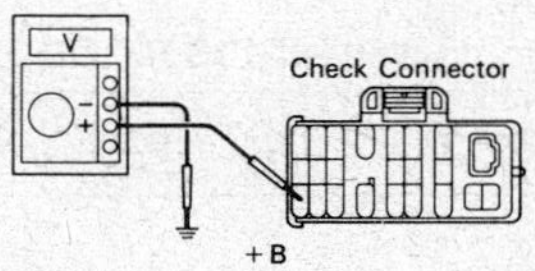

P Turn ignition switch ON.

C Measure voltage between terminal +B of check connector and body ground.

OK **Voltage : 10 – 14 V**

OK → (go to 4)

NG → **Check for ECU power source circuit and check for open in harness and connector between terminal +B of check connector and main relay.**

4 **Check for open and short in harness and connector between terminal FP of check connector, fuel pump and body ground**

OK → Repair or replace fuel pump.

NG ↓

Repair or replace harness or connector.

5 **Check voltage between terminals FPC and E of fuel pump ECU connector.**

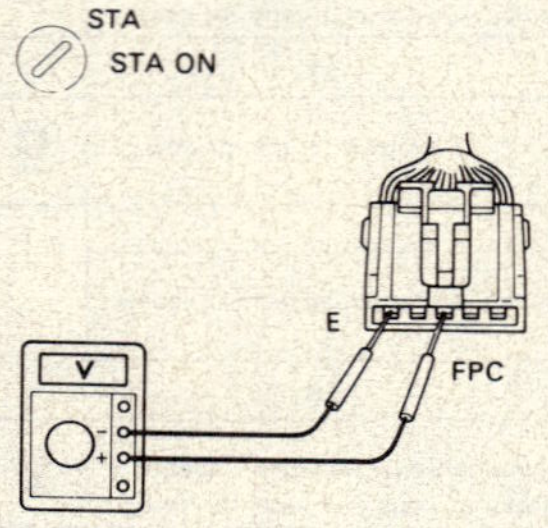

P (1) Remove the LH quarter trim panel.

(2) Disconnect fuel pump ECU connector.

C Measure voltage between terminals FPC and E of fuel pump ECU connector when ignition switch is turned to start.

OK **Voltage: 4 – 6 V**

OK → Replace fuel pump ECU.

NG ↓

6 **Check for open in harness and connector between terminal FPC of engine (& ECT) ECU and terminal FPC of fuel pump ECU, terminal E of fuel pump ECU and body ground**

NG → Repair or replace harness or connector.

OK ↓

7 **Check for open and short in harness and connector between terminal DI of engine (& ECT) ECU and terminal DI of fuel pump ECU**

NG → Repair or replace harness or connector.

OK ↓

Check and replace engine (& ECT) ECU.

Diag. Code 51 — Switch Condition Signal Circuit

CIRCUIT DESCRIPTION

Neutral Start Switch Signal*
The ECU uses the signals from the neutral start switch to determine whether the transmission is in park or neutral, or in some other gear.

Air Conditioner Switch Signal
The ECU uses the output from the air conditioner switch to determine whether or not the air conditioner is operating so that it can increase the idling speed of the engine if necessary.

Throttle Position Sensor IDL Signal
The IDL contacts are mounted in the throttle position sensor, and detects the idle condition.

Code No.	Diagnostic Code Detecting Condition	Trouble Area
51	(1) 3 sec. or more after engine starts idle switch OFF (IDL1). (2)* Neutral start switch OFF (NSW). (Shift position in "R", "D", "2" or "1" ranges). (3) A/C switch ON.	• Throttle position sensor IDL circuit • Accelerator pedal and cable • Neutral start switch circuit • A/C switch circuit • ECU

*: Only vehicles with ECT.
HINT: In this circuit, diagnosis can only be made in the test mode.

DIAGNOSTIC CHART

See next page for DIAGNOSTIC CHART

WIRING DIAGRAM

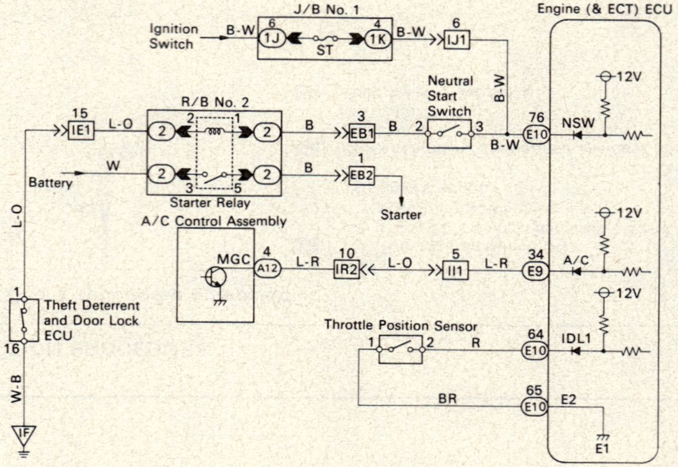

DIAGNOSTIC CHART

1 Check for input signal.
- NG (NSW) → See Neutral start switch circuit
- NG (A/C) → Step 3
- NG (IDL) → Step 2
- OK → Proceed to next circuit inspection shown on matrix chart

2 Check resistance of throttle position sensor.
- NG → Adjust or replace throttle position sensor
- OK → Check and repair harness or connector between sensor and ECU.

3 Check input circuit of A/C.
- OK → Check and replace A/C control assembly.
- NG → Step 4

4 Check for open and short in harness and connector between A/C control assy. and ECU.
- NG → Repair or replace harness or connector.
- OK → Check and replace ECU.

INSPECTION PROCEDURE

1 Check output condition of diag. code 51.

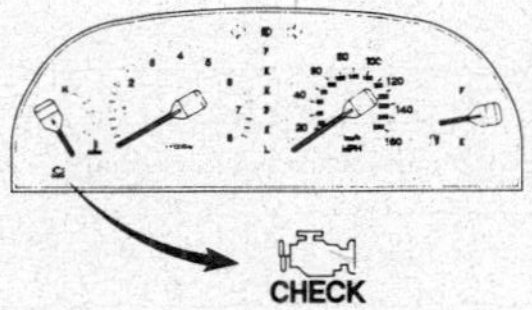

P Setting the test mode.
(1) Turn ignition switch OFF.
(2) Connect terminals TE2 and E1 of TDCL.
(3) Turn ignition switch ON.
(For checking terminal A/C, start the engine.)
(4) Connect terminlas TE1 and E1 of TDCL.

C Check if code "51" is output by the "CHECK" engine warning light.

OK

	Condition	Code
Neutral Start Switch (NSW)	P or N range	Normal*
	R,D,2 or L range	51*
Throttle Position Sensor (IDL1)	Accelerator pedal released	Normal*
	Accelerator pedal depressed	51*
A/C Switch (A/C)	A/C SW ON	51
	A/C SW OFF	Normal

*: Before the STA signal is input (ST is not ON), diagnostic code 43 is also output.

Hint Diag. code 42 is output with vehicle speed 5 km/h (3 mph) or below

NG IDL1 ..... Go to step 2 NSW
A/C ...... Go to step 3

OK Proceed to next circuit inspection shown on matrix chart

2 Check throttle position sensor.

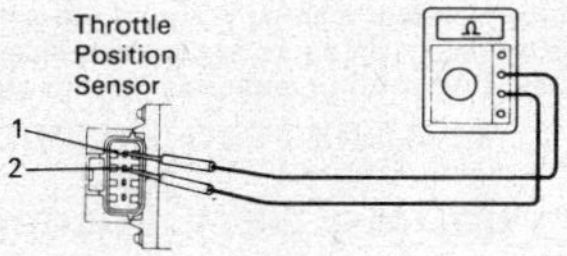

P Disconnect throttle position sensor connector.

C Measure resistance between terminals 2 and 1 of throttle position sensor connector.

OK

Throttle Valve	Resistance
Fully closed	Less than 0.5 kΩ
Opened	1 MΩ or higher

NG Adjust or replace throttle position sensor

OK Check and repair harness or connector between engine (& ECT) ECU and throttle position sensor.

3 **Disconnect A/C control assembly connector, check voltage between terminal MGC of A/C control assembly connector and body ground.**

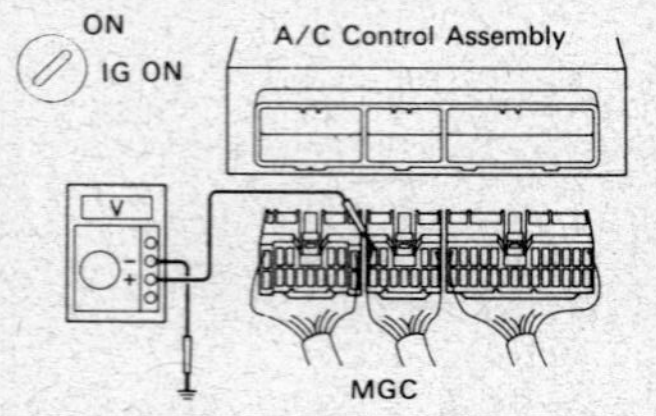

P (1) Remove air conditioner control assembly.
(2) Disconnect air conditioner control assembly connector.
(3) Turn ignition switch on.

C Measure voltage between terminal MGC of air conditioner control assembly connector and body ground.

OK **Voltage: 10 – 14 V**

OK → **Check and replace A/C control assembly.**

NG ↓

4 **Check for open and short in harness and connector between engine (& ECT) ECU and A/C control assembly**

NG → **Repair or replace harness or connector.**

OK ↓

Check and replace engine (& ECT) ECU.

Neutral Start Switch Circuit (Only vehicles with ECT.)

CIRCUIT DESCRIPTION

The neutral start switch goes on when the shift lever is in the N or P shift position. When it goes on the terminal NSW of the ECU is grounded to body ground via the starter relay and theft deterrent ECU, thus the terminal NSW voltage becomes 0V. When the shift lever is in the D, 2, L or R position, the neutral start switch goes off, so the voltage of ECU terminal NSW becomes battery voltage, the voltage of the ECU internal power source.
If the shift lever is moved from the N range to the D range, this signal is used for air-fuel ratio correction and for idle speed control (estimated control), etc.
When the neutral start switch is off, code "51" is output in the test mode diagnosis. (This is not abnormal.)

DIAGNOSTIC CHART

HINT: This diagnosis chart is based on premise that the engine is cranked normally. If the engine is not cranked, proceed to the matrix chart of problem symptoms

1 Check output condition of diag. code 51.
- OK → Proceed to next circuit inspection shown on matrix chart
- NG Type I → 2
- NG Type II → 3

2 Check for open in harness and connector between neutral start switch and ECU.
- OK → Check and replace ECU.
- NG → Repair or replace harness or connector.

3 Check neutral start switch.
- NG → Replace neutral start switch.
- OK → Check and replace ECU.

WIRING DIAGRAM

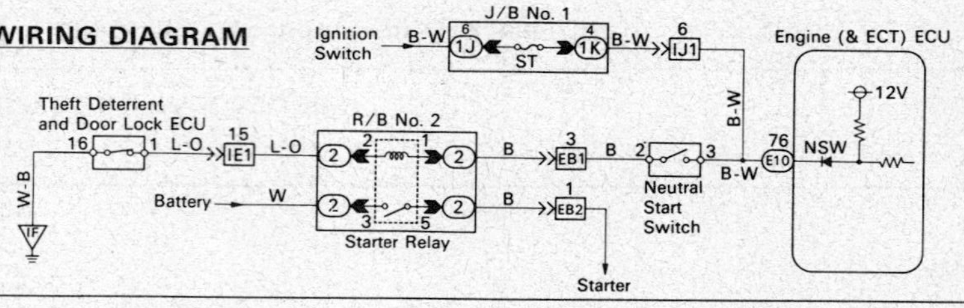

INSPECTION PROCEDURE

1 Check output condition of diag. code 51.

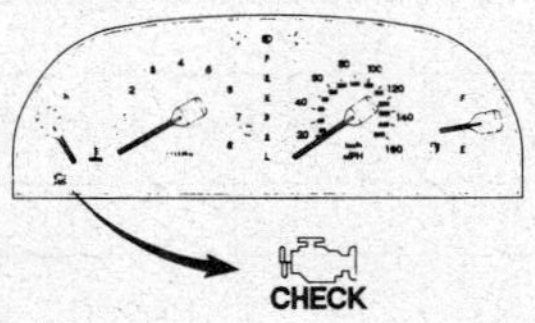

P (1) Connect terminals TE2 and E1 of TDCL.
(2) Turn ignition switch on.
(3) Crank the engine.
(4) Connect terminals TE1 and E1 of TDCL.

C Check if diagnostic code "51" is output when the shift lever is in the P and D shift positions.

Result

Shift Position	Result		
	OK	NG Type I	NG Type II
"P"	Normal Code	Code 51	Normal Code
"D"	Code 51	Code 51	Normal Code

NG Type I → Go to step 2.

NG Type II → Go to step 3.

OK ↓

Proceed to next circuit inspection shown on matrix chart

2 Check for open in harness and connector between engine (& ECT) ECU and neutral start switch

OK → Check and replace engine (& ECT) ECU.

NG ↓

Repair or replace harness or connector.

3 Check neutral start switch

NG → Replace neutral start switch.

OK ↓

Check and replace engine (& ECT) ECU.

ECU Power Source Circuit

CIRCUIT DESCRIPTION

When the ignition switch is turned on, battery voltage is applied to the terminal IGSW of the ECU, and the main relay control circuit in the ECU sends a signal to the terminal M-REL of the ECU, switching on the main relay. This signal causes current to flow to the coil, closing the contacts of the main relay and supplying power to the terminals +B and +B1 of the ECU.
If the ignition switch is turned off, the ECU continues to switch on the main relay for a maximum of 2 seconds for the initial setting of the ISC valve.

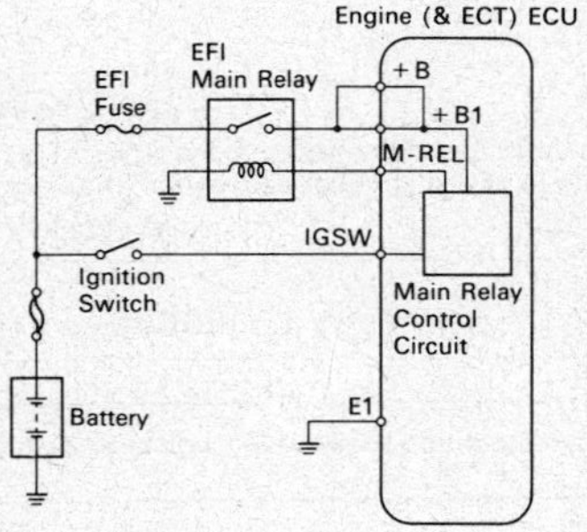

DIAGNOSTIC CHART

See next page for the DIAGNOSTIC CHART.

WIRING DIAGRAM

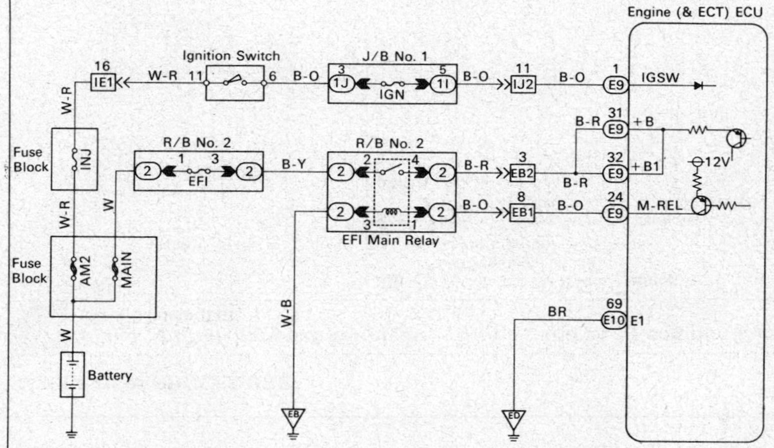

DIAGNOSTIC CHART

Step	Check	Result	Action
1	Check voltage of ECU power source.	OK	Proceed to next circuit inspection shown on matrix chart
		NG	Go to 2
2	Check continuity between terminals E1 and body ground.	NG	Repair or replace harness or connector.
		OK	Go to 3
3	Check voltage of terminal IGSW.	OK	Go to 6
		NG	Go to 4
4	Check IGN fuse.	NG	Check for short in all the harness and components connected to IGN fuse.
		OK	Go to 5
5	Check ignition switch.	NG	Replace ignition switch.
		OK	Check for open in harness and connector between battery and ECU.
6	Check voltage of terminal M-REL.	NG	Check and replace ECU.
		OK	Go to 7
7	Check EFI fuse.	NG	Check for short in all the harness and components connected to EFI fuse.
		OK	Go to 8
8	Check EFI main relay.	NG	Replace EFI main relay.
		OK	Go to 9
9	Check for open and short in harness and connector between terminal M-REL and body ground.	NG	Repair or replace harness or connector.
		OK	Check and repair harness or connector between EFI fuse and battery.

INSPECTION PROCEDURE

1 Check voltage between terminals +B, +B1 and E1 of engine (& ECT) ECU connector.

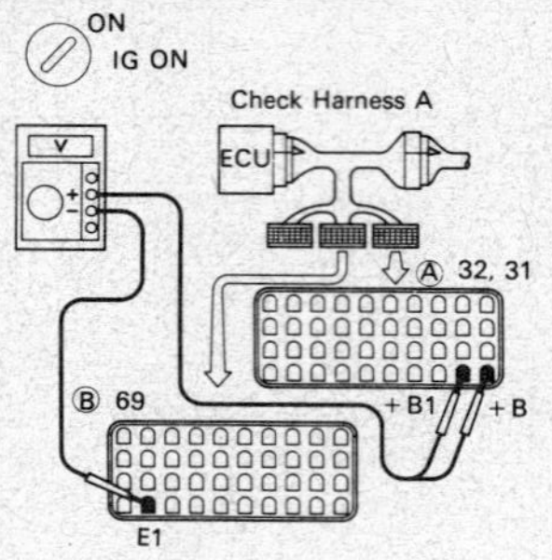

P (1) Connect the Check Harness A.
(See page TR-34)
(2) Turn ignition switch on.

C Measure voltage between terminals +B, +B1 and E1 of engine (& ECT) ECU connector.

OK Voltage: 10 – 14 V

OK Proceed to next circuit inspection shown on matrix chart

NG

2 Check for open in harness and connector between terminal E1 of engine (& ECT) ECU and body ground

NG Repair or replace harness or connector.

OK

3 Check voltage between terminal IGSW of engine (& ECT) ECU connector and body ground.

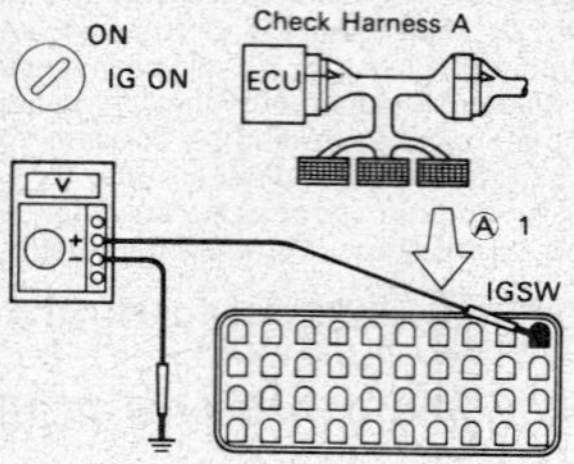

P Turn ignition switch on.

C Measure voltage between terminal IGSW of engine (& ECT) ECU connector and body ground.

OK Voltage: 10 – 14 V

OK Go to step 6.

NG

4 Check IGN fuse.

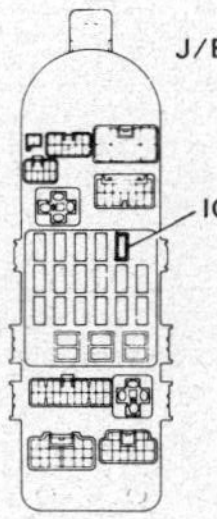

- **P** Remove IGN fuse from J/B No.1.
- **C** Check continuity of IGN fuse.
- **OK** **Continuity**

NG › **Check for short in all the harness and components connected to IGN fuse**

OK ↓

5 Check ignition switch.

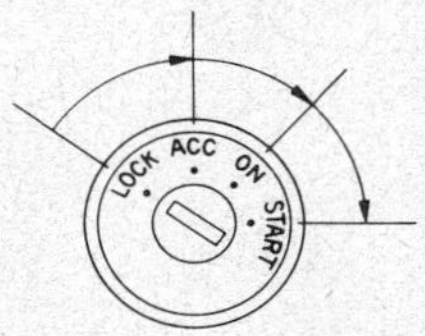

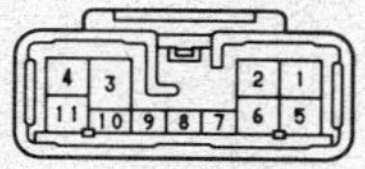

- **P** Remove under cover and finish panel.
- **C** Check continuity between terminals.
- **OK**

○—○ continuity

Terminal / Switch position	2	3	4	5	6	11
LOCK						
ACC		○—	—○			
ON	○—	—○—	—○		○—	—○
START	○—	——	—○	○—	—○—	—○

NG › **Replace ignition switch.**

OK ↓

Check and repair harness and connector between battery and ignition switch, ignition switch and engine (& ECT) ECU.

6 Check voltage between terminal M-REL of engine (& ECT) ECU connector and body ground.

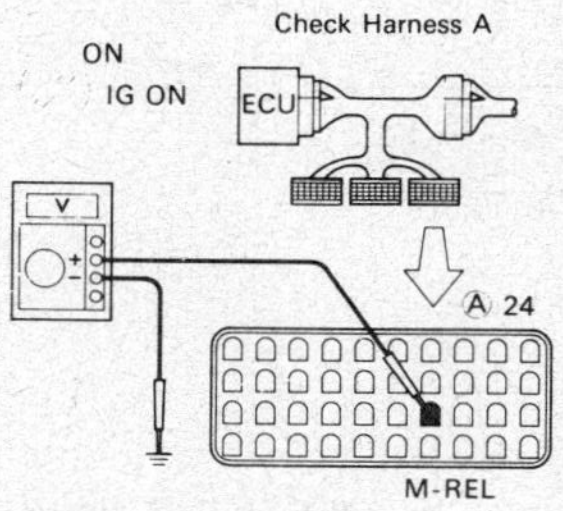

- **P** Turn ignition switch on.
- **C** Measure voltage between terminal M-REL of engine (& ECT) ECU connector and body ground.
- **OK** **Voltage: 10 – 14 V**

NG › **Check and replace engine (& ECT) ECU.**

OK ↓

7 Check EFI fuse.

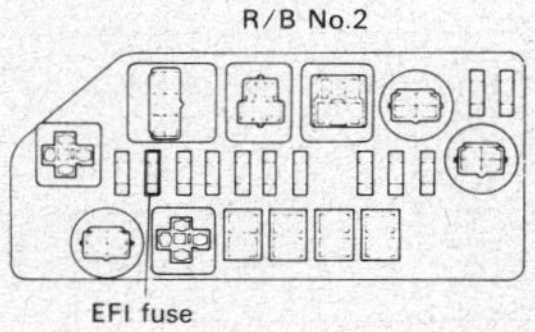

- **P** Remove EFI fuse from R/B No.2.
- **C** Check continuity of EFI fuse.
- **OK** **Continuity**

NG › **Check for short in all the harness and components connected to EFI fuse**

OK ↓

8 **Check EFI main relay.**

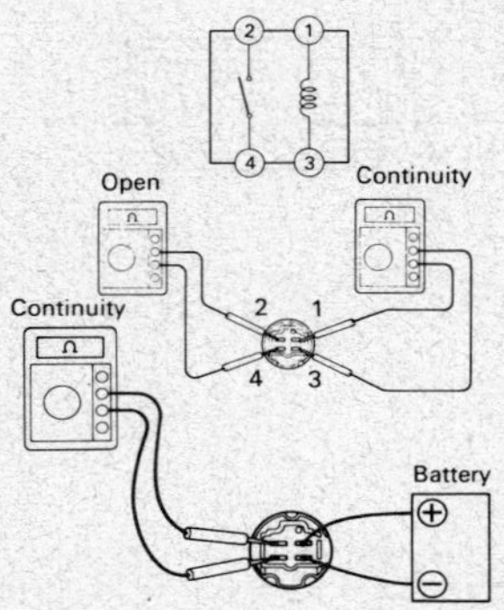

P Remove EFI main relay.

C Check continuity between terminals of EFI main relay shown below.

OK

Terminals 2 and 4	Open
Terminals 1 and 3	Continuity (Reference value 72 Ω)

C (1) Apply battery voltage between terminals 1 and 3.
(2) Check continuity between terminals 2 and 4.

OK

Terminals 2 and 4	Continuity

NG → **Replace EFI main relay.**

OK

9 **Check for open and short in harness and connector between terminal M-REL of engine (& ECT) ECU and body ground**

NG → **Repair or replace harness or connector.**

OK

Check and repair harness or connector between EFI fuse and battery.

Back Up Power Source Circuit

CIRCUIT DESCRIPTION

Battery voltage is supplied to terminal BATT of the ECU even when the ignition switch is off for use by the diagnostic code memory and air-fuel ratio adaptive control value memory, etc.

DIAGNOSTIC CHART

Step	Check	Result	Action
1	Check EFI fuse.	NG	Check for short in all the harness and components connected to EFI fuse.
		OK	↓
2	Check voltage of terminal BATT.	NG	Check and repair harness or connector between battery, EFI fuse and ECU.
		OK	↓
3	Check operation for the back up.	NG	Check and replace ECU.
		OK	Proceed to next circuit inspection shown on matrix chart

WIRING DIAGRAM

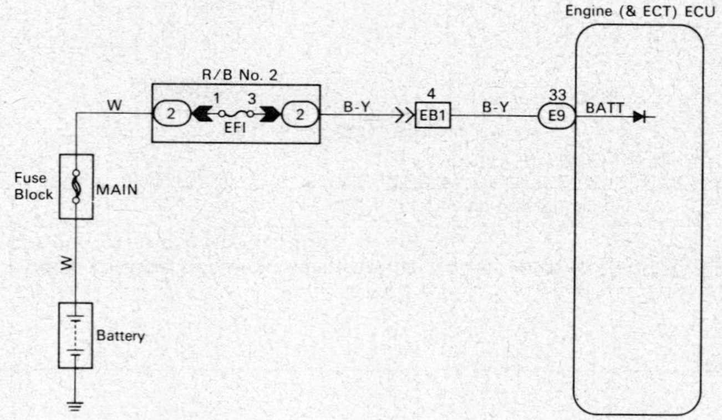

INSPECTION PROCEDURE

1 Check EFI fuse.

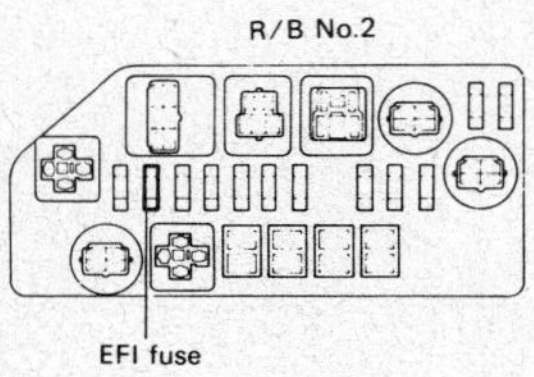

- **P** Remove EFI fuse from R/B No.2.
- **C** Check continuity of EFI fuse.
- **OK** **Continuity**

NG → **Check for short in all the harness and components connected to EFI fuse**

OK ↓

2 Check voltage between terminal BATT of engine (& ECT) ECU connector and body ground.

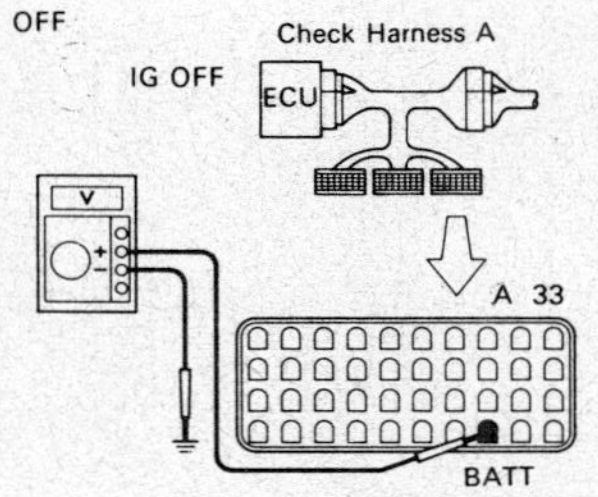

- **P** Connect the Check Harness A.
- **C** Measure voltage between terminal BATT of engine (& ECT) ECU connector and body ground.
- **OK** **Voltage: 10 – 14 V**

NG → **Check and repair harness or connector between engine (& ECT) ECU and EFI fuse, EFI fuse and battery.**

OK ↓

3 Are the diagnostic codes still in the memory when the ignition switch is turned OFF?

NO → **Check and replace engine (& ECT) ECU.**

YES ↓

Proceed to next circuit inspection shown on matrix chart

Injector Circuit

CIRCUIT DESCRIPTION

The injectors are provided to the intake manifold. They inject fuel into the cylinders based on the signals from the engine (& ECT) ECU.

Reference — INSPECTION USING OSCILLOSCOPE

- With the engine idling measure between terminals #10 ~ 60 and E01 of engine (& ECT) ECU.

HINT: The correct waveform appears as shown in the illustration on the below.

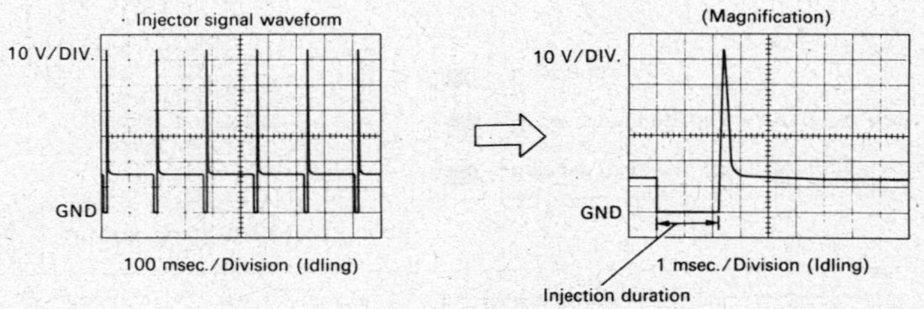

DIAGNOSTIC CHART

1. Check voltage of terminals #10 ~ 60. — OK → go to 3; NG ↓
2. Check AM2 and INJ fuses. — NG → Check for short in all the harness and components connected to AM2 and INJ fuses. OK ↓
   - Check for open in harness and connector between ECU and battery.
3. Check continuity between terminals E01, E02 and body ground. — NG → Repair or replace harness or connector. OK ↓
4. Check operation for injectors. — NG → Replace injector. OK ↓
   - Check and replace ECU.

WIRING DIAGRAM

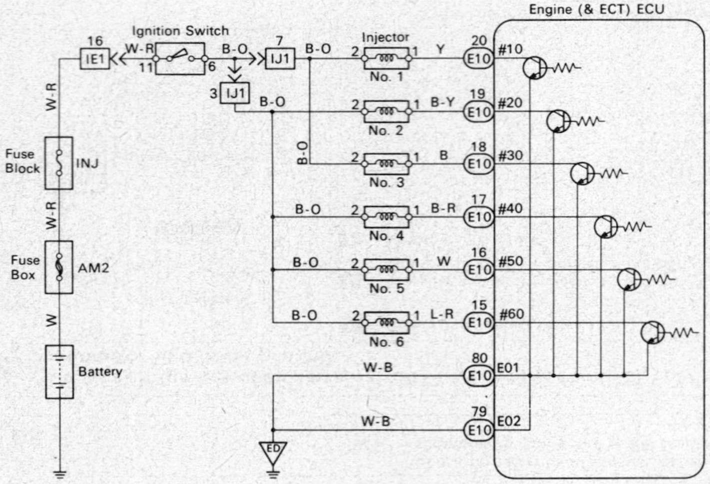

INSPECTION PROCEDURE

1 Check voltage between terminals #10 ~ 60 of engine (& ECT) ECU and body ground.

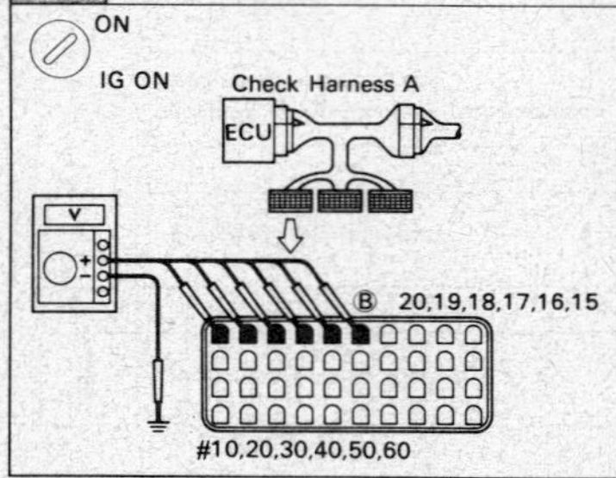

P (1) Connect the Check Harness A.

(2) Turn ignition switch on.

C Measure voltage between terminals #10 ~ 60 of engine (& ECT) ECU and body ground.

OK **Voltage: 10 – 14 V**

OK → **Go to step 3.**

NG ↓

2 Check AM2 and INJ fuses.

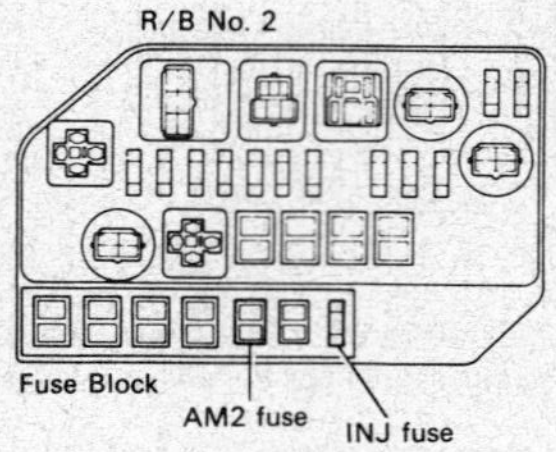

P Remove AM2 and INJ fuses from fuse block.

C Check continuity of AM2 and INJ fuses.

OK **Continuity**

NG → **Check for short in all the harness and components connected to AM2 and INJ fuses.**

OK ↓

Check for open in harness and connector between engine (& ECT) ECU and battery.

3 **Check for open in harness and connector between terminals E01, E02 of engine (& ECT) ECU connector and body ground**

OK | NG: **Repair or replace harness or connector.**

4 **Check injectors.**

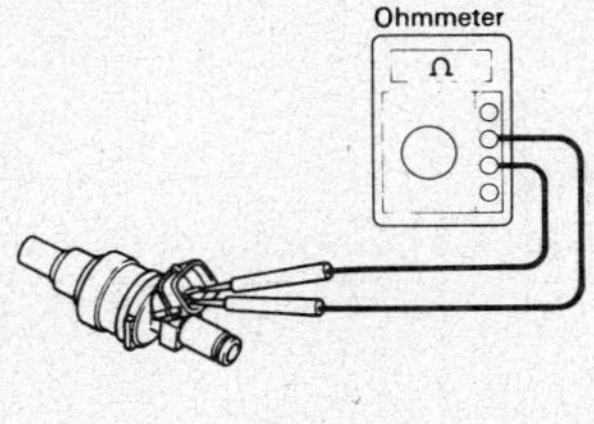

P Disconnect injector connector

C Measure resistance of injector.

OK **Resistance: 13.4 – 14.2 Ω at 20°C (68°F)**

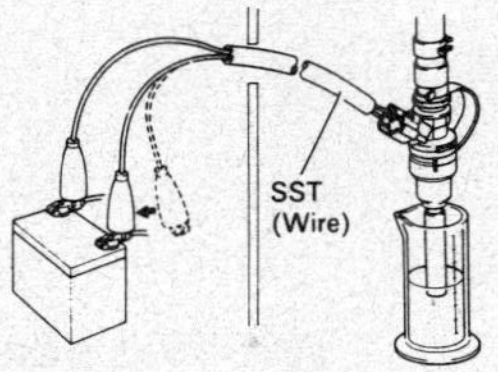

C Check injection volume of injector

OK
- **Injection volume**
  70 – 88 cc/15 sec. (4.3 – 5.4 cu in.)
  Difference between each injector:
  Less than 9 cc (0.5 cu in.)
- **Leakage**
  Fuel drop: One drop or less per minute.

OK | NG: **Replace injector.**

Check and replace engine (& ECT) ECU.

ISC Valve Circuit

CIRCUIT DESCRIPTION

The ISC valve is provided on the intake air chamber and intake air bypassing the throttle valve is directed to the ISC valve through a passage.
A step motor is built into the ISC valve. It consists of four coils, the magnetic rotor, valve shaft and valve. When current flows to the coils due to signals from the ECU, the rotor turns and moves the valve shaft forward or backward, changing the clearance between the valve and the valve seat.
In this way the intake air volume bypassing the throttle valve is regulated, controlling the engine speed.
There are 125 possible positions to which the valve can be opened.

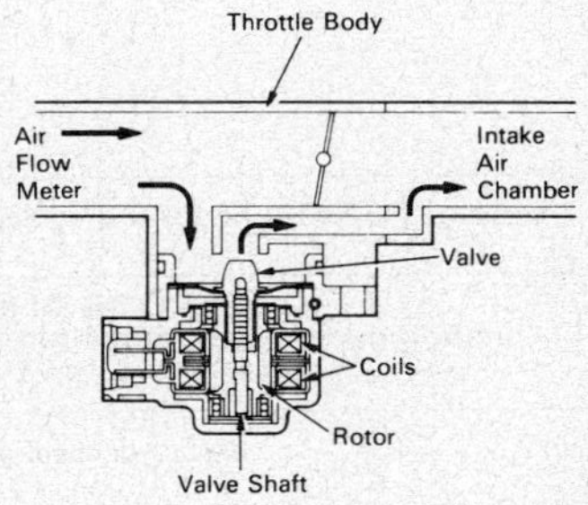

Reference **INSPECTION USING OSCILLOSCOPE**

A/C switch OFF → ON
(ISC valve open)

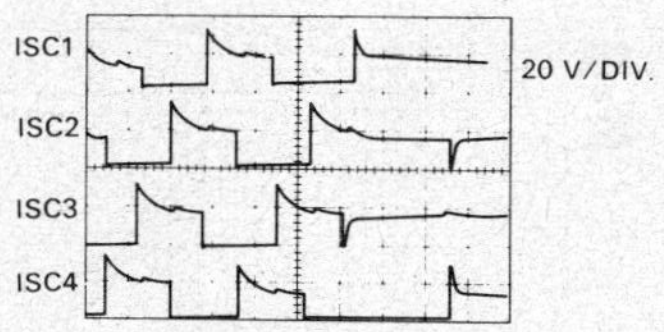

A/C switch ON → OFF
(ISC valve close)

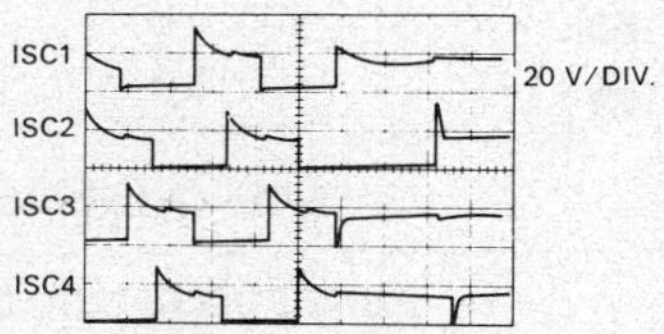

- With the engine idling measure between terminals ISC1, ISC2, ISC3, ISC4 and E01 of engine (& ECT) ECU when A/C switch ON or OFF.

HINT:
The correct waveform appears as shown in the illustration on the left.

DIAGNOSTIC CHART

1 Check operation for ISC valve. — NG → Replace ISC valve.

↓ OK

2 Check harness and connector between EFI main relay, ISC valve and ECU. — NG → Repair or replace harness or connector.

↓ OK

Proceed to next circuit inspection shown on matrix chart

WIRING DIAGRAM

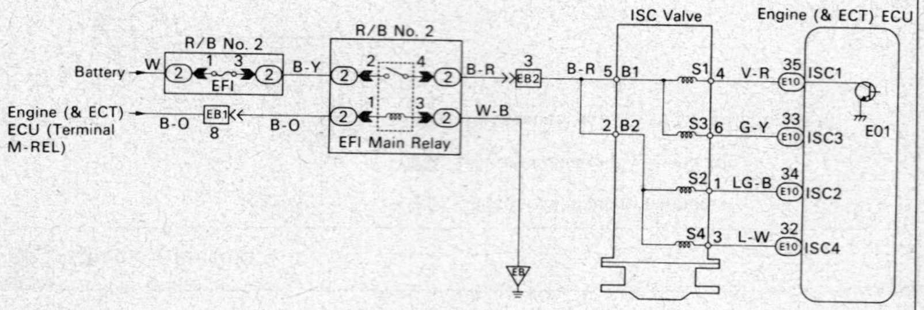

INSPECTION PROCEDURE

1 Check ISC valve.

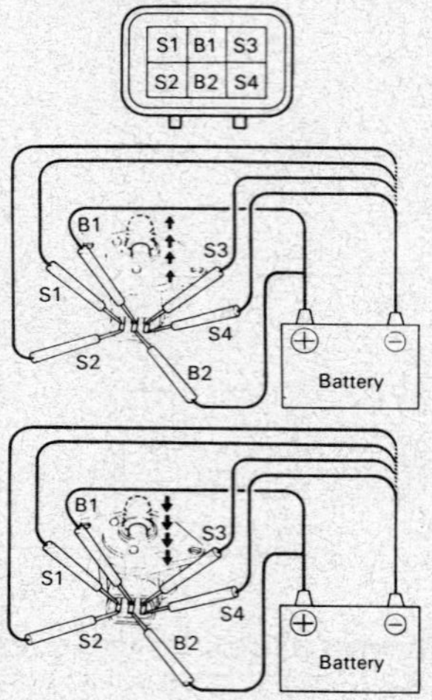

P Disconnect ISC valve connector.

C Measure resistance between terminals shown below.

OK

Terminal	Resistance
B1 – S1	10 Ω – 30 Ω
B1 – S3	10 Ω – 30 Ω
B2 – S2	10 Ω – 30 Ω
B2 – S4	10 Ω – 30 Ω

P Remove ISC Valve.

C (1) Connect the battery positive lead to terminals B1 and B2, and the negative lead to terminals S1-S2-S3-S4 in that order.
(2) Connect the battery positive lead to terminals B1 and B2, and the negative lead to terminals S4-S3-S2-S1 in that order.

OK **(1) The valve moves in the closing direction.**
(2) The valve moves in the opening direction.

NG → **Replace ISC valve.**

OK ↓

2 Check for open and short in harness and connector between EFI main relay and ISC valve, ISC valve and engine (& ECT) ECU

NG → **Repair or replace harness or connector.**

OK ↓

Proceed to next circuit inspection shown on matrix chart

IACV Control VSV Circuit

CIRCUIT DESCRIPTION

The circuit opens and closes the IACV (Intake Air Control Valve) in response to the engine load in order to increase the intake efficiency (ACIS: Acoustic Control Induction System).
When the engine speed is 4,500 rpm or less and throttle valve opening angle is 30° or more, or engine speed is 4,500 rpm or more and throttle valve opening angle is 30° or less, the engine (& ECT) ECU turns the VSV ON and closes the IACV. At all other times, the VSV is OFF, so the IACV is open.

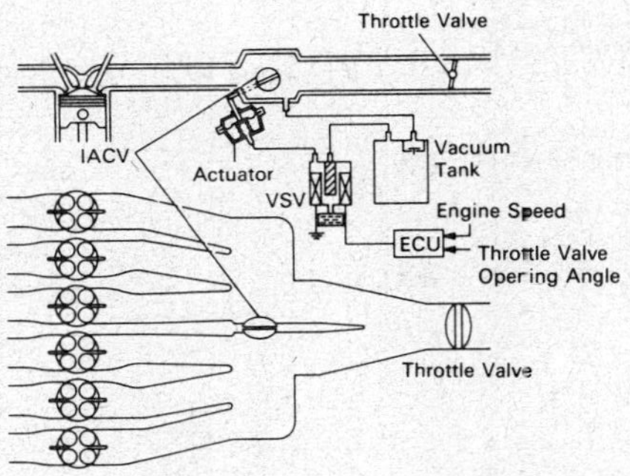

VSV	IACV
ON	closed
OFF	open

DIAGNOSTIC CHART

1. Check operation of the IACV control VSV. — NG → Replace IACV control VSV.
   - OK ↓
2. Check voltage of IACV control VSV power source. — OK → Check for vacuum tank
   - NG ↓
3. Check for open and short in IACV control VSV circuit. — NG → Repair or replace harness or connector.
   - OK ↓

Check and replace ECU.

WIRING DIAGRAM

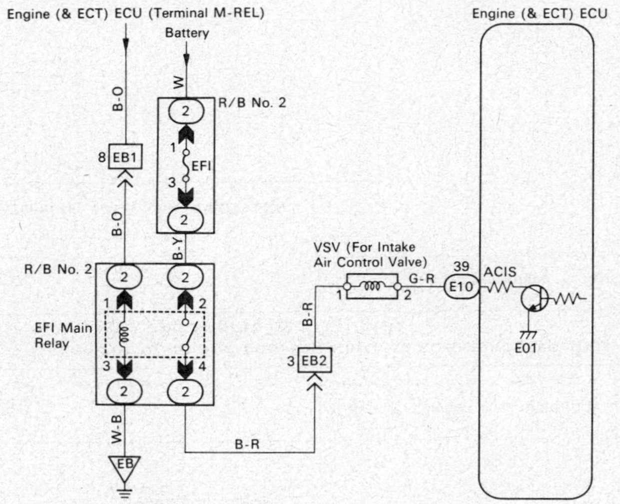

INSPECTION PROCEDURE

1 Check IACV control VSV.

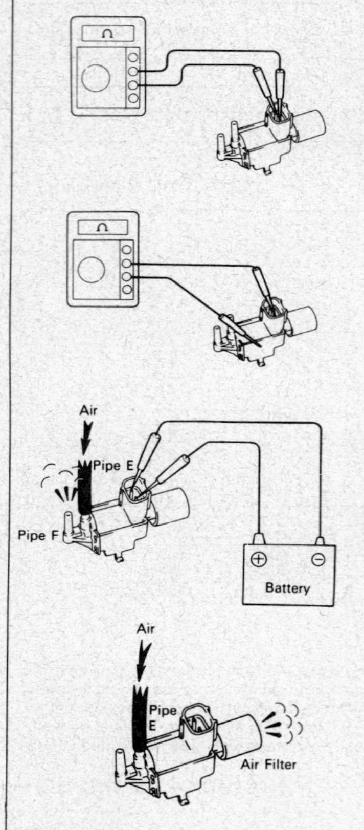

P (1) Remove IACV control VSV.
(2) Disconnect IACV control VSV connector.

C (1) Measure resistance between terminals.
(2) Measure resistance between each terminal and the body.

OK (1) **Resistance: 38.5 – 44.5 Ω at 20°C (68°F)**
(2) **Resistance: 1MΩ or higher**

C Check operation of IACV control VSV when battery voltage is applied to the terminals of IACV control VSV connector or not.

OK **Battery voltage is applied:**
The air from pipe E is flowing out through pipe F.

Battery voltage is not applied:
The air from pipe E is flowing out through the air filter.

NG — **Replace IACV control VSV.**

OK

2 Check voltage between terminal ACIS of engine (& ECT) ECU connector and body ground.

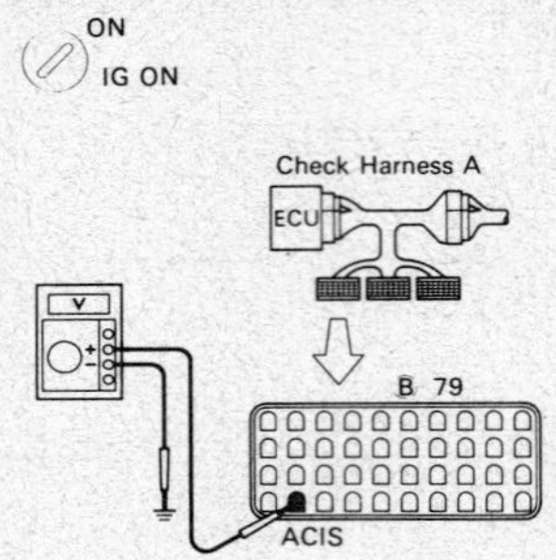

P (1) Connect the Check Harness A.
(2) Turn ignition switch on.

C Measure voltage between terminal ACIS of engine (& ECT) ECU connector and body ground.

OK **Voltage: 10 – 14 V**

OK — **Check for vacuum tank.**

NG

3 Check for open and short in harness and connector between EFI main relay and engine (& ECT) ECU

NG — **Repair or replace harness or connector.**

OK

Check and replace engine (& ECT) ECU.

TE1, TE2 Terminal Circuit

CIRCUIT DESCRIPTION

Terminal TE1 is located in the check connector in the engine compartment and terminals TE1 and TE2 are located in the TDCL in the cabin. When these terminals are connected with the E1 terminal, diagnostic codes in normal mode or test mode can be read from the "CHECK" engine warning light on the combination meter.

DIAGNOSTIC CHART

HINT: If terminals TE1 and TE2 are connected with terminal E1, diagnostic code is not output or test mode is not activated.
Even though terminal TE1 is not connected with terminal E1, the "CHECK" engine warning light blinks.
For the above phenomenon, the likely cause is an open or short in the wire harness, or malfunction inside the ECU.

Step	Check	Result	Action
1	Check voltage of terminals TE1, TE2.	OK	Check and replace ECU.
	↓ NG		
2	Check continuity between terminal E1 and body ground.	NG	Repair or replace harness or connector.
	↓ OK		
3	Check for open and short in harness and connector between check connector, TDCL and ECU.	NG	Repair or replace harness or connector.
	↓ OK		
	Check and replace ECU.		

WIRING DIAGRAM

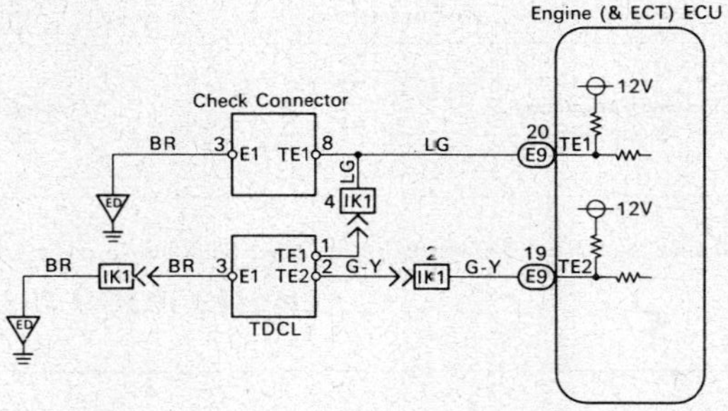

INSPECTION PROCEDURE

1 Check voltage between terminals TE1, TE2 and E1 of check connector, TDCL.

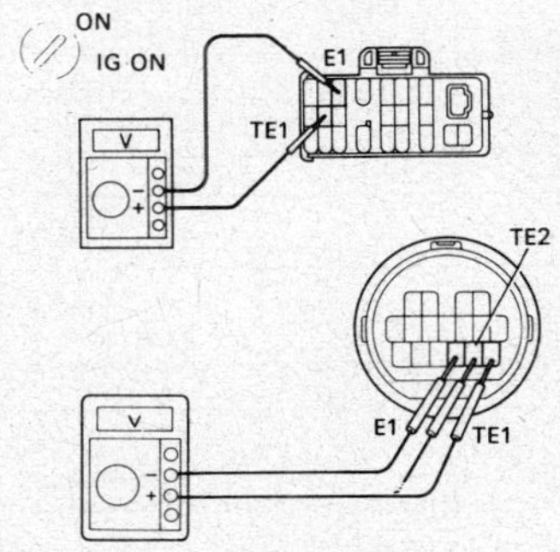

P Turn ignition switch on.
C Measure voltage between terminals TE1, TE2 and E1 of check connector, TDCL.

OK **Voltage: 10 – 14 V**

OK ▷ **Check and replace engine (& ECT) ECU.**

NG ↓

2 Check continuity between terminal E1 of check connector, TDCL and body ground.

NG ▷ **Repair or replace harness or connector.**

OK ↓

3 Check for open and short in harness and connector between engine (& ECT) ECU and check connector, TDCL

NG ▷ **Repair or replace harness or connector.**

OK ↓

Check and replace engine (& ECT) ECU.

DIAGNOSTIC CODE CHART

HINT: Parameters listed in the chart may not be exactly same as your reading due to type of the instruments or other factors.

If a malfunction code is displayed during the diagnostic code check in test mode, check the circuit for that code listed in the table below (Proceed to the page given for that circuit).

Code No.	Number of Check Engine Blinks	Circuit	Diagnostic Code Detecting Condition	Trouble Area	"CHECK" Engine Warning Light*1		Memory*2	
					Normal Mode	Test Mode		
-		Normal	No code is recorded.	——	——	——	——	——
12		RPM Signal	No "NE" or "G1" and "G2" signal to ECU within 2 sec. after cranking.	● Open or short in engine speed sensor, No. 1, No. 2 cam position sensor circuit. ● Engine speed sensor ● No. 1, No. 2 cam position sensor ● Starter ● ECU	ON	N.A.	○	
13		RPM Signal	No NE signal to ECU for 0.1 sec. or more at 1,000 rpm or more.	● Open or short in engine speed sensor circuit. ● Engine speed sensor ● ECU	ON	N.A.	○	
			No 12 pulses of NE to ECU during the interval between G1 and G2 pulses.	● Open or short in engine speed sensor circuit. ● Mechanical system malfunction (skipping teeth of timing belt, belt stretched) ● Engine speed sensor ● ECU	N.A.	ON		
			Deviation in G (G1, G2) and NE signal continues for 1 sec. during idling (throttle fully closed) after engine warmed up.	● Mechanical system malfunction (skipping teeth of timing belt, belt stretched) ● No. 1, No. 2 cam position sensor ● ECU	ON	N.A.		
14		Ignition No. 1 Signal	No IGF1 signal to ECU for 8-11 consecutive IGT1 signal.	● Open or short in IGF1 circuit from igniter No. 1 – ECU. ● Igniter No. 1 ● ECU	ON	N.A.	○	
15		Ignition No. 2 Signal	No IGF2 signal to ECU for 8-11 consecutive IGT2 signal.	● Open or short in IGF2 circuit from igniter No. 2 – ECU. ● Igniter No. 2 ● ECU	ON	N.A.	○	
16		ECT Control Signal	Fault in communications between the engine CPU and ECT CPU in the ECU.	● ECU	ON	N.A.	X	
21		Main Oxygen Sensor Signal (on left bank)	(1) Open or short in heater circuit of main oxygen sensor for 0.5 sec. or more.	● Open or short in heater circuit of main oxygen sensor. ● Main oxygen sensor heater ● ECU	ON	N.A.	○	
			(2) Main oxygen sensor signal voltage is reduced to between 0.35 V and 0.70 V for 60 sec. under conditions (a) ~ (d). (2 trip detection logic) *4 (a) Coolant temp. : Between 70°C (158°F) and 95°C (203°F). (b) Engine speed : 1,500 rpm or more. (c) Load driving (EX. ECT in 4th speed, A/C ON, Flat road, 50 mph <80 km/h>) (d) Main oxygen sensor signal voltage : Alternating above and below 0.45 V.	● Main oxygen sensor circuit ● Main oxygen sensor	ON	ON		

4.0L ENGINE—DIAGNOSTIC CODES—SC400

DIAGNOSTIC CODE CHART (Cont'd)

Code No.	Number of Check Engine Blinks	Circuit	Diagnostic Code Detecting Condition	Trouble Area	"CHECK" Engine Warning Light*1 Normal Mode	"CHECK" Engine Warning Light*1 Test Mode	Memory*2	
22		Water Temp. Sensor Signal	Open or short in water temp. sensor circuit for 0.5 sec. or more.	● Open or short in water temp. sensor circuit. ● Water temp. sensor ● ECU	ON	ON	○	
24		Intake Air Temp. Sensor Signal	Open or short in intake air temp. sensor circuit for 0.5 sec. or more.	● Open or short in intake air temp. sensor circuit. ● Intake air temp. sensor ● ECU	OFF ON*3	ON	○	
25 (26)		Air-Fuel Ratio Lean (Rich) Malfunction	(1) Main oxygen sensor voltage is 0.45 V or less (lean) for 90 sec. under conditions (a) and (b). (2 trip detection logic)*4 (a) Coolant temp. : 60°C (140°F) or more. (b) Engine speed : 1,500 rpm or more.	● Open or short in main oxygen sensor circuit. ● Main oxygen sensor ● Ignition system ● Water temp. sensor	ON	ON	○	
			(2)*3 Main oxygen sensor voltage is alternating above and below 0.45 V at 5 times par second or more under conditions (a) and (b). (2 trip detection logic)*4 (a) Engine speed : Idling (b) Coolant temp. : Between 60°C (140°F) and 95°C (203°F).	● Open or short in injector circuit. ● Fuel line pressure (injector blockage) ● Air flow meter (air intake) ● Engine ground bolt loose ● Foreign object caught in valve				
			(3)*3 Difference of air-fuel ratio feedback compensation value between right and left banks is more than 10 percentage for 30 sec. or more under conditions (a) and (b). (2 trip detection logic)*4 (a) Engine speed : 2,000 rpm or more. (b) Coolant temp. : Between 60°C (140°F) and 95°C (203°F).	● Fuel line pressure (injector leak, blockage) ● Mechanical system malfunction (skipping teeth of timing belt) ● Ignition system				
27		Sub-Oxygen Sensor Signal (on left bank)	(1) Open or short in heater circuit of sub-oxygen sensor for 0.5 sec. or more.	● Open or short in heater circuit of sub-oxygen sensor. ● Sub-oxygen sensor heater. ● ECU	ON	N.A.	○	
			(2) Main oxygen sensor signal is 0.45 V or more and sub-oxygen sensor signal is 0.45 V or less under conditions (a) and (b). (2 trip detection logic)*4 (a) Coolant temp. : 80°C (176°F) or more. (b) Accel-pedal : Full depressed for 2 sec. or more.	● Open or short in sub-oxygen sensor circuit. ● Sub-oxygen sensor ● ECU	ON	ON		
28		Main Oxygen Sensor Signal (on right back)	Same as (1) of Code No. 21	● Same as (1) of Code No. 21	ON	N.A.	○	
			Same as (2) of Code No. 21	● Same as (2) of Code No. 21	ON	ON		
29		Sub-Oxygen Sensor Signal (on right bank)	Same as (1) of Code No. 27	● Same as (1) of Code No. 27	ON	N.A.	○	
			Same as (2) of Code No. 27	● Same as (2) of Code No. 27	ON	ON		

DIAGNOSTIC CODE CHART (Cont'd)

Code No.	Number of Check Engine Blinks	Circuit	Diagnostic Code Detecting Condition	Trouble Area	"CHECK" Engine Warning Light*1 Normal Mode	"CHECK" Engine Warning Light*1 Test Mode	Memory*2	
31	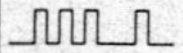	Air Flow Meter Signal	All conditions below are detected. (a) No air-flow meter signal to ECU for 2 sec. when engine speed is above 300 rpm. (b) Engine stall.	● Open or short in air flow meter circuit. ● Air flow meter ● ECU	ON	N.A.	○	
35	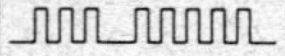	HAC Sensor Signal	Open or short in HAC sensor circuit for 0.5 sec. or more.	● ECU	ON	ON	○	
41	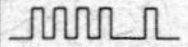	Throttle Position Sensor Signal	(1) Open or short in throttle position sensor circuit (VTA1) for 0.5 sec. or more.	● Open or short in throttle position sensor circuit.	ON	ON	○	
			(2) IDL1 contact is ON and VTA1 output exceeds 1.5 V for 0.5 sec. or more.	● Throttle position sensor ● ECU				
42	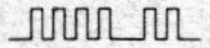	Vehicle Speed Sensor Signal	All conditions below are detected continuously for 8 sec. or more. (a) Vehicle speed signal : 0 mph (b) Engine speed : 1,500 rpm or more (c) Neutral start switch (NSW) : OFF	● No. 1 speed sensor ● Combination meter ● Open or short in No. 1 speed sensor circuit. ● ECU	ON	ON	○	
43	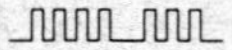	Starter Signal	No starter signal to ECU.	● Open or short in starter signal circuit. ● Open or short in ignition switch or main relay circuit. ● ECU	N.A.	OFF	X	
47	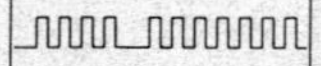	Sub-Throttle Position Sensor Signal	(1) Open or short in sub-throttle position sensor circuit (VTA2) for 0.5 sec. or more.	● Open or short in sub-throttle position sensor circuit.	OFF	ON	○	
			(2) IDL2 contact is ON and VTA2 output exceeds 1.5 V for 0.5 sec. or more.	● Sub-throttle position sensor ● ECU				
52	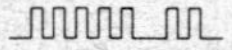	No. 1 Knock Sensor Signal	No No. 1 knock sensor signal to ECU for 3 crank revolutions with engine speed between 1,600 rpm ~ 5,200 rpm.	● Open or short in No. 1 knock sensor circuit. ● No. 1 knock sensor (looseness) ● ECU	ON	N.A.	○	
53	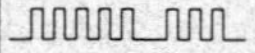	Knock Control Signal	Engine control computer (for knock control) malfunction at engine speed between 650 rpm and 5,200 rpm.	● ECU	ON	N.A.	X	
55	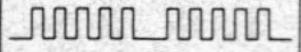	No. 2 Knock Sensor Signal	No No. 2 knock sensor signal to ECU for 3 crank revolutions with engine speed between 1,600 rpm ~ 5,200 rpm.	● Open or short in No. 2 knock sensor circuit. ● No. 2 knock sensor (looseness) ● ECU	ON	N.A.	○	

DIAGNOSTIC CODE CHART (Cont'd)

Code No.	Number of Check Engine Blinks	Circuit	Diagnostic Code Detecting Condition	Trouble Area	"CHECK" Engine Warning Light*1 Normal Mode	"CHECK" Engine Warning Light*1 Test Mode	Memory*2	See Page
71*3		EGR System Malfunction	(1) Open or short in EGR step motor circuit for 1 sec. or more.	● Open or short in EGR step motor circuit. ● ECU	ON	N.A.	○	
			(2) EGR gas temp. is 65°C (149°F) or below for 1 ~ 4 min. under conditions (a) and (b). (2 trip detection logic)*4 (a) Coolant temp. : 65°C (149°F) or more. (b) EGR operation possible (EX. ECT in 3rd speed, A/C ON, 60 mph <96 km/h>, Flat road).	● Open in EGR gas temp. sensor circuit. ● ECU	ON	ON		
78		Fuel Pump Control Signal	(1) Open or short in fuel pump circuit for 1 sec. or more with engine speed 1,000 rpm or less. (2 trip detection logic)*4 (2) Open in input circuit of fuel pump ECU (FPC) with engine speed 1,000 rpm or less. (2 trip detection logic)*4 (3) Open or short in diagnostic signal line (DI) of fuel pump ECU with engine speed 1,000 rpm or less. (2 trip detection logic)*4	● Open or short in fuel pump ECU circuit. ● Fuel pump ECU ● Engine & ECT ECU power source circuit. ● Fuel pump ● Engine & ECT ECU	ON	ON	○	
51		Switch Condition Signal	(1) 3 sec. or more after engine starts, idle switch OFF (IDL1). (2) Neutral start switch OFF (NSW). (Shift position in "R", "D", "2" or "1" ranges). (3) A/C switch ON.	● A/C switch circuit ● Throttle position sensor IDL circuit ● Neutral start switch circuit. ● Accelerator pedal and cable ● ECU	N.A.	OFF	X	

*1: "ON" displayed in the diagnosis mode column indicates that the "CHECK" Engine Warning Light is lighted up when a malfunction is detected. "OFF" indicates that the "CHECK" does not light up during malfunction diagnosis, even if a malfunction is detected. "N.A." indicates that the item is not included in malfunction diagnosis.

*2: "○" in the memory column indicates that a diagnostic code is recorded in the ECU memory when a malfunction occurs. "X" indicates that a diagnostic code is not recorded in the ECU memory even if a malfunction occurs. Accordingly, output of diagnostic results in normal or test mode is performed with the IG switch ON.

*3: Only for USA specification vehicles.

*4: This indicates items for which "2 trip detection logic" is used. With this logic, when a logic malfunction is first detected, the malfunction is temporarily stored in the ECU memory. If the same case is detected again during the second drive test, this second detection causes the "CHECK" Engine Warning Light to light up.
The 2 trip repeats the same mode a 2nd time. (However, the IG switch must be turned OFF between the 1st trip and 2nd trip).
In the Test Mode, the "CHECK" Engine Warning Light lights up the 1st trip a malfunction is detected.

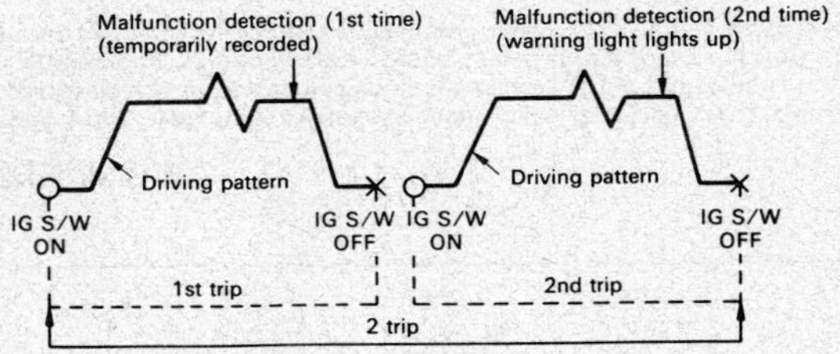

FAIL-SAFE CHART

If any of the following codes is recorded, the ECU enters fail-safe mode.

Code No.	Fail-Safe Operation	Fail-Safe Deactivation Conditions
14	Fuel cut	1 IGF1 detected in consecutive 8 ignitions.
15	Fuel cut	1 IGF2 detected in consecutive 8 ignitions.
16	Torque control prohibited.	Returned to normal condition.
22	THW is fixed at 80°C (176°F).	Returned to normal condition.
24	THA is fixed at 20°C (68°F).	Returned to normal condition.
31	Ignition timing controlled between 10° ~ 30° BTDC, Injection time controlled between 3 msec. ~ 30 m sec., in relation to : ● Engine rpm ● Throttle angle ● ISC step value	KS input 15 times/sec. or more.
35	Atmospheric pressure is fixed at 760 mmHg.	Returned to normal condition.
41	VTA1 is fixed at 0°	The following 1 or 2 must be repeated at least 2 times consecutively. 1 When IDL is ON 0.25 V ≦ VTA1 ≦ 0.95 V 2 When IDL is OFF 0.25 V ≦ VTA1 ≦ 4.9 V
47	Value of VTA1 is used.	The following 1 or 2 must be repeated at least 2 times consecutively. 1 When IDL2 is ON 0.25 V ≦ VTA2 ≦ 0.95 V 2 When IDL2 is OFF 0.25 V ≦ VTA2 ≦ 4.9 V
52	Max. timeing retardation.	IG switch OFF.
53	Max. timeing retardation.	Returned to normal condition.
55	Max. timeing retardation.	IG switch OFF.

Back-Up Function
If there is trouble with the program in the ECU and the ignition signals (IGT) are not output, the ECU controls fuel injection and ignition timing at predetermined levels as a back-up function to make it possible to continue to operate the vehicle.
Furthermore, the injection duration is calculated from the satrting signal (STA) and the throttle position signal (IDL). Also, the ignition timing is fixed at the initial ignition timing, 5° BTDC, without relation to the engine speed.

HINT: If the engine is controlled by the back-up function, the CHECK ENGINE light lights up to warn the driver of the malfunction but the diagnostic code is not output.

BASIC INSPECTION

When the normal code is displayed in the diagnostic code check, troubleshooting should be performed in the order for all possible circuits to be considered as the causes of the problems.
In many cases, by carrying out the basic engine check shown in the following flow chart, the location causing the problem can be found quickly and efficiently. Therefore, use of this check is essential in engine troubleshooting.

1 Is battery voltage 11 V or more when engine is stopped ?

NO → **Charge or replace battery.**

YES ↓

2 Is engine cranked ?

NO → **Proceed to matrix to matrix chart of problem symptoms on**

YES ↓

3 Does engine start ?

NO → **Go to step 7.**

YES ↓

4 Check air filter.

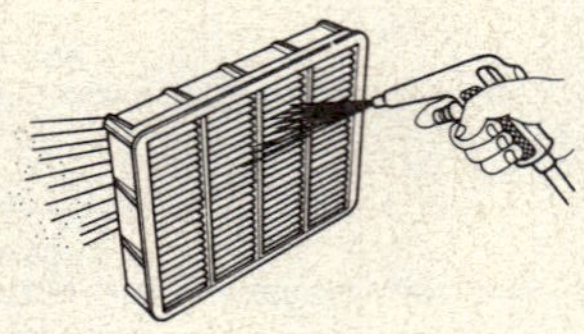

P Remove air filter.

C Visually check that the air cleaner element is not excessively damaged or oily.

Hint If necessary, clean element with compressed air. First blow from inside thoroughly, then blow off outside of element.

NG → **Repair or replace.**

OK ↓

Go to step 5.

5 Check idle speed.

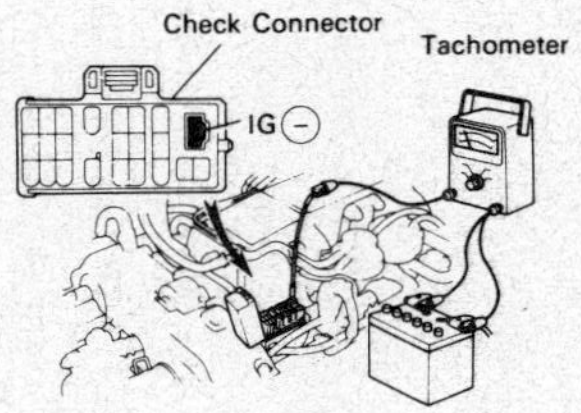

P (1) Warm up engine at normal operating temperature.
(2) Switched off all accessories.
(3) Switched off air conditioner.
(4) Shift transmission into "N" range.
(5) Connect tachometer test probe to terminal IG ⊖ of check connector, and set the tachometer to the 4-cylinder range.

C Check idle speed.

OK **Idle speed: 650 ~ 750 rpm**

Caution
- **NEVER allow tachometer test probe to touch ground as it could result in damage to igniter and/or ignition coil.**
- **As some tachometers are not compatible with this ignition system, we recommended that you confirm compatibility of your unit before use.**

NG 〉 **Proceed to matrix chart of problem symptoms**

OK

6 Check ignition timing.

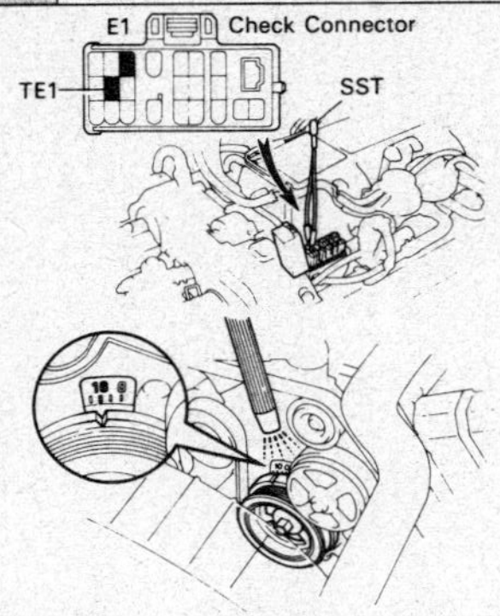

P (1) Warm up engine at normal operating temperature.
(2) Shift transmission into "N" range.
(3) Keep the engine speed at idle.
(4) Using SST, connect terminals TE1 and E1 of check connector.

SST 09843-18020

(5) Using a timing light, connect the tester to No. 6 high-tension cord.

C Check ignition timing.

OK **Ignition timing: 8 – 12° BTDC at idle**

NG 〉 **Proceed to troubleshoot.**

OK

Proceed to matrix chart of problem symptoms

7 Check fuel pressure.

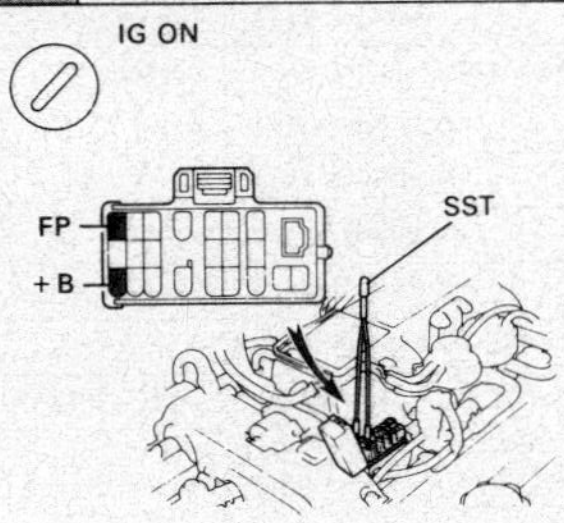

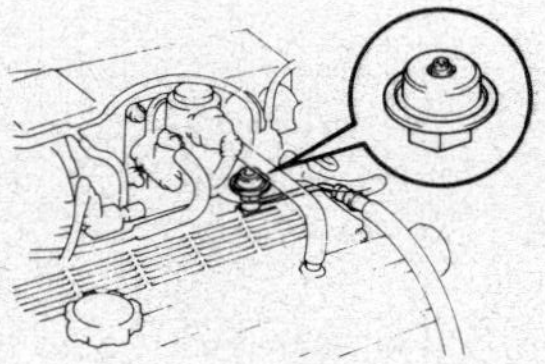

P (1) Be sure that fuel is enough in tank.
(2) Turn ignition switch on.
(3) Using SST, connect terminals FP and +B of check connector.

SST 09843-18020

C Check that pulsation damper screw rises up when teminals are connected.

Caution
Never make a mistake with the terminal connection position as this will cause a malfunction.

NG 〉 **Proceed to troubleshoot.**

OK

8 Check for spark.

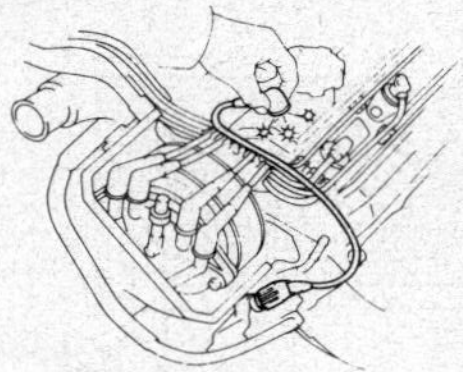

C Disconnect the high-tension cord from the distributor and, hold the end about 12.5 mm (1/2") from the ground, see if spark occurs while the engine is being cranked.

Hint To prevent excessive fuel injected from the injectors during this test, Don't crank the engine for more than 1 – 2 seconds at a time.

NG 〉 **Proceed to troubleshoot.**

OK

Proceed to matrix chart of problem systems

TERMINALS OF ECU

ECU Terminals

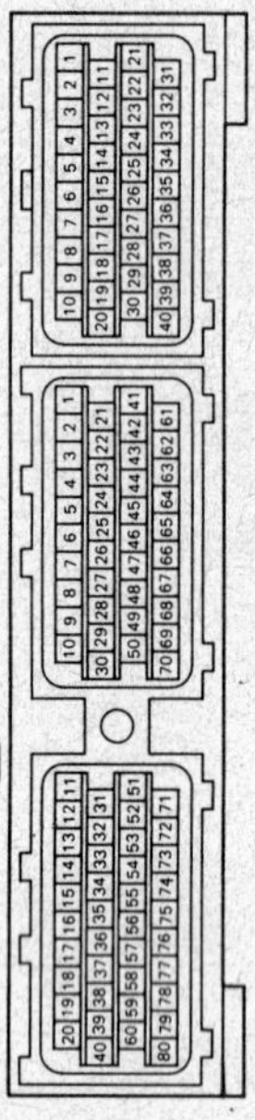

Terminal No.	Symbol	Connection	Terminal No.	Symbol	Connection
E9 – 1	IGSW	Ignition switch	E9 – 31	+B	EFI main relay
2	SPD	Speed sensor No. 1	32	+B1	EFI main relay
3	KD	Kick-down switch	33	BATT	Battery
4	STP	Stop light switch	34	A/C	A/C control assembly
5	—	—	35	—	—
6	W	"CHECK" engine warning light	36	—	—
7	R	Neutral start switch	37	TR	TRAC ECU
8	—	—	38	NEO	TRAC ECU
9	2	Neutral start switch	39	VTO2	TRAC ECU
10	L	Neutral start switch	40	VTO1	TRAC ECU
11	AD	Cruise control ECU	E10 – 1	NCO⊖	T/M input speed sensor
12	OD1	Cruise control ECU	2	—	—
13	—	—	3	SP2⊖	Speed sensor No. 2 (for ECT)
14	—	—	4	E11	ECU ground
15	—	—	5	G2⊖	Cam position sensor
16	—	—	6	G1⊖	Cam position sensor
17	TT	TDCL	7	NE⊖	Engine speed sensor
18	P	ECT pattern select switch	8	—	—
19	TE2	Check connector	9	S2	ECT solenoid
20	TE1	Check connector	10	S1	ECT solenoid
21	DI	Fuel pump ECU	11	—	—
22	FPC	Fuel pump ECU	12	—	—
23	ACMG	A/C magnet clutch relay	13	SLN⊖	ECT solenoid No. 4
24	M-REL	EFI main relay	14	SLU⊖	ECT solenoid No. 3
25	—	—	15	—	—
26	—	—	16	—	—
27	—	—	17	#40	Injectors (No. 3 and No. 5)
28	OD2	O/D main switch	18	#30	Injectors (No. 4 and No. 6)
29	—	—	19	#20	Injectors (No. 2 and No. 8)
30	—	—	20	#10	Injectors (No. 1 and No. 7)

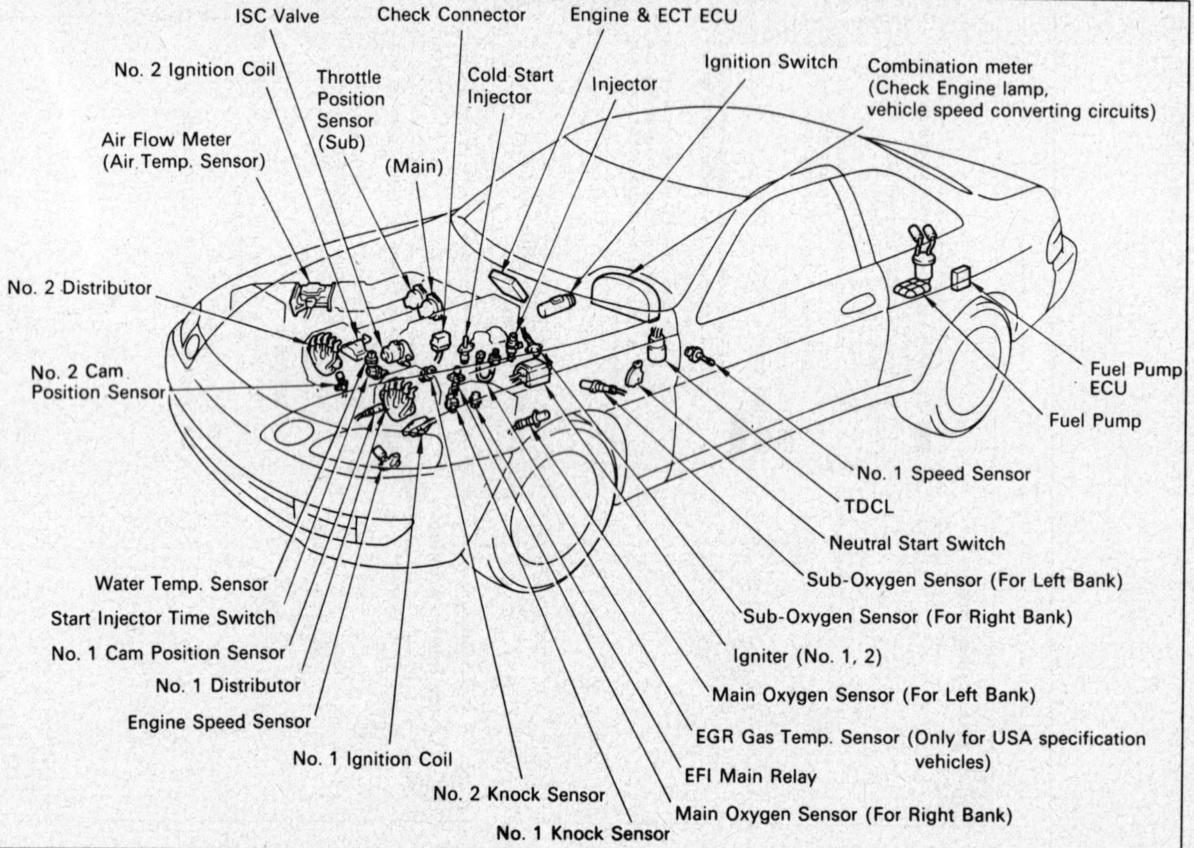

4.0L ENGINE – ECU TERMINALS – SC400

TERMINALS OF ECU (Cont'd)

ECU Terminals

E10 (B) E9 (A)

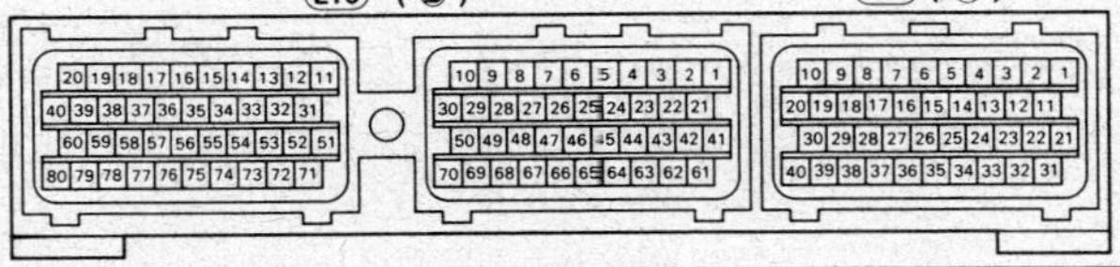

Terminal No.	Symbol	Connection	Terminal No.	Symbol	Connection
E10 – 21	NCO	T/M input speed sensor	E10 – 51	—	———
22	—	———	52	HTL2	Sub-oxygen sensor heater (on left bank)
23	SP2	Speed sensor No. 2 (for ECT)	53	HTR2	Sub-oxygen sensor heater (on right bank)
24	—	———	54	—	———
25	G2	No. 2 cam position sensor	55	—	———
26	G1	No. 1 cam position sensor	56	IGT2	No. 2 igniter
27	NE	Engine speed sensor	57	IGT1	No. 1 igniter
28	VF2	Check connector	58	IGF1	No. 1 igniter
29	VF1	Check connector	59	IGF2	No. 2 igniter
30	—	———	60	—	———
31	—	———	61	—	———
32	ISC4	ISC valve	62	—	———
33	ISC3	ISC valve	63	IDL2	Sub-throttle position sensor
34	ISC2	ISC valve	64	IDL1	Throttle position sensor
35	ISC1	ISC valve	65	E2	Sensor ground
36	EGR4*	EGR valve	66	KS	Air flow meter
37	EGR3*	EGR valve	67	OXR2	Sub-oxygen sensor (on right bank)
38	EGR2*	EGR valve	68	OXL2	Sub-oxygen sensor (on left bank)
39	EGR1	EGR valve (or VSV)	69	E1	ECU ground
40	—	———	70	—	———
41	VC	Throttle position sensor. Air flow meter.	71	HTL1	Main oxygen sensor heater (on left bank)
42	VTA2	Sub-throttle position sensor	72	HTR1	Main oxygen sensor heater (on right bank)
43	VTA1	Throttle position sensor	73	FPU	Fuel pressure up VSV
44	THW	Water temp. sensor	74	PAG	EVAP purge VSV
45	THA	Intake air temp. sensor	75	—	———
46	THG*	EGR gas temp. sensor	76	NSW	Neutral start switch
47	OXR1	Main oxygen sensor (on right bank)	77	STA	Starter relay
48	OXL1	Main oxygen sensor (on left bank)	78	STJ	Cold start injector
49	KNK2	No. 2 knock sensor	79	EO2	Power ground
50	KNK1	No. 1 knock sensor	80	EO1	Power ground

* Only for USA specification vehicles.

4.0L ENGINE – ECU TERMINALS STANDARD VALUES SC400

STANDARD VALUE OF ECU TERMINALS

(※ : A = E9, B = E10)

※ Symbols (Terminals No.)	STD Voltage (V)	Condition
BATT (A33) – E1 (B69)	10 ~ 14	Always
IGSW (A1), +B (A31), +B1 (A32) – E1 (B69)	10 ~ 14	IG switch ON
VC (B41) – E2 (B65)	4.0 ~ 6.0	IG switch ON
IDL1 (B64), IDL2 (B63) – E2 (B65)	Below 1.0	IG switch ON Main or sub throttle valve fully closed.
	10 ~ 14	IG switch ON Main or sub throttle valve fully open.
VTA1 (B43), VTA2 (B42) – E2 (B65)	0.1 ~ 1.0	IG switch ON Throttle valve fully closed.
	3.0 ~ 6.0	IG switch ON Throttle valve fully open.
KS (B66) – E1 (B69)	Pulse generation (0 and 4 ~ 6)	Idling
THA (B45) – E2 (B65)	1.0 ~ 3.0	Idling, Intake air temp. 20°C (68°F).
THW (B44) – E2 (B65)	0.1 ~ 1.0	Idling, Water temp. 80°C (176°F).
STA (B77) – E1 (B69)	6.0 or more	Cranking
#10 (B20), #20 (B19), #30 (B18), #40 (B17) – E01 (B80)	10 ~ 14	IG switch ON
	Pulse generation	Idling
IGT1 (B57), IGT2 (B56) – E1 (B69)	Pulse generation (0 and 4 ~ 6)	Idling
IGF1 (B58), IGF2 (B59) – E1 (B69)	Below 1.0	IG switch ON
	Pulse generation (0 and 4 ~ 6)	Idling
G1 (B26), G2 (B25) – G1⊖ (B6)	Pulse generation (–1.0 ~ +1.0 or higher)	Idling
NE (B27) – NE⊖ (B7)	Pulse generation (–0.8 ~ +0.8 or higher)	Idling
MREL (A24) – E1 (B69)	10 ~ 14	IG switch ON
FPC (A22) – E1 (B69)	4.2 ~ 6.0	Cranking, Sudden racing (6,000 rpm)
	Pulse generation (0 and 4 ~ 6)	Idling
DI (A21) – E1 (B69)	7.5 or more	Idling
FPU (B73) – E1 (B69)	10 ~ 14	IG switch ON
	Below 2.0	Restarting high temp.
PAG (B74) – E1 (B69)	10 ~ 14	Idling

4.0L ENGINE – ECU TERMINALS STANDARD VALUES SC400

STANDARD VALUE OF ECU TERMINALS (Cont'd)

(※: A = E9, B = E10)

※ Symbols (Terminals No.)	STD Voltage (V)	Condition
ISC1 (B35), ISC2 (B34), ISC3 (B33), ISC4 (B32) – E1 (B69)	10 ~ 14	IG switch ON
VF1 (B29), VF2 (B28) – E1 (B69)	1.0 ~ 4.0	Maintain engine speed at 2,500 rpm for two minutes after warming up then return to idling.
OXL1 (B48), OXR1 (B47) – E1 (B69)	Pulse generation (0 ~ 1.0)	Maintain engine speed at 2,500 rpm for two minutes after warming up.
HTL1 (B71), HTR1 (B72), HTL2 (B52), HTR2 (B53) – E01 (B80)	Below 2.0	Idling
	10 ~ 14	IG switch ON
EGR1 (B39), EGR2 (B38), EGR3 (B37), EGR4 (B36) – E1 (B69)*	10 ~ 14	IG switch ON
EGR (B39) – E1 (B69)	10 ~ 14	IG switch ON
KNK1 (B50), KNK2 (B49) – E1 (B69)	Pulse generation (Frequency : 6.6 KHz)	Idling
NSW (B76) – E1 (B69)	10 ~ 14	IG switch ON Other shift position in "P", "N" ranges.
	Below 1.0	IG switch ON Shift position in "P", "N" ranges.
SPD (A2) – E1 (B69)	Pulse generation (0 and 10 ~ 14)	IG switch ON Rotate driving wheel slowly.
TE1 (A20) – E1 (B69)	10 ~ 14	IG switch ON
TE2 (A19) – E1 (B69)	10 ~ 14	IG switch ON
W (A6) – E1 (B69)	10 ~ 14	Idling
	Below 2.0	IG switch ON
AD (A11) – E1 (B69)	10 ~ 14	Idling
ODI (A12) – E1 (B69)	4.0 ~ 6.0	IG switch ON
A/C (A34) – E1 (B69)	Bellow 2.0	A/C switch ON (At idling)
	10 ~ 14	A/C switch OFF
ACMG (A23) – E1 (B69)	Below 2.0	A/C switch ON (At idling)
	10 ~ 14	A/C switch OFF
TR (A37) – E1 (B69)	10 ~ 14	IG switch ON
VTO1 (A40), VTO2 (A39) – E2 (B65)	Below 1.0	IG switch ON Main or sub throttle valve fully closed.
	3.0 ~ 5.5	IG switch ON Main or sub throttle valve fully open.
NEO (A38) – E1 (B69)	4.0 ~ 6.0	IG switch ON

* Only for USA specification vehicles.

4.0L ENGINE – MATRIX CHART OF PROBLEM SYMPTOMS SC400

MATRIX CHART OF PROBLEM SYMPTOMS

When the malfunction code is not confirmed in the diagnostic code check and the problem still can not be confirmed in the basic inspection, then proceed to this step and perform troubleshooting according to the numbered order given in the table below.

The circuits indicated by ※ on the matrix chart can be inspected using the TCCS checker.

Symptom \ Suspect area		※ Air flow meter circuit	※ Starter signal circuit	EGR System	Switch condition signal circuit	※ Neutral start switch circuit	ECU power source circuit	Back up power source circuit	※ Injector circuit	Cold start injector circuit	※ ISC valve circuit	Fuel pump control circuit	Fuel pressure control VSV circuit	A/C signal circuit (Compressor circuit)	Starter and Starter relay	Ignition signal circuit (Spark test)	Spark plug	Ignition coil	Distributor	Compression	ECT faulty	Theft deterrent ECU	Engine & ECT ECU
Does not start	Engine does not crank					2									1							3	
	No initial combustion						1		5	6		3				2	4						
	No complete combustion								6	7		1				5	4	2	3				
Difficult to start	Difficult to start normally		1						8	9	2	3					6	4	5	7			
	Difficult to start cold		1						5	3	2	4					8	6	7				
	Difficult to start hot		1						5	6	2	4	3				9	7	8				
Poor Idling	Incorrect first idle				1						2												
	High engine idle speed				1	5	4	6			2			3									
	Low engine idle speed	7			3	4		8	6		1	5		2									
	Rough idling	3		5	1			12	4	13	2	8				6	11	9	10	7			
	Hunting	3			1		4				2	5											
Poor Driveability	Hesitation/Poor acceleration	2			1				3			4				5	8	6	7		9		
	Muffler explosion (after fire)				1				5								4	2	3				
	Surging				1				5			2					4		3				
Engine Stall	Engine stalls soon after starting	2								4	3	1											
	After acceleration pedal depressed	2			1					3													
	After acceleration pedal released								1	3	2												4
	During A/C operation										1			2									3
	When N to D shift					1				3	2												

4.0L ENGINE—CONNECTOR LOCATIONS—SC400

LOCATION OF CONNECTORS

Location of Connectors in Engine Compartment

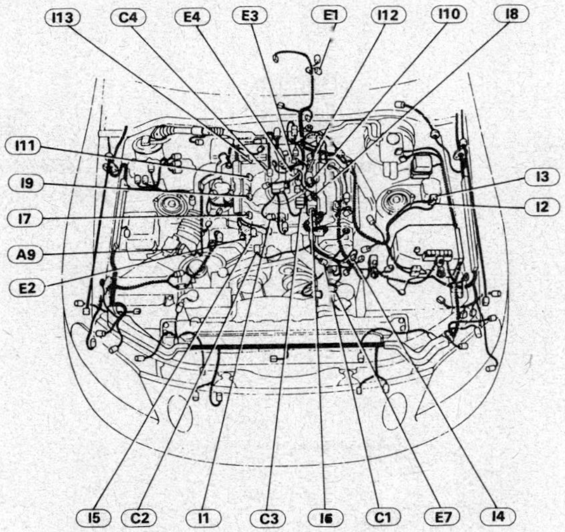

A9
Intake Air Temp. Sensor

IS-5-1

C1
No. 1 Cam Position sensor

le-2-1-D

C2
No. 2 Cam Position Sensor

le-2-1-E

C3
Check Connector

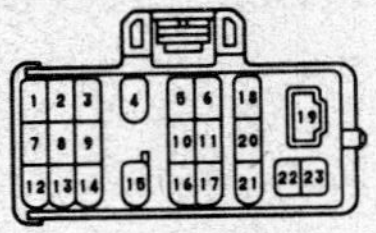

lej-23-1

C4
Cold Start Injector

IS-2-1-I

E1
ECT No. 1, 2, 3, 4 Solenoid Valve

le-6-1

E2
Water Temp. Sensor

V-2-1-C

E3
EGR Gas Temp. Sensor

le-2-1-C

E4
EGR Valve

le-6-1-D

E7
Engine Speed Sensor

le-2-1-b

I1
ISC Valve

le-6-1-D

I2
No. 1 Igniter

le-5-1-B

I3
No. 2 Igniter

le-5-1-B

I4
No. 1 Ignition Coil

IS-2-1-L

I5
No. 2 Ignition Coil

IS-2-1-L

I6 I7 I8 I9
I10 I11 I12 I13
Injector

le-2-1-a

Location of Connectors in Engine Compartment (Cont'd)

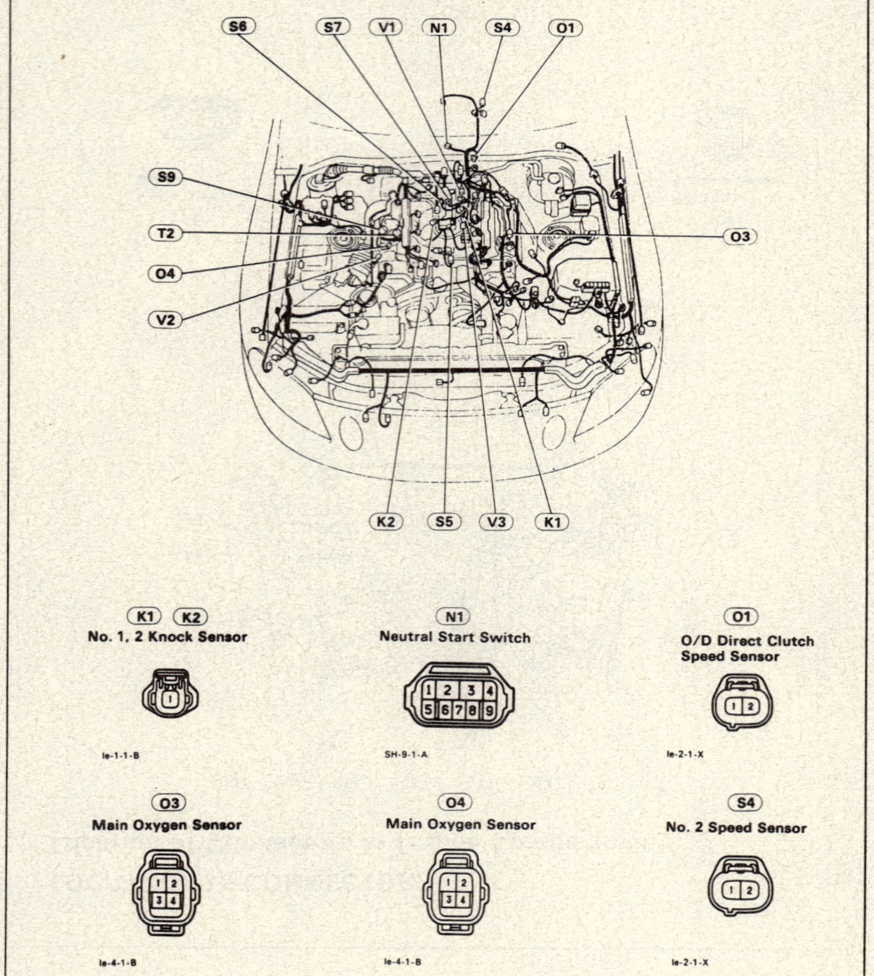

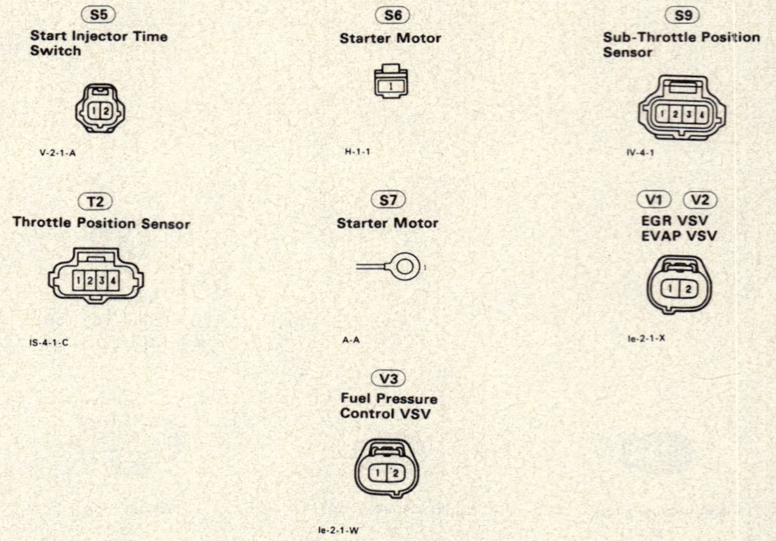

4.0L ENGINE—CONNECTOR LOCATIONS—SC400

Location of Connectors in Engine Compartment (Cont'd)

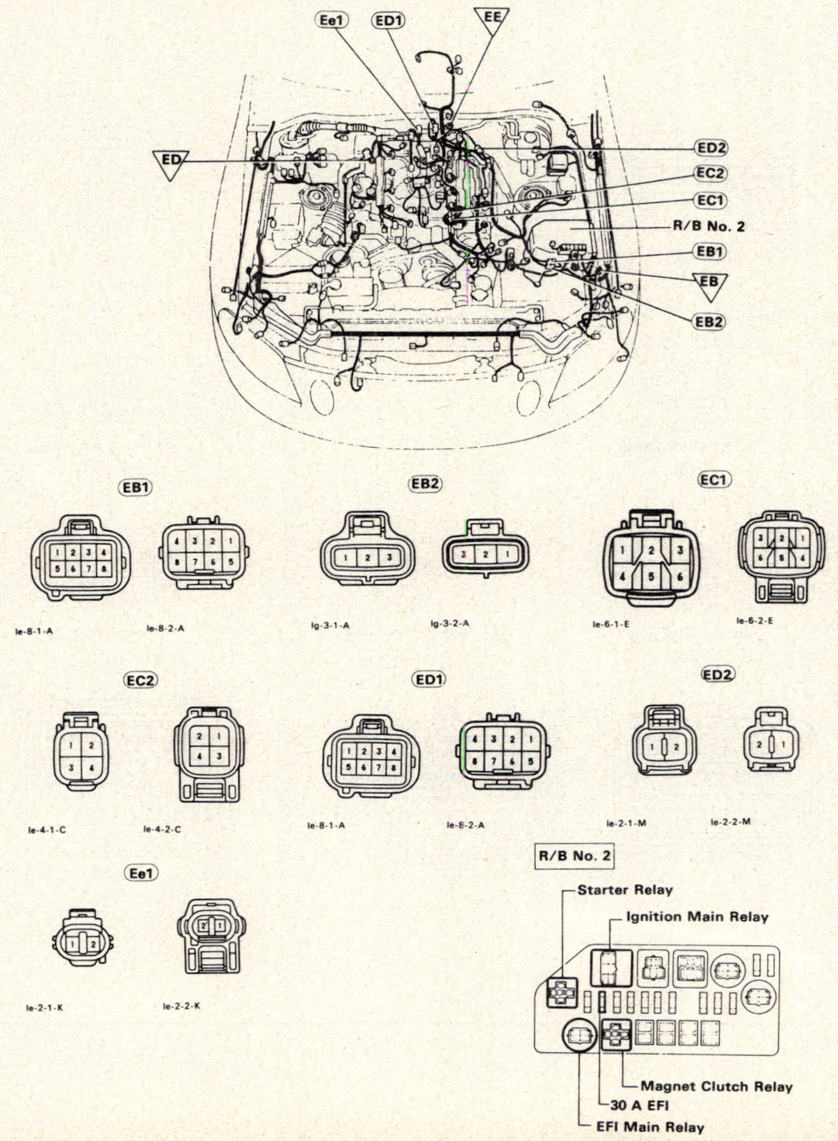

Location of Connectors in Instrument Panel

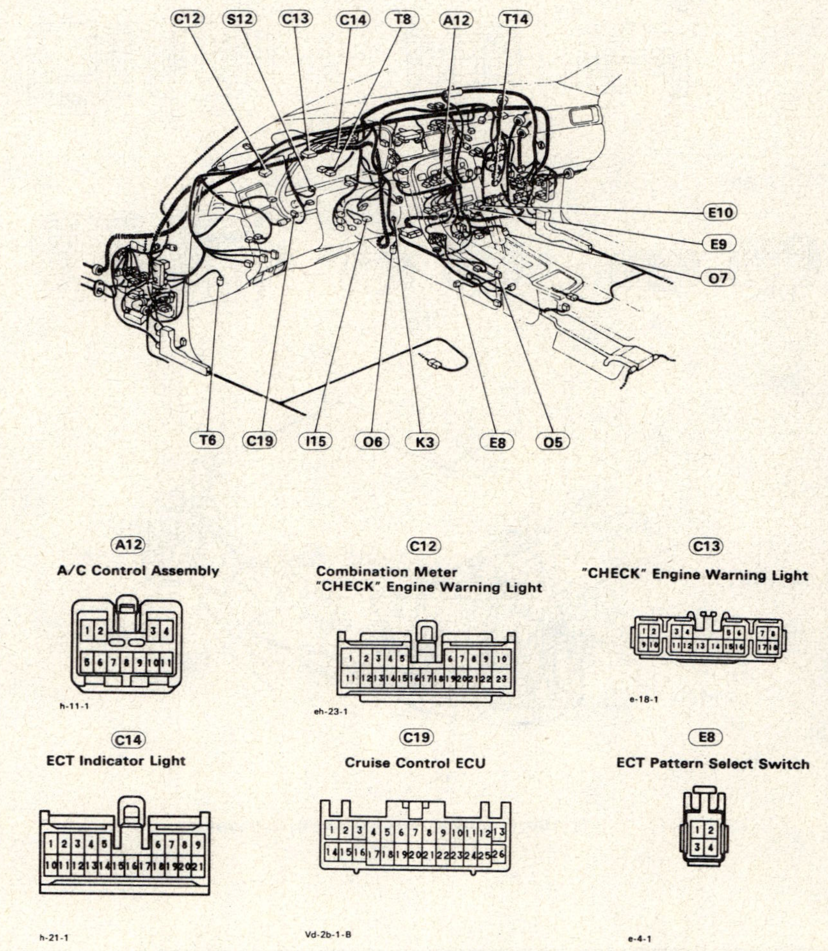

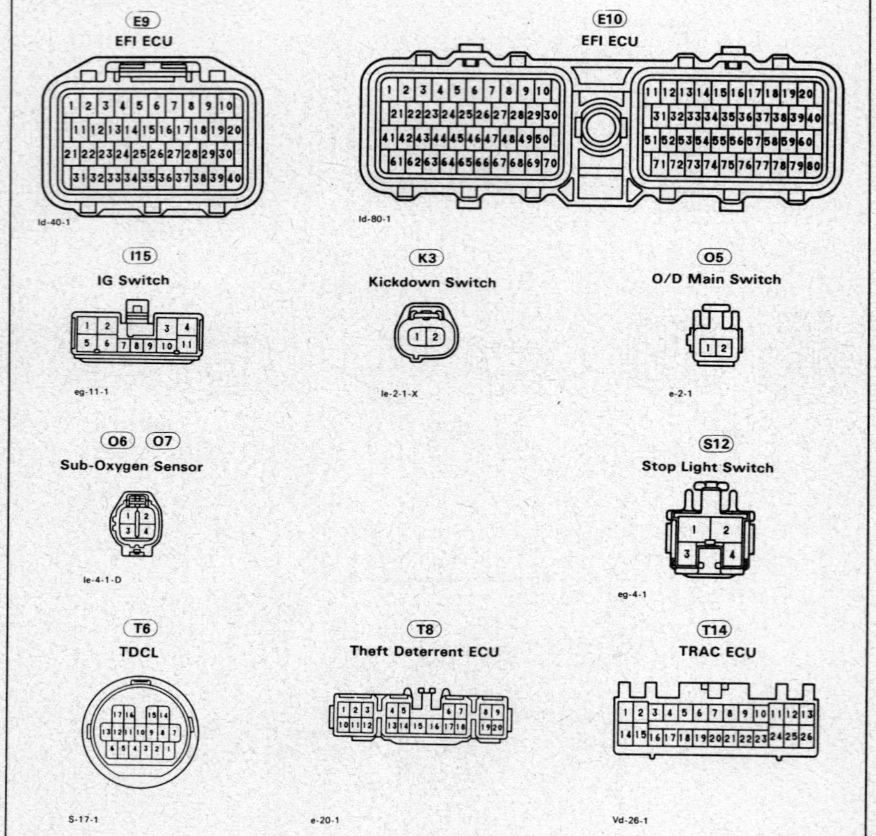

Location of Connectors in Instrument Panel (Cont'd)

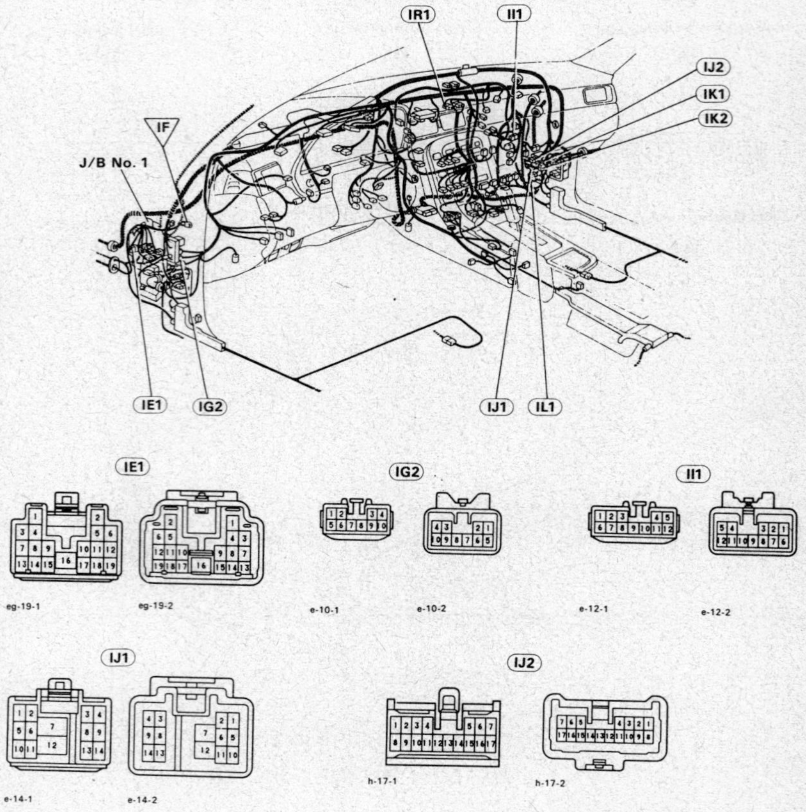

4.0L ENGINE – CONNECTOR LOCATIONS – SC400

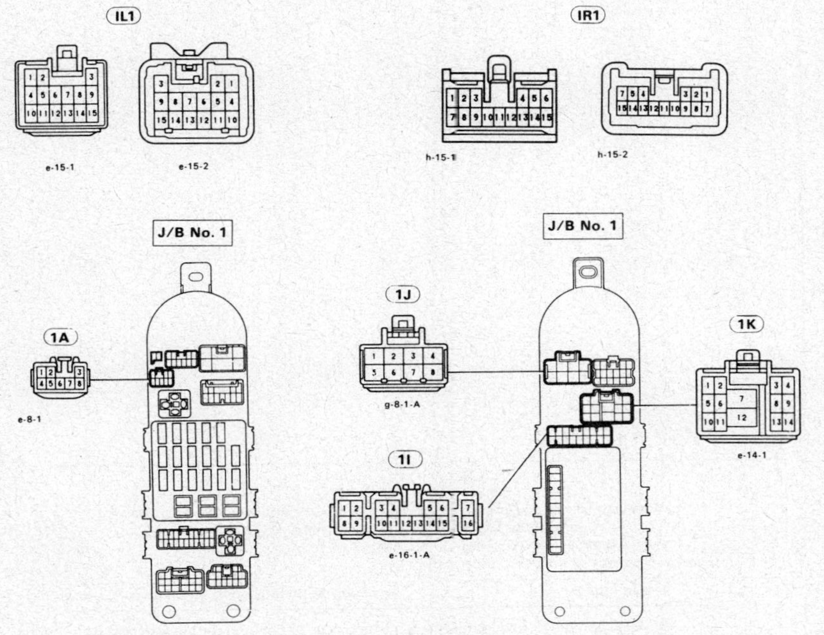

Location of Connectors in Body

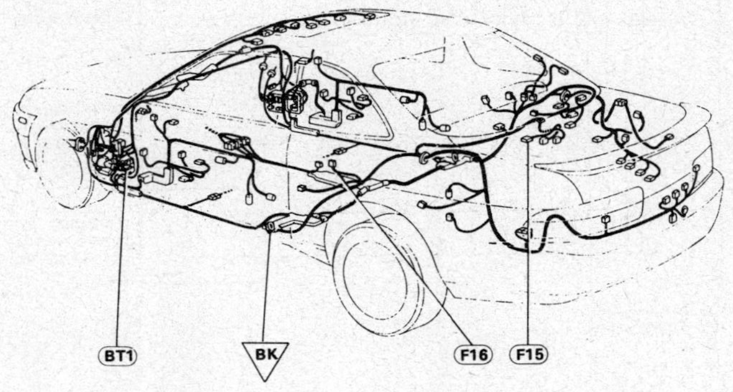

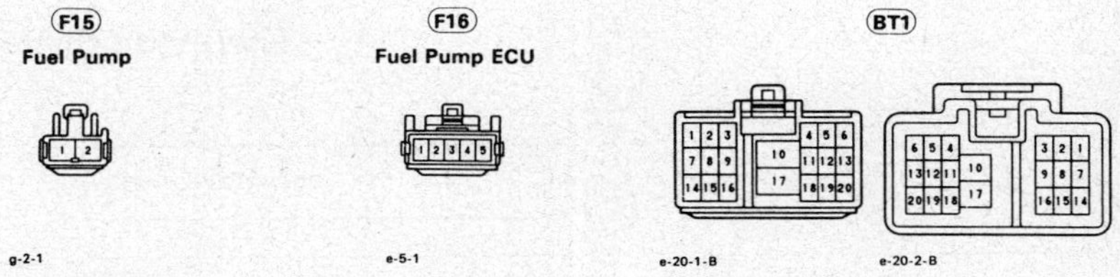

CIRCUIT INSPECTION

Diag. Code	12	RPM Signal Circuit (No. 1)

CIRCUIT DESCRIPTION

Cam position sensors (G1 and G2 signals) and engine speed sensor (NE signal) consist of a signal plate and a pick up coil.
The G1, G2 signal plates have one tooth each on its outer circumference and are mounted on the left and right bank camshafts.
When the camshafts rotate, the protrusion on the signal plate and the air gap on the pick up coil change, causing fluctuations in the magnetic field and generating an electromotive force in the pick up coil.
The NE signal plate has 12 teeth and is mounted on the crankshaft. The NE signal sensor generates 12 NE signals per engine revolution. The ECU detects the standard crankshaft angle based on the G1, G2 signals, and the actual crankshaft angle and the engine speed by the NE signals.

Code No.	Diagnostic Code Detecting Condition	Trouble Area
12	No "NE" or "G1" and "G2" signal to ECU within 2 sec. after cranking.	• Open or short in engine speed sensor, No. 1, No. 2 cam position sensor circuit • Engine speed sensor • No. 1, No. 2 cam position sensor • Starter • ECU

DIAGNOSTIC CHART

Step	Check	Result	Action
1	Check resistance of engine speed sensor, No. 1, No. 2 cam position sensor.	NG	Replace sensor.
		OK	Go to 2
2	Check for open and short in harness and connector between ECU and each sensor.	NG	Repair or replace harness or connector.
		OK	Go to 3
3	Inspect sensor installation and teeth of signal plate.	NG	Tighten the sensor. Replace signal plate.
		OK	Check and replace ECU.

WIRING DIAGRAM

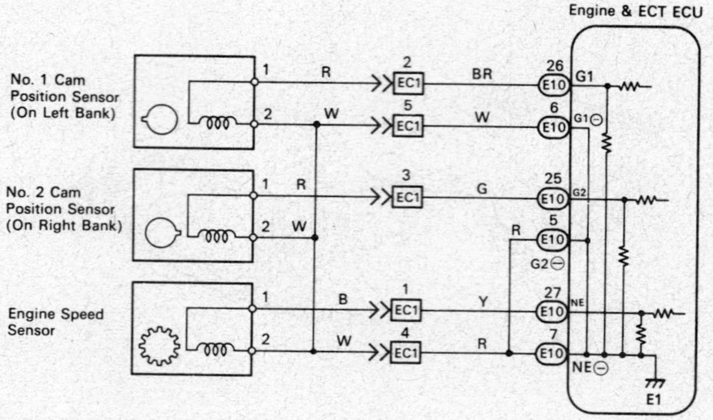

INSPECTION PROCEDURE

1 Check engine speed sensor, No. 1, No. 2 cam position sensor.

No. 1 cam position sensor

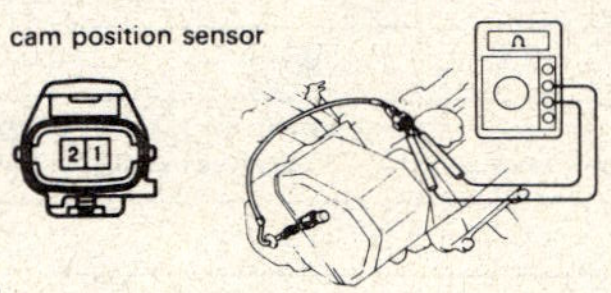

le-2-2-D

No. 2 cam position sensor

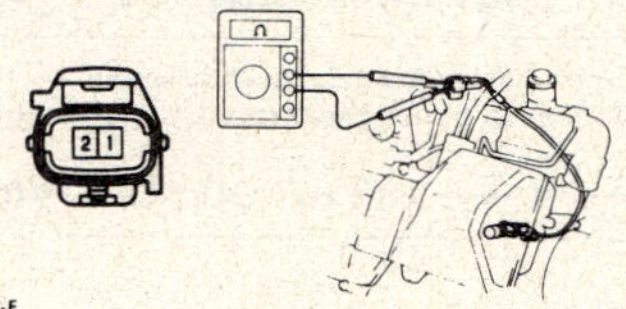

le-2-2-E

Engine speed sensor

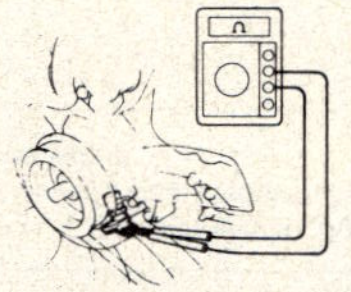

P For engine speed sensor,
(1) Remove engine under cover.
(2) Disconnect engine speed sensor connector.

For No. 1, No. 2 cam position sensor,
(1) Disconnect No. 1, No. 2 cam position sensor connectors.

C Measure resistance of engine speed sensor, No. 1 and No. 2 cam position sensor.

OK

	Resistance
No. 1 cam position sensor	950 ~ 1,250 Ω
No. 2 cam position sensor	950 ~ 1,250 Ω
Engine speed sensor	950 ~ 1,250 Ω

Reference **INSPECTION USING OSCILLOSCOPE**

G, NE signal waveforms

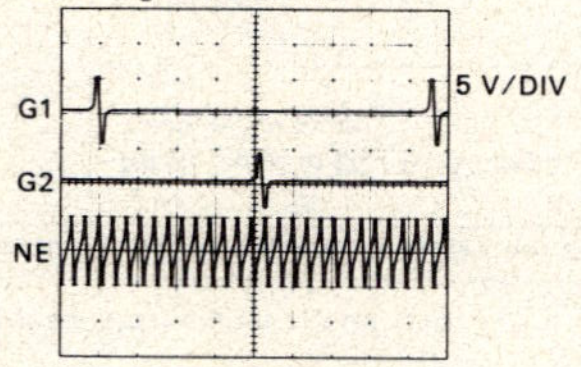

20 m sec/Division (Idling)

- During cranking or idling, check between terminals G1, G2 and G1⊖, NE and NE⊖ of engine & ECT ECU.

HINT: The correct waveforms appear as shown in the illustration on the left.

OK

NG Replace engine speed sensor, No. 1, No. 2 cam position sensor.

2 Check for open and short in harness and connector between engine & ECT ECU and each sensor

OK

NG Repair or replace harness or connector.

3 Inspect sensor installation and teeth of signal plate.

OK

NG Tighten the sensor.
Replace signal plate.

Check and replace engine & ECT ECU.

Diag. Code	13	RPM Signal Circuit (No. 2)

CIRCUIT DESCRIPTION

Refer to RPM signal circuit (No. 1)

Code No.	Diagnostic Code Detecting Condition	Trouble Area
13	No NE signal to ECU for 0.1 sec. or more at 1,000 rpm or more.	• Open or short in engine speed sensor circuit. • Engine speed sensor • ECU
	No 12 pulses of NE to ECU during the interval between G1 and G2 pulses.	• Open or short in engine speed sensor circuit. • Mechanical system malfunction (skipping teeth of timing belt, belt stretched) • Engine speed sensor • ECU
	Deviation in G (G1, G2) and NE signal continues for 1 sec. during idling (throttle fully closed) after engine warmed up.	• Mechanical system malfunction (skipping teeth of timing belt, belt stretched) • No. 1, No. 2 cam position sensor • ECU

DIAGNOSTIC CHART

HINT: Perform troubleshooting of diag. code 12 first. If no trouble is found, troubleshoot the following mechanical systems.

1 Inspect sensor installation. Check if any teeth of NE signal plate are broken. — NG → Tighten the sensor. Replace signal plate.

OK ↓

2 Check valve timing (Check for loose and jumping teeth of timing belt — NG → Adjust valve timing (Repair or replace timing belt).

OK ↓

Check for momentary interruption

Diag. Code	14, 15	Ignition Signal Circuit

CIRCUIT DESCRIPTION

The ECU determines the ignition timing, turns on Tr_1 at a predetermined angle (°CA) before the desired ignition timing and outputs an ignition signal (IGT) "1" to the igniter.
Since the width of the IGT signal is constant, the dwell angle control circuit in the igniter determines the time the control circuit starts primary current flow to the ignition coil based on the engine rpm and ignition timing one revolution ago, that is, the time the Tr_2 turns on.
When it reaches the ignition timing, the ECU turns Tr_1 off and outputs the IGT signal "O".
This turns Tr_2 off, interrupting the primary current flow and generating a high voltage in the secondary coil which causes the spark plug to spark. Also, by the counter electromotive force generated when the primary current is interrupted, the igniter sends an ignition confirmation signal (IGF) to the ECU.
The ECU stops fuel injection as a fail safe function when the IGF signal is not input to the ECU.

Code No.	Diagnostic Code Detecting Condition	Trouble Area
14	No IGF1 signal to ECU for 8-11 consecutive IGT1 signal.	• Open or short in IGF1 and IGT1 circuit from No. 1 igniter – ECU. • No. 1 igniter • ECU
15	No IGF2 signal to ECU for 8-11 consecutive IGT2 signal.	• Open or short in IGF2 and IGT2 circuit from No. 2 igniter – ECU. • No. 2 igniter • ECU

HINT: No. 1 igniter is for left bank and No. 2 igniter is for right bank.

WIRING DIAGRAM

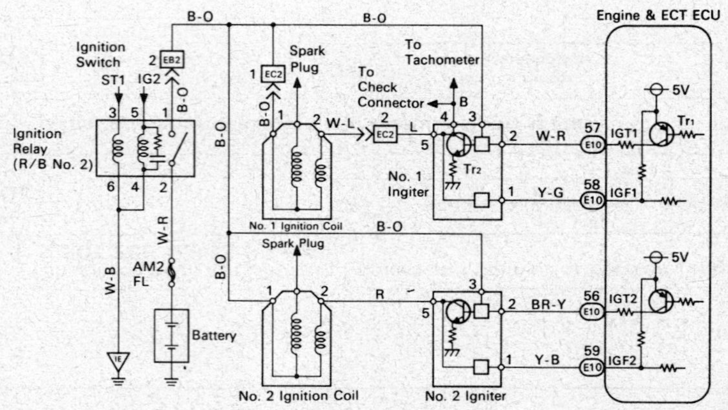

DIAGNOSTIC CHART

Refer to inspection procedure in details.

HINT: If diag. code 14 is displayed, check No. 1 igniter circuit.
If diag. code 15 is displayed, check No. 2 igniter circuit.

Step	Check	Result	Action
1	Check for spark.	NG	Go to 4
		OK	Go to 2
2	Check for open and short in IGF signal circuit between igniter and ECU.	NG	Repair or replace harness or connector.
		OK	Go to 3
3	Check voltage of terminal IGF.	OK	Replace igniter.
		NG	Check and replace ECU.
4	Check voltage of terminal IGT.	NG	Go to 9
		OK	Go to 5
5	Check voltage of igniter power source.	NG	Go to 8
		OK	Go to 6
6	Check for open and short in primary coil circuit.	NG	Repair or replace harness or connector.
		OK	Go to 7
7	Check ignition coil.	NG	Replace ignition coil.
		OK	Replace igniter.
8	Check ignition relay.	NG	Replace ignition relay.
		OK	Check and repair igniter power source circuit.
9	Check voltage of terminal IGT (Disconnect igniter connector).	OK	Replace igniter.
		NG	Go to 10
10	Check for open and short in IGT signal circuit between igniter and ECU.	NG	Repair or replace harness or connector.
		OK	Check and replace ECU.

INSPECTION PROCEDURE

HINT: If diag. code 14 is displayed, check No. 1 igniter circuit.
If diag. code 15 is displayed, check No. 2 igniter circuit.

1 Check for spark.

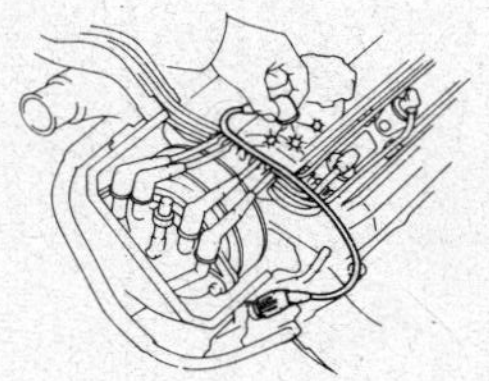

C Disconnect the high-tension cord from the distributor, hold its end about 12.5 mm (1/2") from the ground, see if spark occurs while the engine is being cranked.

OK **Spark should be generated.**

Hint: To prevent excessive fuel injected from the injectors during this check. Don't crank the engine for more than 1 – 2 seconds at a time.

OK → (next step)

NG → **Go to step 4.**

2 Check for open and short in harness and connector in IGF and IGT signal circuit between engine & ECT ECU and igniter

OK → (next step)

NG → **Repair or replace harness or connector.**

3 Disconnect igniter connector and check voltage between terminals IGF1, 2 of engine & ECT ECU connector and body ground.

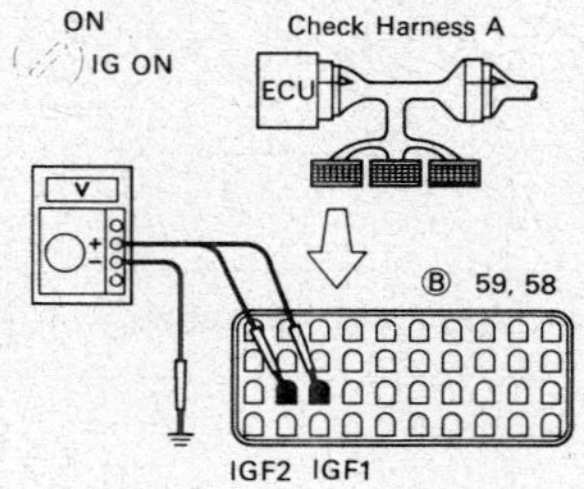

P (1) Disconnect igniter connector.
(2) Connect the Check Harness A.
(3) Turn ignition switch on.

C Measure voltage between terminals IGF1, 2 of engine & ECT ECU connector and body ground.

OK **Voltage: 4 – 6 V**

OK → **Replace igniter.*1**

NG → **Check and replace engine & ECT ECU.**

*1: When diag. code 14 is displayed, replace the igniter with 5 wire harness. (extra wire is for tachometer). When diag. code 15 is displayed, replace the igniter with 4 wire harness.

4 Check voltage between terminals IGT1, 2 of engine & ECT ECU connector and body ground.

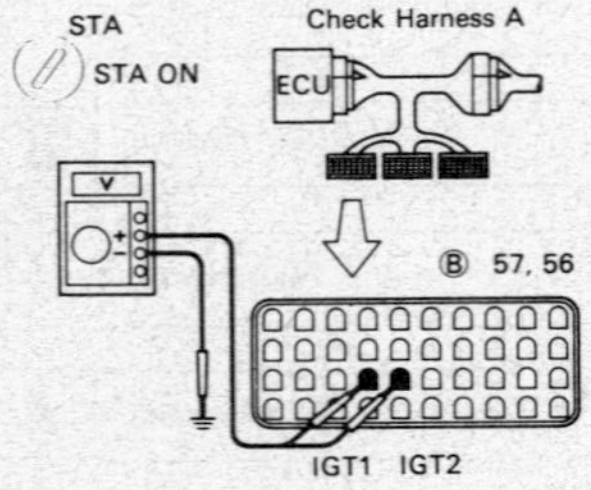

P Connect the Check Harness A.

C Measure voltage between terminals IGT1, 2 of engine & ECT ECU connector and body ground when engine is cranked.

OK **Voltage: 0.5 – 1.0 V**
(Neither 0 V nor 5 V)

Reference INSPECTION USING OSCILLOSCOPE

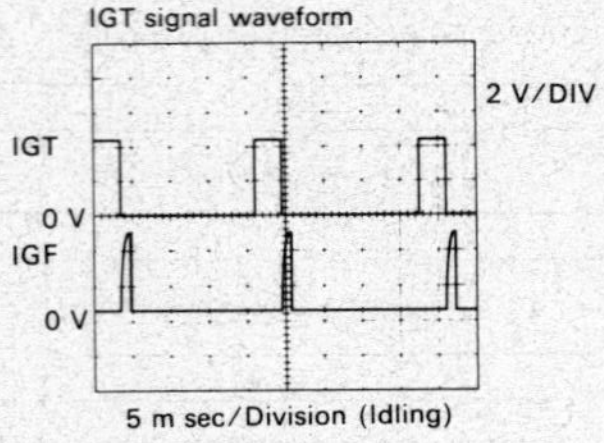

- During cranking or idling, check waveform between terminals IGT1, IGT2 and E1 of engine & ECT ECU.

HINT: The correct waveform appears as shown in the illustration on the left, with rectangle waves.

OK

NG 〉 Go to step **9**.

5 Check voltage between terminal 3 of No. 1, No. 2 igniter connectors and body ground.

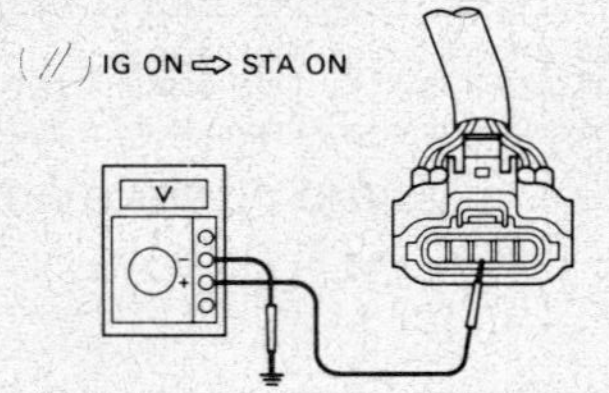

P Disconnect igniter connector.

C Measure voltage between terminal 3 of No. 1, No. 2 igniter connectors and body ground, when ignition switch is turned to "ON" and "STA" position.

OK **Voltage: 10 – 14 V**

OK

NG 〉 Go to step **8**.

6 Check for open and short in harness and connector between ignition relay and ignition coil, ignition coil and igniter

OK

NG 〉 **Repair or replace harness or connector.**

7 Check ignition coil.

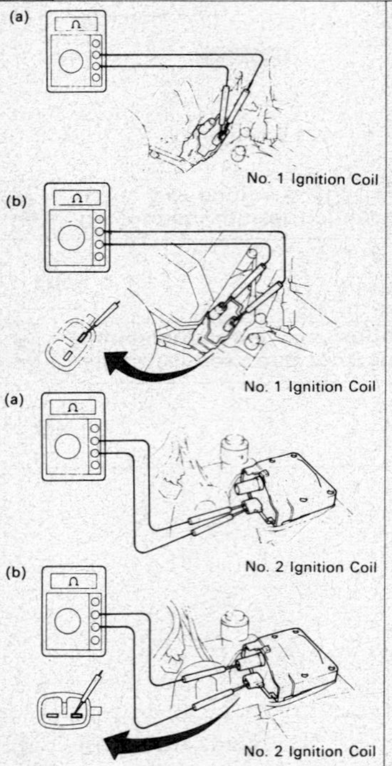

P Disconnect ignition coil connector.
(For No. 1 ignition coil (on left bank), remove the battery)

C (1) Check primary coil.
Measure resistance between terminals of ignition coil connector.
(2) Check secondary coil.
Measure resistance between terminal 1 of ignition coil connector and high-tension terminal.

OK

	Resistance
(a) Primary Coil (Cold)	0.40 – 0.50 Ω
(b) Secondary Coil (Cold)	10.2 – 13.8 kΩ

OK

NG 〉 **Replace ignition coil.**

Replace igniter. *1

*1: When diag. code 14 is displayed, replace the igniter with 5 wire harness. (extra wire is for tachometer). When diag. code 15 is displayed, replace the igniter with 4 wire harness.

8 Check ignition relay.

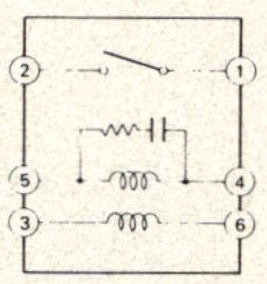

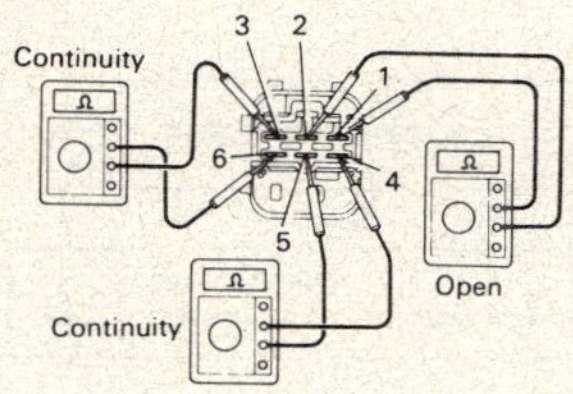

P Remove ignition relay.

C Check continuity between terminals of ignition relay shown below.

OK

Terminals 1 and 2	Open
Terminals 3 and 6	Continuity (Reference value 22 Ω)
Terminals 4 and 5	Continuity (Reference value 107 Ω)

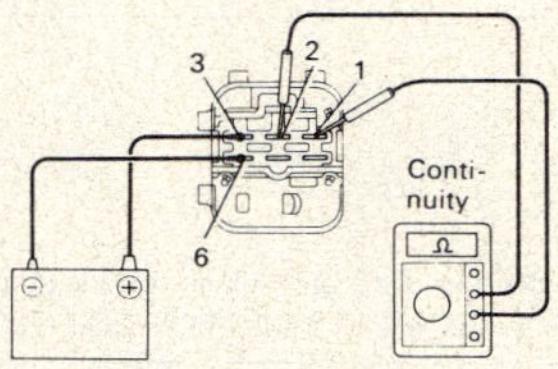

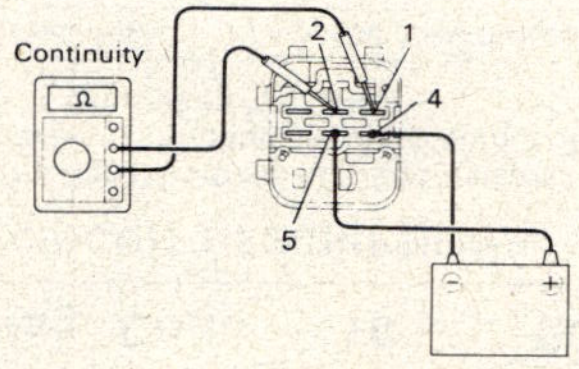

C (1) Apply battery voltage between terminals 3 and 6, 4 and 5.
(2) Check continuity between terminals 1 and 2.

OK

Terminals 1 and 2	Continuity

NG Replace ignition relay.

OK

Check and repair harness or connector between battery and ignition relay, ignition relay and igniter.

9 Disconnect igniter connector and check voltage between terminals IGT1, 2 of engine & ECT ECU connector and body ground.

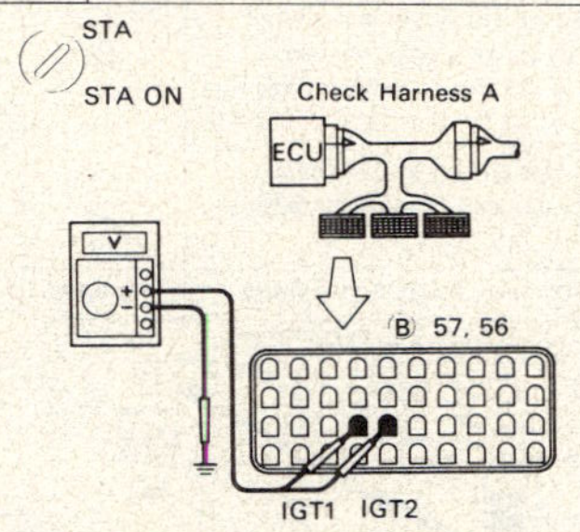

P (1) Disconnect igniter connector.
(2) Connect engine & ECT ECU connector.

C Measure voltage between terminals IGT1, 2 of engine & ECT ECU connector and body ground when engine is cranked.

OK **Voltage: 0.5 – 1.0 V**
(Neither 0 V nor 5 V)

Reference — INSPECTION USING OSCILLOSCOPE

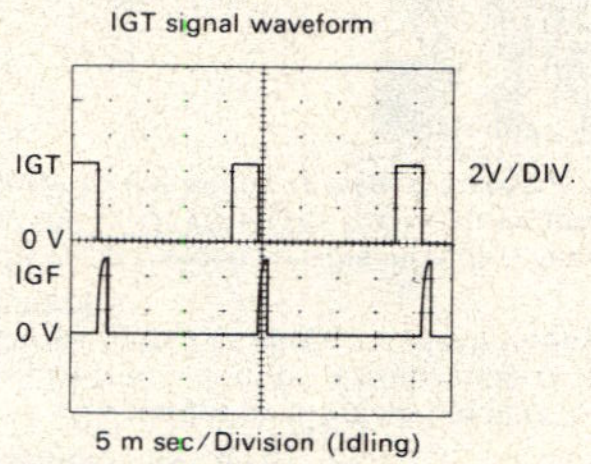

- During cranking or idling, measure between terminals IGT1, IGT2 and E1 of engine & ECT ECU.

HINT: The correct waveform appears as shown in the illustration on the left, with rectangle waves.

OK Replace igniter.*1

NG

10 Check for open and short in harness and connector in IGT signal circuit between engine & ECT ECU and igniter

NG Repair or replace harness or connector.

OK

Check and replace engine & ECT ECU.

*1: When diag. code 14 is displayed, replace the igniter with 5 wire harness. (extra wire is for tachometer). When diag. code 15 is displayed, replace the igniter with 4 wire harness.

Diag. Code 16 — ECT Control Signal Malfunction

CIRCUIT DESCRIPTION

The signal from the ECT CPU retards the ignition timing of the engine during ECT gear shifting, thus momentarily reducing torque output of the engine for smooth clutch operation inside the transmission and reduced shift shock.

Code No.	Diagnostic Code Detecting Condition	Trouble Area
16	Fault in communications between the engine CPU and ECT CPU in the ECU.	• ECU

If the ECU records the diagnostic code "16" in memory, it prohibits the torque control of the ECT which performs smooth gear shifting.

DIAGNOSTIC CHART

1 Are there any other codes (besides Code 16) being output?

- YES → Go to relevant diag. code chart.
- NO → Replace engine & ECT ECU.

Diag. Code 21, 28 — Main Oxygen Sensor Circuit

CIRCUIT DESCRIPTION

To obtain a high purification rate for the CO, HC and NOx components of the exhaust gas, a three-way catalyst is used, but for most efficient use of the three-way catalyst, the air-fuel ratio must be precisely controlled so that it is always close to the stoichiometric air-fuel ratio.

The oxygen sensor has the characteristic whereby its output voltage changes suddenly in the vicinity of the stoichiometric air-fuel ratio. This characteristic is used to detect the oxygen concentration in the exhaust gas and provide feedback to the computer for control of the air-fuel ratio.

When the air-fuel ratio becomes LEAN, the oxygen concentration in the exhaust increases and the oxygen sensor informs the ECU of the LEAN condition (small electromotive force: 0 V).

When the air-fuel ratio is RICHER than the stoichiometric air-fuel ratio the oxygen concentration in the exhaust gas is reduced and the oxygen sensor informs the ECU of the RICH condition (large electromotive force: 1V).

The ECU judges by the electromotive force from the oxygen sensor whether the air-fuel ratio is RICH or LEAN and controls the injection time accordingly. However, if malfunction of the oxygen sensor causes output of abnormal electromotive force, the ECU is unable to perform accurate air-fuel ratio control.

The main oxygen sensors include a heater which heats the Zirconia element. The heater is controlled by the ECU. When the intake air volume is low (the temperature of the exhaust gas is low) current flows to the heater to heat the sensor for accurate oxygen concentration detection.

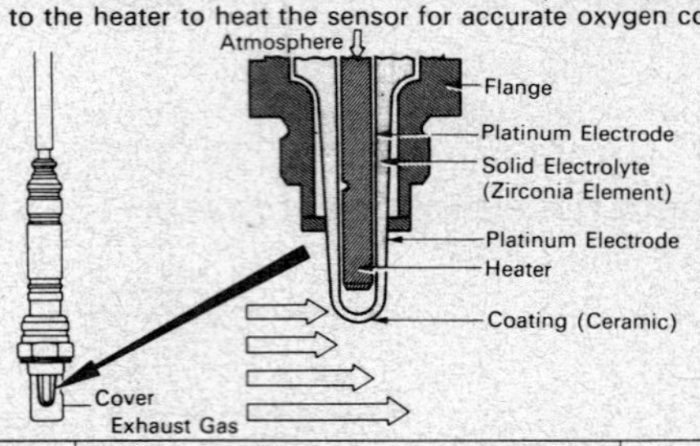

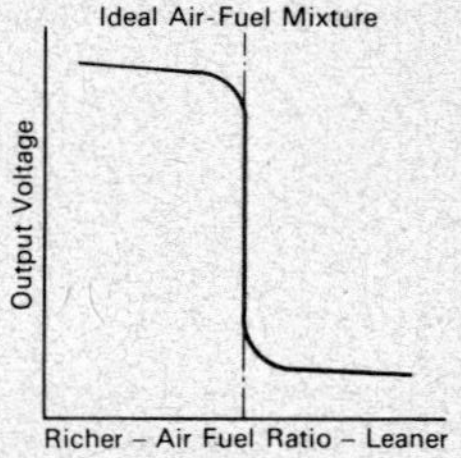

Code No.	Diagnostic Code Detecting Condition	Trouble Area
21 · 28	(1) Open or short in heater circuit of main oxygen sensor for 0.5 sec. or more.	• Open or short in heater circuit of main oxygen sensor. • Main oxygen sensor heater • ECU
	(2) Main oxygen sensor signal voltage is reduced to between 0.35 V and 0.70 V for 60 sec. under conditions (a) ~ (d). (2 trip detection logic) (a) Coolant temp.: Between 70°C (158°F) and 95°C (203°F). (b) Engine speed: 1,500 rpm or more. (c) Load driving (EX. ECT in 4th speed, A/C ON, Flat road, 50 mph (80km/h)). (d) Main oxygen sensor signal voltage: Alternating above and below 0.45 V.	• Main oxygen sensor circuit • Main oxygen sensor

HINT: Diag. code 21 is for the left bank main oxygen sensor circuit. Diag. code 28 is for the right bank main oxygen sensor circuit.

CIRCUIT DESCRIPTION (Cont'd)

DIAGNOSIS CODE DETECTION DRIVING PATTERN

Purpose of the driving pattern.
(a) To simulate diag. code detecting condition after diag. code is recorded.
(b) To check that the malfunction is corrected when the repair is completed confirming that diag. code is no longer detected.

Malfunction: **Main Oxygen Sensor Deterioration**

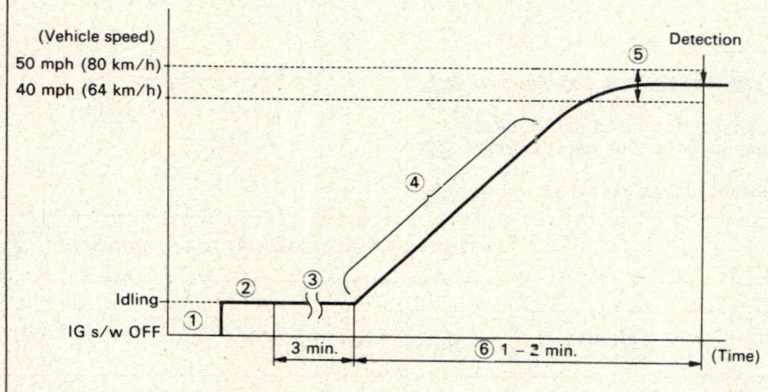

① Initiate test mode
② Start engine and warm up.
③ After engine is warmed up, let it idle for 3 min.
④ With the A/C ON and the transmission in D range (O/D ON), perform gradual acceleration with in the range 1,300 – 1,700 rpm (centered around 1,500 rpm).
(Take care that the engine speed does not fall below 1,200 rpm when shifting. Gradually depress the accelerator pedal and kept it. Steady so that engine braking does not occur.)
⑤ Maintain the vehicle speed at 40 – 50 mph (64 – 80 km/h).
⑥ Keep the vehicle running for 1 – 2 min. after starting acceleration.

HINT: If a malfunction exists, the "CHECK" engine warning light will light up after approx. 60 sec. from the start of acceleration.

NOTICE: If the conditions in this test are not strictly followed, detection of the malfunction will not be possible.

DIAGNOSTIC CHART

HINT: If diag. code "21" is output, check the left bank main oxygen sensor circuit. If diag. code "28" is output, check the right bank main oxygen sensor circuit.

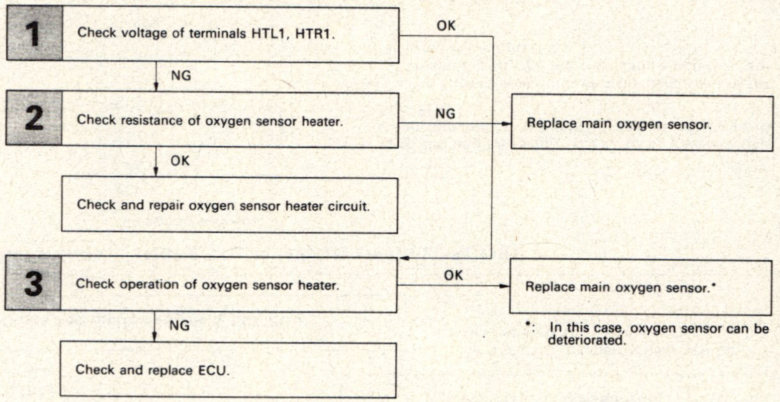

WIRING DIAGRAM

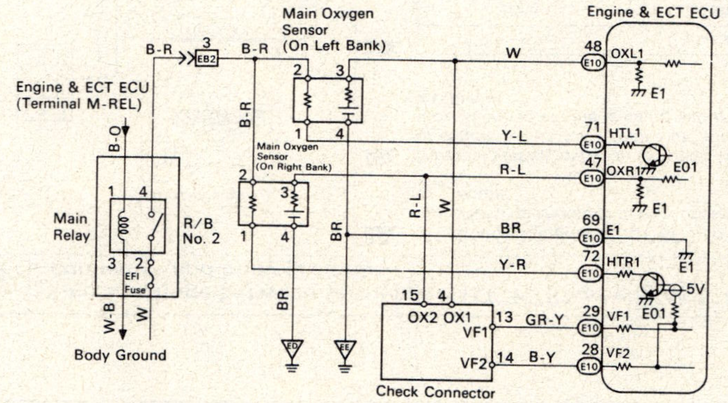

INSPECTION PROCEDURE

1 Disconnect the engine & ECT ECU connector, check voltage between terminals HTL1, HTR1 of engine & ECT ECU connector and body ground.

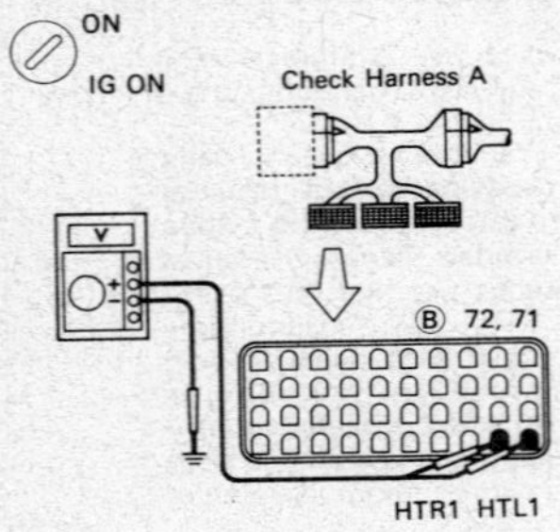

P (1) Connect the Check Harness A.
(2) Disconnect engine & ECT ECU connector.
(3) Turn ignition switch on.

C Measure voltage between terminals HTL1, HTR1 of engine & ECT ECU connector and body ground.

OK **Voltage: 10 – 14 V**

OK → Go to step **3**.

NG ↓

2 Check main oxygen sensor heater.

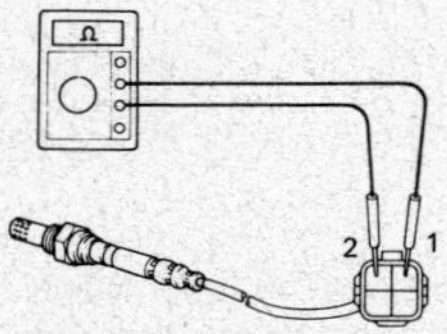

P Disconnect main oxygen sensor connector.

C Measure resistance between terminals 1 and 2 of main oxygen sensor connector.

OK **Resistance: 5.1 – 6.3 Ω at 20°C (68°F)**

NG → **Replace main oxygen sensor.**

OK ↓

Check and repair harness or connector between main relay and main oxygen sensor, main oxygen sensor and engine & ECT ECU.

3 Check voltage between terminals HTL1, HTR1 of engine & ECT ECU connector and body ground.

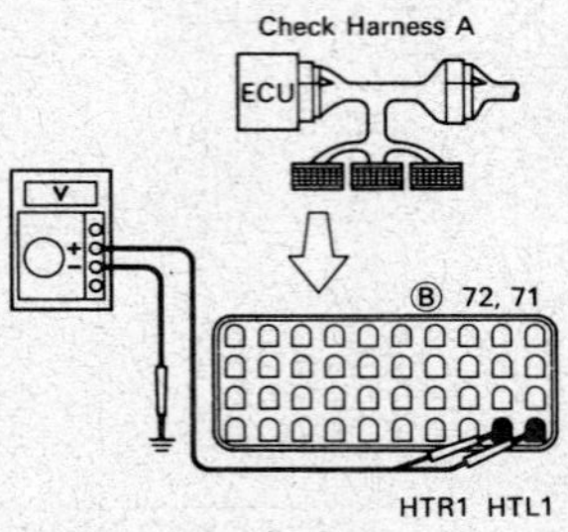

P (1) Connect engine & ECT ECU connector.
(2) Warm up engine to normal operating temperature.

C Measure voltage between terminals HTL1, HTR1 of engine & ECT ECU connector and body ground, when engine is idling and racing at 4,000 rpm.

OK

	Voltage
Idling	0 V
Racing at 4,000 rpm	10 – 14 V

Hint In the 4,000 rpm racing check, continue engine racing at 4,000 rpm for approx. 20 seconds or more.

OK → **Replace main oxygen sensor.***

*: In this case, oxygen sensor can be deteriorated.

NG ↓

Check and replace engine & ECT ECU.

Reference INSPECTION USING OSCILLOSCOPE

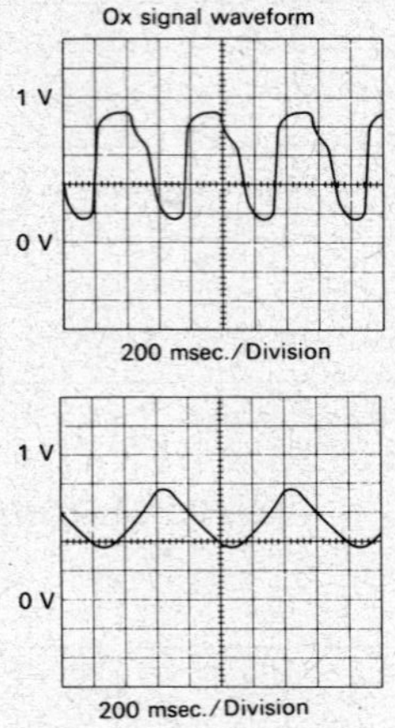

- With the engine racing (4,000 rpm) measure between terminals OXL1, OXR2 and E1 of engine & ECT ECU.

HINT: The correct waveform appears as shown in the illustration on the left, oscillating between approx. 0.1 V and 0.9 V

If the oxygen sensor is deteriorated, the amplitude of the voltage is reduced as shown on the left.

Diag. Code 22 Water Temp. Sensor Circuit

CIRCUIT DESCRIPTION

The water temperature sensor senses the coolant temperature. A thermistor built in the sensor changes the resistance value according to the coolant temperature. The lower the coolant temperature, the greater the thermistor resistance value, and the higher the coolant temperature, the lower the thermistor resistance value (See Fig. 1.).

The water temperature sensor is connected to the ECU (See next page). The 5 V power source voltage in the ECU is applied to the water temperature sensor from the terminal THW via a resistor R. That is, the resistor R and the water temperature sensor are connected in series. When the resistance value of the water temperature sensor changes in accordance with changes in the coolant temperature, the potential at the terminal THW also changes. Based on this signal, the ECU increases the fuel injection volume to improve driveability during cold engine operation. If the ECU records the diagnostic code 22, it operates the fail safe function, keeping the coolant temperature at a constant 80°C (176°F).

(fig-1)

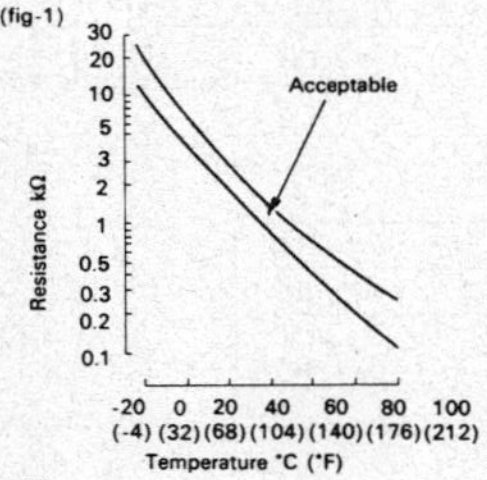

FI4741

< Reference >

Water Temp. °C (°F)	Resistance (kΩ)	Voltage (V)
-20 (-4)	16.0	4.3
0 (32)	5.9	3.4
20 (68)	2.5	2.4
40 (104)	1.2	1.5
60 (140)	0.6	0.9
80 (176)	0.3	0.5
100 (212)	0.2	0.3

Code No.	Diagnostic Code Detecting Condition	Trouble Area
22	Open or short in water temp. sensor circuit for 0.5 sec. or more.	• Open or short in water temp. sensor circuit • Water temp. sensor • ECU

DIAGNOSTIC CHART

HINT: If diagnostic codes "22" (water temperature sensor circuit), "24" (intake air temperature sensor circuit) and "41" (throttle position sensor circuit) are output simultaneously, E2 (sensor ground) may be open.

Step	Check	Result	Action
1	Check voltage of sensor.	OK	Check for momentary interruption
		NG ↓	
2	Check resistance of sensor.	NG	Replace water temp. sensor.
		OK ↓	
3	Check for open and short in harness and connector between ECU and water temp. sensor.	NG	Repair or replace harness or connector.
		OK ↓	
	Check and replace ECU.		

WIRING DIAGRAM

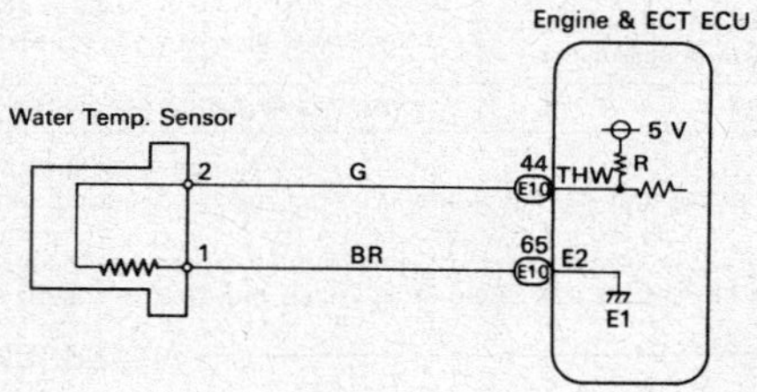

INSPECTION PROCEDURE

HINT: If diagnostic codes "22" (water temperature sensor circuit), "24" (intake air temperature sensor circuit) and "41" (throttle position sensor circuit) are output simultaneously, E2 (sensor ground) may be open.

1 Check voltage between terminals THW and E2 of engine & ECT ECU connector.

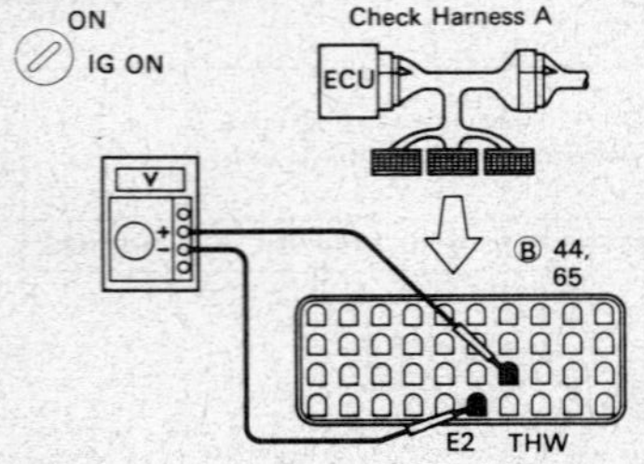

P (1) Connect the Check Harness A.

(2) Turn ignition switch on.

C Measure voltage between terminals THW and E2 of engine & ECT ECU connector.

OK

Water Temp. °C (°F)	Voltage
20 (68) (Engine is cool)	1 – 3 V
80 (176) (Engine is hot)	0.1 – 1.0 V

OK → **Check for momentary imterruption**

NG ↓

2 Check water temp. sensor.

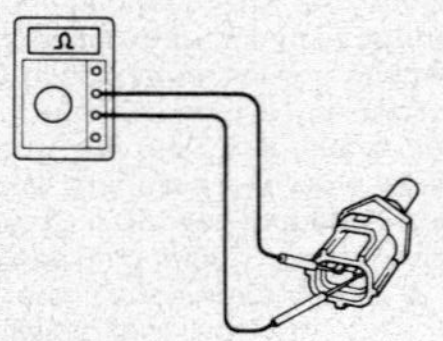

P Disconnect the water temp. sensor connector.

C Measure resistance between terminals.

OK **Resistance is within Acceptable Zone on chart.**

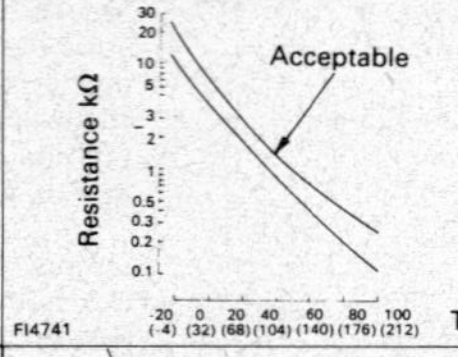

Water temp. °C (°F)	Resistance
20 (68)	2 – 3 kΩ
80 (176)	0.2 – 0.4 kΩ

NG → **Replace water temp. sensor.**

OK ↓

3 Check for open and short in harness and connector between engine & ECT ECU and water temp. sensor

NG → **Repair or replace harness or connector.**

OK ↓

Check and replace engine & ECT ECU.

Diag. Code	24	Intake Air Temp. Sensor Circuit

CIRCUIT DESCRIPTION

The intake air temp. sensor is built into the air flow meter and senses the intake air temperature. The structure of the sensor and connection to the ECU is the same as in the water temp. sensor shown on page TR-68.
If the ECU records the diagnostic code "24", it operates the fail safe function, keeping the intake air temperature at a constant 20°C (68°F).

Code No.	Diagnostic Code Detecting Condition	Trouble Area
24	Open or short in intake air temp. sensor circuit for 0.5 sec. or more.	• Open or short in intake air temp. sensor circuit. • Intake air temp. sensor • ECU

DIAGNOSTIC CHART

HINT: If diagnostic codes "22" (water temperature sensor circuit), "24" (intake air temperature sensor circuit) and "41" (throttle position sensor circuit) are output simultaneously, E2 (sensor ground) may be open.

1 Check voltage of sensor. — OK → Check for momentary interruption
↓ NG
2 Check resistance of sensor. — NG → Replace intake air temp. sensor.
↓ OK
3 Check for open and short in harness and connector between ECU and intake air temp. sensor. — NG → Repair or replace harness or connector.
↓ OK
Check and replace ECU.

WIRING DIAGRAM

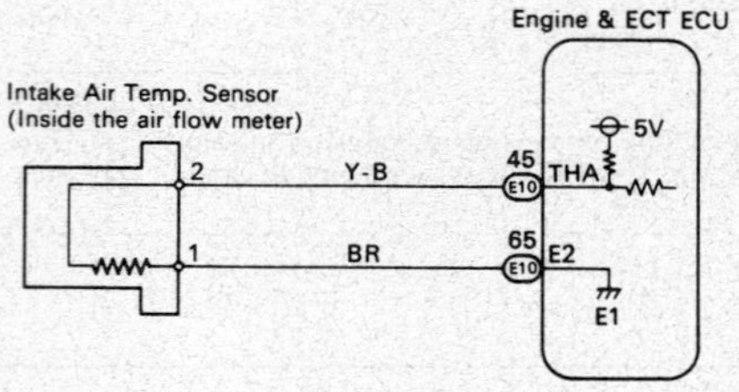

Diag. Code	25	Air-Fuel Ratio Lean Malfunction
	26	Air-Fuel Ratio Rich Malfunction

CIRCUIT DESCRIPTION

Code No.	Diagnostic Code Detecting Condition	Trouble Area
25 · 26	(1) Main oxygen sensor voltage is 0.45 V or less (lean) for 90 sec. under conditions (a) and (b). (2 trip detection logic) (a) Coolant temp. : 60°C (140°F) or more. (b) Engine speed : 1,500 rpm or more.	• Open or short in main oxygen sensor circuit • Main oxygen sensor • Ignition system • Water temp. sensor
	(2) Main oxygen sensor voltage is alternating above and below 0.45 V at 5 times par second or more under conditions (a) and (b). (2 trip detection logic)* (a) Engine speed : Idling (b) Coolant temp. : Between 60°C (140°F) and 95°C (203°F)	• Open or short in injector circuit. • Fuel line pressure (injector blockage) • Air flow meter (air intake) • Engine ground bolt loose • Foreign object caught in valve
	(3) Difference of air-fuel ratio feedback compensation value between right and left banks is more than 10 percentage for 30 sec. or more under conditions (a) and (b). (2 trip detection logic)* (a) Engine speed : 2,000 rpm or more. (b) Coolant temp. : Between 60°C (140°F) and 95°C (203°F)	• Fuel line pressure (injector leak, blockage) • Mechanical system malfunction (skipping teeth of timing belt) • Ignition system

INSPECTION PROCEDURE

1 Check voltage between terminals THA and E2 of engine & ECT ECU connector.

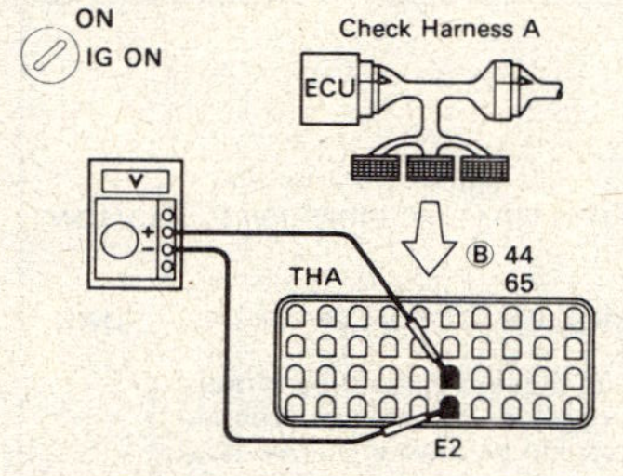

P (1) Connect the Check Harness A.

(2) Turn ignition switch on.

C Measure voltage between terminals THA and E2 of engine & ECT ECU connector.

OK

Intake air temp. °C (°F)	Voltage
20 (68)	1 – 3 V
60 (140)	0.5 – 1.0 V

NG ↓

OK → Check for momentary interruption

2 Check intake air temp. sensor.

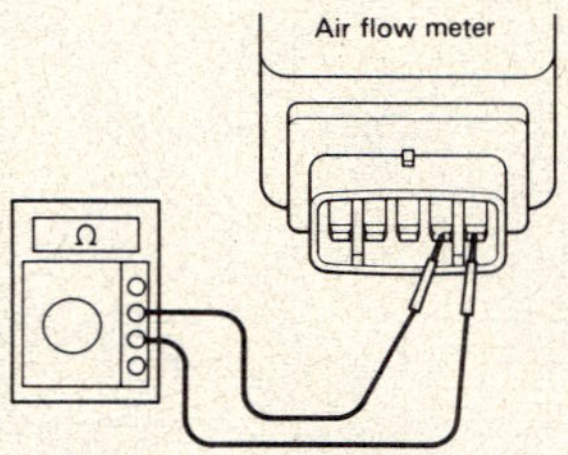

P Disconnect the air flow meter connector.

C Measure resistance between terminals 1 and 2 of air flow meter connector.

OK **Resistance is within Acceptable Zone on chart.**

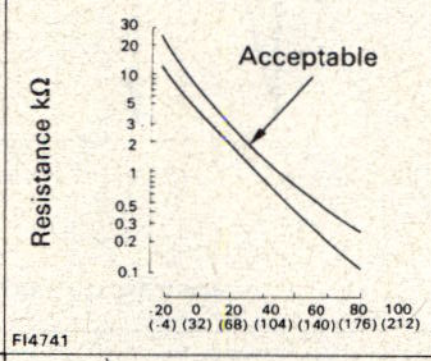

Intake air temp. °C (°F)	Resistance
20 (68)	2 – 3 kΩ
60 (140)	0.4 – 0.7 kΩ

OK ↓

NG → Replace intake air temp. sensor (Replace air flow meter).

3 Check for open and short in harness and connector between engine & ECT ECU and intake air temp. sensor

OK ↓

NG → Repair or replace harness or connector.

Check and replace engine & ECT ECU.

CIRCUIT DESCRIPTION (Cont'd)

DIAGNOSIS CODE DETECTION DRIVING PATTERN

Purpose of the driving pattern.

(a) To simulate diag. code detecting condition after diag. code is recorded.

(b) To check that the malfunction is corrected when the repair is completed confirming that diag. code is no longer detected.

Malfunction: Open or Short in Main Oxygen Sensor

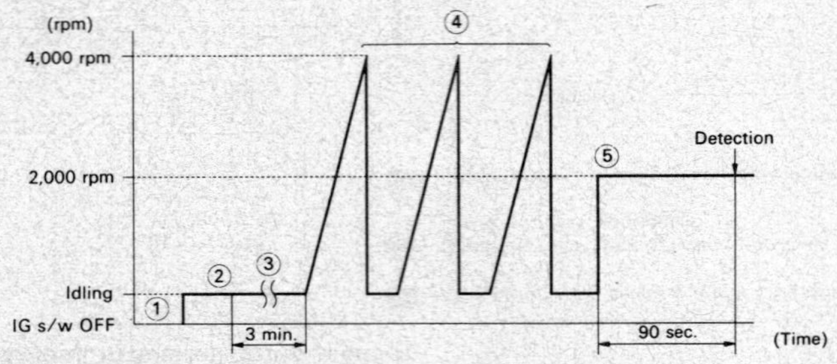

① Initiate test mode (See page TR-12).
② Start engine and warm up.
③ After engine is warmed up, let it idle for 3 min.
④ Perform quick racing to 4,000 rpm three times by accelerator pedal.
⑤ After performing the racing in ④, perform racing at 2,000 rpm for 90 sec.

HINT: If a malfunction exists, the "CHECK" engine warning light will light up after 90 sec. from the start of racing.

NOTICE: If the conditions in this test are not strictly followed, detection of the malfunction will not be possible.

CIRCUIT DESCRIPTION (Cont'd)

DIAGNOSIS CODE DETECTION DRIVING PATTERN

Purpose of the driving pattern.

(a) To simulate diag. code detecting condition after diag. code is recorded.

(b) To check that the malfunction is corrected when the repair is completed confirming that diag. code is no longer detected.

Malfunction: Open or Short in Injector circuit, Injector Leak or Blockage, Loose E/G Earth Bolt.

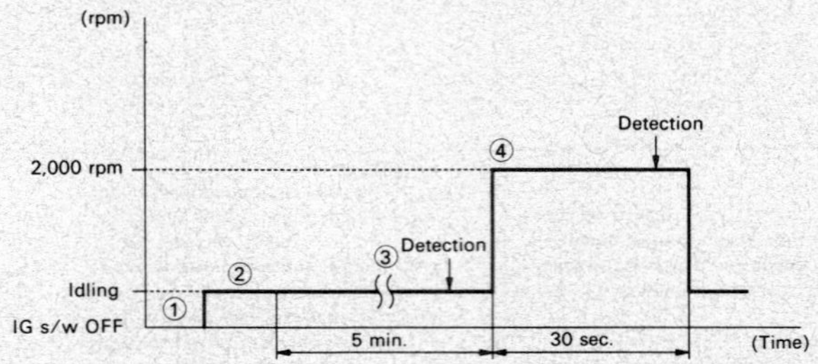

① Initiate test mode
② Start engine and warm up.
③ After engine is warmed up, let it idle for 5 min.
(After the engine is started, do not depress the accelerator pedal.)
④ If the malfunction is not detected during idling, perform racing without any load at approx. 2,000 rpm for 30 sec.

HINT: If a malfunction exists, the "CHECK" engine warning light will light up during the 5 min. idling period or within 30 sec. of starting racing.

NOTICE: If the conditions in this test are not strictly followed, detection of the malfunction will not be possible.

DIAGNOSTIC CHART

1 Check voltage of terminal VF.
- NG Type I → 2
- NG Type II → 7
- (OK → 7)

2 Check voltage of terminal OX.
- NG → 3
- OK → 7

3 Check for open and short in harness and connector.
- NG → Repair or replace harness or connector.
- OK → 4

4 Check each item found to be a possible cause of problem.
- NG → Repair or replace.
- OK → 5

5 Check compression
- NG → Repair or replace.
- OK → 6

6 Does malfunction disappear when a good main oxygen sensor is installed?
- YES → Replace main oxygen sensor.
- NO → Check and replace ECU.

7 Check each item found to be a possible cause of problem.
- NG → Repair or replace.
- OK → 8

8 Check compression
- NG → Repair or replace.
- OK → 9

9 Does malfunction disappear when a good main oxygen sensor is installed?
- YES → Replace main oxygen sensor.
- NO → Check and replace ECU.

INSPECTION PROCEDURE

1 Check voltage between terminals VF1, VF2 and E1 of check connector.

Left Bank

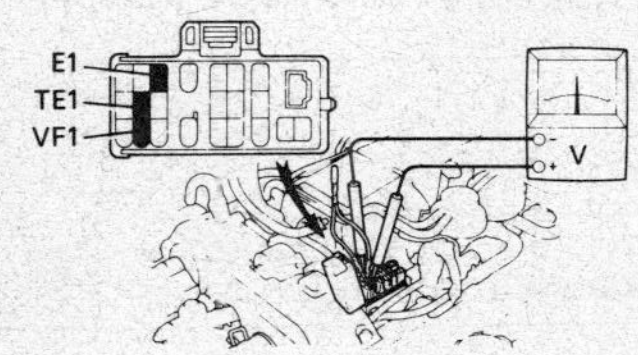

Right Bank

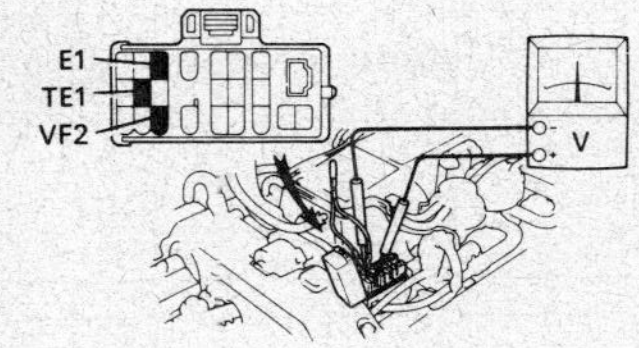

P
(1) Warm up engine at normal operating temperature.
(2) Connect terminals TE1 and E1 of check connector.
(3) Connect positive prove to terminal VF1, VF2 and negative prove to terminal E1 of check connector.

C
(1) Warm up the oxygen sensor by running engine at 2,500 rpm for about 2 minutes.
(2) Then, maintaining engine at 2,500 rpm, count how many times needle of voltmeter fluctuates between 0 and 5 V.

Result

Result	
Needle fluctuates of 8 times for every ten seconds	OK
Continue at 0 V	NG Type I
Continue at 5 V	NG Type II

NG Type II → Go to step 7.

NG Type I ↓

2 Check voltage between terminals OX1, OX2 and E1 of check connector.

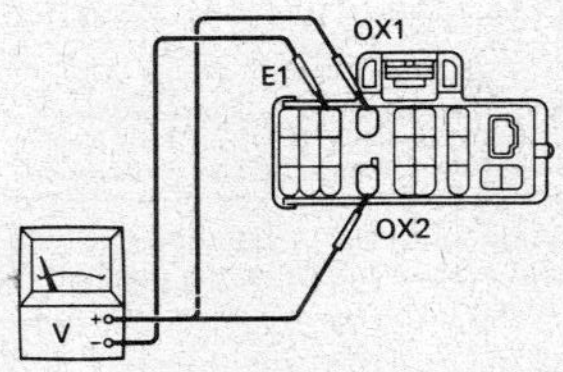

P Warm up engine at normal operating temperature.

C Measure voltage between terminals OX1, OX2 and E1 of check connector when engine is suddenly raced to full throttle.

OK **The voltage should be 0.5 V or higher at least once.**

Hint Perform inspection within 1 second.

OK → Go to step 7.

NG ↓

3 Check for open and short in harness and connector between engine & ECT ECU and main oxygen sensor, engine & ECT ECU and check connector

NG → Repair or replace harness or connector.

OK ↓

4 Check each item found to be a possible cause of problem.

Check each circuit found to be a possible cause of trouble according to the results of the check in 1 or 2. The numbers in the table below show the order in which the checks should be performed.

Main oxygen sensor signal from either side continues at 0 V.	Main oxygen sensor signals from both sides continue at 0 V.	Possible Cause	See page
1		Faulty sensor installation.	–
3		Injector circuit	TR-132
2	3	Misfire	IG-6
4		Valve timing	EM-51
	1	Air leakage	FI-13
	2	Fuel system	TR-114
	6	Characteristics deviation in air flow meter.	TR-88
	4	Characteristics deviation in water temp. sensor.	TR-68
	5	Characteristics deviation in intake air temp. sensor.	TR-72

NG → Repair or replace.

OK ↓

5 Check compression

NG → Repair or replace.

OK ↓

6 Does malfunction disappear when a good main oxygen sensor is installed?

YES → Replace main oxygen sensor.

NO ↓

Check and replace engine & ECT ECU.

7 Check each item found to be a possible cause of problem.

Check each circuit found to be a possible cause of trouble according to the results of the check in 1. The numbers in the table below show the order in which the checks should be performed.

Main oxygen sensor signal from either side continues at 5.0 V.	Main oxygen sensor signals from both sides continue at 5.0 V.	Main oxygen sensor signals from both sides are normal.	Possible Cause	
1		7	Injector circuit	
		3	Misfire	
2		4	Valve timing	
		1	Air leakage	
	1	2	Fuel system	
3	4		Cold start injector circuit	
	5	8	Characteristics deviation in air flow meter.	
	2	5	Characteristics deviation in water temp. sensor.	
	3	6	Characteristics deviation in intake air temp. sensor.	

NG → Repair or replace.

OK ↓

8 Check compression

NG → Repair or replace.

OK ↓

9 Does malfunction disappear when a good main oxygen sensor is installed?

YES → Replace main oxygen sensor.

NO ↓

Check and replace engine & ECT ECU.

Diag. Code	27, 29	Sub-Oxygen Sensor Circuit

CIRCUIT DESCRIPTION

The sub-oxygen sensor is installed on the exhaust pipe. Its construction and operation is the same as the main oxygen sensor

Code No.	Diagnostic Code Detecting Condition	Trouble Area
27 · 29	(1) Open or short in heater circuit of sub-oxygen sensor for 0.5 sec. or more.	• Open or short in heater circuit of sub-oxygen sensor. • Sub-oxygen sensor heater. • ECU
	(2) Main oxygen sensor signal is 0.45 V or more and sub-oxygen sensor signal is 0.45 V or less under conditions (a) and (b). (2 trip detection logic) (a) Coolant temp. : 80°C (176°F) or more. (b) Accel. pedal : Fully depressed for 2 sec. or more.	• Open or short in sub-oxygen sensor circuit. • Sub-oxygen sensor • ECU

HINT: Diag. code 27 is for the left bank sub-oxygen sensor circuit.
Diag. code 29 is for the right bank sub-oxygen sensor circuit.

CIRCUIT DESCRIPTION (Cont'd)

DIAGNOSIS CODE DETECTION DRIVING PATTERN

Purpose of the driving pattern.

(a) To simulate diag. code detecting condition after diag. code is recorded.

(b) To check that the malfunction is corrected when the repair is completed confirming that diag. code is no longer detected.

Malfunction: Open or Short in Sub-Oxygen Sensor

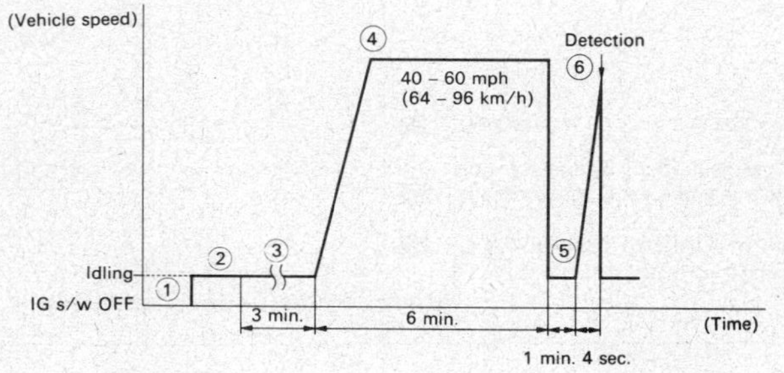

① Initiate test mode
② Start engine and warm up.
③ After engine is warmed up, let it idle for 3 min.
④ With the A/C ON and the transmission in D range (O/D OFF), drive at 40 – 60 mph (64 – 96 km/h) for 6 min.
⑤ After the driving in ④, stop the vehicle.
⑥ Once the vehicle is stopped, within the next minute apply acceleration with the throttle fully open for 4 sec.

HINT: If a malfunction exists, the "CHECK" engine warning light will light up after full acceleration for 4 sec.

NOTICE: **If the conditions in this test are not strictly followed, detection of the malfunction will not be possible.**

DIAGNOSTIC CHART

HINT:
- When other codes are output in addition to 27 and 29 at the same time, check the circuits for other codes first.
- If diag. code 27 is displayed, check left bank sub-oxygen sensor circuit, and if 29 is displayed, check right bank sub-oxygen sensor circuit.

1 Check voltage of terminals HTL2, HTR2.
- OK → go to 3
- NG ↓

2 Check resistance of sub-oxygen sensor heater.
- NG → Replace sub-oxygen sensor.
- OK ↓

Check and repair sub-oxygen sensor heater circuit.

3 Check operation of sub-oxygen sensor heater.
- OK → Replace sub-oxygen sensor.*
- NG ↓

Check and replace ECU.

*: In this case, oxygen sensor can be deteriorated.

WIRING DIAGRAM

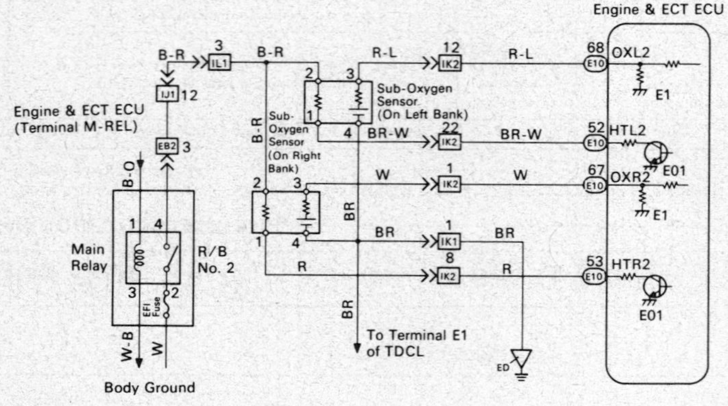

INSPECTION PROCEDURE

1 Disconnect the engine & ECT ECU connector, check voltage between terminals HTL2, HTR2 of engine & ECT ECU connector and body ground.

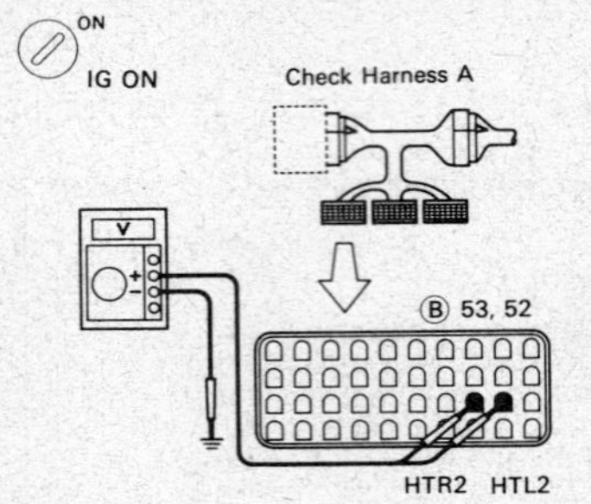

P (1) Connect the Check Harness A.
(2) Disconnect engine & ECT ECU connector.
(3) Turn ignition switch on.

C Measure voltage between terminals HTL2, HTR2 of engine & ECT ECU connector and body ground.

OK **Voltage: 10 – 14 V**

NG

OK **Go to step 3.**

2 Check sub-oxygen sensor heater.

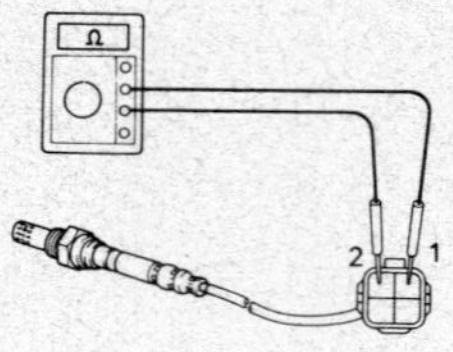

P Disconnect sub-oxygen sensor connector.

C Measure resistance between terminals 1 and 2 of sub-oxygen sensor connector.

OK **Resistance: 5.1 – 6.3 Ω 20°C (68°F)**

NG **Replace sub-oxygen sensor.**

OK

Check and repair harness or connector between main relay and sub-oxygen sensor, sub-oxygen sensor and engine & ECT ECU.

3 Check voltage between terminals HTL2, HTR2 of engine & ECT ECU connector and body ground.

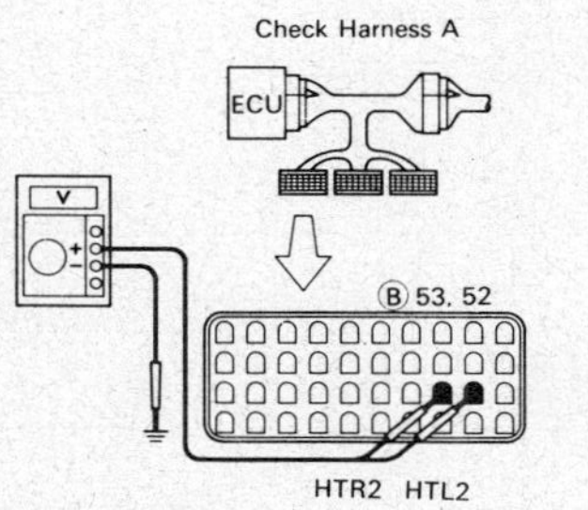

P (1) Connect engine & ECT ECU connector.
(2) Warm up engine to normal operating temperature.

C Measure voltage between terminals HTL2, HTR2 of engine & ECT ECU connector and body ground, when engine is idling and racing at 4,000 rpm.

OK

	Voltage
Idling	0 V
Racing at 4,000 rpm	10 – 14 V

Hint In the 4,000 rpm racing check, continue engine racing at 4,000 rpm for approx. 20 seconds or more.

OK → **Replace sub-oxygen sensor.***

NG → **Check and replace engine & ECT ECU.**

*: **In this case, oxygen sensor can be deteriorated.**

Diag. Code 31 Air Flow Meter Circuit

CIRCUIT DESCRIPTION

As shown in the figure at right, when a pillar (Vortex generating body) is placed in the path of a uniform flow, vortices called Karman-Vortex are generated downstream of the object. Using this principle, a vortex generator is placed inside the air flow meter. By measuring the frequency of the vortices generated, the ECU can determine the volume of air flowing through the air flow meter. The vortices are detected by their exerting pressure on thin metal foil (mirror) surfaces and a light emitting element and light receptor (LED and photo transistor) positioned opposite the mirror which senses the vibrations in the mirror optically. The ECU uses these signals mainly for calculation of the basic injection volume and the basic ignition advance angle.

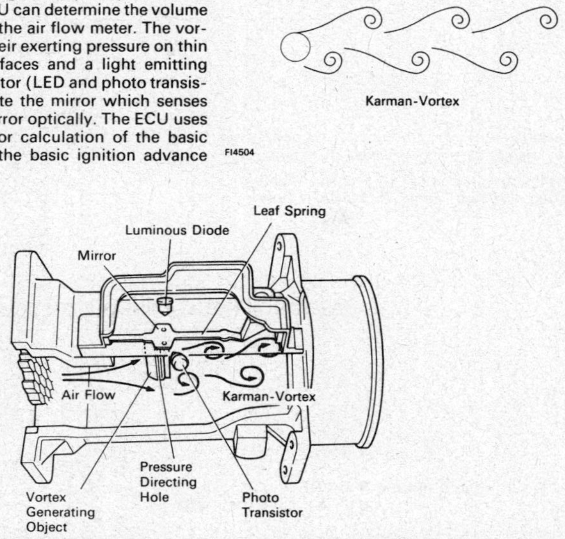

Code No.	Diagnostic Code Detecting Condition	Trouble Area
31	All conditions below are detected. (a) No air-flow meter signal to ECU for 2 sec. when engine speed is above 300 rpm. (b) Engine stall.	• Open or short in air flow meter circuit • Air flow meter • ECU

If the ECU records diagnostic code "31", it enters fail safe mode

DIAGNOSTIC CHART

Step	Check	Result	Action
1	Check voltage of terminal KS.	OK	When diag. code 31 is displayed, check and replace ECU.
		NG ↓	
2	Check for open and short in harness and connector between air flow meter and ECU.	NG	Repair or replace harness or connector.
		OK ↓	
3	Check voltage terminal KS of ECU.	NG	Check and replace ECU.
		OK ↓	
4	Check voltage of air flow meter power source.	OK	Replace air flow meter.
		NG ↓	
	Check and replace ECU.		

WIRING DIAGRAM

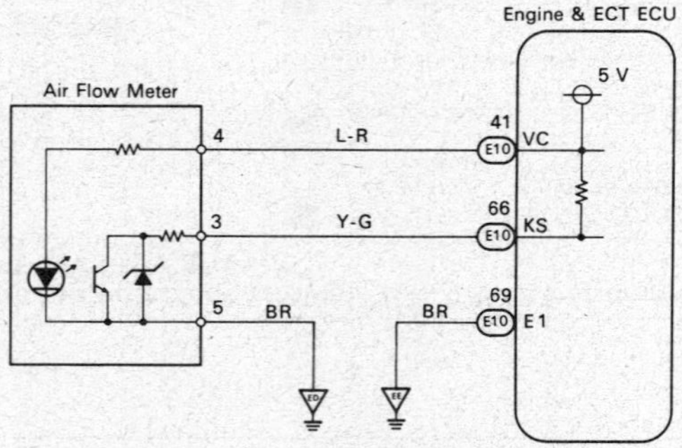

INSPECTION PROCEDURE

1 Check voltage between terminals KS and E1 of engine & ECT ECU connector.

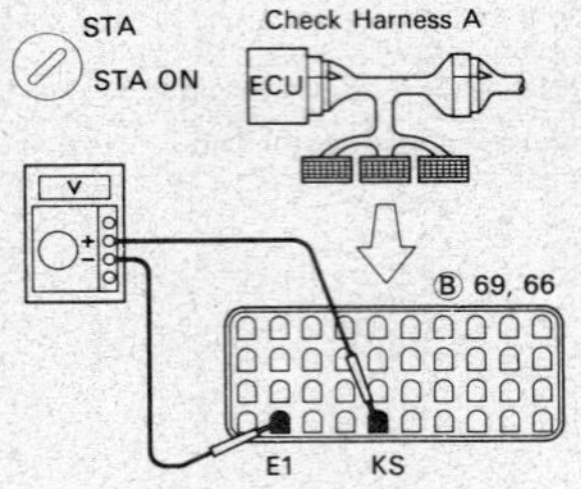

P Connect the check harness A.

C Measure voltage between terminals KS and E1 of engine & ECT ECU connector while engine is cranked.

OK **Voltage: 2.0 – 4.0 V (Neither 0 V nor 5 V)**

Reference **INSPECTION USING OSCILLOSCOPE**

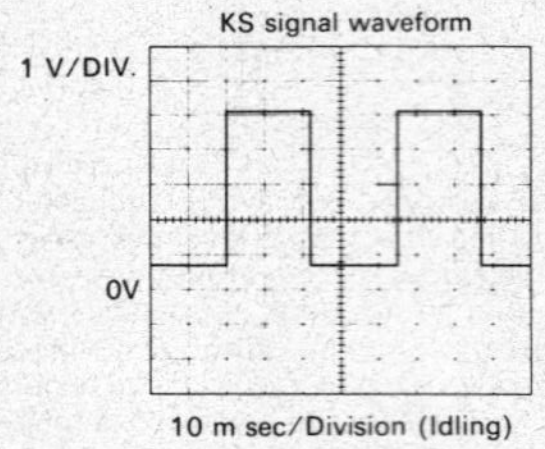

- During cranking or idling, measure between terminals KS and E1 of engine & ECT ECU.

HINT: The correct waveform appears as shown in the illustration on the left, with rectangle waves.

NG ↓

OK → **When diag. code 31 is displayed, check and replace engine & ECT ECU.**

2 Check for open and short in harness and connector between engine & ECT ECU and air flow meter

OK ↓

NG → **Repair or replace harness or connector.**

3 **Disconnect air flow meter connector and check voltage between terminals KS and E1 of engine & ECT ECU.**

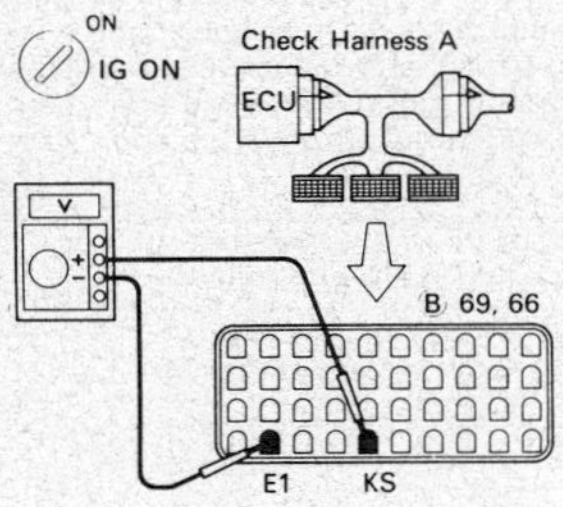

P (1) Disconnect the air flow meter connector.
(2) Turn ignition switch on.

C Measure voltage between terminals KS and E1 of engine & ECT ECU.

OK **Voltage: 4 – 6 V**

NG Check and replace ECU.

OK

4 **Disconnect air flow meter connector and check voltage between terminals VC and E1 of engine & ECT ECU.**

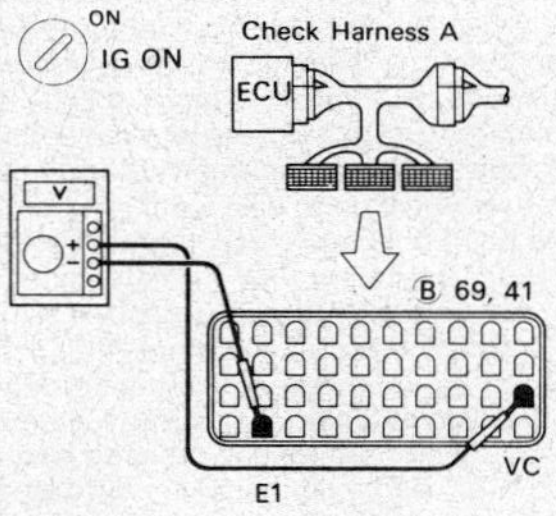

P (1) Disconnect the air flow meter connector.
(2) Turn ignition switch on.

C Measure voltage between terminals VC and E1 of engine & ECT ECU connector.

OK **Voltage: 4 – 6 V**

OK Replace air flow meter.

NG

Check and replace engine & ECT ECU.

Diag. Code	35	High Altitude Compensator Sensor (HAC Sensor) Circuit

CIRCUIT DESCRIPTION

The HAC sensor is built into the ECU. This is a semiconductor pressure sensor with properties which cause its electrical resistance to change when stress is applied to the sensor's crystal (silicon) (piezoelectric effect). This sensor is used to detect the atmospheric (absolute) pressure and outputs corresponding electrical signals. Fluctuations in the air pressure cause changes in the intake air density, which can cause deviations in the air-fuel ratio. The signals from HAC sensor are used to make corrections for these fluctuations. If the ECU records diagnostic code "35", the fail safe function operates and the atmospheric pressure is set at a constant 760 mmHg.

Code No.	Diagnostic Code Detecting Condition	Trouble Area
35	Open or short detected in HAC sensor circuit for 0.5 sec. or more.	• ECU

DIAGNOSTIC CHART

1 Are there any other codes (besides Code 16) being output?

YES → Go to relevant diag. code chart.

NO → Replace engine & ECT ECU.

Diag. Code	41, 47	(Sub-) Throttle Position Sensor Circuit

CIRCUIT DESCRIPTION

The throttle position sensor is mounted in the throttle body and detects the throttle valve opening angle. When the throttle valve is fully closed, the IDL contacts in the throttle position sensor are on, so the voltage at the terminal IDL of the ECU become 0 V. At this time, a voltage of approximately 0.7 V is applied to the terminal VTA of the ECU. When the throttle valve is opened, the IDL contacts go off and thus the power source voltage of approximately 12 V in the ECU is applied to the terminal IDL of the ECU. The voltage applied to the terminal VTA of the ECU increases in proportion to the opening angle of the throttle valve and becomes approximately 3.5 – 5.0 V when the throttle valve is fully opened. The ECU judges the vehicle driving conditions from these signals input from the terminals VTA and IDL, and uses them as one of the conditions for deciding the air-fuel ratio correction, power increase correction and fuel-cut control etc. The sub-throttle position sensor is built and operates in the same way as the main throttle position sensor. This sensor is used for traction control. The sub-throttle valve is opened and closed by the sub-throttle actuator according to signals from the TRAC ECU to control the engine output.

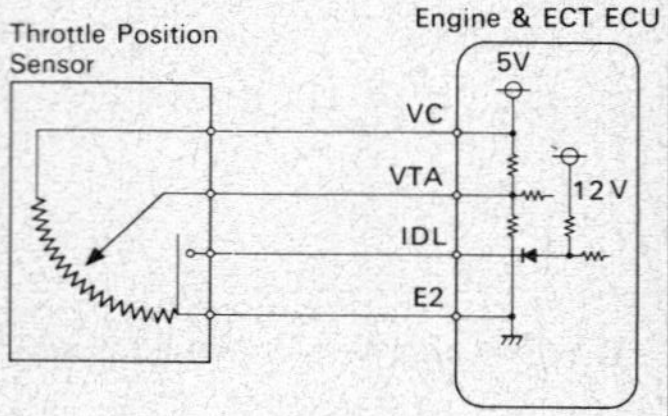

Code No.	Diagnostic Code Detecting Condition	Trouble Area
41	(1) Open or short in throttle position sensor circuit (VTA1) for 0.5 sec. or more.	• Open or short in throttle position sensor circuit.
	(2) IDL1 contact is ON and VTA1 output exceeds 1.5 V for 0.5 sec. or more.	• Throttle position sensor • ECU
47	(1) Open or short in sub-throttle position sensor circuit (VTA2) for 0.5 sec. or more.	• Open or short in sub-throttle position sensor circuit.
	(2) IDL2 contact is ON and VTA2 output exceeds 1.5 V for 0.5 sec. or more.	• Sub-throttle position sensor • ECU

HINT:
Diag. code 41 is for the throttle position sensor circuit.
Diag. code 47 is for the sub-throttle position sensor circuit.
- When the connector for the (sub-) throttle position sensor is disconnected, diagnostic code 41 or 47 is not displayed. Diagnostic code 41 or 47 is displayed only when there is an open or short in the VTA signal circuit of the (sub-) throttle position sensor.
- Signals from the (sub-) throttle position sensor are also input to the TRAC ECU, So when a malfunction occurs on the TRAC side, code 41 or 47 may be displayed.

DIAGNOSTIC CHART

HINT:
- If diag. code 41 is displayed, check throttle position sensor circuit; if diag. code 47 is displayed, check sub-throttle position sensor circuit.
- If diag. codes "22"(water temperature sensor circuit), "24" (intake air temperature sensor circuit) and "41" (throttle position sensor circuit) are output simultaneously, E2 (sensor ground) may be open.

1 Check voltage of throttle position sensor. — OK → Check for momentary interruption

↓ NG

2 Check operation for throttle position sensor. — NG → Adjust or replace throttle position sensor.

↓ OK

3 Check for open and short in harness and connector between throttle position sensor and ECU. — NG → Repair or replace harness or connector.

↓ OK

Check and replace ECU.

WIRING DIAGRAM

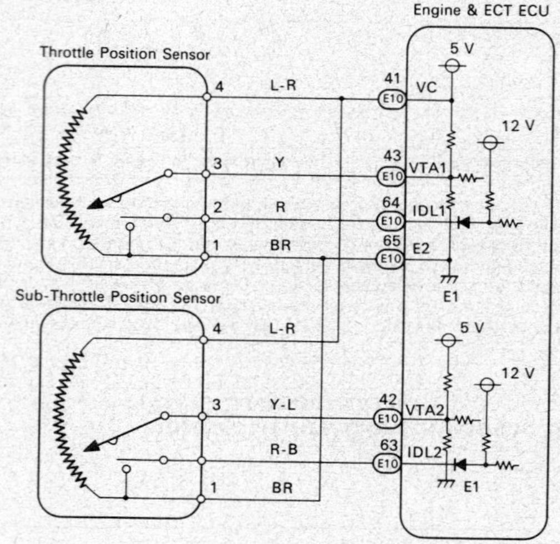

INSPECTION PROCEDURE

HINT:
- If diag. code 41 is displayed, check throttle position sensor circuit. If diag. code 47 is displayed, check sub-throttle position sensor circuit.
- If diag. codes "22" (water temperature sensor circuit), "24" (intake air temperature sensor circuit) and "41" (throttle position sensor circuit) are output simultaneously, E2 (sensor ground) may be open.

1 Check voltage between terminals VTA1, 2, IDL1, 2 and E2 of engine & ECT ECU connector.

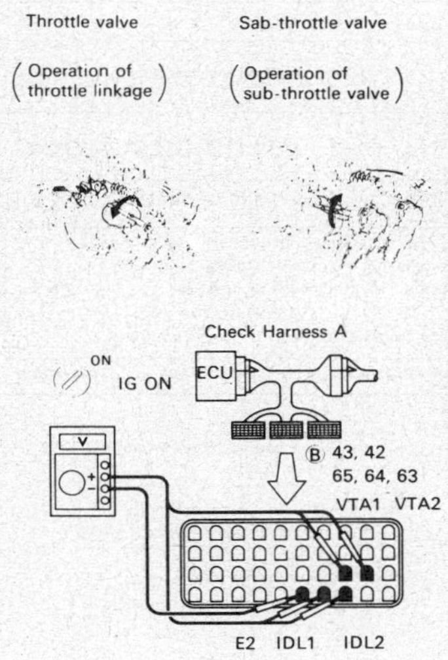

P (1) Connect the Check Harness A.
(2) Turn ignition switch on.
(3) For sub-throttle position sensor, remove intake air duct and disconnect sub-throttle valve step motor connector.

C Measure voltage between terminals VTA1, 2, IDL1, 2 and E2 of engine & ECT ECU connector when the (sub-) throttle valve is opened gradually from the closed condition.

OK

Terminal / Throttle Valve	VTA1, 2	IDL1, 2
Fully Closed	0.1 – 1.0 V	Below 1.0 V
Fully Open	3 – 5 V	10 – 14 V

Hint The voltage should increase steadily in proportion to the throttle valve opening angle.

OK → Check for momentary interruption

NG → Go to step 2.

2 Check (sub-) throttle position sensor.

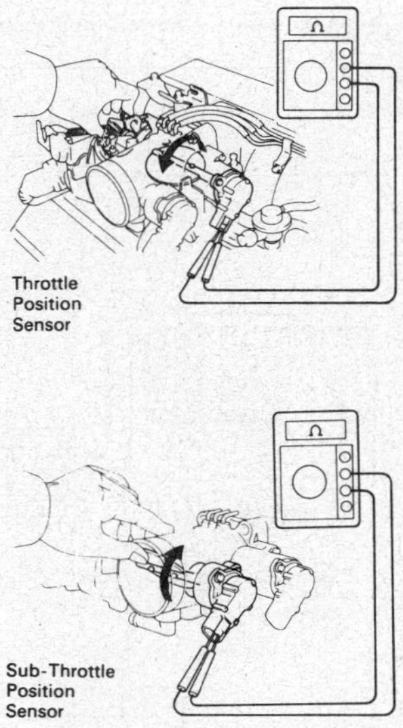

P (1) Remove intake air duct.
(2) Disconnect (sub-) throttle position sensor connector.

C Measure resistance between terminals 3, 2 and 1 of (sub-) throttle position sensor connector when the throttle valve is opened gradually from the closed condition.

OK

Terminal / Throttle Valve	3 – 1	2 – 1
Fully Closed	0.2 – 0.8 kΩ	Less than 0.5 kΩ
Fully Open	2.8 – 8.0 kΩ	1MΩ or higher

Hint Resistance between terminals 3 and 1 should increase gradually in accordance with the throttle valve opening angle.

NG → Adjust or replace (sub-) throttle position sensor

OK ↓

3 Check for open and short in harness and connector between engine & ECT ECU and (sub-) throttle position sensor

NG → Repair or replace harness or connector.

OK ↓

Check and replace engine & ECT ECU.

Diag. Code	42	Vehicle Speed Sensor Signal Circuit

CIRCUIT DESCRIPTION

The No. 1 speed sensor outputs a 20-pulse signal for every revolution of the rotor shaft, which is rotated by the transmission output shaft via the driven gear. This signal is converted in the combination meter to 4-pulse signal by the pulse conversion circuit and then sent to the Engine & ECT ECU. The ECU determines the vehicle speed based on the frequency of these pulse signals.

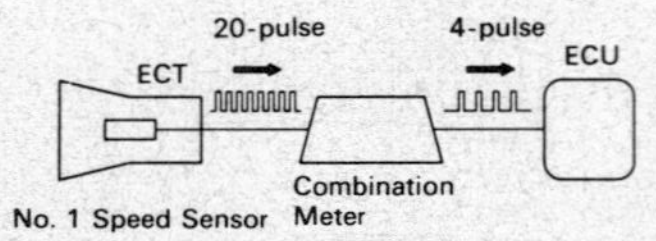

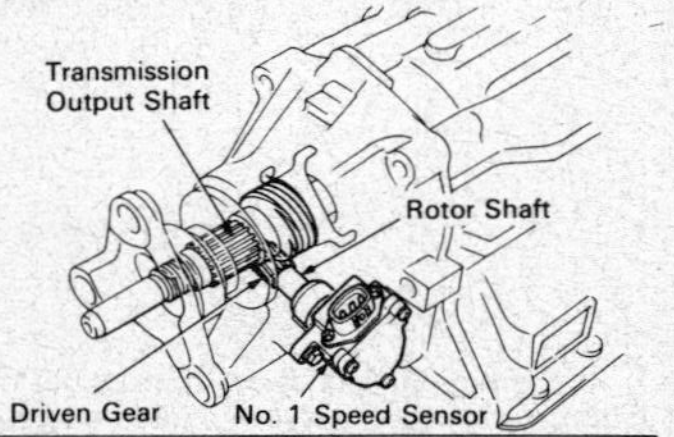

Code No.	Diagnostic Code Detecting Condition	Trouble Area
42	All conditions below are detected continuously for 8 sec. or more. (a) Vehicle speed signal : 0 km/h (0 mph) (b) Engine speed : 1,500 rpm or more (c) Neutral start switch (NSW) : OFF	• No. 1 speed sensor. • Combination meter. • Open or short in No. 1 speed sensor circuit. • ECU

HINT: In test mode, diag. code 42 is output when vehicle speed is 5 km/h (3 mph) or below.

DIAGNOSTIC CHART

1 Check operation of speedometer. — NG → Check speedometer circuit.

↓ OK

2 Check voltage of terminal SPD. — NG → Check speedometer circuit.

↓ OK

Check and replace ECU.

WIRING DIAGRAM

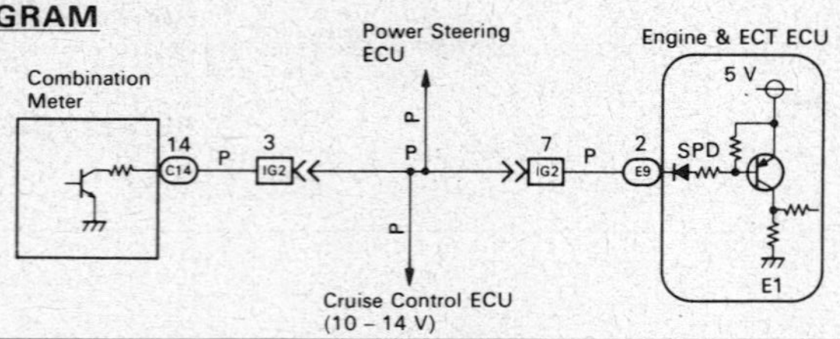

INSPECTION PROCEDURE

1 Check operation of speedometer.

C Drive the vehicle and check if the operation of the speedometer in the combination meter is normal.

Hint The No. 1 speed sensor is operating normally if the speedometer display is normal.

OK ↓ **NG** 〉 **Check speedometer circuit.**

2 Check voltage between terminal SPD of engine & ECT ECU connector and body ground.

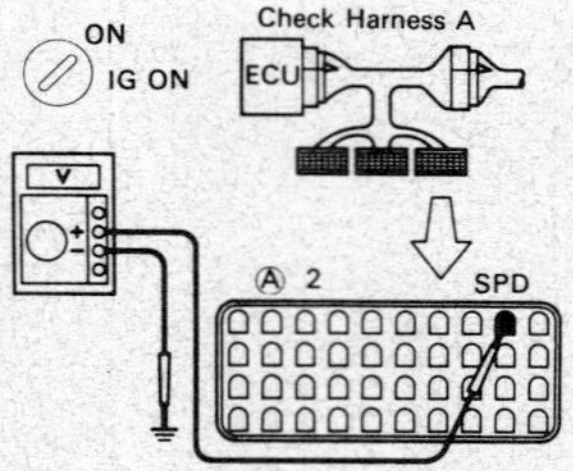

P (1) Shift the shift lever to N range.
(2) Jack up a rear wheel on one side.
(3) Connect the Check Harness A.

(4) Turn ignition switch on.

C Measure voltage between terminal SPD of engine & ECT ECU connector and body ground when the wheel is turned slowly.

OK **Voltage is generated intermittently.**

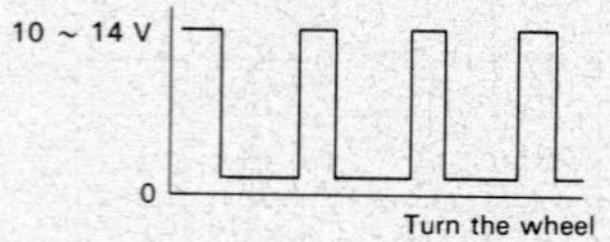

NG 〉 **Check speedometer circuit.**

OK ↓

Check and replace engine & ECT ECU.

Diag. Code	43	Starter Signal Circuit

CIRCUIT DESCRIPTION

When the engine is cranked, the intake air flow is slow, so fuel vaporization is poor. A rich mixture is therefore necessary in order to achieve good startability. While the engine is being cranked, the battery voltage is applied to terminal STA of the ECU. The starter signal is mainly used to increase the fuel injection volume for the starting injection control and after-start injection control.

Code No.	Diagnostic Code Detecting Condition	Trouble Area
43	No starter signal to ECU.	• Open or short in starter signal circuit. • Open or short in ignition switch or main relay circuit. • ECU

DIAGNOSTIC CHART

HINT: This diagnostic chart is based on the premise that the engine is cranked normally. If the engine is not cranked, proceed to the matrix chart of problem symptoms

1 Check for the test mode. — OK → Proceed to next circuit inspection shown on matrix chart

NG ↓

2 Check for open in harness and connectors between ECU and relay. — NG → Repair or replace harness or connector.

OK ↓

Check and replace ECU.

WIRING DIAGRAM

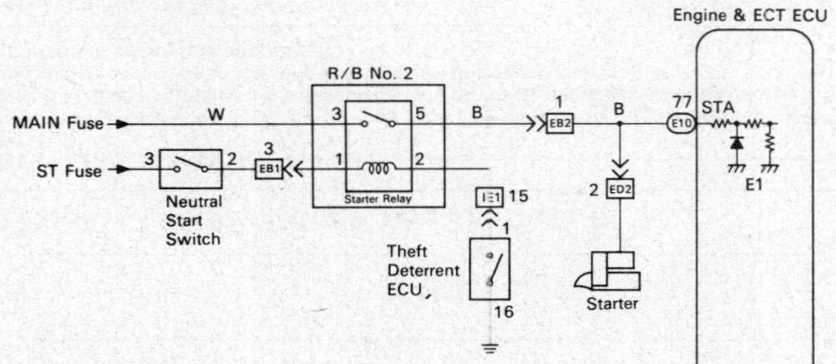

INSPECTION PROCEDURE

1 **Check output condition of diag. code 43.**

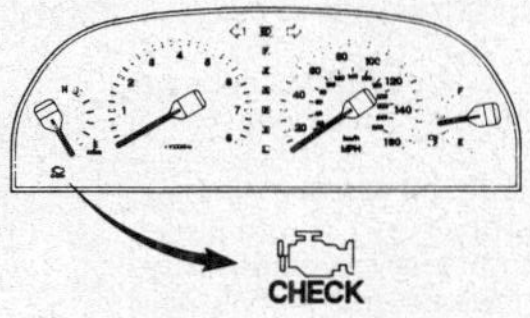

P Setting the test mode.
(1) Turn ignition switch OFF.
(2) Connect terminals TE2 and E1 of TDCL.
(3) Turn ignition switch ON.
(Don't start the engine)
(4) Connect terminals TE1 and E1 of TDCL.

C Check if code "43" is output by the "CHECK" engine warning light.

OK **Code "43" is output.**

C Start the engine.
Check if the code "43" disappear.

OK **Code "43" is not output.**

OK → **Proceed to next circuit inspection shown on matrix chart**

NG ↓

2 **Check for open in harness and connectors between engine & ECT ECU and starter relay**

NG → **Repair or replace harness or connector.**

OK ↓

Check and replace engine & ECT ECU.

Diag. Code 52,53,55 Knock Sensor Circuit

CIRCUIT DESCRIPTION

Knock sensors are fitted one each to the right bank and left bank of the cylinder block to detect engine knocking. This sensor contains a piezoelectric element which generates a voltage when it becomes deformed, which occurs when the cylinder block vibrates due to knocking. If engine knocking occurs, ignition timing is retarded to suppress it.

Code No.	Diagnostic Code Detecting Condition	Trouble Area
52	No No. 1 knock sensor signal to ECU for 3 crank revolutions with engine speed between 1,600 rpm ~ 5,200 rpm.	• Open or short in No. 1 knock sensor circuit. • No. 1 knock sensor (looseness) • ECU
53	Engine control computer (for knock control) malfunction at engine speed between 650 rpm and 5,200 rpm.	• ECU
55	No No. 2 knock sensor signal to ECU for 3 crank revolutions with engine speed between 1,600 rpm ~ 5,200 rpm.	• Open or short in No. 2 knock sensor circuit. • No. 2 knock sensor (looseness) • ECU

If the ECU detects the above diagnosis conditions, it operates the fail safe function in which the corrective retard angle value is set to the maximum value.

CIRCUIT DESCRIPTION (Cont'd)

DIAGNOSIS CODE DETECTION DRIVING PATTERN

Purpose of the driving pattern.
(a) To simulate diag. code detecting condition after diag. code is recorded.
(b) To check that the malfunction is corrected when the repair is completed confirming that diag. code is no longer detected.

Malfunction: Open or Short in Knock Sensor

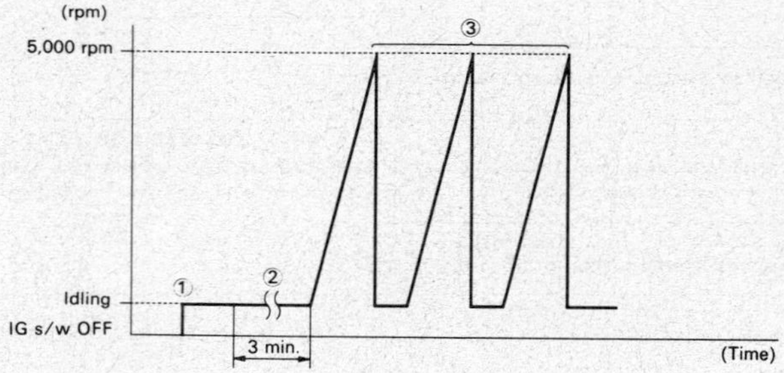

① Start engine and warm up.
② After engine is warmed up, let it idle for 3 min.
③ With the A/C ON, perform quick racing (5,000 rpm) three times. (Rapidly depress the accelerator pedal and suddenly release it.)

HINT: If a malfunction exists, the "CHECK" engine warning light will light up when sudden racing is performed.

NOTICE: If the conditions in this test are not strictly followed, detection of the malfunction will not be possible.

DIAGNOSTIC CHART

HINT: If diag. code 52 is displayed, check No. 1 knock sensor (for left bank) circuit.
If diag. code 55 is displayed, check No. 2 knock sensor (for right bank) circuit.
If diag. code 53 is displayed, replace engine & ECT ECU.

1 Check no continuity of knock sensor circuit. — OK → go to 3; NG ↓

2 Check knock sensor. — NG → Replace knock sensor.; OK ↓

3 Check for open and short in harness and connector between knock sensor and ECU. — NG → Repair or replace harness or connector.; OK ↓

4 Does malfunction disappear when a good knock sensor is installed? — YES → Replace knock sensor.; NO ↓

Check and replace ECU.

WIRING DIAGRAM

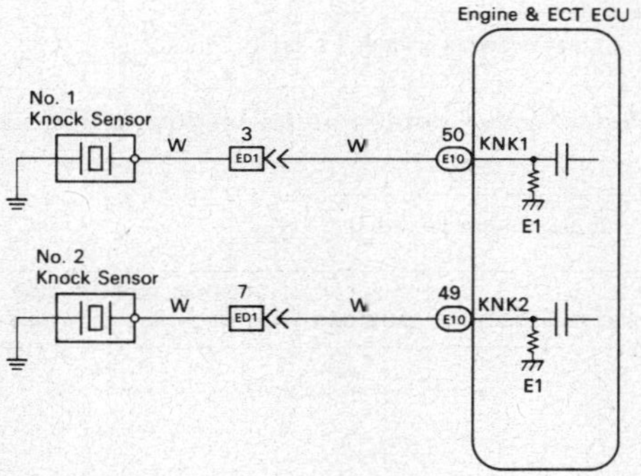

INSPECTION PROCEDURE

1 Check continuity between terminals KNK1, KNK2 of engine & ECT ECU connector and body ground.

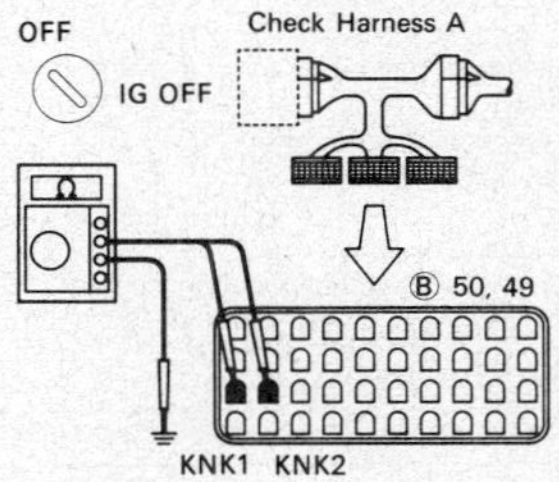

P (1) Connect the Check Harness A.

(2) Disconnect the engine & ECT ECU connector.

C Measure resistance between terminals KNK1, KNK2 of engine & ECT ECU connector and body ground.

OK **Resistance: 1 MΩ or higher**

NG ↓

OK 〉 **Go to step 3.**

2 Check knock sensor.

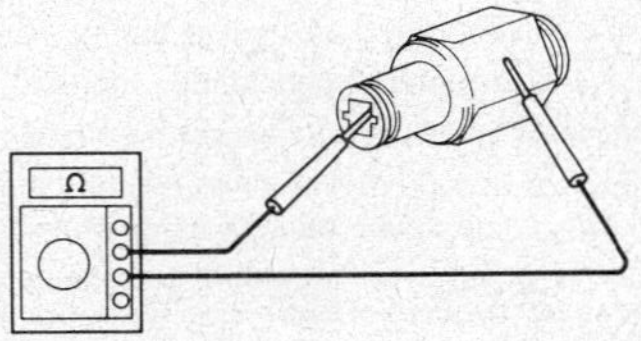

P Disconnect knock sensor connector.

C Measure resistance between the knock sensor terminal and body.

OK **Resistance: 1 MΩ or higher**

OK ↓

NG 〉 **Replace knock sensor**

3	Check for open and short in harness and connectors between engine & ECT ECU and knock sensor

NG — Repair or replace harness or connector.

OK

4	Does malfunction disappear when a good knock sensor is installed?

YES — Replace knock sensor

NO

Check and replace engine & ECT ECU.

Reference INSPECTION USING OSCILLOSCOPE

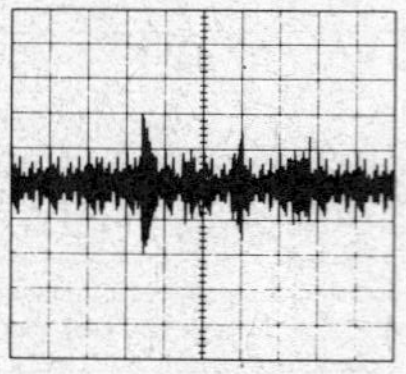

- With the engine racing (4,000 rpm) measure between terminals KNK1, KNK2 of engine & ECT ECU and body ground.

HINT: The correct waveform appears as shown in the illustration on the left.

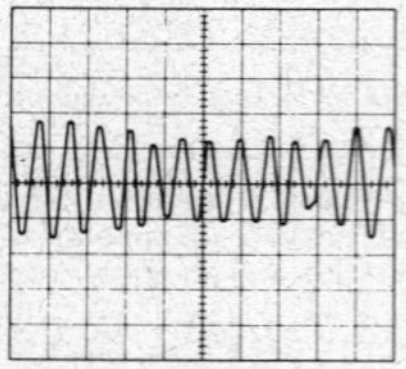

- Spread the time on the horizontal axis, and confirm that period of the wave is 151μ sec. (Normal mode vibration frequency of knock sensor: 6.6 KHz).

HINT: If normal mode vibration frequency is not 6.6 KHz, the sensor is malfunctioning.

Diag. Code	71	EGR System Malfunction (Only for USA spec.)

CIRCUIT DESCRIPTION

The EGR system recirculates exhaust gas, which is controlled to the proper quantity to suit the driving conditions, into the intake air mixture to slow down combustion, reduce the combustion temperature and reduce NOx emissions.

The lift amount of the EGR valve is controlled by a step motor operated by the ECU. The ECU operates the motor only a preprogrammed amount in response to the engine operating conditions (engine rpm, intake air volume to adjust the EGR volume to the target valve.)

Under the following conditions, EGR is cut to maintain driveability.

- Coolant temp. below 53°C (127.4°F)
- During deceleration (throttle valve closed)
- Light engine load (amount of intake air very small)
- Engine speed over 4,000 rpm
- Engine idling
- Neutral start switch on

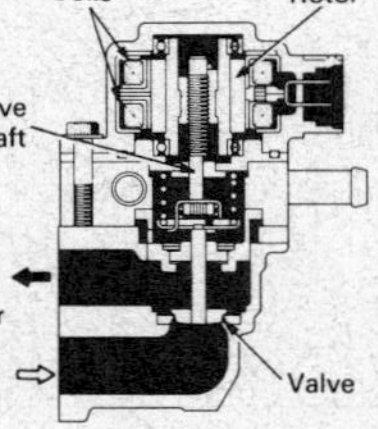

Code No.	Diagnostic Code Detecting Condition	Trouble Area
71	(1) Open or short in EGR step motor circuit for 1 sec. or more.	• Open or short in EGR step motor circuit • ECU
	(2) EGR gas temp. is 65°C (149°F) or below for 1 ~ 4 min. under conditions (a) and (b). (2 trip detection logic) (a) Coolant temp. : 65°C (149°F) or more. (b) EGR operation possible (EX. ECT in 3rd speed, A/C ON, 60 mph (96 km/h), Flat road).	• Open in EGR gas temp. sensor circuit • ECU

CIRCUIT DESCRIPTION (Cont'd)

DIAGNOSIS CODE DETECTION DRIVING PATTERN

Purpose of the driving pattern.

(a) To simulate diag. code detecting condition after diag. code is recorded.

(b) To check that the malfunction is corrected when the repair is completed confirming that diag. code is no longer detected.

Malfunction: Open in EGR gas temp. sensor circuit

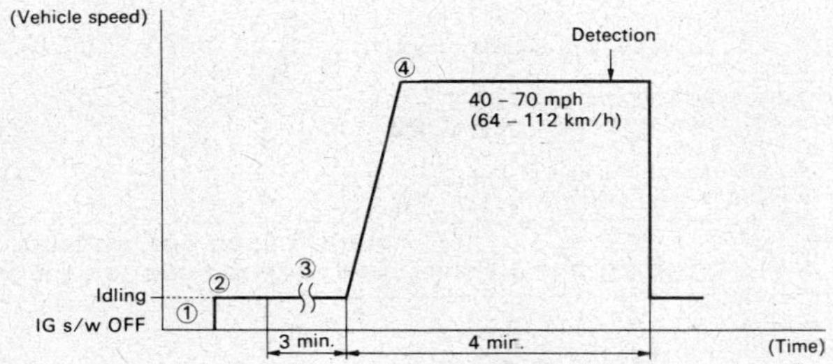

① Initiate test mode (See page TR-12).
② Start engine and warm up.
③ After engine is warmed up, let it idle for 3 min.
④ With the A/C ON and transmission in D range (O/D OFF), drive at 40 – 70 mph for 4 min.

HINT: If a malfunction exists, the "CHECK" engine warning light will light up at 1 – 4 min. of driving at 40 – 70 mph (64 – 112 km/h).

NOTICE: If the conditions in this test are not strictly followed, detection of the malfunction will not be possible.

DIAGNOSTIC CHART

1 Check voltage of terminal EGR 1 ~ 4.
- OK → go to 5
- NG ↓

2 Check voltage of EGR step motor power source.
- NG → Check for ECU power source circuit.
- OK ↓

3 Check resistance of EGR step motor.
- NG → Replace EGR step motor.
- OK ↓

4 Check for open and short in harness and connector between EGR step motor and ECU.
- NG → Repair or replace harness or connector.
- OK ↓

Check and replace ECU.

5 Check operation for EGR step motor.
- NG → Replace EGR step motor.
- OK ↓

6 Check resistance of EGR gas temp. sensor.
- NG → Replace EGR gas temp. sensor.
- OK ↓

7 Check for open and short in harness and connector between EGR gas temp. sensor and ECU.
- NG → Repair or replace harness or connector.
- OK ↓

Check and replace ECU.

WIRING DIAGRAM

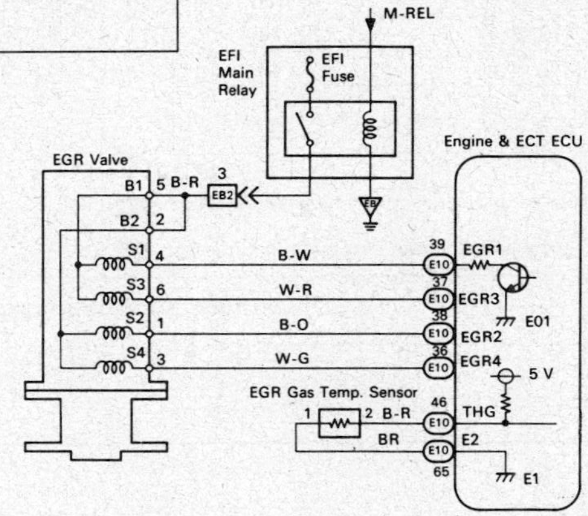

INSPECTION PROCEDURE

1 Check voltage between terminals EGR 1 ~ 4 of engine & ECT ECU connector and body ground.

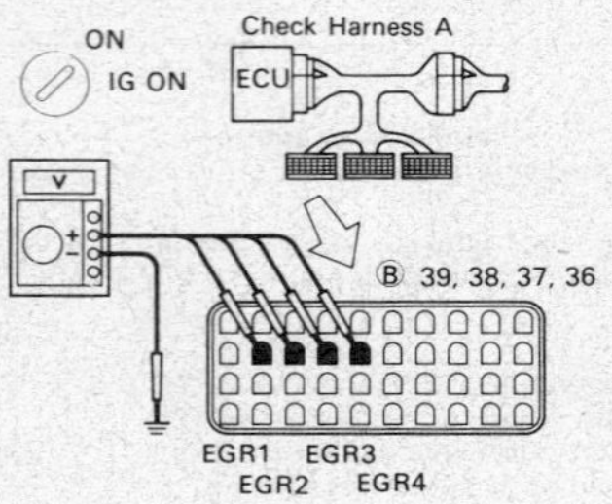

P (1) Connect the Check Harness A.

(2) Turn ignition switch on.

C Measure voltage between terminals EGR 1 ~ 4 of engine & ECT ECU connector and body ground.

OK Voltage: 10 – 14 V

OK › Go to step 5.

NG ↓

2 Check voltage between terminals B1 and B2 of EGR step motor connector and body ground.

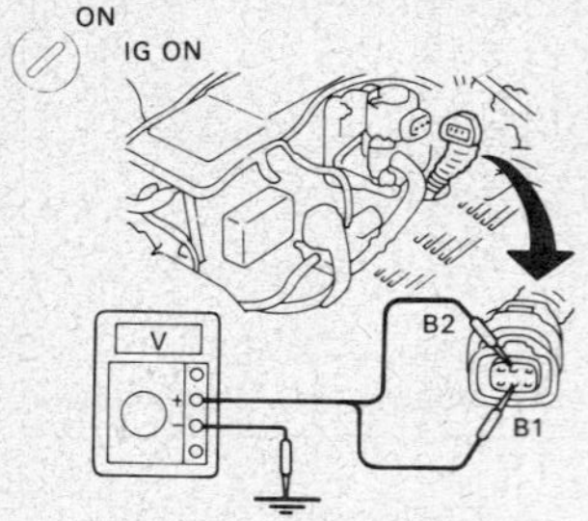

P (1) Disconnect EGR step motor connector.
(2) Turn ignition switch on.

C Measure voltage between terminals B1 and B2 of EGR step motor connector and body ground.

OK Voltage: 10 – 14 V

NG › Check for ECU power source circuit

OK ↓

3 Check resistance of EGR step motor.

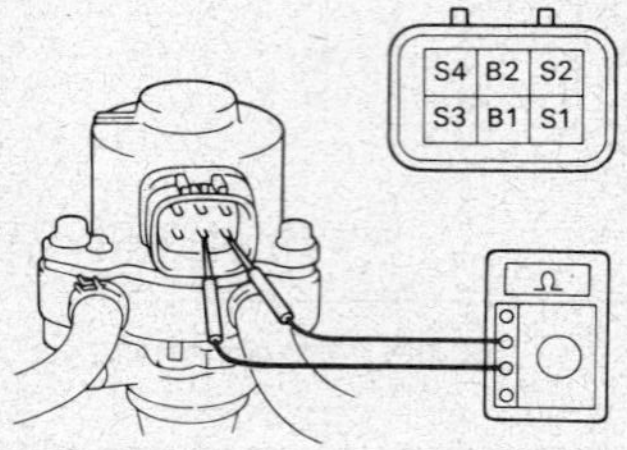

P Disconnect EGR step motor connector.

C Measure resistance between terminals shown below.

OK

Terminal	Resistance
B1 – S1	10 Ω – 30 Ω
B1 – S3	10 Ω – 30 Ω
B2 – S2	10 Ω – 30 Ω
B2 – S4	10 Ω – 30 Ω

NG › Replace EGR step motor.

OK ↓

4 Check for open and short in harness and connector between EGR step motor and engine & ECT ECU

NG › Repair or replace harness or connector.

OK ↓

Check and replace engine & ECT ECU.

5 Check operation for EGR step motor.

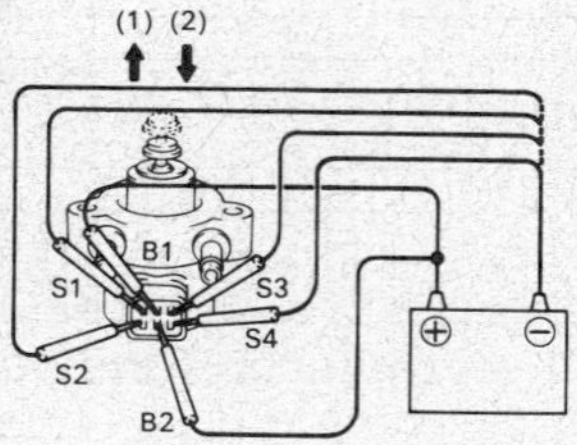

P Remove EGR step motor.

C (1) Connect the battery positive lead to terminals B1 and B2, and the negative lead to terminals S1 – S2 – S3 – S4 in that order.
(2) Connect the battery positive lead to terminals B1 and B2, and the negative lead to terminals S4 – S3 – S2 – S1 in that order.

OK (1) **The valve moves in the closing direction.**
(2) **The valve moves in the opening direction.**

NG › Replace EGR step motor.

OK ↓

6 Check resistance of EGR gas temp. sensor.

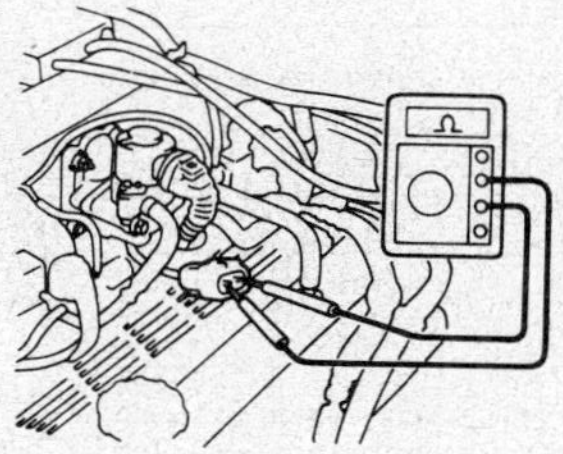

P Disconnect EGR gas temp. sensor connector

C Measure resistance between terminals of EGR gas temp. sensor connector.

OK **Resistance: 69 – 89 kΩ at 50°C (122°F)**
12 – 15 kΩ at 100°C (212°F)
2 – 4 kΩ at 150°C (302°F)

NG → **Replace EGR gas temp. sensor.**

OK ↓

7 Check for open and short in harness and connector between EGR gas temp. sensor and engine & ECT ECU

NG → **Repair or replace harness or connector.**

OK ↓

Check and replace engine & ECT ECU.

Diag. Code 78 Fuel Pump Control Circuit

CIRCUIT DESCRIPTION

The fuel pump speed is controlled at two steps (high speed,low speed) by the condition of the engine (starting, light load, heavy load), when the engine starts (STA ON), the engine & ECT ECU sends a Hi signal to the fuel pump ECU (FPC terminal).

The fuel pump ECU then outputs Hi voltage (battery voltage) to the fuel pump so that the fuel pump operates at high speed.

After the engine starts, during idling or light loads, the engine & ECT ECU outputs a High Low signal to the fuel pump ECU, the fuel pump ECU outputs Lo battery voltage (about 9V) to the fuel pump and causes the fuel pump to operate at low speed.

If the intake air volume increases (high engine load), the engine & ECT ECU sends a Hi signal to the fuel pump ECU and causes the fuel pump to operate at high speed.

Code No.	Diagnostic Code Detecting Condition	Trouble Area
78	(1) Open or short in fuel pump circuit for 1 sec. or more with engine speed 1,000 rpm or less. (2 trip detection logic) (2) Open in input circuit of fuel pump ECU (FPC) with engine speed 1,000 rpm or less. (2 trip detection logic) (3) Open or short in diagnostic signal line (DI) of fuel pump ECU with engine speed 1,000 rpm or less. (2 trip detection logic)	• Open or short in fuel pump ECU circuit. • Fuel pump ECU • Engine & ECT ECU power source circuit. • Fuel pump • Engine & ECT ECU

WIRING DIAGRAM

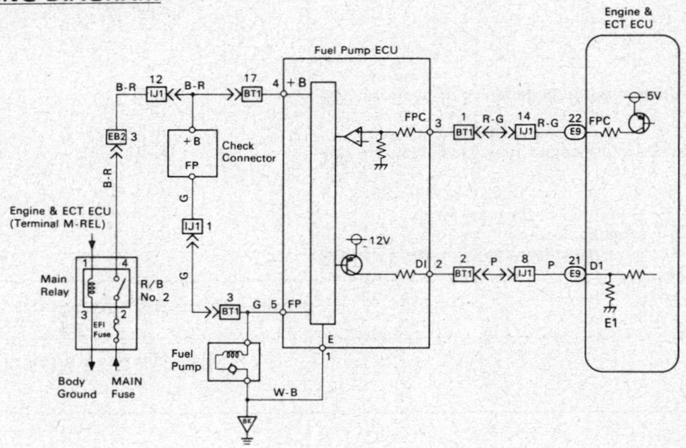

DIAGNOSTIC CHART

1. Check fuel pump operation. — NG → go to 3; OK ↓
2. Check for open and short in harness and connector between check connector and fuel pump ECU. — OK → go to 5; NG ↓
   - Repair or replace harness or connector.
3. Check voltage of terminal + B of check connector. — NG → Check for ECU power source circuit.; OK ↓
4. Check for open and short in harness and connector between check connector and fuel pump. — OK → Repair or replace fuel pump.; NG ↓
   - Repair or replace harness or connector.
5. Check voltage of terminal FPC. — OK → Replace fuel pump ECU.; NG ↓
6. Check for open and short in harness and connector between ECU and fuel pump ECU. (FPC) — NG → Repair or replace harness or connector.; OK ↓
7. Check for open and short in harness and connector between ECU and fuel pump ECU. (DI) — NG → Repair or replace harness or connector.; OK ↓
   - Check and replace ECU.

INSPECTION PROCEDURE

1 Check fuel pump operation.

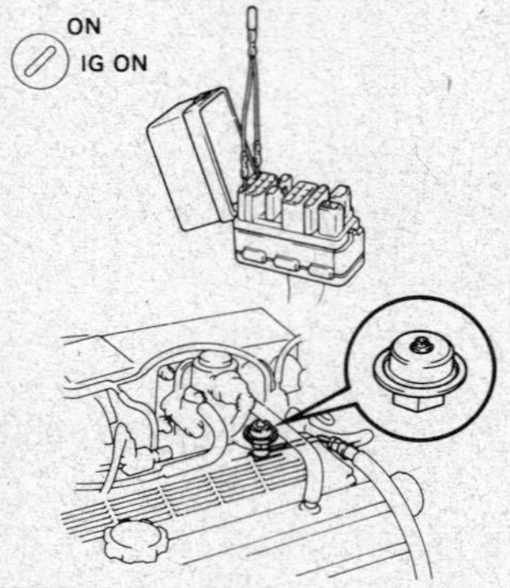

P (1) Turn ignition switch ON.
(2) Using SST, connect terminals + B and FP of check connector.
SST 09843-18020

C Check that there is projection of the head of the pulsation damper screw.

OK **Projection of the head of the pulsation damper screw.**

OK ↓

NG 〉 **Go to step 3.**

2 Check for open and short in harness and connector between terminals + B ↔ + B, FP ↔ FP of the check connector and fuel pump ECU

OK 〉 **Go to step 5.**

NG ↓

Repair or replace harness or connector.

3 Check voltage of terminal + B of check connector.

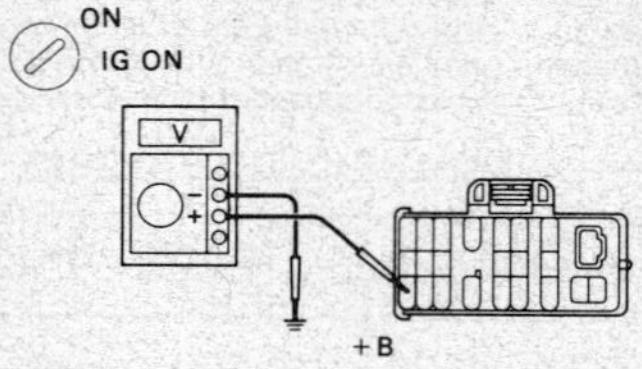

P Turn ignition switch ON.

C Measure voltage between terminal + B of check connector and body ground.

OK **Voltage : 10 – 14 V**

OK ↓

NG 〉 **Check for ECU power source circuit and check for open in harness and connector between terminal + B of check connector and main relay.**

4 Check for open and short in harness and connector between terminal FP of check connector, fuel pump and body ground

OK 〉 **Repair or replace fuel pump.**

NG

Repair or replace harness or connector.

5 Check voltage between terminals FPC and E of fuel pump ECU connector.

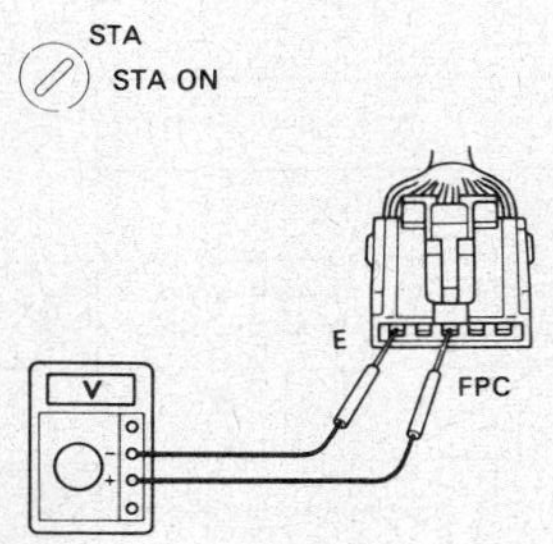

P (1) Remove the LH quarter trim panel.

(2) Disconnect fuel pump ECU connector.

C Measure voltage between terminals FPC and E of fuel pump ECU connector when ignition switch is turned to start.

OK **Voltage: 4 – 6V**

OK 〉 **Replace fuel pump ECU.**

NG

6 Check for open in harness and connector between terminal FPC of engine & ECT ECU and terminal FPC of fuel pump ECU, terminal E of fuel pump ECU and body ground

NG 〉 **Repair or replace harness or connector.**

OK

7 Check for open and short in harness and connector between terminal DI of engine & ECT ECU and terminal DI of fuel pump ECU

NG 〉 **Repair or replace harness or connector.**

OK

Check and replace engine & ECT ECU.

Diag. Code 51 Switch Condition Signal Circuit

CIRCUIT DESCRIPTION

Neutral Start Switch Signal
The ECU uses the signals from the neutral start switch to determine whether the transmission is in park or neutral, or in some other gear.

Air Conditioner Switch Signal
The ECU uses the output from the air conditioner switch to determine whether or not the air conditioner is operating so that it can increase the idling speed of the engine if necessary.

Throttle Position Sensor IDL Signal
The IDL contacts are mounted in the throttle position sensor, and detects the idle condition.

Code No.	Diagnostic Code Detecting Condition	Trouble Area
51	(1) 3 sec. or more after engine starts idle switch OFF (IDL1). (2) Neutral start switch OFF (NSW). (Shift position in "R", "D", "2" or "1" ranges). (3) A/C switch ON.	• Throttle position sensor IDL circuit • Accelerator pedal and cable • Neutral start switch circuit • A/C switch circuit • ECU

HINT: In this circuit, diagnosis can only be made in the test mode.

DIAGNOSTIC CHART

See next page for DIAGNOSTIC CHART

WIRING DIAGRAM

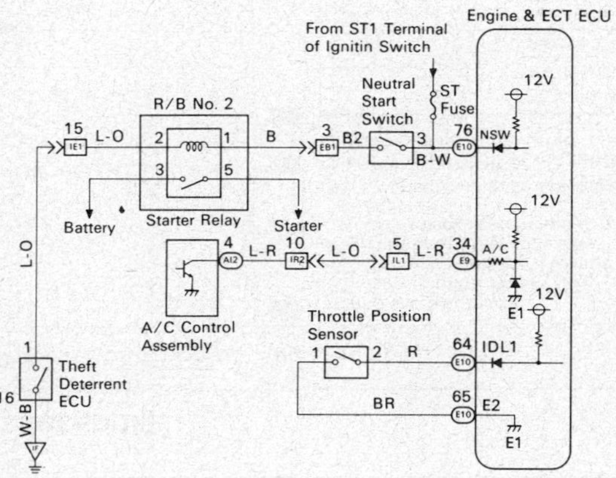

DIAGNOSTIC CHART

1 Check for input signal.
- NG (NSW) → See Neutral start switch circuit
- NG (A/C) → go to 3
- NG (IDL) → go to 2
- OK → Proceed to next circuit inspection shown on matrix chart

2 Check resistance of throttle position sensor.
- NG → Adjust or replace throttle position sensor
- OK → Check and repair harness or connector between sensor and ECU.

3 Check input circuit of A/C.
- OK → Check and replace A/C control assembly.
- NG → go to 4

4 Check for open and short in harness and connector between A/C control assy. and ECU.
- NG → Repair or replace harness or connector.
- OK → Check and replace ECU.

INSPECTION PROCEDURE

1 Check output condition of diag. code 51.

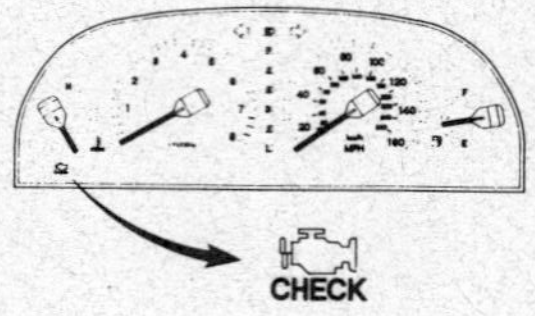

P Setting the test mode.
(1) Turn ignition switch OFF.
(2) Connect terminals TE2 and E1 of TDCL.
(3) Turn ignition switch ON.
(For checking terminal A/C, start the engine.)
(4) Connect terminlas TE1 and E1 of TDCL.

C Check if code "51" is output by the "CHECK" engine warning light.

OK

	Condition	Code
Neutral Start Switch (NSW)	P or N range	Normal*
	R,D,2 or L range	51*
Throttle Position Sensor (IDL1)	Accelerator pedal released	Normal*
	Accelerator pedal depressed	51*
A/C Switch (A/C)	A/C SW ON	51
	A/C SW OFF	Normal

*: Before the STA signal is input (ST is not ON), diagnostic code 43 is also output.

Hint: Diag. code 42 is output with vehicle speed 5 km/h (3 mph) or below

NG IDL1 ..... Go to step 2
A/C ...... Go to step 3

OK Proceed to next circuit inspection shown on matrix chart

2 Check throttle position sensor.

P Disconnect throttle position sensor connector.

C Measure resistance between terminals IDL1 and E2 of throttle position sensor connector.

OK

Throttle Valve	Resistance
Fully closed	Less than 0.5 kΩ
Opened	1 MΩ or higher

NG Adjust or replace throttle position sensor

OK Check and repair harness or connector between engine & ECT ECU and throttle position sensor.

3 Disconnect A/C control assembly connector, check voltage between terminal MGC of A/C control assembly connector and body ground.

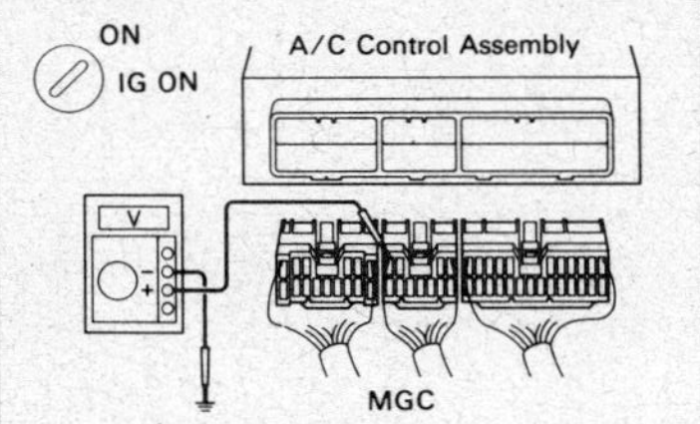

P (1) Remove air conditioner control assembly.
(2) Disconnect air conditioner control assembly connector.
(3) Turn ignition switch on.

C Measure voltage between terminal MGC of air conditioner control assembly connector and body ground.

OK **Voltage: 10 – 14 V**

OK → **Check and replace A/C control assembly.**

NG ↓

4 Check for open and short in harness and connector between engine & ECT ECU and A/C control assembly

NG → **Repair or replace harness or connector.**

OK ↓

Check and replace engine & ECT ECU.

Neutral Start Switch Circuit

CIRCUIT DESCRIPTION

The neutral start switch goes on when the shift lever is in the N or P shift position. When it goes on the terminal NSW of the ECU is grounded to body ground via the starter relay and theft deterrent ECU, thus the terminal NSW voltage becomes 0V. When the shift lever is in the D, 2, L or R position, the neutral start switch goes off, so the voltage of ECU terminal NSW becomes battery voltage, the voltage of the ECU internal power source.
If the shift lever is moved from the N range to the D range, this signal is used for air-fuel ratio correction and for idle speed control (estimated control), etc.
When the neutral start switch is off, code "51" is output in the test mode diagnosis. (This is not abnormal.)

DIAGNOSTIC CHART

HINT: This diagnosis chart is based on premise that the engine is cranked normally. If the engine is not cranked, proceed to the matrix chart of problem symptoms

1 Check output condition of diag. code 51.
- OK → Proceed to next circuit inspection shown on matrix chart
- NG Type I → 2
- NG Type II → 3

2 Check for open in harness and connector between neutral start switch and ECU.
- OK → Check and replace ECU.
- NG → Repair or replace harness or connector.

3 Check neutral start switch.
- NG → Replace neutral start switch.
- OK → Check and replace ECU.

WIRING DIAGRAM

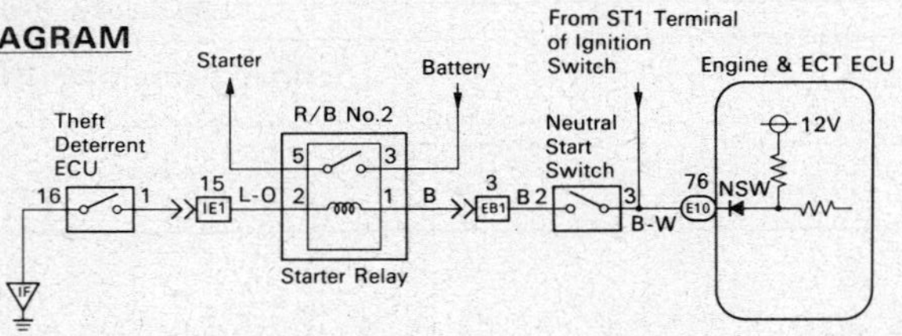

ECU Power Source Circuit

CIRCUIT DESCRIPTION

When the ignition switch is turned on, battery voltage is applied to the terminal IG SW of the ECU, and the main relay control circuit in the ECU sends a signal to the terminal M-REL of the ECU, switching on the main relay. This signal causes current to flow to the coil, closing the contacts of the main relay and supplying power to the terminals +B and +B1 of the ECU.

If the ignition switch is turned off, the ECU continues to switch on the main relay for a maximum of 2 seconds for the initial setting of the ISC valve.

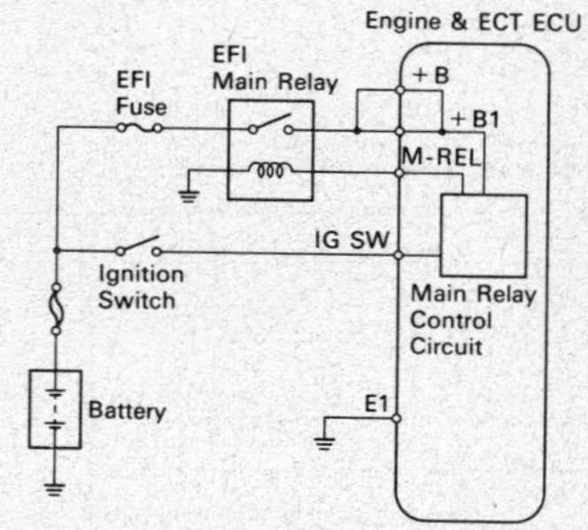

DIAGNOSTIC CHART

See next page for the DIAGNOSTIC CHART.

WIRING DIAGRAM

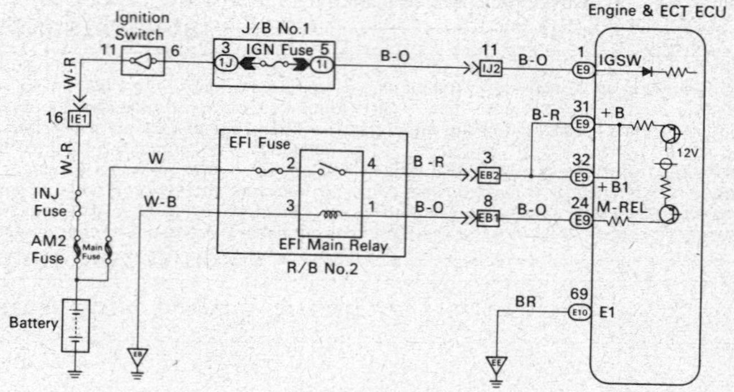

INSPECTION PROCEDURE

1 Check output condition of diag. code 51.

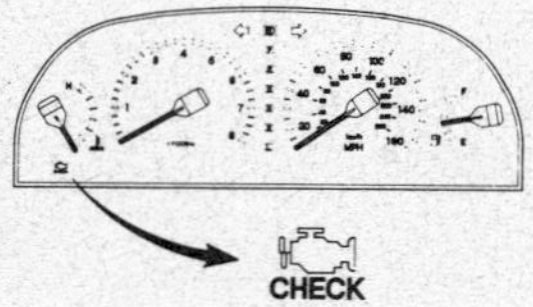

P (1) Connect terminals TE2 and E1 of TDCL.
(2) Turn ignition switch on.
(3) Crank the engine.
(4) Connect terminals TE1 and E1 of TDCL.

C Check if diagnostic code "51" is output when the shift lever is in the P and D shift positions.

Result

Shift Position	Result		
	OK	NG Type I	NG Type II
"P"	Normal Code	Code 51	Normal Code
"D"	Code 51	Code 51	Normal Code

NG Type I → Go to step 2.

NG Type II → Go to step 3.

OK ↓

Proceed to next circuit inspection shown on matrix chart

2 Check for open in harness and connector between engine & ECT ECU and neutral start switch

OK → Check and replace engine & ECT ECU.

NG ↓

Repair or replace harness or connector.

3 Check neutral start switch

NG → Replace neutral start switch.

OK ↓

Check and replace engine & ECT ECU.

DIAGNOSTIC CHART

1 Check voltage of ECU power source.
- OK → Proceed to next circuit inspection shown on matrix chart
- NG ↓

2 Check continuity between terminals E1 and body ground.
- NG → Repair or replace harness or connector.
- OK ↓

3 Check voltage of terminal IGSW.
- OK → (go to 6)
- NG ↓

4 Check IGN fuse.
- NG → Check for short in all the harness and components connected to IGN fuse.
- OK ↓

5 Check ignition switch.
- NG → Replace ignition switch.
- OK ↓

Check for open in harness and connector between battery and ECU.

6 Check voltage of terminal M-REL.
- NG → Check and replace ECU.
- OK ↓

7 Check EFI fuse.
- NG → Check for short in all the harness and components connected to EFI fuse.
- OK ↓

8 Check EFI main relay.
- NG → Replace EFI main relay.
- OK ↓

9 Check for open and short in harness and connector between terminal M-REL and body ground.
- NG → Repair or replace harness or connector.
- OK ↓

Check and repair harness or connector between EFI fuse and battery.

INSPECTION PROCEDURE

1 Check voltage between terminals +B, +B1 and E1 of engine & ECT ECU connector.

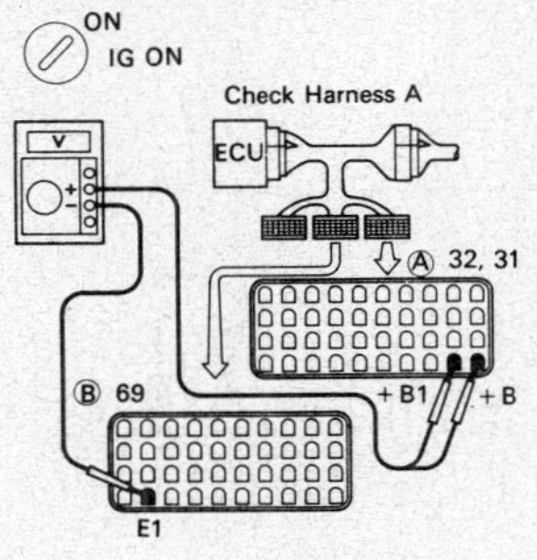

P (1) Connect the Check Harness A.

(2) Turn ignition switch on.

C Measure voltage between terminals +B, +B1 and E1 of engine & ECT ECU connector.

OK **Voltage: 10 – 14 V**

OK → **Proceed to next circuit inspection shown on matrix chart**

NG ↓

2 Check for open in harness and connector between terminal E1 of engine & ECT ECU and body ground

NG → **Repair or replace harness or connector.**

OK ↓

3 Check voltage between terminal IG SW of engine & ECT ECU connector and body ground.

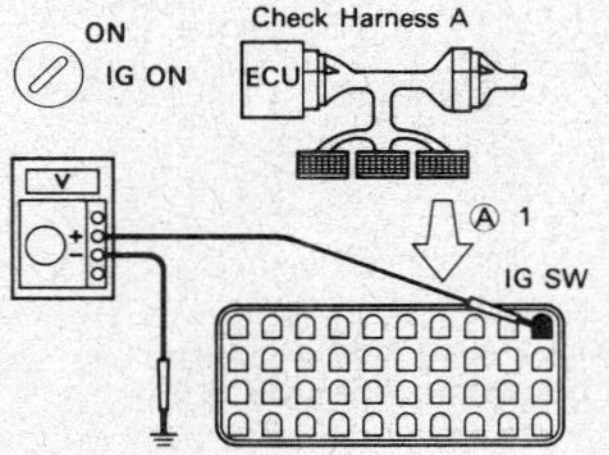

P Turn ignition switch on.

C Measure voltage between terminal IG SW of engine & ECT ECU connector and body ground.

OK **Voltage: 10 – 14V**

OK → **Go to step 6.**

NG ↓

4 Check IGN fuse.

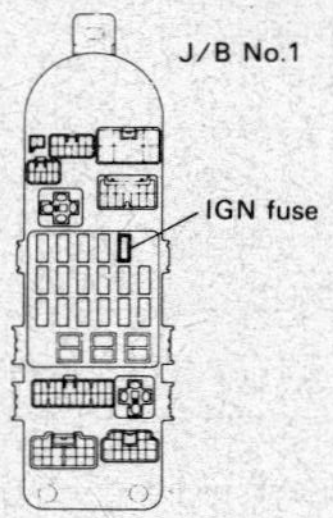

P Remove IGN fuse from J/B No.1.
C Check continuity of IGN fuse.
OK **Continuity**

OK

NG **Check for short in all the harness and components connected to IGN fuse**

5 Check ignition switch.

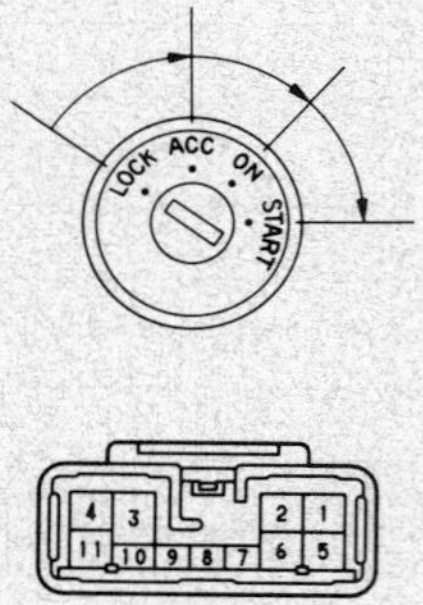

P Remove under cover and finish panel.
C Check continuity between terminals.
OK

O—O continuity

Terminal / Switch position	2	3	4	5	6	11
LOCK						
ACC		O—	—O			
ON	O—	—O—	—O		O—	—O
START	O—		—O	O—	—O—	—O

OK

Check and repair harness and connector between battery and ignition switch, ignition switch and engine & ECT ECU.

NG **Replace ignition switch.**

6 Check voltage between terminal M-REL of engine & ECT ECU connector and body ground.

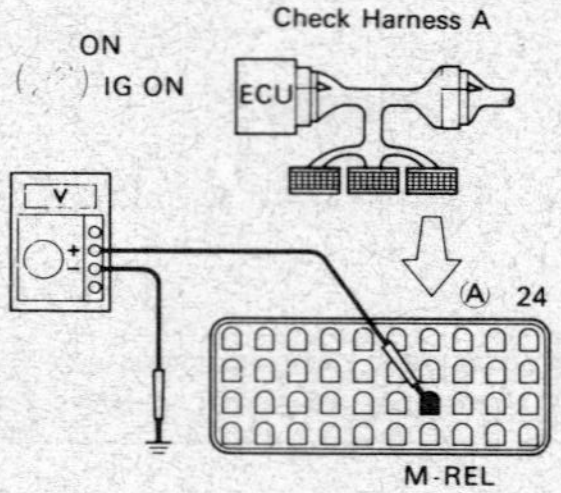

P Turn ignition switch on.
C Measure voltage between terminal M-REL of engine & ECT ECU connector and body ground.
OK **Voltage: 10 – 14V**

OK

NG **Check and replace engine & ECT ECU.**

7 Check EFI fuse.

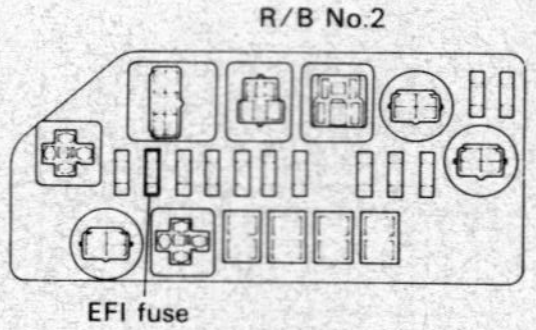

P Remove EFI fuse from R/B No.2.
C Check continuity of EFI fuse.
OK **Continuity**

OK

NG **Check for short in all the harness and components connected to EFI fuse**

8 **Check EFI main relay.**

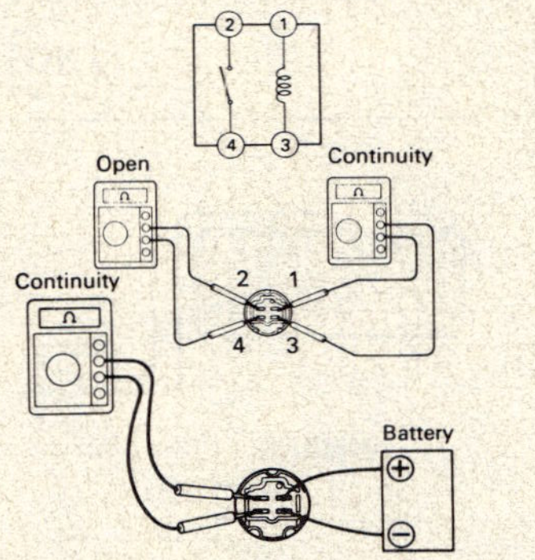

P Remove EFI main relay.

C Check continuity between terminals of EFI main relay shown below.

OK

Terminals 2 and 4	Open
Terminals 1 and 3	Continuity (Reference value 72 Ω)

C (1) Apply battery voltage between terminals 1 and 3.
(2) Check continuity between terminals 2 and 4.

OK

Terminals 2 and 4	Continuity

NG → Replace EFI main relay.

OK ↓

9 **Check for open and short in harness and connector between terminal M-REL of engine & ECT ECU and body ground**

NG → Repair or replace harness or connector.

OK ↓

Check and repair harness or connector between EFI fuse and battery.

Back Up Power Source Circuit

CIRCUIT DESCRIPTION

Battery voltage is supplied to terminal BATT of the ECU even when the ignition switch is off for use by the diagnostic code memory and air-fuel ratio adaptive control value memory, etc.

DIAGNOSTIC CHART

1 Check EFI fuse. — NG → Check for short in all the harness and components connected to EFI fuse.

OK ↓

2 Check voltage of terminal BATT. — NG → Check and repair harness or connector between battery, EFI fuse and ECU.

OK ↓

3 Check operation for the back up. — NG → Check and replace ECU.

OK ↓

Proceed to next circuit inspection shown on matrix chart

WIRING DIAGRAM

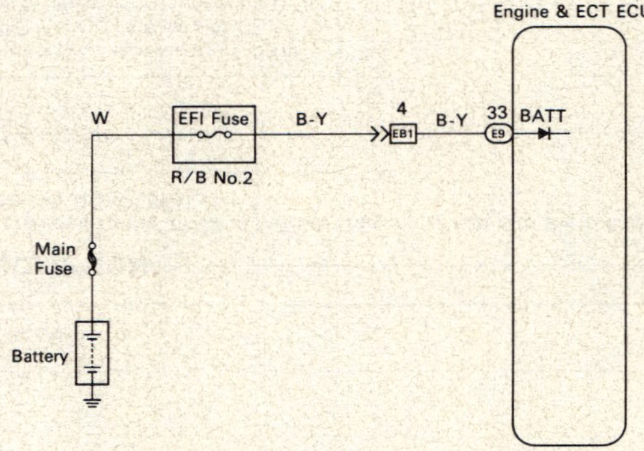

Injector Circuit

CIRCUIT DESCRIPTION

The injectors are provided to the intake manifold. They inject fuel into the cylinders based on the signals from the engine & ECT ECU.

DIAGNOSTIC CHART

- **1** Check voltage between terminals #10 ~ 40 of engine & ECT ECU and body ground.
  - OK → go to 3
  - NG → go to 2
- **2** Check ignition relay.
  - NG → Replace ignition relay.
  - OK → Check and repair harness and connector between ignition relay and battery.
- **3** Check continuity between terminals E01, E02 and body ground.
  - NG → Repair or replace harness or connector.
  - OK → go to 4
- **4** Check for open in harness and connector between ignition relay and ECU.
  - NG → Repair or replace harness or connector.
  - OK → go to 5
- **5** Check operation for injectors.
  - NG → Replace injector.
  - OK → Proceed to next circuit inspection shown on matrix chart

INSPECTION PROCEDURE

1 Check EFI fuse.

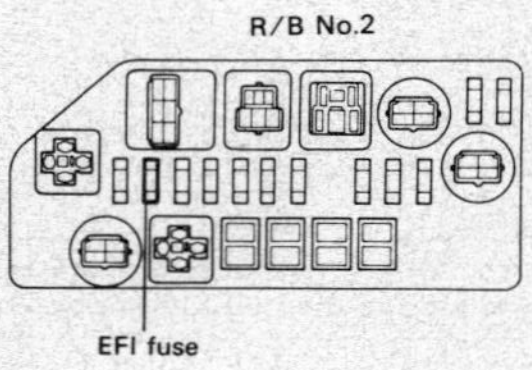

P Remove EFI fuse from R/B No.2.

C Check continuity of EFI fuse.

OK **Continuity**

NG Check for short in all the harness and components connected to EFI fuse

OK

2 Check voltage between terminal BATT of engine & ECT ECU connector and body ground.

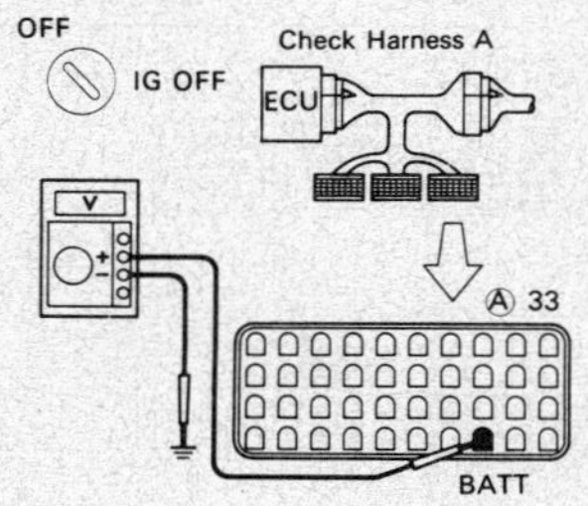

P Connect the Check Harness A.

C Measure voltage between terminal BATT of engine & ECT ECU connector and body ground.

OK **Voltage: 10 – 14V**

NG Check and repair harness or connector between engine & ECT ECU and EFI fuse, EFI fuse and battery.

OK

3 Are the diagnostic codes still in the memory when the ignition switch is turned OFF?

NO Check and replace engine & ECT ECU.

YES

Proceed to next circuit inspection shown on matrix chart

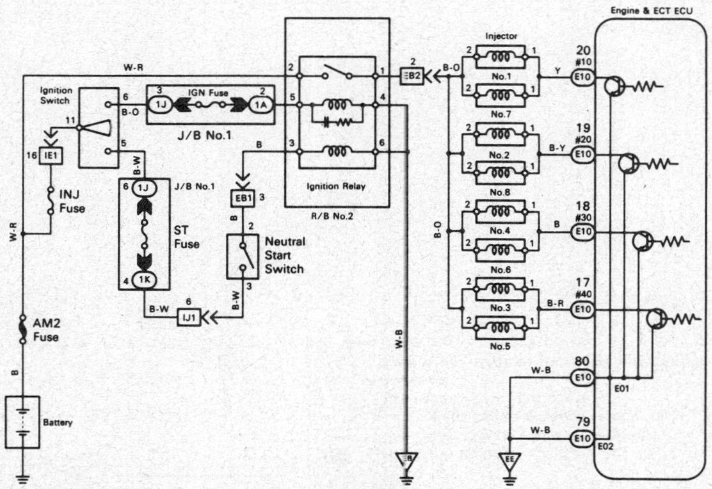

INSPECTION PROCEDURE

1 Check voltage between terminals #10 ~ 40 of engine & ECT ECU and body ground.

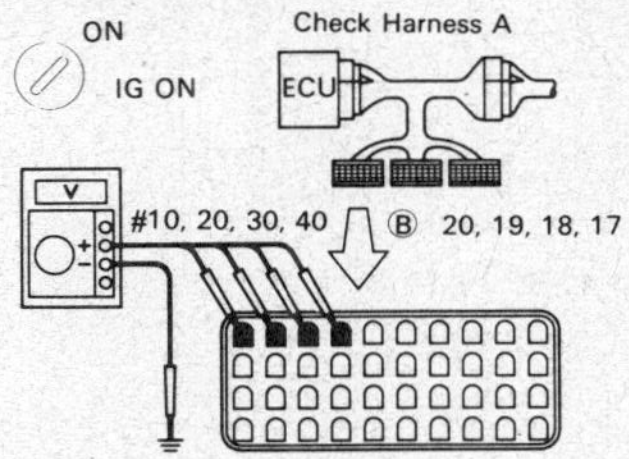

P (1) Connect the Check Harness A.

(2) Turn ignition switch on.

C Measure voltage between terminals #10 ~ 40 of engine & ECT ECU and body ground.

OK **Voltage: 10 – 14V**

NG

OK Go to step **3**.

2 Check ignition relay.

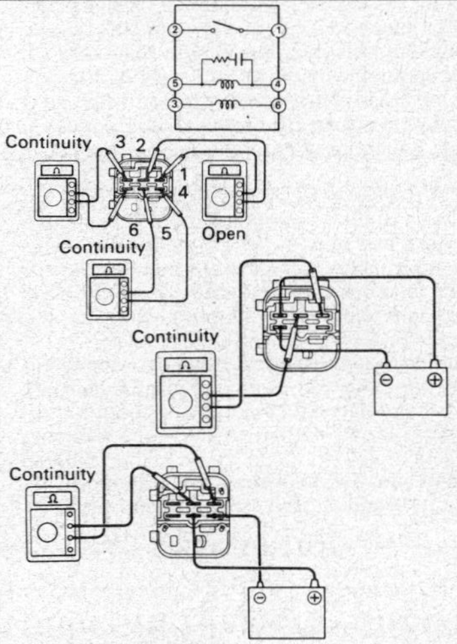

P Remove ignition relay from R/B No.2.

C Check continuity between terminals of ignition relay shown below.

OK

Terminals 1 and 2	Open
Terminals 3 and 6	Continuity (Reference value 22 Ω)
Terminals 4 and 5	Continuity (Reference value 107 Ω)

C (1) Apply battery voltage between terminals 3 and 6, 4 and 5.
(2) Check continuity between terminals 1 and 2.

OK

Terminals 1 and 2	Continuity

NG Replace ignition relay.

OK

Check and repair harness and connector between ignition relay and battery.

3 **Check for open in harness and connector between terminal E01, E02 of engine & ECT ECU connector and body ground**

OK

NG — Repair or replace harness or connector.

4 **Check for open in harness and connector between ignition relay and injector, injector and engine & ECT ECU**

OK

NG — Repair or replace harness or connector.

5 **Check injectors.**

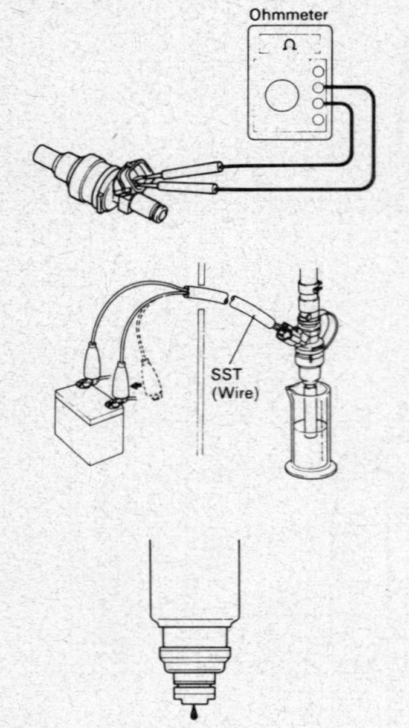

P Disconnect injector connector

C Measure resistance of injector.

OK **Resistance: 13.4 – 14.2 Ω at 20°C (68°F)**

C Check injection volume of injector (See page FI-52).

OK
- **Injection volume**
  55 – 70 cc/15 sec. (3.4 – 4.3 cu in.)
  Difference between each injector:
  Less than 10 cc (0.6 cu in.)
- **Leakage**
  Fuel drop: One drop or less per minute.

OK

NG — Replace injector.

Proceed to next circuit inspection shown on matrix chart

Cold Start Injector Circuit

CIRCUIT DESCRIPTION

The cold start injector is used to maintain the engine startability when it is cold. The injection volume, i.e, the length of time the injector is energized, is controlled by the ECU and the cold start injector time switch.
During a cold start, when the starter turns the contacts in the cold start injector time switch close. Thus current flows to the cold start injector coil, injecting fuel. At the same time, a bimetal in the heat coil is energized and heats up. This soon causes the contacts to open, cutting off the current flow to the injector coil and stopping fuel injection.

The injection duration of the cold start injector is determined by the coolant temperature and the length of time current flows to the heat coil. When the engine is warm, the contacts are opened by the bimetal and the cold start injector does not operate.
When the engine is hard to start and the starter is operated continuously, heat coil (2) heats up the bimetal keeping the contacts open to prevent spark plugs from becoming fouled, which is caused by the cold start injector operation when the open contacts close again.

When the engine is started at a coolant temperature of 22°C (72°F) or lower, the cold start injector operation time is controlled by the cold start injector time switch.
When the coolant temperature is in the normal temperature range 22°C (72°F) or higher, the contacts of the cold start injector time switch are open and the time switch is off, instead, the ECU controls the operating time of the cold start injector.
In this way, the CO and HC levels can be reduced while the engine is being started and the engine startability is maintained. Control by the ECU ends when the coolant temperature reaches 60°C (140°F).

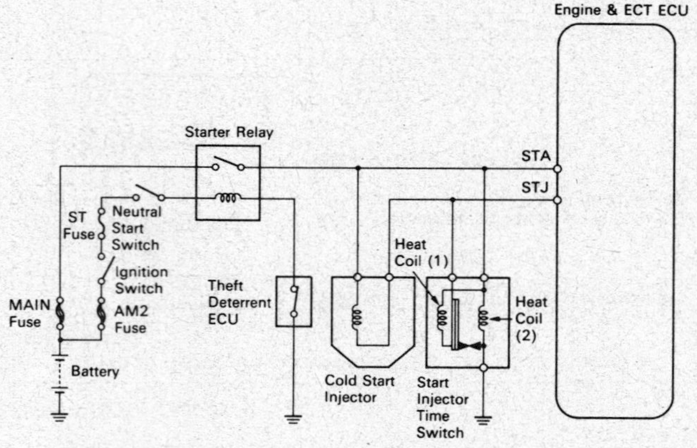

4.0L – CIRCUIT INSPECTION AND TROUBLESHOOTING – SC400

DIAGNOSTIC CHART

1 Check voltage of cold start injector power source.
- OK → go to **3**
- NG ↓

2 Check cold start injector.
- NG → Replace cold start injector.
- OK ↓

Check for open in harness and connector between battery and ECU.

3 Check start injector time switch.
- NG → Replace start injector time switch.
- OK ↓

4 Check for open and short in harness and connector between cold start injector and start injector time switch.
- NG → Repair or replace harness or connector.
- OK ↓

5 Check fuel injection of cold start injector.
- NG → Replace cold start injector.
- OK ↓

Proceed to next circuit inspection shown on matrix chart.

WIRING DIAGRAM

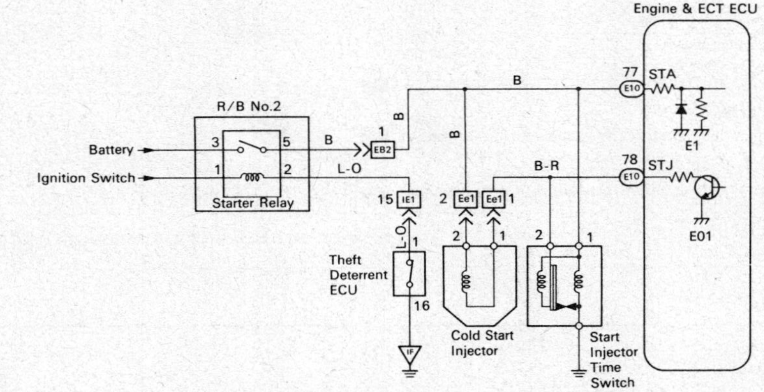

INSPECTION PROCEDURE

1 Check voltage between terminal STJ of engine & ECT ECU connector and body ground.

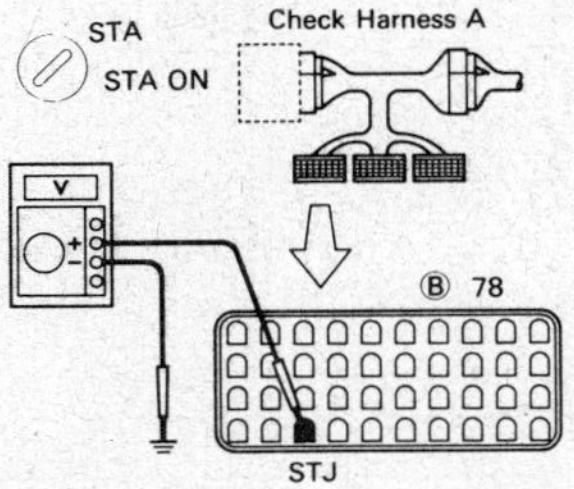

P (1) Connect the Check Harness A.

(2) Disconnect start injector time switch connector.

(3) Disconnect engine & ECT ECU connector.

C Measure voltage between terminal STJ of engine & ECT ECU connector and body ground when ignition switch is turned to STA.

OK **Voltage: 10 – 14V**

NG

OK Go to step **3**.

2 Check resistance of cold start injector.

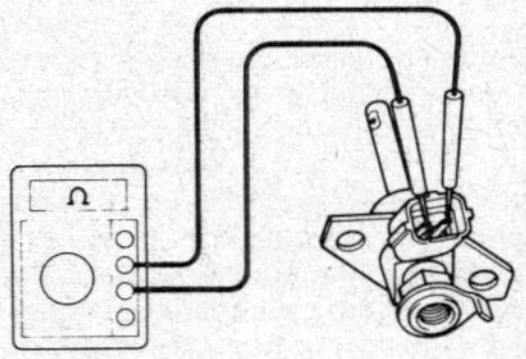

P Remove cold start injector

C Measure resistance between terminals of cold start injector.

OK **Resistance: 2 – 4 Ω at 20°C (68°F)**

NG **Replace cold start injector.**

OK

Check and repair harness and connector between engine & ECT ECU and cold start injector, cold start injector and battery.

3 Check start injector time switch.

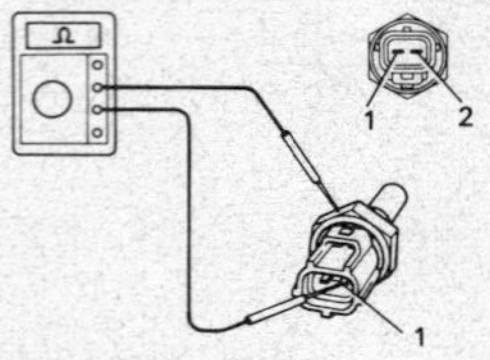

P Remove start injector time switch

C Measure resistance between terminals shown below.

OK

Terminals	Resistance	Coolant temperature
1 – 2	25 – 45 Ω	below 15°C (59°F)
	65 – 85 Ω	above 30°C (86°F)
1-Ground	25 – 85 Ω	—

OK

NG Replace start injector time switch.

4 Check harness and connector between cold start injector and start injector time switch

OK

NG Repair or replace harness or connector.

5 Check fuel injection of cold start injector.

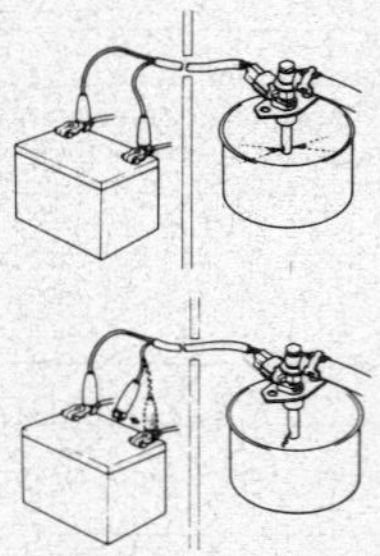

P Remove cold start injector

C (1) Check fuel injection of cold start injector

(2) Check fuel leakage of cold start injector

OK (1) Fuel is injected normally.
(2) Fuel leakage is less than one drop per minute.

Notice Perform this check within the shortest possible time.

OK

NG Repair or replace cold start injector.

Proceed to next circuit inspection shown on matrix chart

ISC Valve Circuit

CIRCUIT DESCRIPTION

The ISC valve is provided on the intake air chamber and intake air bypassing the throttle valve is directed to the ISC valve through a hose.
A step motor is built into the ISC valve. It consists of four coils, the magnetic rotor, valve shaft and valve. When current flows to the coils due to signals from the ECU, the rotor turns and moves the valve shaft forward or backward, changing the clearance between the valve and the valve seat.
In this way the intake air volume bypassing the throttle valve is regulated, controlling the engine speed.
There are 125 possible positions to which the valve can be opened.

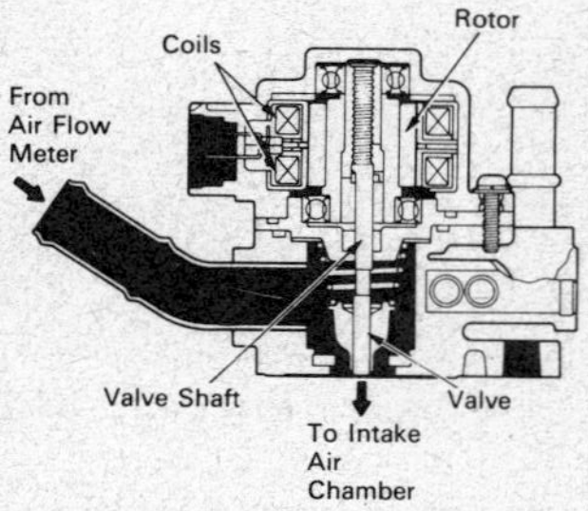

DIAGNOSTIC CHART

1 Check operation for ISC valve. — NG → Replace ISC valve.

OK ↓

2 Check harness and connector between EFI main relay, ISC valve and ECU. — NG → Repair or replace harness or connector.

OK ↓

Proceed to next circuit inspection shown on matrix chart

WIRING DIAGRAM

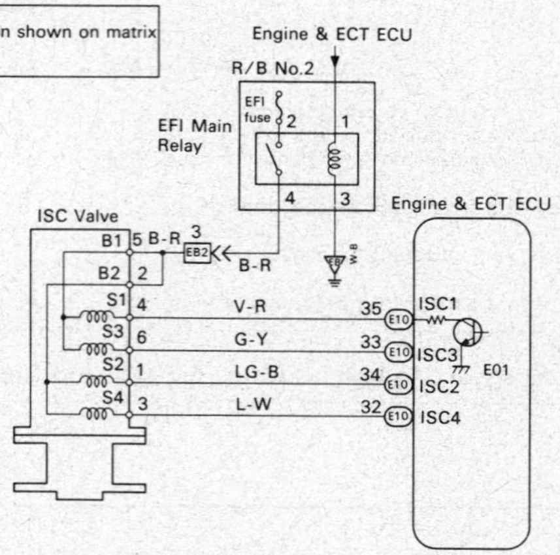

INSPECTION PROCEDURE

1 Check ISC valve.

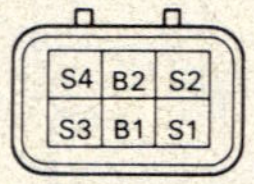

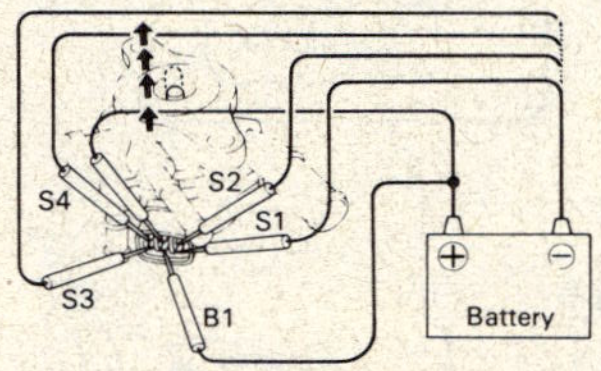

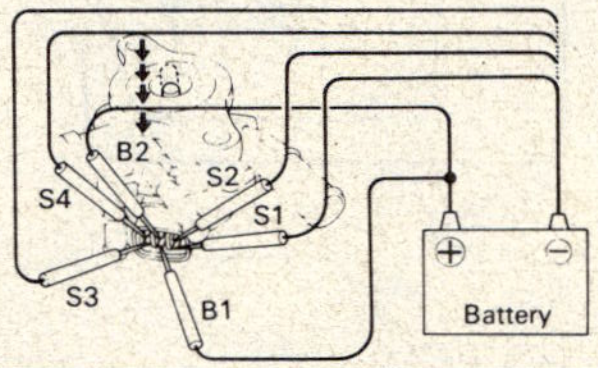

P Disconnect ISC valve connector.

C Measure resistance between terminals shown below.

OK

Terminal	Resistance
B1 – S1	10 Ω – 30 Ω
B1 – S3	10 Ω – 30 Ω
B2 – S2	10 Ω – 30 Ω
B2 – S4	10 Ω – 30 Ω

P Remove ISC Valve.

C (1) Connect the battery positive lead to terminals B1 and B2, and the negative lead to terminals S1-S2-S3-S4 in that order.
(2) Connect the battery positive lead to terminals B1 and B2, and the negative lead to terminals S4-S3-S2-S1 in that order.

OK **(1) The valve moves in the closing direction.**
(2) The valve moves in the opening direction.

NG → **Replace ISC valve.**

OK ↓

2 Check for open and short in harness and connector between EFI main relay and ISC valve, ISC valve and engine & ECT ECU

NG → **Repair or replace harness or connector.**

OK ↓

Proceed to next circuit inspection shown on matrix chart

Fuel Pressure Control VSV Circuit

CIRCUIT DESCRIPTION

The ECU turns on a VSV (Vacuum Switching Valve) to draw the air into the diaphragm chamber of the pressure regulator if it detects that the temperature of the coolant is too high during engine starting.
The air drawn into the chamber increases the fuel pressure to prevent fuel vapor lock at high engine temperature in order to help the engine start when it is warm.
Fuel pressure control ends approx. 100 secs. after the engine is started.

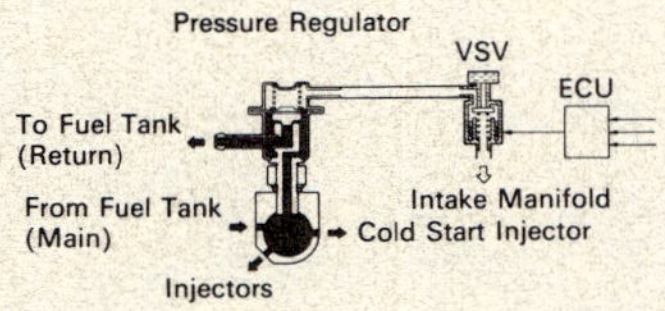

DIAGNOSTIC CHART

1 Check operation for the fuel pressure control VSV. — NG → Replace fuel pressure control VSV.

OK ↓

2 Check voltage of VSV power source. — OK → Proceed to next circuit inspection shown on matrix chart

NG ↓

3 Check for open and short in harness and connector between main relay and ECU. — NG → Repair or replace harness or connector.

OK ↓

Check and replace ECU.

WIRING DIAGRAM

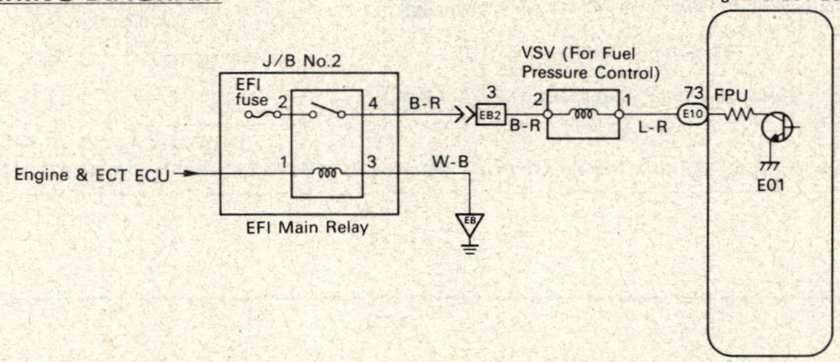

INSPECTION PROCEDURE

1 Check fuel pressure control VSV.

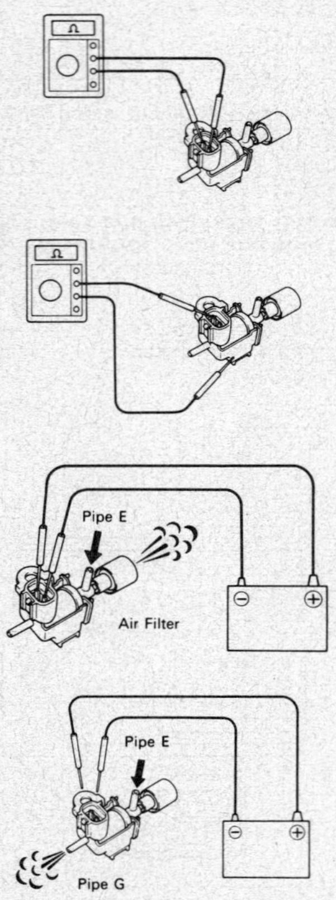

P (1) Remove fuel pressure control VSV.
(2) Disconnect fuel pressure control VSV connector.

C (1) Measure resistance between terminals.
(2) Measure resistance between each terminal and the body.

OK **(1) Resistance: 30 – 50 Ω at 20°C (68°F)**
(2) Resistance: 1MΩ or higher

C Check operation of fuel pressure control VSV when battery voltage is applied to the terminals of fuel pressure control VSV connector or not.

OK **Battery voltage is applied:**
The air from pipe E is flowing out through the air filter.

Battery voltage is not applied:
The air from pipe E is flowing out through pipe G.

NG 〉 **Replace fuel pressure control VSV.**

OK

Go to step 2.

2 Check voltage between terminal FPU of engine & ECT ECU connector and body ground.

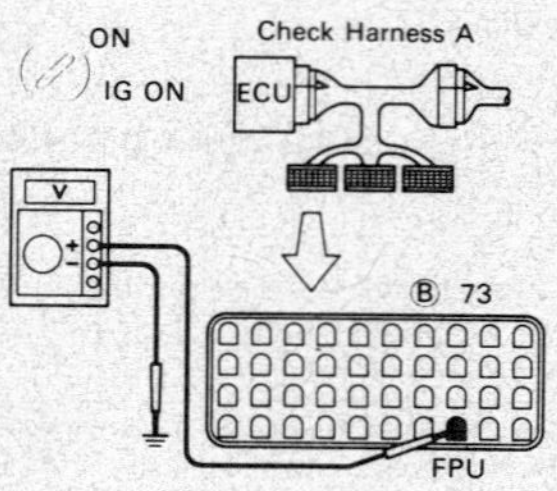

P (1) Connect the Check Harness A.
(2) Turn ignition switch on.

C Measure voltage between terminal FPU of engine & ECT ECU connector and body ground.

OK **Voltage: 10 – 14V**

OK 〉 **Proceed to next circuit inspection shown on matrix chart**

NG

3 Check for open and short in harness and connector between engine & ECT ECU and VSV, VSV and EFI main relay

NG 〉 **Repair or replace harness or connector.**

OK

Check and replace engine & ECT ECU.

TE1, TE2 Terminal Circuit

CIRCUIT DESCRIPTION

Terminal TE1 is located in the check connector in the engine compartment and terminals TE1 and TE2 are located in the TDCL in the cabin. When these terminals are connected with the E1 terminal, diagnostic codes in normal mode or test mode can be read from the "CHECK" engine warning light on the combination meter.

DIAGNOSTIC CHART

HINT: If terminals TE1 and TE2 are connected with terminal E1, diagnostic code is not output or test mode is not activated.
Even though terminal TE1 is not connected with terminal E1, the "CHECK" engine warning light blinks.
For the above phenomenon, the likely cause is an open or short in the wire harness, or malfunction inside the ECU.

Step	Check	Result	Action
1	Check voltage of terminals TE1, TE2.	OK	Check and replace ECU.
		NG ↓	
2	Check continuity between terminal E1 and body ground.	NG	Repair or replace harness or connector.
		OK ↓	
3	Check for open and short in harness and connector between check connector, TDCL and ECU.	NG	Repair or replace harness or connector.
		OK ↓	
	Check and replace ECU.		

WIRING DIAGRAM

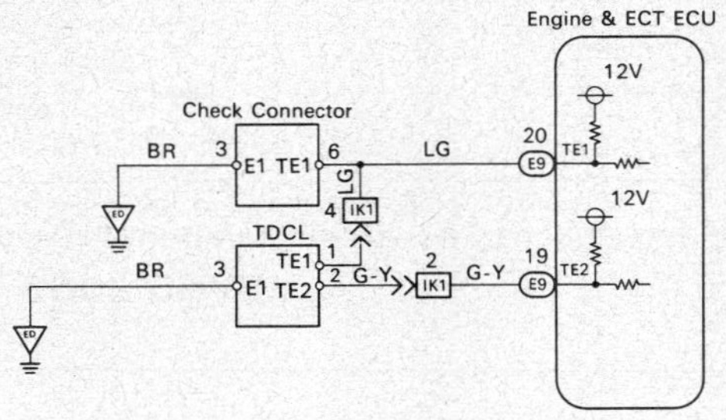

INSPECTION PROCEDURE

1 Check voltage between terminals TE1, TE2 and E1 of check connector, TDCL.

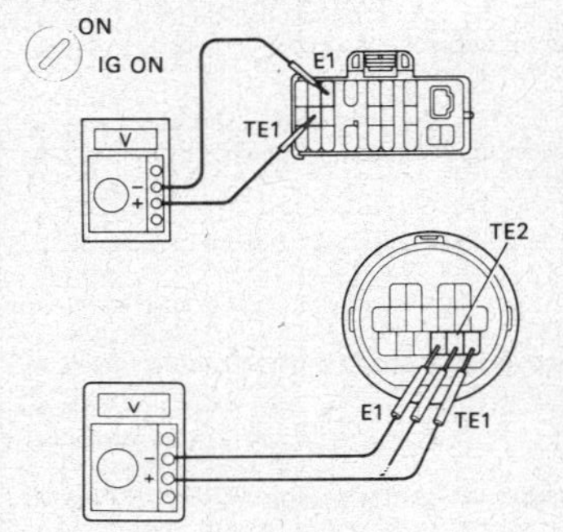

P Turn ignition switch on.

C Measure voltage between terminals TE1, TE2 and E1 of check connector, TDCL.

OK Voltage: 10 – 14V

OK ⟩ Check and replace engine & ECT ECU.

NG ↓

2 Check continuity between terminal E1 of check connector, TDCL and body ground.

NG ⟩ Repair or replace harness or connector.

OK ↓

3 Check for open and short in harness and connector between engine & ECT ECU and check connector, TDCL

NG ⟩ Repair or replace harness or connector.

OK ↓

Check and replace engine & ECT ECU.

DIAGNOSTIC CODE CHART

If a malfunction code is displayed during the diagnostic code check in test mode, check the circuit for that code listed in the table below (Proceed to the page given for that circuit).

Code No.	Number of Check Engine Blinks	Circuit	Diagnosis	Trouble Area	"CHECK" Engine Warning Light*1 Normal Mode	"CHECK" Engine Warning Light*1 Test Mode	Memory*2	
–		Normal	Output when no other code is recorded.	———	—	—	———	
12		RPM Signal	G1, G2 or NE signal is not input to the ECU for 2 sec. or more after STA turns ON.	• Open or short in engine speed sensor, No. 1, No. 2 cam position sensor circuit. • Engine speed sensor • No. 1, No. 2 cam position sensor • Starter • ECU	ON	N.A.	○	
13		RPM Signal	No NE signal input to ECU for 50 msec. or more when engine speed is 1000 rpm or more.	• Open or short in engine speed sensor circuit. • Engine speed sensor • ECU	ON	ON	○	
			Between 2 pulses of G signal, NE signal of other than 12 pulses is input to ECU.	• Open or short in engine speed sensor circuit. • Mechanical system malfunction (skipping teeth of timing belt, belt stretched) • Engine speed sensor • ECU				
			Deviance in G1, G2 and NE signal continues for 1 sec. during idling (throttle fully closed) after engine warmed up.	• Mechanical system malfunction (skipping teeth of timing belt, belt stretched) • No. 1, No. 2 cam position sensor • ECU	ON	N.A.		
14		Ignition No. 1 Signal	IGF1 signal from igniter No. 1 is not input to ECU for 8 consecutive ignition.	• Open or short in IGF1 circuit from igniter No. 1 – ECU. • Igniter No. 1 • ECU	ON	N.A.	○	
15		Ignition No. 2 Signal	IGF2 signal from igniter No. 2 is not input to ECU for 8 consecutive ignition.	• Open or short in IGF2 circuit from igniter No. 2 – ECU. • Igniter No. 2 • ECU	ON	N.A.	○	
16		ECT Control Signal	Normal signal is not output from ECT CPU.	• ECU	ON	N.A.	X	
17		No. 1 Cam Position Sensor Signal	G1 signal is not input to ECU.	• Open or short in No. 1 cam position sensor circuit. • No. 1 cam position sensor • ECU	N.A.	OFF	X	
18		No. 2 Cam Position Sensor Signal	G2 signal is not input to ECU.	• Open or short in No. 2 cam position sensor circuit. • No. 2 cam position sensor • ECU	N.A.	OFF	X	
21		Main Oxygen Sensor Signal (on left bank)	(1) Open or short in heater circuit of main oxygen sensor for 0.5 sec. or more.	• Open or short in heater circuit of main oxygen sensor • Main oxygen sensor heater • ECU	ON	ON	○	
			(2) At normal driving speed (ECT in 4th speed, A/C ON, 40–50 mph), amplitude of oxygen sensor signal is reduced to between 0.35–0.70 V continuously for 60 sec. or more. (2 trip detection logic)	• Main oxygen sensor circuit • Main oxygen sensor				

4.0L ENGINE—DIAGNOSTIC CODES—LS400

DIAGNOSTIC CODE CHART (Cont'd)

Code No.	Number of Check Engine Blinks	Circuit	Diagnosis	Trouble Area	"CHECK" Engine Warning Light*1 Normal Mode	"CHECK" Engine Warning Light*1 Test Mode	Memory*2
22		Water Temp. Sensor Signal	Open or short in water temp. sensor circuit for 0.5 sec. or more.	• Open or short in water temp. sensor circuit. • Water temp. sensor • ECU	ON	ON	○
24		Intake Air Temp. Sensor Signal	Open or short in intake air temp. sensor circuit for 0.5 sec. or more.	• Open or short in intake air temp. circuit. • Intake air temp. sensor • ECU	ON*3	ON	○
25		Air-Fuel Ratio Lean Malfunction	(1) Oxygen sensor output is less than 0.45 V for at least 90 sec. when main oxygen sensor is warmed up (racing at 2000 rpm). (2 trip detection logic)*4	• Open or short in main oxygen sensor circuit • Main oxygen sensor • Ignition system • Water temp. sensor	ON	ON	○
			(2) With engine warmed up and idling, main oxygen sensor signal is centered around 0.45 V with frequency at 5 Hz or above. (2 trip detection logic)*4	• Open or short in injector circuit. • Fuel line pressure (injector blockage) • Air flow meter (air intake) • Engine ground bolt loose • Foreign object caught in valve			
			(3) Deviation between air-fuel ratio feedback compensation values for left and right banks is beyond acceptable limits. (If air-fuel ratio feedback compensation values are deviating, this can be detected when engine is raced at 2000 rpm for 30 sec. when engine is warmed up). (2 trip detection logic)*4	• Fuel line pressure (injector leak, blockage) • Mechanical system malfunction (skipping teeth of timing belt) • Ignition system			
26		Air-Fuel Ratio Rich Malfuntion	Same as (2) and (3) of Code No. 25	Same as (2) and (3) of Code No. 25	ON	ON	○
27		Sub-Oxygen Sensor Signal (on left bank)	When sub-oxygen sensor is warmed up and apply acceleration with the throttle fully open for 4 sec., output of main oxygen sensor is 0.45 V or more (rich) and output of sub-oxygen sensor is 0.45 V or less (lean). (2 trip detection logic)*4 Sub-oxygen sensor warm-up condition: After engine warmed up and idled for 120 sec., ECT in 3rd speed, A/C ON and vehicle driven 45–60 mph for 6 min.	• Short or open in sub-oxygen sensor circuit. • Sub-oxygen sensor • ECU	ON	ON	○

DIAGNOSTIC CODE CHART (Cont'd)

Code No.	Number of Check Engine Blinks	Circuit	Diagnosis	Trouble Area	"CHECK" Engine Warning Light*1		Memory*2	
					Normal Mode	Test Mode		
28		Main Oxygen Sensor Signal (on right bank)	(1) Open or short in heater circuit of main oxygen sensor for 0.5 sec. or more.	• Open or short in heater circuit of main oxygen sensor • Main oxygen sensor heater • ECU	ON	ON	○	
			(2) At normal driving speed (ECT in 4th speed, A/C ON, 40–50 mph), amplitude of oxygen sensor signal is reduced to between 0.35–0.70 V continuously for 60 sec. or more. (2 trip detection logic)*4	• Main oxygen sensor circuit • Main oxygen sensor				
29		Sub-Oxygen Sensor Signal (on right bank)	When sub-oxygen sensor is warmed up and apply acceleration with the throttle fully open for 4 sec., output of main oxygen sensor is 0.45 V or more (rich) and output of sub-oxygen sensor is 0.45 V or less (lean). (2 trip detection logic)*4 Sub-oxygen sensor warm-up condition: After engine warmed up and idled for 120 sec., ECT in 3rd speed, A/C ON and vehicle driven 45–60 mph for 6 min.	• Short or open in sub-oxygen sensor circuit. • Sub-oxygen sensor • ECU	ON	ON	○	
31		Air Flow Meter Signal	Engine speed is above 300 rpm and air flow meter signal is not input to ECU for 2 sec.	• Open or short in air flow meter circuit. • Air flow meter • ECU	ON	N.A.	○	
35		HAC Sensor Signal	Open or short detected in HAC sensor circuit for 0.5 sec. or more.	• ECU	ON	ON	○	
41		Throttle Position Sensor Signal	Open or short detected in throttle position sensor signal (VTA1) for 0.5 sec. or more.	• Open or short in throttle position sensor circuit.	ON*3	ON	○	
			IDL1 contact is ON and VTA1 output exceeds 1.45 V.	• Throttle position sensor • ECU				
43		Starter Signal	Starter signal is not input to ECU even once.	• Open or short in stater signal circuit. • Open or short in IG s/w or main relay circuit. • ECU	N.A.	OFF	×	
47		Sub-Throttle Position Sensor Signal	Open or short detected in sub-throttle position sensor signal (VTA2) for 0.5 sec. or more.	• Open or short in sub-throttle position sensor circuit.	ON*3	ON	○	
			IDL2 contact is ON and VTA2 output exceeds 1.45 V.	• Sub-throttle position sensor • ECU				

DIAGNOSTIC CODE CHART (Cont'd)

Code No.	Number of Check Engine Blinks	Circuit	Diagnosis	Trouble Area	"CHECK" Engine Warning Light*1		Memory*2
					Normal Mode	Test Mode	
52		No. 1 Knock Sensor Signal	No. 1 knock sensor signal is not input to ECU for 3 crank revolution with engine speed between 1600 rpm – 5200 rpm.	• Open or short in No. 1 knock sensor circuit. • No. 1 knock sensor (looseness) • ECU	ON	N.A.	○
53		Knock Control Signal	Engine speed is between 650 rpm and 5200 rpm and engine control computer (for knock control) malfunction is detected.	• ECU	ON	N.A.	×
55		No. 2 Knock Sensor Signal	No. 2 knock sensor signal is not input to ECU for 3 crank revolution with engine speed between 1600 rpm – 5200 rpm.	• Open or short in No. 2 knock sensor circuit. • No. 2 knock sensor (looseness) • ECU	ON	N.A.	○
71		EGR System Malfunction	With engine warmed up and EGR operation possible (ECT in 3rd speed, A/C ON, 45–70 mph), code is displayed when EGR gas temp. sensor is 65°C (149°F) or less when vehicle is driven for 90 sec. (2 trip detection logic)*4	• Open in EGR gas temp. sensor circuit • Short in VSV circuit for EGR • EGR hose disconnected, valve stuck • ECU	ON	ON	○
51		Switch Condition Signal	Displayed when A/C is ON or IDL contact OFF or shift position in "R", "D", "2" or "1" renges.	• A/C switch circuit • Throttle position sensor IDL circuit • Neutral start switch circuit • Accelerator pedal and cable • ECU	N.A.	OFF	×

*1: "ON" displayed in the diagnosis mode column indicates that the "CHECK" Engine Warning Light is lighted up when a malfunction is detected. "OFF" indicates that the "CHECK" does not light up during malfunction diagnosis, even if a malfunction is detected. "N.A." indicates that the item is not included in malfunction diagnosis.

*2: "○" in the memory column indicates that a diagnostic code is recorded in the ECU memory when a malfunction occurs. "×" indicates that a diagnostic code is not recorded in the ECU memory even if a malfunction occurs. Accordingly, output of diagnostic results in normal or test mode is performed with the IG switch ON.

*3: If a code 24, 41 or 47 malfunction occurs in the normal mode, the "CHECK" Engine Warning Light lights up.

*4: This indicates items for which "2 trip detection logic" is used. With this logic, when a logic malfunction is first detected, the malfunction is temporarily stored in the ECU memory. If the same case is detected again during the second drive test, this second detection causes the "CHECK" Engine Warning Light to light up.
The 2 trip repeats the same mode a 2nd time. (However, the IG switch must be turned OFF between the 1st trip and 2nd trip).
In the Test Mode, the "CHECK" Engine Warning Light lights up the 1st trip a malfunction is detected.

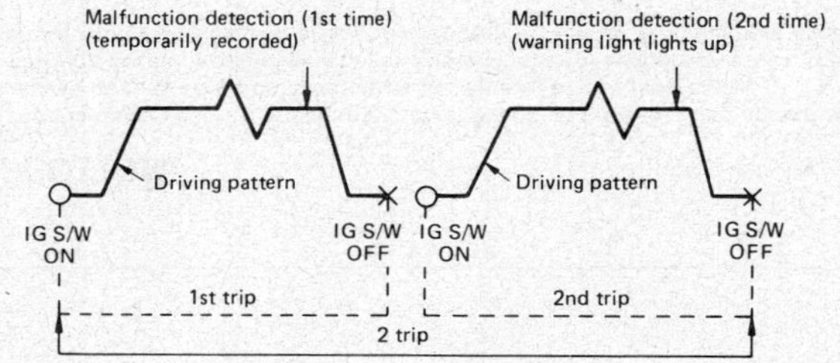

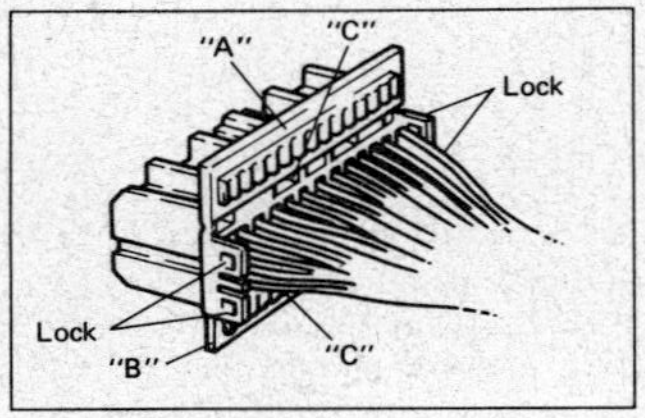

TERMINALS OF ECU

When measuring the voltage or resistance of the connector part of the ECU, remove the locks as required in the following procedure and always insert the test probe into the connector from the wire harness side.

PREPARATION

(a) Turn the ignition switch to LOCK position.

(b) Disconnect the connectors from the ECU.

(c) To remove section "A" and "B", carefully pull outward the lock on both sides as shown in the illustration.

NOTICE: Pay attention to section "C" in the illustration which can be easily broken.

(d) Reconnect the connectors to the ECU.

TERMINALS OF ECU

Terminal No.	Symbol	Connection	Terminal No.	Symbol	Connection	Terminal No.	Symbol	Connection
E8- 1			E9- 5	IGT1	No. 1 ignitor	E10-19	SPD	Speed sensor No. 1
2			6	G2	No. 2 cam position sensor	20	TE2	Check connector
3	NCO+	OD direct clutch speed sensor	7	G1	No.1 cam position sensor	21	SP2–	Speed sensor No. 2 (For ECT)
4	STJ	Cold start injector	8	NE	Engine speed sensor	22	E1	ECU ground
5	ISC4	ISC valve	9	IDL2*1	Sub-throttle position sensor	E11- 1		
6	ISC3	ISC valve	10	IGF2	No. 2 ignitor	2	OD1	Cruise control ECU
7	ISC2	ISC valve	11	KNK2	No. 2 knock sensor	3	OD2	O/D main switch
8	ISC1	ISC valve	12	VF2	Check connector	4	M-REL	EFI main relay
9	SOL2	ECT solenoid No. 2	13	IGT2	No. 2 ignitor	5	W	"CHECK" engine warning light
10	SOL1	ECT solenoid No. 1	14			6	FPR	Fuel pump relay
11	#20	Injectors (No. 2 and No. 8)	15	G–	Cam position sensor	7	ACMG	A/C magnet clutch relay
12	#10	Injectors (No. 1 and No. 7)	16	NE–	Cam position sensor	8	FC	Circuit opening relay
13	E01	Power ground	E10- 1	IGSW	Ignition switch	9	L3	Suspension ECU
14	SLU–	ECT solenoid No. 3	2	STA	Starter relay	10	TR2	ABS and TRAC ECU
15	SLN–	ECT solenoid No. 4	3			11	OXR2	Sub-oxygen sensor (on right bank)
16	NCO–	O/D direct clutch speed sensor	4	VTA1	Throttle position sensor	12	VSH	ABS and TRAC ECU
17	STJ1	Cold start injector	5	THW	Water temp. sensor	13	VTH	ABS and TRAC ECU
18	DG	TDCL	6	OXL1	Main oxygen sensor (on left bank)	14		
19	HT2	Main oxygen sensor (on right bank)	7	THG	EGR gas temp. sensor	15	2	Neutral start switch
20	HT1	Main oxygen sensor (on left bank)	8	KS	Air flow meter	16	L	Nuetral start switch
21	EGR	EGR valve	9	TE1	Check connector	17	P	ECT pattern select switch
22	PR	Fuel pressure control VSV	10	SP2+	Speed sensor No. 2 (For ECT)	18	R	Neutral start switch
23	A/D	Cruise control ECU	11	VC	Throttle position sensor Air flow meter	19	BK	Stop light switch
24	#40	Injectors (No. 3 and No. 5)	12	+B	EFI main relay	20	KD	Kick-down switch
25	#30	Injectors (No. 4 and No. 6)	13	+B1	EFI main relay	21	L2	Suspension ECU
26	E02	Power ground	14	NSW	Neutral start switch	22	L1	Suspension ECU
E9- 1	IDL1	Throttle position sensor	15	VTA2*1	Sub-throttle position sensor	23	AC	A/C control assembly
2	IGF1	No. 1 ignitor	16	THA	Intake air temp. sensor	24	OXL2	Sub-oxygen sensor (on left bank)
3	KNK1	No. 1 knock sensor	17	OXR1	Main oxygen sensor (on right bank)	25	NE0	ABS and TRAC ECU
4	VF1	Check connector	18	E2	Sensor ground	26	BATT	Battery

ECU Terminals

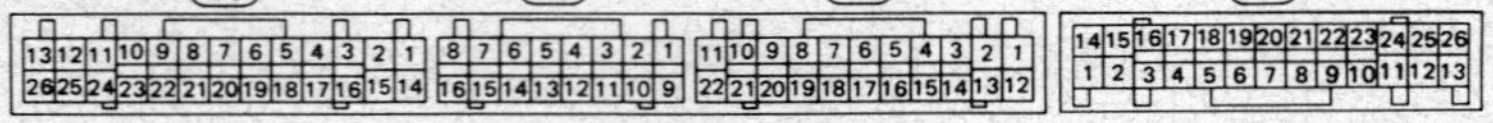

*1: Only w/ TRAC

BASIC INSPECTION

When the normal code is displayed in the diagnostic code check, troubleshooting should be performed in the order for all possible circuits to be considered as the causes of the problems.
In many cases, by carrying out the basic engine check shown in the following flow chart, the location causing the problem can be found quickly and efficiently. Therefore, use of this check is essential in engine troubleshooting.

1 Is battery voltage 11 V or more when engine is stopped ?

YES

NO → **Charge or replace battery.**

2 Is engine cranked ?

YES

NO → **Proceed to matrix chart of problem symptoms**

3 Does engine start ?

YES

NO → **Go to step 7 .**

4 Check air filter.

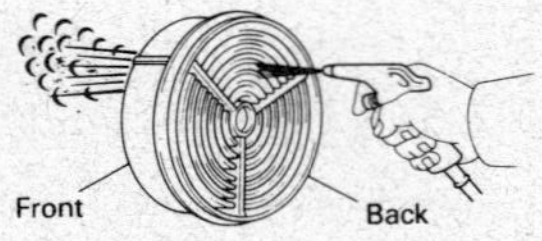

P Remove air filter.

C Visually check that the air cleaner element is not excessively damaged or oily.

Hint If necessary, clean element with compressed air. First blow from inside thoroughly, then blow off outside of element.

OK

NG → **Repair or replace.**

Go to step 5 .

5 Check idle speed.

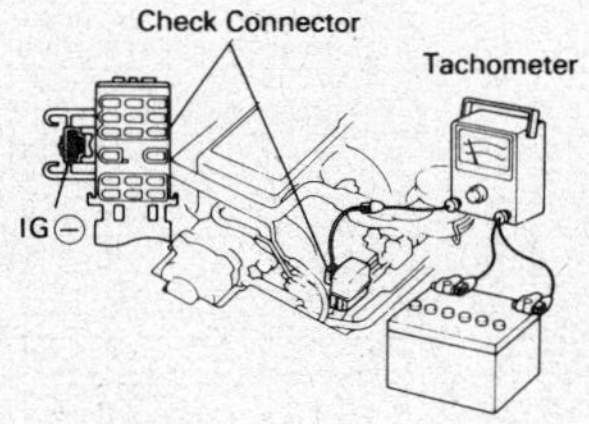

P (1) Warm up engine at normal operating temperature.
(2) Switched off all accessories.
(3) Switched off air conditioner.
(4) Shift transmission into "N" range.
(5) Connect tachometer test probe to terminal IG ⊖ of check connector, and set the tachometer to the 4-cylinder range.

C Check idle speed.

OK **Idle speed: 600 ~ 700 rpm [at approx. 10°C (50°F) or higher]**

Hint Engine rpm increases to 750 ~ 850 rpm when intake air temp. is below approx. 10°C (50°F).

Caution
- **NEVER allow tachometer test probe to touch ground as it could result in damage to igniter and/or ignition coil.**
- **As some tachometers are not compatible with this ignition system, we recommended that you confirm compatibility of your unit before use.**

OK

NG — **Proceed to matrix chart of problem symptoms**

6 Check ignition timing.

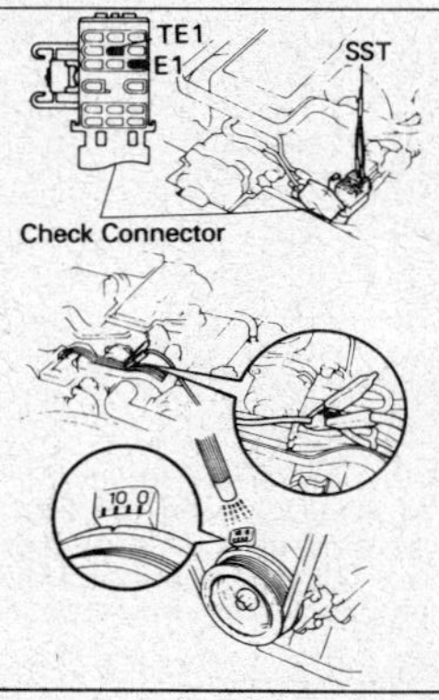

P (1) Warm up engine at normal operating temperature.
(2) Shift transmission into "N" range.
(3) Keep the engine speed at idle.
(4) Using SST, connect terminals TE1 and E1 of check connector.

SST 09843-18020

(5) Using a timing light, connect the tester to No.6 high-tension cord.

C Check ignition timing.

OK **Ignition timing: 8 — 12° BTDC at idle**

OK

NG — **Proceed to troubleshoot.**

Proceed to matrix chart of problem symptoms

7 Check fuel pressure.

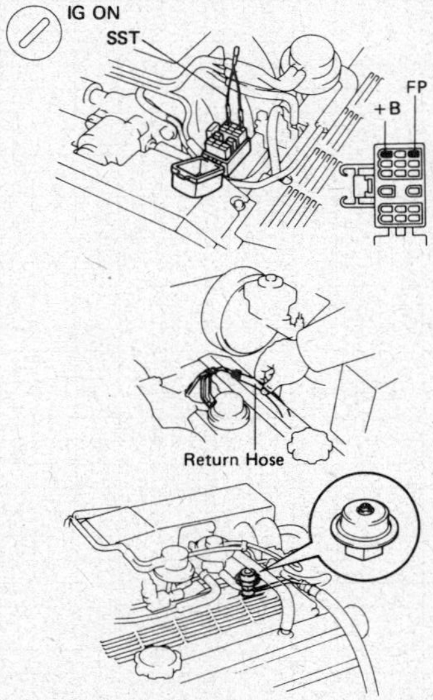

P (1) Be sure that fuel is enough in tank.
(2) Turn ignition switch on.
(3) Using SST, connect terminals FP and +B of check connector.

SST 09843-18020

C Check for fuel pressure in the return hose when it is pinched off.

(Check that pulsation damper screw rises up when teminals are connected.)

Caution
Never make a mistake with the terminal connection position as this will cause a malfunction.

OK

NG — **Proceed to troubleshoot.**

8 Check for spark.

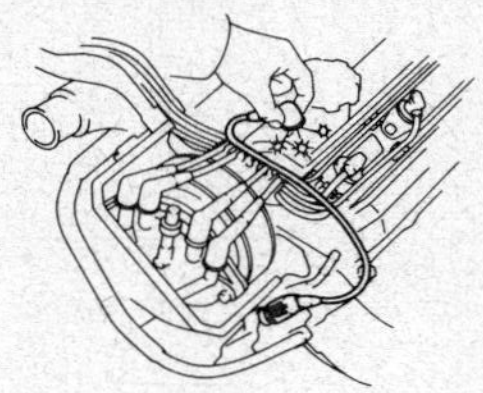

C Disconnect the high-tension cord from the distributor and, hold the end about 12.5 mm (1/2") from the ground, see if spark occurs while the engine is being cranked.

Hint To prevent gasoline from being injected from the injectors during this test, crank the engine for no more than 1 — 2 seconds at a time.

OK

NG — **Proceed to troubleshoot.**

Proceed to matrix chart of problem symptoms

MATRIX CHART OF PROBLEM SYMPTOMS

When the malfunction code is not confirmed in the diagnostic code check and the problem still can not be confirmed in the basic inspection, then proceed to this step and perform troubleshooting according to the numbered order given in the table below. For "Engine Mechanical and Others" items in the suspect area column, details of each check item are provided on the next page.

The circuits indicated by ※ on the matrix chart can be inspected using the TCCS checker.

Symptom	Suspect area	※ RPM signal circuit (No.1)	Ignition signal circuit (Spark test)	※ Main oxygen sensor circuit	※ Water temp. sensor circuit	Intake air temp. sensor circuit	Sub-oxygen sensor circuit	※ Air flow meter circuit	※ Throttle position sensor circuit	※ Starter signal circuit	Knock sensor circuit	EGR System	※ Neutral start switch circuit	ECU power source circuit	Back up power source circuit	※ Injector circuit	Cold start injector circuit	※ ISC valve circuit	※ Fuel system circuit
Does not start	Engine does not crank													3					
	Starter runs-engine does not crank																		
	No initial combustion	9	2											1		6	8		3
	No complete combustion		8		10			7								11	12	3	2
Difficult to start	Engine cranks slowly																		
	Difficult to start normally		11		3	14						5				12	13	4	6
	Difficult to start cold				1	8				2		6				10	7	5	9
	Difficult to start hot				1	8						6				10	7	4	9
Poor Idling	Incorrect first idle				4	5												6	
	High engine idle speed				4	5			10				9	8				6	
	Low engine idle speed				1							4	3			6		2	5
	Rough idling		5	9	3			14				4			13	15	16	7	8
	Misfire		4		8			11				3				13	14	9	10
Poor Driveability	Hesitation/Poor acceleration		12	10	8	9		7	6			5				16	17		11
	Back fire				3	4		6	5			2				8			7
	Muffler explosion (after fire)		7		2	3		5	4							11	12	6	1
	Surging		13	8	3	4			6			5				14	15		7
	Knocking										2	5				4			
Engine Stall	Engine stalls soon after starting				5	6		8				4				9		7	3
	After acceleration pedal depressed							1	3			2							
	After acceleration pedal released							3				2						1	
	During A/C operation																	1	
	When N to D shift												1					2	
	When turning the steering																		
	When starting off or stopping																		
Others	Poor fuel economy			19	6	7	20		8				18			14	16	17	13
	Engine overheat																		
	Engine overcool																		
	Excessive oil consumption																		
	Low oil pressure																		
	High oil pressure																		
	Starter keeps running																		
	Battery often discharges																		

HINT: Before proceeding with troubleshooting using this matrix chart of problem symptoms, basic inspection should be performed first.

Symptom	Suspect area	Fuel pressure control VSV circuit	A/C signal circuit (Compressor circuit)	Fuel quality	Fuel leakage	Coolant leakage	Oil leakage	Vacuum leakage	Starter relay	Neutral start switch	Starter	Spark plug	Ignition coil	Distributor	Accelerator pedal link	Cooling fan system	Dash pot	PS idle-up	Compression	Brakes drag even when released	Sub-throttle valve faulty	ECT faulty	Theft deterrent ECU	Engine mechanical and others	Engine & ECT ECU
Does not start	Engine does not crank								1	4	2												5		
	Starter runs-engine does not crank										1													2	
	No initial combustion											5	4						7					10	11
	No complete combustion							1				6	4	5					9					13	14
Difficult to start	Engine cranks slowly		2								1													3	
	Difficult to start normally			2				1				9	7	8					10					15	16
	Difficult to start cold			4				3				13	11	12											14
	Difficult to start hot	5		3				2				13	11	12											14
Poor Idling	Incorrect first idle														1		2	3							7
	High engine idle speed		7												1		2	3							11
	Low engine idle speed																								7
	Rough idling			2				1				12	10	11					6					17	18
	Misfire			2				1				5	6	7					12					15	16
Poor Driveability	Hesitation/Poor acceleration			4				1				15	13	14	2					3	19	18		20	21
	Back fire							1																9	10
	Muffler explosion (after fire)											10	8	9										13	14
	Surging			2				1				11	9	10					12					16	17
	Knocking			1								3												6	7
Engine Stall	Engine stalls soon after starting			2				1																	10
	After acceleration pedal depressed																								4
	After acceleration pedal released																								4
	During A/C operation		2																						3
	When N to D shift																								3
	When turning the steering																	1							
	When starting off or stopping																					1			
Others	Poor fuel economy	15		2	1							9	10	11	4	5			12	3		21		22	23
	Engine overheat					1										2								3	
	Engine overcool															1								2	
	Excessive oil consumption						1																	2	
	Low oil pressure						1																	2	
	High oil pressure																							1	
	Starter keeps running								1		2														
	Battery often discharges																							1	

CIRCUIT INSPECTION

Diag. Code	12,17,18	RPM Signal Circuit (No.1)

CIRCUIT DESCRIPTION

Cam position sensors (G1 and G2 signals) and engine speed sensor (NE signal) consist of a signal plate and a pick up coil.
The G1, G2 signal plates have one tooth each on its outer circumference and are mounted on the left and right bank camshafts.
When the camshafts rotate, the protrusion on the signal plate and the air gap on the pick up coil change, causing fluctuations in the magnetic field and generating an electromotive force in the pick up coil.
The NE signal plate has 12 teeth and is mounted on the crankshaft. The NE signal sensor generates 12 NE signals per engine revolution. The ECU detects the standard crankshaft angle based on the G1, G2 signals, and the actual crankshaft angle and the engine speed by the NE signals.

Code No.	Diagnosis	Trouble Area
12	G1, G2 or NE signal is not input to the ECU for 2 sec. or more after STA turns ON.	• Open or short in engine speed sensor, No. 1, No. 2 cam position sensor circuit • Engine speed sensor • No. 1, No. 2 cam position sensor • Starter • ECU
17	G1 signal is not input to ECU.	• Open or short in No. 1 cam position sensor circuit • No. 1 cam position sensor • ECU
18	G2 signal is not input to ECU.	• Open or short in No. 2 cam position sensor circuit • No. 2 cam position sensor • ECU

DIAGNOSTIC CHART

HINT: If diagnostic code 17 is displayed, check the No. 1 cam position sensor circuit. If 18 is displayed, check the No. 2 cam position sensor circuit. If 12 is displayed, check the engine speed sensor, No. 1 and No. 2 cam position sensor circuit.

Step	Check	Result	Action
1	Check engine speed sensor, No. 1, No. 2 cam position sensor.	NG	Replace engine speed sensor, No. 1, No. 2 cam position sensor.
		OK	↓
2	Check harness and connector between engine & ECT ECU and each sensor	NG	Repair or replace harness or connector.
		OK	↓
3	Inspect sensor installation and teeth of signal plate.	NG	Tighten the sensor. Replace signal plate.
		OK	↓
	Proceed to next circuit inspection shown on matrix chart *1		

*1: When diag. code 12, 17 or 18 is displayed, check and replace engine and ECT ECU.

WIRING DIAGRAM

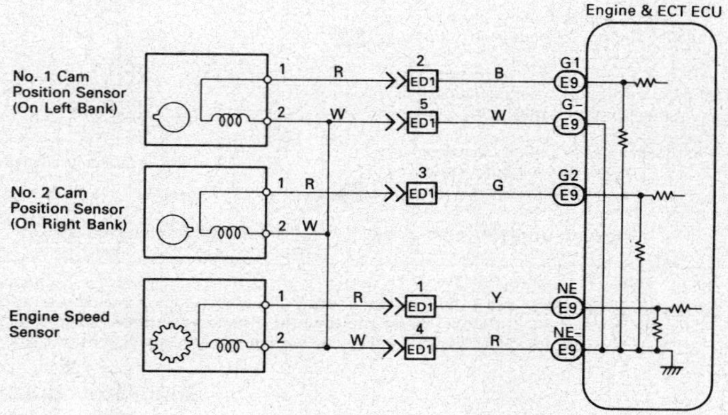

WIRING ROUTING

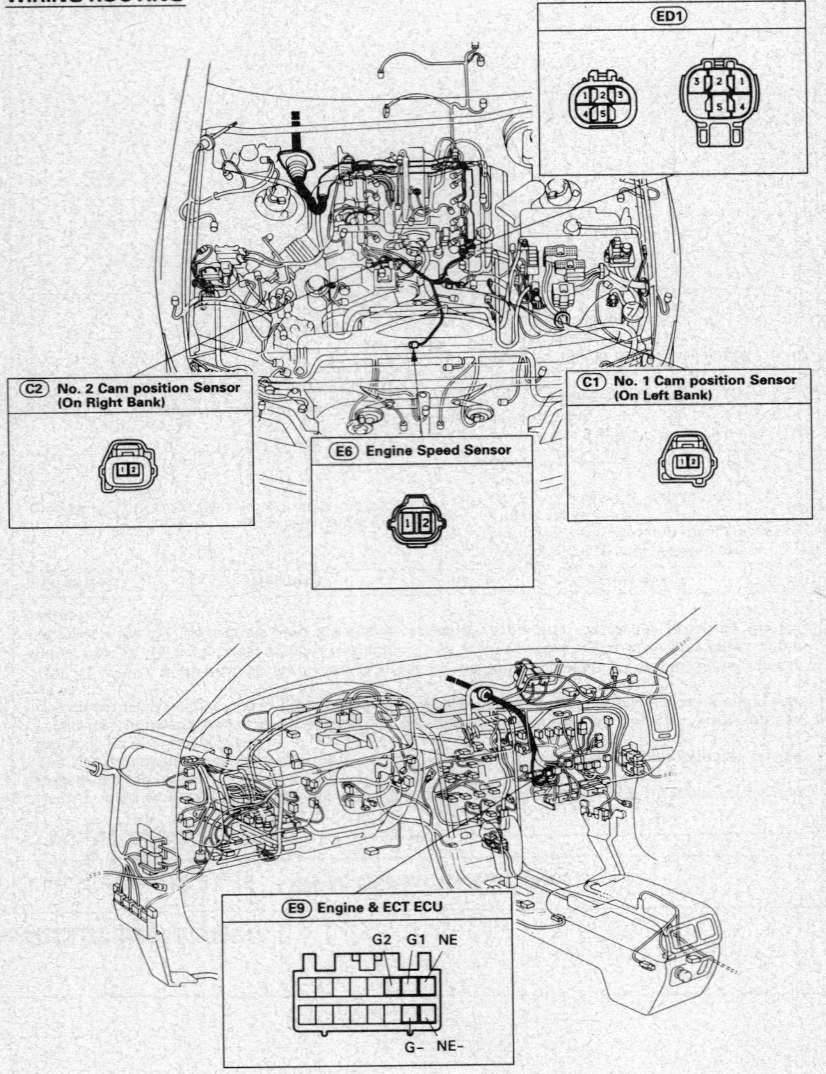

INSPECTION PROCEDURE

HINT: If diag. code 17 is displayed, check the No. 1 cam position sensor circuit.
If 18 is displayed, check the No. 2 cam position sensor circuit.
If 12 is displayed, check the engine speed sensor, No. 1 and No. 2 cam position sensor circuit.

1 Check engine speed sensor, No. 1, No. 2 cam position sensor.

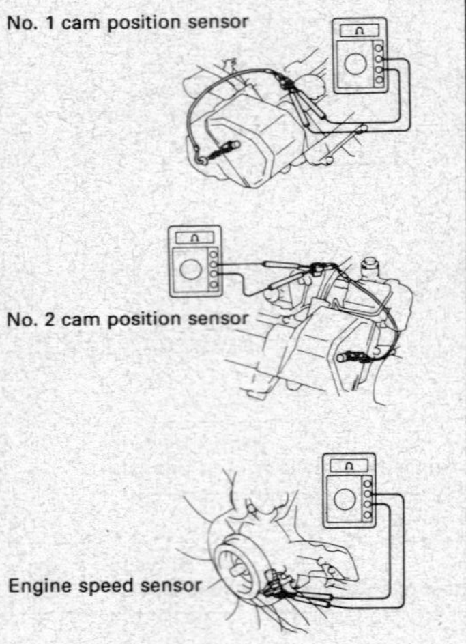

P For engine speed sensor,
(1) Remove engine under cover.
(2) Disconnect engine speed sensor connector.

For No. 1, No. 2 cam position sensor,
(1) Disconnect No. 1, No. 2 cam position sensor connectors.

C Measure resistance of engine speed sensor, No. 1 and No. 2 cam position sensor.

OK

	Resistance
No.1 cam position sensor	950 ~ 1250 Ω
No.2 cam position sensor	950 ~ 1250 Ω
Engine speed sensor	950 ~ 1250 Ω

NG → Replace engine speed sensor, No. 1, No. 2 cam position sensor.

OK ↓

2 Check harness and connector between engine & ECT ECU and each sensor

NG → Repair or replace harness or connector.

OK ↓

3 Inspect sensor installation and teeth of signal plate.

NG → Tighten the sensor. Replace signal plate.

OK ↓

Proceed to next circuit inspection shown on matrix chart *1

*1: However, when diag. code 12, 17 or 18 is displayed, check and replace engine & ECT ECU.

Diag. Code	13	RPM Signal Circuit (No.2)

CIRCUIT DESCRIPTION

Refer to RPM signal circuit (No.1)

Code No.	Diagnosis	Trouble Area
13	No NE signal input to ECU for 50 msec. or more when engine speed is 1000 rpm or more.	• Open or short in engine speed sensor circuit. • Engine speed sensor • ECU
	Between 2 pulses of G signal, NE signal of other than 12 pulses is input to ECU.	• Open or short in engine speed sensor circuit. • Mechanical system malfunction (skipping teeth of timing belt, belt stretched) • Engine speed sensor • ECU
	Deviance in G1, G2 and NE signal continues for 1 sec. during idling (throttle fully closed) after engine warmed up.	• Mechanical system malfunction (skipping teeth of timing belt, belt stretched) • No. 1, No. 2 cam position sensor • ECU

DIAGNOSTIC CHART

HINT: If diagnostic code 12, 17 or 18 is output, perform inspection of RPM signal circuit (No. 1)

1 Inspect sensor installation. Check if any teeth of NE signal plate are broken.
- NG → Tighten the sensor. Replace signal plate.
- OK ↓

2 Check valve timing
Check for loose and jumping teeth of timing belt
- NG → Adjust valve timing (Repair or replace timing belt).
- OK ↓

Check for momentary interruption

Diag. Code	14, 15	Ignition Signal Circuit

CIRCUIT DESCRIPTION

The ECU determines the ignition timing, turns on Tr_1 at a predetermined angle (°CA) before the desired ignition timing and outputs an ignition signal (IGT) "1" to the ignitor.
Since the width of the IGT signal is constant, the dwell angle control circuit in the ignitor determines the time the control circuit starts primary current flow to the ignition coil based on the engine rpm and ignition timing one revolution ago, that is, the time the Tr_2 turns on.
When it reaches the ignition timing, the ECU turns Tr_1 off and outputs the IGT signal "0".
This turns Tr_2 off, interrupting the primary current flow and generating a high voltage in the secondary coil which causes the spark plug to spark. Also, by the counter electromotive force generated when the primary current is interrupted, the ignitor sends an ignition confirmation signal (IGF) to the ECU. The ECU stops fuel injection as a fail safe function when the IGF signal is not input to the ECU.

Code No.	Diagnosis	Trouble Area
14	IGF1 signal from No. 1 igniter is not input to ECU for 8 consecutive ignition.	• Open or short in IGF1 and IGT1 circuit from No. 1 igniter – ECU. • No. 1 igniter • ECU
15	IGF2 signal from No. 2 igniter is not input to ECU for 8 consecutive ignition.	• Open or short in IGF2 and IGT2 circuit from No. 2 igniter – ECU. • No. 2 igniter • ECU

HINT: No.1 ignitor is for left bank and No. 2 ignitor is for right bank.

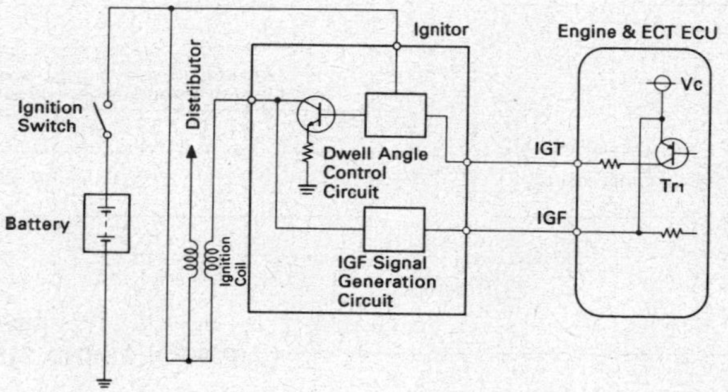

DIAGNOSTIC CHART

HINT: If diag. code 14 is displayed, check No. 1 ignitor circuit.
If diag. code 15 is displayed, check No. 2 ignitor circuit.

1 Check for spark.
- NG → Go to step 4
- OK ↓

2 Check harness and connector in IGF and IGT signal circuit between engine & ECT ECU and ignitor
- NG → Repair or replace harness or connector.
- OK ↓

3 Disconnect ignitor connector and check voltage between terminals IGF1, 2 of engine & ECT ECU connector and body ground.
- OK → Replace ignitor. *1
- NG ↓

Check and replace engine & ECT ECU.

4 Check voltage between terminals IGT1, 2 of engine & ECT ECU connector and body ground.
- NG → Go to step 9
- OK ↓

5 Check voltage between terminal 3 of No. 1, No. 2 ignitor connectors and body ground.
- NG → Go to step 8
- OK ↓

6 Check harness and connector between ignition relay and ignition coil, ignition coil and ignitor
- NG → Repair or replace harness or connector.
- OK ↓

7 Check ignition coil.
- NG → Replace ignition coil.
- OK ↓

Replace ignitor. *1

Go to step 8 | Go to step 9

DIAGNOSTIC CHART (Cont'd)

From step 5 | From step 4

8 Check ignition relay.
- NG → Replace ignition relay.
- OK ↓

Check and repair harness or connector between battery and ignition relay, ignition relay and ignitor.

9 Disconnect ignitor connector and check voltage between terminals IGT1, 2 of engine & ECT ECU connector and body ground.
- OK → Replace ignitor. *1
- NG ↓

10 Check harness and connector in IGT signal circuit between engine & ECT ECU and ignitor
- NG → Repair or replace harness or connector.
- OK ↓

Check and replace engine & ECT ECU.

*1: If diag. code 14 is displayed, replace No. 1 ignitor, connected with harness wrapped with yellow tape.
If diag. code 15 is displayed, replace No. 2 ignitor, connected with harness not wrapped with yellow tape.

WIRING DIAGRAM

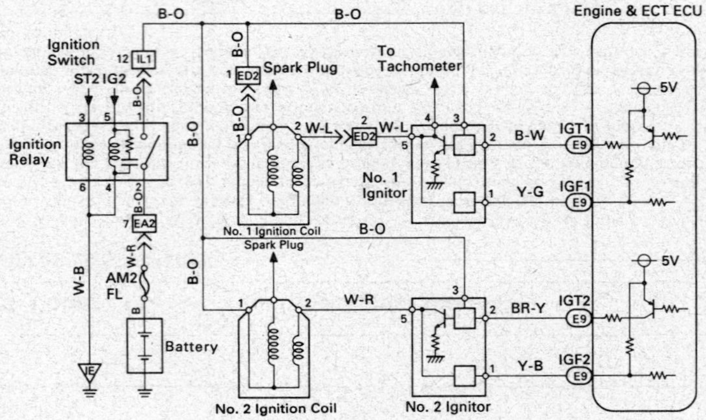

WIRING ROUTING

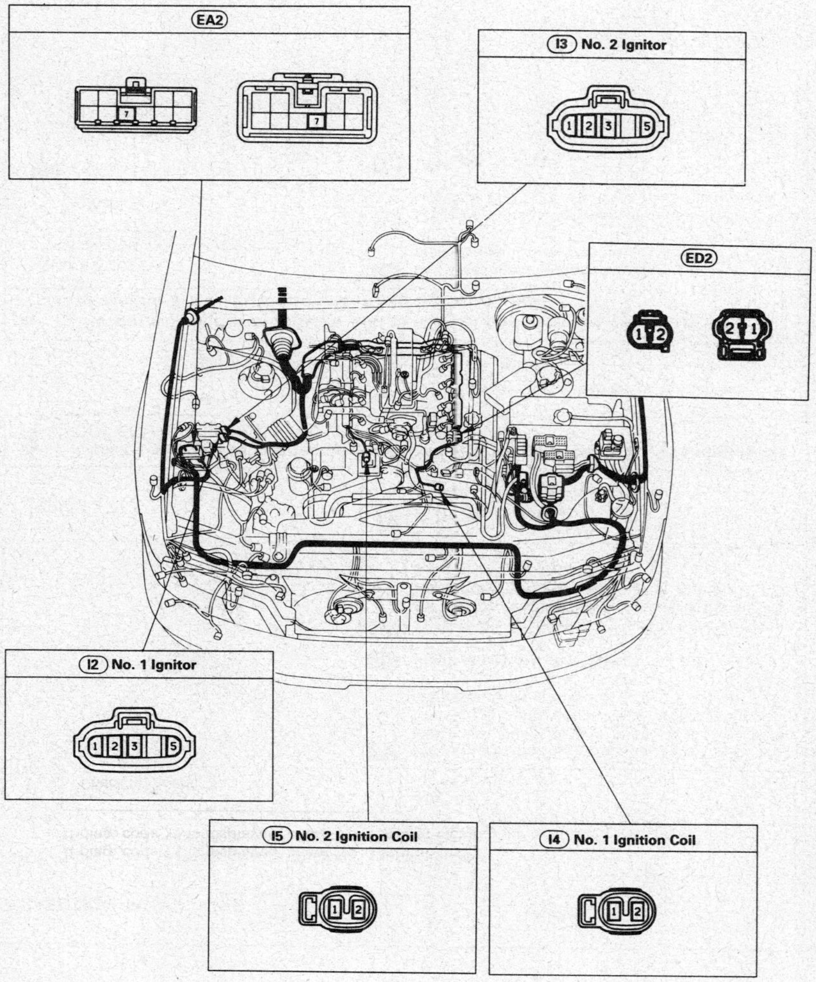

WIRING ROUTING

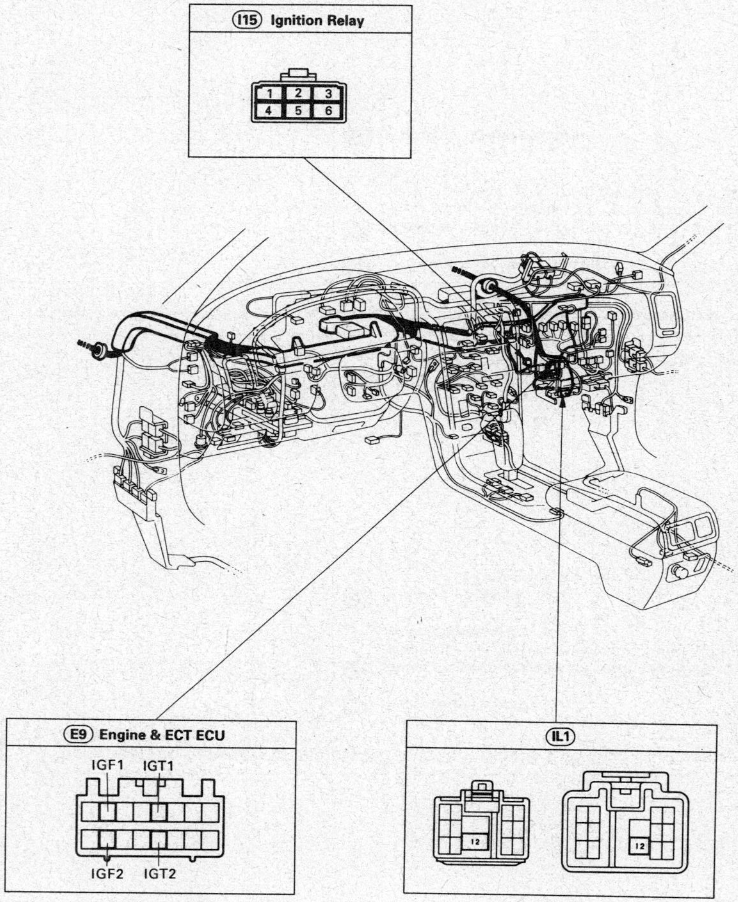

INSPECTION PROCEDURE

HINT: If diag. code 14 is displayed, check No. 1 ignitor circuit.
If diag. code 15 is displayed, check No. 2 ignitor circuit.

1 Check for spark.

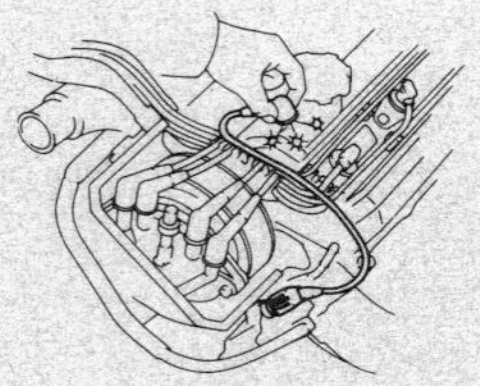

C Disconnect the high-tension cord from the distributor, hold its end about 12.5 mm (1/2") from the ground, see if spark occurs while the engine is being cranked.

OK **Spark should be generated.**

Hint To prevent gasoline from being injected from the injectors during this check, crank the engine for no more than 1 – 2 seconds at a time.

NG Go to step **4**.

OK

2 Check harness and connector in IGF and IGT signal circuit between engine & ECT ECU and ignitor

NG Repair or replace harness or connector.

OK

3 Disconnect ignitor connector and check voltage between terminals IGF1, 2 of engine & ECT ECU connector and body ground.

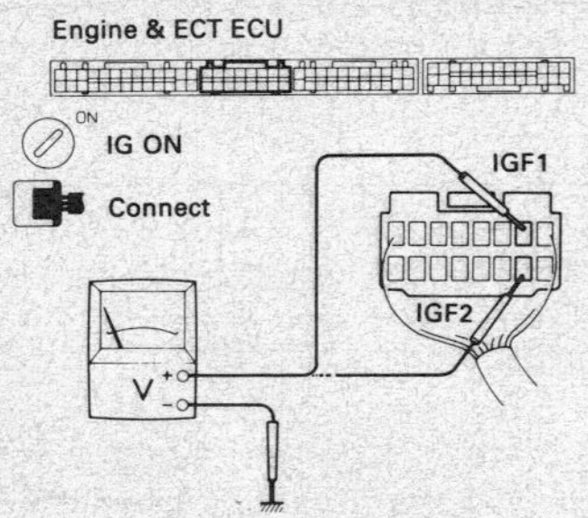

P (1) Disconnect ignitor connector.
(2) Turn ignition switch on.

C Measure voltage between terminals IGF1, 2 of engine & ECT ECU connector and body ground.

OK **Voltage: 4 – 6 V**

OK Replace ignitor. *1

NG

Check and replace engine & ECT ECU.

*1: If diag. code 14 is displayed, replace No. 1 ignitor, connected with harness wrapped with yellow tape.
If diag. code 15 is displayed, replace No. 2 ignitor, connected with harness not wrapped with yellow tape.

4 Check voltage between terminals IGT1, 2 of engine & ECT ECU connector and body ground.

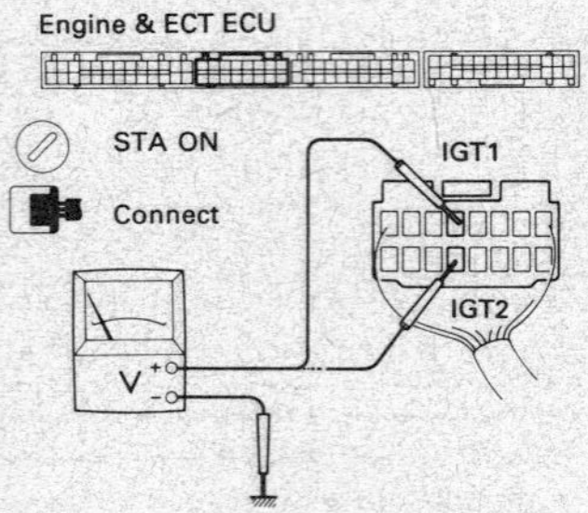

P Remove RH lower pad.

C Measure voltage between terminals IGT1, 2 of engine & ECT ECU connector and body ground when engine is cranked.

OK **Voltage: Approx. 1.2V**
(Neither 0V nor 5V)

NG Go to step **9**.

OK

5 Check voltage between terminal 3 of No. 1, No. 2 ignitor connectors and body ground.

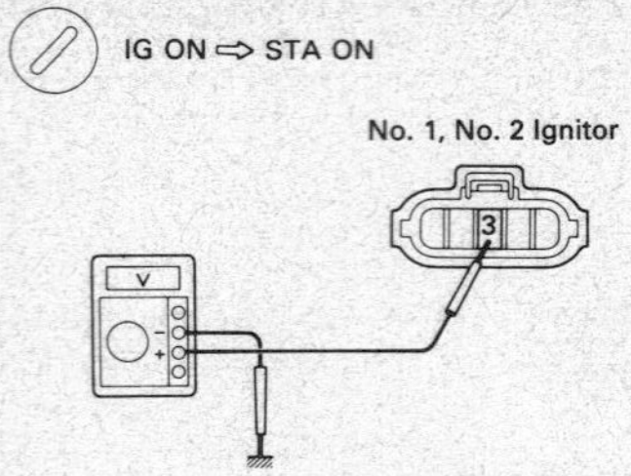

P Disconnect ignitor connector.

C Measure voltage between terminal 3 of No. 1, No. 2 Ignitor connectors and body ground, when ignition switch is turned to "ON" and "STA" position.

OK **Voltage: Battery voltage**

NG Go to step **8**.

OK

6 Check harness and connector between ignition relay and ignition coil, ignition coil and ignitor

NG Repair or replace harness or connector.

OK

7 Check ignition coil.

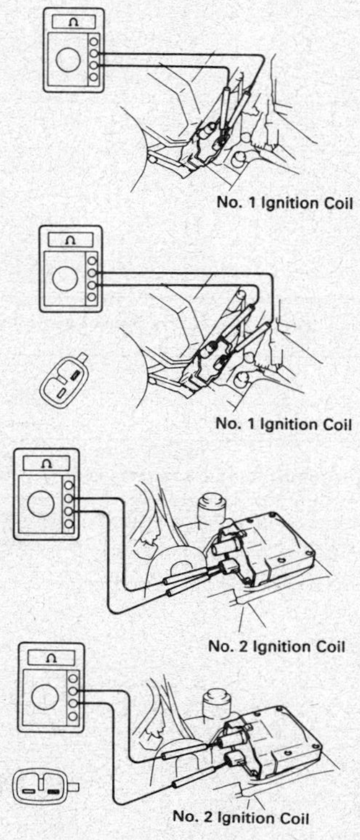

P Disconnect ignition coil connector.
(For No. 1 ignition coil (on left bank), remove the battery)

C
1. Check primary coil.
   Measure resistance between terminals of ignition coil connector.
2. Check secondary coil.
   Measure resistance between terminal 1 of ignition coil connector and high-tension terminal.

OK

	Resistance
Primary Coil (Cold)	0.40 – 0.50 Ω
Secondary Coil (Cold)	8.5 – 11.5 Ω

OK → **Replace ignitor. *1**

NG → **Replace ignition coil.**

*1: If diag. code 14 is displayed, replace No. 1 ignitor, connected with harness wrapped with yellow tape.
If diag. code 15 is displayed, replace No. 2 ignitor, connected with harness not wrapped with yellow tape.

8 Check ignition relay.

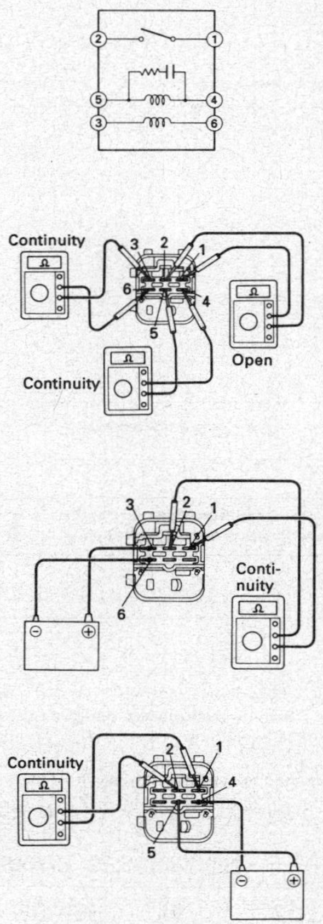

P Remove ignition relay.

C Check continuity between terminals of ignition relay shown below.

OK

Terminals 1 and 2	Open
Terminals 3 and 6	Continuity
Terminals 4 and 5	Continuity

C
(1) Apply battery voltage between terminals 3 and 6, 4 and 5.
(2) Check continuity between terminals 1 and 2.

OK

Terminals 1 and 2	Continuity

OK → **Check and repair harness or connector between battery and ignition relay, ignition relay and ignitor.**

NG → **Replace ignition relay.**

9 **Disconnect ignitor connector and check voltage between terminals IGT1, 2 of engine & ECT ECU connector and body ground.**

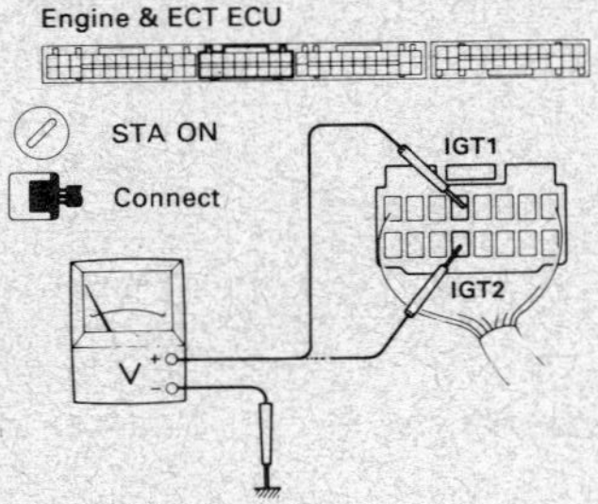

P Disconnect ignitor connector.

C Measure voltage between terminals IGT1, 2 of engine & ECT ECU connector and body ground when engine is cranked.

OK **Voltage: Approx. 1.2 V**
(Neither 0V nor 5V)

OK → **Replace ignitor. *1**

NG ↓

10 **Check harness and connector in IGT signal circuit between engine & ECT ECU and ignitor**

NG → **Repair or replace harness or connector.**

OK ↓

Check and replace engine & ECT ECU.

*1: If diag. code 14 is displayed, replace No. 1 ignitor, connected with harness wrapped with yellow tape.
If diag. code 15 is displayed, replace No. 2 ignitor, connected with harness not wrapped with yellow tape.

Diag.Code 16 ECT Control Signal Malfunction

CIRCUIT DESCRIPTION

Code No.	Diagnosis	Trouble Area
16	Normal signal is not output from ECT CPU.	• ECU

If the ECU records the diagnostic code "16" in memory, it prohibits the torque control of the ECT which performs smooth gear shifting.

DIAGNOSTIC CHART

1 Are there any other codes (besides Code 16) being output?

YES → Go to relevant diag. code chart.

NO ↓

Replace engine & ECT ECU.

Diag. Code	21, 28	Main Oxygen Sensor Circuit

CIRCUIT DESCRIPTION

The main oxygen sensor senses whether the air-fuel ratio is richer or leaner than the stoichiometric air-fuel ratio. The main oxygen sensor is fitted to each exhaust manifold on left and right sides and consists of an element made of zirconium dioxide (ZrO_2, a kind of ceramic material). This element is coated on both the inside and outside with a thin layer of platinum. Ambient air is introduced into the inside of the sensor, and the outside of the sensor is exposed to exhaust gases.

If the oxygen concentration on the inside surface of the ZrO_2 element differs greatly from that on the outside surface at a set temperature or higher, the ZrO_2 element generates a voltage. When the air-fuel mixture is lean, there is a lot of oxygen in the exhaust gas, so there is little difference between the oxygen concentration inside and outside the sensor element. Thus, the voltage generated by the ZrO_2 element is low (close to 0V). Conversely, if the air-fuel mixture is rich, the oxygen in the exhaust gas almost disappears. This creates a large difference in the oxygen concentrations inside and outside the sensor and the voltage generated by the ZrO_2 element is large (approximately 1 V).

The ECU uses this signal to increase or reduce the injection volume to keep the air-fuel ratio at an even value near the stoichiometric air-fuel ratio.

The main oxygen sensors include a heater which heats the ZrO_2 element. The heater is controlled by the ECU. When the intake air volume is low (the temperature of the exhaust gas is low) current flows to the heater to heat the sensor for accurate oxygen concentration detection.

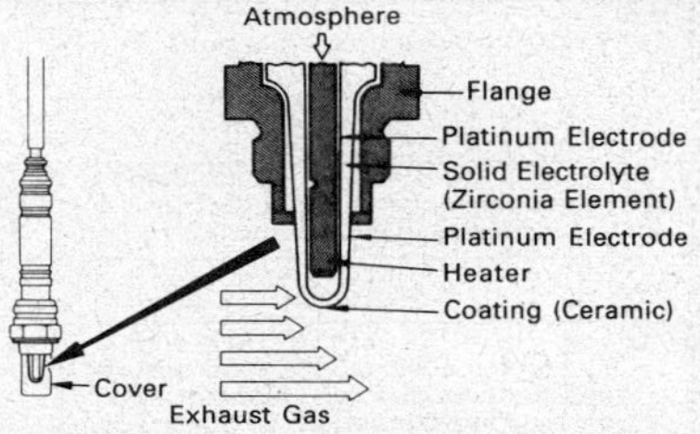

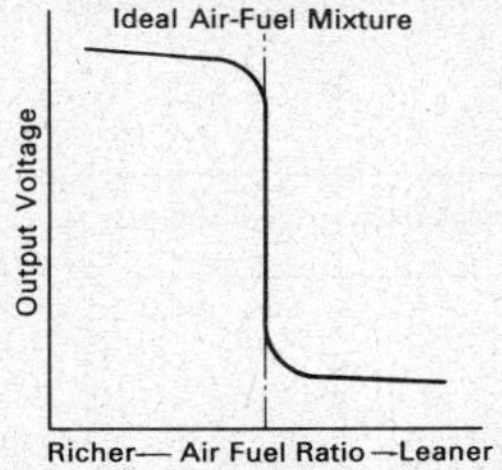

Code No.	Diagnosis	Trouble Area
21 • 28	(1) Open or short in heater circuit of main oxygen sensor for 0.5 sec. or more.	• Open or short in heater circuit of main oxygen sensor. • Main oxygen sensor heater • ECU
	(2) At normal driving speed (ECT in 4th speed, A/C ON, 40 – 50 mph), amplitude of oxygen sensor signal is reduced to between 0.35 – 0.70 V continuously for 60 sec. or more. (2 trip detection logic)	• Main oxygen sensor circuit • Main oxygen sensor

HINT: Diag. code 21 is for the left bank main oxygen sensor circuit. Diag. code 28 is for the right bank main oxygen sensor circuit.

CIRCUIT DESCRIPTION (Cont'd)

DIAGNOSIS CODE DETECTION DRIVING PATTERN

Purpose of the driving pattern.

(a) After the diag. code is erased, the malfunction is reproduced and the diag. code is recorded again.

(b) To check that the malfunction is cleared once the repair is completed (confirm that diag. code is no longer detected).

Malfunction: Deterioration of Main Oxygen Sensor

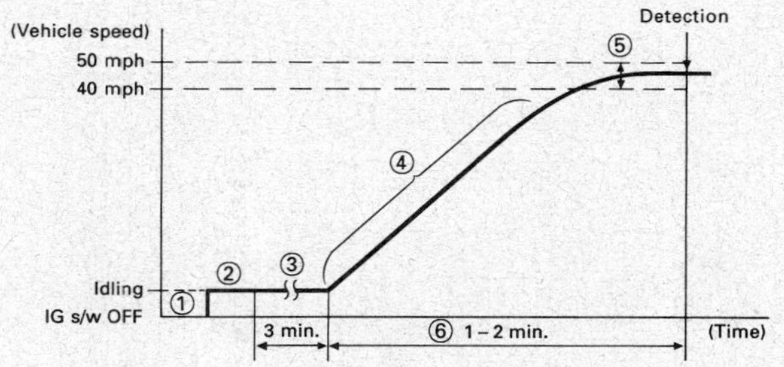

① Initiate test mode
② Start engine and warm engine up.
③ After engine is warmed up, let it idle for 3 min.
④ With the A/C ON and the transmission in D range (O/D ON), perform gradual acceleration with in the range 1300 – 1700 rpm (centered around 1500 rpm). (Take care that the engine speed does not fall below 1200 rpm when shifting. Gradually depress the accelerator pedal and kept it. Steady so that engine braking does not occur.)
⑤ Maintain the vehicle speed at 40 – 50 mph.
⑥ Keep the vehicle running for 1 – 2 min. after starting acceleration.

HINT: If a malfunction exists, the "CHECK" engine warning light will light up after approx. 60 sec. from the start of acceleration.

NOTICE: **If the conditions in this test are not strictly observed, detection of the malfunction will not be possible.**

DIAGNOSTIC CHART

HINT: If diag. code "21" is output, check the left bank main oxygen sensor circuit. If diag. code "28" is output, check the right bank main oxygen sensor circuit.

1 Disconnect the engine & ECT ECU connector, check voltage between terminals HT1, HT2 of engine & ECT ECU connector and body ground.

- OK → go to 3
- NG ↓

2 Check main oxygen sensor heater.

- NG → Replace main oxygen sensor.
- OK ↓

Check and repair harness or connector between main relay and main oxygen sensor, main oxygen sensor and engine & ECT ECU.

3 Check voltage between terminals HT1, HT2 of engine & ECT ECU connector and body ground.

- OK → Replace main oxygen sensor.
- NG ↓

Check and replace engine & ECT ECU.

WIRING DIAGRAM

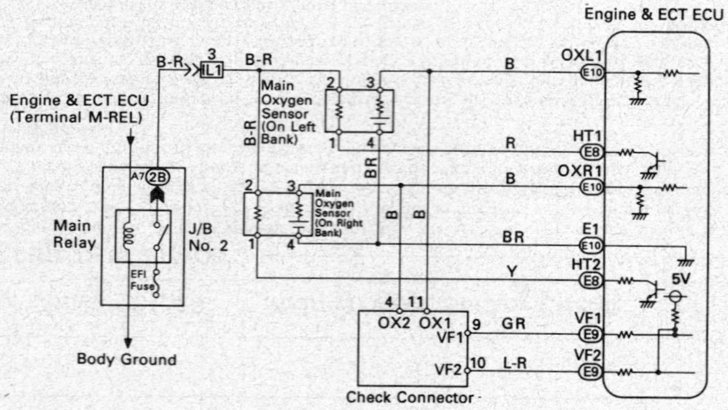

WIRING ROUTING

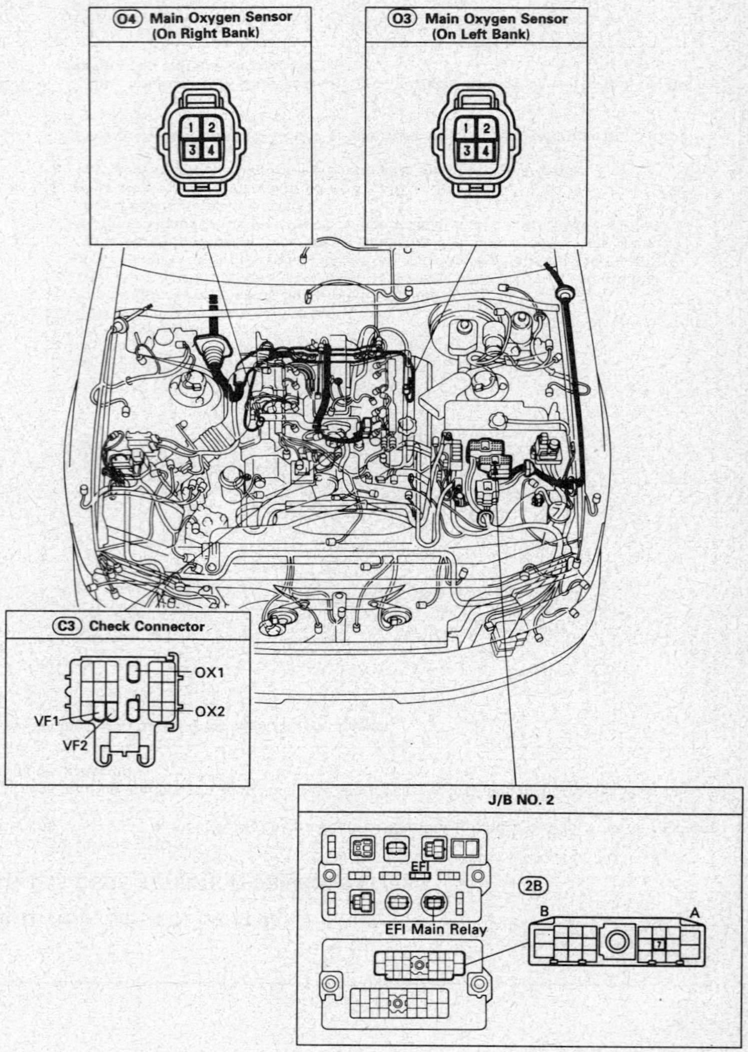

WIRING ROUTING

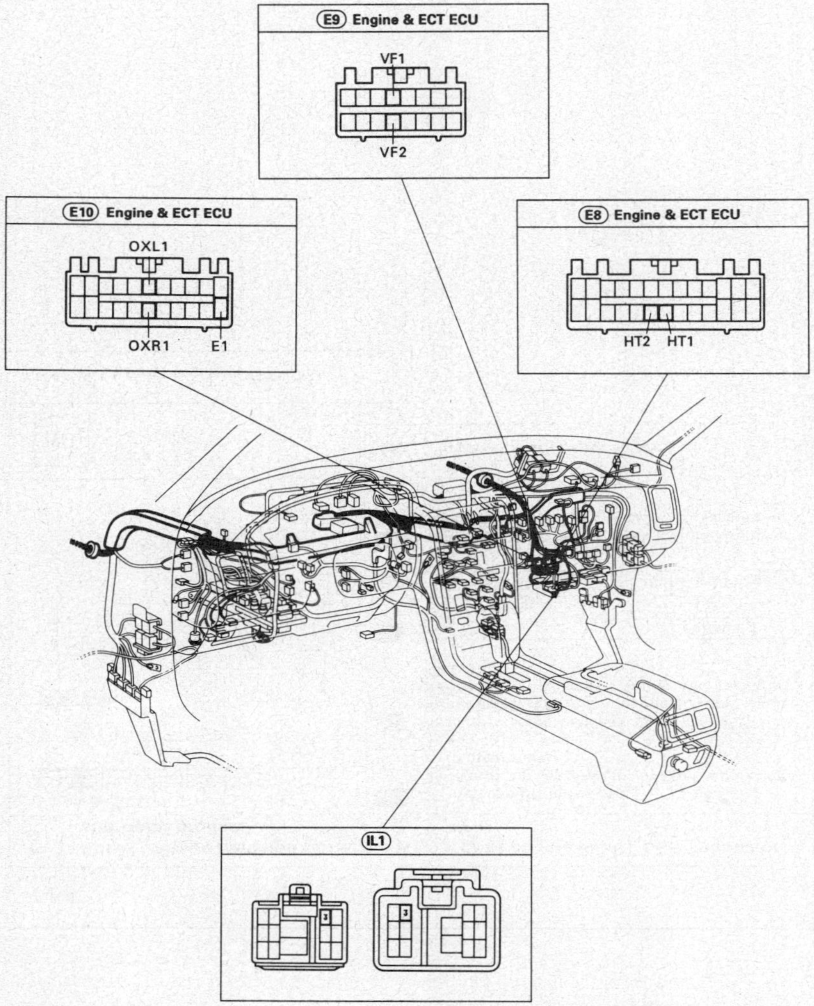

INSPECTION PROCEDURE

HINT: If diag. code "21" is output, check the left bank main oxygen sensor circuit.
If diag. code "28" is output, check the right bank main oxygen sensor circuit.

1 Disconnect the engine & ECT ECU connector, check voltage between terminals HT1, HT2 of engine & ECT ECU connector and body ground.

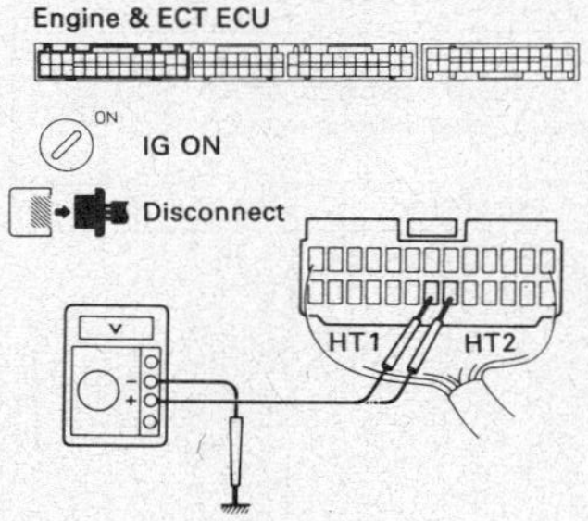

P (1) Remove RH lower pad.
(2) Disconnect engine & ECT ECU connector.
(3) Turn ignition switch on.

C Measure voltage between terminals HT1, HT2 of engine & ECT ECU connector and body ground.

OK **Voltage : Battery Voltage**

OK — Go to step **3**.

NG ↓

2 Check main oxygen sensor heater.

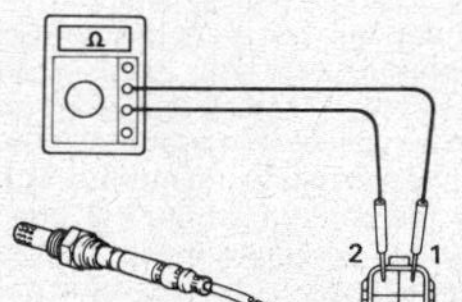

P Disconnect main oxygen sensor connector.

C Measure resistance between terminals 1 and 2 of main oxygen sensor connector.

OK **Resistance : 5.1 — 6.3 Ω at 20°C (68°F)**

NG — **Replace main oxygen sensor.**

OK ↓

Check and repair harness or connector between main relay and main oxygen sensor, main oxygen sensor and engine & ECT ECU.

3	**Check voltage between terminals HT1, HT2 of engine & ECT ECU connector and body ground.**

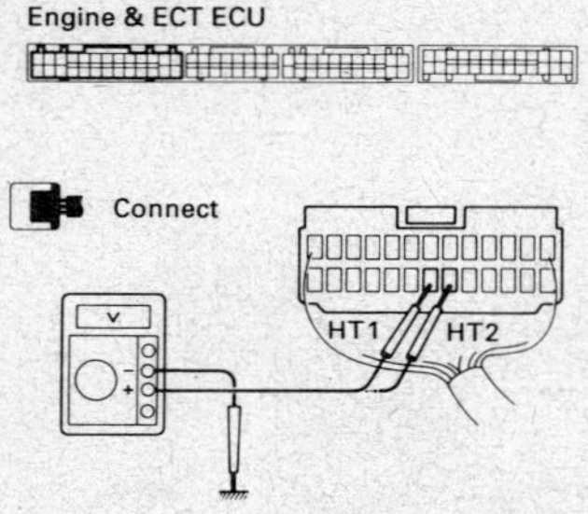

P (1) Remove RH lower pad.
(2) Warm up engine to normal operating temperature.

C Measure voltage between terminals HT1, HT2 of engine & ECT ECU connector and body ground, when engine is idling and racing at 4000 rpm.

OK

	Voltage
Idling	0V
Racing at 4000 rpm	Battery Voltage

Hint In the 4000 rpm racing check, continue engine racing at 4000 rpm for approx. 20 seconds or more.

OK **Replace main oxygen sensor.**

NG

Check and replace engine & ECT ECU.

Diag.Code	22	Water Temp. Sensor Circuit

CIRCUIT DESCRIPTION

The water temperature sensor senses the coolant temperature. A thermistor built in the sensor changes the resistance value according to the coolant temperature. The lower the coolant temperature, the greater the thermistor resistance value, and the higher the coolant temperature, the lower the thermistor resistance value (See Fig. 1.). The water temperature sensor is connected to the ECU as shown in Fig. 2. The 5 V power source voltage in the ECU is applied to the water temperature sensor from the terminal THW via a resistor R. That is, the resistor R and the water temperature sensor are connected in series. When the resistance value of the water temperature sensor changes in accordance with changes in the coolant temperature, the potential at the terminal THW also changes. Based on this signal, the ECU increases the fuel injection volume to improve driveability during cold engine operation. If the ECU records the diagnostic code 22, it operates the fail safe function, keeping the coolant temperature at a constant 80°C (176°F).

(fig-1)

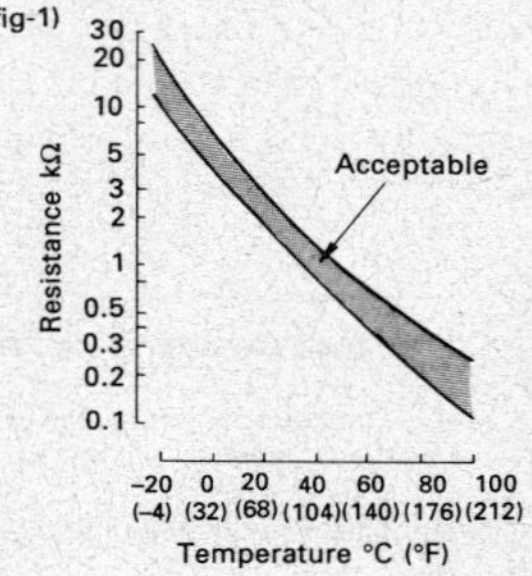

(fig-2)

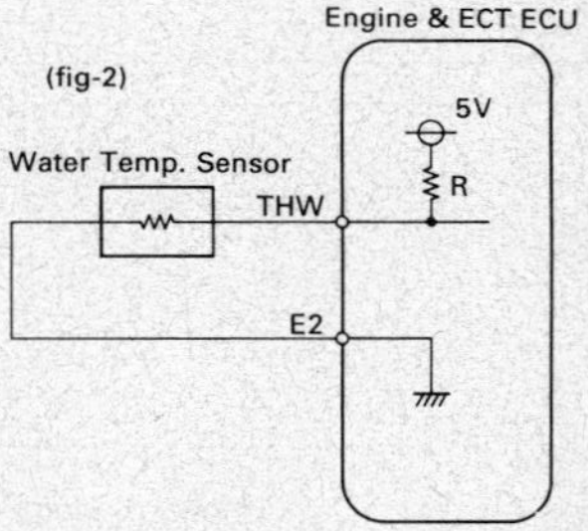

Code No.	Diagnosis	Trouble Area
22	Open or short in water temp. sensor circuit for 0.5 sec. or more.	• Open or short in water temp. sensor circuit. • Water temp. sensor • ECU

DIAGNOSTIC CHART

HINT: If diagnostic codes "22" (water temperature sensor circuit), "24" (intake air temperature sensor circuit) and "41" (throttle position sensor circuit) are output simultaneously, E2 (sensor ground) may be open.

Step	Check	Result	Action
1	Check voltage between terminals THW and E2 of engine & ECT ECU connector.	OK	Proceed to next circuit inspection shown on matrix chart *1
		NG	Go to step 2
2	Check water temp. sensor.	NG	Replace water temp. sensor.
		OK	Go to step 3
3	Check harness and connector between engine & ECT ECU and water temp. sensor	NG	Repair or replace harness or connector.
		OK	Check and replace engine & ECT ECU.

*1: When diag. code 22 is displayed, check for momentary interruption

WIRING DIAGRAM

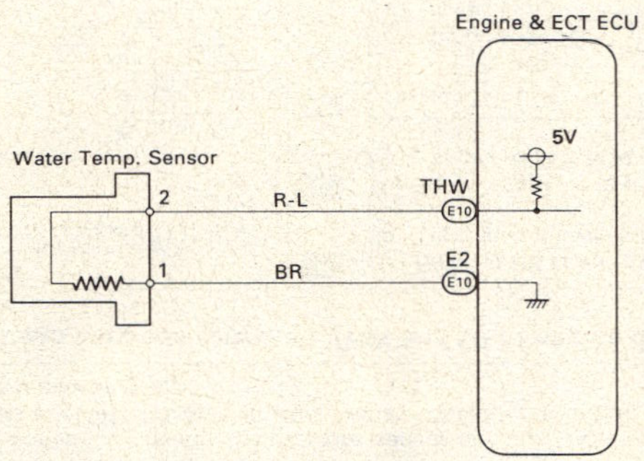

WIRING ROUTING

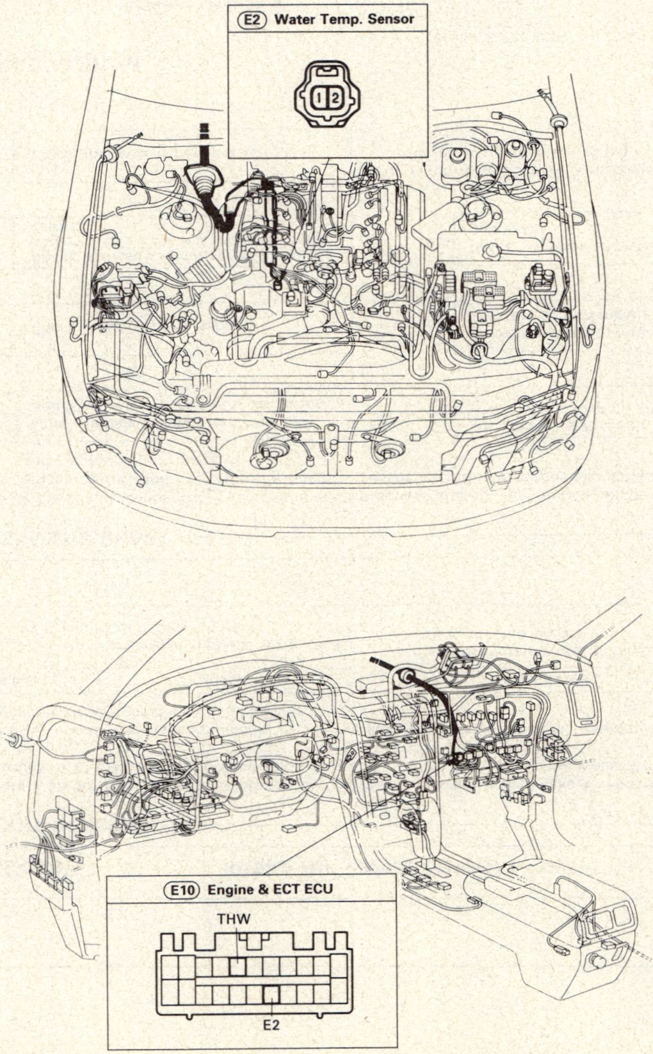

INSPECTION PROCEDURE

HINT: If diagnostic codes "22" (water temperature sensor circuit), "24" (intake air temperature sensor circuit) and "41" (throttle position sensor circuit) are output simultaneously, E2 (sensor ground) may be open.

1 Check voltage between terminals THW and E2 of engine & ECT ECU connector.

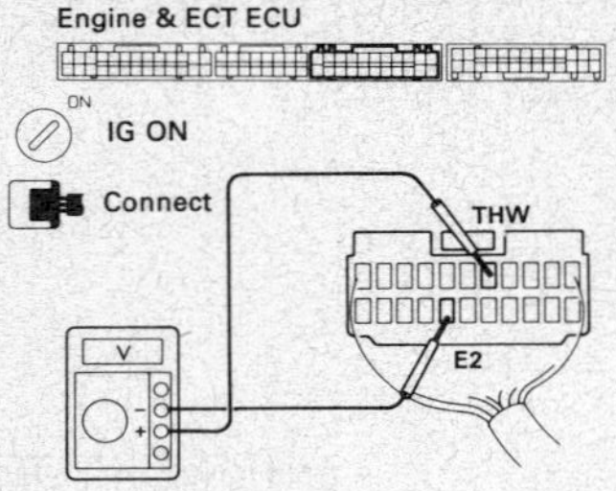

P (1) Remove RH lower pad.
(2) Turn ignition switch on.

C Measure voltage between terminals THW and E2 of engine & ECT ECU connector.

OK

Water Temp. °C (°F)	Voltage
20 (68) (Engine is cool)	1 – 3 V
80 (176) (Engine is hot)	0.1 – 1.0 V

OK → **Proceed to next circuit inspection shown on matrix chart** *1

NG ↓

2 Check water temp. sensor.

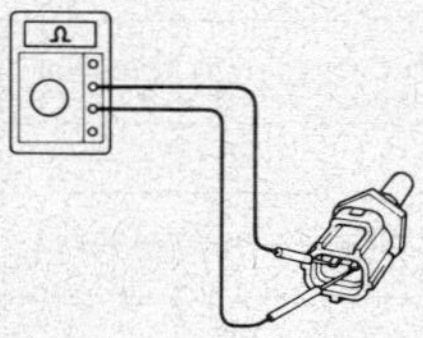

P Disconnect the water temp. sensor connector. (See page FI-68).

C Measure resistance between terminals.

OK **Resistance is within Acceptable zone on chart.**

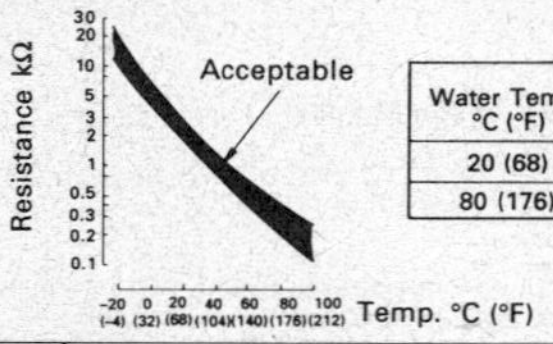

Water Temp. °C (°F)	Resistance
20 (68)	2 – 3 kΩ
80 (176)	0.2 – 0.4 kΩ

NG → **Replace water temp. sensor.**

OK ↓

3 Check harness and connector between engine & ECT ECU and water temp. sensor

NG → **Repair or replace harness or connector.**

OK ↓

Check and replace engine & ECT ECU.

*1: However, when diag. code 22 is displayed, check for momentary interruption

Diag.Code	24	Intake Air Temp. Sensor Circuit

CIRCUIT DESCRIPTION

The intake air temp. sensor is built into the air flow meter and senses the intake air temperature. The structure of the sensor and connection to the ECU is the same as in the water temp. sensor

If the ECU records the diagnostic code "24", it operates the fail safe function, keeping the intake air temperature at a constant 20°C (68°F).

Code No.	Diagnosis	Trouble Area
24	Open or short in intake air temp. sensor circuit for 0.5 sec. or more.	• Open or short in intake air temp. circuit. • Intake air temp. sensor • ECU

DIAGNOSTIC CHART

HINT: If diagnostic codes "22" (water temperature sensor circuit), "24" (intake air temperature sensor circuit) and "41" (throttle position sensor circuit) are output simultaneously, E2 (sensor ground) may be open.

Step	Check	Result	Action
1	Check voltage between terminals THA and E2 of engine & ECT ECU connector.	OK	Proceed to next circuit inspection shown on matrix chart *1
		NG	Go to 2
2	Check intake air temp. sensor.	NG	Replace intake air temp. sensor (Replace air flow meter).
		OK	Go to 3
3	Check harness and connector between engine & ECT ECU and intake air temp. sensor	NG	Repair or replace harness or connector.
		OK	Check and replace engine & ECT ECU.

*1: When diag. code 24 is displayed, check for momentary interruption

WIRING DIAGRAM

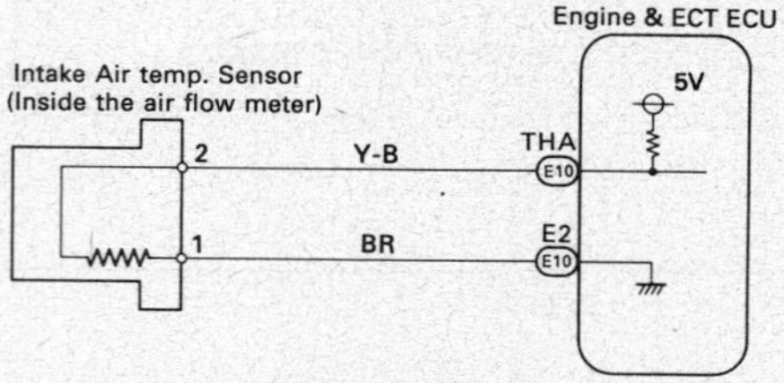

WIRING ROUTING

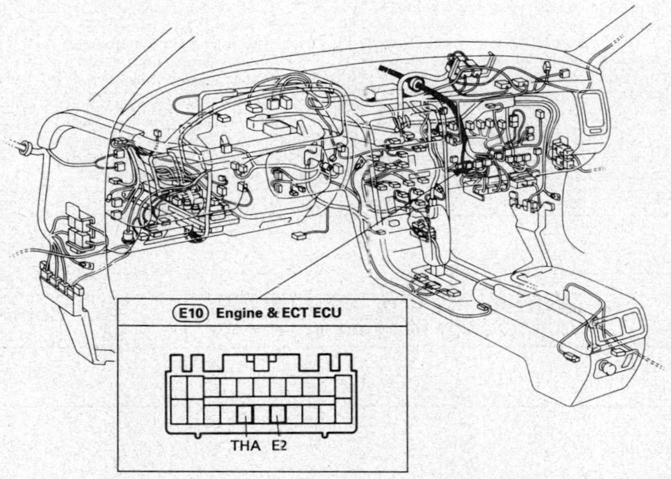

INSPECTION PROCEDURE

HINT: If diagnostic codes "22" (water temperature sensor circuit), "24" (intake air temperature sensor circuit) and "41" (throttle position sensor circuit) are output simultaneously, E2 (sensor ground) may be open.

1 Check voltage between terminals THA and E2 of engine & ECT ECU connector.

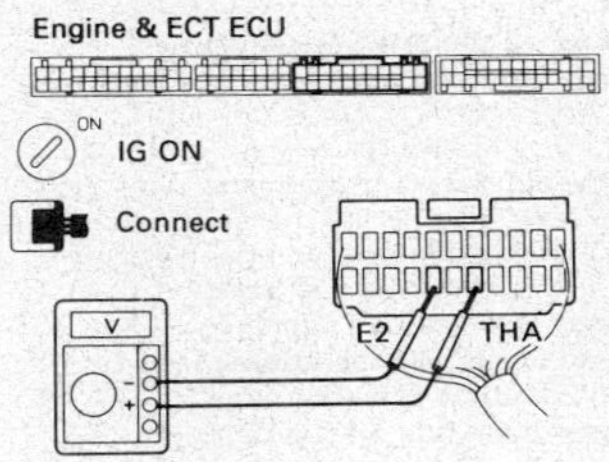

P (1) Remove RH lower pad.
(2) Turn ignition switch on.

C Measure voltage between terminals THA and E2 of engine & ECT ECU connector.

OK

Intake air temp. °C (°F)	Voltage
20 (68)	1 – 3 V
60 (140)	0.5 – 1.0 V

NG

OK → **Proceed to next circuit inspection shown on matrix chart *1**

2 Check intake air temp. sensor.

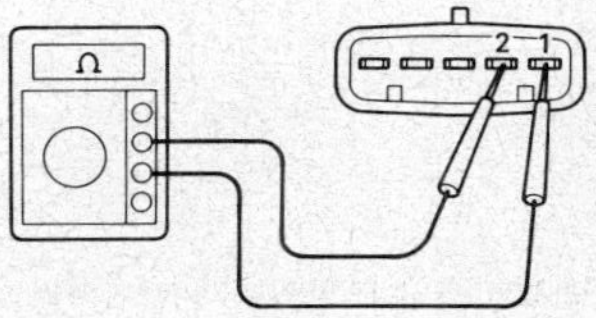

P Disconnect the air flow meter connector.

C Measure resistance between terminals 1 and 2 of air flow meter connector.

OK **Resistance is within Acceptable zone on chart.**

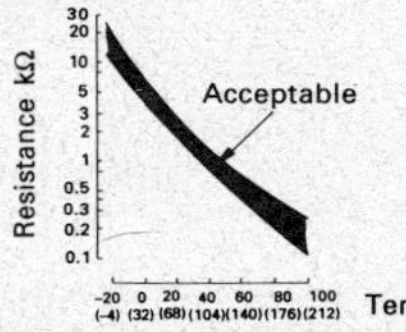

Intake air temp. °C (°F)	Resistance
20 (68)	2 – 3 kΩ
60 (140)	0.4 – 0.7 kΩ

OK

NG → **Replace intake air temp. sensor (Replace air flow meter).**

3 Check harness and connector between engine & ECT ECU and intake air temp. sensor

OK

NG → **Repair or replace harness or connector.**

Check and replace engine & ECT ECU.

*1: When diag. code 24 is displayed, check for momentary interruption

Diag. Code		
Diag. Code	25	**Air-Fuel Ratio Lean Malfunction**
	26	**Air-Fuel Ratio Rich Malfunction**

CIRCUIT DESCRIPTION

Code No.	Diagnosis	Trouble Area
25	(1) Oxygen sensor output is less than 0.45 V for at least 90 sec. when main oxygen sensor is warmed up (racing at 2000 rpm). (2 trip detection logic)	• Open or short in main oxygen sensor circuit • Main oxygen sensor • Ignition system • Water temp. sensor
	(2) With engine warmed up and idling, main oxygen sensor signal is centered around 0.45 V with frequency at 5 Hz or above. (2 trip detection logic)	• Open or short in injector circuit. • Fuel line pressure (injector blockage) • Air flow meter (air intake) • Engine ground bolt loose • Foreign object caught in valve
	(3) Deviation between air-fuel ratio feedback compensation values for left and right banks is beyond acceptable limits. (If air-fuel ratio feedback compensation values are deviating, this can be detected when engine is raced at 2000 rpm for 30 sec. when engine is warmed up). (2 trip detection logic)	• Fuel line pressure (injector leak, blockage) • Mechanical system malfunction (skipping teeth of timing belt) • Ignition system
26	(1) With engine warmed up and idling, main oxygen sensor signal is centered around 0.45 V with frequency at 5 Hz or above. (2 trip detection logic)	• Open or short in main oxygen sensor circuit • Fuel line pressure (injector blockage) • Air flow meter (air intake) • Engine ground bolt loose • Foreign object caught in valve
	(2) Deviation between air-fuel ratio feedback compensation values for left and right banks is beyond acceptable limits. (If air-fuel ratio feedback compensation values are deviating, this can be detected when engine is raced at 2000 rpm for 30 sec. when engine is warmed up). (2 trip detection logic,	• Fuel line pressure (injector leak, blockage) • Mechanical system malfunction (skipping teeth of timing belt) • Ignition system

CIRCUIT DESCRIPTION (Cont'd)

DIAGNOSIS CODE DETECTION DRIVING PATTERN

Purpose of the driving pattern.

(a) After the diag. code is erased, the malfunction is reproduced and the diag. code is recorded again.

(b) To check that the malfunction is cleared once the repair is completed (confirm that diag. code is no longer detected).

Malfunction: **Open or Short in Main Oxygen Sensor**

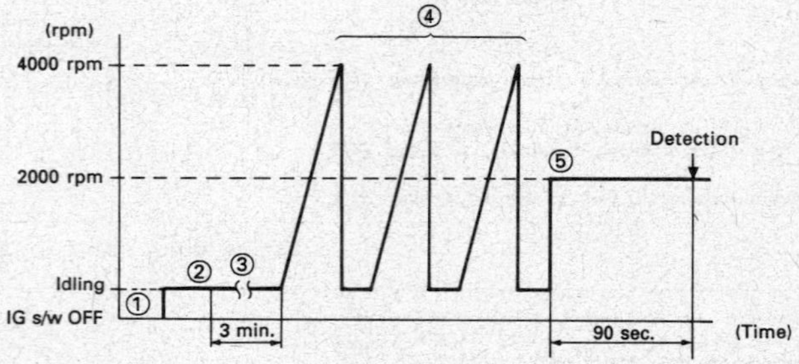

① Initiate test mode
② Start engine and warm engine up.
③ After engine is warmed up, let it idle for 3 min.
④ Perform sudden racing to 4000 rpm three times.
(Rapidly depress the accelerator pedal, then remove your foot from the pedal at 4000 rpm.)
⑤ After performing the racing in ④ , perform racing at 2000 rpm for 90 sec.

HINT: If a malfunction exists, the "CHECK" engine warning light will light up after 90 sec. from the start of racing.

NOTICE: If the conditions in this test are not strictly observed, detection of the malfunction will not be possible.

CIRCUIT DESCRIPTION (Cont'd)

DIAGNOSIS CODE DETECTION DRIVING PATTERN

Purpose of the driving pattern.
(a) After the diag. code is erased, the malfunction is reproduced and the diag. code is recorded again.
(b) To check that the malfunction is cleared once the repair is completed (confirm that diag. code is no longer detected).

Malfunction: Open or Short of Injector, Leak, Blockage, Loose E/G Earth Bolt

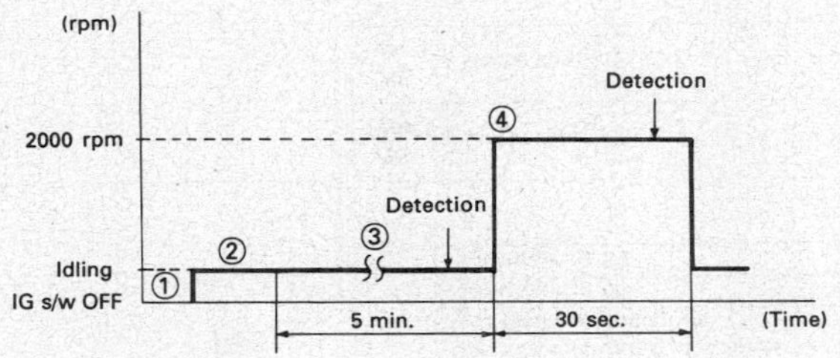

① Initiate test mode
② Start engine and warm engine up.
③ After engine is warmed up, let it idle for 5 min.
(After the engine is started, do not depress the accelerator pedal.)
④ If the malfunction is not detected during idling, perform racing without any load at approx. 2000 rpm for 30 sec.

HINT: If a malfunction exists, the "CHECK" engine warning light will light up during the 5 min. idling perild or within 30 sec. of starting racing.

NOTICE: If the conditions in this test are not strictly abserved, detection of the malfunction will not be possible.

DIAGNOSTIC CHART

1 Check voltage between terminals VF1, VF2 and E1 of check connector.
- NG Type I (Continue at 0V) → go to step 2
- NG Type II (Continue at 5V) → go to step 7
- OK → Proceed to next circuit inspection shown on matrix chart *1

2 Check voltage between terminals OX1, OX2 and E1 of check connector.
- NG → go to step 4
- OK ↓

3 Check voltage between terminals VF1, VF2 and E1 of check connector.
- NG → go to step 5
- OK → Proceed to next circuit inspection shown on matrix chart *1

4 Check harness and connector between engine & ECT ECU and main oxygen sensor, engine & ECT ECU and check connector
- NG → Repair or replace harness or connector.
- OK ↓

5 Check each item found to be a possible cause of problem.
- NG → Repair or replace.
- OK ↓

6 Does malfunction disappear when a known good main oxygen sensor is installed?
- YES → Replace main oxygen sensor.
- NO → Check and replace engine & ECT ECU.

7 Check each item found to be a possible cause of problem.
- NG → Repair or replace.
- OK ↓

8 Does malfunction disappear when a known good main oxygen sensor is installed?
- YES → Replace main oxygen sensor.
- NO → Check and replace engine & ECT ECU.

*1: When diag. code 25 or 26 is displayed, go to step **7**.

INSPECTION PROCEDURE

1 Check voltage between terminals VF1, VF2 and E1 of check connector.

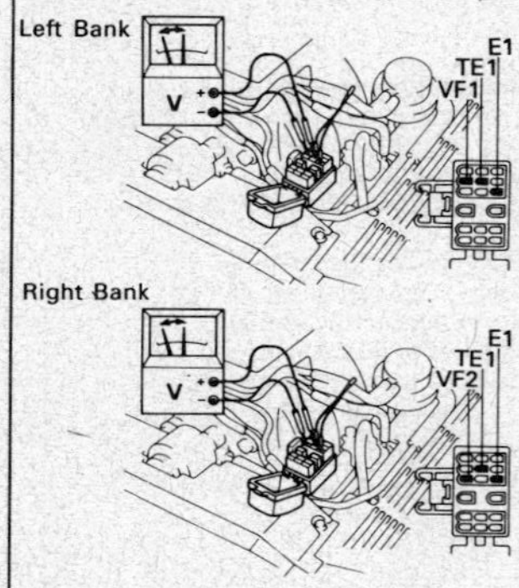

P
(1) Warm up engine at normal operating temperature.
(2) Connect terminals TE1 and E1 of check connector.
(3) Connect positive prove to terminal VF1, VF2 and negative prove to terminal E1 of check connector.

C
(1) Warm up the oxygen sensor by running engine at 2,500 rpm for about 2 minutes.
(2) Then, maintaining engine at 2,500 rpm, count how many times needle of voltmeter fluctuates between 0 and 5V.

Result

Result	
Needle fluctuates of 8 times for every ten seconds	OK
Continue at 0V	NG Type I
Continue at 5V	NG Type II

NG Type I → Go to step **2**.

NG Type II → Go to step **7**.

OK → Proceed to next circuit inspection shown on matrix chart *1

*1: When diag. code 25 or 26 is displayed, go to step **7**.

2 Check voltage between terminals OX1, OX2 and E1 of check connector.

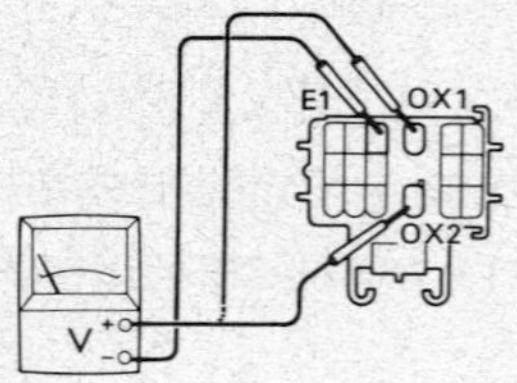

P Warm up engine at normal operating temperature.

C Measure voltage between terminals OX1, OX2 and E1 of check connector when engine is suddenly raced to full throttle.

OK **The voltage should be 0.5 V or higher at least once.**

Hint Perform inspection within 1 second.

NG → Go to step **4**.

OK ↓

3 Check voltage between terminals VF1, VF2 and E1 of check connector (Repeat step 1).

NG → Go to step **5**.

OK → Proceed to next circuit inspection shown on matrix chart *1

*1: When diag. code 25 or 26 is displayed, go to step 7.

4 Check harness and connector between engine & ECT ECU and main oxygen sensor, engine & ECT ECU and check connector

NG → **Repair or replace harness or connector.**

OK ↓

5 Check each item found to be a possible cause of problem.

Check each circuit found to be a possible cause of trouble according to the results of the check in 2 or 3. The numbers in the table below show the order in which the checks should be performed.

Possible Cause	Main oxygen sensor signal from either side continues at 0V.	Main oxygen sensor signals from both sides continue at 0V.
Faulty sensor installation.	1	
Injector circuit	3	
Misfire	2	3
Valve timing	4	
Air leakage		1
Fuel system		2
Characteristics deviation in air flow meter.		6
Characteristics deviation in water temp. sensor.		4
Characteristics deviation in intake air temp. sensor.		5

NG → **Repair or replace.**

OK ↓

6 Does malfunction disappear when a known good main oxygen sensor is installed?

YES → **Replace main oxygen sensor.**

NO ↓

Check and replace engine & ECT ECU.

7 Check each item found to be a possible cause of problem.

Check each circuit found to be a possible cause of trouble according to the results of the check in ■. The numbers in the table below show the order in which the checks should be performed.

	Possible Cause	Main oxygen sensor signal from either side continues at 5.0V.	Main oxygen sensor signals from both sides continue at 5.0V.	Main oxygen sensor signals from both sides are normal.
	Injector circuit	1		7
	Misfire			3
	Valve timing	2		4
	Air leakage			1
	Fuel system		1	2
	Cold start injector circuit	3	4	
	Characteristics deviation in air flow meter.		5	8
	Characteristics deviation in water temp. sensor.		2	5
	Characteristics deviation in intake air temp. sensor.		3	6

NG Repair or replace.

OK

8 Does malfunction disappear when a known good main oxygen sensor is installed?

YES Replace main oxygen sensor.

NO

Check and replace engine & ECT ECU.

Diag. Code 27, 29 Sub-Oxygen Sensor Circuit

CIRCUIT DESCRIPTION

The sub-oxygen sensor is installed on the exhaust pipe. Its construction and operation is the same as the main oxygen sensor

Code No.	Diagnosis	Trouble Area
27 • 29	When sub-oxygen sensor is warmed up and apply acceleration with the throttle fully open for 4 sec. output of main oxygen sensor is 0.45 V or more (rich) and output of sub-oxygen sensor is 0.45 V or less (lean). (2 trip detection logic) Sub-oxygen sensor warm-up condition: After engine warmed up and idled for 120 sec., ECT in 3rd speed, A/C ON and vehicle driven 45 – 60 mph for 6 min.	• Open or short in sub-oxygen sensor circuit • Sub-oxygen sensor • ECU

HINT: Diag.code 27 is for the left bank sub-oxygen sensor circuit.
Diag.code 29 is for the right bank sub-oxygen sensor circuit.

CIRCUIT DESCRIPTION (Cont'd)

DIAGNOSIS CODE DETECTION DRIVING PATTERN

Purpose of the driving pattern.
(a) After the diag. code is erased, the malfunction is reproduced and the diag. code is recorded again.
(b) To check that the malfunction is cleared once the repair is completed (confirm that diag. code is no longer detected).

Malfunction: Open or Short in Sub-Oxygen Sensor

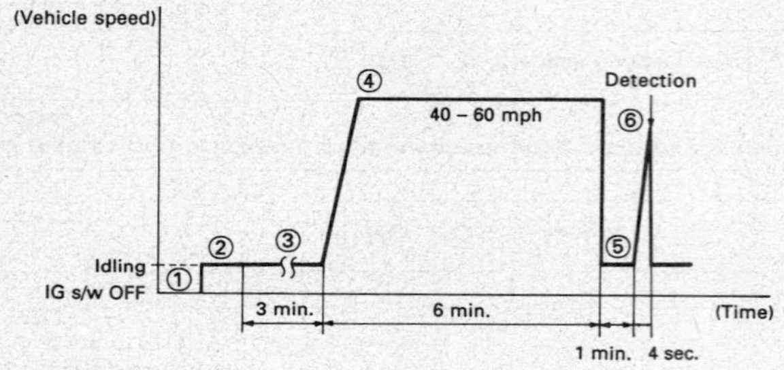

① Initiate test mode
② Start engine and warm engine up.
③ After engine is warmed up, let it idle for 3 min.
④ With the A/C ON and the transmission in D range (O/D OFF), drive at 40 – 60 mph for 6 min.
⑤ After the driving in ④, stop the vehicle.
⑥ Once the vehicle is stopped, within the next minute apply acceleration with the throttle fully open for 4 sec.

HINT: If a malfunction exists, the "CHECK" engine warning light will light up after full acceleration for 4 sec.

NOTICE: If the conditions in this test are not strictly observed, detection of the malfunction will not be possible.

DIAGNOSTIC CHART

HINT:
- When other codes are output in addition to 27 and 29 at the same time, check the circuits for other codes first.
- If diag. code 27 is displayed, check left bank sub-oxygen sensor circuit, and if 29 is displayed, check right bank sub-oxygen sensor circuit.

1 Check for air leakage. — NG → Repair air leakage.
OK ↓

2 Check voltage between terminals OXL2, OXR2 and E2 of engine & ECT ECU connector. — OK → Proceed to next circuit inspeciton shown on matrix chart *1
NG ↓

3 Check harness and connector between engine & ECT ECU, sub-oxygen sensor and body ground — NG → Repair or replace harness or connector.
OK ↓

4 Check fuel system circuit — NG → Repair fuel system circuit.
OK ↓

5 Check injector circuit — NG → Repair or replace.
OK ↓

Replace sub-oxygen sensor.

*1: When diag. code 27 or 29 is displayed, check and replace engine & ECT ECU.

WIRING DIAGRAM

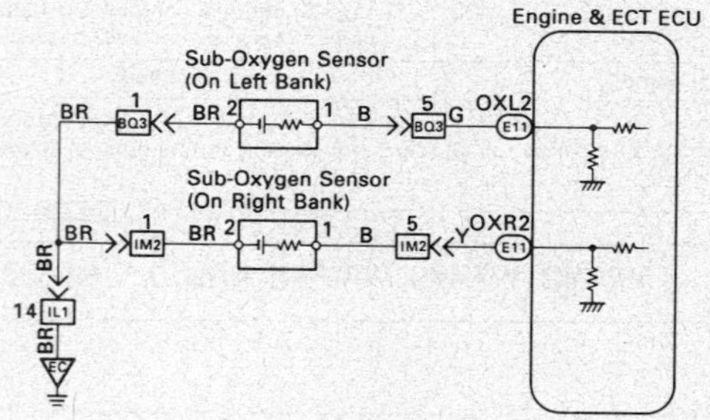

WIRING ROUTING

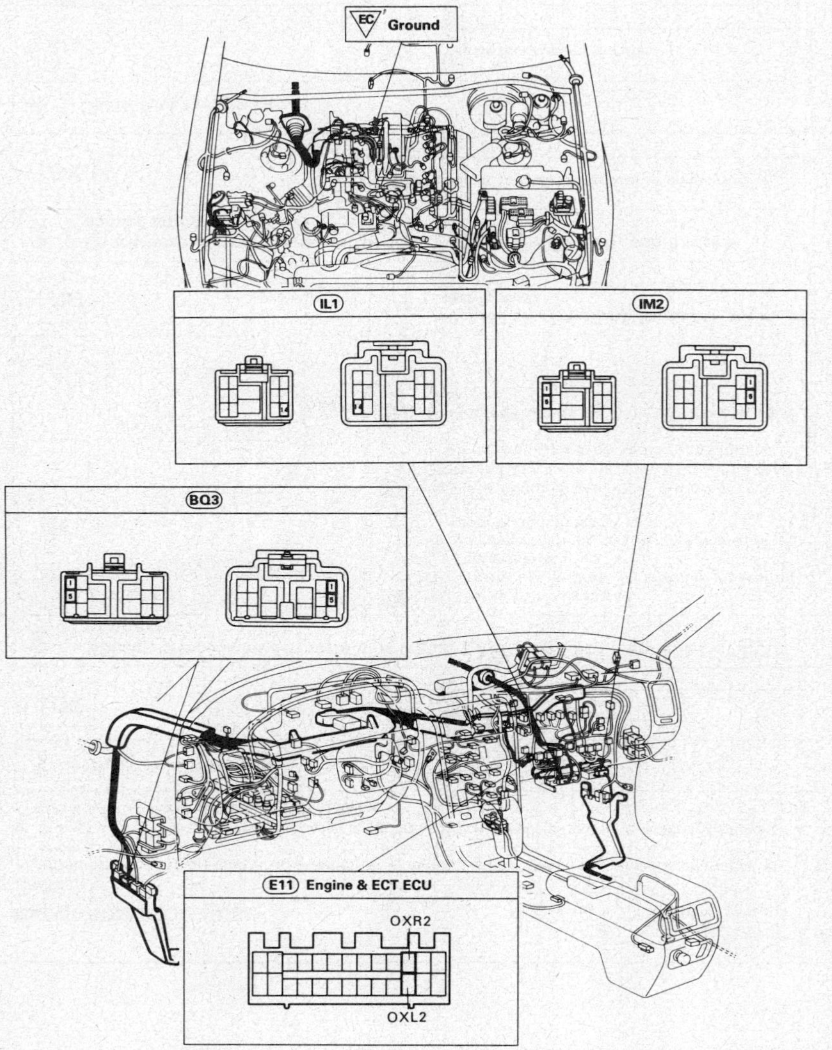

WIRING ROUTING

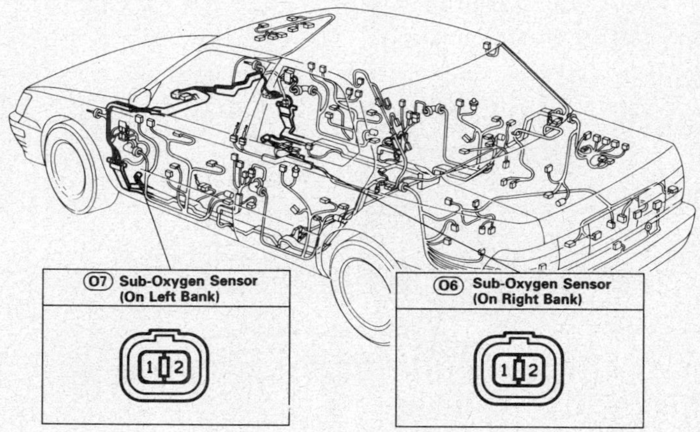

Diag.Code 31 Air Flow Meter Circuit

CIRCUIT DESCRIPTION

As shown in the figure at right, when a pillar (Vortex generating body) is placed in the path of a uniform flow, vortices called Karman-Vortex are generated downstream of the object. Using this principle, a vortex generator is placed inside the air flow meter. By measuring the frequency of the vortices generated, the ECU can determine the volume of air flowing through the air flow meter. The vortices are detected by their exerting pressure on thin metal foil (mirror) surfaces and a light emitting element and light receptor (LED and photo transistor) positioned opposite the mirror which senses the vibrations in the mirror optically. The ECU uses these signals mainly for calculation of the basic injection volume and the basic ignition advance angle.

Karman-Vortex

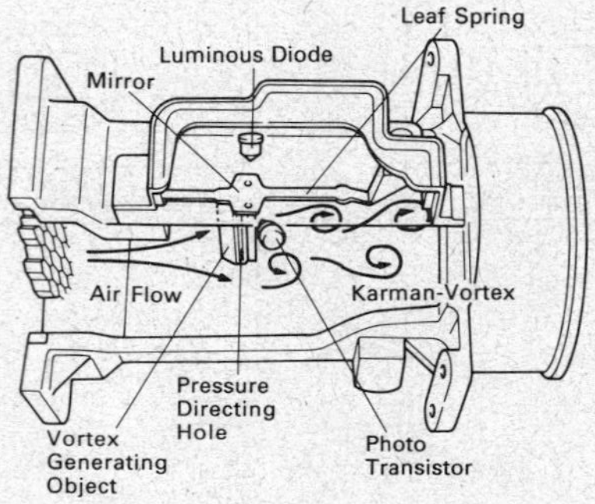

Code No.	Diagnosis	Trouble Area
31	Engine speed is above 300 rpm and air flow meter signal is not input to ECU for 2 seconds.	• Open or short in air flow meter circuit • Air flow meter • ECU

If the ECU records diagnostic code "31", it operates the fail safe function, keeping the ignition timing and fuel injection volume constant and making it possible to drive the vehicle.

INSPECTION PROCEDURE

HINT:
- When other codes are output in addition to 27 and 29 at the same time, check the circuits for other codes first.
- If diag. code 27 is displayed, check left bank sub-oxygen sensor circuit, and if 29 is displayed, check right bank sub-oxygen sensor circuit.

1 Check for air leakage.

NG → **Repair air leakage.**

OK ↓

2 Check voltage between terminals OXL2, OXR2 and E2 of engine & ECT ECU connector.

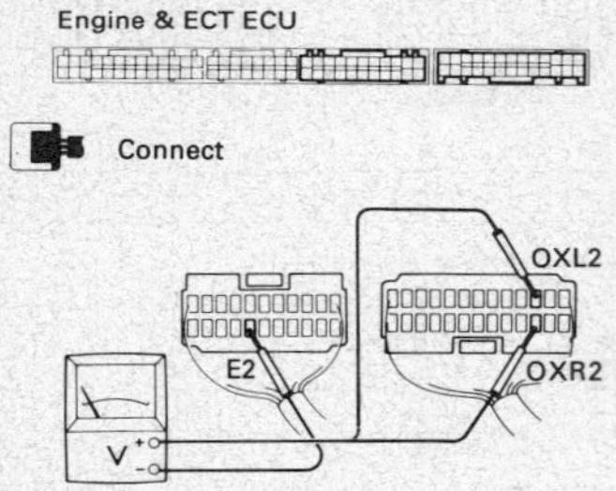

P (1) Remove RH lower pad.
(2) Warm up engine at normal operating temperature.
(3) Engine continue racing at 4,000 rpm to approx. 3 minutes.

C Measure voltage between terminals OXL2, OXR2 and E2 of engine & ECT ECU connector when engine is suddenly raced to full throttle.

OK **The voltage should be 0.5 V or higher at least once.**

OK → **Proceed to next circuit inspection shown on matrix chart** *1

NG ↓

3 Check harness and connector between engine & ECT ECU, sub-oxygen sensor and body ground

NG → **Repair or replace harness or connector.**

OK ↓

4 Check fuel system circuit

NG → **Repair fuel system circuit.**

OK ↓

5 Check injector circuit

NG → **Repair or replace.**

OK ↓

Replace sub-oxygen sensor.

*1: When diag. code 27 or 29 is displayed, check and replace engine & ECT ECU.

DIAGNOSTIC CHART

1 Check voltage between terminals KS and E2 of engine & ECT ECU connector.
- OK → go to 4
- NG → go to 2

2 Check harness and connector between engine & ECT ECU and air flow meter
- NG → Repair or replace harness or connector.
- OK → go to 3

3 Disconnect air flow meter connector and check voltage between terminals VC and E2 of engine & ECT ECU connector.
- OK → Replace air flow meter.
- NG → Check and replace engine & ECT ECU.

4 Is diag. code 31 displayed?
- YES → Check and replace engine & ECT ECU.
- NO → go to 5

5 Does malfunction disappear when a known good air flow meter is installed?
- YES → Replace air flow meter.
- NO → Proceed to next circuit inspection shown on matrix chart

WIRING DIAGRAM

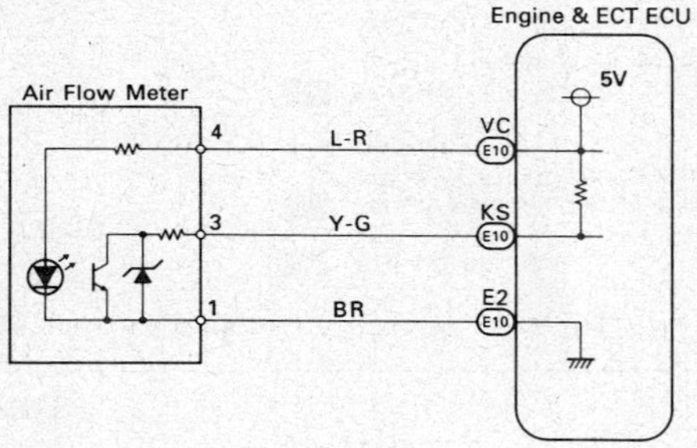

WIRING ROUTING

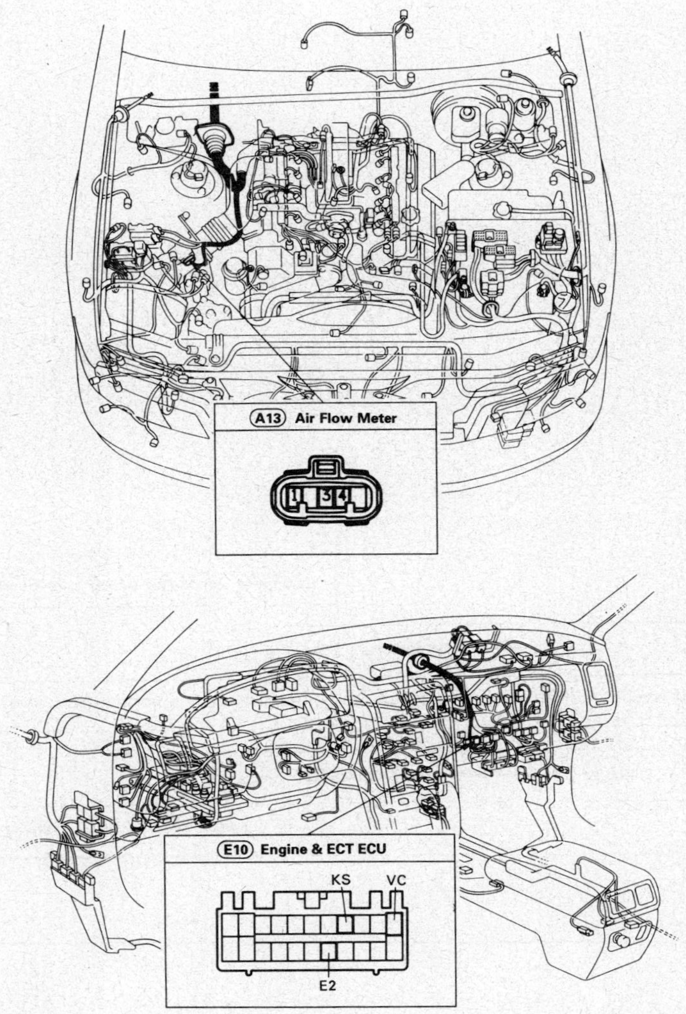

INSPECTION PROCEDURE

1 Check voltage between terminals KS and E2 of engine & ECT ECU connector.

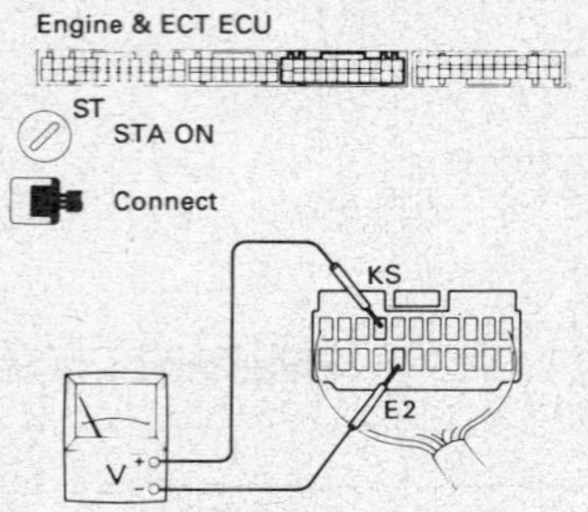

P Remove RH lower pad.

C Measure voltage between terminals KS and E2 of engine & ECT ECU connector while engine is cranked.

OK Voltage: 2.0 – 4.0 V
(Neither 0 V nor 5 V)

OK → Go to step 4 .

NG ↓

2 Check harness and connector between engine & ECT ECU and air flow meter

NG → Repair or replace harness or connector.

OK ↓

3 Disconnect air flow meter connector and check voltage between terminals VC and E2 of engine & ECT ECU connectors.

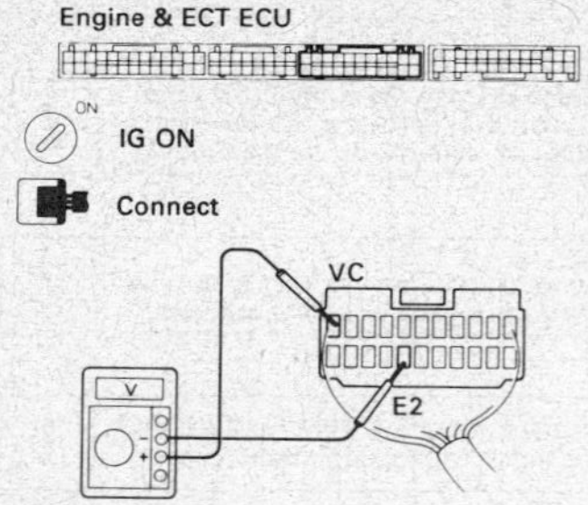

P (1) Disconnect the air flow meter connector.
(2) Turn ignition switch on.

C Measure voltage between terminals VC and E2 of engine & ECT ECU connector.

OK Voltage: 4 – 6 V

OK → Replace air flow meter.

NG ↓

Check and replace engine & ECT ECU.

4 Is diag. code 31 displayed ?

YES → Check and replace engine & ECT ECU.

NO ↓

5 Does malfunction disappear when a known good air flow meter is installed ?

YES → Replace air flow meter.

NO ↓

Proceed to next circuit inspection shown on matrix chart

Diag. Code	35	High Altitude Compensator Sensor (HAC Sensor) Circuit

CIRCUIT DESCRIPTION

The HAC sensor is built into the ECU. This is a semiconductor pressure sensor with properties which cause its electrical resistance to change when stress is applied to the sensor's crystal (silicon) (piezoelectric effect). This sensor is used to detect the atmospheric (absolute) pressure and outputs corresponding electrical signals. Fluctuations in the air pressure cause changes in the intake air density, which can cause deviations in the air-fuel ratio. The signals from HAC sensor are used to make corrections for these fluctuations. If the ECU records diagnostic code "35", the fail safe function operates and the atmospheric pressure is set at a constant 760 mmHg.

Code No.	Diagnosis	Trouble Area
35	Open or short detected in HAC sensor circuit for 0.5 sec. or more.	• ECU

DIAGNOSTIC CHART

1 Are there any other codes (besides Code 16) being output?

- YES → Go to relevant diag. code chart.
- NO → Replace engine & ECT ECU.

Diag.Code	41, 47	(Sub-) Throttle Position Sensor Circuit

CIRCUIT DESCRIPTION

The throttle position sensor is mounted in the throttle body and detects the throttle valve opening angle. When the throttle valve is fully closed, the IDL contacts in the throttle position sensor are on, so the voltage at the terminal IDL of the ECU become 0 V. At this time, a voltage of approximately 0.7 V is applied to the terminal VTA of the ECU. When the throttle valve is opened, the IDL contacts go off and thus the power source voltage of approximately 5 V in the ECU is applied to the terminal IDL of the ECU. The voltage applied to the terminal VTA of the ECU increases in proportion to the opening angle of the throttle valve and becomes approximately 3.5 – 5.0 V when the throttle valve is fully opened. The ECU judges the vehicle driving conditions from these signals input from the terminals VTA and IDL, and uses them as one of the conditions for deciding the air-fuel ratio correction, power increase correction and fuel-cut control etc. The sub-throttle position sensor is built and operates in the same way as the main throttle position sensor. This sensor is used for traction control. The sub-throttle valve is opened and closed by the sub-throttle actuator according to signals from the ABS & TRAC ECU to control the engine output.

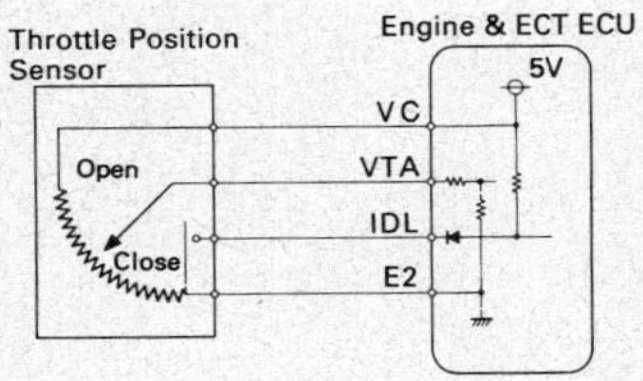

Code No.	Diagnosis	Trouble Area
41	Open or short detected in throttle position sensor signal (VTA1) for 0.5 sec. or more.	• Open or short in throttle position sensor circuit.
	IDL1 contact is ON and VTA1 output exceeds 1.45 V.	• Throttle position sensor • ECU
47	Open or short detected in sub-throttle position sensor signal (VTA2) for 0.5 sec. or more.	• Open or short in sub-throttle position sensor circuit.
	IDL2 contact is ON and VTA2 output exceeds 1.45 V.	• Sub-throttle position sensor • ECU

HINT:
- Diag. code 41 is for the throttle position sensor circuit.
  Diag. code 47 is for the sub-throttle position sensor circuit.
- When the connector for the (sub-) throttle position sensor is disconnected, diagnostic code 41 or 47 is not displayed. Diagnostic code 41 or 47 is displayed only when there is an open or short in the VTA signal circuit of the (sub-) throttle position sensor.

DIAGNOSTIC CHART

HINT:
- If diag. code 41 is displayed, check throttle position sensor circuit; if diag. code 47 is displayed, check sub-throttle position sensor circuit.
- If diag. codes "22" (water temperature sensor circuit), "24" (intake air temperature sensor circuit) and "41" (throttle position sensor circuit) are output simultaneously, E2 (sensor ground) may be open.

1 Check voltage between terminals VTA1, 2, IDL1, 2 and E2 of engine & ECT ECU connector.
- OK → Proceed to next circuit inspection shown on matrix chart *1
- NG ↓

2 Check (sub-) throttle position sensor.
- NG → Adjust or replace (sub-) throttle position sensor
- OK ↓

3 Check harness and connector between engine & ECT ECU and (sub-) throttle position sensor
- NG → Repair or replace harness or connector.
- OK ↓

Check and replace engine & ECT ECU.

*1: When diag. code 41 or 47 is displayed, check for momentary interruption

WIRING DIAGRAM

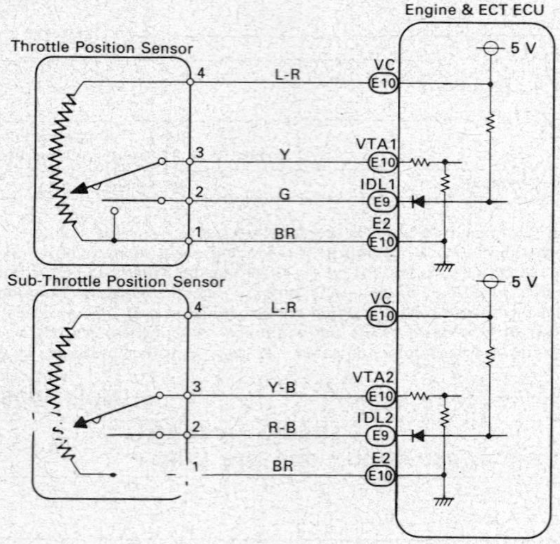

WIRING ROUTING

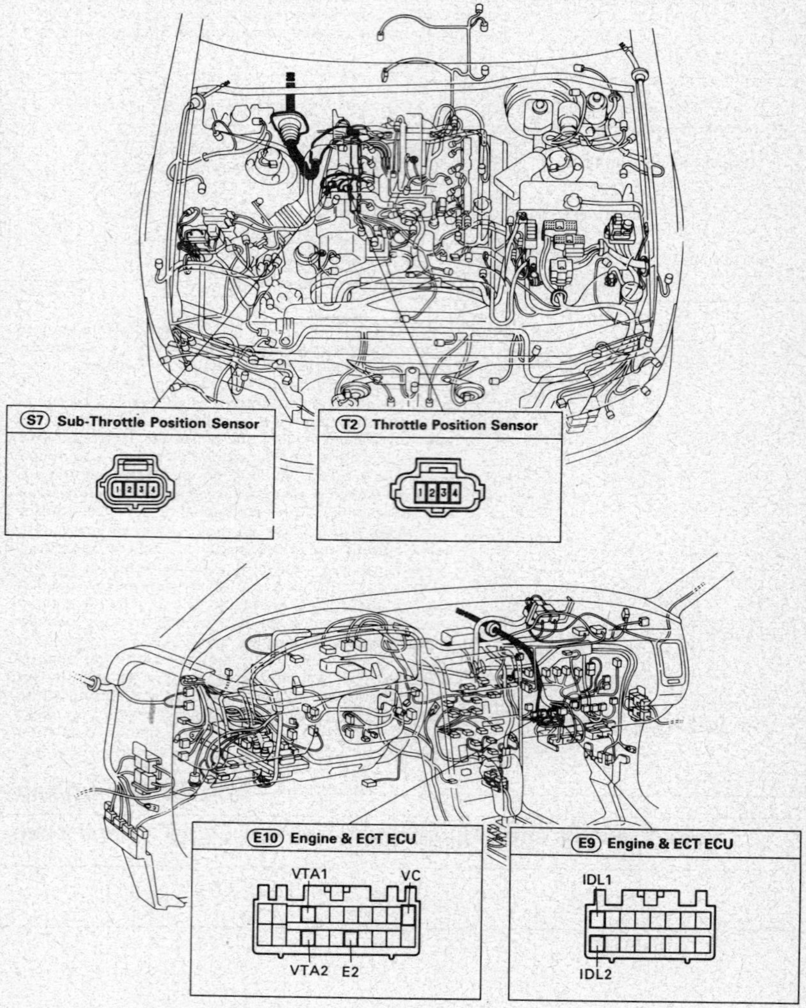

INSPECTION PROCEDURE

HINT:
- If diag. code 41 is displayed, check throttle position sensor circuit. If diag. code 47 is displayed, check sub-throttle position sensor circuit.
- If diag. codes "22" (water temperature sensor circuit), "24" (intake air temperature sensor circuit) and "41" (throttle position sensor circuit) are output simultaneously, E2 (sensor ground) may be open.

1 Check voltage between terminals VTA1, 2, IDL1, 2 and E2 of engine & ECT ECU connector.

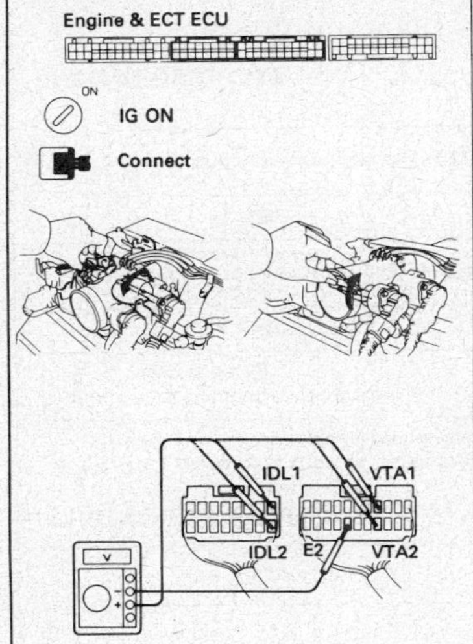

P (1) Remove RH lower pad.
(2) Turn ignition switch on.
(3) For sub-throttle position sensor, remove intake air duct and disconnect sub-throttle valve step motor connector.

C Measure voltage between terminals VTA1, 2, IDL1, 2 and E2 of engine & ECT ECU connector when the (sub-) throttle valve is opened gradually from the closed condition.

OK

Terminal / Throttle Valve	VTA1, 2	IDL1, 2
Fully Closed	0.1 – 1.0 V	Below 1.0 V
Fully Open	3 – 5 V	4 – 6 V

Hint The voltage should increase steadily in proportion to the throttle valve opening angle.

OK → Proceed to next circuit inspection shown on matrix chart *1

*1: When diag. code 41 or 47 is displayed, check for momentary interruption

NG ↓

Go to step **2**.

2 Check (sub-) throttle position sensor.

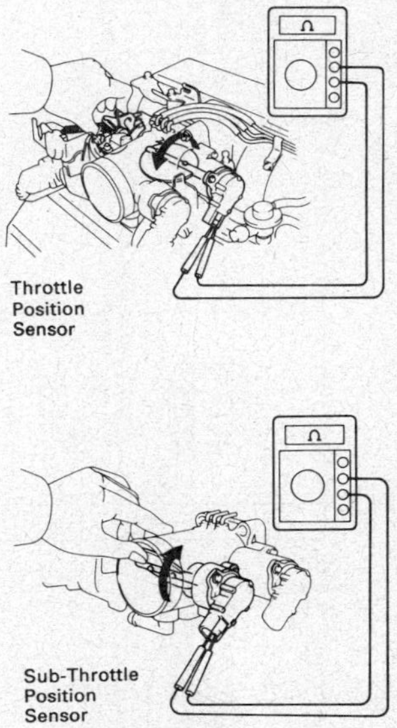

P (1) Remove intake air duct.
(2) Disconnect (sub-) throttle position sensor connector.

C Measure resistance between terminals 3, 2 and 1 of (sub-) throttle position sensor connector when the throttle valve is opened gradually from the closed condition.

OK

Terminal / Throttle Valve	3 – 1	2 – 1
Fully closed	0.2 – 0.8 kΩ	Less than 2.3 kΩ
Fully open	2.8 – 8.0 kΩ	∞ Ω

Hint Resistance between terminals 3 and 1 should increase gradually in accordance with the throttle valve opening angle.

NG → Adjust or replace (sub-) throttle position sensor

OK ↓

3 Check harness and connector between engine & ECT ECU and (sub-) throttle position sensor

NG → Repair or replace harness or connector.

OK ↓

Check and replace engine & ECT ECU.

Diag. Code	43	Starter Signal Circuit

CIRCUIT DESCRIPTION

When the engine is cranked, the intake air flow is slow, so fuel vaporization is poor. A rich mixture is therefore necessary in order to achieve good startability. While the engine is being cranked, the battery voltage is applied to terminal STA of the ECU. The starter signal is mainly used to increase the fuel injection volume for the starting injection control and after-start injection control.

Code No.	Diagnosis	Trouble Area
43	Starter signal is not input to ECU even once.	• Open or short in stater signal circuit. • Open or short in IG S/W or main relay circuit.

DIAGNOSTIC CHART

HINT: This diagnostic chart is based on the premise that the engine is cranked normally. If the engine is not cranked, proceed to the matrix chart of problem symptoms

1 Check output condition of diag. code 43.

- OK → Proceed to next circuit inspection shown on matrix chart
- NG ↓

2 Check for open in harness and connectors between engine & ECT ECU and starter relay

- NG → Repair or replace harness or connector.
- OK ↓

Check and replace engine & ECT ECU.

WIRING DIAGRAM

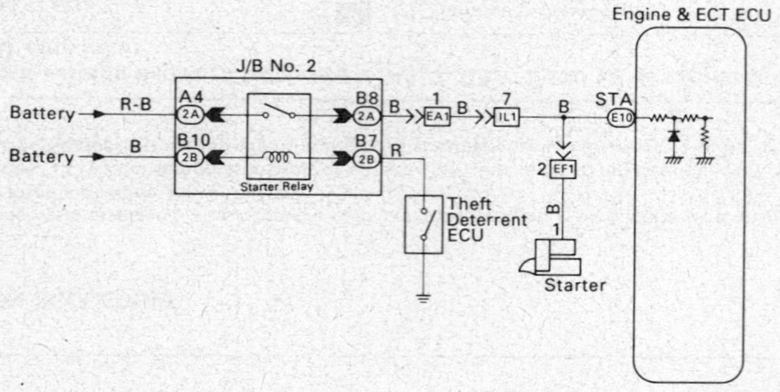

WIRING ROUTING

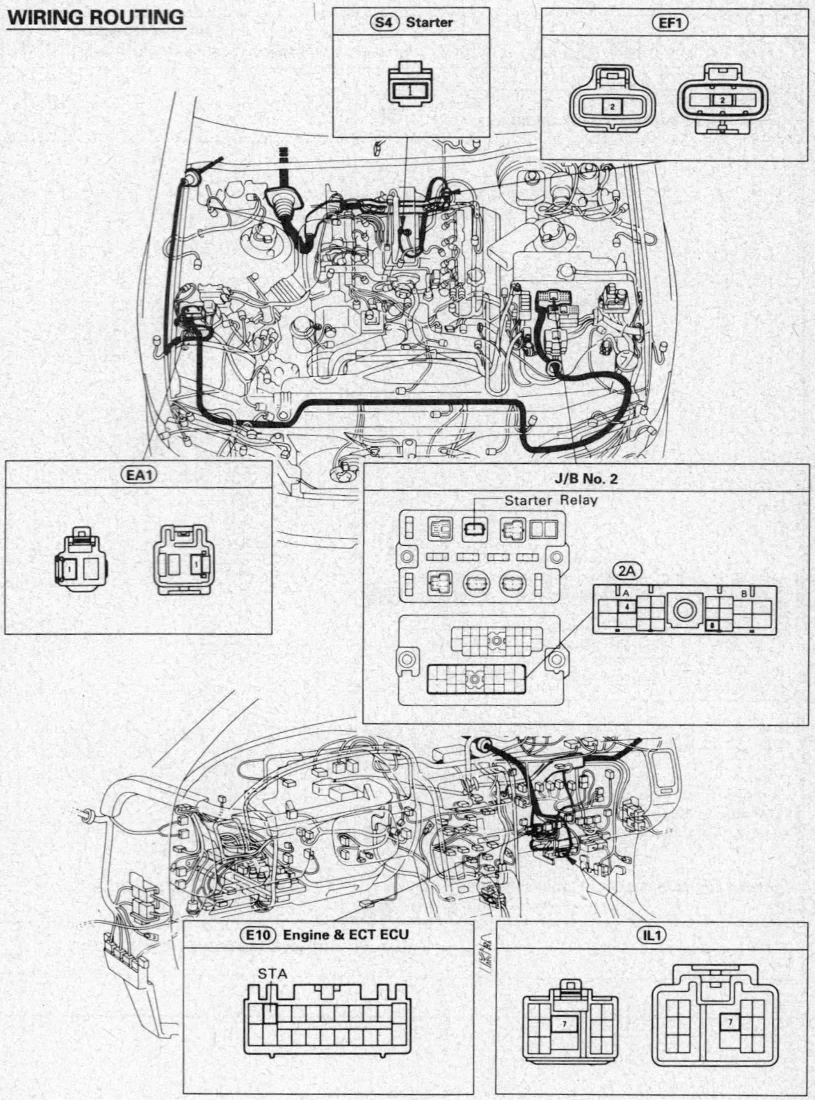

INSPECTION PROCEDURE

HINT: This diagnostic chart is based on the premise that the engine is cranked normally. If the engine is not cranked, proceed to the matrix chart of problem symptoms

1 Check output condition of diag. code 43.

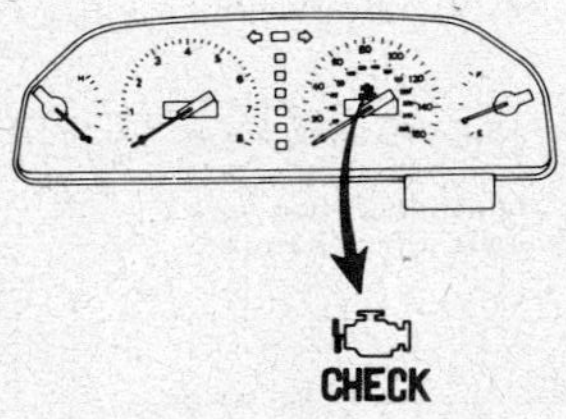

P Setting the test mode.
(1) Turn ignition switch OFF.
(2) Connect terminals TE2 and E1 of TDCL.
(3) Turn ignition switch ON.
(Don't start the engine)
(4) Connect terminals TE1 and E1 of TDCL.

C Check if code "43" is output by the "CHECK" engine warning light.

OK **Code "43" is output.**
(Codes "17" and "18" are also output.)

C Start the engine.
Check if the code "43" disappear.

OK **Code "43" is not output.**
(Neither codes "17" nor "18" are output.)

OK → **Proceed to next circuit inspection shown on matrix chart**

NG ↓

2 Check for open in harness and connectors between engine & ECT ECU and starter relay

NG → **Repair or replace harness or connector.**

OK ↓

Check and replace engine & ECT ECU.

Diag. Code	52,53,55	Knock Sensor Circuit

CIRCUIT DESCRIPTION

Knock sensors are fitted one each to the right bank and left bank of the cylinder block to detect engine knocking. This sensor contains a piezoelectric element which generates a voltage when it becomes deformed, which occurs when the cylinder block vibrates due to knocking. If engine knocking occurs, ignition timing is retarded to suppress it.

Code No.	Diagnosis	Trouble Area
52	No. 1 knock sensor signal is not input to ECU for 3 crank revolution with engine speed between 1600 rpm – 5200 rpm.	• Open or short in No. 1 knock sensor circuit. • No. 1 knock sensor (looseness) • ECU
53	Engine speed is between 650 rpm and 5200 rpm and engine control computer (for knock control) malfunction is detected.	• ECU
55	No. 2 knock sensor signal is not input to ECU for 3 crank revolution with engine speed between 1600 rpm – 5200 rpm.	• Open or short in No. 2 knock sensor circuit. • No. 2 knock sensor (looseness) • ECU

If the ECU detects the above diagnosis conditions, it operates the fail safe function in which the corrective retard angle value is set to the maximum value.

CIRCUIT DESCRIPTION (Cont'd)

DIAGNOSIS CODE DETECTION DRIVING PATTERN

Purpose of the driving pattern.
(a) After the diag. code is erased, the malfunction is reproduced and the diag. code is recorded again.
(b) To check that the malfunction is cleared once the repair is completed (confirm that diag. code is no longer detected).

Malfunction: **Open or Short in Knock Sensor**

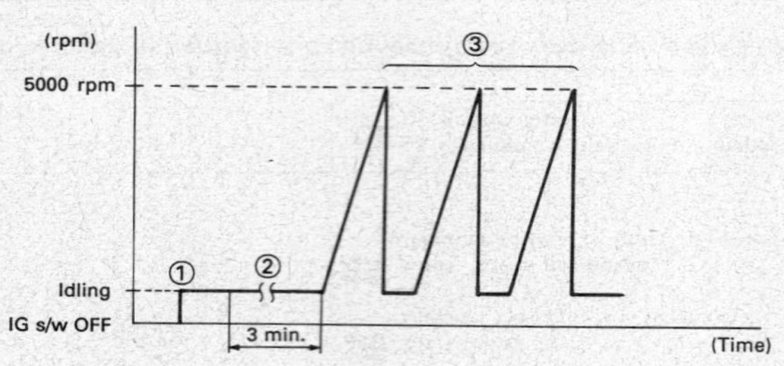

① Start engine and warm engine up.
② After engine is warmed up, let it idle for 3 min.
③ With the A/C ON, perform sudden racing (5000 rpm) three times. (Rapidly depress the accelerator pedal and suddenly release it.)

HINT: If a malfunction exists, the "CHECK" engine warning light will light up when sudden racing is performed.

NOTICE: If the conditions in this test are not strictly observed, detection of the malfunction will not be possible.

DIAGNOSTIC CHART

HINT: If diag. code 52 is displayed, check No.1 knock sensor (for left bank) circuit.
If diag. code 55 is displayed, check No.2 knock sensor (for right bank) circuit.
If diag. code 53 is displayed, replace engine & ECT ECU.

1 Check continuity between terminals KNK1, KNK2 of engine & ECT ECU connector and body ground.
- OK → go to 3 (Check for open and short in harness and connectors)
- NG ↓

2 Check knock sensor.
- NG → Replace knock sensor
- OK ↓

3 Check for open and short in harness and connectors between engine & ECT ECU and knock sensor
- NG → Repair or replace harness or connector.
- OK ↓

4 Does malfunction disappear when a known good knock sensor is installed?
- YES → Replace knock sensor
- NO ↓

Proceed to next circuit inspection shown on matrix chart *1

*1: However, when diag. code 52 or 55 is displayed, check and replace engine & ECT ECU.

WIRING DIAGRAM

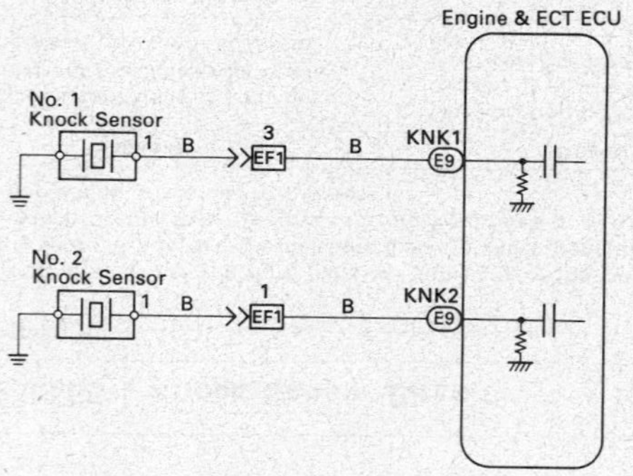

WIRING ROUTING

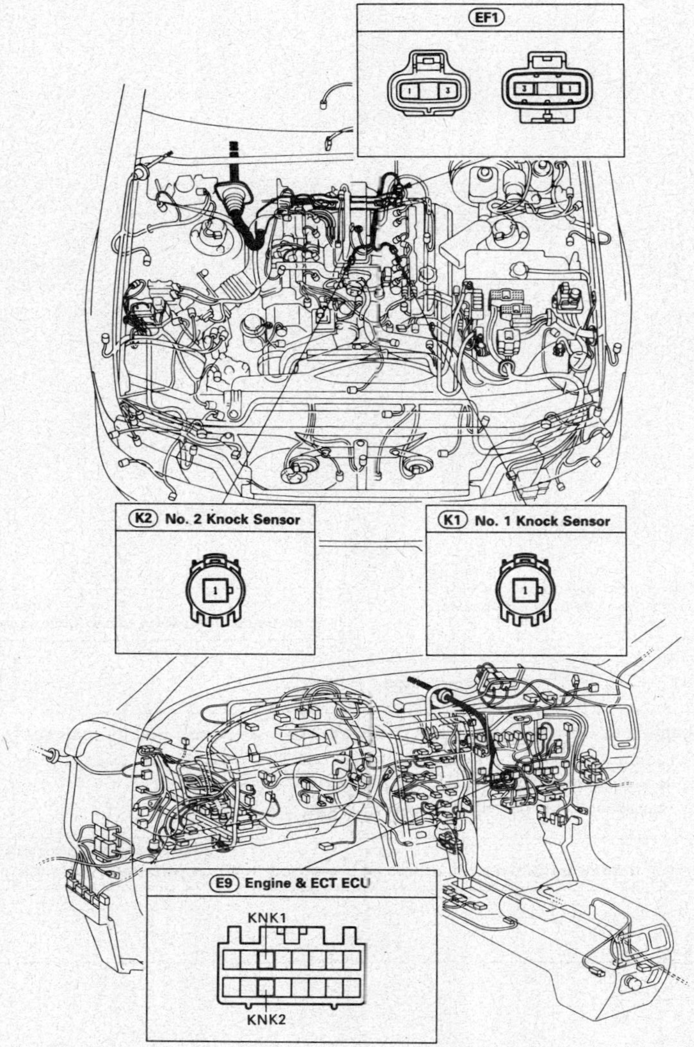

INSPECTION PROCEDURE

HINT: If diag. code 52 is displayed, check No.1 knock sensor (for left bank) circuit.
If diag. code 55 is displayed, check No.2 knock sensor (for right bank) circuit.
If diag. code 53 is displayed, replace engine & ECT ECU.

1 Check continuity between terminals KNK1, KNK2 of engine & ECT ECU connector and body ground.

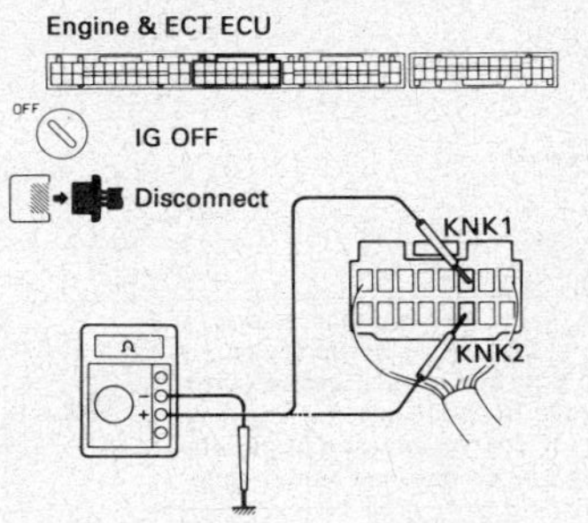

P (1) Remove RH lower pad.
(2) Disconnect the engine & ECT ECU connector.

C Check continuity between terminals KNK1, KNK2 of engine & ECT ECU connector and body ground.

OK **No continuity**

OK Go to step 3.

NG

2 Check knock sensor.

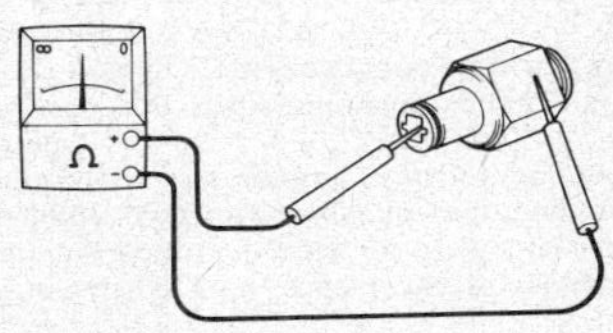

P Disconnect knock sensor connector.

C Measure resistance between the knock sensor terminal and body.

OK **No continuity**

NG Replace knock sensor

OK

3 **Check for open and short in harness and connectors between engine & ECT ECU and knock sensor**

NG — **Repair or replace harness or connector.**

OK

4 **Does malfunction disappear when a known good knock sensor is installed ?**

YES — **Replace knock sensor**

NO

Proceed to next circuit inspection shown on matrix chart *1

*1: However, when diag. code 52 or 55 is displayed, check and replace engine & ECT ECU.

Diag. Code	71	EGR System Malfunction

CIRCUIT DESCRIPTION

The EGR system is designed to recirculate the exhaust gas properly controlled according to the driving condition back into the intake air-fuel mixture. It helps to slow down combustion in the cylinder and thus lower the combustion temperature which, in turn, reduces the amount of NOx emission. The amount of EGR is regulated by the EGR vacuum modulator according to the engine load.

If even one of the following conditions is fulfilled, the VSV is turned on by a signal from the ECU. This results in atmospheric air acting on the EGR valve, closing the EGR valve and shutting off the exhaust gas (EGR cut-off).

- Coolant temperature below predetermined level.
- During deceleration (throttle valve closed).
- Light engine load (amount of intake air very small).
- Engine speed over 4000 rpm.
- Engine racing (neutral start switch turned on).

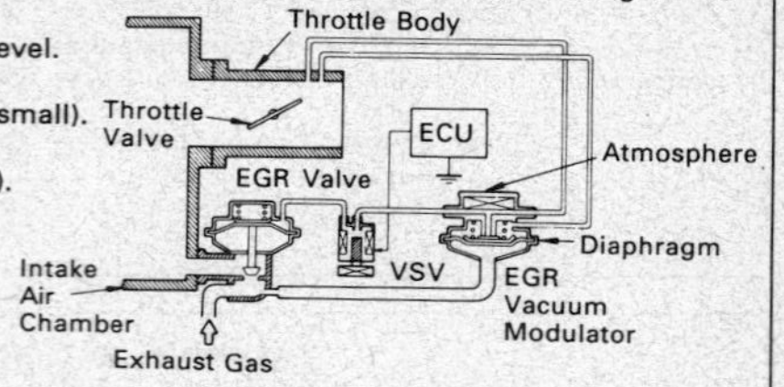

Code No.	Diagnosis	Trouble Area
71	With engine warmed up and EGR operation possible (ECT in 3rd speed, A/C ON, 45 – 70 mph), code is displayed when EGR gas temp. sensor is 65°C or less when vehicle is driven for 90 sec. (2 trip detection logic)	• Open in EGR gas temp. sensor circuit • Short in VSV circuit for EGR • EGR hose disconnected, valve stuck • ECU

CIRCUIT DESCRIPTION (Cont'd)

DIAGNOSIS CODE DETECTION DRIVING PATTERN

Purpose of the driving pattern.

(a) After the diag. code is erased, the malfunction is reproduced and the diag. code is recorded again.

(b) To check that the malfunction is cleared once the repair is completed (confirm that diag. code is no longer detected).

Malfunction: Short in VSV Circuit for EGR, Loose EGR Hose, Valve Stuck

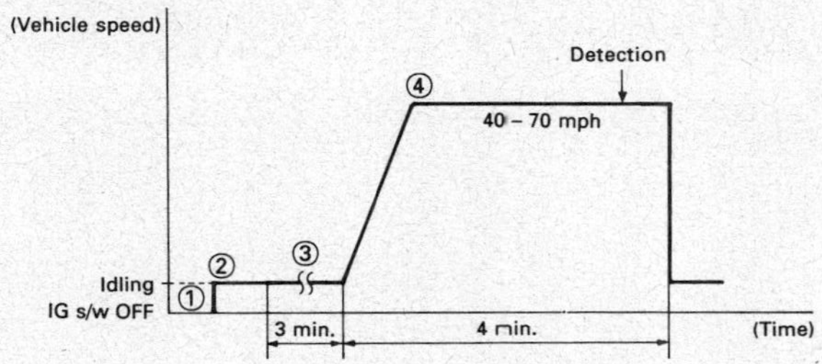

① Initiate test mode
② Start engine and warm engine up.
③ After engine is warmed up, let it idle for 3 min.
④ With the A/C ON and transmission in D range (O/D OFF), drive at 40 – 70 mph for 4 min.

HINT: If a malfunction exists, the "CHECK" engine warning light will light up at 1 – 4 min. of driving at 40 – 70 mph.

NOTICE: If the conditions in this test are not strictly observed, detection of the malfunction will not be possible.

DIAGNOSTIC CHART

1 Check for air leakage in EGR system
- NG → Repair air leakage.
- OK ↓

2 Check EGR vacuum modulator
- NG → Repair or replace EGR vacuum modulator.
- OK ↓

3 Check VSV with cold engine. Check VSV and EGR vacuum modulator with warm engine
- NG → go to step 5
- OK ↓

4 Check EGR valve
- OK → Proceed to next circuit inspection shown on matrix chart *1
- NG ↓ Repair or replace EGR valve.

*1: When diag. code 71 is displayed, go to step **8**.

5 Check VSV for EGR
- NG → Repair or replace VSV for EGR.
- OK ↓

6 Check voltage between terminal EGR of engine & ECT ECU connector and body ground.
- OK → Repair or replace EGR valve.
- NG ↓

7 Check harness and connector between engine & ECT ECU and VSV for EGR, VSV for EGR and EFI main relay
- NG → Repair or replace harness or connector.
- OK ↓ Check and replace engine & ECT ECU.

8 Check EGR gas temp. sensor.
- NG → Replace EGR gas temp. sensor.
- OK ↓ Check and replace engine & ECT ECU.

WIRING DIAGRAM

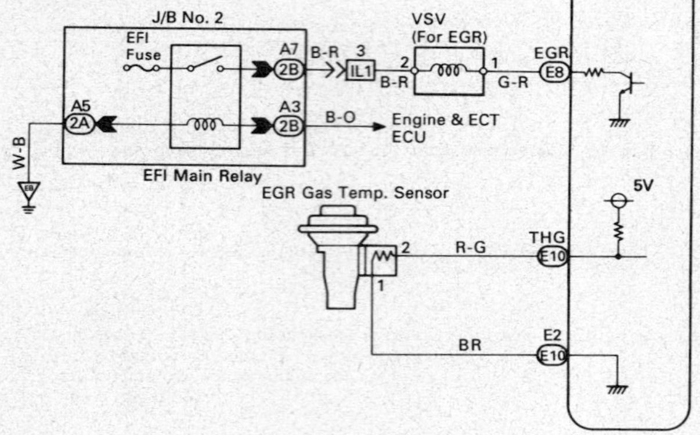

WIRING ROUTING

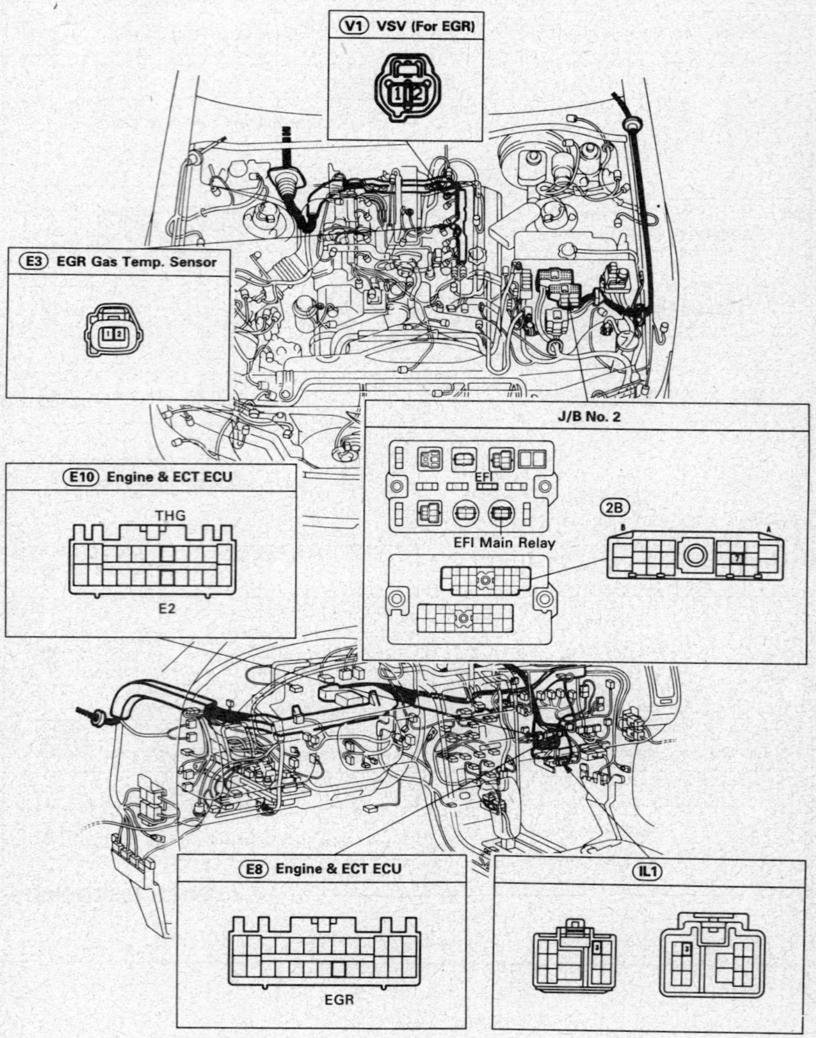

INSPECTION PROCEDURE

1 Check for air leakage in EGR system

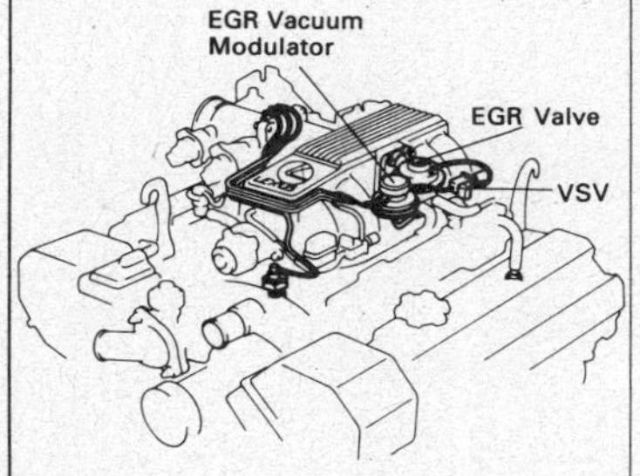

C Check for air leakage in EGR system.

OK ↓

NG → **Repair air leakage.**

2 Check EGR vacuum modulator

OK ↓

NG → **Repair or replace EGR vacuum modulator.**

**3 Check VSV with cold engine.
Check VSV and EGR vacuum modulator with warm engine**

OK ↓

NG → **Go to step 5.**

4 Check EGR valve

OK → **Proceed to next circuit inspection shown on matrix chart** *1

NG ↓

Repair or replace EGR valve.

*1: However, when diag. code 71 is displayed, go to step 8.

5 Check VSV for EGR (See page EC-12).

OK ↓

NG → **Repair or replace VSV for EGR.**

6 Check voltage between terminal EGR of engine & ECT ECU connector and body ground.

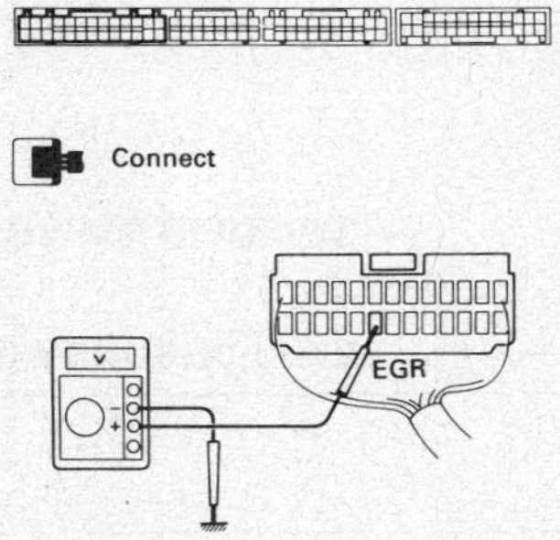

P (1) Remove RH lower pad.
(2) Start the engine.
(3) Warm up engine to normal operating temperature.

C Measure voltage between terminal EGR of engine & ECT ECU connector and body ground, when engine is following conditions.
Condition A : Idling
Condition B : Fully depress the parking brake pedal, and while strongly depressing the brake pedal with your left leg, shift from N range to D range and control the accelerator pedal with your right leg at the engine speed is 1400rpm.

OK

Engine Condition	Voltage
Condition A	Battery voltage
Condition B	Approx. 0V

Notice **Do not perform the operation continuously for longer than 5 seconds.**

OK → **Repair or replace EGR valve.**

NG ↓

7 Check harness and connector between engine & ECT ECU and VSV for EGR, VSV for EGR and EFI main relay

OK ↓

NG → **Repair or replace harness or connector.**

Check and replace engine & ECT ECU.

8 Check EGR gas temp. sensor.

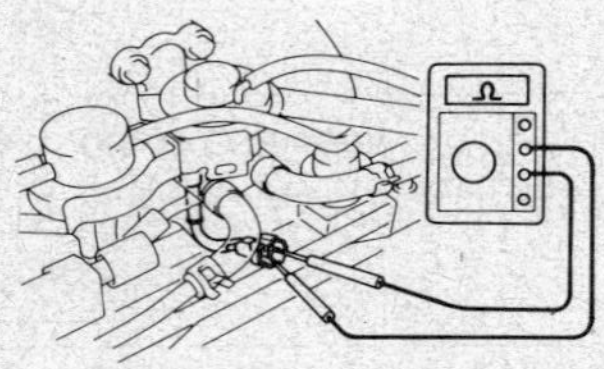

P Disconnect EGR gas temp. sensor connector

C Measure resistance between terminals of EGR gas temp. sensor connector.

OK **Resistance: 69.40 — 88.50 kΩ at 50°C (122°F)**
11.89 — 14.37 kΩ at 100°C (212°F)
2.79 — 3.59 kΩ at 150°C (302°F)

NG **Replace EGR gas temp. sensor.**

OK

Check and replace engine & ECT ECU.

Diag. Code	51	Switch Condition Signal Circuit

CIRCUIT DESCRIPTION

Neutral Start Switch Signal
The ECU uses the signals from the neutral start switch to determine whether the transmission is in park or neutral, or in some other gear.

Air Conditioner Switch Signal
The ECU uses the output from the air conditioner switch to determine whether or not the air conditioner is operating so that it can increase the idling speed of the engine if necessary.

Throttle Position Sensor IDL Signal
The IDL contacts are mounted in the throttle position sensor, and detects the idle condition.

Code No.	Diagnosis	Trouble Area
51	Displayed when A/C is ON, or IDL contact OFF or shift position in "R", "D", "2", or "1" renges.	• A/C switch system • Throttle position sensor IDL circuit • Neutral start switch circuit • Accelerator pedal and cable • ECU

DIAGNOSTIC CHART

WIRING DIAGRAM

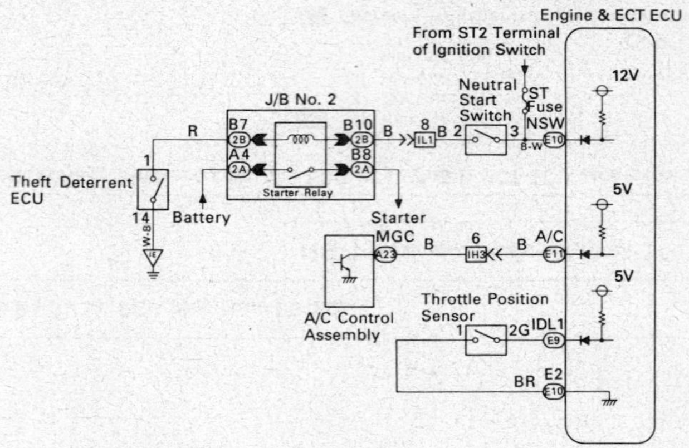

DIAGNOSTIC CHART

1 Check voltage between terminals NSW, IDL1 and A/C of engine & ECT ECU connector and body ground.
- OK → Check and replace engine & ECT ECU.
- NG (IDL) → go to 2
- NG (NSW) → go to 3
- NG (A/C) → go to 5

2 Check throttle position sensor.
- NG → Adjust or replace throttle position sensor
- OK → Check and repair harness or connector between engine & ECT ECU and throttle position sensor.

3 Check operation of starter.
- OK → Check and repair harness or connector between engine & ECT ECU and neutral start switch.
- NG → go to 4

4 Check neutral start switch
- NG → Replace neutral start switch.
- OK → Check and repair harness or connector between engine & ECT ECU and neutral start switch, neutral start switch and starter relay.

5 Disconnect A/C control assembly connector, check voltage between terminal MGC of A/C control assembly connetor and body ground.
- OK → Check and replace A/C control assembly.
- NG → go to 6

6 Check harness and connector between engine & ECT ECU and A/C control assembly
- NG → Repair or replace harness or connector.
- OK → Check and replace engine & ECT ECU.

WIRING ROUTING

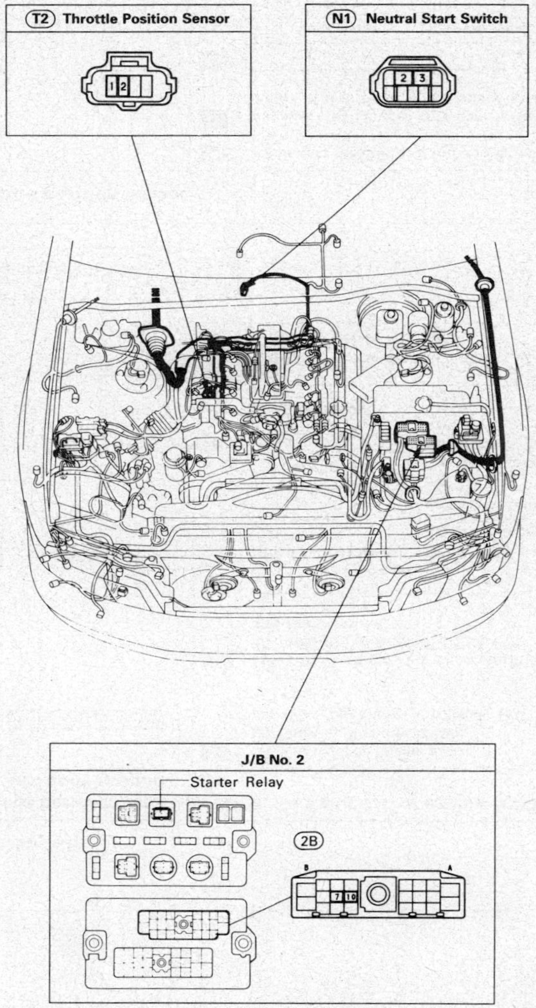

WIRING ROUTING

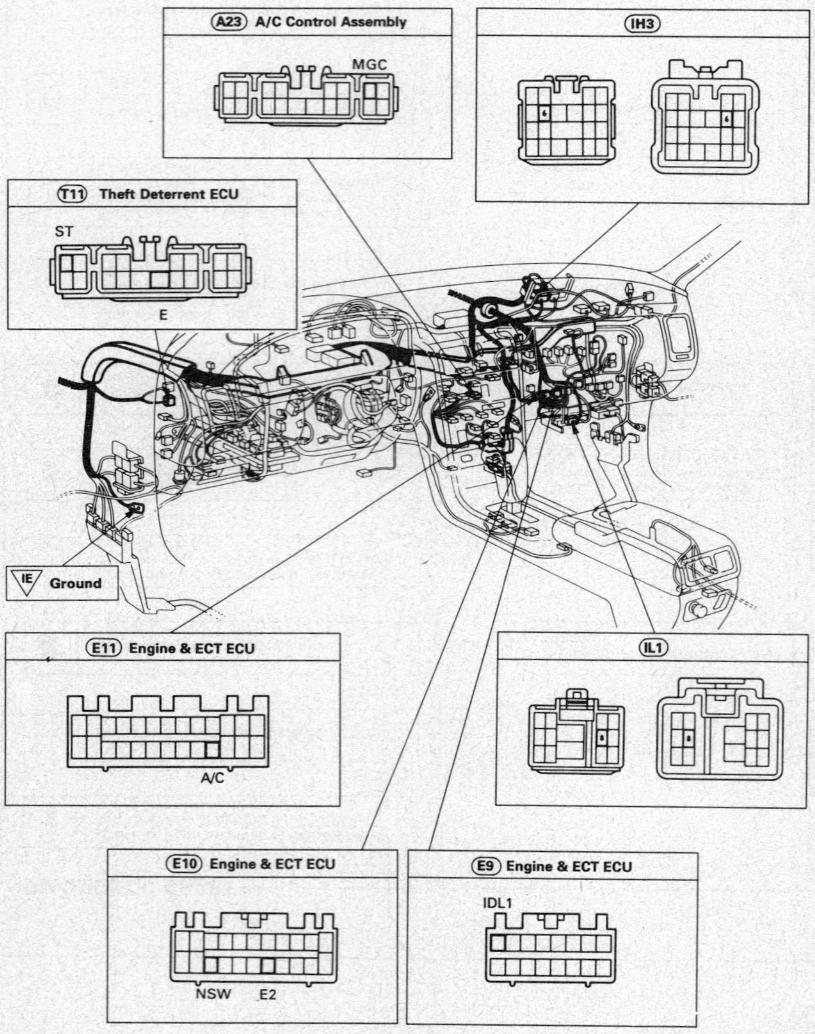

INSPECTION PROCEDURE

1 Check voltage between terminals NSW, IDL1 and A/C of engine & ECT ECU connector and body ground.

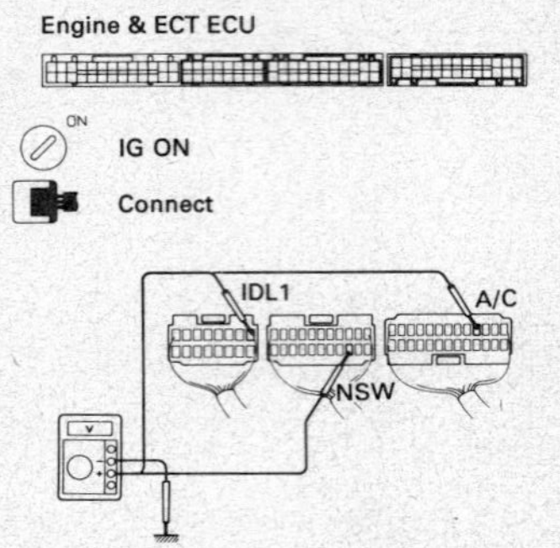

P (1) Remove RH lower pad.
(2) Turn ignition switch on.
(For checking terminal A/C, start the engine.)

C Measure voltage between terminals NSW, IDL1 and A/C of engine & ECT ECU connector and body ground.

OK

	Condition	Voltage
Neutral Start Switch (NSW)	P or N range	Below 1V
	R,D,2 or L range	Battery voltage
Throttle Position Sensor (IDL1)	Accelerator pedal released	Below 1V
	Accelerator pedal depressed	5V
A/C Switch (A/C)	A/C SW ON	Below 2V
	A/C SW OFF	5V

NG IDL1 .... Go to step 2 NSW .... Go to step 3
A/C ..... Go to step 5

OK

Check and replace engine & ECT ECU.

2 Check throttle position sensor.

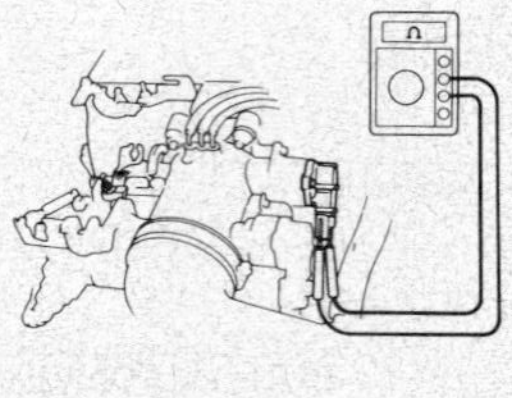

P Disconnect throttle position sensor connector.

C Measure resistance between terminals IDL1 and E2 of throttle position sensor connector.

OK

Throttle Valve	Resistance
Fully closed	Less than 2.3 kΩ
Opened	∞ Ω

NG Adjust or replace throttle position sensor

OK

Check or repair harness or connector between engine & ECT ECU and throttle position sensor.

3 Check operation of starter.

OK → Check and repair harness or connector between engine & ECT ECU and neutral start switch.

NG ↓

4 Check neutral start switch

NG → Replace neutral start switch.

OK ↓

Check and repair harness or connector between engine & ECT ECU and neutral start switch, neutral start switch and starter relay.

5 Disconnect A/C control assembly connector, check voltage between terminal MGC of A/C control assembly connector and body ground.

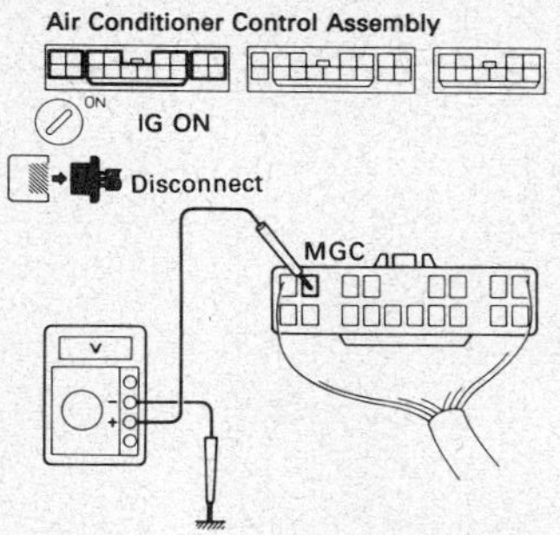

P (1) Remove air conditioner control assembly.
(2) Disconnect air conditioner control assembly connector.
(3) Turn ignition switch on.

C Measure voltage between terminal MGC of air conditioner control assembly connector and body ground.

OK **Voltage: 4 – 6 V**

OK → Check and replace A/C control assembly.

NG ↓

6 Check harness and connector between engine & ECT ECU and A/C control assembly

NG → Repair or replace harness or connector.

OK ↓

Check and replace engine & ECT ECU.

Neutral Start Switch Circuit

CIRCUIT DESCRIPTION

The neutral start switch goes on when the shift lever is in the N or P shift position. When it goes on the terminal NSW of the ECU is grounded to body ground via the starter relay and theft deterrent ECU, thus the terminal NSW voltage becomes 0V. When the shift lever is in the D, 2, L or R position, the neutral start switch goes off, so the voltage of ECU terminal NSW becomes battery voltage, the voltage of the ECU internal power source.
If the shift lever is moved from the N range to the D range, this signal is used for air-fuel ratio correction and for idle speed control (estimated control), etc.
When the neutral start switch is off, code "51" is output in the test mode diagnosis. (This is not abnormal.)

DIAGNOSTIC CHART

HINT: This diagnosis chart is based on premise that the engine is cranked normally. If the engine is not cranked, proceed to the matrix chart of problem symptoms

1 Check output condition of diag. code 51.

- OK → Proceed to next circuit inspection shown on matrix chart
- NG Type I → 2
- NG Type II → 3

2 Check voltage between terminal NSW of engine & ECT ECU connector and body ground.

- OK → Proceed to next circuit inspection shown on matrix chart
- NG ↓

3 Check harness and connector between engine & ECT ECU and neutral start switch

- NG → Repair or replace harness or connector.
- OK ↓

4 Is engine cranked in the D range?

- YES → Replace neutral start switch.
- NO → Check and replace engine & ECT ECU.

WIRING DIAGRAM

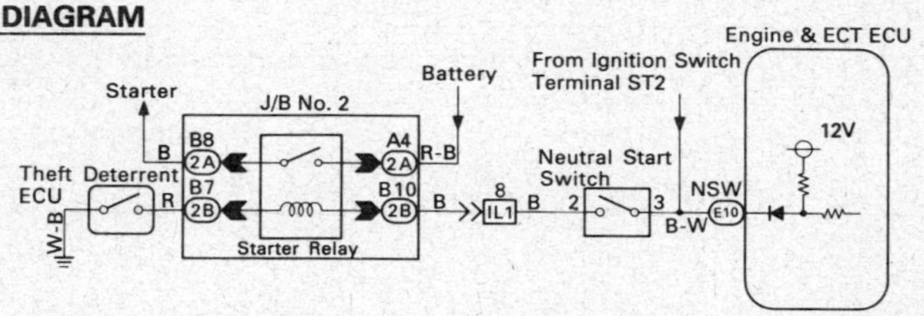

WIRING ROUTING

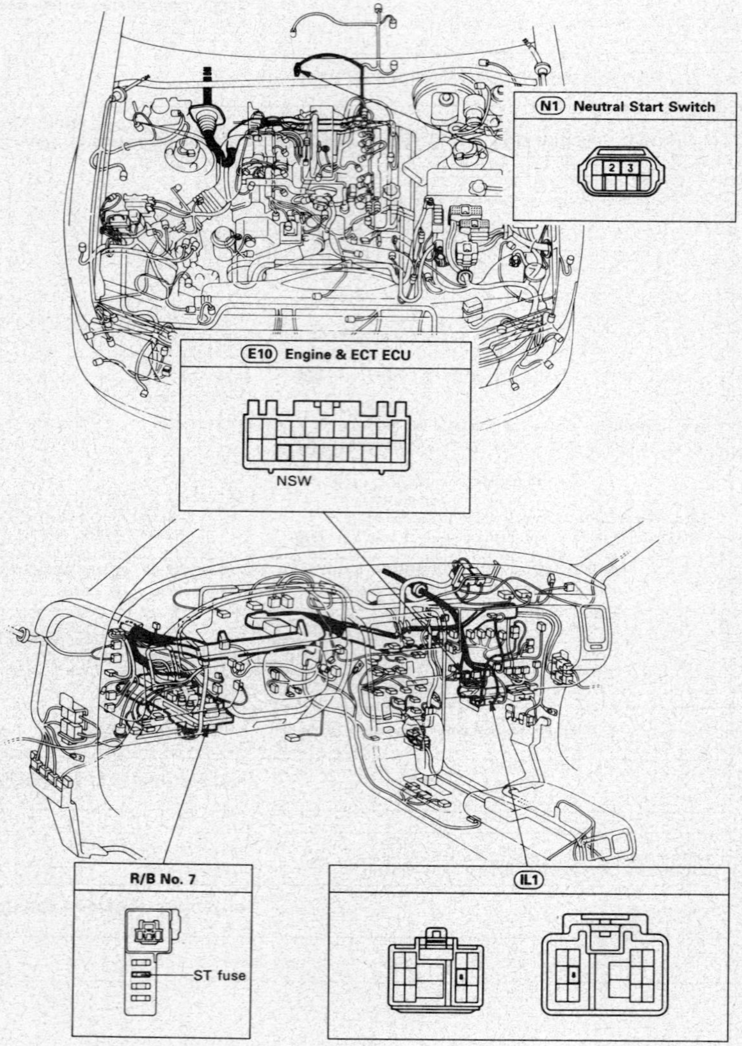

INSPECTION PROCEDURE

1 Check output condition of diag. code 51.

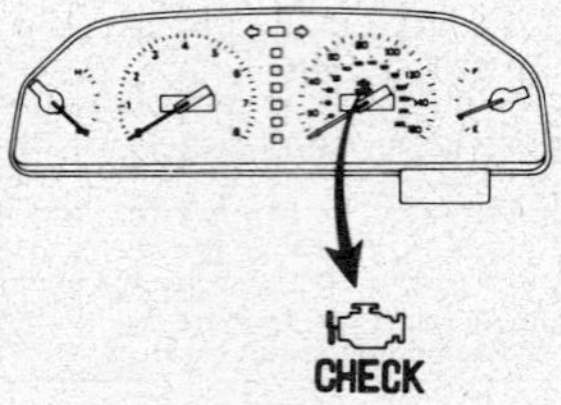

P
(1) Connect terminals TE2 and E1 of TDCL.
(2) Turn ignition switch on.
(3) Crank the engine.
(4) Connect terminals TE1 and E1 of TDCL.

C Check if diagnostic code "51" is output when the shift lever is in the P and D shift positions.

Result

Shift Position	Result: OK	Result: NG Type I	Result: NG Type II
"P"	Normal Code	Code 51	Normal Code
"D"	Code 51	Code 51	Normal Code

NG Type I — Go to step **2**.

NG Type II — Go to step **3**.

OK

Proceed to next circuit inspection shown on matrix chart

2 Check voltage between terminal NSW of engine & ECT ECU connector and body ground.

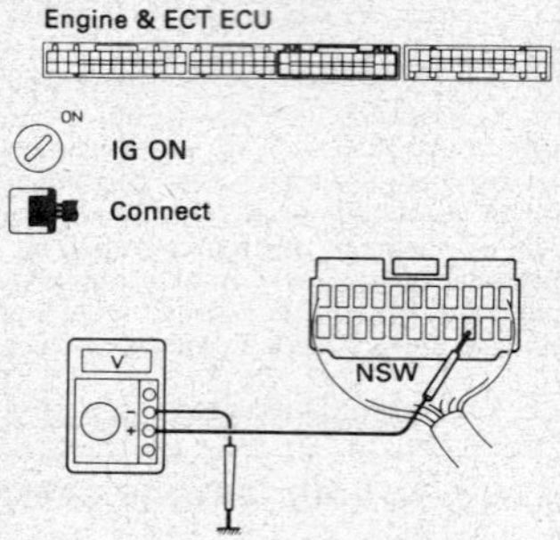

P
(1) Remove RH lower pad.
(2) Turn ignition switch on.

C Measure voltage between terminal NSW of engine & ECT ECU connector and body ground when shift position is in "P" range and "D" range.

OK

Shift position	Voltage
"P"	Below 1V
"D"	Battery voltage

NG

OK — Proceed to next circuit inspection shown on matrix chart

3 **Check harness and connector between engine & ECT ECU and neutral start switch**

NG — **Repair or replace harness or connector.**

OK

4 **Is engine cranked in the D range ?**

YES — **Replace neutral start switch.**

NO

Check and replace engine & ECT ECU.

ECU Power Source Circuit

CIRCUIT DESCRIPTION

When the ignition switch is turned on, battery voltage is applied to the terminal IG SW of the ECU, and the main relay control circuit in the ECU sends a signal to the terminal M-REL of the ECU, switching on the main relay. This signal causes current to flow to the coil, closing the contacts of the main relay and supplying power to the terminals +B and +B1 of the ECU.
If the ignition switch is turned off, the ECU continues to switch on the main relay for a maximum of 2 seconds for the initial setting of the ISC valve.

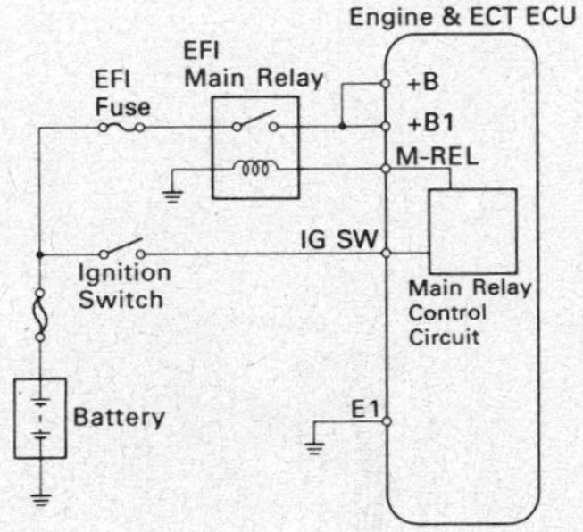

DIAGNOSTIC CHART

WIRING DIAGRAM

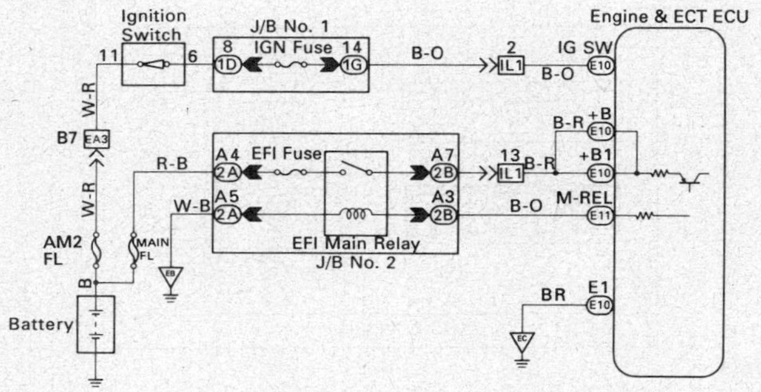

DIAGNOSTIC CHART

1 Check voltage between terminals +B, +B1 and E1 of engine & ECT ECU connector.
- OK → Proceed to next circuit inspection shown on matrix chart
- NG ↓

2 Check for open in harness and connector between terminal E1 of engine & ECT ECU and body ground
- NG → Repair or replace harness or connector.
- OK ↓

3 Check voltage between terminal IG SW of engine & ECT ECU connector and body ground.
- OK → go to step 6
- NG ↓

4 Check IGN fuse.
- NG → Check for short in all the harness and components connected to IGN fuse
- OK ↓

5 Check ignition switch.
- NG → Replace ignition switch.
- OK ↓

Check and repair harness and connector between battery and ignition switch, ignition switch and engine & ECT ECU.

6 Check harness and connector between engine & ECT ECU and EFI main relay
- NG → Repair or replace harness or connector.
- OK ↓

7 Check voltage between terminal M-REL of engine & ECT ECU connector and body ground.
- NG → Check and replace engine & ECT ECU.
- OK ↓

8 Check EFI main relay.
- NG → Replace EFI main relay.
- OK ↓

9 Check EFI fuse.
- NG → Check for short in all the harness and components connected to EFI fuse
- OK ↓

10 Check harness and connector between battery and EFI main relay, EFI main relav and engine & ECT ECU
- NG → Repair or replace harness or connector.
- OK ↓

Check and repair harness or connector between EFI main relay and body ground.

WIRING ROUTING

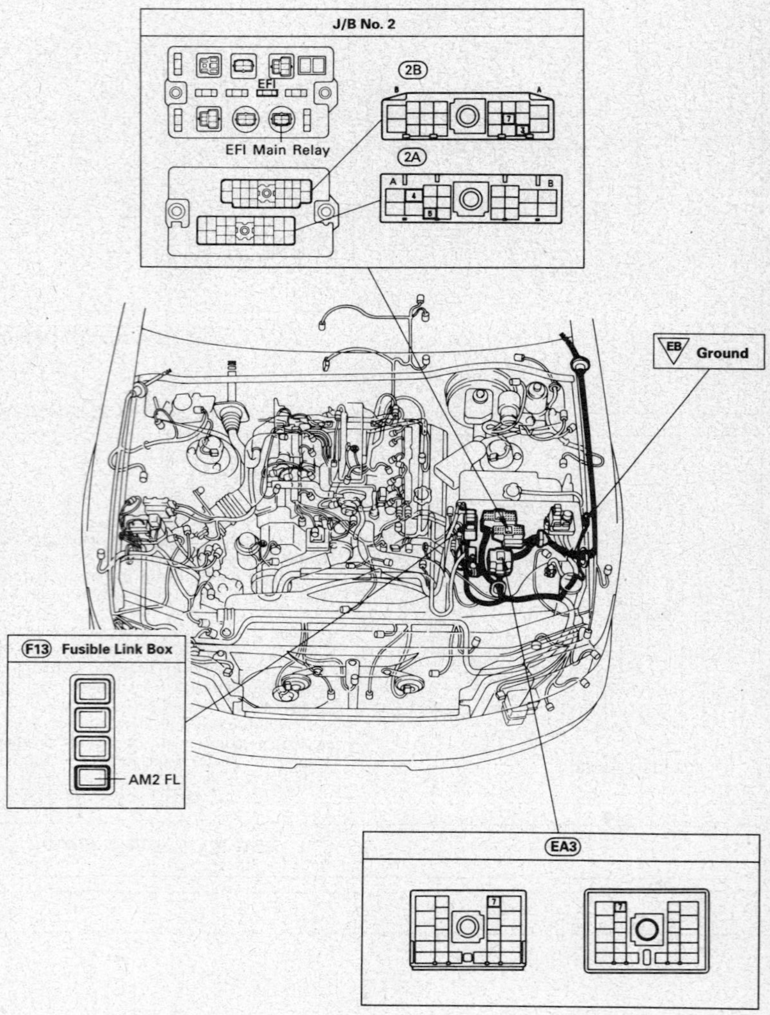

WIRING ROUTING

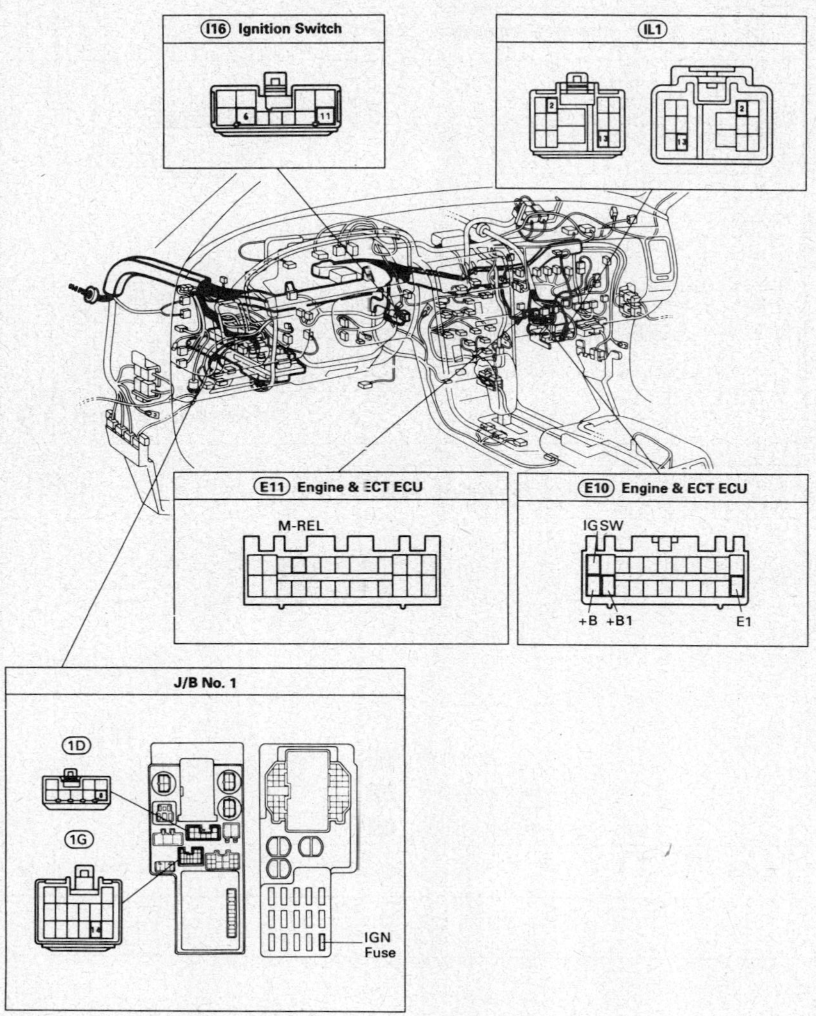

INSPECTION PROCEDURE

1 Check voltage between terminals +B, +B1 and E1 of engine & ECT ECU connector.

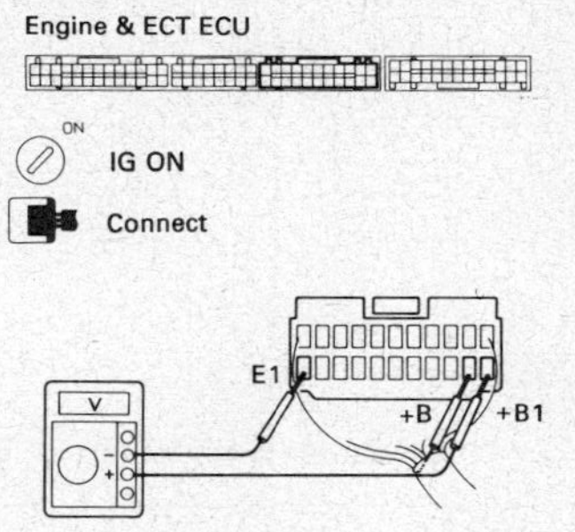

P (1) Remove RH lower pad.
(2) Turn ignition switch on.

C Measure voltage between terminals +B, +B1 and E1 of engine & ECT ECU connector.

OK **Voltage : Battery voltage**

NG

OK ⟩ **Proceed to next circuit inspection shown on matrix chart**

2 Check for open in harness and connector between terminal E1 of engine & ECT ECU and body ground

OK

NG ⟩ **Repair or replace harness or connector.**

3 Check voltage between terminal IG SW of engine & ECT ECU connector and body ground.

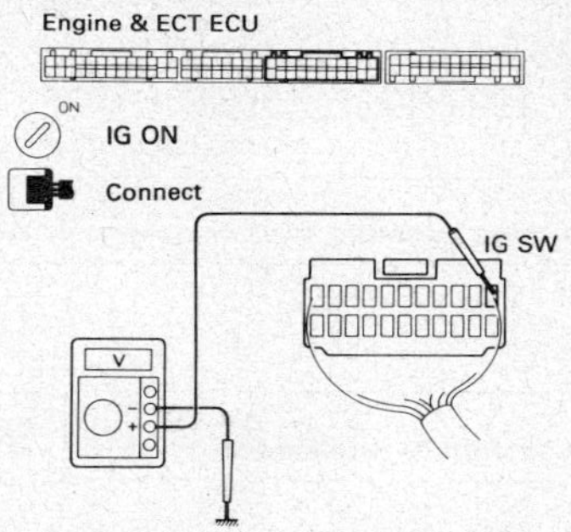

P Turn ignition switch on.

C Measure voltage between terminal IG SW of engine & ECT ECU connector and body ground.

OK **Voltage : Battery voltage**

NG

OK ⟩ **Go to step 6** .

4 Check IGN fuse.

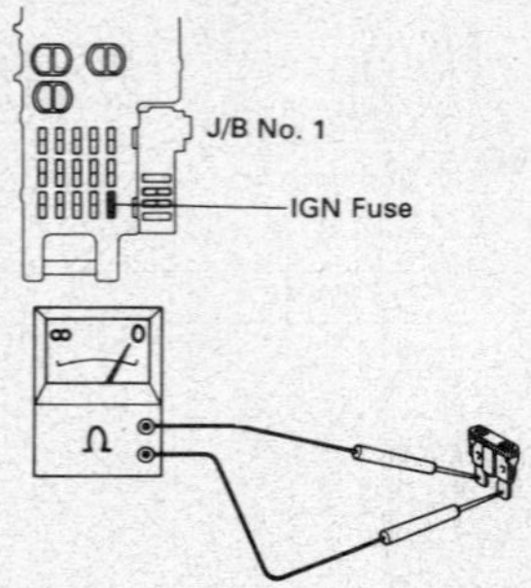

P Remove IGN fuse from J/B No.1.

C Check continuity of IGN fuse.

OK **Continuity**

NG 〉 **Check for short in all the harness and components connected to IGN fuse**

OK ↓

5 Check ignition switch.

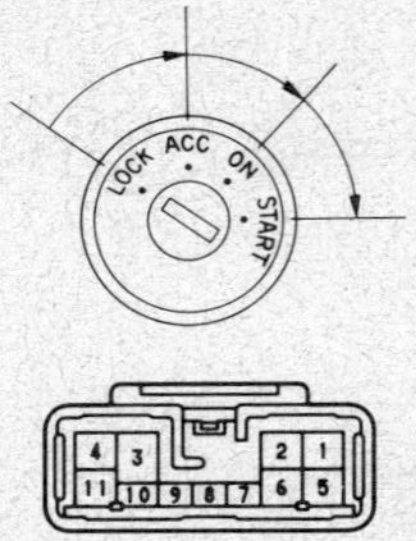

P Remove under cover and finish panel.

C Check continuity between terminals.

OK O—O continuity

Terminal / Switch position	2	3	4	5	6	11
LOCK						
ACC		O—	—O			
ON	O—	—O—	—O		O—	—O
START	O—	—	—O	O—	—O—	—O

NG 〉 **Replace ignition switch.**

OK ↓

Check and repair harness and connector between battery and ignition switch, ignition switch and engine & ECT ECU.

6 Check harness and connector between engine & ECT ECU and EFI main relay

NG 〉 **Repair or replace harness or connector.**

OK ↓

7 Check voltage between terminal M-REL of engine & ECT ECU connector and body ground.

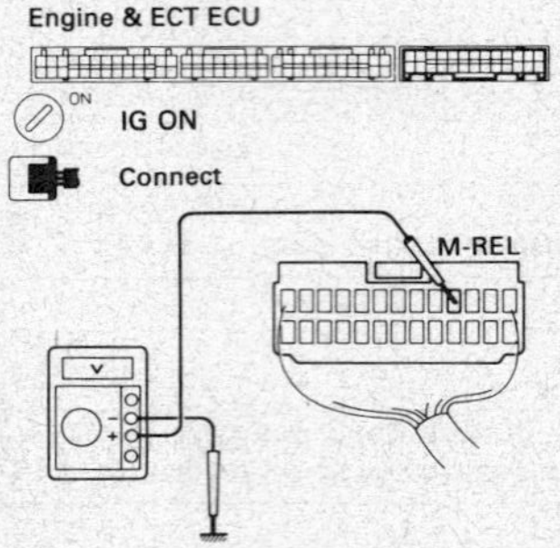

P Turn ignition switch on.

C Measure voltage between terminal M-REL of engine & ECT ECU connector and body ground.

OK **Voltage : Battery voltage**

NG 〉 **Check and replace engine & ECT ECU.**

OK ↓

8 Check EFI main relay.

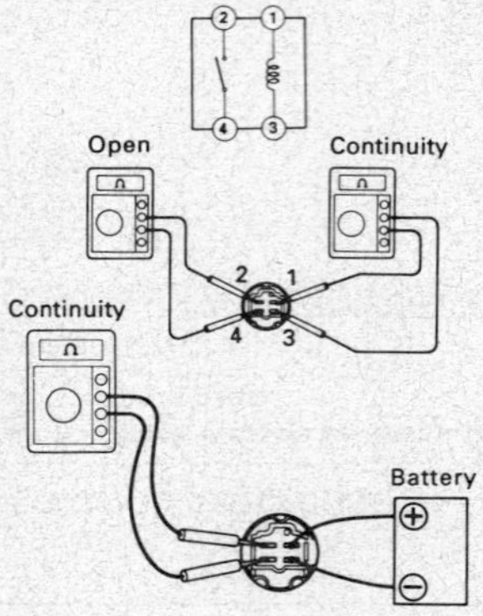

P Remove EFI main relay.

C Check continuity between terminals of EFI main relay shown below.

OK

Terminals 2 and 4	Open
Terminals 1 and 3	Continuity

C
1. Apply battery voltage between terminals 1 and 3.
2. Check continuity between terminals 2 and 4.

OK

Terminals 2 and 4	Continuity

NG 〉 **Replace EFI main relay.**

OK ↓

9 **Check EFI fuse.**

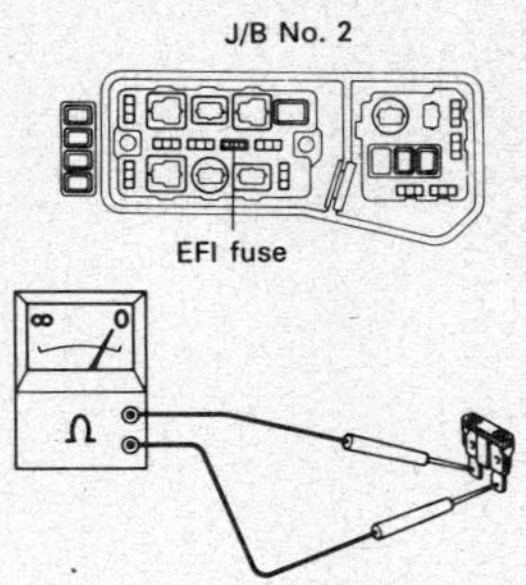

P Remove EFI fuse from J/B No.2.

C Check continuity of EFI fuse.

OK **Continuity**

NG → **Check for short in all the harness and components connected to EFI fuse**

OK ↓

10 **Check harness and connector between battery and EFI main relay, EFI main relay and engine & ECT ECU**

NG → **Repair or relace harness or connector.**

OK ↓

Check and repair harness or connector between EFI main relay and body ground.

Back Up Power Source Circuit

CIRCUIT DESCRIPTION

Battery voltage is supplied to terminal BATT of the ECU even when the ignition switch is off for use by the diagnostic code memory and air-fuel ratio adaptive control value memory, etc.

DIAGNOSTIC CHART

1 Check EFI fuse. — NG → Check for short in all the harness and components connected to EFI fuse

OK ↓

2 Check voltage between terminal BATT of engine & ECT ECU connector and body ground. — OK → Proceed to next circuit inspection shown on matrix chart

NG ↓

Check and repair harness or connector between engine & ECT ECU and EFI fuse, EFI fuse and battery.

WIRING DIAGRAM

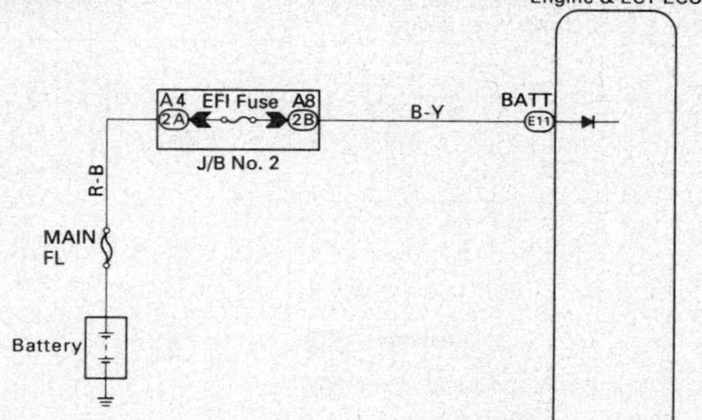

WIRING ROUTING

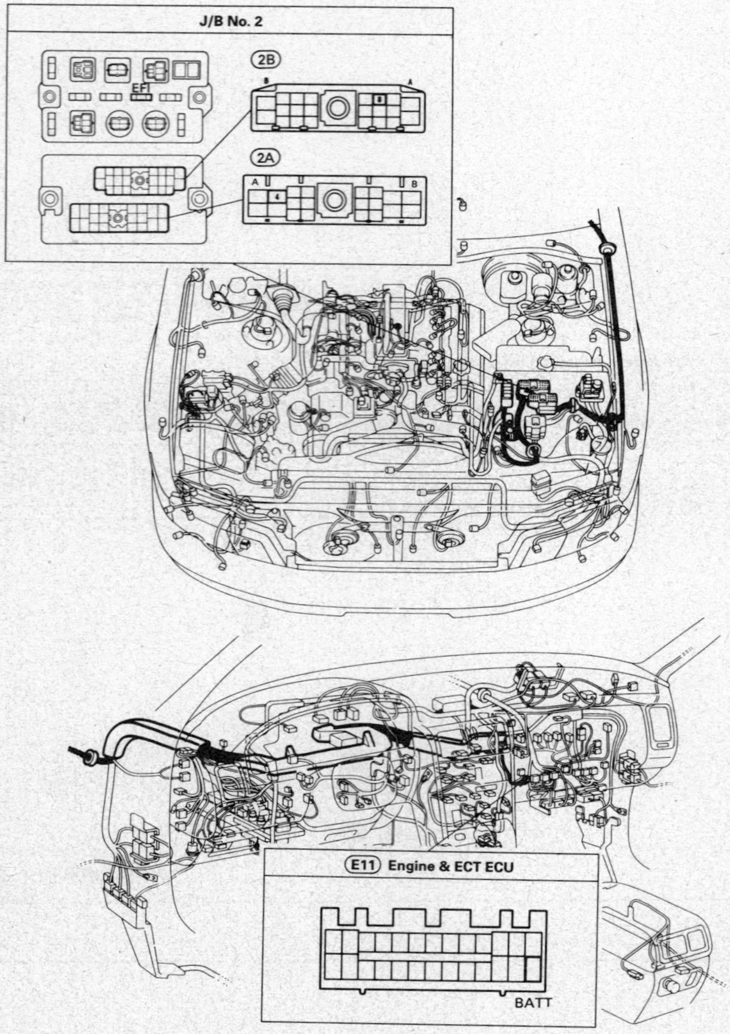

INSPECTION PROCEDURE

1 Check EFI fuse.

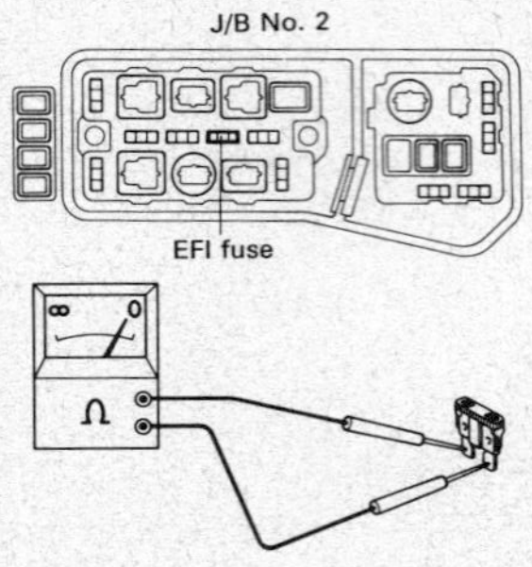

P Remove EFI fuse from J/B No.2.

C Check continuity of EFI fuse.

OK **Continuity**

NG → **Check for short in all the harness and components connected to EFI fuse**

OK ↓

2 Check voltage between terminal BATT of engine & ECT ECU connector and body ground.

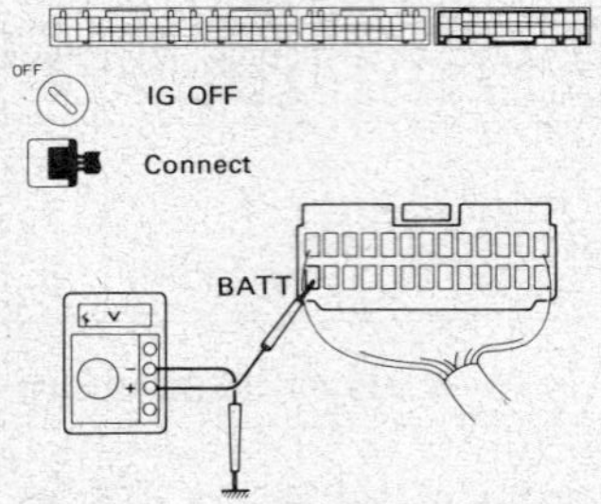

P Remove RH lower pad.

C Measure voltage between terminal BATT of engine & ECT ECU connector and body ground.

OK **Voltage : Battery voltage**

OK → **Proceed to next circuit inspection shown on matrix chart**

NG ↓

Check and repair harness or connector between engine & ECT ECU and EFI fuse, EFI fuse and battery.

Injector Circuit

CIRCUIT DESCRIPTION

The injectors are provided to the intake manifold. They inject fuel into the cylinders based on the signals from the engine & ECT ECU.

DIAGNOSTIC CHART

1 Check voltage between terminal 1 of ignition relay connector and body ground.
- OK → go to 3
- NG → go to 2

2 Check ignition relay.
- NG → Replace ignition relay.
- OK → Check and repair harness and connector between ignition relay and battery.

3 Check for open in harness and connector between terminal E01, E02 of engine & ECT ECU connector and body ground
- NG → Repair or replace harness or connector.
- OK → go to 4

4 Check for open in harness and connector between ignition relay and injector, injector and engine & ECT ECU
- NG → Repair or replace harness or connector.
- OK → go to 5

5 Check injectors.
- NG → Replace injector.
- OK → Proceed to next circuit inspection shown on matrix chart

WIRING DIAGRAM

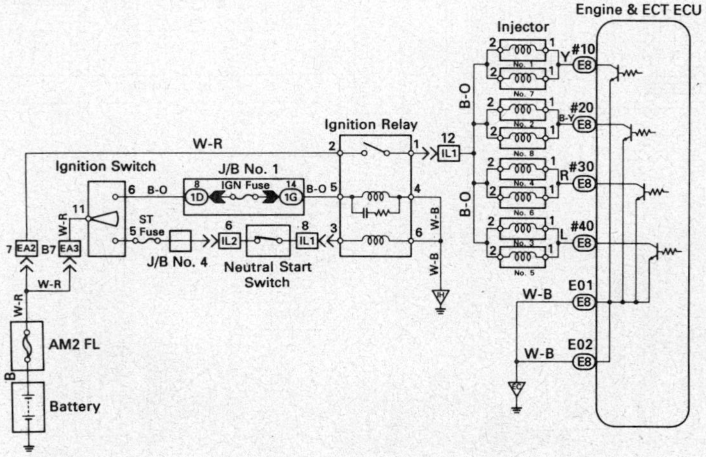

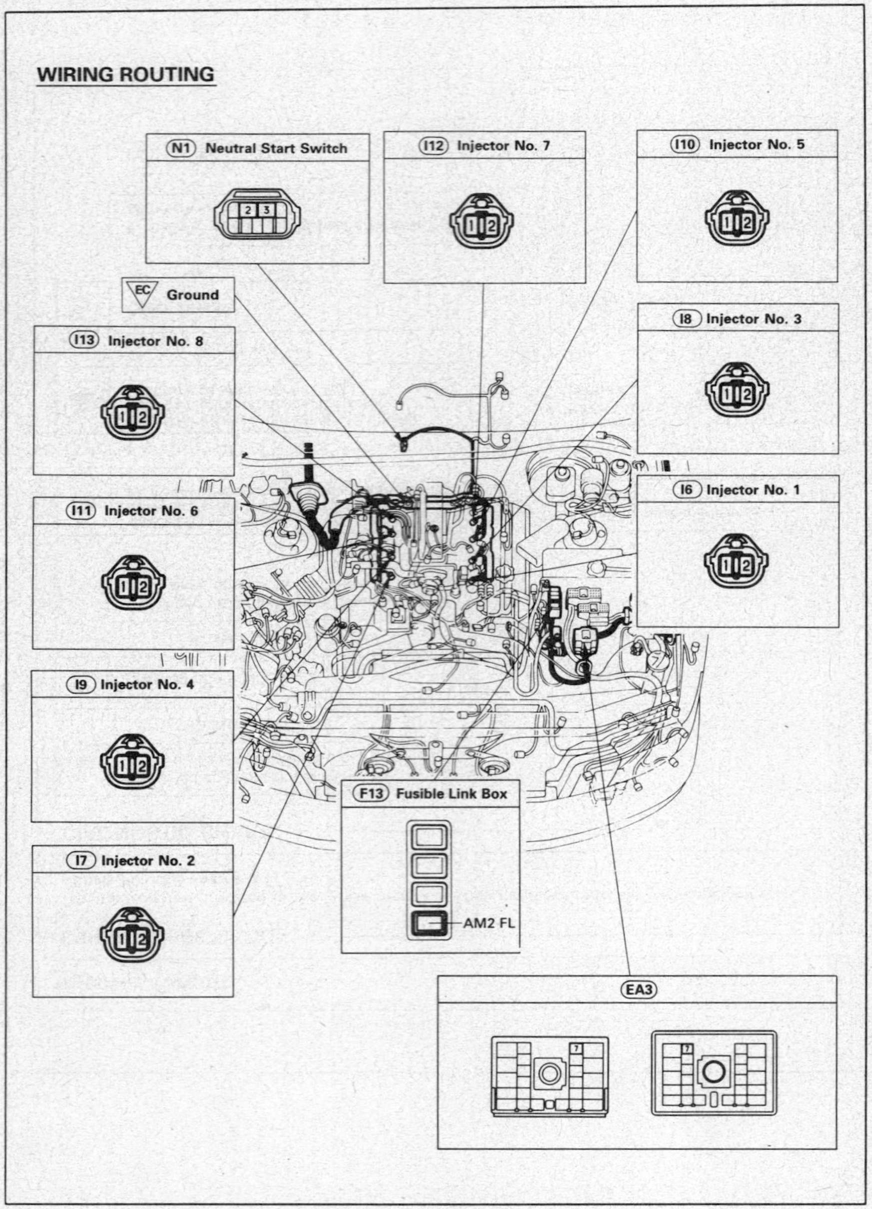
WIRING ROUTING
N1 Neutral Start Switch
I12 Injector No. 7
I10 Injector No. 5
EC Ground
I8 Injector No. 3
I13 Injector No. 8
I6 Injector No. 1
I11 Injector No. 6
I9 Injector No. 4
F13 Fusible Link Box
AM2 FL
I7 Injector No. 2
EA3

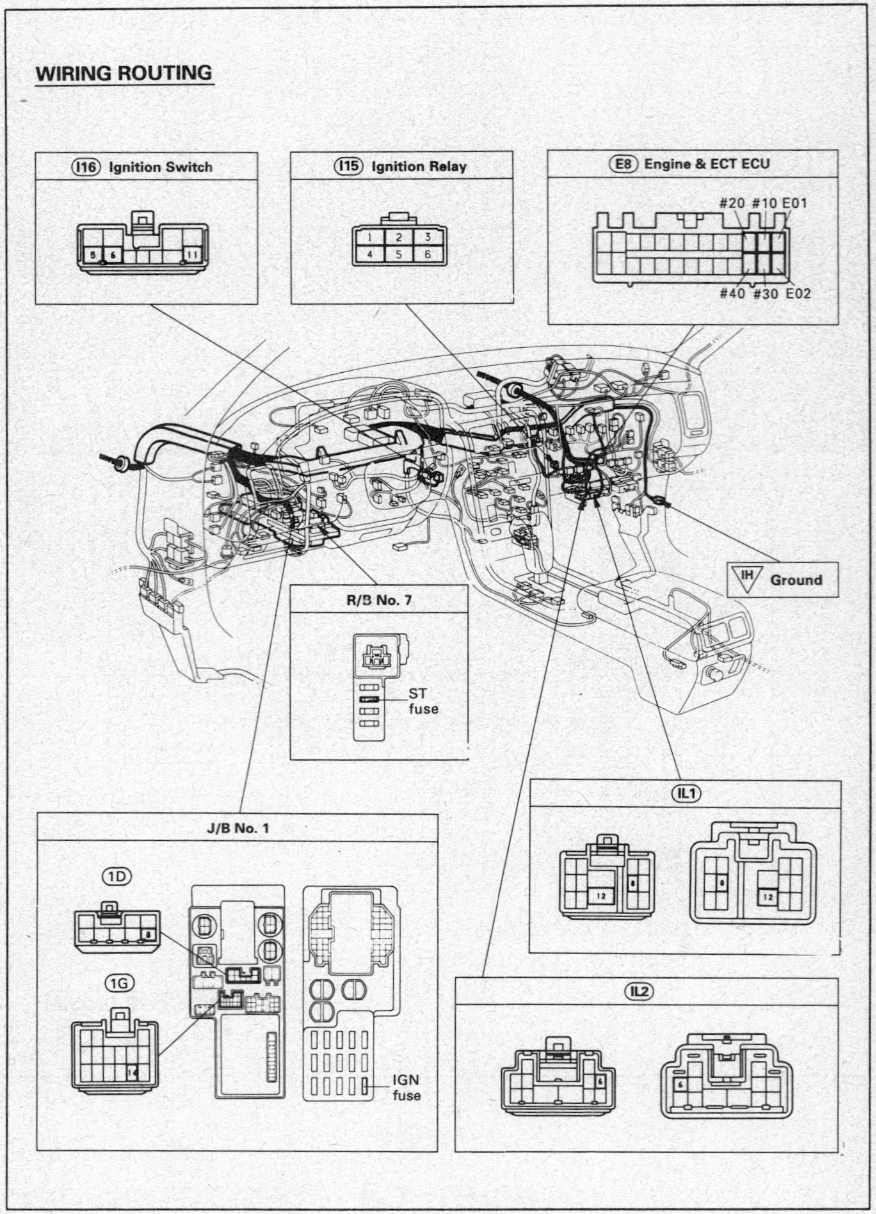
WIRING ROUTING
I16 Ignition Switch
I15 Ignition Relay
E8 Engine & ECT ECU
#20 #10 E01
#40 #30 E02
IH Ground
R/B No. 7
ST fuse
IL1
J/B No. 1
1D
1G
IGN fuse
IL2

INSPECTION PROCEDURE

1 Check voltage between terminal 1 of ignition relay connector and body ground.

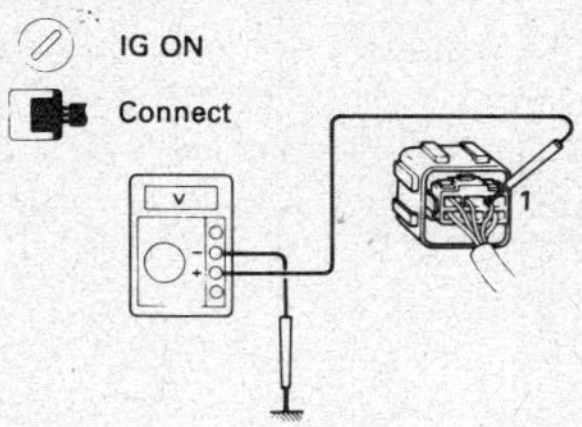

P (1) Remove RH lower pad.
(2) Turn ignition switch on.

C Measure voltage between terminal 1 of ignition relay connector and body ground.

OK **Voltage : Battery voltage**

OK 〉 Go to step **3**.

NG

2 Check ignition relay.

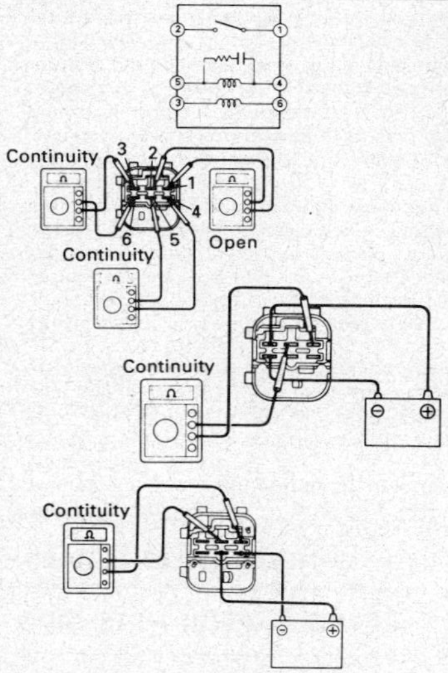

P Remove ignition relay.

C Check continuity between terminals of ignition relay shown below.

OK

Terminals 1 and 2	Open
Terminals 3 and 6	Continuity
Terminals 4 and 5	Continuity

C (1) Apply battery voltage between terminals 3 and 6, 4 and 5.
(2) Check continuity between terminals 1 and 2.

OK

Terminals 1 and 2	Continuity

NG 〉 **Replace ignition relay.**

OK

Check and repair harness and connector between ignition relay and battery.

3 Check for open in harness and connector between terminal E01, E02 of engine & ECT ECU connector and body ground

NG 〉 **Repair or replace harness or connector.**

OK

4 Check for open in harness and connector between ignition relay and injector, injector and engine & ECT ECU

NG 〉 **Repair or replace harness or connector.**

OK

5 Check injectors.

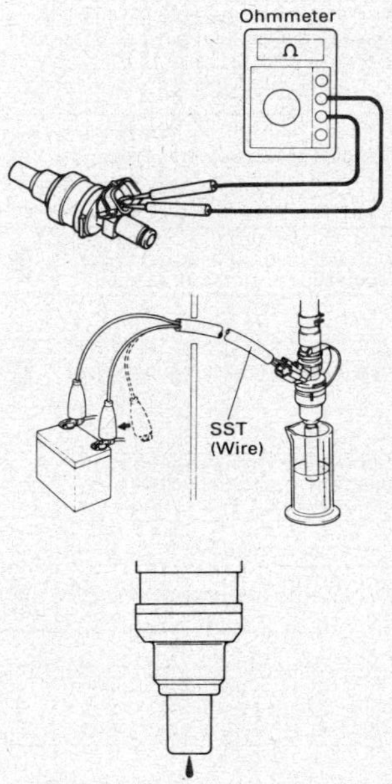

P Disconnect injector connector

C Measure resistance of injector.

OK **Resistance : 13.4 – 14.2 Ω at 20°C (68°F)**

C Check injection volume of injector

OK
- **Injection volume**
  55 – 70 cc/15 sec. (3.4 – 4.3 cu in.)
  Difference between each injector :
  Less than 10 cc (0.6 cu in.)
- **Leakage**
  Fuel drop : One drop or less per minute.

NG 〉 **Replace injector.**

OK

Proceed to next circuit inspection shown on matrix chart

Cold Start Injector Circuit

CIRCUIT DESCRIPTION

The cold start injector is used to maintain the engine startability when it is cold. The injection volume, i,e, the length of time the injector is energized, is controlled by the ECU and the cold start injector time switch.
During a cold start, when the starter turns the contacts in the cold start injector time switch close. Thus current flows to the cold start injector coil, injecting fuel. At the same time, a bimetal in the heat coil is energized and heats up. This soon causes the contacts to open, cutting off the current flow to the injector coil and stopping fuel injection.

The injection duration of the cold start injector is determined by the coolant temperature and the length of time current flows to the heat coil. When the engine is warm, the contacts are opened by the bimetal and the cold start injector does not operate.
When the engine is hard to start and the starter is operated continuously, heat coil (2) heats up the bimetal keeping the contacts open to prevent spark plugs from becoming fouled, which is caused by the cold start injector operation when the open contacts close again.

When the engine is started at a coolant temperature of 22°C (72°F) or lower, the cold start injector operation time is controlled by the cold start injector time switch.
When the coolant temperature is in the normal temperature range 22°C (72°F) or higher, the contacts of the cold start injector time switch are open and the time switch is off, instead, the ECU controls the operating time of the cold start injector.
In this way, the CO and HC levels can be reduced while the engine is being started and the engine startability is maintained. Control by the ECU ends when the coolant temperature reaches 60°C (140°F).

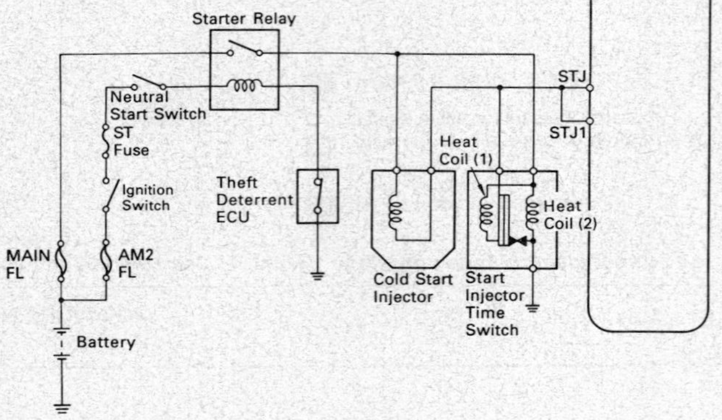

DIAGNOSTIC CHART

1 Check voltage between terminals STJ, STJ1 of engine & ECT ECU connector and body ground.
- OK → go to 3
- NG → go to 2

2 Check resistance of cold start injector.
- NG → Replace cold start injector.
- OK → Check and repair harness and connector between engine & ECT ECU and cold start injector, cold start injector and battery.

3 Check start injector time switch.
- NG → Replace start injector time switch.
- OK ↓

4 Check harness and connector between cold start injector and start injector time switch
- NG → Repair or replace harness or connector.
- OK ↓

5 Check fuel injection of cold start injector.
- NG → Repair or replace cold start injector.
- OK → Proceed to next circuit inspection shown on matrix chart

WIRING DIAGRAM

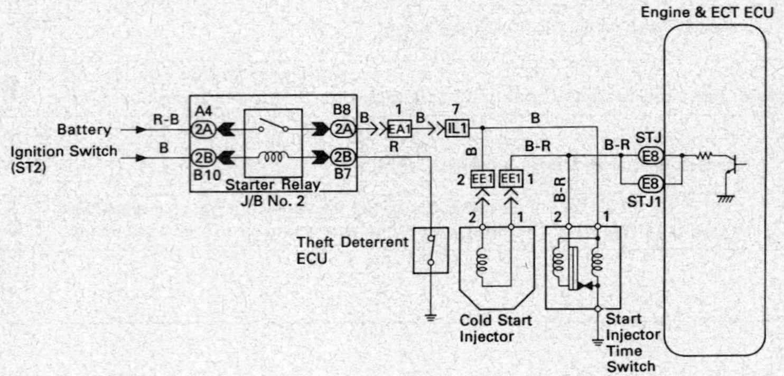

WIRING ROUTING

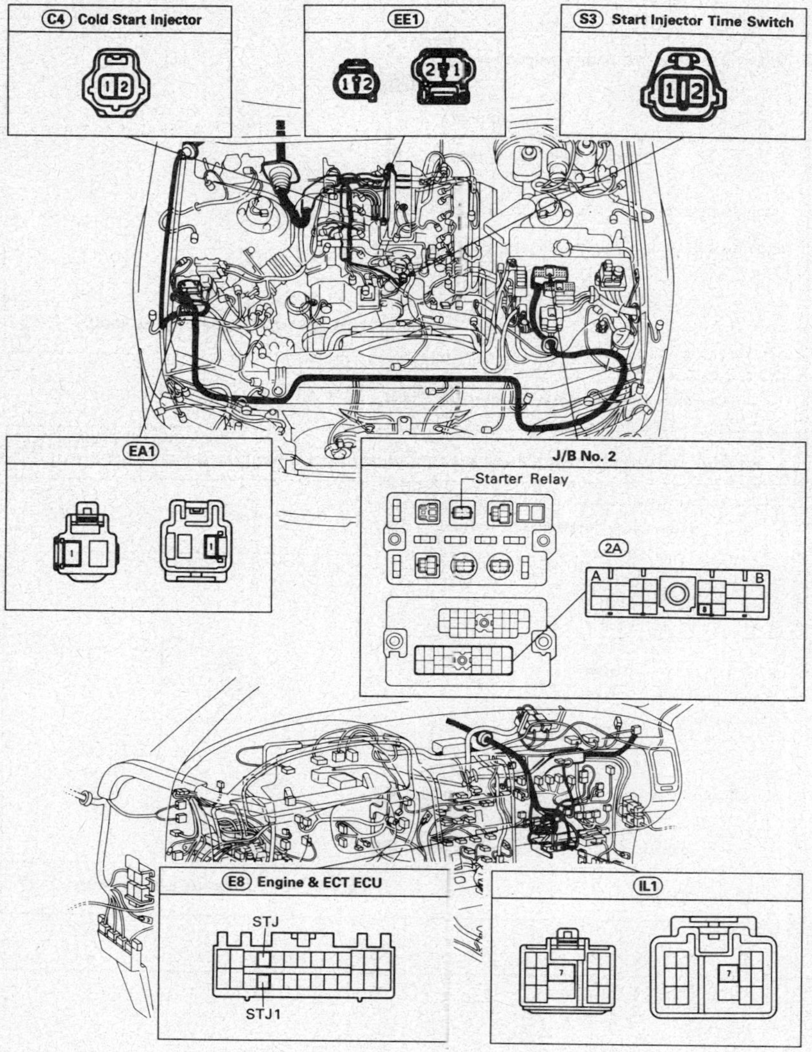

INSPECTION PROCEDURE

1 Check voltage between terminals STJ, STJ1 of engine & ECT ECU connector and body ground.

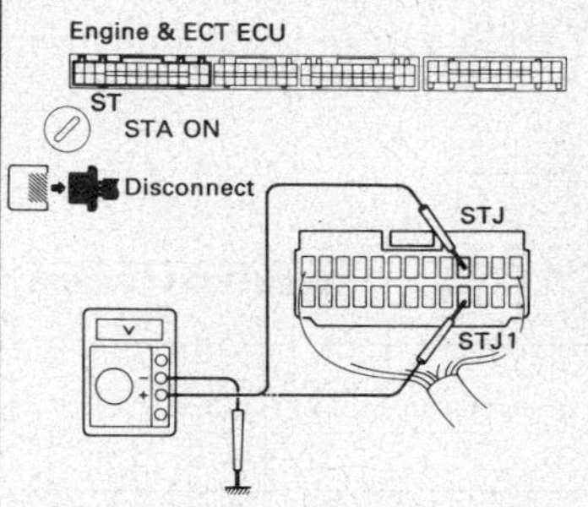

P
(1) Remove RH lower pad.
(2) Disconnect start injector time switch connector.
(3) Disconnect engine & ECT ECU connector.

C Measure voltage between terminals STJ, STJ1 of engine & ECT ECU connector and body ground when ignition switch is turned to STA.

OK **Voltage : Battery voltage**

NG

OK Go to step **3** .

2 Check resistance of cold start injector.

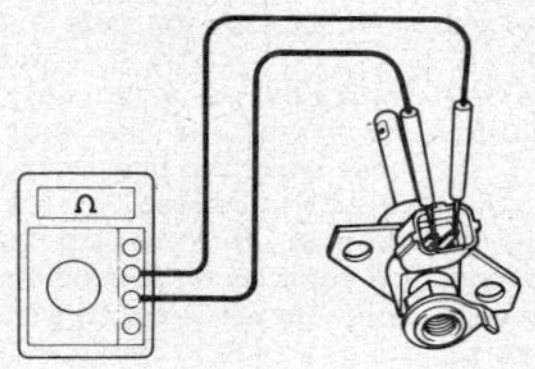

P Remove cold start injector

C Measure resistance between terminals of cold start injector.

OK **Resistance : 2 – 4 Ω at 20°C (68°F)**

OK

NG Replace cold start injector.

Check and repair harness and connector between engine & ECT ECU and cold start injector, cold start injector and battery.

3 **Check start injector time switch.**

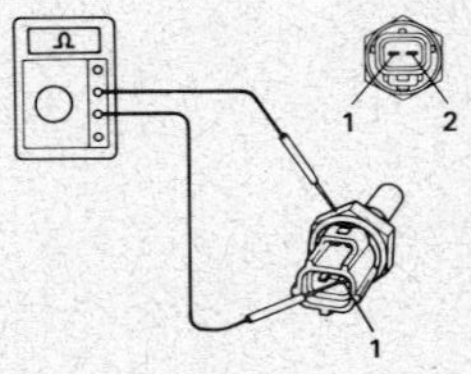

P Remove start injector time switch

C Measure resistance between terminals shown below.

OK

Terminals	Resistance	Coolant temperature
1 – 2	25 – 45 Ω	below 15°C (59°F)
	65 – 85 Ω	above 30°C (86°F)
1-Ground	25 – 85 Ω	—

NG **Replace start injector time switch.**

OK

4 **Check harness and connector between cold start injector and start injector time switch**

NG **Repair or replace harness or connector.**

OK

5 **Check fuel injection of cold start injector.**

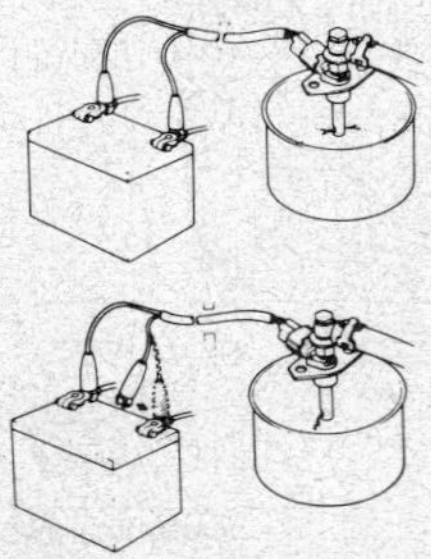

P Remove cold start injector

C (1) Check fuel injection of cold start injector

(2) Check fuel leakage of cold start injector

OK **(1) Fuel is injected normally.**
(2) Fuel leakage is less than one drop per minute.

Notice
Perform this check within the shortest possible time.

NG **Repair or replace cold start injector.**

OK

Proceed to next circuit inspection shown on matrix chart

ISC Valve Circuit

CIRCUIT DESCRIPTION

The ISC valve is provided on the intake air chamber and intake air bypassing the throttle valve is directed to the ISC valve through a hose.
A step moter is built into the ISC valve. It consists of four coils, the magnetic rotor, valve shaft and valve.
When current flows to the coils due to signals from the ECU, the rotor turns and moves the valve shaft forward or backward, changing the clearance between the valve and the valve seat.
In this way the intake air volume bypassing the throttle valve is regulated, controlling the engine speed.
There are 125 possible positions to which the valve can be opened.

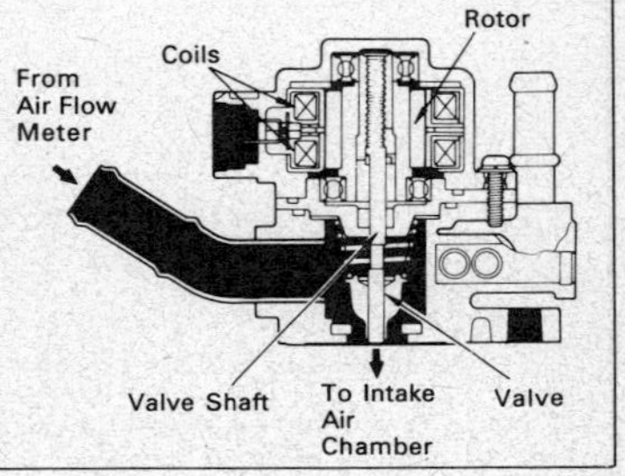

DIAGNOSTIC CHART

1 Check ISC valve. — NG → Replace ISC valve.

↓ OK

2 Check harness and connector between EFI main relay and ISC valve, ISC valve and engine & ECT ECU — NG → Repair or replace harness or connector.

↓ OK

Proceed to next circuit inspection shown on matrix chart

WIRING DIAGRAM

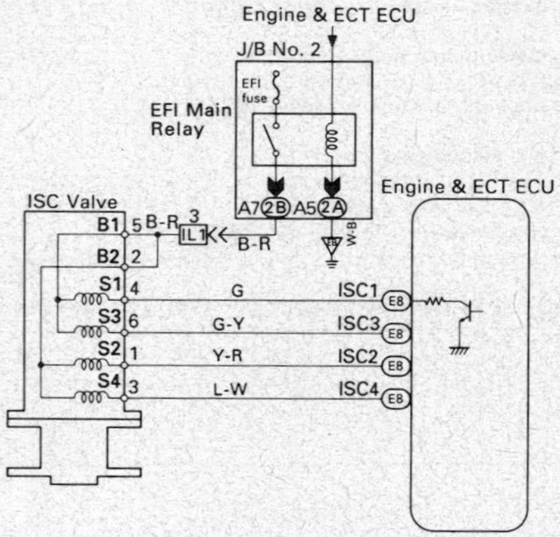

WIRING ROUTING

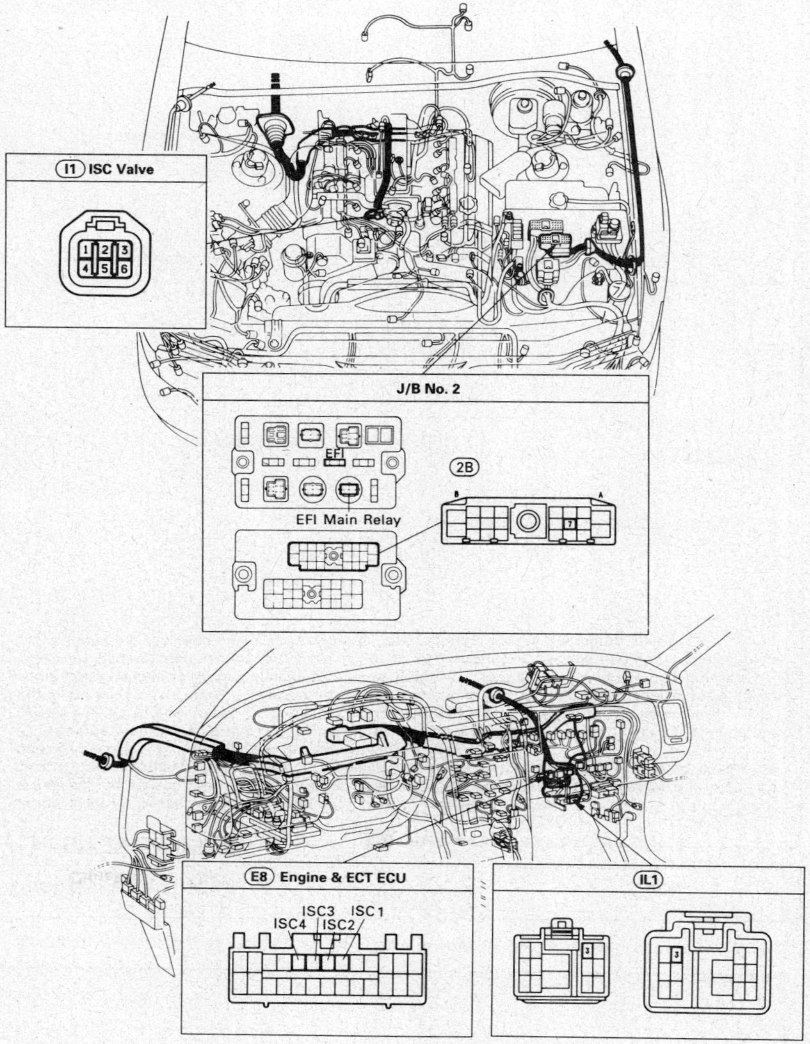

INSPECTION PROCEDURE

1 Check ISC valve.

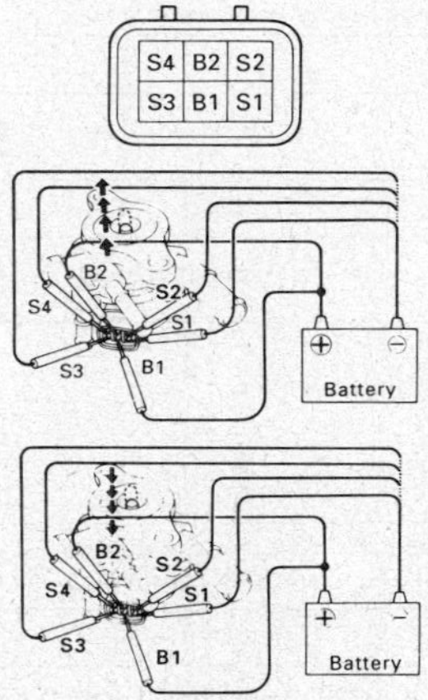

P Disconnect ISC valve connector.

C Measure resistance between terminals shown below.

OK

Terminal	Resistance
B1 – S1	10 Ω – 30 Ω
B1 – S3	10 Ω – 30 Ω
B2 – S2	10 Ω – 30 Ω
B2 – S4	10 Ω – 30 Ω

P Remove ISC Valve.

C (1) Connect the battery positive lead to terminals B1 and B2, and the negative lead to terminals S1-S2-S3-S4 in that order.
(2) Connect the battery positive lead to terminals B1 and B2, and the negative lead to terminals S4-S3-S2-S1 in that order.

OK **(1) The valve moves in the closing direction.**
(2) The valve moves in the opening direction.

NG 〉 **Replace ISC valve.**

OK

2 Check harness and connector between EFI main relay and ISC valve, ISC valve and engine & ECT ECU

NG 〉 **Repair or replace harness or connector**

OK

Proceed to next circuit inspection shown on matrix chart

Fuel System Circuit

CIRCUIT DESCRIPTION

In the diagram below, when the engine is cranked, current flows from terminal ST of the ignition switch to the starter relay coil, the starter relay switches on and current flows to coil L1 of the circuit opening relay. Thus the circuit opening relay switches on, power is supplied to the fuel pump via the fuel pump relay and the fuel pump operates.
At the same time, the engine speed signal NE is input to the ECU. This turns Tr1 in the ECU on and causes current to flow to coil L2 of the circuit opening relay, thus keeping the contacts on.
When the engine is started, Tr2 in the ECU is off, so contact B in the fuel pump control relay closes and battery voltage is applied directly to the fuel pump. The fuel pump speed is thus increased.
When the engine has been started, since Tr2 goes on, current flows to the fuel pump control relay coil, contact A closes and power is supplied to the fuel pump via the fuel pump resistor.

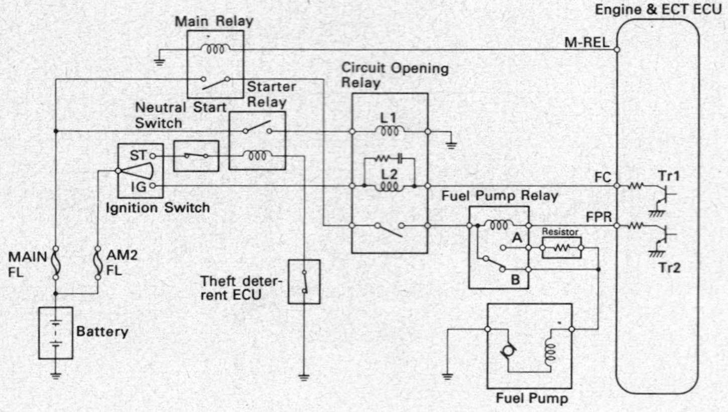

DIAGNOSTIC CHART

1 Check fuel pressure. — OK → Proceed to next circuit inspection shown on matrix chart

NG ↓

① Fuel pressure is zero while cranking.	Go to step **2**
② Fuel pressure gradually drops after engine is started.	Go to step **3**
③ Fuel pressure is above 3.1 kg/cm² (44 psi, 304 kPa)	Go to step **9**
④ Fuel pressure is below 2.7 kg/cm² (38 psi, 265 kPa) when engine is quickly raced to 5000rpm.	Go to srep **10**
⑤ Fuel pressure is above 2.4 kg/cm² (34 psi, 235 kPa) when engine is idling.	Go to step **11**

2 Proceed to "CHECK FUEL PUMP OPERATION" section and check and replace trouble parts.

3 Check IGN fuse (Check operation of discharge warning light). — NG → Proceed to inspection of discharge warning light circuit

OK ↓

4 Check voltage between terminal FC of engine & ECT ECU and body ground. — NG Type I (Constantly 0V) → Go to step **7**; NG Type II (Constantly battery voltage) → Check and replace engine & ECT ECU.

OK ↓

5 Check fuel pump resistor. — NG → Replace fuel pump resistor.

OK ↓

6 Check fuel pump relay. — NG → Replace fuel pump relay.

OK ↓

Check and repair harness and connector between fuel pump relay and fuel pump resistor, fuel pump resistor and fuel pump.

DIAGNOSTIC CHART

From step **4**

7 Check voltage between terminal 5 of circuit opening relay connector and body ground.
- OK → Check and repair harness or connector between circuit opening relay and engine & ECT ECU.
- NG ↓

8 Check circuit opening relay.
- NG → Replace circuit opening relay.
- OK ↓

Check and repair harness or connector between IGN fuse and circuit opening relay.

9 Check fuel return line for clogging.
- NG → Repair or replace fuel return line.
- OK ↓

Replace fuel pressure regulator.

10 Check fuel pressure when flexible hose of fuel return line is pinched off.
- OK → Replace fuel pressure regulator.
- NG ↓

Replace fuel pump.

11 Check vacuum hose connected to pressure regulator.
- NG → Repair or replace vacuum hose.
- OK ↓

12 Check fuel pressure control VSV circuit
- NG → Replace trouble parts.
- OK ↓

13 Check fuel return line for clogging.
- NG → Repair or replace fuel return line.
- OK ↓

Replace fuel pressure regulator.

WIRING DIAGRAM

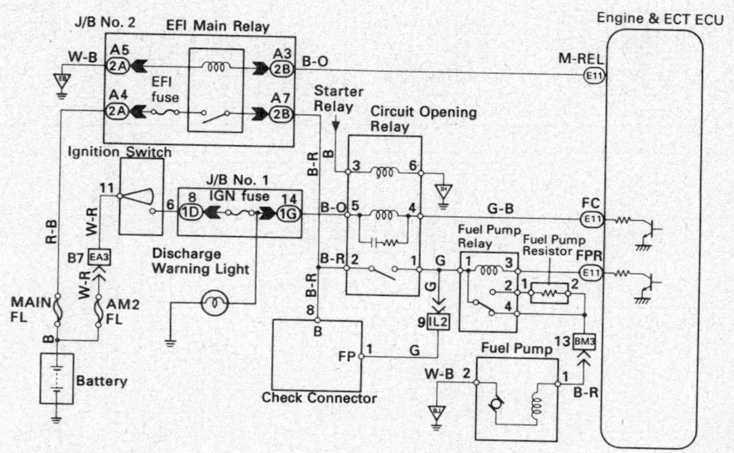

WIRING ROUTING

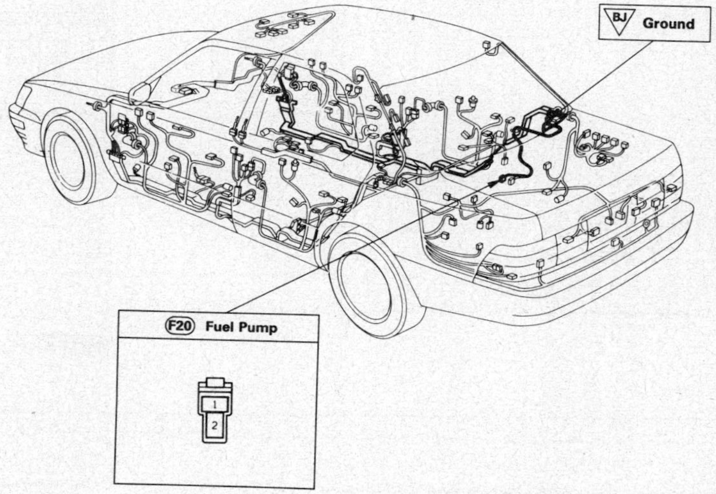

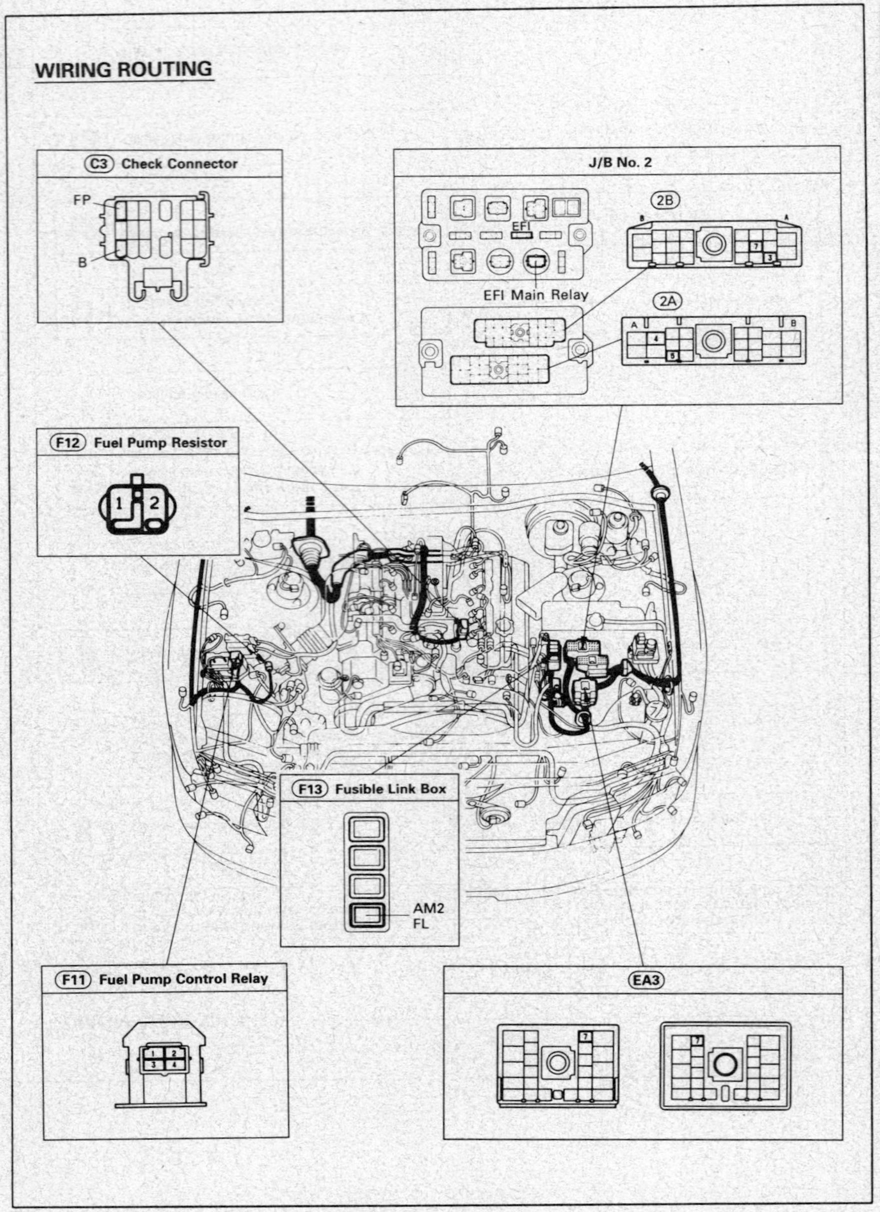
WIRING ROUTING
C3 Check Connector
FP
B
J/B No. 2
EFI
EFI Main Relay
2B
2A
F12 Fuel Pump Resistor
1
2
F13 Fusible Link Box
AM2
FL
F11 Fuel Pump Control Relay
EA3

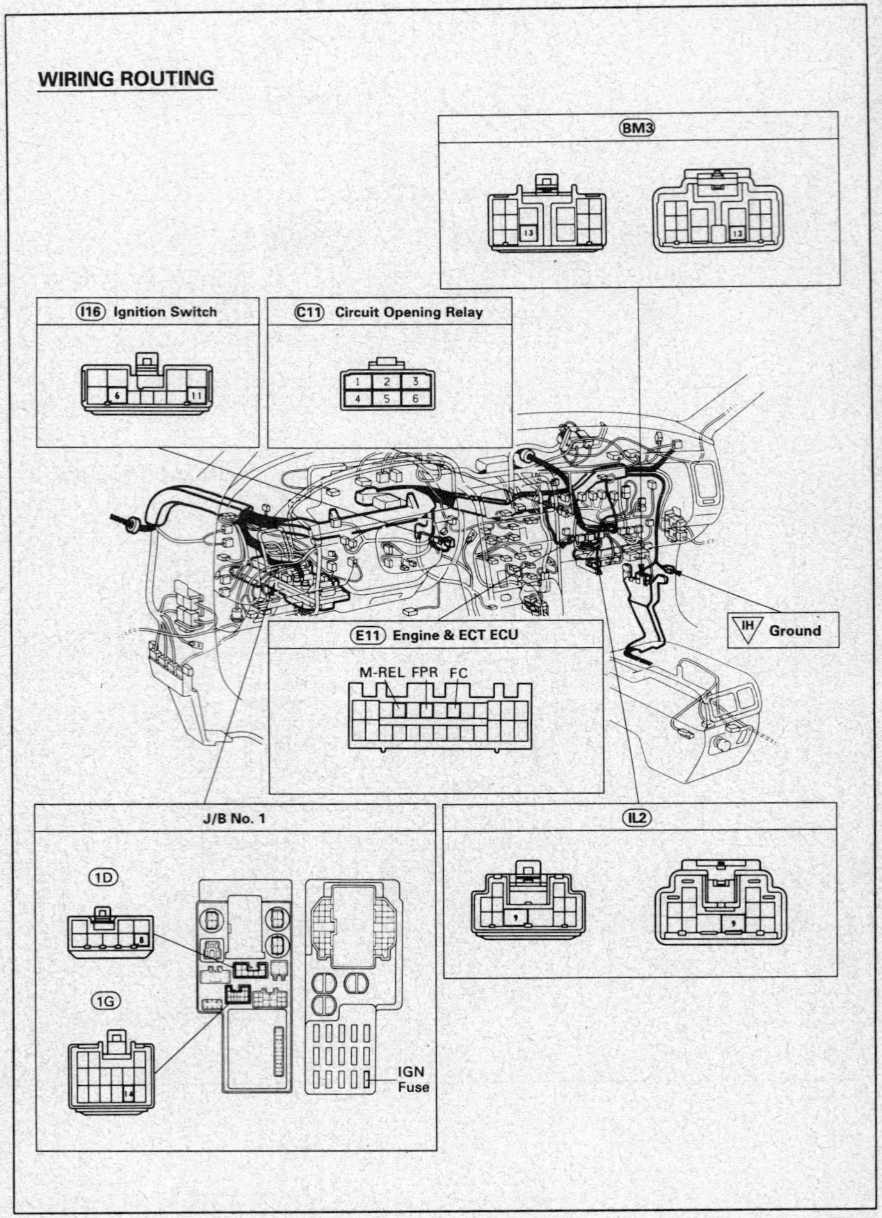
WIRING ROUTING
BM3
I16 Ignition Switch
C11 Circuit Opening Relay
1 2 3
4 5 6
IH Ground
E11 Engine & ECT ECU
M-REL FPR FC
J/B No. 1
1D
1G
IGN
Fuse
IL2

INSPECTION PROCEDURE

1 Check fuel pressure.

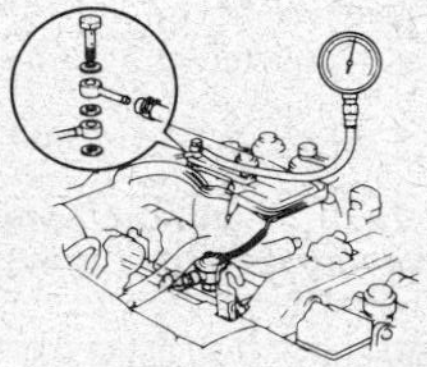

P (a) Enough fuel in tank.
(b) Hook up the fuel pressure gauge (SST 09268-45012)

C Check fuel pressure under the following conditions :
(1) During engine cranking.
(2) When engine is quickly raced from idle to 5,000 rpm.
(3) When engine is idling.

OK **(1), (2) Fuel pressure : 2.7 – 3.1 kg/cm² (38 – 44 psi, 265 – 304 kPa)**
(3) Fuel pressure : 2.0 – 2.4 kg/cm² (28 – 34 psi, 196 – 235 kPa)

Hint When checking the fuel pressure during idling, perform the check after idling has occurred for at least 2 minutes.

OK → **Proceed to next circuit inspection shown on matrix chart**

NG ↓

①	Fuel pressure is zero while cranking.	Go to step 2 .
②	Fuel pressure gradually drops after engine is started.	Go to step 3 .
③	Fuel pressure is above 3.1 kg/cm² (44 psi, 304 kPa)	Go to step 9 .
④	Fuel pressure is below 2.7 kg/cm² (38 psi, 265 kPa) when engine is quickly raced to 5000rpm.	Go to srep 10 .
⑤	Fuel pressure is above 2.4 kg/cm² (34 psi, 235 kPa) when engine is idling.	Go to step 11 .

2 Proceed to "CHECK FUEL PUMP OPERATION" and check and replace trouble parts.

3 Check IGN fuse (Check operation of discharge warning light).

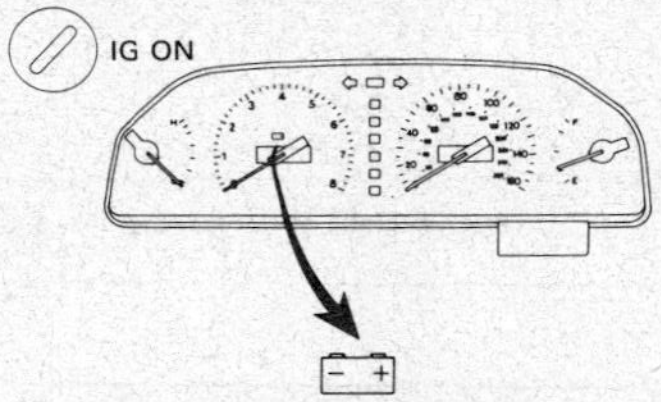

C Check operation of discharge warning light when ignition switch is turned on.

OK **Discharge warning light comes on.**

NG → **Proceed to inspection of discharge warning light circuit**

OK ↓

4 Check voltage between terminal FC of engine & ECT ECU connector and body ground.

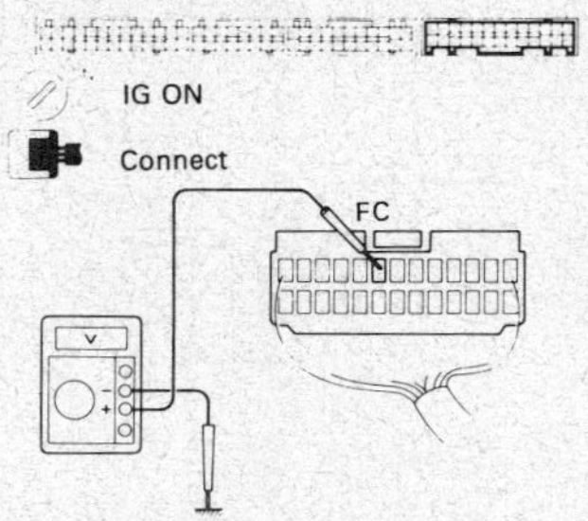

P (1) Turn ignition switch on.
(2) Remove RH lower pad.

C Measure voltage between terminal FC of engine & ECT ECU connector and body ground, when the engine is stopped (with ignition switch on) and when running (or cranking).

Result

Engine condition	Stopped (IG; ON)	Running (or cranking)	Result
Voltage	Battery voltage	Below 1V	OK
	0V	0V	NG Type I
	Battery voltage	Battery voltage	NG Type II

NG Type I → **Go to step 7 .**

NG Type II → **Check and replace engine & ECT ECU.**

OK ↓

5 Check fuel pump resistor.

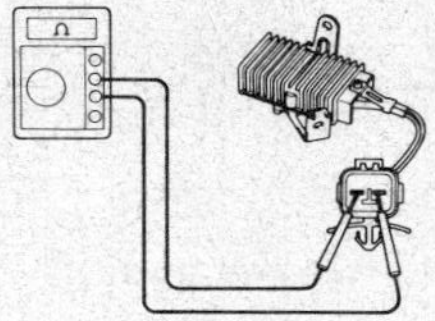

P Disconnect the fuel pump resistor connector.

C Measure resistance between terminals.

OK **Resistance : 0.67– 0.79 Ω**

NG → **Replace fuel pump resistor.**

OK ↓

Go to step 6 .

6 Check fuel pump relay.

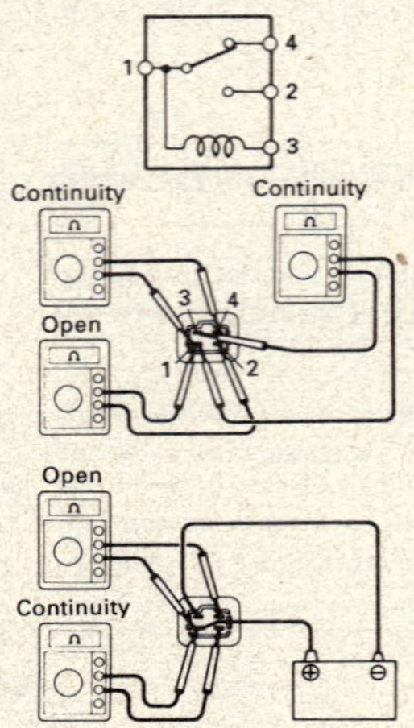

P Remove fuel pump relay.

C Check continuity between terminals of fuel pump relay shown below.

OK

Terminals 1 and 2	Open
Terminals 1 and 3	Continuity
Terminals 1 and 4	Continuity

C (1) Apply battery voltage between terminals 1 and 3.
(2) Check continuity between terminals shown below.

OK

Terminals 1 and 2	Continuity
Terminals 1 and 4	Open

NG Replace fuel pump relay.

OK

Check and repair harness and connectors between fuel pump relay and fuel pump resistor, fuel pump resistor and fuel pump.

7 Check voltage between terminal 5 of circuit opening relay connector and body ground.

IG ON

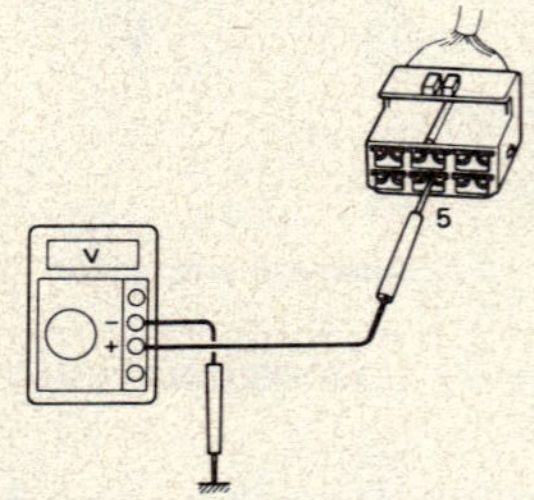

P (1) Remove circuit opening relay.
(2) Disconnect circuit opening relay connector.
(3) Turn Ignition switch on.

C Measure voltage between terminal 5 of circuit opening relay and body ground.

OK **Voltage: Battery voltage**

NG

OK Check and repair harness or connector between circuit opening relay and engine & ECT ECU.

8 Check circuit opening relay.

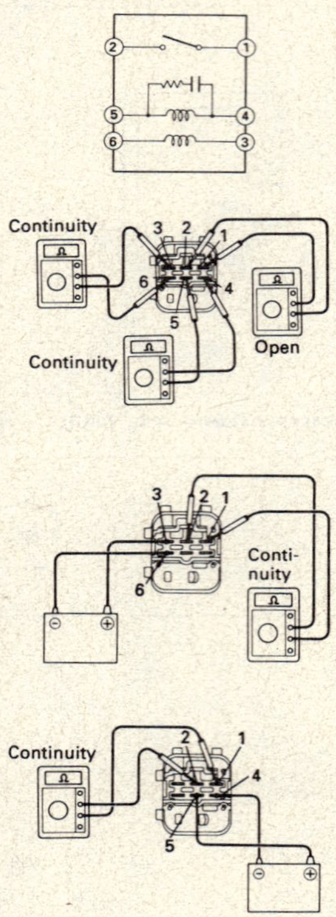

P Remove circuit opening relay.

C Check continuity between terminals of circuit opening relay shown below.

OK

Terminals 1 and 2	Open
Terminals 3 and 6	Continuity
Terminals 4 and 5	Continuity

C (1) Apply battery voltage between terminals 3 and 6, 4 and 5.
(2) Check continuity between terminals 1 and 2.

OK

Terminals 1 and 2	Continuity

NG Replace circuit opening relay.

OK

Check and repair harness or connector between IGN fuse and circuit opening relay.

9 Check fuel return line for clogging.

NG → Repair or replace fuel return line.

OK ↓

Replace fuel pressure regulator.

10 Check fuel pressure when flexible hose of fuel return line is pinched off.

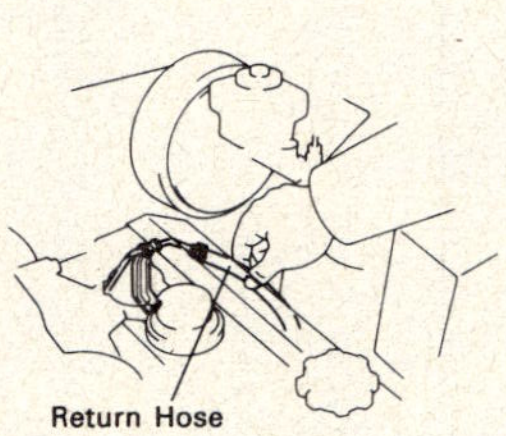

C Check fuel pressure when flexible hose of fuel return line is pinched off with the engine cranking or started.

OK **Fuel pressure is above 2.7 kg/cm² (38 psi, 265 kPa).**

OK → Replace fuel pressure regulator.

NG ↓

Replace fuel pump.

11 Check vacuum hose connected to pressure regulator.

NG → Repair or replace vacuum hose.

OK ↓

12 Check fuel pressure control VSV circuit

NG → Replace trouble parts.

OK ↓

13 Check fuel return line for clogging.

NG → Repair or replace fuel return line.

OK ↓

Replace fuel pressure regulator.

Fuel Pressure Control VSV Circuit

CIRCUIT DESCRIPTION

The ECU turns on a VSV (Vacuum Switching Valve) to draw the air into the diaphragm chamber of the pressure regulator if it detects that the temperature of the coolant is too high during engine starting.
The air drawn into the chamber increases the fuel pressure to prevent fuel vapor lock at high engine temperatures in order to help the engine start when it is warm.
Fuel pressure control ends approx. 100 secs. after the engine is started.

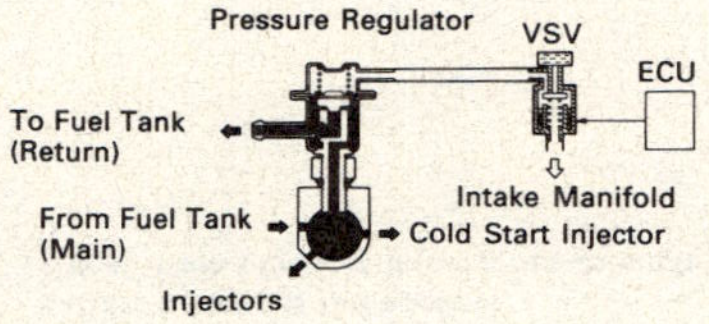

DIAGNOSTIC CHART

1 Check fuel pressure control VSV.
NG → Replace fuel pressure control VSV.

OK ↓

2 Check voltage between terminal PR of engine & ECT ECU connector and body ground.
OK → Proceed to next circuit inspection shown on matrix chart

NG ↓

3 Check harness and connector between engine & ECT ECU and VSV, VSV and EFI main relay
NG → Repair or replace harness or connector.

OK ↓

Check and replace engine & ECT ECU.

WIRING DIAGRAM

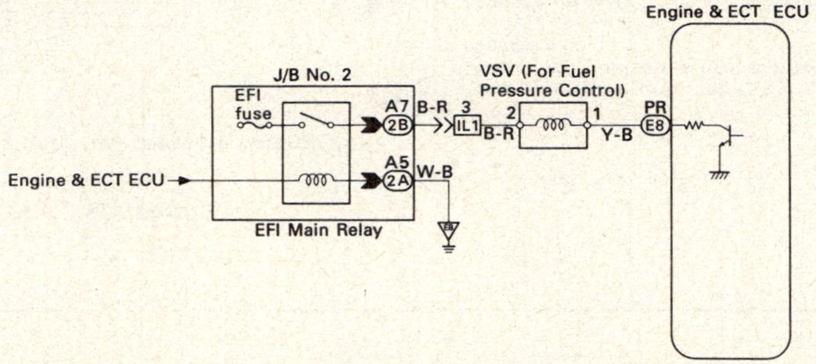

WIRING ROUTING

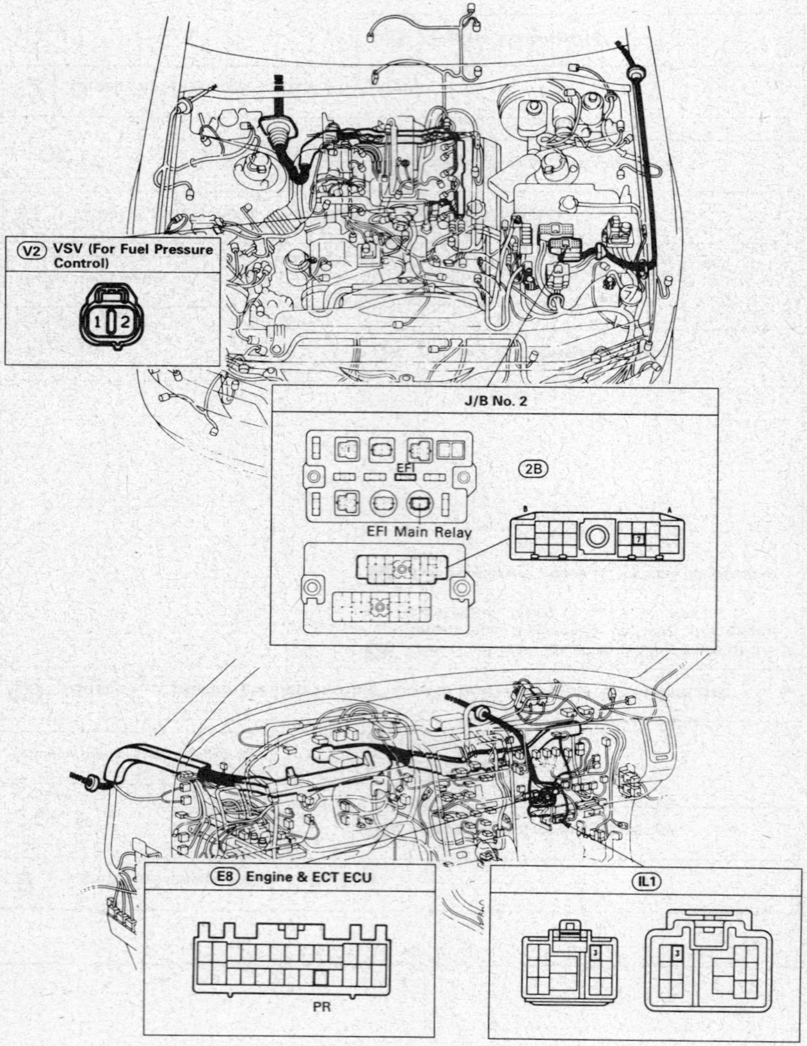

INSPECTION PROCEDURE

1 Check fuel pressure control VSV.

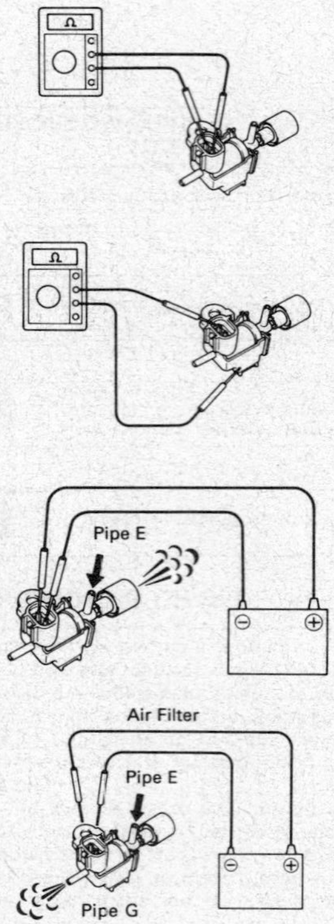

P (1) Remove fuel pressure control VSV.
(2) Disconnect fuel pressure control VSV connector.

C (1) Measure resistance between terminals.
(2) Measure resistance between each terminal and the body.

OK **(1) Resistance : 30 – 50 Ω at 20°C (68°F)**
(2) Open

C Check operation of fuel pressure control VSV when battery voltage is applied to the terminals of fuel pressure control VSV connector or not.

OK **Battery voltage is applied :**
The air from pipe E is flowing out through the air filter.

Battery voltage is not applied :
The air from pipe E is flowing out through pipe G.

OK — Go to step **2**.

NG — **Replace fuel pressure control VSV.**

2 Check voltage between terminal PR of engine & ECT ECU connector and body ground.

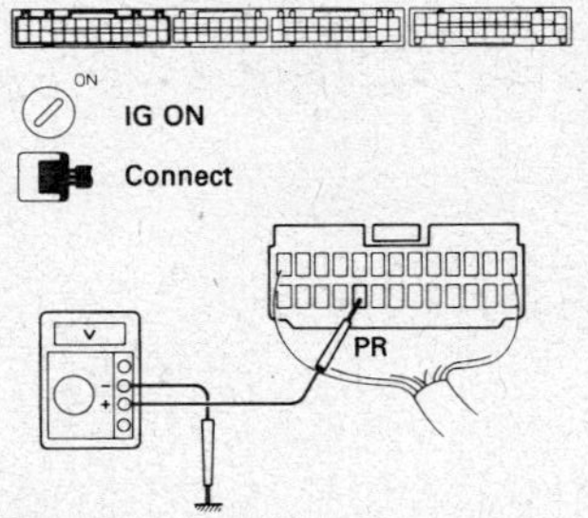

P (1) Remove RH lower pad.
(2) Turn ignition switch on.

C Measure voltage between terminal PR of engine & ECT ECU connector and body ground.

OK **Voltage : Battery voltage**

OK — **Proceed to next circuit inspection shown on matrix chart**

NG ↓

3 Check harness and connector between engine & ECT ECU and VSV, VSV and EFI main relay

NG — **Repair or replace harness or connector.**

OK ↓

Check and replace engine & ECT ECU.

TE1, TE2, Terminal Circuit

CIRCUIT DESCRIPTION

Terminal TE1 is located in the check connector in the engine compartment and terminals TE1 and TE2 are located in the TDCL in the cabin. When these terminals are connected with the E1 terminal, diagnostic codes in normal mode or test mode can be read from the "CHECK" engine warning light on the combination meter.

DIAGNOSTIC CHART

1	Check voltage between terminals TE1, TE2 and E1 of check connector, TDCL.	OK	Check and replace engine & ECT ECU.
	↓ NG		
2	Check continuity between terminal E1 of check connector, TDCL and body ground.	NG	Repair or replace harness or connector.
	↓ OK		
3	Check harness and connector between engine & ECT ECU and check connector, TDCL	NG	Repair or replace harness or connector.
	↓ OK		
	Check and replace engine & ECT ECU.		

WIRING DIAGRAM

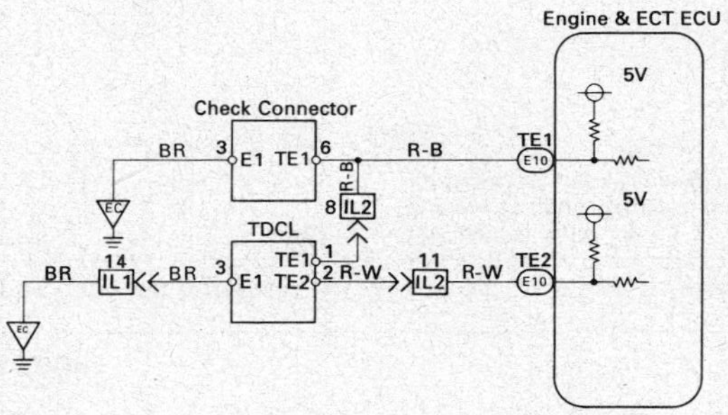

WIRING ROUTING

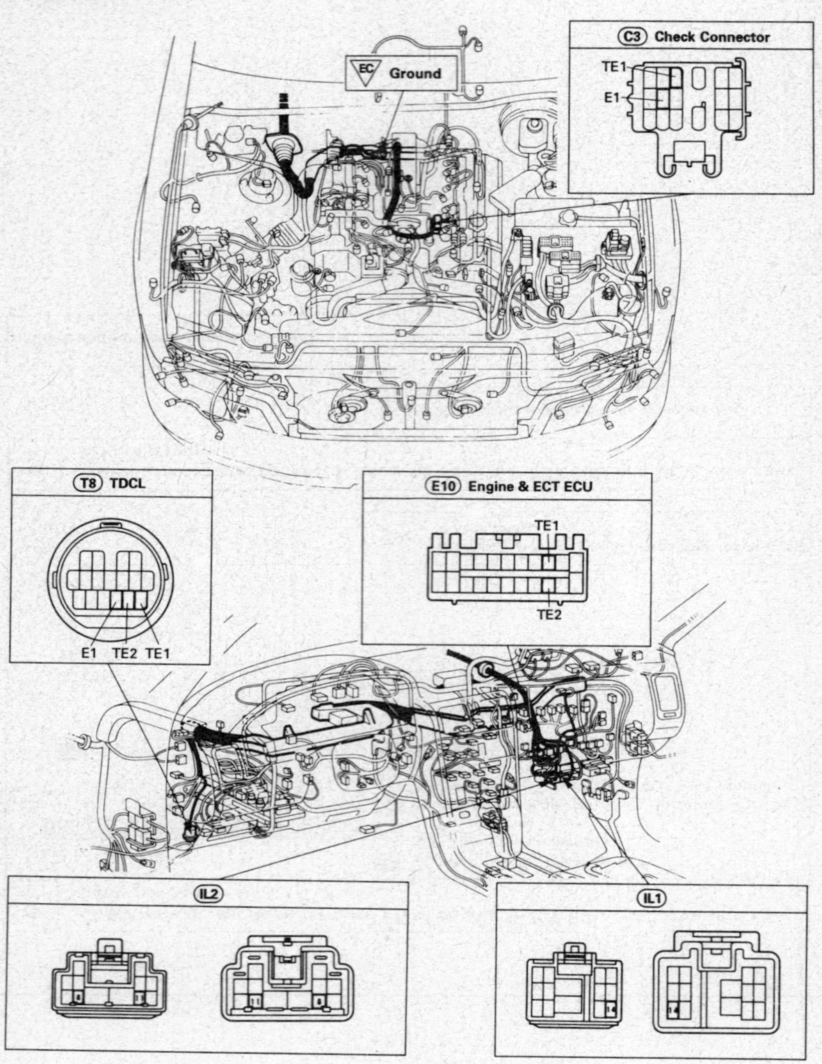

INSPECTION PROCEDURE

1 **Check voltage between terminals TE1, TE2 and E1 of check connector, TDCL.**

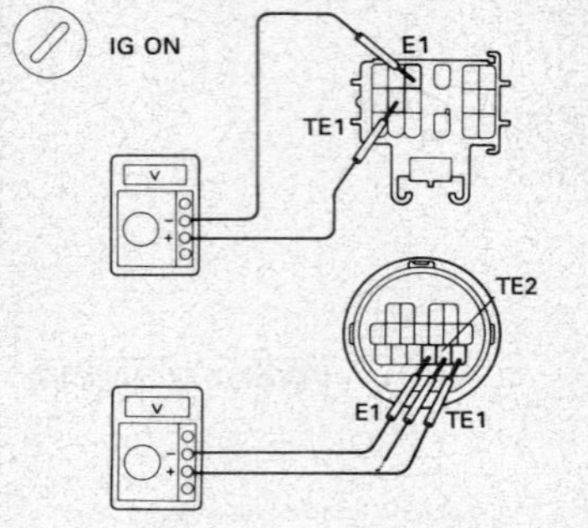

C (1) Turn ignition switch on.
(2) Measure voltage between terminals TE1, TE2 and E1 of check connector, TDCL.

OK **Voltage: 4 – 6 V**

OK → **Check and replace engine & ECT ECU.**

NG ↓

2 **Check continuity between terminal E1 of check connector, TDCL and body ground.**

NG → **Repair or replace harness or connector.**

OK ↓

3 **Check harness and connector between engine & ECT ECU and check connector, TDCL**

NG → **Repair or replace harness or connector.**

OK ↓

Check and replace engine & ECT ECU.

Component Replacement

NOTE: Disconnecting the battery will erase any stored trouble codes and will erase any accessory memories (i.e. radio memory). Read any trouble codes before disconnecting the battery and obtain any radio codes necessary for reprogramming. On air bag equipped vehicles do not start any work on the vehicle until a minimum of 30 seconds has passed from the time the battery was disconnected to prevent accidental airbag deployment.

FUEL PUMP

CAUTION

Do not smoke or have open flame or sparks near during this procedure. Fire and explosion may result in personal injury.

Removal and Installation

ES250

1. Disconnect the negative battery cable.
2. Raise and safely support vehicle. Drain the fuel into a safe container.
3. Remove the fuel tank lines and fill tube. Support the tank and remove the retaining straps. Lower tank partially and disconnect the fuel pump and sender electrical leads. Remove tank.
4. Remove the retaining screws which secure the pump access plate to the fuel tank. Withdraw the plate, gasket and pump assembly.
5. Disconnect the leads and hoses from the fuel pump.
6. Install pump assembly with new gasket and fuel filter.
7. Install the fuel tank lines and leads. Replace the tank and tighten the strap retaining bolts to 29 ft. lbs. (39 Nm).
8. Fill tank with new fuel and connect battery. Check operation of pump.

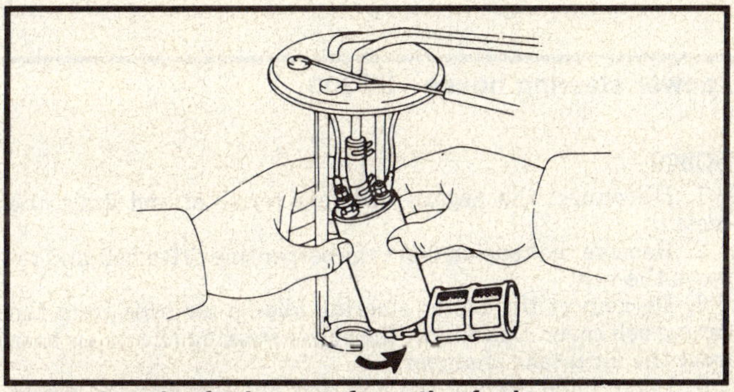

Removing the fuel pump from the fuel pump bracket – ES250

ES300

1. Disconnect the negative battery terminal.
2. Remove the rear seat bottom cushion.
3. Disconnect the fuel pump electrical connector. Remove the 5 screws holding the access hole service plate.
4. Disconnect the fuel pump power lead from the pump plate.
5. Remove the fuel filler cap to relieve pressure and prevent fuel from spilling out the fuel lines.
6. Disconnect the fuel feed line coupler. Use a backup wrench to prevent damage to the fitting. Disconnect the fuel return hose.
7. Remove the 8 bolts holding the fuel pump bracket and remove from the tank opening. Remove the gasket.

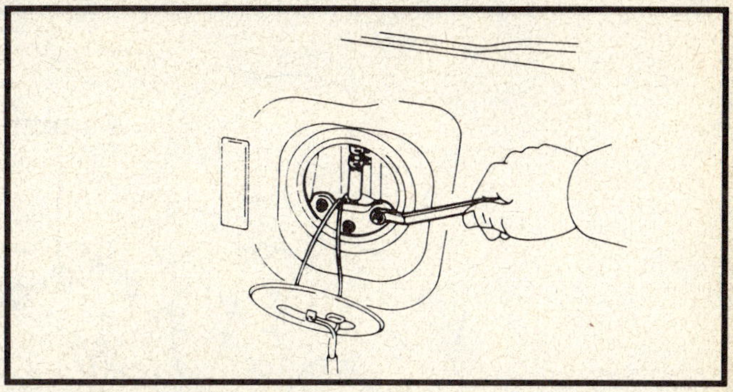

Accessing the fuel pump connections behind the rear seat back – LS400

8. Install a new gasket on the fuel pump bracket and install in the tank.
9. Replace the 8 bolts and torque to 35 inch lbs. (3.9 Nm).
10. Connect the fuel return hose and thread on the fuel feed line. Torque the connection to 22 ft. lbs. (30 Nm).
11. Connect the fuel pump power lead to the pump plate.
12. Before installing the service plate connect the power lead and connect the negative battery terminal. Jumper the +B and FP terminals of the check connector and turn **ON** the ignition switch. This will activate the fuel pump. Check for leaks.
13 Install the access hole service plate with the 5 screws. Install the seat cushion.

LS400, SC400 AND SC300

1. Disconnect the negative battery terminal and drain the fuel from the tank.
2. Remove the front luggage compartment cover and the rear seat back cushion.
3. Remove the plug from the panel partition.
4. Remove the fuel pump connector and the 8 bolts holding the cover in place.
5. Disconnect the fuel hose by moving the 2 clips towards the fuel pump bracket.
6. Remove the 3 bolts holding the fuel pump and remove the 2 clips.
7. Remove the fuel pump from the tank. Remove the nuts and washers from the fuel pump and disconnect the electrical leads at the pump.
8. Pull out the clip retaining the fuel filter.
9. Replace the fuel filter and electrical connections. Replace the pump in the bracket and place the assembly into the tank.
10. Install the 3 bolts and 2 clips holding the pump and line. Torque to 48 inch lbs. (5.4 Nm).
11. Replace the cover and 8 bolts and torque to 26 inch lbs. (2.99 Nm) on the LS400 and 35 inch lbs. (3.9 Nm) on the SC400 and SC300. Attach the electrical leads.
12. Replace the partition plug, seat back and compartment cover. Fill tank with fuel and connect the battery.
13. Check the operation of the fuel pump and check for leaks.

WATER TEMPERATURE SENSOR

Removal and Installation

ES250 AND ES300

1. Disconnect the negative battery terminal. Drain the coolant to below the level of the water outlet pipe where the sensor is located.
2. Disconnect the water temperature sensor connector from the body of the sensor.
3. Unscrew the sensor from the water outlet. Remove the gasket.

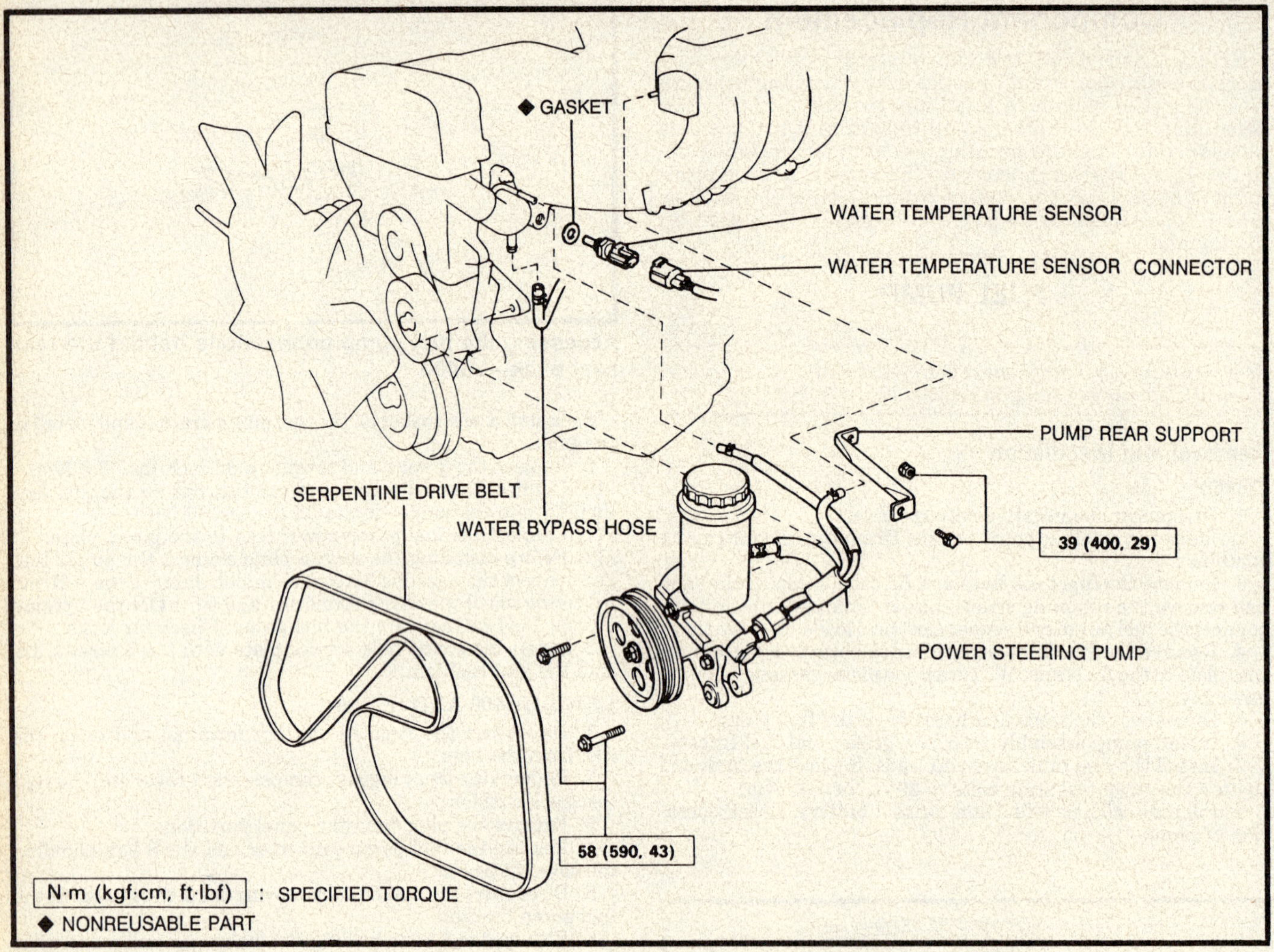

Water temperature sensor removal. Do not disconnect the power steering hoses—SC300

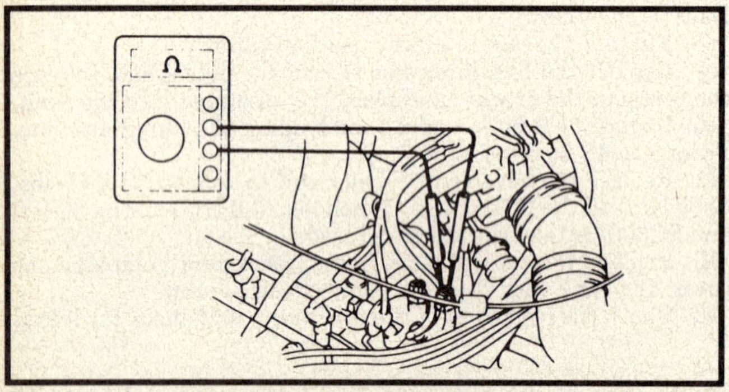

Testing the water temperature sensor. The cold start injector time switch is located behind the water temperature sensor—ES250 and ES300

To install:

4. Replace the sensor gasket and install the sensor.
5. Replace the sensor wiring connector.
6. Fill the cooling system and connect the negative battery terminal. Check for leaks.

SC300

1. Disconnect the negative battery terminal and drain the coolant.
2. Remove the tension from the serpentine drive belt and remove the belt.
3. Disconnect the power steering idle-up air hose from the timing belt cover. Disconnect the power steering idle-up air hose from the air intake chamber.
4. Loosen and remove the bolt and nut securing the power steering pump support. Remove the 2 nuts securing the power steering pump and remove the pump from the engine. Do not disconnect the fluid lines and secure the pump out of the way.
5. Loosen the water bypass hose clamp and disconnect the bypass hose from the water outlet.
6. Disconnect the water temperature sensor connector and remove the sensor from the engine. Remove the gasket.

To install:

7. Install a new gasket and thread in the water temperature sensor. Torque the sensor to 14 ft. lbs. (20 Nm). Connect the sensor electrical connector.
8. Connect the water bypass hose to the water outlet.
9. Install the power steering pump. Torque the nut and bolt on the pump support to 29 ft. lbs. (39 Nm) and the pump bracket bolts to 43 ft. lbs. (58 Nm). Tighten the fasteners in alternating sequence.
10. Connect the 2 power steering idle-up air hoses to their original locations.

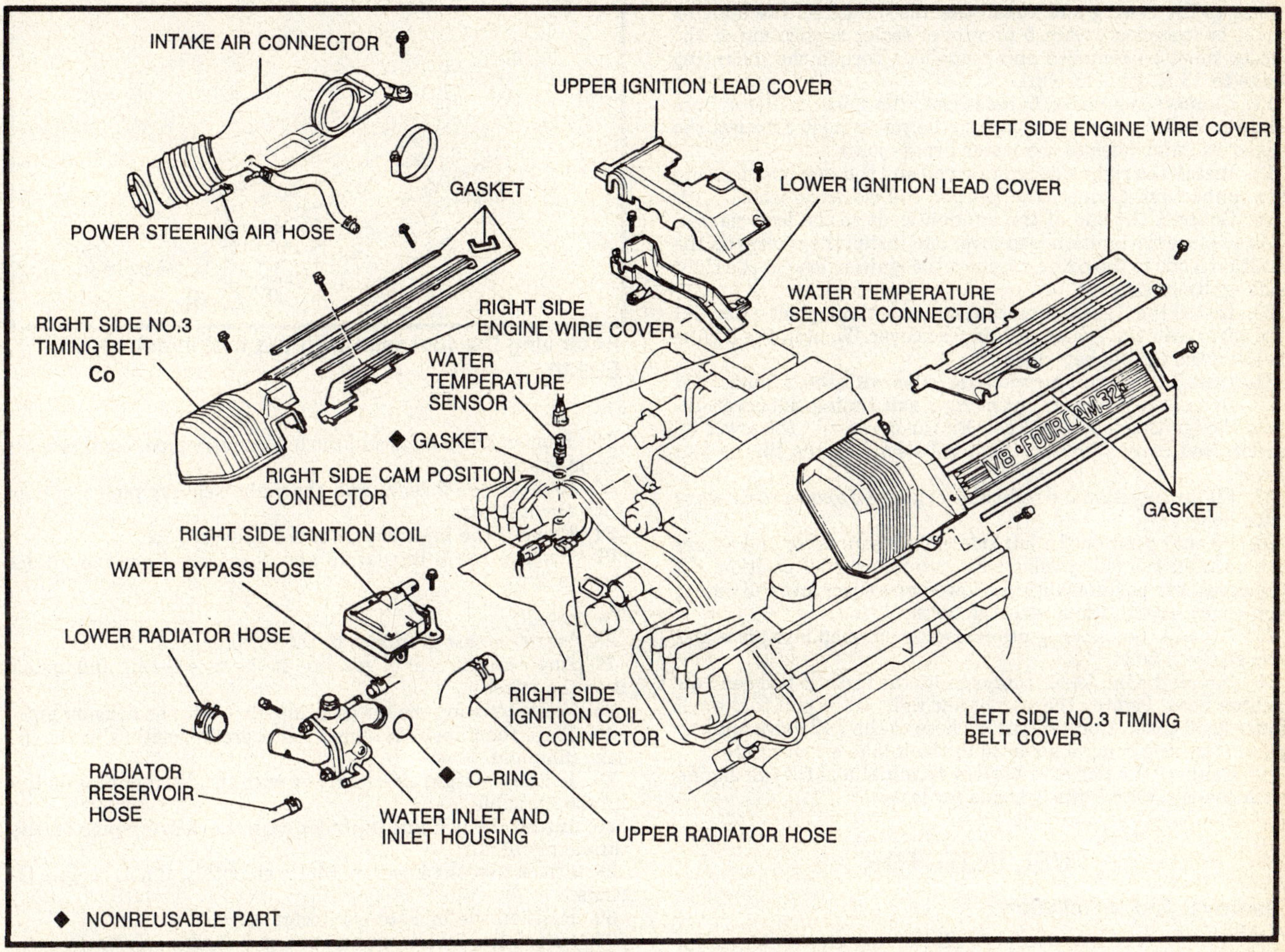

Water temperature sensor removal – LS400 and SC400

11. Install the serpentine drive belt. Fill engine with coolant and connect the negative battery terminal. Check for leaks.

SC400 AND LS400

1. Disconnect the negative battery terminal and drain the coolant.
2. Remove the nut and 2 bolts holding the throttle body cover and remove the cover.
3. Disconnect the accelerator cable, the throttle control cable (automatic transmission) and the cruise control actuator cable from the throttle body.
4. Disconnect the ISC valve air hose and the power steering air control valve air hose. Remove the bolt securing the intake air connector to the cylinder head cover. Loosen the 2 hose clamps and disconnect the air intake connector from the throttle body and air cleaner hose. Remove the intake air connector.
5. Remove the 2 bolts securing the upper ignition lead cover and disengage the clip securing the upper and lower halves of the cover. Separate the halves and remove the upper portion.
6. Remove the bolt securing the right side wire cover and remove the cover. Remove the 2 bolts holding the left side wire cover and remove the cover.
7. Remove the 2 bolts securing the EVAP system vacuum switching valve to the cylinder head and timing belt cover.
8. Remove the 3 timing belt cover bolts from right side and remove the timing belt cover. Remove the 4 timing belt cover bolts from the left side cover and remove the ignition lead grommet from the cover and remove the cover.
9. Disconnect the ignition lead from the right side ignition coil and remove the ignition leads from the lower ignition lead cover. Remove the mounting bolt and remove the lower cover.
10. Disconnect the ignition coil connector from the right side coil. Remove the 2 bolts securing the coil and pull the coil away from the engine. Disconnect the cam position sensor connection from the ignition coil bracket and remove the coil.
11. Disconnect the upper radiator hose from the water bypass joint. Disconnect the lower radiator hose and the radiator reservoir hose from the water inlet.
12. Disconnect the water bypass hose from the water inlet housing. Remove the 2 bolts securing the inlet housing to the water pump and remove the inlet housing. Remove the O-ring from the inlet housing.
13. Disconnect the water temperature sensor electrical connector. Remove the water temperature sensor and gasket.

To install:

14. Install the water temperature sensor with a new gasket and torque to 14 ft. lbs. (20 Nm). Connect the sensor electrical connector.
15. Install a new O-ring on the inlet housing. Remove all the old sealing material from the inlet housing and clean the mating surfaces. Use a non-residue type solvent and take care not to leave any oil or loose material on the mating surfaces.
16. Use sealing material 08826-00100 or equivalent, and apply a 0.08–0.12 in. (2–3 mm) bead to the sealing groove of the inlet housing. Do not use an excessive amount of sealer. Apply soapy

water to the O-ring and install the inlet housing. The housing must be installed within 5 minute of sealer application or the sealer must be removed and reapplied. Torque the mounting bolts to 13 ft. lbs. (18 Nm).
17. Connect the water bypass hose, the lower radiator hose and the radiator reservoir hose to the water inlet. Connect the upper radiator hose to the water bypass joint.
18. Install the right side ignition coil and the cam position sensor connection. Connect the ignition coil electrical lead.
19. Connect the end of the ignition leads to the lead clamps. Bolt in place the ignition lead cover and install the clamps on the ignition leads to the cover. Connect the ignition lead to the right side ignition coil.
20. Install the 3 gaskets on the right side timing belt cover. Fit the cover over the lower ignition lead cover. Tighten the 3 bolts holding the timing belt cover.
21. Install the vacuum switching valve with the 2 bolts.
22. Install the 3 gaskets on the right side timing belt cover. Install the ignition lead grommet. Fit the cover over the lower ignition lead cover. Tighten the 4 bolts holding the timing belt cover.
23. Fit into position the right side engine wire cover and secure with bolt.
24. Fit into position the left side engine wire cover and secure the vacuum switching valve wire into its original position. Fit the cover into position with the lower wire cover and the timing belt cover. Install the 2 securing bolts.
25. Replace the upper ignition lead cover, snap into place and secure the 2 bolts.
26. Install the intake air connector to the throttle body and air cleaner hose. Tighten the clamps and tighten the bolt to the cylinder head cover. Connect the air hose to the ISC valve and the power steering control air hose to the intake air connector.
27. Connect the negative battery terminal and fill the engine with coolant. Check the systems for leaks.

FUEL INJECTOR

Removal and Installation

NOTE: Before removal, tag all the hoses and electrical connections for reference during reassembly.

ES250

1. Disconnect the negative battery cable.
2. Drain the engine coolant.
3. Disconnect the accelerator cable and throttle cable from the throttle body, if equipped.
4. Remove the air cleaner, air flow meter and air cleaner hose.
5. Disconnect PCV, vacuum sensing, emission control and water by-pass hoses.
6. Disconnect ISC, TPS and EGR temperature sensor connectors.
7. Remove the No. 1 right engine mount.
8. Disconnect the cold start injector.
9. Place a towel under the injector tube, slowly remove the union bolt. Remove the injector tube.
10. Disconnect the brake booster hose, power steering vacuum and air hoses and cruise control hose.
11. Disconnect the ground strap and remove the nut to disconnect the wire clamp.
12. Disconnect the EGR pipe.
13. Remove the 2 bolts and the No. 1 engine hanger.
14. Remove the bolt, and disconnect the air intake chamber support from the air intake chamber.
15. Remove the air intake chamber.
16. Disconnect the injector connectors and the water temperature sensor connector.
17. Disconnect the fuel lines.
18. Remove the fuel pipe.

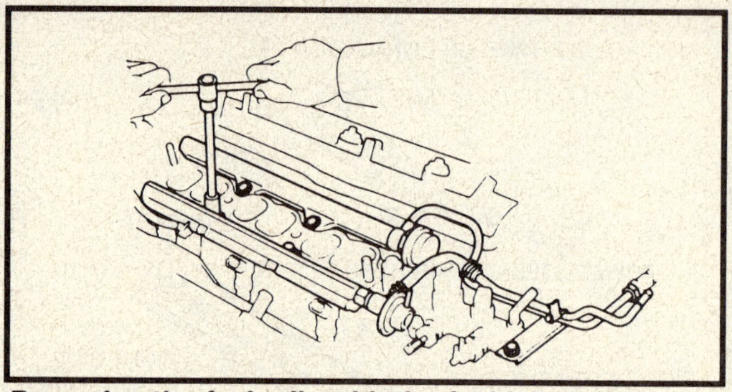

Removing the fuel rails with the fuel injectors—ES250

19. Remove the 2 bolts and the left delivery pipe together with the injectors.
20. Remove the 2 bolts and the right delivery pipe together with the injectors.
21. Remove the injectors from the delivery pipe.
22. Remove the 6 insulators and 4 spacers from the intake manifold.

To install:

23. Install a new grommet to the injector.
24. Apply a light coat of gasoline to the new O-ring and install it to the injector.
25. While twisting the injector, install it into the delivery pipe.
26. Place the 6 insulators and 4 spacers in position on the intake manifold.
27. Install the right side injectors with the delivery pipe on the intake manifold.
28. Install the left side injectors with the delivery pipe on the intake manifold.
29. Check that the injectors rotate smoothly; if not, replace O-rings.
30. Position the injector connector upward.
31. Install the 5 bolts and torque to 9 ft. lbs. (13 Nm).
32. Install the No. 2 fuel pipe with 4 new gaskets and torque bolts to 24 ft. lbs. (32 Nm).
33. Install the fuel lines.
34. Connect the 3 wiring harness clamps to the left delivery pipe.
35. Connect the injector and water sensor connectors.
36. Install the air chamber with new gasket and torque the bolts to 32 ft. lbs. (43 Nm).
37. Connect the EGR pipe and torque to 58 ft. lbs. (78 Nm).
38. Connect the wire clamp with nut.
39. Install the No. 1 engine hanger and air intake support torque the bolts to 27 ft. lbs. (37 Nm).
40. Connect the hoses and electrical connectors.
41. Connect the cold start injector tube and torque to 13 ft. lbs. (18 Nm).
42. Install the No. 1 engine right mount and torque bolts to 38 ft. lbs. (52 Nm).
43. Connect the ISC, TPS and EGR temperature sensor connectors.
44. Connect PCV, vacuum sensing, water by-pass and emission hoses, if not already installed.
45. Install the air cleaner cap, air flow meter and air cleaner hose.
46. Connect the accelerator and throttle cable, if equipped.
47. Fill engine with coolant.
48. Connect the negative battery cable.

ES300

1. Disconnect the negative battery terminal. Drain the engine coolant.

2. Disconnect the accelerator and throttle cables from the throttle linkage.
3. Disconnect the connector from the airflow meter. Disconnect the clamp from the coil wire. Disconnect the air ducting and the air cleaner hose clamp. Release the air cleaner cover clips and remove the air cleaner cover and airflow meter assembly.
4. Unscrew the 2 5mm bolts and remove the V-bank cover.
5. Disconnect the EGR gas temperature sensor connector clamp. Disconnect the 2 vacuum line from the fuel pressure control vacuum switching valve. Disconnect the 2 vacuum lines from the IACV vacuum switching valve. Disconnect the connectors from the 2 vacuum switching valves. Remove the 2 bolts holding the emission control valve set.
6. Disconnect the PCV line, the power steering idle-up air line, the brake booster vacuum line and the IACV vacuum line.
7. Disconnect the 2 ground straps.
8. Disconnect the electrical connector from the cold start injector. Place a container underneath the cold start injector to catch any spilled fuel. Disconnect the cold start injector fuel feed pipe union bolt slowly and remove.
9. Remove 2 bolts each from the engine lifting hook bracket and the intake plenum support. Remove the bracket and support.
10. Loosen the EGR pipe union and the 2 mounting bolts. Remove the EGR pipe.
11. Remove the bolt holding the hydraulic motor pressure pipe from the intake plenum. Move the pipe off the plenum.
12. Disconnect the 3 vacuum lines and 2 water lines from the intake plenum. Disconnect the EGR gas temperature sensor.
13. Disconnect the throttle position sensor connector, the ISC valve connector and air hose and the power steering idle-up air line.
14. Unscrew the 8mm nuts and bolts and remove the plenum and gasket.
15. Disconnect the fuel inlet from the fuel filter and the fuel return from the fuel pipe.
16. Disconnect the fuel injector electrical connectors.
17. Remove the fuel pipe vacuum hose, 2 union bolts, 4 gaskets and the fuel pipe itself.
18. Disconnect the fuel hose from the fuel pipe and remove the 2 bolts holding the left side fuel rail. Carefully lift the fuel rail with the injectors so the injectors do not drop.
19. Remove the injectors from the fuel rail and the 2 spacers from the intake manifold.
20. Remove the air pipe by disconnecting the 2 air hoses and removing the bolts.
21. Remove the 2 bolts holding the right side fuel rail. Carefully lift the fuel rail with the injectors so the injectors do not drop.
22. Remove the injectors from the fuel rail and the 2 spacers from the intake manifold.

To install:

23. Install new grommets and O-rings to the injectors. Apply a light coat of gasoline to the O-rings.
24. Install the injectors into the fuel rails. Turn the injectors right and left to insure seating of the injectors.
25. Install the spacers on the intake manifold and clean the injector bores. Place the fuel rail with injectors and the fuel pipe onto the manifold.
26. Check that the injectors rotate smoothly in their bores and position the injector connectors upwards. Torque the bolts to 9 ft. lbs. (13 Nm).
27. Install the air pipe and torque the bolts to 74 inch lbs. (8.3 Nm). Connect the air hoses.
28. Install the spacers on the intake manifold and clean the injector bores. Place the fuel rail with injectors and the fuel pipe onto the manifold.
29. Check that the injectors rotate smoothly in their bores and position the injector connectors upwards. Torque the bolts to 9 ft. lbs. (13 Nm).
30. Connect the fuel hose to the fuel pipe.
31. Install the second fuel pipe using 4 new gaskets and the union bolts. Torque to 25 ft. lbs. (34 Nm) and connect the IACV vacuum line.
32. Connect the injector electrical connectors.
33. Connect fuel line to the fuel filter and torque the connection to 22 ft. lbs. (29 Nm). Connect the fuel return hose.
34. Install the intake plenum with a new gasket onto the intake manifold and torque the 8mm hardware to 32 ft. lbs. (43 Nm). Connect the throttle position sensor, the ISC valve connection, the ISC valve air hose and the power steering idle-up air hose.
35. Connect the EGR gas temperature sensor, the throttle water bypass hose, the EGR cooler water bypass hose, the EGR valve to EGR BVSV vacuum line, the EGR vacuum modulator to EGR BVSV vacuum hose and the throttle body to EVAP BVSV vacuum hose.
36. Attach the hydraulic motor pressure pipe to the plenum.
37. Install a new gasket and sleeve ball on the EGR pipe and bolt into place. Torque the bolts to 13 ft. lbs. (18 Nm). Torque the union to 58 ft. lbs. (78 Nm).
38. Install the engine lift hook bracket and plenum support and torque the 2 bolts each to 29 ft. lbs. (39 Nm).
39. Connect the cold start injector line to the injector with 2 new gaskets and torque to 11 ft. lbs. (15 Nm).
40. Connect the ground straps and torque to 9 ft. lbs. (13 Nm).
41. Connect the brake booster vacuum line, the power steering idle-up air line, the PCV hose and the IACV vacuum line.
42. Install the emission control valve assembly and bolt in place. Connect the vacuum switching valve connectors and vacuum lines. Connect the 2 vacuum lines to the fuel pressure control vacuum switching valves. Attach the EGR gas temperature sensor clamp to the emission control valve assembly.
43. Install the V-bank cover with the 2 5mm nuts.
44. Install the air cleaner cover and air ducting. Tighten the hose clamp and attach the air hoses. Connect the coil wire clamp and airflow meter connector.
45. Connect the throttle cable and the accelerator cable. Adjust the throttle cable lock nuts so that the stopper on the cable extends from the the outer cable end cavity 0–0.04 in. (0–1mm).
46. Fill the engine with coolant and connect the negative battery terminal. Check for proper operation and leaks.

SC300

1. Disconnect the negative battery terminal and drain the coolant.
2. Disconnect the throttle cable, the accelerator cable and cruise control actuator cable from the throttle body.
3. Remove the power steering air hose from the timing belt cover. Remove the PCV hose from the cylinder head cover.
4. Loosen the EGR pipe union and remove the 2 bolts and gasket. Remove the pipe.
5. Remove the 4 bolts securing the throttle body support bracket.
6. Disconnect the PCV hose, the 3 vacuum hoses, the EGR vacuum switching valve connector, the throttle position sensor connector, the ISC valve connector, the subthrottle position sensor connector if equipped, the subthrottle actuator connector if equipped and the EGR gas temperature sensor connector.
7. Remove the 2 nuts and 4 bolts holding the intake air connector to the intake chamber. Remove the vacuum switching valve nut and disconnect and plug the 2 water bypass hoses from the throttle body. Remove the intake air connector along with the throttle body. Remove the gasket.
8. Slowly loosen the union bolt and disconnect the No.1 fuel pipe from the delivery pipe. Use a rag or container to catch any spilled fuel.
9. Remove the 2 bolts each from the air intake chamber supports and remove the supports. Take notice to their positions.
10. Disconnect the vacuum line from the fuel pressure regulator and disconnect the electrical connectors from the injectors.

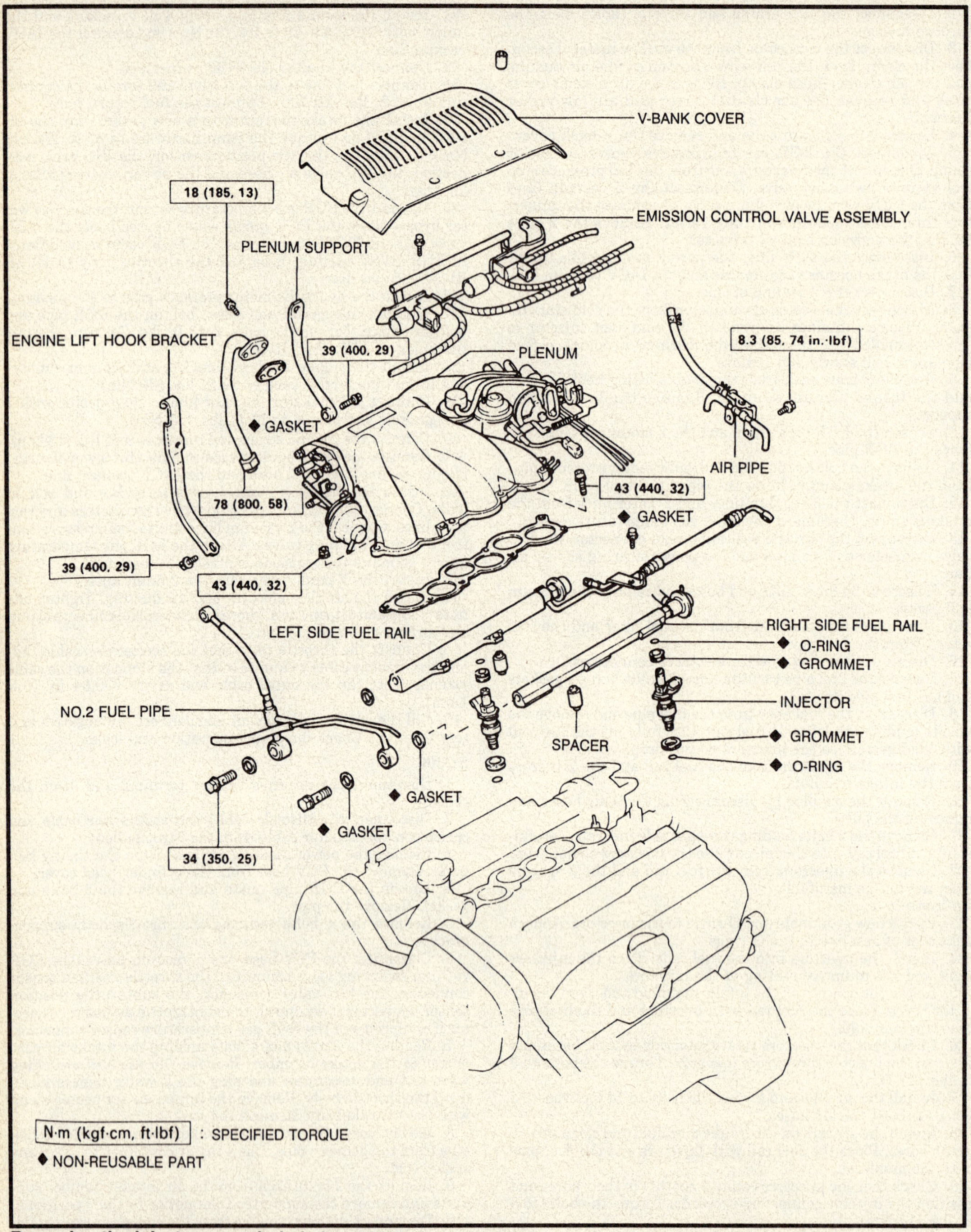

Removing the injectors—exploded view—ES300

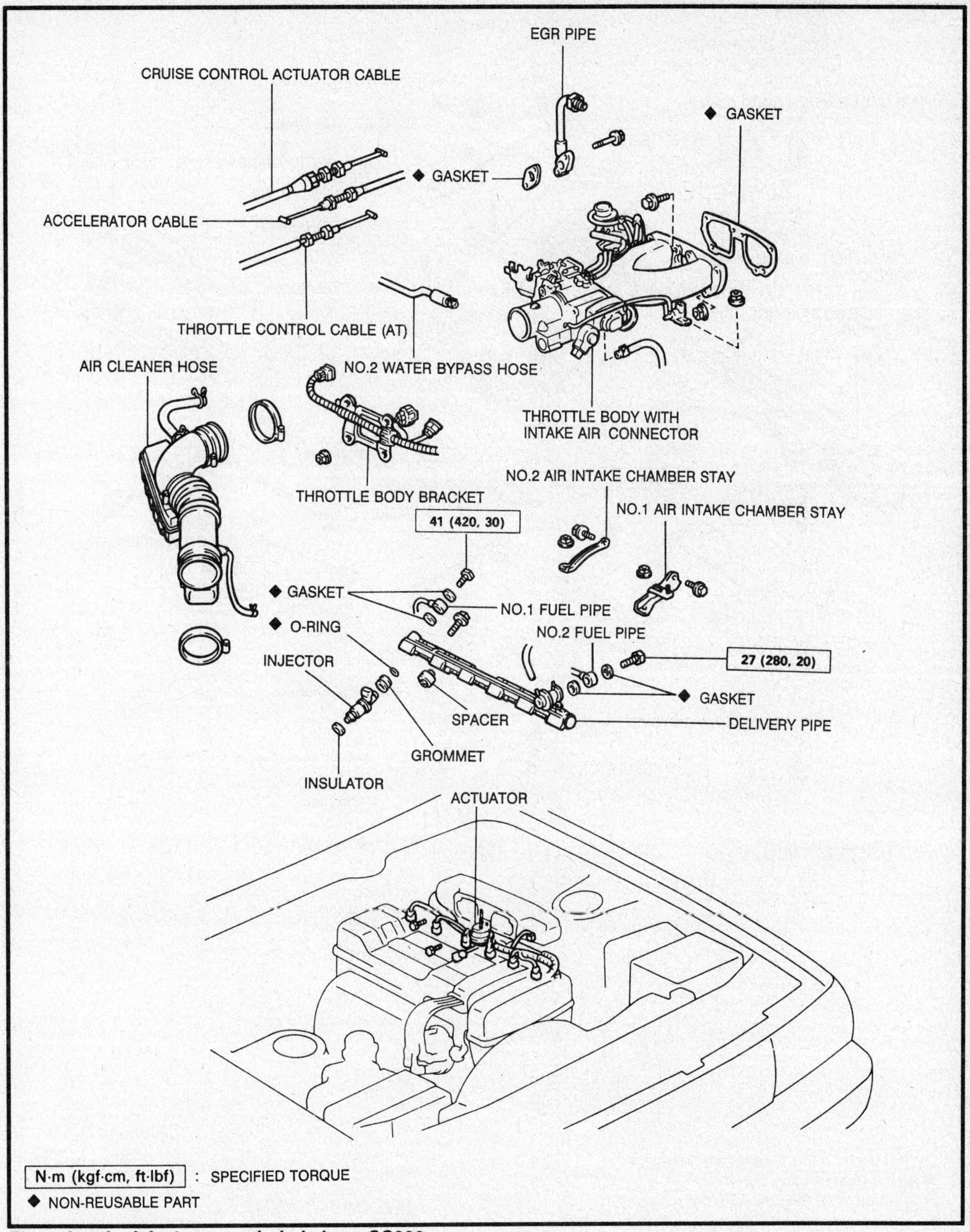

Removing the injectors—exploded view—SC300

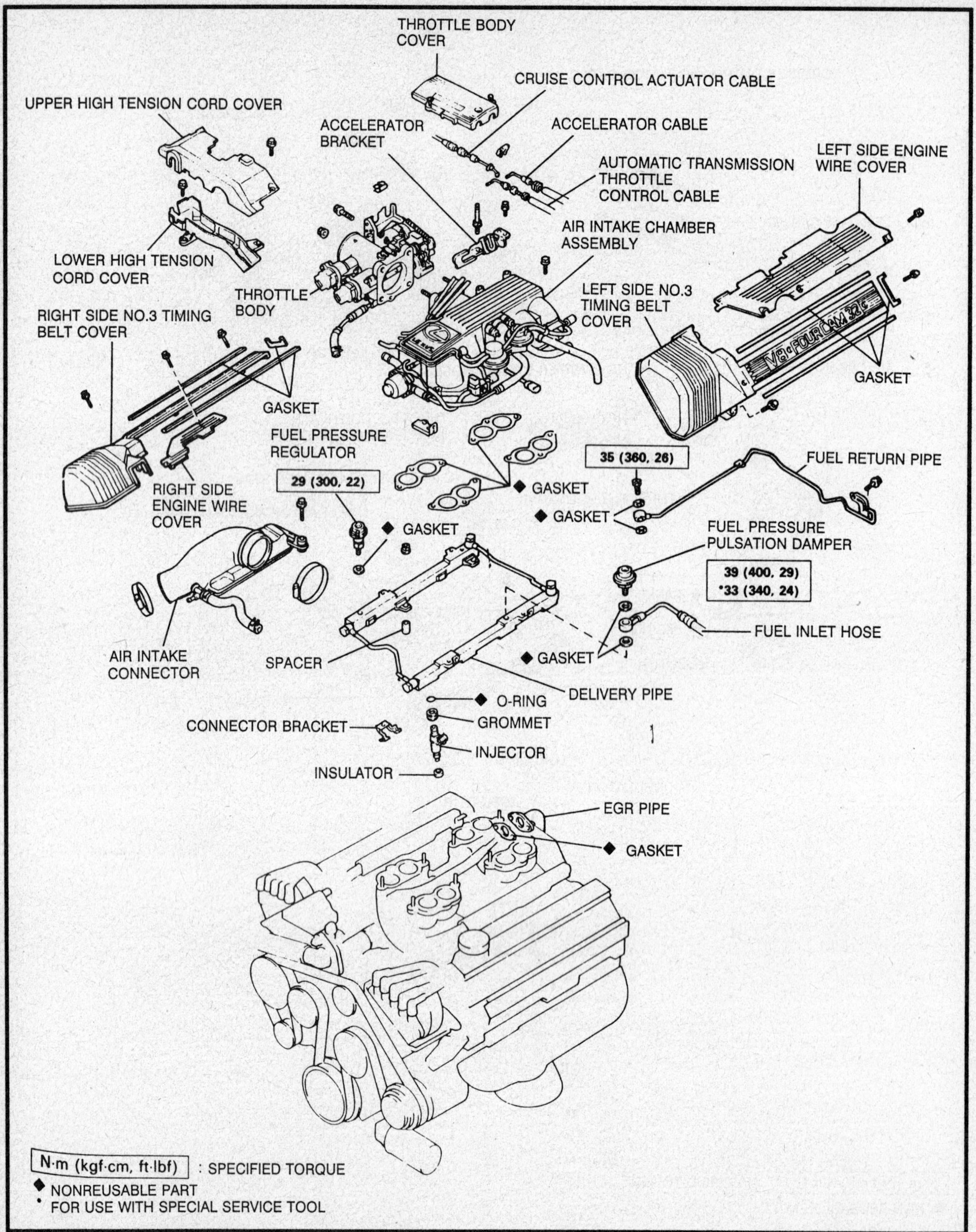

Removing the injectors – exploded view – SC400

11. Remove the 2 bolts from the control valve actuator located above the injectors on the intake chamber and rotate the actuator up. Support the actuator being careful not to bend the control rod.
12. Remove the 3 bolts holding the fuel delivery pipe and lift the pipe with the injectors. Be careful not to drop the injectors. Remove the injectors from the pipe and remove the O-rings and grommets from the injectors.
13. Remove the 6 insulators and 3 spacers from the head.

To install:

14. Install a new O-ring and grommet to the injector and lightly coat the O-ring with gasoline. Using a twisting motion push the injector into the fuel delivery pipe and position the connector outwards.
15. Place the 6 insulators and the 3 spacers into position and lower the injector and delivery pipe assembly onto the cylinder head.
16. Install the 3 bolts loosely and check that all 6 injectors rotate smoothly in their bores. If they do not rotate smoothly the O-rings are installed incorrectly; replace the O-rings.
17. Position the injector connectors upwards and torque the 3 bolts to 15 ft. lbs. (21 Nm).
18. Position the control valve actuator and install the mounting bolts. Torque the bolts to 61 inch lbs. (6.8 Nm). Check the operation of the control rod and IACV valve.
19. Note that the No. 1, 3 and 5 injector connectors are gray and the No. 2, 4 and 6 connectors are black. Connect the injector connectors.
20. Connect the vacuum line to the fuel pressure regulator.
21. Install the intake chamber supports into their original locations; the supports are marked F and R for front and rear. Torque the bolts to 13 ft. lbs. (18 Nm).
22. Use 2 new gaskets and install the No.2 fuel pipe to the delivery pipe. Torque the union bolt to 30 ft. lbs. (42 Nm). Use 2 new gaskets and install the No.1 fuel pipe to the delivery pipe. Torque the union bolt to 30 ft. lbs. (42 Nm).
23. Install a new intake air connector gasket so that the protrusion on the edge of the gasket faces the intake chamber.
24. Connect the water bypass hoses to the throttle body. Install the vacuum switching valve nut and install the air intake connector. Torque the 2 nuts and 6 bolts to 15 ft. lbs. (21 Nm).
25. Connect the PCV hose, the 3 vacuum hoses, the EGR vacuum switching valve connector, the throttle position sensor connector, the ISC valve connector, the subthrottle position sensor connector if equipped, the subthrottle actuator connector if equipped and the EGR gas temperature sensor connector.
26. Install the throttle body support bracket and torque the 4 nuts to 15 ft. lbs. (21 Nm).
27. Install a new gasket on the EGR pipe and install the pipe into place. Torque the 2 bolts to 15 ft. lbs. (21 Nm) and the union nut to 47 ft. lbs. (64 Nm).
28. Install the air cleaner hose hose clamps and connect the power steering air hose to the timing belt cover. Install the PCV hose to the cylinder head cover.
29. Connect the accelerator cable, the automatic transmission throttle control cable and cruise control actuator cable.
30. Fill the engine with coolant and connect the negative battery terminal. Check for proper operation and leaks.

SC400

1. Disconnect the negative battery terminal and drain the engine coolant.
2. Remove the nut and 2 bolts holding the throttle body cover and remove the cover.
3. Disconnect the accelerator cable, the throttle control cable (automatic transmission) and the cruise control actuator cable from the throttle body.
4. Disconnect the ISC valve air hose and the power steering air control valve air hose. Remove the bolt securing the intake air connector to the cylinder head cover. Loosen the 2 hose clamps and disconnect the air intake connector from the throttle body and air cleaner hose. Remove the intake air connector.
5. Remove the 2 bolts securing the upper ignition lead cover and disengage the clip securing the upper and lower halves of the cover. Separate the halves and remove the upper portion.
6. Remove the bolt securing the right side wire cover and remove the cover. Remove the 2 bolts holding the left side wire cover and remove the cover.
7. Remove the 2 bolts securing the EVAP system vacuum switching valve to the cylinder head and timing belt cover.
8. Remove the 3 timing belt cover bolts from right side and remove the timing belt cover. Remove the 4 timing belt cover bolts from the left side cover and remove the ignition lead grommet from the cover and remove the cover.
9. Disconnect the ignition lead from the right side ignition coil and remove the ignition leads from the lower ignition lead cover. Remove the mounting bolt and remove the lower cover.
10. Disconnect the throttle position sensor, the subthrottle position sensor if equipped and the subthrottle actuator connector if equipped. Disconnect the heater hose from the heater valve, the ISC valve water bypass hose, the vacuum hose from the throttle body.
11. Remove the 2 bolts and 2 nuts securing the throttle body to the intake chamber. Remove the throttle body and disconnect the PCV hose and water bypass hose. Remove the throttle body gasket.
12. Disconnect the cold start injector connector, the ISC valve connectors, the EGR valve connector and the fuel pressure control vacuum switching valve.
13. Disconnect the ISC valve water bypass hose from the water inlet housing, the power steering air control valve vacuum hose from the intake chamber, the fuel pressure regulator vacuum hose from the vacuum pipe, the 2 EVAP system vacuum switching valves vacuum lines from the vacuum pipe and the charcoal canister vacuum line from the vacuum pipe.
14. Disconnect the EGR water bypass hose from the water bypass pipe, the brake booster vacuum hose from the intake chamber, the heater valve vacuum switching valve vacuum hose from the intake chamber and the PCV hose from the left side cylinder head cover.
15. Remove the stud bolt and bolt retaining the accelerator bracket. Remove the union bolt and washers connecting the cold start injector tube to the right side delivery pipe. Remove the bolts and gasket from the EGR pipe connection and disconnect the EGR pipe from the intake chamber. Remove the 8 nuts, 4 bolts and the automatic transmission throttle cable bracket. Disconnect the check connector bracket from the air intake chamber.
16. Remove the intake chamber assembly and the 4 gaskets.
17. Remove the pulsation damper and the 2 gaskets and disconnect the inlet hose from the left side delivery pipe. Disconnect the fuel return hose from the fuel return pipe.
18. Disconnect the EGR pipe from the cylinder head by removing the securing bolt.
19. Remove the 2 bolts holding the wiring harness to the engine.
20. Remove the 4 bolts securing the engine wiring harness to the fuel delivery pipes. Remove the 2 bolts securing the engine wiring harness to the rear water bypass joint. Remove the 2 bolts securing the engine wiring harness to the rear side of the right side cylinder head. Remove the 2 wiring harness connectors from the connector bracket on the front side of the left side delivery pipe. Disconnect the injector connections at the injectors.
21. Remove the return pipe to left side cylinder head bolt. Remove the union bolt and gaskets and remove the return pipe.
22. Remove the 4 nuts securing the fuel delivery pipe to the intake manifold and lift the connector bracket with the fuel delivery pipes and the injectors. Be careful not to drop the injectors. Remove the 4 spacers and the 8 insulators.
23. Pull the injectors out of the delivery pipes and remove the O-rings and grommets from the injectors.

To install:

24. Install a new O-ring and grommet to the injector and light-

ly coat the O-ring with gasoline. Using a twisting motion push the injector into the fuel delivery pipe and position the connector outwards.

25. Place the 8 insulators and the 4 spacers into position and lower the injector and delivery pipe assembly onto the cylinder head.

26. Install the 4 bolts loosely and check that all 8 injectors rotate smoothly in their bores. If they do not rotate smoothly the O-rings are installed incorrectly; replace the O-rings.

27. Position the injector connectors upwards and torque the 4 bolts to 13 ft. lbs. (18 Nm).

28. Loosely install the return pipe bolt and the union bolt with 2 new gaskets. Torque the return pipe to pressure regulator union bolt to 26 ft. lbs. (35 Nm). Tighten the bolt securing the return pipe to the cylinder head.

29. Torque the fuel pressure regulator locknut to 22 ft. lbs. (29 Nm).

30. Connect the injector connectors. Install the engine wiring harness connectors to the connector bracket on the left side delivery pipe. Install the engine wiring harness to the right side cylinder head with the 2 bolts. Install the engine wiring harness to the rear water bypass joint with 2 bolts and install the engine wiring harness to the delivery pipes with the 4 bolts.

31. Install the 2 bolts securing the engine wiring harness to the intake manifold.

32. Loosely install the EGR pipe to the right side cylinder head. This connection will be tightened after the intake chamber is installed.

33. Connect the fuel return line to the return pipe.

34. Connect the inlet hose to the pulsation damper with 2 new gaskets. Use SST 09612-24014 or equivalent and torque to 24 ft. lbs. (33 Nm). If not using SST, obtain a torque of 29 ft. lbs. (39 Nm) at the nut. The lower torque rating using the SST takes in account the added length of the SST and the resulting torque multiplication due to the longer moment arm. Using the SST the total torque will equal the higher torque rating at the nut. Use a torque wrench with a length of 11.81 in. (30 cm) between the handle pivot and the drive end.

35. Install 4 new gaskets and install the intake chamber on the intake manifold. Loosely install the EGR pipe bolts using a new gasket. Install the check connector, the throttle cable bracket, the fuel pressure control vacuum switching valve and the EGR system vacuum switching valve. Torque the 8 nuts and 4 bolts to 13 ft. lbs. (18 Nm). Torque the EGR pipe bolts to 13 ft. lbs. (18 Nm).

36. Torque the cold start injector tube to right side delivery pipe union bolt to 11 ft. lbs. (15 Nm). Install the accelerator bracket and tighten the stud bolt and bolt.

37. Connect the EGR valve water bypass hose, the brake booster vacuum hose, the heater water valve vacuum switching valve vacuum hose, the PCV hose, the water inlet water hose, the power steering air control valve vacuum hose, the fuel pressure regulator vacuum hose, the EVAP system vacuum switching valves vacuum lines and the charcoal canister vacuum hose.

38. Connect the cold start injector connector, the ISC valve connector, the EGR valve connector and the vacuum switching valve connector for the fuel pressure control.

39. Connect the PCV hose and water hose to the throttle body. Install the throttle body on the intake chamber with a new gasket and torque the 2 nuts and 2 bolts to 13 ft. lbs. (18 Nm). Install the water bypass pipe to the clamp on the engine wiring harness cover.

40. Connect the water bypass hose to the ISC valve, the vacuum hose to the throttle body and the heater hose to the heater valve. Connect the throttle position sensor connector, the subthrotttle position sensor if equipped and the subthrottle actuator connector if equipped.

41. Connect the end of the ignition leads to the lead clamps. Bolt in place the ignition lead cover and install the clamps on the ignition leads to the cover. Connect the ignition lead to the right side ignition coil.

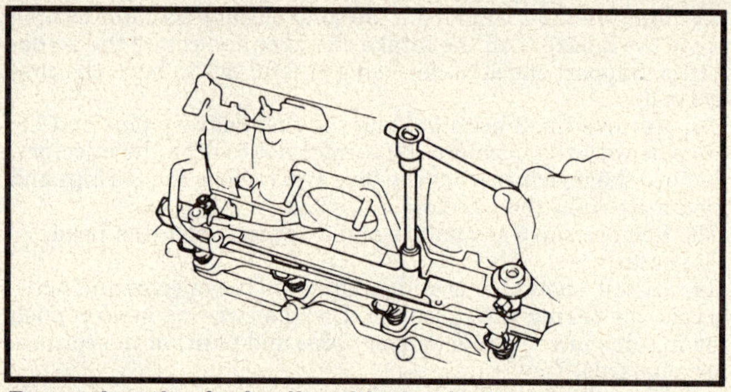

Removing the fuel rails with the fuel injectors—LS400

42. Install the 3 gaskets on the right side timing belt cover. Fit the cover over the lower ignition lead cover. Tighten the 3 bolts holding the timing belt cover.

43. Install the vacuum switching valve with the 2 bolts.

44. Install the 3 gaskets on the right side timing belt cover. Install the ignition lead grommet. Fit the cover over the lower ignition lead cover. Tighten the 4 bolts holding the timing belt cover.

45. Fit into position the right side engine wire cover and secure with bolt.

46. Fit into position the left side engine wire cover and secure the vacuum switching valve wire into its original position. Fit the cover into position with the lower wire cover and the timing belt cover. Install the 2 securing bolts.

47. Replace the upper ignition lead cover, snap into place and secure the 2 bolts.

48. Install the intake air connector to the throttle body and air cleaner hose. Tighten the clamps and tighten the bolt to the cylinder head cover. Connect the air hose to the ISC valve and the power steering control air hose to the intake air connector.

49. Connect the throttle cable, accelerator cable and the cruise control cable to the throttle body. Install the throttle body cover.

50. Connect the negative battery terminal and fill the engine with coolant. Check the systems for leaks.

LS400

1. Disconnect the negative battery cable and drain the coolant.
2. Remove the air cleaner and air flow meter assembly.
3. Remove the throttle body and the engine wire covers. Remove the PCV hose.
4. Remove the timing belt covers. Remove the right side ignition coil and the upper and lower ignition lead covers.
5. Disconnect the 3 vacuum hoses from the vacuum pipe, the 4 vacuum hoses from the air pipe, the 2 vacuum hoses from the EGR VSV and the EGR vacuum modulator hose from the base of the EGR valve. Remove the EGR vacuum modulator and vacuum control valve with the bracket.
6. Remove the 2 vacuum switching valves and the water bypass pipes. Remove the 2 bolts holding the engine wiring to the fuel delivery pipe.
7. Disconnect the water temperature sender connector, the water temperature sensor connector, the cold start switch connector, the distributor connector, the engine speed sensor connector, the EGR gas temperature sensor connector and the injector connectors.
8. Disconnect the fuel return line, the cold start injector tube and the front fuel pipe. Disconnect the fuel inlet hose bolt and gaskets, pulsation damper and remove the hose from the delivery rail.
9. Remove the 2 union bolts and the 4 gaskets and remove the rear pipe. Leave it attached to the vehicle. Disconnect the front fuel pipe.

10. Disconnect the vacuum line from the pressure regulator. Remove the nuts holding the fuel delivery pipe and remove with the injectors. Remove the insulators from the engine.
11. Remove the injectors from the pipe.

To install:

12. Install the injectors onto the rails. Use new O-rings and grommets on the injectors. Lubricate with gasoline before installation. The injectors should rotate smoothly in their bores. If they don't, this is probably caused by incorrect installation of the O-ring.
13. Install the delivery pipes with the injectors on the engine. Temporarily tighten the mounting nuts. Check that the injectors can still rotate. Check and replace the O-ring if they do not rotate smoothly.
14. Temporarily install the fuel pipes and delivery pipes. Torque the delivery pipe to 13 ft. lbs. (18 Nm), second the front fuel pipe to 22 ft. lbs. (29 Nm), then the rear fuel pipe to 22 ft. lbs. (29 Nm).
15. Connect the cold start injector tube and torque to 11 ft. lbs. (15 Nm) and install the fuel return pipe and torque to 22 ft. lbs. (29 Nm).
16. Connect the fuel inlet hose to the delivery pipe with new gaskets and torque to 69 inch lbs. (7.8 Nm) and install the pulsation damper and torque to 22 ft. lbs. (29 Nm).
17. Connect the electrical connectors to all disconnected sensors and injectors. Install the bolts holding the engine wiring and torque to 74 inch lbs. (8.3 Nm).
18. Install the water bypass hoses, the upper and lower ignition lead covers, the right side ignition coil and the timing belts covers. Install the ignition lead covers on both sides. Install the vacuum switching valves and torque to 13 ft. lbs. (18 Nm). Install the check connector, vacuum modulator and vacuum control valve with bracket and bolt.
19. Connect the vacuum hoses to there original positions.
20. Install the throttle body and the PCV hose. Install the air cleaner with the air flow meter.
21. Fill the engine with coolant and connect the battery cable. Check for fuel and coolant leakages and check coolant level.

COLD START INJECTOR

Removal and Installation

NOTE: Before removal, tag all the hoses and electrical connections for reference during reassembly.

ES250

1. Disconnect the negative battery cable.
2. Disconnect the cold start injector.
3. Place a towel under the injector line and slowly remove the union bolt. Remove the injector line.
4. Disconnect the injector line from the clamp.
5. Remove the 2 bolts, cold start injector and gasket.

To install:

6. Install a new gasket, the injector and bolts. Torque the injector bolts to 48 inch lbs. (5.4 Nm).
7. Connect the injector tube to clamp.
8. Connect the injector tube with the 2 new gaskets and the union bolt and torque to 14 ft. lbs. (20 Nm).
9. Connect the cold start injector connector.
10. Connect the negative battery cable.

ES300

1. Disconnect the negative battery terminal. Place a rag or container under the injector fuel connection.
2. Disconnect the injector electrical connector. Slowly loosen the injector union bolt and remove the bolt and gaskets.
3. Loosen and remove the injector mounting nuts. Remove the injector and gasket.

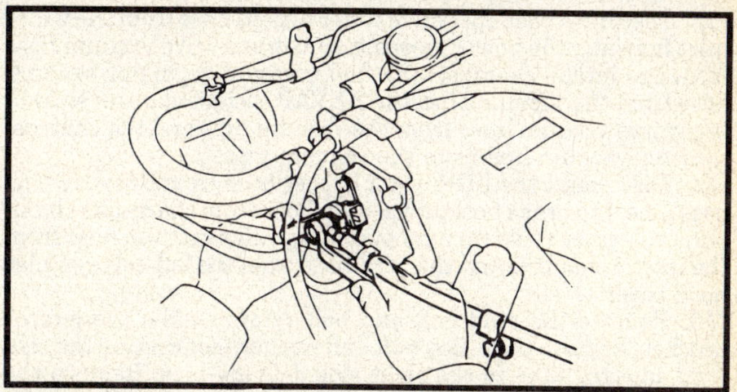

Removing the fuel line to the cold start injector—ES250

To install:

4. Install the injector with a new gasket and torque the nuts to 74 inch lbs. (8.3 Nm).
5. Install the fuel pipe with 2 new gaskets and tighten the union bolt to 11 ft. lbs. (15 Nm).
6. Connect the injector electrical connector. Connect the negative battery terminal and check for fuel leakage.

SC300

The SC300 does not make use of a cold start injector. The fuel injection computer varies the standard injection routine to accommodate cold starts.

SC400

1. Disconnect the negative battery terminal and drain the engine coolant.
2. Remove the nut and 2 bolts holding the throttle body cover and remove the cover.
3. Disconnect the accelerator cable, the throttle control cable (automatic transmission) and the cruise control actuator cable from the throttle body.
4. Disconnect the ISC valve air hose and the power steering air control valve air hose. Remove the bolt securing the intake air connector to the cylinder head cover. Loosen the 2 hose clamps and disconnect the air intake connector from the throttle body and air cleaner hose. Remove the intake air connector.
5. Remove the 2 bolts securing the upper ignition lead cover and disengage the clip securing the upper and lower halves of the cover. Separate the halves and remove the upper portion.
6. Remove the bolt securing the right side wire cover and remove the cover. Remove the 2 bolts holding the left side wire cover and remove the cover.
7. Remove the 2 bolts securing the EVAP system vacuum switching valve to the cylinder head and timing belt cover.
8. Remove the 3 timing belt cover bolts from right side and remove the timing belt cover. Remove the 4 timing belt cover bolts from the left side cover and remove the ignition lead grommet from the cover and remove the cover.
9. Disconnect the ignition lead from the right side ignition coil and remove the ignition leads from the lower ignition lead cover. Remove the mounting bolt and remove the lower cover.
10. Disconnect the throttle position sensor, the subthrottle position sensor if equipped and the subthrottle actuator connector if equipped. Disconnect the heater hose from the heater valve, the ISC valve water bypass hose, the vacuum hose from the throttle body.
11. Remove the 2 bolts and 2 nuts securing the throttle body to the intake chamber. Remove the throttle body and disconnect the PCV hose and water bypass hose. Remove the throttle body gasket.
12. Disconnect the cold start injector connector, the ISC valve connectors, the EGR valve connector and the fuel pressure control vacuum switching valve.

13. Disconnect the ISC valve water bypass hose from the water inlet housing, the power steering air control valve vacuum hose from the intake chamber, the fuel pressure regulator vacuum hose from the vacuum pipe, the 2 EVAP system vacuum switching valves vacuum lines from the vacuum pipe and the charcoal canister vacuum line from the vacuum pipe.
14. Disconnect the EGR water bypass hose from the water bypass pipe, the brake booster vacuum hose from the intake chamber, the heater valve vacuum switching valve vacuum hose from the intake chamber and the PCV hose from the left side cylinder head cover.
15. Remove the stud bolt and bolt retaining the accelerator bracket. Remove the union bolt and washers connecting the cold start injector tube to the right side delivery pipe. Remove the bolts and gasket from the EGR pipe connection and disconnect the EGR pipe from the intake chamber. Remove the 8 nuts, 4 bolts and the automatic transmission throttle cable bracket. Disconnect the check connector bracket from the air intake chamber.
16. Remove the intake chamber assembly and the 4 gaskets.
17. Remove the injector line union bolt and the 2 gaskets. Remove the line from the injector. Disconnect the injector electrical connector. Remove the 2 bolts securing the injector and remove the injector and gasket.

To install:

18. Install the injector with a new gasket and torque the bolts to 69 inch lbs. (7.8 Nm).
19. Install the injector line with 2 new gaskets and the union bolt. Torque the union bolt to 11 ft. lbs. (15 Nm). Connect the injector electrical connector.
20. Install 4 new gaskets and install the intake chamber on the intake manifold. Loosely install the EGR pipe bolts using a new gasket. Install the check connector, the throttle cable bracket, the fuel pressure control vacuum switching valve and the EGR system vacuum switching valve. Torque the 8 nuts and 4 bolts to 13 ft. lbs. (18 Nm). Torque the EGR pipe bolts to 13 ft. lbs. (18 Nm).
21. Torque the cold start injector tube to right side delivery pipe union bolt to 11 ft. lbs. (15 Nm). Install the accelerator bracket and tighten the stud bolt and bolt.
22. Connect the EGR valve water bypass hose, the brake booster vacuum hose, the heater water valve vacuum switching valve vacuum hose, the PCV hose, the water inlet water hose, the power steering air control valve vacuum hose, the fuel pressure regulator vacuum hose, the EVAP system vacuum switching valves vacuum lines and the charcoal canister vacuum hose.
23. Connect the cold start injector connector, the ISC valve connector, the EGR valve connector and the vacuum switching valve connector for the fuel pressure control.
24. Connect the PCV hose and water hose to the throttle body. Install the throttle body on the intake chamber with a new gasket and torque the 2 nuts and 2 bolts to 13 ft. lbs. (18 Nm). Install the water bypass pipe to the clamp on the engine wiring harness cover.
25. Connect the water bypass hose to the ISC valve, the vacuum hose to the throttle body and the heater hose to the heater valve. Connect the throttle position sensor connector, the subthrottle position sensor if equipped and the subthrottle actuator connector if equipped.
26. Connect the end of the ignition leads to the lead clamps. Bolt in place the ignition lead cover and install the clamps on the ignition leads to the cover. Connect the ignition lead to the right side ignition coil.
27. Install the 3 gaskets on the right side timing belt cover. Fit the cover over the lower ignition lead cover. Tighten the 3 bolts holding the timing belt cover.
28. Install the vacuum switching valve with the 2 bolts.
29. Install the 3 gaskets on the right side timing belt cover. Install the ignition lead grommet. Fit the cover over the lower ignition lead cover. Tighten the 4 bolts holding the timing belt cover.

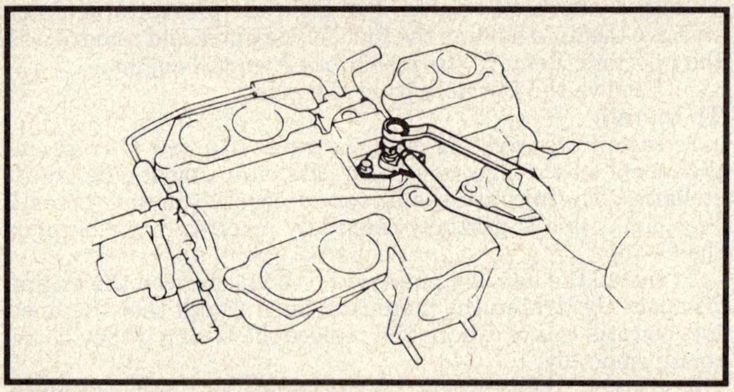

Removing the fuel line to the cold start injector with the air intake chamber removed—LS400

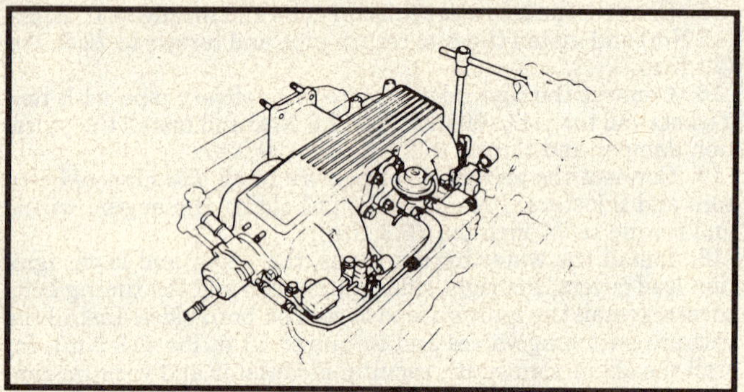

Removing the air intake chamber—LS400

30. Fit into position the right side engine wire cover and secure with bolt.
31. Fit into position the left side engine wire cover and secure the vacuum switching valve wire into its original position. Fit the cover into position with the lower wire cover and the timing belt cover. Install the 2 securing bolts.
32. Replace the upper ignition lead cover, snap into place and secure the 2 bolts.
33. Install the intake air connector to the throttle body and air cleaner hose. Tighten the clamps and tighten the bolt to the cylinder head cover. Connect the air hose to the ISC valve and the power steering control air hose to the intake air connector.
34. Connect the throttle cable, accelerator cable and the cruise control cable to the throttle body. Install the throttle body cover.
35. Connect the negative battery terminal and fill the engine with coolant. Check the systems for leaks.

LS400

1. Disconnect the negative battery cable.
2. Drain the engine coolant.
3. Remove the air cleaner and air flow meter.
4. Disconnect the water by-pass hoses from the throttle body.
5. Disconnect the hose from the ISC valve body.
6. Tag and disconnect the hoses. Remove the left side engine wire cover.
7. Disconnect the accelerator linkage and the cable from the throttle body.
8. Disconnect the cold start injector, TPS, ISC and VSV connectors and related hoses.
9. Disconnect the EGR vacuum modulator and vacuum control valve.
10. Disconnect the 3 bolts holding the air pipe and vacuum pipe. Disconnect the water bypass pipe. Remove the 2 bolts holding the accelerator cable bracket.
11. Disconnect the cold start fuel hose from the delivery pipe.

12. Disconnect the EGR pipe from the air intake chamber by removing the 2 bolts and remove the gasket once it is off.
13. Remove the 4 bolts and 8 nuts, throttle cable clamp, the 2 vacuum switching valves and check connector and remove the air intake chamber. Disconnect the water hose from the water inlet housing. Remove the 4 gaskets from the intake manifold. Remove the cold start injector from the air intake manifold.

To install:

14. Place the new gasket and install the cold start injector and bolts. Torque to 42 inch lbs. (4.9 Nm). Install the cold start injector pipe to the injector and torque to 11 ft. lbs. (15 Nm). Use new gaskets and washers throughout.
15. Install the air chamber with new gaskets. Install the bolts, nuts and accessories. Torque to 13 ft. lbs. (18 Nm).
16. Tighten the EGR connecting bolts to 13 ft. lbs. (18 Nm).
17. Connect the cold start fuel pipe using new washers and torque to 11 ft. lbs. (15 Nm).
18. Install the accelerator control cable bracket and torque to 69 inch lbs. (7.8 Nm).
19. Connect the water bypass hose.
20. Install the air and vacuum pipe and torque the 3 bolts to 69 inch lbs. (7.8 Nm).
21. Install the vacuum modulator and vacuum control valve.
22. Connect the emission control, actuator, brake booster, PCV, ISC, ISC water by-pass hoses and related hoses.
23. Connect the electrical connectors and wiring cover.
24. Install the throttle body and related connections. Install the air cleaner and air flow meter. Fill the cooling system and connect the negative battery cable.
25. Check for coolant and fuel leaks. Check coolant level after running.

AIR FLOW METER

Removal and Installation

ES250 AND ES300

1. Disconnect the negative battery cable and the air flow meter connector.
2. Disconnect the vacuum hoses and air cleaner hose.
3. Remove the air cleaner cap by removing the 2 bolts from the resonator and pry off the lock plate. Remove the bolt, the 4 nuts and plate washers.
4. Remove the air flow meter and gasket from the cap.

To install:

5. Install air flow meter with gasket, lock plate, plate washers and nuts.
6. Pry lock plate over nut.
7. Install the air cleaner cap and air flow meter assembly.
8. Connect the air flow meter wiring.
9. Connect the negative battery cable.

SC300

1. Disconnect the negative battery cable.

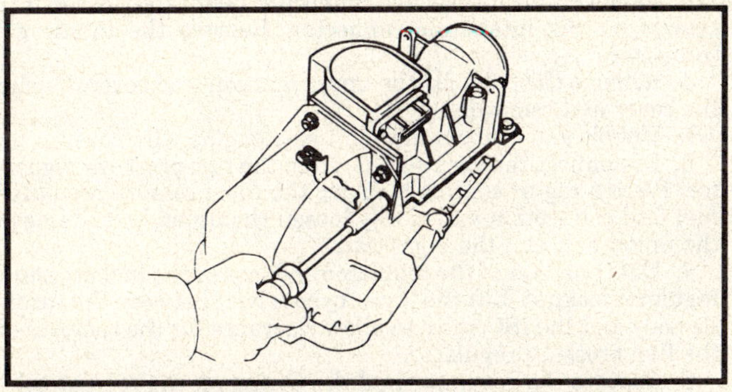
Removing the air flow meter from the air cleaner assembly – ES250

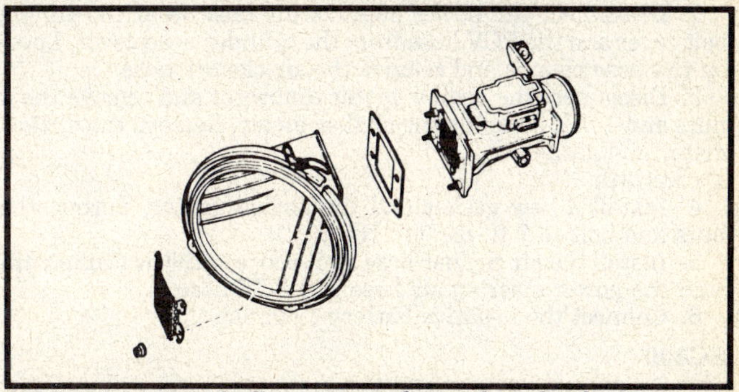
Removing the air flow meter from the air cleaner assembly – LS400

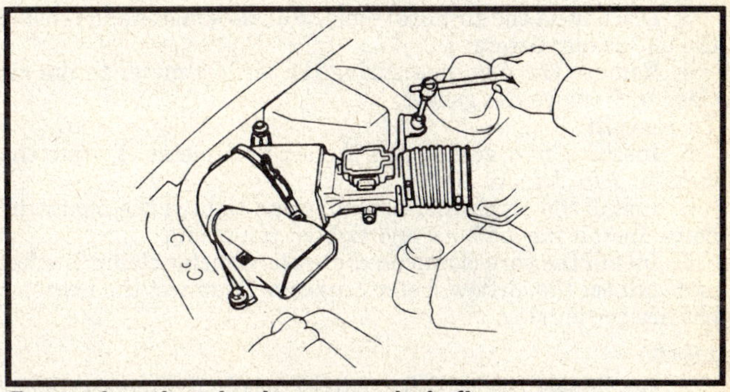
Removing the air cleaner and air flow meter assembly from the vehicle – LS400

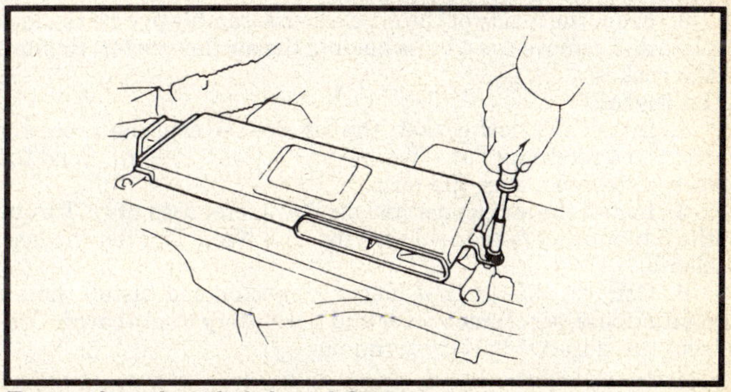
Removing the air inlet – LS400

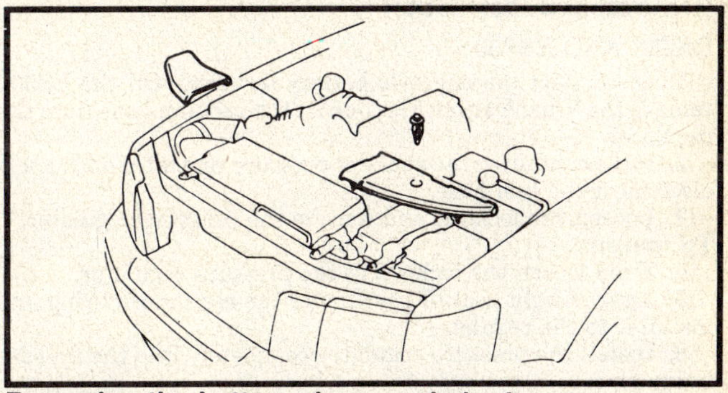
Removing the battery clamp and air cleaner covers – LS400

2. Disconnect the power steering air hose from the timing belt cover and the PCV hose from the cylinder head cover. Loosen the hose clamps and remove the air cleaner hose.
3. Disconnect the airflow meter connector and remove the 4 nuts and 1 bolt securing the airflow meter. Remove the airflow meter and gasket.

To install:

4. Install a new gasket and the airflow meter. Torque the nuts and bolt to 7 ft. lbs. (10 Nm).
5. Install the air cleaner hose and secure with the clamps. Install the power steering air hose and PCV hose.
6. Connect the negative battery terminal.

SC400

1. Disconnect the negative battery cable and the airflow meter connector. Disconnect the wire clamp from the airflow meter clamp bracket.
2. Loosen the air hose clamp and the 4 air cleaner assembly securing bolts.
3. Disconnect the air hose from the airflow meter and remove the air cleaner assembly.
4. Remove the 4 bolts securing the airflow meter to the air cleaner. Remove the gasket.

To install:

5. Install a new gasket and the air flow meter. Torque the bolts to 7 ft. lbs. (10 Nm).
6. Install the air cleaner assembly and tighten the mounting bolts. Install the air hose and tighten the clamp.
7. Install the wire clamp on the air flow meter clamp bracket and connect the airflow meter connector. Connect the negative battery terminal.

LS400

1. Disconnect the negative battery terminal. Remove the air cleaner cover, battery clamp cover and the air cleaner hose.
2. Disconnect the air flow meter connector. Loosen the clamp and remove the 3 bolts retaining the air cleaner and air flow meter assembly. Remove assembly.
3. From the inside of the air cleaner assembly pry off the lock plate and remove the 4 bolts holding the air flow meter. Remove the gasket.

To install:

4. Install the new gasket, the air cleaner cap, the protector and the 4 retaining nuts. Torque to 7 ft. lbs. (10 Nm). Bend the lock plate down over the nuts.
5. Install the air cleaner and air flow meter assembly. Torque the 3 mounting nuts to 43 inch lbs. (4.9 Nm). Tighten the hose clamp.
6. Connect the air flow meter connector and install the air cleaner hose, air cleaner cover and the battery clamp cover. Connect the negative battery terminal.

PRESSURE REGULATOR

Removal and Installation

ES250 AND ES300

1. Disconnect the negative battery terminal. On the ES300 remove the V-bank cover. Disconnect the vacuum hose from the regulator.
2. Put a contain or towel under pressure regulator and slowly disconnect the fuel hose.
3. Loosen the locknut and remove the pressure regulator.

To install:

4. Fully loosen the locknut of the pressure regulator.
5. Apply a light coat of gasoline to the regulator O-ring and install it to the regulator.
6. Install the pressure regulator completely into the fuel delivery pipe.
7. Turn the pressure regulator counterclockwise until the vacuum fitting is vertical.

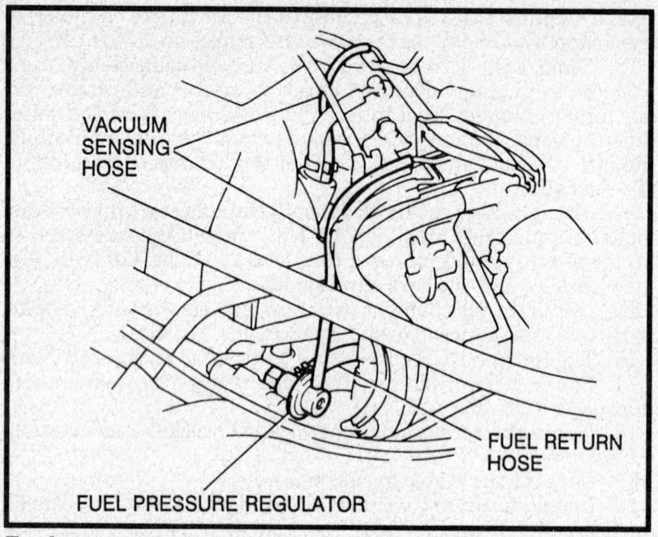

Fuel pressure regulator and related hoses. Note the position of the vacuum line outlet on the regulator—ES250

8. Tighten the locknut to 24 ft. lbs. (32 Nm) on the ES250 and 16 ft. lbs. (22 Nm) on the ES300.
9. Connect fuel hose and vacuum line. Replace the V-bank cover on the ES300. Connect the negative battery terminal. Start the engine and check for fuel leaks.

SC300

1. Disconnect the negative battery terminal. Disconnect the vacuum hose from the fuel pressure regulator.
2. Place a rag or container under the fuel pressure regulator fuel feed connection and slowly loosen the union bolt. Remove the union bolt and the 2 gaskets.
3. Remove the 2 bolts from the fuel pressure regulator and remove the regulator. Remove the O-ring from the regulator.

To install:

4. Install a new O-ring on the fuel pressure regulator and lightly coat with gasoline.
5. Install the regulator and torque the bolts to 78 inch lbs. (8.8 Nm).
6. Connect the fuel feed line with 2 new gaskets and torque the union bolt to 20 ft. lbs. (27 Nm). Connect the vacuum line and the negative battery terminal. Check for leaks.

SC400

1. Disconnect the negative battery terminal.
2. Disconnect the ISC valve air hose and the power steering air control valve air hose from the intake air connector.
3. Remove the intake air connector bolts and loosen the clamps on the intake air connector. Remove the intake air connector.
4. Remove the bolts for the upper ignition lead cover. Unclip the cover and remove.
5. Unbolt and remove the right side engine wire cover.
6. Disconnect the vacuum line from the fuel pressure regulator. Place a rag or container under the fuel pressure regulator fuel feed connection and slowly loosen the union bolt. Remove the union bolt and the 2 gaskets.
7. Note position of the regulator and loosen the fuel pressure regulator locknut. Lift the water bypass hose between the throttle body and the ISC valve to allow clearance for the removal of the fuel pressure regulator.
8. Remove the regulator and the O-ring on the regulator.

To install:

9. Fully loosen the locknut of the pressure regulator.

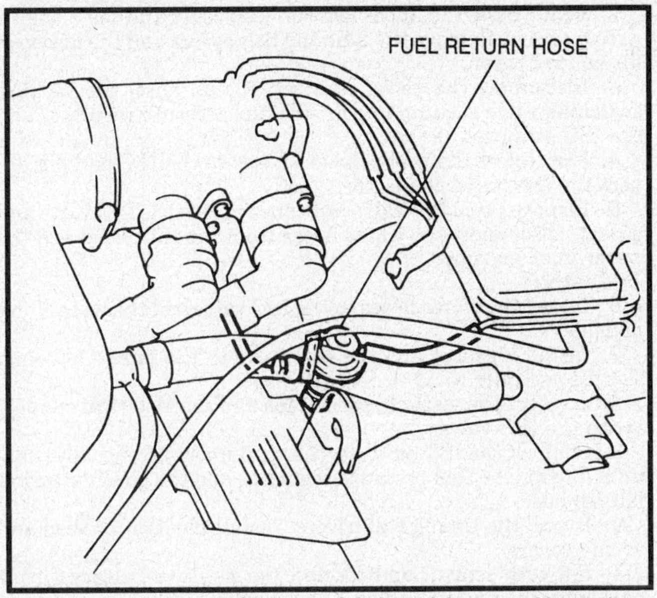

Fuel pressure regulator and related hoses. Note the position of the vacuum line outlet on the regulator—LS400

10. Apply a light coat of gasoline to the regulator O-ring and install it to the regulator.
11. Install the pressure regulator completely into the fuel delivery pipe.
7. Turn the pressure regulator counterclockwise until the regulator is in the original postion.
12. Connect the fuel return line with 2 new gaskets and torque the union bolt to 26 ft. lbs. (35 Nm). Tighten the locknut to 22 ft. lbs. (29 Nm).
13. Connect the vacuum line. Replace the right side wire cover and the upper ignition lead cover.
14. Install the intake air connector. Tighten the clamps and bolt. Connect the air hoses. Connect the negative battery terminal. Start the engine and check for fuel leaks.

LS400

1. Disconnect the negative battery terminal. Disconnect the ISC air hose.
2. Disconnect the vacuum hose from the regulator.
3. Remove the upper ignition lead cover and the right side engine wiring cover.
4. Disconnect the water bypass hose.
5. Place towel or container under the pressure regulator. Remove the union bolt and washers from the return line connection and remove from the pressure regulator body.
6. Note the positioning of the fuel pressure regulator, loosen the locknut and remove the pressure regulator

To install:

7. Fully loosen the locknut of the pressure regulator.
8. Apply a light coat of gasoline to the regulator O-ring and install it to the regulator.
9. Install the pressure regulator completely into the fuel delivery pipe.
10. Turn the pressure regulator counterclockwise until the vacuum fitting is in its original position approximately 20 degrees right of center.
11. Connect the fuel return line with new gaskets and tighten the union bolt to 22 ft. lbs. (29 Nm).
12. Torque the regulator locknut to 22 ft. lbs. (29 Nm).
13. Connect the water bypass hose, the right side engine wiring cover and the upper ignition lead cover.
14. Connect the vacuum hose and ISC air hose. Connect the negative battery terminal.
15. Start the engine and check for fuel leaks.

IDLE SPEED CONTROL VALVE

Removal and Installation

ES250

1. Disconnect the negative battery terminal. Drain engine coolant.
2. Disconnect ISC valve connector.
3. Disconnect air hose and the 2 water by-pass hoses.
4. Remove the nut on the ISC valve and disconnect the wiring harness clamp. Remove the 2 bolts retaining the ISC valve and remove the ISC valve and gasket.

To install:

5. Install the ISC valve and gasket. Torque the bolts to 9 ft. lbs. (13 Nm). Connect the wiring harness clamp with the nut.
6. Connect the air hose and water by-pass hoses.
7. Connect the ISC valve connector.
8. Fill the cooling system. Connect the negative battery terminal.

ES300

1. Disconnect the negative battery terminal. Drain engine coolant.
2. Disconnect ISC valve connector.
3. Disconnect the 2 air hoses and the 2 water by-pass hoses.
4. Disconnect the power steering idle-up vacuum hose. Remove the 2 bolts retaining the ISC valve and remove the ISC valve and gasket.

To install:

5. Install the ISC valve and gasket. Torque the bolts to 9 ft. lbs. (13 Nm). Connect the power steering idle-up vacuum hose.
6. Connect the air hoses and water by-pass hoses.
7. Connect the ISC valve connector.
8. Fill the cooling system. Connect the negative battery terminal.

SC300

1. Disconnect the negative battery terminal and drain the coolant.
2. Disconnect the throttle cable, the accelerator cable and cruise control actuator cable from the throttle body.
3. Remove the power steering air hose from the timing belt cover. Remove the PCV hose from the cylinder head cover.
4. Loosen the air cleaner hose clamps and remove the hose.
5. Remove the 4 bolts securing the throttle body support bracket.
6. Disconnect the 4 vacuum hoses, the throttle position sensor connector, the ISC valve connector, the subthrottle position sensor connector if equipped and the subthrottle actuator connector if equipped.

Removing the ISC valve—ES250

7. Remove the 2 nuts and 2 bolts securing the throttle body and separate the throttle body from the intake air connector. Disconnect the 2 water bypass hoses from the throttle body and remove the throttle body.
8. Remove the 2 screws retaining the ISC valve. Remove the ISC valve and the O-ring on the ISC valve.

To install:

9. Install a new O-ring on the ISC valve and fit the ISC valve on the throttle body. Tighten the 2 mounting screws.
10. Connect the 2 water bypass hoses to the throttle body. Install a new gasket on the throttle body base and replace the throttle body on the intake air connector. Torque the nuts and bolts to 15 ft. lbs. (21 Nm).
11. Connect the subthrottle position sensor connector if equipped, the subthrottle actuator connector if equipped, the throttle position sensor connector, the ISC valve connector and the 4 vacuum hoses.
12. Install the throttle body bracket and torque the 4 nuts to 15 ft. lbs. (21 Nm).
13. Install the air cleaner hose and connect the power steering air hose and the PCV hose. Tighten the clamps.
14. Connect the accelerator cable, the throttle control cable and the cruise control cable to the throttle body.
15. Connect the negative battery terminal and fill the engine with coolant. Check for leaks.

LS400 AND SC400

1. Disconnect the negative battery terminal. Drain the coolant.
2. Remove the the upper ignition lead cover and the 2 engine wiring covers. Remove the 2 timing belt covers and the lower ignition lead cover.
3. Disconnect the power steering idle-up hose, the vacuum switching valve vacuum line for the fuel pressure regulator and the ISC air hose.
4. Disconnect the water bypass hoses to the ISC and disconnect the ISC valve connector.
5. Remove the 2 ISC valve nuts and remove the ISC valve and gasket. Disconnect the water hoses from the ISC valve and the water inlet housing.

To install:

6. Install the water hoses to the ISC valve and the water inlet housing.
7. Install a gasket and the ISC valve to the intake housing. Torque the nuts to 13 ft. lbs. (18 Nm).
8. Connect the water bypass hoses and the electrical connector to the ISC valve.
9. Connect the ISC air hose, the vacuum switching valve vacuum line for the fuel pressure regulator and the power steering idle-up hose.
10. Install the timing belt covers. Install the ignition lead and wiring covers.
11. Fill with coolant and connect the negative battery terminal. Start the engine and check for coolant leaks.

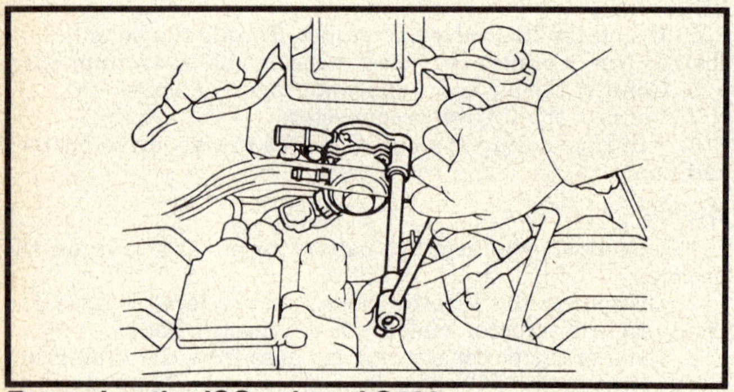

Removing the ISC valve—LS400

THROTTLE BODY

Removal and Installation

ES250

1. Disconnect the negative battery cable.
2. Drain the engine coolant.
3. Disconnect the accelerator cable and throttle cable from the throttle body. the throttle cable will be on automatic transmission equipped vehicles.
4. Remove the air cleaner, air flow meter and air cleaner hose.
5. Disconnect PCV, vacuum sensing, emission control and water by-pass hoses.
6. Disconnect throttle position sensor.
7. Remove the throttle cable bracket on automatic transmission equipped vehicles.
8. Remove the 4—manual transmission or 3—automatic transmission throttle body retaining bolts.
9. Remove the throttle body.

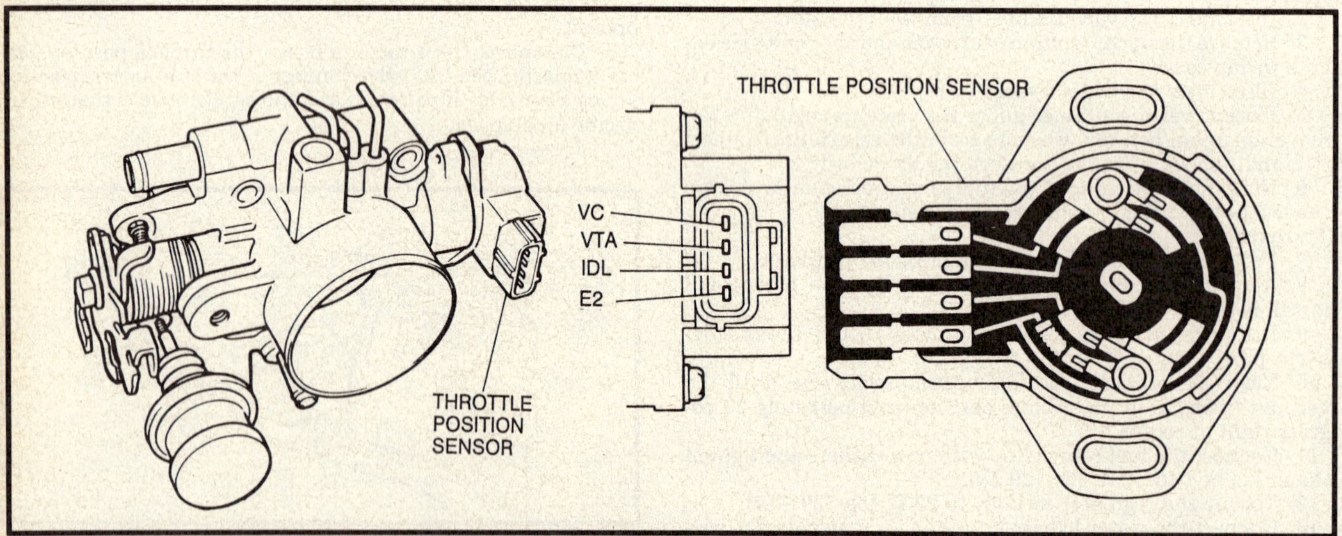

Throttle body with the throttle position sensor— ES250

To install:

10. Install the throttle body and new gasket. Torque to 9 ft. lbs. (13 Nm).
11. Install the the throttle cable bracket, if equipped. Torque to 9 ft. lbs. (13 Nm).
12. Connect the throttle position sensor.
13. Connect the PCV, vacuum sensing, emission control and water by-pass hoses.
14. Install the air cleaner, air flow meter and air cleaner hose.
15. Connect the accelerator cable and throttle cable.
16. Fill the cooling system and connect the negative battery cable.

ES300

1. Disconnect the negative battery terminal. Drain the engine coolant.
2. Disconnect the accelerator and throttle cables from the throttle linkage.
3. Disconnect the connector from the airflow meter. Disconnect the clamp from the coil wire. Disconnect the air ducting and the air cleaner hose clamp. Release the air cleaner cover clips and remove the air cleaner cover and airflow meter assembly.
4. Disconnect the throttle position sensor connector and the water bypass hoses from the throttle body. Disconnect the vacuum lines.
5. Remove the 2 nuts and the 2 bolts securing the throttle body and remove the throttle body and gasket.

To install:

6. Using a new gasket, replace the throttle body and torque the hardware to 9 ft. lbs. (13 Nm).
7. Connect the water bypass hoses, the vacuum lines and the throttle position sensor connector.
8. Install the air cleaner and airflow meter assembly. Install the air cleaner hose. Tighten the clamps and fasten the 4 clips. Connect the airflow meter connector.
9. Connect the throttle cable and the accelerator cable. Fill the engine with coolant and connect the negative battery terminal.

SC300

1. Disconnect the negative battery terminal and drain the coolant.
2. Disconnect the throttle cable, the accelerator cable and cruise control actuator cable from the throttle body.
3. Remove the power steering air hose from the timing belt cover. Remove the PCV hose from the cylinder head cover.
4. Loosen the air cleaner hose clamps and remove the hose.
5. Remove the 4 bolts securing the throttle body support bracket.
6. Disconnect the 4 vacuum hoses, the throttle position sensor connector, the ISC valve connector, the subthrottle position sensor connector if equipped and the subthrottle actuator connector if equipped.
7. Remove the 2 nuts and 2 bolts securing the throttle body and separate the throttle body from the intake air connector. Disconnect the 2 water bypass hoses from the throttle body and remove the throttle body.

To install:

8. Connect the 2 water bypass hoses to the throttle body. Install a new gasket on the throttle body base and replace the throttle body on the intake air connector. Torque the nuts and bolts to 15 ft. lbs. (21 Nm).
9. Connect the subthrottle position sensor connector if equipped, the subthrottle actuator connector if equipped, the throttle position sensor connector, the ISC valve connector and the 4 vacuum hoses.
10. Install the throttle body bracket and torque the 4 nuts to 15 ft. lbs. (21 Nm).
11. Install the air cleaner hose and connect the power steering air hose and the PCV hose. Tighten the clamps.
12. Connect the accelerator cable, the throttle control cable and the cruise control cable to the throttle body.
13. Connect the negative battery terminal and fill the engine with coolant. Check for leaks.

SC400

1. Disconnect the negative battery terminal and drain the engine coolant.
2. Remove the nut and 2 bolts holding the throttle body cover and remove the cover.
3. Disconnect the accelerator cable, the throttle control cable (automatic transmission) and the cruise control actuator cable from the throttle body.
4. Disconnect the ISC valve air hose and the power steering air control valve air hose. Remove the bolt securing the intake air connector to the cylinder head cover. Loosen the 2 hose clamps and disconnect the air intake connector from the throttle body and air cleaner hose. Remove the intake air connector.
5. Remove the 2 bolts securing the upper ignition lead cover and disengage the clip securing the upper and lower halves of the cover. Separate the halves and remove the upper portion.
6. Remove the bolt securing the right side wire cover and remove the cover. Remove the 2 bolts holding the left side wire cover and remove the cover.
7. Remove the 2 bolts securing the EVAP system vacuum switching valve to the cylinder head and timing belt cover.
8. Remove the 3 timing belt cover bolts from right side and remove the timing belt cover. Remove the 4 timing belt cover bolts from the left side cover and remove the ignition lead grommet from the cover and remove the cover.
9. Disconnect the ignition lead from the right side ignition coil and remove the ignition leads from the lower ignition lead cover. Remove the mounting bolt and remove the lower cover.
10. Disconnect the throttle position sensor, the subthrottle position sensor if equipped and the subthrottle actuator connector if equipped. Disconnect the heater hose from the heater valve, the ISC valve water bypass hose, the vacuum hose from the throttle body.
11. Remove the 2 bolts and 2 nuts securing the throttle body to the intake chamber. Remove the throttle body and disconnect the PCV hose and water bypass hose. Remove the throttle body gasket.
12. Disconnect the cold start injector connector, the ISC valve connectors, the EGR valve connector and the fuel pressure control vacuum switching valve.
13. Disconnect the ISC valve water bypass hose from the water inlet housing, the power steering air control valve vacuum hose from the intake chamber, the fuel pressure regulator vacuum hose from the vacuum pipe, the 2 EVAP system vacuum switching valves vacuum lines from the vacuum pipe and the charcoal canister vacuum line from the vacuum pipe.
14. Disconnect the EGR water bypass hose from the water bypass pipe, the brake booster vacuum hose from the intake chamber, the heater valve vacuum switching valve vacuum hose from the intake chamber and the PCV hose from the left side cylinder head cover.
15. Remove the stud bolt and bolt retaining the accelerator bracket. Remove the union bolt and washers connecting the cold start injector tube to the right side delivery pipe. Remove the bolts and gasket from the EGR pipe connection and disconnect the EGR pipe from the intake chamber. Remove the 8 nuts, 4 bolts and the automatic transmission throttle cable bracket. Disconnect the check connector bracket from the air intake chamber.
16. Remove the intake chamber assembly and the 4 gaskets.

To install:

17. Install 4 new gaskets and install the intake chamber on the intake manifold. Loosely install the EGR pipe bolts using a new gasket. Install the check connector, the throttle cable bracket, the fuel pressure control vacuum switching valve and the EGR system vacuum switching valve. Torque the 8 nuts and 4 bolts

to 13 ft. lbs. (18 Nm). Torque the EGR pipe bolts to 13 ft. lbs. (18 Nm).

18. Torque the cold start injector tube to right side delivery pipe union bolt to 11 ft. lbs. (15 Nm). Install the accelerator bracket and tighten the stud bolt and bolt.

19. Connect the EGR valve water bypass hose, the brake booster vacuum hose, the heater water valve vacuum switching valve vacuum hose, the PCV hose, the water inlet water hose, the power steering air control valve vacuum hose, the fuel pressure regulator vacuum hose, the EVAP system vacuum switching valves vacuum lines and the charcoal canister vacuum hose.

20. Connect the cold start injector connector, the ISC valve connector, the EGR valve connector and the vacuum switching valve connector for the fuel pressure control.

21. Connect the PCV hose and water hose to the throttle body. Install the throttle body on the intake chamber with a new gasket and torque the 2 nuts and 2 bolts to 13 ft. lbs. (18 Nm). Install the water bypass pipe to the clamp on the engine wiring harness cover.

22. Connect the water bypass hose to the ISC valve, the vacuum hose to the throttle body and the heater hose to the heater valve. Connect the throttle position sensor connector, the subthrottle position sensor if equipped and the subthrottle actuator connector if equipped.

23. Connect the end of the ignition leads to the lead clamps. Bolt in place the ignition lead cover and install the clamps on the ignition leads to the cover. Connect the ignition lead to the right side ignition coil.

24. Install the 3 gaskets on the right side timing belt cover. Fit the cover over the lower ignition lead cover. Tighten the 3 bolts holding the timing belt cover.

25. Install the vacuum switching valve with the 2 bolts.

26. Install the 3 gaskets on the right side timing belt cover. Install the ignition lead grommet. Fit the cover over the lower ignition lead cover. Tighten the 4 bolts holding the timing belt cover.

27. Fit into position the right side engine wire cover and secure with bolt.

28. Fit into position the left side engine wire cover and secure the vacuum switching valve wire into its original position. Fit the cover into position with the lower wire cover and the timing belt cover. Install the 2 securing bolts.

29. Replace the upper ignition lead cover, snap into place and secure the 2 bolts.

30. Install the intake air connector to the throttle body and air cleaner hose. Tighten the clamps and tighten the bolt to the cylinder head cover. Connect the air hose to the ISC valve and the power steering control air hose to the intake air connector.

31. Connect the throttle cable, accelerator cable and the cruise control cable to the throttle body. Install the throttle body cover.

32. Connect the negative battery terminal and fill the engine with coolant. Check the systems for leaks.

LS400

1. Disconnect the negative battery terminal and drain the coolant.
2. Remove the air cleaner and air flow meter assembly.
3. Disconnect the ISC air hose and the power steering idle-up hose and loosen the clamp on the air intake pipe. Remove the 2 bolts from the air intake pipe and remove the pipe.
4. Remove the throttle body cover and disconnect the accelerator control cable, the cruise control cable and the throttle cable.
5. Disconnect the vacuum hoses from the throttle body and the pressure regulator.
6. Remove the upper ignition lead cover and the water bypass hose from the ISC valve.
7. Disconnect the throttle position sensor, the throttle valve motor and sub-throttle position sensor, as equipped.
8. Remove the 2 nuts and 2 bolts retaining the throttle body and separate from the intake chamber. Remove the gasket.
9. Remove the PCV hose from the valve cover and remove the water bypass hose from the throttle body. Remove throttle body from engine. Remove hoses from the throttle body.

To install:

10. Install the bypass hoses and the PCV hose to the throttle body.
11. Install a new gasket to the air intake chamber and connect the PCV hose to the valve cover. Connect the bypass hose to the throttle body.
12. Place the throttle body in position and install the 2 bolts and 2 nuts retaining the throttle body. Torque the hardware to 13 ft. lbs. (18 Nm).
13. Reconnect the electrical connectors to the throttle position sensors and throttle valve motor. The sub-throttle position sensor connector has a yellow tape stripe.
14. Install the bypass hose to the ISC valve and install the upper ignition lead cover.

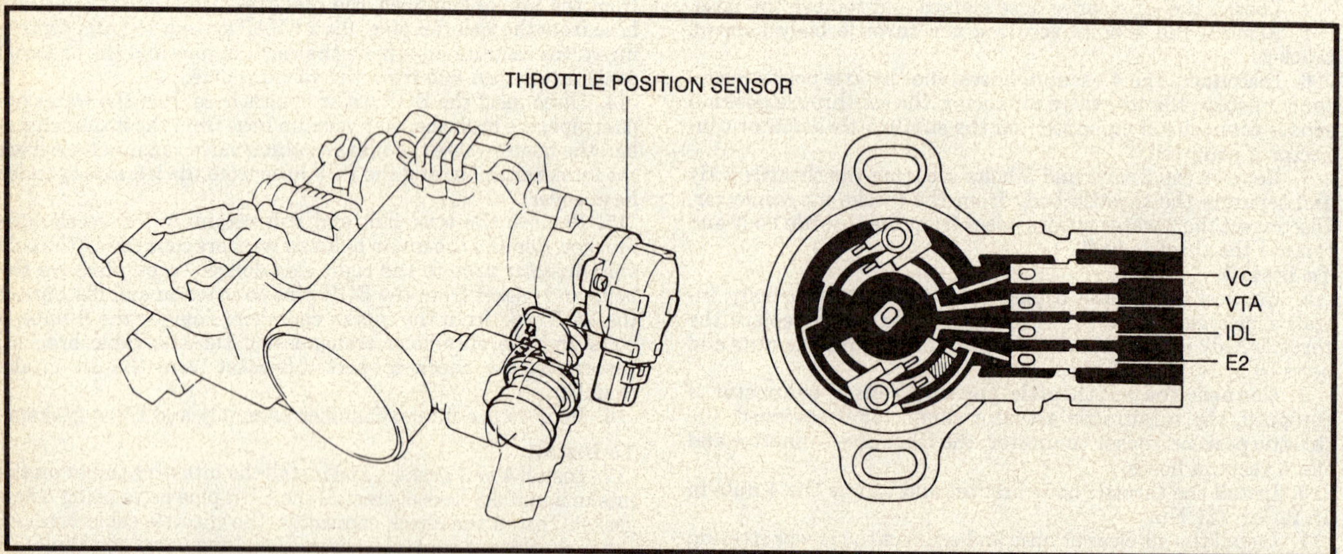

Throttle body with the throttle position sensor— LS400

15. Install the vacuum hoses to the throttle body and the pressure regulator. Connect the throttle cable, cruise control cable and the accelerator control cable.
16. Install the throttle body cover and the intake air pipe. Torque the bolts holding the pipe to 45 inch lbs. (4.9 Nm). Tighten the pipe clamp.
17. Connect the ISC air hose and the power steering idle-up hose.
18. Install the air cleaner and air flow meter assembly. Fill the cooling system with coolant.
19. Run the engine and check for coolant leaks and for proper coolant level.

THROTTLE POSITION SENSOR

Removal and Installation

1. Disconnect the negative battery terminal an the throttle position sensor connector.
2. Remove the screws retaining the throttle position sensor and remove the sensor.

To install:

3. Make sure that the throttle plate is fully closed.
4. Place the throttle position sensor into position rotated approximately 90 degrees to the right of the normal installed position.
5. Rotate the body of the throttle position sensor to the left and install the mounting screws. Adjust the sensor.
6. Connect the throttle position sensor connector and the negative battery terminal.

Adjustment

1. Loosen the sensor screws. On the ES300 apply vacuum to the throttle opener.
2. Insert the appropriate feeler gauge between the throttle stop screw and the stop lever. Correct size feeler gauge is:
   ES250–0.020 in. (0.50mm)
   ES300–0.021 in. (0.54mm)
   SC300–0.018 in. (0.45mm)
   SC400–0.016 in. (0.40mm)
   LS400–0.016 in. (0.40mm)
3. Connect an ohmmeter across terminal IDL and E2 of the sensor.
4. Gradually turn sensor clockwise until the ohmmeter deflects and tighten the screws.
5. With a continuity meter connected across terminals IDL and E2 check the adjustment by inserting a feeler gauge between the throttle stop screw and the lever. There should be continuity. Correct size feeler gauge is:
   ES250–0.012 in. (0.30mm)
   ES300–0.014 in. (0.35mm)
   SC300–0.018 in. (0.45mm)
   SC400–0.016 in. (0.40mm)
   LS400–0.016 in. (0.40mm)
6. Insert a feeler gauge between the throttle stop screw and the lever. There should be no continuity. Correct size feeler gauge is:
   ES250–0.028 in. (0.70mm)
   ES300–0.028 in. (0.70mm)
   SC300–0.022 in. (0.55mm)
   SC400–0.026 in. (0.65mm)
   LS400–0.036 in. (0.85mm)

SUBTHROTTLE SENSOR

Removal and Installation

1. Disconnect the negative battery terminal an the throttle position sensor connector.
2. Remove the screws retaining the throttle position sensor and remove the sensor.

To install:

3. Make sure that the throttle plate is fully closed.
4. Place the throttle position sensor into position rotated approximately 90 degrees to the right of the normal installed position.
5. Rotate the body of the throttle position sensor to the left and install the mounting screws. Adjust the sensor.
6. Connect the throttle position sensor connector and the negative battery terminal.

Adjustment

1. Loosen the sensor screws.
2. Insert the appropriate feeler gauge between the throttle stop screw and the stop lever. Correct size feeler gauge is:
   SC300–0.014 in. (0.35mm)
   SC400–0.013 in. (0.30mm)
3. Connect an ohmmeter across terminal IDL2 and E2 of the sensor; the bottom 2 terminals of the connector.
4. Gradually turn sensor clockwise until the ohmmeter deflects and tighten the screws.
5. With a continuity meter connected across terminals IDL2 and E2 check the adjustment by inserting a feeler gauge between the throttle stop screw and the lever. There should be continuity. Correct size feeler gauge is:
   SC300–0.014 in. (0.35mm)
   SC300–0.012 in. (0.30mm)
6. Insert a feeler gauge between the throttle stop screw and the lever. There should be no continuity. Correct size feeler gauge is:
   SC300–0.020 in. (0.50mm)
   SC400–0.020 in. (0.50mm)

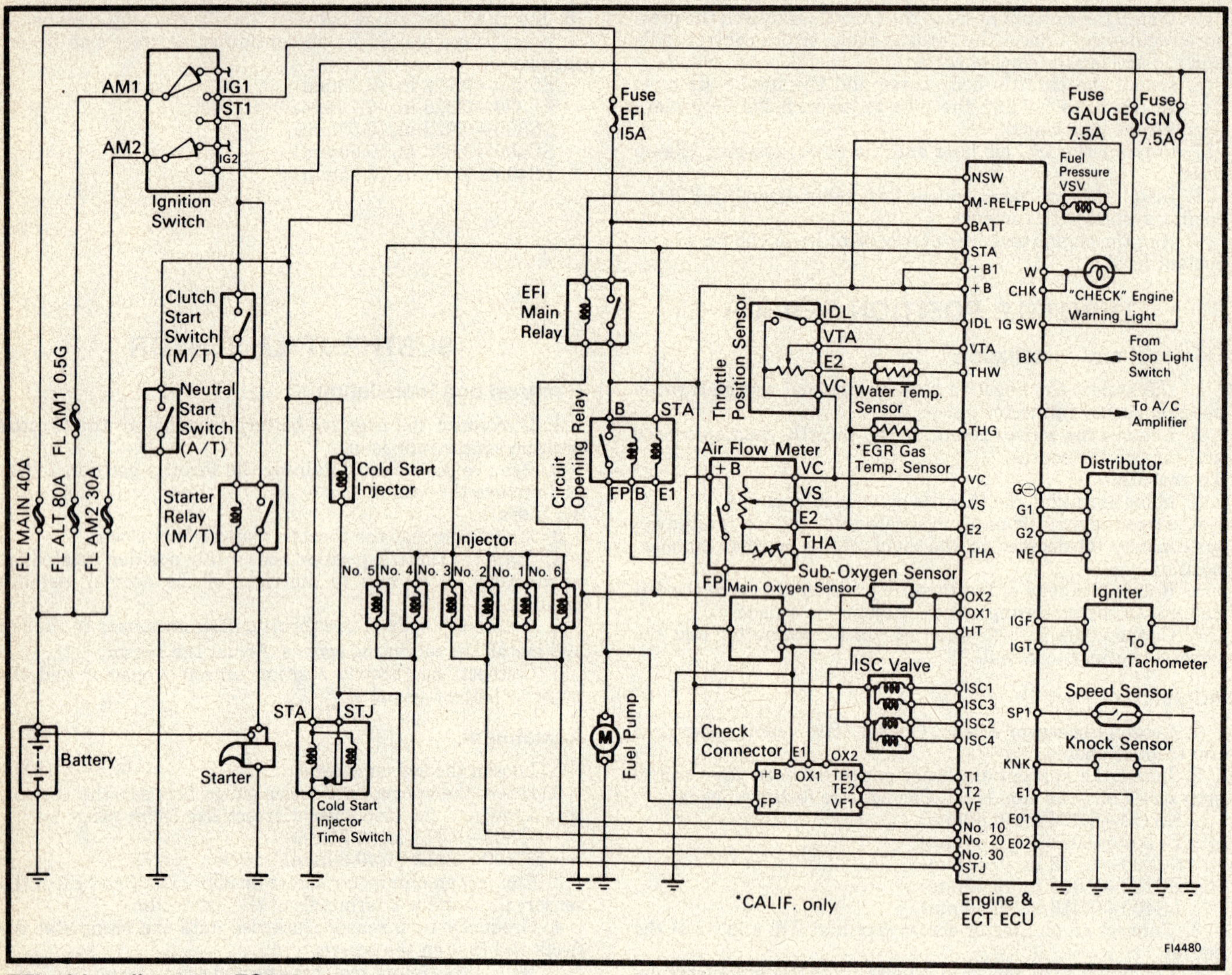

EFI wiring diagram—ES250

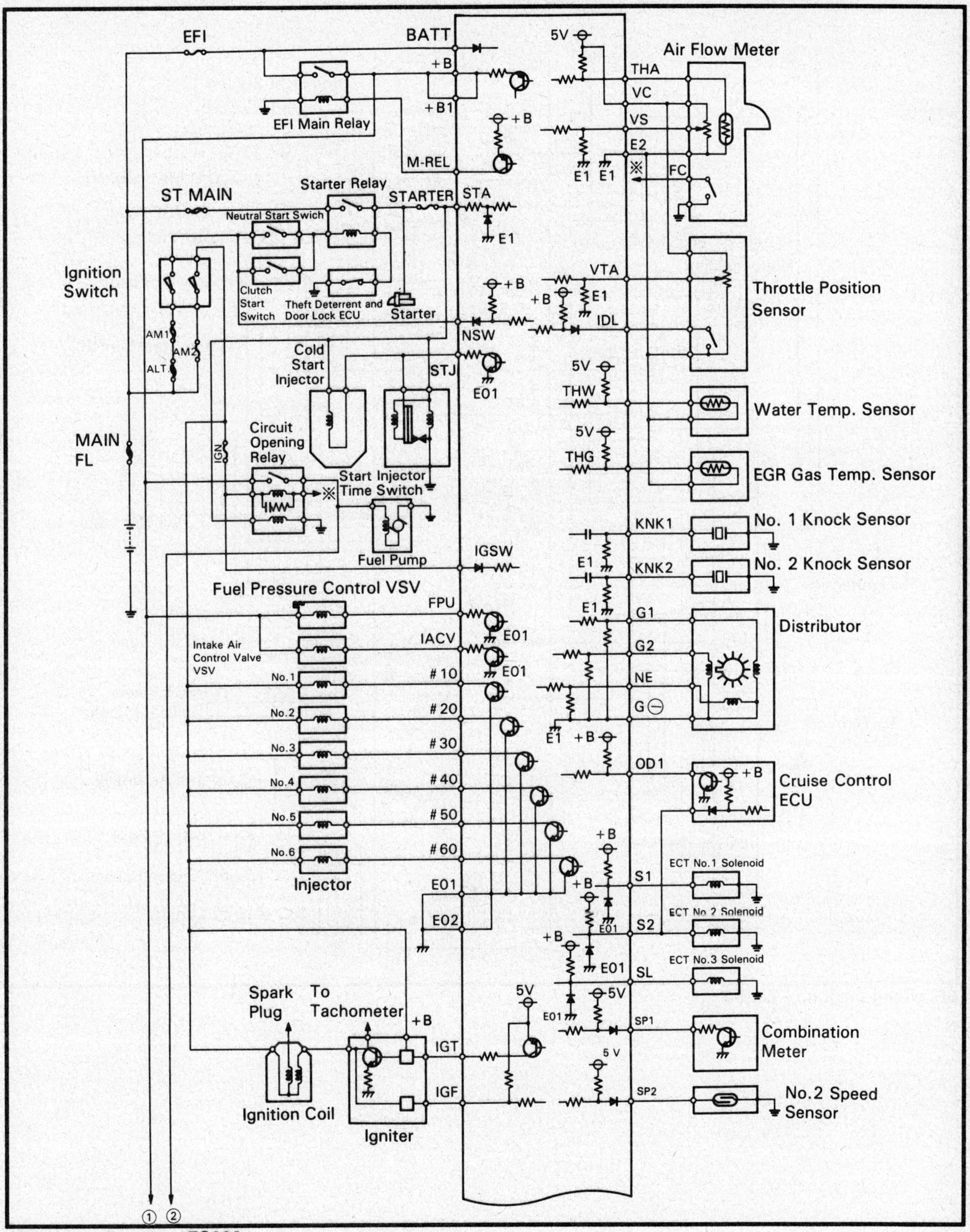

EFI wiring diagram—ES300

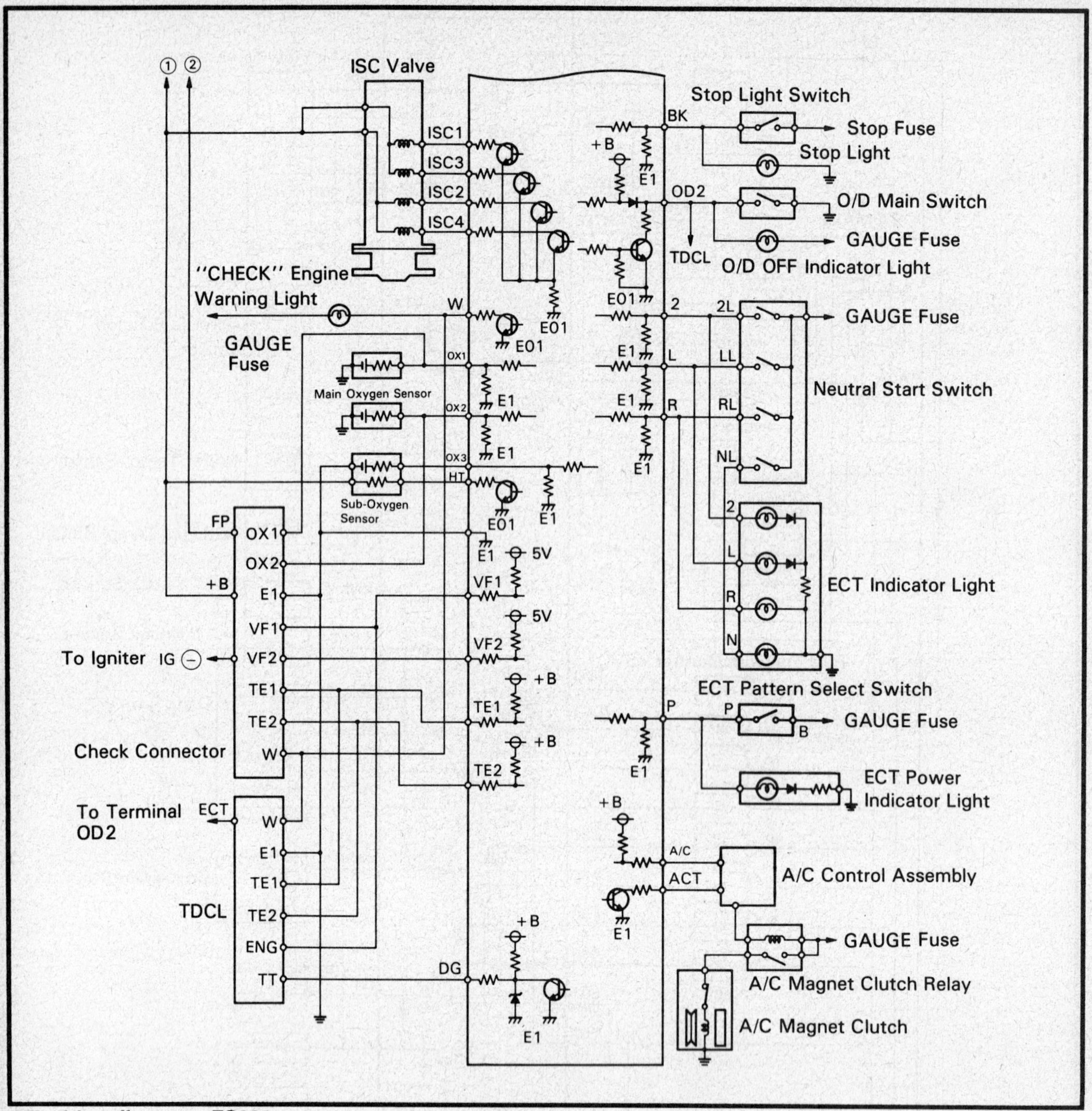

EFI wiring diagram—ES300

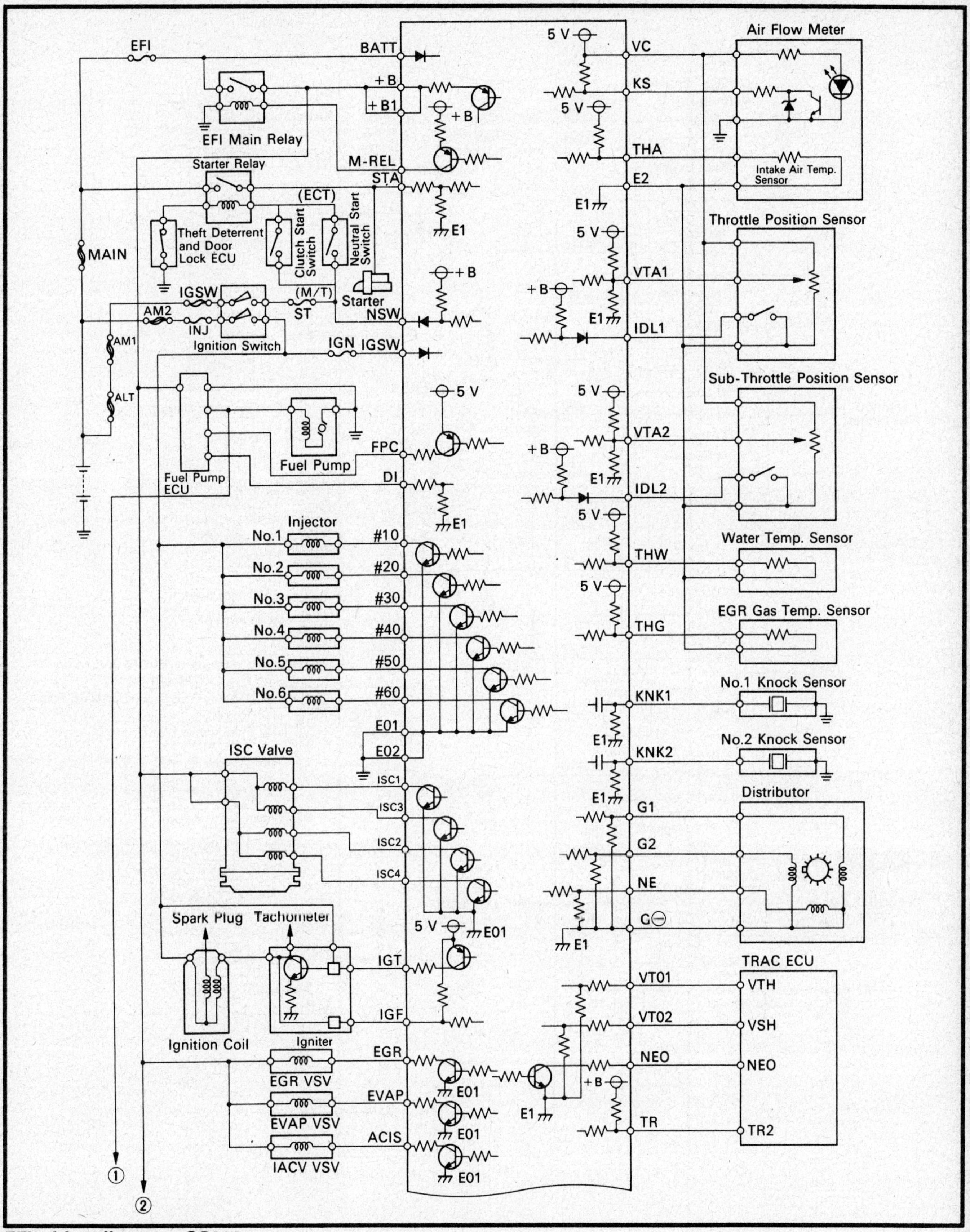

EFI wiring diagram—SC300

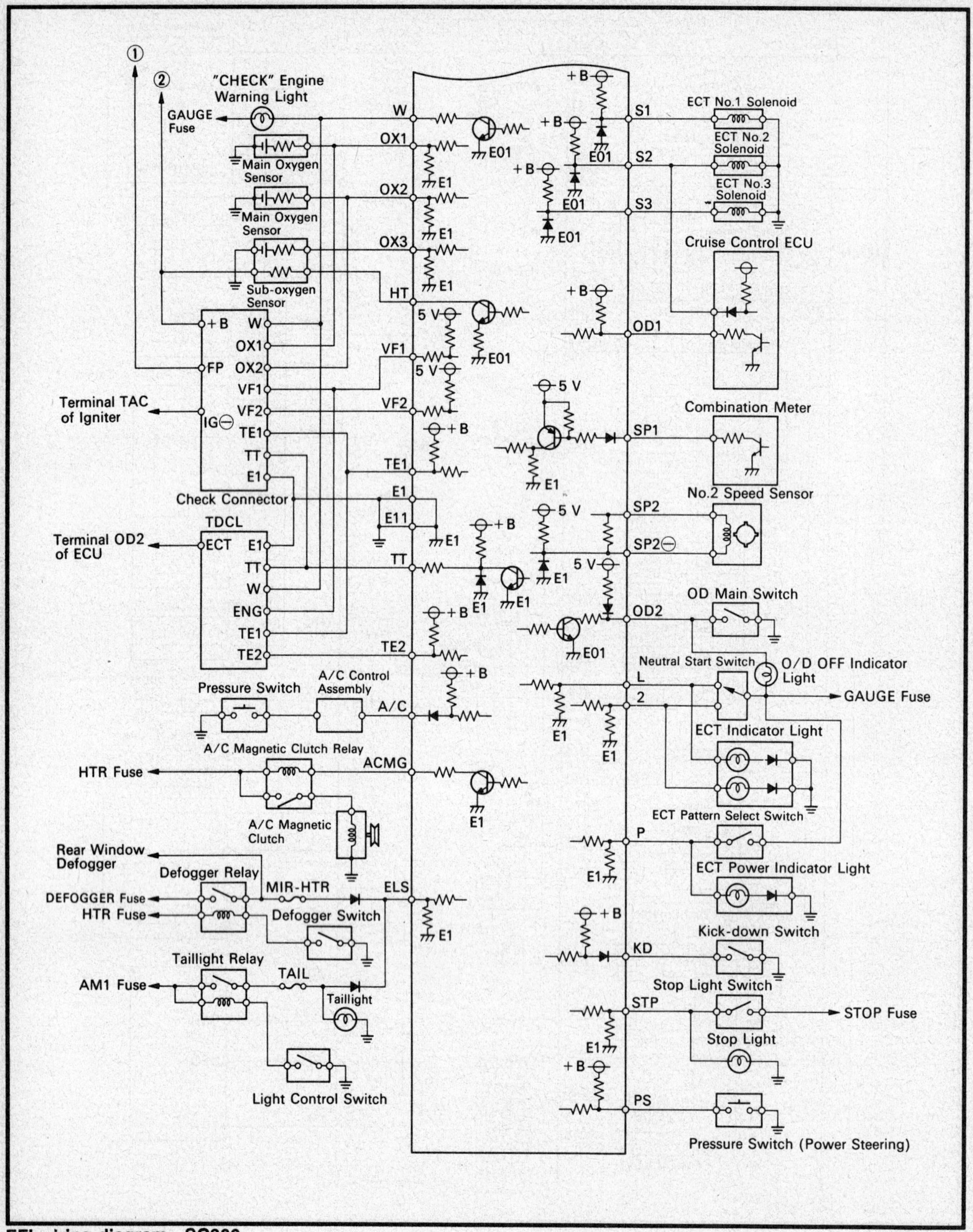

EFI wiring diagram—SC300

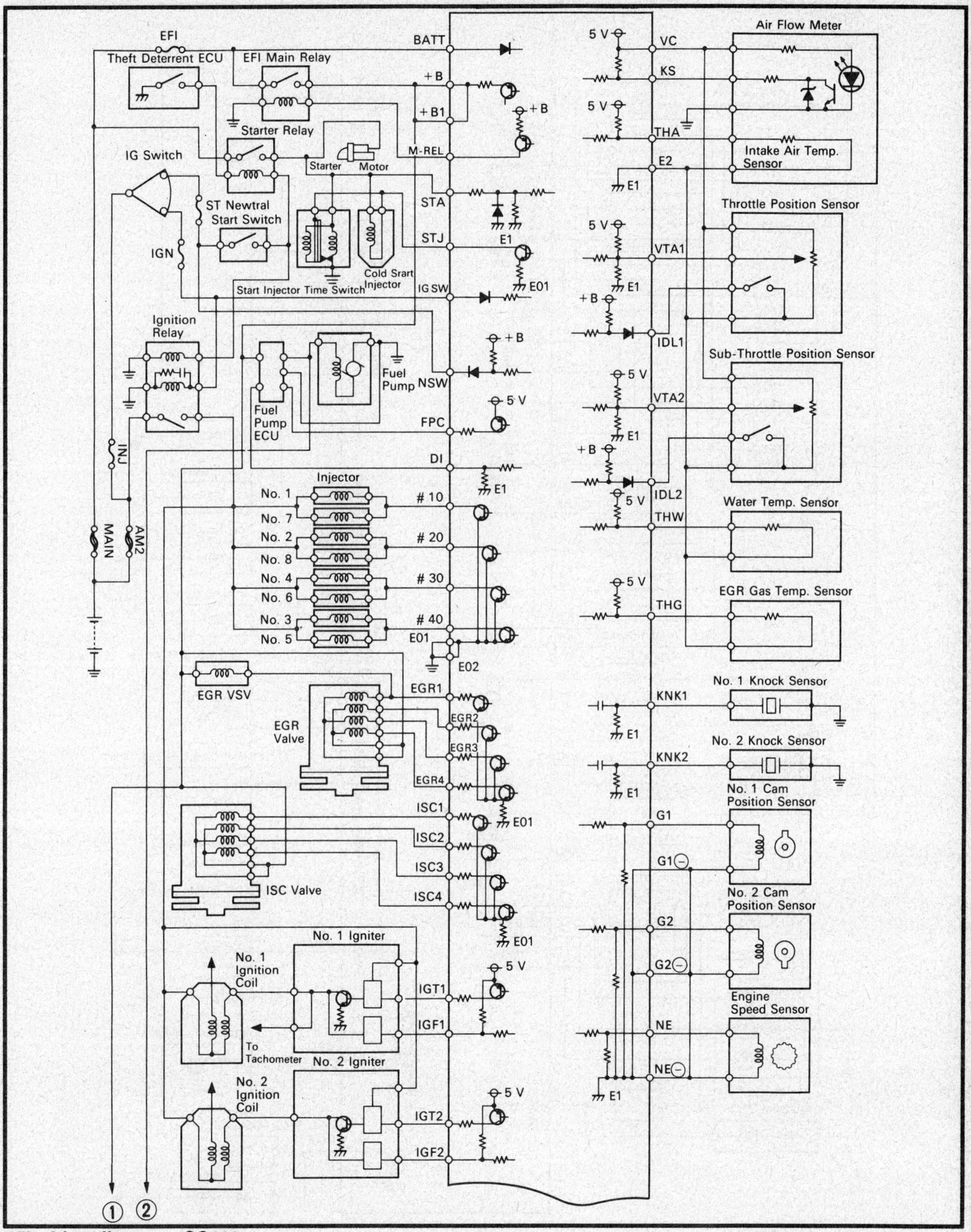

EFI wiring diagram—SC400

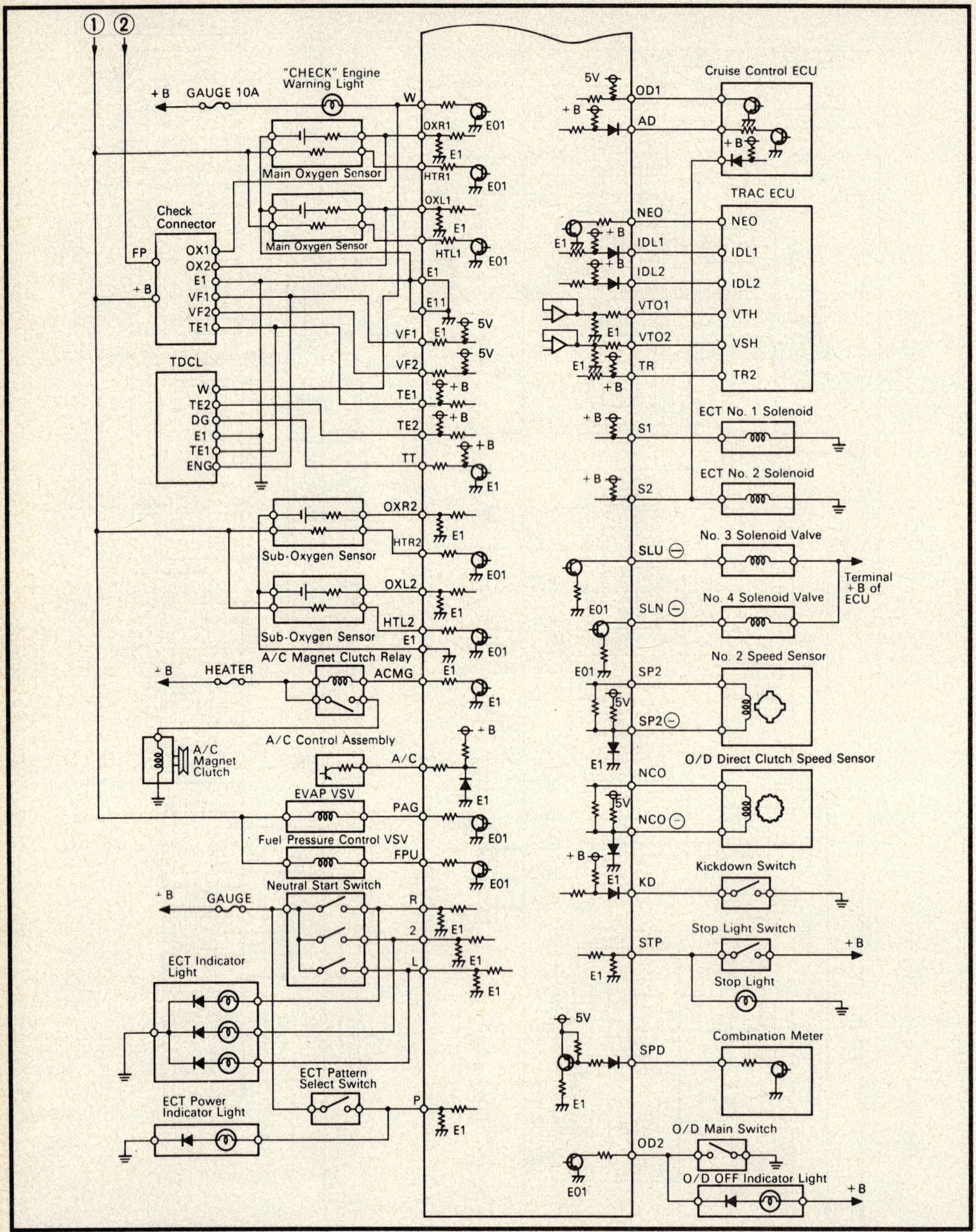

EFI wiring diagram—SC400

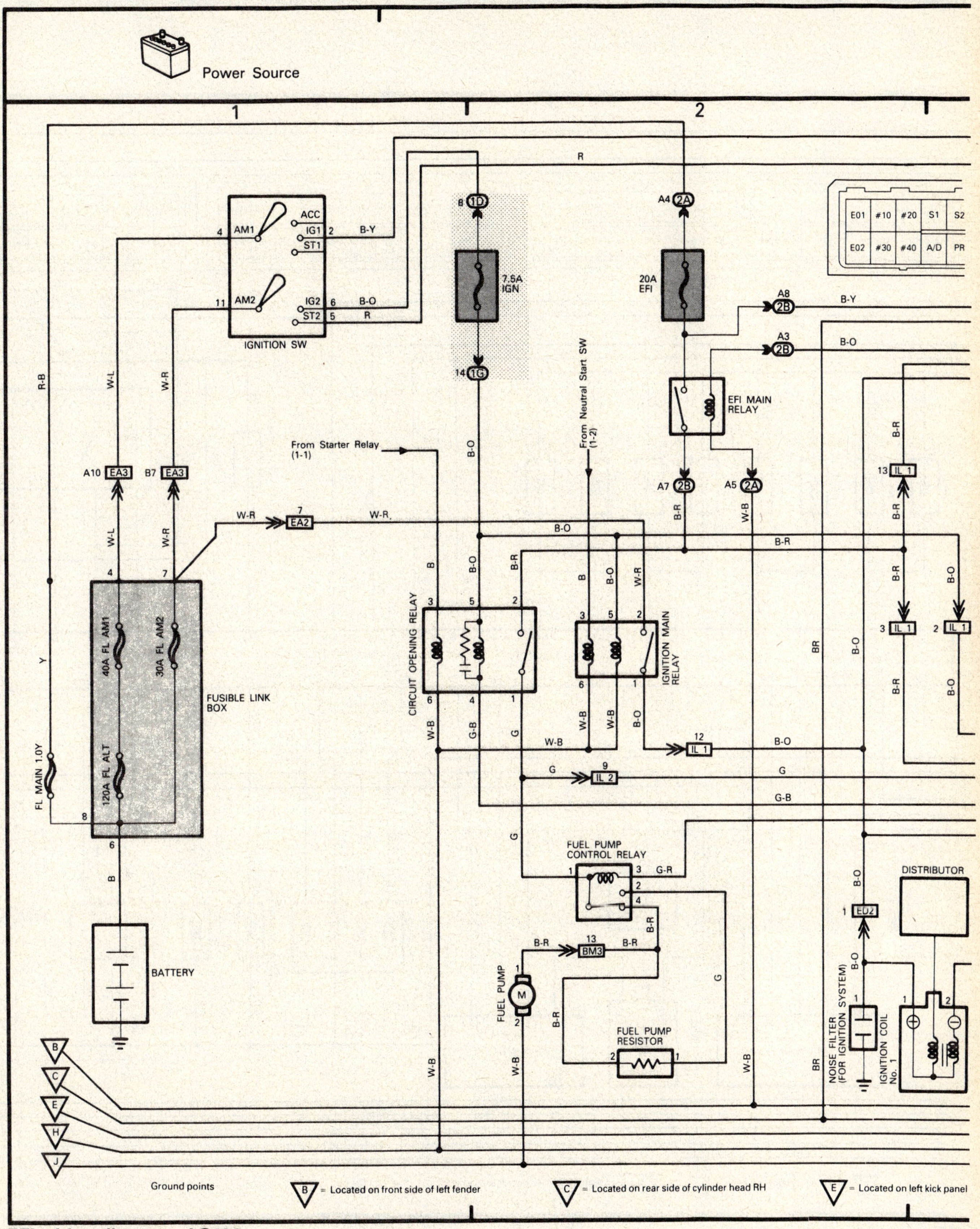

EFI wiring diagram—LS400

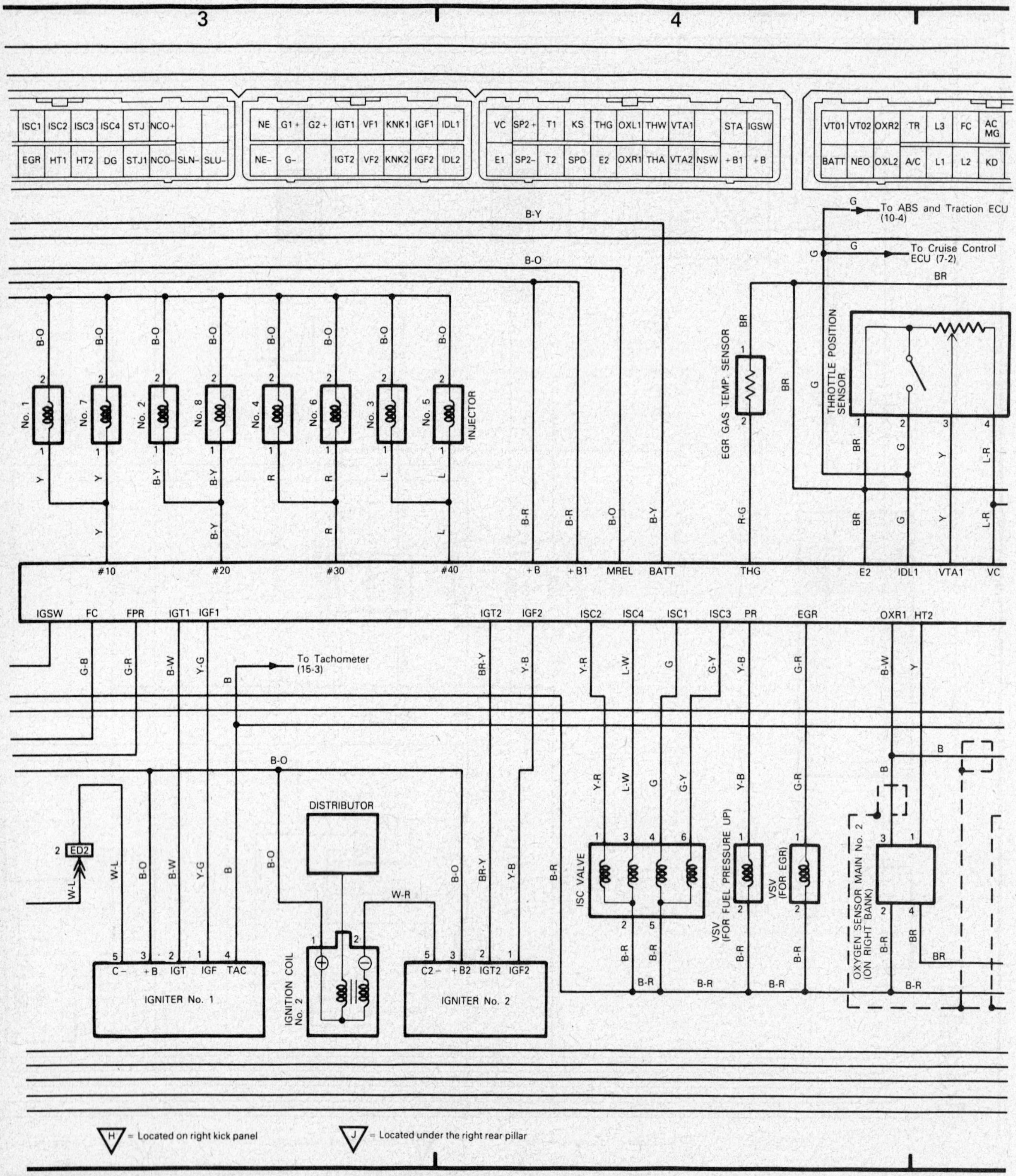

EFI wiring diagram—LS400

Engine Control

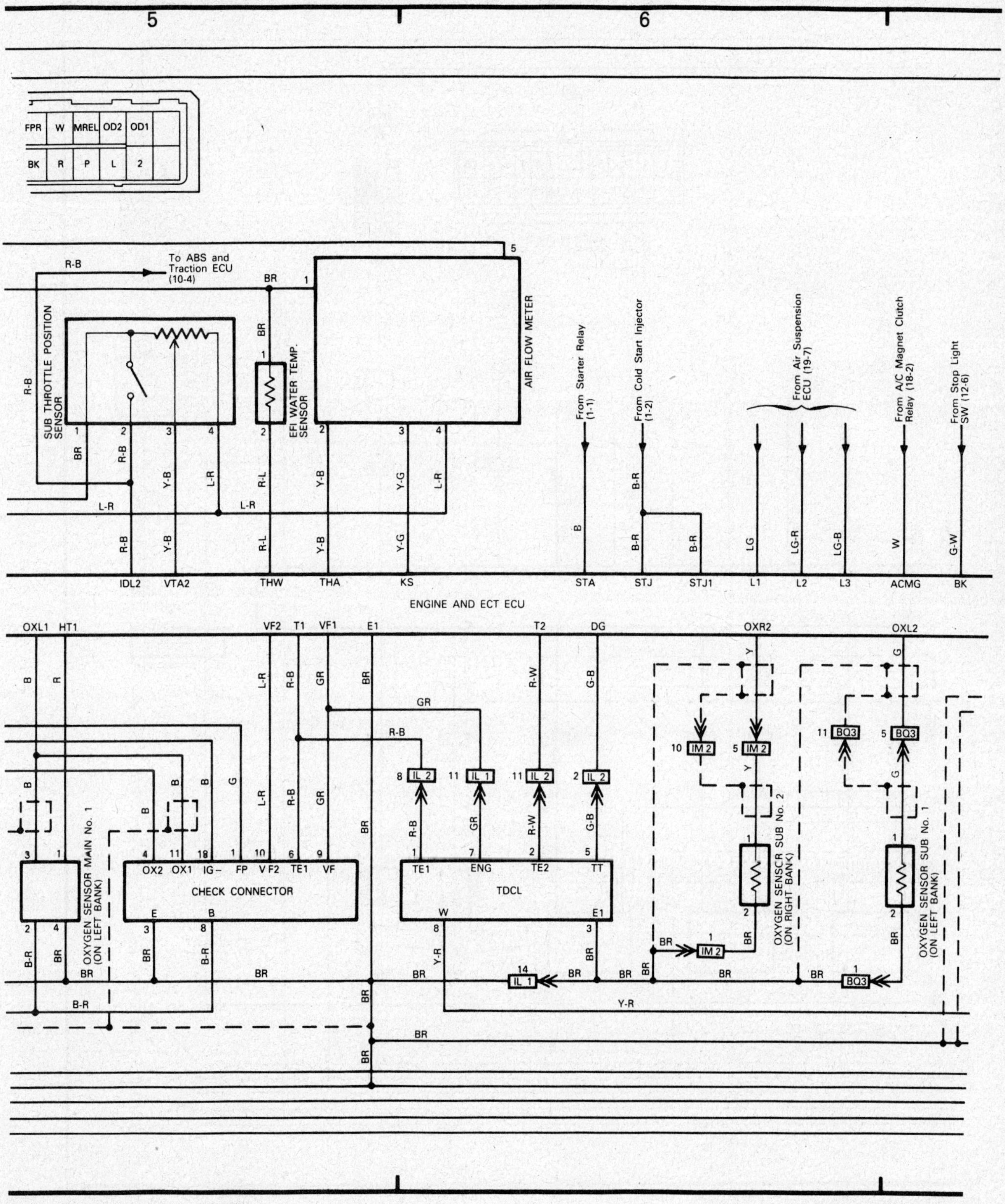

EFI wiring diagram—LS400

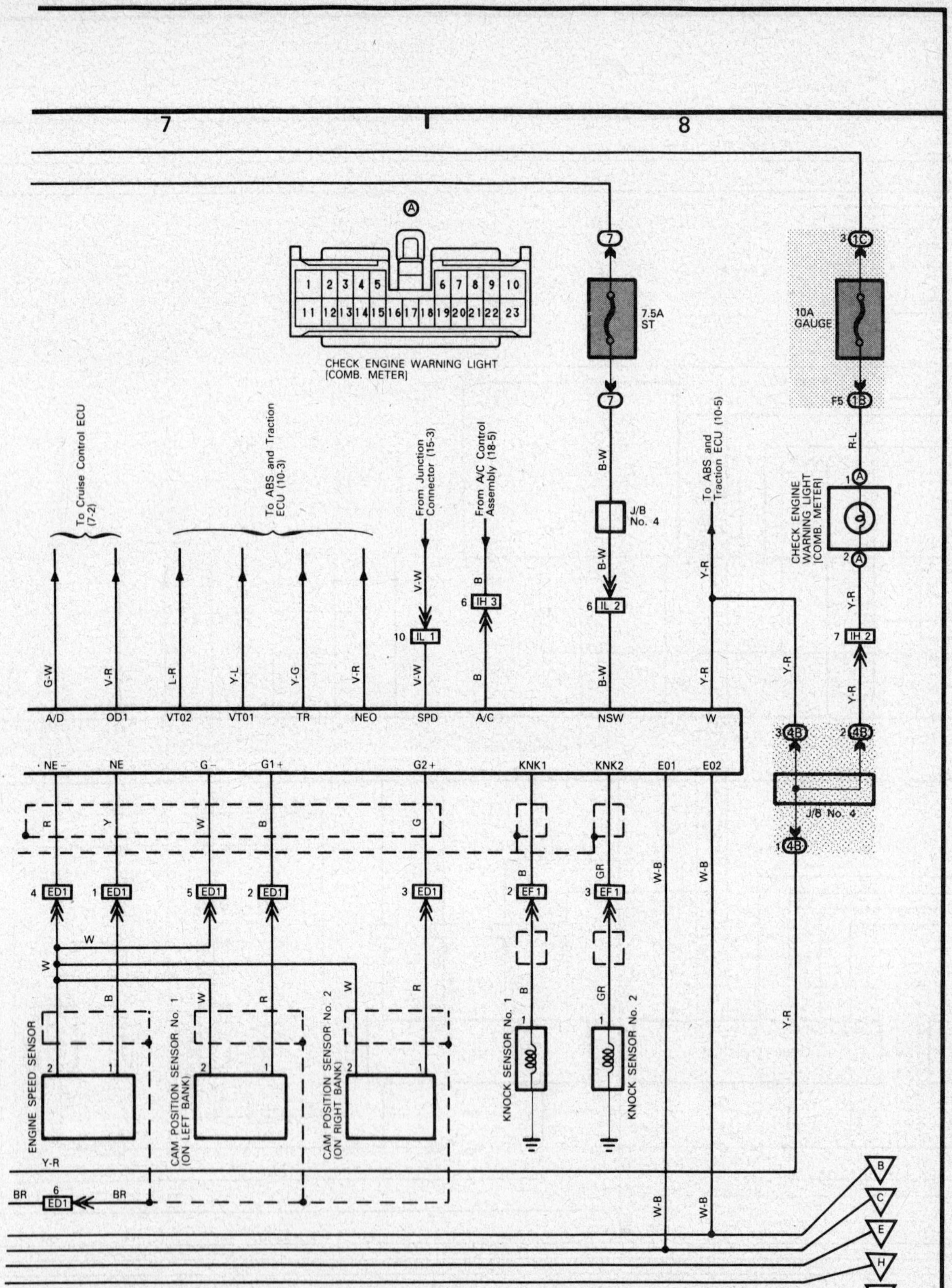

EFI wiring diagram—LS400

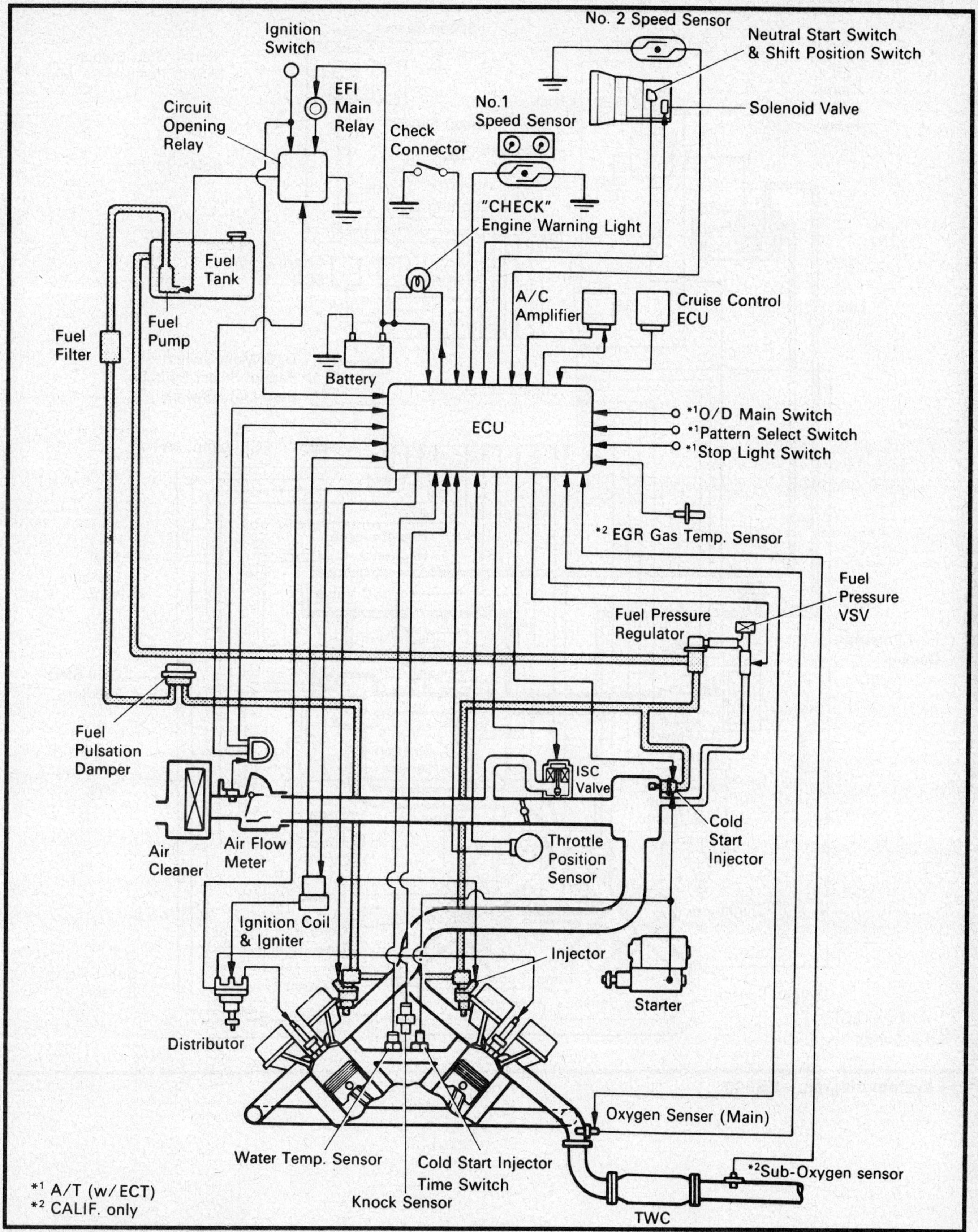

EFI system diagram—ES250

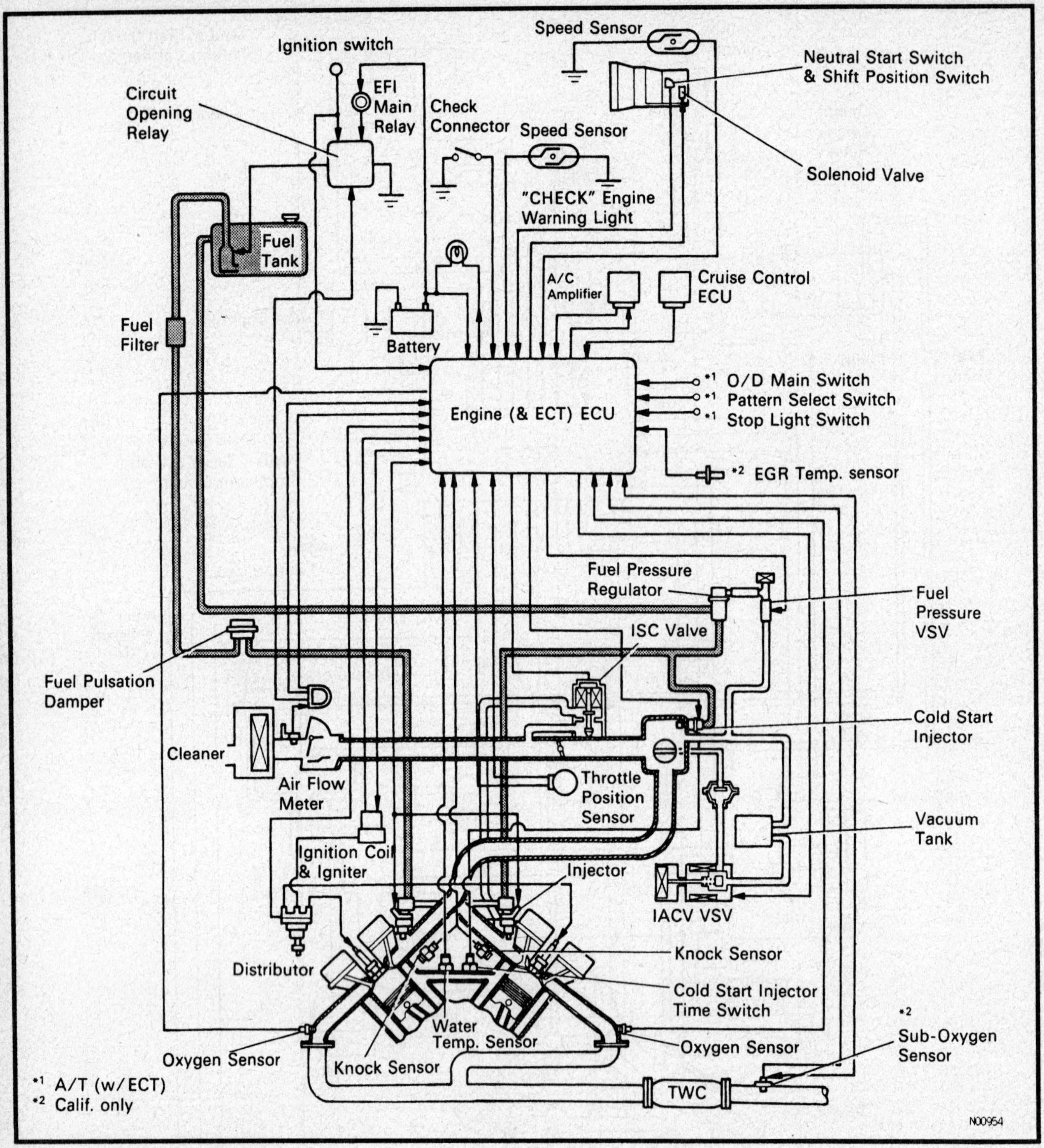

EFI system diagram—ES300

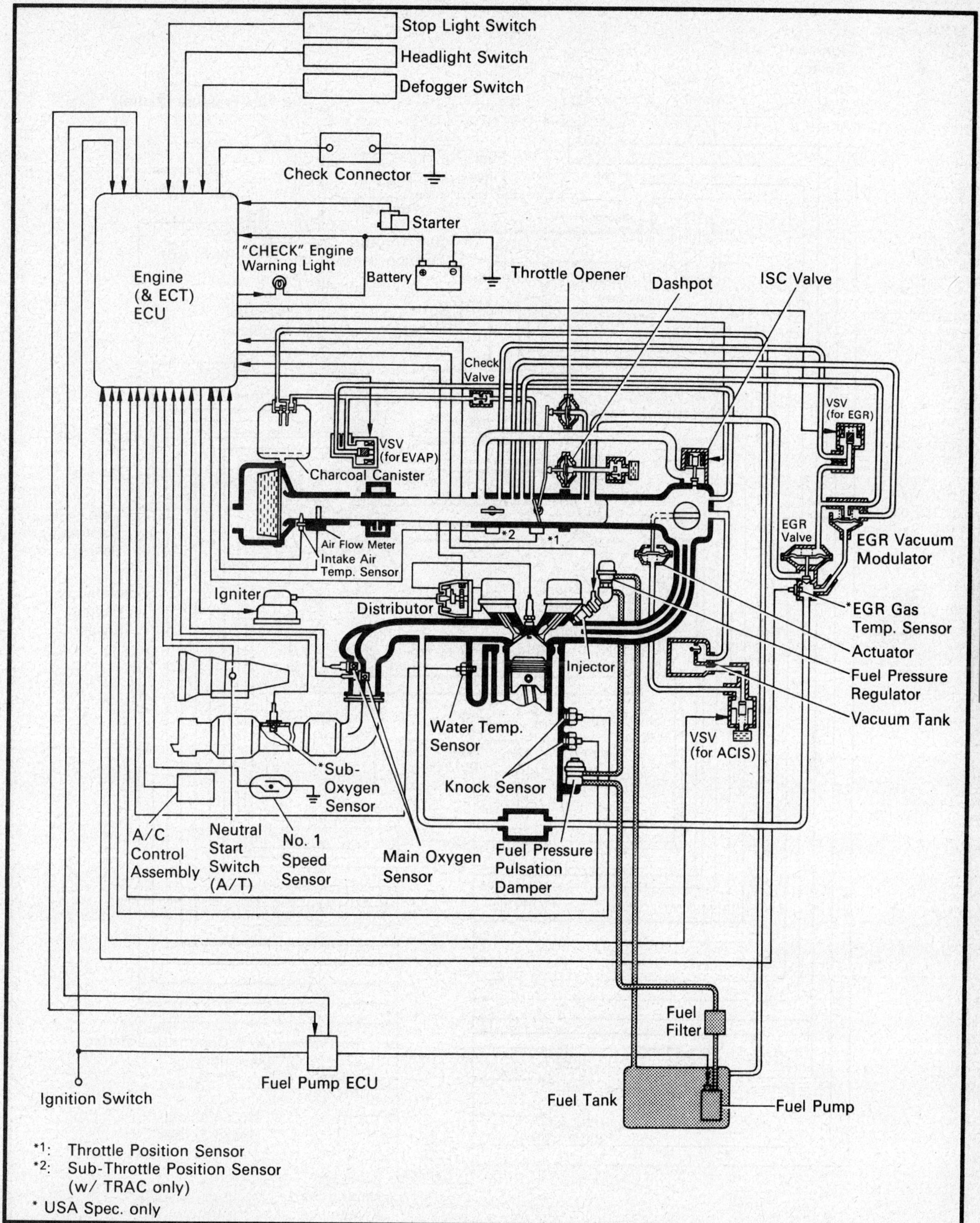

EFI system diagram—SC300

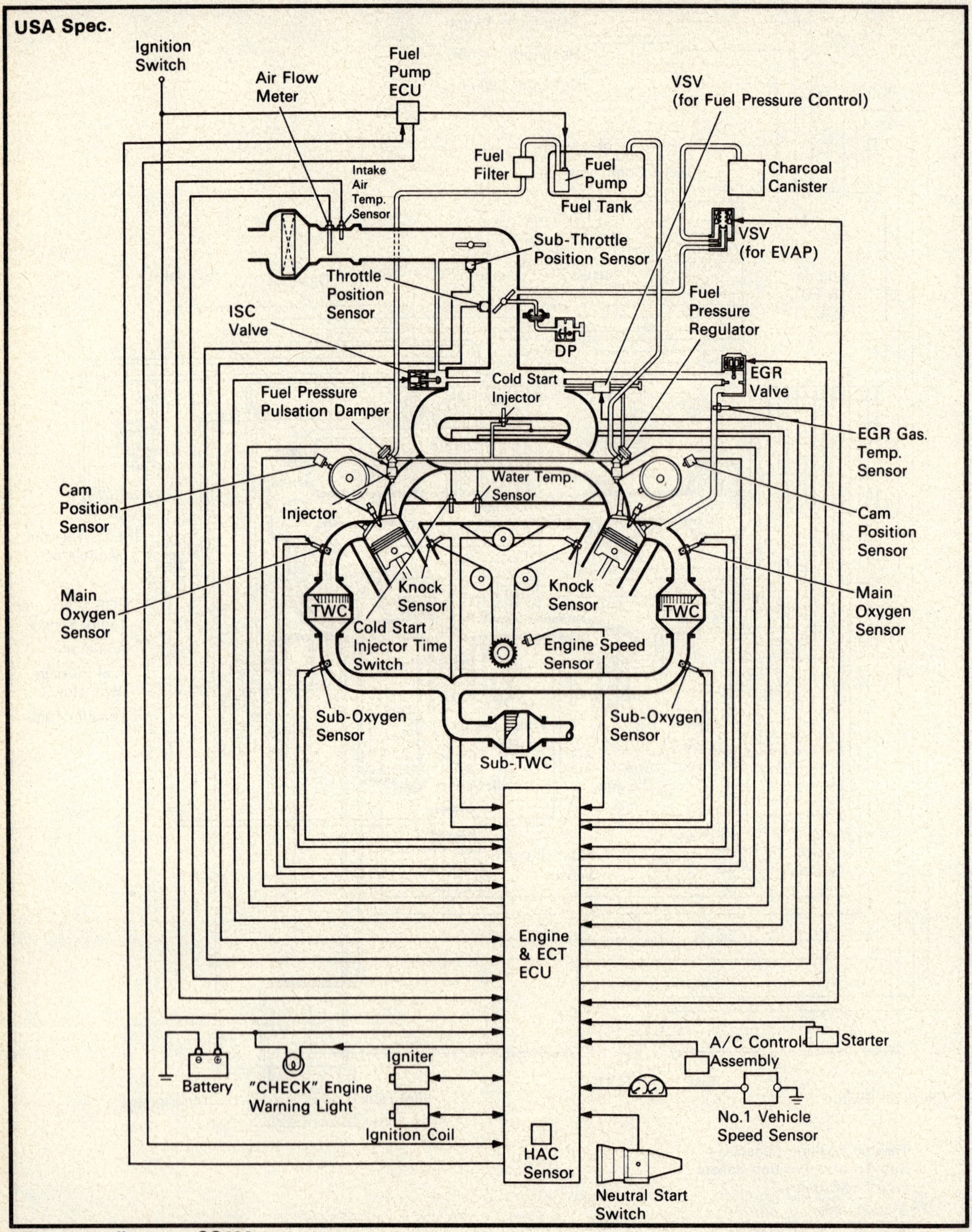

EFI system diagram—SC400

Mazda Fuel Injection

13 SECTION

MAZDA FUEL INJECTION SYSTEM EXCEPT RX-7 AND NAVAJO

General Information

ENGINE CONTROL UNIT (ECU)

The ECU, through various input signals, monitors battery voltage, engine rpm, amount of air intake, cranking signal, intake temperature, coolant temperature, oxygen concentration in the exhaust gases, throttle opening, atmospheric pressure gearshift position, clutch engagement, braking, power steering operation and A/C compressor operation.

The ECU controls the operation of the fuel injection system, idle-up system, evaporative emissions system and ignition timing. The ECU has a built in fail-safe mechanism. If a fault occurs while driving, The ECU will substitute pre-programmed values. Driving performance will be affected, but the vehicle will still be driveable.

Pressure Regulator Control (PRC) System

This system, prevents fuel loss during idle after the engine is hot re-started, by cutting the vacuum to the pressure regulator causing the fuel pressure to increase.

Evaporative Emission Control (EEC) System

This system is controlled by signals sent from the water thermo sensor, intake air thermo sensor, air flow sensor and the engine speed sensor (ignition coil). The ECU calculates the engine operating conditions from the signals, and controls the EEC system by the operation of the solenoid valves for the No. 1 purge control valve and the vacuum switching valve.

Idle Speed Control (ISC) System

The ISC system controls the amount of air detected by the air flow meter, by regulating the amount of bypass air that passes through the throttle body, which helps in maintaining a steady idle speed.

FUEL SYSTEM

The fuel system, supplies the necessary fuel to the injectors to achieve proper combustion at a constant pressure. Fuel is metered and injected into the intake manifold according to the signals from the engine control unit.

The fuel system consists of the fuel pump, fuel filter, distribu-

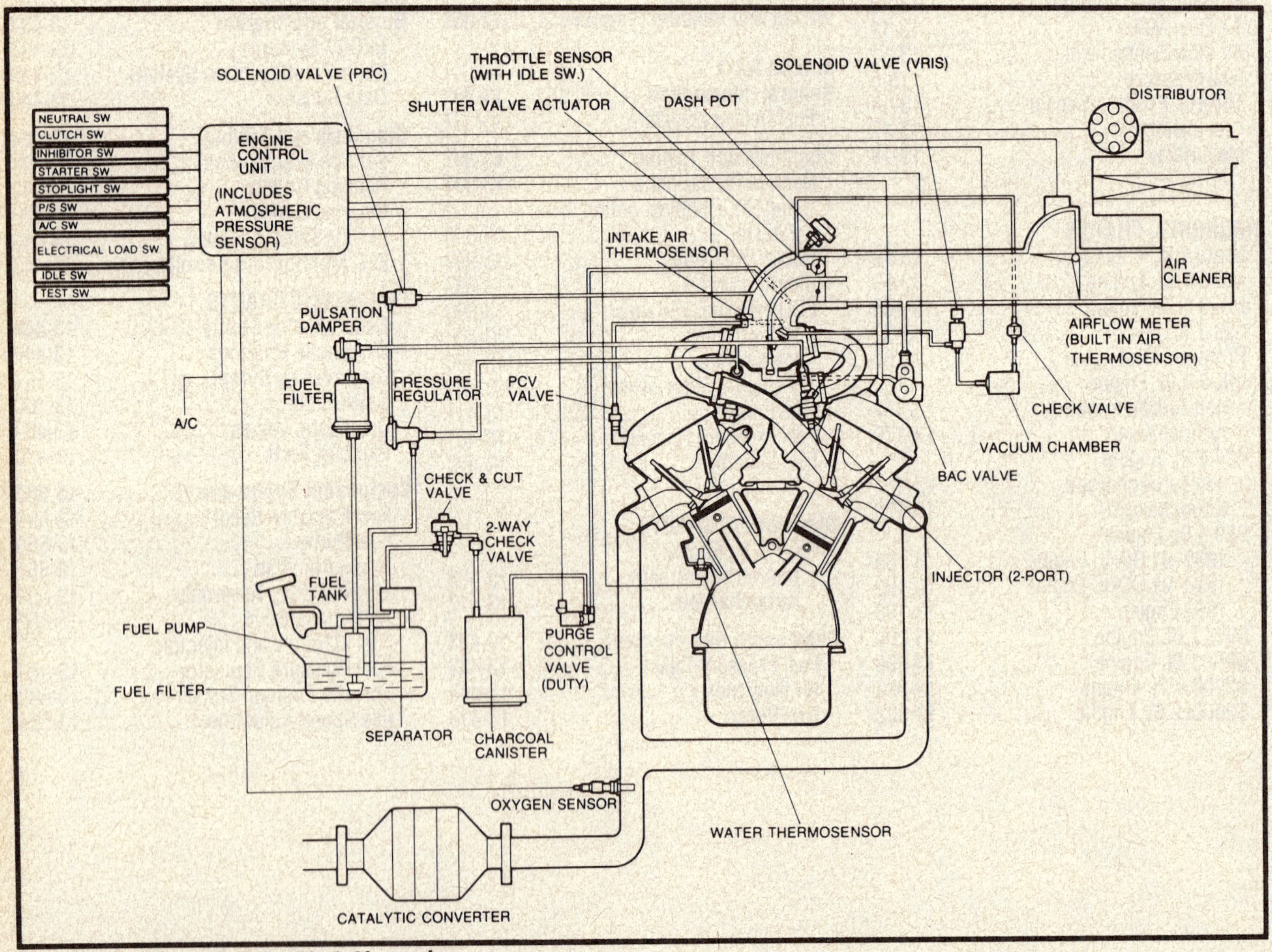

System diagram—MPV with 3.0L engine

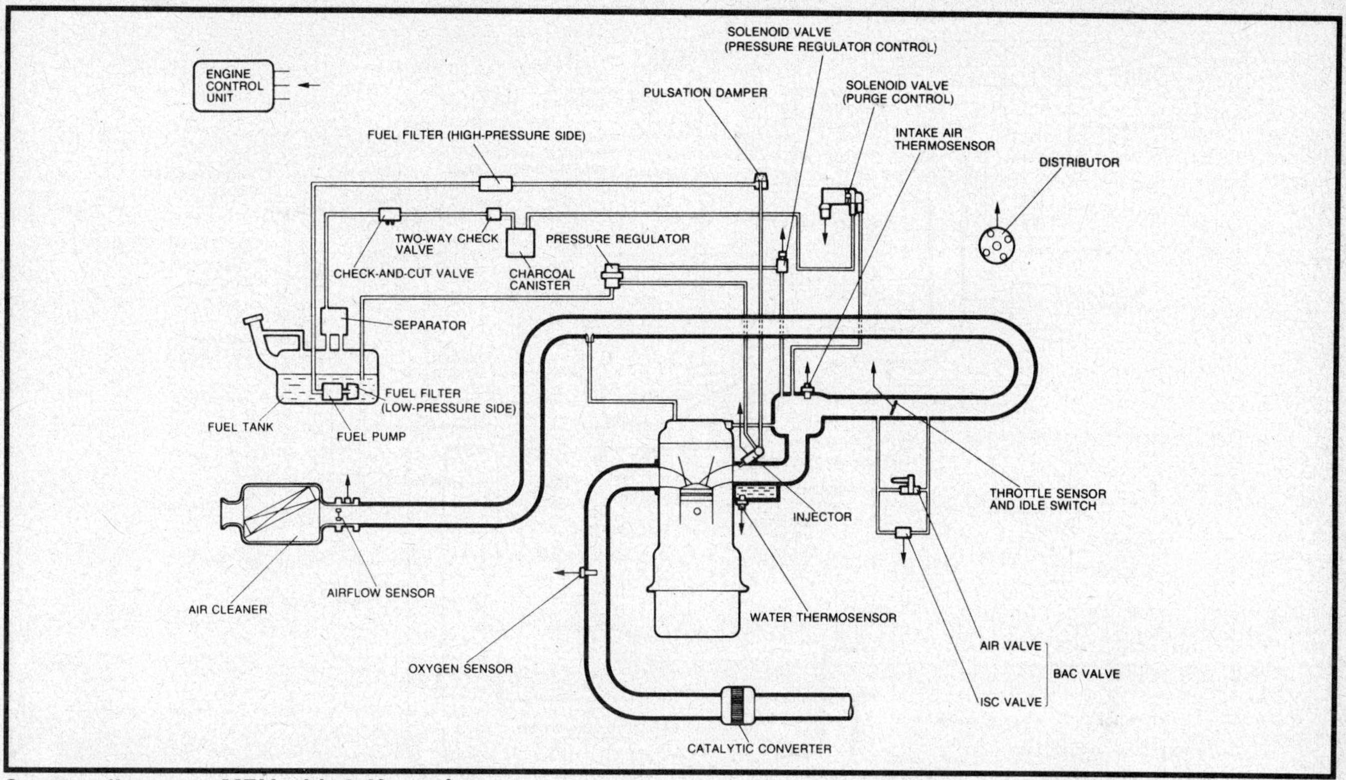

System diagram—MPV with 2.6L engine

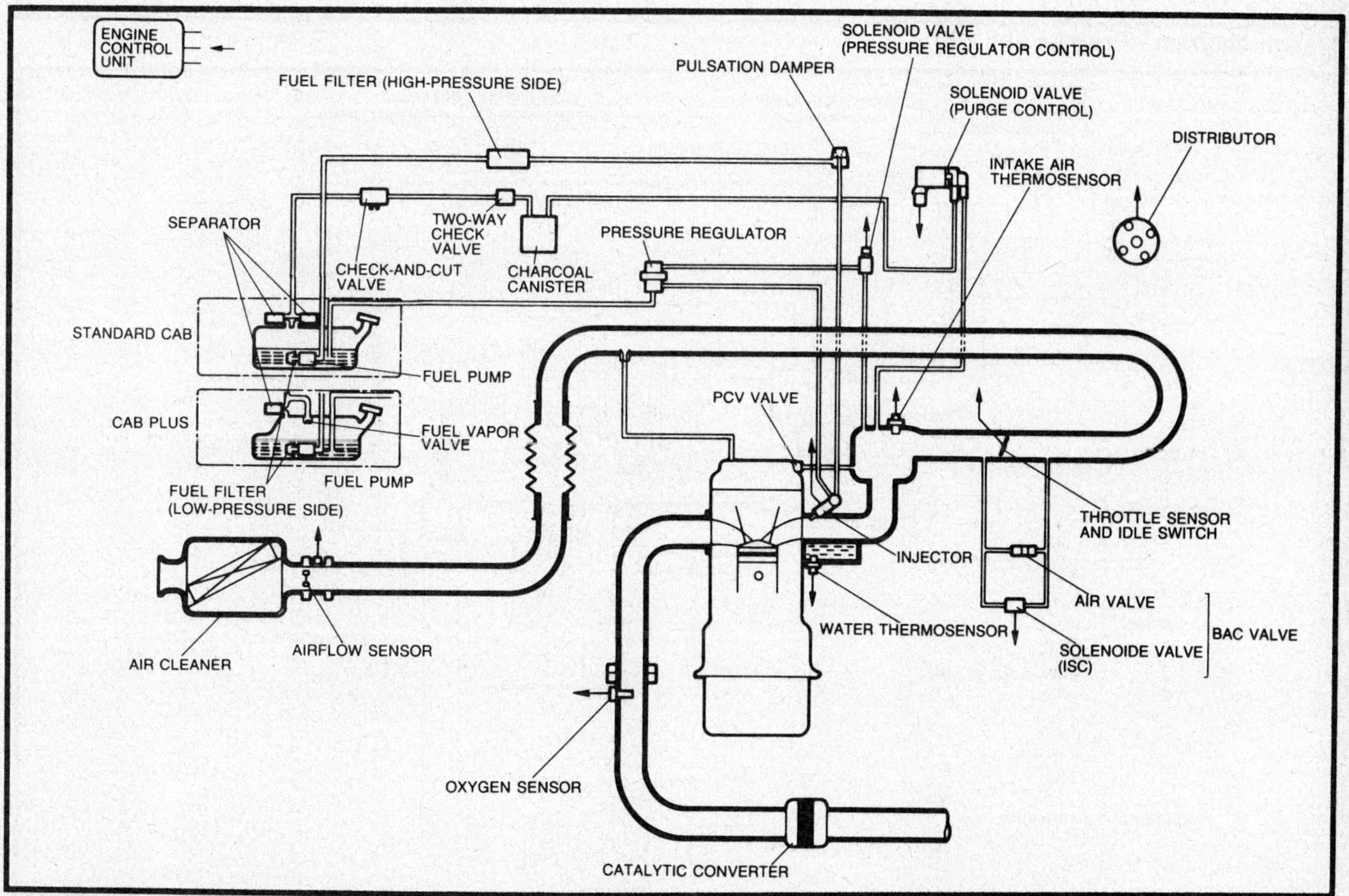

System diagram—B2200, B2600i

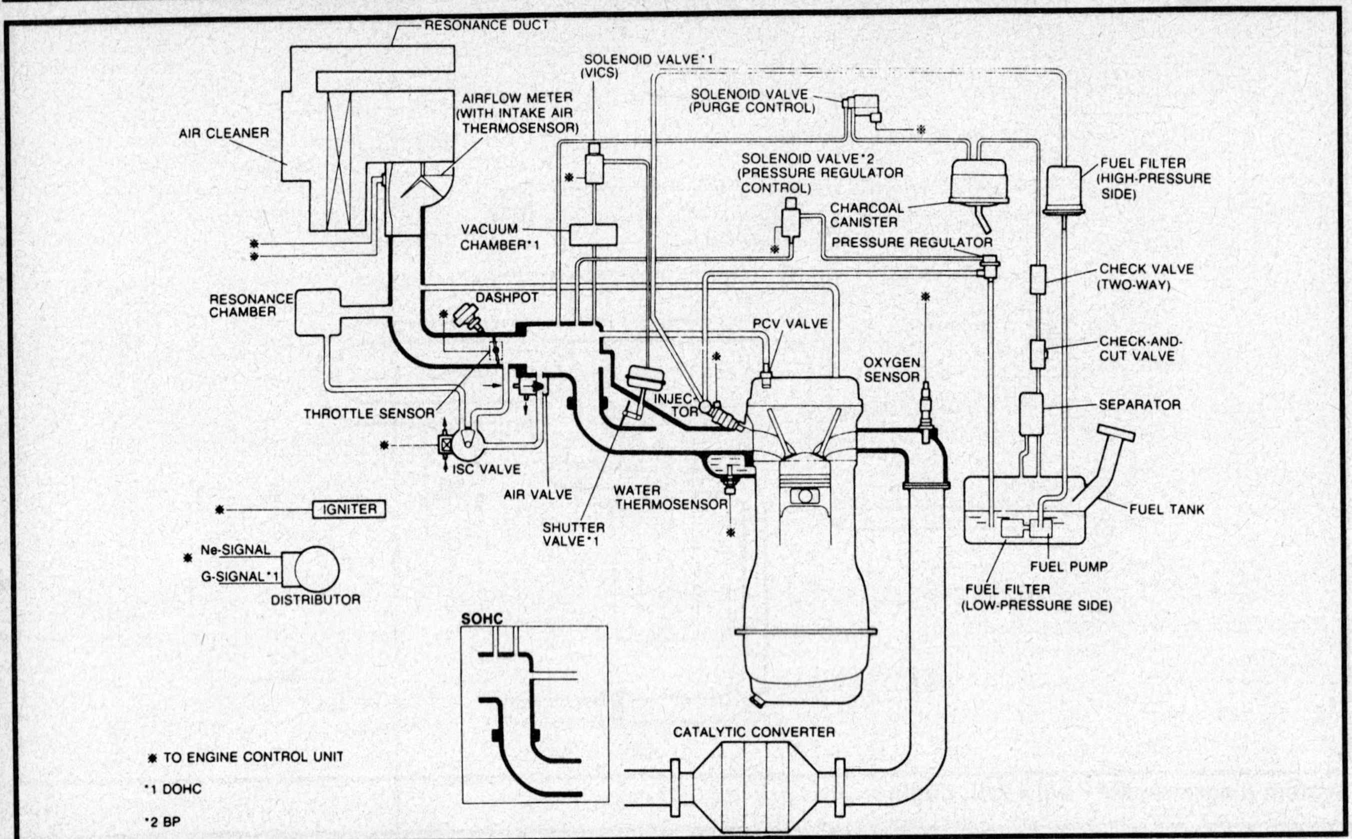

System diagram—Protege and 323

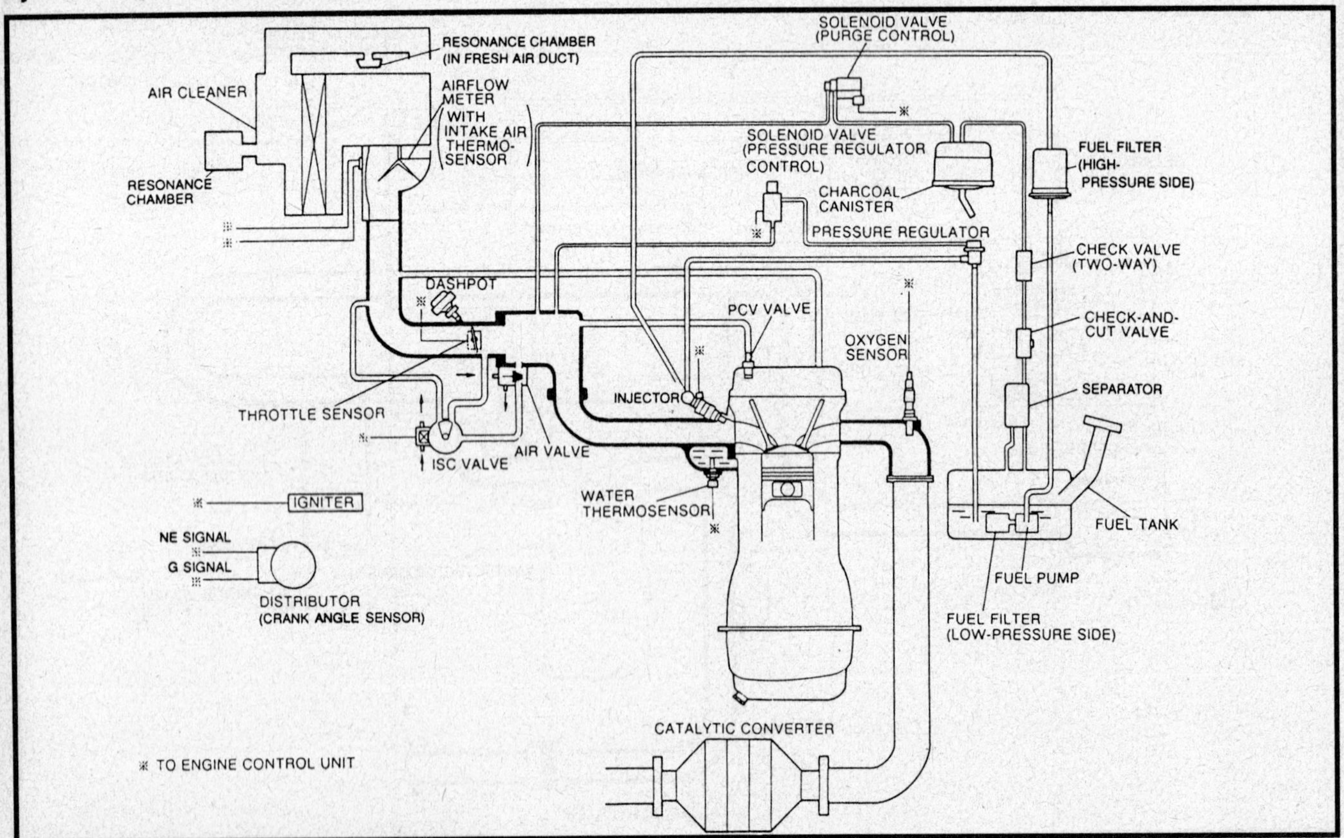

System diagram—MX-3 with 1.6L engine

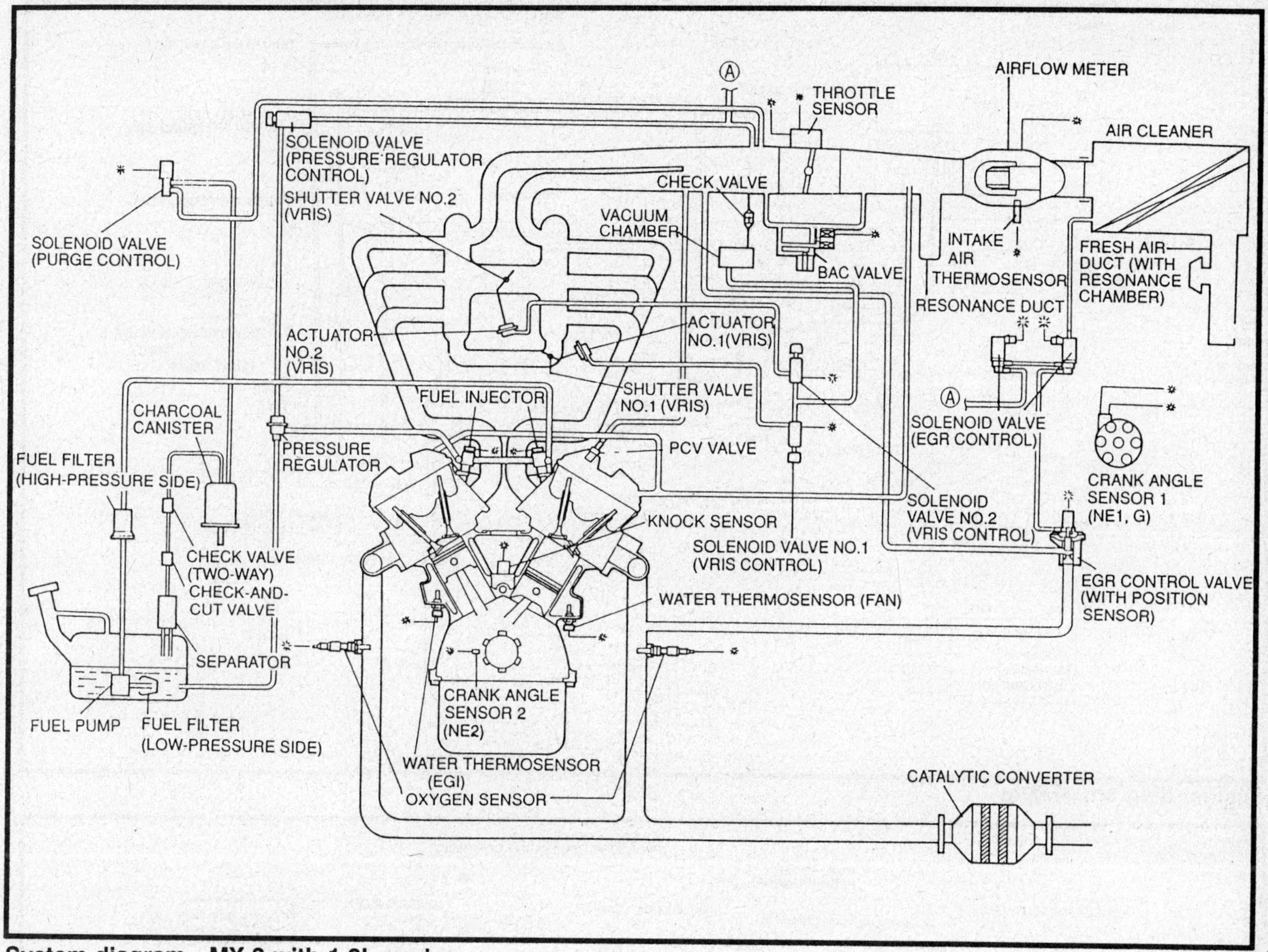

System diagram – MX-3 with 1.8L engine

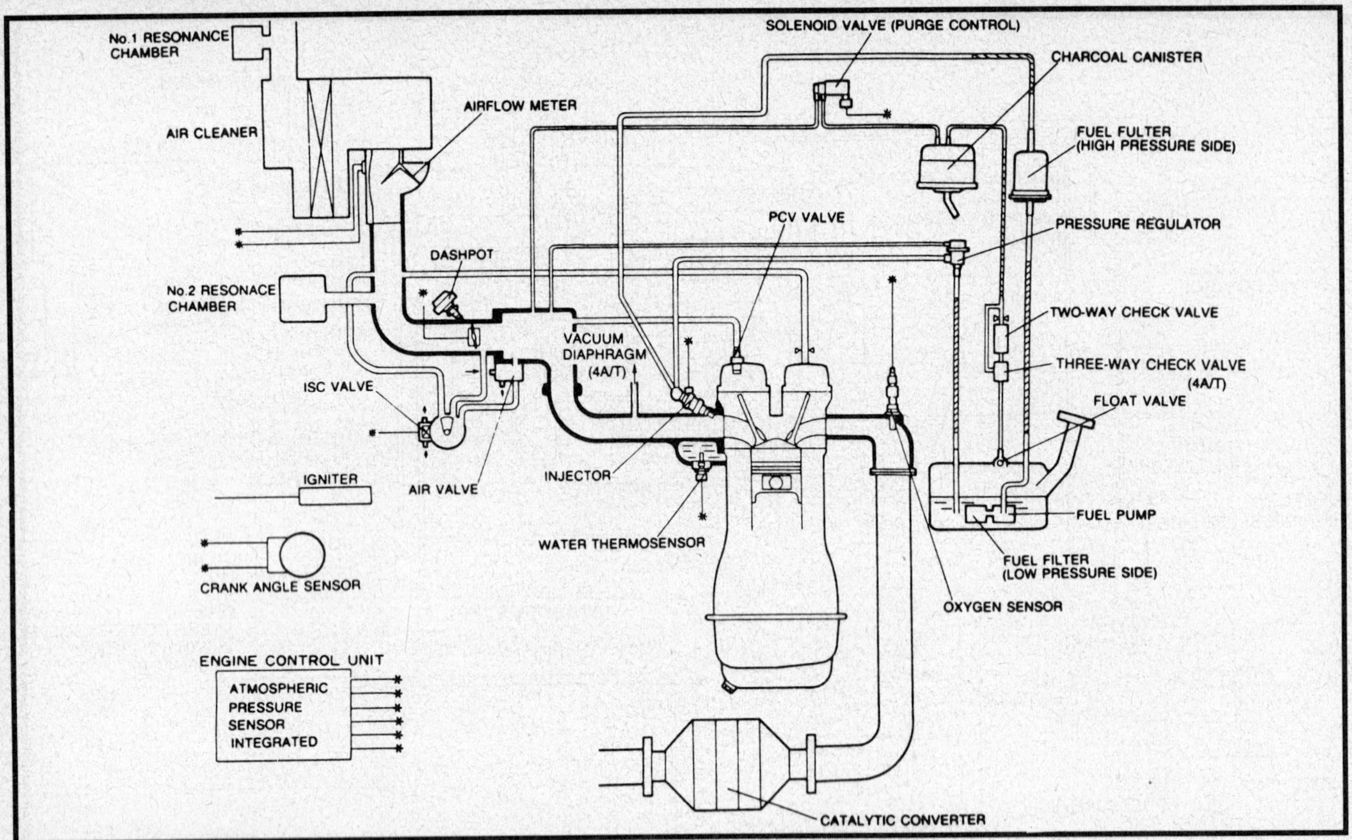

System diagram—Miata

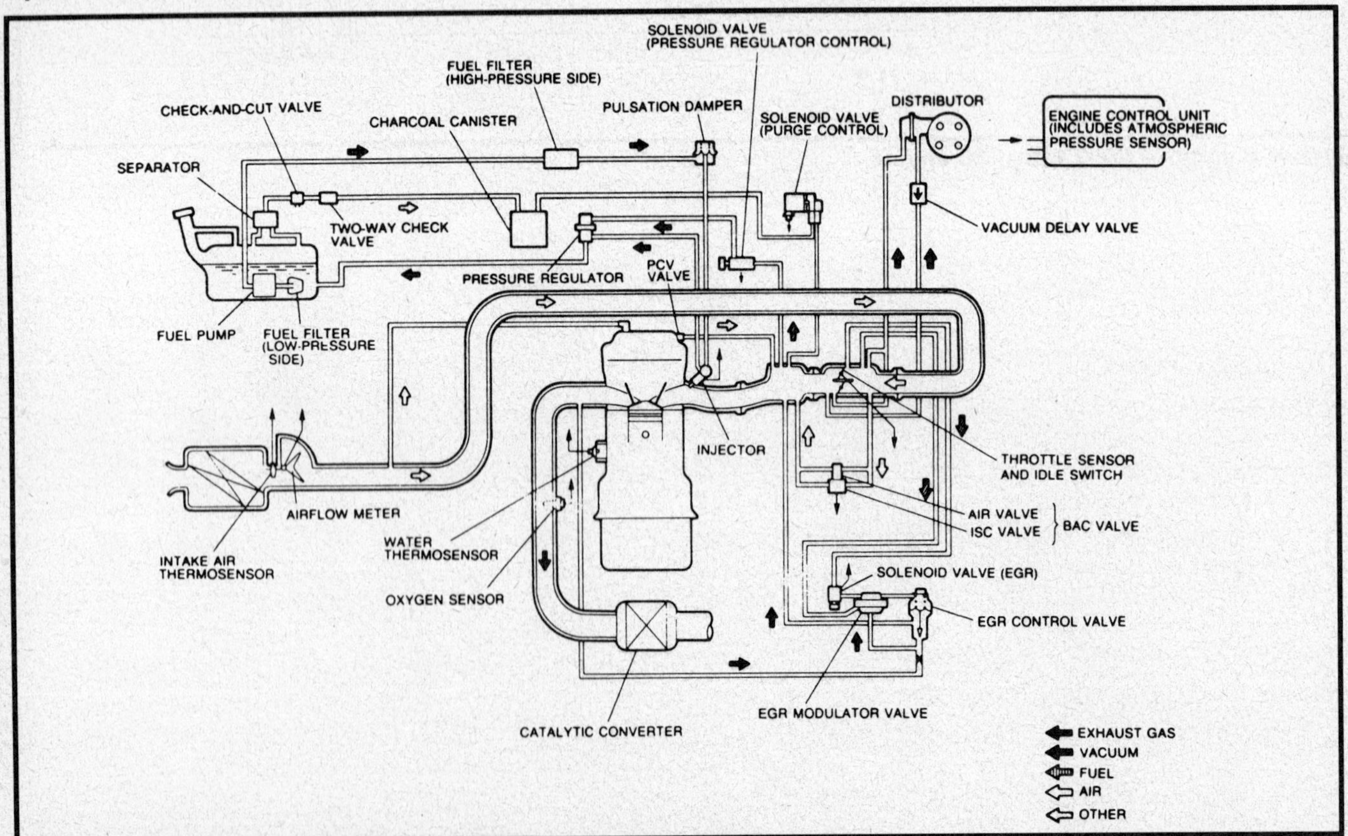

System diagram—626 and MX-6 non-turbo

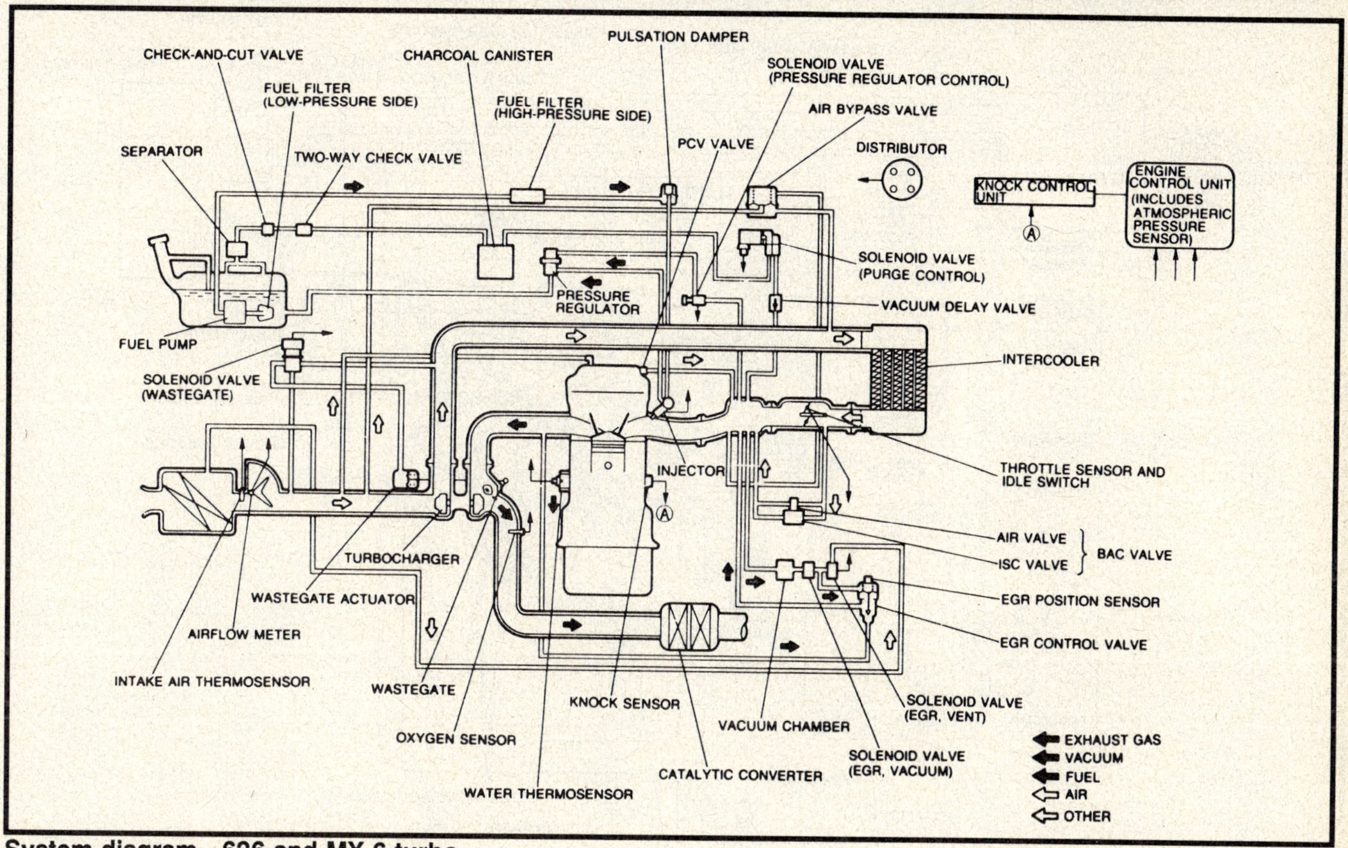

System diagram—626 and MX-6 turbo

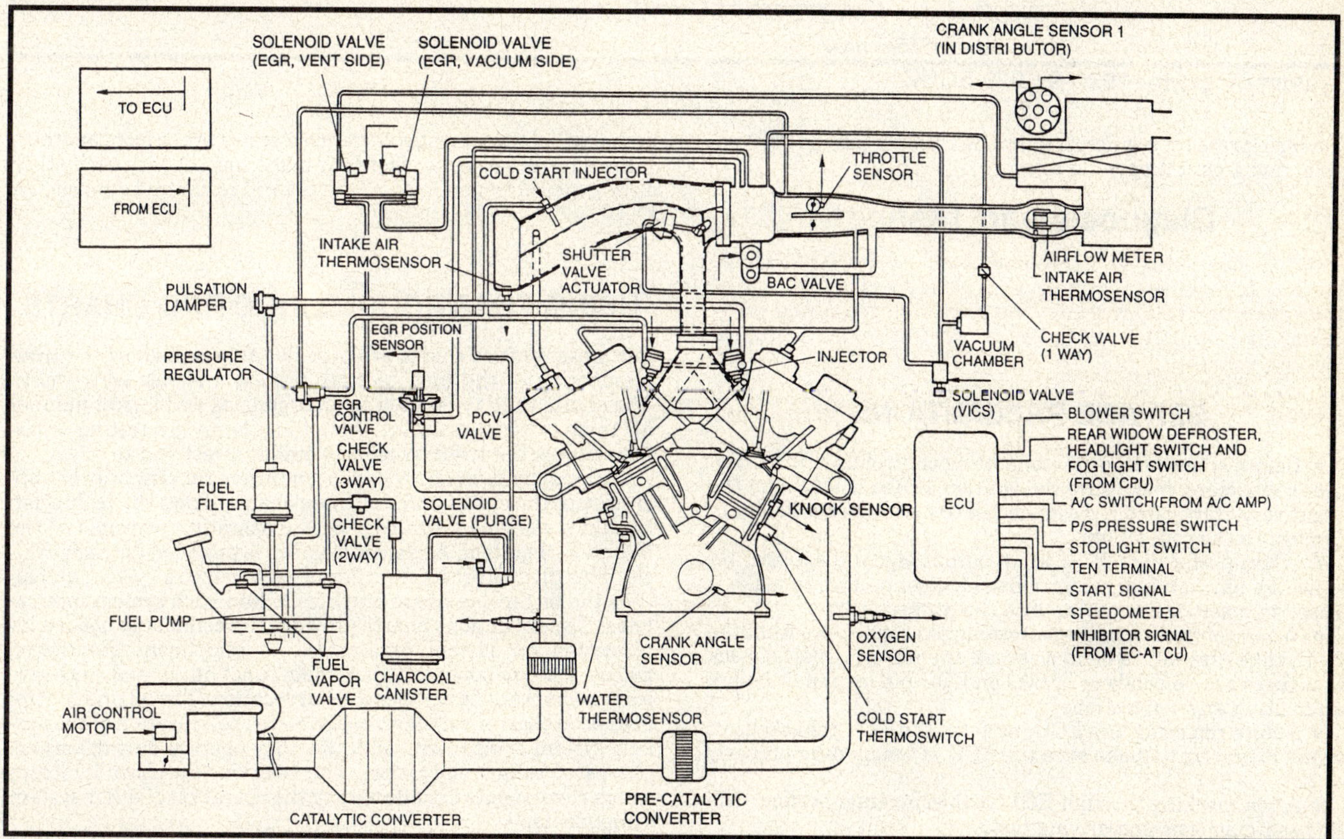

System diagram—1992 929

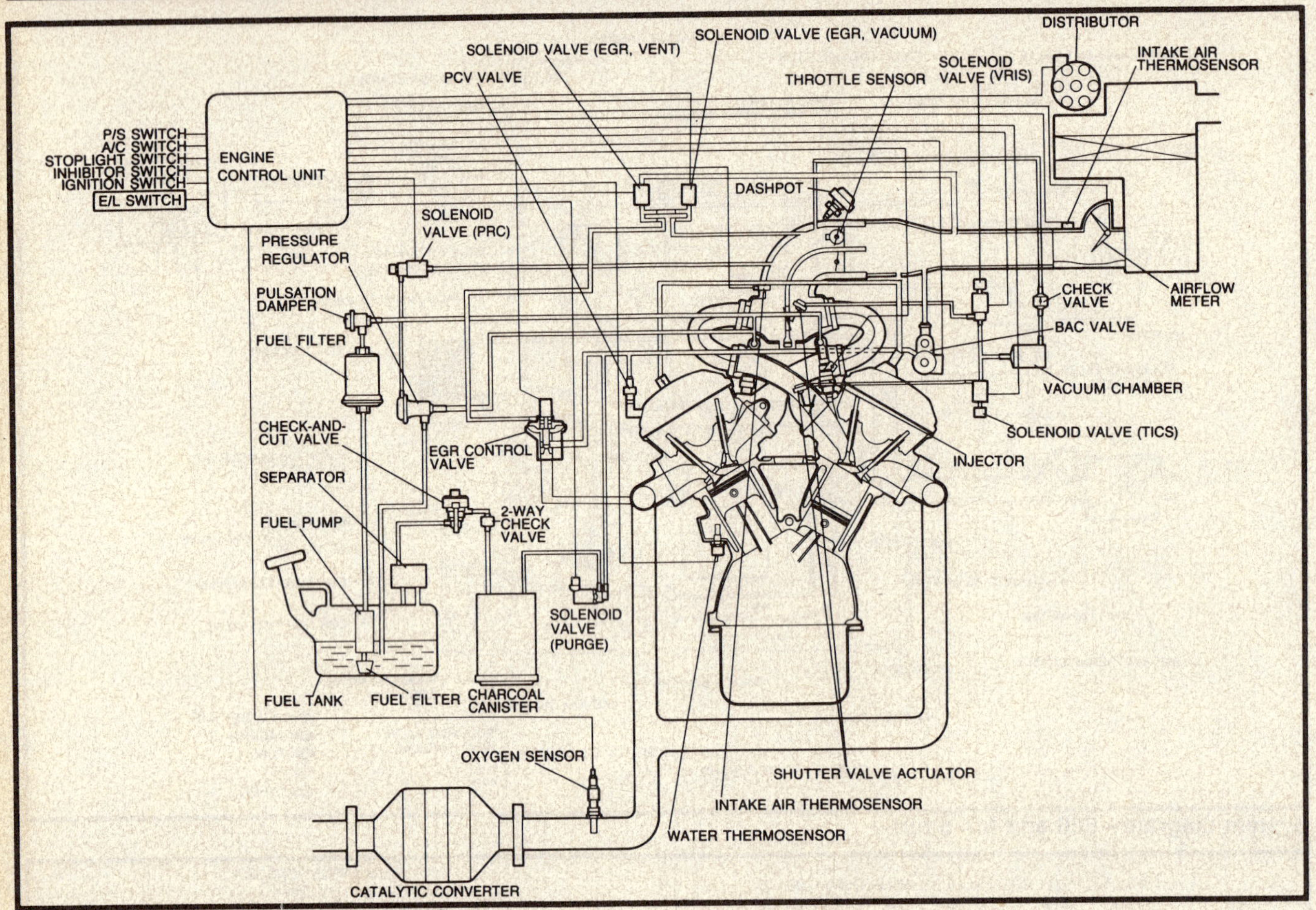

System diagram—1990–91 929 SOHC

tion pipe, pressure regulator, injectors, transfer pump (4WD vehicle) and a circuit opening relay.

Diagnosis and Testing

NOTE: On vehicles equipped with Audio Anti-Theft system, before performing any repairs, the Personal Code Number (P.C.N.) must be obtained from the individual owner of the vehicle, or an authorized Mazda dealer.

SERVICE PRECAUTIONS

- Before connecting or disconnecting control unit ECU harness connectors, make sure the ignition switch is OFF and the negative battery cable is disconnected to avoid the possibility of damage to the control unit.
- When performing ECU input/output signal diagnosis, remove the pin terminal retainer from the connectors to make it easier to insert tester probes into the connector.
- When connecting or disconnecting pin connectors from the ECU, take care not to bend or break any pin terminals. Check that there are no bends or breaks on ECU pin terminals before attempting any connections.
- Before replacing any ECU, perform the ECU input/output signal diagnosis to make sure the ECU is functioning properly or not.
- After checking through EGI troubleshooting, perform the EFI self-diagnosis and driving test.
- When measuring supply voltage of ECU controlled components with a circuit tester, separate 1 tester probe from another. If the 2 tester probes accidentally make contact with each other during measurement, a short circuit will result and damage the ECU.

USING THE TROUBLESHOOTING CHARTS

For most vehicles charts are provided that will allow the troubleshooting of the fuel injection system and related systems. The charts need to be used in an organized and logical manner for them to be most useful. Do not just jump to a testing procedure unless the troubleshooting guides direct you to.

Various charts are provided to give a general view as to the operation of the system. Use of these charts allows the technician to gain a working knowledge of the operating principle of the Mazda EGI system. A system diagram is provided for each vehicle. This gives a broad overview of each vehicles system and allows the technician to see graphically how each system interrelates. A specifications charts provides the commonly needed information on system parameters. A component description names the components, what their function is, and may give some remarks as to its operation or location. The engine control operation chart provides an overview to the operation of the various system components and how they operate during various phases of vehicle operation. The relationship chart indicates which components are affected by the operation of other system components.

To start troubleshooting the system first go to the trouble-

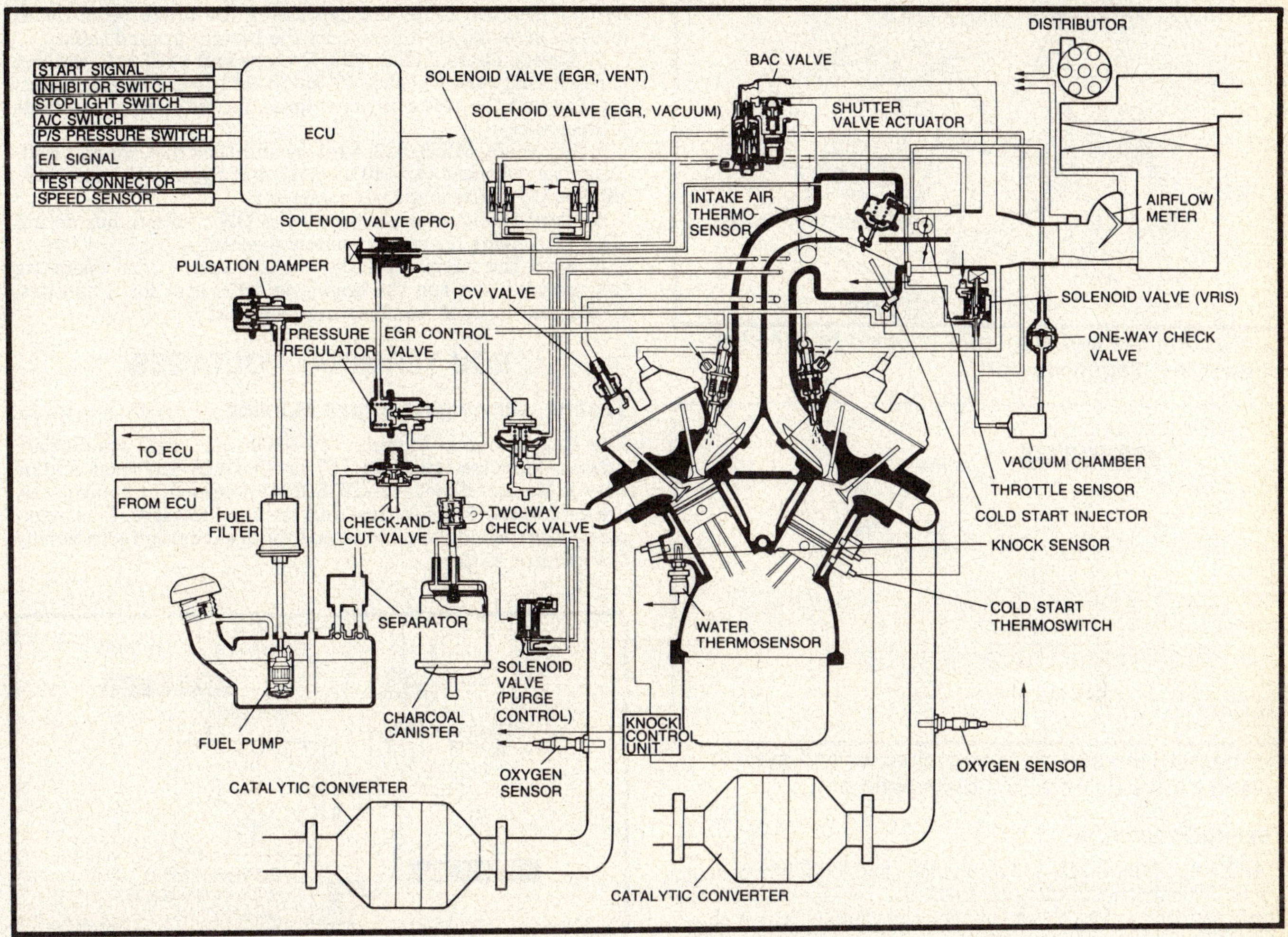

System diagram—1990–91 929 DOHC

shooting charts with the symptom lists. The troubleshooting item (symptom) will be followed by the number of the testing procedure for tracking down the cause. Some of the charts give various possible causes and will give a number of causes in order of priority. The troubleshooting should be followed by the priority list as the most likely causes will have a higher order of priority than the lower causes and this should save the technician time. Depending on the vehicle the troubleshooting procedure will either have the technician pull any trouble codes from the ECU during the procedure or the code pulling procedure may be separate from the troubleshooting procedure. In the cases where the trouble code pulling is done separate from the troubleshooting procedure there will be a chart with the trouble codes and a separate section with trouble code troubleshooting arranged by trouble code number.

The troubleshooting procedure may callout a subsystem to be checked. In these cases there will be individual subsystem troubleshooting guides. These charts will also have an order or priority listing. Following this priority list time will be saved.

RETRIEVING CODES

NOTE: On vehicles equipped with Audio Anti-Theft system, before performing any repairs, the Personal Code Number (P.C.N.) must be obtained from the individual owner of the vehicle, or an authorized Mazda dealer.

With Self-Diagnosis Checker

The self diagnosis checker (49–H018–9A1) and System Selector (49–B019–9A0, used on Miata, MX-3, 323, Protege and 1992 929) are used to retrieve code numbers of malfunctions which have happened and were memorized or are continuing. The malfunction is indicated by the code number and a buzzer.

If there is more than 1 malfunction, the code numbers will display on the self diagnosis checker 1 by 1 in numerical order. In the case of malfunctions, 09, 13, and 01, the code numbers are displayed in a order of 01, 09 and then 13.

The memory of malfunctions is canceled by disconnecting the negative battery cable for at least 20 seconds.

The ECU has a built in fail-safe mechanism for the main input sensors. If a malfunction occurs, the emission control unit will substitute values. This will slightly effect the driving performance, but the vehicle may still be driven.

The ECU continuously checks for malfunctions of the input devices. But, the ECU checks checks for malfunctions of the output devices within 3 seconds after turning the ignition switch to the **ON** position and the test connector is grounded.

The malfunction indicator light indicates a pattern the same as the buzzer of the self-diagnosis checker when the self-diagnosis check connector is grounded. When the self-diagnosis check connector is not grounded, the lamp illuminates steady while malfunction of the main input sensor occurs and goes out if the malfunction recovers. However, the malfunction code is memorized in the emission control unit.

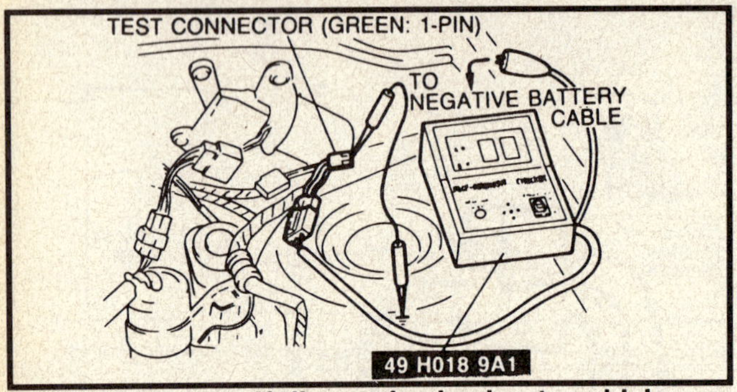

Connecting the self diagnosis checker to vehicles without the diagnostic port

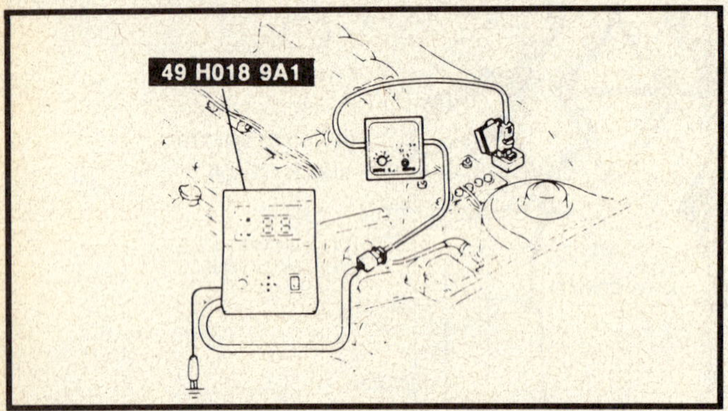

Connecting the self diagnosis checker and system selector to vehicles with a diagnostic port

Testing Procedure

1. Except Miata, MX-3, 323, Protege and 1992 929 connect the Self Diagnosis Checker (49–H018–9A1) or equivalent to the check connector. The check connector is usually located above the right side wheel housing.
2. On Miata, MX-3, 323, Protege and 1992 929, connect Self Diagnosis Checker (49–H018–9A1) and System Selector (49–B019–9A0) to the diagnosis connector and ground.
3. Set the select switch on the Self Diagnosis Checker to the "A" position.
4. Except Miata, MX-3, 323, Protege and 1992 929, ground the green test connector using a suitable jumper wire.
5. On Miata, MX-3, 323, Protege and 1992 929 set the System Selector (49–B019–9A0) to position 1 and the SELF TEST switch to the up position.
6. Turn the ignition switch to the **ON** position. Check that the number "88" flashes on the digital display and the buzzer sounds for 3 seconds after turning the ignition switch **ON**.
7. If the number "88" does not flash, check the main relay, power supply circuit and the check the check connector wiring.
8. Except Miata, MX-3, 323, Protege and 1992 929, if the number "88" flashes and the buzzer sounds continuously for more than 20 seconds, replace the ECU and perform Steps 3 and 4 again.
9. On Miata, MX-3, 323, Protege and 1992 929, if the number "88" flashes and the buzzer sounds continuously for more than 20 seconds, check for a short circuit between the ECU and terminal 1F on the diagnosis connector. Replace the ECU if necessary.
10. Note the code numbers and check the causes, repair as necessary. Be sure to recheck the code numbers after repairing.

CLEARING CODES

1. Cancel the memory of malfunctions by disconnecting the battery negative cable and depressing the brake pedal for at least 20 seconds, then reconnect the battery ground cable.
2. Except Miata, MX-3, 323, Protege and 1992 929, connect the Self-Diagnosis Checker 49–H018–9A1 to the check connector. Ground the test connector (green: 1 pin) using a suitable jumper wire.
3. On Miata, MX-3, 323, Protege and 1992 929, connect Self-Diagnosis Checker (49–H018–9A1) and System Selector (49–B019–9A0) to the diagnosis connector.
4. Turn the ignition switch to the **ON** position, but do not start the engine for approximately 6 seconds.
5. Start the engine and allow it to reach normal operating temperature, the run the engine at 2000 rpm for 2 minutes. Check that no code numbers are displayed.

ECU TERMINAL VOLTAGES

Testing with Engine Signal Monitor

The Engine Signal Monitor (49–9200–162), with appropriate Adapter Harness (49–9200–167 or 49–G018–903) and Engine Digital Monitor Sheet (49–G018–904) if needed, or equivalents, are used to check the control unit terminal voltages. This monitor checks the individual terminal voltages through selection by the monitor switch.

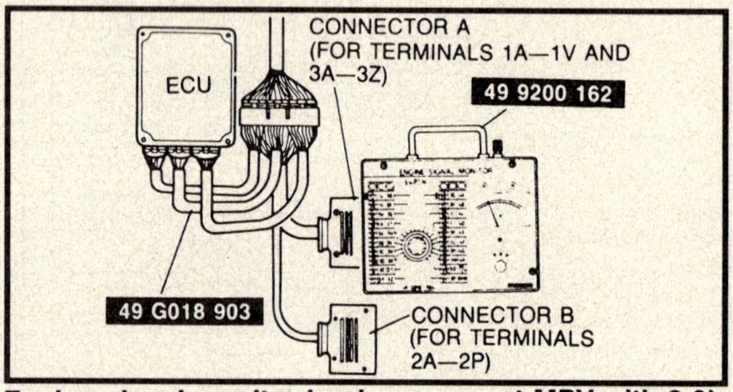

Engine signal monitor hookup—except MPV with 3.0L engine

Testing

1. Start the engine and allow it to reach normal operating temperature. Stop the engine.
2. Disconnect the ECU connector.
2. On MPV with 3.0L engine, connect Engine Signal Monitor (49–9200–162) and Adapter Harness (49–9200–167) between the ECU and the ECU wiring harness.
3. Except MPV with 3.0L engine connect Engine Signal Monitor (49-9200-162) and the Adapter Harness (49-9200-163) between the ECU and the ECU wiring harness. Install Engine Signal Monitor Sheet (49–G018–904) on the Engine Signal Monitor.

NOTE: On 626 and MX-6 equipped with manual transmission, use connector (A) of the Adapter. On vehicles equipped with automatic transmission, use connector (A) of the Adapter to check voltages of terminals 1A through 1V and 3A through 3Z, and use connector (B) to check voltages of terminals 2A through 2P.

4. Position the select switch and the monitor switch on the Engine Signal Monitor to the terminal number to be tested. Check the voltage of each terminal.
5. If any ECU terminal voltage is incorrect, check the input

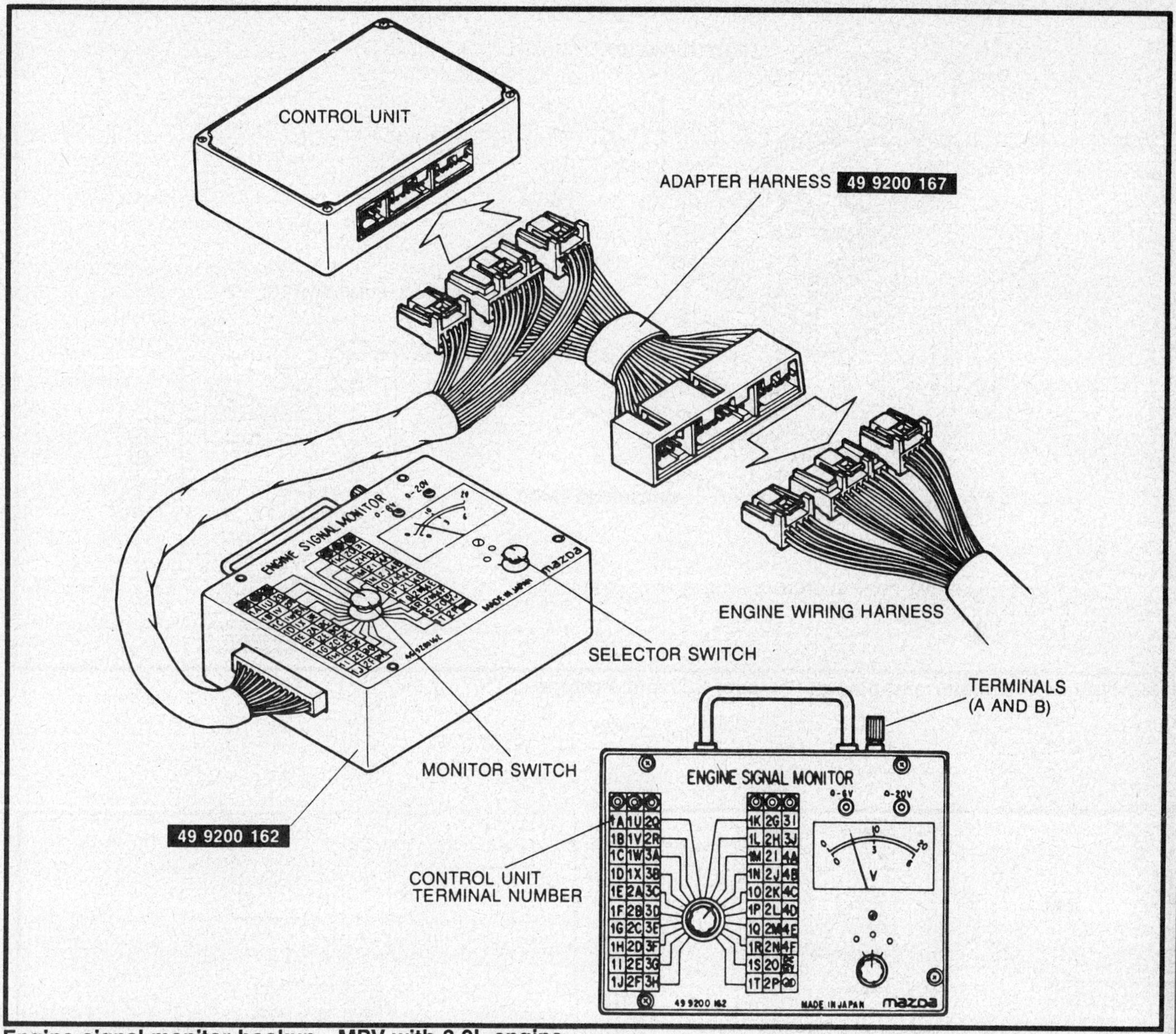

Engine signal monitor hookup—MPV with 3.0L engine

or output component and related wiring. If components and wiring are normal, replace the ECU.

AIR FLOW METER

Testing

EXCEPT B2200, B2600i AND MPV WITH 2.6L ENGINE

1. Disconnect the electrical connector from the airflow meter body and remove the airflow meter from the vehicle.
2. Inspect the air flow meter body for cracks. Using a suitable ohmmeter, check the resistance between the terminals.
3. Using a suitable tool, open the measuring plate or cone. Check for smooth operation.
4. Measure the resistance between the terminals as listed in the specifications chart.
5. On all vehicles, if specifications are not as indicated, replace the air flow meter.

AIR FLOW SENSOR

Testing

B2200, B2600i AND MPV WITH 2.6L ENGINE

1. Slide back the rubber boot from the air flow sensor connector.
2. Using a suitable voltmeter, check the terminal voltages.
3. If specifications are not as indicated, check the wiring harness for an open circuit.
4. If no problem is found in the wiring harness, check the burn-off operation as follows:
   a. Disconnect the battery ground cable for a minimum of 20 seconds, then reconnect it.
   b. Start the engine allow it to reach normal operating temperature. Slide back the rubber boot from the air flow sensor connector.
   c. With the transmission in **N**, operate the engine for approximately 3 minutes at 2000 rpm.

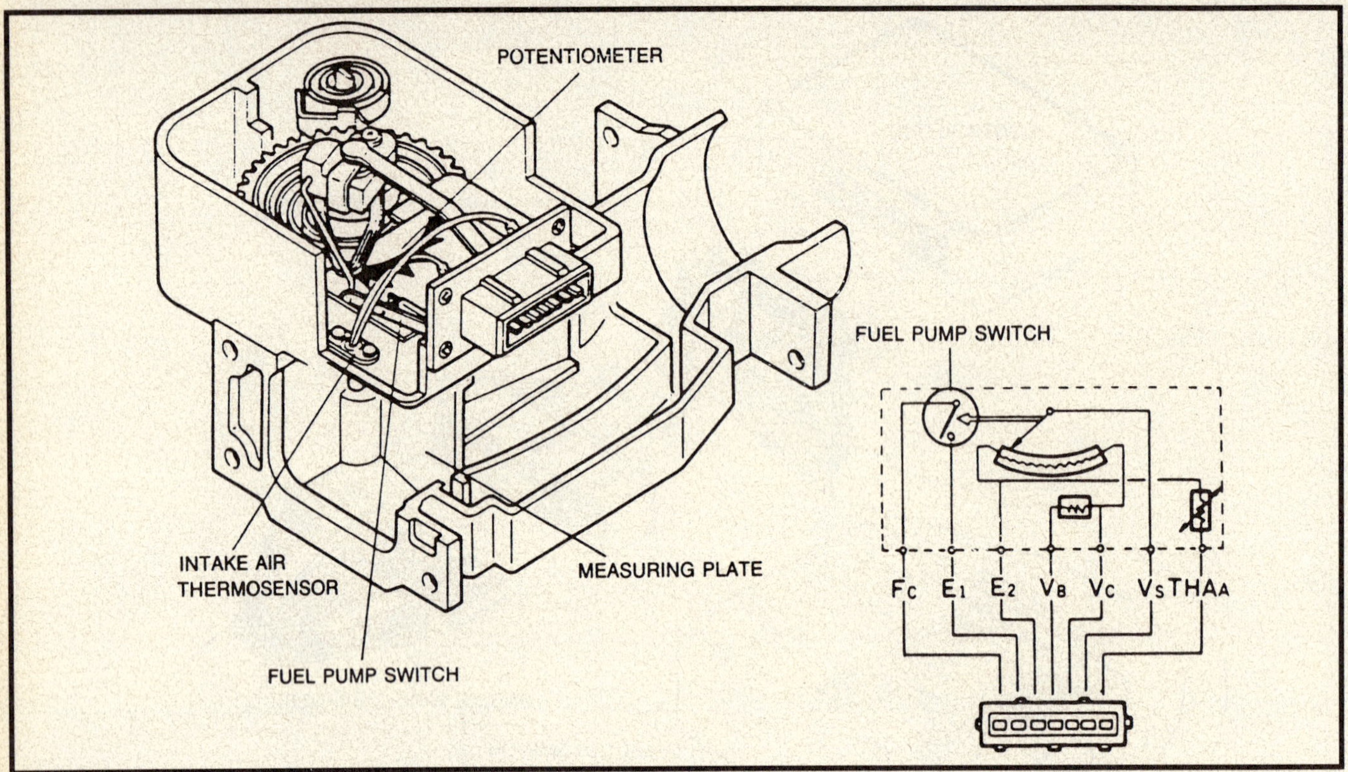

Flap type airflow meter and pinouts—except 323 and Protege

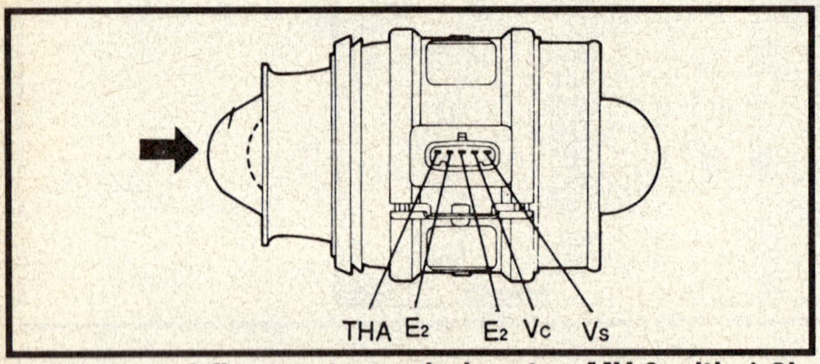

Cone type airflow meter and pinouts—MX-3 with 1.8L engine and 1992 929

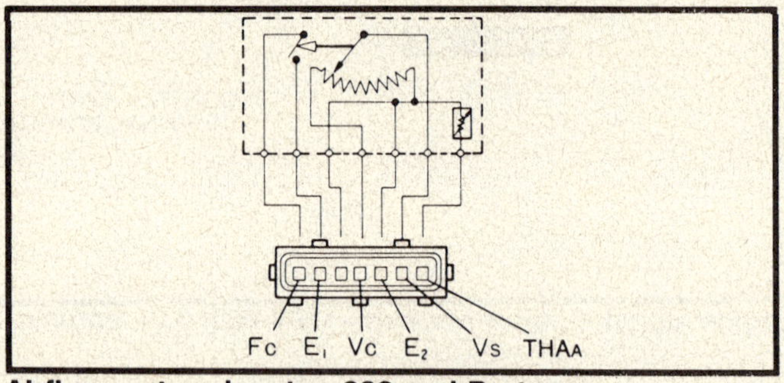

Airflow meter pinouts—323 and Protege

Terminal	Resistance (Ω)	
	Fully closed	Fully open
E2↔Vs	20–400	20–1,000
E2↔Vc	100–400	
E2↔VB	200–400	
E2↔THA (Intake air thermosensor)	-20°C (-4°F) 20°C (68°F) 60°C (140°F)	13.6–18.4 kΩ 2.21–2.69 kΩ 493–667 Ω
E1↔Fc	∞	0

Airflow meter specifications—626, 626 Turbo, MX-6 and MX-6 Turbo

Terminal	Resistance (Ω)	
	Fully closed	Fully open
E2↔Vs	20–400	20–1,000
E2↔Vc	100–300	
E2↔VB	200–400	
E2↔THAA (Intake air thermosensor)	-20°C (-4°F) 20°C (68°F) 60°C (140°F)	13,600–18,400 2,210– 2,690 493– 667
E1↔Fc	∞	0

Airflow meter specifications—MPV with 3.0L engine and 1990–91 929 with SOHC engine

Terminal	Resistance (Ω)	
	Fully closed	Fully open
E2 ↔ Vs	200—600	20—1,200
E2 ↔ Vc	200—400	
E2 ↔ THAA (Intake air thermosensor)	−20°C (−4°F): 13.6—18.4 KΩ 20°C (68°F): 2.21—2.69 kΩ 60°C (140°F): 493—667Ω	
E1 ↔ Fc	∞	0

Airflow meter specifications—323 and Protege

Terminal	Resistance (Ω)	
	Fully closed	Fully open
E2↔Vs	20—600	20—1,000
E2↔Vc	200—400	
E2↔THAA (Intake air thermosensor)	−20°C (−4°F) 20°C (68°F) 60°C (140°F)	13,600—18,400 2,210—2,690 493—667
E1↔Fc	∞	0

Airflow meter specifications—1990–91 929S with DOHC engine, MX-3 with 1.6L engine and Miata

Terminal	Resistance (Ω)
E2 ↔ VS	20—600 (Closed; 20°C (68°F)) 20—1,000 (Open; 20°C (68°F))
E2 ↔ VC	200—400 (Closed ↔ Open; 20°C (68°F))
E2 ↔ THA (Intake air thermosensor)	−20°C (−4°F) 10,000—20,000 20°C (68°F) 2,000—3,000 60°C (140°F) 400—700

Airflow meter specifications—1992 929

Terminal	Resistance (Ω)
E2 ↔ Vs	20—600 (Closed; 20°C (68°F)) 20—1,000 (Open; 20°C (68°F))
E2 ↔ Vc	200—400 (Closed ↔ Open; 20°C (68°F))
E2 ↔ THA (Intake air thermosensor)	−20°C (−4°F) 13,600—18,400 20°C (68°F) 2,210—2,690 60°C (140°F) 493—667

Airflow meter specifications—MX-3 with 1.8L engine

VB: Battery voltage

Terminal wire \ Condition	Ignition switch ON	Engine running
B/W (Power supply)	VB	
G/O (Burn-off)	0V	
G/B (Airflow mass)	1.0—2.0V	1.9—5V
B/W (Ground)	0V	
B/O (Ground)	0V	

Airflow sensor specifications—MPV with 2.6L engine

Terminal wire \ Condition	Ignition switch ON	Engine running
B/Y (Power supply)	Battery voltage	
G/O (Burn-off)	0V	
G/B (Airflow mass)	1.0—2.0V	1.9—5V
G/Y (Ground)	0V	
B/O (Ground)	0V	

Airflow sensor specifications – B2200, B2600i

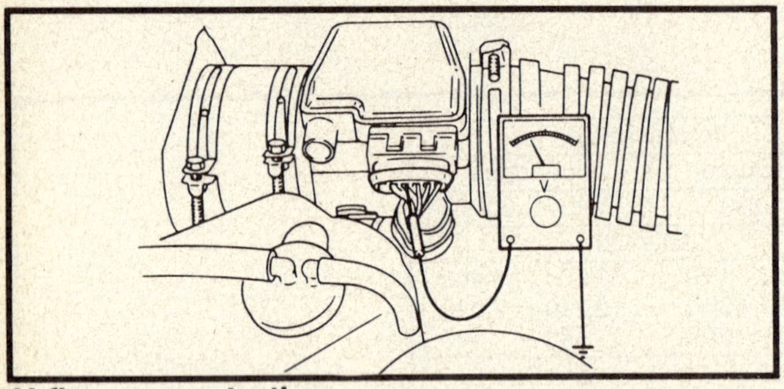

Airflow sensor testing

d. Turn the ignition switch to the **OFF** position.

e. On 1990 MPV with 2.6L engine, using a suitable voltmeter, check the voltage of the air flow sensor wire (G/O) and terminal (2K) on the ECU. Voltmeter should read 0 volts when the ignition switch is turned **OFF** and 8–12 volts momentarily, 2–5 seconds after the ignition switch is turned **OFF**.

f. On B2200, B2600i and 1991–92 MPV with 2.6L engine, using a suitable voltmeter, check the voltage of the air flow sensor wire (G/O) and terminal (2H) on the ECU. Voltmeter should read 0 volts when the ignition switch is turned Off and 8–12 volts momentarily, 2–5 seconds after the ignition switch is turned Off.

g. If specifications are not as indicated, check the voltage of the ECU terminals 2E, 2I and 2J on 1990 MPV with 2.6L engine, then the wiring harness. On 1991–92 MPV with 2.6L engine, check the voltage of the ECU terminals 2O, 2Q and 1L, then the wiring harness. On B2200, B2600is check the voltage of the ECU terminals 2P, 2Q and 1L, then the wiring harness. Replace the air flow sensor if necessary.

THROTTLE BODY

Testing

1. Check the throttle valve operation. Check the free play of the accelerator cable. Free play should be 0.04–0.12 in. (1–3mm). Adjust by using nuts at throttle end of the cable, if necessary.
2. Depress the accelerator pedal to the floor and check that the throttle valve is fully opened. Adjust by using throttle stop bolt at the pedal, if necessary.
3. Check that the throttle bores are clear of deposits. Do not remove the thin sealing coating from the bores or the throttle butterflies.

THROTTLE POSITION SENSOR

Testing

MX-3 WITH 1.6L ENGINE, 323, PROTEGE AND MIATA

Manual Transmission/Transaxle

1. Disconnect the electrical connector from the throttle position sensor.
2. Connect throttle position sensor tester (49-9200-165) and adapter (49-9200-166) or equivalent to the throttle position sensor. If testers are not available, use a suitable ohmmeter.
3. On MX-3, 323 and Protege, insert a feeler gauge of 0.004 in. (0.1mm) between the throttle stop screw and the stop lever. There should be continuity between the IDL and E terminals and no continuity between the POW and E terminals. Insert a 0.039 in. (1.0mm) feeler gauge between the throttle stop screw and the stop lever. There should be no continuity between either pairs of terminals.
4. On 1990 Miata, insert a feeler gauge of 0.016in (0.4mm) between the throttle stop screw and the stop lever. There should be continuity between the IDL and TL terminals and no continuity between the POW and TL terminals. Insert a 0.027 in. (0.7mm) feeler gauge between the throttle stop screw and the stop lever. There should be no continuity between either pairs of terminals.
5. Open the throttle to the wide open position. Their should be continuity between the POW and TL terminals on the Miata and the POW and E terminals on the MX-3.
6. If the throttle position sensor does not meet these parameters, the unit may need adjustment or replacement. If adjustment does not correct the problem replace the sensor.

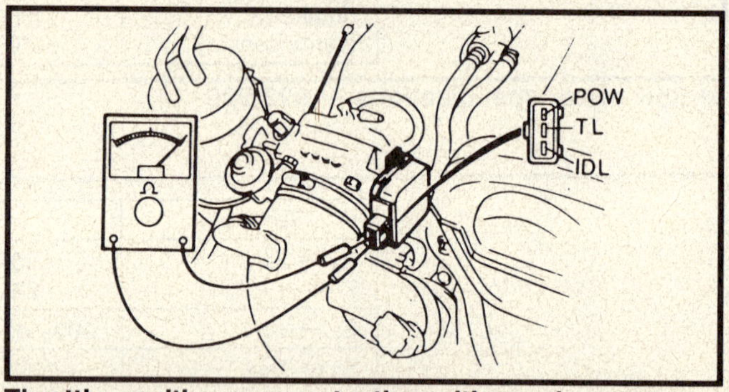

Throttle position sensor testing with an ohm-meter. The TL terminal is also called the E terminal on some vehicles – Miata, MX-3 with 1.6L engine, 323 and Protege with manual transmissions

MX-3 WITH 1.6L ENGINE, 323, PROTEGE AND MIATA

Automatic Transmission/ Transaxle

1. Disconnect the electrical connector from the throttle sensor.
2. Connect a suitable ohmmeter between terminals E and IDL on the throttle sensor.
3. Insert a 0.004 in. (0.1mm) feeler gauge between the throttle stop screw and the stop lever. Continuity should exists between the terminals.
4. Insert a 0.024 in. (0.6mm) feeler gauge between the throttle stop screw and the stop lever. No continuity should exists between the terminals.

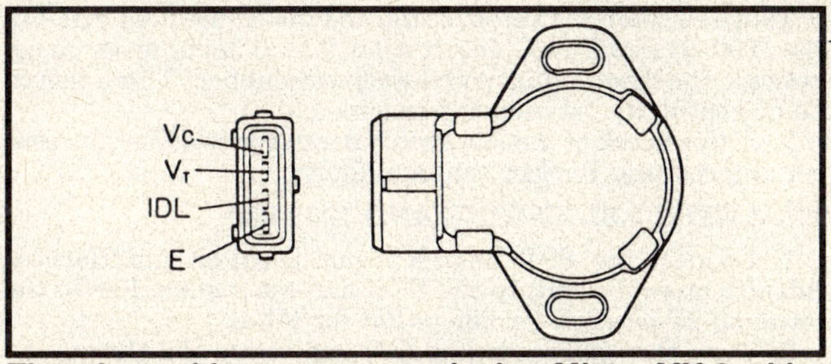

Throttle position sensor terminals—Miata, MX-3 with 1.6L engine, 323 and Protege with automatic transmissions

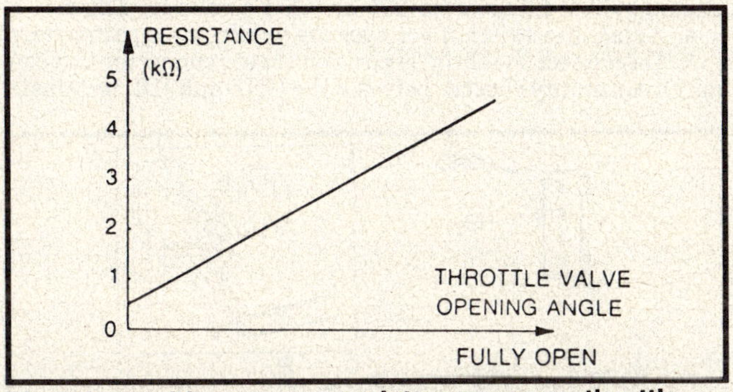

Throttle position sensor resistance versus throttle angle. The resistance must increase linearly as the throttle opens—Miata, MX-3 with 1.6L engine, 323 and Protege with automatic transmissions

5. Connect a suitable ohmmeter between terminals VT and E on the throttle sensor. Ohmmeter should read as follows:
   a. Throttle valve fully closed—Below 1 Kilo-ohm
   b. Throttle valve fully open—Approximately 5 Kilo-ohms

626 AND MX-6, MPV WITH 2.6L ENGINE, B2200 AND B2600i

1. Disconnect the air hose from the throttle body assembly, then disconnect the 3-pin throttle sensor electrical connector.
2. Connect Adapter Harness (49–G018–901) between the throttle sensor and the wiring harness.
3. Turn the ignition switch to the **ON** position. Check that the throttle valve is in the fully closed position.
4. Using a suitable voltmeter, measure the black and red wire voltages. Voltage should be as follows:
   Black wire—Approximately 0 volts
   Red wire—4.5–5.5 volts
5. If specifications are not as indicated, check the battery voltage and the wiring harness. If no problems with the wiring or voltage are indicated, replace the ECU.
6. Note the red wire voltage. Compare the blue wire voltage to the recorded red wire voltage as specified.
7. Hold the throttle valve in the fully opened position. Compare the blue wire voltage to the recorded red wire voltage as specified.
8. Check that the blue wire voltage increases smoothly, when opening the throttle valve from the fully closed to the fully opened positions. If not, replace the throttle sensor. Turn the ignition switch to the **OFF** position.
9. Disconnect the adapter harness (49–G018–901) and reconnect the throttle sensor electrical connector. Disconnect the battery ground cable and depress the brake pedal for at least 5 seconds, to eliminate the ECU malfunction memory.

RED wire voltage (V)	BLUE wire voltage (V)	RED wire voltage (V)	BLUE wire voltage (V)
4.50—4.59	0.37—0.54	5.10—5.19	0.42—0.61
4.60—4.69	0.38—0.55	5.20—5.29	0.43—0.62
4.70—4.79	0.39—0.56	5.30—5.39	0.44—0.63
4.80—4.89	0.40—0.57	5.40—5.49	0.44—0.64
4.90—4.99	0.40—0.58	5.50	0.44—0.66
5.00—5.09	0.41—0.60		

Throttle closed throttle position sensor voltage specifications—B2200, B2600i, 626, MX-6 and MPV with 2.6L engine

RED wire voltage (V)	BLUE wire voltage (V)	RED wire voltage (V)	BLUE wire voltage (V)
4.50—4.59	3.58—4.23	5.10—5.19	4.05—4.79
4.60—4.69	3.66—4.32	5.20—5.29	4.13—4.88
4.70—4.79	3.74—4.41	5.30—5.39	4.21—4.98
4.80—4.89	3.82—4.51	5.40—5.49	4.29—5.07
4.90—4.99	3.90—4.60	5.50	4.29—5.17
5.00—5.09	3.97—4.70		

Throttle open throttle position sensor voltage specifications—B2200, B2600i, 626, MX-6 and MPV with 2.6L engine

MPV WITH 3.0L ENGINE AND 1990–91 929

1. Disconnect the electrical connector from the throttle sensor.
2. Connect a suitable ohmmeter between terminals VREF and GND on the throttle sensor. Ohmmeter should read 3.5–6.5 Kilo-ohms (3.0–6.0 Kilo-ohms for 1992 MPV and 1990–91 929 with DOHC).
3. Connect the ohmmeter between terminals TVO and GND on the throttle sensor. Slowly open the throttle valve and check the resistance. Resistance should be as follows:

Year	Model	Throttle Valve Position	
		Open	Closed
1990–91	929 SOHC	3.5–6.5 kilo-ohms	1 kilo-ohm
1990–91	929 DOHC	3.3–7.0 kilo-ohms	0.2–0.6 kilo-ohms
1990–91	MPV 3.0L	3.5–6.5 kilo-ohms	1 kilo-ohm
1992	MPV 3.0L	3.0–6.0 kilo-ohms	1 kilo-ohm

Throttle position sensor test specifications—1990–991 929 and MPV 3.0L

4. If specifications are not as indicated, check the throttle sensor adjustment.
5. Except 1990–91 929 DOHC, connect the ohmmeter to the IDL and GND terminals and insert a 0.02 in.(0.5mm) feeler gauge between the throttle stop and the adjustment bolt. On the 1990–91 929 DOHC, connect the ohmmeter to the IDL and GND terminals and insert a 0.004 in.(0.1mm) feeler gauge between the throttle stop and the adjustment bolt.There should be continuity between the terminals.
6. Except 1990–91 929 DOHC, insert a 0.028 in.(0.7mm) feeler gauge between the throttle stop and the adjustment bolt. On the 1990–91 929 DOHC, insert a 0.012 in.(0.3mm) feeler gauge between the throttle stop and the adjustment bolt. There should be no continuity between the terminals.
7. If this check of the idle switch doesn't pass adjust the sensor and recheck. Replace sensor if faulty.

MX-3 WITH 1.8L ENGINE AND 1992 929

1. Connect the SST Engine Signal Monitor and Harness Adapter or equivalent, to the ECU and set the monitor to the terminal 2F position for 929 or 2M for MX-3.
2. Turn the ignition switch **ON** and rotate the throttle by hand.
3. With the throttle fully closed the voltage on terminal 2F for 929 or2M for MX-3 should be 0.1–1.1 volts. With the throttle fully open the voltage on terminal 2F for 929 or 2M for MX-3 should be 3.0–4.3 volts for 929 or 3.1–4.4 volts for MX-3.
4. Adjust the switch if voltages are within these parameters.
5. Disconnect the throttle position sensor connector and connect a continuity checker between the GND and IDL terminals.

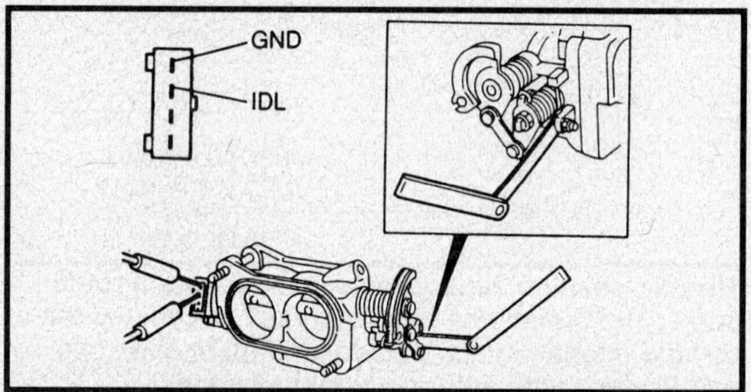

Testing the idle switch portion of the throttle position sensor—1992 929

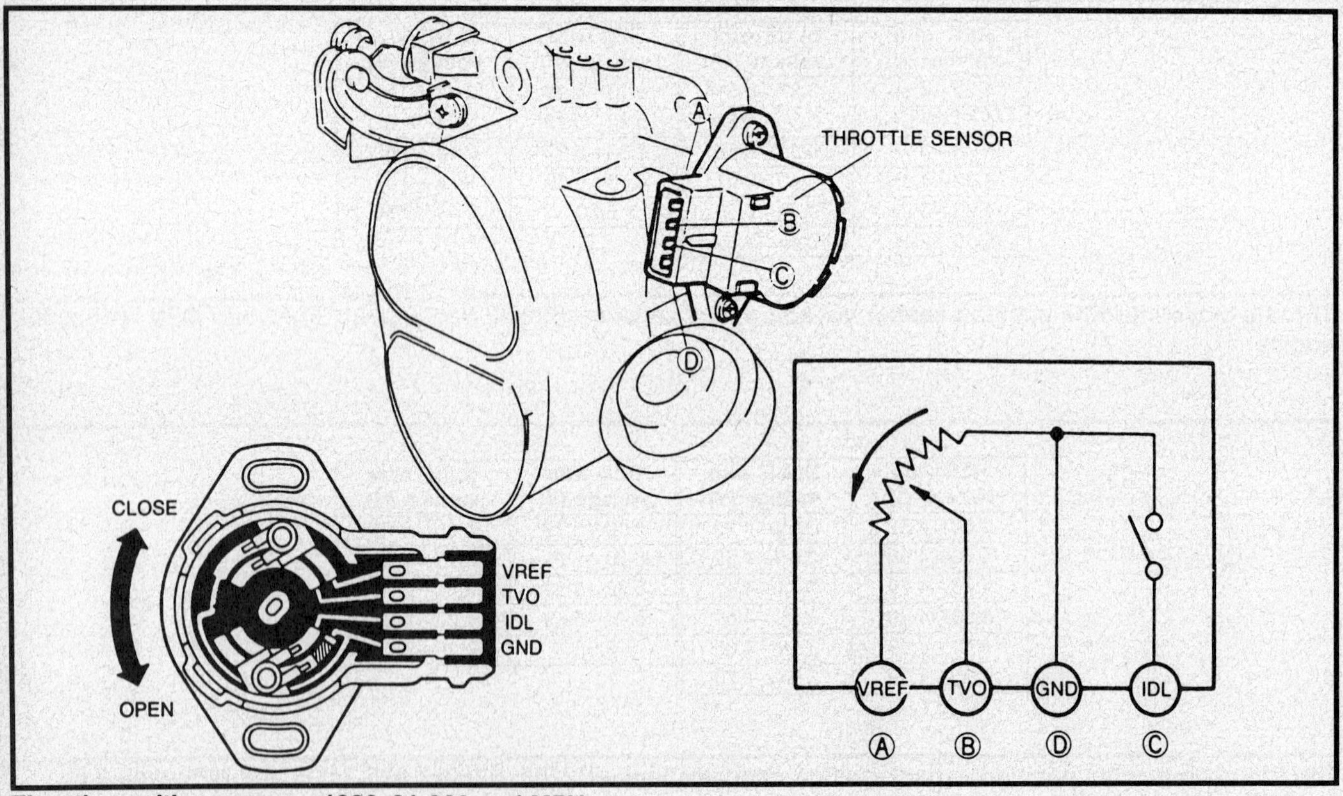

Throttle position sensor—1990–91 929 and MPV

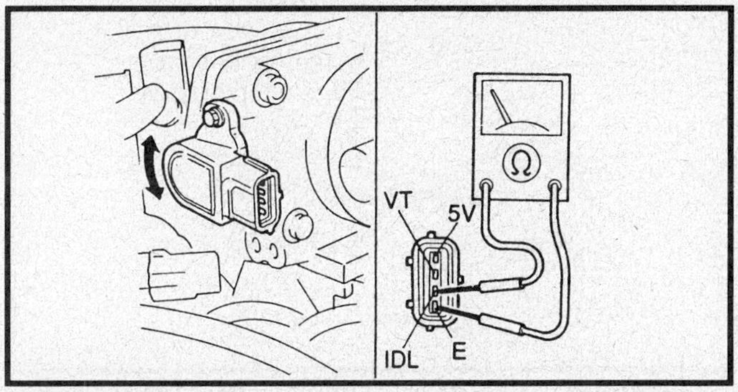

Testing the throttle position idle switch—MX-3 with 1.8L engine

6. Insert a 0.004 in.(0.1mm) feeler gauge for 9299 or 0.006 in. (0.15mm) feeler gauge for MX-3 between the throttle stop screw and the lever. There should be continuity.
7. Insert a 0.008 in.(0.2mm) feeler gauge for 929 or 0.020 in. (0.50mm) feeler gauge for MX-3 between the throttle stop screw and the lever. There should be no continuity.
8. Adjust the sensor if these values are not obtained.

FUEL INJECTORS

On-Vehicle Testing

1. Start the engine and allow it to reach normal operating temperature, and let run at idle.
2. Apply a sound scope or equivalent to the injector and check for operating sound.
3. If no sound is detected, check the wiring harness, injector resistance and the ECU terminal voltages.

Resistance Test

1. Disconnect the electrical connector from the fuel injector.
2. Using a suitable ohmmeter, measure the resistance across the fuel injector terminals. On the MX-3 with 1.8L engine check the resistance at the fuel injector harness connectors. Ohmmeter should read 12–16 ohms.
3. If specifications are not as indicated, replace the fuel injector.
4. Repeat Step 1 through 3 for the remaining fuel injectors.

Operational Tests

MPV, B2200, B2600i AND 1990–91 929

1. Relieve the fuel pressure from the fuel system, the disconnect the battery ground cable.
2. Remove the delivery pipe, injectors and pressure regulator as an assembly.
3. Firmly attach each injector to the distribution pipe with a piece of wire. Ensure that the injectors are not allowed to move or rotate on the distribution pipe.
4. Connect the distribution pipe assembly between the fuel filter and return pipe, then connect the return hose to the pressure regulator.
5. Connect the battery ground cable. Install a jumper across the yellowtest connector.
6. Turn the ignition switch to the **ON** position, and check the injectors for fuel leakage.

NOTE: After approximately 1 minute, a slight amount of fuel leakage is acceptable.

7. If fuel leaks from any injector, replace it.
8. Suspend injector in a graduated container and apply battery voltage to the injector terminal for 15 seconds.
9. Check that the volume flowed in the 15 second period is within specifications.

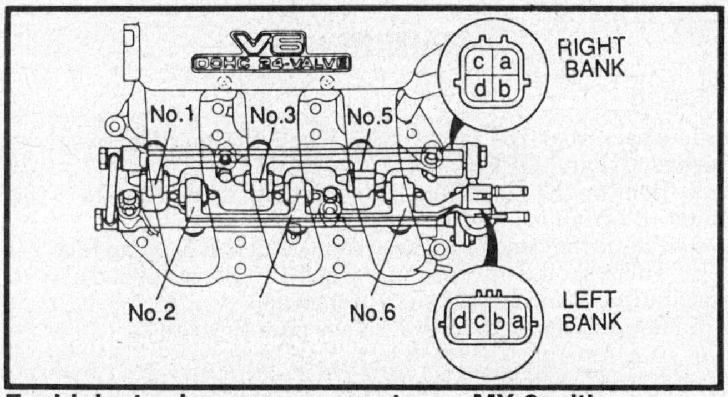

Fuel injector harness connectors—MX-3 with 1.8Lengine

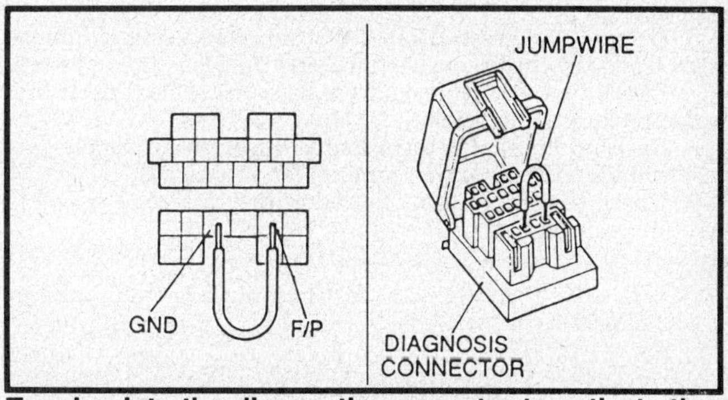

Tapping into the diagnostic connector to activate the fuel pump—323, Protege, Miata, MX-3 and 1992 929

Harness	Terminal	injector No.
Right Bank	a—b	1
	a—c	5
	a—d	3
Left Bank	d—c	2
	d—b	4
	d—a	6

Testing fuel injector resistances at the harness connectors—MX-3 with 1.8L engine

323, PROTEGE, MIATA, MX-3 AND 1992 929

1. Relieve the fuel pressure from the fuel system.
2. On Miata, remove the air valve.
3. Disconnect the electrical connector from each fuel injector.
4. Remove the delivery pipe attaching bolts, then affix the injectors to the to the delivery pipe with suitable wire.

NOTE: Affix the injectors firmly to the distribution pipe so that no movement of the injectors is possible.

5. Install a jumper wire between terminals F/P and GND on the diagnosis connector. Turn the ignition switch to the **ON** position.
6. Tilt the injectors approximately 60 degrees, and check that no fuel leaks from the injector nozzles. If fuel leaks from any injector, replace that injector.

NOTE: After approximately 1 minute, a slight amount of fuel leakage is acceptable.

7. Turn the ignition switch to the **OFF** position and remove the jumper wire.
8. Suspend injector in a graduated container and apply battery voltage to the injector terminal for 15 seconds.
9. Check that the volume flowed in the 15 second period is within specifications.

626 AND MX-6

1. Relieve the fuel pressure from the fuel system.
2. Lift the dynamic chamber upward.
3. Disconnect the electrical connector from each injector, then remove the delivery pipe with the injectors attached.
4. Affix the injectors to the to the delivery pipe with some suitable wire.

NOTE: Affix the injectors firmly to the distribution pipe so that no movement of the injectors is possible.

5. Connect jumper wire between the terminals of the yellow fuel pump test connector. Turn the ignition switch to the **ON** position.
6. Ground the green 1-pin test connector using a jumper wire. Open the throttle valve to release the air in the injectors.
7. Check that no fuel leaks from the injector nozzles. If fuel leaks, replace the injector.
8. Suspend injector in a graduated container and apply battery voltage to the injector terminal for 15 seconds.
9. Check that the volume flowed in the 15 second period is within specifications.

NOTE: After 1 minute, a slight amount of fuel leakage is acceptable.

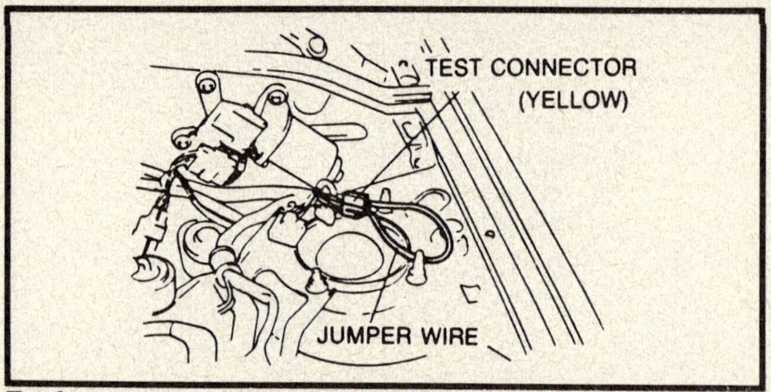

Fuel pump test connector (yellow) – 626, MX-6, MPV, B2200, B2600i and 1990–91 929

Injector test connector (green) – 626 and MX-6

10. Suspend injector in a graduated container and apply battery voltage to the injector terminal for 15 seconds.
11. Check that the volume flowed in the 15 second period is within specifications.

MAIN RELAY

Testing

1. Check that the relay clicks when the ignition switch is turned **ON** and **OFF**.
2. Remove the relay from the socket and apply 12 volts to the A and B terminals of the relay.
3. Check the continuity between the C and D terminals.
4. There should be continuity when 12 volts is applied and no continuity when the 12 volts is removed.
5. Replace the relay if it does not pass these tests.

1.6L ENGINE SPECIFICATIONS CHART—MIATA

Item			Unit	Specification
Idle speed			rpm	850 ± 50*
Ignition timing			BTDC	10° ± 1°* (M/T), 8° ± 1°* (4A/T)
Throttle body				
Type				Horizontal draft
Throat diameter			mm (in)	55 (2.2)
Fuel pump				
Type				Impeller (in-tank)
Output pressure			kPa (kg/cm², psi)	441—589 (4.5—6.0, 64—85)
Fuel filter				
Type	Low-pressure side			Nylon element
	High-pressure side			Paper element
Pressure regulator				
Type				Diaphragm
Regulating pressure			kPa (kg/cm², psi)	265—314 (2.7—3.2, 38—46)
Injector				
Type				High-ohmic
Type of drive				Voltage
Resistance			Ω	12—16 (at 20°C, 68°F)
ISC valve				
Solenoid resistance			Ω	11—13 (at 20°C, 68°F)
Air valve				
Opening temperature				Below 40°C (104°F)
Solenoid valve (Purge control)				
Solenoid resistance			Ω	23—27 (at 20°C, 68°F)
Crank angle sensor				
Type				Optical pickup
Airflow meter				
Resistance Ω	E2↔Vs	Fully closed		200—600
		Fully open		20—1,000
	E2↔Vc			200—400
	E2↔THAA (Intake air thermosensor)	−20°C (−4°F)		13,600—18,400
		20°C (68°F)		2,210—2,690
		60°C (140°F)		493—667
	E1↔Fc	Fully closed		∞
		Fully open		0
Water thermosensor				
Resistance kΩ		−20°C (−4°F)		14.6—17.8
		20°C (68°F)		2.2—2.7
		80°C (179°F)		0.29—0.35
Circuit opening relay				
Resistance Ω		STA—E1		21—43
		B—Fc		109—226
		B—Fp		∞
Fuel tank				
Capacity			liters (US gal, Imp gal)	45 (11.9, 9.9)
Air cleaner				
Element type				Oil permeated
Accelerator cable				
Free play			mm (in)	1—3 (0.039—0.118)
Fuel				
Specification				Unleaded (RON 87 or higher)

*... With System Selector (49 B019 9A0) test switch at SELF TEST.

1.6L ENGINE COMPONENENT DESCRIPTIONS—MIATA

Component	Function	Remark
Air cleaner	Filters air entering throttle body	
Airflow meter	Detects amount of intake air; sends signal to engine control unit	• Intake air thermosensor and fuel pump switch included • Use Vref (5 volt) as power source
Air valve	When cold, supplies bypass air into dynamic chamber	• Engine speed increased to shorten warm-up period • Thermowax type
Atmospheric pressure sensor	Detects atmospheric pressure; sends signal to engine control unit	Built-in ECU
Catalytic converter	Reduces HC, CO, and NOx by chemical reaction	Monolith type
Charcoal canister	Stores gas tank fumes while engine stopped	
Circuit opening relay	Voltage for fuel pump while engine running	
Clutch switch (M/T)	Detects in-gear condition; sends signal to engine control unit	Switch OFF when clutch pedal released
Diagnosis connector	Concentrated service connector Concentrated terminals are: 1. EGI self-diagnosis terminal 2. Initial set terminal 3. Fuel pump check terminal 4. Cooling fan check terminal	21-pin (Black)
Crank angle sensor **G-signal** **Ne-signal**	Detects No.1 cylinder TDC; sends signal to engine control unit Detects crank angle at 180° intervals; sends signal to control unit	
Dynamic chamber	Interconnects all cylinders	
Engine control unit (ECU)	**Detects the following:** 1. A/C operation 2. Air/fuel ratio (Oxygen concentration) 3. Atmospheric pressure 4. Braking signal 5. Cranking signal 6. E/L operation 7. Engine coolant temperature 8. Engine speed 9. In-gear condition 10. Intake air amount 11. Intake air temperature 12. No.1 piston TDC 13. P/S operation 14. Initial set signal 15. Throttle valve fully closed/ fully open condition	1. A/C switch 2. Oxygen sensor 3. Atmospheric pressure sensor 4. Stoplight switch 5. Ignition switch (START position) 6. Cooling fan relay Light and turn relay Blower motor switch 7. Water thermosensor 8. Crank angle sensor (Ne-signal) 9. Neutral and clutch switches 10. Airflow meter 11. Intake air thermosensor 12. Crank angle sensor (G-signal) 13. P/S pressure switch 14. Diagnosis connector 15. Throttle sensor
	Controls operation of the following: 1. A/C (Cut off) 2. Fail-safe function 3. Fuel injection system 4. Idle speed control 5. Ignition timing control system 6. Monitor function 7. Purge control system 8. Lockup control system	1. A/C relay 2. Self-diagnosis checker and MIL 3. Injector 4. ISC valve 5. Igniter 6. Monitor lamp (Self-diagnosis checker) 7. Solenoid valve (Purge control) 8. 4A/T control unit

1.6L ENGINE COMPONENENT DESCRIPTIONS—MIATA

Component	Function	Remark
Fuel vapor valve	Prevents fuel from entering canister during vehicle roll over	
Fuel filter	Filters particles from fuel	
Fuel pump	Provides fuel to injectors	• Operates while engine running • Installed in fuel tank
Igniter	Receives spark signal from engine control unit and generates high voltage in ignition coil Detects high voltage ignition; sends substitute ignition signal to engine control unit	
Ignition switch (START position)	Sends engine cranking signal to engine control unit	
Injector	Injects fuel into intake port	• Controlled by signals from engine control unit • High-ohmic injector • Single port injector nozzle
Intake air thermosensor	Detects intake air temperature; sends signal to engine control unit	Installed in airflow meter
ISC valve	Controls bypass air amount	• Controlled by duty signal from engine control unit • Controls idle-up
Solenoid valve (Purge control)	Controls evaporative fumes from canister to intake manifold	• Controlled by duty signal from engine control unit
Main relay	Supplies electric current to injectors, engine control unit, etc.	
MIL (Malfunction indicator lamp)	Lamp illuminates when input device malfunctions	(TEN) terminal of diagnosis connector
	Lamp flashes to indicate malfunction code number of input and output devices	(TEN) terminal of diagnosis connector
Neutral switch	Detects in-gear condition; sends signal to engine control unit	Switch ON when in neutral
Oxygen sensor	Detects oxygen concentration; sends signal to engine control unit	Zirconia ceramic and platinum coating
PCV valve	Controls blow by gas amount introduced into engine	
P/S pressure switch	Detects P/S operation; sends signal to engine control unit	P/S: ON when steering wheel turned
Pressure regulator	Adjusts fuel pressure supplied to injectors	
Resonance chamber	Improves mid-range torque characteristics	
Stoplight switch	Detects braking operation (deceleration); sends signal to engine control unit	
Throttle body	Controls intake air quantity	Integrated throttle sensor, ISC valve, and dashpot
Throttle sensor	• Detects position of the throttle valve fully closed; sends signal to engine control unit • Detects the throttle valve opening at a preset angle; sends signal to engine control unit	
Two-way check valve	Controls pressure in fuel tank	
Three-way check valve (4A/T)	Controls pressure in fuel tank	
Water thermosensor	Detects coolant temperature; sends signal to engine control unit	

1.6L ENGINE ENGINE CONTROL OPERATION CHART MIATA

INPUT DEVICES / ENGINE CONDITIONS	APPROXIMATE TIME (BASED ON 10—16°C, or 50—60°F AMBIENT)	SENSORS: CRANK ANGLE SENSOR: G-SIGNAL	CRANK ANGLE SENSOR: NE-SIGNAL	AIRFLOW METER: INTAKE AIR THERMOSENSOR	AIRFLOW METER: POTENTIOMETER	WATER THERMOSENSOR	OXYGEN SENSOR	IGF SIGNAL	ATMOSPHERIC PRESSURE SENSOR
CRANKING —COLD ENGINE • COLD AIR • COLD COOLANT	ZERO			SIGNAL HAS NO EFFECT ON ECU	SIGNAL HAS NO EFFECT ON ECU		SIGNAL HAS NO EFFECT ON ECU		
COLD START —FAST IDLE • COLD AIR • COLD COOLANT	ONE MINUTE				LOW VOLUME AIRFLOW: (3.0V)	COOL TO WARM: MEDIUM VOLTAGE (2.3V and DROPPING)	SENSOR COLD: LOW TO HIGH VOLTAGE (0—0.8V)		
COLD DRIVE AWAY —PART THROTTLE • WARM AIR • WARM COOLANT	TWO MINUTES								
WARM DRIVE AWAY —PART THROTTLE • WARM AIR • WARM COOLANT	THREE MINUTES				MODERATE VOLUME AIRFLOW: LOW TO MEDIUM VOLTAGE (1—3.5V)	WARM: MEDIUM VOLTAGE (APPROX. 0.7V AND DROPPING)	SENSOR WARM: HIGH VOLTAGE (ABOVE 0.6V)		
HOT CRUISE • WARM AIR • WARM COOLANT							SENSOR HOT SWITCHING FROM HIGH VOLTAGE (ABOVE 0.6V RICH)		SENDS VOLTAGE SIGNAL TO ECU THAT VARIES WITH ALTITUDE; VOLTAGE (APPROX. 4V)
HOT ACCELERATION —60% THROTTLE		SENDS No.1 CYLINDER TDC SIGNAL TO ECU	SENDS ENGINE SPEED SIGNAL TO ECU	COOL TO WARM: MEDIUM VOLTAGE (4.3—1.5V)	MODERATE TO STRONG VOLUME AIRFLOW: (0.5—1.5V)		TO LOW VOLTAGE (BELOW. 0.4V LEAN)	SENDS IGNITION SPARK SIGNAL TO ECU	
HOT ACCELERATION —WIDE OPEN THROTTLE	MORE THAN FOUR MINUTES					HOT: LOW VOLTAGE (APPROX. 0.4V)	HIGH VOLTAGE (ABOVE 0.6V RICH)		
DECELERATION —CLOSED THROTTLE							LOW VOLTAGE (BELOW 0.4V LEAN)		
HOT CURB IDLE —EXTENDED					LOW VOLUME AIRFLOW (3.0V)		SWITCHING FROM HIGH TO LOW VOLTAGE (0.4—0.6V)		
HOT ENGINE SHUTDOWN	—	OFF	OFF	OFF	OFF	OFF	SENSOR HOT: LOW VOLTAGE (0.1V LEAN) UNTIL SENSOR COOLS DOWN	OFF	OFF

1.6L ENGINE ENGINE CONTROL OPERATION CHART—MIATA

SWITCHES										
THROTTLE SENSOR (4A/T) TVO TERMINAL	THROTTLE SENSOR (M/T) POWER TERMINAL	THROTTLE SENSOR (M/T) IDLE TERMINAL	A/C SWITCH	NEUTRAL AND CLUTCH SWITCHES (M/T)	INHIBITOR SWITCH (4A/T)	STOP LIGHT SWITCH	E/L SIGNAL	P/S PRESSURE SWITCH	IGNITION SWITCH START POSITION	TEST TERMINAL
SEND SIGNAL TO ECU WHEN PEDAL DEPRESSED	SEND SIGNAL TO ECU WHEN PEDAL DEPRESSED	SIGNAL HAS NO EFFECT ON ECU	SIGNAL HAS NO EFFECT ON ECU	SIGNAL HAS NO EFFECT ON ECU	SIGNAL HAS NO EFFECT ON ECU	SIGNAL HAS NO EFFECT ON ECU	SIGNAL HAS NO EFFECT ON ECU	SIGNAL HAS NO EFFECT ON ECU	SENDS SIGNAL TO ECU	SIGNAL HAS NO EFFECT ON ECU
LOW VOLTAGE SIGNAL TO ECU (0.2–0.8)	HIGH VOLTAGE SIGNAL TO ECU (APPROX. 5V)	LOW VOLTAGE SIGNAL TO ECU (BELOW 1.5V)	A/C SWITCH ON: SENDS SIGNAL TO ECU (0V) A/C SWITCH OFF: NO SIGNAL TO ECU (BATTERY VOLTAGE)	IN NEUTRAL: LOW VOLTAGE SIGNAL TO ECU (0V)	IN N OR P RANGE: NO SIGNAL TO ECU (0V)	NO SIGNAL TO ECU (BELOW 1.5V)	HEADLIGHT SWITCH ON: HIGH VOLTAGE SIGNAL TO ECU (BATTERY VOLTAGE) OFF: LOW VOLTAGE SIGNAL TO ECU (BELOW 1.5V) BLOWER CONTROL SWITCH OR COOLING FAN SWITCH ON: LOW VOLTAGE SIGNAL TO ECU (0V) OFF: HIGH VOLTAGE SIGNAL TO ECU (BATTERY VOLTAEG)	STEERING WHEEL TURNED: LOW VOLTAGE SIGNAL TO ECU (0V) STEERING WHEEL STRAIGHT AHEAD: HIGH VOLTAGE TO ECU (BATTERY VOLTAGE)	NO SIGNAL TO ECU (BELOW 1.5V)	TERMINAL NOT GROUNDED: HIGH VOLTAGE TO ECU (BATTERY VOLTAGE)
MEDIUM VOLTAGE SIGNAL TO ECU (0.8–3.0V)	LOW VOLTAGE SIGNAL TO ECU (0V)	HIGH VOLTAGE SIGNAL TO ECU (BATTERY VOLTAGE)		DRIVING IN ANY GEAR: HIGH VOLTAGE SIGNAL TO ECU (BATTERY VOLTAGE)	IN D RANGE: NO SIGNAL TO ECU (0V)	BRAKE PEDAL DEPRESSED: SENDS SIGNAL TO ECU (BATTERY VOLTAGE)				LOW VOLTAGE SIGNAL TO ECU WHEN CONNECTOR GROUNDED (BELOW 1.5V)
HIGH VOLTAGE SIGNAL TO ECU (ABOVE 3.0V)	HIGH VOLTAGE SIGNAL TO ECU (APPROX. 5V)	LOW VOLTAGE SIGNAL TO ECU (0V)		IN NEUTRAL: LOW VOLTAGE SIGNAL TO ECU (0V)	IN N OR P RANGE: SENDS SIGNAL TO ECU (0V)					
LOW VOLTAGE SIGNAL TO ECU (0.2–0.8V)										
OFF	OFF	OFF	OFF	OFF	OFF	OFF	OFF	OFF	OFF	OFF

ENGINE CONDITIONS / OUTPUT DEVICES	APPROXIMATE TIME (BASED ON 10–16°C or 50–60°F AMBIENT)	INJECTOR: INJECTION	INJECTOR: INJECTION TIMING	ISC VALVE	SOLENOID VALVE (PURGE CONTROL)	A/C RELAY (A/C CUT-OFF)	MAIN RELAY	IGNITER
CRANKING —COLD ENGINE • COLD AIR • COLD COOLANT	ZERO		ALL CYLINDERS EACH Ne-SIGNAL			OFF (A/C CUT)		
COLD START —FAST IDLE • COLD AIR • COLD COOLANT	ONE MINUTE	RICH		LARGE AMOUNT OF BYPASS AIR	OFF (PURGE CUT)			
COLD DRIVE AWAY —PART THROTTLE • COLD AIR • COLD COOLANT	TWO MINUTES					ON (A/C ON)		
WARM DRIVE AWAY —PART THROTTLE • WARM AIR • WARM COOLANT	THREE MINUTES	RICH AND LEAN	2-GROUP					
HOT CRUISE • WARM AIR • WARM COOLANT				SMALL AMOUNT OF BYPASS AIR	OPERATES (DUTY VALUES [PURGE GAS AMOUNT] CHANGE)		ON	IGNITION SPARK ADVANCE SIGNAL
HOT ACCELERATION —60% THROTTLE		RICH						
HOT ACCELERATION —WIDE OPEN THROTTLE	MORE THAN FOUR MINUTES					OFF (A/C CUT)		
DECELERATION —CLOSED THROTTLE		FUEL CUT		LARGE AND SMALL AMOUNT OF BYPASS AIR	OFF (PURGE CUT)	ON (A/C ON)		
HOT CURB IDLE —EXTENDED		RICH AND LEAN	2-GROUP	SMALL AMOUNT OF BYPASS AIR				
HOT ENGINE SHUT DOWN	—	DOES NOT INJECT		OFF	OFF	OFF	OFF	OFF

1.6L ENGINE RELATIONSHIP CHART—MIATA

Input Devices \ Output Devices		Injector: Fuel Injection Amount	Injector: Fuel Injection Timing	ISC Valve	Solenoid Valve (Purge Control)	A/C Relay (A/C Cut-Off)	Igniter (Ignition Timing Control)	4A/T Control Unit
Test Terminal		×	×	○	×	×	○	×
Ignition Switch (Start Position)		○	○	○	×	○	○	×
P/S Pressure Switch		×	×	○	×	×	×	×
Cooling Fan Switch		×	×	○	×	×	×	×
Blower Control Switch		×	×	○	×	×	×	×
Headlight Switch		×	×	○	×	×	×	×
Stoplight Switch (M/T)		○	×	×	×	×	×	×
Neutral and Clutch Switches (M/T)		○	×	○	○	×	○	×
Inhibitor switch (4A/T)		○	×	○	○	○	○	×
A/C Switch		×	×	○	×	○	×	○
Throttle Sensor	Idle Terminal (IDL)	○	×	○	○	×	○	×
	Power Terminal (M/T) Tvo Terminal (4A/T)	○	×	×	×	○	×	×
Atmospheric Pressure Sensor		○	×	○	×	×	×	×
IGF Signal		○	×	×	×	×	×	×
Oxygen Sensor		○	×	×	○	×	×	×
Water Thermosensor		○	×	○	○	×	○	×
Airflow Meter	Potentiometer	○	×	×	○	×	○	○
	Intake Air Thermosensor	○	×	○	○	×	×	×
Crank Angle Sensor	Ne-Signal	○	○	○	○	×	○	×
	G-Signal	×	○	×	×	×	○	×

1.6L ENGINE TROUBLESHOOTING GUIDE—MIATA

No.	TROUBLESHOOTING ITEM	DESCRIPTION
1	Will not crank or cranks slowly	Refer to Engine Electrical System
2	Cranks normally but will not start (No combustion)	Engine cranks at normal speed but shows no sign of "firing"
3	Cranks normally but will not start (Partial combustion) — When engine is cold	Engine cranks at normal speed but shows partial combustion and will not continue to run
4	Cranks normally but will not start (Partial combustion) — After warm-up	Engine cranks at normal speed but shows partial combustion and will not continue to run after running and hot soaked
5	Cranks normally but hard to start — Always	Engine cranks at normal speed but requires excessive cranking time (more than 5 sec.) before starting
6	Cranks normally but hard to start — When engine is cold	Same condition as No.5 when engine is cold Restarts OK after warm-up
7	Cranks normally but hard to start — After warm-up	Same condition as No.5 after running and hot soaked Starts normally when cold
8	Rough idle — Always	Engine vibrates excessively at idle in every condition
9	Low idle speed/Rough idle — Before warm-up	Engine speed low or engine vibrates excessively at idle during warm-up
10	Low idle speed/Rough idle — After warm-up	Engine runs normally at idle during warm-up but vibrates excessively after warm-up
11	High idle speed — After warm-up	Engine idle excessive for operation mode
12	Low idle speed — When A/C, P/S, or E/L ON	Engine speed decreases at idle when A/C, P/S, or E/L is ON
13	Rough idle just after starting	Engine starts normally but vibrates excessively only just after starting
14	Idle moves up and down	Engine speed up and down periodically at idle
15	Engine stalls at idle — Always	Engine starts normally but vibrates excessively and stalls at idle in every condition
16	Engine stalls at idle — Before warm-up	Engine starts normally but vibrates excessively and stalls at idle before warm-up
17	Engine stalls at idle — After warm-up	Engine runs normally at idle during warm-up but becomes rough and stalls after warm-up
18	Engine stalls during start-up	Engine unexpectedly stops running while starting
19	Engine stalls on deceleration	Engine unexpectedly stops running while decelerating or after deceleration
20	Engine stalls at idle — When A/C, P/S, or E/L ON	Engine unexpectedly stops running at idle when A/C, P/S, or E/L is ON
21	Engine stalls suddenly (Intermittent)	Engine intermittently stops running
22	Hesitates/Stumbles on acceleration	Flat spot occurs just after accelerator is depressed or mild jerking occurs during acceleration
23	Surges while cruising	Unexpected, usually repetitive change in engine speed
24	Lack of power	Performance is poor under load Maximum speed reduced
25	Poor acceleration	Performance is poor while accelerating
26	Runs rough on deceleration/Afterburn	Engine runs rough while decelerating and abnormal combustion in exhaust system
27	Knocking	Abnormal combustion accompanied by audible "pinging" noise
28	Fuel odor	Gasoline odor in cabin
29	Exhaust sulfur smell	Exhaust gas smells abnormal (rotten egg smell)
30	High oil consumption	Oil consumption excessive
31	Poor fuel economy	Fuel economy unsatisfactory
32	MIL always ON	Self-Diagnosis Checker does not indicate Malfunction Code No. but MIL comes on
33	MIL never ON	Self-Diagnosis Checker indicates Malfunction Code No. of input device but MIL never comes ON
34	A/C does not work	Blower fan operates but no cool air is discharged

1.6L ENGINE TROUBLESHOOTING GUIDE—MIATA

2	CRANKS NORMALLY BUT WILL NOT START (NO COMBUSTION)		
DESCRIPTION	• Engine cranks at normal speed but shows no sign of "firing" • Battery in normal condition • Throttle valve not held fully open while cranking • Fuel in tank		
[TROUBLESHOOTING HINTS] Because of no combustion, possibly no fuel is injected to engine or no ignition at all cylinders ① No spark • Ignition control malfunction • Ignition system component malfunction ② No fuel injection • Fuel pump does not operate • Injector does not operate ③ Low fuel line pressure ④ Low engine compression			
STEP	**INSPECTION**		**ACTION**
1	Check if strong blue spark is visible at disconnected high-tension lead while cranking engine 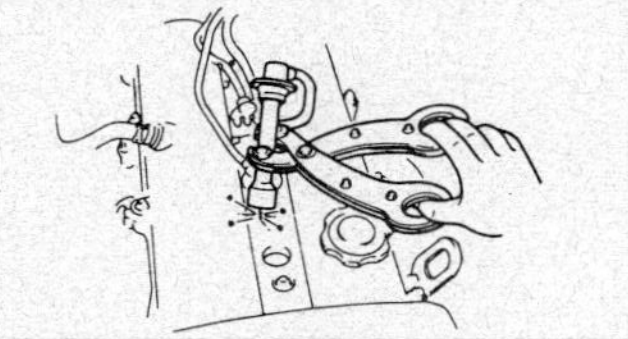	Yes	Go to Step 3
		No	Go to Step 2
2	Check if "00" is displayed on Self-Diagnosis Checker with ignition switch ON 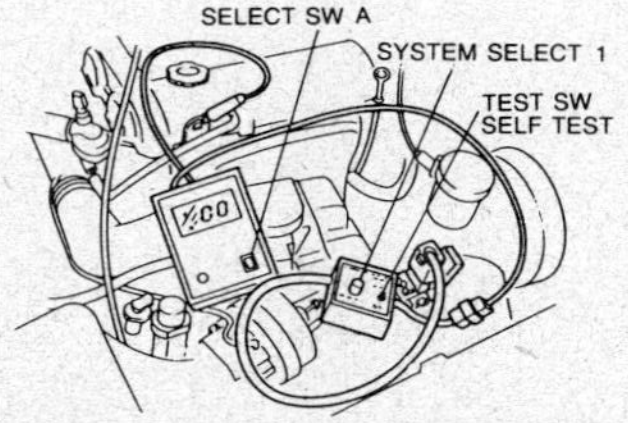	Yes	Check ignition system (Refer to Troubleshooting "Misfire")
		No	**Malfunction Code No. displayed** Check for cause (Refer to specified check sequence)
			"88" flashes Check ECU terminal 1F voltage **Specification: Battery voltage (Ignition switch ON)** ⇨ If OK, replace ECU ⇨ If not OK, check wiring between ECU and Self-Diagnosis Checker
3	Connect diagnosis connector terminals F/P and GND with jumper wire and check for fuel pump operating sound with ignition switch ON 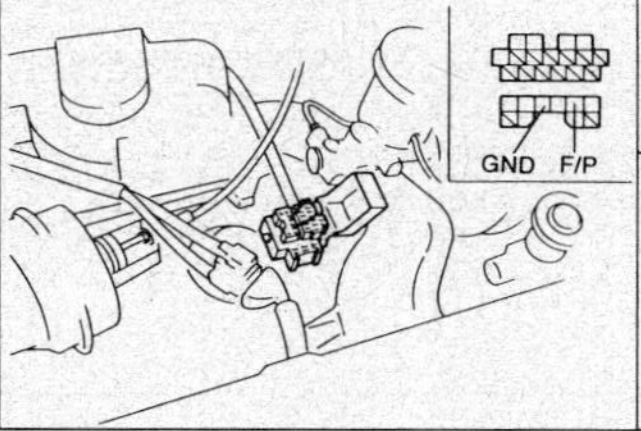	Yes	Check if engine starts in this condition ⇨ If starts, check circuit opening relay ⇨ If does not start, go to Step 5
		No	Go to Step 4

STEP	INSPECTION		ACTION
4	Check if battery voltage exists at fuel pump connector (L/R) wire with jumper wire connected (Step 3) 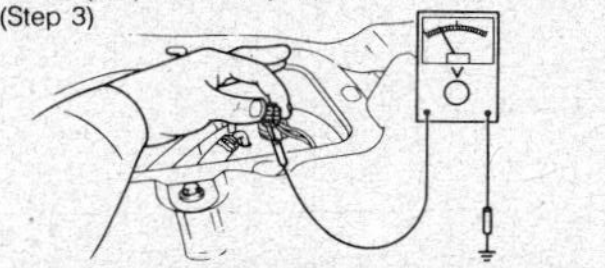	Yes	Check continuity of fuel pump
		No	Check circuit opening relay
5	Check for injector operating sound while cranking engine 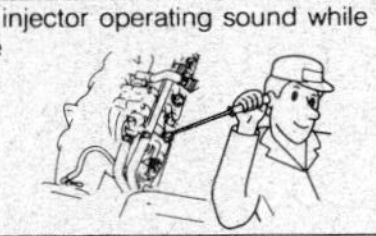	Yes	Go to Step 7
		No	Go to Step 6
6	Check if battery voltage exists at injector connector (W/R) wire with ignition switch ON 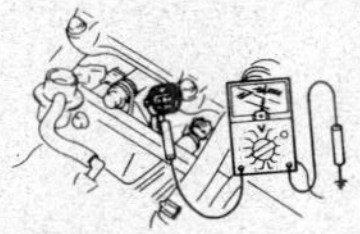	Yes	Check ECU terminals 2A, 2U and 2V voltages
		No	Check for open circuit in wiring between main relay and injector
7	Connect diagnosis connector terminals F/P and GND with jumper wire and check for correct fuel line pressure with ignition switch ON **Fuel Line pressure: 265—314 kPa (2.7—3.2 kg/cm², 38—46 psi)** 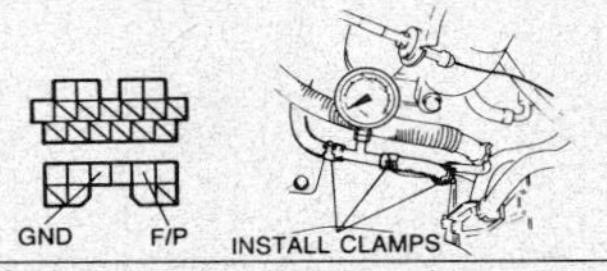	Yes	Go to next step
		No	**Low pressure** Check fuel line pressure while pinching fuel return hose ⇨ If fuel line pressure quickly increases, check pressure regulator ⇨ If fuel line pressure gradually increases, check for clogging between fuel pump and pressure regulator If not clogged, check fuel pump maximum pressure
8	Check for correct engine compression **Engine compression: 1,324—932 kPa (13.5—9.5 kg/cm², 192—135 psi)-300 rpm**	Yes	Go to next step
		No	Check engine condition • Worn piston, piston rings or cylinder wall • Defective cylinder head gasket • Distorted cylinder head • Improper valve seating • Valve sticking in guide
9	Check if spark plugs are OK 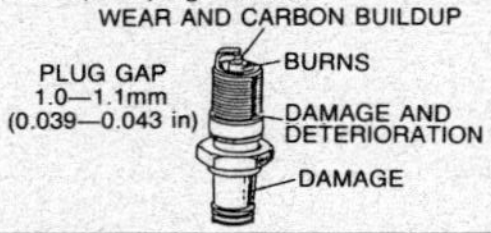	Yes	Go to next step
		No	Repair, clean, or replace
10	Try known good ECU and check if condition improves		

3	CRANKS NORMALLY BUT WILL NOT START (PARTIAL COMBUSTION) — WHEN ENGINE IS COLD		
DESCRIP-TION	• Engine cranks at normal speed but shows partial combustion and will not continue to run • Battery in normal condition • Fuel in tank		
[TROUBLESHOOTING HINTS] ① Air/Fuel mixture too rich • Air cleaner element clogged • Airflow meter stuck ② Air/Fuel mixture too lean • Fuel injection control malfunction (Correction for coolant temperature) • Low fuel line pressure • Air leakage of intake air system ③ Low engine compression			

STEP	INSPECTION		ACTION
1	Check if "00" is displayed on Self-Diagnosis Checker with ignition switch ON 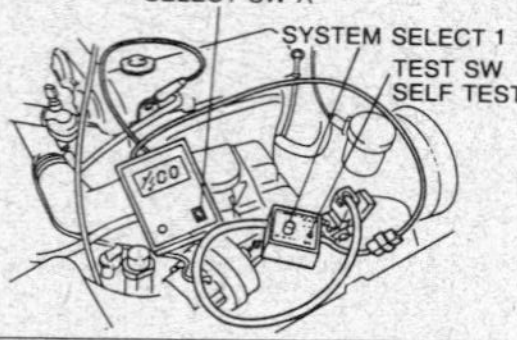	Yes	Go to next step
		No	**Malfunction Code No. displayed** Check for cause (Refer to specified check sequence) **"88" flashes** Check ECU terminal 1F voltage **Specification: Battery voltage (Ignition switch ON)** ⇨ If OK, replace ECU ⇨ If not OK, check wiring between ECU and Self-Diagnosis Checker
2	Check if strong blue spark is visible at each disconnected high-tension lead while cranking engine 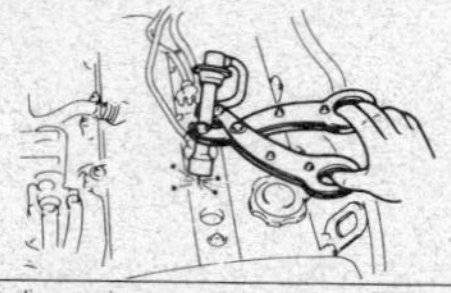	Yes	Go to next step
		No	Check ignition system (Refer to Troubleshooting "Misfire")
3	Connect diagnosis connector terminal F/P and GND with jumper wire and check for correct fuel line pressure with ignition switch ON **Fuel line pressure: 265—314 kPa (2.7—3.2 kg/cm², 38—46 psi)** 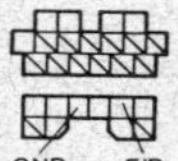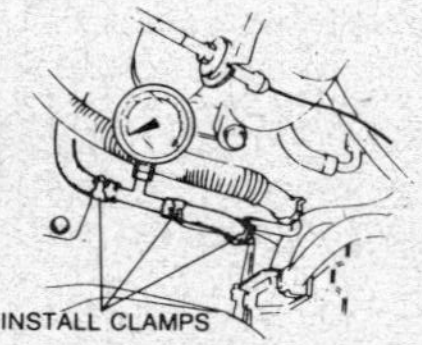	Yes	Go to next step
		No	**Low pressure** Check fuel line pressure while pinching fuel return hose ⇨ If fuel line pressure quickly increases, check pressure regulator ⇨ If fuel line pressure gradually increases, check for clogging between fuel pump and pressure regulator If not clogged, check fuel pump maximum pressure **High pressure** Check if fuel return hose is clogged or restricted ⇨ If OK, replace pressure regulator ⇨ If not OK, repair or replace
4	Check if ECU terminal voltages are OK (2D, 2O and 2Q)	Yes	Go to next step
		No	Check for cause (Refer to "Check Point for Each Terminal")

STEP	INSPECTION		ACTION
5	Check for air leakage of intake air system	Yes	Repair or replace
		No	Go to next step
6	Check if airflow meter vane moves smoothly	Yes	Go to next step
		No	Repair or replace
7	Check for correct engine compression **Engine compression: 1,324—932 kPa (13.5—9.5 kg/cm², 192—135 psi)-300 rpm** 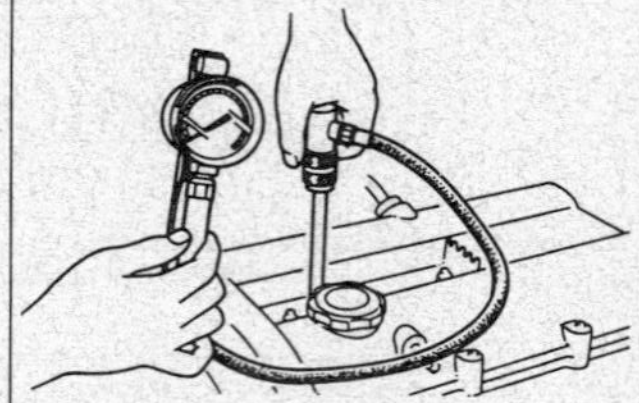	Yes	Go to next step
		No	Check engine condition • Worn piston, piston rings or cylinder wall • Defective cylinder head gasket • Distorted cylinder head • Improper valve seating • Valve sticking in guide
8	Check if spark plugs are OK 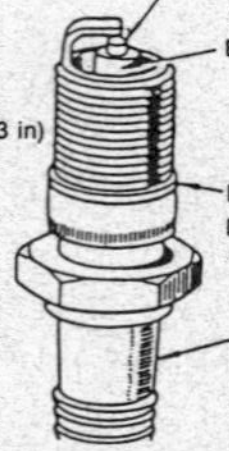	Yes	Go to next step
		No	Repair, clean, or replace
9	Try known good ECU and check if condition improves		

1.6L ENGINE TROUBLESHOOTING GUIDE—MIATA

4	CRANKS NORMALLY BUT WILL NOT START (PARTIAL COMBUSTION) — AFTER WARM-UP
DESCRIP-TION	•Engine cranks at normal speed but shows partial combustion and will not continue to run after running and hot soaked •Battery in normal condition •Engine starts normally when cold

[TROUBLESHOOTING HINTS]

① Air/Fuel mixture too rich
- •Fuel injection control malfunction (Correction for coolant temperature)
- •Injector fuel leakage

② Vapor lock
- •Fuel pressure not held in fuel line after engine stops
- •High RVP (winter) fuel used in warm weather

STEP	INSPECTION		ACTION
1	Check if "00" is displayed on Self-Diagnosis Checker with ignition switch ON 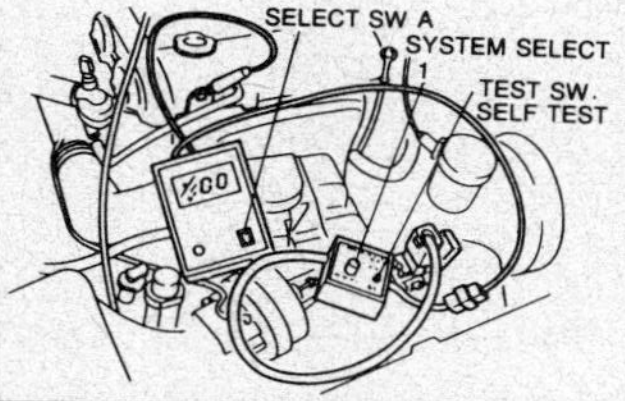	Yes	Go to next step
		No	**Malfunction Code No. displayed** Check for cause (Refer to specified check sequence) **"88" flashes** Check ECU terminal 1F voltage **Specification: Battery voltage (Ignition switch ON)** ⇨ If OK, replace ECU ⇨ If not OK, check wiring between ECU and Self-Diagnosis Checker
2	Check if ECU terminal voltages are OK (2D and 2Q)	Yes	Go to next step
		No	Check for cause (Refer to "Check Point for Each Terminal")
3	Connect diagnosis connector terminals F/P and GND with jumper wire and check for correct fuel line pressure with ignition switch ON **Fuel line pressure:** **265—314 kPa (2.7—3.2 kg/cm², 38—46 psi)** 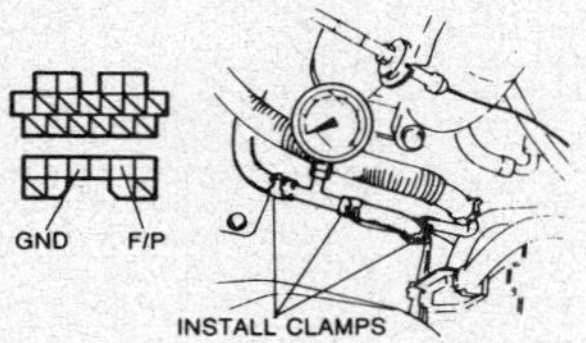	Yes	Go to next step
		No	**Low pressure** Check fuel line pressure while pinching fuel return hose ⇨ If fuel line pressure quickly increases, check pressure regulator ⇨ If fuel line pressure gradually increases, check fuel line and filter for clogging If not clogged, check fuel pump maximum pressure **High pressure** Check if fuel return hose is clogged or restricted ⇨ If OK, replace pressure regulator ⇨ If not OK, repair or replace hose

STEP	INSPECTION		ACTION
4	With condition of step 3, check if fuel line pressure is held after ignition switch is turned OFF **Fuel line pressure: More than 147 kPa (1.5 kg/cm², 21 psi) for 5 min.** 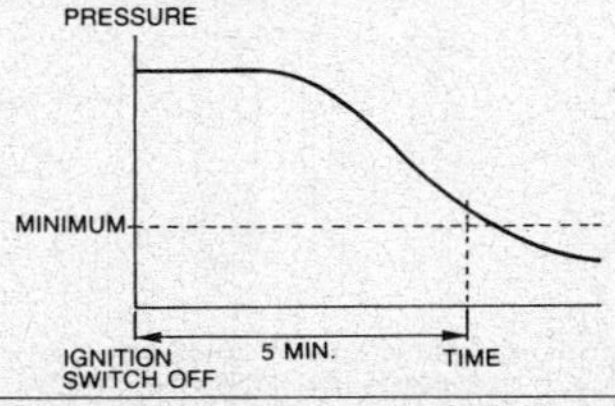	Yes	Go to Step 6
		No	Go to Step 5
5	Check if fuel line pressure is held after ignition switch is turned OFF and blocking outlet of pressure regulator **Fuel line pressure:** **More than 147 kPa (1.5 kg/cm², 21 psi) for 5 min.**	Yes	Replace pressure regulator
		No	Check fuel pump hold pressure ⇨ If OK, check injector for fuel leakage ⇨ If not OK, replace fuel pump
6	Try known good ECU and check if condition improves	Yes	Replace ECU
		No	Change fuel to another brand

5	CRANKS NORMALLY BUT HARD TO START — ALWAYS		
DESCRIPTION	• Engine cranks at normal speed but requires excessive cranking time (more than 5 sec.) before starting • Battery in normal condition		
[TROUBLESHOOTING HINTS]	① Air/Fuel mixture too lean • Fuel injection control malfunction (Correction for coolant temperature) • Low fuel line pressure • Air leakage ② Air/Fuel mixture too rich • Air cleaner element clogged • Airflow meter stuck ③ Poor ignition spark		

STEP	INSPECTION		ACTION
1	Check if "00" is displayed on Self-Diagnosis Checker with ignition switch ON 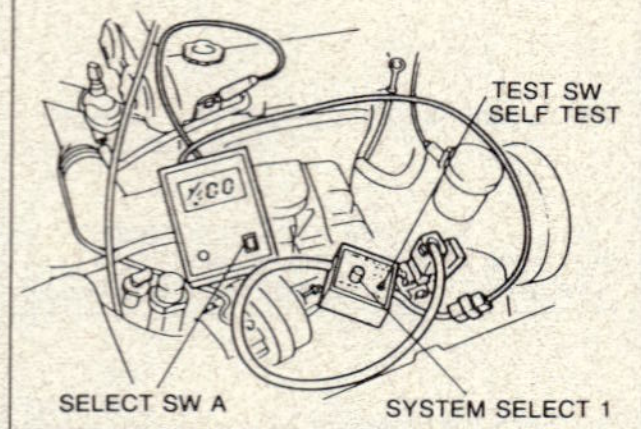 	Yes	Go to next step
		No	**Malfunction Code No. displayed** Check for cause (Refer to specified check sequence) **"88" flashes** Check ECU terminal 1F voltage **Specification: Battery voltage (Ignition switch ON)** ⇨ If OK, replace ECU ⇨ If not OK, check wiring between ECU and Self-Diagnosis Checker
2	Check for correct intake manifold vacuum at idle **Vacuum: More than 450 mmHg (17.7 inHg)** 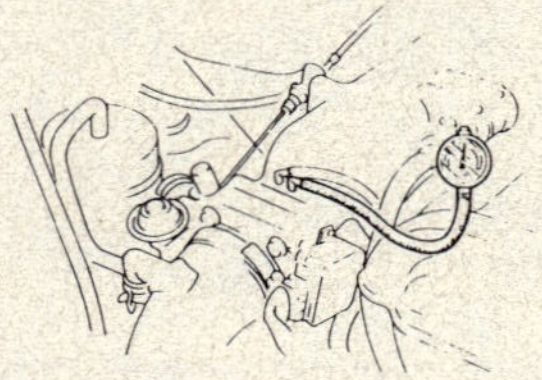	Yes	Go to next step
		No	Check for air leakage of intake air system components
3	Check if air cleaner element is clean	Yes	Go to next step
		No	Replace air cleaner element
4	Check for correct fuel line pressure at idle **Fuel line pressure:** **265—314 kPa (2.7—3.2 kg/cm^2, 38—46 psi)** **(Vacuum hose to pressure regulator disconnected)** 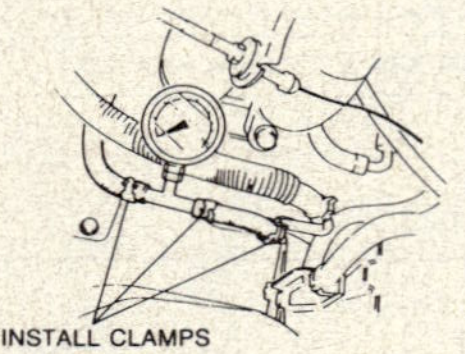 	Yes	Go to next step
		No	**Low pressure** Check fuel line pressure while pinching fuel return hose ⇨ If fuel line pressure quickly increases, check pressure regulator ⇨ If fuel line pressure gradually increases, check for clogging between fuel pump and pressure regulator If not clogged, check fuel pump maximum pressure

STEP	INSPECTION		ACTION
5	Check if ECU terminal voltages are OK (2D, 2O and 2Q)	Yes	Go to next step
		No	Check for cause (Refer to "Check Point for Each Terminal")
6	Check if strong blue spark is visible at each disconnected high-tension lead while cranking engine 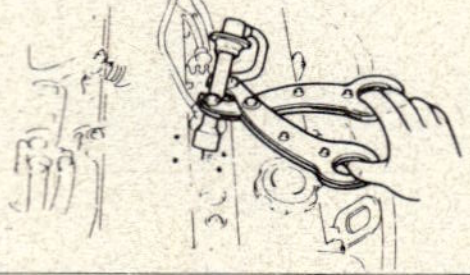	Yes	Go to next step
		No	Check ignition system (Refer to Troubleshooting "Misfire")
7	Check for injector operating sound at each injector at idle 	Yes	Go to Step 9
		No	Go to Step 8
8	Check if battery voltage exists at injector connector (W/R) wire with ignition switch ON 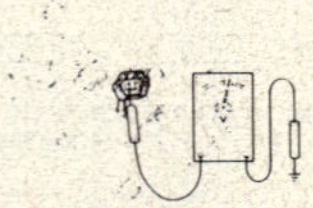	Yes	Check if injector resistance is OK **Resistance: Approx. 14Ω** ⇨ If OK, check wiring between injector and ECU ⇨ If not OK, replace injector
		No	Check wiring between main relay and injector
9	Check for correct engine compression **Engine compression: 1,324—932 kPa (13.5—9.5 kg/cm^2, 192—135 psi) - 300 rpm** 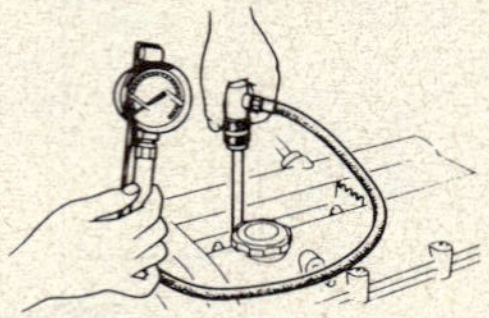	Yes	Go to next step
		No	Check engine condition • Worn piston, piston rings or cylinder wall • Defective cylinder head gasket • Distorted cylinder head • Improper valve seating • Valve sticking in guide
10	Check if spark plugs are OK 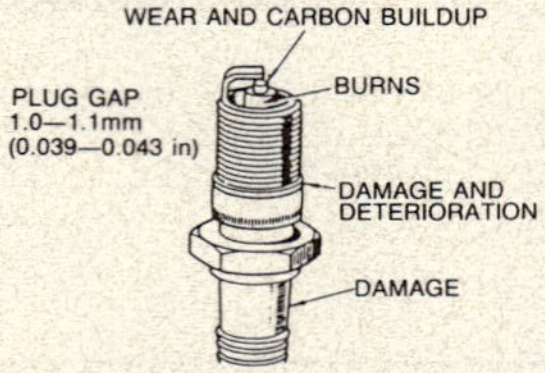 	Yes	Go to next step
		No	Repair, clean, or replace
11	Try known good ECU and check if condition improves		

1.6L ENGINE TROUBLESHOOTING GUIDE—MIATA

6	CRANKS NORMALLY BUT HARD TO START — WHEN ENGINE IS COLD		
DESCRIPTION	• Engine cranks at normal speed but requires excessive cranking time before starting • Battery in normal condition • Restarts OK after warm-up		
[TROUBLESHOOTING HINTS]	① Air/Fuel mixture too rich • Airflow meter stuck • Air cleaner element clogged • Idle speed control malfunction ② Air/Fuel mixture too lean • Fuel injection control malfunction (Correction for coolant temperature) ③ Poor atomization of fuel • Low RVP (summer) fuel used in cold weather		
STEP	**INSPECTION**		**ACTION**
1	Check if "00" is displayed on Self-Diagnosis Checker with ignition switch ON 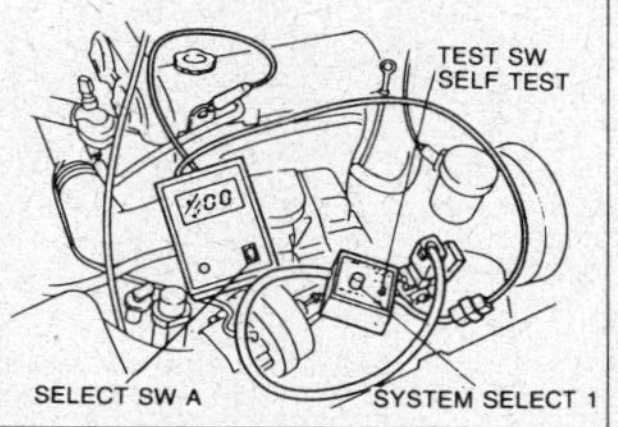 	Yes	Go to next step
		No	**Malfunction Code No. displayed** Check for cause (Refer to specified check sequence)
			"88" flashes Check ECU terminal 1F voltage **Specification: Battery voltage (Ignition switch ON)** ⇨ If OK, replace ECU ⇨ If not OK, check wiring between ECU and Self-Diagnosis Checker
2	Check if ECU terminal voltages are OK (1C, 2D, 2O and 2Q)	Yes	Go to next step
		No	Check for cause (Refer to "Check Point for Each Terminal")
3	Check if engine starts easily when depressing accelerator while cranking	Yes	Check if ISC valve is OK ⇨ If OK, check air valve ⇨ If not OK, replace ISC valve
		No	Go to next step
4	Check for correct intake manifold vacuum at idle **Vacuum: More than 450 mmHg (17.7 inHg)** 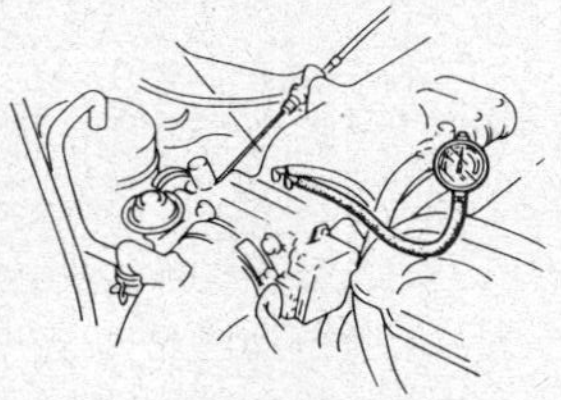	Yes	Go to next step
		No	Check for air leakage of intake air system components
5	Check if air cleaner element is clean	Yes	Go to next step
		No	Replace air cleaner element
6	Try known good ECU and check if condition improves	Yes	Replace ECU
		No	Change fuel to another brand

7	CRANKS NORMALLY BUT HARD TO START — AFTER WARM-UP		
DESCRIPTION	• Engine cranks at normal speed but requires excessive cranking time (more than 5 sec.) before starting after running and hot soaked • Battery in normal condition • Engine starts normally when cold		
[TROUBLESHOOTING HINTS]	① Air/Fuel mixture too rich • Fuel injection control malfunction • Injector fuel leakage ② Vapor lock • Fuel pressure not held in fuel line after engine stops • High RVP (winter) fuel used in warm weather		
STEP	**INSPECTION**		**ACTION**
1	Check if "00" is displayed on Self-Diagnosis Checker with ignition switch ON 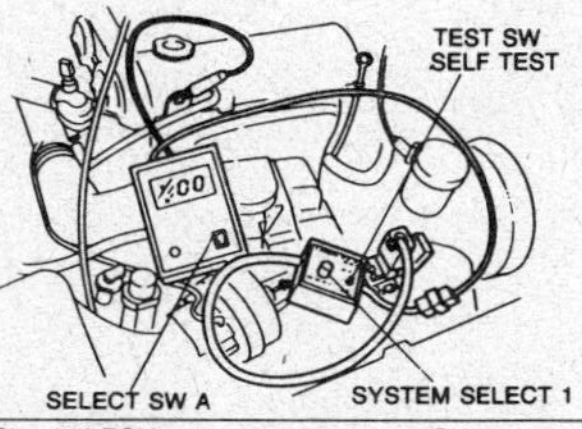 	Yes	Go to next step
		No	**Malfunction Code No. displayed** Check for cause (Refer to specified check sequence)
			"88" flashes Check ECU terminal 1F voltage **Specification: Battery voltage (Ignition switch ON)** ⇨ If OK, replace ECU ⇨ If not OK, check wiring between ECU and Self-Diagnosis Checker
2	Check if ECU terminal voltages are OK (2D and 2Q)	Yes	Go to next step
		No	Check for cause (Refer to "Check Point for Each Terminal")
3	Run engine at idle and check if fuel line pressure is held after ignition switch turned OFF **Fuel line pressure: More than 147 kPa (1.5 kg/cm², 21 psi) for 5 min.** 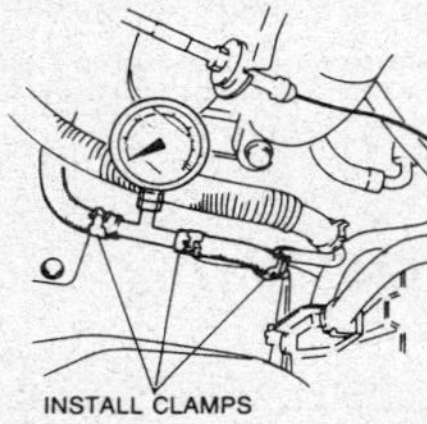 	Yes	Go to next step
		No	Block outlet of pressure regulator and check if fuel line pressure is held after ignition switch is turned OFF ⇨ If OK, replace pressure regulator ⇨ If not OK, check fuel pump hold pressure If fuel pump is OK, check injector for fuel leakage
4	Try known good ECU and check if condition improves	Yes	Replace ECU
		No	Change fuel to another brand

8	ROUGH IDLE — ALWAYS		
DESCRIP-TION	• Engine vibrates excessively at idle in every condition		
[TROUBLESHOOTING HINTS] ① Air/Fuel mixture too lean • Air leakage • Fuel injection control malfunction • Low fuel line pressure ② One or more injectors not operating or clogged ③ One or more spark plugs not sparking ④ Injection timing misadjustment ⑤ Low engine compression			
STEP	INSPECTION		ACTION
1	Check for correct intake manifold vacuum at idle **Vacuum: More than 450 mmHg (17.7 inHg)** 	Yes	Go to next step
		No	Check for air leakage of intake air system components
2	Check if air cleaner element is clean	Yes	Go to next step
		No	Replace air cleaner element
3	Check if "00" is displayed on Self-Diagnosis Checker with ignition switch ON 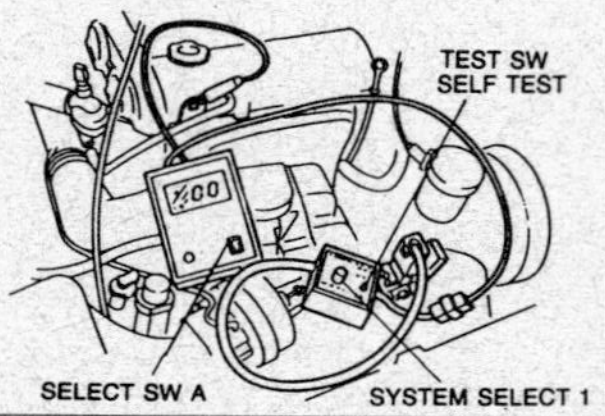	Yes	Go to next step
		No	**Malfunction Code No. displayed** Check for cause (Refer to specified check sequence) **"88" flashes** Check ECU terminal 1F voltage **Specification: Battery voltage (Ignition switch ON)** ⇨ If OK, replace ECU ⇨ If not OK, check wiring between ECU and Self-Diagnosis Checker
4	Check switches for correct operation with Self-Diagnosis Checker Monitor Lamp and ignition switch ON 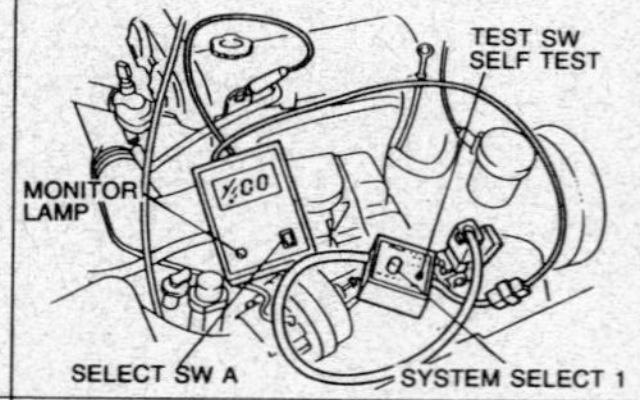	Yes	Go to next step
		No	**Lamp not ON/OFF with specified switch** Check for cause (Refer to specified check sequence) **Lamp always ON** Check wiring between ECU terminal 1D and Self-Diagnosis Checker
5	Check if ECU terminal voltages are OK (2D, 2O and 2Q)	Yes	Go to next step
		No	Check for cause (Refer to "Check Point for Each Terminal")

STEP	INSPECTION		ACTION
6	Check for injector operating sound at idle with sound scope or screwdriver	Yes	Go to Step 8
		No	Go to Step 7
7	Check if battery voltage exists at injector connector (W/R) wire 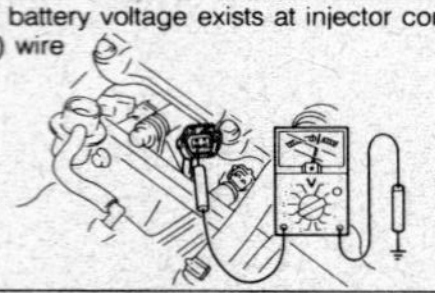	Yes	Check if injector resistance is OK **Resistance: Approx. 14Ω** ⇨ If OK, check wiring between ECU and injector ⇨ If not OK, replace injector
		No	Check wiring between ECU and injector
8	Disconnect each high-tension lead at idle and check if engine speed decreases equally each time 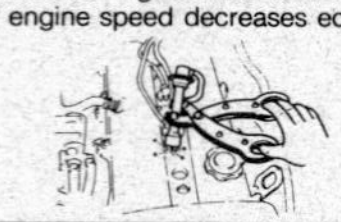	Yes	Disconnect each injector connector at idle and check if engine speed decreases equally each time ⇨ If OK, go to Step 10 ⇨ If not OK, check injector for fuel leakage
		No	Go to step 9
9	Check if spark plugs are OK 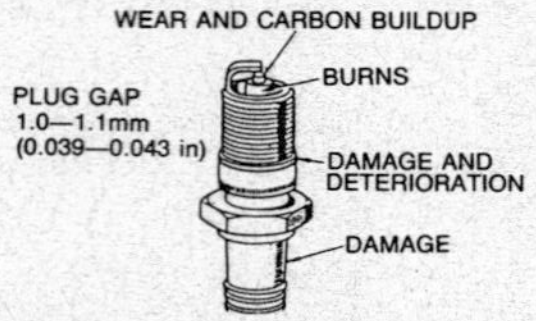	Yes	Check for correct engine compression ⇨ If OK, replace injector ⇨ If not OK, check for cause (Refer to Section B)
		No	Repair, clean, or replace
10	Check for correct ignition timing at idle **Ignition timing: 10° ± 1° BTDC (M/T)** **8° ± 1° BTDC (4A/T)** 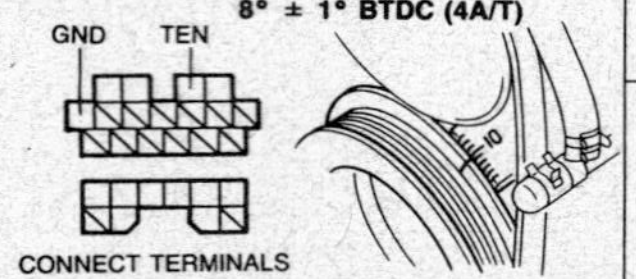	Yes	Check for correct idle speed **Idle speed: 850 ± 50 rpm** ⇨ If OK, go to next step ⇨ If not OK, adjust idle speed
		No	Adjust
11	Check for correct fuel line pressure at idle **Fuel line pressure:** **265—314 kPa (2.7—3.2 kg/cm², 38—48 psi)** **(Vacuum hose to pressure regulator disconnected)** 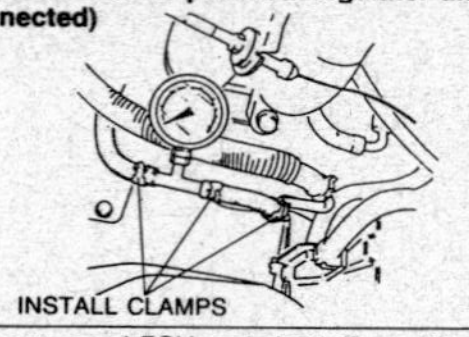	Yes	Go to next step
		No	**Low pressure** Check fuel line pressure while pinching fuel return hose ⇨ If fuel line pressure quickly increases, check pressure regulator ⇨ If fuel line pressure gradually increases, check for clogging between fuel pump and pressure regulator If not clogged, check fuel pump maximum pressure
12	Try known good ECU and check if condition improves		

1.6L ENGINE TROUBLESHOOTING GUIDE—MIATA

9	LOW IDLE SPEED/ROUGH IDLE — BEFORE WARM-UP
DESCRIPTION	• Engine speed low or engine vibrates excessively at idle during warm-up

[TROUBLESHOOTING HINTS]

① Low intake air amount
- • Airflow meter stuck
- • Air cleaner element clogged
- • Idle speed control [Air valve / Correction for coolant temperature]

② Low fuel injection amount
- • Fuel injection control malfunction (Correction for coolant temperature)

③ Poor atomization of fuel
- • Low RVP (summer) fuel used in cold weather

STEP	INSPECTION		ACTION
1	Check if "00" is displayed on Self-Diagnosis Checker with ignition switch ON 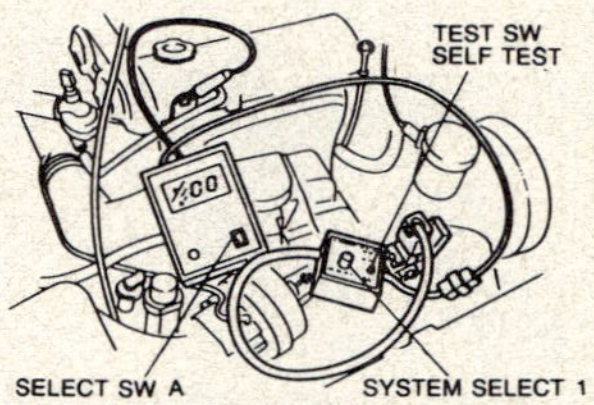	Yes	Go to next step
		No	**Malfunction Code No. displayed** Check for cause (Refer to specified check sequence)
			"88" flashes Check ECU terminal 1F voltage **Specification: Battery voltage (Ignition switch ON)** ⇨ If OK, replace ECU ⇨ If not OK, check wiring between ECU and Self-Diagnosis Checker
2	Check switches for correct operation with Self-Diagnosis Checker Monitor Lamp and ignition switch ON 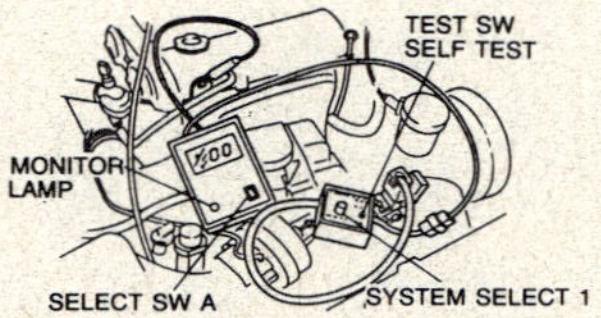	Yes	Go to next step
		No	**Lamp not ON/OFF with specified switch** Check for cause (Refer to specified check sequence)
			Lamp always ON Check wiring between ECU terminal 1D and Self-Diagnosis Checker
3	Check if ECU terminal voltages are OK (2D, 2O and 2Q)	Yes	Go to next step
		No	Check for cause (Refer to "Check Point for Each Terminal")
4	Check for correct intake manifold vacuum at idle **Vacuum: More than 450 mmHg (17.7 inHg)** 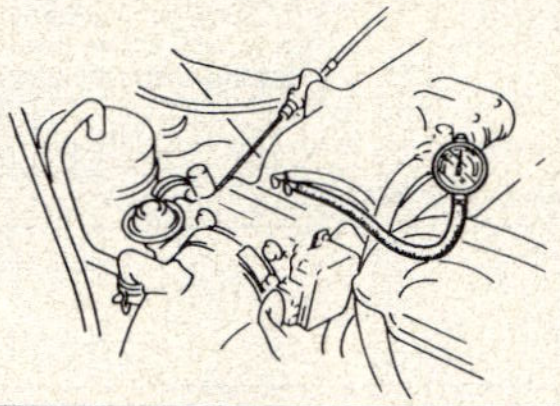	Yes	Go to next step
		No	Check for air leakage of intake air system components
5	Check if air cleaner element is clean	Yes	Go to next step
		No	Replace air cleaner element

STEP	INSPECTION		ACTION
6	Connect System Selector to diagnosis connector and set Test Switch to "SELF TEST" when engine is cold Check if engine speed decreases as engine warms up 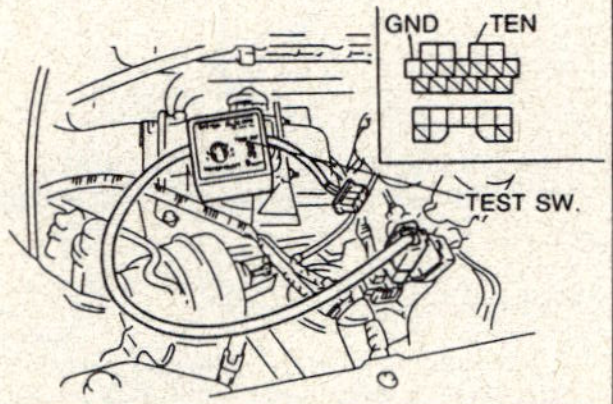	Yes	Go to next step
		No	Check air valve
7	With condition of Step 5 check for correct ignition timing at idle after warm-up **Ignition timing: 10° ± 1° BTDC (M/T) 8° ± 1° BTDC (4A/T)**	Yes	Go to next step
		No	Adjust
8	Try known good ECU and check if condition improves	Yes	Replace ECU
		No	Change fuel to another brand

10	LOW IDLE SPEED/ROUGH IDLE — AFTER WARM-UP
DESCRIP-TION	• Engine runs normally at idle during warm-up but vibrates excessively after warm-up

[TROUBLESHOOTING HINTS]
① Idle speed control malfunction
② Air/Fuel mixture too lean
• Air leakage of intake air system
• Low fuel line pressure
③ Air/Fuel mixture too rich
• Fuel injection control malfunction (Correction for coolant temperature)
④ Poor ignition spark
⑤ Low engine compression

STEP	INSPECTION		ACTION
1	Check if "00" is displayed on Self-Diagnosis Checker with ignition switch ON 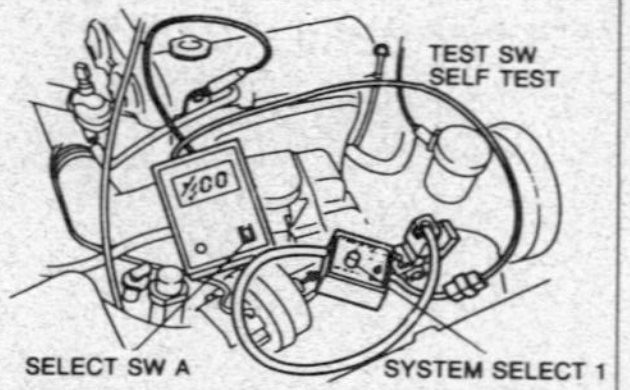	Yes	Go to next step
		No	**Malfunction Code No. displayed** Check for cause (Refer to specified check sequence) **"88" flashes** Check ECU terminal 1F voltage **Specification: Battery voltage (Ignition switch ON)** ⇨ If OK, replace ECU ⇨ If not OK, check wiring between ECU and Self-Diagnosis Checker
2	Check switches for correct operation with Self-Diagnosis Checker Monitor Lamp and ignition switch ON 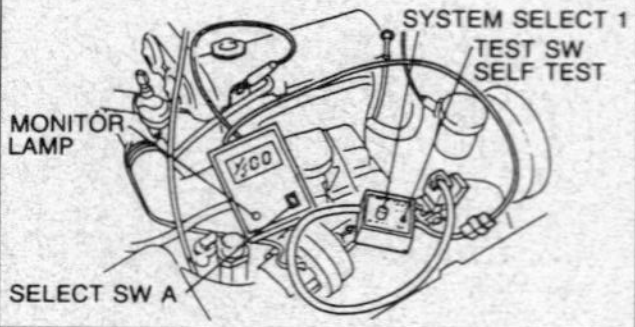	Yes	Go to next step
		No	**Lamp not ON/OFF with specified switch** Check for cause (Refer to specified check sequence) **Lamp always ON** Check wiring between ECU terminal 1D and Self-Diagnosis Checker
3	Disconnect ISC valve connector at idle and check if "clicking" is heard from ISC valve 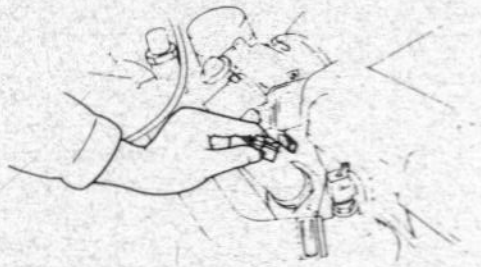	Yes	Go to next step
		No	Replace ISC valve
4	Check for correct intake manifold vacuum at idle **Vacuum: More than 450 mmHg (17.7 inHg)** 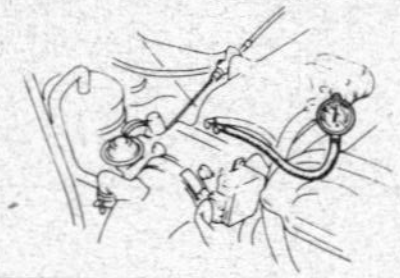	Yes	Go to next step
		No	**Low vacuum** Check for air leakage of intake air system components

STEP	INSPECTION		ACTION
5	Check if ECU terminal voltages are OK (2D, 2O and 2Q)	Yes	Go to next step
		No	Check for cause (Refer to "Check Point for Each Terminal")
6	Check for correct ignition timing at idle **Ignition timing: 10° ± 1° BTDC (M/T) 8° ± 1° BTDC (4A/T)** 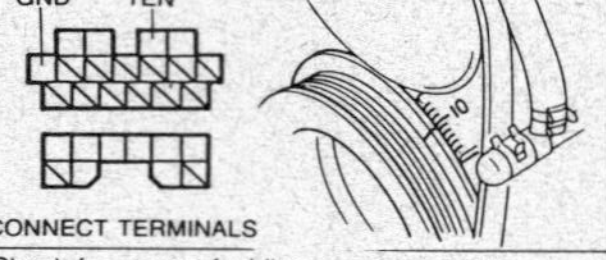	Yes	Check for correct idle speed **Idle speed: 850 ± 50 rpm** ⇨ If OK, go to next step ⇨ If not OK, adjust idle speed
		No	Adjust
7	Check for correct fuel line pressure at idle **Fuel line pressure: 265—314 kPa (2.7—3.2 kg/cm², 38—46 psi) (Vacuum hose to pressure regulator disconnected)** 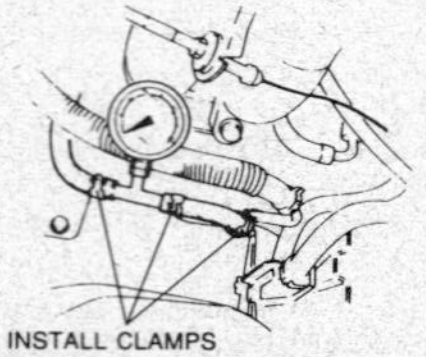	Yes	Go to next step
		No	**Low pressure** Check fuel line pressure while pinching fuel return hose ⇨ If fuel line pressure quickly increases, check pressure regulator ⇨ If fuel line pressure gradually increases, check for clogging between fuel pump and pressure regulator If not clogged, check fuel pump maximum pressure
8	Check if strong blue spark is visible at each disconnected high-tension lead while cranking engine	Yes	Go to next step
		No	Check ignition system (Refer to Troubleshooting "Misfire")
9	Check for correct engine compression **Engine compression: 1,324—932 kPa (13.5—9.5 kg/cm², 192—135)-300 rpm** 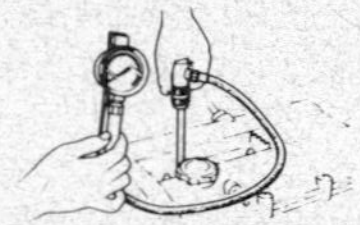	Yes	Go to next step
		No	Check engine condition • Worn piston, piston rings or cylinder wall • Defective cylinder head gasket • Distorted cylinder head • Improper valve seating • Valve sticking in guide
10	Check if spark plugs are OK 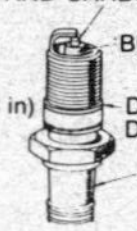	Yes	Go to next step
		No	Repair, clean, or replace
11	Try known good ECU and check if condition improves		

11 HIGH IDLE SPEED — AFTER WARM-UP

[TROUBLESHOOTING HINTS]
Excessive intake air supplied to engine
① Throttle valve not fully closed
② Idle speed control malfunction
- Air valve not closing
- ISC valve stuck
- Incorrect coolant temperature signal

STEP	INSPECTION		ACTION
1	Check if throttle valve is fully closed when accelerator released	Yes	Go to Step 3
		No	Check if throttle linkage is correctly installed and operates freely ⇨ If OK, go to Step 2 ⇨ If not OK, clean, adjust or replace linkage
2	Check if dashpot is correctly adjusted **Dashpot set speed: 2,500 ± 150 rpm**	Yes	Check if throttle valve is contaminated ⇨ If contaminated, clean throttle body ⇨ If not contaminated, replace throttle body
		No	Adjust
3	Check if "00" is displayed on Self-Diagnosis Checker with ignition switch ON 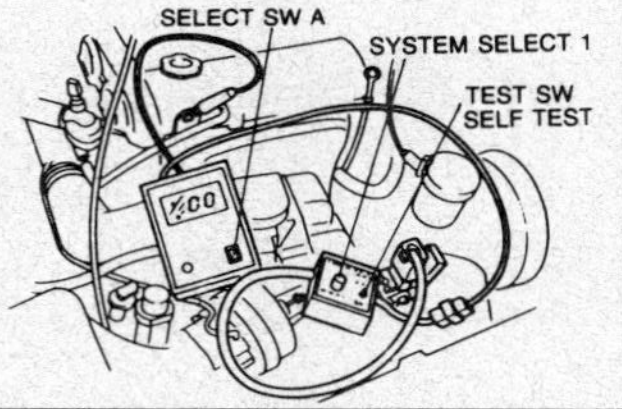	Yes	Go to next step
		No	**Malfunction Code No. displayed** Check for cause (Refer to specified check sequence) **"88" flashes** Check ECU terminal 1F voltage **Specification: Battery voltage (Ignition switch ON)** ⇨ If OK, replace ECU ⇨ If not OK, check wiring between ECU and Self-Diagnosis Checker
4	Check switches for correct operation with Self-Diagnosis Checker Monitor Lamp and ignition switch ON 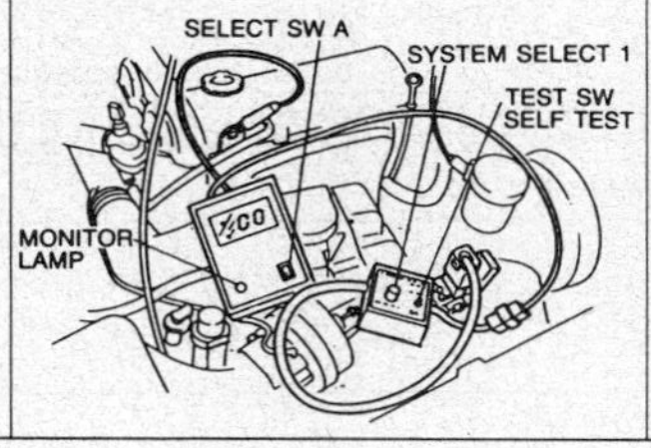	Yes	Go to next step
		No	**Lamp not ON/OFF with specified switch** Check for cause (Refer to specified check sequence) **Lamp always ON** Check wiring between ECU terminal 1D and Self-Diagnosis Checker

STEP	INSPECTION		ACTION
5	Connect System Selector to diagnosis connector and set Test Switch to "SELF TEST" when engine is cold Check if engine speed decreases as engine warms up 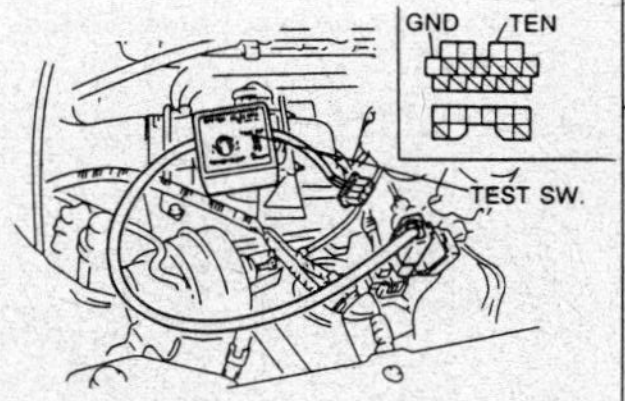	Yes	Go to next step
		No	Check air valve
6	Disconnect ISC valve connector at idle and check if "clicking" is heard from ISC valve 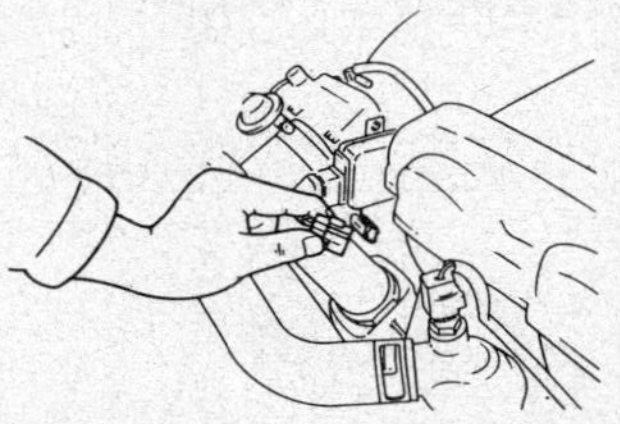	Yes	Go to next step
		No	Check ISC valve
7	Pinch PCV hose with pliers and check if engine speed decreases 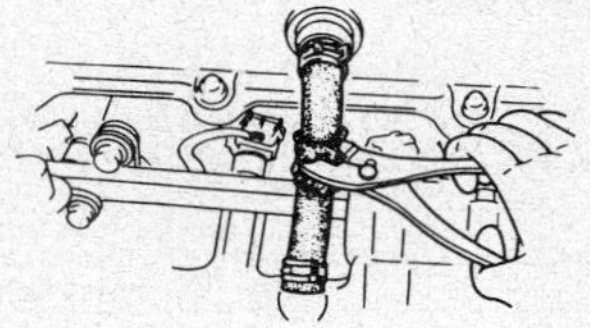	Yes	Check PCV valve
		No	Go to next step
8	Check if ECU terminal voltages are OK (2D, 2O and 2Q)	Yes	Try known good ECU
		No	Check for cause (Refer to "Check Point for Each Terminal")

12	LOW IDLE SPEED — WHEN A/C, P/S, OR E/L ON		
DESCRIPTION	•Engine speed decreases at idle when A/C, P/S, or E/L ON •A/C, P/S, headlights, blower fan and electric cooling fan operate normally		
[TROUBLESHOOTING HINTS]	① Idle speed control malfunction •Engine speed feedback control malfunction •ISC valve stuck		
STEP	INSPECTION		ACTION
1	Check if "00" is displayed on Self-Diagnosis Checker with ignition switch ON 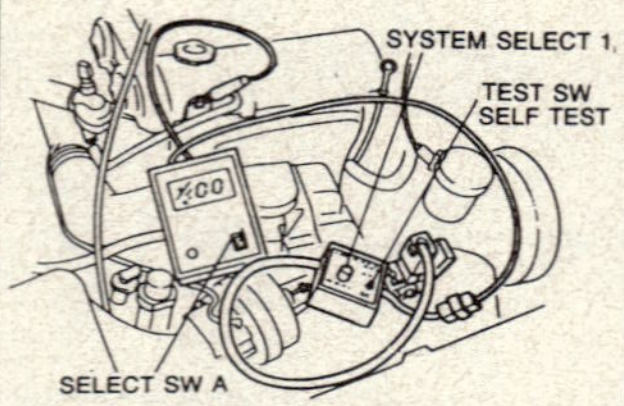	Yes	Go to next step
		No	**Malfunction Code No. displayed** Check for cause (Refer to specified check sequence) **"88" flashes** Check ECU terminal 1F voltage **Specification: Battery voltage (Ignition switch ON)** ⇨ If OK, replace ECU ⇨ If not OK, check wiring between ECU and Self-Diagnosis Checker
2	Check switches for correct operation with Self-Diagnosis Checker Monitor Lamp and ignition switch ON 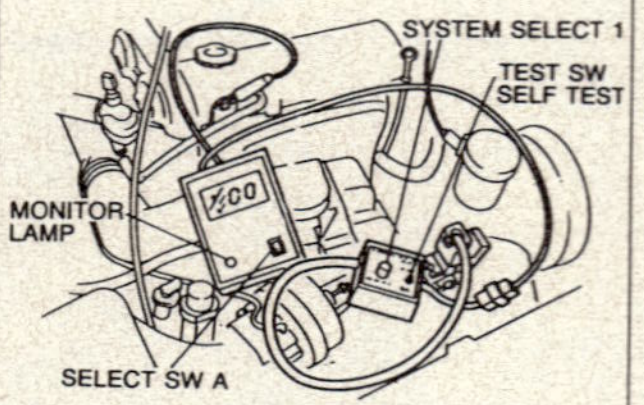	Yes	Go to next step
		No	**Lamp not ON/OFF with specified switch** Check for cause (Refer to specified check sequence) **Lamp always ON** Check wiring between ECU terminal 1D and Self-Diagnosis Checker
3	Check if continuity exists between diagnosis connector terminal TEN and ground 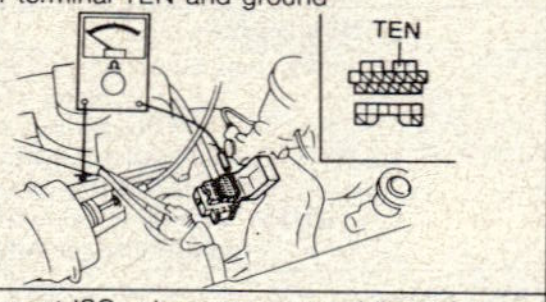	Yes	Check for short circuit in wiring between diagnosis connector terminal TEN and ground
		No	Go to next step
4	Disconnect ISC valve connector at idle and check if "clicking" is heard from ISC valve 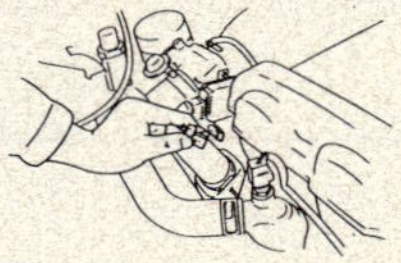	Yes	Try known good ECU
		No	Replace ISC valve

13	ROUGH IDLE JUST AFTER STARTING		
DESCRIPTION	•Engine starts normally but vibrates excessively just after starting		
[TROUBLESHOOTING HINTS]	① Fuel injection control and idle speed control malfunction •Start signal not input to ECU ② Idle speed misadjustment ③ Ignition timing misadjustment		
STEP	INSPECTION		ACTION
1	Check if "00" is displayed on Self-Diagnosis Checker with ignition switch ON 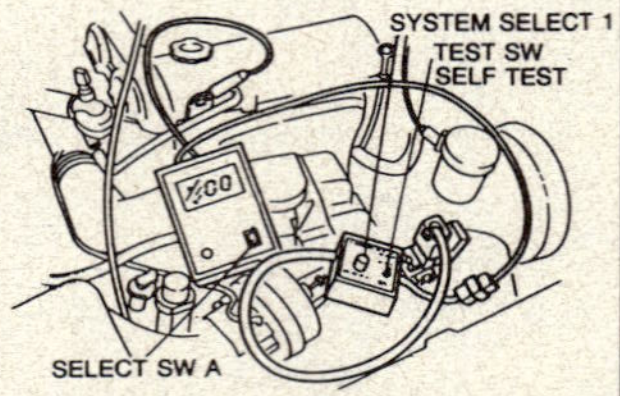	Yes	Go to next step
		No	**Malfunction Code No. displayed** Check for cause (Refer to specified check sequence) **"88" flashes** Check ECU terminal 1F voltage **Specification: Battery voltage (Ignition switch ON)** ⇨ If OK, replace ECU ⇨ If not OK, check wiring between ECU and Self-Diagnosis Checker
2	Check switches for correct operation with Self-Diagnosis Checker Monitor Lamp and ignition switch ON 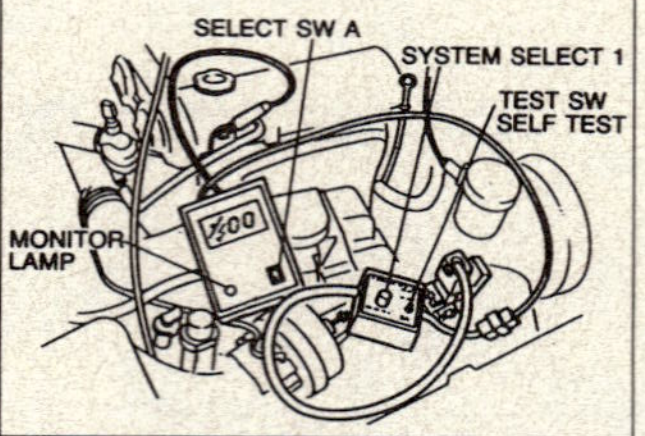	Yes	Go to next step
		No	**Lamp not ON/OFF with specified switch** Check for cause (Refer to specified check sequence) **Lamp always ON** Check wiring between ECU terminal 1D and Self-Diagnosis Checker
3	Check if ECU terminal 1C voltage is OK **Voltage: Approx. 10V (While cranking)**	Yes	Go to next step
		No	Check for cause (Refer to "Check Point for Each Terminal")
4	Check for correct ignition timing at idle **Ignition timing: 10° ± 1° BTDC (M/T)** **8° ± 1° BTDC (4A/T)** 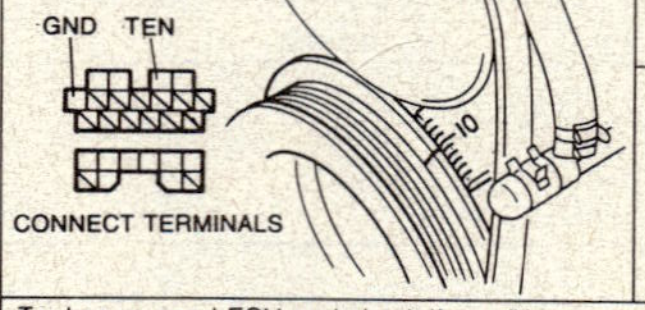	Yes	Check for correct idle speed **Idle speed: 850 ± 50 rpm** ⇨ If OK, go to next step ⇨ If not OK, adjust idle speed
		No	Adjust
5	Try known good ECU and check if condition improves		

1.6L ENGINE TROUBLESHOOTING GUIDE—MIATA

14	IDLE MOVES UP AND DOWN		
DESCRIPTION	• Engine speeds up and down periodically at idle		
[TROUBLESHOOTING HINTS]	① Fuel cut occurs at idle • Air valve not closing after warm-up and idle speed reaches to fuel cut speed ② Fuel injection amount fluctuating • Bad contact point inside airflow meter ③ Air leakage of intake air system ④ Poor ignition spark ⑤ Air/Fuel mixture too rich • Evaporative emission control system malfunction ⑥ Low engine compression		
STEP	INSPECTION		ACTION
1	Check if "00" is displayed on Self-Diagnosis Checker with ignition switch ON 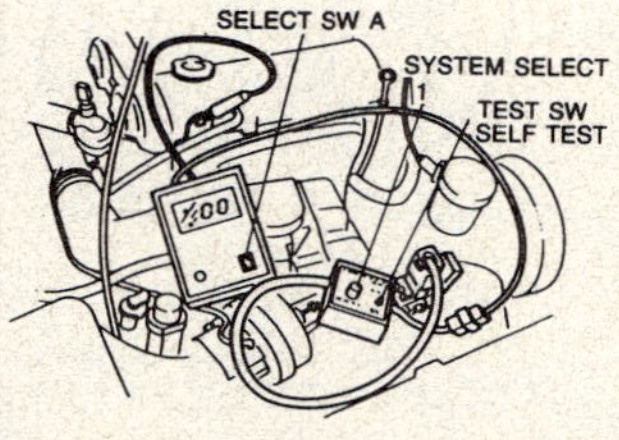	Yes	Go to next step
		No	**Malfunction Code No. displayed** Check for cause (Refer to specified check sequence) **"88" flashes** Check ECU terminal 1F voltage **Specification: Battery voltage (Ignition switch ON)** ⇨ If OK, replace ECU ⇨ If not OK, check wiring between ECU and Self-Diagnosis Checker
2	Check for correct idle speed **Idle speed: 850 ± 50 rpm** 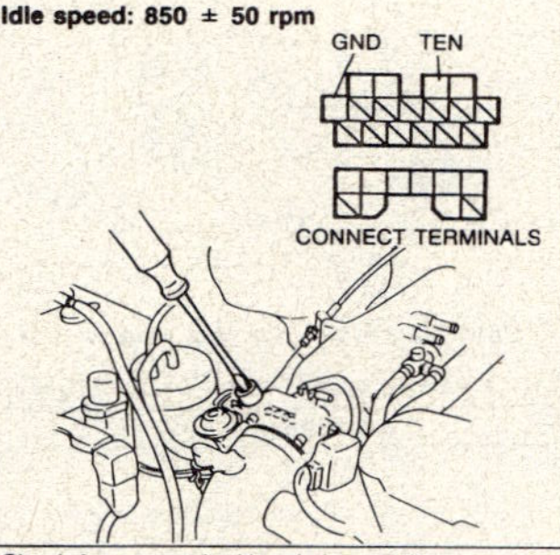	Yes	Go to next step
		No	Check if idle speed can be adjusted by turning air adjust screw ⇨ If OK, adjust idle speed ⇨ If not OK, check air valve
3	Check for correct ignition timing at idle **Ignition timing: 10° ± 1° BTDC (M/T) 8° ± 1° BTDC (4A/T)** 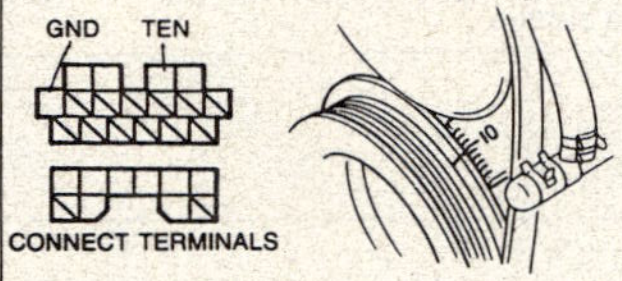	Yes	Go to next step
		No	Adjust

STEP	INSPECTION		ACTION
4	Check for correct intake manifold vacuum at idle **Intake manifold vacuum: More than 450 mmHg (17.7 inHg)** 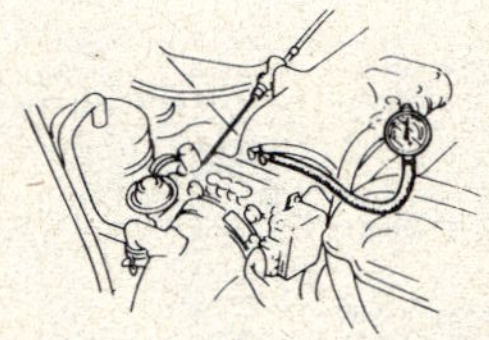	Yes	Go to next step
		No	**Low vacuum** Check for air leakage of intake air system
5	Check for injector operating sound at idle with sound scope or screwdriver	Yes	Go to Step 7
		No	Go to Step 6
6	Check if battery voltage exists at injector connector (W/R) wire 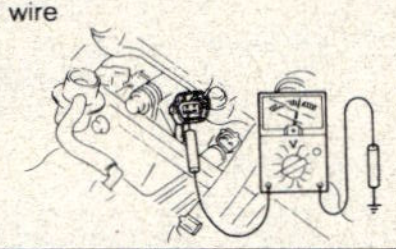	Yes	Check if injector resistance is OK **Resistance: Approx. 14Ω** ⇨ If OK, check wiring between ECU and injector ⇨ If not OK, replace injector
		No	Check wiring between ECU and injector
7	Disconnect each high-tension lead at idle and check if engine speed decreases equally each time 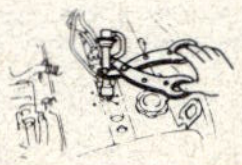	Yes	Disconnect each injector connector at idle and check if engine speed decreases equally each time ⇨ If OK, go to Step 9 ⇨ If not OK, check injector for fuel leakage
		No	Go to Step 8
8	Check if spark plugs are OK 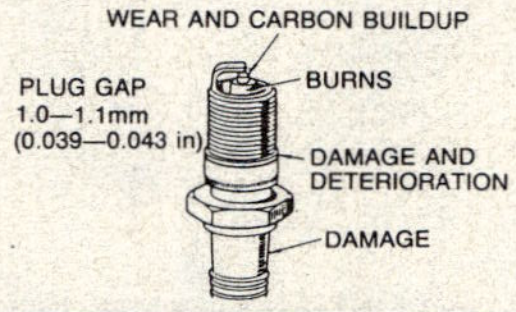	Yes	Check for correct engine compression ⇨ If OK, replace injector ⇨ If not OK, check for cause (Refer to Section B)
		No	Repair, clean, or replace
9	Check if ECU terminal voltages are OK (2D, 2Q and 2X)	Yes	Go to next step
		No	Check for cause (Refer to "Check Point for Each Terminal")
10	Check if vacuum is felt at solenoid valve (purge control) at idle 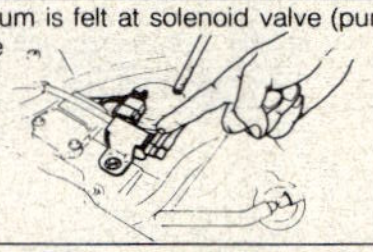	Yes	Check solenoid valve
		No	Go to next step
11	Try known good ECU and check if condition improves		

1.6L ENGINE TROUBLESHOOTING GUIDE—MIATA

15	ENGINE STALLS AT IDLE — ALWAYS		
DESCRIPTION	•Engine starts normally but vibrates excessively and stalls at idle in every condition		
[TROUBLESHOOTING HINTS]	① Incorrect idle speed •Idle speed misadjustment •Idle speed control malfunction ② Air/Fuel mixture too rich or lean •Injector clogged or inoperative •Low fuel line pressure •Low intake air amount or air leakage ③ Poor ignition spark		

STEP	INSPECTION		ACTION
1	Check for correct intake manifold vacuum at idle **Vacuum: More than 450 mmHg (17.7 inHg)** 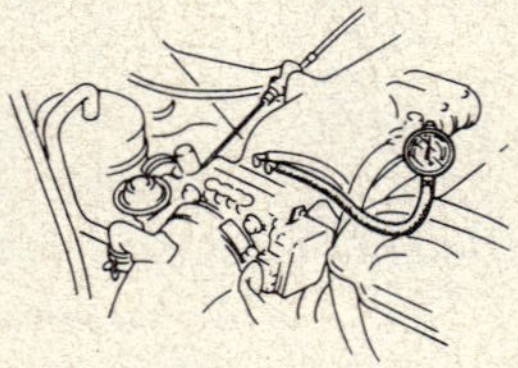	Yes	Go to next step
		No	Check for air leakage of intake air system components
2	Check if air cleaner element is clean	Yes	Go to next step
		No	Replace air cleaner element
3	Check if "00" is displayed on Self-Diagnosis Checker with ignition switch ON 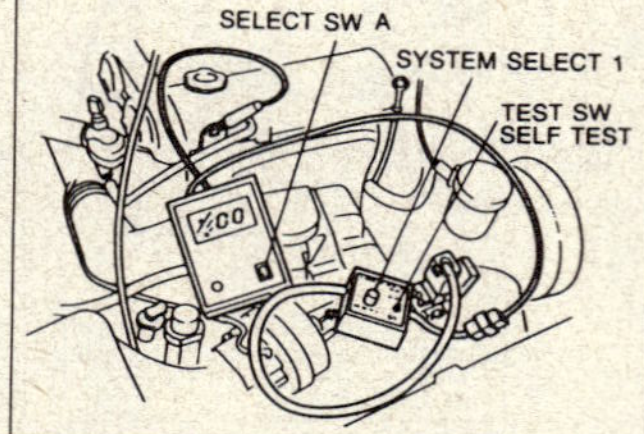	Yes	Go to next step
		No	**Malfunction Code No. displayed** Check for cause (Refer to specified check sequence) **"88" flashes** Check ECU terminal 1F voltage **Specification: Battery voltage (Ignition switch ON)** ⇨ If OK, replace ECU ⇨ If not OK, check wiring between ECU and Self-Diagnosis Checker
4	Check switches for correct operation with Self-Diagnosis Checker Monitor Lamp and ignition switch ON 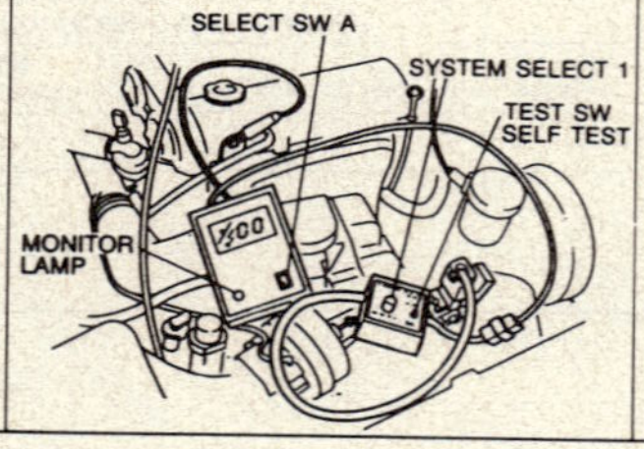	Yes	Go to next step
		No	**Lamp not ON/OFF with specified switch** Check for cause (Refer to specified check sequence) **Lamp always ON** Check wiring between ECU terminal 1D and Self-Diagnosis Checker

STEP	INSPECTION		ACTION
5	Check if ECU terminal voltages are OK (2D, 2O and 2Q)	Yes	Go to next step
		No	Check for cause (Refer to "Check Point for Each Terminal")
6	Check if strong blue spark is visible at each disconnected high-tension lead while cranking engine 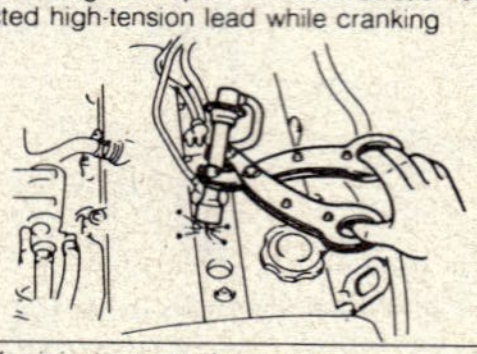	Yes	Go to next step
		No	Check ignition system (Refer to Troubleshooting "Misfire")
7	Check for injector operating sound at each injector at idle	Yes	Go to Step 9
		No	Go to Step 8
8	Check if battery voltage exists at injector connector (W/R) wire with ignition switch ON 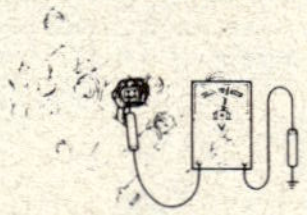	Yes	Check if injector resistance is OK **Resistance: Approx. 14Ω** ⇨ If OK, check wiring between injector and ECU ⇨ If not OK, replace injector
		No	Check wiring between main relay and injector
9	Check if spark plugs are OK 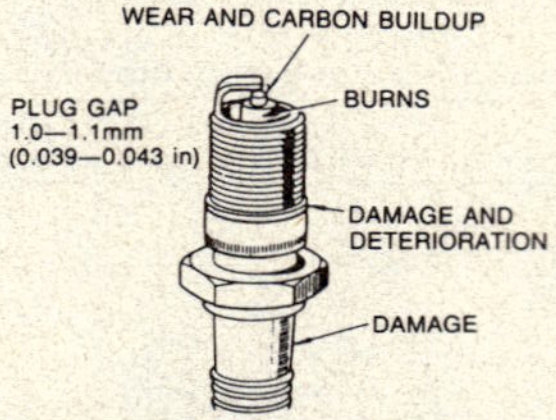	Yes	Go to next step
		No	Repair or replace
10	Connect diagnosis connector terminals F/P and GND with jumper wire and check for correct fuel line pressure with ignition switch ON **Fuel line pressure: 265—314 kPa (2.7—3.2 kg/cm², 38—46 psi)** 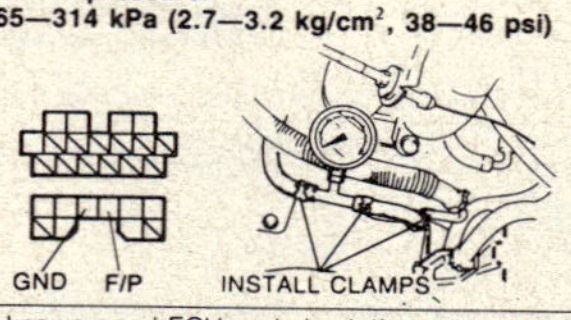	Yes	Go to next step
		No	**Low pressure** Check fuel line pressure while pinching fuel return hose ⇨ If fuel line pressure quickly increases, check pressure regulator ⇨ If fuel line pressure gradually increases, check for clogging between fuel pump to pressure regulator If not clogged, check for fuel pump maximum pressure
11	Try known good ECU and check if condition improves		

16 ENGINE STALLS AT IDLE — BEFORE WARM-UP

DESCRIPTION
- Engine starts normally but vibrates excessively and stalls at idle before warm-up

[TROUBLESHOOTING HINTS]
① Low intake air amount
- Idle speed control malfunction
- Air cleaner element clogged
- Airflow meter stuck

② Air/Fuel mixture too lean
- Air leakage of intake air system

③ Poor atomization of fuel
- Low RVP (summer) fuel used in cold weather

STEP	INSPECTION		ACTION
1	Check if "00" is displayed on Self-Diagnosis Checker with ignition switch ON 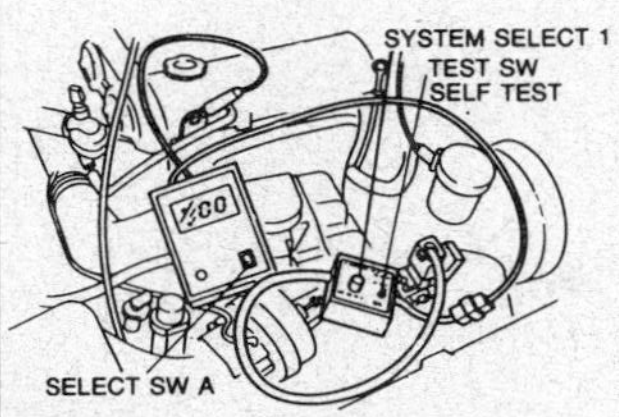	No	Go to next step
		No	**Malfunction Code No. displayed** Check for cause (Refer to specified check sequence)
			"88" flashes Check ECU terminal 1F voltage **Specification: Battery voltage (Ignition switch ON)** ⇨ If OK, replace ECU ⇨ If not OK, check wiring between ECU and Self-Diagnosis Checker
2	Check switches for correct operation with Self-Diagnosis Checker Monitor Lamp with ignition switch ON 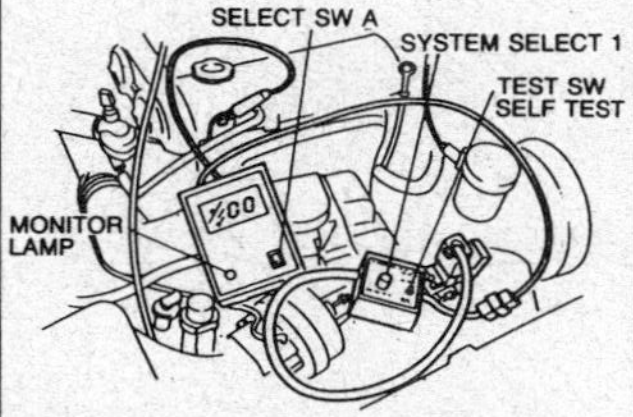	Yes	Go to next step
		No	**Lamp not ON/OFF with specified switch** Check for cause (Refer to specified check sequence)
			Lamp always ON Check wiring between ECU terminal 1D and Self-Diagnosis Checker
3	Check if ECU terminal voltages are OK (2D, 2O and 2Q)	Yes	Go to next step
		No	Check for cause (Refer to "Check Point for Each Terminal")
4	Check for correct intake manifold vacuum at idle **Vacuum: More than 450 mmHg (17.7 inHg)** 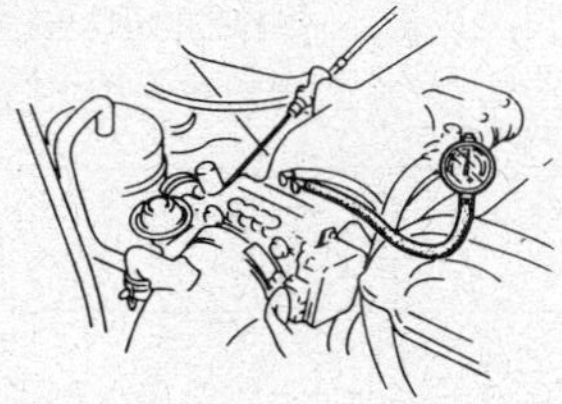	Yes	Go to next step
		No	Check for air leakage of intake air system components

STEP	INSPECTION		ACTION
5	Check if air cleaner element is clean	Yes	Go to next step
		No	Replace air cleaner element
6	Disconnect ISC valve connector when engine is cold and note idle speed Check if engine speed decreases after warm-up 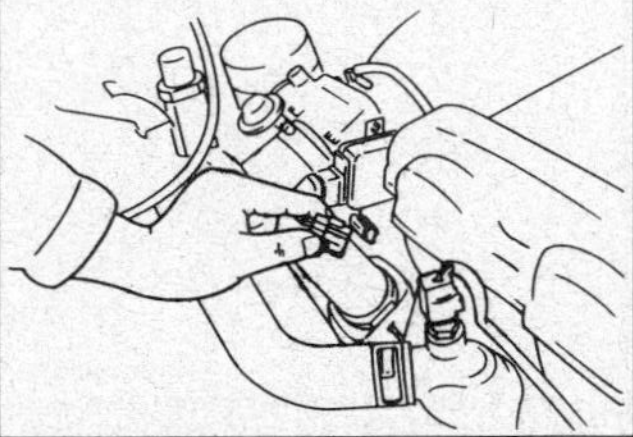	Yes	Go to next step
		No	Check air valve
7	Try known good ECU and check if condition improves	Yes	Replace ECU
		No	Change fuel to another brand

17 ENGINE STALLS AT IDLE — AFTER WARM-UP

DESCRIPTION
- Engine runs normally at idle during warm-up but becomes rough and stalls after warm-up

[TROUBLESHOOTING HINTS]
① Air/Fuel mixture too lean
- Air leakage of intake air system

② Low intake air amount
- Idle speed control malfunction

STEP	INSPECTION		ACTION
1	Check if air is leaked from intake air system component	Yes	Repair or replace
		No	Go to next step
2	Disconnect ISC valve connector at idle and check if "clicking" is heard from ISC Valve 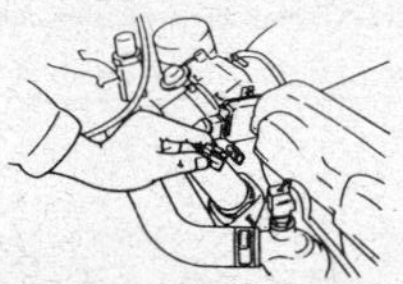	Yes	Go to next step
		No	Replace ISC valve
3	Go to troubleshooting No.10 "LOW IDLE SPEED/ROUGH IDLE—AFTER WARM-UP"		

18	ENGINE STALLS DURING START-UP		
DESCRIPTION	• Engine unexpectedly stops running while starting		
[TROUBLESHOOTING HINTS]	① Misfire occurs when depressing accelerator • Air/Fuel mixture too rich or too lean • Incorrect ignition timing • Weak ignition	② Lack of engine torque for start-up • Air/Fuel mixture too rich or too lean • Low intake air amount • Low engine compression	
STEP	**INSPECTION**		**ACTION**
1	Check if brakes are dragging	Yes	Repair
		No	Go to next step
2	Check if "00" is displayed on Self-Diagnosis Checker with ignition switch ON	No	Go to next step
	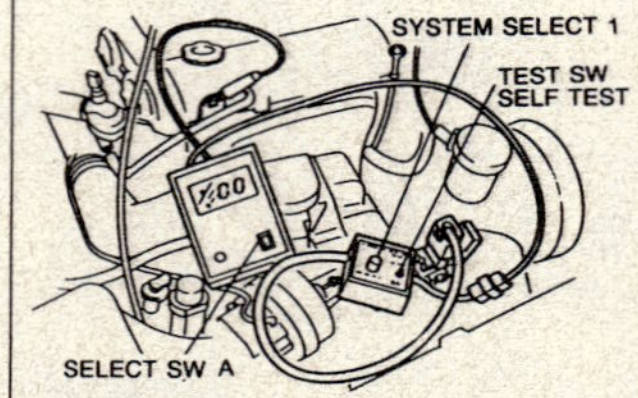	No	**Malfunction Code No. displayed** Check for cause (Refer to specified check sequence)
			"88" flashes Check ECU terminal 1F voltage **Specification: Battery voltage (Ignition switch ON)** ⇨ If OK, replace ECU ⇨ If not OK, check wiring between ECU and Self-Diagnosis Checker
3	Check switches for correct operation with Self-Diagnosis Checker Monitor Lamp and ignition switch ON	Yes	Go to next step
	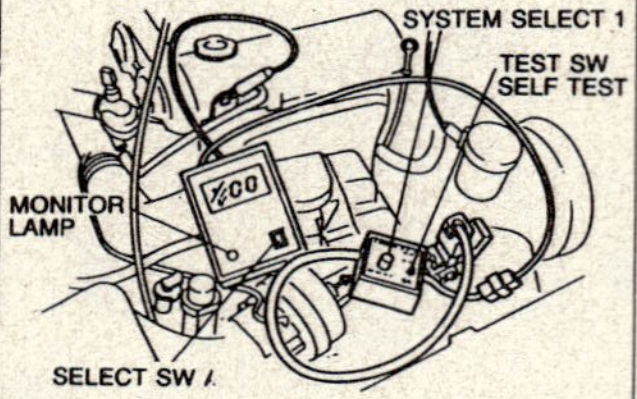	No	**Lamp not ON/OFF with specified switch** Check for cause (Refer to specified check sequence)
			Lamp always ON Check wiring between ECU terminal 1D and Self-Diagnosis Checker
4	Disconnect oxygen sensor connector and check if condition improves 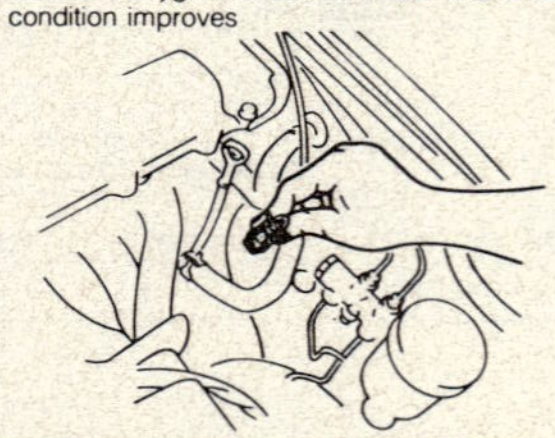	Yes	Check oxygen sensor
		No	Go to next step
5	Check if ECU terminal voltages are OK	Yes	Go to next step
		No	Check for cause (Refer to "Check Point for Each Terminal")

STEP	INSPECTION		ACTION
6	Check if throttle linkage is correctly installed and operates freely	Yes	Go to next step
		No	Correct, clean, or replace as required any binding or damaged linkage and adjust cable deflection at throttle body
7	Check for correct intake manifold vacuum at idle	Yes	Go to next step
	Vacuum: More than 450 mmHg (17.7 inHg) 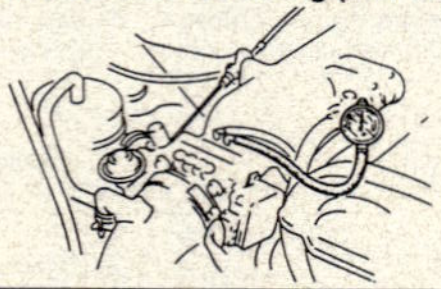	No	Check for air leakage of intake air system components
8	Check if air cleaner element is clean	Yes	Go to next step
		No	Replace air cleaner element
9	Check for correct ignition timing at idle **Ignition timing: 10° ± 1° BTDC (M/T)** **8° ± 1° BTDC (4A/T)** 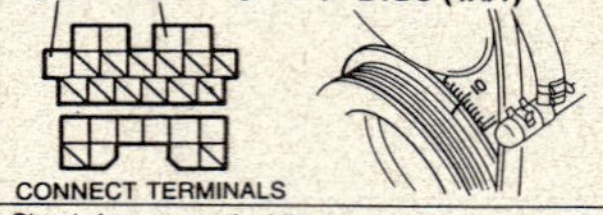	Yes	Check if ignition timing advances when accelerating ⇨ If advances, go to next step ⇨ If no advance, replace ECU
		No	Adjust
10	Check for correct fuel line pressure at idle **Fuel line pressure:** **265—314 kPa (2.7—3.2 kg/cm², 38—46 psi)** **(Vacuum hose to pressure regulator disconnected)** 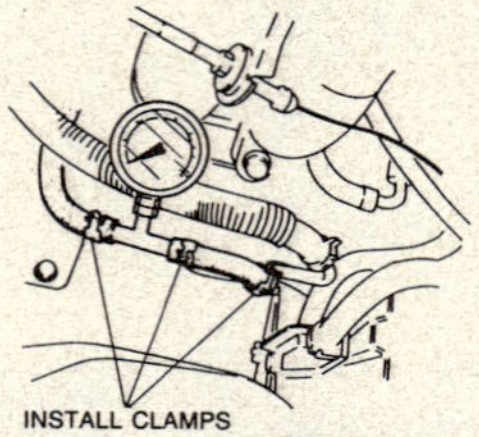	Yes	Check if fuel line pressure decreases when accelerating quickly ⇨ If decreases, check fuel pump maximum pressure If OK, check fuel line and filter for clogging ⇨ If no decrease, go to next step
		No	**Low pressure** Check for fuel line pressure while pinching fuel return hose ⇨ If fuel line pressure quickly increases, check pressure regulator ⇨ If fuel line pressure gradually increases, check for clogging between fuel pump and pressure regulator If not clogged, check fuel pump maximum pressure
			High pressure Check if fuel return line is clogged ⇨ If OK, replace pressure regulator ⇨ If not OK, replace
11	Check for correct engine compression	Yes	Go to next step
	Engine compression: 1,324—932 kPa (13.5—9.5 kg/cm², 192—135 psi)-300 rpm	No	Check engine condition • Worn piston, piston rings or cylinder wall • Defective cylinder head gasket • Distorted cylinder head • Improper valve seating • Valve sticking in guide
12	Try known good ECU and check if condition improves		

19	ENGINE STALLS ON DECELERATION		
DESCRIPTION	• Engine unexpectedly stops running while decelerating or after deceleration		

[TROUBLESHOOTING HINTS]
Engine speed drops too much when releasing accelerator
① Idle speed control malfunction
② Fuel cut control malfunction
③ Engine feedback control malfunction
④ Idle speed misadjustment

STEP	INSPECTION		ACTION
1	Check if "00" is displayed on Self-Diagnosis Checker with ignition switch ON 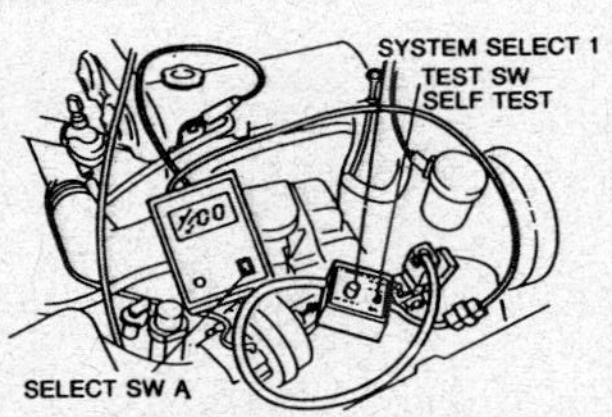	Yes	Go to next step
		No	**Malfunction Code No. displayed** Check for cause (Refer to specified check sequence) **"88" flashes** Check ECU terminal 1F voltage **Specification: Battery voltage (Ignition switch ON)** ⇨ If OK, replace ECU ⇨ If not OK, check wiring between ECU and Self-Diagnosis Checker
2	Check switches for correct operation with Self-Diagnosis Checker Monitor Lamp and ignition switch ON 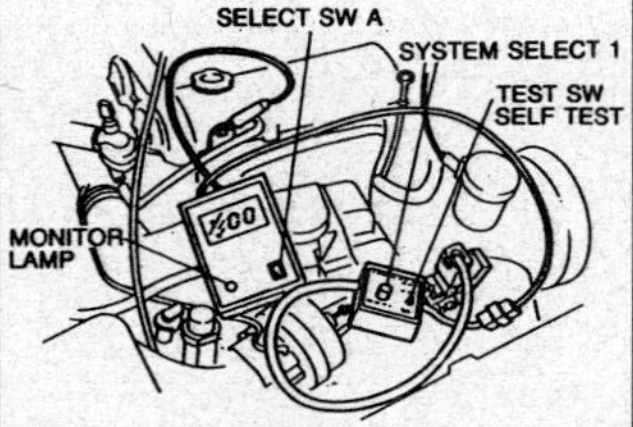	Yes	Go to next step
		No	**Lamp not ON/OFF with specified switch** Check for cause (Refer to specified check sequence) **Lamp always ON** Check wiring between ECU terminal 1D and Self-Diagnosis Checker
3	Disconnect oxygen sensor connector and check if condition improves 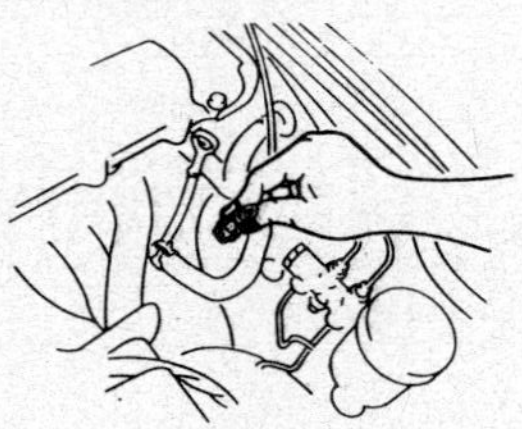	Yes	Check oxygen sensor
		No	Go to next step
4	Check if ECU terminal voltages are OK (2D, 2O, 2U, 2V and 2Q)	Yes	Go to next step
		No	Check for cause (Refer to "Check Point for Each Terminal")

STEP	INSPECTION		ACTION
5	Disconnect ISC valve connector at idle and check if "clicking" is heard from ISC valve 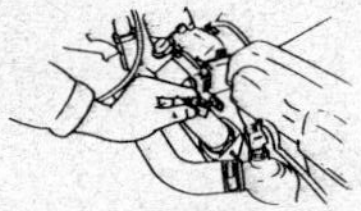	Yes	Go to next step
		No	Replace ISC valve
6	Check for correct idle speed **Idle speed: 850 ± 50 rpm** 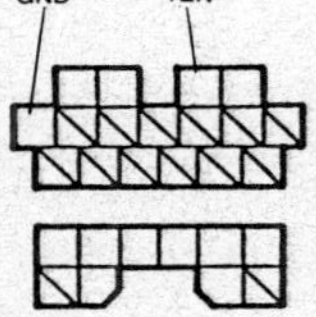 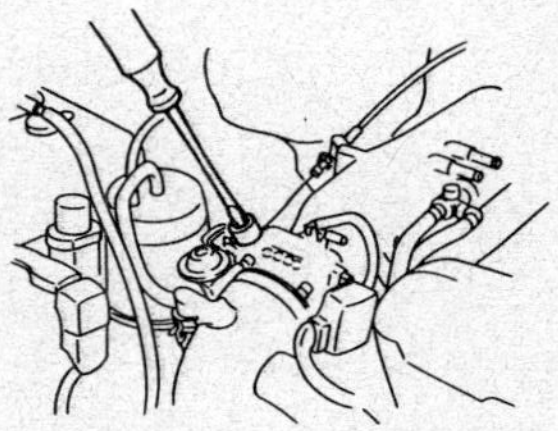	Yes	Go to next step
		No	Adjust
7	Try known good ECU and check if condition improves		

20	ENGINE STALLS AT IDLE — WHEN A/C, P/S, OR E/L ON		
DESCRIPTION	• Engine unexpectedly stops running at idle when A/C, P/S, or E/L ON • A/C, P/S, headlight, blower fan and electric cooling fan OK		
[TROUBLESHOOTING HINTS] ① Idle speed control malfunction • No input signal from switch • Idle speed misadjustment • ISC valve stuck			
STEP	INSPECTION		ACTION
1	Check if "00" is displayed on Self-Diagnosis Checker with ignition switch ON 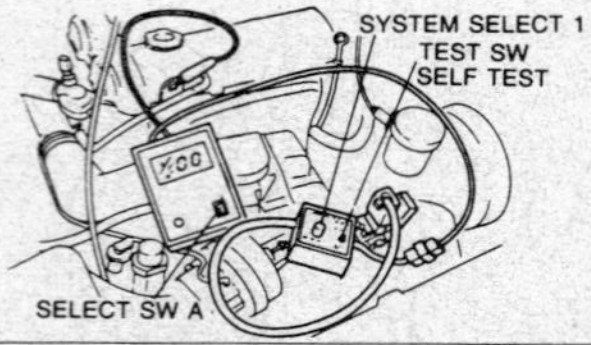 	Yes	Go to next step
		No	**Malfunction Code No. displayed** Check for cause (Refer to specified check sequence)
			"88" flashes Check ECU terminal 1F voltage **Specification: Battery voltage (Ignition switch ON)** ⇨ If OK, replace ECU ⇨ If not OK, check wiring between ECU and Self-Diagnosis Checker
2	Check switches for correct operation with Self-Diagnosis Checker Monitor Lamp and ignition switch ON 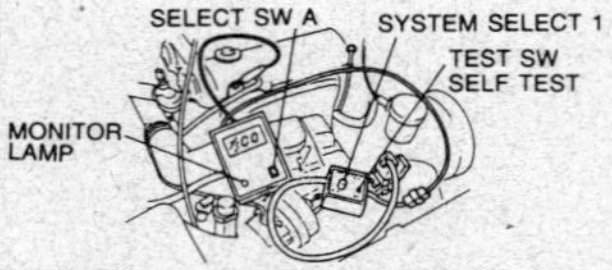 	Yes	Go to next step
		No	**Lamp not ON/OFF with specified switch** Check for cause (Refer to specified check sequence)
			Lamp always ON Check wiring between ECU terminal 1D and Self-Diagnosis Checker
3	Check if ECU terminal voltages are OK (1G, 1P, 1U, 2D, 2Q and 2W)	Yes	Go to next step
		No	Check for cause (Refer to "Check Point for Each Terminal")
4	Check for correct idle speed **Idle speed: 850 ± 50 rpm** 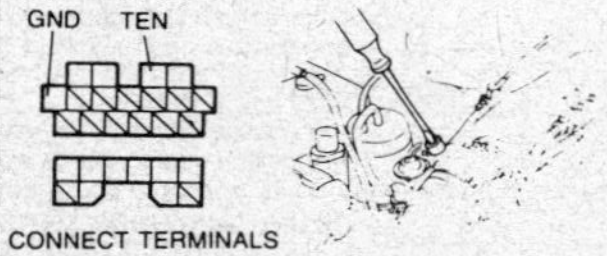 	Yes	Go to next step
		No	Adjust
5	Disconnect ISC valve connector at idle and check if "clicking" is heard from ISC valve 	Yes	Go to next step
		No	Replace ISC valve
6	Try known good ECU and check if condition improves		

21	ENGINE STALLS SUDDENLY (INTERMITTENT)		
DESCRIPTION	• Engine intermittently stops running • Before stalling, engine condition OK		
[TROUBLESHOOTING HINTS] ① Intermittently no spark or no fuel injection • Poor connection in wiring harness			
STEP	INSPECTION		ACTION
1	Check if "00" is displayed on Self-Diagnosis Checker with ignition switch ON 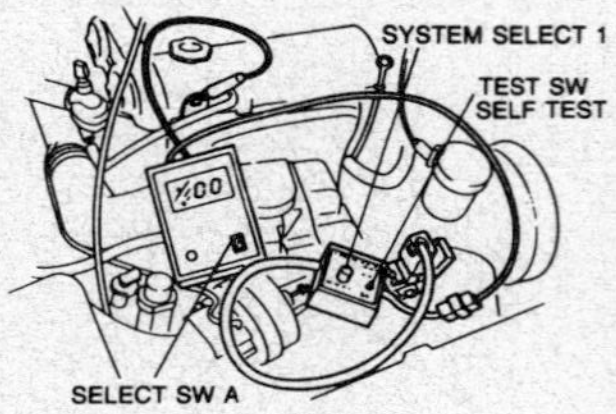 	Yes	Go to next step
		No	**Malfunction Code No. displayed** Check for cause (Refer to specified check sequence) **Note** **When checking wiring harness and connectors, tap, move, and wiggle suspect sensor and/or harness to recreate problem**
			"88" flashes Check ECU terminal 1F voltage **Specification: Battery voltage (Ignition switch ON)** ⇨ If OK, replace ECU ⇨ If not OK, check wiring between ECU and Self-Diagnosis Checker
2	Check if ECU terminal voltages are OK (1B, 2A, 2B and 2C) **Note** **When checking voltages, tap, move, and wiggle harness and connector**	Yes	Go to Troubleshooting No.2 "CRANKS NORMALLY BUT WILL NOT START (NO COMBUSTION)"
		No	Check for cause (Refer to "Check Point for Each Terminal")

22	HESITATES/STUMBLES ON ACCELERATION		
DESCRIPTION	• Flat spot occurs just after accelerator depressed or mild jerking occurs during acceleration		
[TROUBLESHOOTING HINTS] ① Air/Fuel mixture leaning when depressing accelerator • Fuel injection control malfunction (Correction for accelerating condition) • Air leakage of intake air system • Fuel line pressure low • Spark advance control malfunction			
STEP	**INSPECTION**		**ACTION**
1	Check if "00" is displayed on Self-Diagnosis Checker with ignition switch ON 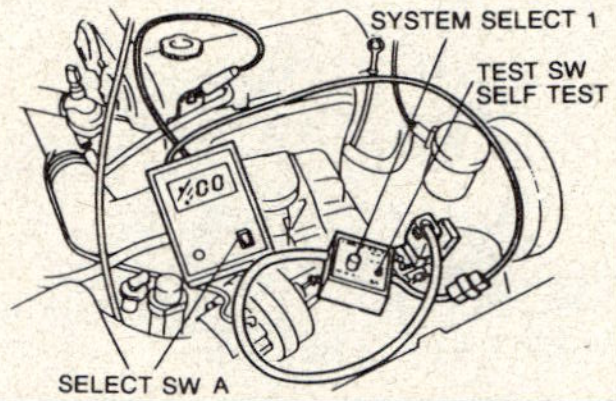	Yes	Go to next step
		No	**Malfunction Code No. displayed** Check for cause (Refer to specified check sequence)
			"88" flashes Check ECU terminal 1F voltage **Specification: Battery voltage (Ignition switch ON)** ⇨ If OK, replace ECU ⇨ If not OK, check wiring between ECU and Self-Diagnosis Checker
2	Check switches for correct operation with Self-Diagnosis Checker Monitor Lamp and ignition switch ON 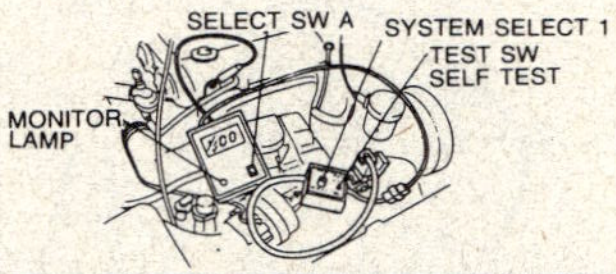	Yes	Go to next step
		No	**Lamp not ON/OFF with specified switch** Check for cause (Refer to specified check sequence)
			Lamp always ON Check wiring between ECU terminal 1D and Self-Diagnosis Checker
3	Disconnect oxygen sensor connector and check if condition improves 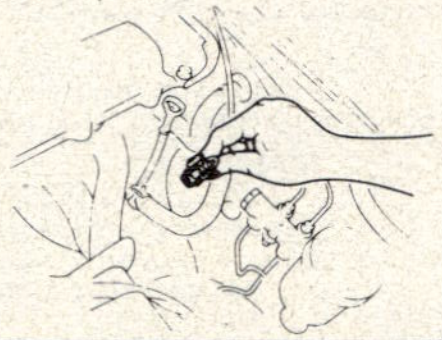	Yes	Check oxygen sensor
		No	Go to next step
4	Check if ECU terminal voltages are OK	Yes	Go to next step
		No	Check for cause (Refer to "Check Point for Each Terminal")
5	Check if throttle linkage is correctly installed and operates freely	Yes	Go to next step
		No	Correct, clean, or replace as required any binding or damaged linkage and adjust cable deflection at throttle body

STEP	INSPECTION		ACTION
6	Check if air duct and air hoses are correctly installed	Yes	Go to next step
		No	Repair
7	Check for correct intake manifold vacuum at idle **Vacuum: More than 450 mmHg (17.7 inHg)** 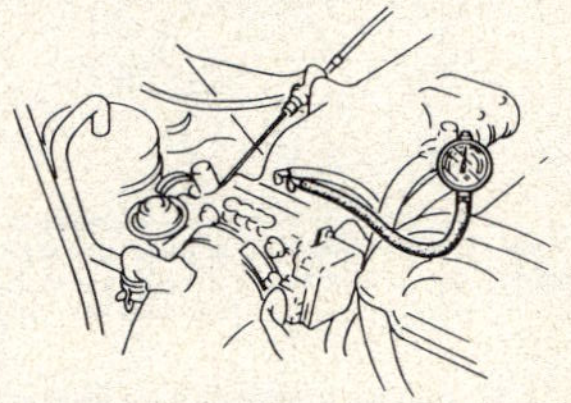	Yes	Go to next step
		No	Check for air leakage of intake air system components
8	Check if air cleaner element is clean	Yes	Go to next step
		No	Replace air cleaner element
9	Check for correct ignition timing at idle **Ignition timing: 10° ± 1° BTDC (M/T)** **8° ± 1° BTDC (4A/T)** 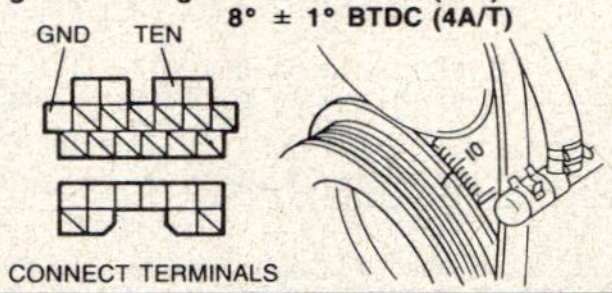	Yes	Check if ignition timing advances when accelerating ⇨ If advances, go to next step ⇨ If no advance, replace ECU
		No	Adjust
10	Check for correct fuel line pressure at idle **Fuel line pressure:** **265—314 kPa (2.7—3.2 kg/cm², 38—46 psi)** **(Vacuum hose to pressure regulator disconnected)** 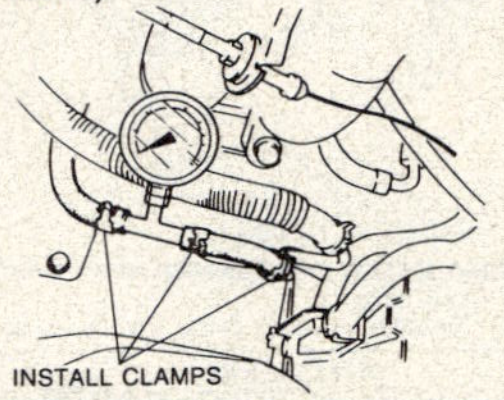	Yes	Check if fuel line pressure decreases when accelerating quickly ⇨ If decreases, check fuel line and filter for clogging ⇨ If no decrease, go to next step
		No	**Low pressure** Check for fuel line pressure while pinching fuel return hose ⇨ If fuel line pressure quickly increases, check pressure regulator ⇨ If fuel line pressure gradually increases, check for clogging between fuel pump and pressure regulator If not clogged, check fuel pump maximum pressure
11	Check if exhaust system is restricted	Yes	Repair or replace
		No	Go to next step
12	Try known good ECU and check if condition improves		

23	SURGES WHILE CRUISING		
DESCRIPTION	• Unexpected change in engine speed which is usually repetitive		
[TROUBLESHOOTING HINTS] ① Air/Fuel mixture too lean or too rich • Fuel injection control malfunction • Air leakage of intake air system • Fuel line pressure low • Evaporative emission control malfunction • Spark advance control malfunction			
STEP	**INSPECTION**		**ACTION**
1	Check if "00" is displayed on Self-Diagnosis Checker with ignition switch ON 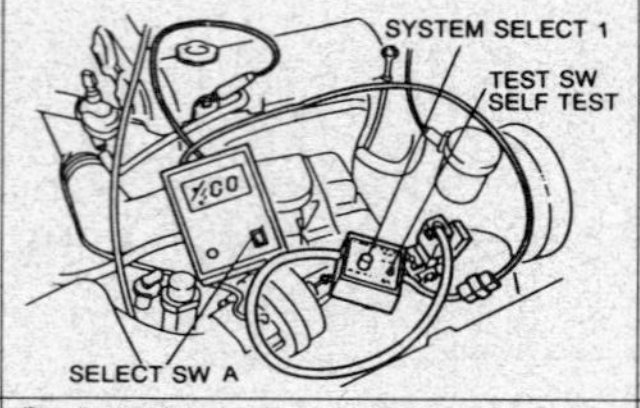	Yes	Go to next step
		No	**Malfunction Code No. displayed** Check for cause (Refer to specified check sequence) **"88" flashes** Check ECU terminal 1F voltage **Specification: Battery voltage (Ignition switch ON)** ⇨ If OK, replace ECU ⇨ If not OK, check wiring between ECU and Self-Diagnosis Checker
2	Check switches for correct operation with Self-Diagnosis Checker Monitor Lamp and ignition switch ON 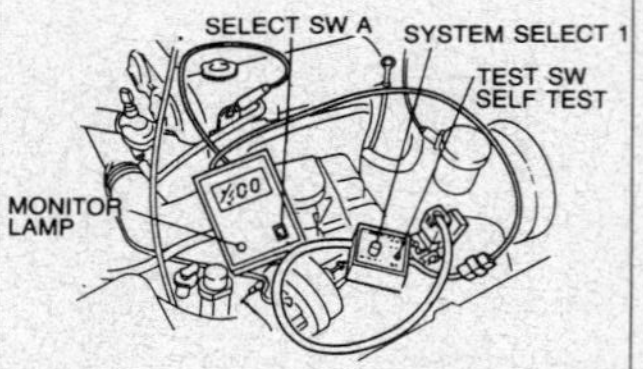	Yes	Go to next step
		No	**Lamp not ON/OFF with specified switch** Check for cause (Refer to specified check sequence) **Lamp always ON** Check wiring between ECU terminal 1D and Self-Diagnosis Checker
3	Disconnect oxygen sensor connector and check if condition improves 	Yes	Check oxygen sensor
		No	Go to next step
4	Check if ECU terminal voltages are OK	Yes	Go to next step
		No	Check for cause (Refer to "Check Point for Each Terminal")
5	Check if throttle linkage is correctly installed and operates freely	Yes	Go to next step
		No	Correct, clean, or replace as required any binding or damaged linkage, and adjust cable deflection at throttle body

STEP	INSPECTION		ACTION
6	Check for correct intake manifold vacuum **Vacuum: More than 450 mmHg (17.7 inHg)** 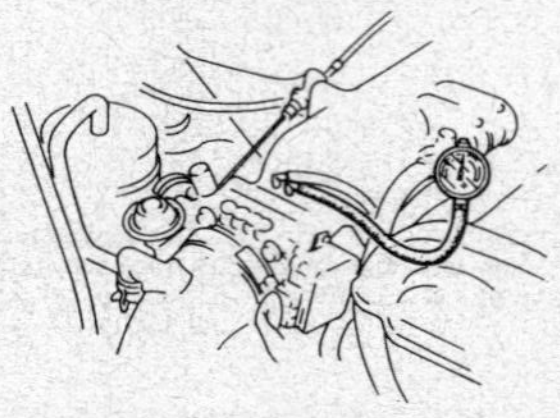	Yes	Go to next step
		No	Check for air leakage of intake air system components
7	Check if air cleaner element is clean	Yes	Go to next step
		No	Replace air cleaner element
8	Check for correct ignition timing at idle **Ignition timing: 10° ± 1° BTDC (M/T)** **8° ± 1° BTDC (4A/T)** 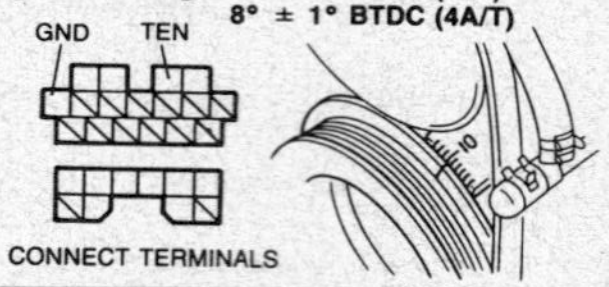	Yes	Check if ignition timing advances when accelerating ⇨ If advances, go to next step ⇨ If no advance, replace ECU
		No	Adjust
9	Check for correct fuel line pressure at idle **Fuel line pressure:** **265—314 kPa (2.7—3.2 kg/cm², 38—46 psi)** **(Vacuum hose to pressure regulator disconnected)** 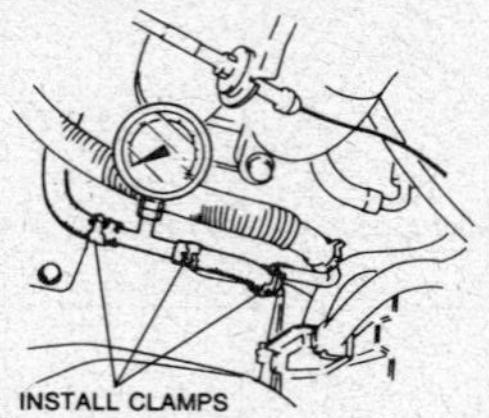	Yes	Check if fuel line pressure decreases when accelerating quickly ⇨ If decreases, check fuel line and filter for clogging ⇨ If no decrease, go to next step
		No	**Low pressure** Check for fuel line pressure while pinching fuel return hose ⇨ If fuel line pressure quickly increases, check pressure regulator ⇨ If fuel line pressure gradually increases, check for clogging between fuel pump and pressure regulator If not clogged, check fuel pump maximum pressure
10	Check if exhaust system is restricted	Yes	Repair or replace
		No	Go to next step
11	Try known good ECU and check if condition improves		

24	LACK OF POWER		
DESCRIP-TION	• Performance poor under load • Reduced maximum speed		
[TROUBLESHOOTING HINTS] ① Factors other than engine malfunction • Clutch slipping • Brake dragging • Low tire pressure • Unrecommended tire size • Overloaded ② Low intake air amount • Throttle valve not open fully • Clogged intake air system ③ Air/Fuel mixture too lean or too rich • Fuel line pressure low or high • Insufficient fuel injection ④ Poor ignition ⑤ Low engine compression			

STEP	INSPECTION		ACTION
1	Check factors other than engine • Clutch slipping • Brake dragging • Low tire pressure • Unrecommended tire size	Yes	Go to next step
		No	Repair
2	Check if throttle valve fully opens when depressing accelerator fully	Yes	Go to next step
		No	Check if accelerator cable is correctly installed ⇨ If OK, check throttle body ⇨ If not OK, install accelerator cable correctly
3	Check if "00" is displayed on Self-Diagnosis Checker with ignition switch ON 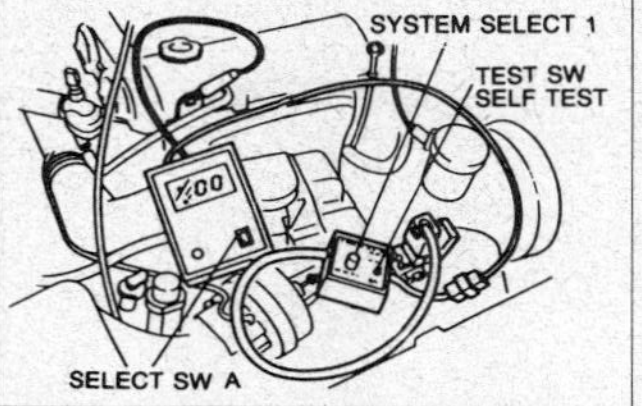	Yes	Go to next step
		No	**Malfunction Code No. displayed** Check for cause (Refer to specified check sequence)
			"88" flashes Check ECU terminal 1F voltage **Specification: Battery voltage (Ignition switch ON)** ⇨ If OK, replace ECU ⇨ If not OK, check wiring between ECU and Self-Diagnosis Checker
4	Check switches for correct operation with Self-Diagnosis Checker Monitor Lamp and ignition switch ON 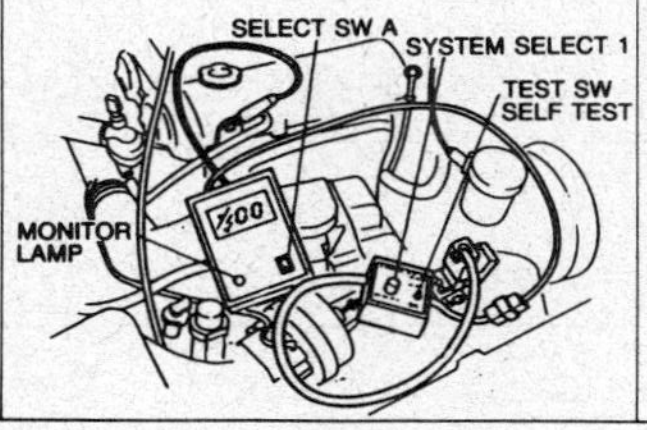	Yes	Go to next step
		No	**Lamp not ON/OFF with specified switch** Check for cause (Refer to specified check sequence)
			Lamp always ON Check wiring between ECU terminal 1D and Self-Diagnosis Checker

STEP	INSPECTION		ACTION
5	Check for correct ignition timing at idle **Ignition timing: 10° ± 1° BTDC (M/T) 8° ± 1° BTDC (4A/T)** 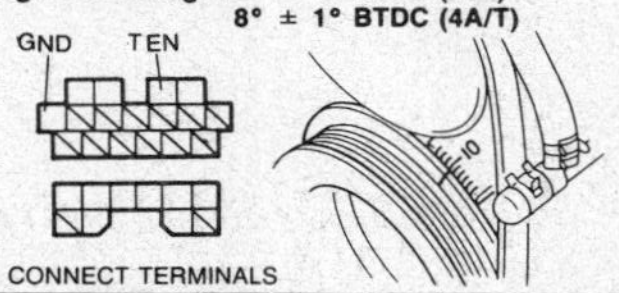	Yes	Check if ignition timing advances when accelerating ⇨ If advances, go to next step ⇨ If no advance, check ECU terminal voltages
		No	Adjust
6	Check if spark plugs are OK 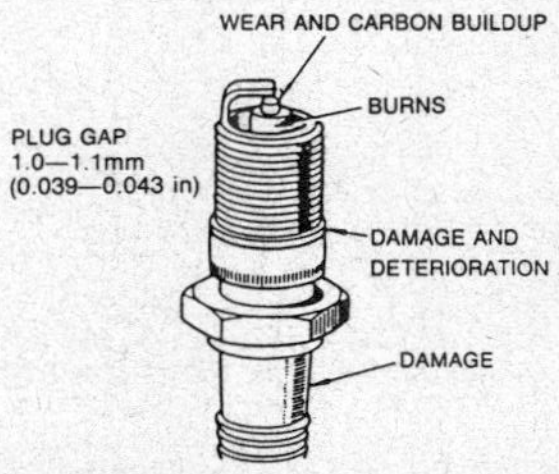	Yes	Go to next step
		No	Repair, clean, or replace
7	Check if resistance of high-tension leads are OK **Resistance: 16 kΩ per 1m (3.28 ft)** 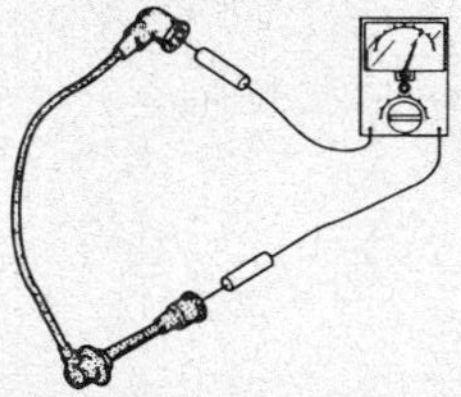	Yes	Go to next step
		No	Replace
8	Check if resistance of ignition coil is OK **Resistance (at 20°C [68°F]): ① Primary coil........ 0.78—0.94Ω ② Secondary coil.... 11.2—15.2 kΩ** 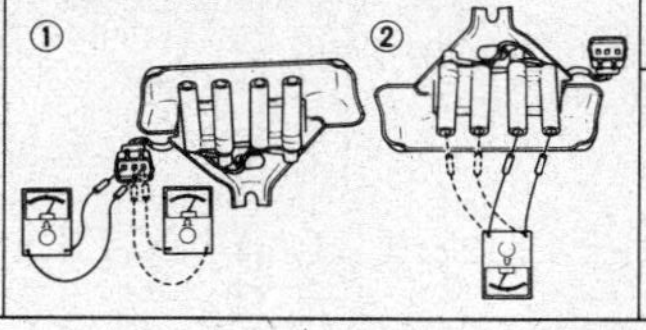	Yes	Go to next step
		No	Replace

STEP	INSPECTION		ACTION
9	Check for correct engine compression **Engine compression: 1,324—932 kPa (13.5—9.5 kg/cm², 192—135 psi)-300 rpm** 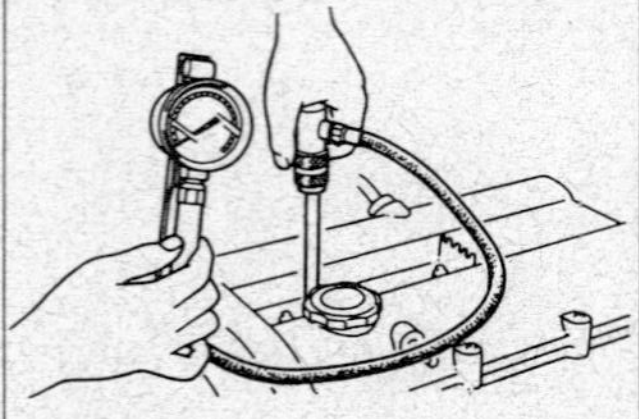	Yes	Go to next step
		No	Check engine condition • Worn piston, piston rings or cylinder wall • Defective cylinder head gasket • Distorted cylinder head • Improper valve seating • Valve sticking in guide
10	Check for correct intake manifold vacuum at idle	Yes	Go to next step
	Intake manifold vacuum: More than 450 mmHg (17.7 inHg) 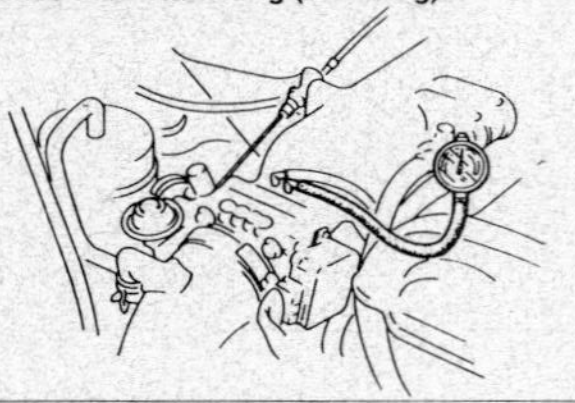	No	Check for air leakage of intake air system components
11	Check if air cleaner element is clean	Yes	Go to next step
		No	Replace air cleaner element
12	Check for injector operating sound at idle	Yes	Go to Step 14
		No	Go to Step 13
13	Check if battery voltage exists at injector connector (W/R) wire with ignition switch ON 	Yes	Check if injector resistance is OK **Resistance: Approx. 14Ω** ⇨ If OK, check wiring between ECU and injector ⇨ If not OK, check injector for fuel leakage
		No	Check wiring between ECU and injector
14	Check if ECU terminal voltages are OK (2D, 2O and 2Q)	Yes	Go to next step
		No	Check for cause (Refer to "Check Point for Each Terminal")

STEP	INSPECTION		ACTION
15	Check for correct fuel line pressure at idle **Fuel line pressure: 265—314 kPa (2.7—3.2 kg/cm², 38—46 psi) (Vacuum hose to pressure regulator disconnected)** 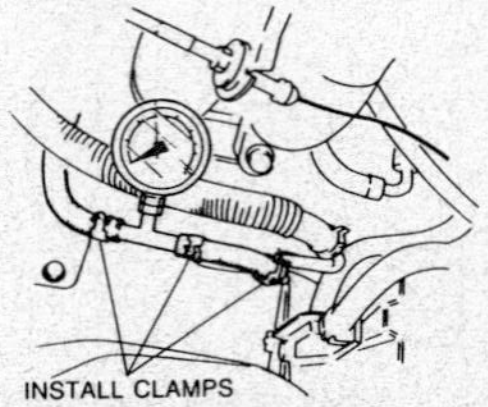	Yes	Check if fuel line pressure decreases when accelerating quickly ⇨ If decreases, check fuel pump maximum pressure If OK, check fuel line and filter for clogging ⇨ If no decreases, go to next step
		No	**Low pressure** Check for fuel line pressure while pinching fuel return hose ⇨ If fuel line pressure quickly increases, check pressure regulator ⇨ If fuel line pressure gradually increases, check for clogging between fuel pump and pressure regulator If not clogged, check fuel pump maximum pressure
16	Run engine several minutes at idle and check if fuel line pressure is held after ignition switch is turned OFF **Fuel line pressure: More than 147 kPa (1.5 kg/cm², 21 psi) for 5 min.** 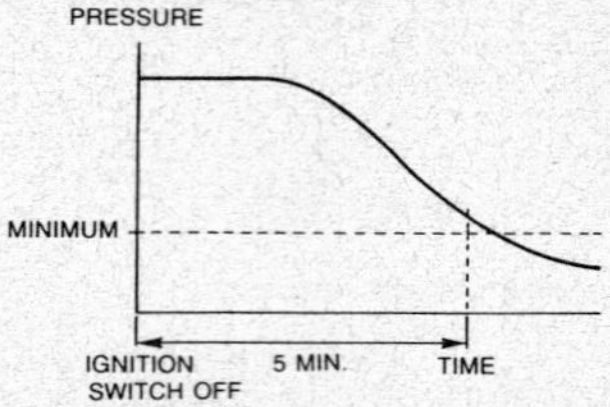	Yes	**High pressure** Check if vacuum hose to pressure regulator is damaged or poorly connected ⇨ If OK, replace pressure regulator ⇨ If not OK, repair or replace hose
		No	Check injector for fuel leakage
17	Disconnect each injector connector at idle and check if engine speed decreases equally each time 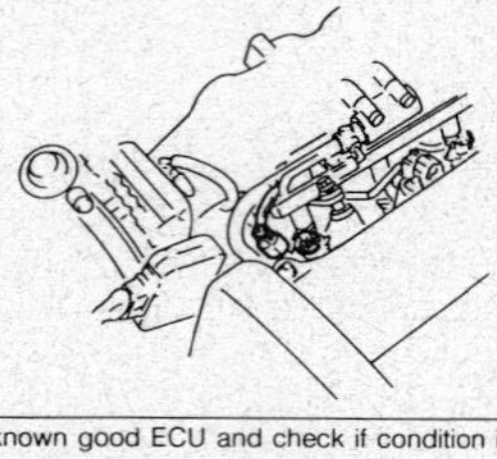	Yes	Go to next step
		No	Replace injector
18	Try known good ECU and check if condition improves		

25	POOR ACCELERATION
DESCRIPTION	• Performance poor while accelerating

[TROUBLESHOOTING HINTS]

① Factors other than engine malfunction
- Clutch slipping
- Brake dragging
- Low tire pressure
- Unrecommended tire size
- Overloaded

② Low intake air amount
- Throttle valve not open fully
- Clogged intake air system

④ Air/Fuel mixture too lean or too rich
- Fuel line pressure low or high
- Insufficient fuel injection

④ Poor ignition

⑤ Low engine compression

STEP	INSPECTION		ACTION
1	Check factors other than engine • Clutch slipping • Brake dragging • Low tire pressure • Unrecommended tire size	Yes	Go to next step
		No	Repair
2	Check if throttle valve fully opens when depressing accelerator fully	Yes	Go to next step
		No	Check if accelerator cable is correctly installed ⇨ If OK, check throttle body ⇨ If not OK, install accelerator cable correctly
3	Check if "00" is displayed on Self-Diagnosis Checker with ignition switch ON 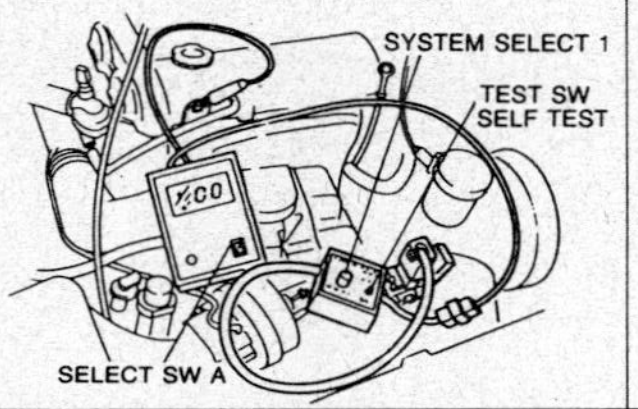	Yes	Go to next step
		No	**Malfunction Code No. displayed** Check for cause (Refer to specified check sequence)
			"88" flashes Check ECU terminal 1F voltage **Specification: Battery voltage (Ignition switch ON)** ⇨ If OK, replace ECU ⇨ If not OK, check wiring between ECU and Self-Diagnosis Checker
4	Check switches for correct operation with Self-Diagnosis Checker Monitor Lamp and ignition switch ON 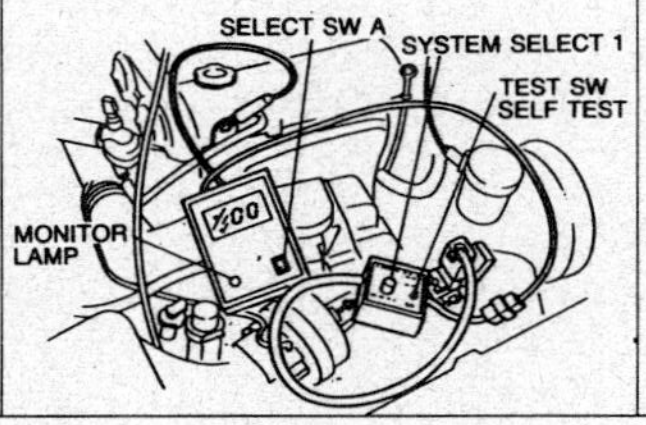	Yes	Go to next step
		No	**Lamp not ON/OFF with specified switch** Check for cause (Refer to specified check sequence)
			Lamp always ON Check wiring between ECU terminal 1D and Self-Diagnosis Checker

STEP	INSPECTION		ACTION
5	Check for correct ignition timing at idle **Ignition timing: 10° ± 1° BTDC (M/T) 8° ± 1° BTDC (4A/T)** 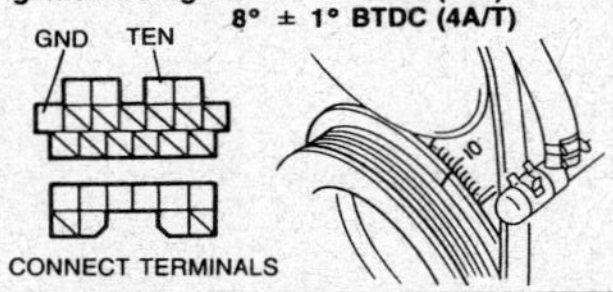	Yes	Check if ignition timing advances when accelerating ⇨ If advances, go to next step ⇨ If no advance, check ECU terminal voltages
		No	Adjust
6	Check if spark plugs are OK 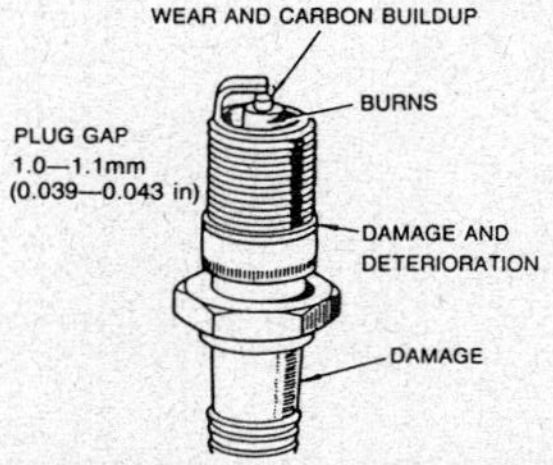	Yes	Go to next step
		No	Repair, clean, or replace
7	Check if resistance of high-tension leads are OK **Resistance: 16 kΩ per 1m (3.28 ft)** 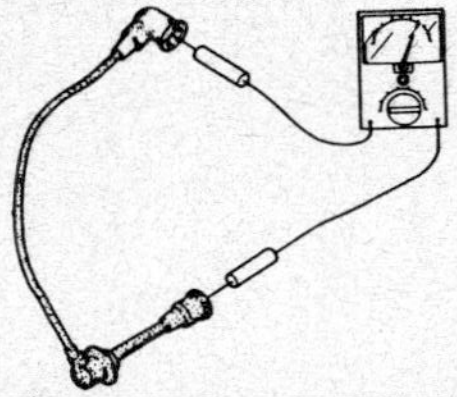	Yes	Go to next step
		No	Replace
8	Check if resistance of ignition coil is OK **Resistance (at 20°C [68°F]): ① Primary coil........ 0.78—0.94Ω ② Secondary coil.... 11.2—15.2 kΩ** 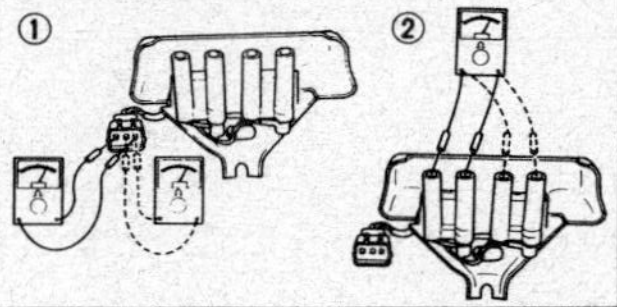	Yes	Go to next step
		No	Replace

STEP	INSPECTION		ACTION
9	Check for correct engine compression **Engine compression: 1,324—932 kPa (13.5—9.5 kg/cm², 192—135 psi)-300 rpm** 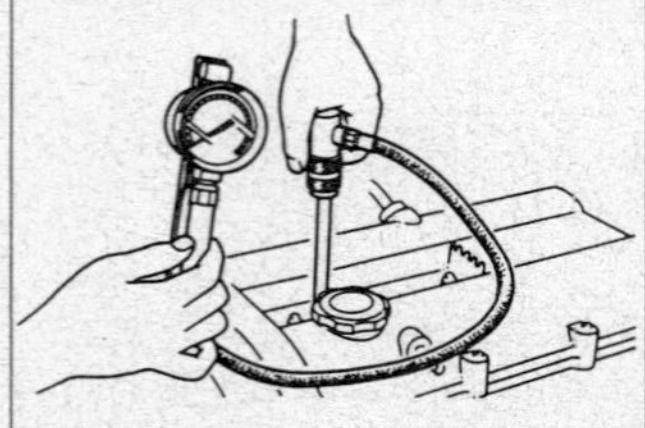	Yes	Go to next step
		No	Check engine condition • Worn piston, piston rings or cylinder wall • Defective cylinder head gasket • Distorted cylinder head • Improper valve seating • Valve sticking in guide
10	Check for correct intake manifold vacuum at idle **Intake manifold vacuum: More than 450 mmHg (17.7 inHg)** 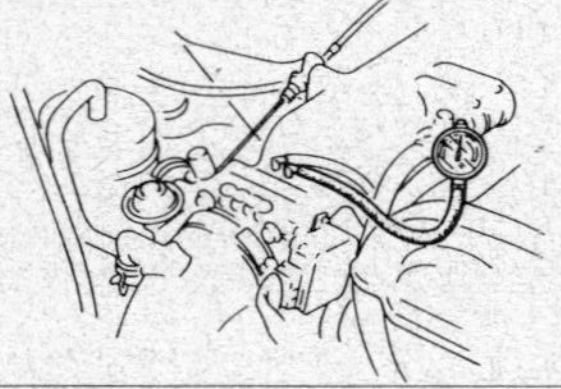	Yes	Go to next step
		No	Check for air leakage of intake air system components
11	Check if air cleaner element is clean	Yes	Go to next step
		No	Replace air cleaner element
12	Check for injector operating sound at idle	Yes	Go to Step 14
		No	Go to Step 13
13	Check if battery voltage exists at injector connector (W/R) wire with ignition switch ON 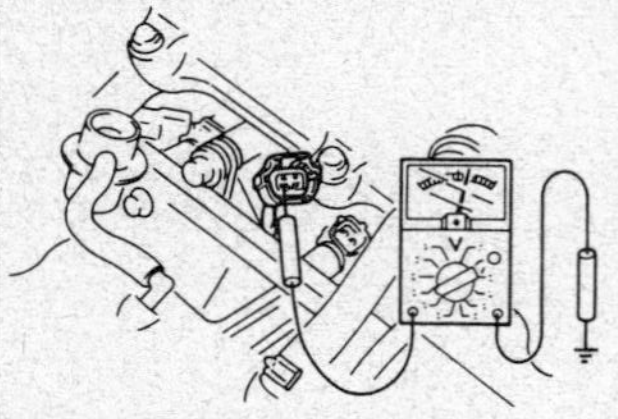	Yes	Check if injector resistance is OK **Resistance: Approx. 14Ω** ⇨ If OK, check wiring between ECU and injector ⇨ If not OK, check injector for fuel leakage
		No	Check wiring between ECU and injector
14	Check if ECU terminal voltages are OK (2D, 2O and 2Q)	Yes	Go to next step
		No	Check for cause (Refer to "Check Point for Each Terminal")

STEP	INSPECTION		ACTION
15	Check for correct fuel line pressure at idle **Fuel line pressure: 265—314 kPa (2.7—3.2 kg/cm², 38—46 psi) (Vacuum hose to pressure regulator disconnected)** 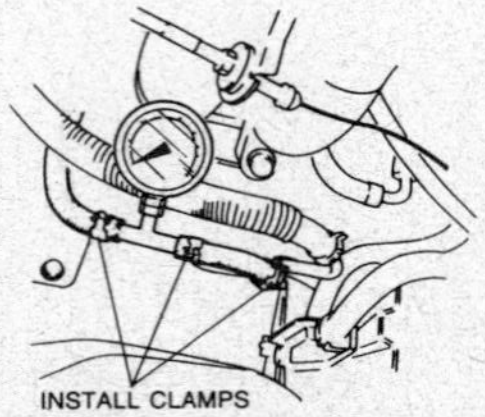	Yes	Check if fuel line pressure decreases when accelerating quickly ⇨ If decreases, check fuel pump maximum pressure If OK, check fuel line and filter for clogging ⇨ If no decreases, go to next step
		No	**Low pressure** Check for fuel line pressure while pinching fuel return hose ⇨ If fuel line pressure quickly increases, check pressure regulator ⇨ If fuel line pressure gradually increases, check for clogging between fuel pump and pressure regulator If not clogged, check fuel pump maximum pressure
16	Run engine several minutes at idle and check if fuel line pressure is held after ignition switch is turned OFF **Fuel line pressure: More than 147 kPa (1.5 kg/cm², 21 psi) for 5 min.** 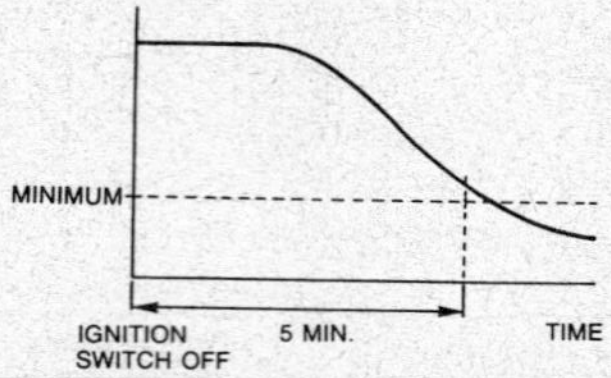	Yes	**High pressure** Check if vacuum hose to pressure regulator is damaged or poorly connected ⇨ If OK, replace pressure regulator ⇨ If not OK, repair or replace hose
		No	Check injector for fuel leakage
17	Disconnect each injector connector at idle and check if engine speed decreases equally each time 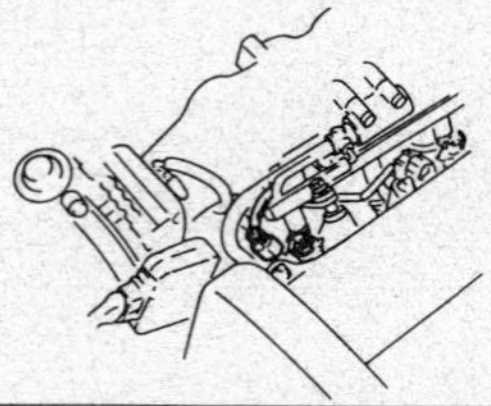	Yes	Go to next step
		No	Replace injector
18	Try known good ECU and check if condition improves		

26	RUNS ROUGH ON DECELERATION/AFTERBURN
DESCRIPTION	• Engine runs rough while decelerating and abnormal combustion occurs in exhaust system

[TROUBLESHOOTING HINTS]
① Air/Fuel mixture too rich
- Air cleaner element clogged
- Fuel injection control malfunction (Fuel cut control)
- Injector fuel leakage
- Ignition timing misadjustment

STEP	INSPECTION		ACTION
1	Check if "00" is displayed on Self-Diagnosis Checker with ignition switch ON 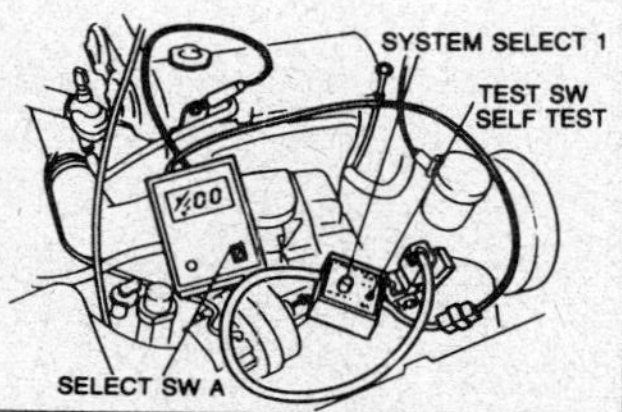	Yes	Go to next step
		No	**Malfunction Code No. displayed** Check for cause (Refer to specified check sequence) **"88" flashes** Check ECU terminal 1F voltage **Specification: Battery voltage (Ignition switch ON)** ⇨ If OK, replace ECU ⇨ If not OK, check wiring between ECU and Self-Diagnosis Checker
2	Check switches for correct operation with Self-Diagnosis Checker Monitor Lamp with ignition switch ON 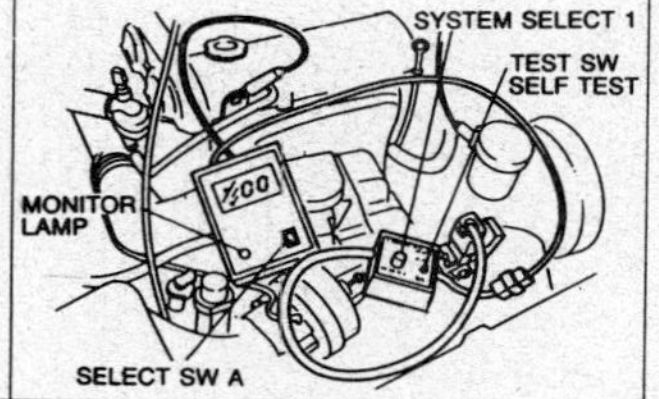	Yes	Go to next step
		No	**Lamp not ON/OFF with specified switch** Check for cause (Refer to specified check sequence) **Lamp always ON** Check wiring between ECU terminal 1D and Self-Diagnosis Checker
3	Check for correct ignition timing at idle **Ignition timing: 10° ± 1° BTDC (M/T) 8° ± 1° BTDC (4A/T)** 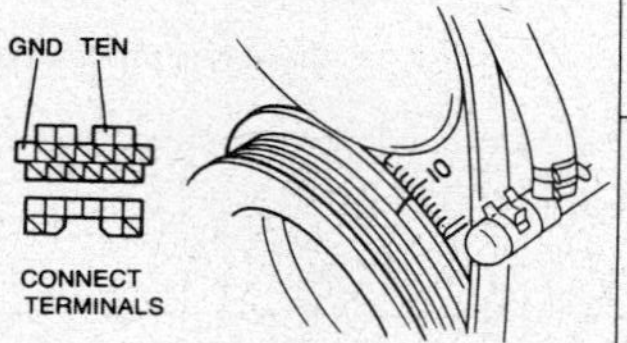	Yes	Go to next step
		No	Adjust

STEP	INSPECTION		ACTION
4	Check if fuel cut operation is OK during deceleration **Fuel cut: Above 1,900 rpm after warm-up** 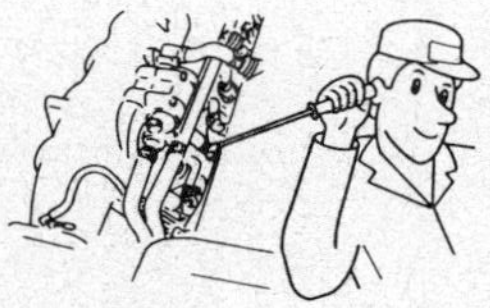	Yes	Go to next step
		No	Try known good ECU
5	Run engine several minutes at idle and check if fuel line pressure is held after ignition switch is turned OFF **Fuel line pressure: More than 147 kPa (1.5 kg/cm², 21 psi) for 5 min.** 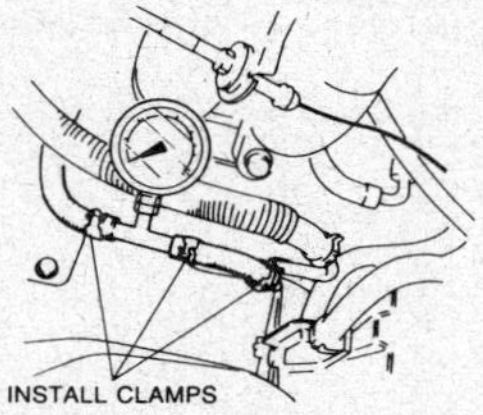	Yes	Go to next step
		No	Check injector for fuel leakage
6	Check if air cleaner element is clean 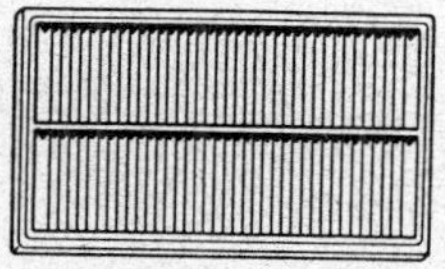	Yes	Go to next step
		No	Replace
7	Try known good ECU and check if condition improves		

27	KNOCKING		
[TROUBLESHOOTING HINTS] ① Air/Fuel mixture too lean • Fuel injection amount incorrect • Fuel line pressure decreases while accelerating ② Incorrect ignition timing (too advance) ③ Overheating ④ Carbon deposits in engine			
STEP	**INSPECTION**		**ACTION**
1	Check if "00" is displayed on Self-Diagnosis Checker with ignition switch ON	Yes	Go to next step
	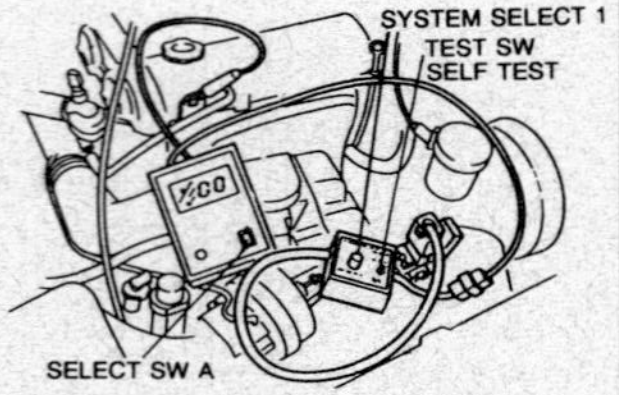	No	**Malfunction Code No. displayed** Check for cause (Refer to specified check sequence)
			"88" flashes Check ECU terminal 1F voltage **Specification: Battery voltage (Ignition switch ON)** ⇨ If OK, replace ECU ⇨ If not OK, check wiring between ECU and Self-Diagnosis Checker
2	Check switches for correct operation with Self-Diagnosis Checker Monitor Lamp and ignition switch ON	Yes	Go to next step
	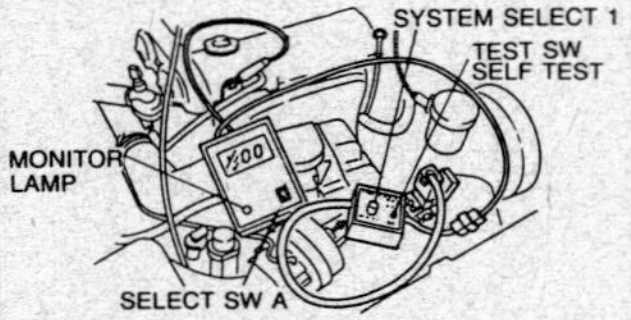	No	**Lamp not ON/OFF with specified switch** Check for cause (Refer to specified check sequence)
			Lamp always ON Check wiring between ECU terminal 1D and Self-Diagnosis Checker
3	Check if ECU terminal voltages are OK (2D, 2O and 2Q)	Yes	Go to next step
		No	Check for cause (Refer to "Check Point for Each Terminal")
4	Check for correct intake manifold vacuum at idle	Yes	Go to next step
	Vacuum: More than 450 mmHg (17.7 inHg) 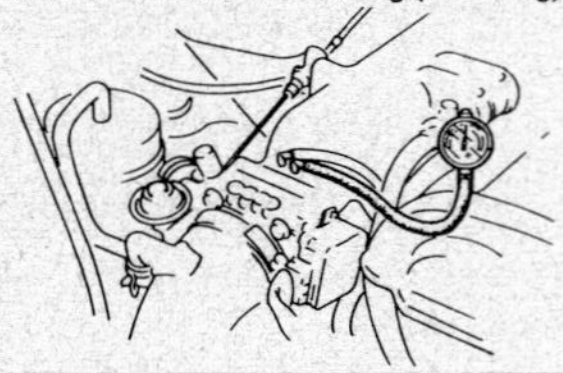	No	Check for air leakage of intake air system components
5	Check if air cleaner element is clean	Yes	Go to next step
		No	Replace air cleaner element
6	Check for correct engine compression	Yes	Go to next step
	Engine compression: 1,324—932 kPa (13.5—9.5 kg/cm², 192—135 psi)-300 rpm	No	**High compression** Check engine condition • Carbon deposits

STEP	INSPECTION		ACTION
7	Check for correct fuel line pressure at idle **Fuel line pressure: 265—314 kPa (2.7—3.2 kg/cm², 38—46 psi) (Vacuum hose to pressure regulator disconnected)**	Yes	Check if fuel line pressure decreases when accelerating quickly ⇨ If decreases, check for clogging between fuel pump and pressure regulator ⇨ If no decrease, go to next step
	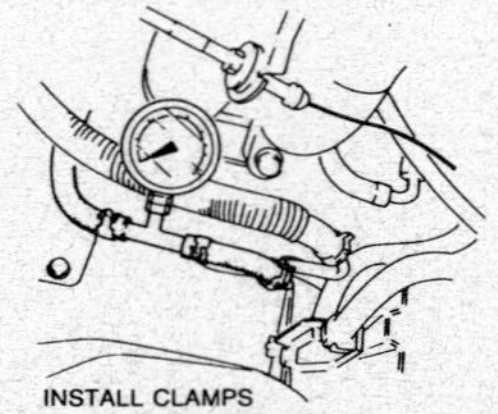	No	**Low pressure** Check for fuel line pressure while pinching fuel return hose ⇨ If fuel line pressure quickly increases, check pressure regulator ⇨ If fuel line pressure gradually increases, check for clogging between fuel pump and pressure regulator If not clogged, check fuel pump maximum pressure
8	Check for correct ignition timing at idle **Ignition timing: 10° ± 1° BTDC (M/T) 8° ± 1° BTDC (4A/T)**	Yes	Check if ignition timing advances when accelerating ⇨ If advances, go to next step ⇨ If no advance, replace ECU
	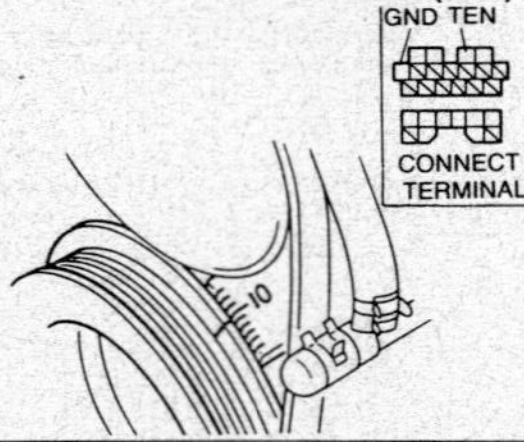	No	Adjust
9	Check if cooling system is OK	Yes	Go to next step
		No	Repair or replace • Thermostat • Electric cooling fan • Radiator
10	Try known good ECU and check if condition improves	Yes	Replace ECU
		No	Change fuel to another brand or use higher octane fuel

28	FUEL ODOR		
DESCRIPTION	•Gasoline odor in cabin		
[TROUBLESHOOTING HINTS] ① Poor connection or damaged fuel system or evaporative emission control system ② Charcoal canister overflow due to evaporative emission control system malfunction			
STEP	INSPECTION		ACTION
1	Check if fuel leak or damage are visible at fuel system and evaporative emission control system	Yes	Repair or replace
		No	Go to next step
2	Check if "00" is displayed on Self-Diagnosis Checker with ignition switch ON 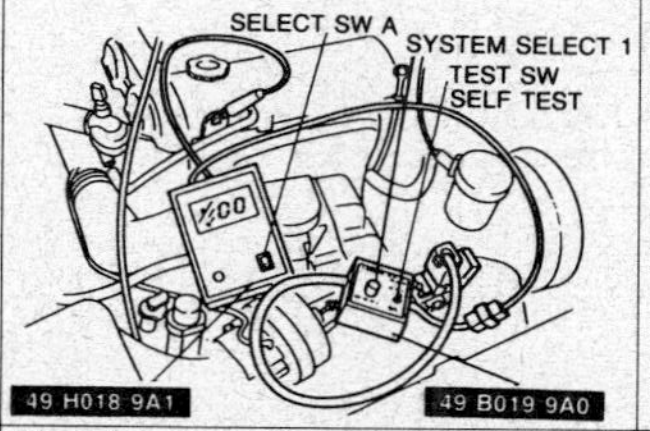	Yes	Go to next step
		No	**Malfunction Code No. displayed** Check for cause (Refer to specified check sequence) **"88" flashes** Check ECU terminal 1F voltage **Specification: Battery voltage (Ignition switch ON)** ⇨ If OK, replace ECU ⇨ If not OK, check wiring between ECU and Self-Diagnosis Checker
3	Check if vacuum is felt at solenoid valve (purge control) with engine running and throttle valve opened (Neutral switch connector disconnected) 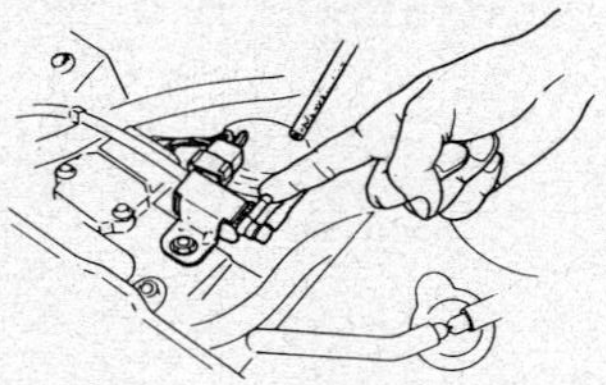	Yes	Go to Step 5
		No	Check for solenoid valve operating sound in this condition ⇨ If OK, check vacuum hoses for clogging ⇨ If not OK, go to next step
4	Apply battery voltage and ground to solenoid valve (purge control) and check if air flows through valve 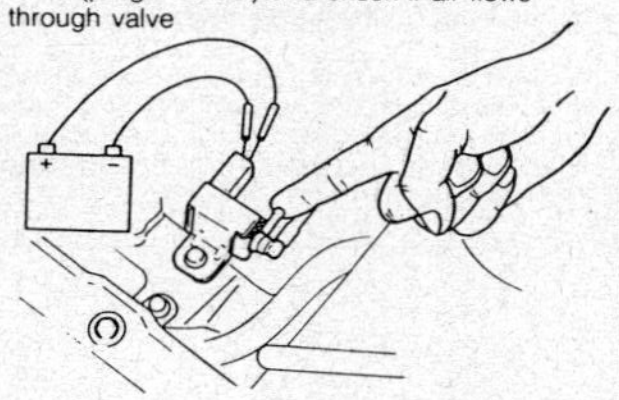	Yes	Check ECU terminal 2X voltage
		No	Replace solenoid valve
5	Try known good ECU		

29	EXHAUST SULFER SMELL	
DESCRIPTION	•Exhaust gas smells abnormal (Rotten egg smell)	
[TROUBLESHOOTING HINTS] High sulfer content fuel used		
STEP	INSPECTION	ACTION
1	Change fuel to another brand	

30	HIGH OIL CONSUMPTION		
[TROUBLESHOOTING HINTS] ① PCV system malfunction ② Engine malfunction (Oil working up, working down, or leakage)			
STEP	INSPECTION		ACTION
1	Check if PCV hose, ventilation hose or their attaching nipples are separated, damaged, clogged, or restricted	Yes	Repair or replace
		No	Go to next step
2	Check if air pressure or oil is present at ventilation hose 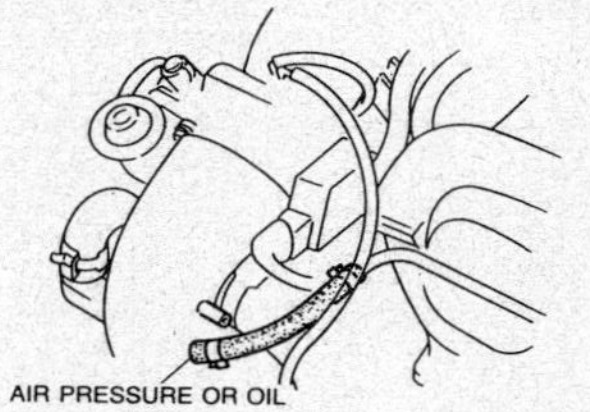	Yes	Go to next step
		No	Check engine condition •Oil leakage •Worn valve seal •Worn valve stem •Worn valve guide
3	Check if vacuum is felt at PCV valve at idle 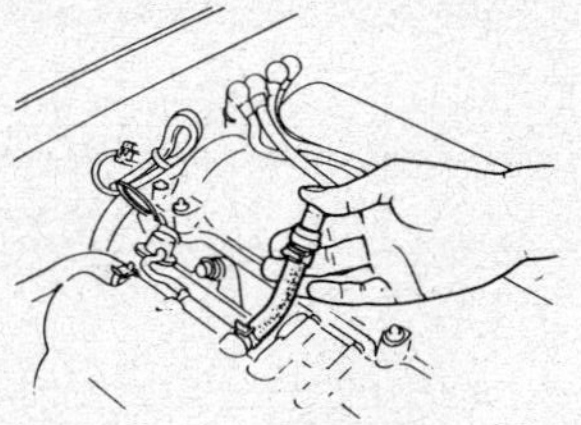	Yes	Check engine condition •Worn piston ring groove •Stuck piston ring •Worn piston or cylinder
		No	Replace PCV valve

31	POOR FUEL ECONOMY

[TROUBLESHOOTING HINTS]
While fuel consumption is drastically increased during city driving, short-run operation, stop and go driving, extended winter warm-up periods, etc., as opposed to "trip" mileage, an attempt should be made to determine these factors when confronted with "poor mileage" conditions. However, since the operator is not always at fault, the following is offered.

① Operator depressing accelerator more than usual due to low engine power
- Poor ignition
- Low intake air amount
- Electric spark advance control malfunction
- Clutch slipping
- Exhaust component restricted

② Air/Fuel mixture too rich
- High fuel line pressure

③ Alcohol blended fuel used
④ High vehicle load
- Low tire pressure
- Unrecommended tire used
- Brake dragging

⑤ Fuel cut control malfunction
⑥ High idle speed

STEP	INSPECTION		ACTION
1	Check factors other than engine • Low tire pressure • Unrecommended tire used • Clutch slipping • Brake dragging • Exhaust component restricted	Yes	Go to next step
		No	Repair
2	Check if air hoses are connected correctly	Yes	Go to next step
		No	Repair
3	Check if air cleaner element is clean	Yes	Go to next step
		No	Replace
4	Check if "00" is displayed on Self-Diagnosis Checker with ignition switch ON 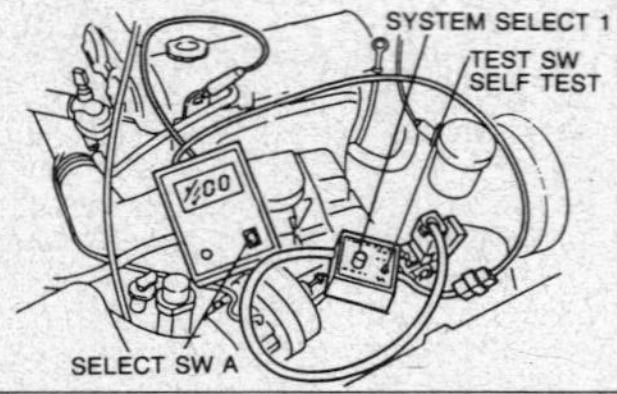	Yes	Go to next step
		No	**Malfunction Code No. displayed** Check for cause (Refer to specified check sequence) **"88" flashes** Check ECU terminal 1F voltage **Specification: Battery voltage (Ignition switch ON)** ⇨ If OK, replace ECU ⇨ If not OK, check wiring between ECU and Self-Diagnosis Checker
5	Check switches for correct operation with Self-Diagnosis Checker Monitor Lamp and ignition switch ON 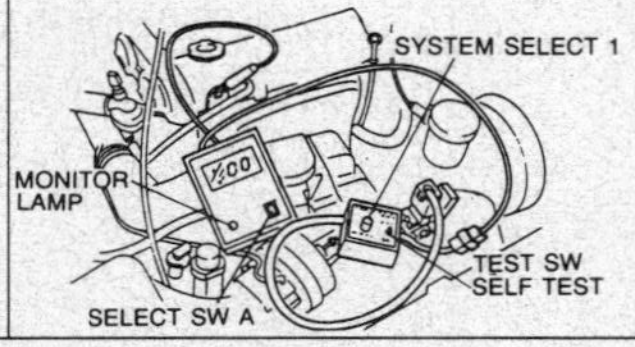	Yes	Go to next step
		No	**Lamp not ON/OFF with specified switch** Check for cause (Refer to specified check sequence) **Lamp always ON** Check wiring between ECU terminal 1F and Self-Diagnosis Checker

STEP	INSPECTION		ACTION
6	Check if ECU terminal voltages are OK (2D, 2N, 2O, 2P, 2Q, 2U and 2V)	Yes	Go to next step
		No	Check for cause (Refer to "Check Point for Each Terminal")
7	Check if fuel cut operation is OK during deceleration **Fuel cut: Above 1,900 rpm after warm-up**	Yes	Go to next step
		No	Try known good ECU
8	Check for correct ignition timing at idle **Ignition timing: 10° ± 1° BTDC (M/T) 8° ± 1° BTDC (4A/T)** 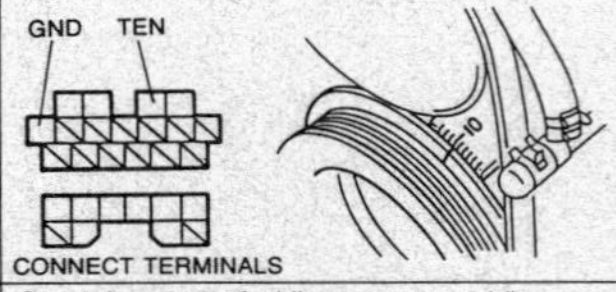	Yes	Go to next step
		No	Adjust
9	Check for correct fuel line pressure at idle **Fuel line pressure: 216—265 kPa (2.2—2.7 kg/cm², 31—38 psi)** 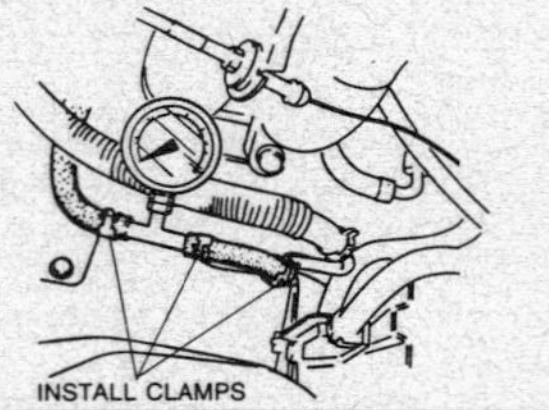	Yes	Go to next step
		No	**High pressure** Check if vacuum hose to pressure regulator is damaged or poorly connected ⇨ If OK, replace pressure regulator ⇨ If not OK, repair or replace hose
10	Run engine several minutes at idle and check if fuel line pressure is held after ignition switch is turned OFF **Fuel line pressure: More than 147 kPa (1.5 kg/cm², 21 psi) for 5 min.** 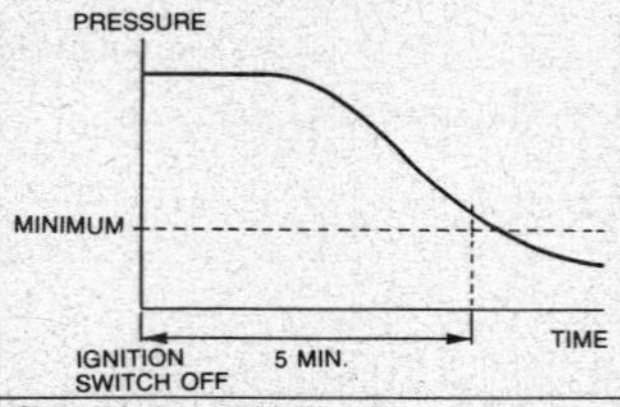	Yes	Go to next step
		No	Check injector for fuel leakage
11	Change fuel to another brand		

32	MIL ALWAYS ON		
DESCRIP-TION	•Self-Diagnosis Checker does not indicate Malfunction Code No. but MIL always ON		
[TROUBLESHOOTING HINTS] •Short circuit in wiring harness •ECU malfunction			
STEP	INSPECTION		ACTION
1	Disconnect (Y/B) wire from ECU and check if MIL goes off 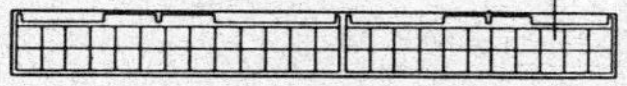 	Yes	Replace ECU
		No	Check for short circuit in wiring between combination meter and ECU

33	MIL NEVER ON		
DESCRIP-TION	•Self-Diagnosis Checker indicates Malfunction code No. o input device but MIL never ON •Other indicator and warning lamps OK		
[TROUBLESHOOTING HINTS] •Bulb burned out •Open circuit in wiring harness •ECU malfunction			
STEP	INSPECTION		ACTION
1	Ground (Y/B) wire at ECU with jumper wire and check if MIL comes on 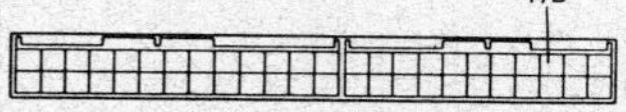 	Yes	Check connection condition of ECU connector ⇨ If OK, replace ECU ⇨ If not OK, repair ECU connector
		No	Check if bulb is OK ⇨ If OK, repair (Y/B) wire between ECU and combination meter ⇨ If not OK, replace bulb

34	A/C DOES NOT WORK		
DESCRIP-TION	•Blower fan operates but cool air not expelled		
[TROUBLESHOOTING HINTS] •Open or short circuit in wiring harness •A/C relay malfunction •A/C switch malfunction •Magnetic clutch malfunction •ECU malfunction			
STEP	INSPECTION		ACTION
1	Ground (L/B) wire at A/C relay with jumper wire and check if condenser fan operates when ignition switch is turned ON 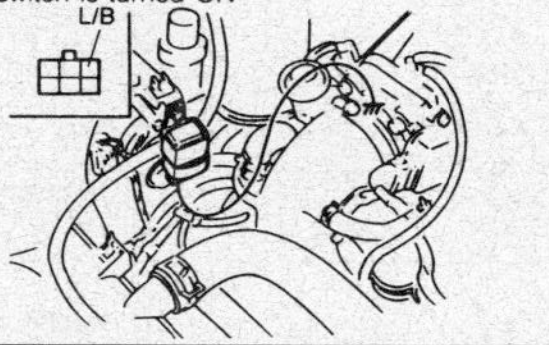 	Yes	Go to next step
		No	Go to A/C system Troubleshooting
2	Ground (L/B) wire at ECU with jumper wire and check if condenser fan operates when ignition switch is turned ON 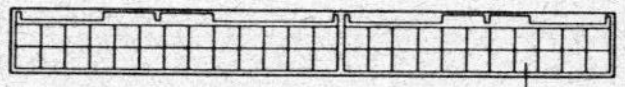 	Yes	Go to next step
		No	Repair (L/B) wire between ECU and A/C relay
3	Check if A/C switch is OK	Yes	Try known good ECU
		No	Check A/C switch and wiring

Malfunction display		Sensor or subsystem	Self-diagnosis	Fail-safe
Code No.	Pattern of output signal (Self-Diagnosis Checker)			
01	ON OFF	Ignition pulse	No IGf-signal	—
02	ON OFF	Ne-signal	No Ne-signal	—
03	ON OFF	G-signal	No G-signal	Cancels 2-group injection and electronic spark distribution
08	ON OFF	Airflow meter	Open or short circuit	Basic fuel injection amount fixed as for two driving modes (1) Idle switch: ON (2) Idle switch: OFF
09	ON OFF	Water thermosensor	Open or short circuit	Maintains constant 20°(68°F) command
10	ON OFF	Intake air thermosensor (Airflow meter)	Open or short circuit	Maintains constant 20°C (68°F) command
12	ON OFF	Throttle sensor	Open or short circuit	Maintains constant command of throttle valve fully open
14	ON OFF	Atmospheric pressure sensor	Open or short circuit	Maintains constant command of sea level pressure
15	ON OFF	Oxygen sensor (Inactivation)	Sensor output continues less than 0.55V 180 sec. after engine exceeds 1,500 rpm	Cancels engine feedback operation
17	ON OFF	Oxygen sensor (Inversion)	Sensor output continues unchanged 20 sec. after engine exceeds 1,500 rpm	Cancels engine feedback operation
26	ON OFF	Solenoid valve (Purge control)	—	—
34	ON OFF	ISC valve		—

Caution
- **If there is more than one failure present, the lowest number malfunction code is displayed first, the remaining codes are displayed in order.**
- **After repairing all failures, turn the ignition switch off, disconnect the negative battery cable for at least 20 seconds to erase the memory of a malfunction code from the engine control unit.**

CODE No.	01 (IGf-SIGNAL)		
STEP	INSPECTION		ACTION
1	Are there any poor connections at ignition coil connectors and igniter connectors?	Yes	Repair or replace connector
		No	Go to next step
2	Does tachometer operate?	Yes	Go to next step
		No	Check for open circuit in wiring from igniter to ECU terminal 2I
3	Is resistance of ignition coil OK? Resistance: Primary 0.78—0.94Ω Secondary 11.2—15.2 kΩ	Yes	Go to next step
		No	Replace ignition coil
4	Is there continuity between ignition coil and igniter? (Ignition coil – Igniter: A (W) – A (W); B (Y) – H (Y))	Yes	Go to next step
		No	Check for open circuit in wiring from ignition coil to igniter
5	Is ignition coil terminal wire (L) voltage OK?	Yes	Go to next step
		No	Check for open circuit in wiring from ignition coil to ignition switch
6	Is igniter terminal wire (L) voltage OK?	Yes	Go to next step
		No	Check for open circuit in wiring from igniter to ignition switch
7	Is there continuity between igniter and ground?	Yes	Go to next step
		No	Check for open circuit in wiring from igniter to ground
8	Check igniter	Yes	Replace ECU
		No	Replace igniter

Circuit Diagram

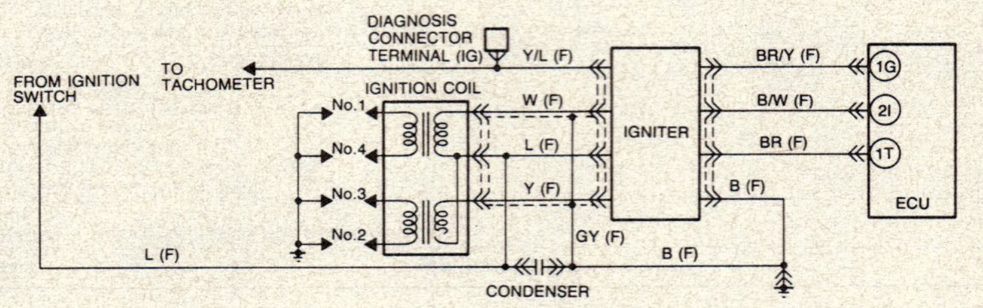

CODE No.	02 (Ne-SIGNAL)		
STEP	INSPECTION		ACTION
1	Are there any poor connections in crank angle sensor circuit?	Yes	Repair or replace connector
		No	Go to next step
2	Is Code No.03 present at same time?	Yes	Go to next step
		No	Go to Step 5
3	Is there continuity between crank angle sensor terminal-wire (B/LG) and ground?	Yes	Go to next step
		No	Check for open circuit in wiring from crank angle sensor to ground
4	Is there battery voltage at crank angle sensor terminal-wire (W/R)?	Yes	Go to next step
		No	Check for open circuit in wiring from crank angle sensor to main relay
5	Is there continuity between crank angle sensor and ECU? (Crank angle sensor C (W) — ECU 2E)	Yes	Go to next step
		No	Check for open circuit in wiring from crank angle sensor to ECU
6	Is there approx. 5V at ECU terminal 2E? (With crank angle sensor connector disconnected)	Yes	Go to next step
		No	Replace ECU
7	Is there approx. 5V at crank angle sensor terminal-wire (W)? (At harness-side connector with connector disconnected)	Yes	Replace crank angle sensor
		No	Check for short circuit in wiring from crank angle sensor to ECU

CODE No.	03 (G-SIGNAL)		
STEP	INSPECTION		ACTION
1	Are there any poor connections in crank angle sensor circuit?	Yes	Repair or replace connector
		No	Go to next step
2	Is Code No.02 also present?	Yes	Go to next step
		No	Go to Step 5
3	Is there continuity between crank angle sensor terminal-wire (B/LG) and ground?	Yes	Go to next step
		No	Check for open circuit in wiring from crank angle sensor to ground
4	Is there battery voltage at crank angle sensor terminal-wire (W/R)?	Yes	Go to next step
		No	Check for open circuit in wiring from crank angle sensor to main relay
5	Is there continuity between crank angle sensor and ECU? (Crank angle sensor D (Y/L) — ECU 2G)	Yes	Go to next step
		No	Check for open circuit in wiring from crank angle sensor to ECU
6	Is there approx. 5V at ECU terminal 2E? (With crank angle sensor connector disconnected)	Yes	Go to next step
		No	Replace ECU
7	Is there approx. 5V at crank angle sensor terminal-wire (Y/L)? (At harness-side connector with connector disconnected)	Yes	Replace crank angle sensor
		No	Check for short circuit in wiring from crank angle sensor to ECU

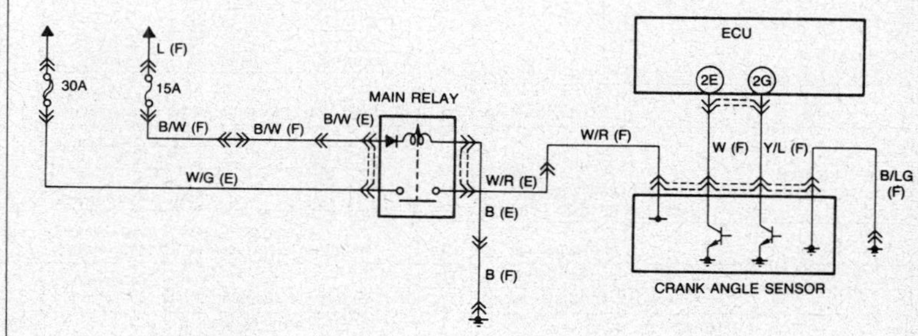

CODE No.	08 (AIRFLOW METER)		
STEP	INSPECTION		ACTION
1	Are there any poor connections in airflow meter circuit?	Yes	Repair or replace connector
		No	Go to next step
2	Is Code No.10 present at same time?	Yes	Check for open circuit in wiring from airflow meter terminal-wire (B/LG) to ground
		No	Go to next step
3	Is resistance of airflow meter OK? Airflow meter / Fully closed Ω / Fully open Ω D (LG/R)—F (R) / 200—600 / 20—1,000 D (LG/R)—C (B/LG) / 200—400	Yes	Go to next step
		No	Replace airflow meter
4	Is there continuity between airflow meter connector and ECU? Airflow meter / ECU D (LG/R) / 2K F (R) / 2O	Yes	Go to next step
		No	Check for open circuit in wiring from airflow meter to ECU
5	Are ECU terminals 2D and 2K voltages OK?	Yes	Replace ECU
		No	Check for short circuit in wiring from airflow meter to ECU

Circuit Diagram

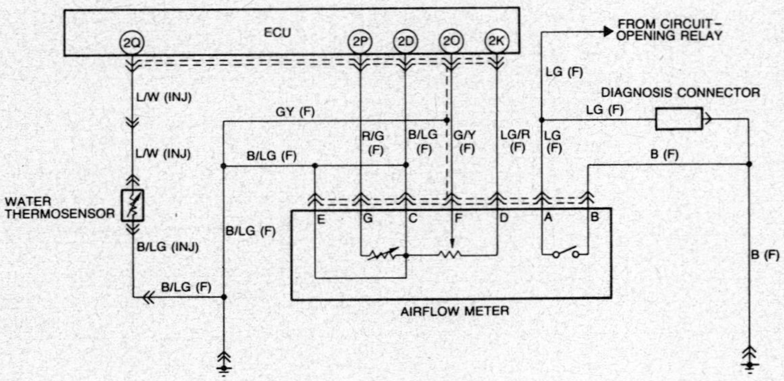

CODE No.	09 (WATER THERMOSENSOR)		
STEP	INSPECTION		ACTION
1	Are there any poor connections in water thermosensor circuit?	Yes	Repair or replace connector
		No	Go to next step
2	Is there continuity between water thermosensor and ECU? Water thermosensor / ECU A (L/W) / 2Q B (B/LG) / 2D	Yes	Go to next step
		No	Check for open circuit in wiring from water thermosensor to ECU
3	Is resistance of water thermosensor OK? Coolant temp. / Resistance −20°C (−4°F) / 14.6—17.8 kΩ 20°C (68°F) / 2.2—2.7 kΩ 80°C (176°F) / 290—350Ω	Yes	Go to next step
		No	Replace water thermosensor
4	Is same Code No. present after performing after-repair procedure?	Yes	Go to next step
		No	Water thermosensor and circuit OK
5	Are ECU terminals 2Q and 2D voltages OK?	Yes	Replace ECU
		No	Check for short circuit in wiring from water thermosensor to ECU

CODE No.	10 (INTAKE AIR THERMOSENSOR-IN AIRFLOW METER)		
STEP	INSPECTION		ACTION
1	Are there any poor connections in intake air thermosensor circuit?	Yes	Repair or replace connector
		No	Go to next step
2	Is there continuity between intake air thermosensor (in airflow meter) and ECU? Intake air thermosensor / ECU C (B/LG) / 2D G (R/G) / 2P	Yes	Go to next step
		No	Check for open circuit in wiring from intake air thermosensor (in airflow meter) to ECU
3	Is resistance of intake air thermosensor (in airflow meter) OK? Terminal / Temperature / Resistance C (B/RG)—G (R/G) / −20°C (−4°F) / 13.6—18.4 kΩ 20°C (68°F) / 2.21—2.69 kΩ 60°C (140°F) / 493—667Ω	Yes	Go to next step
		No	Replace airflow meter
4	Is same Code No. present after performing after-repair procedure?	Yes	Go to next step
		No	Intake air thermosensor and circuit OK
5	Are engine control unit terminals 2P and 2D voltages OK?	Yes	Replace ECU
		No	Check for short circuit in wiring from intake air thermosensor to ECU

1.6L ENGINE TROUBLESHOOTING GUIDE—MIATA

CODE No. 12 (THROTTLE SENSOR-4A/T)

STEP	INSPECTION		ACTION
1	Check throttle sensor circuit for poor connection	Yes	Repair or replace connector
		No	Go to next step
2	Check wire harness between throttle sensor and ECU for continuity Throttle sensor / ECU: A (LG/R) – 2K (LG/R); B (R/B) – 2M (R/B); D (B/LG) – 2D (B/LG)	Yes	Go to next step
		No	Repair or replace
3	Check if resistances between terminals B (R/B) and D (B/LG) are OK Throttle valve / Resistance: Fully closed – Below 1 kΩ; Fully open – Approx. 5 kΩ	Yes	Go to next step
		No	Adjust or replace throttle sensor
4	Check if ECU terminal 2M voltage is OK	Yes	Replace ECU
		No	Check for short circuit in wiring from throttle sensor to ECU

CODE No. 14 (ATMOSPHERIC PRESSURE SENSOR)

Replace ECU

CODE No. 15 (OXYGEN SENSOR)

Note
- **If Code Nos.15 and 17 are both present, first perform the checking procedure for Code No.17**

STEP	INSPECTION		ACTION
1	Are there any poor connections in oxygen sensor circuit?	Yes	Repair or replace connector
		No	Go to next step
2	Is oxygen sensor output voltage OK?	Yes	Go to next step
		No	Replace oxygen sensor
3	Is there continuity between oxygen sensor and ECU terminal 2N?	Yes	Go to next step
		No	Check for open circuit in wiring from oxygen sensor to ECU
4	Is ECU terminal 2N voltage OK?	Yes	Go to next step
		No	Check for short circuit in wiring from oxygen sensor to ECU
5	Is sensitivity of oxygen sensor OK?	Yes	Replace ECU
		No	Replace oxygen sensor

CODE No. 17 (FEEDBACK SYSTEM)

STEP	INSPECTION		ACTION
1	Warm up engine and run it at 2,500—3,000 rpm for 3 min.		
2	Does monitor lamp of Self-Diagnosis Checker illuminate at idle?	Yes	Go to next step
		No	Check for air leak in vacuum hoses or emission components Check for contaminated oxygen sensor Check for insufficient fuel injection
3	Are spark plugs clean?	Yes	Go to next step
		No	Clean or replace spark plugs
4	Is oxygen sensor voltage OK?	Yes	Go to next step
		No	Replace oxygen sensor
5	Is same Code No. present after performing after-repair procedure?	Yes	Go to next step
		No	Check for short circuit in wiring from oxygen sensor to ECU terminal 2N
6	Is there continuity between oxygen sensor and ECU terminal 2N?	Yes	Go to next step
		No	Check for open circuit in wiring from oxygen sensor to ECU
7	Is ECU terminal 2N voltage OK?	Yes	Replace ECU
		No	Check for short circuit in wiring from oxygen sensor to ECU

Circuit Diagram

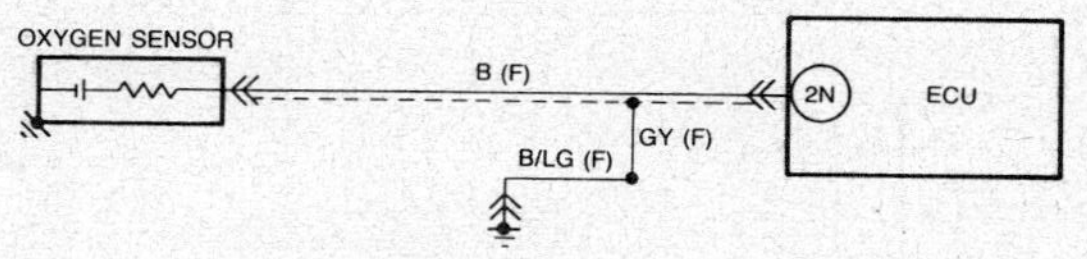

CODE No. 26 (SOLENOID VALVE-PURGE CONTROL)

STEP	INSPECTION		ACTION
1	Are there any poor connections in solenoid valve circuit?	Yes	Repair or replace connector
		No	Go to next step
2	Is resistance of solenoid valve OK? **Resistance: 25 ± 2Ω**	Yes	Go to next step
		No	Replace solenoid valve
3	Is there battery voltage at terminal wire (W/R) of solenoid valve circuit?	Yes	Go to next step
		No	Check for open circuit in wiring from solenoid valve to main relay
4	Is there continuity between solenoid valve and ECU? Solenoid valve / ECU: B (Y/R) – 2X	Yes	Go to next step
		No	Check for open circuit in wiring from solenoid valve to ECU
5	Is ECU terminal (2X) voltage OK?	Yes	Replace ECU
		No	Check for short circuit in wiring from solenoid valve to ECU

CODE No.	34 (ISC VALVE)		
STEP	INSPECTION		ACTION
1	Are there any poor connections in ISC valve circuit?	Yes	Repair or replace connector
		No	Go to next step
2	Is resistance of ISC valve OK? **Resistance: 12 ± 1Ω**	Yes	Go to next step
		No	Replace ISC valve
3	Is there battery voltage at terminal-wire (W/R) of ISC valve circuit?	Yes	Go to next step
		No	Check for open circuit in wiring from ISC valve to main relay
4	Is there continuity between ISC valve and ECU? (ISC valve: B (L/O); ECU: 2W)	Yes	Go to next step
		No	Check for open circuit in wiring from ISC valve to ECU
5	Is ECU terminal 2W voltage OK?	Yes	Replace ECU
		No	Check for short circuit in wiring from ISC valve to ECU

Circuit Diagram

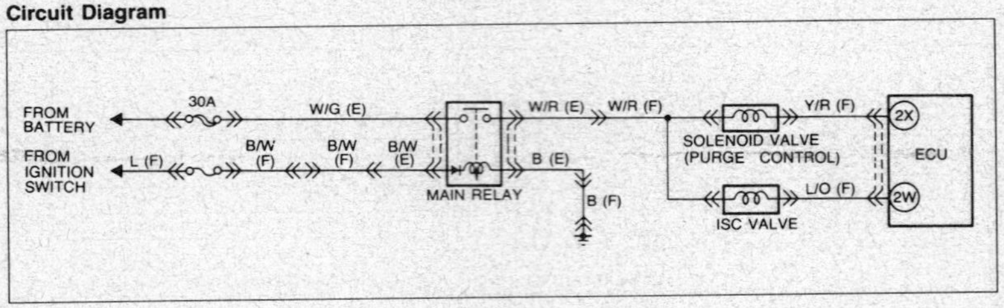

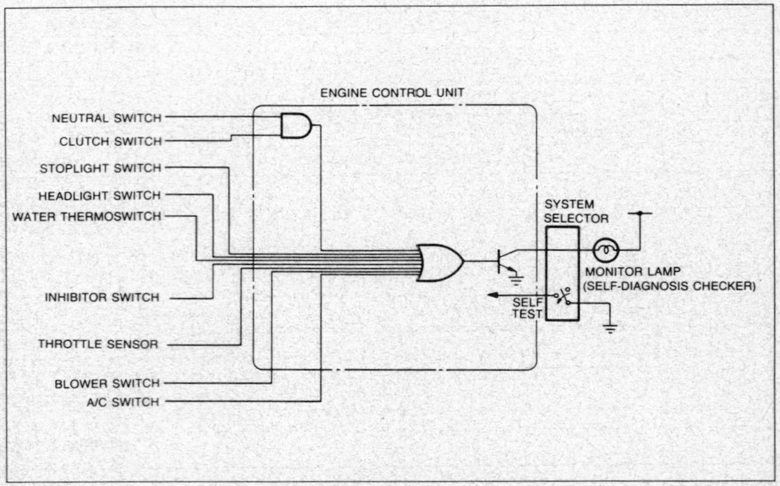

Switch	Self-Diagnosis Checker (Monitor lamp)		Remark
	Light ON	Light OFF	
Clutch switch (M/T)	Pedal released	Pedal depressed	In gear
Neutral switch (M/T)	In gear	Neutral	Clutch pedal released
Inhibitor switch (4A/T)	L, S, D and R range	N and P range	—
Throttle sensor	Pedal depressed (not fully)	Pedal fully depressed	—
Stoplight switch (M/T)	Pedal depressed	Pedal released	—
Headlight switch	ON	OFF	—
Blower switch	ON	OFF	Blower motor position: "medium" "high" or "super high" position
A/C switch	ON	OFF	Fan speed control: Low position
Water thermoswitch	Fan operating	Fan not operating	

1.6L ENGINE TROUBLESHOOTING GUIDE–MIATA

Procedure

Set conditions to deactivate each switch
- All accessories OFF
- Transmission in neutral
- All pedals released

Verify that monitor lamp does not illuminate

NO → Check each switch and related wire harness
- Clutch and Neutral switch
- Idle switch
- Pow switch
- Stoplight switch
- Headlight switch
- Blower switch
- A/C switch
- Water thermoswitch

YES → Check each switch as described

Neutral and Clutch switches (M/T)

Shift transmission into gear
Verify that monitor lamp illuminates with clutch pedal is released

NO → PC:
- Neutral or clutch switch malfunction
- Open circuit in related wire harness
- Engine control terminal 1V malfunction

YES → Depress clutch pedal
Verify that monitor lamp does not illuminate
Return transmission to neutral

NO → PC:
- Clutch switch malfunction

Inhibitor switch (4A/T)

Shift transmission into L, S, D, or P range
Check that monitor lamp illuminates

NO → PC:
- Inhibitor switch malfunction
- Open or short circuit in related wire harness
- ECU terminal 1V voltage incorrect
- ECU malfunction

YES → Inhibitor switch OK
Shift into P or N range

Throttle sensor

Depress accelerator pedal (not fully) and check that monitor lamp illuminates

NO → PC:
- Throttle sensor malfunction
- Short circuit in wire harness from throttle sensor to ECU
- ECU terminal 1N voltage incorrect
- ECU malfunction

YES → Depress acceleration pedal fully and check that monitor lamp does not illuminate

NO → PC:
- Throttle sensor malfunction
- Open circuit in wire harness from throttle sensor to ECU
- ECU terminal 2L (M/T) or 2K (4A/T) voltage incorrect

YES → Throttle sensor OK
Release accelerator pedal

Stoplight switch

Depress brake pedal and verify that monitor lamp illuminates

NO → PC:
- Stoplight switch malfunction
- Open circuit in wire harness from stoplight switch to engine control unit
- Engine control unit terminal 1O malfunction (Refer to page F–128)

YES → Stoplight switch OK
Release brake pedal

Headlight switch

Turn ON headlight switch
Verify that monitor lamp illuminates

NO → PC:
- Headlight switch malfunction
- Open circuit in wire harness from headlight switch to engine control unit
- Engine control unit terminal 1U malfunction

YES → Headlight switch OK
Turn OFF headlight switch

Blower switch

Turn blower switch to "medium" position
Verify that monitor lamp illuminates

NO → PC:
- Blower switch malfunction
- Open circuit in wire harness from blower switch to engine control unit
- Engine control unit terminal 1S malfunction

YES → Blower switch OK
Turn OFF blower switch

A/C switch

Set fan speed control to low position
Turn A/C switch ON, and verify that monitor lamp illuminates

NO → PC:
- A/C switch malfunction
- Blower switch malfunction
- Open circuit in wire harness from A/C switch to blower switch
- Engine control unit terminal 1U malfunction

YES → A/C switch OK
Turn A/C switch and blower switch OFF

Water thermoswitch circuit (not included in switch inspection)

Warning
- **The cooling fan operates when the diagnosis connector terminals TFA and GND are connected with a jumper wire. Use caution.**

Connect diagnosis connector terminals TFA and GND with a jumper wire
Verify that monitor lamp illuminates

NO → PC:
- Relay malfunction
- Open circuit in wire harness from water thermoswitch to engine control unit
- Engine control unit terminal 1R malfunction

YES → Water thermoswitch circuit is OK
Disconnect a jumper wire

VB: Battery voltage

Terminal	Input	Output	Connection to	Test condition	Voltage	Remark
1A	—	—	Battery	Constant	VB	For backup
1B	○		Main relay	Ignition switch OFF	0V	
				Ignition switch ON	VB	
1C	○		Ignition switch (Start position)	While cranking	Approx. 10V	
				Ignition switch ON	0V	
1D		○	Self-Diagnosis Checker (Monitor lamp)	Test switch at "SELF-TEST" Lamp illuminated for 3 sec. after ignition switch OFF→ON	4.5—5.5V	With Self-Diagnosis Checker and System Selector
				Lamp not illuminated after 3 sec.	VB	
				Test switch at "O_2 MONITOR" at idle Monitor lamp illuminated	4.5—5.5V	
				Test switch at "O_2 MONITOR" at idle Monitor lamp not illuminated	VB	
1E		○	Malfunction indicator lamp	Lamp illuminated for 3 sec. after ignition switch OFF→ON	Below 2.5V	With System Selector test switch at "SELF-TEST"
				Lamp not illuminated after 3 sec.	VB	
				Lamp illuminated	Below 2.5V	
				Lamp not illuminated	VB	
1F		○	Self-Diagnosis Checker (Code number)	Buzzer sound for 3 sec. after ignition switch OFF→ON	Below 2.5V	• With Self-Diagnosis Checker and System Selector • With System Selector test switch at "SELF-TEST"
				Buzzer not sounded after 3 sec.	VB	
				Buzzer sounded	Below 2.5V	
				Buzzer not sounded	VB	
1G		○	Igniter	Ignition switch ON	0V	
				Idle	Approx. 0.2V	
1H		○	Igniter	Ignition switch ON	0V	
				Idle	Approx. 0.2V	
1I	—	—	—	—	—	—
1J		○	A/C relay	Ignition switch ON	VB	
				A/C switch ON at idle	Below 2.5V	
				A/C switch OFF at idle	VB	
1K	○		Diagnosis connector	System Selector test switch at "O_2 MONITOR"	VB	
				System Selector test switch at "SELF-TEST"	0V	
1L	—	—	—	—	—	—
1M	—	—	—	—	—	—
1N	○		Throttle sensor (Idle point)	Accelerator pedal released	0V	Ignition switch ON
				Accelerator pedal depressed	VB	
1O	○		Stoplight switch	Brake pedal released	0V	
				Brake pedal depressed	VB	
1P	○		P/S pressure switch	Ignition switch ON	VB	
				P/S ON (at idle)	0V	
				P/S OFF (at idle)	VB	
1Q	○		A/C switch	A/C switch ON (Ignition switch ON)	Below 2.5V	Blower motor ON
				A/C switch OFF (Ignition switch ON)	VB	

VB: Battery voltage

Terminal	Input	Output	Connection to	Test condition	Voltage	Remark
1R	○		Fan switch	Fan operating (Engine coolant temperature over 97°C (207°F) or diagnosis connector terminal TFA grounded)	0V	
				Fan not operating (Idle)	VB	
1S	○		Blower control switch	Blower control switch at mid, high or super high position	Approx. 0V	Ignition switch ON
				Blower control switch OFF or low	Approx. 12V	
1T	—	—	—	—	—	
1U	○		Headlight switch	Headlights ON (Tail, parking, low beam/high beam)	Approx. 12V	
				Headlights OFF	0V	
1V	○		Neutral or clutch switch (M/T)	Neutral position or clutch pedal depressed	0V	
				Other conditions	VB	
			Inhibitor switch (4A/T)	N or P range	0V	
				Other conditions	VB	
2A	—	—	Ground (Injector)	Constant	0V	
2B	—	—	Ground (Output)	Constant	0V	
2C	—	—	Ground (CPU)	Constant	0V	
2D	—	—	Ground (Input)	Constant	0V	
2E	○		Crank angle sensor (Ne-signal)	Ignition switch ON	0V or 5V	
				Idle	Approx. 2V	
2F	—	—	—	—	—	—
2G	○		Crank angle sensor (G-signal)	Ignition switch ON	0V or 5V	
				Idle	Approx. 1.5V	
2H	○		Ground	Constant	0V	
			Open	Federal and Canada spec	Approx. 2V	
2I	○		Igniter	Ignition switch ON	Below 0.5V	
				Idle	Approx. 1V	
2K		○	Airflow meter	Constant	4.5—5.5V	
2L	○		Throttle sensor (Power terminal)	Accelerator pedal released	Approx. 5V	
				Accelerator pedal fully depressed	0V	
2M	○		Throttle sensor (4A/T)	Accelerator pedal released	Approx. 0.5V	
				Accelerator pedal fully depressed	Approx. 4.0V	
2N	○		Oxygen sensor	Ignition switch ON	0V	
				Idle (Cold engine)	0V	
				Idle (After warm-up)	0—1V	
				Increase engine speed (After warm-up)	0.5—1V	
				Deceleration	0—0.4V	
2O	○		Airflow meter	Ignition switch ON	Approx. 3.8V	
				Idle	Approx. 3.3V	
2P	○		Airflow sensor (Intake air thermosensor)	At 20°C (68°F)	Approx. 2.5V	
2Q	○		Water thermosensor	Engine coolant temperature 20°C (68°F)	Approx. 2.5V	
				After warm-up	Approx. 0.4V	
2R	—	—	—	—	—	—

1.6L ENGINE TROUBLESHOOTING GUIDE—MIATA

VB: Battery voltage

Terminal	Input	Output	Connection to	Test condition	Voltage	Remark
2S	—	—	—	—	—	—
2T	—	—	—	—	—	—
2U		○	Injector (Nos.1, 3)	Ignition switch ON	VB	* Engine Signal Monitor: Green and red lights flash
				Idle	Approx. 12V*	
2V		○	(Nos.2, 4)	Deceleration from 3,000 rpm to 1,900 rpm (After warm-up)	Approx. 12V	
2W		○	ISC valve	Ignition switch ON	Approx. 7V	
				Idle	Approx. 9V	
2X		○	Solenoid valve (Purge control)	Ignition switch ON	VB	
				Idle	VB	
2Y	—	—	—	—	—	—
2Z		○	4A/T control unit	Ignition switch ON	0V	

Terminal location

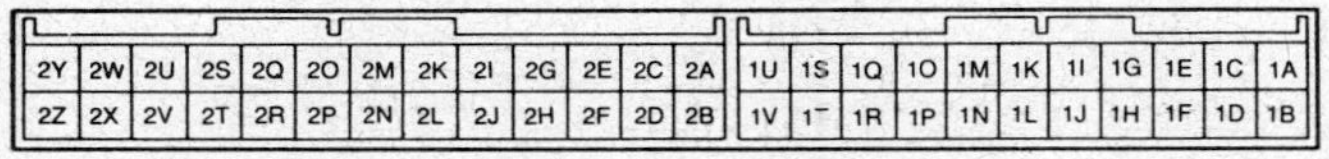

VB: Battery voltage

Terminal	Connection to	Abnormal voltage	Possible cause
1A	Battery	Always 0V (Battery OK)	• ROOM 10A fuse burned • Open circuit in wiring from ROOM 10A fuse to ECU terminal 1A
1B	Main relay	Always 0V	• Main relay malfunction • Open circuit in wiring from main relay to ECU terminal 1B
1C	Ignition switch (Start position)	Always 0V (Starter turns)	• Open circuit in wiring from starter interlock switch to ECU terminal 1C
1D	Self-Diagnosis Checker (Monitor lamp)	Always 0V	• Main relay malfunction • Open circuit in wiring from main relay to diagnosis connector terminal +B • Open or short circuit in wiring from diagnosis connector terminal MEN to ECU terminal 1D
		Always VB	• Poor connection at ECU connector • ECU malfunction
		Always approx. 5V	• ECU malfunction
1E	Malfunction indicator lamp (MIL)	Always below 2.5V (MIL always ON)	• Short circuit in wiring from combination meter to ECU terminal 1E • ECU malfunction
		Always below 2.5V (MIL never ON)	• Open circuit in wiring from combination meter to ECU terminal 1E
		Always VB	• Poor connection at ECU connector • ECU malfunction
1F	Self-Diagnosis Checker (Code No.)	Always below 2.5V (No display on Self-Diagnosis Checker)	• Main relay malfunction • Open circuit in wiring from main relay to diagnosis connector terminal +B
		Always below 2.5V ("88" is displayed and buzzer sounds continuously)	• Open or short circuit in wiring from diagnosis connector terminal FEN to ECU terminal 1F
		Always VB	• Poor connection at ECU connector • ECU malfunction
1G 1H	Igniter	Always 0V	Refer to Code No.01 troubleshooting
1J	A/C relay	Always below 2.5V (A/C does not operate)	• A/C relay malfunction • Open circuit in wiring from main relay to A/C relay • Open circuit in wiring from A/C relay to ECU terminal 1J
		Always below 2.5V (A/C switch OFF but A/C operates)	• Short circuit in wiring from A/C relay to ECU terminal 1J • ECU malfunction
		Always VB	• A/C switch malfunction • Poor connection at ECU connector • ECU malfunction
1K	Diagnosis connector (Terminal TEN)	Always 0V	• Short circuit in wiring from ECU terminal 1K to diagnosis connector terminal TEN
		Always VB	• Open circuit in wiring from ECU terminal 1K to diagnosis connector terminal TEN • Open circuit in wiring from diagnosis connector terminal GND to ground
1N	Throttle sensor (Idle terminal)	Always 0V	• Throttle sensor misadjustment • Short circuit in wiring from ECU terminal 1N to throttle sensor • ECU malfunction
		Always VB	• Throttle sensor misadjustment • Open circuit in wiring from ECU terminal 1N to throttle sensor • Open circuit in wiring from throttle sensor to ground
1O	Stoplight switch	Always 0V (Stoplights OK)	• Open circuit in wiring from stoplight switch to ECU terminal 1O

1.6L ENGINE TROUBLESHOOTING GUIDE—MIATA

V_B: Battery voltage

Terminal	Connection to	Abnormal voltage	Possible cause
1P	P/S pressure switch	Always 0V	• P/S pressure switch malfunction • Short circuit in wiring from ECU terminal 1P to P/S pressure switch • ECU malfunction
		Always V_B	• P/S pressure switch malfunction • Open circuit in wiring from ECU terminal 1P to P/S pressure switch • Open circuit in wiring from P/S pressure switch to ground
1Q	A/C switch	Always 0V (with blower switch ON)	• A/C switch malfunction • Short circuit in wiring from ECU terminal 1Q to A/C switch • Poor connection at ECU connector • ECU malfunction
		Always V_B (with blower switch ON) (Blower fan OK)	• A/C switch malfunction • Open circuit in wiring from ECU terminal 1Q to A/C switch • Open circuit in wiring from A/C switch to blower control switch
1R	Fan switch	Always 0V (Cooling fan OK)	• Open or short circuit in wiring from electric cooling fan relay to ECU terminal 1R • ECU malfunction
1S	Blower control switch	Always 0V (Blower fan OK)	• Short circuit in wiring from blower control switch to ECU terminal 1S • Poor connection at ECU connector • ECU malfunction
		Always V_B (Blower fan OK)	• Open circuit in wiring from blower control switch to ECU terminal 1S
1U	Headlight switch	Always 0V (Headlights OK)	• Open or short circuit in wiring from headlight relay to ECU terminal 1U
1V	Neutral switch Clutch switch (M/T)	Always 0V (M/T)	• Neutral switch malfunction • Clutch switch malfunction • Short circuit in wiring from ECU terminal 1V to neutral or clutch switch
		Always V_B (M/T)	• Neutral switch malfunction • Clutch switch malfunction • Open circuit in wiring from ECU terminal 1V to neutral or clutch switch • Poor connection at ECU connector
	Inhibitor switch (4A/T)	Always 0V (4A/T)	• Inhibitor switch malfunction • Short circuit in wiring from inhibitor switch to ECU terminal 1V
		Always V_B (4A/T)	• Inhibitor switch malfunction • Open circuit in wiring from inhibitor switch to ECU terminal 1V
2A 2B 2C 2D	Ground	More than 0V	• Poor contact at ground terminal • Open circuit in wiring from ECU to ground
2E	Crank angle sensor (Ne-signal)	Always 0V or approx. 5V	• Refer to Code No.02 troubleshooting
2G	Crank angle sensor (G-signal)	Always 0V or approx. 5V	• Refer to Code No.03 troubleshooting
2H	Ground	Approx. 2V	• Open circuit in wiring from ECU terminal 2H to ground
	— (Federal and Canada)	0V	• Short circuit in wiring from ECU terminal 2H to ground

V_B: Battery voltage

Terminal	Connection to	Abnormal voltage	Possible cause
2I	Igniter	Always 0V	• Refer to Code No.01 troubleshooting
2K	Airflow meter	Always 0V	• Short circuit in wiring from ECU terminal 2K to airflow meter • Poor connection at ECU connector • ECU malfunction
		Below 4.5V or above 5.5V	• ECU malfunction
2L	Throttle sensor (Power terminal) (M/T)	Always 0V	• Throttle sensor malfunction • Short circuit in wiring from ECU terminal 2L to throttle sensor • Poor connection at ECU connector • ECU malfunction
		Always approx. 5V	• Throttle sensor misadjustment • Open circuit in wiring from ECU terminal 2L to throttle sensor • Open circuit in wiring from throttle sensor to ground
2M	Throttle sensor (4A/T)	Always constant	• Open circuit in wiring from ECU terminal 2M to throttle sensor • Open circuit in wiring from ECU terminal 2K to throttle sensor • Open circuit in wiring from ECU terminal 2D to throttle sensor
		Always above 1V	• Throttle sensor misadjustment
2N	Oxygen sensor	0V after warm-up	• Refer to Code No.15 troubleshooting
		Always approx. 1V after warm-up	• Refer to Code No.17 troubleshooting
2O	Airflow meter	Always 0V or approx. 5V	• Refer to Code No.08 troubleshooting
2P	Airflow meter (Intake air thermosensor)	Always 0V or approx. 5V	• Refer to Code No.10 troubleshooting
			• Intake air thermosensor malfunction
2Q	Water thermosensor	Always 0V or approx. 5V	• Refer to Code No.09 troubleshooting
			• Water thermosensor malfunction
2U 2V	Injector	Always 0V	• Main relay malfunction Open or short circuit in wiring from injector to ECU terminal 2U or 2V
		Always V_B	• ECU malfunction
2W	ISC valve	Always 0V or V_B	• Refer to Code No.34 troubleshooting
			• ISC valve malfunction
2X	Solenoid valve (Purge control)	Always 0V or V_B	• Refer to Code No.26 troubleshooting
			• Solenoid valve (Purge control) malfunction
2Z	4A/T control unit	Always V_B	• ECU malfunction

ENGINE CONTROL TERMINAL RELATIONSHIP CHART 1992 VERSUS 1990–91 323 AND PROTEGE

Connected to	Terminal New	Terminal Previous	Connected to	Terminal New	Terminal Previous
Headlight switch	1H	1U	Water thermosensor	2E	2Q
Diagnosis connector (TEN terminal)	1I	1K	Throttle sensor	2F	2M
Rear window defroster switch	1J	1T	Throttle sensor Airflow meter	2I	2K
Ground (California) Main relay (Canada)	1K	2H	Distributor (G-signal) [DOHC]	2J	2G
A/C relay	1L	1J	Intake air thermosensor	2K	2P
Vehicle speed sensor	1M	—	Solenoid valve (Purge control)	2O	2X
P/S pressure switch	1N	1P	Ground (Injector)	3A	2A
A/C switch	1O	1Q	Ground (Output)	3B	2B
Blower control switch	1P	1S	Ground (CPU)	3C	2C
Stoplight switch	1Q	1O	Ground (Input)	3D	2D
Cruise control main switch	1S	—	DRL relay (Canada)	3F	1L
Throttle sensor (Idle switch)	1T	1N	Solenoid valve (VICS)	3I	2S
Distributor (Ne-signal)	2A	2E	Solenoid valve (Pressure regulator) [BP]	3M	2T
Airflow meter	2B	2O	ISC valve	3Q	2W
Oxygen sensor	2C	2N	Injector (Nos. 1, 3)	3U	3U
Fan switch	2D	1R	Injector (Nos. 2, 4)	3V	2V

1.6L ENGINE SPECIFICATIONS CHART—MX-3

Item				B6 SOHC
Idle speed			rpm	700 – 800 (750±50)
Ignition timing*			BTDC	9—11° (10±1°)
Throttle body				
Type				Horizontal draft
Throat diameter			mm (in)	50 (2.0)
Fuel pump				
Type				Impeller (in-tank)
Output pressure			kPa (kg/cm², psi)	441—589 (4.5—6.0, 64—85)
Fuel filter				
Type	Low-pressure side			Nylon element (in fuel pump)
	High-pressure side			Paper element
Pressure regulator				
Type				Diaphragm
Regulating pressure			kPa (kg/cm², psi)	265—314 (2.7—3.2, 38—46)
Injector				
Type				High-ohmic, top-feeding
Type of drive				Voltage
Resistance			Ω	12—16 (at 20°C, 68°F)
ISC valve (solenoid valve [idle speed control])				
Solenoid resistance				11—13 (at 20°C, 68°F)
Solenoid valve (purge control)				
Solenoid resistance				23—27 (at 20°C, 68°F)
Crank angle sensor				
Type				Hall element pick-up
Airflow meter				
Type				Linear Slide Type
Resistance Ω	$E_2 \leftrightarrow Vs$	Fully closed		20—600
		Fully open		20—1,000
	$E_2 \leftrightarrow Vc$			200—400
	$E_2 \leftrightarrow THAA$ (Intake air thermosensor)	−20°C (−4°F)		13,600—18,400
		20°C (68°F)		2,210—2,690
		60°C (140°F)		493—667
	$E_1 \leftrightarrow Fc$	Fully closed		∞
		Fully open		0
Water thermosensor				
Resistance kΩ		−20°C (−4°F)		14.6—17.8
		20°C (68°F)		2.2—2.7
		80°C (176°F)		0.29—0.35
Circuit opening relay				
Resistance Ω		STA—E_1		21—43
		B — F_C		109—226
		B — F_P		∞
Fuel Tank				
Capacity			liters (US gal, Imp gal)	50 (13.2, 11.0)
Air cleaner				
Element type				Oil permeated
Accelerator cable				
Free play			mm (in)	1—3 (0.039—0.118)
Fuel				
Specification				Unleaded regular (RON 87 or higher)

* With System Selector (49 B019 9A0) test switch at "SELF TEST."
* TEN terminal of diagnosis connector grounded.

1.6L AND 1.8L ENGINES SPECIFICATIONS CHART PROTEGE AND 323

Item	Engine	B6 SOHC	BP SOHC	BP DOHC
Idle speed*1 *2	rpm	700—800		
Ignition timing*2	BTDC	6°—8°	4°—6°	9°—11°
Fuel pump				
Maximum output pressure	kPa (kg/cm², psi)	441—637 (4.5—6.5, 64—92)		
Fuel filter				
Type	Low-pressure side	Nylon element (in fuel pump)		
	High-pressure side	Paper element		
Pressure regulator				
Regulating pressure	kPa (kg/cm², psi)	265—314 (2.7—3.2, 38—46)		
Injector				
Type		High-ohmic		
Type of drive		Electromechanical		
Resistance	Ω	12—16		
Idle speed control (ISC) valve				
Type		Rotary		
Resistance	Ω	11—13		
Purge control solenoid valve				
Resistance	Ω	23—27		
Water thermosensor				
Resistance kΩ	−20°C (−4°F)	14.6—17.8		
	20°C (68°F)	2.21—2.69		
	40°C (104°F)	1.0—1.3		
	80°C (176°F)	0.29—0.35		
Airflow meter				
Resistance Ω	E2↔Vs Fully closed	20—600		
	E2↔Vs Fully open	20—1,000		
	E2↔Vc	200—400		
	E2↔THAA (Intake air thermosensor) −20°C (−4°F)	13,600—18,400		
	E2↔THAA (Intake air thermosensor) 20°C (68°F)	2,210—2,690		
	E2↔THAA (Intake air thermosensor) 60°C (140°F)	493—667		
	E1↔Fc Fully closed	∞		
	E1↔Fc Fully open	0		
Fuel tank				
Capacity	liters (US gal, Imp gal)	Hatchback 50 (13.2, 11.0), PROTEGE...55 (14.5, 12.1)		
Air cleaner				
Element type		Oil permeated		
Fuel				
Specification		Unleaded regular (RON 91 or higher)		

*1 With parking brake applied (Canada)
*2 TEN terminal of diagnosis connector grounded

1.6L AND 1.8L ENGINES COMPONENT DESCRIPTIONS 323 AND MX-3 WITH 1.6L ENGINE AND PROTEGE

Component	Function	Remark
Air cleaner	Filters air entering throttle body	
Airflow meter	Detects amount of intake air; sends signal to ECU	Intake air thermosensor and fuel pump switch included
Air valve	Supplies bypass air into dynamic chamber when engine is cold	• Engine speed increased to shorten warm-up period • Thermowax type • Installed in dynamic chamber
Atmospheric pressure sensor	Detects atmospheric pressure; sends signal to ECU	Built in ECU
Catalytic converter	Reduces HC, CO, and NOx by chemical reaction	Monolith type
Charcoal canister	Stores fuel tank fumes while engine stopped	
Check valve	Controls pressure in fuel tank	Two-way type
Circuit opening relay	Voltage for fuel pump while engine running	
Clutch switch (MTX)	Detects clutch condition; sends signal to ECU	Switch OFF when clutch pedal released
Diagnosis connector	Concentrated service connector Concentrated terminals: 1. EGI self-diagnostic terminal 2. EC-AT self-diagnostic terminal 3. Test terminal 4. Fuel pump check terminal 5. Cooling fan check terminal 6. Engine rpm output terminal	25-pin (located near left suspension mounting block)
Dynamic chamber	Interconnects all cylinders	
Engine control unit (ECU)	**Detects the following:** 1. A/C operation 2. Air/fuel ratio (Oxygen concentration) 3. ATF thermosensor 4. Atmospheric pressure 5. Braking signal 6. Cranking signal 7. DRL (Daytime Running Light) operation 8. E/L operation 9. Engine coolant temperature 10. Engine speed 11. Ignition ON signal 12. In-gear condition 13. Inhibitor switch (ATX) 14. Intake air amount 15. Intake air temperature 16. No.1 piston TDC of compression 17. P/S operation 18. Shift solenoid (ATX) 19. Test signal (Ignition timing, idle speed, Malfunction code No.) 20. Throttle valve fully closed/fully open condition 21. Throttle valve opening angle 22. Turbine sensor	1. A/C switch 2. Oxygen sensor 4. Atmospheric pressure sensor 5. Stoplight switch (MTX) 6. Ignition switch (START Position) 7. DRL relay (Canada) 8. Blower motor switch, cooling fan relay, headlight switch, and rear window defroster switch 9. Water thermosensor 10. Distributor (Ne-signal) 11. Ignition switch 12. Neutral and clutch switches (MTX) 13. N range, D range, S range, L range switch 14. Airflow meter 15. Intake air thermosensor (In airflow meter) 16. Distributor (G-signal) DOHC) 17. P/S pressure switch 18. 1-2, 2-3, 3-4, Lockup solenoid 19. Diagnosis connector (TEN terminal) 20. Throttle sensor (MTX) 21. Throttle sensor (ATX)

1.6L AND 1.8L ENGINES COMPONENT DESCRIPTIONS 323 AND MX-3 WITH 1.6L ENGINE AND PROTEGE

Component	Function	Remark
Engine control unit (ECU) (Cont'd)	**Controls operation of the following:** 1. A/C (Cut-off) 2. Fail-safe function 3. Fuel injection system 4. Idle speed control 5. Ignition timing control system 6. Monitor function 7. Pressure regulator control system 8. Purge control system 9. VICS	1. A/C relay 2. Self-Diagnosis Checker and MIL 3. Injector 4. Idle speed control (ISC) valve 5. Igniter 6. Monitor lamp (Self-Diagnosis Checker) 7. Solenoid valve (Pressure regulator control) (BP) 8. Purge control solenoid valve 9. Solenoid valve (VICS) (DOHC)
Fuel filter	Filters particles from fuel	
Fuel pump	Provides fuel to injectors	• Operates while engine running • Installed in fuel tank
Igniter	Receives spark signal from ECU and generates high voltage in ignition coil	
Ignition switch (START position)	Engine cranking signal sent to ECU	
Inhibitor switch (ATX)	Detects in-gear condition; sends signal to ECU	Switch ON in N or P range
Injector	Injects fuel into intake port	• Controlled by signals from ECU • High-ohmic injector
Intake air thermosensor	Detects intake air temperature; sends signal to ECU	Installed on dynamic chamber
Idle speed control (ISC) valve	Controls bypass air amount	Controlled by duty signal from ECU
Main relay (FUEL INJ relay)	Supplies electric current to injectors, ECU, etc.	
MIL (Malfunction indicator lamp)	Lamp flashes to indicate malfunction code number of input and output devices	TEN terminal grounded
Neutral switch (MTX)	Detects in-gear condition; sends signal to ECU	Switch ON in neutral
Oxygen sensor	Detects oxygen concentration; sends signal to ECU	Zirconia ceramic and platinum coating
PCV valve	Controls blowby gas introduced into engine	
P/S pressure switch	Detects P/S operation; sends signal to ECU	P/S: ON when steering whee[illegible]ned right or left
Pressure regulator	Adjusts fuel pressure supplied to injectors	
Resonance chamber	Reduces intake air noise	
Resonance duct	Reduces intake air noise	
Separator	Prevents fuel from flowing into charcoal canister	
Stoplight switch	Detects braking operation (deceleration), sends signal to ECU	
Solenoid valve (Purge control)	Controls evaporative fumes from canister to intake manifold	Controlled by duty signal from ECU
Shutter valve actuator (DOHC)	Closes/opens shutter valve to improve torque characteristics	For Variable inertia charging system (VICS*)
Throttle body	Controls intake air quantity	Integrated throttle sensor, dashpot, and ISC valve
Throttle sensor	• Detects throttle valve fully closed/fully opened condition (MTX) • Detects throttle valve opening angle (ATX) • Sends signals to ECU	

Component		Function	Remark
Three-way solenoid valve	**VICS***	Controls vacuum to shutter valve actuator	Cuts vacuum when engine speed above 5,000 rpm
	Pressure regulator control	Controls vacuum to pressure regulator	Cuts vacuum just after starting when engine hot starting
Vacuum chamber (DOHC)		Stores vacuum for use during wide open throttle	For VICS*
Water thermosensor		Detects coolant temperature; sends signal to ECU	

* VICS: Variable Inertia Charging System

1.6L AND 1.8L ENGINES TROUBLESHOOTING GUIDE 323 AND MX-3 WITH 1.6L ENGINE AND PROTEGE

INPUT DEVICES / ENGINE CONDITIONS	APPROXIMATE TIME (BASED ON 10—16°C, 50—60°F AMBIENT)	SENSORS: DISTRIBUTOR (G-SIGNAL)[1] (Ne-SIGNAL)	SENSORS: WATER THERMO-SENSOR	SENSORS: OXYGEN SENSOR	SENSORS: AIRFLOW METER	SENSORS: INTAKE AIR THERMO-SENSOR	THROTTLE SENSOR: V_T TERMINAL (ATX)	THROTTLE SENSOR: POW TERMINAL (MTX)	THROTTLE SENSOR: IDL TERMINAL
CRANKING —COLD ENGINE • COLD AIR • COLD COOLANT	Zero			Signal has no effect on ECU	Signal has no effect on ECU	Signal has no effect on ECU	Signal has no effect on ECU	Signal has no effect on ECU	Signal has no effect on ECU
COLD START —FAST IDLE • COLD AIR • COLD COOLANT	One minute		Cool to warm: Medium voltage (3.5V and dropping)	Sensor cold: Low to high voltage (0—0.9V)	Low volume airflow: High voltage (Above 3.0V)	Cool [Below 20°C (68°F)]: Above 2.3V	Closed throttle: Low voltage (0.2—0.8V)	Closed throttle: Low voltage (4.5—5.5V)	Closed throttle: Low voltage (0V)
COLD DRIVEAWAY —PART THROTTLE • COLD AIR • COLD COOLANT	Two minutes								
WARM DRIVEAWAY —PART THROTTLE • WARM AIR • WARM COOLANT	Three minutes	Ne-SIGNAL: Sends all cylinders TDC signal to ECU	Warm: Medium voltage (Approx 0.7V and dropping)	Sensor hot: switching from high voltage (0.9V)	Moderate volume airflow: Low to medium voltage (1.0—3.5V)		Part throttle: Medium voltage (0.8—3.0V)	Part throttle: Medium voltage (4.5—5.5V)	
HOT CRUISE • WARM AIR • WARM COOLANT		G-SIGNAL[1]: Sends No. 1 cylinder TDC (compression) signal to ECU		to low voltage (0.1V)					Open throttle: High voltage (battery voltage)
HOT ACCELERATION —60% THROTTLE				High voltage (0.9V)	Moderate to strong volume of airflow: Low voltage (0.5—1.5V)	Cool to warm [-20—40°C (-4—104°F)]: 1.5—4.3V			
HOT ACCELERATION —WIDE OPEN THROTTLE	More than four minutes		Hot: Low voltage (Below 0.5V)				Wide open throttle: High voltage (Above 3.0V)	Wide open throttle: High voltage (0V)	
DECELERATION —CLOSED THROTTLE				Low voltage (0V)	Low volume of airflow (Above 3.0V)		Closed throttle: Low voltage (0.2—0.8V)	Closed throttle: Low voltage (4.5—5.5V)	Closed throttle: Low voltage (0V)
HOT CURB IDLE —EXTENDED				Switching from high to low voltage (0.75—0.25V)					
HOT ENGINE SHUTDOWN	—	OFF	OFF	Sensor hot: Low voltage (0.1V) until sensor cools	OFF	OFF	OFF	OFF	OFF

[1] DOHC

ENGINE CONDITIONS	SENSORS: STOP-LIGHT SWITCH	SENSORS: NEUTRAL AND CLUTCH SWITCHES (MTX)	SENSORS: INHIBITOR SWITCH (ATX)	SENSORS: A/C SWITCH	SENSORS: P/S PRESSURE SWITCH	SENSORS: E/L SIGNAL	IGNITION SWITCH: START POSITION	IGNITION SWITCH: ON POSITION	DIAGNOSIS CONNECTOR TEN TERMINAL (TEST TERMINAL)	DRL (Daytime Running Light) CONTROL UNIT (CANADA)
CRANKING —COLD ENGINE • COLD AIR • COLD COOLANT	Signal has no effect on ECU	Signal has no effect on ECU	Signal has no effect on ECU	Signal has no effect on ECU	Signal has no effect on ECU	Signal has no effect on ECU	Sends signal to ECU (Battery voltage)	Signal has no effec on ECU	Signal has no effect on ECU	OFF (Battery voltage)
COLD START —FAST IDLE • COLD AIR • COLD COOLANT	Brake pedal depressed: Sends signal to ECU (Battery voltage)	In neutral: Low voltage signal to ECU (0V)	In N or P range: No signal to ECU (0V)							
COLD DRIVEAWAY —PART THROTTLE • COLD AIR • COLD COOLANT										
WARM DRIVEAWAY —PART THROTTLE • WARM AIR • WARM COOLANT										
HOT CRUISE • WARM AIR • WARM COOLANT	No signal sent to ECU (Below 1.5V)	Driving in any gear: High voltage signal to ECU (Battery voltage)	In D range: No signal to ECU (Battery voltage)	A/C switch ON: Sends signal to ECU (Below 1.5V) A/C switch OFF: No signal to ECU (Battery voltage)	Steering wheel turned: Low voltage signal to ECU (Below 1.5V) Steering wheel straight ahead: High voltage signal to ECU (Battery voltage)	Headlight or rear window defroster switch ON: High voltage signal to ECU (Battery voltage) OFF: Low voltage signal to ECU (Below 1.5V) Blower switch[2] or Electrical fan ON: Low voltage signal to ECU (Below 1.5V) OFF: High voltage signal to ECU (Battery voltage)	No signal to ECU (Below 1.5V)	Sends signal to ECU (Battery voltage)	Terminal not grounded: High voltage Signal to ECU (Battery voltage) Terminal grounded: Low voltage signal to ECU (Below 1.5V)	DRL ON: (0V) DRL OFF: (Battery voltage)
HOT ACCELERATION —60% THROTTLE										
HOT ACCELERATION —WIDE OPEN THROTTLE										
DECELERATION —CLOSED THROTTLE	Brake pedal depressed: Sends signal to ECU (Battery voltage)									
HOT CURB IDLE —EXTENDED		In neutral: Low voltage signal to ECU (0V)	In N or P range: Sends signal to ECU (0V)							
HOT ENGINE SHUTDOWN	OFF	OFF	OFF	OFF	OFF	OFF	OFF	OFF	OFF	OFF

[2] **Blower switch: 2nd position or more**

1.6L AND 1.8L ENGINES TROUBLESHOOTING GUIDE—323 AND MX-3 WITH 1.6L ENGINE AND PROTEGE

OUTPUT DEVICES / ENGINE CONDITIONS	APPROXIMATE TIME (BASED ON 10—16°C, 50—60°F AMBIENT)	INJECTOR: INJECTION	INJECTOR: INJECTION TIMING	AIR VALVE	ISC VALVE	SOLENOID VALVE (PURGE CONTROL)
CRANKING —COLD ENGINE • COLD AIR • COLD COOLANT	Zero		All cylinders each ignition pulse			
COLD START —FAST IDLE • COLD AIR • COLD COOLANT	One minute	Rich			Large amount of bypass air	OFF (Purge cut)
COLD DRIVEAWAY —PART THROTTLE • COLD AIR • COLD COOLANT	Two minutes			Open: coolant temperature: Below 40°C (104°F)		
WARM DRIVEAWAY —PART THROTTLE • WARM AIR • WARM COOLANT	Three minutes	Rich and lean	2-group*1 1-group*2			
HOT CRUISE • WARM AIR • WARM COOLANT						Operates (Duty values [purge gas amount] change)
HOT ACCELERATION —60% THROTTLE		Rich			Small amount of bypass air	
HOT ACCELERATION —WIDE OPEN THROTTLE	More than four minutes					
DECELERATION —CLOSED THROTTLE		Fuel cut		Closed		
HOT CURB IDLE —EXTENDED		Rich and lean	2-group*1 1-group*2			OFF (Purge cut)
HOT ENGINE SHUTDOWN	—	Does not inject			OFF	OFF

*1 DOHC, *2 SOHC, *3 BP

OUTPUT DEVICES / ENGINE CONDITIONS	THREE-WAY SOLENOID VALVE (PRESSURE REGULATOR CONTROL)*3	A/C RELAY	IGNITER	THREE-WAY SOLENOID VALVE (VICS)*1
CRANKING —COLD ENGINE • COLD AIR • COLD COOLANT		OFF (A/C OFF)	IGNITION TIMING: BTDC 6°–8° (B6) 4°–6° (BP) 9°–11° (DOHC)	
COLD START —FAST IDLE • COLD AIR • COLD COOLANT		OFF (A/C OFF: Approx. 5 sec [DOHC] 10 sec. [SOHC])		ON (Vacuum to shutter valve actuator)
COLD DRIVEAWAY —PART THROTTLE • COLD AIR • COLD COOLANT				
WARM DRIVEAWAY —PART THROTTLE • WARM AIR • WARM COOLANT	OFF (Vacuum to pressure regulator)	ON (A/C ON)		
HOT CRUISE • WARM AIR • WARM COOLANT			IGNITION SPARK ADVANCE SIGNAL (Ignition timing: ATDC 10°-BTDC 50°)	
HOT ACCELERATION —60% THROTTLE				Engine speed below 5,000 rpm: OFF (Vacuum cut)
HOT ACCELERATION —WIDE OPEN THROTTLE		OFF (A/C cut)		
DECELERATION —CLOSED THROTTLE		ON (A/C ON)		
HOT CURB IDLE —EXTENDED	After starting: ON during hot start only (Vacuum cut)			ON (Vacuum to shutter valve actuator)
HOT ENGINE SHUTDOWN	OFF	OFF	OFF	OFF

*1 DOHC, *2 SOHC, *3 BP

1.6L AND 1.8L ENGINES RELATIONSHIPS CHART
323 AND MX-3 WITH 1.6L ENGINE AND PROTEGE

INPUT DEVICE \ OUTPUT DEVICE		INJECTOR: FUEL INJECTION AMOUNT	INJECTOR: FUEL INJECTION TIMING	ISC VALVE	SOLENOID VALVE (PURGE CONTROL)	A/C RELAY (A/C CUT-OFF)	SOLENOID VALVE (LOCK-UP CONTROL)	IGNITER (IGNITION TIMING CONTROL)	THREE-WAY SOLENOID VALVE: PRESSURE REGULATOR*1	THREE-WAY SOLENOID VALVE: VICS*2	SELF-DIAGNOSIS CHECKER/ MIL*3 (MALFUNCTION CODE)	SELF-DIAGNOSIS CHECKER (MONITOR LAMP)
DIAGNOSIS CONNECTOR (TEN TERMINAL)				○				○			○	○
IGNITION SWITCH (START POSITION)		○	○	○		○		○	○			
E/L SIGNAL*4				○								○
P/S PRESSURE SWITCH				○		○						
A/C SWITCH				○		○						○
INHIBITOR SWITCH (ATX)		○		○	○	○		○	○			○
NEUTRAL AND CLUTCH SWITCHES (MTX)		○		○	○	○		○	○			○
STOPLIGHT SWITCH		○										○
ATMOSPHERIC PRESSURE SENSOR (IN ECU)		○		○	○	○					○	
THROTTLE SENSOR		○	○	○	○	○		○	○		○ (ATX)	○
INTAKE AIR THERMOSENSOR		○		○	○				○		○	
AIRFLOW METER		○	○		○			○			○	
OXYGEN SENSOR		○			○						○	○
WATER THERMOSENSOR		○		○	○		○	○	○		○	
DISTRIBUTOR	G-SIGNAL*2		○								○	
DISTRIBUTOR	Ne-SIGNAL	○	○	○	○	○		○		○	○	

*1 BP
*2 DOHC
*3 MIL: Malfunction Indicator Lamp
*4 E/L SIGNAL: Blower fan control switch second position or higher, cooling fan operating, headlights ON, or rear window defroster switch ON

1.6L AND 1.8L ENGINES TROUBLESHOOTING GUIDE
323 AND MX-3 WITH 1.6L ENGINE AND PROTEGE

No.	TROUBLESHOOTING ITEM		DESCRIPTION
1	Will not crank or cranks slowly		Refer to Engine Electrical System
2	Cranks normally but will not start	No combustion	Engine cranks at normal speed but shows no sign of firing
3		Partial combustion—When engine cold	Engine cranks at normal speed but shows only partial combustion and will not continue to run
4		Partial combustion—After warm-up	Engine starts normally when cold, but will not start after running and hot soaked
5	Cranks normally but hard to start	Always	Engine cranks at normal speed but requires excessive cranking time before starting
6		When engine is cold	Same condition as No.5 when engine cold Restarts normally after warm-up
7		After warm-up	Same condition as No.5 after running and hot soaked Starts normally when cold
8	Rough idle (Low idle speed)/ Engine stalls at idle	Always	Engine stalls or vibrates excessively at idle
9		Before warm-up	Engine stalls or vibrates excessively at idle during warm-up
10		After warm-up	Engine runs normally at idle during warm-up but vibrates excessively or stalls after warm-up
11		When A/C, P/S, or E/L ON	Engine stalls or vibrates excessively at idle when A/C, P/S, or E/L ON
12	Rough idle/Engine stalls just after starting		Engine stalls or vibrates excessively only just after starting (acceleration from idle)
13	High idle speed after warm-up		Idle speed excessive after warm-up
14	Idle moves up and down/Idle hunting		Engine speeds up and down periodically at idle
15	Engine stalls on deceleration		Engine unexpectedly stops running while decelerating or after deceleration
16	Engine stalls suddenly (Intermittent)		Engine intermittently stops running
17	Hesitates/Stumbles on acceleration		Flat spot occurs just after accelerator depressed or mild jerking occurs during acceleration
18	Surges while cruising		Unexpected change in engine speed which is usually repetitive
19	Lack of power		Performance poor under load when throttle valve wide open Maximum speed reduced
20	Poor acceleration		Performance poor while accelerating
21	Runs rough on deceleration/Backfire		Engine runs rough while decelerating and abnormal combustion occurs in exhaust system
22	Knocking		Abnormal combustion accompanied by audible "pinging" noise
23	Fuel odor		Gasoline odor in cabin
24	Exhaust sulfur smell		Exhaust gas smells abnormal
25	High oil consumption		Oil consumption excessive
26	Poor fuel economy		Fuel economy unsatisfactory
27	MIL always ON		Self-Diagnosis Checker does not indicate malfunction Code No. but MIL always ON
28	MIL never ON		Self-Diagnosis Checker indicates malfunction
29	A/C does not work		Blower fan operates but magnetic clutch does not operate

1.6L AND 1.8L ENGINES TROUBLESHOOTING GUIDE—323 AND MX-3 WITH 1.6L ENGINE AND PROTEGE

2	CRANKS NORMALLY BUT WILL NOT START (NO COMBUSTION)
DESCRIPTION	• Engine cranks at normal speed but shows no sign of "firing" • Battery in normal condition • Throttle valve not held fully open while cranking (Dechoke system not operate) • Fuel in tank

[TROUBLESHOOTING HINTS]
Because of no combustion, possibly no fuel is injected to engine or no ignition at all cylinders

1. No spark
   - Ignition control malfunction
   - Ignition system component malfunction
2. No fuel injection
   - Fuel pump inoperative
   - Injectors inoperative
3. Low fuel line pressure
4. Low engine compression

STEP	INSPECTION		ACTION
1	Check if strong blue spark is visible at disconnected high tension lead 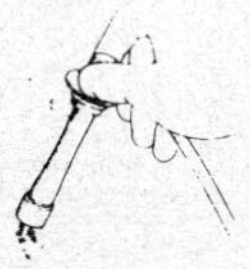	Yes	Go to Step 3
		No	Go to Step 2
2	Check if "00" is displayed on Self-Diagnosis Checker with ignition switch ON 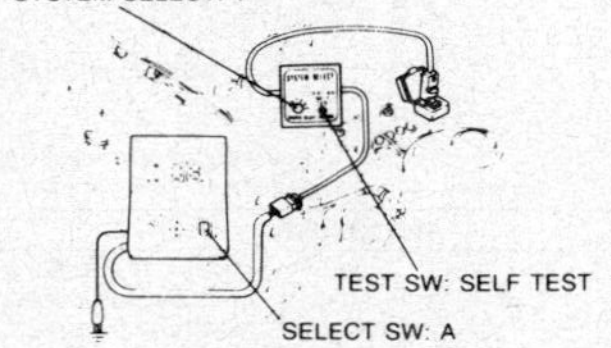	Yes	Check ignition system (Refer to Troubleshooting "Misfire")
		No	**Service Code No. displayed** Check for cause (Refer to specified check sequence)
			"88" flashes Check ECU terminal 1F voltage **Specification: Battery voltage (Ignition switch ON)** ⇨ If OK, replace ECU ⇨ If not OK, check wiring between ECU and Self-Diagnosis Checker
3	Connect diagnosis connector terminals F/P and GND with jumper wire and check for fuel pump operating sound with ignition switch ON 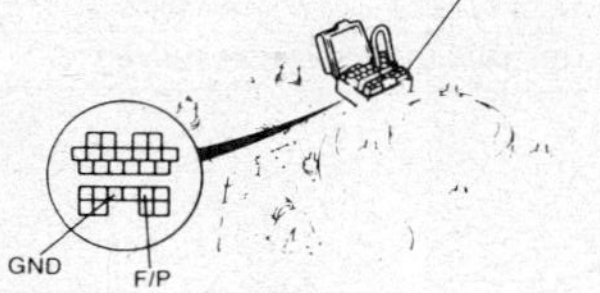	Yes	Check if engine starts in this condition ⇨ If starts, check circuit opening relay ⇨ If does not start, go to Step 5
		No	Go to next step

STEP	INSPECTION		ACTION
4	Check if battery voltage exists at fuel pump connector B/P wire with ignition switch ON 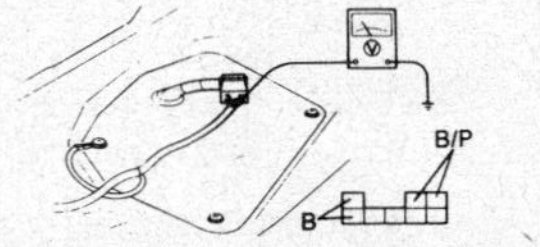	Yes	Check continuity of fuel pump (Between terminals B/P and B)
		No	Check circuit opening relay
5	Check for injector operating sound while cranking engine	Yes	Go to Step 7
		No	Go to next step
6	Check if battery voltage exists at injector connector (W/R) wire with ignition switch ON 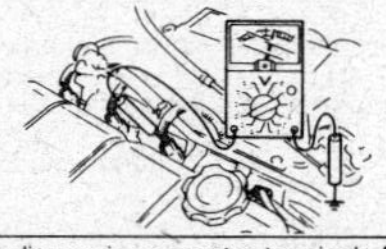	Yes	Check ECU terminal voltages MTX...1N, 2A, 2U, 2V ATX... 1T, 3A, 3U, 3V
		No	Check for open circuit in wiring between main relay (FUEL INJ relay) and injector
7	Connect diagnosis connector terminals F/P and GND with jumper wire and check for correct fuel line pressure with ignition switch ON **Fuel Line pressure:** **265—314 kPa (2.7—3.2 kg/cm², 38—46 psi)** 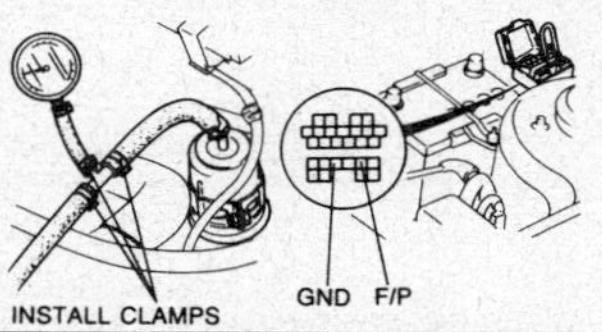	Yes	Go to next step
		No	**Low pressure** Check fuel line pressure while pinching fuel return hose ⇨ If fuel line pressure quickly increases, check pressure regulator ⇨ If fuel line pressure gradually increases, check for clogging between fuel pump and pressure regulator If not clogged, check fuel pump maximum pressure
			High pressure Check if fuel return hose is clogged or restricted ⇨ If OK, replace pressure regulator ⇨ If not OK, repair or replace
8	Check for correct engine compression **Engine compression:** • **BP SOHC** **834 kPa (8.5 kg/cm², 121 psi)-300 rpm** • **BP DOHC** **883 kPa (9.0 kg/cm², 128 psi)-300 rpm** • **B6** **932 kPa (9.5 kg/cm², 135 psi)-300 rpm**	Yes	Go to next step
		No	Check engine condition • Worn piston, piston rings or cylinder wall • Defective cylinder head gasket • Distorted cylinder head • Improper valve seating • Valve sticking in guide
9	Check if all spark plugs are OK WEAR AND CARBON BUILDUP ☞ page G–18 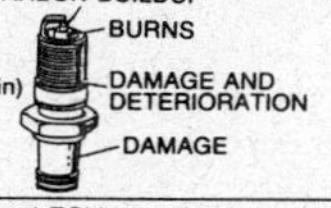	Yes	Go to next step
		No	Clean or replace
10	Try known good ECU and check if condition improves		

3	CRANKS NORMALLY BUT WILL NOT START (PARTIAL COMBUSTION) — WHEN ENGINE COLD
DESCRIPTION	• Engine cranks at normal speed but shows only partial combustion and will not continue to run • Battery in normal condition • Fuel in tank

[TROUBLESHOOTING HINTS]
① Air/Fuel mixture too rich
- Air cleaner element clogged
- Airflow meter stuck

② Air/Fuel mixture too lean
- Fuel injection control malfunction (Correction for coolant temperature)
- Low fuel line pressure
- Air leakage of intake air system

③ Low engine compression

STEP	INSPECTION		ACTION
1	Check if "00" is displayed on Self-Diagnosis Checker with ignition switch ON 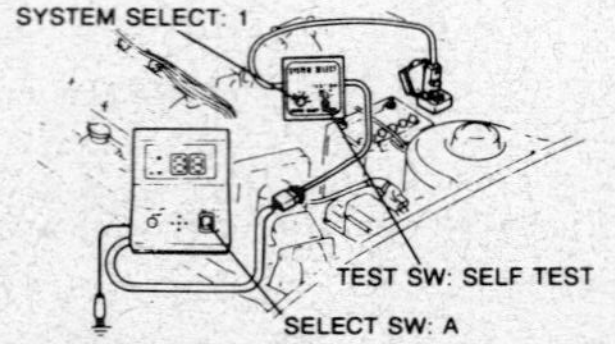	Yes	Go to next step
		No	**Service Code No. displayed** Check for cause (Refer to specified check sequence)
			"88" flashes Check ECU terminal 1F voltage **Specification: Battery voltage (Ignition switch ON)** ⇨ If OK, replace ECU ⇨ If not OK, check wiring between ECU and Self-Diagnosis Checker
2	Check if strong blue spark is visible at each disconnected high-tension lead 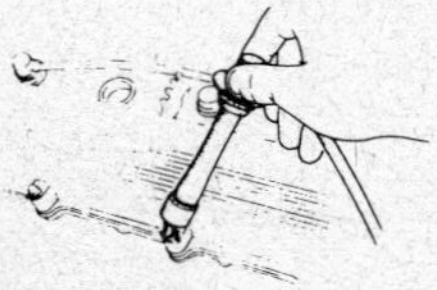	Yes	Go to next step
		No	Check distributor cap and rotor
3	Connect diagnosis connector terminals F/P and GND with jumper wire and check for correct fuel line pressure with ignition switch ON **Fuel line pressure:** **265—314 kPa (2.7—3.2 kg/cm², 38—46 psi)** 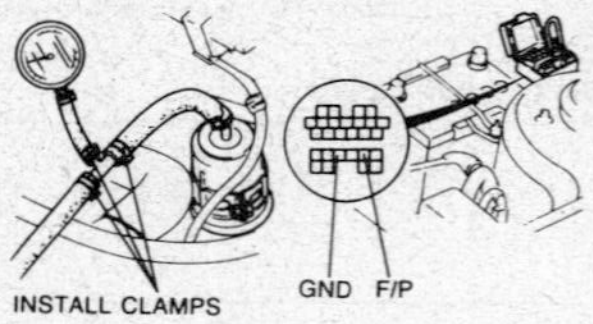	Yes	Go to next step
		No	**Low pressure** Check fuel line pressure while pinching fuel return hose ⇨ If fuel line pressure quickly increases, check pressure regulator ⇨ If fuel line pressure gradually increases, check for clogging between fuel pump and pressure regulator If not clogged, check fuel pump maximum pressure **High pressure** Check if fuel return hose is clogged or restricted ⇨ If OK, replace pressure regulator ⇨ If not OK, repair or replace

STEP	INSPECTION		ACTION
4	Check if ECU terminal voltages are OK MTX...Especially 1C, 2D and 2Q ATX... Especially 1C, 2E and 3D	Yes	Go to next step
		No	Check for cause
5	Check for air leakage of intake air system components	Yes	Repair or replace
		No	Go to next step
6	Check if engine starts when water thermosensor is disconnected	Yes	Check water thermosensor
		No	Go to next step
7	Manually check if airflow meter moves smoothly from fully closed to fully open 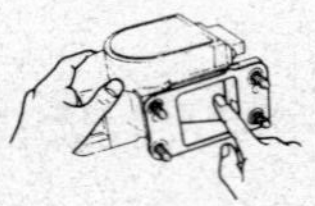	Yes	Go to next step
		No	Repair or replace
8	Check for correct engine compression **Engine compression:** • **BP SOHC** **834 kPa (8.5 kg/cm², 121 psi)-300 rpm** • **BP DOHC** **883 kPa (9.0 kg/cm², 128 psi)-300 rpm** • **B6** **932 kPa (9.5 kg/cm², 135 psi)-300 rpm**	Yes	Go to next step
		No	Check engine condition • Worn piston, piston rings or cylinder wall • Defective cylinder head gasket • Distorted cylinder head • Improper valve seating • Valve sticking in guide
9	Check if spark plugs are OK 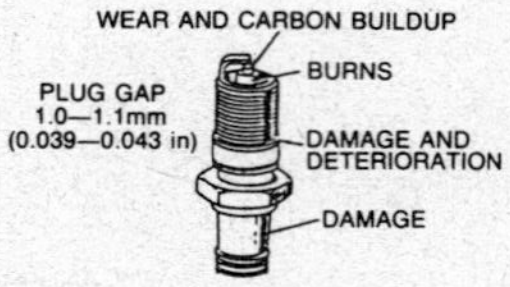	Yes	Go to next step
		No	Clean, or replace
10	Try known good ECU and check if condition improves		

4	CRANKS NORMALLY BUT WILL NOT START (PARTIAL COMBUSTION) — AFTER WARM-UP
DESCRIPTION	• Engine cranks at normal speed and shows partial combustion but will not continue to run after running and hot soaked • Battery in normal condition • Engine starts normally when cold

[TROUBLESHOOTING HINTS]

① Air/Fuel mixture too rich
- Insufficient fuel injection control (Correction for coolant temperature)
- Injector fuel leakage

② Vapor lock
- Fuel pressure not held in fuel line after engine stopped
- High RVP (winter) fuel used in warm weather

③ Pressure regulator control system malfunction [BP]

STEP	INSPECTION		ACTION
1	Warm-up engine to normal operating temperature and stop it Connect diagnosis connector terminals F/P and GND with jumper wire for 3 minutes with ignition switch ON, then check if engine starts 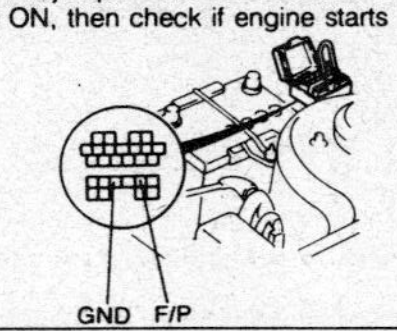	Yes	Change fuel to another brand
		No	**BP engine** Go to next step **B6 engine** Go to Step 3
2	Remove vacuum hose from pressure regulator and plug it Check if engine starts 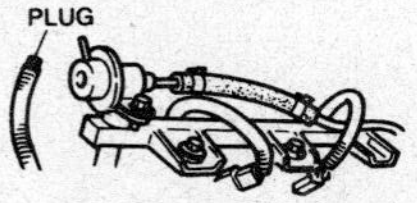	Yes	Check pressure regulator control system
		No	Go to next step
3	Connect diagnosis connector terminals F/P and GND with jumper wire and check for correct fuel line pressure with ignition switch ON **Fuel line pressure:** **265—314 kPa (2.7—3.2 kg/cm², 38—46 psi)** 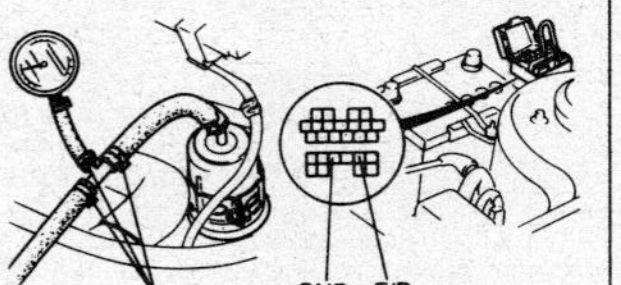	Yes	Go to next step
		No	**Low pressure** Check fuel line pressure while pinching fuel return hose ⇨ If fuel line pressure quickly increases, check pressure regulator ⇨ If fuel line pressure gradually increases, check fuel line and filter for clogging If not clogged, check fuel pump maximum pressure **High pressure** Check if fuel return hose is clogged or restricted ⇨ If OK, replace pressure regulator ⇨ If not OK, repair or replace
4	In same condition as Step 2, check if fuel line pressure is held after ignition switch is turned OFF **Fuel line pressure: More than 147 kPa (1.5 kg/cm², 21 psi) for 5 min.**	Yes	Go to Step 6
		No	Go to next step

STEP	INSPECTION		ACTION
5	Plug outlet of pressure regulator and check if fuel line pressure is held after ignition switch is turned OFF **Fuel line pressure:** **More than 147 kPa (1.5 kg/cm², 21 psi) for 5 min.**	Yes	Replace pressure regulator
		No	Check fuel pump hold pressure ⇨ If OK, check injector for fuel leakage ⇨ If not OK, replace fuel pump
6	Disconnect water thermosensor connector and check if engine starts 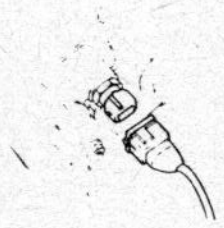	Yes	Check water thermosensor ⇨ If OK, check connections of water thermosensor and ECU 2Q terminal (MTX) ⇨ If OK, check connections of water thermosensor and ECU 2E terminal (ATX) ⇨ If not OK, replace water thermosensor
		No	Go to next step
7	Check if "00" is displayed on Self-Diagnosis Checker with ignition switch ON 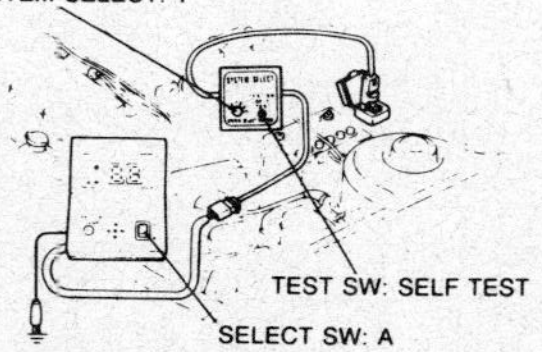	Yes	Go to next step
		No	**Service Code No. displayed** Check for cause (Refer to specified check sequence) **"88" flashes** Check ECU terminal 1F voltage **Specification: Battery voltage (Ignition switch ON)** ⇨ If OK, replace ECU ⇨ If not OK, check wiring between ECU and Self-Diagnosis Checker
8	Check if ECU terminal voltages are OK MTX...Especially 1C, 2D and 2Q ATX...Especially 1C, 2E and 3D	Yes	Go to next step
		No	Check for cause
9	Try known good ECU and check if condition improves	Yes	Replace ECU
		No	Change fuel to another brand

5	CRANKS NORMALLY BUT HARD TO START — ALWAYS
DESCRIP-TION	• Engine cranks at normal speed but requires excessive cranking time before starting • Battery in normal condition • Engine runs normally at idle [If idle condition not OK, refer to "Rough idle" (Nos. 8—12)]

[TROUBLESHOOTING HINTS]

① Air/Fuel mixture too lean
- Fuel injection control malfunction (Correction for coolant temperature)
- Low fuel line pressure
- Air leakage

② Air/Fuel mixture too rich
- Air cleaner element clogged
- Airflow meter stuck

③ Poor ignition spark

STEP	INSPECTION		ACTION
1	Check if "00" is displayed on Self-Diagnosis Checker with ignition switch ON 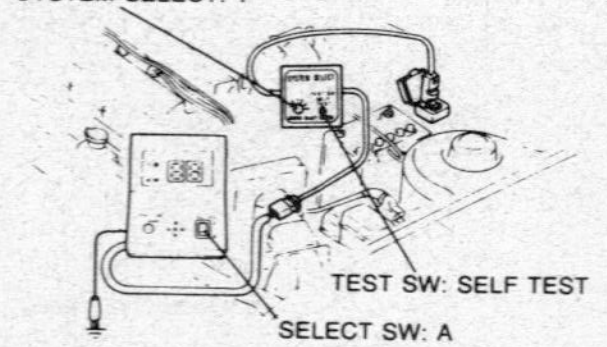	Yes	Go to next step
		No	**Service Code No. displayed** Check for cause (Refer to specified check sequence) **"88" flashes** Check ECU terminal 1F voltage **Specification: Battery voltage (Ignition switch ON)** ⇨ If OK, replace ECU ⇨ If not OK, check wiring between ECU and Self-Diagnosis Checker
2	Check for correct intake manifold vacuum at idle **Vacuum: More than 450 mmHg (17.7 inHg)**	Yes	Go to next step
		No	Check for air leakage of intake air system components
3	Check if air cleaner element is clean	Yes	Go to next step
		No	Replace air cleaner element
4	Check if engine starts easily when throttle valve quarter open	Yes	Check throttle valve for carbon deposit and then go to Step 6
		No	Go to next step
5	Check for correct fuel line pressure at idle **Fuel line pressure:** **265—314 kPa (2.7—3.2 kg/cm², 38—46 psi)** **(Vacuum hose to pressure regulator disconnected)** 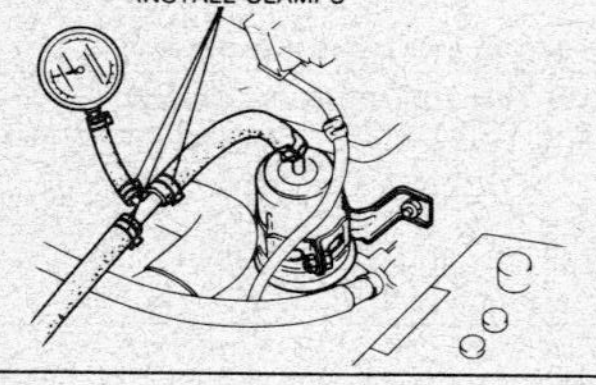	Yes	Go to next step
		No	**Low pressure** Check fuel line pressure while pinching fuel return hose ⇨ If fuel line pressure quickly increases, check pressure regulator ⇨ If fuel line pressure gradually increases, check for clogging between fuel pump and pressure regulator If not clogged, check fuel pump maximum pressure

STEP	INSPECTION		ACTION
6	Disconnect ISC valve connector at idle and check if engine speed increases 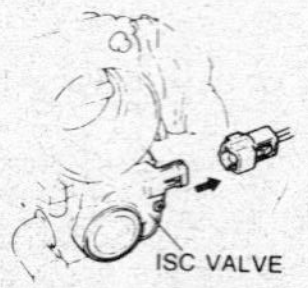	Yes	Go to next step
		No	Replace ISC valve
7	Check if ECU terminal voltages are OK MTX...Especially 1C, 2D and 2Q ATX... Especially 1C, 2E and 3D	Yes	Go to next step
		No	Check for cause
8	Connect diagnosis connector terminals F/P and GND with jumper wire and check if engine starts normally 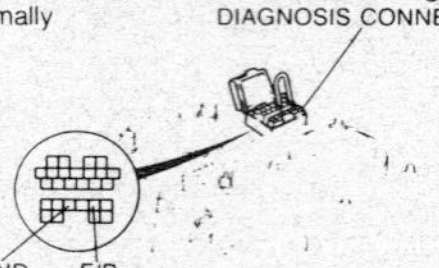	Yes	Check circuit opening relay ⇨ If relay OK, repair or replace wire harness ⇨ If relay not OK, replace relay
		No	Go to next step
9	Connect System Selector to diagnosis connector and set Test Switch to "SELF TEST" and check for correct ignition timing at idle after warm-up **Ignition timing (BTDC)** **BP DOHC: 9°—11°** **BP SOHC: 4°—6°** **B6 (323) : 6°—8°** **B6 (MX-3): 9°—11°** 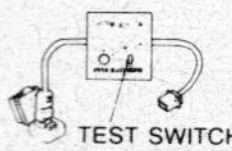	Yes	Go to next step
		No	Adjust ignition timing
10	Check for correct engine compression **Engine compression (Minimum):** • **BP DOHC** **883 kPa (9.0 kg/cm², 128 psi)-300 rpm** • **BP SOHC** **834 kPa (8.5 kg/cm², 121 psi)-300 rpm** • **B6** **932 kPa (9.5 kg/cm², 135 psi)-300 rpm**	Yes	Go to next step
		No	Check engine condition • Worn piston, piston rings or cylinder wall • Defective cylinder head gasket • Distorted cylinder head • Improper valve seating • Valve sticking in guide
11	Try known good ECU and check if condition improves		

1.6L AND 1.8L ENGINES TROUBLESHOOTING GUIDE—323 AND MX-3 WITH 1.6L ENGINE AND PROTEGE

6	CRANKS NORMALLY BUT HARD TO START — WHEN ENGINE COLD		
DESCRIP-TION	• Engine cranks at normal speed but requires excessive cranking time before starting • Battery in normal condition • Restarts OK after warm-up • Engine runs normally at idle [If idle condition is not OK, refer to "Rough idle" (Nos. 8—12)]		
[TROUBLESHOOTING HINTS]	① Air/Fuel mixture too rich • Airflow meter sticking • Air cleaner element clogged • Idle speed control malfunction ② Air/Fuel mixture too lean • Fuel injection control malfunction (Correction for coolant temperature) ③ Poor atomization • Low RVP (summer) fuel is used in cold weather		
STEP	INSPECTION		ACTION
1	Check if "00" is displayed on Self-Diagnosis Checker with ignition switch ON 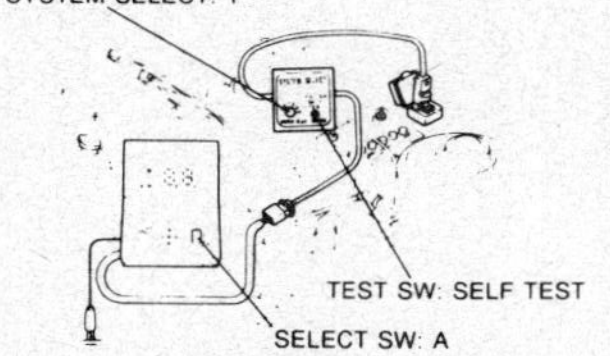 	Yes	Go to next step
		No	**Service Code No. displayed** Check for cause (Refer to specified check sequence)
			"88" flashes Check ECU terminal 1F voltage **Specification: Battery voltage (Ignition switch ON)** ⇨ If OK, replace ECU ⇨ If not OK, check wiring between ECU and Self-Diagnosis Checker
2	Check if ECU terminal voltages are OK MTX...Especially 1C, 2D and 2Q ATX... Especially 1C, 2E and 3D	Yes	Go to next step
		No	Check for cause
3	Check if engine starts easily when throttle valve quarter open 	Yes	Check if ISC valve is OK ⇨ If OK, check air valve ⇨ If not OK, replace ISC valve
		No	Go to next step
4	Check for correct intake manifold vacuum at idle **Vacuum: More than 450 mmHg (17.7 inHg)**	Yes	Go to next step
		No	Check for air leakage of intake air system components
5	Check if air cleaner element is clean	Yes	Go to next step
		No	Replace air cleaner element
6	Disconnect ISC valve connector at idle and check if engine speed increases  	Yes	Go to next step
		No	Replace ISC valve

7	CRANKS NORMALLY BUT HARD TO START — AFTER WARM-UP		
DESCRIP-TION	• Engine cranks at normal speed but requires excessive cranking time before starting after running and hot soaked • Battery in normal condition • Engine starts normally when cold • Engine runs normally at idle [If idle condition not OK, refer to "Rough idle" (Nos. 8—12)]		
[TROUBLESHOOTING HINTS]	① Air/Fuel mixture too rich • Fuel injection control malfunction • Injector fuel leakage ② Vapor lock • Fuel pressure not held in fuel line after engine stopped • High RVP (winter) fuel used in warm weather • Pressure regulator control system malfunction [BP]		
STEP	INSPECTION		ACTION
1	Check if "00" is displayed on Self-Diagnosis Checker with ignition switch ON 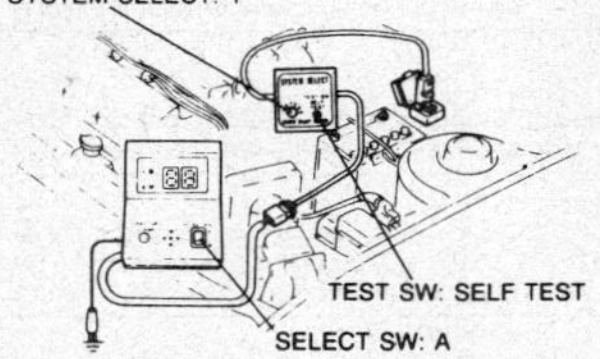 	Yes	**BP engine** Go to next step
			B6 engine Go to Step 3
		No	**Service Code No. displayed** Check for cause (Refer to specified check sequence)
			"88" flashes Check ECU terminal 1F voltage **Specification: Battery voltage (Ignition switch ON)** ⇨ If OK, replace ECU ⇨ If not OK, check wiring between ECU and Self-Diagnosis Checker
2	Remove vacuum hose from pressure regulator and plug hose Check if engine starts normally 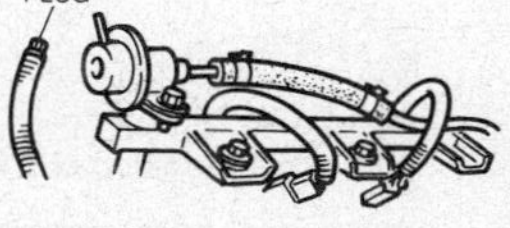 	Yes	Check pressure regulator control system
		No	Go to next step
3	Check if ECU terminal voltages are OK MTX...Especially 1C, 2D, 2Q, and 2T [BP] ATX... Especially 1C, 2E, 3D, and 3M [BP]	Yes	Go to next step
		No	Check for cause
4	Run engine at idle and check if fuel line pressure is held after ignition switch turned OFF **Fuel line pressure: More than 147 kPa (1.5 kg/cm², 21 psi) for 5 min.** 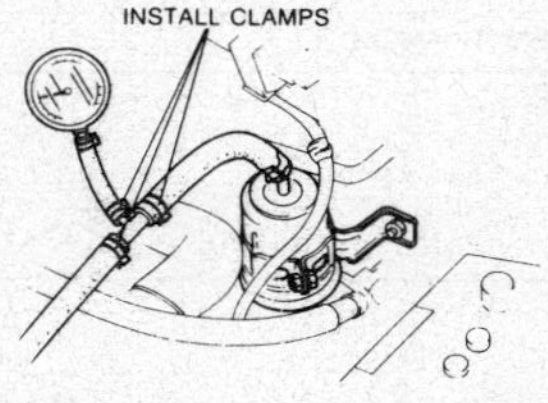 	Yes	Go to next step
		No	Plug outlet of pressure regulator and check if fuel line pressure is held after ignition switch is turned OFF ⇨ If OK, replace pressure regulator ⇨ If not OK, check fuel pump hold pressure If fuel pump is OK, check injectors for fuel leakage

STEP	INSPECTION		ACTION
5	Warm-up engine to normal operating temperature and stop it Connect diagnosis connector terminals F/P and GND with jumper wire for 3 minutes with ignition switch ON, then check if engine starts easily 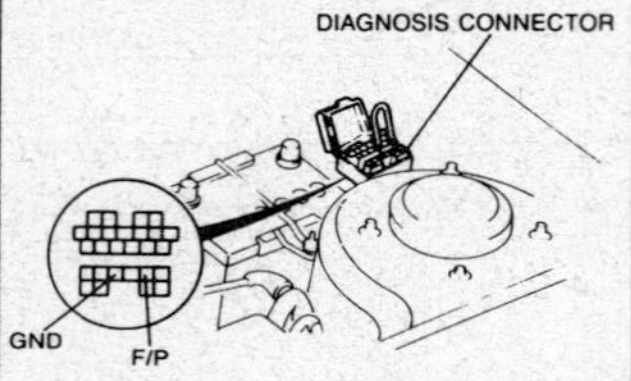 	Yes	Change fuel to another brand
		No	Go to next step
6	Try known good ECU and check if condition improves	Yes	Replace ECU
		No	Change fuel to another brand

8	ROUGH IDLE/ENGINE STALLS AT IDLE — ALWAYS		
DESCRIPTION	• Engine starts normally but stalls or vibrates excessively at idle in every condition		
[TROUBLESHOOTING HINTS] ① Air/Fuel mixture too lean • Air leakage • Fuel injection control malfunction • Low fuel line pressure ② One or more injectors clogged or not operating ③ Low intake air amount	④ Incorrect idle speed • Idle speed misadjustment • Idle speed control malfunction ⑤ One or more spark plugs not firing ⑥ Low engine compression ⑦ Airflow meter stuck		
STEP	**INSPECTION**		**ACTION**
1	Check if "00" is displayed on Self-Diagnosis Checker with ignition switch ON 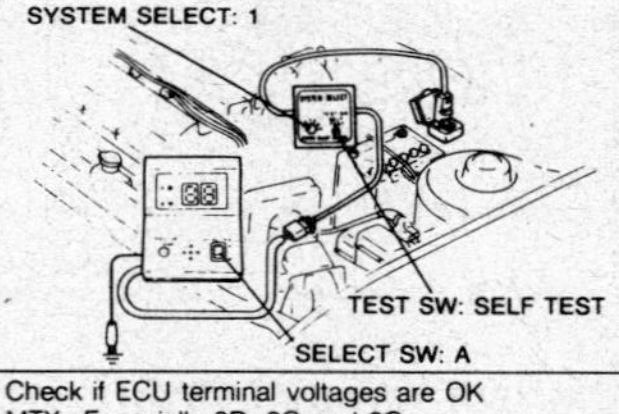 	Yes	Go to next step
		No	**Service Code No. displayed** Check for cause (Refer to specified check sequence) **"88" flashes** Check ECU terminal 1F voltage **Specification: Battery voltage (Ignition switch ON)** ⇨ If OK, replace ECU ⇨ If not OK, check wiring between ECU and Self-Diagnosis Checker
2	Check if ECU terminal voltages are OK MTX...Especially 2D, 2O and 2Q ATX... Especially 2B, 2E and 3D	Yes	Go to next step
		No	Check for cause
3	Disconnect high-tension lead at idle and check if engine speed decreases equally at each cylinder	Yes	Go to Step 8
		No	Go to next step
4	Check for injector operating sound at idle	Yes	Go to Step 6
		No	Go to Step 5
5	Check if battery voltage exists at injector connector (W/R) wire 	Yes	Check if injector resistance is OK **Resistance: Approx. 12—16Ω** ⇨ If OK, check wiring between ECU and injector ⇨ If not OK, replace injector
		No	Check wiring between ECU and injector
6	Check if strong blue spark is visible at disconnected high-tension lead 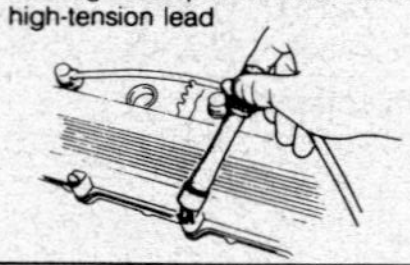	Yes	Go to next step
		No	Check high-tension lead ⇨ If OK, check distributor cap and rotor ⇨ If not OK, replace high-tension lead

STEP	INSPECTION		ACTION
7	Check if spark plugs are OK 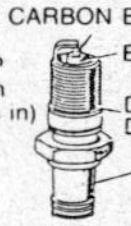 	Yes	Check for correct engine compression ⇨ If OK, replace injector ⇨ If not OK, check for cause
		No	Clean or replace
8	Check for correct fuel line pressure at idle **Fuel line pressure:** **265—314 kPa (2.7—3.2 kg/cm², 38—48 psi)** **(Vacuum hose to pressure regulator disconnected)** 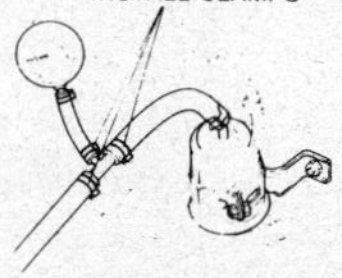 	Yes	Go to next step
		No	**Low pressure** Check fuel line pressure while pinching fuel return hose ⇨ If fuel line pressure quickly increases, check pressure regulator ⇨ If fuel line pressure gradually increases, check for clogging between fuel pump and pressure regulator. If not clogged, check fuel pump maximum pressure
9	Manually check if airflow meter moves smoothly from fully closed to fully open 	Yes	Go to next step
		No	Repair or replace
10	Check for air leakage at intake air system components	Yes	Repair or replace
		No	Go to next step
11	Check for correct engine compression **Engine compression (Minimum):** • **BP SOHC** **834 kPa (8.5 kg/cm², 121 psi)-300 rpm** • **BP DOHC** **883 kPa (9.0 kg/cm², 128 psi)-300 rpm** • **B6** **932 kPa (9.5 kg/cm², 135 psi)-300 rpm**	Yes	Go to next step
		No	Check for cause
12	Connect System Selector to diagnosis connector and set Test Switch to "SELF TEST" and check for correct ignition timing at idle after warm-up **Ignition timing (BTDC)** **BP DOHC: 9°—11°** **BP SOHC: 4°—6°** **B6 (323) : 6°—8°** **B6 (MX-3): 9°—11°**	Yes	Try known good ECU and check if condition improves
		No	Adjust ignition timing

9	ROUGH IDLE/ENGINE STALLS AT IDLE — BEFORE WARM-UP
DESCRIPTION	• Engine speed low or engine stalls or vibrates excessively at idle during warm-up

[TROUBLESHOOTING HINTS]

① Low intake air amount
- Airflow meter stuck
- Air cleaner element clogged
- Air valve

② Low fuel injection control
- Fuel injection control malfunction (Correction for coolant temperature)

③ Poor atomization
- Low RVP (summer) fuel used in cold weather

STEP	INSPECTION		ACTION
1	Check if "00" is displayed on Self-Diagnosis Checker with ignition switch ON 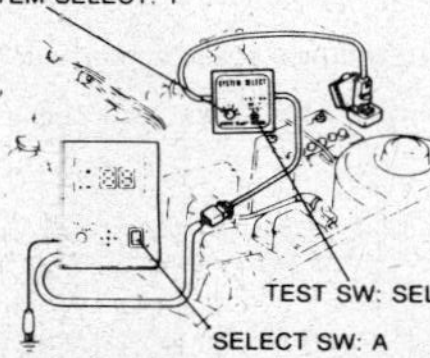 	Yes	Go to next step
		No	**Service Code No. displayed** Check for cause (Refer to specified check sequence) **"88" flashes** Check ECU terminal 1F voltage **Specification: Battery voltage (Ignition switch ON)** ⇨ If OK, replace ECU ⇨ If not OK, check wiring between ECU and Self-Diagnosis Checker
2	Check for correct intake manifold vacuum at idle after warm-up **Vacuum: More than 450 mmHg (17.7 inHg)**	Yes	Go to next step
		No	Check for air leakage of intake air system components
3	Check if air cleaner element is clean	Yes	Go to next step
		No	Replace air cleaner element
4	Check if ECU terminal voltages are OK MTX...Especially 2D, 2O and 2Q ATX...Especially 2B, 2E and 3D	Yes	Go to next step
		No	Check for cause
5	Disconnect ISC valve connector and check if engine speed increases after warm-up 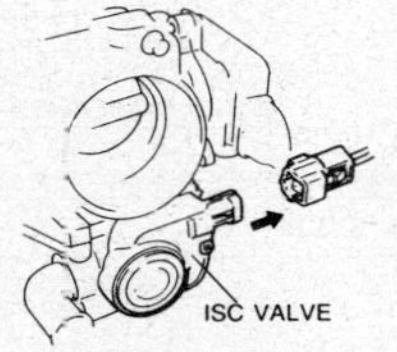 	Yes	Go to next step
		No	Replace ISC valve

STEP	INSPECTION		ACTION
6	Check for correct water thermosensor resistance 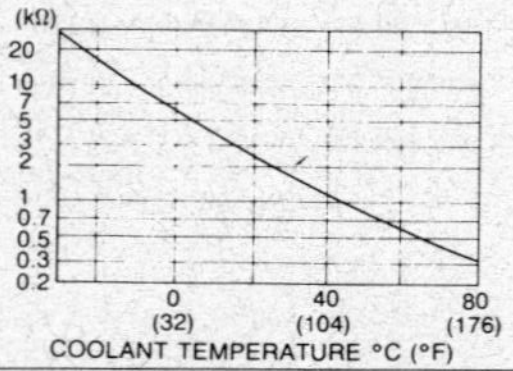	Yes	Go to next step
		No	Replace water thermosensor
7	Connect System Selector to diagnosis connector and set Test Switch to "SELF TEST" when engine is cold Check if engine speed decreases as engine warms up 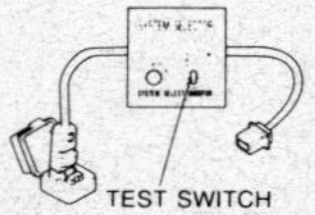	Yes	Go to next step
		No	Check air valve
8	Connect System Selector to diagnosis connector and set Test Switch to "SELF TEST" and check for correct ignition timing at idle after warm-up **Ignition timing (BTDC)** **BP DOHC: 9°—11°** **BP SOHC: 4°—6°** **B6 (323) : 6°—8°** **B6 (MX-3): 9°—11°** 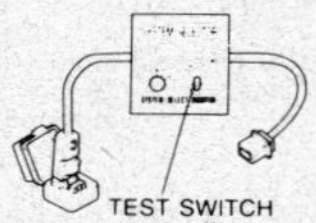	Yes	Go to next step
		No	Adjust ignition timing
9	Try known good ECU and check if condition improves	Yes	Replace ECU
		No	Change fuel to another brand

10	ROUGH IDLE/ENGINE STALLS AT IDLE — AFTER WARM-UP
DESCRIPTION	• Engine runs normally at idle during warm-up but engine stalls or vibrates excessively after warm-up

[TROUBLESHOOTING HINTS]

① Idle speed control system malfunction
② Air/Fuel mixture too lean
- Air leakage
- Low fuel line pressure

③ Air/Fuel mixture too rich
- Fuel injection control malfunction (Correction for coolant temperature)

④ Poor ignition
⑤ Low engine compression

STEP	INSPECTION		ACTION
1	Check if "00" is displayed on Self-Diagnosis Checker with ignition switch ON 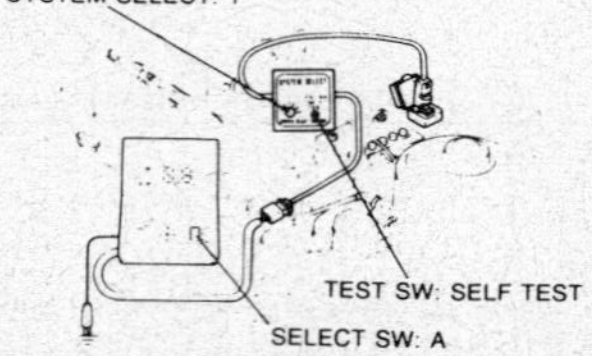	Yes	Go to next step
		No	**Service Code No. displayed** Check for cause (Refer to specified check sequence) **"88" flashes** Check ECU terminal 1F voltage **Specification: Battery voltage (Ignition switch ON)** ⇨ If OK, replace ECU ⇨ If not OK, check wiring between ECU and Self-Diagnosis Checker
2	Check if ECU terminal voltages are OK MTX...Especially 2D, 2O and 2Q ATX...Especially 2B, 2E and 3D	Yes	Go to next step
		No	Check for cause
3	Disconnect high-tension lead at idle and check if engine speed decreases equally at all cylinders	Yes	Go to next step
		No	Go to Step 10
4	Check for correct intake manifold vacuum at idle **Vacuum: More than 450 mmHg (17.7 inHg)**	Yes	Go to next step
		No	Check for air leakage of intake air system components
5	Check if air cleaner element is clean	Yes	Go to next step
		No	Replace air cleaner element
6	Disconnect ISC valve connector at idle and check if engine speed increases 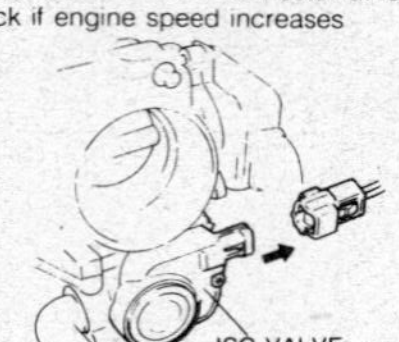	Yes	Go to next step
		No	Replace ISC valve

1.6L AND 1.8L ENGINES TROUBLESHOOTING GUIDE—323 AND MX-3 WITH 1.6L ENGINE AND PROTEGE

STEP	INSPECTION		ACTION
7	Check for correct fuel line pressure at idle	Yes	Go to next step
	Fuel line pressure: **265—314 kPa (2.7—3.2 kg/cm², 38—46 psi) (Vacuum hose to pressure regulator disconnected)** 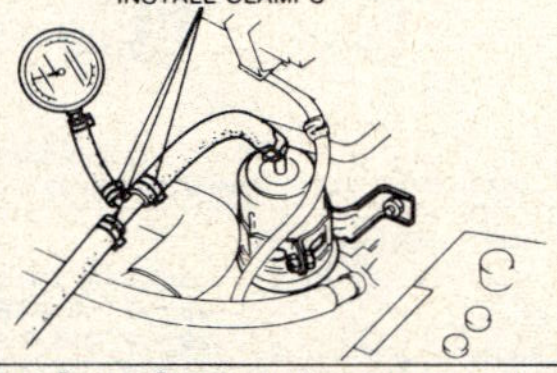	No	**Low pressure** Check fuel line pressure while pinching fuel return hose ⇨ If fuel line pressure quickly increases, check pressure regulator ⇨ If fuel line pressure gradually increases, check for clogging between fuel pump and pressure regulator If not clogged, check fuel pump maximum pressure
8	Connect System Selector to diagnosis connector and set Test Switch to "SELF TEST" and check for correct ignition timing at idle after warm-up	Yes	Go to next step
	Ignition timing (BTDC) **BP DOHC : 9°—11°** **BP SOHC : 4°—6°** **B6 (323) : 6°—8°** **B6 (MX-3): 9°—11°**	No	Adjust ignition system
9	Disconnect water thermosensor connector and check if engine condition improves	Yes	Replace water thermosensor
		No	Try known good ECU and check if condition improves
10	Check for injector operating sound at idle	Yes	Go to next step
		No	Check if injector resistance is OK **Resistance: 12—16Ω** ⇨ If OK, check wiring between ECU and injector ⇨ If not OK, replace injector
11	Check for correct engine compression	Yes	Go to next step
	Engine compression (Minimum): • **BP SOHC :** **834 kPa (8.5 kg/cm², 121 psi)-300 rpm** • **BP DOHC :** **883 kPa (9.0 kg/cm², 128 psi)-300 rpm** • **B6 :** **932 kPa (9.5 kg/cm², 135 psi)-300 rpm**	No	Check engine
12	Check if strong blue spark is visible at disconnected high-tension lead	Yes	Go to next step
		No	Check high-tension lead ⇨ If OK, check distributor cap and rotor ⇨ If not OK, replace high-tension lead
13	Check if spark plugs are OK	Yes	Try known good ECU and check if condition improves
		No	Repair or replace

11	ROUGH IDLE/ENGINE STALLS AT IDLE — WHEN A/C, P/S, OR E/L ON
DESCRIPTION	• Engine stalls or vibrates excessively at idle when A/C, P/S, or E/L ON • A/C, P/S, daytime running lights (Canada), headlights, blower fan and electric cooling fan operate normally • Idle condition is normal when A/C, P/S, and E/L is OFF

[TROUBLESHOOTING HINTS]
① Idle speed control system malfunction
• Engine speed feedback control inoperative

STEP	INSPECTION		ACTION
1	Check if "00" is displayed on Self-Diagnosis Checker with ignition switch ON	Yes	Go to next step
	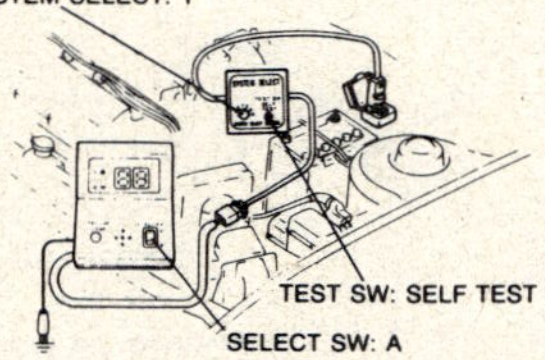	No	**Service Code No. displayed** Check for cause (Refer to specified check sequence) **"88" flashes** Check ECU terminal 1F voltage **Specification: Battery voltage (Ignition switch ON)** ⇨ If OK, replace ECU ⇨ If not OK, check wiring between ECU and Self-Diagnosis Checker
2	Check switches for correct operation by Self-Diagnosis Checker Monitor Lamp with ignition switch ON	Yes	Go to next step
	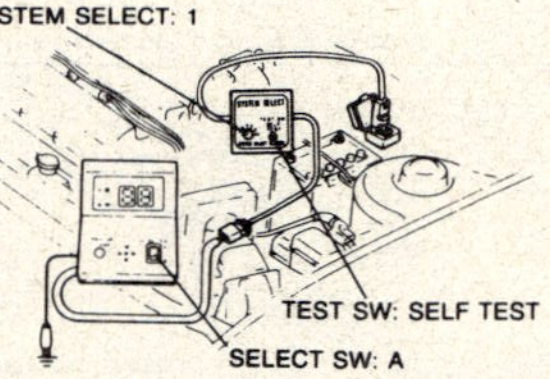	No	**Lamp not ON or OFF with specified switch** Check for cause (Refer to specified check sequence) **Lamp always ON** Check wiring between ECU terminal 1D and Self-Diagnosis Checker
3	Check if ECU terminal 1K (MTX) 1I (ATX) voltage is OK	Yes	Try known good ECU and check if condition improves
	Specification: Battery voltage (at idle) 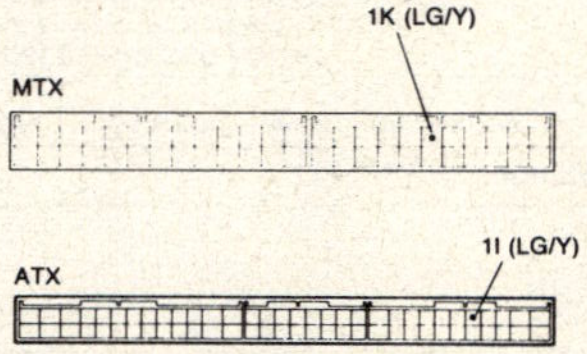	No	Check for short circuit in wiring between diagnosis connector terminal TEN and ground

12	ROUGH IDLE/ENGINE STALLS JUST AFTER STARTING		
DESCRIPTION	• Engine starts normally but vibrates excessively or stalls just after starting (acceleration from idle) • Idle conditions are normal in the other conditions		
[TROUBLESHOOTING HINTS]	① Fuel injection control system or idle speed control system malfunction • Start signal not input to ECU ② Idle speed misadjustment ③ Ignition timing misadjustment		
STEP	INSPECTION		ACTION
1	Check if "00" is displayed on Self-Diagnosis Checker with ignition switch ON 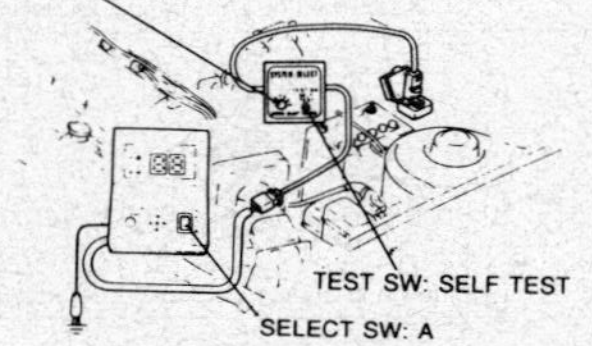	Yes	Go to next step
		No	**Service Code No. displayed** Check for cause (Refer to specified check sequence)
			"88" flashes Check ECU terminal 1F voltage **Specification: Battery voltage (Ignition switch ON)** ⇨ If OK, replace ECU ⇨ If not OK, check wiring between ECU and Self-Diagnosis Checker
2	Check if ECU terminal voltages are OK MTX...Especially 1N ATX... Especially 1T	Yes	Go to next step
		No	Check for cause
3	Connect System Selector to diagnosis connector and set Test Switch to "SELF TEST" and check for correct ignition timing at idle after warm-up **Ignition timing (BTDC)** **BP DOHC: 9°—11°** **BP SOHC: 4°—6°** **B6 (323) : 6°—8°** **B6 (MX-3): 9°—11°** 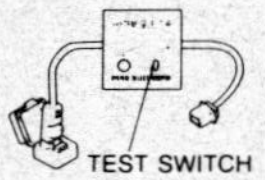	Yes	In same condition as Step 3 inspection and apply parking brake Check for correct idle speed **Idle speed: 700—800 rpm** ⇨ If OK, go to next step ⇨ If not OK, adjust idle speed
		No	Adjust
4	Check continuity between throttle sensor terminals IDL and E 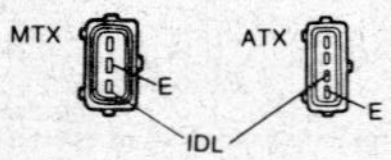	Yes	Go to next step
	Clearance between throttle lever and stopper / Continuity: MTX: 0.1mm (0.004 in) — Yes; 1.0mm (0.039 in) — No ATX: 0.1mm (0.004 in) — Yes; 0.6mm (0.024 in) — No	No	Adjust
5	Try known good ECU and check if condition improves		

Clearance between throttle lever and stopper		Continuity
MTX	0.1mm (0.004 in)	Yes
	1.0mm (0.039 in)	No
ATX	0.1mm (0.004 in)	Yes
	0.6mm (0.024 in)	No

13	HIGH IDLE SPEED AFTER WARM-UP		
DESCRIPTION	• Idle speed excessive after warm-up		
[TROUBLESHOOTING HINTS]	Excessive intake air supplied to engine ① Throttle valve not fully closed ② Idle speed control malfunction • Air valve not closed • ISC valve stuck • ISC valve connector disconnected • A/C, P/S, or E/L signal sent to ECU • Incorrect coolant temperature signal		
STEP	INSPECTION		ACTION
1	Check if throttle valve is fully closed when accelerator released 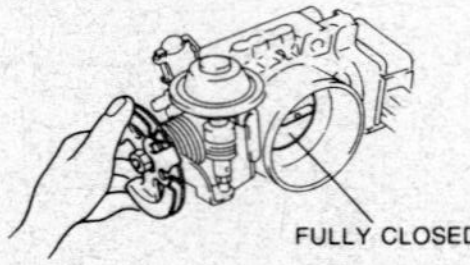	Yes	Go to Step 3
		No	Check if throttle linkage is correctly installed and operates freely ⇨ If OK, go to Step 2 ⇨ If not OK, clean, adjust or replace linkage
2	Check if dashpot is correctly adjusted **Dashpot set speed:** **B6.............................. Approx. 3,000 rpm** **BP SOHC..................... Approx. 2,700 rpm** **BP DOHC..................... Approx. 3,500 rpm**	Yes	Check if throttle valve is contaminated ⇨ If contaminated, clean throttle body ⇨ If not contaminated, replace throttle body
		No	Adjust
3	Check if "00" is displayed on Self-Diagnosis Checker with ignition switch ON 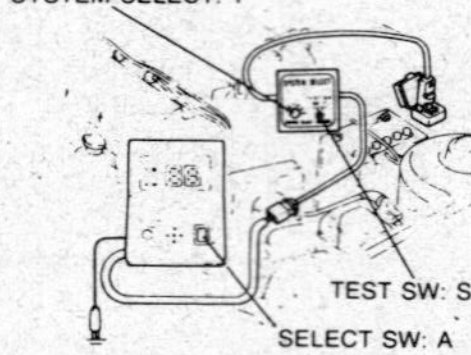	Yes	Go to next step
		No	**Service Code No. displayed** Check for cause (Refer to specified check sequence)
			"88" flashes Check ECU terminal 1F voltage **Specification: Battery voltage (Ignition switch ON)** ⇨ If OK, replace ECU ⇨ If not OK, check wiring between ECU and Self-Diagnosis Checker
4	Check switches for correct operation by Self-Diagnosis Checker Monitor Lamp with ignition switch ON 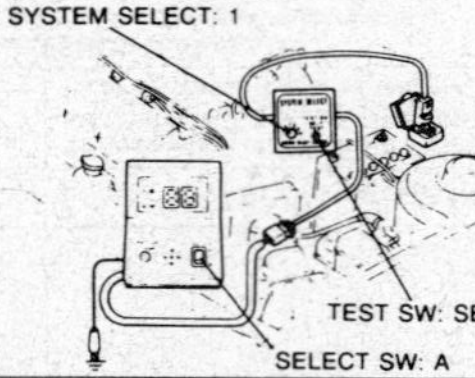	Yes	Go to next step
		No	**Lamp not ON or OFF with specified switch** Check for cause (Refer to specified check sequence)
			Lamp always ON Check wiring between ECU terminal 1D and Self-Diagnosis Checker
5	Check if ECU terminal voltages are OK MTX...Especially 2Q ATX... Especially 2E	Yes	Try known good ECU and check if condition improves
		No	Check for cause

STEP	INSPECTION		ACTION
6	Connect System Selector to diagnosis connector and set Test Switch to "SELF TEST" when engine is cold Check if engine speed decreases as engine warms up 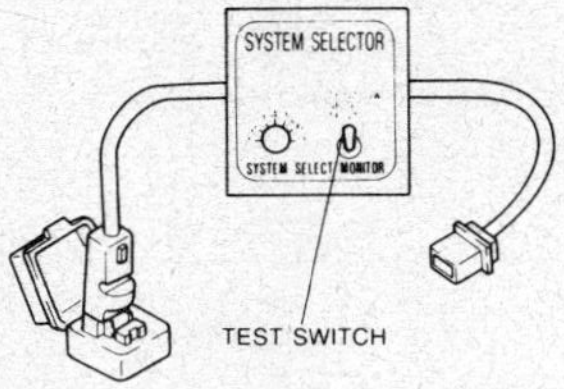 SYSTEM SELECTOR TEST SWITCH	Yes	Go to next step
		No	Check air valve
7	Disconnect ISC valve connector at idle and check if idle speed increases 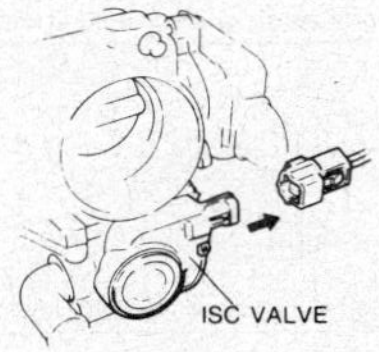 ISC VALVE	Yes	Go to next step
		No	Check ISC valve
8	Pinch PCV hose with pliers and check if engine speed decreases 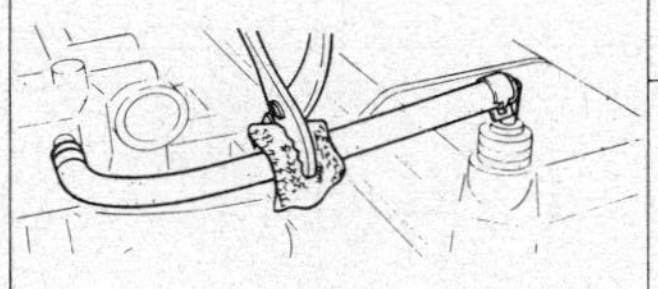	Yes	Check PCV valve
		No	Go to next step
9	Check if ECU terminal voltages are OK MTX...Especially 2D, 2O and 2Q ATX... Especially 2B, 2E and 3D	Yes	Try known good ECU and check if condition improves
		No	Check for cause

14	IDLE MOVES UP AND DOWN/IDLE HUNTING		
DESCRIPTION	• Engine speeds up and down periodically at idle		
[TROUBLESHOOTING HINTS] ① Idle switch (built in throttle sensor) OFF at idle ② Air leakage ③ Fuel injection amount inconstant • Poor contact point inside airflow meter ④ Poor ignition			
STEP	**INSPECTION**		**ACTION**
1	Check if "00" is displayed on Self-Diagnosis Checker with ignition switch ON SYSTEM SELECT: 1 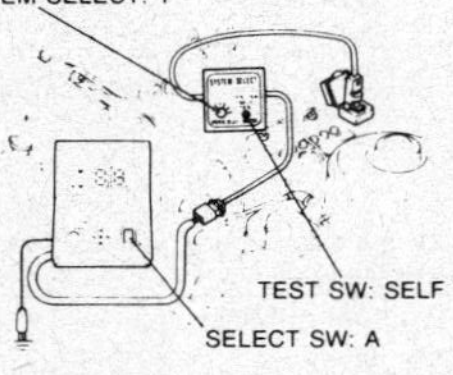 TEST SW: SELF TEST SELECT SW: A	Yes	Go to next step
		No	**Service Code No. displayed** Check for cause (Refer to specified check sequence) **"88" flashes** Check ECU terminal 1F voltage **Specification: Battery voltage (Ignition switch ON)** ⇨ If OK, replace ECU ⇨ If not OK, check wiring between ECU and Self-Diagnosis Checker
2	Check for correct intake manifold vacuum at idle **Intake manifold vacuum:** **More than 450 mmHg (17.7 inHg)**	Yes	Go to next step
		No	**Low vacuum** Check for air leakage at intake air system
3	Check if air cleaner element is clean	Yes	Go to next step
		No	Replace air cleaner element
4	Disconnect high-tension lead at idle and check if engine speed decreases equally at each cylinder	Yes	Go to next step
		No	Go to Step 9
5	Check if ECU terminal voltages are OK	Yes	Go to next step
		No	Check for cause
6	Connect System Selector to diagnosis connector and set Test Switch to "SELF TEST" and check for correct ignition timing at idle after warm-up **Ignition timing (BTDC)** **BP DOHC: 9°—11°** **BP SOHC: 4°—6°** **B6 (323) : 6°—8°** **B6 (MX-3): 9°—11°** 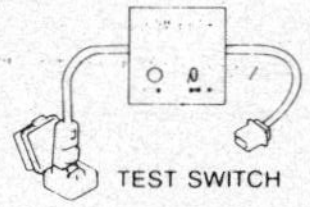 TEST SWITCH	Yes	Go to next step
		No	Adjust ignition timing

STEP	INSPECTION		ACTION
7	Connect System Selector to diagnosis connector and set Test Switch to " SELF TEST" and check if idle is normal 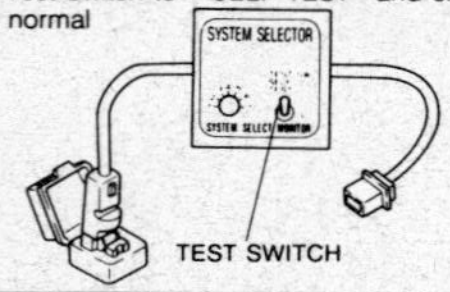	Yes	Try known good ECU
		No	Go to next step
8	Check if airflow meter is OK	Yes	Go to Step 14
		No	Replace airflow meter
9	Check for injector operating sound at idle	Yes	Go to Step 11
		No	Go to next Step
10	Check if battery voltage exists at injector connector (W/R) wire 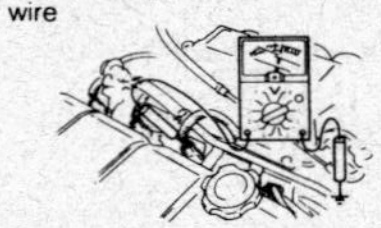	Yes	Check if injector resistance is OK **Resistance: 12—16Ω** ⇨ If OK, check wiring between ECU and injector ⇨ If not OK, replace injector
		No	Check wiring between ECU and injector
11	Check if strong blue spark is visible at disconnected high-tension lead 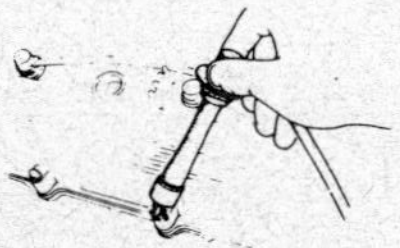	Yes	Go to next step
		No	Check high-tension lead ⇨ If OK, check distributor cap and rotor ⇨ If not OK, replace high-tension lead
12	Check if spark plugs are OK 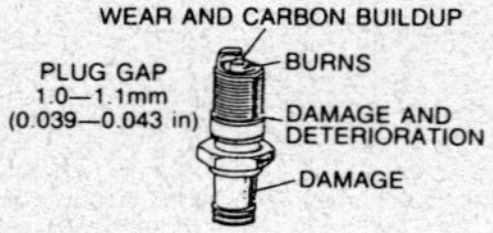	Yes	Check for correct engine compression **Compression (Minimum):** • **BP SOHC** **834 kPa (8.5 kg/cm², 121 psi)-300 rpm** • **BP DOHC** **883 kPa (9.0 kg/cm², 128 psi)-300 rpm** • **B6** **932 kPa (9.5 kg/cm², 135 psi)-300 rpm** ⇨ If OK, go to next step ⇨ If not OK, check for cause
		No	Clean or replace
13	Check for injector leakage	Yes	Replace injector
		No	Go to next step
14	Try known good ECU and check if condition improves		

15	ENGINE STALLS ON DECELERATION		
DESCRIPTION	• Engine unexpectedly stops running while decelerating or after deceleration • Idle condition normal		
[TROUBLESHOOTING HINTS] Engine speed drops too much when releasing accelerator or poor connector connection disconnected by deceleration force ① Idle speed control malfunction ③ Idle speed misadjustment ② Fuel cut control malfunction			
STEP	**INSPECTION**		**ACTION**
1	Check if idle condition is normal	Yes	Go to next step
		No	Adjust or perform troubleshooting Nos. 8—11 "ROUGH IDLE"
2	Check if "00" is displayed on Self-Diagnosis Checker with ignition switch ON SYSTEM SELECT: 1 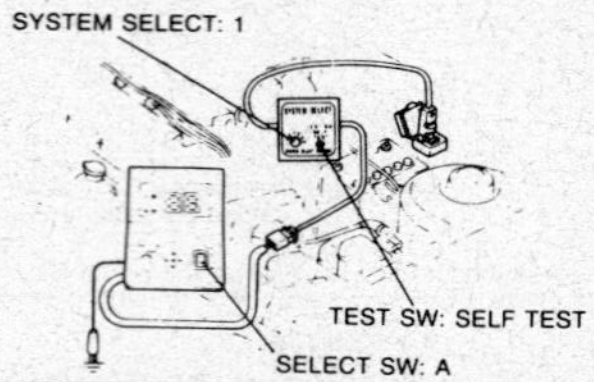	Yes	Go to next step
		No	**Service Code No. displayed** Check for cause (Refer to specified check sequence)
			"88" flashes Check ECU terminal 1F voltage **Specification: Battery voltage (Ignition switch ON)** ⇨ If OK, replace ECU ⇨ If not OK, check wiring between ECU and Self-Diagnosis Checker
3	Check switches for correct operation with Self-Diagnosis Checker Monitor Lamp with ignition switch ON SYSTEM SELECT: 1 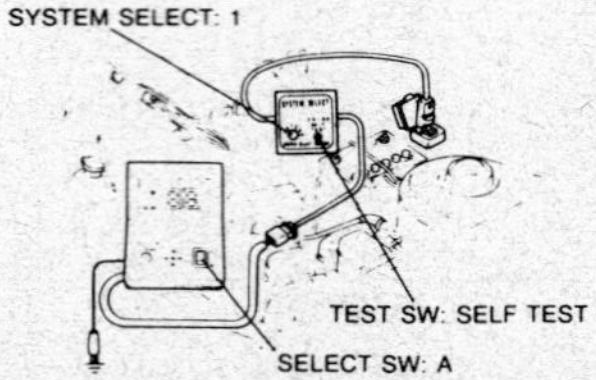	Yes	Go to next step
		No	**Lamp not ON or OFF with specified switch** Check for cause (Refer to specified check sequence)
			Lamp always ON Check wiring between ECU terminal 1D and Self-Diagnosis Checker
4	Check if ECU terminal voltages are OK MTX...Especially 1V, 2D, 2O, 2U, 2V and 2Q ATX...Especially 2B, 2E, 3D, 3U, and 3V	Yes	Go to next step
		No	Check for cause
5	Check for poor connection of following parts • Ignition coil • Igniter • Distributor • High-tension lead • Injector • Circuit opening relay • ECU	Yes	Repair or replace
		No	Go to next step

STEP	INSPECTION		ACTION
6	Check if dashpot is correctly adjusted **Dashpot set speed:** **B6............................ Approx. 3,000 rpm** **BP SOHC...................... Approx. 2,700 rpm** **BP DOHC...................... Approx. 3,500 rpm**	Yes	Go to next step
		No	Adjust dashpot
7	Disconnect ISC valve connector at idle and check if engine speed increases 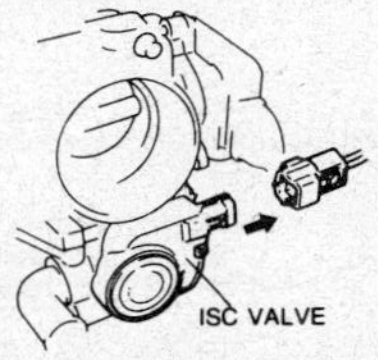 	Yes	Go to next step
		No	Replace ISC valve
8	Try known good ECU and check if condition improves		

16	ENGINE STALLS SUDDENLY (INTERMITTENT)		
DESCRIPTION	• Engine intermittently stops running • Before stalling, engine condition OK		
[TROUBLESHOOTING HINTS] ① Intermittently no spark or no fuel injection caused by vehicle vibration, acceleration, or deceleration • Poor connection in wire harness			
STEP	**INSPECTION**		**ACTION**
1	Check if "00" is displayed on Self-Diagnosis Checker with ignition switch ON 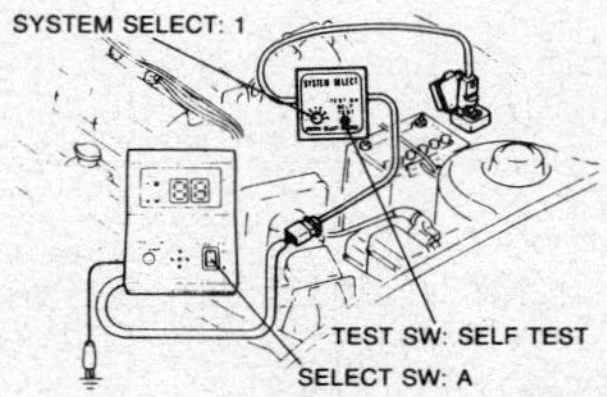 	Yes	Go to next step
		No	**Service Code No. displayed** Check for cause (Refer to specified check sequence) **Note** • **When checking wiring harness and connectors, tap, move, and wiggle suspect sensor and/or harness to recreate problem**
			"88" flashes Check ECU terminal 1F voltage **Specification: Battery voltage (Ignition switch ON)** ⇨ If OK, replace ECU ⇨ If not OK, check wiring between ECU and Self-Diagnosis Checker
2	Check for poor connection of following parts • Ignition coil • Igniter • Distributor • High-tension lead • Injector • Circuit opening relay • ECU	Yes	Repair or replace
		No	Go to next step
3	Check if ECU terminal voltages are OK MTX...Especially 1B, 2A, 2B and 2C ATX... Especially 1B, 3A, 3B and 3C **Note** • **When checking voltages, tap, move, and wiggle harness and connector to recreate problem**	Yes	Go to Troubleshooting No.2 "CRANKS NORMALLY BUT WILL NOT START (NO COMBUSTION)"
		No	Check for cause

17 HESITATES/STUMBLES ON ACCELERATION

DESCRIPTION
- Flat spot occurs just after accelerator depressed or mild jerking occurs during acceleration

[TROUBLESHOOTING HINTS]
① Air/Fuel mixture becomes lean when depressing accelerator
- Fuel injection control malfunction (Correction for accelerating condition)
- Air leakage
- Fuel line pressure decreases
- Spark advance control malfunction

STEP	INSPECTION		ACTION
1	Check if "00" is displayed on Self-Diagnosis Checker with ignition switch ON SYSTEM SELECT: 1 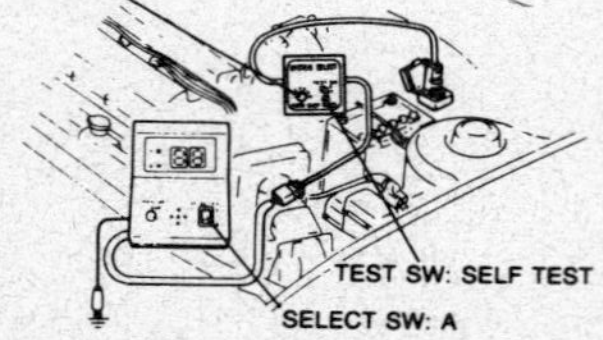 TEST SW: SELF TEST SELECT SW: A	Yes	Go to next step
		No	**Service Code No. displayed** Check for cause (Refer to specified check sequence) **"88" flashes** Check ECU terminal 1F voltage **Specification: Battery voltage (Ignition switch ON)** ⇨ If OK, replace ECU ⇨ If not OK, check wiring between ECU and Self-Diagnosis Checker
2	Check for correct intake manifold vacuum at idle **Vacuum: More than 450 mmHg (17.7 inHg)**	Yes	Go to next step
		No	Check for air leakage at intake air system components
3	Check if air cleaner element is clean	Yes	Go to next step
		No	Replace air cleaner element
4	Check if ECU terminal voltages are OK MTX...Especially 2L ATX... Especially 1T, 2F, and 2I	Yes	Go to next step
		No	Check for cause
5	Check if throttle linkage is correctly installed and operates freely	Yes	Go to next step
		No	Correct, clean, or replace as required any binding or damaged linkage and adjust cable deflection at throttle body
6	Check continuity between throttle sensor terminals IDL and E 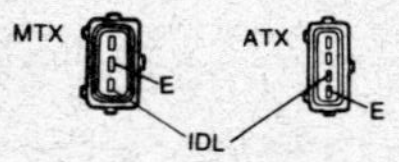	Yes	Go to next step
		No	Adjust

Clearance between throttle lever and stopper		Continuity
MTX	0.1mm (0.004 in)	Yes
	1.0mm (0.039 in)	No
ATX	0.1mm (0.004 in)	Yes
	0.6mm (0.024 in)	No

STEP	INSPECTION		ACTION
7	Check for correct fuel line pressure at idle **Fuel line pressure:** **265—314 kPa (2.7—3.2 kg/cm², 38—46 psi)** **(Vacuum hose to pressure regulator disconnected)** INSTALL CLAMPS 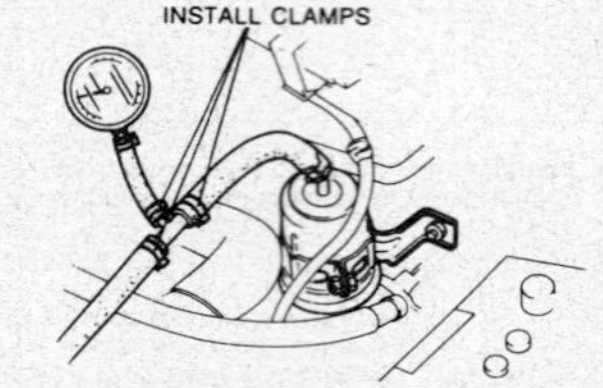	Yes	Check if fuel line pressure decreases when accelerating quickly ⇨ If decreases, check fuel line and filter for clogging ⇨ If no decrease, go to next step
		No	**Low pressure** Check fuel line pressure while pinching fuel return hose ⇨ If fuel line pressure quickly increases, check pressure regulator ⇨ If fuel line pressure gradually increases, check for clogging between fuel pump and pressure regulator If not clogged, check fuel pump maximum pressure
8	Connect System Selector to diagnosis connector and set Test Switch to "SELF TEST" and check for correct ignition timing at idle after warm-up **Ignition timing (BTDC)** **BP DOHC: 9°—11°** **BP SOHC: 4°—6°** **B6 (323) : 6°—8°** **B6 (MX-3): 9°—11°** 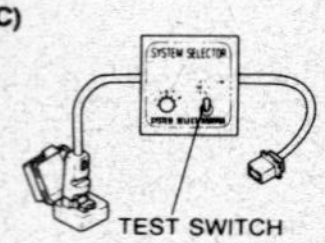 TEST SWITCH	Yes	Check if ignition timing advances when accelerating ⇨ If advances, go to next step ⇨ If not advance, replace ECU
		No	Adjust
9	Check if air duct and air hoses are correctly installed	Yes	Go to next step
		No	Repair
10	Check if exhaust system is restricted	Yes	Repair or replace
		No	**SOHC** Go to Step 12 **DOHC** Go to next step
11	Check if variable inertia control system (VICS) is OK	Yes	Go to next step
		No	Repair or replace
12	Try known good ECU and check if condition improves		

1.6L AND 1.8L ENGINES TROUBLESHOOTING GUIDE—323 AND MX-3 WITH 1.6L ENGINE AND PROTEGE

18	SURGES WHILE CRUISING		
DESCRIP-TION	• Unexpected change in engine speed which is usually repetitive		
[TROUBLESHOOTING HINTS] ① Air/Fuel mixture too lean ② Misfire ③ Poor connection in wiring harness			
STEP	**INSPECTION**		**ACTION**
1	Check if "00" is displayed on Self-Diagnosis Checker with ignition switch ON 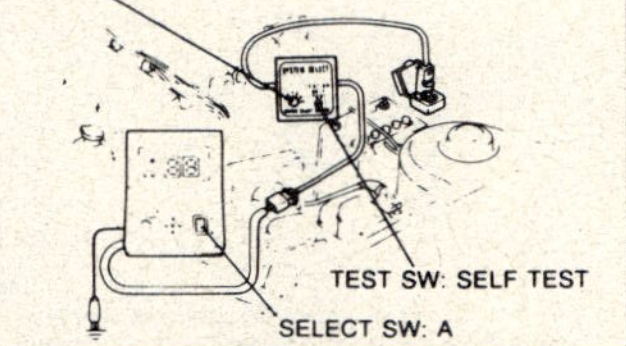	Yes	Go to next step
		No	**Service Code No. displayed** Check for cause (Refer to specified check sequence) **"88" flashes** Check ECU terminal 1F voltage **Specification: Battery voltage (Ignition switch ON)** ⇨ If OK, replace ECU ⇨ If not OK, check wiring between ECU and Self-Diagnosis Checker
2	Check switches for correct operation by Self-Diagnosis Checker Monitor Lamp with ignition switch ON 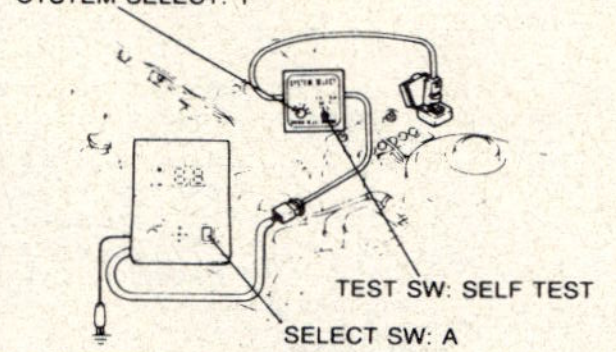	Yes	Go to next step
		No	**Lamp not ON or OFF with specified input switch** Check for cause (Refer to specified check sequence) **Lamp always ON** Check ECU terminal 1D voltage
3	Check if throttle sensor is OK	Yes	Go to next step
		No	Adjust
4	Disconnect oxygen sensor connector and check if condition improves 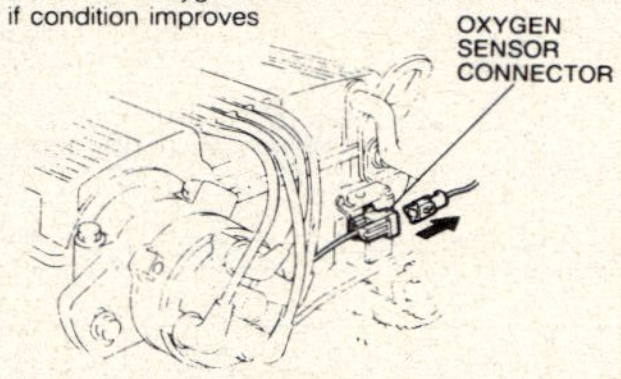	Yes	Check oxygen sensor
		No	Go to next step
5	Check if ECU terminal voltages are OK	Yes	Go to next step
		No	Check for cause

STEP	INSPECTION		ACTION
6	Check if throttle linkage is correctly installed and operates freely	Yes	Go to next step
		No	Correct, clean, or replace as required any binding or damaged linkage, and adjust cable deflection at throttle body
7	Check for correct intake manifold vacuum at idle **Vacuum: More than 450 mmHg (17.7 inHg)**	Yes	Go to next step
		No	Check for air leakage of intake air system components
8	Check if air cleaner element is clean	Yes	Go to next step
		No	Replace air cleaner element
9	Connect System Selector to diagnosis connector and set Test Switch to "SELF TEST" and check for correct ignition timing at idle after warm-up **Ignition timing (BTDC)** **BP DOHC: 9°—11°** **BP SOHC: 4°—6°** **B6 (323) : 6°—8°** **B6 (MX-3): 9°—11°** 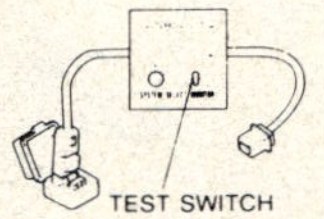	Yes	Check if ignition timing advances when accelerating ⇨ If advances, go to next step ⇨ If not advance, replace ECU
		No	Adjust
10	Check for correct fuel line pressure at idle **Fuel line pressure:** **265—314 kPa (2.7—3.2 kg/cm², 38—46 psi)** **(Vacuum hose to pressure regulator disconnected)** 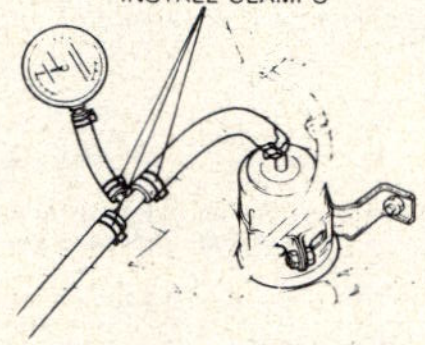	Yes	Check if fuel line pressure decreases when accelerating quickly ⇨ If decreases, check for fuel pump maximum pressure If OK, check fuel line and filter for clogging ⇨ If no decrease, go to next step
		No	**Low pressure** Check fuel line pressure by pinching fuel return hose ⇨ If fuel line pressure quickly increases, check pressure regulator ⇨ If fuel line pressure gradually increases, check for clogging between fuel pump and pressure regulator If not clogged, check fuel pump maximum pressure **High pressure** Check if fuel return line is clogged ⇨ If OK, replace pressure regulator ⇨ If not OK, replace
11	Check if exhaust system is restricted	Yes	Repair or replace
		No	Go to next step
12	Try known good ECU and check if condition improves		

19	LACK OF POWER
DESCRIPTION	• Performance poor under load when throttle valve wide open • Reduced maximum speed • Idle condition normal

[TROUBLESHOOTING HINTS]

1 Factors other than engine malfunction
- Clutch slipping
- ATX slipping
- Brake dragging
- Low tire pressure
- Incorrect tire size
- Overloaded

2 Low intake air amount
- Throttle valve not open fully
- Clogged intake air system

③ Air/Fuel mixture too lean
- Fuel line pressure decreases
- Fuel injection malfunction

④ Poor ignition
⑤ Low engine compression
⑥ Alcohol blended fuel used

STEP	INSPECTION		ACTION
1	Check factors other than engine • Clutch slipping • ATX slipping • Brake dragging • Low tire pressure • Incorrect tire size	Yes	Go to next step
		No	Repair
2	Check if throttle valve fully opened when accelerator depressed fully	Yes	Go to next step
		No	Check if accelerator cable is correctly installed ⇨ If OK, check throttle body ⇨ If not OK, install accelerator cable correctly
3	Check if "00" is displayed on Self-Diagnosis Checker with ignition switch ON 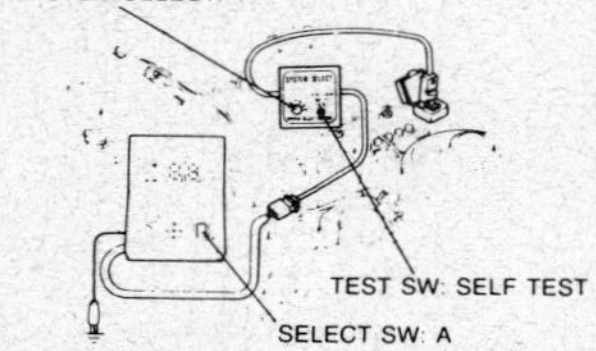	Yes	Go to next step
		No	**Service Code No. displayed** Check for cause (Refer to specified check sequence)
			"88" flashes Check ECU terminal 1F voltage **Specification: Battery voltage (Ignition switch ON)** ⇨ If OK, replace ECU ⇨ If not OK, check wiring between ECU and Self-Diagnosis Checker
4	Connect System Selector to diagnosis connector and set Test Switch to "SELF TEST" and check for correct ignition timing at idle after warm-up **Ignition timing (BTDC)** **BP DOHC: 9°—11°** **BP SOHC: 4°—6°** **B6 (323) : 6°—8°** **B6 (MX-3): 9°—11°** 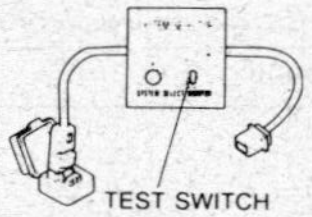	Yes	Check if ignition timing advances when accelerating ⇨ If advances, go to next step ⇨ If not advance, check ECU terminal voltages
		No	Adjust
5	Check if ECU terminal voltages are OK MTX...Especially 1K, 1N, 2L, 2K, and 2S (BPD) ATX... Especially 1I, 1T, 2F, 2I and 3I (BPD)	Yes	Go to next step
		No	Check for cause

STEP	INSPECTION		ACTION
6	Check for correct intake manifold vacuum at idle **Intake manifold vacuum: More than 450 mmHg (17.7 inHg)**	Yes	Go to next step
		No	Check for air leakage of intake air system components
7	Check if air cleaner element is clean	Yes	Go to next step
		No	Replace air cleaner element
8	Check for correct fuel line pressure at idle **Fuel line pressure: 265—314 kPa (2.7—3.2 kg/cm², 38—46 psi) (Vacuum hose to pressure regulator disconnected)** 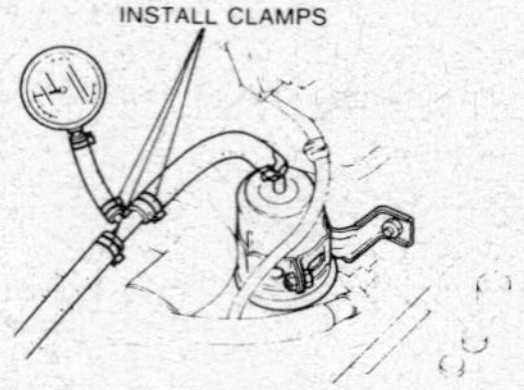	Yes	Check if fuel line pressure decreases when accelerating quickly ⇨ If decreases, check fuel pump maximum pressure If OK, check fuel line and filter for clogging ⇨ If not decrease, go to next step
		No	**Low pressure** Check fuel line pressure while pinching fuel return hose ⇨ If fuel line pressure quickly increases, check pressure regulator ⇨ If fuel line pressure gradually increases, check for clogging between fuel pump and pressure regulator If not clogged, check fuel pump maximum pressure
			High pressure Check if fuel return line is clogged ⇨ If OK, replace pressure regulator ⇨ If not OK, replace
9	Check if airflow meter is OK I. Check if measuring plate moves smoothly 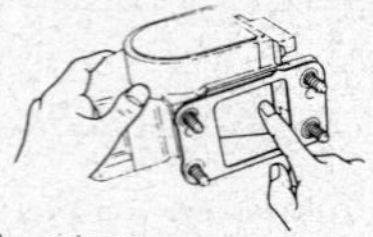	Yes	Go to next step
	II. Check resistance 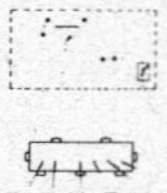	No	Repair or replace

Terminal	Resistance (Ω) Fully closed	Resistance (Ω) Fully open
E2↔Vs	20—600	20—1,000
E2↔Vc	200—400	
E1↔Fc	∞	0

STEP	INSPECTION		ACTION
10	Check if spark plugs are OK 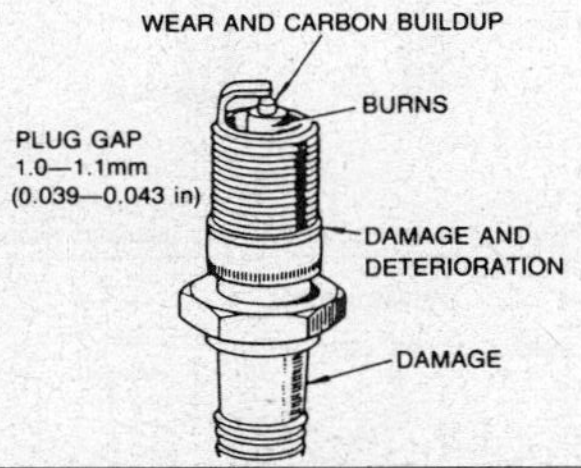	Yes	Go to next step
		No	Clean or replace
11	Check if resistance of high-tension leads are OK **Resistance: 16 kΩ per 1 m (3.28 ft)** 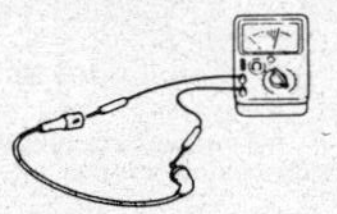	Yes	Go to next step
		No	Replace
12	Check if resistance of ignition coil is OK **Resistance (at 20°C [68°F]):** **Primary coil winding........ 0.81—0.99Ω** **Secondary coil winding.... 10—16 kΩ** 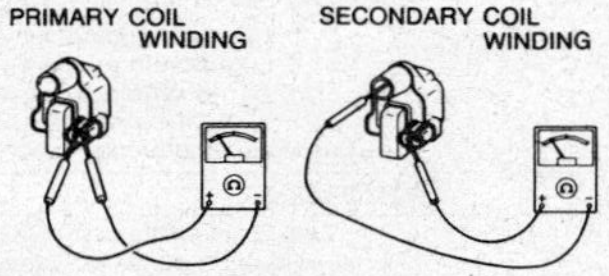	Yes	Go to next step
		No	Replace
13	Check for correct engine compression **Engine compression:** • **BP DOHC** **883 kPa (9.0 kg/cm², 128 psi)-300 rpm** • **BP SOHC** **834 kPa (8.5 kg/cm², 121 psi)-300 rpm** • **B6** **932 kPa (9.5 kg/cm², 135 psi)-300 rpm**	Yes	Go to next step
		No	Check engine condition • Worn piston, piston rings or cylinder wall • Defective cylinder head gasket • Distorted cylinder head • Improper valve seating • Valve sticking in guide
14	Change fuel and check if condition improves	Yes	Change fuel to another brand
		No	**SOHC** Go to Step 16
			DOHC Go to next step

STEP	INSPECTION		ACTION
15	Check if variable inertia charging system (VICS) is OK (BPD)	Yes	Go to next step
		No	Repair or replace
16	Try known good ECU and check if condition improves		

20	POOR ACCELERATION
DESCRIPTION	• Performance poor while accelerating • Idle condition normal

[TROUBLESHOOTING HINTS]

① Factors other than engine malfunction
- Clutch slipping
- ATX slipping
- Brake dragging
- Low tire pressure
- Incorrect tire size
- Over-loaded

② Low intake air amount
- Throttle valve not open fully
- Clogging in intake air system

③ Air/Fuel mixture too lean
- Fuel line pressure decreases
- Fuel injection malfunction

④ Poor ignition

⑤ Low engine compression

⑥ Alcohol blended fuel used

STEP	INSPECTION		ACTION
1	Check factors other than engine • Clutch slipping • Brake dragging • Low tire pressure • Incorrect tire size • ATX slipping	Yes	Go to next step
		No	Repair
2	Check if throttle valve fully opens when depressing accelerator fully	Yes	Go to next step
		No	Check if accelerator cable is correctly installed ⇨ If OK, check throttle body ⇨ If not OK, install accelerator cable correctly
3	Check if "00" is displayed on Self-Diagnosis Checker with ignition switch ON 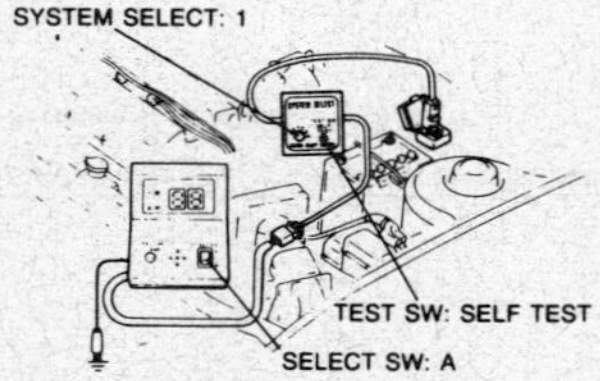 	Yes	Go to next step
		No	**Service Code No. displayed** Check for cause (Refer to specified check sequence)
			"88" flashes Check ECU terminal 1F voltage **Specification: Battery voltage (Ignition switch ON)** ⇨ If OK, replace ECU ⇨ If not OK, check wiring between ECU and Self-Diagnosis Checker
4	Check continuity between throttle sensor terminals IDL and E 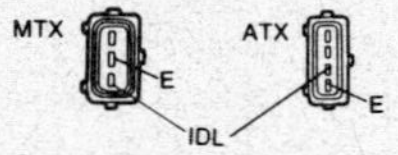 	Yes	Go to next step
		No	Adjust

Clearance between throttle lever and stopper		Continuity
MTX	0.1mm (0.004 in)	Yes
	1.0mm (0.039 in)	No
ATX	0.1mm (0.004 in)	Yes
	0.6mm (0.024 in)	No

STEP	INSPECTION		ACTION
5	Connect System Selector to diagnosis connector and set Test Switch to "SELF TEST" and check for correct ignition timing at idle after warm-up **Ignition timing (BTDC)** **BP DOHC: 9°—11°** **BP SOHC: 4°—6°** **B6 (323) : 6°—8°** **B6 (MX-3): 9°—11°** 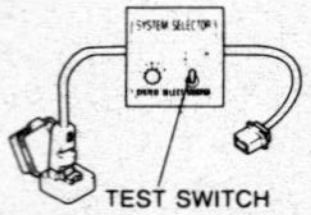 	Yes	Check if ignition timing advances when accelerating ⇨ If advances, go to next step ⇨ If not advance, check ECU terminal voltages
		No	Adjust
6	Check if ECU terminal voltages are OK	Yes	Go to next step
		No	Check for cause
7	Check for correct intake manifold vacuum at idle **Intake manifold vacuum:** **More than 450 mmHg (17.7 inHg)**	Yes	Go to next step
		No	Check for air leakage of intake air system components
8	Check for correct fuel line pressure at idle **Fuel line pressure:** **265—314 kPa (2.7—3.2 kg/cm², 38—46 psi)** **(Vacuum hose to pressure regulator disconnected)** 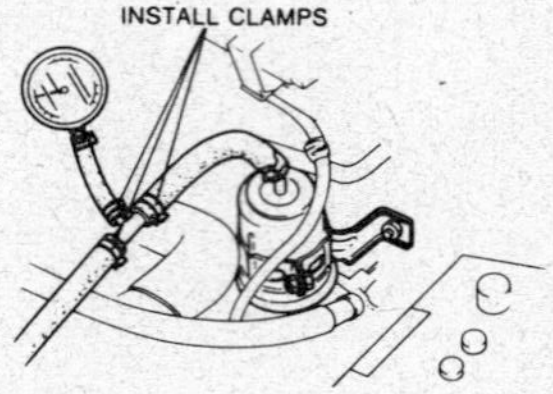 	Yes	Check if fuel line pressure decreases when accelerating quickly ⇨ If decreases, check fuel pump maximum pressure If OK, check fuel line and filter for clogging ⇨ If not decrease, go to next step
		No	**Low pressure** Check for fuel line pressure while pinching fuel return hose ⇨ If fuel line pressure quickly increases, check pressure regulator ⇨ If fuel line pressure gradually increases, check for clogging between fuel pump and pressure regulator If not clogged, check fuel pump maximum pressure
			High pressure Check if fuel line is clogged ⇨ If OK, replace pressure regulator ⇨ If not OK, replace

STEP	INSPECTION		ACTION
9	Check if airflow meter is OK I. Check if measuring plate moves smoothly 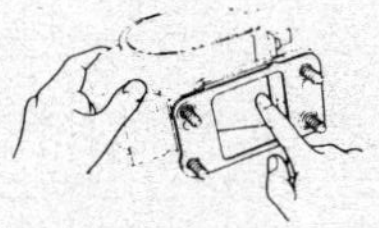	Yes	Go to next step
	II. Check resistance 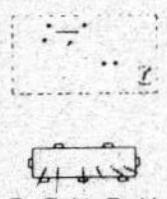	No	Repair or replace
10	Check if spark plugs are OK 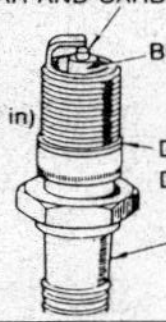	Yes	Go to next step
		No	Clean or replace
11	Check if resistance of ignition coil is OK **Resistance (at 20°C [68°F]):** **Primary coil winding........ 0.81—0.99Ω** **Secondary coil winding.... 10—16 kΩ** 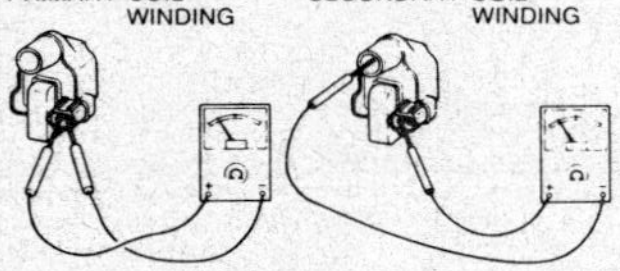	Yes	Go to next step
		No	Replace
12	Check for correct engine compression	Yes	Go to next step
	Engine compression (Minimum): • **BP DOHC** **883 kPa (9.0 kg/cm², 128 psi)-300 rpm** • **BP SOHC** **834 kPa (8.5 kg/cm², 121 psi)-300 rpm** • **B6** **932 kPa (9.5 kg/cm², 135 psi)-300 rpm**	No	Check engine condition • Worn piston, piston rings or cylinder wall • Defective cylinder head gasket • Distorted cylinder head • Improper valve seating • Valve sticking in guide

Step 9 resistance table:

Terminal	Resistance (Ω) Fully closed	Resistance (Ω) Fully open
E2↔Vs	20—600	20—1,000
E2↔Vc	200—400	200—400
E1↔Fc	∞	0

STEP	INSPECTION		ACTION
13	Change fuel and check if acceleration improves	Yes	Change fuel to another brand
		No	Go to next step
14	Check if A/C cut-off control system is OK	Yes	**SOHC** Go to Step 16 **DOHC** Go to next step
		No	Repair or replace
15	Check if variable inertia charging system (VICS) is OK (BPD)	Yes	Go to next step
		No	Repair or replace
16	Try known good ECU and check if condition improves		

21	RUNS ROUGH ON DECELERATION/BACKFIRE
DESCRIPTION	• Engine runs rough while decelerating and abnormal combustion occurs in exhaust system • Transaxle in normal condition

[TROUBLESHOOTING HINTS]

1 Air/Fuel mixture too rich
- Air cleaner element clogged
- Fuel injection control malfunction (Fuel cut control)
- Injector fuel leakage
- Throttle sensor (IDL terminal) malfunction
- Dashpot malfunction

STEP	INSPECTION		ACTION
1	Check if "00" is displayed on Self-Diagnosis Checker with ignition switch ON	Yes	Go to next step
	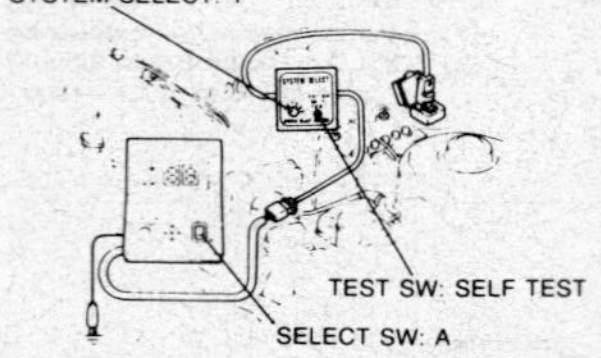	No	**Service Code No. displayed** Check for cause (Refer to specified check sequence) **"88" flashes** Check ECU terminal 1F voltage **Specification: Battery voltage (Ignition switch ON)** ⇨ If OK, replace ECU ⇨ If not OK, check wiring between ECU and Self-Diagnosis Checker
2	Check switches for correct operation by Self-Diagnosis Checker Monitor Lamp with ignition switch ON	Yes	Go to next step
	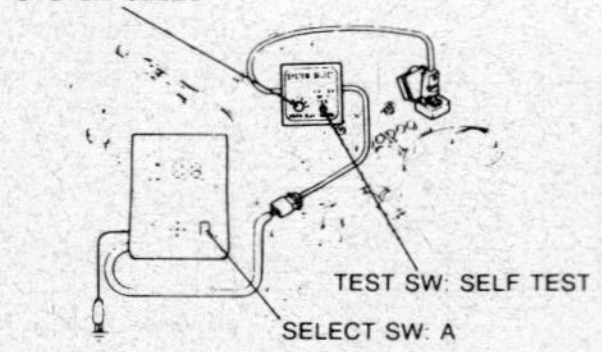	No	**Lamp not ON or OFF with specified switch** Check for cause (Refer to specified check sequence) **Lamp always ON** Check wiring between ECU terminal 1D and Self-Diagnosis Checker
3	Check for correct intake manifold vacuum at idle **Vacuum: More than 450 mmHg (17.7 inHg)**	Yes	Go to next step
		No	Check for air leakage at intake air system components
4	Check if air cleaner element is clean	Yes	Go to next step
		No	Replace air cleaner element
5	Check if ECU terminal voltages are OK MTX...Especially 1N, 2A, 2U and 2V ATX...Especially 1T, 3A, 3U and 3V	Yes	Go to next step
		No	Check for cause

STEP	INSPECTION		ACTION
6	Check if fuel cut operation is OK	Yes	Go to next step
		No	Try known good ECU and check if condition improves
7	Check if airflow meter is OK I. Check if measuring plate moves smoothly 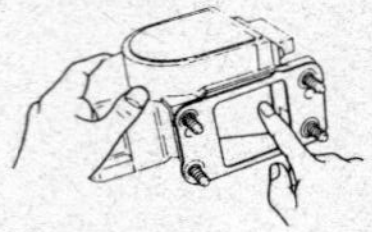	Yes	Go to next step
	II. Check resistance 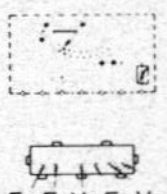	No	Repair or replace
8	Check if throttle sensor IDL terminal is OK	Yes	Go to next step
		No	Adjust
9	Check if dashpot is correctly adjusted **Dashpot set speed:** **B6.................. Approx. 3,000 rpm** **BP SOHC.............. Approx. 2,700 rpm** **BP DOHC.............. Approx. 3,500 rpm**	Yes	Check if throttle valve is contaminated ⇨ If contaminated, clean throttle body ⇨ If not contaminated, replace throttle body
		No	Adjust
10	Check injector for fuel leakage ☞ **page F–132** 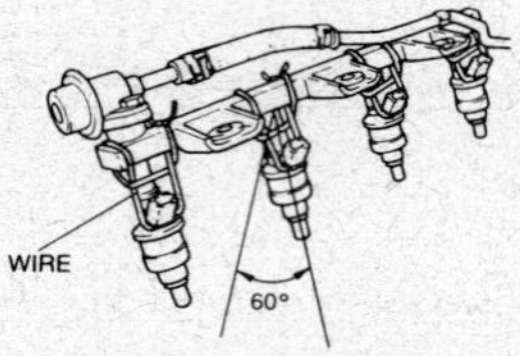	Yes	Replace
		No	Go to next step
11	Try known good ECU and check if condition improves		

Step 7 resistance check:

Terminal	Resistance (Ω) Fully closed	Resistance (Ω) Fully open
E2↔Vs	20—600	20—1,000
E2↔Vc	200—400	
E1↔Fc	∞	0

22	KNOCKING		
DESCRIPTION	• Abnormal combustion accompanied by audible "pinging" noise		
[TROUBLESHOOTING HINTS] ① Incorrect ignition timing (Too advanced) ② Carbon deposits in cylinder ③ Overheating	④ Air/Fuel mixture too lean • Fuel injection amount not correct • Fuel line pressure decreases while accelerating		
STEP	INSPECTION		ACTION
1	Connect System Selector to diagnosis connector and set Test Switch to "SELF TEST" and check for correct ignition timing at idle after warm-up **Ignition timing (BTDC)** **BP DOHC: 9°—11°** **BP SOHC: 4°—6°** **B6 (323) : 6°—8°** **B6 (MX-3): 9°—11°** 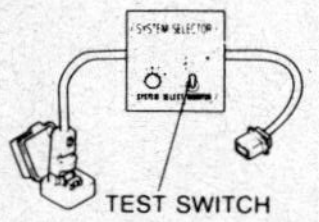 	Yes	Go to next step
		No	Adjust
2	Check if "00" is displayed on Self-Diagnosis Checker with ignition switch ON 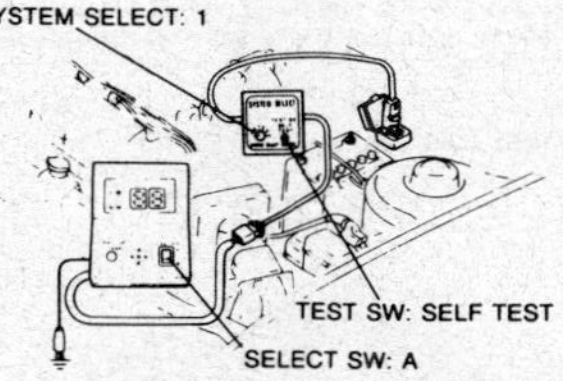 	Yes	Go to next step
		No	**Service Code No. displayed** Check for cause (Refer to specified check sequence) **"88" flashes** Check ECU terminal 1F voltage **Specification: Battery voltage (Ignition switch ON)** ⇨ If OK, replace ECU ⇨ If not OK, check wiring between ECU and Self-Diagnosis Checker
3	Check for correct intake manifold vacuum at idle **Vacuum: More than 450 mmHg (17.7 inHg)**	Yes	Go to next step
		No	Check for air leakage of intake air system components
4	Check for correct engine compression **Engine compression (Minimum):** • **BP DOHC** **883 kPa (9.0 kg/cm², 128 psi)-300 rpm** • **BP SOHC** **834 kPa (8.5 kg/cm², 121 psi)-300 rpm** • **B6** **932 kPa (9.5 kg/cm², 135 psi)-300 rpm**	Yes	Go to next step
		No	**High compression** Check engine condition • Carbon deposits

STEP	INSPECTION		ACTION
5	Check for correct fuel line pressure at idle **Fuel line pressure:** **265—314 kPa (2.7—3.2 kg/cm², 38—46 psi)** **(Vacuum hose to pressure regulator disconnected)** 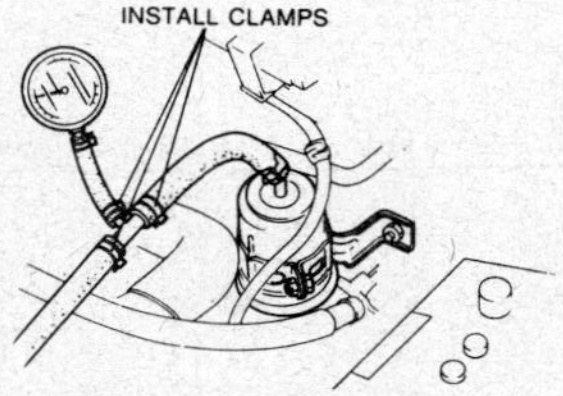 	Yes	Check if fuel line pressure decreases when accelerating quickly ⇨ If it decreases, check for clogging between fuel pump and pressure regulator ⇨ If not decrease, go to next step
		No	**Low pressure** Check fuel line pressure while pinching fuel return hose ⇨ If fuel line pressure quickly increases, check pressure regulator ⇨ If fuel line pressure gradually increases, check for clogging between fuel pump and pressure regulator If not clogged, check fuel pump maximum pressure
6	Check if cooling system is OK	Yes	Go to next step
		No	Repair or replace • Thermostat • Electric cooling fan • Radiator
7	Try known good ECU and check if condition improves	Yes	Replace ECU
		No	Change fuel to another brand or use higher octane fuel

23	FUEL ODOR		
DESCRIPTION	• Gasoline odor in cabin		
[TROUBLESHOOTING HINTS] ① Poor connection or damaged at fuel system or evaporative emission control system ② Charcoal canister overflow due to evaporative emission control system malfunction			
STEP	INSPECTION		ACTION
1	Check if fuel leakage or damage of fuel system and evaporative emission control system	Yes	Repair or replace
		No	Go to next step
2	Check if "00" is displayed on Self-Diagnosis Checker with ignition switch ON 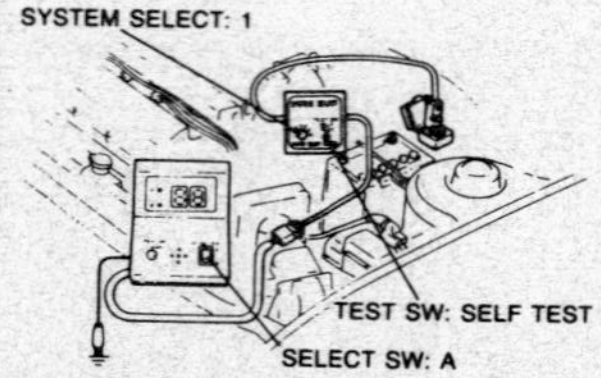 	Yes	Go to next step
		No	**Service Code No. displayed** Check for cause (Refer to specified check sequence) **"88" flashes** Check ECU terminal 1F voltage **Specification: Battery voltage (Ignition switch ON)** ⇨ If OK, replace ECU ⇨ If not OK, check wiring between ECU and Self-Diagnosis Checker
3	Check if vacuum is felt at solenoid valve (purge control) with engine running and throttle valve opened (Neutral switch connector disconnected) 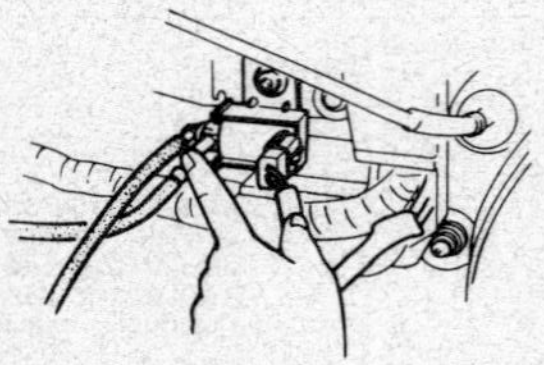	Yes	Go to Step 5
		No	Check for solenoid valve operating sound ⇨ If OK, check vacuum hoses for clogging ⇨ If not OK, go to next step
4	Apply battery voltage and ground to solenoid valve (purge control) and check if vacuum is felt at solenoid valve at idle 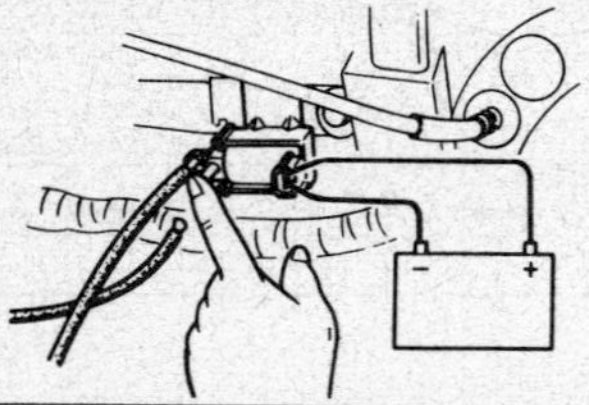	Yes	Check ECU terminal voltages
		No	Replace solenoid valve
5	Try known good ECU and check if condition improves		

24	EXHAUST SULFUR SMELL	
DESCRIPTION	• Exhaust gas smells abnormally	
[TROUBLESHOOTING HINTS] High sulfur content fuel used		
STEP	INSPECTION	ACTION
1	Change fuel to another brand	

25	HIGH OIL CONSUMPTION		
DESCRIPTION	• Oil consumption excessive		
[TROUBLESHOOTING HINTS] ① PCV system malfunction ② Engine malfunction (Oil working up, working down, or leakage)			
STEP	INSPECTION		ACTION
1	Check if PCV hose, ventilation hose or their attaching nipples are separated, damaged, clogged, or restricted	Yes	Repair or replace
		No	Go to next step
2	Check if air pressure or oil is present at ventilation hose 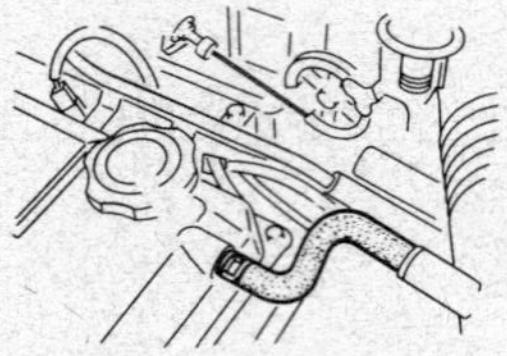	Yes	Go to next step
		No	Check engine condition • Oil leakage • Worn valve seal • Worn valve stem • Worn valve guide
3	Check if vacuum is felt at PCV valve at idle 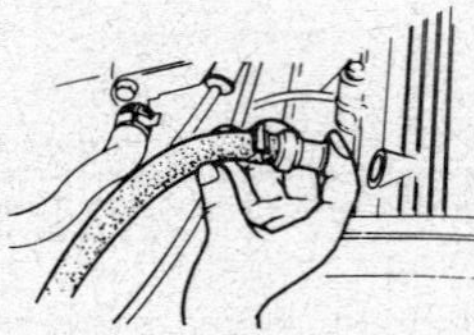	Yes	Check engine condition • Worn piston ring groove • Stuck piston rings • Worn piston or cylinder
		No	Replace PCV valve

1.6L AND 1.8L ENGINES TROUBLESHOOTING GUIDE—323 AND MX-3 WITH 1.6L ENGINE AND PROTEGE

26	POOR FUEL ECONOMY
DESCRIPTION	• Fuel economy unsatisfactory

[TROUBLESHOOTING HINTS]
While fuel consumption is drastically increased by city driving, short-run operation, stop and go driving, extended winter warm-up periods, etc., as opposed to "trip" mileage, an attempt should be made to determine these factors when confronted with "poor mileage" conditions. However, since the operator is not always at fault, the following is offered.

1. Operator depressing accelerator more than usual due to low engine power
   - Poor ignition
   - Low intake air amount
   - Electronic spark advance control system malfunction
   - Clutch slipping/ATX slipping
   - Exhaust component restricted
2. Air/Fuel mixture too rich
   - High fuel line pressure
3. Alcohol blended fuel used
4. High vehicle load
   - Low tire pressure
   - Incorrect tire size
   - Brake dragging

STEP	INSPECTION		ACTION
1	Check factors other than engine • Low tire pressure • Unrecommended tire size • Clutch slipping • Brake dragging • Exhaust component restricted	Yes	Go to next step
		No	Repair
2	Check if air hoses are connected correctly	Yes	Go to next step
		No	Repair
3	Check if "00" is displayed on Self-Diagnosis Checker with ignition switch ON 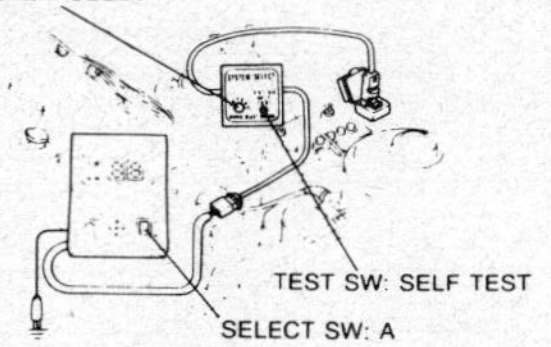	Yes	Go to next step
		No	**Service Code No. displayed** Check for cause (Refer to specified check sequence) **"88" flashes** Check ECU terminal 1F voltage **Specification: Battery voltage (Ignition switch ON)** ⇨ If OK, replace ECU ⇨ If not OK, check wiring between ECU and Self-Diagnosis Checker
4	Check for correct intake manifold vacuum at idle **Vacuum: More than 450 mmHg (17.7 inHg)**	Yes	Go to next step
		No	Check for air leakage at intake air system components
5	Check if air cleaner element is clean	Yes	Go to next step
		No	Replace air cleaner element

STEP	INSPECTION		ACTION
6	Check if ECU terminal voltages are OK MTX...Especially 2D, 2N, 2O, 2Q, 2U and 2V ATX...Especially 2B, 2C, 2E, 3D, 3U and 3V	Yes	Go to next step
		No	Check for cause
7	Connect System Selector to diagnosis connector and set Test Switch to "SELF TEST" and check for correct ignition timing at idle after warm-up **Ignition timing (BTDC)** **BP DOHC: 9°—11°** **BP SOHC: 4°—6°** **B6 (323) : 6°—8°** **B6 (MX-3): 9°—11°** 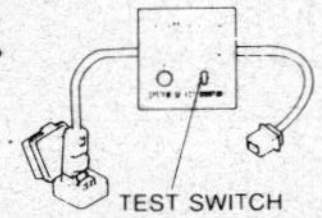	Yes	Go to next step
		No	Adjust
8	Check for correct fuel line pressure at idle **Fuel line pressure:** **216—265 kPa (2.2—2.7 kg/cm², 31—38 psi)** 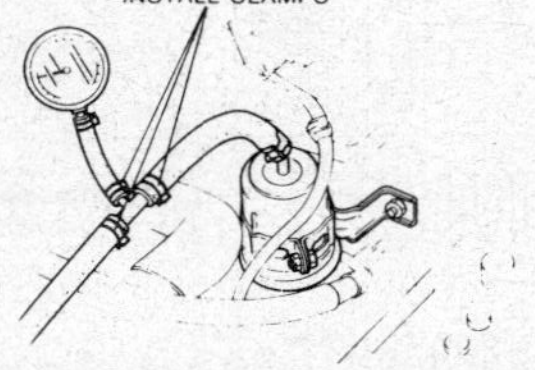	Yes	Go to next step
		No	**High pressure** Check if vacuum hose to pressure regulator is damaged or poorly connected ⇨ If OK, replace pressure regulator ⇨ If not OK, repair or replace hose
9	Run engine at idle and check if fuel line pressure is held after ignition switch turned OFF **Fuel line pressure: More than 147 kPa (1.5 kg/cm², 21 psi) for 5 min.** 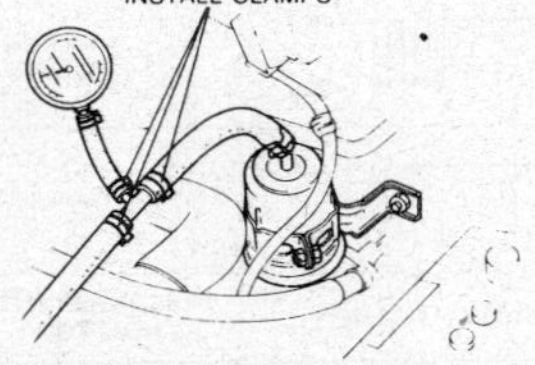	Yes	Go to next step
		No	Check injectors for fuel leakage
10	Change fuel to another brand		

1.6L AND 1.8L ENGINES TROUBLESHOOTING GUIDE
323 AND MX-3 WITH 1.6L ENGINE AND PROTEGE

27	MIL ALWAYS ON		
DESCRIPTION	• Self-Diagnosis Checker does not indicate Malfunction Code No. but MIL always ON		
[TROUBLESHOOTING HINTS] • Short circuit in wiring harness • ECU malfunction			
STEP	INSPECTION		ACTION
1	Disconnect (Y/B) wire from ECU and check if MIL goes off 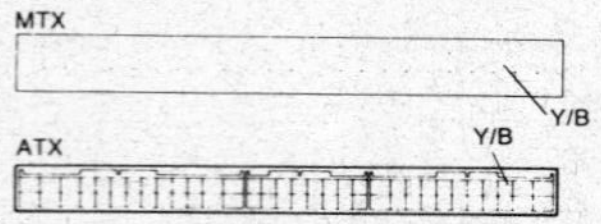	Yes	Replace ECU
		No	Check for short circuit in wiring between instrument cluster and ECU

28	MIL NEVER ON		
DESCRIPTION	• Self-Diagnosis Checker indicates Malfunction Code No. of input device but MIL never ON • Other indicator and warning lamps OK		
[TROUBLESHOOTING HINTS] • Bulb burned • Open circuit in wiring harness • ECU malfunction			
STEP	INSPECTION		ACTION
1	Ground (Y/B) wire at ECU with jumper wire and check if MIL comes on 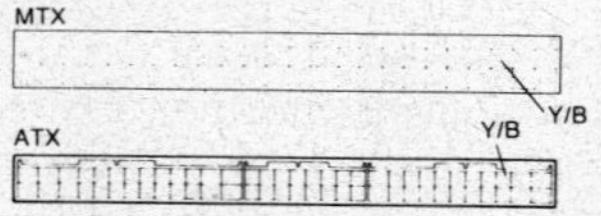	Yes	Check connection of ECU connector ⇨ If OK, replace ECU ⇨ If not OK, repair ECU connector
		No	Check if bulb is OK ⇨ If OK, repair (Y/B) wire between ECU and instrument cluster ⇨ If not OK, replace bulb

29	A/C DOES NOT WORK		
DESCRIPTION	• Blower fan operates but magnet clutch does not operate		
[TROUBLESHOOTING HINTS] • Open or short circuit in wiring harness • A/C relay, A/C switch, or magnetic clutch malfunction • ECU malfunction			
STEP	INSPECTION		ACTION
1	Check ECU terminal 1Q (MTX) 1O (ATX) voltage OK 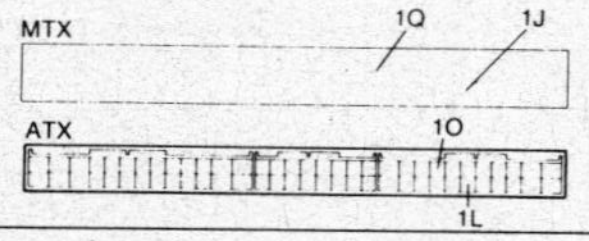	Yes	Check ECU terminal 1J (MTX) 1L (ATX) voltage OK ⇨ If OK, check A/C system ⇨ If not OK, replace ECU
		No	Check for cause

1.6L AND 1.8L ENGINES TROUBLE CODES
323 AND MX-3 WITH 1.6L ENGINE AND PROTEGE

CODE NO.	LOCATION OF MALFUNCTION	OUTPUT SIGNAL PATTERN	SELF-DIAGNOSIS	FAIL-SAFE
02	Ne-signal		No Ne-signal	—
03	G-signal (DOHC)		No G-signal	Cancels 2-group injection
06	Vehicle speed sensor		No input signal from vehicle speed sensor	Shifting performed normally
08	Airflow meter		Open or short circuit	Basic fuel injection amount fixed as for two driving modes (1) Idle switch ON (2) Idle switch OFF
09	Water thermosensor	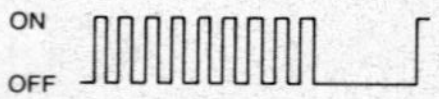		Maintains constant 20°C (68°F) command
10	Intake air thermosensor			Maintains constant 20°C (68°F) command
12	Throttle sensor			Throttle opening judged as full stroke lockup not provided
14	Atmospheric pressure sensor (in ECU)			Maintains constant command of sea level pressure
15	Oxygen sensor (Inactivation)	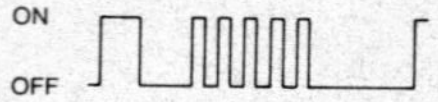	Sensor output continues less than 0.55V 95 sec. after engine starts (1500 rpm)	Cancels engine feedback operation
17	Oxygen sensor (Inversion)	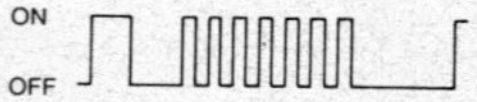	Sensor output continues uncharged 50 sec. after engine exceeds 1500 rpm	Cancels engine feedback operation
25	Solenoid valve (pressure regulator)	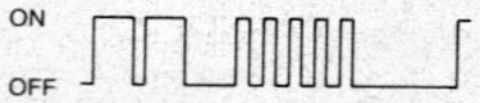	Open or short circuit	—

1.6L AND 1.8L ENGINES TROUBLE CODES 323 AND MX-3 WITH 1.6L ENGINE AND PROTEGE

CODE NO.	LOCATION OF MALFUNCTION	OUTPUT SIGNAL PATTERN	SELF-DIAGNOSIS	FAIL-SAFE
26	Solenoid valve (purge control)		Open or short circuit	—
34	ISC valve			—
41	Solenoid valve (VICS) [DOHC]	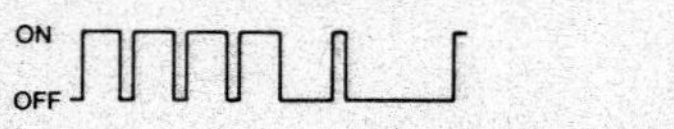		—
55	Pulse generator		No input signal from pulse generator	Shifting performed in accordance with signals from vehicle speed sensor
60	1-2 shift solenoid valve		Open or short circuit	Solenoid valve(s) performs the shifting with as little interference as possible with driving performance Lockup not provided
61	2-3 shift solenoid valve			
62	3-4 shift solenoid valve			
63	Lockup solenoid valve	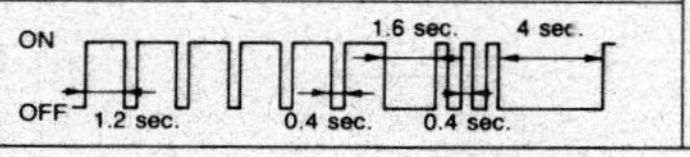		Shifting performed normally Lockup not provided

Note
- **The HOLD indicator does not flash the service code numbers.**
- **If there is more than one failure present, the code numbers will be indicated in numerical order, lowest number first.**
- **After repairing a problem, turn the ignition switch OFF and disconnect the negative battery cable at least 20 seconds to erase the service code numbers from the engine control unit (ECU) memory.**

1.6L AND 1.8L ENGINES TROUBLE CODE TROUBLESHOOTING—323 AND MX-3 WITH 1.6L ENGINE AND PROTEGE

CODE No.	02 (DISTRIBUTOR Ne-SIGNAL)		
STEP	INSPECTION		ACTION
1	Check distributor circuit for poor connection	Yes	Repair or replace connector
		No	**SOHC** Go to Step 3
			DOHC Go to next step
2	Check if Code No.03 is also present	Yes	Go to next step
		No	Go to Step 5
3	Check terminal-wire for continuity	Yes	Go to next step
		No	Repair or replace
4	Check if battery voltage exists at distributor terminal-wire	Yes	Go to next step
		No	Check for open circuit in wiring from distributor to main relay (FUEL INJ relay)
5	Check terminal-wire between distributor and ECU terminal 2E (MTX) 2A (ATX) for continuity and ground	Yes	Go to next step
		No	Repair or replace
6	Check if ECU terminal 2E (MTX) 2A (ATX) voltage is OK	Yes	Replace ECU
		No	Go to next step
7	Check if 0V or 4.5V—5.5V exists at distributor terminal-wire (W)	Yes	Replace distributor
		No	Go to next step
8	Check if 4.5V—5.5V exists at ECU terminal 2E (MTX) 2A (ATX) (With distributor connector disconnected)	Yes	Check for short circuit in wiring from distributor to ECU
		No	Replace ECU

MTX

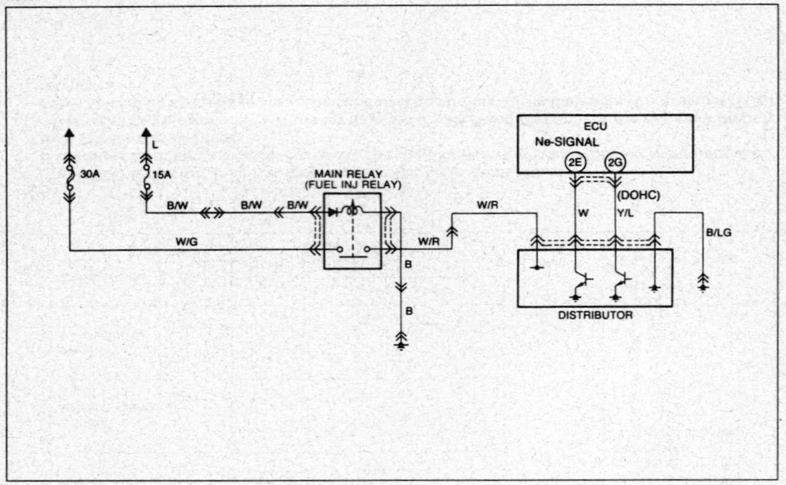

ATX

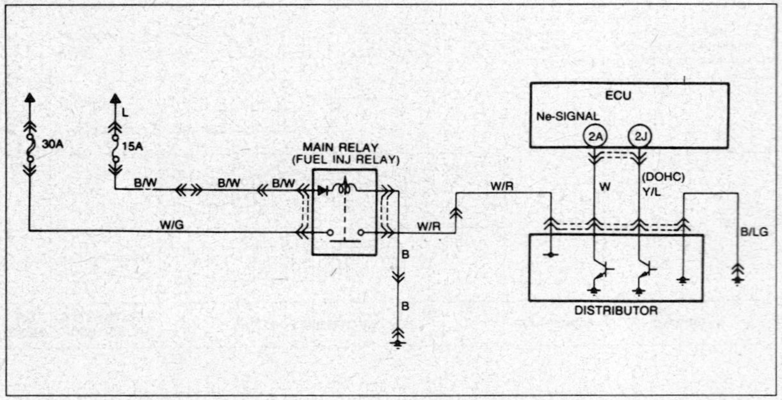

CODE No.	03 (DISTRIBUTOR G-SIGNAL) — DOHC		
STEP	INSPECTION		ACTION
1	Check distributor circuit for poor connection	Yes	Repair or replace connector
		No	Go to next step
2	Check if Code No.02 is also present	Yes	Go to next step
		No	Go to Step 5
3	Check terminal-wire for continuity	Yes	Go to next step
		No	Repair or replace
4	Check if battery voltage exists at distributor terminal-wire	Yes	Go to next step
		No	Check for open circuit in wiring from distributor to main relay
5	Check terminal-wire between distributor and ECU terminal 2G (MTX) 2J (ATX) for continuity and ground	Yes	Go to next step
		No	Repair or replace
6	Check if ECU terminal 2G (MTX) 2J (ATX) voltage is OK	Yes	Replace ECU
		No	Go to next step
7	Check if 0V or 4.5V—5.5V exists at distributor terminal-wire	Yes	Replace distributor
		No	Go to next step
8	Check if 4.5V—5.5V exists at ECU terminal 2G (MTX) 2J (ATX) (With distributor connector disconnected)	Yes	Check for short circuit in wiring from distributor to ECU
		No	Replace ECU

CODE No.	08 (AIRFLOW METER)		
STEP	INSPECTION	ACTION	
1	Check airflow meter circuit for poor connection	Yes	Repair or replace connector
		No	Go to next step
2	Check if Code No.10 is also present	Yes	Check for open circuit in wiring from airflow meter terminal-wire (B/LG) to ground
		No	Go to next step
3	Check if resistance of airflow meter is OK Airflow meter / Fully closed (Ω) / Fully open (Ω) D(LG/R)—F(R) / 20—600 / 20—1,000 D(LG/R)—C(B/LG) / 200—400	Yes	Go to next step
		No	Replace airflow meter
4	Check wire harness between airflow meter and ECU for continuity Airflow meter / ECU D (LG/R) / 2K (MTX) 2I (ATX) F (R) / 2O (MTX) 2B (ATX)	Yes	Go to next step
		No	Repair or replace wire harness
5	Check if ECU terminals voltages are OK MTX...2D, 2K and 2O ATX... 2B, 2I and 3D	Yes	Replace ECU
		No	Check for short circuit in wiring from airflow meter to ECU

MTX

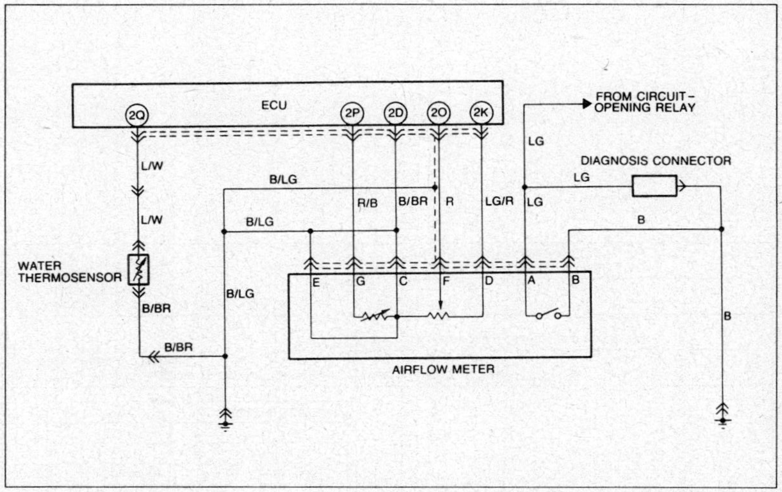

ATX

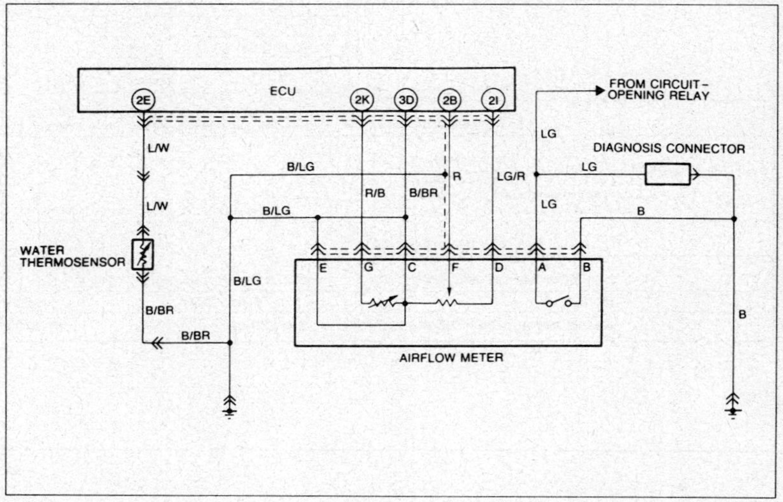

<table>
<tr><th>CODE No.</th><th colspan="3">09 (WATER THERMOSENSOR)</th></tr>
<tr><th>STEP</th><th>INSPECTION</th><th colspan="2">ACTION</th></tr>
<tr><td rowspan="2">1</td><td rowspan="2">Check water thermosensor circuit for poor connection</td><td>Yes</td><td>Repair or replace connector</td></tr>
<tr><td>No</td><td>Go to next step</td></tr>
<tr><td rowspan="2">2</td><td rowspan="2">Check wire harness between water thermosensor and ECU for continuity
<table>
<tr><th>Water thermosensor</th><th>ECU</th></tr>
<tr><td>A (L/W)</td><td>2Q (MTX) 2E (ATX)</td></tr>
<tr><td>B (B/BR)</td><td>2D (MTX) 3D (ATX)</td></tr>
</table></td><td>Yes</td><td>Go to next step</td></tr>
<tr><td>No</td><td>Repair or replace</td></tr>
<tr><td rowspan="2">3</td><td rowspan="2">Check if resistance of water thermosensor is OK
<table>
<tr><th>Coolant temp.</th><th>Resistance (kΩ)</th></tr>
<tr><td>−20°C (−4°F)</td><td>14.6—17.8</td></tr>
<tr><td>20°C (68°F)</td><td>2.21—2.69</td></tr>
<tr><td>80°C (176°F)</td><td>0.29—0.35</td></tr>
</table></td><td>Yes</td><td>Go to next step</td></tr>
<tr><td>No</td><td>Replace water thermosensor</td></tr>
<tr><td rowspan="2">4</td><td rowspan="2">Check if same Code No. is present following after-repair procedure</td><td>Yes</td><td>Go to next step</td></tr>
<tr><td>No</td><td>Water thermosensor and circuit OK</td></tr>
<tr><td rowspan="2">5</td><td rowspan="2">Check if ECU terminals voltages are OK
MTX...2D and 2Q
ATX...2E and 3D</td><td>Yes</td><td>Replace ECU</td></tr>
<tr><td>No</td><td>Check for short circuit in wiring from water thermosensor to ECU</td></tr>
</table>

<table>
<tr><th>CODE No.</th><th colspan="3">10 (INTAKE AIR THERMOSENSOR — IN AIRFLOW METER)</th></tr>
<tr><th>STEP</th><th>INSPECTION</th><th colspan="2">ACTION</th></tr>
<tr><td rowspan="2">1</td><td rowspan="2">Check intake air thermosensor circuit for poor connection</td><td>Yes</td><td>Repair or replace connector</td></tr>
<tr><td>No</td><td>Go to next step</td></tr>
<tr><td rowspan="2">2</td><td rowspan="2">Check if Code No.08 is also present</td><td>Yes</td><td>Check for open circuit in wiring from airflow meter terminal wires (B/LG) and (B/BR) to ground</td></tr>
<tr><td>No</td><td>Go to next step</td></tr>
<tr><td rowspan="2">3</td><td rowspan="2">Check wire harness between intake air thermosensor and ECU for continuity
<table>
<tr><th>Intake air thermosensor (In airflow meter)</th><th>ECU</th></tr>
<tr><td>C (B/BR)</td><td>2D (MTX) 3D (ATX)</td></tr>
<tr><td>G (R/B)</td><td>2P (MTX) 2K (ATX)</td></tr>
</table></td><td>Yes</td><td>Go to next step</td></tr>
<tr><td>No</td><td>Repair or replace</td></tr>
<tr><td rowspan="2">4</td><td rowspan="2">Check if resistance of airflow meter between terminals C (B/BR) and G (R/B) is OK
<table>
<tr><th>Temperature</th><th>Resistance (kΩ)</th></tr>
<tr><td>−20°C</td><td>13.6—18.4</td></tr>
<tr><td>20°C</td><td>2.21—2.69</td></tr>
<tr><td>60°C</td><td>0.493—0.667</td></tr>
</table></td><td>Yes</td><td>Go to next step</td></tr>
<tr><td>No</td><td>Replace airflow meter</td></tr>
<tr><td rowspan="2">5</td><td rowspan="2">Check if same Code No. is present following after-repair procedure</td><td>Yes</td><td>Go to next step</td></tr>
<tr><td>No</td><td>Intake air thermosensor and circuit OK</td></tr>
<tr><td rowspan="2">6</td><td rowspan="2">Check if ECU terminals voltages are OK
MTX...2D and 2P
ATX... 2K and 3D</td><td>Yes</td><td>Replace ECU</td></tr>
<tr><td>No</td><td>Check for short circuit in wiring from intake air thermosensor to ECU</td></tr>
</table>

CODE No.	12 (THROTTLE SENSOR — ATX)		
STEP	INSPECTION		ACTION
1	Check throttle sensor circuit for poor connection	Yes	Repair or replace connector
		No	Go to next step
2	Check wire harness between throttle sensor and ECU for continuity Throttle sensor / ECU: A (LG/R) — 2K (MTX) 2I (ATX) B (LG/W) — 2M (MTX) 2F (ATX) D (B/BR) — 2D (MTX) 3D (ATX)	Yes	Go to next step
		No	Repair or replace
3	Check if resistances between terminals B (LG/W) and D (B/BR) are OK Throttle valve / Resistance: Fully closed — Below 1 kΩ Fully open — Approx. 5 kΩ	Yes	Go to next step
		No	Adjust or replace throttle sensor
4	Check if ECU terminal 2M (MTX) 2F (ATX) voltage is OK	Yes	Replace ECU
		No	Check for short circuit in wiring from throttle sensor to ECU

ATX

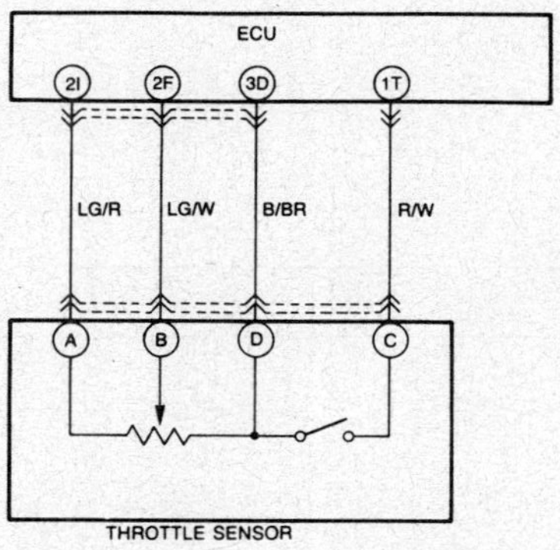

CODE No.	14 (ATMOSPHERIC PRESSURE SENSOR — IN ECU)
STEP	ACTION
1	Replace ECU

CODE No.	15 (OXYGEN SENSOR — INACTIVATION)		
	Note •If Code Nos. 15 and 17 are both present, first perform the checking procedure for Code No.17		
STEP	INSPECTION		ACTION
1	Check oxygen sensor circuit for poor connection	Yes	Repair or replace connector
		No	Go to next step
2	Check if oxygen sensor output voltage is OK	Yes	Go to next step
		No	Replace oxygen sensor
3	Check wire harness between oxygen sensor and ECU terminal 2N (MTX) 2C (ATX) for continuity	Yes	Go to next step
		No	Repair or replace
4	Check if ECU terminal 2N (MTX) 2C (ATX) voltage is OK	Yes	Go to next step
		No	Check for short circuit in wiring from oxygen sensor to ECU
5	Check if sensitivity of oxygen sensor is OK	Yes	Replace ECU
		No	Try known good oxygen sensor and check if condition improves

MTX

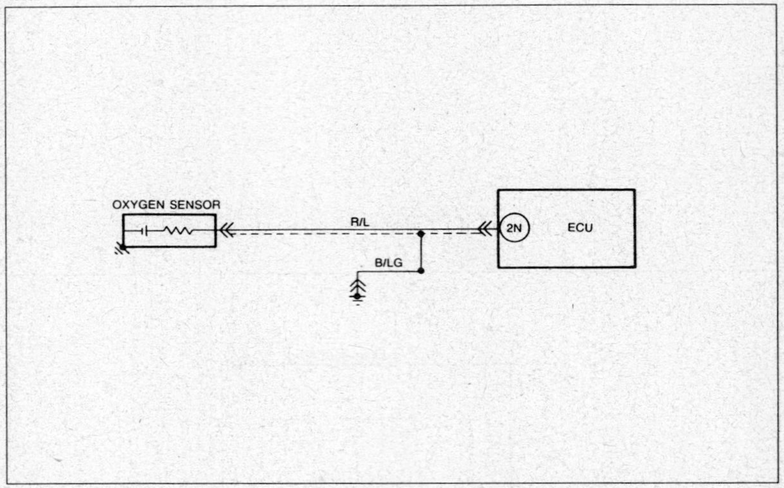

ATX

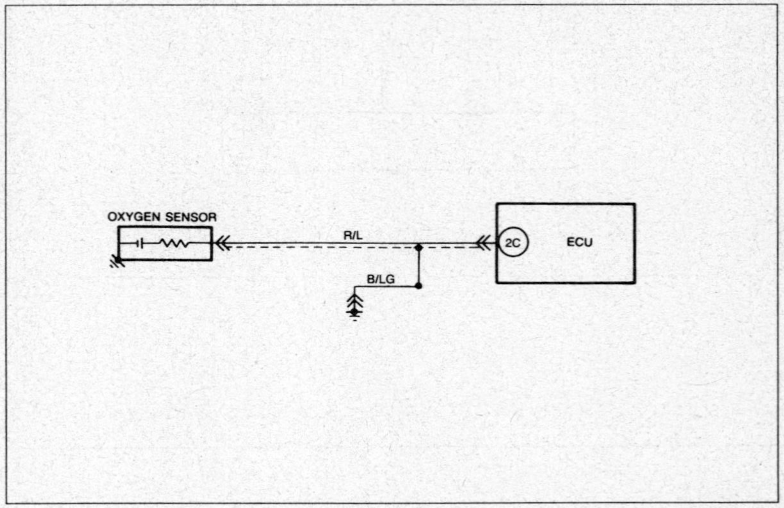

CODE No.	17 (Oxygen sensor—Inversion)		
STEP	INSPECTION		ACTION
1	Check if same Code No. is present after performing after-repair procedure	Yes	Go to next step
		No	Check oxygen sensor circuit for poor connection ⇨ If OK, perform troubleshooting Code No.15 ⇨ If not OK, repair or replace connector
2	Check monitor lamp of Self-Diagnosis Checker illuminate at idle after warming-up engine and running it at **2,500—3,000 rpm** for **3 minutes**	Yes	Go to next step **Note** • **A/F mixture rich**
		No	Go to Step 5 **Note** • **A/F mixture is lean or misfire occurring**
3	Check for correct fuel line pressure at idle **Fuel line pressure:** **265—314 kPa (2.7—3.2 kg/cm², 38—46 psi)** **(Vacuum hose to pressure regulator disconnected)**	Yes	Go to next step
		No	**High pressure** Check if fuel return hose is clogged or restricted ⇨ If OK, replace pressure regulator ⇨ If not OK, repair or replace
4	Check injectors for fuel leakage	Yes	Replace injector
		No	Check if water thermosensor is OK ⇨ If OK, Go to Step 11 ⇨ If not OK, replace water thermosensor
5	Disconnect each high-tension lead at idle and check if engine speed decreases equally at each cylinder	Yes	Go to next step
		No	Go to Step 8

Circuit Diagram
MTX

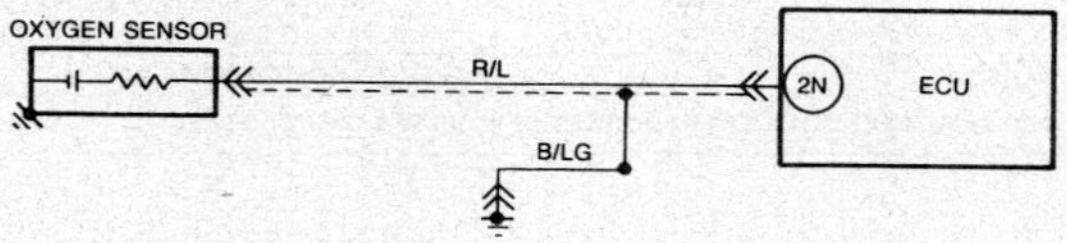

STEP	INSPECTION		ACTION
6	Check for correct fuel line pressure at idle **Fuel line pressure:** **265—314 kPa (2.7—3.2 kg/cm², 38—46 psi)** **(Vacuum hose to pressure regulator disconnected)**	Yes	Go to next step
		No	**Low pressure** Check fuel line pressure while pinching fuel return hose ⇨ If fuel line pressure quickly increases, check pressure regulator ⇨ If fuel line pressure gradually increases, check for clogging between fuel pump and pressure regulator If not clogged, check fuel pump maximum pressure
7	Check intake air system components for air leakage	Yes	Go to Step 11
		No	Replace
8	Check for misfire of dead cylinder from Step 5 inspection	Yes	Repair or replace ignition system
		No	Go to next step
9	Check for injector operating sound at idle of dead cylinder from Step 5 inspection	Yes	Go to next step
		No	Check for battery voltage at injector connector terminal-wire (W/R) ⇨ If Yes, replace injector ⇨ If No, check wire harness for short or open
10	Replace injector at dead cylinder from Step 5 inspection Check if same Code No. is present following performing after-repair procedure	Yes	Go to next step
		No	Removed injector at fault
11	Try known good ECU and check if condition improves		

Circuit Diagram
ATX

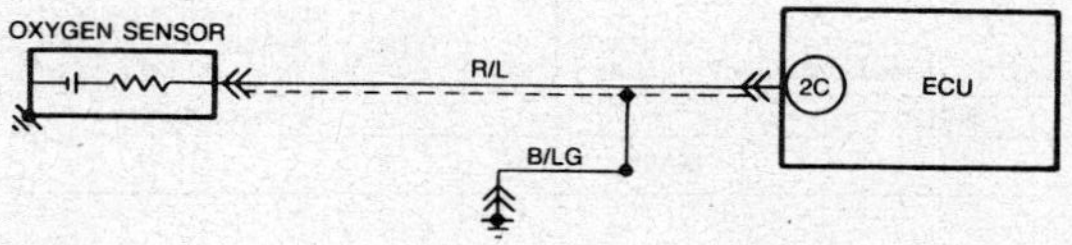

CODE No.	25 (SOLENOID VALVE — PRESSURE REGULATOR CONTROL)—BP ENGINE		
STEP	INSPECTION		ACTION
1	Disconnect connector from ECU and check if battery voltage exists at ECU terminal 2T (MTX) 3M (ATX) wire (G/O) with ignition switch ON	Yes	Check ECU terminal connector for poor connection ⇨ If OK, replace ECU ⇨ If not OK, repair or replace connector
		No	Go to next step
2	In same condition as Step 1, check if battery voltage exists at solenoid valve connector terminal-wire (G/O)	Yes	Repair or replace wire (G/O)
		No	Go to next step
3	Check if solenoid valve is OK	Yes	Check for short or open circuit in wiring from main relay (FUEL INJ relay) to solenoid valve
		No	Replace solenoid valve

CODE No.	26 (SOLENOID VALVE — PURGE CONTROL)		
STEP	INSPECTION		ACTION
1	Disconnect connector from ECU and check if battery voltage exists at ECU terminal 2X (MTX) 2O (ATX) wire harness (W/L) with ignition switch ON	Yes	Check ECU terminal connector for poor connection ⇨ If OK, replace ECU ⇨ If not OK, repair or replace connector
		No	Go to next step
2	In same condition as Step 1, check if battery voltage exists at solenoid valve connector terminal-wire (W/L)	Yes	Repair or replace wire harness (W/L)
		No	Go to next step
3	Check if solenoid valve is OK	Yes	Check for short or open circuit in wiring from main relay (FUEL INJ relay) to solenoid valve
		No	Replace solenoid valve

Circuit Diagram
MTX

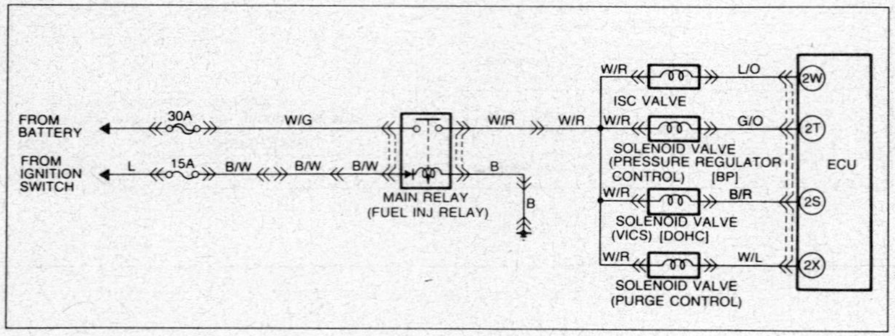

CODE No.	34 (ISC VALVE)		
STEP	INSPECTION		ACTION
1	Disconnect connector from ECU and check if battery voltage exists at ECU terminal 2W (MTX) 3Q (ATX) wire (L/O) with ignition switch ON	Yes	Check ECU terminal connector for poor connection ⇨ If OK, replace ECU ⇨ If not OK, repair or replace connector
		No	Go to next step
2	In same condition as Step 1, check if battery voltage exists at ISC valve connector terminal-wire (L/O)	Yes	Repair or replace wire (L/O)
		No	Go to next step
3	Check ISC valve for correct resistance **Resistance: 11—13Ω [at 20°C(68°F)]**	Yes	Check for short or open circuit in wiring from main relay (FUEL INJ relay) to ISC valve
		No	Replace ISC valve

CODE No.	41 (SOLENOID VALVE — VICS)—DOHC		
STEP	INSPECTION		ACTION
1	Disconnect connector from ECU and check if battery voltage exists at ECU terminal 2S (MTX) 3I (ATX) wire (B/R) with ignition switch ON	Yes	Check ECU terminal connector for poor connection ⇨ If OK, replace ECU ⇨ If not OK, repair or replace connector
		No	Go to next step
2	In same condition as Step 1, check if battery voltage exists at solenoid valve connector terminal-wire (B/R)	Yes	Repair or replace wire (B/R)
		No	Go to next step
3	Check if solenoid valve is OK	Yes	Check for short or open circuit in wiring from main relay to solenoid valve
		No	Replace solenoid valve

Circuit Diagram
ATX

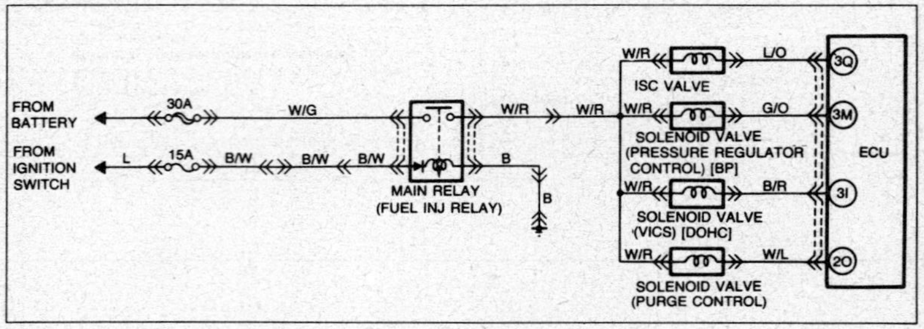

1.6L AND 1.8L ENGINES SWITCH MONITOR FUNCTION 323 AND MX-3 WITH 1.6L ENGINE AND PROTEGE

SWITCH MONITOR FUNCTION
Individual switches can be inspected by the **SST** (Self-Diagnosis checker).

Note
- **The TEN terminal of the diagnosis connector must be grounded and the ignition switch ON (engine stopped).**
- **If either of the switches remain activated, the monitor lamp will be illuminated.**

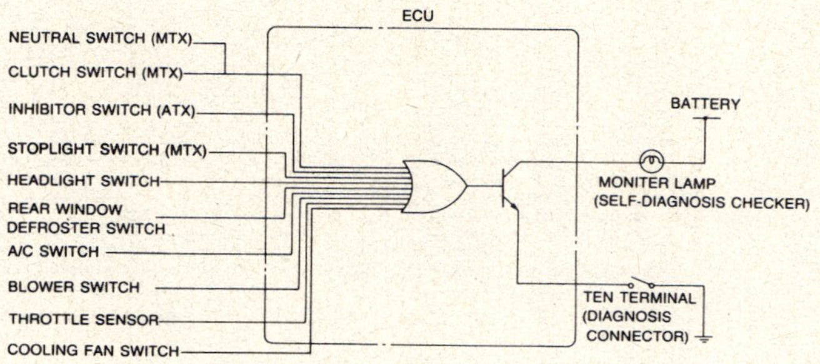

Switch		Self-Diagnosis Checker (Monitor lamp)		Remark
		Light ON	Light OFF	
Clutch switch	(MTX)	Pedal released	Pedal depressed	In gear
Neutral switch	(MTX)	In gear	Neutral	Clutch pedal released
Inhibitor switch	(ATX)	L, S, D or R range	N or P range	—
Throttle sensor		Pedal depressed (Not fully)	Pedal released/fully depressed	—
Stoplight switch		Pedal depressed	Pedal released	—
Headlight switch		ON	OFF	Headlights/parking lights: ON
Blower switch		ON	OFF	Blower switch at 2nd or above position
A/C switch		ON	OFF	Fan speed control at 1st position
Rear window defroster switch		ON	OFF	—
Water thermoswitch		Fan operating	Fan not operating	—

1.6L AND 1.8L ENGINES SWITCH MONITOR FUNCTION TROUBLESHOOTING—323 AND MX-3 WITH 1.6L ENGINE AND PROTEGE

Procedure

Set conditions to deactivate each switch
- All accessories OFF
- Transaxle in neutral
- All pedals released

Verify that monitor lamp does not illuminate

YES → Check each switch as described

NO → Check each switch and related wire harness
- Clutch and Neutral switch
- Throttle sensor
- Stoplight switch
- Headlight switch
- Rear defroster switch
- Blower switch
- Inhibitor switch
- A/C switch
- Water thermoswitch

Neutral and Clutch switch (MTX)

Shift transaxle into gear
Check that monitor lamp illuminates with clutch pedal released

NO → PC:
- Neutral switch malfunction
- Clutch switch malfunction
- Open circuit in related wire harness
- ECU terminal 1V voltage incorrect
- ECU malfunction

YES → Depress clutch pedal
Check that monitor lamp does not illuminate
Return transmission to neutral

NO → PC: • Clutch switch malfunction

SWITCH MONITOR FUNCTION TROUBLESHOOTING—323 AND MX-3 WITH 1.6L ENGINE AND PROTEGE

Headlight switch

Turn headlight switch ON
Check that monitor lamp illuminates

- NO → PC: • Headlight switch malfunction • Open circuit in wire harness from headlight switch to ECU • ECU terminal 1U (MTX) 1H (ATX) voltage incorrect • ECU malfunction
- YES → Headlight switch OK
Turn headlight switch OFF

Rear window defroster switch

Turn rear window defroster switch ON
Check that monitor lamp illuminates

- NO → PC: • Rear window defroster switch malfunction • Open circuit in wire harness from rear window defroster switch to ECU • ECU terminal 1T (MTX) 1J (ATX) voltage incorrect • ECU malfunction
- YES → Rear defroster switch OK
Turn rear window defroster switch OFF

Inhibitor switch (ATX)

Shift transaxle into L, S, D, or P range
Check that monitor lamp illuminates

- NO → PC: • Inhibitor switch malfunction • Open or short circuit in related wire harness • ECU terminal 1V voltage incorrect • ECU malfunction
- YES → Inhibitor switch OK
Shift into P or N range

Throttle sensor

Depress accelerator pedal (not fully) and check that monitor lamp illuminates

- NO → PC: • Throttle sensor malfunction • Short circuit in wire harness from throttle sensor to ECU • ECU terminal 1N (MTX) 1T (ATX) voltage incorrect • ECU malfunction
- YES → Depress acceleration pedal fully and check that monitor lamp does not illuminate
  - NO → PC: • Throttle sensor malfunction • Open circuit in wire harness from throttle sensor to ECU • ECU terminal 2L (MTX) or 2I (ATX) voltage incorrect
  - YES → Throttle sensor OK
Release accelerator pedal

Stoplight switch

Depress brake pedal
Check that monitor lamp illuminates

- NO → PC: • Stoplight switch malfunction • Open circuit in wire harness from stoplight switch to ECU • ECU terminal 1O (MTX) 1Q (ATX) voltage incorrect • ECU malfunction
- YES → Stoplight switch OK
Release brake pedal

Blower switch

Turn blower switch to 2nd or higher position
Check that monitor lamp illuminates

- NO → PC: • Blower switch malfunction • Open circuit in wire harness from blower switch to ECU • ECU terminal 1S (MTX) 1P (ATX) voltage incorrect • ECU malfunction
- YES → Blower switch OK

1.6L AND 1.8L ENGINES ENGINE CONTROL UNIT TERMINAL VOLTAGES—323 AND MX-3 WITH 1.6L ENGINE AND PROTEGE

V_B: Battery voltage

Terminal	Input	Output	Connection to	Test condition		Correct voltage	Remark
1A	—	—	Battery	Constant		V_B	For backup
1B	○		Main relay (FUEL INJ relay)	Ignition switch	OFF	0V	—
					ON	V_B	
1C	○		Ignition switch (START)	While cranking		Approx. 10V	—
				Ignition switch ON		0V	
1D		○	Self-Diagnosis Checker (Monitor lamp)	Test switch at "SELF-TEST" Lamp illuminated for 3 sec. after ignition switch OFF→ON		4.5—5.5V	With Self-Diagnosis Checker and System Selector
				Lamp not illuminated after 3 sec.		V_B	
				Test switch at "O_2 MONITOR" at idle Monitor lamp illuminated		4.5—5.5V	
				Test switch at "O_2 MONITOR" at idle Monitor lamp not illuminated		V_B	
1E		○	Malfunction indicator lamp (MIL)	Lamp illuminated for 3 sec. after ignition switch OFF→ON		Below 2.5V	With System Selector test switch at "SELF-TEST"
				Lamp not illuminated after 3 sec.		V_B	
				Lamp illuminated		Below 2.5V	
				Lamp not illuminated		V_B	
1F		○	Self-Diagnosis Checker (Code Number)	Buzzer sounded for 3 sec. after ignition switch OFF→ON		Below 2.5V	• With Self-Diagnosis Checker and System Selector • With System Selector test switch at "SELF-TEST"
				Buzzer not sounded after 3 sec.		V_B	
				Buzzer sounded		Below 2.5V	
				Buzzer not sounded		V_B	
1G		○	Igniter	Ignition switch ON		0V	—
				Idle		Approx. 0.2V	
1H	—	—	—	—		—	—
1I	—	—	—	—		—	—
1J		○	A/C relay	Ignition switch ON		V_B	—
				A/C switch ON at idle		Below 2.5V	
				A/C switch OFF at idle		V_B	
1K	○		Diagnosis connector (TEN terminal)	System Selector test switch at "O_2 MONITOR"		V_B	Ignition switch ON
				System Selector test switch at "SELF-TEST"		0V	
1L	○		DRL relay (Canada)	Parking brake pulled with ignition switch ON (DRL OFF)		V_B	• DRL: Daytime Running Lights
				Idle (DRL ON)		Below 2.5V	

2Y	2W	2U	2S	2Q	2O	2M	2K	2I	2G	2E	2C	2A	1U	1S	1Q	1O	1M	1K	1I	1G	1E	1C	1A
2Z	2X	2V	2T	2R	2P	2N	2L	2J	2H	2F	2D	2B	1V	1T	1R	1P	1N	1L	1J	1H	1F	1D	1B

V_B: Battery voltage

Incorrect voltage		Possible cause
Always 0V		• ROOM 10A fuse burned • Open circuit in wiring from ROOM 10A fuse to ECU terminal 1A
Always 0V		• Main relay malfunction • Open or short circuit in wiring from main relay to ECU terminal 1B
Always 0V (Starter turns)		• Open or short circuit in wiring from starter interlock switch (USA), ignition switch (CANADA), to ECU terminal 1C
Always 0V		• Main relay (FUEL INJ relay) malfunction • Open circuit in wiring from main relay to diagnosis connector terminal +B • Open or short circuit in wiring from diagnosis connector terminal MEN to ECU terminal 1D
Always V_B		• Poor connection at ECU connector • ECU malfunction
Always approx. 5V		• ECU malfunction
Always below 2.5V	MIL always ON	• Short circuit in wiring from combination meter to ECU terminal 1E • ECU malfunction
	MIL never ON	• Open circuit in wiring from combination meter to ECU terminal 1E
Always V_B		• Poor connection at ECU connector • ECU malfunction
Always below 2.5V	No display on Self-Diagnosis Checker	• Main relay (FUEL INJ relay) malfunction • Open circuit in wiring from main relay to diagnosis connector terminal +B
	"88" displayed and buzzer sounds continuously	• Open or short circuit in wiring from diagnosis connector terminal FEN to ECU terminal 1F
Always V_B		• Poor connection at ECU connector • ECU malfunction
Always 0V		• Short circuit in wiring from igniter to ECU terminal 1G
—		—
—		—
Always below 2.5V	A/C does not operate	• A/C relay malfunction • Open circuit in wiring from ignition switch to A/C relay • Open circuit in wiring from A/C relay to ECU terminal 1J
	A/C switch OFF but A/C operates	• Short circuit in wiring from A/C relay to ECU terminal 1J • ECU malfunction
Always V_B		• Poor connection at ECU connector • ECU malfunction
Always below 1.0V		• Short circuit in wiring from diagnosis connector terminal TEN to ECU terminal 1K
Always V_B		• Open circuit in wiring from diagnosis connector terminal TEN to ECU terminal 1K • Open circuit in wiring from diagnosis connector terminal GND to ground
Always below 2.5V	DRL ON when ignition switch ON	• Short circuit in wiring from DRL relay to ECU terminal 1L • Short circuit in wiring from DRL relay to DRL unit
	DRL never ON	• Open circuit in wiring from DRL relay to ignition switch
Always V_B		• Parking brake switch always ON • DRL unit malfunction

1.6L AND 1.8L ENGINES ENGINE CONTROL UNIT TERMINAL VOLTAGES—323 AND MX-3 WITH 1.6L ENGINE AND PROTEGE

VB: Battery voltage

Terminal	Input	Output	Connection to	Test condition	Correct voltage	Remark
1M	—	—	—	—	—	—
1N	○		Throttle sensor (Idle switch)	Accelerator pedal released	0V	Ignition switch ON
				Accelerator pedal depressed	VB	
1O	○		Stoplight switch	Brake pedal released	0V	—
				Brake pedal depressed	VB	
1P	○		P/S pressure switch	Ignition switch ON	VB	—
				P/S ON at idle	0V	
				P/S OFF at idle	VB	
1Q	○		A/C switch	A/C switch ON	Below 2.5V	Ignition switch ON and blower motor ON
				A/C switch OFF	VB	
1R	○		Fan switch	Fan operating (Engine coolant temperature over 97°C (207°F) or diagnosis connector terminal TFA grounded)	0V	—
				Fan not operating (Idle)	VB	
1S	○		Blower control switch	Blower control switch OFF or 1st position	VB	Ignition switch ON
				Blower control switch 2nd or higher position	0V	
1T	○		Rear window defroster switch	Rear window defroster switch OFF	0V	Ignition switch ON
				Rear window defroster switch ON	VB	
1U	○		Headlight switch	Headlights ON	VB	—
				Headlights OFF	0V	
1V	○		Neutral/Clutch switches	Neutral position or clutch pedal depressed	0V	—
				Others	VB	

2Y	2W	2U	2S	2O	2O	2M	2K	2I	2G	2E	2C	2A	1U	1S	1Q	1O	1M	1K	1I	1G	1E	1C	1A
2Z	2X	2V	2T	2R	2P	2N	2L	2J	2H	2F	2D	2B	1V	1T	1R	1P	1N	1L	1J	1H	1F	1D	1B

VB: Battery voltage

Incorrect voltage		Possible cause
—		—
Always below 1.0V		• Throttle sensor misadjustment • Short circuit in wiring from throttle sensor to ECU terminal 1N • ECU malfunction
Always VB		• Throttle sensor misadjustment • Open circuit in wiring from throttle sensor to ECU terminal 1N • Open circuit in wiring from throttle sensor to ECU terminal 2D
Always below 1.0V (Stoplight OK)		• Open circuit in wiring from stoplight switch to ECU terminal 1O
Always below 1.0V		• P/S pressure switch malfunction • Short circuit in wiring from P/S pressure switch to ECU terminal 1P • ECU malfunction
Always VB		• P/S pressure switch malfunction • Open circuit in wiring from P/S pressure switch to ECU terminal 1P • Open circuit in wiring from P/S pressure switch to ground
Always below 2.5V (Blower fan OK)		• A/C switch malfunction • Short circuit in wiring from A/C switch to ECU terminal 1Q • Poor connection at ECU connector • ECU malfunction
Always VB (Blower fan OK)		• A/C switch malfunction • Open circuit in wiring from A/C switch to ECU terminal 1Q • Open circuit in wiring from A/C switch to blower control switch
Always below 1.0V (Electrical cooling fan OK)		• Open or short circuit in wiring from fan relay to ECU terminal 1R • ECU malfunction
Always below 1.0V (Blower fan OK)		• Short circuit in wiring from blower control switch to ECU terminal 1S • Poor connection at ECU connector • ECU malfunction
Always VB (Blower fan OK)		• Open circuit in wiring from blower control switch to ECU terminal 1S
Always below 1.0V	Illumination lamp ON when rear window defroster switch ON	• Open or short circuit in wiring from rear window defroster switch to ECU terminal 1T
	Illumination lamp never ON	• Open circuit in wiring from ignition switch to rear window defroster switch • Rear window defroster switch malfunction
Always below 1.0V (Headlights OK)		• Open or short circuit in wiring from headlight relay to ECU terminal 1U
Always below 1.0V		• Neutral switch malfunction • Clutch switch malfunction • Short circuit in wiring from ECU terminal 1V to neutral or clutch switch
Always VB		• Neutral switch malfunction • Clutch switch malfunction • Open circuit in wiring from ECU terminal 1V to neutral or clutch switch • Poor connection at ECU connector

1.6L AND 1.8L ENGINES ENGINE CONTROL UNIT TERMINAL VOLTAGES 323 AND MX-3 WITH 1.6L ENGINE AND PROTEGE

VB: Battery voltage

Terminal	Input	Output	Connection to	Test condition	Correct voltage	Remark	Incorrect voltage	Possible cause
2A	—	—	Ground (Injector)	Constant	0V	—	Above 0V	• Poor contact at ground terminal • Open circuit in wiring from ECU to ground
2B	—	—	Ground (Output)	Constant	0V	—		
2C	—	—	Ground (CPU)	Constant	0V	—		
2D	—	—	Ground (Input)	Constant	0V	—		
2E	O		Distributor (Ne-signal)	Ignition switch ON	Approx. 0V or 4.5—5.5V	—	Always approx. 0V or approx. 2V	• Refer to Code No.02 Troubleshooting
				Idle	Approx. 2V			
2F	—	—	—	—	—	—	—	—
2G	O		Distributor (G-signal) [DOHC]	Ignition switch ON	Approx. 0V or 4.5—5.5V		Always approx. 0V or approx. 1.5V	• Refer to Code No.03 Troubleshooting
				Idle	Approx. 1.5V			
2H	O		Ground (California)	Constant	0V	—	Approx. 4.5—5.5V (California)	• Open circuit in wiring from ECU terminal 2H to ground
			Main relay (Canada)	Ignition switch ON	VB		Always 0V (Canada)	• Short circuit in wiring from ECU terminal 2H to main relay (FUEL INJ relay)
2I	—	—	—	—	—	—	—	—
2J	O		Open	Constant	Approx. 1—2.5V	—	Approx. 4.5—5.5V	• ECU malfunction
2K	O		Throttle sensor (ATX)/Airflow meter	Constant	4.5—5.5V	—	Always 0V	• Short circuit in wiring from ECU terminal 2K to throttle sensor, or airflow meter • Poor connection at ECU connector • ECU malfunction
							Below 4.5V or above 5.5V	• ECU malfunction
2L	O		Throttle sensor (Power switch)	Accelerator pedal released	Approx. 4.5—5.5V	—	Always 0V	• Throttle sensor malfunction • Short circuit in wiring from ECU terminal 2L to throttle sensor • Poor connection at ECU connector • ECU malfunction
				Accelerator pedal fully depressed	0V		Always 4.5—5.5V	• Throttle sensor misadjustment • Open circuit in wiring from ECU terminal 2L to throttle sensor • Open circuit in wiring from ECU terminal 2L to ECU terminal 2D
2M	O		Throttle sensor	Accelerator pedal released	Approx. 0.5V	—	Always constant	• Open circuit in wiring from ECU terminal 2M to throttle sensor • Open circuit in wiring from ECU terminal 2K to throttle sensor • Open circuit in wiring from ECU terminal 2D to throttle sensor
				Accelerator pedal fully depressed	Approx. 4.0V		Always above 1V	• Throttle sensor misadjustment
2N	O		Oxygen sensor	Ignition switch ON	0V	—	0V after warm-up	• Refer to Code No.15 Troubleshooting
				Idle (Cold engine)	0V		Always approx. 1V after warm-up	• Refer to Code No.17 Troubleshooting
				Idle (After warm-up)	0—1.0V			
				Increasing engine speed (After warm-up)	0.5—1.0V			
				Deceleration	0—0.4V			
2O	O		Airflow meter	Ignition switch ON	Approx. 3.8V	—	Always 0V or 4.5—5.5V	• Refer to Code No.08 Troubleshooting
				Idle	Approx. 3.3V			
2P	O		Intake air thermosensor	Ambient air temperature 20°C (68°F)	Approx. 2.5V	Built in airflow meter	Always 0V or 4.5—5.5V	• Refer to Code No.10 Troubleshooting
2Q	O		Water thermosensor	Engine coolant temperature 20°C (68°F)	Approx. 2.5V	—	Always 0V or 4.5—5.5V	• Refer to Code No.09 Troubleshooting
				After warm-up	Below 0.5V			
2R	—	—	—	—	—	—	—	—

2Y	2W	2U	2S	2Q	2O	2M	2K	2I	2G	2E	2C	2A	1U	1S	1Q	1O	1M	1K	1I	1G	1E	1C	1A		
2Z	2X	2V	2T	2R	2P	2N	2L	2J	2H	2F	2D	2B	1V	1T	1R	1P	1N	1L	1J	1H	1F	1D	1B		

1.6L AND 1.8L ENGINES ENGINE CONTROL UNIT TERMINAL VOLTAGES—323 AND MX-3 WITH 1.6L ENGINE AND PROTEGE

VB: Battery voltage

Terminal	Input	Output	Connection to	Test condition	Correct voltage	Remark
2S		○	Solenoid valve (VICS)	Engine speed below 5,000 rpm	0V	VICS: Variable Inertia Charging System [DOHC]
				Engine speed above 5,000 rpm	VB	
2T		○	Solenoid valve (Pressure regulator) [BP]	60 [DOHC]/120 [SOHC] seconds after engine started when engine coolant temperature above 90°C (194°F) and intake air temperature above 58°C (136°F) [DOHC]/50°C (122°F) [SOHC]	0V	—
				Other condition at idle	VB	
2U		○	Injector (Nos. 1, 3)	Ignition switch ON	VB	* Engine Signal Monitor: Green and red lamps flash
				Idle	VB	
				Engine speed above 2,000 rpm on deceleration (After warm-up)	VB	
2V		○	Injector (Nos. 2, 4)	Ignition switch at idle	VB	
				Idle	VB	
				Engine speed above 2,000 rpm on deceleration (After warm-up)	VB	
2W		○	ISC valve	Ignition switch ON	Approx. 7V	—
				Idle	Approx. 9V	
2X		○	Solenoid valve (Purge control)	Ignition switch ON	VB	—
				Idle	VB	
2Y	—	—	—	—	—	—

2Y	2W	2U	2S	2Q	2O	2M	2K	2I	2G	2E	2C	2A	1U	1S	1Q	1O	1M	1K	1I	1G	1E	1C	1A
2Z	2X	2V	2T	2R	2P	2N	2L	2J	2H	2F	2D	2B	1V	1T	1R	1P	1N	1L	1J	1H	1F	1D	1B

VB: Battery voltage

Incorrect voltage	Possible cause
Always 0V or VB	• Refer to Code No.41 Troubleshooting
Always 0V or VB	• Refer to Code No.25 Troubleshooting
Always 0V	• Main relay (FUEL INJ relay) malfunction • Open or short circuit in wiring from injector to ECU terminal 2U or 2V
Always VB	• ECU malfunction
Always 0V or VB	• Refer to Code No.34 Troubleshooting
Always 0V or VB	• Refer to Code No.26 Troubleshooting
—	—

1.6L AND 1.8L ENGINES ENGINE CONTROL UNIT TERMINAL VOLTAGES—323 AND MX-3 WITH 1.6L ENGINE AND PROTEGE

VB: Battery voltage

Terminal	Input	Output	Connection to	Test condition		Correct voltage	Remark
1A	—	—	Battery	Constant		VB	For backup
1B			Main relay (FUEL INJ relay)	Ignition switch	OFF	0V	—
					ON	VB	
1C			Ignition switch (START)	While cranking		Approx. 10V	—
				Ignition switch ON		0V	
1D			Self-Diagnosis Checker (Monitor lamp)	Test switch at "SELF-TEST" Lamp illuminated for 3 sec. after ignition switch OFF→ON		4.5—5.5V	With Self-Diagnosis Checker and System Selector
				Lamp not illuminated after 3 sec.		VB	
				Test switch at "O2 MONITOR" at idle Monitor lamp illuminated		4.5—5.5V	
				Test switch at "O2 MONITOR" at idle Monitor lamp not illuminated		VB	
1E		○	Malfunction indicator lamp (MIL)	Lamp illuminated for 3 sec. after ignition switch OFF→ON		Below 2.5V	With System Selector test switch at "SELF-TEST"
				Lamp not illuminated after 3 sec.		VB	
				Lamp illuminated		Below 2.5V	
				Lamp not illuminated		VB	
1F		○	Self-Diagnosis Checker (Code Number)	Buzzer sounded for 3 sec. after ignition switch OFF→ON		Below 2.5V	• With Self-Diagnosis Checker and System Selector • With System Selector test switch at "SELF-TEST"
				Buzzer not sounded after 3 sec.		VB	
				Buzzer sounded		Below 2.5V	
				Buzzer not sounded		VB	
1G		○	Igniter	Ignition switch ON		0V	—
				Idle		Approx. 0.2V	
1H	○		Headlight switch	Headlights ON		VB	—
				Headlights OFF		Below 1.0V	
1I	○		Diagnosis connector (TEN terminal)	System Selector test switch at "O2 MONITOR"		VB	Ignition switch ON
				System Selector test switch at "SELF-TEST"		0V	
1J	○		Rear window defroster switch	Rear window defroster switch OFF		0V	Ignition switch ON
				Rear window defroster switch ON		VB	
1K	○		Ground (California)	Constant		0V	—
			Main relay (Canada)	Ignition switch ON		VB	

3Y	3W	3U	3S	3Q	3O	3M	3K	3I	3G	3E	3C	3A	2O	2M	2K	2I	2G	2E	2C	2A	1U	1S	1Q	1O	1M	1K	1I	1G	1E	1C	1A
3Z	3X	3V	3T	3R	3P	3N	3L	3J	3H	3F	3D	3B	2P	2N	2L	2J	2H	2F	2D	2B	1V	1T	1R	1P	1N	1L	1J	1H	1F	1D	1B

VB: Battery voltage

Incorrect voltage		Possible cause
Always 0V		• ROOM 10A fuse burned • Open circuit in wiring from ROOM 10A fuse to ECU terminal 1A
Always 0V		• Main relay malfunction • Open or short circuit in wiring from main relay to ECU terminal 1B
Always 0V (Starter turns)		• Open or short circuit in wiring from or inhibitor switch (ATX) to ECU terminal 1C
Always 0V		• Main relay (FUEL INJ relay) malfunction • Open circuit in wiring from main relay to diagnosis connector terminal +B • Open or short circuit in wiring from diagnosis connector terminal MEN to ECU terminal 1D
Always VB		• Poor connection at ECU connector • ECU malfunction
Always 4.5—5.5V		• ECU malfunction
Always below 2.5V	MIL always ON	• Short circuit in wiring from combination meter to ECU terminal 1E • ECU malfunction
	MIL never ON	• Open circuit in wiring from combination meter to ECU terminal 1E
Always VB		• Poor connection at ECU connector • ECU malfunction
Always below 2.5V	No display on Self-Diagnosis Checker	• Main relay (FUEL INJ relay) malfunction • Open circuit in wiring from main relay to diagnosis connector terminal +B
	"88" displayed and buzzer sounds continuously	• Open or short circuit in wiring from diagnosis connector terminal FEN to ECU terminal 1F
Always VB		• Poor connection at ECU connector • ECU malfunction
Always 0V		• Short circuit in wiring from igniter to ECU terminal 1G
Always below 1.0V (Headlights OK)		• Open or short circuit in wiring from headlight relay to ECU terminal 1H
Always 0V		• Short circuit in wiring from diagnosis connector terminal TEN to ECU terminal 1I
Always VB		• Open circuit in wiring from diagnosis connector terminal TEN to ECU terminal 1I • Open circuit in wiring from diagnosis connector terminal GND to ground
Always 0V	Illumination lamp ON when rear window defroster switch ON	• Open or short circuit in wiring from rear window defroster switch to ECU terminal 1J
	Illumination lamp never ON	• Open circuit in wiring from ignition switch to rear window defroster switch • Rear window defroster switch malfunction
Approx. 5V (California)		• Open circuit in wiring from ECU terminal 1K to ground
Always 0V (Canada)		• Short circuit in wiring from ECU terminal 1K to main relay (FUEL INJ relay)

1.6L AND 1.8L ENGINES ENGINE CONTROL UNIT TERMINAL VOLTAGES—323 AND MX-3 WITH 1.6L ENGINE AND PROTEGE

VB: Battery voltage

Terminal	Input	Output	Connection to	Test condition	Correct voltage	Remark
1L		○	A/C relay	Ignition switch ON	VB	—
				A/C switch ON at idle	Below 2.5V	
				A/C switch OFF at idle	VB	
1M	○		Vehicle speed sensor	While driving	4—5V	—
				Vehicle stopped	0 or 8V	
1N	○		P/S pressure switch	Ignition switch ON	VB	—
				P/S ON at idle	0V	
				P/S OFF at idle	VB	
1O	○		A/C switch	A/C switch ON	Below 2.5V	Ignition switch ON and blower motor ON
				A/C switch OFF	VB	
1P	○		Blower control switch	Blower control switch OFF or 1st position	VB	Ignition switch ON
				Blower control switch 2nd or higher position	Below 1.0V	
1Q	○		Stoplight switch	Brake pedal released	Below 1.0V	—
				Brake pedal depressed	VB	
1R	○		Inhibitor switch (P, N ranges)	N or P range	0V	Ignition switch ON
				Other range	VB	
1S	○		Cruise control main switch	Normal condition	VB	Ignition switch ON
				Set or Resume switch ON or vehicle speed 8 km/h (5 mph) lower than preset speed (Driving vehicle cruise control operation)	Below 1.0V	
1T	○		Throttle sensor (Idle switch)	Accelerator pedal released	Below 1.0V	Ignition switch ON
				Accelerator pedal depressed	VB	
1U	—	—	—	—	—	—
1V	—	—	—	—	—	—
2A	○		Distributor (Ne-signal)	Ignition switch ON	Approx. 0V or 5V	—
				Idle	Approx. 2V	

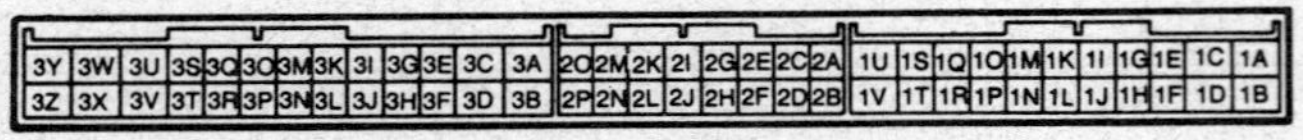

VB: Battery voltage

Incorrect voltage		Possible cause
Always below 2.5V	A/C does not operate	• A/C relay malfunction • Open circuit in wiring from ignition switch to A/C relay • Open circuit in wiring from A/C relay to ECU terminal 1L
	A/C switch OFF but A/C operates	• Short circuit in wiring from A/C relay to ECU terminal 1L • ECU malfunction
Always VB		• Poor connection at ECU connector • ECU malfunction
Always 0V		• Vehicle speed sensor malfunction • Short circuit in wiring from vehicle speed sensor to ECU terminal 1M
		• Vehicle speed sensor malfunction • Open circuit in wiring from vehicle speed sensor to ECU terminal 1M
Always 0V		• P/S pressure switch malfunction • Short circuit in wiring from P/S pressure switch to ECU terminal 1P • ECU malfunction
Always VB		• P/S pressure switch malfunction • Open circuit in wiring from P/S pressure switch to ECU terminal 1P • Open circuit in wiring from P/S pressure switch to ground
Always below 2.5V (Blower fan OK)		• A/C switch malfunction • Short circuit in wiring from A/C switch to ECU terminal 1Q • Poor connection at ECU connector • ECU malfunction
Always VB (Blower fan OK)		• A/C switch malfunction • Open circuit in wiring from A/C switch to ECU terminal 1Q • Open circuit in wiring from A/C switch to blower control switch
Always 0V (Blower fan OK)		• Short circuit in wiring from blower control switch to ECU terminal 1S • Poor connection at ECU connector • ECU malfunction
Always VB (Blower fan OK)		• Open circuit in wiring from blower control switch to ECU terminal 1S
Always below 1.0V (Stoplight OK)		• Open circuit in wiring from stoplight switch to ECU terminal 1O
Always 0V		• Inhibitor switch malfunction • Short circuit in wiring from inhibitor switch to ECU terminal 1R
Always VB		• Inhibitor switch malfunction • Open circuit in wiring from inhibitor switch to ECU terminal 1R
Always 0V		• Cruise switch malfunction • Short circuit in wiring from cruise switch to ECU terminal 1S
		• Cruise switch malfunction • Open circuit in wiring from cruise switch to ECU terminal 1S
Always 0V		• Throttle sensor misadjustment • Short circuit in wiring from throttle sensor to ECU terminal 1T • ECU malfunction
Always VB		• Throttle sensor misadjustment • Open circuit in wiring from throttle sensor to ECU terminal 1T
—		—
—		—
Always approx. 0V or approx. 2V		• Refer to Code No.02 Troubleshooting

1.6L AND 1.8L ENGINES ENGINE CONTROL UNIT TERMINAL VOLTAGES—323 AND MX-3 WITH 1.6L ENGINE AND PROTEGE

VB: Battery voltage

Terminal	Input	Output	Connection to	Test condition	Correct voltage	Remark	Incorrect voltage	Possible cause
2B	○		Airflow meter	Ignition switch ON	Approx. 3.8V	—	Always approx. 0V or approx. 5V	• Refer to Code No.08 Troubleshooting
				Idle	Approx. 3.3V			
2C	○		Oxygen sensor	Ignition switch ON	0V	—	0V after warm-up	• Refer to Code No.15 Troubleshooting
				Idle (Cold engine)	0V		Always approx. 1V after warm-up	• Refer to Code No.17 Troubleshooting
				Idle (After warm-up)	0—1.0V			
				Increasing engine speed (After warm-up)	0.5—1.0V			
				Deceleration	0—0.4V			
2D	○		Fan switch	Fan operating (Engine coolant temperature over 97°C (207°F) or diagnosis connector terminal TFA grounded)	0V	—	Always 0V (Electrical cooling fan OK)	• Open or short circuit in wiring from fan relay to ECU terminal 2D • ECU malfunction
				Fan not operating (Idle	VB			
2E	○		Water thermosensor	Engine coolant temperature 20°C (68°F)	Approx. 2.5V	Ignition switch ON	Always approx. 0V or approx. 5V	• Refer to Code No.09 Troubleshooting
				After warm-up	Below 0.5V			
2F	○		Throttle sensor	Accelerator pedal released	Approx. 0.5V	—	Always constant	• Open circuit in wiring from ECU terminal 2F to throttle sensor • Open circuit in wiring from ECU terminal 2I to throttle sensor • Open circuit in wiring from ECU terminal 3D to throttle sensor
				Accelerator pedal fully depressed	Approx. 4.0V		Always above 1V	• Throttle sensor misadjustment
2G	○		ATF thermosensor	ATF Temp −30°C (−22°F) –150°C (302°F)	Approx. 4.95—1.12V	Ignition switch ON	Always constant	• ATF termosensor malfunction • Open or short circuit in wiring from ATF termosensor to ECU terminal 2G
				ATF Temp 20°C (68°F)	Approx. 4.6V			
				ATF Temp 130°C (266°F)	Approx. 1.54V			
2H	○		Hold switch	Switch depressed	0V	Ignition switch ON	Always 0V	• Hold switch malfunction • Short circuit in wiring from hold switch to ECU terminal 2H
				Switch released	VB		Always VB	• Hold switch malfunction • Open circuit in wiring from hold switch to ECU terminal 2H
2I	○		Throttle sensor	Constant	4.5—5.5V	Ignition switch ON	Always 0V	• Short circuit in wiring from ECU terminal 2I to throttle sensor airflow meter, or water thermosensor • Poor connection at ECU connector • ECU malfunction
							Below 4.5V or above 5.5V	• ECU malfunction
2J	○		Distributor (G-signal) [DOHC]	Ignition switch ON	Approx. 0V or 5V	—	Always approx. 0V or approx. 1.5V	• Refer to Code No.03 Troubleshooting
				Idle	Approx. 1.5V			
2K	○		Intake air thermosensor	Ambient air temperature 20°C (68°F)	Approx. 2.5V	Built in airflow meter	Always approx. 0V or approx. 5V	• Refer to Code No.10 Troubleshooting
2L	—	—	—	—	—	—	—	—
2M	○		Pulse generator (IN)	Engine running at idle (N range)	Above 0.6V	—	Always constant	• Pulse genrotor malfunction • Open or short circuit in wiring from pulse generator to ECU terminal 2M
				Engine stopped (Ignition switch ON)	0V			
2N			Pulse generator ground	Constant	0V	—	Above 0V	• Poor connection at ground terminal • Open circuit in wiring from ECU terminal 2N to ground
2O		○	Solenoid valve (Purge control)	Ignition switch ON	VB	—	Always 0V or VB	• Refer to Code No.26 Troubleshooting
				Idle	VB			
2P		○	Hold indicator	Hold mode	Below 2V	Ignition switch ON	Always 0V	• Hold indicator or malfunction • Short circuit in wiring from hold indicator to ECU terminal 2P
				Normal mode	VB		Always VB	• Hold indicator malfunction • Open circuit in wiring from hold indicator to ECU terminal 2P

3Y	3W	3U	3S	3Q	3O	3M	3K	3I	3G	3E	3C	3A	2O	2M	2K	2I	2G	2E	2C	2A	1U	1S	1Q	1O	1M	1K	1I	1G	1E	1C	1A
3Z	3X	3V	3T	3R	3P	3N	3L	3J	3H	3F	3D	3B	2P	2N	2L	2J	2H	2F	2D	2B	1V	1T	1R	1P	1N	1L	1J	1H	1F	1D	1B

1.6L AND 1.8L ENGINES ENGINE CONTROL UNIT TERMINAL VOLTAGES—323 AND MX-3 WITH 1.6L ENGINE AND PROTEGE

VB: Battery voltage

Terminal	Input	Output	Connection to	Test condition	Correct voltage	Remark
3A	—	—	Ground (Injector)	Constant	0V	—
3B	—	—	Ground (Output)	Constant	0V	—
3C	—	—	Ground (CPU)	Constant	0V	—
3D	—	—	Ground (Input)	Constant	0V	—
3E	○		Inhibitor switch (D range)	D range	VB	Ignition switch ON
				Other ranges	0V	
3F	○		DRL relay (Canada)	Parking brake pulled with ignition switch ON (DRL OFF)	VB	• DRL: Daytime Running Lights
				Idle (DRL ON)	Below 2.5V	
3G	○		Inhibitor switch (L range)	L range	VB	Ignition switch ON
				Other ranges	0V	
3H	○		Inhibitor switch (S range)	S range	VB	Ignition switch ON
				Other ranges	0V	
3I		○	Solenoid valve (VICS)	Engine speed below 5,000 rpm	Below 1.5V	VICS: Variable Inertia Charging System [DOHC]
				Engine speed above 5,000 rpm	VB	
3J	—	—	—	—	—	—
3K	—	—	—	—	—	—
3L	—	—	—	—	—	—
3M		○	Solenoid valve (Pressure regulator) [BP]	60 [DOHC]/120 [SOHC] seconds after engine started when engine coolant temperature above 90°C (194°F) and intake air temperature above 58°C (136°F) [DOHC]/50°C (122°F) [SOHC]	Below 1.5V	—
				Other condition at idle	VB	
3N	—	—	—	—	—	—
3O	—	—	—	—	—	—
3P	—	—	—	—	—	—
3Q		○	ISC valve	Ignition switch ON	VB	• Engine signal Monitor: Green and red lamps flash
				Idle	Approx. 10V	
3R	—	—	—	—	—	—
3S	—	—	—	—	—	—
3T	—	—	—	—	—	—

3Y	3W	3U	3S	3Q	3O	3M	3K	3I	3G	3E	3C	3A	2O	2M	2K	2I	2G	2E	2C	2A	1U	1S	1Q	1O	1M	1K	1I	1G	1E	1C	1A
3Z	3X	3V	3T	3R	3P	3N	3L	3J	3H	3F	3D	3B	2P	2N	2L	2J	2H	2F	2D	2B	1V	1T	1R	1P	1N	1L	1J	1H	1F	1D	1B

VB: Battery voltage

Incorrect voltage		Possible cause
Above 0V		• Poor connect at ground terminal • Open circuit in wiring from ECU to ground
Always 0V		• Inhibitor switch malfunction • Short circuit in wiring from inhibitor switch to ECU terminal 3E
Always VB		• Inhibitor switch malfunction • Open circuit in wiring from inhibitor switch to ECU terminal 3E
Always below 2.5V	DRL ON when ignition switch ON	• Short circuit in wiring from DRL relay to ECU terminal 1L • Short circuit in wiring from DRL relay to DRL unit
	DRL never ON	• Open circuit in wiring from DRL relay to ignition switch
Always VB		• Parking brake switch always ON • DRL unit malfunction
Always 0V		• Inhibitor switch malfunction • Short circuit in wiring from inhibitor switch to ECU terminal 3G
Always VB		• Inhibitor switch malfunction • Short circuit in wiring from inhibitor switch to ECU terminal 3G
Always 0V		• Inhibitor switch malfunction • Short circuit in wiring from inhibitor switch to ECU terminal 3H
Always VB		• Inhibitor switch malfunction • Open circuit in wiring from inhibitor switch to ECU terminal 2H
Always approx. 0V or VB		• Refer to Code No.41 Troubleshooting
—		—
—		—
—		—
Always 0V or VB		• Refer to Code No.25 Troubleshooting
—		—
—		—
—		—
Always 0V or VB		• Refer to Code No.34 Troubleshooting
—		—
—		—
—		—

1.6L AND 1.8L ENGINES ENGINE CONTROL UNIT TERMINAL VOLTAGES 323 AND MX-3 WITH 1.6L ENGINE AND PROTEGE

V_B: Battery voltage

Terminal	Input	Output	Connection to	Test condition	Correct voltage	Remark
3U		○	Injector (Nos. 1, 3)	Ignition switch ON	V_B	* Engine Signal Monitor: Green and red lamps flash
				Idle	V_B	
				Engine speed above 2,000 rpm on deceleration (After warm-up)	V_B	
3V		○	Injector (Nos. 2, 4)	Ignition switch at idle	V_B	
				Idle	V_B	
				Engine speed above 2,000 rpm on deceleration (After warm-up)	V_B	
3W		○	1-2 shift solenoid valve	Solenoid valve ON	V_B	AT running
				Solenoid valve OFF	0V	
3X		○	2-3 shift solenoid valve	Solenoid valve ON	V_B	AT running
				Solenoid valve OFF	0V	
3Y		○	3-4 shift solenoid valve	Solenoid valve ON	V_B	AT running
				Solenoid valve OFF	0V	
3Z		○	Lockup solenoid valve	Solenoid valve ON	V_B	AT running
				Solenoid valve OFF	0V	

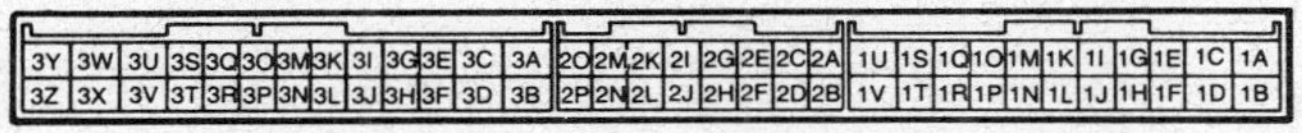

V_B: Battery voltage

Incorrect voltage	Possible cause
Always 0V	• Main relay (FUEL INJ relay) malfunction • Open or short circuit in wiring from injector to ECU terminal 3U or 3V
Always V_B	• ECU malfunction
Always 0V	• 1-2 shift solenoid valve malfunction • Short circuit in wiring from solenoid valve to ECU terminal 3W
Always V_B	• 1-2 shift solenoid valve malfunction • Open circuit in wiring from solenoid valve to ECU terminal 3W
Always 0V	• 2-3 shift solenoid valve malfunction • Short circuit in wiring from solenoid valve to ECU terminal 3X
Always V_B	• 2-3 shift solenoid valve malfunction • Open circuit in wiring from solenoid valve to ECU terminal 3X
Always 0V	• 3-4 shift solenoid valve malfunction • Short circuit in wiring from solenoid valve to ECU terminal 3Y
Always V_B	• 3-4 shift solenoid valve malfunction • Open circuit in wiring from solenoid valve to ECU terminal 3Y
Always 0V	• Lockup solenoid valve malfunction • Short circuit in wiring from solenoid valve to ECU terminal 3Z
Always V_B	• Lockup solenoid valve malfunction • Open circuit in wiring from solenoid valve to ECU terminal 3Z

1.8L ENGINE SPECIFICATIONS CHART – MX-3

Item			Engine	K8 DOHC
Idle speed			rpm	640—700* (670 ± 30)
Ignition timing			BTDC	9—11°* (10 ± 1°)
Throttle body				
Type				Horizontal draft
Throat diameter			mm (in)	55 (2.2)
Fuel pump				
Type				Impeller (in-tank)
Output pressure			kPa (kg/cm², psi)	638(6.5, 92)
Fuel filter				
Type	Low-pressure side			Nylon element (in fuel pump)
	High-pressure side			Paper element
Pressure regulator				
Type				Diaphragm
Regulating pressure			kPa (kg/cm², psi)	284 (2.9, 42)
Injector				
Type				High-ohmic, side feed
Type of drive				Voltage
Resistance			Ω	12—16 (at 20°C, 68°F)
ISC valve (solenoid valve [Idle speed control])				
Solenoid resistance				10.7—12.3 (at 20°C, 68°F)
Solenoid valve (Purge control)				
Solenoid resistance				23—27 (at 20°C, 68°F)
Crank angle sensor 1 (NE1, G signal)				
Type				Hall element pick-up
Airflow meter				
Type				Measuring core type
Resistance Ω	E2 ↔ Vs	Fully closed		20—600
		Fully open		20—1,000
	E2 ↔ Vc			200—400
	E2 ↔ THA (Intake air thermosensor)	−20°C (−4°F)		13600—18400
		20°C (68°F)		2210—2690
		60°C (140°F)		493—667
Water thermosensor (EGI)				
Resistance kΩ		−20°C (−4°F)		14.6—17.8
		20°C (68°F)		2.2—2.7
		80°C (176°F)		0.29—0.35
Water thermosensor (FAN)				
Resistance kΩ		91°C (196°F)		1.70—1.84
		97°C (207°F)		1.42—1.53
		108°C (22°F)		1.03—1.11
Fuel Tank				
Capacity			liters (US gal, Imp gal)	50 (13.2, 11.0)
Air cleaner				
Element type				Oil permeated
Accelerator cable				
Free play			mm (in)	1—3 (0.039—0.118)
Fuel				
Specification				Unleaded regular (RON 87 or higher)

* TEN terminal of diagnosis connector grounded.

1.8L ENGINE COMPONENT DESCRIPTION – MX-3

Component	Function	Remarks
A/C relay	Controls A/C compressor magnet clutch	
Air cleaner	Filters air entering throttle body	
Airflow meter	Detects amount of intake air; sends signal to ECU	Intake air thermosensor included
Air valve	Supplies bypass air into intake manifold (engine cold)	• Engine speed increased to shorten warm-up period • Thermowax type • Installed in BAC valve
Atmospheric pressure sensor	Detects atmospheric pressure; sends signal to ECU	In ECU
Bypass air control (BAC) valve	Supplies bypass air into intake manifold	Consist of air valve and ISC valve
Catalytic converter	Reduces HC, CO, and NOx by chemical reaction	Monolith type
Charcoal canister	Stores fuel tank fumes (engine stopped)	
Check valve (two-way)	Controls pressure in fuel tank	Two-way type
Crank angle sensor (In distributor) **G signal** **NE1 signal**	Detects No.1 and No.4 cylinder TDC; sends signal to ECU (G signal) Detects crank angle at 60° intervals; sends signal to ECU (NE1 signal)	Installed in distributor
Crank angle sensor NE2 signal	Detects crank angle 60° intervals; sends signal to ECU	Installed in engine
Check and cut valve	Prevents fuel from flowing into charcoal canister	
Fuel pump relay	Voltage for fuel pump while engine running	
Diagnosis connector	Concentrate service connector terminals; 1. EGI self - diagnostic terminal 2. Initial set terminal 3. Fuel pump check terminal 4. Cooling fan check terminal 5. Engine speed output terminal	25-pin (located near left suspension mounting block)
Engine control unit (ECU)	**Detects the following:** 1. A/C operation 2. Air/Fuel ratio (oxygen concentration) 3. Atmospheric pressure 4. Braking signal 5. Cranking Signal 6. E/L operation 7. Engine coolant temperature 8. Engine speed 9. Ignition ON signal 10. In-gear condition 11. Intake air amount 12. Intake air temperature 13. EGR valve position 14. No.1 piston TDC 15. P/S operation 16. Test mode signal (ignition timing, idle speed, malfunction code No.) 17. Throttle valve opening angle 18. Throttle valve fully closed 19. Vehicle speed 20. Knocking	1. A/C switch 2. Oxygen sensor 3. Atmospheric pressure sensor 4. Stoplight switch 5. Ignition switch (START position) 6. Blower motor switch, cooling fan relay, headlight switch, and rear window defroster switch 7. Water thermosensor 8. Crank angle sensor (NE2 signal) 9. Ignition switch 10. Neutral and clutch switches 11. Airflow meter 12. Intake air thermosensor (in airflow meter) 13. EGR position sensor 14. Crank angle sensor (G signal) 15. P/S pressure switch 16. Diagnosis connector (TEN terminal) 17. Throttle sensor 18. Idle switch (in throttle sensor) 19. Speed sensor 20. Knock sensor

1.8L ENGINE COMPONENT DESCRIPTION—MX-3

Component	Function	Remarks
Engine control unit (ECU) cont'd	**Controls operation of following:** 1. A/C (Cut-off) 2. Self-diagnosis function 3. Fuel injection system 4. Idle speed control 5. Ignition system (Electronic spark advance) 6. Monitor function 7. Purge control system 8. VRIS control system 9. EGR control system 10. Pressure regulator control system 11. Electric cooling fan control system	1. A/C relay 2. Self-diagnosis Checker 3. Injector 4. Idle speed control (ISC) valve 5. Ignitor 6. Monitor lamp (Self-Diagnosis Checker) 7. Solenoid valve (Purge) 8. Solenoid valve (VRIS) 9. Solenoid valve (EGR) 10. Solenoid valve (PRCV) 11. Electric cooling fan/condenser fan
Fuel filter	Filters particles from fuel	
Fuel pump	Provides fuel to injectors	• Operates while engine running • Installed in fuel tank
Igniter	Receives spark signal from ECU and generates high voltage in ignition coil	Installed in distributor
Ignition switch (start position)	Sends engine cranking signal to ECU	
Inhibitor switch	Detect in-gear condition; send signal to ECU	Switch ON in N and P ranges
Injector	Injects fuel into intake port	Controlled by signals from ECU (side feed type)
Intake manifold	Supplies intake air to all cylinders	
Intake air thermosensor	Detects intake air temperature; sends signal to ECU	Installed in airflow meter
Knock sensor	Detects intake air knocking; sends signal to ECU	
Main relay	Supplies current to output devices and ECU	
Neutral/Clutch switch	Detects in gear condition; sends signal to ECU	Switch ON in terminal
One way check valve	Prevents misoperation of shutter valve (VRIS) and EGR valve during heavy-load operations	For VRIS and EGR
Oxygen sensor	Detects oxygen concentration; sends signal to ECU	• Zirconic and platinum coating • With heater
PCV valve	Controls blow by gas introduced into engine	
P/S pressure switch	Detects P/S operation	P/S ON when steering wheel turned
Pressure regulator	Adjusts fuel pressure supplied to injectors	
Resonance chamber	Reduces intake air noise	
Separator	Prevents fuel from flowing into charcoal canister	
Shutter valve actuator	Closes/opens shutter valve to improve torque characteristics	For variable resonance induction system (VRIS)
Solenoid valve (ISC)	Supplies bypass air into intake manifold	• Controlled by duty signal from engine control unit • Works independent of idle-up • Installed in BAC valve
Solenoid value (EGR vent)	Controls vacuum to EGR control value (Open air side)	Controlled by duty signal from ECU
Solenoid value (EGR vacuum)	Controls vacuum to EGR control value (Vacuum side)	Controlled by duty signal form ECU

Component	Function	Remarks
Solenoid valve (PRCV)	Controls vacuum to pressure regulator control valve	
Solenoid valve (Purge)	Controls evaporative fumes from canister to intake manifold	Controlled by duty signal from ECU
Solenoid valve (VRIS)	Controls vacuum to shutter valve actuator	
Speedometer sensor	Detects vehicle speed; sends signal to ECU	Installed on instrument cluster
Stoplights switch	Detects braking; sends signal to ECU	
Throttle body	Control intake air amount	
Water thermosensor	Detects coolant temperature; sends signal to ECU	Installed on engine (for EGI and fan)

1.8L ENGINE ENGINE CONTROLS OPERATION CHART — MX-3

ENGINE CONDITIONS \ INPUT DEVICES	APPROXIMATE TIME (BASED ON 10—16°C OR 50—60°F AMBIENT)	SENSORS: CRANK ANGLE SENSOR 1 (Ne1, SIGNALG SIGNAL)	CRANK ANGLE SENSOR 2 (Ne2 SIGNAL)	THROTTLE SENSOR	WATER THERMOSENSOR (EGI)	WATER THERMOSENSOR (FAN)	INTAKE AIR THERMOSENSOR	OXYGEN SENSRO	EGR POSITION SENSOR
CRANKING —COLD ENGINE • COLD AIR • COLD COOLANT	Zero			Closed throttle: low voltage (0.1–1.1V)			Signal has no effect on ECU	Signal has no effect on ECU	
COLD START —FAST IDLE • COLD AIR • COLD COOLANT	One minutes				Cool to warm: medium voltae (3.5V and dropping)	Cool to warm: medium voltage (4.5V and dropping)		Sensor cold: low to high voltage (0–0.9V)	
COLD DRIVEAWAY —PART THROTTLE • COLD AIR • COLD COOLANT	Two minutes								EGR stopped: low voltage (Approx. 0.8V)
WARM DRIVEAWAY —PART THROTTLE • WARM AIR • WARM COOLANT	Three minutes	Ne 1 SIGNAL Sends engine speed signal to ECU		Part throttle: medium voltage (1–3.1V)	Warm: medium voltage (Approx. 0.7V and dropping)	Warm: medium voltage (Approx. 4.0V and dropping)		Sensor warm: high voltage (0.9V)	
HOT CRUISE • WARM AIR • WARM COOLANT		G SIGNAL Sends No.1 and No.4 cylinder TDC signal to ECU	Sends engine speed signal to ECU				Cool to warm: medium voltage (1.4–3.4V)	Sensor hot: switching from high voltage (0.9V)	EGR operate: meduim voltage (0.8–4.5V)
HOT ACCELERATION —60%THROTTLE								to low volt (0.1V)	
HOT ACCELERATION —WIDE OPEN THROTTLE	More than four minutes			Wide open throttle: high voltage (Approx. 4.0V)	Hot: low voltage (Approx. 0.5V)	hot: low voltage (Approx. 3.2V)		High voltage (0.9V)	
DECELERATION —CLOSED THROTTLE				Closed throttle: low voltage (0.1–1.1V)				Low voltage (0V)	EGR stopped: low voltage (Approx. 0.8V)
HOT CURB IDLE —EXTENDED								Switching from high to low voltage (0.75–0.25V)	
HOT ENGINE SHUTDOWN	—	OFF	OFF	OFF	OFF	OFF	OFF	Sensor hot: low voltage (0.1V) until sensor cools	OFF

ENGINE CONDITIONS	SWITCHES: IDLE SWITCH	STOP-LIGHT SWITCH	NEUTRAL AND CLUTCH SWITCHES (MTX)	INHIBITOR SIGNAL (ATX)	A/C AMPLIFIER	P/S PRESSURE SWITCH	IGNITION SWITCH: START POSITION	IGNITION SWITCH: ON POSITION	TEST CONNECTOR
CRANKING	Signal has no effect on ECU	Signal has no effect on ECU	Signal has no effect on ECU	Signal has no effect on ECU	Signal has no effet on ECU	Signal has no effect on ECU	Sends signal to ECU (10V)	Signal has no effect on ECU	Signal has no effect on ECU
COLD START	Low voltage signal to ECU (below 1.0V)	Brake pedal depressed: sends signal to ECU (approx. 12V)	In neutral: low voltage signal to ECU (below 1.0V)	In N or P range: sends signal to ECU (below 1.0V)					
COLD DRIVEAWAY									
WARM DRIVEAWAY					A/C switch ON: sends signal to ECU (1.5–3.5V)	Steering wheel turned: low voltage signal to ECU (below 1.0V)			
HOT CRUISE	High voltage signal to ECU (V_B)	No signal send to ECU (below 1.0V)	Driving in any gear: high voltage signal to ECU (V_B)	In D range: no voltage signal to ECU (V_B)	A/C switch OFF: Sends signal to ECU (4.5–5.5V)	Steering wheel straight ahead: high voltage signal to ECU (V_B)	No signal to ECU (below 1.0V)	Sends signal to ECU (V_B)	Connector not grounded: high voltage signal to ECU (V_B)
HOT ACCELERATION —60%THROTTLE									
HOT ACCELERATION —WIDE OPEN THROTTLE									
DECELERATION	Low voltage signal to ECU (below 1.0V)	Brake pedal depressed: sends signal to ECU (approx. 12V)							
HOT CURB IDLE			In neutral : low voltage signal to ECU (below 1.0V)	In N or P range: sends signal to ECU (below 1.0V)					Low voltage signal to ECU when connector grounded (below 1.0V)
HOT ENGINE SHUTDOWN	OFF	OFF	OFF	OFF	OFF	OFF	OFF	OFF	OFF

1.8L ENGINE ENGINE CONTROLS OPERATION CHART—MX-3

ENGINE CONDITIONS / OUTPUT DEVICES	APPROXIMATE TIME (BASED ON 10–16°C or 50–60°F AMBIENT)	INJECTOR		BAC VALVE		SOLENOID VALVE (PURGE CONTROL)	SOLENOID VALVE (PRESSURE REGULATOR CONTROL)	A/C RELAY	IGNITER
		INJECTION	INJECTION TIMING	AIR VALVE	ISC VALVE				
CRANKING —COLD ENGINE • COLD AIR • COLD COOLANT	Zero		All cylinders each ignition pulse			OFF (Purge cut		OFF (A/C CUT)	Ignition timing BTDC 6—8°
COLD START —FAST IDLE • COLD AIR • COLD COOLANT	One minute	Rich		Open (coolant temperature: below 50°C 122°F)	Large amount of bypass air			ON (A/C CUT: approx. 5 sec.)	
COLD DRIVEAWAY —PART THROTTLE • COLD AIR • COLD COOLANT	Two minutes								
WARM DRIVEAWAY —PART THROTTLE • WARM AIR • WARM COOLANT	Three minutes	Rich and Lean	Sequential			Operates (Duty values [purge gas amount] change)	OFF (Vacuum to pressure regulator	ON (A/C ON)	
HOT CRUISE • WARM AIR • WARM COOLANT					Small amount of bypass air				Ignition spark advance signal
HOT ACCELERATION —60% THROTTLE								ON (A/C cut: approx. 5 sec.)	
HOT ACCELERATION —WIDE OPEN THROTTLE	More than four minutes	Rich		Closed					
DECELERATION —CLOSED THROTTLE		Fuel cut			Large and small amount of bypass air	OFF (Purge cut)			
HOT CURB IDLE —EXTENDED		Rich and Lean	Sequential		Small amount of bypass air		After starting: ON during hot start only (Vacuum cut)	ON (A/C ON)	
HOT ENGINE SHUTDOWN	—	Does not inject			OFF	OFF	OFF	OFF	OFF

SOLENOID VALVE (EGR)			
EGR VENT	EGR VACUUM	VRIS NO.1	VRIS NO.2
OFF (No vacuum to EGR valve)	OFF (Atmospheric pressure to EGR valve)	OFF	OFF
		ON (Engin speed 1900–2700 rpm or 3900–6300 rpm	ON (Engine speed 1900–2700 rpm or 4700–6300 rpm)
System operates (amount of EGR changes)			
OFF	OFF		
		OFF	OFF

1.8L ENGINE TROUBLESHOOTING GUIDE—MX-3

NO.	TROUBLESHOOTING ITEM		DESCRIPTION
1	Will not crank or cranks slowly		Refer to Engine Electrical System
2	Cranks normally but will not start	(No combustion)	Engine cranks at normal speed but shows no sign of firing
3		(Partial combustion) — When engine cold	Engine cranks at normal speed but shows only partial combustion and will not continue to run
4		(Partial combustion) — After warm-up	Engine starts normally when cold but will not start after running and being hot soaked
5	Cranks normally but had to start	Always	Engine cranks at normal speed but requires excessive cranking time before starting
6		When engine cold	Same condition as No.5 when engine cold Restarts normally after warm-up
7		After warm-up	Same condition as No.5 after running and being hot soaked Starts normally when cold
8	Rough idle Low idle speed Engine stalls at idle	Always	Engine stalls or vibrates excessively at idle
9		Before warm-up	Engine stalls or vibrates excessively at idle during warm-up
10		After warm-up	Engine runs normally at idle during warm-up but stalls or vibrates excessively after warm-up
11		When A/C, P/S, or E/L ON	Engine stalls or vibrates excessively at idle when A/C, P/S, or E/L ON
12	Rough idle/Engine stalls just after starting		Engine stalls or vibrates excessively just after starting (acceleration from idle)
13	High idle speed after warm-up		Idle speed excessively high after warm-up
14	Idle moves up and down/idle hunting		Engine speeds up and down periodically at idle
15	Engine stalls on deceleration		Engine unexpectedly stops running during or after deceleration
16	Engine stalls suddenly (Intermittent)		Engine intermittently stops running
17	Hesitates/Stumbles on acceleration		Flat spot occurs just after accelerator is depressed, or mild jerking occurs during acceleration
18	Surges while cruising		Unexpected change in engine speed is usually repetitive
19	Lack of power		Performance poor under load when throttle valve is wide open Maximum speed reduced
20	Poor acceleration		Performance poor during acceleration
21	Runs rough during deceleration/Backfire		Engine runs rough during deceleration, and abnormal combustion occurs in exhaust system
22	Knocking		Abnormal combustion accompanied by audible "pinging" noise
23	Fuel odor		Gasoline odor in cabin
24	Exhaust sulfur odor		Exhaust gas has an unusual odor
25	High oil consumption		Oil consumption excessive
26	Poor fuel economy		Fuel economy unsatisfactory
27	A/C does not work		Blower fan operates, but magnetic clutch does not

1.8L ENGINE RELATIONSHIPS CHART—MX-3

INPUT DEVICE \ OUTPUT DEVICE		INJECTOR: FUEL INJECTION AMOUNT	INJECTOR: FUEL INJECTIO TIMING	CIRCUIT OPENING RELAY	IGNITER	A/C RELAY (A/C CUT-OFF) AND CONDENSER FAN LOW RELAY	BAC VALVE (ISC VALVE)	ELECTRIC COOLING FAN RELAY	CONDENSER FAN HIGH RELAY NO.1 NO.2	SOLENOID VALVE: PURGE CONTROL	SOLENOID VALVE: VRIS (NO.1 AND NO.2) CONTROL	SOLENOID VALVE: EGR CONTROL	SOLENOID VALVE: PRESSURE REGULATOR CONTROL	EC-AT CU (TORQUE REDUCED SIGNAL/WATER THERMO-SIGNAL)	SELF-DIAGNOSIS CHECKER (SERVICE CODE)	SELF-DIAGNOSIS CHECKER (MONITOR LAMP)
CRANK ANGLE SENSOR 1	NE1 SIGNAL	○	○	○	○										○	
	G SIGNAL		○		○										○	
CRANK ANGLE SENSOR 2 (NE2 SIGNAL)		○	○	○	○		○			○	○	○	○		○	
THROTTLE SENSOR	IDLE SWITCH	○	○		○		○	○		○		○				○
	THROTTLE SENSOR	○			○	○	○				○	○	○	○	○	
WATER THERMOSENSOR (EGI)		○	○		○		○	○	○	○		○		○	○	
WATERR THERMOSENSOR (FAN)						○		○	○				○		○	
INTAKE AIR THERMOSENSOR		○					○			○			○		○	
OXYGEN SENSOR		○								○					○	○
KNOCK SENSOR					○									○	○	
A/C AMPLIFIER						○	○	○								○
P/S PRESSURE SWITCH						○	○									
IGNITION SWITCH (START SIGNAL)		○	○	○	○	○	○	○	○				○			
STOPLIGHT SWITCH		○									○					○
E/L LOAD*							○									○
NEUTRAL AND CLUTCH SWITCHES		○			○	○	○			○			○			○
DIAGNOSIS CONNECTOR (TEN TERMINAL)					○		○	○					○		○	○
EGR POSITION SENSOR					○							○			○	
ATMOSPHERIC PRESSURE SENSOR		○					○			○					○	
SPEEDOMETER SENSOR							○				○					
AIRFLOW METER		○			○					○		○		○	○	
INHIVITOR SIGNAL (EC-AT CU)		○			○	○				○			○			○
REDUCE TORQUE SIGNAL 1, 2 (EC-AT CU)		○			○									○		

* E/L Load: Blower fan control switch 2nd position or higher, headlights ON, fog lights ON, or rear window defroster switch ON.

2	CRANKS NORMALLY BUT WILL NOT START (NO COMBUSTION)
DESCRIP-TION	• Engine cranks at normal speed but shows no sign of firing • Battery in normal condition • Throttle valve not held fully open while cranking (dechoke system not operating) • Fuel in tank

[TROUBLESHOOTING HINTS]
Because of no combustion, possibly no fuel is injected to engine or ignition at all cylinders

① No spark
- Ignition control malfunction
- Ignition system component malfunction

② No fuel injection
- Fuel pump inoperative
- Injectors inoperative

③ Low fuel line pressure
④ Low engine compression
⑤ ECU power supply circuit open
⑥ ECU ground circuit open

STEP	INSPECTION		ACTION
1	Is strong blue spark visible at disconnected high-tension lead? 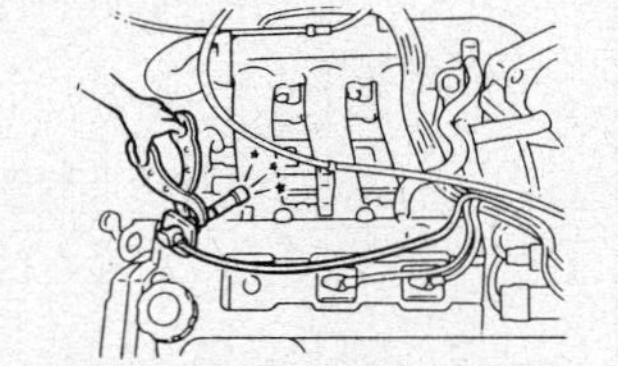	Yes	Go to Step 3
		No	Go to next step
2	Is "00" displayed on Self-Diagnosis Checker with ignition switch ON? 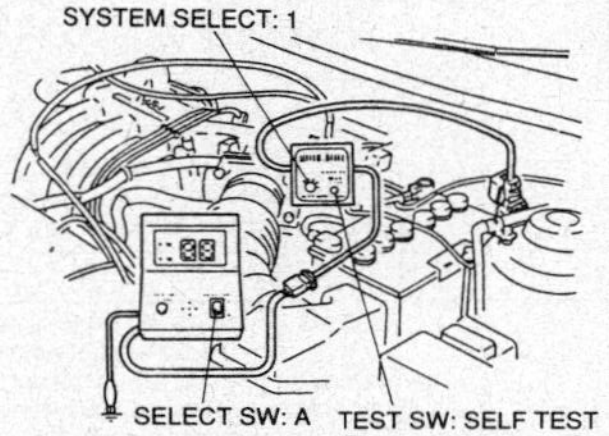 	Yes	Check ignition system (Refer to Troubleshooting "Misfire")
		No	**Service Code No. displayed** Check for cause (Refer to specified check sequence)
3	Connect diagnosis connector terminals F/P and GND with jumper wire; fuel pump operating sound with ignition switch ON? 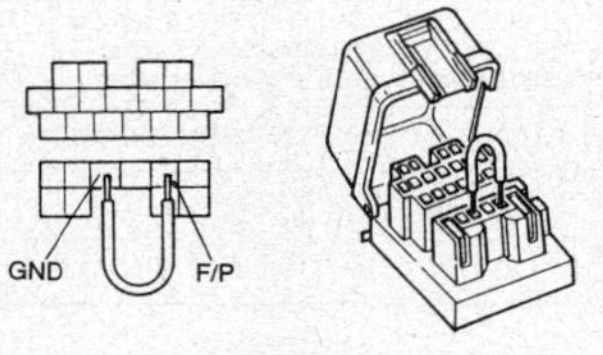 	Yes	Does engine start in this condition? ⇨ If it does, check the circuit-opening relay ⇨ If it does not, go to Step 5
		No	Go to next step

STEP	INSPECTION		ACTION
4	Does V_B exist at the fuel pump connector (B/P) wire with ignition switch ON? 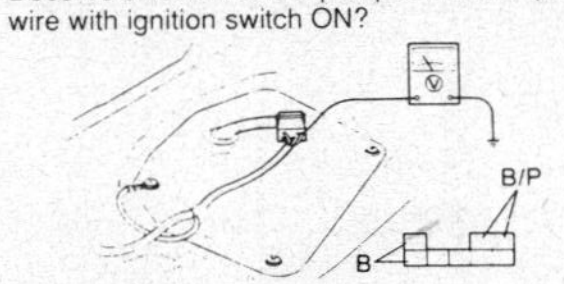 	Yes	Check continuity of fuel pump (Between terminals B/P and B)
		No	Check circuit opening relay
5	Is there an injector operating sound while the engine is being cranked?	Yes	Go to Step 7
		No	Go to next step
6	Does V_B exist at the injector connector with the ignition switch ON?	Yes	Check ECU terminals 3U, 3V, 3W, 3X, 3Y and 3Z voltages
		No	Check for open circuit in wiring between main relay (FUEL INJ rely) and injector
7	Connect diagnosis connector terminals F/P and GND with jumper wire; is the fuel line pressure correct with the ignition switch ON? **Fuel Line Pressure:** **255—314 kPa (2.6—3.2 kg/cm², 37—46 psi)** 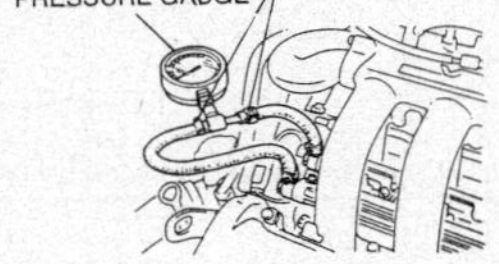 	Yes	Go to next step
		No	**Low pressure** Check fuel line pressure while pinching fuel return hose ⇨ If pressure **quickly** increases, check pressure regulator ⇨ If pressure **gradually** increase, check for clogging between fuel pump and pressure regulator If hose is not clogged, check fuel pump maxumum pressure
			High pressure Is fuel return hose clogged or restricted? ⇨ If it is not, replace pressure regulator ⇨ If it is, repair or replace it
8	Is the engine compression correct? **Engine compression:** **981 kPa (10.0 kg/cm², 142 psi)-300 rpm**	Yes	Go to next step
		No	Check engine condition • Worn piston, piston rings, or cylinder wall • Defective cylinder head gasket • Distorted cylinder head • Improper valve seating • Valve sticking in guide
9	Are all spark plugs OK? 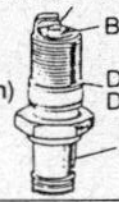 	Yes	Go to next step
		No	Clean or replace
10	Try a known good ECU; does condition improve?		

3	CRANKS NORMALLY BUT WILL NOT START (PARTIAL COMBUSTION) — WHEN ENGINE COLD
DESCRIPTION	• Engine cranks at normal speed but shows only partial combustion and will not continue to run • Battery in normal condition • Fuel in tank

[TROUBLESHOOTING HINTS]

① Air/Fuel mixture too rich
- Air cleaner element clogged
- Airflow meter stuck

② Air/Fuel mixture too lean
- Fuel injection control malfunction (correction for coolant temperature)
- Low fuel line pressure
- Air leakage of intake air system

③ Low engine compression

STEP	INSPECTION		ACTION
1	Is "00" displayed on Self-Diagnosis Checker with ignition switch ON? 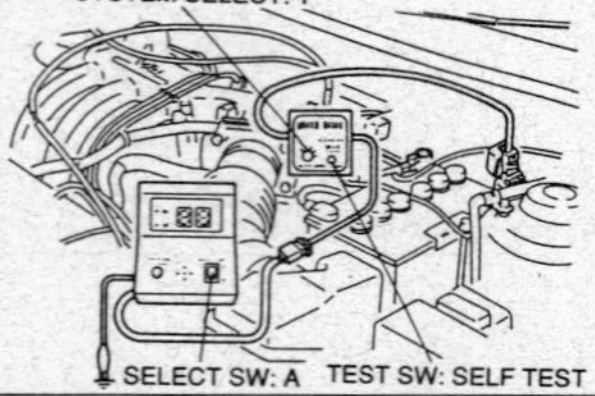	Yes	Go to next step
		No	**Service Code No. displayed** Check for cause (Refer to specified check sequence)
2	Is strong blue spark visible at each disconnected high-tension lead? 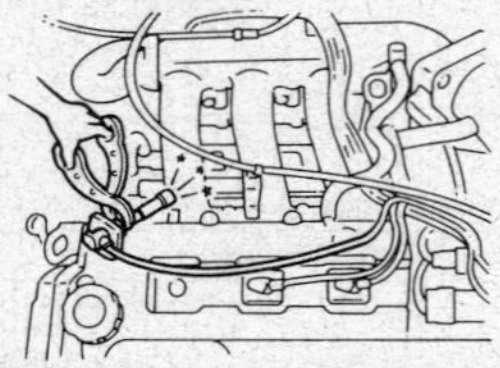	Yes	Go to next step
		No	Check distributor cap and rotor
3	Connect diagnosis connector terminals F/P and GND with jumper wire; is fuel line pressure correct with ignition switch ON?	Yes	Go to next step
	Fuel line pressure: **255—314 kPa (2.6—3.2 kg/cm², 37—46 psi)** PRESSURE GAUGE 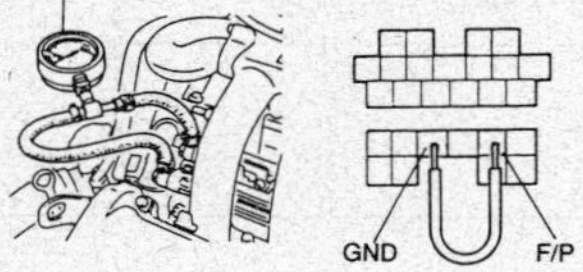	No	**Low pressure** Check fuel line pressure while pinching fuel return hose ➪ If pressure **quickly** increases, check pressure regulator ➪ If pressure **gradually** increases, check for clogging between fuel pump and pressure regulator If not clogged, check fuel pump maximum pressure **High pressure** Is fuel return hose clogged or restricted? ➪ If it is not, replace pressure regulator ➪ If it is, repair or replace it

STEP	INSPECTION		ACTION
4	Are ECU terminal voltage OK? (especially 1C, 2D)	Yes	Go to next step
		No	Checke for cause
5	Is there air leakage of intake air system components?	Yes	Repair or replace
		No	Go to next step
6	Does engine starts when water thermosensor is disconnected?	Yes	Check water thermosensor
		No	Go to next step
7	Does airflow meter moves smoothly from fully dosed to fully open? 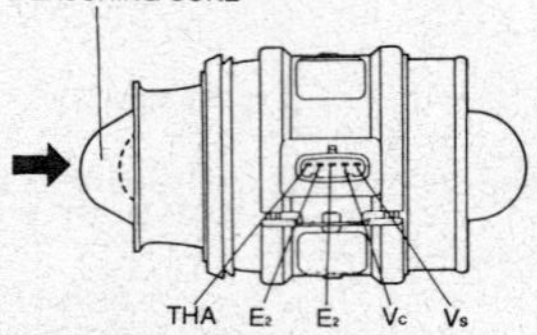	Yes	Go to next step
		No	Repair or replace
8	Is engine compression correct?	Yes	Go to next step
	Engine compression (Minimum): **981 kPa (10.0 kg/cm², 142 psi)-300 rpm**	No	Check engine condition • Worn piston, piston rings, or cylinder wall • Defective cylinder head gasket • Distorted cylinder head • Improper valve seating • Valve sticking in guide
9	Are spark plugs OK? 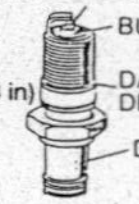	Yes	Go to next step
		No	Clean or replace
10	Try a known good ECU: does condition improve?		

4	CRANKS NORMALLY BUT WILL NOT START (PARTIAL COMBUSTION) — AFTER WARM-UP
DESCRIP-TION	• Engine cranks at normal speed and shows partial combustion but will not continue hot soaked • Battery is in normal condition • Engine starts normally when cold

[TROUBLESHOOTING HINTS]
① Air/Fuel mixture too rich
- Insufficient fuel injection control (correction for coolant temperature)
- Injector fuel leakage

② Vapor lock
- Fuel pressure does not hold in fuel line after engine
- Pressure regulator control system malfunction

STEP	INSPECTION		ACTION
1	Warm-up engine to normal operating temperature and stop it Connect diagnosis connector terminals F/P and GND with jumper wire for 3 minutes with ignition switch ON. Does engine starts? 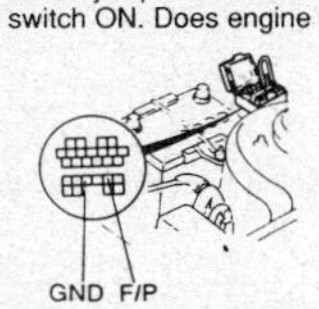	Yes	Change to another brand of fuel
		No	Go to next step
2	Remove vacuum hose from pressure regulator and plug it; Does engine start? 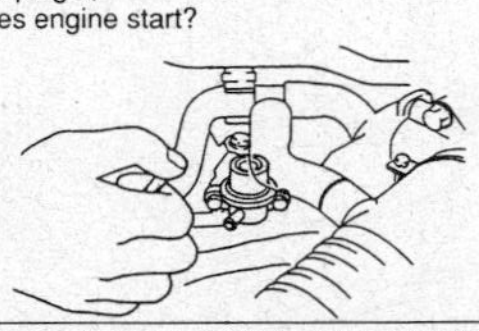	Yes	Check pressure regulator control system
		No	Go to next step
3	Connect diagnosis connector terminals F/P and GND with jumper wire; is fuel line pressure correct with ignition switch ON? **Fuel line pressure:** **255—314 kPa (2.6—3.2 kg/cm², 37—46 psi)** 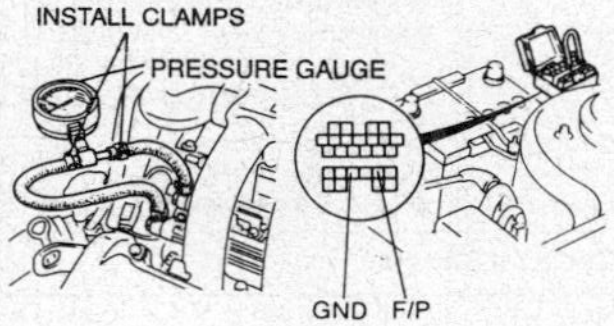	Yes	Go to next step
		No	**Low pressure** Check fuel line pressure while pinching fuel return hose ➪ If pressure **quickly** increases, check pressure regulator ➪ If pressure **gradually** increases, check fuel line and filter for clogging If hose is not clogged, check fuel pump maximum pressure **High pressure** Is fuel return hose clogged or restricted? ➪ If it is not, replace pressure regulator ➪ If it is, repair or replace it
4	In the same condition as Step 2, does fuel line pressure hold after ignition switch is turned OFF? **Fuel line pressure: More than 147 kPa (1.5 kg/cm², 21 psi) for 5 min.**	Yes	Go to Step 6
		No	Go to next step

STEP	INSPECTION		ACTION
5	Plug outlet of pressure regulator; does fuel line pressure hold after ignition switch is turned OFF? **Fuel line pressure:** **More then 147 kPa (1.5 kg/cm², 21 psi) for 5 min.** 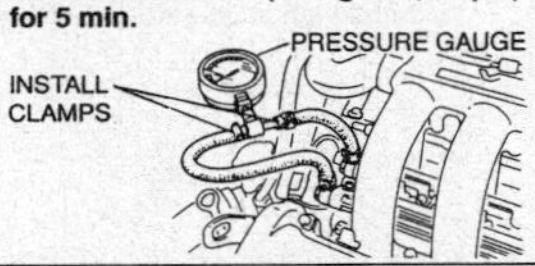	Yes	Replace pressure regulator
		No	Check fuel pump hold pressure ➪ If OK, check injector for fuel leakage ➪ If not OK, replace fuel pump
6	Disconnect water thermosensor connector; does engine starts? 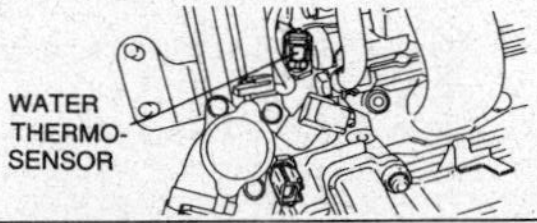	Yes	Check water thermosensor ➪ If OK, check connectors of water thermosensor and ECU 2Q terminal ➪ If not OK, replace water thermosensor
		No	Go to next step
7	Is "00" displayed on Self-Diagnosis Checker with ignition switch ON? 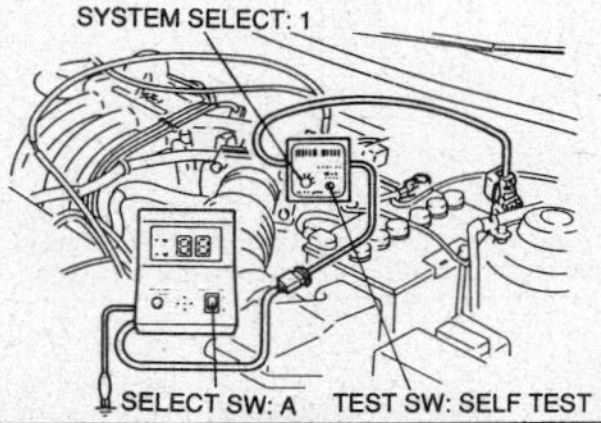	Yes	Go to next step
		No	**Service Code No. displayed** Check for cause (Refer to specified check sequence)
8	Are ECU terminal voltage OK? (especially 1B, 2B and 2E)	Yes	Go to next step
		No	Check for cause
9	Try a known good ECU; does condition improve?	Yes	Replace ECU
		No	Change to another brand of fuel

5	CRANKS NORMALLY BUT HARD TO START- ALWAYS		
DESCRIP-TION	• Engine cranks at normal speed but requires excessive cranking time before starting • Battery is in normal condition • Engine runs normally at idle (If idle condition not OK, refer to "Rough Idle" [Nos. 8-12])		
[TROUBLESHOOTING HINTS] ① Air/Fuel mixture too lean • Fuel injection control malfunction (correct for coolant temperature) • Low fuel line pressure • Air keakage ② Air/Fuel mixture too rich • Air cleaner element clogged • Airflow meter stuck ③ Poor ignition speak			
STEP	**INSPECTION**		**ACTION**
1	Is "00" displayed on Self-Diagnosis Checker with ignition switch ON? 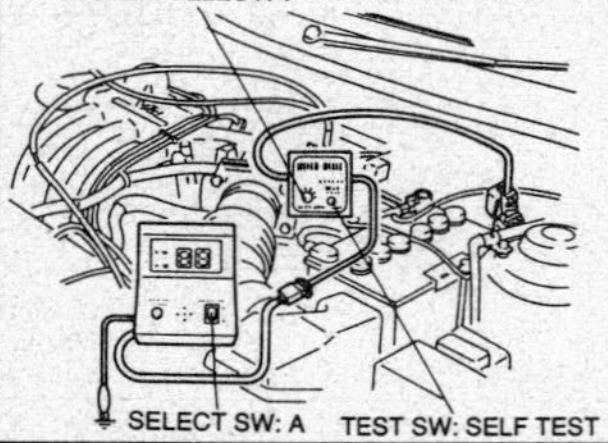	Yes	Go to next step
		No	**Service Code No. displayed** Check for cause (Refer to specified check sequence)
2	Is intake manifold vacuum at idle? **Vacuum: More than 450 mmHg (17.7 inHg)**	Yes	Go to next step
		No	Check for air leakage of intake air system components
3	Is air cleaner element clean?	Yes	Go to next step
		No	Replace air cleaner element
4	Does engine start easily when throttle valve quarter is open?	Yes	Check throttle valve for carbon deposit then go to Step 6
		No	Go to next step
5	Is fuel line pressure at idle correct? **Fuel line pressure: 255—314 kPa (2.6—3.2 kg/cm², 37—46 psi) (Vacuum hose to pressure regulator disconnected)** 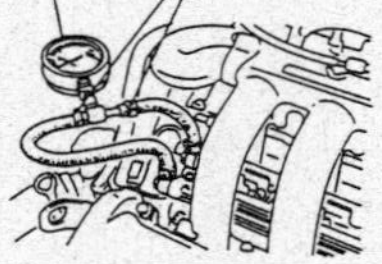	Yes	Go to next step
		No	**Low pressure** Check fuel line pressure while pinching fuel return hose ➪ If pressure **quickly** increases, check pressure regulator ➪ If pressure **gradually** increases, check for clogging between fuel pump and pressure regulator If hose is not clogged, check fuel pump maximum pressure

STEP	INSPECTION		ACTION
6	Disconnect ISC valve connector at idle: Does engine speed increase? 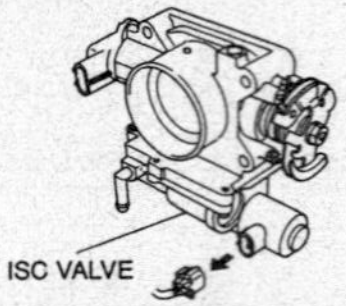	Yes	Go to next step
		No	Replace ISC value
7	**Are ECU terminal voltages OK (especially 1C, 2E, and 3D)?**	Yes	Go to next step
		No	Check for cause
8	Connect diagnosis connector terminals F/P and GND with jumper wire; does engine start normally? 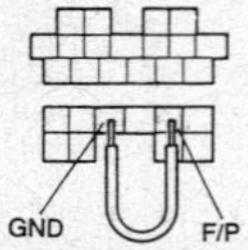	Yes	Check circuit-opening relay ➪ If OK, repair or replace wire harness ➪ If not OK, replace relay
		No	Go to next step
9	Connect System Selector to diagnosis connector and set Test Switch to "SELF TEST"; is ignition timing at idle correct after warm-up? **Ignition timing (BTDC): 9—11° (10 ± 1°)** 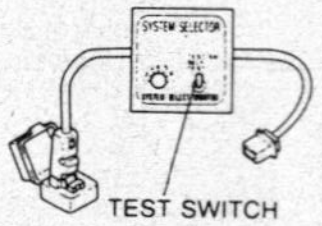	Yes	Go to next step
		No	Adjust ignition timing
10	Is engine compression correct? **Engine compression (Minimum): 981 kPa (10.0 kg/cm², 142 psi)-300 rpm**	Yes	Go to next step
		No	Check engine condition • Worn piston, piston rings, or cylinder wall • Defective cylinder head gasket • Distorted cylinder head • Improper valve seating • Valve sticking in guide
11	Try a known good ECU; does condition improve?		

6	CRANKS NORMALLY BUT HARD TO START — WHEN ENGINE COLD
DESCRIP-TION	• Engine cranks at normal speed but requires excessive cranking time before starting • Battery in normal condition • Restarts OK after warm-up • Engine runs normally at idle (If idle condition is not OK, refer to "Rough idle" [Nos. 8—12])

[TROUBLESHOOTING HINTS]

① Air/Fuel mixture too rich
- Airflow meter sticking
- Air cleaner element clogged
- Idle-speed control malfunction

② Air/Fuel mixture too lean
- Fuel injection control malfunction (correction for coolant temperature)

③ Poor atomization

STEP	INSPECTION		ACTION
1	Is "00" displayed on Self-Diagnosis Checker ignition switch ON? 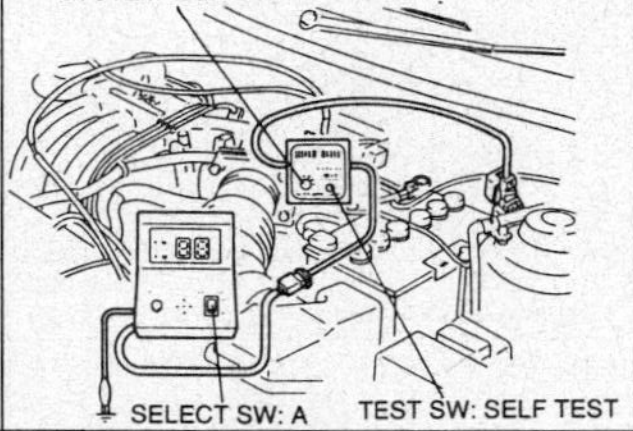	Yes	Go to next step
		No	**Service Code No. displayed** Check for cause (Refer to specified check sequence)
2	Are ECU terminal voltages OK? (especially 1C, 2E and 3D)?	Yes	Go to next step
		No	Check for cause
3	Does engine start easily when throttle valve is quarterly open? 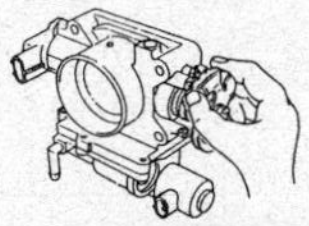	Yes	Is ISC valve OK? ➪ If it is, check air valve ➪ If it is not, replace ISC valve
		No	Go to next step
4	Is intake manifold vacuum correct at idle? **Vacuum: More than 450 mmHg (17.7 inHg)**	Yes	Go to next step
		No	Check for air leakage of intake air system components
5	Is air cleaner element clean?	Yes	Go to next step
		No	Repair air cleaner element
6	Disconnect ISC valve connector at idle: Does engine speed increase? 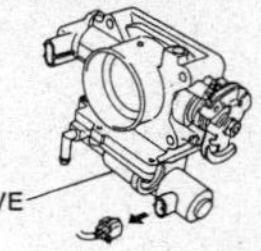	Yes	Go to next step
		No	Replace ISC valve

STEP	INSPECTION		ACTION
7	Connect diagnosis connector terminals F/P and GND with jumper wire; is fuel line pressure correct with ignition switch ON? **Fuel line pressure:** **255—314 kPa (2.6—3.2 kg/cm^2, 37—46 psi)** 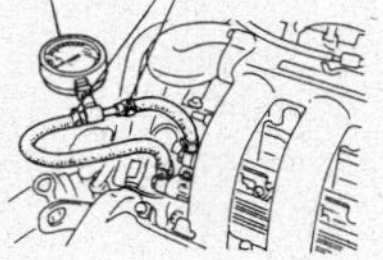	Yes	Go to next stop
		No	**Low pressure** Check fuel line pressure while pinching fuel return hose ➪ If pressure **quickly** increases, check pressure regulator ➪ If pressure **gradually** increases, check fuel line for clogging between fuel pump and pressure regulator If hose is not clogged, check fuel pump maximum pressure
			High pressure Is fuel return hose clogged or restricted? ➪ If it is not, replace pressure regulator ➪ If it is, repair or replace it
8	Try a known good ECU; does condition improves?	Yes	Replace ECU
		No	Change to another brand of fuel

7	CRANKS NORMALLY BUT HARD TO START — AFTER WARM-UP
DESCRIPTION	• Engine cranks at normal speed but requires excessive cranking time before starting after running and being hot soaked • Battery in normal condition • Engine starts normally when cold • Engine runs normally at idle (If idle condition is not OK, refer to "Rough Idle" [Nos. 8—12])

[TROUBLESHOOTING HINTS]

① Air/Fuel mixture too rich
- Fuel injection control malfunction
- Injector fuel leakage

② Vapor lock
- Fuel pressure not held in fuel line after engine stopped
- Pressure regulator control system malfunction

STEP	INSPECTION		ACTION
1	Is "00" displayed on Self-Diagnosis Checker with ignition switch ON? 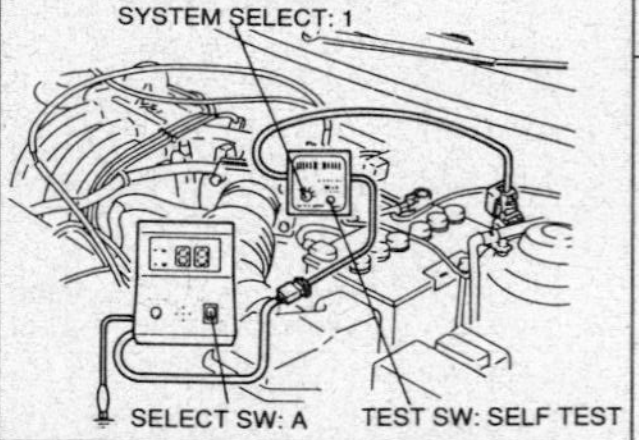	Yes	Go to next step
		No	**Service Code No. displayed** Check for cause (Refer to specified check sequence)
2	Remove vacuum hose from pressure regulator and plug it. Does engine start normally? 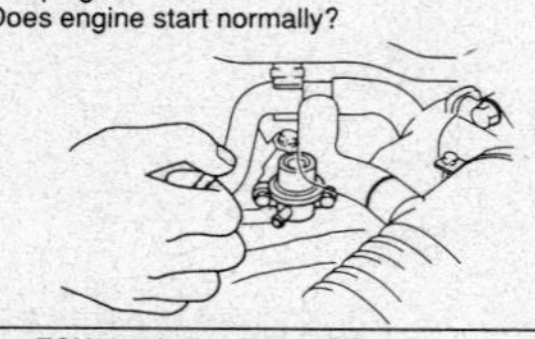	Yes	Check pressure regulator control system
		No	Go to next step
3	Are ECU terminal voltages OK (especially 1C, 2E and 3D)?	Yes	Go to next step
		No	Check for cause
4	Run engine at idle; does fuel line pressure hold after ignition switch is turned OFF? **Fuel line pressure: More than 147 kPa (1.5 kg/cm², 21 psi) for 5 min.** 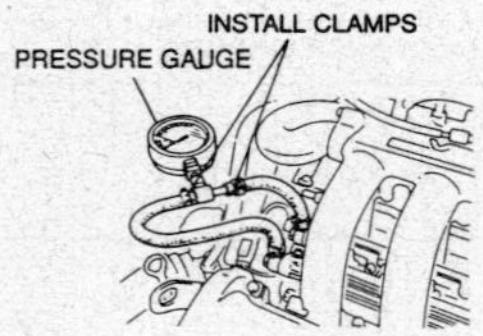	Yes	Go to next step
		No	Plug outlet of pressure regulator; does fuel line pressure hold after ignition switch is turned OFF? ➪ If it does, replace pressure regulator ➪ If it does not, check fuel pump hold pressure If fuel pump is OK, check injectors for fuel leakage

STEP	INSPECTION		ACTION
5	Warm-up engine to normal operating temperature and stop it Connect diagnosis connector terminals F/P and GND with jumper wire for 3 minutes with ignition switch ON; does engine start easily? 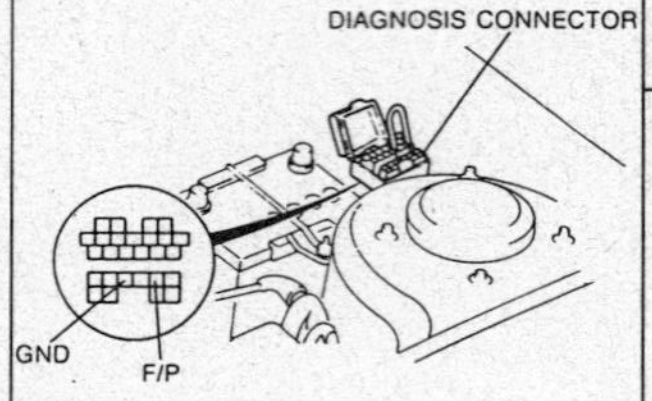	Yes	Change to another brand of fuel
		No	Go to next step
6	Try a known good ECU; does condition improve?	Yes	Replace ECU
		No	Change to another brand of fuel

8	ROUGH IDLE/ENGINE STALLS AT IDLE — ALWAYS
DESCRIPTION	• Engine starts normally but stalls or vibrates excessively at idle in every condition

[TROUBLESHOOTING HINTS]
① Air/Fuel mixture too lean
- Air leakage
- Fuel injection control malfunction
- Low fuel line pressure

② One or more injectors clogged or not operating
③ Low intake air amount
④ Incorrect idle speed
- Idle speed misadjustment
- Idle speed control malfunction

⑤ One or more spark plugs not firing
⑥ Low engine compression
⑦ Airflow meter stuck

STEP	INSPECTION		ACTION
1	Is "00" displayed on Self-Diagnosis Checker with ignition switch ON? 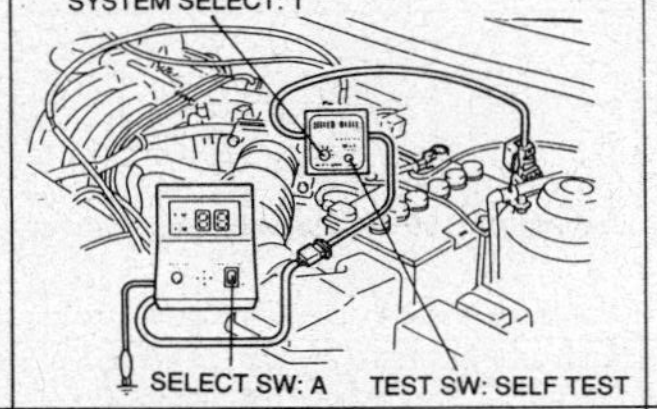	Yes	Go to next step
		No	**Service Code No. displayed** Check for cause (Refer to specified check sequence)
2	Are ECU terminal voltages OK (especially 2B, 2E, and 3D)?	Yes	Go to next step
		No	Check for cause
3	Disconnect high-tension lead at idle; does engine speed decreases equally at each cylinder?	Yes	Go to Step 8
		No	Go to next step
4	Is there an injector operating sound at idle?	Yes	Go to Step 6
		No	Go to next step
5	Does approximately 12V exist at injector connector (W/R) wire?	Yes	Is injector resistance OK? **Resistance: 12—16Ω (20°C [68°F])** ➪ If it is, check wiring between ECU and injector ➪ If it is not, replace injector
		No	Check wiring between ECU and injector
6	Is strong blue spark visible at disconnected high-tension lead? 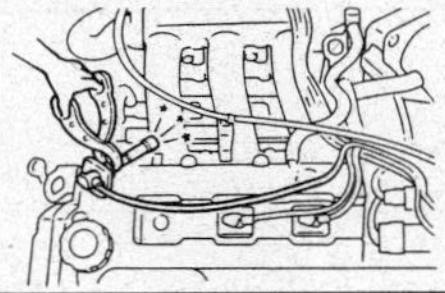	Yes	Go to next step
		No	Check high-tension lead ➪ If OK, check distributor cap and rotor ➪ If not OK, replace high-tension lead

STEP	INSPECTION		ACTION
7	Are spark plugs OK? 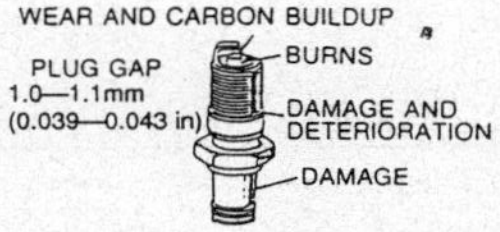	Yes	Is engine compression correct? ➪ If OK, replace injector ➪ If not OK, check for cause
		No	Clean or replace
8	Is fuel line pressure correct at idle? **Fuel line pressure: 255—314 kPa (2.6—3.2 kg/cm², 37—48 psi) (Vacuum hose to pressure regulator disconnected)** 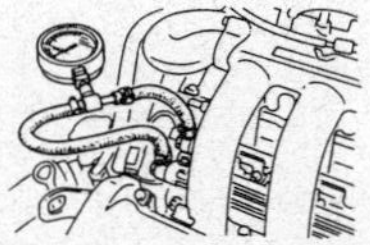	Yes	Go to next step
		No	**Low pressure** Check fuel line pressure while pinching fuel return hose ➪ If pressure **quickly** increases, check pressure regulator ➪ If pressure **gradually** increases, check for clogging between fuel pump and pressure regulator ➪ If hose is not clogged, check fuel pump maximum pressure
9	Does airflow meter moves smoothly from fully closed to fully open? 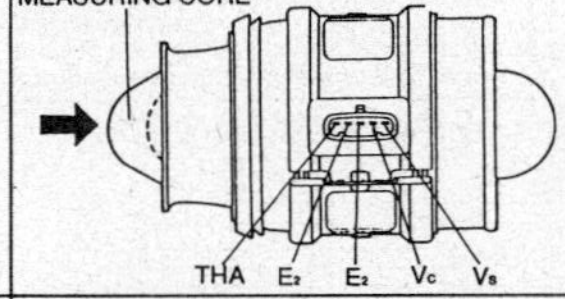	Yes	Go to next step
		No	Repair or replace
10	Is air leakage at intake air system components?	Yes	Repair or replace
		No	Go to next step
11	Is engine compression correct? **Engine compression (Minimum): 981 kPa (10.0 kg/cm², 142 psi)-300 rpm**	Yes	Go to next step
		No	Check for cause
12	Connect System Selector to diagnosis connector and set Test Switch to "SELF TEST"; is ignition timing correct at idle after warm-up? **Ignition Timing (BTDC): 9—11° (10 ± 1°)**	Yes	Try a known good ECU; does condition improve?
		No	Adjust ignition timing

9	ROUGH IDLE/ENGINE STALLS AT IDLE — BEFORE WARM-UP		
DESCRIP-TION	• Engine speed low or engine stalls or vibrates excessively at idle during warm-up		
[TROUBLESHOOTING HINTS] ① Low intake air amount • Airflow meter stuck • Air cleaner element clogged • Air valve ② Low fuel injection control • Fuel injection control malfunction (Correction for coolant temperature) ③ Poor atomization			
STEP	**INSPECTION**		**ACTION**
1	Is "00" displayed on Self-Diagnosis Checker with ignition switch ON? 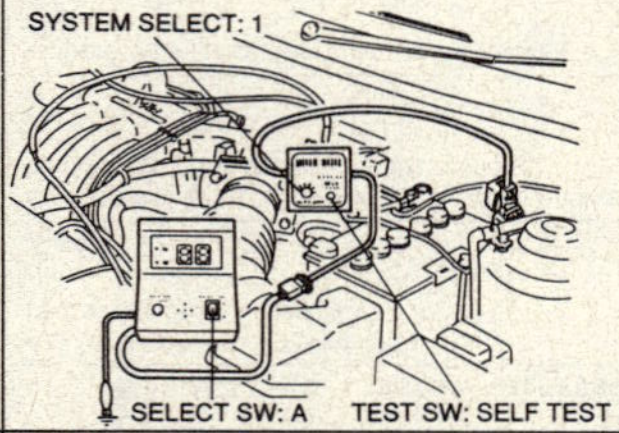	Yes	Go to next step
		No	**Service Code No. displayed** Check for cause (Refer to specified check sequence)
2	Is intake manifold vacuum correct at idle after warm-up? **Vacuum: More than 450 mmHg (17.7 inHg)**	Yes	Go to next step
		No	Check for air leakage of intake air system components
3	Is air cleaner element clean?	Yes	Go to next step
		No	Repair air cleaner element
4	Are ECU terminal voltages OK (especially 2B, 2E, and 3D)?	Yes	Go to next step
		No	Check for cause
5	Disconnect ISC valve connector Does engine speed increase after warm-up? 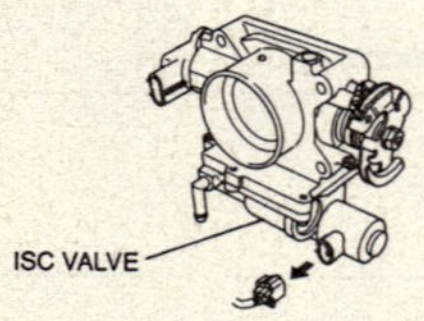	Yes	Go to next step
		No	Replace ISC valve

STEP	INSPECTION		ACTION
6	Is water thermosensor resistance correct? 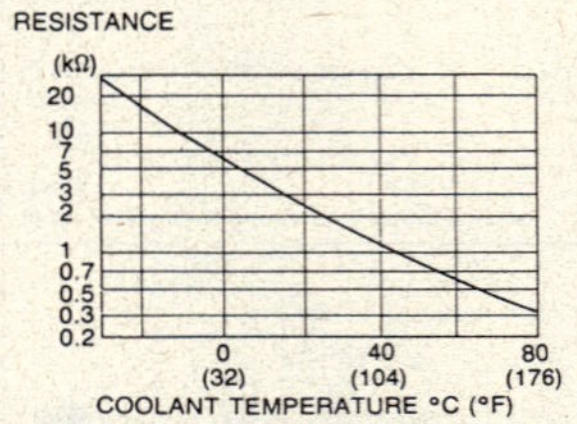	Yes	Go to next step
		No	Replace water thermosensor
7	Connect System Selector to diagnosis connector and set Test switch to "SELF TEST" when engine is cold. Does engine speed decreases as engine warms up?	Yes	Go to next step
		No	Check air valve
8	Connect System Selector to diagnosis connector and set Test Switch to "SELF TEST"; is ignition timing at idle correct after warm-up? **Ignition timing (BTDC): 9—11° (10 ± 1°)** 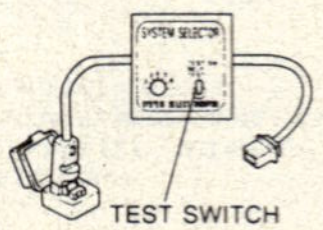	Yes	Go to next step
		No	Adjust ignition timing
9	Try a known good ECU; does condition improve?	Yes	Replace ECU
		No	Change to another brand of fuel

1.8L ENGINE TROUBLESHOOTING GUIDE—MX-3

10	ROUGH IDLE/ENGINE STALLS AT IDLE — AFTER WARM-UP
DESCRIP-TION	• Engine runs normally at idle during warm-up but stalls or vibrates excessively after warm-up

[TROUBLESHOOTING HINTS]
① Idle-speed control system malfunction
② Air/Fuel mixture too lean
- Air leakage
- Low fuel line pressure

③ Air/Fuel mixture too rich
- Fuel injection control malfunction (correction for coolant temperature)

④ Poor ignition
⑤ Low engine compression

STEP	INSPECTION		ACTION
1	Is "00" displayed on Self-Diagnosis Checker with ignition switch ON? 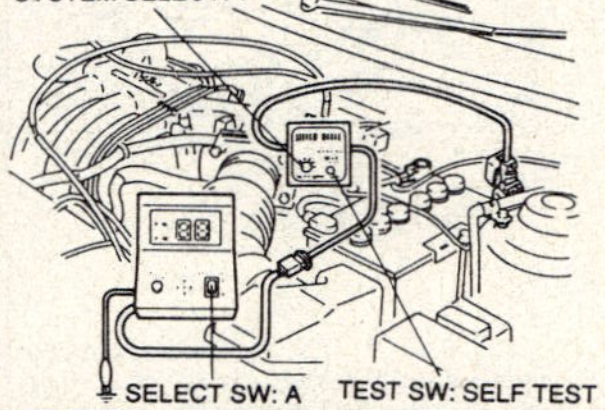	Yes	Go to next step
		No	**Service Code No. displayed** Check for cause (Refer to specified check sequence)
2	Are ECU terminal voltages OK (especially 2B, 2E, and 3D)?	Yes	Go to next step
		No	Check for cause
3	Disconnect high-tension lead idle; does engine speed decrease equally at all cylinders?	Yes	Go to next step
		No	Go to Step 9
4	Is intake manifold vacuum at idle correct? **Vacuum: More than 450 mmHg (17.7 inHg)**	Yes	Go to next step
		No	Check for air leakage of intake air system components
5	Is air cleaner element clean?	Yes	Go to next step
		No	Replace air cleaner element
6	Disconnect ISC valve connector at idle: Does engine speed increase? 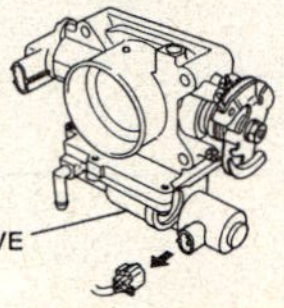	Yes	Go to next step
		No	Replace ISC valve

STEP	INSPECTION		ACTION
7	Is fuel line pressure correct at idle? **Fuel line pressure: 255—314 kPa (2.6—3.2 kg/cm², 37—46 psi) (Vacuum hose to pressure regulator disconnected)** 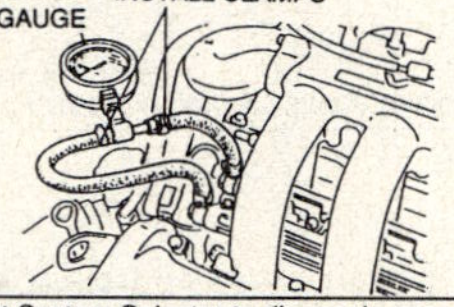	Yes	Go to next step
		No	**Low pressure** Check fuel line pressure while pinching fuel return hose ⇨ If pressure **quickly** increases, check pressure regulator ⇨ If pressure **gradually** increases, check for clogging between fuel pump and pressure regulator ⇨ If hose is not clogged, check fuel pump maximum pressure
8	Connect System Selector to diagnosis connector and set Test Switch to "SELF TEST"; is ignition timing at idle correct after warm-up? **Ignition timing (BTDC): 9—11° (10 ± 1°)**	Yes	Go to next step
		No	Adjust iginition system
9	Disconnect water thermosensor connector; does engine condition improve?	Yes	Replace water thermosensor
		No	Try a known good ECU; does condition improve?
10	Is there an injector operating sound at idle? 	Yes	Go to next step
		No	Is injector resistance OK? **Resistance: 12—16Ω (20°C [68°F])** ⇨ If it is, check wiring between ECU and injector ⇨ If it is not, replace injector
11	Is engine compression correct? **Engine compression (Minimum): 981 kPa (10.0 kg/cm², 142 psi)-300 rpm**	Yes	Go to next step
		No	Check engine
12	Is strong blue spark visible at disconnected high-tension lead? 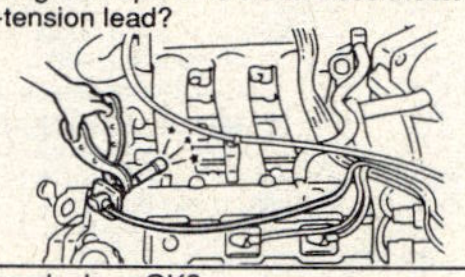	Yse	Go to next step
		No	Check high-tension lead ⇨ If OK, check distributor cap and rotor ⇨ If not OK, replace high-tension lead
13	Are spark plugs OK?	Yes	Try a known good ECU; does condition improve?
		No	Repair or replace

11	ROUGH IDLE/ENGINE STALLS AT IDLE — WHEN A/C, P/S, OR E/L ON		
DESCRIP-TION	• Engine stalls or vibrates excessively at idle when A/C, P/S, or E/L ON • A/C, P/S, headlights, blower fan, and electric cooling fan operate normally • Idle condition is normal when A/C, P/S, and E/L are OFF		
[TROUBLESHOOTING HINTS] ① Idle-speed control system malfunction			
STEP	INSPECTION		ACTION
1	Is "00" displayed on Self-Diagnosis Checker with ignition switch ON?	Yes	Go to next step
	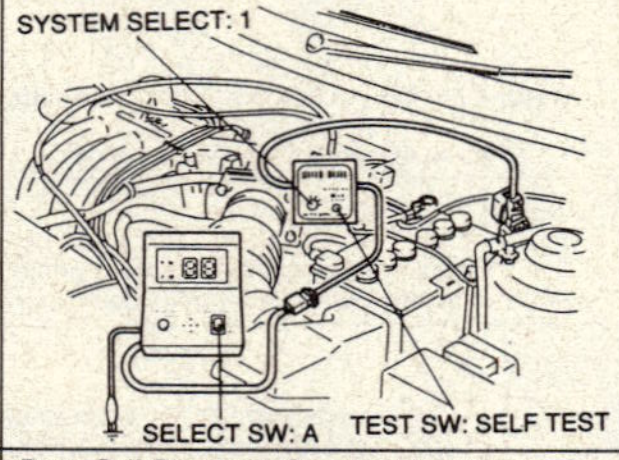	No	**Service Code No. displayed** Check for cause (Refer to specified check sequence)
2	Does Self-Diagnosis Checker Monitor Lamp with ignition switch ON show that switches operate correctly?	Yes	Go to next step
	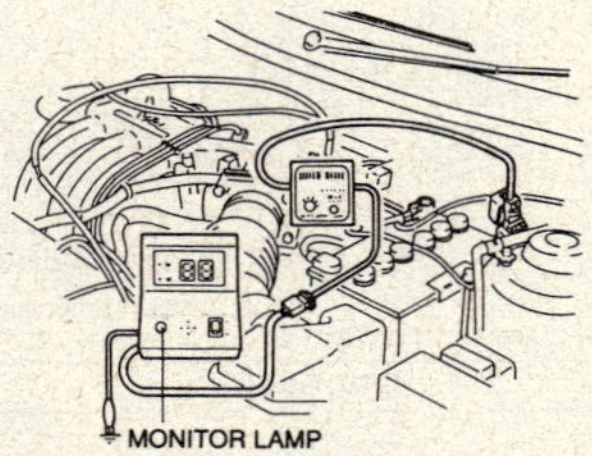	No	**Lamp not ON or OFF with specified switch** Check for cause (Refer to specified check sequence) **Lamp always ON** Check wiring between ECU terminal 1D and Self-Diagnosis Checker
3	Is ECU terminal 1I voltage OK? **Voltage: V_B (at idle)**	Yes	Try known good ECU; does condition improve?
	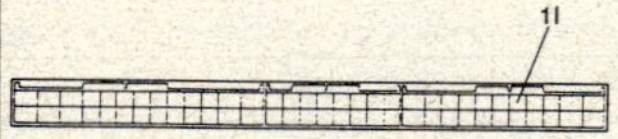	No	Check for short circuit in wiring between diagnosis connector terminal TEN and Ground

12	ROUGH IDLE/ENGINE STALLS JUST AFTER STARTING		
DESCRIP-TION	• Engine starts normally, but vibrates excessively or stalls just after starting (acceleration from idle) • Idle conditions are normal in the other conditions		
[TROUBLESHOOTING HINTS] ① Fuel injection control system • Start signal not input to ECU ② Ignition timing misadjustment			
STEP	INSPECTION		ACTION
1	Is "00" displayed on Self-Diagnosis Checker with ignition switch ON?	Yes	Go to next step
	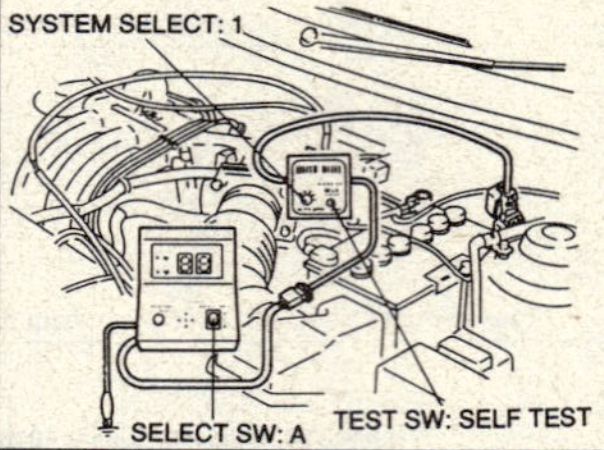	No	**Service Code No. displayed** Check for cause (Refer to specified check sequence)
2	Are ECU terminal voltages OK? (especially 1N)?	Yes	Go to next step
		No	Check for cause
3	Connect System Selector to diagnosis connector and set Test Switch to "SELF TEST"; is the ignition timing at idle correct after warm-up? **Ignition timing (BTDC): 9—11° (10 ± 1°)**	Yes	In same condition as Step 3 inspection and apply parking brake. Is the idle break correct? **Idle speed: 640—700 (670 ± 30) rpm** ⇨ If OK, go to next step ⇨ If not OK, adjust idle speed
		No	Adjust
4	Check for continuity between throttle sensor terminals IDL and E 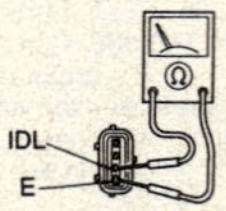	Yse	Go to next step
	Clearance between throttle lever and stopper / Continuity: 0.15mm (0.006 in) — Yes 0.50mm (0.0320 in) — No	No	Adjust
5	Try a known good ECU; does condition improve?		

1.8L ENGINE TROUBLESHOOTING GUIDE – MX-3

13	HIGH IDLE SPEED AFTER WARM-UP		
DESCRIP-TION	• Idle speed excessively high after warm-up		
[TROUBLESHOOTING HINTS] Excessive intake air supplied to engine ① Throttle valve not fully closed ② Idle-speed control malfunction • All valve not closed • Idle-up solenoid valve stuck • A/C E/L signal sent to ECU • P/S pressure switch always ON			
STEP	**INSPECTION**		**ACTION**
1	Is throttle valve fully closed when accelerator is released? 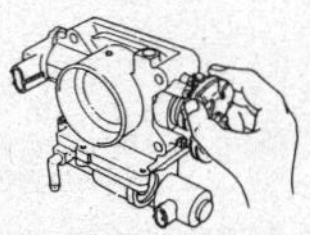	Yes	Go to Step 3
		No	Is throttle linkage correctly installed and does it operate freely? ➪ If OK, go to next step ➪ If not OK, clean, adjust, or replace linkage
2	Is "00" displayed on Self-Diagnosis Checker with ignition switch ON? 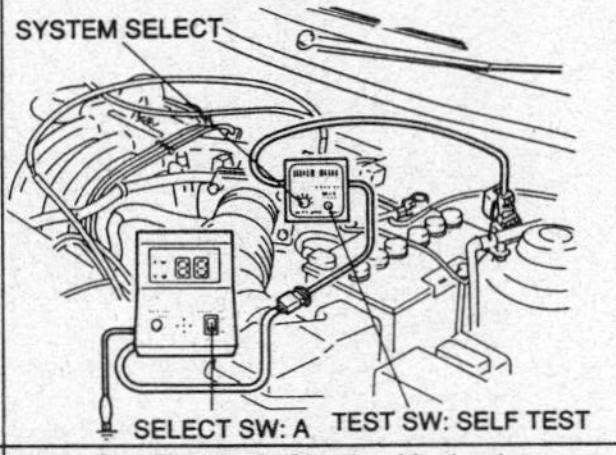 	Yes	Go to next step
		No	**Service Code No. displayed** Check for cause (Refer to specified check sequence)
3	Does Self-Diagnosis Checker Monitor Lamp, with ignition switch ON, show that switches operate correctly? 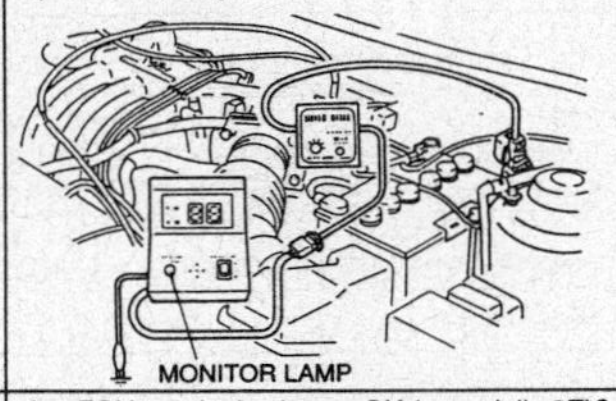 	Yes	Go to next step
		No	**Lamp not ON or OFF with specified switch** Check for cause (Refer to specified check sequence) **Lamp always ON** Check wiring between ECU terminal 1D and Self-Diagnosis Checker
4	Are ECU terminal voltages OK (especially 2E)?	Yes	Try a known ECU; does condition improve?
		No	Check for cause

STEP	INSPECTION		ACTION
5	Does engine speed decrease as engine warms up?	Yes	Go to next step
		No	Check air valve
6	Is ISC valve OK?	Yes	Go to next step
		No	Repair or replace
7	Pinch PCV hose with pliers; does engine speed decrease?	Yes	Check PCV valve
		No	Go to next step
8	Are ECU terminal voltages OK (especially 2B, 2E, and 3D)?	Yse	Try a known good ECU; does condition improve?
		No	Check for cause

14	IDLE MOVES UP AND DOWN/IDLE HUNTING		
DESCRIP-TION	• Engine speeds up and down at idle		
[TROUBLESHOOTING HINTS] ① Idle switch (built-in throttle sensor) OFF at idle ② Air leakage ③ Fuel injection amount inconstant • Poor contact point inside airflow meter ④ Poor ignition			
STEP	INSPECTION		ACTION
1	Is "00" displayed on Self-Diagnosis Checker with ignition switch ON? 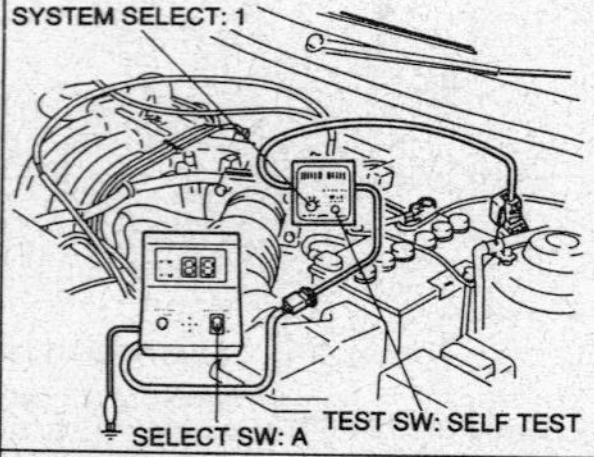 	Yes	Go to next step
		No	**Service Code No. displayed** Check for cause (Refer to specified check sequence)
2	Is intake manifold vacuum correct at idle? **Intake manifold vacuum:** **More than 450 mmHg (17.7 inHg)**	Yes	Go to next step
		No	**Low vacuum** Check for air leakage at intake air system
3	Is air cleaner element clean?	Yes	Go to next step
		No	Replace air cleaner element
4	Disconnect high-tension lead at idle; does engine speed decrease equally at each cylin-der?	Yes	Go to next step
		No	Go to Step 8
5	Are ECU terminal voltages OK?	Yes	Go to next step
		No	Check for cause
6	Connect System Selector to diagnosis connec-tor and set Test Switch to "SELF TEST"; is ignition timing at idle correct after warm-up? **Ignition timing (BTDC): 9—11° (10 ± 1°)**	Yes	Go to next step
		No	Adjust ignition timing

STEP	INSPECTION		ACTION
7	Is airflow meter OK? 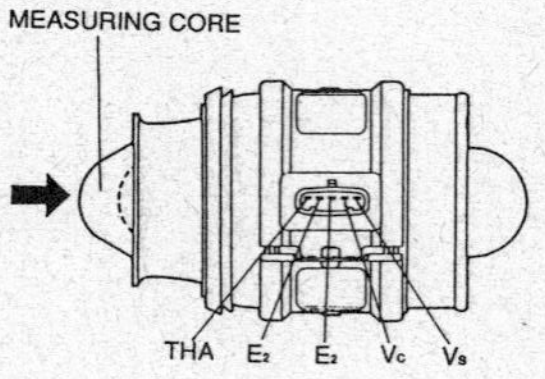 	Yes	Go to Step 13
		No	Replace airflow meter
8	Is there an injector operating sound at idle?	Yes	Go to Step 10
		No	Go to next step
9	Does approximately 12V exist at the injector connector (W/R) wire?	Yes	Is injector resistance OK? **Resistance: 12—16Ω (20°C[68°F])** ⇨ If it is, check wiring between ECU and injector ⇨ If it is not, replace injector
		No	Check wiring between ECU and injector
10	Is strong blue spark visible at disconnected high-tension lead? 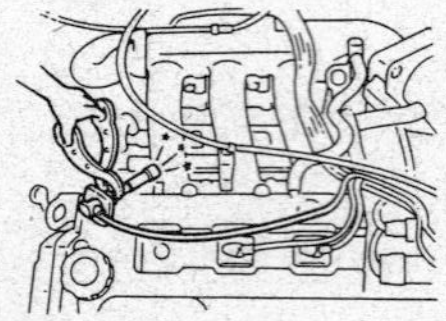	Yes	Go to next step
		No	Check high-tension lead ⇨ If OK, check distributor cap and rotor ⇨ If not OK, replace high-tension lead
11	Are spark plugs OK? 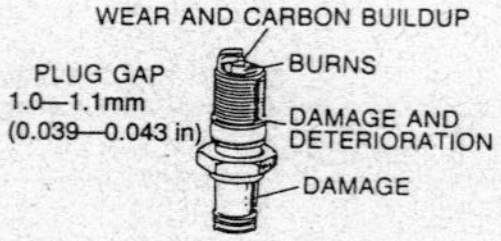 	Yse	Is engine compression correct? **Compression (Minimum):** **981 kPa (10.0 kg/cm², 142 psi)-300 rpm** ⇨ If OK, go to next step ⇨ If not OK, check for cause
		No	Clean or replace
12	Check for injector leakage	Yes	Replace injector
		No	Go to next step
13	Try a known good ECU; does condition improve?		

1.8L ENGINE TROUBLESHOOTING GUIDE—MX-3

15	ENGINE/STALLS ON DECELERATION		
DESCRIPTION	• Engine unexpectedly stops running during or after deceleration • Idle condition normal		
[TROUBLESHOOTING HINTS] Engine speed drops too much when accelerator is released or because of a poor connector connection that has been disconnected by the force of deceleration ① Idle speed misadjustment ② Fuel cut control malfunction			
STEP	**INSPECTION**		**ACTION**
1	Is idle condition normal? **Idle speed: 640—700 rpm (670 ± 30) rpm**	Yes	Go to next step
		No	Adjust or perform troubleshooting Nos 8—11 "ROUGH IDLE"
2	Is "00" displayed on Self-Diagnosis Checker with ignition switch ON? 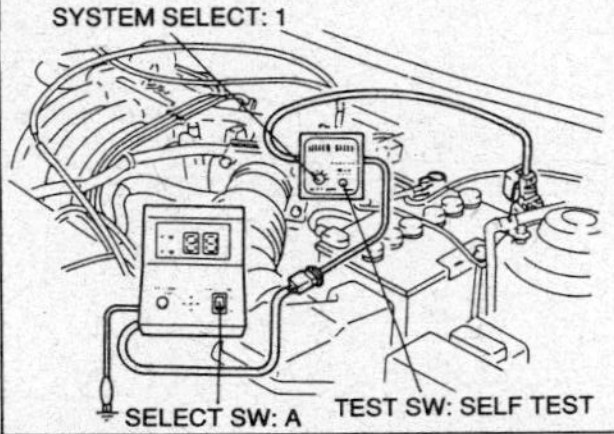	Yes	Go to next step
		No	**Service Code No. displayed** Check for cause (Refer to specified check sequence)
3	Does-Self Diagnosis Checker Monitor Lamp, with ignition switch ON, show that switches operate correctly? 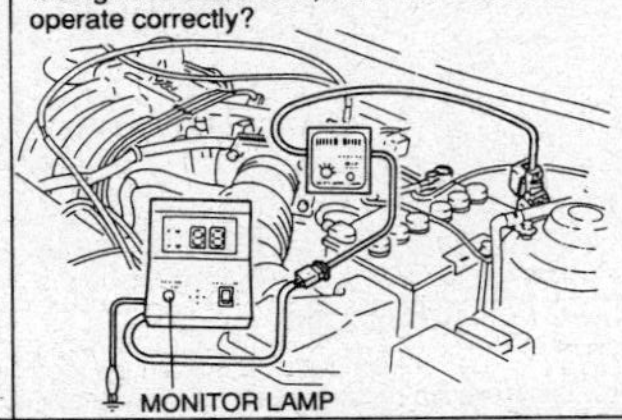	Yes	Go to next step
		No	**Lamp not ON or OFF with specified switch** Check for cause (Refer to specified check sequence)
			Lamp always ON Check wiring between ECU terminal 1D and Self-Diagnosis Checker
4	Are ECU terminal voltages OK (especially 1R, 2B, 3D, 3U, 3V, 3W, 3X, 3Y, and 3Z)?	Yes	Go to next step
		No	Check for cause
5	Check for poor connection of the following parts • Ignition coil • Igniter • Distributor • High-tension lead • Injector • Circuit-opening relay • ECU	Yes	Repair or replace
		No	Go to next step

STEP	INSPECTION		ACTION
6	Disconnect ISC valve connector at idle: Dose engine speed increase? 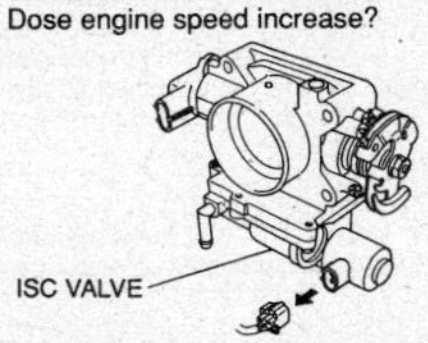	Yes	Go to next step
		No	Replace ISC valve
7	Try a known good ECU; does condition improve?		

16	ENGINE STALLS SUDDENLY (INTERMITTENT)		
DESCRIPTION	• Engine intermittently stops running • Before stalling, is engine condition is OK		

[TROUBLESHOOTING HINTS]
① Intermittently no spark or no fuel injection caused by vehicle vibration, acceleration, or deceleration
- Poor connection in wire harness

STEP	INSPECTION		ACTION
1	Is "00" displayed on Self-Diagnosis Checker with ignition switch ON? 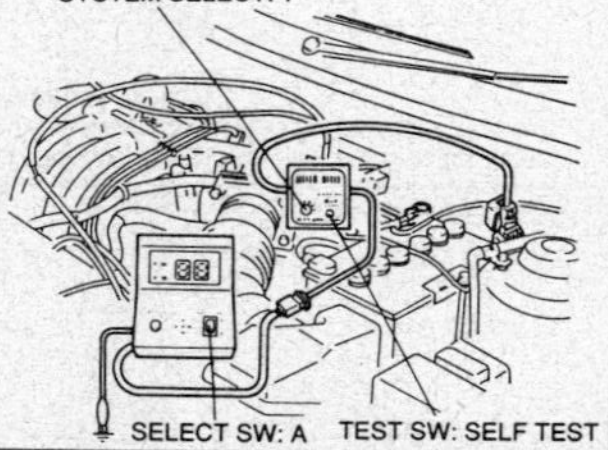	Yes	Go to next step
		No	**Service Code No. displayed** Check for cause (Refer to specified check sequence) **Note** **• When checking wiring harness and connectors, tap, move, and wiggle suspect sensor and harness to recreate problem**
2	Check for poor connection of the following parts • Ignition coil • Igniter • Distributor • High-tension lead • Injector • Circuit-opening relay • ECU	Yes	Repair or replace
		No	Go to next step
3	Are ECU terminal voltages OK (especially 1B, 3A, 3B, and 3C)? **Note** **• When checking voltages, tap, move, and wiggle harness and connector to recreate problem**	Yes	Go to Troubleshooting No.2: "CRANKS NORMALLY BUT WILL NOT START (NO COMBUSTION)"
		No	Check for cause

17	HESITATES/STUMBLES ON ACCELERATION		
DESCRIPTION	• Flat spot occurs just after accelerator depressed of mild jerking occurs during acceleration		

[TROUBLESHOOTING HINTS]
① Air/Fuel mixture becomes lean when depressing accelerator
- Fuel injection control malfunction (Correction for accelerating condition)
- Air leakage
- Fuel line pressure decreases
- Sparks advance control malfunction

STEP	INSPECTION		ACTION
1	Is "00" displayed on Self-Diagnosis Checker ignition switch ON? 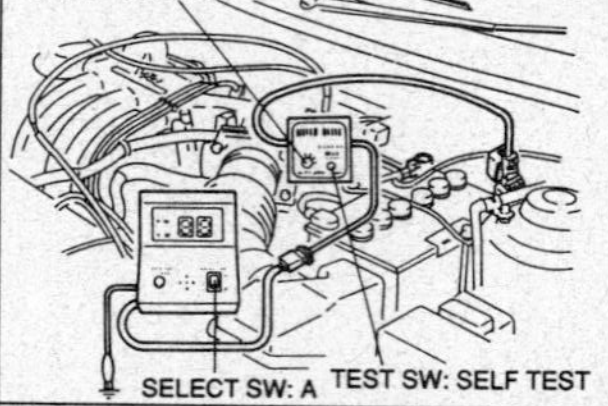	Yes	Go to next step
		No	**Service Code No. displayed** Check for cause (Refer to specified check sequence)
2	Is intake manifold vacuum correct at idle? **Vacuum: More than 450 mmHg (17.7 inHg)**	Yes	Go to next step
		No	Check for air leakage at intake air system components
3	Is air cleaner element is clean?	Yes	Go to next step
		No	Replace air cleaner element
4	Are ECU terminal voltages OK (especially 2F, 2I)?	Yes	Go to next step
		No	Check for cause
5	Is throttle linkage correctly installed and does it operate freely?	Yes	Go to next step
		No	As required, correct, clean, or replace any binding or damaged linkage, and adjust cable deflection at throttle body
6	Check for continuity between throttle sensor terminals IDL and E	Yes	Go to next step
		No	Adjust

Clearance between throttle lever and stopper	Continuity
0.15mm (0.006 in)	Yes
0.50mm (0.020 in)	No

STEP	INSPECTION		ACTION
7	Is fuel line pressure correct at idle? **Fuel line pressure:** **255—314 kPa (2.6—3.2 kg/cm², 37—48 psi)** **(vacuum hose to pressure regulator disconnected)** 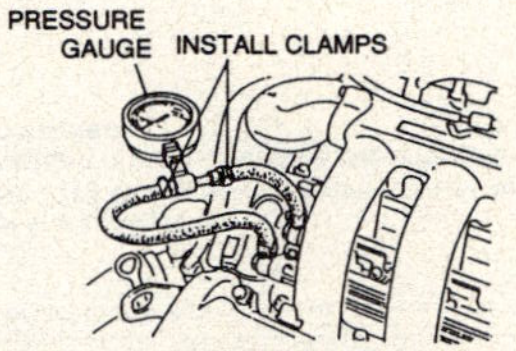	Yes	Does fuel line pressure decrease quickly acceleration? ⇨ If it does check fuel line and filter for clogging ⇨ If it does not, go to next step
		No	**Low pressure** Check fuel line pressure while pinching fuel return hose ⇨ If pressure **quickly** increases, check pressure regulator ⇨ If pressure **gradually** increases, check for clogging between fuel pump and pressure regulator If hose is not clogged, check fuel pump maximum pressur
8	Connect System Selector to diagnosis connector and set Test Switch to "SELF TEST"; is for ignition timing at idle correct after warm-up? **Ignition timing (BTDC):** **9—11° (10 ± 1°)** 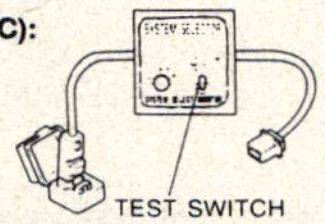	Yes	Does ignition timing advances during acceleration? ⇨ If it does, go to next step ⇨ If not, replace the ECU
		No	Adjust
9	Are air duct and air hoses are correctly installed?	Yes	Go to next step
		No	Repair
10	Is exhaust system restricted?	Yes	Repair or replace
		No	Go to next step
11	Is variable resonance induction system (VRIS) OK?	Yes	Go to next step
		No	Repair or replace
12	Try a known good ECU; condition improve?		

18	SURGES WHILE CRUISING		
DESCRIPTION	• Unexpected change in engine speed that is usually repetitive		
	[TROUBLESHOOTING HINTS] ① Air/Fuel mixture too lean ② Misfire ③ Poor connection in wiring harness		
STEP	**INSPECTION**		**ACTION**
1	Is "00" displayed on Self-Diagnosis Checker with ignition switch ON? 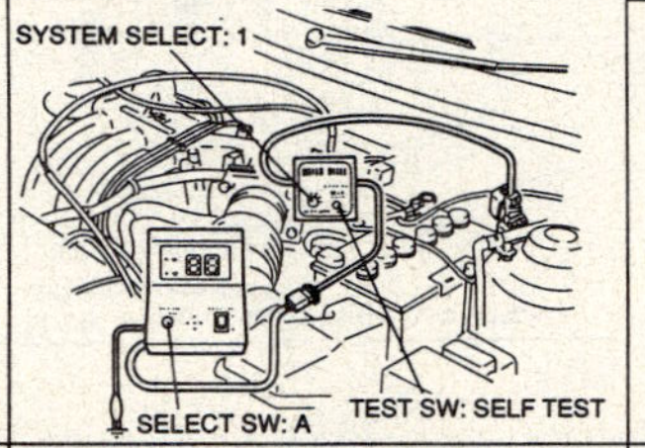	Yes	Go to next step
		No	**Service Code No. displayed** Check for cause (Refer to specified check sequence)
2	Does Self-Diagnosis Checker Monitor Lamp with ignition switch ON show that switches operate correction? 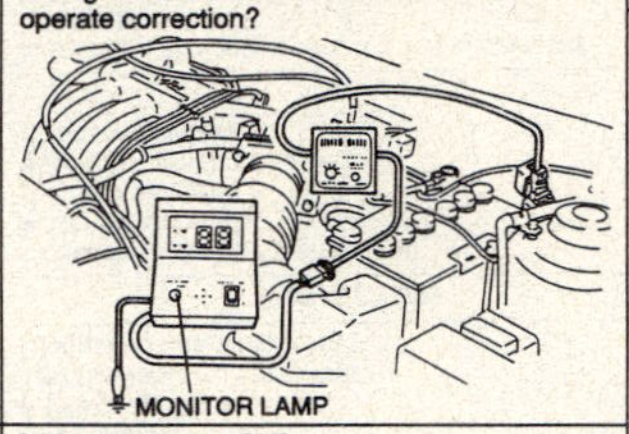	Yes	Go to next step
		No	**Lamp does not illuminate ON or OFF with specified input switch** Check for cause (Refer to specified check sequence) **Lamp always ON** Check ECU terminal 1D voltage
3	Is throttle sensor OK?	Yes	Go to next step
		No	Adjust
4	Disconnect oxygen sensor connector does condition improve?	Yes	Check oxygen sensor
		No	Go to next step
5	Are ECU terminal 2C and 2D voltages OK?	Yes	Go to next step
		No	Check for cause

STEP	INSPECTION		ACTION
6	Is throttle linkage correctly installed and does it operates freely?	Yes	Go to next step
		No	As required, correct, clean, or replace any binding or damaged linkage, and adjust cable deflection at throttle body
7	Is intake manifold vacuum correct at idle? **Vacuum: More than 450 mmHg (17.7 inHg)**	Yes	Go to next step
		No	Check for air leakage of intake air system components
8	Is air cleaner element clean?	Yes	Go to next step
		No	Replace air cleaner element
9	Connect System Selector to diagnosis connector and set Test Switch to "SELF TEST"; is ignition timing at idle after warm-up? **Ignition timing (BTDC): 9—11° (10 ± 1°)** 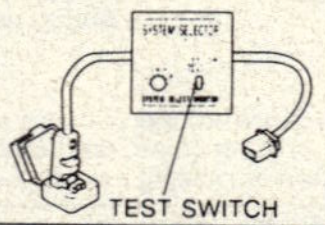 	Yes	Does ignition timing advance during acceleration? ⇨ If it does, go to next step ⇨ If it does not, replace the ECU
		No	Adjust
10	Is correct fuel line pressure at idle **Fuel line pressure: 255—314 kPa (2.6—3.2kg/cm², 37—46 psi) (vacuum hose to pressure regulator disconnected)** 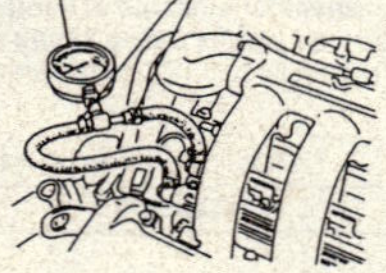 	Yes	Does the fuel line pressure decrease during quick acceleration ⇨ If it does, check for fuel pump maximum pressure If pressure is OK, check the fuel line and filter for clogging ⇨ If it does not decrease, go to next step
		No	**Low pressure** Check fuel line pressure while pinching fuel return hose ⇨ If pressure **quickly** increases, check pressure regulator ⇨ If pressure **gradually** increases, check for clogging between fuel pump and pressure regulator If hose is not clogged, check fuel pump maximum pressure **High pressure** Is fuel return line clogged? ⇨ If it is not, replace pressure regulator ⇨ If it is, replace it
11	Check if exhaust system is restricted	Yes	Repair or replace
		No	Go to next step
12	Try a known good ECU; does condition improve?		

19	LACK OF POWER		
DESCRIPTION	• Perform poor under load when throttle valve is wide open • Reduced maximum speed • Idle condition normal		

[TROUBLESHOOTING HINTS]

① Factors other than engine malfunction
- Clutch slipping
- Brake drugging
- Low tire pressure
- Incorrect tire size
- Over-loaded

② Low intake air amount
- Throttle valve not open fully
- Clogged intake air system

③ Air/Fuel mixture too lean
- Fuel line pressure decreases
- Fuel injection malfunction

④ Poor ignition
④ Low engine compression
④ Alcohol blended fuel used

STEP	INSPECTION		ACTION
1	Check factors other than engine • Clutch slipping • Brake dragging • Low tire pressure • Incorrect tire size	Yes	Go to next step
		No	Repair
2	Is throttle valve fully opened when accelerator is fully depressed?	Yes	Go to next step
		No	Is accelerator cable correctly installed? ⇨ If it is, check throttle body ⇨ If it is not, install it correctly
3	Is "00" displayed on Self-Diagnosis Checker with ignition switch ON 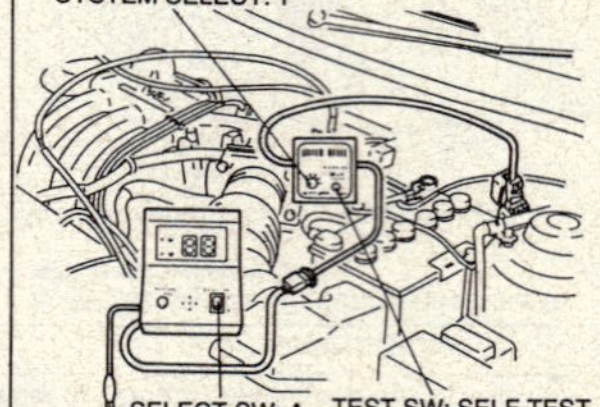 	Yes	Go to next step
		No	**Service Code No. displayed** Check for cause (Refer to specified check sequence)
4	Connect System Selector to diagnosis connector and set Test Switch to "SELF TEST"; is ignition timing at idle correct after warm-up? **Ignition timing (BTDC): 9—11° (10 ± 1°)** 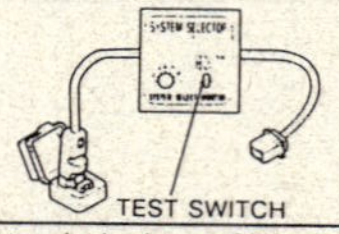 	Yes	Ignition timing advances when accelerating ⇨ If advances, go to next step ⇨ If not advances, check ECU terminal voltages
		No	Adjust
5	Are ECU terminal voltages OK?	Yes	Go to next step
		No	Check for cause

1.8L ENGINE TROUBLESHOOTING GUIDE—MX-3

STEP	INSPECTION		ACTION
6	Ils intake manifold vacuum correct at idle? **Intake manifold vacuum: More than 450 mmHg (17.7 inHg)**	Yes	Go to next step
		No	Check for air leakage of intake air system components
7	Is air cleaner element clean?	Yes	Go to next step
		No	Replace sir cleaner element
8	Is fuel line pressure correct at idle? **Fuel line pressure: 255—314 kPa (2.6—3.2kg/cm², 37—46 psi) (vacuum hose to pressure regulator disconnected)** 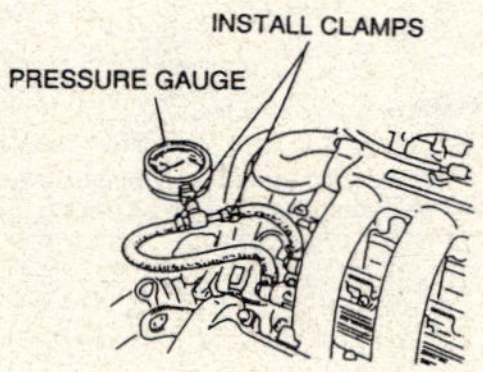	Yes	Does fuel line pressure decrease during quick acceleration? ⇨ If it does, check fuel pump maximum pressure ⇨ If pressure is OK, check fuel line and filter for clogging ⇨ If it does not decrease, go to next step
		No	**Low pressure** Check fuel line pressure while pinching fuel return hose ⇨ If pressure **quickly** increases, check pressure regulator ⇨ If pressure **gradually** increases, check for clogging between fuel pump and pressure regulator If hose is not clogged, check fuel pump maximum pressure
			High pressure Is fuel return line clogged? ⇨ If it is not, replace pressure regulator ⇨ If it is, replace it
9	Is airflow meter OK? I. Does measuring core move smoothly? 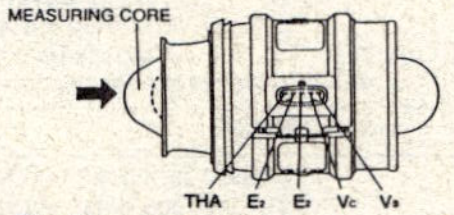	Yes	Go to next step
	II. Check resistance 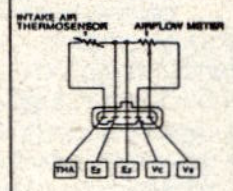 (see resistance table below)	No	Repair or replace

Terminal	Resistance (Ω)	
	Fully closed	Fully open
E2 ↔ Vs	200—600	20—1,000
E2 ↔ Vc	200—400	

STEP	INSPECTION		ACTION
10	Are spark plugs OK? 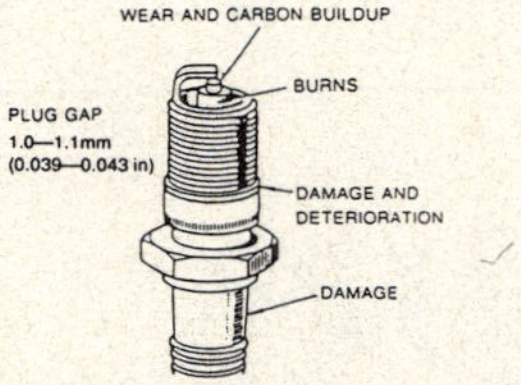	Yes	Go to next step
		No	Clean or replace
11	Is resistance of high-tension leads OK? **Resistance: 16 kΩ per 1 m (3.28 ft)** 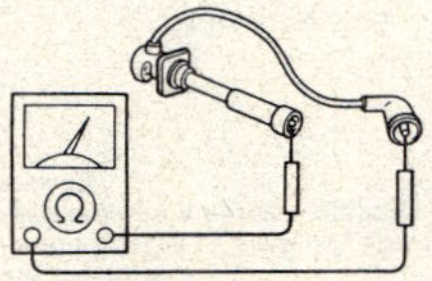	Yes	Go to next step
		No	Replace
12	Is resistance of ignition coil OK? **Resistance (at 20°C [68°F]): Primary coil winding.......... 0.58—0.86Ω Secondary coil winding...... 11.5—18.5 kΩ** 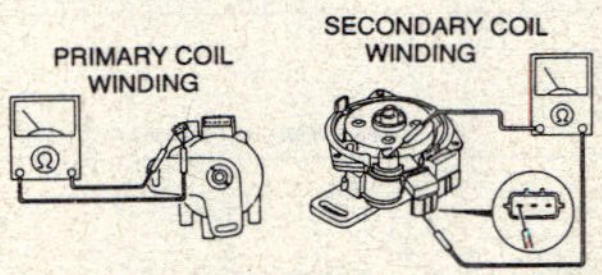	Yes	Go to next step
		No	Replace
13	Is engine compression correct? **Engine compression: 981 kPa (10.0 kg/cm², 142 psi)-300 rpm**	Yes	Go to next step
		No	Check engine condition • Worn piston, piston rings, or cylinder wall • Defective cylinder head gasket • Distorted cylinder head • Improper valve seating • Valve sticking in guide
14	Change to another brand of fuel; does condition improve?	Yes	Change to another brand of fuel
		No	Go to next step
15	Is variable resonance induction system (VIRS) OK?	Yes	Go to next step
		No	Repair or replace
16	Try a known good ECU; condition improve?		

20 POOR ACCELERATION

DESCRIPTION	• Performance poor during acceleration • Idle condition normal

[TROUBLESHOOTING HINTS]

① Factors other than engine malfunction
- Clutch slipping
- Brake drugging
- Low tire pressure
- Incorrect tire size
- Over-loaded

② Low intake air amount
- Throttle valve not open fully
- Clogged intake air system

③ Air/Fuel mixture too lean
- Fuel line pressure decreases
- Fuel injection malfunction

④ Poor ignition

④ Low engine compression

④ Alcohol blended fuel used

STEP	INSPECTION		ACTION
1	Check factors other than engine • Clutch slipping • Brake dragging • Low tire pressure • Incorrect tire size	Yes	Go to next step
		No	Repair
2	Does throttle valve fully open when the accelerator is fully depressed?	Yes	Go to next step
		No	Is accelerator cable correctly installed? ⇨ If it is, check throttle body ⇨ If it is not, install it correctly
3	Is "00" displayed on the Self-Diagnosis Checker with ignition switch ON? 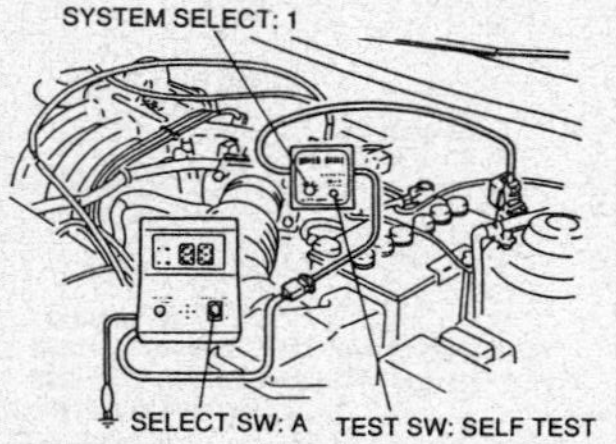	Yes	Go to next step
		No	**Service Code No. displayed** Check for cause (Refer to specified check sequence)
4	Check continuity between throttle sensor terminals IDL and E 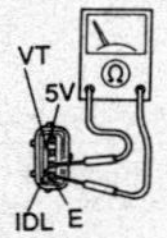	Yes	Go to next step
		No	Adjust

Clearance between throttle lever and stopper	Continuity
0.15mm (0.006 in)	Yes
0.50mm (0.020 in)	No

STEP	INSPECTION		ACTION
5	Connect System Selector to diagnosis connector and set Test Switch to "SELF TEST" Is ignition timing at idle after warm-up? **Ignition timing (BTDC): 9—11° (10 ± 1°)** 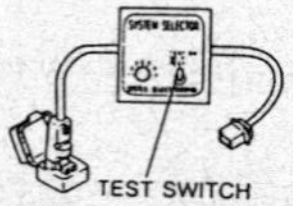	Yes	Does ignition timing advance during acceleration? ⇨ If it does, go to next step ⇨ If it does not, check the ECU terminal voltages
		No	Adjust
6	Are ECU terminal voltages OK?	Yes	Go to next step
		No	Check for cause
7	Is intake manifold vacuum correct at idle? **Intake manifold vacuum: More than 450 mmHg (17.7 inHg)**	Yes	Go to next step
		No	Check for air leakage of intake air system components
8	Is fuel line pressure correct at idle? **Fuel line pressure: 255—314 kPa (2.6—3.2kg/cm², 37—46 psi) (vacuum hose to pressure regulator disconnected)** 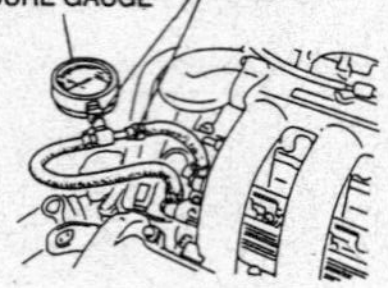	Yes	Does fuel line pressure decrease during quick acceleration? ⇨ If it does, check fuel pump maximum pressure ⇨ If pressure is OK, check fuel line and filter for clogging ⇨ If it does not decrease, go to next step
		No	**Low pressure** Check fuel line pressure while pinching fuel return hose ⇨ If pressure **quickly** increase, check pressure regulator ⇨ If pressure **gradually** increases, check for clogging between fuel pump and pressure regulator. If hose is not clogged, check fuel pump maximum pressure **High pressure** Is fuel line clogged? ⇨ If it is not, replace pressure regulator ⇨ If it is, replace it

1.8L ENGINE TROUBLESHOOTING GUIDE – MX-3

STEP	INSPECTION		ACTION
9	Is airflow meter OK? I. Does measuring core move smoothly? 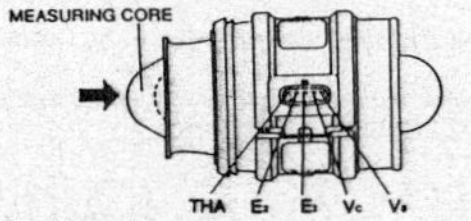	Yes	Go to next step
	II. Check resistance 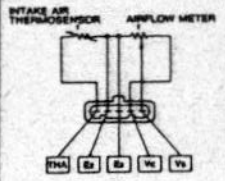	No	Repair or replace
10	Are spark plugs OK? 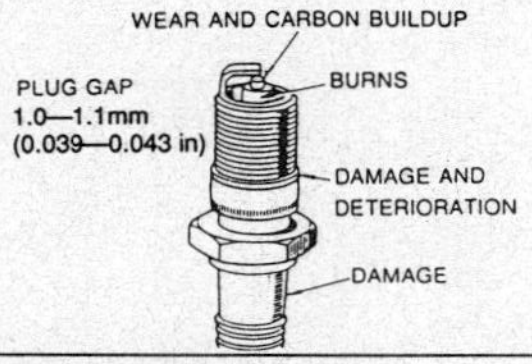	Yes	Go to next step
		No	Clean or replace
11	Is resistance of ignition coil OK? **Resistance (at 20°C [68°F]:** **Primary coil winding........... 0.58—0.86Ω** **Secondary coil winding...... 11.5—18.5 kΩ**	Yes	Go to next step
	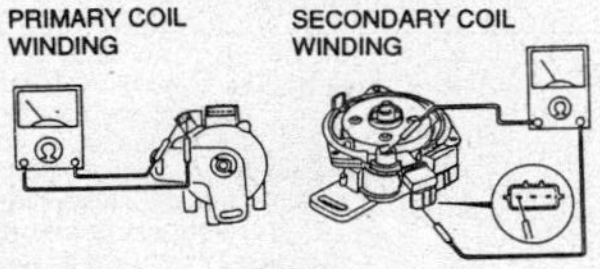	No	Replace
12	Is engine compression correct?	Yes	Go to next step
	Engine compression (minimum): **981 kPa (10.0 kg/cm², 142 psi)-300 rpm**	No	Check engine condition • Worn piston, piston rings, or cylinder wall • Defective cylinder head gasket • Distorted cylinder head • Improper valve seating • Valve sticking in guide

Resistance table for step 9 (II. Check resistance):

Terminal	Resistance (Ω) Fully closed	Resistance (Ω) Fully open
E2 ↔ Vs	200—600	20—1,000
E2 ↔ Vc	200—400	

STEP	INSPECTION		ACTION
13	Change to another brand fuel; does acceleration improve?	Yes	Change to another brand of fuel
		No	Go to next step
14	Is A/C cut-off control system OK?	Yes	Go to next step
		No	Repair or replace
15	Is variable resonance induction system (VRIS) OK?	Yes	Go to next step
		No	Repair or replace
16	Try a known good ECU; does condition improve?		

21	RUNS ROUGH DURING DECELERATION/BACKFIRE
DESCRIP-TION	• Engine runs rough during deceleration, and abnormal combustion occurs in exhaust system • Transaxle in normal condition

[TROUBLESHOOTING HINTS]
① Air/Fuel mixture too rich
- Air cleaner element clogged
- Fuel injection control malfunction (Fuel-cut control)
- Injector fuel leakage

② Throttle sensor (IDL terminal) malfunction

STEP	INSPECTION		ACTION
1	Is "00" displayed on Self-Diagnosis Checker with ignition switch ON? 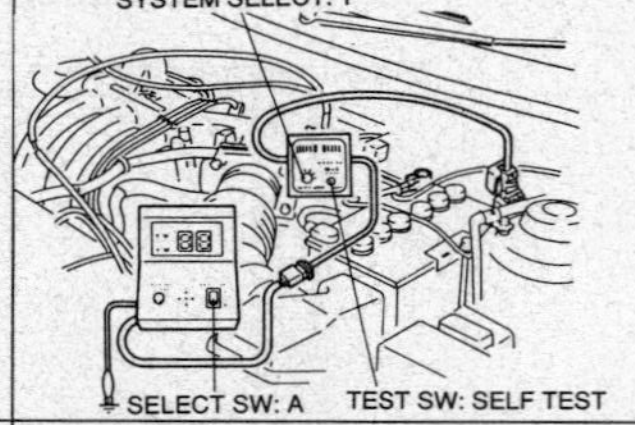	Yes	Go to next step
		No	**Service Code No. displayed** Check for cause (Refer to specified check sequence)
2	Does Self-Diagnosis Checker Monitor Lamp with ignition switch ON show that switches operate correctly? 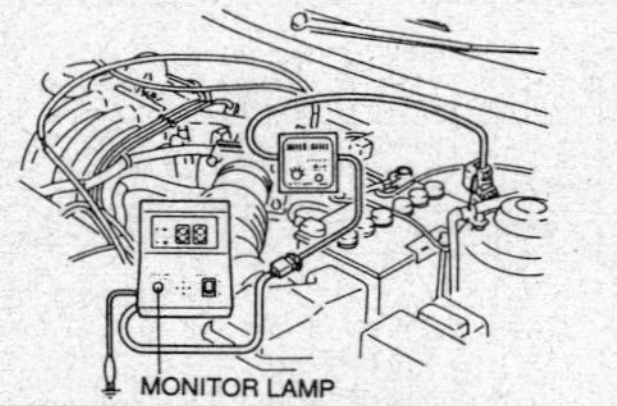	Yes	Go to next step
		No	**Lamp not ON or OFF with specified switch** Check for cause (Refer to specified check sequence)
			Lamp always ON Check wiring between ECU terminal 1D and Self-Diagnosis Checker
3	Is intake manifold vacuum correct at idle? **Vacuum: More than 450 mmHg (17.7 inHg)**	Yes	Go to next step
		No	Check for air leakage at intake air system components
4	Is air cleaner element clean?	Yes	Go to next step
		No	Replace air cleaner element
5	Are ECU terminal voltages OK (especially 2F, 3A, 3U, 3V, 3W, 3X, 3Y and 3Z)?	Yes	Go to next step
		No	Check for cause

STEP	INSPECTION		ACTION
6	Is fuel cut operation OK?	Yes	Go to next step
		No	Try a known good ECU; does condition improve?
7	Is airflow meter OK? I. Does measuring core move smoothly? 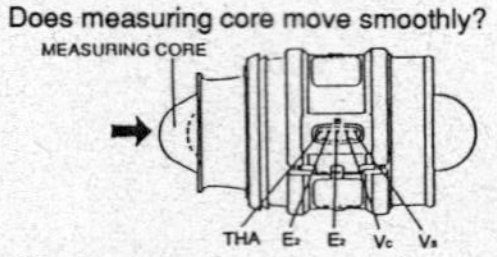	Yes	Go to next step
	II. Check resistance 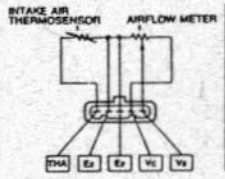	No	Repair or replace
8	Is throttle sensor IDL terminal OK?	Yes	Go to next step
		No	Adjust
9	Check injector for fuel leakage 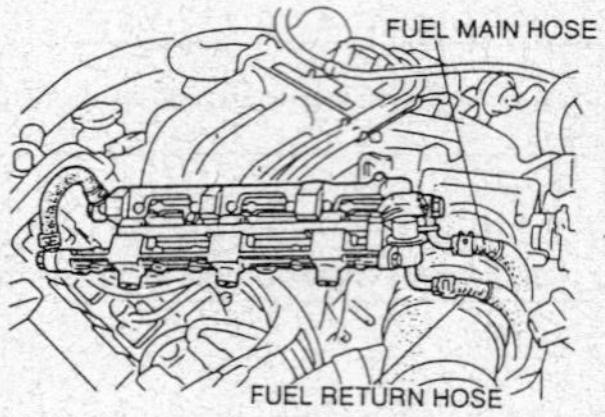	Yes	Replace
		No	Go to next step
10	Try a known good ECU; does condition improve?		

Resistance check (step 7 II):

Terminal	Resistance (Ω) Fully closed	Resistance (Ω) Fully open
E₂ ↔ Vs	200—600	20—1,000
E₂ ↔ Vc	200—400	

22	KNOCKING		
DESCRIPTION	• Abnormal combustion accompanied by audible "pinging" noise		
[TROUBLESHOOTING HINTS] ① Incorrect ignition timing (Too advanced) ② Carbon deposits in cylinder ③ Overheating ④ Air/Fuel mixture too lean • Fuel injection amount not correct • Fuel line pressure decreases during acceleration			
STEP	**INSPECTION**		**ACTION**
1	Connect System Selector to diagnosis connector and set Test Switch to "SELF TEST"; is ignition timing at idle correct after warm-up? **Ignition timing (BTDC): 9—11° (10 ± 1°)** 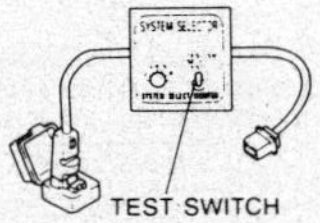	Yes	Go to next step
		No	Adjust
2	Is "00" displayed on Self-Diagnosis Checker with ignition switch ON? 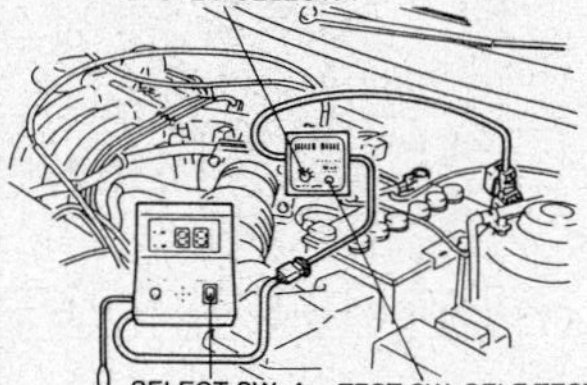	Yes	Go to next step
		No	**Service Code No. displayed** Check for cause (Refer to specified check sequence)
3	Is intake manifold vacuum correct at idle? **Vacuum: More than 450 mmHg 17.7 inHg)**	Yes	Go to next step
		No	Check for air leakage of intake air system components
4	Is engine compression correct? **Engine compression (minimum): 981 kPa (10.0 kg/cm², 142 psi)-300 rpm**	Yes	Go to next step
		No	High compression Check engine condition • Carbon deposits

STEP	INSPECTION		ACTION
5	Is fuel line pressure correct at idle? **Fuel line pressure: 255—314 kPa (2.6—3.2 kg/cm², 37—48 psi) (vacuum hose from pressure regulator disconnected)** 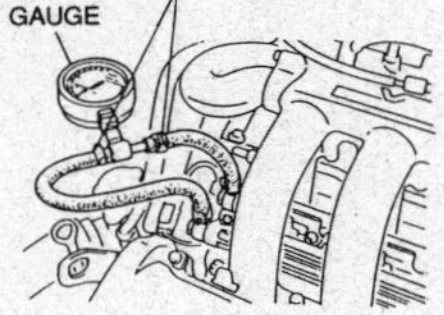	Yes	Does fuel line pressure decrease during quickly acceleration? ⇨ If it does, check for clogging between fuel pump and pressure regulator ⇨ If it does not, go to next step
		No	**Low pressure** Check fuel line pressure while pinching fuel return hose ⇨ If pressure **quickly** increases, check pressure regulator ⇨ If pressure **gradually** increases, check for clogging between fuel pump and pressure regulator If hose is not clogged, check fuel pump maximum pressure
6	Is cooling system OK?	Yes	Go to next step
		No	Repair or replace • Thermostat • Electric cooling fan • Radiator
7	Try a known good ECU; does condition improve?	Yes	Replace ECU
		No	Change another brand of fuel or use a higher-octane fuel

23	FUEL ODOR		
DESCRIP-TION	• Gasoline odor in cabin		
[TROUBLESHOOTING HINTS] ① Poor connection or damaged fuel system ② Chacoal canister overflow due to evapolative emission control system malfunction			
STEP	INSPECTION		ACTION
1	Is there fuel leakage or damage of fuel system and evapolative emission control system?	Yes	Repair or replace
		No	Go to next step
2	Is "00" displayed on Self-Diagnosis Checker with ignition switch ON? 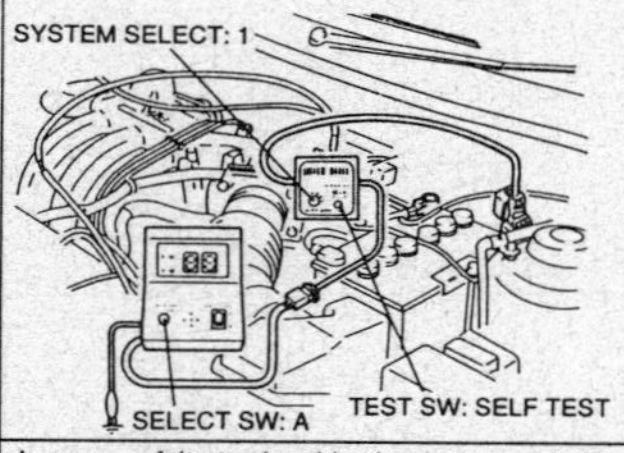 	Yes	Go to next step
		No	**Service Code No. displayed** Check for cause (Refer to specified check sequence)
3	Is vacuum felt at solenoid valve (purge control) with engine running and throttle valve opened? (Neutral switch connector disconnected) 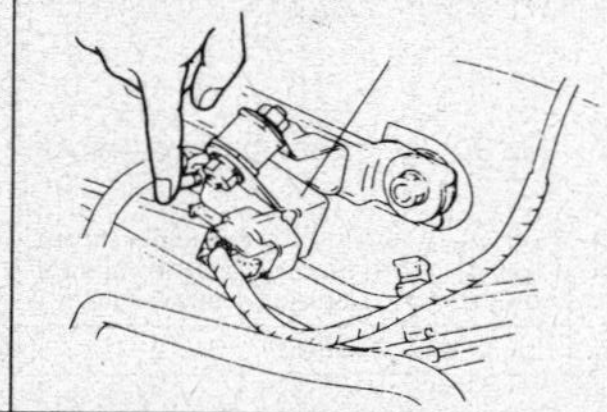	Yes	Go to Step 5
		No	Check for solenoid valve opening sound ➪ If OK, check vacuum hoses for clogging ➪ If not OK, go to next step
4	Apply 12V and ground to solenoid valve (purge control) Is vacuum felt at solenoid valve at idle? 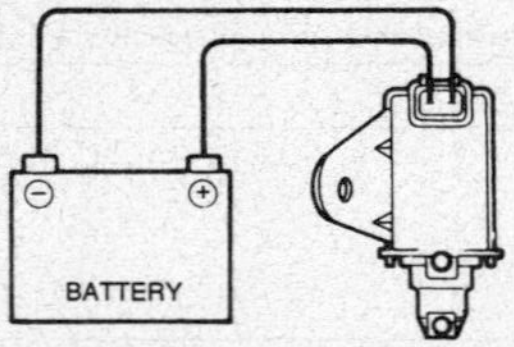 	Yes	Check ECU terminal voltages
		No	Replace solenoid valve
5	Try a known good ECU; does condition improve?		

24	EXHAUST SULFUR ODOR	
DESCRIP-TION	• Exhaust gas has an unusual odor	
[TROUBLESHOOTING HINTS] High sulfur content fuel used		
STEP	INSPECTION	ACTION
1	Change to another brand of fuel	

25	HIGH OIL CONSUMPTION		
DESCRIP-TION	• Oil consumption excessive		
[TROUBLESHOOTING HINTS] ① PCV system malfunction ① Engine malfunction (Oil working up, working down, or leaking)			
STEP	INSPECTION		ACTION
1	Is the PCV hose, the ventilation hose, or the attaching nipples separated, damaged, clogged, or restricted?	Yes	Repair or replace
		No	Go to next step
2	Is there air pressure or is there oil at the venti-lation hose? 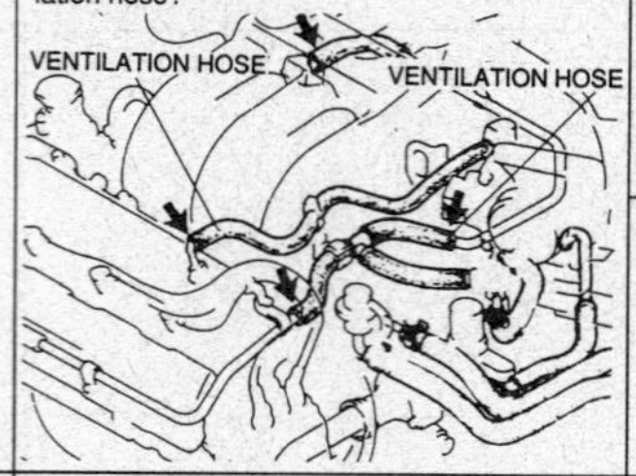 	Yes	Go to next step
		No	Check engine codition • Oil leakage • Worn valve seal • Worn valve steam • Worn valve guide
3	Is vacuum at the PCV valve felt at idle? 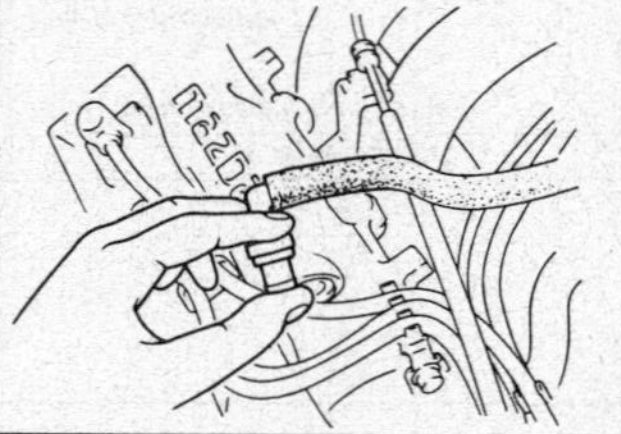	Yes	Check engine condition • Worn piston ring groove • Stuck piston rings • Worn piston or cylinder
		No	Replace PCV valve

2PU0F2-043

26	POOR FUEL ECONOMY
DESCRIPTION	• Fuel economy unsatisfactory

[TROUBLESHOOTING HINTS]
Although fuel consumption is drastically increased by such operation as city driving, short-runs, stop and go driving, andextended winter worn-up periods, as opposed to "trip" mileage an attempt should be made to determine if any these factors were an influence when confronted with "poor mileage". However, since the operator and driving conditions are not always at fault, the following suggestions are offered as possible other reasons for poor mileage.

① Operator depressing accelerator more than usual because of low engine power
- Poor ignition
- Low intake air amount
- Electronic spark advance control system malfunction
- Clutch slipping
- Exhaust component restricted

② Air/fuel mixture too rich
- High fuel line pressure

③ Alcohol-blended fuel used

④ High vehicle load
- Low tire pressure
- Incorrect tire size
- Brake dragging

STEP	INSPECTION		ACTION
1	Check factors other than engine • Low tire pressure • Unrecommended tire size • Clutch slipping • Brake dragging • Exhaust component restricted	Yes	Go to next step
		No	Repair
2	Are air hoses connected correctly?	Yes	Go to next step
		No	Repair
3	Is "00" displayed on Self-Diagnosis Checker with ignition switch ON? 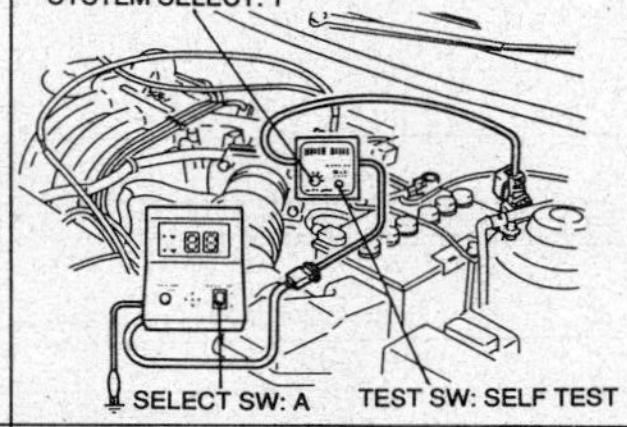	Yes	Go to next step
		No	**Service Code No. displayed** Check for cause (Refer to specified check sequence)
4	Is intake manifold vacuume correct? **Vacuum: More than 450 mmHg**	Yes	Go to next step
		No	Check fro air leakage at intake air system components
5	Air element clean ?	Yse	Go to next step
		No	Replace air cleaner element

STEP	INSPECTION		ACTION
6	Are ECU terminal voltage OK? (especially [2B, 2C, 2D 2E, 3U, 3V, 3W, 3X, 3Y, and 3Z])	Yes	Go to next step
		No	Check for cause
7	Connect System Selection to diagnosis connector and set Test Switch to "SELF-TEST"; is ignition timing at idle correct after warm-up? **Ignition timing (BTDC): 9—11° (10 ± 1°)** 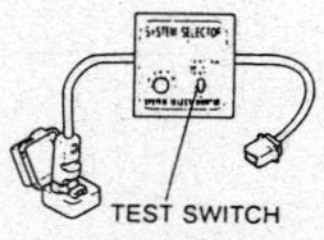	Yes	Go to next step
		No	Adjust
8	Is fuel line pressure correct at idle? **Fuel line pressure:** **206—265 kPa (2.1—2.7 kg/cm², 30—38 psi)** 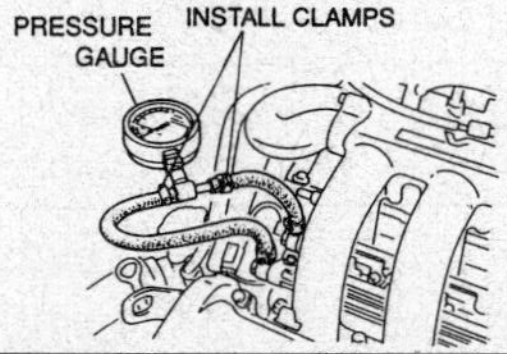	Yes	Go to next step
		No	**High pressure** Is vacuum hose to pressure regulation damaged or poorly connected? ⇨ If it is not, replace pressure regulator ⇨ If it is, repair or replace it
9	Run engine at idle; does fuel line pressure hold after ignition switch is turned off? **Fuel line pressure: More than 147 kPa (1.5 kg/cm², 21 psi) for 5 min.** 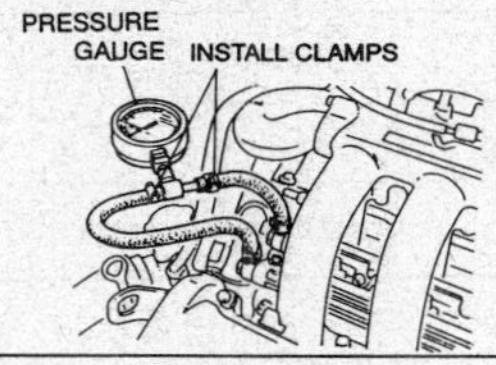	Yes	Go to next step
		No	Check injectors for fuel leakage
10	Change to another brand of fuel		

27	A/C DOES NOT WORK		
DESCRIPTION	• Blower fan operates, but magnetic clutch does not		
[TROUBLESHOOTING HINTS] • Open or short circuit in wiring harness • A/C relay, A/C switch, or magnetic clutch malfunction • ECU malfunction			
STEP	INSPECTION		ACTION
1	Is ECU 1O terminal voltage OK? 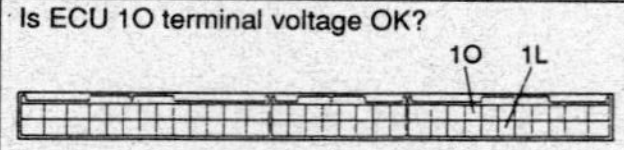 	Yes	Is ECU 1L terminal voltage OK? ⇨ If it is, check A/C system ⇨ If it is not, replace ECU
		No	Check for cause

No.	Indicator flashing pattern	Diagnosed circuit	Condition	Point	Memorized
02	ON / OFF	NE2 signal crank angle sensor 2	No NE2 signal	• Crank angle sensor 2 connector • Wiring from crank angle sensor 2 to ECU • Crank angle sensor 2	Yes
03	ON / OFF	G signal crank angle sensor 1	No. G signal input while a specified number of Ne1 or Ne2 signal pulses are input	• Distributor connector • Wiring from main relay to distributor • Wiring from distributor to ECU • Crank angle sensor 1	Yes
04	ON / OFF	NE1 signal crank angle sensor	No NE1 signal		Yes
05	ON / OFF	Knock sensor	Open or short circuit	• Knock sensor connector • Wiring from knock sensor to ECU • Knock sensor	Yes
08	ON / OFF	Airflow meter	Short circuit	• Airflow meter connector • Wiring from airflow meter to ECU • Airflow meter	Yes
09	ON / OFF	Water thermosensor	Open or short circuit	• Water thermosensor connector • Wiring from water thermosensor to ECU • Water thermosensor resistance	Yes
10	ON / OFF	Intake air thermosensor		• Airflow meter connector • Wiring from airflow meter to ECU • Intake air thermosensor resistance	Yes
12	ON / OFF	Throttle sensor		• Throttle sensor connector • Wiring from throttle sensor to ECU • Throttle sensor	Yes
14	ON / OFF	Atmospheric pressure sensor		• ECU	Yes
15	ON / OFF	Oxygen sensor (Left side)	Sensor output continues less than 0.55V 100 sec. after engine exceeds 1,500 rpm	• Oxygen sensor connector • Wiring from oxygen sensor to ECU • Oxygen sensor	Yes
16	ON / OFF	EGR position sensor	Open or short circuit	• EGR position sensor connector • Wiring from EGR position sensor to ECU • EGR position sensor	Yes
17	ON / OFF	Feed Back system (Left side)	Sensor output not changed 50 sec. after engine exceeds 1,500 rpm	• Fuel pressure • Injection fuel leakage • Ignition system • Air leakage • ECU	Yes

1.8L ENGINE TROUBLE CODES—MX-3

No.	Indicator flashing pattern	Diagnosed circuit	Condition	Point	Memo-rized
23	ON / OFF	Oxygen sensor (Right side)	Sensor output continues less than 0.55V 100 sec. after engine exceeds 1,500 rpm	• Oxygen sensor connector • Wiring from oxygen sensor to ECU • Oxygen sensor	Yes
24	ON / OFF	Feed Back system (Right side)	Sensor output not changed 50 sec. after engine exceeds 1,500 rpm	• Fuel pressure • Injection fuel leakage • Ignition system • Air leakage • ECU	Yes
25	ON / OFF	Solenoid valve (pressure regulator control)		• Solenoid valve connection • Wiring from solenoid valve to ECU • Solenoid valve continuity	No
26	ON / OFF	Solenoid valve (purge control)		• Solenoid valve connection • Wiring from solenoid valve to ECU • Solenoid valve continuity	No
28	ON / OFF	Solenoid valve (EGR-Vaccume)		• Solenoid valve connection • Wiring from solenoid valve to ECU • Solenoid valve continuity	No
29	ON / OFF	Solenoid valve (EGR-VENT)		• Solenoid valve connection • Wiring from solenoid valve to ECU • Solenoid valve continuity	No
34	ON / OFF	Idle speed control (ISC valve)	Open or short circuit	• ISC valve connector • Wiring from ISC valve to ECU • ISC valve continuity	No
41	ON / OFF	Solenoid valve (VRIS 1)		• Solenoid valve connection • Wiring from solenoid valve to ECU • Solenoid valve continuity	No
46	ON / OFF	Solenoid valve (VRIS 2)		• Solenoid valve connection • Wiring from solenoid valve to ECU • Solenoid valve continuity	No
67	ON / OFF	Electrical Cooling fan relay No.2		• Fan relay connector • Wiring from relay to ECU • Fan relay	No
69	ON / OFF	Water thermosensor (fan) *		• Water thermosensor connector • Wiring from water thermosensor to ECU • Water thermosensor resistance	Yes

* Detects malfunction of water thermosensor (fan), when coolant temerature above 60°C (140°F)

1.8L ENGINE TROUBLE CODE TROUBLESHOOTING MX-3

CODE No.	02 (CRANK ANGLE SENSOR No2 SIGNAL)		
STEP	INSPECTION		ACTION
1	Is there poor connection in crank angle sensor circuit?	Yes	Repair or replace
		No	Go to next step
2	Disconnect connector from ECU; Is resistance of connection between 3F (L/Y) and 3H (L/G) terminal OK? **Resistance: 0.95—1.25 KΩ**	Yes	Go to next step
		No	Go to Step 4
3	Is there continuity of connector between ground and 3F or ground and 3H terminal? (At harness side)	Yes	Check for short circuit in wiring from crank angle sensor to 3F or 3H terminal
		No	Replace ECU
4	Is resistance of crank angle sensor OK? **Resistance: 0.95—1.25 KΩ**	Yes	Check for open circuit in wiring from ECU to crank angle sensor
		No	Replace crank angle sensor

Circuit Diagram

[ ] . . . EC-AT

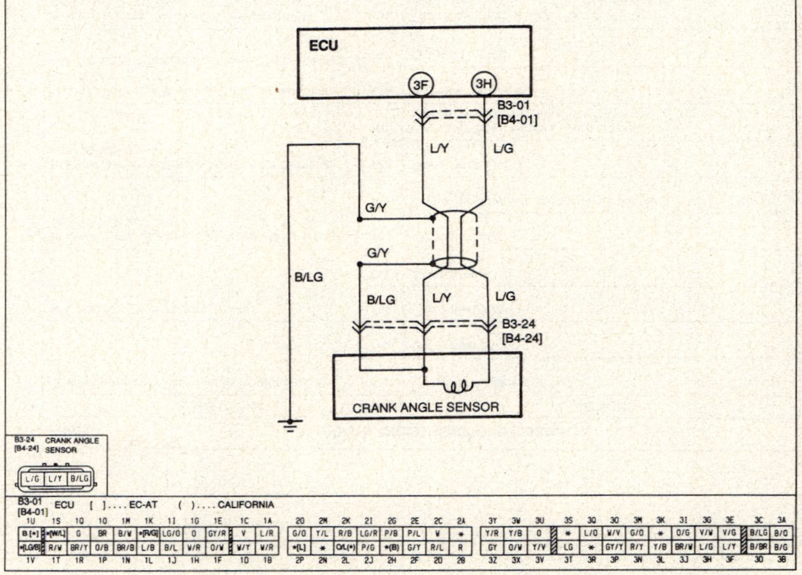

CODE No.	03 (CRANK ANGLE SENSOR G SIGNAL)		
STEP	INSPECTION		ACTION
1	Is there poor connection in distributor circuit?	Yes	Repair or replace connector
		No	Go to next step
2	Is Code No. 04 also present?	Yes	Go to next step
		No	Go to Step 5
3	Is there continuity between distributor terminal-wire (B/LG) and ground?	Yes	Go to next step
		No	Check for open circuit in wiring from distributor to ground
4	Is there battery voltage at distributor terminal-wire (B/W)?	Yes	Go to next step
		No	Check for open circuit in wiring from distributor to main relay
5	Is there approx. 5V at distributor terminal-wire (L)? (At harness-side connector with disconnected)	Yes	Replace distributor
		No	Go to next step
6	Is there approx. 5V at ECU terminal 3G? (With distributor connector disconnected)	Yes	Check for short on open circuit in wiring from distributor to ECU terminal 3G
		No	Replace ECU

Circuit Diagram

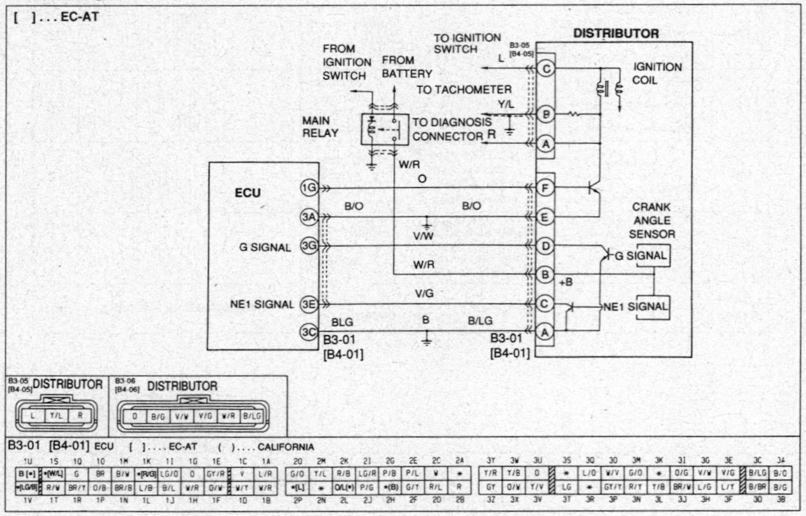

CODE No.	04 (CRANK ANGLE SENSOR Ne1 SIGNAL)		
STEP	INSPECTION		ACTION
1	Is there poor connection in distributor circuit?	Yes	Repair or replace connector
		No	Go to next step
2	Is Code No. 04 also present?	Yes	Go to next step
		No	Go to Step 5
3	Is there continuity between distributor terminal-wire (B/LG) and ground?	Yes	Go to next step
		No	Check for open circuit in wiring from distributor to ground
4	Is there battery voltage at distributor terminal-wire (B/W)?	Yes	Go to next step
		No	Check for open circuit in wiring from distributor to main relay
5	Is there approx. 5V at distributor terminal-wire (G/W)? (AT harness-side connector with disconnected)	Yes	Replace distributor
		No	Go to next step
6	Is there approx. 5V at ECU terminal 3G? (With distributor connector disconnected)	Yes	Check for short on open circuit in wiring from distributor to ECU terminal 3E
		No	Replace ECU

Circuit Diagram

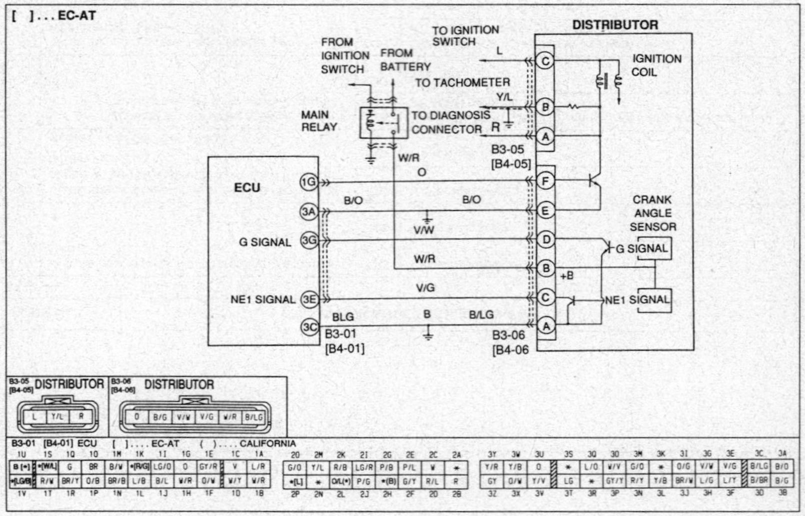

1.8L ENGINE TROUBLE CODE TROUBLESHOOTING—MX-3

CODE No.	05 (KNOCK SENSOR)		
STEP	INSPECTION		ACTION
1	Does knock sensor circuit have a poor connection?	Yes	Repair or replace connector
		No	Go to next step
2	Does the wire harness between the knock sensor and ECU have a continuity?	Yes	Repair or replace harness
		No	Check continuity between ECU terminal 2M and ground ⇨ If continue, repair or replace harness ⇨ If does not continue, go to next step
3	Try known good knock sensor does same Code No. present?	Yes	Replace ECU
		No	Replace knock sensor

Circuit Diagram

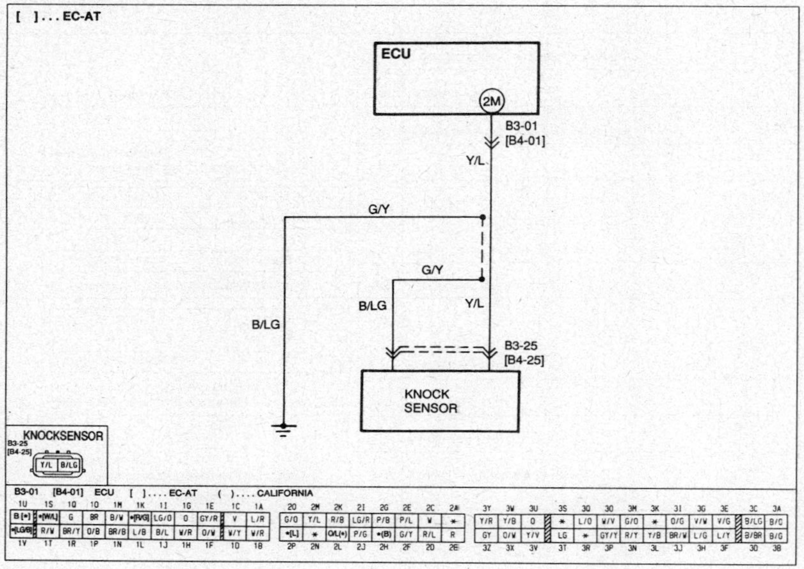

CODE No.	08 (AIRFLOW METER)		
STEP	INSPECTION		ACTION
1	Does airflow meter circuit have a poor connection?	Yes	Repair or replace connector
		No	Go to next step
2	Is Code No. 10 also present?	Yes	Look for open circuit in wiring from airflow meter wire-terminal (B/LG) to ground
		No	Go to next step
3	Is resistance of airflow meter OK? (see table below)	Yes	Go to next step
		No	Replace airflow meter
4	Does wire harness between airflow meter and ECU have continuity? (see table below)	Yes	Go to next step
		No	Repair or replace wire harness
5	Are ECU terminal 3D, 2I, and 2B voltages OK?	Yes	Replace ECU
		No	Look for a short circuit in wiring from airflow meter to ECU

Step 3:

Airflow meter	Fully closed (Ω)	Fully open (Ω)
B (LG/R)—A (R)	20—600	20—1,000
B (LG/R)—D (B/BR)	200—400	

Step 4:

Airflow meter	ECU
B (LG/R)	2I (LG/R)
A (R)	2B (R)

Circuit Diagram

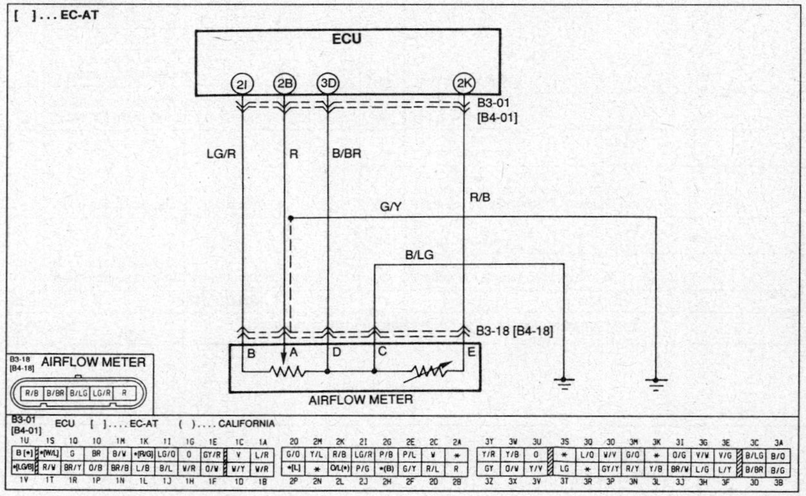

1.8L ENGINE TROUBLE CODE TROUBLESHOOTING—MX-3

CODE No.	09 (WATER THERMOSENSOR)		
STEP	INSPECTION		ACTION
1	Does the water thermosensor circuit have a poor connection?	Yes	Repair or replace connector
		No	Go to next step
2	Does the wire harness between the water thermosensor and ECU have continuity?	Yes	Go to next step
		No	Repair or replace
3	Is resistance of water thermosensor OK?	Yes	Go to next step
		No	Replace water thermosensor
4	Is the same Code No. present following after-repair procedures?	Yes	Go to next step
		No	Water thermosensor and circuit OK
5	Are ECU terminal 2E and 3D voltage OK?	Yes	Replace ECU
		No	Look for a short circuit in wiring from water thermosensor to ECU

Step 2:

Water thermosensor	ECU
A (P/L)	2E (P/L)
B (B/BR)	3D (B/BR)

Step 3:

Coolant temp.	Resistance (kΩ)
−20°C (−4°F)	14.6—17.8
20°C (68°F)	2.21—2.69
80°C (176°F)	0.29—0.35

Circuit Diagram

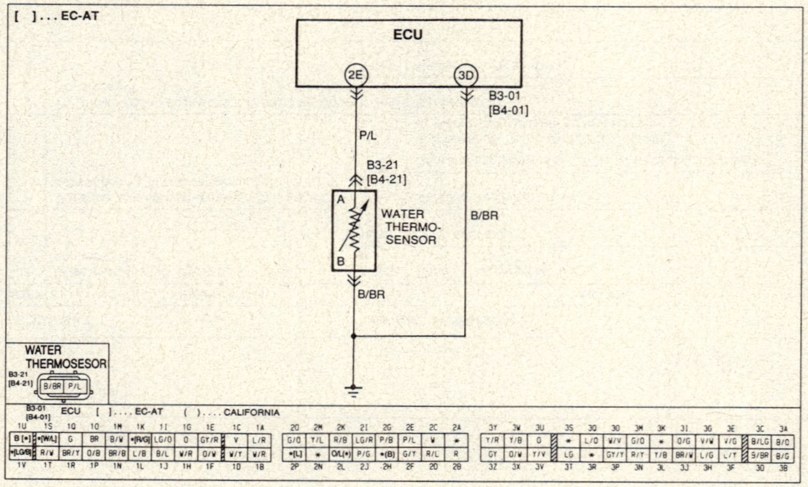

CODE No.	10 (INTAKE AIR THERMOSENSOR — IN AIRFLOW METER)		
STEP	INSPECTION		ACTION
1	Does intake air thermosensor circuit have a poor connection?	Yes	Repair or replace connector
		No	Go to next step
2	Does the Code No. 08 also present?	Yes	Look for an open circuit in wiring from airflow meter terminal wires (B/LG) and (B/BR) to ground
		No	Go to next step
3	Does the wire harness between intake air thermosensor and ECU have continuity?	Yes	Go to next step
		No	Repair or replace
4	Is resistance of airflow meter between terminals C (B/BR) and G (R/B) is OK?	Yes	Go to next step
		No	Replace airflow meter
5	Is same Code No. present following after-repair procedure?	Yes	Go to next step
		No	Intake air thermosensor and circuit OK
6	Are ECU terminals 2K and 3D voltages OK?	Yes	Replace ECU
		No	Look for a short circuit wiring from intake air thermosensor to ECU

Step 3:

Intake air thermosensor (In airflow meter)	ECU
D (B/BR)	3D (B/BR)
E (R/B)	2K (R/B)

Step 4:

Temperature	Resistance (kΩ)
−20°C	13.6—18.4
20°C	2.21—2.69
60°C	0.493—0.667

Circuit Diagram

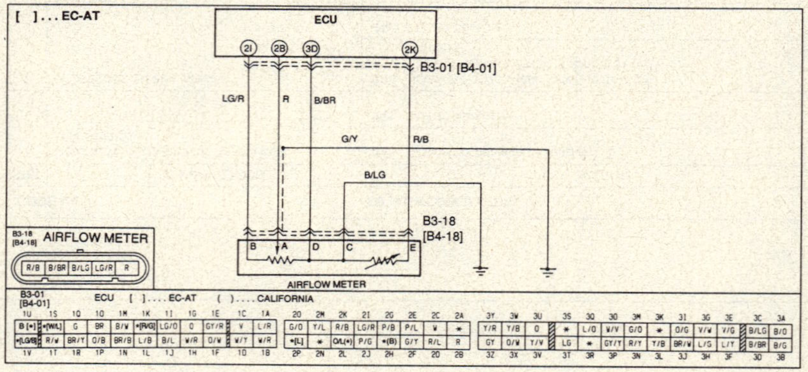

1.8L ENGINE TROUBLE CODE TROUBLESHOOTING—MX-3

CODE No.	12 (THROTTLE SENSOR)		
STEP	INSPECTION	ACTION	
1	Does throttle sensor circuit have a poor connection?	Yes	Repair or replace connector
		No	Go to next step
2	Does wire harness between throttle sensor and ECU have continuity? Throttle sensor — ECU A (LG/R) — 2I (LG/R) B (G/Y) — 2M (G/Y) D (B/BR) — 2D (B/BR)	Yes	Go to next step
		No	Repair or replace
3	Are resistances between terminals B (G/Y) and D (B/BR) OK? Throttle valve — Resistance Fully closed — Below 1 kΩ Fully open — Approx. 5 kΩ	Yes	Go to next step
		No	Adjust or replace throttle sensor
4	Is ECU terminal 2F voltage OK?	Yes	Replace ECU
		No	Look for short circuit in wiring from throttle sensor to ECU

Circuit Diagram

[ ] . . . EC-AT

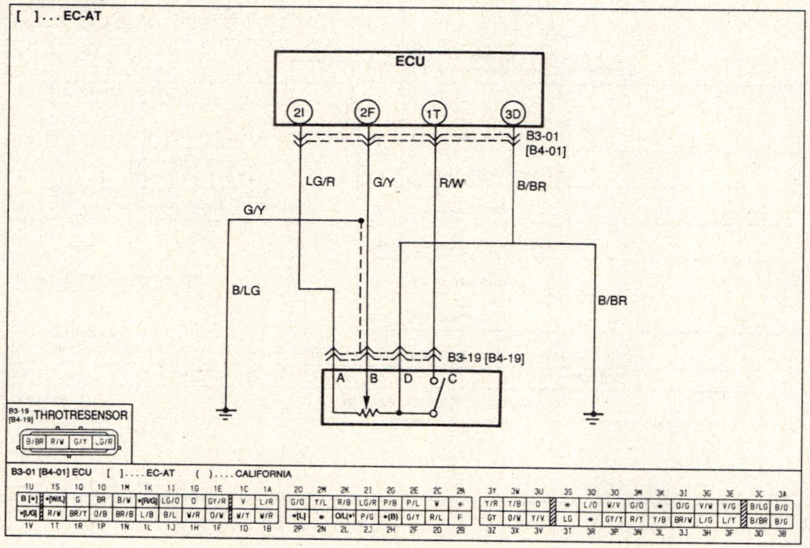

CODE No.	14 (ATMOSPHERIC PRESSURE SENSOR — IN ECU)
STEP	ACTION
1	Replace the ECU

CODE No.	15 and/or 23 (OXYGEN SENSOR — INACTIVATION)		
Note • If Code no. 15 and 17 are both present, first perform the checking procedure for Code No. 17.			
STEP	INSPECTION	ACTION	
1	Does oxygen sensor circuit have a poor connection?	Yes	Repair or replace connector
		No	Go to next step
2	Is oxygen sensor output voltage OK?	Yes	Go to next step
		No	Replace oxygen sensor
3	Does the wire harness between oxygen sensor and ECU terminal (left bank 2D, right bank 2d) have continuity?	Yes	Go to next step
		No	Repair or replace
4	Is ECU terminal (left bank 2D, right bank 2d) voltage OK?	Yes	Go to next step
		No	Look for a short circuit in wiring from oxygen sensor to ECU
5	Is sensitivity of oxygen sensor OK?	Yes	Replace ECU
		No	Try a known good oxygen sensor; does condition improve?

Circuit Diagram

[ ] . . . EC-AT

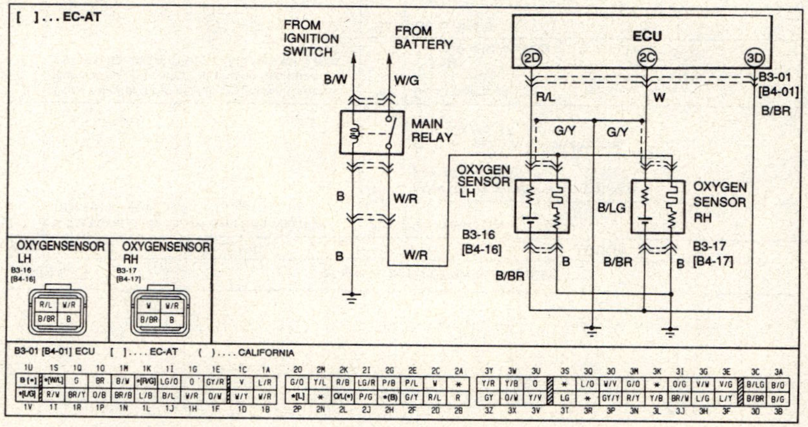

CODE No.	16 (EGR POSITION SENSOR)		
STEP	INSPECTION		ACTION
1	Does EGR position sensor circuit have a poor connection?	Yes	Repair or replace connector
		No	Go to next step
2	Is EGR control valve OK?	Yes	Go to next step
		No	Replace EGR control valve
3	Is resistance of EGR position sensor OK? Resistance: (B) ↔ (C) Approx. 5 kΩ (A) ↔ (C) 0.7—5 kΩ (A) ↔ (B) 0.7—5 kΩ	Yes	Go to next step
		No	Replace EGR control valve
4	Is there continuity between EGR position and ECU? EGR position sensor / ECU (A) (P/G wire) — 2J (B) (B/BR wire) — 3D (C) (LG/R wire) — 2I	Yes	Go to next step
		No	Check for open circuit in wiring from EGR position sensor to ECU
5	Is there 4.5—5.5V at C terminal of EGR position sensor connector?	Yes	Go to next step
		No	Check for short circuit in wiring from EGR position sensor termial C to ECU
6	Is ECU terminal 2J voltage OK?	Yes	Replace ECU
		No	Short circuit in wiring harness EGR position sensor terminal A to ECU

Circuit Diagram

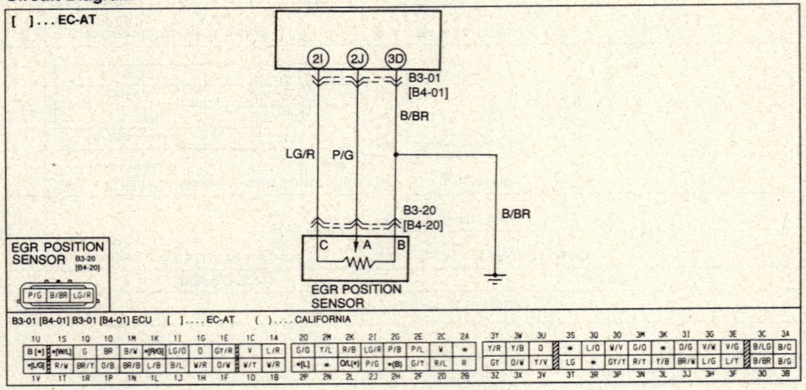

CODE No.	17 and/or 24 (FEEDBACK SYSTEM)		
STEP	INSPECTION		ACTION
1	Is the same Code No. present following after-repair procedure?	Yes	Go to next step
		No	Does the oxygen sensor circuit have a poor connection? ⇨ If not, perform troubleshooting Code No. 15
2	Does monitor lamp of Self-Diagnosis Checker illuminate at idle after the engine has been warmed up and at **2500—3000 rpm** for **3 min**	Yes	Go to next step **Note** • **A/F mixture rich**
		No	Go to Step 5 **Note** • **A/F mixture is lean or misfire is occurring**
3	Is fuel line pressure correct at idle? **Fuel line pressure:** **255—314 kPa (2.6—3.2 Kg/cm², 37—46 psi)** **(vacuum hose to pressure regulator disconnected)**	Yes	Go to next step
		No	**High pressure** Is fuel return hose clogged or restricted? ⇨ If not, replace pressure regulator
4	Check injector for fuel leakage?	Yes	Replace injector
		No	Is water thermosensor OK? ⇨ If it is, replace ECU ⇨ If it is not, replace it
5	Disconnect each high tension lead at idle; does engine speed decrease equally at each cylinder?	Yes	Go to next step
		No	Go to Step 8

Circuit Diagram

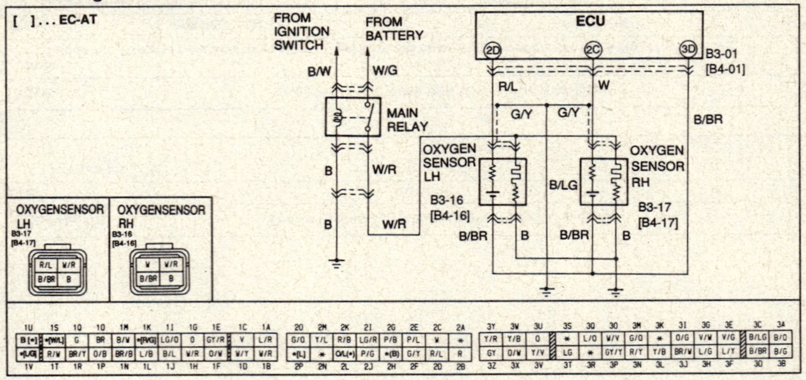

STEP	INSPECTION		ACTION
6	Is fuel line pressure correct at idle?	Yes	Go to next step
	Fuel line pressure: **255—314 kPa (2.6—3.2kg/cm², 37—46 psi) (vacuum hose from pressure regulator disconnected)**	No	**Low pressure** Check fuel line pressure while pinching fuel return hose ⇨ If it **quickly** increases, check pressure regulator ⇨ If it **gradually** increases, check for clogging between fuel pump and pressure regulator If hose is not clogged, check fuel pump maximum pressure
7	Check intake air system components for air leakage	Yes	Go to Step 11
		No	Replace
8	Is there a misfire of a dead cylinder from Step 5 inspection?	Yes	Repair or replace ignition system
		No	Go to next step
9	Is there an injector operating sound at idle of dead cylinder from Step 5 inspection?	Yes	Go to next step
		No	Is there V_B at the injector connector? ⇨ If there is, replace injector ⇨ If there is not, look for a short or open circuit in the wire harness
10	Replace injector at dead cylinder from Step 5 inspection; Is the same Code No. present following after-repair procedure?	Yes	Go to next step
		No	Removed injector at fault
11	Try a known good ECU; does condition improves?		

CODE No.	25 (SOLENOID VALVE — PRESSURE REGULATOR CONTROL)		
STEP	INSPECTION		ACTION
1	Disconnect connector from ECU; is battery voltage at ECU terminal 3M wire (G/O) with ignition switch ON?	Yes	Does the ECU terminal connector have a poor connection? ⇨ If not, repair or replace the ECU ⇨ If it does, repair or replace the con nector
		No	Go to next step
2	In the same condition as step 1, is there battery voltage at solenoid valve connector terminal wire (W/R)?	Yes	Repair or replace wire
		No	Go to next step
3	Is solenoid valve OK?	Yes	Look for a short or open circuit in wiring from main relay (FUEL INJ relay) to solenoid valve
		No	Replace solenoid valve

CODE No.	26 (SOLENOID VALVE — PURGE CONTROL)		
STEP	INSPECTION		ACTION
1	Disconnect connector from ECU; Is battery voltage at ECU terminal 2O wire (G/O) with ignition switch ON?	Yes	Does the ECU terminal connector have a poor connection ⇨ If not, replace the ECU ⇨ If it does, repair or replace the connector
		No	Go to next step
2	In the same condition as Step 1; is there battery voltage at solenoid valve connector terminal wire (W/R)	Yes	Repair or replace wire
		No	Go to next step
3	Is solenoid valve OK?	Yes	Look for a short or open circuit in wiring from the main relay (FUEL INJ relay) to the solenoid valve
		No	Replace solenoid valve

Circuit Diagram

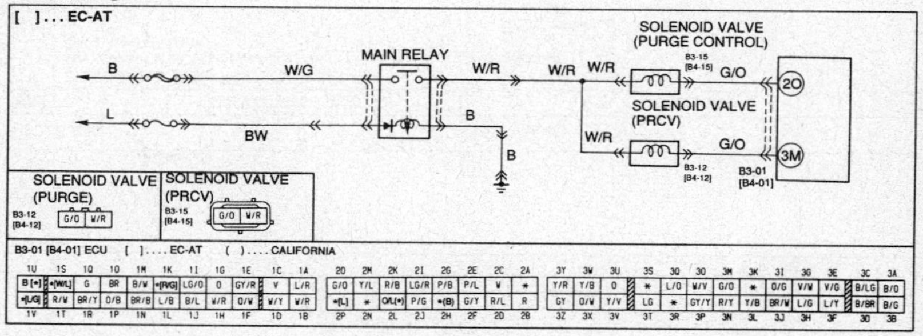

CODE No.	28 (SOLENOID VALVE — EGR — VACUUM)		
STEP	INSPECTION		ACTION
1	Disconnect from ECU; is there battery voltage at ECU terminal 3O wire (W/V) with ignition switch ON?	Yes	Does the ECU terminal connector have a poor connection? ⇨ If not, replace the ECU ⇨ If it does, repair or replace the connector
		No	Go to next step
2	In thew same condition as Step 1, is there battery voltage at solenoid valve con nector terminal wire?	Yes	Repair or replace wire
		No	Go to next step
3	Is solenoid valve OK?	Yes	Look for a short or open circuit in wiring from main relay (FUEL INJ relay) to solenoid valve
		No	Replace solenoid valve

CODE No.	29 (SOLENOID VALVE — EGR — VENT)		
STEP	INSPECTION		ACTION
1	Disconnect from ECU; is there battery voltage at ECU terminal 3P wire (GY/Y) with ignition switch ON?	Yes	Does the ECU terminal have a poor connection? ⇨ If not, replace the ECU ⇨ If it does, repair or replace the connector
		No	Go to next step
2	In the same condition as Step 1 is there battery voltage at solenoid valve connector terminal wire?	Yes	Repair or replace wire
		No	Go to next step
3	Is solenoid valve OK?	Yes	Look for a short or open circuit in wiring from main relay (FUEL INJ relay) to solenoid valve
		No	Replace solenoid valve

Circuit Diagram

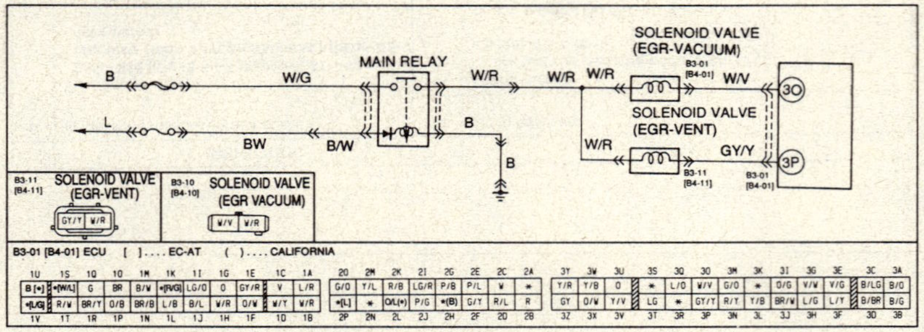

CODE No.	34 (ISC VALVE)		
STEP	INSPECTION		ACTION
1	Disconnect connector from ECU; is there battery voltage at ECU terminal 3Q wire (L/O) with ignition switch ON?	Yes	Check ECU terminal for poor connection ⇨ If OK, replace ECU ⇨ If not OK, repair or replace connector
		No	Go to next step
2	In the same condition as Step 1, is there battery voltage at idle-up solenoid valve connector terminal-wires (L/O) and (L/O)?	Yes	Repair or replace wire (L/O)
		No	Go to next step
3	Does ISC valve have correct resistance? **Resistance: 10.7—12.3 [at 20°C (68°F)]**	Yes	Look for a short or open circuit in wiring from main relay (FUEL INJ relay) to ISC valve
		No	Replace ISC valve

Circuit Diagram

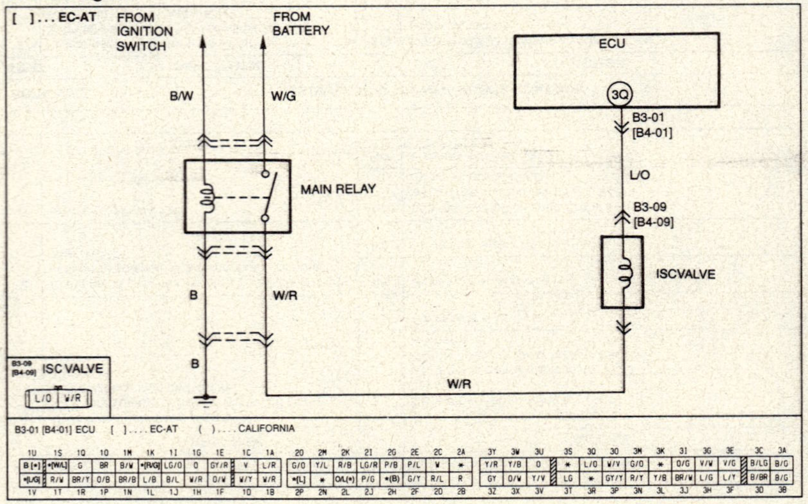

CODE No.	41 (SOLENOID VALVE — VRIS 1)		
STEP	INSPECTION		ACTION
1	Disconnect connector from ECU; is there battery voltage at ECU terminal 3I wire (O/G) with ignition switch ON?	Yes	Check ECU terminal connector for poor connection ⇨ If OK, replace ECU ⇨ If not OK, repair or replace connector
		No	Go to next step
2	In same condition as Step 1, is there battery voltage at solenoid valve connector terminal wire (W/R)?	Yes	Repair or replace wire (B/R)
		No	Go to next step
3	Is solenoid valve OK?	Yes	Look for a short or open circuit in wiring from main relay to solenoid valve
		No	Replace solenoid valve

CODE No.	46 (SOLENOID VALVE - VRIS 2)		
STEP	INSPECTION		ACTION
1	Disconnect connector from ECU; is there battery voltage at ECU terminal 3J wire (BR/W) with ignition switch ON?	Yes	Check ECU terminal connector for poor connection ⇨ If OK, replace ECU ⇨ If not OK, repair or replace connector
		No	Go to next step
2	In same condition as Step 1, is there battery voltage at solenoid valve connector terminal wire (W/R)?	Yes	Repair or replace wire
		No	Go to next step
3	Is solenoid valve OK?	Yes	Look for a short or open circuit in wiring from main relay (FUEL INJ relay) to solenoid valve
		No	Replace solenoid valve

Circuit Diagram

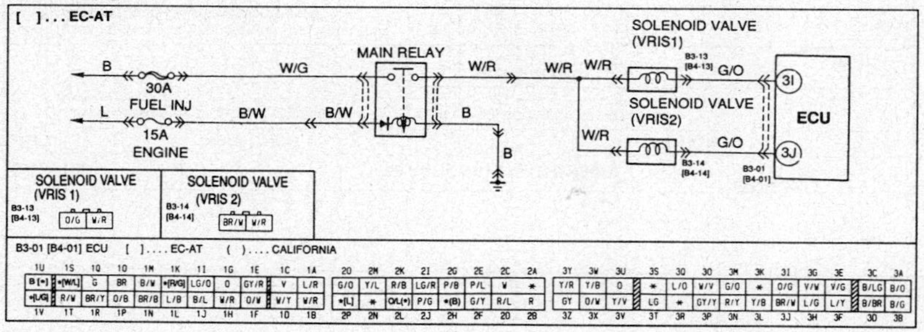

CODE No.	67 (ELECTRICAL COOLING FAN RELAY)		
STEP	INSPECTION		ACTION
	Disconnect connector from ECU; is there battery voltage at ECU termianl 3L wire with ignition switch ON?	Yes	Check ECU terminal connector for poor connection ⇨ If OK, replace ECU ⇨ If not OK, repair replace connector
		No	Go to next step
	In same condition as Step 1, is there battery voltage at solenoid valve connector terminal wire Y/B?	Yes	Repair or replace wire Y/B
		No	Go to next step
	In same condition as Step 1, is there battery voltage at solenoid valve connector terminal wire W/R?	Yes	Go to next step
		No	Look fan shoot circuit in wiring from main relay to electrical cooling fan relay
	There continuity at cooling fan relay connector (Y/B) and (B/W)?	Yes	Look for a short circuit in wiring from electrical cooling fan relay to ECU
		No	Replace electrical cooling fan relay

Circuit Diagram

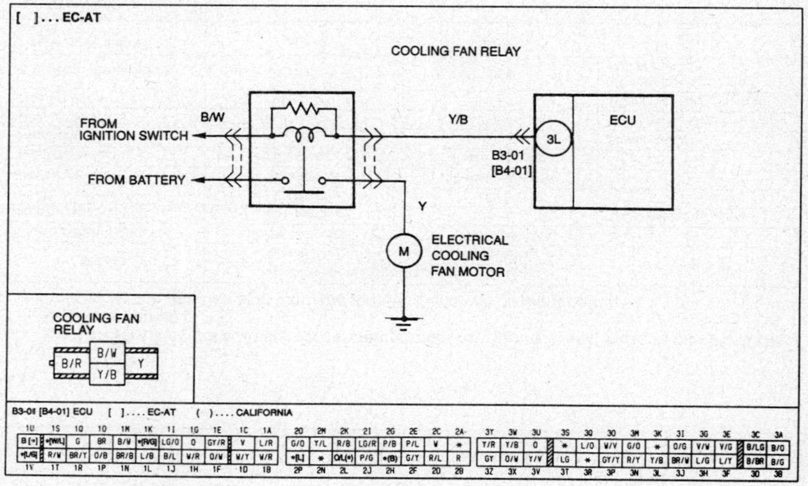

CODE No.	69 (WATER THERMOSENSOR)		
STEP	INSPECTION		ACTION
1	Does the water thermosensor circuit have a poor connection?	Yes	Repair or replace connector
		No	Go to next step
2	Does the wire harness between the water thermosensor and ECU have continuity? Water thermosensor / ECU A (P/L) / 2E (P/L) B (B/BR) / 3D (B/BR)	Yes	Go to next step
		No	Repair or replace
3	Is resistance of water thermosensor OK? Coolant temp. / Resistance (kΩ) 91°C (195°F) / 1.70—1.84 97°C (207°F) / 1.42—1.53 108°C (226°F) / 1.03—1.11	Yes	Go to next step
		No	Replace water thermosensor
4	Is the same Code No. present following after-repair procedures?	Yes	Go to next step
		No	Water thermosensor and circuit OK
5	Are ECU terminal 2E and 3D voltage OK?	Yes	Replace ECU
		No	Look for a short circuit in wiring from water thermosensor to ECU

Circuit Diagram

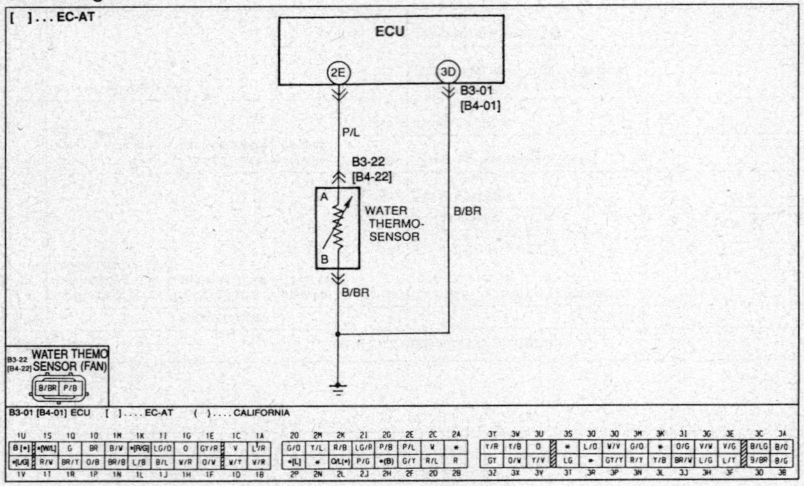

Note

- **The TEN terminal of the diagnosis connector must be grounded and the ignition switch ON (engine stopped).**
- **If either switch remains activated, the monitor lamp will be illuminated.**

Switch	Self-Diagnosis Checker (Monitor lamp)		Remarks
	Light ON	Light OFF	
Clutch switch (MTX)	Pedal released	Pedal depressed	In gear
Neutral switch (MTX)	In gear	Neutral	Clutch pedal released
Inhibitor switch (ATX)	ON	OFF	—
Idle switch	Pedal depressed	Pedal released	—
Stoplight switch	Pedal depressed	Pedal released	—
Headlight switch	ON	OFF	Headlight ON
DRL (CANADA)	ON	OFF	ON; parking lever released
Blower switch	ON	OFF	At 2nd or higher position
A/C switch (if equipped)	ON	OFF	Blower switch at 1st or higher position
Rear window defroster switch	ON	OFF	—

Procedure

Set conditions to deactivate each switch
- All accessories OFF
- Transaxle in neutral
- All pedals released

Does monitor lamp fail to illuminate?

- YES → Check each switch as described
- NO → Check each switch and related wire harness
  - Clutch and neutral switch (MTX)
  - Inhibitor switch
  - Idle switch
  - Stoplight switch
  - Headlight switch
  - Rear window defroster switch
  - Blower switch
  - A/C switch
  - Foglight switch

Neutral and Clutch switches (MTX)

Shift transaxle into gear
Does monitor lamp illuminate with clutch pedal released?

- NO → PC:
  - Neutral switch malfunction
  - Clutch switch malfunction
  - Open circuit in related wire harness
  - ECU terminal 1R voltage incorrect
  - ECU malfunction
- YES → Depress clutch pedal
  Does monitor lamp fail to illuminate?
  Return transmission to neutral
  - NO → PC: • Clutch switch malfunction

Inhibitor switch (ATX)

Shift tranmission into except P and N range
Is monitor lamp illumunated?

- YES → Inhibitor switch OK
  Shift into P or N range
- NO →
  - Inhibitor switch malfunction
  - Open or short circuit in related wire harness
  - ECU terminal (1R) malfunction

Idle switch (in throttle sensor)

Depress accelerator pedal
Is monitor lamp illuminated?

- YES → Idle switch OK
  Release accelerator pedal
- NO →
  - Idle switch malfunction
  - Open or short circuit in wire harness from throttle sensor to ECU
  - ECU terminal 1T or 2I voltage incorrect

Stoplight switch

Depress brake pedal
Does monitor lamp illuminate?

- YES → Stoplight switch OK
  Release brake pedal
- NO → PC:
  - Stoplight switch malfunction
  - Open circuit in wire harness from stoplight switch to ECU
  - ECU terminal 1Q voltage incorrect
  - ECU malfunction

Headlight switch

Turn headlight switch ON
Does monitor lamp illuminate?

- YES → Headlight switch OK
  Turn headlight switch OFF
- NO → PC:
  - Headlight switch malfunction
  - Open circuit in wire harness from headlight switch to ECU
  - ECU terminal 1H voltage incorrect
  - ECU malfunction

DRL

Release parking brake
Does monitor lamp illuminate?

- YES → DRL OK
  Pull parking brake
- NO → PC:
  - DRL malfunction
  - Open circuit in wire harness from DRL relay to ECU
  - ECU terminal 2L voltage incorrect
  - ECU malfunction

Blower switch

Turn blower switch to 2nd or higher position
Does monitor lamp illuminate?

- YES → Blower switch OK
- NO → PC:
  - Blower switch malfunction
  - Open circuit in wire harness from blower switch to ECU
  - ECU terminal 1P voltage incorrect
  - ECU malfunction

A/C switch

Turn A/C switch to 1st position
Turn A/C switch ON
Does monitor lamp illuminate?

- YES → A/C switch OK
- NO → PC:
  - A/C switch malfunction
  - Open circuit in wire harness from A/C switch to blower switch
  - ECU terminal 1O voltage incorrect
  - ECU malfunction

Rear window defroster switch

Turn rear window defroster switch ON
Does monitor lamp illuminate?

- YES → Rear window defroster switch OK
  Turn rear window defroster switch OFF
- NO → PC:
  - Rear window defroster switch malfunction
  - Open circuit in wire harness from rear window defroster switch to ECU
  - ECU terminal 1J voltage incorrect
  - ECU malfunction

1.8L ENGINE ENGINE CONTROL UNIT TERMINAL VOLTAGES—MX-3

Terminal	Input	Output	Connection to	Test condition		Correct voltage	Remarks
1A	—	—	Battery	Constant		V_B	For backup
1B	○		Main relay (FUEL INJ relay)	Ignition switch	OFF	0V	—
					ON	V_B	
1C	○		Ignition switch (START)	While cranking		10V	—
				Ignition switch ON		Below 1.0V	
1D		○	Self-Diagnosis Checker (monitor lamp)	Test switch at "SELF-TEST" Lamp illuminated for 3 sec. after ignition switch OFF→ ON		4.5—5.5V	With Self-Diagnosis Checker and System Selector
				Lamp not illuminated after 3 sec.		V_B	
				Test switch at "O_2 MONITOR" at idle Monitor lamp illuminated		4.5—5.5V	
				Test switch at "O_2 MONITOR" at idle Monitor lamp not illuminated		V_B	
1E		○	Malfubction indicator lamp (MIL)	Lamp illuminated for 3 sec. after ignition switch OFF→ON		Below 2.5V	With System selector test Switch at "SELF-TEST"
				Lamp not illuminated after 3 sec.		V_B	
				Lamp illuminated		Below 2.5V	
				Lamp not illuminated		V_B	
1F		○	Self-Diagnosis Checker (Code Number)	Buzzer sounded for 3 sec. after ignition switch OFF→ ON		Below 2.5V	• With Self-Diagnosis Cheker and System Selector • With System Selector test switch at "SELF-TEST"
				Buzzer not sounded after 3 sec.		V_B	
				Buzzer sounded		Below 2.5V	
				Buzzer not sounded		V_B	
1G		○	Igniter	Ignition switch ON		Below 1.0V	—
				Idle			
1H	○		Head light switch	Head light ON		V_B	—
				Head light OFF		0V	
1I	○		Diagnosis connector (TEN terminal)	System Selector test switch at "O_2 MONITOR"		V_B	• With Self-Diagnosis Checker and System Selector • Ignition switch ON
				System Selector test switch at "SELF-TEST"		Below 1.0V	
1J	○		Rear window defroster relay	Electrical load OFF		V_B	Ignition switch ON
				Electrical load ON		Below 1.5V	
1K		○	EC-AT CU	Engine coolant temperature below 60°C (140°F)		Below 1.0V	Ignition switch ON
				Engine coolant temperature above 60°C (140°F)		V_B	
1L		○	A/C relay	A/C switch ON		Below 2.5V	Ignition switch ON
				A/C switch OFF		V_B	

Incorrect voltage		Possible cause
Always 0V		• ROOM 15A fuse burned • Open circuit in wiring from ROOM 15A fuse to ECU terminal 1A
Always 0V		• Main relay malfunction • Open or short circuit in wiring from main relay to ECU terminal 1B
Always 0V (starter turns)		• Open or short circuit in wiring from ignition switch to ECU terminal 1C
Always 0V		• Main relay (FUEL INJ relay) malfunction • Open circuit in wiring from ignition switch to diagnosis connector terminal +B • Open or short circuit in wiring from diagnosis connector terminal MEN to ECU terminal 1D
Always approx. 12V		• Poor connection at ECU connector • ECU malfunction
Always approx. 5V		• ECU malfunction
Always below 2.5V	MIL always ON	• Short circuit in wiring from combination meter to ECU terminal 1E • ECU malfunction
	MIL never ON	• Open circuit in wiring from combination meter to ECU terminal 1E
Always approx. V_B		• Poor connection at ECU connector • ECU malfunciton
Always below 2.5V	No display on Self-Diagnosis Checker	• Main relay (FUEL INJ relay) malfunction • Open circuit in wiring from ignition switch to diagnosis connector terminal +B
	"88" displayed and buzzer sounds continuously	• Open or short circuit in wiring from diagnosis connector terminal FEN to ECU terminal 1F
Always approx. 12V		• Poor connection at ECU connector • ECU malfunction
Always 0V		• Short circuit in wiring from igniter to ECU terminal 1G
Always below 1.0 (Headlights OK)		• Opener short circuit in wiring from headlight relay to ECU terminal 1H.
Always below 1.0V		• Short circuit in wiring from diagnosis connector terminal TEN to ECU terminal 1I
Always approx 12V		• Open circuit in wiring from diagnosis connector terminal TEN to ECU terminal 1I • Open circuit in wiring from diagnosis connector terminal GND to grpund
Always 0V		• Short circuit in wiring from rear defroster switch to ECU terminal 1J
Always V_B		• Open circuit in wiring from rear defroster switch to ECU terminal 1J
Always below 1.0V		• Open or short circuit in wiring from ECU terminal 1K to EC-AT CU terminal 1N
Always below 2.5V	A/C does not operate	• A/C relay malfunction • Open circuit in wiring from ignition switch to A/C relay • Open circuit in wiring from A/C relay to ECU terminal 1L
	A/C switch OFF but A/C operates	• Short circuit in wiring from A/C relay to ECU terminal 1L • ECU malfunction
Always V_B		• Poor connection at ECU connector • ECU malfunction

1.8L ENGINE ENGINE CONTROL UNIT TERMINAL VOLTAGES—MX-3

Terminal	Input	Output	Connection to	Test condition	Correct voltage	Remarks	Incorrect voltage	Possible cause
1M	○		Speed meter	Ignition switch ON Ignition switch OFF	4.5—5.5V 0V	—	Always 0V	• Open or short circuit in wiring from speed meter to ECU terminal 1M
1N	○		P/S pressure switch	Ignition switch ON P/S ON at idle P/S OFF at idle	V_B Below 1.0V V_B	—	Always below 1.0V Always V_B	• P/S pressure switch malfunction • Short circuit in wiring from P/S pressure switch to ECU terminal 1N • ECU malfunction • P/S pressure switch malfunction • Open circuit in wiring from P/S pressure switch to ECU terminal 1N • Open circuit in wiring from P/S pressure switch to ground
1O	○		A/C amplifier	A/C switch ON A/C switch OFF	1.5—3.5V 4.5—5.5V	Ignition switch ON and blower	Always below 1.5V Always 4.5—5.5V	• Short circuit in wiring from A/C amplifier to ECU terminal 1O • A/C amplifier malfunction • ECU malfunction • Open circuit in wiring from A/C amplifier to ECU terminal 1O • A/C amplifier malfunction
1P	○		A/C amplifier	Fan speed control 4th position Fan Speed control 1st—3rd position	Below 1.0V V_B	Ignition switch ON	Always below 1.0V Always V_B	• Short circuit in wiring from A/C amplifier to ECU terminal 1P • A/C amplifier malfunction • ECU malfunction • Open circuit in wiring from A/C amplifier to ECU terminal 1P • A/C amplifier malfunction
1Q	○		Stoplight switch	Brake pedal released Brake pedal depressed	Below 1.0V V_B	—	Always below 1.0V (Stoplight OK)	• Open circuit in wiring from stoplight switch to ECU terminal 1Q
1R	○		Neutral/Clutch switch (MTX)	Neutral position or clutch pedal depressed Others	Below 1.0V V_B	Ignition switch ON	Always below 1.0 V (Neutral/clutch switch OK) Always V_B	• Short circuit in wiring from Neutral/clutch switch to ECU terminal 1R • Open circuit in wiring from Neutral/clutch switch to ECU terminal 1R
1R	○		EC-AT CU	N or P range Others	Below 1.0V V_B	—	Always below 1.0V (Inhibitor switch OK) Always V_B	• Short circuit in wiring from ECU terminal 1R to EC-AT CU terminal 1D • Open circuit in wiring from ECU terminal 1R to EC-AT CU terminal 1D
1S	○		EC-AT CU	Shifting up performed Others	V_B Approx.2.5V	—	Always 0V	• Open or short circuit in wiring from ECU terminal 1S to EC-AT CU terminal 2M
1T	○		Throttle sensor (Idle switch)	Accelerator pedal released Accelerator pedal depressed	Below 1.0V V_B	Ignition switch ON	Always below 1.0 V Always V_B	• Short circuit in wiring from throttle sensor to ECU terminal 1S • ECU malfunction • Throttle sensor misadjustment • Open circuit in wiring from throttle sensor to ECU terminal 1S • ECU malfunction
1U	—	—	Ground (MTX)	Constant	0V	—	—	—
1V	○		EC-AT CU	Shifting down performed Others	0V V_B	—	Always V_B Always 0V	• Open circuit in wiring from ECU terminal 1V to EC-AT CU terminal 1A • Short circuit in wiring from ECU terminal 1V to EC-AT CU terminal 1A
2A	—	—	—	—	—	—	—	—
2B	○		Airflow meter	Ignition switch ON Idle	Approx. 4V Approx. 2.6V	—	Always below 2.0 V or Approx. 4V	• Refer to Code No. 08 Troubleshooting
2C	○		Oxygen sensor (Right side)	Ignition switch ON Idle (Cold engine) Idle (After warm-up)	0V 0V 0—1.0V	—	0V after warm-up	• Refer to Code No. 23 Troubleshooting
2D	○		Oxygen sensor (Left side)	Increasing engine speed (After warm-up) Deceleration	0.5—1.0V 0—0.4V	—	Always approx. 1V after warm-up 0V after warm-up	• Refer to Code No. 24 Troubleshooting • Refer to Code No. 15 Troubleshooting
2E	○		Water thermo-sensor	Engine coolant temperature 20°C (68°F) After warm-up	Approx. 2.5V Below 0.5V	Ignition switch ON	Always approx. 1V after warm-up Always approx. 0V or approx. 5V	• Refer to Code No. 17 Troubleshooting • Refer to Code No. 09 Troubleshooting
2F	○		Throttle sensor	Accelerator pedal released Accelerator pedal fully depressed	Approx. 0.5V Approx. 3.7V	Ignition switch ON	Always constant Always above 1V	• Open circuit in wiring from ECU terminal to throttle sensor • Throttle sensor misadjustment
2G	○		Water thermo-sensor (FAN)	Engine coolant temperature 20°C (68°F) After warm-up	Approx.4.5V Approx.3.2V	Ignition switch ON	Always approx. 0V or approx. 5V	• Refer to Code No. 67 Troubleshooting

1.8L ENGINE ENGINE CONTROL UNIT TERMINAL VOLTAGES—MX-3

Terminal	Input	Output	Connection to	Test condition	Correct voltage	Remarks	Incorrect voltage	Possible cause
2H	○		Ground (Carifornia)	Constant	0V	—	—	—
2I	○		Throttle sensor	Constant	4.5—5.5V		Always approx. 0V or approx. 5V	• Refer to Code No. 12 Troubleshooting
2J	○		EGR position sensor	Ignition switch ON	Approx. 0.8V	—	Always approx. 0V or approx. 5V	• Refer to Code No. 16 Troubleshooting
				Engine running	Approx. 0.8—4.5V			
2K	○		Intake air thermosensor (Airflow meter)	Ambient air temperature 20°C (68°F)	Approx. 2.5V	Ignotion switch ON	Always 0V	• Refer to Code No. 10 Troubleshooting
2L	○		DRL relay (Canada)	Parking brake pulled with ignition switch ON (DRL OFF)	V_B	DRL: Daytime Running Lights	Always 0V (DRL SYSTEM OK)	• Open or short circuit in wiring from DRL relay to ECU terminal 2L
				Idle (DRL ON)	Below 2.5V			
2M	○		Knock sensor	Ignition switch ON	0V	—	Always 0V or V_B	• Refer to Code No. 05 Troubleshooting
				Idle	0.7V			
2N	—	—	—	—	—	—	—	—
2O		○	Solenoid valve (Purge control)	Ignition switch ON or idle			Always 0V or approx. 5V	• Refer to Code No. 26 Troubleshooting
2P		○	Cooling fan relay	Engine coolant temp. 108° (226°F)	Below 2V	—	Always below 2V or V_B	• Open or short circuit in wiring from cooling fan relay to ECU terminal 2P • ECU malfunction
				Others	V_B			
3A	—	—	Ground (Output)	Constant	0V	—	Above 0V	• Poor contact at ground terminal • Open circuit in wiring from ECU
3B	—	—	Ground (Injector)	Constant	0V	—		
3C	—	—	Ground (CPU)	Constant	0V	—		
3D	—	—	Ground (Input)	Constant	0V	—		
3E	○		Crank angle sensor (in distributor) (Ne1 signal)	Ignition switch ON	Approx. 0V or 5V	—	Always approx. 0V or approx. 2V	• Refer to Code No. 04 Troubleshooting
				Idle	Approx. 2V			
3F	○		Crank angle sensor (Crankshaft pulley) (Ground)	Constant	0V	—	Above 0V	• Poor connection at ground terminal • Open circuit in wiring from ECU
3G	○		Crank angle sensor (in distributor) (G signal)	Ignition switch ON	Approx. 0V or 5V	—	Always approx. 0V or approx. 5V	• Refer to Code No. 03 Troubleshooting
				Idle	Approx. 2.5V			
3H	○		Crank angle sensor (Crankshaft pulley) (Ne2 signal)	Ignition switch ON	0V	* Engine signal Monitor: Green and red lamps flash	Always approx. 0V or approx. 2V	• Refer to Code No. 02 Troubleshooting
				Idle*	Approx. 0V			
3I		○	Solenoid valve (VRIS1)	Engine speed 3,900—6,300 rpm	Approx. 1.2V	—	Always approx. 0V or V_B	• Refer to Code No. 41 Troubleshooting
				Others	V_B			
3J		○	Solenoid valve (VRIS2)	Engine speed 4,700—6,300 rpm	Approx. 1.2V	—	Always approx. 0V or V_B	• Refer to Code No. 46 Troubleshooting
				Others	V_B			
3K	—	—	—	—	—	—	—	—
3L		○	Cooling fan relay	Engine coolant temp. 100°C (212°F)	Below 1.0V	—	Always below 1.0V or V_B	• Open or short circuit in wiring from Cooling fan relay to ECU terminal 3L • ECU malfunction
				Others	V_B			
3M		○	Solenoid valve (Purge control)	Hot condition: Engine coolant temp. above 70°C and intake air temp. above 75°C for 120 sec. after engine starting	Approx. 1.2V	—	Always below 1.0V or V_B	• Refer to Code No. 26 Troubleshooting
				Others	V_B			
3N		○	Condensor fan relay	Ignition switch ON	V_B	—	Always below 2.0V	• Open or short circuit in wiring from condenser fan relay to ECU terminal 3N • ECU malfunction
				Idle: Engine coolant temp. 108°C (226°F)	Below 2.0V			
				Idle: Others	V_B			

1.8L ENGINE ENGINE CONTROL UNIT TERMINAL VOLTAGES – MX-3

Terminal	Input	Output	Connection to	Test condition	Correct voltage	Remarks	Incorrect voltage	Possible cause
3O		○	Solenoid valve (EGR vent)	Idle	V_B	* Initial acceleration Engine Signal Monitor Green and headlamps flash	Always 0V	• Refer to Code No. 28 Troubleshooting
				Initial acceleration	V_B*			
3P		○	Solenoid valve (EGR vaccume)	Idle	V_B		Always 0V	• Refer to Code No. 29 Troubleshooting
				Initial acceleration	V_B*			
3Q		○	ISC valve	Ignition switch ON	Approx. 7V	—	Always approx. 0V or V_B	• Refer to Code No. 34 Troubleshooting
				Idle	Approx. 9V			
3R	—	—	—	—	—	—	—	—
3S	—	—	—	—	—	—	—	—
3T		○	Circuit opening relay	Ignition switch ON	V_B	—	Always below 1.0V	• Open or short circuit in wiring from circuit opening relay to ECU terminal 3T
				Idle	Below 1.0V			
3U		○	Injector No.1	Ignition switch ON or Idle*	V_B	* Engine Signal Monitor: Green and red lamps flash	Always 0V	• Open or short circuit in wiring from injector to ECU terminal 3U, 3V, 3W, 3X, 3Y or 3Z • Main relay malfunction
3V		○	Injector No.2					
3W		○	Injector No.3					
3X		○	Injector No.4					
3Y		○	Injector No.5					
3Z		○	Injector No.6					

Control Unit Connector (Control Unit Side)

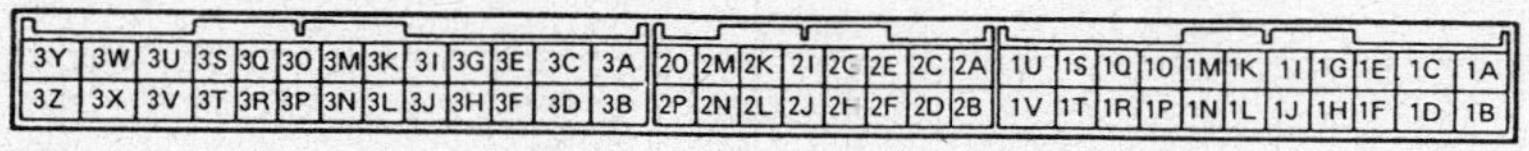

2.2L ENGINE SPECIFICATIONS CHART 626 AND MX-6 NON-TURBO

Item / Engine			Non-Turbo Engine
Idle speed	rpm		750 ± 25 (ATX: P range)*
Throttle body			
Type			Horizontal draft (2-barrel)
Throat diameter	mm (in)	No.1	MTX: 40 (1.6), ATX: 46 (1.8)
		No.2	MTX: 46 (1.8), ATX: 40 (1.6)
Airflow meter			
Resistor	Ω	E2—Vs	Fully closed: 20—400 Fully open: 20—1,000
		E2—Vc	100—400
		E2—VB	200—400
		E2—THA	−20°C (−4°F) 13,600—18,400 20°C (68°F) 2,210— 2,690 60°C (140°F) 493— 667
Fuel pump			
Type			Impeller (in tank)
Output pressure	kPa (kg/cm², psi)		441—588 (4.5—6.0, 64—85)
Feeding capacity	cc (cu in)/10 sec.		220 (13.4) min
Fuel filter			
Type	Low-pressure side		Nylon element
	High-pressure side		Paper element
Pressure regulator			
Type			Diaphragm
Regulating pressure	kPa (kg/cm², psi)		235—275 (2.4—2.8, 34—40)
Injector			
Type			High-ohmic
Type of drive			Voltage
Resistance	Ω		12—16
Injection amount	cc (cu in)/15 sec.		44—61 (2.68—3.72)
Idle speed control valve			
Solenoid resistance	Ω		6.3—9.9
Fuel tank			
Capacity	liters (US gal, Imp gal)		60 (15.9, 13.2)
Air cleaner			
Element type			Oil permeated
Fuel			
Specification			Unleaded regular (RON 87 or higher)

* With test connector grounded.

2.2L ENGINE TROUBLESHOOTING GUIDE 626 AND MX-6 NON-TURBO

Symptom / Possible cause		Ignition pulse	Airflow meter	Water thermosensor	Intake air thermosensor	Throttle sensor	Atmospheric pressure sensor	Oxygen sensor	EGR position sensor (California only)	Feedback system	Solenoid valve (Pressure regulator)	Solenoid valve (Purge)	Solenoid valve (EGR)	Solenoid valve (Idle speed control)
		Input sensors and switches									Output solenoid valves			
1	Fault indicated by SST Code No.	01	08	09	10	12	14	15	16	17	25	26	28	34
2	Hard start or won't start (Cranks OK)													
3	Engine stalls — While warming up													
3	Engine stalls — After warming up													
4	Rough idle — While warming up													
4	Rough idle — After warming up													
5	High idle speed after warming up													
6	Poor acceleration, hesitation or lack of power													
7	Runs rough on deceleration													
8	Afterburn in exhaust system													
9	Poor fuel consumption													
10	Engine stalls or rough after hot starting													

TROUBLESHOOTING PROCEDURE

Note
- **Step 1 under symptom is to quickly determine what system or unit may be at fault by use of the SST. (Self-Diagnosis Checker 49 H018 9A1)**

1st: Check input sensors and output solenoid valves with the **SST**.

2nd: Check other switches with the **SST**.

3rd: Check the following items:

Electrical system
1) Battery condition
2) Fuses

Ignition system
1) Ignition spark
2) Ignition timing

Fuel system
1) Fuel level
2) Fuel leakage
3) Fuel filter
4) Idle speed (with test connector grounded)

Intake air system
1) Air cleaner element
2) Vacuum or air leakage
3) Vacuum hose routing
4) Accelerator cable

Engine
1) Compression
2) Overheating

Others
1) Clutch slippage
2) Brake dragging

4th: Check Fuel and Emission Control Systems.

2.2L ENGINE TROUBLESHOOTING GUIDE 626 AND MX-6 NON-TURBO

Fuel and Emission Control Systems

Symptom	Intake Air System (Poor connection of components, throttle body)	Fuel System (Fuel injection, Fuel pressure)	Pressure Regulator Control System	Idle Speed Control (ISC) System (Air valve, Idle speed control solenoid malfunction)	EGR System (EGR control valve stuck and open)	Evaporative Emission Control system (Solenoid valve [Purge control] malfunction)	PCV System (System clogged)	Deceleration System (Fuel cut operation malfunction)	Exhaust system (System clogged)
2	2	1							
3	4	3		1	2				
3	5	4		2	3		1		
4	5	4		1	3		2		
4	6	5		2	3	4	1		
5	2			1					
6	3	4			1	2			5
7		3		2				1	
8	3	4		1				2	
9		2			3			1	4
10		2	1						

The numbers of the list show the priorities of inspections from the most probable to that with the lowest probability.
These were determined on the following basis:
- Ease of inspection
- Most probable system
- Most probable point in system

2.2L ENGINE TROUBLE CODES 626 AND MX-6 NON-TURBO

Malfunction display: Code No.	Malfunction display: MIL output signal pattern	Sensor or subsystem	Self-diagnosis	Fail-safe
01	ON / OFF	Ignition pulse	No ignition signal	—
08	ON / OFF	Airflow meter	Open or short circuit	Maintains basic signal at preset value
09	ON / OFF	Water thermosensor	Open or short circuit	Maintains constant command • 40°C (104°F) for EGI • 50°C (122°F) for ISC control use
10	ON / OFF	Intake air thermosensor (airflow meter)	Open or short circuit	Maintains constant 20°C (68°F) command
12	ON / OFF	Throttle sensor	Open or short circuit	Maintains constant command of throttle valve fully open
14	ON / OFF	Atmospheric pressure sensor	Open or short circuit	Maintains constant command of sea level pressure
15	ON / OFF	Oxygen sensor	Sensor output continues less than 0.55V 120 sec. after engine starts (1,500 rpm)	Cancels EGI feedback operation
16	ON / OFF	EGR position sensor (California only)	Open short circuit	Cuts off EGR
17	ON / OFF	Feedback system	Sensor output not changed 20 sec. after engine exceeds 1,500 rpm	Cancels EGI feedback operation
25	ON / OFF	Solenoid valve (pressure regulator)	Open or short circuit	—
26	ON / OFF	Solenoid valve (purge control)		—
28	ON / OFF	Solenoid valve (EGR)		—
34	ON / OFF	ISC valve		—

Caution
- **If there is more than one failure present, the lowest number malfunction code is displayed first, the remaining codes are displayed sequentially.**
- **After repairing a failure, turn off the ignition switch and disconnect the negative battery cable and depress the brake pedal for at least 5 seconds to erase the memory of a malfunction code.**

2.2L ENGINE TROUBLE CODE TROUBLESHOOTING—626 AND MX-6 NON-TURBO

Code No.01 (Ignition pulse)

PC: Possible Cause

1. Are there poor connections in ignition coil circuit?
   - **YES** → Repair or replace connector
   - **NO** → go to step 2
2. Is resistance of ignition coil OK?
   Resistance: Primary 0.77—0.95Ω
   Secondary 10.2—15.2 kΩ
   - **NO** → Replace ignition coil
   - **YES** → go to step 3
3. Is there continuity between ignition coil (–) terminal wire and engine control unit 2I (MTX) or 1V (ATX) terminal?
   - **NO** → PC: Open circuit in wiring harness from ignition coil to engine control unit 2I (MTX) or 1V (ATX) terminal
   - **YES** → go to step 4
4. Is same Code No. present after performing after-repair procedure?
   - **NO** → Ignition pulse and circuit OK
   - **YES** → go to step 5
5. Is engine control unit 2I (MTX) or 1V (ATX) terminal voltage OK?
   - **NO** → PC: • No power supply to ignition coil • Short circuit in wiring harness
   - **YES** → PC: Engine control unit malfunction

Code No.08 (Airflow meter)

PC: Possible Cause

1. Are there poor connections in airflow meter circuit?
   - **YES** → Repair or replace connector
   - **NO** → go to step 2
2. Is there continuity between airflow meter and engine control unit?

Airflow meter	Control unit
V_B (R/B wire)	1B
V_C (R/W wire)	2J (MTX) 2A (ATX)
V_S (R/B wire)	2O (MTX) 2B (ATX)
E_2 (LG/Y wire)	2D (MTX) 3D (ATX)

   - **NO** → PC: Open circuit in wiring harness from airflow meter to engine control unit
   - **YES** → go to step 3
3. Is resistance of airflow meter OK?
   Resistance:

Terminal	Fully closed (Ω)	Fully open (Ω)
$E_2 \leftrightarrow V_S$	20—400	20—1,000
$E_2 \leftrightarrow V_C$	100—400	
$E_2 \leftrightarrow V_B$	200—400	

   - **NO** → Replace airflow meter
   - **YES** → go to step 4
4. Is same Code No. present after performing after-repair procedure?
   - **NO** → Airflow meter and circuit OK
   - **YES** → go to step 5
5. Are engine control unit terminal voltages OK?
   Check terminal: 1B,2D,2J,2O (MTX) 1B,2A,2B,3D (ATX)
   - **NO** → PC: Short circuit in wiring harness
   - **YES** → PC: Engine control unit malfunction

2.2L ENGINE TROUBLE CODE TROUBLESHOOTING—626 AND MX-6 NON-TURBO

Code No.09 (Water thermosensor)

PC: Possible Cause

1. Are there poor connections at water thermosensor circuit?
   - YES → Repair or replace connector
   - NO → go to 2
2. Is there continuity between water thermosensor and control unit?

Water thermosensor	Control unit
A (Y/B wire)	2Q (MTX), 2E (ATX)
B (LG/Y wire)	2D (MTX), 3D (ATX)

   - NO → PC: Open circuit in wiring harness from water thermosensor to engine control unit
   - YES → go to 3
3. Is resistance of the water thermosensor OK?
   Resistance:

Coolant temp	Resistance
−20°C (−4°F)	14.6—17.8 kΩ
20°C (68°F)	2.2— 2.7 kΩ
40°C (104°F)	1.0— 1.3 kΩ
60°C (140°F)	500—650Ω
80°C (176°F)	290—350Ω

   - NO → Replace water thermosensor
   - YES → go to 4
4. Is same Code No. present after performing after-repair procedure?
   - NO → Water thermosensor and circuit OK
   - YES → go to 5
5. Are engine control unit terminal voltages OK?
   Check terminal: 2D, 2Q (MTX) 2E, 3D (ATX)
   - NO → PC: Engine short circuit in wiring harness
   - YES → PC: Engine control unit malfunction

Code No.10 (Intake air thermosensor)

PC: Possible Cause

1. Are there poor connections in airflow meter circuit?
   - YES → Repair or replace connector
   - NO → go to 2
2. Is there continuity between intake air thermosensor (in airflow meter) and engine control unit?

Intake air thermosensor (in airflow meter)	Control unit
THA (R wire)	2P (MTX), 2K (ATX)
E_2 (LG/Y wire)	2D (MTX), 3D (ATX)

   - NO → PC: Open circuit in wiring harness from intake air thermosensor (in airflow meter) to engine control unit
   - YES → go to 3
3. Is resistance of intake air thermosensor (in airflow meter) OK?
   Resistance:

Terminal	Resistance
E_2 ↔ THA	−20°C (−4°F): 13.6—18.4 kΩ 20°C (68°F): 2.21—2.69 kΩ 60°C (140°F): 493 —667Ω

   - NO → Replace airflow meter
   - YES → go to 4
4. Is same Code No. present after performing after-repair procedure?
   - NO → Intake air thermosensor and circuit OK
   - YES → go to 5
5. Are engine control unit terminal voltages OK?
   Check terminal: 2D, 2P (MTX) 2K, 3D (ATX)
   - NO → PC: Short circuit in wiring harness
   - YES → PC: Engine control unit malfunction

Code No.12 (Throttle sensor)

PC: Possible cause

Check	Result	Action / PC: Possible cause
Are there poor connections in throttle sensor circuit?	YES	Repair or replace connector terminal
↓ NO		
Is there 4.5—5.5 V at A terminal of throttle sensor connector?	NO	PC: • Open or short circuit in wiring harness from A terminal to 2K (MTX) or 2I (ATX) terminal of control unit • Control unit malfunction
↓ YES		
Is C terminal of throttle sensor connector grounded?	NO	PC: Open circuit in wiring harness from C terminal to ground
↓ YES		
Is throttle sensor adjusted correctly?	NO	Adjust throttle sensor
↓ YES		
Is same Code No. present after performing after-repair procedure?	NO	Throttle sensor and circuit OK
↓ YES		
Is engine control unit 2M (MTX) or 2F (ATX) terminal voltage OK?	NO	PC: Open or short circuit in wiring harness from B terminal of throttle sensor to 2M (MTX) or 2F (ATX) terminal of engine control unit
	YES	PC: Engine control unit malfunction

Code No.14 (Atmospheric pressure sensor)

Check	Result	Action
Is same Code No. present after performing after-repair procedure?	NO	Atmospheric pressure sensor OK
↓ YES		
Replace engine control unit		

Code No.15 (Oxygen sensor)

PC: Possible Cause

Note
- **When Codes No.15 and 17 are present at the same time, first perform the checking procedure for Code No.17.**

Check	Result	Action / PC: Possible Cause
Are there poor connections in oxygen sensor circuit?	YES	Repair or replace connector
↓ NO		
Is oxygen sensor output voltage OK?	NO	Replace oxygen sensor
↓ YES		
Is sensitivity of oxygen sensor OK?	NO	PC: • Oxygen sensor malfunction • Engine control unit malfunction
↓ YES		
Is same Code No. present after performing after-repair procedure?	NO	Oxygen sensor and circuit OK
↓ YES		
Is engine control unit 2N (MTX) or 2C (ATX) terminal voltage OK?	NO	PC: Open or short circuit in wiring harness between oxygen sensor and engine control unit 2N (MTX) or 2C (ATX) terminal
	YES	PC: Engine control unit malfunction

2.2L ENGINE TROUBLE CODE TROUBLESHOOTING—626 AND MX-6 NON-TURBO

Code No.16 (EGR position sensor) (California only)

PC: Possible Cause

Note
- **Inspect the vacuum hose to the EGR control valve for air leakage, blockage and damage if the MIL illuminates only during cruising.**

Check	Result	Action / Possible Cause
Are there poor connections in sensor circuit?	YES	Repair or replace connector
	NO	Go to next step
Is EGR control valve OK?	NO	Replace EGR control valve
	YES	Go to next step
Is resistance of EGR position sensor OK?	NO	Replace EGR control valve
	YES	Go to next step
Is there continuity between EGR position sensor and engine control unit? (see table below)	NO	PC: Open circuit in wiring harness from EGR position sensor to engine control unit
	YES	Go to next step
Is there 4.5—5.5V at A terminal of EGR position sensor connector?	NO	PC: Short circuit in wiring harness from EGR position sensor A terminal to engine control unit 2K (MTX) or 2I (ATX) terminal •Engine control unit malfunction
	YES	Go to next step
Is same Code No. present after performing after-repair procedure?	NO	EGR position sensor and circuit OK
	YES	Go to next step
Is engine control unit 2L (MTX) or 2J (ATX) terminal voltage OK?	NO	PC Short circuit in wiring harness from EGR position sensor C terminal to engine control unit 2L (MTX) or 2J (ATX) terminal
	YES	Replace engine control unit

EGR position sensor	Engine control unit
(C) (Y/L wire)	2L (MTX), 2J (ATX)
(B) (LG/Y wire)	2D (MTX), 3D (ATX)
(A) (LG/R wire)	2K (MTX), 2I (ATX)

Code No.17 (Feedback system)

PC: Possible Cause

Check	Result	Action / Possible Cause
Warm up engine and run it at 2,500—3,000 rpm for three minutes. Does monitor lamp of Self-Diagnosis Checker illuminate at idle?	NO	PC: •Air leak in vacuum hoses or emission component •Contaminated oxygen sensor •Insufficient fuel injection
	YES	Go to next step
Are spark plugs clean?	NO	Clean or replace spark plugs
	YES	Go to next step
Is oxygen sensor voltage OK?	NO	PC: Oxygen sensor malfunction
	YES	Go to next step
Is same Code No. present after performing after-repair procedure?	NO	Feedback system OK
	YES	Go to next step
Is engine control unit 2N (MTX) or 2C (ATX) terminal voltage OK?	NO	PC: Open or short circuit in wiring harness between oxygen sensor and engine control unit 2N (MTX) or 2C (ATX) terminal
	YES	PC: Engine control unit malfunction

PC: Possible Cause

Code No.25 (Solenoid valve — Pressure regulator)

Check	Result	Action / Next step
Are there poor connections in solenoid valve circuit?	YES	Repair or replace connector
	NO	Go to next check
Is there continuity of solenoid valve?	NO	Replace solenoid valve
	YES	Go to next check
Is there battery voltage at (R/B) wire of solenoid valve connector?	NO	PC: Open circuit in wiring harness from (R/B) wire to main relay (for engine control unit)
	YES	Go to next check
Is there continuity between solenoid valve and engine control unit? (Solenoid valve B (W/R wire) – Control unit 2T (MTX), 3M (ATX))	NO	PC: Open circuit in wiring harness from solenoid valve to engine control unit
	YES	PC: • Short circuit in wiring harness • Engine control unit malfunction

Solenoid valve	Control unit
B (W/R wire)	2T (MTX), 3M (ATX)

Code No.26 (Solenoid valve — Purge)

Check	Result	Action / Next step
Are there poor connections in solenoid valve circuit?	YES	Repair or replace connector
	NO	Go to next check
Is there continuity of solenoid valve?	NO	Replace solenoid valve
	YES	Go to next check
Is there battery voltage at (R/B) wire of solenoid valve circuit?	NO	PC: Open circuit in wiring harness from (R/B) wire to main relay (for engine control unit)
	YES	Go to next check
Is there continuity between solenoid valve and engine control unit? (Solenoid valve B (W/B wire) – Control unit 2X (MTX), 2O (ATX))	NO	PC: Open circuit in wiring harness from solenoid valve to engine control unit
	YES	PC: • Short circuit in wiring harness • Engine control unit malfunction

Solenoid valve	Control unit
B (W/B wire)	2X (MTX), 2O (ATX)

Code No.28 (Solenoid valve — EGR)

Check	Result	Action / Next step
Are there poor connections in solenoid valve circuit?	YES	Repair or replace connector
	NO	Go to next check
Is there continuity of solenoid valve?	NO	Replace solenoid valve
	YES	Go to next check
Is there battery voltage at (R/B) wire of solenoid valve connector?	NO	PC: Open circuit in wiring harness from (R/B) wire to main relay (for engine control unit)
	YES	Go to next check
Is there continuity between solenoid valve and engine control unit? (Solenoid valve B (W/L wire) – Control unit 2Y (MTX), 3O (ATX))	NO	PC: Open circuit in wiring harness from solenoid valve to engine control unit
	YES	PC: • Short circuit in wiring harness • Engine control unit malfunction

Solenoid valve	Control unit
B (W/L wire)	2Y (MTX), 3O (ATX)

Code No.34 (ISC valve)

Check	Result	Action / Next step
Are there poor connections in ISC valve circuit?	YES	Repair or replace connector
	NO	Go to next check
Is resistance of ISC valve OK? Resistance: 6.3—9.9Ω	NO	Replace ISC valve
	YES	Go to next check
Is there battery voltage at (R/B) wire of ISC valve connector?	NO	PC: Open or short circuit in wiring harness from (R/B) wire to main relay (for engine control unit)
	YES	Go to next check
Is there continuity between ISC valve and engine control unit? (ISC valve B (W wire) – Control unit 2W (MTX), 3Q (ATX))	NO	PC: Open circuit in wiring harness from ISC valve to engine control unit
	YES	PC: • Short circuit in wiring harness • Engine control unit malfunction

ISC valve	Control unit
B (W wire)	2W (MTX), 3Q (ATX)

2.2L ENGINE SWITCH MONITOR FUNCTION 626 AND MX-6 NON-TURBO

PREPARATION
SST

49 H018 9A1 Self-Diagnosis Checker	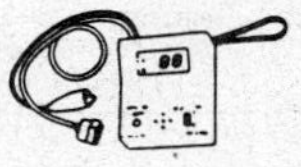	For diagnosis

Individual switches can be monitored by the **SST**.

Note
- **The test connector must be grounded and the ignition switch ON (engine stopped).**

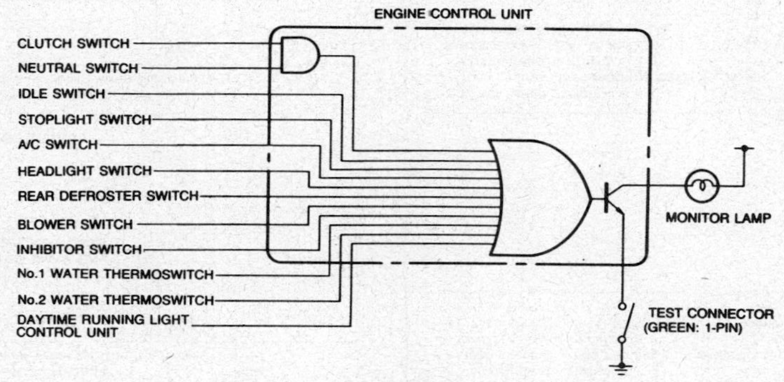

Switch	Self-Diagnosis Checker (Monitor lamp) Light ON	Light OFF	Remarks
Clutch switch	Pedal released	Pedal depressed	Gear: IN
Neutral switch	In gear	Neutral	Clutch pedal released
Idle switch	Pedal depressed	Pedal released	—
Stoplight switch	Pedal depressed	Pedal released	—
A/C switch	ON	OFF	Blower motor position: "1" position
Headlight switch	ON	OFF	—
Rear defroster switch	ON	OFF	—
Blower switch	ON	OFF	Blower motor position: "3" or "4" position
Inhibitor switch	D, 1, 2, and R ranges	P and N ranges	—
No.1 water thermoswitch (Electrical fan)	Check connector (for electrical fan) (B/G) terminal grounded	Check connector (for electrical fan) (B/G) terminal not grounded	While fan not operating
No.2 water thermoswitch (Electrical fan) (ATX)	Check connector (for electrical fan) (L/B) terminal grounded	Check connector (for electrical fan) (L/B) terminal not grounded	While fan not operating
Daytime running light control unit (Canada only)	Parking brake lever released	Parking brake lever pulled up	—

2.2L ENGINE SWITCH MONITOR FUNCTION TROUBLESHOOTING—626 AND MX-6 NON-TURBO

Procedure

Set conditions to deactivate each switch
- All accessories OFF
- Transmission in neutral
- All pedals released
- Parking brake lever pulled

Verify that monitor lamp does not illuminate

NO → Check each switch and related wiring harness
- Clutch and Neutral switch
- Idle switch
- Stoplight switch
- A/C switch
- Headlight switch
- Rear defroster switch
- Blower switch
- Inhibitor switch
- No.1 water thermoswitch
- No.2 water thermoswitch
- Parking brake switch

YES → Check each switch as described

Neutral and Clutch switch (MTX)

Shift transmission into gear
Check that monitor lamp illuminates with clutch pedal released

NO → PC. • Neutral or clutch switch malfunction
- Open circuit in related wiring harness
- Engine control 1V terminal malfunction

YES ↓

Depress clutch pedal
Check that monitor lamp does not illuminate
Return transmission to neutral

NO → PC • Clutch switch malfunction

Idle switch

Depress accelerator pedal, and check that the monitor lamp illuminates

- **NO** → PC: •Idle switch malfunction
  •Short circuit in wiring harness from idle switch to engine control unit
  •Engine control unit 1N (MTX) or 1T (ATX) terminal malfunction
- **YES** → Idle switch OK
  Release accelerator pedal

Stoplight switch

Depress brake pedal
Check that monitor lamp illuminates

- **NO** → PC: •Stoplight switch malfunction
  •Open circuit in wiring harness from brake light switch to engine control unit
  •Engine control unit 1O (MTX) or 1Q (ATX) terminal malfunction
- **YES** → Stoplight switch OK
  Release brake pedal

Blower switch

Turn blower switch to "3rd" position
Check that monitor lamp illuminates

- **NO** → PC: •Blower switch malfunction
  •Open circuit in the wiring harness from blower switch to engine control unit
  •Engine control unit 1S (MTX) or 1P (ATX) terminal malfunction (Refer to page F1–87)
- **YES** → Blower switch OK

A/C switch

Set blower switch to 1st position
Turn A/C switch ON
Check that monitor lamp illuminates

- **NO** → PC: •A/C switch malfunction
  •Blower switch malfunction
  •Open circuit in wiring harness from A/C switch to engine control unit through E/L unit
  •Engine control unit 1Q (MTX) or 1O (ATX) terminal malfunction
- **YES** → A/C switch OK
  Turn OFF A/C switch and blower switch

Headlight switch

Turn ON headlight switch
Check that monitor lamp illuminates

- **NO** → PC: •Headlight switch malfunction
  •Open circuit in wiring harness from headlight switch to engine control unit
  •Engine control unit 1U (MTX) or 1H (ATX) terminal malfunction
- **YES** → Headlight switch OK
  Turn OFF headlight switch

Rear defroster switch

Turn ON rear defroster switch
Check that monitor lamp illuminates

- **NO** → PC: •Rear defroster switch malfunction
  •Open circuit in wiring harness from rear defroster switch to engine control unit
  •Engine control unit 1T (MTX) or 1J (ATX) terminal malfunction
- **YES** → Rear defroster switch OK
  Turn OFF rear defroster switch

Inhibitor switch (ATX)

Shift transmission into gear
Check that monitor lamp illuminates

- **NO** → PC: •Inhibitor switch malfunction
  •Open or short circuit in related wiring harness
  •Control unit 3E, 3G and 3H terminals malfunction
- **YES** → Inhibitor switch OK
  Shift into "P" or "N" range

No.1 water thermoswitch circuit (not included in switch inspection)

Warning
- **The electrical fan operates when the check connector (for electrical fan) (B/G) terminal is grounded. Use caution.**

Ground check connector (for electrical fan) (B/G) terminal with jumper wire
Check that monitor lamp illuminates

- **NO** → PC: •Open circuit wiring harness from No.1 water thermoswitch to engine control unit
  •Engine control unit 1R (MTX) or 2D (ATX) terminal malfunction
- **YES** → No.1 water thermoswitch circuit OK
  Remove jumper wire

2.2L ENGINE SWITCH MONITOR FUNCTION TROUBLESHOOTING—626 AND MX-6 NON-TURBO

No.2 water thermoswitch circuit (not included in switch inspection) (ATX)

Warning
- **The electrical fan operates when the check connector (for electrical fan) (L/B) terminal is grounded. Use caution.**

Ground check connector (for electrical fan) (L/B) terminal with jumper wire
Check that monitor lamp illuminates

NO → PC: • Open circuit wiring harness from No.2 water thermoswitch to engine control unit
• Engine control unit 2G terminal malfunction

YES → No.2 water thermoswitch circuit OK
Remove jumper wire

Daytime running light control unit (Canada only)

Release parking brake lever
Check that monitor lamp illuminates

NO → PC: • Daytime running light control unit malfunction
• Parking brake switch malfunction
• Open circuit wiring harness from daytime running light control unit to engine control unit
• Engine control unit 1M (MTX) or 3F (ATX) terminal malfunction

YES → Daytime running light control unit circuit OK
Pull up parking brake lever

2.2L ENGINE ISC SYSTEM TROUBLESHOOTING IN ORDER OF PRIORITY—626 AND MX-6 NON-TURBO

Note
- **Make the system inspection first. If no problem is found, continue with the next system inspection of the Troubleshooting Guide.**

Symptom \ Possible cause		Air valve	ISC valve	Engine control unit terminal							System in-spection
				1K (MTX) 1I (ATX)	1P (MTX) 1N (ATX)	1R (MTX) 2D (ATX)	1S (MTX) 1P (ATX)	1T (MTX) 1J (ATX)	2G (ATX)	2W (MTX) 3Q (ATX)	
Engine stalls	While warming up	4	1	2						3	
	After warming up		1	7	2	3	4	5	6	8	
Rough idle	While warming up	5	2	3						4	1
	After warming up		2	8	3	4	5	6	7	9	1
High idle speed after warming up		8	2	8	3	4	5	6	7	9	1
Runs rough on deceleration			2	3						4	1
Afterburn in exhaust system		5	2	3						4	1

2.2L ENGINE FUEL SYSTEM TROUBLESHOOTING IN ORDER OF PRIORITY—626 AND MX-6 NON-TURBO

Symptom		Airflow meter	Oxygen sensor	Throttle sensor	Water thermosensor	Fuel pump	Injector	Fuel pressure	Engine control unit terminal: 2U and 2V (MTX) 3U and 3V (ATX)	Engine control unit terminal: 1C
Hard start or won't start (Cranks OK)					4	1	3			2
Engine stalls	While warming up	4					2	1	5	
	After warming up						3	2	4	
Rough idle	While warming up	4			3		2	1		
	After warming up	1					3	2		
Poor acceleration, hesitation, or lack of power		1		3			4	2		
Runs rough on deceleration		1					2			
Afterburn on deceleration		1					2			
Poor fuel consumption		5		4		3		2	1	
Engine stalls or rough after hot starting		1					3	2		

2.2L ENGINE PRESSURE REGULATOR SYSTEM TROUBLESHOOTING IN ORDER OF PRIORITY 626 AND MX-6 NON-TURBO

Note
- Make the system inspection first. If no problem is found, continue with the next system inspection of the Troubleshooting Guide.

Symptom / Possible cause	Solenoid valve (Pressure regulator control)	Water thermosensor	Intake air thermosensor	Throttle sensor	Engine control unit terminal 2T (MTX) 3M (ATX)	System inspection
Engine stalls or rough after hot starting	2	3	4	5	6	1

2.2L ENGINE DECELERATION CONTROL SYSTEM TROUBLESHOOTING IN ORDER OF PRIORITY 626 AND MX-6 NON-TURBO

COMPONENT DESCRIPTIONS

Component	Function	Remarks
Stoplight switch	Detects braking operation (deceleration); sends signal to control unit	
Clutch switch (MTX)	Detects in-gear condition; sends signal to control unit	Switch ON when clutch pedal released
Engine control unit	Detects signals from input sensors and switches; cuts fuel injection	
Idle switch	Detects when throttle valve fully closed; sends signal to control unit	Installed on throttle body
Ignition coil (–) terminal	Detects engine speed; sends signal to control unit	
Inhibitor switch (ATX)	Detects in-gear condition; sends signal to engine control unit	Switch ON in "N" or "P" range
Neutral switch (MTX)	Detects in-gear condition; sends signal to control unit	Switch ON when in-gear
Water thermosensor	Detects coolant temperature; sends signal to control unit	

TROUBLESHOOTING

Check the condition of the wiring harness and connectors before checking the sensor or switches below.

Note

- **Make the system inspection first. If no problem is found, continue with the next system inspection of the Troubleshooting Guide.**

Possible cause	Water thermosensor	Engine control unit
		2U or 2V (MTX) 3U or 3V (ATX)
Checking order	2	1

2.2L ENGINE EGR SYSTEM TROUBLESHOOTING IN ORDER OF PRIORITY—626 AND MX-6 NON-TURBO

Note

- **Make the system inspection first. If no problem is found, continue with the next system inspection of the Troubleshooting Guide.**

Possible cause	Solenoid valve (EGR)	EGR modulator valve	EGR control valve	Water thermo-sensor	EGR position sensor (California)	Engine control unit terminal	System inspection
						2Y (MTX) 30 (ATX)	
Page							
Checking order	3	2	4	6	5	7	1

2.2L ENGINE EVAP SYSTEM TROUBLESHOOTING IN ORDER OF PRIORITY—626 AND MX-6 NON-TURBO

COMPONENT DESCRIPTIONS

Component	Function	Remarks
Airflow meter	Detects amount of intake air; sends signal to control unit	Intake air temp sensor and fuel pump switch are integrated
Charcoal canister	Stores gas tank fumes when engine stopped	
Check-and-cut valve	Releases excessive pressure or vacuum in fuel tank to atmosphere	
Clutch switch	Detects in-gear condition; sends signal to control unit	Switch ON when clutch pedal released
Engine control unit	Detects signals from input sensors and switches; controls solenoid valve (Purge control)	
Idle switch	Detects when throttle valve fully closed; sends signal to control unit	Installed on throttle body
Ignition coil (–) terminal	Detects engine speed; sends signal to control unit	
Inhibitor switch	Detects in-gear condition; sends signal to control unit	Switch On in "N" or "P" range
Neutral switch	Detects in-gear condition; sends signal to control unit	Switch ON when in-gear
Oxygen sensor	Detects Oxygen concentration; sends signal to control unit	Zirconia ceramic and platinum coating
Separator	Prevents fuel from flowing into charcoal canister	
Solenoid valve (Purge control)	Controls vacuum line to vacuum switch valve	
Two-way check valve	Controls pressure in fuel tank	
Water thermosensor	Detects coolant temperature; sends signal to control unit	

TROUBLESHOOTING

Check the condition of the wiring harness or connectors before checking the sensors or switches.

Note

- **Make the system inspection first. If no problem is found, continue with the next system inspection of the Troubleshooting Guide.**

Possible cause	Solenoid valve (Purge control)	Two-way check valve	Check-and-cut valve	Separator	Engine control unit
					2X (MTX) 2O (ATX)
Page					
Checking order	1	3	4	5	2

2.2L ENGINE RELATIONSHIP CHART 626 AND MX-6 NON-TURBO

O: Related ×: Not related

INPUT DEVICES \ OUTPUT DEVICES	INJECTOR: FUEL INJECTION AMOUNT	INJECTOR: FUEL INJECTION TIMING	BAC VALVE: AIR VALVE	BAC VALVE: ISC VALVE	SOLENOID VALVE (EGR)	SOLENOID VALVE (PURGE)	SOLENOID VALVE (PRESSURE REGULATOR)
TEST CONNECTOR	×	×	×	O	×	×	×
STOPLIGHT SWITCH	O	×	×	×	×	×	×
VEHICLE SPEED SWITCH	O	×	×	×	×	×	×
ELECTRICAL LOAD	×	×	×	O	×	×	×
P/S PRESSURE SWITCH	×	×	×	O	×	×	×
A/C SWITCH	O	×	×	O	×	×	×
IGNITION SWITCH (STA POSITION)	O	O	×	×	×	×	O
INHIBITOR SWITCH	O	×	×	O	O	O	×
NEUTRAL AND CLUTCH SWITCH	O	×	×	O	×	O	×
OXYGEN SENSOR	O	×	×	×	×	O	×
ATMOSPHERIC PRESSURE SENSOR	O	×	×	O	×	O	×
INTAKE AIR THERMOSENSOR	O	×	×	O	×	O	O
WATER THERMOSENSOR	O	×	×	O	O	O	O
IDLE SWITCH	O	O	×	O	×	×	×
THROTTLE SENSOR	O	O	×	O	O	O	O
AIRFLOW METER	O	×	×	×	×	O	×
IGNITION COIL	O	O	×	O	O	O	O

2.2L ENGINE ENGINE CONTROL OPERATION CHART 626 AND MX-6 NON-TURBO

Output Devices and Engine Conditions

OUTPUT DEVICES \ ENGINE CONDITIONS		CRANKING (COLD ENGINE)	WARMING UP (DURING IDLE)	MEDIUM LOAD: COLD	MEDIUM LOAD: WARM	ACCEL-ERATION	HEAVY LOAD	DECEL-ERATION	IDLE (THROT-TLE VALVE FULLY CLOSED)	IGN: ON (ENGINE NOT RUNNING)	REMARKS
INJECTOR	INJECTION	Rich			Rich and lean	Rich		Fuel cut	Rich and lean	No injection	
	INJECTION TIMING	1 group (once per revolution)				1 group (once per revolution) (once per two revolutions)*			1 group (once per revolution)		Above 6,300 rpm: fuel cut *Above 4,500 rpm
BAC VALVE	AIR VALVE	Open*			Close						*Coolant temp: below 50°C (122°F)
	ISC VALVE	Large amount of bypass air	Large amount of bypass air*	Small amount of bypass air						No bypass	*In extreme cold condition
SOLENOID VALVE (EGR)		ON (EGR cut)			OFF (EGR)	ON (EGR cut)	OFF (EGR)	ON (EGR cut)	ON (EGR cut)*	ON	*ATX D range: OFF (NO EGR due to no vacuum to system)
SOLENOID VALVE (PURGE)		OFF (Purge cut)			Operates (Duty valves [purge gas amount] change*)			OFF (Purge cut)			*Depends on engine condition
SOLENOID VALVE (PRESSURE REGULATOR CONTROL)		OFF (Vacuum to pressure regulator)							After starting* ON (Vacuum cut)	OFF	*During hot start only

2.2L ENGINE ENGINE CONTROL UNIT TERMINAL VOLTAGES—626 AND MX-6 NON-TURBO WITH MANUAL TRANSMISSION

Terminal	Input	Output	Connected to	Voltage (After warming-up) IGN: ON	Voltage (After warming-up) Idle	Remark
1A	—	—	Battery	Battery voltage		For backup
1B	—	—	Main relay	Battery voltage		—
1C	○		Ign. switch (START)	Below 2.5V		While cranking: Battery voltage
1D		○	Self-Diagnosis Checker (Monitor lamp)	Test connector grounded • For 3 sec. after ign. switch OFF→ON: Below 6.2V (lamp illuminates) • After 3 sec.: Battery voltage (lamp does not illuminate)	Test connector not grounded • Lamp illuminates: Below 6.2V • Lamp does not illuminate: Battery voltage Test connector grounded: Approx. 5V	Using Self-Diagnosis Checker
1E		○	Malfunction indicator lamp (MIL)	• For 3 sec. after ign. switch OFF→ON: Below 4.8V (lamp illuminates) • After 3 sec. : Battery voltage (lamp does not illuminate)		• Test connector grounded • Lamp illuminates: Below 4.8V • Lamp does not illuminate: Battery voltage
1F		○	Self-Diagnosis Checker (Code number)	• For 3 sec. after ign. switch OFF→ON: Below 6.2V (Buzzer sounds) • After 3 sec. : Battery voltage (Buzzer does not sound)		• Using Self-Diagnosis Checker and test connector grounded • Buzzer sounds: Below 6.2V • Buzzer does not sound: Battery voltage
1G	—	—	—	—		—
1H	—	—	—	—		—
1I	—	—	—	—		—
1J		○	A/C relay	Battery voltage	• A/C switch ON: Below 2.5V • A/C switch OFF: Battery voltage	Blower motor ON
1K	○		Test connector	• Test connector grounded: Below 1.5V • Test connector not grounded: Above 10.5V		Test connector: 1-pin, Green connector
1L	—	—	—	—		—
1M	○		Daytime running light control unit (Canada only)	• Parking brake lever pulled up: Battery voltage • Parking brake lever released: Below 1.5V		
1N	○		Idle switch	• Accelerator pedal released: Below 0.5V • Accelerator pedal depressed: Above 7.7V		—
1O	○		Stoplight switch	• Brake peoal released: Below 3.6V • Brake pedal depressed: Above 10.0V		
1P	○		Power steering pressure switch	Above 10.5V	• P/S ON: Below 1.5V • P/S OFF: Above 10.5V	—
1Q			A/C switch	• A/C switch ON: Below 1.5V • A/C switch OFF: Above 10.0V		Blower motor ON
1R	○		Electrical fan (Water thermoswitch)	Battery voltage		Coolant temp. : Below 97C° (207°F)
				Below 1.5V		Coolant temp. : Above 97C° (207°F)

2.2L ENGINE ENGINE CONTROL UNIT TERMINAL VOLTAGES—626 AND MX-6 NON- TURBO WITH MANUAL TRANSMISSION

Terminal	Input	Output	Connected to	Voltage (After warming-up)		Remark
				IGN: ON	Idle	
1S	○		Blower fan switch	• Switch less than 2nd position: Battery voltage • Switch 3rd or 4th position: Below 1.5V		
1T	○		Rear window defroster switch	• Switch OFF: Battery voltage • Switch ON: Below 1.5V		—
1U	○		Headlight switch	• Headlight OFF: Below 1.5V • Headlight ON: Battery voltage		—
1V	○		Neutral and clutch switch	In-gear condition • Clutch pedal depressed: Below 0.5V • Clutch pedal released: Battery voltage		Neutral: Below 0.5V
2A	—	—	Ground (EO1)	0V		—
2B	—	—	Ground (EO2)	0V		—
2C	—	—	Ground (E1)	0V		—
2D	—	—	Ground (E2)	0V		—
2E	—	—	—	—		—
2F	—	—	—	—		—
2G	○		Speedometer	Battery voltage		Above 113 mph (180 km/h): Below 1.0V
2H	—	—	—	—		—
2I	○		Ignition coil-terminal	Battery voltage	*Battery voltage	*Engine signal monitor green and red lamp flash
2J	○		Airflow meter (Vc)	7—9V		—
2K	—	—	Vref	4.5—5.5V		—
2L	○		EGR position sensor (California only)	0.25—0.95V		
2M	○		Throttle sensor	Accelerator pedal released: Approx. 0.5V (depends on terminal voltage)		Throttle valve fully open: 4.3V
2N	○		Oxygen sensor	0V	0—1.0V	• Cold engine at idle: 0V • After warming-up Acceleration: 0.5—1.0V Deceleration: 0—0.4V
2O	○		Airflow meter (Vs)	Approx. 1.7V	Approx. 3—5V	Increase engine speed: voltage increase
2P	○		Air flow meter (Intake air thermo-sensor)	Approx. 2.5V at 20°C (68°F)		—
2Q	○		Water thermosensor	0.3—0.6V		Coolant temp. 20°C (68°F): Approx. 2.5V
2R	—	—	—	—		—
2S	—	—	—	—		—
2T		○	Solenoid valve (Pressure regulator control)	For 120 sec. after ign. switch OFF ON: Below 3.5V	For 120 sec. after starting: Below 3.5V	Coolant temp. above 70°C (158°F) and intake air temp. above 20°C (68°F)
2U		○	Injector (No.1 and No.3)	Battery voltage	*Battery voltage	*Engine signal monitor green and red lamps flash
2V		○	Injector (No.2 and No.4)	Battery voltage	*Battery voltage	*Engine signal monitor green and red lamps flash

Terminal	Input	Output	Connected to	Voltage (After warming-up)		Remark
				IGN: ON	Idle	
2W		○	ISC valve	Engine signal monitor green and red lamps flash		—
2X		○	Solenoid valve (Purge control)	Battery voltage		—
2Y		○	Solenoid valve (EGR)	Below 3.5V		• Engine coolant temp. —below 50°C Below 3.5V • Engine speed above approx. 1,500 rpm: Battery voltage
2Z	—	—	—	—		—

Terminal locations

2Y	2W	2U	2S	2Q	2O	2M	2K	2I	2G	2E	2C	2A	1U	1S	1Q	1O	1M	1K	1I	1G	1E	1C	1A
2Z	2X	2V	2T	2R	2P	2N	2L	2J	2H	2F	2D	2B	1V	1T	1R	1P	1N	1L	1J	1H	1F	1D	1B

2.2L ENGINE ENGINE CONTROL UNIT TERMINAL VOLTAGES—626 AND MX-6 NON- TURBO WITH AUTOMATIC TRANSMISSION

Terminal	Input	Output	Connected to	Voltage (After warming-up)		Remark
				IGN: ON	Idle	
1A	—	—	Battery	Battery voltage		For back-up
1B	—	—	Main relay	Battery voltage		—
1C	○		Inhibitor switch (ATX)	Below 2.5V		While cranking: Battery voltage
1D		○	Self-Diagnosis Checker (Monitor lamp)	Test connect grounded • For 3 sec. after ign. switch OFF→ON: Below 6.2V (lamp illuminates) • After 3 sec.: Battery voltage (lamp does not illuminate)	Test connector not grounded • Lamp illuminates: Below 6.2V • Lamp does not illuminate: Battery voltage Test connector grounded: Approx. 5V	Using Self-Diagnosis Checker
1E		○	Malfunction indicator lamp (MIL)	• For 3 sec. after ign. switch OFF→ON: Below 4.8V (lamp illuminates) • After 3 sec.: Battery voltage (lamp not illuminate)		• Test connector grounded • Lamp illuminates: Below 4.8V • Lamp does not illuminate: Battery voltage
1F		○	Self-Diagnosis Checker (Code number)	• For 3 sec. after ign. switch OFF→ON: Below 6.2V (Buzzer sounds) • After 3 sec.: Battery voltage (Buzzer not sound)		• Using Self-Diagnosis Checker and test connector grounded • Buzzer sounds: Below 6.2V • Buzzer not sound: Battery voltage
1G	—	—	—	—		—
1H	○		Headlight switch	• Headlight OFF: Below 1.5V • Headlight ON: Battery voltage		—
1I	○		Test connector	• Test connector grounded: Below 1.5V • Test connector not grounded: Above 10.5V		Test connector: 1-pin, Green connector
1J	○		Rear window defroster switch	• Switch OFF: Battery voltage • Switch ON: Below 1.5V		—
1K	—	—	—	—		—
1L		○	A/C relay	Battery voltage	• A/C switch ON: Below 2.5V • A/C switch OFF: Battery voltage	Blower motor ON
1M	○		Vehicle speed sensor	Approx. 4.5V or below 1.5V		During driving: Approx. 4.5V
1N	○		Power steering pressure switch	Above 10.5V	• P/S ON: Below 1.5V • P/S OFF: Above 10.5V	—
1O			A/C switch	• A/C switch ON: Below 1.5V • A/C switch OFF: above 10.0V		Blower motor ON
1P	○		Blower fan switch	• Switch less than 2nd position: Battery voltage • Switch 3rd or 4th position: Below 1.5V		
1Q	○		Stoplight switch	• Brake pedal released: Below 3.6V • Brake pedal depressed: Above 10.0V		
1R	○		Inhibitor switch (N and P range)	• N or P range: Below 1.5V • Others: Battery voltage		
1S	○		Blower fan switch	• Switch less than 2nd position: Battery voltage • Switch 3rd or 4th position: Below 1.5V		
1T	○		Idle switch	• Accelerator pedal released: Below 0.5V • Accelerator pedal depressed: Above 7.7V		—
1U	—	—	Ignition switch (IG1)	Battery voltage		For EC-AT shift-solenoid valves
1V	○		Ignition coil ⊖ terminal	Battery voltage	*Battery voltage	*Engine signal monitor: green and red lamp flash
2A	○		Airflow meter (Vc)	7—9V		—
2B	○		Airflow meter (Vs)	Approx. 1.7V	Approx. 3—5V	Increase engine speed: voltage increase
2C	○		Oxygen sensor	0V	0—10V	• Cold engine at idle: 0V • After warming-up Acceleration: 0.5—1.0V Deceleration: 0—0.4V
2D	○		Electrical fan [Low] (No.1 water thermoswitch)	Battery voltage		Coolant temp.: Below 97°C (207°F)
				Below 1.5V		Coolant temp.: Above 97°C (207°F)
2E	○		Water thermosensor	0.3—0.6V		Coolant temp. 20°C (68°F): Approx. 2.5V
2F	○		Throttle sensor	Accelerator pedal released: Approx. 0.5V (depends on 2I terminal voltage)		Throttle valve fully open: 4.3V
2G	○		Electrical fan [High] (No.2 thermoswitch)	Battery voltage		Coolant temp.: Below 108°C (226°F)
				Below 1.5V		Coolant temp.: Above 108°C (226F°)
2H	○		Hold switch	• Switch depressed: Battery voltage • Switch released: Below 1.5V		—
2I	—	—	Vref	4.5—5.5V		—
2J	○		EGR position sensor (California only)	0.25—0.95V		
2K	○		Airflow meter (Intake air thermosensor)	Approx. 2.5V at 20°C (68°F)		—
2L	○		Mode switch (Power side)	• POWER mode: Below 1.5V • ECONOMY mode or HOLD mode: Battery voltage		
2M	○		Pulse generator	Below 1.5V	*Battery voltage	*P or N range
2N	—	—	Pulse generator	Below 1.5V		Ground
2O		○	Solenoid valve (Purge control)	Battery voltage		—
2P		○	Hold indicator	• Hold mode: Below 1.5V • Other modes: Battery voltage		—
3A	—	—	Ground (EO1)	0V		—
3B	—	—	Ground (EO2)	0V		—
3C	—	—	Ground (E1)	0V		—
3D	—	—	Ground (E2)	0V		—
3E	○		Inhibitor switch (D range)	• D range: Battery voltage • Other range: Below 1.5V		—

2.2L ENGINE ENGINE CONTROL UNIT TERMINAL VOLTAGES—626 AND MX-6 NON- TURBO WITH AUTOMATIC TRANSMISSION

Terminal	Input	Output	Connected to	Voltage (After warming-up) IGN: ON	Idle	Remark
3F	○		Daytime running light control unit (Canada only)	• Parking brake lever pulled up: Battery voltage • Parking brake lever released: Below 1.5V		
3G	○		Inhibitor switch (L range)	• L range: Battery to voltage • Other range: Below 1.5V		—
3H	○		Inhibitor switch (S range)	• S range: Battery voltage • Other range: Below 1.5V		—
3I	—	—	—	—		—
3J	—	—	—	—		—
3K	—	—	—	—		—
3L		○	Mode indicator	• HOLD mode: Battery voltage • POWER or ECONOMY mode: Below 1.5V		
3M		○	Solenoid valve (Pressure regulator control)	For 120 sec. after ign. Switch OFF→ON: Below 3.5V	For 120 sec. after starting: Below 3.5V	Coolant temp. above 70°C (158°F) and intake air temp. above 20°C (63°F)
3N	○		Fluid thermoswitch	• Fluid temp. below 143°C (389°F): Approx.10—12V • Fluid temp. above 150°C (302°F): Below 1.5V		—
3O		○	Solenoid valve (EGR)	Below 3.5V		• Engine coolant temp. —below 50°C Below 3.5V • Engine speed above approx. 1,500 rpm: Battery voltage
3P	—	—	—	—		—
3Q		○	ISC valve	Engine signal monitor green and red lamps flash		—
3R	—	—	—	—		—
3S	—	—	—	—		—
3T	—	—	—	—		—
3U		○	Injector (No.1 and No.3)	Battery voltage	*Battery voltage	*Engine signal monitor green and red lamps flash
3V		○	Injector (No.2 and No.4)	Battery voltage	*Battery voltage	*Engine signal monitor green and red lamps flash
3W		○	1—2 shift solenoid valve	• Solenoid valve ON: Battery voltage • Solenoid valve OFF: Below 1.5V		
3X		○	2—3 shift solenoid valve	• Solenoid valve ON: Battery voltage • Solenoid valve OFF: Below 1.5V		
3Y		○	3—4 shift solenoid valve	• Solenoid valve ON: Battery voltage • Solenoid valve OFF: Below 1.5V		
3Z		○	Lockup solenoid valve	• Lockup: Battery voltage • No lockup: Below voltage		

Terminal locations

3Y	3W	3U	3S	3Q	3O	3M	3K	3I	3G	3E	3C	3A	2O	2M	2K	2I	2G	2E	2C	2A	1U	1S	1Q	1O	1M	1K	1I	1G	1E	1C	1A
3Z	3X	3V	3T	3R	3P	3N	3L	3J	3H	3F	3D	3B	2P	2N	2L	2J	2H	2F	2D	2B	1V	1T	1R	1P	1N	1L	1J	1H	1F	1D	1B

2.2L ENGINE SOLENOID VALVE OPERATION CHART 626 AND MX-6 NON-TURBO WITH AUTOMATIC TRANSMISSION

RANGE		GEAR			SOLENOID VALVES 1-2	2-3	3-4	Lock-up
P		Non					ON	
R		Reverse			ON			
N		—	Below approx. 6 km/h (3.7 mph)				ON	
			Above approx. 6 km/h (3.7 mph)		ON			
D		1st				ON	ON	
		2nd			ON	ON	ON	
		3rd	Below approx. 40 km/h (25 mph)					
			Above approx. 40 km/h (25 mph)	Lock-up OFF	ON			
				Lock-up ON	ON			ON
		OD	Lock-up OFF		ON		ON	
			Lock-up ON		ON		ON	ON
S		1st				ON	ON	
		2nd			ON	ON	ON	
		3rd	Below approx. 40 km/h (25 mph)					
			Above approx. 40 km/h (25 mph)		ON			
L		1st				ON	ON	
		2nd	Below approx. 110 km/h (68 mph)		ON	ON		
			Above approx. 110 km/h (68 mph)		ON			
HOLD	D	2nd			ON	ON	ON	
		3rd	Below approx. 40 km/h (25 mph)					
			Above approx. 40 km/h (25 mph)		ON			
	S	2nd			ON	ON		
		3rd	Below approx. 40 km/h (25 mph)					
			Above approx. 40 km/h (25 mph)		ON			
	L	1st				ON		
		2nd	Below approx. 110 km/h (68 mph)		ON	ON		
			Above approx. 110 km/h (68 mph)		ON			

2.2L ENGINE SPECIFICATIONS CHART 626 AND MX-6 TURBO

Item		Engine type	Turbo Engine
Idle speed		rpm	750 ± 25 (ATX: P range)*
Throttle body			
Type			Horizontal draft (2-barrel)
Throat diameter	mm (in)	No.1	MTX: 40 (1.6), ATX: 46 (1.8)
		No.2	MTX: 46 (1.8), ATX: 40 (1.6)
Airflow meter			
Resistor	Ω	E_2—Vs	Fully closed: 20—400 Fully open: 20—1,000
		E_2—Vc	100—400
		E_2—V_B	200—400
		E_2—THA	−20°C (−4°F) 13,600—18,400 20°C (68°F) 2,210— 2,690 60°C (140°F) 493— 667
Fuel pump			
Type			Impeller (in tank)
Output pressure		kPa (kg/cm², psi)	Main pump: 441—588 (4.5—6.0, 64—85) Transfer pump: 39 (0.4, 5.7) max.
Feeding capacity		cc (cu in)/10 seconds	Main pump: 220 (13.4) min. Transfer pump: 190 (11.6) min.
Fuel filter			
Type	Low-pressure side		Nylon element
	High-pressure side		Paper element
Pressure regulator			
Type			Diaphragm
Regulating pressure		kPa (kg/cm², psi)	235—275 (2.4—2.8, 34—40)
Injector			
Type			High-ohmic
Type of drive			Voltage
Resistance		Ω	12—16
Injection amount		cc (cu in)/15 seconds	73—90 (4.45—5.49)
Idle speed control valve			
Solenoid resistance		Ω	6.3—9.9
Turbocharger			
Cooling method			Engine coolant
Lubrication method			Engine oil
Boost pressure (Maximum)		kPa (kg/cm², psi)	60 (0.61, 8.7): Solenoid duty value 100% 45 (0.46, 6.5): Solenoid duty value 0%
Fuel tank			
Capacity		liters (US gal, Imp gal)	60 (15.9, 13.2)
Air cleaner			
Element type			Oil permeated
Fuel			
Specification			Unleaded premium (Unleaded regular)

* With test connector grounded.

2.2L ENGINE TROUBLESHOOTING GUIDE 626 AND MX-6 TURBO

	Symptom		Possible cause
1	Fault indicated by SST Code No.		See codes below
2	Hard start or won't start (Cranks OK)		
3	Engine stalls	While warming up	
		After warming up	
4	Rough idle	While warming up	
		After warming up	
5	High idle speed after warming up		
6	Poor acceleration, hesitation or lack of power		
7	Runs rough on deceleration		
8	Afterburn in exhaust system		
9	Poor fuel consumption		
10	Excessive oil consumption		
11	Engine stalls or rough after hot starting		
12	Knocking		
13	Abnormal noise or vibration		

Possible cause		SST Code No.
Input sensors and switches	Ignition pulse	01
	Distributor (Ne signal)	02
	Distributor (G1 signal)	03
	Distributor (G2 signal)	04
	Knock sensor	05
	Airflow meter	08
	Water thermosensor	09
	Intake air thermosensor	10
	Throttle sensor	12
	Atmospheric pressure sensor	14
	Oxygen sensor	15
	EGR position sensor	16
	Feedback system	17
Output solenoid valves	Solenoid valve (Pressure regulator)	25
	Solenoid valve (Purge)	26
	Solenoid valve (EGR, vacuum side)	28
	Solenoid valve (EGR, vent side)	29
	ISC valve	34
	Solenoid valve (Waste gate)	42

TROUBLESHOOTING PROCEDURE

Note

- **Step 1 under symptom is to quickly determine what system or unit may be at fault by use of the SST. (Self-Diagnosis Checker 49 H018 9A1)**

1st: Check input sensors and output solenoid valves with the **SST**.

2nd: Check other switches with the **SST**.

3rd: Check the following items:

Electrical system
1) Battery condition
2) Fuses

Ignition system
1) Ignition spark
2) Ignition timing (with test connector grounded)

Fuel system
1) Fuel level
2) Fuel leakage
3) Fuel filter
4) Idle speed (with test connector grounded)

Intake air system
1) Air cleaner element
2) Vacuum or air leakage
3) Vacuum hose routing
4) Accelerator cable

Engine
1) Compression
2) Overheating

Others
1) Clutch slippage
2) Brake dragging

4th: Check Fuel and Emission Control Systems.

2.2L ENGINE TROUBLE CODES 626 AND MX-6 TURBO

Malfunction display: Code No.	Malfunction display: MIL output signal pattern	Sensor or subsystem	Self-diagnosis	Fail-safe
01	ON / OFF	Ignition pulse	No ignition signal	—
02	ON / OFF	Ne signal	No Ne signal from crank angle sensor	—
03	ON / OFF	G_1 signal	No G_1 signal	Neither G_1 nor G_2 signal: Engine stopped
04	ON / OFF	G_2 signal	No G_2 signal	
05	ON / OFF	Knock sensor and knock control unit	Open or short circuit	• Retards ignition timing 6° in heavy-load condition • Waste gate opens earlier
08	ON / OFF	Airflow meter	Open or short circuit	Maintains basic signal at preset value
09	ON / OFF	Water thermosensor	Open or short circuit	Maintains constant command • 40°C (104°F) for EGI • 50°C (122°F) for ISC control use
10	ON / OFF	Intake air thermosensor (airflow meter)	Open or short circuit	Maintains constant 20°C (68°F) command
12	ON / OFF	Throttle sensor	Open or short circuit	Maintains constant command of throttle valve fully open
14	ON / OFF	Atmospheric pressure sensor	Open or short circuit	Maintains constant command of sea level pressure
15	ON / OFF	Oxygen sensor	Sensor output continues less than 0.55V 120 sec. after engine starts (1,500 rpm)	Cancels EGI feedback operation
16	ON / OFF	EGR position sensor	Open or short circuit	Cuts off EGR
			Sensor output does not match target value (incorrect output)	—
17	ON / OFF	Feedback system	Sensor output not changed 20 sec. after engine exceeds 1,500 rpm	Cancels EGI feedback operation

2.2L ENGINE TROUBLESHOOTING GUIDE 626 AND MX-6 TURBO

Possible cause — Fuel and Emission Control Systems

Symptom	Intake Air System (Poor connection of components, throttle body)	Fuel System (Fuel injection, Fuel pressure)	Pressure Regulator Control System	Idle Speed Control (ISC) System (Air valve, ISC valve malfunction)	Turbocharging System (Oil & water passage, Turbine and compressor wheels malfunction)	Electronic Spark Advance (ESA) System (Knock control system)	EGR System (EGR control valve stuck and open)	Evaporative Emission Control System (Solenoid valve [Purge control] malfunction)	PCV System (System clogged)	Deceleration System (Fuel cut operation malfunction)	Exhaust System (System clogged)
2	3	2				1					
3	4	3		1			2				
	5	4		2			3		1		
4	5	4		1			3		2		
	6	5		2			3	4	1		
5	2			1							
6	3	4			6		1	2			5
7		3		2						1	
8	3	4		1						2	
9		2					3			1	4
10					1						
11		2	1								
12					2	1					
13					1						

The numbers of the list show the priorities of inspections from the most probable to that with the lowest probability.
These were determined on the following basis:

- Ease of inspection
- Most probable system
- Most probable point in system

2.2L ENGINE TROUBLE CODES 626 AND MX-6 TURBO

Malfunction display		Sensor or subsystem	Self-diagnosis	Fail-safe
Code No.	MIL output signal pattern			
25	ON / OFF 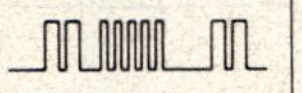	Solenoid valve (pressure regulator)	Open or short circuit	—
26	ON / OFF 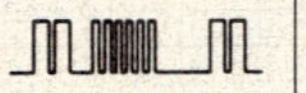	Solenoid valve (purge control)		—
28	ON / OFF 	Solenoid valve (EGR-vacuum)		—
29	ON / OFF 	Solenoid valve (EGR-vent)		—
34	ON / OFF 	ISC valve		—
42	ON / OFF 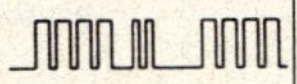	Solenoid valve (waste gate)		—

Caution
- **If there is more than one failure present, the lowest number malfunction code is displayed first, the remaining codes are displayed sequentially.**
- **After repairing a failure, turn off the ignition switch and disconnect the negative battery cable and depress the brake pedal for at least 5 seconds to erase the memory of a malfunction code.**

2.2L ENGINE TROUBLE CODE TROUBLESHOOTING 626 AND MX-6 TURBO

Code No.01 (Ignition pulse) — **PC: Possible Cause**

1. Are there poor connections in igniter circuit?
   - YES → Repair or replace connector
   - NO → next step
2. Is there battery voltage at igniter terminal WR?
   - NO → PC: Open circuit in wiring harness from main relay to igniter
   - YES → next step
3. Is igniter OK?
   - NO → Replace igniter
   - YES → next step
4. Is engine control unit 1G terminal OK?
   - NO → PC: • Malfunction of IGt signal or engine control unit • Open circuit in related wiring harness
   - YES → next step
5. Is same Code No. present after performing after-repair procedure?
   - NO → Ignition pulse and circuit OK
   - YES → next step
6. Is engine control unit 1V terminal OK?
   - NO → PC: Open circuit in wiring harness between engine control unit 1V terminal and igniter
   - YES → PC: Engine control unit malfunction

Code No.02 (Ne signal) — **PC: Possible Cause**

1. Are there poor connections in distributor circuit?
   - YES → Repair or replace connector
   - NO → next step
2. Is resistance between A and B terminal OK?
   - NO → Relace distributor
   - YES → next step
3. Is there continuity between distributor and engine control unit?

Distributor	Control unit
A	3E
B	3F

   - NO → PC: Open circuit in wiring harness from distributor to engine control unit
   - YES → next step
4. Is same Code No. present after performing after-repair procedure?
   - NO → Ne signal and circuit OK
   - YES → next step
5. Are control unit 3E and 3F terminal voltages OK?
   - NO → PC: Open or short circuit in wiring harness
   - YES → PC: Engine control unit malfunction

Code No.03 (G1 signal) **PC: Possible Cause**

- Are there poor connections in distributor circuit?
  - YES → Repair or replace connector
  - NO ↓
- Is resistance between B and D terminal OK? (Refer to page F2–80)
  - NO → Replace distributor
  - YES ↓
- Is there continuity between distributor and engine control unit?

Distributor	Control unit
B	3F
D	3G

  - NO → PC: Open circuit in wiring harness from distributor to engine control unit
  - YES ↓
- Is same Code No. present after performing after-repair procedure?
  - NO → G1 signal and circuit OK
  - YES ↓
- Are engine control unit 3F and 3G terminal voltages OK?
  - NO → PC: Short circuit in wiring harness
  - YES → PC: Engine control unit malfunction

Code No.04 (G2 signal) **PC: Possible Cause**

- Are there poor connections in distributor circuit?
  - YES → Repair or replace connector
  - NO ↓
- Is resistance between B and C terminal OK?
  - NO → Replace distributor
  - YES ↓
- Is there continuity between distributor and engine control unit?

Distributor	Control unit
B	3F
C	3H

  - NO → PC: Open circuit in wiring harness from distributor to engine control unit
  - YES ↓
- Is same Code No. present after performing after-repair procedure?
  - NO → G2 signal and circuit OK
  - YES ↓
- Are engine control unit 3F and 3H terminal voltages OK?
  - NO → PC: Short circuit in wiring harness
  - YES → PC: Engine control unit malfunction

Code No.05 (Knock sensor and Knock control unit) **PC: Possible Cause**

- Are there poor connections in relative circuit?
  - YES → Repair or replace connector
  - NO ↓
- Are knock control unit terminal voltages correct?

Terminal	Voltage
R/B wire	Approx. 12V
R/Y wire	3.3—5.0V
Y/L wire	
B/LG wire	0V

  - NO (all wires) → PC: • Open or short circuit in wiring harness between main relay and knock control unit • Open circuit in wiring harness between knock control unit and ground
  - (only RY wire) → PC: Knock sensor, knock control unit, or related harness malfunction
  - (only YL wire) → PC: Open circuit in wiring harness between knock control unit and knock sensor
  - YES ↓
- Is same Code No. present after performing after-repair procedure?
  - NO → Knock control unit, knock sensor, and circuit OK
  - YES ↓
- Is engine control unit 2M terminal voltage OK?
  - NO → PC: • Open or short circuit in wiring harness • Knock control unit malfunction
  - YES → PC: Engine control unit malfunction

2.2L ENGINE TROUBLE CODE TROUBLESHOOTING—626 AND MX-6 TURBO

Code No.08 (Airflow meter)

PC: Possible Cause

Are there poor connections in airflow meter circuit?
- YES → Repair or replace connector
- NO ↓

Is there continuity between airflow meter and engine control unit?

Airflow meter	Control unit
V_B (R/B wire)	1B
Vc (R/W wire)	2A
Vs (R/B wire)	2B
E_2 (LG/Y wire)	3D

- NO → PC: Open circuit in wiring harness from airflow meter to engine control unit
- YES ↓

Is resistance of airflow meter OK?
Resistance:

Terminal	Fully closed (Ω)	Fully open (Ω)
E_2 ↔ Vs	20—400	20—1,000
E_2 ↔ Vc	100—400	
E_2 ↔ V_B	200—400	

- NO → Repair airflow meter
- YES ↓

Is same Code. No. present after performing after-repair procedure?
- NO → Airflow meter and circuit OK
- YES ↓

Are engine control unit terminal voltages OK?
Check terminal: 1B, 2A, 2B, 3D
- NO → PC: Short circuit in wiring harness
- YES → PC: Engine control unit malfunction

Code No.09 (Water thermosensor)

PC: Possible Cause

Are there poor connections at water thermosensor circuit?
- YES → Repair or replace connector
- NO ↓

Is there continuity between water thermosensor and control unit?

Water thermosensor	Control unit
A (Y/B wire)	2E
B (LG/Y wire)	3D

- NO → PC: Open circuit in wiring harness from water thermosensor to engine control unit
- YES ↓

Is resistance of the water thermosensor OK?
Resistance:

Coolant temp	Resistance
−20°C (−4°F)	14.6—17.8 kΩ
20°C (68°F)	2.2— 2.7 kΩ
40°C (104°F)	1.0— 1.3 kΩ
60°C (140°F)	500—650Ω
80°C (176°F)	290—350Ω

- NO → Replace water thermosensor
- YES ↓

Is same Code No. present after performing after-repair procedure?
- NO → Water thermosensor and circuit OK
- YES ↓

Are engine control unit 2E and 3D terminal voltages OK?
- NO → PC: Engine short circuit in wiring harness
- YES → PC: Engine control unit malfunction

Code No.10 (Intake air thermosensor)

PC: Possible Cause

1. Are there poor connections in airflow meter circuit?
   - YES → Repair or replace connector
   - NO → go to 2
2. Is there continuity between intake air thermosensor (in airflow meter) and engine control unit?

Intake air thermosensor (in airflow meter)	Control unit
THA(R wire)	2K
E_2 (LG/Y wire)	3D

   - NO → PC: Open circuit in wiring harness from intake air thermosensor (in airflow meter) to engine control unit
   - YES → go to 3
3. Is resistance of intake air thermosensor (in airflow meter) OK?
   Resistance:

Terminal	Resistance
E_2 ↔ THA	−20°C (−4°F) : 13.6—18.4 kΩ 20°C (68°F) : 2.21—2.69 kΩ 60°C (140°F) : 493—667Ω

   - NO → Replace airflow meter
   - YES → go to 4
4. Is same Code No. present after performing after-repair procedure?
   - NO → Intake air thermosensor and circuit OK
   - YES → go to 5
5. Are engine control unit 2K and 3D terminal voltages OK?
   - NO → PC: Short circuit in wiring harness
   - YES → PC: Engine control unit malfunction

Code No.12 (Throttle sensor)

PC: Possible cause

1. Are there poor connections in throttle sensor circuit?
   - YES → Repair or replace connector terminal
   - NO → go to 2
2. Is there 4.5—5.5 V at A terminal of throttle sensor connector?
   - NO → PC: • Open or short circuit in wiring harness from A terminal to 2I terminal of control unit • Control unit malfunction
   - YES → go to 3
3. Is C terminal of throttle sensor connector grounded?
   - NO → PC: Open circuit in wiring harness from C terminal to ground
   - YES → go to 4
4. Is throttle sensor adjusted correctly?
   - NO → Adjust throttle sensor
   - YES → go to 5
5. Is same Code No. present after performing after-repair procedure?
   - NO → Throttle sensor and circuit OK
   - YES → go to 6
6. Is engine control unit 2F terminal voltage OK?
   - NO → PC: Open or short circuit in wiring harness from B terminal of throttle sensor to 2F terminal of engine control unit
   - YES → PC: Engine control unit malfunction

Code No.14 (Atmospheric pressure sensor)

1. Is same Code No. present after performing after-repair procedure?
   - NO → Atmospheric pressure sensor OK
   - YES → Replace engine control unit

2.2L ENGINE TROUBLE CODE TROUBLESHOOTING—626 AND MX-6 TURBO

Code No.15 (Oxygen sensor)

PC: Possible Cause

Note
- **When Codes No.15 and 17 are present at the same time, first perform the checking procedure for Code No.17.**

Step	Check	Result	Action
1	Are there poor connections in oxygen sensor circuit?	YES	Repair or replace connector
		NO	Go to next step
2	Is oxygen sensor output voltage OK?	NO	Replace oxygen sensor
		YES	Go to next step
3	Is sensitivity of oxygen sensor OK?	NO	PC: • Oxygen sensor malfunction • Engine control unit malfunction
		YES	Go to next step
4	Is same Code No. present after performing after-repair procedure?	NO	Oxygen sensor and circuit OK
		YES	Go to next step
5	Is engine control unit 2C terminal voltage OK?	NO	PC: Open or short circuit in wiring harness between oxygen sensor and engine control unit 2C terminal
		YES	PC Engine control unit malfunction

Code No.16 (EGR position sensor)

PC: Possible Cause

Note
- **Inspect the vacuum hose to the EGR control valve for air leakage, blockage and damage if MIL illuminates only during cruising.**

Step	Check	Result	Action
1	Are there poor connections in sensor circuit?	YES	Repair or replace connector
		NO	Go to next step
2	Is EGR control valve OK?	NO	Replace EGR control valve
		YES	Go to next step
3	Is resistance of EGR position sensor OK?	NO	Replace EGR control valve
		YES	Go to next step
4	Is there continuity between EGR position sensor and engine control unit? (see table below)	NO	PC: Open circuit in wiring harness from EGR position sensor to engine control unit
		YES	Go to next step
5	Is there 4.5—5.5V at C terminal of EGR position sensor connector?	NO	PC: Short circuit in wiring harness from EGR position sensor C terminal to engine control unit 2I terminal • Engine control unit malfunction
		YES	Go to next step
6	Is same Code No. present after performing after-repair procedure?	NO	EGR position sensor and circuit OK
		YES	Go to next step
7	Is engine control unit 2J terminal voltage OK?	NO	PC: Short circuit in wiring harness from EGR position sensor A terminal to engine control unit 2J terminal
		YES	Replace engine control unit

EGR position sensor	Engine control unit
(A) (Y/L wire)	2J
(B) (LG/Y wire)	3D
(C) (LG/R wire)	2I

Code No.17 (Feedback system)

PC: Possible Cause

Check	Result	Action / Possible Cause
Warm up engine and run it at 2,500—3,000 rpm for three minutes. Does monitor lamp of Self-Diagnosis Checker illuminate at idle?	NO	PC: • Air leak in vacuum hoses or emission component • Contaminated oxygen sensor • Insufficient fuel injection
	YES	↓
Are spark plugs clean?	NO	Clean or replace spark plugs
	YES	↓
Is oxygen sensor voltage OK?	NO	PC: Oxygen sensor malfunction
	YES	↓
Is same Code No. present after performing after-repair procedure?	NO	Feedback system OK
	YES	↓
Is engine control unit 2C terminal voltage OK?	NO	PC: Open or short circuit in wiring harness between oxygen sensor and engine control unit 2C terminal
	YES	PC: Engine control unit malfunction

Code No.25 (Solenoid valve — Pressure regulator)

PC: Possible Cause

Check	Result	Action / Possible Cause
Are there poor connections in solenoid valve circuit?	YES	Repair or replace connector
	NO	↓
Is there continuity of solenoid valve?	NO	Replace solenoid valve
	YES	↓
Is there battery voltage at (R/B) wire of solenoid valve connector?	NO	PC: Open circuit in wiring harness from (R/B) wire to main relay (for engine control unit)
	YES	↓
Is there continuity between solenoid valve and engine control unit? (Solenoid valve B (W/R wire) — Control unit 2O)	NO	PC: Open circuit in wiring harness from solenoid valve to engine control unit
	YES	PC: • Short circuit in wiring harness • Engine control unit malfunction

Solenoid valve	Control unit
B (W/R wire)	2O

Code No.26 (Solenoid valve — Purge)

Check	Result	Action / Possible Cause
Are there poor connections in solenoid valve circuit?	YES	Repair or replace connector
	NO	↓
Is there continuity of solenoid valve?	NO	Replace solenoid valve
	YES	↓
Is there battery voltage at (R/B) wire of solenoid valve circuit?	NO	PC: Open circuit in wiring harness from (R/B) wire to main relay (for engine control unit)
	YES	↓
Is there continuity between solenoid valve and engine control unit? (Solenoid valve B (W/B wire) — Control unit 3M)	NO	PC: Open circuit in wiring harness from solenoid valve to engine control unit
	YES	PC: • Short circuit in wiring harness • Engine control unit malfunction

Solenoid valve	Control unit
B (W/B wire)	3M

Code No.28 (Solenoid valve — EGR, vacuum side)

PC: Possible Cause

Check	Result	Action / Possible Cause
Are there poor connections in solenoid valve circuit?	YES	Repair or replace connector
	NO	↓
Is there continuity of solenoid valve?	NO	Replace solenoid valve
	YES	↓
Is there battery voltage at R/B wire of solenoid valve connector?	NO	PC: Open circuit in wiring harness from R/B wire to main relay (for engine control unit)
	YES	↓
Is there continuity between solenoid valve and engine control unit? (Solenoid valve B (W/L wire) — Control unit 3P)	NO	PC: Open circuit in wiring harness from solenoid valve to engine control unit
	YES	PC: • Short circuit in wiring harness • Engine control unit malfunction

Solenoid valve	Control unit
B (W/L wire)	3P

2.2L ENGINE TROUBLE CODE TROUBLESHOOTING—626 AND MX-6 TURBO

Code No.29 (Solenoid valve — EGR, vent side)

1. Are there poor connections in solenoid valve circuit?
   - YES → Repair or replace connector
   - NO → go to next step
2. Is there continuity of solenoid valve?
   - NO → Replace solenoid valve
   - YES → go to next step
3. Is there battery voltage of R/B wire of solenoid valve connector?
   - NO → PC: Open circuit in wiring harness from R/B wire to main relay (for engine control unit)
   - YES → go to next step
4. Is there continuity between solenoid valve and engine control unit?

Solenoid valve	Control unit
B (L/W wire)	3O

   - NO → PC: Open circuit in wiring harness from solenoid valve to engine control unit
   - YES → PC: • Short circuit in wiring harness • Engine control unit malfunction

Code No.34 (ISC valve)

PC: Possible Cause

1. Are there poor connections in ISC valve circuit?
   - YES → Repair or replace connector
   - NO → go to next step
2. Is resistance of ISC valve OK? Resistance: 6.3—9.9Ω
   - NO → Replace ISC valve
   - YES → go to next step
3. Is there battery voltage at (R/B) wire of ISC valve connector?
   - NO → PC: Open or short circuit in wiring harness from (R/B) wire to main relay (for engine control unit)
   - YES → go to next step
4. Is there continuity between ISC valve and engine control unit?

ISC valve	Control unit
B (W wire)	3Q

   - NO → PC: Open circuit in wiring harness from ISC valve to engine control unit
   - YES → PC: • Short circuit in wiring harness • Engine control unit malfunction

Code No.42 (Solenoid valve — Waste gate)

1. Are there poor connections in solenoid valve circuit?
   - YES → Repair or replace connector
   - NO → go to next step
2. Is there continuity of solenoid valve?
   - NO → Repair solenoid valve
   - YES → go to next step
3. Is there battery voltage of R/B wire of solenoid valve connector?
   - NO → PC: Open circuit in wiring harness from R/B wire to main relay (for engine control unit)
   - YES → go to next step
4. Is there continuity between solenoid valve and engine control unit?

Solenoid valve	Control unit
B (BR/Y wire)	3R

   - NO → PC: Open circuit in wiring harness from solenoid valve to engine control unit
   - YES → PC: • Short circuit in wiring harness • Engine control unit malfunction

2.2L ENGINE SWITCH MONITOR FUNCTION 626 AND MX-6 TURBO

PREPARATION
SST

49 H018 9A1 Self-Diagnosis Checker		For diagnosis

Individual switches can be monitored by the **SST**.

Note
- **The test connector must be grounded and the ignition switch ON (engine stopped).**

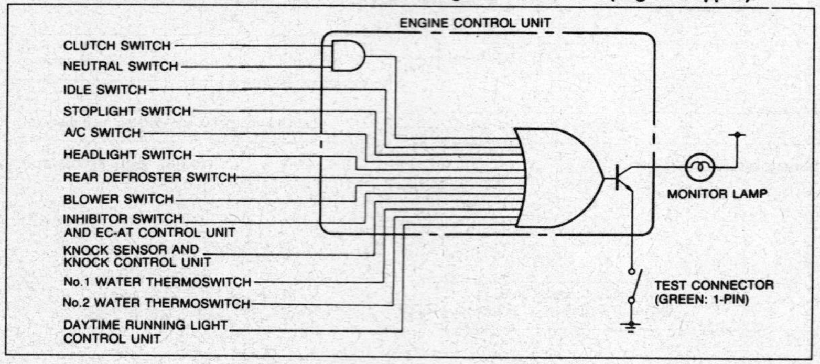

Switch	Self-Diagnosis Checker (Monitor lamp)		Remarks
	Light ON	Light OFF	
Clutch switch	Pedal released	Pedal depressed	Gear: IN
Neutral switch	In gear	Neutral	Clutch pedal released
Idle switch	Pedal depressed	Pedal released	—
Stoplight switch (MTX)	Pedal depressed	Pedal released	—
A/C switch	ON	OFF	Blower motor position: "1" position
Headlight switch	ON	OFF	—
Rear defroster switch	ON	OFF	—
Blower switch	ON	OFF	Blower motor position: "3" or "4" position
Inhibitor switch and EC-AT control unit	D, 1, 2, and R ranges	P and N ranges	—
No.1 water thermoswitch (Electrical fan)	Check connector (for electrical fan) (B/L) terminal grounded	Check connector (for electrical fan) (B/L) terminal not grounded	While fan not operating
Knock sensor and knock control unit	While tapping engine hanger	—	With test connector ground
Daytime running light control unit (Canada only)	Parking brake lever released	Parking brake lever pulled up	—
No.2 water thermoswitch (Electrical fan) (ATX)	Check connector (for electrical fan) (L/R) terminal grounded	Check connector (for electrical fan) (L/R) terminal not grounded	While fan not operating

2.2L ENGINE SWITCH MONITOR FUNCTION TROUBLESHOOTING—626 AND MX-6 TURBO

Procedure

Set conditions to deactivate each switch
- All accessories OFF
- Transmission in neutral
- All pedals released

Verify that monitor lamp does not illuminate

NO → Check each switch and related wiring harness
- Clutch and Neutral switch
- Idle switch
- Stoplight switch
- A/C switch
- Headlight switch
- Rear defroster switch
- Blower switch
- Inhibitor switch
- No.1 water thermoswitch
- No.2 water thermoswitch
- Parking brake switch

YES → Check each switch as described

Neutral and Clutch switch (MTX)

Shift transmission into gear
Check that monitor lamp illuminates with clutch pedal released

NO → PC: • Neutral or clutch switch malfunction
- Open circuit in related wiring harness
- Engine control unit (1R) terminal malfunction

YES → Depress clutch pedal
Check that monitor lamp does not illuminate
Return transmission to neutral

NO → PC: • Clutch switch malfunction

2.2L ENGINE SWITCH MONITOR FUNCTION TROUBLESHOOTING—626 AND MX-6 TURBO

Idle switch

Depress accelerator pedal, and check that the monitor lamp illuminates

- YES → Idle switch OK
  Release accelerator pedal
- NO → PC:
  - Idle switch malfunction
  - Short circuit in wiring harness from idle switch to engine control unit
  - Engine control unit (1T) terminal malfunction

Stoplight switch (MTX)

Depress brake pedal
Check that monitor lamp illuminates

- YES → Stoplight switch OK
  Release brake pedal
- NO → PC:
  - Stoplight switch malfunction
  - Open circuit in wiring harness from stoplight switch to engine control unit
  - Engine control unit (1Q) terminal malfunction

Blower switch

Turn blower switch to "3rd" position
Check that monitor lamp illuminates

- YES → Blower switch OK
- NO → PC:
  - Blower switch malfunction
  - Open circuit in the wiring harness from blower switch to engine control unit
  - Engine control unit (1P) terminal malfunction

A/C switch

Set blower switch to 1st position
Turn A/C switch ON
Check that monitor lamp illuminates

- YES → A/C switch OK
  Turn OFF A/C switch and blower switch
- NO → PC:
  - A/C switch malfunction
  - Blower switch malfunction
  - Open circuit in wiring harness from A/C switch to engine control unit
  - Engine control unit (1O) terminal malfunction

Headlight switch

Turn ON headlight switch
Check that monitor lamp illuminates

- YES → Headlight switch OK
  Turn OFF headlight switch
- NO → PC:
  - Headlight switch malfunction
  - Open circuit in wiring harness from headlight switch to engine control unit
  - Engine control unit (1H) terminal malfunction

Rear defroster switch

Turn ON rear defroster switch
Check that monitor lamp illuminates

- YES → Rear defroster switch OK
  Turn OFF rear defroster switch
- NO → PC:
  - Rear defroster switch malfunction
  - Open circuit in wiring harness from rear defroster switch to engine control unit
  - Engine control unit (1J) terminal malfunction

Inhibitor switch (ATX)

Shift transmission into gear
Check that monitor lamp illuminates

- YES → Inhibitor switch and EC-AT unit OK
  Shift into "P" or "N" range
- NO → PC:
  - Inhibitor switch malfunction
  - Open or short circuit in related wiring harness
  - Engine control unit (1R) terminal malfunction

No.1 water thermoswitch circuit (not included in switch inspection)

Warning
- **The electrical fan operates when the check connector (for electrical fan) (B/L) terminal is grounded. Use caution.**

Ground check connector (for electrical fan) (B/L) terminal with jumper wire
Check that monitor lamp illuminates

- YES → No.1 water thermoswitch circuit OK
  Remove jumper wire
- NO → PC:
  - Open circuit wiring harness from No.1 water thermoswitch to engine control unit
  - Engine control unit (2D) terminal malfunction

2.2L ENGINE SWITCH MONITOR FUNCTION TROUBLESHOOTING—626 AND MX-6 TURBO

No.2 water thermoswitch circuit (not included in switch inspection) (ATX)

Warning
- **The electrical fan operates when the check connector (for electrical fan) (L/R) terminal is grounded. Use caution.**

Ground check connector (for electrical fan) (L/R) terminal with jumper wire
Check that monitor lamp illuminates

NO → PC:
- Open circuit wiring harness from No.2 water thermoswitch to engine control unit
- Engine control unit (2G) terminal malfunction

YES → No.2 water thermoswitch circuit OK
Remove jumper wire

Daytime running light control unit (Canada only)

Release parking brake lever
Check that monitor lamp illuminates

NO → PC:
- Daytime running light control unit malfunction
- Parking brake switch malfunction
- Open circuit wiring harness from daytime running light control unit to engine control unit
- Engine control unit (1S) terminal malfunction

YES → Daytime running light control unit circuit OK
Pull up parking brake lever

2.2L ENGINE ISC SYSTEM TROUBLESHOOTING IN ORDER OF PRIORITY—626 AND MX-6 TURBO

TROUBLESHOOTING
Check the condition of the wiring harness and connectors before checking the sensors or switches.

Note
- **Make the system inspection first. If no problem is found, continue with the next system inspection of the Troubleshooting Guide.**

Symptom \ Possible cause		Air valve	ISC valve	Engine control unit terminal							System inspection
				1J	1N	1I	1P	2D	2G (ATX)	3Q	
Engine stalls	While warming up	4	1			2				3	
	After warming up		1	5	2	7	4	3	6	8	
Rough idle	While warming up	5	2			3				4	1
	After warming up		2	6	3	8	5	4	7	9	1
High idle speed after warming up		7	2	6	3	8	5	4	7	9	1
Runs rough on deceleration			2			3				4	1
Afterburn in exhaust system		5	2			3				4	1

2.2L ENGINE TURBOCHARGER SYSTEM TROUBLESHOOTING IN ORDER OF PRIORITY 626 AND MX-6 TURBO

Symptom \ Possible cause	Turbocharger	Air bypass valve	Intercooler	Boost pressure	Waste gate valve	Overboost warning
Poor acceleration, hesitation or lack of power	4	3	6	1	2	5
Excessive oil consumption	1					
Knocking	3			2	1	
Abnormal noise or vibration	2	1				

2.2L ENGINE PRESSURE REGULATOR SYSTEM TROUBLESHOOTING IN ORDER OF PRIORITY 626 AND MX-6 TURBO

Note
- Make the system inspection first. If no problem is found, continue with the next system inspection of the Troubleshooting Guide.

Symptom \ Possible cause	Solenoid valve (Pressure regulator control)	Water thermosensor	Intake air thermosensor (Airflow meter)	Throttle sensor	Engine control unit terminal	System inspection
					3M	
Engine stalls or rough after hot starting	2	3	4	5	6	1

2.2L ENGINE FUEL SYSTEM TROUBLESHOOTING IN ORDER OF PRIORITY—626 AND MX-6 TURBO

Symptom \ Possible cause		Airflow meter	Oxygen sensor	Throttle sensor	Water thermosensor	Fuel pump	Injector	Fuel pressure	Engine control unit terminal		
									1C	3T	3U,3V
Hard start or won't start (Cranks OK)					4	1	3		2		
Engine stalls	While warming up	4			3		2	1			5
	After warming up	1					3	2			4
Rough idle	While warming up	4			3		2	1			
	After warming up	1					3	2			
Poor acceleration, hesitation, or lack of power		1		3			4	2			
Runs rough on deceleration		1					3			2	
Afterburn on deceleration		1					2				
Poor fuel consumption		5	4		3		2	1			
Engine stalls or rough after hot starting		1					3	2			

2.2L ENGINE DECELERATION CONTROL SYSTEM TROUBLESHOOTING IN ORDER OF PRIORITY 626 AND MX-6 TURBO

Note
- Make the system inspection first. If no problem is found, continue with the next system inspection of the Troubleshooting Guide.

Possible cause	Water thermosensor	Engine control unit
		3U, 3V
Checking order	2	1

2.2L ENGINE RELATIONSHIP CHART 626 AND MX-6 TURBO

○: Related ×: Not related

INPUT DEVICES \ OUTPUT DEVICES	INJECTOR: FUEL INJECTION AMOUNT	INJECTOR: FUEL INJECTION TIMING	BAC VALVE: AIR VALVE	BAC VALVE: ISC VALVE	SOLENOID VALVE (EGR Vent side)	SOLENOID VALVE (EGR Vacuum side)	SOLENOID VALVE (PURGE)	SOLENOID VALVE (PRESSURE REGULATOR)	SOLENOID VALVE (WASTEGATE)	IGNITER
TEST CONNECTOR	×	×	×	○	×	×	×	×	×	○
KNOCK SENSOR	×	×	×	×	×	×	×	×	○	○
STOPLIGHT SWITCH	○	×	×	×	×	×	×	×	×	×
EGR POSITION SENSOR	×	×	×	×	○	○	×	×	×	○
ELECTRICAL LOAD	×	×	×	○	×	×	×	×	×	×
P/S PRESSURE SWITCH	×	×	×	○	×	×	×	×	×	×
A/C SWITCH	○	×	×	○	×	×	×	×	×	×
IGNITION SWITCH (STA POSITION)	○	○	×	×	×	×	×	○	×	○
INHIBITOR SWITCH and EC-AT CONTROL UNIT	○	×	×	○	×	×	○	×	×	○
NEUTRAL AND CLUTCH SWITCH	○	×	×	○	×	×	○	×	×	○
OXYGEN SENSOR	○	×	×	×	×	×	○	×	×	×
ATMOSPHERIC PRESSURE SENSOR	○	×	×	○	×	×	○	×	×	×
INTAKE AIR THERMOSENSOR	○	×	×	○	×	×	○	○	○	×
WATER THERMOSENSOR	○	○	×	○	○	○	○	○	○	○
IDLE SWITCH	○	○	×	○	○	○	×	×	×	○
THROTTLE SENSOR	○	○	×	○	○	○	○	○	×	○
AIRFLOW METER	○	×	×	×	○	○	○	×	○	○
Ne SIGNAL	○	○	×	○	○	○	○	○	○	○
G2 SIGNAL	×	○	×	×	×	×	×	×	×	○
G1 SIGNAL	×	○	×	×	×	×	×	×	×	○

2.2L ENGINE ESA SYSTEM TROUBLESHOOTING IN ORDER OF PRIORITY—626 AND MX-6 TURBO

Symptom \ Possible cause	Distributor	Igniter	Engine control unit terminal 3E, 3F, 3G, and 3H	Knock control system
Hard start or won't start (Cranks OK)	1	2	3	
Knocking				1

2.2L ENGINE EGR SYSTEM TROUBLESHOOTING IN ORDER OF PRIORITY—626 AND MX-6 TURBO

Note
- Make the system inspection first. If no problem is found, continue with the next system inspection of the Troubleshooting Guide.

Possible cause	Water thermo-sensor	EGR control valve	EGR position sensor	Solenoid valve		Throttle sensor	Vacuum chamber	Engine control unit terminal		System in-spection
				Vent	Vac.			2J	3O, 3P	
Checking order	7	3	4	2		8	6	9	10	1

2.2L ENGINE ENGINE CONTROL UNIT TERMINAL VOLTAGES—626 AND MX-6 TURBO

Terminal	Input	Output	Connected to	Voltage (After warming-up) IGN: ON	Voltage (After warming-up) Idle	Remark
1A	—	—	Battery	Battery voltage		For backup
1B	—	—	Main relay	Battery voltage		—
1C	○		Ign. switch [START] (MTX) Inhibitor switch (ATX)	Below 2.5V		While cranking: Battery voltage
1D		○	Self-Diagnosis Checker (Monitor lamp)	Test connect grounded • For 3 sec. after ign. switch OFF→ON: Below 6.2V (lamp illuminates) • After 3 sec.: Battery voltage (lamp does not illuminate)	Test connector not grounded • Lamp illuminates: Below 6.2V • Lamp does not illuminate: Battery voltage Test connector grounded: Approx. 5V	Using Self-Diagnosis Checker
1E		○	Malfunction indicator lamp (MIL)	• For 3 sec. after ign. switch OFF→ON: Below 4.8V (lamp illuminates) • After 3 sec.: Battery voltage (lamp does not illuminate)		• Test connector grounded • Lamp illuminates: Below 4.8V • Lamp does not illuminate: Battery voltage
1F		○	Self-Diagnosis Checker (Code number)	• For 3 sec. after ign. switch OFF→ON: Below 6.2V (Buzzer sounds) • After 3 sec.: Battery voltage (Buzzer does not sound)		• Using Self-Diagnosis Checker and test connector grounded • Buzzer sounds: Below 6.2V • Buzzer does not sound: Battery voltage
1G		○	Igniter	0V	Approx. 0.6—0.8V	—
1H	○		Headlight switch	• Headlight OFF: Below 1.5V • Headlight ON : Battery voltage		—
1I	○		Test connector	• Test connector grounded: Below 1.5V • Test connector not grounded: Above 10.5V		Test connector: 1-pin, Green connector
1J	○		Rear window defroster switch	• Switch OFF: Battery voltage • Switch ON : Below 1.5V		—
1K	—	—	—	—		—
1L		○	A/C relay	Battery voltage	• A/C switch ON: Below 2.5V • A/C switch OFF: Battery voltage	Blower motor ON
1M	○		Vehicle speed switch	Batter voltage		Above 130 mph (210 km/h): Below 1.0V
1N	○		Power steering pressure switch	Above 10.5V	• P/S ON: Below 1.5V • P/S OFF: Above 10.5V	—
1O			A/C switch	• A/C switch ON : Below 1.5V • A/C switch OFF: above 10.0V		Blower motor ON
1P	○		Blower fan switch	• Switch less than 2nd position: Battery voltage • Switch 3rd or 4th position: Below 1.5V		

2.2L ENGINE ENGINE CONTROL OPERATION CHART 626 AND MX-6 TURBO

OUTPUT DEVICES \ ENGINE CONDITIONS		CRANKING (COLD ENGINE)	WARMING UP (DURING IDLE)	MEDIUM LOAD COLD	MEDIUM LOAD WARM	ACCELE-RATION	HEAVY LOAD	DECELE-RATION	IDLE (THROTTLE VALVE FULLY CLOSED)	IGN: ON (ENGINE NOT RUNNING)	REMARKS
INJECTOR	INJECTION	Rich			Rich and lean	Rich		Fuel cut	Rich and lean	No injection	
	INJECTION TIMING	1 group (once per revolution)	2 group (once per two revolutions)						2 group (once per two revolutions)		Above 6,300 rpm: fuel cut
BAC VALVE	AIR VALVE	Open*			Close						*Coolant temp: below 50°C (122°F)
	ISC VALVE	Large amount of bypass air	Large amount of bypass air*	Small amount of bypass air						No bypass	*In extremely cold condition
SOLENOID VALVE (EGR Vent side)		OFF (Duty value 0%) (Atmospheric pressure to EGR valve) — EGR cut —			Operate [Duty values (EGR amount) change*]		OFF (Duty value 0%) (Atmospheric pressure to EGR valve) — EGR cut —			Does not operate	*Depends on engine condition
SOLENOID VALVE (EGR Vacuum side)		OFF (No vacuum to EGR valve)					OFF (No vacuum to EGR valve)				
SOLENOID VALVE (PURGE)		OFF (Purge cut)			Operate [Duty values (purge amount) change*]			OFF (Purge cut)			*Depends on engine condition
SOLENOID VALVE (PRESSURE REGULATOR CONTROL)		OFF (Vacuum to pressure regulator)							After starting: ON* (Vacuum cut)	OFF	*During hot start only
SOLENOID VALVE (WASTEGATE)		OFF (Boost pressure not released)		ON (Duty value 100%)		ON (Duty value changes)* (Boost pressure controlled)		OFF (Boost pressure not released)		OFF	*When knocking occurs
IGNITER (Ignition timing)		Fixed at BTDC 6°	Advanced: depends on engine condition			Retarded: depends on intensity of knocking		Advanced: depends on engine speed		—	

2.2L ENGINE ENGINE CONTROL UNIT TERMINAL VOLTAGES—626 AND MX-6 TURBO

Terminal	Input	Output	Connected to	Voltage (After warming-up) IGN: ON	Voltage (After warming-up) Idle	Remark
1Q	○		Stoplight switch (MTX)	• Brake pedal released: Below 3.6V • Brake pedal depressed: Above 10.0V		
			EC-AT control unit (ATX)	Below 1.5V		3-2, 2-1 shifting and throttle valve more than 5/8 open: Battery voltage
1R	○		Inhibitor switch through EC-AT unit	"N" or "P" range: below 2.5V Other ranges: Battery voltage		ATX
			Neutral and clutch switch	In-gear condition Clutch pedal depressed: below 0.5V Clutch pedal released: Battery voltage		MTX (Neutral: Below 0.5V)
1S	○		Daytime running light control unit (Canada only)	• Parking brake lever pulled up: Battery voltage • Parking brake lever released: Below 1.5V		—
1T	○		Idle switch	• Accelerator pedal released: Below 0.5V • Accelerator pedal depressed: Above 7.7V		—
1U	○		—	Constant below 1.5V		MTX
				Constant above 10.5V		ATX
1V	○		Igniter	Below 1.0V	0.1—1.8V	
2A	○		Airflow meter (Vc)	7—9V		—
2B	○		Airflow meter (Vs)	Approx. 1.7V	Approx. 3—5V	Increase engine speed: voltage increase
2C	○		Oxygen sensor	0V	0—1.0V	• Cold engine at idle: 0V • After warming-up Acceleration: 0.5V—1.0V Deceleration: 0—0.4V
2D	○		Electrical fan [Low] (No.1 water thermo-switch)	Battery voltage		Coolant temp.: Below 97°C (207°F)
				Below 1.5V		Coolant temp.: Above 97°C (207°F)
2E	○		Water thermosensor	0.3—0.6V		Coolant temp.: 20°C (68°F): Approx. 2.5V
2F	○		Throttle sensor	Accelerator pedal released: Approx. 0.5V (depends on 21 terminal voltage)		Throttle valve fully open: 4.3V
2G	○		Electrical fan [High] (No.2 water thermo-switch) (ATX)	Battery voltage		Coolant temp.: Below 108°C (226°F)
				Below 1.5V		Coolant temp.: Above 108°C (226°F)
2H	—	—	—	—		—
2I	—	—	Vref	4.5—5.5V		—
2J	○		ECR position sensor	0.25—0.95V		—
2K	○		Airflow meter (Intake air thermo-sensor)	Approx. 2.5V at 20C° (68°F)		—
2L	—	—	—	—		—
2M	○		Knock control unit	3.3—5.0V		Knocking: 1.3—2.6V
2N	—	—	—	—		—
2O		○	Solenoid valve (Purge control)	Battery voltage		—
2P	—	—	—	—		—
3A	—	—	Ground (E01)	0V		—

Terminal	Input	Output	Connected to	Voltage (After warming-up) IGN: ON	Voltage (After warming-up) Idle	Remark
3B	—	—	Ground (E02)	0V		—
3C	—	—	Ground (E1)	0V		—
3D	—	—	Ground (E2)	0V		—
3E	○		Distributor (Ne ⊕signal)	Approx. 0.6—0.8V		
3F	○		Distributor (Ne, G1, G2, ⊖ signal)	Approx. 0.6—0.8V		
3G	○		Distributor (G1 ⊕signal)	Approx. 0.6—0.8V		
3H	○		Distributor (G2 ⊕signal)	Approx. 0.6—0.8V		
3I	—	—	—	—		—
3J	—	—	—	—		—
3K	—	—	—	—		—
3L		○	Warning buzzer Overboost	Type A[*1]: Approx. 0.08V Type B[*2]: Approx. 12V		Buzzer sounds: Type A: 0V Type B: Below 1V
3M		○	Solenoid valve (Pressure regulator control)	For 120 sec. after ign. switch OFF → ON: Below 3.5V	For 120 sec. after staring: Below 3.5V	Coolant temp. above 70°C (158°F) and intake air temp. above 20°C (63°F)
3N	—	—	—	—		—
3O		○	Solenoid valve (EGR-vent side)	Battery voltage		
3P		○	Solenoid valve (EGR-vacuum side)	Battery voltage		• Voltages change depending on driving condition (EGR amount) • Cold engine: battery voltage Engine coolant temp-below 40°C (104°F)
3Q		○	ISC valve	Engine signal monitor green and red lamps flash		
3R		○	Solenoid valve (Waste gate)	Battery voltage		• Suddenly increase engine speed to above 4,500 rpm: Below 3.5V
3S	—	—	—	—		—
3T		○	Circuit opening relay	Battery voltage	Below 3.5V	
3U		○	Injector (No.1 and No.3)	Battery voltage	Battery voltage	• Engine signal monitor green and red lamps flash
3V		○	Injector (No.2 and No.4)			
3W	—	—	—	—		—
3X	—	—	—	—		—
3Y	—	—	—	—		—
3Z	—	—	—	—		—

*1 Type A: Manufacturer for CPU of body electrical system is "NEC".
*2 Type B: Manufacturer for CPU of body electrical system is "YAZAKI or U-shin".

3.0L ENGINE COMPONENT DESCRIPTIONS—1990–91 929 SOHC

No.	Component	Function	Remark
1	**Air cleaner**	Filters air entering throttle body	—
2	**Airflow meter**	Detects amount of intake air; sends signal to engine control unit	Intake air thermosensor and fuel pump switch included
3	**Atmospheric pressure sensor**	Detects atmospheric pressure; sends signal to engine control unit	Built-in engine control unit
4	**Air valve**	Supplies bypass air into dynamic chamber (engine cold)	• Engine speed increased to shorten warm-up period • Thermowax type • Installed in BAC valve
	BAC valve	Supplies bypass air into dynamic chamber	Consists of air valve and ISC valve
	ISC valve	Supplies bypass air into dynamic chamber	• Controlled by duty signal from engine control unit • Works independent of idle-up • Installed in BAC valve
5	**Brake light switch**	Detects braking (deceleration); sends signal to engine control unit	—
6	**Catalytic converter**	Reduces HC, CO, and NOx by chemical reaction	Monolith type
7	**Charcoal canister**	Stores fuel tank fumes (engine stopped)	—
8	**Check connectors**	For Self-Diagnosis Checker and fuel pump	• 6-pin connector (Green); Self-Diagnosis Checker • 2-pin connector (Yellow); Fuel pump
9	**Check valve**	Prevents misoperation of shutter valve or swirl control valve during heavy load operation	—
10	**Check-and-cut valve**	Controls pressure in fuel tank	—
11	**Circuit opening relay**	Voltage for fuel pump (engine running)	—
12	**Control unit (Engine control unit)**	Detects the following: 1. Engine speed 2. No.1 and No.4 piston TDC 3. Intake air amount 4. Engine coolant temperature 5. Atmospheric pressure 6. Throttle valve opening angle 7. Throttle valve fully closed 8. Air/fuel ratio (oxygen concentration) 9. In-gear condition 10. Intake air temperature 11. Engine air temperature 12. A/C operation 13. P/S operation 14. E/L operation 15. Cranking signal 16. Initial set signal 17. EGR operation signal	1. Ne rotor (in distributor) 2. G rotors (in distributor) 3. Airflow meter 4. Water thermosensor 5. Atmospheric pressure sensor 6. Throttle sensor 7. Idle switch (in throttle sensor) 8. Oxygen sensor 9. Inhibitor switch 10. Intake air thermosensor (in airflow meter) 11. Intake air thermosensor (on dynamic chamber) 12. A/C switch 13. P/S switch 14. E/L switches 15. Ignition switch 16. Test connector 17. EGR position sensor
		Controls operation of the following: 1. Fuel injection system 2. Idle speed control system 3. Pressure regulator control system 4. Variable resonance induction system 5. Triple induction control system 6. Ignition system (Electronic spark advance) 7. Purge control system 8. EGR control system 9. Fail-safe system 10. Switch monitor function	1. Injector 2. BAC valve 3. Solenoid valve (Pressure regulator control) 4. Solenoid valve (VRIS) 5. Solenoid valve (TICS) 6. Igniter 7. Solenoid valve (Purge control) 8. Solenoid valves (EGR) 9. Self-Diagnosis Checker and MIL 10. Monitor lamp (Self-Diagnosis Checker)

No.	Component	Function	Remark
13	**Distributor**	Detects crank angle at 30° intervals; No.1 cylinder TDC and No.4 cylinder TDC; sends signal to engine control unit	• Ne (Number of engine speed) signal • G1 (Group 1) signal • G2 (Group 2) signal
	G1 coil	Detects No.4 cylinder TDC; sends signal to engine control unit	For determining fuel injection timing
	G2 coil	Detects No.1 cylinder TDC; sends signal to engine control unit	For determining fuel injection timing
14	**Dashpot**	Causes gradual throttle valve closing during deceleration	Adjustment speed; 3,500 ± 300 rpm
15	**Dynamic chamber**	Interconnects all cylinders	Contains shutter valve
16	**EC-AT control unit**	Detect N or P range	—
17	**Exhaust gas recirculation (EGR) position sensor**	Detects EGR control valve opening; sends signal to engine control unit	Installed on EGR control valve
18	**Exhaust gas recirculation control valve**	Recirculates portion of exhaust gas	Contains EGR position sensor
19	**Fuel filter**	Filters fuel	—
20	**Fuel pump**	Provides fuel to injectors	• Operates only while engine running • Installed in fuel tank
21	**Idle switch**	Detects when throttle valve fully closed; sends signal to engine control unit	Installed in throttle sensor Switch closed when throttle valve fully closed
	Throttle sensor	Detects throttle valve opening angle; sends signal to engine control unit	• Idle switch included • Installed on throttle body
22	**Igniter**	Receives spark signal from engine control unit; generates high voltage in ignition coil	High voltage distributed by rotor in distributor
23	**Ignition switch**	Detects engine cranking; sends signal to engine control unit	—
24	**Inhibitor switch**	Detects in-gear condition; sends signal through EC-AT control unit to engine control unit	Switch closed in N or P range (A/T)
25	**Injector**	Injects fuel into intake port	• Controlled by signals from engine control unit • High-ohmic injector
26	**Intake manifold**	Supplies intake air to all cylinders	Contains swirl control valve
27	**Intake air thermosensor**	Detects intake air temperature; sends signal to engine control unit	Installed in airflow meter
		Detects engine air temperature; sends signal to engine control unit	Installed on dynamic chamber
28	**Main relay**	Supplies current to injectors and engine control unit	—
29	**MIL (Malfunction indicator light)**	Light illuminates when input device malfunctions	Installed in instrument cluster
30	**Oxygen sensor**	Detects oxygen concentration; sends signal to engine control unit	Zirconia ceramic and platinum coating
31	**PCV valve**	Controls blowby gas amount pulled into engine	—
32	**P/S switch**	Detects P/S operation; sends signal to engine control unit	ON when steering wheel turned right or left
33	**Pressure regulator**	Adjusts fuel pressure supplied to injectors	—
34	**Pulsation damper**	Absorbs fuel pulsation	—

3.0L ENGINE COMPONENT DESCRIPTIONS
1990–91 929 SOHC

No.		Component	Function	Remark
35		**Separator**	Prevents fuel from flowing into charcoal canister	—
36		**Shutter valve actuator**	Controls shutter valve operation	For variable resonance induction system (VRIS)
37		**Swirl control valve actuator**	Controls swirl control valve operation	For triple induction control system (TICS)
38	Solenoid valve	**EGR (Vent side, Vacuum side)**	Controls EGR control valve operation	• Exhaust gas recirculation • Duty solenoid valve
39	Solenoid valve	**Purge control**	Controls evaporative fumes from canister	Controlled by duty signal from ECU
40	Solenoid valve	**TICS**	Controls vacuum to swirl control valve actuator	Triple induction control system (TICS)
41	Solenoid valve	**VRIS**	Controls vacuum to shutter valve actuator	Variable resonance induction system (VRIS)
42	Solenoid valve	**Pressure regulator control**	Blocks vacuum from intake manifold to pressure regulator	Hot engine start only
43		**Test connector**	For Self-Diagnosis Checker and idle speed/ignition timing adjustment	1-pin connector (Green)
44		**Throttle body**	Controls intake air amount	Integrated throttle sensor and dashpot incorporated
45		**Water thermosensor**	Detects engine coolant temperature; sends signal to engine control unit	—

3.0L ENGINE SPECIFICATIONS CHART
1990–91 929 SOHC

Item			JE SOHC
Idle speed (with test connector grounded) rpm			650 ± 20 (A/T: P range)
Air cleaner	Element type		Wet
Throttle body	Throat diameter mm (in)		45 (1.772) x 2
Dashpot	Adjustment speed rpm		3,500 ± 300
Throttle sensor	Idle switch (adjustment clearance) mm (in)		0.5 (0.020) (Continuity) 0.7 (0.028) (No continuity)
Throttle sensor	Resistance kΩ		Ⓐ — Ⓓ 3.5—6.5 Ⓑ — Ⓓ below 1 ↔ 3.5—6.5
Airflow meter	Resistance Ω	E_2 ↔ Vs	20—400 (Measuring plate fully closed), 20—1,000 (fully open)
Airflow meter	Resistance Ω	E_2 ↔ Vc	100—300
Airflow meter	Resistance Ω	E_2 ↔ VB	200—400
Airflow meter	Resistance Ω	E_2 ↔ THA	−20°C (−4°F) 13,600—18,400 20°C (68°F) 2,210— 2,690 60°C (140°F) 493— 667
Airflow meter	Resistance Ω	E_1 ↔ Fc	Fully closed; ∞ Fully open: 0
Fuel tank	Capacity liters (US gal, Imp gal)		70 (18.5, 15.4)
Fuel filter	Low pressure		Nylon 6 (250 mesh)
Fuel filter	High pressure		Filter paper
Fuel pump	Type		Impeller (in tank)
Fuel pump	Fuel pressure kPa (kg/cm², psi)		441—588 (4.5—6.0, 64.0—85.3)
Fuel pump	Feeding capacity liters (US gal, Imp gal)/min		1.3—2.3 (0.34—0.61, 0.29—0.51)
Pressure regulator	Type		Diaphragm
Pressure regulator	Regulated pressure kPa (kg/cm², psi)		264—314 (2.7—3.2, 38.4—45.5)...Vacuum cut
Injector	Drive		Voltage
Injector	Type		High-ohmic injector
Injector	Resistance Ω		12—16
Injector	Injection amount cc (cu in)/15 sec		44—65 (2.68—3.97)
Ignition timing (with test connector grounded)			BTDC 15 ± 1°
BAC valve	Resistance Ω		10.8—12.3
Fuel specification			Unleaded gasoline

3.0L ENGINE TROUBLESHOOTING GUIDE—1990–91 929 SOHC

No.	Symptoms \ Possible cause		Input sensors: Ignition pulse	Distributor (Ne signal)	Distributor (G1 signal)	Distributor (G2 signal)	Airflow meter	Water thermosensor	Intake air thermosensor (Airflow meter)	Intake air thermosensor (Dynamic chamber)	Throttle sensor	Atmospheric pressure sensor	Oxygen sensor	EGR position sensor	Fuel feedback system	Output solenoid valves: Solenoid valve (Pressure regulator control)	Solenoid valve (Purge control)
1	Fault indicated by SST malfunction Code No.		01	02	03	04	08	09	10	11	12	14	15	16	17	25	26
2	Hard starts or won't start (cranks OK)																
3	Engine stalls	While warming up															
		After warming up															
4	Rough idle	While warming up															
		After warming up															
5	High speed idle after warming up																
6	Poor acceleration, hesitation, or lack of power																
7	Runs rough on deceleration																
8	Afterburn in exhaust system																
9	Poor fuel consumption																
10	Excessive oil consumption																
11	Engine stalls or runs rough after hot starting																
12	Fails emission test																

TROUBLESHOOTING PROCEDURE

Note
- **Symptom 1 shows the Code No. to quickly determine what system or unit may be at fault by use of the SST. (SST: Self-Diagnosis Checker 49 H018 9A1)**

1st. Check the electrical system
1) Battery condition
2) Main relays
3) Main fuses
4) Fuses

2nd. Check input sensors and output solenoid valves diagnosed with the **SST**. (49 H018 9A1)

3rd. Check input switches with **SST**. (49 H018 9A1)

4th. Check the following items:
Ignition system
1) Ignition spark
2) Ignition timing (with test connector grounded)
Intake air system
1) Air cleaner element
2) Vacuum or air leakage
3) Vacuum hose and air hose routing
4) Accelerator cable
Fuel system
1) Fuel amount
2) Fuel leakage
3) Fuel filter
4) Idle speed (with test connector grounded)
Engine
1) Compression
2) Overheating
Others
1) Clutch slippage
2) Brake dragging

5th. Check Fuel and Emission Control Systems.

Output solenoid valves: Solenoid valve (EGR, vacuum side)	Solenoid valve (EGR, vent side)	Idle speed control valve	Solenoid valve (TICS) or Oxygen sensor heater	Solenoid valve (VRIS)	Possible cause (Symptom / Page)	Fuel and Emission Control Systems: Intake air system (Poor connection of components, Throttle body)	Idle speed control system (Air valve, Idle speed control valve malfunction)	VRIS (Variable resonance induction system)	TICS (Triple induction control system)	Fuel system (Fuel injection, Fuel pressure)	Pressure regulator control system	Electronic spark advance control system	Deceleration control system (Fuel cut operation malfunction)	EGR system (EGR control valve open)	Evaporative emission control system	PCV system (System clogged)	Exhaust system (System clogged)
28	29	34	40	41													
					2	3				2		1		4			
					3	4	1			3				2			
					3	5	2			4				3		1	
					4	6	1			5				3	4	2	
					4	6	2			5				3	4	1	
					5	3	2			4			1				
					6	3		5	6	4				1	2		7
					7		2			3			1				
					8	3	2			4			1				
					9				2	4	5		1	3			6
					10											1	
					11		2			3	1						
					12	6	5			7	8		3	2	4		1

The numbers of the list show the priorities of inspections from the most possible to that with the lowest possibility.
These were determined on the following basis:
- Ease of inspection
- Most possible system
- Most possible point in system

Input Devices

Mal-function Code No.	Input devices	Malfunction	Fail-safe function	Pattern of output signals (Self-Diagnosis Checker or MIL)
01	Ignition pulse (Igniter, Ignition coil)	Broken wire, short circuit	—	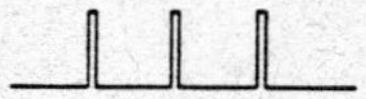
02	Distributor (Ne signal)	Ne signal not input for 1.5 sec. during cranking	—	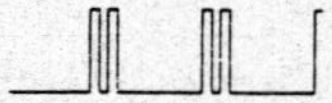
03	Distributor (G1 signal)	Broken wire, short circuit	—	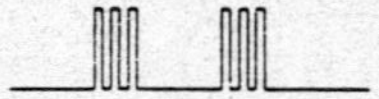
04	Distributor (G2 signal)	Broken wire, short circuit	—	
08	Airflow meter	Broken wire, short circuit	Basic fuel injection amount fixed as for 2 driving modes 1) Idle switch: ON 2) Idle switch: OFF	
09	Water thermosensor	Broken wire, short circuit	Coolant temp. input fixed at 80°C (176°F)	
10	Intake air thermosensor (Airflow meter)	Broken wire, short circuit	Intake air temp. input fixed at 20°C (68°F)	
11	Intake air thermosensor (Dynamic chamber)	Broken wire, short circuit	Intake air temp. input fixed at 20°C (68°F)	
12	Throttle sensor	Broken wire, short circuit	Throttle valve opening angle signal input fixed at fully open	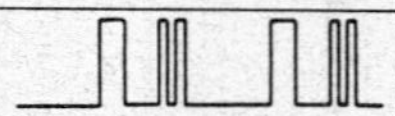
14	Atmospheric pressure sensor	ECU	Atmospheric pressure input fixed at 760 mmHg (29.9 in Hg)	
15	Oxygen sensor	Oxygen sensor output below 0.55V 120 sec. after engine at 1,500 rpm	Feedback system canceled (for EGI)	
16	EGR position sensor	Broken wire, short circuit	EGR position signal input fixed at fully closed	
		Sensor output does not match target value (incorrect output)	—	
17	Feedback system	Oxygen sensor output does not change from 0.55V 60 sec. after engine at 1,500 rpm	Feedback system canceled (for EGI)	

Output Devices

Malfunction code	Output devices	Pattern of output signals (Self-Diagnosis Checker or MIL)
25	Solenoid valve (Pressure regulator control)	
26	Solenoid valve (Purge control)	
28	Solenoid valve (EGR, vacuum side)	
29	Solenoid valve (EGR, vent side)	
34	Idle speed control valve (ISC valve)	
40	Solenoid valve (Triple induction control system) and oxygen sensor heater	
41	Solenoid valve (Variable resonance induction system)	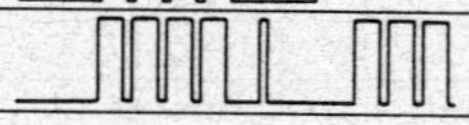

3.0L ENGINE TROUBLE CODE TROUBLESHOOTING—1990–91 929 SOHC

PC: Possible Cause

No.01 Code

- Are there poor connections at ignition coil connectors?
  - YES → Repair or replace connector
  - NO ↓
- Is resistance of ignition coil OK ?
  Resistance: Primary 0.72—0.88Ω
  Secondary 10—30 kΩ
  - NO → Replace ignition coil
  - YES ↓
- Is operation of igniter OK ?
  - NO → Replace igniter
  - YES ↓
- Is there continuity between IGf terminal of igniter connector (wiring harness side) and engine control unit (1G) terminal ?
  - NO → PC: Open circuit in wiring harness from igniter to engine control unit (1G) terminal
  - YES ↓
- Is engine control unit (1M) terminal voltage OK?
  - NO → PC: Short circuit wiring harness
  - YES → Replace engine control unit

No.02 Code

- Are there poor connections at distributor connectors?
  - YES → Repair or replace connector
  - NO ↓
- Is resistance of distributor OK?
  - NO → Replace distributor
  - YES ↓
- Is there continuity between distributor connector (wiring harness side) and engine control unit?

Distributor	Engine control unit
Ne ① W	2E
COM ② G	2F

  - NO → PC: Open circuit in wiring harness
  - YES ↓
- Are engine control unit (2E) and (2F) terminal voltages OK?
  - NO → PC: Short circuit in wiring harness
  - YES → Replace engine control unit

No.03 Code

PC: Possible Cause

- Are there poor connections at distributor connectors?
  - YES → Repair or replace connector
  - NO ↓
- Is resistance of distributor OK?
  - NO → Replace distributor
  - YES ↓
- Is there continuity between distributor connector (wiring harness side) and engine control unit?

Distributor	Engine control unit
G1 ① L	2G

  - NO → PC: Open circuit in wiring harness
  - YES ↓
- Are engine control unit (2G) terminal voltages OK?
  - NO → PC: Short circuit in wiring harness
  - YES → Replace engine control unit

No.04 Code

- Are there poor connections at distributor connectors?
  - YES → Repair or replace connector
  - NO ↓
- Is resistance of distributor OK?
  - NO → Replace distributor
  - YES ↓
- Is there continuity between distributor connector (wiring harness side) and engine control unit?

Distributor	Engine control unit
G2 ① R	2H

  - NO → PC: Open circuit in wiring harness
  - YES ↓
- Are engine control unit (2H) terminal voltages OK?
  - NO → PC: Short circuit in wiring harness
  - YES → Replace engine control unit

No.08 Code

PC: Possible Cause

Are there poor connections at airflow meter connectors
- YES → Repair or replace connector
- NO ↓

Is resistance of airflow meter OK?
Resistance:

Terminal	Fully closed (Ω)	Fully open (Ω)
E_2 ↔ Vs	20—400	20—1,000
E_2 ↔ Vc	100—300	
E_2 ↔ V_B	200—400	

- NO → Repair airflow meter
- YES ↓

Is there continuity between airflow meter connector (wiring harness side) and engine control unit?

Airflow meter	Control unit
V_B (B/W wire)	1B
Vc (BR/Y wire)	2J
Vs (Y/L wire)	2O
E_2 (R/B wire)	2D

- NO → PC: Open circuit in wiring harness from airflow meter to engine control unit
- YES ↓

Are engine control unit terminal voltages OK? Check terminal 2J, 2O, 2D, 1B
- NO → PC: Short circuit in wiring harness
- YES → Replace engine control unit

No.09 Code

PC: Possible Cause

Are there poor connections at water thermosensor connectors?
- YES → Repair or replace connectors
- NO ↓

Is resistance of water thermosensor OK?
Resistance:

Coolant temp	Resistance
−20°C (−4° F)	14.5—17.8 kΩ
20°C (68°F)	2.2—2.7 kΩ
40°C (104°F)	1.0—1.3 kΩ
60°C (140°F)	0.5—0.64 kΩ
80°C (176°F)	0.28—0.35 kΩ

- NO → Replace water thermosensor
- YES ↓

Is there continuity between water thermosensor connector (wiring harness side) and engine control unit?

Water thermosensor	Engine control unit
A (G/W wire)	2Q
B (R/B wire)	2D

- NO → PC: Open circuit in wiring harness from water thermosensor to engine control unit
- YES ↓

Are engine control unit (2Q) and (2D) terminal voltages OK?
- NO → PC: Short circuit in wiring harness
- YES → Replace engine control unit

No.10 Code

PC: Possible Cause

Are there poor connections at airflow meter connectors?
- YES → Repair or replace connector
- NO ↓

Is resistance of intake air thermosensor (in airflow meter) OK?
Resistance:

Terminal	Resistance
E_2↔THA_A	−20°C (−4°F): 13.6—18.4 kΩ 20°C (68°F): 2.21—2.69 kΩ 60°C (140°F): 0.493—0.667 kΩ

- NO → Replace airflow meter
- YES ↓

Is there continuity between intake air thermosensor (in airflow meter) connector (wiring harness side) and engine control unit?

Intake air thermosensor (in airflow meter)	Engine control unit
THA_A (G/O wire)	2P
E_2 (R/B wire)	2D

- NO → PC: Open circuit in wiring harness from intake air thermosensor (in airflow meter) to engine control unit
- YES ↓

Are engine control unit (2P) and (2D) terminal voltages OK?
- NO → PC: Short circuit in wiring harness
- YES → Replace engine control unit

No.11 Code

PC: Possible Cause

Are there poor connections at intake air thermosensor connectors?
- YES → Repair or replace connector
- NO ↓

Is resistance of intake air thermosensor (dynamic chamber) OK?
Resistance:

Temperature	Resistance
20°C (68°F)	37.3—45.6 kΩ
50°C (122°F)	10.5—12.9 kΩ
85°C (185°F)	3.1—3.8 kΩ

- NO → Replace intake air thermosensor (dynamic chamber)
- YES ↓

Is there continuity between intake air thermosensor (dynamic chamber) connector (wiring harness side) and engine control unit?

Intake air thermosensor (dynamic chamber)	Engine control unit
A (L/Y wire)	2L
B (R/B wire)	2D

- NO → PC: Open circuit in wiring harness from intake air thermosensor (dynamic chamber) to engine control unit
- YES ↓

Are engine control unit (2L) and (2D) terminal voltages OK?
- NO → PC: Short circuit in wiring harness
- YES → Replace engine control unit

3.0L ENGINE TROUBLE CODE TROUBLESHOOTING—1990–91 929 SOHC

No.12 Code

PC: Possible Cause

1. Are there poor connections at throttle sensor connectors?
   - YES → Repair or replace connector
   - NO → go to step 2
2. Is throttle sensor adjusted correctly?
   - NO → Adjust throttle sensor (for idle switch)
   - YES → go to step 3
3. Is resistance of throttle sensor OK?
   Resistance: (A) ↔ (D) 3.5—6.5 kΩ
   (B) ↔ (D) Below 1 kΩ .....Throttle valve fully closed
   Approx. 3.5—6.5 kΩ .....Throttle valve fully open
   - NO → Replace throttle sensor
   - YES → go to step 4
4. Is there continuity between throttle sensor connector (wiring harness side) and engine control unit?

Throttle sensor	Engine control unit
A (BR/W wire)	2K
B (R/W wire)	2M
D (R/B wire)	2D

   - NO → PC: Open circuit in wiring harness from throttle sensor to engine control unit
   - YES → go to step 5
5. Are engine control unit terminal voltages OK?
   Check terminal: 2K, 2M
   - NO → PC: Short circuit in wiring harness
   - (2K) terminal voltage not within specification → Replace engine control unit
   - YES → Replace engine control unit

No.14 Code

Replace engine control unit

No.15 Code

PC: Possible Cause

Note
- **When the malfunction Code Nos 15 and 17 are present at the same time, first perform the checking procedure for the malfunction Code No.17.**

1. Are there any poor connections at oxygen sensor connector?
   - YES → Repair or replace connector
   - NO → go to step 2
2. Is operation of oxygen sensor heater OK?
   - NO → Repair or replace related parts
   - YES → go to step 3
3. Is oxygen sensor output voltage OK?
   - NO → Replace oxygen sensor
   - YES → go to step 4
4. Is sensitivity of oxygen sensor OK?
   - NO → PC: • Oxygen sensor malfunction • Engine control unit malfunction
   - YES → go to step 5
5. Is there continuity between oxygen sensor connector (wiring harness side) and engine control unit (2N) terminal?
   - NO → PC: Open circuit in wiring harness from oxygen sensor to engine control unit (2N) terminal
   - YES → go to step 6
6. Is (2N) terminal voltage of engine control unit OK?
   - NO → PC: Short circuit in wiring harness between oxygen sensor and engine control unit (2N) terminal
   - YES → Replace engine control unit

No.16 Code

1. Are there any poor connections at EGR position sensor connector?
   - YES → Repair or replace connector
   - NO → go to step 2
2. Is EGR control valve OK?
   - NO → Replace EGR control valve
   - YES → go to step 3
3. Is resistance of EGR position sensor OK?
   Resistance: (A) ↔ (B) Approx. 5 kΩ
   (A) ↔ (C) 0.7—5 kΩ
   (B) ↔ (C) 0.7—5 kΩ
   - NO → Replace EGR control valve
   - YES → go to step 4
4. Is there continuity between EGR position sensor connector (wiring harness side) and engine control unit?

EGR position sensor	Engine control unit
A (Y)	1M
B (B/W)	2D
C (B/L)	2K

   - NO → PC: Open circuit in wiring harness from EGR position sensor to engine control unit
   - YES → Cont'd

Cont'd

Check	Result	Action
Is there 4.5—5.5V at (A) terminal of EGR position sensor connector?	NO	PC: • Short circuit in wiring harness from EGR position sensor (A) terminal to engine control unit (1M) terminal • Engine control unit malfunction
	YES	Go to next step
Is engine control unit (2K) terminal voltage OK?	NO	PC: Short circuit in wiring harness EGR position sensor (C) terminal to engine control unit (2K) terminal
	YES	Replace engine control unit

No.17 Code

PC: Possible Cause

Check	Result	Action
Warm up engine and run it at 2,500—3,000 rpm for 3 minutes		Go to next step
Does monitor lamp of **SST** illuminate on at idle without test connector (Green: 1-pin) grounded? **SST:** Self-Diagnosis Checker	NO	PC: • Air leak in vacuum hoses or intake air system • Contaminated oxygen sensor • Insufficient fuel injection
	YES	Go to next step
Are spark plugs clean?	NO	Clean or replace spark plugs
	YES	Go to next step
Is oxygen sensor operation OK?	NO	Replace oxygen sensor
	YES	Go to next step
Is same Code No. present after performing after-repair procedure?	NO	Feedback system OK
	YES	Go to next step
Is (2N) terminal voltage of engine control unit OK?	NO	PC: Short circuit in wiring harness from oxygen sensor connector to engine control unit
	YES	Replace engine control unit

No.25 Code

PC: Possible Cause

Check	Result	Action
Are there poor connections at solenoid valve (PRC) connectors?	YES	Repair or replace connector
	NO	Go to next step
Is there continuity of solenoid valve (PRC)?	NO	Replace solenoid valve
	YES	Go to next step
Is there battery voltage at (B/W) wire of solenoid valve (PRC) connector?	NO	PC: Open or short circuit in wiring harness from (B/W) wire to main relay (for engine control unit)
	YES	Go to next step
Is there continuity between solenoid valve (PRC) connector (wiring harness side) and engine control unit? (see table below)	NO	PC: Open circuit in wiring harness from solenoid valve (PRC) to engine control unit
	YES	PC: • Engine control unit malfunction • Short circuit in wiring harness

Solenoid valve (PRC)	Engine control unit
B (L/B wire)	2T

No.26 Code

Check	Result	Action
Are there poor connections at solenoid valve (purge control) connectors?	YES	Repair or replace connector
	NO	Go to next step
Is there continuity of solenoid valve (Purge control)?	NO	Replace solenoid valve
	YES	Go to next step
Is there battery voltage at (B/W) wire of solenoid valve (Purge control) connector?	NO	PC: Open or short circuit in wiring harness from (B/W) wire to main relay (for engine control unit)
	YES	Go to next step
Is there continuity between solenoid valve (Purge control) connector (wiring harness side) and control unit? (see table below)	NO	PC: Open circuit in wiring harness from solenoid valve (Purge control) to engine control unit
	YES	PC: • Engine control unit malfunction • Short circuit in wiring harness

Solenoid valve	Engine control unit
B (BR/L wire)	1H

3.0L ENGINE TROUBLE CODE TROUBLESHOOTING—1990–91 929 SOHC

PC: Possible Cause

No.28 Code

- Are there any poor connections at solenoid valve (EGR, vacuum) connectors?
  - YES → Repair or replace connector
  - NO ↓
- Is there continuity of solenoid valve (EGR, vacuum)?
  - NO → Replace solenoid valve
  - YES ↓
- Is there battery voltage at (B/W) wire of solenoid valve (EGR, vacuum) connector?
  - NO → PC: Open or short circuit in wiring harness from (B/W) wire to main relay (for engine control unit)
  - YES ↓
- Is there continuity between solenoid valve (EGR, vacuum) connector (wiring harness side) and engine control unit?

Solenoid valve	Engine control unit
C (Y/R wire)	2Z

  - NO → PC: Open circuit in wiring harness from solenoid valve (EGR, vacuum) to engine control unit
  - YES → PC: • Engine control unit malfunction • Short circuit in wiring harness

No.29 Code

- Are there any poor connections at solenoid valve (EGR, vent) connectors?
  - YES → Repair or replace connector
  - NO ↓
- Is there continuity of solenoid valve (EGR, vent)?
  - NO → Replace solenoid valve
  - YES ↓
- Is there battery voltage at (B/W) wire of solenoid valve (EGR, vent) connector?
  - NO → PC: Open or short circuit in wiring harness from (B/W) wire to main relay (for engine control unit)
  - YES ↓
- Is there continuity between solenoid valve (EGR, vent) connector (wiring harness side) and engine control unit?

Solenoid valve	Engine control unit
D (V/W wire)	2Y

  - NO → PC: Open circuit in wiring harness from solenoid valve (EGR, vent) to engine control unit
  - YES → PC: • Engine control unit malfunction • Short circuit in wiring harness

No.34 Code

- Are there poor connectors at BAC valve connectors?
  - YES → Repair or replace connector
  - NO ↓
- Is resistance of BAC valve OK? Resistance: 10.8—12.3Ω
  - NO → Replace BAC valve
  - YES ↓
- Is there battery voltage at (B/W) wire of BAC valve connector?
  - NO → PC: Open or short circuit in wiring harness from (B/W) wire to main relay (for engine control unit)
  - YES ↓
- Is there continuity between BAC valve connector (wiring harness side) and engine control unit?

BAC valve	Engine control unit
B (L/O wire)	2W

  - NO → PC: Open circuit in wiring harness from BAC valve to engine control unit
  - YES → PC: • Engine control unit malfunction • Short circuit in wiring harness

No.40 Code
For solenoid valve (TICS)

- Are there any poor connections at solenoid valve (TICS) connectors?
  - YES → Repair or replace connector
  - NO ↓
- Is there continuity of solenoid valve (TICS)?
  - NO → Replace solenoid valve
  - YES ↓
- Is there battery voltage at (B/W) wire of solenoid valve (TICS) connector?
  - NO → PC: Open or short circuit in wiring harness from (B/W) wire to main relay (for engine control unit)
  - YES ↓
- Is there continuity between solenoid valve (TICS) connector (wiring harness side) and engine control unit?

Solenoid valve (TICS)	Engine control unit
B (L/R wire)	2R

  - NO → PC: Open circuit in wiring harness from solenoid valve (TICS) to engine control unit
  - YES → PC: • Engine control unit malfunction • Short circuit in wiring harness

3.0L ENGINE TROUBLE CODE TROUBLESHOOTING 1990–91 929 SOHC

For oxygen sensor heater

PC: Possible Cause

1. Are there any poor connections at oxygen sensor heater connectors?
   - YES → Repair or replace connector
   - NO → go to next step
2. Is oxygen sensor heater operation OK?
   - NO → Replace oxygen sensor
   - YES → go to next step
3. Is there continuity between oxygen sensor heater connector and main relay?

Oxygen sensor	Main relay
C (B/W wire)	D (B/W wire)

   - NO → PC: Open circuit in wiring harness from oxygen sensor to main relay
   - YES → go to next step
4. Is there continuity between oxygen sensor heater connector and engine control unit?

Oxygen sensor	Engine control unit
B (L/R wire)	2R

   - NO → PC: Open circuit in wiring harness from oxygen sensor to engine control unit
   - YES → PC: • Engine control unit malfunction • Short circuit in wiring harness

No.41 Code

1. Are there any poor connections at solenoid valve (VRIS) connectors?
   - YES → Repair or replace connector
   - NO → go to next step
2. Is there continuity of solenoid valve (VRIS)?
   - NO → Replace solenoid valve
   - YES → go to next step
3. Is there battery voltage at (B/W) wire of solenoid valve (VRIS) connector?
   - NO → PC: Open or short circuit in wiring harness from (B/W) wire to main relay (for engine control unit)
   - YES → go to next step
4. Is there continuity between solenoid valve (VRIS) connector (wiring harness side) and engine control unit?

Solenoid valve (VRIS)	Engine control unit
B (V wire)	2S

   - NO → PC: Open circuit in wiring harness from solenoid valve (VRIS) to engine control unit
   - YES → PC: • Engine control unit malfunction • Short circuit in wiring harness

3.0L ENGINE SWITCH MONITOR FUNCTION 1990–91 929 SOHC

MONITOR SWITCH FUNCTION

The operation of individual switches can be checked by the monitor lamp of the Self-Diagnosis Checker.

Note
- **The test connector must be grounded and the ignition switch ON (engine stopped).**

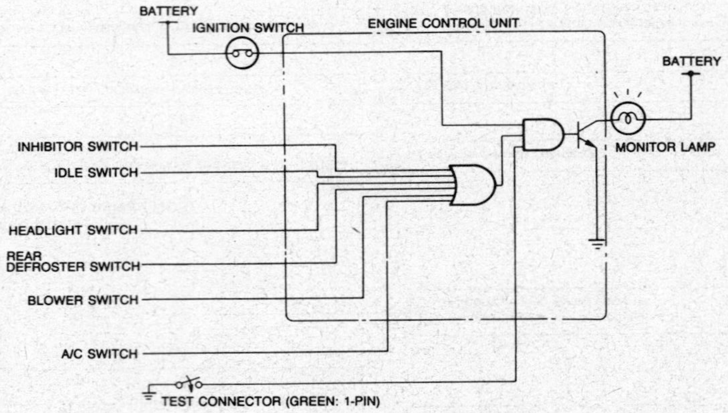

Switch	Self-Diagnosis Checker (Monitor Lamp)		Remark
	Light ON	Light OFF	
Idle switch	Pedal depressed	Pedal released	
A/C switch	ON	OFF	Fan speed control: 1st—4th position
Headlight switch	ON	OFF	
Rear defroster switch	ON	OFF	
Blower switch	ON	OFF	Fan speed control: 5th—8th position
Inhibitor switch	Except "N" and "P" range	"N", "P" range	

Caution
- **If any of the switches remains activated, the monitor lamp will stay on.**

3.0L ENGINE SWITCH MONITOR FUNCTION TROUBLESHOOTING—1990–91 929 SOHC

PC: Possible Cause

Procedure

Set conditions to deactivate each switch
- All accessories OFF
- Transmission in neutral
- All pedals released

Verify that monitor lamp does not illuminate

YES → Check each switch as described

NO → Check each switch and related wiring harness
- Idle switch
- A/C switch
- Headlight switch
- Rear defroster switch
- Blower switch
- Inhibitor switch

Idle switch

Depress accelerator pedal, and check that monitor lamp illuminates

YES → Idle switch OK
Release accelerator pedal

NO → PC:
- Maladjustment of idle switch
- Idle switch malfunction
- Short circuit in wiring harness from idle switch to engine control unit
- Engine control unit (1N) terminal malfunction

Blower switch

Set fan speed control to "5th—8th" position, and check that monitor lamp illuminates

YES → Blower switch OK

NO → PC:
- Fan speed control switch malfunction
- Open circuit in wiring harness from fan speed control switch to engine control unit

A/C switch

Set fan speed control to 1st position
Turn ON A/C switch, and check that monitor lamp illuminates

YES → A/C switch OK
Turn off A/C switch and blower switch

NO → PC:
- A/C switch malfunction
- Fan speed control switch malfunction
- Open circuit in wiring harness from A/C switch to engine control unit

PC: Possible Cause

Headlight switch

Turn ON headlight switch, and check that monitor lamp illuminates

YES → Headlight switch OK
Turn OFF headlight switch

NO → PC:
- Headlight switch malfunction
- Open circuit in wiring harness from headlight switch to engine control unit

Rear defroster switch

Turn ON rear defroster switch, and check that monitor lamp illuminates

YES → Rear defroster switch OK
Turn OFF rear defroster switch

NO → PC:
- Rear defroster switch malfunction
- Open circuit in wiring harness from rear defroster switch to engine control unit

Inhibitor switch

Shift transmission into gear
Check that monitor lamp illuminates

YES → Inhibitor switch OK
Return transmission to neutral

NO → PC:
- EC-AT control unit or Inhibitor switch malfunction
- Open or short circuit in related wiring harness

3.0L ENGINE ENGINE CONTROL OPERATIONS CHART 1990–91 929 SOHC

<table>
<tr><th colspan="2" rowspan="2">ENGINE CONDITION / OUTPUT DEVICE</th><th rowspan="2">CRANKING (COLD ENGINE)</th><th rowspan="2">WARM-UP (DURING IDLE)</th><th colspan="2">MEDIUM LOAD</th><th rowspan="2">ACCEL-ERATION</th><th rowspan="2">HEAVY LOAD</th><th rowspan="2">DECEL-ERATION</th><th rowspan="2">IDLE (THROTTLE VALVE FULLY CLOSED)</th><th rowspan="2">IG: ON (ENGINE NOT RUNNING)</th><th rowspan="2">REMARKS</th></tr>
<tr><th>COLD</th><th>WARM</th></tr>
<tr><td rowspan="4">INJECTOR</td><td>No.1, No.2</td><td colspan="3" rowspan="3">Rich</td><td rowspan="3">Rich and lean (Fuel feedback operation) *1</td><td colspan="2" rowspan="3">Rich</td><td rowspan="4">Fuel cut</td><td rowspan="3">Rich and lean (Fuel feedback operation) *1</td><td rowspan="4">Does not inject</td><td rowspan="3">*1 Operated by oxygen sensor</td></tr>
<tr><td>No.3, No.4</td></tr>
<tr><td>No.5, No.6</td></tr>
<tr><td>INJECTION TIMING</td><td>1 group</td><td colspan="5">3 group</td><td>3 group</td><td>Fuel cut: above 6,500 rpm</td></tr>
<tr><td colspan="2">SOLENOID VALVE (PRC)</td><td colspan="7">OFF (Vacuum to pressure regulator)</td><td>After start ON (Vacuum cut)*2</td><td>Does not operate</td><td>*2 During hot condition only</td></tr>
<tr><td rowspan="2">BAC VALVE</td><td>AIR VALVE</td><td colspan="3">Open*3</td><td colspan="6">Close</td><td>*3 Coolant temp.: below 60°C (140°F)</td></tr>
<tr><td>ISC VALVE</td><td colspan="2">Large amount of bypass air*4</td><td colspan="5">Large and small amount of bypass air</td><td>Large and small amount of bypass air *4</td><td>Does not operate</td><td>*4 Engine speed feedback system operates when the test connector open</td></tr>
<tr><td colspan="2">SOLENOID VALVE (VRIS)</td><td colspan="2">OFF (Shutter valve closed)</td><td colspan="4">ON*5 (Shutter valve open)</td><td colspan="2">OFF (Shutter valve closed)</td><td>Does not operate</td><td>*5 ON: approx. 3,700—5,750 rpm</td></tr>
<tr><td colspan="2">SOLENOID VALVE (TICS)</td><td colspan="2">ON (Swirl control valve closed)</td><td colspan="4">OFF (Swirl control valve open)</td><td colspan="2">ON (Swirl control valve closed)</td><td>Does not operate</td><td></td></tr>
<tr><td colspan="2">IGNITER</td><td>BTDC 7°</td><td>Max.; BTDC 42° Min.; BTDC 11°</td><td>Advance</td><td colspan="3">Max.; BTDC 42° Min.; BTDC −6°</td><td>Gradually returned to initial timing</td><td>Max.; BTDC 32° Min.; BTDC 11°</td><td>Does not operate</td><td></td></tr>
<tr><td colspan="2">SOLENOID VALVE (PURGE CONTROL)</td><td colspan="3">OFF (Purge cut)</td><td colspan="3">Operate [Duty values (purge amount) change]*6</td><td colspan="3">OFF (Purge cut)</td><td>*6 Depends on engine condition</td></tr>
<tr><td rowspan="2">EGR CONTROL</td><td>SOLENOID VALVE (VENT SIDE)</td><td colspan="3">OFF (No vacuum to EGR valve)
— EGR cut —</td><td colspan="3" rowspan="2">— System operates*7 —
(Amount of EGR changes)</td><td colspan="2">OFF (No vacuum to EGR valve)
— EGR cut —</td><td rowspan="2">Does not operate</td><td rowspan="2">*7 Depends on engine condition</td></tr>
<tr><td>SOLENOID VALVE (VACUUM SIDE)</td><td colspan="3">OFF (Duty value 100%) (Atmospheric pressure to EGR valve)</td><td colspan="2">OFF (Duty value 100%) (Atmospheric pressure to EGR valve)</td></tr>
</table>

3.0L ENGINE RELATIONSHIPS CHART 1990–91 929 SOHC

○: Related
X: Not related

OUTPUT DEVICE / INPUT DEVICE	P/S SWITCH	A/C SWITCH	INHIBITOR SWITCH	TEST CONNECTOR	EGR POSITION SENSOR	ATMOSPHERIC PRESSURE SENSOR	OXYGEN SENSOR	INTAKE AIR THERMOSENSOR: DYNAMIC CHAMBER	INTAKE AIR THERMOSENSOR: AIRFLOW METER	WATER THERMOSENSOR	THROTTLE SENSOR: THROTTLE SENSOR	THROTTLE SENSOR: IDLE SWITCH	AIRFLOW METER	DISTRIBUTOR: G ROTOR	DISTRIBUTOR: Ne ROTOR	IG COIL
EGR: VACUUM SIDE	X	X	X	X	○	X	X	X	X	○	○	○	○	X	○	X
EGR: VENT SIDE	X	X	X	X	○	X	X	X	X	○	○	○	○	X	○	X
SOLENOID VALVE (PURGE CONTROL)	X	○	○	X	X	○	○	X	X	○	X	○	○	X	○	○
IGNITER	X	○	○	○	○	X	X	X	X	○	○	○	○	○	○	X
SOLENOID VALVE (TICS)	X	X	X	X	X	X	X	X	X	X	X	X	○	X	○	X
SOLENOID VALVE (VRIS)	X	X	X	X	X	X	X	X	X	X	X	X	○	X	○	X
BAC VALVE: ISC VALVE	○	○	○	○	X	X	X	○	○	○	○	○	X	X	○	X
BAC VALVE: AIR VALVE	X	X	X	X	X	X	X	X	X	X	X	X	X	X	X	X
PRESSURE REGULATOR	X	X	X	X	X	X	X	○	X	○	○	X	X	X	○	X
INJECTOR: FUEL INJECTION TIMING	X	X	X	X	X	X	X	X	X	X	○	○	X	○	○	X
INJECTOR: FUEL INJECTION AMOUNT	X	○	○	X	X	○	○	X	○	○	○	○	○	X	○	X

3.0L ENGINE ENGINE CONTROL UNIT TERMINAL VOLTAGES—1990–91 929 SOHC

Terminal	Input	Output	Connection to	Voltage (after warm-up) Ignition switch: ON	Voltage (after warm-up) Idle	Remarks
1A	—	—	Battery	Approx. 12V		For back-up
1B	○		Main relay	Approx. 12V		
1C	○		Ignition switch (START position)	Below 2.5V		While cranking 6—11V
1D		○	Self-Diagnosis Checker (Monitor lamp)	<Test connector grounded> For 3 sec after ignition switch OFF→ON: Approx. 5V (lamp illuminates) After 3 sec: Approx. 12V	<Test connector grounded> Approx. 5V <Test connector is not grounded> Monitor lamp ON: Below 6V Monitor lamp OFF: Approx. 12V	With Self-Diagnosis Checker
1E		○	MIL (Malfunction Indicator Light)	For 3 sec. after ignition switch OFF→ON: Below 2.5V (light illuminates) After 3 sec: Approx. 12V		• Test connector grounded • Light illuminates: Below 2.5V • Light does not illuminates: Approx. 12V
1F		○	Self-Diagnosis Checker (for Code No.)	For 3 sec. after ignition switch OFF→OFF: Below 2.5V (Buzzer sounds) After 3 sec: Approx. 12V (Buzzer does not sound)		• With Self-Diagnosis Checker and Test connector grounded • Buzzer sounds: Below 2.5V • Buzzer does not sound: Approx. 12V
1G		○	Igniter	0V	Approx. 0.8V	
1H		○	Solenoid valve (Purge control)	Approx. 12V		Driving in-gear 5—15V (Engine Signal Monitor: green and red lights flash)
1I	—	—	—	For 180 sec after ignition switch OFF →ON: Approx. 12V	For 180 sec after starting: Approx. 12V	During hot condition: Coolant temp.: above 90°C (194°F) and engine air temp.: above 85°C (185°F)
				Below 2.5V		Other condition
1J		○	A/C relay	A/C switch ON: Below 2.5V A/C switch OFF: Approx. 12V		Blower motor: ON
1K	○		Test connector	Test connector: Grounded 0V Test connector: Not grounded approx. 12V		Green connector: 1-pin
1L	○		—	Below 1.5V		M/T
				Approx. 12V		EC-AT
1M	○		EGR position sensor	Approx. 0.25—0.9V		
1N	○		Idle switch	Pedal depressed: Approx. 10V Pedal released: 0V		
1O	—	—	—	—		—
1P	○		Power steering (P/S) switch	Constant approx. 12V	P/S: ON: 0V P/S: OFF: Approx. 12V	
1Q	○		A/C switch	A/C switch ON: Below 2.5V A/C switch OFF: Approx. 12V		Blower motor: ON
1R	—	—	—	—		—
1S	○		Blow motor switch	Blow motor switch ON: Below 2.5V Blow motor switch OFF: Approx. 10—12V		

Terminal	Input	Output	Connection to	Voltage (after warm-up) Ignition switch: ON	Voltage (after warm-up) Idle	Remark
1T	○		Rear defroster switch	Rear defroster switch ON: Below 2.5V Rear defroster switch OFF: Approx. 10—12V		
1U	○		Headlight switch	Headlight switch ON: below 2.5V Headlight switch OFF: Approx. 10—12V		
1V	○		Neutral/clutch switch	Clutch pedal depressed: Approx. 12V Clutch pedal released: 0V		Neutral condition
2A	—	—	Ground (E_{01})	0V		
2B	—	—	Ground (E_{02})	0V		
2C	—	—	Ground (E_1)	0V		—
2D	—	—	Ground (E_2)	0V		
2E	○		Distributor (Ne signal)	Approx. 0.6—0.8V		Ne: Number of engine speed
2F	○		Distributor (COM)	Approx. 0.6—0.8V		
2G	○		Distributor (G_1 signal)	Approx. 0.6—0.8V		G_1: Group 1
2H	○		Distributor (G_2 signal)	Approx. 0.6—0.8V		G_2: Group 2
2I	○		Igniter	0V	Approx. 0.6—0.8V	Engine Signal Monitor green and red lights flash
2J	○		Airflow meter (Vc)	7—9V		
2K		○	Vref	4.5—5.5V		
2L	○		Intake air thermosensor (Dynamic chamber)	Approx. 1—2V at 80°C (176°F)		
2M	○		Throttle sensor	Throttle valve fully closed: Approx. 0.3V Throttle valve fully open: Approx. 4V		
2N	○		Oxygen sensor	0V	0—1V	• Cold engine 0V at idle • After warming-up: Acceleration: 0.7—1.0V Deceleration: 0—0.2V
2O	○		Airflow meter (Vs)	Approx. 2.0V	Approx. 3.5—5V	
2P	○		Intake air thermosensor (airflow meter)	Approx. 2—3V at 20°C (68°F)		
2Q	○		Water thermosensor	Approx. 0.3—0.6V		
2R		○	Solenoid valve (Trip induction control system (TICS)) and oxygen sensor heater	Below 2.5V		Above approx. 4,000 rpm: Approx. 12V
2S		○	Solenoid valve (Variable resonance induction system (VRIS))	Approx. 12V		• 3,700—6,500 rpm: Below 2.0V • Other conditions: Approx. 12V
2T		○	Solenoid valve (Pressure regulator control)	For 240 sec after ignition switch OFF→ON Below 2.5V	For 240 sec after starting: Below 2.5V	During hot condition: coolant temp.: Above 80°C (176°F) and engine air temp.: above 75°C (167°F)
				Approx. 12V		Other conditions
2U		○	Injector (No.1 and No.2)	Approx. 12V		Engine Signal Monitor green and red lights flash

3.0L ENGINE SPECIFICATIONS CHART 1990–91 929 DOHC

Item			JE DOHC
Idle speed (Test connector grounded) rpm			700 ± 20 (P range)
Ignition timing (Test connector grounded)			BTDC 8 ± 1°
Air cleaner	Element type		Oil permeated
Throttle body	Throat diameter mm (in)		46 (1.811) x 2
Throttle sensor	Idle switch (adjustment clearance) mm (in)		0.1 (0.004) (Continuity) 0.3 (0.012) (No continuity)
	Resistance kΩ	Vref ↔ GND	3–7
		TVO ↔ GND	0.2–6.0 (Fully closed) 3.3–7.0 (Fully open)
Airflow meter	Resistance Ω	E_2 ↔ Vs	20–600 (Measuring plate fully closed) 20–1,000 (Fully open)
		E_2 ↔ Vc	200–400
		E_1 ↔ Fc	0 (Fully open) ∞ (Fully closed)
		E_2 ↔ THA	−20°C (−4°F) 13,600–18,400 20°C (68°F) 2,210–2,690 60°C (140°F) 493–667
Fuel tank	Capacity liters (US gal, Imp gal)		70 (18.5)
Fuel filter	Low-pressure		Nylon 6 (250 mesh)
	High-pressure		Filter paper
Fuel pump	Type		Impeller (in tank)
	Fuel pressure kPa (kg/cm², psi)		441–588 (4.5–6.0, 64.0–85.3)
	Feeding capacity liters (US gal, Imp gal)/min.		1.3–2.3 (0.34–0.61, 0.29–0.51)
Pressure regulator	Type		Diaphragm
	Fuel pressure kPa (kg/cm², psi)		264–314 (2.7–3.2, 38.4–45.5)... Vacuum cut
Injector	Drive		Voltage
	Type		High-ohmic injector
	Resistance Ω		12–16
	Injection amount cc (cu in)/15 sec		60–65 (3.66–3.97)
Cold start injector	Resistance Ω		3–5
	Injection amount cc (cu in)/15 sec		32–40 (1.95–2.44)
BAC valve	Resistance Ω		10–13
Solenoid valve	Purge control	Resistance Ω	29–35
	EGR vacuum side	Resistance Ω	29–35
	EGR vent side	Resistance Ω	29–35
Fuel specification			Unleaded gasoline (RON 91 or higher)

3.0L ENGINE ENGINE CONTROL UNIT TERMINAL VOLTAGES – 1990–91 929 SOHC

Terminal	Input	Output	Connection to	Voltage (after warm-up) Ignition switch: ON	Voltage (after warm-up) Idle	Remarks
2V		○	Injector (No.3 and No.4)	Approx. 12V		Engine Signal Monitor green and red lights flash
2W		○	Idle speed control (ISC) valve	Approx. 2–5V	Approx. 9–11V	Engine Signal Monitor green and red lights flash
2X		○	Injector (No.5 and No.6)	Approx. 12V		Engine Signal Monitor green and red lights flash
2Y		○	Solenoid valve (EGR, vent side)	Approx. 12V		Initial acceleration: Engine Signal Monitor green and red lights flash
2Z		○	Solenoid valve (EGR, vacuum side)	Approx. 12V		Initial acceleration: Engine Signal Monitor green and red lights flash

Control unit connector (Control unit side)

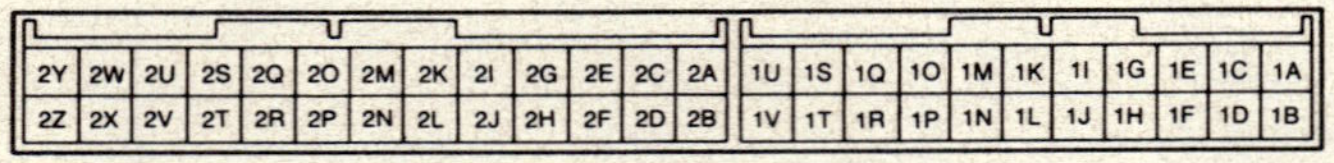

3.0L ENGINE COMPONENT DESCRIPTIONS—1990–91 929 DOHC

Component	Function	Remark
Air cleaner	Filters air entering throttle body	—
Airflow meter	Detects amount of intake air; sends signal to engine control unit	Intake air thermosensor and fuel pump switch included
Air valve	Supplies bypass air into dynamic chamber (engine cold)	• Engine speed increased to shorten warm-up period • Thermowax type • Installed in BAC valve
Bypass air control (BAC) valve	Supplies bypass air into dynamic chamber	Consists of air valve and ISC valve
Idle speed control (ISC) valve	Supplies bypass air into dynamic chamber	• Controlled by duty signal from engine control unit • Works independent of idle-up • Installed in BAC valve
Brake light switch	Detects braking; sends signal to engine control unit	—
Catalytic converter	Reduces HC, CO, and NOx by chemical reaction	Monolith type
Charcoal canister	Stores fuel tank fumes (engine stopped)	—
Check connectors	For Self-Diagnosis Checker and fuel pump	• 6-pin connector (Green); Self-Diagnosis Checker • 2-pin connector (Yellow); Fuel pump
Check-and-cut valve	Controls pressure in fuel tank	—
Circuit opening relay	Voltage for fuel pump (engine running)	—
Engine control unit (ECU)	Detects the following: 1. Engine speed 2. No.1 and No.4 piston TDC 3. Intake air amount 4. Engine coolant temperature 5. Atmospheric pressure 6. Throttle valve opening angle 7. Throttle valve fully closed 8. Air/fuel ratio (oxygen concentration) 9. In-gear condition 10. Intake air temperature 11. Engine air temperature 12. A/C operation 13. P/S operation 14. E/L operation 15. Cranking signal 16. Initial set signal 17. Braking signal 18. EGR operation signal 19. Knocking signal Controls operation of the following: 1. Fuel injection system 2. Idle speed control system 3. Pressure regulator control system 4. Variable resonance induction system 5. Ignition system (Electronic spark advance) 6. Purge control system 7. EGR control system 8 Fail-safe system 9. Switch monitor function	1. Ne rotor (in distributor) 2. G rotor (in distributor) 3. Airflow meter 4. Water thermosensor 5. Atmospheric pressure sensor (in ECU) 6. Throttle sensor 7. Idle switch (in throttle sensor) 8. Oxygen sensor 9. EC-AT control unit 10. Intake air thermosensor (in airflow meter) 11. Intake air thermosensor (on dynamic chamber) 12. A/C switch 13. P/S switch 14. E/L switch 15. Ignition switch 16. Test connector 17. Brake light switch 18. EGR position sensor 19. Knock control unit 1. Injector 2. ISC valve 3. Solenoid valve (PRC) 4. Solenoid valve (VRIS) 5. Igniter 6. Solenoid valve (Purge control) 7. Solenoid valves (EGR) 8. Self-Diagnosis Checker and MIL 9. Monitor lamp (Self-Diagnosis Checker)
Cold start injector	Injects fuel into throttle body (engine cold)	—
Cold start thermoswitch	Controls cold start injector	—

Component	Function	Remark
Distributor	Detects crank angle at 30° intervals; No.1 cylinder TDC and No.4 cylinder TDC; sends signal to ECU	• Ne (Engine speed) signal • G1 signal • G2 signal
G1 coil	Detects No.4 cylinder TDC; sends signal to ECU	For determining fuel injection timing
G2 coil	Detects No.1 cylinder TDC; sends signal to ECU	For determining fuel injection timing
Dynamic chamber	Interconnects all cylinders	Contains shutter valve
EC-AT control unit	Detects N or P range; sends signal to ECU	—
Exhaust gas recirculation (EGR) position sensor	Detects EGR control valve opening; sends signal to ECU	Installed on EGR control valve
Exhaust gas recirculation (EGR) control valve	Recirculates portion of exhaust gas	Contains EGR position sensor
Fuel filter	Filters fuel	—
Fuel pump	Provides fuel to injectors	• Operates only while engine running • Installed in fuel tank
Idle switch	Detects when throttle valve fully closed; sends signal to ECU	Installed in throttle sensor Switch closed when throttle valve fully closed
Throttle sensor	Detects throttle valve opening angle; sends signal to ECU	• Idle switch included • Installed on throttle body
Igniter	Receives spark signal from ECU; generates high voltage in ignition coil	High voltage distributed by rotor in distributor
Ignition switch	Detects engine cranking; sends signal to ECU	—
Inhibitor switch	Detects in-gear condition; sends signal to EC-AT control unit	Switch closed in N or P range
Injector	Injects fuel into intake port	• Controlled by signals from ECU • High-ohmic injector
Intake manifold	Supplies intake air to all cylinders	—
Intake air thermosensor	Detects intake air temperature; sends signal to ECU	Installed in airflow meter
	Detects engine air temperature; sends signal to ECU	Installed on dynamic chamber
Knock control unit	Receives knock signal from knock sensor; sends signal to ECU	—
Knock sensor	Detects engine knocking; sends signal to knock control unit	—
Main relay	Supplies current to injectors and ECU	—
MIL (Malfunction indicator lamp)	Light illuminates if input device malfunctions	Installed in instrument cluster
One-way check valve	Prevents misoperation of shutter valve during heavy-load operations	—
Oxygen sensor	Detects oxygen concentration; sends signal to ECU	Zirconia ceramic and platinum coating
Oxygen sensor relay	Supplies current to oxygen sensor heater	—

3.0L ENGINE RELATIONSHIP CHART
1990–91 929 DOHC

O: Related
X : Not related

OUTPUT DEVICE \ INPUT DEVICE		DISTRIBUTOR		AIR-FLOW METER	THROTTLE SENSOR		WATER THERMO-SENSOR	INTAKE AIR THERMO-SENSOR		OXYGEN SENSOR	ATMOS-PHERIC PRESSURE SENSOR	EGR POSITION SENSOR	EC-AT CONTROL UNIT	KNOCK CONTROL UNIT	A/C SWITCH	P/S SWITCH	IGNITION SWITCH (START SIGNAL)	BRAKE LIGHT SWITCH	E/L SWITCH	TEST CON-NECTOR
		Ne ROTOR	G ROTOR		IDLE SWITCH	THROT-TLE SENSOR		AIR-FLOW METER	DYNAMIC CHAMBER											
IN-JECTOR	FUEL INJECTION AMOUNT	O	X	O	O	O	O	O	O	O	O	X	O	X	O	X	O	X	X	X
	FUEL INJECTION TIMING	O	O	X	X	O	X	X	X	X	X	X	X	X	X	X	O	X	X	X
IGNITER		O	O	O	O	O	O	X	X	X	O	O	O	O	O	X	O	X	X	O
BAC VALVE (ISC VALVE)		O	X	X	O	O	O	O	O	X	X	X	O	X	O	O	X	X	O	O
SOLENOID VALVE (PRC)		O	X	X	X	O	O	X	O	X	X	X	X	X	X	X	O	X	X	X
SOLENOID VALVE (VRIS)		O	X	X	X	O	X	X	X	X	X	X	X	X	X	X	O	X	X	X
SOLENOID VALVE (PURGE CONTROL)		O	X	O	O	O	O	O	X	O	O	X	O	X	X	X	X	X	X	X
SOLENOID VALVE (EGR)	VENT SIDE	O	X	O	O	O	O	X	X	X	O	X	X	X	X	X	X	X	X	X
	VACUUM SIDE	O	X	O	O	O	O	X	X	X	O	X	X	X	X	X	X	X	X	X
EC-AT CONTROL UNIT (LOCKUP CONTROL)		O	X	X	X	X	O	X	X	X	X	X	X	X	X	X	O	X	X	X
OXYGEN SENSOR HEATER RELAY		O	X	O	X	X	O	X	X	X	X	X	X	X	X	X	O	X	X	X
A/C RELAY		O	X	X	X	O	X	X	X	X	X	X	O	X	O	X	O	X	X	X

3.0L ENGINE COMPONENT DESCRIPTIONS
1990–91 929 DOHC

Component		Function	Remark
PCV valve		Controls blowby gas amount pulled into engine	—
Power steering (P/S) switch		Detects P/S operation; sends signal to ECU	ON when steering wheel turned right or left
Pressure regulator		Adjusts fuel pressure supplied to injectors	—
Pulsation damper		Absorbs fuel pulsations	—
Separator		Prevents fuel from flowing into charcoal canister	—
Shutter valve actuator		Controls shutter valve operation	For VRIS
Solenoid valve	EGR (Vent side, Vacuum side)	Controls EGR control valve operation	• Exhaust gas recirculation • Duty solenoid valve
	Purge control	Controls evaporative fumes from canister to intake manifold	Controlled by duty signal from ECU
	Variable resonance induction system (VRIS)	Controls vacuum to shutter valve actuator	—
	Pressure regulator control (PRC)	Blocks vacuum from intake manifold to pressure regulator	Hot engine restart only
Test connector		For Self-Diagnosis Checker and idle speed/ignition timing adjustment	Green: 1-pin connector
Throttle body		Controls intake air amount	Integrated throttle sensor and cold start injector incorporated
Water thermosensor		Detects engine coolant temperature; sends signal to ECU	—

3.0L ENGINE TROUBLESHOOTING GUIDE
1990–91 929 DOHC

Symptoms \ Possible cause (Input sensors)	Ignition pulse	Distributor (Ne-signal)	Distributor (G1 signal)	Distributor (G2 signal)	Knock sensor and knock control unit	Airflow meter	Water thermosensor	Intake air thermosensor (Airflow meter)	Intake air thermosensor (Dynamic chamber)	Throttle sensor	Atmospheric pressure sensor	Left side oxygen sensor	EGR position sensor	Fuel feedback system (Left side)	Right side oxygen sensor	Fuel feedback system (Right side)
1 Fault indicated by SST Malfunction Code No.	01	02	03	04	05	08	09	10	11	12	14	15	16	17	23	24

Symptoms:

1. Fault indicated by SST Malfunction Code No.
2. Hard starts or won't start (cranks OK)
3. Engine stalls — While warming up; After warming up
4. Rough idle — While warming up; After warming up
5. High speed idle after warming up
6. Poor acceleration, hesitation, or lack of power
7. Runs rough on deceleration
8. Afterburn in exhaust system
9. Poor fuel economy
10. Excessive oil consumption
11. Engine stalls or runs rough after hot starting
12. Fails emission test

TROUBLESHOOTING PROCEDURE

Note
- **Symptom 1 shows the Code No. to quickly determine what system or unit may be at fault by use of the SST. (SST: Self-Diagnosis Checker 49 H018 9A1)**

1st. Check the electrical system.
1) Battery condition
2) Main relays
3) Main fuses
4) Fuses

2nd. Check input sensors and output solenoid valves diagnosed with the **SST**.

3rd. Check input switches with **SST**.

4th. Check the following items:
Ignition system
1) Ignition spark
2) Ignition timing (with test connector grounded)
Intake air system
1) Air cleaner element
2) Vacuum or air leakage
3) Vacuum hose and air hose routing
4) Accelerator cable
Fuel system
1) Fuel amount
2) Fuel leakage
3) Fuel filter
4) Idle speed (with test connector grounded)
Engine
1) Compression
2) Overheating

5th. Check Fuel and Emission Control Systems.

3.0L ENGINE ENGINE CONTROL OPERATIONS CHART
1990–91 929 DOHC

OUTPUT DEVICE \ ENGINE CONDITION		CRANKING (COLD ENGINE)	WARM-UP (DURING IDLE)	MEDIUM LOAD COLD	MEDIUM LOAD WARM	ACCELE-RATION	HEAVY LOAD	DECEL-ERATION	IDLE (THROTTLE VALVE FULLY CLOSED)	IGN: ON (ENGINE NOT RUNNING)	REMARK
INJECTOR	No.1–No.6	Rich			Rich and lean (Fuel feedback operation)*1	Rich		Fuel cut	Rich and lean (Fuel feedback operation)*1	Does not inject	*1 Operated by oxygen sensor
	INJECTION TIMING	Sequential injection							Sequential injection		Fuel cut: Above 6,500 rpm
IGNITER		Fixed at BTDC 7°	Advanced: Depends on engine condition			Retarded: Depends on intensity of knocking		Advanced: Depends on engine condition		—	
BAC VALVE	AIR VALVE	Open*2			Closed						*2 Coolant temp: Below 60°C (140°F)
	ISC VALVE	Large amount of bypass air			Small amount of bypass air			Large and small amount of bypass air	Large and small amount of bypass air*3	Does not operate	*3 Engine speed feedback system operates when test connector open
SOLENOID VALVE (PRC)		OFF (Vacuum to pressure regulator)							After start ON (Vacuum cut)*4	Does not operate	*4 During hot condition only
SOLENOID VALVE (VRIS)		OFF (Shutter valve closed)		ON (Shutter valve open)*5				OFF (Shutter valve closed)		Dose not operate	*5 ON: Below 3,800 rpm
SOLENOID VALVE (PURGE CONTROL)		OFF (Purge cut)			Operates (Duty values [purge gas amount] change)				OFF (Purge cut)	Does not operate	
SOLENOID VALVE (EGR)	VENT SIDE	OFF (No vacuum to EGR valve) — EGR cut —			System operates*6			OFF (No vacuum to EGR valve) — EGR cut —		Does not operate	*6 Depends on engine condition
	VACUUM SIDE	OFF (Duty value 100%) (Atmospheric pressure to EGR valve)			(Amount of EGR changes)			OFF (Duty value 100%) (Atmospheric pressure to EGR valve)			
OXYGEN SENSOR HEATER RELAY		OFF	ON				OFF	ON		Does not operate	
A/C RELAY		OFF (A/C OFF)	ON (A/C ON)				OFF (A/C OFF)	ON (A/C ON)		ON (A/C OFF)	

3.0L ENGINE TROUBLE CODES—1990–91 929 DOHC

Mal-function Code No.	Input devices	Malfunction	Fail-safe function	Pattern of output signals (Self-Diagnosis Checker or MIL)
01	Ignition pulse (Igniter, Ignition coil)	Broken wire, short circuit	—	
02	Distributor (Ne-signal)	Ne-signal not input for 1.5 sec. during cranking	—	
03	Distributor (G1 signal)	Broken wire, short circuit	—	
04	Distributor (G2 signal)	Broken wire, short circuit	—	
05	Knock sensor and knock control unit	Broken wire, short circuit	—	
08	Airflow meter	Broken wire, short circuit	Basic fuel injection amount fixed as for 2 driving modes 1) Idle switch: ON 2) Idle switch: OFF	
09	Water thermosensor	Broken wire, short circuit	Coolant temp. input fixed at 80°C (176°F)	
10	Intake air thermosensor (Airflow meter)	Broken wire, short circuit	Intake air temp. input fixed at 20°C (68°F)	
11	Intake air thermosensor (Dynamic chamber)	Broken wire, short circuit	Intake air temp. input fixed at 20°C (68°F)	
12	Throttle sensor	Broken wire, short circuit	Throttle valve opening angle signal input fixed at fully open	
14	Atmospheric pressure sensor	Short circuit	Atmospheric pressure input fixed at 760 mmHg (29.9 in Hg)	
15	Left side oxygen sensor	Oxygen sensor output below 0.55V 120 sec. after engine at 1,500 rpm	Feedback system canceled (for EGI)	
16	EGR position sensor	Broken wire, short circuit	EGR position signal input fixed at fully closed	
		Sensor output does not match target value (incorrect output)	—	
17	Feedback system (Left side)	Oxygen sensor output does not change from 0.55V 60 sec. after engine at 1,500 rpm	Feedback system canceled (for EGI)	

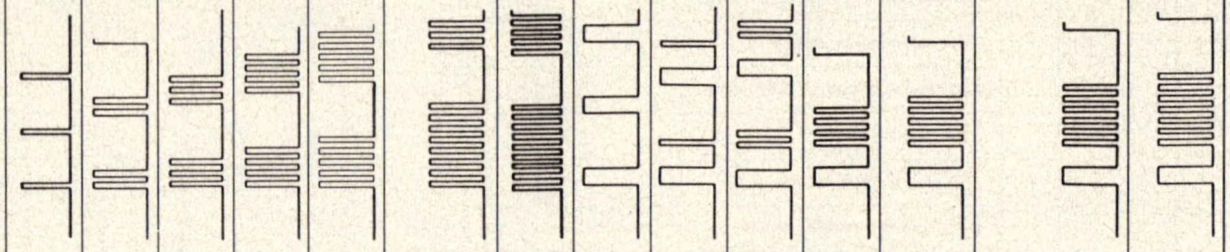

3.0L ENGINE TROUBLESHOOTING GUIDE 1990–91 929 DOHC

Possible cause	Output solenoid valves							Fuel and Emission Control Systems										
	Solenoid valve (PRC)	Solenoid valve (Purge control)	Solenoid valve (EGR, vacuum side)	Solenoid valve (EGR, vent side)	Idle speed control valve	Oxygen sensor heater relay	Solenoid valve (VRIS)	Intake air system (Poor connection of components, Throttle body)	Idle speed control system (Air valve, Idle speed control valve malfunction)	Variable resonance induction system (VRIS)	Fuel system (Fuel injection, Fuel pressure)	Pressure regulator control (PRC) system	Electronic spark advance (ESA) control system	Deceleration control system (Fuel cut operation malfunction)	EGR system (EGR control valve open)	Evaporative emission control system	PCV system (System clogged)	Exhaust system (System clogged)
Symptom / Page	25	26	28	29	34	40	41											
2								3			2		1		4			
3								4	1		3				2			
								5	2		4				3		1	
4								6	1		5				3	4	2	
								6	2		5				3	4	1	
5								2	1		3							
6								3		5	4				1	2		6
7									2		3			1				
8								3	2		4			1				
9											3	4		1	2			5
10																	1	
11									2		3	1						
12								6	5		7	8		3	2	4		1

The numbers of the list show the priorities of inspections from the most possible to that with the lowest possibility. These were determined on the following basis:

- Ease of inspection
- Most possible system
- Most possible point in system

3.0L ENGINE TROUBLE CODES—1990–91 929 DOHC

Mal-function Code No.	Input devices	Malfunction	Fail-safe function	Pattern of output signals (Self-Diagnosis Checker or MIL)
23	Right side oxygen sensor	Oxygen sensor output below 0.55V 120 sec. after engine at 1,500 rpm	Feedback system canceled (for EGI)	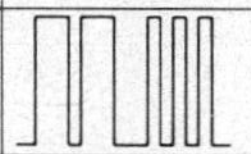
24	Feedback system (Right side)	Oxygen sensor output does not change from 0.55V 60 sec. after engine at 1,500 rpm	Feedback system canceled (for EGI)	

Output Devices

Malfunction code	Output devices	Pattern of output signals (Self-Diagnosis Checker or MIL)
25	Solenoid valve (PRC)	
26	Solenoid valve (Purge control)	
28	Solenoid valve (EGR, vacuum side)	
29	Solenoid valve (EGR, vent side)	
34	ISC valve	
40	Oxygen sensor heater relay	
41	Solenoid valve (VRIS)	

Caution

- If there is more than one failure present, the lowest number malfunction code is displayed first, any subsequent malfunction codes light up in order.
- After repairing all failures, turn off the ignition switch and disconnect the negative battery cable and depress the brake pedal for at least five seconds to erase the memory of a malfunction code.

3.0L ENGINE TROUBLE CODE TROUBLESHOOTING 1990–91 929 DOHC

No.01 Code

- Are there poor connections at ignition coil connectors? — YES → Repair or replace connector
- NO ↓ Is resistance of ignition coil OK ? Resistance: Primary 0.72—0.88Ω, Secondary 10—30 kΩ — NO → Replace ignition coil
- YES ↓ Is operation of igniter OK ? — NO → Replace igniter
- YES ↓ Is there continuity between IGf terminal of igniter connector (wiring harness side) and engine control unit (1V) terminal? — NO → Open circuit in wiring harness from igniter to engine control unit (1V) terminal
- YES ↓ Is engine control unit (1V) terminal voltage OK? — NO → Short circuit in wiring harness
- YES → Replace engine control unit

No.02 Code

- Are there poor connections at distributor connectors? — YES → Repair or replace connector
- NO ↓ Is resistance of distributor OK? — NO → Replace distributor
- YES ↓ Is there continuity between distributor connector (wiring harness side) and engine control unit? — NO → Open circuit in wiring harness

Distributor	Engine control unit
Ne ① W	3E
Ne ② G	3F

- YES ↓ Are engine control unit (3E) and (3F) terminal voltages OK? — NO → Short circuit in wiring harness
- YES → Replace engine control unit

No.03 Code

Are there poor connections at distributor connectors?
- YES → Repair or replace connector
- NO ↓

Is resistance of distributor OK?
- NO → Replace distributor
- YES ↓

Is there continuity between distributor connector (wiring harness side) and engine control unit?

Distributor	Engine control unit
G1 L	3G

- NO → Open circuit in wiring harness
- YES ↓

Is engine control unit 3G terminal voltage OK?
- NO → Short circuit in wiring harness
- YES → Replace engine control unit

No.04 Code

Are there poor connections at distributor connectors?
- YES → Repair or replace connector
- NO ↓

Is resistance of distributor OK?
- NO → Replace distributor
- YES ↓

Is there continuity between distributor connector (wiring harness side) and engine control unit?

Distributor	Engine control unit
G2 ① R	3H

- NO → Open circuit in wiring harness
- YES ↓

Is engine control unit 3H terminal voltage OK?
- NO → Short circuit in wiring harness
- YES → Replace engine control unit

No.05 Code

Are there poor connections in relative circuit?
- YES → Repair or replace connector
- NO ↓

Are knock control unit terminal voltages correct?

Terminal	Voltage
B/W wire (G)	Approx. 12V
G/W wire	3.3—5.0V
Y and W wires	3.3—5.0V
B/LG wire (E)	0V

- NO (all wires) → • Open or short circuit in wiring harness between main relay and knock control unit • Open circuit in wiring harness between knock control unit and ground
- (only G/W wire) → Knock sensor, knock control unit, or related harness malfunction
- (only Y and W wires) → Open circuit in wiring harness between knock control unit and knock sensor
- YES ↓

Is same Code No. present after performing after-repair procedure?
- NO → Knock control unit, knock sensor, and circuit OK
- YES ↓

Is engine control unit 2M terminal voltage OK?
- NO → • Open or short circuit in wiring harness • Knock control unit malfunction
- YES → Engine control unit malfunction

No.08 Code

Are there poor connections at airflow meter connectors
- YES → Repair or replace connector
- NO ↓

Is resistance of airflow meter OK?
Resistance:

Terminal	Fully closed (Ω)	Fully open (Ω)
$E_2 \leftrightarrow V_S$	20—600	20—1,000
$E_2 \leftrightarrow V_C$	200—400	

- NO → Repair airflow meter
- YES ↓

Is there continuity between airflow meter connector (wiring harness side) and engine control unit?

Airflow meter	Control unit
V_B (BR/W wire)	2I
V_S (Y/L wire)	2B
E_1, E_2 (R/B wire)	3D

- NO → Open circuit in wiring harness from airflow meter to engine control unit
- YES ↓

Are engine control unit terminal voltages OK?
Check terminal 2B, 2I, 3D
- NO → Short circuit in wiring harness
- YES → Replace engine control unit

3.0L ENGINE TROUBLE CODE TROUBLESHOOTING – 1990–91 929 DOHC

No.10 Code

Are there poor connections at airflow meter connectors?
- YES → Repair or replace connector
- NO ↓

Is resistance of intake air thermosensor (in airflow meter) OK?
Resistance:

Terminal	Resistance
$E_2 \leftrightarrow THA_A$	−20°C (−4°F): 13.6—18.4 kΩ 20°C (68°F): 2.21—2.69 kΩ 60°C (140°F): 0.493—0.667 kΩ

- NO → Replace airflow meter
- YES ↓

Is there continuity between intake air thermosensor (in airflow meter) connector (wiring harness side) and engine control unit?

Intake air thermosensor (in airflow meter)	Engine control unit
THA_A (G/O wire)	2K
E_2 (R/B wire)	3D

- NO → Open circuit in wiring harness from intake air thermosensor (in airflow meter) to engine control unit
- YES ↓

Are engine control unit (2K) and (3D) terminal voltages OK?
- NO → Short circuit in wiring harness
- YES → Replace engine control unit

No.09 Code

Are there poor connections at water thermosensor connectors?
- YES → Repair or replace connectors
- NO ↓

Is resistance of water thermosensor OK?
Resistance:

Coolant temp	Resistance
−20°C (−4°F)	14.5—17.8 kΩ
20°C (68°F)	2.2—2.7 kΩ
40°C (104°F)	1.0—1.3 kΩ
60°C (140°F)	0.5—0.64 kΩ
80°C (176°F)	0.28—0.35 kΩ

- NO → Replace water thermosensor
- YES ↓

Is there continuity between water thermosensor connector (wiring harness side) and engine control unit?

Water thermosensor	Engine control unit
A (G/W wire)	2E
B (R/B wire)	3D

- NO → Open circuit in wiring harness from water thermosensor to engine control unit
- YES ↓

Are engine control unit (2E) and (3D) terminal voltages
- NO → Short circuit in wiring harness
- YES → Replace engine control unit

No.11 Code

1. Are there poor connections at intake air thermosensor connectors?
   - YES → Repair or replace connector
   - NO → go to step 2
2. Is resistance of intake air thermosensor (dynamic chamber) OK?
   Resistance:

Temperature	Resistance
20°C (68°F)	37.3—45.6 kΩ
50°C (122°F)	10.5—12.9 kΩ
85°C (185°F)	3.1—3.8 kΩ

   - NO → Replace intake air thermosensor (dynamic chamber)
   - YES → go to step 3
3. Is there continuity between intake air thermosensor (dynamic chamber) connector (wiring harness side) and engine control unit?

Intake air thermosensor (dynamic chamber)	Engine control unit
A (L/Y wire)	2L
B (R/B wire)	3D

   - NO → Open circuit in wiring harness from intake air thermosensor (dynamic chamber) to engine control unit
   - YES → go to step 4
4. Are engine control unit (2L) and (3D) terminal voltages OK?
   - NO → Short circuit in wiring harness
   - YES → Replace engine control unit

No.12 Code

1. Are there poor connections at throttle sensor connectors?
   - YES → Repair or replace connector
   - NO → go to step 2
2. Is throttle sensor adjusted correctly?
   - NO → Adjust throttle sensor (for idle switch)
   - YES → go to step 3
3. Is resistance of throttle sensor OK?
   Resistance: (A) ↔ (D) 3—7 kΩ
   (B) ↔ (D) Below 6 kΩ
   .....Throttle valve fully closed
   Approx. 3.3—7.0 kΩ
   .....Throttle valve fully open
   - NO → Replace throttle sensor
   - YES → go to step 4
4. Is there continuity between throttle sensor connector (wiring harness side) and engine control unit?

Throttle sensor	Engine control unit
A (BR/W wire)	2I
B (G/Y wire)	2F
D (R/B wire)	3D

   - NO → Open circuit in wiring harness from throttle sensor to engine control unit
   - YES → go to step 5
5. Are engine control unit terminal voltages OK?
   Check terminal: 2F, 2I
   - NO → Short circuit in wiring harness
   - (2I) terminal voltage not within specification → Replace engine control unit
   - YES → Replace engine control unit

No.14 Code

Replace engine control unit

3.0L ENGINE TROUBLE CODE TROUBLESHOOTING—1990–91 929 DOHC

No.15 Code and/or No.23 Code

Note
- **When Malfunction Code No.s 15 and 17 are present at the same time, first perform the checking procedure for Code No.17.**

- Are there any poor connections at oxygen sensor connector?
  - YES → Repair or replace connector
  - NO ↓
- Is operation of oxygen sensor heater OK?
  - NO → Repair or replace related parts
  - YES ↓
- Is oxygen sensor output voltage OK?
  - NO → Replace oxygen sensor
  - YES ↓
- Is sensitivity of oxygen sensor OK?
  - NO → •Oxygen sensor malfunction •Engine control unit malfunction
  - YES ↓
- Is there continuity between oxygen sensor connector (wiring harness side) and engine control unit (2C and/or 2D) terminal?
  - NO → Open circuit in wiring harness from oxygen sensor to engine control unit (2C and/or 2D) terminal
  - YES ↓
- Is (2C and/or 2D) terminal voltage of engine control unit OK?
  - NO → Short circuit in wiring harness between oxygen sensor and engine control unit (2C and/or 2D) terminal
  - YES → Replace engine control unit

No.16 Code

- Are there poor connections at EGR position sensor connector?
  - YES → Repair or replace connector
  - NO ↓
- Is EGR control valve OK?
  - NO → Replace EGR control valve
  - YES ↓
- Is resistance of EGR position sensor OK?
  Resistance: (A) ↔ (B) Approx. 5 kΩ
  (A) ↔ (C) 0.7—5 kΩ
  (B) ↔ (C) 0.7—5 kΩ
  - NO → Replace EGR control valve
  - YES ↓
- Is there continuity between EGR position sensor connector (wiring harness side) and engine control unit?

EGR position sensor	Engine control unit
A (Y)	2J
B (R/B)	3D
C (BR/W)	2I

  - NO → Open circuit in wiring harness from EGR position sensor to engine control unit
  - YES ↓ Cont'd

Cont'd

- Is there 4.5—5.5V at (A) terminal of EGR position sensor connector?
  - NO → •Short circuit in wiring harness from EGR position sensor (A) terminal to engine control unit (2J) terminal •Engine control unit malfunction
  - YES ↓
- Is engine control unit (2I) terminal voltage OK?
  - NO → Short circuit in wiring harness EGR position sensor (C) terminal to engine control unit (2I) terminal
  - YES → Replace engine control unit

No.17 Code and/or No.24 Code

- Warm up engine and run it at 2,500—3,000 rpm for 3 minutes
  ↓
- Does monitor lamp of **SST** illuminate at idle without test connector (Green: 1-pin) grounded?
  SST: Self-Diagnosis Checker
  - NO → •Air leak in vacuum hoses or intake air system •Contaminated oxygen sensor •Insufficient fuel injection
  - YES ↓
- Are spark plugs clean?
  - NO → Clean or replace spark plugs
  - YES ↓
- Is oxygen sensor operation OK?
  - NO → Replace oxygen sensor
  - YES ↓
- Is same Code No. present after performing after-repair procedure
  - NO → Feedback system OK
  - YES ↓
- Is (2C and/or 2D) terminal voltage of engine control unit OK?
  - NO → Short circuit in wiring harness from oxygen sensor connector to engine control unit
  - YES → Replace engine control unit

No.25 Code

- Are there poor connections at solenoid valve (PRC) connectors?
  - YES → Repair or replace connector
  - NO ↓
- Is there continuity of solenoid valve (PRC)?
  - NO → Replace solenoid valve
  - YES ↓
- Is there battery voltage at (B/W) wire of solenoid valve (PRC) connector?
  - NO → Open or short circuit in wiring harness from (B/W) wire to main relay (for engine control unit)
  - YES ↓
- Is there continuity between solenoid valve (PRC) connector (wiring harness side) and engine control unit?

Solenoid valve (PRC)	Engine control unit
B (L/B wire)	3M

  - NO → Open circuit in wiring harness from solenoid valve (PRC) to engine control unit
  - YES → • Engine control unit malfunction • Short circuit in wiring harness

No.26 Code

- Is there poor connection at solenoid valve (Purge control) connectors?
  - YES → Repair or replace connector
  - NO ↓
- Is there continuity of solenoid valve (Purge control)?
  - NO → Replace solenoid valve
  - YES ↓
- Is there battery voltage at (B/W) wire of solenoid valve (Purge control) connector?
  - NO → Open or short circuit in wiring harness from (B/W) wire to main relay (for engine control unit)
  - YES ↓
- Is there continuity between solenoid valve (Purge control) connector (wiring harness side) and control unit?

Solenoid valve	Engine control unit
B (BR/L wire)	2O

  - NO → Open circuit in wiring harness from solenoid valve (Purge control) to engine control unit
  - YES → • Engine control unit malfunction • Short circuit in wiring harness

No.28 Code

- Are there any poor connections at solenoid valve (EGR, vacuum) connectors?
  - YES → Repair or replace connector
  - NO ↓
- Is there continuity of solenoid valve (EGR, vacuum)?
  - NO → Replace solenoid valve
  - YES ↓
- Is there battery voltage at (B/W) wire of solenoid valve (EGR, vacuum) connector?
  - NO → Open or short circuit in wiring harness from (B/W) wire to main relay (for engine control unit)
  - YES ↓
- Is there continuity between solenoid valve (EGR, vacuum) connector (wiring harness side) and engine control unit?

Solenoid valve	Engine control unit
B (Y/R wire)	3P

  - No → Open circuit in wiring harness from solenoid valve (EGR, vacuum) to engine control unit
  - YES → • Engine control unit malfunction • Short circuit in wiring harness

No.29 Code

- Are there any poor connections at solenoid valve (EGR, vent) connectors?
  - YES → Repair or replace connector
  - NO ↓
- Is there continuity of solenoid valve (EGR, vent)?
  - NO → Replace solenoid valve
  - YES ↓
- Is there battery voltage at (B/W) wire of solenoid valve (EGR, vent) connector?
  - NO → Open or short circuit in wiring harness from (B/W) wire to main relay (for engine control unit)
  - YES ↓
- Is there continuity between solenoid valve (EGR, vent) connector (wiring harness side) and engine control unit?

Solenoid valve	Engine control unit
B (V/W wire)	3O

  - No → Open circuit in wiring harness from solenoid valve (EGR, vent) to engine control unit
  - YES → • Engine control unit malfunction • Short circuit in wiring harness

3.0L ENGINE TROUBLE CODE TROUBLESHOOTING—1990–91 929 DOHC

No.34 Code

- Are there poor connectors at BAC valve connectors?
  - YES → Repair or replace connector
  - NO ↓
- Is resistance of BAC valve OK? Resistance: 10—13Ω
  - NO → Replace BAC valve
  - YES ↓
- Is there battery voltage at (B/W) wire of BAC valve connector?
  - NO → Open or short circuit in wiring harness from (B/W) wire to main relay (for engine control unit)
  - YES ↓
- Is there continuity between BAC valve connector (wiring harness side) and engine control unit?

BAC valve	Engine control unit
B (L/O wire)	3Q

  - NO → Open circuit in wiring harness from BAC valve to engine control unit
  - YES → • Engine control unit malfunction • Short circuit in wiring harness

No.40 Code

- Are there poor connections at oxygen sensor heater relay connectors?
  - YES → Repair or replace connector
  - NO ↓
- Is oxygen sensor heater relay operation OK?
  - NO → Replace oxygen sensor relay
  - YES ↓
- Is there continuity between oxygen sensor heater relay connector and main relay?

Oxygen sensor relay	Main relay
A (B/W wire)	D (B/W wire)

  - NO → Open circuit in wiring harness from oxygen sensor heater relay to main relay
  - YES ↓
- Is there continuity between oxygen sensor heater relay connector and engine control unit?

Oxygen sensor	Engine control unit
B (R/Y wire)	3K

  - NO → Open circuit in wiring harness from oxygen sensor heater relay to engine control unit
  - YES → • Engine control unit malfunction • Short circuit in wiring harness

No.41 Code

- Are there poor connections at solenoid valve (VRIS) connectors?
  - YES → Repair or replace connector
  - NO ↓
- Is there continuity of solenoid valve (VRIS)?
  - NO → Replace solenoid valve
  - YES ↓
- Is there battery voltage at (B/W) wire of solenoid valve (VRIS) connector?
  - NO → Open or short circuit in wiring harness from (B/W) wire to main relay (for engine control unit)
  - YES ↓
- Is there continuity between solenoid valve (VRIS) connector (wiring harness side) and engine control unit?

Solenoid valve (VRIS)	Engine control unit
B (V wire)	3J

  - NO → Open circuit in wiring harness from solenoid valve (VRIS) to engine control unit
  - YES → • Engine control unit malfunction • Short circuit in wiring harness

3.0L ENGINE SWITCH MONITOR FUNCTION 1990–91 929 DOHC

Note
- **The test connector must be grounded and the ignition switch ON (engine stopped).**

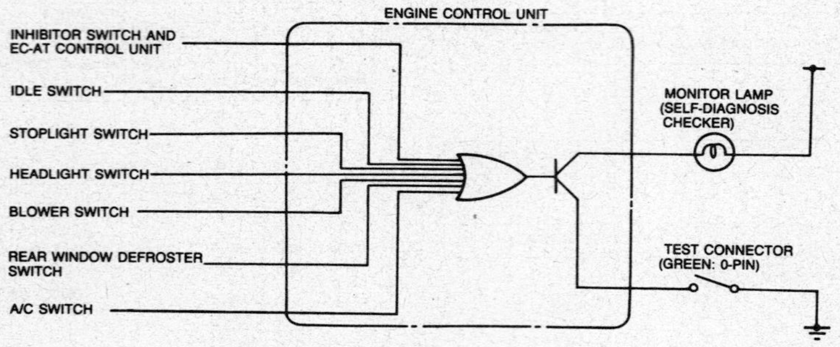

Switch	Self-Diagnosis Checker (Monitor Lamp)		Remark
	Light ON	Light OFF	
Idle switch	Pedal depressed	Pedal released	
Brake light switch	Pedal depressed	Pedal released	
A/C switch	ON	OFF	Fan speed control: 1st—4th position
Headlight switch	ON	OFF	
Rear defroster switch	ON	OFF	
Blower switch	ON	OFF	Fan speed control: 5th—8th position
Inhibitor switch	Except N and P range	N, P range	

Caution
- **If any of the switches remains activated, the monitor lamp will stay on.**

3.0L ENGINE SWITCH MONITOR FUNCTION TROUBLESHOOTING—1990–91 929 DOHC

Procedure

Set conditions to deactivate each switch
- All accessories OFF
- All pedals released

Verify that monitor lamp not illuminated

NO → Check each switch and related wiring harness
- Idle switch
- Brake light switch
- A/C switch
- Headlight switch
- Rear defroster switch
- Blower switch
- Inhibitor switch

YES → Check each switch as described

Idle switch

Depress accelerator pedal and verify that monitor lamp illuminates

NO →
- Misadjustment of idle switch
- Idle switch malfunction
- Short circuit in wiring harness from idle switch to engine control unit
- Engine control unit (1T) terminal malfunction

YES → Idle switch OK
Release accelerator pedal

3.0L ENGINE SWITCH MONITOR FUNCTION TROUBLESHOOTING—1990–91 929 DOHC

Brake light switch

Depress brake pedal and verify that monitor lamp illuminates

- **YES** → Brake light switch OK
  Release brake pedal
- **NO** →
  - Brake light switch malfunction
  - Open circuit in wiring harness from brake light switch to engine control unit
  - Engine control unit (1Q) terminal malfunction

Blower switch

Set fan speed control to 5th—8th position and verify that monitor lamp illuminates

- **YES** → Blower switch OK
- **NO** →
  - Fan speed control switch malfunction
  - Open circuit in wiring harness from fan speed control switch to engine control unit
  - Engine control unit (1P) terminal malfunction

A/C switch

Set fan speed control to 1st position
Turn ON A/C switch, and verify that monitor lamp illuminates

- **YES** → A/C switch OK
  Turn off A/C switch and blower switch
- **NO** →
  - A/C switch malfunction
  - Fan speed control switch malfunction
  - Open circuit in wiring harness from A/C switch to engine control unit
  - Engine control unit (1O) terminal malfunction

Headlight switch

Turn ON headlight switch and verify that monitor lamp illuminates

- **YES** → Headlight switch OK
  Turn OFF headlight switch
- **NO** →
  - Headlight switch malfunction
  - Open circuit in wiring harness from headlight switch to engine control unit
  - Engine control unit (1H) terminal malfunction

Rear defroster switch

Turn ON rear defroster switch and verify that monitor lamp illuminates

- **YES** → Rear defroster switch OK
  Turn OFF rear defroster switch
- **NO** →
  - Rear defroster switch malfunction
  - Open circuit in wiring harness from rear defroster switch to engine control unit
  - Engine control unit (1J) terminal malfunction

Inhibitor switch

Shift transmission into gear and verify that monitor lamp illuminates

- **YES** → Inhibitor switch OK
  Return transmission to neutral
- **NO** →
  - Inhibitor switch malfunction
  - Open or short circuit in related wiring harness
  - Control unit (1R) terminal malfunction

3.0L ENGINE ISC SYSTEM TROUBLESHOOTING IN ORDER OF PRIORITY—1990–91 929 DOHC

Note
- When the System Inspection is completed and no problem is found, return to Troubleshooting Guide and inspect the next system.

Symptom \ Possible cause		System operation	Air valve	ISC valve	Idle switch	Water thermosensor	Intake air thermosensor (Airflow meter)	Intake air thermosensor (Dynamic chamber)	P/S pressure switch	Engine control unit 1I	Engine control unit 3Q
Engine stalls	While warming up	1	2	3	6	7				4	5
	After warming up	1		2	5				*1 6	3	4
Rough idle	While warming up	1	2	3	6	7				4	5
	After warming up	1		2	5		6		*1 7	3	4
High idle speed after warming up		1	3	2	6	5					4
Runs rough on deceleration		1	6	2	5					3	4
Afterburn in exhaust system		1	6	2	5					3	4
Engine stalls or rough after hot starting		1		2	7	5		6		3	4
Fails emission test		1	8	2	5	7	6			3	4

*1: Inspect P/S pressure switch if rough idle or engine stall occurs only when P/S is operated.

3.0L ENGINE VARIABLE RESONANCE INDUCTION SYSTEM TROUBLESHOOTING IN ORDER OF PRIORITY—1990–91 929 DOHC

Note
- Numbers show order of inspection.
- When the System Inspection is completed and no problem is found, return to Troubleshooting Guide and inspect the next system.
- Be careful not to confuse the YES and NO results when making a vacuum check.

Symptom \ Possible cause	System inspection	Vacuum signal	Vacuum signal OK: Shutter valve	Vacuum signal OK: Shutter valve actuator	Vacuum signal NG: Solenoid valve (VRIS)	Vacuum signal NG: Check valve	Vacuum signal NG: Vacuum chamber	Vacuum signal NG: Engine control unit (3J) terminal
Poor acceleration, hesitation, or lack of power	1	2	4	3	4	5	6	3

3.0L ENGINE FUEL SYSTEM TROUBLESHOOTING IN ORDER OF PRIORITY—1990–91 929 DOHC

Symptom \ Possible cause		Airflow meter	Intake air thermosensor (Airflow meter)	Throttle sensor	Water thermosensor	Oxygen sensor	Oxygen sensor heater	Injector	Fuel pump	Pulsation damper	Circuit opening relay	Fuel pressure	Engine control unit	
													3U 3V 3W 3X 3Y 3Z	1C
Hard start					5			4	7		1	6	3	2
Won't start (cranks OK)					6			5	2		1	7	3	4
Engine stalls	While warming up	4			3			2				1	5	
	After warming up	3						2				1	4	
Rough idle	While warming up	4			3			2				1	5	
	After warming up	3	7	6		4	5	2				1	8	
High speed idle after warming up		2			3							1		
Poor acceleration, hesitation, or lack of power		2		3				4				1		
Runs rough on deceleration				1				2					3	
Afterburn in exhaust system		1	3		2			4					5	
Poor fuel economy		5			3	4		2				1		
Engine stalls or runs rough after hot starting		2		3				4				1	5	
Fails emission test			6		5	1	2	7			3	4		

3.0L ENGINE PRESSURE REGULATOR CONTROL SYSTEM TROUBLESHOOTING IN ORDER OF PRIORITY—1990–91 929 DOHC

Note
- Numbers show order of inspection.
- When the System Inspection is completed and no problem is found, return to Troubleshooting Guide and inspect the next system.
- Be careful not to confuse the YES and NO results when making a vacuum check.

Symptom \ Possible cause	Vacuum signal	Vacuum signal OK	Vacuum signal NG				
		Pressure regulator	Solenoid valve (PRC)	Water thermosensor	Intake air thermosensor (Dynamic chamber)	Throttle sensor	Engine control unit 3M
Poor fuel economy	1	2	2				3
Engine stalls or runs rough after hot starting	1	2	2	3	4	5	6
Fails emission test	1	2	2				3

3.0L ENGINE EGR SYSTEM TROUBLESHOOTING IN ORDER OF PRIORITY—1990–91 929 DOHC

Note
- When the System Inspection is completed and no problem is found, return to Troubleshooting Guide and inspect the next system.

Symptom \ Possible cause		System inspection	Solenoid valve: EGR, vacuum side	Solenoid valve: EGR, vent side	Airflow meter	Intake air thermosensor (Dynamic chamber)	Throttle sensor	Water thermosensor	EGR position sensor	EGR control valve	Engine control unit terminal: 3P	Engine control unit terminal: 3O
Hard starts		1	3	4						2	5	6
Won't start (cranks OK)			1							4	2	3
Engine stalls	While warming up	1	3	4		6	7	5		2	8	9
Engine stalls	After warming up	1	3	4			5	6		2	7	8
Rough idle	While warming up	1	3	4		6	7	5		2	8	9
Rough idle	After warming up	1	3	4			5	6		2	7	8
Poor acceleration, hesitation, or lack of power		1	3	4	5		6		7	2	8	9
Poor fuel economy		1	3	4		6	7	5		2	8	9
Fails emission test		1	3	4		6	8	5	7	2	9	10

3.0L ENGINE EVAP SYSTEM TROUBLESHOOTING IN ORDER OF PRIORITY—1990–91 929 DOHC

Note
- Numbers show order of inspection.
- When the System Inspection is completed and no problem is found, return to Troubleshooting Guide and inspect the next system.
- Be careful not to confuse the YES and NO results when making electrical checks.

Symptom \ Possible cause		System inspection	Solenoid valve (Purge control)	Charcoal canister	Airflow meter	Water thermosensor	Engine control unit terminal: 2O
Rough idle	While warming up	1	2	4		3	5
Rough idle	After warming up	1	2	3			4
Poor acceleration, hesitation, or lack of power		1			2		3
Fails emission test		1	2	5	4	3	6

3.0L ENGINE FUEL CUT SYSTEM TROUBLESHOOTING IN ORDER OF PRIORITY—1990–91 929 DOHC

Note

- **When the System Inspection is completed and no problem is found, return to Troubleshooting Guide and inspect the next system.**

Symptom \ Possible cause	System inspection	Water thermosensor	Throttle sensor
Runs rough on deceleration	1	2	3
Afterburn in exhaust system	1	2	3
Poor fuel economy	1	2	3
Fails emission test	1	2	3

3.0L ENGINE ENGINE CONTROL UNIT TERMINAL VOLTAGES—1990–91 929 DOHC

Terminal	Input	Output	Connected to	Voltage (after warm-up): Ignition switch: ON	Voltage (after warm-up): Idle	Remark
1A	○		Battery	Approx. 12V		For backup
1B	○		Main relay	Approx. 12V		—
1C	○		Ignition switch (START position)	Below 2.5V ("P" or "N" range)		While cranking: 6—11V
1D		○	Self-Diagnosis Checker (Monitor lamp)	<Test connector grounded> For 3 sec. after ignition switch OFF → ON: Approx. 5V (lamp illuminates) After 3 sec. approx. 12V	<Test connector grounded> Approx. 5V <Test connector not grounded> Monitor lamp ON; Below 6V Monitor lamp OFF; Approx. 12V	With Self-Diagnosis Checker
1E		○	MIL (Malfunction Indicator Lamp)	For 3 sec. after ignition switch OFF → ON: Below 2.5V (light illumination) After 3 sec.: approx. 12V		• Test connector grounded • Lamp illuminates: Below 2.5V • Lamp does not illuminate: Approx. 12V
1F		○	Self-Diagnosis Checker (For Code No.)	For 3 sec. after ignition switch OFF → ON: Below 2.5V (Buzzer sounds) After 3 sec.: approx. 12V (Buzzer does not sound)		• With Self-Diagnosis Checker and test connector grounded • Buzzer sounds: Below 2.5V • Buzzer does not sound: Approx. 12V
1G		○	Igniter	0V	Approx. 0.8V	—
1H	○		Headlight switch	Headlight switch ON: Approx. 10—12V Headlight switch OFF: Below 2.5V		—
1I	○		Test connector	Test connector grounded: 0V Test connector not grounded: Approx. 12V		Green connector: 1-pin
1J	○		Rear defroster switch	Rear defroster switch ON: Below 2.5V Rear defroster switch OFF: Approx. 12V		—
1K		○	EC-AT control unit	Approx. 12V		—
1L		○	A/C relay	A/C switch ON: Below 2.5V A/C switch OFF: Approx. 12V		Blower motor: ON
1M	○		Speed sensor	0V or approx. 5V		—
1N	○		Power steering (P/S) switch	Constant approx. 12V	P/S: ON: 0V P/S: OFF: Approx. 12V	—
1O	○		A/C switch	A/C switch ON: Below 2.5V A/C switch OFF: Approx. 12V		Blower motor: ON
1P	○		Blower motor switch	Fan speed control ON (5th—8th position): Below 1.5V Others: Approx. 12V		—
1Q	○		Brake light switch	Brake pedal depressed: Approx. 12V Brake pedal released: 0V		—
1R	○		EC-AT control unit	P or N range: Below 2.5V Other ranges: Approx. 12V		—
1S	—	—	—	—		—
1T	○		Idle switch	Accelerator pedal depressed: Approx. 12V Accelerator pedal released: 0V.		—
1U	—	—	—	—		—
1V	○		Igniter	0V	Approx. 0.8—1.2V	Engine Signal Monitor: Green and red lights flash

Terminal	Input	Output	Connected to	Voltage (after warm-up)		Remark
				Ignition switch: ON	Idle	
2A	—	—	—	—	—	—
2B	○		Airflow meter (Vs)	Approx. 4V	Approx. 3.5—5V	—
2C	○		Oxygen sensor (Right side)	0V	0 ↔ 1V	• Cold engine: 0V at idle • After warming-up: Acceleration: 0.7—1.0V Deceleration: 0—0.2V
2D	○		Oxygen sensor (Left side)			
2E	○		Water thermosensor	Approx. 0.3—0.6V		—
2F	○		Throttle sensor	Throttle valve fully closed: Approx. 0.3V Throttle valve fully open: Approx. 4V		—
2G	—	—	—	—		—
2H	—	—	—	—		—
2I		○	Vref	4.5—5.5V		—
2J	○		EGR position sensor	Approx. 0.25—0.9V		—
2K	○		Intake air thermosensor (Airflow meter)	Approx. 2—3V at 20°C (68°F)		—
2L	○		Intake air thermosensor (Dynamic chamber)	Approx. 1—2V at 80°C (176°F)		—
2M	○		Knock control unit	Approx. 4—5V		—
2N	—	—	—	—		—
2O		○	Solenoid valve (Purge control)	Approx. 12V		Engine Signal Monitor: Green and red lights flash
2P	—	—	—	—		—
3A	—	—	Ground	0V		—
3B	—	—	Ground	0V		—
3C	—	—	Ground	0V		—
3D	—	—	Ground	0V		—
3E	○		Distributor (Ne-signal)	0V	Approx. 0.6—0.8V	Ne: Number of engine speed (RPM)
3F	—	—	Distributor (Ground)	0V		—
3G	○		Distributor (G1 signal)	0V	Approx. 0.6—0.8V	—
3H	○		Distributor (G2 signal)	0V	Approx. 0.6—0.8V	—
3I	—	—	—	—		—
3J		○	Solenoid valve (VRIS)	Approx. 12V	Approx. 13—14V	—
3K		○	Oxygen sensor heater relay	Below 2.5V		Above approx. 4,000 rpm: approx. 12V
3L	—	—	—	—		—
3M		○	Solenoid valve (PRC)	For 240 sec. after ignition switch OFF → ON: Below 2.5V	For 240 sec. after starting: Below 2.5V	During hot condition: Coolant temp.: Above 80°C (176°F) and Engine air temp.: Above 75°C (167°F)
				Approx. 12V		Other conditions
3N	—	—	—	—		—
3O		○	Solenoid valve (EGR, Vent side)	Approx. 12V		Initial acceleration: Engine Signal Monitor; green and red lights flash

Terminal	Input	Output	Connection to	Voltage (after warm-up)		Remark
				Ignition switch: ON	Idle	
3P		○	Solenoid valve (EGR, Vacuum side)	Approx. 12V		Initial acceleration: Engine Signal Monitor; green and red lights flash
3Q		○	Idle Speed Control (ISC) valve	Approx. 12V	Approx. 9—11V	Engine Signal Monitor: Green and red lights flash
3R	—	—	—	—		—
3S	—	—	—	—		—
3T	—	—	—	—		—
3U		○	Injector No.1	Approx. 12V		Engine Signal Monitor: Green and red lights flash
3V		○	Injector No.2			
3W		○	Injector No.3			
3X		○	Injector No.4			
3Y		○	Injector No.5			
3Z		○	Injector No.6			

Control Unit Connector (Control Unit Side)

3.0L ENGINE SPECIFICATIONS CHART—1992 929

Item		Engine	JE DOHC
Idle speed		rpm	*680—720 (Prange)
Ignition timing		BTDC	*11—13°
Fuel pump			
Maximum output pressure		kPa (kg/cm², psi)	Below 638 (6.5, 92)
Fuel filter			
Type	Low-pressure side		Nylon element (in fuel pump)
	High-pressure side		Paper element
Pressure regulator			
Regulating pressure		kPa (kg/cm², psi)	255—314 (2.6—3.2, 37—46)
Injector			
Type			Side-feeding
Type of drive			Voltage
Resistance		Ω	12—16 (20°C [68°F])
BAC valve			
Resistance		Ω	10.7—12.3 (20°C [68°F])
Solenoid valve (purge)			
Resistance		Ω	30—34 (20°C [68°F])
Solenoid valve (EGR)			
Resistance (vacuum side)		Ω	Max. 30 (20°C [68°F])
Resistance (vent side)		Ω	Max. 30 (20°C [68°F])
Water thermosensor			
Resistance	kΩ	-20°C (-4°F)	14.6—17.8
		20°C (68°F)	2.2—2.7
		80°C (176°F)	0.2—0.4
Fuel tank		20°C (68°F)	
Capacity		liters (US gal, Imp gal)	70 (18.5, 15.4)
Air cleaner			
Element type			Dry type
Fuel			
Specification			Premium unleaded (RON 91 or higher)

* TEN terminal of diagnosis connector grounded.

3.0L ENGINE COMPONENT DESCRIPTIONS—1992 929

Component	Function	Remark
Air cleaner	Filters air entering throttle body	
Air control motor	Closes/opens variable exhaust shutter valve	
Airflow meter	Detects amount of intake air; sends signal to ECU	Intake air thermosensor included
Air valve	Supplies bypass air into dynamic chamber(engine cold)	• Engine speed increased to shorten warm-up period • Thermowax type • Installed in BAC valve
Atmospheric pressure sensor	Detects atmospheric pressure	In ECU
Bypass air control (BAC) valve	Supplies bypass air into dynamic chamber	Consists of air valve and solenoid valve (ISC)
Catalytic converter	Reduces HC, CO, and NOx by chemical reaction	Monolith type
Charcoal canister	Stores fuel tank fumes (engine stopped)	
Check valve (three-way)	Releases excessive pressure or vacuum in fuel tank to atmosphere	
Check valve (two-way)	Controls pressure in fuel tank	
Circuit opening relay	Voltage for fuel pump while engine running	
Crank angle sensor 1 (Ne 1 signal)	Detects crank angle at 120° intervals; sends signal to ECU	
Crank angle sensor 2 (Ne 2 signal)	Detects crank angle at 60° intervals; sends signal to ECU	
Diagnosis connector	Concentrated connector 1. EGI self-diagnosis 2. EC-AT self-diagnosis 3. A/C self-diagnosis 4. Cruise control self-diagnosis 5. Initial set 6. Fuel pump check 7. Engine speed output	18-pin (located near air cleaner)
Dynamic chamber	Interconnects all cylinders	
EGR control valve	Recircurates portion of exhaust gas	
EGR control valve position sensor	Detects EGR valve position	Installed in EGR control valve
Engine control unit (ECU)	**Detects following:** 1. A/C operation 2. Air/fuel ratio (oxygen concentration) 3. Atmospheric pressure 4. Braking signal 5. Cranking signal 6. E/L operation 7. Engine coolant temperature 8. Engine speed 9. Ignition ON signal 10. In-gear condition 11. Intake air amount 12. Intake air temperature 13. Engine air temperature 14. No.1 and No. 4 piston TDC 15. P/S operation 16. Test mode signal (ignition timing, idle speed, service code no.) 17. Throttle valve opening angle 18. Throttle valve fully closed 19. Shift condition 20. EGR control valve	1. A/C switch 2. Oxygen sensor 3. Atmospheric pressure sensor 4. Stoplight switch 5. Ignition switch (START position) 6. Blower motor switch, headlight switch, fog light switch and rear window defroster switch 7. Water thermosensor 8. Crank angle sensor 1 (Ne 1 signal) Crank angle sensor 2 (Ne 2 signal) 9. Ignition switch 10. Inhibitor switch 11. Airflow meter 12. Intake air thermosensor (in airflow meter) 13. Intake air thermosensor (on dynamic chamber) 14. Crank angle sensor (G signal) 15. P/S pressure switch 16. Diagnosis connector (TEN terminal) 17. Throttle sensor 18. Idle switch (in throttle sensor) 19. EC-AT control unit 20. EGR position sensor

3.0L ENGINE COMPONENT DESCRIPTIONS—1992 929

Component	Function	Remark
Engine control unit (ECU)-cont'd	**Controls operation of the following:** 1. A/C (cut-off) 2. Self-diagnosis function 3. Fuel injection system 4. Idle speed control 5. Ignition system (electronic spark advance) 6. Monitor function 7. Purge control system 8. VICS 9. Variable exhaust system 10. EGR control system	1. A/C relay 2. Self-Diagnosis Checker 3. Injector 4. Solenoid valve (ISC) 5. Igniter 6. Monitor lamp (Self-Diagnosis Checker) 7. Solenoid valve (purge control) 8. Solenoid valve (VICS) 9. Air control motor 10. Solenoid valves (EGR)
Fuel filter	Filters particles from fuel	
Fuel pump	Provides fuel to injectors	• Operates while engine running • Installed in fuel tank
Fuel vapor valve	Prevent liquid fuel from entering charcoal canister	
Igniter	Receives spark signal from ECU and generates high voltage in ignition coil	
Ignition switch	Engine cranking signal sends to ECU	START position
Inhibitor switch	Detects in-gear condition; sends signal to ECU	Switch ON in N and P ranges
Injector	Injects fuel into intake port	Controlled by signals from ECU
Intake manifold	Supplies intake air to all cylinders	
Intake air thermosensor	Detects intake air temperature; sends signal to ECU	Installed in airflow meter
	Detects engine air temperature; sends signal to ECU	Installed on dynamic chamber
Knock sensor	Detects engine knocking; sends signal to ECU	
Main relay	Supplies current to output devices and ECU	
MIL (malfunction indicator lamp)	Lamp illuminates if input device malfunctions	Installed in instrument cluster
One-way check valve	Prevents misoperation of shutter valve (for VICS) during heavy-load operations	
Oxygen sensor	Detects oxygen concentration; sends signal to ECU	• Zirconia ceramic and platinum coating • With heater
PCV valve	Controls blowby gas introduced into engine	
Pressure regulator	Adjusts fuel pressure supplied to injectors	
P/S pressure switch	Detects front P/S operation	P/S: ON when steering wheel turned
Pulsation damper	Absorbs fuel pulsations	
Resonance chamber	Reduces intake air noise	
Shutter valve actuator	Closes/opens shutter valve to improve torque characteristics	For variable inertia charging system (VICS)
Solenoid valves (EGR) vent side, vacuum side	Controls EGR control valve operation	Controlled by duty signal from ECU
Solenoid valve (ISC)	Supplies bypass air into dynamic chamber	• Controlled by duty signal from ECU • Works independent of idle-up • Installed in BAC valve
Solenoid valve (purge control)	Controls evaporative fumes from charcoal canister to intake manifold	Controlled by duty signal from ECU
Solenoid valve (VICS)	Controls vacuum to shutter valve actuator	
Speedometer	Detects vehicle speed; sends signal to ECU	Installed in instrument cluster
Stoplight	Detects braking; sends signal to ECU	
Throttle body	Controls intake air amount	
Water thermosensor	Detects coolant temperature; sends signal to ECU	Installed on intake manifold

3.0L ENGINE RELATIONSHIP CHART—1992 929

INPUT DEVICE \ OUTPUT DEVICE		COLD START INJECTOR	AIR CONTROL MOTOR	INJECTOR: FUEL INJECTION AMOUNT	INJECTOR: FUEL INJECTION TIMING	CIRCUIT OPENING RELAY	IGNITER	A/C RELAY (A/C CUT-OFF)	SOLENOID VALVE: ISC (BAC VALVE)	SOLENOID VALVE: PURGE	SOLENOID VALVE: VICS	SOLENOID VALVE: EGR (VENT VACUUM)	SELF-DIAGNOSIS CHECKER (SERVICE CODE)	SELF-DIAGNOSIS CHECKER (MONITOR LAMP)
CRANK ANGLE SENSOR (IN DISTRIBUTOR)	Ne 1-SIGNAL		○	○	○	○	○		○	○	○	○	○	
	G-SIGNAL				○		○						○	
CRANK ANGLE SENSOR	Ne 2-SIGNAL		○	○	○	○	○		○	○	○	○	○	
THROTTLE SENSOR	IDLE SWITCH			○	○		○		○	○		○		○
	THROTTLE SENSOR			○	○		○	○	○		○	○	○	
WATER THERMOSENSOR		○		○			○		○	○		○	○	
INTAKE AIR THERMOSENSOR	AIRFLOW METER			○					○	○			○	
	DYNAMIC CHAMBER			○			○						○	
OXYGEN SENSOR				○						○			○	○
A/C SIGNAL (A/C AMPLIFIER) *1							○	○	○					○
P/S PRESSURE SWITCH									○					
START SIGNAL (IGNITION SWITCH)		○		○		○	○	○	○		○			
STOPLIGHT SWITCH				○										○
E/L SIGNAL *2									○					○
DIAGNOSIS CONNECTOR (TEN TERMINAL)							○		○				○	○
REDUCE TORQUE SIGNAL (EC-AT CU)							○							
INHIBITOR SIGNAL (EC-AT CU)				○			○	○	○	○	○	○		○
ATMOSPHERIC PRESSURE SENSOR				○			○		○	○			○	
SPEEDOMETER				○			○		○	○			○	
AIRFLOW METER			○	○			○			○		○	○	
KNOCK SENSOR							○						○	
EGR POSITION SENSOR							○					○	○	

*1 Based on A/C load, ECU receives high, low, or OFF signal input.
*2 E/L SIGNAL: Blower fan control switch maximum position, headlight switch ON, rear window defroster switch ON, or fog light switch ON.

3.0L ENGINE ENGINE CONTROL OPERATION CHART—1992 929

Note
- **The data in this chart is for reference only.**

Input Device and Engine Conditions

ENGINE CONDITIONS \ INPUT DEVICES	APPROXIMATE TIME (BASED ON 10—16°C OR 50—60°F AMBIENT)	SENSORS: CRANK ANGLE SENSOR 1 (NE 1, G SIGNAL)	SENSORS: CRANK ANGLE SENSOR 2 (NE 2 SIGNAL)	SENSORS: WATER THERMOSENSOR	SENSORS: OXYGEN SENSOR	SENSORS: AIRFLOW METER	SENSORS: INTAKE AIR THERMOSENSOR (IN AIRFLOW METER)	THROTTLE SENSOR: VT TERMINAL	THROTTLE SENSOR: IDL TERMINAL
CRANKING —COLD ENGINE • COLD AIR • COLD COOLANT	ZERO	↕	↕	↕	Signal has no effect on ECU	Signal has no effect on ECU	Signal has no effect on ECU	Signal has on effect on ECU	Signal has no effect on ECU
COLD START —FAST IDLE • COLD AIR • COLD COOLANT	ONE MINUTE	↕	↕	Cool to warm: Medium voltage (2.5V and dropping)	↕	Low volume airflow: High voltage (Above. 2 V)	↕	Closed throttle: Low voltage (0.1-1.1V)	Closed throttle: Low voltage (Bellow 0.1V)
COLD DRIVEAWAY —PART THROTTLE • WARM AIR • WARM COOLANT	TWO MINUTES	↕	↕	↕	Sensor cold: Low to high voltage (0–1V)	↕	Cool [Below 20°C (68°F)]: Above 2.5V	↕	↕
WARM DIRVEAWAY —PART THROTTLE • WARM AIR • WARM COOLANT	THREE MINUTES	Ne 1 signal (Sends crank angle signal to ECU)	↕	↕	Sensor hot: switching from high voltage (1V)	↕	↕	↕	↕
HOT CRUISE • WARM AIR • WARM COOLANT	↕	G signal (Sends No. 1 and No. 4 cylinders TDC signal to ECU)	Sends crank angle signal to ECU	↕	to low voltage (0.1V)	Moderate volume airflow: Low to medium voltage (0.5-2.5V)	↕	Part throttle: Medium voltage (0.5-4.0V)	Open throttle: High voltage (VB)
HOT ACCELERATION —60%THROTTLE	↕	↕	↕	↕	↕	↕	Cool to warm [−20-40°C (−4-104°F)] 1.5-4.3V	↕	↕
HOT ACCELERATION —WIDE OPEN THROTTLE	MORE THAN FOUR MINUTES	↕	↕	Hot: Low voltage (Below 0.5V)	High voltage (1V)	↕	↕	Wide open throttle: High voltage (Above 4V)	↕
DECELERATION —CLOSED THROTTLE	↕	↕	↕	↕	Low voltage (Bellow 0.4V)	Low volume airflow: (Above 2V)	↕	Closed throttle: Low voltage (0.1-1.1V)	Closed throttle: Low voltage (Bellow 0.1V)
HOT CURB IDLE —EXTENDED	↕	↕	↕	↕	Switching from high to low voltage (0.75-0.25V)	↕	↕	↕	↕
HOT ENGINE SHUTDOWN		OFF	OFF	OFF	Sensor hot: Low voltage (0.1V) until sensor cools	OFF	OFF	OFF	OFF

VB: Battery voltage

VB: Battery voltage

A/C SWITCH	P/S PRESSURE SWITCH	E/L SIGNAL	IGNITION SWITCH: START POSITION	IGNITION SWITCH: ON POSITION	DIAGNOSIS CONNECTOR TEN TERMINAL (TEST TERMINAL)	EC-AT CU: INHIBITOR SWITCH	EC-AT CU: REDUCE TORQUE SIGNAL
Signal has no effect on ECU	Signal has no effect on ECU	Signal has no effect on ECU	Sends signal to ECU (VB)	Signal has no effect on ECU	Signal has no effect on ECU	Signal has no effect on ECU	OFF (VB)
A/C switch ON: Sends signal to ECU [Below (1.5-3.5V) A/C switch OFF: No signal to ECU (4.5-5.5V)	Steering wheel turned: Low voltage signal to ECU (Below 1V) Steering wheel straight ahead: High voltage signal to ECU (VB)	Headlight, fog light, rear window defroster or blower switch* ON: Low voltage signal to ECU (Below 1V) OFF: High voltage signal to ECU (VB)	No signal to ECU	Sends signal to ECU (VB)	Terminal not grounded: High voltage signal to ECU (VB) Terminal grounded: Low voltage signal to ECU (Bellow 1V)	In N or P ranges: No signal to ECU (Bellow 1V) In D range: No signal to ECU (VB) In N or P ranges: Sends signal to ECU (Bellow 1V)	Downshift or upshift ON: (Bellow 1V) OFF
OFF	OFF	OFF	OFF	OFF	OFF	OFF	↕

*1 Blower switch : maximum position

3.0L ENGINE ENGINE CONTROL OPERATION CHART – 1992 929

Output Devices and Engine Conditions

ENGINE CONDITIONS / OUTPUT DEVICES	APPROXI-MATE TIME (BASED ON 10–16°C OR 50–60°F AMBIENT)	INJECTOR		SOLENOID VALVE		BAC VALVE		A/C RELAY (A/C CUT-OFF)	IGNITER	SOLENOID VALVE (EGR)		AIR CONTROL MOTOR (VARIABLE EXHAST SYSTEM)
		INJECTION	INJECTION TIMING	PURGE	VICS	ISC	AIR VALVE			VENT SIDE	VACUUM SIDE	
CRANKING —COLD ENGINE • COLD AIR • COLD COOLANT	ZERO				OFF (Shutter valve close)			OFF (A/C cut)	Ignition timing BTDC 7°			
COLD START —FAST IDLE • COLD AIR • COLD COOLANT	ONE MINUTE	Rich		OFF (Purge cut)	ON (Shutter valve open)	Large amount of by-pass air	Open coolant tempera-ture: bellow 40°C (104°F)	ON (A/C cut: approx. 5sec.)		OFF (No vacuum to EGR valve)	OFF (Atmos-pheric pres-sure to EGR valve)	
COLD DRIVEAWAY —PART THROTTLE • COLD AIR • COLD COOLANT	TWO MINUTES											Shutter valve close
WARM DRIVEAWAY —PART THROTTLE • WARM AIR • WARM COOLANT	THREE MINUTES	Rich and Lean	Sequential					ON (A/C ON)				
HOT CRUISE • WARM AIR • WARM COOLANT				Operates (Duty signal [Purge gas amount] change)	ON or OFF (OFF: Engine speed above 1,100rpm. bellow 4,800rpm. and throttle open)	Small amount of by-pass air			Ignition spark advance signal	System operates (amount of EGR changes)		
HOT ACCELERATION —60% THROTTLE		Rich										
HOT ACCELERATION —WIDE OPEN THROTTLE	MORE THAN FOUR MINUTES						Closed	ON (A/C cut: approx. 5sec.)				Shutter valve open
DECELERATION —CLOSED THROTTLE		Fuel cut				Large and small amount of by-pass air						
HOT CURB IDLE —EXTENDED		Rich and Lean	Sequential	OFF (Purge cut)	ON	Small aount of by pass air		ON (A/C ON)		OFF	OFF	Shutter valve close
HOT ENGINE SHUT-DOWN	—	Does not inject		OFF	OFF	OFF		OFF	OFF	Does no operate		

3.0L ENGINE TROUBLESHOOTING GUIDE—1992 929

DIAGNOSTIC INDEX

NO.	TROUBLESHOOTING ITEM		DESCRIPTION
1	Will not crank or cranks slowly		Refer to Engine Electrical System
2	Cranks normally but will not start	No combustion	Engine cranks at normal speed but shows no sign of firing
3		Partial combustion — When engine cold	Engine cranks at normal speed but shows only partial combustion and will not continue to run
4		Partial combustion — After warm-up	Engine starts normally when cold, but will not start after running and being hot soaked
5	Cranks normally but had to start	Always	Engine cranks at normal speed but requires excessive cranking time before starting
6		When engine is cold	Same condition as No.5 when engine cold Restarts normally after warm-up
7		After warm-up	Same condition as No.5 after running and being hot soaked Starts normally when cold
8	Rough idle Low idle speed Engine stalls at idle	Always	Engine stalls or vibrates excessively at idle
9		During warm-up	Engine stalls or vibrates excessively at idle during warm-up
10		After warm-up	Engine runs normally at idle during warm-up but stalls or vibrates excessively after warm-up
11		When A/C, P/S, or E/L ON	Engine stalls or vibrates excessively at idle when A/C, P/S, or E/L ON
12	Rough idle/Engine stalls just after starting		Engine stalls or vibrates excessively just after starting (acceleration from idle)
13	High idle speed after warm-up		Idle speed excessively high after warm-up
14	Idle moves up and down/idle hunting		Engine speeds up and down periodically at idle
15	Engine stalls on deceleration		Engine unexpectedly stops running during or after deceleration
16	Hesitates/Stumbles on acceleration		Flat spot occurs just after accelerator is depressed, or mild jerking occurs during acceleration
17	Surges while cruising		Unexpected change in engine speed that is usually repetitive
18	Lack of power or poor acceleration		Performance poor under load when throttle valve is wide open Maximum speed reduced Performance poor during acceleration
19	Runs rough during deceleration/Backfire		Engine runs rough during deceleration, and abnormal combustion occurs in exhaust system
20	Knocking		Abnormal combustion accompanied by audible "pinging" noise
21	Fuel odor		Gasoline odor in cabin
22	A/C does not work		Blower fan operates but magnetic clutch does not operate

2 CRANKS NORMALLY BUT WILL NOT START (NO COMBUSTION)

DESCRIPTION
- Engine cranks at normal speed but shows no sign of "firing"
- Battery in normal condition
- Throttle valve not held fully open while cranking (dechoke system does not operated)
- Fuel in tank

[TROUBLESHOOTING HINTS]

① No spark
- Ignition control malfunction
- Ignition system component malfunction

② No fuel injection
- Fuel pump inoperative
- injectors inoperative

③ Low fuel line pressure
④ Low engine compression

STEP	INSPECTION		ACTION
1	Is a strong blue spark visible at disconnected high-tension lead? 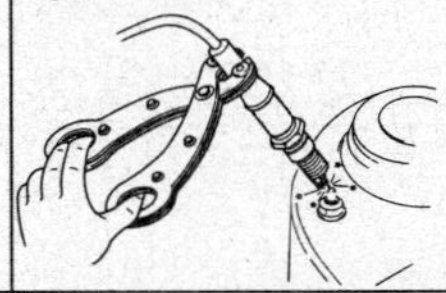	Yes	Go to Step 3
		No	Go to Step 2
2	Is "00" displayed on Self-Diagnosis Checker while ignition switch ON? 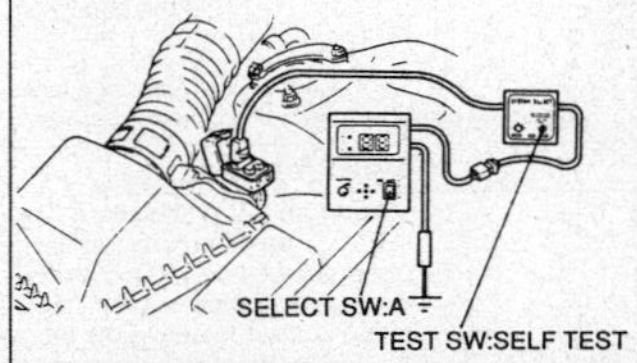	Yes	Check ignition system (Refer to Troubleshooting "Misfire")
		No	**Service Code No. displayed** Check for cause (Refer to specified check sequence)
3	Connect diagnosis connector terminals F/P and GND with jumper wire; is there a fuel pump operating sound while ignition switch ON? 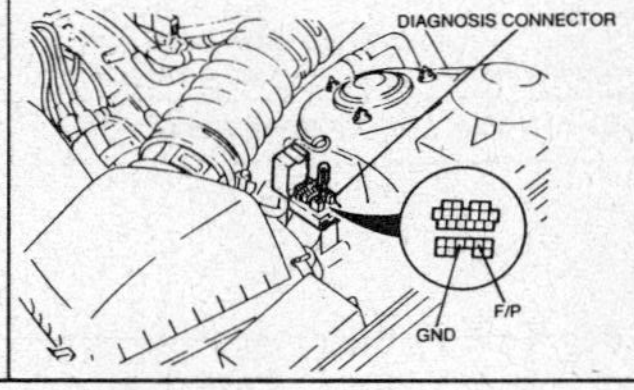	Yes	Check if engine starts in this condition? ➪ If it does, check ECU terminal 3T ➪ If it does not, go to Step 5
		No	Go to next step

STEP	INSPECTION		ACTION
4	In same condition as Step 3, is there approx. 12V at fuel pump connector L/O wire with ignition switch ON? 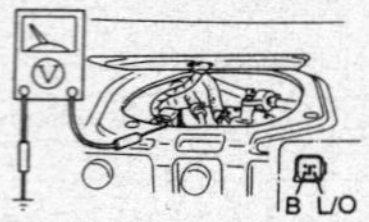	Yes	Check continuity of fuel pump between terminals L/O and B
		No	Check circuit-opening relay
5	Is there injector operating sounds while engine is being cranked?	Yes	Go to Step 7
		No	Go to next step
6	Is there approx. 12V at ECU terminals 3U, 3V, 3W, 3X, 3Y, and 3Z while ignition switch ON? 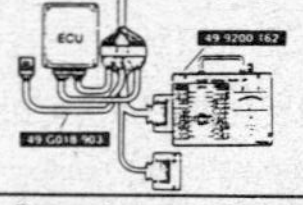	Yes	Check injectors ➪ If OK, go to Step 10 ➪ If not OK, replace injectors
		No	Check for open circuit in wiring between main relay (EGI relay) and ECU terminals
7	Connect diagnosis connector terminals F/P and GND with jumper wire; is fuel line pressure correct while ignition switch ON? **Fuel Line pressure:** **255—314 kPa (2.6—3.2 kg/cm², 37—46 psi)** 	Yes	Go to next step
		No	**Low pressure** Check fuel line pressure while pinching fuel return hose ➪ If pressure **quickly** increases, check pressure regulator ➪ If pressure **gradually** increases, check for clogging between fuel pump and pressure regulator If hose is not clogged, check fuel pump maximum pressure
			High pressure Check if fuel return hose is clogged or restricted? ➪ If not, replace pressure regulator ➪ If it is, repair or replace hose
8	Is engine compression correct? **Engine compression:** **1079 kPa (11.0 kg/cm², 156 psi)-270 rpm**	Yes	Go to next step
		No	Check engine condition • Worn pistons, piston rings, or cylinder walls • Damaged cylinder head gasket • Distorted cylinder head • Improper valve seating • Valve sticking in guide
9	Are all spark plugs OK? 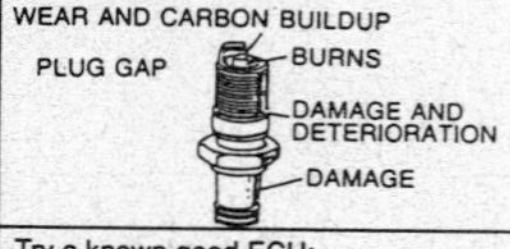	Yes	Go to next step
		No	Go to next step
10	Try a known good ECU; does condition improve?		

3	CRANKS NORMALLY BUT WILL NOT START (PARTIAL COMBUSTION) — WHEN ENGINE COLD
DESCRIPTION	• Engine cranks at normal speed but shows only partial combustion and will not continue to run • Battery in normal condition • Fuel in tank

[TROUBLESHOOTING HINTS]
① Air/Fuel mixture too rich
- Air cleaner element clogged
- Airflow meter stuck

② Air/Fuel mixture too lean
- Cold start injector system malfunction
- Fuel injection control malfunction (correction for coolant temperature)
- Low fuel line pressure
- Air leakage in intake air system

③ Low engine compression

STEP	INSPECTION		ACTION
1	Is "00" displayed on Self-Diagnosis Checker while ignition switch ON? SYSTEM SELECT:1 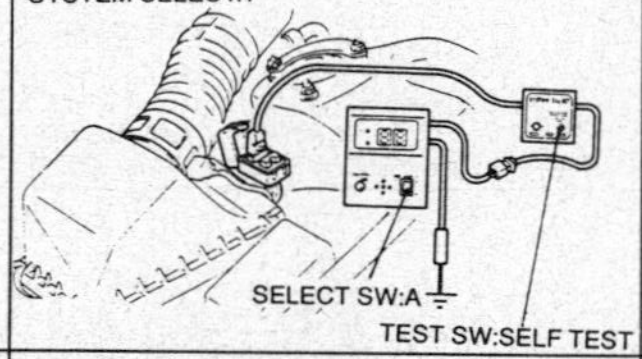	Yes	Go to next step
		No	**Service Code No. displayed** Check for cause (Refer to specified check sequence)
2	Is a strong blue spark visible at each disconnected high-tension lead? 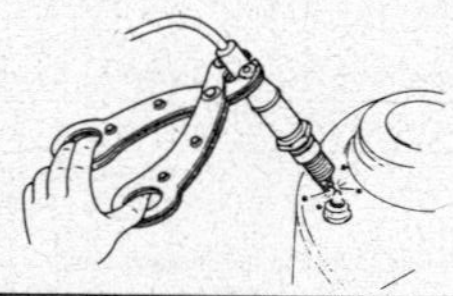	Yes	Go to next step
		No	Check distributor cap and rotor
3	Connect diagnosis connector terminals F/P and GND with a jumper wire; is fuel line pressure correct with ignition switch ON **Fuel line pressure:** **255—314 kPa (2.6—3.2 kg/cm², 37—46psi)** 	Yes	Go to next step
		No	**Low pressure** Check fuel line pressure while pinching fuel return hose ➪ If pressure **quickly** increases, check pressure regulator ➪ If pressure **gradually** increases, check for clogging between fuel pump and pressure regulator If hose is not clogged, check fuel pump maximum pressure
			High pressure Check if fuel return hose is clogged or restricted ➪ If not, replace pressure regulator ➪ If it is, repair or replace hose

3.0L ENGINE TROUBLESHOOTING GUIDE—1992 929

STEP	INSPECTION		ACTION
4	Are ECU terminal voltages OK? (Especially 1C and 2E)	Yes	Go to next step
		No	Check for cause
5	Is there air leakage of intake air system components?	Yes	Repair or replace
		No	Go to next step
6	Does airflow meter move smoothly?	Yes	Go to next step
		No	Repair or replace
7	Is engine compression correct?	Yes	Go to next step
	Engine compression: **981 kPa (10.0 kg/cm², 142 psi)-300 rpm**	No	Check engine condition • Worn pistons, piston rings, or cylinder walls • Damaged cylinder head gasket • Distorted cylinder head • Improper valve seating • Valve sticking in guide
8	Are all spark plugs OK?	Yes	Go to next step
		No	Clean or replace
9	Is there approx. 10V at cold start injector connector (B/L) wire while cranking engine?	Yes	Go to next step
		No	Check for open circuit in wiring between cold start injector and ignition switch
10	Is there approx. 10V at cold start thermoswitch connector (B/O) wire while cranking engine with cold start thermoswitch connector disconnected? (Engine coolant temperature: Below 0°C [32°F])	Yes	Check cold start thermoswitch ⇨ If OK, go to next step ⇨ If not OK, replace thermoswitch
		No	Check cold start injector relay ⇨ If not OK, replace relay ⇨ If OK, check ECU terminal voltage (3I) If ECU terminal voltage is OK, go to next step
11	Is cold start injector OK?	Yes	Go to next step
		No	Replace
12	Try a known good ECU; does condition improved?		

4	CRANKS NORMALLY BUT WILL NOT START (PARTIAL COMBUSTION) — AFTER WARM-UP		
DESCRIPTION	• Engine cranks at normal speed and shows partial combustion but will not continue running after being hot soaked • Battery in normal condition • Engine starts normally when cold		
[TROUBLESHOOTING HINTS]	① Air/Fuel mixture too rich • Insufficient fuel injection control (correction for coolant temperature) • Injector fuel leakage ② Vapor lock • Fuel pressure not held in fuel line after engine stopped		
STEP	**INSPECTION**		**ACTION**
1	Warmup engine to normal operating temperature and stop it Connect diagnosis connector terminals F/P and GND with jumper wire for 3 minutes with ignition switch ON; will engine start? 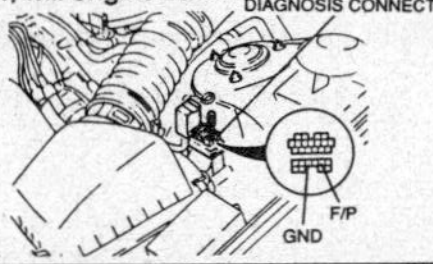 	Yes	Change to another brand of fuel
		No	Go to next step
2	In same condition as Step 1; is fuel line pressure correct with ignition switch ON? **Fuel line pressure:** **255—314 kPa (2.6—3.2 kg/cm², 37—46 psi)** 	Yes	Go to next step
		No	**Low pressure** Check fuel line pressure while pinching fuel return hose ⇨ If pressure **quickly** increases, check pressure regulator ⇨ If pressure **gradually** increases, check fuel line and filter for clogging If hose is not clogged, check fuel pump maximum pressure **High pressure** Check if fuel return hose is clogged or restricted ⇨ If OK, replace pressure regulator ⇨ If not OK, repair or replace hose
3	In same condition as Step 2; is fuel line pressure held after ignition switch is turned OFF? **Fuel line pressure: More than 147 kPa (1.5 kg/cm², 21 psi) for 5 min.**	Yes	Go to Step 5
		No	Go to next step

STEP	INSPECTION		ACTION
4	Plug outlet of pressure regulator; is fuel line pressure held after ignition switch is turned OFF? **Fuel line pressure: More than 147 kPa (1.5 kg/cm², 21 psi) for 5 min.**	Yes	Replace pressure regulator
		No	Check fuel pump hold pressure ⇨ If OK, check injector for fuel leakage ⇨ If not OK, replace fuel pump
5	Disconnect water thermosensor connector; does engine start? 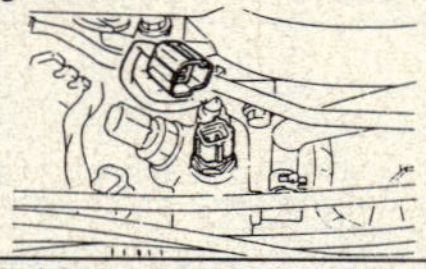	Yes	Check water thermosensor ⇨ If OK, check connections of water thermosensor and ECU terminal 2Q ⇨ If not OK, replace water thermosensor
		No	Go to next step
6	Is "00" displayed on Self-Diagnosis Checker while ignition switch ON? SYSTEM SELECT:1 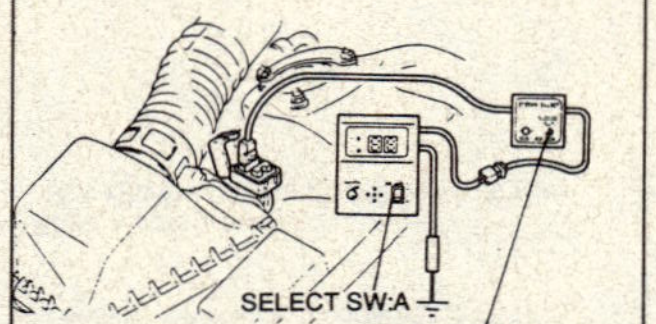 SELECT SW:A / TEST SW:SELF TEST	Yes	Go to next step
		No	**Service Code No. displayed** Check for cause (Refer to specified check sequence)
7	Are ECU terminal voltages OK? (Especially 1C, and 2E)	Yes	Go to next step
		NO	Check for cause
8	Try a known good ECU; does condition improve?	Yes	Replace ECU
		No	Change to another brand of fuel

5	CRANKS NORMALLY BUT HARD TO START — ALWAYS		
DESCRIPTION	• Engine cranks at normal speed but requires excessive cranking time before starting • Battery in normal condition • Engine runs normally at idle (if idle condition not OK, refer to "Rough idle" [Nos. 8—12])		
[TROUBLESHOOTING HINTS]	① Air/Fuel mixture too lean • Fuel injection control malfunction (correction for coolant temperature) • Low fuel pressure • Air leakage in intake air system ② Air/Fuel mixture too rich • Air cleaner element clogged • Airflow meter stuck ③ Low engine compression		
STEP	INSPECTION		ACTION
1	Is "00" displayed on Self-Diagnosis Checker while ignition switch ON? SYSTEM SELECT:1 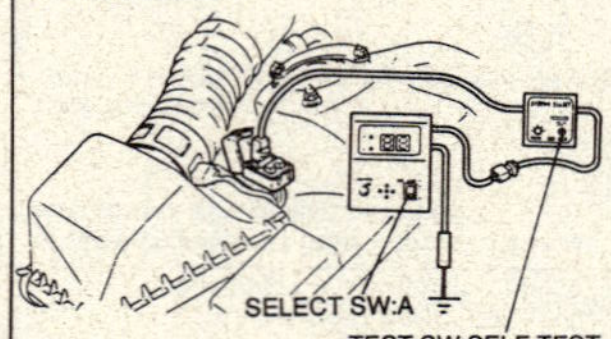 SELECT SW:A / TEST SW:SELF TEST	Yes	Go to next step
		No	**Service Code No. displayed** Check for cause (Refer to specified check sequence)
2	Is intake manifold vacuum correct at idle? **Vacuum: More than 500 mmHg (19.7 inHg)**	Yes	Go to next step
		No	Check for air leakage of intake air system components
3	Is air cleaner element clean?	Yes	Go to next step
		No	Replace air cleaner element
4	Does engine start easily when throttle valve is one-quarter open?	Yes	Check throttle valve for carbon deposits; then go to Step 6
		No	Go to next step
5	Is fuel line pressure correct at idle? **Fuel line pressure: 255—314 kPa (2.6—3.2 kg/cm², 37—46 psi) (vacuum hose to pressure regulator dis connected)** 	Yes	Go to next step
		No	**Low pressure** Check fuel line pressure while pinching fuel return hose ⇨ If pressure **quickly** increases, check pressure regulator ⇨ If pressure **gradually** increases, check for clogging between fuel pump and pressure regulator If hose is not clogged, check fuel pump maximum pressure

STEP	INSPECTION		ACTION
6	Disconnect ISC valve connector before warm-up; does engine speed decrease? 	Yes	Go to next step
		No	Check ISC valve
7	Are ECU terminal voltage OK? (Especially 1C and 2E)	Yes	Go to next step
		No	Check for cause
8	Connect diagnosis connector terminals F/P and GND with jumper wire; does engine start normally? 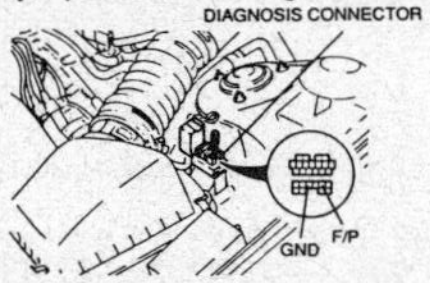	Yes	Check circuit-opening relay ➪ If relay OK, repair or replace wire harness ➪ If relay not OK, replace relay
		No	Go to next step
9	Connect System Selector to diagnosis connector and set Test Switch to SELF TEST; Is ignition timing correct at idle after warm up?	Yes	Go to next step
	Ignition timing (BTDC): 11—13° 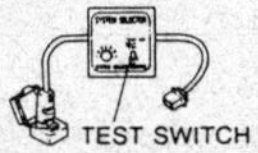	No	Adjust ignition timing
10	Is engine compression correct?	Yes	Go to next step
	Engine compression (Minimum): 1079 kPa (11.0 kg/cm^2, 156 psi)-270 rpm	No	Check engine condition • Worn pistons, piston rings, or cylinder walls • Defective cylinder head gasket • Distorted cylinder head • Improper valve seating • Valve sticking in guide
11	Try a known good ECU; does condition improve?		

6	CRANKS NORMALLY BUT HARD TO START — WHEN ENGINE COLD
DESCRIPTION	• Engine cranks at normal speed but requires excessive cranking time before starting • Battery in normal condition • Restarts OK after warm up • Engine runs normally at idle (if idle condition not OK, refer to "Rough idle" [Nos. 8—12])

[TROUBLESHOOTING HINTS]

① Air/Fuel mixture too rich
- Airflow meter sticking
- Air cleaner element clogged
- Idle speed control malfunction

② Air/Fuel mixture too rich
- Cold start injector system malfunction
- Fuel injection control malfunction (correction for coolant temperature)

③ Poor atomization
- Low RVP (summer) fuel used in cold weather

STEP	INSPECTION		ACTION
1	Is "00" displayed on Self-Diagnosis Checker while ignition switch ON? SYSTEM SELECT:1 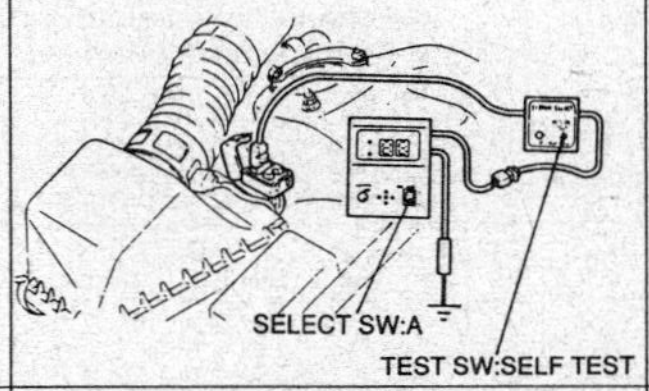	Yes	Go to next step
		No	**Malfunction Code No. displayed** Check for cause (Refer to specified check sequence)
2	Are ECU terminal voltages OK? (Especially 1C and 2E)	Yes	Go to next step
		No	Check for cause
3	Does engine start easily when throttle valve is one-quarter open? 	Yes	Is ISC valve OK? ➪ If OK, check air valve ➪ If not OK, replace BAC valve
		No	Go to next step
4	Is intake manifold vacuum correct at idle?	Yes	Go to next step
	Vacuum: More than 500 mmHg (19.7 inHg)	No	Check for air leakage of intake air system components
5	Is air cleaner element clean?	Yes	Go to next step
		No	Replace air cleaner element
6	Disconnect ISC valve connector before warm-up; does engine speed decrease 	Yes	Go to next step
		No	Check ISC valve

STEP	INSPECTION		ACTION
7	Connect diagnosis connector terminals F/P and GND with jumper wire; is fuel line pressure correct while ignition switch ON? **Fuel line pressure:** **255—314 kPa (2.6—3.2 kg/cm², 37—46 psi)** 	Yes	Go to next step
		No	**Low pressure** Check fuel line pressure while pinching fuel return hose ➪ If pressure **quickly** increases, check pressure regulator ➪ If pressure **gradually** increases, check fuel line for clogging between fuel pump and pressure regulator If hose is not clogged, check fuel pump maximum pressure **High pressure** Check if fuel return hose is clogged or restricted ➪ If OK, replace pressure regulator ➪ If not OK, repair or replace hose
8	Is there approx. 10V exists at cold start injector connector (B/L) wire while cranking engine?	Yes	Go to next step
		No	Check for open circuit in wiring between cold start injector and ignition switch
9	Disconnect cold start thermoswitch connector; is there approx. 10V exists at cold start thermoswitch connector wire (B/O) while cranking engine? (Engine coolant temperature; Below 0°C [32°F])	Yes	Check the cold start thermoswitch ➪ If OK, go to next step ➪ If not OK, replace thermoswitch
		No	Check the cold start injector relay ➪ If not OK, replace relay ➪ If OK, check ECU terminal voltage (3I) If ECU terminal voltage is OK, go to next step
10	Is cold start injector OK?	Yes	Go to next step
		No	Replace injector
11	Try a known good ECU; does condition improve?	Yes	Replace ECU
		No	Change fuel to another brand

7	CRANKS NORMALLY BUT HARD TO START— AFTER WARM-UP
DESCRIPTION	• Engine cranks at normal speed but requires excessive cranking time before starting after running and hot soaked • Battery in normal condition • Engine starts normally when cold • Engine runs normally at idle [if idle condition not OK, refer to "Rough idle" (Nos 8–12)]

[TROUBLESHOOTING HINTS]
① Air/Fuel mixture too rich
- Fuel injection control malfunction
- Injector fuel leakage

② Vapor lock
- Fuel pressure not held in fuel line after engine stopped

STEP	INSPECTION		ACTION
1	Is "00" displayed on Self-Diagnosis Checker while ignition switch ON? 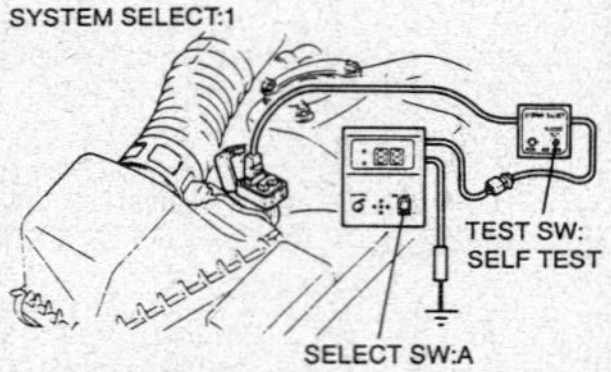	Yes	Go to next step
		No	**Service Code No. displayed** Check for cause (Refer to specified check sequence)
2	Are ECU terminal voltages are OK? (Especially 1C, 2D, 2Q, and 2T)	Yes	Go to next step
		No	Check for cause
3	Run engine at idle; is fuel line pressure held after ignition switch turned OFF? **Fuel line pressure: More than 147 kPa (1.5 kg/cm², 21 psi) for 5 min.**	Yes	Go to next step
		No	Plug outlet of pressure regulator and check if fuel line pressure is held after ignition switch turned OFF ➪ If OK, replace pressure regulator ➪ If not OK, check fuel pump hold pressure If fuel pump is OK, check injectors for fuel leakage
4	Warm up engine to normal operating temperature and stop it Connect diagnosis connector terminals F/P and GND with jumper wire for 3 minutes with ignition switch ON; does engine start easily? 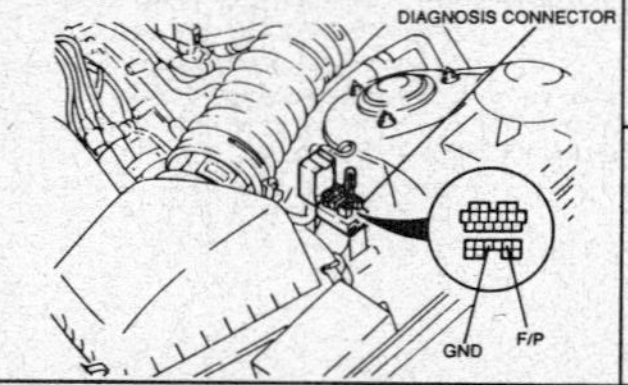	Yes	Change fuel to another brand
		No	Go to next step
5	Try known good ECU; does condition improve?	Yes	Replace ECU
		No	Change fuel to another brand

8	ROUGH IDLE/ENGINE STALLS AT IDLE — ALWAYS
DESCRIPTION	• Engine starts normally but stalls or vibrates excessively at idle (every condition)

[TROUBLESHOOTING HINTS]
① Air/Fuel mixture too lean
- Air leakage
- Fuel injection control malfunction
- Low fuel line pressure

② Injector(s) clogged or not operating
③ Low intake air amount
④ Incorrect idle speed
- Idle speed misadjustment
- Idle speed control malfunction

⑤ Spark plug(s) not firing
⑥ Low engine compression
⑦ Airflow meter stuck
⑧ EGR malfunction

STEP	INSPECTION		ACTION
1	Is "00" displayed on Self-Diagnosis Checker while ignition switch ON? 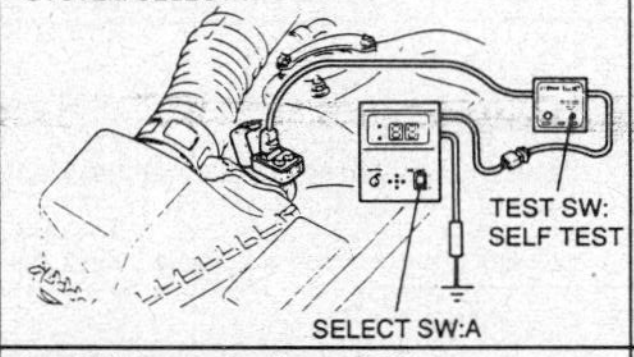	Yes	Go to next step
		No	**Service Code No. displayed** Check for cause (Refer to specified check sequence)
2	Are ECU terminal voltages OK? (Especially 2B, 2E, and 3Q)	Yes	Go to next step
		No	Check for cause
3	Disconnect each high-tension lead at idle; does engine speed decrease equally at each cylinder?	Yes	Go to step 8
		No	Go to next step
4	Is there injector operating sound at idle?	Yes	Go to step 6
		No	Go to step 5
5	Is there approx. 12V at ECU terminals 3U, 3V, 3W, 3X, 3Y, and 3Z? 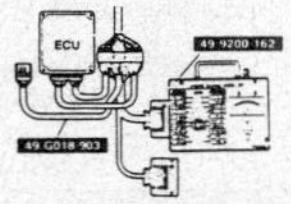	Yes	Check injectors ➪ If OK, try known good ECU ➪ If not OK, replace injector(s)
		No	Check wiring between ECU and injector Check for open circuit in wiring between main relay and ECU terminals
6	Is strong blue spark visible at disconnected high-tension lead? 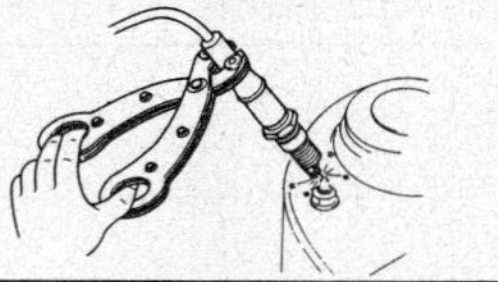	Yes	Go to next step
		No	Check high-tension lead ➪ If OK, check distributor cap and rotor ➪ If not OK, replace high-tension lead

STEP	INSPECTION		ACTION
7	Are spark plugs OK? 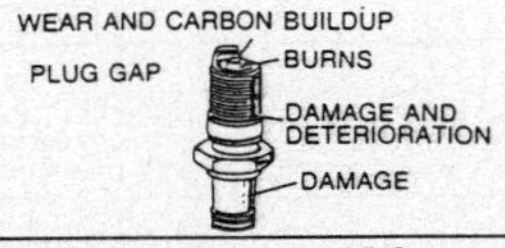	Yes	Check for correct engine compression ➪ If OK, replace injector(s) ➪ If not OK, check for cause
		No	Clean or replace
8	Is fuel line pressure correct at idle? **Fuel line pressure: 255—314 kPa (2.6—3.2 kg/cm², 37—48 psi) (vacuum hose to pressure regulator disconnected)** 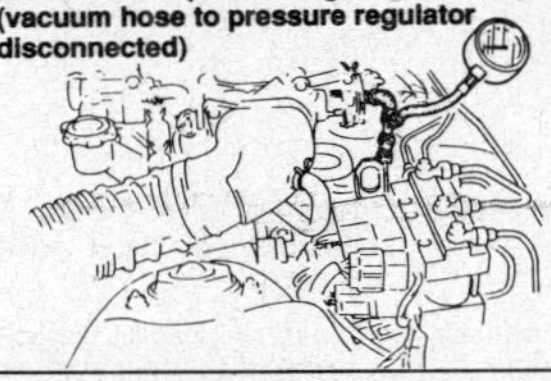	Yes	Go to next step
		No	**Low pressure** Check fuel line pressure while pinching fuel return hose ➪ If pressure **quickly** increases, check pressure regulator ➪ If pressure **gradually** increases, check for clogging between fuel pump and pressure regulator If hose not clogged, check fuel pump maximum pressure
9	Does airflow meter move smoothly? 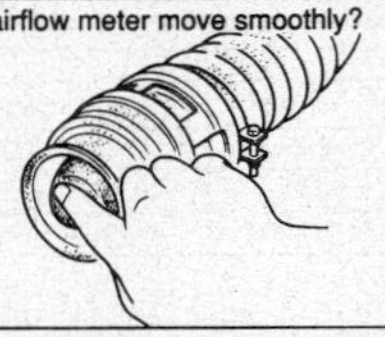	Yes	Go to next step
		No	Repair or replace
10	Is there air leakage at intake air system components?	Yes	Repair or replace
		No	Go to next step
11	Is engine compression correct? **Engine compression (Minimum): 1079 kPa (11.0 kg/cm², 156 psi)-270 rpm**	Yes	Go to next step
		No	Check for cause
12	Connect System Selector to diagnosis connector and set Test Switch to SELF TEST; is ignition timing correct at idle after warm-up? **Ignition timing (BTDC): 11—13°**	Yes	Try known good ECU and check if condition improves
		No	Adjust ignition timing

9	ROUGH IDLE/ENGINE STALLS AT IDLE—DURING WARM-UP
DESCRIPTION	• Engine speed low or engine stalls or vibrates excessively at idle during warm-up

[TROUBLESHOOTING HINTS]
① Low intake air amount
- Airflow meter stuck
- Air cleaner element clogged
- Air valve

② Idle speed control system malfunction
③ Low fuel injection control
- Fuel-injection control malfunction (correction for coolant temperature)

④ Poor atomization
- Low RVP (summer) fuel used in cold weather

STEP	INSPECTION		ACTION
1	Is "00" displayed on Self-Diagnosis Checker while ignition switch ON? SYSTEM SELECT:1 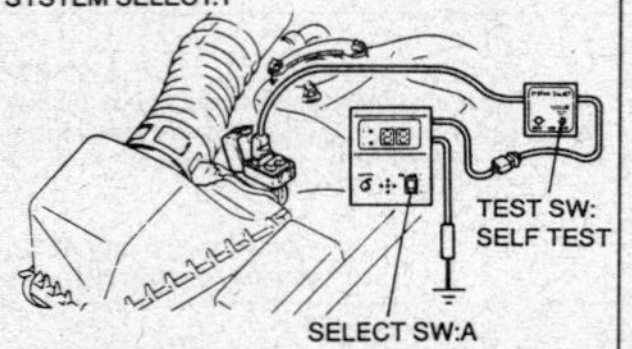	Yes	Go to next step
		No	**Service Code No. displayed** Check for cause (Refer to specified check sequence)
2	Is intake manifold vacuum correct at idle after warm-up? **Vacuum: More than 500 mmHg (19.7 inHg)**	Yes	Go to next step
		No	Check for air leakage of intake air system components
3	Is air cleaner element clean?	Yes	Go to next step
		No	Replace air cleaner element
4	Are ECU terminal voltages OK? (Especially 2B, 2E, and 3Q)	Yes	Go to next step
		No	Check for cause
5	Blow through the BAC valve from port; is air flow OK? **Cold engine: Air flows** **Warm engine: Air does not flow** 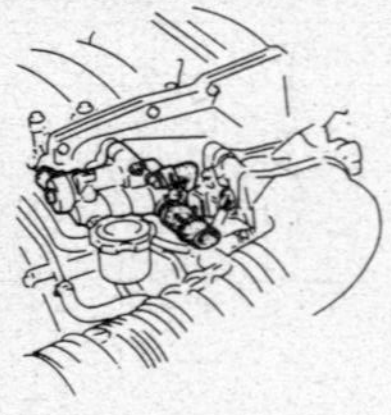	Yes	Go to next step
		No	Replace BAC valve

STEP	INSPECTION		ACTION
6	Disconnect ISC valve connecter before warm-up; does engine speed decrease? 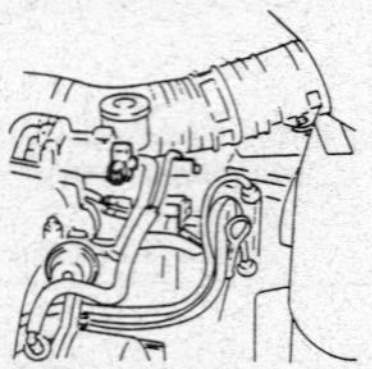	Yes	Go to next step
		No	Replace BAC valve
7	Is water thermosensor resistance OK? Coolant temp. / Resistance (KΩ): −20°C (−4°F) 14.6—17.8; 20°C (68°F) 2.2—2.7; 80°C (176°F) 0.2—0.4	Yes	Go to next step
		No	Replace water thermosensor
8	Connect System Selector to diagnosis connector and set Test Switch to "SELF TEST"; is ignition timing at idle correct after warm-up? **Ignition timing (BTDC): 11—13°** 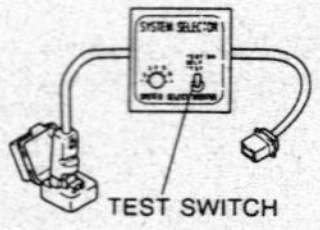	Yes	Go to next step
		No	Adjust ignition timing
9	Try a known good ECU; does condition improve?	Yes	Replace ECU
		No	Change to another brand of fuel

Coolant temp.	Resistance (KΩ)
−20°C (−4°F)	14.6—17.8
20°C (68°F)	2.2—2.7
80°C (176°F)	0.2—0.4

10	ROUGH IDLE/ENGINE STALLS AT IDLE—AFTER WARM-UP		
DESCRIPTION	• Engine runs normally at idle during warm-up; but stalls or vibrates excessively after warm-up		

[TROUBLESHOOTING HINTS]
① Air/Fuel mixture too lean
- Air leakage
- Low fuel line pressure

② Air/Fuel mixture too rich
- Fuel injection control malfunction (correction for coolant temperature)

③ Poor ignition
④ Low engine compression

STEP	INSPECTION		ACTION
1	Is "00" displayed on Self-Diagnosis Checker while ignition switch ON? 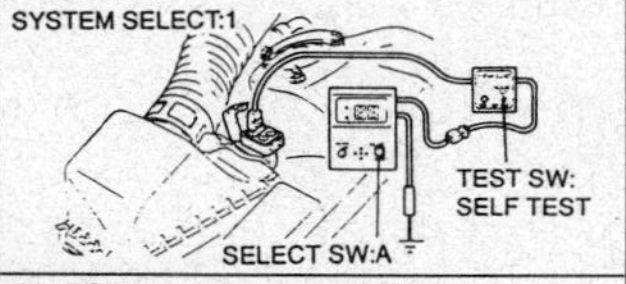 	Yes	Go to next step
		No	**Service Code No. displayed** Check for cause (Refer to specified check sequence)
2	Are ECU terminal voltages OK? (Especially 2B, 2E, and 3Q)	Yes	Go to next step
		No	Check for cause
3	Disconnect each high-tension lead at idle; does engine speed decrease equally at each cylinder?	Yes	Go to next step
		No	Go to Step 9
4	Check for correct intake manifold vacuum at idle **Vacuum: More than 500 mmHg (19.7 inHg)**	Yes	Go to next step
		No	Check for air leakage of intake air system components
5	Is air cleaner element clean	Yes	Go to next step
		No	Replace air cleaner element

STEP	INSPECTION		ACTION
6	Is fuel line pressure correct at idle? **Fuel line pressure:** **255—314 kPa (2.6—3.2 kg/cm², 37—46 psi)** **(Vacuum hose to pressure regulator disconnected)** 	Yes	Go to next step
		No	**Low pressure** Check fuel line pressure while pinching fuel return hose ⇨ If fuel line pressure **quickly** increases, check pressure regulator ⇨ If fuel line pressure **gradually** increases, check for clogging between fuel pump and pressure regulator If hose is not clogged, check fuel pump maximum pressure
7	Connect System Selector to diagnosis connector and set Test Switch to SELF TEST; is ignition timing correct at idle after warm-up? **Ignition timing (BTDC): 11—13°**	Yes	Go to next step
		No	Adjust ignition system
8	Disconnect water thermosensor connector; does engine condition improve?	Yes	Replace water thermosensor
		No	Try known good ECU and check if condition improve
9	Is there injector operating sound at idle? 	Yes	Go to next step
		No	Check injector resistance **Resistance: 12—16Ω** ⇨ If OK, check wiring between ECU and injector ⇨ If not OK, replace injector
10	Is engine compression correct? **Engine compression (Minimum):** **1079 kPa (11.0 kg/cm², 156 psi)-270 rpm**	Yes	Go to next step
		No	Check engine
11	Is strong blue spark visible at disconnected high-tension lead? 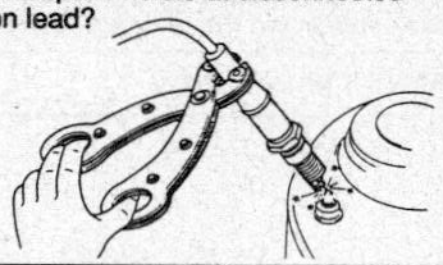	Yes	Go to next step
		No	Check high-tension lead ⇨ If OK, check distributor cap and rotor ⇨ If not OK, replace high-tension lead
12	Are spark plugs OK?	Yes	Try known good ECU and check if condition improves
		No	Clean or replace spark plag(s)

11	ROUGH IDLE/ENGINE STALLS AT IDLE — WHEN A/C, P/S, OR E/L ON		
DESCRIP-TION	• Engine stalls or vibrates excessively at idle when A/C, P/S, or E/L ON • A/C, P/S, headlights, and blower fan operate normally • Idle condition is normal when A/C, P/S, and E/L is OFF		
[TROUBLESHOOTING HINTS] ① Idle speed control system malfunction • Engine speed feedback control inoperative			
STEP	INSPECTION		ACTION
1	Is "00" displayed on Self-Diagnosis Checker while ignition switch ON? SYSTEM SELECT:1 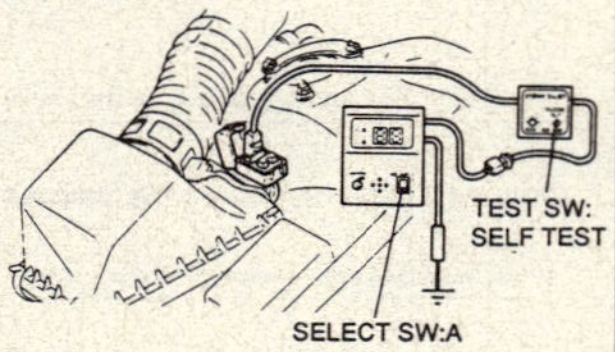 	Yes	Go to next step
		No	**Service Code No. displayed** Check for cause (Refer to specified check sequence)
2	Are switches correct when checked by using Self-Diagnosis Checker Monitor Lamp while ignition switch ON? SYSTEM SELECT:1 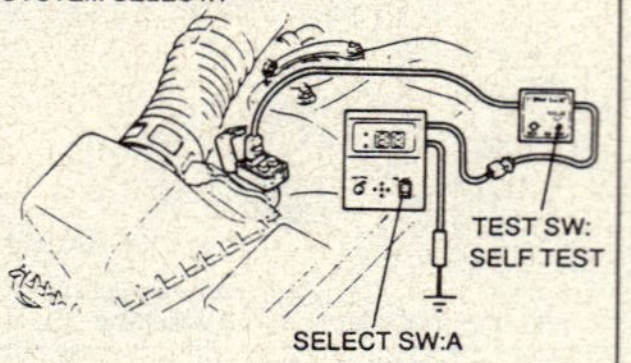 	Yes	Go to next step
		No	**Lamp not ON or OFF with specified switch** Check for cause (Refer to specified check sequence) **Lamp always ON** Check wiring between ECU terminal 1D and Self-Diagnosis Checker
3	Is ECU terminal 1I voltage OK? **Voltage: Approx. 12V (at idle)** 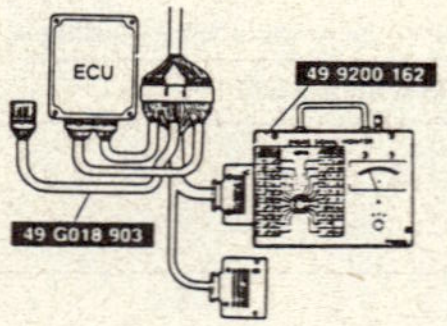 	Yes	Try known good ECU and check if condition improves
		No	Check for short circuit in wiring between diagnosis connector terminal TEN and ground

12	ROUGH IDLE/ENGINE STALLS JUST AFTER STARTING		
DESCRIP-TION	• Engine starts normally but vibrates excessively or stalls just after starting (acceleration from idle) • Idle conditions are normal in the other conditions		
[TROUBLESHOOTING HINTS] ① Fuel injection control system or idle speed control system malfunction • Start signal not input to ECU ② Idle speed misadjustment ③ Ignition timing misadjustment			
STEP	INSPECTION		ACTION
1	Check if "00" is displayed on Self-Diagnosis Checker while ignition switch ON SYSTEM SELECT:1 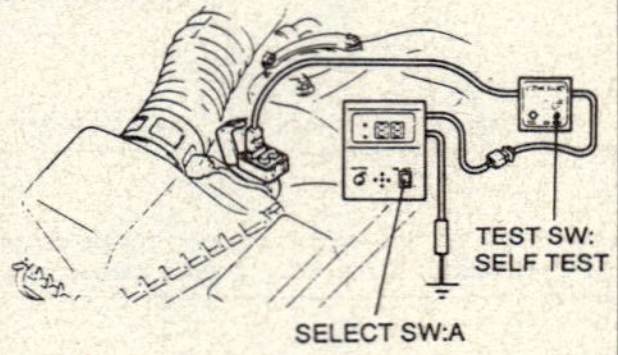 	Yes	Go to next step
		No	**Service Code No. displayed** Check for cause (Refer to specified check sequence)
2	Are ECU terminal voltages are OK? (Especially 1T)	Yes	Go to next step
		No	Check for cause
3	Connect System Selector to diagnosis connector and set Test Switch to SELF TEST; is ignition timing correct at idle after warm-up? **Ignition timing (BTDC): 11—13°** 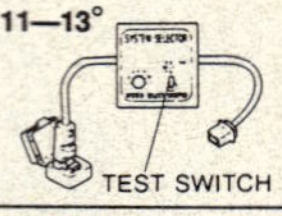 	Yes	In same condition as Step 3 inspection, check for correct idle speed **Idle speed: 680—720 rpm** ⇨ If OK, go to next step ⇨ If not OK, adjust idle speed
		No	Adjust
4	Is throttle sensor adjustment OK? 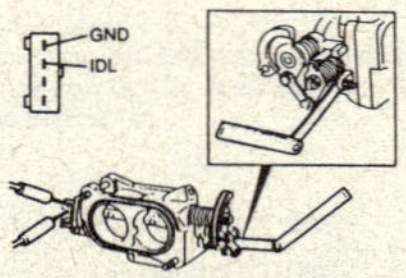 	Yes	Go to next step
		No	Adjust
5	Try known good ECU; does condition improve?		

Clearance between throttle lever and stopper	Continuity
0.1mm (0.004 in)	Yes
0.2mm (0.008 in)	No

13	HIGH IDLE SPEED AFTER WARM-UP		
DESCRIP-TION	• Idle speed excessive after warm-up		
[TROUBLESHOOTING HINTS] Excessive intake air supplied to engine ① Throttle valve not fully closed ② Idle speed control malfunction • Air valve not closed • ISC valve stuck • A/C, P/S, or E/L signal sent to ECU • Incorrect coolant temperature signal			
STEP	**INSPECTION**		**ACTION**
1	Is throttle valve fully closed when accelerator released? 	Yes	Go to Step 3
		No	Check if throttle linkage is correctly installed and operates freely ⇨ If OK, go to Step 2 ⇨ If not OK, clean, adjust, or replace linkage
2	Is "00" displayed on Self-Diagnosis Checker while ignition switch ON? SYSTEM SELECT:1 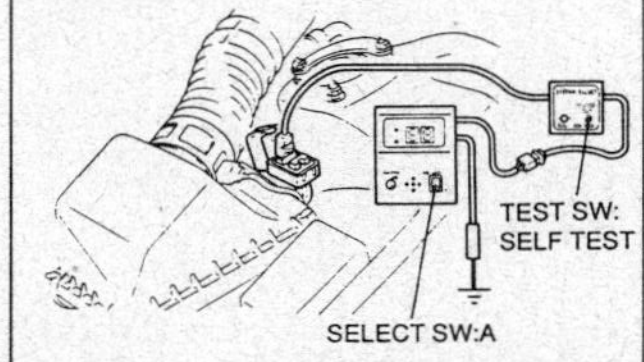	Yes	Go to next step
		No	**Service Code No. displayed** Check for cause (Refer to specified check sequence)
3	Are switches correct when checked by using Self-Diagnosis Checker Monitor Lamp with ignition switch ON? SYSTEM SELECT:1 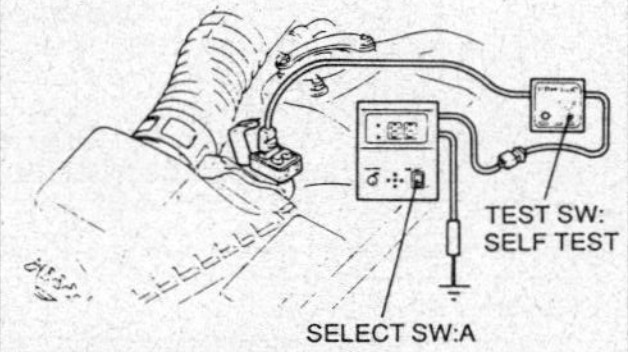	Yes	Go to next step
		No	**Lamp not ON or OFF with specified switch** Check for cause (Refer to specified check sequence)
			Lamp always ON Check wiring between ECU terminal 1D and Self-Diagnosis Checker
4	Are ECU terminal voltages OK? (Especially 2E)	Yes	Go to next step
		No	Check for cause

STEP	INSPECTION		ACTION
5	Blow through the BAC valve; is air flow OK? **Cold engine: Air flows** **Warm engine: Air does not flow** 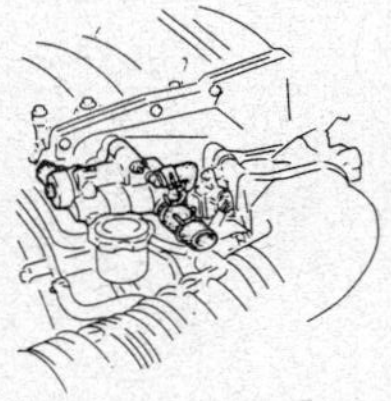	Yes	Go to next step
		No	Check air valve
6	Disconnect ISC valve connector before warm-up; does engine speed decrease? 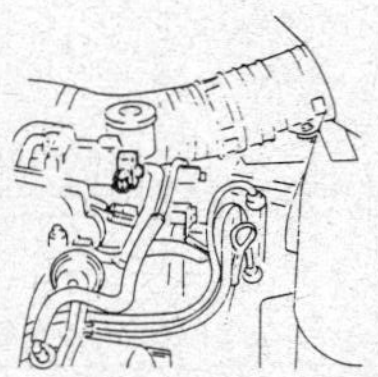	Yes	Go to next step
		No	Check ISC valve
7	Pinch PCV hose with pliers; does engine speed decreases? 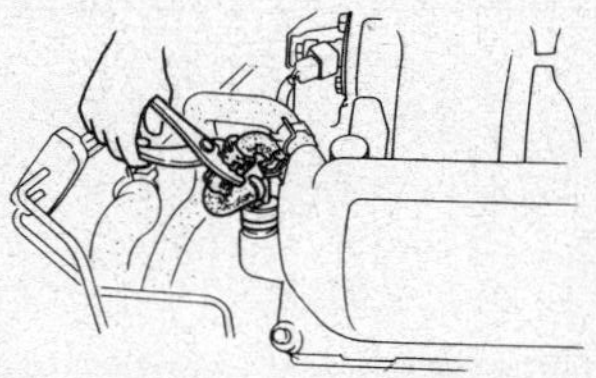	Yes	Check PCV valve
		No	Go to next step
8	Check if ECU terminal voltages are OK (Especially 2B and 2E)	Yes	Try known good ECU and check if condition improves
		No	Check for cause

14	IDLE MOVES UP AND DOWN/IDLE HUNTING		
DESCRIPTION	• Engine speeds up and down periodically at idle		
[TROUBLESHOOTING HINTS] ① Idle switch (in throttle sensor) OFF at idle ② Air leakage in intake air system ③ Fuel injection amount inconstant • Poor contact point inside airflow meter ④ Poor ignition ⑤ Idle speed control malfunction • BAC valve stuck • Short circuit in wiring from BAC valve to ECU • Incorrect coolant temperature signal			
STEP	**INSPECTION**		**ACTION**
1	Is "00" displayed on Self-Diagnosis Checker while ignition switch ON? SYSTEM SELECT:1 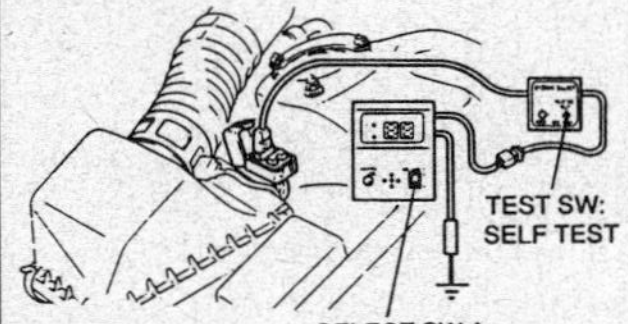 	Yes	Go to next step
		No	**Service Code No. displayed** Check for cause (Refer to specified check sequence)
2	Is intake manifold vacuum correct at idle? **Intake manifold vacuum:** **More than 500 mmHg (19.7 inHg)**	Yes	Go to next step
		No	**Low vacuum** Check for air leakage at intake air system
3	Is air cleaner element clean?	Yes	Go to next step
		No	Replace air cleaner element
4	Disconnect each high-tension lead at idle; does engine speed decrease equally at each cylinder?	Yes	Go to next step
		No	Go to Step 9
5	Are ECU terminal voltages OK? (Especially 2B, 2E, and 3Q)	Yes	Go to next step
		No	Check for cause
6	Connect System Selector to diagnosis connector and set Test Switch to "SELF TEST"; is ignition timing correct at idle after warm-up? **Ignition timing (BTDC): 11—13°** 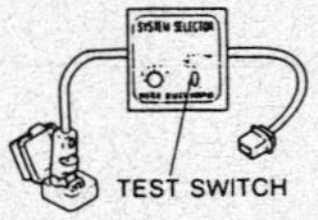 	Yes	Go to next step
		No	Adjust ignition timing

STEP	INSPECTION		ACTION
7	With same condition as Step 6; is idle normal? 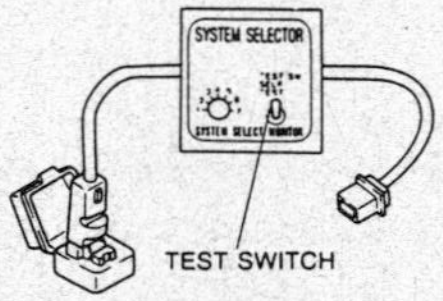 	Yes	Try known good ECU and check if condition improves
		No	Go to next step
8	Is airflow meter OK?	Yes	Go to Step 14
		No	Replace airflow meter
9	Is there injector operating sound at idle?	Yes	Go to Step 11
		No	Go to next step
10	Is there approx. 12V at ECU terminals 3U, 3V, 3W, 3X, 3Y, and 3Z? 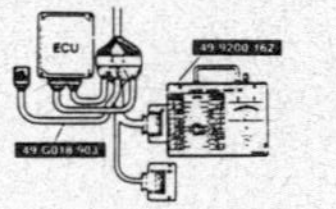 	Yes	Are injector(s) OK? ➪ If OK, try known good ECU ➪ If not OK, replace injector
		No	Check for open circuit in wiring between main relay and ECU terminals
11	Is strong blue spark visible at disconnected high-tension lead? 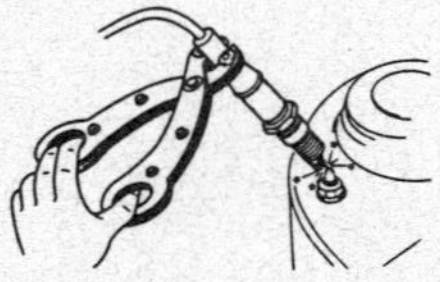	Yes	Go to next step
		No	Check high-tension lead ➪ If OK, check distributor cap and rotor ➪ If not OK, replace high-tension lead
12	Are spark plugs OK? 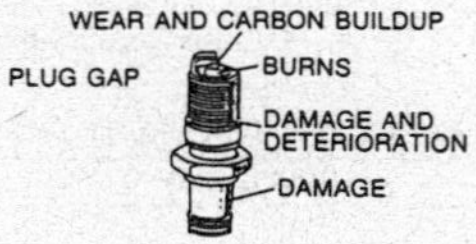 	Yes	Check for correct engine compression **Compression (Minimum):** **1079 kPa (11.0 kg/cm^2, 156 psi)-270 rpm** ➪ If OK, go to next step ➪ If not OK, check for cause
		No	Clean or replace spark plug(s)
13	Try known good ECU and Check if condition improves		

15	ENGINE STALLS ON DECELERATION		
DESCRIPTION	• Engine unexpectedly stops running during or after deceleration • Idle condition normal		
[TROUBLESHOOTING HINTS] Engine speed drops too much when releasing accelerator ① Idle speed control malfunction ② Fuel cut control malfunction ③ Idle speed misadjustment			
STEP	INSPECTION		ACTION
1	Is idle condition normal?	Yes	Go to next step
		No	Adjust or perform troubleshooting Nos. 8—12 "ROUGH IDLE"
2	Is "00" displayed on Self-Diagnosis Checker while ignition switch ON? SYSTEM SELECT:1 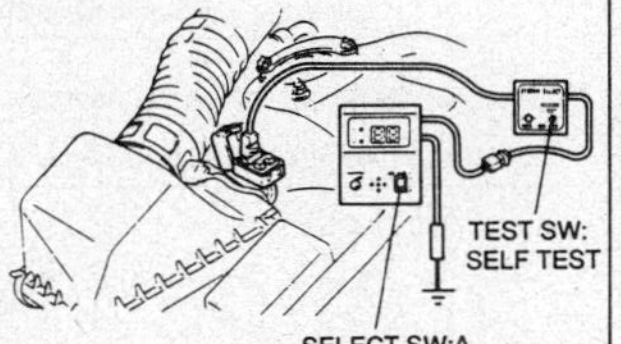	Yes	Go to next step
		No	**Service Code No. displayed** Check for cause (Refer to specified check sequence)
3	Are switches correct when checked by using Self-Diagnosis Checker Monitor Lamp while ignition switch ON? SYSTEM SELECT:1 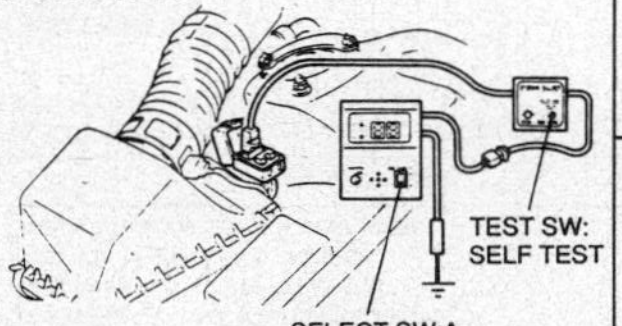	Yes	Go to next step
		No	**Lamp not ON or OFF with specified switch** Check for cause (Refer to specified check sequence)
			Lamp always ON Check wiring between ECU terminal 1D and Self-Diagnosis Checker
4	Are ECU terminal voltages OK? (Especially 2B, 2E, 2F, 2R, 3Q, 3U, 3V, 3W, 3X, 3Y, and 3Z)	Yes	Go to next step
		No	Check for cause
5	Check for poor connection of: • Ignition coil • Igniter • Distributor • High-tension leads • Injectors • Circuit opening relay • ECU	Yes	Repair or replace
		No	Go to next step

STEP	INSPECTION		ACTION
6	Disconnect ISC valve connector before warm-up; does engine speed decrease? 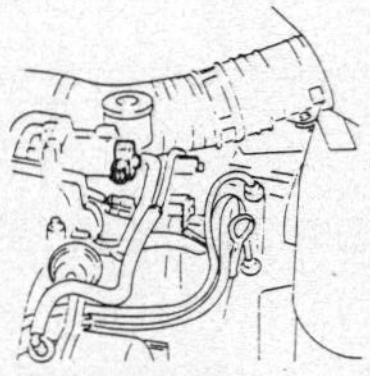	Yes	Go to next step
		No	Check ISC valve
7	Try known good ECU; does condition improves?		

16 — HESITATES/STUMBLES ON ACCELERATION

DESCRIPTION: • Flat spot occurs just after accelerator depressed or mild jerking occurs during acceleration

[TROUBLESHOOTING HINTS]

① Air/Fuel mixture becomes lean when depressing accelerator
- Fuel injection control malfunction (correction for accelerating condition)
- Air leakage in intake air system
- Fuel line pressure decreases
- Spark advance control malfunction

STEP	INSPECTION		ACTION
1	Is "00" displayed on Self-Diagnosis Checker with ignition switch ON? 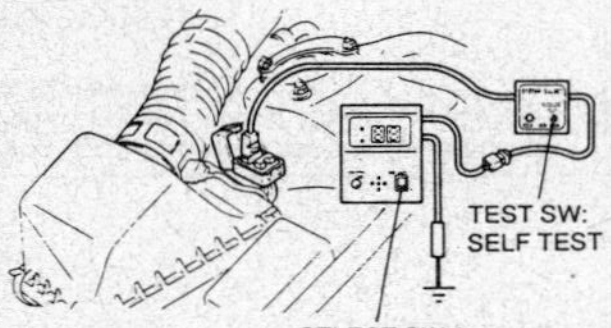	Yes	Go to next step
		No	**Service Code No. displayed** Check for cause (Refer to specified check sequence)
2	Is intake manifold vacuum correct at idle? **Vacuum: More than 500 mmHg (19.7 inHg)**	Yes	Go to next step
		No	Check for air leakage at intake air system components
3	Is air cleaner element clean?	Yes	Go to next step
		No	Replace air cleaner element
4	Are ECU terminal voltages OK? (Especially 1T, 2F, and 2I)	Yes	Go to next step
		No	Check for cause
5	Is throttle linkage correctly installed and operating freely?	Yes	Go to next step
		No	Correct, clean, or replace as required any binding or damaged linkage and adjust cable deflection at throttle body
6	Is throttle sensor adjustment OK? 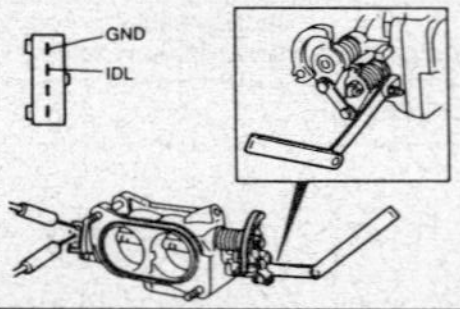	Yes	Go to next step
		No	Adjust

Clearance between throttle lever and stopper	Continuity
0.1mm (0.004 in)	Yes
0.2mm (0.008 in)	No

STEP	INSPECTION		ACTION
7	Is the fuel line pressure correct at idle? **Fuel line pressure: 255—314 kPa (2.6—3.2 kg/cm^2, 37—46 psi) (vacuum hose to pressure regulator disconnected)** 	Yes	Does fuel line pressure decrease when accelerating quickly? ⇨ If decreases, check fuel line and filter for clogging ⇨ If no decrease, go to next step
		No	**Low pressure** Check fuel line pressure while pinching fuel return hose ⇨ If pressure **quickly** increases, check pressure regulator ⇨ If pressure **gradually** increases, check for clogging between fuel pump and pressure regulator If hose is not clogged, check fuel pump maximum pressure
8	Connect System Selector to diagnosis connector and set Test Switch to "SELF TEST"; is ignition timing correct at idle after warm-up? **Ignition timing (BTDC): 11—13°** 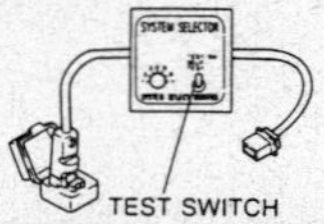	Yes	Check if ignition timing advances when accelerating ⇨ If advances, go to next step ⇨ If not advance, replace ECU
		No	Adjust
9	Are air duct and air hoses correctly installed?	Yes	Go to next step
		No	Repair
10	Is exhaust system restricted?	Yes	Repair or replace
		No	Go to next step
11	Is variable inertia control system (VICS) OK?	Yes	Go to next step
		No	Repair or replace
12	Try known good ECU; does if condition improve?		

17	SURGES WHILE CRUISING		
DESCRIPTION	• Unexpected change in engine speed which is usually repetitive		
[TROUBLESHOOTING HINTS] ① Air/Fuel mixture too lean ② Misfire ③ Poor connection in wiring harness			
STEP	INSPECTION		ACTION
1	Is "00" displayed on Self-Diagnosis Checker while ignition switch ON? SYSTEM SELECT:1 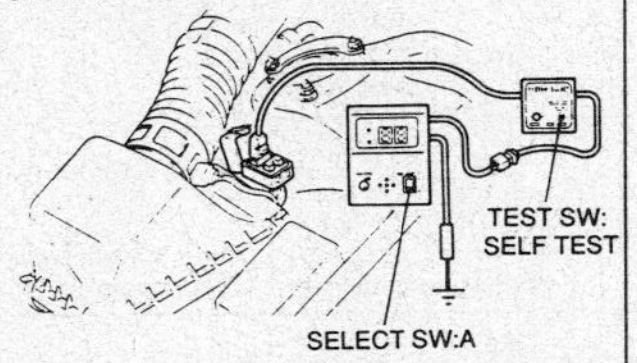 	Yes	Go to next step
		No	**Service Code No. displayed** Check for cause (Refer to specified check sequence)
2	Are switches correct as checked by using Self-Diagnosis Checker Monitor Lamp while ignition switch ON? SYSTEM SELECT:1 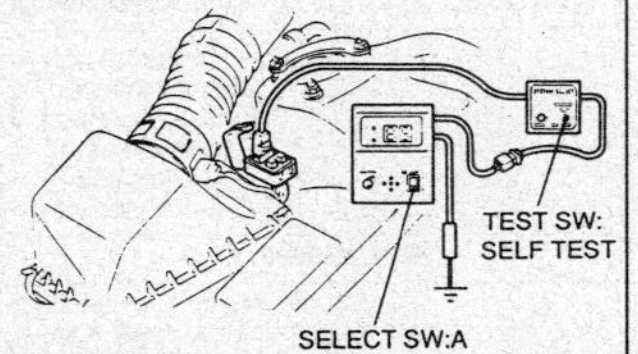 	Yes	Go to next step
		No	**Lamp not ON or OFF with specified input switch** Check for cause (Refer to specified check sequence) **Lamp always ON** Check ECU terminal 1D voltage
3	Is throttle sensor OK?	Yes	Go to next step
		No	Adjust
4	Disconnect oxygen sensor connector; does condition improve? 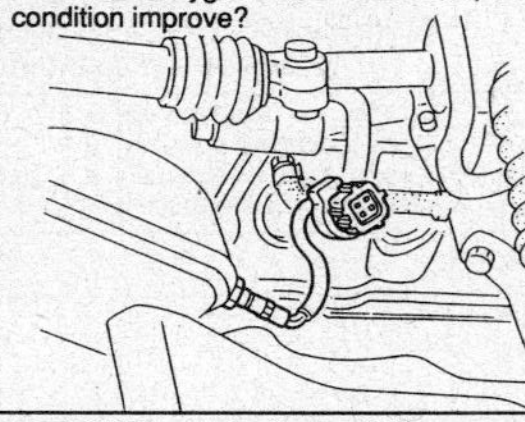	Yes	Check oxygen sensor
		No	Go to next step
5	Are ECU terminal voltages OK?	Yes	Go to next step
		No	Check for cause

STEP	INSPECTION		ACTION
6	Is throttle linkage correctly installed and operating freely?	Yes	Go to next step
		No	Correct, clean, or replace as required any binding or damaged linkage, and adjust cable deflection at throttle body
7	Check for correct intake manifold vacuum at idle **Vacuum: More than 500 mmHg (19.7 inHg)**	Yes	Go to next step
		No	Check for air leakage of intake air system components
8	Is air cleaner element clean?	Yes	Go to next step
		No	Replace air cleaner element
9	Connect System Selector to diagnosis connector and set Test Switch to SELF TEST; is ignition timing correct at idle after warm-up? **Ignition timing (BTDC): 11—13°** 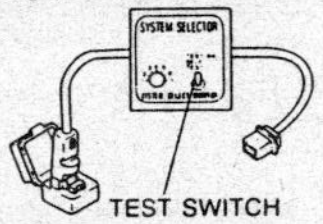 	Yes	Check if ignition timing advances when accelerating ➪ If advances, go to next step ➪ If not advance, replace ECU
		No	Adjust
10	Is fuel line pressure correct at idle? **Fuel line pressure:** **255—314 kPa (2.6—3.2 kg/cm², 37—46 psi)** **(vacuum hose to pressure regulator disconnected)** 	Yes	Does fuel line pressure decrease when accelerating quickly? ➪ If decreases, check for fuel pump maximum pressure If OK, check fuel line and filter for clogging ➪ If no decrease, go to next step
		No	**Low pressure** Check fuel line pressure by pinching fuel return hose ➪ If fuel line pressure **quickly** increases, check pressure regulator ➪ If fuel line pressure **gradually** increases, check for clogging between fuel pump and pressure regulator If not clogged, check fuel pump maximum pressure **High pressure** Is fuel return line clogged? ➪ If OK, replace pressure regulator ➪ If not OK, replace
11	Is exhaust system restricted?	Yes	Repair or replace
		No	Go to next step
12	Try known good ECU; does condition improve?		

18	LACK OF POWER, POOR ACCELERATION
DESCRIPTION	• Performance poor under load when throttle valve wide open • Performance poor while accelerating • Reduced maximum speed • Idle condition normal

[TROUBLESHOOTING HINTS]

① Factors other than engine malfunction
- Clutch slipping
- AT slipping
- Brake dragging
- Low tire pressure
- Incorrect tire size
- Overloaded vihicle

② Low intake air amount
- Throttle valve not open fully
- Clogging in intake air system

③ Air/Fuel mixture too lean
- Fuel line pressure decrease
- Fuel injection malfunction

④ Poor ignition

⑤ Low engine compression

⑥ Alcohol blended fuel used

STEP	INSPECTION		ACTION
1	Check factors other than engine • Clutch slipping • Brake dragging • Low tire pressure • Incorrect tire size • AT slipping	Yes	Go to next step
		No	Repair
2	Does throttle valve fully open when depressing accelerator fully?	Yes	Go to next step
		No	Check if accelerator cable is correctly installed ⇨ If OK, check throttle body ⇨ If not OK, install accelerator cable correctly
3	Is "00" displayed on Self-Diagnosis Checker while ignition switch ON?	Yes	Go to next step
	SYSTEM SELECT:1 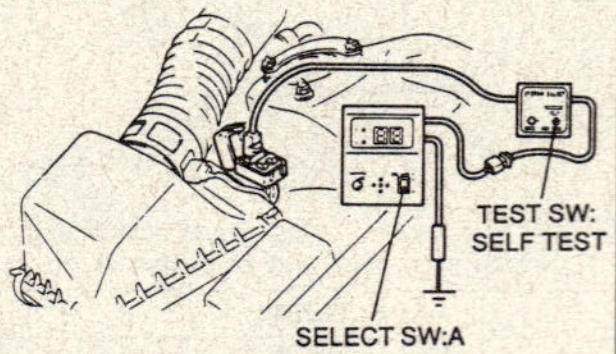	No	**Service Code No. displayed** Check for cause (Refer to specified check sequence)
4	Is throttle sensor adjustment OK? 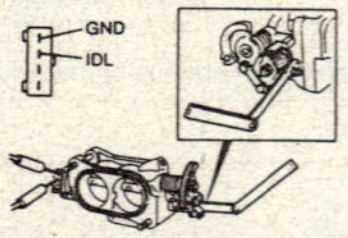	Yes	Go to next step
		No	Adjust

Clearance between throttle lever and stopper	Continuity
0.1mm (0.004 in)	Yes
0.2mm (0.008 in)	No

STEP	INSPECTION		ACTION
5	Connect System Selector to diagnosis connector and set Test Switch to SELF TEST; is for ignition timing at idle correct after warm-up? **Ignition timing (BTDC): 11—13°** 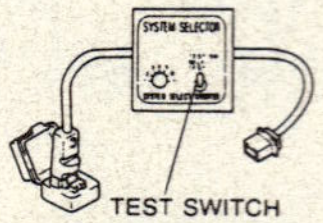	Yes	Check if ignition timing advances during acceleration ⇨ If no advances, go to next step ⇨ If no advance, check the ECU terminal voltages
		No	Adjust
6	Are ECU terminal voltages OK? (Especially 1I, 1T, 2I, and 3J)	Yes	Go to next step
		No	Check for cause
7	Is intake manifold vacuum correct at idle? **Intake manifold vacuum: More than 500 mmHg (19.7 inHg)**	Yes	Go to next step
		No	Check for air leakage of intake air system components
8	Is fuel line pressure correct at idle? **Fuel line pressure: 255—314 kPa (2.6—3.2 kg/cm², 37—46 psi) (vacuum hose to pressure regulator disconnected)**	Yes	Check if fuel line pressure decreases when accelerating quickly ⇨ If it does, check fuel pump maximum pressure If pressure is OK, check fuel line and filter for clogging ⇨ If it does not decrease, go to next step
		No	**Low pressure** Check for fuel line pressure while pinching fuel return hose ⇨ If pressure **quickly** increases, check pressure regulator ⇨ If pressure **gradually** increases, check for clogging between fuel pump and pressure regulator ⇨ If hose is not clogged, check fuel pump maximum pressure
		No	**High pressure** Is fuel line clogged? ⇨ If it is not, replace pressure regulator ⇨ If it is, replace it
9	Are ECU terminal voltages OK?	Yes	Go to next step
		No	Replace air cleaner element
10	Is airflow meter OK?	Yes	Go to next step
		No	Replace

STEP	INSPECTION		ACTION
11	Are spark plugs OK? 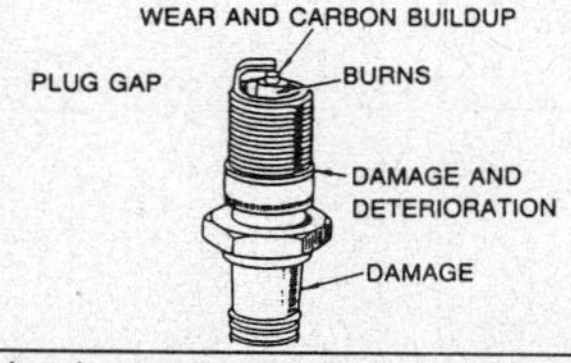	Yes	Go to next step
		No	Clean or replace
12	Is resistance of high-tension leads OK? **Resistance: 16 kΩ per 1 m (3.28 ft)** 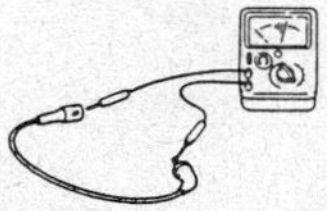	Yes	Go to next step
		No	Replace
13	Is resistance of ignition coil OK? **Resistance (at 20°C [68°F]):** **Primary coil winding ....0.7—1.0 Ω** **Secondary coil winding .....9—17 kΩ**	Yes	Go to next step
		No	Replace
14	Is engine compression correct? **Engine compression (Minimum):** **1079 kPa (11.0 kg/cm², 156 psi)-270 rpm**	Yes	Go to next step
		No	Check engine condition • Worn pistons, piston rings, or cylinder walls • Defective cylinder head gasket • Distorted cylinder head • Improper valve seating • Valve sticking in guide
15	Change fuel and check; does acceleration improve?	Yes	Change fuel to another brand
		No	Go to next step
16	Is A/C cut-off control system OK?	Yes	Go to next step
		No	Repair or replace
17	Is variable inertia charging system (VICS) OK?	Yes	Go to next step
		No	Repair or replace
18	Is variable exhaust system OK?	Yes	Go to next step
		No	Repair or replace
19	Try known good ECU; does condition improve?		

19	RUNS ROUGH ON DECELERATION/AFTER BURN
DESCRIP-TION	• Engine runs rough while decelerating and abnormal combustion occurs in exhaust system • Transmission in normal condition

[TROUBLESHOOTING HINTS]

① Air/Fuel mixture too rich
- Air cleaner element clogged
- Fuel injection control malfunction (Fuel cut control)
- Injector fuel leakage
- Throttle sensor (IDL terminal) malfunction
- Idle speed control malfunction

STEP	INSPECTION		ACTION
1	Is "00" displayed on Self-Diagnosis Checker while ignition switch ON? SYSTEM SELECT:1 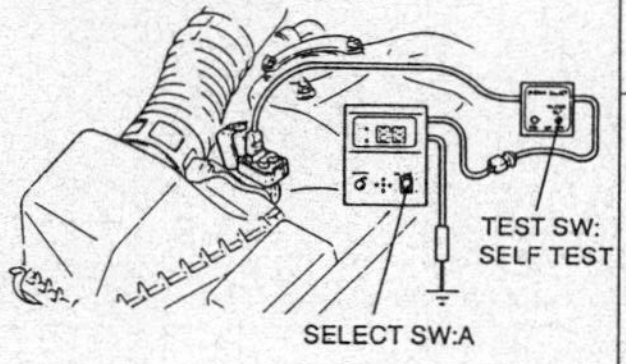	Yes	Go to next step
		No	**Service Code No. displayed** Check for cause (Refer to specified check sequence)
2	Are switches correct when checked by using Self-Diagnosis Checker Monitor Lamp while ignition switch ON? SYSTEM SELECT:1 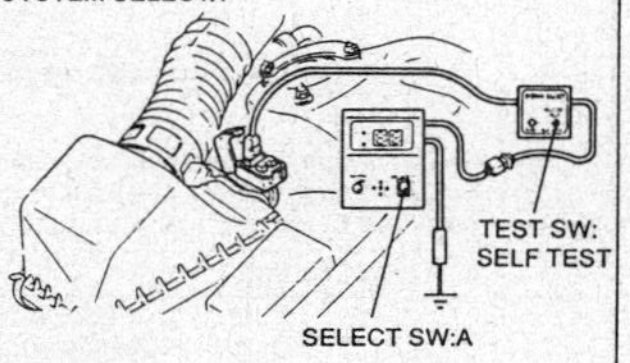	Yes	Go to next step
		No	**Lamp not ON or OFF with specified switch** Check for cause (Refer to specified check sequence) **Lamp always ON** Check wiring between ECU terminal 1D and Self-Diagnosis Checker
3	Is intake manifold vacuum correct at idle? **Vacuum: More than 500 mmHg (19.7 inHg)**	Yes	Go to next step
		No	Check for air leakage of intake air system components
4	Is air cleaner element clean?	Yes	Go to next step
		No	Replace air cleaner element
5	Are ECU terminal voltages OK? (Especially 1T, 2F, 3B, 3Q, 3U, 3V, 3W, 3X, 3Y, and 3Z)	Yes	Go to next step
		No	Check for cause

STEP	INSPECTION		ACTION
6	Is fuel cut operation OK?	Yes	Go to next step
		No	Try known good ECU and check if condition improves
7	Is airflow meter OK?	Yes	Go to next step
		No	Repair or replace
8	Is throttle sensor TVO terminal voltage OK?	Yes	Change to another brand of fuel
		No	Adjust or replace
9	Is if throttle sensor IDL terminal voltave OK?	Yes	Go to next step
		No	Adjust
10	Disconnect ISC valve connector before warm-up; does engine speed decrease? 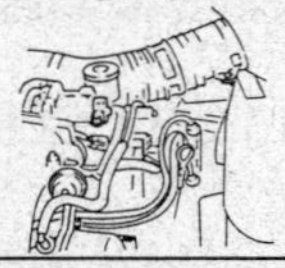	Yes	Go to next step
		No	Check BAC valve
11	Is there injector fuel leakage? 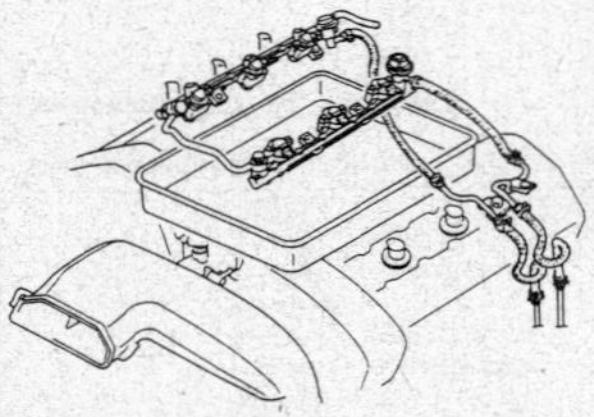	Yes	Replace
		No	Go to next step
12	Try known good ECU; does condition improve?		

20	KNOCKING		
DESCRIP-TION	• Abnormal combustion accompanied by audible "pinging" noise		
[TROUBLESHOOTING HINTS] ① Incorrect ignition timing (advanced) ② Carbon deposits in cylinder ③ Overheating	④ Air/Fuel mixture too lean • Fuel injection amount not correct • Fuel line pressure decrease while accelerating		
STEP	INSPECTION		ACTION
1	Connect System Selector to diagnosis connector and set Test Switch to SELF TEST; is ignition timing correct at idle after warm-up? **Ignition timing (BTDC): 11—13°** 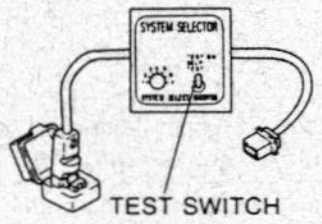	Yes	Go to next step
		No	Adjust
2	Is "00" displayed on Self-Diagnosis Checker with ignition switch ON? SYSTEM SELECT:1 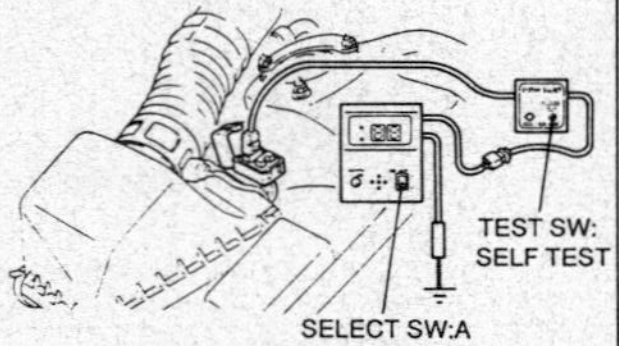	Yes	Go to next step
		No	**Service Code No. displayed** Check for cause (Refer to specified check sequence)
3	Is intake manifold vacuum correct at idle	Yes	Go to next step
	Vacuum: More than 500 mmHg (19.7 inHg)	No	Check for air leakage of intake air system components
4	Is engine compression correct? **Engine compression (Minimum):** **1079 kPa (11.0 kg/cm², 156 psi)-270 rpm**	Yes	Go to next step
		No	**High compression** Check engine condition • Carbon deposits

STEP	INSPECTION		ACTION
5	Is the fuel line pressure correct at idle? **Fuel line pressure:** **255—314 kPa (2.6—3.2 kg/cm², 37—46 psi)** **(vacuum hose to pressure regulator disconnected)** 	Yes	Check if fuel line pressure decreases when accelerating quickly? ⇨ If decreases, check for clogging between fuel pump and pressure regulator ⇨ If no decrease, go to next step
		No	**Low pressure** Check fuel line pressure while pinching fuel return hose ⇨ If pressure **quickly** increases, check pressure regulator ⇨ If pressure **gradually** increases, check for clogging between fuel pump and pressure regulator If hose is not clogged, check fuel pump maximum pressure
6	Is cooling system OK?	Yes	Go to next step
		No	Repair or replace • Thermostat • Electric cooling fan • Radiator
7	Try known good ECU; does condition improve?	Yes	Replace ECU
		No	Change fuel to another brand or use higher octane fuel

21	FUEL ODOR
DESCRIPTION	• Gasoline odor in cabin

[TROUBLESHOOTING HINTS]
① Poor connection or damaged at fuel system or evaporative emission control system
② Charcoal canister overflow due to evaporative emission control system malfunction

STEP	INSPECTION		ACTION
1	Are there fuel leakage or damage of fuel system and evaporative emission control system?	Yes	Repair or replace
		No	Go to next step
2	Is "00" displayed on Self-Diagnosis Checker while ignition switch ON? SYSTEM SELECT:1 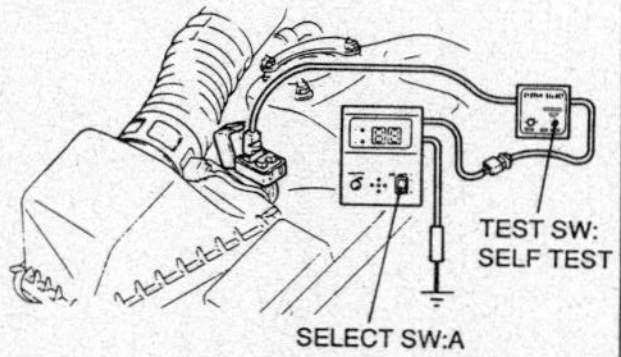	Yes	Go to next step
		No	**Service Code No. displayed** Check for cause (Refer to specified check sequence)

STEP	INSPECTION		ACTION
3	Is vacuum felt at solenoid valve (purge control) while driving vehicle on chassis roller 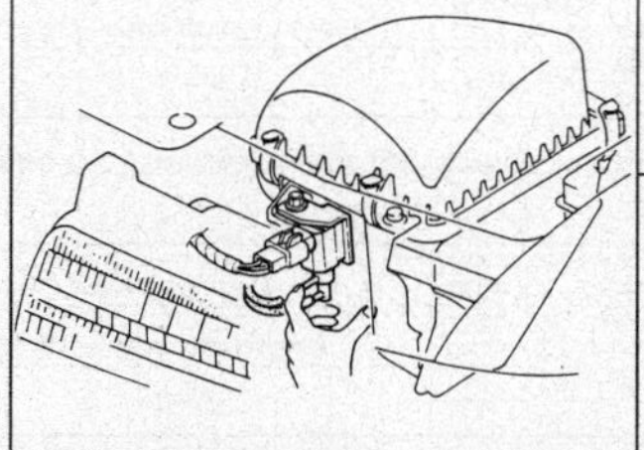	Yes	Go to Step 5
		No	Check for solenoid valve operating sound ⇨ If OK, check vacuum hoses for clogging ⇨ If not OK, go to next step
4	Apply battery voltage and a ground to solenoid valve (purge control) Is vacuum felt at solenoid valve at idle?	Yes	Check ECU terminal voltages
		No	Replace solenoid valve
5	Try known good ECU; does condition improve?		

22	A/C DOES NOT WORK
DESCRIPTION	• Blower fan operates but magnetic clutch does not

[TROUBLESHOOTING HINTS]
- Open or short circuit in wiring harness
- A/C relay, A/C switch, or magnetic clutch malfunction
- ECU malfunction

STEP	INSPECTION		ACTION
1	Is ECU 1O terminal voltage OK? 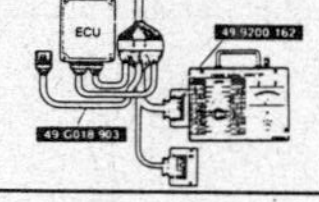	Yes	Check ECU 1L terminal voltage ⇨ If OK, check A/C system ⇨ If not OK, check for cause
		No	Check for cause

3.0L ENGINE ELECTRICAL DIAGNOSIS SERVICE POINTS—1992 929

Main Relay (Battery Power)

Circuit	Condition		
	Open circuit	Short circuit	Poor ground
ECU (1B)—Main relay	Engine will not start	EGI fuse (30A) burns out when ignition switch ON	NA

Room Fuse (Memory Power)

Circuit	Condition		
	Open circuit	Short circuit	Poor ground
ECU (1A)—Room fuse	No symptom	Room fuse (15A) burns out	NA

Ground (Output Device, Injector, System, Analogue)

Circuit	Condition		
	Open circuit	Short circuit	Poor ground
ECU (3A)—Ground (Output device)	(One side open circuit) No symptom (Both sides open circuit) Engine will not start	NA	Engine may not start Hard starting
ECU (3B)—Ground (Injector)			
ECU (3C)—Ground (System)	No symptom		
ECU (3D)—Ground (Analogue)			

Diagnosis Connector (TEN terminal)

Circuit	Condition		
	Open circuit	Short circuit	Poor ground
ECU (1I)—Diagnosis connector	Cannot perform service code checks and switch monitor checks	Hard starting Rough idle Poor acceleration	NA

Diagnosis Connector (MEN terminal)

Circuit	Condition		
	Open circuit	Short circuit	Poor ground
ECU (1D)—Diagnosis connector	Self-Diagnosis Checker monitor lamp will not illuminate	Self-Diagnosis Checker monitor lamp will stay ON	NA

Diagnosis Connector (FEN terminal)

Circuit	Condition		
	Open circuit	Short circuit	Poor ground
ECU (1F)—Diagnosis connector	Self-Diagnosis Checker buzzer will no sound during service code check	Code "88" will keep flashing and buzzer will continue sounding during service code check	NA

Malfunction Indicator Lamp (MIL)

Circuit	Condition		
	Open circuit	Short circuit	Poor ground
ECU (1E)—MIL	MIL will not illuminate with ignition switch ON	MIL will not go off after engine is started	NA

NA: Not applicable

[Input Device]
Airflow Meter (Intake Air Thermosensor Included)

Circuit	Condition		
	Open circuit	Short circuit	Poor ground
ECI (2B)—Airflow meter	Code No. 08 output Lack of power Poor acceleration	Code No. 08 output Lack of power Poor acceleration	NA
ECU (2I)—Airflow meter			
ECU (2K)—Airflow meter	Code No. 10 output No symptom	Code No. 10 output No symptom	
FCU (3D)—Airflow meter	(One side open circuit) No symptom (Both sides open circuit) Code No. 08, 10 output Poor acceleration	NA	Poor acceleration
Airflow meter—Ground			

Crank Angle Sensor (Ne 1, Ne 2, G signal)

Circuit	Condition		
	Open circuit	Short circuit	Poor ground
ECU (3H)—Crank angle sensor	Code No. 02 output No symptom	Code No. 02 output No symptom	NA
ECU (3F)—Crank angle sensor	Code No. 02 output No symptom	NA	
ECU (3E)—Crank position sensor	Code No. 04 output No symptom	Code No. 04 output No symptom	
FCU (3G)—Crank position sensor	Code No. 03, 04 output Lack of power Poor acceleration	Code No. 03, 04 output Lack of power Poor acceleration	
Main relay—Crank position sensor	Code No. 03, 04 output Engine will not start	Code No. 03, 04 output EGI fuse (30A) burns out when ignition switch ON	
Crank position sensor—Ground	Code No. 03, 04 output Engine will not start	NA	Engine may not start

Symptom is different with combination of Ne 1, Ne 2, and G signal malfunction.

Water Thermosensor

Circuit	Condition		
	Open circuit	Short circuit	Poor ground
ECU (2E)—Water thermosensor	Code No. 09 output Rough idle, Hard starting	Code No. 09 output Rough idle, Hard starting	NA
Water thermosensor—Ground	Code No. 09 output Rough idel, Hard starting	NA	Rough idle Hard starting

Intake Air Thermosensor (Dynamic Chamber)

Circuit	Condition		
	Open circuit	Short circuit	Poor ground
ECU (2L)—Intake air thermosensor	Code No. 11 output No symptom	Code No. 11 output No symptom	NA
Intake air thermosensor—Ground	Code No. 11 output No symptom	NA	No symptom

NA: Not applicable

3.0L ENGINE ELECTRICAL DIAGNOSIS SERVICE POINTS—1992 929

Throttle Sensor (Idle Switch Included)

Circuit	Condition		
	Open circuit	Short circuit	Poor ground
ECU (2I)—Throttle sensor	Code No. 12 output Poor acceleration Strong shift shock	Code No. 12 output Poor acceleration Strong shift shock	NA
ECU (2F)—Throttle sensor	Code No. 12 output Poor acceleration Strong shift shock	Code No. 12 output Poor acceleration Strong shift shock	NA
ECU (1T)—Throttle sensor	Code No. 12 may output Rough idle	Fuel cut when accelerator pedal is pressed, causing hunting	NA
Throttle sensor—Ground	Code No. 12 output Poor acceleration Strong shift shock	NA	Poor acceleration Strong shift shock

Oxygen Sensor

Circuit	Condition		
	Open circuit	Short circuit	Poor ground
ECU (2C)—Oxygen sensor (right bank)	Code No. 17 output No symptom	Code No. 17 output No symptom	NA
Oxygen sensor (right bank)—Ground	Code No. 17 output No symptom	NA	No symptom
ECU (2D)—Oxygen sensor (left bank)	Code No. 15 output No symptom	Code No. 15 output No symptom	NA
Oxygen sensor (left bank)—Ground	Code No. 15 output No symptom	NA	No symptom

Knock Sensor

Circuit	Condition		
	Open circuit	Short circuit	Poor ground
ECU (2M)—Knock sensor (right bank)	Code No. 07 output Poor acceleration	Code No. 07 output Poor acceleration	NA
ECU (2N)—Knock sensor (left bank)	Code No. 05 output Poor accelaration	Code No. 05 output Poor acceleration	NA

EGR Position Sensor (EGR Control Valve)

Circuit	Condition		
	Open circuit	Short circuit	Poor ground
ECU (2I)—EGR position sensor	Code No. 16 output No symptom	Code No. 16 output No symptom	NA
ECU (2J)—EGR position sensor	Code No. 16 output No symptom	Code No. 16 output No symptom	NA
EGR position sensor—ground	Code No. 16 output No symptom	NA	No symptom

Starter Signal (Ignition Switch)

Circuit	Condition		
	Open circuit	Short circuit	Poor ground
ECU (1C)—Ignition switch	Hard starting	Starter signal fuse (10A) burns out when cranking	NA

P/S Pressure Switch

Circuit	Condition		
	Open circuit	Short circuit	Poor ground
ECU (1N)—P/S pressure switch (front)	Engine speed may momentarily drop during steering	Engine speed may momentarily drop during steering	Engine speed may momentarily drop during steering

NA: Not applicable

CPU

Circuit	Condition		
	Open circuit	Short circuit	Poor ground
ECU (1J)—CPU	Engine speed may momentarily drop when headlights, fog lights, or rear defroster is turned ON	Engine speed may momentarily drop when headlights, fog lights, or rear defroster is turned ON	NA

A/C Signal (A/C Amplifier)

Circuit	Condition		
	Open circuit	Short circuit	Poor ground
ECU (1O)—A/C amplifier	Air conditioner will not inoperate	Air conditioner will constantly operate with blower ON	NA

Blower Signal (A/C Amplifier)

Circuit	Condition		
	Open circuit	Short circuit	Poor ground
ECU (1P)—A/C amplifier	Engine speed may momentarily drop in 3rd or higher position	Engine speed may momentarily drop in 3rd or higher position	NA

Stoplight Signal (Stoplight Switch)

Circuit	Condition		
	Open circuit	Short circuit	Poor ground
ECU (1Q)—Stoplight switch	No symptom	No symptom	NA

Vehicle Speed signal (Speedometer)

Circuit	Condition		
	Open circuit	Short circuit	Poor ground
ECU (1M)—Speedometer	Code No. 06 output Solenoid valve (purge control) will not operate	Code No. 06 output Solenoid valve (purge control) will not operate	NA

Inhibitor Signal (EC-AT CU)

Circuit	Condition		
	Open circuit	Short circuit	Poor ground
ECU (1R)—EC-AT CU	Engine speed will be slightly lower in P and N range	Engine speed will be slightly higher in R, D, S, and L ranges	NA

Reduce Torque Signal (EC-AT CU)

Circuit	Condition		
	Open circuit	Short circuit	Poor ground
ECU (1S)—EC-AT CU	Shift shock may be slightly stronger	Shift shock may be slightly stronger	NA

NA: Not applicable

Shift Solenoid A, B

Circuit	Condition		
	Open circuit	Short circuit	Poor ground
ECU (1U)—Shift solenoid A	Shift shock may be slightly stronger	Shift shock may be slightly stronger	NA
Shift solenoid A—Ground		NA	Shifting may be abnormal
ECU (1V)—Shift solenoid B	Shift shock may be slightly stronger	Shift shock may be slightly stronger	NA
Shift solenoid B—Ground		NA	Shifting may be abnormal

[Output Device] Injector

Circuit	Condition		
	Open circuit	Short circuit	Poor ground
ECU (3U, 3V, 3W, 3X, 3Y, 3Z)—Injector	Rough idle, Misfiring	Rough idle, Misfiring	NA

IGT Signal (Igniter)

Circuit	Condition		
	Open circuit	Short circuit	Poor ground
ECU (1G)—Igniter	Engine will not start	Engine will not start	NA
Igniter—Ignition coil			
Igniter—Ground		NA	Engine will not start

ISC Solenoid Valve (BAC Valve)

Circuit	Condition		
	Open circuit	Short circuit	Poor ground
ECU (3Q)—ISC solenoid valve	Code No. 34 output Rough idle Engine stalling	Code No. 34 output Hunting	NA
ISC solenoid valve—Main relay		EGI fuse (30A) burns out when ignition switch ON	

Circuit Opening Relay

Circuit	Condition		
	Open circuit	Short circuit	Poor ground
ECU (3T)—Circuit opening relay	Engine will not start	Fuel pump will continue operating when ignition switch ON	NA

Solenoid Valve (Purge Control)

Circuit	Condition		
	Open circuit	Short circuit	Poor ground
ECU (2O)—Solenoid valve (purge control)	Code No. 26 output Solenoid valve will not operate	Code No. 26 output Rough idle, Hard starting	NA
Solenoid valve (purge control)—Main relay		EGI fuse (30A) burns out when ignition switch ON	

NA: Not applicable

Solenoid Valve (VICS)

Circuit	Condition		
	Open circuit	Short circuit	Poor ground
ECU (3J)—Solenoid valve (VICS)	Code No. 41 output Power drop in high-speed range	Code No. 41 output Power drop in mid-speed range	NA
Solenoid valve (VICS)—Main relay		EGI fuse (30A) burns out when ignition switch ON	

Solenoid Valve (EGR—Vacuum Side)

Circuit	Condition		
	Open circuit	Short circuit	Poor ground
ECU (3P)—Solenoid valve	Code No. 28 output No symptom	Code No. 28 output Engine will not start	NA
Solenoid valve—Main relay	Code No. 28 output No symptom	EGI fuse (30A) burns out when ignition switch ON	

Solenoid Valve (EGR—Vent Side)

Circuit	Condition		
	Open circuit	Short circuit	Poor ground
ECU (3O)—Solenoid valve	Code No. 29 output No symptom	Code No. 29 output No symptom	NA
Solenoid valve—Main relay	Code No. 29 output No symptom	EGI fuse (30A) burns out when ignition switch ON	

Air Control Motor (Variable Exhaust System)

Circuit	Condition		
	Open circuit	Short circuit	Poor ground
ECU (3R)—Air control motor	Power drop in high-speed heavy-load range	Exhaust noise in the low-speed range will be greather than usual	NA
ECU (3S)—Air control motor		Power drop in high-speed, heavy-load range	

Oxygen Sensor Heater

Circuit	Condition		
	Open circuit	Short circuit	Poor ground
ECU (3K)—Oxygen sensor heater (right bank)	Code No. 36 output No symptom	Code No. 36 output No symptom	NA
ECU (3N)—Oxygen sensor heater (left bank)	Code No. 37 output No symptom	Code No. 37 output No symptom	

A/C Signal (A/C Relay)

Circuit	Condition		
	Open circuit	Short circuit	Poor ground
ECU (1L)—A/C relay	Air conditioner will not operate	Air conditioner will remain operating with blower ON	NA

Cold Start Injector (CSI) Relay

Circuit	Condition		
	Open circuit	Short circuit	Poor ground
ECU (3I)—CSI relay	Code No. 30 output No symptom	Code No. 30 output Hard starting when cold	NA

Atmospheric pressure sensor (EC-AT CU)

Circuit	Condition		
	Open circuit	Short circuit	Poor ground
ECU (2G)—EC-AT CU	Shift shock will be strong at high altitudes	Shift shock will be stronger at high altitudes	NA

NA: Not applicable

3.0L ENGINE ELECTRICAL DIAGNOSIS SERVICE POINTS 1992 929

A/C Signal (EC-AT CU)

Circuit	Condition		
	Open circuit	Short circuit	Poor ground
ECU (1K)—EC-AT CU	Code No. 65 output The air conditioning will be in the ON mode in D range (NORMAL), and slip lock up will be prohibited	Code No. 65 output The air conditioning will be in the OFF mode in D range (NORMAL)	NA

Torque Reduced Signal (EC-AT CU)

Circuit	Condition		
	Open circuit	Short circuit	Poor ground
ECU (2P)—EC-AT CU	Shift shock may be slightly stronger	Shift shock may be slightly stronger	NA

NA: Not applicable

3.0L ENGINE TROUBLE CODES—1992 929

Code Numbers

Service Code No.	Sensor or Subsystem	Condition	Fail-safe function	Pattern of output signals (Self-Diagnosis Checker or MIL)
02	Ne 2 signal	No Ne 2 signal	(see table below)	ON / OFF
03	G signal	No G signal input while a certain number of Ne 1 or Ne 2 signal pulses are input.	(see table below)	ON / OFF
04	Ne 1 signal	No Ne 1 signal	(see table below)	ON / OFF
05	Left side knock sensor	Open or short circuit	Ignition timing retard	ON / OFF
06	Speed signal	No speed signal input while driving vehicle at more than 40 km/h (25 mph) in D, S or L range for 10 sec	Maintains constant 0 km/h (0 mph) command	ON / OFF
07	Right side knock sensor	Open or short circuit	Ignition timing retard	ON / OFF
08	Airflow meter	Short circuit	Basic fuel injection amount fixed as for 3 driving modes 1) Idle switch ON 2) Slight throttle angle 3) Large throttle angle.	ON / OFF
09	Water thermosensor	Open or short circuit	Maintains constant 80°C(176°F) command	ON / OFF
10	Intake air thermosensor (in airflow meter)	Open or short circuit	Maintains constant 20°C(68°F) command	ON / OFF
11	Intake airthermosensor (on dynamic chamber)	Open or short circuit	Maintains constant 20°C(68°F) command	ON / OFF
12	Throttle sensor	Input voltage is less than specified when idle switch OFF, or input voltage is more than specified.	Maintains constant command of throttle valve fully open	ON / OFF
14	Atmospheric pressure sensor	Open or short circuit	Maintains constant command of sea level pressure	ON / OFF
15	Left side oxygen sensor	Sensor output continues less than 0.55V 130 sec. after engine exceeds1,500 rpm	Cancels engine feedback operation	ON / OFF
16	EGR position sensor	Open or short circuit	Maintains constant command of EGR valve fully open	ON / OFF

Fail-safe function (codes 02, 03, 04):

Ne 2	G	Ne 1	✻
NG	OK	OK	A
OK	OK	NG	A
OK	NG	OK	B
NG	NG	OK	B
OK	NG	NG	C
NG	OK	NG	C
NG	NG	NG	C

✻ Fail safe operation
A: sequential injection (normal injection)
B: simultaneous injection to all cylinders
C: fuel injection and ignition stopped

3.0L ENGINE TROUBLE CODES—1992 929

Service Code No.	Sensor or subsystem	Condition	Fail-safe function	Pattern of output signals (Self-Diagnosis checker or MIL)
17	Left side feedback system	Sensor output continues unchanged 50sec. after engine exceeds 1,500 rpm	Cancels engine feedback operation	ON / OFF
23	Right side oxygen sensor	Sensor output continues less than 0.55V 130 sec. after engine exceeds 1,500 rpm		ON / OFF
24	Right side feedback system	Sensor output continues unchanged 50sec. after engine exceeds 1,500 rpm		ON / OFF
26	Solenoid valve (purge control)	Open or short circuit	—	ON / OFF
28	Solenoid valve (EGR—vacuum side)		—	ON / OFF
29	Solenoid valve (EGR—vent side)		—	ON / OFF
30	Relay (Cold start injector)		—	ON / OFF
34	Solenoid valve (ISC)		—	ON / OFF
36	Right side oxygen sensor heater		—	ON / OFF
37	Left side oxygen sensor heater		—	ON / OFF
41	Solenoid valve (VICS)		—	ON / OFF
65	A/C signal (EC-AT CU)		—	ON / OFF

Caution

- **If there is more than one failure present, the code numbers will be indicated in numerical order lowest number first.**
- **After repairing all failures, turn the ignition switch OFF, disconnect the negative battery cable for twenty seconds to erase the service code from the memory of the ECU.**

3.0L ENGINE TROUBLE CODE TROUBLESHOOTING 1992 929

Troubleshooting

If a service code number is shown on the **SST**, check for the cause by using the chart related to the code number shown.

CODE No.	02 (CRANK ANGLE SENSOR — Ne 2 SIGNAL)		
STEP	**INSPECTION**		**ACTION**
1	Is there poor connection in clank angle sensor circuit?	Yes	Repair or replace connector
		No	Go to next step
2	Disconnect connector from ECU Is resistance of connector between 3F (L/Y) and 3H (L/G) terminal OK? **Resistance: 0.95—1.25 kΩ**	Yes	Go to next step
		No	Go to Step 4
3	Is there continuity of connector between ground and 3F or ground and 3H terminal? (At harness side)	Yes	Check for short circuit in wiring from crank angle sensor to 3F or 3H terminal
		No	Replace ECU
4	Is resistance of crank angle sensor OK? **Resistance: 0.95—1.25 kΩ**	Yes	Check for open circuit in wiring from ECU to crank angle sensor
		No	Replace crank angle sensor

Circuit Diagram

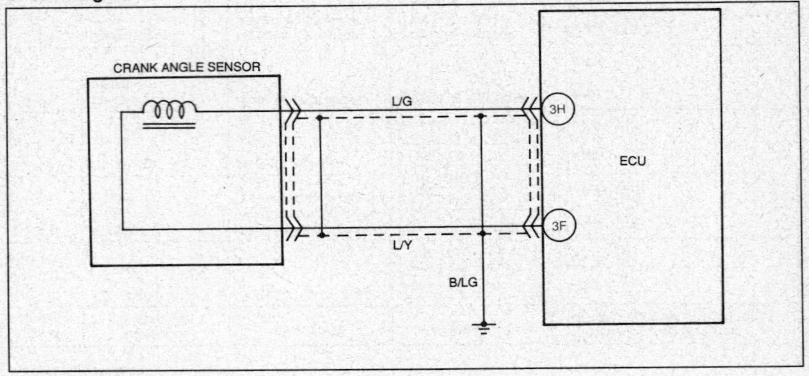

CODE No.	03 (DISTRIBUTOR G SIGNAL)		
STEP	INSPECTION		ACTION
1	Is there poor connection in distributor circuit?	Yes	Repair or replace connector
		No	Go to next step
2	Is Code No. 04 also present?	Yes	Go to next step
		No	Go to Step 5
3	Is there continuity between distributor terminal-wire (B/LG) and ground?	Yes	Go to next step
		No	Check for open circuit in wiring from distributor to ground
4	Is there battery voltage at distributor terminal-wire (B/W)?	Yes	Go to next step
		No	Check for open circuit in wiring from distributor to main relay
5	Is there approx. 5V at distributor terminal-wire (L)? (With AT harness-side connector with disconnected)	Yes	Replace distributor
		No	Go to next step
6	Is there approx. 5V at ECU terminal 3G? (With distributor connector disconnected)	Yes	Check for short or open circuit in wiring from distributor to ECU terminal 3G
		No	Go to next step

Circuit Diagram

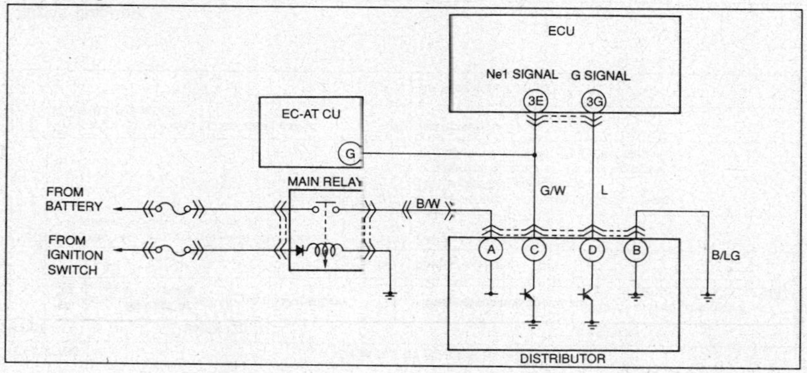

CODE No.	04 (DISTRIBUTOR Ne 1 SIGNAL)		
STEP	INSPECTION		ACTION
1	Is there poor connection in distributor circuit?	Yes	Repair or replace connector
		No	Go to next step
2	Is Code No. 03 also present?	Yes	Go to next step
		No	Go to Step 5
3	Is there continuity between distributor terminal-wire (B/LG) and ground?	Yes	Go to next step
		No	Check for open circuit in wiring from distributor to ground
4	Is there battery voltage at distributor terminal-wire (B/W)?	Yes	Go to next step
		No	Check for open circuit in wiring from distributor to main relay
5	Is there approx. 5V at distributor terminal-wire (G/W)? (AT harness-side connector with disconnected)	Yes	Replace distributor
		No	Go to next step
6	Is there approx. 5V at ECU terminal 3E? (With distributor connector disconnected)	Yes	Check for short or open circuit in wiring from distributor to ECU terminal 3E
		No	Replace ECU

Circuit Diagram

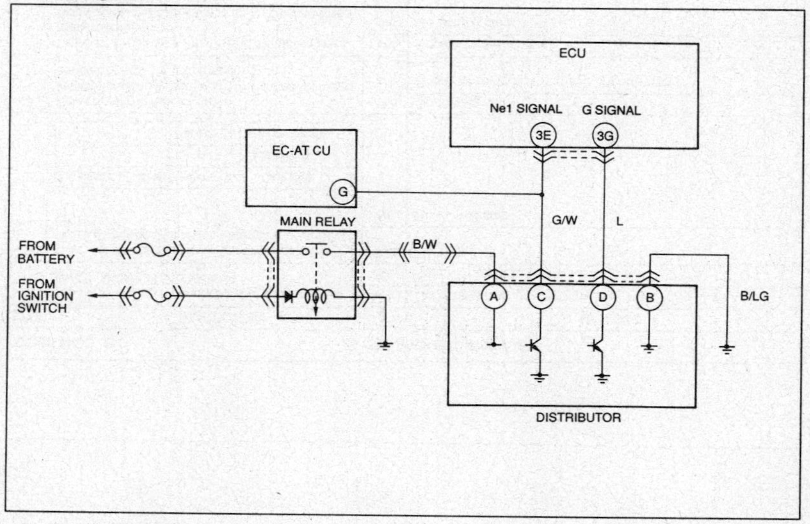

CODE NO.	05 and/or 07 (KNOCK SENSOR)		
STEP	INSPECTION		ASCTION
1	Does knock sensor circuit have a poor connection?	Yes	Repair or replace connector
		No	Go to next step
2	Does the wire harness between the knock sensor and ECU have a continuity?	Yes	Repair or replace harness
		No	Check continuity between ECU terminals (2M and/or 2N) and ground ⇨ If continue, repair or replace harness ⇨ If does not contiuie, go to next step
3	Try known good knock sensor does same Code No. present?	Yes	Replace ECU
		No	Replace knock sensor

Circuit Diagram

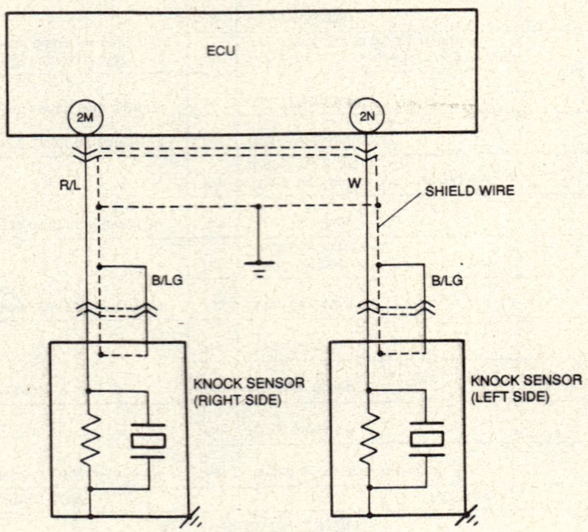

CODE No.	06 (SPEEDOMETER SENSOR)		
STEP	INSPECTION		ACTION
1	Are there any poor connections at speedometer sensor and ECU connectors?	Yes	Go to next step
		No	Repair or replace connector
2	Connect a circuit tester to terminals as shown Is input voltage of speedometer sensor at ECU OK? (+) term.: 1M; (−) term.: Ground; Voltage (V): 2–3 — Condition: While driving; 0 or 4.5–5.5 — Condition: Vehicle stopped	Yes	Go to Step 8
		No	Go to next step
3	Remove combination meter Is there continuity between 1M terminal of meter connector and 1M terminal of ECU	Yes	Go to next step
		No	Repair or replace wiring and/or connector
4	Connect circuit tester to 1K and 1L terminals of meter connector Does pointer of circuit tester move slightly when rear wheels slowly turned?	Yes	Replace speedometer
		No	Go to next step
5	Remove speedometer sensor Is resistance felt when turning speedometer driven gear by hand?	Yes	Go to next step
		No	Replace speedometer sensor
6	Disconnect speedometer sensor connector and connect circuit tester Does pointer of circuit tester moves slightly when driven gear slowly turned?	Yes	Go to next step
		No	Replace speedometer sensor
7	Disconnect speedometer sensor connector Is continuity of sensor OK? **Resistance: Approx. 290 Ω (20°C [68°F]); reference**	Yes	Check wiring and connectors from speedometer sensor to speedometer If OK, go to next step If not OK, repair wiring and/or connector
8	Disconnect negative battery cable for at least 20 seconds Connect battery cable and recheck for service code Is service code displayed?	Yes	Replace ECU
		No	Intermittent poor connection Check for cause

Circuit Diagram

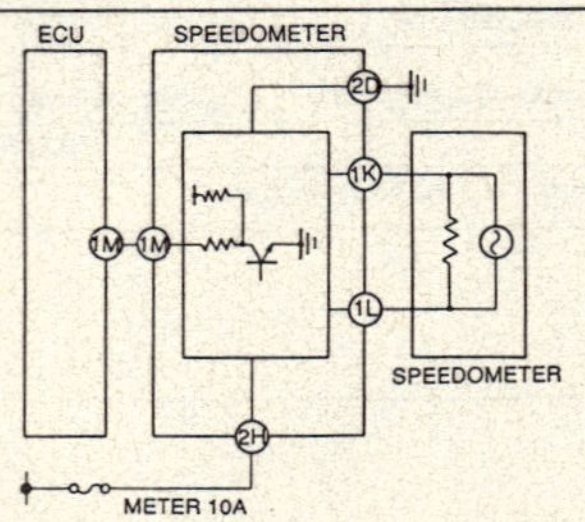

CODE No.	08 (AIRFLOW METER)		
STEP	INSPECTION		ACTION
1	Is there poor connection in airflow sensor circuit?	Yes	Repair or replace connector
		No	Go to next step
2	Is there continuity between airflow meter and ECU?	Yes	Go to next step
		No	Check for open circuit in wiring from airflow meter to ECU
3	Is resistance of airflow meter OK?	Yes	Go to next step
		No	Replace airflow meter
4	Are ECU terminals 2B, 2I, and 3D voltages OK?	Yes	Replace ECU
		No	Check for short circuit in wiring from airflow meter to ECU ➪ If OK, replace ECU ➪ If not OK, repair or replace wiring

Airflow meter	ECU
A (W)	2B (W)
B (BR/W)	2I (BR/W)
C (B)	3D (B)

Circuit Diagram

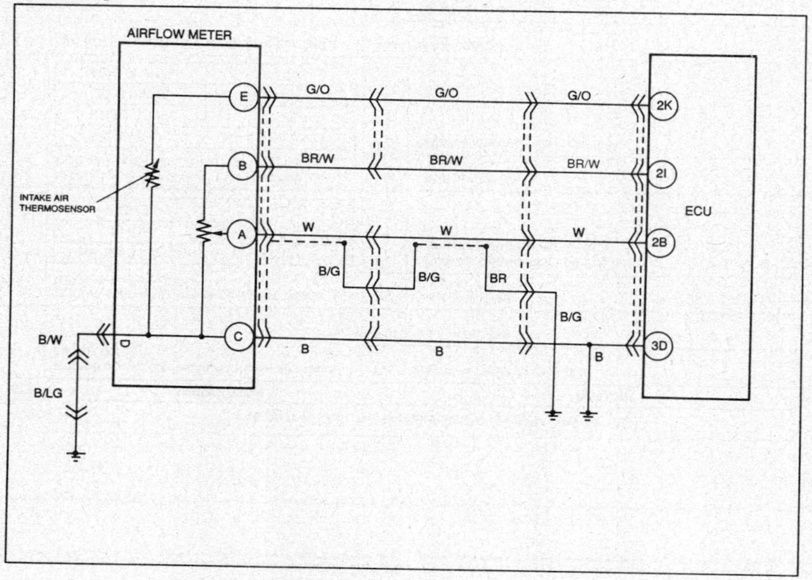

CODE No.	09 (WATER THERMOSENSOR)		
STEP	INSPECTION		ACTION
1	Is there a poor connection in water thermosensor circuit?	Yes	Repair or replace connector
		No	Go to next step
2	Is there continuity between water thermosensor and ECU?	Yes	Go to next step
		No	Check for open circuit wiring from water thermosensor to ECU
3	Is resistance of water thermosensor correct?	Yes	Go to next step
		No	Replace water thermosensor
4	Are ECU terminal 2E and 3D voltages correct?	Yes	Replace ECU
		No	Check for short circuit in wiring from water thermosensor to ECU

Water thermosensor	ECU
A (G/W)	2E (G/W)

Coolant temp	Resistance (kΩ)
–20°C (–4°F)	14.6—17.8
20°C (68°F)	2.2—2.7
80°C (176°F)	0.2—0.4

Circuit Diagram

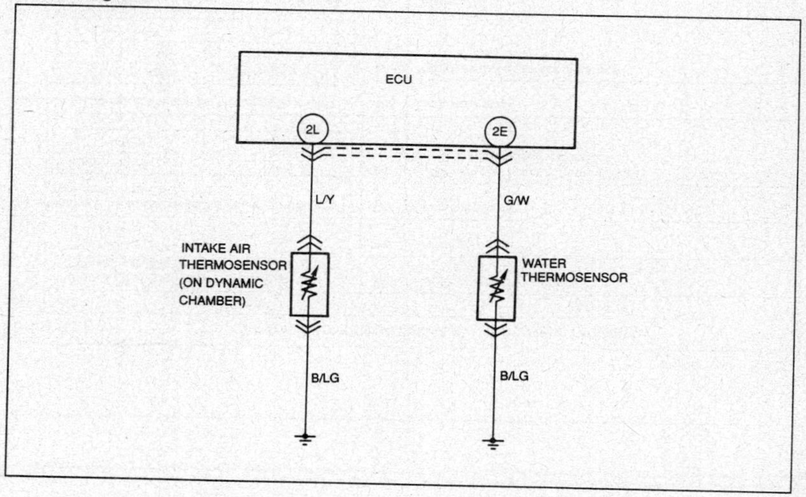

CODE No.	10 (INTAKE AIR THERMOSENSOR — IN AIRFLOW METER)		
STEP	INSPECTION		ACTION
1	Is there a poor connection in intake air thermosensor circuit?	Yes	Repair or replace connector
		No	Go to next step
2	Is there continuity between intake air thermosensor and ECU? Intake air thermosensor / ECU C (B) / 3D (B) E (G/O) / 2K (G/O)	Yes	Go to next step
		No	Check for open circuit in wiring from intake air thermosensor to ECU
3	Is resistance of intake air thermosensor between terminals C(B) and E(G/O) OK? Coolant temp / Resistance (kΩ) −20°C (−4°F) / 10—20 20°C (68°F) / 2—3 60°C (140°F) / 0.4—0.7	Yes	Go to next step
		No	Replace intake air thermosensor
4	Are ECU terminal 2K and 3D voltages OK?	Yes	Replace ECU
		No	Check for short circuit in wiring from intake air thermosensor to ECU

Circuit Diagram

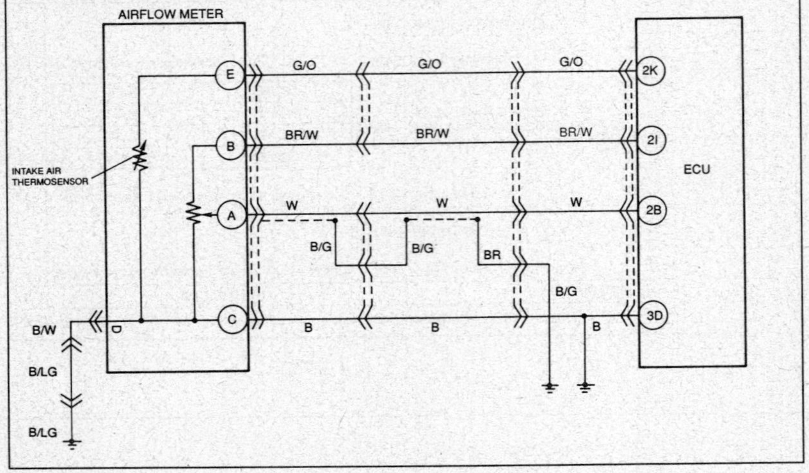

CODE No.	11 (INTAKE AIR THERMOSENSOR — IN DYNAMIC CHAMBER)		
STEP	INSPECTION		ACTION
1	Is there a poor connection in intake air thermosensor circuit?	Yes	Repair or replace connector
		No	Go to next step
2	Is there continuity between intake air thermosensor and ECU? Intake air thermosensor / ECU A (L/Y) / 2L (L/Y)	Yes	Go to next step
		No	Check for open circuit in wiring from intake air thermosensor and ECU
3	Is resistance of intake air thermosensor OK? Coolant temp / Resistance (kΩ) 25°C (77°F) / 29.7—36.3 85°C (185°F) / 3.3—3.7	Yes	Go to next step
		No	Replace intake air thermosensor
4	Is there continuity between intake air termosensor and ground?	Yes	Replace ECU
		No	Check for open circuit in wiring from intake air thermosensor to ground

Circuit Diagram

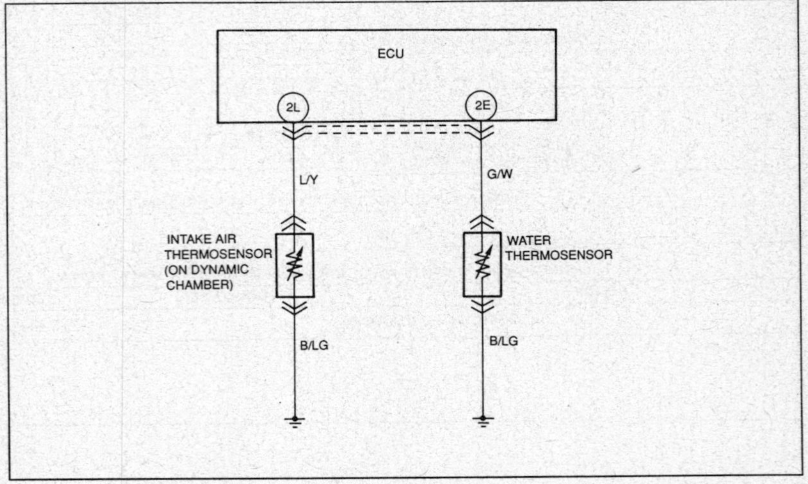

CODE No.	12 (THROTTLE SENSOR)		
STEP	INSPECTION		ACTION
1	Is there a poor connection in throttle sensor circuit?	Yes	Repair or replace connector
		No	Go to next step
2	Is there continuity between throttle sensor and ECU Throttle sensor / ECU: A (B/LG) — ground C (L) — 2F (L) D (BR/W) — 2I (BR/W)	Yes	Go to next step
		No	Check for open circuit in wiring from throttle sensor and ECU
3	Is idle switch of throttle sensor OK? Check following switch monitor function test	Yes	Go to next step
		No	Replace throttle sensor
4	Is ECU terminal 2F voltage correct?	Yes	Replace ECU
		No	Check for short circuit in wiring from throttle sensor to ECU

Circuit Diagram

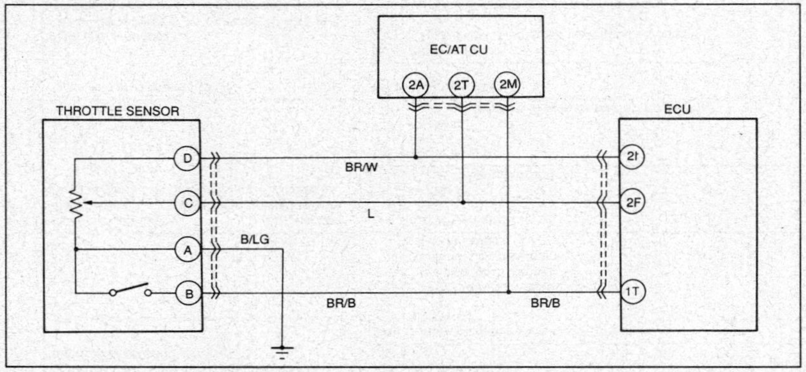

CODE No.	14 (ATMOSPHERIC PRESSURE SENSOR — IN ECU)
STEP	ACTION
1	Replace ECU

CODE No.	15 and/or 23 (OXYGEN SENSOR — INACTIVATION)		
Note: • If Code Nos. 15 and 17, and/or 23 and 24 are both present, first perform the checking procedure for Code No. 17 and/or 24			
STEP	INSPECTION		ACTION
1	Are there are poor connections in oxygen sensor circuit?	Yes	Repair or replace connector
		No	Go to next step
2	Is oxygen sensor output voltage correct?	Yes	Go to next step
		No	Replace oxygen sensor
3	Is there continuity between oxygen sensor (left and/or right) and ECU terminals (2D and/or 2C)?	Yes	Go to next step
		No	Check for open circuit in wiring from oxygen sensor to ECU
4	Are ECU terminals (2D and/or 2C) voltage correct?	Yes	Go to next step
		No	Check for short circuit in wiring from oxygen sensor to ECU
5	Is sensitivity of oxygen sensor correct?	Yes	Replace ECU
		No	Try known good oxygen sensor and check if condition improves

Circuit Diagram

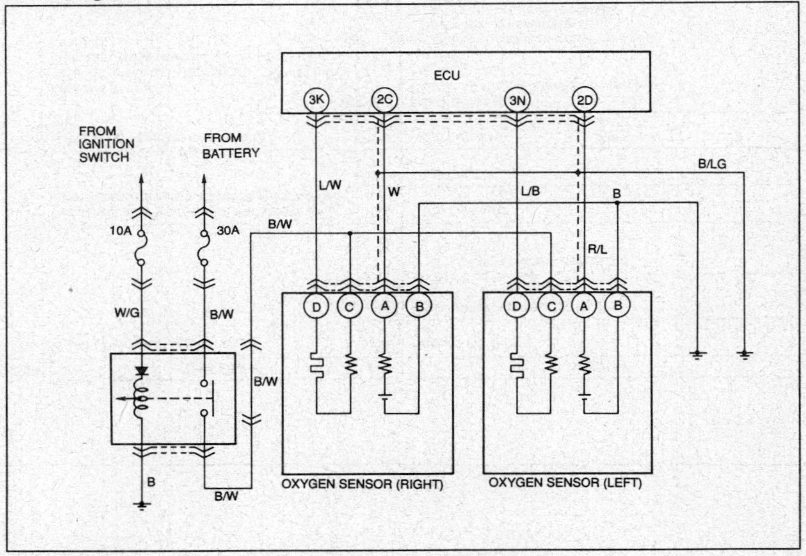

CODE No.	17 and/or 24 (FEEDBACK SYSTEM)		
STEP	INSPECTION		ACTION
1	Is same Code No. present after performing after-repair procedure?	Yes	Go to next step
		No	Check oxygen sensor circuit for poor connection ⇨ If OK, perform troubleshooting Code No. 15 and/or 23 ⇨ If not OK, repair or replace connector
2	Does monitor lamp of Self-Diagnosis Checker illuminate at idle after warming-up engine and running it at 2,500-3,000 rpm for 3 minutes?	Yes	Go to next step **Note** **• A/F mixture is rich**
		No	Go to Step 5 **Note** **• A/F mixture is lean or misfire occurring**
3	Is fuel line pressure at idle OK? **Fuel line pressure:** **255—314 kPa (2.6—3.2 kg/cm², 37—45 psi) (vacuum hose to pressure regulator disconnected)**	Yes	Go to next step
		No	**High pressure** Check if fuel return hose is clogged or restricted? ⇨ If OK, replace pressure regulator ⇨ If not OK, repair or replace hose

Circuit Diagram

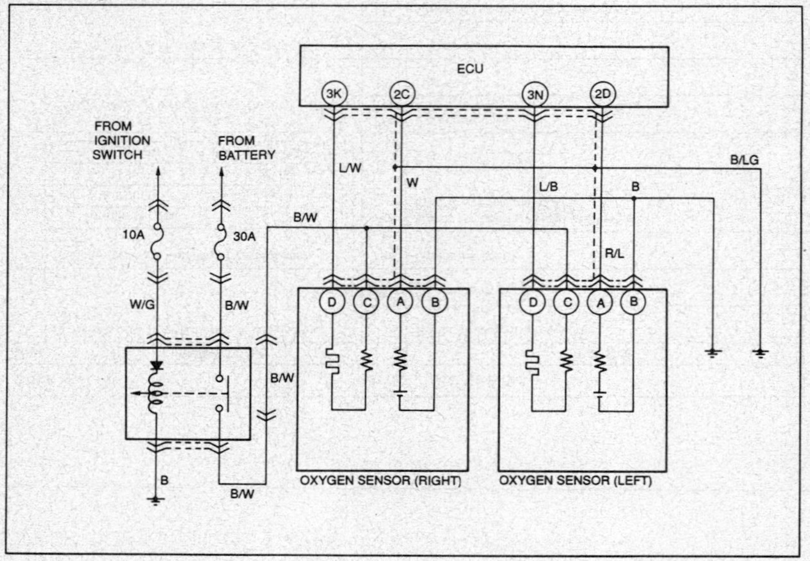

STEP	INSPECTION		ACTION
4	Is there leakage of fuel injector?	Yes	Replace injector
		No	Is water thermosensor OK? ⇨ If OK, go to Step 7 ⇨ If not OK, replace water thermosensor
5	Disconnect each high-tension lead at idle; does engine speed decrease equally at each cylinder?	Yes	Go to next step
		No	Go to Step 8
6	Is fuel line pressure at idle correct? **Fuel line pressure:** **255—314 kPa (2.6—3.2 kg/cm², 37—45 psi) (vacuum hose to pressure regulator disconnected)**	Yes	Go to next step
		No	**Low pressure** Check fuel line pressure while pinching fuel return hose ⇨ If fuel line pressure **quickly** increases, check pressure regulator ⇨ If fuel line pressure **gradually** increases, check for clogging between fuel pump and pressure regulator If hose is not clogged, check fuel pump maximum pressure
7	Is there air leakage in intake air system components?	Yes	Go to Step 10
		No	Repair or replace
8	Was misfire found in Step 5 of inspection?	Yes	Repair or replace ignition system
		No	Go to next step
9	Replace injector Is same Code No. present following performing after-repair procedure?	Yes	Go to next step
		No	Replace injector
10	Try known good ECU; does condition improve?		

CODE No.	16 (EGR POSITION SENSOR)		
STEP	INSPECTION		ACTION
1	Does EGR position sensor circuit have a poor connection?	Yes	Repair or replace connector
		No	Go to next step
2	Is EGR control valve OK?	Yes	Go to next step
		No	Replace EGR control valve
3	Is resistance of EGR position sensor OK? Resistance: (B) ↔ (C) Approx. 5kΩ (A) ↔ (C) 0.7—5 kΩ (A) ↔ (B) 0.7—5 kΩ	Yes	Go to next step
		No	Replace EGR control valve
4	Is there continuity between EGR position and ECU? EGR position sensor — ECU (A) (Y wire) — 2J (C) (BR/W wire) — 2I	Yes	Replace ECU
		No	Check for open circuit in wiring from EGR position sensor to ECU
5	Is there 4.5—5.5V at C terminal of EGR position sensor connector?	Yes	Go to next step
		No	Check for short circuit in wiring from EGR position sensor terminal C to ECU
6	Is ECU terminal 2J voltage OK?	Yes	Replace ECU
		No	Short circuit in wiring harness EGR position sensor terminal A to ECU

Circuit Diagram

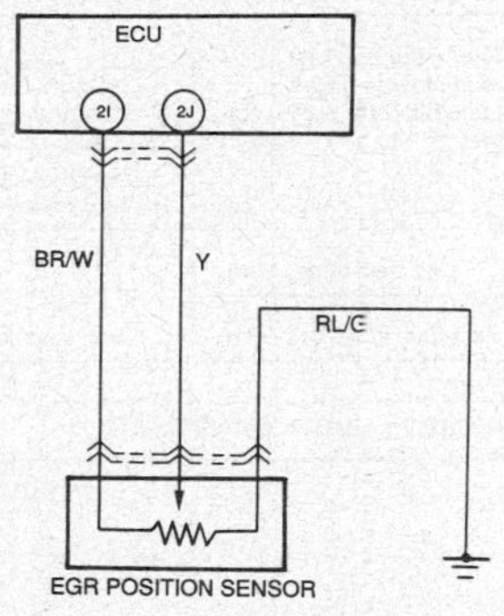

CODE No.	26 (SOLENOID VALVE — PURGE CONTROL)		
STEP	INSPECTION		ACTION
1	Is there a poor connection in solenoid valve circuit?	Yes	Repair or replace connector
		No	Go to next step
2	Disconnect connector from solenoid valve Is there continuity of solenoid valve between terminals?	Yes	Go to next step
		No	Replace solenoid valve
3	Is there battery voltage at solenoid valve terminal-wire (B/W)?	Yes	Go to next step
		No	Check for open circuit in wiring from main relay to solenoid valve terminal-wire (B/W)
4	Is there continuity between solenoid valve and ECU terminal 2O?	Yes	Check for short circuit in wiring from solenoid valve terminal-wire (BR/L) to ECU terminal 2O
		No	Check for open circuit in wiring from solenoid valve terminal-wire (BR/L) to ECU terminal 2O

CODE No.	30 (COLD START INJECTOR RELAY)		
STEP	INSPECTION		ACTION
1	Is there a poor connection in CSI relay circuit?	Yes	Repair or replace connector
		No	Go to next step
2	Disconnect connector from CSI relay Is there continuity of CSI relay terminals between A (B/W) and B (P)?	Yes	Go to next step
		No	Replace solenoid valve
3	Is there battery voltage at CSI relay terminal-wire (B/W)?	Yes	Go to next step
		No	Check for open circuit in wiring from main relay to CSI relay terminal-wire (B/W)
4	Is there continuity between CSI relay and ECU terminal 3I?	Yes	Check for short circuit in wiring from CSI relay terminal-wire (P) to ECU terminal 3I
		No	Check for open circuit in wiring from CSI relay terminal-wire (P) to ECU terminal 3I

Circuit Diagram

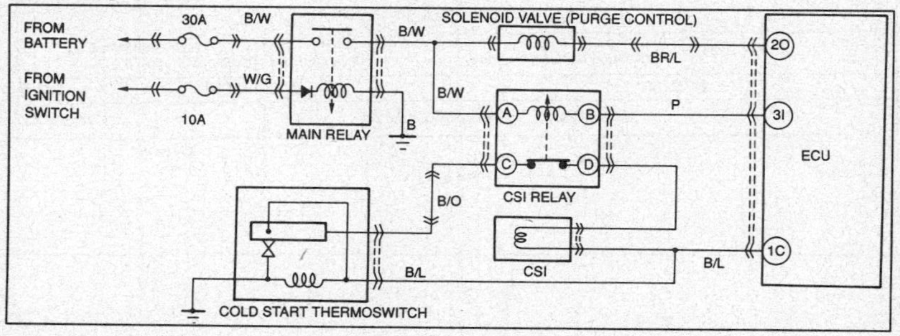

CODE No.	28 (SOLENOID VALVE — EGR VACUUM)		
STEP	INSPECTION		ACTION
1	Is there a poor connection in solenoid valve (EGR vacuum) circuit?	Yes	Repair or replace connector
		No	Go to next step
2	Is connector terminal A (B/W) voltages with solenoid valve connector disconnected OK? Condition: IG switch ON — Voltage: Battery voltage	Yes	Go to next step
		No	Check for open or short circuit in wiring from solenoid valve terminal A (B/W) to main relay terminal
3	Is continuity between solenoid valve terminal B(L/W) and ECU terminal 3P OK?	Yes	Check for short circuit in wiring from solenoid valve terminal B (L/W) to ECU terminal 3P ➪ If OK, go to next step ➪ If not OK, repair or replace wire harness
		No	Repair or replace wire harness
4	Is solenoid valve OK?	Yes	Replace ECU
		No	Replace solenoid valve

Circuit Diagram

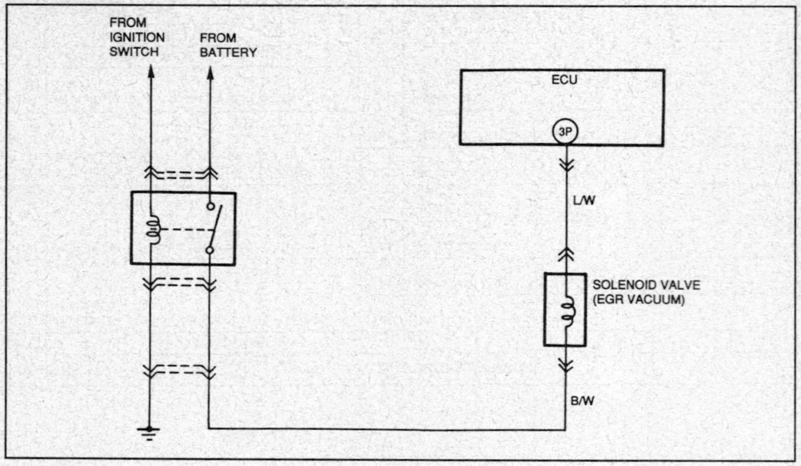

CODE No.	29 (SOLENOID VALVE — EGR VENT)		
STEP	INSPECTION		ACTION
1	Is there a poor connection in solenoid valve (EGR vent) circuit?	Yes	Repair or replace connector
		No	Go to next step
2	Is connector terminal A (B/W) voltages with solenoid valve connector disconnected OK? Condition: IG switch ON — Voltage: Battery voltage	Yes	Go to next step
		No	Check for open or short circuit in wiring from solenoid valve terminal A (B/W) to main relay terminal
3	Is continuity between solenoid valve terminal B (O/L) and ECU terminal 30 OK?	Yes	Check for short circuit in wiring from solenoid valve terminal B (O/L) to ECU terminal 3O ➪ If OK, go to next step ➪ If not OK, repair or replace wire harness
		No	Repair or replace wire harness
4	Is solenoid valve OK?	Yes	Replace ECU
		No	Replace solenoid valve

Circuit Diagram

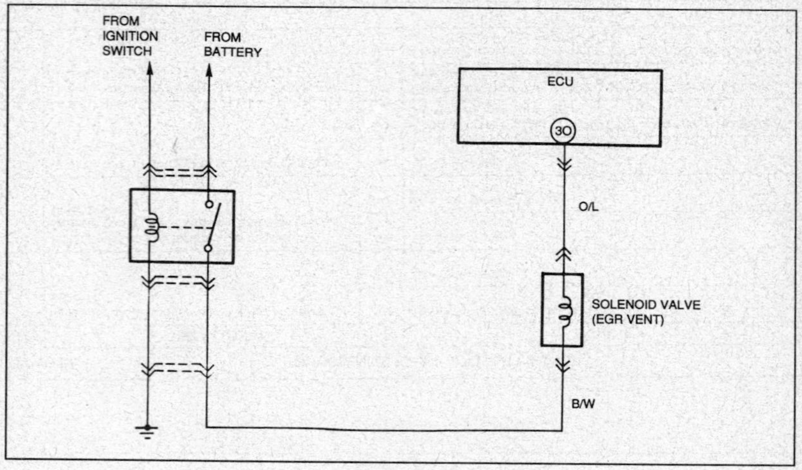

CODE No.	34 (SOLENOID VALVE-ISC) — In BAC valve		
STEP	INSPECTION		ACTION
1	Is there a poor connection in solenoid valve circuit?	Yes	Repair or replace connector
		No	Go to next step
2	Disconnect connector from solenoid valve Is there continuity of solenoid valve between terminals?	Yes	Go to next step
		No	Replace solenoid valve
3	Is there battery voltage at solenoid valve terminal-wire (B/W)?	Yes	Go to next step
		No	Check for open circuit in wiring from main relay to solenoid valve terminal-wire (B/W)
4	Is there continuity between solenoid valve and ECU terminal 3Q?	Yes	Check for short circuit in wiring from solenoid valve terminal-wire (L/O) to ECU terminal 3Q
		No	Check for open circuit in wiring from solenoid valve terminal-wire (L/O) to ECU terminal 3Q

Circuit Diagram

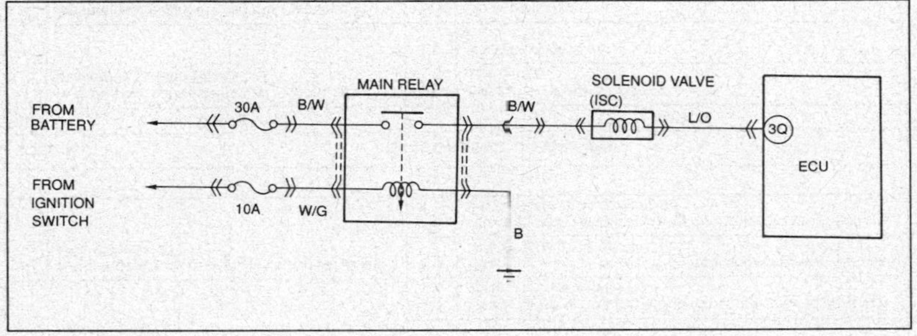

CODE No.	36 or 37 (HEATER-OXYGEN SENSOR RIGHT or LEFT)		
STEP	INSPECTION		ACTION
1	Is there a poor connection in solenoid valve circuit?	Yes	Repair or replace connector
		No	Go to next step
2	Disconnect connector from oxygen sensor Is there continuity of oxygen sensor heater between terminal (B/W) and (L/W) or (B/W) and (L/B)?	Yes	Go to next step
		No	Replace oxygen sensor
3	Is there battery voltage at oxygen sensor terminal-wire (B/W)?	Yes	Go to next step
		No	Check for open circuit in wiring from main relay to oxygen sensor terminal-wire (B/W)
4	Is there continuity between oxygen sensor and ECU terminal 3K and 3N?	Yes	Check for short circuit in wiring from oxygen sensor terminal-wire (L/W) or (L/B) to ECU terminal
		No	Check for open circuit in wiring from oxygen sensor terminal-wire (L/W) or (L/B) to ECU terminal

Circuit Diagram

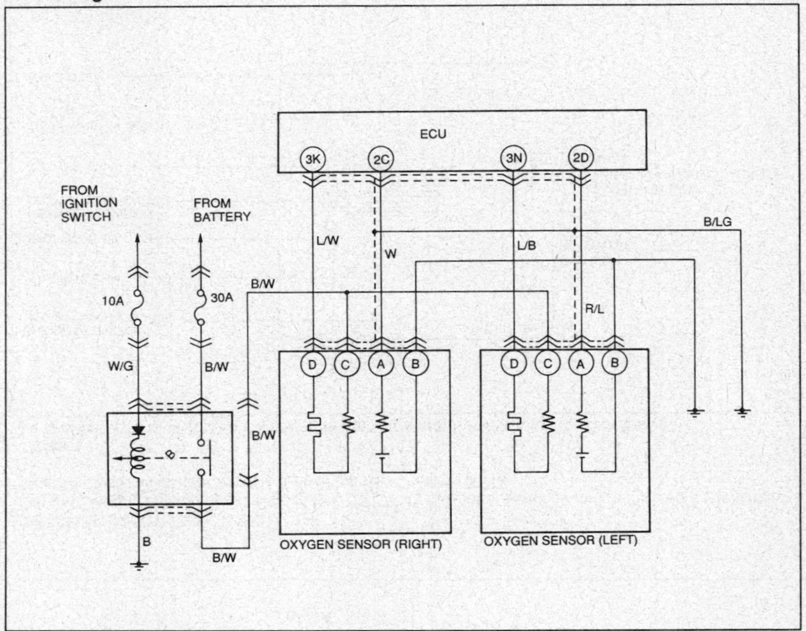

3.0L ENGINE TROUBLE CODE TROUBLESHOOTING 1992 929

CODE No.	41 (SOLENOID VALVE-VICS)		
STEP	INSPECTION		ACTION
1	Is there poor connection in solenoid valve circuit?	Yes	Repair or replace connector
		No	Go to next step
2	Disconnect connector from solenoid valve Is there continuity of solenoid valve between terminals?	Yes	Go to next step
		No	Replace solenoid valve
3	Is there battery voltage at solenoid valve terminal-wire (B/W)?	Yes	Go to next step
		No	Check for open circuit in wiring from main relay to solenoid valve terminal-wire (B/W)
4	Is there continuity between solenoid valve and ECU terminal 3J?	Yes	Check for short circuit in wiring from solenoid valve terminal-wire (Y/R) to ECU terminal 3J
		No	Check for open circuit in wiring from solenoid valve terminal-wire (Y/R) to ECU terminal 3J

CODE No.	65 (A/C-SIGNAL — EC-AT CU)		
STEP	INSPECTION		ACTION
1	Is there poor connection in A/C-signal circuit between ECU and EC-AT CU?	Yes	Repair or replace connector
		No	Go to next step
2	Is there continuity between ECU terminal 1K and EC-AT CU terminal L?	Yes	Go to next step
		No	Check for open circuit in wiring from EC-AT CU to ECU
3	Is EC-AT CU terminal L voltage OK?	Yes	Go to next step
		No	Check for cause
4	Is ECU terminal 1K voltage OK?	Yes	Replace ECU
		No	Check for short circuit in wiring from EC-AT CU to ECU

Circuit Diagram

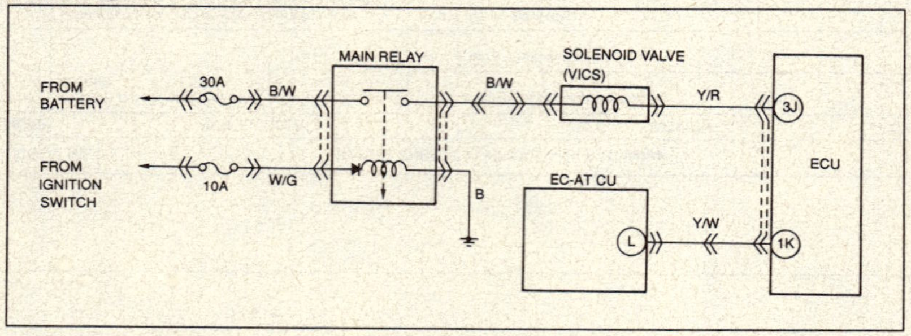

3.0L ENGINE SWITCH MONITOR FUNCTION – 1992 929

SWITCH MONITOR FUNCTION

With the System Selector test switch set to "Self Test", individual switches are monitored as described using the System Selector and the Self-Diagnosis Checker, **SSTs.**

Note

- **If either of the switches remain activated, the monitor lamp will be illustrated.**

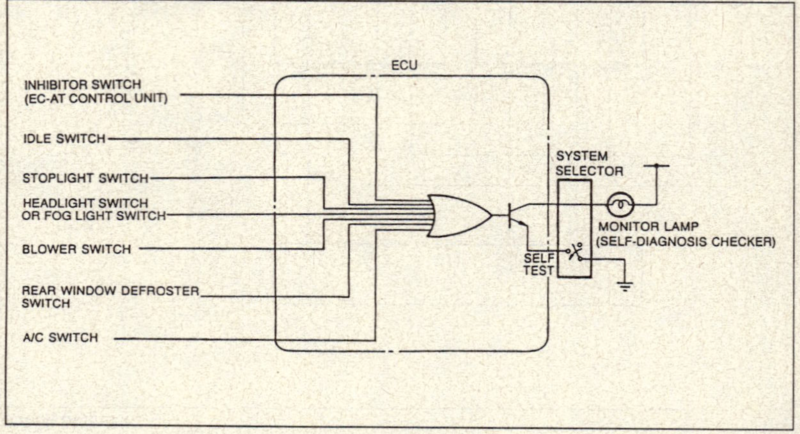

Switch	Self-Diagnosis Checker (Monitor lamp)		Remarks
	Light ON	Light OFF	
Idle switch	Pedal depressed	Pedal released	in throttle sensor
Stoplight switch	Pedal depressed	Pedal released	
Headlight switch	ON	OFF	
Rear window defroster switch	ON	OFF	
Inhibitor switch	Except N and P range	N, P range	
Blower control switch	Blower control switch max. position	Blower control switch 1st position	A/C switch OFF
A/C switch	ON	OFF	Blower control switch 1st position
Fog light switch	ON	OFF	

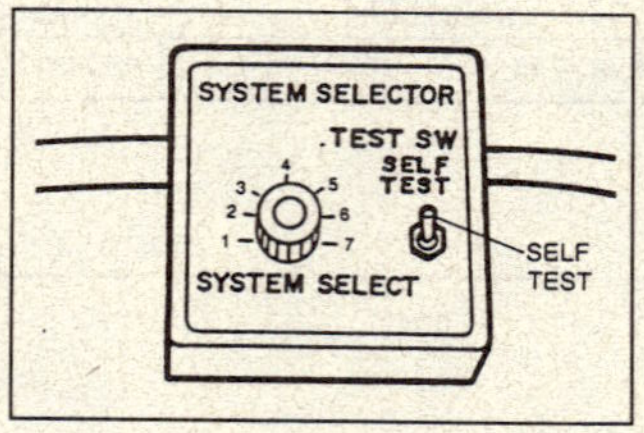

Inspection Procedure

1. Connect the **SST** (System Selector) to the diagnosis connector.
2. Set switch A to position 1.
3. Set TEST SW to SELF-TEST.

3.0L ENGINE SWITCH MONITOR FUNCTION TROUBLESHOOTING – 1992 929

Procedure

Set conditions to deactivate each switch
- All accessories OFF
- All pedals released

Is monitor lamp illuminated?

YES → Check each switch and related wire harness
- Idle switch
- Stoplight switch
- A/C switch
- Headlight switch
- Fog light switch
- Rear defroster switch
- Blower control switch
- Inhibitor switch

NO → Check each switch as described

Idle switch (in throttle sensor)

Depress accelerator pedal
Is monitor lamp illuminated?

NO →
- Idle switch malfunction
- Open or short circuit in wiring harness from throttle sensor to ECU
- ECU terminals 1T or 2I voltage incorrect

YES → Idle switch OK
Release accelerator pedal

Stoplight switch

Depress brake pedal
Is monitor lamp illuminated?

NO →
- Stoplight switch malfunction
- Open circuit in wiring harness from stoplight switch to ECU
- ECU terminal (1Q) malfunction

YES → Stoplight switch OK
Release brake pedal

Headlight switch

Turn ON headlight switch
Is monitor lamp illuminated?

NO →
- Headlight switch malfunction
- Open circuit in wiring garness from CPU to ECU terminal (1J)
- ECU terminal (1J) malfunction

YES → Headlight switch OK
Turn OFF headlight switch

Rear window defroster switch

Turn ON rear window defroster switch
Is monitor lamp illuminated?

NO →
- Rear window defroster switch malfunction
- Open circuit in wiring farness from CPU to ECU terminal (1J)
- ECU terminal (1J) malfunction

YES → Rear window defroster switch OK
Turn OFF rear window defroster switch

Inhibitor switch

Shift tranmission into except P and N range
Is monitor lamp illuminated?

NO →
- Inhibitor switch malfunction
- Open or short circuit in related wire harness
- ECU terminal (1R) malfunction

YES → Inhibitor switch OK
Shift into P or N range

Blower control switch

Blower control switch to max. position
Is monitor lamp illuminated?

NO →
- Blower control switch malfunction
- Open circuit in the wiring harness from A/C amplifier to ECU terminal (1P)
- ECU terminal (1P) malfunction

YES → Blower control switch OK

A/C switch

Set blower contrl switch 1st position
Turn ON A/C switch
Is monitor lamp illuminated?

NO →
- A/C switch malfunction
- Blower control switch malfunction
- Open circuit in the wiring harness from A/C amplifier to ECU terminal (1O)
- ECU terminal (1O) malfunction

YES → A/C switch OK
Turn off A/C switch and blower control switch

Fog light switch

Turn ON fog light switch
Is monitor lamp illuminated?

NO →
- Fog light switch malfunction
- Open circuit in wiring harness from CPU to ECU terminal (1J)
- ECU terminal (1J) mulfunction

YES → Fog light switch OK
Turn OFF fog light switch

Terminal voltage

V_B: Battery voltage

Terminal	Input	Output	Connection to	Test condition		Correct voltage	Remarks
1A	—	—	Battery	Constant		V_B	For backup
1B	○		Main relay (EGI relay)	Ignition switch	OFF	0V	—
					ON	V_B	
1C	○		Ignition switch (START)	While cranking		10V	—
				Ignition switch ON		Below 1.0V	
1D		○	Self-Diagnosis Checker (monitor lamp)	Test switch at "SELF-TEST" Lamp illuminated for 3 sec. after ignition switch OFF→ON		Approx. 5V	With Self-Diagnosis Checker and System Selector
				Lamp not illuminated after 3 sec.		V_B	
				Test switch at "O_2 MONITOR" at idle Monitor lamp illuminated		Approx. 5V	
				Test switch at "O_2 MONITOR" at idle Monitor lamp not illuminated		V_B	
1E		○	Malfubction indicator lamp (MIL)	Lamp illuminated for 3 sec. after ignition switch OFF→ON		Below 2.5V	With System Selector test Switch at "SELF-TEST"
				Lamp not illuminated after 3 sec		V_B	
				Lamp illuminated		Below 2.5V	
				Lamp not illuminated		V_B	
1F		○	Self-Diagnosis Checker (Code Number)	Buzzer sounded for 3 sec. after ignition switch OFF→ON		Below 2.5V	• With Self-Diagnosis Cheker and System Selector • With System Selector test switch at "SELF-TEST"
				Buzzer not sounded after 3 sec.		V_B	
				Buzzer sounded		Below 2.5V	
				Buzzer not sounded		V_B	
1G		○	Igniter	Ignition switch ON		Below 1.0V	—
				Idle			
1H	—	—	—	—		—	—
1I	○		Diagnosis connector (TEN terminal)	System Selector test switch at "O_2 MONITOR"		V_B	• With Self-Diagnosis Checker and System Selector • Ignition switch ON
				System Selector test switch at "SELF-TEST"		Below 1.0V	
1J	○		Central processing unit	Electrical load* OFF		V_B	Ignition switch ON
				Electrical load ON		Below 1.5V	
1K		○	EC/AT control unit	A/C switch ON		V_B	Ignition switch ON
				A/C switch OFF		Below 1.0V	
1L		○	A/C relay	A/C switch ON		Below 1.0V	Ignition switch ON
				A/C switch OFF		V_B	
1M	○		Speedometer	Ignition switch ON		4.5—5.5V	—
				Ignition switch OFF		0V	
1N	○		P/S pressure switch	Ignition switch ON		V_B	—
				P/S ON at idle		Below 1.0V	
				P/S OFF at idle		V_B	

* Electrical load: Blower fan control switch third position or higher, headlights ON, rear window defroster swith ON, or fog light switch ON.

Incorrect voltage		Possible cause
Always 0V		• ROOM 15A fuse burned • Open circuit in wiring from ROOM 15A fuse to ECU terminal 1A
Always 0V		• Main relay malfunction • Open or short circuit in wiring from main relay to ECU terminal 1B
Always 0V (starter turns)		• Open or short circuit in wiring from ignition switch to ECU terminal 1C
Always 0V		• Main relay (EGI relay) malfunction • Open circuit in wiring from ignition switch to diagnosis connector terminal +B • Open or short circuit in wiring from diagnosis connector terminal MEN to ECU terminal 1D
Always approx. V_B		• Poor connection at ECU connector • ECU malfunction
Always approx. 5V		• ECU malfunction
Always below 2.5V	MIL always ON	• Short circuit in wiring from combination meter to ECU terminal 1E • ECU malfunction
	MIL never ON	• Open circuit in wiring from combination meter to ECU terminal 1E
Always approx. V_B		• Poor connection at ECU connector • ECU malfunction
Always below 2.5V	No display on Self-Diagnosis Checker	• Main relay (EGI relay) malfunction • Open circuit in wiring from ignition switch to diagnosis connector terminal +B
	"88" displayed and buzzer sounds continuously	• Open or short circuit in wiring from diagnosis connector terminal FEN to ECU terminal 1F
Always approx. 12V		• Poor connection at ECU connector • ECU malfunction
Always 0V		• Short circuit in wiring from igniter to ECU terminal 1G
—		—
Always below 1.0V		• Short circuit in wiring from diagnosis connector terminal TEN to ECU terminal 1I
Always approx V_B		• Open circuit in wiring from diagnosis connector terminal TEN to ECU terminal 1I • Open circuit in wiring from diagnosis connector terminal GND to ground
Always below 1.5V		• Short circuit in wiring from CPU to ECU terminal 1J
Always V_B		• Open circuit in wiring from CPU to ECU terminal 1J
Always below 1.0V		• Open or short circuit in wiring from EC-AT control unit to ECU terminal 1K
Always below 1.0V	A/C does not operator	• A/C relay malfunction • Open circuit in wiring from ignition switch to A/C relay • Open circuit in wiring from A/C relay to ECU terminal 1L
	A/C switch OFF but A/C operates	• Short circuit in wiring from A/C relay to ECU terminal 1L • ECU malfunction
Always V_B		• Poor connection at ECU connector • ECU malfunction
Always 0V		• Refer to Code No. 06 Troubleshooting
Always below 1.0V		• P/S pressure switch malfunction • Short circuit in wiring from P/S pressure switch to ECU terminal 1N • ECU malfunction
Always V_B		• P/S pressure switch malfunction • Open circuit in wiring from P/S pressure switch to ECU terminal 1N • Open circuit in wiring from P/S pressure switch to ground

3.0L ENGINE ENGINE CONTROL UNIT TERMINAL VOLTAGES—1992 929

V_B: Battery voltage

Terminal	Input	Output	Connection to	Test condition	Correct voltage	Remarks	Incorrect voltage	Possible cause
1O	○		A/C amplifier	A/C switch ON	1.5—3.5V	Ignition switch ON	Always below 1.5V	• Short circuit in wiring from A/C amplifier to ECU terminal 1O • A/C amplifier malfunction • ECU malfunction
				A/C switch OFF	4.5—5.5V		Always 4.5—5.5V	• Open circuit in wiring from A/C amplifier to ECU terminal 1O • A/C amplifier malfunction
1P	○		A/C amplifier	Fan speed control 4th position	Below 1.0V	Ignition switch ON and blower motor ON	Always below 1.0V	• Short circuit in wiring from A/C amplifier to ECU terminal 1P • A/C amplifier malfunction • ECU malfunction
				Fan Speed control 1st position	V_B		Always V_B	• Open circuit in wiring from A/C amplifier to ECU terminal 1P • A/C amplifier malfunction
1Q	○		Stoplight switch	Brake pedal released	Below 1.0V	—	Always below 1.0V (Stoplight OK)	• Open circuit in wiring from stoplight switch to ECU terminal 1Q
				Brake pedal depressed	V_B			
1R	○		EC-AT control unit	N or P range	Below 1.0V	Ignition switch ON	Always below 1.0V	• Inhibitor switch malfunction • Short circuit in wiring from EC-AT control unit to ECU terminal 1R
				Others	V_B		Always V_B	• Inhibitor switch malfunction • Open circuit in wiring from EC-AT control unit to ECU terminal 1R
1S	○		EC-AT control unit	Shifting performed	Below 1.0V	—	Always below 1.0V	• Short circuit in wiring from EC-AT control unit to ECU terminal 1S • ECU malfunction
				Others	V_B		Always V_B	• Open circuit in wiring from EC-AT control unit to ECU terminal 1S • EC-AT control unit malfunction
1T	○		Throttle sensor (Idle switch)	Accelerator pedal released	Below 1.0V	Ignition switch ON	Always below 1.0V	• Throttle sensor misadjustment • Short circuit in wiring from throttle sensor to ECU terminal 1T • ECU malfunction
				Accelerator pedal depressed	V_B		Always approx. 12V	• Throttle sensor misadjustment • Open circuit in wiring from throttle sensor to ECU terminal 1T • Open circuit in wiring from throttle sensor to ECU terminal 3D
1U	○		EC-AT control unit	P, R, or N range and 1st, 4th gear	V_B	—	Always below 1.0V	• Short circuit in wiring from EC-AT control unit to ECU terminal 1U • ECU malfunction
				2nd, 3rd gear	Below 1.0V		Always V_B	• Open circuit in wiring from EC-AT control unit to ECU terminal 1U • EC-AT control unit malfunction
1V	○		EC-AT control unit	P, R, or N range and 1st, 2nd gear	V_B	—	Always below 1.0V	• Short circuit in wiring from EC-AT control unit to ECU terminal 1V • ECU malfunction
				3rd, 4th gear	Below 1.0V		Always V_B	• Open circuit in wiring from EC-AT control unit to E • EC-AT control unit malfunction
2A	—	—	—	—	—	—	—	—
2B	○		Airflow meter	Ignition switch ON	Approx. 4V	—	Always approx. 0V or approx. 5V	• Refer to Code No.08 Troubleshooting
				Idle	Approx. 2V			
2C	○		Oxygen sensor (Right side)	Ignition switch ON	0V	—	0V after warm-up	• Refer to Code No.23 Troubleshooting
				Idle (Cold engine)	0V		Always approx. 1V after warm-up	• Refer to Code No.24 Troubleshooting
2D	○		Oxygen sensor (Left side)	Idle (After warm-up)	0—1.0V		0V after warm-up	• Refer to Code No. 15 Troubleshooting
				Increasing engine speed (After warm-up)	0.5—1.0V		Always approx. 1V after warm-up	• Refer to Code No. 17 Troubleshooting
				Deceleration	0—0.4V			
2E	○		Water thermo-sensor	Engine coolant temperature 20°C (68°F)	Approx. 2.5V	Ignition switch ON	Always approx. 0V or approx. 5V	• Refer to Code No.09 Troubleshooting
				After warm-up	Below 0.5V			
2F	○		Throttle sensor	Accelerator pedal released	0.1—1.1V	Ignition switch ON	Always approx. 0V	• Open circuit in wiring from ECU terminal 2F to throttle sensor • Open circuit in wiring from ECU terminal 2I to throttle sensor • Open circuit in wiring from ECU terminal 3D to throttle sensor
				Accelerator pedal fully depressed	3.0—4.3V		Always above 1V	• Throttle sensor misadjustment
2G	○		Atmospheric pres-suresensor	Atmospheric pressure above 725 mmHg (28.5 in-Hg) which is approx. below 400 m (1312 ft)	Approx. above 3.8V	Ignition switch ON	Always 0V	• Refer to Code No.14 Troubleshooting
				Atmospheric pressure below 600 mmHg (26.0 in-Hg) which is approx. above 1200 m (3937 ft)	Approx. below3.3V			

3.0L ENGINE ENGINE CONTROL UNIT TERMINAL VOLTAGES—1992 929

V_B: Battery voltage

Terminal	Input	Output	Connection to	Test condition	Correct voltage	Remarks	Incorrect voltage	Possible cause
2H	—	—	—	—	—	—	—	—
2I	○		Throttle sensor/ EC-AT control/ Airflow meter	Constant	4.5—5.5V	—	Always 0V	• Short circuit in wiring from ECU terminal 2I to throttle sensor, EC-AT control unit, or airflow meter • Poor connection at ECU connector • ECU malfunction
2J	○		EGR position sensor	Ignitio switch ON	Approx. 0.25—0.9V	—	Always approx. 0V	• Refer to Code No.16 Troubleshooting
				Idle				
2K	○		Intake air thermosensor (Airflow meter)	Ambient air temperature 20°C (68°F)	Approx. 2.5V	Ignition switch ON	Below 4.5V or above 5.5 V Always approx. 0V or approx. 5V	• ECU malfunction • Refer to Code No.10 Troubleshooting
2L	○		Intake air thermosensor (Dynamic chamber)	Ambient air temperature 20°C (68°F)	Approx. 2.5V	Ignition switch ON	Always approx. 0V or approx. 5V	• Refer to Code No.11 Troubleshooting
2M	○		Knock sensor (Right side)	Ignition switch ON	Below 1.0V	—	Always 0V	• Refer to Code No.07 Troubleshooting
2N	○		Knock sensor (Left side)	Ignition switch ON	Below 1.0V	—	Always 0V	• Refer to Code No.05 Troubleshooting
2O		○	Solenoid valve (Purge control)	Ignition switch ON	V_B	—	Always approx. 0V or V_B	• Refer to Code No.26 Troubleshooting
				Idle	V_B			
2P		○	EC-AT control unit	Engine coolant temperature below 55°C (131°F)	Below 1.0V	Ignition switch	Always approx. 0V	• Open or short circuit in wiring from ECU terminal 2P to EC-AT control unit terminal 2H
				Engine coolant temperature below 55°C (131°F)	V_B			
3A	—	—	Ground (Injector)	Constant	0V	—	Above 0V	• Poor contact at ground terminal • Open circuit in wiring from ECU to ground
3B	—	—	Ground (Output)	Constant	0V	—		
3C	—	—	Ground (CPU)	Constant	0V	—		
3D	—	—	Ground (Input)	Constant	0V	—		
3E	○		Crank angle sensor 1 (in distributor) (Ne 1 signal)	Ignition switch ON	Approx. 0V or 5V	—	Always approx. 0V or approx. 2V	• Refer to Troubleshooting, Code No.4
				Idle	Approx. 2V			
3F	○		Crank angle sensor 2 (Crankshaft pulley) (Ground)	Constant	0V	—	Above 0V	• Poor contact at ground terminal • Open circuit in wiring from ECU to ground
3G	○		Crank angle sensor 1 (in distributor) (G signal)	Ignition switch ON	Approx. 0V or 5V	—	Always approx. 0V or approx. 2V	• Refer to troubleshooting, Code No.03
				Idle	Approx. 2V			
3H	○		Crank angle sensor 2 (Crankshaft pulley) (Ne 2 signal)	Ignition switch ON	0V	*Engine Signal Monitor: Green and red lamps flash	Always approx. 0V or approx. 2V	• Refer to Troubleshooting, Code No.02
				Idle*	Approx. 0V			
3I		○	CSI relay	Engine coolant temperature below 0°C (32°F)	V_B	—	Always below 1V	• Open or short circuit in wiring from CSI relay to ECU terminal 3I
				Engine coolant temperature above 5°C (41°F)	Below 1.0V			
3J		○	Solenoid valve (VICS)	While cranking	V_B	—	Always approx. 0V	• Refer to Code No.41 Troubleshooting
				Idle	Below 2V			
3K		○	Oxygen heater (Right side)	While cranking or High load	V_B	—	Always approx. 0V or approx. 12V	• Refer to Code No. 36 Troubleshooting
				Others	Below 2V			
3M	—	—	—	—	—	—	—	—
3N		○	Oxygen heater (Left side)	While cranking or High load	V_B	—	Always approx. 0V or approx. 12V	• Refer to Code No.37 Troubleshooting
				Others	Below 2V			

3.0L ENGINE ENGINE CONTROL UNIT TERMINAL VOLTAGES – 1992 929

V_B: Battery voltage

Terminal	Input	Output	Connection to	Test condition	Correct voltage	Remarks	Incorrect voltage	Possible cause
3O		○	Solenoid valve (EGR, Vent side)	Ignition switch ON	V_B	Initial acceleration after warm up: Engine Signal Monitor; green and red lights flash	Always approx. 0V or V_B	• Refer to Code NO.29 Troubleshooting
				Idle				
3P		○	Solenoid valve (EGR, Vacuum side)	Ignition switch ON	V_B		Always approx. 0V or V_B	• Refer to Code NO. 28 Troubleshooting
				Idle				
3Q		○	ISC valve	Ignition switch ON	V_B	—	Always approx. 0V or approx. 12V	• Refer to Code NO. 34 Troubleshooting
				Idle	Approx. 9V			
3R		○	Air control motor	Idle	Below 1V	—	Always V_B	• ECU malfunction
				Engine speed above 3,400 rpm and high load for one second	V_B			
3S		○	Air control motor	Idle	Below 1V			
				Engine speed above 3 400 rpm and high load for one second	V_B			
3T		○	Circuit opening relay	Ignition switch ON	V_B	—	Always below 1V	• Open or short circuit in wiring from circuit opening relay to ECU terminal 3T
				Idle	Below 1V			
3U		○	Injector No.1	Ignition switch ON or Idle*	V_B	*Engine Signal Monitor: Green and red lamps flash	Always 0V	• Open or short circuit in wiring from injector to ECU terminal 3U, 3V, 3W, 3Y, or 3Z • Main relay malfunction
3V		○	Injector No.2					
3W		○	Injector No.3					
3X		○	Injector No.4					
3Y		○	Injector No.5					
3Z		○	Injector No.6					

Control Unit Connector (Control Unit Side)

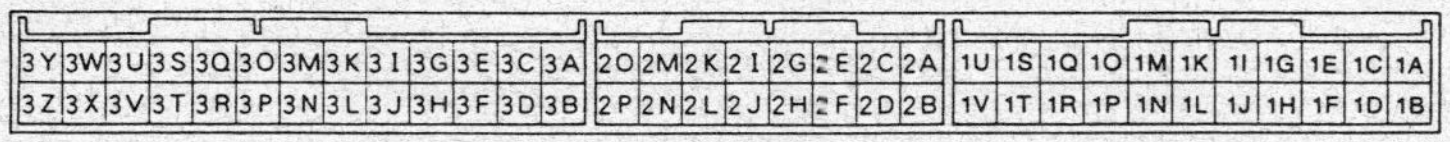

2.6L ENGINE SPECIFICATION CHART—MPV

Item			Specifications
Idle speed		rpm	A/T: 750—790 (P range)*
Ignition timing		BTDC	4—6° (A/T: P range)*
Throttle body			
Type			Horizontal draft
Throat diameter		mm (in)	55 (2.2)
Fuel pump			
Type			Impeller (in-tank)
Output pressure		kPa (kg/cm², psi)	441—588 (4.5—6.0, 64—85)
Fuel filter			
Type	Low-pressure side		Nylon element
	High-pressure side		Paper element
Pressure regulator			
Type			Diaphragm
Regulating pressure		kPa (kg/cm², psi)	265—314 (2.7—3.2, 38—46)
Injector			
Type			High-ohmic
Type of drive			Voltage
Resistance		Ω	12—16 (at 20°C, 68°F)
BAC valve (Solenoid valve [Idle speed control])			
Solenoid resistance		Ω	7.7—9.3 (at 23°C, 73°F)
Solenoid valve (Purge control)			
Solenoid resistance		Ω	23—27 (at 20°C, 68°F)
Water thermosensor			
Resistance kΩ	−20°C (−4°F)		14.5—17.8
	20°C (68°F)		2.2—2.7
	80°C (176°F)		0.28—0.35
Intake air thermosensor			
Resistance kΩ	25°C (77°F)		29.7—36.3
	85°C (185°F)		3.3—3.7
Circuit opening relay			
Resistance Ω	STA—E₁		21—43
	B—Fc		109—226
	B—Fp		∞
Fuel tank			
Capacity		liters (US gal, Imp gal)	60 (15.9, 13.2)
Air cleaner			
Element type			Dry
Accelerator cable			
Free play		mm (in)	1—3 (0.039—0.118)
Fuel			
Specification			Unleaded regular (RON 87 or higher)

*...Test connector grounded.

2.6L ENGINE COMPONENT DESCRIPTIONS—MPV

Component	Function	Remarks
Air cleaner	Filters air entering throttle body	
Airflow sensor	Detects amount of intake air; sends signal to engine control unit	
Air valve	When cold, supplies bypass air into dynamic chamber	• Engine speed increased to shorten warm-up period • Thermowax type • Installed in BAC valve
Atmospheric pressure sensor	Detects atmospheric pressure; sends signal to engine control unit	Built in ECU
BAC valve	Supplies bypass air into dynamic chamber	Consists of air valve and ISC valve
Catalytic converter	Reduces HC, CO, and NOx by chemical reaction	Monolith type
Central processing unit (CPU)	Detects that E/L is being applied; sends singal to engine control unit	
Charcoal canister	Stores gas tank fumes when engine stopped	
Check connector	For Self-Diagnosis Checker	6-pin connector (Green)
Check-and-cut valve	Releases excessive pressure or vacuum in fuel tank to atmosphere	
Circuit opening relay	Voltage for fuel pump while engine running	
Clutch switch (M/T)	Detects in-gear condition; sends signal to engine control unit	Switch ON when clutch pedal depressed
Dynamic chamber	Interconnects all cylinders	
Engine control unit	Detects following: 1. Engine speed 2. No.1 piston TDC 3. Intake air amount 4. Engine coolant temperature 5. Ignition ON signal 6. Throttle valve opening angle 7. Throttle valve fully closed 8. Air/fuel ratio (Oxygen concentration) 9. In-gear condition 10. Intake air temperature 11. Atmospheric pressure 12. A/C operation 13. P/S operation 14. E/L operation 15. Cranking signal 16. Test signal (idle speed, malfunction code No.) 17. Braking signal Controls operation of the following: 1. Fuel injection system 2. Idle speed control 3. Pressure regulator control system 4. Purge control system 5. Fail-safe function 6. Monitor function 7. Burn-off system 8. Lockup inhibition control 9. Electronic spark advance control system 10. Fuel pump 11. A/C (cut off) 12. Main relay control	1. Ignition coil 2. G-signal 3. Airflow sensor 4. Water thermosensor 5. Ignition switch 6. Throttle sensor 7. Idle switch 8. Oxygen sensor 9. Inhibitor switch (A/T) 10. Intake air thermosensor (on dynamic chamber) 11. Atmospheric pressure sensor 12. A/C switch 13. P/S pressure switch 14. Central processing unit (CPU) 15. Ignition switch (START position) 16. Test connector 17. Stoplight switch 1. Injector 2. Solenoid valve (Idle speed control) 3. Solenoid valve (Pressure regulator control) 4. Solenoid valve (Purge control) 5. Self-Diagnosis Checker and MIL 6. Monitor lamp (Self-Diagnosis Checker) 7. Airflow sensor 8. 4 A/T control unit 9. Igniter 10. Circuit opening relay 11. A/C relay 12. Main relay

2.6L ENGINE COMPONENT DESCRIPTIONS—MPV

Component	Function	Remarks
Fuel filter	Filters particles from fuel	
Fuel pump	Provides fuel to injectors	• Operates while engine running • Installed in fuel tank
G signal (crank angle sensor)	Detects No.1 cylinder TDC; sends signal to engine control unit	For determining fuel injection timing
Idle switch	Detects when throttle valve fully closed; sends signal to engine control unit	Installed on throttle body
Igniter	Receives spark signal from signal rotor and generates high voltage to ignition coil	
Ignition switch (START position)	Sends engine cranking signal to engine control unit	
Inhibitor switch (A/T)	Detects in-gear condition; sends signal to engine control unit	Switch ON in "N" or "P" range
Injector	Injects fuel into intake port	• Controlled by signals from engine control unit • High-ohmic injector • Two port injector nozzle
Intake air thermosensor	Detects intake air temperature; sends signal to engine control unit	Installed on dynamic chamber
Main relay	Supplies electric current to injectors and engine control unit.	
MIL (Malfunction indicator lamp)	Light illuminates when input device malfunctions and oxygen sensor replacement time comes	Test connector not grounded
	Light flashes to indicate malfunction code No. of input and output devices	Test connector grounded
	Lamp stays ON when oxygen sensor replacement time comes	
Ne signal (crank angle sensor)	Detects crank angle at 180° intervals; sends signal to engine control unit	For determining fuel injection timing and ignition timing
Neutral switch (M/T)	Detects in-gear condition; sends signal to engine control unit	Switch ON when neutral
Oxygen sensor	Detects oxygen concentration; sends signal to engine control unit	Zirconia ceramic and platinum coating
PCV valve	Controls amount of blowby gas introduced into engine	
P/S pressure switch	Detects P/S operation; sends signal to engine control unit	P/S: ON when steering wheel turned right or left
Pressure regulator	Adjusts fuel pressure supplied to injectors	
Resonance chamber	Improves mid-range torque characteristics	
Separator	Prevents fuel from flowing into charcoal canister	
Solenoid valve: **Idle speed control**	Controls bypass air amount	• Controlled by duty signal from engine control unit • With integrated air valve • Controls idle-up
Solenoid valve: **Pressure regulator control**	Controls vacuum to pressure regulator	Cuts vacuum passage when hot
Solenoid valve: **Purge control**	Controls evaporative fumes from canister to intake manifold	
Stoplight switch	Detects braking operation (deceleration); sends signal to engine control unit	
Test connector	For Self-Diagnosis Checker and idle speed adjustment	1-pin connector (Green)
Throttle body	Controls intake air quantity	Integrated throttle sensor and idle switch
Throttle sensor	Detects throttle valve opening angle; sends signal to engine control unit	Installed on throttle body
Two-way check valve	Controls pressure in fuel tank	
Water thermosensor	Detects coolant temperature; sends signal to engine control unit	

2.6L ENGINE TROUBLESHOOTING GUIDE—MPV

INPUT DEVICES \ OUTPUT DEVICES	INJECTOR: FUEL INJECTION AMOUNT	INJECTOR: FUEL INJECTION TIMING	BAC VALVE: AIR VALVE	BAC VALVE: ISC VALVE	SOLENOID VALVE (PURGE CONTROL)	SOLENOID VALVE (PRESSURE REGULATOR CONTROL)	A/C RELAY (A/C CUT-OFF)	4A/T CONTROL UNIT (LOCKUP INHIBITION CONTROL)	AIRFLOW SENSOR (BURN-OFF)	CIRCUIT OPENING RELAY (FUEL PUMP CONTROL)	IGNITER (ELECTRONIC SPARK ADVANCE CONTROL)
TEST CONNECTOR	×	×	×	○	×	×	×	×	×	×	○
IGNITION SWITCH (ON POSITION)	×	×	×	×	×	×	×	×	○	×	×
IGNITION SWITCH (START POSITION)	○	○	×	○	×	○	○	×	×	×	○
CENTRAL PROCESSING UNIT (CPU)	×	×	×	○	×	×	×	×	×	×	×
P/S PRESSURE SWITCH	○	×	×	○	×	×	×	×	×	×	×
A/C SWITCH	○	×	×	○	×	×	○	×	×	×	×
INHIBITOR SWITCH (A/T)	○	×	×	○	○	×	○	○	×	×	×
STOPLIGHT SWITCH	○	×	×	×	×	×	×	×	×	×	×
IDLE SWITCH	○	×	×	○	○	○	×	×	×	×	○
SPEED SENSOR (A/T) (IN COMBINATION METER)	○	×	×	○	×	×	×	×	×	×	○
ATMOSPHERIC PRESSURE SENSOR	○	×	×	○	×	×	×	×	×	×	×
THROTTLE SENSOR	○	×	×	×	×	○	○	×	×	×	×
INTAKE AIR THERMOSENSOR	○	×	×	×	×	○	×	×	×	×	×
AIRFLOW SENSOR	○	×	×	○	○	×	×	○	○	×	○
OXYGEN SENSOR	○	×	×	×	○	×	×	×	×	×	×
WATER THERMOSENSOR	○	×	×	○	○	○	×	×	○	×	○
DISTRIBUTOR: Ne-SIGNAL	○	○	×	○	○	○	×	×	○	○	○
DISTRIBUTOR: G-SIGNAL	×	○	×	×	×	×	×	×	×	×	×

2.6L ENGINE ENGINE CONTROL OPERATION CHART—MPV

Input Devices and Engine Conditions

Note
• The data in this chart is for reference only.

ENGINE CONDITIONS \ INPUT DEVICES	APPROXIMATE TIME (BASED ON 10—16°C or 50—60°F AMBIENT)	SENSORS: DISTRIBUTOR (G-SIGNAL)	DISTRIBUTOR (Ne-SIGNAL)	WATER THERMOSENSOR	OXYGEN SENSOR	AIRFLOW SENSOR	INTAKE AIR THERMOSENSOR	THROTTLE SENSOR	ATMOSPHERIC PRESSURE SENSOR (IN ECU)
CRANKING —COLD ENGINE • COLD AIR • COLD COOLANT	Zero				Signal has no effect on ECU	Signal has no effect on ECU	Signal has no effect on ECU	Signal has no effect on ECU	
COLD START —FAST IDLE • COLD AIR • COLD COOLANT	One minutes			Cool to warm: medium voltage (3.5V and dropping)	Sensor cold: low to high voltage (0—0.9V)	Low volume airflow: low to high voltage (2.4—2.6V)		Closed throttle: low voltage (0.3—0.7V)	
COLD DRIVEAWAY —PART THROTTLE • COLD AIR • COLD COOLANT	Two minutes								
WARM DRIVEAWAY —PART THROTTLE • WARM AIR • WARM COOLANT	Three minutes			Warm: medium voltage (Approx. 0.7V and dropping)	Sensor warm: high voltage (0.9V)	Moderate volume airflow: low to medium voltage (3.0V)		Part throttle: medium voltage (1—3.5V)	
HOT CRUISE • WARM AIR • WARM COOLANT		Sends No.1 cylinder TDC signal to ECU	Sends engine speed signal to ECU		Sensor hot: switching from high voltage (0.9V) to low voltage (0.1V)		Cool to warm: medium voltage (1.4—3.4V)		Sends voltage signal to ECU that varies with altitude: voltage (approx. 4V at sea level)
HOT ACCELERATION —60% THROTTLE						Moderate to strong volume of airflow: (3.8V)			
HOT ACCELERATION —WIDE OPEN THROTTLE	More than four minutes			Hot: low voltage (Approx. 0.4V)	High voltage (0.9V)	Strong volume of airflow: (4.0V)		Wide open throttle: high voltage (Approx. 4.0V)	
DECELERATION —CLOSED THROTTLE					Low voltage (0V)	Low volume of airflow: (2.4V)		Closed throttle: low voltage (0.3—0.7V)	
HOT CURB IDLE —EXTENDED					Switching from high to low voltage (0.75—0.25V)				
HOT ENGINE SHUTDOWN	—	OFF	OFF	OFF	Sensor hot: low voltage (0.1V) until sensor cools	OFF	OFF	OFF	OFF

SWITCHES: IDLE SWITCH	STOP-LIGHT SWITCH	INHIBITOR SWITCH (A/T)	A/C SWITCH	P/S PRESSURE SWITCH	CPU (E/L SIGNAL)	IGNITION SWITCH: START POSITION	IGNITION SWITCH: ON POSITION	TEST CONNECTOR
Signal has no effect on ECU	Signal has no effect on ECU	Signal has no effect on ECU	Signal has no effect on ECU	Signal has no effect on ECU	Signal has no effect on ECU	Sends signal to ECU (VB)	Signal has no effect on ECU	Signal has no effect on ECU
Low voltage signal to ECU (below 1.5V)	Brake pedal depressed: sends signal to ECU (VB)	In N or P range: sends signal to ECV (VB)						
High voltage signal to ECU (VB)	No signal send to ECU (below 1.5V)	In D range: no voltage signal to ECU (below 1.5V)	A/C switch ON: sends signal to ECU (VB) A/C switch OFF: no signal to ECU (below 1.5V)	Steering wheel turned: low voltage signal to ECU (below 1.5V) Steering wheel straight ahead: high voltage signal to ECU (VB)	Headlight or blower or rear window defroster switch ON: low voltage signal to ECU (below 1.5V) Headlight and blower and rear window defroster switches OFF: high voltage signal to ECU (VB)	No signal to ECU (below 1.5V)	Sends signal to ECU (VB)	Connector not grounded: high voltage signal to ECU (VB)
Low voltage signal to ECU (below 1.5V)	Brake pedal depressed: sends signal to ECU (VB)							
		In N or P range: sends signal to ECU (VB)						Low voltage signal to ECU when connector grounded (below 1.5V)
OFF	OFF	OFF	OFF	OFF	OFF	OFF	OFF	OFF

2.6L ENGINE ENGINE CONTROL OPERATION CHART—MPV

Output Devices and Engine Conditions

Note
• **The data in this chart is for reference only.**

ENGINE CONDITIONS / OUTPUT DEVICES	APPROXIMATE TIME (BASED ON 10—16°C or 50—60°F AMBIENT)	INJECTOR: INJECTION	INJECTOR: INJECTION TIMING	BAC VALVE: AIR VALVE	BAC VALVE: ISC VALVE	SOLENOID VALVE (PURGE CONTROL)	SOLENOID VALVE (PRESSURE REGULATOR CONTROL)	A/C RELAY (A/C CUT-OFF)	4 A/T CONTROL UNIT (LOCK UP INHIBITOR CONTROL)	AIRFLOW SENSOR (BURN-OFF)	CIRCUIT OPENING RELAY	MAIN RELAY	IGNITER
CRANKING —COLD ENGINE • COLD AIR • COLD COOLANT	Zero		All cylinders each ignition pulse			OFF (Purge cut)		OFF (A/C ON)			OFF		Ignition timing BTDC 6 ± 1°
COLD START —FAST IDLE • COLD AIR • COLD COOLANT	One minute	Rich		Open (coolant temperature: below 50°C 122°F)	Large amount of bypass air			ON (A/C OFF: approx. 5 sec.)					
COLD DRIVEAWAY —PART THROTTLE • COLD AIR • COLD COOLANT	Two minutes								OFF (Lock up)				
WARM DRIVEAWAY —PART THROTTLE • WARM AIR • WARM COOLANT	Three minutes	Rich and lean	2-group			Operates (Duty values [purge gas amount] change)	OFF (Vacuum to pressure regulator)	OFF (A/C ON)					
HOT CRUISE • WARM AIR • WARM COOLANT					Small amount of bypass air					OFF	ON (Engine speed above 50 rpm)	ON	Ignition spark advance signal
HOT ACCELERATION —60% THROTTLE								ON (A/C CUT)					
HOT ACCELERATION —WIDE OPEN THROTTLE	More than four minutes	Rich		Closed					ON (Lock up inhibit)				
DECELERATION —CLOSED THROTTLE		Fuel cut	Fuel cut		Large and small amount of bypass air	OFF (Purge cut)		OFF (A/C ON)	OFF (Lock up)				
HOT CURB IDLE —EXTENDED		Rich and lean	2-group		Small amount of bypass air		After starting: ON during hot start only (Vacuum cut)						
HOT ENGINE SHUTDOWN	—	Does not inject	Does not inject		OFF	OFF	OFF	OFF	OFF	ON (Burn-off)	OFF	OFF	OFF

Cranks normally but won't start (No combustion)

STEP	QUICK INSPECTION		ACTION		POSSIBLE CAUSE AND DETAILED INSPECTION
1	Check for malfunction code (01) with SST [Ign ON, Test connector (Green: 1-pin) grounded]	Yes	Check for cause by referring to check sequence		
		No	Go to Next Step		
2	Check for spark by disconnecting high-tension lead while cranking	Yes	Go to Next Step		
		No	Check ignition system (Refer to ignition system troubleshooting)		
3	Check for fuel pump operating sound from fuel filler port [Ign ON, Test connector (Yellow: 2-pin) connected]	Yes	Check if engine starts in this condition	Yes	Check circuit opening relay switching operation
					Check circuit opening relay circuit (Ign: START)
				No	Go to Next Step
		No	Check circuit opening relay switching operation	Yes	Check circuit opening relay circuit
					Check fuel pump circuit
					Check fuel pump operation
				No	Replace circuit opening relay
4	Check fuel line pressure [IGN ON, Test connector (Yellow: 2-pin) connected] **Fuel line pressure: 265—314 kPa (2.7—3.2 kg/cm², 38—46 psi)**	Yes	Go to Next Step		
		No	Check fuel pump maximum pressure **Fuel pump maximum pressure: 441—588 kPa (4.5—6.0 kg/cm², 64—85 psi)**	Yes	Replace pressure regulator
				No	Replace fuel pump
5	Check for injector operating sound while cranking	Yes	Go to Next Step		
		No	Check voltage at ECU (2U) and (2V) terminals with SST **Voltage: VB (Ign ON)**	Yes	Check throttle sensor
					Replace ECU
				No	Check wiring for short or open
					Poor ground circuit from ECU (2A) terminal (Check terminal voltage with SST)
6	Check continuity between check connector [for tachometer (White: 1-pin)] and ground (Ign ON)	Yes			Wiring short to ground
		No	Go to Next Step		
7	Substitute a well-known ECU Check if the condition improves	Yes			ECU malfunction
		No	Check ground circuit from ECU (2B) terminal with SST **Voltage: 0V (Ign ON)**	Yes	Go to Next Step
				No	Poor ground circuit
8					Low compression

Cranks normally but hard to start (Always)

STEP	QUICK INSPECTION		ACTION		POSSIBLE CAUSE AND DETAILED INSPECTION
1	Check ignition timing at idle after warm up **Ignition timing: 4—6° BTDC** [Test connector (Green: 1-pin) grounded]	Yes	Go to Next Step		
		No	Adjust ignition timing		
2	Disconnect high-tension lead of each cylinder at idle Check if engine condition changes	Yes	Go to Next Step		
		No	Check ignition system	Yes	Replace injector (If step 3 OK)
				No	Check spark plug
					Check high-tension lead
					Check distributor cap
3	Check for injector operating sound at idle	Yes	Go to Next Step		
		No	Check resistance at injector harness connector (X-12) Terminals: (B/Y)—(LG/B), (B/Y)—(LG/R); Resistance: 6—8Ω	Yes	Check wiring short or open
				No	Check injector resistance
					Check wiring short or open
4	Check fuel line pressure [Ign ON, Test connector (Yellow: 2-pin) connected] **Fuel line pressure: 265—314 kPa (2.7—3.2 kg/cm², 38—46 psi)**	Yes	Go to Next Step		
		No	Substitute a good fuel filter and retest	Yes	Replace fuel filter
			Check fuel pump maximum pressure **Fuel pump maximum pressure: 441—588 kPa (4.5—6.0 kg/cm², 64—85 psi)**	Yes	Replace pressure regulator
				No	Replace fuel pump
5	Operate fuel pump [Ign ON, test connector (Yellow: 2-pin) connected] Turn ignition switch OFF and observe fuel pressure **for 5 minutes** **Fuel pressure: More than 147 kPa (1.5 kg/cm², 21 psi)**	Yes	Go to Next Step		
		No	Check fuel pump pressure drop	No	Replace fuel pump
			Check pressure regulator pressure drop	Yes	Check injector fuel leakage
				No	Replace pressure regulator
6					Check compression
7					ECU malfunction

2.6L ENGINE SYMPTOM TROUBLESHOOTING – MPV

Cranks normally but hard to start (Only when engine is cold)

STEP	QUICK INSPECTION		ACTION		POSSIBLE CAUSE AND DETAILED INSPECTION
1	Check specific gravity of battery using a hydrometer **Specific gravity: Above 1.200**	Yes	Go to Next Step		
		No	Recharge battery		
2	Check for malfunction code (09) (26) with SST [Ign ON, Test connector (Green: 1-pin) grounded]	Yes	Check for cause by referring to check sequence		
		No	Go to Next Step		
3	Disconnect high-tension lead of each cylinder at idle Check if engine condition changes	Yes	Go to Next Step		
			Check ignition system		Check spark plug
					Check high-tension lead
					Check distributor cap
4	Check fuel line pressure [Ign ON, Test connector (Yellow: 2-pin) connected] **Fuel line pressure: 265—314 kPa (2.7—3.2 kg/cm², 38—46 psi)**	Yes	Go to Next Step		
		No	Check for fuel leaks		
			Substitute a good fuel filter and retest	Yes	Replace fuel filter
			Check fuel pump maximum pressure **Fuel pump maximum pressure: 441—588 kPa (4.5—6.0 kg/cm², 64—85 psi)**	Yes	Replace pressure regulator
				No	Replace fuel pump
5	Disconnect ISC valve connector when engine is cold Check if idle speed decreases during warm up	Yes	Go to Next Step		
		No			Check if BAC valve (air valve) opens when cold
6					ECU malfunction

Cranks normally but hard to start (Only when engine is warm)

STEP	QUICK INSPECTION		ACTION		POSSIBLE CAUSE AND DETAILED INSPECTION
1	Check for malfunction code (26) with SST [Ign ON, Test connector (Green: 1-pin) grounded]	Yes	Check for cause by referring to check sequence		
		No	Go to Next Step		
2	Disconnect water thermosensor connector Check if condition improves	Yes	Check water thermosensor connector condition as follows: 1. Shake connector and check if condition changes 2. Check condition of terminal (burned or damaged) 3. Connect a good terminal to harness side connector and check for looseness	Yes	Check water thermosensor
				No	Poor contact of water thermosensor connector
		No	Go to Next Step		
3	Operate fuel pump [Ign ON, Test connector (Yellow: 2-pin) connected] Turn ignition switch OFF and observe fuel pressure **for 5 minutes** **Fuel pressure: More than 147 kPa (1.5 kg/cm², 21 psi)**	Yes	Go to Next Step		
		No	Check fuel pump pressure drop	No	Replace fuel pump
			Check pressure regulator pressure drop	Yes	Check injector fuel leakage
				No	Replace pressure regulator
4					ECU malfunction

Cranks normally but hard to start (Only after heat soak)

STEP	QUICK INSPECTION		ACTION		POSSIBLE CAUSE AND DETAILED INSPECTION
1	Check for malfunction code (09), (11), (25), (26) with SST [Ign ON, Test connector (Green: 1-pin) grounded]	Yes	Check for cause by referring to check sequence		
		No	Go to Next Step		
2	Circulate fuel by operating fuel pump **for 20 seconds** [Ign ON, Test connector (Yellow: 2-pin) connected] Check if condition improves	Yes	Go to Step 3		
		No	Go to Step 4		
3	Disconnect vacuum hose from pressure regulator Check if condition improves	Yes	Check the components related to pressure regulator control system		Check water thermosensor
					Check intake air thermosensor
					Check solenoid valve (PRC)
					ECU malfunction (Check (2T) terminal voltage)
		No	Go to Next Step		
4	Operate fuel pump [Ign ON, Test connector (Yellow: 2-pin) connected] Turn ignition switch OFF and observe fuel pressure **for 5 minutes** **Fuel pressure: More than 147 kPa (1.5 kg/cm², 21 psi)**	Yes	Go to Next Step		
		No	Check fuel pump pressure drop	No	Replace fuel pump
			Check pressure regulator pressure drop	Yes	Check injector fuel leakage
				No	Replace pressure regulator
5	Change fuel with specified one Check if condition improves	Yes			Poor fuel quality
		No	Go to Next Step		
6					ECU malfunction

Cranks normally but won't start (Intermittent)

STEP	QUICK INSPECTION		ACTION	POSSIBLE CAUSE AND DETAILED INSPECTION
1	Shake connector of ignition coil, main relay and ECU while cranking Check if the engine starts	Yes	There may be a poor contact of the connector. Repair or replace the wiring	
		No	Go to troubleshooting "Cranks normally but hard to start (Always)"	

2.6L ENGINE SYMPTOM TROUBLESHOOTING—MPV

Rough idle (Always)

STEP	QUICK INSPECTION		ACTION		POSSIBLE CAUSE AND DETAILED INSPECTION
1	Check for malfunction code (08), (09), (14), (17), (26), (34) with SST [Ign ON, Test connector (Green: 1-pin) grounded]	Yes	Check for cause by referring to the check sequence		
		No	**"88" flashing** Check voltage at ECU (3G) terminal with SST **Voltage: 0V (Ign ON)**	Yes	Replace ECU
				No	Poor ground circuit
			"00" Go to Next Step		
2	Check ignition at idle after warm up **Ignition timing: 4—6° BTDC** [Test connector (Green: 1-pin) grounded]	Yes	Go to Next Step		
		No	Adjust ignition timing (If possible)		
3	Disconnect high-tension lead of each cylinder at idle Check if engine condition changes	Yes	Go to Next Step		
		No	Check ignition system	Yes	Replace injector (If Step 3 OK)
				No	Check spark plug
					Check high-tension lead
					Check distributor cap
4	Check idle speed after warm up **Idle speed: 750—790 rpm (A/T)** [Test connector (Green: 1-pin) grounded]	Yes	Go to Next Step		
		No	Adjust idle speed (If possible)		
5	Check for injector operating sound at idle	Yes	Go to Next Step		
		No	Check resistance at injector harness connector (X-12) Terminals: (B/Y)—(LG/B), (B/Y)—(LG/R); Resistance: 6—8Ω	Yes	Check wiring short or open
				No	Check injector resistance
					Check wiring short or open
6	Check fuel line pressure [Ign ON, Test connector (Yellow: 2-pin) connected] **Fuel line pressure: 265—314 kPa (2.7—3.2 kg/cm², 38—46 psi)**	Yes	Go to Next Step		
		No	Check for fuel leakage		
			Substitute a good fuel filter and retest	Yes	Replace fuel filter
			Check fuel pump maximum pressure **Fuel pump maximum pressure: 441—588 kPa (4.5—6.0 kg/cm², 64—85 psi)**	Yes	Replace pressure regulator
				No	Replace fuel pump
7	Check intake manifold vacuum at idle **Vacuum: 480 ± 20 mmHg (18.9 ± 0.79 inHg)**	Yes	Go to Next Step		
		No	Check for air leaks	Yes	Intake air system components damaged
					Vacuum and intake air hoses loose or damaged
					Bolts or nuts loose
					Gaskets damaged
				No	Check throttle valve closing condition
8	Substitute a well-known ECU Check if condition improves	Yes			ECU malfunction
		No	Check voltage at ECU (2C) terminal with SST **Voltage: 0V (Ign ON)**	Yes	Go to Next Step
				No	Poor ground circuit
9					Check compression

Rough idle (Only when engine is cold)

STEP	QUICK INSPECTION		ACTION		POSSIBLE CAUSE AND DETAILED INSPECTION
1	Check for malfunction code (08), (09), (26) with SST [Ign ON, Test connector (Green: 1-pin) grounded]	Yes	Check for cause by referring to check sequence		
		No	Go to Next Step		
2	Disconnect high-tension lead of each cylinder at idle Check if engine condition changes	Yes	Go to Next Step		
		No	Check ignition system [Refer to ignition system troubleshooting (Misfire)]	Yes	Replace injector (If step 3 OK)
				No	Check spark plug
					Check high-tension lead
					Check distributor cap
3	Check for injector operating sound at idle	Yes	Go to Next Step		
		No	Check resistance at injector harness connector (X-12) Terminals: (B/Y)—(LG/B), (B/Y)—(LG/R); Resistance: 6—8Ω	Yes	Check wiring short or open
				No	Injector malfunction (Check resistance)
					Wiring short or open
4	Disconnect ISC valve connector at idle when engine is cold Check if idle speed decreases during warm up	Yes	Go to Next Step		
		No			Check if BAC valve (air valve) opens when cold
5	Check if specified engine oil is used	Yes	Go to Next Step		
		No	Change engine oil to specified oil		
6	Subsitute a well-known ECU Check if condition improves	Yes			ECU malfunction
		No			Check airflow sensor

2.6L ENGINE SYMPTOM TROUBLESHOOTING—MPV

Rough idle (Only when engine is warm)

STEP	QUICK INSPECTION		ACTION		POSSIBLE CAUSE AND DETAILED INSPECTION
1	Run engine at **2,000 rpm** for more than **20 seconds** Check for malfunction code (08), (09), (15), (17), (26), (34) with SST [Ign ON, Test connector (Green: 1-pin) grounded]	Yes	Check for cause by referring to check sequence		
		No	Go to Next Step		
2	Check idle speed after warm up **Idle speed: 750—790 rpm (A/T)** [Test connector (Green: 1-pin) grounded]	Yes	Go to Next Step		
		No	Adjust idle speed (if possible)		
3	Check for flashing of SST monitor lamp after warm up **Monitor lamp: Flashes more than 8 times/10 seconds at 2,000—3,000 rpm** [Test connector (Green: 1-pin) not grounded]	Yes	Go to Next Step		
		No			Replace oxygen sensor
4	Disconnect ISC valve connector after warm up Check if engine speed drops	Yes	Go to Next Step		
		No			Check ISC valve
5	Disconnect water thermosensor connector Check if condition improves	Yes	Check water thermosensor connector condition as follows: 1. Shake connector and check if condition changes 2. Check condition of terminal (burned or damaged) 3. Connect a good terminal to harness side connector and check for looseness	Yes	Check water thermosensor
				No	Poor contact of water thermosensor connector
		No	Go to Next Step		
6	Disconnect high-tension lead of each cylinder at idle Check if engine condition changes	Yes	Go to Next Step		
		No	Check ignition system	Yes	Replace injector (If step 7 OK)
				No	Check spark plug
					Check high-tension lead
					Check distributor cap
			Note: If spark plug is wet, injector may be leaking		
7	Check for injector operating sound at idle	Yes	Go to Next Step		
		No	Check resistance at injector harness connector (X-12) Terminals: (B/Y)—(LG/B), (B/Y)—(LGR); Resistance: 6—8Ω	Yes	Check wiring short or open
				No	Check injector resistance
					Check wiring short or open
8	Check for air leaks by listening for sucking noise	Yes	Go to Next Step		
		No			Intake air system components damaged
					Vacuum and intake air hoses loose or damaged
					Bolts or nuts loose
					Gaskets damaged
9					ECU malfunction

Rough idle (Only after heat soak)

STEP	QUICK INSPECTION		ACTION		POSSIBLE CAUSE AND DETAILED INSPECTION
1	Run engine at **2,000 rpm** for more than **20 seconds** Check for malfunction code (09), (11), (15), (17), (25), (26) with SST [Ign ON, Test connector (Green: 1-pin) grounded]	Yes	Check for cause by referring to check sequence		
		No	Go to Next Step		
2	Check switches with SST Neutral switch (M/T) Clutch switch (M/T) Inhibitor switch (A/T) [Ign ON, Test connector (Green: 1-pin) grounded]	Yes	Go to Next Step		
		No	Check for cause by referring to check sequence		
3	Check for flashing of SST monitor lamp after warm up **Monitor lamp: Flashes more than 8 times 10 seconds at 2,000—3,000 rpm** [Test connector (Green: 1-pin) not grounded]	Yes	Go to Next Step		
		No			Replace oxygen sensor
4	Disconnect vacuum hose from pressure regulator Check if condition improves	Yes	Check components related to pressure regulator control system		Check water thermosensor
					Check intake air thermosensor
					Check solenoid valve (PRC)
					ECU malfunction (Check (2T) terminal voltage)
		No	Go to Next Step		
5	Run engine at idle and stop it Observe fuel pressure **for 5 minutes** **Fuel pressure: More than 147 kPa (1.5 kg/cm^2, 21 psi)**	Yes	Go to Next Step		
		No	Check fuel pump pressure drop	No	Replace fuel pump malfunction
			Check pressure regulator pressure drop	Yes	Check injector fuel leakage
				No	Replace pressure regulator
6	Disconnect high-tension lead of each cylinder at idle Check if engine condition changes	Yes	Go to Next Step		
		No	Check ignition system	Yes	Replace injector (If step 3 OK)
				No	Check spark plug
					Check high-tension lead
					Check distributor cap

Rough idle just after starting

STEP	QUICK INSPECTION		ACTION		POSSIBLE CAUSE AND DETAILED INSPECTION
1	Check for malfunction code (09), (34) with SST [Ign ON, Test connector (Green: 1-pin) grounded]	Yes	Check for cause by referring to check sequence		
		No	Go to Next Step		
2	Check idle switch with SST [Ign ON, Test connector (Green: 1-pin) grounded]	Yes	Go to Next Step		
		No	Check for cause by referring to check sequence		
3	Check ignition timing at idle after warm up **Ignition timing: 4—6° BTDC** [Test connector (Green: 1-pin) grounded]	Yes	Go to Next Step		
		No	Adjust ignition timing		
4	Check idle speed after warm up **Idle speed: 750—790 rpm (A/T)** [Test connector (Green: 1-pin) grounded]	Yes	Go to Next Step		
		No	Try to adjust idle speed	Yes	Idle-speed misadjustment
				No	Check accelerator cable free play
					Check ISC valve (Stuck closed)
					Check throttle body
5	Substitute a well-known ECU Check if condition improves	Yes			ECU malfunction
		No	Check voltage at ECU (1C) terminal with SST **Voltage: Approx. 10V (While cranking)**	Yes	Go to Next Step
				No	Check starter interlock switch (M/T)
					Check inhibitor switch (A/T)
					Check related wiring
6					Poor quality engine oil
7	Check for injector operating sound at idle	Yes	Go to Next Step		
		No	Check resistance at injector harness connector (X-12) Terminals: (B/Y)—(LG/B), (B/Y)—(LG/R); Resistance: 6—8Ω	Yes	Check wiring short or open
				No	Check injector resistance
					Check wiring short or open
8	Change fuel to specified grade Check if condition improves	Yes			Poor fuel quality
		No	Go to Next Step		
9					ECU malfunction

Low idle speed (When A/C, P/S, E/L ON)

STEP	QUICK INSPECTION		ACTION	POSSIBLE CAUSE AND DETAILED INSPECTION
1	Check for malfunction code (34) with SST [Ign ON, Test connector (Green: 1-pin) grounded]	Yes	Check for cause by referring to check sequence	
		No	Go to Next Step	
2	Disconnect ISC valve connector at idle Check if the condition does not change	Yes	Go to Next Step	
		No		Check coolant level
				Check engine oil
3	Check switches with SST Idle switch Inhibitor switch (A/T) [Ign ON, Test connector (Green: 1-pin) grounded]	Yes	Go to Next Step	
		No	Check for cause by referring to check sequence	
4	Check continuity between test connector (Green: 1-pin) and ground			Wiring short to ground

Note

- **In case of low idle speed with A/C ON, if the problem cannot be solved by the above steps, it may be an A/C compressor malfunction.**

2.6L ENGINE SYMPTOM TROUBLESHOOTING—MPV

High idle speed after warm up

STEP	QUICK INSPECTION		ACTION		POSSIBLE CAUSE AND DETAILED INSPECTION
1	Check for malfunction code (09), (26), (34) with SST [Ign ON, Test connector (Green: 1-pin) grounded]	Yes	Check for cause by referring to check sequence		
		No	Go to Next Step		
2	Check ignition timing at idle after warm up **Ignition timing: 4—6° BTDC** [Test connector (Green: 1-pin) grounded]	Yes	Go to Next Step		
		No	Adjust ignition timing		
3	Check intake manifold vacuum at idle	Yes	Go to Next Step		
		No	Check for air leaks	Yes	Intake air system components damaged
					Vacuum and intake air hoses loose or damaged
					Bolts or nuts loose
					Gaskets damaged
				No	Check vacuum hose routing
					Check throttle body
4	Check idle speed after warm up **Idle speed: 750—790 rpm** [Test connector (Green: 1-pin) grounded]	Yes			Check ISC valve
		No	Try to adjust idle speed	Yes	Idle speed misadjustment
				No	Go to Next Step
5	Disconnect ISC valve connector at idle when engine is cold Check if idle speed decreases during warm up	Yes	Go to Next Step		
		No			Check air valve
6	Disconnect water thermosensor connector and check if condition improves	Yes	Check water thermosensor connector condition as follows: 1. Shake connector and check if condition changes 2. Check condition of terminal (burned or damaged) 3. Connect a good terminal to harness side connector and check for looseness	Yes	Check water thermo-sensor
				No	Poor contact of water thermo-sensor connector
		No	Go to Next Step		
7					ECU malfunction

Idle hunting or surging

STEP	QUICK INSPECTION		ACTION		POSSIBLE CAUSE AND DETAILED INSPECTION
1	(If trouble occurs only at warm condition) Run engine at **2,000 rpm** for more than **20 seconds** Check for malfunction code (09), (15), (17), (26), (34) with SST [Ign ON, Test connector (Green: 1-pin) grounded]	Yes	Check for cause by referring to check sequence		
		No	Go to Next Step		
2	(If trouble occurs only at warm condition) Check for flashing of SST monitor lamp after warm up **Monitor lamp:** **Flashes more than 8 times/10 seconds at 2,000—3,000 rpm** [Test connector (Green: 1-pin) not grounded]	Yes	Go to Next Step		
		No			Replace oxygen sensor
3	Check intake manifold vacuum at idle **Vacuum: 460—500 mmHg (18.1—19.7 inHg)**	Yes	Go to Next Step		
		No	Check for air leaks		Intake air system components damaged
					Vacuum and air intake hoses loose or damaged
					Bolts or nuts loose
					Gaskets damaged
4	Pinch PCV hose Check if condition improves	Yes			Check PCV valve
		No	Go to Next Step		
5	Check fuel line pressure [Ign ON, Test connector (Yellow: 2-pin) connected] **Fuel line pressure:** **265—314 kPa (2.7—3.2 kg/cm², 38—46 psi)**	Yes	Go to Next Step		
		No	Check for fuel leaks		
			Substitute a good fuel filter and retest	Yes	Replace fuel filter
			Check fuel pump maximum pressure **Fuel pump maximum pressure: 441—588 kPa (4.5—6.0 kg/cm², 64—85 psi)**	Yes	Replace pressure regulator
				No	Replace fuel pump
6					ECU malfunction

Engine stall at idle (Always)

STEP	QUICK INSPECTION		ACTION			POSSIBLE CAUSE AND DETAILED INSPECTION
1	Check for malfunction code (08), (26) with SST [Ign ON, Test connector (Green: 1-pin) grounded]	Yes	Check for cause by referring to the check sequence			
		No	Go to Next Step			
2	Slightly open throttle valve and disconnect high-tension lead of each cylinder Check if engine condition changes	Yes	Go to Next Step			
		No	Check ignition system			Check spark plug
						Check high-tension lead
						Check distributor cap
3	Check fuel line pressure [Ign ON, Test connector (Yellow: 2-pin) connected] **Fuel line pressure: 265—314 kPa (2.7—3.2 kg/cm², 38—46 psi)**	Yes	Go to Next Step			
		No	Check for fuel leaks			
			Substitute a good fuel filter and retest		Yes	Replace fuel filter
			Check fuel pump maximum pressure **Fuel pump maximum pressure: 441—588 kPa (4.5—6.0 kg/cm², 64—85 psi)**		Yes	Replace pressure regulator
					No	Replace fuel pump
4	Check if vacuum hoses and the air hoses are connected correctly	Yes	Go to Next Step			
		No	Connect correctly			
5						Airflow sensor
6						ECU malfunction

Engine stall at idle (Only when engine is cold)

STEP	QUICK INSPECTION		ACTION	POSSIBLE CAUSE AND DETAILED INSPECTION
1	Check for malfunction code (09), (26) with SST [Ign ON, Test connector (Green: 1-pin) grounded]	Yes	Check for cause by referring to check sequence	
		No	Go to Next Step	
2				Check BAC valve (air valve)
3				ECU malfunction

Engine stall at idle (Only when engine is warm)

STEP	QUICK INSPECTION		ACTION		POSSIBLE CAUSE AND DETAILED INSPECTION
1	Check for malfunction code (09), (26) with SST [Ign ON, Test connector (Green: 1-pin) grounded]	Yes	Check for cause by referring to check sequence		
		No	Go to Next Step		
2	Disconnect water thermosensor connector Check if condition improves	Yes	Check water thermosensor connector as follows: 1. Shake connector and check if condition changes 2. Check condition of terminal (burned or damaged) 3. Connect a good terminal to harness side connector and check for looseness	Yes	Check water thermosensor
				No	Poor contact of water thermosensor connector
		No	Go to Next Step		
3					ECU malfunction

Engine stall at idle (Only after heat soak)

STEP	QUICK INSPECTION	ACTION	POSSIBLE CAUSE AND DETAILED INSPECTION
1			ECU malfunction

Engine stall at idle (When A/C, P/S, E/L is ON)

STEP	QUICK INSPECTION		ACTION	POSSIBLE CAUSE AND DETAILED INSPECTION
1	Check for malfunction code (34) with SST [Ign ON, Test connector (Green: 1-pin) grounded]	Yes	Check for cause by referring to check sequence	
		No	Go to Next Step	
2	Check switches with SST • Headlight switch • Rear defroster switch • Blower switch [Ign ON, Test connector (Green: 1-pin) grounded]	Yes	Go to Next Step	
		No	Check for cause by referring to check sequence	
3	Disconnect ISC valve connector at idle [Test connector (Green: 1-pin) grounded] Check if the condition does not change	Yes	Go to Next Step	
		No		Check ISC valve
				Check engine oil
4	Check idle speed after warm up **Idle speed: 750—790 rpm (A/T)** [Test connector (Green: 1-pin) grounded]	Yes	Go to Next Step	
		No	Adjust idle speed	
5				ECU malfunction

Note
- **Engine stalls at idle with A/C ON, if the trouble can not be fixed after checking above steps, it may be A/C compressor malfunction**

Engine stall at idle (in "D" range)

STEP	QUICK INSPECTION		ACTION	POSSIBLE CAUSE AND DETAILED INSPECTION
1	Check for malfunction code (34) with SST [Ign ON, Test connector (Green: 1-pin) grounded]	Yes	Check for cause by referring to check sequence	
		No	Go to Next Step	
2	Check inhibitor switch with SST [Ign ON, Test connector (Green: 1-pin) grounded]	Yes	Go to Next Step	
		No	Check for cause by referring to check sequence	
3				ECU malfunction

Engine stall during start up

STEP	QUICK INSPECTION		ACTION		POSSIBLE CAUSE AND DETAILED INSPECTION
1	Check for malfunction code (08) with SST [Ign ON, Test connector (Green: 1-pin) grounded]	Yes	Check for cause by referring to check sequence		
		No	Go to Next Step		
2	Check idle speed after warm up **Idle speed: 750—790 rpm (A/T)** [Test connector (Green: 1-pin) grounded]	Yes	Go to Next Step		
		No	Adjust idle speed		
3	Check for injector operating sound at idle	Yes	Go to Next Step		
		No	Check resistance at injector harness connector (X-12) Terminal: (B/Y)—(LG/B), (B/Y)—(LG/R); Resistance: 6—8Ω	Yes	Check wiring short or open
				No	Cheek injector resistance
					Check wiring
4					ECU malfunction

Engine stall on deceleration (Only when engine is cold)

STEP	QUICK INSPECTION		ACTION	POSSIBLE CAUSE AND DETAILED INSPECTION
1	Check for malfunction code (08), (09) with SST [Ign ON, Test connector (Green: 1-pin) grounded]	Yes	Check for cause by referring to check sequence	
		No	Go to Next Step	
2				ECU malfunction

Engine stall on deceleration (Only after warm up)

STEP	QUICK INSPECTION		ACTION	POSSIBLE CAUSE AND DETAILED INSPECTION
1	Check for malfunction code (08), (09) with SST [Ign ON, Test connector (Green: 1-pin) grounded]	Yes	Check for cause by referring to check sequence	
		No	Go to Next Step	
2	Check stoplight switch with SST [Ign ON, Test connector (Green: 1-pin) grounded]	Yes	Go to Next Step	
		No	Check for cause by referring to check sequence	
3				ECU malfunction

Engine stall at idle (Intermittent)

STEP	QUICK INSPECTION		ACTION	POSSIBLE CAUSE AND DETAILED INSPECTION
1	Shake connector of ignition coil, main relay and ECU while cranking Check if the engine starts	Yes	There may be a poor contact at the connector. Repair or replace the wiring	
		No	Go to troubleshooting "Engine stall at idle (Always)"	

Hesitates/Stumbles on acceleration

QUICK	INSPECTION		ACTION	POSSIBLE CAUSE AND DETAILED INSPECTION
1	Run engine at **2,000 rpm** for **20 seconds** and stop it Check for malfunction code (09), (15), (17) with SST [Ign ON, Test connector (Green: 1-pin) grounded]	Yes	Check for cause by referring to check sequence	
		No	Go to Next Step	
2	Check idle switch with SST [Ign ON, Test connector (Green: 1-pin) grounded]	Yes	Go to Next Step	
		No	Check for cause by referring to check sequence	
3	Disconnect oxygen sensor connector Check if condition improves	Yes		Check oxygen Sensor
		No	Go to Next Step	
4	Check fuel line pressure while accelerating (Vacuum hose to pressure regulator disconnected) **Fuel line pressure:** **Keeps 265—314 kPa** **2.7—3.2 kg/cm², 38—46 psi)**	Yes	Go to Next Step	
		No	Substitute a good fuel filter and retest — Yes	Replace fuel filter
				Pressure regulator Replace
5	Check for air leaks with throttle valve open by listening for sucking noise	Yes		Intake air system components damaged
				Vacuum and intake air hoses loose or damaged
				Bolts or nuts loose
				Gaskets damaged
		No	Go to Next Step	
6	Substitute a well-known ECU Check if condition improves	Yes		ECU malfunction
		No		Check airflow sensor
				Check spark plug
7	Check other systems			Clutch slipping
				AT kick-down control malfunction

Hesitates at steady speed

STEP	QUICK INSPECTION		ACTION	POSSIBLE CAUSE AND DETAILED INSPECTION
1	Run engine at **2,000 rpm** for **20 seconds** and stop it Check for malfunction code (08), (09), (12), (14), (15), (17) with SST [Ign ON, Test connector (Green: 1-pin) grounded]	Yes	Check for cause by referring to check sequence	
		No	Go to Next Step	
2	Disconnect oxygen sensor connector Check if condition improves	Yes		Check oxygen sensor
		No	Go to Next Step	
3	Check for air leaks with throttle valve open by listening for sucking noise	Yes	Go to Next Step	
		No		Intake air system components damaged
				Vacuum and intake air hoses loose or damaged
				Nuts or bolts loose
				Gasket damaged
4	Check fuel line pressure while accelerating (Vacuum hose to pressure regulator disconnected) **Fuel line pressure:** **Keeps 265—314 kPa** **(2.7—3.2 kg/cm², 38—46 psi)**	Yes	Go to Next Step	
		No	Substitute a good fuel filter and retest — Yes	Replace fuel filter
				Replace pressure regulator
5	Check condition of ignition coil and airflow meter connectors (Burned or damaged)	Yes		Poor contact
		No	Go to Next Step	
6	Gradually open throttle valve Check if engine speed increases smoothly	Yes	Go to Next Step	
		No		Check airflow sensor
				Check throttle sensor
7				Check spark plug
8	Change fuel to specified grade Check if condition improves	Yes		Poor fuel quality
		No	Go to Next Step	
9				ECU malfunction

2.6L ENGINE SYMPTOM TROUBLESHOOTING – MPV

Jerking on acceleration

STEP	QUICK INSPECTION		ACTION	POSSIBLE CAUSE AND DETAILED INSPECTION
1	Run engine at **2,000 rpm** for **20 seconds** and stop it Check for malfunction code (09), (15), (17) with SST [Ign ON, Test connector (Green: 1-pin) grounded]	Yes	Check for cause by referring to check sequence	
		No	Go to Next Step	
2	Check idle switch with SST [Ign ON, Test connector (Green: 1-pin) grounded]	Yes	Go to Next Step	
		No	Check for cause by referring to check sequence	
3	Disconnect oxygen sensor connector Check if condition improves	Yes		Check oxygen Sensor
		No	Go to Next Step	
4	Check fuel line pressure while accelerating (Vacuum hose to pressure regulator disconnected) **Fuel line pressure: Keeps 265—314 kPa 2.7—3.2 kg/cm², 38—46 psi)**	Yes	Go to Next Step	
		No	Substitute a good fuel filter and retest — Yes	Replace fuel filter
				Pressure regulator Replace
5	Check for air leaks with throttle valve open by listening for sucking noise	Yes		Intake air system components damaged
				Vacuum and intake air hoses loose or damaged
				Bolts or nuts loose
				Gaskets damaged
		No	Go to Next Step	
6	Substitute a well-known ECU Check if condition improves	Yes		ECU malfunction
		No		Check airflow sensor
				Check throttle body
				Check spark plug
7	Check other systems			Clutch slipping
				AT kick-down control malfunction

Knocking

STEP	QUICK INSPECTION		ACTION			POSSIBLE CAUSE AND DETAILED INSPECTION
1	Check malfunction code (09) with SST [Ign ON, Test connector (Green: 1-pin) grounded]	Yes	Check for cause by referring to the check sequence			
		No	Go to Next Step			
2	Check ignition timing after warm up **Ignition timing: 4—6° BTDC** [Test connector (Green: 1-pin) grounded]	Yes	Go to Next Step			
		No				Ignition timing misadjust
3	Disconnect water thermosensor connector Check if condition improves	Yes				Check water thermo-sensor
		No	Go to Next Step			
4	Observe fuel line pressure while accelerating from idle **Fuel line pressure: Keeps 265—314 kPa (2.7—3.2 kg/cm², 38—46 psi)** (Vacuum hose to pressure regulator disconnected)	Yes	Go to Next Step			
		No	Check fuel pump maximum pressure **Fuel pump maximum pressure: 441—588 kPa (4.5—6.0 kg/cm², 64—85 psi)**		Yes	Replace fuel filter
					No	Replace fuel pump
5						Check airflow sensor
6						Check spark plug
7	Change fuel to specified grade Check if condition improves	Yes				Poor fuel quality
		No	Go to Next Step			
8	Check cooling system					Thermostat
						Radiator
9						ECU malfunction

Poor acceleration

STEP	QUICK INSPECTION		ACTION		POSSIBLE CAUSE AND DETAILED INSPECTION
1	Check for malfunction code (08), (09), (12), (14), (26) with SST [Ign ON, Test connector (Green: 1-pin) grounded]	Yes	Check for cause by referring to check sequence		
		No	Go to Next Step		
2	Check idle switch with SST [Ign ON, Test connector (Green: 1-pin) grounded]	Yes	Go to Next Step		
		No	Check for cause by referring to check sequence		
3	Disconnect high—tension lead of each cylinder at idle. Check if engine condition changes [ISC valve connector disconnected]	Yes	Go to Next Step		
		No	Check ignition system [Refer to ignition system troubleshooting (Misfire)]	Yes	Replace injector
				No	Check spark plug
					Check high-tension
					Check distributor cup
4	Check for air leaks by listening for sucking noise	Yes			Intake air system components damaged
					Vacuum and air intake hoses loose or damaged
					Nuts or bolts loose
					Gasket damaged
		No	Go to Next Step		
5	Observe fuel line pressure while accelerating from idle **Fuel line pressure: Keeps 265—314 kPa (2.7—3.2 kg/cm², 38—46 psi)** [Vacuum hose to pressure regulator disconnected]	Yes	Go to Next Step		
		No	Substitute a good fuel filter and retest	Yes	Replace fuel filter
				No	Replace pressure regulator
6	Gradually depress accelerator from idle Check if engine speed increases smoothly	Yes	Go to Next Step		
		No	Check accelerator cable free play	Yes	Check airflow sensor
					Check throttle body
				No	Adjust
7	Check fuel to specified grade Check if condition improves	Yes			Poor fuel quality
		No	Go to Next Step		
8	Substitute a well-known ECU Check if condition improves	Yes			ECU malfunction
		No	Go to Next Step		
9	Check other systems				Clutch slipping
					Transmission (M/T)
					Transmission (A/T)
					Brake
					Belt tension

Lack of power

STEP	QUICK INSPECTION		ACTION		POSSIBLE CAUSE AND DETAILED INSPECTION
1	Check for malfunction code (08) and (14) (only high-altitude) with SST [Ign ON, Test connector (Green:1-pin) grounded]	Yes	Check for cause by referring to check sequence		
		No	Go to Next Step		
2	(Only at high-altitude) Check ignition timing at idle **Ignition timing: 4—6° BTDC** [Test connector (Green: 1-pin) grounded]	Yes	Go to Next Step		
		No			Adjust ignition timing
3	Disconnect ISC valve connector and the high-tension lead of each cylinder Check if condition changes	Yes	Go to Next Step		
		No	Check ignition system	Yes	Replace injector (If step 4 OK)
				No	Check high-tension lead
					Check distributor cap
					Check spark plug
4	Check for injector operating sound at idle	Yes	Go to Next Step		
		No	Check resistance at injector harness connector (X-12) Terminals: (B/Y)—(LG/B), (B/Y)—(LG/R); Resistance: 6—8Ω	Yes	Check wiring short or open
				No	Check injector resistance
					Check wiring short or open
5	Check air cleaner element for clogging	Yes	Go to Next Step		
		No	Clean air cleaner element		
6	Check for air leaks by listening for sucking noises • At idle • When throttle valve is open	Yes			Intake air system
					Components damaged
					Vacuum and air intake hoses loose or damaged
					Nuts or bolts loose
					Gasket damaged

2.6L ENGINE SYMPTOM TROUBLESHOOTING—MPV

Lack of power (Cont'd)

STEP	QUICK INSPECTION		ACTION		POSSIBLE CAUSE AND DETAILED INSPECTION
7	Check fuel line pressure [Ign ON, Test connector (Yellow: 2-pin) connected] **Fuel line pressure: 265—314 kPa (2.7—3.2 kg/cm^2, 38—46 psi)**	Yes	Go to Next Step		
		No	Check for fuel leakage		
			Substitute a good fuel filter and retest	Yes	Replace fuel filter
			Check fuel pump maximum pressure **Fuel pump maximum pressure: 441—588 kPa (4.5—6.0 kg/cm^2, 64—85 psi)**	Yes	Replace pressure regulator
				No	Replace fuel pump
8	Check fuel line pressure at idle **Fuel line pressure: 206—255 kPa (2.1—2.6 kg/cm^2, 30—37 psi)**	Yes	Go to Next Step		
		No			Replace pressure regulator
9	Check if fuel line pressure drops while accelerating (Vacuum hose disconnected)	Yes	Substitute a good fuel filter and retest	Yes	Replace fuel filter
		No	Go to Next Step		
10	Check exhaust system for damage	Yes	Go to Next Step		
		No	Repair or replace		
11	Check A/C, P/S and alternator belts tensions	Yes	Go to Next Step		
		No	Adjust belt tension		
12	Check if accelerator can be depressed fully	Yes	Go to Next Step		
		No	Check accelerator cable	Yes	Throttle body
				No	Accelerator cable
13	Substitute a well-known ECU Check if condition improves	Yes			ECU malfunction
		No			Check air flow sensor
					Check throttle sensor
					Go to Next Step
14	Substitute a specified fuel Check if condition improves	Yes			Poor fuel quality
		No	Go to Next Step		
15	Check other systems				Brake
					Clutch
					A/T
					Engine

Bucking at high speed

STEP	QUICK INSPECTION		ACTION		POSSIBLE CAUSE AND DETAILED INSPECTION
1	Run engine at **2,000 rpm** for more than **20 seconds** Check for malfunction code (08), (09), (12), (14), (15), (17) with SST [Ign ON, Test connector (Green: 1-pin) grounded]	Yes	Check for cause by referring to check sequence		
		No	Go to Next Step		
2	Disconnect oxygen sensor connector Check if condition improves	Yes			Check oxygen sensor
		No	Go to Next Step		
3	Observe fuel line pressure while accelerating from idle **Fuel line pressure: Keeps 265—314 kPa (2.7—3.2 kg/cm^2, 38—46 psi)** [Vacuum hose to pressure regulator disconnected]	Yes	Go to Next Step		
		No	Substitute a good fuel filter and retest	Yes	Replace fuel filter
					Replace pressure regulator
4	Check for air leaks by listening sucking noise	Yes	Go to Next Step		
		No			Intake air system components damaged
					Vacuum and air intake hoses loose or damaged
					Nuts or bolts loose
					Gasket damaged
5	Gradually open throttle valve from idle check if engine speed increases smoothly	Yes	Go to Next Step		
		No			Check airflow sensor
6					Check spark plug
7					ECU malfunction

Bucking on deceleration

STEP	QUICK INSPECTION		ACTION	POSSIBLE CAUSE AND DETAILED INSPECTION
1	Check for malfunction code (08), (09), (12), (26), (34) with SST [Ign ON, Test connector (Green: 1-pin) grounded]	Yes	Check for cause by referring to the check sequence	
		No	Go to Next Step	
2	Check switches with SST [Ign ON, Test connector (Green: 1-pin) grounded] • Idle switch • Stoplight switch	Yes	Go to Next Step	
		No	Check for cause by referring to the check sequence	
3	Substitute a well-known ECU Check if condition improves	Yes		ECU malfunction
		No		Check throttle sensor
				Go to Next Step
4				Check spark plug
5				Check clutch slipping
6				Check compression between cylinders

Poor fuel economy

STEP	QUICK INSPECTION		ACTION		POSSIBLE CAUSE AND DETAILED INSPECTION
1	Run the engine at **2,000 rpm** for more than **20 seconds** after warm up and stop it Check for malfunction code (15) (17) with SST [Ign ON, Test connector (Green: 1-pin) grounded]	Yes	Check for cause by referring to check sequence		
		No	Go to Next Step		
2	Check idle switch with SST [Ign ON, Test connector (Green: 1-pin) grounded]	Yes	Go to Next Step		
		No	Check for cause by referring to check sequence		
3	Check for flashing of monitor lamp after warm up **Monitor lamp: Flashes more than 8 times /10 seconds at 2,000—3,000 rpm** [Test connector (Green: 1-pin) grounded]	Yes	Go to Next Step		
		No			Replace oxygen sensor
4	Check fuel line pressure at idle **Fuel line pressure: 206—255 kPa (2.1—2.6 kg/cm², 30—37 psi)**	Yes	Go to Next Step		
		No	Check vacuum line to pressure regulator for clogging or air leakage	Yes	Vacuum line clogging or damaged
				No	Check solenoid valve (PRC)
					ECU malfunction (Check (2T) terminal voltage)
					Replace pressure regulator
5	Check for fuel cut operation during deceleration **Fuel cut: Above 1,600 rpm after warm up**	Yes	Go to Next Step		
		No	Check water thermosensor	Yes	Replace ECU
				No	Replace water thermosensor
6	Check other systems				Clutch slipping
					A/T malfunction
					Brake
					Tire air pressure

2.6L ENGINE SYMPTOM TROUBLESHOOTING – MPV

High oil consumption/White exhaust smoke

STEP	QUICK INSPECTION		ACTION		POSSIBLE CAUSE AND DETAILED INSPECTION
1	Check for oil leak from engine	Yes	Repair or replace		
		No	Go to Next Step		
2	Disconnect PCV valve from engine Check if vacuum is felt at idle	Yes	Go to Next Step		
		No	Check PCV valve	Yes	PCV hose clogging
				No	Replace PCV valve
3	Check that ventilation hose is installed correctly	Yes	Go to Next Step		
		No	Install ventilation hose correctly		
4	Possible malfunction of engine				

Afterburn on deceleration

STEP	QUICK INSPECTION		ACTION		POSSIBLE CAUSE AND DETAILED INSPECTION
1	Check malfunction code (34) with SST [Ign ON, Test connector (Green: 1-pin) grounded]	Yes	Check for cause by referring to the check sequence		
		No	Go to Next Step		
2	Check idle switch with SST [Ign ON, Test connector (Green: 1-pin) grounded]	Yes	Go to Next Step		
		No	Check for cause by referring to the check sequence		
3	Check air cleaner element for clogging	Yes	Go to Next Step		
		No	Clean air cleaner element		
4	Check fuel cut operation during deceleration **Fuel cut: Above 1,600 rpm after warm up**	Yes	Go to Next Step		
		No	Check water thermosensor	Yes	ECU malfunction Check (2Q) terminal voltage
				No	Replace water thermosensor
5	Run engine at idle and stop it (Ign OFF) Observe fuel pressure for **5 minutes** **Fuel pressure: More than 147 kPa (1.5 kg/cm², 21 psi)**	Yes	Go to Next Step		
		No	Check fuel pump for pressure drop	No	Replace fuel pump
			Check pressure regulator for pressure drop	Yes	Check injector fuel leakage
				No	Replace pressure regulator
6					Check compression
					Check valve timing

Rotten egg smell

STEP	QUICK INSPECTION	ACTION	POSSIBLE CAUSE AND DETAILED INSPECTION
1	Change fuel to specified grade Check if condition improves		Poor fuel quality

Black exhaust smoke

STEP	QUICK INSPECTION		ACTION		POSSIBLE CAUSE AND DETAILED INSPECTION
1	Run engine at **2,000 rpm** for more then **20 seconds** Check for malfunction code (17) with SST [Ign ON, Test connector (Green: 1-pin grounded]	Yes	Check for cause by referring to the check sequence		
		No	Go to Next Step		
2	Check fuel line pressure [Ign ON, Test connector (Yellow: 2-pin) connected] **Fuel line pressure: 265—314 kPa (2.7—3.2 kg/cm², 38—46 psi)**	Yes	Go to Next Step		
		No (high)			Replace pressure regulator
3	Disconnect water thermosensor connector Check if condition improves	Yes	Check water thermosensor connector condition as follows: 1. Shake connector and check if condition changes 2. Check condition of terminal (burned or damaged) 3. Connect a good terminal to harness side connector and check for looseness	Yes	Check water thermosensor
				No	Poor contact of water thermosensor connector
		No	Go to Step 4 [at high-altitude (**1,000 m, 3,280 ft**)] Go to Step 5 (at sea level)		
4					Check airflow sensor
5					ECU malfunction

Fails emission test (Engine condition OK)

STEP	QUICK INSPECTION		ACTION			POSSIBLE CAUSE AND DETAILED INSPECTION
1	Run engine at **2,000 rpm** for **20 seconds** and stop it Check for malfunction code (08), (09), (14), (15), (17), (26) with SST [Ign ON, Test connector (Green: 1-pin) grounded]	Yes	Check for cause by referring to the check sequence			
		No	Go to Next Step			
2	Check idle switch with SST [Ign ON, Test connector (Green: 1-pin) grounded]	Yes	Go to Next Step			
		No	Check for cause by referring to the check sequence			
3	Check for air leaks by listening for sucking noise • At idle • Throttle valve open	Yes	Go to Next Step			
		No				
4	Check for flashing of SST monitor lamp after warm up **Monitor lamp:** **Flashes more than 8 times/10 seconds at 2,000—3,000 rpm** [Test connector (Green: 1-pin) not grounded]	Yes	Go to Next Step			
		No				Replace oxygen sensor
5	Check fuel line pressure [Ign ON, Test connector (Yellow: 2-pin) connected] **Fuel line pressure:** **265—314 kPa (2.7—3.2 kg/cm², 38—46 psi)**	Yes	Go to Next Step			
		No (low)	Check for fuel leakage			
			Substitute a good fuel filter and retest		Yes	Replace fuel filter
			Check fuel pump maximum pressure **Fuel pump maximum pressure:** **441—588 kPa (4.5—6.0 kg/cm², 64—85 psi)**		Yes	Replace pressure regulator
					No	Replace fuel pump
		No (high)				Replace pressure regulator
6	Check fuel line pressure at idle **Fuel line pressure:** **206—255 kPa (2.1—2.6 kg/cm², 30—37 psi)**	Yes	Go to Next Step			
		No	Check vacuum line to pressure regulator for clogging or air leakage		Yes	Vacuum line
					No	Check solenoid valve (PRC)
						ECU malfunction
						Replace pressure regulator
7	Check fuel cut operation while decelerating **Fuel cut:** **Above 1,600 rpm after warm up**	Yes	Go to Next Step			
		No	Check water thermosensor		Yes	Replace ECU
					No	Replace water thermosensor
8	Check for fuel fumes	Yes	Check for cause by referring to troubleshooting "Gasoline fumes"			
		No	Go to Next Step			
9						Check PCV valve
10						Check Spark plugs
11						Poor fuel quality

Gasoline fumes

STEP	QUICK INSPECTION		ACTION		POSSIBLE CAUSE AND DETAILED INSPECTION
1	Check for leaks	Yes	Replace		
		No	Go to Next Step		
2	Check if fumes are emitted from check-and-cut valve	Yes	Check check-and cut valve	Yes	Check two-way check valve
					Purge line clogging
				No	Replace check-and-cut valve
		No	Go to Next Step		
3	Check for malfunction code (08), (15), (09), (17), (26) with SST [Ign ON, Test connector (Green: 1-pin) grounded]	Yes	Check for cause by referring to the check sequence		
		No	Go to Next Step		
4	Check switches with SST • Idle switch • Inhibitor switch (A/T) [Ign ON, Test connector (Green: 1-pin) grounded]	Yes	Go to Next Step		
		No	Check for cause by referring to the check sequence		
5	Run engine at idle. Ground the solenoid valve (Purge control) terminal-wire (L/Y) and disconnect vacuum hose (white) from solenoid valve. Check for vacuum at solenoid valve	Yes			ECU malfunction Check (2X) terminal voltage
		No			Replace solenoid valve (Purge control)

2.6L ENGINE SYMPTOM TROUBLESHOOTING—MPV

MIL always ON (Engine condition is OK)

STEP	QUICK INSPECTION		ACTION		POSSIBLE CAUSE AND DETAILED INSPECTION
1	Check for malfunction code with SST [Ign ON, Test connector (Green: 1-pin) grounded]		**"88"** Replace ECU		
			"00" (Except California) Go to Step 2		
			"00" (California only) Wiring between ECU (1E) terminal and MIL short to ground		
2	(Except California) Check if oxygen sensor replacement time has come **Oxygen sensor replacement schedule: Every 80,000 miles (128,000 km)**	Yes	Check if MIL has been reset by exchanging MIL set screw position	Yes	Go to next action
				No	Reset the MIL
			Check continuity between MIL and ground	Yes	Wiring between MIL and ECU is short to ground
				No	Replace mileage sensor
		No	Check continuity between MIL and ground	Yes	Wiring between MIL and ECU shorted
				No	Replace mileage sensor

MIL never ON (Engine condition is not OK)

STEP	QUICK INSPECTION		ACTION		POSSIBLE CAUSE AND DETAILED INSPECTION
1	Check if other indicator lamps illuminate	Yes	Go to Next Step		
		No	Check power supply circuit to combination meter		
2	Check bulb of the MIL	Yes	Ground ECU (1E) terminal Check if MIL illuminates	Yes	Replace ECU
				No	Wiring between ECU and MIL open
		No	Replace		
3	(Except California) Check if the MIL illuminated at oxygen sensor replacement time **Oxygen sensor replacement time: Every 80,000 miles (120,800 km)**				(Except California) MIL set screw loose or left out
					(Except California) Replace mileage sensor

A/C does not work

STEP	QUICK INSPECTION		ACTION			POSSIBLE CAUSE AND DETAILED INSPECTION
1	Check if condenser fan operates when grounding A/C relay terminal-wire (R/W) (Ign ON)	Yes	Check voltage at ECU (1Q) terminal with SST Voltage at idle after warm up: 0V (A/C and blower switches ON)	**F2–179**	Yes	ECU malfunction (Check (1J) terminal voltage)
						Wiring between ECU (1J) and A/C relay open
					No	A/C system malfunction
		No	Check A/C system			

2.6L ENGINE TROUBLE CODES—MPV

Code No.	Malfunction display: Pattern of output signal (Self-Diagnosis Checker)	Sensor or subsystem	Self-diagnosis	Fail-safe
02	ON / OFF	Ne-signal	No Ne-signal	—
03	ON / OFF	G signal	No G signal	Cancels 2-group injection
08	ON / OFF	Airflow sensor	Open or short circuit	Basic fuel injection amount fixed as for two driving modes (1) Idle switch: ON (2) Idle switch: OFF
09	ON / OFF	Water thermosensor	Open or short circuit	Maintains constant 80°(176°F) command
11	ON / OFF	Intake air thermosensor (dynamic chamber)	Open or short circuit	Maintains constant 20°C (68°F) command
12	ON / OFF	Throttle sensor	Open or short circuit	Maintains constant command of throttle valve fully open
14	ON / OFF	Atmospheric pressure sensor (Engine control unit)	Open or short circuit	Maintains constant command of sea level pressure
15	ON / OFF	Oxygen sensor (Inactivation)	Sensor output continues less than 0.45V 180 sec. after engine exceeds 1,500 rpm	Cancels engine feedback operation
17	ON / OFF	Feedback system (Inversion)	Sensor output not changed 20 sec. after engine exceeds 1,500 rpm	Cancels engine feedback operation
25	ON / OFF	Solenoid valve (pressure regulator control)	Open or short circuit	—
26	ON / OFF	Solenoid valve (purge control)		—
34	ON / OFF	Solenoid valve (idle speed control)		—

Caution
- **If there is more than one failure present, the lowest number malfunction code is displayed first, the remaining codes are displayed in order.**
- **After repairing all failures, turn off the ignition switch, disconnect the negative battery cable, for at least 20 sec. to erase the memory of a malfunction code from the engine control unit.**

2.6L ENGINE TROUBLE CODE TROUBLESHOOTING MPV

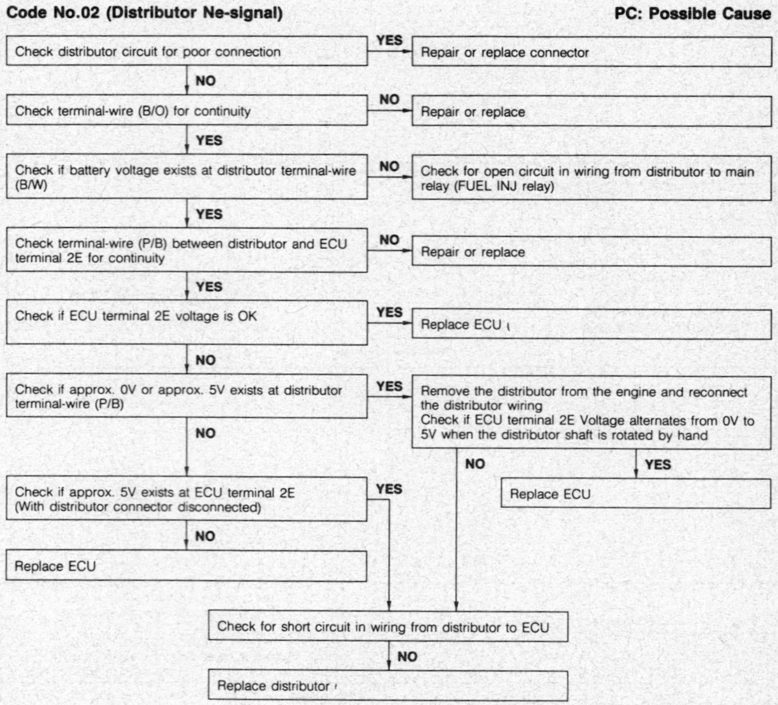

2.6L ENGINE TROUBLE CODE TROUBLESHOOTING—MPV

Code No.03 (Distributor G-signal)

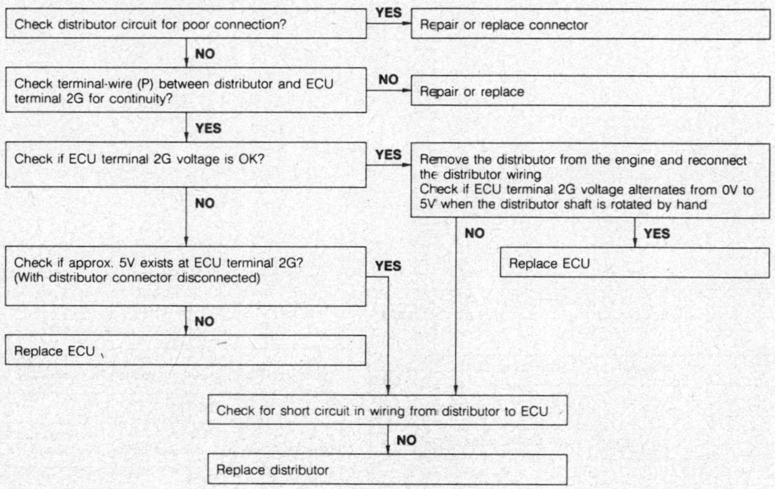

Code No.08 (Airflow sensor)

PC: Possible Cause

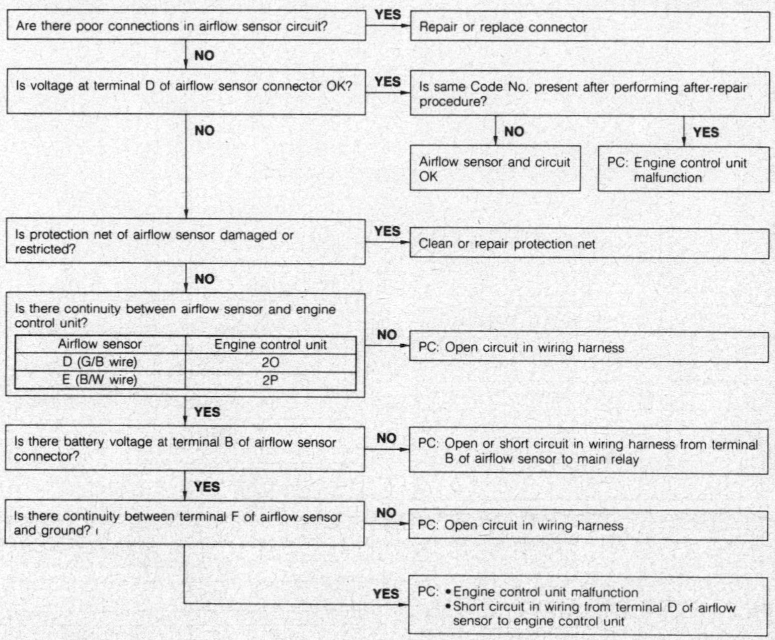

Code No.09 (Water thermosensor)

PC: Possible Cause

Are there poor connections at water thermosensor circuit?
- YES → Repair or replace connector
- NO ↓

Is there continuity between water thermosensor and engine control unit?

Water thermosensor	Engine control unit
A (G/W wire)	2Q
B (BR/B wire)	2D

- NO → PC: Open circuit in wiring harness from water thermosensor to engine control unit
- YES ↓

Is resistance of the water thermosensor OK?
Resistance:

Coolant temp	Resistance
−20°C (−4°F)	14.5— 17.8 kΩ
20°C (68°F)	2.2— 2.7 kΩ
80°C (176°F)	280 —350Ω

- NO → Replace water thermosensor
- YES ↓

Is same Code No. present after performing after-repair procedure?
- NO → Water thermosensor and circuit OK
- YES ↓

Are engine control unit 2Q and 2D terminal voltages OK?
- NO → PC: Engine short circuit in wiring harness
- YES → PC: Engine control unit malfunction

No.11 Code (Intake air thermosensor)

PC: Possible Cause

Are there poor connections at intake air thermosensor connectors?
- YES → Repair or replace connector
- NO ↓

Is there continuity between intake air thermosensor (dynamic chamber) and engine control unit?

Intake air thermosensor (dynamic chamber)	Engine control unit
A (G wire)	2L
B (BR/B wire)	2D

- NO → PC: Open circuit in wiring harness
- YES ↓

Is resistance of intake air thermosensor (dynamic chamber) OK?
Resistance:

Temperature	Resistance
25°C (77°F)	29.7—36.3 kΩ
85°C (185°F)	3.3—3.7 kΩ

- NO → Replace intake air thermosensor (dynamic chamber)
- YES ↓

Is same Code No. present after performing after-repair procedure?
- No → Intake air thermosensor and circuit OK
- YES ↓

Are engine control unit 2L and 2D terminal voltages OK?
- NO → PC: Short circuit in wiring harness
- YES → PC: Engine control unit malfunction

2.6L ENGINE TROUBLE CODE TROUBLESHOOTING—MPV

Code No.12 (Throttle sensor)

PC: Possible cause

Check	Result	Action / Possible cause	Otherwise
Are there poor connections in throttle sensor circuit?	YES	Repair or replace connector terminal	NO → next
Is there 4.5—5.5 V at A terminal of throttle sensor connector?	NO	FC: • Open or short circuit in wiring harness from A terminal of throttle sensor to 2K terminal of engine control unit • Engine control unit malfunction	YES → next
Is C terminal of throttle sensor connector grounded?	NO	PC: Open circuit in wiring harness from C terminal of throttle sensor to ground	YES → next
Is throttle sensor adjusted correctly?	NO	Adjust throttle sensor	YES → next
Is same Code No. present after performing after-repair procedure?	NO	Throttle sensor and circuit OK	YES → next
Is engine control unit 2M terminal voltage OK?	NO	PC: Open or short circuit in wiring harness from B terminal of throttle sensor to 2M terminal of engine control unit	
	YES	PC: Engine control unit malfunction	

Code No.14 (Atmospheric pressure sensor in ECU)

Replace ECU

Code No.15 (Oxygen sensor)

PC: Possible Cause

Note
- **When Codes No.15 and 17 are present at the same time, first perform the checking procedure for Code No.17.**

Check	Result	Action / Possible cause	Otherwise
Are there poor connections in oxygen sensor circuit?	YES	Repair or replace connector	NO → next
Is oxygen sensor output voltage OK?	NO	Replace oxygen sensor	YES → next
Is sensitivity of oxygen sensor OK?	NO	PC: • Oxygen sensor malfunction • Engine control unit malfunction	YES → next
Is same Code No. present after performing after-repair procedure?	NO	Oxygen sensor and circuit OK	YES → next
Is engine control unit 2N terminal voltage OK?	NO	PC: Open or short circuit in wiring harness between oxygen sensor and engine control unit 2N terminal	
	YES	PC: Engine control unit malfunction	

Code No.17 (Feedback system)

PC: Possible Cause

Warm up engine and run it at 2,500—3,000 rpm for three minutes. Does monitor lamp of Self-Diagnosis Checker illuminate at idle?	NO	PC: • Air leak in vacuum hoses or emission component • Contaminated oxygen sensor • Insufficient fuel injection
	YES	Go to next step
Are spark plugs clean?	NO	Clean or replace spark plugs
	YES	Go to next step
Is oxygen sensor voltage OK?	NO	PC: Oxygen sensor malfunction
	YES	Go to next step
Is same Code No. present after performing after-repair procedure?	NO	Feedback system OK
	YES	Go to next step
Is engine control unit 2N terminal voltage OK?	NO	PC: Open or short circuit in wiring harness between oxygen sensor and engine control unit 2N terminal
	YES	PC: Engine control unit malfunction

Code No.25 (Solenoid valve—Pressure regulator control)

PC: Possible Cause

Are there poor connections in solenoid valve circuit?	YES	Repair or replace connector
	NO	Go to next step
Is there continuity of solenoid valve?	NO	Replace solenoid valve
	YES	Go to next step
Is there battery voltage at (B/W) wire of solenoid valve connector?	NO	PC: Open circuit in wiring harness from (B/W) wire to main relay
	YES	Go to next step
Is there continuity between solenoid valve and engine control unit? (Solenoid valve B (L/B wire) — Engine control unit 2T)	NO	PC: Open circuit in wiring harness from solenoid valve to engine control unit
	YES	PC: • Short circuit in wiring harness • Engine control unit malfunction

Solenoid valve	Engine control unit
B (L/B wire)	2T

Code No.26 (Solenoid valve—Purge control)

Are there poor connections in solenoid valve circuit?	YES	Repair or replace connector
	NO	Go to next step
Is there continuity of solenoid valve?	NO	Replace solenoid valve
	YES	Go to next step
Is there battery voltage at (B/W) wire of solenoid valve circuit?	NO	PC: Open circuit in wiring harness from (B/W) wire to main relay
	YES	Go to next step
Is there continuity between solenoid valve and engine control unit? (Solenoid valve B (L/Y wire) — Engine control unit 2X)	NO	PC: Open circuit in wiring harness from solenoid valve to engine control unit
	YES	PC: • Short circuit in wiring harness • Engine control unit malfunction

Solenoid valve	Engine control unit
B (L/Y wire)	2X

2.6L ENGINE TROUBLE CODE TROUBLESHOOTING MPV

Code No.34 (Solenoid valve—Idle speed control (ISC))

PC: Possible Cause

1. Are there poor connections in solenoid valve circuit?
   - **YES** → Repair or replace connector
   - **NO** → go to 2
2. Is resistance of solenoid valve OK? Resistance: 7.7—9.3Ω
   - **NO** → Replace BAC valve
   - **YES** → go to 3
3. Is there battery voltage at (B/W) wire of solenoid valve connector?
   - **NO** → PC: Open or short circuit in wiring harness from (B/W) wire to main relay
   - **YES** → go to 4
4. Is there continuity between solenoid valve and engine control unit?

Solenoid valve	Engine control unit
B (L/G wire)	2W

   - **NO** → PC: Open circuit in wiring harness from solenoid valve to engine control unit
   - **YES** → PC: • Short circuit in wiring harness • Engine control unit malfunction

2.6L ENGINE SWITCH MONITOR FUNCTION—MPV

PREPARATION
SST

49 H018 9A1

Checker, Self-diagnosis

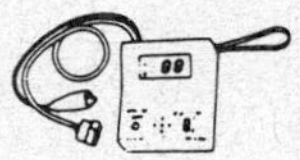

Individual switches can be monitored by the **SST**.

Note
- **The test connector must be grounded and the ignition switch ON (engine stopped).**

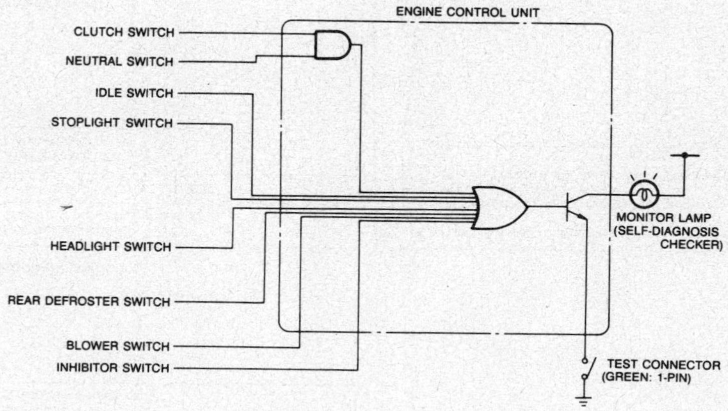

Switch	Self-Diagnosis Checker (Monitor lamp)		Remarks
	Light ON	Light OFF	
Idle switch	Pedal depressed	Pedal released	—
Stoplight switch	Pedal depressed	Pedal released	—
Headlight switch	ON	OFF	—
Rear defroster switch	ON	OFF	—
Blower switch	ON	OFF	Blower motor position: "high" or "super high" position
Inhibitor switch	D, 1, 2, and R ranges	P and N ranges	

2.6L ENGINE SWITCH MONITOR FUNCTION TROUBLESHOOTING – MPV

Procedure

Set conditions to deactivate each switch
- All accessories OFF
- Transmission in neutral
- All pedals released

Verify that monitor lamp does not illuminate

- **YES** → Check each switch as described
- **NO** → Check each switch and related wiring harness
  - Clutch and Neutral switch
  - Idle switch
  - Stoplight switch
  - Headlight switch
  - Rear defroster switch
  - Blower switch
  - Inhibitor switch

Neutral and Clutch switch (M/T)

Shift transmission into gear
Check that monitor lamp illuminates with clutch pedal released

- **NO** → PC: • Neutral or clutch switch malfunction
  - Open circuit in related wiring harness
  - Engine control (1V) terminal malfunction
- **YES** → Depress clutch pedal
  Check that monitor lamp does not illuminate
  Return transmission to neutral
  - **NO** → PC: Clutch switch malfunction

Idle switch

Depress accelerator pedal, and check that the monitor lamp illuminates

- **NO** → PC: • Idle switch malfunction
  - Short circuit in wiring harness from idle switch to engine control unit
  - Engine control unit (1N) terminal malfunction
- **YES** → Idle switch OK
  Release accelerator pedal

Stoplight switch

Depress brake pedal
Check that monitor lamp illuminates

- **NO** → PC: • Stoplight switch malfunction
  - Open circuit in wiring harness from stoplight switch to engine control unit
  - Engine control unit (1O) terminal malfunction
- **YES** → Stoplight switch OK
  Release brake pedal

Blower switch

Turn blower switch to "high" position
Check that monitor lamp illuminates

- **NO** → PC: • CPU malfunction
  - Blower switch malfunction
  - Open circuit in the wiring harness from blower switch to engine control unit through CPU
  - Engine control unit (1S) terminal malfunction
- **YES** → Blower switch OK

2.6L ENGINE SWITCH MONITOR FUNCTION TROUBLESHOOTING – MPV

Headlight switch

Turn ON headlight switch
Check that monitor lamp illuminates

- NO → PC: • CPU malfunction
  • Headlight switch malfunction
  • Open circuit in wiring harness from headlight switch to engine control unit through CPU
  • Engine control unit (1S) terminal malfunction
- YES → Headlight switch OK
  Turn OFF headlight switch

Rear defroster switch

Turn ON rear defroster switch
Check that monitor lamp illuminates

- NO → PC: • CPU malfunction
  • Rear defroster switch malfunction
  • Open circuit in wiring harness from rear defroster switch to engine control unit through CPU
  • Engine control unit (1S) terminal malfunction
- YES → Rear defroster switch OK
  Turn OFF rear defroster switch

Inhibitor switch (A/T)

Shift transmission into gear
Check that monitor lamp illuminates

- NO → PC: • Inhibitor switch malfunction
  • Open or short circuit in related wiring harness
  • Control unit (1V) terminal malfunction
- YES → Inhibitor switch OK
  Shift into "P" or "N" range

2.6L ENGINE ENGINE CONTROL UNIT TERMINAL VOLTAGES – MPV

Terminal voltage

V_B: Battery voltage

Terminal	Input	Output	Connection to	Test condition	Voltage	Remarks
1A	—	—	Battery	Constant	V_B	For backup
1B	○		Main relay	Ignition switch OFF	0V	
				Ignition switch ON	V_B	
				During burn-off (airflow sensor)		
1C	○		Ignition switch (Start position)	While cranking	Approx. 10V	
				Ignition switch ON	0V	
1D		○	Self-Diagnosis Checker (Monitor lamp)	Test connector (Green: 1-pin) grounded For 3 seconds after ignition switch OFF→ON (Lamp illuminates)	Approx. 5V	With Self-Diagnosis Checker
				After 3 seconds (Lamp does not illuminated)	V_B	
				Test connector (Green: 1-pin) not grounded at idle. Monitor lamp ON	Approx. 5V	
				Test connector (Green: 1-pin) not grounded at idle. Monitor lamp OFF	V_B	
1E		○	Malfunction indicator lamp	For 3 seconds after ignition switch OFF→ON (Lamp illuminates)	Below 2.5V	Test connector (Green: 1-pin) grounded
				After 3 seconds (Lamp dose not illuminates)	V_B	
				Lamp illuminates	Below 2.5V	
				Lamp not illuminate	V_B	
1F		○	Self-Diagnosis checker (Code number)	For 3 seconds after ignition switch OFF→ON (Buzzer sounds)	Below 2.5V	• With Self-Diagnosis Checker • Test connector (Green: 1-pin) grounded
				After 3 seconds (Buzzer does not sounded)	V_B	
				Buzzer sounds	Below 2.5V	
				Buzzer not sounded	V_B	
1G		○	Main relay	Ignition switch OFF	V_B	
				During burn-off (airflow sensor) Ignition switch ON	0V	
1H		○	Circuit opening relay	Ignition switch ON	V_B	
				During cranking or at idle	Below 2.5V	
1I	○		Ignition switch (ON position)	Ignition switch OFF	0V	
				Ignition switch ON	V_B	
1J		○	A/C relay	Ignition switch ON	V_B	Blower motor: ON
				For 10 seconds After fully depressing accelerator pedal with A/C switch ON (A/C does not operate) (in-gear, ignition switch ON)	V_B	
				After 10 seconds	Below 2.5V	
				For 5 seconds after cranking with A/C switch ON (A/C does not operate)	V_B	
				After 5 seconds (A/C operates)	Below 2.5V	
				A/C switch ON at idle	Below 2.5V	
				A/C switch OFF at idle	V_B	
1K	○		Test connector	Test connector (Green: 1-pin) not grounded	V_B	Ignition switch ON
				Test connector (Green: 1-pin) grounded	0V	
1L	○		Open (A/T)	Ignition switch ON	V_B	
1M	○		Speed sensor (A/T)	Ignition switch ON	0 or 4.5V	
				Idle	Approx 4.5V	
1N	○		Idle switch	Accelerator pedal released	0V	Ignition switch ON
				Accelerator pedal depressed	V_B	
1O	○		Stoplight switch	Brake pedal released	0V	Ignition switch ON
				Brake pedal depressed	V_B	
1P	○		P/S pressure switch	Ignition switch ON	V_B	
				P/S ON (at idle)	0V	
				P/S OFF (at idle)	V_B	
1Q	○		A/C switch	A/C switch ON (Ignition switch ON)	Below 2.5V	Blower motor: ON
				A/C switch OFF (Ignition switch ON)	V_B	

2.6L ENGINE ENGINE CONTROL UNIT TERMINAL VOLTAGES—MPV

Terminal voltage

V_B: Battery voltage

Terminal	Input	Output	Connection to	Test condition	Voltage	Remarks
1R	—	—	—	—	—	—
1S	○		Central processing unit	Electrical load OFF	V_B	Ignition switch ON
				Electrical load ON	Below 1.5V	
1T	—	—	—	—	—	—
1U	—	—	—	—	—	—
1V	○		Neutral or clutch switch (Inhibitor switch)	Neutral or clutch pedal depressed (P or N ranges)	0V	Ignition switch ON
				Other condition	V_B	
2A	—	—	Ground (E01)	Constant	0V	
2B	—	—	Ground (E02)	Constant	0V	
2C	—	—	Ground (E1)	Constant	0V	
2D	—	—	Ground (E2)	Constant	0V	
2E		○	Distributor (Ne-signal)	Ignition switch ON	0 or 5V	Ne-Signal
				Idle	2V	
2F		○	Igniter	Ignition switch ON	1—2V	Ignition-timing signal
				Idle	Approx. 0.5V	
2G	○		Distributor (G-signal)	Ignition switch ON	0 or 5V	G-Signal
				Idle	V_B	
2H		○	Airflow sensor (Burn-off)	Just after ignition switch OFF	0V	Burn-off functions momentarily
				Burn off (2-5 seconds after ignition switch OFF) (Refer to page F2–170)	8—V_B	
2I	—	—	—	—	—	—
2J	—	—	—	—	—	—
2K		○	Vref	Ignition switch ON	4.5—5.5V	
2L	○		Intake air thermosensor (Dynamic chamber)	At 20°C (68°F)	Approx. 2.5V	
2M	○		Throttle sensor	Accelerator pedal released	Approx. 0.5V	Ignition switch ON
				Accelerator pedal fully depressed	Approx. 4.3V	
2N	○		Oxygen sensor	Ignition switch ON	0V	
				Idle (Cold engine)	0V	
				Idle (After warm up)	0—1.0V	Needle moves from 0V to 1V
				Increase engine speed (After warm up)	0.5—1.0V	
				Deceleration	0—0.4V	
2O	○		Airflow sensor (Intake air mass)	Ignition switch ON	1.0—2.0V	
				Idle (After warm up)	1.9—2.6V	
				Increase engine speed (After warm up)	2—5V	
2P	○		Airflow sensor (Ground)	Constant	0V	
2Q	○		Water thermosensor	Engine coolant termperautre 20°C (68°F)	Approx. 2.5V	Ignition switch ON
				After warm up	Approx. 0.4V	
2R	—	—	—	—	—	—
2S	—	—	—	—	—	—
2T		○	Solenoid valve (PRC)	For 120 seconds after ignition switch OFF→ON	Below 2.5V	During hot condition. Coolant temp. above 90°C (194°F) Intake air temp. above 65°C (149°F)
				For 120 seconds after starting	Below 2.5V	
				Ignition switch ON	V_B	Other conditions
2U		○	Injector (No.1, 3)	Ignition switch ON	V_B	* Engine Signal Monitor: Green and red lights flash
				Idle	V_B	

Terminal voltage

V_B: Battery voltage

Terminal	Input	Output	Connection to	Test condition	Voltage	Remarks
2V		○	Injector (No.2, 4)	Ignition switch ON	V_B	* Engine Signal Monitor: Green and red lights flash
				Idle	V_B	
2W		○	Solenoid valve (Idle speed control)	Ignition switch ON	Approx. 11V	Engine signal monitor: Green and red lights flash
				Idle	Approx. 10V	
2X		○	Solenoid valve (Purge control)	Ignition switch ON	V_B	
				Idle	V_B	* Engine signal monitor: Green and red lights flash
				Driving in gear	5—15V*	
2Y		○	HAT control unit	Ignition switch ON	V_B	
2Z	—	—	—	—	—	—

Terminal location

2Y	2W	2U	2S	2Q	2O	2M	2K	2I	2G	2E	2C	2A	1U	1S	1Q	1O	1M	1K	1I	1G	1E	1C	1A
2Z	2X	2V	2T	2R	2P	2N	2L	2J	2H	2F	2D	2B	1V	1T	1R	1P	1N	1L	1J	1H	1F	1D	1B

2.6L ENGINE CENTRAL PROCESSING UNIT—MPV

CENTRAL PROCESSING UNIT (CPU)

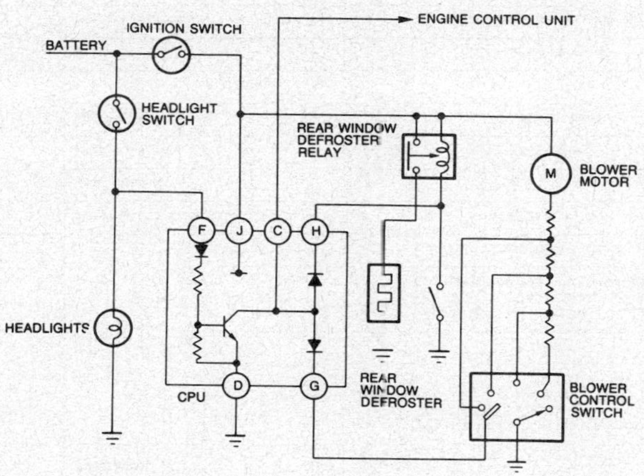

Operation
This CPU detects that an electrical load is being applied and signals the engine control unit to stabilize the idle. This CPU detects the electrical loads below.

Load	Condition
Headlights	Headlights ON (low or high beam)
Rear defroster	Defroster ON
Blower motor	Fan switch high or super high position

2.6L ENGINE CENTRAL PROCESSING UNIT TERMINAL VOLTAGES—MPV

VB: Battery voltage

Terminal	Connected to	Test condition	Specification	To correct
b	Battery (through ROOM 10A fuse)	Constant	VB	Check ROOM 10A fuse and wiring harness (Refer to Section T)
c	Engine control unit	Electrical load OFF (Ignition switch ON)	VB	Check engine control unit and wiring harness (Refer to page F2–180)
		Electrical load ON (Ignition switch ON)	Below 1.5V	
d	Ground	Constant	0V	Check wiring harness
f	Headlight switch	Headlight switch ON	VB	Check headlight switch and wiring harness (Refer to Section T)
g	Blower fan switch	Blower fan switch High or Super high position	0V	Check blower fan switch and wiring harness (Refer to Section U)
h	Rear window defroster switch	Rear window defroster switch ON	0V	Check rear window defroster switch and wiring harness (Refer to Section T)
j	Ignition switch	Ignition switch ON	VB	Check ignition switch and wiring harness (Refer to Section T)

3.0L ENGINE SPECIFICATION CHART—MPV

Item			Specification
Idle speed (test connector grounded) rpm			800 ± 20 (M/T: in Neutral) (A/T: in P range)
Air cleaner	Element type		Oil permeated
Throttle body	Throat diameter mm (in)		45 (1.772) x 2
Dashpot	Adjustment speed rpm		3,500 ± 300
Throttle sensor	Idle switch (adjustment clearance) mm (in)		0.5 (0.020) (Continuity) 0.7 (0.028) (No continuity)
	Resistance kΩ		Ⓐ— Ⓓ 3.5—6.5 Ⓑ— Ⓓ below 1 ↔ 3.0—6.0
Airflow meter	Resistance Ω	$E_2 \leftrightarrow V_S$	20—400 (Measuring plate fully closed), 20—1,000 (fully open)
		$E_2 \leftrightarrow V_C$	100—300
		$E_2 \leftrightarrow V_B$	200—400
		$E_2 \leftrightarrow THA$	−20°C (−4°F) 13,600—18,400 20°C (68°F) 2,210— 2,690 60°C (140°F) 493— 667
		$E_1 \leftrightarrow F_C$	Fully closed: ∞ Fully open: 0
Fuel tank	Capacity liters (US gal, Imp gal)		74 (19.6, 16.3)
Fuel filter	Low pressure		Nylon 6 (250 mesh)
	High pressure		Filter paper
Fuel pump	Type		Impeller (in tank)
	Output pressure kPa (kg/cm², psi)		441—588 (4.5—6.0, 64—85)
Pressure regulator	Type		Diaphragm
	Regulated pressure kPa (kg/cm², psi)		265—314 (2.7—3.2, 38—46)
Injector	Drive		Voltage
	Type		High-ohmic injector
	Resistance Ω		12—16
Ignition timing (test connector grounded)			10—12° BTDC (M/T: in Neutral) (A/T: in P range)
BAC valve	Resistance Ω		10.7—12.3
Fuel specification			Unleaded gasoline (RON 87 or higher)

3.0L ENGINE COMPONENT DESCRIPTIONS—MPV

Component	Function	Remark
Air cleaner	Filters air entering throttle body	
Airflow meter	Detects amount of intake air; sends signal to engine control unit	Intake air thermosensor and fuel pump switch included
Air valve	Supplies bypass air into dynamic chamber (engine cold)	• Engine speed increased to shorten warm-up period • Thermo wax type • Installed in BAC valve
Atmospheric pressure sensor	Detects atmospheric pressure; sends signal to engine control unit	Built in ECU
BAC valve	Supplies bypass air into dynamic chamber	Consists of air valve and ISC valve
Catalytic converter	Reduce HC, CO and NOx by chemical reaction	Monolith type
Central processing unit (CPU) (E/L control function)	Detects electrical load operation; sends signal to engine control unit	Electrical load • Rear defroster • Blower motor • Headlights
Charcoal canister	Store fuel tank fumes	
Check connectors	For Self-Diagnosis Checker and fuel pump	• 6-pin connector (Green); Self-Diagnosis Checker • 2-pin connector (Yellow); Fuel pump
Check valve	Prevents misoperation of shutter valve during heavy-load operation	
Check-and-cut valve	Controls pressure in fuel tank	
Circuit opening relay	Voltage for fuel pump (engine running)	
Clutch switch (M/T)	Detects clutch condition; sends signal to engine control unit	Switch opened when clutch pedal released
Engine control unit (ECU)	Detects the following; 1. A/C operation 2. Air/fuel ratio (oxygen concentration) 3. Atmospheric pressure 4. Braking signal 5. Cranking signal 6. E/L operation 7. Engine air temperature 8. Engine coolant temperature 9. Engine speed 10. In-gear condition 11. Initial set signal 12. Intake air amount 13. Intake air temperature 14. No.1 piston TDC 15. P/S operation 16. Throttle valve fully closed 17. Throttle valve opening angle Controls operation of the following: 1. Fail-safe system 2. Fuel injection system 3. Idle speed control system 4. Ignition system (atmospheric pressure sensor) 5. Pressure regulator control system 6. Purge control system 7. Switch monitor function 8. Variable resonance induction system	1. A/C switch 2. Oxygen sensor 3. Atmospheric pressure sensor 4. Stoplight switch 5. Ignition switch 6. E/L switch, central processing unit (CPU) 7. Intake air thermosensor (on dynamic chamber) 8. Water thermosensor 9. Ignition coil 10. Neutral, clutch, and inhibitor switches 11. Test connector 12. Airflow meter 13. Intake air thermosensor (in airflow meter) 14. G rotors (in distributor) 15. P/S pressure switch 16. Idle switch (in throttle sensor) 17. Throttle sensor 1. Self-Diagnosis Checker and MIL 2. Injector 3. BAC valve 4. Igniter 5. Solenoid valve (pressure regulator control) 6. Solenoid valve (purge control) 7. Monitor lamp (Self-Diagnosis Checker) 8. Solenoid valve (VRIS)

3.0L ENGINE COMPONENT DESCRIPTIONS—MPV

Component	Function	Remark
Distributor	Detects crank angle at 30° intervals; No.1 cylinder TDC and No.4 cylinder TDC; sends signal to engine control unit	G-signal
Dashpot	Causes gradual throttle valve closing during deceleration	Adjustment speed; 3,500 ± 300 rpm
Dynamic chamber	Interconnects all cylinders	Includes shutter valve
Fuel filter	Filters fuel	
Fuel pump	Provides fuel to injectors	• Operates only while engine running • Installed in fuel tank
G-coil	Detects No.1 cylinder TDC; sends signal to engine control unit	For determining fuel injection timing
Idle switch	Detects when throttle valve fully closed; sends signal to engine control unit	Installed in throttle sensor. Switch closed when throttle valve fully closed
Igniter	Receives spark signal of signal rotor and controls primary current flow in ignition coil	Controls spark advance by signal of ATP sensor.
Ignition switch	Detects engine cranking; sends signal to engine control unit	
Inhibitor switch (A/T)	Detects in-gear condition; sends signal to engine control unit	Switch closed in N or P range (EC-AT)
Injector	Injects fuel into intake port	• Controlled by signals from engine control unit • High-ohmic injector
Intake manifold	Supplies intake air to all cylinders	
Intake air thermosensor	Detects intake air temperature; sends signal to engine control unit	Installed in airflow meter
	Detects engine air temperature; sends signal to engine control unit	Installed on dynamic chamber
Main relay	Supplies current to injectors and engine control unit	
MIL (Malfunction indicator lamp)	Illuminates when input device malfunctions	Installed in instrument cluster
Neutral switch (M/T)	Detects in-gear condition; sends signal to engine control unit	Switch open when in-gear
Oxygen sensor	Detects oxygen concentration; sends signal to engine control unit	Zirconia ceramic and platinum coating
PCV valve	Controls blowby gas amount pulled into engine	
P/S pressure switch	Detects P/S operation; sends signal to engine control unit	ON when steering wheel turned right or left
Pressure regulator	Adjusts fuel pressure supplied to injectors	
Pulsation damper	Absorbs fuel pulsations	
Separator	Prevents fuel from flowing into charcoal canister	
Shutter valve actuator	Controls shutter valve operation	For variable resonance induction system (VRIS)
Solenoid valve — Idle speed control	Supplies bypass air into dynamic chamber	• Controlled by duty signal from engine control unit • Works independent of idle-up • Installed in BAC valve
Solenoid valve — Pressure regulator control	Blocks vacuum from intake manifold to pressure regulator	Hot engine only
Solenoid valve — Purge control	Controls evaporative fumes from canister to intake manifold	
Solenoid valve — VRIS	Cuts vacuum to shutter valve actuator	Variable resonance induction system (VRIS)
Stoplight switch (M/T)	Detects braking (deceleration); sends signal to engine control unit	
Test connector	For Self-Diagnosis Checker and idle speed adjustment	1-pin connector (Green)
Throttle body	Controls intake air amount	Integrated throttle sensor and dashpot incorporated
Throttle sensor	Detects throttle valve opening angle; sends signal to engine control unit	• Idle switch included • Installed on throttle body
Water thermosensor	Detects coolant temperature; sends signal to engine control unit	

3.0L ENGINE ENGINE CONTROL OPERATION CHART MPV

ENGINE CONDITIONS \ INPUT DEVICES	APPROXIMATE TIME (BASED ON 10–16°C, 50–60°F AMBIENT TEMPERATURE)	IGNITION COIL	THROTTLE SENSOR: IDLE SWITCH	THROTTLE SENSOR: THROTTLE POSITION SENSOR	WATER THERMO-SENSOR	OXYGEN SENSOR	AIRFLOW METER: AIRFLOW SENSOR	AIRFLOW METER: INTAKE AIR THERMO-SENSOR	ATMOSPHERIC PRESSURE SENSOR
CRANKING COLD ENGINE · COLD AIR · COLD COOLANT	ZERO		SIGNAL HAS NO EFFECT ON ECU	SIGNAL HAS NO EFFECT ON ECU		SIGNAL HAS NO EFFECT ON ECU	SIGNAL HAS NO EFFECT ON ECU	SIGNAL HAS NO EFFECT ON ECU	
COLD START FAST IDLE · COLD AIR · COLD COOLANT	ONE MINUTE		LOW VOLTAGE SIGNAL TO ECU (BELOW 1.5V)	CLOSED THROTTLE: LOW VOLTAGE (0.3—0.7V)	COOL TO WARM: MEDIUM VOLTAGE (3V AND DROPPING)	SENSOR COLD: LOW TO HIGH VOLTAGE (0—0.8V)	LOW VOLUME AIRFLOW: LOW-TO MEDIUM VOLTAGE (4.0—7.0V)		
COLD DRIVE AWAY PART THROTTLE · COLD AIR · COLD COOLANT	TWO MINUTES								
WARM DRIVE AWAY PART THROTTLE (25%) · WARM INTAKE AIR · WARM COOLANT	THREE MINUTES				WARM: MEDIUM VOLTAGE (APPROX. 0.7V AND DROPPING)	SENSOR WARM: HIGH VOLTAGE (ABOVE 0.6V)	MODERATE VOLUME AIR FLOW: LOW-TO-MEDIUM VOLTAGE (6.1—8.5V)		SENDS VOLTAGE SIGNAL TO ECU THAT VARIES WITH ALTITUDE: VOLTAGE (APPROX. 4V)
HOT CRUISE · WARM INTAKE AIR · HOT COOLANT		SENDS ENGINE RPM SIGNAL TO ECU	HIGH: VOLTAGE SIGNAL TO ECU (VB)	PART THROTTLE: MEDIUM VOLTAGE (1—2.5V)		SENSOR HOT: SWITCHING FROM HIGH VOLTAGE (ABOVE 0.6V RICH) TO LOW VOLTAGE (BELOW 0.4V LEAN)		COOL TO WARM: MEDIUM VOLTAGE (1.4—3V)	
HOT ACCELERATION · 60% THROTTLE							MODERATE-TO-HIGH VOLUME OF AIR FLOW: (8.1—8.7V)		
HOT ACCELERATION · WIDE OPEN THROTTLE	MORE THAN FOUR MINUTES			WIDE OPEN THROTTLE (APPROX. 4.0V)	HOT: LOW VOLTAGE (APPROX. 0.5V)	HIGH VOLTAGE (ABOVE 0.6V RICH)	HIGH VOLUME OF AIR FLOW: (8.4—9.0V)		
DECELERATION · CLOSED THROTTLE			LOW VOLTAGE SIGNAL TO ECU (BELOW 1.5V)	CLOSED THROTTLE: LOW VOLTAGE (0.3—0.7V)		LOW VOLTAGE (BELOW 0.4V LEAN)	LOW VOLUME AIR FLOW: (4.0—5.0V)		
HOT CURB IDLE (EXTENDED)						SWITCHING FROM HIGH TO LOW VOLTAGE (0.4—0.6V)			
HOT ENGINE SHUTDOWN		OFF	OFF	OFF	OFF	SENSOR HOT: LOW VOLTAGE (0.1V LEAN) UNTIL SENSOR COOLS DOWN	OFF	OFF	OFF

3.0L ENGINE TROUBLESHOOTING GUIDE—MPV

OUTPUT DEVICE \ INPUT DEVICE		DISTRIBUTOR: NE SIGNAL	DISTRIBUTOR: G1, G2 SIGNAL	AIR-FLOW METER	THROTTLE SENSOR: IDLE SWITCH	THROTTLE SENSOR: THROTTLE SENSOR	WATER THERMO SENSOR	IGN SW (START)	INTAKE AIR THERMOSENSOR: AIR-FLOW METER	INTAKE AIR THERMOSENSOR: DYNAMIC CHAMBER	OXYGEN SENSOR	ATP SENSOR (IN-ECU)	STOP-LIGHT SW (M/T)	TEST CONNECTOR	NEUTRAL CLUTCH (M/T) INHIBITOR (A/T) SW	A/C SW	P/S PRESSURE SW	E/L SIGNAL
INJECTOR	FUEL INJECTION AMOUNT	○	X	○	○	○	○	○	○	○	○	○	○	○	○	X	X	X
	FUEL INJECTION TIMING	○	○	X	X	X	X	○	X	X	X	X	X	X	X	X	X	X
PRESSURE REGULATOR		○	X	X	X	○	○	○	X	○	X	X	X	X	○	X	X	X
BAC VALVE	AIR VALVE	X	X	X	X	X	X	X	X	X	X	X	X	X	X	X	X	X
	ISC VALVE	○	X	X	○	○	○	○	○	○	X	○	○	○	○	○	○	○
SOLENOID VALVE (VRIS)		○	X	X	X	X	X	X	X	X	X	X	X	X	X	X	X	X
IGNITER		1	1	1	X	X	1	1	X	X	X	○	X	○	1	X	X	X
SOLENOID VALVE (PURGE)		○	X	○	○	○	○	X	○	X	○	○	X	X	○	X	X	X
OXYGEN SENSOR HEATER		○	X	○	X	X	○	○	X	X	X	X	X	X	X	X	X	X
A/C RELAY		X	X	X	X	○	X	○	X	X	X	X	X	X	○	○	X	X

3.0L ENGINE ENGINE CONTROL OPERATION CHART—MPV

INPUT DEVICES

DISTRIBUTOR (G-SIGNAL)	NEUTRAL AND CLUTCH SWITCH (M/T)	INHIBITOR SWITCH (A/T)	STOPLIGHT SWITCH (M/T)	A/C SWITCH	P/S PRESSURE SWITCH	CPU (E/L SIGNAL)	IGNITION SWITCH: START POSITION	IGNITION SWITCH: ON POSITION	TEST CONNECTOR
↕	SIGNAL HAS NO EFFECT ON ECU	SIGNAL HAS NO EFFECT ON ECU	SIGNAL HAS NO EFFECT ON ECU	SIGNAL HAS NO EFFECT ON ECU	SIGNAL HAS NO EFFECT ON ECU	SIGNAL HAS NO EFFECT ON ECU	SENDS SIGNAL TO ECU	SIGNAL HAS NO EFFECT ON ECU	SIGNAL HAS NO EFFECT ON ECU
↕	IN NEUTRAL: SENDS SIGNAL TO ECU (VB)	IN N or P SENDS SIGNAL TO ECU (VB)	BRAKE PEDAL DEPRESSED: SENDS SIGNAL TO ECU (VB)	↕	↕	↕	↕	↕	↕
SENDS NO.1 CYLINDER TDC SIGNAL TO ECU	DRIVING IN ANY GEAR: NO VOLTAGE SEND SIGNAL TO ECU (BELOW 1.5V)	IN D,R SECOND AND LOW: NO VOLTAGE SIGNAL TO ECU (BELOW 1.5V)	NO SIGNAL TO ECU (BELOW 1.5V)	A/C SWITCH ON: SENDS SIGNAL TO ECU (VB) A/C SWITCH OFF: NO SIGNAL TO ECU (BELOW 1.5V)	STEERING WHEEL TURNED: LOW VOLTAGE SIGNAL TO ECU (BELOW 1.5V) STEERING WHEEL STRAIGHT AHEAD: HIGH VOLTAGE SIGNAL TO ECU (VB)	HEADLIGHTS, BLOWER OR REAR DEFOGER ON: LOW VOLTAGE SIGNAL TO ECU (BELOW 2.5V) HEADLIGHTS, BLOWER, AND REAR DEFOGER OFF: HIGH VOLTAGE SIGNAL TO ECU (VB)	NO SIGNAL TO ECU (BELOW 1.5V)	SENDS SIGNAL TO ECU (VB)	WHEN CONNECTOR NOT GROUNDED: HIGH VOLTAGE SIGNAL TO ECU (VB)
↕	IN NEUTRAL: SENDS SIGNAL TO ECU (VB)	IN N OR P RANGE; SENDS SIGNAL TO ECU (BELOW 1.5V)	BRAKE PEDAL DEPRESSED: SENDS SIGNAL TO ECU (VB)	↕	↕	↕	↕	↕	WHEN CONNECTOR GROUNDED: LOW VOLTAGE SIGNAL TO ECU (BELOW 1.5V)
OFF	OFF	OFF	OFF	OFF	OFF	OFF	OFF	OFF	OFF

OUTPUT DEVICES

ENGINE CONDITIONS / OUTPUT DEVICES	APPROXIMATE TIME BASED ON 10–16°C or 50–60°F AMBIENT TEMPERATURE	INJECTOR: INJECTION	INJECTOR: INJECTION TIMING	BAC VALVE: AIR VALVE	BAC VALVE: ISC VALVE	A/C CUT RELAY	SOLENOID VALVE (PURGE CONTROL)	SOLENOID VALVE (PRESSURE REGULATOR CONTROL)	MAIN RELAY
CRANKING COLD ENGINE · COLD AIR · COLD COOLANT	ZERO	↕	1-Group	↕	↕	OFF (A/C OFF)	↕	↕	↕
COLD START FAST IDLE · COLD AIR · COLD COOLANT	ONE MINUTE	RICH	↕	OPEN (COOLANT TEMP: BELOW 60°C [140°F])	LARGE AMOUNT OF BY-PASS AIR	↕	OFF (PURGE CUT)	↕	↕
COLD DRIVEAWAY PART THROTTLE · COLD AIR · COLD COOLANT	TWO MINUTES	↕	↕	↕	↕	↕	↕	↕	↕
WARM DRIVEAWAY PART THROTTLE (25%) · WARM INLET AIR · WARM COOLANT	THREE MINUTES	RICH and LEAN	↕	↕	↕	ON (A/C ON)	↕	↕	↕
HOT CRUISE · WARM INLET AIR · HOT COOLANT	↕	↕	3-Group	↕	↕	↕	OPERATES (DUTY VALUES) (PURGE GAS AMOUNT CHANGES)	OFF (VACUUM TO PRESSURE REGULATOR)	↕
HOT ACCELERATION · 60% THROTTLE	↕	RICH	↕	↕	SMALL AMOUNT OF BY-PASS AIR	↕	↕	↕	ON
HOT ACCELERATION · WIDE OPEN THROTTLE	MORE THAN FOUR MINUTES	↕	↕	↕	↕	OFF (A/C OFF)	OFF (PURGE CUT)	↕	↕
DECELERATION · CLOSED THROTTLE	↕	FUEL CUT	FUEL CUT	CLOSE	LARGE AND SMALL AMOUNT OF BY-PASS AIR	↕	ON (SMALL)	↕	↕
HOT CURB IDLE (EXTENDED)	↕	RICH and LEAN	3-Group	↕	SMALL AMOUNT OF BY-PASS AIR	ON (A/C ON)	OFF (PURGE CUT)	AFTER STARTING ON (DURING HOT START ONLY) (VACUUM CUT)	↕
HOT ENGINE SHUTDOWN		NOT INJECTION	NOT INJECTION	↕	OFF	ON	OFF	OFF	OFF

3.0L ENGINE ENGINE CONTROL OPERATION CHART MPV

<table>
<tr><th rowspan="2">ENGINE CONDITIONS \ OUTPUT DEVICES</th><th colspan="3">OUTPUT DEVICES</th></tr>
<tr><th>IGNITER</th><th>SOLENOID VALVE (V.R.I.S)</th><th>OXYGEN SENSOR HEATER</th></tr>
<tr><td>CRANKING
COLD ENGINE
· COLD AIR
· COLD COOLANT</td><td rowspan="9">ON (MORE THAN 1,000 m (3,280 ft) ABOVE SEA LEVEL)</td><td rowspan="2">OFF (V.R.I.S OFF)</td><td>OFF</td></tr>
<tr><td>COLD START
FAST IDLE
· COLD AIR
· COLD COOLANT</td><td rowspan="2">ON (BELOW 2,250 rpm)</td></tr>
<tr><td>COLD DRIVEAWAY
PART THROTTLE
· COLD AIR
· COLD COOLANT</td><td rowspan="2">ON (V.R.I.S ON, 3,600–5,650 rpm ONLY)</td></tr>
<tr><td>WARM DRIVEAWAY
PART THROTTLE (25%)
· WARM INLET AIR
· WARM COOLANT</td><td rowspan="6">OFF</td></tr>
<tr><td>HOT CRUISE
· WARM INLET AIR
· HOT COOLANT</td><td rowspan="3">OFF (V.R.I.S OFF)</td></tr>
<tr><td>HOT ACCELERATION
· 60% THROTTLE</td></tr>
<tr><td>HOT ACCELERATION
· WIDE OPEN THROTTLE</td></tr>
<tr><td>DECELERATION
· CLOSED THROTTLE</td><td>ON (V.R.I.S ON, 3,600–5,650 rpm ONLY)</td></tr>
<tr><td>HOT CURB IDLE (EXTENDED)</td><td>OFF (V.R.I.S OFF)</td></tr>
<tr><td>HOT ENGINE SHUTDOWN</td><td>OFF</td><td>OFF</td><td>OFF</td></tr>
</table>

3.0L ENGINE SYMPTOM TROUBLESHOOTING—MPV

Cranks normally but won't start (No combustion)

STEP	QUICK INSPECTION		ACTION		POSSIBLE CAUSE AND DETAILED INSPECTION
1	Check malfunction code (01) with SST [IG ON, Test connector (Green: 1-pin) grounded]	Yes	Go to Next Step		
		No	Check for the cause by referring to the check sequence		
2	Check for spark by disconnecting high-tension leads while cranking	Yes	Go to Next Step		
		No	Check ignition system (Refer to ignition system troubleshooting)		
3	Check for fuel pump operating sound from fuel filler port [IG ON, Test connector (Yellow: 2-pin) connected]	Yes	Check if the engine starts with this condition	Yes	Circuit opening relay and wiring (Starting circuit)
					Go to Next Step
		No	Check circuit opening relay switching operation	Yes	Wiring open (Check relay circuit)
					Fuel pump malfunction
					Wiring open (Check fuel pump circuit)
				No	Circuit opening relay malfunction (Replace)
4	Check fuel line pressure [IG ON, Test connector (Yellow: 2-pin) connected] **Fuel line pressure: 265—314 kPa (2.7—3.2 kg/cm^2, 38—46 psi)**	Yes	Go to Next Step		
		No	Check fuel pump maximum pressure **Fuel pump max. pressure: 441—589 kPa (4.5—6.0 kg/cm^2, 64—85 psi)**	Yes	Pressure regulator malfunction (Replace)
				No	Fuel pump malfunction (Replace)
5	Check for injector operating sound while cranking	Yes	Go to Next Step		
		No	Check voltage at ECU 3C, 3E and 3F terminals with SST **Voltage: Vв (IG ON)**	Yes	Throttle sensor malfunction
					ECU malfunction (Replace)
				No	Wiring open or short
					Poor ground circuit from ECU 3A terminal
6	Check continuity between check connector [for tachometer (Black: 1-pin)] and ground (Ign ON)	Yes			Wiring short to ground
		No	Go to Next Step		
7	Substitute a well-known ECU Check if the condition becomes better	Yes			ECU malfunction
		No	Check ground circuit from ECU 2R terminal with SST **Voltage: 0V (Ign ON)**	Yes	Go to Next Step
				No	Poor ground circuit
8					Low compression

Cranks normally but hard to start (Always)

STEP	QUICK INSPECTION		ACTION		POSSIBLE CAUSE AND DETAILED INSPECTION
1	Check ignition timing at idle after warm up **Ignition timing: 10—12° BTDC** [Test connector (Green: 1-pin) grounded]	Yes	Go to Next Step		
		No	Adjust ignition timing		
	Note: At high-altitude area **[1,000 m (3,280 ft)]**, ignition timing advances to **approx. 20° BTDC** when test connector (Green: 1-pin) is not grounded				
2	Disconnect high-tension lead of each cylinder at idle Check if the engine condition changes	Yes	Go to Next Step		
		No	Check ignition system	Yes	Injector clogging (Replace)
				No	Spark plug
					High-tension lead
					Distributor cap
3	Check injector operating sound at idle	Yes	Go to Next Step		
		No	Check injector resistance	Yes	Wiring short or open
				No	Injector malfunction (Replace)
4	Check fuel line pressure [IG ON, Test connector (Yellow: 2-pin) connected] **Fuel line pressure:** **265—314 kPa (2.7—3.2 kg/cm², 38—46 psi)**	Yes	Go to Next Step		
		No	Check fuel pump max. pressure **Fuel pump max. pressure:** **441—589 kPa (4.5—6.0 kg/cm², 64—85 psi)**	Yes	Pressure regulator malfunction (Replace)
				No	Fuel pump malfunction (Replace)
5	Operate the fuel pump [IG ON, Test connector (Yellow: 2-pin) connected] Turn the ignition switch OFF and observe fuel pressure for 5 minutes **Fuel pressure:** **More than 147 kPa (1.5 kg/cm², 21 psi)**	Yes	Go to Next Step		
		No	Check fuel pump pressure drop	No	Fuel pump malfunction (Replace)
			Check pressure regulator pressure drop	Yes	Injector fuel leaks
				No	Pressure regulator malfunction (Replace)
6					Low compression
7	Substitute a well-known ECU				

Cranks normally but hard to start (only when engine cold)

STEP	QUICK INSPECTION		ACTION		POSSIBLE CAUSE AND DETAILED INSPECTION
1	Check gravity of battery by using a hydrometer **Specific gravity: Above 1.200**	Yes	Go to Next Step		
		No	Recharge the battery		
2	Check malfunction codes (09) (26) with SST [IG ON, Test connector (Green: 1-pin) grounded]	Yes	Go to Next Step		
		No	Check for the cause by referring to the check sequence		
3	Disconnect high-tension lead of each cylinder at idle Check if the engine condition changes	Yes	Go to Next Step		
		No	Check ignition system	Yes	Injector clogging (Replace)
				No	Spark plug
					High-tension lead
					Distributor cap
4	Check fuel line pressure [IG ON, Test connector (Yellow: 2-pin) connected] **Fuel line pressure:** **265—314 kPa (2.7—3.2 kg/cm², 38—46 psi)**	Yes	Go to Next Step		
		No	Check fuel pump max. pressure **Fuel pump max. pressure:** **441—589 kPa (4.5—6.0 kg/cm², 64—85 psi)**	Yes	Pressure regulator malfunction (Replace)
				No	Fuel pump malfunction (Replace)
5	Disconnect the ISC valve connector Check idle speed during warm up	Yes	Go to Next Step		
		No			BAC valve stuck close
6	Substitute a well-known ECU				

Cranks normally but hard to start (only when engine warm)

STEP	QUICK INSPECTION		ACTION		POSSIBLE CAUSE AND DETAILED INSPECTION
1	Check malfunction code (26) with SST [IG ON, Test connector (Green: 1-pin) grounded]	Yes	Go to Next Step		
		No	Check for the cause by referring to the check sequence		
2	Disconnect water thermosensor connector Check if the condition becomes better	Yes	Check water thermosensor connector for connecting condition as follows 1. Shake connector and check if the condition changes 2. Check condition of terminal (burning or damaged) 3. Connect a male terminal to the connector and check for connection condition	Yes	Water thermosensor malfunction
				No	Poor contact at water thermosensor connector
		No	Go to Next Step		
3	Operate the fuel pump [IG ON, Test connector (Yellow: 2-pin) connected] Turn the ignition switch OFF and observe fuel pressure for 5 minutes **Fuel pressure:** **More than 147 kPa (1.5 kg/cm², 21 psi)**	Yes	Go to Next Step		
		No	Check fuel pump pressure drop	No	Fuel pump malfunction (Replace)
			Check pressure regulator pressure drop	Yes	Injector fuel leaks
				No	Pressure regulator malfunction (Replace)
4	Substitute a well-known ECU				

Cranks normally but hard to start (only after heat soak)

STEP	QUICK INSPECTION		ACTION		POSSIBLE CAUSE AND DETAILED INSPECTION
1	Check malfunction code (09) (11) (25) (26) with SST [IG ON, Test connector (Green: 1-pin) grounded]	Yes	Go to Next Step		
		No	Check for the cause by referring to the check sequence		
2	Circulate the fuel by operating fuel pump **for 20 sec.** [IG ON, Test connector (Yellow: 2-pin) connected] Check if the condition becomes better	Yes	Go to Step 4		
		No	Go to Next Step		
3	Disconnect vacuum hose from pressure regulator Check if the condition becomes better	Yes	Check the components related to pressure regulator control system		Water thermosensor malfunction
					Intake air thermo-sensor malfunction
					Solenoid valve (PRC) malfunction
					ECU malfunction (Check 2N terminal voltage)
		No	Go to Next Step		
4	Operate the fuel pump [IG ON, Test connector (Yellow: 2-pin) connected] Turn the ignition switch OFF and observe fuel pressure **for 5 min.** **Fuel pressure:** **More than** **147 kPa (1.5 kg/cm², 21 psi)**	Yes	Go to Next Step		
		No	Check fuel pump pressure drop	No	Fuel pump malfunction (Replace)
			Check pressure regulator pressure drop	Yes	Injector fuel leaks
				No	Pressure regulator malfunction (Replace)
5	Change fuel according to specifications Check if the condition becomes better	Yes			Poor fuel quality
		No			
6	Substitute a well-known ECU				

Cranks normally but won't start (Intermittent)

STEP	QUICK INSPECTION		ACTION	POSSIBLE CAUSE AND DETAILED INSPECTION
1	Shake connector of ignition coil, main relay and ECU while cranking Check if the engine starts	Yes	There may be a poor contact at the connector Repair or replace the wiring	
		No	Go to troubleshooting "Cranks normally but hard to start (always)"	

Rough idle (Always)

STEP	QUICK INSPECTION		ACTION		POSSIBLE CAUSE AND DETAILED INSPECTION
1	Check malfunction code (08) (09) (14) (17) (26) (34) with SST [IG ON, Test connector (Green: 1-pin) grounded]	Yes	Check for the cause by referring to the check sequence		
		No	"88" flashing Check voltage at ECU 3G terminal with SST **Voltage:** **0V (IG ON)**	Yes	ECU malfunction (Replace)
				No	Poor ground circuit
			"00" Go to Next Step		
2	Check ignition timing at idle after warm up **Ignition timing: 10—12° BTDC** [Test connector (Green: 1-pin) grounded]	Yes	Go to Next Step		
		No	Adjust ignition timing (If possible)		
3	Disconnect high-tension lead of each cylinder at idle Check if the condition changes	Yes	Go to Next Step		
		No	Check ignition system	Yes	Injector clogging (Replace)
				No	Spark plug
					High-tension lead
					Distributor cap
4	Check idle speed after warm up **Idle speed: 780—820 rpm** **(M/T: in Neutral)** **(A/T: in P range)** [Test connector (Green: 1-pin) grounded]	Yes	Go to Next Step		
		No	Adjust idle speed (If possible)		
5	Check injector operating sound at idle	Yes	Go to Next Step		
		No	Check injector resistance	Yes	Wiring short or open
				No	Injector malfunction (Check resistance)
					Wiring short or open
6	Check fuel line pressure [IG ON, Test connector (Yellow: 2-pin) connected] **Fuel line pressure:** **265—314 kPa (2.7—3.2 kg/cm², 38—46 psi)**	Yes	Go to Next Step		
		No			Fuel leaks
					Fuel filter clogging
			Check fuel pump max. pressure **Fuel pump max. pressure:** **441—589 kPa (4.5—6.0 kg/cm², 64—85 psi)**	Yes	Pressure regulator malfunction (Replace)
				No	Fuel pump malfunction (Replace)
7	Check intake manifold vacuum at idle **Vacuum: 460—500 mmHg (18.1—19.7 inHg)**	Yes	Go to Next Step		
		No	Check for air leaks	Yes	Throttle valve
				No	Vacuum hoses
					Gaskets
					Bolts or nuts
					Component broken
8	Substitute a well-known ECU Check if the condition becomes better	Yes			ECU malfunction
		No	Check voltage at ECU 3G terminal with SST **Voltage:** **0V (Ign ON)**	Yes	Check compression
				No	Poor ground circuit

Rough idle (only when engine is cold)

STEP	QUICK INSPECTION		ACTION		POSSIBLE CAUSE AND DETAILED INSPECTION
1	Check malfunction code (08) (09) (26) with SST [IG ON, Test connector (Green: 1-pin) grounded]	Yes	Go to Next Step		
		No	Check for the cause by referring to the check sequence		
2	Check ignition timing advance (Centrifugal advance)	Yes	Go to Next Step		
		No			Distributor malfunction
3	Disconnect high-tension lead of each cylinder at idle Check if the engine condition changes	Yes	Go to Next Step		
		No	Check ignition system	Yes	Injector clogging
				No	Spark plug
					High-tension lead
					Distributor cap
4	Check injector operating sound at idle	Yes	Go to Next Step		
		No	Check injector resistance	Yes	Wiring short or open
				No	Injector malfunction (Check resistance)
					Wiring short or open
5	Disconnect ISC valve connector and check idle speed during warm up	Yes	Go to Next Step		
		No			BAC valve stuck close
6	Disconnect water thermosensor connector Check if the condition becomes better	Yes	Check water thermosensor connector for connecting condition as follows 1. Shake connector and check if the condition changes 2. Check condition of terminal (burning or damaged) 3. Connect a male terminal to the connector and check for connection condition	Yes	Water thermosensor malfunction
				No	Poor contact at water thermosensor connector
		No	Go to Next Step		
7	Check if the specified engine oil is used	Yes	Go to Next Step		
		No	Change engine oil with specified one		
8	Substitute a well-known ECU Check if the condition becomes better	Yes			ECU malfunction
		No			Airflow meter malfunction

Rough idle (only when engine is warm)

STEP	QUICK INSPECTION		ACTION		POSSIBLE CAUSE AND DETAILED INSPECTION
1	Run the engine at **2,000 rpm for more than 20 sec.** Check malfunction code (08) (09) (15) (17) (26) with SST [IG ON, Test connector (Green: 1-pin) grounded]	Yes	Go to Next Step		
		No	Check for the cause by referring to the check sequence		
2	Check idle speed after warm up **Idle speed: 780—820 rpm (M/T: in Neutral) (A/T: in P range)** [Test connector (Green: 1-pin) grounded]	Yes	Go to Next Step		
		No	Adjust idle speed (if possible)		
3	Check monitor lamp flashing with SST after warm up **Monitor lamp: Flashes ON and OFF more than 8 times/10 seconds at 2,000—3,000 rpm** [Test connector (Green: 1-pin) not grounded]	Yes	Go to Next Step		
		No			Oxygen sensor deterioration
4	Disconnect ISC valve connector Check if the engine speed drops	Yes	Go to Next Step		
		No			ISC valve malfunction
5	Disconnect water thermosensor connector Check if the condition becomes better	Yes	Check water thermosensor connector for connecting condition as follows 1. Shake connector and check if the condition changes 2. Check condition of terminal (burning or damaged) 3. Connect a male terminal to the connector and check for connection condition	Yes	Water thermosensor malfunction
				No	Poor contact at water thermosensor connector
		No	Go to Next Step		
6	Disconnect high-tension lead of each cylinder at idle Check if the engine condition changes	Yes	Go to Next Step		
		No	Check ignition system	Yes	Injector clogging
				No	Spark plug
					High-tension lead
					Distributor cap
			Note: If spark plug is wet, it may be from injector fuel leaks		
7	Check for injector operating sound at idle	Yes	Go to Next Step		
		No	Check injector resistance	Yes	Wiring short or open
				No	Injector malfunction (Check resistance)
					Wiring short or open
8	Check intake manifold vacuum at idle **Vacuum: 460—500 mmHg (18.1—19.7 inHg)**	Yes	Go to Next Step		
		No	Check for air leaks	Yes	Throttle valve
				No	Vacuum hoses
					Gaskets
					Bolts or nuts loosen
					Components damaged
9	Substitute a well-known ECU				ECU malfunction

Rough idle (only after heat soak)

STEP	QUICK INSPECTION		ACTION		POSSIBLE CAUSE AND DETAILED INSPECTION
1	Run the engine at **2,000 rpm for more than 20 sec.** Check malfunction code (09) (11) (15) (17) (25) (26) with SST [IG ON, Test connector (Green: 1-pin) grounded]	Yes	Go to Next Step		
		No	Check for the cause by referring to the check sequence		
2	Check switches with SST Neutral switch (M/T) Clutch switch (M/T) Inhibitor switch (A/T) [IG ON, Test connector (Green: 1-pin) grounded]	Yes	Go to Next Step		
		No	Check for the cause by referring to the check sequence		
3	Check monitor lamp flashing with SST after warm up **Monitor lamp:** **Flashes ON and OFF more than 8 times/10 seconds at 2,000—3,000 rpm** [Test connector (Green: 1-pin) not grounded]	Yes	Go to Next Step		
		No			Oxygen sensor deterioration
4	Disconnect vacuum hose from pressure regulator Check if the condition becomes better	Yes	Check components related to pressure regulator control system		Water thermosensor malfunction
					Intake air thermo-sensor malfunction
					Solenoid valve (PRC) malfunction
					ECU malfunction (Check 2N terminal voltage)
		No	Go to Next Step		
5	Run the engine at idle and stop it Observe fuel pressure **for 5 min.** **Fuel pressure:** **More than 147 kPa (1.5 kg/cm², 21 psi)**	Yes	Go to Next Step		
		No	Check fuel pump pressure drop	No	Fuel pump malfunction (Replace)
			Check pressure regulator pressure drop	Yes	Injector fuel leaks
				No	Pressure regulator malfunction
6	Disconnect high-tension lead of each cylinder at idle Check if the condition changes	Yes	Go to Next Step		
		No	Check ignition system	Yes	Injector clogging/Replace (If step 3 OK)
				No	Spark plug
					High-tension lead
					Distributor cap
7	Check for injector operating sound at idle	Yes	Go to Next Step		
		No	Check injector resistance	Yes	Wiring short or open
				No	Injector malfunction (Check resistance)
8	Change fuel to specified one				Poor fuel quality

Rough idle just after starting

STEP	QUICK INSPECTION		ACTION		POSSIBLE CAUSE AND DETAILED INSPECTION
1	Check malfunction code (09) (34) with SST [IG ON, Test connector (Green: 1-pin) grounded]	Yes	Go to Next Step		
		No	Check for the cause by referring to the check sequence		
2	Check idle switch with SST [IG ON, Test connector (Green: 1-pin) grounded]	Yes	Go to Next Step		
		No	Check for the cause by referring to the check sequence		
3	Check ignition timing at idle after warm up **Ignition timing:** **10—12° BTDC at sea level** **Approx. 20° BTDC at high altitude** **[1,000 m (3,280 ft)]** [Test connector (Green: 1-pin) grounded]	Yes	Go to Next Step		
		No	Check ignition timing after warm up **Ignition timing:** **10—12° BTDC** [Test connector (Green: 1-pin) grounded]	Yes	Atmospheric pressure sensor (ECU Replace)
				No	Adjust ignition timing
4	Check idle speed after warm up **Idle speed: 780—820 rpm** **(M/T: In Neutral)** **(A/T: In P range)** [Test connector (Green: 1-pin) grounded]	Yes	Go to Next Step		
		No	Try to adjust idle speed	Yes	Idle speed misadjustment
				No	Accelerator cable free play
					ISC valve stuck close
					Throttle valve
5	Substitute a well-known ECU Check if the condition becomes better	Yes			ECU malfunction
		No	Check voltage at ECU 3B terminal with SST **Voltage:** **Approx. 6—11V** **(While cranking)**	Yes	Go to Section H
				No	Inhibitor switch (A/T)
					Interlock switch
					Replated wiring open
6					Poor quality engine oil

Low idle speed (when A/C, P/S, E/L is ON)

STEP	QUICK INSPECTION		ACTION	POSSIBLE CAUSE AND DETAILED INSPECTION
1	Check malfunction code (34) with SST [Ign ON, Test connector (Green: 1-pin) grounded]	Yes	Go to Next Step	
		No	Check for the cause by referring to the check sequence	
2	Disconnect ISC valve connected at idle Check if the condition does not change	Yes	Go to Next Step	
		No		Low coolant level
				Engine oil poor quality
3	Check switches with SST Idle switch Neutral switch (M/T) Clutch switch (M/T) Inhibitor switch (A/T) [Ign ON, Test connector (Green: 1-pin) grounded]	Yes	Go to Next Step	
		No	Check for the cause by referring to the check sequence	
4	Check the continuity between test connector (Green: 1-pin) and ground			Wiring short to ground

Note

- **In case of low idle speed with A/C ON, if the problem can not be solved by the above steps, it may be an A/C compressor malfunction.**

High idle speed after warm up

STEP	QUICK INSPECTION		ACTION		POSSIBLE CAUSE AND DETAILED INSPECTION
1	Check malfunction code (09) (26) (34) with SST [IG ON, Test connector (Green: 1-pin) grounded]	Yes	Go to Next Step		
		No	Check for the cause by referring to the check sequence		
2	Check ignition timing at idle after warm up **Ignition timing: 10—12° BTDC** [Test connector (Green: 1-pin) grounded]	Yes	Go to Next Step		
		No	Adjust ignition timing		
3	Check intake manifold vacuum at idle	Yes	Go to Next Step		
		No	Check for air leak	Yes	Vacuum hose mis-routing
					Throttle valve open
				No	Vacuum hoses
					Gaskets
					Bolts or nuts loose
					Intake air system components
4	Check idle speed **Idle speed: 780—820 rpm** **(M/T: in Neutral)** **(A/T: in P range)** [Test connector (Green: 1-pin) grounded]	Yes			ISC valve malfunction
		No	Adjust idle speed (If possible)		
5	Disconnect ISC valve connector and check idle speed during warm up	Yes	Go to Next Step		
		No			Air valve stuck open
6	Disconnect water thermosensor connector and check if the condition becomes better	Yes	Check for water thermosensor connector condition as follows 1. Shake the connector and check if the engine condition become better 2. Check pin condition (burning or broken) 3. Connect the male pin to the connector and check for connector condition	Yes	Water thermosensor malfunction
				No	Poor contact at water thermo-sensor connector
		No	Go to Next Step		
7	Substitute a well-known ECU				

Idle hunting or surging

STEP	QUICK INSPECTION		ACTION		POSSIBLE CAUSE
1	(If trouble occurs at warm condition) Check malfunction code (15) (34) with SST [Ign ON, Test connector (Green: 1-pin) grounded]	Yes	Check for the cause by referring to the check sequence		
		No	Go to Next Step		
2	(If trouble occurs at warm condition) Check monitor lamp flashing with SST after warm up **Monitor lamp: Flashes ON and OFF more than 8 times/10 seconds at 2,000—3,000 rpm** [Test connector (Green: 1-pin) not grounded]	Yes	Go to Next Step		
		No			Oxygen sensor deterioration (Replace)
3	Check for intake manifold vacuum at idle **Vacuum: 460—500 mmHg (18.1—19.7 inHg)**	Yes	Go to Next Step		
		No			Air leakage
4	Pinch PCV hose check if the engine condition changes	Yes			PCV valve malfunction
		No	Go to Next Step		
5	Disconnect the water thermosensor connector Check if the engine condition changes	Yes			Water thermosensor malfunction
		No	Go to Next Step		
6	Connect the test connector (Yellow: 2-pin) with a jumper wire and disconnect the airflow meter connector Check if the hunting stops	Yes			Airflow meter malfunction
		No	Go to Next Step		
7	Check fuel line pressure at idle (Vacuum hose to pressure regulator disconnected) **Fuel line pressure: 265—314 kPa (2.7—3.2 kg/cm², 38—46 psi)**	Yes	Go to Next Step		
		No	Fuel line pressure does not drop when connecting vacuum hose to pressure regulator: Check for vacuum at vacuum hose	Yes	Pressure regulator malfunction. (Replace)
				No	Vacuum hose clogging (Replace)
			Fuel line pressure drops but not with in specification when connecting vacuum hose to pressure regulator **Fuel line pressure: 265—314 kPa (2.7—3.2 kg/cm², 38—46 psi)**		Pressure regulator malfunction (Replace)
8	Substitute a well-known ECU				ECU malfunction
9	Substitute a well-known pulsation damper				Pulsation damper malfunction

Engine stall at idle (Always)

STEP	QUICK INSPECTION		ACTION		POSSIBLE CAUSE
1	Check malfunction code with SST [Ign: ON, Test connector (Green: 1-pin) grounded]	Yes	Check the cause by referring to the check sequence		
		No	Go to Next Step		
2	Disconnect high-tension lead of each cylinder (Engine speed: 1,000 rpm) Is engine condition changed?	Yes	Go to Next Step		
		No	Check ignition system	Yes	Injector malfunction
				No	Ignition system malfunction
3	Check fuel line pressure [Ign ON, Test connector (Yellow: 2-pin) connected] **Fuel line pressure: 265—314 kPa (2.7—3.2 kg/cm², 38—46 psi)**	Yes	Go to Next Step		
		No			Fuel leaks
					Fuel filter clogging
			Check fuel pump maximum pressure **Fuel pump maximum pressure: 441—589 kPa (4.5—6.0 kg/cm², 64—85 psi)**	Yes	Pressure regulator malfunction (Replace)
				No	Fuel pump malfunction (Replace)
4	Check the vacuum hoses and the air funnel for sucking air	Yes	Shut the air leak part		
		No	Go to Next Step		
5	Connect terminals of test connector (Yellow: 2-pin) Check if engine stalls	Yes	Go to Next Step		
		No	Replace airflow meter		
6					ECU malfunction

3.0L ENGINE SYMPTOM TROUBLESHOOTING—MPV

Engine stall at idle (Only when engine is cold)

STEP	QUICK INSPECTION		ACTION	POSSIBLE CAUSE
1	Check 2I terminal of ECU	Yes	Go to Next Step	
		No	Replace water thermosensor	Water thermosensor malfunction
2	Blow the air valve (cold condition) Check the airflow	Yes	Go to Next Step	
		No	Replace the BAC valve	BAC valve malfunction (Replace)
3	Substitute a well-known ECU			ECU malfunction (Replace)

Engine stall at idle (only when engine is warm)

STEP	QUICK INSPECTION		ACTION	POSSIBLE CAUSE
1	Check 2I terminal of ECU	Yes	Go to Next Step	
		No	Replace water thermosensor	Water thermosensor malfunction
2	Substitute a well-known ECU			ECU malfunction

Engine stall at idle (Only after heat soak)

STEP	QUICK INSPECTION		ACTION		POSSIBLE CAUSE AND DETAILED INSPECTION
1	Run the engine at **2,000 rpm for more than 20 sec.** Check malfunction code (09) (11) (15) (17) (25) (26) with SST [Ign ON, Test connector (Green: 1-pin) grounded]	Yes	Go to Next Step		
		No	Check for the cause by referring to the check sequence		
2	Check switches with SST Neutral switch (M/T) Clutch switch (M/T) Inhibitor switch (A/T) [Ign ON, Test connector (Green: 1-pin) grounded]	Yes	Go to Next Step		
		No	Check for the cause by referring to the check sequence		
3	Check monitor lamp flashing with SST after warm up **Monitor lamp:** **Flashes ON and OFF more than 8 times/10 seconds at 2,000—3,000 rpm** [Test connector (Green: 1-pin) not grounded]	Yes	Go to Next Step		
		No			Oxygen sensor deterioration
4	Disconnect vacuum hose from pressure regulator Check if the condition becomes better	Yes	Check components related to pressure regulator control system		Water thermosensor malfunction
					Intake air thermo-sensor malfunction
					Solenoid valve (PRC) malfunction
					ECU malfunction (Check 2N terminal voltage)
		No	Go to Next Step		
5	Run the engine at idle and stop it Observe fuel pressure **for 5 minutes** **Fuel pressure:** **More than 147 kPa (1.5 kg/cm², 21 psi)**	Yes	Go to Next Step		
		No	Check fuel pump pressure drop	No	Fuel pump malfunction (Replace)
			Check pressure regulator pressure drop	Yes	Injector fuel leaks
				No	Pressure regulator malfunction
6	Disconnect high-tension lead of each cylinder at idle Check if the condition changes	Yes	Go to Next Step		
		No	Check ignition system	Yes	Injector clogging Replace (If step 3 OK)
				No	Spark plug
					High-tension lead
					Distributor cap

Engine stall at idle (When A/C, P/S, E/L (CPU) is ON)

STEP	QUICK INSPECTION		ACTION	POSSIBLE CAUSE AND DETAILED INSPECTION
1	Check malfunction code (34) with SST [Ign ON, Test connector (Green: 1-pin) grounded]	Yes	Go to Next Step	
		No	Check for the cause by referring to the check sequence	
2	Disconnect ISC valve connected at idle Check if the condition does not change	Yes	Go to Next Step	
		No		Low coolant level
				Engine oil poor quality
3	Check switches with SST Idle switch Neutral switch (M/T) Clutch switch (M/T) Inhibitor switch (A/T)	Yes	Go to Next Step	
		No	Check for the cause by referring to the check sequence	
4	Check the continuity between test connector (Green: 1-pin) and ground			Wiring short to ground

Note
- **Engine stall at idle with A/C ON, if the trouble can not be fixed after checking above steps, it may be A/C compressor malfunction.**

Engine stall at idle (in "D" range)

STEP	QUICK INSPECTION		ACTION	POSSIBLE CAUSE
1	Check malfunction code with SST [IG ON, Test connector (Green: 1-pin) grounded]	Yes	Check the cause by referring to the check sequence	
		No	Go to Next Step	
2	Check switches with SST Inhibitor switch [IG ON, Test connector (Green: 1-pin) grounded]	Yes	Go to Next Step	
		No	Adjustment or replacement	Inhibitor switch malfunction
3	Substitute a well-known ECU			ECU malfunction (Replace)

Engine stall during start up

STEP	QUICK INSPECTION		ACTION	POSSIBLE CAUSE
1	Check malfunction code (08) with SST [IG ON, Test connector (Green: 1-pin) grounded]	Yes	Check the cause by referring to the check sequence	
		No	Go to Next Step	
2	Check idle switch with SST [IG ON, Test connector (Green: 1-pin) grounded]	Yes	Go to Next Step	
		No	Check for the cause by referring to the check sequence	
3	Check the idle speed **Idle speed: 780—820 rpm (M/T: in Neutral) (A/T: in P range)** [Test connector (Green: 1-pin) grounded]	Yes	Go to Next Step	
		No	Adjust the idle speed	
4	Check for injector operating sound at idle	Yes	Go to Next Step	
5	Check for correct ignition timing advance	Yes	Go to Next Step	
		No	Adjust or replace	Distributor malfunction
6	Substitute a well-known ECU			ECU malfunction

3.0L ENGINE SYMPTOM TROUBLESHOOTING—MPV

Engine stall on deceleration (only when engine is cold)

STEP	QUICK INSPECTION		ACTION	POSSIBLE CAUSE
1	Check malfunction code with SST [IG ON, Test connector (Green: 1-pin) grounded]	Yes	Check the cause by referring to the check sequence	
		No	Go to Next Step	
2	Check 2I terminal of ECU	Yes	Go to Next Step	
		No	Replace the water thermosensor	Water thermosensor malfunction
3	Blow the air valve (cold condition) check the air flow	Yes	Go to Next Step	
		No	Replace the BAC valve	BAC valve malfunction (Replace)
4	Check the dashpot touch rpm and the operation	Yes	Go to Next Step	
		No	Adjust or replace the dashpot	Dashpot malfunction
5	Connect terminals of test connector (Yellow: 2-pin) Check the engine stalls	Yes	Go to Next Step	
		No	Replace the circuit opening relay	Circuit opening relay malfunction
6	Substitute a well-known ECU			ECU malfunction

Engine stall on deceleration (Only after warm up)

STEP	QUICK INSPECTION		ACTION	POSSIBLE CAUSE
1	Check malfunction code (08) with SST [IG ON, Test connector (Green: 1-pin) grounded	Yes	Check the cause by referring to the check sequence	
		No	Go to Next Step	
2	Substitute a well-known ECU			ECU malfunction (Replace)

Engine stall at idle (Intermittent)

STEP	QUICK INSPECTION		ACTION	POSSIBLE CAUSE AND DETAILED INSPECTION
1	Shake connector of ignition coil, main relay and ECU while cranking Check if the engine starts	Yes	There may be a poor contact at the connector. Repair or replace the wiring	
		No	Go to troubleshooting "Engine stall at idle (always)"	

Hesitates/Stumbles on acceleration

STEP	QUICK INSPECTION		ACTION		POSSIBLE CAUSE AND DETAILED INSPECTION
1	Run the engine at **2,000 rpm for 20 sec.** and stop it Check malfunction code (08) (09) (15) (17) with SST [IG ON, Test connector (Green: 1-pin) grounded]	Yes	Go to Next Step		
		No	Check for the cause by referring to the check sequence		
2	Check idle switch with SST [IG ON, Test connector (Green: 1-pin) grounded]	Yes	Go to Next Step		
		No	Check for the cause by referring to the check sequence		
3	Disconnect oxygen sensor connector Check if the condition becomes better	Yes			Oxygen sensor
		No	Go to Next Step		
4	Check fuel line pressure while accelerating (Vacuum hose to pressure regulator disconnected) **Fuel line pressure: Keeps 265—314 kPa (2.7—3.2 kg/cm², 38—46 psi)**	Yes	Go to Next Step		
		No			Fuel line clogging
					Fuel filter clogging
					Pressure regulator malfunction/(Replace)
5	Check for air leaks with throttle valve open by listening for sucking noise	Yes	Go to Next Step		
		No			Vacuum hoses
					Gaskets
					Nuts or bolts loose
					Damaged components
6	Disconnect water thermosensor connector Check if the condition improves	Yes	Check water thermosensor connector for connecting condition as follows 1. Shake connector and check if the condition changes 2. Check condition of terminal (burning or damaged) 3. Connect a male terminal to the connector and check for connection condition	Yes	Water thermosensor malfunction
				No	Poor contact at water thermosensor connector
		No	Go to Next Step		
7	Substitute a well-known ECU Check if the condition becomes better	Yes			ECU malfunction
		No			Airflow meter
					Throttle valve
					Spark plug
8	Check other systems				Clutch slipping

Hesitates at steady speed

STEP	QUICK INSPECTION		ACTION	POSSIBLE CAUSE AND DETAILED INSPECTION
1	Run the engine at **2,000 rpm for 20 sec.** and stop it Check malfunction code (08) (09) (12) (14) (15) (17) with SST [IG ON, Test connector (Green: 1-pin) grounded]	Yes	Go to Next Step	
		No	Check for the cause by referring to the check sequence	
2	Disconnect oxygen sensor connector Check if the condition becomes better	Yes		Oxygen sensor deterioration
		No	Go to Next Step	
3	Check for air leaks with throttle valve open by listening for sucking noise	Yes	Go to Next Step	
		No		Vacuum hoses
				Gaskets
				Nuts or bolts loosen
				Damaged components
4	Check fuel line pressure while accelerating (Vacuum hose to pressure regulator disconnected) **Fuel line pressure: Keeps 265—314 kPa (2.7—3.2 kg/cm², 38—46 psi)**	Yes	Go to Next Step	
		No		Pressure regulator malfunction (Replace)
5	Check condition ignition coil and airflow meter connectors terminal (Burning or damaged)	Yes	Go to Next Step	
		No		Poor contact
6	Gradually open the throttle valve Check if engine speed increases smoothly	Yes	Go to Next Step	
		No		Airflow meter
				Throttle valve
				Throttle sensor
7				Spark plug • Gap • Damage
8	Change fuel with specified one Check if the condition becomes better	Yes		Poor fuel quality
		No	Go to Next Step	
9	Substitute a well-known ECU			

Hesitates/Stumbles on acceleration

STEP	QUICK INSPECTION		ACTION	POSSIBLE CAUSE AND DETAILED INSPECTION
1	Run the engine at **2,000 rpm for 20 sec.** and stop it Check malfunction code (08) (09) (15) (17) with SST [IG ON, Test connector (Green: 1-pin) grounded]	Yes	Go to Next Step	
		No	Check for the cause by referring to the check sequence	
2	Check idle switch with SST [IG ON, Test connector (Green: 1-pin) grounded]	Yes	Go to Next Step	
		No	Check for the cause by referring to the check sequence	
3	Disconnect oxygen sensor connector Check if the condition becomes better	Yes		Oxygen sensor
		No	Go to Next Step	
4	Check fuel line pressure while accelerating (Vacuum hose to pressure regulator disconnected) **Fuel line pressure:** **Keeps 265—314 kPa (2.7—3.2 kg/cm², 38—46 psi)**	Yes	Go to Next Step	
		No		Fuel line clogging Fuel filter clogging Pressure regulator malfunction/(Replace)
5	Check for air leaks with throttle valve open by listening for sucking noise	Yes	Go to Next step	
		No		Vacuum hoses Gaskets Nuts or bolts loosen Damaged components
6	Disconnect water thermosensor connector Check if the condition improves	Yes	Check water thermosensor connector for connecting condition as follows 1. Shake connector and check if the condition changes 2. Check condition of terminal (burning or damaged) 3. Connect a male terminal to the connector and check for connection condition — Yes: Water thermosensor malfunction; No: Poor contact at water thermosensor connector	
		No	Go to Next Step	
7	Substitute a well-known ECU Check if the condition becomes better	Yes		ECU malfunction
		No		Airflow meter Throttle valve Spark plug
8	Check other systems			Clutch slipping

Knocking

STEP	QUICK INSPECTION		ACTION	POSSIBLE CAUSE AND DETAILED INSPECTION
1	Check malfunction code (09) with SST [IG ON, Test connector (Green: 1-pin) grounded]	Yes	Check for cause by referring to check sequence	
		No	Go to Step 2 (at sea level) Go to Step 3 [at high altitude (1,000 m, 3,280 ft)]	
2	Check ignition timing after warm up **Ignition timing 10—12° BTDC** [Test connector (Green: 1-pin) grounded]	Yes	Go to Next Step	
		No	Adjust ignition timing	
3	Check fuel line pressure at idle (Disconnecting vacuum hose to pressure regulator) **Fuel line pressure:** **Keeps 265—314 kPa (2.7—3.2 kg/cm², 38—46 psi)**	Yes	Go to Next Step	
		No	Fuel pressure does not drop when connecting vacuum hose to pressure regulator: **Fuel line pressure 206—255 kPa (2.1—2.6 kg/cm², 30—37 psi)** Check for vacuum at vacuum hose — Yes: Pressure regulator malfunction (Replace); No: Vacuum hose clogging (Replace)	
			Fuel pressure drops but not within specification when connecting vacuum hose to pressure regulator **Fuel line pressure 206—255 kPa (2.1—2.6 kg/cm², 30—37 psi)**	Pressure regulator malfunction (Replace)
4	Check the airflow meter	Yes	Go to Next Step	
		No	Replace the airflow meter	Airflow meter malfunction
5	Change the fuel with specified one Check if the condition becomes better **Fuel spec: RON87 or higher**	Yes		Poor fuel quality
		No	Go to Next Step	
6	Substitute a well-known ECU			ECU malfunction

Poor acceleration

STEP	QUICK INSPECTION		ACTION		POSSIBLE CAUSE
1	Check malfunction code with SST [IG ON, Test connector (Green: 1-pin) grounded	Yes	Check the cause by referring to the check sequence		
		No	Go to Next Step		
2	Check switches with SST Idle switch	Yes	Check the cause by referring to the check sequence		Idle switch malfunction (Adjust or replace)
		No	Go to Next Step		
3	Disconnect high-tension lead of each cylinder at idle IS engine condition changed?	Yes	Go to Next Step		
		No	Check ignition system	Yes	Injector malfunction
				No	Ignition system malfunction
4	Check air cleaner element for clogging	Yes	Go to Next Step		
		No	Clean air cleaner element		
5	Check ignition timing at idle after warm up **Ignition timing: 10—12° BTDC** [Test connector (Green: 1-pin) grounded]	Yes	Go to Next Step		
		No	Adjust ignition timing		
	Note: At high-altitude area **[1,000 m (3,280 ft)]**, ignition timing advances to **approx. 20° BTDC** when test connector (Green: 1-pin) is not grounded)				
6	Check the vacuum hoses and the air funnel for sucking air	Yes	Shut the air leak part		
		No	Go to Next Step		
7	Check fuel line pressure at idle (vacuum hose to pressure regulator disconnected) **Fuel line pressure: 265—314 kPa (2.7—3.2 kg/cm², 38—46 psi)**	Yes	Go to Next Step		
		No	Fuel line pressure does not drop when connecting vacuum hose to pressure regulator Check for vacuum at vacuum hose	Yes	Pressure regulator malfunction (Replace)
				No	Vacuum hose clogging (Replace)
			Fuel line pressure drops but not within specification when connecting vacuum hose to pressure regulator		Pressure regulator malfunction (Replace) Fuel pump malfunction (Replace)
8	When the accelerator pedal is depressed, engine speed increases smoothly	Yes	Go to Next Step		
		No	Airflow meter malfunction		
9	Change the fuel with specified one Check if the condition becomes better **Fuel spec: RON87 or higher**	Yes			Poor fuel quality
		No	Go to Next Step		
10	Substitute a well-known ECU				ECU malfunction

Bucking at high speed

STEP	QUICK INSPECTION		ACTION		POSSIBLE CAUSE
1	Check malfunction code with SST [IG ON, Test connector (Green: 1-pin) grounded]	Yes	Check for the cause by referring to the check sequence		
		No	Go to Next Step		
2	Disconnect the oxygen sensor connector, check if the engine condition better	Yes	Replace the oxygen sensor		
		No	Go to Next Step		
3	Check fuel line pressure at idle (Vacuum hose to pressure regulator disconnected) **Fuel line pressure: 265—314 kPa (2.7—3.2 kg/cm², 38—46 psi)**	Yes	Go to Next Step		
		No	Fuel line pressure does not drop when connecting vacuum hose to pressure regulator: **Fuel line pressure: 206—255 kPa (2.1—2.6 kg/cm², 30—37 psi)** Check for vacuum at vacuum hose	Yes	Pressure regulator malfunction (Replace)
				No	Vacuum hose clogging (Replace)
			Fuel line pressure drops but not within specification when connecting vacuum hose to pressure regulator		Pressure regulator malfunction (Replace)
4	Check the vacuum hoses and the air funnel for sucking air	Yes	Shut the air leak part		
		No	Go to Next Step		
5	Check ignition timing at idle after warm up **Ignition timing: 10—12° BTDC** [Test connector (Green: 1-pin) grounded]	Yes	Go to Next Step		
		No	Adjust ignition timing		
	Note: At high-altitude area **[1,000 m (3,280 ft)]**, ignition timing advances to **approx. 20° BTDC** when test connector (Green: 1-pin) is not grounded				
6	When the accelerator pedal is depressed, engine speed increases smoothly	Yes	Go to Next Step		
		No	Airflow meter malfunction		
7	Substitute a well-known ECU				ECU malfunction

3.0L ENGINE SYMPTOM TROUBLESHOOTING—MPV

Bucking on deceleration

STEP	QUICK INSPECTION		ACTION	POSSIBLE CAUSE AND DETAILED INSPECTION
1	Check malfunction code (08) (09) (12) (26) (34) with SST [IG ON, Test connector (Green: 1-pin) grounded]	Yes	Go to Next Step	
		No	Check for the cause by referring to the check sequence	
2	Check switches with SST [IG ON, Test connector (Green: 1-pin) grounded] • Idle switch • Stoplight switch	Yes	Go to Next Step	
		No	Check for the cause by referring to the check sequence	
3	Check the dashpot operation	Yes	Go to Next Step	
		No	Adjust or replace the dashpot	Dashpot malfunction
4	Substitute a well-known ECU Check if the condition becomes better	Yes		ECU malfunction
		No		Throttle sensor
				Go to Next Step
5				Spark plug gap
6				Clutch slipping
7				Compression differs between cylinders

Poor fuel economy

STEP	QUICK INSPECTION		ACTION		POSSIBLE CAUSE AND DETAILED INSPECTION
1	Run the engine at **2,000—3,000 rpm for more than 20 sec.** after warm up and stop it Check malfunction code (15) (17) with SST [IG ON, Test connector (Green: 1-pin) grounded]	Yes	Go to Next Step		
		No	Check for the cause by referring to the check sequence		
2	Check idle switch with SST [IG ON, Test connector (Green: 1-pin) grounded]	Yes	Go to Next Step		
		No	Check for the cause by referring to the check sequence		
3	Check monitor lamp flashing with SST after warm up **Monitor lamp: Flashes ON and OFF more than 8 times/10 seconds at 2,000—3,000 rpm** [Test connector (Green: 1-pin) not grounded]	Yes	Go to Next Step		
		No			Oxygen sensor deterioration
4	Check fuel line pressure at idle **Fuel line pressure: 206—255 kPa (2.1—2.6 kg/cm^2, 30—37 psi)**	Yes	Go to Next Step		
		No	Check vacuum line to pressure regulator for clogging or air leak	Yes	Vacuum line
				No	Solenoid valve (PRC) malfunction
					ECU malfunction (Check 2N terminal voltage)
					Pressure regulator malfunction/Replace
5	Check for fuel cut operation while deceleration **Fuel cut:** **Until approx. 1,800 rpm (In-gear and braking condition-M/T Only)** **Until approx. 2,600 rpm (During A/C operation)** **Until approx. 2,300—4,600 rpm (Other conditions)** **After warm up**	Yes	Go to Next Step		
		No	Check water thermosensor	Yes	ECU malfunction
				No	Water thermosensor malfunction/Replace
			Check stoplight switch		
			Check neutral switch		
6	Check ignition timing advance	Yes	Go to Next Step		
		No			Centrifugal advance bad: Distributor malfunction
			Vacuum advance bad: Check for vacuum routing and delay valve direction		Distributor malfunction
7	Check other systems				Clutch plate slipping
					A/T malfunction
					Brake wear
					Tire air pressure

High oil consumption/White exhaust smoke

STEP	QUICK INSPECTION		ACTION		POSSIBLE CAUSE AND DETAILED INSPECTION
1	Check for oil leak from engine	Yes	Repair or replace		
		No	Go to Next Step		
2	Disconnect PCV valve from engine Check if the vacuum is felt at idle	Yes	Go to Next Step		
		No	Check PCV valve	Yes	PCV hose clogging
				No	PCV valve malfunction
3	Check that the ventilation hose is installed correctly	Yes	Go to Next Step		
		No	Install ventilation hose correctly		
4	It may be a malfunction of the engine Check for the cause				

Afterburn on deceleration

STEP	QUICK INSPECTION		ACTION		POSSIBLE CAUSE AND DETAILED INSPECTION
1	Check malfunction code (34) with SST [IG ON, Test connector (Green: 1-pin) grounded]	Yes	Go to Next Step		
		No	Check for the cause by referring to the check sequence		
2	Check idle switch with SST [Test connector (Green: 1-pin) grounded]	Yes	Go to Next Step		
		No	Check for the cause by referring to the check sequence		
3	Check ignition timing at idle after warm up **Ignition timing: 10—12° BTDC** [Test connector (Green: 1-pin) grounded]	Yes	Go to Next Step		
		No			Centrifugal advance bad: Distributor malfunction
			Vacuum advance bad: Check for vacuum routing.		Distributor malfunction
4	Check air cleaner element for clogging	Yes	Go to Next Step		
		No	Clean air cleaner element		
5	Check fuel cut operation while decelerating **Fuel cut: Above approx. 1,800 rpm after warm up**	Yes	Go to Next Step		
		No	Check water thermosensor	Yes	ECU malfunction 2I terminal
				No	Water thermosensor (Replace)
6	Run the engine at idle and stop the engine (IG OFF) Check fuel pressure drop	Yes	Go to Next Step		
		No	Check fuel pump for pressure drop	Yes	Check pressure regulator for pressure drop
				No	Fuel pump malfunction (Replace)
			Check pressure regulator for pressure drop	Yes	Injector leak
				No	Pressure regulator malfunction (Replace)
7					Low compression
					Valve timing

Rotten egg smell

STEP	QUICK INSPECTION	ACTION	POSSIBLE CAUSE AND DETAILED INSPECTION
1	Change fuel with specified one		Poor fuel quality

Black exhaust smoke

STEP	QUICK INSPECTION		ACTION		POSSIBLE CAUSE AND DETAILED INSPECTION
1	Run the engine at **2,000 rpm for more than 20 sec.** check for malfunction code (17) with SST [IG ON, Test connector (Green: 1-pin) grounded]	Yes	Go to Next Step		
		No	Check for the cause by referring to the check sequence		
2	Check fuel line pressure [IG ON, Test connector (Yellow: 2-pin) connected] **Fuel line pressure: 265—314 kPa (2.7—3.2 kg/cm², 38—46 psi)**	Yes	Go to Next Step		
		No (high)			Pressure regulator malfunction (Replace)
3	Disconnect water thermosensor connector Check if the condition improves	Yes	Check water thermosensor connector for connecting condition as follows 1. Shake connector and check if the condition changes 2. Check condition of terminal (burning or damaged) 3. Connect a male terminal to the connector and check for connection condition	Yes	Water thermosensor malfunction
				No	Poor contact at water thermosensor connector
		No	Go to step 4 **[at high-altitude (1,000 m, 3,280 ft)]** Go to step 5 (at sea level)		
4					Atmospheric pressure sensor malfunction (Replace the ECU)
5					Airflow sensor malfunction
6	Substitute a well-known ECU				ECU malfunction

Fails emission test (Engine condition OK)

STEP	QUICK INSPECTION		ACT ON		POSSIBLE CAUSE AND DETAILED INSPECTION
1	Run the engine at **2,000 rpm for 20 sec.** and stop it Check malfunction code (08) (09) (14) (15) (17) (26) with SST [IG ON, Test connector (Green: 1-pin) grounded]	Yes	Go to Next Step		
		No	Check for the cause by referring to the check sequence		
2	Check idle switch with SST [IG ON, Test connector (Green: 1-pin) grounded]	Yes	Go to Next Step		
		No	Check for the cause by referring to the check sequence		
3	Check for air leaks by hearing sucking noise • At idle • Throttle valve open	Yes	Go to Next Step		
		No			
4	Check monitor lamp flashing with SST after warm up **Monitor lamp:** **Flashes ON and OFF more than 8 times/10 seconds at 2,000—3,000 rpm** [Test connector (Green: 1-pin) not grounded]	Yes	Go to Next Step		
		No			Oxygen sensor deterioration (Replace)
5	Check fuel line pressure [IG ON, Test connector (Yellow: 2-pin) connected] **Fuel line pressure:** **265—314 kPa (2.7—3.2 kg/cm², 38—46 psi)**	Yes	Go to Next Step		
		No (Low)			Fuel leaks
					Fuel filter clogging
			Check fuel pump maximum pressure **Fuel pump maximum pressure:** **441—589 kPa (4.5—6.0 kg/cm², 64—85 psi)**	Yes	Pressure regulator malfunction (Replace)
				No	Fuel pump malfunction (Replace)
		No (High)			Pressure regulator malfunction (Replace)
6	Check fuel line pressure at idle **Fuel line pressure:** **206—255 kPa (2.1—2.6 kg/cm², 30—37 psi)**	Yes	Go to Next Step		
		No	Check vacuum line to pressure regulator for clogging or air leak	Yes	Vacuum line
				No	Solenoid valve (PRC) malfunction
					ECU malfunction (2N)
					Pressure regulator malfunction (Replace)
7	Check fuel cut operation while decelerating **Fuel cut:** **Above approx. 1,800 rpm after warm up**	Yes	Go to Next Step		
		No	Check water thermosensor	Yes	ECU malfunction (Replace)
				No	Water thermosensor malfunction/Replace
8	Check for fuel fumes	Yes	Check for the cause by referring to troubleshooting "Gasoline fumes"		
		No	Go to Next Step		
9					PCV valve
					PCV hose
10					Spark plug • Gap • Damage
11					Poor fuel quality

Gasoline fumes

STEP	QUICK INSPECTION		ACTION		POSSIBLE CAUSE AND DETAILED INSPECTION
1	Check for fuel leaks	Yes	Go to Next Step		
		No	Replace		
2	Check if the evaporative fume is emitted from the check-and-cut valve	Yes	Check check-and-cut valve	Yes	Two-way check valve
					Purge line clogging
				No	Check-and-cut valve malfunction (Replace)
		No	Go to Next Step		
3	Check malfunction code (08) (15) (09) (17) (26) with SST [IG ON, Test connector (Green: 1-pin) grounded]	Yes	Go to Next Step		
		No	Check for the cause by referring to the check sequence		
4	Check switches with SST • Idle switch • Neutral switch (M/T) • Clutch switch (M/T) • Inhibitor switch (A/T) [IG ON, Test connector (Green: 1-pin) grounded]	Yes	Go to Next Step		
		No	Check for the cause by referring to the check sequence		
5	Ground the solenoid valve (Purge control) connector (L/Y) terminal and disconnect vacuum hose from charcoal canister and check for vacuum	Yes			ECU malfunction (Check 2K) terminal voltage
		No			Solenoid valve (Purge control) malfunction (Replace)

STEP 1

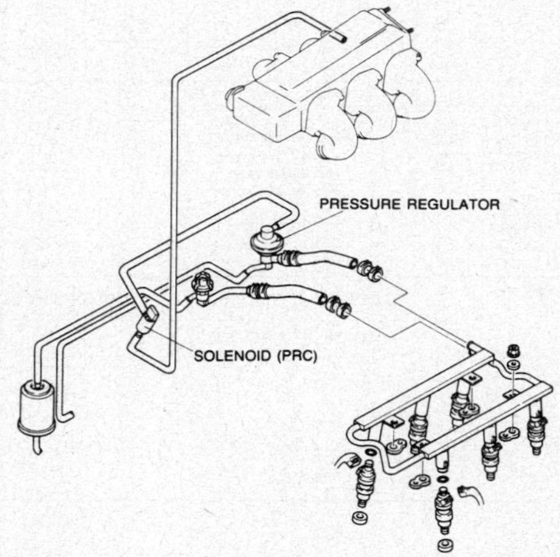

MIL always ON (Engine condition is OK)

STEP	QUICK INSPECTION		ACTION		POSSIBLE CAUSE AND DETAILED INSPECTION
1	Check malfunction code with SST [IG ON, Test connector (Green: 1-pin) grounded]		"88" ECU malfunction (Replace)		
			"00" (Except california) Go to Next Step		
			"00" (California only) Wiring between ECU 1A terminal and MIL short to ground		
2	(Except California) Check if the oxygen sensor replacement time has come **Oxygen sensor replacement time: Every 80,000 miles (128,000 km)**	Yes	Check if MIL was reset by ex-changing MIL set screw position	Yes	Go to next action
				No	Reset the MIL
			Check continuity between MIL and ground	Yes	Wiring between MIL and ECU is short to ground
				No	Mileage sensor mal-function
		No	Check continuity between MIL and ground	Yes	Wiring between MIL and ECU is short to ground
				No	Mileage sensor malfunction

MIL never ON (Engine condition is not OK)

STEP	QUICK INSPECTION		ACTION		POSSIBLE CAUSE AND DETAILED INSPECTION
1	Check if other indicator lamps illuminate	Yes	Go to Next Step		
		No	Check power supply circuit to combination meter		
2	Check bulb of the MIL	Yes	Ground the ECU 1A terminal check if the MIL illuminates	Yes	ECU malfunction
				No	Wiring between ECU and MIL open
		No	Replace		
3	(Except California) Check if the MIL illuminated at oxygen sensor replacement time **Oxygen sensor replacement time: Every 80,000 miles (128,000 km)**				MIL set screw loose or left
					Mileage sensor malfunction

A/C does not work

STEP	QUICK INSPECTION		ACTION		POSSIBLE CAUSE AND DETAILED INSPECTION
1	Check if the condenser fan oper-ates when grounding A/C relay connector (R/W) terminal (IG ON)	Yes	Check the volt-age at ECU 1L terminal with SST **Voltage at idle after warm up: 0V (A/C and blower switches ON)**	Yes	ECU malfunction (Check 1C terminal voltage)
					Wiring between ECU 1C and A/C relay open
				No	A/C system mal-function
		No	Check A/C system		

3.0L ENGINE TROUBLE CODES—MPV

Malfunction Code No.	Input devices	Malfunction	Fail-safe function	Pattern of output signals (Self-Diagnosis Checker or MIL)
01	Ignition pulse (Igniter, Ignition coil)	Broken wire, short circuit	—	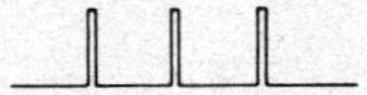
02	Distributor (Ne signal)	Ne signal not input for 1.5 sec. during cranking	—	
03	Distributor (G1 signal)	Broken wire, short circuit	—	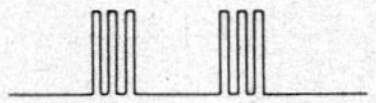
04	Distributor (G2 signal)	Broken wire, short circuit	—	
08	Airflow meter	Broken wire, short circuit	Basic fuel injection amount fixed as for 2 driving modes 1) Idle switch: ON 2) Idle switch: OFF	
09	Water thermosensor	Broken wire, short circuit	Coolant temp. input fixed at 80°C (176°F)	
10	Intake air thermosensor (Airflow meter)	Broken wire, short circuit	Intake air temp. input fixed at 20°C (68°F)	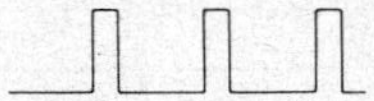
11	Intake air thermosensor (Dynamic chamber)	Broken wire, short circuit	Intake air temp. input fixed at 20°C (68°F)	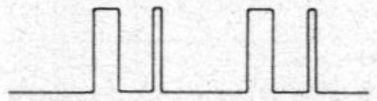
12	Throttle sensor	Broken wire, short circuit	Throttle valve opening angle signal input fixed at fully open	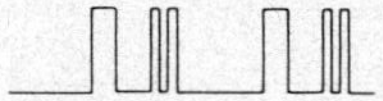
14	Atmospheric pressure sensor	ECU	Atmospheric pressure input fixed at 760 mmHg (29.9 in Hg)	
15	Oxygen sensor	Oxygen sensor output below 0.55V 120 sec. after engine at 1,500 rpm	Feedback system canceled (for EGI)	
17	Feedback system	Oxygen sensor output does not change from 0.55V 60 sec. after engine at 1,500 rpm	Feedback system canceled (for EGI)	

Code No.	Output devices	Output signal pattern (Self-Diagnosis Checker or MIL)
25	Solenoid valve (Pressure regulator control)	
26	Solenoid valve (Purge control)	
34	Idle speed control valve (BAC valve)	
36	Oxygen sensor heater	
41	Solenoid valve (Variable resonance induction system)	

Caution
- **If there is more than one failure present, the lowest number malfunction code is displayed first, the subsequent malfunction codes appear in order.**
- **After repairing all failures, turn the ignition switch OFF, disconnect the negative battery cable, for approx. 20 seconds to erase the malfunction code memory.**

3.0L ENGINE TROUBLE CODE TROUBLESHOOTING – MPV

PC: Possible Cause

No.01 Code

Check	Result	Action
Are there poor connections at ignition coil connectors?	YES	Repair or replace connector
	NO	Go to next step
Is resistance of ignition coil OK ? Resistance: Primary 0.72—0.88Ω Secondary 10—30 kΩ	NO	Replace ignition coil
	YES	Go to next step
Is operation of igniter OK ?	NO	Replace igniter
	YES	Go to next step
Is there continuity between IGf terminal of igniter connector (wiring harness side) and engine control unit (1G) terminal ?	NO	PC: Open circuit in wiring harness from igniter to engine control unit (1G) terminal
	YES	Go to next step
Is engine control unit (1M) terminal voltage OK?	NO	PC: Short circuit wiring harness
	YES	Replace engine control unit

No.02 Code

Check	Result	Action
Are there poor connections at distributor connectors?	YES	Repair or replace connector
	NO	Go to next step
Is resistance of distributor OK?	NO	Replace distributor
	YES	Go to next step
Is there continuity between distributor connector (wiring harness side) and engine control unit?	NO	PC: Open circuit in wiring harness
	YES	Go to next step
Are engine control unit (2E) and (2F) terminal voltages OK?	NO	PC: Short circuit in wiring harness
	YES	Replace engine control unit

Distributor	Engine control unit
Ne ① W	2E
COM ② G	2F

No.03 Code

Check	Result	Action
Are there poor connections at distributor connectors?	YES	Repair or replace connector
	NO	Go to next step
Is resistance of distributor OK?	NO	Replace distributor
	YES	Go to next step
Is there continuity between distributor connector (wiring harness side) and engine control unit?	NO	PC: Open circuit in wiring harness
	YES	Go to next step
Are engine control unit (2G) terminal voltages OK?	NO	PC: Short circuit in wiring harness
	YES	Replace engine control unit

Distributor	Engine control unit
G1 ① L	2G

No.04 Code

Check	Result	Action
Are there poor connections at distributor connectors?	YES	Repair or replace connector
	NO	Go to next step
Is resistance of distributor OK?	NO	Replace distributor
	YES	Go to next step
Is there continuity between distributor connector (wiring harness side) and engine control unit?	NO	PC: Open circuit in wiring harness
	YES	Go to next step
Are engine control unit (2H) terminal voltages OK?	NO	PC: Short circuit in wiring harness
	YES	Replace engine control unit

Distributor	Engine control unit
G2 ① R	2H

Code No.08 (Airflow meter)

PC: Possible Cause

Are there poor connections at airflow meter connectors?
- YES → Repair or replace connector
- NO ↓

Is resistance of airflow meter OK?
Resistance:

Terminal	Fully closed (Ω)	Fully open (Ω)
E_2 ↔ Vs	20—400	20—1,000
E_2 ↔ Vc	100—300	
E_2 ↔ V_B	200—400	

- NO → Repair airflow meter
- YES ↓

Is there continuity between airflow meter connector (wiring harness side) and engine control unit?

Airflow meter	Control unit
V_B (B/W wire)	3I
Vc (B/W wire)	2B
Vs (G/B wire)	2E
E_2 (BR/B wire)	2C

- NO → PC: Open circuit in wiring harness from airflow meter to engine control unit
- YES ↓

Are engine control unit terminal voltages OK?
Check terminals 2B, 2C, 2E, 3I
- NO → PC: Short circuit in wiring harness
- YES → Replace engine control unit

Code No.09 (Water thermosensor)

Are there poor connections at water thermosensor connectors?
- YES → Repair or replace connectors
- NO ↓

Is resistance of water thermosensor OK?
Resistance:

Coolant temp	Resistance
−20°C (−4°F)	14.5—17.8 kΩ
20°C (68°F)	2.2— 2.7 kΩ
40°C (104°F)	1.0— 1.3 kΩ
60°C (140°F)	0.5— 0.7 kΩ
80°C (176°F)	0.28— 0.35 kΩ

- NO → Replace water thermosensor
- YES ↓

Is there continuity between water thermosensor connector (wiring harness side) and engine control unit?

Water thermosensor	Engine control unit
(G/W wire)	2I

- NO → PC: Open circuit in wiring harness from water thermosensor to engine control unit
- YES ↓

Are engine control unit (2I) terminal voltages OK?
- NO → PC: Short circuit in wiring harness
- YES → Replace engine control unit

Code No.10 (Intake air thermosensor—Airflow meter)

PC: Possible Cause

Are there poor connections at airflow meter connectors?
- YES → Repair or replace connector
- NO ↓

Is resistance of intake air thermosensor (in airflow meter) OK?
Resistance:

Terminal	Resistance
E_2↔THA_A	−20°C (−4°F): 13.6—18.4 kΩ 20°C (68°F): 2.21—2.69 kΩ 60°C (140°F): 0.493—0.667 kΩ

- NO → Replace airflow meter
- YES ↓

Is there continuity between intake air thermosensor (in airflow meter) connector (wiring harness side) and engine control unit?

Intake air thermosensor (in airflow meter)	Engine control unit
THA_A (G/O wire)	2J
E_2 (BR/B wire)	2C

- NO → PC: Open circuit in wiring harness from intake air thermosensor (in airflow meter) to engine control unit
- YES ↓

Are engine control unit (2J) and (2C) terminal voltages OK ?
- NO → PC: Short circuit in wiring harness
- YES → Replace engine control unit

Code No.11 (Intake air thermosensor—Dynamic chamber)

Are there poor connections at intake air thermosensor connectors?
- YES → Repair or replace connector
- NO ↓

Is resistance of intake air thermosensor (dynamic chamber) OK?
Resistance:

Temperature	Resistance
25°C (77°F)	29.7—36.3 kΩ
85°C (185°F)	3.3— 3.7 kΩ

- NO → Replace intake air thermosensor (dynamic chamber)
- YES ↓

Is there continuity between intake air thermosensor (dynamic chamber) connector (wiring harness side) and engine control unit?

Intake air thermosensor (dynamic chamber)	Engine control unit
A (G wire)	2L
B (BR/B wire)	2C

- NO → PC: Open circuit in wiring harness from intake air thermosensor (dynamic chamber) to engine control unit
- YES ↓

Are engine control unit (2L) and (2C) terminal voltages OK?
- NO → PC: Short circuit in wiring harness
- YES → Replace engine control unit

Code No.12 (Throttle sensor)

PC: Possible Cause

Check	Result	Action
Are there poor connections at throttle sensor connectors?	YES	Repair or replace connector
(NO →) Is throttle sensor adjusted correctly?	NO	Adjust throttle sensor (for idle switch)
(YES →) Is resistance of throttle sensor OK? Resistance: (A) ↔ (C) 3.5—6.5 kΩ (B) ↔ (C) Below 1 kΩ .....Throttle valve fully closed Approx. 3.5—6.5 kΩ .....Throttle valve fully open	NO	Replace throttle sensor
(YES →) Is there continuity between throttle sensor connector (wiring harness side) and engine control unit?	NO	PC: Open circuit in wiring harness from throttle sensor to engine control unit
(YES →) Are engine control unit terminal voltages OK? Check terminal: 2A, 1E, 2G	NO	PC: Short circuit in wiring harness
	(2A) terminal voltage not within specification	Replace engine control unit
	YES	Replace engine control unit

Throttle sensor	Engine control unit
A (BR/W wire)	2A
B (R/L wire)	2G
C (G/W wire)	1E

Code No.14 (Atmospheric pressure sensor)

Replace engine control unit

Code No.15 (Oxygen sensor)

PC: Possible Cause

Note

- **When the malfunction code No.15 and 17 are present at the same time, first perform the checking procedure for the malfunction code No.17. (Refer to page F1–125.)**

Check	Result	Action
Are there any poor connections at oxygen sensor connector?	YES	Repair or replace connector
(NO →) Is operation of oxygen sensor heater OK?	NO	Repair or replace related parts
(YES →) Is oxygen sensor output voltage OK?	NO	Replace oxygen sensor
(YES →) Is sensitivity of oxygen sensor OK?	NO	PC: • Oxygen sensor malfunction • Engine control unit malfunction
(YES →) Is there continuity between oxygen sensor connector (wiring harness side) and engine control unit (2D) terminal?	NO	PC: Open circuit in wiring harness from oxygen sensor to engine control unit (2D) terminal
(YES →) Is (2D) terminal voltage of engine control unit OK?	NO	PC: Short circuit in wiring harness between oxygen sensor and engine control unit (2D) terminal
	YES	Replace engine control unit

Code No.17 (Feedback system)

Check	Result	Action
Warm up engine and run it at **2,500—3,000 rpm** for 3 minutes		
Does monitor lamp of **SST** illuminate at idle without test connector (Green: 1-pin) grounded? **SST:** Self-Diagnosis Checker	NO	PC: • Air leak in vacuum hoses or intake air system • Contaminated oxygen sensor • Insufficient fuel injection
(YES →) Are spark plugs clean?	NO	Clean or replace spark plugs
(YES →) Is oxygen sensor operation OK?	NO	Replace oxygen sensor
(YES →) Is same code No. present after performing after repair procedure?	NO	Feedback system OK
(YES →) Is (2D) terminal voltage of engine control unit OK?	NO	PC: Short circuit in wiring harness from oxygen sensor connector to engine control unit
	YES	Replace engine control unit

3.0L ENGINE TROUBLE CODE TROUBLESHOOTING—MPV

Code No.25 (Solenoid valve (Pressure regulator control))

PC: Possible Cause

- Are there poor connections at solenoid valve (PRC) connectors?
  - YES → Repair or replace connector
  - NO ↓
- Is there continuity of solenoid valve (PRC)?
  - NO → Replace solenoid valve
  - YES ↓
- Is there battery voltage at (B/W) wire of solenoid valve (PRC) connector?
  - NO → PC: Open or short circuit in wiring harness from (B/W) wire to main relay (for engine control unit)
  - YES ↓
- Is there continuity between solenoid valve (PRC) connector (wiring harness side) and engine control unit?

Solenoid valve (PRC)	Engine control unit
L/B wire	2N

  - NO → PC: Open circuit in wiring harness from solenoid valve (PRC) to engine control unit
  - YES → PC: • Engine control unit malfunction • Short circuit in wiring harness

Code No.26 (Solenoid valve (Purge control))

- Are there poor connections at solenoid valve (purge control) connectors?
  - YES → Repair or replace connector
  - NO ↓
- Is there continuity of solenoid valve (purge control)?
  - NO → Replace solenoid valve
  - YES ↓
- Is there battery voltage at (B/W) wire of solenoid valve (purge control) connector?
  - NO → PC: Open or short circuit in wiring harness from (B/W) wire to main relay (for engine control unit)
  - YES ↓
- Is there continuity between solenoid valve (purge control) connector (wiring harness side) and engine control unit?

Solenoid valve	Engine control unit
L/O wire	2K

  - NO → PC: Open circuit in wiring harness from solenoid valve (purge control) to engine control unit
  - YES → PC: • Engine control unit malfunction • Short circuit in wiring harness

Code No.34 (Idle-speed control valve (BAC valve))

PC: Possible Cause

- Are there poor connectors at BAC valve connectors?
  - YES → Repair or replace connector
  - NO ↓
- Is resistance of BAC valve OK? Resistance: 10.7—12.3Ω
  - NO → Replace BAC valve
  - YES ↓
- Is there battery voltage at (B/W) wire of BAC valve connector?
  - NO → PC: Open or short circuit in wiring harness from (B/W) wire to main relay (for engine control unit)
  - YES ↓
- Is there continuity between BAC valve connector (wiring harness side) and engine control unit?

BAC valve	Engine control unit
L/G	2Q

  - NO → PC: Open circuit in wiring harness from BAC valve to engine control unit
  - YES → PC: • Engine control unit malfunction • Short circuit in wiring harness

Code No.36 (Heater-oxygen sensor)

- Is there a poor connection in solenoid valve circuit?
  - YES → Repair or replace connector
  - NO ↓
- Disonnect connector from oxygen sensor is there continuity of oxygen sensor heater between terminal (BR) and (W) or (B/W) and (R/B)?
  - NO → Replace oxygen sensor
  - YES ↓
- Is there battery voltage at oxygen sensor terminal-wire (B/W)?
  - NO → Check for open circuit in wiring from main relay to oxygen sensor terminal-wire (B/W)
  - YES ↓
- Is there continuity between oxygen sensor and ECU terminal 2N and 2R?
  - NO → Check for open circuit in wiring from oxygen sensor terminal-wire (W) or (R/B) to ECU terminal
  - YES → Check for short circuit in wiring from oxygen sensor terminal-wire (W) or (R/B) to ECU terminal

Code No.41 (Solenoid valve (Variable resonance induction system))

Are there any poor connections at solenoid valve (VRIS) connectors?
- YES → Repair or replace connector
- NO ↓

Is there continuity of solenoid valve (VRIS)?
- NO → Replace solenoid valve
- YES ↓

Is there battery voltage at (B/W) wire of solenoid valve (VRIS) connector?
- NO → PC: Open or short circuit in wiring harness from (B/W) wire to main relay (for engine control unit)
- YES ↓

Is there continuity between solenoid valve (VRIS) connector (wiring harness side) and engine control unit?

Solenoid valve (VRIS)	Engine control unit
Y/G	2O

- NO → PC: Open circuit in wiring harness from solenoid valve (VRIS) to engine control unit
- YES → PC: • Engine control unit malfunction • Short circuit in wiring harness

SWITCH MONITOR FUNCTION

PREPARATION
SST

49 H018 9A1

Checker, Self-diagnosis

Individual switches can be monitored by the **SST**.

Note
- **The test connector must be grounded and the ignition switch ON (engine stopped).**

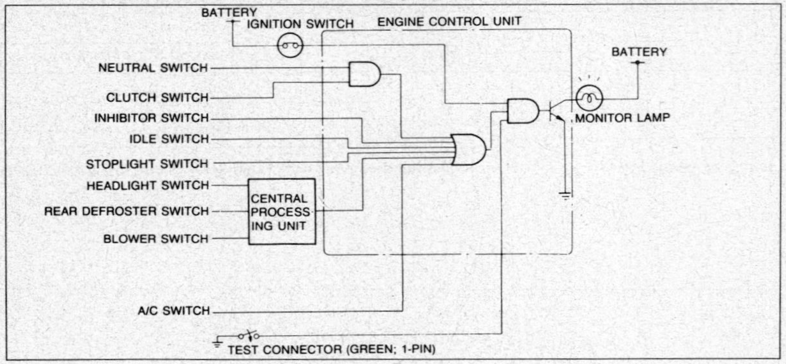

Switch	Self-Diagnosis Checker (Monitor Lamp)		Remarks
	Light ON	Light OFF	
Clutch switch (M/T)	Pedal released	Pedal depressed	In gear
Neutral switch (M/T)	In gear	Neutral	Clutch pedal released
Idle switch	Pedal depressed	Pedal released	
Stoplight switch (M/T)	Pedal depressed	Pedal released	
A/C switch	ON	OFF	Fan speed control: Low, Mid position
Headlight switch	ON	OFF	
Rear defroster switch	ON	OFF	
Blower switch	ON	OFF	Fan speed control: High, Super-high position
Inhibitor switch (A/T)	Except N and P range	N and P range	

Caution
- **When any of the switch is activated, the monitor lamp will illuminate.**

3.0L ENGINE SWITCH MONITOR FUNCTION TROUBLESHOOTING—MPV

Procedure

Set conditions to deactivate each switch
- All accessories OFF
- Transmission in neutral
- All pedals released

Verify that monitor lamp does not illuminate

- NO → Check each switch and related wiring harness
  - Clutch and neutral switch
  - Idle switch
  - Stoplight switch
  - A/C switch
  - Headlight switch
  - Rear defroster switch
  - Blower switch
  - Inhibitor switch
- YES → Check each switch as described

Clutch and neutral switch (M/T)

Shift transmission to 1st gear
Check that monitor lamp illuminates with clutch pedal released

- NO → PC: • Clutch or neutral switch malfunction
  - Open circuit in wiring harness from clutch switch and neutral switch to engine control unit
  - Engine control unit (1G) terminal malfunction
- YES ↓

Depress clutch pedal
Check that monitor lamp does not illuminate

- NO → PC: • Clutch switch malfunction (Refer to page F1–182)
  - Short circuit in wiring harness from clutch switch to engine control unit
- YES → Clutch and neutral switch OK
  Return transmission to neutral and released clutch pedal

Idle switch

Depress accelerator pedal, and check that monitor lamp illuminates

- NO → PC: • Misadjustment of idle switch
  - Idle switch malfunction
  - Short circuit in wiring harness from idle switch to engine control unit
  - Engine control unit (1E) terminal malfunction
- YES → Idle switch OK
  Release accelerator pedal

Stoplight switch (M/T)

Depress brake pedal, and check that monitor lamp illuminates

- NO → PC: • Stoplight switch malfunction
  - Open circuit in wiring harness from stoplight switch to engine control unit
  - Engine control unit (1J) terminal malfunction
- YES → Stoplight switch OK
  Release brake pedal

Blower switch

Set fan speed control to "High—Super high" position, and check that monitor lamp illuminates

- NO → PC: • CPU malfunction
  - Fan speed control switch malfunction
  - Open circuit in the wiring harness from fan speed control switch to E/L control unit
  - Engine control unit (1H) terminal malfunction
- YES → Blower switch OK

A/C switch

Set fan speed control to Low position
Turn ON A/C switch, and check that monitor lamp illuminates

- NO → PC: • CPU malfunction
  - A/C switch malfunction
  - Fan speed control switch malfunction
  - Open circuit in wiring harness from A/C switch to CPU
  - Engine control unit (1L) terminal malfunction
- YES → A/C switch OK
  Turn off A/C switch and blower switch

Headlight switch

Turn ON headlight switch, and check that monitor lamp illuminates

- NO → PC: • CPU malfunction
  - Headlight switch malfunction
  - Open circuit in wiring harness from headlight switch to CPU
  - Engine control unit (1H) terminal malfunction
- YES → Headlight switch OK
  Turn OFF headlight switch

Rear defroster switch

Turn ON rear defroster switch, and check that monitor lamp illuminates

- NO → PC: • CPU malfunction
  - Rear defroster switch malfunction
  - Open circuit in wiring harness from rear defroster switch to CPU
  - Engine control unit (1H) terminal malfunction
- YES → Rear defroster switch OK
  Turn OFF rear defroster switch

Inhibitor switch (A/T)

Shift transmission into gear
Check that monitor lamp illuminates

- NO → PC: • Inhibitor switch malfunction
  - Open or short circuit in related wiring harness
  - Control unit (3D) terminal malfunction
- YES → Inhibitor switch OK
  Return transmission to neutral

Engine Control Unit Terminal voltage

V_B: Battery voltage

Terminal	Input	Output	Connection to	Voltage (after warm-up) Ignition switch: ON	Voltage (after warm-up) Idle	Remarks
1A	—	—	Battery	V_B		For back-up
1B	○		Main relay	V_B		
1C	○		Ignition switch (START position)	Below 2.5V		While cranking 6—11V
1D		○	Self-Diagnosis Checker (Monitor lamp)	<Test connector grounded> For 3 sec after ignition switch OFF→ON: Approx. 5V (lamp illuminates) After 3 sec: V_B	<Test connector grounded> Approx. 5V <Test connector is not grounded> Monitor lamp ON: Below 6V Monitor lamp OFF: V_B	With Self-Diagnosis Checker
1E		○	MIL (Malfunction Indicator Light)	For 3 sec. after ignition switch OFF→ON: Below 2.5V (light illuminates) After 3 sec: V_B		• Test connector grounded • Light illuminates: Approx. 5V • Light does not illuminates: V_B
1F		○	Self-Diagnosis Checker (for Code No.)	For 3 sec. after ignition switch OFF→OFF: Below 2.5V (Buzzer sounds) After 3 sec: V_B (Buzzer does not sound)		• With Self-Diagnosis Checker and Test connector grounded • Buzzer sounds: Below 2.5V • Buzzer does not sound: V_B
1G		○	Igniter	0V	Approx. 0.8V	
1H		○	Solenoid valve (Purge control)	V_B		Driving in-gear 5—15V (Engine Signal Monitor: green and red lights flash)
1I	—	—	—	—		—
1J		○	A/C relay	A/C switch ON: Below 2.5V A/C switch OFF: V_B		Blower motor: ON Ignition switch: ON
1K	○		Test connector	Test connector: Grounded 0V Test connector: Not grounded V_B		Green connector: 1-pin
1L	○		—	0V		M/T
				V_B		EC-AT
1M	○		Vehicle speed sensor	Ignition switch ON	4.5—5.5V	—
				Ignition switch OFF	0V	
1N	○		Idle switch	Pedal depressed: Approx. 10V Pedal released: 0V		
1O	—	—	—	—		—
1P	○		Power steering (P/S) switch	Constant V_B	P/S: ON: 0V P/S: OFF: V_B	
1Q	○		A/C switch	A/C switch ON: Below 2.5V A/C switch OFF: V_B		Blower motor: ON
1R	—	—	—	—		—
1S	—	—	—	—		—

V_B: Battery voltage

Terminal	Input	Output	Connection to	Voltage (after warm-up) Ignition switch: ON	Voltage (after warm-up) Idle	Remark
1T	○		CPU	Electric load ON: Below 2.5V Electric load OFF: Approx. 10—12V		E/L: Blower switch 3rd or higher Rear window defroster switch: ON Headlights: ON
1U	—		—	—		—
1V	○		Neutral/clutch switch	Clutch pedal depressed: V_B Clutch pedal released: 0V		Neutral condition
2A	—	—	Ground (E_{01})	0V		
2B	—	—	Ground (E_{02})	0V		
2C	—	—	Ground (E_1)	0V		—
2D	—	—	Ground (E_2)	0V		
2E	○		Distributor (Ne signal)	Approx. 0.6—0.8V		Ne: Number of engine speed
2F	○		Distributor (COM)	Approx. 0.6—0.8V		
2G	○		Distributor (G_1 signal)	Approx. 0.6—0.8V		G_1: Group 1
2H	○		Distributor (G_2 signal)	Approx. 0.6—0.8V		G_2: Group 2
2I	○		Igniter	0V	Approx. 0.6—0.8V	Engine Signal Monitor green and red lights flash
2J	○		Airflow meter (Vc)	7—9V		
2K		○	Vref	4.5—5.5V		
2L	○		Intake air thermo-sensor (Dynamic chamber)	Approx. 1—2V at 80°C (176°F)		
2M	○		Throttle sensor	Throttle valve fully closed: Approx. 0.3V Throttle valve fully open: Approx. 4V		
2N	○		Oxygen sensor	0V	0—1V	• Cold engine 0V at idle • After warming-up: Acceleration: 0.7—1.0V Deceleration: 0—0.2V
2O	○		Airflow meter (Vs)	Approx. 2.0V	Approx. 3.5—5V	
2P	○		Intake air thermo-sensor (airflow meter)	Approx. 2—3V at 20°C (68°F)		
2Q	○		Water thermosensor	Approx. 0.3—0.6V Approx. 2—3V at engine coolant temp. 20°C (68°F)		
2R		○	Oxygen sensor heater	Below 2.5V		Above approx. 4,000 rpm: Approx. V_B
2S		○	Solenoid valve (Variable resonance induction system (VRIS))	V_B		• 3,600—5,700 rpm: Below 2.0V • Other conditions: V_B
2T		○	Solenoid valve (Pressure regulator control)	For 180 sec after ignition switch OFF→ON Below 2.5V	For 180 sec after starting: Below 2.5V	During hot condition: coolant temp.: Above 80°C (176°F) and engine air temp.: above 75°C (167°F)
				V_B		Other conditions
2U		○	Injector (No.1 and No.2)	V_B		Engine Signal Monitor green and red lights flash

3.0L ENGINE ENGINE CONTROL UNIT TERMINAL VOLTAGES—1992 MPV

VB: Battery voltage

Terminal	Input	Output	Connection to	Voltage (after warm-up) Ignition switch: ON	Voltage (after warm-up) Idle	Remarks
2V		○	Injector (No.3 and No.4)	VB		Engine Signal Monitor green and red lights flash
2W		○	Idle speed control (ISC) valve	Above: 8V	Approx. 9—11V	Engine Signal Monitor green and red lights flash
2X		○	Injector (No.5 and No.6)	VB		Engine Signal Monitor green and red lights flash
2Y	—	—	—	—		—
2Z		○	EC-AT control unit	Sea level: VB high altitude (1,000 m [3,280 ft]: 0V)		Ignition switch: ON

Control unit connector (Control unit side)

2Y	2W	2U	2S	2Q	2O	2M	2K	2I	2G	2E	2C	2A	1U	1S	1Q	1O	1M	1K	1I	1G	1E	1C	1A
2Z	2X	2V	2T	2R	2P	2N	2L	2J	2H	2F	2D	2B	1V	1T	1R	1P	1N	1L	1J	1H	1F	1D	1B

3.0L ENGINE ENGINE CONTROL UNIT TERMINAL VOLTAGES—1990–91 MPV

Terminal	Input	Output	Connection to	Test condition	Voltage	Remarks
1A		○	Malfunction indicator light	For 3 seconds after ignition switch OFF→ON (Light illuminates)	Below 4.8V	• Test connector grounded
				After 3 seconds (Light does not illuminate)	Approx. 12V	
				Light illuminates	Below 4.8V	
				Light does not illuminate	Approx. 12V	
1B		○	Self-Diagnosis Checker (Code number)	For 3 seconds after ignition switch OFF→ON (Buzzer sounds)	Below 6.2V	• Using Self-Diagnosis Checker and test connector grounded
				After 3 seconds (Buzzer does not sound)	Approx. 12V	
				Buzzer sounds	Below 6.2V	
				Buzzer does not sound	Approx. 12V	
1C		○	A/C relay	Ignition switch ON	Approx. 12V	Blower motor: ON
				For 10 seconds after cranking with A/C switch ON (A/C does not operate)	Approx. 12V	
				After 10 seconds (A/C operates)	Below 2.5V	
				A/C switch ON at idle	Below 2.5V	
				A/C switch OFF at idle	Approx. 12V	
1D		○	Self-Diagnosis Checker (Monitor lamp)	Test connector grounded For 3 seconds after ignition switch OFF→ON (Light illuminates)	Below 6.2V	With Self-Diagnosis Checker
				After 3 seconds (Light does not illuminate)	Approx. 12V	
				Test connector not grounded at idle Monitor lamp ON	Below 6.2V	
				Test connector not grounded at idle Monitor lamp OFF	Approx. 12V	
1E	○		Idle switch	Accelerator pedal released	Approx. 0V	Ignition switch ON
				Accelerator pedal depressed	Approx. 12V	
1F	—	—	—	—	—	—
1G	○		M/T Switch	M/T	Approx. 0V	Ignition switch ON
				A/T	Approx. 12V	
1H	○		Central processing unit	Electrical load ON (Ign ON)	Below 3V	Electrical load: Rear defroster Headlight Blower motor (3rd & 4th positions)
				Electrical load OFF (Ign ON)	Above 7.3V	
1I			—	—	—	—
1J	○		Stoplight switch (M/T)	Brake pedal released	Approx. 0V	Ignition switch ON
				Brake pedal depressed	Approx. 12V	
1K	○		P/S pressure switch	P/S ON (at idle)	Approx. 0V	
				P/S OFF (at idle)	Approx. 12V	
1L	○		A/C switch	A/C switch ON (Ignition switch ON)	Approx. 0V	Blower motor: ON
				A/C switch OFF (Ignition switch ON)	Approx. 12V	
1M	○		Ignition coil ⊖ terminal	Ignition switch ON	Approx. 12V	* Engine Signal Monitor: green and red lights flash
				Idle	Approx. 12V*	
1N	○		Distributor (G signal)	Ignition switch ON	Approx. 0V	* Engine Signal Monitor: green and red lights flash
				Idle	Approx. 1V*	

3.0L ENGINE ENGINE CONTROL UNIT TERMINAL VOLTAGES—1990–91 MPV

Terminal	Input	Output	Connection to	Test condition	Voltage	Remarks
2A		○	Vref	Ignition switch ON	4.5—5.5V	
2B	○		Airflow meter (Vc) (Ground)	Constant	7—9V	
2C	—	—	Ground (E2)	Constant	0V	
2D	○		Oxygen sensor	Ignition switch ON	0V	
				Idle (Cold engine)	0V	
				Idle (After warm up)	0—1.0V	
				Increase engine speed (After warm up)	0.5—1.0V	
				Deceleration	0—0.4V	
2E	○		Airflow meter (Vs) (Intake air mass)	Ignition switch ON	Approx. 2V	
				Idle (After warm up)	Approx. 3.5—5V	
				Increase engine speed (After warm up)	5—9V	
2F	○		Test connector	Test connector (Green: 1-pin) not grounded	Approx. 12V	Ignition switch ON
				Test connector (Green: 1pin) grounded	0V	
2G	○		Throttle sensor	Accelerator pedal released	Approx. 0.3V	Ignition switch ON
				Accelerator pedal fully depressed	Approx. 4V	
2H			—	—	—	—
2I	○		Water thermo-sensor	Engine coolant temperature 20°C (68°F)	Approx. 2.5V	Ignition switch ON
				After warm up	Approx. 0.4V	
2J	○		Intake air thermosensor (Airflow meter)	At 20°C (68°F)	Approx. 2—3V	
2K		○	Solenoid valve (Purge control)	Ignition switch ON	Approx. 12V	*Engine Signal Monitor: green and red lights flash
				Idle	Approx. 12V	
				Driving in-gear	5—15V*	
2L	○		Intake air thermo-sensor (Dynamic chamber)	At 20°C (68°F)	Approx. 2.5V	
2M		○	Igniter	At sea level	Approx. 12V	Ignition switch ON
				At high altitude (1000 m [3280 ft])	Approx. 0V	
2N		○	Solenoid valve (PRC)	For 120 seconds after ignition switch OFF→ON	Below 2.5V	During hot condition. Coolant temp. above 80°C (194°F) Intake air temp. above 75°C (167°F)
				For 120 seconds after starting	Below 2.5V	
				Ignition switch ON	Approx. 12V	Other conditions
2O		○	Solenoid valve (VRIS)	3,500—5,600 rpm	Below 2.5V	
				Other conditions	Approx. 12V	
2P		○	Solenoid valve (Equip only kit)	—	—	—
2Q		○	Solenoid valve (Idle speed control)	Ignition switch ON	Above 8V	Engine Signal Monitor: green and red light flash
				Idle	Approx. 9V	
2R	—	—	Ground (Eo2)	Constant	0V	

Terminal	Input	Output	Connection to	Test condition	Voltage	Remarks
3A	—	—	Ground (Eo1)	Constant	0V	
3B	○		Ignition switch (Start position) (M/T)	While cranking	6—11V	
				Ignition switch ON	Below 2.5V	
3C		○	Injector (No.3 and No.4)	Ignition switch ON	Approx. 12V	*Engine Signal Monitor: green and red lights flash
				Idle	Approx. 12V*	
3D	○		Neutral or clutch switch (M/T)	Neutral or clutch pedal depressed	Approx. 0V	Ignition switch ON
				Other conditions	Approx. 12V	
			Inhibitor switch	"N" or "P" range	Approx. 0V	
				Other ranges	Approx. 12V	
3E		○	Injector (No.1 and No.2)	Ignition switch ON	Approx. 12V	*Engine Signal Monitor: green and red lights flash
				Idle	Approx. 12V*	
3F		○	Injector (No.5 and No.6)	Ignition switch ON	Approx. 12V	*Engine Signal Monitor: green and red lights flash
				Idle	Approx. 12V*	
3G	—	—	Ground (E1)	Constant	0V	
3H		○	Oxygen sensor	Above 2,250 rpm	Approx. 12V	
				Other condition	Below 2.5V	
3I	○		Main relay	Ignition switch OFF	Approx. 0V	
				Ignition switch ON	Approx. 12V	
3J	—	—	Battery	Constant	Approx. 12V	For backup

Control Unit Connector (Control unit side)

Control unit connector

3I	3G	3E	3C	3A	2Q	2O	2M	2K	2I	2G	2E	2C	2A	1M	1K	1I	1G	1E	1C	1A
3J	3H	3F	3D	3B	2R	2P	2N	2L	2J	2H	2F	2D	2B	1N	1L	1J	1H	1F	1D	1B

3.0L ENGINE CENTRAL PROCESSING UNIT—MPV

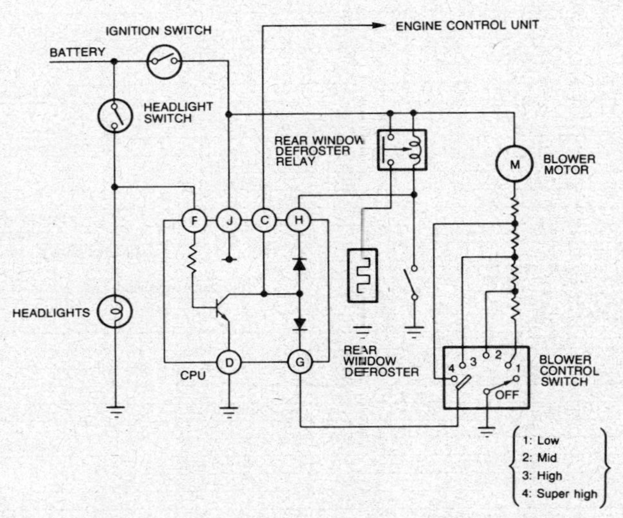

Operation

This CPU detects that an electrical load is being applied and signals the engine control unit to stabilize the idle. The CPU detects the electrical loads below.

Load	Condition of detection	Voltage (1H terminal of ECU)
Headlights	Headlights ON (Low or high beam)	0V
Rear defroster	Rear defroster ON	0V
Blower motor	Fan switch High & Super-high position	0V

3.0L ENGINE CENTRAL PROCESSING UNIT TERMINAL VOLTAGES—MPV

VB: Battery voltage

Terminal	Connected to	Test condition	Specification	To correct
b	Battery (through ROOM 10A fuse)	Constant	VB	Check ROOM 10A fuse and wiring harness
c	Engine control unit	Ignition switch ON	VB	Check engine control unit and wiring harness
d	Ground	Constant	0V	Check wiring harness
f	Headlight switch	Headlight switch ON	VB	Check headlight switch and wiring harness
g	Blower fan switch	Blower fan switch High or Super-high position	0V	Check blower fan switch and wiring harness
h	Rear window defroster switch	Rear window defroster switch ON	0V	Check rear window defroster switch and wiring harness
j	Ignition switch	Ignition switch ON	VB	Check ignition switch and wiring harness

SPECIFICATION CHART—B2200, B2600i

Item		Unit	Specification
Idle speed*1		rpm	M/T: 730—770, A/T: 750—790 (P range)
Ignition timing*1		BTDC	**G6**: 4—6°, **F2**: 5—7°
Throttle body			
Type			Horizontal draft (1 barrel)
Throat diameter	No.1 No.2	mm (in)	**G6**: 55 (2.2) / **F2**: 50 (2.0)
Fuel pump			
Type			Impeller (in-tank)
Output pressure		kPa (kg/cm², psi)	441—589 (4.5—6.0, 64—85)
Fuel filter			
Type	Low-pressure side		Nylon element
	High-pressure side		Paper element
Pressure regulator			
Type			Diaphragm
Regulating pressure		kPa (kg/cm², psi)	265—314 (2.7—3.2, 38—46)
Injector			
Type			High-ohmic
Type of drive			Voltage
Resistance		Ω	12—16 (at 20°C, 68°F)
Volume			**G6**: 74—89 cc (4.51—5.43 cu in)/15 sec. **F2**: 50—62 cc (3.05—3.78 cu in)/15 sec.
BAC valve (solenoid valve [Idle speed control])			
Solenoid resistance		Ω	7.7—9.3 (at 23°C, 73°F)
Solenoid valve (Purge control)			
Solenoid resistance		Ω	30—34 (at 20°C, 68°F)
Water thermosensor			
Resistance	−20°C (−4°F)	kΩ	14.5—17.8
	20°C (68°F)	kΩ	2.2—2.7
	80°C (176°F)	kΩ	0.28—0.35
Intake air thermosensor			
Resistance	25°C (77°F)	kΩ	29.7—36.3
	85°C (185°F)	kΩ	3.3—3.7
Circuit opening relay			
Resistance	STA—E1	Ω	21—43
	B—Fc	Ω	109—226
	B—Fp	Ω	∞
Fuel tank			
Capacity		liters (US gal, Imp gal)	56 (14.8, 12.3)
Air cleaner			
Element type			Dry
Accelerator cable			
Free play		mm (in)	1—3 (0.039—0.118)
Fuel			
Specification			Unleaded regular (RON 87 or higher)

COMPONENT DESCRIPTIONS—B2200, B2600i

Component	Function	Remarks
Air cleaner	Filters air entering throttle body	
Airflow sensor	Detects amount of intake air; sends signal to engine control unit	
Air valve	When cold, supplies bypass air into dynamic chamber	• Engine speed increased to shorten warm-up period • Thermowax type • Installed in BAC valve
Atmospheric pressure sensor	Detects atmospheric pressure	In ECU
BAC valve	Supplies bypass air into dynamic chamber	Consists of air valve and ISC valve
Catalytic converter	Reduces HC, CO, and NOx by chemical reaction	Monolith type
Charcoal canister	Stores gas tank fumes when engine stopped	
Check connector	For Self-Diagnosis Checker	6-pin connector (Green)
Check-and-cut valve	Releases excessive pressure or vacuum in fuel tank to atmosphere	
Circuit opening relay	Voltage for fuel pump while engine running	
Clutch switch	Detects in-gear condition; sends signal to engine control unit	Switch ON when clutch pedal depressed
Crank angle sensor (In distributor)	1. Detects No.1 cylinder TDC; sends signal to engine control unit 2. Detects engine speed; sends signal to engine	For determining fuel injection timing
Dynamic chamber	Interconnects all cylinders	
Engine control unit	Detects following: 1. Engine speed 2. No.1 piston TDC 3. Intake air amount 4. Engine coolant temperature 5. Ignition ON signal 6. Throttle valve opening angle 7. Throttle valve fully closed 8. Air/fuel ratio (Oxygen concentration) 9. In-gear condition 10. Intake air temperature 11. Atmospheric pressure 12. A/C operation 13. P/S operation 14. E/L operation 15. Cranking signal 16. Test signal (idle speed, malfunction code No.) 17. Braking signal Controls operation of the following: 1. Fuel injection system 2. Idle speed control 3. Pressure regulator control system 4. Purge control system 5. Fail-safe function 6. Monitor function 7. Burn-off system 8. Ignition timing control system 9. Fuel pump 10. A/C (cut off) 11. Main relay control	 1. Ne-Signal 2. G-signal 3. Airflow sensor 4. Water thermosensor 5. Ignition switch 6. Throttle sensor 7. Idle switch 8. Oxygen sensor 9. Neutral and clutch switches 10. Intake air thermosensor (on dynamic chamber) 11. Atmospheric pressure sensor (In ECU) 12. A/C switch 13. P/S pressure switch 14. Headlight and blower switches 15. Ignition switch (START position) 16. Test connector 17. Stoplight switch 1. Injector 2. Solenoid valve (Idle speed control) 3. Solenoid valve (Pressure regulator control) 4. Solenoid valve (Purge control) 5. Self-Diagnosis Checker and MIL 6. Monitor lamp (Self-Diagnosis Checker) 7. Airflow sensor 8. Igniter 9. Circuit opening relay 10. A/C relay 11. Main relay

COMPONENT DESCRIPTIONS—B2200, B2600i

Component		Function	Remarks
Fuel filter		Filters particles from fuel	
Fuel pump		Provides fuel to injectors	• Operates while engine running • Installed in fuel tank
Fuel vapor valve		Prevents fuel from flowing into charcoal canister	
Idle switch		Detects when throttle valve fully closed; sends signal to engine control unit	Installed on throttle body
Igniter		Receives spark signal from signal ECU and generates high voltage to ignition coil	
Ignition switch (START position)		Sends engine cranking signal to engine control unit	
Injector		Injects fuel into intake port	• Controlled by signals from engine control unit • High-ohmic injector • Two port injector nozzle **(G6)**
Intake air thermosensor		Detects intake air temperature; sends signal to engine control unit	Installed on dynamic chamber
Main relay		Supplies electric current to injectors and engine control unit.	
MIL (Malfunction indicator lamp)		(For Federal and Canada) Lamp illuminates to indicate the maintenance schedule for the emission control system	Every 60,000 and 80,000 miles (Federal) or 90,000 and 130,000 km (Canada)
		(For California) Lamp illuminates when input device malfunctions	Test connector not grounded
		(For California) Lamp flashers to indicate malfunction code No. of input and output devices	Test connector grounded
Neutral switch		Detects in-gear condition; sends signal to engine control unit	Switch ON when neutral
Oxygen sensor		Detects oxygen concentration; sends signal to engine control unit	Zirconia ceramic and platinum coating
PCV valve		Controls amount of blowby gas introduced into engine	
P/S pressure switch		Detects P/S operation; sends signal to engine control unit	P/S: ON when steering wheel turned right or left
Pressure regulator		Adjusts fuel pressure supplied to injectors	
Resonance chamber (G6)		Improves mid-range torque characteristics	
Separator		Prevents fuel from flowing into charcoal canister	
Solenoid valve	Idle speed control	Controls bypass air amount	• Controlled by duty signal from engine control unit • With integrated air valve • Controls idle-up
Solenoid valve	Pressure regulator control	Controls vacuum to pressure regulator	Cuts vacuum passage when hot
Solenoid valve	Purge control	Controls evaporative fumes from canister to intake manifold	
Stoplight switch		Detects braking operation (deceleration); sends signal to engine control unit	
Test connector		For Self-Diagnosis Checker and idle speed ignition timing adjustment	1-pin connector (Green)
Throttle body		Controls intake air quantity	Integrated throttle sensor and idle switch
Throttle sensor		Detects throttle valve opening angle; sends signal to engine control unit	Installed on throttle body
Two-way check valve		Controls pressure in fuel tank	
Water thermosensor		Detects coolant temperature; sends signal to engine control unit	

TROUBLESHOOTING GUIDE—B2200, B2600i

INPUT DEVICES \ OUTPUT DEVICES		INJECTOR: FUEL INJECTION AMOUNT	INJECTOR: FUEL INJECTION TIMING	BAC VALVE: AIR VALVE	BAC VALVE: ISC VALVE	SOLENOID VALVE (PURGE CONTROL)	SOLENOID VALVE (PRESSURE REGULATOR CONTROL)	A/C RELAY (A/C CUT-OFF)	AIRFLOW SENSOR (BURN-OFF)	CIRCUIT OPENING RELAY (FUEL PUMP CONTROL)	IGNITER (IGNITION TIMING CONTROL)
TEST CONNECTOR		×	×	×	○	×	×	×	×	×	○
IGNITION SWITCH (ON POSITION)		×	×	×	×	×	×	×	○	×	×
IGNITION SWITCH (START POSITION)		○	○	×	○	×	○	○	×	×	○
HEADLIGHT AND BLOWER SWITCH		×	×	×	○	×	×	×	×	×	×
P/S PRESSURE SWITCH		×	×	×	○	×	×	×	×	×	×
A/C SWITCH		×	×	×	○	×	×	○	×	×	×
NEUTRAL AND CLUTCH SWITCH		○	×	×	○	○	×	○	×	×	×
STOPLIGHT SWITCH		○	×	×	×	×	×	×	×	×	×
IDLE SWITCH		○	×	×	○	○	○	×	×	×	○
ATMOSPHERIC PRESSURE SENSOR		○	×	×	F2 × G6 ○	×	×	×	×	×	×
THROTTLE SENSOR		○	×	×	×	×	○	○	×	×	×
INTAKE AIR THERMOSENSOR		○	×	×	×	×	○	×	×	×	×
AIRFLOW SENSOR		○	×	×	○	○	×	×	○	×	○
OXYGEN SENSOR		○	×	×	×	○	×	×	×	×	×
WATER THERMOSENSOR		○	×	×	○	○	○	×	○	×	○
DISTRIBUTOR	(Ne-SIGNAL)	○	○	×	○	○	○	×	○	○	○
DISTRIBUTOR	(G-SIGNAL)	×	○	×	×	×	×	×	×	×	×

ENGINE CONTROL OPERATION CHART—B2200, B2600i

INPUT DEVICES / ENGINE CONDITIONS	APPROXIMATE TIME (BASED ON 10—16°C or 50—60°F AMBIENT)	SENSORS: DISTRIBUTOR (G-SIGNAL)	SENSORS: DISTRIBUTOR (Ne-SIGNAL)	SENSORS: WATER THERMOSENSOR	SENSORS: OXYGEN SENSOR	SENSORS: AIRFLOW SENSOR	SENSORS: INTAKE AIR THERMOSENSOR	SENSORS: THROTTLE SENSOR	SENSORS: ATMOSPHERIC PRESSURE SENSOR (IN ECU)
CRANKING —COLD ENGINE • COLD AIR • COLD COOLANT	Zero				Signal has no effect on ECU	Signal has no effect on ECU	Signal has no effect on ECU	Signal has no effect on ECU	
COLD START —FAST IDLE • COLD AIR • COLD COOLANT	One minutes			Cool to warm: medium voltage (3.5V and dropping)	Sensor cold: low to high voltage (0—0.9V)	Low volume airflow: low to high voltage (2.4—2.6V)		Closed throttle: low voltage (0.3—0.7V)	
COLD DRIVEAWAY —PART THROTTLE • COLD AIR • COLD COOLANT	Two minutes								
WARM DRIVEAWAY —PART THROTTLE • WARM AIR • WARM COOLANT	Three minutes			Warm: medium voltage (Approx. 0.7V and dropping)	Sensor warm: high voltage (0.9V)	Moderate volume airflow: low to medium voltage (3.0V)		Part throttle: medium voltage (1—3.5V)	
HOT CRUISE • WARM AIR • WARM COOLANT		Sends No.1 cylinder TDC signal to ECU	Sends engine speed signal to ECU		Sensor hot: switching from high voltage (0.9V) to low voltage (0.1V)		Cool to warm: medium voltage (1.4—3.4V)		Sends voltage signal to ECU that varies with altitude: voltage (approx. 4V at sea level)
HOT ACCELERATION —60% THROTTLE						Moderate to strong volume of airflow: (3.8V)			
HOT ACCELERATION —WIDE OPEN THROTTLE	More than four minutes			Hot: low voltage (Approx. 0.4V)	High voltage (0.9V)	Strong volume of airflow: (4.0V)		Wide open throttle: high voltage (Approx. 4.0V)	
DECELERATION —CLOSED THROTTLE					Low voltage (0V)	Low volume of airflow: (2.4V)		Closed throttle: low voltage (0.3—0.7V)	
HOT CURB IDLE —EXTENDED					Switching from high to low voltage (0.75—0.25V)				
HOT ENGINE SHUTDOWN	—	OFF	OFF	OFF	Sensor hot: low voltage (0.1V) until sensor cools	OFF	OFF	OFF	OFF

ENGINE CONDITIONS	SWITCHES: IDLE SWITCH	SWITCHES: STOP-LIGHT SWITCH	SWITCHES: NEUTRAL AND CLUTCH SWITCHES	SWITCHES: A/C SWITCH	SWITCHES: P/S PRESSURE SWITCH	SWITCHES: HEAD-LIGHT SWITCH	SWITCHES: BLOWER SWITCH	SWITCHES: IGNITION SWITCH START POSITION	SWITCHES: IGNITION SWITCH ON POSITION	SWITCHES: TEST CONNECTOR
CRANKING	Signal has no effect on ECU	Signal has no effect on ECU	Signal has no effect on ECU	Signal has no effect on ECU	Signal has no effect on ECU	Signal has no effect on ECU	Signal has no effect on ECU	Sends signal to ECU (approx. 12V)	Signal has no effect on ECU	Signal has no effect on ECU
COLD START	Low voltage signal to ECU (below 1.5V)	Brake pedal depressed: sends signal to ECU (approx. 12V)	In neutral: low voltage signal to ECU (approx. 0V)							
COLD DRIVEAWAY										
WARM DRIVEAWAY										
HOT CRUISE	High voltage signal to ECU (battery voltage)	No signal send to ECU (below 1.5V)	Driving in any gear: high voltage signal to ECU (battery voltage)	A/C switch ON: sends signal to ECU (battery voltage)	Steering wheel turned: low voltage signal to ECU (below 1.5V)	Headlight switch ON: low voltage signal to ECU (below 1.5V)	Blower switch ON: low voltage signal to ECU (below 1.5V)	No signal to ECU (below 1.5V)	Sends signal to ECU (battery voltage)	Connector not grounded: high voltage signal to ECU (battery voltage)
HOT ACCELERATION —60% THROTTLE				A/C switch OFF: no signal to ECU (below 1.5V)	Steering wheel straight ahead: high voltage signal to ECU (battery voltage)	Headlight switch OFF: high voltage signal to ECU (battery voltage)	Blower switch OFF: high voltage signal to ECU (battery voltage)			
HOT ACCELERATION —WIDE OPEN THROTTLE										
DECELERATION	Low voltage signal to ECU (below 1.5V)	Brake pedal depressed: sends signal to ECU (approx. 12V)								
HOT CURB IDLE			In neutral: low voltage signal to ECU (approx. 0V)							Low voltage signal to ECU when connector grounded (below 1.5V)
HOT ENGINE SHUTDOWN	OFF	OFF	OFF	OFF	OFF	OFF	OFF	OFF	OFF	OFF

ENGINE CONTROL OPERATION CHART—B2200, B2600i

ENGINE CONDITIONS \ OUTPUT DEVICES	APPROXI-MATE TIME (BASED ON 10–16°C or 50–60°F AMBIENT)	INJECTOR: INJECTION	INJECTOR: INJECTION TIMING	BAC VALVE: AIR VALVE	BAC VALVE: ISC VALVE	SOLENOID VALVE (PURGE CONTROL)	SOLENOID VALVE (PRESSURE REGULATOR CONTROL)	A/C RELAY (A/C CUT-OFF)	AIRFLOW SENSOR (BURN-OFF)
CRANKING —COLD ENGINE • COLD AIR • COLD COOLANT	Zero		All cylinders each ignition pulse			OFF (Purge cut)		OFF (A/C ON)	
COLD START —FAST IDLE • COLD AIR • COLD COOLANT	One minute	Rich		Open (coolant tempera-ture: below 50°C 122°F)	Large amount of bypass air			ON (A/C OFF: approx. 5 sec.)	
COLD DRIVEAWAY —PART THROTTLE • COLD AIR • COLD COOLANT	Tow minutes								
WARM DRIVEAWAY —PART THROTTLE • WARM AIR • WARM COOLANT	Three minutes	Rich and lean	2-group			Operates (Duty values [purge gas amount] change)	OFF (Vacuum to pressure regulator)	OFF (A/C ON)	
HOT CRUISE • WARM AIR • WARM COOLANT					Small amount of bypass air				OFF
HOT ACCELERATION —60% THROTTLE		Rich						ON (A/C CUT)	
HOT ACCELERATION —WIDE OPEN THROTTLE	More than four minutes			Closed					
DECELERATION —CLOSED THROTTLE		Fuel cut			Large and small amount of bypass air	OFF (Purge cut)		OFF (A/C ON)	
HOT CURB IDLE —EXTENDED		Rich and lean	2-group		Small amount of bypass air		After starting: ON during hot start only (Vacuum cut)		
HOT ENGINE SHUTDOWN	—	Does not inject			OFF	OFF	OFF	OFF	ON (Burn-off)

ENGINE CONDITIONS \ OUTPUT DEVICES	CIRCUIT OPENING RELAY	MAIN RELAY	IGNITER
CRANKING —COLD ENGINE • COLD AIR • COLD COOLANT	OFF		
COLD START —FAST IDLE • COLD AIR • COLD COOLANT			
COLD DRIVEAWAY —PART THROTTLE • COLD AIR • COLD COOLANT			
WARM DRIVEAWAY —PART THROTTLE • WARM AIR • WARM COOLANT			
HOT CRUISE • WARM AIR • WARM COOLANT	ON (Engine speed above 50 rpm)	ON	Operation
HOT ACCELERATION —60% THROTTLE			
HOT ACCELERATION —WIDE OPEN THROTTLE			
DECELERATION —CLOSED THROTTLE			
HOT CURB IDLE —EXTENDED			
HOT ENGINE SHUTDOWN	OFF	OFF	OFF

SYMPTON TROUBLESHOOTING—B2200, B2600i

Cranks normally but won't start (No combustion)

STEP	QUICK INSPECTION		ACTION		POSSIBLE CAUSE AND DETAILED INSPECTION
1	Check for malfunction code (02) with SST [IGN ON, Test connector (Green: 1-pin) grounded]	Yes	Check for cause by referring to check sequence		
		No	Go to Next Step		
2	Check for spark by disconnecting high-tension lead while cranking	Yes	Go to Next Step		
		No	Check ignition system (Refer to ignition system troubleshooting)		
3	Check for fuel pump operating sound from fuel filler port [IGN ON, Test connector (Yellow: 2-pin) connected]	Yes	Check if engine starts in this condition	Yes	Check circuit opening relay switching operation
					Check circuit opening relay circuit (IGN: START)
				No	Go to Next Step
		No	Check circuit opening relay switching operation	Yes	Check circuit opening relay circuit
					Check fuel pump operation
				No	Replace circuit opening relay
4	Check fuel line pressure [IGN ON, Test connector (Yellow: 2-pin) connected] **Fuel line pressure: 265—314 kPa (2.7—3.2 kg/cm², 38—46 psi)**	Yes	Go to Next Step		
		No	Check fuel pump maximum pressure **Fuel pump maximum pressure: 441—588 kPa (4.5—6.0 kg/cm², 64—85 psi)**	Yes	Replace pressure regulator
				No	Replace fuel pump
5	Check for injector operating sound while cranking	Yes	Go to Next Step		
		No	Check voltage at ECU (2U) and (2V) terminals with SST **Voltage: Approx. 12V (IGN ON)**	Yes	Check throttle sensor
					Replace ECU
				No	Check wiring for short or open
					Poor ground circuit from ECU (2A) terminal (Check terminal voltage with SST)
6	Substitute a well-known ECU. Check if the condition improves	Yes			ECU malfunction
		No	Check ground circuit from ECU (2B) terminal with SST **Voltage: 0V (IGN ON)**	Yes	Go to Next Step
				No	Poor ground circuit
7					Low compression

Cranks normally but hard to start (Always)

STEP	QUICK INSPECTION		ACTION		POSSIBLE CAUSE AND DETAILED INSPECTION
1	Check if vacuum hoses and the air hoses are connected correctly	Yes	Go to Next Step		
		No	Connect correctly		
2	Check air cleaner element for clogging	Yes	Go to Next Step		
		No	Clean air cleaner element		
3	Check ignition timing at idle after warm up **Ignition timing: BTDC 4—6° (G6) 5—7° (F2)** [Test connector (Green: 1-pin) grounded]	Yes	Go to Next Step		
		No	Adjust ignition timing		
4	Disconnect high-tension lead of each cylinder at idle. Check if engine condition changes	Yes	Go to Next Step		
		No	Check ignition	Yes	Replace injector (If step 3 OK)
				No	Check spark plug
					Check high-tension lead
					Check distributor cap
5	Check for injector operating sound at idle	Yes	Go to Next Step		
		No	Check resistance at injector harness connector (EMINJ-01) Terminals: (B/Y)—(LG/B), (B/Y)—(LG/R); Resistance: 6—8Ω	Yes	Check wiring short or open
				No	Check injector resistance
					Check wiring short or open
6	Check fuel line pressure [IGN ON, Test connector (Yellow: 2-pin) connected] **Fuel line pressure: 265—314 kPa (2.7—3.2 kg/cm², 38—46 psi)**	Yes	Go to Next Step		
		No	Check if fuel filter has been replaced according to maintenance schedule	Yes	Check fuel line for clogging
				No	Replace fuel filter
			Check fuel pump maximum pressure **Fuel pump maximum pressure: 441—588 kPa (4.5—6.0 kg/cm², 64—85 psi)**	Yes	Replace pressure regulator
				No	Replace fuel pump
7	Operate fuel pump [IGN ON, Test connector (Yellow: 2-pin) connected] Turn ignition switch OFF and observe fuel pressure **for 5 minutes** **Fuel pressure: More than 147 kPa (1.5 kg/cm², 21 psi)**	Yes	Go to Next Step		
		No	Check fuel pump pressure drop	No	Replace fuel pump
			Check pressure regulator pressure drop	Yes	Check injector fuel leakage
				No	Replace pressure regulator
8					Check compression

SYMPTON TROUBLESHOOTING—B2200, B2600i

Cranks normally but hard to start (Only when engine is cold)

STEP	QUICK INSPECTION		ACTION		POSSIBLE CAUSE AND DETAILED INSPECTION
1	Check specific gravity of battery using a hydrometer **Specific gravity: Above 1.200**	Yes	Go to Next Step		
		No	Recharge battery		
2	Check for malfunction code (09) (26) with SST [IGN ON, Test connector (Green: 1-pin) grounded]	Yes	Check for cause by referring to check sequence		
		No	Go to Next Step		
3	Disconnect high-tension lead of each cylinder at idle Check if engine condition changes	Yes	Go to Next Step		
			Check ignition system		Check spark plug
					Check high-tension lead
					Check distributor cap
4	Check fuel line pressure [IGN ON, Test connector (Yellow: 2-pin) connected] **Fuel line pressure: 265—314 kPa (2.7—3.2 kg/cm², 38—46 psi)**	Yes	Go to Next Step		
		No	Check for fuel leaks		
			Check if fuel filter has been replaced according to maintenance schedule	Yes	Check fuel line for clogging
				No	Replace fuel filter
			Check fuel pump maximum pressure **Fuel pump maximum pressure: 441—588 kPa (4.5—6.0 kg/cm², 64—85 psi)**	Yes	Replace pressure regulator
				No	Replace fuel pump
5	Disconnect ISC valve connector when engine is cold Check if idle speed decreases during warm up	Yes	Go to Next Step		
		No			Check if BAC valve (air valve) opens when cold
6	Check voltage at ECU (1C) terminal with SST **Voltage: Approx. 10V (while cranking)**	Yes	Go to Next Step		
		No	Check starter interlock switch	Yes	Check related wiring
				No	Replace switch
7	Check voltage at ECU (2Q) terminal with SST **Voltage: Approx. 2.5V (IGN ON, Engine coolant temperature 20°C [68°F])**	Yes	Go to Next Step		
		No			Check water thermosensor
8					Check compression

Cranks normally but hard to start (Only when engine is warm)

STEP	QUICK INSPECTION		ACTION		POSSIBLE CAUSE AND DETAILED INSPECTION
1	Check for malfunction code with SST [IGN ON, Test connector (Green: 1-pin) grounded]	Yes	Check for cause by referring to check sequence		
		No	Go to Next Step		
2	Disconnect water thermosensor connector Check if condition improves	Yes	Check water thermosensor connector condition as follows: 1. Shake connector and check if condition changes 2. Check condition of terminal (burned or damaged) 3. Connect a good terminal to harness side connector and check for looseness	Yes	Check water thermosensor
				No	Poor contact of water thermosensor connector
		No	Go to Next Step		
3	Operate fuel pump [IGN ON, Test connector (Yellow: 2-pin) connected] Turn ignition switch OFF and observe fuel pressure **for 5 minutes** **Fuel pressure: More than 147 kPa (1.5 kg/cm², 21 psi)**	Yes	Go to Next Step		
		No	Check fuel pump pressure drop	No	Replace fuel pump
			Check pressure regulator pressure drop	Yes	Check injector fuel leakage
				No	Replace pressure regulator
4					ECU malfunction

SYMPTON TROUBLESHOOTING—B2200, B2600i

Cranks normally but hard to start (Only after heat soak)

STEP	QUICK INSPECTION		ACTION		POSSIBLE CAUSE AND DETAILED INSPECTION
1	Check for malfunction code with SST [IGN ON, Test connector (Green: 1-pin) grounded]	Yes	Check for cause by referring to check sequence		
		No	Go to Next Step		
2	Circulate fuel by operating fuel pump **for 20 seconds** [IGN ON, Test connector (Yellow: 2-pin) connected] Check if condition improves	Yes	Go to Step 3		
		No	Go to Step 4		
3	Disconnect vacuum hose from pressure regulator Check if condition improves	Yes	Check the components related to pressure regulator control system		Check water thermo-sensor
					Check intake air thermosensor
					Check solenoid valve (PRC)
					ECU malfunction (Check (2T) terminal voltage)
		No	Go to Next Step		
4	Operate fuel pump [IGN ON, Test connector (Yellow: 2-pin) connected] Turn ignition switch OFF and observe fuel pressure **for 5 minutes** **Fuel pressure: More than 147 kPa (1.5 kg/cm², 21 psi)**	Yes	Go to Next Step		
		No	Check fuel pump pressure drop	No	Replace fuel pump
			Check pressure regulator pressure drop	Yes	Check injector fuel leakage
				No	Replace pressure regulator
5	Change fuel with specified one Check if condition improves	Yes			Poor fuel quality
		No	Go to Next Step		
6					ECU malfunction

Cranks normally but won't start (Intermittent)

STEP	QUICK INSPECTION		ACTION	POSSIBLE CAUSE AND DETAILED INSPECTION
1	Shake connector of ignition coil, main relay and ECU while cranking Check if the engine starts	Yes	There may be a poor contact of the connector. Repair or replace the wiring	
		No	Go to troubleshooting "Cranks normally but hard to start (Always)"	

Rough idle (Always)

STEP	QUICK INSPECTION		ACTION		POSSIBLE CAUSE AND DETAILED INSPECTION
1	Check for malfunction code with SST [IGN ON, Test connector (Green: 1-pin) grounded]	Yes	Check for cause by referring to the check sequence		
		No	**"88" flashing** Check voltage at ECU (2C) terminal with SST **Voltage: 0V (IG ON)**	Yes	Replace ECU
				No	Poor ground circuit
			"00" Go to Next Step		
2	Check ignition at idle after warm up **Ignition timing: BTDC 4—6° (G6) 5—7° (F2)** [Test connector (Green: 1-pin) grounded]	Yes	Go to Next Step		
		No	Adjust ignition timing (If possible)		
3	Disconnect high-tension lead of each cylinder at idle Check if engine condition changes	Yes	Go to Next Step		
		No	Check ignition system	Yes	Replace injector (If Step 3 OK)
				No	Check spark plug
					Check high-tension lead
					Check distributor cap
4	Check idle speed after warm up **Idle speed: 730—770 rpm (M/T) 750—790 rpm (A/T, P range)** [Test connector (Green: 1-pin) grounded]	Yes	Go to Next Step		
		No	Adjust idle speed (If possible)		
5	Check for injector operating sound at idle	Yes	Go to Next Step		
		No	Check resistance at injector harness connector (EMINJ-01) Terminals / Resistance: (B/Y)—(LG/B), (BY)—(LG/R) / 6—8Ω	Yes	Check wiring short or open
				No	Check injector resistance
					Check wiring short or open
6	Check fuel line pressure [IGN ON, Test connector (Yellow: 2-pin) connected] **Fuel line pressure: 265—314 kPa (2.7—3.2 kg/cm², 38—46 psi)**	Yes	Go to Next Step		
		No	Check for fuel leakage		
			Substitute a good fuel filter and retest	Yes	Replace fuel filter
			Check fuel pump maximum pressure **Fuel pump maximum pressure: 441—588 kPa (4.5—6.0 kg/cm², 64—85 psi)**	Yes	Replace pressure regulator
				No	Replace fuel pump
7	Check intake manifold vacuum at idle **Vacuum: 500—540 mmHg (19.7—21.3 inHg)**	Yes	Go to Next Step		
		No	Check for air leaks	Yes	Intake air system components damaged
					Vacuum and intake air hoses loose or damaged
					Bolts or nuts loose
					Gaskets damaged
				No	Check throttle valve closing condition
8	Substitute a well-known ECU Check if condition improves	Yes			ECU malfunction
		No	Check voltage at ECU (2C) terminal with SST **Voltage: 0V (IGN ON)**	Yes	Go to Next Step
				No	Poor ground circuit
9					Check compression

Rough idle (Only when engine is cold)

STEP	QUICK INSPECTION		ACTION	POSSIBLE CAUSE AND DETAILED INSPECTION
1	Check for malfunction code with SST [IGN ON, Test connector (Green: 1-pin) grounded]	Yes	Check for cause by referring to check sequence	
		No	Go to Next Step	
2	Check ignition at idle after warm up **Ignition timing:** **BTDC 4—6° (G6)** **5—7° (F2)** [Test connector (Green: 1-pin) grounded]	Yes	Go to Next Step	
		No	Adjust ignition timing (If possible)	
3	Disconnect high-tension lead of each cylinder at idle Check if engine condition changes	Yes	Go to Next Step	
		No	Check ignition system	Yes: Replace injector (If step 4 OK) No: Check spark plug Check high-tension lead Check distributor cap
4	Check for injector operating sound at idle	Yes	Go to Next Step	
		No	Check resistance at injector harness connector (EMINJ-01) **Terminals** (B/Y)—(LG/B), (B/Y)—(LG/R) — **Resistance** 6—8Ω	Yes: Check wiring short or open No: Injector malfunction (Check resistance) Wiring short or open
5	Disconnect ISC valve connector at idle when engine is cold Check if idle speed decreases during warm up	Yes	Go to Next Step	
		No		Check if BAC valve (air valve) opens when cold
6	Check if specified engine oil is used	Yes	Go to Next Step	
		No	Change engine oil to specified oil	
7	Subsitute a well-known ECU Check if condition improves	Yes		ECU malfunction
		No		Check airflow sensor

Rough idle (Only when engine is warm)

STEP	QUICK INSPECTION		ACTION	POSSIBLE CAUSE AND DETAILED INSPECTION
1	Run engine at **2,000 rpm** for more than **20 seconds** Check for malfunction code with SST [IGN ON, Test connector (Green: 1-pin) grounded]	Yes	Check for cause by referring to check sequence	
		No	Go to Next Step	
2	Check idle speed after warm up **Idle speed: 730—770 rpm (M/T)** **750—790 rpm (A/T, P range)** [Test connector (Green: 1-pin) grounded]	Yes	Go to Next Step	
		No	Adjust idle speed (If possible)	
3	Check for flashing of SST monitor lamp after warm up **Monitor lamp: Flashes more than 8 times/10 seconds at 2,000—3,000 rpm** [Test connector (Green: 1-pin) not grounded]	Yes	Go to Next Step	
		No		Replace oxygen sensor
4	Disconnect ISC valve connector after warm up Check if engine speed drops	Yes	Go to Next Step	
		No		Check ISC valve
5	Disconnect water thermosensor connector Check if condition improves	Yes	Check water thermosensor connector condition as follows: 1. Shake connector and check if condition changes 2. Check condition of terminal (burned or damaged) 3. Connect a good terminal to harness side connector and check for looseness	Yes: Check water thermosensor No: Poor contact of water thermosensor connector
		No	Go to Next Step	
6	Disconnect high-tension lead of each cylinder at idle Check if engine condition changes	Yes	Go to Next Step	
		No	Check ignition system [Refer to ignition system troubleshooting (Misfire)] Note: If spark plug is wet, injector may be leaking	Yes: Replace injector (If step 7 OK) No: Check spark plug Check high-tension lead Check distributor cap
7	Check for injector operating sound at idle	Yes	Go to Next Step	
		No	Check resistance at injector harness connector (EMINJ-01) **Terminals** (B/Y)—(LG/B), (B/Y)—(LGR) — **Resistance** 6—8Ω	Yes: Check wiring short or open No: Check injector resistance Check wiring short or open
8	Check for air leaks by listening for sucking noise	Yes	Go to Next Step	
		No		Intake air system components damaged Vacuum and intake air hoses loose or damaged Bolts or nuts loose Gaskets damaged
9				Check compression

SYMPTON TROUBLESHOOTING—B2200, B2600i

Rough idle (Only after heat soak)

STEP	QUICK INSPECTION		ACTION		POSSIBLE CAUSE AND DETAILED INSPECTION
1	Run engine at **2,000 rpm** for more than **20 seconds** Check for malfunction code with SST [IGN ON, Test connector (Green: 1-pin) grounded]	Yes	Check for cause by referring to check sequence		
		No	Go to Next Step		
2	Check switches with SST Neutral switch Clutch switch [IGN ON, Test connector (Green: 1-pin) grounded]	Yes	Go to Next Step		
		No	Check for cause by referring to check sequence		
3	Check for flashing of SST monitor lamp after warm up **Monitor lamp:** **Flashes more than 8 times 10 seconds at 2,000—3,000 rpm** [Test connector (Green: 1-pin) not grounded]	Yes	Go to Next Step		
		No			Replace oxygen sensor
4	Disconnect vacuum hose from pressure regulator Check if condition improve	Yes	Check components related to pressure regulator control system		Check water thermosensor
					Check intake air thermosensor
					Check solenoid valve (PRC)
					ECU malfunction (Check (2T) terminal voltage)
		No	Go to Next Step		
5	Run engine at idle and stop it Observe fuel pressure **for 5 minutes** **Fuel pressure:** **More than** **147 kPa (1.5 kg/cm², 21 psi)**	Yes	Go to Next Step		
		No	Check fuel pump pressure drop	No	Replace fuel pump malfunction
			Check pressure regulator pressure drop	Yes	Check injector fuel leakage
				No	Replace pressure regulator
6	Disconnect high-tension lead of each cylinder at idle Check if engine condition changes	Yes	Go to Next Step		
		No	Check ignition system	Yes	Replace injector (If step 3 OK)
				No	Check spark plug
					Check high-tension lead
					Check distributor cap
7	Check for injector operating sound at idle	Yes	Go to Next Step		
		No	Check resistance at injector harness connector (EMINJ-01) Terminals: (B/Y)—(LG/B), (B/Y)—(LG/R); Resistance: 6—8Ω	Yes	Check wiring short or open
				No	Check injector resistance
					Check wiring short or open
8	Change fuel to specified grade Check if condition improves	Yes			Poor fuel quality
		No	Go to Next Step		
9					ECU malfunction

Rough idle just after starting

STEP	QUICK INSPECTION		ACTION		POSSIBLE CAUSE AND DETAILED INSPECTION
1	Check for malfunction code with SST [IGN ON, Test connector (Green: 1-pin) grounded]	Yes	Check for cause by referring to check sequence		
		No	Go to Next Step		
2	Check idle switch with SST [IGN ON, Test connector (Green: 1-pin) grounded]	Yes	Go to Next Step		
		No	Check for cause by referring to check sequence		
3	Check ignition timing at idle after warm up **Ignition timing:** **BTDC 4—6° (G6)** **5—7° (F2)** [Test connector (Green: 1-pin) not grounded]	Yes	Go to Next Step		
		No	Adjust ignition timing		
4	Check idle speed after warm up **Idle speed: 730—770 rpm (M/T)** **750—790 rpm (A/T, P range)** [Test connector (Green: 1-pin) grounded]	Yes	Go to Next Step		
		No	Try to adjust idle speed	Yes	Idle-speed misadjustment
				No	Check accelerator cable free play
					Check ISC valve (Stuck closed)
					Check throttle body
5	Substitute a well-known ECU Check if condition improves	Yes			ECU malfunction
		No	Check voltage at ECU (1C) terminal with SST **Voltage:** **Approx. 10V** **(While cranking)**	Yes	Go to Next Step
				No	Check starter interlock switch
					Check related wiring
6					Poor quality engine oil

Low idle speed (When A/C, P/S, E/L is ON)

STEP	QUICK INSPECTION		ACTION	POSSIBLE CAUSE AND DETAILED INSPECTION
1	Check for malfunction code with SST [IGN ON, Test connector (Green: 1-pin) grounded]	Yes	Check for cause by referring to check sequence	
		No	Go to Next Step	
2	Disconnect ISC valve connector at idle Check if the condition does not change	Yes	Go to Next Step	
		No		Check coolant level
				Check engine oil
3	Check switches with SST Idle switch Neutral switch Clutch switch [IGN ON, Test connector (Green: 1-pin) grounded]	Yes	Go to Next Step	
		No	Check for cause by referring to check sequence	
4	Check continuity between test connector (Green: 1-pin) and ground			Wiring short to ground

Note:
In case of low idle speed with A/C ON, if the problem cannot be solved by the above steps, it may be an A/C compressor malfunction. (Refer to Section U.)

SYMPTON TROUBLESHOOTING—B2200, B2600i

High idle speed after warm up

STEP	QUICK INSPECTION		ACTION		POSSIBLE CAUSE AND DETAILED INSPECTION
1	Check for malfunction code with SST [IGN ON, Test connector (Green: 1-pin) grounded]	Yes	Check for cause by referring to check sequence		
		No	Go to Next Step		
2	Check ignition timing at idle after warm up **Ignition timing:** **BTDC 4—6° (G6)** **5—7° (F2)** [Test connector (Green: 1-pin) grounded]	Yes	Go to Next Step		
		No	Adjust ignition timing		
3	Check if throttle valve is fully closed when accelerator released	Yes	Go to Next Step		
		No	Check if throttle linkage is correctly installed and operates freely		
4	Check idle speed after warm up **Idle speed: 730—770 rpm (M/T)** **Idle speed: 750—790 rpm (A/T, P range)** [Test connector (Green: 1-pin) grounded]	Yes			Check ISC valve
		No	Try to adjust idle speed	Yes	Idle speed misadjustment
				No	Go to Next Step
5	Disconnect ISC valve connector at idle when engine is cold Check if idle speed decreases during warm up	Yes	Go to Next Step		
		No			Check air valve
6	Disconnect water thermosensor connector and check if condition improves	Yes	Check water thermosensor connector conditon as follows: 1. Shake connector and check if condition changes 2. Check condition of terminal (burned or damaged) 3. Connect a good terminal to harness side connector and check for looseness	Yes	Check water thermo-sensor
				No	Poor contact of water thermo-sensor connector
		No	Go to Next Step		
7					ECU malfunction

Idle hunting or surging

STEP	QUICK INSPECTION		ACTION		POSSIBLE CAUSE AND DETAILED INSPECTION
1	(If trouble occurs only at warm condition) Run engine at **2,000 rpm** for more than **20 seconds** Check for malfunction code with SST [IGN ON, Test connector (Green: 1-pin) grounded]	Yes	Check for cause by referring to check sequence		
		No	Go to Next Step		
2	(If trouble occurs only at warm condition) Check for flashing of SST monitor lamp after warm up **Monitor lamp:** **Flashes more than 8 times 10 seconds at 2,000—3,000 rpm** [Test connector (Green: 1-pin) not grounded]	Yes	Go to Next Step		
		No			Replace oxygen sensor
3	Check intake manifold vacuum at idle **Vacuum: G6 500—540 mmHg (19.7—21.3 inHg)** **F2 510—550 mmHg (20.1—21.7inHg)**	Yes	Go to Next Step		
		No	Check for air leaks	Yes	Intake air system components damaged
					Vacuum and air intake hoses loose or damaged
					Bolts or nuts loose
					Gaskets damaged
				No	Check throttle body
4	Pinch PCV hose Check if condition improves	Yes			Check PCV valve
		No	Go to Next Step		
5	Check fuel line pressure [IGN ON, Test connector (Yellow: 2-pin) connected] **Fuel line pressure:** **265—314 kPa (2.7—3.2 kg/cm², 38—46 psi)**	Yes	Go to Next Step		
		No	Check for fuel leaks		
			Substitute a good fuel filter and retest	Yes	Replace fuel filter
			Check fuel pump maximum pressure **Fuel pump maximum pressure: 441—588 kPa (4.5—6.0 kg/cm², 64—85 psi)**	Yes	Replace pressure regulator
				No	Replace fuel pump
6					ECU malfunction

Engine stall at idle (Always)

STEP	QUICK INSPECTION		ACTION		POSSIBLE CAUSE AND DETAILED INSPECTION
1	Check for malfunction code with SST [IGN ON, Test connector (Green: 1-pin) grounded]	Yes	Check for cause by referring to the check sequence		
		No	Go to Next Step		
2	Check if strong blue spark is visible at spark plug while cranking	Yes	Go to Next Step		
		No	Check ignition system		Check spark plug Check high-tension lead Check distributor cap
3	Check fuel line pressure [IGN ON, Test connector (Yellow: 2-pin) connected] **Fuel line pressure: 265—314 kPa (2.7—3.2 kg/cm², 38—46 psi)**	Yes	Go to Next Step		
		No	Check for fuel leaks		
			Check if fuel filter has been replaced according to maintenance schedule	Yes	Check fuel line for clogging
				No	Replace fuel filter
			Check fuel pump maximum pressure **Fuel pump maximum pressure: 441—588 kPa (4.5—6.0 kg/cm², 64—85 psi)**	Yes	Replace pressure regulator
				No	Replace fuel pump
4	Check if vacuum hoses and the air hoses are connected correctly	Yes	Go to Next Step		
		No	Connect correctly		
5					Airflow sensor
6					ECU malfunction

Engine stall at idle (Only when engine is cold)

STEP	QUICK INSPECTION		ACTION	POSSIBLE CAUSE AND DETAILED INSPECTION
1	Check for malfunction code with SST [IGN ON, Test connector (Green: 1-pin) grounded]	Yes	Check for cause by referring to check sequence	
		No	Go to Next Step	
2				Check BAC valve (air valve)
3				ECU malfunction

Engine stall at idle (Only when engine is warm)

STEP	QUICK INSPECTION		ACTION		POSSIBLE CAUSE AND DETAILED INSPECTION
1	Check for malfunction code with SST [IGN ON, Test connector (Green: 1-pin) grounded]	Yes	Check for cause by referring to check sequence		
		No	Go to Next Step		
2	Disconnect water thermosensor connector Check if condition improves	Yes	Check water thermosensor connector as follows: 1. Shake connector and check if condition changes 2. Check condition of terminal (burned or damaged) 3. Connect a good terminal to harness side connector and check for looseness	Yes	Check water thermosensor
				No	Poor contact of water thermosensor connector
		No	Go to Next Step		
3					ECU malfunction

Engine stall at idle (When A/C, P/S, E/L is ON)

STEP	QUICK INSPECTION		ACTION	POSSIBLE CAUSE AND DETAILED INSPECTION
1	Check for malfunction code with SST [IGN ON, Test connector (Green: 1-pin) grounded]	Yes	Check for cause by referring to check sequence	
		No	Go to Next Step	
2	Check switches with SST • Headlight switch • Blower switch [IGN ON, Test connector (Green: 1-pin) grounded]	Yes	Go to Next Step	
		No	Check for cause by referring to check sequence	
3	Disconnect ISC valve connector at idle [Test connector (Green: 1-pin) grounded] Check if the condition does not change	Yes	Go to Next Step	
		No		Check ISC valve Check engine oil
4	Check idle speed after warm up **Idle speed: 730—770 rpm (M/T) 750—790 rpm (A/T, P range)** [Test connector (Green: 1-pin) grounded]	Yes	Go to Next Step	
		No	Adjust idle speed	
5				ECU malfunction

Note:
Engine stalls at idle with A/C ON, if the trouble cannot be fixed after checking above steps, it may be A/C compression malfunction

SYMPTON TROUBLESHOOTING – B2200, B2600i

Engine stall during start up

STEP	QUICK INSPECTION		ACTION	POSSIBLE CAUSE AND DETAILED INSPECTION
1	Check for malfunction code with SST [IGN ON, Test connector (Green: 1-pin) grounded]	Yes	Check for cause by referring to check sequence	
		No	Go to Next Step	
2	Check idle speed after warm up **Idle speed: 730—770 rpm (M/T) 750—790 rpm (A/T, P range)** [Test connector (Green: 1-pin) grounded]	Yes	Go to Next Step	
		No	Adjust idle speed	
3	Check for injector operating sound at idle	Yes	Go to Next Step	
		No	Check resistance at injector harness connector (EMINJ-01) Terminal: (B/Y)—(LG/B), (B/Y)—(LG/R); Resistance: 6—8Ω Yes: Check wiring short or open No: Check injector resistance	Check wiring
4	Check ignition timing at idle after warm up **Ignition timing: BTDC 4—6° (G6) 5—7° (F2)** [Test connector (Green: 1-pin) grounded]	Yes	Go to Next Step	
		No	Adjust ignition timing	
5				ECU malfunction

Engine stall on deceleration

STEP	QUICK INSPECTION		ACTION	POSSIBLE CAUSE AND DETAILED INSPECTION
1	Check for malfunction code with SST [IG ON Test connector (Green: 1-pin) grounded]	Yes	Check for cause by referring to check sequence	
		No	Go to Next Step	
2	Check idle switch and stoplight switch with SST [IGN ON, Test connector (Green: 1-pin) grounded]	Yes	Go to Next Step	
		No	Check for cause by referring to check sequence	
3	Check for flashing of monitor lamp after warm up **Monitor lamp: Flashes more than 8 times/10 seconds at 2,000—3,000 rpm** [The connector (Green: 1-pin) not grounded]	Yes	Go to Next Step	
		No		Replace oxygen sensor
4	Check for fuel cut operation during deceleration **Fuel cut: after warm up Above 1,600 rpm (G6) Above 1,900 rpm (F2)**	Yes	Go to Next Step	
		No	Check water thermosensor Yes: Replace ECU No: Replace water thermosensor	
5	Check idle speed after warm up **Idle speed: 730—770 rpm (M/T) 750—790 rpm (A/T, P range)** [Test connector (Green: 1-pin) grounded]	Yes	Go to Next Step	
		No	Adjust idle speed (If possible)	
6	Check ignition timing at idle after warm up **Ignition timing: BTDC 4—6° (G6) 5—7° (F2)** [Test connector (Green: 1-pin) not grounded]	Yes	Go to Next Step	
		No	Adjust ignition timing	
7				Check ISC valve

SYMPTON TROUBLESHOOTING—B2200, B2600i

Engine stall at idle (Intermittent)

STEP	QUICK INSPECTION		ACTION	POSSIBLE CAUSE AND DETAILED INSPECTION
1	Shake connector of ignition coil, main relay and ECU while cranking Check if the engine starts	Yes	There may be a poor contact at the connector. Repair or replace the wiring	
		No	Go to troubleshooting "Engine stall at idle (Always)"	

Hesitates/Stumbles on acceleration

QUICK	INSPECTION		ACTION		POSSIBLE CAUSE AND DETAILED INSPECTION
1	Run engine at **2,000 rpm** for **20 seconds** and stop it Check for malfunction code with SST [IGN ON, Test connector (Green: 1-pin) grounded]	Yes	Check for cause by referring to check sequence		
		No	Go to Next Step		
2	Check idle switch with SST [IGN ON, Test connector (Green: 1-pin) grounded]	Yes	Go to Next Step		
		No	Check for cause by referring to check sequence		
3	Disconnect oxygen sensor connector Check if condition improves	Yes			Check oxygen Sensor
		No	Go to Next Step		
4	Check fuel line pressure while accelerating (Vacuum hose to pressure regulator disconnected) **Fuel line pressure: Keeps 265—314 kPa 2.7—3.2 kg/cm^2, 38—46 psi)**	Yes	Go to Next Step		
		No	Check if fuel filter has been replaced according to maintenance schedule	Yes	Check fuel line for clogging
				No	Replace fuel filter
					Replace pressure regulator
5	Check for air leaks with throttle valve open by listening for sucking noise	Yes			Intake air system components damaged
					Vacuum and intake air hoses loose or damaged
					Bolts or nuts loose
					Gaskets damaged
		No	Go to Next Step		
6	Substitute a well-known ECU Check if condition improves	Yes			ECU malfunction
		No			Check airflow sensor
					Check throttle body
					Check spark plug
7	Check other systems				Clutch slipping

Hesitates at steady speed

STEP	QUICK INSPECTION		ACTION		POSSIBLE CAUSE AND DETALED INSPECTION
1	Run engine at **2,000 rpm** for **20 seconds** and stop it Check for malfunction code with SST [IGN ON, Test connector (Green: 1-pin) grounded]	Yes	Check for cause by referring to check sequence		
		No	Go to Next Step		
2	Disconnect oxygen sensor connector Check if condition improves	Yes			Check oxygen sensor
		No	Go to Next Step		
3	Check for air leaks with throttle valve open by listening for sucking noise	Yes	Go to Next Step		
		No			Intake air system components damaged
					Vacuum and intake air hoses loose or damaged
					Nuts or bolts loose
					Gasket damaged
4	Check fuel line pressure while accelerating (Vacuum hose to pressure regulator disconnected) **Fuel line pressure: Keeps 265—314 kPa (2.7—3.2 kg/cm^2, 38—46 psi)**	Yes	Go to Next Step		
		No	Check if fuel filter has been replaced according to maintenance schedule	Yes	Check fuel line for clogging
				No	Replace fuel filter
					Replace pressure regulator
5	Check condition of ignition coil and airflow meter connectors (Burned or damaged)	Yes			Poor contact
		No	Go to Next Step		
6	Gradually open throttle valve Check if engine speed increases smoothly	Yes	Go to Next Step		
		No			Check airflow sensor
					Check throttle body
					Check throttle sensor
7					Check spark plug
8	Change fuel to specified grade Check if condition improves	Yes			Poor fuel quality
		No	Go to Next Step		
9					ECU malfunction

SYMPTON TROUBLESHOOTING—B2200, B2600i

Jerking on acceleration

STEP	QUICK INSPECTION		ACTION		POSSIBLE CAUSE AND DETAILED INSPECTION
1	Run engine at **2,000 rpm** for **20 seconds** and stop it Check for malfunction code with SST [IGN ON, Test connector (Green: 1-pin) grounded]	Yes	Check for cause by referring to check sequence		
		No	Go to Next Step		
2	Check idle switch with SST [IGN ON, Test connector (Green: 1-pin) grounded]	Yes	Go to Next Step		
		No	Check for cause by referring to check sequence		
3	Disconnect oxygen sensor connector Check if condition improves	Yes			Check oxygen Sensor
		No	Go to Next Step		
4	Check fuel line pressure while accelerating (Vacuum hose to pressure regulator disconnected) **Fuel line pressure: Keeps 265—314 kPa 2.7—3.2 kg/cm², 38—46 psi)**	Yes	Go to Next Step		
		No	Check if fuel filter has been replaced according to maintenance schedule	Yes	Check fuel line for clogging
				No	Replace fuel filter
					Replace pressure regulator
5	Check for air leaks with throttle valve open by listening for sucking noise	Yes			Intake air system components damaged
					Vacuum and intake air hoses loose or damaged
					Bolts or nuts loose
					Gaskets damaged
		No	Go to Next Step		
6	Substitute a well-known ECU Check if condition improves	Yes			ECU malfunction
		No			Check airflow sensor
					Check throttle body
					Check spark plug
7	Check other systems				Clutch slipping

Knocking

STEP	QUICK INSPECTION		ACTION		POSSIBLE CAUSE AND DETAILED INSPECTION
1	Check malfunction code with SST [IGN ON, Test connector (Green: 1-pin) grounded]	Yes	Check for cause by referring to the check sequence		
		No	Go to Step 2		
2	Check ignition timing at idle after warm up **Ignition timing: BTDC 4—6° (M/T) 5—7° (A/T, P range)** [Test connector (Green: 1-pin) not grounded]	Yes	Go to Next Step		
		No	Adjust ignition timing		
3	Disconnect water thermosensor connector Check if condition improves	Yes			Check water thermosensor
		No	Go to Next Step		
4	Check vacuum routing	Yes	Go to Next Step		
		No			Vacuum hose
5	Observe fuel line pressure while accelerating from idle **Fuel line pressure: Keeps 265—314 kPa (2.7—3.2 kg/cm², 38—46 psi)** (Vacuum hose to pressure regulator disconnected)	Yes	Go to Next Step		
		No	Check fuel pump maximum pressure **Fuel pump maximum pressure: 441—588 kPa (4.5—6.0 kg/cm², 64—85 psi)**	Yes	Replace fuel filter
				No	Replace fuel pump
6					Check airflow sensor
7					Check spark plug
8	Change fuel to specified grade Check if condition improves	Yes			Poor fuel quality
		No	Go to Next Step		
9	Check cooling system.				Thermostat
					Radiator
10					ECU malfunction

SYMPTON TROUBLESHOOTING—B2200, B2600i

Poor acceleration

STEP	QUICK INSPECTION		ACTION		POSSIBLE CAUSE AND DETAILED INSPECTION
1	Check for malfunction code with SST [IGN ON, Test connector (Green: 1-pin) grounded]	Yes	Check for cause by referring to check sequence		
		No	Go to Next Step		
2	Check idle switch with SST [IGN ON, Test connector (Green: 1-pin) grounded]	Yes	Go to Next Step		
		No	Check for cause by referring to check sequence		
3	Disconnect high—tension lead of each cylinder at idle. Check if engine condition changes [ISC valve connector disconnected]	Yes	Go to Next Step		
		No	Check ignition system [Refer to ignition system troubleshooting (Misfire)]	Yes	Replace injector
				No	Check spark plug
					Check high-tension
					Check distributor cup
4	Check ignition at idle after warm up **Ignition timing:** **BTDC 4—6° (G6)** **5—7° (F2)** [Test connector (Green: 1-pin) grounded]	Yes	Go to Next Step		
		No	Adjust ignition timing		
5	Check for air leaks by listening for sucking noise	Yes			Intake air system components damaged
					Vacuum and air intake hoses loose or damaged
					Nuts or bolts loose
					Gasket damaged
		No	Go to Next Step		
6	Observe fuel line pressure while accelerating from idle **Fuel line pressure:** **Keeps 265—314 kPa (2.7—3.2 kg/cm², 38—46 psi)** [Vacuum hose to pressure regulator disconnected]	Yes	Go to Next Step		
		No	Check if fuel filter has been replaced according to maintenance schedule	No	Replace pressure regulator
				Yes	Replace fuel filter
7	Gradually depress accelerator from idle Check if engine speed increases smoothly	Yes	Go to Next Step		
		No	Check accelerator cable free play	Yes	Check airflow sensor
					Check throttle body
				No	Adjust
8	Check fuel to specified grade Check if condition improves	Yes			Poor fuel quality
		No	Go to Next Step		
9	Substitute a well-known ECU Check if condition improves	Yes			ECU malfunction
		No	Go to Next Step		
10	Check other systems				Clutch slipping
					Transmission (M/T)
					Brake dragging
					Belt tension

Lack of power

STEP	QUICK INSPECTION		ACTION		POSSIBLE CAUSE AND DETAILED INSPECTION
1	Check for malfunction code and (only high-altitude) with SST [IGN ON, Test connector (Green:1-pin) grounded]	Yes	Check for cause by referring to check sequence		
		No	Go to Step 2 (High-altitude) Go to Step 3 (Others)		
2	Check ignition timing at idle after warm up **Ignition timing:** **BTDC 4—6° (G6)** **5—7° (F2)** [Test connector (Green: 1-pin) grounded]	Yes	Go to Next Step		
		No	Adjust ignition timing		
3	Disconnect ISC valve connector and the high-tension lead of each cylinder Check if condition changes	Yes	Go to Next Step		
		No	Check ignition system	Yes	Replace injector (If step 4 OK)
				No	Check high-tension lead
					Check distributor cap
					Check spark plug
4	Check for injector operating sound at idle	Yes	Go to Next Step		
		No	Check resistance at injector harness connector (EMINJ-01) **Terminals / Resistance** (B/Y)—(LG/B), (B/Y)—(LG/R): 6—8Ω	Yes	Check wiring short or open
				No	Check injector resistance
					Check wiring short or open
5	Check air cleaner element for clogging	Yes	Go to Next Step		
		No	Clean air cleaner element		
6	Check for air leaks by listening for sucking noises • At idle • When throttle valve is open	Yes			Intake air system
					Components damaged
					Vacuum and air intake hoses loose or damaged
					Nuts or bolts loose
					Gasket damaged

SYMPTON TROUBLESHOOTING—B2200, B2600i

Lack of power (Con't'd)

STEP	QUICK INSPECTION		ACTION		POSSIBLE CAUSE AND DETAILED INSPECTION
7	Check fuel line pressure [IGN ON, Test connector (Yellow: 2-pin) connected] **Fuel line pressure: 265—314 kPa (2.7—3.2 kg/cm², 38—46 psi)**	Yes	Go to Next Step		
		No	Check for fuel leakage		
			Substitute a good fuel filter and retest	Yes	Replace fuel filter
			Check fuel pump maximum pressure **Fuel pump maximum pressure: 441—588 kPa (4.5—6.0 kg/cm² 64—85 psi)**	Yes	Replace pressure regulator
				No	Replace fuel pump
8	Check fuel line pressure at idle **Fuel line pressure: 216—264 kPa (2.2—2.7 kg/cm², 31—38 psi)**	Yes	Go to Next Step		
		No			Replace pressure regulator
9	Check if fuel line pressure drops while accelerating (Vacuum hose disconnected)	Yes	Check if fuel filter has been replaced according to maintenance schedule	Yes	Check fuel line for clogging
				No	Replace fuel filter
		No	Go to Next Step		
10	Check exhaust system for damage	Yes	Go to Next Step		
		No	Repair or replace		
11	Check A/C, P/S and alternator belts tensions	Yes	Go to Next Step		
		No	Adjust belt tension		
12	Check if accelerator can be depressed fully	Yes	Go to Next Step		
		No	Check accelerator cable	Yes	Throttle body
				No	Accelerator cable
13	Substitute a well-known ECU Check if condition improves	Yes			ECU malfunction
		No			Check airflow sensor
					Check throttle sensor
					Go to Next Step
14	Substitute a specified fuel Check if condition improves	Yes			Poor fuel quality
		No	Go to Next Step		
15	Check other systems				Brake
					Clutch
					Engine

Bucking at high speed

STEP	QUICK INSPECTION		ACTION		POSSIBLE CAUSE AND DETAILED INSPECTION
1	Run engine at **2,000 rpm** for more than **20 seconds**. Check for malfunction code with SST [IGN ON, Test connector (Green: 1-pin) grounded]	Yes	Check for cause by referring to check sequence		
		No	Go to Next Step		
2	Disconnect oxygen sensor connector Check if condition improves	Yes			Check oxygen sensor
		No	Go to Next Step		
3	Observe fuel line pressure while accelerating from idle **Fuel line pressure: Keeps 265—314 kPa (2.7—3.2 kg/cm², 38—46 psi)** [Vacuum hose to pressure regulator disconnected]	Yes	Go to Next Step		
		No	Check if fuel filter has been replaced according to maintenance schedule	Yes	Check fuel line for clogging
				No	Replace fuel filter
					Replace pressure regulator
4	Check for air leaks by listening sucking noise	Yes	Go to Next Step		
		No			Intake air system components damaged
					Vacuum and air intake hoses loose or damaged
					Nuts or bolts loose
					Gasket damaged
5	Check ignition timing at idle after warm up **Ignition timing: BTDC 4—6° (G6) 5—7° (F2)** [Test connector (Green: 1-pin) grounded]	Yes	Go to Next Step		
		No	Adjust ignition timing		
6	Gradually open throttle valve from idle check if engine speed increases smoothly	Yes	Go to Next Step		
		No			Check airflow sensor
7					Check spark plug
8					ECU malfunction

SYMPTON TROUBLESHOOTING—B2200, B2600i

Bucking on deceleration

STEP	QUICK INSPECTION		ACTION	POSSIBLE CAUSE AND DETAILED INSPECTION
1	Check for malfunction code with SST [IGN ON, Test connector (Green: 1-pin) grounded]	Yes	Check for cause by referring to the check sequence	
		No	Go to Next Step	
2	Check switches with SST [IGN ON, Test connector (Green: 1-pin) grounded] • Idle switch • Stoplight switch	Yes	Go to Next Step	
		No	Check for cause by referring to the check sequence	
3	Substitute a well-known ECU Check if condition improves	Yes		ECU malfunction
		No		Check throttle sensor
				Go to Next Step
4				Check spark plug
5				Check clutch slipping
6				Check compression between cylinders

Poor fuel economy

STEP	QUICK INSPECTION		ACTION		POSSIBLE CAUSE AND DETAILED INSPECTION
1	Run the engine at **2,000 rpm** for more than **20 seconds** after warm up and stop it Check for malfunction code with SST [IGN ON, Test connector (Green: 1-pin) grounded]	Yes	Check for cause by referring to check sequence		
		No	Go to Next Step		
2	Check idle switch with SST [IGN ON, Test connector (Green: 1-pin) grounded]	Yes	Go to Next Step		
		No	Check for cause by referring to check sequence		
3	Check for flashing of monitor lamp after warm up **Monitor lamp: Flashes more than 8 times /10 seconds at 2,000—3,000 rpm** [Test connector (Green: 1-pin) not grounded]	Yes	Go to Next Step		
		No			Replace oxygen sensor
4	Check fuel line pressure at idle **Fuel line pressure: 196—255 kPa (2.0—2.6 kg/cm², 28—37 psi)**	Yes	Go to Next Step		
		No	Check vacuum line to pressure regulator for clogging or air leakage	Yes	Vacuum line clogging or damaged
				No	Check solenoid valve (PRC)
					ECU malfunction (Check (2T) terminal voltage)
					Replace pressure regulator
5	Check for fuel cut operation during deceleration **Fuel cut: after warm up Above 1,600 rpm (G6) Above 1,900 rpm (F2)**	Yes	Go to Next Step		
		No	Check water thermosensor	Yes	Replace ECU
				No	Replace water thermosensor
6	Check ignition timing at idle after warm up **Ignition timing: BTDC 4—6° (G6) 5—7° (F2)** [Test connector (Green: 1-pin) grounded]	Yes	Go to Next Step		
		No	Adjust ignition timing		
7	Check other systems				Clutch slipping
					Brake
					Tire air pressure

SYMPTON TROUBLESHOOTING—B2200, B2600i

High oil consumption/White exhaust smoke

STEP	QUICK INSPECTION		ACTION		POSSIBLE CAUSE AND DETAILED INSPECTION
1	Check for oil leak from engine	Yes	Repair or replace		
		No	Go to Next Step		
2	Disconnect PCV valve from engine Check if vacuum is felt at idle	Yes	Go to Next Step		
		No	Check PCV valve	Yes	PCV hose clogging
				No	Replace PCV valve
3	Check that ventilation hose is installed correctly	Yes	Go to Next Step		
		No	Install ventilation hose correctly		
4	Possible malfunction of engine Check for cause by referring to the check sequence				

Afterburn on deceleration

STEP	QUICK INSPECTION		ACTION		POSSIBLE CAUSE AND DETAILED INSPECTION
1	Check malfunction code with SST [IGN ON, Test connector (Green: 1-pin) grounded]	Yes	Check for cause by referring to the check sequence		
		No	Go to Next Step		
2	Check idle switch with SST [IGN ON, Test connector (Green: 1-pin) grounded]	Yes	Go to Next Step		
		No	Check for cause by referring to the check sequence		
3	Check ignition timing at idle after warm up **Ignition timing:** **BTDC 4—6° (G6)** **5—7° (F2)** [Test connector (Green: 1-pin) grounded]	Yes	Go to Next Step		
		No	Adjust ignition timing		
4	Check air cleaner element for clogging	Yes	Go to Next Step		
		No	Clean air cleaner element		
5	Check fuel cut operation during deceleration **Fuel cut: after warm up** **Above 1,600 rpm (G6)** **Above 1,900 rpm (F2)**	Yes	Go to Next Step		
		No	Check water thermosensor	Yes	ECU malfunction Check (2Q) terminal voltage
				No	Replace water thermosensor
6	Run engine at idle and stop it (IG OFF) Observe fuel pressure for **5 minutes** **Fuel pressure:** **More than** **147 kPa (1.5 kg/cm², 21 psi)**	Yes	Go to Next Step		
		No	Check fuel pump for pressure drop	No	Replace fuel pump
			Check pressure regulator for pressure drop	Yes	Check injector fuel leakage
				No	Replace pressure regulator
7					Check compression
					Check valve timing

Rotten egg smell

STEP	QUICK INSPECTION	ACTION	POSSIBLE CAUSE AND DETAILED INSPECTION
1	Change fuel to specified grade Check if condition improves		Poor fuel quality

Gasoline fumes

STEP	QUICK INSPECTION		ACTION		POSSIBLE CAUSE AND DETAILED INSPECTION
1	Check for leaks	Yes	Replace		
		No	Go to Next Step		
2	Check if fumes are emitted from check-and-cut valve	Yes	Check check-and cut valve	Yes	Check two-way check valve
					Purge line clogging
				No	Replace check-and-cut valve
		No	Go to Next Step		
3	Check for malfunction code with SST [IGN ON, Test connector (Green: 1-pin) grounded]	Yes	Check for cause by referring to the check sequence		
		No	Go to Next Step		
4	Check switches with SST • Idle switch • Neutral switch • Clutch switch [IGN ON, Test connector (Green: 1-pin) grounded]	Yes	Go to Next Step		
		No	Check for cause by referring to the check sequence		
5	Run engine at idle. Ground the solenoid valve (Purge control) terminal-wire (L/Y) and disconnect vacuum hose (white) from solenoid valve. Check for vacuum at solenoid valve	Yes			ECU malfunction Check (2X) terminal voltage
		No			Replace solenoid valve (Purge control)

STEP 1

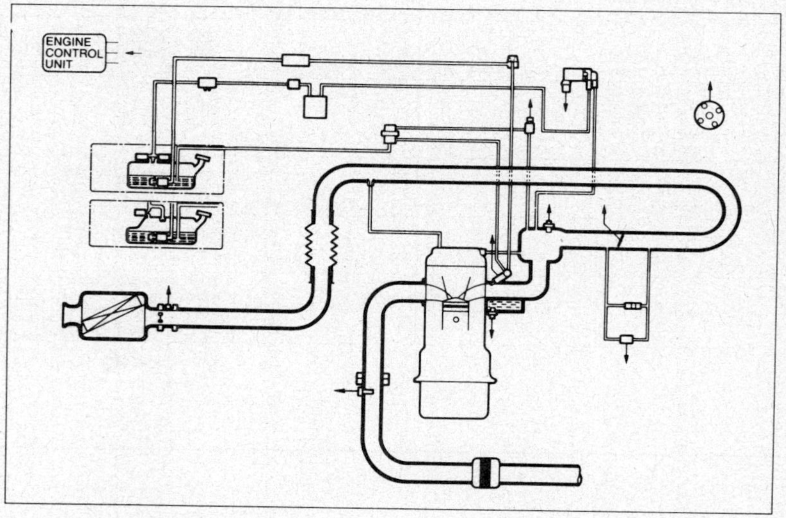

SYMPTON TROUBLESHOOTING—B2200, B2600i

MIL always ON

STEP	QUICK INSPECTION		ACTION		POSSIBLE CAUSE AND DETAILED INSPECTION
1	(California) Check for malfunction code with SST [IGN ON, Test connector (Green: 1-pin) grounded]		**"88"** Replace ECU		
			"00" Wiring between ECU (1E) terminal and MIL short to ground		
2	(Federal and Canada) Check if emission system parts replacement time has come	Yes	Check if MIL has been reset by exchanging MIL set connector	Yes	Replace mileage sensor
				No	Reset the MIL
	Emission system parts replacement schedule: Every 60,000 and 80,000 miles (Federal) or 90,000 and 130,000 km (Canada)	No			Replace mileage sensor

MIL never ON

STEP	QUICK INSPECTION		ACTION		POSSIBLE CAUSE AND DETAILED INSPECTION
1	Check if other indicator lamps illuminate	Yes	Go to Next Step		
		No	Check power supply circuit to combination meter		
2	Check bulb of the MIL	Yes	(California only) Ground ECU (1E) terminal Check if MIL illuminates	Yes	Replace ECU
				No	Wiring between ECU and MIL open
					(Federal and Canada) MIL set connector loose or disconnected
					(Federal and Canada) Replace mileage sensor
		No	Replace		

A/C does not work

STEP	QUICK INSPECTION		ACTION		POSSIBLE CAUSE AND DETAILED INSPECTION
1	Check if condenser fan operates when grounding A/C relay terminal-wire (L/W) (IGN ON)	Yes	Check voltage at ECU (1Q) terminal with SST Voltage at idle after warm up: 0V (A/C and blower switches ON)	Yes	ECU malfunction (Check (1J) terminal voltage)
					Wiring between ECU (1J) and A/C relay open
				No	A/C system malfunction
		No	Check A/C system		

TROUBLE CODES—B2200, B2600i

Code No.	Malfunction display: Pattern of output signal (Self-Diagnosis Checker or MIL (California only))	Sensor or subsystem	Self-diagnosis	Fail-safe
02	ON / OFF	Ne signal	No Ne signal	—
03	ON / OFF	G signal	No G signal	Cancels 2-group injection
08	ON / OFF	Airflow sensor	Open or short circuit	Basic fuel injection amount fixed as for two driving modes (1) Idle switch: ON (2) Idle switch: OFF
09	ON / OFF	Water thermosensor	Open or short circuit	Maintains constant 20°(68°F) command
11	ON / OFF	Intake air thermosensor (dynamic chamber)	Open or short circuit	Maintains constant 20°C (68°F) command
12	ON / OFF	Throttle sensor	Open or short circuit	Maintains constant command of throttle valve fully open
14	ON / OFF	Atmospheric pressure sensor	Open or short circuit	Maintains constant command of sea level pressure
15	ON / OFF	Oxygen sensor (Inactivation)	Sensor output continues less than 0.45V 180 sec. after engine exceeds 1,500 rpm	Cancels engine feedback operation
17	ON / OFF	Oxygen sensor (Inversion)	Sensor output not changed 20 sec. after engine exceeds 1,500 rpm	Cancels engine feedback operation
25	ON / OFF	Solenoid valve (pressure regulator control)	Open or short circuit	—
26	ON / OFF	Solenoid valve (purge control)		—
34	ON / OFF	Solenoid valve (Idle speed control)		—

Caution

a) If there is more than one failure present, the lowest number malfunction code is displayed first, the remaining codes are displayed in order.

b) After repairing all failures, turn off the ignition switch, disconnect the negative battery cable for more than 20 seconds to erase the memory of a malfunction code from the engine control unit.

Code No.02 (Distributor Ne-signal)

PC: Possible Cause

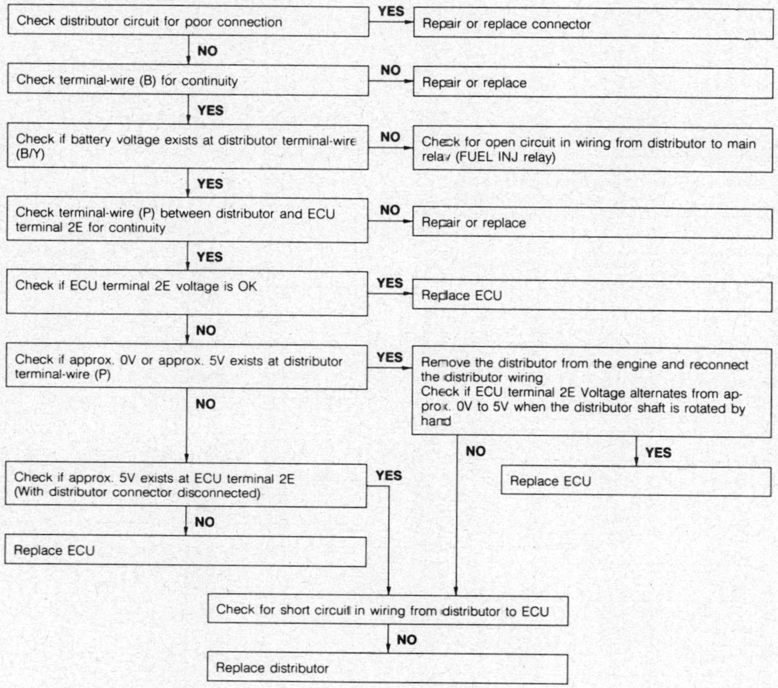

Code No.03 (Distributor G-signal)

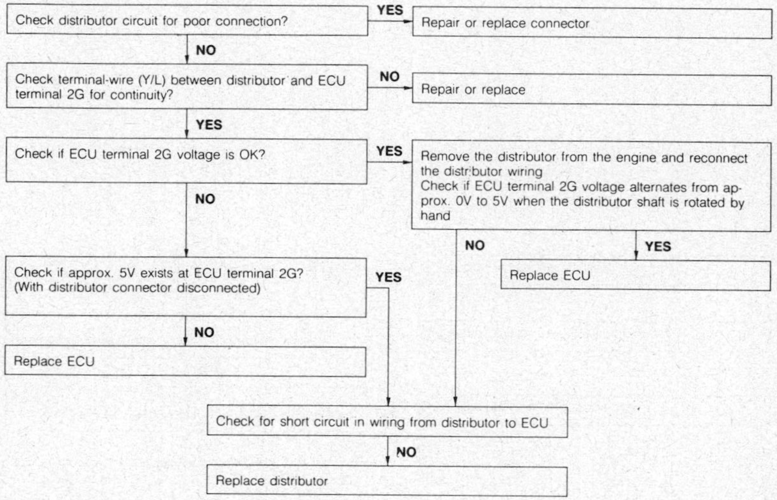

Code No.08 (Airflow sensor)

PC: Possible Cause

- Are there poor connections in airflow sensor circuit?
  - YES → Repair or replace connector
  - NO ↓
- Is voltage at terminal D of airflow sensor connector OK?
  - YES → Is same Code No. present after performing after-repair procedure?
    - NO → Airflow sensor and circuit OK
    - YES → PC: Engine control unit malfunction
  - NO ↓
- Is protection net of airflow sensor damaged or restricted?
  - YES → Clean or repair protection net
  - NO ↓
- Is there continuity between airflow sensor and engine control unit?

Airflow sensor	Engine control unit
D (G/B wire)	2O
E (G/Y wire)	2P

  - NO → PC: Open circuit in wiring harness
  - YES ↓
- Is there battery voltage at terminal B of airflow sensor connector?
  - NO → PC: Open or short circuit in wiring harness from terminal B of airflow sensor to main relay
  - YES ↓
- Is there continuity between terminal F of airflow sensor and ground?
  - NO → PC: Open circuit in wiring harness
  - YES → PC: • Engine control unit malfunction
    • Short circuit in wiring from terminal D of airflow sensor to engine control unit

Code No.09 (Water thermosensor)

PC: Possible Cause

- Are there poor connections at water thermosensor circuit?
  - YES → Repair or replace connector
  - NO ↓
- Is there continuity between water thermosensor and engine control unit?

Water thermosensor	Engine control unit
A (G/W wire)	2Q
B (BR/B wire)	2D

  - NO → PC: Open circuit in wiring harness from water thermosensor to engine control unit
  - YES ↓
- Is resistance of the water thermosensor OK?
  Resistance:

Coolant temp	Resistance
−20°C (−4°F)	14.5— 17.8 kΩ
20°C (68°F)	2.2— 2.7 kΩ
80°C (176°F)	280 —350Ω

  - NO → Replace water thermosensor
  - YES ↓
- Is same Code No. present after performing after-repair procedure?
  - NO → Water thermosensor and circuit OK
  - YES ↓
- Are engine control unit 2Q and 2D terminal voltages OK?
  - NO → PC: Engine short circuit in wiring harness
  - YES → PC: Engine control unit malfunction

No.11 Code (Intake air thermosensor)

PC: Possible Cause

1. Are there poor connections at intake air thermosensor connectors?
   - YES → Repair or replace connector
   - NO → go to 2
2. Is there continuity between intake air thermosensor (dynamic chamber) and engine control unit?

Intake air thermosensor (dynamic chamber)	Engine control unit
A (G wire)	2L
B (BR/B wire)	2D

   - NO → PC Open circuit in wiring harness
   - YES → go to 3
3. Is resistance of intake air thermosensor (dynamic chamber) OK?
   Resistance:

Temperature	Resistance
25°C (77°F)	29.7—36.3 kΩ
85°C (185°F)	3.3—3.7 kΩ

   - NO → Replace intake air thermosensor (dynamic chamber)
   - YES → go to 4
4. Is same Code No. present after performing after-repair procedure?
   - No → Intake air thermosensor and circuit OK
   - YES → go to 5
5. Are engine control unit 2L and 2D terminal voltages OK?
   - NO → PC Short circuit in wiring harness
   - YES → PC Engine control unit malfunction

Code No.12 (Throttle sensor)

PC: Possible cause

1. Are there poor connections in throttle sensor circuit?
   - YES → Repair or replace connector terminal
   - NO → go to 2
2. Is there 4.5—5.5 V at A terminal of throttle sensor connector?
   - NO → PC: • Open or short circuit in wiring harness from A terminal of throttle sensor to 2K terminal of engine control unit
     • Engine control unit malfunction
   - YES → go to 3
3. Is C terminal of throttle sensor connector grounded?
   - NO → PC: Open circuit in wiring harness from C terminal of throttle sensor to ground
   - YES → go to 4
4. Is throttle sensor adjusted correctly?
   - NO → Adjust throttle sensor
   - YES → go to 5
5. Is same Code No. present after performing after-repair procedure?
   - NO → Throttle sensor and circuit OK
   - YES → go to 6
6. Is engine control unit 2M terminal voltage OK?
   - NO → PC: Open or short circuit in wiring harness from B terminal of throttle sensor to 2M terminal of engine control unit
   - YES → PC: Engine control unit malfunction

Code No.14 (Atmospheric pressure sensor in ECU)

Replace ECU

TROUBLE CODE TROUBLESHOOTING—B2200, B2600i

Code No.15 (Oxygen sensor—Inactivation)

PC: Possible Cause

Note
When Codes No.15 and 17 are present at the same time, first perform the checking procedure for Code No.17.

- Are there poor connections in oxygen sensor circuit?
  - YES → Repair or replace connector
  - NO ↓
- Is oxygen sensor output voltage OK?
  - NO → Replace oxygen sensor
  - YES ↓
- Is sensitivity of oxygen sensor OK?
  - NO → PC: • Oxygen sensor malfunction • Engine control unit malfunction
  - YES ↓
- Is same Code No. present after performing after-repair procedure?
  - NO → Oxygen sensor and circuit OK
  - YES ↓
- Is engine control unit 2N terminal voltage OK?
  - NO → PC: Open or short circuit in wiring harness between oxygen sensor and engine control unit 2N terminal
  - YES → PC: Engine control unit malfunction

Code No.17 (Oxygen sensor—Inversion)

PC: Possible Cause

- Warm up engine and run it at 2,500–3,000 rpm for three minutes. Does monitor lamp of Self-Diagnosis Checker illuminate at idle?
  - NO → PC: • Air leak in vacuum hoses or emission component • Contaminated oxygen sensor • Insufficient fuel injection
  - YES ↓
- Are spark plugs clean?
  - NO → Clean or replace spark plugs
  - YES ↓
- Is oxygen sensor voltage OK?
  - NO → PC: Oxygen sensor malfunction
  - YES ↓
- Is same Code No. present after performing after-repair procedure?
  - NO → Feedback system OK
  - YES ↓
- Is engine control unit 2N terminal voltage OK?
  - NO → PC: Open or short circuit in wiring harness between oxygen sensor and engine control unit 2N terminal
  - YES → PC: Engine control unit malfunction

Code No.25 (Solenoid valve—Pressure regulator control)

PC: Possible Cause

- Are there poor connections in solenoid valve circuit?
  - YES → Repair or replace connector
  - NO ↓
- Is there continuity of solenoid valve?
  - NO → Replace solenoid valve
  - YES ↓
- Is there battery voltage at (B/Y) wire of solenoid valve connector?
  - NO → PC: Open circuit in wiring harness from (B/Y) wire to main relay
  - YES ↓
- Is there continuity between solenoid valve and engine control unit?

Solenoid valve	Engine control unit
B (L/B wire)	2T

  - NO → PC: Open circuit in wiring harness from solenoid valve to engine control unit
  - YES → PC: • Short circuit in wiring harness • Engine control unit malfunction

Code No.26 (Solenoid valve—Purge control)

- Are there poor connections in solenoid valve circuit?
  - YES → Repair or replace connector
  - NO ↓
- Is there continuity of solenoid valve?
  - NO → Replace solenoid valve
  - YES ↓
- Is there battery voltage at (B/Y) wire of solenoid valve circuit?
  - NO → PC: Open circuit in wiring harness from (B/Y) wire to main relay
  - YES ↓
- Is there continuity between solenoid valve and engine control unit?

Solenoid valve	Engine control unit
B (L/Y wire)	2X

  - NO → PC: Open circuit in wiring harness from solenoid valve to engine control unit
  - YES → PC: • Short circuit in wiring harness • Engine control unit malfunction

TROUBLE CODE TROUBLESHOOTING—B2200, B2600i

Code No.34 (Solenoid valve—Idle speed control (ISC)) — **PC: Possible Cause**

Step	Check	Result	Action
1	Are there poor connections in solenoid valve circuit?	YES	Repair or replace connector
		NO	Go to step 2
2	Is resistance of solenoid valve OK? Resistance 7.7–9.3Ω	NO	Replace BAC valve
		YES	Go to step 3
3	Is there battery voltage at (B/Y) wire of solenoid valve connector?	NO	PC: Open or short circuit in wiring harness from (B/Y) wire to main relay
		YES	Go to step 4
4	Is there continuity between solenoid valve and engine control unit?	NO	PC: Open circuit in wiring harness from solenoid valve to engine control unit
		YES	PC: • Short circuit in wiring harness • Engine control unit malfunction

Solenoid valve	Engine control unit
B (L/G wire)	2W

SWITCH MONITOR FUNCTION—B2200, B2600i

SWITCH MONITOR FUNCTION

PREPARATION
SST

49 H018 9A1
Self-diagnosis checker

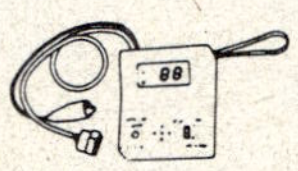

Individual switches can be monitored by the **SST**.

Note
The test connector must be grounded and the ignition switch ON (engine stopped).

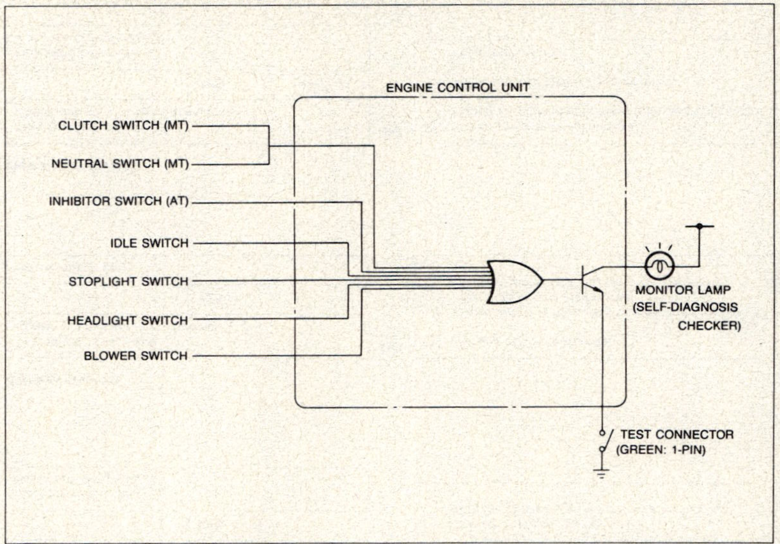

Switch		Self-Diagnosis Checker (Monitor lamp)		Remark
		Light ON	Light OFF	
Clutch switch	(MT)	Pedal released	Pedal depressed	In gear
Neutral switch	(MT)	In gear	Neutral	Clutch pedal released
Inhibitor switch	(AT)	L, S, D or R range	N or P range	—
Idle switch		Pedal depressed	Pedal released	—
Stoplight switch		Pedal depressed	Pedal released	—
Headlight switch		ON	OFF	Headlights/small lights: ON
Blower switch		ON	OFF	Blower motor ON

SWITCH MONITOR FUNCTION TROUBLESHOOTING—B2200, B2600i

Procedure

Set conditions to deactivate each switch
- All accessories OFF
- Transmission in neutral
- All pedals released

Verify that monitor lamp does not illuminate

NO → Check each switch and related wiring harness
- Clutch and Neutral switch
- Idle switch
- Stoplight switch
- Headlight switch
- Blower switch
- Inhibitor switch

YES → Check each switch as described

Neutral and Clutch switch (M/T)

Shift transmission into gear
Check that monitor lamp illuminates with clutch pedal released

NO → PC:
- Neutral or clutch switch malfunction
- Open circuit in related wiring harness
- Engine control (1V) terminal malfunction

YES → Depress clutch pedal
Check that monitor lamp does not illuminate
Return transmission to neutral

NO → PC:
- Clutch switch malfunction

Depress accelerator pedal, and check that the monitor lamp illuminates

NO → PC:
- Idle switch malfunction
- Short circuit in wiring harness from idle switch to engine control unit
- Engine control unit (1N) terminal malfunction

YES → Idle switch OK
Release accelerator pedal

Stoplight switch

Depress brake pedal
Check that monitor lamp illuminates

NO → PC:
- Stoplight switch malfunction
- Open circuit in wiring harness from stoplight switch to engine control unit
- Engine control unit (1O) terminal malfunction

YES → Stoplight switch OK
Release brake pedal

Blower switch

Turn blower switch ON
Check that monitor lamp illuminates

NO → PC:
- Blower switch malfunction
- Open circuit in the wiring harness from blower switch to engine control unit
- Engine control unit (1S) terminal malfunction

YES → Blower switch OK

Headlight switch

Turn headlight switch ON
Check that monitor lamp illuminates

NO → PC:
- Headlight switch malfunction
- Open circuit in wire harness from headlight switch to ECU
- ECU terminal (1U) voltage incorrect
- ECU malfunction

YES → Headlight switch OK
Turn headlight switch OFF

SWITCH MONITOR FUNCTION TROUBLESHOOTING
B2200, B2600i

Inhibitor switch (AT)

Shift transaxle into D, or P range
Check that monitor lamp illuminates

NO → PC: • Inhibitor switch malfunction
- Open or short circuit in related wire harness
- ECU terminal (1V) voltage incorrect
- ECU malfunction

YES ↓

Inhibitor switch OK
Shift into P or N range

Headlight switch

Turn ON headlight switch
Check that monitor lamp illuminates

NO → PC: • Headlight switch malfunction
- Open circuit in wiring harness from headlight switch to engine control unit
- Engine control unit (1U) terminal malfunction

YES ↓

Headlight switch OK
Turn OFF headlight switch

ENGINE CONTROL UNIT TERMINAL VOLTAGES
B2200, B2600i

Terminal voltage

VB: Battery voltage

Terminal	Input	Output	Connection to	Test condition	Voltage	Remarks
1A	—	—	Battery	Constant	VB	For backup
1B	○		Main relay	Ignition switch OFF	0V	
				Ignition switch ON During burn-off (airflow sensor)	VB	
1C	○		Ignition switch (Start position)	While cranking	0V	
				Ignition switch ON	0V	
1D		○	Self-Diagnosis Checker (Monitor lamp)	Test connector (Green: 1-pin) grounded For 3 seconds after ignition switch OFF→ON (Lamp illuminates)	4.5—5.5V	With Self-Diagnosis Checker
				After 3 seconds (Lamp does not illuminated)	VB	
				Test connector (Green: 1-pin) not grounded at idle. Monitor lamp ON	4.5—5.5V	
				Test connector (Green: 1-pin) not grounded at idle. Monitor lamp OFF	VB	
1E		○	Malfunction indicator lamp (California only)	For 3 seconds after ignition switch OFF→ON (Lamp illuminates)	Below 2.5V	Test connector (Green: 1-pin) grounded
				After 3 seconds (Lamp dose not illuminates)	VB	
				Lamp illuminates	Below 2.5V	
				Lamp not illuminate	VB	
1F		○	Self-Diagnosis checker (Code number)	For 3 seconds after ignition switch OFF→ON (Buzzer sounds)	Below 2.5V	• With Self-Diagnosis Checker • Test connector (Green: 1-pin) grounded
				After 3 seconds (Buzzer does not sounded)	VB	
				Buzzer sounds	Below 2.5V	
				Buzzer not sounded	VB	
1G		○	Main relay	Ignition switch OFF	VB	
				During burn-off (airflow sensor) Ignition switch ON	0V	
1H		○	Circuit opening relay	Ignition switch ON	VB	
				During cranking or at idle	Below 2.5V	
1I	○		Ignition switch (ON position)	Ignition switch OFF	0V	
				Ignition switch ON	VB	
1J		○	A/C relay	Ignition switch ON	VB	Blower motor: ON
				For 10 seconds After fully depressing accelerator pedal with A/C switch ON (A/C does not operate) (in-gear, ignition switch ON)	VB	
				After 10 seconds	Below 2.5V	
				For 5 seconds after cranking with A/C switch ON (A/C does not operate)	VB	
				After 5 seconds (A/C operates)	Below 2.5V	
				A/C switch ON at idle	Below 2.5V	
				A/C switch OFF at idle	VB	
1K	○		Test connector	Test connector (Green: 1-pin) not grounded	VB	Ignition switch ON
				Test connector (Green: 1-pin) grounded	0V	
1L	○		Ground (M/T)	Ignition switch ON	0V	
			Open (A/T)	Ignition switch ON	VB	
1M	○		Speed sensor (A/T)	Ignition switch ON	0 or 4.5V—5.5V	
				Idle	4.5—5.5V	
1N	○		Idle switch	Accelerator pedal released	0V	Ignition switch ON
				Accelerator pedal depressed	VB	
1O	○		Stoplight switch	Brake pedal released	0V	Ignition switch ON
				Brake pedal depressed	VB	
1P	○		P/S pressure switch	Ignition switch ON	VB	
				P/S ON (at idle)	0V	
				P/S OFF (at idle)	VB	
1Q	○		A/C switch	A/C switch ON (Ignition switch ON)	Below 2.5V	Blower motor: ON
				A/C switch OFF (Ignition switch ON)	VB	

ENGINE CONTROL UNIT TERMINAL VOLTAGES—B2200, B2600i

Terminal voltage

VB: Battery voltage

Terminal	Input	Output	Connection to	Test condition	Voltage	Remarks
1R	○		Ground (EC-AT)	Ignition switch ON	0V	For G6
			Open (M/T, HAT)	Ignition switch ON	VB	
1S	○		Blower switch	Blower OFF	VB	Ignition switch ON
				Blower ON	Below 1.5V	
1T	—	—	—	—	—	—
1U	○		Headlight switch	Headlight ON	VB	
				Headlight OFF	Below 1.5V	
1V	○		Neutral or clutch switch (Inhibitor switch)	Neutral or clutch pedal depressed (P or N ranges)	0V	Ignition switch ON
				Other condition	VB	
2A	—	—	Ground (E01)	Constant	0V	
2B	—	—	Ground (E02)	Constant	0V	
2C	—	—	Ground (E1)	Constant	0V	
2D	—	—	Ground (E2)	Constant	0V	
2E		○	Distributor	Ignition switch ON	0 or 5V	Ne-Signal
				Idle	2V	
2F		○	Igniter	Ignition switch ON	0 or 5V	Ignition-timing signal
				Idle	Approx. 0.5V	
2G	○		Distributor	Ignition switch ON	0 or 5V	G-Signal
				Idle	Approx. 1.2V	
2H		○	Airflow sensor (Burn-off)	Just after ignition switch OFF	0V	Burn-off functions momentarily
				Burn off (2-5 seconds after ignition switch OFF) (Refer to page F2–174)	8—12V	
2I	—	—	—	—	—	—
2J	—	—	—	—	—	—
2K		○	Vref	Ignition switch ON	4.5—5.5V	
2L	○		Intake air thermosensor (Dynamic chamber)	At 20°C (68°F)	Approx. 2.5V	
2M	○		Throttle sensor	Accelerator pedal released	Approx. 0.5V	Ignition switch ON
				Accelerator pedal fully depressed	Approx. 4.3V	
2N	○		Oxygen sensor	Ignition switch ON	0V	
				Idle (Cold engine)	0V	
				Idle (After warm up)	0—1.0V	Needle moves from 0V to 1V
				Increase engine speed (After warm up)	0.5—1.0V	
				Deceleration	0—0.4V	
2O	○		Airflow sensor (Intake air mass)	Ignition switch ON	1.0—2.0V	
				Idle (After warm up)	1.9—2.6V	
				Increase engine speed (After warm up)	2—5V	
2P	○		Airflow sensor (Ground)	Constant	0V	
2Q	○		Water thermosensor	Engine coolant temperature 20°C (68°F)	Approx. 2.5V	Ignition switch ON
				After warm up	Approx. 0.4V	
2R	—	—	—	—	—	—
2S	—	—	—	—	—	—
2T		○	Solenoid valve (PRC)	For 120 seconds after ignition switch OFF→ON	Below 2.5V	During hot condition. Coolant temp. above 90°C (194°F) Intake air temp. above 75°C (167°F)
				For 120 seconds after starting	Below 2.5V	
				Ignition switch ON	VB	Other conditions
2U		○	Injector G6 (No.3, 4) F2 (No.1, 3)	Ignition switch ON	VB	* Engine Signal Monitor: Green and red lights flash
				Idle	VB	

Terminal voltage

VB: Battery voltage

Terminal	Input	Output	Connection to	Test condition	Voltage	Remarks
2V		○	Injector G6 (No.1, 2) F2 (No.2, 4)	Ignition switch ON	VB	* Engine Signal Monitor: Green and red lights flash
				Idle	VB	
2W		○	Solenoid valve (Idle speed control)	Ignition switch ON	Approx. 11V	Engine signal monitor: Green and red lights flash
				Idle	Approx. 10V	
2X		○	Solenoid valve (Purge control)	Ignition switch ON	VB	
				Idle	VB	* Engine signal monitor: Green and red lights flash
				Driving in gear	5—1.5V*	
2Y		○	HAT control unit	Ignition switch ON	VB	For G6 HAT
				Accelerator for pedal fully depressed	0V	
2Y		○	EC-AT control unit	At sea level	VB	For G6 EC-AT Ignition switch ON
				At high altitude (800 m [2,624 ft])	0V	
2Z	—	—	—	—	—	—

Terminal location

2Y	2W	2U	2S	2Q	2O	2M	2K	2I	2G	2E	2C	2A	U	S	Q	O	M	K	I	G	E	C	A
2Z	2X	2V	2T	2R	2P	2N	2L	2J	2H	2F	2D	2B	V	T	R	P	N	L	J	H	F	D	B

Component Replacement

NOTE: On vehicles equipped with Audio Anti-Theft system, before performing any repairs, the Personal Code Number (P.C.N.) must be obtained from the individual owner of the vehicle, or an authorized Mazda dealer.

SERVICE PRECAUTIONS

- Do not operate the fuel pump when the fuel lines are empty.
- Do not reuse fuel hose clamps.
- Make sure all EGI harness connectors are fastened securely. A poor connection can cause an extremely high surge voltage in the coil and condenser and result in damage to integrated circuits.
- Keep the EGI harness at least 4 in. away from adjacent harnesses to prevent an EGI system malfunction due to external electronic "noise."
- Keep EGI all parts and harnesses dry during service.
- Before attempting to remove any parts, turn off the ignition switch and disconnect the battery ground cable.
- Always use a 12 volt battery as a power source.
- Do not attempt to disconnect the battery cables with the engine running.

up the engine immediately after starting or just prior to shutdown.

- Do not attempt to disassemble the EGI control unit under any circumstances.
- If installing a 2-way or CB radio, keep the antenna as far as possible away from the electronic control unit. Keep the antenna feeder line at least 8 in. away from the EGI harness and do not let the 2 run parallel for a long distance. Be sure to ground the radio to the vehicle body.
- Do not apply battery power directly to injectors.
- Handle air flow meter carefully to avoid damage.
- Do not disassemble air flow meter or clean meter with any type of detergent.

FUEL PRESSURE RELIEF

The fuel in the fuel system remains under high pressure even when the engine is not running. So before disconnecting any fuel lines, release the fuel pressure to reduce the risk of injury or fire. Relieve the fuel pressure as follows:

Except MPV with 2.6L engine, 1990–91 929, 323, Protege and MX-3

1. Start the engine.

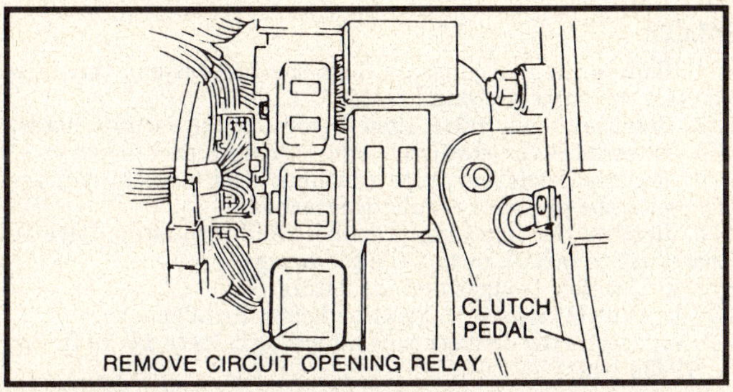

To release fuel pressure remove the circuit opening relay located under the left side of the dashboard—626 and MX-6

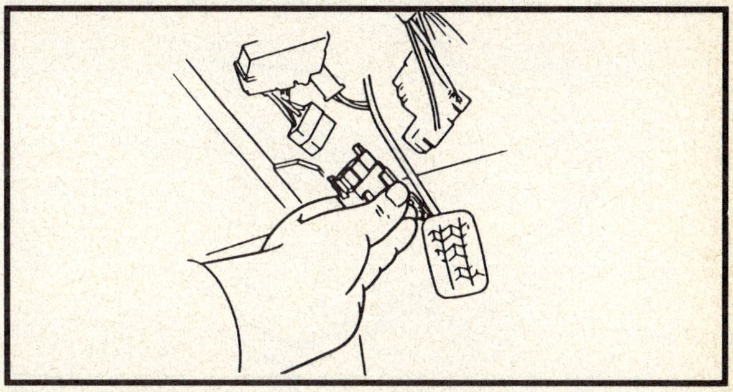

Disconnecting the circuit opening relay to relieve fuel pressure–Miata, B2200 and B2600i

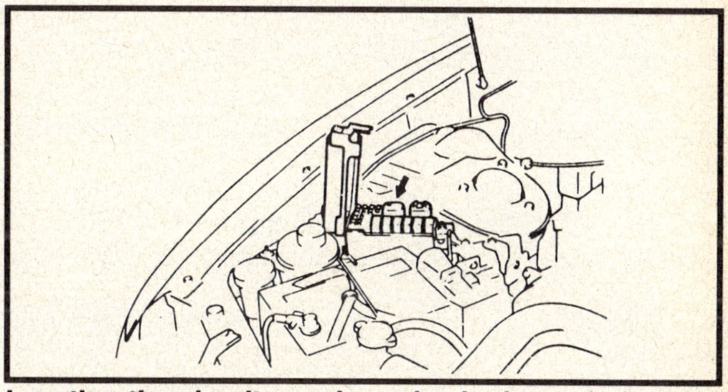

Locating the circuit opening relay in the underhood relay box—1992 929

2. Except MPV with 3.0L engine, disconnect the circuit opening relay connector, with the engine running. The circuit opening relay controls power to the fuel pump and is located in the underhood fuse and relay panel in the 1992 929 and under the dash near the kick panel on the drivers side on other vehicles. Some relays have a yellow harness connector.
3. On MPV with 3.0L engine, disconnect the air flow meter connector.
4. After the engine stalls, turn the ignition switch to the **OFF** position and reconnect the circuit opening relay connector and/or the air flow meter connector.
5. Use a shop rag to cover the fuel lines when disconnecting. Plug all fuel lines after they have been disconnected.

MPV with 2.6L engine, 1990–91 929, 323, Protege and MX-3

1. Start the engine.

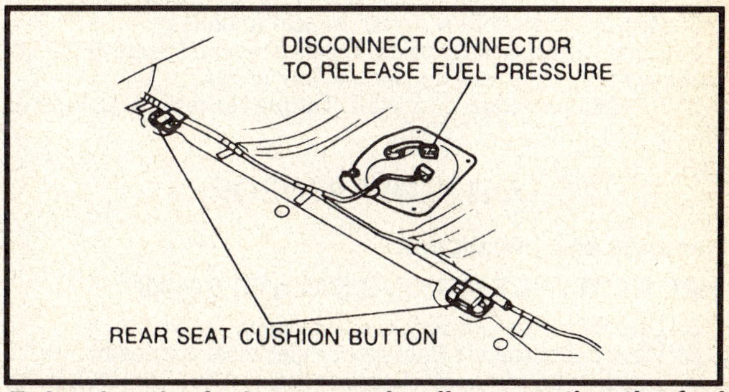

Releasing the fuel pressure by disconnecting the fuel pump—323, Protege and MX-3

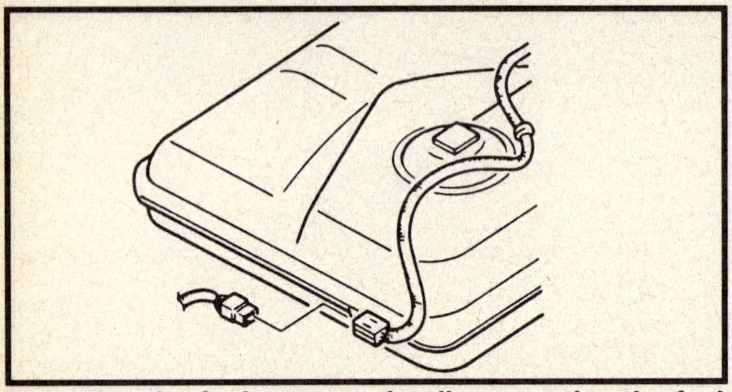

Releasing the fuel pressure by disconnecting the fuel pump—MPV with 2.6L engine

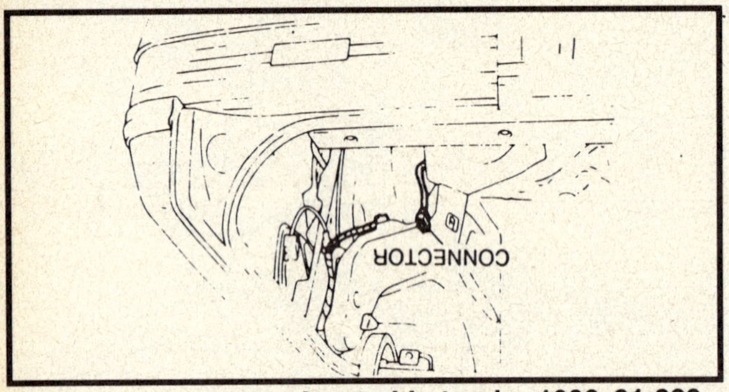

Fuel pump connector located in trunk—1990–91 929

2. On the 323, Protege and MX-3 remove the rear cushion, then disconnect the fuel pump electrical connector. On the MPV locate the fuel pump lead coming from the fuel tank and disconnect the fuel pump electrical connector. On the 1990–91 929 disconnect the fuel pump lead located within the trunk on the right side near the fuel tank.
3. After the engine stalls, turn the ignition switch to the **OFF** position and reconnect the fuel pump electrical connector and install the rear seat cushion.
4. Use a shop rag to cover the fuel lines when disconnecting. Plug all fuel lines after they have been disconnected.

AIR FLOW METER

Removal and Installation

EXCEPT MPV WITH 2.6L ENGINE, B2200 AND B2600i

1. Disconnect the battery ground cable, then the high tension leads and connectors.
2. Loosen the hose band, then remove the intake hose.
3. Remove the air flow meter attaching bolts.
4. Turn the air cleaner cover upside down and remove the attaching nuts, then remove the air flow meter.
5. Reverse procedure to install. Torque the mounting nuts to 69–95 inch lbs. (7.8–11 Nm).

AIR FLOW SENSOR

Removal and Installation

MPV WITH 2.6L ENGINE, B2200 AND B2600i

1. Disconnect the battery ground cable.
2. Disconnect the air flow sensor electrical connector.
3. Loosen the air hose clamps, then remove the mounting bolts.
4. Remove the air flow sensor. Reverse procedure to install.

FUEL INJECTORS

Removal and Installation

MPV WITH 3.0L ENGINE AND 929

1. Relieve the fuel pressure from the fuel system, then disconnect the battery ground cable. Remove the dynamic chamber (intake plenum) cover.
2. Remove the dynamic chamber, then disconnect the fuel hoses and vacuum hoses, then the electrical connectors from each injector.
3. Remove the delivery pipe with the pressure regulator, then the injectors.
4. Reverse procedure to install, noting the following:
   a. Use new O-rings.
   b. Coat O-ring with gasoline before installing injector.
   c. Torque delivery pipe attaching bolts to 14–19 ft. lbs. (19–25 Nm).

MX-3 WITH 1.6L ENGINE, 323 AND PROTEGE

1. Relieve the fuel pressure from the fuel system, then disconnect the battery ground cable.
2. Disconnect the electrical connector from each injector.
3. Remove the injector harness from the delivery pipe, then the delivery pipe attaching bolts from the intake manifold.
4. Remove the delivery pipe assembly, with the injectors and pressure regulator attached.
5. Remove the injectors, then the grommets and insulators. Reverse procedure to install, noting the following:
   a. Use new O-rings and insulators.
   b. Coat O-rings with gasoline before installing.
   c. Torque delivery pipe attaching bolts to 14–19 ft. lbs. (19–25 Nm).

MX-3 WITH 1.8L ENGINE

1. Relieve the fuel pressure from the system. Disconnect the negative battery connector.
2. Remove the air intake hose and the fuel injector harness electrical connectors.
3. Remove the fuel hose to the fuel distribution rails. Remove the bolts holding the fuel distribution rails.
4. Remove the fuel distribution rails being careful not to misplace any of the spacers. Note the position of the spacers for replacement.
5. Remove the injectors and insulators from the rails.
6. Lightly oil the new O–rings and install on the injectors.
7. Install the injectors into the bores by twisting the injector. Align the notch on the injector with the notch on the rail.
8. Install the bolts and torque to 22–31 inch lbs. (2.5–3.5 Nm).
9. Install the fuel hose, harness, and intake hose. Replace the negative battery cable and check for fuel leaks.

MIATA

1. Relieve the fuel pressure from the fuel system, then disconnect the battery ground cable.
2. Remove the air valve, then disconnect the vacuum hoses.
3. Disconnect the electrical connector from each injector.
4. Remove the delivery pipe attaching bolts, then the delivery pipe with the pressure regulator attached.
5. Remove the injectors, grommets and insulators. Reverse procedure to install, noting the following:
   a. Use new O-rings and insulators.
   b. Coat O-rings with gasoline before installing.
   c. Torque the delivery pipe attaching bolts to 14–19 ft. lbs. (19–25 Nm).

626 AND MX-6

1. Relieve the fuel pressure from the fuel system, then disconnect the battery ground cable.

2. Remove the wiring harness bracket attaching bolts, then the wiring harness bracket.
3. On non-turbocharged vehicles, remove the EGR modulator valve, then the EGR modulator valve mounting bracket.
4. Disconnect the air hose from the throttle body. Remove the engine hanger attaching bolts, then the engine hanger.
5. Remove the dynamic chamber mounting bolts and nuts, then lift the dynamic chamber assembly.
6. Disconnect the fuel return line bracket from the intake manifold, then disconnect the injector connectors.
7. Remove the delivery pipe attaching bolts, then the delivery pipe with the injectors, pressure regulator and pulsation damper.
8. Remove the injectors from the delivery pipe.
9. Reverse procedure to install, noting the following:
   a. Use new O-rings.
   b. Coat O-rings with gasoline before installing.
   c. Torque delivery pipe, dynamic chamber and engine hanger nuts and bolts to 14–19 ft. lbs. (19–25 Nm).

MPV WITH 2.6L ENGINE, B2200 AND B2600i

1. Relieve the fuel pressure from the fuel system.
2. Remove the dynamic chamber attaching nuts and bolts, then the dynamic chamber.
3. Disconnect the vacuum and fuel hoses. Remove the delivery pipe attaching bolts, then remove the delivery pipe with the pressure regulator attached.
4. Disconnect the electrical connectors from the injectors. Remove the grommets, injectors and insulators.
5. Reverse procedure to install, noting the following:
   a. Use new O-rings.
   b. Coat O-rings with gasoline before installing.
   c. Torque the delivery pipe and dynamic chamber attaching nuts and bolts to 14–19 ft. lbs. (19–25 Nm).

THROTTLE BODY

Removal and Installation

1. Relieve the fuel system pressure. Disconnect the battery ground cable.
2. Disconnect the accelerator cable from the throttle linkage.
3. Disconnect the air funnel, then the air hoses and lines from the throttle body. Remove the Bypass Air Control valve (BAC), if equipped.
4. On the 1990–91 929 SOHC remove the fuel line at the cold start injector.
5. On the B2200, B2600i, MPV, 1990–91 929 SOHC, 626 and MX-6 drain engine coolant to a level below the throttle body. Remove the water lines on water heated throttle bodies.
6. Disconnect all electrical connectors from the throttle body assembly. Remove the throttle body attaching screws, then the throttle body assembly.
7. Reverse procedure to install. Connect the electrical connectors, vacuum lines and water lines. Torque the retaining nuts to 14–19 ft. lbs.(19–25 Nm). Refill the cooling system.

THROTTLE POSITION SENSOR

Removal and Installation

1. Disconnect the battery ground cable.
2. Disconnect the throttle position electrical connector.
3. Remove the 2 screws fastening the throttle position sensor to the throttle body.
4. Remove the sensor from the throttle body.
5. Replace the throttle sensor on the throttle body. On some vehicles it may be necessary to rotate the sensor to align the tabs on the sensor and throttle body and then rotate the sensor back to align the mounting holes.
6. Affix the sensor with the mounting screws. Connect the throttle sensor electrical lead.
7. Adjust the sensor and check its operation. Connect the negative battery terminal.

Adjustment

MX-3 WITH 1.6L ENGINE, 323, PROTEGE AND MIATA

Manual Transmission/Transaxle

1. Disconnect the electrical connector from the throttle sensor.
2. Connect a suitable ohmmeter to terminals E and IDL on the throttle sensor.
3. Insert a 0.016 in. (0.4mm) feeler gauge between the throttle stop screw and the stop lever.
4. Loosen the throttle sensor attaching screws and rotate the throttle sensor clockwise approximately 30 degrees, then rotate it back counterclockwise until continuity exists.
5. Replace the feeler gauge with a 0.027 in. (0.7mm) feeler gauge. Check that continuity does not exists. If continuity exists, repeat Steps 3 through 5.
6. Tighten the throttle sensor attaching screws.

NOTE: Do not move the throttle sensor while tighten the screws.

7. Operate the throttle valve and recheck the adjustment of the throttle sensor.

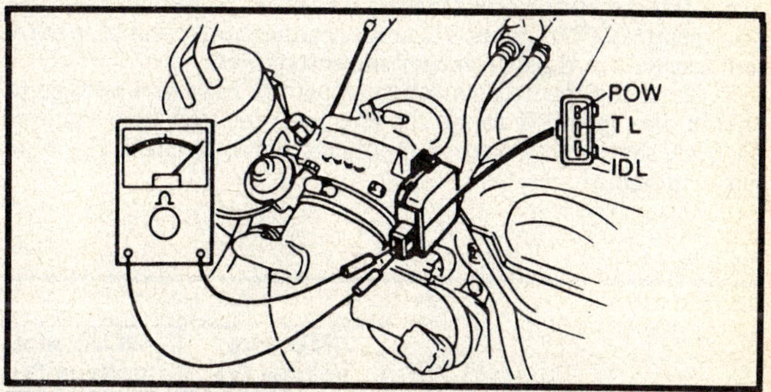

Throttle position sensor testing with an ohm-meter. The TL terminal is also called the E terminal on some vehicles—Miata, MX-3 with 1.6L engine), 323 and Protege with manual transmissions

MX-3 WITH 1.6L ENGINE, 323, PROTEGE AND MIATA

Automatic Transmission/Transaxle

1. Disconnect the electrical connector from the throttle sensor.
2. Connect a suitable ohmmeter between terminals E and IDL on the throttle sensor.
3. Insert a 0.01 in. (0.25mm) feeler gauge between the throttle stop screw and the stop lever.
4. Loosen the throttle sensor attaching screws and rotate the sensor approximately 30 degrees, then rotate it back counterclockwise until continuity exists.
5. Replace the feeler gauge with a 0.016 in. (0.4mm) feeler gauge and check that continuity does not exists. If continuity exists, repeat Steps 3 through 5.
6. Tighten the throttle sensor attaching screws.

NOTE: Do not move the throttle sensor while tighten the screws.

7. Open the throttle valve and check that the resistance between the throttle sensor terminals E and VT using a suitable ohmmeter. Ohmmeter should read approximately 5 Kilo-ohms.

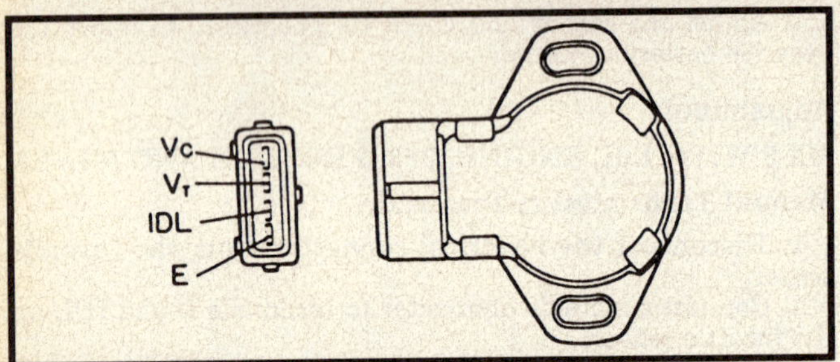

Throttle position sensor terminals – Miata, MX-3 with 1.6L engine, 323 and Protege with automatic transmissions

MPV WITH 2.6L ENGINE, 626, MX-6, B2200 AND B2600i

1. Disconnect the air hose from throttle body assembly, then disconnect the throttle sensor electrical connector (3-pin).
2. Connect Adapter Harness (49–G018–901) between the throttle sensor and the wiring harness.
3. Turn the ignition switch to the **ON** position. Check that the throttle valve is in the fully closed position.
4. Using a suitable voltmeter, measure the red wire voltage and record it. Change the voltmeter connection to the blue wire, and loosen the throttle sensor mounting bolts.
5. Turn the throttle sensor to adjust the blue wire voltage to within the specified range for the recorded red wire voltage. Tighten the throttle sensor mounting bolts. Recheck the blue wire voltage.
6. Hold the throttle valve in the fully opened position, and check that the blue wire voltage is within the specified range.
7. Check that the blue wire voltage increases smoothly when opening the throttle valve from its closed to fully opened position. If not, replace the throttle sensor.
8. Turn the ignition switch to the **OFF** position. Disconnect the Adapter Harness (49–G018–901) and reconnect the throttle sensor electrical connector.
9. Disconnect the battery ground cable and depress the brake pedal for at least 5 seconds to eliminate the ECU malfunction memory.

MPV WITH 3.0L ENGINE AND 1990–91 929

1. Disconnect the throttle sensor electrical connector from the throttle sensor.
2. Connect a suitable ohmmeter between terminals C and D on the throttle sensor.
3. Except 1990–91 929 with DOHC engine, insert a 0.020 in. (0.5mm) feeler gauge between the throttle stopper screw and the throttle lever. On 1990–91 929 with DOHC engine, insert a 0.004 in. (0.1mm) feeler gauge between the throttle stopper screw and the throttle lever. Continuity should exists between terminals C and D.
4. Except 1990–91 929 with DOHC engine, insert a 0.028 in. (0.7mm) feeler gauge between the throttle stopper screw and the throttle lever. On 1990–91 929 with DOHC engine, insert a 0.012 in. (0.3mm) feeler gauge between the Throttler stopper screw and the throttle lever. No continuity should exists between terminals C and D.
5. If specifications in Steps 3 and 4 are not as indicated, loosen the throttle sensor attaching screws and turn the throttle sensor until proper specifications are obtained.
6. Reconnect the throttle sensor electrical connector.

RED wire voltage (V)	BLUE wire voltage (V)	RED wire voltage (V)	BLUE wire voltage (V)
4.50—4.59	0.37—0.54	5.10—5.19	0.42—0.61
4.60—4.69	0.38—0.55	5.20—5.29	0.43—0.62
4.70—4.79	0.39—0.56	5.30—5.39	0.44—0.63
4.80—4.89	0.40—0.57	5.40—5.49	0.44—0.64
4.90—4.99	0.40—0.58	5.50	0.44—0.66
5.00—5.09	0.41—0.60		

Throttle closed throttle position sensor voltage specifications – B2200, B2600i, 626, MX-6 and MPV with 2.6L engine

RED wire voltage (V)	BLUE wire voltage (V)	RED wire voltage (V)	BLUE wire voltage (V)
4.50—4.59	3.58—4.23	5.10—5.19	4.05—4.79
4.60—4.69	3.66—4.32	5.20—5.29	4.13—4.88
4.70—4.79	3.74—4.41	5.30—5.39	4.21—4.98
4.80—4.89	3.82—4.51	5.40—5.49	4.29—5.07
4.90—4.99	3.90—4.60	5.50	4.29—5.17
5.00—5.09	3.97—4.70		

Throttle open throttle position sensor voltage specifications – B2200, B2600i, 626, MX-6 and MPV with 2.6L engine

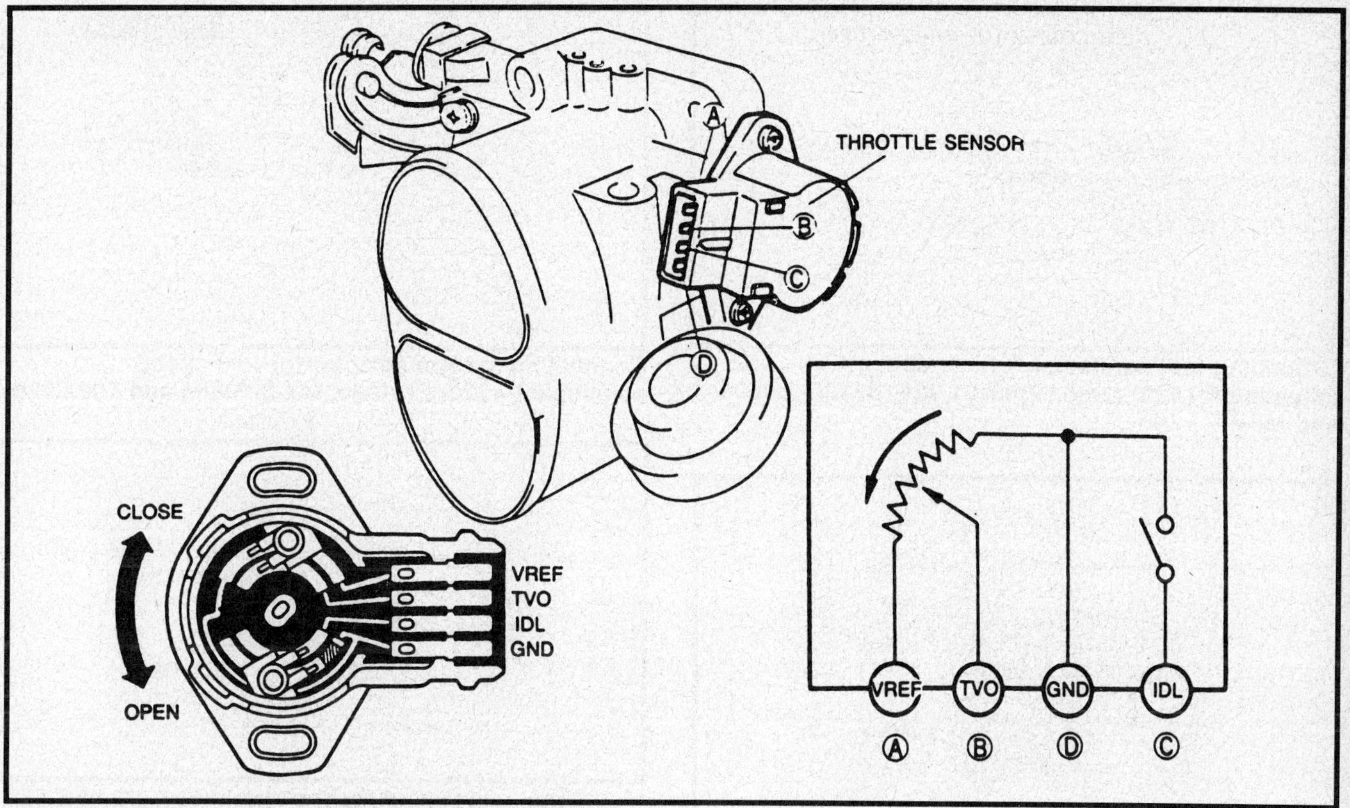

Throttle position sensor—1990–91 929 and MPV

MX-3 WITH 1.8L ENGINE AND 1992 929

1. Connect the SST Engine Signal Monitor and Harness Adapter or equivalent, to the ECU and set the monitor to the terminal 2F position for 929 or 2M for MX-3.
2. Turn the ignition switch **ON** and rotate the throttle by hand.
3. With the throttle fully closed the voltage on terminal 2F for 929 or 2M for MX-3 should be 0.1–1.1 volts. With the throttle fully open the voltage on terminal 2F for 929 or 2M for MX-3 should be 3.0–4.3 volts for 929 or 3.1–4.4 volts for MX-3.
4. Adjust the switch if voltages are within these parameters.
5. Disconnect the throttle position sensor connector and connect a continuity checker between the GND and IDL terminals.
6. Insert a 0.004 in.(0.1mm) feeler gauge for 929 or 0.006 in. (0.15mm) feeler gauge for MX-3 between the throttle stop screw and the lever. There should be continuity.

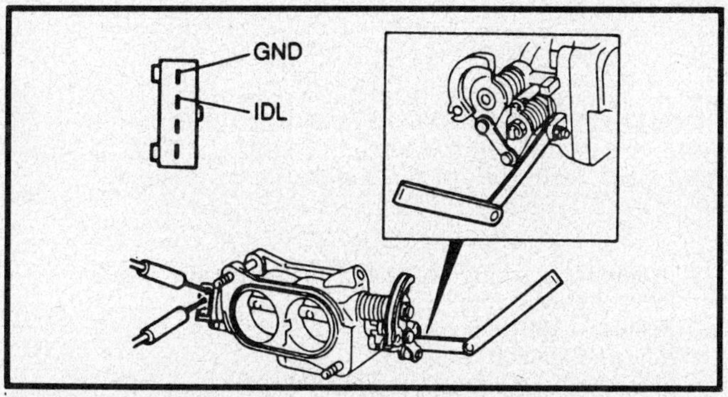

Testing the idle switch portion of the throttle position sensor—1992 929

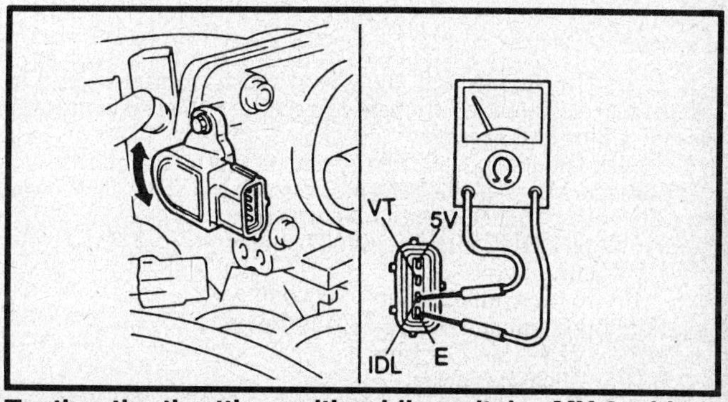

Testing the throttle position idle switch—MX-3 with 1.8L engine

7. Insert a 0.008 in.(0.2mm) feeler gauge for 929 or 0.020 in. (0.50mm) feeler gauge for MX-3 between the throttle stop screw and the lever. There should be no continuity.
8. Adjust the sensor if these values are not obtained. Loosen the mounting screws and rotate the body of the throttle position sensor to obtain the proper adjustment.

IDLE SPEED

Adjustment

626, MX-6, 1990–91 929, B2200, B2600i AND MPV

Before adjusting the idle speed, check that the ignition timing, spark plugs, etc., are in normal operating condition. Turn off all lights and other unnecessary loads.

1. Connect a suitable tachometer to the check connector (white 1 pin).

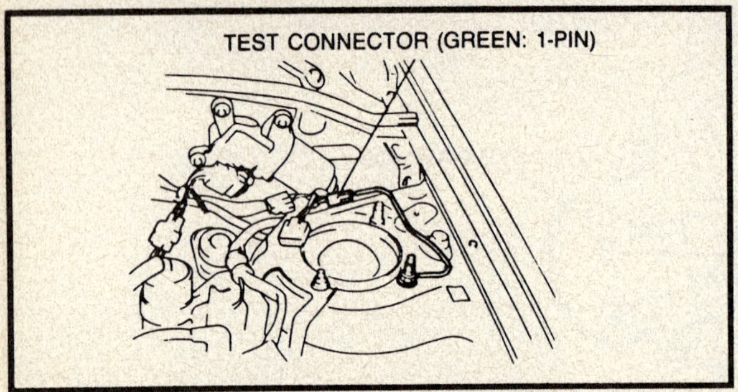

Grounding test connector for idle speed adjustment – 626, MX-6, 1990–91 929, B2200, B2600i and MPV

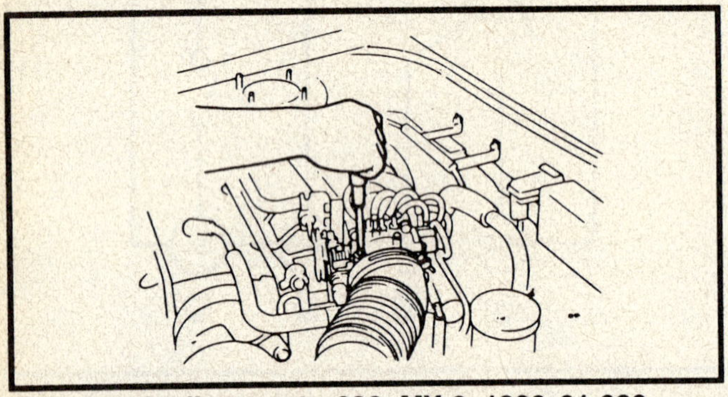

Idle speed adjustment – 626, MX-6, 1990–91 929, B2200, B2600i and MPV

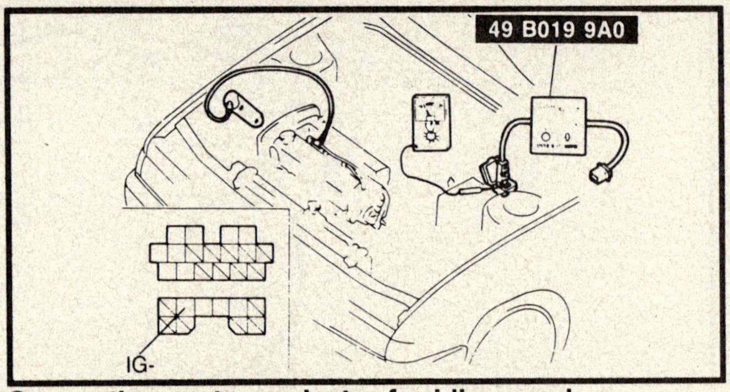

Connecting system selector for idle speed adjustment – 323, Protege, MX-3, Miata and 1992 929

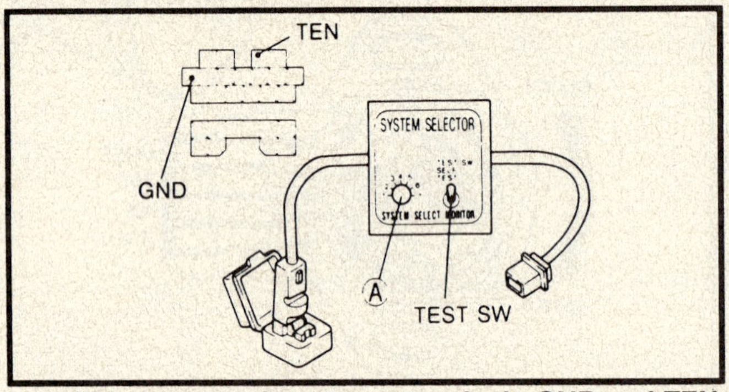

Installing jumper wire across terminals GND and TEN on the diagnosis connector when system selector is not available – 323, Protege, MX-3, Miata and 1992 929

Idle speed adjustment – 323, Protege, MX-3, Miata and 1992 929

2. Start the engine and allow it to reach normal operating temperature. Connect a jumper wire between the test connector (green: 1 pin) and ground.
3. Check the idle speed. Idle speed should be as follows:
   626 and MX-6 – 725–775 rpm
   929 with SOHC engine – 630–670 rpm
   929 with DOHC engine – 680–720 rpm
   MPV 2.6L engine
      automatic transmission – 730–770 rpm
      manual transmission – 750–790 rpm
   MPV 3.0L – 730–770 rpm
   B2200, B2600i trucks
   automatic transmission – 750–790 rpm
      manual transmission – 730–770 rpm
4. If specifications are not as indicated, remove the protector cap from the air adjust screw, and adjust the idle speed by turning the air adjust screw.
5. After adjustment, install the protector cap and disconnect the jumper wire from the test connector. After adjusting the idle speed, the dashpot should be adjusted.

323, PROTEGE, MX-3, MIATA AND 1992 929

1. Start the engine and allow it to reach normal operating temperature. Turn off all lights and other unnecessary loads. When checking the idle speed, ensure that the electric cooling fan is not operating.
2. Connect System Selector 49-B019-9AO to the diagnosis connector. Connect a suitable tachometer to the IG terminal on the diagnosis connector.
3. Set switch (A) on the System Selector to position 1, then set the TEST SW to SELF-TEST.

NOTE: If System Selector 49-B019-9A0 is not available, connect a jumper wire between the TEN terminal and GND terminal on the diagnosis connector.

4. Apply the parking brake and check the idle speed:
   323 – 700–800 rpm
   Protege – 700–800 rpm
   Miata – 800–900 rpm
   MX-3 with 1.6L engine – 800–900 rpm
   MX-3 with 1.8L engine – 640–700 rpm
   1992 929 – 680–720 rpm

NOTE: On 323 and Protege Canadian vehicles equipped with automatic transmission, when the parking brake is not applied, the idle speed should be approximately 800 rpm.

5. If specifications are not as specified, adjust the idle by turning the air adjusting screw. Disconnect the System Selector.

DASHPOT

Adjustment

The MX-3 with 1.6L engine, MPV with 3.0L engine, 323, Protege, Miata and 1990–91 929 SOHC are equipped with a dashpot control. The dashpot controls the action of the throttle to prevent the throttle from snapping shut and boosting emissions.

1. With the engine not running, open the throttle valve fully. Push lightly dashpot rod and check that the rod goes into the dashpot slowly. Release the dashpot rod, and check that the rod comes out quickly. If not, replace the dashpot.
2. Start the engine and allow it to reach normal operating temperature. Stop the engine. On vehicles equipped with a diagnostic connector, connect a suitable tachometer to the IG terminal on the diagnosis connector. On other vehicles, connect a suitable tachometer to the check connector.
3. Start the engine, and slowly increase engine speed to 4000 rpm.
4. Slowly decrease engine rpm, and check that the dashpot rod touches the lever at the specified rpm.

Protege with DOHC engine and MX-3 with 1.6L engine—Approximately 3000 rpm

323 and Protege with SOHC engine—Approximately 2700 rpm

929 with SOHC engine and MPV—3200–3800 rpm

Miata—Approximately 2500 rpm

5. If dashpot actuation speeds not as specified, loosen the locknut and turn the dashpot to adjust.

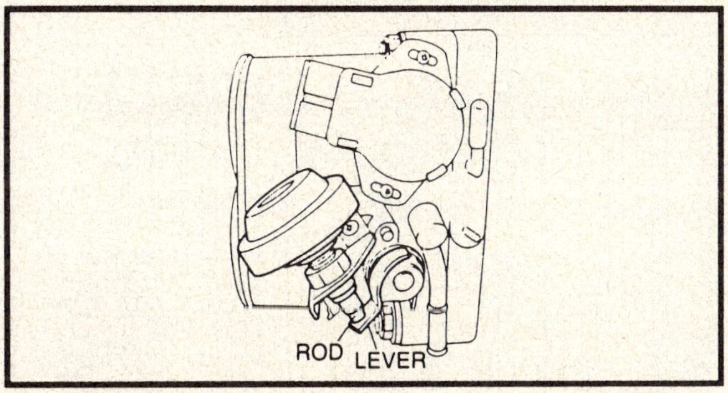

Dashpot adjustment

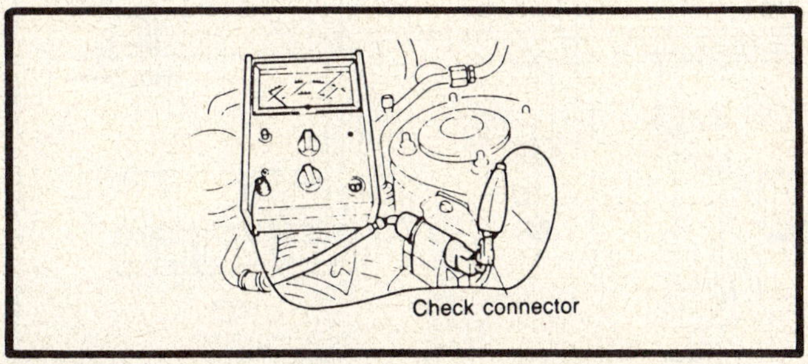

Connecting tachometer to check connector for dashpot adjustment

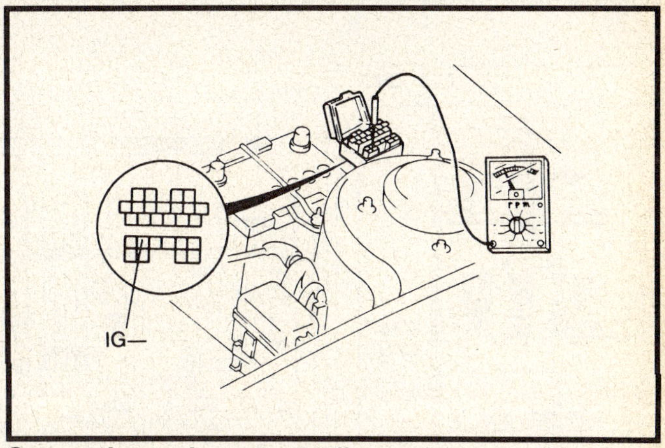

Connecting tachometer to diagnosis connector terminal IG for dashpot adjustment

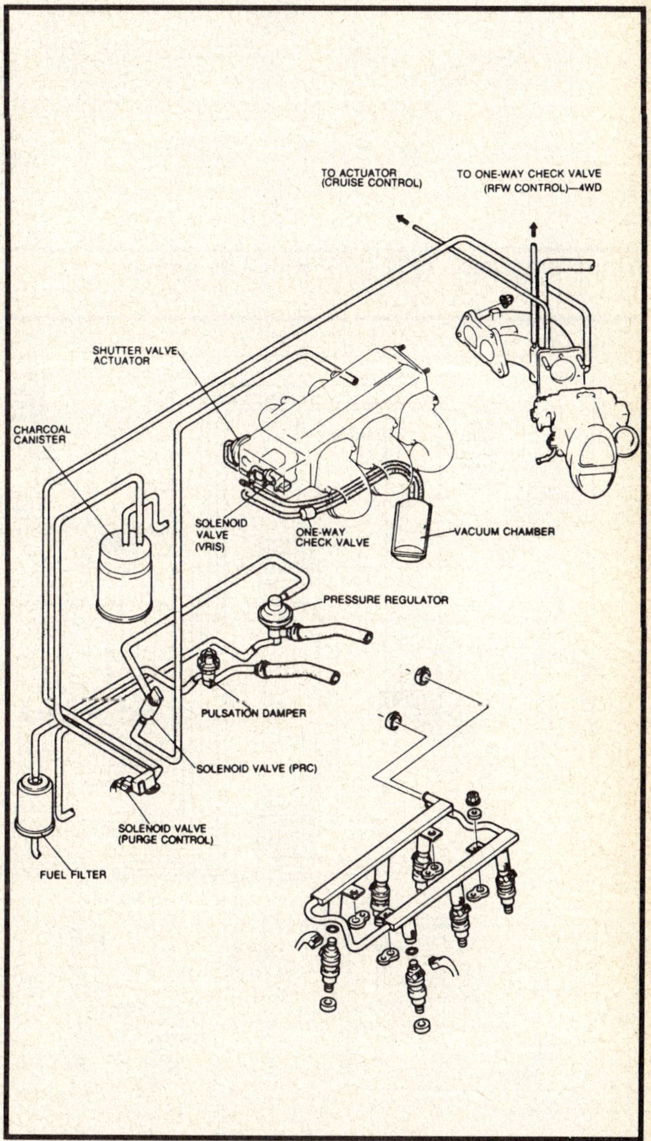

Vacuum hose routing diagram—MPV with 3.0L engine

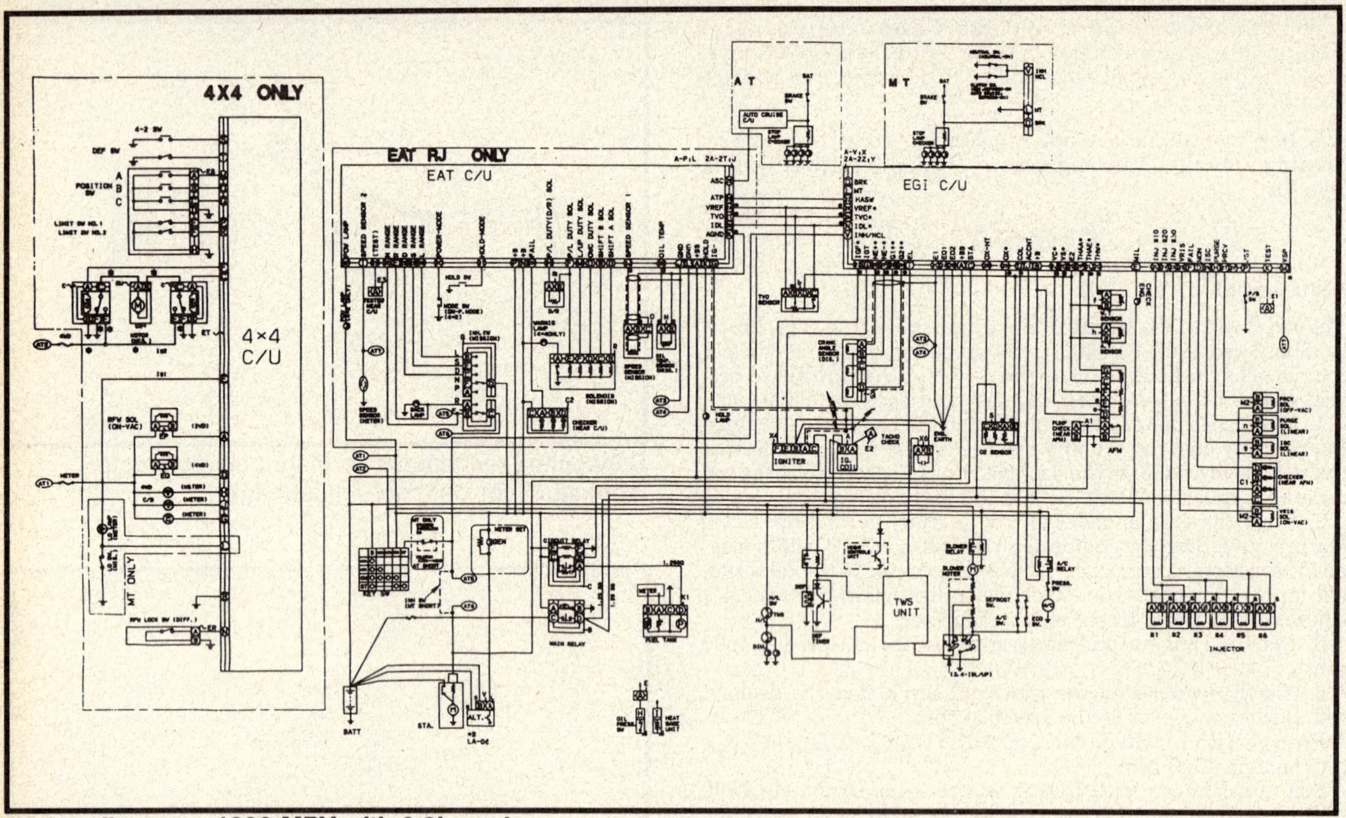

Wiring diagram—1992 MPV with 3.0L engine

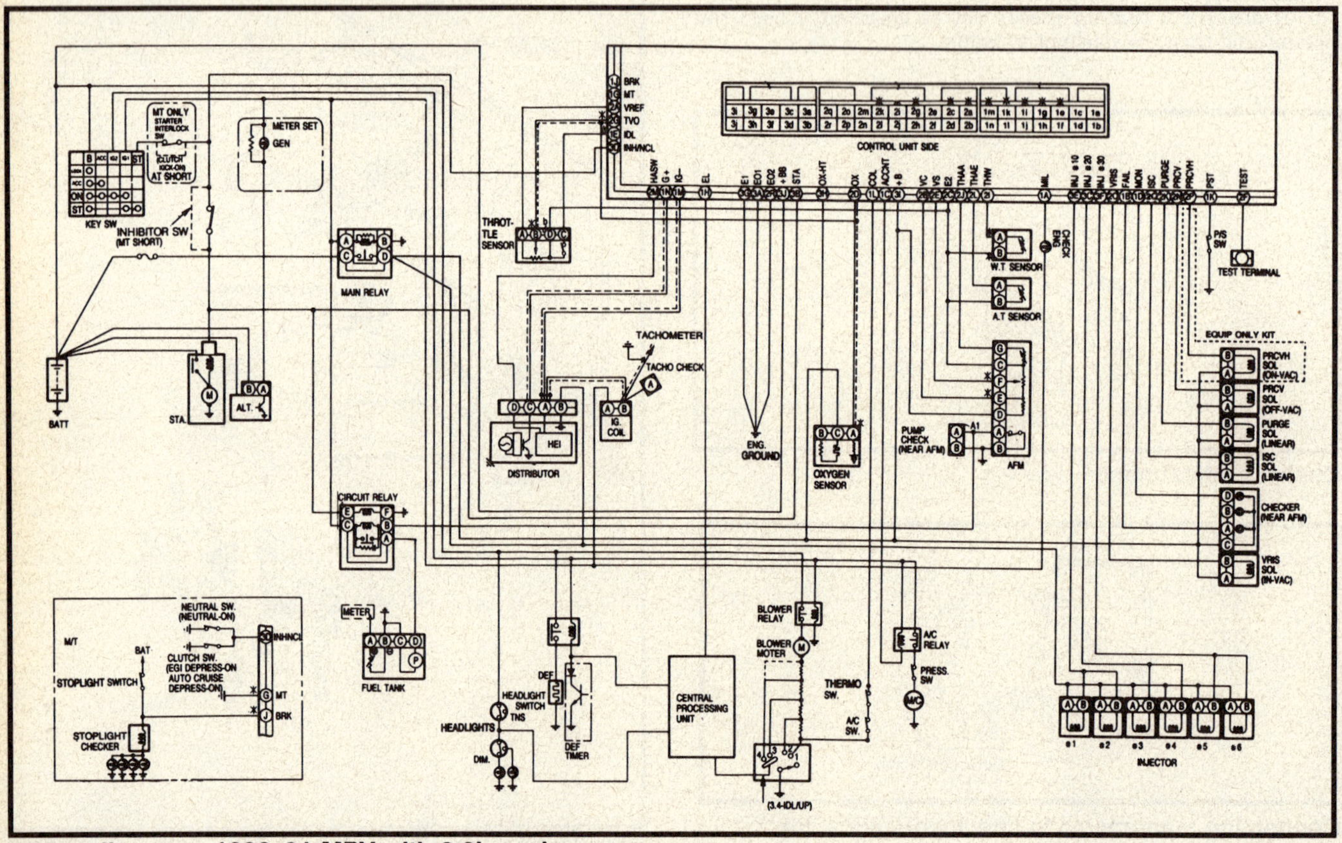

Wiring diagram—1990–91 MPV with 3.0L engine

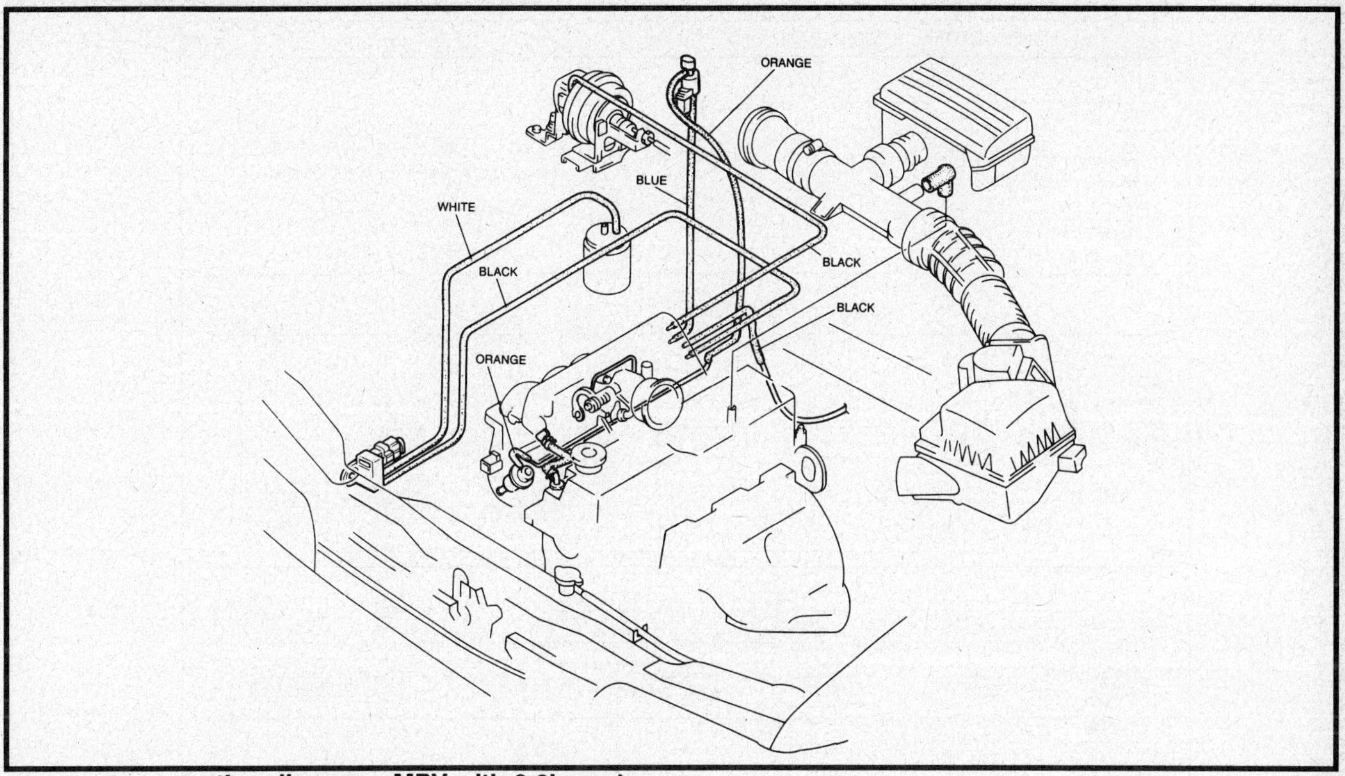

Vacuum hose routing diagram—MPV with 2.6L engine

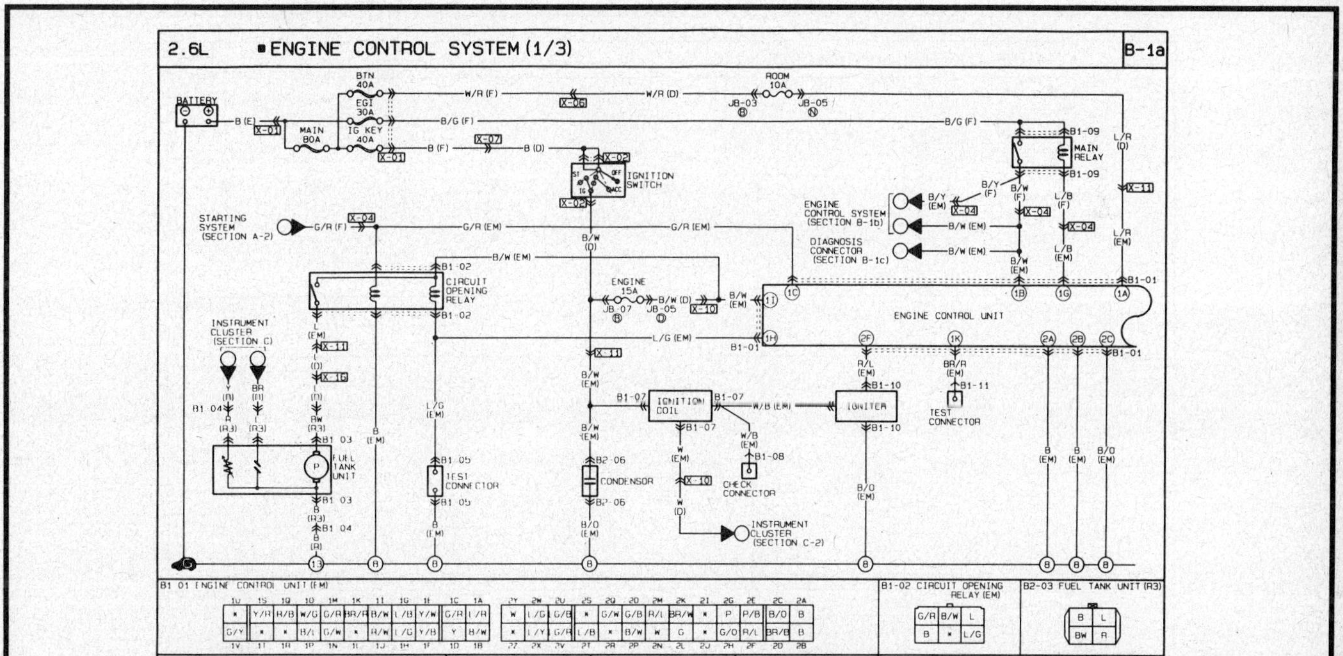

Wiring diagram—MPV with 2.6L engine

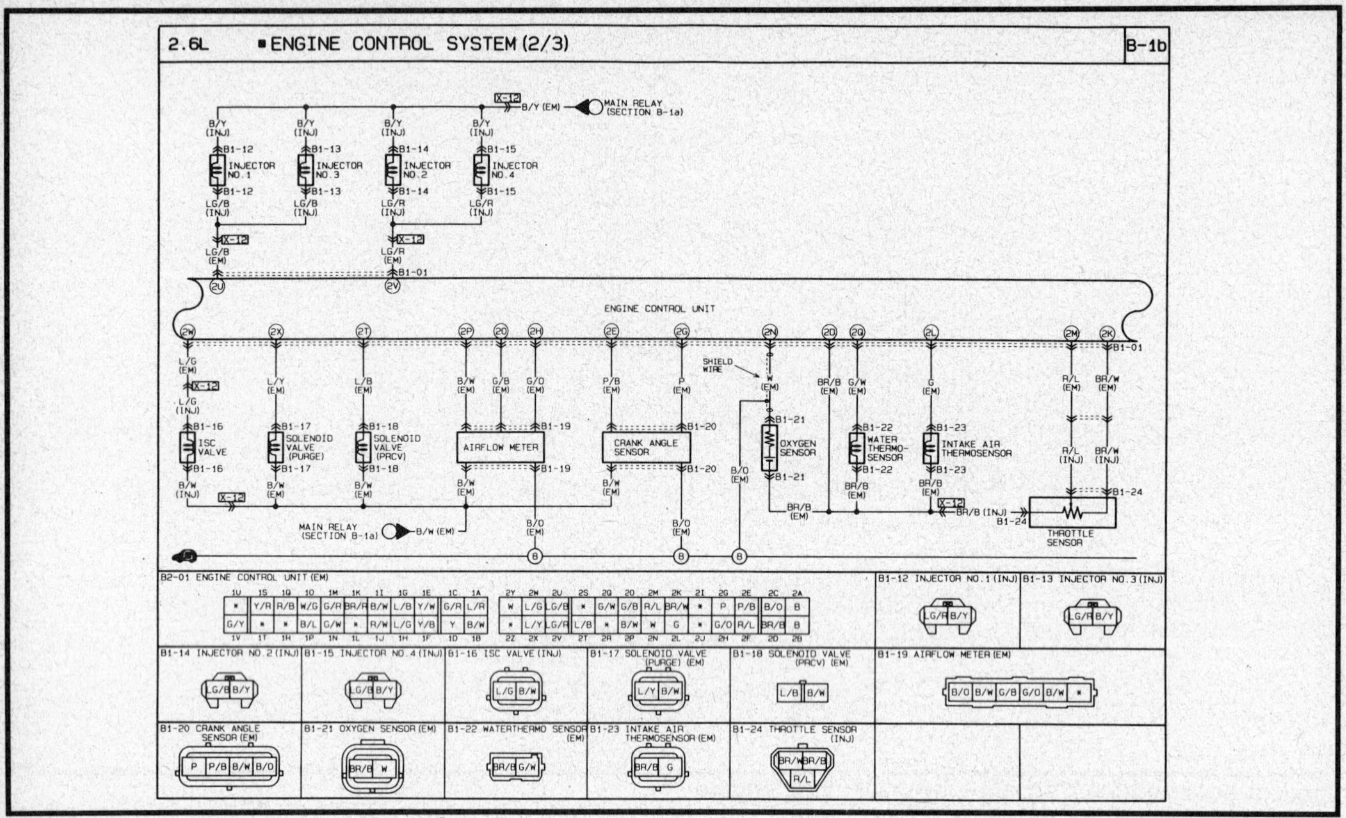

Wiring diagram—MPV with 2.6L engine

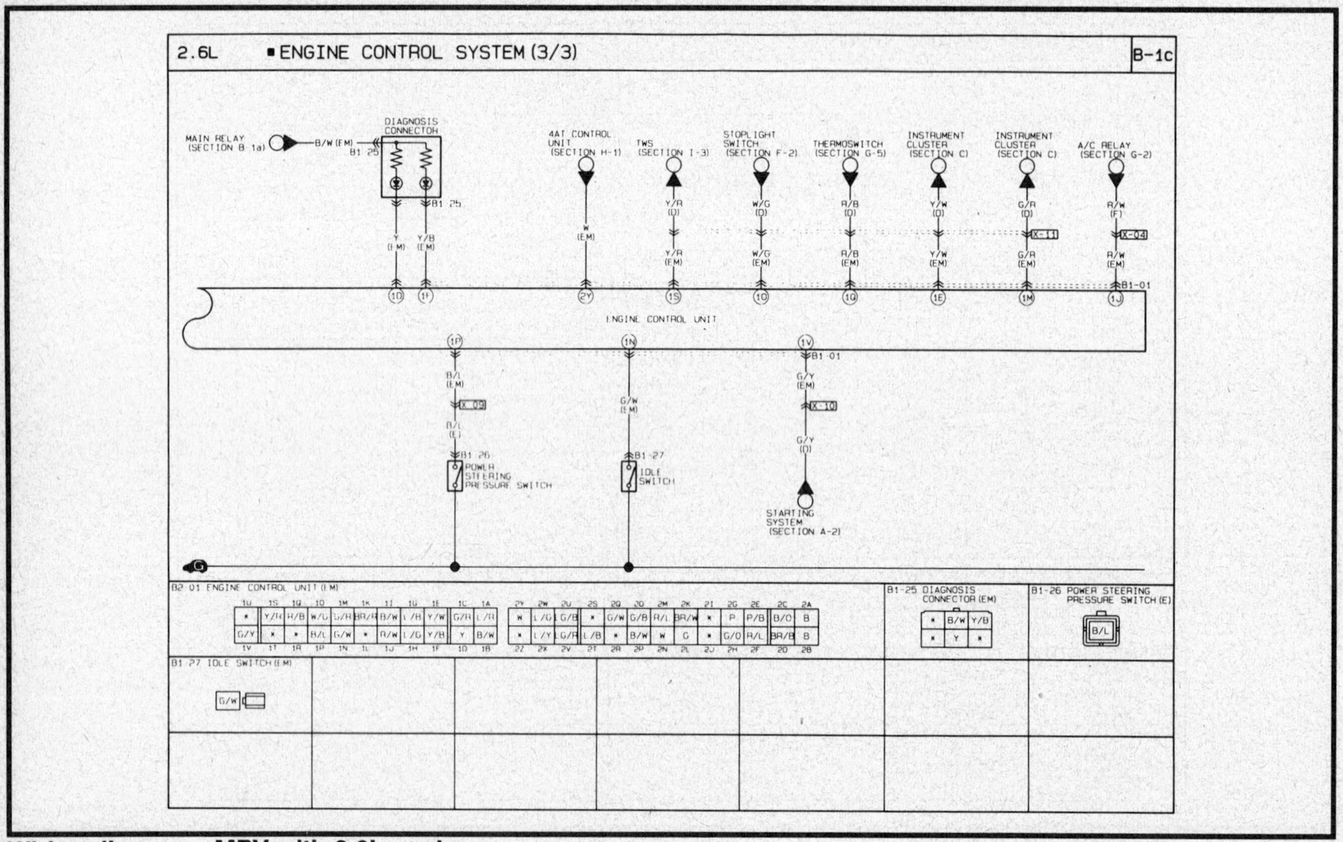

Wiring diagram—MPV with 2.6L engine

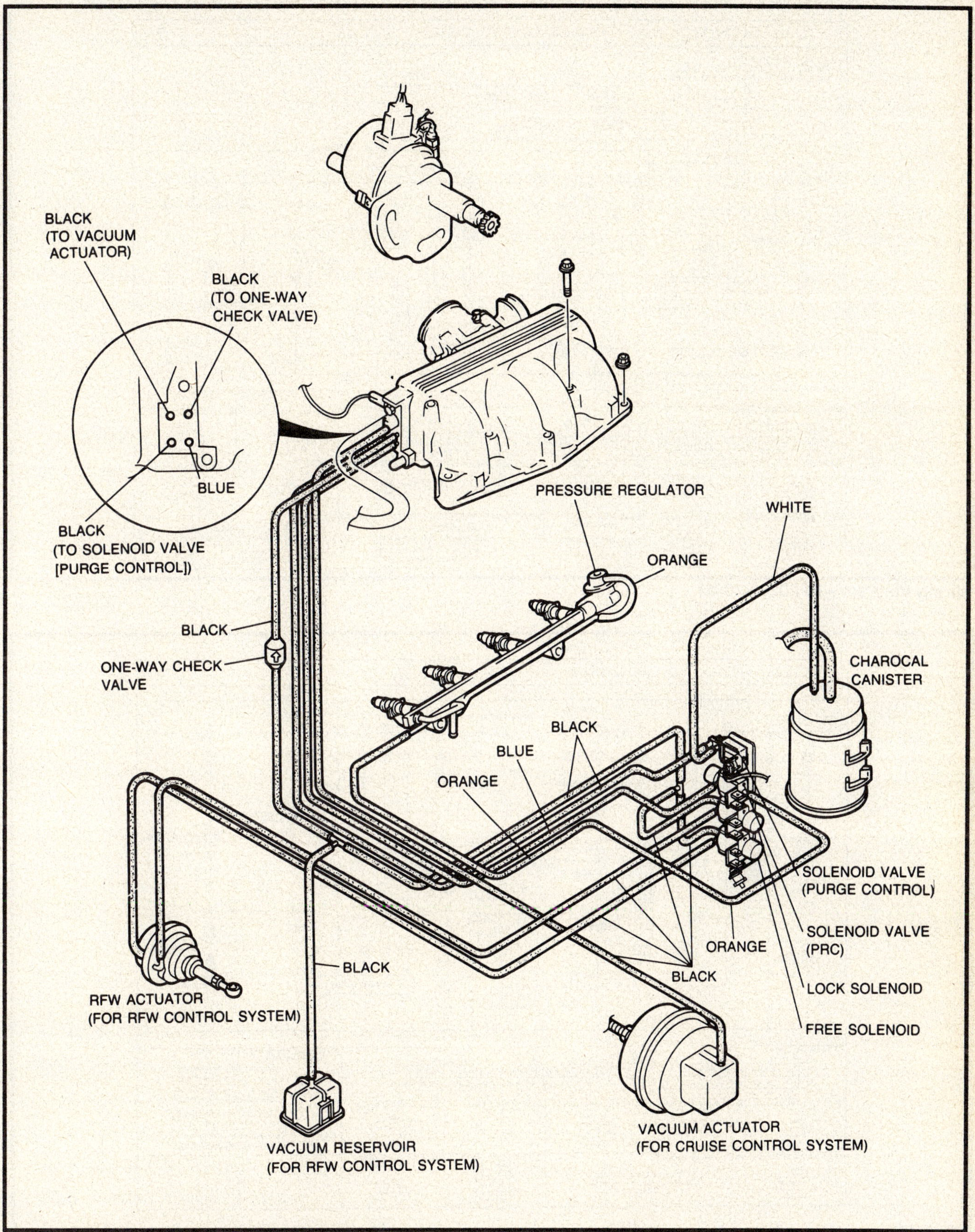

Vacuum hose routing diagram—B2200, B2600i

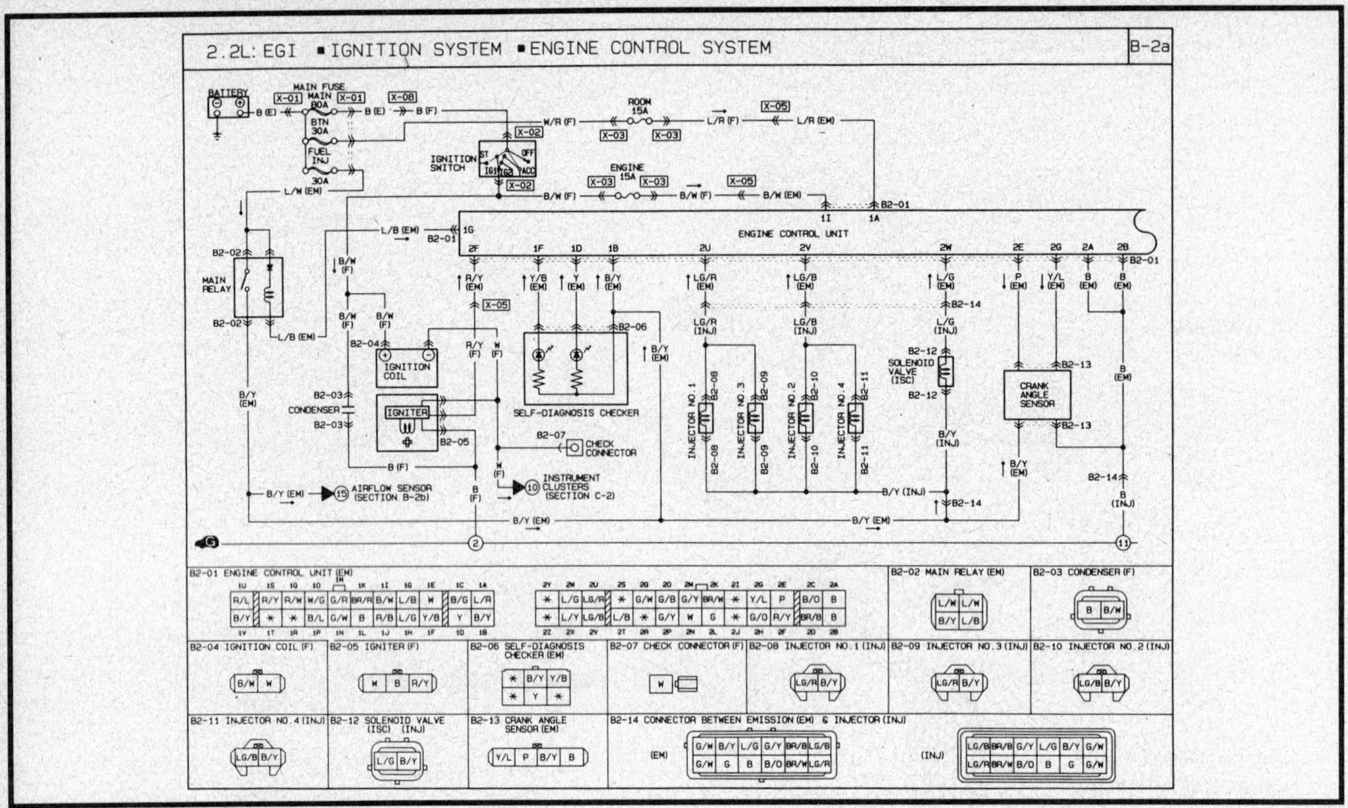

Wiring diagram—B2200, B2600i

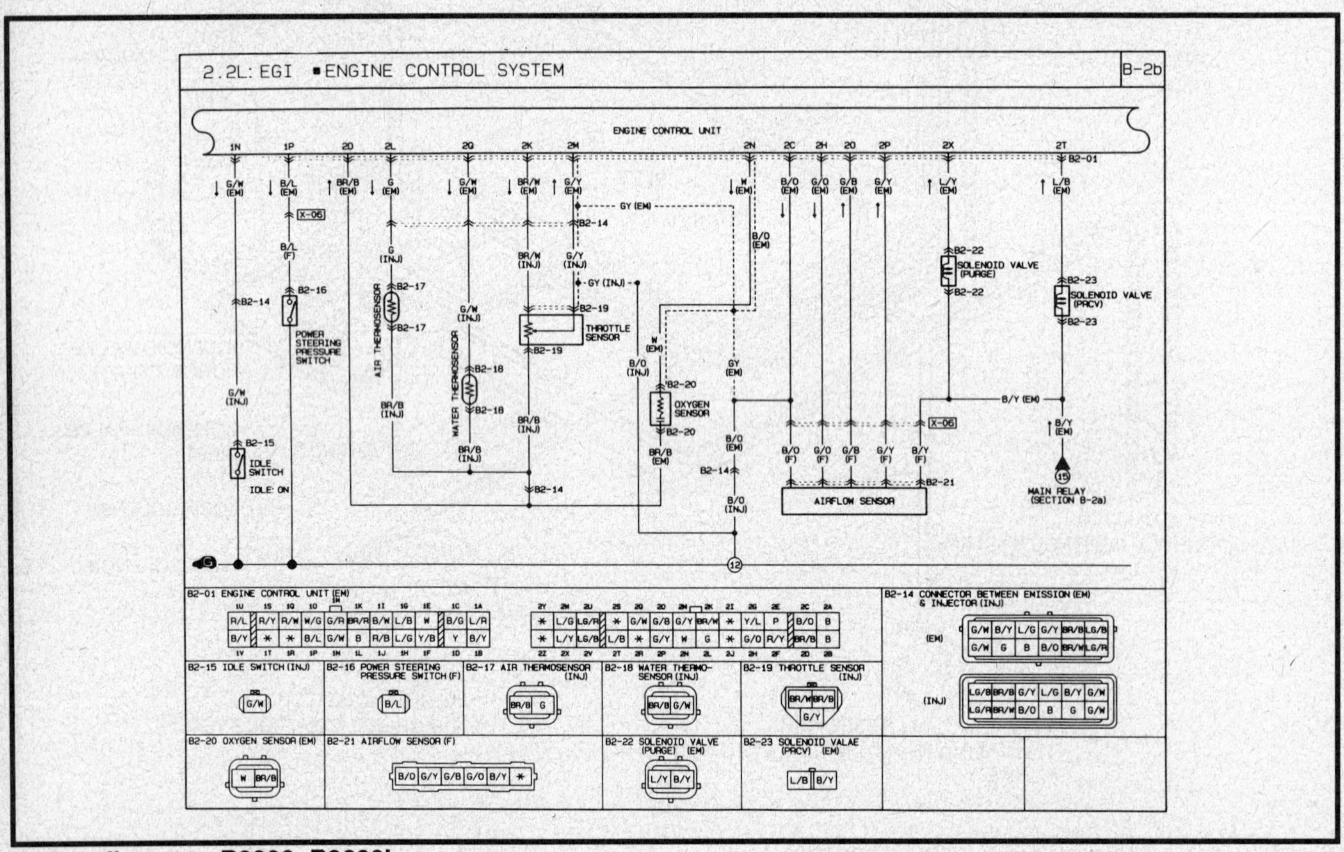

Wiring diagram—B2200, B2600i

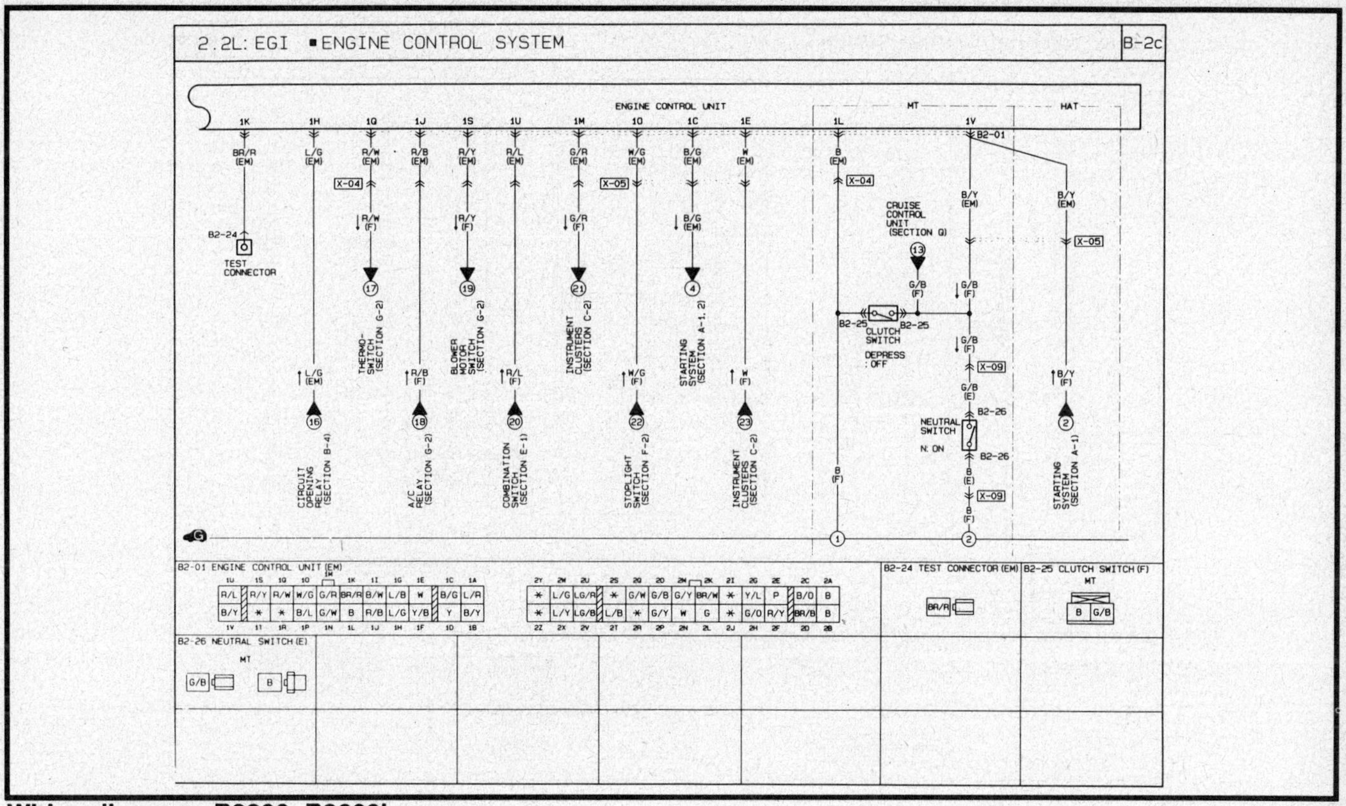

Wiring diagram—B2200, B2600i

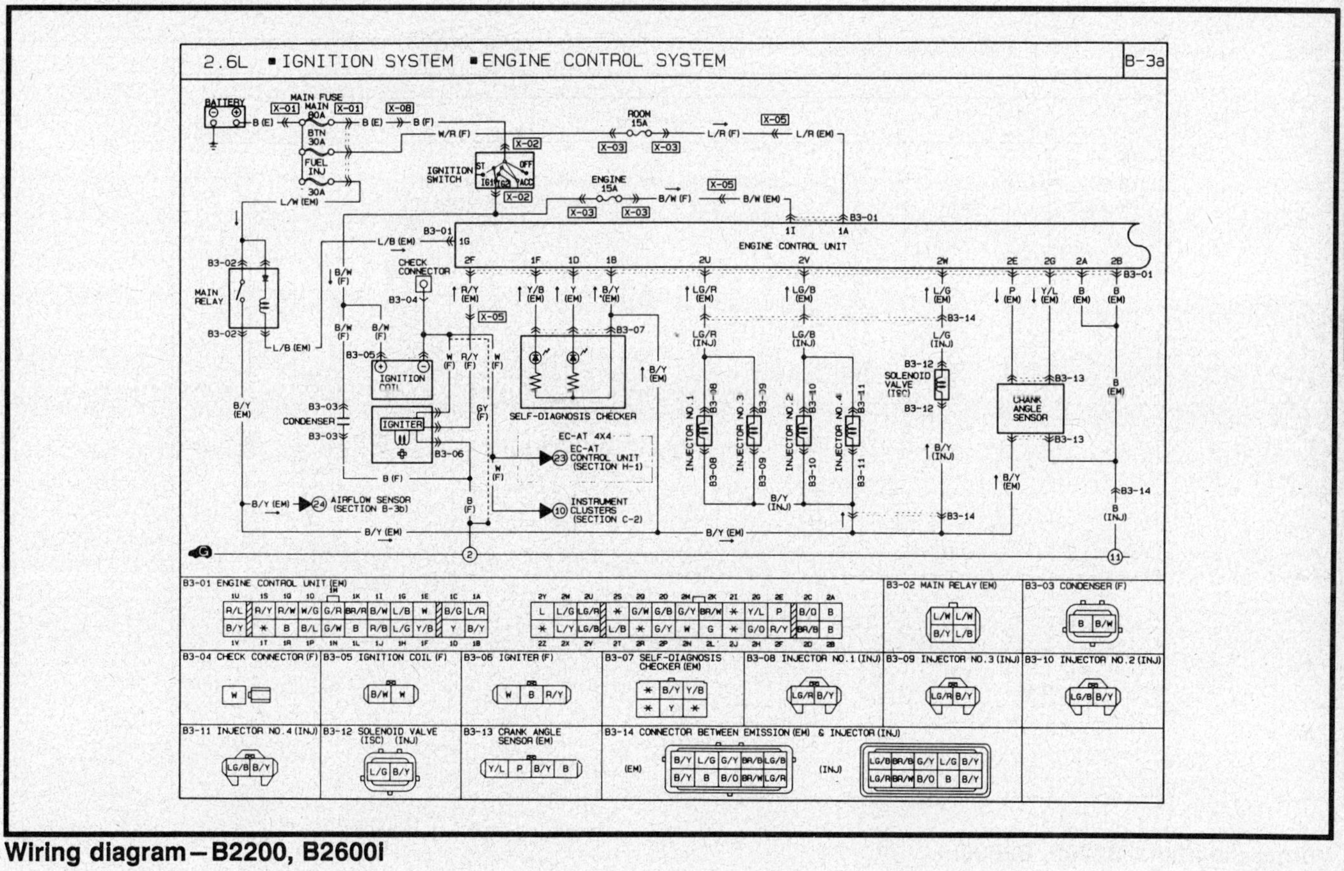

Wiring diagram—B2200, B2600i

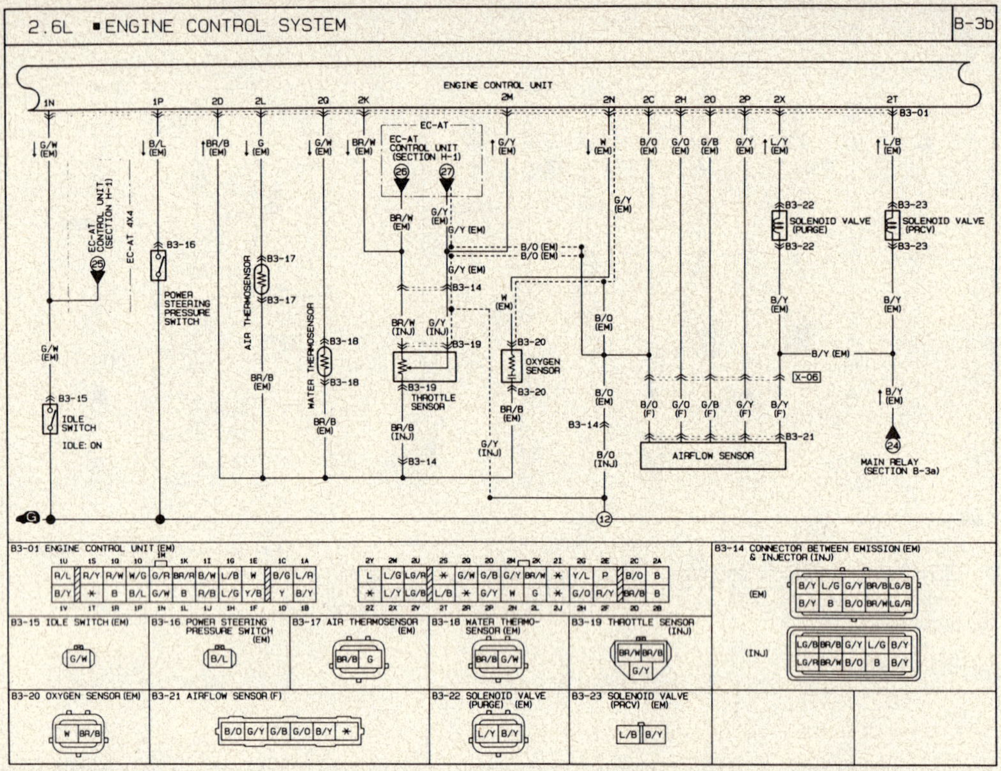

Wiring diagram—B2200, B2600i

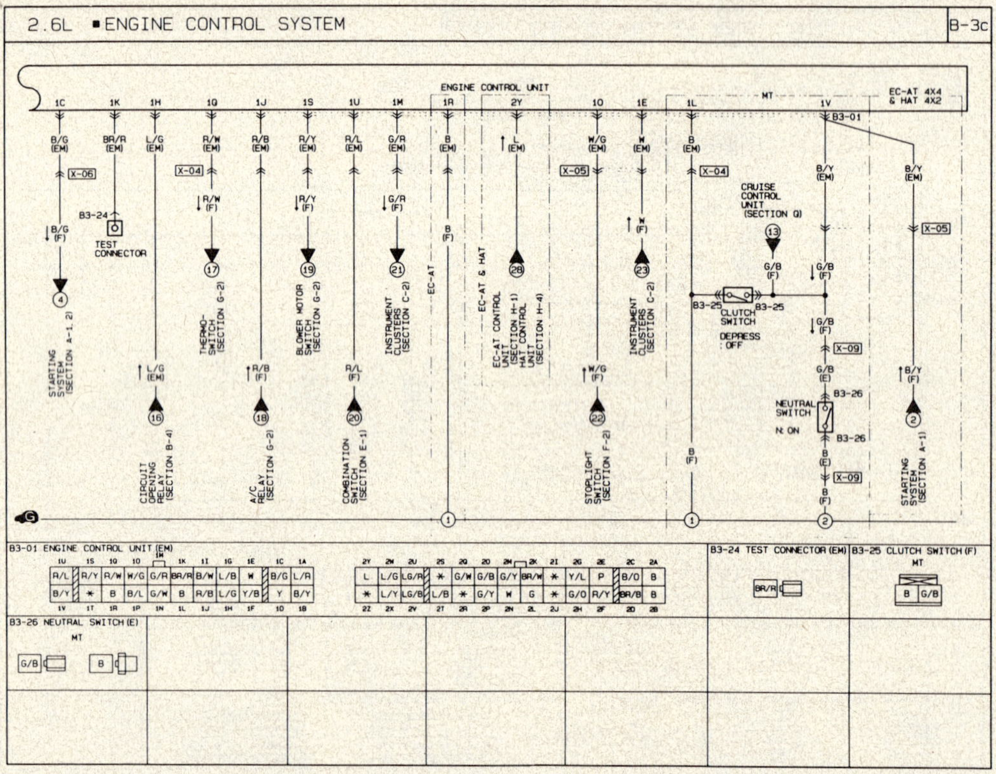

Wiring diagram—B2200, B2600i

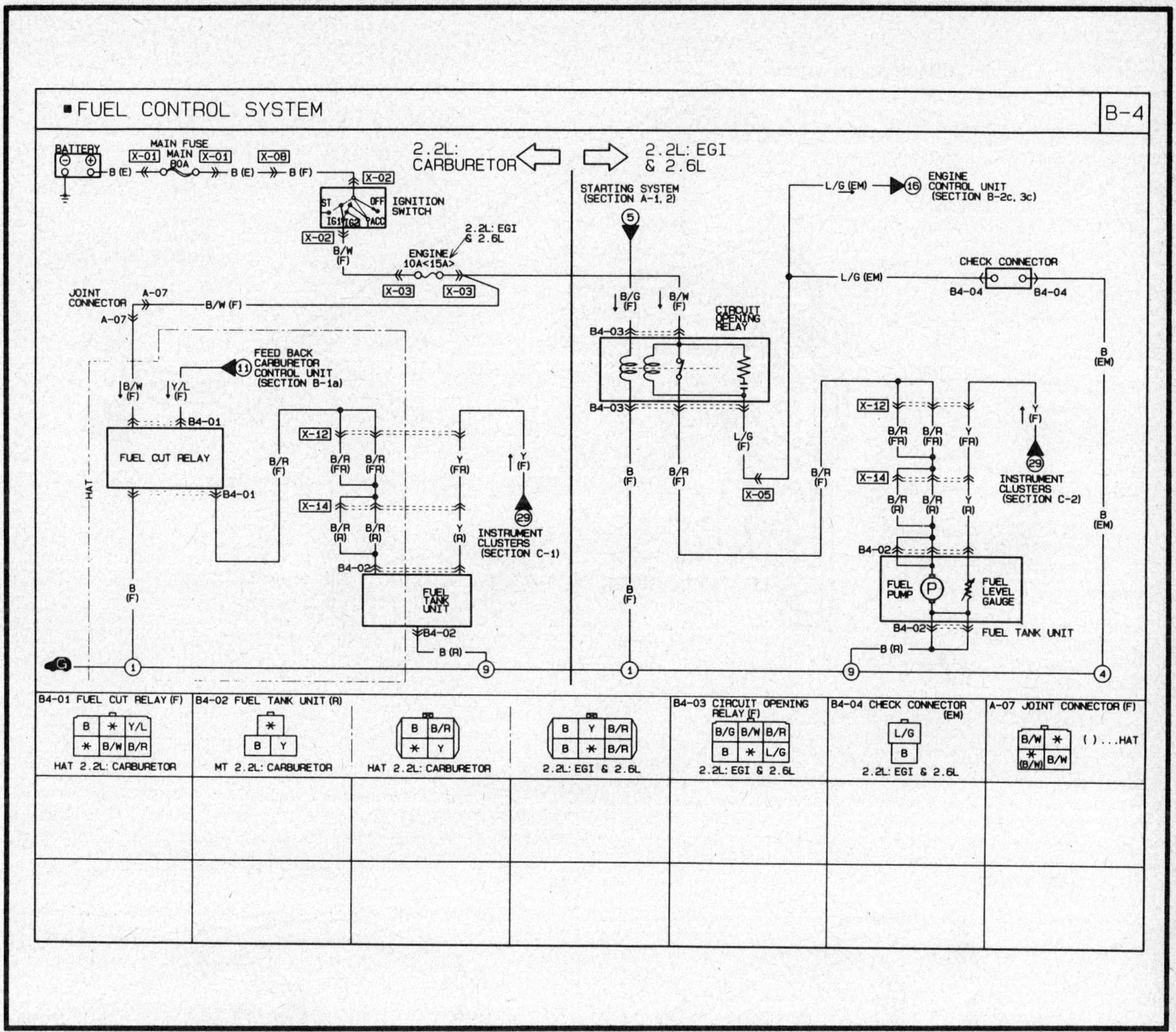

Wiring diagram—B2200, B2600i

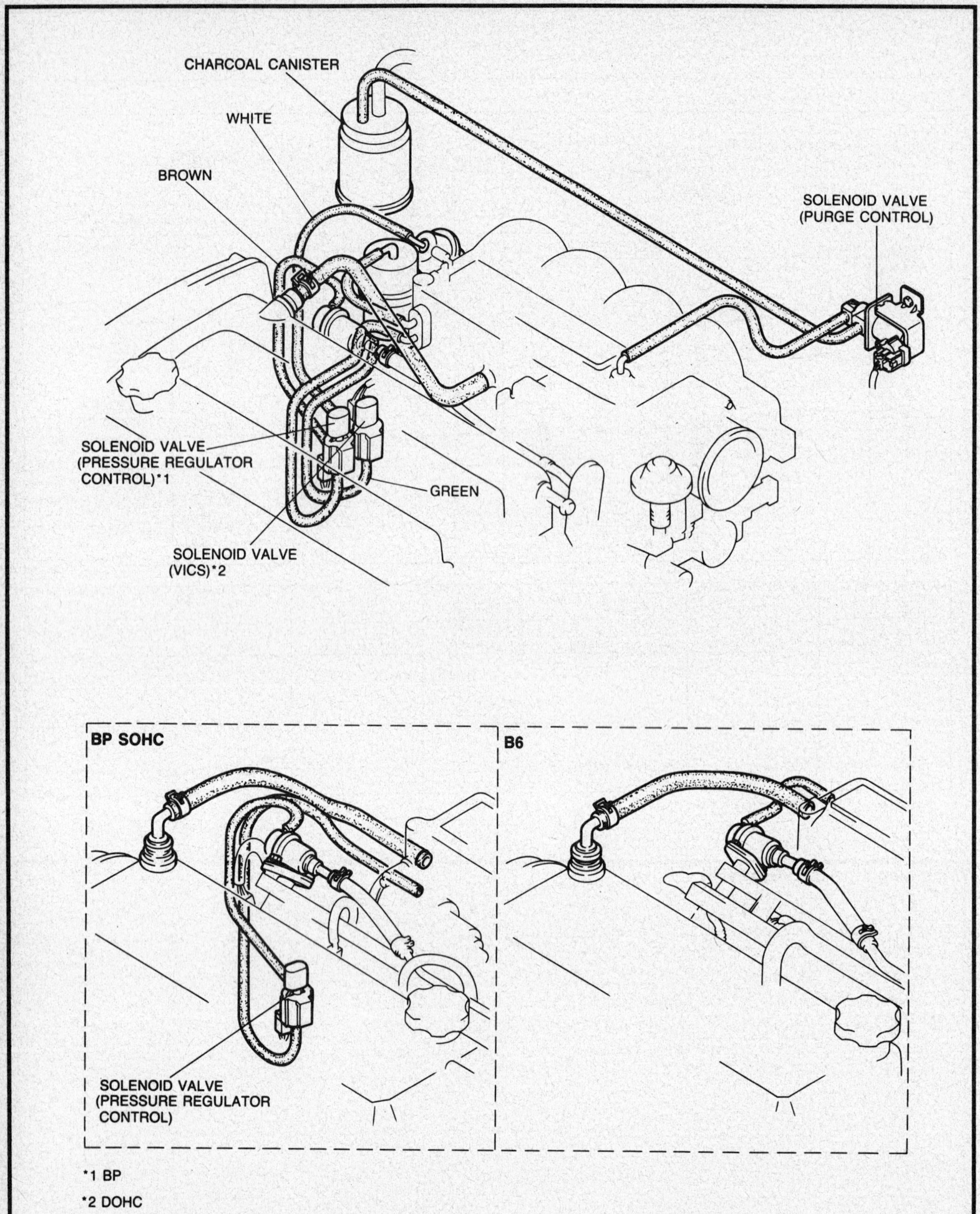

Vacuum hose routing diagram—Protege and 323

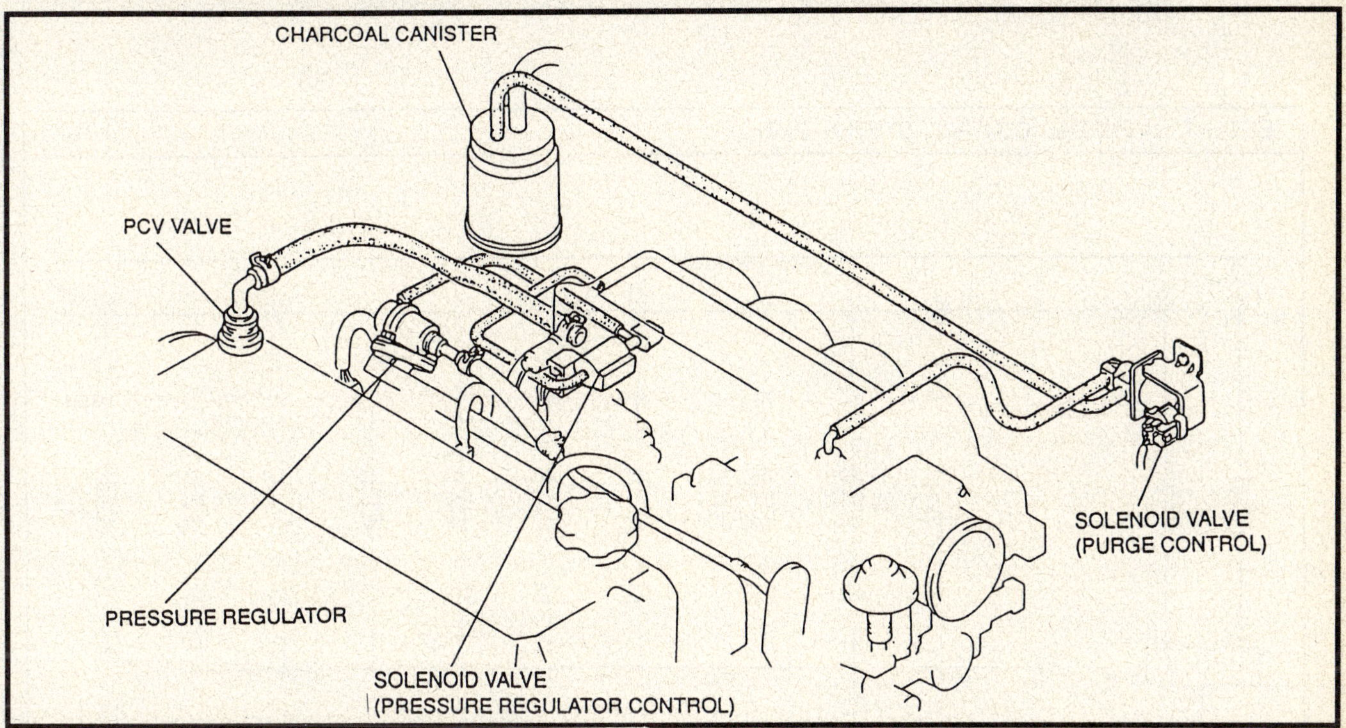

Vacuum hose routing diagram—MX-3 with 1.6L engine

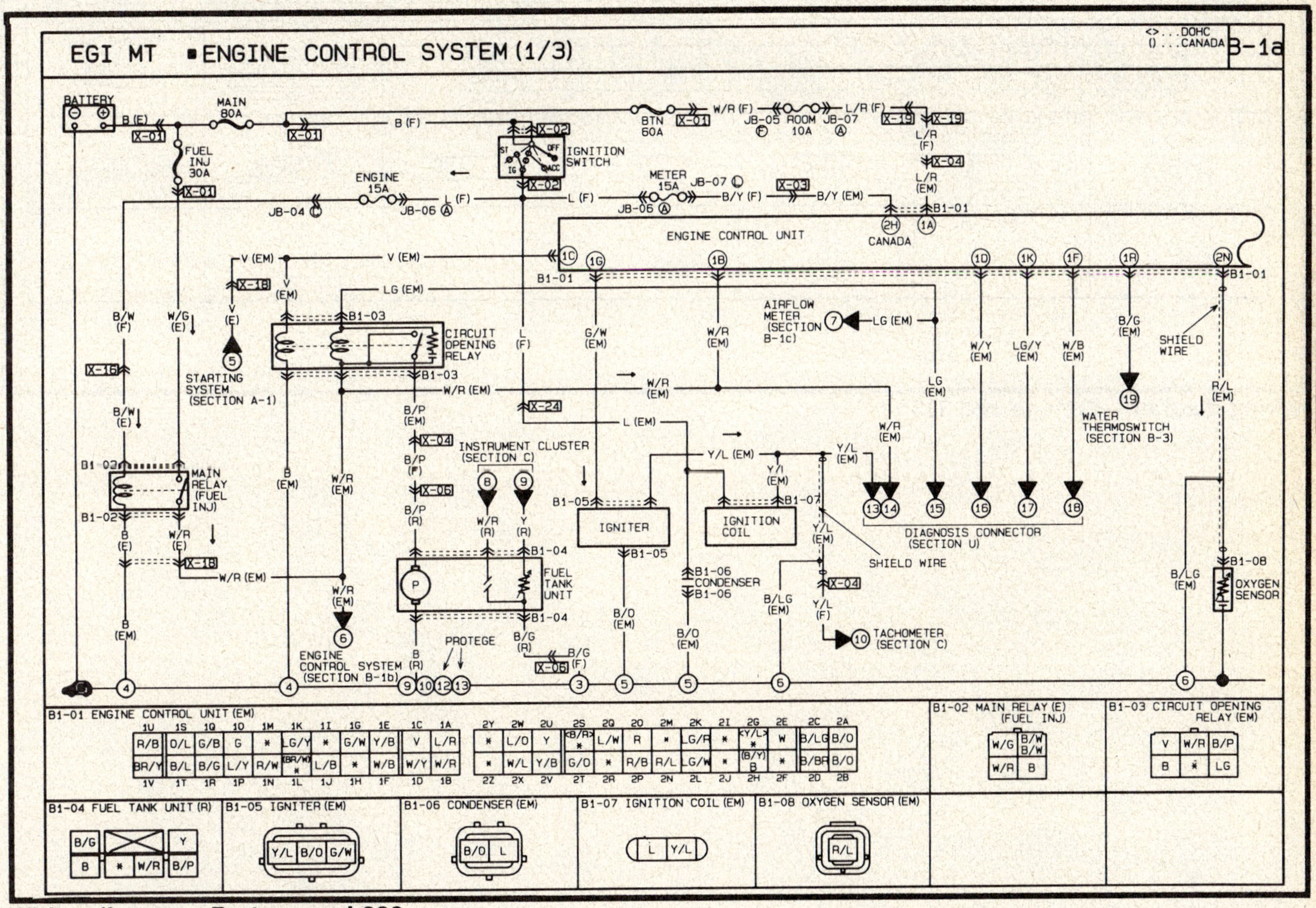

Wiring diagram—Protege and 323

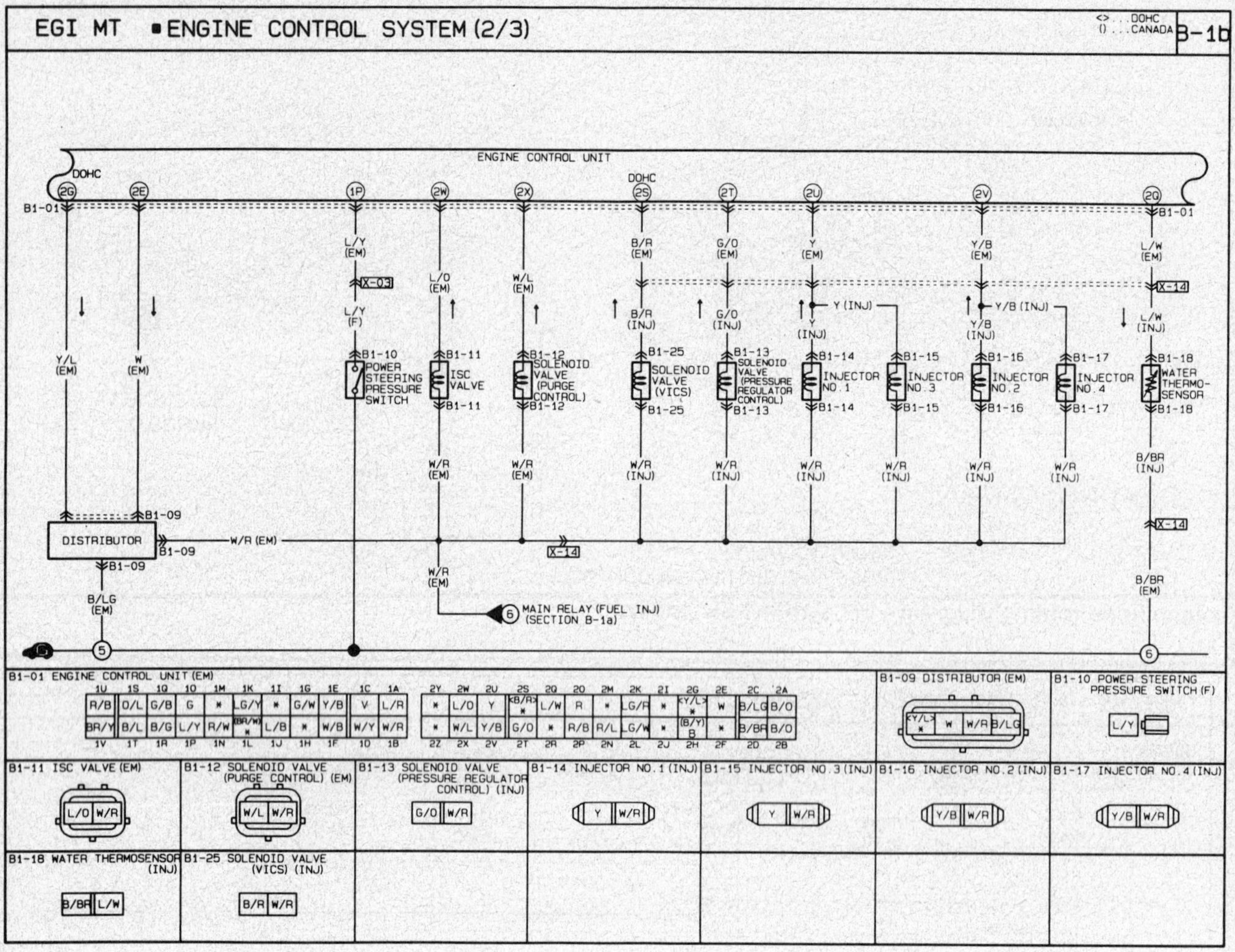

Wiring diagram—Protege and 323

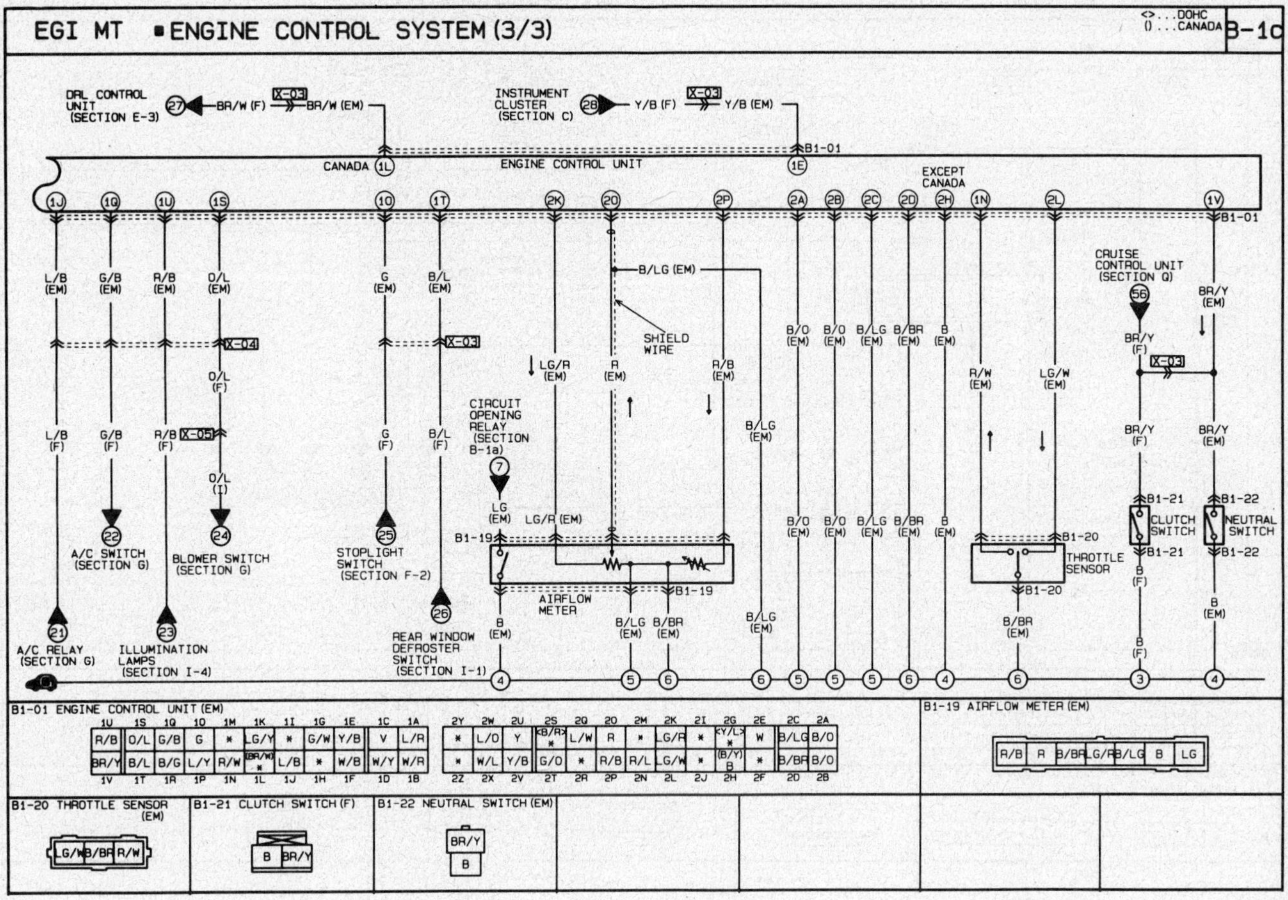

Wiring diagram—Protege and 323

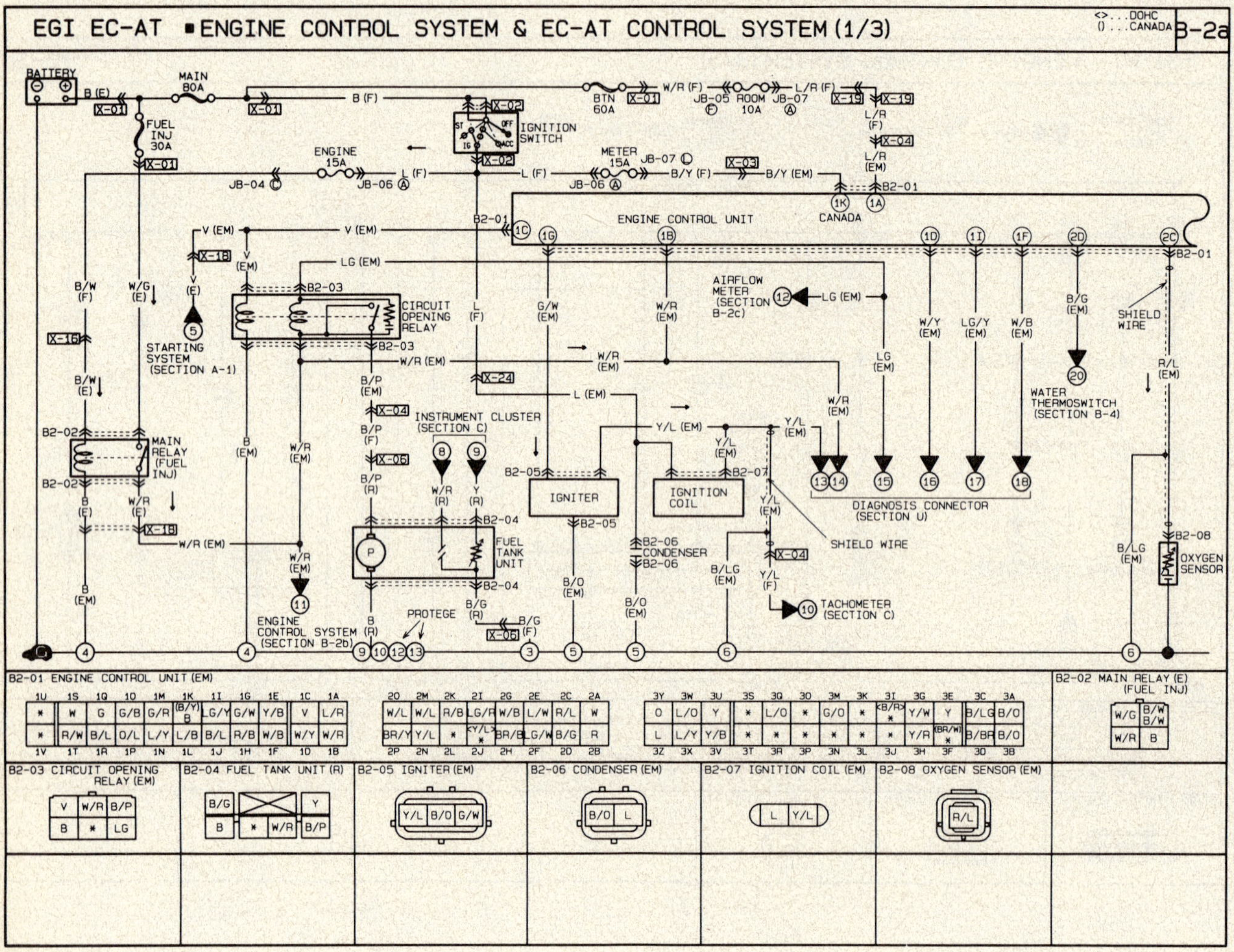

Wiring diagram—Protege and 323

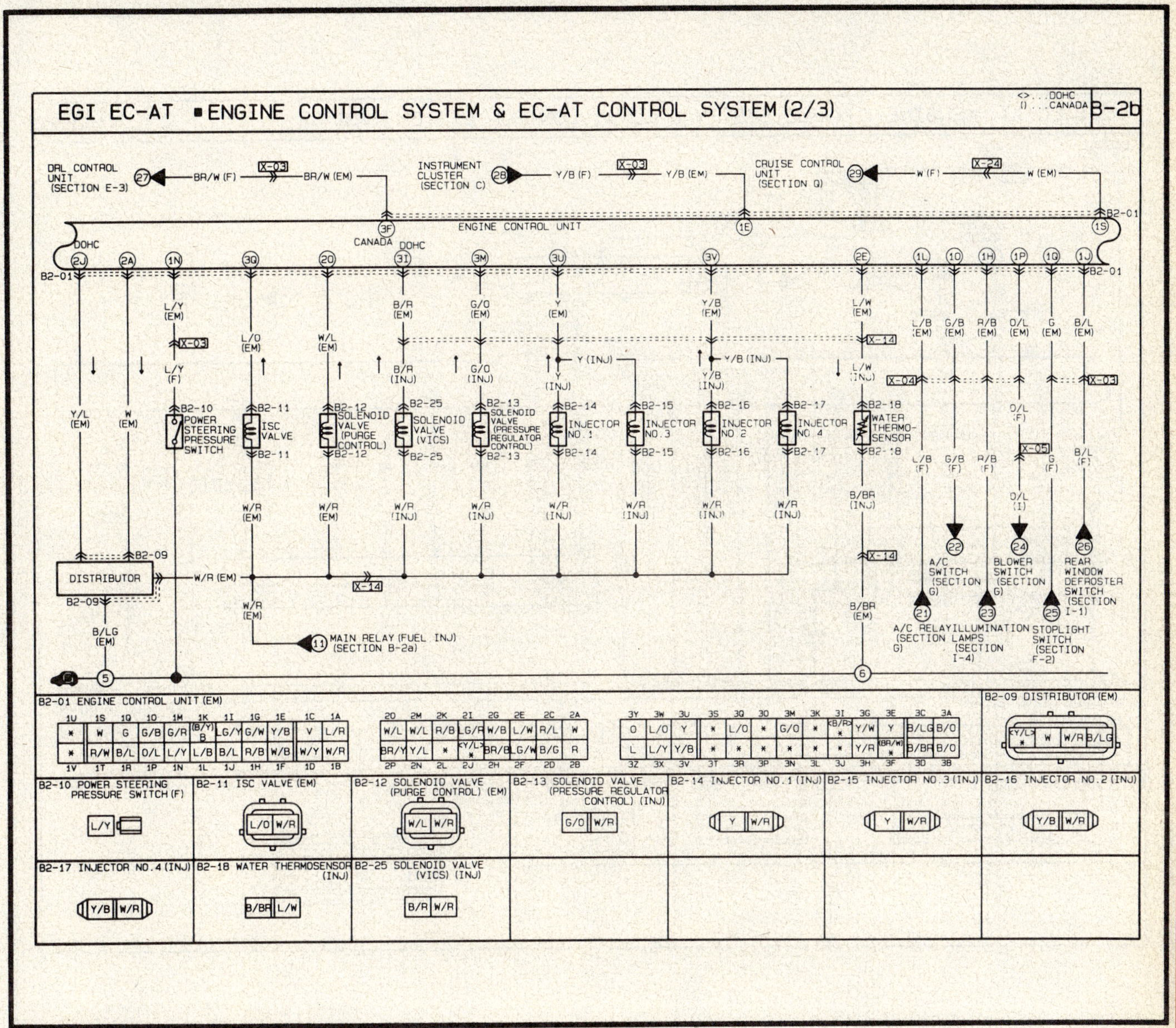

Wiring diagram—Protege and 323

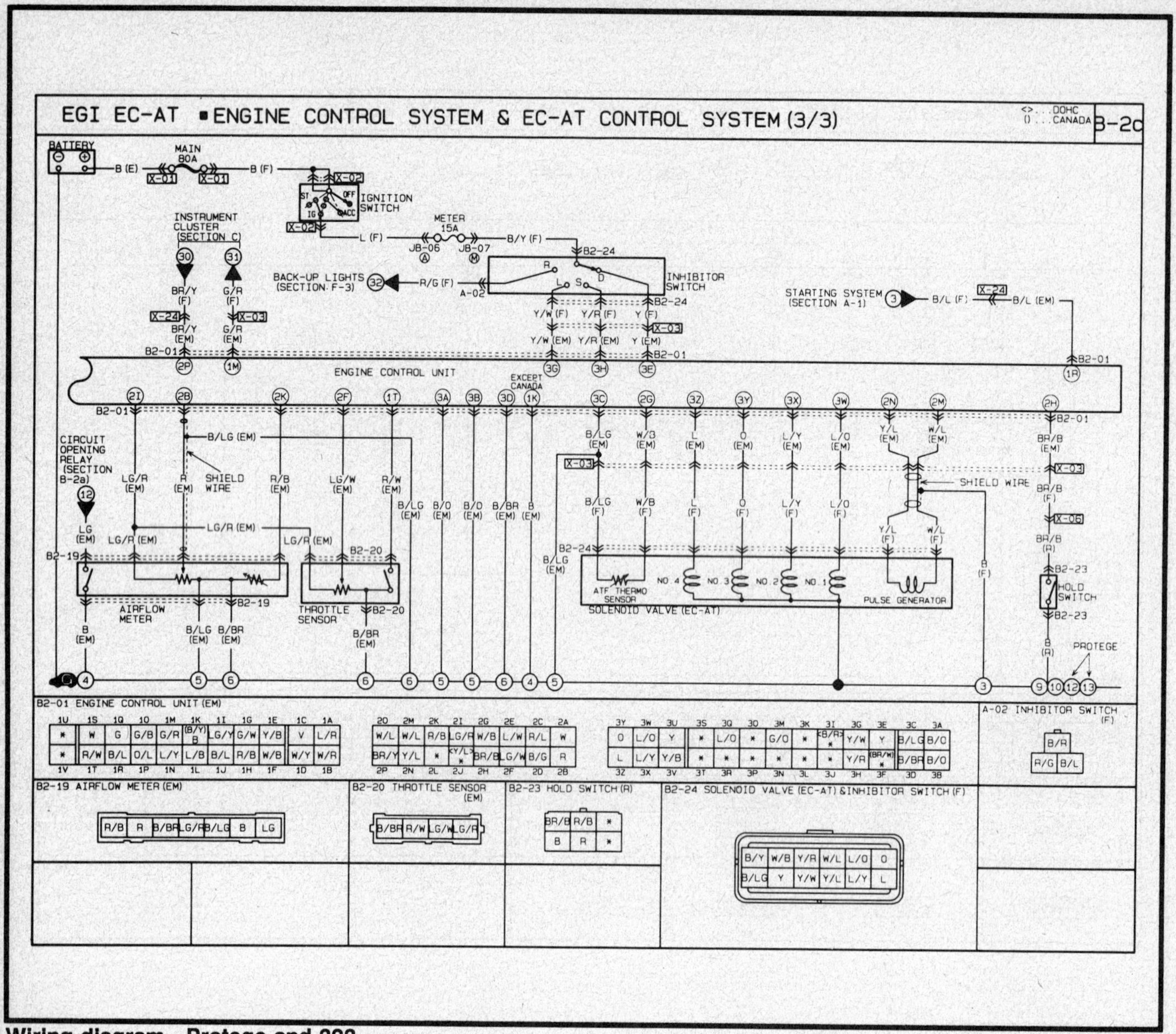

Wiring diagram—Protege and 323

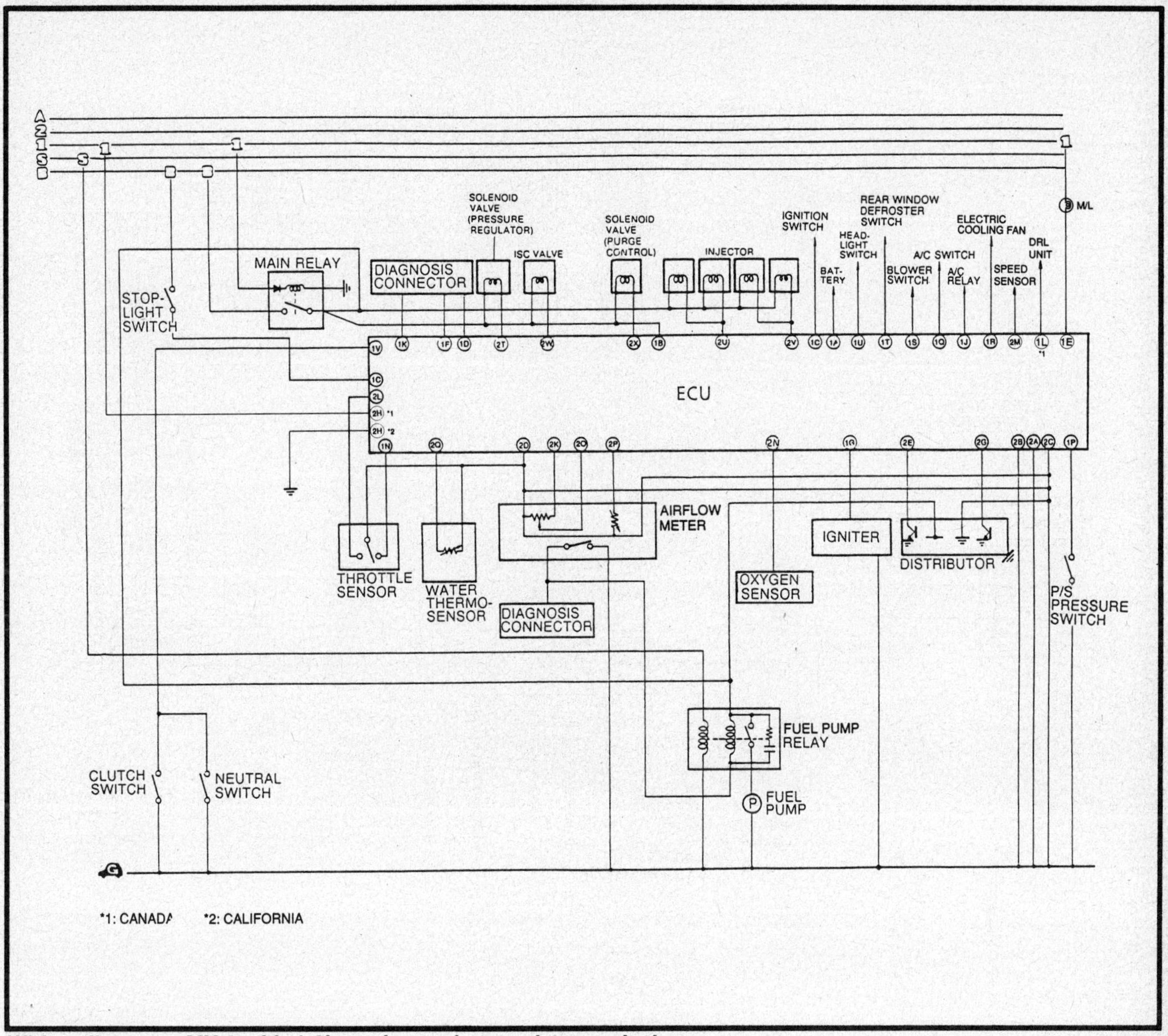

Wiring diagram—MX-3 with 1.6L engine and manual transmission

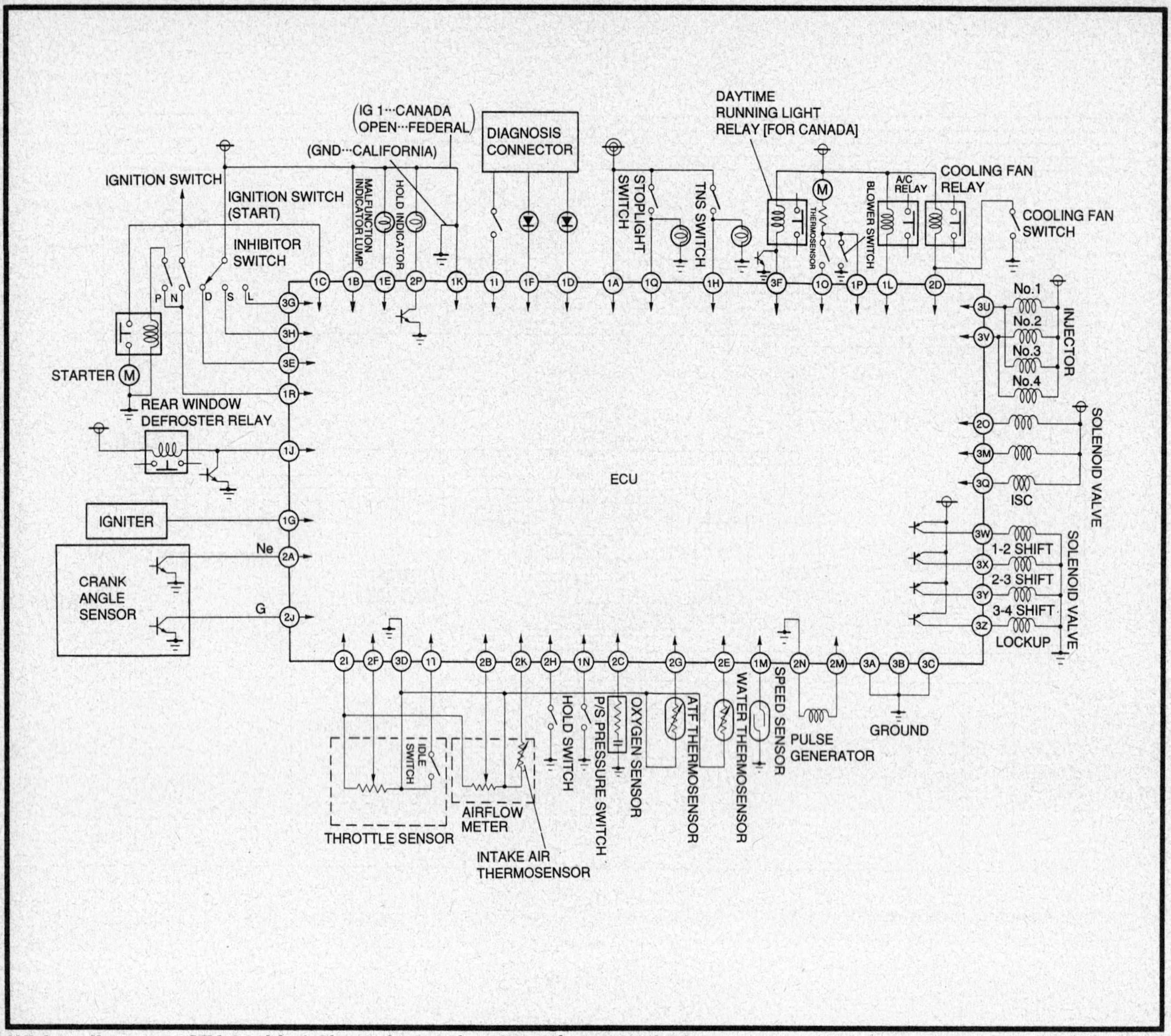

Wiring diagram—MX-3 with 1.6L engine and automatic transmission

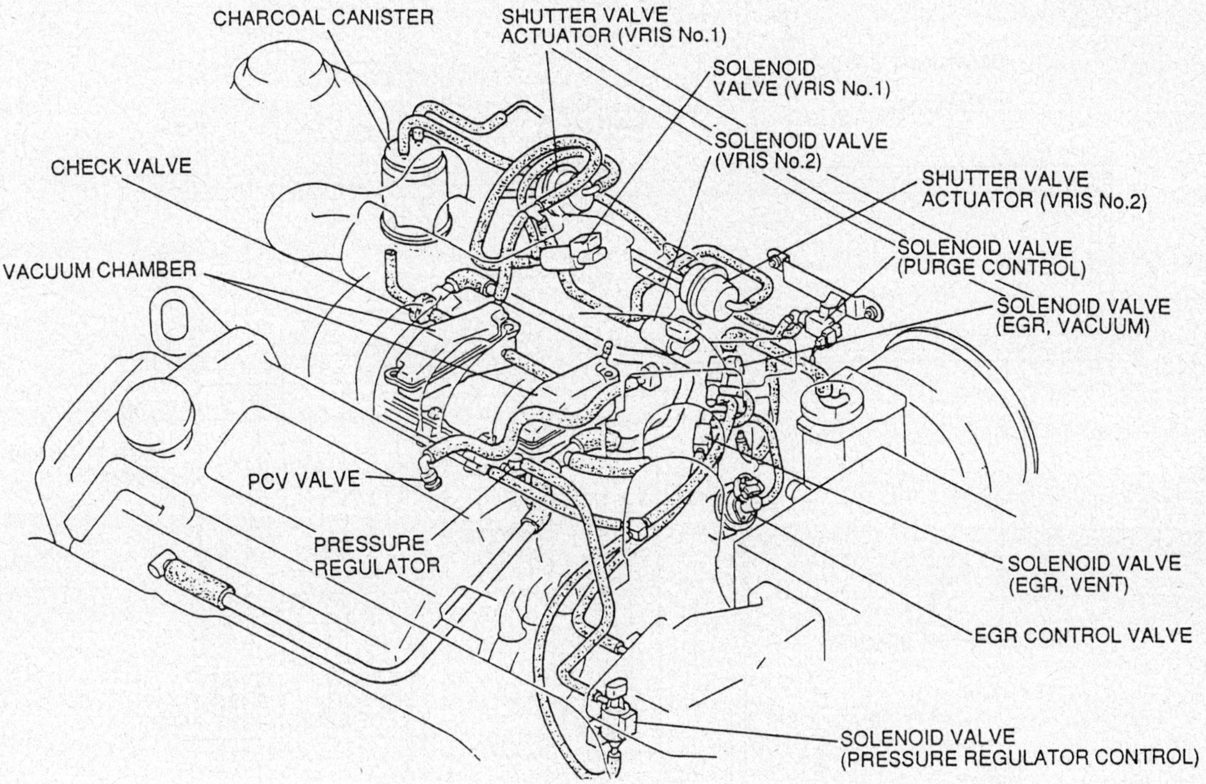

Vacuum hose routing diagram—MX-3 with 1.8L engine

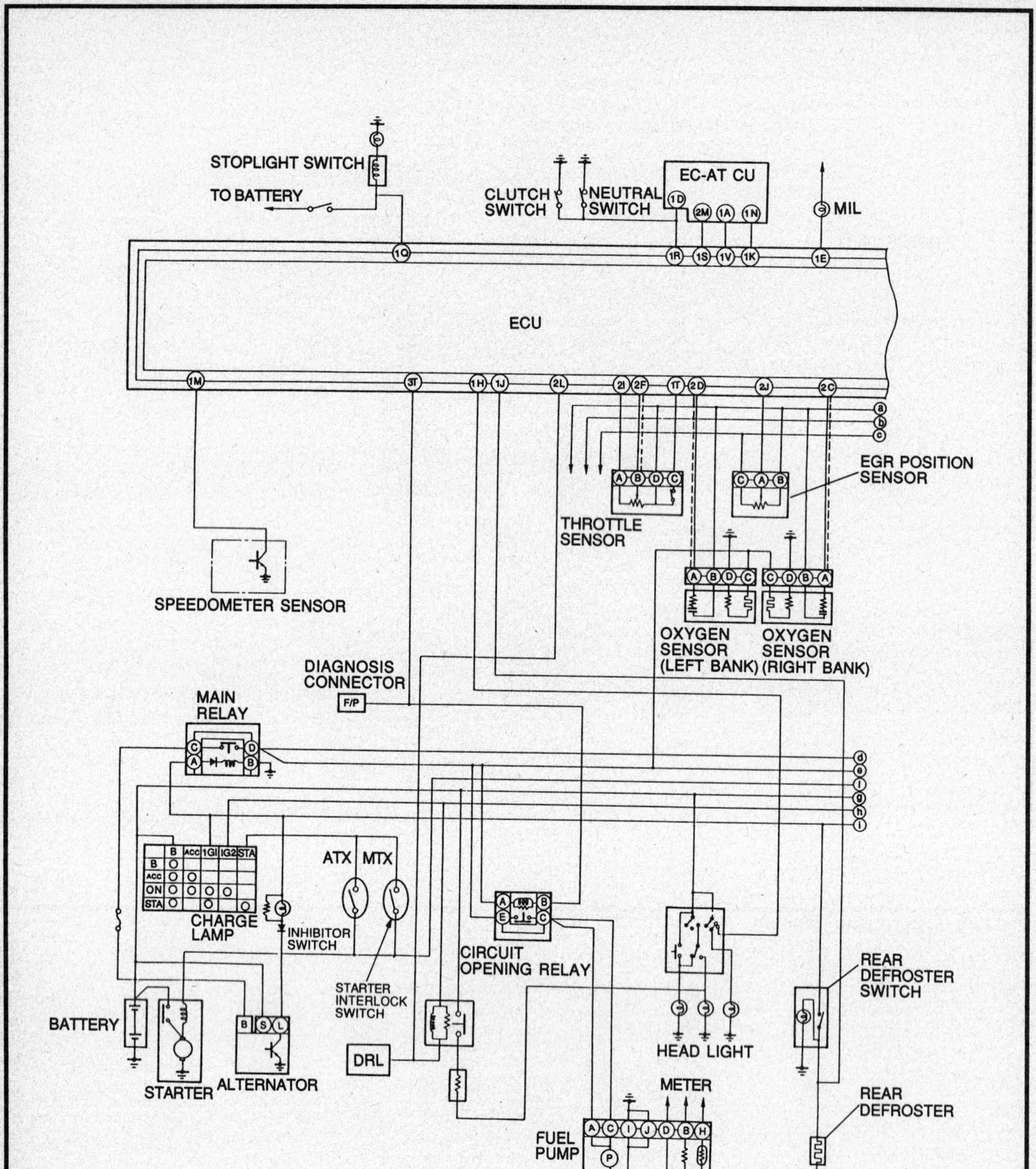

Wiring diagram—MX-3 with 1.8L engine

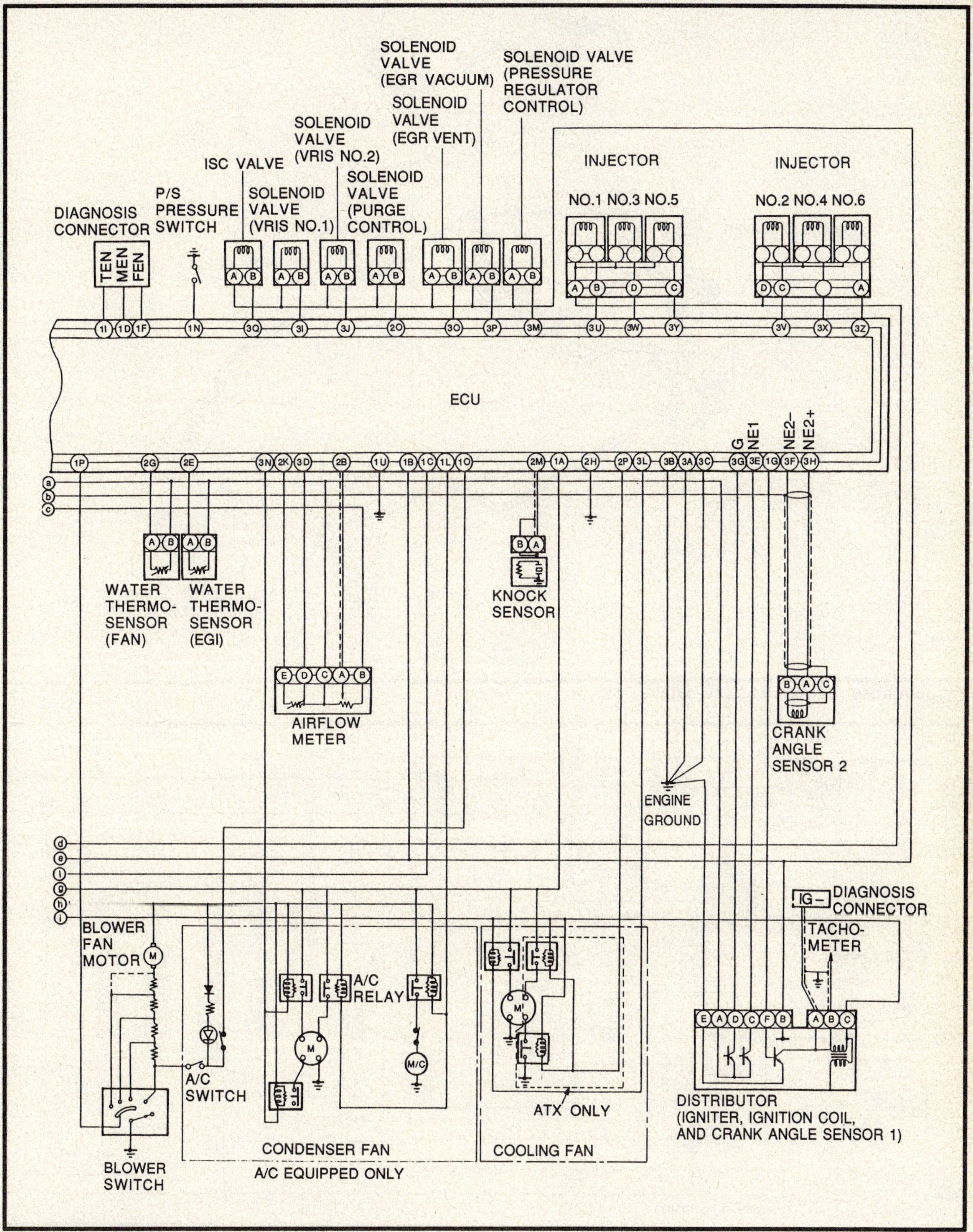

Wiring diagram—MX-3 with 1.8L engine

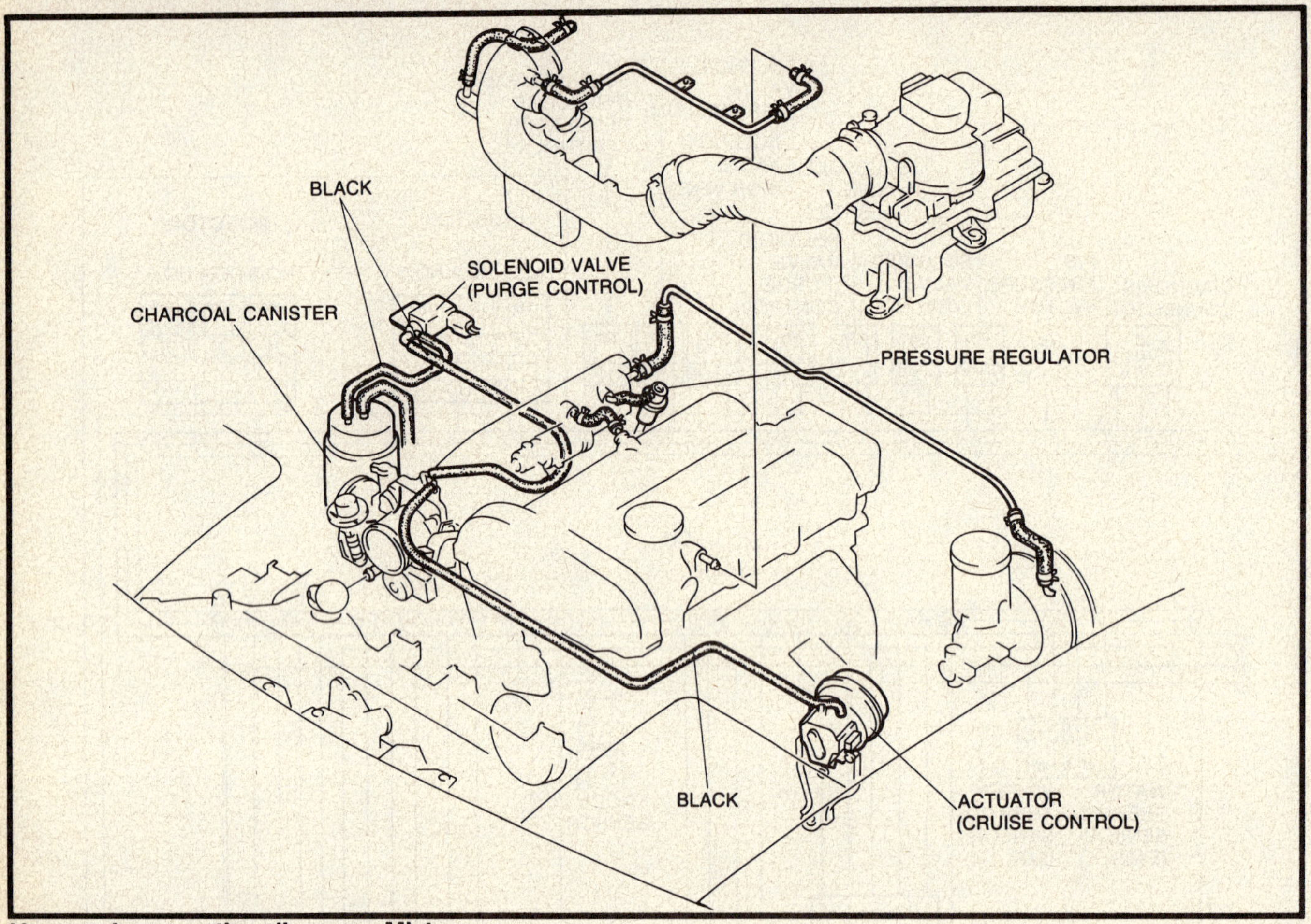

Vacuum hose routing diagram – Miata

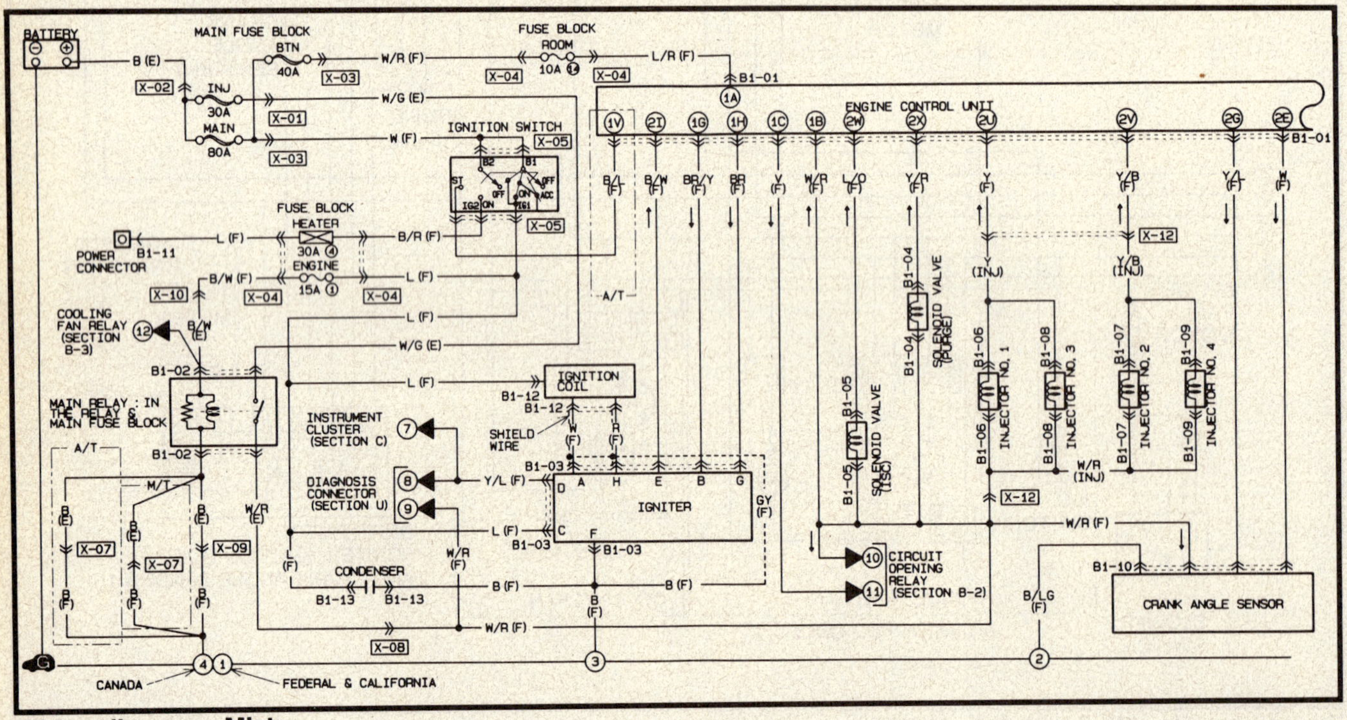

Wiring diagram – Miata

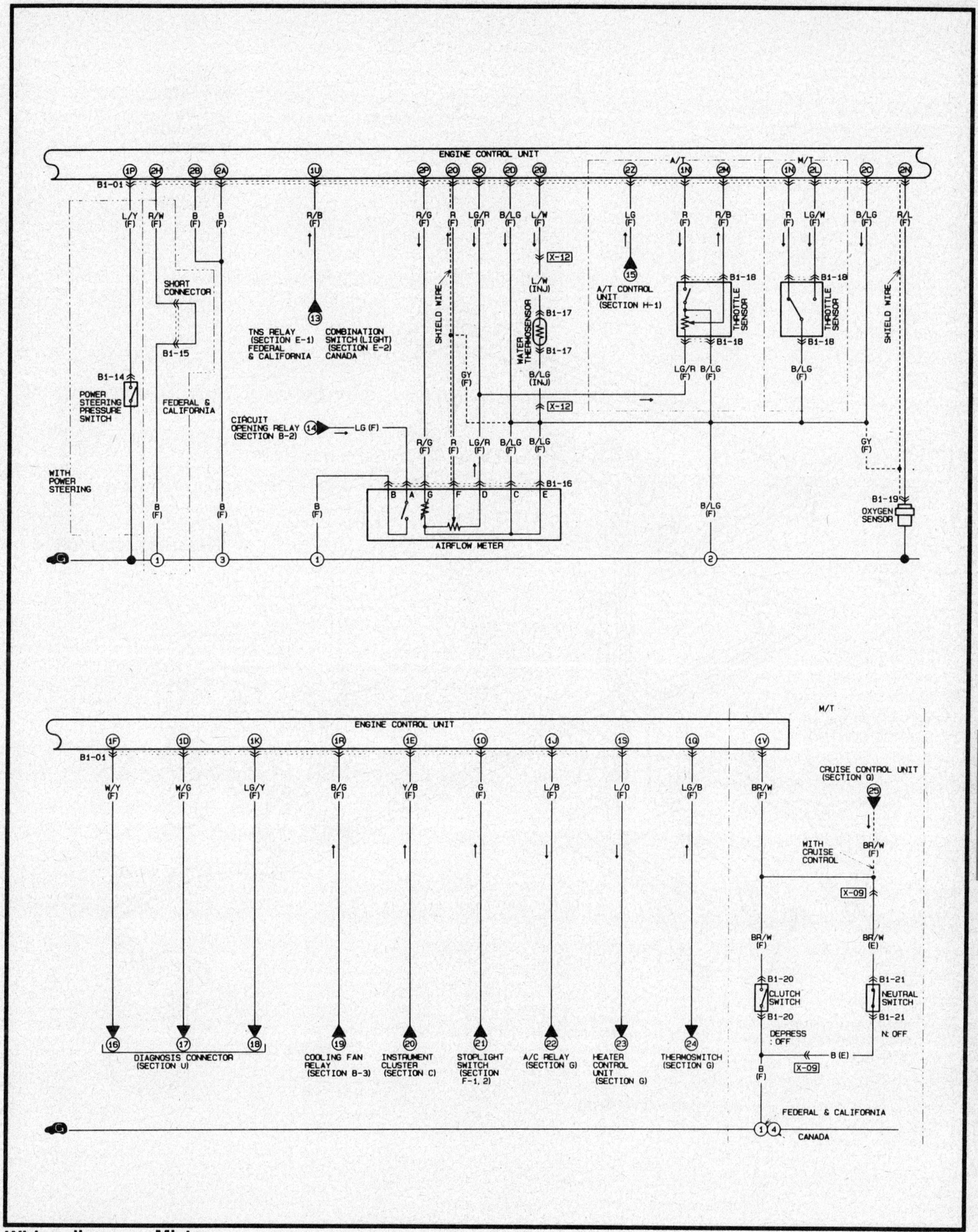

Wiring diagram – Miata

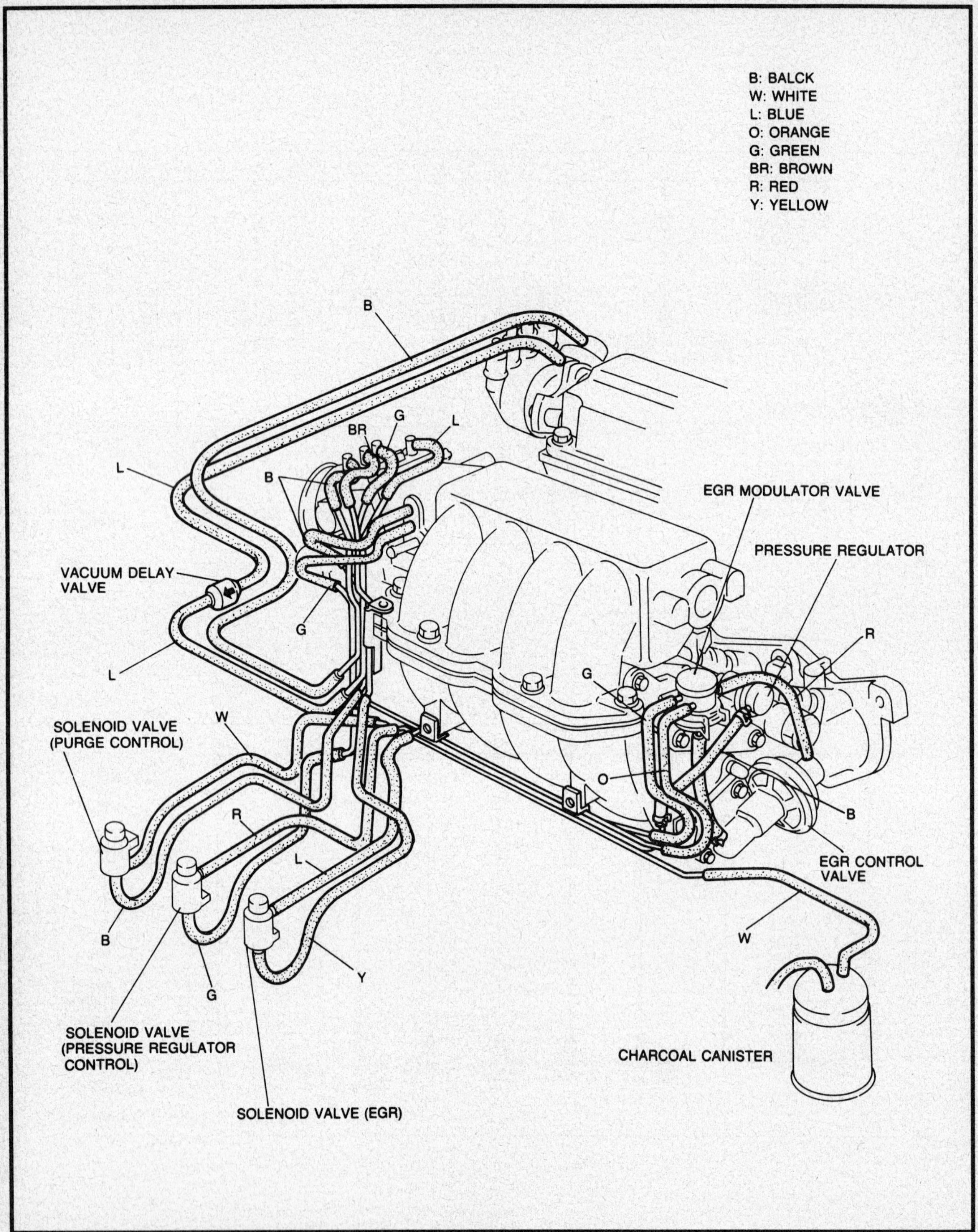

Vacuum hose routing diagram—626 and MX-6 non- turbo

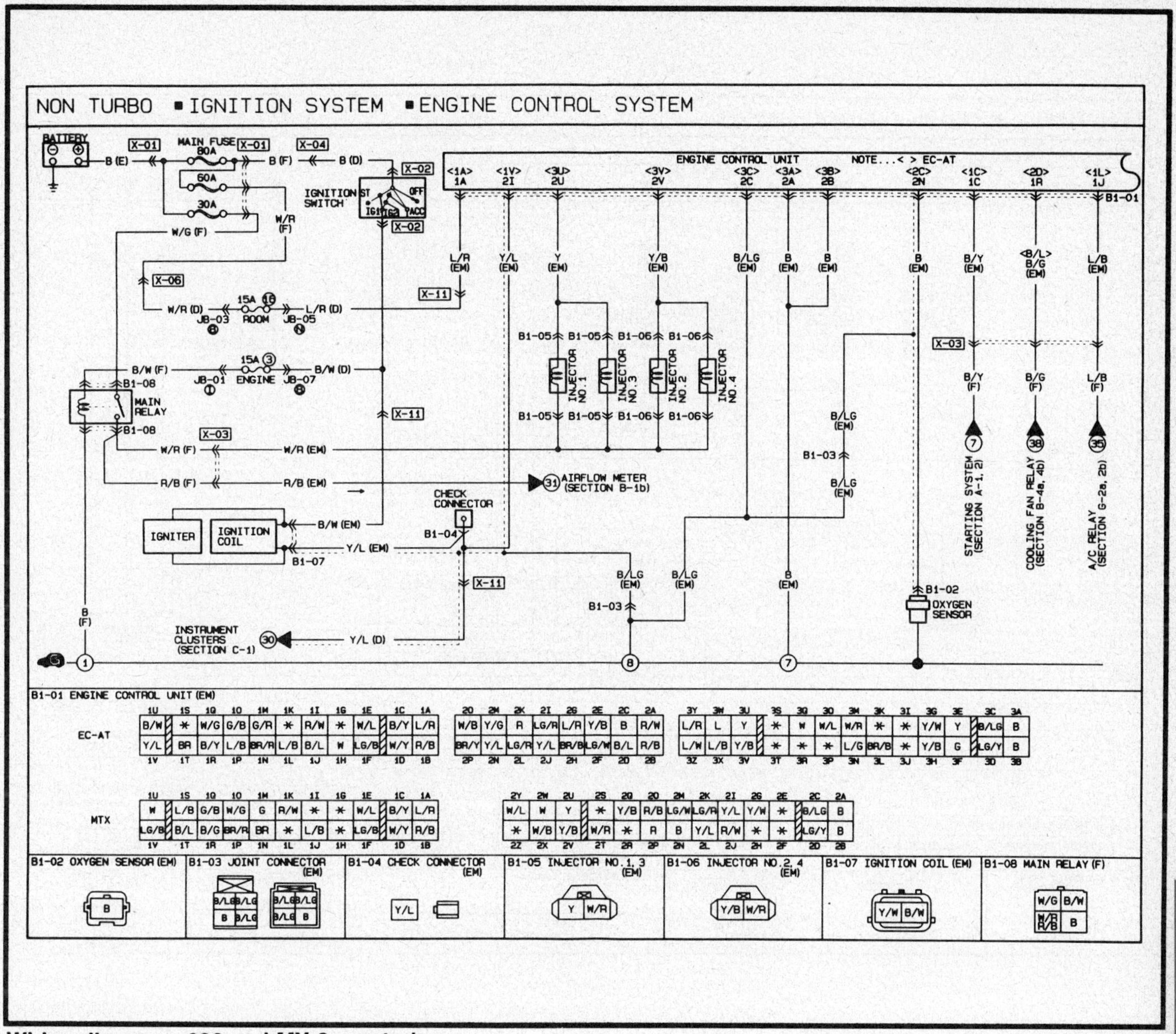

Wiring diagram – 626 and MX-6 non-turbo

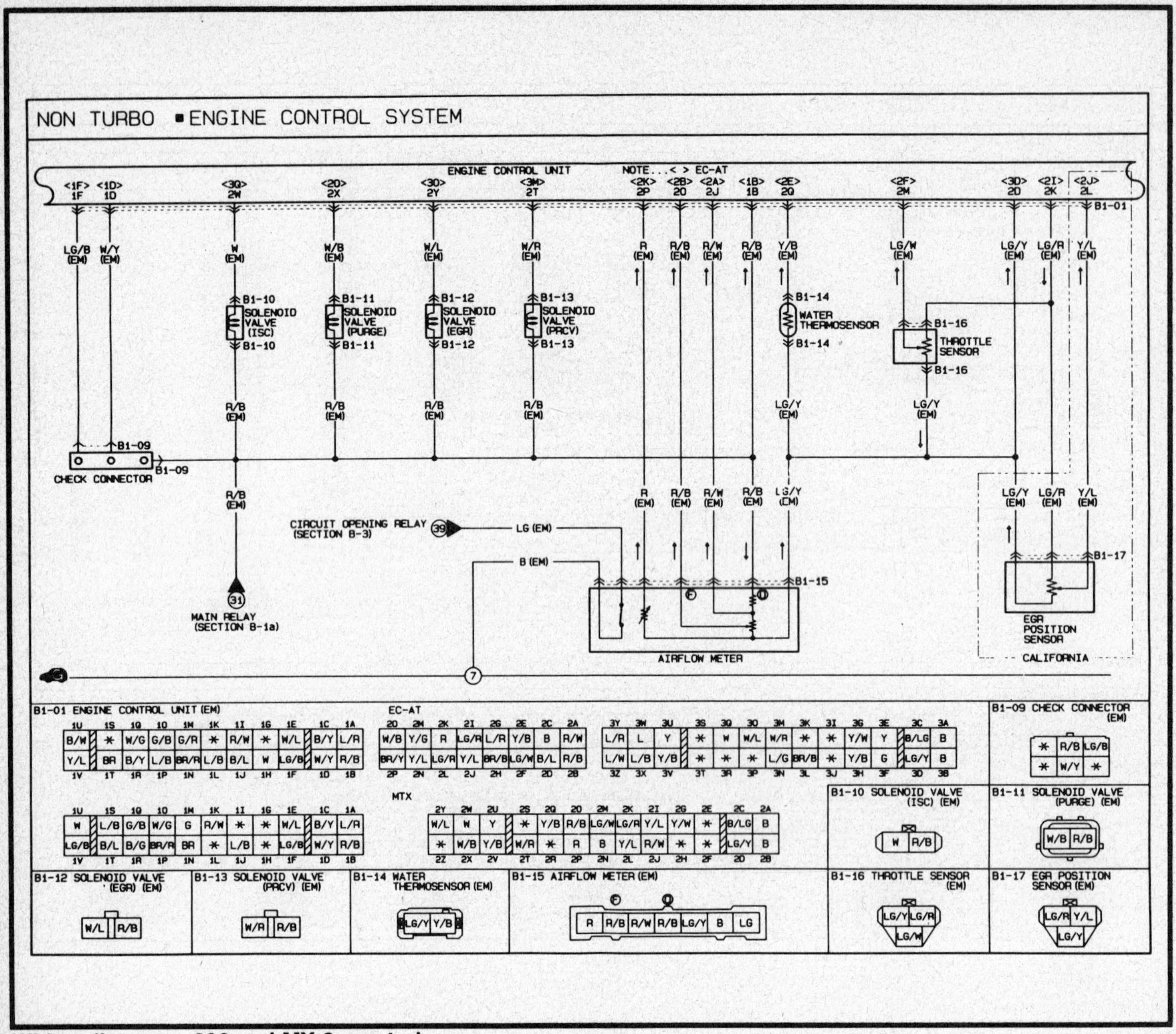

Wiring diagram—626 and MX-6 non-turbo

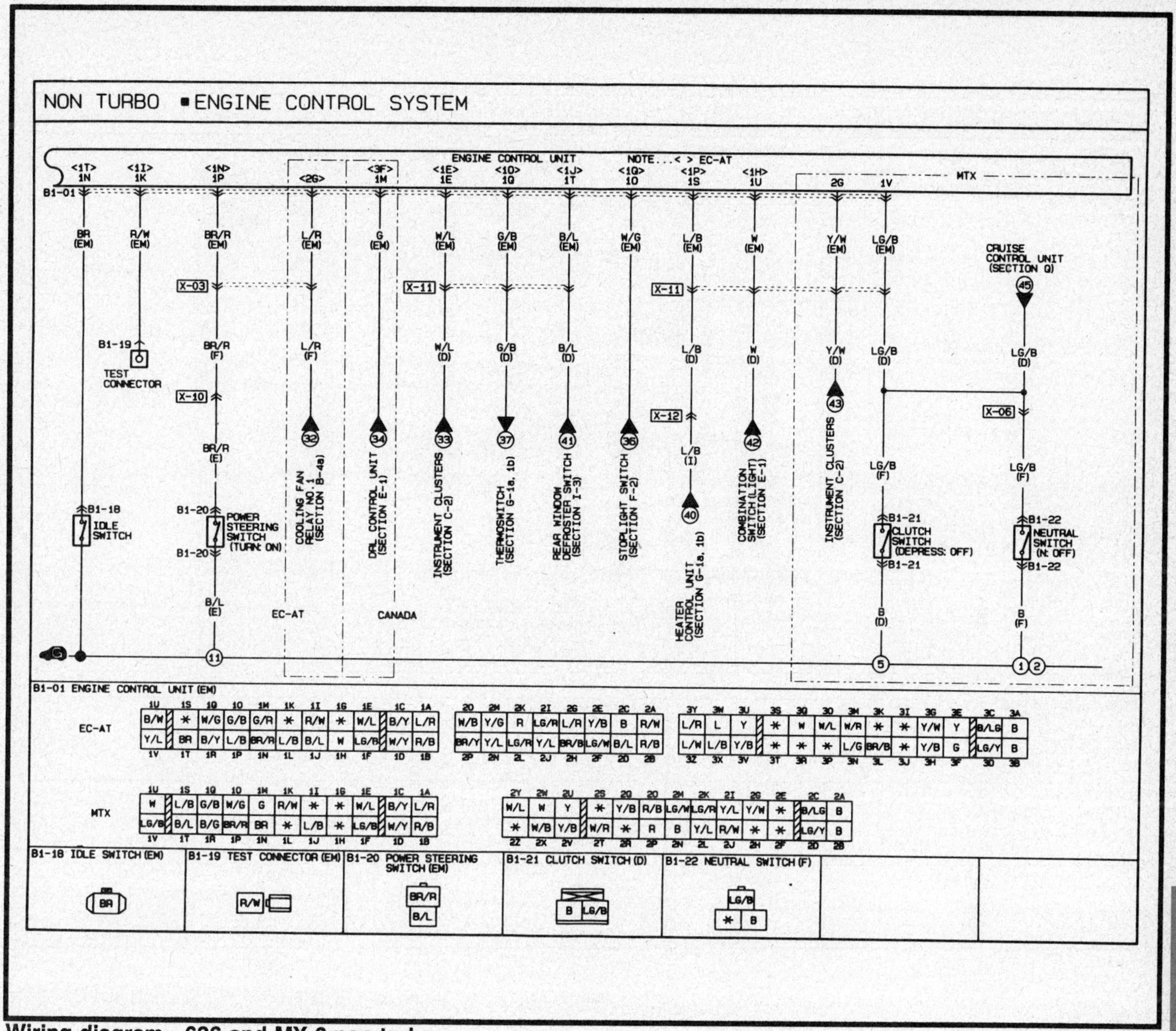

Wiring diagram—626 and MX-6 non-turbo

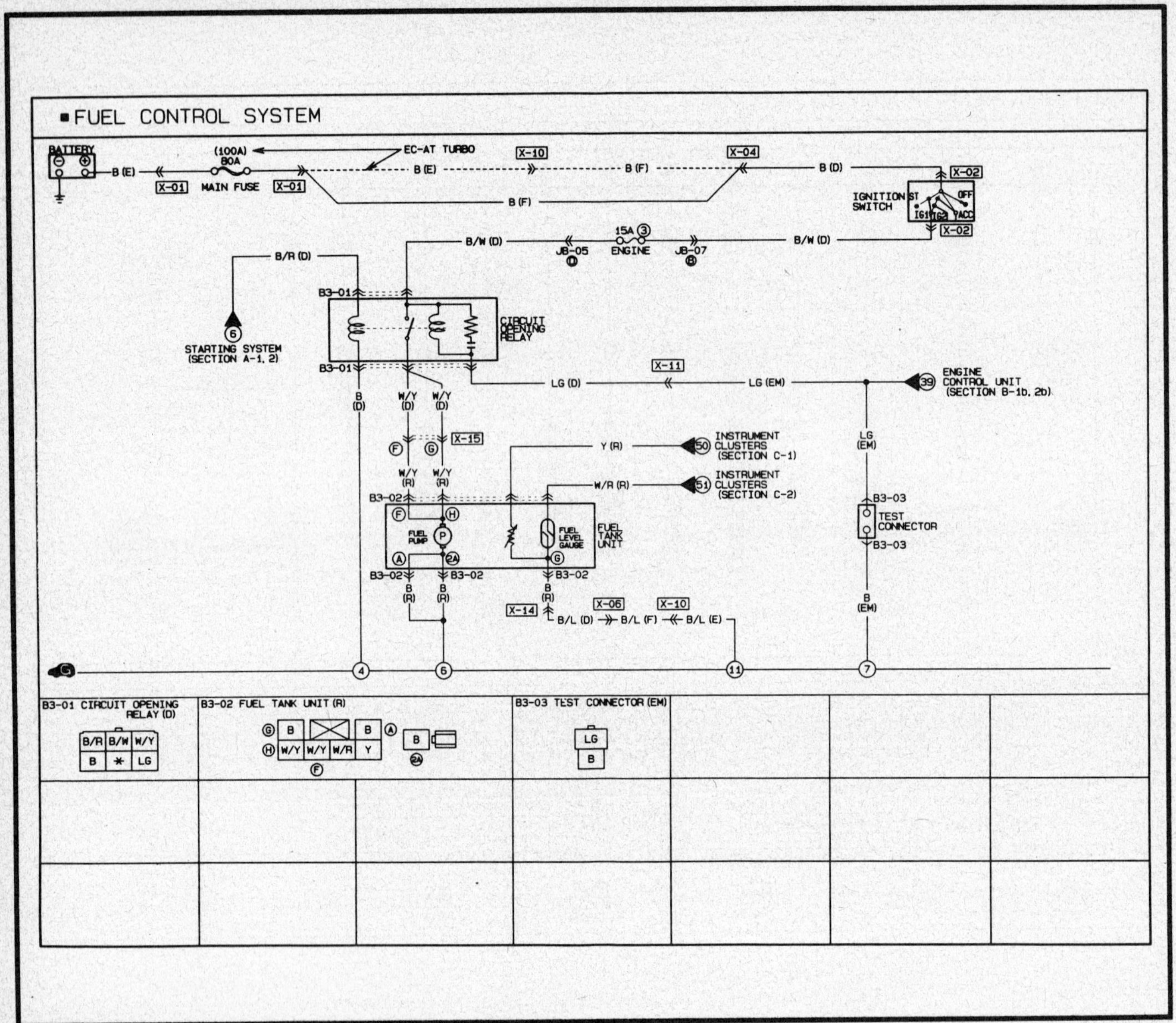

Wiring diagram—626 and MX-6 non-turbo

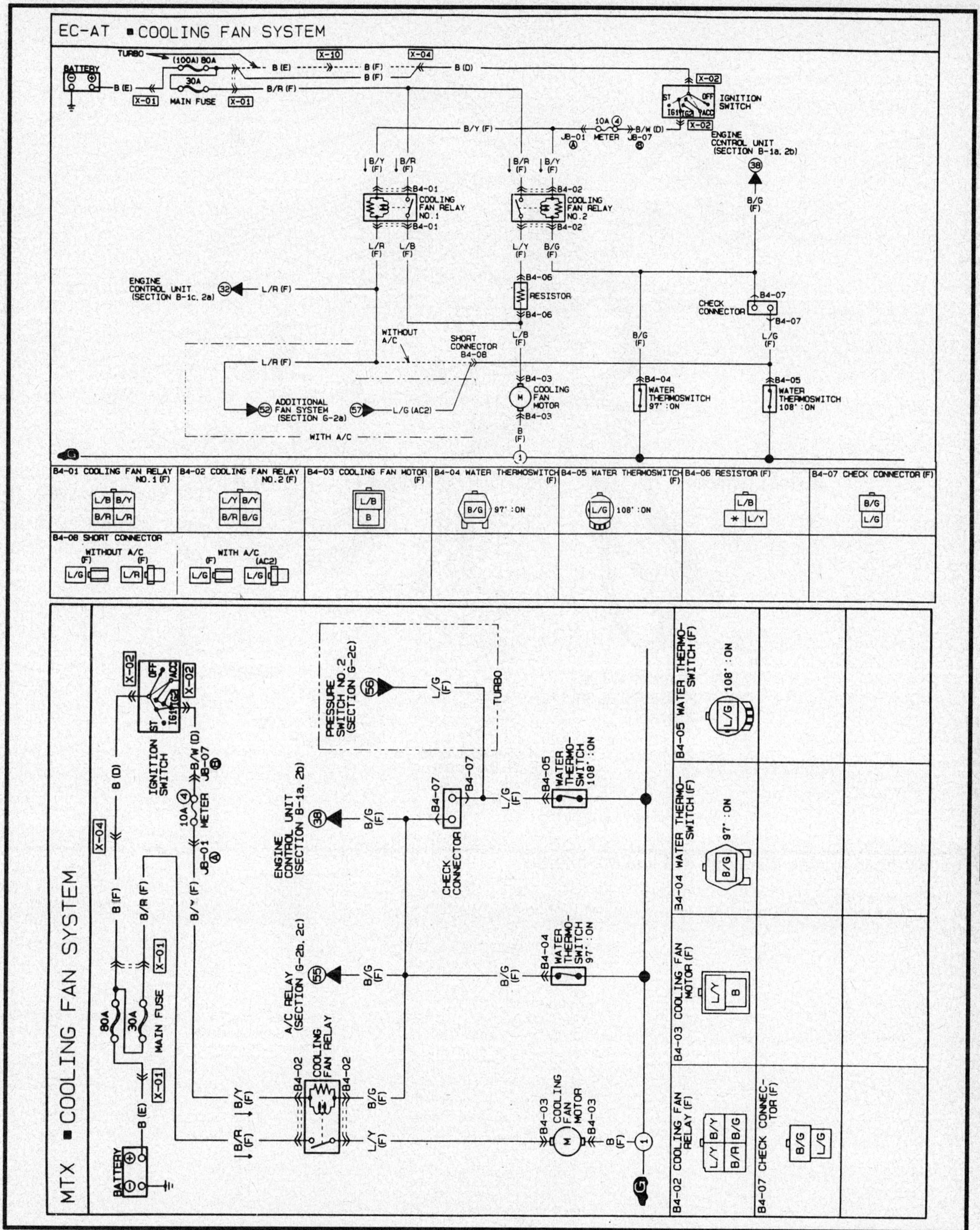

Wiring diagram—626 and MX-6 non-turbo

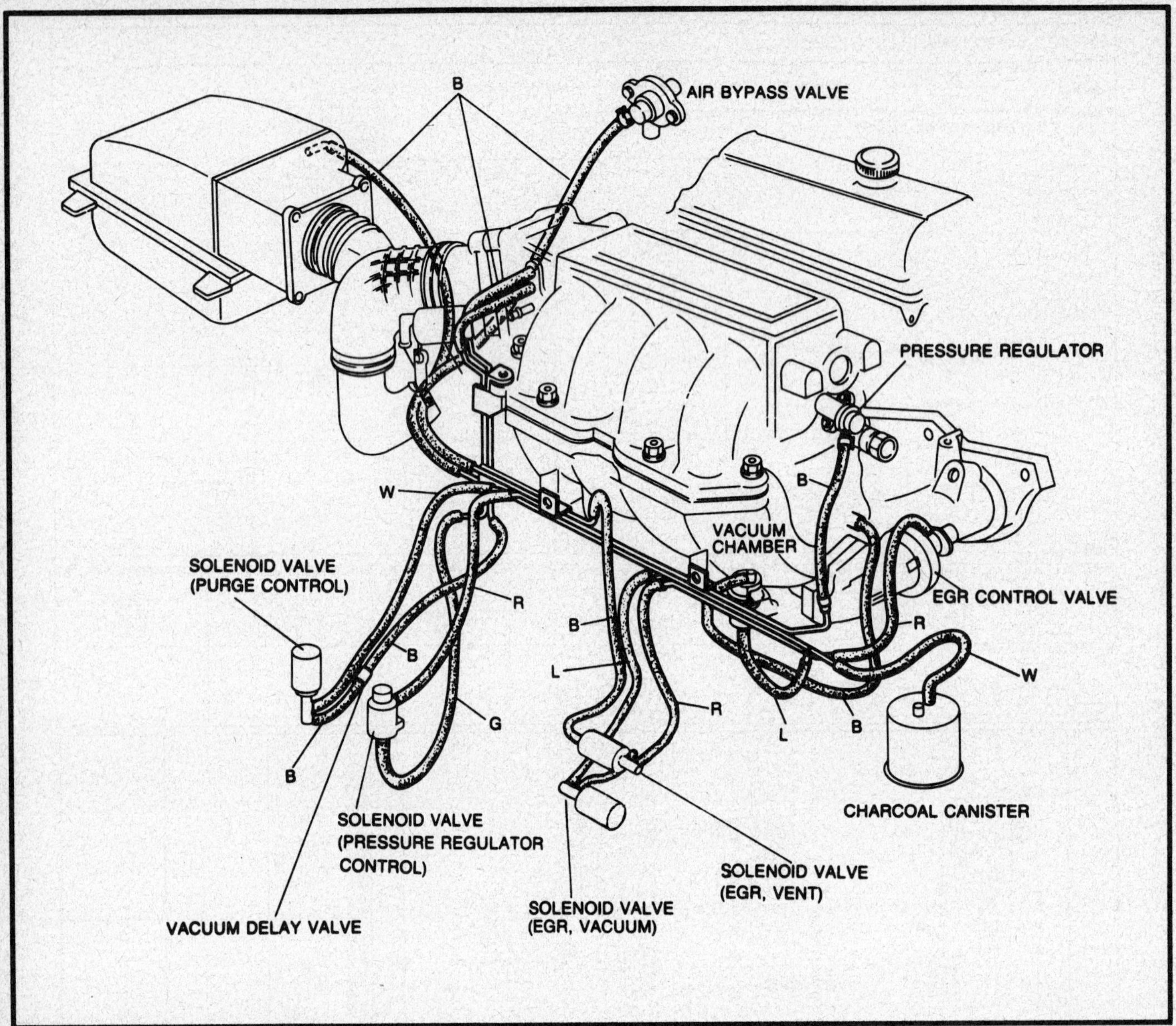

Vacuum hose routing diagram—626 and MX-6 turbo

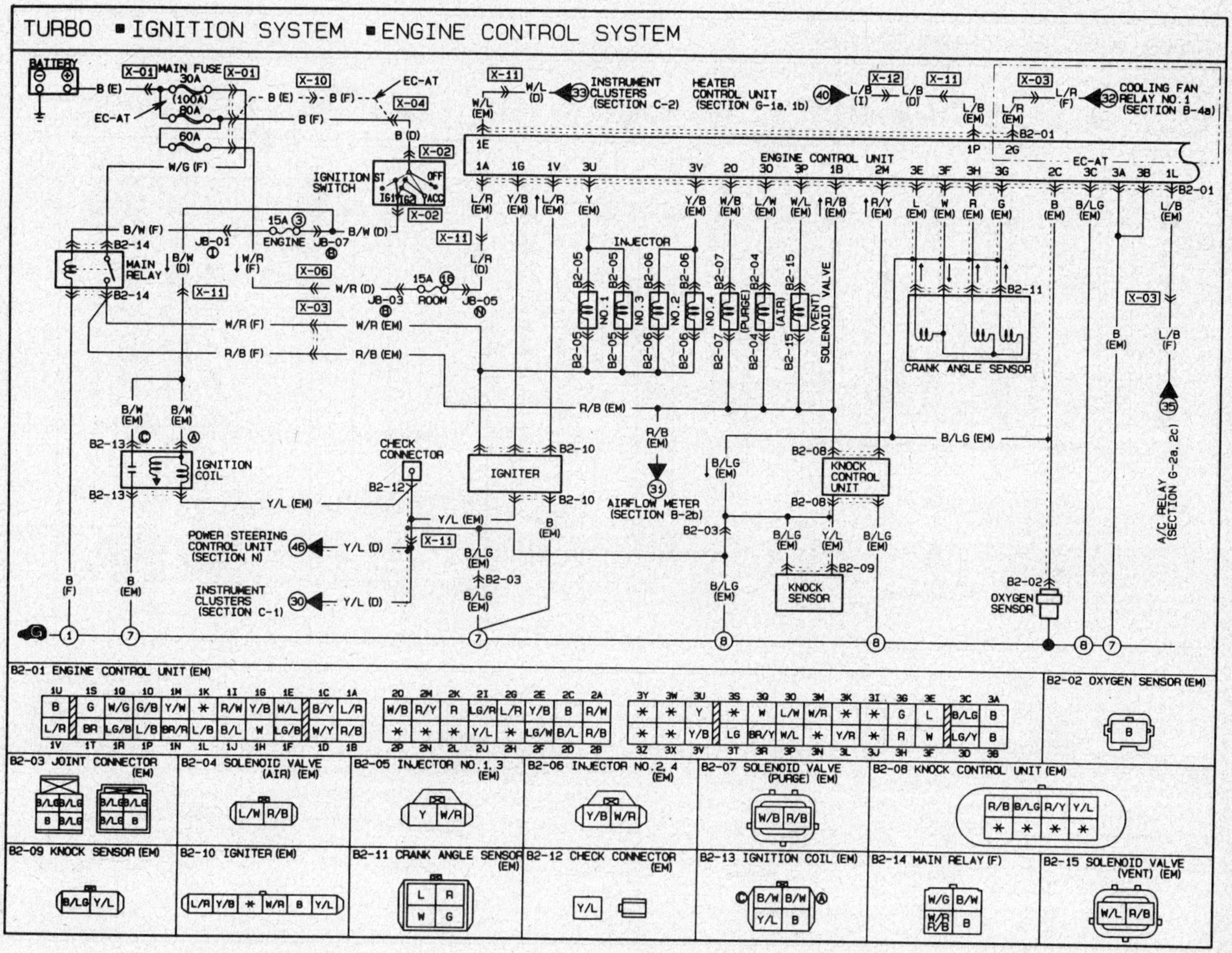

Wiring diagram—626 and MX-6 turbo

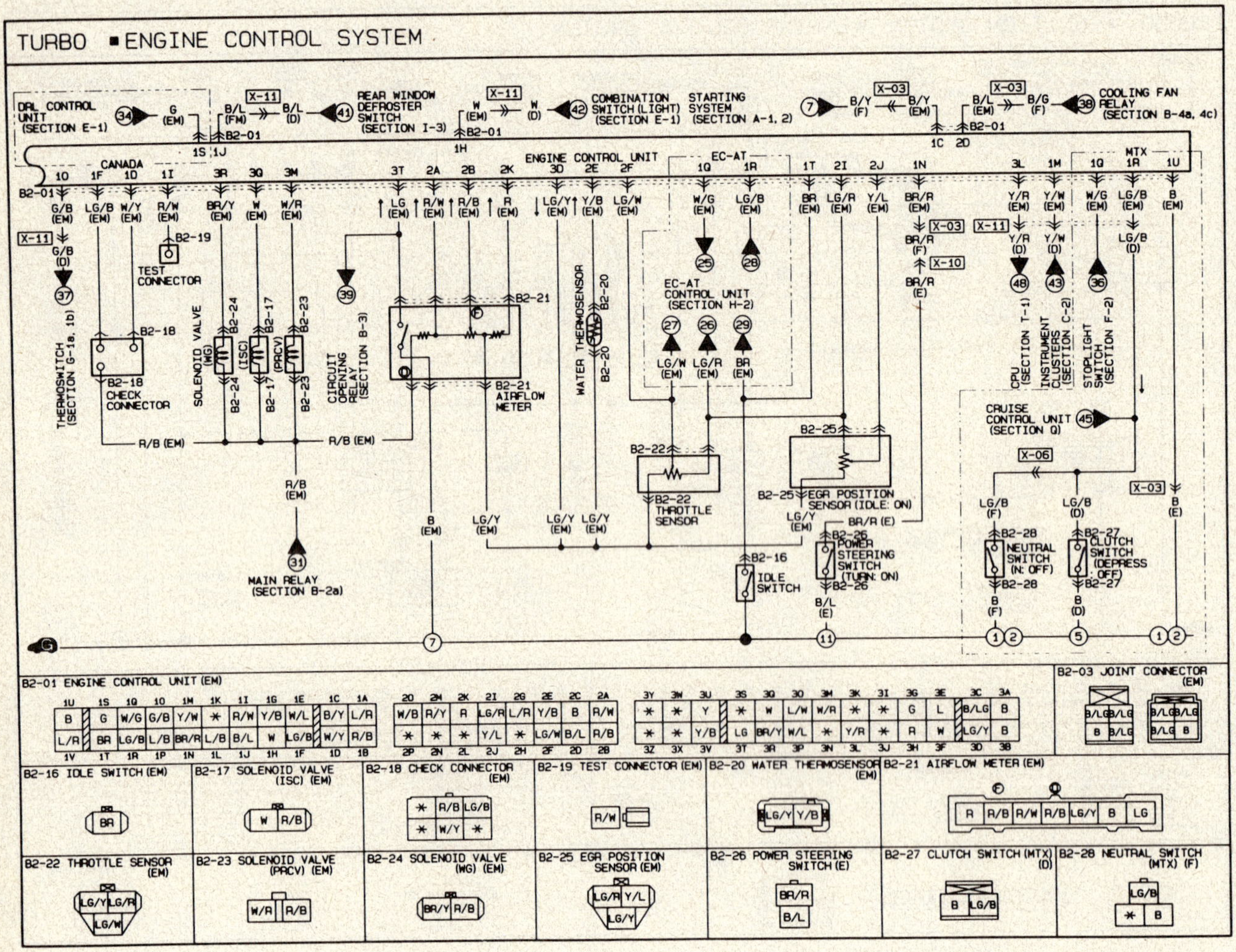

Wiring diagram—626 and MX-6 turbo

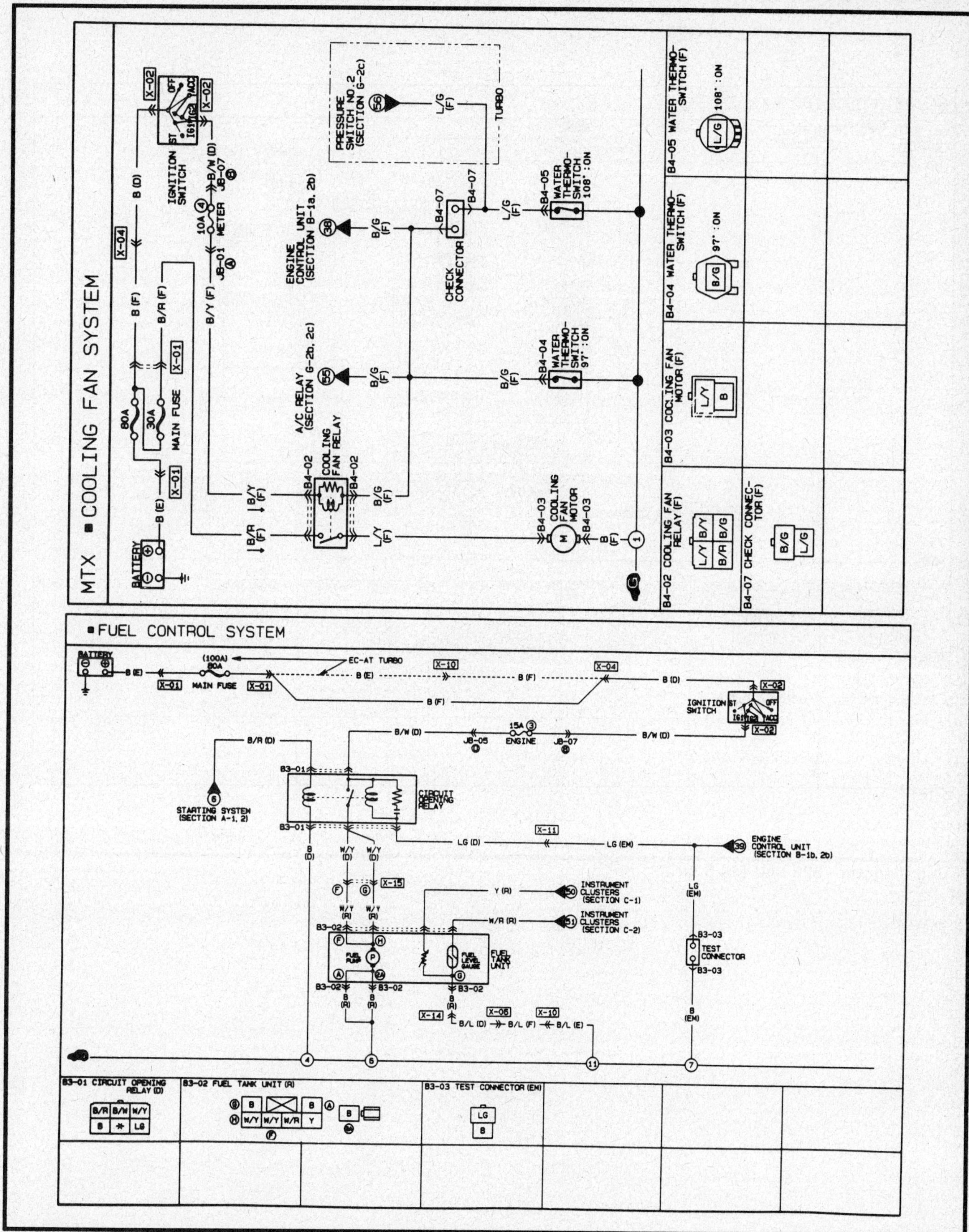

Wiring diagram—626 and MX-6 turbo

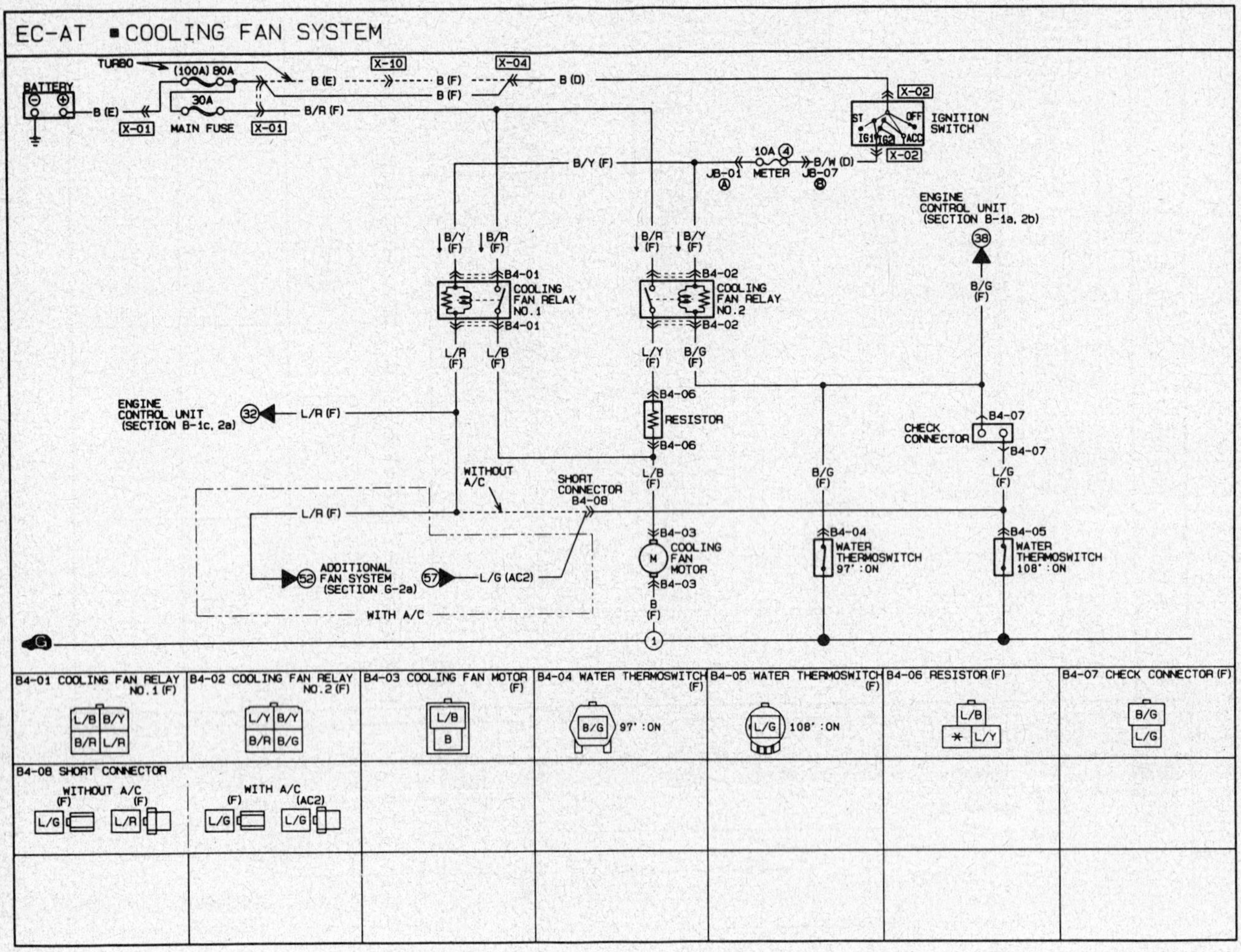

Wiring diagram – 626 and MX-6 turbo

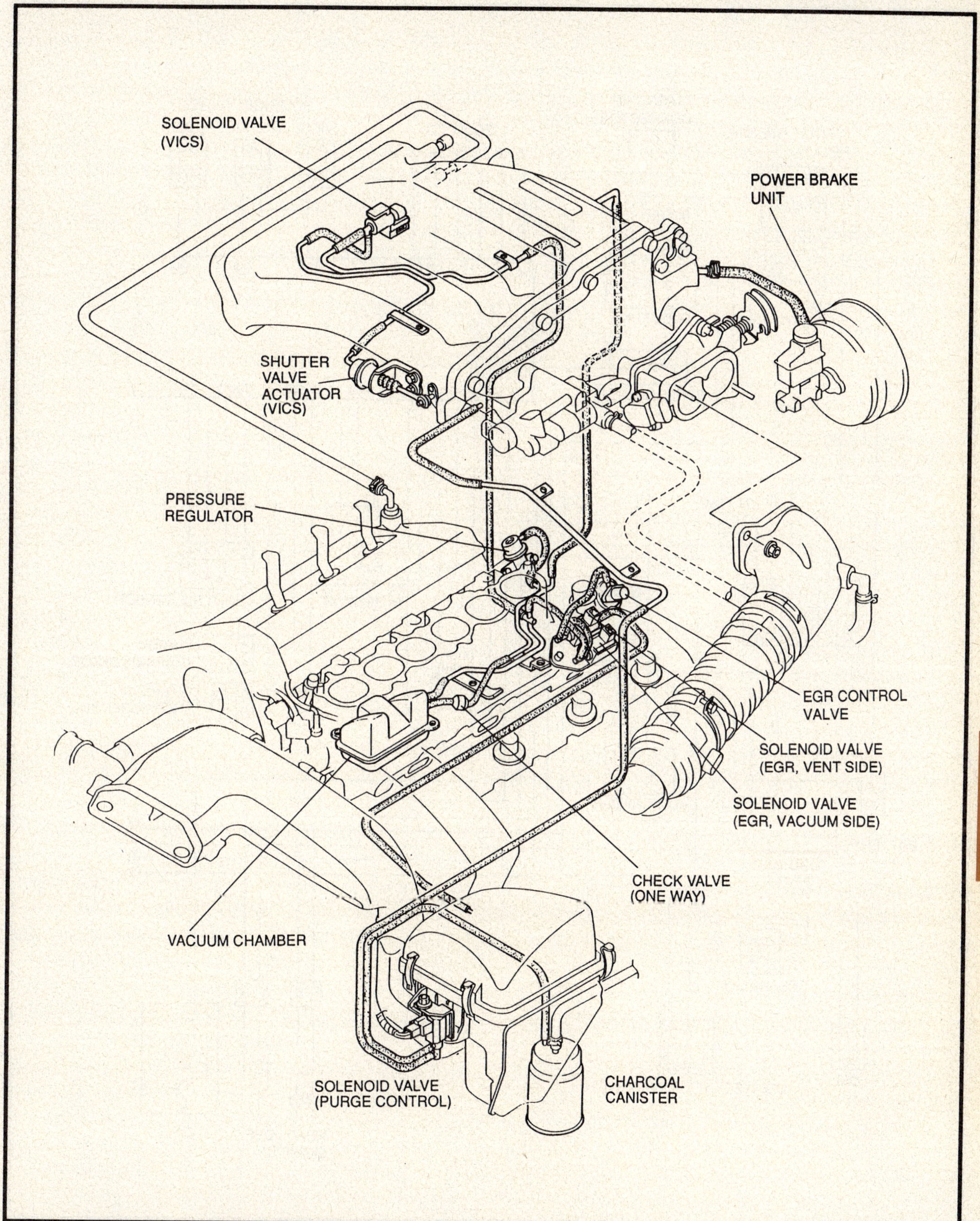

Vacuum hose routing diagram—1992 929

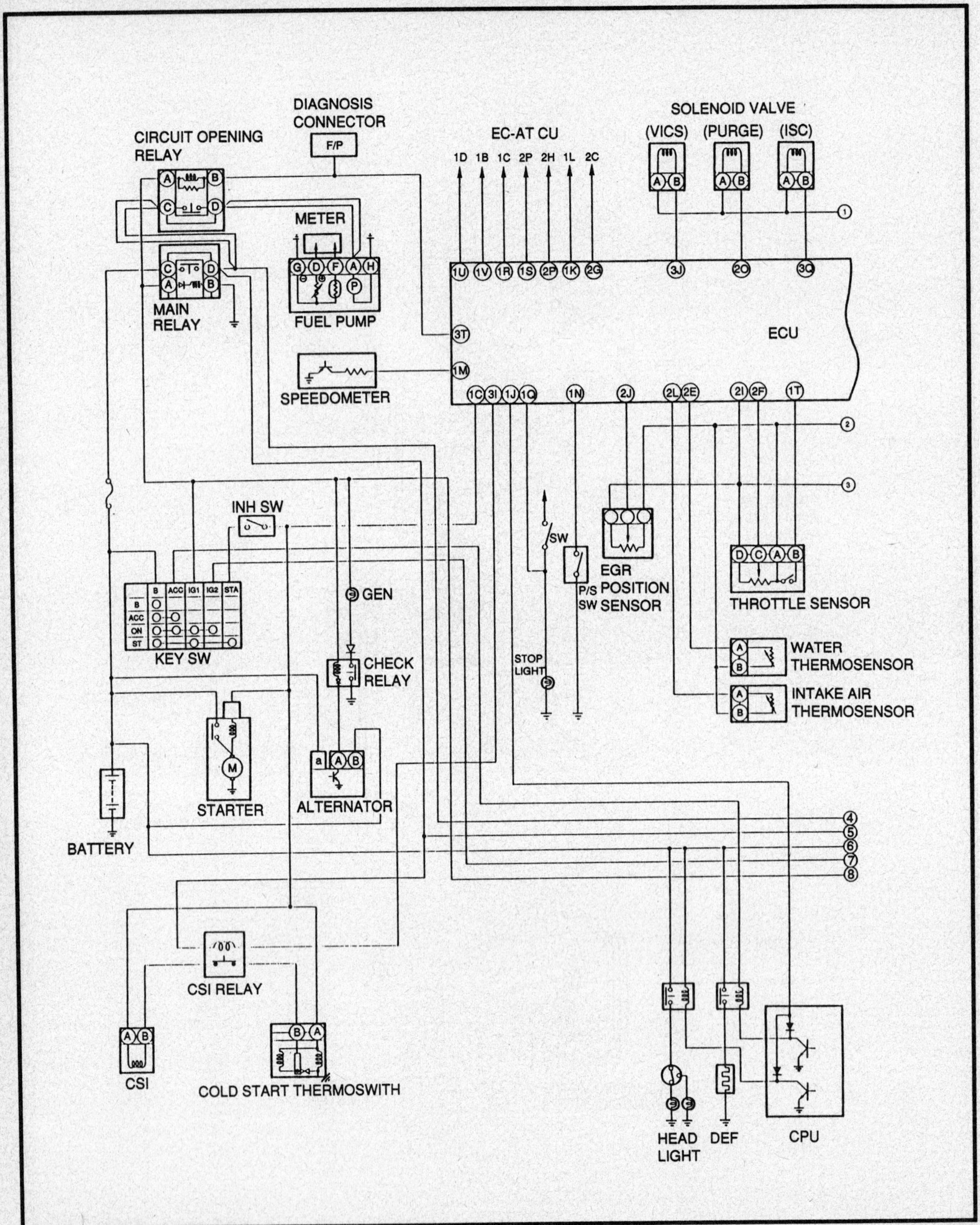

Wiring diagram—1992 929

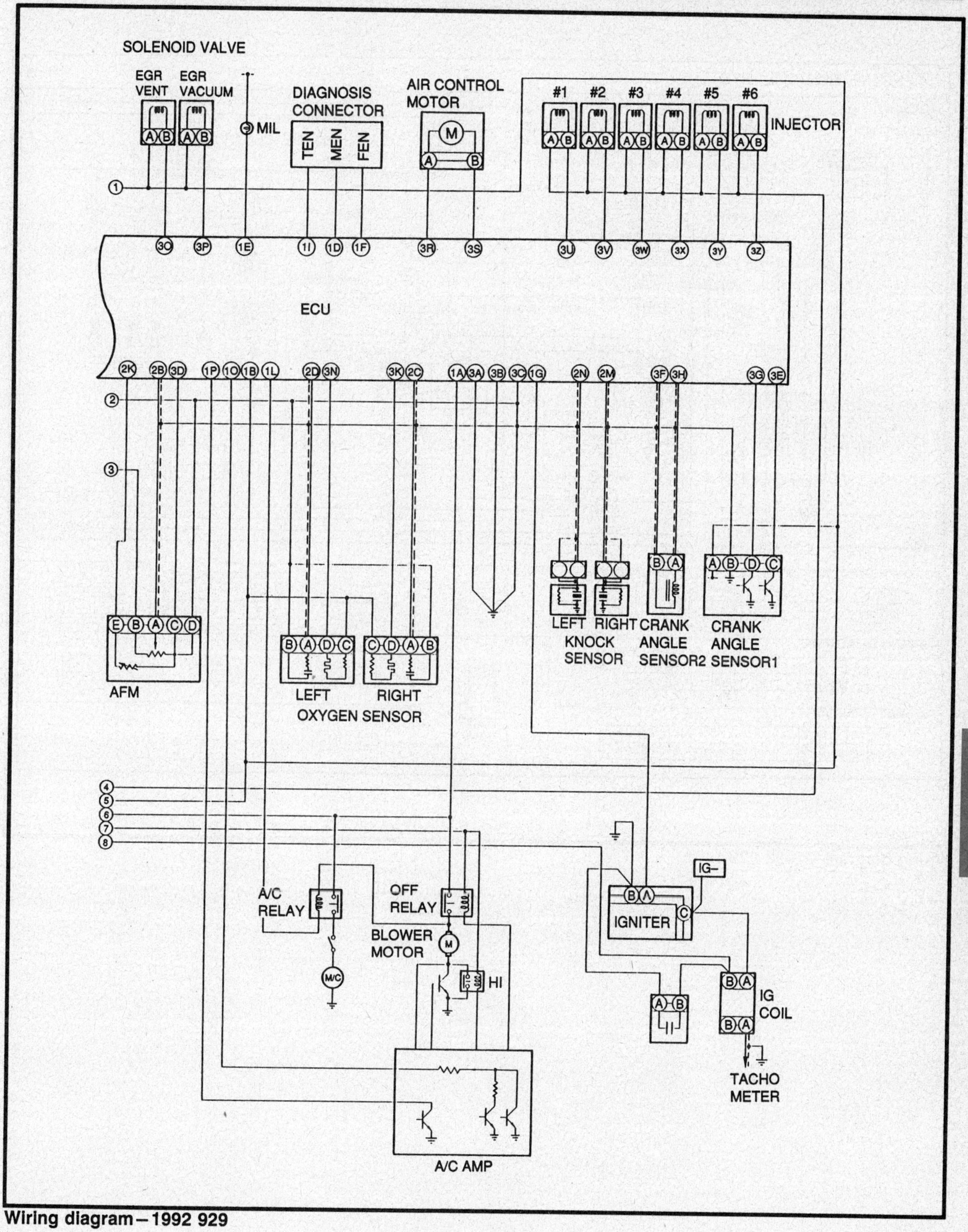

Wiring diagram—1992 929

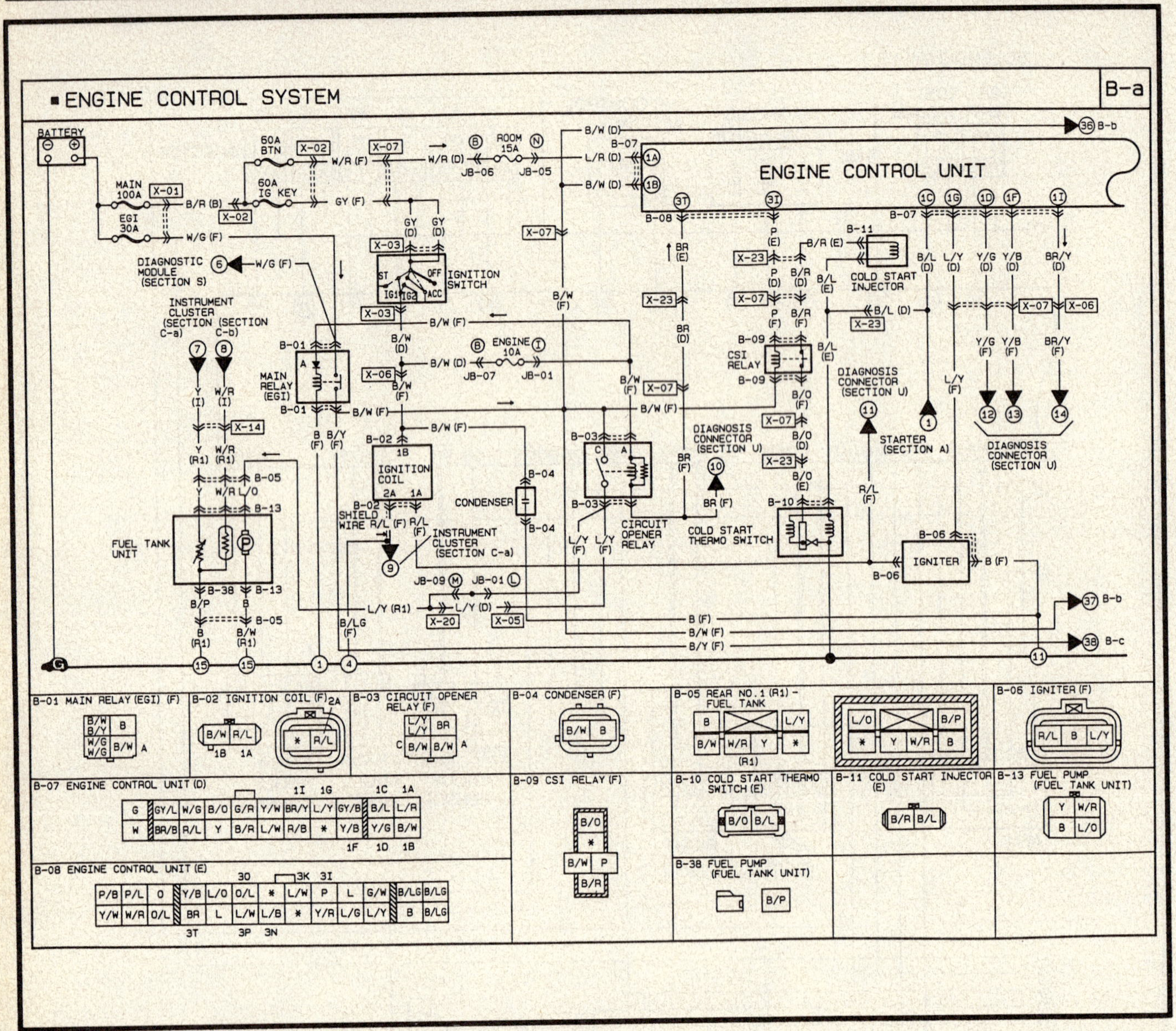

Wiring diagram—1992 929

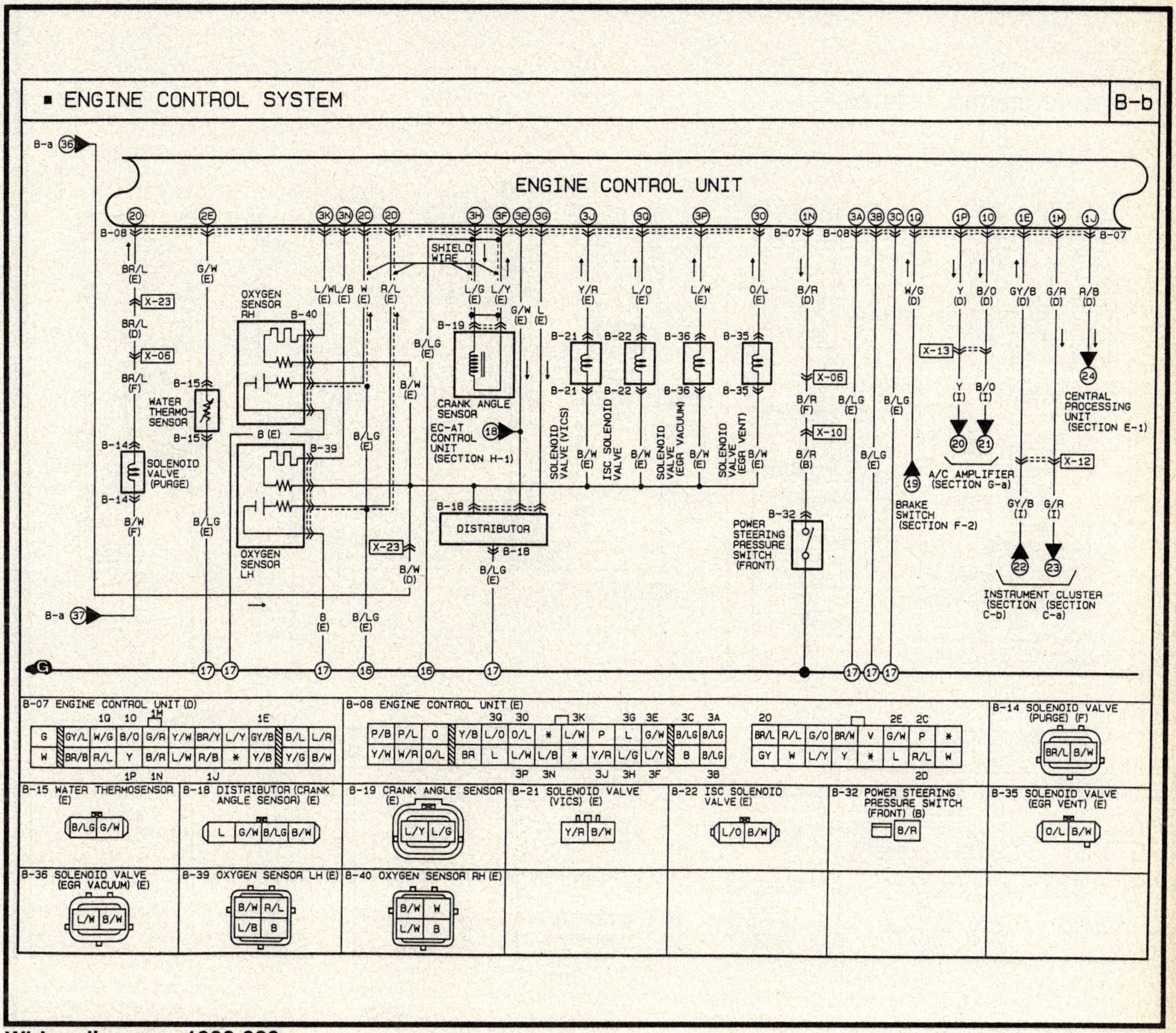

Wiring diagram—1992 929

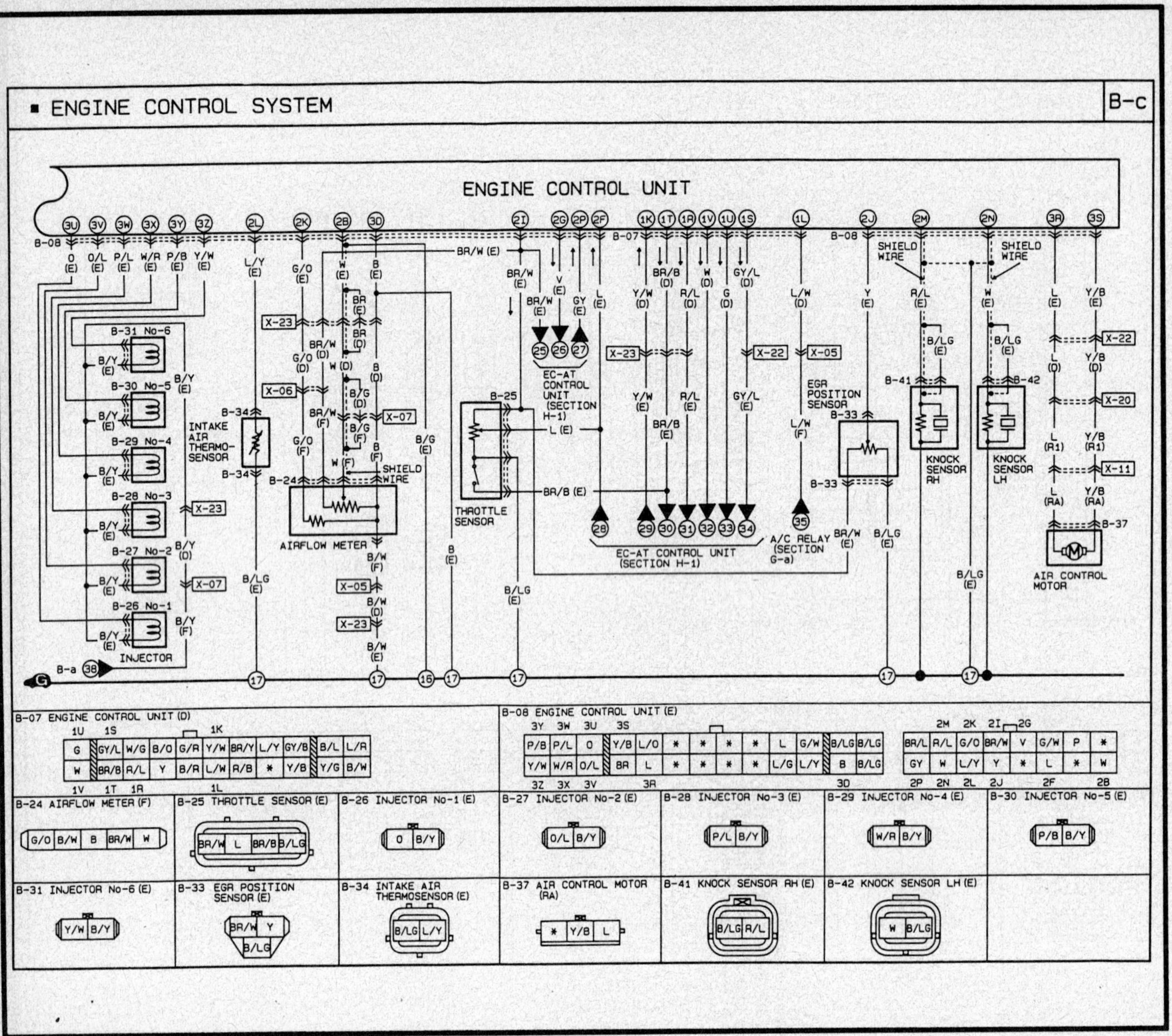

Wiring diagram—1992 929

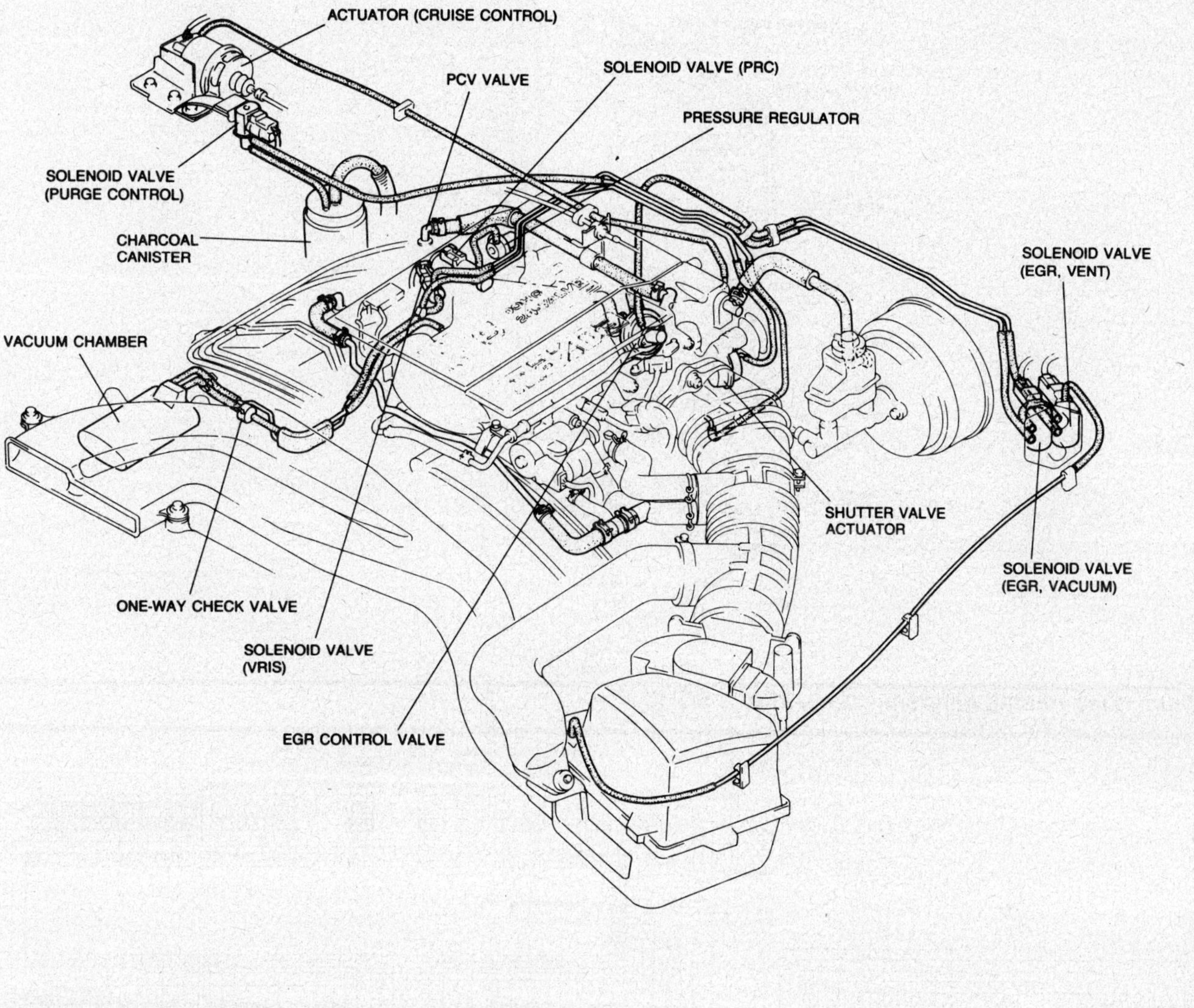

Vacuum hose routing diagram—1990–91 929 DOHC

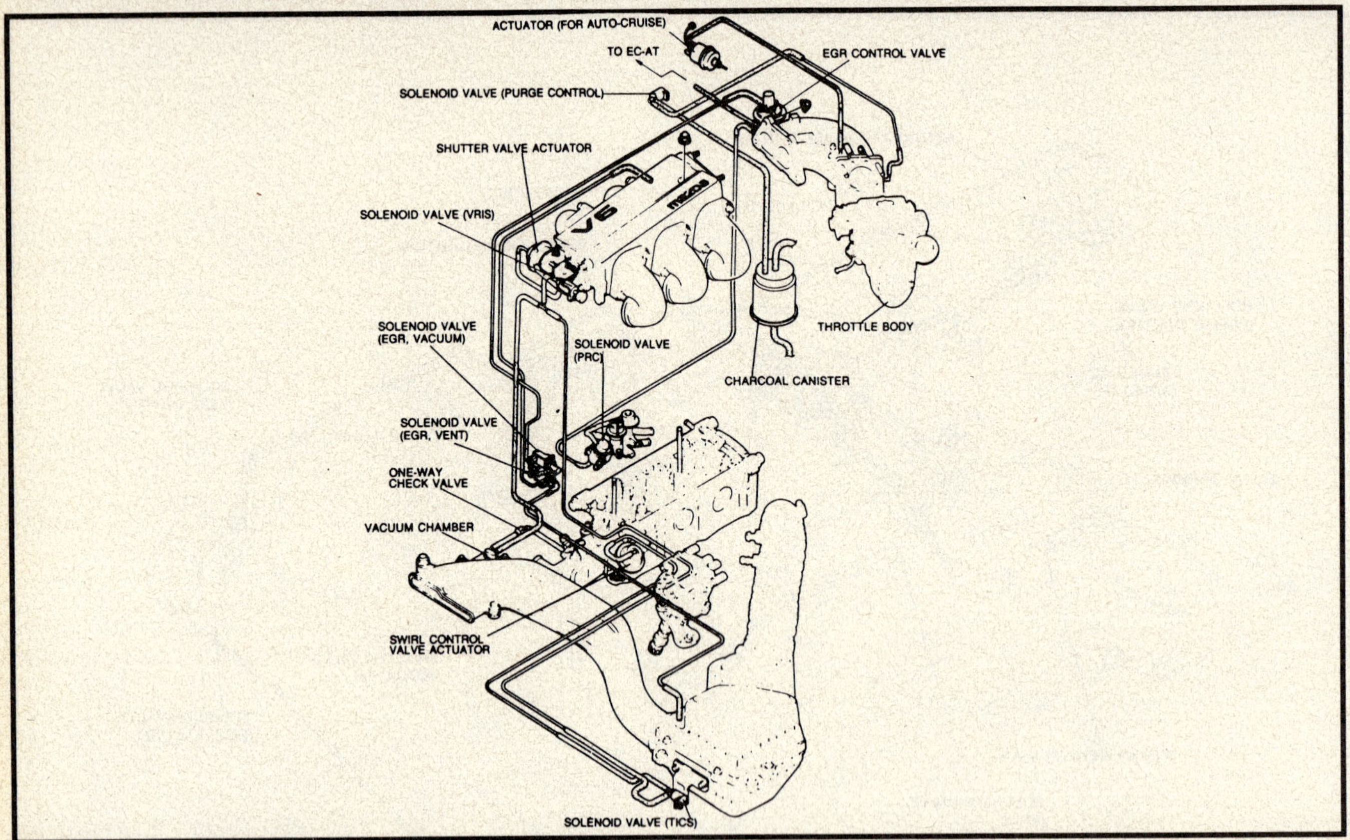

Vacuum hose routing diagram—1990–91 929 SOHC

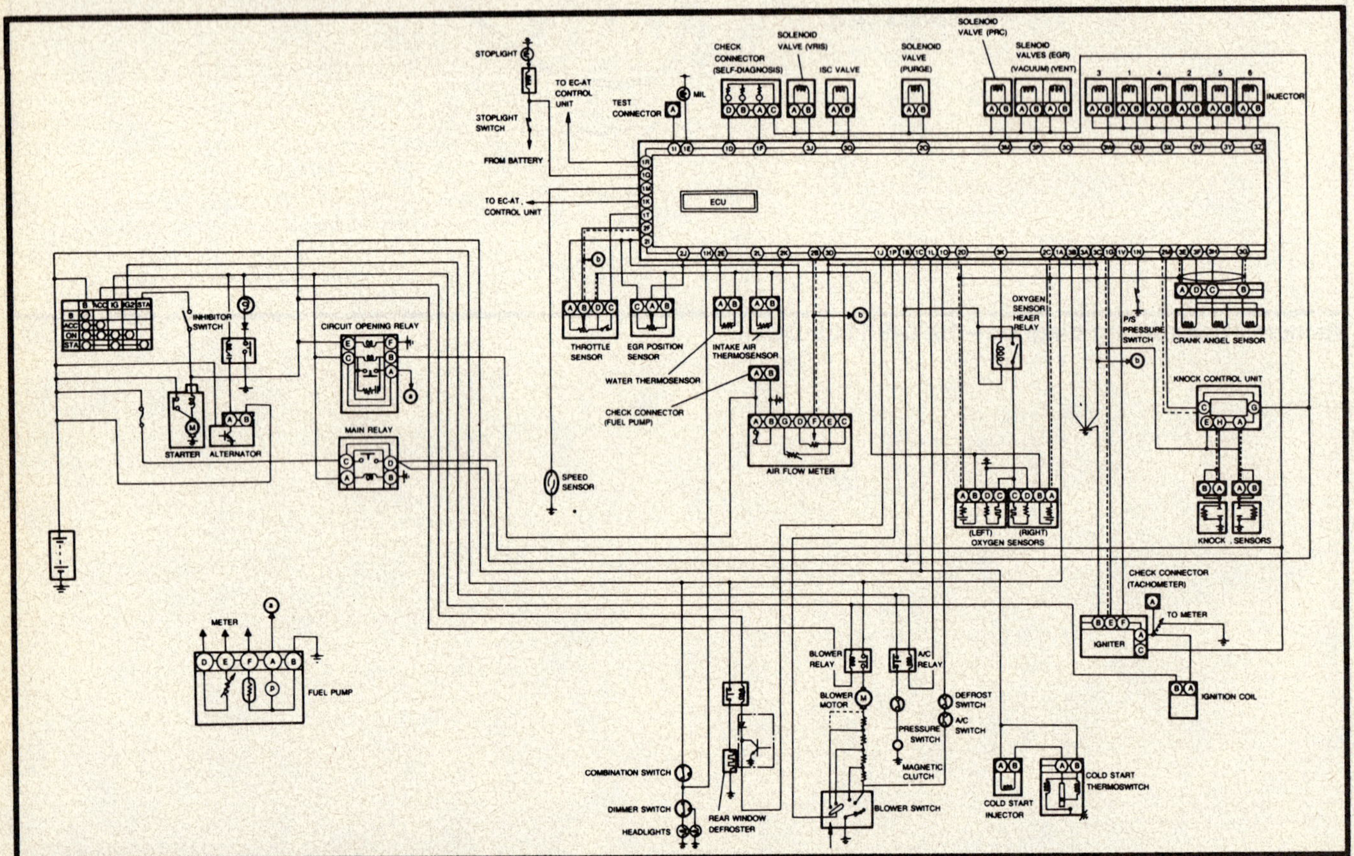

Wiring diagram—1990–91 929 DOHC

MAZDA RX-7 ELECTRONIC GASOLINE INJECTION (EGI) SYSTEM

General Information

This system is broken down into major systems; the Fuel System and Electronic Control System.

ELECTRONIC CONTROL SYSTEM

The EGI control unit (ECU), through various input sensors, monitors battery voltage, engine rpm, amount of air intake, cranking signal, intake temperature, coolant temperature, oxygen concentration in the exhaust gases, throttle opening, atmospheric pressure, gearshift position, clutch engagement, braking, power steering operation and A/C compressor operation.

The ECU controls operation of the fuel injection system, idle-up system , fuel evaporation system and ignition timing. The ECU has a built in fail-safe mechanism. If a fault occurs while driving, the ECU will substitute pre-programmed values. Driving performance will be affected, but the vehicle will still be driveable.

FUEL SYSTEM

The fuel injection system supplies the necessary fuel to the injectors to achieve combustion at constant pressure. Fuel is metered and injected into the intake manifold according to the signals from the EGI control unit. The injection system consists of the fuel pump, fuel filter, distribution pipe, pulsation dampener, pressure regulator, injectors, air flow meter, air valve, and solenoid resistor. The air induction system provides sufficient air for the engine operation.

Diagnosis and Testing

SERVICE PRECAUTION

- Before connecting or disconnecting control unit ECU harness connectors, make sure the ignition switch is OFF and the negative battery cable is disconnected to avoid the possibility of damage to the control unit.

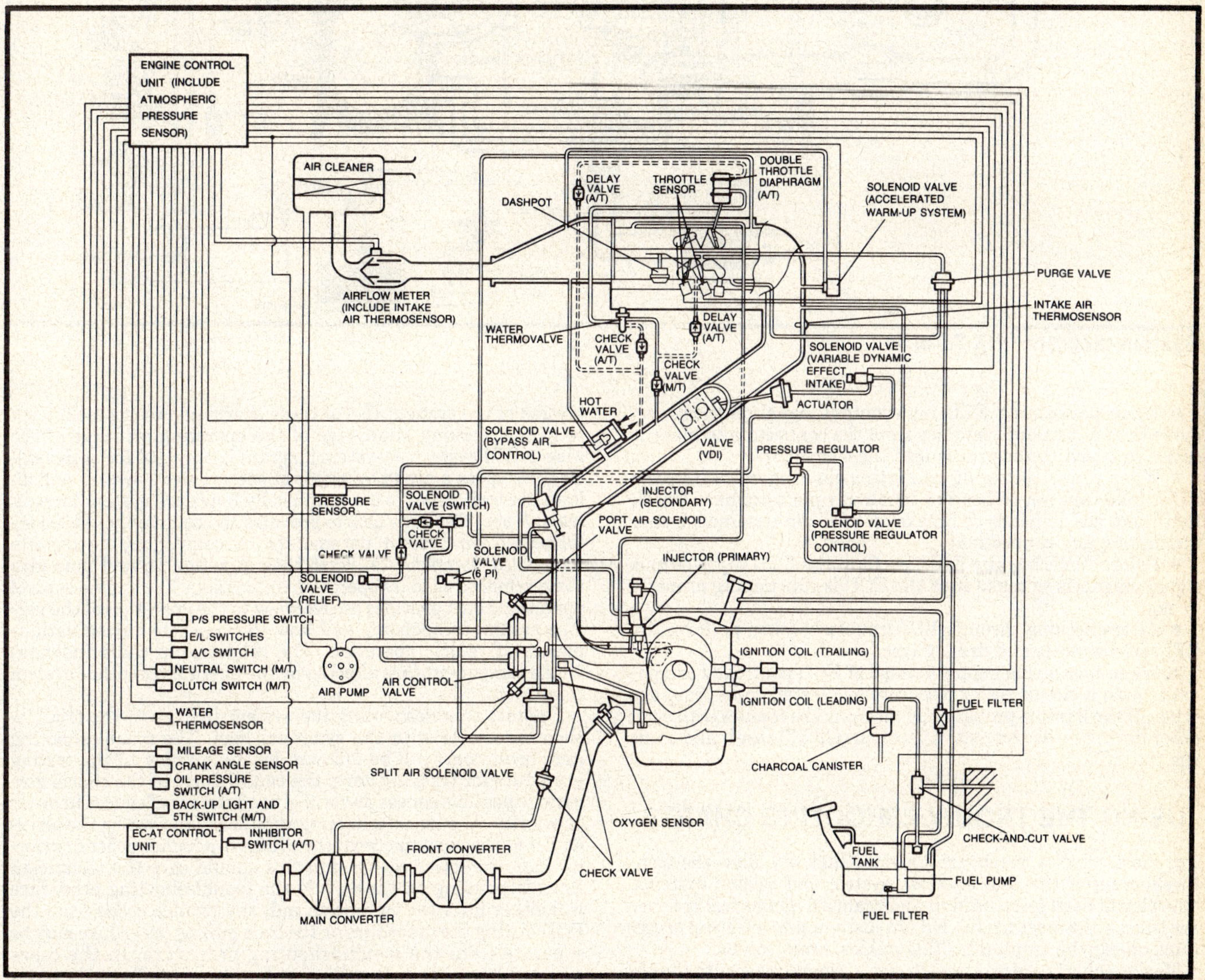

System diagram—RX-7 non-turbo

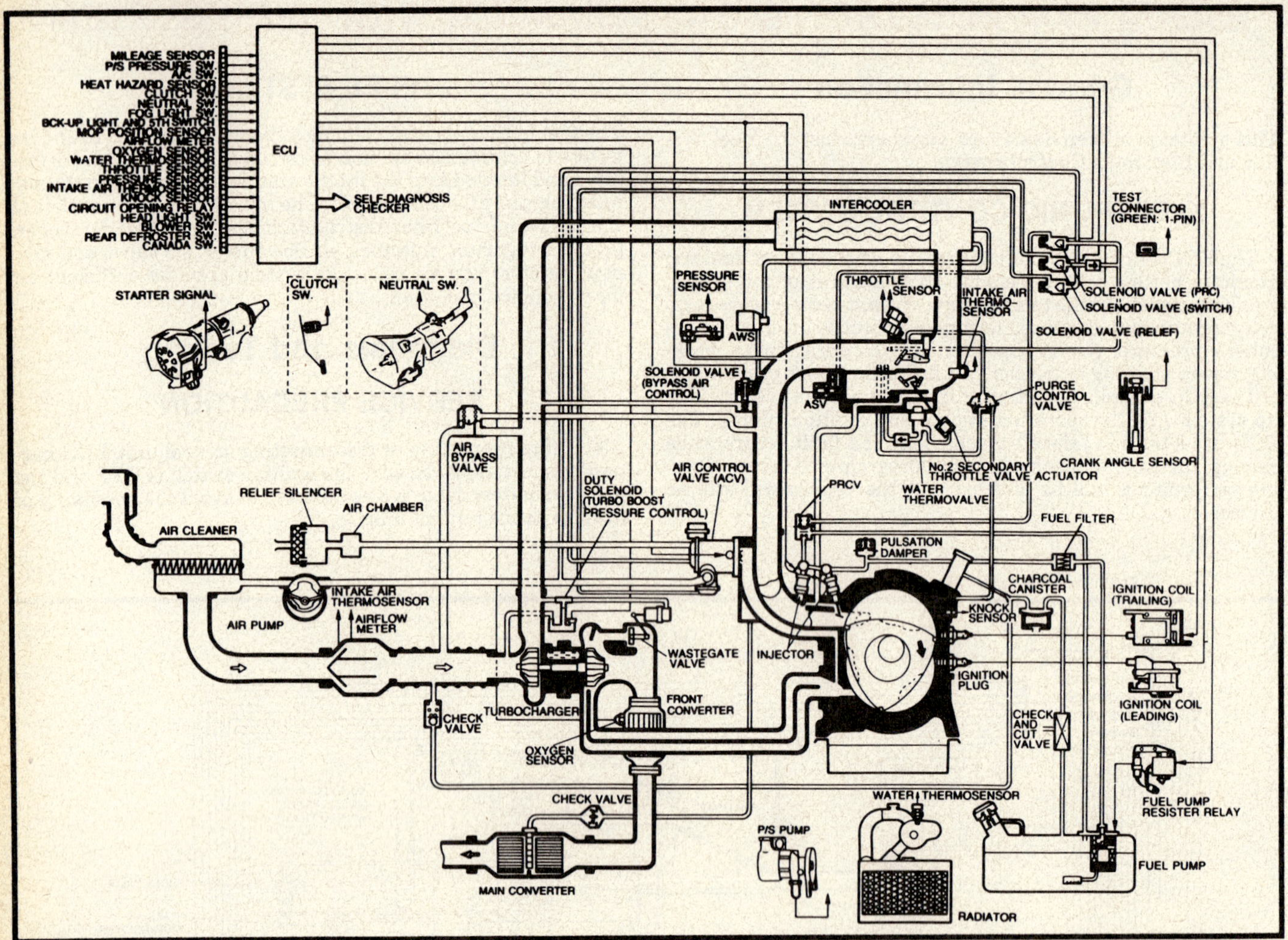

System diagram—RX-7 Turbo

• When performing ECU input/output signal diagnosis, remove the pin terminal retainer from the connectors to make it easier to insert tester probes into the connector.

• When connecting or disconnecting pin connectors from the ECU, take care not to bend or break any pin terminals. Check that there are no bends or breaks on ECU pin terminals before attempting any connections.

• Before replacing any ECU, perform the ECU input/output signal diagnosis to make sure the ECU is functioning properly or not.

• After checking through EGI troubleshooting, perform the EFI self-diagnosis and driving test.

• When measuring supply voltage of ECU controlled components with a circuit tester, separate one tester probe from another. If the 2 tester probes accidentally make contact with each other during measurement, a short circuit will result and damage the ECU.

USING THE TROUBLESHOOTING CHARTS

For most vehicles charts are provided that will allow the troubleshooting of the fuel injection system and related systems. The charts need to be used in an organized and logical manner for them to be most useful. Do not just jump to a testing procedure unless the troubleshooting guides direct you to.

Various charts are provided to give a general view as to the operation of the system. Use of these charts allows the technician to gain a working knowledge of the operating principle of the Mazda EGI system. A system diagram is provided for each vehicle. This gives a broad overview of each vehicles system and allows the technician to see graphically how each system interrelates. A specifications charts provides the commonly needed information on system parameters. A component description names the components, what their function is, and may give some remarks as to its operation or location. The engine control operation chart provides an overview to the operation of the various system components and how they operate during various phases of vehicle operation. The relationship chart indicates which components are affected by the operation of other system components.

To start troubleshooting the system first go to the troubleshooting charts with the symptom lists. The troubleshooting item (symptom) will be followed by the number of the testing procedure for tracking down the cause. Some of the charts give various possible causes and will give a number of causes in order of priority. The troubleshooting should be followed by the priority list as the most likely causes will have a higher order of priority than the lower causes and this should save the technician time. Depending on the vehicle the troubleshooting procedure will either have the technician pull any trouble codes from the ECU during the procedure or the code pulling procedure may be separate from the troubleshooting procedure. In the cases where the trouble code pulling is done separate from the troubleshooting procedure there will be a chart with the trouble codes and a separate section with trouble code troubleshooting arranged by trouble code number.

The troubleshooting procedure may callout a subsystem to be checked. In these cases there will be individual subsystem troubleshooting guides. These charts will also have an order or priority listing. Following this priority list time will be saved.

RETRIEVING CODES

NOTE: On vehicles equipped with Audio Anti-Theft system, before performing any repairs, the Personal Code Number (P.C.N.) must be obtained from the individual owner of the vehicle, or an authorized Mazda dealer.

With Self-Diagnosis Checker

The self diagnosis checker (49–H018–9A1) is used to retrieve code numbers of malfunctions which have happened and were memorized or are continuing. The malfunction is indicated by the code number and a buzzer.

If there is more than 1 malfunction, the code numbers will display on the self diagnosis checker 1 by 1 in numerical order. In the case of malfunctions, 08, 13, and 01, the code numbers are displayed in a order of 01, 08 and then 13.

The memory of malfunctions is canceled by disconnecting the negative battery cable for at least 20 seconds.

The ECU has a built in fail-safe mechanism for the main input sensors. If a malfunction occurs, the emission control unit will substitute values. This will slightly effect the driving performance, but the vehicle may still be driven.

The ECU continuously checks for malfunctions of the input devices. But, the ECU checks checks for malfunctions of the output devices within 3 seconds after turning the ignition switch to the **ON** position and the test connector is grounded.

The malfunction indicator light indicates a pattern the same as the buzzer of the self-diagnosis checker when the self-diagnosis check connector is grounded. When the self-diagnosis check connector is not grounded, the lamp illuminates steady while malfunction of the main input sensor occurs and goes out if the malfunction recovers. However, the malfunction code is memorized in the emission control unit.

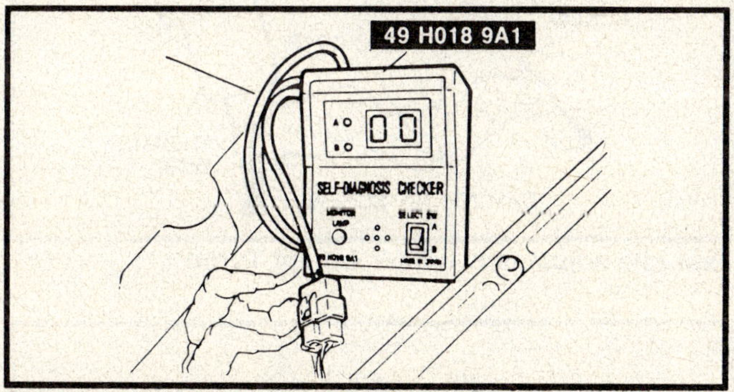

Self-diagnosis checker 49–H018–9A1

Testing Procedure

1. Start and allow the engine to reach normal operating temperature. Stop the engine.
2. Connect the Self Diagnosis Checker (49–H018–9A1) or equivalent to the check connector and the battery ground cable.
3. Set the select switch on the Self Diagnosis Checker to the "A" position.
4. Connect a jumper wire between the test connector and ground.
5. Turn the ignition switch to the **ON** position. Check that the number "88" flashes on the digital display and the buzzer sounds for 3 seconds after turning the ignition switch **ON**.
6. If the number "88" does not flash, check the check connector wiring.
7. If the number "88" flashes and the buzzer sounds continuously for more than 20 seconds, check for a short circuit between terminal 1F on the ECU and the check connector. Check ECU terminals 3X and 3Z voltages. Replace the ECU if necessary.
8. Note the code numbers and check the causes, repair as necessary. Be sure to recheck the code numbers after repairing.

CLEARING CODES

1. Cancel the memory of malfunctions by disconnecting the battery negative cable and depressing the brake pedal for at least 5 seconds, then reconnect the battery ground cable.
2. Connect the Self-Diagnosis Checker 49–H018–9A1 to the check connector. Ground the test connector (green: 1 pin) using a suitable jumper wire.
3. Turn the ignition switch to the **ON** position, but do not start the engine for approximately 6 seconds.
4. Start the engine and allow it to reach normal operating temperature, the run the engine at 2000 rpm for 2 minutes. Check that no code numbers are displayed.

ECU TERMINAL VOLTAGES

Testing with Engine Signal Monitor

The Engine Signal Monitor 49–9200–162, Adapter Harness 49–G018–903 and Engine Digital Monitor Sheet 49–G018–903 are used to check the control unit terminal voltages. This monitor checks the individual terminal voltages through selection by the monitor switch.

Testing

1. Start the engine and allow it to reach normal operating temperature. Stop the engine.

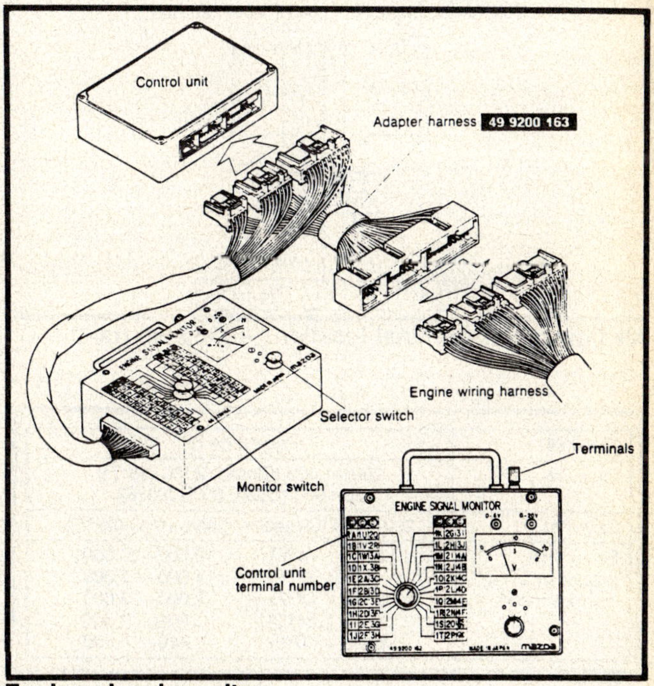

Engine signal monitor

2. Lift up the floor mat on the passenger's side, and remove the ECU protective cover. Disconnect the ECU connector.

4. Connect Engine Signal Monitor (49–9200–162), and Adapter Harness (49–G018–903) between the ECU and the ECU wiring harness, then install Engine Signal Monitor Sheet (49–G018–904) on the Engine Signal Monitor.

NOTE: Use connector (A) of the Adapter Harness to check voltages of terminals 1A through 1V and 3A through 3Z, and use connector (B) to check voltages of terminals 2A through 2P.

5. Position the select switch and the monitor switch on the Engine Signal Monitor to the terminal number to be tested. Check the voltage of each terminal.

6. If any ECU terminal voltage is incorrect, check the input or output component and related wiring. If components and wiring are normal, replace the ECU.

AIR FLOW METER

Testing

1. Inspect the air flow meter body for cracks.
2. Using a suitable ohmmeter, check the resistance between the terminals and check the resistance on the charts below.

Operational Check

1. Disconnect the negative battery terminal and disconnect the airflow meter electrical connector.
2. Remove the airhose to the airpump. Loosen the bolt holding the air funnel and remove the air cleaner and airflow meter. Remove the airflow meter from the air cleaner.
3. Using a suitable tool press open the measuring plate. Measure the resistance between the air flow meter terminals.
4. The resistance should be within the specifications shown in the following charts.
5. If the air flow meter fails to meet these specifications, replace it.
6. Replace the airflow meter in reverse order.

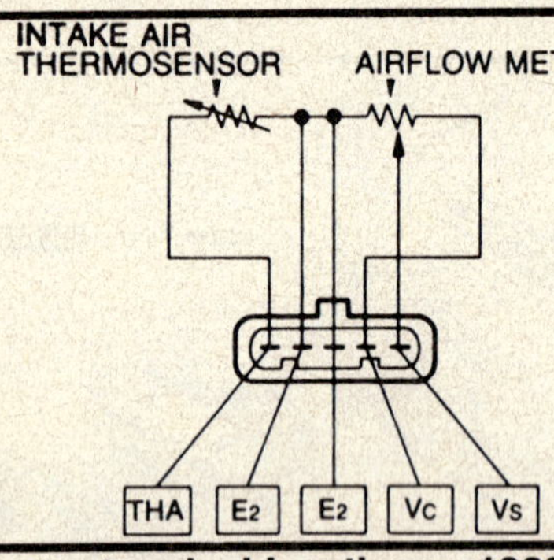

Air flow meter terminal locations—1989-90 RX-7

Terminal	Resistance (Ω)	
E2 ↔ Vs	200—1,000 (Closed; 20°C (68°F)) 20—800 (Open; 20°C (68°F))	
E2 ↔ Vc	200—400 (Closed ↔ Open; 20°C (68°F))	
E2 ↔ THA (Intake air thermosensor)	−20°C (−4°F) 0°C (32°F) 20°C (68°F) 40°C (104°C) 60°C (140°F)	10,000—20,000 4,000— 7,000 2,000— 3,000 900— 1,300 400— 700

Air flow meter resistance values—RX-7

THROTTLE BODY

Checking the No. 1 Secondary Throttle Valve

1. The No. 1 secondary throttle valve starts to open when the primary throttle valve opens to 15 degrees for the Turbo, 12 degrees manual transmission equipped non-Turbo, 8 degrees automatic transmission equipped non-Turbo and completely opens at the same time when the primary throttle valve fully opens.
2. Check the clearance between between the primary throttle valve and the wall of the throttle bore when the No. 1 secondary throttle valve starts to open.
3. If the clearance is not within the specification shown below, bend the tab until the proper clearance is obtained:
   RX-7 except turbocharged
   manual transmission 0.02–0.03 in. (0.5–0.7mm)
   automatic transmission 0.03–0.06 in. (0.8–1.4mm)
   RX-7 turbocharged—0.04–0.07 in. (1.1–1.7mm)

Checking fast idle operation

In order to perform this operation properly the vehicle and the throttle chamber must be at 77°F (25°C).

1. For proper fast idle operation the matching mark on the fast idle cam must be aligned with the center of the cam roller.
2. If the matching mark and the center of the cam roller do not align, turn the cam roller do not align, turn the cam adjusting screw until the proper alignment is obtained.

NOTE: Fast idle adjustment is unnecessary unless it has been tampered with.

3. Once the correct matching mark aligns with the center of the cam roller check the clearance, 0.016–0.02 in. (0.4–0.5mm), (throttle chamber—primary throttle valve), turn the fast idle adjusting screw to the correct specification.

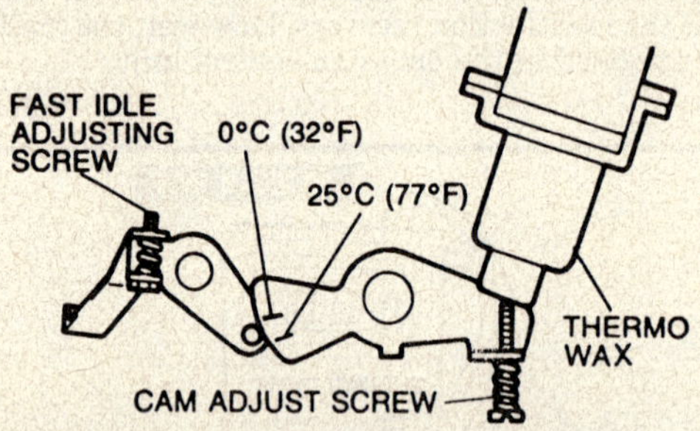

Fast idle adjustment—RX-7 except Turbo

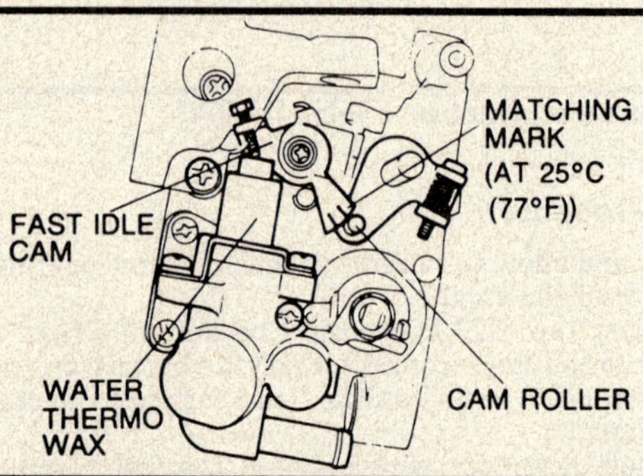

Fast idle adjustment—RX-7 Turbo

THROTTLE POSITION SENSOR

Testing

1. Start the engine and allow it to reach normal operating temperature. Stop the engine.
2. Connect Engine Signal Monitor (49–9200–162) and Adapter Harness (49–G018–903) between the ECU and the ECU wiring harness.

NOTE: When testing the throttle sensor, ensure to use connector B of the Adapter Harness (49–G018–903).

3. Place Engine Signal Monitor Sheet (49–G018–904) on the Engine Signal Monitor. Check the voltage between terminals 2F and 2G of the ECU. Specifications should be as follows:

Terminal 2F
- Throttle closed: 0.75–1.25 volts
- Throttle fully opened: Approximately 5 volts.

Terminal 2G
- Throttle closed: 0.25–1.25 volts
- Throttle fully opened: 4.1–4.44 volts

4. If specifications are not as indicated, adjust or replace the throttle sensor as necessary.

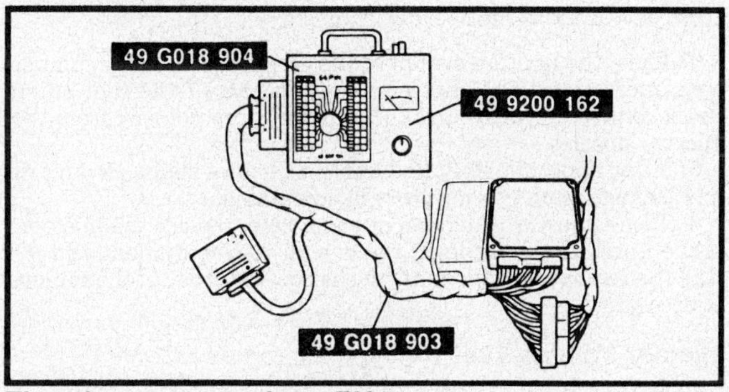

Throttle sensor testing – RX-7

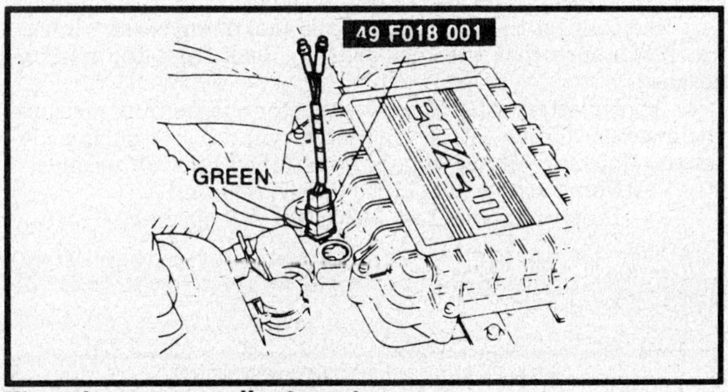

Throttle sensor adjustment

FUEL PUMP

Testing

1. Relieve the fuel system pressure.
2. Disconnect the negative battery cable.
3. Disconnect the main fuel hose from the main fuel line.
4. Connect a suitable fuel pressure gauge.
5. Connect the negative battery cable.
6. Connect the terminals of the check connector with a jumper wire. Turn on the ignition switch to operate the fuel pump.

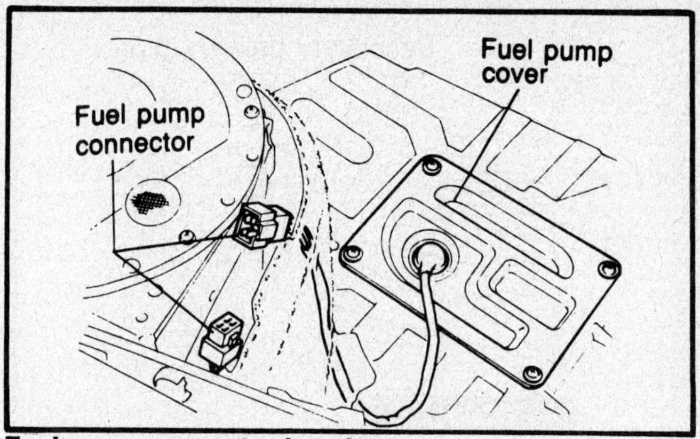

Fuel pump connector location

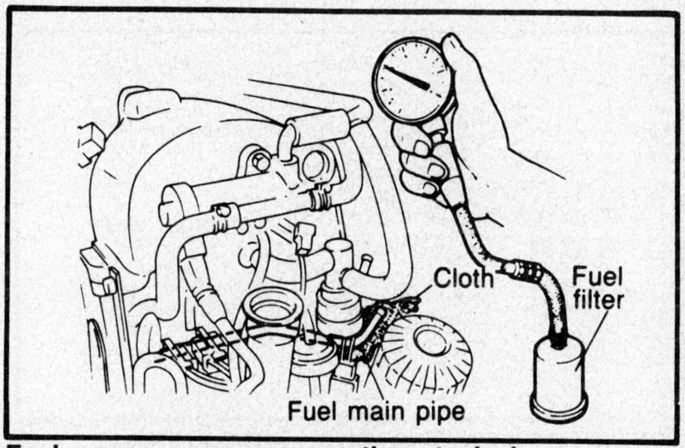

Fuel pressure gauge connection – typical

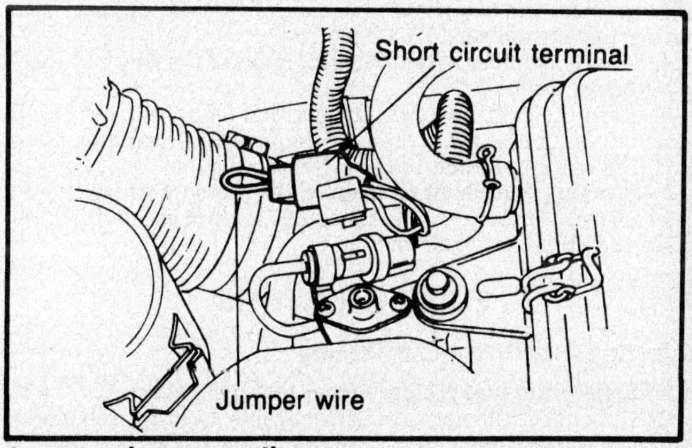

Jumper wire connection

7. Observe that the fuel pressure is within the following specifications:
   RX-7 except Turbo – 64–85.3 psi (441–588 kPa)
   RX-7 Turbo – 71.1–92.4 psi (490–637 kPa)

PRESSURE REGULATOR

Fuel Line Pressure Testing

1. Relieve the fuel pressure from the fuel system.
2. Disconnect the battery ground cable. Connect a suitable pressure gauge between the fuel filter and the pulsation damper.

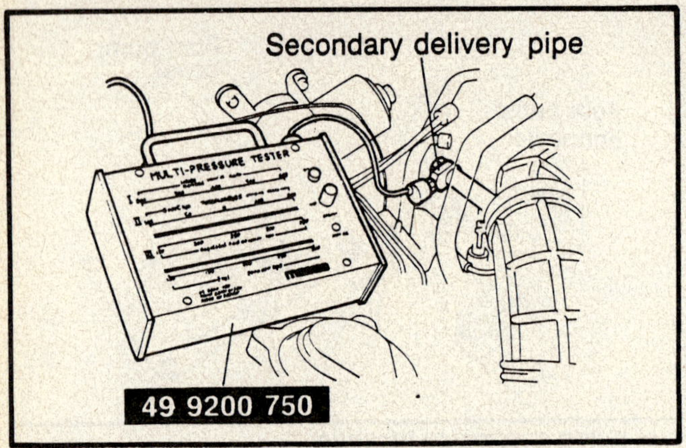

Fuel system pressure tester No. 49 9200 750

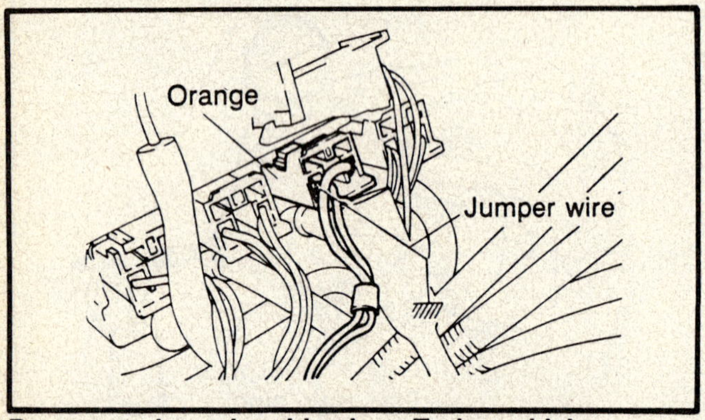

Pressure wire solenoid valve—Turbo vehicles

3. Connect the battery ground cable. Start the engine. Pressure gauge should read 27–33 psi.
4. Install a jumper wire to the terminals of the check connector (yellow: 2-pin).
5. Turn the ignition to the **ON** position for approximately 10 seconds. Pinch the outlet hose of the pressure regulator, then turn the ignition switch to the **OFF** position.
6. After approximately 5 minutes, the pressure gauge should read 21 psi. If pressure gauge readings are not as specified, replace the pressure regulator.

FUEL INJECTORS

Injection Fuel Pressure Testing

1. Relieve the fuel pressure from the fuel system.
2. Disconnect the negative battery cable. Connect the multi-pressure tester (49-9200-750) or equivalent to the fuel line.
3. Reconnect the negative battery cable. Start the engine and measure the fuel pressure at the "III" range of the multi pressure tester. The fuel pressure (injection pressure) should be approximately35.6–37.0 psi.

NOTE: The fuel pressure is indicated on the lower LED line (fuel pressure line) at any ranges.

4. If the injection pressure is lower than specified, check the following:
   a. Fuel pump outlet pressure.
   b. Fuel line pressure.
   c. Fuel filter clog.
5. If the injection pressure is higher than specifications, check the following:
   a. Fuel return pipe clogged.
   b. Fuel line pressure.

Injector Testing

Check the operating sound of the injector, using a sound scope, check that operating sounds are produced from each injector at idle and at acceleration.

1. If the injectors do not operate, check the following:
   a. Check the fuel injection system main fuse (40 amp).
   b. Turn the ignition switch **ON** and **OFF**, verify that the main relay clicks.
2. If the clicking sound is not heard at the main relay, when the ignition switch is turned **ON**, use a voltmeter and check for 12 volts at the main relay connector (BW) terminal wire.

Injector Fuel Leak Test

1. Relieve the fuel pressure from the fuel system.
2. Remove the throttle body unit.
3. Loosen the distribution pipe attaching bolts. Affix the injectors to the distribution pipe with some suitable wire.

NOTE: Affix the injectors firmly to the distribution pipe so no movement of the injectors is possible.

4. Turn the ignition switch to the **ON** position. Use a jumper wire and connect the shot circuit terminals of the fuel pump check connector. Make sure that the fuel does not leak from the injector nozzles.
5. After approximately five minutes, a very slight amount of fuel leakage from the injectors is acceptable.
6. If the injector leaks fuel at a fast rate, replace the injector. If the injectors do not leak, remove all test equipment and install the removed components in the reverse order of the removal procedure.

Injector Volume Test

1. Relieve the fuel pressure from the fuel system.
2. Connect a suitable hose to the injector. Connect injector checker (49–B092–953) or equivalent to the injector.
3. Turn the ignition switch to the **ON** position. Use a jumper wire and connect the terminals of the fuel pump check connector. Make sure that the fuel does not leak from the injector nozzles.
4. Apply battery voltage to the injector checker and measure the injector volume with a graduated cylinder (measuring container). Injector volume should be as follows per 15 seconds:
   Non-Turbo—111–118cc (6.8–7.2 cu. in.)
   Turbocharged—133–142cc (8.1–8.7 cu. in.)

NOTE: Do not apply battery voltage directly to the injector. The injector checker must be used or the injector will be damaged.

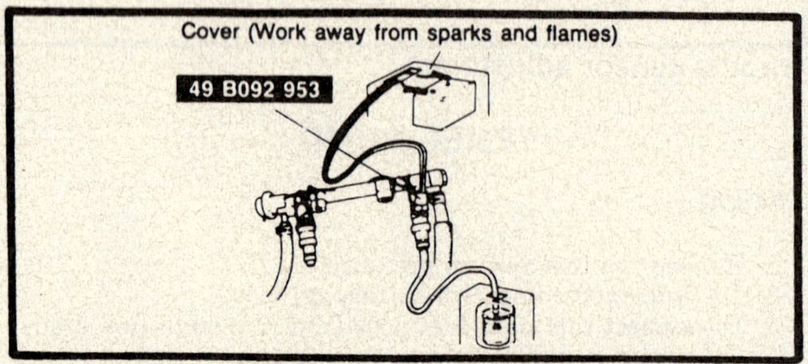

Fuel injector volume test

SPECIFICATION CHART – RX-7 NON-TURBO

SPECIFICATIONS

Item			Model: 13B EGI engine
Idle speed (Test connector grounded)		rpm	750 ± 25 (for A/T...P Range)
Air cleaner	Element type		Long life wet
Throttle body	Type		Horizontal — draft (2 stage — 3 barrel)
	Throat diameter	Primary mm (in)	45 (1.772)
		Secondary mm (in)	45 (1.772) x 2
	Water thermovalve	Operation temp. °C (°F)	M/T: 67—77 (153—171) or more A/T: 60—70 (140—158) or more
Dashpot	Adjustment speed	rpm	2,700—3,100
Fuel tank	Capacity	liters (US gal, Imp gal)	70 (18.5, 15.4)
Fuel filter	Low pressure		Nylon 6 (164 and 45 mesh)
	High pressure		Filter paper
Pressure regulator	Type		Diaphragm
	Regulated pressure	kPa (kg/cm², psi)	235—275 (2.4—2.8, 34.1—39.8)
Fuel pump	Type		Impeller (intank)
	Outlet pressure	kPa (kg/cm², psi)	441—588 (4.5—6.0, 64.0—85.3)
Injector (Primary and Secondary)	Drive		Voltage drive
	Injection volume	cc (cu in)/15 sec.	111—118 (6.8—7.2)
Heat hazard sensor	Operation temperature	°C (°F)	105—115 (221—239)
Main silencer	Capacity	cc (cu in)	M/T: 10,300 (628.3) x 2, A/T: 12,000 (732) x 2
Ignition timing (Test connector grounded)			Leading: 5° ± 1° ATDC Trailing : 20° ± 2° ATDC
Distribution	Type		Control unit
Spark advance	Type		Control unit
Idle-up system	A/C	rpm	M/T: 875 A/T: 800
	"D" range	rpm	750 (at warm engine)
Anti-afterburn valve	Operation time	sec.	M/T: 1.60—2.20 A/T: 0.52—0.92

COMPONENT DESCRIPTIONS – RX-7 NON-TURBO

COMPONENT DESCRIPTION

Component	Function	Remarks
Accelerated Warm-up System (AWS)	Supplies bypass air into dynamic chamber	Controlled by duty signal from control unit
Anti Afterburn Valve	Supplies fresh air into rear port during deceleration	Included in air control valve
Air Bleed Socket	Supplies fresh air into injector hole	
Air Cleaner	Filters air into throttle chamber	
Air Control Valve	Directs air to one of three locations; exhaust port, main converter or relief air silencer	Consists of 3 valves; Relief valve Switching valve Anti afterburn valve
Air Flow Meter	Detects amount of intake air; sends signal to control unit	
Atmospheric Pressure Sensor	Detects atmospheric pressure; sends signal to control unit	Built in ECU
Air Pump	Supplies secondary air to air control valve	
Catalytic Converter	Reduces HC, CO and NOx	
Charcoal Canister	Stores gas tank fumes when engine stopped	Vented to atmosphere through charcoal and filter
Check-and-cut Valve	Controls pressure in fuel tank	
Coil with Igniter	Generates high voltage	Leading; ignite simultaneously Trailing; ignite individually
Crank Angle Sensor	Detects eccentric shaft angle at 30° intervals and front rotor position; sends signal to control unit	
Dashpot	Gradually closes throttle valve during deceleration	
Double Throttle System	Gradually opens the No.2 secondary throttle valve when No.1 secondary throttle valve suddenly opens	
Dynamic Chamber	Connects front and rear ports	Primary and secondary separated
Engine Control Unit	Detects the following: 1. Engine speed 2. Intake air amount 3. Engine coolant temperature 4. Throttle opening 5. Intake manifold vacuum 6. O₂ concentration 7. In-gear condition 8. Intake air temperature 9. Floor temperature 10. A/C operation 11. Cranking signal 12. Atmospheric pressure 13. Initial set signal 14. Position of transmission gear 15. P/S operation 16. Metering oil pump (MOP) position signal 17. Electric load (E/L) condition 18. Mileage Controls operation of the following: 1. Fuel injection system 2. Ignition control system 3. ISC system 4. Pressure regulator control system 5. Secondary air injection control system 6. Variable dynamic effect intake (VDI) system 7. 6-port induction (6PI) system	 Crank angle sensor Air flow meter Water thermosensor Throttle sensors Pressure sensor Oxygen (O₂) sensor Neutral switch and clutch switch (Inhibitor switch) Intake air thermosensor Heat hazard sensor A/C switch Starter switch Atmospheric pressure sensor Test connector (Green: 1-pin) Back-up light and 5th switch, oil pressure switch P/S switch MOP position sensor Headlight switch, Blower switch, Rear defroster switch, Fog light switch Mileage sensor

COMPONENT DESCRIPTIONS—RX-7 NON-TURBO

Component	Function	Remarks
Fast Idle System	Opens primary throttle valve slightly at idle	Only during cold condition
Fuel Filter	Filters particles from fuel	
Fuel Pump	Provides fuel to injectors	Operates while engine is running Installed in fuel tank
Heat Hazard Sensor	Detects floor temperature; sends signal to control unit	Heat hazard sensor turned ON; relieves secondary air
Initial Set Coupler	Sends initial set signal to control unit	While adjustment of idle speed, idle mixture; coupler is shorted
Injector	Injects fuel into intermediate housing and secondary intake manifold	Controlled by signals from control unit
Intake Air Thermosensor	Detects intake air temperature and temperature into the engine; sends signal to control unit	Located on the air flow meter and air intake pipe Thermistor
Mileage Sensor	Detects vehicle mileage sends signal to control unit	Above 20,000 miles; mileage switch ON
Oxygen (O_2) Sensor	Detects O_2 concentration; sends signal to control unit	Zirconia ceramic and platinum coating
Pressure Regulator	Adjusts fuel pressure supplied to injectors	
Pressure Sensor	Detects intake manifold pressure sends signal to control unit	
Pulsation Damper	Absorbs fuel pulsation	
Purge Control Valve	Regulates evaporative fumes from gas tank and canister to intake manifold	
Solenoid Valve (BAC)	Supplies bypass air into intake manifold	Controlled by duty signal from control unit
Solenoid Valve (Relief)	Controls relief valve	
Solenoid Valve (Switch)	Controls switching valve	
Solenoid Valve (6PI)	Controls auxiliary port valve	
Solenoid Valve (PRC)	Shuts vacuum passage between dynamic chamber and pressure regulator	Orange
Solenoid Valve (VDI)	Controls variable dynamic effect intake valve	
Test Connector (Green: 1-pin)	Sends initial set signal to control unit	During adjustment of idle speed, ignition timing; connector grounded
Throttle Body	Controls intake air quantity	
Throttle Sensor	Detects primary throttle valve opening angle; sends signal to control unit	
VDI Valve	Change the pressure wave line	
Water Thermosensor	Detects engine coolant temperature; sends signal to control unit	Thermistor

TROUBLESHOOTING GUIDE—RX-7 NON-TURBO

TROUBLESHOOTING GUIDE

This troubleshooting guide shows the malfunction code numbers retrieved by the **SST** and symptoms of various failures. Perform troubleshooting as described below.

TROUBLESHOOTING PROCEDURE
Troubleshooting With SST
Troubleshooting with the **SST** (Self-Diagnosis Checker 49 H018 9A1) is done to quickly determine what system or unit may be at fault.

1st: Check input sensors and output devices with the **SST**.
2nd: Check other switches with the **SST**.
3rd: Check the following items:

- Electrical system : Battery conditions, fuses
- Ignition system : Ignition spark, ignition timing (with test connector grounded)
- Fuel system : Fuel level, fuel leakage, fuel filter, idle speed (with test connector grounded)
- Intake air system : Air cleaner element, vacuum or air leakage, vacuum hose routing, accelerator cable
- Engine : Compression, overheating
- Others : Clutch slippage, brake dragging

4th: Check fuel and emission control systems.

Malfunction Code No.

Code No.	Input device
01	Ignition coil (Trailing side)
02	Crank angle sensor (Ne-signal)
03	Crank angle sensor (G-signal)
08	Airflow meter (AFM)
09	Water thermosensor
10	Intake air thermosensor (AFM)
11	Intake air thermosensor (Engine)
12	Throttle sensor (Full range)
13	Pressure sensor
14	Atmospheric pressure sensor (Built in ECU)
15	Oxygen sensor
17	Feedback system
18	Throttle sensor (Narrow range)
20	Metering oil pump position sensor
27	Metering oil pump
37	Metering oil pump

Code No.	Output device
25	Solenoid valve (Pressure regulator control (PRC))
26	Stepping motor (Metering oil pump)
30	Split air solenoid valve
31	Solenoid valve (Relief)
32	Solenoid valve (Switch)
33	Port air solenoid valve
34	Solenoid valve (Bypass air control (BAC))
38	Solenoid valve (Accelerated warm-up system (AWS))
40	Auxiliary port valve
41	Solenoid valve (Variable dynamic effect intake (VDI) control)
51	Fuel pump resistor relay
71	Injector (Front secondary)
73	Injector (Rear secondary)

TROUBLESHOOTING GUIDE – RX-7 NON-TURBO

Troubleshooting of Each System
The troubleshooting guide lists the most likely causes to a given symptom. After finding the systems to check, refer to the pages shown for detailed guides.
The numbers of the list show the priorities of inspections from the most probable to that with the lowest probability.
These were determined on the following basis:

- Ease of inspection
- Most possible system
- Most possible point in system

Possible cause \ Symptom	Hard start or No start (Cranks OK)	Rough idle	High idle speed (At normal operating temperature)	Engine does not run smoothly (At normal operating temperature)	Engine does not run smoothly (Only when engine is cold)	Engine does not run smoothly (Only when engine is hot)	Lack of power, poor acceleration or hesitation	Afterburn	Runs rough on deceleration	Fails emission test	Knocking
INTAKE AIR SYSTEM	4	1	2	1	3		1	2		3	
IDLE SPEED CONTROL (ISC) SYSTEM		4	1		1	4		5	2	7	
VARIABLE DYNAMIC-EFFECT INTAKE (VDI) SYSTEM							3				
6-PORT INDUCTION SYSTEM (6PI)		3					2				
FUEL SYSTEM	2	5		4	5		5	7	3	6	
PRESSURE REGULATOR CONTROL (PRC) SYSTEM	3		4			3					
EXHAUST SYSTEM				5	6		7			5	
SECONDARY AIR INJECTION SYSTEM	5				7			6		4	
CRANKCASE AND EVAPORATIVE EMISSION CONTROL SYSTEM										2	
DECELERATION CONTROL SYSTEM					8			4	1	8	
CATALYTIC CONVERTER										1	
A/C CUT-OUT CONTROL SYSTEM							8				
CONTROL SYSTEM	6	6	3	3	2	2	6	3	5	9	2
ELECTRONIC SPARK ADVANCE (ESA) SYSTEM	1	2		2	4	1	4	1	4	10	1

TROUBLE CODES – RX-7 NON-TURBO

Code number

Caution

a) If there is more than one failure present, the lowest number malfunction code is displayed first, the subsequent malfunction codes appear in order.

b) After repairing all failures, turn the ignition switch OFF, disconnect the negative battery cable for at least 5 seconds to erase the malfunction code memory.

Input devices

Code No.	Input devices	Malfunction	Fail-safe function	Output signal pattern (Self-Diagnosis Checker or MIL)
01	Ignition coil (Trailing side)	Malfunction of spark plug broken wire, short circuit	Trailing-side ignition pulse cut	
02	Crank angle sensor (Ne signal)	Broken wire, short circuit	Fuel injection and ignition cut	
03	Crank angle sensor (G signal)	Broken wire, short circuit	Fuel injection and ignition cut	
08	Airflow meter (AFM)	Broken wire, short circuit	Basic fuel injection amount and ignition timing fixed	
09	Water thermosensor	Broken wire, short circuit	Coolant temp. input fixed at 80°C (176°F)	
10	Intake air thermosensor (AFM)	Broken wire, short circuit	Intake air temp. input fixed at 20°C (68°F)	
11	Intake air thermosensor (Engine)	Broken wire, short circuit	Intake air temp. input fixed at 20°C (68°F)	
12	Throttle sensor (Full range)	Broken wire, short circuit	Throttle valve opening angle input signal fixed at 20% open	
13	Pressure sensor (Intake manifold pressure)	Broken wire, short circuit	Intake manifold pressure input signal fixed at 760 mmHg (29.9 inHg)	
14	Atmospheric pressure sensor (ATP)	Malfunctioning ECU	Atmospheric pressure input signal fixed at 760 mmHg (29.9 inHg)	
15	Oxygen sensor	Oxygen sensor output remains below 0.55V 80 sec. after F/B system operation beginning	Feedback system canceled (For EGI)	
17	Feedback system	Oxygen sensor output remains 0.55V 10 sec. after F/B system operation beginning	Feedback system canceled (For EGI)	
18	Throttle sensor (Narrow range)	Broken wire, short circuit	Throttle valve opening angle input signal fixed at full open	
20	Metering oil pump position sensor	Broken wire, short circuit	MOP fixed smallest open Basic fuel injection amount and ignition timing fixed	
27	Metering oil pump (MOP)	Malfunctioning MOP, step motors, broken wire, short circuit or malfunctioning ECU	MOP fixed smallest open Basic fuel injection amount and ignition timing fixed	
37	Metering oil pump (MOP)	Malfunction MOP, step motors, broken wire, short circuit, malfunctioning ECU, alternator or battery	Basic fuel injection amount and ignition timing fixed	

Output devices

Code No.	Output devices	Output signal pattern (Self-Diagnosis Checker or MIL)
25	Solenoid valve (Pressure regulator control (PRC))	
26	Step motor (Metering oil pump)	
30	Split air solenoid valve	
31	Solenoid valve (Relief)	
32	Solenoid valve (Switch)	
33	Port air solenoid valve	
34	Solenoid valve (Bypass air control (BAC))	
38	Solenoid valve (Accelerated warm-up system (AWS))	
40	Auxiliary port valve	
41	Solenoid valve (Variable dynamic effect intake control (VDI))	
51	Fuel pump resistor relay	
71	Injector (Front secondary)	
73	Injector (Rear secondary)	

Troubleshooting chart

If a malfunction code number is shown on the **SST**, check by using the following chart and the wiring diagram.

Input device
No.1 Code (Ignition coil (Trailing side))

- Are there poor connections at ignition coil connectors? — **YES** → Repair or replace connector
  - **NO** ↓
- Is (B/W) terminal voltage of coil (Trailing side) OK? **Voltage: Approx. 12V** — **NO** → Repair or replace wire harness from main relay
  - **YES** ↓
- Is coil (Trailing side) operation OK? — **NO** → Replace coil (Trailing side)
  - **YES** ↓
- Is there continuity between coil (Trailing side) and ECU **1V** and **1G** terminal? — **NO** → Repair or replace wire harness
  - **YES** → Replace ECU

No.2 Code (Crank angle sensor (Ne signal))

- Are there poor connections at crank angle sensor connectors? — **YES** → Repair or replace connector
  - **NO** ↓
- Is resistance of Ne pick-up coil OK? — **NO** → Replace crank angle sensor
  - **YES** ↓
- Is there continuity between crank angle sensor (Ne pick-up coil) and ECU **3E** terminal? — **NO** → Repair or replace wire harness
  - **YES** → Replace ECU

No.3 Code (Crank angle sensor (G signal))

- Are there poor connections, at crank angle sensor connectors? — **YES** → Repair or replace connector
  - **NO** ↓
- Is resistance of G pick-up coil OK? — **NO** → Replace crank angle sensor
  - **YES** ↓
- Is there continuity between crank angle sensor (G pick-up coil) and ECU **3G** and **3H** terminals — **NO** → Repair or replace wire harness
  - **YES** → Replace ECU

TROUBLE CODE TROUBLESHOOTING—RX-7 NON-TURBO

No.8 Code (Airflow meter)

Are there poor connections at airflow meter connectors?
- YES → Repair or replace connector
- NO ↓

Is resistance of airflow meter OK?
- NO → Replace airflow meter
- YES ↓

Is there continuity between airflow meter and ECU?

Airflow meter	ECU
Vs (GY)	2B
E_1 (BR/B)	3C and 3D
VREF (BR/W)	2I

- NO → Repair or replace wire harness
- YES → Replace ECU

No.10 Code (Intake air thermosensor (AFM))

Are there poor connections at thermosensor connectors?
- YES → Repair or replace connector
- NO ↓

Is resistance of intake air thermosensor OK?
- NO → Replace airflow meter
- YES ↓

Is there continuity between intake air thermosensor and ECU?

Intake air thermosensor	ECU
Signal side (G/O)	2K
Ground (BR/B)	3C

- NO → Repair or replace wire harness
- YES → Replace ECU

No.11 Code (Intake air thermosensor (Engine))

Are there poor connections at thermosensor connectors?
- YES → Repair or replace connector
- NO ↓

Is resistance of intake air thermosensor OK?
- NO → Replace intake air thermosensor
- YES ↓

Is there continuity between intake air thermosensor and ECU?

Intake air thermosensor	ECU
Signal side (G)	2L
Ground (BR/B)	3C

- NO → Repair or replace wire harness
- YES → Replace ECU

No.12 Code (Throttle sensor (Full range))

Are there poor connections at throttle sensor connectors?
- YES → Repair or replace connector
- NO ↓

Is voltage of ECU 2G terminal OK?

Throttle opening	Specification voltage
Idle position	0.25—1.25V
Full opening	4.1—4.4V

- NO → Adjust or replace throttle sensor
- YES ↓

Is there continuity between throttle sensor and ECU?

Throttle sensor	ECU
Vref (BR/W)	2I
Ground (BR/B)	3D
Signal (B/G)	2G

- NO → Repair or replace wire harness
- YES → Replace ECU

No.13 Code (Pressure sensor)

Are there poor connections at pressure sensor connectors?
- YES → Repair or replace connector
- NO ↓

Is there continuity between pressure sensor and ECU?

Pressure sensor	ECU
Vref (BR/W)	2I
Ground (BR/B)	3D
Signal (G/Y)	2H

- NO → Repair or replace wire harness
- YES ↓

Is voltage of pressure sensor OK?
When applying **100 mmHg (3.9 inHg)** vacuum

Pressure sensor	Voltage
Vref (BR/W)	Approx. 5V
Ground (BR/B)	Approx. 0V
Signal (G/Y)	2.8—3.2V

- NO → Vref or Ground NG: Replace ECU; Signal NG: Replace pressure sensor
- YES → Replace ECU

No.14 Code (Atmospheric pressure (ATP) sensor)

Because of the ATP sensor built in ECU, if No.14 Code flash, it's mean malfunction ECU
- YES → Replace ECU (ATP sensor in ECU)

TROUBLE CODE TROUBLESHOOTING—RX-7 NON-TURBO

No.15 Code (Oxygen sensor)

Note
If malfunction codes No.15 and 17 are both present, perform the checking procedure for malfunction Code No.17 first.

- Are there poor connections at connectors?
  - YES → Repair or replace connector
  - NO ↓
- Is oxygen sensor output voltage OK?
  - NO → Replace oxygen sensor
  - YES ↓
- Is oxygen sensor sensitivity OK?
  - NO → Replace oxygen sensor
  - YES ↓
- Is there continuity between oxygen sensor and ECU **2C** terminal?
  - NO → Repair or replace wire harness
  - YES → Replace ECU

No.17 Code (Feedback system)

- Are there poor connections at connectors?
  - YES → Repair or replace connector
  - NO ↓
- Is oxygen sensor sensitivity OK?
  - NO → Possible cause:
    - Air leak in vacuum hoses or intake air system
    - Contaminated oxygen sensor
    - Insufficient fuel injection
  - YES ↓
- Are spark plugs OK?
  - NO → Clean or replace spark plugs
  - YES ↓
- Is there continuity between oxygen sensor and ECU **2C** terminal?
  - NO → Repair or replace wire harness
  - YES → Replace ECU

No.18 Code (Throttle sensor (Narrow range))

- Are there poor connections at throttle sensor connectors?
  - YES → Repair or replace connector
  - NO ↓
- Is voltage of ECU 2F terminal OK?

Throttle opening	Specification voltage
Idle position	0.75—1.25V
Full opening	Approx. 5V

  - NO → Adjust or replace throttle sensor
  - YES ↓
- Is there continuity between throttle sensor and ECU?

Throttle sensor	ECU
Vref (BR/W)	2I
Ground (BR/B)	3D
Signal (G/R)	2F

  - NO → Repair or replace wire harness
  - YES → Replace ECU

No.20 Code (Metering oil pump position sensor)

Metering oil pump control system

No.27 Code (Metering oil pump)

Metering oil pump control system

No.37 Code (Metering oil pump)

Metering oil pump control system

TROUBLE CODE TROUBLESHOOTING – RX-7 NON-TURBO

Output device
No.25 Code (Solenoid valve (PRC))

- Are there poor connections at solenoid valve connectors?
  - YES → Repair or replace connector
  - NO ↓
- Is there continuity of solenoid valve?
  - NO → Replace solenoid valve
  - YES ↓
- Is there continuity between solenoid valve and ECU?

Solenoid valve	ECU
Battery (B/W)	1B
Signal side (L/O)	3M

  - NO → Repair or replace wire harness
  - YES → Replace ECU

No.26 Code (Stepping motor (Metering oil pump))

- Metering oil pump control system

No.30 Code (Split air solenoid valve)

- Are there poor connections at solenoid valve connectors?
  - YES → Repair or replace connector
  - NO ↓
- Is there continuity of solenoid valve?
  - NO → Replace solenoid valve
  - YES ↓
- Is there continuity between solenoid valve and ECU?

Solenoid valve	ECU
Battery (B/W)	1B
Signal side (L/B)	3I

  - NO → Repair or replace wire harness
  - YES → Replace ECU

No.31 Code (Solenoid valve (Relief))

- Are there poor connections at solenoid valve connectors?
  - YES → Repair or replace connector
  - NO ↓
- Is there continuity of solenoid valve?
  - NO → Replace solenoid valve
  - YES ↓
- Is there continuity between solenoid valve and ECU?

Solenoid valve	ECU
Battery (B/W)	1B
Solenoid valve (L/Y)	2P

  - NO → Repair or replace wire harness
  - YES → Replace ECU

No.32 Code (Solenoid valve (Switch))

- Are there poor connections at solenoid valve connectors?
  - YES → Repair or replace connector
  - NO ↓
- Is there continuity of solenoid valve?
  - NO → Replace solenoid valve
  - YES ↓
- Is there continuity between solenoid valve and ECU?

Solenoid valve	ECU
Battery (B/W)	1B
Signal side (L/R)	2Q

  - NO → Repair or replace wire harness
  - YES → Replace ECU

No.33 Code (Port air solenoid valve)

- Are there poor connections at solenoid valve connectors?
  - YES → Repair or replace connector
  - NO ↓
- Is there continuity of solenoid valve?
  - NO → Replace solenoid valve
  - YES ↓
- Is there continuity between solenoid valve and ECU?

Solenoid valve	ECU
Battery (B/W)	1B
Signal side (L)	2N

  - NO → Repair or replace wire harness
  - YES → Replace ECU

No.34 Code (Solenoid valve (BAC))

- Are there poor connections at solenoid valve connectors?
  - YES → Repair or replace connector
  - NO ↓
- Is there continuity of solenoid valve?
  - NO → Replace solenoid valve
  - YES ↓
- Is there continuity between solenoid valve and ECU?

Solenoid valve	ECU
Battery (B/W)	1B
Signal side (L/G)	3Q

  - NO → Repair or replace wire harness
  - YES → Replace ECU

No.38 Code (Solenoid valve (AWS))

- Are there poor connections at solenoid valve connectors?
  - YES → Repair or replace connector
  - NO ↓ Next page

TROUBLE CODE TROUBLESHOOTING—RX-7 NON-TURBO

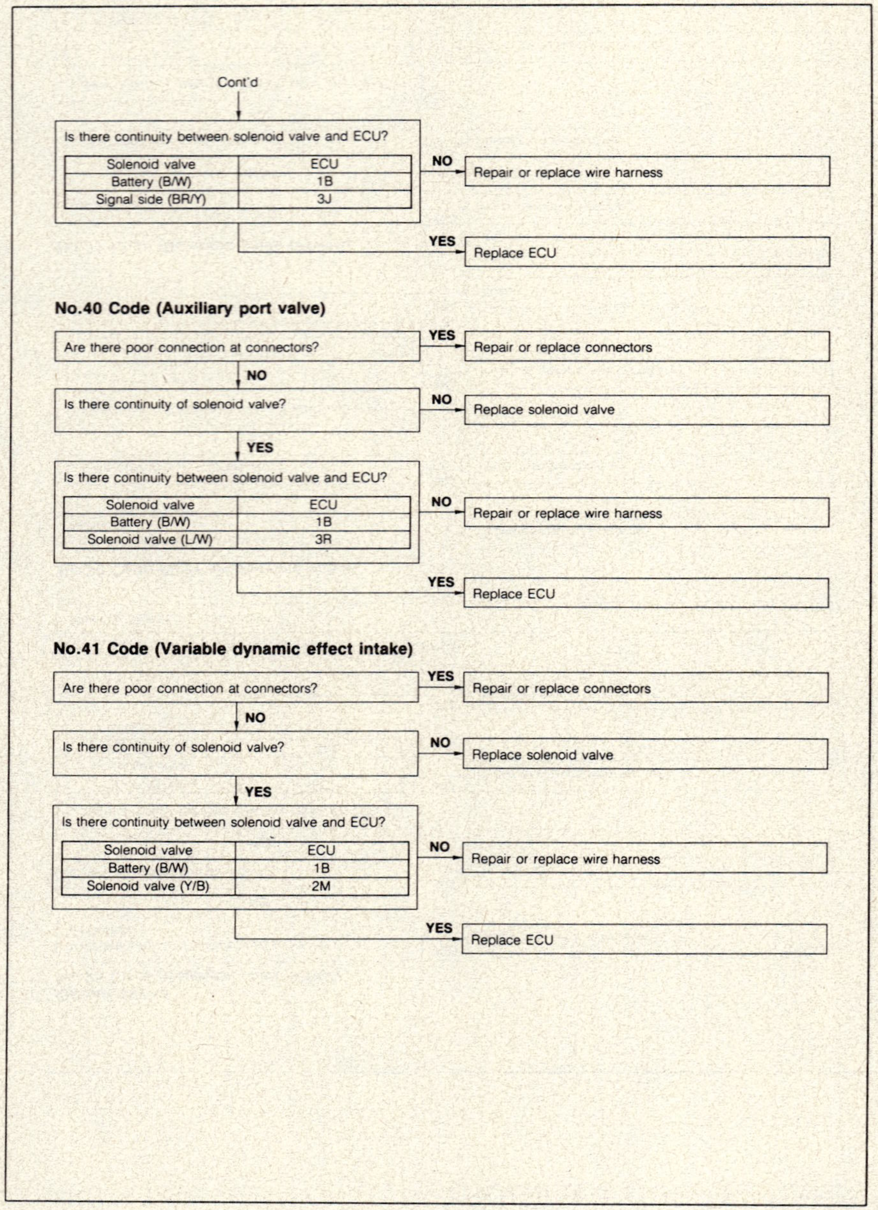

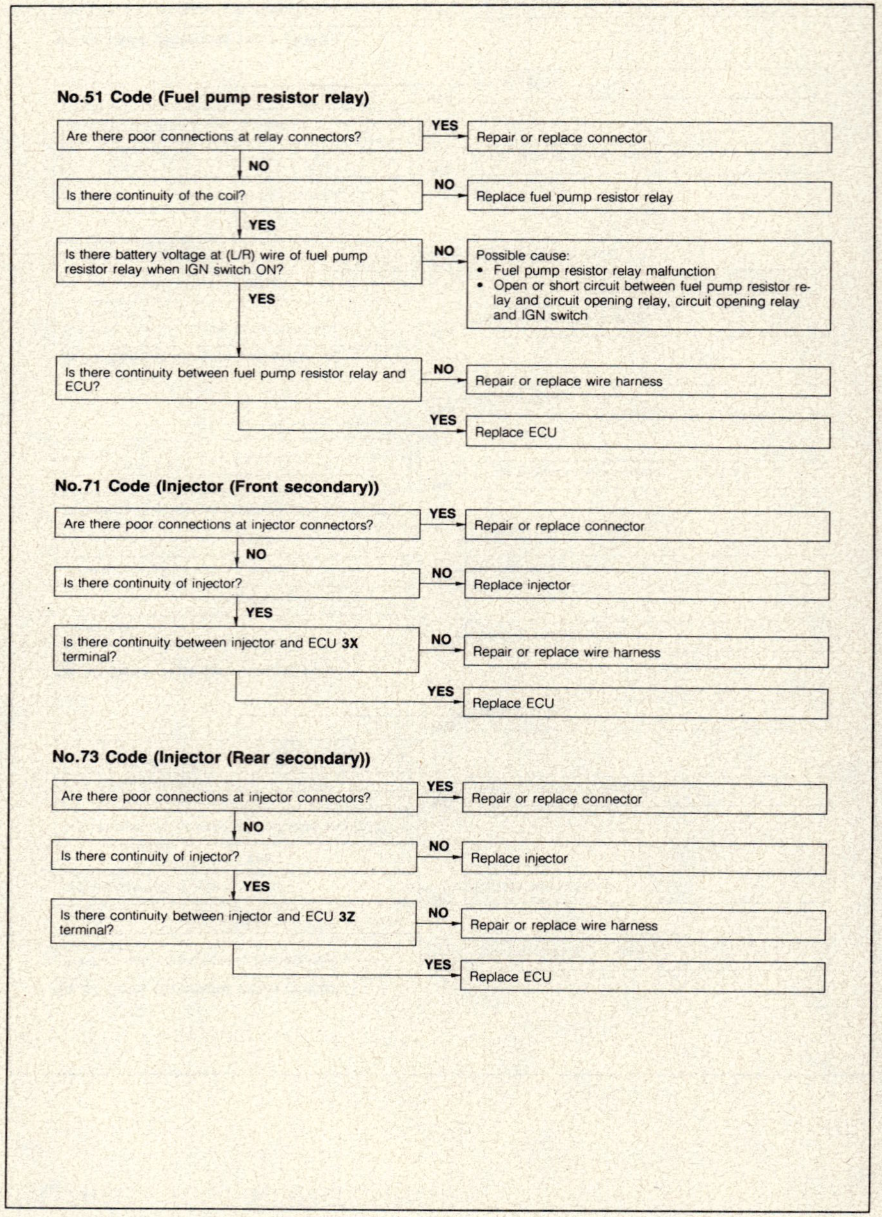

SWITCH MONITOR FUNCTION—RX-7 NON-TURBO

SWITCH MONITOR FUNCTION
Preparation
SST

49 H018 9A1 Self-Diagnosis Checker		For diagnosis

Individual switches can be monitored by the **SST**.

Note
The test connector must be grounded and the ignition switch ON (engine stopped).

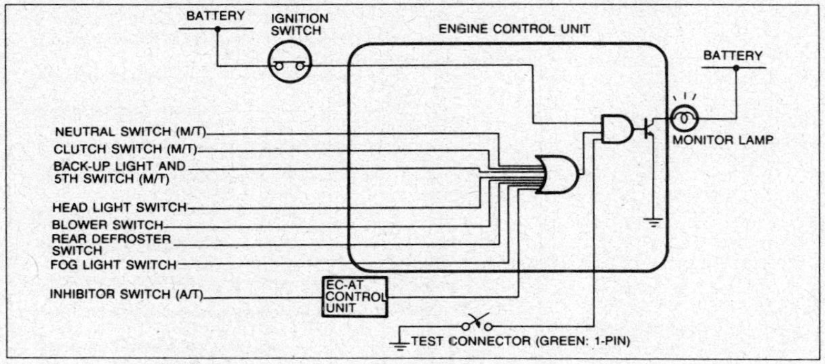

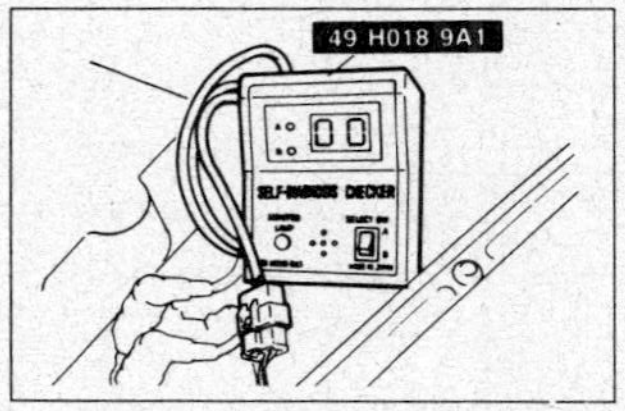

Inspection Procedure
1. Warm up the engine to normal operating temperature and stop it.
2. Connect the **SST** to the check connector (Green: 6-pin) and the negative battery terminal.

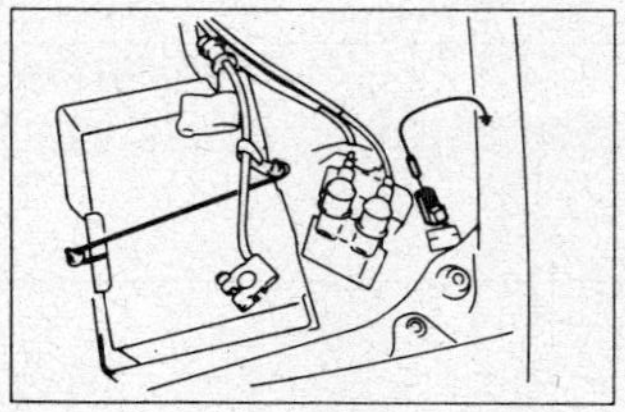

3. Connect a jumper wire between the test connector (Green: 1-pin) and a ground.
4. Turn the ignition switch ON, then check that the monitor lamp illuminates when each switch is made to function according to the table below.

Caution
a) When any one of the switches are activated, the monitor lamp will be on.
b) Do not start the engine.

SWITCH MONITOR FUNCTION TROUBLESHOOTING RX-7 NON-TURBO

Function Table

Switch	Self-Diagnosis Checker (Monitor lamp) Lamp-ON	Lamp-OFF	Possible cause (When incorrect)
Neutral switch (M/T)	Shift transmission to 1, 2, 3 or 4	Neutral position	• Neutral switch malfunction • Open circuit between neutral switch and ECU 1R terminal, neutral switch and ground
Back-up light and 5th switch (M/T)	Shift transmission 5th gear	Neutral position	• Back-up light and 5th switch malfunction • Open circuit between Back-up light and 5th switch and ECU 1T terminal, overdrive and ground
Clutch switch (M/T)	Pedal depressed	Pedal released	• Clutch switch malfunction • Open circuit between clutch switch and ECU 1Q terminal, clutch switch and ground, clutch switch and ACC circuit
Inhibitor switch (A/T)	Except N and P range	N and P range	• Inhibitor switch malfunction • EC-AT control unit malfunction • Open circuit between EC-AT control unit and ECU.
Headlight switch	Switch ON	Switch OFF	• Headlight switch malfunction • Open circuit between headlight switch and ECU 3L terminal, headlight switch and battery line
Blower switch	Switch ON (At any fan speed OK)	Switch OFF	• Blower switch malfunction • Heater relay malfunction • Heater control unit malfunction • Open circuit between heater control unit and ECU 3O terminal
Rear defroster switch	Switch ON	Switch OFF	• Rear defroster switch malfunction • Open circuit between rear defroster switch and ECU 3P terminal, rear defroster switch and ground
Fog light switch (If equipped)	Switch ON (Only equip fog light)	Switch OFF	• Fog light switch malfunction • Open circuit between fog light switch and ECU 1S terminal, fog light switch and battery

ISC SYSTEM TROUBLESHOOTING IN ORDER OF PRIORITY—RX-7 NON-TURBO

IDLE SPEED CONTROL (ISC) SYSTEM

DESCRIPTION

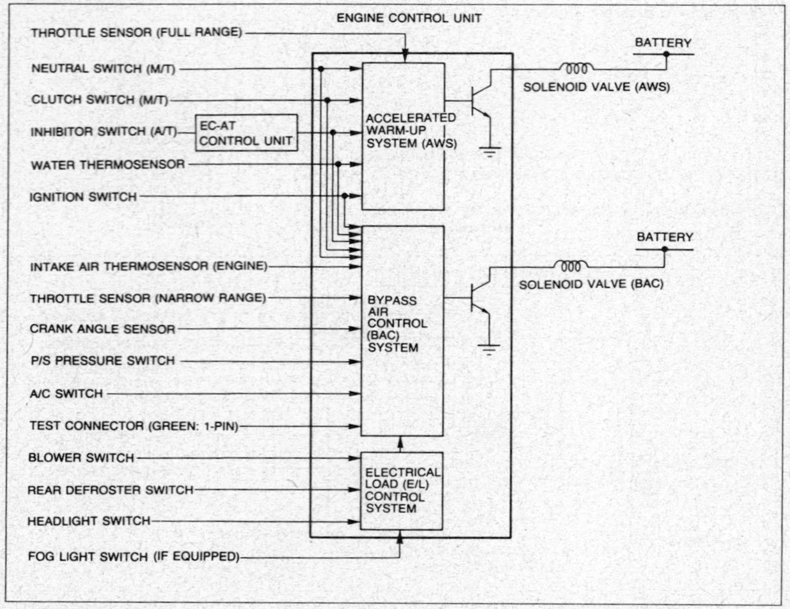

TROUBLESHOOTING

Check the condition of the wiring harness and connectors before checking the sensors or switches.

Symptom \ Possible cause	System inspection	Solenoid valve (BAC)	Solenoid valve (AWS)	Engine control unit terminal 1B	3J	3Q
Hard start or no start (Cranks OK)	1	3	4	2	5	2
Rough idle	1	2	4	3	5	3
High idle speed (At normal operating temperature)	1	2	3	5	4	5
Engine does not run smoothly (Only when engine is cold)	1	2	4	3	5	3
Afterburn	1	2	—	3	1	3
Runs rough on deceleration	1	2	—	3	1	3
Fails emission test	1	4	2	3	3	3

FUEL SYSTEM TROUBLESHOOTING IN ORDER OF PRIORITY—RX-7 NON-TURBO

TROUBLESHOOTING

Check the condition of the wiring harness and connectors before checking the sensors or switches.

Symptom \ Possible cause	Fuel pump	Fuel filter	Injector	Hold pressure (Inspection)	Fuel line pressure (Inspection)	Pressure regulator	Pulsation damper	Fuel pump control system	Engine control unit terminal 1K	3W	3X	3Y	3Z
Hard start or no start (Cranks OK)	5	6	7	1	2	8		3	4	4	4	4	4
Rough idle	4		1		2		3						
Engine does not run smoothly (At normal operating temp.)	6	4	2		1		5			3	3	3	3
Engine does not run smoothly (When engine is cold)					1					2	2	2	2
Lack of power, poor acceleration or hesitation	2	3	4		1	6				5	5	5	5
Afterburn			1							2	2	2	2
Runs rough on deceleration			1							2	2	2	2
Fails emission test			3		1	2				4	4	4	4

Warning
Before release the fuel pressure from the fuel system to reduce the possibility of injury or fire.

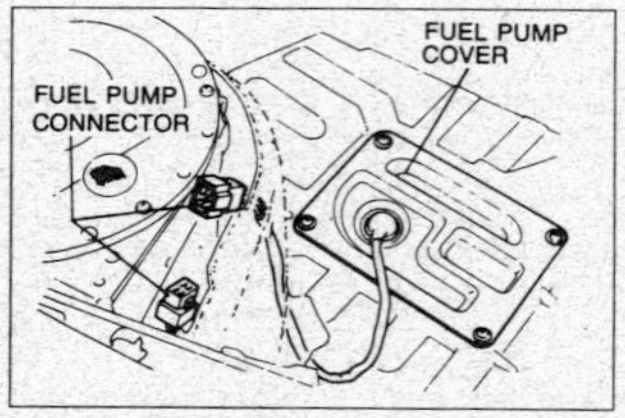

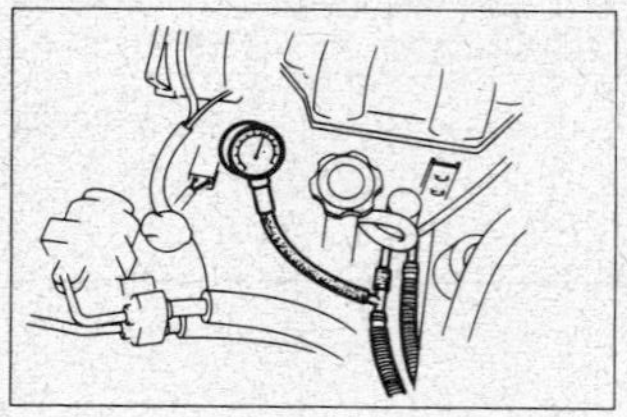

PRECAUTION
Fuel Pressure Release and Fuel System
Fuel in the fuel system remains under high pressure even when the engine is not running.

a) Before disconnecting any fuel line, release the fuel pressure from the fuel system to reduce the possibility of injury or fire.
   1. Start the engine.
   2. Disconnect the fuel pump connector.
   3. Allow the engine to stalls. Turn off the ignition switch.
   4. Reconnect the fuel pump connector.

b) Use a rag as protection from fuel spray when disconnecting the hoses.
c) Plug all disconnected hoses.
d) Inspect the fuel system with a suitable pressure gauge.

Caution
Install hose clamps to secure the fuel pressure gauge to the fuel filter and the fuel main hose to prevent fuel leakage.

SECONDARY AIR SYSTEM TROUBLESHOOTING IN ORDER OF PRIORITY—RX-7 NON-TURBO

DESCRIPTION
This system helps to clean to exhaust gas by introducing fresh air into the exhaust port and main converter.

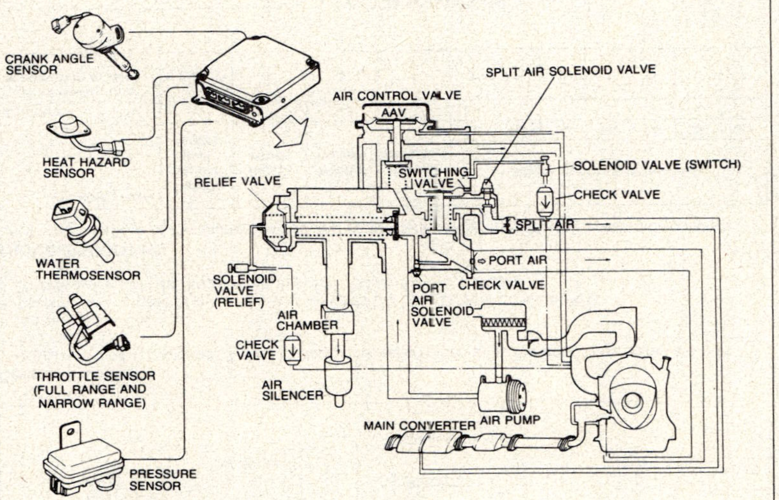

TROUBLESHOOTING
Check the condition of the wiring harness vacuum hoses, and check connectors before checking the sensors or switches.

Symptom \ Possible cause	System inspection	Air pump	Check valves (in intake manifold and in intake manifold to catalytic converter)	Air control valve	Solenoid valve (Switch)	Solenoid valve (Relief)	Heat hazard sensor	Split air solenoid valve	Port air solenoid valve	Engine control unit terminal 3I	2O	2P	2N
Rough idle	1	3	8	4	5	6	7	—	—	—	2	2	—
Afterburn	1	3	8	4	5	6	7	—	—	—	2	2	—
Fails emission test	1	3	10	4	7	8	9	5	6	2	2	2	2

DECELERATION CONTROL SYSTEM TROUBLESHOOTING IN ORDER OF PRIORITY—RX-7 NON-TURBO

TROUBLESHOOTING
Check the condition of the wiring harness, vacuum hoses and connectors before checking the sensors or switches.

Symptom \ Possible cause	Dashpot	Anti-afterburn valve
Afterburn	2	1
Runs rough on deceleration	1	2
Fails emission test	1	2

A/C CUTOUT CONTROL SYSTEM TROUBLESHOOTING IN ORDER OF PRIORITY—RX-7 NON-TURBO

DESCRIPTION

An A/C cut-out system is employed to provide smoother acceleration while the air conditioning is in use.

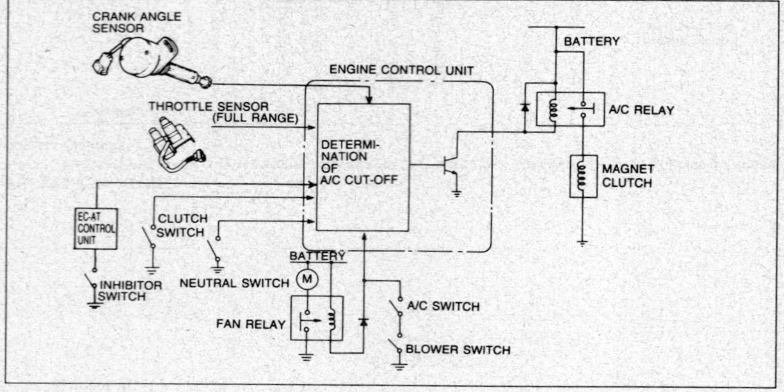

Operate Condition

The A/C is cut-off for **approx. 7 sec.** during acceleration when all of the conditions below are met.

1. A/C switch ON.
2. Blower switch ON.
3. Transmission in gear and clutch pedal released (M/T), in D, S or L range (A/T).
4. Throttle valve opening angle more than 50°. (More than half throttle)

TROUBLESHOOTING

Check the condition of the wiring harness and connectors before checking the sensors or switches.

Possible cause / Symptom	Throttle sensor (Full range)	Clutch switch	Neutral switch	System inspection	Engine control unit terminals						
					1L	1O	1Q	1R	2G	2I	3D
Lack of power, poor acceleration or hesitation	3	4	5	1	2	2	2	2	2	2	2

SYSTEM INSPECTION

1. Turn the A/C and blower switches ON.
2. Turn the ignition switch ON (engine not running).
3. Shift the transmission into gear (M/T), in D, S or L range (A/T).
4. Open the throttle valve fully and verify that the magnetic clutch disengages (Click is heard), then reengages after **approx. 7 sec.**.
5. If not as specified, check the throttle sensor and engine control unit 1L, 1O, 1Q, 1R, 2G, 2I and 3D terminal voltages.

RELATIONSHIP CHART—RX-7 NON-TURBO

RELATIONSHIP CHART
Output Devices and Input Devices

INPUT DEVICES / OUTPUT DEVICES	FUEL INJECTION AMOUNT	FUEL INJECTION TIMING	FUEL PUMP RESISTOR RELAY	A/C RELAY	PORT AIR SOLENOID VALVE	SPLIT AIR SOLENOID VALVE	SOLENOID VALVE (SWITCH)	SOLENOID VALVE (RELIEF)	SOLENOID VALVE (AWS)	SOLENOID VALVE (PRESSURE REGULATOR)	SOLENOID VALVE (BYPASS AIR CONTROL)	SOLENOID VALVE (VARIABLE DYNAMIC EFFECT INTAKE)	SOLENOID VALVE (6 PORT INJECTION)	IGNITER (TRAILING IGNITION TIMING)	IGNITER (LEADING IGNITION TIMING)
TEST CONNECTOR	○	×	×	×	×	×	×	×	×	×	○	×	×	○	○
IGNITION SWITCH	○	×	○	×	×	×	○	○	○	○	○	×	×	○	○
POWER STEERING PRESSURE SWITCH	×	×	×	×	×	×	×	×	×	×	○	×	×	○	○
A/C SWITCH	○	×	×	○	×	×	×	×	×	×	○	×	×	○	○
FOG LIGHT SWITCH (IF EQUIPPED)	○	×	×	×	×	×	×	×	×	×	○	×	×	○	○
HEAD LIGHT SWITCH	○	×	×	×	×	×	×	×	×	×	○	×	×	○	○
BLOWER SWITCH	○	×	×	×	×	×	×	×	×	×	○	×	×	○	○
REAR DEFROSTER SWITCH	○	×	×	×	×	×	×	×	×	×	○	×	×	○	○
CLUTCH SWITCH (M/T)	○	×	○	○	×	○	○	○	○	○	○	×	○	○	○
NEUTRAL SWITCH (M/T)	○	×	○	○	×	○	○	○	○	○	○	×	○	○	○
EC-AT CONTROL UNIT (A/T) (INHIBITOR SWITCH)	○	×	×	○	×	○	○	○	○	○	○	×	○	○	○
BACK-UP LIGHT AND 5TH SWITCH (M/T)	×	×	×	×	×	○	○	○	×	×	○	×	×	○	○
OIL PRESSURE SWITCH (A/T)	○	×	×	×	×	○	○	○	×	×	○	×	×	○	○
MILEAGE SENSOR	○	×	×	×	○	×	○	×	×	×	×	×	×	○	○
HEAT HAZARD SENSOR	×	×	×	×	×	×	×	○	×	×	×	×	×	×	×
OXYGEN SENSOR	○	×	×	×	×	×	×	×	×	×	×	×	×	×	×
THROTTLE SENSOR (NARROW RANGE)	○	×	○	×	○	×	○	○	×	○	○	×	×	○	○
THROTTLE SENSOR (FULL RANGE)	○	×	×	○	○	×	×	○	○	×	×	×	×	○	○
PRESSURE SENSOR	○	×	×	×	○	○	○	○	×	×	×	×	×	○	○
CRANK ANGLE SENSOR (NE)	○	○	○	○	×	×	○	○	○	×	○	○	○	○	○
CRANK ANGLE SENSOR (G+, G–)	○	○	○	×	×	×	×	×	×	○	×	×	×	○	○
WATER THERMOSENSOR	○	×	○	×	×	×	○	○	○	○	○	×	○	○	○
INTAKE AIR THERMOSENSOR (AFM)	○	×	○	×	×	×	×	×	×	×	×	×	×	×	×
INTAKE AIR THERMOSENSOR (ENGINE)	○	×	○	×	×	×	×	×	×	○	○	×	×	○	○
AIRFLOW METER (AFM)	○	×	○	×	×	×	×	×	×	×	×	○	○	○	○

ENGINE CONTROLS TROUBLESHOOTING RX-7 NON-TURBO

TROUBLESHOOTING

Check the condition of the wiring harness and connectors before checking the sensors or switches.

Symptom \ Possible cause	Airflow meter	Intake air thermosensor (Engine)	Throttle sensor (Full and Narrow range)	Oxygen sensor	Pressure sensor	Heat hazard sensor	Main relay	Circuit opening relay	Neutral switch (M/T)	Clutch switch (M/T)	Inhibitor switch (A/T)	Power steering pressure switch	Back-up light and 5th switch (M/T)	Oil pressure switch (A/T)	Water thermosensor	Mileage sensor	Engine control unit terminal
Hard start or no start (Cranks OK)	7	6	4		5		1	2							3		8
Rough idle	1	4	2						6	7	5				3	8	9
High idle speed (At normal operating temperature)	1		2									4			3		5
Engine does not run smoothly (Only when engine is cold)	1	3				4									2		5
Engine does not run smoothly (Only when engine is hot)	1	3				4									2		5
Lack of power, poor acceleration or hesitation	1		2			9			5	6	4	7	10	10	8	3	11
Afterburn	1		2		5					3		4					6
Fails emission test	2		3	1		4									5		6
Knocking	2	4	3						5	6	5				1		7

ENGINE CONTROLS OPERATION CHART RX-7 NON-TURBO

OUTPUT DEVICES \ ENGINE CONDITION	CRANKING (WHEN ENGINE IS COLD)	IDLE (DURING WARM-UP)	MEDIUM LOAD COLD	MEDIUM LOAD WARM	ACCELERATION	HEAVY LOAD	DECELERATION (THROTTLE VALVE FULLY CLOSED)	IDLE (WHEN ENGINE IS WARM)	IGNITION: ON (NOT ENGINE RUN)	REMARKS
FUEL INJECTION AMOUNT — PRIMARY INJECTOR	↔	Rich	↔	Rich and Lean	Rich	↔	Fuel cut	Rich	OFF	Rear side; Above 8,500 rpm; Fuel cut
FUEL INJECTION AMOUNT — SECONDARY INJECTOR	↔	OFF	↔	↔	Rich	↔	↔	OFF	↔	
FUEL PUMP RESISTOR RELAY	12V drive	↔	9—10V drive	↔	12V drive	↔	↔	9—10V drive	OFF	Hot start (at idle) 12V drive
A/C RELAY	↔	ON	↔	↔	*1 OFF	↔	↔	ON	↔	*1 Only 7 sec.
PORT AIR SOLENOID VALVE	↔	Above 20,000 mile (34,000 km): ON Below 20,000 mile (34,000 km): OFF	↔	↔	↔	↔	↔	ON	↔	
SPLIT AIR SOLENOID VALVE	↔	O/D switch ON or back-up light and 5th switch ON and above 20,000 mile (34,000 km): ON	↔	↔	↔	↔	↔	↔	OFF	Always ON at high altitude
SOLENOID VALVE (SWITCH)	↔	OFF (Port air)	↔	↔	ON (Split air)	↔	↔	OFF (Port air)	OFF	
SOLENOID VALVE (RELIEF)	ON (Air injection)	↔	OFF (Relief)	ON (Air injection)	OFF (Relief)	↔	↔	ON (Air injection)	ON	
SOLENOID VALVE (AWS)	ON	For 17 sec. after start: ON	↔	↔	OFF	↔	↔	↔	↔	
SOLENOID VALVE (PRESSURE REGULATOR)	↔	↔	↔	OFF	↔	↔	↔	After start; *2 ON	ON	*2 For 50 sec.—70 sec.
SOLENOID VALVE (BYPASS AIR CONTROL)	Large quantity of air	Feed back duty	↔	↔	Fixed duty	↔	↔	Feed back duty	Fixed duty	Test connector (Green: 1-pin) ground; Stop the feedback control
SOLENOID VALVE (VARIABLE DYNAMIC EFFECT INTAKE)	↔	↔	OFF	↔	↔	ON	↔	OFF	↔	
SOLENOID VALVE (6 PORT INJECTION)	↔	OFF	↔	↔	*3 ON (Above 3,850 rpm)	↔	↔	OFF	↔	*3 Only load condition
IGNITER (TRAILING IGNITION TIMING)	BTDC 5°	Retard	↔	↔	Advanced	↔	Retard	ATDC 20°	OFF	
IGNITER (LEADING IGNITION TIMING)	BTDC 5°	Retard	↔	↔	Advanced	↔	Retard	ATDC 5°	OFF	

ENGINE CONTROL UNIT TERMINAL VOLTAGES—RX-7 NON-TURBO

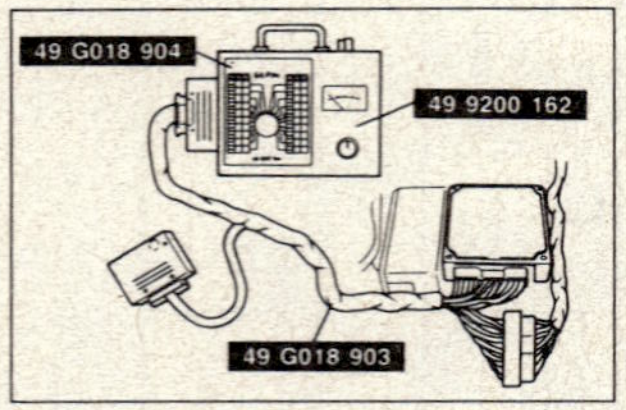

ENGINE CONTROL UNIT

Inspection

1. Lift up the floor mat in front of the passenger's seat.
2. Remove the protector cover.
3. Connect the **SST** (Engine Signal Monitor) between the ECU and wire harness using the **SST** (Adapter harness).

Note

Use connector A of the adapter harness to check voltages at terminals 1A through 1V and 3A through 3Z, and use connector B to check voltages at terminals 2A through 2P.

4. Place the **SST** (Sheet: 64-pin side) on the **SST** (Engine Signal Monitor).
5. Measure the voltage at each terminal.

Engine control unit terminal (unit side)

3Y	3W	3U	3S	3Q	3O	3M	3K	3I	3G	3E	3C	3A	2O	2M	2K	2I	2G	2E	2C	2A	1U	1S	1Q	1O	1M	1K	1I	1G	1E	1C	1A
3Z	3X	3V	3T	3R	3P	3N	3L	3J	3H	3F	3D	3B	2P	2N	2L	2J	2H	2F	2D	2B	1V	1T	1R	1P	1N	1L	1J	1H	1F	1D	1B

Terminal	Input	Output	Connection to	Test condition	Voltage	Remark
1A	○		Battery	Constant	Approx. 12V	Backup
1B	○		Main relay	Ignition switch ON	Approx. 12V	—
				Ignition switch OFF	Approx. 0V	
1C	○		Ignition switch	Ignition switch START (Cranking)	Approx. 12V	—
				Ignition switch ON	Approx. 0V	
				Ignition switch OFF	Approx. 0V	
1D		○	Self-Diagnosis Checker (Monitor lamp)	Test connector grounded For 3 sec. after ignition switch OFF→ON (Lamp illuminates)	Below 6.2V	With Self-Diagnosis checker
				After 3 sec. (Light does not illuminate)	Approx. 12V	
				Test connector grounded at idle	Approx. 12V	
				Test connector grounded (Monitor lamp ON)	Below 6.2V	
				Test connector grounded (Monitor lamp OFF)	Approx. 12V	
1E		○	Malfunction indicator light (MIL) lamp	For 3 sec. after ignition switch OFF→ON (Lamp illuminates)	Below 4.8V	Test connector grounded
				After 3 sec. (Lamp does not illuminate)	Approx. 12V	
				Lamp illuminates	Below 4.8V	
				Lamp does not illuminates	Approx. 12V	
1F		○	Self-Diagnosis Checker (Malfunction code number)	For 3 sec. after ignition switch OFF→ON (Buzzer sounds)	Below 6.2V	With Self-Diagnosis Checker and test connector grounded
				After 3 sec. (Buzzer does not sound)	Approx. 12V	
				Buzzer sounds	Below 6.2V	
				Buzzer does not sound	Approx. 12V	
1G		○	Ignition coil (Trailing)	Ignition switch ON	Approx. 0V	IGt-T (Ignition timing signal)
				Idle	Approx. 0.8V	
1H		○	Ignition coil (Leading)	Ignition switch ON	Approx. 0V	IGt-L (Ignition timing signal)
				Idle	Approx. 0.8V	
1I	○		Test connector (Green: 1-pin)	Test connector grounded	Approx. 0V	Ignition switch ON
				Test connector not grounded	Approx. 12V	
1J		○	Ignition coil (Trailing)	Ignition switch ON	Approx. 4.4V	IGs-T (Select signal)
				Idle	Approx. 2.2V	
1K		○	Fuel pump resistor relay	Cranking	Approx. 12V	—
				Idle (More than 90 sec. after cranking)	Below 2.0V	
1L		○	A/C relay	A/C switch ON	Below 2.5V	Ignition switch ON
				A/C switch OFF	Approx. 12V	Blower switch ON

Terminal	Input	Output	Connection to	Test condition	Voltage	Remark
1M	○		Mileage sensor	Under 20,000 miles (34,000 km)	Approx. 12V	There is an error more or less
				Over 20,000 miles (34,000 km)	Below 1.5V	
1N	○		Power steering (P/S) pressure switch	Ignition switch ON	Approx. 12V	P/S ON: Turning P/S OFF: Straight ahead
				P/S ON (Idle)	Approx. 0V	
				P/S OFF (Idle)	Approx. 12V	
1O	○		A/C switch	A/C switch ON (Idle)	Below 2.5V	Blower switch ON
				A/C switch OFF (idle)	Approx. 12V	
1P	○		Heat hazard sensor	Ignition switch ON	Below 1.5V	—
				Idle (Floor temp.: Below 110°C (230°F))	Approx. 12V	
				Idle (Floor temp.: Above 110°C (230°F))	Below 1.5V	
1Q	○		Clutch switch (M/T)	Clutch pedal: released	Approx. 12V	Ignition switch ON
				Clutch pedal: depressed	Below 2.0V	
1R	○		Neutral switch (M/T)	Neutral	Below 2.0V	Ignition switch ON
				In gear	Approx. 12V	
	○		EC-AT control unit (Inhibitor signal) (A/T)	N, P range	Below 2.0V	Ignition switch ON
				Others	Approx. 12V	
1S	○		Fog light switch	Fog light ON (Idle)	Approx. 12V	If equipped
				Fog light OFF (Idle)	Approx. 0V	
1T	○		Back-up light and 5th switch (M/T)	5th gear	Approx. 12V	Ignition switch ON
				1th — 4th gear	Below 2.0V	
	○		Oil pressure switch (A/T)	Overdrive	Below 2.0V	Can not check (Load condition only)
				Others	Approx. 12V	
1U	○		AT switch	A/T	Approx. 0V	Ignition switch ON
				M/T	Approx. 12V	
1V	○		Ignition coil (Trailing)	Ignition switch ON	Below 2.0V	IGt-T (Ignition confirmation signal)
				Idle	Approx. 1.4V	
2A	○		Metering oil pump (MOP) position sensor	Ignition switch OFF	0V	Refer to Section D
				Idle	Approx. 1.0V	
2B	○		Airflow meter (Vs)	Ignition switch ON	Approx. 4.0V	—
				Idle	2.5V—3.5V	
2C	○		Oxygen sensor	Idle	Below 1.0V	—
				Acceleration	0.5V—1.0V	
				Deceleration	0V—0.4V	
2D	—	—	—	—	—	—
2E	○		Water thermosensor	Idle (Engine hot)	0.3V—1.0V	—
				Water temperature; 20°C (68°F)	Approx. 2.4V	
2F	○		Throttle sensor (Narrow range)	Ignition switch ON (Idle position)	Approx. 1.0V	After warm-up
				Ignition switch ON (Full throttle)	Approx. 5.0V	
				Idle	Approx. 1.0V	
2G	○		Throttle sensor (Full range)	Ignition switch ON (Idle position)	Approx. 0.8V	After warm-up
				Ignition switch ON (Full throttle)	Approx. 4.3V	
				Idle	Approx. 0.8V	
2H	○		Pressure sensor	Vacuum hose disconnected and plugged	3.4V—3.6V	Ignition switch ON
				100 mmHg (3.9 inHg) vacuum applied to pressure sensor	2.8V—3.2V	
2I		○	Sensors	Ignition switch ON	4.5V—5.5V	Vref (Power supply)
				Ignition switch OFF	0V	
2J	○		Ground or open	Canada (Ground)	0V	—
				Except for Canada (Open)	Approx. 12V	
2K	○		Intake air thermosensor (Airflow meter)	Idle (At 20°C (68°F))	2V—3V	—
2L	○		Intake air thermosensor (Engine)	Idle (At 80°C (176°F))	1V—2V	—
2M		○	Solenoid valve (Variable dynamic effect intake control)	Above 5,200 rpm	Below 2.0V	Only while driving
				Below 5,200 rpm	Approx. 12V	—

ENGINE CONTROL UNIT TERMINAL VOLTAGES RX-7 NON-TURBO

Terminal	Input	Output	Connection to	Test condition	Voltage	Remark
2N		○	Port air solenoid valve	Except idle (Below 20,000 miles (34,000 km))	Approx. 12V	—
				Idle or above 20,000 miles (34,000 km)	Below 2.0V	
2O		○	Solenoid valve (Switch)	Half throttle	Below 2.0V	Ignition switch ON
				Idle	Approx. 12V	
2P		○	Solenoid valve (Relief)	Idle	Below 2.0V	—
				Engine speed; above 3,300 rpm	Approx. 12V	
3A	—	—	Ground	Constant	0V	Power
3B	—	—	Ground	Constant	0V	Power
3C	—	—	Ground	Constant	0V	System
3D	—	—	Ground	Constant	0V	Analog
3E	○		Crank angle sensor (Ne)	Ignition switch ON	Below 1.0V	Red
				Idle	Below 1.0V	
3F	—	—	—	—	—	—
3G	○		Crank angle sensor (G+)	Ignition switch ON	Below 1.0V	Black
				Idle	Below 1.0V	
3H	○		Crank angle sensor (G−)	Ignition switch ON	Below 1.0V	White
				Idle	Below 1.0V	
3I		○	Split air solenoid valve	5th gear (M/T), Over drive (A/T)	Below 2.5V	Refer to page F1–65
				Others	Approx. 12V	
3J		○	Solenoid valve (Accelerated warm-up system)	Ignition switch OFF	0V	Engine coolant temperature: 15°C (59°F)–35°C (95°F)
				Ignition switch ON	Approx. 12V	
				Idle (Less than 17 sec. after cranking)	Below 2.0V	
				Idle (More than 17 sec. after cranking)	Approx. 12V	
3K		○	Circuit opening relay	Ignition switch OFF	0V	—
				Ignition switch ON	Approx. 12V	
				Idle	Below 2.0V	
3L	○		Headlight switch	Headlight switch ON	Approx. 12V	—
				Headlight switch OFF	0V	
3M		○	Solenoid valve (Pressure regulator control)	Ignition switch ON	Below 2.0V	—
				Cranking	Below 2.0V	Hot condition only
				Idle (Less than 20 sec. after cranking)	Below 2.0V	
				Idle (More than 90 sec. after cranking)	Approx. 12V	
3N		○	EC-AT control unit	Idle (After warm-up)	Below 2.0V	
				Others (After warm-up)	Approx. 12V	
3O	○		Blower switch	Blower switch ON	Below 2.0V	Ignition switch ON
				Blower switch OFF	Approx. 12V	
3P	○		Rear defroster switch	Rear defroster switch ON	Below 2.0V	Ignition switch ON
				Rear defroster switch OFF	Approx. 12V	
3Q		○	Solenoid valve (Bypass air control)	Ignition switch OFF	0V	Duty pulse
				Ignition switch ON	Approx. 9V	
				Idle	Approx. 9V	
3R		○	6-port induction (6PI) system	Above 3,850 rpm	Below 2.0V	Cannot check (Warm-up and load condition only)
				Below 3,850 rpm	Approx. 12V	
3S 3T 3U 3V		○	Stepping motor (Metering oil pump)	—	—	Can not check with circuit tester (Refer to Section D)
3W		○	Injector (Front primary)	Ignition switch ON	Approx. 12V	Ground time is very short
				Idle	Approx. 12V	
3X		○	Injector (Front secondary)	Ignition switch ON	Approx. 12V	Ground time is very short
				Idle	Approx. 12V	
3Y		○	Injector (Rear primary)	Ignition switch ON	Approx. 12V	Ground time is very short
				Idle	Approx. 12V	
3Z		○	Injector (Rear secondary)	Ignition switch ON	Approx. 12V	Ground time is very short
				Idle	Approx. 12V	

SPECIFICATION CHART—RX-7 TURBO

SPECIFICATIONS

Item			Model: 13B Turbocharged engine
Idle speed (Test connector grounded) rpm			750 ± 25
Air cleaner	Element type		Long life wet
Throttle body	Type		Horizontal draft (2 stage-3 barrel)
	Throat diameter	Primary mm (in)	45 (1.772)
		Secondary mm (in)	45 (1.772) x 2
	Water thermovalve	Operation temp. °C (°F)	55–65 (131–149) or more
Dashpot	Adjustment (Throttle sensor (narrow range) resistance Ⓐ–Ⓑ) kΩ		1.8–3.8
Turbocharger	Type		Water cooled
	Boost pressure kPa (kg/cm², psi)		57.0 (0.58, 8.25)
	Lubrication		Engine oil
Waste gate valve			Incorporated with turbocharger
Fuel tank	Capacity liters (US gal, Imp gal)		70 (18.5, 15.4)
Fuel filter	Type	Low pressure	Nylon 6 (164 & 45 mesh)
		High pressure	Filter paper
Pressure regulator	Type		Diaphragm
	Regulated pressure kPa (kg/cm², psi)		235–275 (2.4–2.8, 34.1–39.8)
Fuel pump	Type		Impeller (in-tank)
	Output pressure kPa (kg/cm², psi)		490–637 (5.0–6.5, 71.1–92.4)
Injector (Primary and Secondary)	Drive		Voltage drive
	Injection volume cc (cu in)/15 sec		133–142 (8.1–8.7)
Heat hazard sensor	Operation temperature °C (°F)		105–115 (221–239)
Main silencer	Capacity cc (cu in)		12,000 (732) x 2
Ignition timing (Test connector grounded)			Leading: 5° ± 1° ATDC Trailing: 20° ± 2° ATDC
Distribution	Type		Control unit
Spark advance	Type		Control unit
Idle up system	A/C rpm		875
Anti-afterburn valve	Operating time sec		1.60–2.20
Intercooler	Type		Air cooled

COMPONENT DESCRIPTIONS – RX-7 TURBO

COMPONENT DESCRIPTIONS

Component	Function	Remarks
Accelerated Warm-up System (AWS)	Supplies bypass air into dynamic chamber	Controlled by duty signal from control unit
Anti-afterburn Valve	Supplies fresh air into rear port during deceleration	Included in air control valve
Air Bleed Socket	Supplies fresh air into primary injector hole	
Air Bypass Valve	Reduces sound of intake air from turbocharger relieved through air flow meter during deceleration	
Air Cleaner	Filters air into throttle chamber	
Air Control Valve	Directs air to one of three locations: exhaust port, main converter, or relief air silencer	Consists of 3 valves: Relief valve Switching valve Anti-afterburn valve
Airflow Meter	Detects amount of intake air; sends signal to control unit	
Atmospheric Pressure Sensor	Detects atmospheric pressure; sends signal to control unit	Built in ECU
Air Pump	Supplies secondary air to air control valve	
Catalytic Converter	Reduces HC, CO, and NOx	
Check Valve	Supplies the blowby gas and evaporative emission gas into the turbocharger when the intake manifold vacuum becomes positive pressure	For evaporative emission control system
Charcoal Canister	Stores fuel tank fumes when engine stops	Vented to atmosphere through charcoal and filter
Check-and-cut Valve	Controls pressure in fuel tank	
Coil with Igniter	Generates high voltage	Leading; ignite simultaneously Trailing; ignite individually
Crank Angle Sensor	Detects eccentric shaft angle at 30° intervals and front rotor position; sends signal to control unit	
Dashpot	Gradually closes throttle valve during deceleration	
Double Throttle System	Gradually opens the No.2 secondary throttle valve when No.1 secondary throttle valve suddenly opens	
Duty Solenoid Valve (Turbo boost pressure control)	Controls turbo boost pressure	
Dynamic Chamber	Connects front and rear ports	Primary and secondary separated
Engine Control Unit	Detects the following: 1. Engine speed 2. Amount of intake air 3. Engine coolant temperature 4. Throttle opening 5. Intake manifold pressure 6. O_2 concentration 7. In-gear condition 8. Intake air temperature 9. Floor temperature 10. A/C operation 11. Cranking signal 12. Atmospheric pressure 13. Knocking signal 14. Initial set signal 15. Position of transmission gear 16. Metering oil pump (MOP) position signal 17. Electric load (E/L) condition 18. Mileage Controls operation of the following: 1. Fuel injection system 2. Ignition control system	 Crank angle sensor Airflow meter Water thermosensor Throttle sensors Pressure sensor Oxygen (O_2) sensor Neutral switch and clutch switch Intake air thermosensor Heat hazard sensor A/C switch Starter switch Atmospheric pressure sensor Knock sensor and knock control unit Test connector (Green: 1-pin) Back-up light and 5th switch MOP position sensor Headlight switch, Blower switch, Rear defroster switch, Fog light switch Mileage sensor (No.1 and No.2)

Component	Function	Remarks
Engine Control Unit (Cont'd)	3. ISC system 4. Pressure regulator control system 5. Secondary air injection control system 6. Turbo boost pressure control 7. Fuel pump control system	
Fast Idle System	Opens primary throttle valve slightly at idle	Only when engine is cold
Fuel Filter	Filters fuel	
Fuel Pump	Provides fuel to injectors	Operates while engine is running Installed in fuel tank
Fuel Pump Resistor Relay	Controls voltage for fuel pump	
Heat Hazard Sensor	Detects floor temperature; sends signal to control unit	Heat hazard sensor turned ON; relieves secondary air
Injector	Injects fuel into intermediate housing and secondary intake manifold	Controlled by signals from control unit
Intake Air Thermosensor	Detects intake air temperature and temperature into the engine; sends signal to control unit	Located on the airflow meter and intake air pipe Thermistor
Intercooler	Prevents increase of intake air temperature	Air cooled type
Knock Sensor	Detects engine knock; sends signal to engine control unit	
Mileage Sensor (No.1 and No.2)	Detects vehicle mileage; sends signal to control unit	Above 20,000 miles (34,000 km); mileage sensor No.1 is ON Below 600 miles (1,000 km); mileage sensor No.2 is ON
Oxygen Sensor	Detects O_2 concentration; sends signal to control unit	Zirconia ceramic and platinum coating
Pressure Regulator	Adjusts fuel pressure supplied to injectors	
Pulsation Damper	Absorbs fuel pulsations	
Pressure sensor	Detects intake manifold pressure, sends signal to control unit	
Purge Control Valve	Regulates evaporative fumes from fuel tank and canister to intake manifold	
Solenoid Valve (Bypass air control)	Supplies bypass air into intake manifold	Controlled by duty signal from control unit
Solenoid Valve (PRC)	Shuts vacuum passage between dynamic chamber and pressure regulator	Orange
Solenoid Valve (Switch)	Controls switching valve of the air control valve	Gray
Solenoid Valve (Relief)	Controls relief valve	Blue
Solenoid Valve (AWS)	Controls accelerated warm-up system (AWS)	
Solenoid Valve (Air supply valve)	Supplies bypass air into dynamic chamber	During AWS operation
Split Air Solenoid Valve	Control split air	
Test Connector (Green: 1-pin)	Sends initial set signal to control unit	During adjustment of idle speed, ignition timing and knock system; connector grounded
Throttle Body	Controls intake air quantity	
Throttle Sensors	Detects primary throttle valve opening angle; sends signal to control unit	
Turbocharger	Pressurizes intake air utilizing exhaust gas flow	Twin-scroll turbocharger
Wastegate Valve	Controls amount of exhaust gas bypassing exhaust turbine to control intake air boost pressure	
Water Thermosensor	Detects engine coolant temperature; sends signal to control unit	Thermistor

TROUBLESHOOTING GUIDE—RX-7 TURBO

TROUBLESHOOTING GUIDE

This troubleshooting guide shows the malfunction code numbers retrieved by the **SST** and symptoms of various failures. Perform troubleshooting as described below.

TROUBLESHOOTING PROCEDURE

Troubleshooting With SST

Troubleshooting with the **SST** (Self-Diagnosis Checker 49 H018 9A1) is done to quickly determine what system or unit may be at fault.

1st: Check input sensors and output devices with the **SST**.
2nd: Check other switches with the **SST**.
3rd: Check the following items:

Electrical system	:	Battery conditions, fuses
Ignition system	:	Ignition spark, ignition timing (with test connector grounded)
Fuel system	:	Fuel level, fuel leakage, fuel filter, idle speed (with test connector grounded)
Intake air system	:	Air cleaner element, vacuum or air leakage, vacuum hose routing, accelerator cable
Engine	:	Compression, overheating
Others	:	Clutch slippage, brake dragging

4th: Check fuel and emission control systems.

Malfunction Code No.

Code No.	Input device
01	Ignition coil (Trailing side)
02	Crank angle sensor (Ne-signal)
03	Crank angle sensor (G-signal)
05	Knock sensor
08	Airflow meter (AFM)
09	Water thermosensor
10	Intake air thermosensor (AFM)
11	Intake air thermosensor (Engine)
12	Throttle sensor (Full range)
13	Pressure sensor
14	Atmospheric pressure sensor (Built in ECU)
15	Oxygen sensor
17	Feedback system
18	Throttle sensor (Narrow range)
20	Metering oil pump position sensor
27	Metering oil pump
37	Metering oil pump

Code No.	Output device
25	Solenoid valve (Pressure regulator control (PRC))
26	Step motor (Metering oil pump)
30	Split air solenoid valve
31	Solenoid valve (Relief)
32	Solenoid valve (Switching)
33	Port air solenoid valve
34	Solenoid valve (Bypass air control (BAC))
38	Solenoid valve (Accelerated warm-up system (AWS) and Air supply valve (ASV))
42	Solenoid valve (Waste gate control)
51	Fuel pump resistor relay
71	Injector (Front secondary)
73	Injector (Rear secondary)

Troubleshooting of Each System

The troubleshooting guide lists the most likely causes to a given symptom. After finding the systems to check, refer to the pages shown for detailed guides.
The numbers of the list show the priorities of inspections from the most probable to that with the lowest probability.
These were determined on the following basis:
- Ease of inspection
- Most possible system
- Most possible point in system

Possible cause \ Symptom	Hard start or No start	Rough idle	High idle speed (At normal operating temperature)	Engine does not run smoothly (At normal operating temperature)	Engine does not run smoothly (Only when engine is cold)	Engine does not run smoothly (Only when engine is hot)	Lack of power, poor acceleration or hesitation	Afterburn	Runs rough on deceleration	Fails emission test	Knocking
INTAKE AIR SYSTEM	5	1	2	1	3		1	2		3	
TURBOCHARGER CONTROL SYSTEM							7				
IDLE SPEED CONTROL (ISC) SYSTEM	4	2	1	5	1	3		5	2	7	
FUEL SYSTEM	2	3		4	5	4	2	6	3	6	
PRESSURE REGULATOR CONTROL (PRC) SYSTEM	3		4								
EXHAUST SYSTEM					6		5	8		5	
SECONDARY AIR INJECTION SYSTEM					7			7		4	
CRANKCASE AND EVAPORATIVE EMISSION CONTROL SYSTEM	7									2	
DECELERATION CONTROL SYSTEM				6				4	1	8	
CATALYTIC CONVERTER										1	
A/C CUT-OUT CONTROL SYSTEM											
KNOCK CONTROL SYSTEM							6				1
CONTROL SYSTEM	6	4	3	3	2	2	4	3	4	9	3
ELECTRONIC SPARK ADVANCE (ESA) SYSTEM	1	5		2	4	1	3	1	5	10	2

TROUBLE CODES—RX-7 TURBO

Code number

Caution

a) If there is more than one failure present, the lowest number malfunction code is displayed first, the subsequent malfunction codes appear in order.

b) After repairing all failures, turn the ignition switch OFF, disconnect the negative battery cable for at least 5 seconds to erase the malfunction code memory.

Input devices

Code No.	Input devices	Malfunction	Fail-safe function	Output signal pattern (Self-Diagnosis Checker or MIL)
01	Ignition coil (Trailing side)	Malfunction of spark plug, broken wire, short circuit	Trailing-side ignition pulse cut	
02	Crank angle sensor (Ne signal)	Broken wire, short circuit	Fuel injection and ignition cut	
03	Crank angle sensor (G signal)	Broken wire, short circuit	Fuel injection and ignition cut	
05	Knock sensor	Broken wire, short circuit	Ignition timing retarded	
08	Airflow meter (AFM)	Broken wire, short circuit	Basic fuel injection amount and ignition timing fixed	
09	Water thermosensor	Broken wire, short circuit	Coolant temp. input fixed at 80°C (176°F)	
10	Intake air thermosensor (AFM)	Broken wire, short circuit	Intake air temp. input fixed at 20°C (68°F)	
11	Intake air thermosensor (Engine)	Broken wire, short circuit	Intake air temp. input fixed at 20°C (68°F)	
12	Throttle sensor (Full range)	Broken wire, short circuit	Throttle valve opening angle input signal fixed at 20% open	
13	Pressure sensor (Intake manifold pressure)	Broken wire, short circuit	Intake manifold pressure input signal fixed at 960 mmHg (37.8 inHg)	
14	Atmospheric pressure sensor (ATP)	Malfunctioning ECU	Atmospheric pressure input signal fixed at 760 mmHg (29.9 inHg)	
15	Oxygen sensor	Oxygen sensor output remains below 0.55V 80 sec. after F/B system operation beginning	Feedback system canceled (For EGI)	
17	Feedback system	Oxygen sensor output remains 0.55V 10 sec. after F/B system operation beginning	Feedback system canceled (For EGI)	
18	Throttle sensor (Narrow range)	Broken wire, short circuit	Throttle valve opening angle input signal fixed at full open	
20	Metering oil pump position sensor	Broken wire, short circuit	MOP fixed smallest open Basic fuel injection amount and ignition timing fixed	
27	Metering oil pump (MOP)	Malfunctioning MOP, step motors, broken wire, short circuit, or malfunctioning ECU	MOP fixed smallest open Basic fuel injection amount and ignition timing fixed	
37	Metering oil pump (MOP)	Malfunction MOP, step motors, broken wire, short circuit, malfunctioning ECU, alternator or battery	Basic fuel injection amount and ignition timing fixed	

Output devices

Code No.	Output devices	Output signal pattern (Self-Diagnosis Checker or MIL)
25	Solenoid valve (Pressure regulator control (PRC))	
26	Step motor (Metering oil pump)	
30	Split air solenoid valve	
31	Solenoid valve (Relief)	
32	Solenoid valve (Switch)	
33	Port air solenoid valve	
34	Solenoid valve (Bypass air control (BAC))	
38	Solenoid valve (Accelerated warm-up system (AWS) and air supply valve (ASV))	
42	Duty solenoid (Turbo boost pressure control)	
51	Fuel pump resistor relay	
71	Injector (Front secondary)	
73	Injector (Rear secondary)	

TROUBLE CODE TROUBLESHOOTING – RX-7 TURBO

Troubleshooting chart

If a malfunction code number is shown on the **SST**, check by using the following chart and the wiring diagram.

Input device

No.1 Code (Ignition coil (Trailing side))

- Are there poor connections at ignition coil connectors?
  - YES → Repair or replace connector
  - NO ↓
- Is (B/W) terminal voltage of coil (Trailing side) OK? **Voltage: Approx. 12V**
  - NO → Repair or replace wire harness from main relay
  - YES ↓
- Is coil (Trailing side) operation OK?
  - NO → Replace coil (Trailing side)
  - YES ↓
- Is there continuity between coil (Trailing side) and ECU **1V** and **1G** terminal?
  - NO → Repair or replace wire harness
  - YES → Replace ECU

No.2 Code (Crank angle sensor (Ne signal))

- Are there poor connections at crank angle sensor connectors?
  - YES → Repair or replace connector
  - NO ↓
- Is resistance of Ne pick-up coil OK?
  - NO → Replace crank angle sensor
  - YES ↓
- Is there continuity between crank angle sensor (Ne pick-up coil) and ECU **3E** terminal?
  - NO → Repair or replace wire harness
  - YES → Replace ECU

No.3 Code (Crank angle sensor (G signal))

- Are there poor connections, at crank angle sensor connectors?
  - YES → Repair or replace connector
  - NO ↓
- Is resistance of G pick-up coil OK?
  - NO → Replace crank angle sensor
  - YES ↓
- Is there continuity between crank angle sensor (G pick-up coil) and ECU **3G** and **3H** terminals
  - NO → Repair or replace wire harness
  - YES → Replace ECU

No.5 Code (Knock sensor)

- Are there poor connections at knock sensor connectors?
  - YES → Repair or replace connector
  - NO ↓
- Check continuity between knock sensor and ECU **2M** terminal
  - NO → Repair or replace wire harness
  - YES → Replace knock sensor or ECU

No.8 Code (Airflow meter)

- Are there poor connections at airflow meter connectors?
  - YES → Repair or replace connector
  - NO ↓
- Is resistance of airflow meter OK?
  - NO → Replace airflow meter
  - YES ↓
- Is there continuity between airflow meter and ECU?

Airflow meter	ECU
Vs (GY)	2B
E_1 (BR/B)	3C and 3D
VREF (BR/W)	2I

  - NO → Repair or replace wire harness
  - YES → Replace ECU

No.10 Code (Intake air thermosensor (AFM))

- Are there poor connections at thermosensor connectors?
  - YES → Repair or replace connector
  - NO ↓
- Is resistance of intake air thermosensor OK?
  - NO → Replace airflow meter
  - YES ↓
- Is there continuity between intake air thermosensor and ECU?

Intake air thermosensor	ECU
Signal side (G/O)	2K
Ground (BR/B)	3C

  - NO → Repair or replace wire harness
  - YES → Replace ECU

TROUBLE CODE TROUBLESHOOTING—RX-7 TURBO

No.11 Code (Intake air thermosensor (Engine))

Are there poor connections at thermosensor connectors?
- YES → Repair or replace connector
- NO ↓

Is resistance of intake air thermosensor OK?
- NO → Replace intake air thermosensor
- YES ↓

Is there continuity between intake air thermosensor and ECU?

Intake air thermosensor	ECU
Signal side (G)	2L
Ground (BR/B)	3C

- NO → Repair or replace wire harness
- YES → Replace ECU

No.12 Code (Throttle sensor (Full range))

Are there poor connections at throttle sensor connectors?
- YES → Repair or replace connector
- NO ↓

Is voltage of ECU 2G terminal OK?

Throttle opening	Specification voltage
Idle position	0.25—1.25V
Full opening	4.1—4.4V

- NO → Adjust or replace throttle sensor
- YES ↓

Is there continuity between throttle sensor and ECU?

Throttle sensor	ECU
Vref (BR/W)	2I
Ground (BR/B)	3D
Signal (B/G)	2G

- NO → Repair or replace wire harness
- YES → Replace ECU

No.13 Code (Pressure sensor)

Are there poor connections at pressure sensor connectors?
- YES → Repair or replace connector
- NO ↓

Is there continuity between pressure sensor and ECU?

Pressure sensor	ECU
Vref (BR/W)	2I
Ground (BR/B)	3D
Signal (G/Y)	2H

- NO → Repair or replace wire harness
- YES ↓

Is voltage of pressure sensor OK?
When applying **100 mmHg (3.9 inHg)** vacuum

Pressure sensor	Voltage
Vref (BR/W)	Approx. 5V
Ground (BR/B)	Approx. 0V
Signal (G/Y)	2.8—3.2V

- NO → Vref or Ground NG: Replace ECU
  Signal NG: Replace pressure sensor
- YES → Replace ECU

No.14 Code (Atmospheric pressure (ATP) sensor)

Because of the ATP sensor built in ECU, if No.14 Code flash, it's mean malfunction ECU
- YES → Replace ECU (ATP sensor in ECU)

No.15 Code (Oxygen sensor)

Note
If malfunction codes No.15 and 17 are both present, perform the checking procedure for malfunction Code No.17 first.

Are there poor connections at connectors?
- YES → Repair or replace connector
- NO ↓

Is oxygen sensor output voltage OK?
- NO → Replace oxygen sensor
- YES ↓

Is oxygen sensor sensitivity OK?
- NO → Replace oxygen sensor
- YES ↓

Is there continuity between oxygen sensor and ECU **2C** terminal?
- NO → Repair or replace wire harness
- YES → Replace ECU

TROUBLE CODE TROUBLESHOOTING—RX-7 TURBO

No.17 Code (Feedback system)

- Are there poor connections at connectors?
  - YES → Repair or replace connector
  - NO → Is oxygen sensor sensitivity OK?
    - NO → Possible cause:
      - Air leak in vacuum hoses or intake air system
      - Contaminated oxygen sensor
      - Insufficient fuel injection
    - YES → Are spark plugs OK?
      - NO → Clean or replace spark plugs
      - YES → Is there continuity between oxygen sensor and ECU **2C** terminal?
        - NO → Repair or replace wire harness
        - YES → Replace ECU

No.18 Code (Throttle sensor (Narrow range))

- Are there poor connections at throttle sensor connectors?
  - YES → Repair or replace connector
  - NO → Is voltage of ECU 2F terminal OK?

Throttle opening	Specification voltage
Idle position	0.75—1.25V
Full opening	Approx. 5V

  - NO → Adjust or replace throttle sensor
  - YES → Is there continuity between throttle sensor and ECU?

Throttle sensor	ECU
Vref (BR/W)	2I
Ground (BR/B)	3D
Signal (G/R)	2F

    - NO → Repair or replace wire harness
    - YES → Replace ECU

No.20 Code (Metering oil pump position sensor)

Metering oil pump control system

No.27 Code (Metering oil pump)

Metering oil pump control system

No.37 Code (Metering oil pump)

Metering oil pump control system

Output device
No.25 Code (Solenoid valve (PRC))

- Are there poor connections at solenoid valve connectors?
  - YES → Repair or replace connector
  - NO → Is there continuity of solenoid valve?
    - NO → Replace solenoid valve
    - YES → Is there continuity between solenoid valve and ECU?

Solenoid valve	ECU
Battery (B/W)	1B
Signal side (L/O)	3M

      - NO → Repair or replace wire harness
      - YES → Replace ECU

No.26 Code (Stepping motor (Metering oil pump))

Metering oil pump control system

No.30 Code (Split air solenoid valve)

- Are there poor connections at solenoid valve connectors?
  - YES → Repair or replace connector
  - NO → Is there continuity of solenoid valve?
    - NO → Replace solenoid valve
    - YES → Is there continuity between solenoid valve and ECU?

Solenoid valve	ECU
Battery (B/W)	1B
Signal side (L/B)	3I

      - NO → Repair or replace wire harness
      - YES → Replace ECU

No.31 Code (Solenoid valve (Relief))

- Are there poor connections at solenoid valve connectors?
  - YES → Repair or replace connector
  - NO → Is there continuity of solenoid valve?
    - NO → Replace solenoid valve
    - YES → Is there continuity between solenoid valve and ECU?

Solenoid valve	ECU
Battery (B/W)	1B
Solenoid valve (L/Y)	2P

      - NO → Repair or replace wire harness
      - YES → Replace ECU

TROUBLE CODE TROUBLESHOOTING – RX-7 TURBO

No.32 Code (Solenoid valve (Switch))

Are there poor connections at solenoid valve connectors?
- YES → Repair or replace connector
- NO ↓

Is there continuity of solenoid valve?
- NO → Replace solenoid valve
- YES ↓

Is there continuity between solenoid valve and ECU?

Solenoid valve	ECU
Battery (B/W)	1B
Signal side (L/R)	2Q

- NO → Repair or replace wire harness
- YES → Replace ECU

No.33 Code (Port air solenoid valve)

Are there poor connections at solenoid valve connectors?
- YES → Repair or replace connector
- NO ↓

Is there continuity of solenoid valve?
- NO → Replace solenoid valve
- YES ↓

Is there continuity between solenoid valve and ECU?

Solenoid valve	ECU
Battery (B/W)	1B
Signal side (L)	2N

- NO → Repair or replace wire harness
- YES → Replace ECU

No.34 Code (Solenoid valve (BAC))

Are there poor connections at solenoid valve connectors?
- YES → Repair or replace connector
- NO ↓

Is there continuity of solenoid valve?
- NO → Replace solenoid valve
- YES ↓

Is there continuity between solenoid valve and ECU?

Solenoid valve	ECU
Battery (B/W)	1B
Signal side (L/G)	3Q

- NO → Repair or replace wire harness
- YES → Replace ECU

No.38 Code (Solenoid valve (AWS and ASV))

Are there poor connections at solenoid valve connectors?
- YES → Repair or replace connector
- NO ↓

Is there continuity between solenoid valve and ECU?

Solenoid valve	ECU
Battery (B/W)	1B
Signal side (BR/Y)	3J

- NO → Repair or replace wire harness
- YES → Replace ECU

Note
The AWS and ASV solenoid valves are connected in parallel. (Refer to the figure.)
If only AWS or ASV circuit is bad code No.38 will not flash.

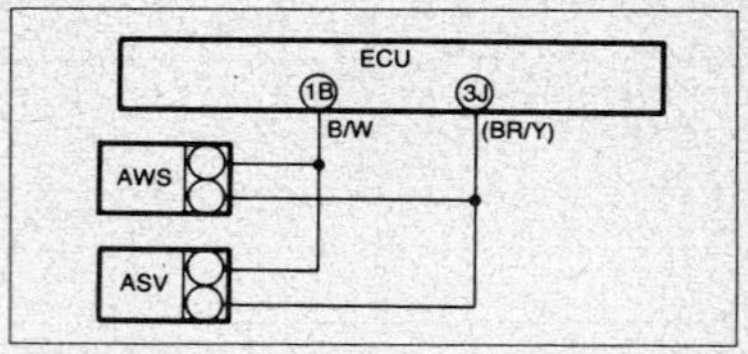

No.42 Code (Duty solenoid (Turbo boost pressure control))

Are there poor connections at solenoid valve connectors?
- YES → Repair or replace connector
- NO ↓

Is there continuity of coil?
- NO → Replace solenoid valve
- YES ↓

Is there continuity between solenoid valve and ECU?

Solenoid valve	ECU
Battery (B/W)	1B
Signal side (L/W)	3R

- NO → Repair or replace wire harness
- YES → Replace ECU

TROUBLE CODE TROUBLESHOOTING—RX-7 TURBO

No.51 Code (Fuel pump resistor relay)

Question	Answer	Action
Are there poor connections at relay connectors?	YES	Repair or replace connector
	NO	Go to next step
Is there continuity of the coil?	NO	Replace fuel pump resistor relay
	YES	Go to next step
Is there battery voltage at (L/R) wire of fuel pump resistor relay when IGN switch ON?	NO	Possible cause: • Fuel pump resistor relay malfunction • Open or short circuit between fuel pump resistor relay and circuit opening relay, circuit opening relay and IGN switch
	YES	Go to next step
Is there continuity between fuel pump resistor relay and ECU?	NO	Repair or replace wire harness
	YES	Replace ECU

No.71 Code (Injector (Front secondary))

Question	Answer	Action
Are there poor connections at injector connectors?	YES	Repair or replace connector
	NO	Go to next step
Is there continuity of injector?	NO	Replace injector
	YES	Go to next step
Is there continuity between injector and ECU **3X** terminal?	NO	Repair or replace wire harness
	YES	Replace ECU

No.73 Code (Injector (Rear secondary))

Question	Answer	Action
Are there poor connections at injector connectors?	YES	Repair or replace connector
	NO	Go to next step
Is there continuity of injector?	NO	Replace injector
	YES	Go to next step
Is there continuity between injector and ECU **3Z** terminal?	NO	Repair or replace wire harness
	YES	Replace ECU

SWITCH MONITOR FUNCTION—RX-7 TURBO

SWITCH MONITOR FUNCTION
Preparation
SST

49 H018 9A1 Self-Diagnosis Checker		For diagnosis

Individual switches can be monitored by the **SST**.

Note
The test connector must be grounded and the ignition switch ON (engine stopped).

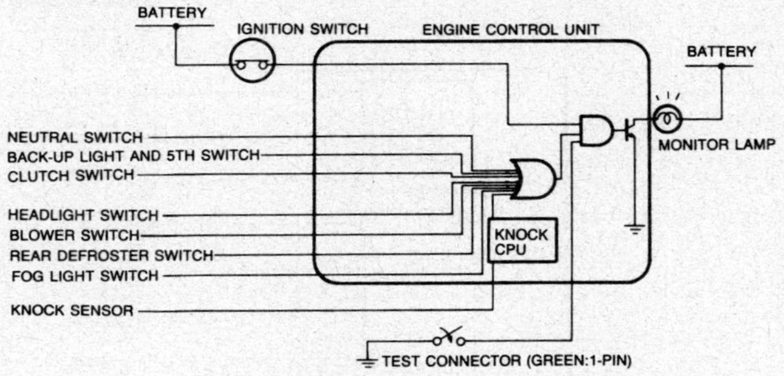

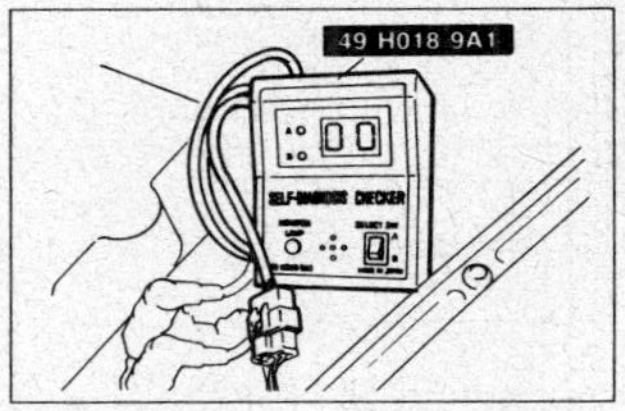

Inspection Procedure
1. Warm up the engine to normal operating temperature and stop it.
2. Connect the **SST** to the check connector (Green: 6-pin) and the negative battery terminal.

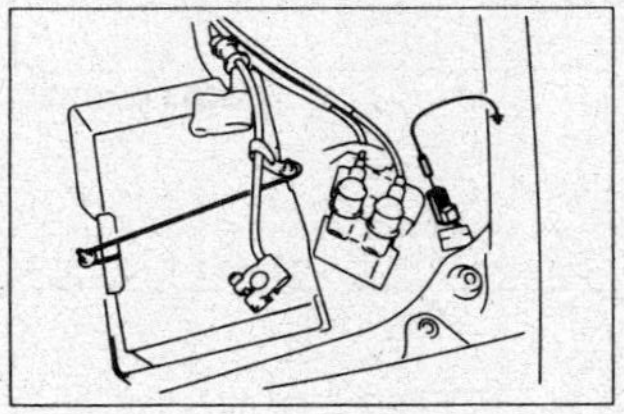

3. Connect a jumper wire between the test connector (Green: 1-pin) and a ground.
4. Turn the ignition switch ON, then check that the monitor lamp illuminates when each switch is made to function according to the table below.

Caution
a) When any one of the switches are activated, the monitor lamp will be on.
b) Do not start the engine.

SWITCH MONITOR FUNCTION TROUBLESHOOTING RX-7 TURBO

Switch	Self-Diagnosis Checker (Monitor lamp) Lamp-ON	Lamp-OFF	Possible cause (When incorrect)
Neutral switch	Shift transmission to 1, 2, 3 or 4	Neutral position	• Neutral switch malfunction • Open circuit between neutral switch and ECU 1R terminal, neutral switch and ground
Back-up light and 5th switch	Shift transmission 5th gear	Neutral position	• Back-up light and 5th switch malfunction • Open circuit between back-up light and 5th switch and ECU 1T terminal, overdrive and ground
Clutch switch	Pedal depressed	Pedal released	• Clutch switch malfunction • Open circuit between clutch switch and ECU 1Q terminal, clutch switch and ground, clutch switch and ACC circuit
Headlight switch	Switch ON	Switch OFF	• Headlight switch malfunction • Open circuit between headlight switch and ECU 3L terminal, headlight switch and battery line
Blower switch	Switch ON (At any fan speed OK)	Switch OFF	• Blower switch malfunction • Heater relay malfunction • Heater control unit malfunction • Open circuit between heater control unit and ECU 3O terminal
Rear defroster switch	Switch ON	Switch OFF	• Rear defroster switch malfunction • Open circuit between rear defroster switch and ECU 3P terminal, rear defroster switch and ground
Fog light switch (If equipped)	Switch ON (Only equip fog light)	Switch OFF	• Fog light switch malfunction • Open circuit between fog light switch and ECU 1S terminal, fog light switch and battery
Knock sensor	While knocking Lamp ON 0.5 sec. (IGN switch ON: hit engine hanger with hammer)	Other	• Knock sensor malfunction • Open circuit between knock sensor and ECU 2M terminal

ISC SYSTEM TROUBLESHOOTING IN ORDER OF PRIORITY—RX-7 TURBO

DESCRIPTION

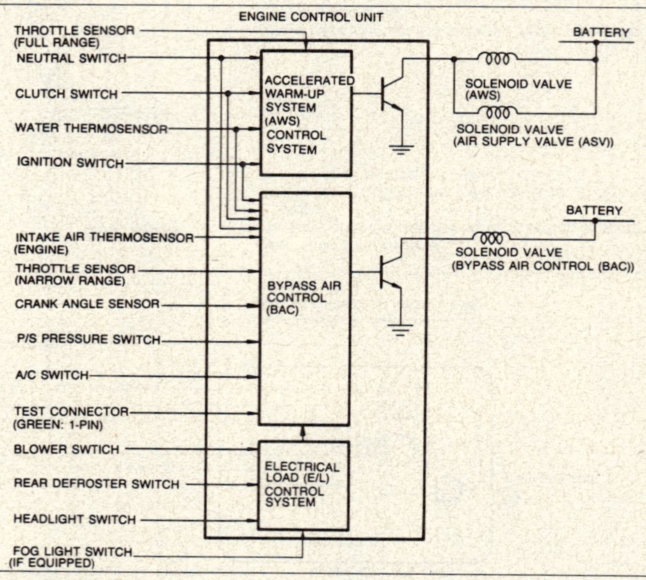

TROUBLESHOOTING

Check the condition of the wiring harness and connectors before checking the sensors or switches.

Symptom \ Possible cause	System inspection	Solenoid valve (BAC)	Solenoid valve (AWS)	Solenoid valve (ASV)	Engine control unit terminal 1B	3J	3Q
Hard start or No start (Cranks OK)	1	3	4	4	2	5	2
Rough idle	1	2	4	4	3	5	3
High idle speed (At normal operating temperature)	1	2	3	3	5	4	5
Engine does not run smoothly (Only when engine is cold)	1	2	4	4	3	5	3
Afterburn	1	2	—	—	3	—	3
Runs rough on deceleration	1	2	—	—	3	—	3
Fails emission test	1	4	2	2	5	3	5

TURBO CONTROL SYSTEM TROUBLESHOOTING IN ORDER OF PRIORITY—RX-7 TURBO

DESCRIPTION
This system utilizes the energy of escaping exhaust gas to pressurize the intake air. This system therefore, supplies more than the normal amount of air into the combustion chamber. As a result of the more fully charged combustion chamber, higher engine output and torque is obtained.
The maximum turbocharger pressure is controlled by the solenoid valve (Turbo boost pressure control) for engine safety.

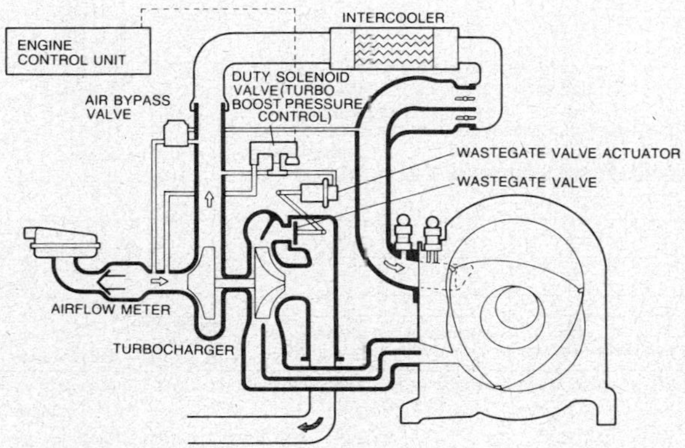

PREPARATION
SST

49 H080 740 Pressure tester	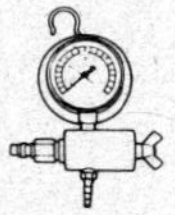	For inspection of turbocharger

TROUBLESHOOTING
Check the condition of the wiring harness and connectors before checking the sensor or switches.

Possible cause / Symptom	Turbo-charger	Air bypass valve	Inter-cooler	Boost pressure	Duty solenoid valve (Turbo boost pressure control)	Waste-gate valve actuator	Engine control unit terminal 1B	Engine control unit terminal 3R
Engine does not run smoothly (At normal operating temp.)	1	3	2					
Lack of power, poor acceleration or hesitation	5		6	1	2	3	4	4

FUEL SYSTEM TROUBLESHOOTING IN ORDER OF PRIORITY—RX-7 TURBO

TROUBLESHOOTING
Check the condition of the wiring harness and connectors before checking the sensors or switches.

Possible cause / Symptom	Fuel pump	Fuel filter	Injector	Hold pressure (Inspection)	Fuel line pressure (Inspection)	Pressure regulator	Pulsation damper	Fuel pump control system	Engine control unit terminal 1K	3W	3X	3Y	3Z
Hard start or no start (Cranks OK)	5	6	7	1	2	8		3	4	4	4	4	4
Rough idle	4		1		2		3						
Engine does not run smoothly (At normal operating temp.)	6	4	2		1		5			3	3	3	3
Engine does not run smoothly (When engine is cold)					1					2	2	2	2
Lack of power, poor acceleration or hesitation	2	3	4		1	6				5	5	5	5
Afterburn			1							2	2	2	2
Runs rough on deceleration			1							2	2	2	2
Fails emission test			3		1	2				4	4	4	4

Warning
Before release the fuel pressure from the fuel system to reduce the possibility of injury or fire.

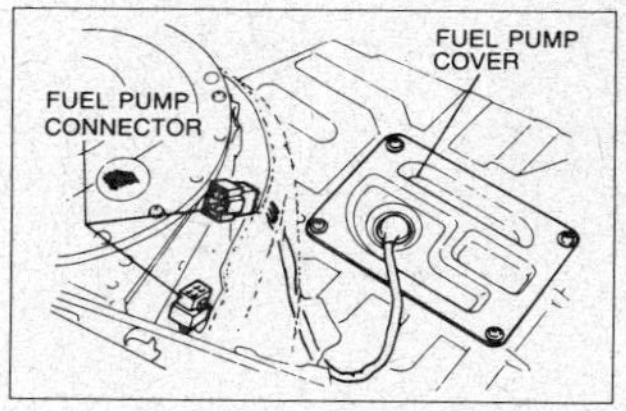

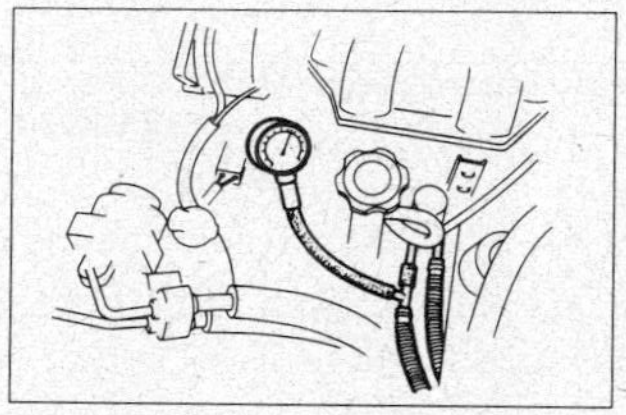

PRECAUTION
Fuel Pressure Release and Fuel System
Fuel in the fuel system remains under high pressure even when the engine is not running.

a) Before disconnecting any fuel line, release the fuel pressure from the fuel system to reduce the possibility of injury or fire.
   1. Start the engine.
   2. Disconnect the fuel pump connector.
   3. Allow the engine to stalls. Turn off the ignition switch.
   4. Reconnect the fuel pump connector.

b) Use a rag as protection from fuel spray when disconnecting the hoses.
c) Plug all disconnected hoses.
d) Inspect the fuel system with a suitable pressure gauge.

Caution
Install hose clamps to secure the fuel pressure gauge to the fuel filter and the fuel main hose to prevent fuel leakage.

PRC SYSTEM TROUBLESHOOTING IN ORDER OF PRIORITY—RX-7 TURBO

DESCRIPTION

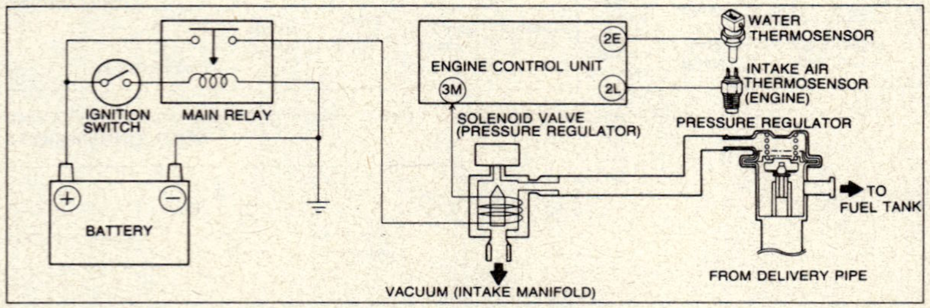

To prevent percolation of the fuel during idle after the engine is restarted, vacuum is cut to the pressure regulator increasing the fuel pressure.

SYSTEM OPERATION

Specified operation time: For above **50 sec.** after cranking (Coolant temperature: Above **65°C**, Intake air temperature (Engine): Above **60°C**)

TROUBLESHOOTING

Check the condition of the wiring harness and connectors before checking the sensors or switches.

Possible cause / Symptom	Solenoid valve (Pressure regulator control)	Water thermosensor	Intake air thermosensor	Engine control unit terminal	
				3R	1B
Hard start or No start (Cranks OK)	1	3	4	2	2
High idle speed (At normal operating temperature)	1	3	4	2	2

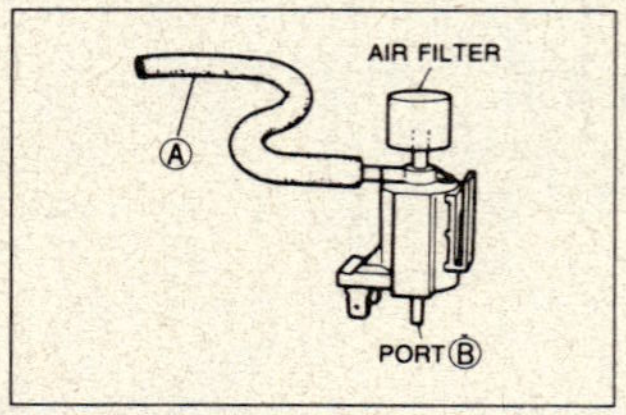

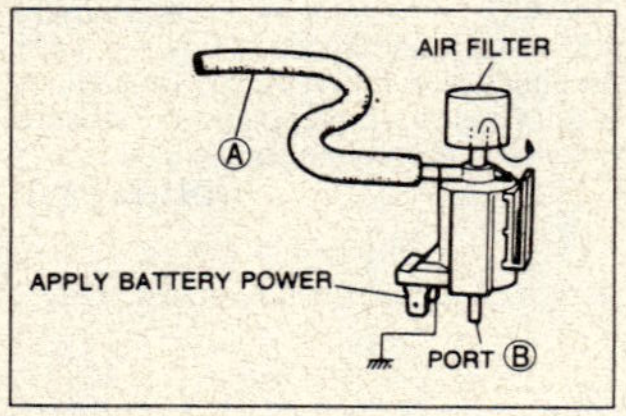

SOLENOID VALVE (PRESSURE REGULATOR CONTROL)

Inspection

1. Disconnect the vacuum hose from the solenoid valve and vacuum pipe.
2. Blow through the solenoid valve from Ⓐ.
3. Verify that air flows from port Ⓑ.
4. Disconnect the solenoid valve connector.
5. Connect 12V and a ground to the terminals of the solenoid valve.
6. Blow through the solenoid valve from the vacuum hose Ⓐ.
7. Verify that air flows from the valve air filter.

SECONDARY AIR SYSTEM TROUBLESHOOTING IN ORDER OF PRIORITY—RX-7 TURBO

DESCRIPTION

This system helps to clean the exhaust gas by introducing fresh air into the exhaust port and main converter.

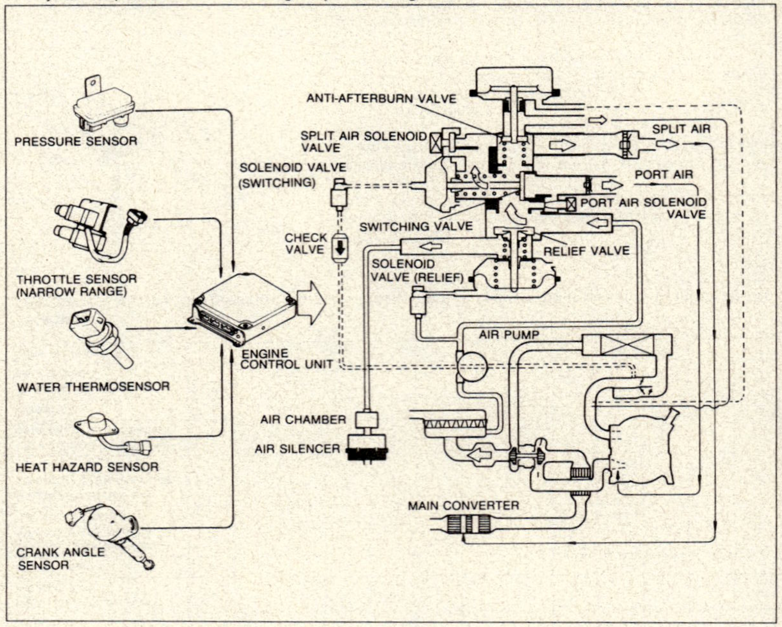

TROUBLESHOOTING

Check the condition of the wiring harness, vacuum hoses, and connectors before checking the sensors or switches.

Possible cause / Symptom	System inspection	Air pump	Check valves (in intake manifold) and (intake manifold to catalytic converter)	Air control valve	Solenoid valve (Switch)	Solenoid valve (Relief)	Heat hazard sensor	Split air solenoid valve	Port air solenoid valve	Engine control unit terminal			
										3I	2O	2P	2M
Rough idle	1	3	8	4	5	6	7	—	—	—	2	2	—
Afterburn	1	3	8	4	5	6	7	—	—	—	2	2	—
Fails emission test	1	3	10	4	7	8	9	5	6	2	2	2	2

DECELERATION CONTROL SYSTEM TROUBLESHOOTING IN ORDER OF PRIORITY – RX-7 TURBO

TROUBLESHOOTING
Check the condition of the wiring harness, vacuum hoses and connectors before checking the sensors or switches.

Symptom \ Possible cause	Dashpot	Anti-afterburn valve	Air bypass valve
Afterburn	2	1	
Runs rough on deceleration	2	3	1
Fails emission test	1	3	2

A/C CUTOUT CONTROL SYSTEM TROUBLESHOOTING IN ORDER OF PRIORITY – RX-7 TURBO

DESCRIPTION
An A/C cut-out system is employed to provide smoother acceleration while the air conditioning is in use.

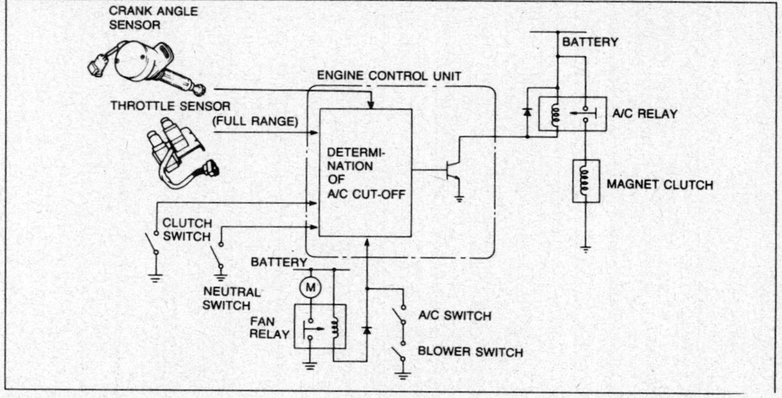

Operate Condition
The A/C is cut-off for **approx. 7 sec.** during acceleration when all of the conditions below are met.
1. A/C switch ON.
2. Blower switch ON.
3. Transmission in gear and clutch pedal released.
4. Throttle valve opening angle more than 50°. (More than half throttle)

TROUBLESHOOTING
Check the condition of the wiring harness and connectors before checking the sensors or switches.

Symptom \ Possible cause	Throttle sensor (Full range)	Clutch switch	Neutral switch	System inspection	Engine control unit terminals						
					1L	1O	1Q	1R	2G	2I	3D
Lack of power, poor acceleration or hesitation	3	4	5	1	2	2	2	2	2	2	2

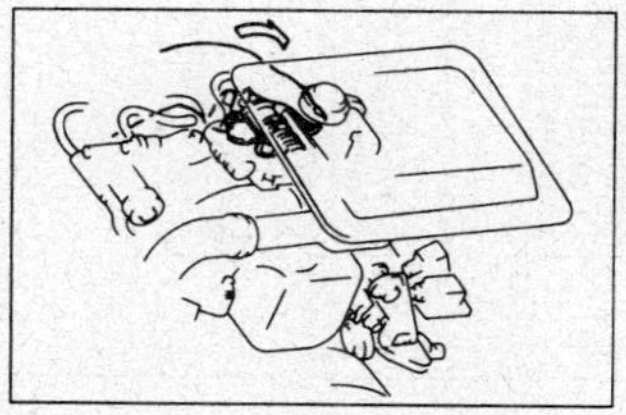

SYSTEM INSPECTION
1. Turn the A/C and blower switches ON.
2. Turn the ignition switch ON (engine not running).
3. Shift the transmission into gear.
4. Open the throttle valve fully and verify that the magnetic clutch disengages (Clicks is heard), then reengages after **approx. 7 sec.**

ENGINE CONTROLS OPERATION CHART – RX-7 TURBO

OUTPUT DEVICES \ ENGINE CONDITION		CRANKING (WHEN ENGINE IS COLD)	IDLE (DURING WARM-UP)	MEDIUM LOAD		ACCELERATION	HEAVY LOAD	DECELERATION (THROTTLE VALVE FULLY CLOSED)	IDLE (WHEN ENGINE IS WARM)	IGNITION: ON (NOT ENGINE RUN)	REMARKS
				COLD	WARM						
FUEL INJECTION AMOUNT	PRIMARY INJECTOR	← Rich →			Rich and Lean	← Rich →		Fuel cut	Rich	OFF	Rear side; Above 8,500 rpm; Fuel cut
	SECONDARY INJECTOR	← OFF →						← OFF →			
FUEL PUMP RESISTOR RELAY		12V drive	← 9—10V drive →			← 12V drive →		← 9—10V drive →		OFF	Hot start (at idle) 12V drive
A/C RELAY		← ON →				*1OFF	← ON →				*1Only 7 sec.
PORT AIR SOLENOID VALVE		← Above 20,000 mile (34,000 km); ON Below 20,000 mile (34,000 km; OFF) →					OFF	← OFF (Above 2,000 mile (34,000 km) ON) →			
SPLIT AIR SOLENOID VALVE		← Back-up light and 5th switch ON or high altitude; ON →									
SOLENOID VALVE (SWITCH)		← OFF (Port air) →			← ON (Split air) →			← OFF (Port air) →			
SOLENOID VALVE (RELIEF)		← ON (Air injection) →		OFF (Relief)	ON (Air injection)	← OFF (Relief) →		← ON (Air injection) →		ON	
SOLENOID VALVE (AWS AND ASV)		ON	For 17 sec. after start: ON	← OFF →							
SOLENOID VALVE (PRESSURE REGULATOR)		← OFF →							After start; *2ON	ON	*2For 50 sec.—90 sec.
SOLENOID VALVE (BYPASS AIR CONTROL)		Large quantity of air	Feed back duty	← Fixed duty →					Feed back duty	Fixed duty	Test connector (Green: 1-pin) ground; Stop the feedback control
DUTY SOLENOID VALVE (TURBO BOOST PRESSURE)		← Feed back duty →									
IGNITER (TRAILING IGNITION TIMING)		BTDC 5°	Retard	Advanced (If occuring knocking, retards ignition timing max. 5°)				Retard	ATDC 20°	OFF	Test connector (Green 1-pin): Ground (Only ignition ON, (engine does not run)) Inspect the knock control system
IGNITER (LEADING IGNITION TIMING)									ATDC 5°		

RELATIONSHIP CHART – RX-7 TURBO

RELATIONSHIP CHART
Output Devices and Input Devices

OUTPUT DEVICES \ INPUT DEVICES	TEST CONNECTOR	IGNITION SWITCH	POWER STEERING PRESSURE SWITCH	A/C SWITCH	FOG LIGHT SWITCH (IF EQUIPPED)	HEAD LIGHT SWITCH	BLOWER SWITCH	REAR DEFROSTER SWITCH	CLUTCH SWITCH	NEUTRAL SWITCH	BACK-UP LIGHT AND 5TH SWITCH	MILEAGE SENSOR No.1	MILEAGE SENSOR No.2	HEAT HAZARD SENSOR	OXYGEN SENSOR	THROTTLE SENSOR (NARROW RANGE)	THROTTLE SENSOR (FULL RANGE)	PRESSURE SENSOR	KNOCK SENSOR	CRANK ANGLE SENSOR (NE)	CRANK ANGLE SENSOR (G+, G–)	WATER THERMOSENSOR	INTAKE AIR THERMOSENSOR (AFM)	INTAKE AIR THERMOSENSOR (ENGINE)	AIRFLOW METER (AFM)
IGNITER (LEADING IGNITION TIMING)	O	O	X	O	O	O	O	O	O	O	O	O	O	X	X	O	O	O	O	O	O	O	X	X	O
IGNITER (TRAILING IGNITION TIMING)	O	O	X	O	O	O	O	O	O	O	O	O	O	X	X	O	O	O	O	O	O	O	X	X	O
DUTY SOLENOID VALVE (TURBO BOOST PRESSURE)	X	X	X	X	X	X	X	X	X	X	X	X	O	X	X	X	O	O	X	O	X	X	X	X	X
SOLENOID VALVE (BYPASS AIR CONTROL)	O	O	O	O	O	O	O	O	O	O	O	X	X	X	X	O	X	X	X	O	X	O	X	O	X
SOLENOID VALVE (PRESSURE REGULATOR)	X	O	X	X	X	X	X	X	O	O	X	X	X	X	X	O	X	X	X	X	O	O	X	O	X
SOLENOID VALVE (AWS AND ASV)	X	O	X	X	X	X	X	X	O	O	X	X	X	X	X	X	O	X	X	O	X	O	X	X	X
SOLENOID VALVE (RELIEF)	X	O	X	X	X	X	X	X	O	O	O	X	X	O	X	O	O	O	X	O	X	O	X	X	X
SOLENOID VALVE (SWITCH)	X	O	X	X	X	X	X	X	O	O	O	X	X	X	X	O	X	O	X	O	X	O	X	X	X
SPLIT AIR SOLENOID VALVE	X	X	X	X	X	X	X	X	X	X	O	X	X	X	X	X	X	X	X	X	X	X	X	X	X
PORT AIR SOLENOID VALVE	X	X	X	X	X	X	X	X	X	X	X	O	X	X	X	X	X	O	X	O	X	X	X	X	X
A/C RELAY	X	X	X	O	X	X	X	X	O	O	X	X	X	X	X	X	O	X	X	O	X	X	X	X	X
FUEL PUMP RESISTOR RELAY	X	O	X	X	X	X	X	X	O	O	X	X	X	X	X	O	X	X	X	O	O	O	O	O	O
FUEL INJECTION TIMING	X	X	X	X	X	X	X	X	X	X	X	X	X	X	X	X	X	X	X	O	O	X	X	X	X
FUEL INJECTION AMOUNT	O	O	X	O	O	O	O	O	O	O	O	O	O	X	O	O	O	O	X	O	O	O	O	O	O

ENGINE CONTROLS TROUBLESHOOTING—RX-7 TURBO

TROUBLESHOOTING
Check the condition of the wiring harness and connectors before checking the sensors or switches.

Symptom \ Possible cause	Airflow meter	Intake air thermosensor (Engine)	Throttle sensor (Full and Narrow range)	Oxygen sensor	Knock sensor	Pressure sensor	Heat hazard sensor	Main relay	Circuit opening relay	Neutral switch	Clutch switch	Power steering pressure switch	Back-up light and 5th switch	Water thermosensor	Mileage sensor (No.1 and No.2)	Engine control unit terminal
Hard start or no start (Cranks OK)	7	6	4			5		1	2					3		8
Rough idle	1	4	2		5					6	7			3	8	9
High idle speed (At normal operating temperature)	1		2									4		3		5
Engine does not run smoothly (Only when engine is cold)	1	3					4							2		5
Engine does not run smoothly (Only when engine is hot)	1	3					4							2		5
Lack of power, poor acceleration or hesitation	1		2		4		9			5	6	7	10	8	3	11
Afterburn	1		2			5					3	4				6
Fails emission test	2		3	1			4							5		6
Knocking	3	5	4		1					6	7			2		8

ENGINE CONTROL UNIT TERMINAL VOLTAGES RX-7 TURBO

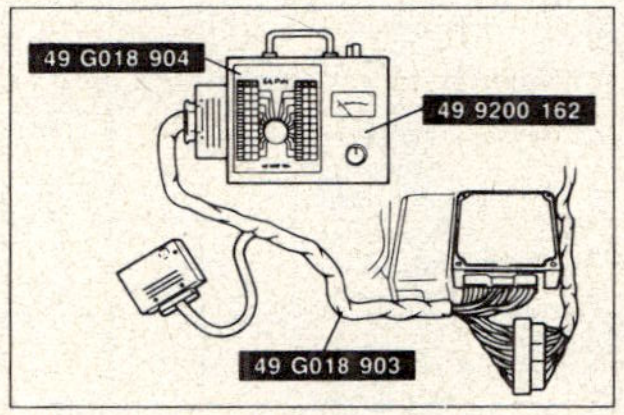

ENGINE CONTROL UNIT
Inspection
1. Lift up the floor mat in front of the passenger's seat.
2. Remove the protector cover.
3. Connect the **SST** (Engine Signal Monitor) between the ECU and wire harness using the **SST** (Adapter harness).

Note
Use connector A of the adapter harness to check voltages at terminals 1A through 1V and 3A through 3Z, and use connector B to check voltages at terminals 2A through 2P.

4. Place the **SST** (Sheet: 64-pin side) on the **SST** (Engine Signal Monitor).
5. Measure the voltage at each terminal.

Engine control unit terminal (unit side)

3Y	3W	3U	3S	3Q	3O	3M	3K	3I	3G	3E	3C	3A	2O	2M	2K	2I	2G	2E	2C	2A	1U	1S	1Q	1O	1M	1K	1I	1G	1E	1C	1A
3Z	3X	3V	3T	3R	3P	3N	3L	3J	3H	3F	3D	3B	2P	2N	2L	2J	2H	2F	2D	2B	1V	1T	1R	1P	1N	1L	1J	1H	1F	1D	1B

Terminal	Input	Output	Connection to	Test condition	Voltage	Remark
1A	○		Battery	Constant	Approx. 12V	Backup
1B	○		Main relay	Ignition switch ON	Approx. 12V	—
				Ignition switch OFF	Approx. 0V	
1C	○		Ignition switch	Ignition switch START (Cranking)	Approx. 12V	—
				Ignition switch ON	Approx. 0V	
				Ignition switch OFF	Approx. 0V	
1D		○	Self-Diagnosis Checker (Monitor lamp)	Test connector grounded For 3 sec. after ignition switch OFF→ON (Lamp illuminates)	Below 6.2V	With Self-Diagnosis checker
				After 3 sec. (Light does not illuminate)	Approx. 12V	
				Test connector grounded at idle	Approx. 12V	
				Test connector grounded (Monitor lamp ON)	Below 6.2V	
				Test connector grounded (Monitor lamp OFF)	Approx. 12V	
1E		○	Malfunction indicator light (MIL) lamp	For 3 sec. after ignition switch OFF→ON (Lamp illuminates)	Below 4.8V	Test connector grounded
				After 3 sec. (Lamp does not illuminate)	Approx. 12V	
				Lamp illuminates	Below 4.8V	
				Lamp does not illuminates	Approx. 12V	
1F		○	Self-Diagnosis Checker (Malfunction code number)	For 3 sec. after ignition switch OFF→ON (Buzzer sounds)	Below 6.2V	With Self-Diagnosis Checker and test connector grounded
				After 3 sec. (Buzzer does not sound)	Approx. 12V	
				Buzzer sounds	Below 6.2V	
				Buzzer does not sound	Approx. 12V	
1G		○	Ignition coil (Trailing)	Ignition switch ON	Approx. 0V	IGt-T (Ignition timing signal)
				Idle	Approx. 0.8V	
1H		○	Ignition coil (Leading)	Ignition switch ON	Approx. 0V	IGt-L (Ignition timing signal)
				Idle	Approx. 0.8V	
1I	○		Test connector (Green: 1-pin)	Test connector grounded	Approx. 0V	Ignition switch ON
				Test connector not grounded	Approx. 12V	
1J		○	Ignition coil (Trailing)	Ignition switch ON	Approx. 4.4V	IGs-T (Select signal)
				Idle	Approx. 2.2V	

ENGINE CONTROL UNIT TERMINAL VOLTAGES—RX-7 TURBO

Terminal	Input	Output	Connection to	Test condition	Voltage	Remark
1K		○	Fuel pump resistor relay	Cranking	Approx. 12V	—
				Idle (More than 90 sec. after cranking)	Below 2.0V	
1L		○	A/C relay	A/C switch ON	Below 2.5V	Ignition switch ON
				A/C switch OFF	Approx. 12V	Blower switch ON
1M	○		Mileage sensor No.1	Under 20,000 miles (34,000 km)	Approx. 12V	There is an error more or less
				Over 20,000 miles (34,000 km)	Below 1.5V	
1N	○		Power steering (P/S) pressure switch	Ignition switch ON	Approx. 12V	P/S ON: Turning
				P/S ON (Idle)	Approx. 0V	P/S OFF: Straight ahead
				P/S OFF (Idle)	Approx. 12V	
1O	○		A/C switch	A/C switch ON (Idle)	Below 2.5V	Blower switch ON
				A/C switch OFF (idle)	Approx. 12V	
1P	○		Heat hazard sensor	Ignition switch ON	Below 1.5V	—
				Idle (Floor temp.: Below 110°C (230°F))	Approx. 12V	
				Idle (Floor temp.: Above 110°C (230°F))	Below 1.5V	
1Q	○		Clutch switch	Clutch pedal: released	Approx. 12V	Ignition switch ON
				Clutch pedal: depressed	Below 2.0V	
1R	○		Neutral switch	Neutral	Below 2.0V	Ignition switch ON
				In gear	Approx. 12V	
1S	○		Fog light switch	Fog light ON (Idle)	Approx. 12V	If equipped
				Fog light OFF (Idle)	Approx. 0V	
1T	○		Back-up light and 5th switch	5th gear	Below 2.0V	Ignition switch ON
				1st—4th gear	Approx. 12V	
1U	○		Mileage sensor No.2	Under 600 miles (1,000 km)	Approx. 12V	There is an error more or less
				Over 600 miles (1,000 km)	Below 2.0V	
1V	○		Ignition coil (Trailing)	Ignition switch ON	Below 2.0V	IGf-T (Ignition confirmation signal)
				Idle	Approx. 1.4V	
2A	○		Metering oil pump (MOP) position sensor	Ignition switch OFF	0V	Refer to Section D
				Idle	Approx. 1.0V	
2B	○		Airflow meter (Vs)	Ignition switch ON	Approx. 4.0V	—
				Idle	2.5V—3.5V	
2C	○		Oxygen sensor	Idle	Below 1.0V	—
				Acceleration	0.5V—1.0V	
				Deceleration	0V—0.4V	
2D	—	—	—	—	—	—
2E	○		Water thermosensor	Idle (Engine hot)	0.3—1.0V	—
				Water temperature: 20°C (68°F)	2.0—3.0V	
2F	○		Throttle sensor (Narrow range)	Ignition switch ON (Idle position)	Approx. 1.0V	After warm-up
				Ignition switch ON (Full throttle)	Approx. 5.0V	
				Idle	Approx. 1.0V	
2G	○		Throttle sensor (Full range)	Ignition switch ON (Idle position)	Approx. 0.8V	After warm-up
				Ignition switch ON (Full throttle)	Approx. 4.3V	
				Idle	Approx. 0.8V	
2H	○		Pressure sensor	Vacuum hose disconnected and plugged	2.2—2.5V	Ignition switch ON
				100 mmHg (3.9 inHg) vacuum applied to pressure sensor	1.8—2.2V	
2I		○	Sensors	Ignition switch ON	4.5V—5.5V	Vref (Power supply)
				Ignition switch OFF	0V	
2J	○		Ground or open	Canada (Ground)	0V	—
				Except for Canada (Open)	Approx. 12V	
2K	○		Intake air thermosensor (Airflow meter)	Idle (At 20°C (68°F))	2V—3V	—
2L	○		Intake air thermosensor (Engine)	Idle (At 80°C (176°F))	1V—2V	—
2M	○		Knock sensor	Ignition switch ON	Approx. 0V	Very low voltage at any condition
				Idle	Approx. 0V	
				Knocking	Approx. 0V	

Terminal	Input	Output	Connection to	Test condition	Voltage	Remark
2N		○	Port air solenoid valve	Idle (Below 20,000 miles (34,000 km))	Approx. 12V	There is an error more or less
				Idle (Above 20,000 miles (34,000 km))	Below 2.0V	
				Engine speed: above 3,500 rpm	Approx. 12V	—
2O		○	Solenoid valve (Switch)	Idle	Approx. 12V	Ignition switch ON
				Half throttle	Below 2.0V	
2P		○	Solenoid valve (Relief)	Idle	Below 2.0V	—
				Engine speed: above 4,000 rpm	Approx. 12V	
3A	—	—	Ground	Constant	0V	Power
3B	—	—	Ground	Constant	0V	Power
3C	—	—	Ground	Constant	0V	System
3D	—	—	Ground	Constant	0V	Analog
3E	○		Crank angle sensor (Ne)	Ignition switch ON	Below 1.0V	Red
				Idle	Below 1.0V	
3F	—	—	—	—	—	—
3G	○		Crank angle sensor (G+)	Ignition switch ON	Below 1.0V	Black
				Idle	Below 1.0V	
3H	○		Crank angle sensor (G−)	Ignition switch ON	Below 1.0V	White
				Idle	Below 1.0V	
3I		○	Split air solenoid valve	5th gear	Below 2.0V	Refer to page F2–60
				Others	Approx. 12V	
3J		○	Solenoid valve (Accelerated warm-up system and air supply valve)	Ignition switch OFF	0V	Engine coolant temperature: 15°C (59°F)—35°C (95°F)
				Ignition switch ON	Approx. 12V	
				Idle (Less than 17 sec. after cranking)	Below 2.0V	
				Idle (More than 17 sec. after cranking)	Approx. 12V	
3K		○	Circuit opening relay	Ignition switch OFF	0V	—
				Ignition switch ON	Approx. 12V	
				Idle	Below 2.0V	
3L	○		Headlight switch	Headlight switch ON	Approx. 12V	—
				Headlight switch OFF	0V	
3M		○	Solenoid valve (Pressure regulator control)	Ignition switch ON	Below 2.0V	—
				Cranking	Below 2.0V	Hot condition only
				Idle (Less than 20 sec. after cranking)	Below 2.0V	
				Idle (More than 90 sec. after cranking)	Approx. 12V	—
3N	—	—	—	—	—	—
3O	○		Blower switch	Blower switch ON	Below 2.0V	Ignition switch ON
				Blower switch OFF	Approx. 12V	
3P	○		Rear defroster switch	Rear defroster switch ON	Below 2.0V	Ignition switch ON
				Rear defroster switch OFF	Approx. 12V	
3Q		○	Solenoid valve (Bypass air control)	Ignition switch OFF	0V	Duty pulse
				Ignition switch ON	Approx. 9V	
				Idle	Approx. 9V	
3R		○	Duty solenoid valve (Turbo boost pressure control)	Ignition switch OFF	0V	Duty pulse
				Idle	Below 2.0V	
3S		○	Stepping motor (Metering oil pump)	—	—	Can not check with circuit tester (Refer to Section D)
3T						
3U						
3V						
3W		○	Injector (Front primary)	Ignition switch ON	Approx. 12V	Ground time is very short
				Idle	Approx. 12V	
3X		○	Injector (Front secondary)	Ignition switch ON	Approx. 12V	Ground time is very short
				Idle	Approx. 12V	
3Y		○	Injector (Rear primary)	Ignition switch ON	Approx. 12V	Ground time is very short
				Idle	Approx. 12V	
3Z		○	Injector (Rear secondary)	Ignition switch ON	Approx. 12V	Ground time is very short
				Idle	Approx. 12V	

Component Replacement

NOTE: On vehicles equipped with Audio Anti-Theft system, before performing any repairs, the Personal Code Number (P.C.N.) must be obtained from the individual owner of the vehicle, or an authorized Mazda dealer.

RELIEVING FUEL PRESSURE

The fuel in the fuel system remains under high pressure even when the engine is not running. So before disconnecting any fuel lines, release the fuel pressure to reduce the risk of injury or fire. Relieve the fuel pressure as follows:

1. Start the engine.
2. Disconnect the fuel pump connector with the engine running.
3. After the engine stalls from lack of fuel, turn the ignition switch off.
4. Use a shop rag to cover the fuel lines when disconnecting and plug all fuel lines after they have been disconnected.

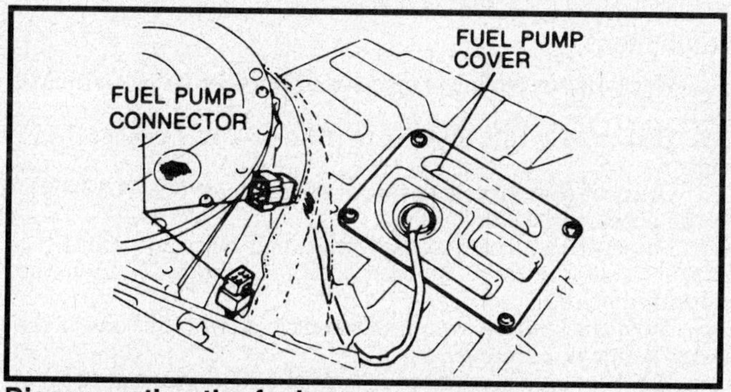

Disconnecting the fuel pump connector in the hatch area to relieve fuel pressure—RX-7

SERVICE PRECAUTIONS

- Do not operate the fuel pump when the fuel lines are empty.
- Do not reuse fuel hose clamps.
- Make sure all EGI harness connectors are fastened securely. A poor connection can cause an extremely high surge voltage in the coil and condenser and result in damage to integrated circuits.
- Keep the EGI harness at least 4 in. away from adjacent harnesses to prevent an EGI system malfunction due to external electronic "noise."
- Keep EGI all parts and harnesses dry during service.
- Before attempting to remove any parts, turn off the ignition switch and disconnect the battery ground cable.
- Always use a 12 volt battery as a power source.
- Do not attempt to disconnect the battery cables with the engine running.

up the engine immediately after starting or just prior to shutdown.

- Do not attempt to disassemble the EGI control unit under any circumstances.
- If installing a 2-way or CB radio, keep the antenna as far as possible away from the electronic control unit. Keep the antenna feeder line at least 8 in. away from the EGI harness and do not let the 2 run parallel for a long distance. Be sure to ground the radio to the vehicle body.
- Do not apply battery power directly to injectors.
- Handle air flow meter carefully to avoid damage.
- Do not disassemble air flow meter or clean meter with any type of detergent.

AIR FLOW METER

Removal and Installation

1. Disconnect the negative battery cable.
2. Loosen the hose band and remove the intake hose.
3. Remove the air flow meter attaching bolts.
4. Turn the air cleaner cover upside down and remove the attaching nuts, remove the air flow meter.
5. Installation is the reverse order of the removal procedure.

FUEL PUMP

Removal and Installation

1. Relieve the fuel pressure from the fuel system.
2. Lift up the rear mat and remove the fuel pump cover.
3. Disconnect the fuel pump wiring connector.
4. Disconnect and plug the fuel main and return lines from the fuel pump.
5. Remove the fuel pump and fuel tank gauge as an assembly.
6. Replace the fuel pump.
7. Installation is the reverse order of the removal procedure.

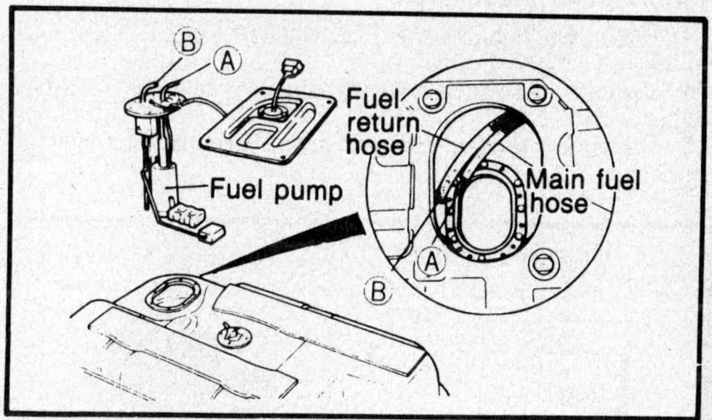

Fuel pump removal

PRESSURE REGULATOR

Removal and Installation

1. Relieve the fuel pressure from the fuel system. Disconnect the negative battery cable.
2. Remove the throttle chamber (dynamic chamber).
3. Disconnect the fuel return hose.
4. Remove the pressure regulator.
5. Installation is the reverse order of the removal procedure. Be sure to replace the O-ring(s) and after installation check for fuel leaks with fuel pressure applied.

FUEL FILTER

Removal and Installation

The fuel filter should be replaced every 30,000 miles. Replace the filter as follows:

1. Relieve the fuel pressure from the fuel system. Disconnect the negative battery cable.
2. Raise the front of the front of the car and support it with jack stands.
2. Disconnect and plug the fuel lines. Remove the fuel filter with bracket as an assembly.
3. Install a new filter and reconnect the fuel lines.

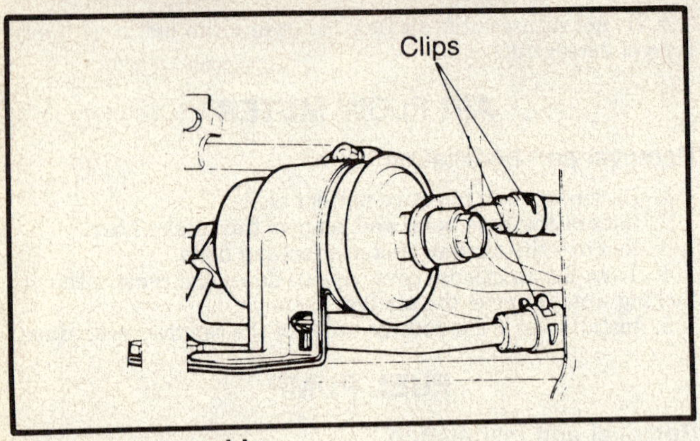

Fuel filter assembly

4. Replace the copper washer with a new one and torque it to 18–25 ft.lbs. When installing the fuel filter make sure that the fuel lines are pushed on as far up as possible. Secure the hose with clamps.

FUEL INJECTORS

Removal and Installation

1. Relieve the fuel pressure from the fuel system. Disconnect the negative battery cable.
2. Remove the throttle body (dynamic chamber) as described earlier
3. Disconnect the fuel hose and pipe. Disconnect the injector connector.

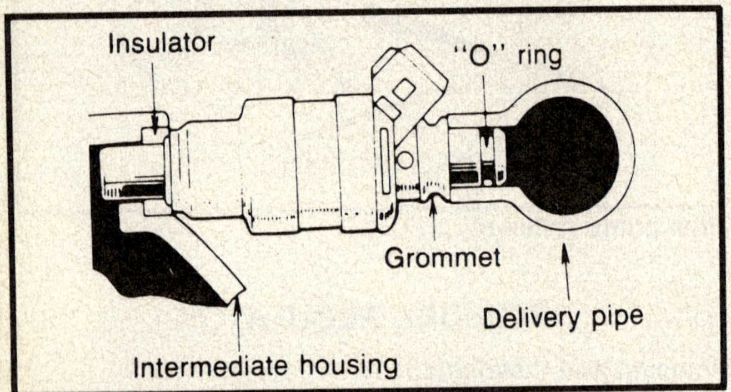

Side view of injector mounting

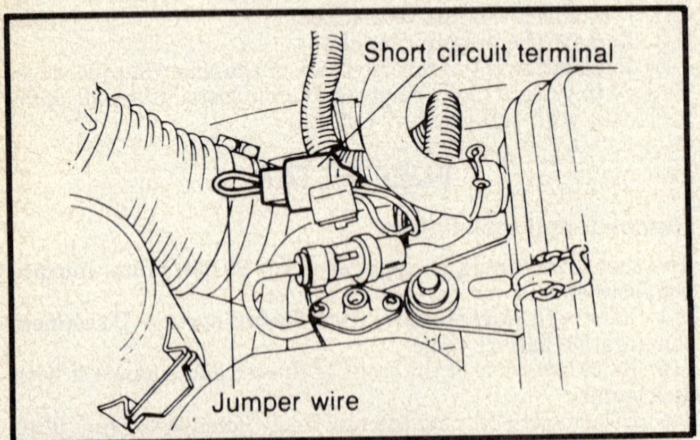

Fuel pump short-circuit connector

4. Remove the pressure regulator and distribution pipe.
5. Remove the injector. Once the injector is removed, use an ohmmeter and measure the resistance at the injector connector (on the injector). The resistance should be 12–16 ohms.
6. Installation is the reverse order of the removal procedure. Be sure to replace the O-ring(s) and after installation check for fuel leaks with fuel pressure applied.

THROTTLE BODY

Removal and Installation

1. Disconnect the negative battery cable. Disconnect the accelerator cable from the throttle linkage.
2. Disconnect the cruise control cable, if equipped.
3. Disconnect the air funnel. Disconnect the hoses and tubes from the throttle body.
4. Disconnect the throttle position sensor connector. Remove the throttle body retaining screws and remove the throttle body assembly.
5. Installation is the reverse order of the removal procedure.

THROTTLE POSITION SENSOR

Adjustment

1. Warm up the engine to operating temperature, then turn it **OFF**.
2. Disconnect the connector from the throttle position sensor.
3. Connect the throttle position sensor tester (49–F018–001) to the check connector (green).
4. Loosen the throttle sensor mounting bolts and move the sensor to set the correct closed position voltage. Tighten the mounting bolts.
5. Turn the ignition switch **ON** and check whether one of the tester lamps illuminates.
6. If both lamps illuminate or if neither does, turn the throttle sensor adjusting screw until one of the lamps illuminates.
   a. If both lamps illuminate turn the adjusting screw counter-clockwise.
   b. If both lamps do not illuminate turn the adjusting screw clockwise.
7. Reinstall the cap on the adjusting screw after adjustment.

NOTE: Do not apply excessive pressure on the adjusting screw, as it may cause incorrect adjustment.

IDLE SYSTEM

Idle Speed Adjustment

Before adjusting the idle speed and idle mixture, be sure the ig-

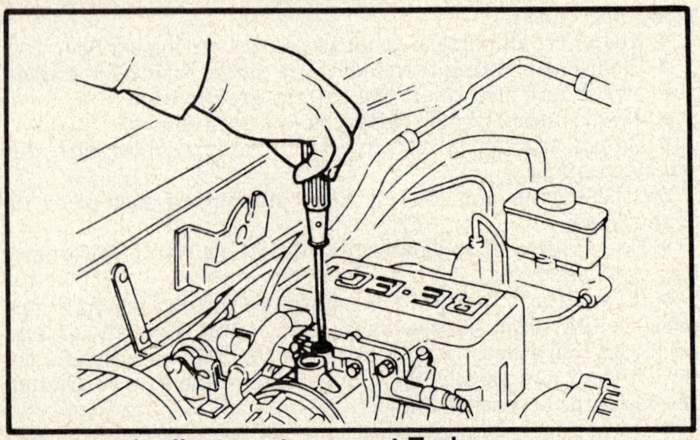

Idle speed adjustment—except Turbo

nition timing, spark plugs, etc., are all in normal operating condition. Turn off all lights and other unnecessary electrical loads.

1. Connect a suitable tachometer to the check connector (Black), located beside the trailing side coil. If the tachometer does not operate properly, reconnect it to the leading side coil.
2. Ground the test connector (green: 1-pin).
3. Warm up the engine.
4. Check the idle speed. If necessary, turn the throttle adjust screw and set the idle speed to 725–775 rpm.

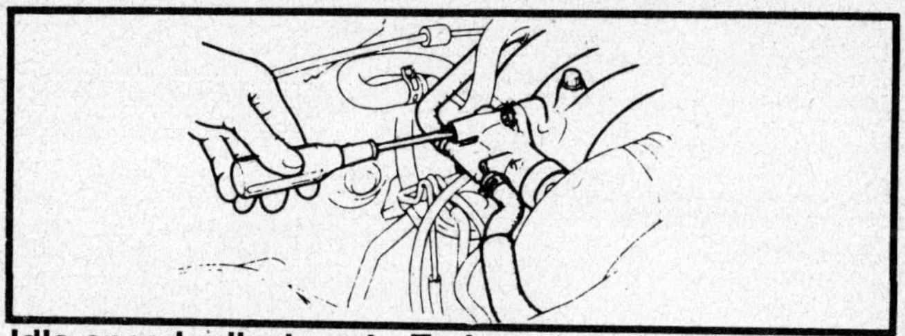

Idle speed adjustment—Turbo

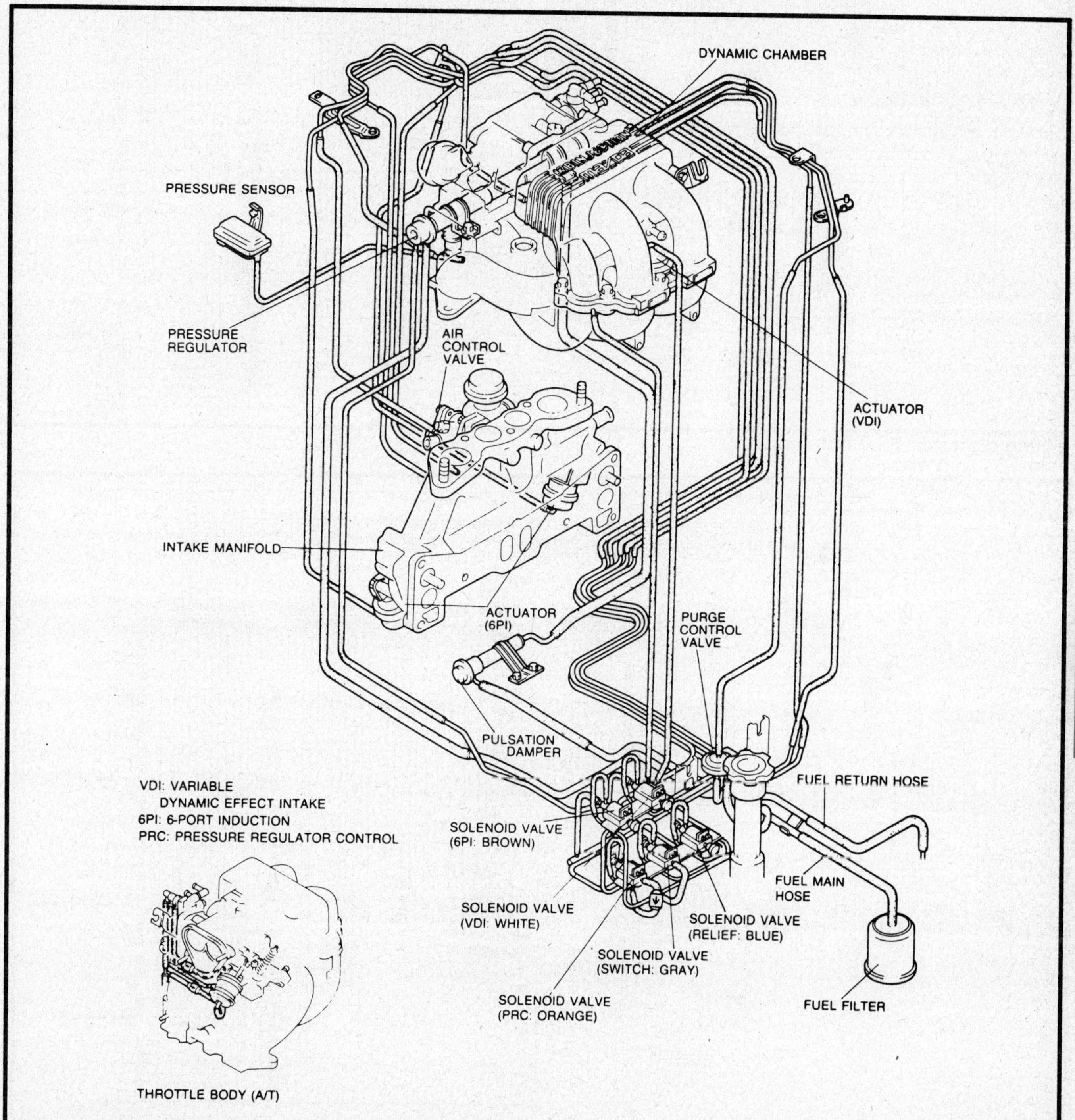

Vacuum hose routing diagram—RX-7 non-turbo

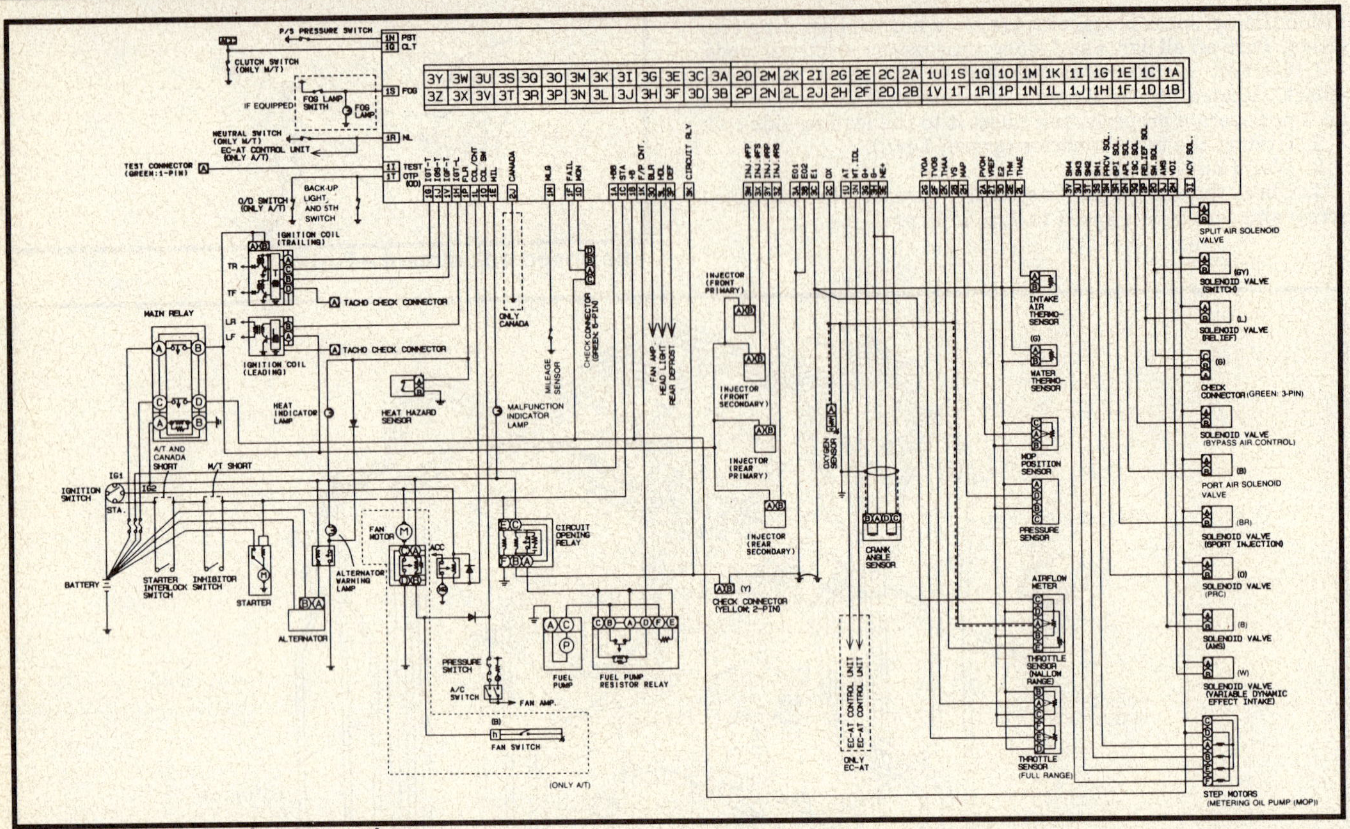

Wiring diagram—RX-7 non-turbo

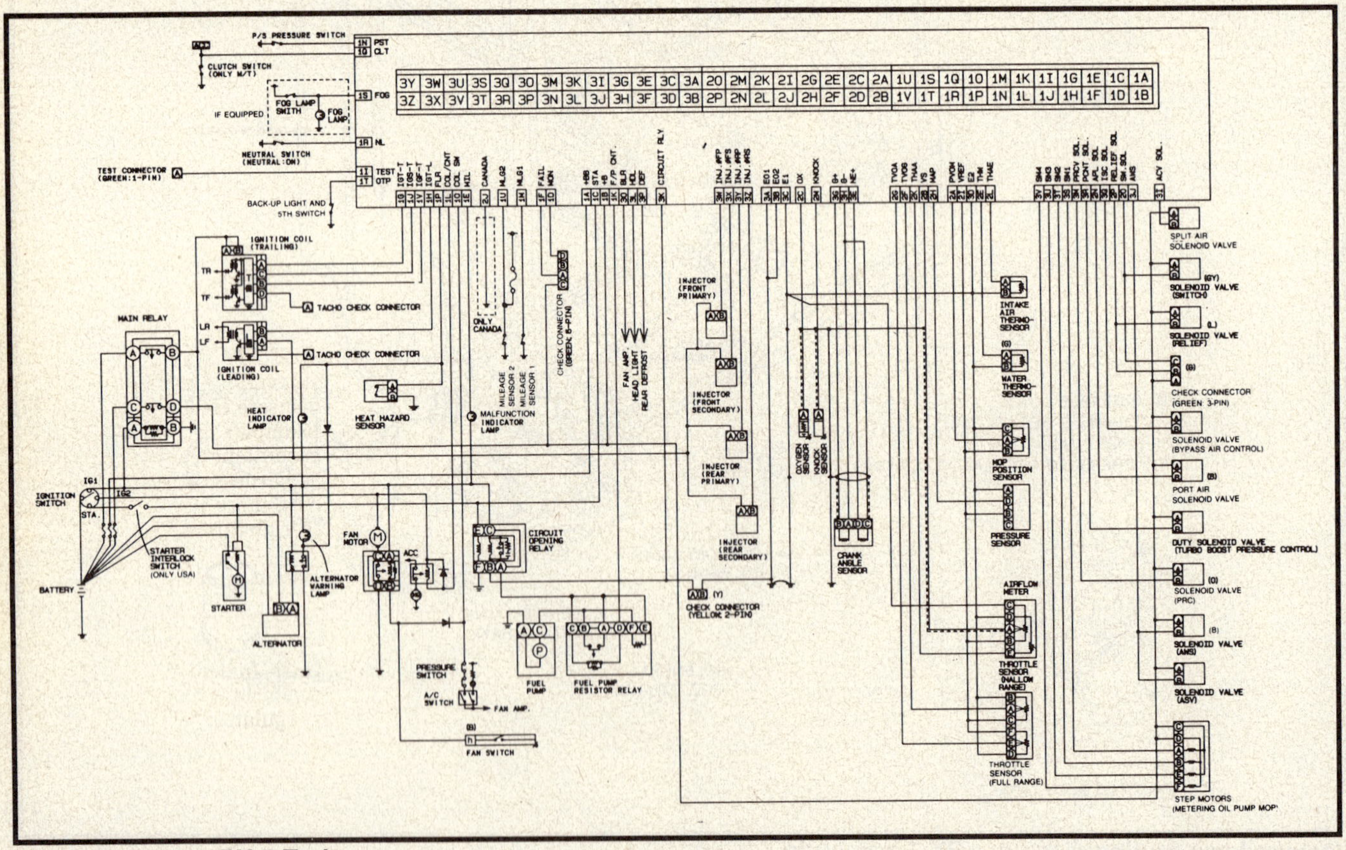

Wiring diagram—RX-7 Turbo

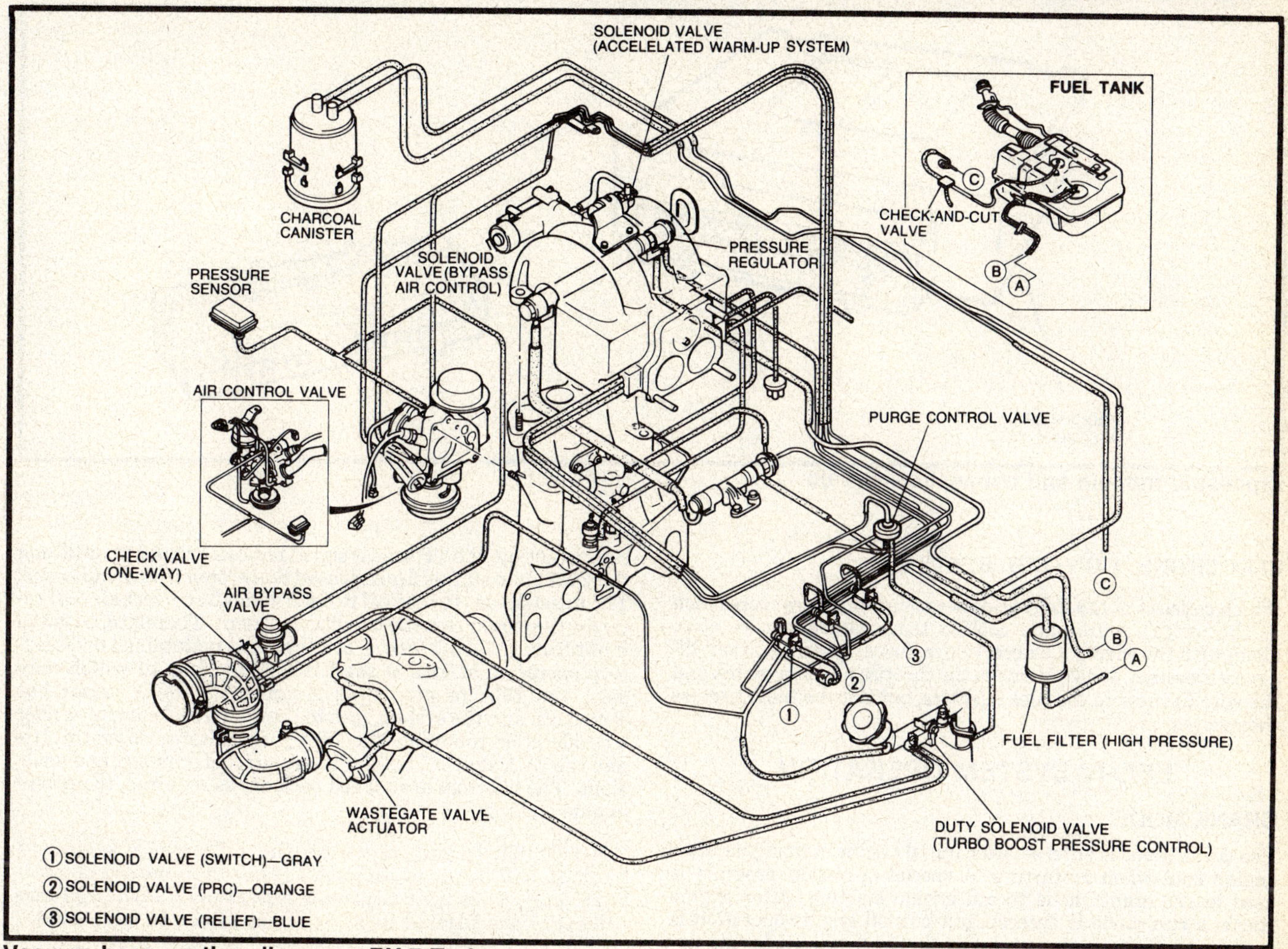

Vacuum hose routing diagram—RX-7 Turbo

MAZDA NAVAJO EEC-IV ENGINE CONTROL SYSTEM

General Information

The EEC-IV engine control system heart is a microcomputer called an Electronic Control Assembly (ECA). The ECA receives data (system inputs) from sensors, switches, relays and other electronic components and issues command signals (system outputs) to various devices in order to control engine operation under a variety of loads and ambient conditions. The ECA is calibrated according to the powertrain, axle ratio and gross vehicle weight (GVW) to optimize fuel economy and driveability while minimizing harmful emissions.

The EEC-IV system electronically controls the fuel injectors to maintain a 14.7:1 air/fuel ratio under all driving conditions. This 14.7:1 air/fuel mixture allows the 3-way catalytic converter to operate at peak efficiency while getting the most performance and economy from the engine. Ignition spark timing, deceleration fuel cut-off, curb and fast idle speed, evaporative emissions purge, air conditioning cut-off during wide open throttle, cold engine start and enrichment, electric fuel pump and self-test engine diagnostics are also controlled by the ECA to maintain consistent driveability across a wide range of operating conditions, temperatures and altitudes. The EEC-IV system is self-adjusting for operation at high altitude elevations (over 4000 ft. above sea level) and can actually compensate for engine component wear (a worn timing chain, for example).

The ECA in the EEC-IV system contains the calibration module within the ECA assembly. The harness connectors are edge-card type contacts which provide a more positive connection. The ECA is mounted behind the right front (passenger) kick panel.

The EEC-IV engine control system is divided into 3 major systems:

FUEL DELIVERY SYSTEM

This includes the fuel tank and lines, fuel pump and fuel injection components. On fuel injected vehicles, this includes the fuel supply manifold, fuel pressure regulator, injectors and fuel filter.

AIR INDUCTION SYSTEM

This includes the air cleaner and ducts, intake manifold, fuel injection throttle body, throttle air bypass valve, vane airflow meter and related vacuum hoses and air ducts.

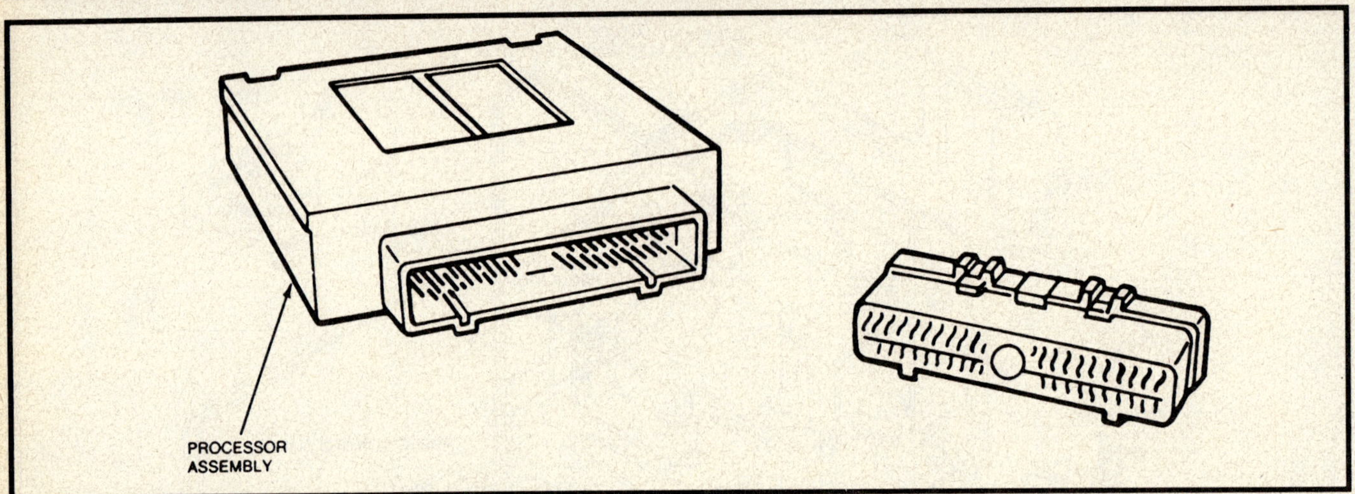

Processor module and connector—Navajo

ELECTRONIC CONTROL SYSTEM

Which consists of the ECA and its various engine sensors (such as the oxygen sensor and coolant temperature sensor, etc.), along with the wiring harness, relays, fuses, battery and self-diagnostic system. It also includes the throttle position switch and the wire harness to the fuel injectors, but not the fuel injectors themselves.

EEC-IV SYSTEM OPERATION

CRANK MODE

The crank mode is entered after initial engine starting, or after engine stall when restarting. A special operation program is used in the crank mode to aid engine starting. After engine starts, a run mode is entered and normal engine operation is performed. If the engine stumbles during a run mode, the underspeed mode is entered to help it recover from the stumble and prevent stalling. When cranking the engine, the fuel control is in the open loop mode (no feedback to the ECA) of operation and the ECA sets engine timing at 10–15 degrees BTDC.

The injectors fire in a simultaneous, double-fire manner (EFI—fires twice every crankshaft revolution), to provide the base crank air/fuel control. The throttle air bypass valve solenoid is set to open the bypass valve to provide the fast idle/no-touch start.

UNDERSPEED MODE

Operation in the underspeed mode (under 500 rpm) is similar to that previously described for the crank mode. The system switches from the underspeed mode to the normal run mode when the required rpm is reached. The underspeed mode is used to provide a good pulse width to the injectors and ignores any signal from the air flow meter.

CLOSED THROTTLE MODE (IDLE OR DECELERATION) MODE

In the closed throttle mode, the air/fuel ratio is trimmed by either varying the pulse width of the output from the ECA to the injectors to obtain the desired mixture. To calculate what this output signal should be, the ECA evaluates inputs from the ECT sensor, the vane meter, the TP sensor, the EGO sensor, the PIP sensor and the air conditioning clutch. These sensors inform the ECA of the various conditions that must be evaluated in order for the ECA to detect the EGO sensor has cooled off, and the system goes to open-loop fuel control. Under a deceleration condition, the TP sensor signal indicates closed throttle and the ECA shut-off fuel for improved fuel economy and emissions. The injectors are turned off to determine the correct air/fuel ratio for the closed-throttle condition present. Therefore, with the input from the EGO sensor, the system is maintained in closed-loop operation at idle. If the EGO sensor fails to switch rich/lean, the ECA programming assumes the EGO sensor has cooled off, and the system goes to open-loop fuel control. Under a deceleration condition, the TP signal indicates closed throttle and the ECA shut-off fuel for improved fuel economy and emissions. The injectors are turned back on, as required, to prevent engine stalling.

NOTE: The point at which the injectors are turned back on will occur at different rpm's, depending on calibration factors and engine temperature, although the injectors are turned back on if the throttle is opened.

Ignition timing is also determined by ECA using these same inputs. The ECA has a series of tables programmed into the assembly at the factory. These tables provide the ECA with a reference of desired ignition timing for the various operating conditions reflected by the sensor inputs. The throttle air bypass valve position is determined by the ECA as a function of RPM, ECT, air conditioning ON or OFF, throttle mode and time since start-up inputs.

PART THROTTLE MODE (CRUISE)

The air/fuel mixture ratio and ignition timing are calculated in the same manner as previously described for the closed throttle mode. The fuel control system remains in closed-loop during part throttle operation, as long as the EGO sensor is operational. In part throttle operation, the throttle air bypass valve is positioned to provide an electronic dashpot function in the event the throttle is closed. Again, as in the closed throttle mode, the ECA makes this determination based on the inputs from the applicable sensors. The TP sensor provides the throttle plate position signal to the ECA.

WIDE OPEN THROTTLE (WOT) MODE

Control of the air/fuel ratio in WOT mode is the same as in part, or closed throttle situations, except that fuel control switches to open-loop, and the fuel injector pulse width is increased to provide additional fuel enrichment. This pulse width increase is applied as a result of the WOT signal from the TP sensor to the ECA. More spark advance is added in WOT for improved performance.

COLD OR HOT ENGINE OPERATION

This modified operation changes the normal engine operation output signals, as required, to adjust for uncommon engine operating conditions. These include cold or excessively hot engine.

LIMITED OPERATION STRATEGY (LOS)

In this operation, the ECA provides the necessary output signals to allow the vehicle to "limp home" when an electronic malfunction occurs, sometimes referred to as limp-in mode. The air bypass valve goes to a fixed voltage, timing is locked at the fixed timing (depends on calibration), and the injector pulse width is constant.

MULTI-POINT FUEL INJECTION SYSTEM

The Multi-Point (EFI) Fuel Injection sub system include a high pressure tank-mounted fuel pump, fuel charging manifold, pressure regulator, fuel filter and both solid and flexible fuel lines. The fuel charging manifold includes 6 electronically controlled fuel injectors, each mounted directly above an intake port in the lower intake manifold. The injectors are energized in 2 banks of 3, once each crankshaft revolution.

The fuel pressure regulator maintains a constant pressure drop across the injector nozzles. The regulator is referenced to intake manifold vacuum and is connected parallel to the fuel injectors and positioned on the far end of the fuel rail. Any excess fuel supplied by the pump passes through the regulator and is returned to the fuel tank via a return line.

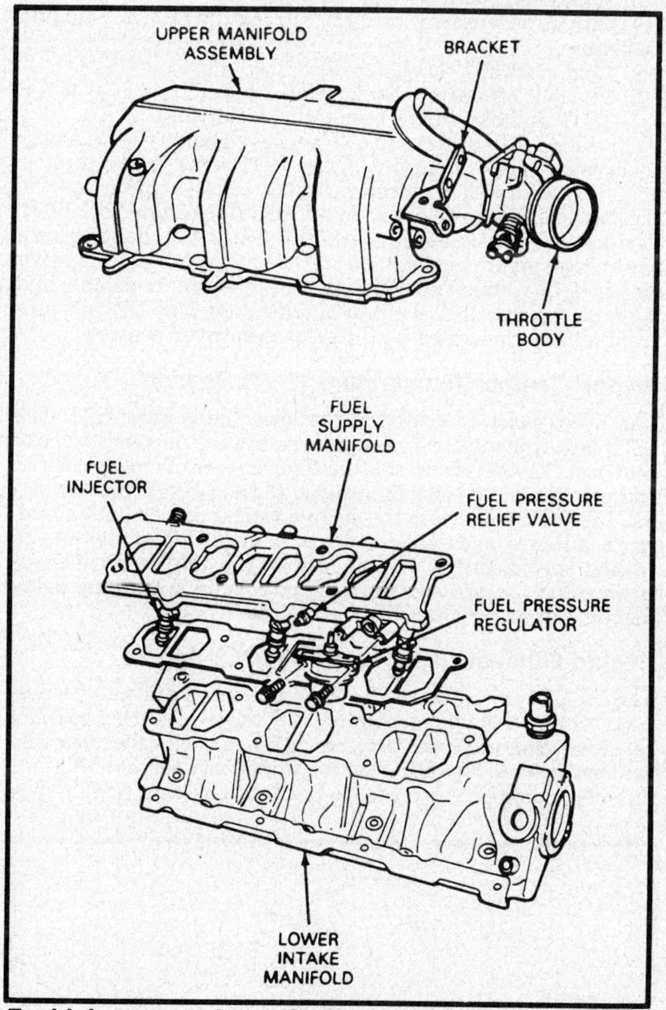

Fuel injectors and manifolding on the 4.0L engine—Navajo

NOTE: The pressure regulator reduces fuel pressure to 39–40 psi under normal operating conditions. At idle or high manifold vacuum condition, fuel pressure is reduced to approximately 30 psi.

The fuel pressure regulator is a diaphragm operated relief valve in which the inside of the diaphragm senses fuel pressure and the other side senses manifold vacuum. Normal fuel pressure is established by a spring preload applied to the diaphragm. Control of the fuel system is maintained through the EEC power relay and the EEC-IV control unit, although electrical power is routed through the fuel pump relay and an inertia switch. The fuel pump relay is located under the hood in the relay box on the passenger side fenderwall and the inertia switch is located under the dash on the passenger side near the transmission hump. The fuel pump is located in-tank.

The inertia switch opens the power circuit to the fuel pump in the event of a collision. Once tripped, the switch must be reset manually by pushing the reset button on the assembly. Check that the inertia switch is reset before diagnosing power supply problems to the fuel pump circuit.

Fuel Injectors

The fuel injectors used with the EFI system are electromechanical (solenoid) type designed to meter and atomize fuel delivered to the intake ports of the engine. The injectors are mounted in the lower intake manifold and positioned so that their spray nozzles direct the fuel charge in front of the intake valves. The injector body consists of a solenoid actuated pintle and needle valve assembly. The control unit sends an electrical impulse that activates the solenoid, causing the pintle to move inward off the seat and allow the fuel to flow. The amount of fuel delivered is controlled by the length of time the injector is energized (pulse width), since the fuel flow orifice is fixed and the fuel pressure drop across the injector tip is constant. Correct atomization is achieved by contouring the pintle at the point where the fuel enters the pintle chamber.

NOTE: Exercise care when handling fuel injectors during service. Be careful not to lose the pintle cap and replace O-rings to assure a tight seal. Never apply direct battery voltage to test a fuel injector.

The injectors receive high pressure fuel from the fuel manifold (fuel rail) assembly. The complete assembly includes a single, preformed tube with 6 injector connectors, mounting flange for the pressure regulator, mounting attachments to locate the manifold and provide the fuel injector retainers and a Schrader® quick-disconnect fitting used to perform fuel pressure tests.

The fuel manifold is normally removed with fuel injectors and pressure regulator attached. Fuel injector electrical connectors are plastic and have locking tabs that must be released when disconnecting when disconnecting the wiring harness.

Throttle Air Bypass Valve

The throttle air bypass valve is an electro-mechanical (solenoid) device whose operation is controlled by the EEC-IV control unit. A variable air metering valve controls both cold and warm idle air flow in response to commands from the control unit. The valve operates by bypassing a regulated amount of air around the throttle plate; the higher the voltage signal from the control unit, the more air is bypassed through the valve. In this manner, additional air can be added to the fuel mixture without moving the throttle plate. At curb idle, the valve provides smooth idle for various engine coolant temperatures, compensates for air conditioning load and compensates for transaxle load and no-load conditions. The valve also provides fast idle for

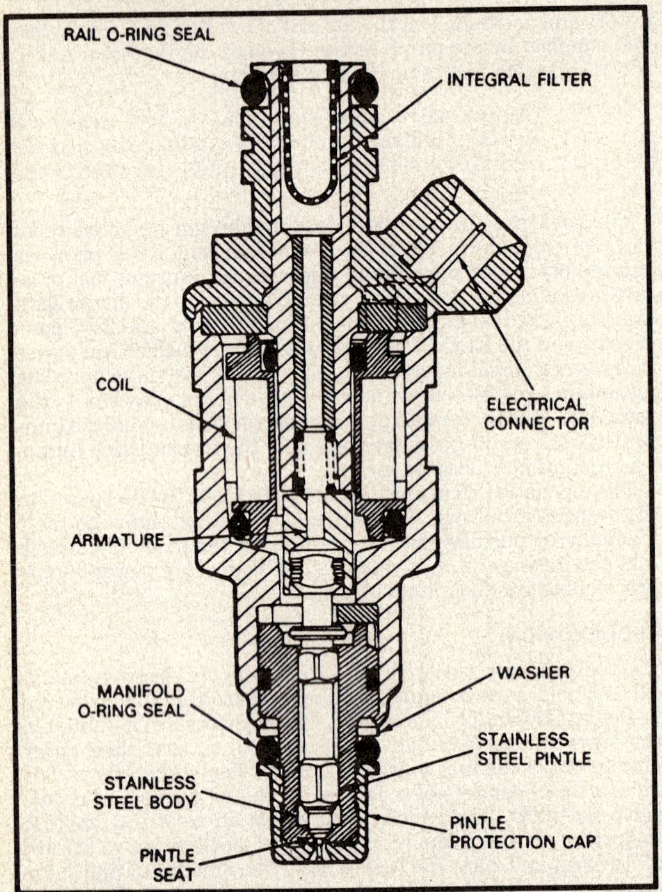

Fuel injector cross section – Navajo

start-up, replacing the fast idle cam, throttle kicker and anti-dieseling solenoid common to previous methods of idle control.

There are no curb idle or fast idle adjustments. As in curb idle operation, the fast idle speed is proportional to engine coolant temperature. Fast idle kick-down will occur when the throttle is kicked. A time-out feature in the ECA will also automatically kick-down fast idle to curb idle after a time period of approximately 15–25 seconds; after coolant has reached approximately 71°C (160°F). The signal duty cycle from the ECA to the valve will be at 100% (maximum current) during the crank to provide maximum air flow to allow no touch starting at any time (engine cold or hot).

DATA SENSORS

The electronic engine control sub system consists of the ECA and various sensors and actuators. The ECA reads inputs from engine sensors, then outputs a voltage signal to various components (actuators) to control engine functions. The period of time that the injectors are energized ("ON" time or "pulse width") determines the amount of fuel delivered to each cylinder. The longer the pulse width, the richer the fuel mixture.

NOTE: The operating reference voltage (Vref) between the ECA and its sensors and actuators is 5 volts. This allows these components to work during the crank operation even though the battery voltage drops.

In order for the ECA to properly control engine operation, it must first receive current status reports on various operating conditions. The control unit constantly monitors crankshaft position, throttle plate position, engine coolant temperature, exhaust gas oxygen level, air intake mass and temperature, air conditioning (On/Off) and barometric pressure.

Distributorless Ignition System (DIS)

The DIS ignition systems eliminates the need for a distributor by using multiple coils, which fire 2 spark plugs at the same time. This system uses a dual function crank sensor. The dual function crank sensor provides both Cylinder Identification (CID), used to choose which coil to fire and Profile Ignition Pick-up (PIP) for spark timing, signals to the DIS module. The DIS module functions include:

- Selection of Coil(s)
- Drives the coils
- Drives the tachometer
- PIP (crankshaft position to ECA)
- CID (cylinder identification)
- LOS (limited operation strategy – base timing)

Throttle Position (TP) Sensor

The TP sensor is mounted on the throttle body. This sensor provides the ECA with a signal that indicates the opening angle of the throttle plate. The sensor output signal uses a 5 volt reference voltage (Vref). From this input, the ECA controls:

1. Operating modes, which are wide open throttle (WOT), part throttle (PT) and closed throttle (CT).
2. Fuel enrichment at WOT.
3. Additional spark advance at WOT.
4. Air conditioning cut off during WOT (30 seconds maximum).
5. Cold start kick-down.
6. Fuel cut off during deceleration.
7. WOT dechoke during crank mode (starting).

On the EEC-IV system, the TP sensor signal to the ECA only changes the spark timing during the WOT mode. As the throttle plate rotates, the TP sensor varies its voltage output. As the throttle plate moves from a closed throttle position to a WOT Position, the voltage output of the TP sensor will change from a low voltage (approximately 1.0 volt) to a high voltage (approximately 4.75 volts). The TP sensor used is not adjustable and must be replaced if it is out of specification. The EEC-IV programming compensates for differences between sensors.

Engine Coolant Temperature (ECT) Sensor

The ECT sensor is located in the lower intake manifold. The ECT is a thermistor (changes resistance as temperature changes). The sensor detects the temperature of engine coolant and provides a corresponding signal to the ECA. From this signal, the ECA will modify the air/fuel ratio (mixture), idle speed, spark advance and canister purge control. When the engine coolant is cold, the ECT signal causes the ECA to provide enrichment to the air/fuel ratio for good cold drive away as engine coolant warms up, the voltage will drop.

Heated Exhaust Gas Oxygen (HEGO) Sensor

The HEGO sensor on the EEC-IV system is a little different from others used and is mounted in its own mounting boss, located between the 2 downstream tubes in the header near the exhaust system. The EGO sensor works between 0–1 volt output, depending on the presence (lean) or absence (rich) of oxy-

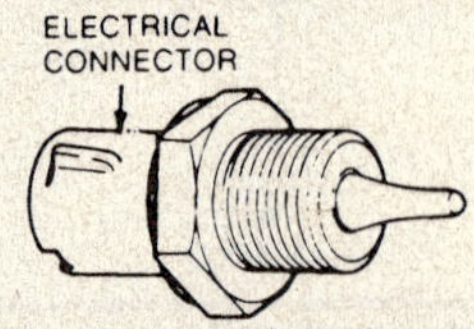

Coolant temperature sensor – Navajo

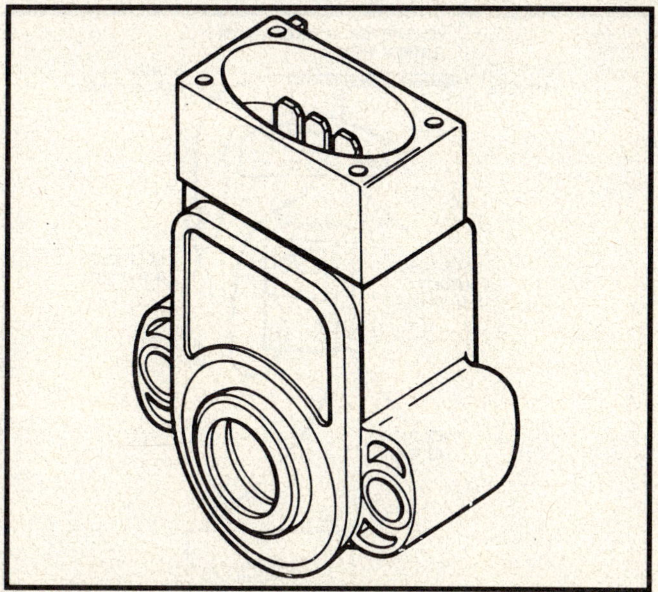

Throttle position sensor—Navajo

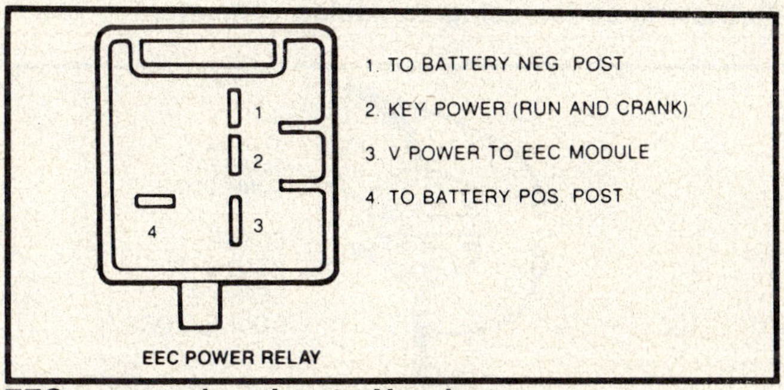

EEC power relay pinout—Navajo

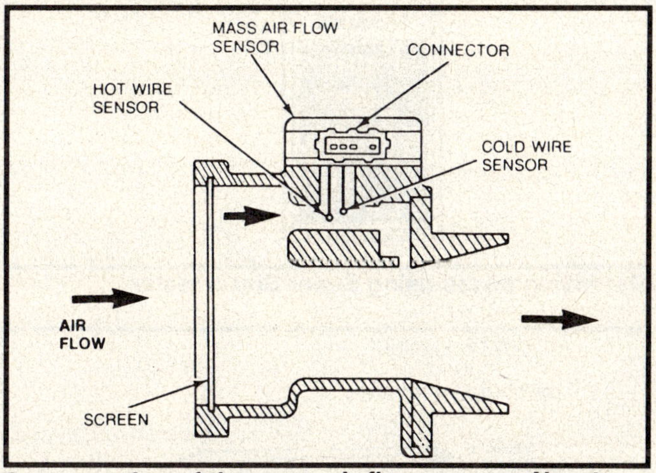

Cross section of the mass air flow sensor—Navajo

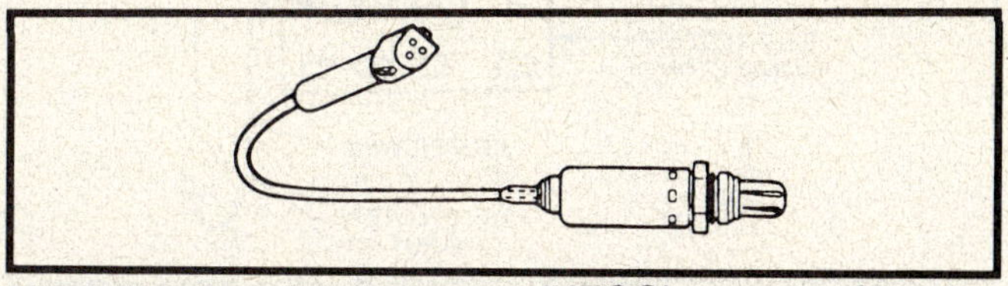

Heated exhaust gas oxygen (HEGO) sensor—Navajo

gen in the exhaust gas. A voltage reading greater than 0.6 volts indicates a rich air/fuel ratio, while a reading of less than 0.4 volts indicates a lean air/fuel ratio.

NOTE: Never apply voltage to the EGO sensor because it could destroy the sensor's calibration. This includes the use of an ohmmeter. Before connecting and using a voltmeter, make sure it has a high-input impedance (at least 10 megohms) and is set on the proper resistance range. Any attempt to use a powered voltmeter to measure the EGO voltage output directly will damage or destroy the sensor.

The EGO sensors incorporate a heating element (HEGO), to bring the sensor up to operating temperature more quickly and keep it there during extended idle periods to prevent the sensor from cooling off and placing the system into open loop operation.

Mass Air Flow Sensor

As opposed to vane type air flow meters, the mass air flow meter on the Navajo directly measures the mass of the air coming into the engine. The signal output from the sensor is an analog voltage ranging from 0.5 volts to 5.0 volts. The ECA uses this information to calculate the injector pulse width to vary the injection amounts. The element used to sense the air flow is a platinum wire wrapped around a ceramic rod and coated with glass. The wire is heated electrically and a second sensing cold wire measures the temperature of the hot wire. The hot wire is maintained at 200° C above the ambient air temperature. The power required to maintain this temperature is proportional to the mass of the air flowing through the sensor.

Air Conditioning Clutch Compressor (ACC) Signal

Anytime battery voltage is applied to the A/C clutch, the same signal is also applied to the ECA. The ECA then maintains the engine idle speed with the throttle air bypass valve control solenoid (fuel injection) to compensate for the added load created by the A/C clutch operation. Shutting down the A/C clutch will have a reverse effect. The ECA will maintain the engine idle speed.

Barometric (BAP) Sensor

The barometric sensor is used to compensate for altitude variations. From this signal, the ECA modifies the air/fuel ratio, spark timing and idle speed. The barometric sensor is a design that produces a frequency based on atmospheric pressure (altitude). The barometric sensor is mounted on the rear of the engine compartment near the firewall.

Diagnosis and Testing

SERVICE PRECAUTIONS

- Do not operate the fuel pump when the fuel lines are empty.
- Do not reuse fuel hose clamps.
- Make sure all ECA harness connectors are fastened securely. A poor connection can cause an extremely high surge voltage in the coil and condenser and result in damage to integrated circuits.
- Keep the EGI harness at least 4 in. away from adjacent harnesses to prevent an EGI system malfunction due to external electronic "noise."
- Keep all ECA related parts and harnesses dry during service.
- Before attempting to remove any parts, turn off the ignition switch and disconnect the battery ground cable.
- Always use a 12 volt battery as a power source.

- Do not attempt to disconnect the battery cables with the engine running.
- Do not attempt to disassemble the ECA control unit under any circumstances.
- If installing a 2-way or CB radio, keep the antenna as far as possible away from the electronic control unit. Keep the antenna feeder line at least 8 in. away from the EGI harness and do not let the 2 run parallel for a long distance. Be sure to ground the radio to the vehicle body.
- Do not apply battery power directly to injectors.
- Handle air flow meter carefully to avoid damage.
- Do not disassemble air flow meter or clean meter with any type of detergent.

READING CODES

With Super Star II Tester

KEY ON ENGINE OFF (KOEO)

1. Specific instructions for the use of the tester are included with the tester.
2. Plug the tester connectors into the vehicle self test connectors.
3. Set the switches on the tester to the appropriate settings for type of EEC system and for fast or slow mode. The same information is contained in fast or slow mode, but the fast mode is 100 times faster than the slow and is for use with the Super Star Tester.
4. Turn on the tester power and depress the test button. Turn **ON** the vehicle ignition key.
5. The tester will read the codes as they exist.

KEY ON ENGINE RUNNING (KOER)

1. Connect the tester as in KOEO test and start and warm the engine to normal operating temperature.
2. Turn the engine off and depress the test button so it stays in the test position.
3. Restart the engine and the tester will read any existing codes. The tester will display an engine I.D. code, then on some vehicles a Dynamic Response code. The final display will be the service codes.

With Analog Voltmeter

1. Turn the vehicle ignition switch **OFF**. Set the range of the analog voltmeter to 0–15 volts.
2. Connect the positive lead of an analog voltmeter to the positive terminal of the battery.
3. Connect the negative lead of the voltmeter to the Self Test Output pin of the Self Test connector.
4. Connect a jumper wire between the Self Test Input connector and the Signal Return pin of the Self Test connector.
5. The needle will sweep 3 times for a code digit of 3, for example. If the code is 123, the needle will sweep once, pause for 2 seconds, sweep twice, pause for 2 seconds, then sweep three times. There will be a 4 second pause between codes. The method of readout for 2 digit codes is similar to the three digit code readout, but only 2 sets of sweeps will occur.
4. The continuous memory codes will be displayed first, then there will be a 6 second pause, a single sweep, another 6 second pause then the KOEO codes will be displayed.

WITH MALFUNCTION INDICATOR OR CHECK ENGINE LIGHT

1. Connect a jumper wire between the Self Test Input (STI) connector and the Signal Return pin of the Self Test connector.
2. This will output the codes to the MIL or Check Engine light on the instrument panel.
3. The light will flash 3 times for a code digit of 3, for example. If the code is 123, the light will flash once, pause for 2 sec-

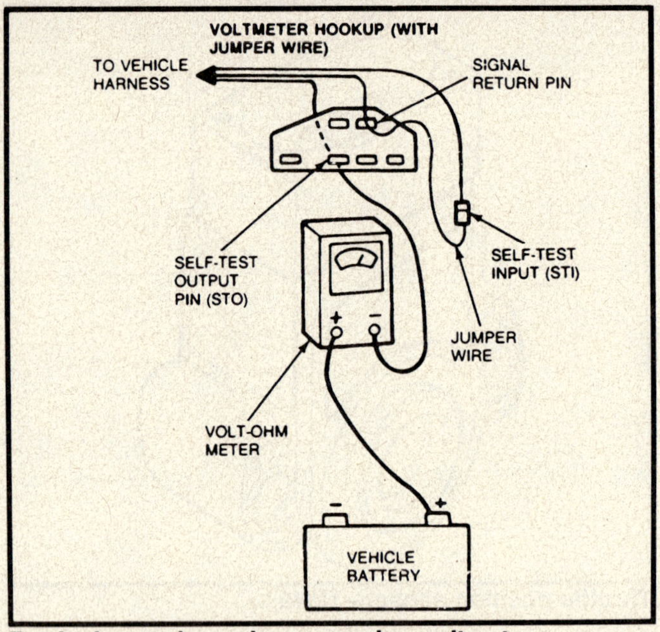

Retrieving codes using an analog voltmeter

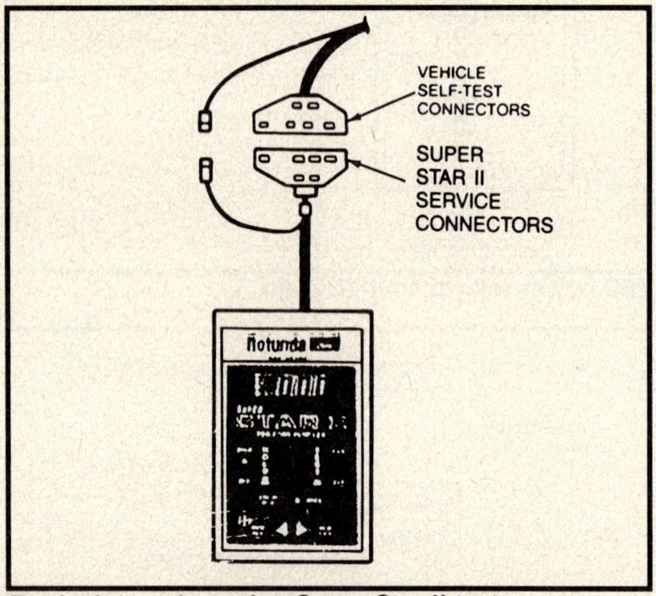

Retrieving codes using Super Star II tester

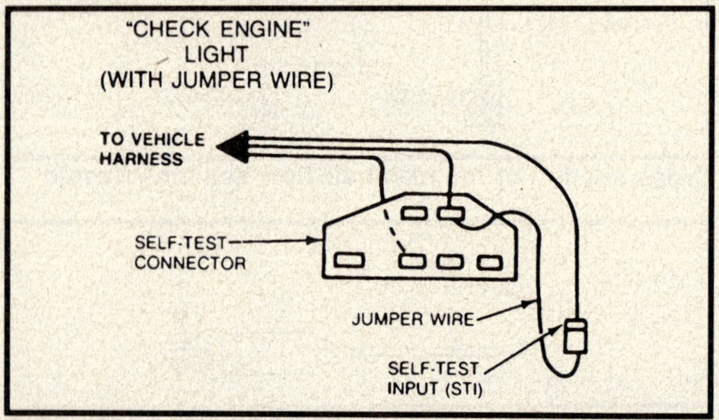

Retrieving codes using the instrument panel light

onds, flash twice, pause for 2 seconds, then flash three times. There will be a 4 second pause between codes. The method of readout for 2 digit codes is similar to the three digit code readout, but only 2 sets of sweeps will occur.

4. The continuous memory codes will be displayed first, then there will be a 6 second pause, a single flash, another 6 second pause then the KOEO codes will be displayed.

CLEARING CODES

Continuous Codes

1. Run the Self Test for the output system being used; Super Star II, analog voltmeter, or dash light.
2. When the service codes start to be displayed the codes will be cleared when the Test button is unlatched on the Super Star II tester. For the other retrieval methods the codes will be cleared when the jumper is removed between the Self Test Input connector and the Signal Return pin of the Self Test connector.
3. Disconnect the Super Star II tester or voltmeter.

Keep Alive Memory

The Keep Alive Memory will be erased when the negative battery cable is disconnected. Keep the cable off for a minimum of 5 minutes. When the cable is connected and the vehicle driven, it may exhibit some drivability problems for the first 10 miles or so. This is due to the relearning process where the ECA memorizes values needed for optimum drivability.

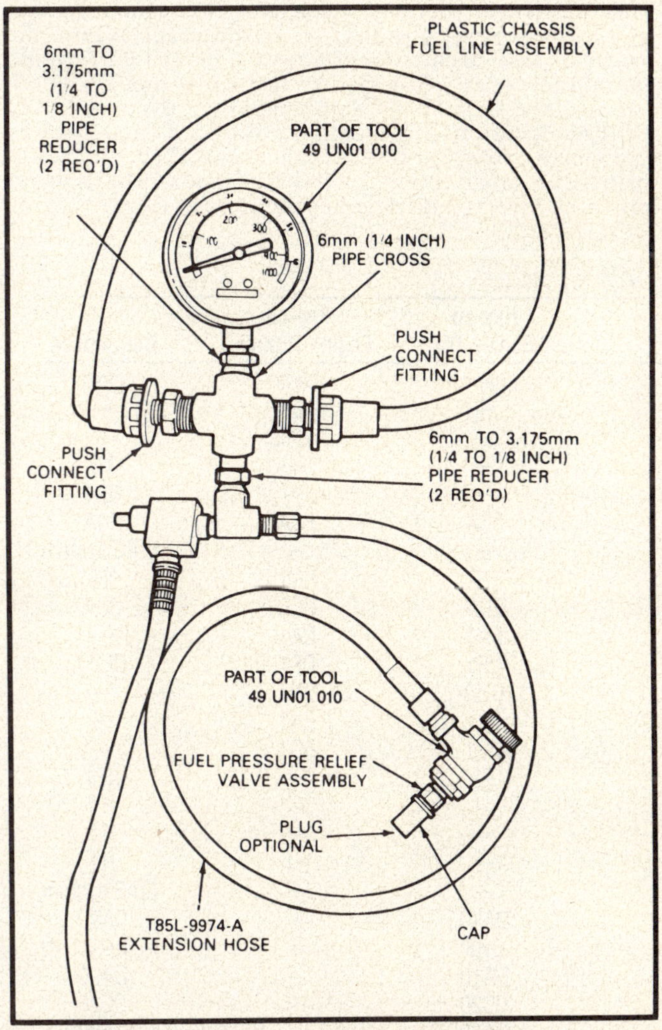

Fuel pressure testing equipment—Navajo

ELECTRIC FUEL DELIVERY SYSTEMS

CAUTION

Fuel pressure must be relieved before attempting to disconnect any fuel lines or personal injury may result.

Pressure Tests

The diagnostic pressure valve (Schrader type) is located on the fuel rail on multi-point systems. This valve provides a convenient point for service personnel to monitor fuel pressure, release the system pressure prior to maintenance, and to bleed out air which may become trapped in the system during filter replacement. A pressure gauge with an adapter is required to perform pressure tests.

If the pressure tap is not installed or an adapter is not available, use a T-fitting to install the pressure gauge between the fuel filter line and the throttle body fuel inlet or fuel rail.

Testing fuel pressure requires the use of a special pressure gauge (49 UN01 010 or equivalent) that attaches to the diagnostic pressure tap fitting. Depressurize the fuel system before disconnecting any lines.

EEC-IV SYSTEM TESTING

As in any service procedure, a routine inspection of the EEC-IV system for loose connections, broken wires or obvious damage is the best way to start. Perform the system Quick Test outlined below before going any further. Check all vacuum connections and secondary ignition wiring before assuming that the problem lies with the EEC-IV system. A self-diagnosis capability is built into the EEC-IV system to aid in troubleshooting. The primary tool necessary to read the trouble codes stored in the system is an analog voltmeter or special Super Self Test Automatic Readout (SUPER STAR II) tester (499 UN10 056, or equivalent). The Malfunction Indicator Light (MIL) can also be used to retrieve the codes. While the self-test is not conclusive by itself, when activated it checks the EEC-IV system by testing its memory integrity and processing capability. The self-test also verifies that all sensors and actuators are connected and working properly.

EEC MONITOR FOR INTERMITTENTS

The EEC monitor and recorder act as a scan tool into the ECA system. They view the same sensor and actuator values that the ECA uses. The monitor displays these values for both static and dynamic conditions. The purpose of this monitor is to let the technician see the same information that the processor sees and observe how the processor reacts to the information.

Connecting the EEC Monitor requires disconnecting the ECA and connecting special harness adapter to the ECA, monitor and engine control harness. The monitor can then view all circuits going to or from the ECA.

The EEC monitor comes with special overlays for each vehicle to be tested. This manual will provide the data reference charts for use with the EEC monitor, or equivalent. But, since the monitor itself comes with special overlays and instruction that only work for the EEC Monitor those are not included. All diagnostic reference valve for all engines are included after the pinpoint tests called "Monitor Intermittents and Diagnostic Values".

EEV IV System Quick Test

Correct test results for the quick test are dependent on the correct operation of related non-EEC components, such as ignition wires, battery, etc. It may be necessary to correct defects in these areas before the EEC-IV system will pass the quick test. Before connecting any test equipment to check the EEC system, make the following checks:

1. Check the air cleaner and intake ducts for leaks or restrictions. Replace the air cleaner if dirty or clogged.
2. Check all engine vacuum hoses for proper routing according to the vacuum schematic on the underhood sticker. Check for proper connections and repair any broken, cracked or pinched hoses or fittings.
3. Check the EEC system wiring harness connectors for tight fit, loose or detached terminals, corrosion, broken or frayed wires, short circuits to metal in the engine compartment or melted insulation exposing bare wire.

NOTE: It may be necessary to disconnect or disassemble the connector to check for terminal damage or corrosion and perform some of the inspections. Note the location of each pin in the connector before disassembly. When doing continuity checks to make sure there are no breaks in the wire, shake or wiggle the harness and connector during testing to check for looseness or intermittent contact.

4. Check the control module, sensors and actuators for obvious physical damage.
5. Turn **OFF** all electrical loads when testing and make sure the doors are closed whenever readings are made. Do not disconnect any electrical connector with key **ON**. Turn the key **OFF** to disconnect or reconnect the wiring harness to any sensor or the control unit.
6. Make sure the engine coolant and oil are at the proper level.
7. Check for leaks around the exhaust manifold, oxygen sensor and vacuum hoses connections with the engine idling at normal operating temperature.
8. Check the ignition and fuel systems to make sure there is fuel and spark. Check the distributor cap, rotor and internal components for damage, corrosion or signs of excessive wear.
9. Check battery for full charge and cable connections for tightness.
10. Inspect the relay connector and make sure the ECA power relay is securely attached and making a good ground connection.
11. Twist the oil filler cap on turbocharged engines to make sure it is tight. A loose oil filler cap will cause turbocharged engines to run rough at idle.
12. Only after all the above checks have been performed should the voltmeter be connected to read the trouble codes. If not, the self-diagnosis system may indicate a failed component when all that is wrong is a loose or broken connection.

QUICK TEST STEPS

1. Visually check of hoses, wiring, sensors, actuators and fuel system. Check battery for proper charge.
2. Warm engine to normal operating temperature.
3. Attach test equipment.
4. Perform "Key ON Engine OFF Self-Test".
5. Perform "Computed Timing Check".
6. Perform "Engine RUNNING Self-Test".
7. Perform "Continuous Self-Test".
8. Perform "Diagnostic by Symptom".
9. Proceed to pinpoint test only as directed by Quick-Test or Self-Test codes.
10. Replace components only when instructed in pinpoint test.

NOTE: Never measure voltage or resistance directly at the processor connector. Disconnect solenoids, switches and circuits from harness, before performing individual testing. For EEC-IV test codes and code definitions, see Section 3 "Self-Diagnostic Systems". Any ECA pin locations given are for reference only. Probing any connectors with DVOM, unless specified, will result in permanent damage to the pin connector. Always probe as directed using a breakout box.

Navajo 2 Digit Engine Codes

Code and Description	Key-on Engine Off	Key-on Engine Running	Continuous
Code 11—System pass:	pass	pass	pass
Code 12—Cannot control rpm during high rpm check:	—	KE1	—
Code 13—Cannot control rpm during low rpm check:	—	KE15	—
Code 14—PIP circuit failure:	—	—	N1
Code 15—ECA keep alive only memory (KAM) test failed:	—	—	QB1
Code 15—ECA read only memory (ROM) test failed:	ECA	ECA	ECA
Code 16—IDM signal not received:	—	—	EDIS system
Code 18—Loss of IDM input or SAW circuit open:	—	PC1	N3
Code 19—Failure in ECA internal voltage:	ECA	ECA	ECA
Code 21—ECT out of range:	DA1	DA1	—
Code 22—Barometric Pressure (BP) sensor:	DF1	DF1	DF90
Code 23—TP sensor out of self-test range:	DH2	DH1	—
Code 24—ACT sensor out of self-test range:	DA1	DA1	—
Code 26—MAF out of self-test range:	DC1	DC1	—
Code 29—Insufficient input from VSS:	—	—	DP1
Code 41—Fuel system at adaptive limits, no HEGO switch:	—	—	H1
Code 41—Lack of HEGO switches, indicates lean:	—	H1	H1
Code 42—Lack of HEGO switches, indicates rich:	—	H1	H1
Code 45—Coil 1, 2 or 3 failure:	—	—	EDIS system
Code 51—ECT indicated −40°F open circuit:	DA10	—	DA90
Code 53—TP sensor circuit above maximum voltage	DH3	—	DH90
Code 54—ACT indicated −40°F open circuit:	DA10	—	DA90
Code 56—MAF circuit above maximum voltage:	DC20	—	DC20
Code 61—ECT indicated 254°F grounded circuit:	DA20	—	DA90

Navajo 2 Digit Engine Codes

Code and Description	Key-on Engine Off	Key-on Engine Running	Continuous
Code 63—TP sensor circuit below minimum voltage:	DH10	—	DH94
Code 64—ACT indicated 254°F grounded circuit:	DA20	—	DA90
Code 66—MAF below minimum voltage:	—	—	DC10
Code 67—Clutch switch circuit failure:	—	—	FA1
Code 67—Vehicle not in Park during KOEO:	FA1	—	—
Code 72—Insufficient MAF change during test:	—	DC10	—
Code 73—Insufficient TP change dynamic rseponse test:	—	DH20	—
Code 74—BOO circuit failure/not actuated during self-test:	—	FD1	FD90
Code 77—Operator error during test:	—	M1	—
Code 79—A/C ON/ Defroster ON/ during self-test:	KM40	—	—
Code 86—Shift solenoid circuit failure	TB1	—	—
Code 87—Primary fuel pump circuit faiure:	J7	—	J95
Code 89—Clutch Converter Override (CCO) circuit failure:	TB1	—	—
Code 95—Fuel pump circuit open-ECA to motor ground:	J20	—	J20
Code 96—Fuel pump circuit open-ECA to battery:	J30	—	J30
Code 98—Hard fault present:	TB1	Quick test	Quick test
Code not listed:	QA1	QA1	QA1

Navajo 3 Digit Engine Codes

Code and Description	Key-on Engine Off	Key-on Engine Running	Continuous
Code 111—System pass:	pass	pass	pass
Code 112—ACT indicated 254°F grounded circuit:	DA20	—	DA90
Code 113—ACT indicated −40°F open circuit:	DA10	—	DA90
Code 114—ACT sensor out of self-test range:	DA1	DA1	—
Code 116—ECT out of range:	DA1	DA1	—
Code 117—ECT indicated 254°F grounded circuit:	DA20	—	DA90
Code 118—ECT indicated −40°F open circuit:	DA10	—	DA90
Code 121—TP sensor out of self-test range:	DH2	DH1	—
Code 122—TP sensor circuit below minimum voltage:	DH10	—	DH94
Code 123—TP sensor circuit above maximum voltage	DH3	—	DH90
Code 126—Barometric Pressure (BP) sensor:	DF1	DF1	DF90
Code 128—Barometric Pressure (BP) sensor:	—	—	DF11
Code 129—Insufficient MAF change during test:	—	DC10	—
Code 136—Fuel control:	—	H1	—
Code 137—Fuel control:	—	H1	—
Code 139—Fuel control:	—	—	H1
Code 144—Fuel control:	—	—	H1
Code 157—MAF below minimum voltage:	—	—	DC10
Code 158—MAF circuit above maximum voltage:	DC20	—	DC20
Code 159—MAF out of self-test range:	DC1	DC1	—
Code 167—Insufficient TP change dynamic response test:	—	DH20	—
Code 171—Fuel system at adaptive limits, no HEGO switch:	—	—	H1
Code 172—Lack of HEGO switches, indicates lean:	—	H1	H1
Code 173—Lack of HEGO switches, indicates rich:	—	H1	H1
Code 174—Fuel control:	—	—	H1
Code 175—Fuel control:	—	—	H1
Code 176—Fuel control:	—	—	H1
Code 177—Fuel control:	—	—	H1
Code 178—Fuel control:	—	—	H1
Code 179—Lean adaptive limit at part throttle, system rich:	—	—	H15
Code 181—Rich adaptive limit at part throttle, system lean:	—	—	H1
Code 182—Fuel control:	—	—	H15
Code 183—Fuel control:	—	—	H1

Navajo 3 Digit Engine Codes

Code and Description	Key-on Engine Off	Key-on Engine Running	Continuous
Code 188—Fuel control:	—	—	H15
Code 189—Fuel control:	—	—	H1
Code 191—Fuel control:	—	—	H15
Code 192—Fuel control:	—	—	H1
Code 194—Fuel control:	—	—	H5
Code 195—Fuel control:	—	—	H5
Code 211—PIP circuit failure:	—	—	NC1
Code 212—Loss of IDM input or SAW circuit grounded:	—	—	NC3
Code 213—SAW circuit open:	—	P1	—
Code 215—Ignition system:	—	—	EDIS system
Code 216—Ignition system:	—	—	EDIS system
Code 217—Ignition system:	—	—	EDIS system
Code 218—Ignition system:	—	—	EDIS system
Code 222—Ignition system:	—	—	EDIS system
Code 223—IDM signal not received:	—	—	EDIS system
Code 224—IDM signal not received:	—	—	EDIS system
Code 226—IDM signal:	NC2	—	—
Code 232—Coil 1, 2 or 3 failure:	—	—	EDIS system
Code 238—Ignition system:	—	—	EDIS system
Code 338—Temperature sensor:	—	—	DA100
Code 339—Temperature sensor:	—	—	DA101
Code 341—Octane adjust circuit open:	KP1	—	—
Code 411—Cannot control rpm during low rpm check:	—	KE15	—
Code 412—Cannot control rpm during high rpm check:	—	KE1	—
Code 452—Insufficient input from VSS:	—	—	DP1
Code 511—ECA read only memory (ROM) test failed:	ECA	ECA	ECA
Code 512—ECA keep alive only memory (KAM) test failed:	—	—	QB1
Code 513—Failure in ECA internal voltage:	ECA	ECA	ECA
Code 522—Vehicle not in P or N during KOEO:	FA1	—	—
Code 525—Vehicle not in P or N during KOEO:	FA1	—	FA1
Code 528—Clutch switch circuit failure:	—	—	FA1
Code 529—Check Engine light:	—	—	ML25
Code 533—Check Engine light:	—	—	ML25
Code 536—BOO circuit failure/not actuated during self-test:	—	FD1	FD90
Code 538—Operator error during test:	—	M1	—
Code 539—A/C ON/ Defroster ON/ during self-test:	KM40	—	—
Code 542—Fuel pump circuit open-ECA to motor ground:	J20	—	J20
Code 542—Fuel pump circuit open-ECA to battery:	J30	—	J93
Code 556—Primary fuel pump circuit faiure:	J7	—	J95
Code 565—Canister purge (CANP) circuit failure:	KD6	—	—
Code 566—Shift solenoid circuit failure:	TB1	—	—
Code 569—Canister purge (CANP) circuit failure:	KD6	—	—
Code 629—Clutch Converter Override (CCO) circuit failure:	TB1	—	—
Code 998—Hard fault present:	TC10	Quick test	Quick test
Code not listed:	QA1	QA1	QA1
600 Series codes are transmission codes			

DIAGNOSTIC ROUTINES—NAVAJO

Diagnostic Routine Index

GASOLINE ENGINES

Title	Routine
Cranks Normally But Won't Start	201
Starts Normally But Won't Run (Stalls)	202
Cranks Normally But Slow to Start	203
Rough Idle	204
Misses Under Load	205
Low Idle (Stalls on Deceleration or Quick Stop)	206
Hesitates or Stalls on Acceleration	207
Backfire (Induction or Exhaust)	208
Lack of Power	209
Surges at Steady Speed	210
High Idle (Engine Diesels)	211
Engine Noise	212
Poor Fuel Economy	213
High Oil Consumption	214
Spark Knock/Pinging	215
Engine Vibrates at Normal/High Speeds	216
Engine Runs Cold	217
Engine Runs Hot	218
Exhaust Smoke	219
Gas Smell	220
"CHECK ENGINE" Light Always On or Never On	221
State Emission Test Failure	222
Improper Shift	223

Diagnostic Routines

201 CRANKS NORMALLY BUT WON'T START

NOTE: Extended cranking, because of a "No Start" condition, can load the exhaust system with raw fuel, which can ruin the catalytic converter after the engine starts. After the "No Start" condition has been repaired, remove the fuel relay fuse, run the engine until surplus fuel is used up and reinstall the fuel relay fuse.

System	Component
EEC	Quick Test
Ignition	Electrical Connections Secondary Ignition Wires Spark Plugs Fouled Ignition Switch Hall Crankshaft Sensor EDIS Ignition Module Coil Pack
Fuel Delivery	Filter Pump Water/Dirt/Rust Contamination in Fuel Lines Tank (Fuel Supply) Sender Filter Fuel Pressure Regulator Injectors Inertia Switch
Fuel Charging Assy./Throttle Body	Electrical Connections Throttle Linkages
Basic Engine	Camshaft Timing Compression
Exhaust	Catalyst Exhaust

Diagnostic Routines

202 STARTS NORMALLY BUT WON'T RUN (STALLS)

System	Component
Fuel Charging Assy/Throttle Body	Idle Speed Electrical and Vacuum Connections
EEC	Quick Test
Fuel Delivery	Filter Pump Water/Dirt/Rust Contamination in Fuel Lines Tank (Fuel Supply) Sender Filter Fuel Pressure Regulator Injectors Inertia Switch
Ignition	Electrical Connections Secondary Ignition Wires Ignition Switch Hall Crankshaft Sensor EDIS Ignition Module Coil Pack
Exhaust	Component (Restricted)
Basic Engine	Camshaft and Valve Train

Diagnostic Routines

203 CRANKS NORMALLY BUT SLOW TO START

NOTE: It is a good practice to confirm that the correct starting procedure was being used by the customer before proceeding with diagnosis.

System	Component
Ignition	Scope Engine for: Spark Plugs, Coil, Secondary Ignition Wires Spark Plugs Fouled Hall Crankshaft Sensor EDIS Ignition Module Coil Pack(s)
EEC	Quick Test
Fuel Delivery	Filter Pump Water/Dirt/Rust Contamination in Fuel Lines Fuel Pressure Regulator Sender Filter Injectors
Fuel Charging Assy/Throttle Body	Idle Speed Electrical and Vacuum Connections
Exhaust	Component (Restricted)
Induction and Vacuum Distribution	Vacuum Leaks Air Cleaner Element Restricted
PCV	Valve
EVAP	Components

DIAGNOSTIC ROUTINES—NAVAJO

Diagnostic Routines

204 ROUGH IDLE

System	Component
Cooling	Fan (Loose or Cracked)
Ignition	Scope Engine For: Spark Plug, Coil, Secondary Wires Ignition Timing
EEC	Quick Test
Fuel Charging Assy/Throttle Body	Idle Speed Electrical and Vacuum Connections
Fuel Delivery	Fuel Pressure Regulator Injectors Fuel Rail
Vacuum Distribution	Vacuum Leaks
Cooling	Thermostat
Basic Engine	Compression Valve Train Camshaft Intake Manifold Gaskets
PCV	Valve
EVAP	Components
Exhaust	Pipes, Muffler, Catalyst, Resonator

Diagnostic Routines

205 MISSES UNDER LOAD

System	Component
Ignition	Scope Engine For: Spark Plug, Coil, Secondary Wires, Timing
EEC	Quick Test
Fuel Delivery	Filter Pump Lines Fuel Pressure Regulator Sender Filter Injectors
Fuel Charging Assy/Throttle Body	Electrical and Vacuum Connections

206 LOW IDLE (STALLS ON DECEL OR QUICK STOP)

System	Component
Fuel Charging Assy/Throttle Body	Curb or Fast Idle Speed Electrical and Vacuum Connections
EEC	Quick Test

Diagnostic Routines

207 HESITATES OR STALLS ON ACCELERATION

System	Component
EEC	Quick Test
Ignition	Scope Engine For: Spark Plug, Coil and Secondary Wires Ignition Timing
Fuel Charging Assy/Throttle Body.	Electrical & Vacuum Connections Idle Speeds
Fuel Delivery	Filter Pump Water/Dirt/Rust Contamination in Fuel Lines Fuel Pressure Regulator: Sender Filter Injectors
Vacuum Distribution	Vacuum Leaks
Induction	Air Cleaner Duct, Stove Pipe, and Valve
PCV	Valve
Base Transmission	Converter Clutch Override Converter Clutch

Diagnostic Routines

208 BACKFIRE (INDUCTION OR EXHAUST)

System	Component
Ignition	Scope Engine For: Spark Plug, Coil, Secondary Wires, Crossed Wires Ignition Timing
Vacuum Distribution	Vacuum Hoses, or Connections Leak(s)
EEC	Quick Test
Basic Engine	Intake Manifold Gaskets Compression Check Camshaft Valves
Exhaust	Components (Restricted)
Fuel Delivery	Filter Pump Water, Dirt, Rust, Contamination in Fuel Lines Fuel Pressure Regulator Injectors Sender Filter

Diagnostic Routines

209 LACK OF POWER

System	Component
Ignition	Timing
EEC	Quick Test
Fuel Delivery	Filter Pump Lines Fuel Pressure Regulator Sender Filter Injectors
Fuel Charging Assy/Throttle Body	Electrical and Vacuum Connections Idle Speed
Ignition	Scope Engine For: Spark Plugs, Coil, etc.
Exhaust	Components (Restricted)
Cooling	Thermostat
Vacuum Distribution	Vacuum Leaks
Induction	Air Cleaner Duct and Valve and Element
Basic Engine	Compression Check Camshaft Valves
Drive Train	Clutch, Automatic Transmission, Brakes

Diagnostic Routines

210 SURGE AT STEADY SPEED

System	Component
EEC	Quick Test
Fuel Delivery	Filter Pump Lines Fuel Pressure Regulator Sender Filter
Fuel Charging Assy/Throttle Body	Electrical & Vacuum Connections Idle Speed
Ignition	Scope Engine For: Spark Plugs, Coil, Secondary Ignition Wires, etc., Timing
Vacuum Distribution	Vacuum Leaks
Induction	Air Cleaner Duct and Element
EVAP	Components
Basic Engine	Valve Train and Camshaft Intake Manifold Gaskets

Diagnostic Routines

211 HIGH IDLE (ENGINE DIESELS)

NOTE: If engine idles smoothly after key is turned off, trouble is likely to be in the ignition switch, ignition harness, starter solenoid "IGN" tap, or EEC relay.

System	Component
Fuel Charging Assy/Throttle Body	Idle Speed Electrical and Vacuum Connections Throttle Plate and Linkage Speed Control Chain
Vacuum Distribution	Vacuum Leaks
EEC	Quick Test
Induction	Vacuum Leaks
Cooling	Overheating

Diagnostic Routines

212 ENGINE NOISE

System	Component
Squeal, Click, or Chirp	Oil Level (low) Valve Train Drive Belts (loose) Belt Driven Components EEC Solenoids
Rumble, Grind	Belt Driven Components
Rattle	Component (loose)
Hiss	Vacuum Distribution System (leak) Induction System (leak) Spark Plug (loose)
	Cooling System (leak)
	EVAP System (leak)
Snap	Secondary Ignition
Rap, Roar	Exhaust System (leak)
Knock	Connecting Rod Bearing (worn) Main Bearing (worn) Piston Pin (loose) Piston to Bore Clearance (cold engine)
	Detonation

DIAGNOSTIC ROUTINES—NAVAJO

Diagnostic Routines

213 POOR FUEL ECONOMY

NOTE: Since fuel consumption is drastically increased for city driving, short-run operation, stop and go driving, trailer towing, extended winter warm-up periods, etc., as opposed to "trip" mileage, an attempt should be made to determine these factors when confronted with "poor mileage" conditions. However, since the operator is not always at fault, the following is appended:

System	Component
Fuel Charging Assy/Throttle Body	Electrical & Vacuum Connections
Fuel Delivery	Fuel Return Line Blocked Fuel Pressure Regulator
Induction	Air Cleaner Duct and Valve Air Cleaner Element (Restricted)
Ignition	Scope Engine For: Spark Plug, Coil, Secondary Wires Ignition Timing
EEC	Quick Test
Cooling	Thermostat
Factors External to the Engine	Tire Pressure & Type Clutch Operation Converter Clutch Override Automatic Transmission Shift Pattern and Fluid Level Brake Drag Exhaust System Speedometer/Odometer Gear Ratio Axle Ratio Vehicle Load Road & Weather Conditions Aftermarket Add-Ons

Diagnostic Routines

214 HIGH OIL CONSUMPTION

NOTE: If the condition cannot be verified, clean engine, if necessary, change oil and filter (at customer's expense), seal and have customer drive 500 miles (804.5 Km) or enough distance to consume two quarts before returning for re-examination.

System	Component
External Leaks	Rocker Cover Gasket, Crankshaft Seals, Engine Assembly
Proper Dipstick	Overfilling (due to improper reading of dipstick).
Induction	Air Cleaner Element (Sealing)
PCV	Valve
Internal Leaks (blue smoke from tailpipe)	Valve Guides Valve Stem Seals Intake Manifold and Gasket Cylinder Head Drain Passages Piston Rings

215 SPARK KNOCK/PINGING

NOTE: If the above fails to correct the condition, it is recommended that the owner change his source of fuel and use higher octane fuel.

System	Component
Ignition	Timing
EEC	Quick Test
Vacuum Distribution	Vacuum Leaks
Cooling	Overheating
Basic Engine	Oil Level Compression Check Intake Manifold Gasket
Fuel Delivery	Injectors
PCV	Valve
Induction	Air Cleaner Duct and Valve Assembly

Diagnostic Routines

216 ENGINE VIBRATES AT NORMAL SPEEDS

System	Component
Engine Accessories	Fan Belt Driven Components Engine Mounts Engine Vibration Damper
Otherwise	Non-Engine Components: Drive Line, Tires, Wheel Balance

217 ENGINE RUNS COLD

System	Component
Gauge System	Gauge, Sender
Cooling	Thermostat

218 ENGINE RUNS HOT

System	Component
Cooling	Thermostat Coolant Level Radiator or A/C Condenser Pressure Cap and Overflow System External Leaks Belts and Belt Tension Fan and Fan Clutch
Gauge System	Gauge, Sender
EEC	Quick Test
Ignition	Timing
Basic Engine	Oil Level Internal Leak(s) Core Sand in Head/Block Water Pump
Brake	Brakes (dragging)

Diagnostic Routines

219 EXHAUST SMOKE

NOTE: White Smoke is normal during warm-up.

System	Component
Black Smoke (rich mixture)	Air Cleaner Element (Restricted)
	Fuel Pressure Regulator Injectors
	EEC Components
Blue Smoke (burning oil)	PCV Valve
	Valve Guides/Stems/Seals Oil Drain Passages in Head
	Rings (not seated, seized, gummed up, worn) Cylinder bores (scuffed)
White Smoke (coolant in combustion)	Intake Manifold (cracked/porous) Cylinder Head/Gasket (leaks) Block (cracked/porous)

Diagnostic Routines

220 GAS SMELL

System	Component
Fuel Delivery	Fuel Filter (leaks) Injectors (leaking) Fuel Line, Pump to Tank (leaks) Fuel Tank (leaks) Fuel Tank Filler Neck/Cap (leaks) Fuel Return Line (Blocked) Fuel Pressure Regulator
EVAP	Carbon Canister, Solenoid, Hoses (leaks)

221 "CHECK ENGINE" LIGHT ALWAYS ON OR NEVER ON

System	Component
EEC	Quick Test

Diagnostic Routines

222 STATE EMISSION TEST FAILURE

NOTE: Canada and some states or metropolitan areas in the United States require periodic Idle Emission Tests. All Mazda products have been designed to pass these tests. If a Mazda product fails an Idle Emission Test, it is probable that 1) The engine temperature was not warm and stabilized prior to the test. 2) The vehicle had idled excessively long prior to the test.

Prior to starting any repairs, complaints of Idle Emission Test failure should be verified.

The following example encompasses most of the emissions measurement modes of the current state idle test procedures:

- Ensure that the engine is at normal operating temperature and that all accessories are turned off.
- Read emissions at idle.
- Run engine at 2500 ± 300 rpm.
- Read emissions within 30 seconds.
- Return engine speed to idle.
- Read emissions within 30 seconds.

If any emission components are replaced, Keep Alive Memory (KAM) should be cleared before repeating State Emission Test procedure. Refer to Quick Test Appendix.

Diagnostic Routines

222 STATE EMISSION TEST FAILURE (CONTINUED)

System	Component
EEC	Quick Test
Ignition	Scope Engine For: Spark Plug, Coil, Secondary Wires Timing
Vacuum Distribution	Vacuum Leaks/Blockage
Fuel Charging Assy/Throttle Body	Idle Speed Electrical and Vacuum Connections
Fuel Delivery	Injectors Fuel Rail
PCV	Valve
EVAP	Carbon Canister, Purge Solenoid
Exhaust	Pipes, Muffler, Catalyst Resonator
Cooling	Unstabilized Engine Temperature
Basic Engine	Scheduled Maintenance Compression Valve Train Camshaft Intake Manifold Gaskets

Diagnostic Routines

223 IMPROPER SHIFT

System	Component
Base Transmission	Converter Clutch Control Solenoid Converter Clutch
EEC	Quick Test

DIAGNOSIS BY SYMPTOM—NAVAJO

EFI System Diagnosis By Symptom

SYMPTOM	POSSIBLE SOURCE	ACTION
• No Start — Cold • Backfire (Induction or Exhaust)	• Inoperative ISC motor.	• Go to ECU Diagnostics.
	• TP sensor stuck at WOT	• Crank engine with TP sensor disconnected.
	• Plugged, leaking, or inoperative injector • Insufficient fuel pressure.	• Electric Fuel Delivery Systems. • EFI — Go to Fuel Injector Testing/Cleaning procedure,
	• Throttle body ISC-BPA contaminated.	• Clean Idle Air Bypass Valve. Refer to Idle Air Bypass Valve Cleaning in this section.
	• Throttle plate stop screw backed out.	• Go to adjustment procedure in this section.
		Return to Routine 201,

SYMPTOM	POSSIBLE SOURCE	ACTION
• Hard Start — Cold • Starts Normally But Won't Run (Stalls) — Cold • Cranks Normally But Slow to Start — Cold • Backfire (Induction or Exhaust)	• ISC inoperative.	• Go to ECU Diagnostics.
	• Plugged, leaking, or inoperative injector.	• Refer to Electric Fuel Delivery Systems. • Fuel Injector Testing/ Cleaning
	• Throttle body, ISC-BPA contaminated, icing.	• Clean Idle Air Bypass Valve.
	• Throttle plate stop screw backed out.	• adjustment procedure
		Return to Routine 203.

SYMPTOM	POSSIBLE SOURCE	ACTION
• Rough Idle — Cold • Emission Test Failure	• Plugged, leaking or inoperative injector. • Throttle body, ISC — BPA contaminated, icing.	• Refer to . Electric Fuel Delivery Systems. • Fuel Injector Testing/ Cleaning procedure • Idle Air Bypass Valve
		Return to Routine 204,

EFI System Diagnosis By Symptom

SYMPTOM	POSSIBLE SOURCE	ACTION
• Stall, Stumble, Hesitation — Cold • Hesitates or Stalls on Acceleration • Backfire (Induction or Exhaust)	• TP sensor failure.	• Go to ECU Diagnostics.
	• Plugged, leaking, or inoperative injectors.	• Refer to : Electric Fuel Delivery Systems. • Go to Fuel Injector Testing/ Cleaning procedure in this Section.
	• Throttle body ISC-BPA contaminated or icing.	• Clean Idle Air Bypass Valve.
	• Throttle plate stop screw backed out.	• adjustment procedure
		Return to Routine 202 or 207,

SYMPTOM	POSSIBLE SOURCE	ACTION
• No Start — Hot • Cranks Normally But Won't Start — Hot	• Inoperative ISC motor. • TP sensor stuck at WOT.	• Go to ECU Diagnostics.
	• Plugged, leaking, or inoperative injector. • Insufficient fuel pressure.	• Refer to Electric Fuel Delivery Systems. • Fuel Injector Testing/ Cleaning
		Return to Routine 201,

SYMPTOM	POSSIBLE SOURCE	ACTION
• Hard Start, Hot • Starts Normally But Won't Run (Stalls) — Hot • Cranks Normally But Slow to Start Hot • Backfire (Induction or Exhaust)	• Excessive fuel pressure.	• Clean, service or replace: — Fuel return line. — Fuel pressure regulator. .
	• Contaminated fuel pressure regulator valve and seat.	• Clean, service or replace fuel pressure regulator. • Check fuel pressure bleed down after engine has been turned off.
	• Plugged, leaking, or inoperative injector. • Injector O-ring seal leaking.	• Refer to · Electric Fuel Delivery Systems. • Fuel Injector Testing/ Cleaning
	• ISC inoperative.	• Go to ECU Diagnostics.
		Return to Routine 203,

EFI System Diagnosis By Symptom

SYMPTOM	POSSIBLE SOURCE	ACTION
• Rough Idle — Hot • Emission Test Failure	• Plugged, leaking or inoperative injector. • Injector O-ring seal leaking.	• Refer to ' Electric Fuel Delivery Systems. • Fuel Injector Testing/ Cleaning
	• Throttle body ISC-BPA contaminated.	• Clean Idle Air Bypass Valve.
	Return to Routine 204,	

SYMPTOM	POSSIBLE SOURCE	ACTION
• Stall, Stumble, Hesitation — Hot • Hesitates or Stalls on a Acceleration-Hot	• TP Sensor Malfunction.	• Go to ECU Diagnostics.
	• Plugged, leaking, or inoperative injector.	• Refer to Electric Fuel Delivery Systems. • Fuel Injector Testing/ Cleaning
	• Throttle body ISC-BPA contaminated.	• Clean Idle Air Bypass Valve.
	Return to Routine 202 or 207,	

EFI System Diagnosis By Symptom

SYMPTOM	POSSIBLE SOURCE	ACTION
• Low Idle Speed, Stalls on Decel or Quick Stop	• ISC inoperative. • TP Sensor malfunction.	• Go to ECU Diagnostics.
	• Base idle adjustment.	• adjustment
	Return to Routine 206,	

SYMPTOM	POSSIBLE SOURCE	ACTION
• Lack of Power • Misses Under Load	• TP sensor malfunction	• Go to ECU Diagnostics.
	• Plugged, leaking, or inoperative injectors. • Fuel pressure regulator failure.	• Refer to Electric Fuel Delivery Systems. • Fuel Injector Testing/ Cleaning
	Return to Routine 209,	

DIAGNOSIS BY SYMPTOM—NAVAJO

EFI System Diagnosis By Symptom

SYMPTOM	POSSIBLE SOURCE	ACTION
• Poor Mileage • Misses Under Load	• Injector O-ring seal leaking	• injector leakage test.
		Return to Routines 213, 219,

SYMPTOM	POSSIBLE SOURCE	ACTION
• Reduced Top Speed/Power • Misses Under Load • Lack of Power	• Plugged, leaking, or inoperative injectors. • Fuel pressure failure.	• Refer to Electric Fuel Delivery Systems. • Fuel Injector Testing/ Cleaning
	• TP Sensor malfunction.	• Go to ECU Diagnostics.
		Return to Routine 209,

EFI System Diagnosis By Symptom

SYMPTOM	POSSIBLE SOURCE	ACTION
• Surge at Cruise • Surges at Steady Speed	• Plugged or leaking fuel injectors.	• Refer to Electric Fuel Delivery Systems. • Fuel Injector Testing/ Cleaning
		Return to Routine 210,

SYMPTOM	POSSIBLE SOURCE	ACTION
• Flooding, Black Exhaust Smoke • Gas Smell • Emission Test Failure	• Excessive fuel pressure. • Injector stuck open, or O-ring seal leaking.	• Refer to Electric Fuel Delivery Systems. • Fuel Injector Testing/ Cleaning

SYMPTOM	SYSTEM	POSSIBLE SOURCE	ACTION
• High Idle Speed Engine Diesels or Idles too Fast		• Incorrect idle base adjustment.	• Perform all idle adjustments.
		• Vacuum leaks.	• Check all vacuum lines and connections.
		• Throttle plate, ISC-BPA or linkage sticking.	• Visual check for proper operation. • Clean Idle Air Bypass Valve.
			Return to Routine 211,

EEC Quick Test — Procedures and Appendix

TEST STEP	RESULT ▶	ACTION TO TAKE
1.0 VISUAL CHECK		
• Check vacuum hoses for damage and proper connection. • Check EEC system wiring harness for proper connections. • Are all visual checks OK?	Yes ▶ No ▶	GO to Quick Test Step [2.0]. SERVICE as necessary.
2.0 VEHICLE PREPARATION AND EQUIPMENT HOOKUP		
• Apply parking brake. • Place shift lever firmly in PARK (NEUTRAL for manual transmissions). • Block drive wheels. • Turn off all electrical loads. • Connect appropriate test equipment to the Self-Test connector in order to run Self-Test. • Is vehicle prepared to run Self-Test?	Yes ▶ No ▶	GO to Quick Test Step [3.0]. REPEAT this test step.

EEC Quick Test — Procedures and Appendix

TEST STEP	RESULT ▶	ACTION TO TAKE
3.0 PERFORM KOEO SELF-TEST		
• Start engine and idle until vehicle is at operating temperature. **NOTE: If engine does not start or stalls after starting, continue to perform KOEO Self-Test.** • Key off, wait 10 seconds. • Activate Self-Test. • Key on, engine off. • Record all KOEO and Continuous Memory Codes received. • **Is a PASS code "11" or "111" present in KOEO? (11, 111-10-any code)** **NOTE: The first two are PASS codes from KOEO. The "10" is considered the separator code between KOEO and Continuous Memory.**	Yes ▶	**If "CHECK ENGINE" light is on:** GO to Three-Digit Service Code Charts [6.0], before addressing any other symptoms. If engine runs rough or idles rough: GO to Quick Test Step [3.1]. If engine is a no start: GO to Pinpoint Test Step [AA1]. If these symptoms are not present: GO to Quick Test Step [4.0].
	No ▶	**If KOEO service codes are received:** GO to Three-Digit Service Code Charts [6.0]. **If no service codes are received:** GO to Pinpoint Test Step [QA1].

QUICK TEST PROCEDURES—NAVAJO

EEC Quick Test — Procedures and Appendix

TEST STEP	RESULT ▶	ACTION TO TAKE
3.1 CHECK FOR CONTINUOUS MEMORY SERVICE CODES 45, 46, 48, 215, 216, 217, 232 or 238		
• Was Continuous Memory service code 45, 46, 48, 215, 216, 217, 232 or 238 received during KOEO Self-Test?	Yes ▶	GO to Pinpoint Test Step [N15].
	No ▶	GO to Pinpoint Test Step [S2].
4.0 CHECK BASE TIMING		
• Key off, wait 10 seconds. • Activate Engine Running Self-Test. • Start engine. **NOTE:** **If engine starts but stalls or stalls during Self-Test, GO to Pinpoint Test Step [S1].** **Engine Running Service Code 98 indicates vehicle is in Failure Mode Effects Management (FMEM) and DID NOT PASS KOEO On-Demand/Continuous Memory Self-Test. Engine Running Self-Test cannot be performed while in FMEM. Go to Continuous Memory code charts, Quick Test Step [8.0].** • Check computed timing after the last service code has been displayed. The timing will remain fixed for two minutes, unless Self-Test is deactivated. **NOTE: Computed timing is equal to base timing plus 20 degrees BTDC with 3 degrees tolerance. See vehicle decal for correct base timing.** • Is computed timing within specs?	Yes ▶ No ▶	Go to Quick Test Step [5.0]. GO to Pinpoint Test Step [P2].

EEC Quick Test — Procedures and Appendix

TEST STEP	RESULT ▶	ACTION TO TAKE
5.0 PERFORM ENGINE RUNNING SELF-TEST		
• Deactivate Self-Test. • Start engine and idle until vehicle is at operating temperature. • Activate Engine Running Self-Test. • Turn engine off. • Start engine. • Record all service codes displayed. **NOTE: Engine Running Service Code 98/998 indicates vehicle is in Failure Mode Effects Management (FMEM) and DID NOT PASS KOEO On-Demand/Continuous Memory Self-Test. Engine Running Self-Test cannot be performed while in FMEM.** • **Is a PASS Code (11 or 111) received during Engine Running Self-Test?** • Rerun KOEO Self-Test Quick Test Step [3.0].	Yes ▶	**If any Continuous Memory service codes were received in Quick Test Step [3.0]:** GO to Three-Digit Service Code Charts [6.0]. **If a Continuous Memory PASS code (11 or 111) was received in Quick Test Step [3.0] and no symptoms are present:** EEC diagnostic testing is complete. **If a Continuous Memory PASS Code (11 or 111) was received in Quick Test Step [3.0] and symptoms are present:** GO to Diagnosis By Symptom chart Quick Test Step [9.0].
	No ▶	**If Engine Running service codes are received:** GO to Three-Digit Service Code Charts [6.0]. **If no service codes are received:** GO to Pinpoint Test Step [QA1].

QUICK TEST STEP: Three-Digit Service Code Charts — 6.0

Service Code	Pinpoint Test Step Go to Direction: KOEO	ER	Cont
No Codes	QA1	QA1	QA1
111	Pass	Pass	Pass
112	DA20	—	DA90
113	DA10	—	DA90
114	DA1	DA1	—
116	DA1	DA1	—
117	DA20	—	DA90
118	DA10	—	DA90
121	DH2	DH1	—
122	DH10	—	DH94
123	DH3	—	DH90
126	DF1	DF1	DF90
128	—	—	DF11
129	—	DC10	—
136	—	H1	—
137	—	H1	—
139	—	—	H1
144	—	—	H1
157	—	—	DC10
158	DC20	—	DC20
159	DC1	DC1	—
167	—	DH20	—
171	—	—	H1
172	—	H1	H1
173	—	H1	H1
174	—	—	H1
175	—	—	H1
176	—	—	H1
177	—	—	H1
178	—	—	H1
179	—	—	HA15
181	—	—	HA1
182	—	—	HA15
183	—	—	HA1
188	—	—	HA15
189	—	—	HA1
191	—	—	HA15
192	—	—	HA1
194	—	—	H5
195	—	—	H5
211	—	—	NC1
212	—	—	NC3
213	—	PC1	—
215	—	—	**EDIS System**
216	—	—	**EDIS System**

(Continued)

QUICK TEST STEP: Three-Digit Service Code Charts — 6.0

Service Code	Pinpoint Test Step Go to Direction: KOEO	ER	Cont
217	—	—	**EDIS System**
218	—	—	**EDIS System**
222	—	—	**EDIS System**
223	—	—	**EDIS System**
224	—	—	**EDIS System**
225	—	DG1	—
226	NC2	—	—
232	—	—	**EDIS System**
238	—	—	**EDIS System**
338	—	—	DA100
339	—	—	DA101
341	KP1	—	—
411	—	KE15	—
412	—	KE1	—
452	—	—	DP1
511		Replace Processor	
512	—	—	QB1
513		Replace Processor	
522	TA1	—	—
525	TA1	—	TA1
528	—	—	TA1
529	—	—	ML25
533	—	—	ML25
536	—	FD1	FD90
538	—	M1	—
539	KM40	—	—
543	J30	—	J93
556	J7	—	J95
565	KD6	—	—
566	TC10	—	—
569	KD6	—	—
617	—	—	TG90
618	—	—	TG90
619	—	—	TG90
621	TC1	—	—
622	TC1	—	—
624	TC10	—	TG90
625	TC10	—	TG90
626	TC1	—	—
627	TC1	—	—
628	—	—	TG90
629	TC10	—	—
631	TB1	—	—
632	—	TB1	—
633	TB1	—	—

(Continued)

QUICK TEST PROCEDURES—NAVAJO

QUICK TEST STEP: Three-Digit Service Code Charts — 6.0

Service Code	Pinpoint Test Step Go to Direction KOEO	ER	Cont
634	TD1	—	TG90
636	TE1	TE1	—
637	TE10	—	TG91
638	TE20	—	TG91
639	—	TF1	TF90
641	TC1	—	—
643	TC20	—	TC30
645	—	—	TG90
646	—	—	TG90
647	—	—	TG90
648	—	—	TG90
649	—	—	TG90
651	—	—	TG90
652	TC1	—	—
654	TD1	—	—
656	—	—	TG95
998	TC10	Quick Test Step 3.0	
Codes Not Listed	QA1	QA1	QA1

QUICK TEST STEP: Diagnosis By Symptom — 9.0

9.0 DIAGNOSIS BY SYMPTOM

SPECIAL NOTES:

- Verify that a **Pass Code 111** was received in Key On Engine Off, Engine Running and Continuous Self-Tests before continuing with this test.

QUICK TEST STEP: Diagnosis By Symptom — 9.0

9.0 DIAGNOSIS BY SYMPTOM	
Symptom	**Pinpoint Test Step Direction**
Engine runs rough or misses Surges Lack of Power Rough idle Erratic rpm	S 2
Engine Stalls Stalls in Self-Test	S 1
A/C not functioning	K M 1
Low idle with A/C On	F A 1 1
"CHECK ENGINE" light always on	M L 1
"CHECK ENGINE" light never on	M L 5
"CHECK ENGINE" light on intermittently	M L 1 0
"CHECK ENGINE" light flashing with erratic idle	M L 1 5
Stumble after hot restart	H 2 0
Gasoline fumes under hood	K D 1
Exhaust Odor Black Smoke	H 2 9

APPENDIX: Visual Check

SPECIAL NOTES:

- Correct results of the QUICK TEST are dependent on the proper operation of related non-EEC components.
- It may be necessary to disconnect or disassemble harness connector assemblies to do some of the inspections. Pin locations should be noted before disassembly.

VISUAL CHECK

1. Inspect the air cleaner and inlet ducting.
2. Check all engine vacuum hoses for damage, leaks, cracks, blockage, proper routing, etc.
3. Check EEC system wiring harness for proper connections, bent or broken pins, corrosion, loose wires, proper routing, etc.
4. Check the processor, sensors and actuators for physical damage.
5. Check the engine coolant for proper level.
6. Check the transmission fluid level and quality.
7. Make all necessary repairs before continuing with QUICK TEST.

Diagnostic Sensor/Actuator Reference Values

NOTE:
- The chart below contains typical component values.
- Values measured in the field may differ slightly from those shown.
- Breakout box pin number assignments differ from vehicle to vehicle. Refer to the Pin Usage Chart applicable to the vehicle being serviced for the correct pin assignments.
- Each vehicle application will not have all components listed below.
- Pre-condition the engine in the following manner before recording any values:
  - —The engine should be at a normal operating condition.
  - —Start and run engine at 2000 rpm for two minutes.

INPUTS	Breakout Box Pin Numbers (+)	(−)	KEY OFF Throttle Closed (ohms)	Key On Eng. Off (volts)	Hot Idle (volts)
Coil Tach Signal (IDM)	4	16	21.8 K	8.0 to 10.0	8.0 to 12.0
* Engine Coolant Temperature (ECT) ①	7	46	1.7 K to 3.6 K	—	0.31 to 1.0
(Heated) Exhaust Gas Oxygen (H) EGO	29	46	> 1.5 M	< 0.4	0.0 to 0.9
* Barometric Pressure (BP) ②	45	46	—	—	—
* Mass Air Flow (MAF) ③	50	40/60	—	—	—
* Neutral/Drive Switch (NDS) ④	30	40	—	—	—
Reference Voltage (VREF)	26	46	—	5.0 ± 5%	5.0 ± 5%
Throttle Position (TP)	47	46	0.5 K to 1.2 K	0.7 to 1.3	0.7 to 1.3
* Vehicle Speed Sensor (VSS) ⑤	3	6	190.0 to 240.0	—	—
OUTPUTS					
Canister Purge (CANP)	37	31	60.0 to 70.0	0.0	0.0
Fuel Pump Relay (FP)	37	22	—	0.0 (OFF)	VBAT (ON)
Idle Speed Bypass (IBA)	37	21	9.5 to 11.5	0.0 (OFF)	2.0 to 5.5
* Injectors Bank #1 ⑥	37	58	—	0.0	—
* Injectors Bank #2 ⑥	37	59	—	0.0	—
Spark Angle Word (SAW)	36	16	—	—	0.5 to 0.7

* **Numbers to the right of these inputs/outputs refer to explanations**

Diagnostic Sensor/Actuator Reference Values

Explanations —

① Engine coolant temperature must be within the 180-240°F (82-116°C) temperature range before measurements are taken.

② Refer to the Altitude vs. Voltage Chart in Pinpoint Test DF, for proper operating values.

③ Refer to the MAF Sensor data chart in Pinpoint Test DC for proper operating values.

④ The NDS is open in any gear, but closed in neutral or park.

⑤ If vehicle is not moving, the speed sensor output to the processor will be zero. The vehicle must be moving for the speed sensor to provide information to the processor.

⑥ Refer to the Injector Resistance Chart in Pinpoint Test H, for the proper resistance values.

Electrical Schematic

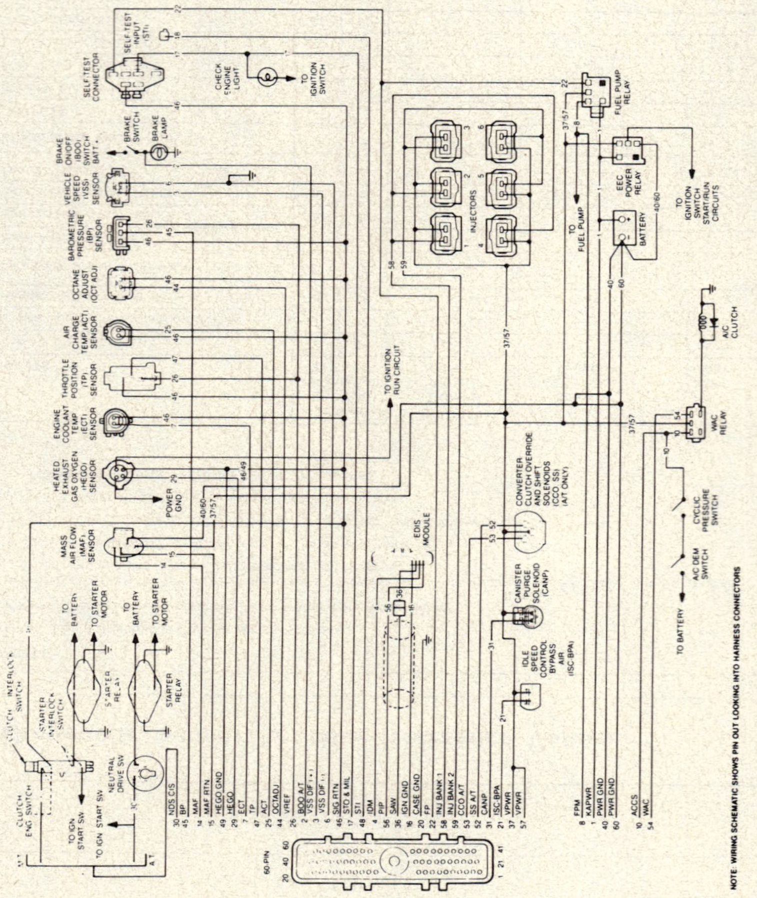

EEC Processor Connector Pin Usage

Pin	Circuit	Wire Color	Application	Abbreviations
1	37	Y	Keep Alive Power	KAPWR
2	511	LG	Brake On/Off	BOO
3	679	DG/W	Vehicle Speed Sensor (+)	VSS (+)
4	11	T/Y	Ignition Diagnostic Monitor	IDM
6	676	BK/Y	Vehicle Speed Sensor (−)	VSS (−)
7	354	LG/R	Engine Coolant Temperature Sensor	ECT
8	238	P/B	Fuel Pump Monitor	FPM
9	915	B/O	Data Communications Link (−)	DATA (−)
10	198	DG/O	A/C Cycling Switch	ACCS
14	967	LB/R	Mass Air Flow Sensor	MAF
15	968	T/LB	Mass Air Flow Return	MAF RTN
16	259	O/R	Ignition Ground	IGN GND
17	658	P/LG	Self-Test Output and Malfunction Indicator Light	STO and MIL
20	57	B	Case Ground	CASE GND
21	264	W/LB	Idle Speed Control Solenoid	ISC
22	926	LB/O	Fuel Pump	FP
25	743	GY	Air Charge Temperature Sensor	ACT
26	351	BR/W	Reference Voltage	VREF
28	914	T/O	Data Communications Link (+)	DATA (+)
29	74	GY/LB	Heated Exhaust Gas Oxygen Sensor	HEGO
30	199	LB/Y	Clutch Interlock Switch Neutral Gear Switch (M/T Only)	CIS/NGS
30	199	LB/Y	Neutral Drive Switch (A/T Only)	NDS
31	101	GY/Y	Canister Purge	CANP
34	305	LB/P	Data Output Link	DOL
36	929	P	Spark Angle Width	SAW
37	361	R	Vehicle Power	VPWR
40	570	B/W	Power Ground	PWR GND
44	242	DG	Octane Adjust	OCT ADJ
46	359	GY/R	Signal Return	SIG RTN
47	355	GY/W	Throttle Position Sensor	TP
48	209	W/P	Self-Test Input	STI
49	89	O	Heated Exhaust Gas Oxygen Sensor Ground	HEGO GND
52	237	O/Y	Shift Solenoid (A/T Only)	SS
53	480	P/Y	Clutch Converter Override (A/T Only)	CCO
54	331	P/Y	WOT A/C Cut-Off	WAC
56	395	GY/O	Profile Ignition Pick-Up	PIP
57	361	R	Vehicle Power	VPWR
58	555	T	Injector (Bank 1 — Controls Engine Cylinder Numbers 1, 2, and 3)	INJ BANK 1
59	556	W	Injector (Bank 2 — Controls Engine Cylinder Numbers 4, 5, and 6)	INJ BANK 2
60	570	B/LG	Power Ground	PWR GND

Pin locations given for reference only. Probing 60 pin connector with DVOM probe will result in permanent damage to the pin connectors. Always probe as directed, using the Breakout Box.

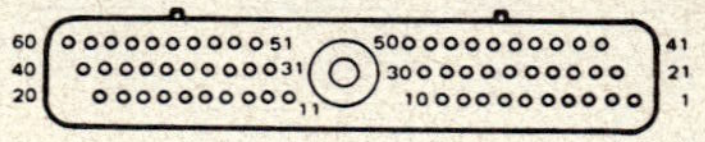

Quick Test Codes and Code Definition

SERVICE CODE		SERVICE CODE DEFINITION
111 orc	▶	System PASS
412 r	▶	Cannot control rpm during Self-Test high rpm check
411 r	▶	Cannot control rpm during Self-Test low rpm check
211 c	▶	PIP circuit failure
511 o	▶	ECA Read Only Memory (ROM) test failed
512 c	▶	ECA Keep Alive Memory (KAM) test failed
226 o	▶	Ignition Diagnostic Monitor (IDM) signal not received
213 r	▶	SAW circuit open
212 c	▶	Loss of IDM input to processor saw circuit grounded
513 o	▶	Failure in ECA internal voltage
116 or	▶	Engine Coolant Temp (ECT) out of Self-Test range
121 or	▶	Throttle Position (TP) out of Self-Test range
114 or	▶	Air Charge Temp (ACT) out of Self-Test range
159 or	▶	Mass Air Flow (MAF) out of Self-Test range
452 c	▶	Insufficient input from the Vehicle Speed Sensor (VSS)
232 c	▶	Coil 1, 2 or 3 failure
118 oc	▶	ECT indicated −40°F (−40°C)/Circuit Open
123 oc	▶	TP circuit above maximum voltage
113 oc	▶	ACT indicated −40°F (−40°C)/circuit open
158 oc	▶	MAF circuit above maximum voltage
341 o	▶	Octane adjust circuit open
117 oc	▶	ECT indicated 254°F (123°C)/circuit grounded
122 oc	▶	TP circuit below minimum voltage
112 oc	▶	ACT indicated 254°F/circuit grounded
157 c	▶	MAF circuit below minimum voltage
522 o	▶	Vehicle not in PARK or neutral during KOEO
528 c	▶	Clutch Switch circuit failure
129 r	▶	Insufficient MAF change during Dynamic Response Test
167 r	▶	Insufficient TP change during Dynamic Response Test
536 r	▶	Brake On/Off (BOO) circuit failure/not actuated during Self-Test
538 r	▶	Operator error (Dynamic Response/Cylinder Balance Tests)
539	▶	A/C ON/Defrost ON during Self-Test
566 o	▶	Shift Solenoid (SS) circuit failure
556 oc	▶	Primary Fuel Pump circuit failure
629 o	▶	Clutch Converter Override (CCO) circuit failure
542 oc	▶	Fuel Pump circuit open-ECA to motor ground
543 oc	▶	Fuel Pump circuit open-battery to ECA
998 r	▶	Hard fault present
124 c	▶	Throttle Position (TP) Sensor voltage higher than expected
125 c	▶	Throttle Position (TP) Sensor voltage lower than expected
171 c	▶	Fuel system at adaptive limits, Oxygen Sensor (HEGO) unable to switch
172 rc	▶	Lack of Oxygen Sensor (HEGO) switches, indicates lean
173 rc	▶	Lack of Oxygen Sensor (HEGO) switches, indicates rich
179 c	▶	Fuel system at lean adaptive limit at part throttle, system rich
181 c	▶	Fuel system at rich adaptive limit at part throttle, system lean
184 c	▶	Mass Air flow (MAF) higher than expected
185 c	▶	Mass Air Flow (MAF) lower than expected
186 c	▶	Injector pulse width higher than expected
187 c	▶	Injector pulse width lower than expected
565 o	▶	Canister Purge (CANP) circuit failure
NO CODES	▶	Unable to initiate Self-Test or unable to output Self-Test codes
CODES NOT LISTED	▶	Service codes displayed are not applicable to the vehicle being tested

KEY: o=Key On Engine Off (KOEO) r=Engine Running (ER) c=Continuous Memory

EEC Diagnostic Reference Values

INPUTS/ OUTPUTS	SIGNAL PIN #	KOEO	HOT IDLE	30 MPH	55 MPH
KAPWR	1	>10.5V	>10.5V	>10.5V	>10.5V
VPWR	37/57	>10.5V	>10.5V	>10.5V	>10.5V
VREF	26	4.5-5.5V	4.5-5.5V	4.5-5.5V	4.5-5.5V
CSE GND	20	0 ± 0.5V	0 ± 0.5V	0 ± 0.5V	0 ± 0.5V
IGN GND	16	0 ± 0.5V	0 ± 0.5V	0 ± 0.5V	0 ± 0.5V
PWR GDN	40/60	0 ± 0.3V	0 ± 0.3V	0 ± 0.3V	0 ± 0.3V
SIG RTN	46	0 ± 0.5V	0 ± 0.5V	0 ± 0.5V	0 ± 0.5V
HEGO GND	49	0 ± 0.5V	0 ± 0.5V	0 ± 0.5V	0 ± 0.5V
MAF RTN	15	0 ± 0.5V	0 ± 0.5V	0 ± 0.5V	0 ± 0.5V
OCTADJ	44	0V	0V	0V	0V
MAF	14	0V	0.6V	1.3-1/4V	1.7-2.0V
TP	47	1.0V	1.0V	1.2-1.3V	1.4-1.6
ECT	7	0.6V	0.6V	0.6V	0.6V
ACT	25	2.2V	2.6V	2.7V	2.9V
Bp	45	158 Hz	158 Hz	158 Hz	158 Hz
IDM	4	0 rpm	750-850 rpm	1500-1700 rpm	1800-2000 rpm
Pip	56	0 rpm	750-850 rpm	1500-1700 rpm	1800-2000 rpm
HEGO	29	0V	Switching	Switching	Switching
BOO	2	0V	0/VBAT①	0V	0V
FPm	8	VBAT	VBAT	VBAT	VBAT
ACCS	10	0V	0/VBAT②	0V	0V
VSS+	3	0MPH	0MPH	30MPH	55MPH
NDS	30	0V	0V	5.0V	5.0V
STI	48	5.0V	5.0V	5.0V	5.0V
INJ BANK 2	59	VBAT	3.0-3.2mS	3.8-4.4mS	5.0-7.0mS
INJ BANK 1	58	VBAT	3.0-3.2mS	3.8-4.4mS	5.0-7.0mS
SS 3/4 (A/T)	52	VBAT	VBAT	VBAT	0V
STO/MIL	17	0.6V	VBAT	VBAT	VBAT
WAC	54	0V	0/VBAT②	0V	0V
ISC-BPA	21	VBAT	8.0-12.0V	7.3-8.4V	5.5-6.8V
FP	22	VBAT	0.1V	0.1V	0.1V
SAW	36	0rpm	750-850rpm	1500-1700rpm	1800-2000rpm
CCO (A/T)	53	VBAT	VBAT	VBAT	0V

NOTES: ① — Brake pedal depressed.
② — A/C on.
Reference values shown may vary ±20% depending on operating conditions and other factors.

EEC Graphs and Charts

ACT Sensor Graph

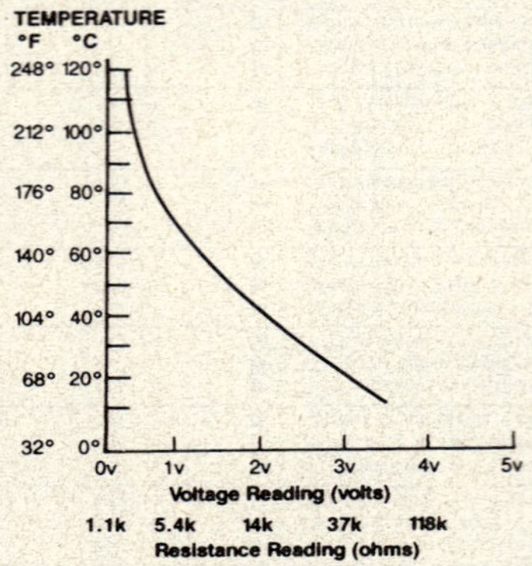

ACT SENSOR DATA* Voltage values calculated for VREF=5.0v (These values may vary ± 15% due to sensor and VREF variations).

TEMPERATURE °F	°C	VOLTAGE* Volts	RESISTANCE K ohms
248	120	0.28	1.18
230	110	0.36	1.55
212	100	.47	2.07
194	90	.61	2.80
176	80	.80	3.84
158	70	1.04	5.37
140	60	1.35	7.60
122	50	1.72	10.97
104	40	2.16	16.15
86	30	2.62	24.27
68	20	3.06	37.30
50	10	3.52	58.75

ECT Sensor Graph

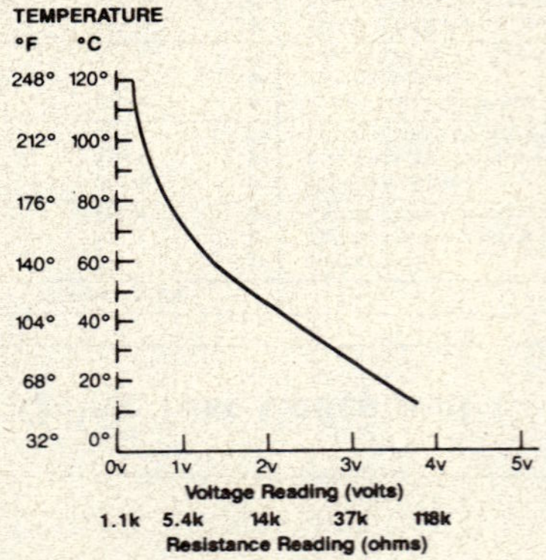

ECT SENSOR DATA* Voltage values calculated for VREF=5.0v (These values may vary ± 15% due to sensor and VREF variations).

TEMPERATURE °F	°C	VOLTAGE* Volts	RESISTANCE K ohms
248	120	0.28	1.18
230	110	0.36	1.55
212	100	.47	2.07
194	90	.61	2.80
176	80	.80	3.84
158	70	1.04	5.37
140	60	1.35	7.60
122	50	1.72	10.97
104	40	12.16	16.15
86	30	12.62	24.27
68	20	13.06	37.30
50	10	13.52	58.75

EEC Graphs and Charts

FUEL PRESSURE SPECIFICATION TABLE

ENGINE RUNNING	30-45 PSI 210-310 kPa
KEY ON ENGINE OFF	35-45 PSI 240-310 kPa

MAF Sensor Graph

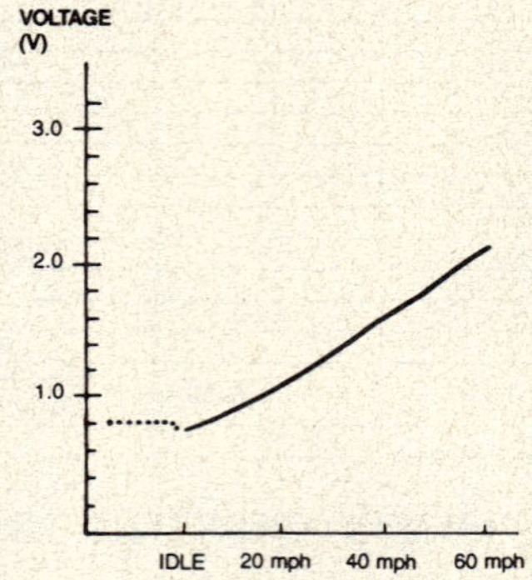

MAF SENSOR DATA* Engine at Normal Operating Temperature.

ENGINE CONDITION	VOLTAGE VOLTS*
IDLE	.8
20 MPH	1.0
40 MPH	1.7
60 MPH	2.1

EEC Graphs and Charts

BP Sensor Graph

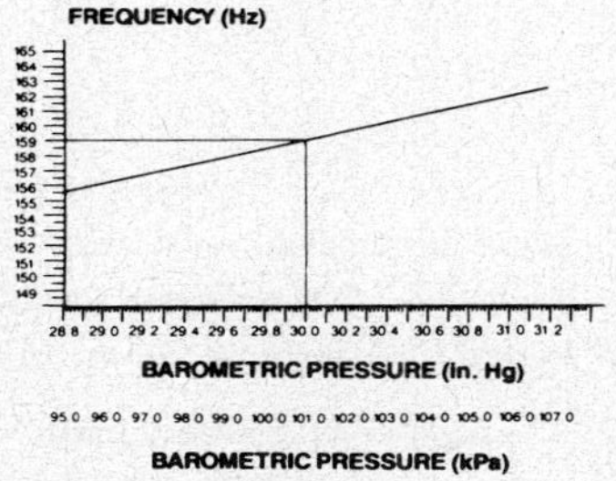

BP SENSOR DATA:
OUTPUT FREQUENCY AT KOEO
Frequency may vary plus or minus 6 Hz of these values due to sensor variations.

BAROMETRIC PRESSURE		FREQUENCY
in. (Hg)	kPa	Hz *
17.1	58	122.4
18.3	62	125.5
19.5	66	128.7
20.7	70	131.9
21.8	74	135.1
23.0	78	138.3
24.2	82	141.8
25.4	86	145.4
26.6	90	148.9
27.7	94	152.5
28.9	98	156.1
30.1	102	159.6
31.0	105	162.4

EEC Graphs and Charts

Throttle Angle

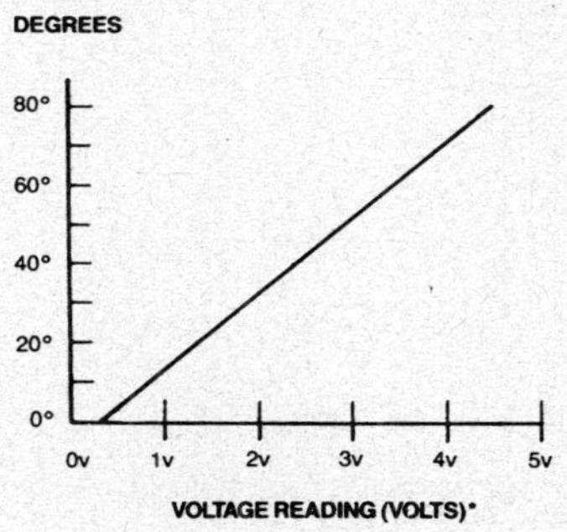

TP SENSOR DATA * Voltage values calculated for VREF=5.0v (These values may vary ± 15% due to sensor and VREF variations)

THROTTLE ANGLE DEGREES	VOLTAGE* VOLTS
0	0.50
10	0.97
20	1.44
30	1.90
40	2.37
50	2.84
60	3.31
70	3.78
80	4.24

EEC No Start (EDIS)	Pinpoint Test	AA

Note

You should enter this Pinpoint Test only when Quick Test has been successfully completed and the engine is still a no start or when directed here

Remember

To prevent the replacement of good components, be aware that the following non-EEC areas may be at fault:

- Fuel: quantity and quality
- Ignition: general condition, moisture, cracks, damage, etc.
- Engine: internal, valves, timing belt, camshaft
- Starter and battery circuit
- Variable Reluctance sensor (VRS)
- EDIS module
- Coil Packs

This Pinpoint Test is intended to diagnose only the following:

- Spark (as related to EEC)
- Harness circuits: PIP, IGN GND, VPWR
- ECU assembly

EEC No Start (EDIS)	Pinpoint Test	AA

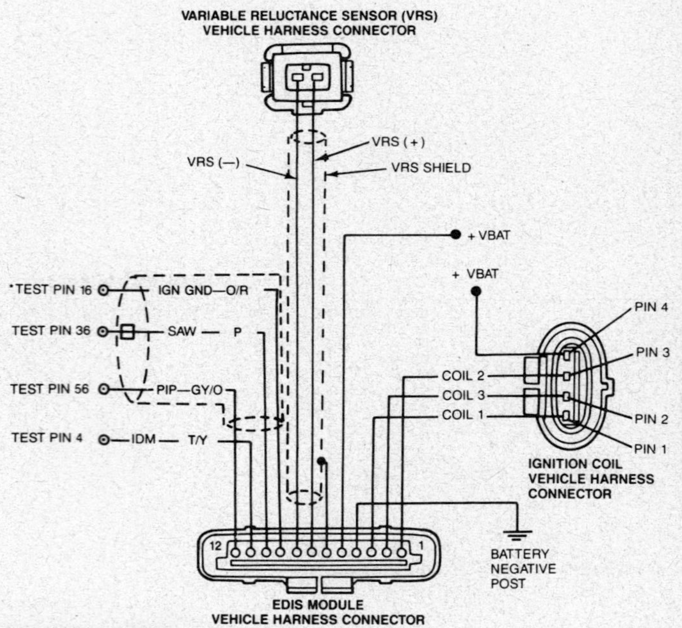

*TEST PINS LOCATED ON BREAKOUT BOX.
ALL HARNESS CONNECTORS VIEWED INTO MATING SURFACE.

PINPOINT TESTS—NAVAJO

EEC No Start (EDIS)	Pinpoint Test	AA

WARNING

STOP THIS TEST AT THE FIRST SIGN OF A FUEL LEAK AND SERVICE AS REQUIRED.

CAUTION

No open flame — No smoking during fuel delivery checks.

TEST STEP	RESULT ▶	ACTION TO TAKE
AA1 ATTEMPT TO CRANK ENGINE		
NOTE: Verify fuel pump inertia switch is set (button pushed in). Refer to Owner's Guide for location. • **Does engine crank?**	Yes ▶ No ▶	GO to AA2. REFER to starting system
AA2 CHECK FOR VREF AT THROTTLE POSITION SENSOR		
• Key off, wait 10 seconds. • Voltmeter on 20 volt scale. • Disconnect TP sensor. • Key on, engine off. • Measure voltage between VREF circuit and SIG RTN circuit at the TP sensor vehicle harness connector. • **Is voltage between 4.0 volts and 6.0 volts?**	Yes ▶ No ▶	RECONNECT TP sensor. GO to AA3. GO to Pinpoint Test Step C1.
AA3 CHECK FOR SPARK AT PLUGS		
• Disconnect the spark plug wire to any accessible cylinder. • Connect spark tester between spark plug wire and engine ground. • Crank engine and check for spark. • Reconnect the spark plug wire to the spark plug. • **Was spark present and consistent?**	Yes ▶ No ▶	GO to AA4. REFER to EDIS diagnosis.

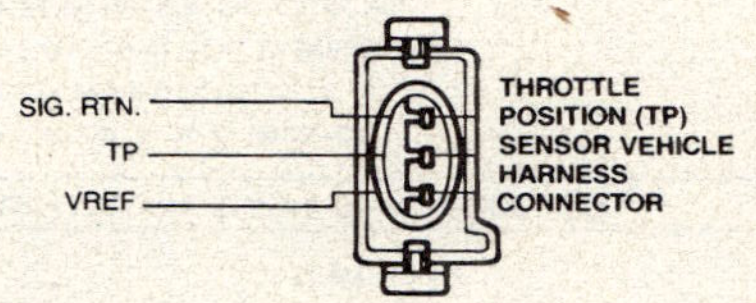

EEC No Start (EDIS)	Pinpoint Test	AA

TEST STEP	RESULT ▶	ACTION TO TAKE
AA4 CHECK CONTINUITY OF IGNITION GROUND CIRCUIT		
• Key off, wait 10 seconds. • Disconnect ECU 60 pin connector. Inspect for damaged or pushed out pins, corrosion, loose wires, etc. Service as necessary. • Install breakout box, leave ECU disconnected. • Ohmmeter on 200 ohm scale. • Disconnect EDIS module. • Measure resistance between Test Pin 16 at the breakout box and IGN GND circuit at the EDIS module vehicle harness connector. • **Is resistance less than 5.0 ohms?**	Yes ▶ No ▶	GO to AA5. SERVICE open circuit. REMOVE breakout box. RECONNECT all components. RERUN Quick Test.
AA5 CHECK PIP CIRCUIT FOR SHORT TO POWER		
• Key off, wait 10 seconds. • Breakout box installed, ECU disconnected. • EDIS module disconnected. • Ohmmeter on 200,000 ohm scale. • Measure resistance between Test Pin 56 (PIP) at the breakout box and battery negative post. • **Is voltage greater than 7.0 volts?**	Yes ▶ No ▶	GO to AA6. SERVICE short circuit. REMOVE breakout box. RECONNECT all components. RERUN Quick Test.
AA6 CHECK PIP CIRCUIT CONTINUITY		
• Key off. • Breakout box installed, ECU disconnected. • EDIS module disconnected. • Measure resistance between Test Pin 56 at the breakout box and PIP circuit at the EDIS module vehicle harness connector. • **Is resistance less than 5.0 ohms?**	Yes ▶ No ▶	GO to AA7. SERVICE open circuit. REMOVE breakout box. RECONNECT all components. RERUN Quick Test.
AA7 CHECK PIP SIGNAL CIRCUIT FOR SHORTS TO GROUND		
• Key off. • Breakout box installed, ECU disconnected. • EDIS module disconnected. • Measure resistance between Test Pin 56 (PIP) and Test Pins 16, 40, 46, 60 (short to GROUND). • **Is each resistance greater than 10,000 ohms?**	Yes ▶ No ▶	GO to AA8. SERVICE short circuit. REMOVE breakout box. RECONNECT all components. RERUN Quick Test. If vehicle does not start. GO to AA8.

EEC No Start (EDIS) — Pinpoint Test AA

TEST STEP	RESULT ▶	ACTION TO TAKE
AA8 ISOLATE SHORT(S) IN PROCESSOR		
• Key off, wait 10 seconds. • Breakout box installed. • Reconnect ECU to breakout box. • EDIS module disconnected. • Ohmmeter on 200,000 ohm scale. • Measure resistance between Test Pin 56 (PIP) and Test Pins 37, 57 (short to POWER), 40, and 60 (short to GROUND) at the breakout box. • **Is each resistance between 20,000 and 150,000 ohms?**	Yes ▶ No ▶	RECONNECT all components. GO to AA9. REPLACE ECU. REMOVE breakout box. RECONNECT all components. RERUN Quick Test.
AA9 CHECK PIP SIGNAL		
• Key off. • Breakout box installed, ECU connected. • Voltmeter to 20 volt scale. • Reconnect EDIS module. • Measure voltage between Test Pin 56 and Test Pin 16 at the breakout box. • Crank engine, record reading. • **Is voltage between 3.0 and 7.0 volts?**	Yes ▶ No ▶	GO to AA10. GO to AA6.
AA10 CHECK FUEL PUMP		
• **No smoking nearby.** • Connect fuel pressure gauge. • Note initial pressure reading. • Observe pressure gauge as you pressurize fuel system. (Turn key to RUN for 1 second, then turn key to OFF. Wait 10 seconds. Repeat 5 times.) • **Does fuel pressure increase?** **WARNING: If fuel starts leaking, turn key off immediately. No smoking.**	Yes ▶ No ▶	GO to Pinpoint Test Step S1. GO check for mechanical problems in fuel system.

Vehicle Battery — Pinpoint Test B

Note

You should enter this Pinpoint Test only when directed here from Pinpoint Tests C, J or P.

Remember

To prevent the replacement of good components, be aware that the following non-EEC areas may be at fault:

- Ignition Switch
- Battery Cables
- Alternator
- Voltage Regulator
- Ground Straps

This Pinpoint Test is intended to diagnose only the following:

- ECU
- Harness circuits: SIG RTN, PWR GND, VPWR, KAPWR, IGNITION Switch.
- Battery Voltage
- Power Relay

Vehicle Battery	Pinpoint Test	B

Pinpoint Test Schematic

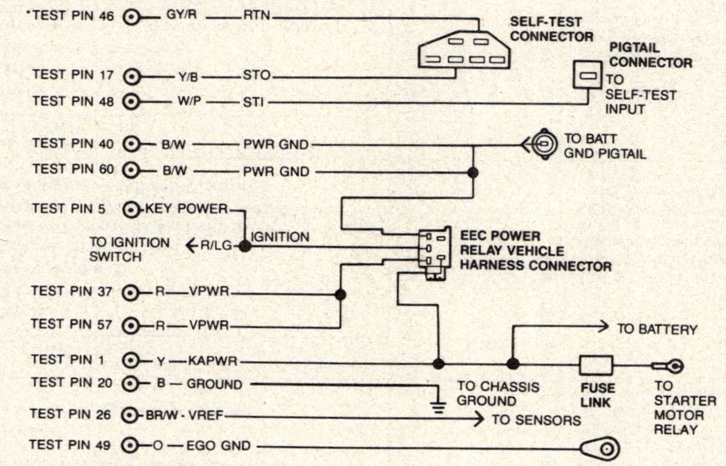

*TEST PINS LOCATED ON BREAKOUT BOX.
ALL HARNESS CONNECTORS VIEWED INTO MATING SURFACE.

Vehicle Battery	Pinpoint Test	B

TEST STEP	RESULT ▶	ACTION TO TAKE
B1 CHECK BATTERY VOLTAGE		
• Key on, engine off. • Voltmeter on 20 volt scale. • Measure voltage across battery terminals. • **Is voltage greater than 10.5 volts?**	Yes ▶ No ▶	GO to B2. SERVICE discharged battery.
B2 CHECK PWR GND CIRCUIT CONTINUITY		
• Key off, wait 10 seconds. • Disconnect ECU 60 pin connector. Inspect for damaged or pushed out pins, corrosion, loose wires, etc. Service as necessary. • Install breakout box and connect ECU to breakout box. • Ohmmeter on 200 ohm scale. • Measure resistance between battery negative post and Test Pins 40 and 60 at the breakout box. • **Is each resistance less than 5 ohms?**	Yes ▶ No ▶	GO to B3. SERVICE open in PWR GND circuit. REMOVE breakout box. RECONNECT ECU. RERUN Quick Test.
B3 CHECK FOR OPEN BETWEEN SIG RTN AND PWR GND CIRCUITS AT PROCESSOR		
• Key off, wait 10 seconds. • Breakout box installed, ECU connected. • Ohmmeter on 200 ohm scale. • Measure resistance between Test Pin 46 and Test Pins 40 and 60 at the breakout box. • **Is each resistance less than 5 ohms?**	Yes ▶ No ▶	GO to B4. REPLACE ECU. REMOVE breakout box. RERUN Quick Test.

Vehicle Battery — Pinpoint Test B

TEST STEP	RESULT ▶	ACTION TO TAKE
B4 CHECK SIG RTN CIRCUIT CONTINUITY • Key off, wait 10 seconds. • Breakout box installed, ECU connected. • Ohmmeter on 200 ohm scale. • Measure resistance between Test Pin 46 at the breakout box and SIG RTN circuit in the Self-Test connector. • **Is resistance less than 5.0 ohms?**	Yes ▶	REMOVE breakout box. RECONNECT ECU. GO to B5.
	No ▶	SERVICE open in SIG RTN circuit. REMOVE breakout box. RECONNECT ECU. RERUN Quick Test.
B5 CHECK KAPWR CIRCUIT VOLTAGE AT EEC POWER RELAY • Key off. • Disconnect EEC power relay. • Key on, engine off. • Voltmeter on 20 volt scale. • Measure voltage between KAPWR circuit at the EEC power relay connector and battery negative post. • **Is voltage greater than 10.5 volts?**	Yes ▶	GO to B6.
	No ▶	SERVICE open in KAPWR circuit between EEC power relay and battery positive post. RECONNECT all components. RERUN Quick Test.
B6 CHECK IGNITION CIRCUIT VOLTAGE AT EEC POWER RELAY • Key on, engine off. • Disconnect EEC power relay. • Voltmeter on 20 volt scale. • Measure voltage between the battery negative post and ignition switch circuit at the EEC power relay connector. • **Is voltage greater than 10.5 volts?**	Yes ▶	GO to B7.
	No ▶	SERVICE open in ignition switch circuits. RECONNECT all components. RERUN Quick Test.

Vehicle Battery — Pinpoint Test B

TEST STEP	RESULT ▶	ACTION TO TAKE
B7 CHECK PWR GND CIRCUIT CONTINUITY • Key off, wait 10 seconds. • Disconnect ECU. • Ohmmeter on 200 ohm scale. • Measure resistance between PWR GND circuit at the EEC power relay connector and battery negative post. • **Is the resistance less than 10 ohms?**	Yes ▶	GO to B8.
	No ▶	SERVICE open circuit. RECONNECT ECU. RERUN Quick Test.
B8 CHECK VPWR CIRCUIT CONTINUITY • Key off. • Breakout box installed, ECU connected. • EEC power relay removed. • Measure resistance between test pins 37/57 at the breakout box and the VPWR terminal of the EEC power relay connector. • Is resistance less than 5 ohms?	Yes ▶	GO to B9.
	No ▶	SERVICE open in VPWR circuit between EEC power relay connector and ECU. Remove breakout box. RECONNECT all components. RERUN Quick Test.
B9 CHECK VPWR CIRCUIT VOLTAGE AT EEC POWER RELAY • Key off. • Breakout box installed, ECU connected, EEC power relay connected. • Key on, engine off. • Voltmeter on 20 volt scale. • Measure voltage between test pin 37/57 and test pins 40/60 at the breakout box. • **Is voltage greater than 10.5 volts?**	Yes ▶	SERVICE open or short to ground in VPWR circuit between ECU and EEC power relay. REMOVE breakout box. RERUN Quick Test.
	No ▶	REPLACE EEC power relay. REMOVE breakout box. RERUN Quick Test.

Reference Voltage — Pinpoint Test C

Note

You should enter this Pinpoint Test only when a check for VREF has failed in the sensor Pinpoint Tests (D-Series) or Pinpoint Tests [A] or [QA].

Remember

This Pinpoint Test is intended to diagnose only the following:

- ECU assembly
- Sensor harness circuits: SIG RTN, VREF
- Three-wire sensors: TP, PFE, BP.

Description

Reference Voltage (VREF) is a positive voltage (about 5.0V) that is output by the ECU. This consistent voltage is used by all three-wire sensors.

Signal Return (SIG RTN) is a dedicated ground used by most EEC sensors and some other inputs.

Pinpoint Test Schematic

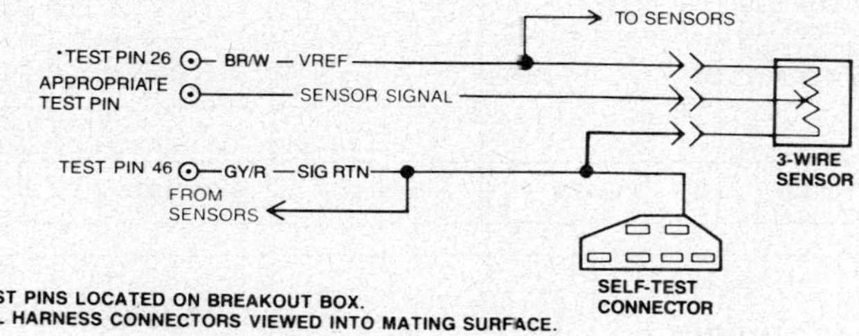

***TEST PINS LOCATED ON BREAKOUT BOX.**
ALL HARNESS CONNECTORS VIEWED INTO MATING SURFACE.

Reference Voltage — Pinpoint Test C

TEST STEP	RESULT ▶	ACTION TO TAKE
C1 CHECK VEHICLE BATTERY POWER CIRCUIT		
• Key off, wait 10 seconds. • Disconnect 60 pin connector. Inspect for damaged or pushed out pins, corrosion, loose wires, etc. Service as necessary. • Install breakout box and connect ECU to breakout box. • Voltmeter on 20 volt scale. • Key on, engine off. • Measure voltage between Test Pin 37 at the breakout box and SIG RTN/PWR GND circuit in the Self-Test connector. • **Are both voltages greater than 10.5 volts, and are both voltages within 1.0 volts of each other?**	Yes ▶ No ▶	GO to [C2]. RECONNECT sensor (if applicable). GO to [B1].
C2 CHECK VREF VOLTAGE		
• Key on, engine off. • Breakout box installed, ECU connected. • Voltmeter on 20 volt scale. • Measure voltage between Test Pin 26 and Test Pin 46 at the breakout box. • **What is the voltage?**	Greater than 6.0 volts ▶ Less than 4.0 volts ▶ Between 4.0 volts and 6.0 volts ▶	GO to [C4]. GO to [C5]. GO to [C3].
C3 CHECK VREF AND SIG RTN CIRCUITS FOR CONTINUITY		
• Key off. • Sensor that sent you here disconnected. • Breakout box installed. • Disconnect ECU. • Ohmmeter on 200 ohm scale. • Measure resistance between Test Pin 26 at breakout box and VREF circuit at vehicle harness connector of the sensor that sent you here. • Measure resistance between Test Pin 46 at breakout box and SIG RTN circuit at vehicle harness connector of the sensor that sent you here. • **Is each resistance less than 5.0 ohms?**	Yes ▶ No ▶	Reference voltage OK. REMOVE breakout box. RECONNECT sensor. RERUN Quick Test. SERVICE open in VREF or SIG RTN circuits. REMOVE breakout box. RECONNECT all components. RERUN Quick Test.

PINPOINT TESTS—NAVAJO

Reference Voltage — Pinpoint Test C

TEST STEP	RESULT ▶	ACTION TO TAKE
C4 CHECK FOR EXCESS VOLTAGE ON VREF CIRCUIT		
• Key off, wait 10 seconds. • Breakout box installed. • Disconnect ECU. • Disconnect SUPER STAR II Tester (if applicable.) **NOTE: For proper results of this test, the SUPER STAR II must be disconnected. Due to the circuitry of the SUPER STAR II Tester and the vehicle, voltage can be fed to the VREF circuit giving the false indication of a short to power.** • Key on, engine off. • Voltmeter on 20 volt scale. • Measure voltage between Test Pin 26 at the breakout box and battery ground. • **Is voltage less than 0.5 volts?**	Yes ▶	REPLACE ECU. REMOVE breakout box RECONNECT sensor. RERUN Quick Test.
	No ▶	SERVICE short to battery power in vehicle harness. REMOVE breakout box. RECONNECT ECU and sensor. RERUN Quick Test. If condition persists, REPLACE ECU.
C5 CHECK FOR SHORTED THROTTLE POSITION SENSOR		
• Key off, wait 10 seconds. • Breakout box installed, ECU connected. • Disconnect Throttle Position (TP) sensor from vehicle harness. • Key on, engine off. • Voltmeter on 20 volt scale. • Measure voltage between Test Pin 26 and Test Pin 46 at the breakout box. • **Is voltage less than 4.0 volts?**	Yes ▶	Key off. RECONNECT TP sensor. GO to C7.
	No ▶	REPLACE TP sensor. REMOVE breakout box. RECONNECT all components. RERUN Quick Test.

Reference Voltage — Pinpoint Test C

TEST STEP	RESULT ▶	ACTION TO TAKE
C7 CHECK FOR SHORTED BP SENSOR		
• Key off, wait 10 seconds. • Breakout box installed, ECU connected. • Disconnect BP sensor. • Key on, engine off. • Voltmeter on 20 volt scale. • Measure voltage between Test Pin 26 and Test Pin 46 at the breakout box. • **Is voltage less than 4.0 volts?**	Yes ▶	Key off. RECONNECT BP sensor. GO to C10.
	No ▶	REPLACE BP sensor. REMOVE breakout box. RECONNECT ECU and sensor(s). RERUN Quick Test.

Reference Voltage

Pinpoint Test **C**

TEST STEP	RESULT ▶	ACTION TO TAKE
C10 CHECK FOR SHORT TO GROUND IN VREF		
• Key off, wait 10 seconds. • Breakout box installed. • Disconnect ECU. • TP sensor disconnected. • BP sensor disconnected. • Ohmmeter on 200 ohm scale. • Measure resistance between Test Pin 26 and Test Pins 20, 40, 46 and 60 at the breakout box. • **Is any resistance less than 5 ohms?**	Yes ▶	SERVICE short to ground. REMOVE breakout box. RECONNECT all components. CONNECT all sensors. RERUN Quick Test. If original condition still exists, REPLACE ECU.
	No ▶	REPLACE ECU. REMOVE breakout box. RECONNECT all components. RERUN Quick Test.

Temperature Sensor Test ACT and ECT

Pinpoint Test **DA**

Note

You should enter this Pinpoint Test only when a Service Code 21, 24, 51, 54, 61 or 64 is received during Quick Test.

Remember

To prevent the replacement of good components, be aware that the following non-EEC areas may be at fault:

- Coolant level
- Cooling System
- Water pump drive belt
- Engine operating temperature
- Engine oil level
- Thermostat
- Heat stove duct valve operation
- Air cleaner duct
- Ambient temperature

This Pinpoint Test is intended to diagnose the following:

- ACT sensor
- ECT sensor
- Harness circuits: ACT, ECT and SIG RTN
- ECU assembly

Description

The Air Charge Temperature (ACT) and Engine Coolant Temperature (ECT) sensors change resistance in response to temperature. ACT and ECT sensor resistance decreases as the surrounding temperature increases, providing a signal to the ECU that indicates the temperature of either the incoming air charge or the engine coolant temperature.

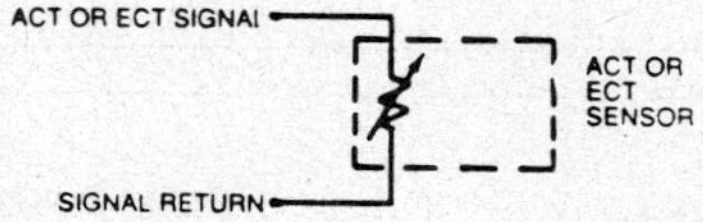

Temperature Sensor Test ACT and ECT	Pinpoint Test	DA

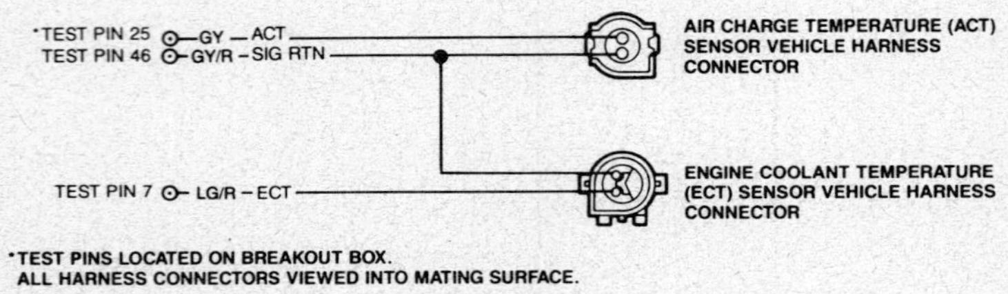

*TEST PINS LOCATED ON BREAKOUT BOX.
ALL HARNESS CONNECTORS VIEWED INTO MATING SURFACE.

Temperature Sensor Test ACT and ECT	Pinpoint Test	DA

Notes:

Engine coolant temperature must be greater than 50°F to pass the KOEO test and greater than 180°F to pass the KOER test. To accomplish this, the engine should be at normal operating temperature.

Ambient temperature should be above 50°F to receive acceptable input from the Air Charge Temperature sensor.

Voltage values were calculated for VREF=5.0 volts. These values may vary 15 percent due to sensor and VREF variations.

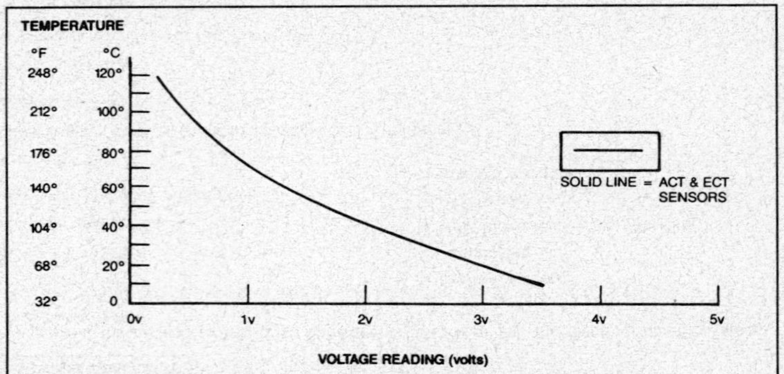

Temperature		Engine Coolant/Air Charge Temperature Sensor Values	
°F	°C	Voltage (volts)	Resistance (K ohms)
248	120	.27	1.18
230	110	.35	1.55
212	100	.46	2.07
194	90	.60	2.80
176	80	.78	3.84
158	70	1.02	5.37
140	60	1.33	7.70
122	50	1.70	10.97
104	40	2.13	16.15
86	30	2.60	24.27
68	20	3.07	37.30
50	10	3.51	58.75

Temperature Sensor Test ACT and ECT — Pinpoint Test DA

TEST STEP	RESULT ▶	ACTION TO TAKE
DA1 SERVICE CODE 21/116 OR 24/114: CHECK OPERATION, INSTALLATION OF TEMPERATURE SENSOR		
Service Code 21/116 (ECT) or 24/114 (ACT) indicates that the corresponding sensor is out of Self-Test range. Correct range of measure is 0.3 to 3.7 volts. Possible causes: — Low coolant level (ECT). — Ambient temperature below 50°F (ACT). — Faulty harness connector. — Faulty sensor. • Run engine for 2 minutes at 2000 rpm. — **For NO STARTS:** GO to DA3. — **For vehicle STALLS:** GO to S1. • Check that upper radiator hose is hot and pressurized. • Rerun Quick Test. • **Are codes 21, 24, 114 or 116 present?**	Yes ▶ No ▶	GO to DA2. SERVICE other codes as necessary.
DA2 CHECK VREF CIRCUIT VOLTAGE AT THROTTLE POSITION SENSOR		
• Refer to schematic in Pinpoint Test DH. • Key off, wait 10 seconds. • Disconnect TP sensor. • Voltmeter on 20 volt scale. • Key on, engine off. • Measure voltage between VREF circuit and SIG RTN circuit at the TP sensor vehicle harness connector. • **Is voltage between 4.0 volts and 6.0 volts?**	Yes ▶ No ▶	RECONNECT TP sensor. GO to DA3. GO to Pinpoint Test Step C1.

Temperature Sensor Test ACT and ECT — Pinpoint Test DA

TEST STEP	RESULT ▶	ACTION TO TAKE
DA3 CHECK RESISTANCE OF TEMPERATURE SENSOR WITH ENGINE OFF		
• Key off, wait 10 seconds. • Disconnect suspect temperature sensor. • Ohmmeter on 200,000 ohm scale. • Measure resistance between sensor signal circuit and SIG RTN circuit at the temperature sensor. Refer to the corresponding chart at the beginning of this Pinpoint Test for resistance specifications. • **Is resistance within specifications?**	Yes ▶ No ▶	GO to DA4. **For ECT sensor with a NO START:** Do not Service Code 21/116 at this time, GO to A1. REPLACE suspect sensor. RECONNECT vehicle harness. RERUN Quick Test.

Temperature Sensor Test ACT and ECT | Pinpoint Test | DA

TEST STEP	RESULT ▶	ACTION TO TAKE
DA4 CHECK RESISTANCE OF TEMPERATURE SENSOR WITH ENGINE RUNNING		
NOTE: Engine may have cooled down. Always warm engine before taking ECT sensor resistance measurements. Check for open thermostat. • Key off, wait 10 seconds. • Suspect temperature sensor disconnected. • Ohmmeter on 200,000 ohm scale. • Run engine for 2 minutes at 2000 rpm. • Measure resistance between sensor signal circuit and SIG RTN circuit at the temperature sensor. Refer to the corresponding chart at the beginning of this Pinpoint Test for resistance specifications. • **Is resistance within specification?**	Yes ▶	REPLACE ECU. RECONNECT vehicle harness. RERUN Quick Test.
	No ▶	REPLACE suspect sensor. RECONNECT vehicle harness. RERUN Quick Test.
DA10 SERVICE CODE 51/118 OR 54/113 INDUCE OPPOSITE CODE 61/117 OR 64/112		
Service Codes 51/118 (ECT) and 54/113 (ACT) indicate that the corresponding sensor signal is greater than the Self-Test maximum. The maximum for ECT and ACT sensors is 4.6 volts. Possible causes: — Open in harness (ACT or ECT). — Faulty connection. — Faulty sensor. — Faulty ECU. • Key off, wait 10 seconds. • Disconnect suspect temperature sensor. • Connect a jumper wire between the sensor signal circuit and SIG RTN circuit at the temperature sensor vehicle harness connector. • Run Key On Engine Off Self-Test. • **Is Code 61, 64, 112 or 117 present?**	Yes ▶	REPLACE suspect sensor. REMOVE jumper wire. RECONNECT vehicle harness. RERUN Quick Test.
	No ▶	REMOVE jumper wire. GO to DA11 .

Temperature Sensor Test ACT and ECT | Pinpoint Test | DA

TEST STEP	RESULT ▶	ACTION TO TAKE
DA11 CHECK CONTINUITY OF SENSOR SIGNAL AND SIG RTN CIRCUITS		
• Key off, wait 10 seconds. • Suspect temperature sensor disconnected. • Disconnect ECU 60 Pin connector. Inspect for damaged or pushed out pins, corrosion, loose wires, etc. Service as necessary. • Install breakout box, leave ECU disconnected. • Ohmmeter on 200 ohm scale. • Measure resistance between sensor signal circuit at the temperature sensor vehicle harness connector and Test Pin 7 (ECT) or 25 (ACT) at the breakout box. • Measure resistance between SIG RTN circuit at the temperature sensor vehicle harness connector and Test Pin 46 at the breakout box. • **Is each resistance less than 5 ohms?**	Yes ▶	REPLACE ECU. REMOVE breakout box. RECONNECT all components. RERUN Quick Test.
	No ▶	SERVICE open circuits. REMOVE breakout box. RECONNECT all components. RERUN Quick Test.
DA20 SERVICE CODE 61/117 OR 64/112: INDUCE OPPOSITE CODE 51/118 OR 54/113		
Service Codes 61/117 (ECT) or 64/112 (ACT) indicate that the corresponding sensor's signal is less than the Self-Test minimum. The ACT and ECT sensor minimum is 0.2 volts. Possible causes: — Grounded circuit in harness. — Faulty sensor. — Faulty ECU. — Faulty connection. • Key off, wait 10 seconds. • Disconnect vehicle harness from suspect sensor. Inspect for damaged, corroded, pushed out pins of loose wires, etc. Service as necessary. • Run Key On Engine Off Self-Test. • **Is Code 51, 54, 113 or 118 present?**	Yes ▶	REPLACE sensor. RECONNECT harness. RERUN Quick Test.
	No ▶	GO to DA21 .

Temperature Sensor Test ACT and ECT — Pinpoint Test DA

TEST STEP	RESULT ▶	ACTION TO TAKE
DA21 CHECK VREF CIRCUIT VOLTAGE AT THROTTLE POSITION SENSOR		
• Refer to schematic in Pinpoint Test DH. • Key off, wait 10 seconds. • Suspect temperature sensor disconnected. • Disconnect TP sensor. • Voltmeter on 20 volt scale. • Key on, engine off. • Measure voltage between VREF circuit and SIG RTN circuit at the TP sensor vehicle sensor connector. • **Is voltage between 4.0 volts and 6.0 volts?**	Yes ▶ No ▶	RECONNECT TP sensor, GO to DA22. GO to Pinpoint Test Step C1.
DA22 CHECK TEMPERATURE SENSOR SIGNAL CIRCUIT FOR SHORT TO GROUND		
• Key off, wait 10 seconds. • Suspect temperature sensor disconnected. • Disconnect ECU 60 Pin connector. Inspect for damaged or pushed out pins, corrosion, loose wires, etc. Service as necessary. • Install breakout box, leave ECU disconnected. • Ohmmeter on 200,000 ohm scale. • Measure resistance between Test Pin 7 (ECT) or 25 (ACT) and Test Pins 40, 46 and 60 at the breakout box. • **Is each resistance greater than 10,000 ohms?**	Yes ▶ No ▶	REPLACE ECU. REMOVE breakout box. RECONNECT all components. RERUN Quick Test. SERVICE short circuit. REMOVE breakout box. RECONNECT all components. RERUN Quick Test.

Temperature Sensor Test ACT and ECT — Pinpoint Test DA

TEST STEP	RESULT ▶	ACTION TO TAKE
DA90 CONTINUOUS MEMORY CODE 51/118, 54/113, 61/117 OR 64/112: CHECK SENSOR		
Continuous Memory Codes 51/118 and 54/113 indicate that the sensor signal was greater than the Self-Test maximum of 4.6 volts. The code was generated under normal driving conditions. Continuous Memory Codes 61/117 and 64/112, indicate that the sensor signal was less than the Self-Test minimum of 0.2 volts. The code was generated under normal driving conditions. (see table below) Possible causes: — Faulty sensor. — Open circuit in harness. — Grounded circuit in harness. — Faulty ECU. • Enter Key On Engine Off continuous monitor mode. Refer to Quick Test Appendix. • Observe Voltmeter or SUPER STAR II for indication of a fault while performing the following: — Tap on the sensor to simulate road shock. — Wiggle the sensor connector. • **Is a fault indicated?**	Yes ▶ No ▶	DISCONNECT and inspect connectors. If OK, REPLACE the sensor. CLEAR Continuous Memory RERUN Quick Test. GO to DA91.

Sensors	Continuous Memory Codes
ACT	54/113 and 64/112
ECT	51/118 and 61/117

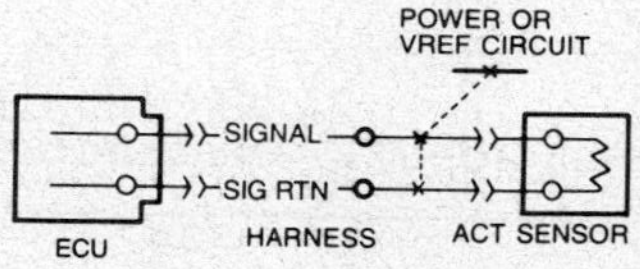

PINPOINT TESTS—NAVAJO

Temperature Sensor Test ACT and ECT — Pinpoint Test DA

TEST STEP	RESULT ▶	ACTION TO TAKE
DA91 CHECK EEC VEHICLE HARNESS • Still in Key On Engine Off continuous monitor mode. • Observe Voltmeter or SUPER STAR II for fault indication while performing the following: — Refer to the illustration in Step DA90. Grasp the vehicle harness close to the sensor connector. Wiggle, shake, or bend a small section of the EEC system vehicle harness while working your way to the dash panel. Also wiggle, shake, or bend the EEC system vehicle harness from the dash panel to the ECU. • **Is a fault indicated?**	Yes ▶ No ▶	ISOLATE fault and SERVICE as necessary. CLEAR Continuous Memory RERUN Quick Test. GO to DA92.
DA92 CHECK ECU AND VEHICLE HARNESS CONNECTORS • Key off, wait 10 seconds. • Disconnect ECU 60 Pin connector. Disconnect sensor connector. Inspect for damage, loose or pushed out pins, loose or poorly crimped wires. • **Are connectors and terminals OK?**	Yes ▶ No ▶	CLEAR Continuous Memory RERUN Quick Test. SERVICE as necessary. CLEAR Continuous Memory RERUN Quick Test.
DA100 CONTINUOUS MEMORY CODE 338 Service Code 338 indicates the engine had not reached the normal operating temperature or the system is not heating. The cooling system is not properly controlling the engine temperature. Possible causes: — Thermostat stuck open. — Water outlet gasket leak. — Water pump gasket leak. — Head gasket leak. — Heater hose leak. • Go directly to the Cooling System Group in the service manual for further diagnosis.		

Temperature Sensor Test ACT and ECT — Pinpoint Test DA

TEST STEP	RESULT ▶	ACTION TO TAKE
DA101 CONTINUOUS MEMORY CODE 339 Service Code 339 indicates the engine had exceeded the high temperature limit. The cooling system is not properly controlling the engine temperature. Possible causes: — Thermostat stuck closed. — Water passages clogged. — Faulty water pump. — Faulty cooling fan. — Low coolant level. — Faulty radiator cap. — Radiator fins clogged. — Coolant leakage. • Go directly to the Cooling System Group in the service manual for further diagnosis.		

Mass Air Flow (MAF) Sensor — Pinpoint Test — DC

Note

You should enter this Pinpoint Test only when a Service Code 26/159, 56/158, or 66/157 is received during Quick Test or you have been directed here from Diagnosis By Symptom.

Remember

To prevent the replacement of good components, be aware that the following non-EEC areas may be at fault:

- Air cleaner element
- Inlet air duct
- Throttle body

This Pinpoint Test is intended to diagnose only the following:

- Mass Air Flow sensor
- Harness circuits: VPWR, PWR GND, MAF and MAF RTN
- ECU assembly

Description

The Mass Air Flow (MAF) sensor is located between the air cleaner and the throttle body. The MAF sensor uses a hot wire sensing element to measure the amount of air entering the engine. Air passing over the hot wire causes it to cool. The MAF sensor then outputs an analog voltage signal to the electronic control unit (ECU) to determine the intake air mass. The ECU will then calculate the required fuel injector pulse width in order to provide the desired air/fuel ratio.

Pinpoint Test Schematic

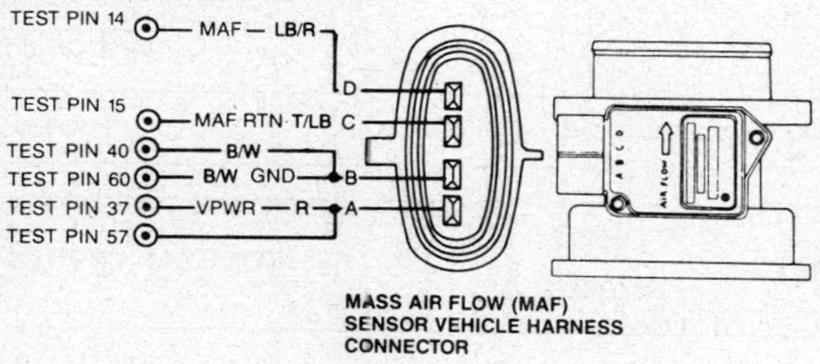

*TEST PINS LOCATED ON THE BREAKOUT BOX.
NOTE: ALL HARNESS CONNECTORS VIEWED INTO MATING SURFACE.

Mass Air Flow (MAF) Sensor — Pinpoint Test — DC

NOTE: MAF Signal Voltage vs. engine operation condition while engine is at normal operating temperature. These values are typical for automatics, but may vary based on vehicle load and temperature.

MAF Sensor Graph

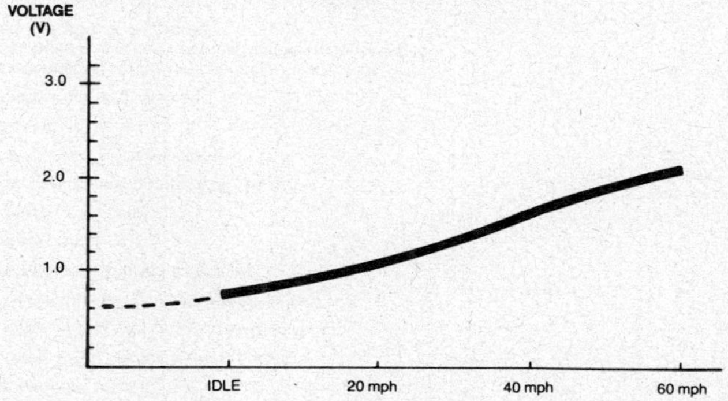

MAF Sensor Data

Engine Condition	MAF Signal Voltage
IDLE	.60
20 mph	1.10
40 mph	1.70
60 mph	2.10

Mass Air Flow (MAF) Sensor — Pinpoint Test DC

TEST STEP	RESULT ▶	ACTION TO TAKE
DC1 SERVICE CODE 26/159: CHECK VPWR CIRCUIT VOLTAGE		
Service Code 26/159 indicates that the Mass Air Flow (MAF) sensor is out of Self-Test range. Key On Engine Off Service Code 26/159 indicates that the MAF Signal was greater than .70 volts during Key On, Engine Off Self-Test. Key On Engine Running Service Code 26/159 indicates that the MAF Signal was not between .20 and 1.50 volts during Key On Engine Running Self-Test. **NOTE: Service Code 26/159 could be generated by the garage exhaust ventilation system. Remove ventilation system and properly vent to outside atmosphere before rerunning Self-Test.** Possible causes: — Faulty MAF sensor. — Faulty ECU. • Key off. • Disconnect MAF sensor. • Voltmeter on 20 volt scale. • Key on, engine off. • Measure voltage between VPWR circuit at the MAF sensor vehicle harness connector and battery negative post. • **Is voltage greater than 10.5 volts?**	Yes ▶ No ▶	GO to DC2. SERVICE open in VPWR circuit. RERUN Quick Test.
DC2 CHECK MAF SENSOR GROUND		
• Key on, engine off. • MAF sensor disconnected. • Voltmeter on 20 volt scale. • Measure voltage between VPWR circuit and PWR GND circuit at the MAF sensor vehicle harness connector. • **Is voltage greater than 10.5 volts?**	Yes ▶ No ▶	GO to DC10. SERVICE open PWR GND circuit. RECONNECT MAF sensor. RERUN Quick Test.

Mass Air Flow (MAF) Sensor — Pinpoint Test DC

TEST STEP	RESULT ▶	ACTION TO TAKE
DC10 SERVICE CODE 66/157 OR 72/129: CHECK CONTINUITY OF MAF AND VPWR CIRCUITS		
Service Code 66/157 indicates that the Mass Air Flow (MAF) sensor signal went below .40 volts sometime during last 80 warm-up cycles. Engine Running Service Code 72/129 indicates insufficient MAF change during Dynamic Response Test. Possible causes: — Open MAF circuit. — Open VPWR circuit to MAF sensor. — Open PWR GND circuit to MAF sensor. — Open MAF RTN circuit to MAF sensor. — MAF circuit shorted to ground. — Faulty ECU. — Faulty MAF sensor. — Air leak before or after MAF sensor. — MAF sensor disconnected. • Key off. • Disconnect MAF sensor. • Disconnect ECU 60 Pin connector. Inspect for damaged or pushed out pins, corrosion, loose wires, etc. Service as necessary. • Install breakout box, leave ECU disconnected. • Ohmmeter on 200 ohm scale. • Measure resistance between VPWR circuit at the MAF sensor vehicle harness connector and Test Pins 37 and 57 at the breakout box. • Measure resistance between MAF circuit at the MAF sensor vehicle harness connector and Test Pin 14 at the breakout box. • **Is each resistance less than 5 ohms?**	Yes ▶ No ▶	GO to DC11. SERVICE open circuit. REMOVE breakout box. RECONNECT all components. RERUN Quick Test.

PINPOINT TESTS—NAVAJO

Mass Air Flow (MAF) Sensor — Pinpoint Test DC

TEST STEP	RESULT ▶	ACTION TO TAKE
DC11 CHECK MAF CIRCUIT FOR SHORTS TO GROUND AND MAF RTN CIRCUIT • Key off. • MAF sensor disconnected. • Breakout box installed, ECU disconnected. • Ohmmeter on 200,000 ohm scale. • Measure resistance between Test Pin 14 and Test Pins 15, 40 and 60 at the breakout box. • **Is each resistance greater than 10,000 ohms?**	Yes ▶	GO to DC12.
	No ▶	SERVICE short circuit(s). REMOVE breakout box. RECONNECT all components. RERUN Quick Test.
DC12 CHECK CONTINUITY OF PWR GND CIRCUIT • Key off. • MAF sensor disconnected. • Breakout box installed, ECU disconnected. • Ohmmeter on 200 ohm scale. • Measure resistance between PWR GND circuit at the MAF sensor vehicle harness connector and battery negative post. • **Is resistance less than 10 ohms?**	Yes ▶	GO to DC13.
	No ▶	SERVICE open circuit. REMOVE breakout box. RECONNECT all components. RERUN Quick Test.
DC13 CHECK CONTINUITY OF MAF RTN CIRCUIT • Key off. • MAF sensor disconnected. • Breakout box installed, ECU disconnected. • Ohmmeter on 200 ohm scale. • Measure resistance between MAF RTN circuit at the MAF sensor vehicle harness connector and Test Pin 15 at the breakout box. • **Is resistance less than 5 ohms?**	Yes ▶	GO to DC14.
	No ▶	SERVICE open circuit. REMOVE breakout box. RECONNECT all components. RERUN Quick Test.

Mass Air Flow (MAF) Sensor — Pinpoint Test DC

TEST STEP	RESULT ▶	ACTION TO TAKE
DC14 CHECK MAF CIRCUIT FOR SHORT TO GROUND • Key off. • MAF sensor disconnected. • Breakout box installed. • Connect ECU to breakout box. • Ohmmeter on 200,000 ohm scale. • Measure resistance between Test Pin 14 and Test Pins 15, 40 and 60 at the breakout box. • **Is each resistance greater than 10,000 ohms?**	Yes ▶	GO to DC15.
	No ▶	REPLACE ECU. REMOVE breakout box. RECONNECT MAF sensor. RERUN Quick Test.
DC15 CHECK MAF CIRCUIT OUTPUT • Key off. • Reconnect MAF sensor. • Breakout box installed, ECU connected. • Voltmeter on 20 volt scale. • Key on, engine running. • Measure voltage between Test Pin 14 at the breakout box and battery negative post. • **Is voltage between .36 and 1.50 volts?**	Yes ▶	REPLACE ECU. REMOVE breakout box. RERUN Quick Test.
	No ▶	REPLACE MAF sensor. REMOVE breakout box. RECONNECT ECU. RERUN Quick Test.
DC16 CHECK MAF CIRCUIT OUTPUT • Key off. • Reconnect MAF sensor. • Breakout box installed, ECU connected. • Key on, engine running. • Measure voltage between Test Pin 14 and Test Pin 15 at the breakout box. • **Is voltage between .36 and 1.50 volts?**	Yes ▶	REPLACE ECU. REMOVE breakout box. RERUN Quick Test.
	No ▶	REPLACE MAF sensor. REMOVE breakout box. RECONNECT ECU. RERUN Quick Test.

Mass Air Flow (MAF) Sensor — Pinpoint Test DC

TEST STEP	RESULT ▶	ACTION TO TAKE
DC20 SERVICE CODE 56/158: RERUN SELF-TEST WITH MAF SENSOR DISCONNECTED		
Service Code 56/158 indicates that the Mass Air Flow (MAF) sensor signal went above 4.5 volts during normal engine operation (continuous) or during Self-Test. • Key off. • Disconnect MAF sensor. • Start engine, idle one minute. • Key off. • Rerun Key On Engine Off Self-Test. • **Is Service Code 66 or 157 present?**	Yes ▶ No ▶	REPLACE MAF sensor. RERUN Quick Test. GO to DC21 .
DC21 CHECK MAF CIRCUIT FOR SHORT TO VPWR		
• Key off. • MAF sensor disconnected. • Disconnect ECU 60 Pin connector. Inspect for damaged or pushed out pins, corrosion, loose wires, etc. Service as necessary. • Install breakout box, leave ECU disconnected. • Ohmmeter on 200,000 ohm scale. • Measure resistance between MAF circuit and VPWR circuit at the MAF sensor vehicle harness connector. • **Is resistance greater than 10,000 ohms?**	Yes ▶ No ▶	REPLACE ECU. REMOVE breakout box. RECONNECT MAF sensor. RERUN Quick Test. SERVICE short circuit. REMOVE breakout box. RECONNECT all components. RERUN Quick Test.

Barometric Pressure (BP) Sensor — Pinpoint Test DF

Note

You should enter this Pinpoint Test only when a Service Code 22/126 or 72/129 is received during Quick Test or if you have been directed here from Pinpoint Test S or Diagnosis By Symptom.

Remember

To prevent the replacement of good components, be aware that the following non-EEC areas may be at fault:

- Unusually high/low barometric pressure.
- Basic engine (valves, vacuum leaks, timing, etc.).

This Pinpoint Test is intended to diagnose only the following:

- BP sensor
- Harness circuits: VREF, BP SIG, and SIG RTN
- ECU assembly.

Description

The Barometric Pressure (BP) sensor operates as a piezoelectric (pressure sensing) disc. However, rather than generating a voltage, its output is a frequency change. The BP sensor is used to sense the changes in barometric pressure, allowing the electronic control unit (ECU) to sense the altitude at which the vehicle is being operated. Its signal affects air/fuel ratio and spark for altitude adjustment.

Barometric Pressure (BP) Sensor	Pinpoint Test	DF

Pinpoint Test Schematic

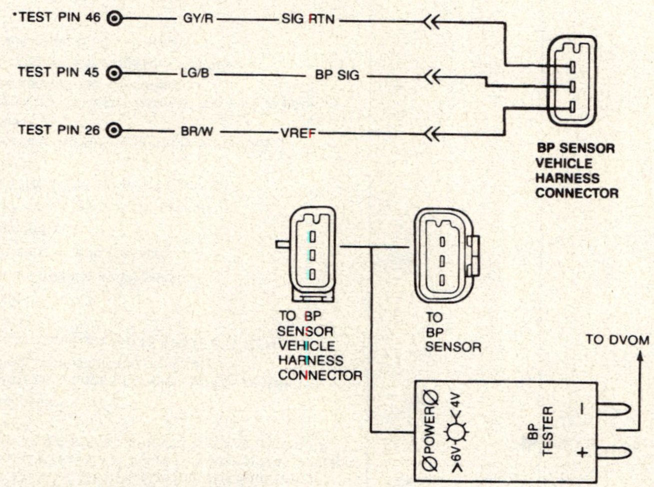

*TEST PINS LOCATED ON BREAKOUT BOX.
ALL HARNESS CONNECTORS VIEWED INTO MATING SURFACE.

Barometric Pressure (BP) Sensor	Pinpoint Test	DF

BP Sensor Graph (KOEO)

NOTE: Frequency may vary plus or minus 3 Hz from the values given due to sensor variations.

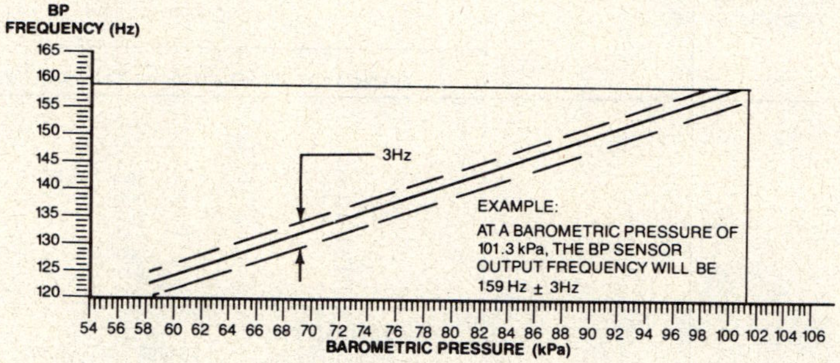

BP Sensor Data

Barometric Pressure in-Hg	Barometric Pressure kPa	Frequency Hz
17.1	58	122.4
18.3	62	125.5
19.5	66	128.7
20.7	70	131.9
21.8	74	135.1
23.0	78	138.3
24.2	82	141.8
25.4	86	145.4
26.6	90	148.9
27.7	94	152.5
28.9	98	156.1
30.1	102	159.6
31.0	105	162.4

Barometric Pressure (BP) Sensor — Pinpoint Test DF

TEST STEP	RESULT ▶	ACTION TO TAKE
DF1 SERVICE CODE 22/126: CHECK FOR POWER TO BP SENSOR Service Code 22/126 indicates that the Barometric Pressure (BP) sensor is out of Self-Test range. Correct BP tester range of measurement is typically from 1.4 to 1.6 volts. Possible causes: — BP SIG circuit open between sensor vehicle harness connector and processor. — BP SIG circuit shorted to VREF, SIG RTN, or GND. Faulty BP sensor. — Vacuum trapped at BP sensor. — High atmospheric pressure. — Damaged ECU. — VREF circuit open at BP sensor. — SIG RTN circuit open at BP sensor. • Key off. • Disconnect the BP sensor from the vehicle harness. • Connect the BP tester between the vehicle harness connector and the BP sensor. • Insert BP banana plugs into Voltmeter. • Set Voltmeter to 20 volt scale. **NOTE: Green light on tester indicates VREF is OK (4-6v). Red light (or no light) indicates VREF is either too low or too high.** • Key on. • **Is green light on?**	Yes ▶ No ▶	GO to DF3. GO to DF2.
DF2 CHECK FOR POWER AT SENSOR VEHICLE HARNESS CONNECTOR **NOTE: Green light reaffirms that VREF is OK (4-6v). Red light (or no light) indicates VREF is either too low or too high.** • Key on. • BP tester connected. • Voltmeter connected to BP tester. • Disconnect Bp sensor. • **Is green light on?**	Yes ▶ No ▶	REPLACE BP sensor. RERUN Quick Test. REMOVE BP tester. RECONNECT the BP sensor. GO to Pinpoint Test Step C1.

Barometric Pressure (BP) Sensor — Pinpoint Test DF

TEST STEP	RESULT ▶	ACTION TO TAKE
DF3 CHECK BP SENSOR OUTPUT **NOTE: Measure several known good BP sensors on available vehicles. The measured voltage will be typical for your location on the day of testing.** • Key on. • BP tester connected. • Voltmeter connected to BP tester. • Measure BP sensor voltage on customer vehicle. • **Is voltage in range for your altitude?**	Yes ▶ No ▶	REMOVE BP tester. GO to DF4. REMOVE BP tester. GO to DF5.
DF4 CHECK CONTINUITY OF BP SIG CIRCUIT • Key off, wait 10 seconds. • BP sensor disconnected. • Disconnect ECU 60 pin connector. Inspect for damaged or pushed out pins, corrosion, loose wires, etc. Service as necessary. • Install breakout box, leave processor disconnected. • Ohmmeter on 200 ohm scale • Measure resistance between BP SIG circuit at the BP sensor vehicle harness connector and Test Pin 45 at the breakout box. • **Is resistance less than 5.0 ohms?**	Yes ▶ No ▶	REPLACE ECU. REMOVE breakout box. RECONNECT BP sensor. RERUN Quick Test. SERVICE open circuit. REMOVE breakout box. RECONNECT all components. RERUN Quick Test.

Altitude table for DF3:

Approximate Altitude (Ft.)	Voltage Output (+/−.04 Volts)
0	1.59
1000	1.56
2000	1.53
3000	1.50
4000	1.47
5000	1.44
6000	1.41
7000	1.39

Barometric Pressure (BP) Sensor — Pinpoint Test DF

TEST STEP	RESULT ▶	ACTION TO TAKE
DF5 CHECK BP SIG CIRCUIT FOR SHORTS TO VREF, SIG RTN AND GROUND		
• Key off, wait 10 seconds. • BP sensor disconnected. • Disconnect ECU 60 pin connector. Inspect for damaged or pushed out pins, corrosion, loose wires, etc. Service as necessary. • Install breakout box, leave ECU disconnected. • Ohmmeter on 200,000 ohm scale. • Measure resistance between Test Pin 45 and Test Pins 26, 46, 40 and 60 at the breakout box. • **Is each resistance greater than 10,000 ohms?**	Yes ▶	REPLACE BP sensor. REMOVE breakout box. RECONNECT ECU. RERUN Quick Test.
	No ▶	SERVICE short circuit. REMOVE breakout box. RECONNECT all components. RERUN Quick Test.

Barometric Pressure (BP) Sensor — Pinpoint Test DF

TEST STEP	RESULT ▶	ACTION TO TAKE
DF90 CONTINUOUS MEMORY CODE 22/126: EXERCISE BP SENSOR		
Continuous Memory Service Code 22/126 indicates the Barometric Pressure (BP) sensor was out of self-test range. The code was set during normal driving conditions. Correct range of measurement is typically from 1.4 to 1.6 volts. Possible causes: — Faulty BP sensor. — Faulty EEC vehicle harness. — Faulty BP sensor vehicle harness connectors and/or terminals. — Unusually high/low barometric pressure. • Using Key On, Engine Off Continuous Monitor Mode, observe Voltmeter or SUPER STAR II for indication of a fault while performing the following: — Connect a vacuum pump to the BP sensor. — Slowly apply 84 kPa (25 in-Hg) vacuum to the BP sensor. — Slowly bleed vacuum off the BP sensor. — Lightly tap on BP sensor (simulate road shock). — Wiggle BP connector. • **Is fault indicated?**	Yes ▶	DISCONNECT and INSPECT connectors. If connector and terminals are good, REPLACE BP sensor, RERUN Quick Test.
	No ▶	GO to DF91.

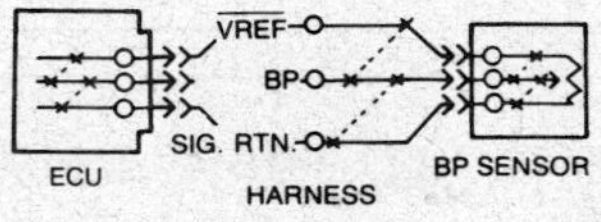

Barometric Pressure (BP) Sensor — Pinpoint Test DF

TEST STEP	RESULT ▶	ACTION TO TAKE
DF91 CHECK EEC VEHICLE HARNESS		
• Remain in Key On, Engine Off Continuous Monitor Mode. • Observe Voltmeter or SUPER STAR II for a fault indication while performing the following: — Referring to the illustration in Step DF90, grasp the vehicle harness close to the sensor connector. Wiggle, shake or bend a small section of the EEC system vehicle harness while working your way to the dash panel. Also wiggle, shake or bend the EEC vehicle harness from the dash panel to the ECU. • **Is a fault indicated?**	Yes ▶	ISOLATE fault and SERVICE as necessary. CLEAR Continuous Memory Code RERUN Quick Test.
	No ▶	GO to DF10.
DF92 CHECK ECU AND VEHICLE HARNESS CONNECTORS		
• Key off, wait 10 seconds. • Disconnect ECU 60 pin connector. • Inspect connectors and connector terminals for obvious damage or faults. • **Are connectors and terminals OK?**	No ▶	SERVICE as necessary. RERUN Quick Test.
	Yes ▶	CLEAR Continuous Memory RERUN Quick Test.

Throttle Position (TP) Sensor — Pinpoint Test DH

Note

You should enter this Pinpoint Test only when a Service Code 23/121, 53/123, 63/122 or 73/167 is received during Quick Test.

Remember

To prevent the replacement of good components, be aware that the following non-EEC areas may be at fault:

- Idle speeds/throttle stop adjustment
- Binding throttle shaft/linkage or speed control linkage.
- TP sensor may not be seated properly (tightened down)

This Pinpoint Test is intended to diagnose only the following:

- Throttle Position (TP) sensor
- Sensor harness circuits: VREF, TP, and SIG RTN
- ECU assembly.

Description

The Throttle Position (TP) sensor is a potentiometer that provides a signal to the electronic control unit (ECU) that is directly proportional to throttle plate position.

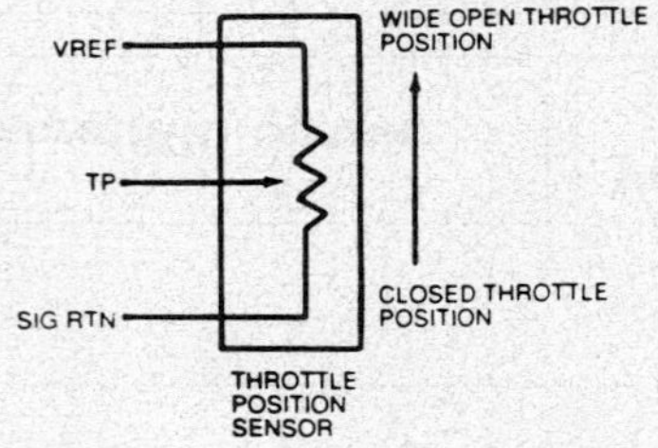

Throttle Position (TP) Sensor	Pinpoint Test	DH

Pinpoint Test Schematic

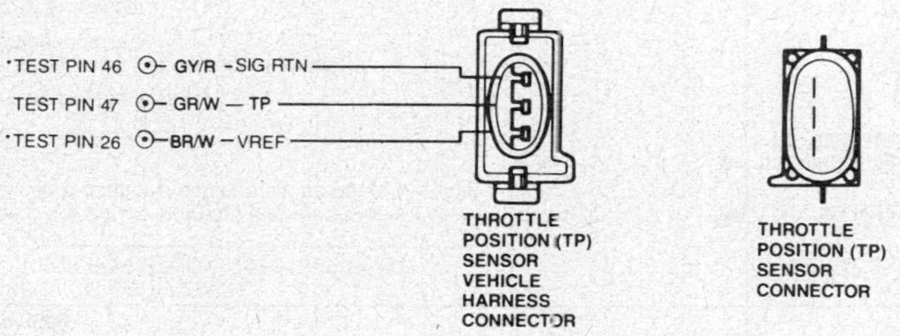

*TEST PINS LOCATED ON BREAKOUT BOX.
ALL HARNESS CONNECTORS VIEWED INTO MATING SURFACE.

Throttle Position (TP) Sensor	Pinpoint Test	DH

NOTE: The normal range of the throttle angle measurement for the Throttle Position (TP) sensor is from 0 to 85 degrees.

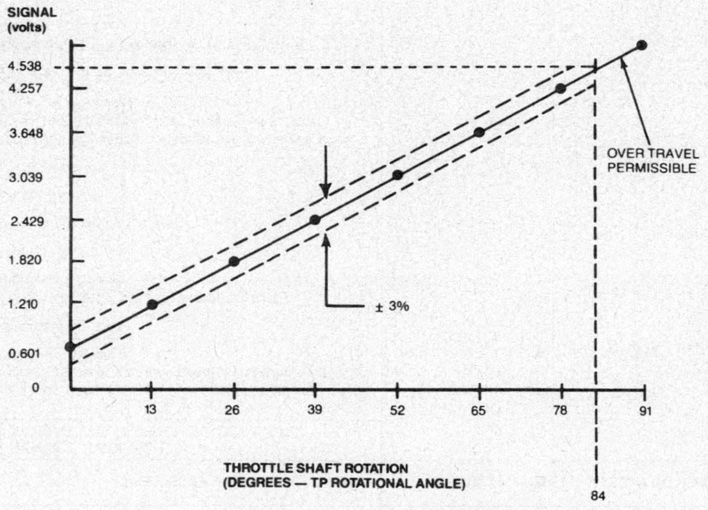

Throttle Position (TP) Sensor — Pinpoint Test DH

TEST STEP	RESULT ▶	ACTION TO TAKE
DH1 ENGINE RUNNING SERVICE CODE 23/121: CHECK FOR OTHER SERVICE CODES		
Service Code 23/121 indicates that the Throttle Position (TP) sensor's rotational setting may be out of Self-Test range. Possible causes are: — Binding throttle linkage. — TP sensor may not be seated properly (tightened down). — Faulty TP sensor. — Faulty ECU. • Check for Codes 31/327 in Key On Engine Running Self-Test. • **Are either of the above codes present with Code 23 or 121?**	Yes ▶ No ▶	**RETURN to the Key On Engine Running Service Code chart and PROCEED as directed with Code 31/327.** GO to DH2.
DC2 SERVICE CODE 23/121: CHECK FOR STUCK THROTTLE PLATE		
Service Code 23/121 indicates that the Throttle Position (TP) sensor's rotational setting may be out of Self-Test range. Possible causes are: — Binding throttle linkage. — TP sensor may not be seated properly (tightened down). — Faulty TP sensor. — Faulty ECU. • Visually inspect throttle body and throttle linkage for binding or sticking. • Verify the throttle linkage is at mechanical/closed throttle. Check for: binding throttle linkage, speed control linkage, vacuum line/electrical harness interference, etc. • **Does throttle move freely and return to closed throttle position?**	Yes ▶ No ▶	GO to DH3. SERVICE as necessary. RERUN Quick Test.

Throttle Position (TP) Sensor — Pinpoint Test DH

TEST STEP	RESULT ▶	ACTION TO TAKE
DH3 SERVICE CODE 53/123: ATTEMPT TO GENERATE CODE 63/122		
Service Code 53/153 indicates that the Throttle Position (TP) sensor signal is greater than the Self-Test maximum value. Possible causes are: — TP sensor may not be seated properly (tightened down). — Faulty TP sensor. — Short to power in harness. — Faulty ECU. • Key off, wait 10 seconds. • Disconnect TP sensor. Inspect for damaged or pushed out pins, corrosion, loose wires, etc. Service as necessary. • Rerun Key On Engine Off Self-Test. • **Is Code 63 or 122 present (ignore all other codes)?**	Yes ▶ No ▶	GO to DH4. GO to DH6.
DH4 CHECK VREF CIRCUIT VOLTAGE		
• Key off, wait 10 seconds. • TP sensor disconnected. • Voltmeter on 20 volt scale. • Key on, engine off. • Measure voltage between VREF circuit and SIG RTN circuit at the TP sensor vehicle harness connector. • **Is voltage between 4.0 and 6.0 volts?**	Yes ▶ No ▶	REPLACE TP sensor. RERUN Quick Test. RECONNECT TP sensor. GO to Pinpoint Test Step C1.

PINPOINT TESTS—NAVAJO

Throttle Position (TP) Sensor — Pinpoint Test DH

TEST STEP	RESULT ▶	ACTION TO TAKE
DH6 CHECK TP CIRCUIT FOR SHORTS TO POWER		
• Key off, wait 10 seconds. • TP sensor disconnected. • Disconnect ECU 60 pin connector. Inspect for damaged or pushed out pins, corrosion, loose wires, etc. Service as necessary. • Install breakout box, leave ECU disconnected. • Ohmmeter on 200,000 ohm scale. • Measure resistance between Test Pin 47 and Test Pins 26 and 57 at the breakout box. • **Is each resistance greater than 10,000 ohms?**	Yes ▶	REPLACE ECU. REMOVE breakout box. RECONNECT TP sensor. RERUN Quick Test.
	No ▶	SERVICE short circuit. REMOVE breakout box. RECONNECT all components. RERUN Quick Test.

Throttle Position (TP) Sensor — Pinpoint Test DH

TEST STEP	RESULT ▶	ACTION TO TAKE
DH10 SERVICE CODE 63/122: ATTEMPT TO GENERATE CODE 53/123 OR 23/121		
Service Code 63/122 indicates that the Throttle Position (TP) sensor signal is less than the Self-Test minimum value. Possible causes are: — TP sensor may not be seated properly (tightened down). — Faulty TP sensor. — Open harness. — Grounded harness. — Faulty ECU. • Key off, wait 10 seconds. • Disconnect TP sensor. Inspect for damaged or pushed out pins, corrosion, loose wires, etc. Service as necessary. • Jumper VREF circuit to TP circuit at TP sensor vehicle harness connector. • Perform Key On Engine Off Self-Test. **NOTE: If no codes are generated, immediately remove jumper and go directly to DH14.** • **Is Code 53/123 or 23/121 present (ignore all other codes)?**	Yes ▶	REPLACE TP sensor. REMOVE jumper. RERUN Quick Test.
	No ▶	REMOVE jumper. GO to DH11.
DH11 CHECK VREF CIRCUIT VOLTAGE		
• Key off, wait 10 seconds. • TP sensor disconnected. • Voltmeter on 20 volt scale. • Key on engine off. • Measure voltage between VREF circuit and SIG RTN circuit at the TP sensor vehicle harness connector. • **Is voltage between 4.0 and 6.0 volts?**	Yes ▶	GO to DH12.
	No ▶	RECONNECT all components. GO to Pinpoint Test Step C1.

Throttle Position (TP) Sensor — Pinpoint Test DH

TEST STEP	RESULT ▶	ACTION TO TAKE
DH12 CHECK TP CIRCUIT CONTINUITY		
• Key off, wait 10 seconds. • TP sensor disconnected. • Disconnect ECU 60 pin connector. Inspect for damaged or pushed out pins, corrosion, loose wires, etc. Service as necessary. • Install breakout box, and leave ECU disconnected. • Ohmmeter on 200 ohm scale. • Measure resistance between TP circuit at the TP sensor vehicle harness connector and Test Pin 47 at the breakout box. • **Is the resistance less than 5.0 ohms?**	Yes ▶	GO to DH14.
	No ▶	SERVICE open circuit. REMOVE breakout box. RECONNECT all components. RERUN Quick Test.
DH14 CHECK TP CIRCUIT FOR SHORTS TO GROUND		
• Key off, wait 10 seconds. • TP sensor disconnected. • Disconnect ECU 60 pin connector. Inspect for damaged or pushed out pins, corrosion, loose wires, etc. Service as necessary. • Install breakout box, leave ECU disconnected. • Ohmmeter on 200,000 ohm scale. • Measure resistance between Test Pin 47 and Test Pins 40, 46, and 60 at the breakout box. • **Is each resistance greater than 10,000 ohms?**	Yes ▶	REPLACE ECU. REMOVE breakout box. RECONNECT all components. RERUN Quick Test.
	No ▶	SERVICE short circuit. REMOVE breakout box. RECONNECT all components. RERUN Quick Test.

Throttle Position (TP) Sensor — Pinpoint Test DH

TEST STEP	RESULT ▶	ACTION TO TAKE
DH20 ENGINE RUNNING SERVICE CODE 73/167: PERFORM PROPER DYNAMIC RESPONSE TEST AT WIDE OPEN THROTTLE		
NOTE: Engine Running Service Code 73/167 indicates the TP sensor did not exceed 25 percent of its rotation during the Dynamic Reponse Test. A complete wide open throttle must be performed during the Dynamic Response portion of the test. • Run Key On Engine Running Self-Test. Be sure a complete WOT is performed during the Dynamic Response portion of the test. • **Is Code 73 or 167 still present?**	Yes ▶	GO to DH21.
	No ▶	Unable to duplicate Code 73/167 at this time. GO to Quick Test Step 5.0 to service other Engine Running Codes. Otherwise, testing completed.
DH21 CHECK TP SENSOR MOVEMENT DURING DYNAMIC RESPONSE TEST		
• Key off. • Disconnect ECU 60 pin connector. Inspect for damaged or pushed out pins, corrosion, loose wires, etc. Service as necessary. • Install breakout box and connect ECU to breakout box. • Voltmeter on 20 volt scale. • Connect Voltmeter to Test Pin 47 and Test Pin 46 at the breakout box. • Rerun Engine Running Self-Test with a proper WOT Dynamic Response portion of the test. • **Does voltage increase to greater than 3.5 volts during the Dynamic Response Test?**	Yes ▶	REPLACE ECU. REMOVE breakout box. RERUN Quick Test.
	No ▶	VERIFY TP sensor is properly installed to throttle body. If OK, REPLACE TP sensor. RERUN Quick Test.

PINPOINT TESTS—NAVAJO

Throttle Position (TP) Sensor — Pinpoint Test DH

TEST STEP	RESULT ▶	ACTION TO TAKE
DH90 CONTINUOUS MEMORY CODE 53/123: MONITOR TP CIRCUIT UNDER SIMULATED ROAD SHOCK		
• Enter Key On Engine Off Continuous Monitor mode. Refer to Quick Test Appendix. • Observe Voltmeter or SUPER STAR II for indication of a fault while performing the following: — Move throttle slowly to WOT position. — Release throttle slowly to closed position and lightly tap on TP sensor (simulate road shock). — Wiggle TP harness connector. • **Does Voltmeter or SUPER STAR II indicate a fault?** 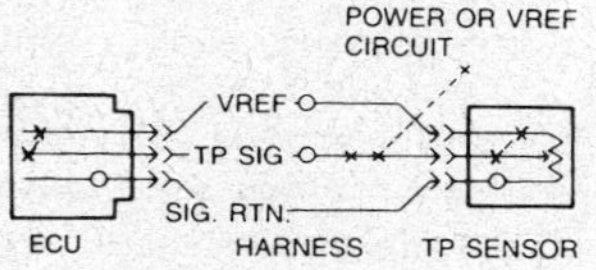 	Yes ▶ No ▶	GO to DH91. GO to DH92.
DH91 MEASURE THROTTLE POSITION SIGNAL VOLTAGE WHILE EXERCISING TP SENSOR		
• Key off, wait 10 seconds. • Disconnect ECU 60 pin connector. Inspect for damaged or pushed out pins, corrosion, loose wires, etc. Service as necessary. • Install breakout box and connect ECU to breakout box. • Voltmeter or SUPER STAR II still connected to STO as in previous step. • Connect a VOM from Test Pin 47 to Test Pin 46 at the breakout box. • Voltmeter on 20 volt scale. • Key on, engine off. • While observing VOM, repeat step DH90. • **Does the fault occur below 4.25 volts?**	Yes ▶ No ▶	DISCONNECT and INSPECT connectors. If connector and terminals are good, REPLACE TP sensor CLEAR Continuous Memory RERUN Quick Test. Throttle position sensor overtravel may have caused the Continuous Memory Code 53/123. VERIFY vehicle harness integrity, GO to DH92.

Throttle Position (TP) Sensor — Pinpoint Test DH

TEST STEP	RESULT ▶	ACTION TO TAKE
DH92 CHECK EEC VEHICLE HARNESS		
• Still in Key On Engine Off Continuous Monitor mode. • Observe Voltmeter or SUPER STAR II for a fault indication while performing the following: — Referring to the illustration in Step DH90, grasp the vehicle harness close to the sensor connector. Wiggle, shake or bend a small section of the EEC system vehicle harness while working your way to the dash panel. Also wiggle, shake or bend the EEC vehicle harness from the dash panel to the ECU. • **Does Voltmeter or SUPER STAR II indicate a fault?**	Yes ▶ No ▶	ISOLATE fault. SERVICE as necessary. REFER to appropriate figure. CLEAR Continuous Memory RERUN Quick Test. GO to DH93.
DH93 CHECK ECU AND HARNESS CONNECTORS		
• Key off, wait 10 seconds. • Disconnect ECU 60 pin connector. Inspect for damaged or pushed out pins, corrosion, loose wires, etc. Service as necessary. • **Are connectors and terminals OK?**	No ▶ Yes ▶	SERVICE as necessary. CLEAR Continuous Memory RERUN Quick Test. CLEAR Continuous Memory RERUN Quick Test.

Throttle Position (TP) Sensor — Pinpoint Test DH

TEST STEP	RESULT ▶	ACTION TO TAKE
DH94 CONTINUOUS MEMORY CODE 63/122: MONITOR TP CIRCUIT UNDER SIMULATED ROAD SHOCK		
• Enter Key On Engine Off Continuous Monitor mode. Refer to Quick Test Appendix. • Observe Voltmeter or SUPER STAR II for indication of a fault while performing the following: — Move throttle slowly to WOT position. — Release throttle slowly to closed position. — Lightly tap on TP sensor (simulate road shock). — Wiggle TP harness connector. • **Does Voltmeter or SUPER STAR II indicate a fault?**	Yes ▶	INSPECT connectors. If connectors and terminals are good, REPLACE TP sensor. CLEAR Continuous Memory RERUN Quick Test.
	No ▶	GO to DH95.

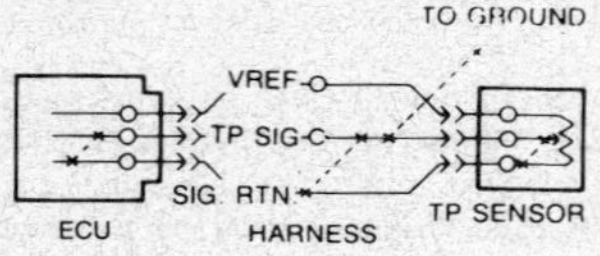

TEST STEP	RESULT ▶	ACTION TO TAKE
DH95 CHECK EEC VEHICLE HARNESS		
• Still in Key On Engine Off Continuous Monitor mode. • Observe Voltmeter or SUPER STAR II for a fault indication while performing the following: — Referring to the illustration in Step DH94, grasp the vehicle harness close to the sensor connector. Wiggle, shake or bend a small section of the EEC system vehicle harness while working your way to the dash panel. Also wiggle, shake or bend the EEC vehicle harness from the dash panel to the ECU. • **Does Voltmeter or SUPER STAR II indicate a fault?**	Yes ▶	ISOLATE fault. SERVICE as necessary. REFER to appropriate figure. CLEAR Continuous Memory RERUN Quick Test.
	No ▶	GO to DH96.

Throttle Position (TP) Sensor — Pinpoint Test DH

TEST STEP	RESULT ▶	ACTION TO TAKE
DH96 CHECK ECU AND HARNESS CONNECTORS		
• Key off, wait 10 seconds. • Disconnect ECU 60 pin connector. Inspect for damaged or pushed out pins, corrosion, loose wires, etc. • **Are connectors and terminals OK?**	No ▶	SERVICE as necessary. CLEAR Continuous Memory. RERUN Quick Test.
	Yes ▶	CLEAR Continuous Memory RERUN Quick Test.

Vehicle Speed Sensor (VSS) | Pinpoint Test | DP

Note

You should enter this Pinpoint Test only when Service Code 29 is received during Quick Test or when you have been directed here from Pinpoint Test TG.

Remember

This Pinpoint Test is intended to diagnose only the following:

- VSS Harness Circuits
- Vehicle Speed Sensor
- ECU Assembly

Description

The Vehicle Speed Sensor (VSS) is a variable reluctance sensor that generates a waveform with a frequency that is proportional to vehicle speed. When the vehicle is moving slowly, the sensor produces a low frequency signal. As the vehicle speed increases, the sensor produces a higher frequency signal. The ECU uses this signal to control fuel njection, ignition timing, and transmission/transaxle shift points.

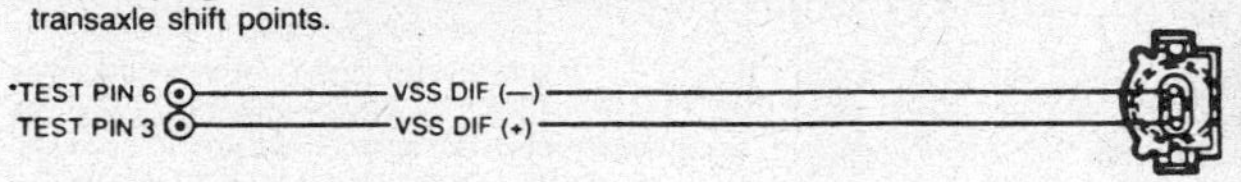

Pinpoint Test Schematic

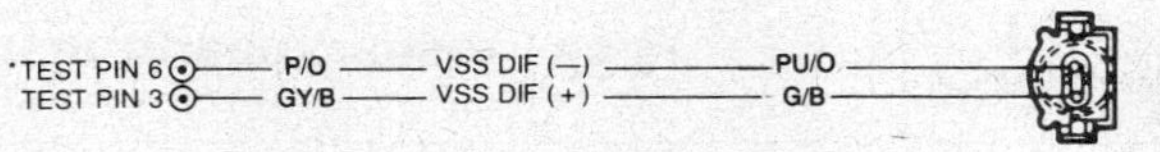

*TEST PINS LOCATED ON BREAKOUT BOX.
ALL HARNESS CONNECTORS VIEWED INTO MATING SURFACE.

Vehicle Speed Sensor (VSS) | Pinpoint Test | DP

VEHICLE SPEED SENSOR (VSS) DRIVE CYCLE

- Record and clear EEC Continuous Memory Codes.
- Warm engine to operating temperature.
- Perform the drive cycle below as appropriate for the vehicle being tested.

Automatic Transmission

Place the gear selector in DRIVE range and heavily accelerate to 35 mph, then coast down to an idle and stop the vehicle. Shut engine off.

Run Key On Engine Off Self-Test and record Continuous Memory Codes.

Manual Transmission

Starting in first gear, shift to second gear and moderately accelerate to 40 mph, then coast down to an idle and stop vehicle. Shut engine off.

Run Key On Engine Off Self-Test and record Continuous Memory Codes.

TEST STEP	RESULT ▶	ACTION TO TAKE
DP1 CONTINUOUS MEMORY CODE 29/452: COMPLETE VSS DRIVE CYCLE: VERIFY CODE OR DRIVE COMPLAINT		
Continuous Memory Code 29/452 indicates that sometime during the last 40 to 80 warm-up cycles, the processor detected an error in the VSS output signal. Possible causes are: — Faulty VSS. — Open or shorted circuit(s). — Faulty ECU. • Perform Vehicle Speed Sensor Drive Cycle at least three times as outlined below. • **VEHICLE SPEED SENSOR (VSS) DRIVE CYCLE:** — Record and clear continuous memory codes. — Warm engine to operating temperature. **AUTOMATIC TRANSMISSIONS:** — Place gear selector in DRIVE range. — Obey all local traffic laws. Accelerate heavily to 35 mph. — Coast down to an idle and stop the vehicle. — Shut the engine off. — After the drive cycle is completed, run Key On Engine Off Self-Test and record the continuous memory service codes displayed. **MANUAL TRANSMISSIONS:** — From first gear, shift to second. — Obey all local traffic laws. Accelerate moderately to 40 mph. — Coast down to an idle and stop the vehicle. — Shut the engine off. — After the drive cycle is completed, run Key On Engine Off Self-Test and record the continuous memory service codes displayed. • **Did Continuous Memory Code 29 or 452 repeat?**	Yes ▶ No ▶	GO to DP2. CLEAR Continuous Memory

Vehicle Speed Sensor (VSS) | Pinpoint Test | DP

TEST STEP	RESULT ▶	ACTION TO TAKE
DP2 CHECK CONTINUITY OF VEHICLE SPEED SENSOR HARNESS CIRCUITS		
• Key off. • Disconnect ECU 60 pin connector. Inspect for damaged or pushed out pins, corrosion, loose wires, etc. Service as necessary. • Install breakout box, ECU disconnected. • Disconnect vehicle speed sensor. • Measure resistance between Test Pin 3 at the Breakout Box and VSSDIF (+) circuit at the vehicle speed sensor vehicle harness connector. • Measure resistance between Test Pin 6 at the breakout box and VSSDIF (−) circuit at the vehicle speed sensor vehicle harness connector. • **Is each resistance less than 5.0 ohms?**	Yes ▶	GO to DP3.
	No ▶	SERVICE open circuit. REMOVE breakout box. RECONNECT all components. REPEAT Test Step DP1 to verify elimination of code or drive complaint.
DP3 CHECK VEHICLE SPEED SENSOR SIGNAL HARNESS CIRCUITS FOR SHORTS TO POWER OR GROUND		
• Key off. • Vehicle speed sensor disconnected. • Breakout box installed, ECU disconnected. • Measure resistance between Test Pin 3 and Test Pins 6, 37 and 40 at the breakout box. • **Is each resistance greater than 500 ohms?**	Yes ▶	RECONNECT the ECU. GO to DP4.
	No ▶	SERVICE short circuit. REMOVE breakout box. RECONNECT all components. REPEAT Test Step DP1 to verify elimination of code or drive complaint.
DP4 CHECK VEHICLE SPEED SENSOR RESISTANCE		
• Key off. • Vehicle speed sensor disconnected. • Measure the resistance of the vehicle speed sensor. • **Is resistance between 190 and 250 ohms?**	Yes ▶	REMOVE breakout box. REPLACE ECU. RECONNECT the vehicle speed sensor. REPEAT Test Step DP1 to verify elimination of code or drive complaint.
	No ▶	REPLACE the vehicle speed sensor. REMOVE breakout box. REPEAT Test Step DP1 to verify elimination of code or drive complaint.

Neutral Drive Input | Pinpoint Test | FA

Note

You should enter this Pinpoint Test only when a Service Code 67/522, 525 or 528, or 79 is received during Quick Test or you have been directed here from Diagnosis By Symptom.

Remember

This Pinpoint Test is intended to diagnose only the following:

- Clutch Engage/Interlock Switch
- Neutral Drive Switch
- Neutral Gear Switch
- ECU
- Harness Circuits: CES, CIS, NDS, NGS, and SIG RTN

Description

The Neutral Drive Input gives an indication of the transmission load to the ECU. This information is required for adjusting fuel/air ratio and idle speed. If the Neutral Drive Switch (NDS), Neutral Gear Switch (NGS) or Clutch Engage Switch (CES) is closed, the circuit from Test Pin 30 will be grounded.

Neutral Drive Input	Pinpoint Test	FA

Pinpoint Test Schematics

Manual Transmission Applications

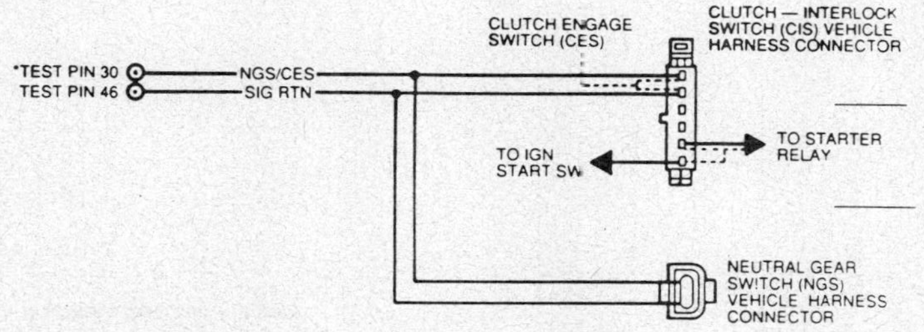

Automatic Transmission Applications

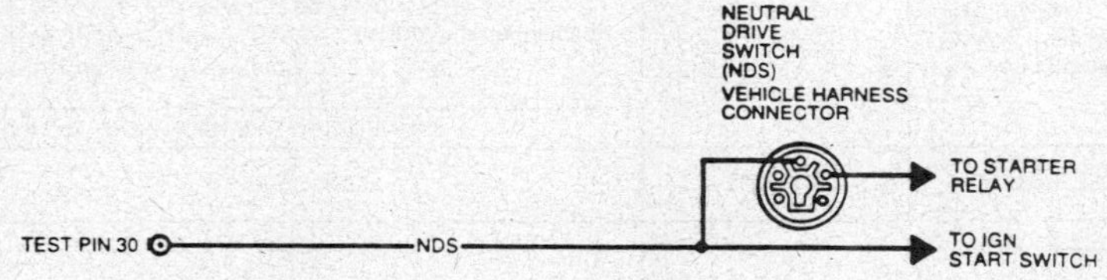

Neutral Drive Input	Pinpoint Test	FA

TEST STEP	RESULT ▶	ACTION TO TAKE
FA1 CODE 67/522, 525 or 528 SYSTEM IDENTIFICATION		
A Code 67/522, 525 or 528 resulted from voltage being high at either: — Pin 10 = A/C input — Pin 30 = Neutral drive while cranking the engine or during KOEO test. Possible causes are: — A/C circuit shorted to power — Clutch engage/interlock circuits open — Neutral drive/gear switch open — Damaged ECU — Starter relay disconnected during Self-Test.	▶	GO to FA8.

Neutral Drive Input | Pinpoint Test | FA

TEST STEP	RESULT ▶	ACTION TO TAKE
FA8 CHECK NEUTRAL DRIVE INPUT		
• Key off, wait 10 seconds. • Verify A/C control is in OFF position, if so equipped. • Verify transmission is in NEUTRAL or PARK. • Disconnect ECU 60 pin connector. Inspect for damaged or pushed out pins, corrosion, loose wires, etc. Service as necessary. • Install breakout box, leave ECU connected. • Key on, engine off. • Voltmeter on 20 volt scale. • Measure voltage between Test Pin 30 at the breakout box and chassis ground. • **Is voltage less than 1.0 volt?**	Yes ▶	**For vehicles with A/C:** Perform Pinpoint Test FA10. **For all others:** REPLACE ECU.
	No ▶	GO to FA9.

Neutral Drive Input | Pinpoint Test | FA

TEST STEP	RESULT ▶	ACTION TO TAKE
FA9 CHECK NEUTRAL DRIVE SWITCH		
• Key off, wait 10 seconds. • Breakout box installed, ECU disconnected. • Ohmmeter on 200 ohm scale. • Locate the Neutral Drive switch. • Disconnect vehicle harness from the Neutral Drive switch and measure resistance across the switch. • **Is resistance less than 5 ohms?**	Yes ▶	SERVICE open in vehicle harness Neutral Drive circuit. REMOVE breakout box. RECONNECT all components. RERUN Quick Test.
	No ▶	REPLACE Neutral Drive switch. REMOVE breakout box. RECONNECT all components. RERUN Quick Test.
FA10 CHECK A/C INPUT		
NOTE: Before entering this test, verify A/C selector is off. If A/C was on, rerun Quick Test. If code 67/522, 525 or 528, or 79 is present, continue with this test. • Disconnect ECU 60 pin connector. Inspect for damaged or pushed out pins, corrosion, loose wires, etc. Service as necesary. • Install breakout box, leave ECU disconnected. • Key on, engine off. • Voltmeter on 20 volt scale. • Measure voltage between Test Pin 10 at the breakout box and chassis ground. • **Is voltage greater than 1.0 volt?**	Yes ▶	SERVICE short to power in A/C circuit. REMOVE breakout box. RECONNECT all components. RERUN Quick Test.
	No ▶	REPLACE ECU. REMOVE breakout box. RECONNECT all components. RERUN Quick Test.

Neutral Drive Input | Pinpoint Test | FA

TEST STEP	RESULT ▶	ACTION TO TAKE
FA11 CHECK A/C INPUT CIRCUIT		
NOTE: A low idle with A/C on could be the result of the ECU not receiving, or recognizing the A/C input on Pin 10. • Key off, wait 10 seconds. • Disconnect ECU 60 pin connector. Inspect for damaged or pushed out pins, corrosion, loose wires, etc. Service as necessary. • Install breakout box, leave ECU disconnected. • Voltmeter on 20 volt scale. • Key on, engine off. • A/C on. • Measure voltage between Test Pin 10 and Test Pin 40. • **Is voltage greater than 10.5 volts?**	Yes ▶	REPLACE ECU. REMOVE breakout box. RECONNECT all components. RERUN Quick Test.
	No ▶	SERVICE open in A/C circuit. REMOVE breakout box. RECONNECT all components. RERUN Quick Test.

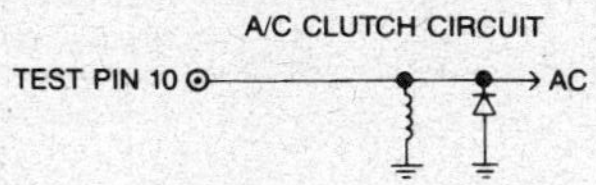

Brake On/Off (BOO) | Pinpoint Test | FD

Note

You should enter this Pinpoint Test only when a Service Code 74/536 or 75/531 is received during Quick Test.

Remember

This Pinpoint Test is intended to diagnose only the following:

- BOO circuit
- ECU assembly

Description

The Brake On/Off (BOO) input is wired to the stoplamp circuit and informs the ECU when the brake is applied. The BOO input is used primarily by the converter clutch lock/unlock strategy.

Pinpoint Test Schematic

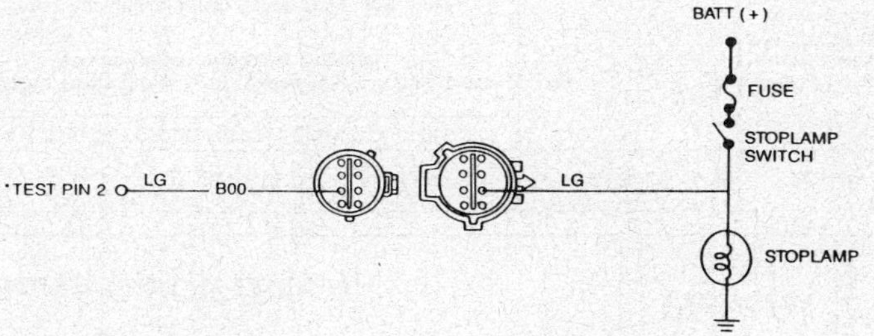

*** TEST PINS LOCATED ON BREAKOUT BOX.**
ALL HARNESS CONNECTORS VIEWED INTO MATING SURFACES.

PINPOINT TESTS—NAVAJO

Brake On/Off (BOO) — Pinpoint Test FD

* TEST PINS LOCATED ON BREAKOUT BOX.
ALL HARNESS CONNECTORS VIEWED INTO MATING SURFACES.

TEST STEP	RESULT ▶	ACTION TO TAKE
FD1 SERVICE CODE 74/536 VERIFY BRAKE WAS PRESSED		
Service Code 74/536 indicates that when the brake pedal was depressed and released during the Engine Running Self-Test, the BOO signal did not cycle high and low. Possible causes: — Brake pedal not depressed and released during the Engine Running Self-Test. — Brake pedal depressed during entire Engine Running Self-Test. — Open BOO/stoplamp circuit. — Short to GROUND or POWER. — Damaged brake switch. — Damaged processor. • **Did you press brake during the Engine Running Self-Test?** **NOTE: On some vehicles it is necessary to depress and release the brake after the Dynamic Response Code 1(0) but before the brief WOT.**	Yes ▶ No ▶	GO to FD2. RERUN Engine Running Self-Test. PRESS brake once during test.
FD2 CHECK OPERATION OF STOPLAMPS		
• Key on. • Check stoplamp operation.	Stoplamps operate normally ▶ Stoplamps never on ▶ Stoplamps always on ▶	GO to FD3. GO to FD4. GO to FD5.
FD3 CHECK FOR BOO CIRCUIT CYCLING		
• Key off. • Disconnect ECU 60 pin connector. Inspect for damaged or pushed out pins, corrosion, loose wires, etc. Service as necessary. • Install breakout box, leave ECU disconnected. • DVOM on 20 volt scale. • Measure voltage between BOO Test Pin (refer to chart on FD cover page) and Test Pin 40 at the breakout box while depressing and releasing brake. • **Does the voltage cycle?**	Yes ▶ No ▶	REPLACE ECU. REMOVE breakout box. RERUN Quick Test. SERVICE open in BOO circuit between ECU and BOO connection to stoplamp circuit. RERUN Quick Test.

Brake On/Off (BOO) — Pinpoint Test FD

TEST STEP	RESULT ▶	ACTION TO TAKE
FD4 CHECK FOR POWER TO BRAKE SWITCH		
NOTE: Verify integrity of related fuses in fuse panel and condition of stoplamp bulbs. • Key off. • Disconnect brake switch (located on brake pedal). • Measure voltage between BATT (+) input to brake switch and chassis ground. • **Is voltage greater than 10 volts?**	Yes ▶ No ▶	VERIFY operation of brake switch. If OK, SERVICE open circuit between brake switch and stoplamp ground. RECONNECT brake switch connector. RERUN Quick Test. SERVICE open BATT (+) circuit to brake switch. RECONNECT brake switch connector. RERUN Quick Test.
FD5 VERIFY BRAKE SWITCH IS NOT ALWAYS CLOSED		
• Key off. • Disconnect brake switch (located on brake pedal). • Key on, engine off. • **Are stoplamps still on?**	Yes ▶ No ▶	GO to FD6. VERIFY proper installation of brake switch. If OK, REPLACE brake switch. RECONNECT harness connector. RERUN Quick Test.
FD6 CHECK FOR SHORT TO POWER IN ECU		
• Key off. • Brake switch disconnected. • Disconnect ECU. • Key on, engine off. • **Are stoplamps still on?**	Yes ▶ No ▶	SERVICE short to power in BOO/Stoplamp circuit. RECONNECT all components. RERUN Quick Test. REPLACE ECU. RECONNECT brake switch. RERUN Quick Test.

PINPOINT TESTS—NAVAJO

Brake On/Off (BOO) — Pinpoint Test FD

TEST STEP	RESULT ▶	ACTION TO TAKE
FD10 SERVICE CODE 75/531: CHECK OPERATION OF BRAKE LAMPS		
Service Code 75/531 indicates that while the brake pedal was released during the Engine Running Self-Test, the BOO signal was high. Possible causes: — Brake pedal depressed during entire Engine Running Self-Test. — Open BOO/stoplamp circuit (between ECU and stoplamp ground). — Short to POWER. — Damaged brake switch. — Damaged processor. • Key on. • Check stoplamp operation.	Stoplamps operate normally ▶	GO to FD11.
	Stoplamps always on ▶	GO to FD5.
	Stoplamps never on ▶	VERIFY condition of stoplamp bulbs. If OK, SERVICE open circuit between BOO connection to stoplamp circuit and stoplamp ground. RERUN Quick Test.
FD11 CHECK BOO CIRCUIT CONTINUITY		
• Key off. • Disconnect ECU 60 pin connector. Inspect for damaged or pushed out pins, corrosion, loose wires, etc. Service as necessary. • Install breakout box, leave ECU disconnected. • Disconnect brake switch (located on brake pedal). • Measure resistance between Test Pin 2 at the breakout box and circuit to stoplamps pin at the brake switch vehicle harness connector. • **Is resistance less than 5.0 ohms?**	Yes ▶	REPLACE ECU. RECONNECT brake switch. REMOVE breakout box. RERUN Quick Test.
	No ▶	SERVICE open BOO circuit between ECU and BOO connection to stoplamp circuit. REMOVE Breakout Box. RECONNECT ECU and brake switch. RERUN Quick Test.

Brake On/Off (BOO) — Pinpoint Test FD

TEST STEP	RESULT ▶	ACTION TO TAKE
FD90 CONTINUOUS MEMORY CODE 536: CHECK FOR PROPER STOPLAMP SWITCH INSTALLATION		
Continuous Memory Code 536 indicates a BOO circuit failure. If the BOO input does not cycle after a predetermined number of transitions from 0 mph to a specific speed, the BOO input is assumed to be damaged and Continuous Memory Code 536 is set. Possible causes: — Stoplamp switch improperly installed. — Open stoplamp/BOO circuit. — Stoplamp/BOO circuit. — Stoplamp/BOO circuit shorted to power. — Damaged stoplamp switch. — Damaged stoplamp ground connection. • Check stoplamp switch for proper installation (alignment with pedal), corrosion, frayed wires, etc. • **Is stoplamp switch in good condition and properly installed?**	Yes ▶	GO to FD91.
	No ▶	SERVICE as necessary. CLEAR Continuous Memory. RERUN Quick Test.
FD91 CHECK STOPLAMP GROUND		
• Check stoplamp ground connection for corrosion or other damage. • Check stoplamp connector and wires for corrosion or other damage. • **Are stoplamp wires, connector, and ground connection OK?**	Yes ▶	GO to FD92.
	No ▶	SERVICE as necessary. CLEAR Continuous Memory. RERUN Quick Test.
FD92 CHECK STOPLAMP/BOO CIRCUITS FOR SHORT TO POWER		
• Key on, engine off. • Brake pedal NOT depressed. • Wiggle stoplamp/BOO circuit wires and connectors while observing stoplamps. • **Do stoplamps flash on while wiggling?**	Yes ▶	ISOLATE short to power and SERVICE as necessary. CLEAR Continuous Memory. RERUN Quick Test.
	No ▶	GO to FD93.

Brake On/Off (BOO) — Pinpoint Test FD

TEST STEP	RESULT ▶	ACTION TO TAKE
FD93 CHECK STOPLAMP CIRCUIT CONTINUITY		
• Key off. • Depress brake pedal and hold. • Wiggle stoplamp circuit wires and connectors while observing stoplamps. • Lightly tap stoplamp switch (simulate road shock) while observing stoplamps. • **Do stoplamps ever go off?**	Yes ▶	ISOLATE open in stoplamp circuit and SERVICE as necessary. CLEAR Continuous Memory RERUN Quick Test.
	No ▶	GO to FD94.
FD94 CHECK BOO CIRCUIT CONTINUITY		
• Key off. • Release brake pedal. • Disconnect ECU 60 pin connector. Inspect for damaged or pushed out pins, corrosion, loose wires, etc. Service as necessary. • Install breakout box, leave ECU disconnected. • Connect DVOM between BOO Test Pin (refer to schematic) at the breakout box and stoplamp circuit at the stoplamp switch. • DVOM on 200 ohm scale. • Wiggle BOO circuit wires and connectors while observing DVOM. • **Is resistance ever greater than 5.0 ohms while wiggling?**	Yes ▶	ISOLATE open in BOO circuit and SERVICE as necessary. REMOVE breakout box. RECONNECT ECU. RERUN Quick Test.
	No ▶	GO to FD99.

Brake On/Off (BOO) — Pinpoint Test FD

Note

A basic working knowledge of the EEC system is critical to effectively analyze road test data.

WARNING: THIS ROAD TEST IS A SUGGESTED BUT OPTIONAL PROCEDURE. ALL APPLICABLE SAFETY PROCEDURES AND TRAFFIC LAWS MUST BE FOLLOWED. IN ORDER FOR A ROAD TEST TO BE PERFORMED IT IS REQUIRED THAT ANOTHER PERSON ACCOMPANY THE DRIVER. THE ACCOMPANYING PERSON CAN MAKE MEASUREMENTS, OBSERVE CHANGES AND RECORD NOTES.

Prepare Vehicle for a Road Test

- Breakout box installed, ECU connected.
- Install fuel pressure gauge and MAP/BP tester (optional).
- Other materials needed; DVOM, pencil, paper, appropriate schematic/pin usage sheet (from Section 3A or 4A).

Preliminary Power/Ground Checks

- With the key ON and a DVOM referenced to the battery negative post, check the following signals for correct values.

POWERS:

— KAPWR > 10.5V (Pin 1)

— VPWR > 10.5V (Pins 37/57)

— VREF 5 ± 1V (pin 26)

GROUNDS (all = 0 ± .5V

— PWR GND (Pins 40/60)

— SIG RTN (Pin 46)

— IGN GND (Pin 16)

OPTIONAL GROUNDS:

— HEGO GND (Pin 49)

— CSE GND (Pin 20)

— MAF RTN (Pin 9 or 15)

Brake On/Off (BOO) — Pinpoint Test FD

FD99 Road Test

The purpose of the road test is to identify an area of concern by monitoring certain controlled parameters while trying to re-create a driveability or MIL symptom.

Obtaining Other Needed Information and Materials Before the Road Test

- Refer to the Symptom Charts , looking at the chart(s) that most resembles the vehicle's driveability or MIL symptom. Before the road test perform the Visual/Mechanical Checks that are listed. Next, list the EEC sensors and actuators in the order given. These circuits, along with the BOO signal, are the main circuits that will be monitored.
- Refer to the proper Diagnostic Reference Value Sheet Although these charts were designed for use with the EEC monitor box. most of the values can be read using the breakout box and a DVOM (with the DVOM referenced to ground all values in DCV units can be used; other values may also be helpful, ex. MAP Hz using the BP tester). Also refer to Section F-16 Appendix.
- The use of test lamps may also aid diagnosis. For example, a test lamp could be connected at the stop lamp switch between BATT (+) and ground, with another connected between the "power to bulbs" pin and ground. The BATT (+) lamp should always stay on, with the "power to bulbs" lamp on only when the brake is depressed. Also, with a DVOM connected between the BOO Test Pin and Test Pin 40 at the breakout box, if 6-7 volts is shown with the brake pedal released, this could indicate an open circuit between the ECU and stop lamp ground (the ECU supplies 6-7 volts on the BOO circuit, but this voltage is not "strong" enough to light the stop lamps or most test lamps).

Road Test

- After starting the engine for the road test, enter Engine Running Continuous Monitor Mode
- Drive the vehicle to create the conditions so that the symptom will occur. If the Customer Information Worksheet has been completed, this information may help when trying to re-create the symptom.
- When the symptom occurs, the accompanying passenger should observe changes in listed EEC signals. Information about the symptom, operating condition value of the EEC signal or other notes should be recorded onto paper.
- If you are unable to duplicate the symptom, it may still be helpful to verify that the EEC values are in the expected range.

Analyzing the Data

- Once the road test is completed, the results need to be analyzed to locate and service the exact fault which caused the symptom.
- If no problem is identified, return for other possible causes of the symptom.

Fuel Control — Pinpoint Test H

Note

You should enter this Pinpoint Test only when a Service Code 172 or 173 is received during Quick Test or when directed here from Diagnosis By Symptom or Pinpoint Test [S].

Remember

To prevent the replacement of good components, be aware that the following non-EEC areas may be at fault:

- Ignition Coil
- Fouled Spark Plugs
- Spark Plug Wires
- CANP Problems
- PCV Valves (see note below)
- Distributorless Ignition System
- Air Filter
- Fuel Contamination, Engine Oil
- Poor Power Ground
- Manifold Leaks, Intake/Exhaust Leaks
- Engine Not at Normal Operating Temperatures

This Pinpoint Test is intended to diagnose only the following:

- HEGO Sensor
- HEGO Signal and Ground Circuit
- HEGO Sensor Connection
- Vacuum Systems
- Fuel Injector
- ECU Assembly
- Harness circuits: HEGO GRD, HEGO, INJ. 1-6, VPWR and SIG RTN

NOTE: Fuel contaminated engine oil may affect 172 and 173 Service Codes, so if it is suspected, remove the PCV from the valve cover, and rerun Quick Test. If the problem is corrected, then change the engine oil and filter.

Description

The Heated Exhaust Gas Oxygen (HEGO) sensor reacts with the oxygen in the exhaust gases and generates a voltage based on this reaction. A low voltage indicates too much oxygen, or a lean condition. A high voltage indicates not enough oxygen, or a rich condition.

The fuel injector is a typical on/off device. The amount of fuel delivered is determined by how long the injector is held open.

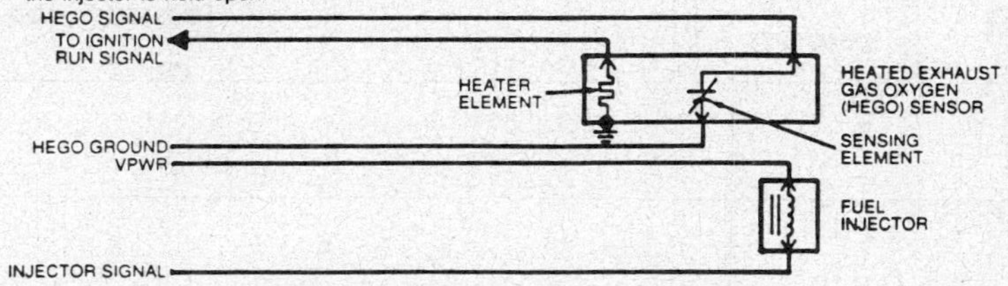

Fuel Control	Pinpoint Test	H

Pinpoint Test Schematic

4.0L EFI

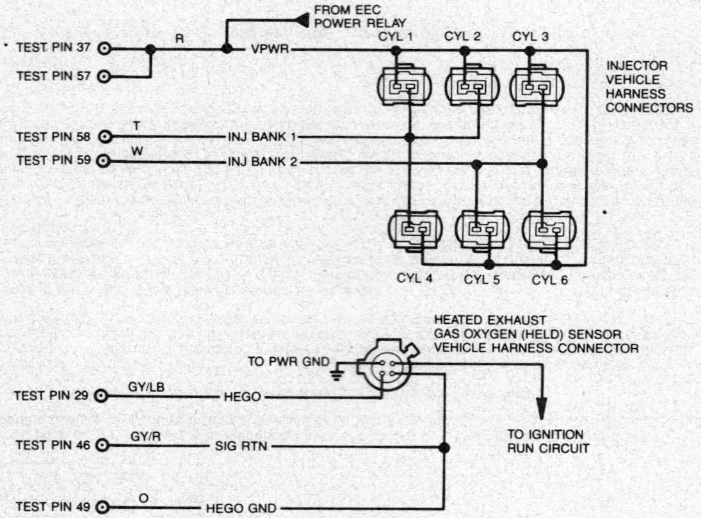

* TEST PINS LOCATED ON BREAKOUT BOX.
ALL HARNESS CONNECTORS VIEWED INTO MATING SURFACE.

Fuel Control	Pinpoint Test	H

TEST STEP	RESULT ▶	ACTION TO TAKE
H1 CHECK FUEL PRESSURE		
HEGO Engine Running code 41/172 indicates the system is always lean. HEGO Engine running code 42/173 indicates the system is always rich. • Key off, wait 10 seconds. • Install fuel pressure gauge. • Verify that manifold vacuum is connected to the fuel pressure regulator if applicable. • Start and run engine at idle. • Is fuel pressure within specification for the engine being tested? **For No Starts:** • If engine will not run, cycle the key off to on several times. • **Is fuel pressure within specification for the engine being tested?**	Yes ▶ No ▶	GO to H2 . GO to Section for Electric Fuel pump and fuel pressure regulator checks.

ENGINE RUNNING	30-45 PSI 210-310 kPa
KEY ON ENGINE OFF	35-45 PSI 240-310 kPa

NOTE: Maximum fuel pressure is obtainable at WOT or with the vacuum hose removed from the fuel pressure regulator.

INJECTOR BANK RESISTANCE SPECIFICATION
4.0 TO 5.5 ohms

SINGLE INJECTOR RESISTANCE SPECIFICATION
13.0 to 16.0 ohms

PINPOINT TESTS – NAVAJO

Fuel Control — Pinpoint Test H

TEST STEP	RESULT ▶	ACTION TO TAKE
H2 CHECK SYSTEM'S ABILITY TO HOLD FUEL PRESSURE		
• Key on, engine off. • **Does fuel pressure remain at specification for 60 seconds?**	Yes ▶	**For No Starts:** GO to H3. **For Runs Rough, Misses or Fuel Service Code (41, 42):** REMOVE fuel pressure gauge. GO to H4.
	No ▶	GO to H6.
H3 FUEL DELIVERY TEST		
NOTE: Verify fuel quality; air and/or water will also pressurize and look like acceptable fuel delivery. • Key off. • Fuel pressure gauge installed. • Pressurize fuel system per step H1. • Locate and disconnect the inertia switch. • Crank engine for 5 seconds. • **Does pressure drop greater than 5 psi. (34 kPa.) by the end of the 5 second crank cycle?**	Yes ▶	The EEC system is not the cause of the No Start. REMOVE the fuel pressure gauge. RECONNECT the inertia switch.
	No ▶	REMOVE fuel pressure gauge. RECONNECT inertia switch. GO to H4.

Fuel Control — Pinpoint Test H

TEST STEP	RESULT ▶	ACTION TO TAKE
H4 CHECK RESISTANCE OF INJECTOR(S) AND HARNESS		
• Key off, wait 10 seconds. • Disconnect ECU 60 pin connector. Inspect for damaged or pushed out pins, corrosion, loose wires, etc. Service as necessary. • Install breakout box, leave ECU disconnected. • Ohmmeter on 200 ohm scale. • Measure resistance of INJECTOR BANK 1 between Test Pin 37 and Test Pin 58 at the breakout box. Record resistance. • Measure resistance of INJECTOR BANK 2 between Test Pin 37 and Test Pin 59 at the breakout box. Record resistance. • **Is each resistance within specifications?**	Yes ▶	GO to H6.
	No ▶	GO to H5. **For NO START:** SERVICE open in VPWR circuit.

Fuel Control — Pinpoint Test H

TEST STEP	RESULT ▶	ACTION TO TAKE
H5 ISOLATE FAULTY INJECTOR CIRCUIT • Key off. • Breakout box installed, ECU disconnected. • Disconnect all injectors on suspect bank. • Ohmmeter on 200 ohm scale. • Connect one injector and measure resistance between Test Pin 37 and either Test Pin 58 or 59 as appropriate. • Disconnect that injector and repeat process for each of the remaining injectors. • **Is each resistance within 13-16 ohms?**	Yes ▶	GO to H6 .
	No ▶	SERVICE open or short in VPWR or injector circuit of the suspect injector(s). If OK, REPLACE injector. REMOVE breakout box. RECONNECT ECU and injectors. DRIVE vehicle five miles at 55 mph. RERUN Quick Test.
H6 CHECK INJECTOR DRIVER SIGNAL Requires standard non-powered 12 volt test lamp. • Key off. • Connect ECU to breakout box • Connect test lamp between Test Pin 37 and Test Pin 58 at the breakout box. • Connect test lamp between Test Pin 37 and 59 at the breakout box. • Crank or start engine. **NOTE: Properly operating systems will show a dim glow on the lamp.** • **Is glow on lamp dim?**	Yes ▶	GO To H7 .
	No ▶	**No light:** VERIFY 12 volt at Test Pins 37 and 57. **Bright light:** CHECK injector circuit for shorts to ground. If system checks OK, REPLACE ECU. REMOVE breakout box. RERUN Quick Test.

Fuel Control — Pinpoint Test H

TEST STEP	RESULT ▶	ACTION TO TAKE
H7 CHECK EXTERNAL SOURCE FOR FUEL PRESSURE PROBLEM • Key off. • Pressurize fuel system per Test Step H1 . — Visually look for fuel leaking at fuel injector O-rings, fuel pressure regulator, and fuel rails. • **Is there a visible leak?**	Yes ▶	REMOVE pressure gauge. SERVICE as necessary. After servicing leak, RERUN Quick Test.
	No ▶	GO to H8 .
H8 INJECTOR BALANCE TEST • Connect tachometer to engine. Run engine at idle. • Disconnect and reconnect the injectors one at a time: Note rpm drop for each injector. • **Does each injector produce at least a 100 rpm momentary drop?** **NOTE: ISC will attempt to re-establish rpm.**	Yes ▶	Fuel delivery OK. Problem is in an area common to all cylinders i.e. air/vacuum leak, fuel contamination. SERVICE as necessary.
	No ▶	GO to injector testing and cleaning instructions. After any servicing, RERUN Quick Test.

Fuel Control — Pinpoint Test H

TEST STEP	RESULT ▶	ACTION TO TAKE
H11 SERVICE CODE 41/172 FUEL CONTROL ALWAYS LEAN		
HEGO Engine Running code 41 indicates the system is always lean. • Run engine at 2000 rpm for 2 minutes. • Key off, wait 10 seconds. • Rerun Engine Running Self-Test. • **Is Code 41 or 172 present?**	Yes ▶	GO to H14.
	No ▶	GO to H20.

Fuel Control — Pinpoint Test H

TEST STEP	RESULT ▶	ACTION TO TAKE
H14 CHECK HEGO SENSOR		
NOTE: The purpose of this test is to verify the HEGO sensor can generate greater than 0.5 volts during Engine Running Self-Test. Any Vacuum/air leaks in non-EEC areas could also cause Code 41/172 or 91/136. Check for: — Leaking vacuum actuator (e.g. A/C control motor). — Engine sealing. — PCV system. — Unmetered air leak between Mass Air Flow sensor and throttle body. — Lead contaminated HEGO sensor. • Key off. • Disconnect appropriate HEGO sensor from vehicle harness. • Connect voltmeter to HEGO SIGNAL and HEGO GND at the HEGO sensor connector. • Voltmeter on 20 volt scale. • Run engine at approximately 2000 rpm for two minutes. • Rerun Engine Running Self-Test and monitor HEGO sensor voltage. • **Is the voltage greater than 0.5 volts at the end of Self-Test?**	Yes ▶	GO to H15.
	No ▶	REPLACE HEGO sensor. RERUN Quick Test.

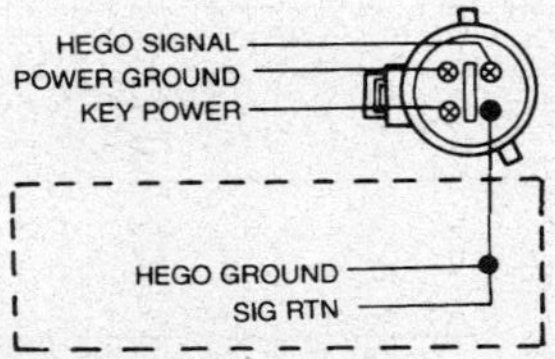

HEGO SENSOR CONNECTOR

PINPOINT TESTS—NAVAJO

Fuel Control — Pinpoint Test H

TEST STEP	RESULT ▶	ACTION TO TAKE
H15 CHECK CONTINUITY OF HEGO SIGNAL AND HEGO GROUND CIRCUITS		
• Key off. • Disconnect ECU 60 pin connector. Inspect for damaged or pushed out pins, corrosion, loose wires, etc. Service as necessary. • Install breakout box, leave ECU disconnected. • HEGO disconnected. • Ohmmeter on 200 ohm scale. • Measure resistance between Test Pin 49 and Test Pin 46 at the breakout box and HEGO GND at the HEGO vehicle harness connector. • **Is each resistance less than 5.0 ohms?**	Yes ▶	GO to H16.
	No ▶	SERVICE open circuit. REMOVE breakout box. RECONNECT ECU, HEGO sensor, and any other components that are disconnected or removed. RERUN Quick Test.

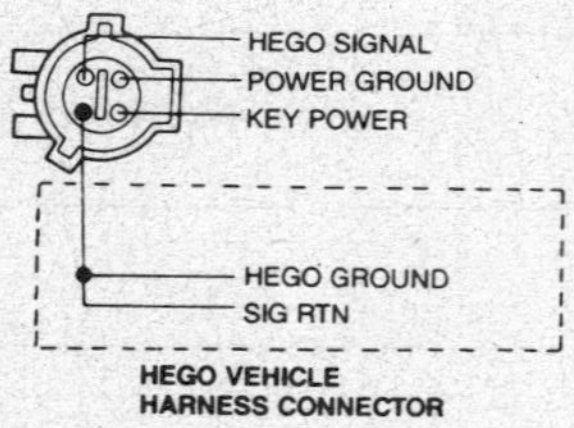

Fuel Control — Pinpoint Test H

TEST STEP	RESULT ▶	ACTION TO TAKE
H16 CHECK HEGO CIRCUIT FOR SHORT TO GROUND		
• Key off. • Breakout box installed, ECU disconnected. • HEGO disconnected. • Ohmmeter on 200,000 ohm scale. • Measure resistance between Test Pin 29 and Test Pins 40 and 49 at the breakout box. • Measure resistance between Test Pin 29 and Test Pins 40 and 46 at the breakout box. • **Is each resistance greater than 10,000 ohms?**	Yes ▶	GO to H17.
	No ▶	SERVICE short circuit. RECONNECT ECU and HEGO sensor. RERUN Quick Test.
H17 CHECK HEGO SENSOR FOR SHORT TO GROUND		
• Key off. • Breakout box installed, ECU disconnected. • HEGO disconnected. • Ohmmeter on 200,000 ohm scale. • Measure resistance between PWR GND and HEGO SIGNAL at the HEGO sensor connector. • Measure resistance between HEGO GND and SIG RTN at the HEGO sensor connector. • **Is resistance greater than 10,000 ohms?**	Yes ▶	REMOVE breakout box. RECONNECT HEGO sensor. REPLACE ECU. RERUN Quick Test.
	No ▶	REPLACE HEGO sensor. RECONNECT ECU. RERUN Quick Test.

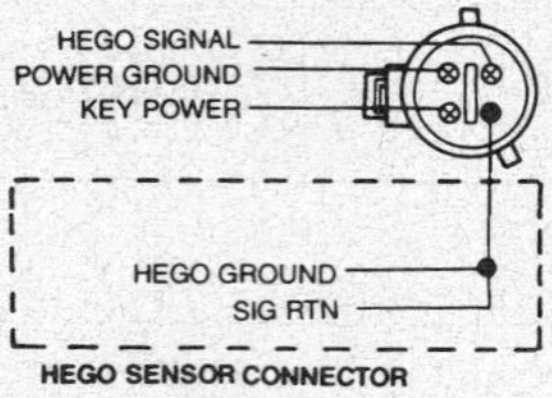

PINPOINT TESTS—NAVAJO

Fuel Control — Pinpoint Test H

TEST STEP	RESULT ▶	ACTION TO TAKE
H20 CHECK RESISTANCE OF HEATER ELEMENT ON HEGO		
• Key off. • Disconnect HEGO. • Ohmmeter on 200 ohm scale. • Measure resistance between KPWR circuit and PWR GND circuit at HEGO sensor connector. • Hot to warm resistance specification is 5.0 to 30.0 ohms. • Room temperature resistance specification is 2.0 to 5.0 ohms. • **Is resistance within specification?**	Yes ▶	GO to H21.
	No ▶	REPLACE HEGO sensor. RERUN Quick Test.

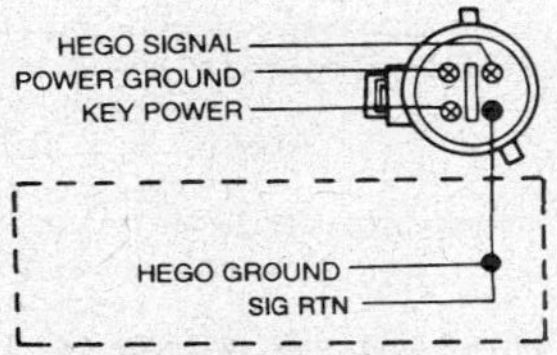

HEGO SENSOR CONNECTOR

Fuel Control — Pinpoint Test H

TEST STEP	RESULT ▶	ACTION TO TAKE
H21 CHECK FOR POWER AT HEGO HARNESS CONNECTOR		
• Key on, engine off. • HEGO disconnected. • Voltmeter on 20 volt scale. • Measure voltage between KEY POWER circuit and PWR GND circuit at the HEGO vehicle harness connector. • **Is voltage greater than 10.5 volts?**	Yes ▶	RECONNECT HEGO sensor. HEGO sensor system OK. GO to H1.
	No ▶	GO to H22.
H22 CHECK CONTINUITY OF POWER GROUND CIRCUIT		
• Key off, wait 10 seconds. • HEGO disconnected. • Ohmmeter on 200 ohm scale. • Measure resistance between PWR GND circuit at the HEGO vehicle harness connector and battery negative post. • **Is resistance less than 5.0 ohms?**	Yes ▶	SERVICE open in KEY POWER circuit. RECONNECT HEGO sensor. RERUN Quick Test.
	No ▶	SERVICE open in PWR GND circuit. RECONNECT HEGO sensor. RERUN Quick Test.

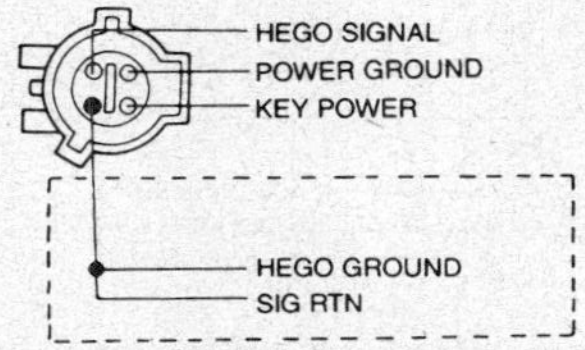

HEGO VEHICLE HARNESS CONNECTOR

Fuel Control	Pinpoint Test	H

TEST STEP	RESULT ▶	ACTION TO TAKE
H23 SERVICE CODE 42/173, 92/137, 177: FUEL CONTROL ALWAYS RICH; CHECK HEGO SIGNAL FOR SHORT TO POWER		
With dual HEGO, code 42/173 refers to right or rear HEGO, code 92/137 or 177 refers to left or front HEGO. HEGO Engine Running Code 42/173 indicates the system is always rich. • Key off, wait 10 seconds. • Disconnect the appropriate HEGO sensor for Code 42/173 or 92/137 or 177. • Voltmeter on 20 volt scale. • Key on, engine off. • Measure voltage between HEGO SIG and PWR GND at the HEGO vehicle harness connector. • **Is voltage less than 0.5 volts?**	Yes ▶	GO to H24 .
	No ▶	SERVICE HEGO circuit short to power. RECONNECT HEGO sensor. RERUN Quick Test.

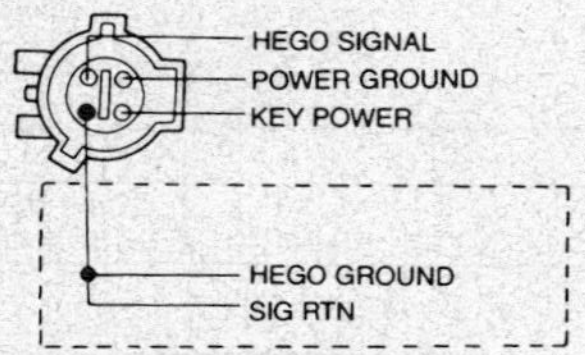

HEGO VEHICLE HARNESS CONNECTOR

Fuel Control	Pinpoint Test	H

TEST STEP	RESULT ▶	ACTION TO TAKE
H24 CHECK HEGO SENSOR FOR SHORT TO IGNITION RUN CIRCUIT		
• Key off. • HEGO disconnected. • Ohmmeter on 200,000 ohm scale. • Measure resistance between KPWR circuit and HEGO SIG circuit at the HEGO sensor connector. • **Is resistance greater than 10,000 ohms?**	Yes ▶	GO to H25 .
	No ▶	REPLACE HEGO sensor. RERUN Quick Test.

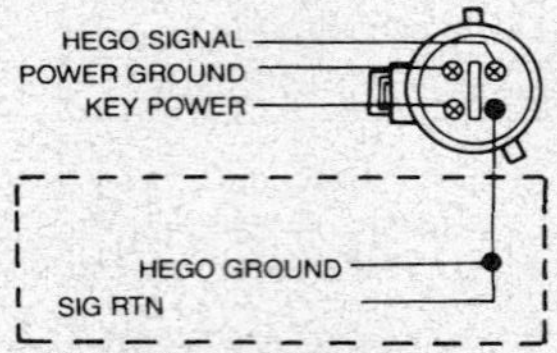

HEGO SENSOR CONNECTOR

TEST STEP	RESULT ▶	ACTION TO TAKE
H25 ATTEMPT TO GENERATE CODE 41/172, 91/136 or 176		
Non-EEC areas could cause a Service Code 42/92. Check for: — Fuel contaminated engine oil — Ignition caused misfire (fouled spark plug) — CANP problems • Key off, wait 10 seconds. • HEGO disconnected. • Jumper HEGO SIG circuit at the HEGO vehicle harness connector to battery negative post. • Rerun Engine Running Self-Test. • **Is Code 41/172, 91/136 or 176 present?**	Yes ▶	REMOVE jumper. GO to H28 .
	No ▶	REMOVE jumper. RECONNECT HEGO sensor. DISCONNECT ECU 60 pin connector. Inspect for damaged or pushed out pins, corrosion, loose wires, etc. SERVICE as necessary. If OK REPLACE ECU. RERUN Quick Test.

Fuel Control | Pinpoint Test | H

TEST STEP	RESULT ▶	ACTION TO TAKE
H28 HEGO SENSOR CHECK		
• Key off, wait 10 seconds. • HEGO sensor disconnected. • Connect VOM to HEGO SIGNAL at the HEGO sensor connector and to battery negative post. • Voltmeter on 20 volt scale. • Create a vacuum leak to cause HEGO sensor to go lean. • Disconnect any vacuum hose from the manifold vacuum tree. • Disconnect the PCV valve hose from the PCV valve. • Start engine and run at approximately 2000 rpm. • **Does the Voltmeter indicate less than 0.4 volts within 30 seconds?**	Yes ▶	RECONNECT HEGO sensor and vacuum lines. HEGO sensor is OK. GO to H1.
	No ▶	REPLACE HEGO sensor. RECONNECT vacuum hoses. RERUN Quick Test.

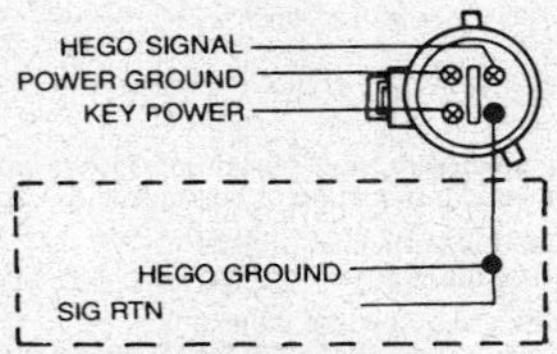

HEGO SENSOR CONNECTOR

Fuel Control | Pinpoint Test | H

H29 CONTINUOUS MEMORY CODE 41/172 EXHAUST ODOR, SMOKE:

CODE 41/172 — Indicates that a HEGO circuit has not switched during closed loop fuel control.

CONTINUOUS MEMORY CODE 41/172 EXHAUST ODOR, SMOKE:

NOTE: In this situation, Code 41/172 does not necessarily indicate a lean condition.

Before attempting to service a Continuous Memory Code 41/172, DIAGNOSE all other driveability complaints first (e.g. rough idle, misses, etc.) using Diagnosis by Symptom.

Exhaust odor, smoke is possible to appear without code 41/172, it is an intermittent fault and could be due to loose HEGO SIG or HEGO GND connection or poor HEGO harness insulation near the exhaust manifold. The ECU thinks the system is lean and therefore leads to an over-rich fuel condition.

Some areas to check are:

- HEGO SIG and GND connections.
- HEGO harness insulation near the exhaust manifold.
- Unmetered Air (vacuum leaks/intake air leaks):
  - — Canister purge system.
  - — PCV system.
  - — Engine sealing.
  - — Crimped fuel lines.
  - — Plugged fuel filter.
  - — Fouled fuel injectors.
  - — Air leaks between mass air flow sensor and air outlet tube to throttle body.
- HEGO Fuel Fouled:

  Whenever an over-rich fuel condition has been experienced (fuel fouled spark plugs), make a thorough check of the ignition system. If a HEGO sensor is suspected of being fuel fouled (low output or slow response), run the vehicle at sustained high speed (within legal limits) followed by a few hard accels. This will burn off the HEGO contamination and restore proper HEGO operation.
- Ignition System:

  If engine is always in DEFAULT spark (base timing) refer to Quick Test Step 4.0.
- Improper Fueling:

  Lead fouled HEGO sensor.
- Fuel Pressure:

  Perform Pinpoint Test Steps H1 and H2.
- TP Sensor:

  Turn key to RUN position. While moving throttle slowly toward wide-open position, measure voltage between Test Pins 47 and 46 at the breakout box. If the voltage does not increase with the increase of throttle opening, replace TP sensor or linkage as necessary.
- If at this point the driveability concern is still present, perform Pinpoint Test Steps H3 through H6.

Fuel Control	Pinpoint Test	H

H30 CONTINUOUS MEMORY CODE 41 or 42:

Code 41/172 — HEGO indicated the fuel system was lean for more than 15 seconds when the fuel system should have been in closed loop fuel control.

Code 42/173 — HEGO indicated the fuel system was rich for more than 15 seconds when the fuel system should have been in closed loop fuel control.

- Before attempting to service a Continuous Memory Code 41/172 or 42/173 DIAGNOSE all other driveability complaints first. Examples: rough idle, misses, etc. using Diagnosis By Symptom.

- Whenever an over-rich fuel condition has been experienced (fuel fouled spark plugs), make a thorough check of the ignition system. If a HEGO sensor is suspected of being fuel fouled (low output or slow response), after the vehicle service, run the vehicle at sustained high speed (within legal limits) followed by a few hard accels. This will burn off the HEGO contamination and restore proper HEGO operation.
- The fuel Service Code may help to isolate the cause of the fuel control problem. Some areas to check are:

For Code 41/172:

- Intermittent HEGO circuit (SIG or GND).
- Low fuel pressure at WOT.
  - — Low-pressure fuel pump.
  - — Restricted fuel supply (crimped fuel lines or plugged fuel filter or screen at the Fuel Pressure Regulator).
- Low fuel flow at WOT with correct fuel pressure.
  - — Clogged fuel injectors.
  - — Low battery voltage less than 11 volts.

Fuel Control	Pinpoint Test	H

H30 (CONTINUED)

For Code 42/173:

- Intermittent HEGO circuit (SIGNAL or GROUND).
- Excessive fuel pressure. Check for fuel pressure regulator vacuum line disconnected or kinked fuel return line.
- Excessive fuel flow. Check for damaged or stuck open fuel injector(s).

For Both Codes 41/172 and 42/173:

- Intermittent HEGO circuit (SIGNAL or GROUND).
- Contaminated HEGO sensor (lead or silicone fouled).
- Improper fuel pressure. Check fuel pump and fuel pressure regulator.

Fuel Pump Circuit	Pinpoint Test	J

Note

You should enter this Pinpoint Test only when a Service Code 87/556, 95/542 or 96/543 is received during Quick Test or if you have been directed here from Pinpoint Test [A] or Diagnosis By Symptom.

Remember

This Pinpoint Test is intended to diagnose only the following:

- Fuel Pump Relay
- Inertia Switch
- Harness Circuits: BATT (+), VPWR, F.P., GND and Power-To-Pump
- ECU Assembly

Description

The Fuel Pump Relay is a normally open relay that is used to supply voltage to the electric fuel pump. When the ignition key is turned to the START or RUN position the ECU grounds Pin 22 (FP), which activates the relay (closes the contacts) and sends voltage to the fuel pump. If, within 1-2 seconds, the ECU does not receive an ignition PIP signal (indicating the engine is not turning), Pin 22 will be un-grounded and the fuel pump will turn off.

The Inertia Switch is a safety device that is wired into either the primary circuit (coil side) or secondary circuit (contact side). In the event of a collision, the inertia switch will open and voltage will be cut off to the fuel pump.

The Fuel Pump Monitor circuit is wired into the power-to-pump circuit and is used by the ECU to monitor the fuel pump secondary circuit.

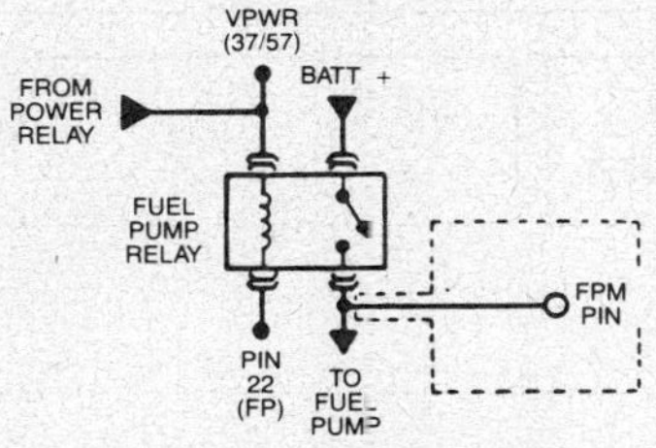

Fuel Pump Circuit	Pinpoint Test	J

Pinpoint Test Schematic

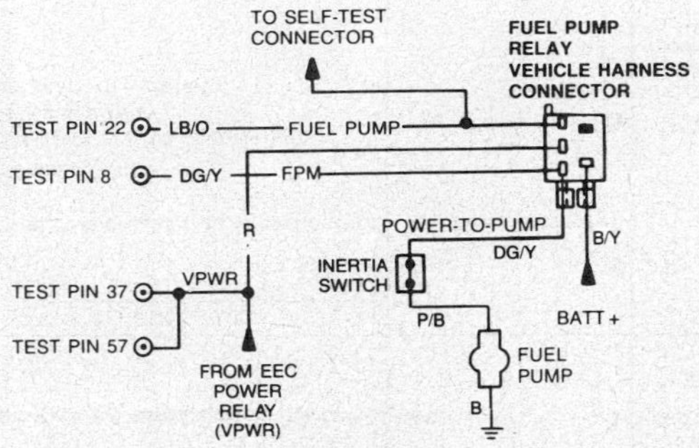

Fuel Pump Circuit — Pinpoint Test J

TEST STEP	RESULT ▶	ACTION TO TAKE
J1 NO FUEL PUMP PRESSURE: CHECK FOR FUEL PUMP ELECTRICAL OPERATION		
• To check if fuel pump runs, turn key from Off to Run position, repeat several times. (Do not turn to start position.) • **Does fuel pump run briefly each time the key is turned to RUN position?**	Yes ▶ No ▶	GO to fuel pump Section GO to J2.
J2 CHECK FOR VPWR TO ECU		
• Key off. • Breakout box installed, ECU connected. • Voltmeter on 20 volt scale. • Key on, engine off. • Measure voltage between Test Pin 37 and Test Pin 40 at the breakout box and between Test Pin 57 and Test Pin 60 at the breakout box. • **Are both voltages greater than 10.5 volts?**	Yes ▶ No ▶	GO to J5. GO to B1.

Fuel Pump Circuit — Pinpoint Test J

TEST STEP	RESULT ▶	ACTION TO TAKE
J5 CHECK FOR VOLTAGE TO POWER-TO-PUMP CIRCUIT		
• Key on, engine off. • Breakout box installed, ECU connected. • Voltmeter on 20 volt scale. • Measure voltage between chassis ground and Power-To-Pump circuit at fuel pump relay while cranking engine. • **Is voltage greater than 8.0 volts during crank?**	Yes ▶ No ▶	GO to . open in Power-To-Pump circuit, Fuel pump GND, open in pump, etc. GO to J6.
J6 CHECK FOR BATT+ TO FUEL PUMP RELAY		
• Key on, engine off. • Breakout box installed (if applicable), ECU connected. • Voltmeter on 20 volt scale. • Measure voltage between chassis ground and BATT(+) at the fuel pump relay. • **Is voltage greater than 10.5 volts?**	Yes ▶ No ▶	GO to J11. SERVICE open in BATT (+) between fuel pump relay and battery positive post (for relay block applications, between fuel pump relay and EEC power relay). REMOVE breakout box. RECONNECT ECU. RERUN Quick Test.

PINPOINT TESTS – NAVAJO

Fuel Pump Circuit — Pinpoint Test J

TEST STEP	RESULT	ACTION TO TAKE
J11 CHECK FOR VOLTAGE AT POWER-TO-PUMP CIRCUIT		
• Key off. • Breakout box installed, ECU disconnected. • Connect jumper wire from Test Pin 22 to Test Pin 40 or 60 at the breakout box. • Voltmeter on 20 volt scale. • Key on, engine off. • Measure voltage between chassis ground and Power-To-Pump circuit at fuel pump relay. • **Is voltage greater than 10.5 volts?**	Yes ▶	REPLACE ECU. REMOVE breakout box.. RERUN Quick Test.
	No ▶	REPLACE fuel pump relay. REMOVE breakout box. RECONNECT ECU. RERUN Quick Test.
J20 SERVICE CODE 95/542: DOES ENGINE START		
Service Code 95/542 indicates that one of the following has occurred: No Start: — Inertia switch not reset or electrically open (if in secondary circuit). — Open circuit in or between the fuel pump and FPM circuit connection to the Power-To-Pump circuit. — Poor fuel pump ground Vehicle Starts: — Fuel pump secondary circuit short to power. — Fuel pump relay contacts always closed. — Open in FPM circuit between ECU and connection to the Power-To-Pump circuit. — Faulty ECU. • **Does the engine start?**	Yes ▶	GO to J21.
	No ▶	GO to J25.
J21 VERIFY THAT FUEL PUMP IS OFF		
• Key on. Wait five seconds. • Listen for motor noise from fuel pump. • **Is fuel pump off?**	Yes ▶	GO to J23.
	No ▶	GO to J22.

Fuel Pump Circuit — Pinpoint Test J

TEST STEP	RESULT	ACTION TO TAKE
J22 CHECK FOR FUEL PUMP RELAY ALWAYS CLOSED		
• Key off. • Locate and disconnect fuel pump relay. • Key on. • **Does fuel pump shut off when relay is disconnected?**	Yes ▶	REPLACE fuel pump relay. RERUN Quick Test.
	No ▶	SERVICE short to power in POWER-TO-PUMP/ FPM circuit. RECONNECT fuel pump relay. RERUN Quick Test.
J23 CHECK FPM CIRCUIT CONTINUITY		
• Key off. • Disconnect ECU 60 pin connector. Inspect for damaged or pushed out pins, corrosion, loose wires, etc. Service as necessary. • Install breakout box, leave ECU disconnected. • Disconnect fuel pump relay. • Ohmmeter on 200 ohm scale. • Measure resistance between Test Pin 8 at the breakout box and Power-To-Pump circuit at the fuel pump relay vehicle harness connector. • **Is resistance less than 5.0 ohms?**	Yes ▶	REPLACE ECU. REMOVE breakout box. RECONNECT fuel pump relay. RERUN Quick Test.
	No ▶	SERVICE open circuit. REMOVE breakout box. RECONNECT all components. RERUN Quick Test.

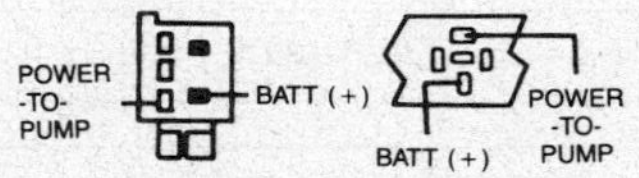

FUEL PUMP RELAY VEHICLE HARNESS CONNECTOR (USE APPLICABLE DRAWING)

Fuel Pump Circuit — Pinpoint Test J

TEST STEP	RESULT ▶	ACTION TO TAKE
J24 CHECK FUEL PUMP PRIMARY CIRCUIT FOR SHORT TO GROUND • Key off. • Breakout box installed, ECU disconnected. • Fuel pump relay disconnected. • Ohmmeter on 200,000 ohm scale. • Measure resistance between Test Pin 22 and Test Pin 40 at the breakout box. • **Is resistance greater than 10,000 ohms?**	Yes ▶ No ▶	REPLACE ECU. REMOVE breakout box. RECONNECT fuel pump relay. RERUN Quick Test. SERVICE short circuit. REMOVE breakout box. RECONNECT all components. RERUN Quick Test.
J25 CHECK INERTIA SWITCH • Key off, wait 10 seconds. • Locate and disconnect fuel pump inertia switch (verify that switch is reset). • Ohmmeter on 200 ohm scale. • Measure resistance of the fuel pump inertia switch. • **Is resistance less than 5.0 ohms?**	Yes ▶ No ▶	RECONNECT inertia switch. GO to J26. REPLACE or RESET inertia switch. RERUN Quick Test.

Fuel Pump Circuit — Pinpoint Test J

TEST STEP	RESULT ▶	ACTION TO TAKE
J26 CHECK FOR CONTINUITY BETWEEN FPM CIRCUIT AND GROUND • Key off. • Breakout box installed, ECU disconnected. • Fuel pump relay disconnected. • Ohmmeter on 200 ohm scale. • Measure resistance between Test Pin 8 at the breakout box and battery negative post. • **Is resistance less than 10.0 ohms?**	Yes ▶ No ▶	REPLACE ECU. REMOVE breakout box. RECONNECT fuel pump relay. RERUN Quick Test. REMOVE breakout box. RECONNECT all components. GO to Section , Electric Fuel Pump, for open in Power-To-Pump circuit, poor fuel pump ground, open in fuel pump, etc.
J30 SERVICE CODE 96/543: DOES ENGINE START? Service Code 96/543 indicates a fuel pump secondary circuit failure between the BATT(+) supply and the FPM connection to the Power-To-Pump circuit. Possible causes are: — Open circuit. — Faulty fuel pump relay. — Faulty ECU. • Key off, wait 10 seconds. • Locate fuel pump relay. • Voltmeter on 20 volt scale. • Measure voltage between BATT(+) circuit at the fuel pump relay and battery negative post. • **Is voltage greater than 10.5 volts?**	Yes ▶ No ▶	GO to J31. SERVICE open in BATT(+) circuit. RERUN Quick Test.

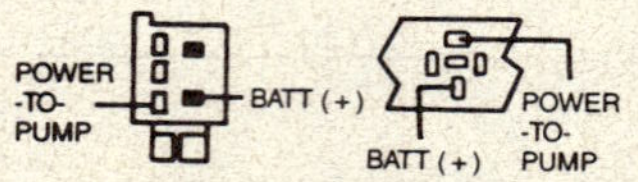

FUEL PUMP RELAY VEHICLE HARNESS CONNECTOR (USE APPLICABLE DRAWING)

PINPOINT TESTS—NAVAJO

Fuel Pump Circuit — Pinpoint Test J

TEST STEP	RESULT ▶	ACTION TO TAKE
J31 CHECK FOR VOLTAGE AT POWER-TO-PUMP CIRCUIT TO VERIFY FUEL PUMP RELAY OPERATION		
• Key off. • Fuel pump relay connected. • Voltmeter on 20 volt scale. • Connect Voltmeter between Power-To-Pump circuit at the fuel pump relay and battery negative post. • Observe Voltmeter as you activate fuel pump relay (turn key to run for 1 second, then to off for 10 seconds. Repeat 5 times.) • **Does voltage measure greater than 10.5 volts for about 1 second after key is turned to RUN position during test?**	Yes ▶ No ▶	REPLACE ECU. RERUN Quick Test. REPLACE fuel pump relay. RERUN Quick Test.

Fuel Pump Circuit — Pinpoint Test J

TEST STEP	RESULT ▶	ACTION TO TAKE
J90 CONTINUOUS MEMORY CODE 95/542: CHECK EEC HARNESS		
A Continuous Memory Code 95/542 indicates that one of the following intermittent conditions has occurred: — Fuel pump circuit activated when ECU expected circuit to be off (i.e., fuel system test or prime procedure). — Open circuit in or between the fuel pump and FPM circuit at the ECU (refer to schematic). — Poor fuel pump ground. — FPM or Power-To-Pump circuit short to power. — Fuel pump relay contacts stuck closed. — Left front HEGO circuit short to power. • Start engine. • Check for engine stall/stumble while performing the following (also, if possible, listen for fuel pump turning off.) — Shake, wiggle, bend the Power-To-Pump circuit between the Power-To-Pump pin at the fuel pump relay and the fuel pump. — Shake, wiggle, bend the fuel pump ground circuit from the fuel pump to ground. — Lightly tap the fuel pump to simulate road shock. — For vehicles with the inertia switch in the Power-To-Pump circuit (refer to schematic), lightly tap inertia switch to simulate road shock. • Key off. • Inspect the fuel pump vehicle harness connector and the fuel pump ground for corrosion, damaged pins, etc. • **Is fault indicated/found?**	Yes ▶ No ▶	ISOLATE fault and SERVICE as necessary. CLEAR Continuous Memory RERUN Quick Test. GO to J91.

Fuel Pump Circuit — Pinpoint Test J

TEST STEP	RESULT ▶	ACTION TO TAKE
J91 CHECK FPM CIRCUIT		
• Key off. • Disconnect ECU 60 pin connector. Inspect for damaged or pushed out pins, corrosion, loose wires, etc. Service as necessary. • Install breakout box, leave ECU disconnected. • Key on, engine off. • Connect a TEST LAMP between Test Pin 8 and Test Pin 37 at the breakout box. • Observe test lamp for an indication of a fault while performing the following (The light will go out when a fault is found, indicating an open): — Shake, wiggle, bend the fuel pump monitor circuit between the fuel pump relay (or splice if applicable, see schematic) and the processor. • **Is fault indicated?**	Yes ▶ No ▶	ISOLATE fault and SERVICE as necessary. REMOVE breakout box. CLEAR Continuous Memory RERUN Quick Test. GO to J92.
J92 CHECK FOR SHORTS TO POWER		
• Key on, engine off. • Breakout box installed, ECU disconnected. • Connect a TEST LAMP between Test Pin 8 and Test Pin 40. • Observe test lamp for an indication of a fault while performing the following (The light will turn on when a fault is detected, indicating a short to power. Also, if possible, listen for fuel pump turning on.): — Shake, wiggle, bend the fuel pump monitor circuit and Power-To-Pump circuit, especially where they may be in the vicinity of a power circuit. • Lightly tap the fuel pump relay (to simulate road shock). • **Is fault indicated?**	Yes ▶ No ▶	ISOLATE fault and SERVICE as necessary. REMOVE breakout box. RECONNECT ECU. CLEAR Continuous Memory RERUN Quick Test. CLEAR Continuous Memory RERUN Quick Test.

Fuel Pump Circuit — Pinpoint Test J

TEST STEP	RESULT ▶	ACTION TO TAKE
J93 CONTINUOUS MEMORY CODE 96/543: CHECK FOR CONTINUOUS MEMORY CODE 87/556		
• **Is Continuous Memory Code 87 or 556 also present?**	Yes ▶ No ▶	GO to J95. GO to J94.
J94 CHECK EEC HARNESS		
A Continuous Memory Code 96/543, without the presence of a Continuous Memory Code 87/556, indicates that during vehicle operation, one of the following has occurred: — Open in the BATT(+) circuit between BATT(+) and the fuel pump relay. — Fuel pump relay contacts opened. — Open in the Power-To-Pump circuit from the fuel pump relay to the FPM splice, if applicable (see schematic). • Start engine. • Check for engine stall/stumble while performing the following (also, if possible, listen for fuel pump turning off): — Shake, wiggle, bend the BATT(+) circuit from BATT(+) to the fuel pump relay. — Lightly tap the fuel pump relay (to simulate road shock). — Shake, wiggle, bend the Power-To-Pump circuit from the fuel pump relay to the FPM splice, if applicable (See schematic). • Key off. • Inspect the fuel pump relay connectors and BATT(+) connector terminal for corrosion, damaged pins, etc. • **Is fault indicated/found?**	Yes ▶ No ▶	ISOLATE fault and SERVICE as necessary. CLEAR Continuous Memory RERUN Quick Test. CLEAR Continuous Memory RERUN Quick Test.

Fuel Pump Circuit — Pinpoint Test J

TEST STEP	RESULT ▶	ACTION TO TAKE
J95 CONTINUOUS MEMORY CODE 87/556: CHECK EEC HARNESS		
A Continuous Memory Code 87/556 indicates that a fuel pump primary circuit failure has occurred during vehicle operation. Possible causes are: — Open in VPWR circuit between the EEC power relay and the fuel pump relay. — Open coil in fuel pump relay. — Open in fuel pump circuit (Pin 22). — Damaged inertia switch. • Start engine. • Check for engine stall/stumble while performing the following (also, if possible, listen for fuel pump turning off): — Shake, wiggle, bend the VPWR circuit between the EEC power relay and the fuel pump relay. For vehicles with the inertia switch in the VPWR circuit (refer to schematic), lightly tap the inertia switch to simulate road shock. — Shake, wiggle, bend the EEC vehicle harness fuel pump circuit (Test Pin 22) between the ECU and the fuel pump relay. — Lightly tap the fuel pump relay to simulate road shock. • Key off. • Inspect the ECU 60 pin connector and the fuel pump relay connectors for corrosion, damaged pins, etc. damaged pins, etc. • **Is fault indicated/found?**	Yes ▶	ISOLATE fault and SERVICE as necessary. CLEAR Continuous Memory RERUN Quick Test.
	No ▶	CLEAR Continuous Memory RERUN Quick Test.

Canister Purge (CANP) — Pinpoint Test KD

Note

You should enter this Pinpoint Test only when a Service Code 85/565 is received during Quick Test or you have been directed here from Diagnosis By Symptom.

Remember

This Pinpoint Test is intended to diagnose only the following:

- CANP Solenoid
- Harness Circuits: CANP and VPWR
- ECU Assembly

Description

The Canister Purge (CANP) Solenoid is an output control which allows venting of the evaporative purge carbon canister back into the fuel (intake) system. This output may be an on-off or constant frequency/varying duty cycle type control.

Pinpoint Test Schematic

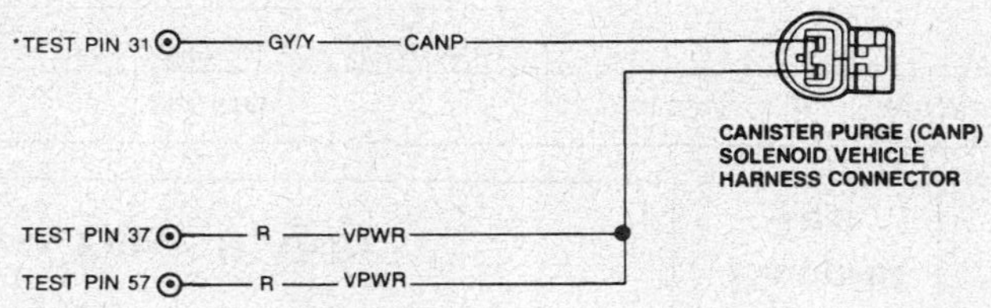

*TEST PINS LOCATED ON BREAKOUT BOX.
ALL HARNESS CONNECTORS VIEWED INTO MATING SURFACE.

Canister Purge (CANP) — Pinpoint Test — KD

TEST STEP	RESULT ▶	ACTION TO TAKE
KD1 ENTER OUTPUT STATE CHECK (REFER TO QUICK TEST APPENDIX)		
NOTE: Do not use SUPER STAR II tester for this step, use VOM/DVOM. • Key off, wait 10 seconds. • Disconnect electrical connector on the speed control servo; if equipped. • Voltmeter on 20 volt scale. • Connect VOM negative test lead to STO circuit at Self-Test connector and positive test lead to battery positive. • Jumper STI circuit to SIG RTN at the Self-Test connector. • Perform Key On Engine Off Self-Test until the completion of the Continuous Test Codes. • VOM will indicate less than 1.0 volt when test is completed. • Depress and release the throttle. • **Does voltage increase?**	Yes ▶	REMAIN in Output State Check. GO to KD2.
	No ▶	DEPRESS throttle to WOT and release. If STO voltage does not go high, leave equipment hooked up and GO to Pinpoint Test Step QC1
KD2 CHECK CANISTER PURGE (CANP) SOLENOID ELECTRICAL OPERATION		
• Key on, engine off. • Disconnect CANP solenoid. • Connect voltmeter positive test lead to VPWR circuit and negative test lead to CANP circuit at the CANP solenoid vehicle harness connector. • Voltmeter on 20 volt scale. • While observing VOM depress and release the throttle several times to cycle output. • **Does CANP circuit cycle 0.5 volts or greater?**	Yes ▶	GO to KD3.
	No ▶	REMOVE jumper. GO to KD6.
KD3 CHECK CANP SOLENOID FOR VACUUM LEAKS		
• Key on, engine off. • CANP solenoid disconnected. • Disconnect vacuum hose at CANP solenoid on manifold vacuum side of CANP solenoid. • Apply 16 in-Hg (54 kPa) of vacuum to manifold vacuum side of CANP solenoid. • **Does CANP solenoid hold vacuum for 20 seconds?**	Yes ▶	REMAIN in Output State Check. Leave vacuum pump setup in place. GO to KD4.
	No ▶	REPLACE CANP solenoid. RERUN Quick Test. If symptom is still present, GO to Evaporative Emission Systems diagnosis.

Canister Purge (CANP) — Pinpoint Test — KD

TEST STEP	RESULT ▶	ACTION TO TAKE
KD4 CHECK CANP SOLENOID FOR MECHANICAL OPERATION		
• While remaining in output state check, reconnect CANP solenoid. • Apply 16 in-Hg (53 kPa) of vacuum to manifold vacuum side of CANP solenoid. • Depress and release throttle. • **Is vacuum released?**	Yes ▶	CHECK hose from CANP solenoid to canister for cracks, leaks, etc. If OK, REMOVE Jumper from STI to SIGNAL RETURN. GO to KD5.
	No ▶	CHECK hose from CANP solenoid to canister for blockage or kinks. If OK, REPLACE CANP solenoid. RERUN Quick Test.
KD5 CHECK FOR VACUUM TO CANP SOLENOID		
• Disconnect vacuum hose at CANP solenoid at manifold vacuum side. • Start engine. • **Is vacuum present at engine vacuum hose?**	Yes ▶	EEC system OK. If a symptom is still present, GO to Evaporative Emission System diagnosis.
	No ▶	CHECK vacuum line for proper routing, kinks, leak or blockage. If OK, probable subjects affecting engine vacuum.

PINPOINT TESTS – NAVAJO

Canister Purge (CANP) — Pinpoint Test KD

TEST STEP	RESULT ▶	ACTION TO TAKE
KD6 MEASURE CANP SOLENOID RESISTANCE		
Service Code 85/565 indicates a failure in the Canister Purge solenoid (CANP) circuit. Possible causes are: — Faulty CANP solenoid — Open harness — Shorted (Power or Ground) harness — Faulty ECU. • Key off, wait 10 seconds. • Disconnect CANP solenoid. • Ohmmeter on 200 ohm scale. • Measure CANP solenoid resistance. • **Is resistance between 40 and 90 ohms?**	Yes ▶ No ▶	GO to KD7. REPLACE CANP solenoid. RERUN Quick Test.
KD7 CHECK VOLTAGE OF VPWR CIRCUIT		
• Key on, engine off. • CANP solenoid disconnected. • Voltmeter on 20 volt scale. • Measure voltage between VPWR at the CANP solenoid vehicle harness connector and battery ground. • **Is voltage greater than 10.5 volts?**	Yes ▶ No ▶	GO to KD8. SERVICE open harness circuit. RECONNECT CANP solenoid. RERUN Quick Test.
KD8 CHECK CONTINUITY OF CANP CIRCUIT		
• Key off, wait 10 seconds. • CANP solenoid disconnected. • Disconnect ECU 60 pin connectors. Inspect for damaged or pushed out pins, corrosion, loose wires, etc. Service as necessary. • Install breakout box, leave ECU disconnected. • Ohmmeter on 200 ohm scale. • Measure resistance between Test Pin 31 at the breakout box and CANP circuit at the CANP solenoid vehicle harness connector. • **Is resistance less than 5 ohms?**	Yes ▶ No ▶	GO to KD9. SERVICE open circuit. REMOVE breakout box. RECONNECT all components. RERUN Quick Test.

Canister Purge (CANP) — Pinpoint Test KD

TEST STEP	RESULT ▶	ACTION TO TAKE
KD9 CHECK CANP CIRCUIT FOR SHORT TO GROUND		
• Key off, wait 10 seconds. • CANP solenoid disconnected. • Breakout box installed, ECU disconnected. • Ohmmeter on 200,000 ohm scale. • Measure resistance between Test Pin 31 and Test Pins 40, 46 and 60 at the breakout box. • **Is each resistance greater than 10,000 ohms?**	Yes ▶ No ▶	GO to KD10. SERVICE short to ground. REMOVE breakout box. RECONNECT all components. RERUN Quick Test.
KD10 CHECK CANP CIRCUIT FOR SHORT TO POWER		
• Key off, wait 10 seconds. • CANP solenoid disconnected. • Breakout box installed, ECU disconnected. • Ohmmeter on 200,000 ohm scale. • Measure resistance between Test Pin 31 and Test Pins 37 and 57 at the breakout box. • **Is each resistance greater than 10,000 ohms?**	Yes ▶ No ▶	REPLACE ECU. RECONNECT CANP solenoid. REMOVE breakout box. RERUN Quick Test. SERVICE short to power. REMOVE breakout box. RECONNECT all components. REPEAT Quick Test. If code is repeated, REPLACE ECU. RERUN Quick Test.

Idle Speed Control (Bypass Air) | Pinpoint Test | KE

Note

You should enter this Pinpoint Test only when a Service Code 12/412, 13/411 16, or 19 is received during Quick Test or when directed here from Pinpoint Test [S].

Remember

To prevent the replacement of good components, be aware that the following non-EEC areas may be at fault:

- Engine not up to operating temperature
- Engine over operating temperature
- A/C Input (electrical problem)
- Throttle Speed Control Linkage
- Throttle sticking or linkage binding

This Pinpoint Test is intended to diagnose only the following:

- Rpm in Self-Test only
- ISC Solenoid
- Harness Circuits: ISC and VPWR
- ECU Assembly

Description

The Idle Speed Control (ISC) solenoid is used to control engine idle speed and dashpot functions. The ISC solenoid is mounted on the throttle body and allows air to bypass the throttle plate. The amount of air allowed to bypass is determined by the ECU and controlled by a duty cycle signal.

Idle Speed Control (Bypass Air) | Pinpoint Test | KE

Pinpoint Test Schematic

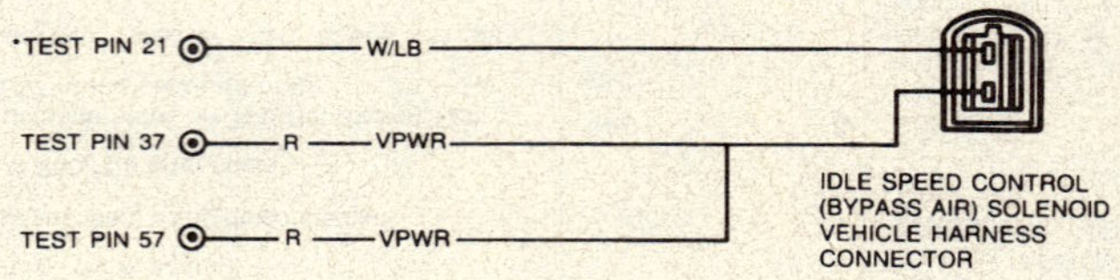

***TEST PINS LOCATED ON BREAKOUT BOX.**
ALL HARNESS CONNECTORS VIEWED INTO MATING SURFACE.

Idle Speed Control (Bypass Air) — Pinpoint Test KE

TEST STEP	RESULT ▶	ACTION TO TAKE
KE1 SERVICE CODE 12/412, 16: CHECK FOR RPM DROP		
Service Code 12/412 indicates that during Engine Running Self-Test, engine rpm could not be controlled within the Self-Test upper limit band. Service Code 16 indicates that during Engine Running Self-Test, engine rpm was too low to perform the HEGO Test. Possible causes are: — Open or shorted circuit — Throttle linkage binding — Improper idle airflow set — Throttle body/ISC solenoid contamination — Items external to Idle Speed Control system that could affect engine rpm. — Faulty ISC solenoid — Faulty ECU • Key off. • Connect engine tachometer. • Start engine. • Disconnect ISC harness connector. • **Does rpm drop or stall?**	Yes ▶ No ▶	GO to KE3. GO to KE4.

Idle Speed Control (Bypass Air) — Pinpoint Test KE

TEST STEP	RESULT ▶	ACTION TO TAKE
KE3 CHECK FOR OTHER EEC CODES		
• **Are Service Codes 22/126, 41/172, 42/173, 91/136 or 92/137 present?**	Yes ▶	RECONNECT ISC solenoid. GO to Quick Test Step 7.0 for appropriate Pinpoint Test.
	No ▶	GO to KE4.
KE4 MEASURE ISC SOLENOID RESISTANCE		
• Key off. • ISC solenoid disconnected. • Ohmmeter on 200 ohm scale. • Measure solenoid resistance. **NOTE: Due to diode in solenoid, place voltmeter (+) lead on VPWR pin and (−) lead on ISC pin** • **Is resistance between 6.0 and 13.0 ohms?**	Yes ▶ No ▶	GO to KE5. REPLACE ISC solenoid. RERUN Quick Test.
KE5 CHECK FOR INTERNAL SHORT TO ISC SOLENOID CASE		
• Key off. • ISC solenoid disconnected. • Ohmmeter on 200,000 ohm scale. • Measure resistance from either ISC solenoid pin to ISC housing. • **Is resistance greater than 10,000 ohms?**	Yes ▶ No ▶	GO to KE6. REPLACE ISC solenoid. RERUN Quick Test.

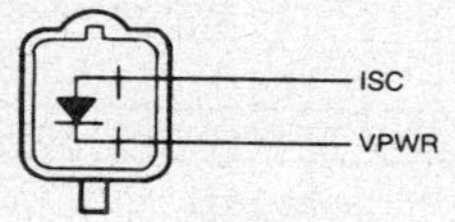

ISC SOLENOID CONNECTOR

PINPOINT TESTS—NAVAJO

Idle Speed Control (Bypass Air) — Pinpoint Test — KE

TEST STEP	RESULT ▶	ACTION TO TAKE
KE6 CHECK VPWR CIRCUIT VOLTAGE		
• Key on, engine off. • ISC solenoid disconnected. • Voltmeter on 20 volt scale. • Measure voltage between VPWR circuit at the ISC solenoid vehicle harness connector and battery ground. • **Is voltage greater than 10.5 volts?**	Yes ▶ No ▶	GO to KE7. SERVICE open circuit. RERUN Quick Test.
KE7 CHECK ISC CIRCUIT CONTINUITY		
• Key off. • ISC solenoid disconnected. • Disconnect ECU and inspect both 60 pin connectors for damaged or pushed out pins, corrosion, loose wires, etc. Service as necessary. • Install breakout box, leave ECU disconnected. • Ohmmeter on 200 ohm scale. • Measure resistance between Test Pin 21 at the breakout box and ISC circuit at ISC solenoid vehicle harness connector. • **Is resistance less than 5 ohms?**	Yes ▶ No ▶	GO to KE8. SERVICE open circuit. REMOVE breakout box. RECONNECT all components. RERUN Quick Test.
KE8 CHECK ISC CIRCUIT FOR SHORT TO GROUND		
• Key off. • Breakout box installed, ECU disconnected. • ISC solenoid disconnected. • Ohmmeter on 200,000 ohm scale. • Measure resistance between Test Pin 21 and Test Pins 40, 46 and 60 at the breakout box. • **Is each resistance greater than 10,000 ohms?**	Yes ▶ No ▶	GO to KE9. SERVICE short circuit. REMOVE breakout box. RECONNECT all components. RERUN Quick Test.
KE9 CHECK ISC CIRCUIT FOR SHORT TO POWER		
• Key off. • Breakout box installed, ECU disconnected. • ISC solenoid disconnected. • Key on. • Measure voltage between breakout box and chassis ground. • **Is voltage less than 1.0 volt?**	Yes ▶ No ▶	GO to KE10. SERVICE short circuit. REMOVE breakout box. RECONNECT all components. RERUN Quick Test. If code or symptom is still present, REPLACE ECU.

Idle Speed Control (Bypass Air) — Pinpoint Test — KE

TEST STEP	RESULT ▶	ACTION TO TAKE
KE10 CHECK FOR ISC SIGNAL FROM THE ECU		
• Key off. • Breakout box installed. • Reconnect ECU to breakout box. • Reconnect ISC solenoid. • Voltmeter on 20 volt scale. • Connect voltmeter between Test Pin 21 and Test Pin 40. • Start engine. • Slowly increase rpm to 3000 rpm then release. • **Does voltage vary?**	Yes ▶ No ▶	GO to KE11. REMOVE ISC solenoid and verify that it is not stuck open. If OK REPLACE ECU. REMOVE breakout box. RERUN Quick Test.
KE11 CHECK IDLE		
• **Is idle speed within specification?**	Yes ▶	Remove ISC solenoid and INSPECT for contamination. If contamination is present, determine if solenoid is type that can be cleaned. **No contamination or solenoid cannot be cleaned** — REPLACE ISC solenoid. RERUN Quick Test. **Solenoid is contaminated or can be cleaned** — CLEAN as necessary. RERUN Quick Test. If code/symptom is still present, REPLACE ISC solenoid.
	No ▶	RESET idle to specification. REFER to idle set procedure. RERUN Quick Test. If UNABLE to RESET idle to specification, GO to KE12.

Idle Speed Control (Bypass Air) — Pinpoint Test KE

TEST STEP	RESULT ▶	ACTION TO TAKE
KE12 CHECK FOR PROBLEMS AFFECTING PROPER ENGINE SPEED		
• Check throttle linkage and/or speed control linkage for binding. • Inspect throttle body for contamination. • Check engine vacuum hoses. Refer to VECI decal. • Check for leaks around ISC solenoid (e.g. mounting gasket, etc.). • **Are all the above checks OK?**	Yes ▶	Remove ISC solenoid and INSPECT for contamination. If contamination is present, determine if solenoid is type that can be cleaned. **No contamination or solenoid cannot be cleaned** — REPLACE ISC solenoid. RERUN Quick Test. **Solenoid is contaminated or can be cleaned** — CLEAN as necessary . RERUN Quick Test. If code/symptom is still present, REPLACE ISC solenoid.
	No ▶	SERVICE as necessary. REMOVE breakout box. RECONNECT processor. RERUN Quick Test.

Idle Speed Control (Bypass Air) — Pinpoint Test KE

TEST STEP	RESULT ▶	ACTION TO TAKE
KE15 SERVICE CODE 13/411: VERIFY IDLE SPEED IS WITHIN SPECIFICATION		
Service Code 13/411 indicates that during Engine Running Self-Test, engine rpm could not be controlled within the Self-Test lower limit band. Possible causes are: — Improper idle airflow set. — Vacuum leaks. — Throttle linkage binding. — Throttle plates open. — Throttle body/ISC solenoid contamination. — ISC circuit short to ground. — Damaged ISC solenoid. • **Is idle speed within specification?**	Yes ▶	Remove ISC solenoid and INSPECT for contamination. If contamination is present, determine if solenoid is type that can be cleaned. **No contamination or solenoid cannot be cleaned** — REPLACE ISC solenoid. RERUN Quick Test. **Solenoid is contaminated or can be cleaned** — CLEAN as necessary . RERUN Quick Test. If code/symptom is still present, REPLACE ISC solenoid.
	No ▶	RESET idle airflow to specification. REFER to : idle airflow set adjust procedure. If UNABLE to RESET idle to specification, GO to KE16 .

Idle Speed Control (Bypass Air) — Pinpoint Test KE

TEST STEP	RESULT ▶	ACTION TO TAKE
KE16 CHECK FOR CONDITIONS AFFECTING IDLE • Check engine vacuum hoses for leaks. Refer to VECI decal. • Check throttle linkage and/or speed control linkage for binding. • Check that throttle plates are closed. • Check for induction system leaks. (e.g. ISC solenoid to throttle body gasket, loose ISC, etc.) • Check throttle body for contamination. • Check that CANP solenoid is not stuck open. • **Are all the above checks OK?**	Yes ▶ No ▶	GO to KE17. SERVICE as necessary. RERUN Quick Test.
KE17 CHECK FOR INTERNAL SHORT TO ISC SOLENOID CASE • Key off. • Disconnect ISC solenoid. • Ohmmeter on 200,000 ohm scale. • Measure resistance from either ISC solenoid pin to ISC housing. • **Is resistance greater than 10,000 ohms?**	Yes ▶ No ▶	GO to KE18. REPLACE ISC solenoid. RERUN Quick Test.
KE18 CHECK ISC CIRCUIT FOR SHORT TO GROUND • Key off. • ISC solenoid disconnected. • Disconnect the ECU 60 pin connector. Inspect for damaged or pushed out pins, corrosion, loose wires, etc. Service as necessary. • Install breakout box, leave ECU disconnected. • Ohmmeter on 200,000 ohm scale. • Measure resistance between Test Pin 21 and Test Pins 40, 46 and 60 at the breakout box. • **Are all resistances greater than 10,000 ohms?**	Yes ▶ No ▶	GO to KE19. SERVICE short circuit. REMOVE breakout box. RECONNECT all components. RERUN Quick Test.

Idle Speed Control (Bypass Air) — Pinpoint Test KE

TEST STEP	RESULT ▶	ACTION TO TAKE
KE19 CHECK ECU OUTPUT • Key off. • Breakout box installed. • Reconnect ECU to breakout box. • Reconnect ISC solenoid. • Voltmeter on 20 volt scale. • Connect Voltmeter between Test Pin 21 and Test Pin 40. • Start engine. • Slowly increase rpm to 3000. • **Is voltage between 3.0 and 11.5 volts?**	Yes ▶	Remove ISC solenoid and INSPECT for contamination. If contamination is present, determine if solenoid is type that can be cleaned. **No contamination or solenoid cannot be cleaned** — REPLACE ISC solenoid. RERUN Quick Test. **Solenoid is contaminated or can be cleaned** — CLEAN as necessary. RERUN Quick Test. If code/symptom is still present, REPLACE ISC solenoid.
	No ▶	REPLACE ECU. REMOVE breakout box. RERUN Quick Test.

Idle Speed Control (Bypass Air) — Pinpoint Test KE

TEST STEP	RESULT ▶	ACTION TO TAKE
KE22 SERVICE CODE 16: HIGH ISC RPM		
Service Code 16 indicates that with the ISC off, engine rpm was above a Self-Test limit. Possible causes are: — Improper idle set — Purge solenoid — Air/vacuum leaks • **Is Code 48 present?**	Yes ▶	RESET throttle plate. REFER to VECI decal for curb idle set procedure. RERUN Quick Test. If Code 48 is still present, GO to KE21.
	No ▶	GO to KD15.
KE25 SERVICE CODE 19: LOW ISC RPM		
Service Code 19 indicates that with the ISC off, engine rpm dropped below a Self-Test limit (usually around 600 rpm). Possible causes are: — Engine not at operating temperature. — Throttle body/air inlet contamination. — Improper idle set. • Key off. • Deactivate Self-Test. • Run engine at 2000 rpm for 2 minutes or until inlet radiator hose is hot and pressurized. • Key off. • Rerun Engine Running Self-Test. • **Does engine stumble and/or is code 19 still present?**	Yes ▶	INSPECT throttle body and air inlet for contamination. SERVICE as necessary. If OK, ADJUST base idle RERUN Quick Test.
	No ▶	SERVICE other codes as necessary.

WOT A/C Cutout (WAC) A/C Demand — Pinpoint Test KM

Note

You should enter this Pinpoint Test only when directed here from Diagnosis By Symptom.

Remember

To prevent the replacement of good components, be aware that the following non-EEC areas may be at fault:

- Refrigerant charge
- Ambient temperature less than 45°

This Pinpoint Test is intended to diagnose only the following:

- Harness circuits: WAC, VPWR, GND, POWER-TO-CLUTCH, ACD
- WAC relay
- ECU assembly

WOT A/C Cutout (WAC) A/C Demand — Pinpoint Test KM

Description

WAC

When the air conditioning demand switch is turned on and the cyclic pressure switch/low pressure switch is closed, voltage passes through the normally closed WAC relay to the A/C clutch. Under certain conditions (e.g. engine cranking, wide open throttle) the ECU can interrupt this voltage, turning off the A/C. To do this, the ECU will ground the WAC output (Pin 54), which will energize the coil in the WAC relay. This will result in the normally closed contacts opening, interrupting the voltage supply to the A/C clutch.

ACC/S

The ECU uses the ACC/S input to determine when the A/C demand switch is on and the cyclic pressure switch/low pressure (LP) switch is closed. This information is used primarily for idle control.

Pinpoint Test Schematic

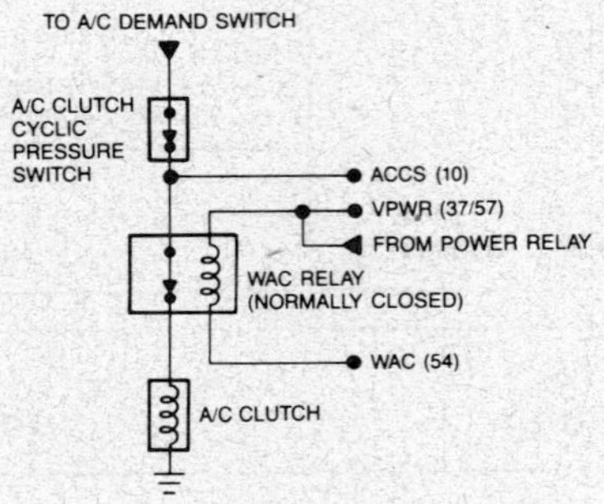

WOT A/C Cutout (WAC) A/C Demand — Pinpoint Test KM

TEST STEP	RESULT ▶	ACTION TO TAKE
KM1 NO A/C: CHECK FOR VOLTAGE AT A/C CLUTCH		
NOTE: Before proceeding with "NO A/C" diagnostics, verify integrity of related fuses in fuse panel. • Key off. • Disconnect A/C clutch. • A/C demand switch to A/C. • Voltmeter on 20 volt scale. • Start engine, wait 10 seconds. • Measure voltage between the power side of the A/C clutch vehicle harness connector and battery negative post (allow time for normal on/off cycling of the ACC/S or L P switch). • **Is voltage greater than 10.5 volts?**	Yes ▶ No ▶	EEC system OK. REFER to check for poor ground, open circuit in clutch, mechanical problems, etc. GO to KM2.
KM2 CHECK POWER-TO-CLUTCH CIRCUIT CONTINUITY		
• Key off. • A/C clutch disconnected. • Disconnect WAC relay. • Ohmmeter on 200 ohm scale. • Measure resistance between power side of the A/C clutch vehicle harness connector and POWER-TO-CLUTCH pin at the WAC relay vehicle harness connector. • **Is resistance less than 5.0 ohms?**	Yes ▶ No ▶	RECONNECT A/C clutch. GO to KM3. SERVICE open circuit. RECONNECT all components. RE-EVALUATE symptom.

PINPOINT TESTS—NAVAJO

WOT A/C Cutout (WAC) A/C Demand — Pinpoint Test KM

TEST STEP	RESULT ▶	ACTION TO TAKE
KM3 CHECK FOR POWER ON A/C DEMAND CIRCUIT		
• Key on, engine off. • WAC relay disconnected. • A/C demand switch to A/C. • Voltmeter on 20 volt scale. • Measure voltage between A/C demand input pin at WAC relay vehicle harness connector and chassis ground. • **Is voltage greater than 10.5 volts?**	Yes ▶	GO to KM6 .
	No ▶	VERIFY operation of A/C clutch cyclic pressure switch and A/C demand switch. . If OK, SERVICE open circuit. RECONNECT all components. RE-EVALUATE symptom.

WOT A/C Cutout (WAC) A/C Demand — Pinpoint Test KM

TEST STEP	RESULT ▶	ACTION TO TAKE
KM6 CHECK WAC CIRCUIT FOR SHORT TO GROUND		
• Key off. • WAC relay disconnected. • Disconnect ECU 60 pin connector. Inspect for damaged or pushed out pins, corrosion, loose wires, etc. Service as necessary. Leave processor disconnected. • Ohmmeter on 200,000 ohm scale. • Measure resistance between WAC circuit at the WAC relay or A/C fan controller harness connector and chassis ground. • **Is resistance greater than 10,000 ohms?**	Yes ▶	GO to KM9 .
	No ▶	SERVICE short circuit. RECONNECT all components. RE-EVALUATE symptom.

WOT A/C Cutout (WAC) A/C Demand — Pinpoint Test KM

TEST STEP	RESULT ▶	ACTION TO TAKE
KM9 CHECK WAC RELAY • Key off. • ECU disconnected. • Reconnect WAC relay • Disconnect A/C clutch. • Voltmeter on 20 volt scale. • Key on, engine off. • A/C demand switch to A/C. • Measure voltage between the power side of the A/C clutch vehicle harness connector and the battery negative post. • **Is voltage greater than 10.5 volts?**	Yes ▶	REPLACE ECU. RECONNECT A/C clutch. RE-EVALUATE symptom.
	No ▶	REPLACE WAC relay. RECONNECT all components. RE-EVALUATE symptom.

WOT A/C Cutout (WAC) A/C Demand — Pinpoint Test KM

TEST STEP	RESULT ▶	ACTION TO TAKE
KM15 NO A/C CUTOUT AT WOT: ENTER OUTPUT STATE CHECK (REFER TO QUICK TEST APPENDIX) **NOTE: Do not use SUPER STAR II tester for this Step, use Voltmeter.** • Key off, wait 10 seconds. • Voltmeter on 20 volt scale. • Disconnect electrical connector on the Speed Control servo; if equipped. • Connect Voltmeter negative test lead to STO at the Self-Test connector and positive test lead to battery positive post. • Jumper STI to SIGNAL RETURN at the Self-Test connector. • Perform Key On Engine Off Self-Test until the completion of the Continuous Memory Codes. • Voltmeter will indicate less than 1.0 volt when test complete. • Depress and release the throttle. • **Does voltage increase to greater than 10.5 volts?**	Yes ▶	REMAIN in Output State Check. GO to KM17.
	No ▶	DEPRESS throttle to WOT and RELEASE. If STO voltage does not go high, leave equipment hooked up and GO to Pinpoint Test Step QC2.
KM17 CHECK FOR VPWR TO RELAY • Still in Output State Check. • Disconnect harness from WAC relay. • Voltmeter on 20 volt scale. • Measure voltage between VPWR circuit at the WAC relay vehicle harness connector and chassis ground. • **Is voltage greater than 10.5 volts?**	Yes ▶	GO to KM18.
	No ▶	SERVICE open in VPWR circuit between power relay and WAC relay. RECONNECT all components and REMOVE jumper. RE-EVALUATE symptom.

PINPOINT TESTS—NAVAJO

WOT A/C Cutout (WAC) A/C Demand — Pinpoint Test KM

TEST STEP	RESULT ▶	ACTION TO TAKE
KM18 CHECK FOR WAC CYCLING		
• Still in Output State Check. • WAC relay disconnected. • Voltmeter on 20 volt scale. • Connect Voltmeter positive test lead to the VPWR circuit and the negative test lead to the WAC circuit at the WAC relay vehicle harness connector. • While observing Voltmeter, depress and release throttle several times (to cycle output on and off). • **Does voltage cycle high and low?**	Yes ▶	REPLACE WAC relay. REMOVE jumper. RE-EVALUATE symptom.
	No ▶	REMOVE jumper. RECONNECT Speed Control servo. GO to KM19 .
KM19 CHECK WAC CIRCUIT CONTINUITY		
• Key off. • Disconnect ECU 60 pin connector. Inspect for damaged or pushed out pins, corrosion, loose wires, etc. Service as necessary. • Install breakout box, leave ECU disconnected. • Disconnect WAC relay. • Measure resistance between Test Pin 54 at the breakout box and WAC circuit at the WAC relay vehicle harness connector. • **Is resistance less than 5.0 ohms?**	Yes ▶	GO to KM20 .
	No ▶	SERVICE open circuit. REMOVE breakout box. RECONNECT all components. RE-EVALUATE symptom.
KM20 CHECK WAC CIRCUIT FOR SHORT TO POWER		
• Key off. • Breakout box installed, ECU disconnected. • WAC relay disconnected. • Key on, engine off. • Measure voltage between Test Pin 54 and chassis ground. • **Is voltage less than 1.0 volt?**	Yes ▶	REPLACE ECU. RECONNECT WAC relay. RE-EVALUATE symptom.
	No ▶	SERVICE short circuit. REMOVE breakout box. RECONNECT all components. RE-EVALUATE symptom. If symptom is still present, REPLACE ECU.

WOT A/C Cutout (WAC) A/C Demand — Pinpoint Test KM

TEST STEP	RESULT ▶	ACTION TO TAKE
KM30 CYCLE A/C DEMAND SWITCH		
• Key off, wait 10 seconds. • Disconnect ECU 60 pin connector. Inspect for damaged or pushed out pins, corrosion, loose wires, etc. Service as necessary. • Install breakout box, leave ECU disconnected. • Voltmeter on 20 volt scale. • Key on, engine off. • Connect Voltmeter positive test lead to Test Pin 43 and negative test lead to Test Pin 40. • **Does voltage cycle high and low when A/C demand switch is cycled?**	Yes ▶	REPLACE ECU. REMOVE breakout box. RERUN Quick Test.
	No ▶	GO to KM31 .
KM31 CHECK ACD CIRCUIT CONTINUITY		
• Key off, wait 10 seconds. • Breakout box installed, ECU disconnected. • Ohmmeter on 200 ohm scale. • Measure resistance between Test Pin 43 at the breakout box and A/C demand switch. • **Is resistance greater than 5 ohms?**	Yes ▶	SERVICE open in ACD circuit. REMOVE breakout box. RECONNECT ECU. RERUN Quick Test.
	No ▶	EEC system OK. REMOVE breakout box. RECONNECT ECU.
KM35 CHECK ACD CIRCUIT FOR SHORT TO POWER		
• Key off. • Disconnect ECU 60 pin connector. Inspect for damaged or pushed out pins, corrosion, loose wires, etc. Service as necessary. • Install breakout box, leave ECU disconnected. • Disconnect WAC relay (TRK). • A/C demand switch "OFF". • Voltmeter on 20 volt scale. • Key on, engine off. • Measure voltage between Test Pin 43 at the breakout box and chassis ground. • **Is voltage less than 1.0 volt?**	Yes ▶	EEC system OK. REMOVE breakout box. RECONNECT all components.
	No ▶	VERIFY operation of A/C demand switch. IF OK, SERVICE short circuit. REMOVE breakout box. RECONNECT all components. RE-EVALUATE symptom.

WOT A/C Cutout (WAC) A/C Demand — Pinpoint Test KM

TEST STEP	RESULT ▶	ACTION TO TAKE
KM50 ATTEMPT TO GENERATE CODE 539		
• Key off. • A/C demand switch to A/C. • Perform KOEO Self-Test. • **Is Code 539 present during Self-Test?**	Yes ▶ No ▶	GO to KM51. GO to KM60.
KM51 CHECK FOR VOLTAGE TO A/C RELAY		
• Key off. • Disconnect A/C relay. • Key on. • Measure voltage between IGN RUN circuit at A/C relay vehicle harness connector and chassis ground. • Measure voltage between BATT(+) circuit at A/C relay vehicle harness connector and chassis ground. • **Are both voltages greater than 10.5 volts?**	Yes ▶ No ▶	GO to KM52. SERVICE open circuit. RECONNECT all components. RE-EVALUATE symptom.

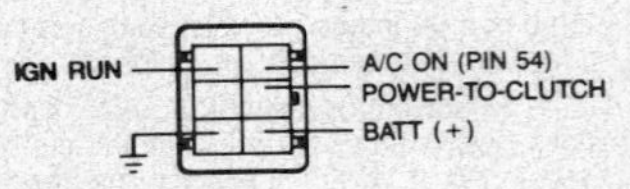

A/C RELAY
VEHICLE HARNESS CONNECTOR

TEST STEP	RESULT ▶	ACTION TO TAKE
KM52 CHECK POWER-TO-CLUTCH CIRCUIT CONTINUITY		
• Key off. • A/C relay disconnected. • A/C clutch disconnected. • Measure resistance between power side of A/C clutch vehicle harness connector and Power-To-Clutch circuit at A/C relay vehicle harness connector. • **Is resistance less than 5.0 ohms?**	Yes ▶ No ▶	GO to KM53. SERVICE open circuit. RECONNECT all components. RE-EVALUATE symptom.

WOT A/C Cutout (WAC) A/C Demand — Pinpoint Test KM

TEST STEP	RESULT ▶	ACTION TO TAKE
KM53 CHECK A/C RELAY		
• Key off. • A/C clutch disconnected. • Reconnect A/C relay. • Jumper A/C ON circuit at A/C relay vehicle harness connector to battery ground. • Key on. • Measure voltage between power side of A/C clutch vehicle harness connector and chassis ground. • **Is voltage greater than 10.5 volts?**	Yes ▶ No ▶	Key off. REMOVE jumper. GO to KM54. REPLACE A/C relay. RECONNECT all components. RE-EVALUATE symptom.
KM54 CHECK A/C ON CIRCUIT CONTINUITY		
• Key off. • Disconnect ECU 60 pin connector. Inspect for damaged or pushed out pins, corrosion, loose wires, etc. Service as necessary. • Install breakout box, leave ECU disconnected. • Disconnect A/C relay. • Measure resistance between Test Pin 54 at breakout box and A/C ON circuit at A/C relay vehicle harness connector. • **Is resistance less than 5.0 ohms?**	Yes ▶ No ▶	REPLACE ECU. REMOVE breakout box. RECONNECT all components. RE-EVALUATE symptom. SERVICE open circuit. RECONNECT all components. RE-EVALUATE symptom.
KM60 CHECK ACD CIRCUIT CONTINUITY TO LOW PRESSURE (LP) SWITCH		
• Key off. • A/C demand switch to A/C • Disconnect LP switch. • Measure resistance between A/C demand switch side of LP switch vehicle harness connector and chassis ground. • **Is resistance less than 5.0 ohms?**	Yes ▶ No ▶	GO to KM61. VERIFY operation of A/C demand switch. REFER to Ventilation/Climate Control Group in Service Manual. If OK, SERVICE open circuit. RECONNECT all components. RE-EVALUATE symptom.

PINPOINT TESTS—NAVAJO

WOT A/C Cutout (WAC) A/C Demand — Pinpoint Test — KM

TEST STEP	RESULT ▶	ACTION TO TAKE
KM61 MEASURE RESISTANCE OF LP SWITCH		
• Key off. • LP switch disconnected. • Measure resistance of LP switch. • **Is resistance less than 5.0 ohms?**	Yes ▶	GO to KM62.
	No ▶	REFER to Ventilation/Climate Control Group Service Manual. RECONNECT all components. RE-EVALUATE symptom.
KM62 CHECK ACD CIRCUIT CONTINUITY FROM LP SWITCH TO PROCESSOR		
• Key off. • LP switch disconnected. • Disconnect ECU 60 pin connector. Inspect for damaged or pushed out pins, corrosion, loose wires, etc. Service as necessary. • Install breakout box, leave ECU disconnected. • Measure resistance between Test Pin 23 at breakout box and ECU side of LP switch vehicle harness connector. • **Is resistance less than 5.0 ohms?**	Yes ▶	REPLACE ECU. REMOVE breakout box. RECONNECT all components. RE-EVALUATE SYMPTOM.
	No ▶	SERVICE open circuit. RECONNECT all components. RE-EVALUATE symptom.
KM70 SERVICE CODE 539: CHECK A/C INPUT		
Service Code 539 indicates that the A/C demand input to the ECU was low (A/C on) during Self-Test. **NOTE: Before entering this test, verify A/C selector is off. If A/C was on, rerun Quick Test. If Code 539 is present, continue with this test.** • Key off. • A/C off. • Disconnect LP switch. • Measure resistance between A/C demand switch side of LP switch vehicle harness connector and chassis ground. • **Is resistance greater than 10,000 ohms?**	Yes ▶	GO to KM71.
	No ▶	VERIFY operation of A/C demand switch. REFER to Ventilation/Climate Control Group in Service Manual. If OK, SERVICE short to ground. RECONNECT all components. RERUN Quick Test.
KM71 CHECK LOW PRESSURE (LP) SWITCH		
• Key off. • LP switch disconnected. • Measure resistance between chassis ground and both pins of LP switch. • **Are both resistances greater than 10,000 ohms?**	Yes ▶	GO to KM72.
	No ▶	REPLACE LP switch. RECONNECT all components. RERUN Quick Test.

WOT A/C Cutout (WAC) A/C Demand — Pinpoint Test — KM

TEST STEP	RESULT ▶	ACTION TO TAKE
KM72 CHECK ACD CIRCUIT TO ECU FOR SHORT TO GROUND		
• Key off. • LP switch disconnected. • Disconnect ECU 60 pin connector. Inspect for damaged or pushed out pins, corrosion, loose wires, etc. Service as necessary. • Measure resistance between ECU side of LP switch vehicle harness connector and chassis ground. • **Is resistance greater than 10,000 ohms?**	Yes ▶	REPLACE ECU. RECONNECT all components. RERUN Quick Test.
	No ▶	SERVICE short to ground. RECONNECT all components. RERUN Quick Test.
KM80 A/C ALWAYS ON: CHECK POWER-TO-CLUTCH CIRCUIT FOR SHORT TO POWER		
NOTE: Before entering this test, verify Code 539 was not received during Self-Test. If Code 539 is present, refer to proper Pinpoint Test direction. • Key off. • Disconnect A/C relay. • Disconnect A/C clutch. • Key on. • Measure voltage between power side of A/C clutch vehicle harness connector and battery negative post. • **Is voltage less than 1.0 volt?**	Yes ▶	GO to KM81.
	No ▶	SERVICE short to power. RECONNECT all components. RE-EVALUATE symptom.
KM81 CHECK A/C RELAY		
• Key off. • A/C relay disconnected. • Measure resistance between Power-To-Clutch Pin and both the IGN RUN and BATT(+) Pins at the A/C relay. • Measure resistance between A/C ON Pin and GROUND Pin at A/C relay. • **Are all resistances greater than 10,000 ohms?**	Yes ▶	GO to KM82.
	No ▶	REPLACE A/C relay. RECONNECT all components. RE-EVALUATE symptom.

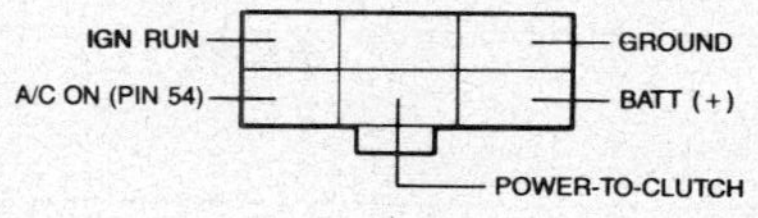

A/C RELAY

WOT A/C Cutout (WAC) A/C Demand — Pinpoint Test KM

TEST STEP	RESULT ▶	ACTION TO TAKE
KM82 CHECK A/C ON CIRCUIT FOR SHORT TO GROUND • Key off. • A/C relay disconnected. • Disconnect ECU 60 pin connector. Inspect for damaged or pushed out pins, corrosion, loose wires, etc. Service as necessary. • Measure resistance between A/C ON circuit at A/C relay vehicle harness connector and chassis ground. • **Is resistance greater than 10,000 ohms?**	Yes ▶	REPLACE ECU. RECONNECT all components. RE-EVALUATE symptom.
	No ▶	SERVICE short to ground. RECONNECT all components. RE-EVALUATE symptom.

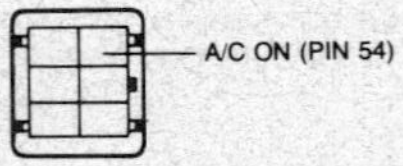

A/C RELAY
VEHICLE HARNESS CONNECTOR

Fuel Pressure Regulator Control (FPRC) Solenoid — Pinpoint Test KN

Note

You should enter this Pinpoint Test only when you have been directed here from Quick Test.

Remember

To prevent replacement of good components, be aware that the following non-EEC area may be at fault:

- Damaged fuel pressure regulator
- Damaged vacuum hoses
- Engine operating temperatures
- Base engine (valves, vacuum leaks, timing, etc.)

This Pinpoint Test is intended to diagnose only the following:

- FPRC solenoid
- Harness circuits: FPRC, VPWR and SIG RTN
- ECU assembly

Description

The Fuel Pressure Regulator Control (FPRC) solenoid controls the vacuum applied to the fuel pressure regulator. The FPRC solenoid contains a valve with two ports. One port is connected to the manifold vacuum. The second port is connected to the fuel pressure regulator. The FPRC solenoid also has an air filter cap (vent) on it. If the solenoid is energized, then the solenoid valve closes the vacuum path to the second port which leads to the fuel pressure regulator. This results in a higher fuel pressure during crank, idle, neutral and wide open throttle drive modes for a predetermined time and engine temperature. If the solenoid is not energized, then the solenoid valve opens the vacuum to the second port, allowing the manifold vacuum signal to be applied as input to the fuel pressure regulator.

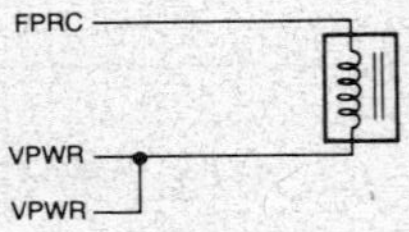

FUEL PRESSURE REGULATOR CONTROL (FPRC) SOLENOID

***Test pins located on breakout box. All harness connectors viewed into mating surface.**

PINPOINT TESTS—NAVAJO

Fuel Pressure Regulator Control (FPRC) Solenoid — Pinpoint Test KN

<table>
<tr><th>TEST STEP</th><th>RESULT ▶</th><th>ACTION TO TAKE</th></tr>
<tr><td colspan="3">KN1 KOEO SERVICE CODE 554: ENTER OUTPUT STATE CHECK (REFER TO QUICK TEST APPENDIX)</td></tr>
<tr><td rowspan="2">Service Code 554 indicates a failure in the Fuel Pressure Regulator Control (FPRC) solenoid circuit.
Possible causes:
— FPRC solenoid
— Damaged vacuum hose(s)
— Open in harness
— Short (to power or ground) in harness
— Faulty ECU.
NOTE: Do not use STAR tester for this step. Use a VOM/DVOM.
• Key off.
• Disconnect electrical connector on the speed control servo, if equipped.
• DVOM on 20 volt scale.
• Connect DVOM negative test lead to STO circuit at Self-Test Connector and positive test lead to battery positive.
• Jumper STI circuit to SIG RTN at Self-Test connector.
• Perform Key On Engine Off Self-Test until completion of Continuous Test codes.
• DVOM will indicate less than 1.0 volt when test is completed.
• Depress and release throttle.
• Does voltage increase?</td><td>Yes ▶</td><td>REMAIN in Output State Check. GO to KN2.</td></tr>
<tr><td>No ▶</td><td>DEPRESS throttle to WOT and release. If STO voltage does not go high leave equipment hooked up and PERFORM Pinpoint Test QC.</td></tr>
<tr><td colspan="3">KN2 CHECK FUEL PRESSURE REGULATOR CONTROL SOLENOID ELECTRICAL OPERATION</td></tr>
<tr><td rowspan="2">• Key on, engine off.
• Disconnect FPRC solenoid.
• Connect DVOM positive test lead to VPWR pin and negative test lead to FPRC signal circuit at the FPRC solenoid vehicle harness connector.
• DVOM on 20 volt scale.
• While observing DVOM depress and release throttle several times to cycle solenoid output.
• Does FPRC solenoid output voltage change greater than 1.0 volt?</td><td>Yes ▶</td><td>RECONNECT FPRC solenoid. REMOVE jumper. GO to KN3.</td></tr>
<tr><td>No ▶</td><td>REMOVE jumper. GO to KN5.</td></tr>
</table>

Fuel Pressure Regulator Control (FPRC) Solenoid — Pinpoint Test KN

<table>
<tr><th>TEST STEP</th><th>RESULT ▶</th><th>ACTION TO TAKE</th></tr>
<tr><td colspan="3">KN3 CHECK VACUUM SIGNAL BETWEEN FUEL PRESSURE REGULATOR AND FPRC SOLENOID</td></tr>
<tr><td rowspan="2">• Key off.
• Disconnect and inspect vacuum hose between fuel pressure regulator and FPRC solenoid for blockage, kinks, etc. Service as necessary.
• Start and idle engine for about five minutes to allow it to reach normal operating temperatures.
• Key off.
• Tee in a standard vacuum gauge between fuel pressure regulator and FPRC solenoid.
• Start and idle engine while observing vacuum gauge needle. Observe vacuum gauge needle for about five minutes.
• Did vacuum signal appear after about two minutes of engine idle?</td><td>Yes ▶</td><td>REMOVE vacuum gauge. GO to Section for Fuel Pressure Regulator diagnostics.</td></tr>
<tr><td>No ▶</td><td>REMOVE vacuum gauge. GO to KN4.</td></tr>
<tr><td colspan="3">KN4 CHECK VACUUM SIGNAL BETWEEN FPRC SOLENOID AND MANIFOLD VACUUM SOURCE</td></tr>
<tr><td rowspan="2">• Key off.
• Tee in a standard vacuum gauge between the FPRC solenoid and manifold vacuum source.
• Start and idle engine while observing the vacuum gauge needle.
• Was any vacuum signal present.</td><td>Yes ▶</td><td>REPLACE FPRC solenoid. REMOVE vacuum gauge. RERUN Quick Test and/or verify elimination of drive symptom.</td></tr>
<tr><td>No ▶</td><td>REPLACE vacuum hose between FPRC solenoid and source vacuum. REMOVE vacuum gauge. RERUN Quick Test and/or verify elimination of drive symptom.</td></tr>
<tr><td colspan="3">KN5 CHECK FPRC SOLENOID VPWR</td></tr>
<tr><td rowspan="2">• Key off.
• FPRC solenoid disconnected.
• Key on, engine off.
• Measure voltage between VPWR at FPRC solenoid vehicle harness connector and battery ground.
• Is voltage greater than 10.5 volts?</td><td>Yes ▶</td><td>GO to KN6.</td></tr>
<tr><td>No ▶</td><td>SERVICE open in VPWR harness circuit. RECONNECT FPRC solenoid. RERUN Quick Test and/or verify elimination of drive symptom.</td></tr>
</table>

Fuel Pressure Regulator Control (FPRC) Solenoid — Pinpoint Test KN

TEST STEP	RESULT ▶	ACTION TO TAKE
KN6 CHECK FPRC CIRCUIT CONTINUITY • Key off. • FPRC solenoid disconnected • Disconnect ECU 60 pin connector. Inspect for damaged or pushed out pins, corrosion, loose wires, etc. Service as necessary. • Install breakout box, leave ECU disconnected. • Measure resistance between Test Pin 13 at breakout box and FPRC circuit at solenoid vehicle harness connector. • **Is resistance less than 5.0 ohms?**	Yes ▶	GO to KN7 .
	No ▶	SERVICE open circuit. REMOVE breakout box. RECONNECT all components. RERUN Quick Test and/or verify elimination of drive symptom.
KN7 CHECK FPRC CIRCUIT FOR SHORT TO POWER AND GROUND • Key off. • FPRC solenoid disconnected. • Breakout box installed, ECU disconnected. • Measure resistance betwen Test Pin 13 and Test Pins 37 and 57 at Breakout Box. • Measure resistance between Test Pin 13 and Test Pins 40, 46 and 60 at breakout box. • **Is each resistance greater than 10,000 ohms?**	Yes ▶	REPLACE ECU. REMOVE breakout box. RECONNECT all components. RERUN Quick Test and/or verify elimination of drive symptom.
	No ▶	SERVICE short circuit. REMOVE breakout box. RECONNECT all components. RERUN Quick Test and/or verify elimination of drive symptom.

Octane Adjust — Pinpoint Test KP

Note

You should enter this Pinpoint Test only when you have been directed here from Diagnosis By Symptom.

Remember

This Pinpoint Test is intended to diagnose only the following:

- Harness circuits: SIG RTN, Octane Adjust (OCT ADJ)
- Octane shorting bar connector

Pinpoint Test Schematic

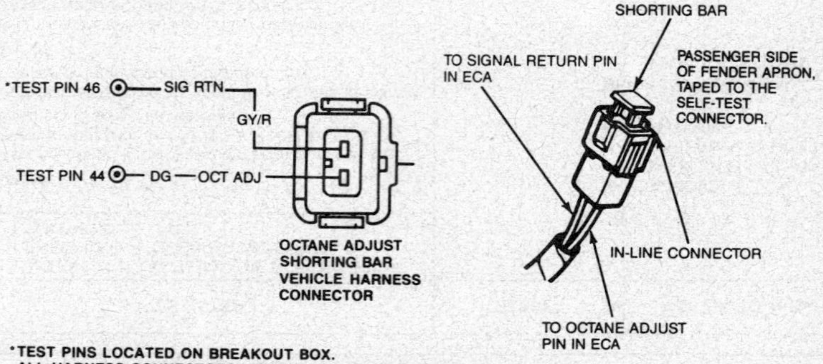

***TEST PINS LOCATED ON BREAKOUT BOX.**
ALL HARNESS CONNECTORS VIEWED INTO MATING SURFACE.

NOTE: The Octane Adjust (OCT ADJ) shorting bar is similar in shape to the SPOUT in-line connector.

PINPOINT TESTS—NAVAJO

Octane Adjust — Pinpoint Test KP

TEST STEP	RESULT	ACTION TO TAKE
KP1 VISUALLY INSPECT OCTANE ADJUST IN-LINE CONNECTOR SERVICE CODE 57/341		
Service Code 57/341 indicates OCT ADJ shorting bar is not in or the OCT ADJ circuit is open. • Key off. • Visually inspect in-line connector. • **Is shorting bar removed?**	Yes ▶ No ▶	GO to KP2. GO to KP4.
KP2 CHECK FOR MODIFICATION DECAL		
• **Is there a modification decal attached to vehicle indicating that OCT ADJ shorting bar was removed?**	Yes ▶ No ▶	Testing complete. GO TO KP3.
KP3 CHECK FOR SERVICE CODE 57/341		
• Replace OCT ADJ shorting bar. • Key off. • Activate self-test. • Key on, engine off. • **Is Service Code 57/341 present?**	Yes ▶ No ▶	GO to KP4. Testing complete. EEC system OK.
KP4 CHECK CONTINUITY OF OCTANE ADJUST CIRCUIT		
NOTE: There should be continuity from OCT ADJ circuit through the in-line connector and shorting bar to SIG RTN circuit. • Key off. • Disconnect ECU 60 pin connector. Inspect for damaged or pushed out pins, corrosion, loose wires, etc. Service as necessary. • Install breakout box, leave ECU disconnected. • Measure resistance between Test Pin 46 and OCT ADJ Test Pin at breakout box (refer to schematic for appropriate OCT ADJ circuit test pin). • **Is the resistance less than 5.0 ohms?**	Yes ▶ No ▶	REPLACE ECU. REMOVE Breakout Box. RERUN Quick Test. SERVICE open OCT ADJ circuit or shorting bar or SIG RTN circuit. REMOVE breakout box. RERUN Quick Test.

Octane Adjust — Pinpoint Test KP

TEST STEP	RESULT	ACTION TO TAKE
KP5 CHECK FOR SERVICE CODE 11/111		
• Start engine and idle vehicle at operating temperature. • Key off. • Activate self-test. • Key on, engine off. • **Is Service Code 11 or 111 present?**	Yes ▶ No ▶	GO to KP6. **For service code 57/341:** GO to KP1. **For all other codes:** RETURN to Quick Test.
KP6 VERIFY IN-LINE CONNECTOR SHORTING BAR IS INSTALLED		
• Key off. • Visually inspect OCT ADJ in-line connector. • **Is shorting bar installed?**	Yes ▶ No ▶	GO to KP8. GO to KP7.
KP7 CHECK FOR MODIFICATION DECAL		
• **Is there a modification decal attached to vehicle indicating that OCT ADJ shorting bar was removed?**	Yes ▶ No ▶	GO to KP10. REPLACE shorting bar. GO address spark knock concern.
KP8 CHECK FOR TSB		
• **Is there a Technical Service Bulletin (TSB) authorizing removal of the OCT ADJ shorting bar?**	Yes ▶ No ▶	GO to KP9. GO address spark knock concern.
KP9 REMOVE OCT ADJ SHORTING BAR		
• Remove OCT ADJ shorting bar only if there is a TSB authorization. • Drive vehicle to verify spark knock. • **Is spark knock present?**	Yes ▶ No ▶	GO to KP10. Testing complete. EEC System OK.

Octane Adjust — Pinpoint Test KP

TEST STEP	RESULT ▶	ACTION TO TAKE
KP10 CHECK OCTANE ADJUST CIRCUIT FOR A SHORT TO GROUND		
• Key off. • Breakout box installed, ECU disconnected. • Measure resistance between OCT ADJ circuit at in-line connector and Test Pins 40, 46 and 60 at breakout box. • **Is each resistance greater than 10,000 ohms?**	Yes ▶ No ▶	GO to KP11. SERVICE short circuit. REMOVE breakout hox.
KP11 CHECK PROCESSOR INTEGRITY		
• Key off. • OCT ADJ shorting bar removed because of TSB authorization. • Breakout box installed, ECU connected. • Key on. • Measure voltage between OCT ADJ circuit and Test Pin 40/60 at breakout box. • **Is voltage greater than 4.0 volts?**	Yes ▶ No ▶	REMOVE breakout box. REPLACE ECU. REMOVE breakout box.

Dynamic Response Test — Pinpont Test M

Note

You should enter this Pinpoint Test only when a Service Code 77/538 is received during Quick Test.

Remember

This Pinpoint Test is intended to diagnose only the following:

- Throttle movement (greater than 3/4 throttle)
- Rpm increase (greater than 2000 rpm)

Pinpoint Test Schematic

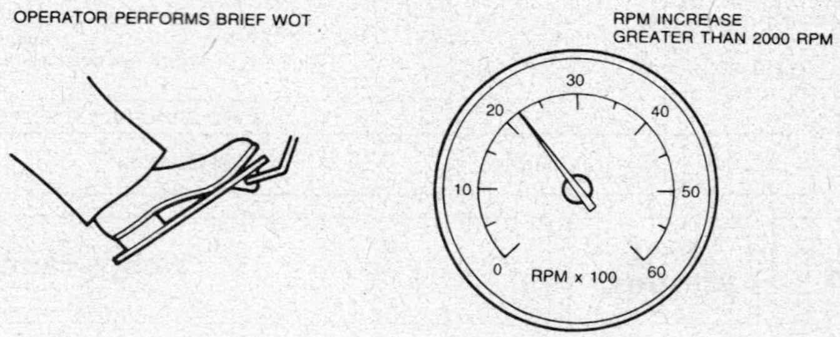

Dynamic Response Test — Pinpoint Test M

TEST STEP	RESULT ▶	ACTION TO TAKE
M1 SERVICE CODE 77/538 SYSTEM FAILED TO RECOGNIZE BRIEF WOT		
NOTE: A brief snap of the throttle may not be sufficient to pass this test. Be sure to go to WOT and return. • Rerun Engine Running Self-Test. Be sure operator is familiar with the Engine Running format which proceeds as follows: — Start engine. — Activate Self-Test. — ID Code 2 (0); start of test. — Dynamic Response code 1 (0); perform brief WOT. — Testing over. — Service code output begins. • **Is Code 77 or 538 still present?**	Yes ▶ No ▶	GO to M2. Dynamic Response Test passed. SERVICE any other service code(s) received as necessary.
M2 DID ENGINE ACHIEVE GREATER THAN 2000 RPM?		
• **During the WOT in the Dynamic response test, did the engine achieve greater than 2000 rpm?**	Yes ▶ No ▶	REPLACE ECU RERUN Quick Test. CHECK for conditions that would present engine from achieving greater than 2000 rpm (binding throttle linkage, etc). Refer to other possible causes of the vehicles driveability symptom.

"CHECK ENGINE" Light — Pinpoint Test ML

Note

You should enter this Pinpoint Test only when directed here from Pinpoint Test QA or Diagnosis By Symptom.

Remember

To prevent the replacement of good components, be aware that the following non-EEC areas may be at fault:

- Fuse, bulb or socket.

This Pinpoint Test is intended to diagnose only the following:

- STO/MIL circuit
- ECU assembly

Description

The "Check Engine" or Malfunction Indicator Light (MIL) is intended to alert the driver of certain malfunctions in the EEC system. The MIL output is turned on when the strategy detects a fault in input/output circuits. The light will remain on as long as the fault causing it is present. Regulations governing this light also require that the Self-Test fault codes be displayed by the flashing of this light.

"CHECK ENGINE" Light — Pinpoint Test — ML

Pinpoint Test Schematic

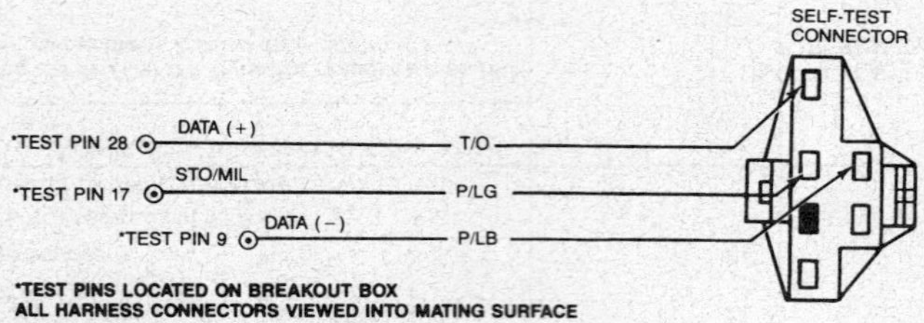

*TEST PINS LOCATED ON BREAKOUT BOX
ALL HARNESS CONNECTORS VIEWED INTO MATING SURFACE

"CHECK ENGINE" Light — Pinpoint Test — ML

TEST STEP	RESULT ▶	ACTION TO TAKE
ML1 "CHECK ENGINE" LIGHT ALWAYS ON: CHECK STO/MIL CIRCUIT FOR SHORTS TO GROUND		
NOTE: If vehicle will not start go to Pinpoint Test Step AC1. • **If any Key On Engine Off service codes or Continuous Memory Codes are present, service before proceeding. If no codes are output, continue with this Test Step.** • Key off. • Disconnect ECU 60 pin connector. Inspect for damaged or pushed out pins, corrosion, loose wires, etc. Service as necessary. • Install breakout box, leave ECU disconnected. • Ohmmeter on 200 ohm scale. • Measure resistance between Test Pin 17 and Test Pin 40 at the breakout box. • **Is resistance less than 5 ohms?**	Yes ▶ No ▶	SERVICE short circuit between Test Pin 17 and "CHECK ENGINE" Light, or between Test Pin 17 and Self-Test Connector. REMOVE breakout box. RECONNECT ECU. RERUN Quick Test. REPLACE ECU. REMOVE breakout box. RERUN Quick Test.
ML5 "CHECK ENGINE" LIGHT NEVER ON: CHECK CONTINUITY OF STO/MIL CIRCUIT		
NOTE: If vehicle will not start perform Pinpoint Test AC1. **Refer to Quick Test Appendix for a detailed description of how the "CHECK ENGINE" light (malfunction indicator light) operates.** • Key off. • Disconnect ECU 60 pin connector. Inspect for damaged or pushed out pins, corrosion, loose wires, etc. Service as necessary. • Install breakout box, leave ECU disconnected. • Ohmmeter on 200 ohm scale. • Measure resistance between Test Pin 17 and the "CHECK ENGINE" light. • **Is resistance less than 5 ohms?**	Yes ▶ No ▶	GO to ML6. SERVICE open circuit. REMOVE breakout box. RECONNECT ECU. RERUN Quick Test.

PINPOINT TESTS – NAVAJO

"CHECK ENGINE" Light — Pinpoint Test ML

TEST STEP	RESULT ▶	ACTION TO TAKE
ML6 CHECK FOR VOLTAGE TO BULB		
• Key off. • Voltmeter on 20 volt scale. • Key on, engine off. • Measure voltage from battery negative post to the "hot" side of the "CHECK ENGINE" bulb. • **Is the voltage greater than 10.5 volts?**	Yes ▶ No ▶	GO to ML7 . GO to ML8 .
ML7 CHECK FOR FAULTY BULB/SOCKET		
• Install known good bulb/socket. • Turn key to "RUN" position. • Is "CHECK ENGINE" light on?	Yes ▶ No ▶	RECONNECT ECU. Bulb/socket was faulty. REPLACE ECU. VERIFY repair by turning key to "RUN" position.
ML8 CHECK FOR VOLTAGE AT FUSE		
• Key off. • Voltmeter on 20 volt scale. • Key on, engine off. • Measure voltage from battery negative post to the "hot" side of the fuse. • Is the voltage greater than 10.5 volts?	Yes ▶ No ▶	SERVICE fuse or open in circuit. RECONNECT ECU. VERIFY repair by turning key to "RUN" position. SERVICE VBAT circuit. RECONNECT ECU. VERIFY repair by turning key to "RUN" position.

"CHECK ENGINE" Light — Pinpoint Test ML

TEST STEP	RESULT ▶	ACTION TO TAKE
ML10 "CHECK ENGINE" LIGHT INTERMITTENTLY ON: CHECK FOR INTERMITTENT STO SHORT TO GROUND		
NOTE: If vehicle will not start go to Pinpoint Test Step AC1 . **The "CHECK ENGINE" light will come ON when there is a Continuous Memory Code present. Service any Continuous Memory Codes before proceeding.** (See Quick Test Appendix for description of "CHECK ENGINE" light function.) If no codes are outputted, continue with this Test Step. • Enter Key On Engine Off Continuous Monitor Mode. Refer to Quick Test Appendix. • Observe Voltmeter or SUPER STAR II for indication of a fault while you wiggle, shake or bend a small section of the EEC system harness in the following locations: — Harness closest to Self-Test connector to the dash panel — Dash panel to the ECU — Dash panel to the "CHECK ENGINE" light • **Is a fault indicated?**	Yes ▶ No ▶	SERVICE short to ground. RERUN Quick Test. UNABLE to duplicate fault at this time. Testing complete.

"CHECK ENGINE" Light — Pinpoint Test ML

TEST STEP	RESULT ▶	ACTION TO TAKE
ML15 "CHECK ENGINE" LIGHT FLASHING WITH ERRATIC IDLE: CHECK FOR STI SHORT TO GROUND		
NOTE: Vehicle symptoms indicate that STI is grounded and the vehicle is actually performing Self-Test without a tester installed. • Key off. • Disconnect ECU 60 pin connector. Inspect for damaged or pushed out pins, corrosion, loose wires, etc. Service as necessary. • Ohmmeter on 200,000 ohm scale. • Measure resistance between the Self-Test Input (STI) connector and engine block ground. • **Is resistance less than 10,000 ohms?**	Yes ▶	SERVICE short circuit. RECONNECT ECU. VERIFY symptom eliminated.
	No ▶	RECONNECT ECU. VERIFY symptom eliminated. REFER to rough idle routines.

Ignition/Erratic Diagnostic Monitor (IDM) — Pinpoint Test N

Note

You should enter this Pinpoint Test only when a Service Code 14, 16, 18, 19 or 45 is received during Quick Test, when directed here from Diagnosis By Symptom

Remember

To prevent the replacement of good components, be aware that the following non-EEC areas may be at fault:

- EDIS ignition module
- EDIS coil packs
- Spark plugs and high tension cables
- Arcing of secondary ignition components

This Pinpoint Test is intended to diagnose only the following:

- Harness circuits: IGN GND, SPOUT, PIP, IDM, DPI, SAW
- Processor assembly

Description

The Ignition Diagnostic Monitor (IDM) is an input signal to the processor that verifies spark function. This signal is transmitted from the EDIS module to the ECU and consists of a single pulse for each successful spark. Lack of an IDM pulse is used as an indication of intermittant and/or missing spark.

PINPOINT TESTS—NAVAJO

Ignition/Erratic Diagnostic Monitor (IDM) — Pinpoint Test N

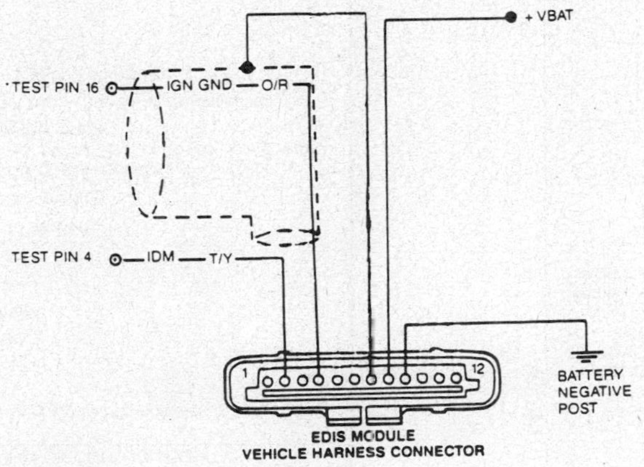

*TEST PINS LOCATED ON BREAKOUT BOX.
ALL HARNESS CONNECTORS VIEWED INTO MATING SURFACE.

Ignition/Erratic Diagnostic Monitor (IDM) — Pinpoint Test N

TEST STEP	RESULT ▶	ACTION TO TAKE
N1 CONTINUOUS MEMORY CODE 211: ERRATIC IGNITION		
Code 211 indicates two successive erratic profile ignition pickup (PIP) pulses occurred, resulting in a possible engine miss or stall. Possible causes: — Loose wires/connectors. — Arcing secondary ignition components (coil, wires, plugs, etc.). — On-board transmitter (2-way radio).* • **Are any of the above present?** *Verify all 2-way radio installations. Carefully follow manufacturer's installation instructions regarding the routing of antenna and power leads.	Yes ▶ No ▶	SERVICE as necessary. CLEAR Continuous Memory Code 211. RERUN Quick Test. GO to N4.
N2 SERVICE CODE 226: IDM CIRCUIT FAILURE (EDIS)		
• **Does engine start?**	Yes ▶ No ▶	GO to N4. GO to EDIS Diagnosis.
N3 CONTINUOUS MEMORY CODE 212: CHECK FOR OTHER EEC CODES		
Code 212 indicates a loss of IDM Processor input. Possible causes: — Open harness. — Shorted harness. — DIS module. — Processor. • **Is Continous Memory Service Code 215, 216 or 217 present?**	Yes ▶ No ▶	GO to N15. GO to N4.

Ignition/Erratic Diagnostic Monitor (IDM) — Pinpoint Test N

TEST STEP	RESULT ▶	ACTION TO TAKE
N4 CONTINUOUS MEMORY CODE 212 CHECK IDM CIRCUIT CONTINUITY		
Continuous Memory Code 212 indicates a loss of IDM ECU input. Possible causes: — Open harness. — Shorted harness. — EDIS module. — ECU. • Key off, wait 10 seconds. • Disconnect EDIS module. • Disconnect ECU 60 pin connector. Inspect for damaged or pushed out pins, corrosion, loose wires, etc. Service as necessary. • Install breakout box, leave ECU disconnected. • Ohmmeter on 200,000 ohm scale.	Yes ▶	GO to N5.
	No ▶	SERVICE open circuit. REMOVE breakout box. RECONNECT ECU. CLEAR Continuous Memory Codes. RERUN Quick Test.

Ignition/Erratic Diagnostic Monitor (IDM) — Pinpoint Test N

TEST STEP	RESULT ▶	ACTION TO TAKE
N4 CONTINUOUS MEMORY CODE 212: CHECK CONTINUITY OF IDM CIRCUIT (Continued)		
• Measure resistance between Test Pin 4 at the breakout box and EDIS module 2 pin harness connector. • **Is resistance less than 5.0 ohms?**		
N5 CHECK IDM CIRCUIT FOR SHORT TO GROUND		
• Key off, wait 10 seconds. • EDIS module disconnected. • Breakout box installed, ECU disconnected. • Ohmmeter on 200,000 ohm scale. NOTE: During this check when the four-wire HEGO is connected to vehicle harness, a short to SIG RTN (Pin 46) may be indicated in conjunction with an actual PWR GND short. • Measure resistance between Test Pin 4 and Test Pins 40, 46 and 60 at the breakout box. • **Is each resistance above 10,000 ohms?**	Yes ▶	GO to N6.
	No ▶	REMOVE breakout box. SERVICE short to ground in IDM circuit. RECONNECT all components. CLEAR Continuous Memory Code. RERUN Quick Test.
N6 CHECK ECU FOR SHORT TO GROUND		
• Key off. • Breakout box installed. • Connect ECU to breakout box. • EDIS module disconnected. • Ohmmeter on 200,000 ohm scale. • Measure resistance between Test Pin 4 and Test Pin 16, 20, 40, 46, and 60 at the breakout box. • **Is each resistance greater than 10,000 ohms?**	Yes ▶	REMOVE breakout box. RECONNECT all components.
	No ▶	REPLACE ECU. REMOVE breakout box. RECONNECT all components. RERUN Quick Test.

PINPOINT TESTS—NAVAJO

Ignition/Erratic Diagnostic Monitor (IDM) — Pinpoint Test N

TEST STEP	RESULT ▶	ACTION TO TAKE
N7 CHECK EDIS MODULE		
• Key off. • Breakout box installed. • Connect ECU and EDIS module to breakout box. • Connect DVOM between Test Pin 4 and Test Pin 16 at breakout box. • Start engine. • Observe DVOM when voltage is allowed to stabilize. • Lightly tap on EDIS ignition components simulate road shock. • Wiggle both EDIS module connectors. • A sudden change in voltage indicates a fault. • **Is a fault indicated?**	Yes ▶ No ▶	DISCONNECT and INSPECT connectors. If connector and terminals are good, REMOVE breakout box, RECONNECT all components and GO to Ignition Diagnostics GO to N8.
N8 CHECK EEC HARNESS		
• DVOM still connected between Test Pin 4 and Test Pin 16 at breakout box. • Key on, engine running. • While observing a voltage change like in N7, do the following: — Grasp vehicle harness closest to EDIS module connector. Wiggle, shake and bend a small section of EEC harness while working your way to other ignition system components and to dash panel. Also wiggle, shake and bend EEC harness from dash panel to ECU. • **Is a fault indicated?**	Yes ▶ No ▶	ISOLATE fault and service as necessary. REMOVE breakout box. RECONNECT all components. CLEAR Continuous Memory Code. RERUN Quick Test. GO to N9.
N9 CHECK ECU AND HARNESS CONNECTORS		
• Key off, wait 10 seconds. • Disconnect ECU 60 pin connector. Inspect for damaged or pushed out pins, corrosion, loose wires, etc. • **Are connectors and terminals OK?**	Yes ▶ No ▶	GO to N6. SERVICE as necessary. REMOVE breakout box. RECONNECT all components. CLEAR Continuous Memory Codes. RERUN Quick Test.

Ignition/Erratic Diagnostic Monitor (IDM) — Pinpoint Test N

TEST STEP	RESULT ▶	ACTION TO TAKE
N15 CONTINUOUS MEMORY CODES 45: CHECK CONTINUITY OF COIL CIRCUITS FROM EDIS TO COIL PACK		
Service Code 45 indicates a fault has been detected by the ECU in one of the three coils contained in the ignition coil pack. **NOTE: Codes 45 refer to faults detected in the circuits related to coils 3, 1 or 2 respectively. Coil 1 provides voltage for cylinder three and four spark plugs. Coil 2 provides voltage for cylinder two and six spark plugs. Coil 3 provides voltage for cylinder one and five spark plugs. The IDM pulse train contains a corresponding pulse for each operating coil. In the event that a coil fails, the corresponding pulse will be absent from the IDM pulse train.** Possible causes: — Open in coil circuit from EDIS module to coil pack. — Shorts to coil circuit GND or PWR in EDIS module. • Key off, wait 10 seconds. • Disconnect DIS module. • Ohmmeter on 200 ohm scale. • Measure resistance between pin 10 at the EDIS module and the coil 1 pin at the ignition coil pack. • Measure resistance between pin 11 at the EDIS module and the coil 2 pin at the ignition coil pack. • Measure resistance between pin 12 at the EDIS module and the coil 3 pin at the ignition coil pack. • **Is each resistance less than 5 ohms?**	Yes ▶ No ▶	GO to N16. SERVICE open circuit. RECONNECT all components. RERUN Quick Test.

Ignition/Erratic Diagnostic Monitor (IDM) — Pinpoint Test N

TEST STEP	RESULT ▶	ACTION TO TAKE
N16 CHECK COIL PACK CIRCUIT FOR SHORTS TO GROUND AND POWER		
• Key off. • EDIS module disconnected pins 1-12. • Disconnect coil pack. • Ohmmeter on 200,000 ohm scale. • Measure resistance between pin 4 and pins 10, 11 and 12 at the EDIS module connector. • Measure resistance between pin 8 and pins 10, 11 and 12 at the EDIS module connector. • **Is each resistance greater than 10,000 ohms?**	Yes ▶ No ▶	REFER to EDIS Diagnosis. SERVICE short circuit. RECONNECT all components. RERUN Quick Test.

Spark Timing Check — Pinpoint Test P

Note

You should enter this Pinpoint Test only when checking computed timing. A Service Code 18/213 is received during Quick Test.

Remember

To prevent the replacement of good components, be aware that the following non-EEC areas may be at fault:

- Base Engine
- EDIS Module
- Variable Reluctance Sensor (VRS)

This Pinpoint Test is intended to diagnose only the following:

- Harness circuit: SAW
- Base Timing
- ECU Assembly

Description

Spark Angle Width (SAW) is a digital signal generated by the ECU which supplies spark timing information to the EDIS module. Normally, there will be one and only one SAW pulse for each PIP pulse.

PINPOINT TESTS—NAVAJO

Spark Timing Check — Pinpoint Test P

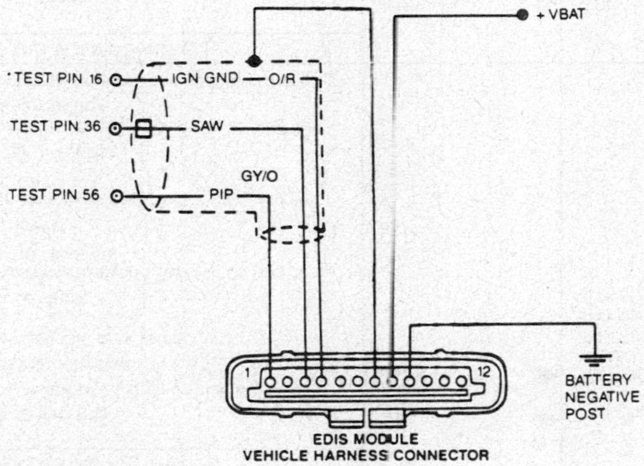

*TEST PINS LOCATED ON BREAKOUT BOX.
ALL HARNESS CONNECTORS VIEWED INTO MATING SURFACE.

Test 36 SAW

Application	Wire Color
4.0L NAVAJO	PK

Spark Timing Check — Pinpoint Test P

TEST STEP	RESULT ▶	ACTION TO TAKE
P1 CHECK FOR POWER TO ECU		
NOTE: Before proceeding with this Pinpoint Test, verify that the base timing check in Quick Test Step 4.0 has been performed. • Key off, wait 10 seconds. • Disconnect ECU 60 pin connector. Inspect for damaged or pushed out pins, corrosion, loose wires. Service as necessary. • Install breakout box, leave ECU disconnected. • Key on, engine off. • Voltmeter on 20 volt scale. • Measure voltage between Test Pin 37 and Test Pin 40 at the breakout box. • Measure voltage between Test Pin 57 and Test Pin 60 at the breakout box. • **Is each voltage greater than 10.5 volts?**	Yes ▶ No ▶	GO to P2. GO to Pinpoint Test Step B1.
P2 CHECK SAW CIRCUIT CONTINUITY		
• Key off, wait 10 seconds. • Breakout box installed, ECU disconnected. • Disconnect EDIS module. • Ohmmeter on 200 Scale. • Measure resistance between Test Pin 36 at the breakout box and SAW circuit at the EDIS module vehicle harness connector. • **Is resistance less than 5 ohms?**	Yes ▶ No ▶	GO to P3. SERVICE open circuit. REMOVE breakout box. RECONNECT all components. CHECK timing per Quick Test Step 4.0.

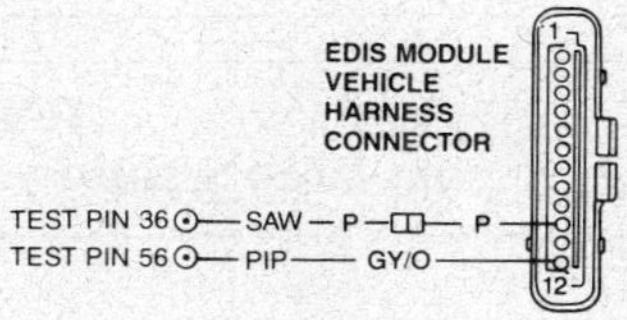

Spark Timing Check | Pinpoint Test | P

TEST STEP	RESULT ▶	ACTION TO TAKE
P3 CHECK SAW CIRCUIT FOR SHORTS TO POWER, GROUND AND PIP		
• Key off, wait 10 seconds. • Breakout box installed, ECU disconnected. • EDIS module disconnected. • Ohmmeter on 200,000 ohm scale. **For Shorts To Ground:** — Measure resistance between Test Pin 36 and Test Pins 16, 20, 40, 46 & 60. **For Shorts To Power:** — Measure resistance between Test Pin 36 and Test Pins 26, 37 & 57. **For Short To PIP Circuit:** — Measure resistance between Test Pin 36 and Test Pin 56 at Breakout Box. • **Is each resistance greater than 10,000 ohms?**	Yes ▶ No ▶	GO to P4. SERVICE short circuit. REMOVE breakout box. RECONNECT all components. RERUN Quick Test.
P4 CHECK SAW VOLTAGE AT ECU		
• Key off, wait 10 seconds. • Breakout box installed. • Connect ECU to breakout box. • Reconnect EDIS module. • Timing switch to "DIST" position on breakout box. • Volmeter on 20 volt scale. • Measure voltage between Test Pin 36 at the breakout box and battery negative post during Engine Running Self-Test. • **Is voltage between 0.01 and 1.89 volts?**	Yes ▶ No ▶	EEC System OK. REMOVE breakout box. RECONNECT ECU. GO to Ignition System Diagnosis. REPLACE ECU. REMOVE breakout box. RERUN Quick Test.

No Codes/Codes Not Listed | Pinpoint Test | QA

Note

You should enter this Pinpoint Test only when you have been directed here from Quick Test.

Remember

This Pinpoint Test is intended to diagnose only the following:

- ECU
- EEC Power Relay
- Harness Circuits: SIG RTN, STO, STI, VPWR, VREF, HEGO

Pinpoint Test Schematic

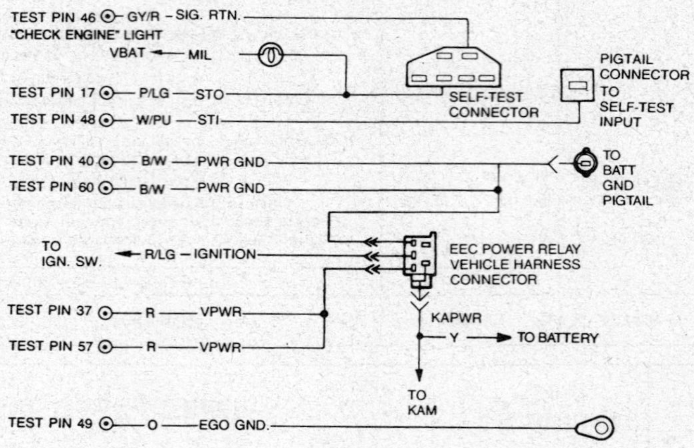

*TEST PINS LOCATED ON BREAKOUT BOX.
ALL HARNESS CONNECTORS VIEWED INTO MATING SURFACE.

PINPOINT TESTS—NAVAJO

No Codes/Codes Not Listed | **Pinpoint Test** | **QA**

TEST STEP	RESULT ▶	ACTION TO TAKE
QA1 CHECK VREF VOLTAGE AT SELF-TEST CONNECTOR		
NOTE: Aftermarket devices such as an alarm system may cause self-test to abort if wiring is connected to certain EEC components. If such a device is installed, disconnect it completely from EEC system, restore EEC circuits to original, and rerun Quick Test before proceeding with the Pinpoint Test. **NOTE: If using a STAR tester to run Self-Test, verify that correct procedure is used for your application.** • Key off, wait 10 seconds. • Disconnect ECU 60 pin connector. Inspect for damaged or pushed out pins, corrosion, loose wires, etc. Service as necessary. • Install breakout box and connect ECU to breakout box. • Voltmeter on 20 volt scale. • Key on, engine off. • Measure voltage between Test Pin 26 at the breakout box and SIG RTN in the Self-Test connector. • **Is voltage between 4.0 and 6.0 volts?**	Yes ▶ No ▶	GO to QA2. GO to Pinpoint Test Step C1.
QA2 CHECK SIG RTN CIRCUIT CONTINUITY		
• Key off. • Breakout box installed, ECU disconnected. • Measure resistance between SIG RTN circuit in Self-Test connector and Test Pin 46 at breakout box. • **Is resistance less than 5 ohms?**	Yes ▶ No ▶	GO to Pinpoint Test Step C1. SERVICE open circuit. REMOVE breakout box. RECONNECT ECU. RERUN Quick Test.
QA3 CHECK STI CIRCUIT CONTINUITY		
• Key off, wait 10 seconds. • Breakout box installed. • Disconnect ECU. • Ohmmeter on 200 ohm scale. • Measure resistance between STI in the Self-Test single pin pigtail connector and Test Pin 48 at the breakout box. • **Is resistance less than 5 ohms?**	Yes ▶ No ▶	GO to QA4. SERVICE open circuit. REMOVE breakout box. RECONNECT ECU. RERUN Quick Test.

No Codes/Codes Not Listed | **Pinpoint Test** | **QA**

TEST STEP	RESULT ▶	ACTION TO TAKE
QA4 CHECK STO CIRCUIT CONTINUITY		
• Key off. • Breakout box installed, ECU disconnected. • Ohmmeter on 200 ohm scale. • Measure resistance between STO in the Self-Test connector and Test Pin 17 at the breakout box. • **Is resistance less than 5 ohms?**	Yes ▶ No ▶	GO to QA5. SERVICE short circuit. REMOVE breakout box. RECONNECT ECU. RERUN Quick Test.
QA5 CHECK HEGO SIGNAL FOR SHORT TO POWER		
NOTE: Due to internal circuitry of ECU a right/rear HEGO signal short to power could cause EEC strategy NOT to enter on-demand Self-Test. • Breakout box installed, ECU disconnected. • Key on, engine off. • Measure voltage between HEGO Signal Test Pin 29 or 44 (refer to Pinpoint Test H schematic) and Test Pin 40 or 60 at breakout box. • **Is voltage greater than 2.0 volts?**	Yes ▶ No ▶	GO to QA6. GO to QA7.
QA6 ISOLATE SHORT TO HARNESS OR HEGO SENSOR		
• Breakout box installed, ECU disconnected. • Disconnect right/rear HEGO sensor. • Key on, engine off. • Measure voltage between HEGO Signal Test Pin 29 or 44 (refer to Pinpoint Test H schematic) and Test Pin 40 or 60 at breakout box. • **Is voltage greater than 2.0 volts?**	Yes ▶ No ▶	SERVICE HEGO Signal circuit for shorts to power. REMOVE breakout box. RECONNECT all components. RERUN Quick Test. REPLACE right/rear HEGO sensor. REMOVE breakout box. RECONNECT all components. RERUN Quick Test.

No Codes/Codes Not Listed — Pinpoint Test QA

TEST STEP	RESULT ▶	ACTION TO TAKE
QA7 CHECK STO CIRCUIT FOR SHORT TO GROUND		
• Key off. • Breakout box installed, ECU disconnected. • Ohmmeter on 200 ohm scale. • Measure resistance between STO in Self-Test connector and engine block ground. • **Is resistance less than 5 ohms?**	Yes ▶	SERVICE STO or MIL circuit for short to ground. REMOVE breakout box. RECONNECT ECU. RERUN Quick Test.
	No ▶	GO to QA8.
QA8 CHECK IF POWER RELAY IS ALWAYS ON		
• Key off. • Breakout box installed, ECU disconnected. • Connect Voltmeter to Test Pin 37 or 57 and to Test Pin 40 or 60 at the breakout box. • Voltmeter on 20 volt scale. • Turn key on and then off. Wait 10 seconds. • **Does voltage change from greater than 10.5 volts to less than 1.0 volt?**	Yes ▶	**For vehicles equipped with MIL (malfunction indicator light) displayed as "CHECK ENGINE":** GO to QA9.
	No ▶	GO to QA7.

No Codes/Codes Not Listed — Pinpoint Test QA

TEST STEP	RESULT ▶	ACTION TO TAKE
QA9 CHECK VPWR CIRCUIT FOR SHORT TO POWER		
• Key off. • Breakout box installed, ECU disconnected. • EEC Power Relay. • Voltmeter on 20 volts scale. • Measure voltage between Test Pin 37 or 57 and Test Pin 40 or 60 at the breakout box. • **Is voltage greater than 1.0 volts?**	Yes ▶	SERVICE VPWR circuit short ot power. REMOVE breakout box. RECONNECT all components. RERUN Quick Test.
	No ▶	REPLACE EEC Power Relay or Integrated Relay Controller. REMOVE breakout box. RECONNECT ECU. RERUN Quick Test.
QA10 MIL EQUIPPED VEHICLES ("CHECK ENGINE" LAMP)		
• **Are any of these conditions present?** • Malfunction indicator light: — Always ON	▶	Go to ML1.
— Always OFF	▶	GO to ML5.
• **Malfunction indicator light functioning normally**		REPLACE the ECU. RERUN Quick Test.

Continuous Memory Service Code 15/512 — Pinpoint Test QB

Note

You should enter this Pinpoint Test only when you have been directed here from Quick Test.

Remember

This Pinpoint Test is intended to diagnose only the following:

- ECU
- Harness circuits: PWR GND, VPWR, KAPWR, IGNITION

Pinpoint Test Schematic

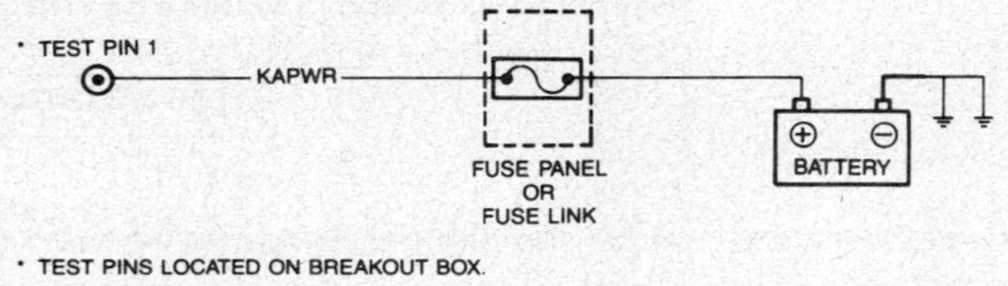

***TEST PINS LOCATED ON BREAKOUT BOX**
ALL HARNESS CONNECTORS VIEWED INTO MATING SURFACE

Continuous Memory Service Code 15/512 — Pinpoint Test QB

TEST STEP	RESULT ▶	ACTION TO TAKE
QB1 CHECK KAPWR TO ECU		
Continuous Code 15/512 indicates that ECU has experienced a power interrupt in its Keep Alive Memory (KAM). **NOTE: If KAPWR is interrupted to the ECU, for example when installing a breakout box or when battery is disconnected, Code 15/512 may be set in continuous memory.** • Key off. • Disconnect ECU 60 pin connector. Inspect for damaged or pushed out pins, corrosion, loose wires, etc. Service as necessary. • Install breakout box, leave ECU disconnected. • Measure voltage between Test Pin 1 and Test Pin 40 or 60 at breakout box. • While observing DVOM, grasp EEC harness and wiggle, shake or bend a small section while working your way from ECU to dash panel. • **Does DVOM indicate less than 10.5 volts?**	Yes ▶ No ▶	Service open circuit. REMOVE breakout box. RECONNECT ECU. RERUN Quick Test. RECONNECT ECU. GO to QB2.
QB2 INSPECT ENGINE COMPARTMENT WIRING FOR PROPER ROUTING		
• Inspect EEC wiring for closeness to ignition components or wires. If EEC wiring is close, reroute as necessary. • Clear continuous Memory codes (refer to Quick Test Appendix). • Wait five minutes to allow Code 15/512 to reset. • Rerun Key On Engine Off Self-Test: • **Is Code 15 or 512 present on retest?**	Yes ▶ No ▶	REPLACE ECU. REMOVE breakout box. RERUN Quick Test. REMOVE breakout box. RECONNECT ECU. SERVICE other codes as necessary. If none, testing is complete.

Output State Check Not Functioning | Pinpoint Test | QC

Note

You should enter this Pinpoint Test only when you have been directed here from other Pinpoint Tests.

Remember

This Pinpoint Test is intended to diagnosis only the following:

- Test conditions
- Throttle Plate Linkage

Output State Check Not Functioning | Pinpoint Test | QC

TEST STEP	RESULT ▶	ACTION TO TAKE
QC1 CHECK FOR CODES 23, 53, 63, 121, 122 or 123		
NOTE: If vehicle is equipped with speed control, the speed control servo must be electrically disconnected. This prevents the speed control system from affecting the throttle plate movement during Output State Check. Rerun Quick Test or Output State Check. • Key off, wait 10 seconds. • Perform Key On Engine Off Self-Test. **NOTE: If additional information is required on Output State Check, refer to diagnostic aids in the Quick Test Appendix.** • **Are any of these codes (23, 53, 63, 121, 122 or 123) present?**	Yes ▶ Code 11 or 111 ▶ No Codes ▶	GO to Quick Test and SERVICE appropriate code as instructed. GO to QC2. GO to QA1.
QC2 CHECK THROTTLE LINKAGE		
• Check throttle and throttle linkages for sticking and binding. • **Is throttle OK?**	Yes ▶ No ▶	REPLACE TP sensor. RERUN Quick Test. SERVICE as necessary. RERUN Quick Test.

Re-Initialization Check | Pinpoint Test | QD

Note

You should enter this Pinpoint Test only when Service Code 19/513 is received during Quick Test or you have been directed here from Diagnosis By Symptom.

Remember

This Pinpoint Test is intended to diagnose only the following:

- ECU
- EEC Power Relay
- Harness circuits: GND, VPWR, IGNITION

Pinpoint Test Schematic

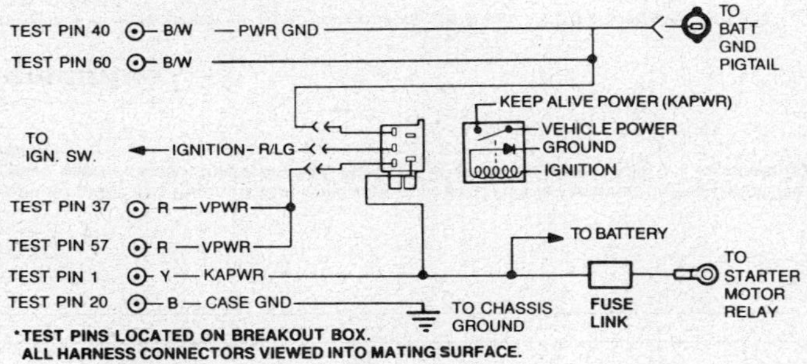

Re-Initialization Check | Pinpoint Test | QD

TEST STEP	RESULT ▶	ACTION TO TAKE
QD5 SERVICE CODE 19/513: INTERNAL VOLTAGE		
This Service Code 19/513 indicates the ECU's voltage regulator inability to maintain proper internal voltage, which is necessary for the ECU to accurately compute data. • Rerun Key On Engine Off Self-Test. • **Is 19/513-10-11/111 present?**	Yes ▶	REPLACE EEC ECU. RERUN Quick Test.
	No ▶	SERVICE other codes. If none, testing complete.

System Check	Pinpoint Test	S

Note

You should enter this Pinpoint Test only after a Code 11/111 is received during Quick Test and you have been directed here from EEC No-Start Pinpoint Test Step A20 or Diagnosis By Symptom.

Remember

To prevent the replacement of good components, be aware that the following Non-EEC areas may be at fault:

- Poor power/ground connections
- Ignition system wires, coil, plugs
- Base engine valves, cam timing, compression, etc.

This Pinpoint Test is intended only as a Quick Check for the basic functioning of the following:

- ISC Bypass Air System
- MAF System

System Check	Pinpoint Test	S

Pinpoint Test Flow Chart

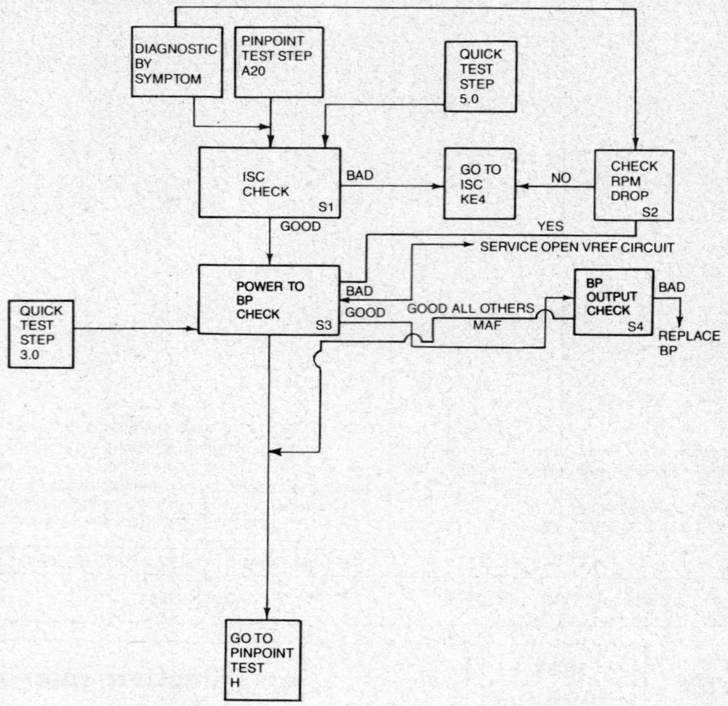

PINPOINT TESTS—NAVAJO

System Check	Pinpoint Test	S

TEST STEP	RESULT ▶	ACTION TO TAKE
S1 ISC-BPA CHECK		
NOTE: This Test Step is for vehicles with stalls and or no starts. • Attempt to start engine at part throttle. • **Will engine run smooth at part throttle?**	Yes ▶ No ▶	GO to KE4 . GO to S3 .
S2 CHECK FOR RPM DROP		
• Key off. • Connect engine tachometer. • Start engine. • Disconnect ISC solenoid. • **Does rpm drop or stall?**	Yes ▶ No ▶	RECONNECT ISC solenoid. GO to S3 . GO to KE4 .
S3 POWER TO BP SENSOR TEST		
NOTE: If the vehicle being checked is not equipped with BP system hardware go directly to S8 . Green light on tester indicates VREF is OK. Red Light (or no light) indicates VREF is either too low or too high. • Key off. • Disconnect the BP sensor from the vehicle harness. • Connect the BP tester between the vehicle harness and the BP sensor, (refer to schematic below). • Insert BP tester banana plugs into DVOM. • Set DVOM to 20 volt scale. • Key on, engine off. • **Is green light on?**	Yes ▶ No ▶	GO to S4 . SERVICE open VREF circuit. REMOVE BP tester. RECONNECT BP. RE-EVALUATE symptom.

System Check	Pinpoint Test	S

TEST STEP	RESULT ▶	ACTION TO TAKE
S3 POWER TO BP SENSOR TEST (Continued)		

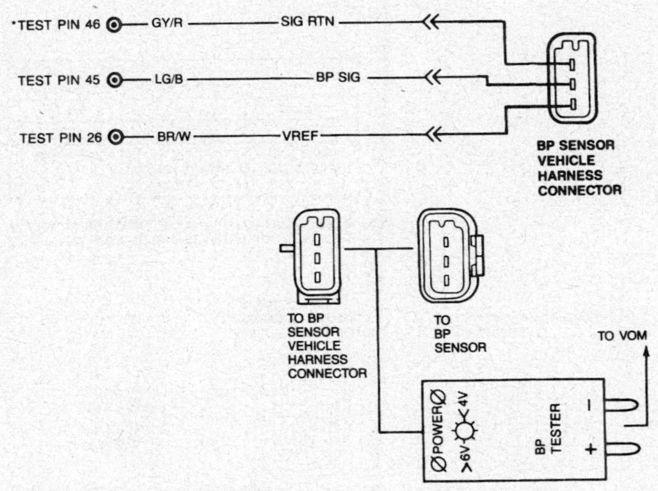

***TEST PINS LOCATED ON BREAKOUT BOX. ALL HARNESS CONNECTORS VIEWED INTO MATING SURFACE.**

System Check | Pinpoint Test | S

TEST STEP	RESULT ▶	ACTION TO TAKE
S4 BP TESTER OUTPUT READING		
NOTE: Measure several known good BP sensors on available vehicles. The measured voltage will be typical for your location on the day of testing. • BP Tester and VOM connected. • Voltmeter or 20 volt scale. • Key on.	Yes ▶	GO to H1.
• Approximate Altitude (Ft.) / Voltage Output (+/−.04 Volts): 0 — 1.59 1000 — 1.56 2000 — 1.53 3000 — 1.50 4000 — 1.47 5000 — 1.44 6000 — 1.41 7000 — 1.39 **• Is voltage in range for your altitude?**	No (Sensor output is out-of-range) ▶	REPLACE BP sensor.

Transmission—A4LD | Pinpoint Test | TB

Pinpoint Test Schematic

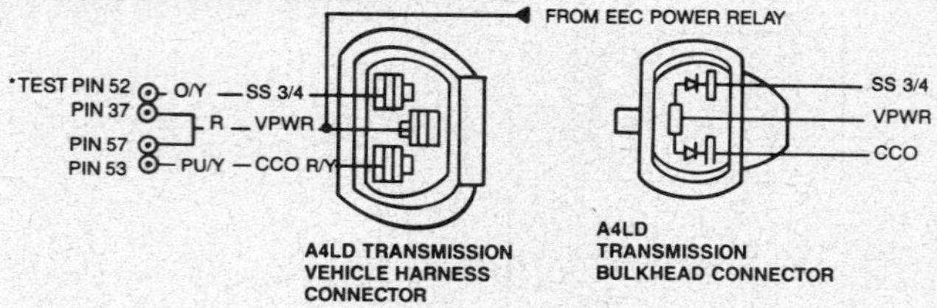

***TEST PINS LOCATED ON BREAKOUT BOX.**
ALL HARNESS CONNECTORS VIEWED INTO MATING SURFACE.

Solenoid	Processor Signal Output Pin	KOEO Self-Test Code
CCO	53	89/629
SS 3/4	52	86/566

CCO - Converter Clutch Override
SS 3/4 Shift Solenoid 3/4-4/3

PINPOINT TESTS—NAVAJO

Transmission — A4LD — Pinpoint Test TB

TEST STEP	RESULT ▶	ACTION TO TAKE
TB2 CHECK CONTINUITY OF VEHICLE HARNESS CIRCUITS • Key off. • Breakout box installed, ECU disconnected. • A4LD transmission bulkhead connector disconnected. • Ohmmeter on 200 ohm scale. • Measure the resistance between Test Pins 37/57 at the breakout box and VPWR circuit at the A4LD transmission vehicle harness connector. • Measure the resistance between the ECU signal output pin (refer to table in TB1) at the breakout box and the same circuit at the A4LD transmission vehicle harness connector. • **Is each resistance less than 5 ohms?**	Yes ▶ No ▶	GO to TB3. SERVICE open circuit. REMOVE breakout box. RECONNECT all components. RERUN Quick Test.
TB3 CHECK A4LD TRANSMISSION VEHICLE HARNESS CIRCUITS • Key off. • Breakout box installed, ECU disconnected. • A4LD transmission bulkhead connector disconnected. • Ohmmeter on 200,000 ohm scale. • Measure the resistance between the ECU signal output pin (refer to table in TB1) and Test Pins 37/57 at the breakout box. • Measure the resistance between the ECU signal output pin (refer to table in TB1) and Test Pins 40, 60 and 46 at the breakout box and chassis ground. • **Is each resistance greater than 10,000 ohms?**	Yes ▶ No ▶	GO to TB4. SERVICE short circuit(s). REMOVE breakout box. RECONNECT all components. RERUN Quick Test.

Transmission — A4LD — Pinpoint Test TB

TEST STEP	RESULT ▶	ACTION TO TAKE
TB4 CHECK VPWR VOLTAGE TO A4LD TRANSMISSION SOLENOIDS • Key off. • Breakout box installed. • Connect ECU to breakout box. • A4LD transmission bulkhead connector disconnected. • Voltmeter on 20 volt scale. • Key on, engine off. • Measure voltage between Test Pins 37/57 at the A4LD transmission vehicle harness connector and chassis ground/battery ground. • **Is voltage greater than 10.5 volts?**	Yes ▶ No ▶	GO to TB5. REPLACE ECU. REMOVE Breakout box. RECONNECT solenoid(s). RERUN Quick Test.
TB5 CHECK OUTPUT DRIVER SIGNAL • Key off. • Breakout box installed, ECU connected. • A4LD transmission bulkhead connector disconnected. • Connect test light between Test Pin 37/57 and signal output pin (refer to table in TB1 at the breakout box. • Enter Key On, Engine Off Self-Test. • **Does test lamp flash momentarily then stay off?**	Yes ▶ No ▶	Problem is internal to the transmission. REPLACE ECU. REMOVE breakout box. RECONNECT solenoid(s). RERUN Quick Test.

Component Replacement

RELIEVING FUEL SYSTEM PRESSURE

The fuel injected engine is equipped with a pressure relief valve located on the fuel supply manifold. Remove the fuel tank cap and attach fuel pressure gauge T80L–9974–A, or equivalent, to the valve to release the fuel pressure. If a suitable pressure gauge is not available, disconnect the vacuum hose from the fuel pressure regulator and attach a hand vacuum pump. Apply about 25 in. Hg (84 kPa) of vacuum to the regulator to vent the fuel system pressure into the fuel tank through the fuel return hose. Note that this procedure will remove the fuel pressure from the lines, but not the fuel itself. Take precautions to avoid the risk of fire and use clean rags to soak up any spilled fuel when the lines are disconnected.

QUICK CONNECT FUEL LINE FITTINGS

Removal and Installation

NOTE: Quick Connect (push) type fuel line fittings must be disconnected using proper procedures or the fitting may be damaged. There are 2 types of retainers are used on the push connect fittings. Line sizes of 3/8 and 5/16 in. use a hairpin clip retainer. In addition, some engines use spring lock connections secured by a garter spring which requires a special tool for removal.

HAIRPIN CLIP

1. Relieve the fuel pressure before disconnecting the line. Clean all dirt and/or grease from the fitting. Spread the 2 clip legs about 1/8 in. (3mm) each to disengage from the fitting and pull the clip outward from the fitting. Use finger pressure only, do not use any tools.

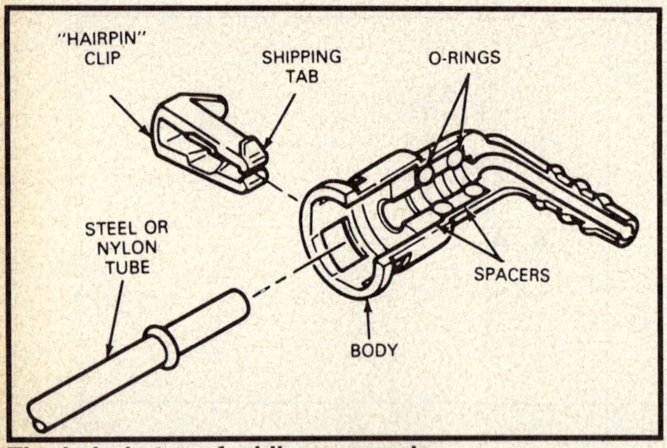

The hairpin type fuel line connector

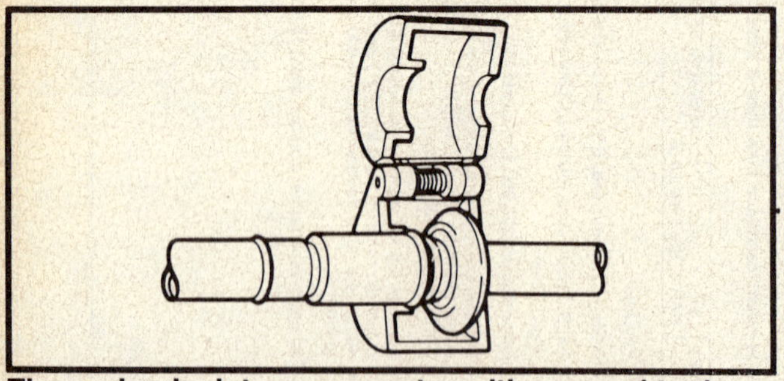

The spring lock type connector with removal tool

2. Grasp the fitting and hose assembly and pull away from the steel line. Twist the fitting and hose assembly slightly while pulling, if necessary, when a sticking condition exists.
3. Inspect the hairpin clip for damage, replace the clip if necessary. Reinstall the clip in position on the fitting.
4. Inspect the fitting and inside of the connector to insure freedom of dirt or obstruction. Install fitting into the connector and push together. A click will be heard when the hairpin snaps into proper connection. Pull on the line to insure full engagement.

SPRING LOCK CONNECTOR

1. Relieve the fuel pressure before disconnecting the line.
2. Assemble the removal tool around the connector (49 UN01 051 for 3/8 in. line or 49 UN01 052 for 1/2 in. line, or equivalent).
3. Pull the tool into the connector to release the spring.
4. Pull the line out of the connector and remove the tool.
5. Clean the connector and apply refrigerant oil to the connection.
6. Press the connector together to install. Pull the connector to ensure the connector is fully engaged.

FUEL PUMP

CAUTION

The fuel supply lines will remain pressurized for long periods of time after the engine is turned off. Relieve the fuel system pressure before perorming any repairs or personal injury my result.

Removal and Installation

IN-TANK PUMP

1. It is necessary to remove the fuel tank.
2. Depressurize the fuel system.
3. Remove fuel from the fuel tank by pumping out through the filter tube.

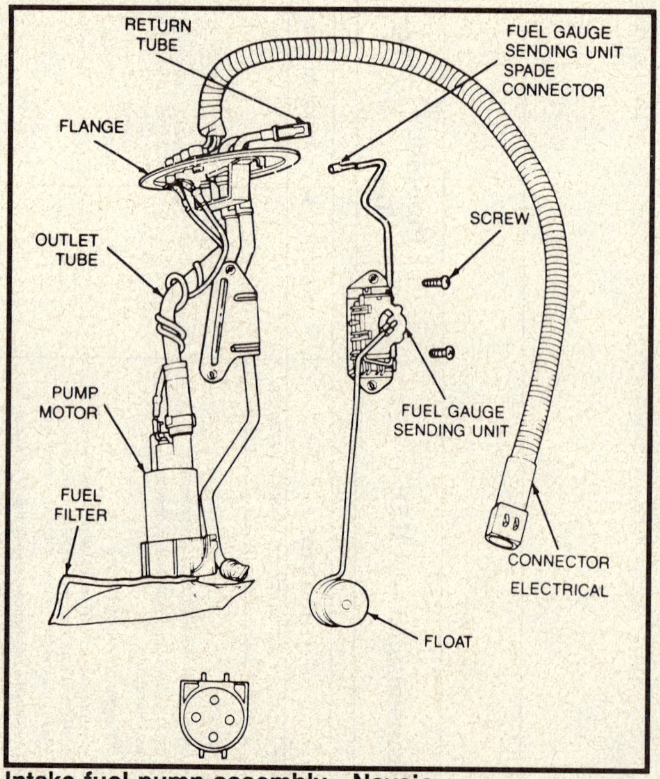

Intake fuel pump assembly—Navajo

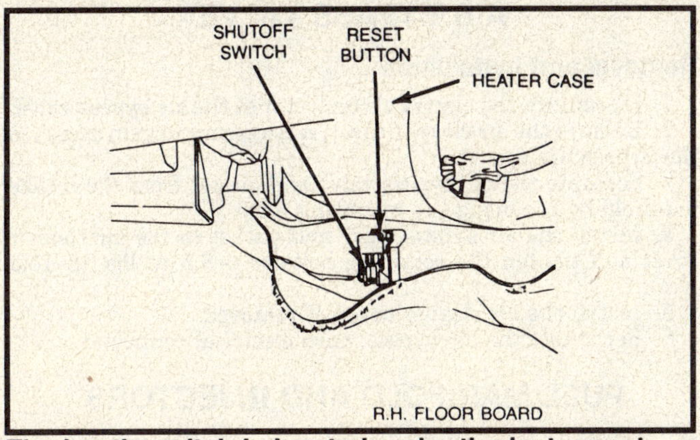

The inertia switch is located under the instrument panel on the right of the transmission hump—Navajo

4. Disconnect the supply and return line fittings and the vent line.
5. Disconnect and remove the fuel filter tube.
6. Disconnect the electrical connections to both the fuel sender and the fuel pump wiring harness.
7. Remove the fuel tank support straps and remove the fuel tank.
8. Remove any dirt that has accumulated around the fuel pump attaching flange, to prevent it from entering the tank during service procedures.
9. Turn the fuel pump locking ring counter-clockwise with the necessary tool and remove the locking ring.
10. Remove the fuel pump and bracket assembly.
11. Remove the seal gasket and discard.

To install:

12. Put a light coating of heavy grease on a new seal ring to hold it in place during assembly. Install it in fuel tank ring groove.
13. Install the tank in the vehicle.
14. Install the electrical connector.
15. Install the fuel line fittings and tighten to 30–40 ft. lbs (40–54 Nm).
16. Install a minimum of 10 gallons of fuel and inspect for leaks.
17. Install pressure gauge on valve on throttle body and turn ignition to **ON** position for 3 seconds. Turn ignition key **OFF** and back on for 3 seconds repeatedly, 5 to 10 times, until pressure gauge shows at least 30 psi. Re-inspect for leaks at fittings.
18. Remove pressure gauge. Start engine and re-inspect for leaks.

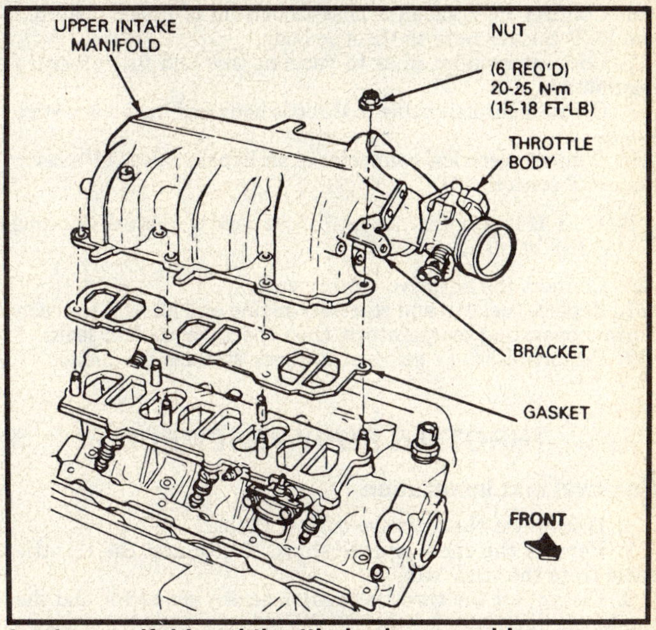

Intake manifold and throttle body assembly removal—Navajo

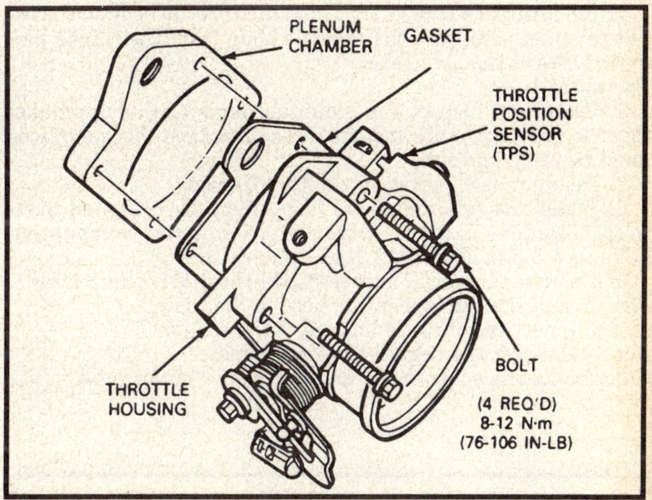

Throttle body assembly—Navajo

INTAKE MANIFOLD AND THROTTLE BODY ASSEMBLY

Removal and Installation

1. Disconnect negative battery cable and all electrical connectors at air bypass valve, throttle position sensor and Air Charge Temperature (ACT) sensor.
2. Remove air inlet tube from air cleaner to throttle body.
3. Remove snow/ice shield to expose throttle linkage. Disconnect throttle cable from ball stud.
4. Disconnect upper intake manifold vacuum connectors; both the front and rear fittings including the vacuum line to fuel pressure regulator.
5. Disconnect PCV closure tube from under throttle body and disconnect PCV vacuum tube from under the manifold.
6. Remove canister purge line from fitting near power steering pump.
7. Loosen bolt which retains A/C line at the upper rear of the upper manifold and disengage retainer.
8. Remove 6 upper intake manifold retaining bolts.
9. Remove upper intake and throttle body as an assembly from lower intake manifold.

To install:

10. Clean and inspect mounting faces of the lower and upper intake manifold.
11. Position new gasket on lower intake mounting face. The use of alignment studs may be helpful.
12. Install upper intake manifold and throttle body assembly to lower manifold making sure gasket remains in place (if alignment studs are not used).
13. Install 6 upper intake manifold retaining bolts. Tighten to 11–15 ft. lbs. (16–20 Nm).
14. Engage A/C line retainer cup and tighten bolt to specification.
15. Install canister purge line to fitting.

16. Connect PCV vacuum hose to bottom of upper manifold and PCV closure hose to throttle body.
17. Connect vacuum lines to vacuum tree and fuel pressure regulator.
18. Connect throttle cable to throttle body and install snow/ice shield.
19. Connect electrical connector at air bypass valve, TP sensor and ACT sensor.

NOTE: If lower intake manifold was removed, fill and bleed the cooling system.

20. Connect the negative battery cable.
21. Replace fuel cap and start the engine and allow it to reach normal operating temperature, then check for coolant leaks.
22. Perform EEC-IV self-test to check systems functions.

THROTTLE BODY ASSEMBLY

Removal and Installation

1. Disconnect the negative battery cable.
2. Remove the snow and ice shield. Disconnect the throttle cable from the stud ball.
3. Disconnect the throttle position sensor connector and the intake air duct.
4. Disconnect the canister purge hose from the throttle body. There is a fitting on the underside of the throttle body that the hose connects to.
5. Remove the 4 screws that hold the throttle body to the intake plenum. Separate the throttle body from the intake plenum. Discard the old gasket.

To install:

6. Clean and inspect the mounting faces of the air intake/throttle body assembly and the intake manifold. Both surfaces must be clean and flat.
7. Install a new gasket on the throttle body.
8. Install the retaining bolts finger tight, then tighten them to specifications in a criss-cross pattern. Torque the mounting bolts to 6–8.5 ft. lbs. (8–12 Nm).
9. Connect the wiring harness to the throttle position sensor.
10. Install the canister purge hose to the fitting.
11. Connect the throttle linkage to the stud ball.
12. Reconnect the negative battery cable.
13. Install the snow shield and air cleaner outlet tube.

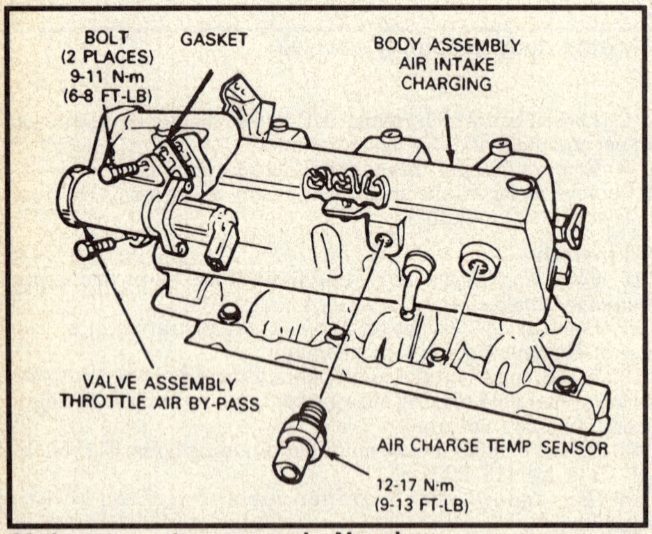

Air bypass valve removal – Navajo

AIR BYPASS VALVE

Removal and Installation

1. Disconnect the electrical connector at the air bypass valve.
2. Remove the air cleaner cover, if necessary to gain access to the bypass valve.
3. Separate the air bypass valve and gasket from the intake manifold by removing the mounting bolts.
4. Install the air bypass valve and gasket to the air cleaner cover and tighten the retaining bolts to 6–8.5 ft. lbs. (8–11.5 Nm).
5. Install the air cleaner cover, if removed.
6. Reconnect the air bypass valve electrical connector.

FUEL MANIFOLD AND INJECTORS

Removal and Installation

1. Remove the air intake and throttle body assembly. Be sure to depressurize the fuel system before disconnecting any fuel lines.
2. Carefully disconnect the fuel supply line from the fitting on the fuel manifold..
3. Disconnect the fuel return line from the fuel pressure regulator. Unlock the locking tabs on the connector retainer and split the halves. Clean the connection to prevent dirt from entering. Push the fitting towards the regulator and insert the the fingers of the fuel line coupling key 49 UN01 006, or equivalent, into the coupling slots. Pull the fitting off the regulator.
4. Remove the 6 Torx fuel injector manifold retaining bolts.
5. Remove the fuel injectors from the supply manifold by lifting while gently rocking from side to side. Place all removed components on a clean surface to prevent contamination by dirt or grease.

NOTE: Injectors and fuel rail must be handles with extreme care to prevent damage to sealing areas and sensitive fuel metering orifices.

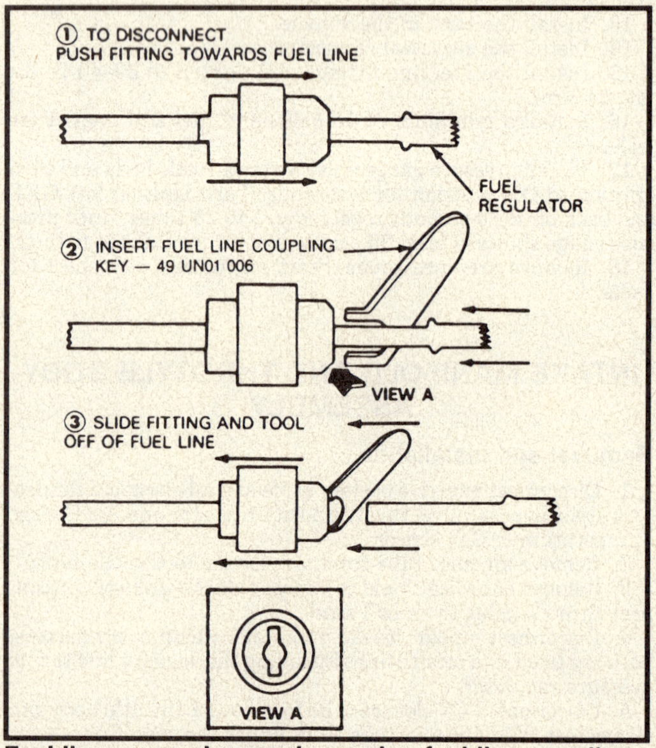

Fuel line removal procedure using fuel line coupling key 49 UN01 006 – Navajo

7. Examine the injector O-rings for deterioration or damage and install new O-rings, as required (2 per injector).
8. Make sure the injector caps are clean and free from contamination or damage.

To install:

9. Lubricate all O-rings with clean engine oil, then install the injectors in the manifold using a light twisting/pushing motion.

NOTE: Never use silicone grease, it will clog the injector.

10. Carefully install the fuel rail assembly onto the lower intake manifold, pushing down on the fuel rail to make sure the O-rings are seated.
11. Hold the fuel rail assembly in place and install the retaining bolts finger tight. Tighten the retaining bolts to 7–10 ft. lbs. (10–14 Nm).
12. Connect the fuel supply and return lines.
13. Connect the fuel injector wiring harness at the injectors.
14. Connect the vacuum line to the fuel pressure regulator.
15. Install the air intake and throttle body assembly.

FUEL PRESSURE REGULATOR

Removal and Installation

1. Depressurize the fuel system, remove shielding as needed.
2. Remove the vacuum line at the pressure regulator.
3. Disconnect the fuel return line from the fuel pressure regulator. Unlock the locking tabs on the connector retainer and split the halves. Clean the connection to prevent dirt from entering. Push the fitting towards the regulator and insert the the fingers of the fuel line coupling key 49 UN01 006, or equivalent, into the coupling slots. Pull the fitting off the regulator.
4. Remove the 2 retaining screws from the regulator housing.
4. Remove the pressure regulator assembly, gasket and O-ring. Discard the gasket and check the O-ring for signs of cracks or deterioration.
5. Clean the gasket mating surfaces. If scraping is necessary, be careful not to damage the fuel pressure regulator or supply line gasket mating surfaces.
6. Lubricate the pressure regulator O-ring with with light engine oil. Do not use silicone grease; it will clog the injectors.
7. Install the O-ring and a new gasket on the pressure regulator.
8. Install the pressure regulator on the fuel manifold and tighten the retaining screws to 6–8 ft. lbs. (7.5–10.5 Nm).
9. Install the vacuum line at the pressure regulator. Install the fuel line to the regulator by pushing the line on.
10. Build up fuel pressure by turning the ignition switch **ON** and **OFF** at least 6 times, leaving the ignition **ON** for at least 5 seconds each time.
11. Check for system for fuel leaks.

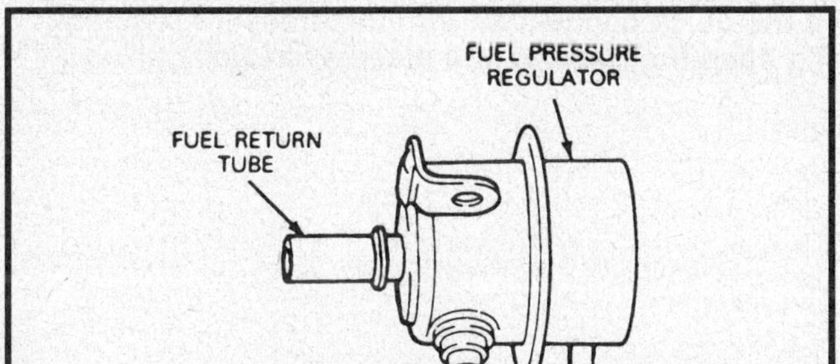

Fuel pressure regulator—Navajo

THROTTLE POSITION (TP) SENSOR

Adjustment is not possible with this style of throttle position sensor. Use of alignment marks when replacing the throttle position sensor will ensure proper operation.

Removal and Installation

1. Disconnect the throttle position sensor electrical connector.
2. Remove the screw retaining the TP sensor electrical connector to the air throttle body, if equipped.
3. Scribe alignment marks on the air throttle body and TP sensor to indicate proper alignment during installation.
4. Remove the TP sensor retaining screws, then remove the TP sensor and gasket from the throttle body.

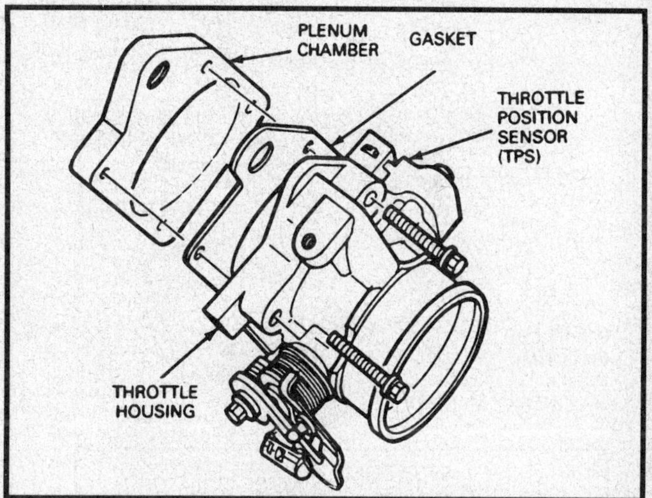

Throttle body with throttle position sensor—Navajo

To install:

5. Position the TP sensor so that the wiring harness is parallel to the venturi bores. Place the TP sensor and gasket on the throttle body, making sure the rotary tangs on the sensor are aligned with the throttle shaft blade. Slide the rotary tangs into position over the throttle shaft blade, then rotate the throttle position sensor clockwise only to its installed position (align the scribe marks made earlier).

NOTE: Slide the rotary tangs into position over the throttle shaft blade, then rotate the throttle position sensor clockwise only its installed position. Failure to install the TP sensor in this manner may result in excessive idle speeds.

6. Once the scribe marks are aligned, install the TP sensor retaining screws and tighten them to 11–16 inch lbs. (1.2–1.8 Nm)
7. Position the electrical connector over the locating dimple, if equipped, then secure to the throttle body with the retaining screw.
8. Inspect the connector terminals for corrosion, damage and bent pins and repair as necessary. Replace any cracked or burned connector terminals. Reconnect the TP sensor electrical connector, start the engine and check the idle speed.

IDLE SPEED ADJUSTMENT

The idle speed is controlled by the EEC processor and is not normally adjusted manually. If the idle speed changes it is indicative of a problem existing in the fuel system. The problems that may cause a change in the idle speed include contamination of the throttle bore, HEGO or idle speed control device. Also if the throttle is sticking or there are vacuum leaks, the idle will be affected. If the engine is not fully warmed, the idle speed will be off spec.

Adjustment

1. Place the transmission in **N** or **PARK**. Apply the parking brake and chock the wheels. The cooling system must be filled and the engine at operating temperature. The heater and accessories must be OFF. Check that the throttle lever is resting on the throttle plate stop screw. The ignition timing should be on spec.
2. Read any engine codes and correct problems. With the engine OFF, disconnect the negative battery terminal for a minimum of 5 minutes, then reconnect it.
3. Start the engine and let it run and stabilize for 2 minutes. Raise the engine RPM and allow the engine to return to idle.
4. If the idle is not correct shut off the engine and disconnect the idle speed control air bypass solenoid.
5. Run the engine at 2500 RPM for 30 seconds and then let engine idle for 2 minutes. Check the idle speed. If the idle speed needs adjustment, turn off the engine and make the estimated adjustment. Start the engine and repeat steps 3, 4 and 5.
6. Shut off the engine and disconnect the negative battery terminal for 5 minutes and reconnect the idle speed control air bypass solenoid. Check that the throttle and linkage moves freely.
7. Start the engine and run at idle for 2 minutes. Blip the throttle and let it return to idle. Lightly depress the throttle and let it return to idle. Check the idle. If problems still exist, check for additional problems.
8. If the engine idle oscillates the throttle plates may be open enough to allow purge flow. Disconnect the carbon canister purge line and plug it. If there is purge flow present, close the throttle plate until oscillations stop. Reconnect the purge line. Check idle speed.

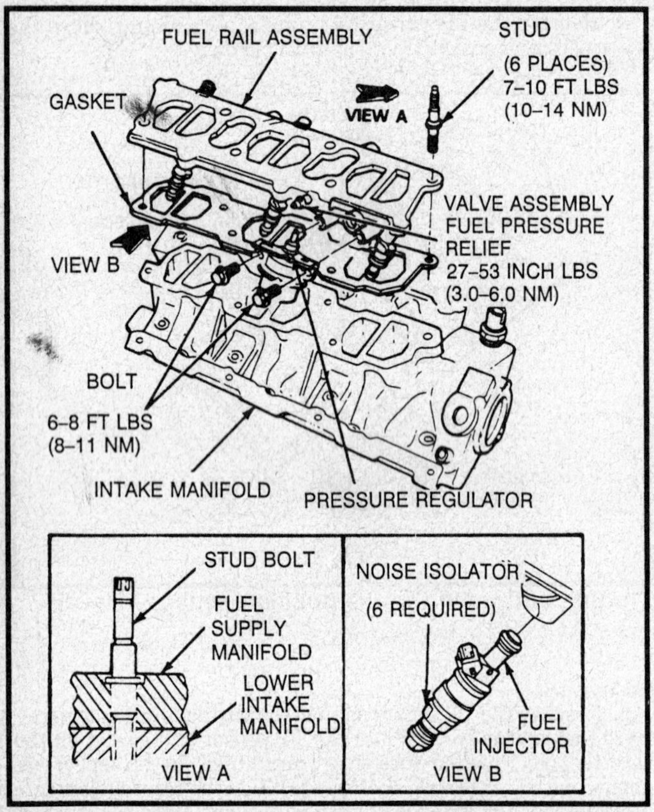

Fitting the fuel rail assembly on the intake manifold – Navajo

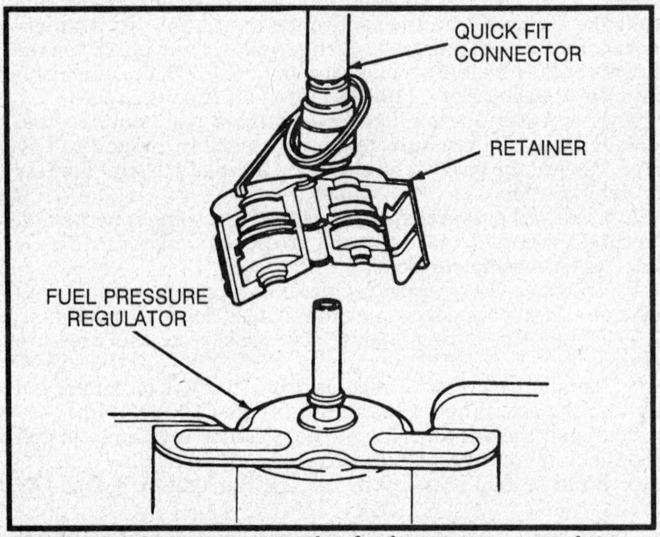

Fit the quick connect on the fuel pressure regulator and snap the retainer into place – Navajo

Mitsubishi Fuel Injection

14 SECTION

MITSUBISHI FUEL INJECTION SYSTEM

General Information

The Electronic Fuel Injection (EFI) system, used on Mitsubishi vehicles, is classified as a Multi-Point Injection (MPI) system. The MPI system controls the fuel flow, idle speed, and ignition timing. The basic function of the MPI system is to control the air/fuel ratio in accordance with all engine operating conditions. An Electronic Control Unit (ECU) is the heart of the MPI system. Based on data from various sensors, the ECU computes the desired air/fuel ratio.

The MPI system can be divided into 3 basic areas:

Fuel System

The fuel is pressure-fed through an in-tank fuel pump assembly and is distributed to the respective injectors via a fuel filter and delivery pipe. The fuel pressure applied to the injectors is at a continuous higher pressure than that of the intake manifold. The fuel pressure is controlled by a pressure regulator. Excess fuel is returned to the fuel tank through the fuel return pipe.

Base on ECU signals, the injectors inject fuel into the cylinder manifold ports. When an electrical current is applied to the injector, the injector valve if fully opened to inject fuel. Since the fuel pressure is a continuous constant pressure, the amount of fuel injected is increased or decreased in proportion to the time the electrical current flows.

Intake System

The flow rate of air drawn through the air cleaner is measured by an Air Flow Sensor (AFS). The air enters the air intake plenum through the throttle body an is distributed to each cylinder manifold. Inside the manifold, the air is mixed with the fuel from the injectors and is drawn into the cylinder. The air flow rate is controlled according to the degree of the throttle valve. In addition to throttle valve opening, air flow rate may be controlled by a stepper motor opening and/or Idle Speed Control (ISC) servo.

Control System

The control system is composed of the sensors, which monitors the engine conditions and the Electronic Control Unit (ECU), which calculates the injection timing and rate according to the signals from the sensors.

The sensors converts engine operating conditions such as intake air temperature, exhaust oxygen contents, air flow rate, coolant temperature and rpm into electrical signals, which is sent to the ECU. The ECU analyzes these signal and calculates the desired fuel to meet the engine operating conditions.

During idling, the ISC servo is driven according to the load to ensure stable idling.

DATA SENSORS

Air Flow Sensor (AFS)

The AFS measures the intake air volume. The ECU uses this intake air volume signal to decide the basic fuel injection duration.

Intake Air Temperature Sensor

The air temperature sensor is a resistor based sensor for detecting the intake air temperature. The ECU provides fuel injection control based on this information. The air temperature sensor is in located on the AFS.

Engine Coolant Temperature Sensor

The coolant temperature sensor is located in the coolant passage of the intake manifold. The ECU uses this signal to determine the base fuel enrichment for cold and warm engine operation.

Barometric Pressure Sensor

The barometric pressure sensor signal is used by the ECU to compute the altitude of the vehicle and so the ECU can correct the ignition timing and air-fuel ratio. The atmosphere pressure sensor is contain in the AFS.

Throttle Position Sensor (TPS)

The TPS is a rotating type variable resistor that rotates with the throttle body shaft to sense the throttle valve opening. Based on TPS voltage signals, the ECU computes the throttle valve opening and accordingly corrects fuel for engine acceleration.

Idle Switch

The idle switch is a contact type switch. The switch is installed at the tip of the ISC. This switch provides the ECU with idle or off idle signal.

Motor Position Sensor (MPS)

The MPS is a variable resistor type sensor and is installed in the ISC servo. The MPS senses the ISC servo plunger position and sends the signal to the ECU. The ECU controls the valve opening, and consequently the idle speed by using the MPS signal, idle signal, engine coolant temperature signal, load signals and vehicle speed sensor.

Cylinder TDC and Crankshaft Angle Sensor

The No. 1 TDC sensor and the crankshaft angle sensor are composed of a disc and unit assembly in the distributor. The No. 1 cylinder TDC is detected by the signal obtained through the single inner slit of the disc. The ECU, based upon this signal, determines the fuel injection cylinder. The crankshaft angle sensor signal comes from the the 4 slits at the outer circumference of the disc serve to detect the position of the crankshaft. The ECU, based on this signal, determines the fuel injection timing, and also calculates the amount of intake air, the ignition timing, etc. for each stroke of engine.

Oxygen Sensor

The oxygen sensor, by detecting the oxygen content in the exhaust gas, maintains the stoichiometric air/fuel ratio. In operation, the ECU receives the signals from the sensor and changes the duration during which fuel is injected. The oxygen sensor is located in the exhaust manifold.

Vehicle Speed Sensor

The vehicle speed sensor uses a reed type switch. The speed sensor is built into the speedometer and converts the speedometer gear revolution into pulse signals, which are sent to the ECU.

Accelerator Pedal Position Sensor

The accelerator pedal position sensor, vehicles equipped with Traction Control (TCL) converts the amount the accelerator pedal is depressed into a voltage and inputs this to the traction control unit. The traction control unit uses this signal for traction control.

Detonation Sensor

The detonation sensor generates a voltage proportional to the magnitude of cylinder block vibration due to knocking and inputs it to the ECU. Based on this signal, the ECU provides retard control of the ignition timing.

Neutral (Inhibitor) Switch

The ECU based on this signal senses when the automatic transaxle is in **N** or **P** and operates the ISC servo to keep the idle speed correct.

Idle Speed Control (ISC) Servo

The ISC servo consists of a motor, worm gear, worm wheel and plunger. The MPS is used to detect plunger position and an idle switch to detect closed throttle position.

As the motor rotates, according to signals from the ECU, the plunger extends or retracts depending on direction of the motor rotation. This actuates the throttle valve via the ISC lever. The idle speed is controlled by changing the throttle open by the ECU through this function.

FUEL DELIVERY SYSTEM

Fuel Pump

The fuel pump is a compact impeller design and is installed inside the fuel tank. The pump assembly consists of the impeller (driven by the motor), the pump casing (which forms the pumping chamber), and cover of the pump. This pump is called a wet-type pump, because the inside is also filled with fuel. Never operate this type of pump when removed from the vehicle or explosion will result, due to fuel fumes, electric sparks and fresh air, which is not available in the fuel tank.

Fuel Injectors

The injectors are solenoid valves. When the solenoid coil is energized, the plunger is retracted. The needle valve that is attached to the plunger is pulled to the full open position.

Diagnosis and Testing

SERVICE PRECAUTIONS

- Do not apply battery power directly to injectors.
- To avoid damage to the ECU, always turn the ignition switch **OFF** before disconnecting the battery cable.
- Do not attempt to disconnect the battery cables with the engine running.
- Do not disconnect wiring connector with the engine running, unless instructed to do so.
- Before removing or installing any part(s), always check the self-diagnosis system.
- After completion of repairs, when the battery was disconnected, operate the engine at idle for approximately 5 minutes in order to stabilize the engine control conditions (Turn all accessories **OFF**).

SELF-DIAGNOSIS SYSTEM

The Engine Control Unit (ECU) is capable of monitoring both the input and output signals which it performed. When the ECU noticed that an irregularity has continued for a specified time or longer, it memorizes a trouble code and outputs the signal to the self-diagnosis output terminal. There are 16 diagnosis items, including the normal state. If more than 1 item is activated, they will be displayed sequentially from the smallest code number to the largest. The trouble codes can be read out with a voltmeter or multi-tester.

Among the self-diagnosis items, a Malfunction Indicator Light (MIL) comes ON to notify the driver of the emission control items when an irregularity is detected.

Generally, the MIL will come ON for 5 seconds when the ignition switch is turned **ON**.

NOTE: Always check that the MIL illuminates for approximately 5 seconds, when the ignition switch is turned ON. If the light does not illuminate, check for an open circuit harness, blown fuse or blown bulb.

READING CODES

Using Voltmeter

1. Connect an analog voltmeter to the self-diagnostic connector, across the MPI diagnosis and ground terminal.
2. Turn the ignition switch **ON**. The ECU diagnostics memory will immediately start.
3. Read the diagnosis output pattern displayed by the voltmeter needle deflections and record it.
4. Compare the readings with the diagnosis code(s) and repair the fault.
5. After repairing the fault, erase the ECU memory.

Using Multi-Unit Tester (MUT)

1. Turn the ignition switch **OFF**.
2. Connect the Multi-Unit Tester (MUT), (Tool 09391–33200 or equivalent), to the diagnosis connector in the fuse box.
3. Connect the power-source terminal of the MUT to the cigarette lighter switch.

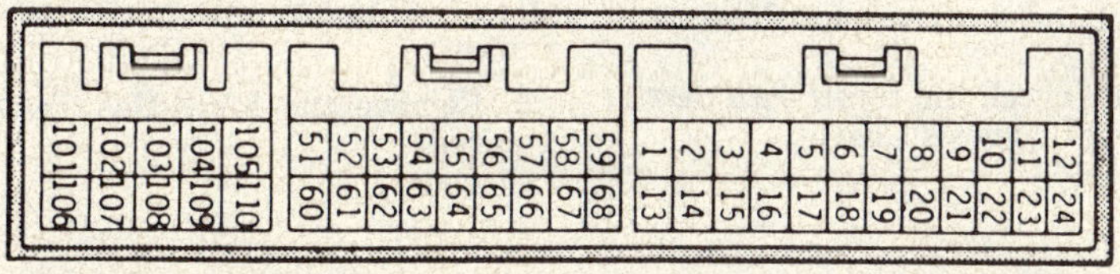

ECU connector terminals identification—1.5L (VIN A), 1.6L (VIN Y), 1.8L (VIN T), 2.0L (VIN R, U and V), 2.4L (VIN W) and Trucks with 3.0L engines

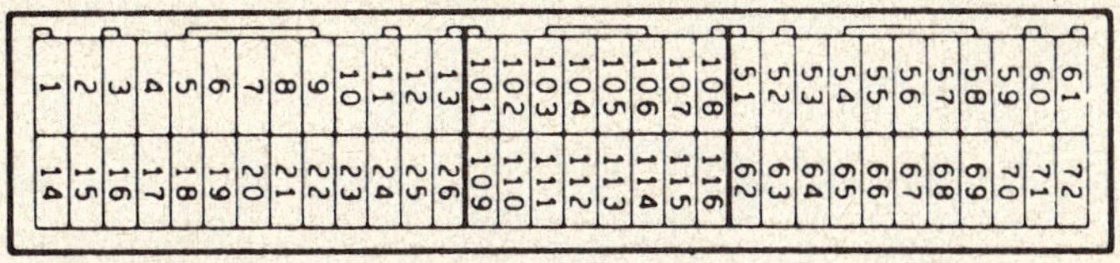

ECU connector terminals identification—1.8L (VIN D) and cars with 3.0L (VIN B, C AND S) engines

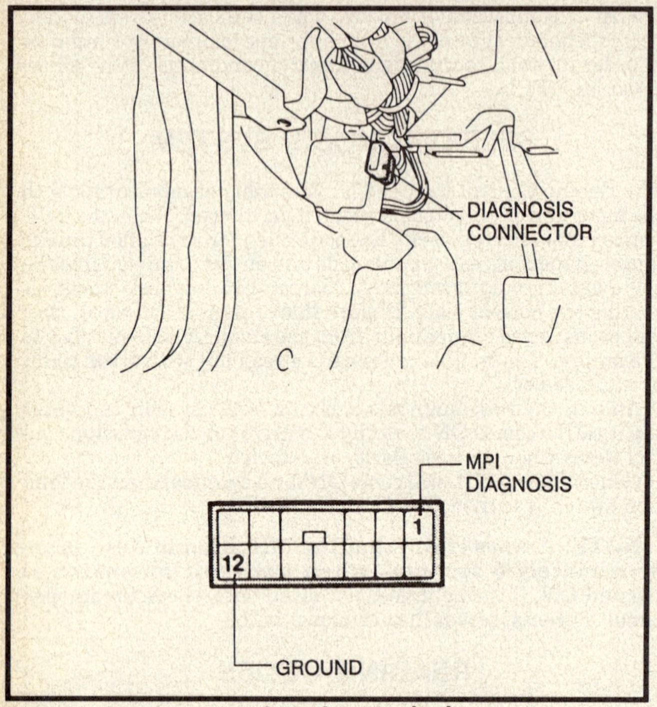

Diagnosis connector location, typical—car

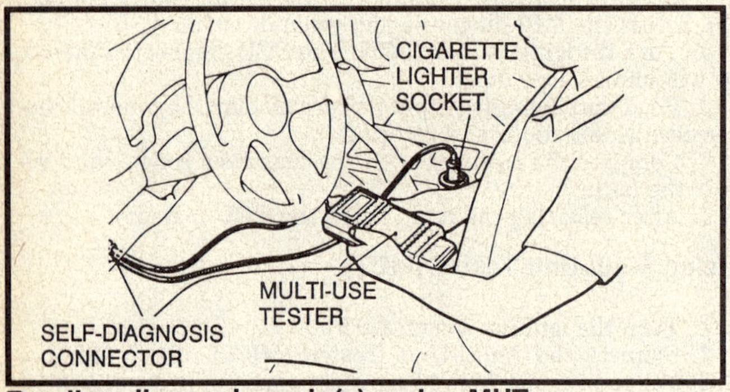

Reading diagnosis code(s), using MUT

4. Turn the ignition switch **ON** and the ECU memory contents will immediately start.
5. Read the self-diagnosis output and check output of correct code.
6. After recording the fault codes, check and repair each part according to the check items in the Diagnosis Charts.

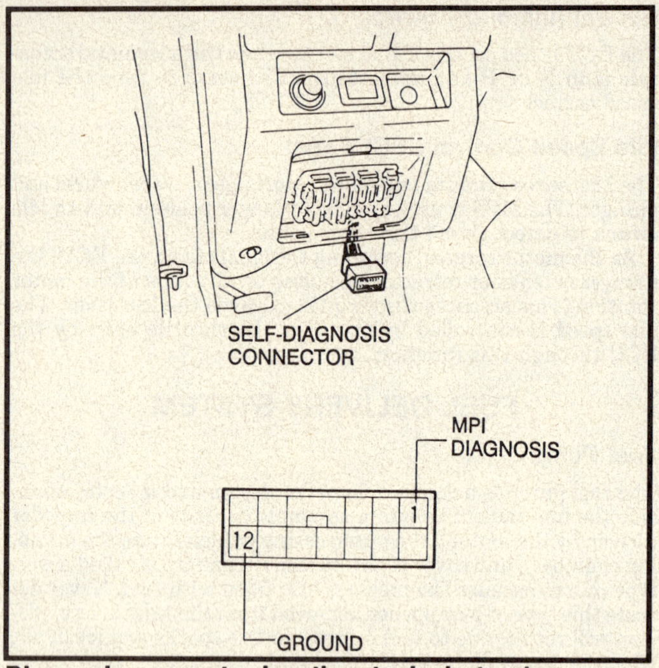

Diagnosis connector location, typical—trucks

NOTE: If the battery voltage is too low, trouble codes cannot be read. If the battery or the ECU connector is disconnected, the diagnosis memory will be erased.

When the ignition key is in the **START** position, the cigarette lighter power is OFF. If it is necessary to perform tests, using the MUT in the crank mode, an auxiliary power supply will be necessary. Use the battery clamp harness provided.

CLEARING CODES

After completion of the repair or correction of the problem:

1. Turn the ignition switch **OFF**.
2. Disconnect the negative battery cable for approximately 10 seconds.
3. Reconnect the negative battery cable.
4. Check that no trouble codes are displayed.

NOTE: It is not necessary to disconnect the battery terminal, when clearing codes, if the MUT is being used. The MUT is capable of clearing the ECU self-diagnosis memory.

5. Remove the Multi-Use Tester (MUT) or voltmeter, as required.

1.5L ENGINE—SELF-DIAGNOSIS MALFUNCTION CODES—MIRAGE

DIAGNOSIS CHART (FAULT TREE)

Output preference order	Diagnosis item	Diagnosis code			Check item (Remedy)
		Output signal pattern	No.	Memory	
1	Engine control unit	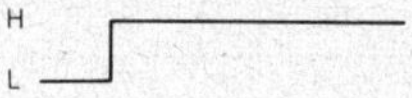	—		(Replace engine control unit)
2	Oxygen sensor	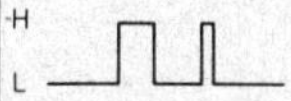	11	Retained	• Harness and connector • Oxygen sensor • Fuel pressure • Injectors (Replace if defective.) • Intake air leaks
3	Air flow sensor	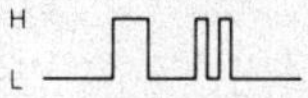	12	Retained	• Harness and connector (If harness and connector are normal, replace air flow sensor assembly.)
4	Intake air temperature sensor		13	Retained	• Harness and connector • Intake air temperature sensor
5	Throttle position sensor	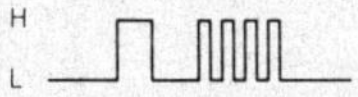	14	Retained	• Harness and connector • Throttle position sensor • Idle position switch
6	Motor position sensor		15	Retained	• Harness and connector • Motor position sensor • Throttle position sensor
7	Engine coolant temperature sensor		21	Retained	• Harness and connector • Engine coolant temperature sensor
8	Crank angle sensor	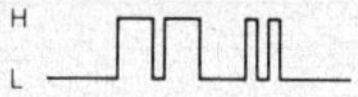	22	Retained	• Harness and connector (If harness and connector are normal, replace distributor assembly.)
9	Top dead center sensor (No. 1 cylinder)	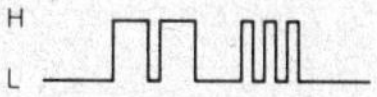	23	Retained	• Harness and connector (If harness and connector are normal, replace distributor assembly.)
10	Vehicle speed sensor (reed switch)	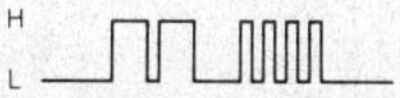	24	Retained	• Harness and connector • Vehicle speed sensor (reed switch)
11	Barometric pressure sensor	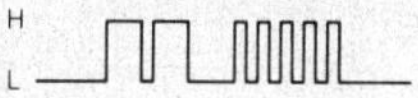	25	Retained	• Harness and connector (If harness and connector are normal, replace barometric pressure sensor assembly.)
12	Ignition timing adjustment signal		36		• Harness and connector
13	Injector		41	Retained	• Harness and connector • Injector coil resistance
14	Fuel pump		42	Retained	• Harness and connector • Control relay
15	EGR <California>		43	Retained	• Harness and connector • EGR temperature sensor • EGR valve • EGR control solenoid valve • EGR valve control vacuum
16	Normal state	 (Continue)			

NOTE
Replace the engine control unit if a malfunction code is output although the inspection reveals that there is no problem with the check items.

Output preference order	Diagnosis item	Trouble code: Output signal pattern	Trouble code: No.	Trouble code: Memory	Check item (Remedy)
1	Electronic control unit		—	—	(Replace electronic control unit)
2	Oxygen sensor	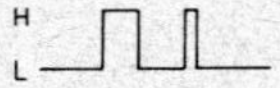	11	Retained	o Harness and connector o Fuel pressure o Injectors (Replace if defective) o Intake air leaks o Oxygen sensor
3	Air flow sensor	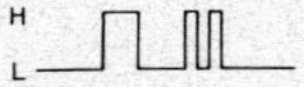	12	Retained	o Harness and connector (If harness and connector are normal, replace air flow sensor assembly)
4	Air temperature sensor	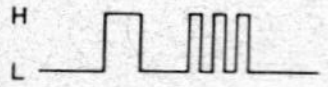	13	Retained	o Harness and connector o Air temperature sensor
5	Throttle position sensor	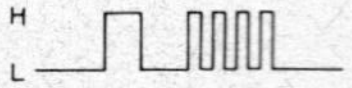	14	Retained	o Harness and connector o Throttle position sensor o Idle position switch
6	Motor position sensor		15	Retained	o Harness and connector o Motor position sensor
7	Engine coolant temperature sensor	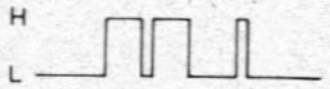	21	Retained	o Harness and connector o Engine coolant temperature sensor
8	Crank angle sensor		22	Retained	o Harness and connector (If harness and connector are normal, replace the distributor assembly.)
9	No.1 cylinder top dead center sensor	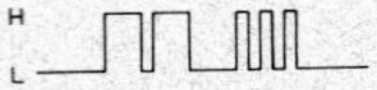	23	Retained	o Harness and connector (If harness and connector are normal, replace the distributor assembly.)
10	Vehicle-speed sensor (reed switch)		24	Retained	o Harness and connector o Vehicle-speed sensor (reed switch)
11	Barometric pressure sensor	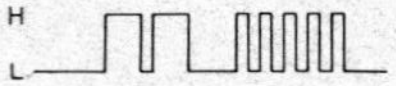	25	Retained	o Harness and connector (If harness and connector are normal, replace the barometric pressure sensor assembly.)
12	Injector		41	Retained	o Harness and connector o Injector coil resistance
13	Fuel pump		42	Retained	o Harness and connector o Control relay
14	EGR*		43	Retained	o Harness and connector o EGR temperature sensor o EGR valve o EGR control solenoid valve o EGR valve control vacuum
15	Normal state	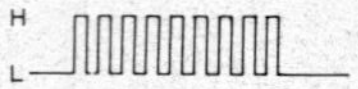	—	—	—

NOTE
1. Replace the ECU only when all other possible causes for a malfunction have been explored.
2. The diagnostic item marked * is applicable to the California vehicles only.

1.6L ENGINE—SELF-DIAGNOSIS MALFUNCTION CODES—MIRAGE

DIAGNOSIS CHART (FAULT TREE)

Output preference order	Diagnosis item	Diagnosis code			Check item (Remedy)
		Output signal pattern	No.	Memory	
1	Engine control unit	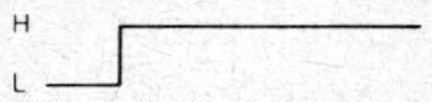	–	–	(Replace engine control unit)
2	Oxygen sensor	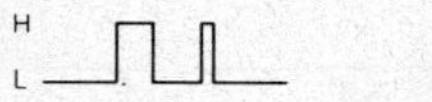	11	Retained	• Harness and connector • Oxygen sensor • Fuel pressure • Injectors (Replace if defective.) • Intake air leaks
3	Air flow sensor	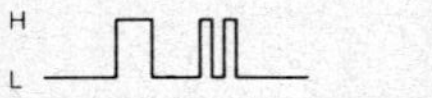	12	Retained	• Harness and connector (If harness and connector are normal, replace air flow sensor assembly.)
4	Intake air temperature sensor	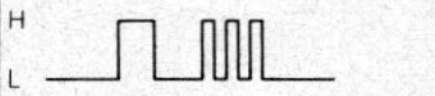	13	Retained	• Harness and connector • Intake air temperature sensor
5	Throttle position sensor	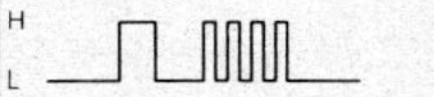	14	Retained	• Harness and connector • Throttle position sensor • Idle position switch
6	Coolant temperature sensor	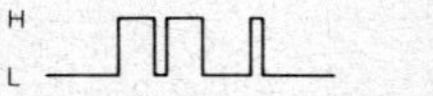	21	Retained	• Harness and connector • Coolant temperature sensor
7	Crank angle sensor	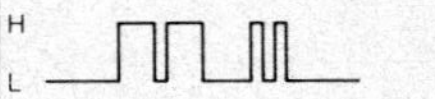	22	Retained	• Harness and connector (If harness and connector are normal, replace crank angle sensor assembly.)
8	Top dead center sensor (No. 1 and No. 4 cylinder sensor)		23	Retained	• Harness and connector (If harness and connector are normal, replace crank angle sensor assembly.)
9	Vehicle speed sensor (reed switch)	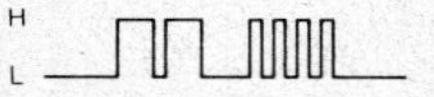	24	Retained	• Harness and connector • Vehicle speed sensor (reed switch)
10	Barometric pressure sensor	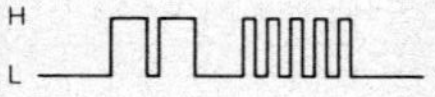	25	Retained	• Harness and connector (If harness and connector are normal, replace barometric pressure sensor assembly.)
11	Injector		41	Retained	• Harness and connector • Injector coil resistance

NOTE
Replace the engine control unit if a malfunction code is output although the inspection reveals that there is no problem with the check items.

Output preference order	Diagnosis item	Diagnosis code			Check item (Remedy)
		Output signal pattern	No.	Memory	
12	Fuel pump	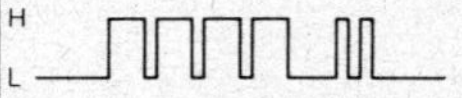	42	Retained	• Harness and connector • Control relay
13	EGR <California>		43	Retained	• Harness and connector • EGR thermo-sensor • EGR valve • EGR control solenoid valve • EGR valve control vacuum
14	Ignition coil, Power transistor unit	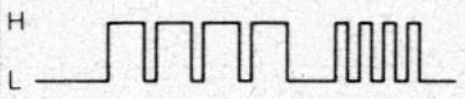	44	Retained	• Harness and connector • Ignition coil • Power transistor unit
15	Normal state	 (Continue)	–	–	–

NOTE
Replace the engine control unit if a malfunction code is output although the inspection reveals that there is no problem with the check items.

1.8L ENGINE—SELF-DIAGNOSIS MALFUNCTION CODES—ECLIPSE

DIAGNOSIS CHART (FAULT TREE)

Output preference order	Diagnosis item	Diagnosis code			Check item (Remedy)
		Output signal pattern	No.	Memory	
1	Engine control unit		–	–	(Replace engine control unit)
2	Oxygen sensor		11	Retained	• Harness and connector • Oxygen sensor • Fuel pressure • Injectors (Replace if defective.) • Intake air leaks
3	Air flow sensor	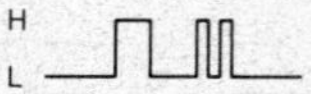	12	Retained	• Harness and connector (If harness and connector are normal, replace air flow sensor assembly.)
4	Intake air temperature sensor		13	Retained	• Harness and connector • Intake air temperature sensor
5	Throttle position sensor		14	Retained	• Harness and connector • Throttle position sensor • Idle position switch
6	Motor position sensor	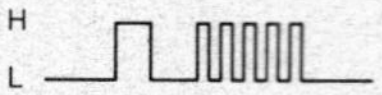	15	Retained	• Harness and connector • Motor position sensor • Throttle position sensor
7	Engine coolant temperature sensor	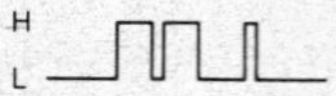	21	Retained	• Harness and connector • Engine coolant temperature sensor
8	Crank angle sensor		22	Retained	• Harness and connector (If harness and connector are normal, replace distributor assembly.)
9	Top dead center sensor (No. 1 cylinder)		23	Retained	• Harness and connector (If harness and connector are normal, replace distributor assembly.)
10	Vehicle speed sensor (reed switch)		24	Retained	• Harness and connector • Vehicle speed sensor (reed switch)
11	Barometric pressure sensor	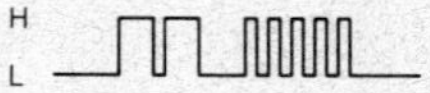	25	Retained	• Harness and connector (If harness and connector are normal, replace barometric pressure sensor assembly.)
12	Ignition timing adjustment signal	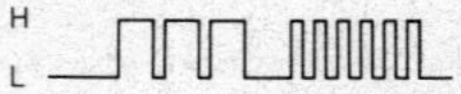	36	–	• Harness and connector
13	Injector		41	Retained	• Harness and connector • Injector coil resistance
14	Fuel pump		42	Retained	• Harness and connector • Control relay
15	EGR <California>		43	Retained	• Harness and connector • EGR temperature sensor • EGR valve • EGR control solenoid valve • EGR valve control vacuum
16	Normal state		–	–	–

NOTE
1. Replace the engine control unit if a malfunction code is output although the inspection reveals that there is no problem with the check items.

2.0L ENGINE—SELF-DIAGNOSIS MALFUNCTION CODES—ECLIPSE

DIAGNOSIS CHART (FAULT FREE)

Output preference order	Diagnosis item	Diagnosis code: Output signal pattern	Diagnosis code: No.	Diagnosis code: Memory	Check item (Remedy)
1	Engine control unit		–	–	(Replace engine control unit)
2	Oxygen sensor		11	Retained	• Harness and connector • Oxygen sensor • Fuel pressure • Injectors (Replace if defective.) • Intake air leaks
3	Air flow sensor		12	Retained	• Harness and connector (If harness and connector are normal, replace air flow sensor assembly.)
4	Intake air temperature sensor	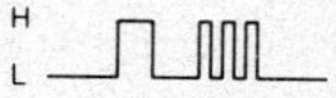	13	Retained	• Harness and connector • Intake air temperature sensor
5	Throttle position sensor	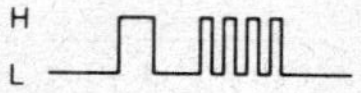	14	Retained	• Harness and connector • Throttle position sensor • Idle position switch
6	Engine coolant temperature sensor		21	Retained	• Harness and connector • Coolant temperature sensor
7	Crank angle sensor		22	Retained	• Harness and connector (If harness and connector are normal, replace crank angle sensor assembly.)
8	Top dead center sensor (No. 1 and No. 4 cylinder)	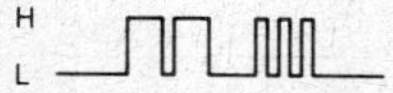	23	Retained	• Harness and connector (If harness and connector are normal, replace crank angle sensor assembly.)

Output preference order	Diagnosis item	Diagnosis code: Output signal pattern	Diagnosis code: No.	Diagnosis code: Memory	Check item (Remedy)
9	Vehicle speed sensor (reed switch)		24	Retained	• Harness and connector • Vehicle speed sensor (reed switch)
10	Barometric pressure sensor	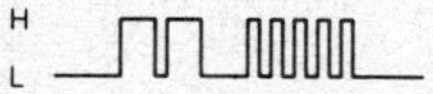	25	Retained	• Harness and connector (If harness and connector are normal, replace barometric pressure sensor assembly.)
11	Detonation sensor <Turbo>		31	Retained	• Harness and connector (If harness and connector are normal, replace detonation sensor.)
12	Injector		41	Retained	• Harness and connector • Injector coil resistance
13	Fuel pump	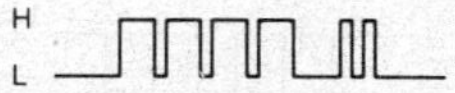	42	Retained	• Harness and connector • Control relay
14	EGR <California>	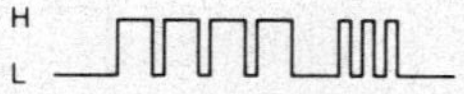	43	Retained	• Harness and connector • EGR temperature sensor • EGR valve • EGR control solenoid valve • EGR valve control vacuum
15	Ignition coil	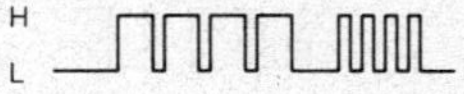	44	Retained	• Harness and connector • Ignition coil • Power transistor
16	Normal state		–	–	–

NOTE
1. Replace the engine control unit if a malfunction code is output although the inspection reveals that there is no problem with the check items.

DIAGNOSIS CHART (FAULT TREE)

Output preference order	Diagnosis item	Diagnosis code: Output signal pattern	Diagnosis code: No.	Diagnosis code: Memory	Check item (Remedy)
1	Engine control unit		–	–	(Replace engine control unit)
2	Oxygen sensor	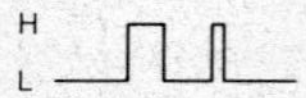	11	Retained	• Harness and connector • Oxygen sensor • Fuel pressure • Injectors (Replace if defective.) • Intake air leaks
3	Air flow sensor	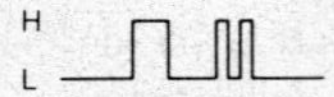	12	Retained	• Harness and connector (If harness and connector are normal, replace air flow sensor assembly.)
4	Intake air temperature sensor	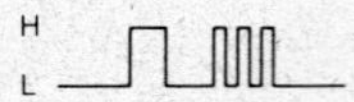	13	Retained	• Harness and connector • Intake air temperature sensor
5	Throttle position sensor	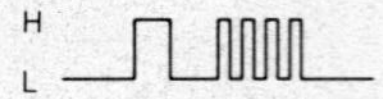	14	Retained	• Harness and connector • Throttle position sensor • Idle position switch
6	Motor position sensor		15	Retained	• Harness and connector • Motor position sensor • Throttle position sensor
7	Engine coolant temperature sensor	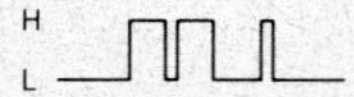	21	Retained	• Harness and connector • Engine coolant temperature sensor
8	Crank angle sensor	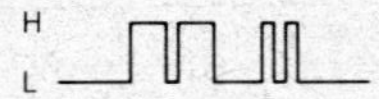	22	Retained	• Harness and connector (If harness and connector are normal, replace distributor assembly.)
9	Top dead center sensor (No.1 cylinder)	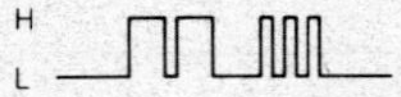	23	Retained	• Harness and connector (If harness and connector are normal, replace distributor assembly.)
10	Vehicle speed sensor (reed switch)	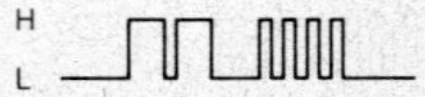	24	Retained	• Harness and connector • Vehicle speed sensor (reed switch)
11	Barometric pressure sensor	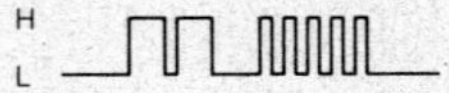	25	Retained	• Harness and connector (If harness and connector are normal, replace barometric pressure sensor assembly.)
12	Injector	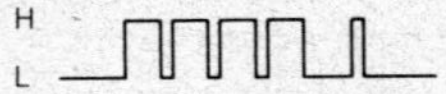	41	Retained	• Harness and connector • Injector coil resistance
13	Fuel pump		42	Retained	• Harness and connector • Control relay
14	EGR <California>		43	Retained	• Harness and connector • EGR temperature sensor • EGR valve • EGR control solenoid valve • EGR valve control vacuum
15	Normal state		–	–	–

NOTE
Replace the engine control unit if a malfunction code is output although the inspection reveals that there is no problem with the check items.

2.4L ENGINE – SELF-DIAGNOSIS MALFUNCTION CODES – EXPO/LRV

DIAGNOSIS CHART (FAULT TREE)

Output preference order	Diagnosis item	Malfunction code			Check items (Remedy)
		Output signal pattern	No.	Memory	
1	Engine control unit	H L	–	–	(Replace engine control unit)
2	Oxygen sensor	H L	11	Retained	• Harness and connector • Fuel pressure • Injectors (Replace if defective.) • Intake air leaks • Oxygen sensor
3	Air flow sensor	H L	12	Retained	• Harness and connector (If harness and connector are normal, rplace air flow sensor assembly)
4	Intake air temperature sensor	H L	13	Retained	• Harness and connector • Intake air temperature sensor
5	Throttle position sensor	H L	14	Retained	• Harness and connector • Throttle position sensor • Idle position switch
6	Engine coolant temperature sensor	H L	21	Retained	• Harness and connector • Engine coolant temperature sensor
7	Crank angle sensor	H L	22	Retained	• Harness and connector (If harness and connector are normal, replace distributor assembly.)
8	No.1 cylinder top dead center sensor	H L	23	Retained	• Harness and connector (If harness and connector are normal, replace distributor assembly.)
9	Vehicle speed sensor (reed switch)	H L	24	Retained	• Harness and connector • Vehicle speed sensor (reed switch)
10	Barometric pressure sensor	H L	25	Retained	• Harness and connector (If harness and connector are normal, replace barometric pressure sensor assembly.)
11	Ignition timing adjustment signal	H L	36	–	• Harness and connector

Output preference order	Diagnosis item	Malfunction code			Check items (Remedy)
		Output signal pattern	No.	Memory	
12	Injector	H L	41	Retained	• Harness and connector • Injector coil resistance
13	Fuel pump	H L	42	Retained	• Harness and connector • Control relay
14	EGR <California>	H L	43	Retained	• Harness and connector • EGR valve • EGR control solenoid valve • EGR valve control vacuum • EGR temperature sensor
15	Servo valve position sensor	H L	55	Retained	• Harness and connector • Servo valve movement (If harness, connector and servo valve movement are nomal, replace Idle speed control servo assembly)
16	Normal state	H L	–	–	–

NOTE
1. Replace the engine control unit if a malfunction code is output although the inspection reveals that there is no problem with the check items.
2. The code numbers will be displayed in order, starting from the lowest.

PROBLEM DIAGNOSIS CONTENT CHART

Malfunction code No.	Diagnosis item	Diagnosis contents	Probable cause	Remark (Trouble symptom, etc.)
–	Engine control unit	Abnormality in engine control unit	–	• Engine stalls • Starting is impossible
11	Oxygen sensor	Oxygen sensor signal voltage does not change (lean/rich) even when air/fuel ratio feedback control (close loop control) operates.	(1) Oxygen sensor malfunction (2) Open circuit or short circuit in oxygen sensor, or connector contact is defective	• Reduction in exhaust gas purification efficiency *
–			(3) Inappropriate fuel pressure (4) Injector malfunction (5) Air leaking in through clearance in gasket (6) Engine control unit malfunction	• Reduction in exhaust gas purification * • Poor starting • Unstable idling • Poor acceleration
12	Air flow sensor	Frequency of air flow sensor signal is 10Hz or less, even though engine is running.	(1) Air flow sensor malfunction (2) Open circuit or short circuit in air flow sensor, or connector contact is defective. (3) Engine control unit malfunction.	• Poor acceleration * • Inappropriate idle speed * • Unstable idling *
13	Intake air temperature sensor	(1) Voltage of intake air temperature sensor signal is 4.5V or more. (2) Voltage of intake air temperature sensor signal is 0.27V or less.	(1) Intake air temperature sensor malfunction (2) Open circuit or short circuit in the intake air temperature sensor, or connector contact is defective (3) Engine control unit malfunction	• Slightly poor driveability * • At high temperatures: (a) Poor starting * (b) Unstable idling *
14	Throttle position sensor	(1) Voltage of throttle position sensor signal is 0.2V or less. (2) Voltage of throttle position sensor signal is 2V or more, even though idle position switch is ON.	(1) Throttle position sensor malfunction, or adjustment is defective (2) Open circuit or short circuit in throttle position sensor, or connector contact is defective.	• Slightly poor acceleration <M/T> • Poor driveability <A/T> • Engine stalls
			(3) Idle position switch ON malfunction (4) Short circuit in idle position switch signal wire (5) Engine control unit malfunction	• Engine stalls • Racing is impossible

NOTE
*: Fail-Safe/backup function is operating.

Malfunction code No.	Diagnosis item	Diagnosis contents	Probable cause	Remark (Trouble symptom, etc.)
21	Engine coolant temperature sensor	(1) Voltage of engine coolant temperature sensor signal is 4.6V or more. (2) Voltage of engine coolant temperature sensor signal is 0.11V or less. (3) While engine is warming up, engine coolant temperature sensor signal shows a drop in engine coolant temperature.	(1) Engine coolant temperature malfunction (2) Open circuit or short circuit in engine coolant temperature sensor, or connector contact is defective (3) Engine control unit malfunction	When engine is cold: • Poor starting * • Unstable idling * • Poor acceleration *
22	Crank angle sensor	Voltage of crank angle sensor signal does not change (high/low), even though engine has been cranking for 4 seconds or more.	(1) Crank angle sensor malfunction (2) Open circuit or short circuit in crank angle sensor, or connector contact is defective (3) Engine control unit malfunction	• Engine stalls • Starting is impossible
23	Top dead center sensor	Voltage of top dead center sensor signal does not change (high/low), even though engine is running.	(1) Top dead center sensor malfunction (2) Open circuit or short circuit in top dead center sensor, or connector contact is defective. (3) Engine control unit malfunction	• Unstable idling * • Poor acceleration *
24	Vehicle speed sensor (reed switch)	Voltage of vehicle speed sensor does not change (high/low) even though vehicle is accelerating at an engine speed of 3000 rpm or more.	(1) Vehicle speed sensor malfunction (2) Open circuit or short circuit in vehicle speed sensor, or connector contact is defective (3) Engine control unit malfunction	Engine stalls when vehicle stops after decelerating
25	Barometric pressure sensor	(1) Voltage of barometric pressure sensor signal is 4.5V or more. (2) Voltage of barometric pressure sensor signal is 0.2V or less	(1) Barometric pressure sensor malfunction (2) Open circuit or short circuit in barometric pressure sensor, or connector contact is defective (3) Engine control unit malfunction	• Poor starting * • Unstable idling * • Poor acceleration *
36	Ignition timing adjustment signal	Ignition timing adjustment signal wire is short-circuited to earth.	(1) Ignition timing adjustment signal wire is short-circuited to earth (2) Engine control unit malfunction	• Poor acceleration • Engine overheats

2.4L ENGINE
MALFUNCTION CODES DIAGNOSIS CONTINUED
EXPO/LRV

Mulfunction code No.	Diagnosis item	Diagnosis contents	Probable cause	Remark (Trouble symptom, etc.)
41	Injector	Injectors do not operate for a continuous 4 second period while engine is cranking or idling.	(1) Injection malfunction (2) Open circuit or short circuit in injector, or connector contact is defective (3) Engine control unit malfunction.	• Poor starting • Unstable idling • Poor acceleration
42	Fuel pump	Though engine is revolving, fuel pump is not supplied with power to drive motor.	(1) Fuel pump power circuit is broken, short-circuited or poor contact of connector. (2) Control relay is defective. (3) Engine control unit malfunction.	• Incapable to start • Engine stall
43	EGR <California>	During driving after engine warm-up; (1) Volume of EGR is scarce. (EGR temperature sensor signal voltage is too hight.) (2) EGR temperature sensor signal voltage is below 0.1V.	(1) EGR valve does not open. (2) EGR valve control negative pressure is too low. (3) EGR control solenoid valve is defective. (4) EGR temperature sensor is defective. (5) EGR temperature sensor circuit is broken, short-circuited or poor contact of connector. (6) Engine control unit malfunction.	• Reduction is exhaust gas purfication efficiency
55	Servo valve position sensor	Servo valve does not move to the intended position (opening angle), even though idle speed control servo motor operates many times.	(1) Servo valve position sensor malfunction (2) Open circuit or short circuit in servo valve position sensor, or connector contact is defective (3) Idle speed control servo motor (DC motor) malfunction (4) Open circuit or short circuit in idle speed control servo motor (DC motor), or connector contact is defective (5) Engine control unit malfunction.	• Inappropriate idle speed * • Engine stops * • Unstable idling *

2.4L ENGINE
SELF-DIAGNOSIS MALFUNCTION CODES
TRUCK

Output preference order	Diagnosis item	Diagnosis code			Check item (Remedy)
		Output signal pattern	No.	Memory	
1	Engine control unit	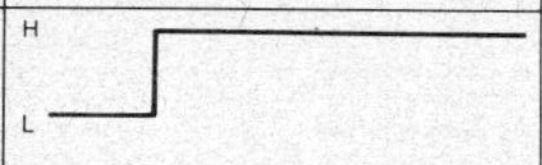	–	–	(Replace engine control unit)
2	Oxygen sensor	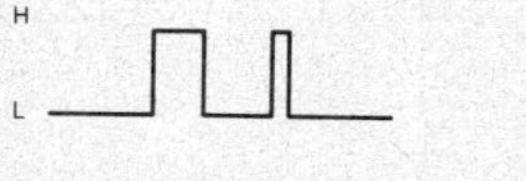	11	Retained	• Harness and connector • Oxygen sensor • Fuel pressure • Injectors (Replace if defective.) • Intake air leaks
3	Air flow sensor		12	Retained	• Harness and connector (If harness and connector are normal, replace air flow sensor assembly.)
4	Intake air temperature sensor		13	Retained	• Harness and connector • Intake air temperature sensor
5	Throttle position sensor		14	Retained	• Harness and connector • Throttle position sensor • Idle position switch
6	Motor position sensor		15	Retained	• Harness and connector • Motor position sensor • Throttle position sensor
7	Engine coolant temperature sensor		21	Retained	• Harness and connector • Engine coolant temperature sensor
8	Crank angle sensor	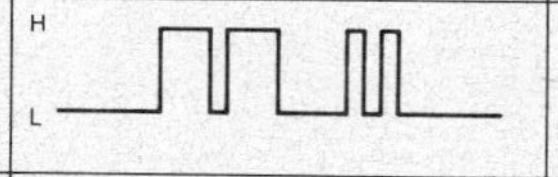	22	Retained	• Harness and connector (If harness and connector are normal, replace distributor assembly.)
9	Top dead center sensor (No. 1 cylinder)	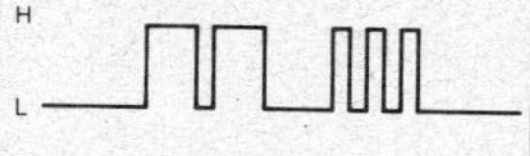	23	Retained	• Harness and connector (If harness and connector are normal, replace distributor assembly.)
10	Vehicle speed sensor (reed switch)	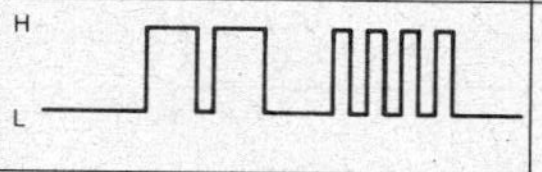	24	Retained	• Harness and connector • Vehicle speed sensor (reed switch)
11	Barometric pressure sensor	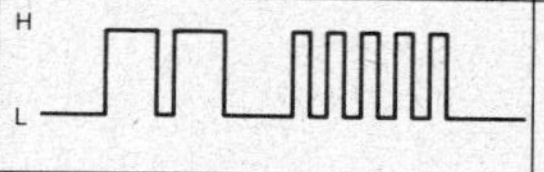	25	Retained	• Harness and connector (If harness and connector are normal, replace barometric pressure sensor assembly.)
12	Ignition timing adjustment signal	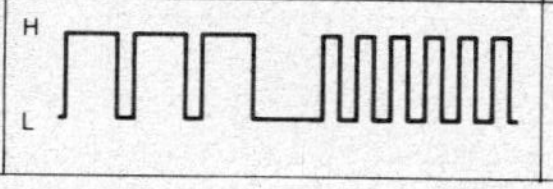	36	–	• Harness and connector

2.4L ENGINE SELF-DIAGNOSIS MALFUNCTION CODES CONTINUED TRUCK

Output preference order	Diagnosis item	Diagnosis code			Check item (Remedy)
		Output signal pattern	No.	Memory	
13	Injector	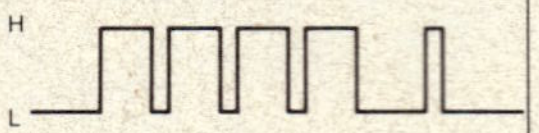	41	Retained	• Harness and connector • Injector coil resistance
14	Fuel pump	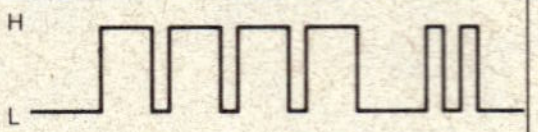	42	Retained	• Harness and connector • Control relay
15	EGR <California>		43	Retained	• Harness and connector • EGR temperature sensor • EGR valve • EGR control solenoid valve • EGR valve control vacuum
16	Normal state		–	–	–

NOTE
Replace the engine control unit if a malfunction code is output although the inspection reveals that there is no problem with the check items.

2.4L ENGINE MALFUNCTION CODES DIAGNOSIS TRUCK

DIAGNOSIS CHART

NOTE
*: The fail safe/back up functions are operating.

Malfunction code No.	Diagnosis item	Trouble symptom	Main cause	Note (malfunction symptom,etc.)
–	Engine control unit	Malfunction of engine control unit itself	–	• Engine stall • No starting
11	Oxygen sensor	The oxygen sensor signal voltage does not change(lean/rich), even though the air fuel ratio feed-back control(closed-loop control) is being carried out	(1) Malfunction of oxygen sensor (2) Open-circuit, shortcircuit of oxygen sensor circuit or poor connection of connector	• Failure of exhaust gas purification function*
			(3) Inadequate fuel pressure (4) Malfunction of injector (5) Air absorption from gasket space (6) Malfunction of engine control unit	• Failure of exhaust gas purification function* • Poor starting • Unstable idling • Poor acceleration
12	Air flow sensor	The air flow sensor signal frequency is less than 10Hz,even though the engine is operating	(1) Malfunction of air flow sensor (2) Open-circuit, shortcircuit of air flow-sensor circuit or poor connection of connector (3) Malfunction of engine control unit	• Poor acceleration* • Unstable idling*
13	Intake air temperature sensor	(1) The intake air temperature sensor signal voltage is more than 4.5V. (2) The intake air temperature sensor signal voltage is less than 0.27V.	(1) Malfunction of intake air temperature sensor (2) Open-circuit, shortcircuit of intake air temperature sensor circuit or poor connection of connector (3) Malfunction of engine control unit	• Slightly poor drivability* • In high temperature: (a) Poor starting* (b) Unstable idling*
14	Throttle position sensor	(1) The throttle position sensor signal voltage is less than 0.2V. (2) The throttle position sensor signal voltage is more than 2V, even though the idle position switch is "ON"	(1) Malfunction or maladjustment of throttle position sensor (2) Open-circuit, shortcircuit of throttle position sensor circuit or poor connection of connrctor	• Slightly poor acceleration<M/T> • Poor drivability<A/T> • Engine stall
			(3) Malfunction at idle position switch "ON" (4) Shortcircuit of idle position switch signal line (5) Malfunction of engine control unit	• Engine stall • No racing
15	Motor position sensor	(1) The motor position sensor signal voltage is more than 4.8V. (2) The motor position sensor signal voltage is less than 0.2V. (3) During idling, the motor position sensor signal voltage is too low for the opening of the throttle valve.	(1) Malfunction of motor position sensor (2) Open-circuit, shortcircuit of motor position sensor circuit or poor connection of connector (3) Malfunction ormaladjustment of throttle position sensor (4) Open-circuit, shortcircuit of throttle position sensor circuit or poor connection of connector (5) Malfunction of engine control unit	• Engine stall*

2.4L ENGINE—MALFUNCTION CODES DIAGNOSIS CONTINUED—TRUCK

Malfunction code No.	Diagnosis item	Trouble symptom	Main cause	Note (malfunction symptom,etc.)
21	Engine coolant temperature sensor	(1) The engine coolant temperature sensor signal voltage is more than 4.6V (2) The engine coolant temperature sensor signal voltage is less than 0.11V (3) During the engine warming-up, the engine coolant temperature sensor signal has indicated dropping of engine cooling temperature	(1) Malfunction of engine coolant temperature sensor (2) Open circuit, shortcircuit of engine coolant temperature sensor circuit or poor connection of connector (3) Malfunction of engine control unit	In cold engine condition; • No starting* • Unstable idling* • Poor acceleration*
22	Crank angle sensor	The crank angle sensor signal voltage does not change(High/Low), even though the engine is cranking more than 4 seconds	(1) Malfunction of crank angle sensor (2) Open circuit, shortcircuit of crank angle sensor circuit or poor connection of connector (3) Malfunction of engine control unit	• Engine stall • No starting
23	Top dead center sensor	The top dead center sensor signal voltage does not change(High/Low), even thought the engine is operating.	(1) Malfunction of top dead center sensor (2) Open circuit, shortcircuit of top dead center senser circuit or poor connection of connector (3) Malfunction of engine control unit	• Unstable idling* • Poor acceleration*
24	Vehicle speed sensor (reed switch)	The vehicle speed sensor signal voltage does not change(High/Low), when the vehicle is accelerated with engine speed more than 3000rpm.	(1) Malfunction of vehicle speed sensor (2) Open circuit, shortcircuit of vehicle speed sensor or poor connection of connector (3) Malfunction of engine control unit	Engine may stall when the vehicle is stopped from reduced speed driving.
25	Barometric pressure sensor	(1) The barometric pressure sensor signal voltage is more than 4.5V. (2) The air pressure sensor signal voltage is less than 0.2V.	(1) Malfunction of barometric pressure sensor (2) Open circuit, shortcircuit of barometric pressure sensor or poor connection of connector (3) Malfunction of engine control unit	• Unstable idling* • Poor acceleration* • Poor starting*

Malfunction code No.	Diagnosis item	Trouble symptom	Main cause	Note (malfunction symptom,etc.)
36	Ignition timing adjustment signal	The ignition timing adjustment signal line to ground is shortcircuited.	(1) Shortcircuit of ignition timing adjustment signal line to ground (2) Malfunction of engine control unit	• Poor acceleration • Engine overheat
41	Injector	During engine cranking or idling, the injector does not continue to drive more than 4 seconds.	(1) Malfunction of injector (2) Open circuit, shortcircuit of injector circuit or poor connection of connector (3) Malfunction of engine control unit	• Unstable idling • Poor acceleration • Poor starting
42	Fuel pump	The fuel pump is not provided with motor driving power, even though the engine is operating.	(1) Open-circuit, shortcircuit of fuel pump power circuit or poor connection of connector (2) Malfunction of control relay (3) Malfunction of engine control unit	• No starting • Engine stall
43	EGR <California>	During operation after engine warm-up; (1) The EGR volume is scarce. (EGR temperature sensorsignal voltage is too high.) (2) The EGR temperature sensor signal voltage is less than 0.1V.	(1) EGR valve does not open. (2) EGR valve control negative pressure is too low. (3) Malfunction of EGR control solenoid valve (4) Malfunction of EGR temperature sensor (5) Open-circuit, shortcircuit of EGR temperature sensor circuit or poor connection of connector (6) Malfunction of engine control unit	• Failure of exhaust gas purification function

DIAGNOSIS CHART (FAULT TREE)

Output preference order	Diagnosis item	Diagnosis code			Check item (Remedy)
		Output signal pattern	No.	Memory	
1	Engine control unit		—	—	(Replace engine control unit)
2	Oxygen sensor		11	Retained	• Harness and connector • Oxygen sensor • Fuel pressure • Injectors (Replace if defective.) • Intake air leaks
3	Air flow sensor		12	Retained	• Harness and connector (If harness and connector are normal, replace air flow sensor assembly.)
4	Intake air temperature sensor		13	Retained	• Harness and connector • Intake air temperature sensor
5	Throttle position sensor		14	Retained	• Harness and connector • Throttle position sensor • Idle position switch
6	Engine coolant temperature sensor		21	Retained	• Harness and connector • Engine coolant temperature sensor
7	Crank angle sensor		22	Retained	• Harness and connector (If harness and connector are normal, replace distributor assembly.)
8	Top dead center sensor	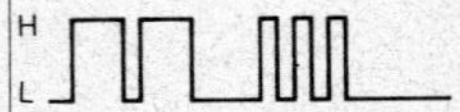	23	Retained	• Harness and connector (If harness and connector are normal, replace distributor assembly.)
9	Vehicle speed sensor (reed switch)		24	Retained	• Harness and connector • Vehicle speed sensor (reed switch)
10	Barometric pressure sensor		25	Retained	• Harness and connector (If harness and connector are normal, replace barometric pressure sensor assembly.)
11	Injector		41	Retained	• Harness and connector • Injector coil resistance
12	Fuel pump		42	Retained	• Harness and connector • Control relay
13	Normal state		—	—	—

NOTE
Replace the engine control unit if a malfunction code is output although the inspection reveals that there is no problem with the check items.

3.0L ENGINE—SELF-DIAGNOSIS MALFUNCTION CODES—DIAMANTE

DIAGNOSIS CHART (FAULT TREE)

Output preference order	Diagnosis item	Diagnosis code: Output signal pattern	Diagnosis code: No.	Diagnosis code: Memory	Check item (Remedy)
1	Engine control unit		–	–	(Replace engine control unit)
2	Oxygen sensor		11	Retained	• Harness and connector • Oxygen sensor • Fuel pressure • Injectors (Replace if defective.) • Intake air leaks
3	Air flow sensor		12	Retained	• Harness and connector (If harness and connector are normal, replace air flow sensor assembly.)
4	Intake air temperature sensor	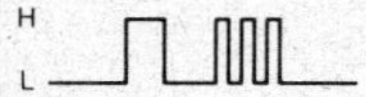	13	Retained	• Harness and connector • Intake air temperature sensor
5	Throttle position sensor		14	Retained	• Harness and connector • Throttle position sensor • Idle position switch
6	Coolant temperature sensor	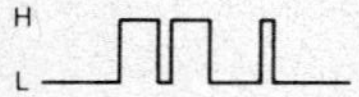	21	Retained	• Harness and connector • Coolant temperature sensor
7	Crank angle sensor	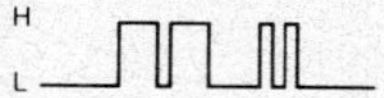	22	Retained	• Harness and connector (If harness and connector are normal, replace crank angle sensor or distributor assembly.)
8	Top dead center sensor	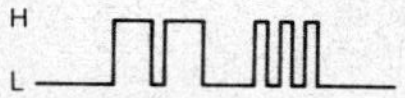	23	Retained	• Harness and connector (If harness and connector are normal, replace crank angle sensor or distributor assembly.)
9	Vehicle speed sensor (reed switch)	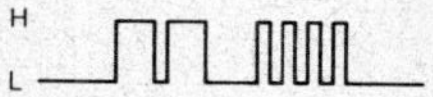	24	Retained	• Harness and connector • Vehicle speed sensor (reed switch)
10	Barometric pressure sensor	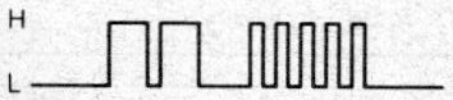	25	Retained	• Harness and connector (If harness and connector are normal, replace barometric pressure sensor assembly.)
11	Detonation sensor	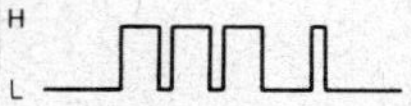	31	Retained	• Harness and connector (If harness and connector are normal, replace detonation sensor.)

NOTE
Replace the engine control unit if a malfunction code is output although the inspection reveals that there is no problem with the check items.

Output preference order	Diagnosis item	Diagnosis code: Output signal pattern	Diagnosis code: No.	Diagnosis code: Memory	Check item (Remedy)
12	Ignition timing adjustment signal	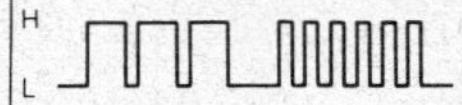	36	–	• Harness and connector
13	Injector		41	Retained	• Harness and connector • Injector coil resistance
14	EGR <California>	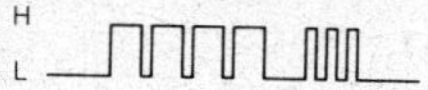	43	Retained	• Harness and connector • EGR temperature sensor • EGR valve • EGR control solenoid valve • EGR valve control vacuum
15	Ignition coil, Power transistor unit (No. 1–4 cylinder)	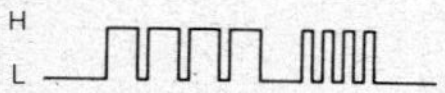	44	Retained	• Harness and connector • Ignition coil • Power transistor unit
16	Ignition coil, Power transistor unit (No. 2–5 cylinder)		52	Retained	• Harness and connector • Ignition coil • Power transistor unit
17	Ignition coil, Power transistor unit (No. 3–6 cylinder)	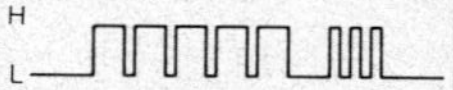	53	Retained	• Harness and connector • Ignition coil • Power transistor unit
18	A cable from transaxle control unit, for transmission of torque reduction signal <DOHC>	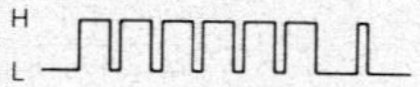	61	Retained	• Harness and connector (If harness and connector are normal, replace only transaxle control unit.)
19	Induction control valve position sensor	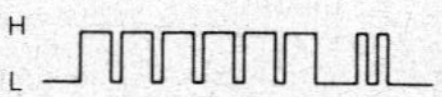	62	Retained	• Harness and connector (If harness, connector and induction control valve are normal, replace air intake plenum assembly.)
20	Vacuum control solenoid valve <Vehicles with TCL>		71	Retained	• Harness and connector • Vacuum solenoid valve
21	Ventilation control solenoid valve <Vehicles with TCL>		72	Retained	• Harness and connector • Ventilation solenoid valve
22	Normal state	 (Continue)	–	–	–

NOTE
Replace the engine control unit if a malfunction code is output although the inspection reveals that there is no problem with the check items.

TROUBLESHOOTING TABLE

NOTE
*: The failsafe/back-up function is in operation.

Trouble code no.	Diagnosis Item	Diagnosis content	Major cause	Remarks (troube phenomenon, etc.)
–	Engine control unit	The engine control unit itself is abnormal.	–	• Engine stop • Impossible start
11	Oxygen sensor	Though the air/fuel mixture ratio feed-back control (closed loop control) is operated, the signal voltage of the oxygen sensor does not vary (to be lean/rich).	(1) The oxygen sensor is troubled. (2) Open-circuit, short-circuit or improper connector contact occurs in the oxygen sensor circuit.	• Deterioration of exhaust gas purifying performance*
			(3) The fuel pressure is improper. (4) The injector is troubled. (5) Air is sucked through the gasket clearance, etc. (6) The engine control unit is troubled.	• Deterioration of exhaust gas purifying performance* • Abnormal start • Unstable idling • Abnormal acceleration
12	Air flow sensor	Though the engine is running, the signal frequency of the air flow sensor is 10Hz or less.	(1) Air flow sensor is troubled. (2) Open-circuit, short-circuit or improper connector contact occurs in the air flow sensor circuit. (3) The engine control unit is troubled.	• Abnormal acceleration* • Improper idling revolution speed* • Unstable idling*
13	Intake air thermo-sensor	(1) The signal voltage of the intake air temperature sensor is 4.5V or higher. (2) The signal voltage of the intake air temperature sensor is 0.27V or lower.	(1) The intake air temperature sensor is troubled. (2) Open-circuit, short-circuit or improper connector contact occurs in the air intake temperature sensor circuit. (3) The engine control unit is troubled.	• Slightly improper driveability* • At high temperatures, (a) Improper start* (b) Unstable idling*
14	Throttle position sensor	(1) The signal voltage of the throttle position sensor is 0.2V or lower. (2) Though the idle position switch is on, the signal voltage of the throttle position sensor is 2V or higher.	(1) The throttle position sensor is troubled or improperly adjusted. (2) Open-circuit, short-circuit or improper connector contact occurs in the throttle position sensor circuit.	• Improper driveability • Engine stop
			(3) The idle position switch ON is troubled. (4) The idle position switch signal line is short-circuited. (5) The engine control unit is troubled.	• Engine stop • Racing is impossible
21	Coolant thermosensor	(1) The signal voltage of the coolant thermosensor is 4.6V or higher. higher. (2) The signal voltage of the coolant thermosensor is 0.11V or lower. (3) During engine warming-up, the coolant thermosensor signal indicates that the engine coolant temperature drops.	(1) The coolant thermosensor is troubled. (2) Open-circuit, short-circuit or improper connector contact occurs in the coolant thermosensor circuit. (3) The engine control unit is troubled.	During cold weather, • Starting is impossible. • Unstable idling • Improper acceleration
22	Crank angle sensor	(1) Though the engine is cranking 4 seconds or more, the signal voltage of the crank angle sensor does not vary (to be high/low). (2) Crank angle sensor signal pattern is abnormal. <DOHC>	(1) The crank angle sensor is troubled. (2) Open-circuit, short-circuit or improper connector contact occurs in the crank angle sensor circuit. (3) The TDC sensor is troubed or connector contact is improper. <DOHC> (4) The engine control unit is troubled.	• Engine stop • Starting is impossible.
23	TDC sensor	(1) Though the engine is running, the signal voltage of the TDC sensor does not vary (to be high/low). (2) The TDC sensor signal pattern is abnormal. <DOHC>	(1) The TDC sensor is troubled. (2) Open-circuit, short-circuit or improper connector contact occurs in the TDC sensor circuit. (3) The crank angle sensor is troubled or connector contact is improper. <DOHC> (4) The engine control unit is troubled.	• Engine stop* <DOHC> • Unstable idling* <SOHC> • Improper acceleration* <SOHC>
24	Vehicle speed sensor (reed switch)	During acceleration of engine revolution speed of 3,000 rpm or more, the signal voltage of the vehicle speed sensor does not vary (to be high/low).	(1) The vehicle speed sensor is troubled. (2) Open-circuit, short-circuit or improper connector contact occurs in the vehicle speed sensor circuit. (3) The engine control unit is troubled.	The engine sometimes cuts off when stopping after deceleration.
25	Open-air pressure sensor	(1) The signal voltage of the open air pressure sensor is 4.5V or higher. (2) The signal voltage of the open air pressure sensor is 0.2V or lower.	(1) The open-air pressure sensor is troubled. (2) Open-circuit, short-circuit or improper connector contact occurs in the open-air pressure sensor circuit. (3) The engine control unit is troubled.	• Unstable idling* • Improper acceleration* • Improper start*
31	Detonation sensor	Signal voltage is abnormal.	(1) The detonation sensor is troubled. (2) Open-circuit, short-circuit or improper connector contact occurs in the detonation sensor circuit. (3) The engine control unit is troubled.	• Improper acceleration*

3.0L ENGINE – MALFUNCTION CODES DIAGNOSIS CONTINUED – DIAMANTE

Trouble code no.	Diagnosis Item	Diagnosis content	Major cause	Remarks (troube phenomenon, etc.)
36	Ignition timing adjusting signal	The ignition timing adjusting signal line is short-circuited to the ground.	(1) The ignition timing adjusting signal line is short-circuited to the ground. (2) The engine control unit is troubled.	• Improper acceleration • Engine overheat
41	Injector	Injector is not continuously driven for 4 seconds during engine cranking or idle operation.	(1) The injector is troubled. (2) Open-circuit, short-circuit or improper connector contact occurs in the injector sensor circuit. (3) The engine control unit is troubled.	• Unstable idling • Improper acceleration • Improper start
43	EGR <California>	During engine running after warming-up, (1) The EGR amount is small. (The signal voltage of the EGR thermosensor is excessively high.) (2) The signal voltage of the EGR thermosensor is 0.1V or less.	(1) The EGR valve is not opened. (2) The negative pressure of the EGR valve control is excessively low. (3) The EGR control solenoid valve is troubled. (4) The EGR thermosensor is troubled. (5) Open-circuit, short-circuit or improper connector contact occurs in the EGR thermosensor circuit. (6) The engine control unit is troubled.	• Deterioration of exhaust gas purifying performance
44	Ignition coil/power transistor unit for cylinders 1 and 4 <DOHC>	During engine running, the ignition signal is not input. However, cases in which ignition signals are input to all cylinders are excluded.	(1) The ignition coil is troubled. (2) Open-circuit, short-circuit or improper connector contact occurs in the ignition primary circuit (3) The power transistor unit is troubled. (4) The engine control unit is troubled.	• Unstable idling* • Improper acceleration* • Improper start*
52	Ignition coil/power transistor unit for cylinders 2 and 5 <DOHC>	During engine running, the ignition signal is not input. However, cases in which ignition signals are input to all cylinders are excluded.	(1) The ignition coil is troubled. (2) Open-circuit, short-circuit or improper connector contact occurs in the ignition primary circuit (3) The power transistor unit is troubled. (4) The engine control unit is troubled.	• Unstable idling* • Improper acceleration* • Improper start*
53	Ignition coil/power transistor unit for cylinders 3 and 6 <DOHC>	During engine running, the ignition signal is not input. However, cases in which ignition signals are input to all cylinders are excluded.	(1) The ignition coil is troubled. (2) Open-circuit, short-circuit or improper connector contact occurs in the ignition primary circuit (3) The power transistor unit is troubled. (4) The engine control unit is troubled.	• Unstable idling* • Improper acceleration* • Improper start*

Trouble code no.	Diagnosis Item	Diagnosis content	Major cause	Remarks (troube phenomenon, etc.)
61	A cable from transaxle control unit, for transmission of torque reduction signal <DOHC>	During engine running, the signal voltage of torque reduction request from the transaxle control unit is kept low for several seconds.	(1) The signal line of the torque reduction request is short-circuited. (2) The transaxle control unit is troubled. (3) The engine control unit is troubled.	• Speed-change feeling deteriorates slightly.*
62	Intake air control valve position sensor	Even if the intake air control servo motor is driven several times, the intake air control valve will not achieve its target position (opening).	(1) The intake air control valve position sensor is troubled. (2) Open-circuit, short-circuit or improper connector contact occurs in the intake air control valve position sensor circuit. (3) The intake air control servo motor (DC motor) is troubled. (4) Open-circuit, short-circuit or improper connector contact occurs in the intake air control servo motor (DC motor) circuit. (5) The engine control unit is troubled.	• Improper acceleration
71	Vacuum control solenoid valve <Vehicles with TCL>	(1) Though the vacuum control solenoid valve is driven, the solenoid is not energized. (2) Though the vacuum control solenoid valve is not driven, the solenoid is energized.	(1) The vacuum control solenoid valve is troubled. (2) Open-circuit, short-circuit or improper connector contact occurs in the vacuum control solenoid valve circuit. (3) The engine control unit is troubled.	• Even if the acceleration pedal is pressed, the engine revolution speed does not rise. • Traction control does not work.*
72	Ventilation control solenoid valve <Vehicles with TCL>	(1) Though the ventilation control solenoid valve is driven, the solenoid is not energized. (2) Though the ventilation control solenoid valve is not driven, the solenoid is energized.	(1) The ventilation control solenoid valve is troubled. (2) Open-circuit, short-circuit or improper connector contact occurs in the ventilation control solenoid valve circuit. (3) The engine control unit is troubled.	• Even if the acceleration pedal is pressed, the engine revolution speed does not rise. • Traction control does not work.*

DIAGNOSIS CHART (FAULT TREE)

Output preference order	Diagnosis item	Diagnosis code			Check item (Remedy)
		Output signal pattern	No.	Memory	
1	Engine control unit	H L 	–	–	(Replace engine control unit)
2	Oxygen sensor	H L 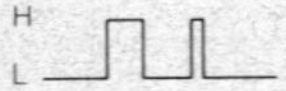	11	Retained	• Harness and connector • Fuel pressure • Injectors (Replace if defective.) • Intake air leaks • Oxygen sensor
3	Air flow sensor	H L 	12	Retained	• Harness and connector (If harness and connector are normal, replace air flow sensor assembly.)
4	Intake air tempera-ture sensor	H L 	13	Retained	• Harness and connector • Intake air temperature sensor
5	Throttle position sensor	H L 	14	Retained	• Harness and connector • Throttle position sensor • Idle position switch
6	Coolant tempera-ture sensor	H L 	21	Retained	• Harness and connector • Coolant temperature sensor
7	Crank angle sensor	H L 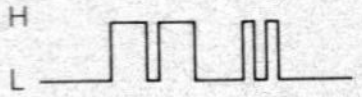	22	Retained	• Harness and connector (If harness and connector are normal, replace distributor assembly.)
8	Top dead center sensor	H L 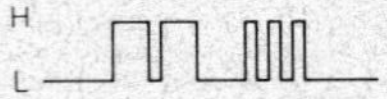	23	Retained	• Harness and connector (If harness and connector are normal, replace distributor assembly.)
9	Vehicle-speed sensor (reed switch)	H L 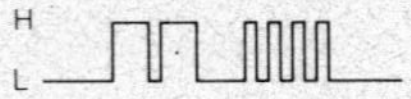	24	Retained	• Harness and connector • Vehicle-speed sensor (reed switch)
10	Barometric pressure sensor	H L 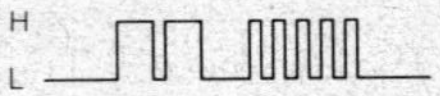	25	Retained	• Harness and connector (If harness and connecor are normal, replace barometric pressure sensor assembly.)
11	Injector	H L 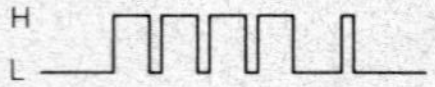	41	Retained	• Harness and connector • Injector coil resistance
12	Fuel pump	H L 	42	Retained	• Harness and connector • Control relay
13	EGR <Califor-nia>	H L 	43	Retained	• Harness and connector • EGR temperature sensor • EGR valve • EGR control solenoid valve • EGR valve control vacuum
14	Normal state	H L 	–	–	–

NOTE
Replace the engine control unit if a malfunction code is output although the inspection reveals that there is no problem with the check items.

3.0L ENGINE—SELF-DIAGNOSIS MALFUNCTION CODES—3000GT

DIAGNOSIS CHART (FAULT TREE)

Output preference order	Diagnosis item	Diagnosis code			Check item (Remedy)
		Output signal pattern	No.	Memory	
1	Engine control unit	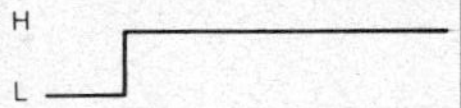	–	–	(Replace engine control unit)
2	Oxygen sensor (Turbo: Rear bank)		11	Retained	• Harness and connector • Oxygen sensor • Fuel pressure • Injectors (Replace if defective.) • Intake air leaks
3	Air flow sensor		12	Retained	• Harness and connector (If harness and connector are normal, replace air flow sensor assembly.)
4	Intake air temperature sensor	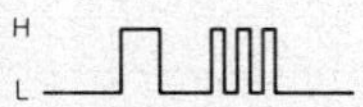	13	Retained	• Harness and connector • Intake air temperature sensor
5	Throttle position sensor		14	Retained	• Harness and connector • Throttle position sensor • Idle position switch
6	Coolant temperature sensor	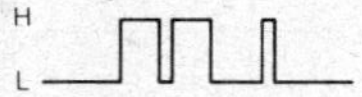	21	Retained	• Harness and connector • Coolant temperature sensor
7	Crank angle sensor	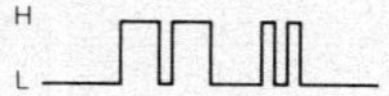	22	Retained	• Harness and connector (If harness and connector are normal, replace crank angle sensor assembly.)
8	Top dead center sensor	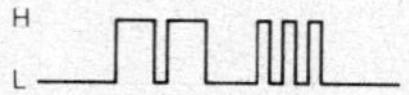	23	Retained	• Harness and connector (If harness and connector are normal, replace crank angle sensor assembly.)
9	Vehicle speed sensor (reed switch)		24	Retained	• Harness and connector • Vehicle speed sensor (reed switch)
10	Barometric pressure sensor		25	Retained	• Harness and connector (If harness and connector are normal, replace barometric pressure sensor assembly.)
11	Detonation sensor	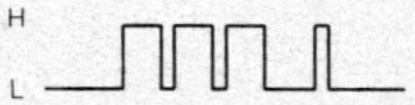	31	Retained	• Harness and connector (If harness and connector are normal, replace detonation sensor.)

NOTE
Replace the engine control unit if a malfunction code is output although the inspection reveals that there is no problem with the check items.

Output preference order	Diagnosis item	Diagnosis code			Check item (Remedy)
		Output signal pattern	No.	Memory	
12	Ignition timing adjustment signal	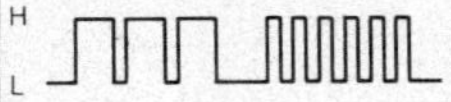	36	–	• Harness and connector
13	Oxygen sensor (Turbo: Front bank)	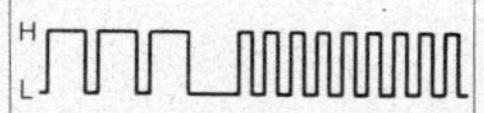	39	Retained	• Harness and connector • Oxygen sensor • Fuel pressure • Injectors (Replace if defective) • Intake air leaks
14	Injector		41	Retained	• Harness and connector • Injector coil resistance
15	EGR <California>	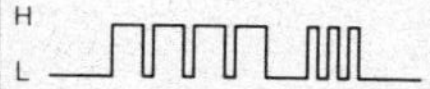	43	Retained	• Harness and connector • EGR thermo-sensor • EGR valve • EGR control solenoid valve • EGR valve control vacuum
16	Ignition coil, Power transistor unit (No. 1–4 cylinder)		44	Retained	• Harness and connector • Ignition coil • Power transistor unit
17	Ignition coil, Power transistor unit (No. 2–5 cylinder)		52	Retained	• Harness and connector • Ignition coil • Power transistor unit
18	Ignition coil, Power transistor unit (No. 3–6 cylinder)	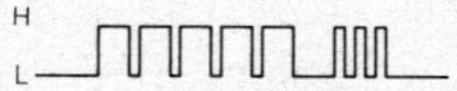	53	Retained	• Harness and connector • Ignition coil • Power transistor unit
19	A cable from trans-axle control unit, for transmission of torque reduction signal <A/T>		61	Retained	• Harness and connector (If harness and connector are normal, replace only transaxle control unit.)
20	Induction control valve position sensor <Non Turbo>	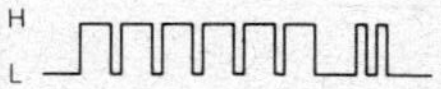	62	Retained	• Harness and connector (If harness, connector and induction control valve are normal, replace air intake plenum assembly.)
21	Normal state		–	–	–

NOTE
Replace the engine control unit if a malfunction code is output although the inspection reveals that there is no problem with the check items.

DIAGNOSIS CHART

Output preference order	Diagnosis item	Diagnosis code			Check item (Remedy)
		Output signal pattern	No.	Memory	
1	Engine control unit		–	–	(Replace engine control unit)
2	Oxygen sensor		11	Retained	• Harness and connector • Oxygen sensor • Fuel pressure • Injectors (Replace if defective.) • Intake air leaks
3	Air flow sensor		12	Retained	• Harness and connector (If harness and connector are normal, replace air flow sensor assembly.)
4	Intake air temperature sensor		13	Retained	• Harness and connector • Intake air temperature sensor
5	Throttle position sensor		14	Retained	• Harness and connector • Throttle position sensor • Idle position switch
6	Engine coolant temperature sensor		21	Retained	• Harness and connector • Engine coolant temperature sensor
7	Crank angle sensor		22	Retained	• Harness and connector (If harness and connector are normal, replace distributor assembly.)
8	Top dead center sensor	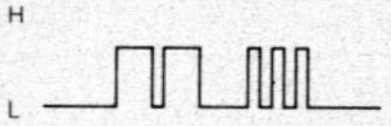	23	Retained	• Harness and connector (If harness and connector are normal, replace distributor assembly.)
9	Vehicle speed sensor (reed switch)	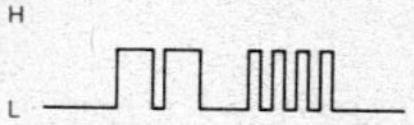	24	Retained	• Harness and connector • Vehicle speed sensor (reed switch)
10	Barometric pressure sensor		25	Retained	• Harness and connector (If harness and connector are normal, replace barometric pressure sensor assembly.)
11	Injector	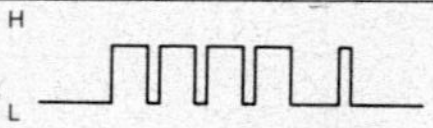	41	Retained	• Harness and connector • Injector coil resistance
12	EGR <California>		43	Retained	• Harness and connector • EGR thermo sensor • EGR valve • EGR valve control solenoid valve • EGR valve control vacuum
13	Normal state		–	–	–

NOTE
Replace the engine control unit if a malfunction code is output although the inspection reveals that there is no problem with the check items.

3.0L ENGINE—MALFUNCTION CODES DIAGNOSIS—TRUCK

DIAGNOSIS CHART

NOTE
*: The fail safe/back up functions are operating.

Malfunction code No	Diagnosis item	Trouble symptom	Main cause	Note (malfunction symptom,etc.)
–	Engine control unit	Malfunction of engine control unit itself	–	• Engine stall • No starting
11	Oxygen sensor	The oxygen sensor signal voltage does not change (lean/rich), even though the air/fuel ratio feed-back control (closedloop control) is being carried out	(1) Malfunction of oxygen sensor (2) Open circuit, shortcircuit of oxygen sensor circuit or poor connection of connector	• Failure of exhaust gas purification function*
			(3) Inadequate fuel pressure (4) Malfunction of injector (5) Air absorption from gasket space (6) Malfunction of engine control unit	• Failure of exhaust gas purification function* • Poor starting • Unstable idling • Poor acceleration
12	Air flow sensor	The air flow sensor signal frequency is less than 10Hz,even though the engine is operating	(1) Malfunction of air flow sensor (2) Open circuit, shortcircuit of air flow-sensor circuit or poor connection of connector (3) Malfunction of engine control unit	• Poor acceleration* • Improper idling speed* • Unstable idling*
13	Intake air temperature sensor	(1) The intake air temperature signal voltage is more than 4.5V. (2) The intake air temperature signal voltage is less than 0.27V.	(1) Malfunction of intake air temperature sensor (2) Open circuit, shortcircuit of intake air temperature sensor circuit or poor connection of connector (3) Malfunction of engine control unit	• Slightly poor drivability* • In high temperature: (a) Poor starting* (b) Unstable idling*
14	Throttle position sensor	(1) The throttle position sensor signal voltage is less than 0.2V. (2) The throttle position sensor signal voltage is more than 2V, even though the idle position switch is "ON".	(1) Malfunction or maladjustment of throttle position sensor (2) Open circuit, shortcircuit of throttle position sensor circuit or poor connection of connrctor	• Slightly poor acceleration • Engine stall
			(3) Malfunction at idle position switch "ON" (4) Shortcircuit of idle position switch signal line (5) Malfunction of engine control unit	• Engine stall • No racing
21	Engine coolant temperature sensor	(1) The engine coolant temperature sensor signal voltage is more than 4.6V. (2) The engine coolant temperature sensor signal voltage is less than 0.11V. (3) During the engine warming-up, the engine coolant temperature sensor signal has indicated dropping of engine coolant water-temperature	(1) Malfunction of engine coolant temperature sensor (2) Open circuit, shorecircuit of engine coolant temperature sensor circuit or poor connection of connector (3) Malfunction of engine control unit	In cold engine condition; • No starting* • Unstable idling* • Poor acceleration*

Malfunction code No	Diagnosis item	Trouble symptom	Main cause	Note (malfunction symptom,etc.)
22	Crank angle sensor	The crank angle sensor signal voltage does not change(High/Low), even though the engine is cranking more than 4 seconds	(1) Malfunction of crank angle sensor (2) Open circuit, shortcircuit of crank angle sensor circuit or poor connection of connectror (3) Malfunction of engine control unit	• Engine stall • No starting
23	Top dead center sensor	The top dead center signal voltage does not change(High/Low), even though the engine is operating.	(1) Malfunction of top dead center sensor (2) Open circuit, shortcircuit of top dead center senser circuit or poor connection of connector (3) Malfunction of engine control unit	• Unstable idling* • Poor acceleration*
24	Vehicle speed sensor (reed switch)	The vehicle speed sensor signal voltage does not change(High/Low), when the vehicle is accelerated with engine speed more than 3000rpm.	(1) Malfunction of vehicle speed sensor (2) Open circuit, shortcircuit of vehicle speed sensor or poor connection of connector (3) Malfunction of engine control unit	Engine may stall when the vehicle is stopped from reduced speed driving.
25	Barometric pressure sensor	(1) The barometric pressure sensor signal voltage is more than 4.5V. (2) The barometric pressure sensor signal voltage is less than 0.2V.	(1) Malfunction of barometric pressure sensor (2) Open circuit, shortcircuit of barometric pressure sensor or poor connection of connector (3) Malfunction of engine control unit	• Unstable idling* • Poor acceleration* • Poor starting*
41	Injector	During engine cranking or idling, the injector does not continue to drive more than 4 seconds.	(1) Malfunction of injector (2) Open circuit, shortcircuit of injector circuit or poor connection of connector (3) Malfunction of engine control unit	• Unstable idling • Poor acceleration • Poor starting
43	EGR <California>	During operation after engine warm-up; (1) The EGR volume is scarce. (EGR temperature sensorsignal voltage is too high.) (2) The EGR temperature sensor signal voltage is less than 0.1V.	(1) EGR valve does not open. (2) EGR valve control negative pressure is too low. (3) Malfunction of EGR control solenoid valve (4) Malfunction of EGR temperature sensor (5) Open-circuit, shortcircuit of EGR temperature sensor circuit or poor connection of connector (6) Malfunction of engine control unit	• Failure of exhaust gas purification function

FUEL PUMP

Operational Check

1. Turn the ignition switch **OFF**.
2. Apply battery voltage to the fuel pump drive connector and check that the fuel pump operates.

NOTE: The fuel pump is located inside the fuel tank and cannot be readily heard. Remove the fuel tank cap and listen for the fuel pump operating sound through the filler cap.

3. Pinch the high pressure hose and note if fuel pressure is felt.

FUEL PRESSURE

Relieving

1. Disconnect the fuel pump harness at the rear of the tank.
2. Start engine and allow to idle.
3. After engine stalls, fuel pressure is reduced.
4. Disconnect the negative battery cable and reconnect the fuel pump wiring.
5. Remove the fuel tank cap.

NOTE: When disconnecting any fuel system hoses, always place a shop towel to prevent residual fuel from spaying.

Testing

1. Relieve fuel system pressure.
2. Turn ignition switch **OFF** and disconnect the negative battery cable.

NOTE: When disconnecting any fuel system hoses, always place a shop towel to prevent residual fuel from spaying.

3. Disconnect the high pressure hose at the delivery side. Using a fuel pressure gauge and adapter (MD998742-01 or equivalent), install between the delivery pipe and high pressure hose.
4. Connect the negative battery cable.
5. Activate the fuel pump, by applying battery voltage to the fuel pump drive (check terminal) terminal. Check that there is no leakage from pressure gauge or connections.
6. Operate the engine at curb idle speed.
7. Disconnect the vacuum hose from the pressure regulator and plug. Record the fuel pressure. Fuel pressure, at curb idle, should be as follows:
   1.5L (VIN A AND J) engines – 47–50 psi (330–350 kPa).
   1.6L (VIN Y) engine – 47–50 psi (330–350 kPa).
   1.8L (VIN D and T) engines – 47–50 psi (330–350 kPa).

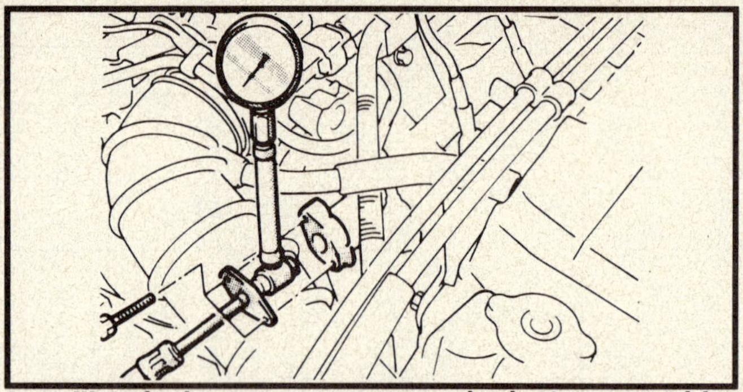

Installing fuel pressure gauge and adapter assembly

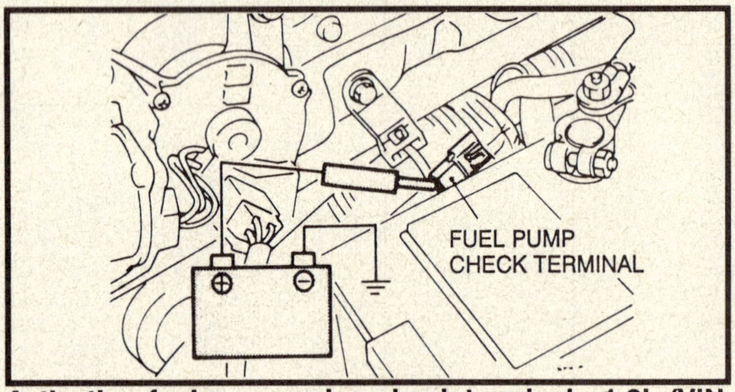

Activating fuel pump using check terminal – 1.8L (VIN T) and 2.0L (VIN R AND U) engines

FUEL PRESSURE DIAGNOSIS CHART

Condition	Probable cause	Remedy
Fuel pressure too low	a. Clogged fuel filter. b. Fuel leakage to the return side, caused by poor seating of the fuel-pressure regulator. c. Low discharge pressure of the fuel pump	a. Replace fuel filter b. Replace fuel pressure regulator. c. Check the in-tank fuel hose for leakage or replace the fuel pump.
Fuel pressure too high	a. Sticking fuel-pressure regulator b. Clogged or bent fuel return hose or pipe.	a. Replace fuel pressure regulator b. Repair or replace hose or pipe.
There is no difference in fuel pressure when the vacuum hose is connected and when it is not.	a. Clogging, or damaged vacuum hose or the nipple. b. Sticking or poor seating of the fuel-pressure regulator.	a. Repair or replace the vacuum hose or the nipple b. Repair or replace hose or pipe.
Fuel pressure drops slowly after engine is stopped	a. Injector leakage	a. Replace injector
Fuel pressure drops immediately after engine is stopped	a. The check valve within the fuel pump is open	a. Replace fuel pump

2.0L (VIN R and V) non-turbocharged engines – 47–50 psi (330–350 kPa).
2.0L (VIN U) turbocharged engine with manual transaxle – 36–38 psi (250–270 kPa).
2.0L (VIN U) turbocharged engine with automatic transaxle – 41–46 psi (290–320 kPa).
2.4L (VIN W) engine – 47–50 psi (330–350 kPa).
3.0L (VIN B AND S) engines – 47–50 psi (330–350 kPa).
3.0L (VIN C) turbocharged engine – 43–45 psi (295–315 kPa).
8. Reconnect the vacuum hose to the pressure regulator. With the engine at curb idle, fuel pressure should be as follows:
1.5L (VIN A AND J) engines – 38 psi (270 kPa).
1.6L (VIN Y) engine – 38 psi (270 kPa).
1.8L (VIN D and T) engines – 38 psi (270 kPa).
2.0L (VIN R and V) non-turbocharged engines – 38 psi (270 kPa).
2.0L (VIN U) turbocharged engine with manual transaxle – 27 psi (190 kPa).
2.0L (VIN U) turbocharged engine with automatic transaxle – 33 psi (230 kPa).
2.4L (VIN W) engine – 38 psi (270 kPa).
3.0L (VIN B AND S) engines – 38 psi (270 kPa).
3.0L (VIN C) turbocharged engine – 34 psi (235 kPa).
9. If the specified values are not met, make the necessary repairs, as indicated in the diagnosis chart.
10. Shut the engine OFF. Check that the fuel pressure is maintained at the value indicated in Step 8, for approximately 5 minutes.
11. Relieve the fuel pressure.
12. Remove the fuel pressure gauge and reconnect the high pressure hose, using a new O-ring.
13. Start the engine and check for leaks.

CONTROL RELAY

Failure of the control relay will interrupt power supply to the fuel pump, injectors and the ECU, resulting in a No Start condition.

Testing

1.5L (VIN J) ENGINE

1. Turn ignition switch **OFF** and disconnect the negative battery cable.
2. Disconnect the control relay harness connector and remove the control relay.
3. Check the control relay L_1 and L_2 coils (relay de-energized), as follows:
   a. Terminals 1 and 4. Should be non-conductive (infinite).
   b. Terminals 3 and 8. Should indicate a value of approximately 95 ohms.
   c. Terminals 6 and 7. Should indicate a value of approximately 35 ohms.
4. Connect an ohmmeter between terminals 1 and 4 of the control relay. Using jumper wires, connect the control relay terminal 7 to battery (+). Note the ohmmeter reading, when terminal 6 is connected and disconnected to battery (−). There should be continuity when connected and no continuity when disconnected.

NOTE: Be very careful when making these connections. Any mistakes and the relay will be damaged.

5. Check the control relay L_3 coil (relay de-energized), as follows:
   a. Terminals 1 and 2. Should be non-conductive (infinite).
   b. Terminals 7 and 5. Should be non-conductive (infinite) in 1 direction and conductive (0 ohms) in the opposite direction.
6. Connect an ohmmeter between terminals 1 and 2 of the control relay. Using jumper wires, connect the control relay terminal 5 to battery (+). Note the ohmmeter reading, when terminal 7 is connected and disconnected to battery (−). There should be continuity when connected and no continuity when disconnected.
7. If the results are not as specified, replace the control relay.

1.8L (VIN D) ENGINE

1. Turn ignition switch **OFF** and disconnect the negative battery cable.
2. Disconnect the control relay harness connector and remove the control relay.
3. Check for continuity between the relay terminals, as follows:
   a. Terminals 5 and 7. Should indicate a continuity.
   b. Terminals 6 and 8. Should indicate continuity in 1 direction only.
4. Connect a voltmeter from terminals 1 of the control relay to ground. Using jumper wires, connect the control relay terminal 7 to battery (+). Note the voltmeter reading, when terminal 5 is connected and disconnected to battery (−). There should be battery voltage when connected and no voltage when disconnected.

NOTE: Be very careful when making these connections. Any mistakes and the relay will be damaged.

5. Connect an ohmmeter between terminals 2 and 4; also at terminals 3 and 4 of the control relay. Using jumper wires, connect the control relay terminal 8 to battery (+). Note the ohmmeter reading, when terminal 6 is connected and disconnected to battery (−). There should be continuity when connected and no continuity when disconnected, between terminals 2 and 4; also at terminals 3 and 4 of the control relay.
6. If the results are not as specified, replace the control relay.

1.5L (VIN A), 1.8L (VIN T) AND 2.0L (VIN V) ENGINES

1. Turn ignition switch **OFF** and disconnect the negative battery cable.
2. Disconnect the control relay harness connector and remove the control relay.
3. Check for continuity between the relay terminals, as follows:
   a. Terminals 3 and 5. Should indicate a value of approximately 95 ohms.
   b. Terminals 2 and 5. Should indicate a value of approximately 95 ohms.
   c. Terminals 6 and 7. Should indicate a value of approximately 35 ohms.
   d. Terminals 6 and 8. Should indicate continuity in 1 direction only.
4. Connect an ohmmeter between terminals 1 and 4 of the control relay. Using jumper wires, connect the control relay terminal 7 to battery (+). Note the ohmmeter reading, when terminal 6 is connected and disconnected to battery (−). There should be continuity when connected and no continuity when disconnected.

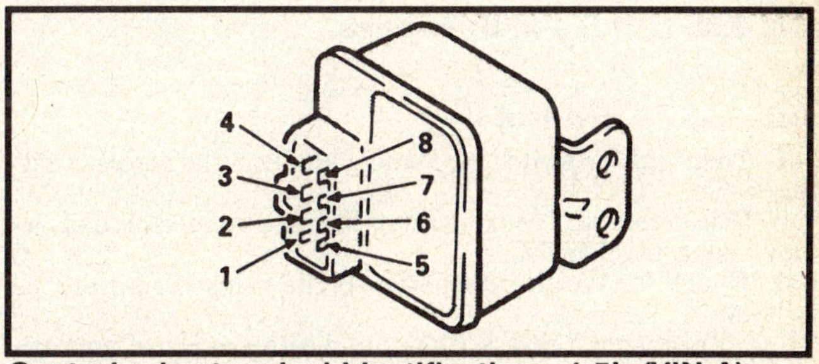

Control relay terminal identification – 1.5L (VIN A), 1.8L (VIN T AND D), 2.0L (VIN V) and 2.4L (VIN W) engines

NOTE: Be very careful when making these connections. Any mistakes and the relay will be damaged.

5. Connect an ohmmeter between terminals 2 and 4 of the control relay. Using jumper wires, connect the control relay terminal 8 to battery (+). Note the ohmmeter reading, when terminal 6 is connected and disconnected to battery (−). There should be continuity when connected and no continuity when disconnected.
6. If the results are not as specified, replace the control relay.

1.6L (VIN Y), 2.0L (VIN R AND U) AND 3.0L (VIN B, C AND S) ENGINES

1. Turn ignition switch **OFF** and disconnect the negative battery cable.
2. Disconnect the control relay harness connector and remove the control relay.

NOTE: Be very careful when making the following connections, as damage to the control relay may occur.

3. Connect a voltmeter across terminals 4 and 5 of the control relay. Using jumper wires, apply 12 volts (+) power supply to terminal 10 of the control relay. Note the voltage when the power supply negative (−) lead is connected and disconnected from terminal 8. Should indicated 12 volts when connected and 0 volt when disconnected.
4. Connect a voltmeter across terminals 2 and 3 of the control relay. Using jumper wires, apply 12 volts (+) power supply to terminal 9 of the control relay. Note the voltage when the power supply negative (−) lead is connected and disconnected from terminal 6. Should indicated 12 volts when connected and 0 volt when disconnected.
5. Connect a voltmeter between terminals 2 of the control relay and ground. Using jumper wires, apply 12 volts (+) power supply to terminal 3 of the control relay. Note the voltage when the power supply negative (−) lead is connected and disconnected from terminal 7. Should indicated 12 volts when connected and 0 volt when disconnected.
6. If the results is not satisfactory, replace the control relay.

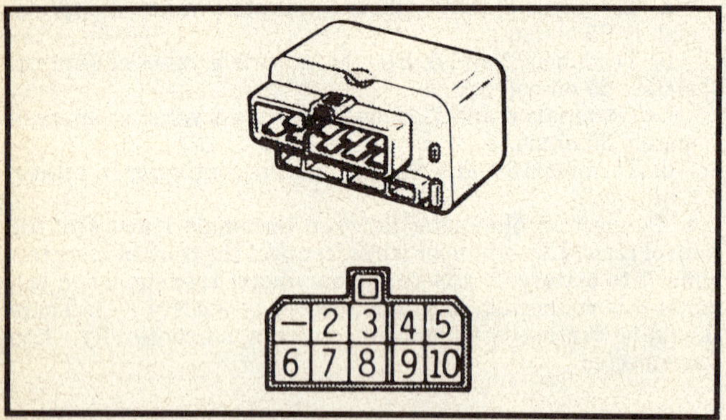

Control relay terminals identification—1.6L (VIN Y) and 2.0L (VIN R and U) engines

2.4L (VIN W) ENGINE

1. Turn ignition switch **OFF** and disconnect the negative battery cable.
2. Disconnect the control relay harness connector and remove the control relay.
3. Check for continuity between the relay terminals, as follows:
   a. Terminals 3 and 5. Should indicate a value of approximately 95 ohms.
   b. Terminals 2 and 5. Should indicate a value of approximately 95 ohms.
   c. Terminals 6 and 7. Should indicate a value of approximately 35 ohms.
   d. Terminals 6 and 8. Should indicate continuity in 1 direction only.
4. Connect an ohmmeter between terminals 1 and 4 of the control relay. Using jumper wires, connect the control relay terminal 7 to battery (+). Note the ohmmeter reading, when terminal 6 is connected and disconnected to battery (−). There should be continuity when connected and no continuity when disconnected.

NOTE: Be very careful when making these connections. Any mistakes and the relay will be damaged.

5. Connect an ohmmeter between terminals 1 and 4 of the control relay. Using jumper wires, connect the control relay terminal 2 to battery (+). Note the ohmmeter reading, when terminal 5 is connected and disconnected to battery (−). There should be continuity when connected and no continuity when disconnected.
6. If the results are not as specified, replace the control relay.

AIR FLOW SENSOR (AFS)

Testing

The AFS pulse signals must be checked, using the Multi-Use Tester (MUT) or an oscilloscope.

EXCEPT 1.5L (VIN J) ENGINE

Using The Multi-Use Tester (MUT)

1. Run the engine until operating temperature is reached.
2. Turn OFF all accessories: Lights, cooling fan, etc.
3. Set the front wheel in straight forward position.
4. Position the gear selector in **P** (automatic transaxle) or **N** (manual transaxle).
5. Note the AFS output frequency under the following engine conditions:

NOTE: With the engine racing, the AFS output frequency will increase proportionately with rpm.

1.5L (VIN A) Engine
750 rpm—30–45 Hz
2000 rpm—80–110 Hz
1.6L (VIN Y) Engine
750 rpm—27–33 Hz
2000 rpm—60–80 Hz
1.8L (VIN D) Engine
750 rpm—23–49 Hz
2000 rpm—51–91 Hz
1.8L (VIN T) Engine
700 rpm—25–40 Hz
2000 rpm—67–88 Hz
2.0L (VIN R and U) Engines
700 rpm—25–50 Hz
2000 rpm—60–85 Hz Turbocharged
2000 rpm—70–90 Hz Non-turbocharged
2.0L (VIN V) Engine
700 rpm—25–50 Hz
2000 rpm—70–130 Hz
2.4L Engine
750 rpm—34–60 Hz
2000 rpm—70–110 Hz
3.0L (VIN B, C AND S) Engines
700 rpm—25–50 Hz
2000 rpm—70–100 Hz Turbocharged
2000 rpm—70–90 Hz Non-turbocharged

NOTE: The AFS output frequency may be approximately 10% higher than indicated, when the vehicle is new (driven 300 miles or less).

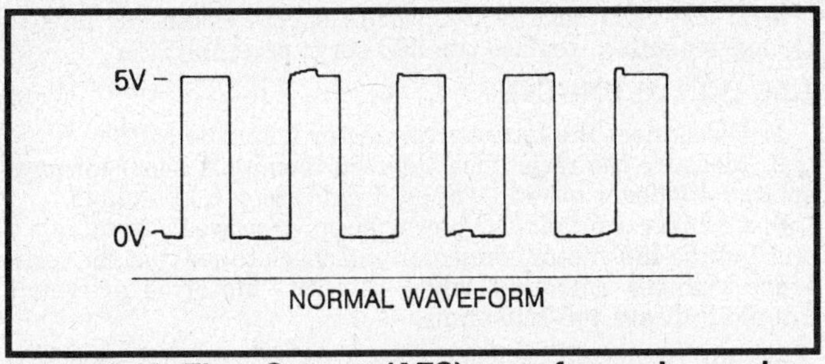

Correct Air Flow Sensor (AFS) waveform when using an oscilloscope

6. If the AFS fails to meet these specifications, replace the AFS.

Using An Oscilloscope

1. Run the engine until operating temperature is reached.
2. Turn **OFF** all accessories: Lights, cooling fan, etc.
3. Set the front wheel in straight forward position.
4. Position the gear selector in **P** (automatic transaxle) or **N** (manual transaxle).
5. Operate the engine at idle speed.
6. Connect the probe to the oscilloscope pick-up point and compare the scope pattern with diagram shown.

1.5L (VIN J) ENGINE

1. Connect a voltmeter between terminals 6 (sensor ground) and 3 (output) of the AFS connector.
2. Operate the engine until normal operating temperature is reached.
3. Check the output voltage. The output voltage should be 2.7–3.2 volts at idle and at 3,000 rpm.

NOTE: If the AFS fails, the intake air volume cannot be measured and as a result, normal fuel injection control is no longer available. During this condition, the vehicle will operate at a preset map value.

INTAKE AIR TEMPERATURE SENSOR

Testing

1. Disconnect the harness connector from the sensor.
2. Measure the resistance between terminals 4 and 6 (sensor side connector). The readings should be as follows:
   32°F (0°C) – 5.4–6.0 ohms
   68°F (20°C) – 2.3–2.7 ohms
   176°F (80°C) – 0.31–0.43 ohms
3. Measure the sensor resistance, while heating the sensor with a hair drier. As the temperature increase, the resistance should decrease.

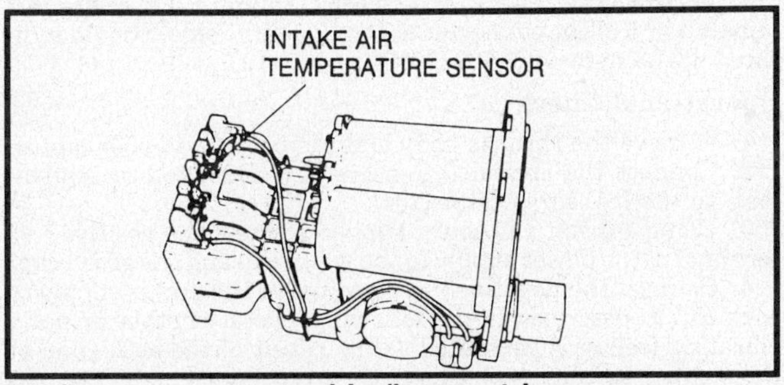

Air flow sensor assembly (barometric pressure sensor with intake air temperature sensor contained) – typical

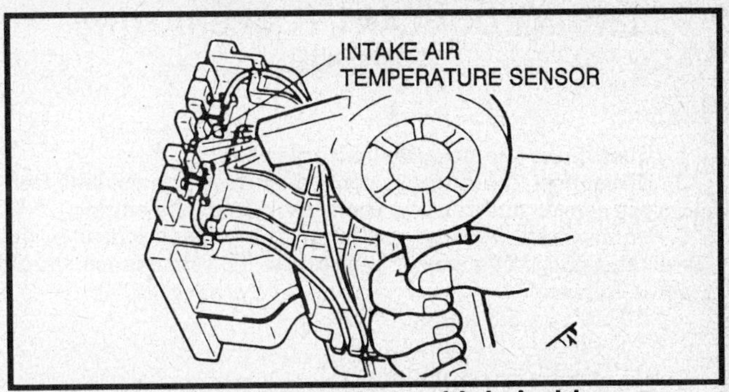

Heating intake air temperature with hair drier

4. If the sensor resistance deviates from the specified values, replace the air flow sensor assembly.

BAROMETRIC PRESSURE SENSOR

Testing

EXCEPT 1.5L (VIN J) ENGINE

The following procedure requires the use of the Multi-Use Tester (MUT) MB991341 or equivalent.

1. Connect the MUT or equivalent to the diagnosis connector.
2. Turn the ignition switch **ON**.
3. Select item 25 (sensor pressure) and check the sensor readings at the following altitudes:
   0 ft. (0 m) – 29.92 in.Hg (760 mmHg)
   1,969 ft. (600 m) – 27.95 in.Hg (710 mmHg)
   3,937 ft. (1,200 m) – 25.98 in.Hg (660 mmHg)
   5,906 ft. (1,800 m) – 24.02 in.Hg (610 mmHg)
4. If the sensor resistance deviates from the specified values, replace the sensor.

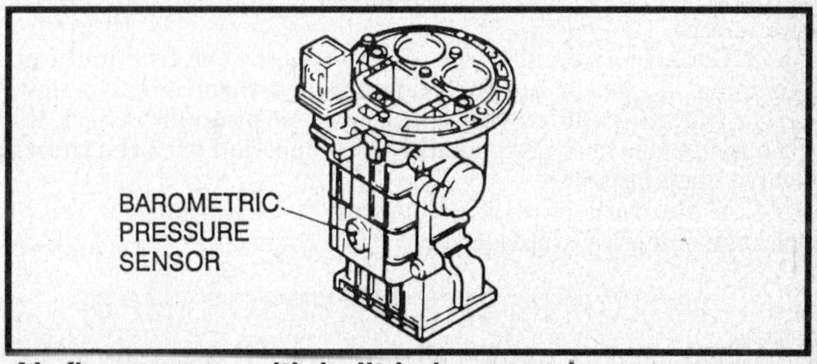

Air flow sensor with built-in barometric pressure sensor – typical

1.5L (VIN J) ENGINE

1. Connect a voltmeter between terminals 6 (sensor ground) and 5 (output) of the pressure sensor connector.
2. Operate the engine until normal operating temperature is reached.
3. With the engine at normal idle, gradually cover approximately ½ of the air cleaner air intake, while checking for a change in voltage. The voltage should decrease and the pressure decrease.
4. Use the following as a reference:
   2.9 psi (150mmHg) – 0.79 volts
   6.9 psi (350mmHg) – 1.84 volts
   15 psi (760mmHg) – 4.0 volts
5. Replace the AFS assembly, if necessary.

ENGINE COOLANT TEMPERATURE SENSOR

Testing

1. Disconnect the negative battery cable.
2. Disconnect the harness connector from the coolant temperature sensor and remove the sensor from the engine.
3. Immerse the sensing portion of the coolant sensor in hot water and check the sensor resistance. The resistance should read as follows:
   32°F (0°C) – 5.8 ohms
   68°F (20°C) – 2.4 ohms
   104°F (40°C) – 1.1 ohms
   176°F (80°C) – 0.3 ohms
4. If the sensor resistance deviates from the specified values, replace the sensor.

THROTTLE POSITION SENSOR (TPS)

Testing

**1.5L (VIN A), 1.6L (VIN Y)
2.0L (VIN R, U AND V) AND
3.0L (VIN B, C AND S) ENGINES**

1. Disconnect the harness connector from the TPS.
2. Measure the resistance between terminal 4 (sensor ground) and terminal 1 (sensor power). Should indicate 3.5–6.5 kilo-ohms.
3. Connect an analog type ohmmeter between terminal 4 and terminal 2 (sensor output). Operate the throttle valve slowly from the idle position to the full open position and check that the resistance changes smoothly in proportion with the throttle valve opening angle.
4. If the resistance is not within specifications, or fails to change smoothly, replace the TPS.

1.5L (VIN J) AND 1.8L (VIN D) ENGINES

1. Disconnect the harness connector from the TPS.
2. Measure the resistance between terminal 1 (sensor ground) and terminal 4 (sensor power). Should indicate 3.5–6.5 kilo-ohms.
3. Connect an analog type ohmmeter between terminal 1 and terminal 3 (sensor output). Operate the throttle valve slowly from the idle position to the full open position and check that the resistance changes smoothly in proportion with the throttle valve opening angle.
4. If the resistance is not within specifications, or fails to change smoothly, replace the TPS.

MOTOR POSITION SENSOR (MPS)

The MPS may be referred to as a Servo Valve Position Sensor on other vehicle.

Testing

1.5L (VIN A), 1.8L (VIN T) AND 2.0L (VIN V) ENGINES

1. Disconnect the harness connector from the MPS.
2. Measure the resistance between terminals 2 and 3. Should indicate 4–6 kilo-ohms.
3. Disconnect the ISC servo connector.
4. Using a 6 volt DC power supply, connect terminal 1 of the ISC servo connector to negative (−) and terminal 2 to positive (+). Then, measure the resistance between terminals 3 and 5 of the motor position sensor connector when the ISC servo is activated. The resistance should smoothly increase and decrease in accordance with plunger movement (extend and retract).

NOTE: Do not use voltage higher than 6 volts DC. Applying higher voltage could cause locking of the servo gears.

5. If the resistance is not within specifications, or fails to change smoothly, replace the ISC servo assembly.

1.5L (VIN J) ENGINE

1. Disconnect the harness connector from the MPS.
2. Measure the resistance between terminals 1 (sensor power) and 4 (sensor ground). Should indicate 4–6 kilo-ohms.
3. Connect a 6 volt DC power supply, between terminals 2 and 1 of the ISC motor connector and check to see that the resistance changes smoothly when the ISC motor is activated. Should indicate 4–6 kilo-ohms.

NOTE: Do not use voltage higher than 6 volts DC. Applying higher voltage could cause locking of the servo gears.

4. If the resistance is not within specifications, or fails to change smoothly, replace the ISC servo assembly.

IDLE SPEED CONTROL SERVO (DC MOTOR)

Testing

2.0L (VIN V) ENGINE

Actuator Inspection

1. Disconnect the harness connector from the ISC servo.
2. Check for continuity of the ISC servo coil across the actuator terminals.
3. Using a 6 volt DC power supply, connect terminal 1 of the ISC servo connector to negative (−) and terminal 2 to positive (+). Then, check that the ISC servo operates.

NOTE: Do not use voltage higher than 6 volts DC. Applying higher voltage could cause locking of the servo gears.

4. If the ISC servo does not operate, replace the ISC servo as an assembly.

IDLE SPEED CONTROL SERVO (STEPPER MOTOR)

Testing

2.0L (VIN R AND U) ENGINES

Coil Resistance

1. Disconnect the harness connector from the ISC servo and connect the special test harness (MD998464 or equivalent).
2. Measure the resistance between terminal 2 and either terminals 1 or 3 of the connector at the ISC servo side. Should indicated 28–33 ohms at 68°F (20°C).
3. Measure the resistance between terminal 5 and either terminals 4 or 6 of the connector at the ISC servo side. Should indicated 28–33 ohms at 68°F (20°C).

Operational Check

1. Remove the throttle body and remove the stepper motor.
2. Connect the special test harness (MD998463 or equivalent) to the ISC servo connector.
3. Using a 6 volt DC power supply, connect the positive (+) terminal of the power supply to the white clip and the green clip.
4. Connect the negative (−) terminal of the power supply to each clip as described in the following, check whether or not a vibrating feeling is generated as the result of the activation of the stepper motor.
   a. Negative (−) terminal to the red and black clip.
   b. Negative (−) terminal to the blue and black clip.
   c. Negative (−) terminal to the blue and yellow clip.

d. Negative (−) terminal to the red and yellow clip.
e. Negative (−) terminal to the red and black clip.
f. Repeat the tests in sequence from Step a to e.
5. If the results are satisfactory, the stepper motor is operating normal.

SERVO VALVE POSITION SENSOR

The servo valve position sensor may be referred to as a MPS on other vehicle.

Testing

1.8L (VIN D) AND 2.4L ENGINES

1. Disconnect the harness connector from the MPS.
2. Measure the voltage to the sensor terminals 1 to ground, with the ignition switch **ON**. Should be 4.8–5.2 volts. If not, repair the harness.
3. Measure the voltage to the sensor terminals 2 and 4 to ground, with the ignition switch **ON**. Should be 4.8–5.2 volts. If not, repair the harness.
4. Check for continuity, at the harness side connector, between terminal 3 (ground circuit) and a good engine ground. Should indicated continuity. If not, repair the harness.
5. On 1.8L engine, check for an open circuit or shorted circuit to ground, at the ECU terminals 61, 5 and 18 and the servo valve position sensor terminals 1, 2, and 4 (sensor disconnected).
6. On 2.4L engine, check for an open circuit or shorted circuit to ground, at the ECU terminals 23, 13, 67 and 68 and the servo valve position sensor terminals 1, 2, and 4 (sensor disconnected).
7. If the correct results are obtained and the condition still exist, replace the servo position sensor.

IDLE POSITION SWITCH (TPS INCORPORATED)

Testing

1. With the accelerator pedal released, check to be sure that the throttle valve lever or the fixed SAS is pushed.

NOTE: if it is not pushed, adjust the fixed SAS.

2. Disconnect the TPS connector.
3. Check for continuity across the TPS connector terminal 4 (sensor ground) and terminal 3 (idle position switch).
   a. With the accelerator pedal depressed, the switch should be non-conductive (infinity).
   b. With the accelerator pedal released, the switch should be conductive (0 ohms).

NOTE: If there is no continuity when the accelerator pedal is returned, loosen the TPS retaining screws, turn the TPS completely counter-clockwise and check again.

4. Replace the TPS (idle position switch incorporated), if the results are not satisfactory.

TOP DEAD CENTER (TDC) SENSOR

Testing

EXCEPT 1.5L (VIN J) ENGINE

Using Oscilloscope

1. Operate the engine until normal operating temperature is reached.
2. Connect the probe to the oscilloscope pick-up point.
3. Check and compare the waveform pattern.
4. Replace the TDC sensor, if the results are not as indicated.

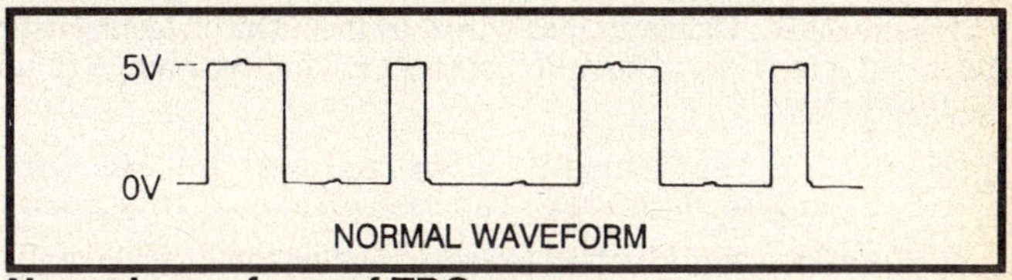

Normal waveform of TDC sensor

CRANK ANGLE SENSOR

Testing

EXCEPT 1.5L (VIN J) ENGINE

Using Oscilloscope

1. Operate the engine until normal operating temperature is reached.
2. Connect the probe to the oscilloscope pick-up point.
3. Check and compare the waveform pattern.
4. Replace the crank angle sensor, if the results are not as indicated.

Using Multi-Use Tester

1. Connect the multi-use tester to the diagnosis connector.
2. Connect a tachometer to the engine.
3. Crank the engine and compare the cranking rpm. Both readings should agree.

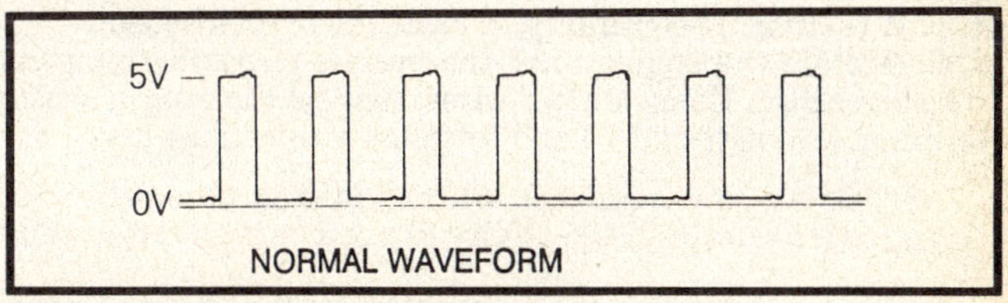

Normal waveform of crank angle sensor

TDC SENSOR AND CRANKSHAFT ANGLE SENSOR

Testing

1.5L (VIN J) ENGINE

1. Connect a voltmeter between terminals 1 (sensor ground) and 2 (crank angle signal), and between terminals 1 and 3 (No. 1 TDC signal).
2. Measure the output voltage, while cranking the engine. Output voltage should be as follows:
   a. No. 1 cylinder TDC sensor – 0.2–1.2 volts (needle fluctuates).
   b. Crankshaft angle sensor – 1.8–2.5 volts.
3. If the voltage is abnormal, check the sensor power and ground circuit. If okay, disassemble the distributor and service, as necessary.

OXYGEN SENSOR

Testing

1.6L (VIN Y) AND 2.0L (VIN R AND U) ENGINES

1. Disconnect the harness connector from the oxygen sensor.
2. Connect the special harness connector (MD998460 or equivalent) to the oxygen sensor side connector.
3. Check for continuity between the oxygen sensor connector terminal 3 (red clip of special tool) and terminal 4 (blue clip). Should read approximately 12 ohms at 68°F (20°C).
4. If there is no continuity, replace the oxygen sensor.
5. Operate the engine until the normal operating tempera-

ture is reached. Using jumper wires, connect the oxygen sensor terminal 3 (red clip of tool) to battery (+) and terminal 4 (blue clip) to battery (−).

NOTE: Be very careful when making the following connections, as damage to the oxygen sensor may occur.

6. Connect a digital voltmeter between terminal 1 (yellow clip of tool) and terminal 2 (black clip).
7. While repeatedly increasing the engine rpm, measure the output voltage of the oxygen sensor. Should indicated 0.6–1.0 volts.
8. Replace the oxygen sensor, if it is faulty.

1.5L (VIN A AND J), 1.8L (VIN D AND T), 2.0L (VIN V) AND 2.4L (VIN W) ENGINES

1. Operate the engine until the normal operating temperature is reached.
2. Disconnect the harness connector from the oxygen sensor.
3. Connect a digital voltmeter between the sensor connector and ground. While repeatedly increasing the engine rpm, measure the output voltage of the oxygen sensor. Should indicated 0.6–1.0 volts.
4. Replace the oxygen sensor, if it is faulty.

3.0L (VIN B, C AND S) ENGINES

1. Disconnect the harness connector from the oxygen sensor.
2. Check for continuity between the oxygen sensor connector terminal 3 and terminal 4. Should read approximately 20 ohms at 68°F (20°C).
3. If there is no continuity, replace the oxygen sensor.
4. Operate the engine until the normal operating temperature is reached. Using jumper wires, connect the oxygen sensor terminal 3 to battery (+) and terminal 4 to battery (−).

NOTE: Be very careful when making the following connections, as damage to the oxygen sensor may occur.

5. Connect a digital voltmeter between terminal 1 and terminal 2.
6. While repeatedly increasing the engine rpm, measure the output voltage of the oxygen sensor. Should indicated 0.6–1.0 volts.
7. Replace the oxygen sensor, if it is faulty.

INJECTORS

Testing

1. Measure the resistance across the injector terminals.
2. Readings should be as follows:
   Except 2.0L turbocharged and 3.0L turbocharged engines – 13–16 ohms at 68°F (20°C)
   2.0L turbocharged and 3.0L turbocharged engines – 2–3 ohms at 68°F (20°C)
3. If the resistance is not within specifications, replace the injector.

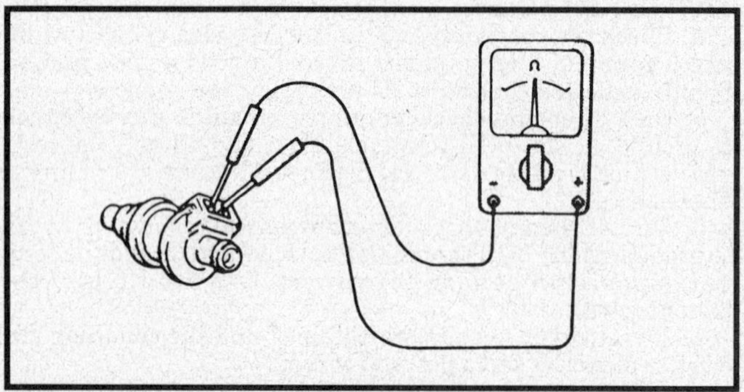

Measuring injector resistance

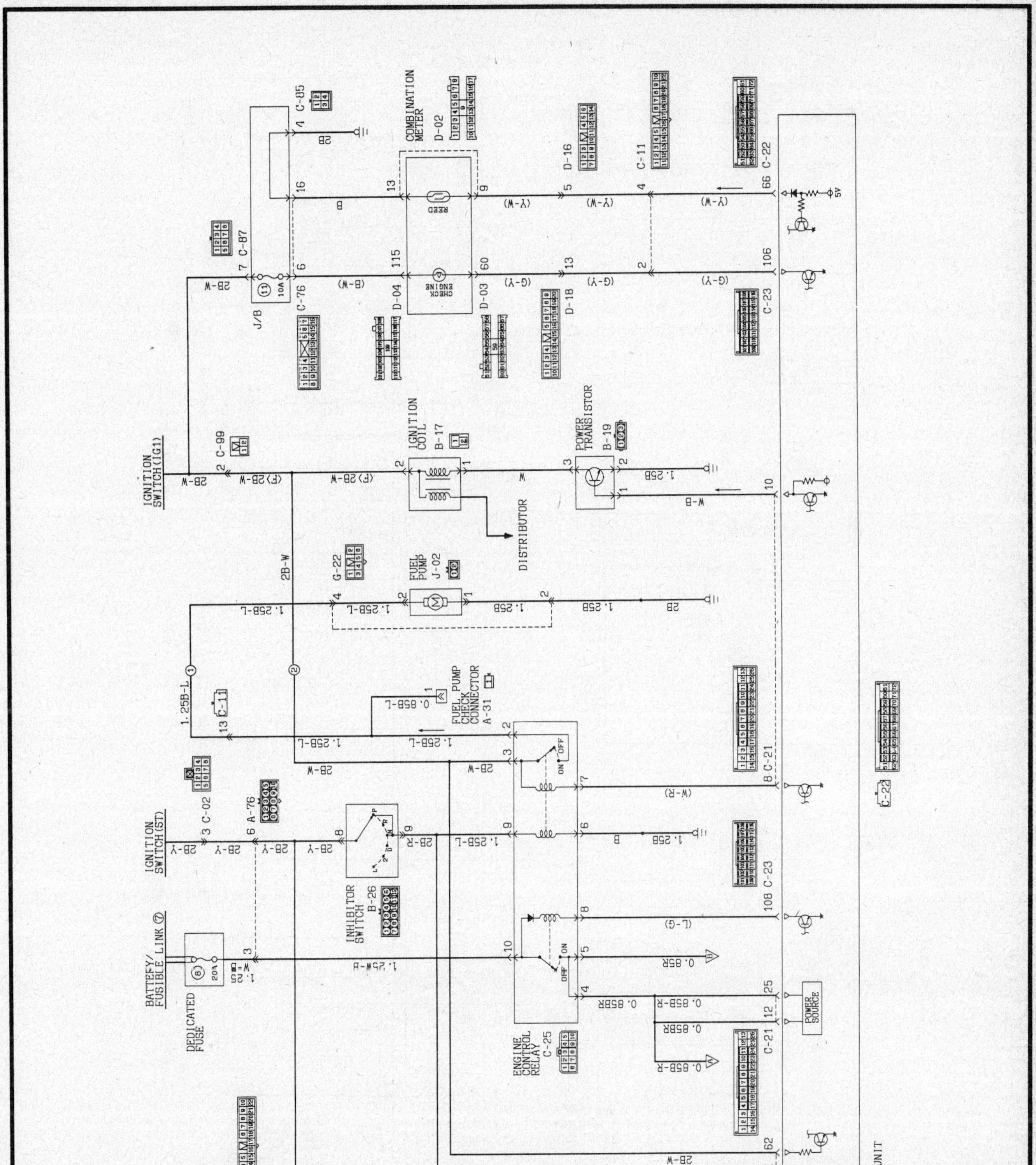

Fuel system electrical diagram—Diamante with 3.0L (VIN S) engine

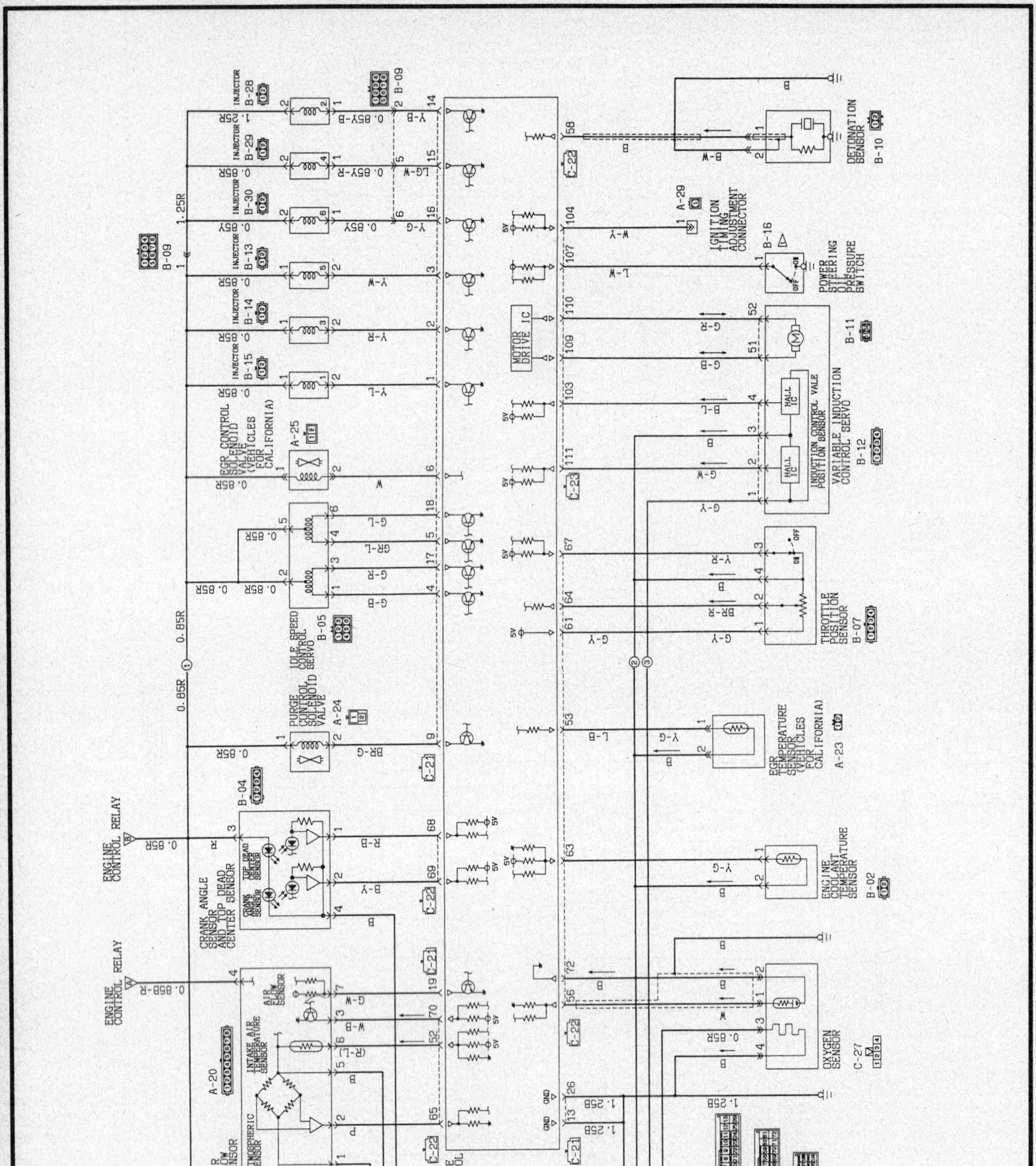

Fuel system electrical diagram, continued – Diamante with 3.0L (VIN S) engine

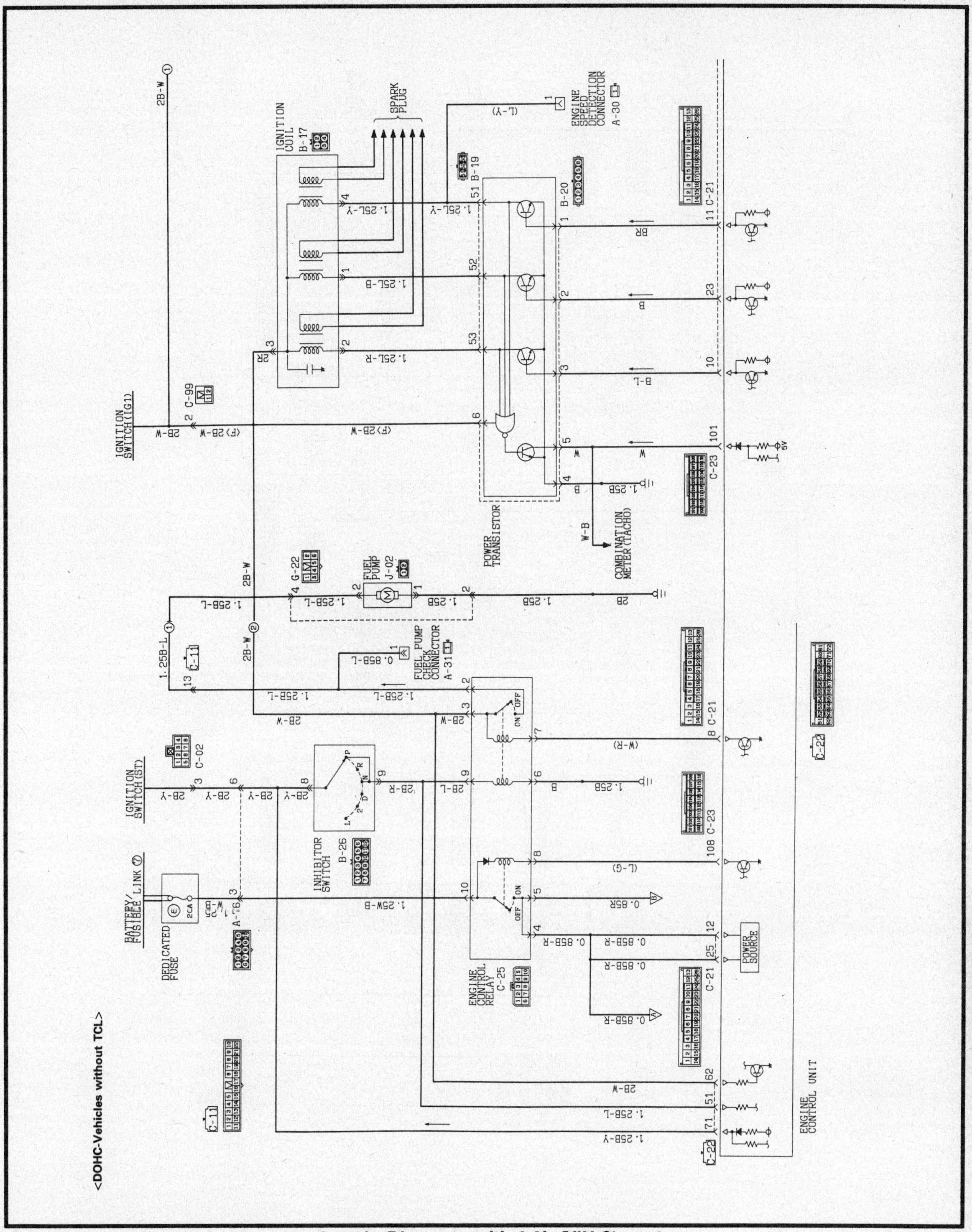

Fuel system electrical diagram, continued—Diamante with 3.0L (VIN S) engine

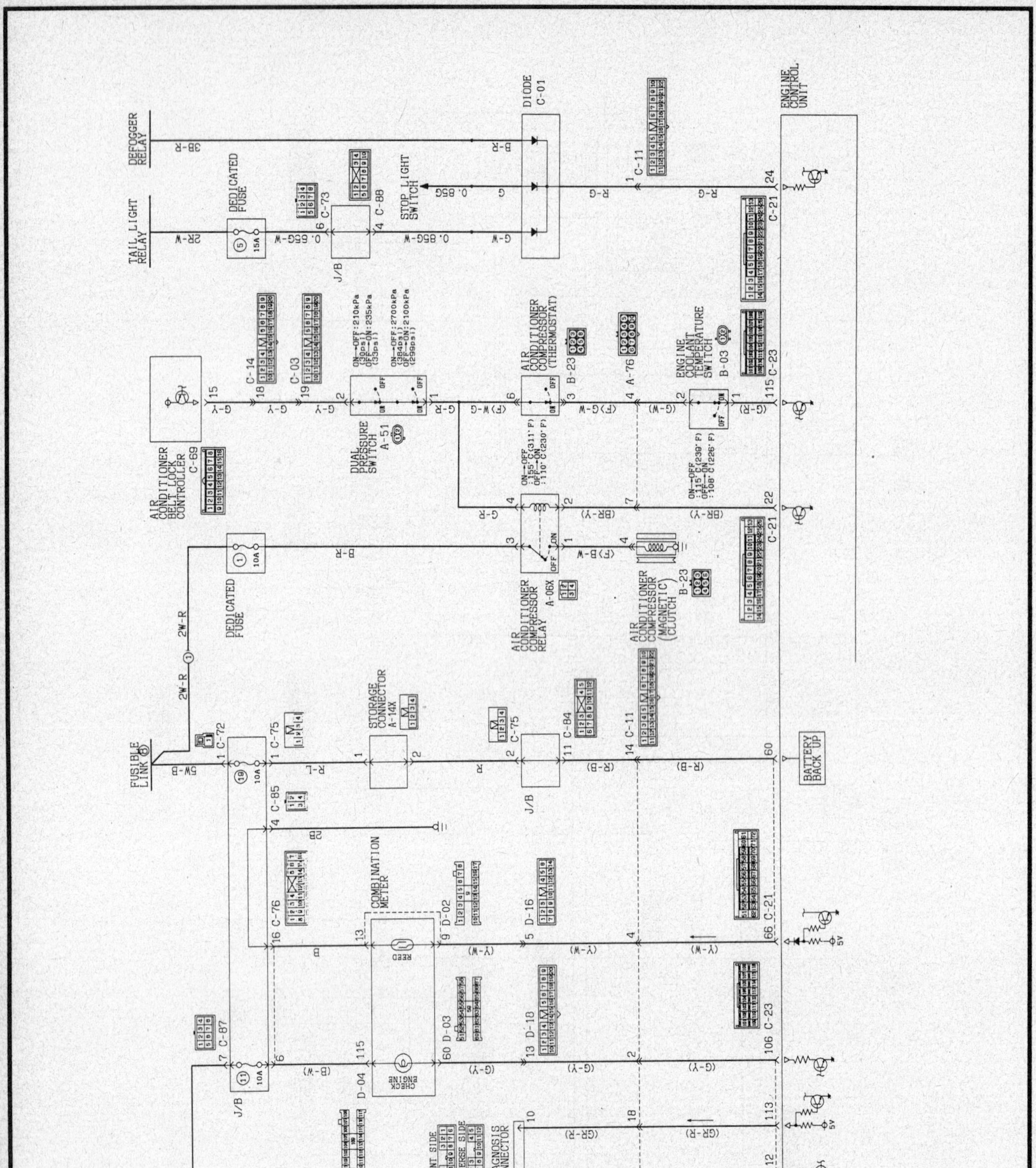

Fuel system electrical diagram, continued – Diamante with 3.0L (VIN S) engine

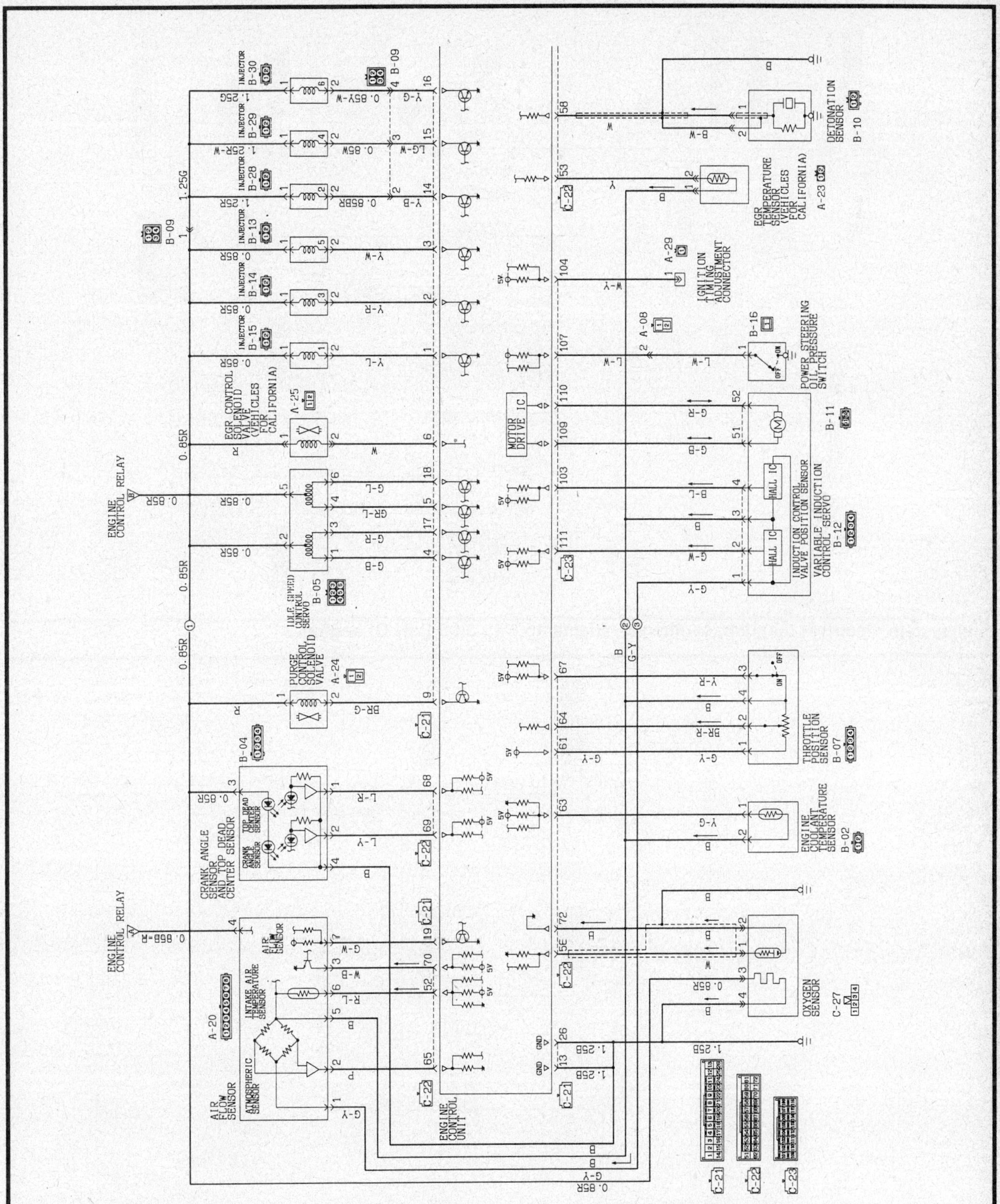

Fuel system electrical diagram, continued—Diamante with 3.0L (VIN S) engine

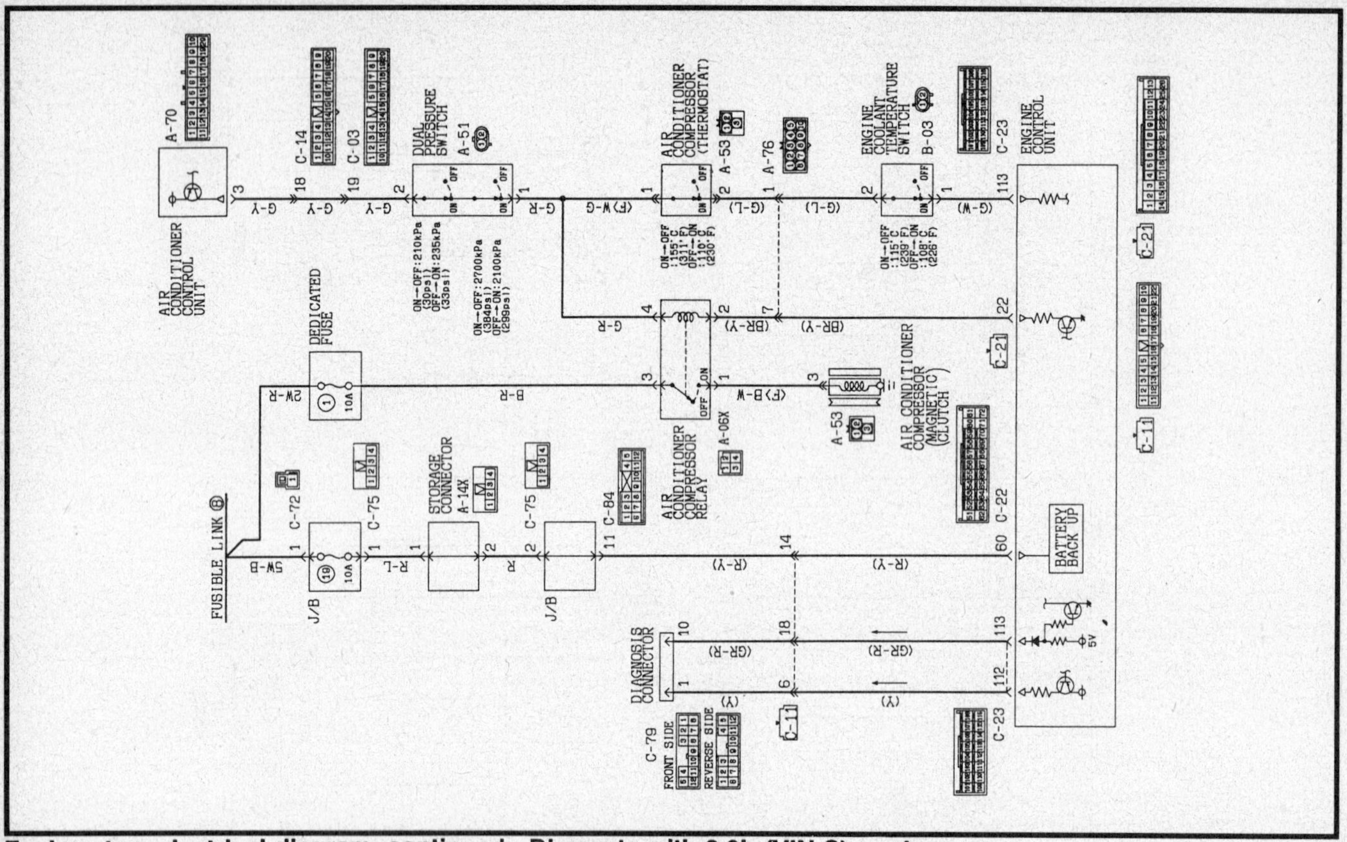

Fuel system electrical diagram, continued—Diamante with 3.0L (VIN S) engine

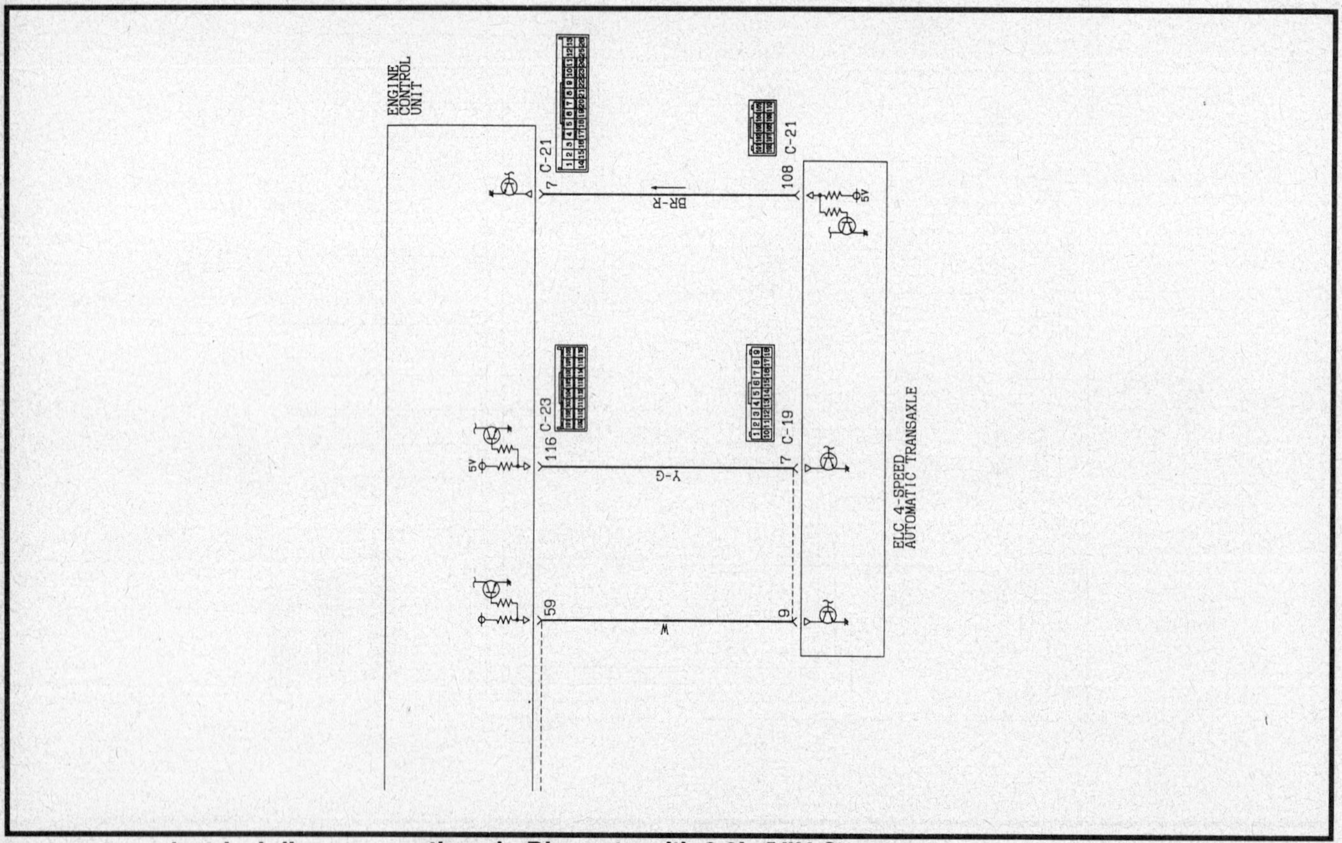

Fuel system electrical diagram, continued—Diamante with 3.0L (VIN S) engine

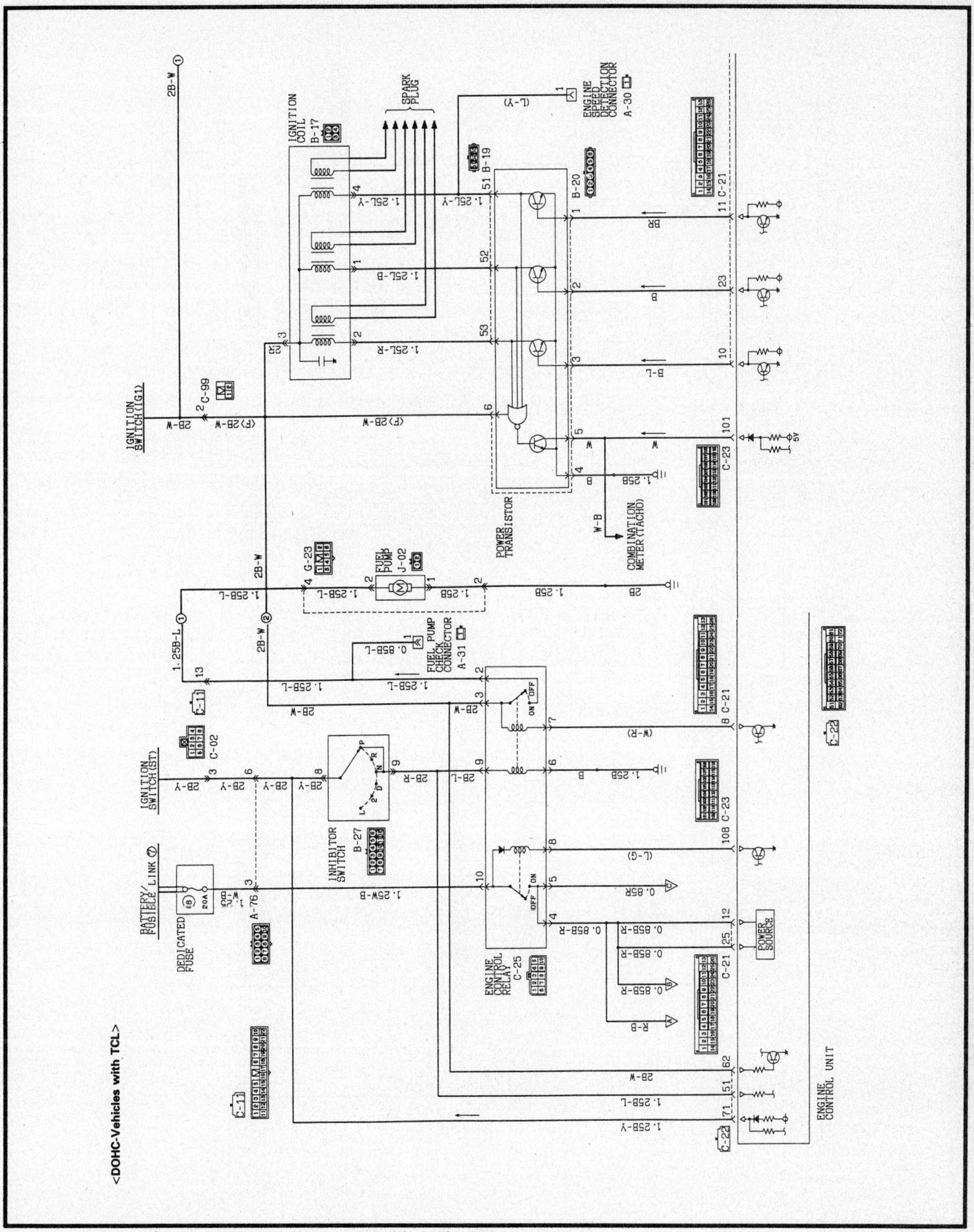

Fuel system electrical diagram – Diamante with 3.0L (VIN B) engine

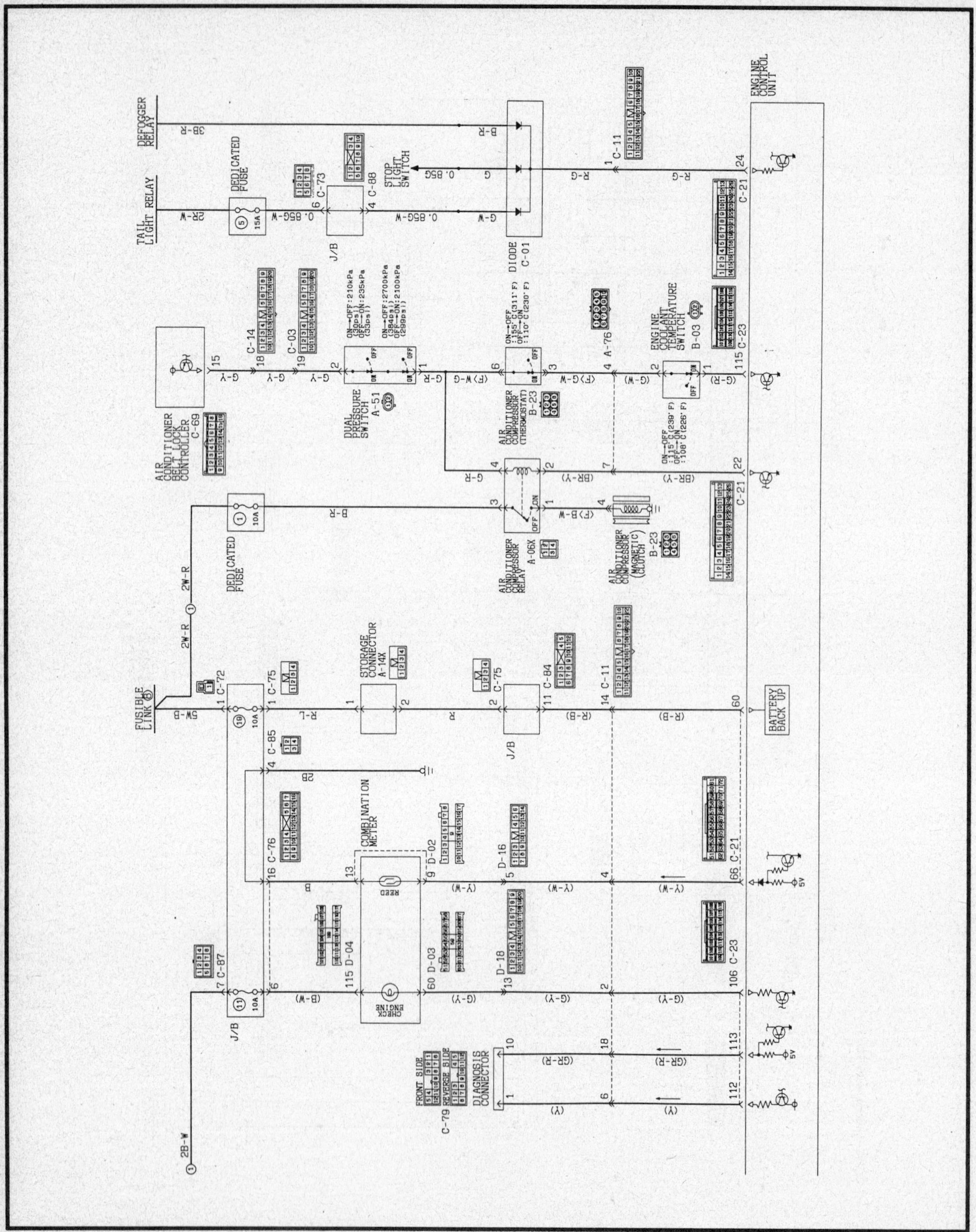

Fuel system electrical diagram, continued – Diamante with 3.0L (VIN B) engine

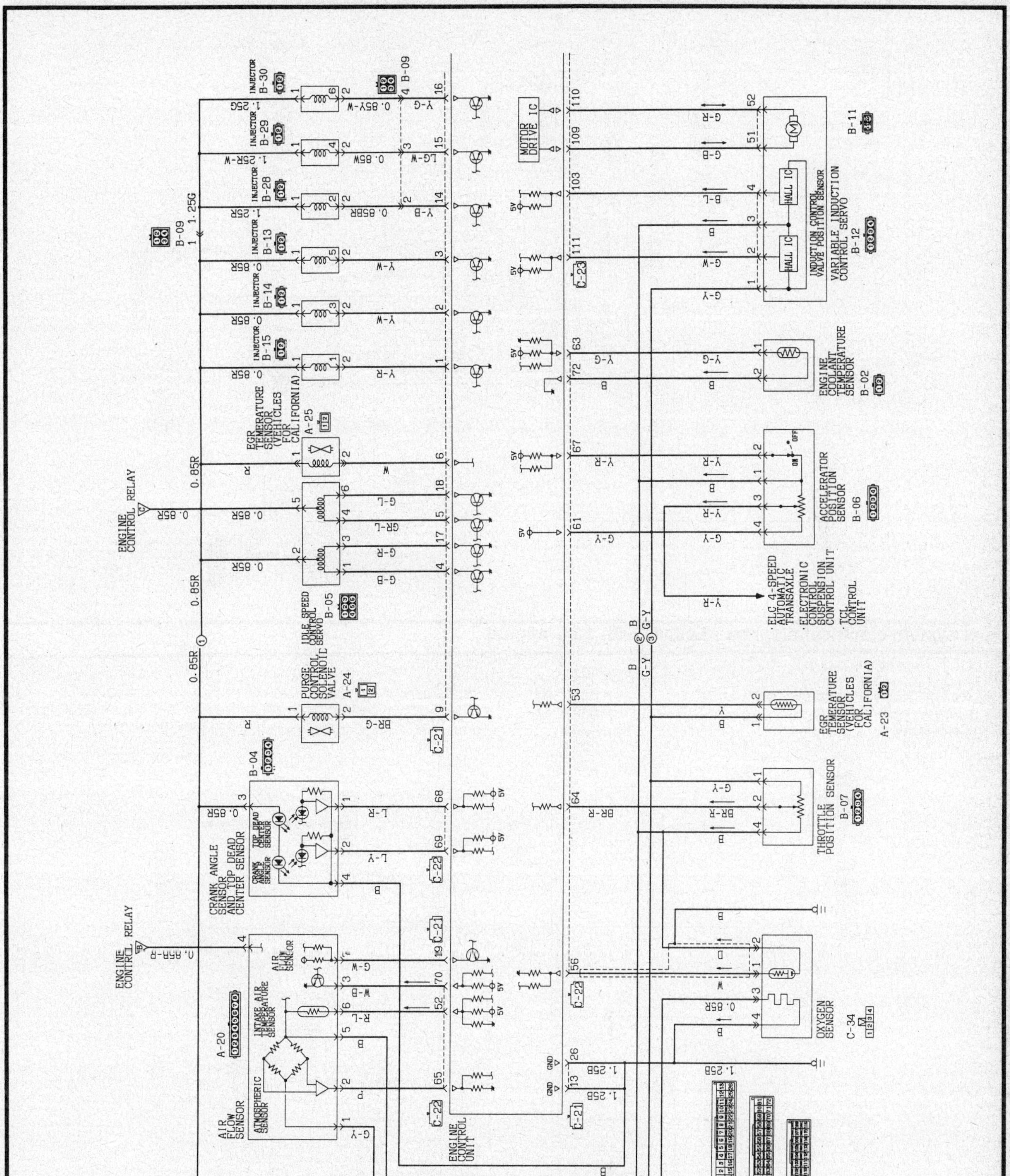

Fuel system electrical diagram, continued – Diamante with 3.0L (VIN B) engine

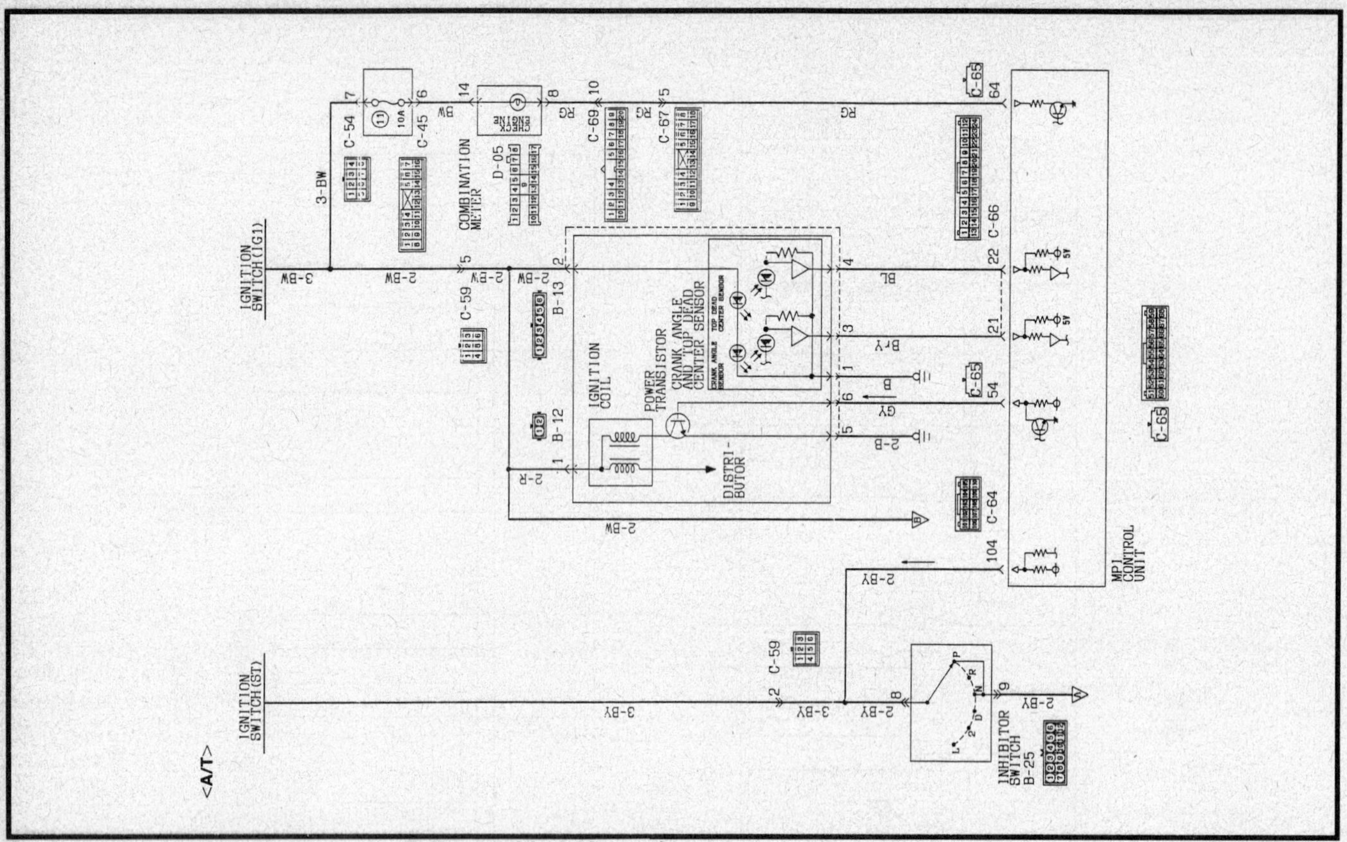

Fuel system electrical diagram – Eclipse with 1.8L engine

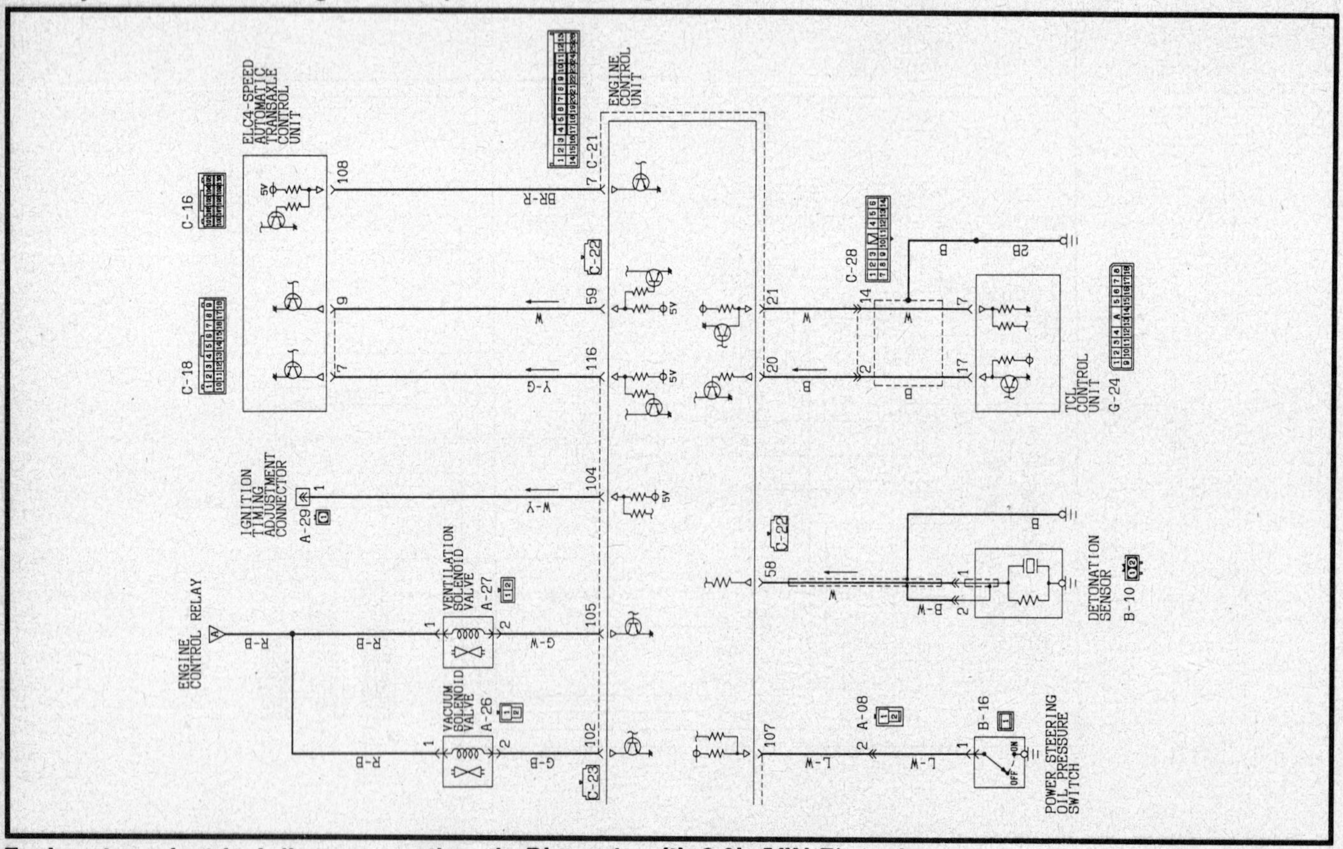

Fuel system electrical diagram, continued – Diamante with 3.0L (VIN B) engine

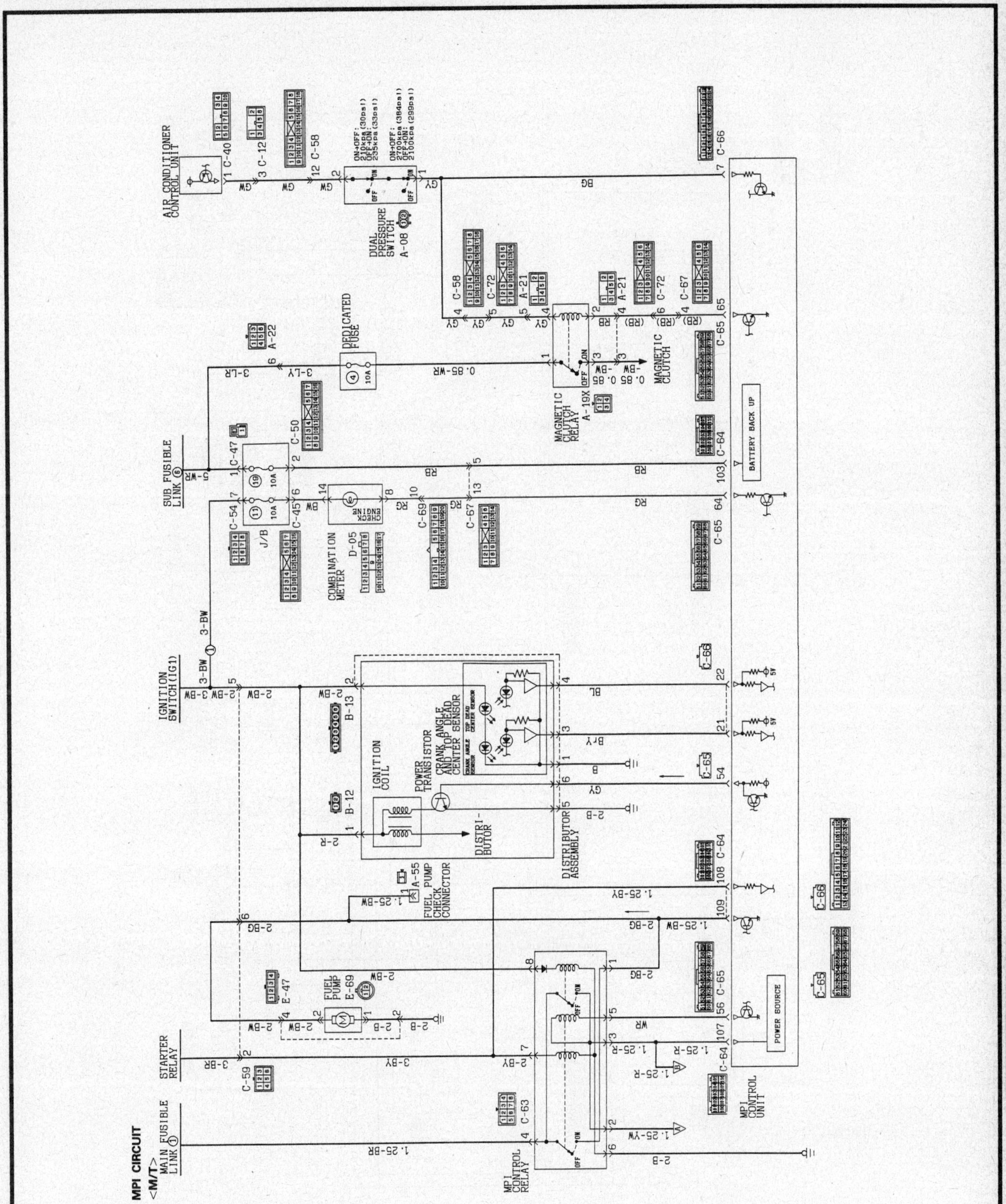

Fuel system electrical diagram, continued—Eclipse with 1.8L engine

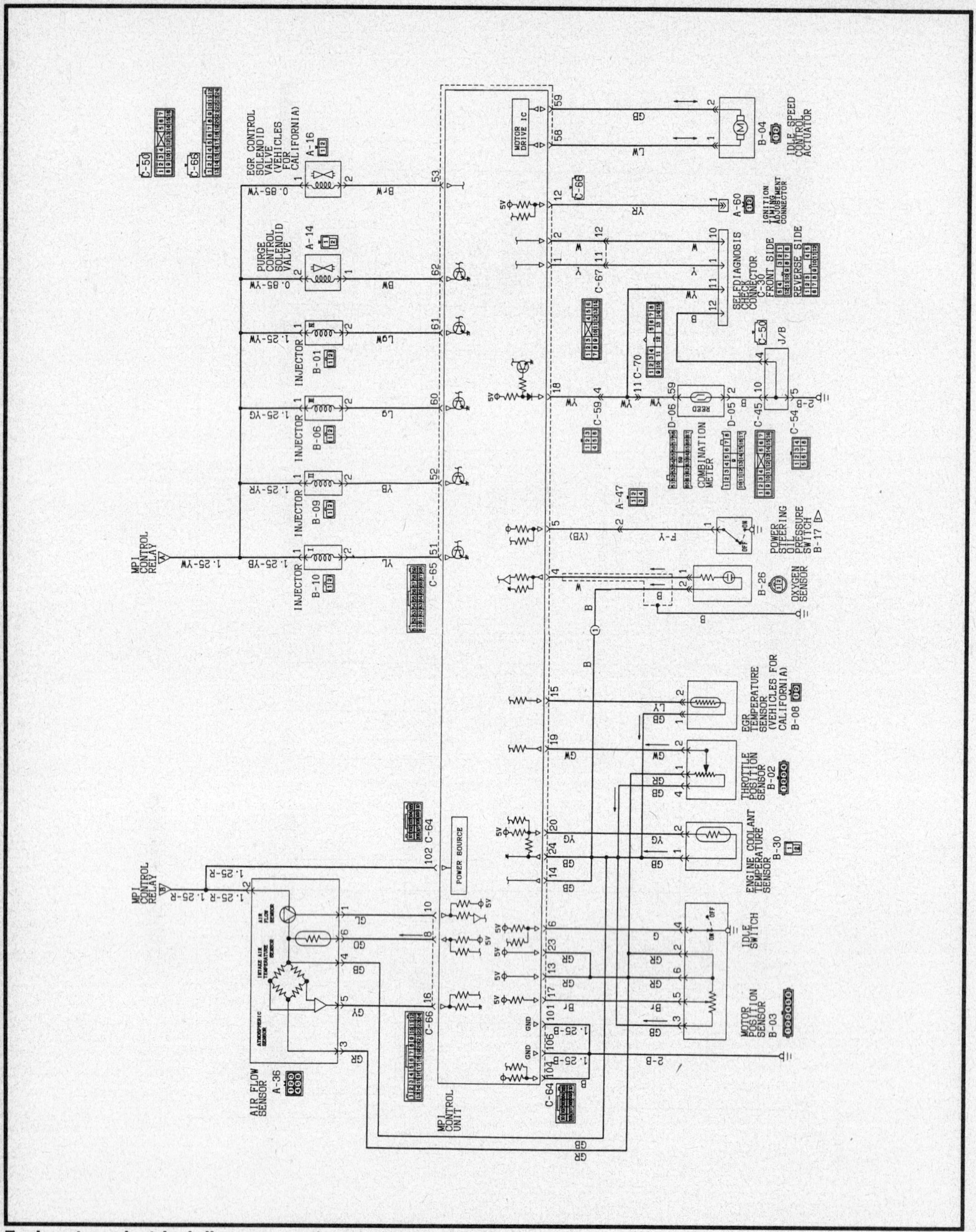

Fuel system electrical diagram, continued – Eclipse with 1.8L engine

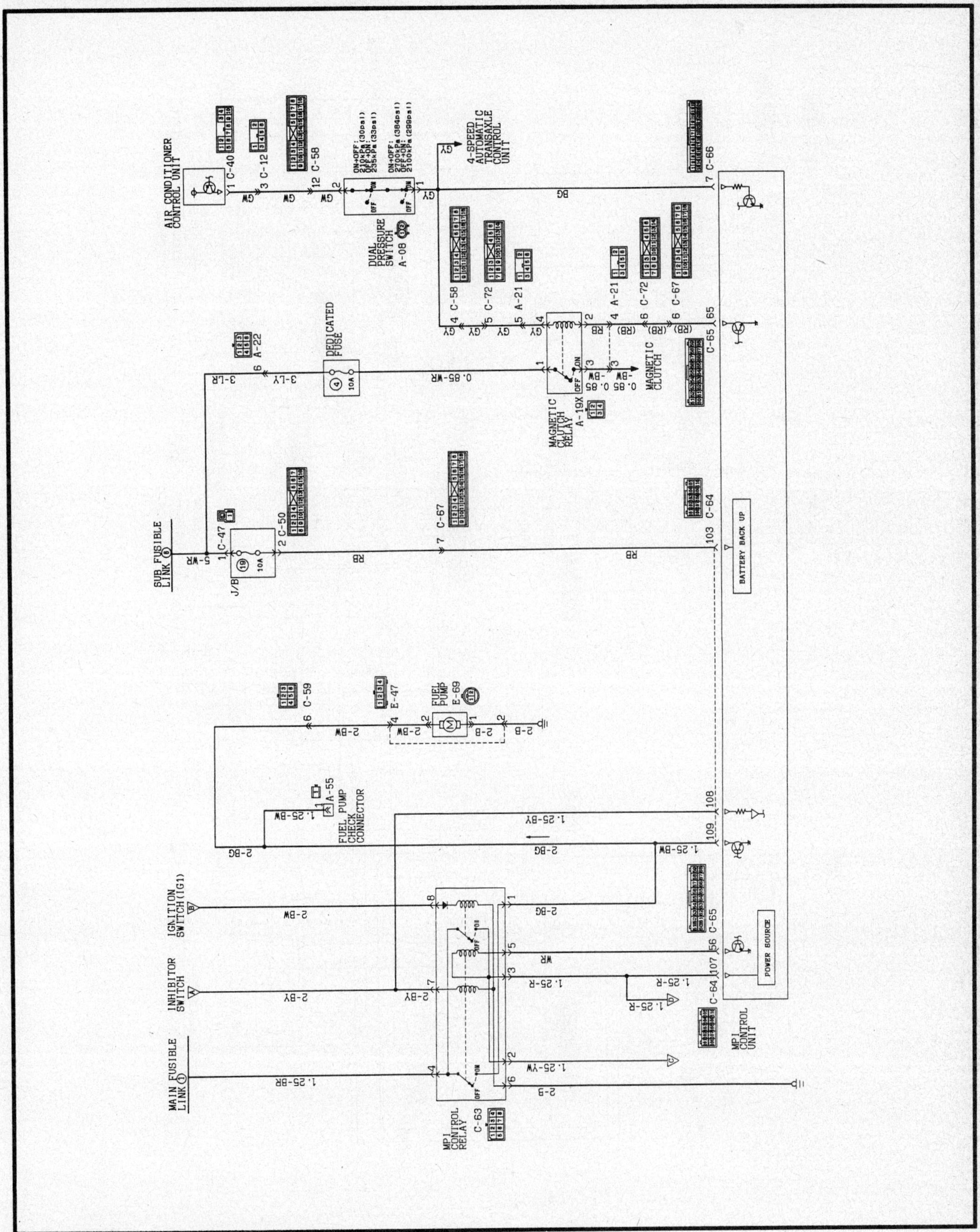

Fuel system electrical diagram, continued—Eclipse with 1.8L engine

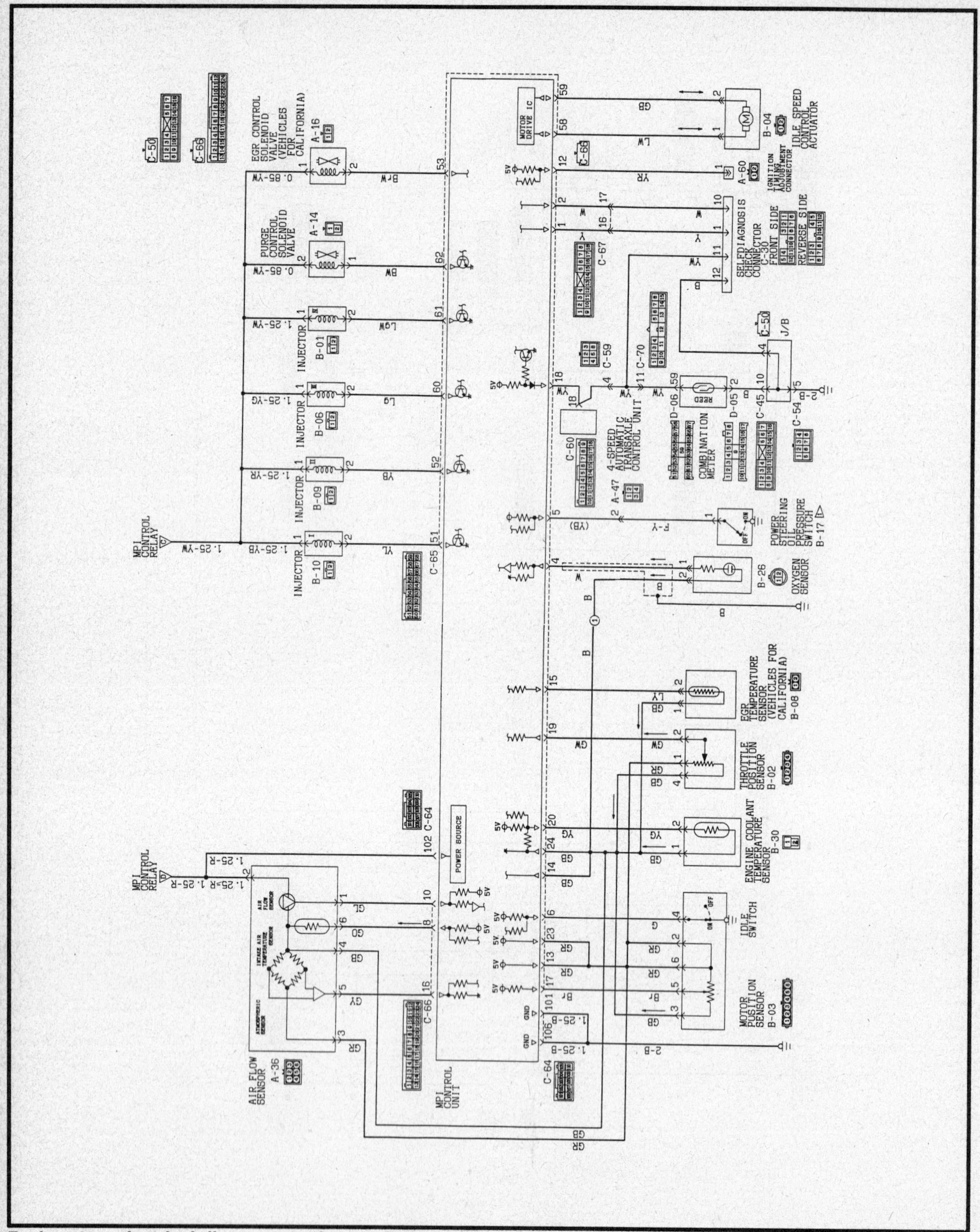

Fuel system electrical diagram, continued – Eclipse with 1.8L engine

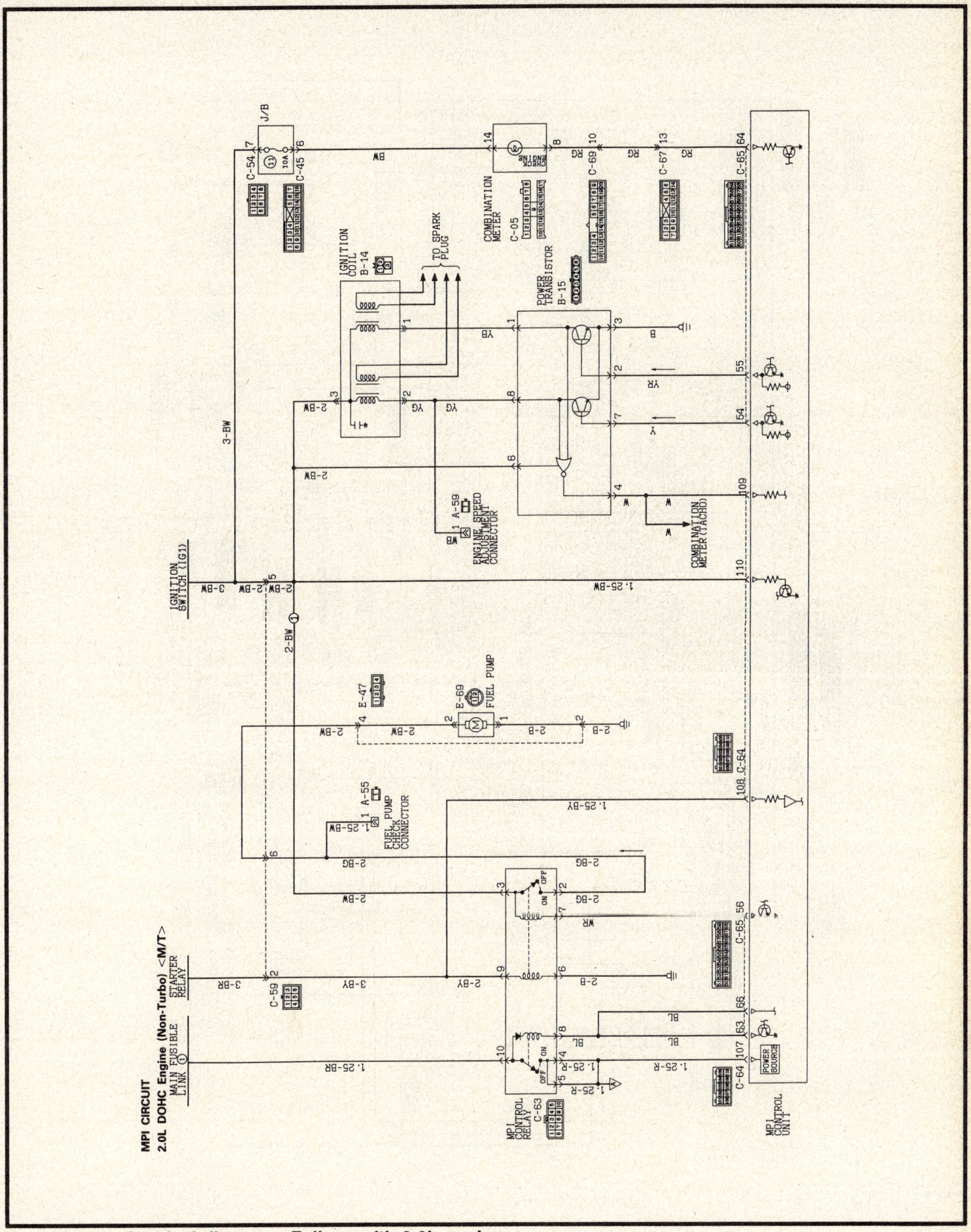

Fuel system electrical diagram—Eclipse with 2.0L engine

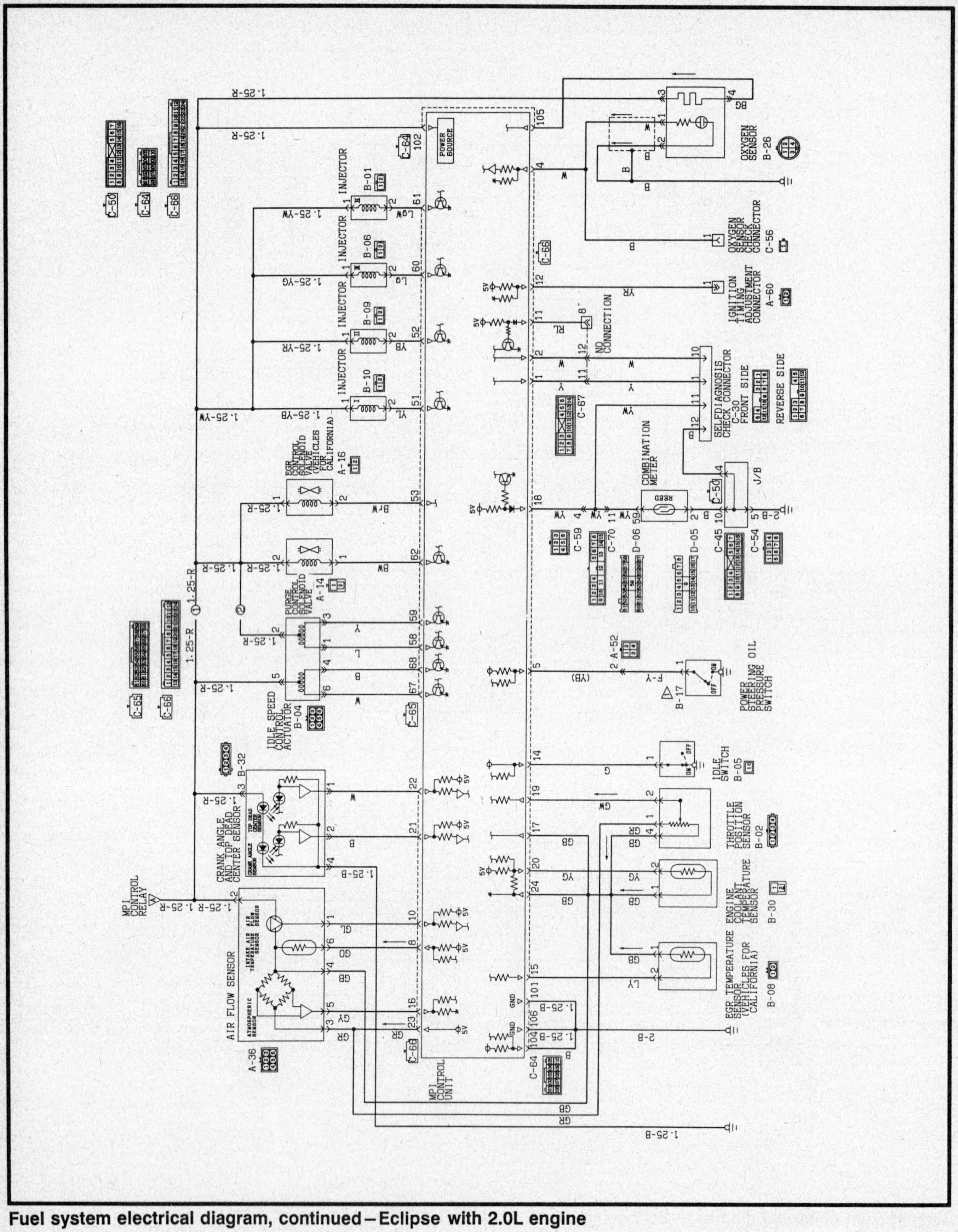

Fuel system electrical diagram, continued – Eclipse with 2.0L engine

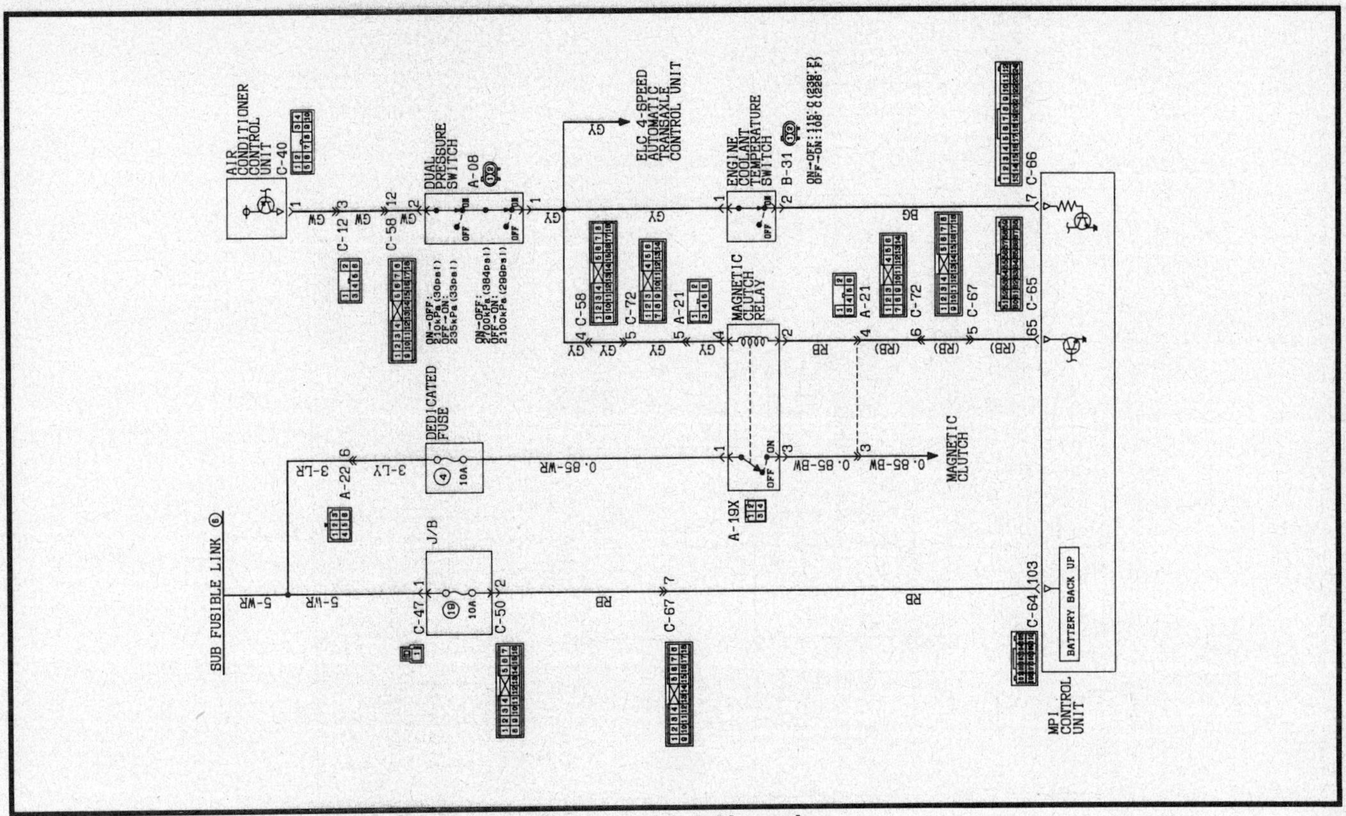

Fuel system electrical diagram, continued – Eclipse with 2.0L engine

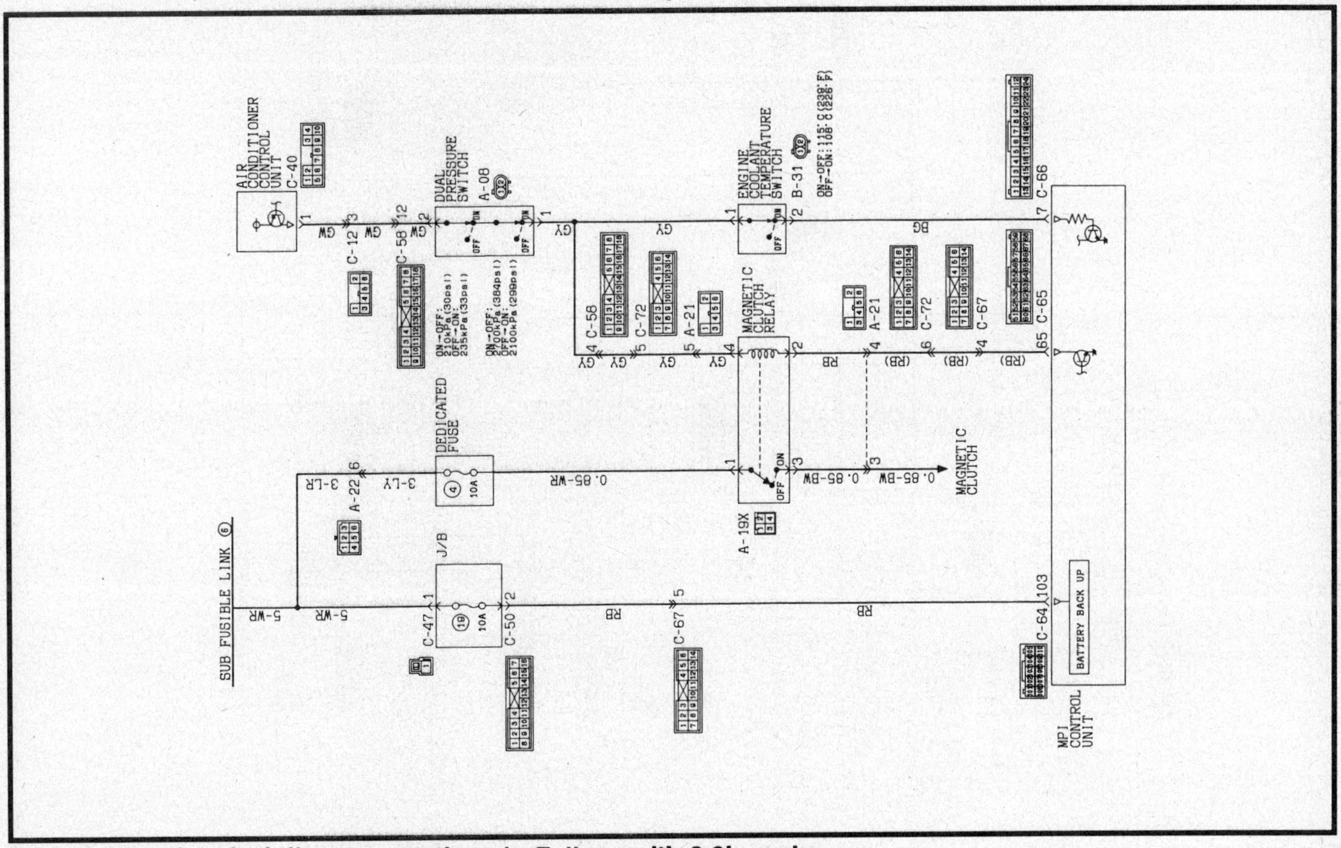

Fuel system electrical diagram, continued – Eclipse with 2.0L engine

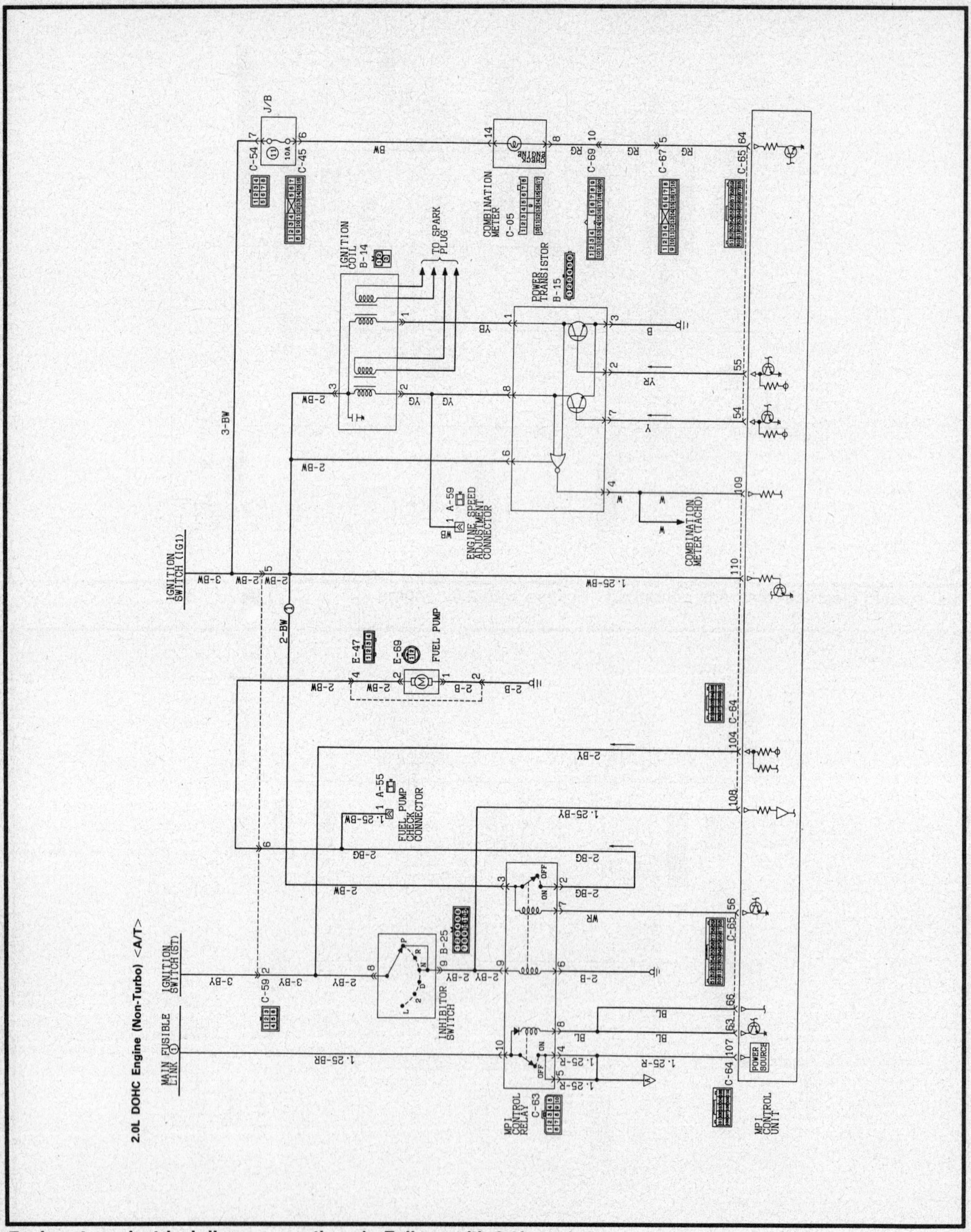

Fuel system electrical diagram, continued—Eclipse with 2.0L engine

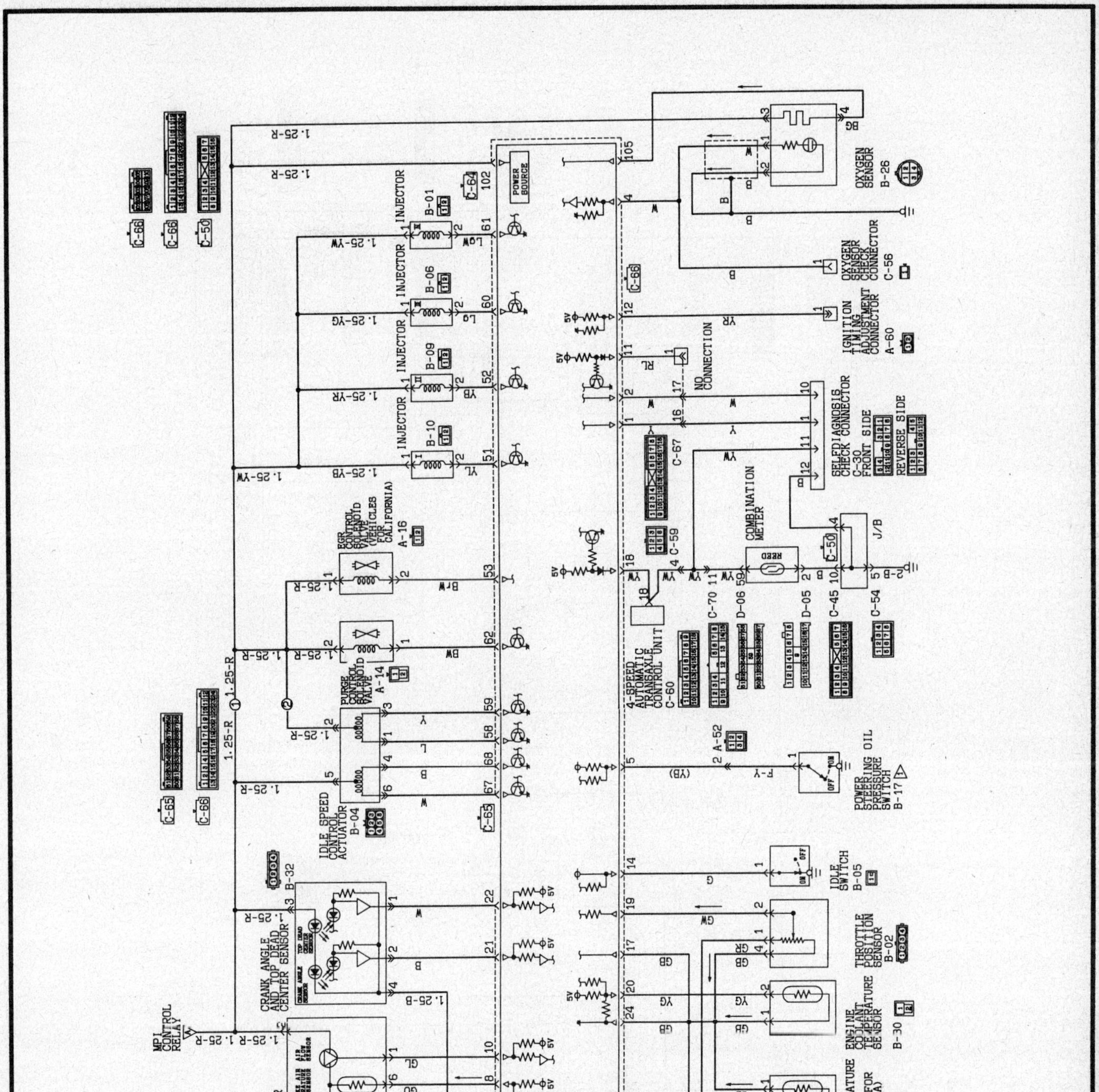

Fuel system electrical diagram, continued—Eclipse with 2.0L engine

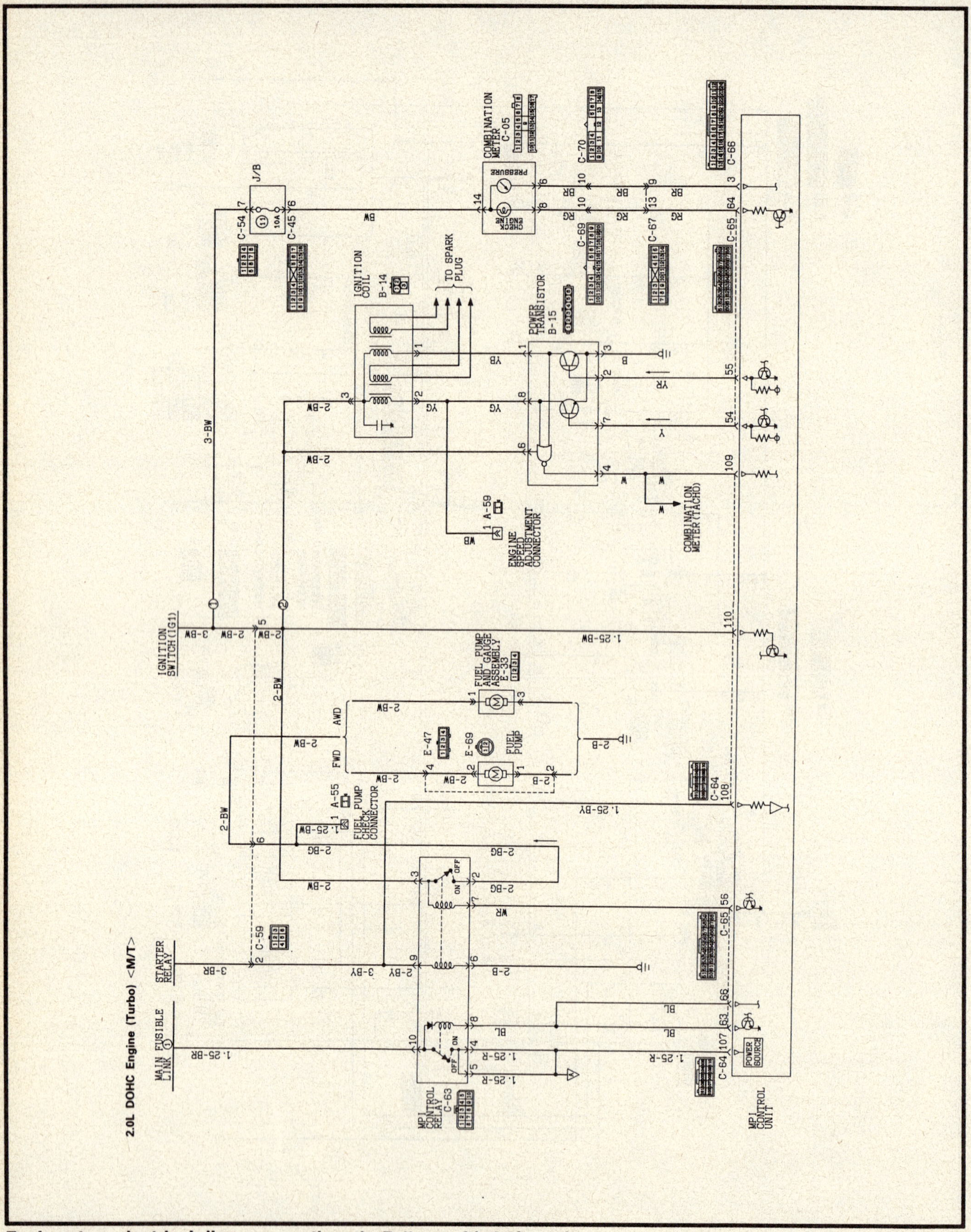

Fuel system electrical diagram, continued—Eclipse with 2.0L engine

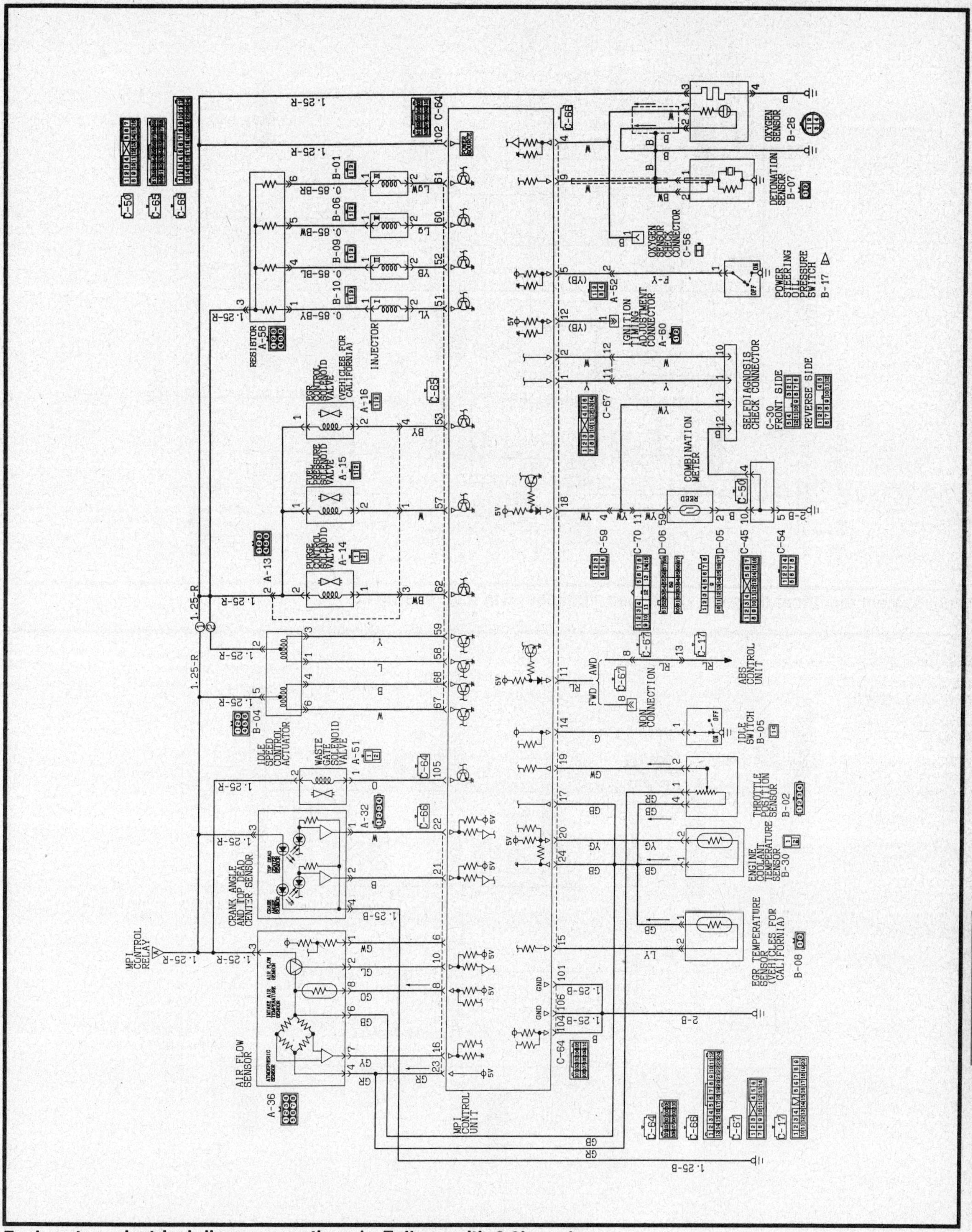

Fuel system electrical diagram, continued—Eclipse with 2.0L engine

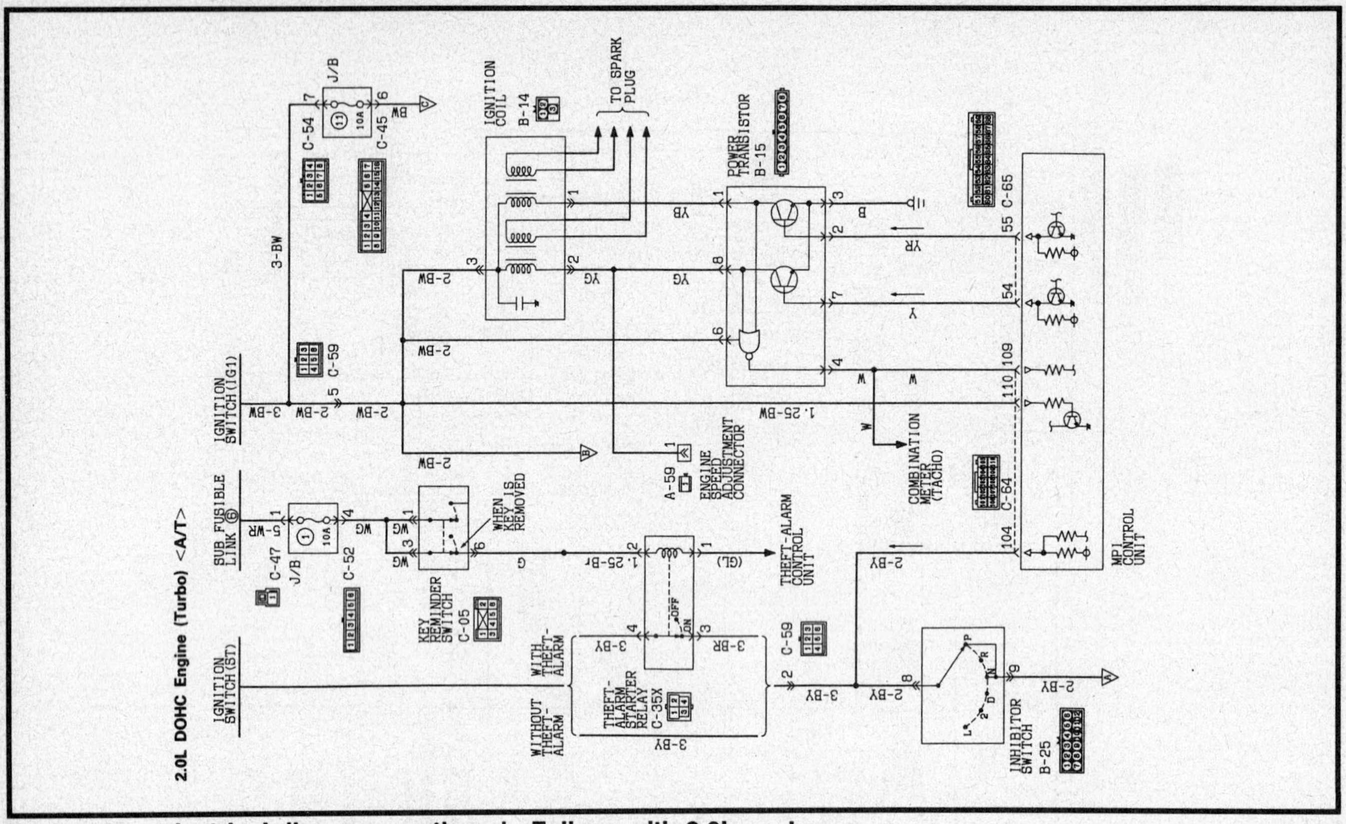

Fuel system electrical diagram, continued–Eclipse with 2.0L engine

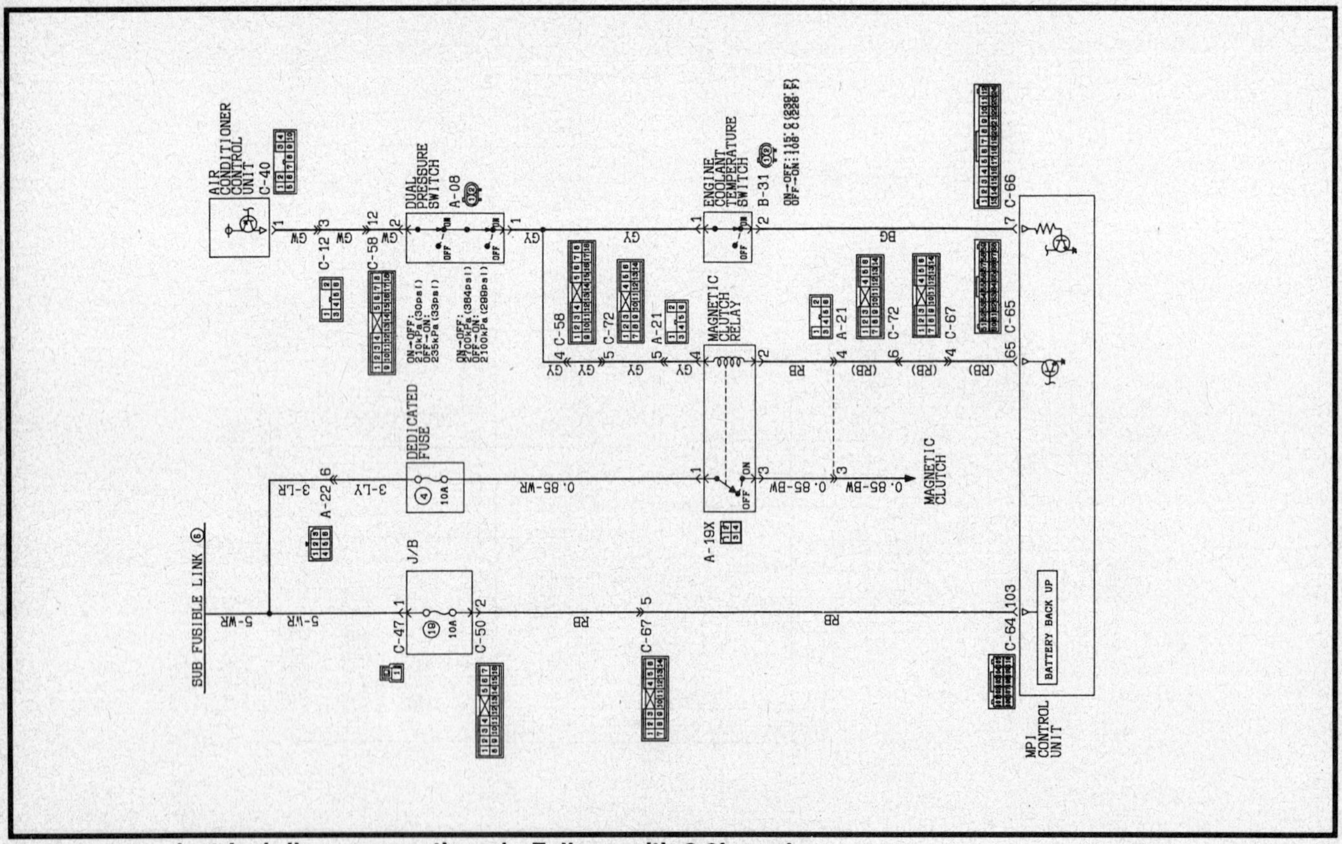

Fuel system electrical diagram, continued–Eclipse with 2.0L engine

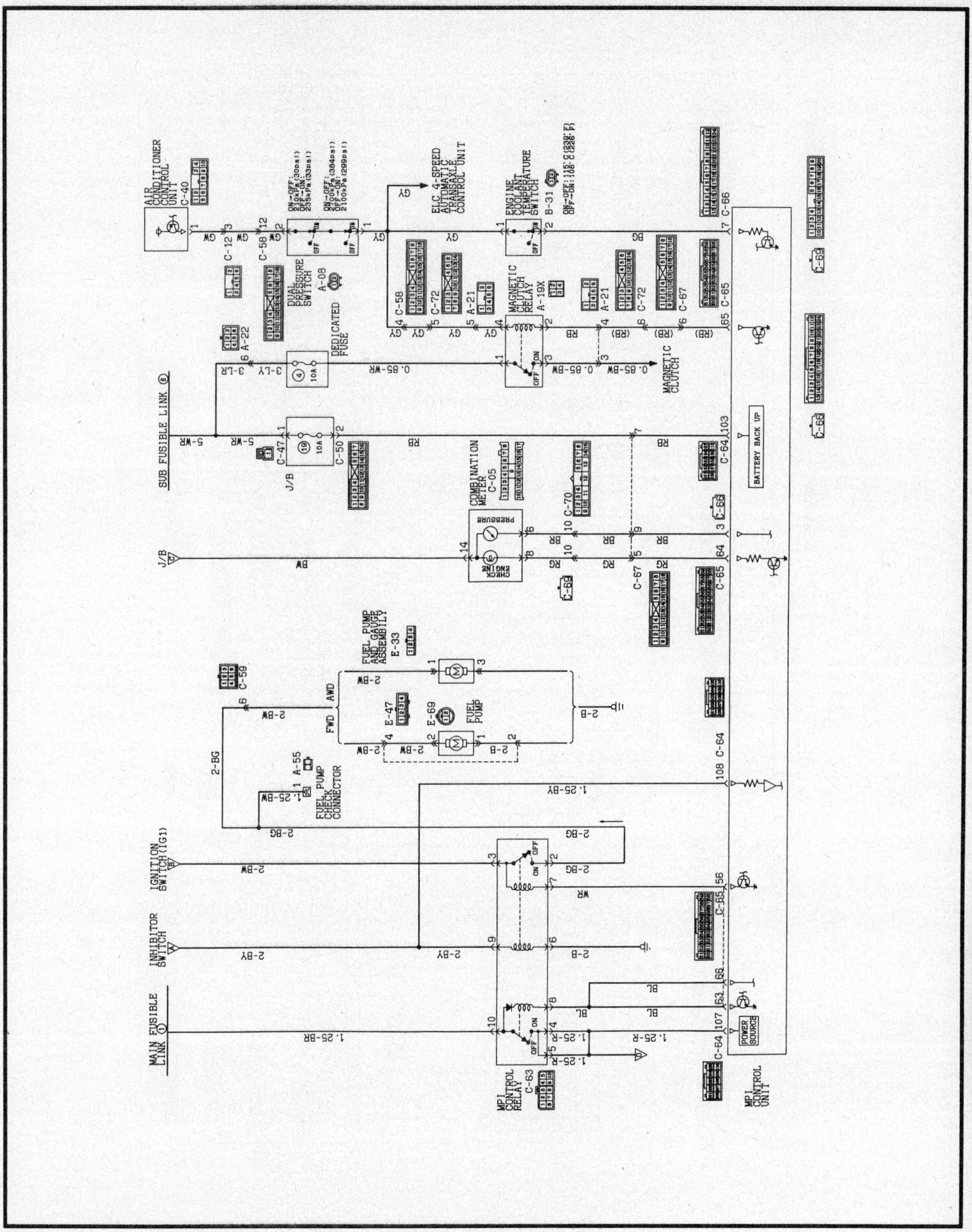

Fuel system electrical diagram, continued – Eclipse with 2.0L engine

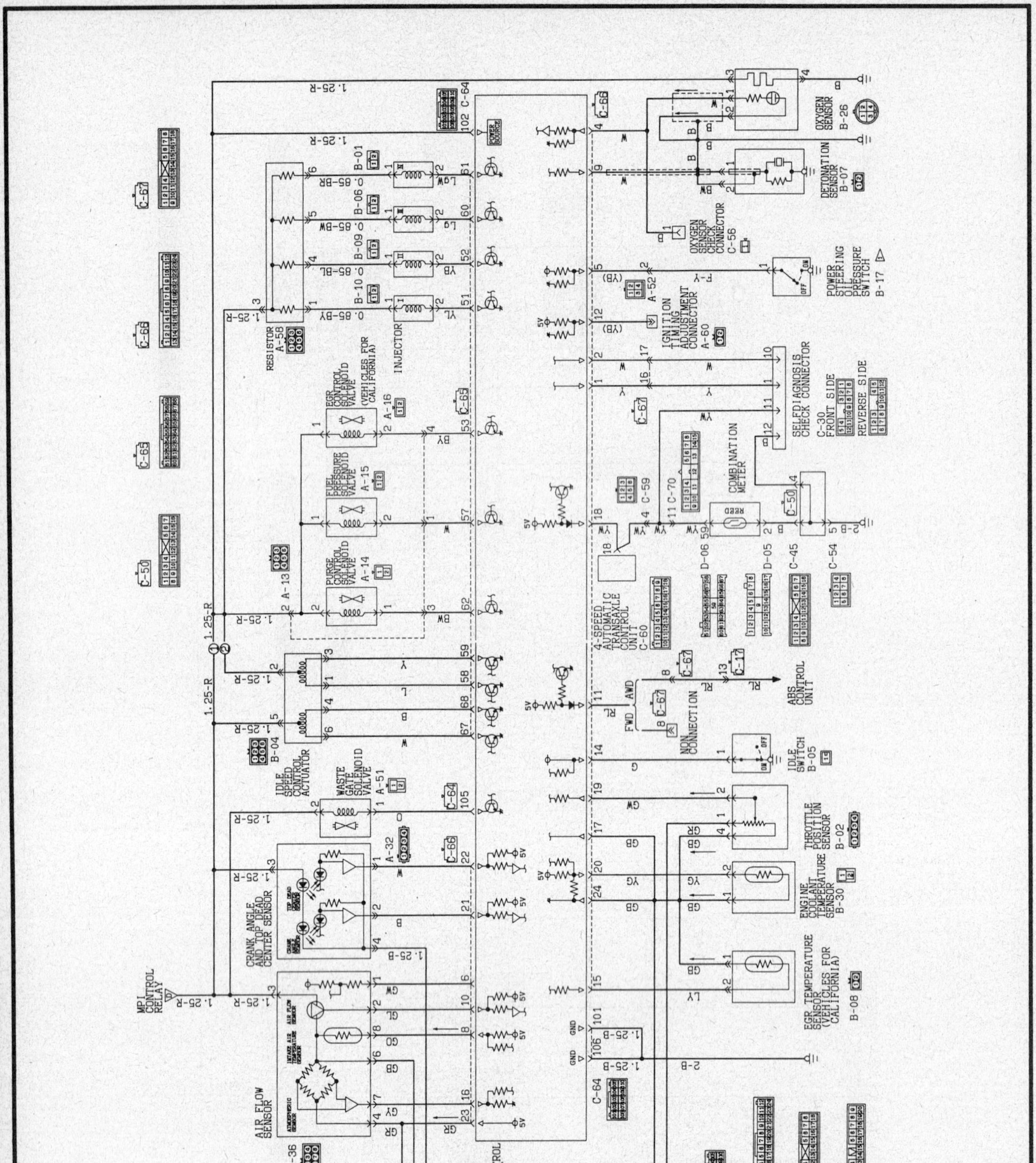

Fuel system electrical diagram, continued – Eclipse with 2.0L engine

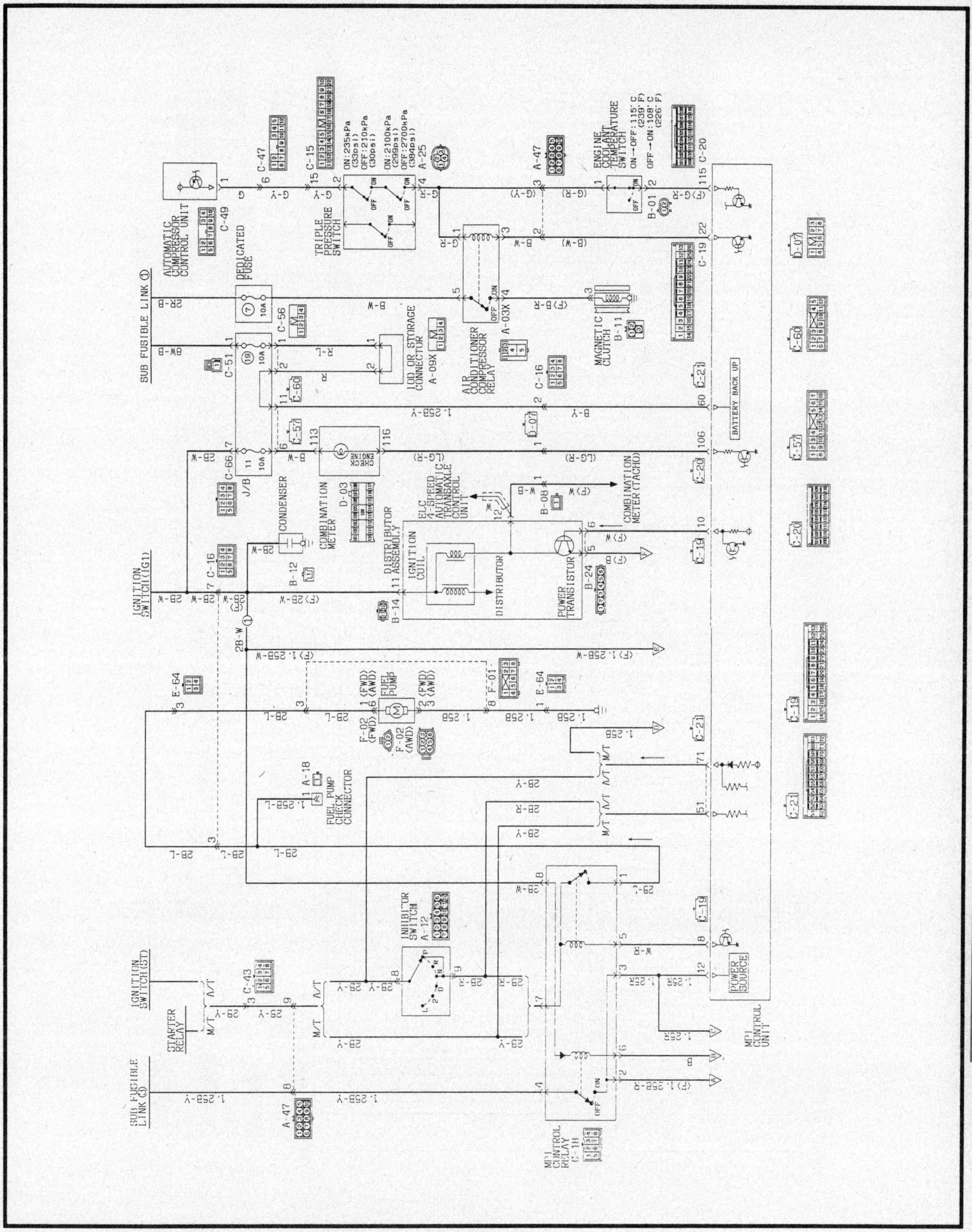

Fuel system electrical diagram—Exp/LRV with 1.8L engine

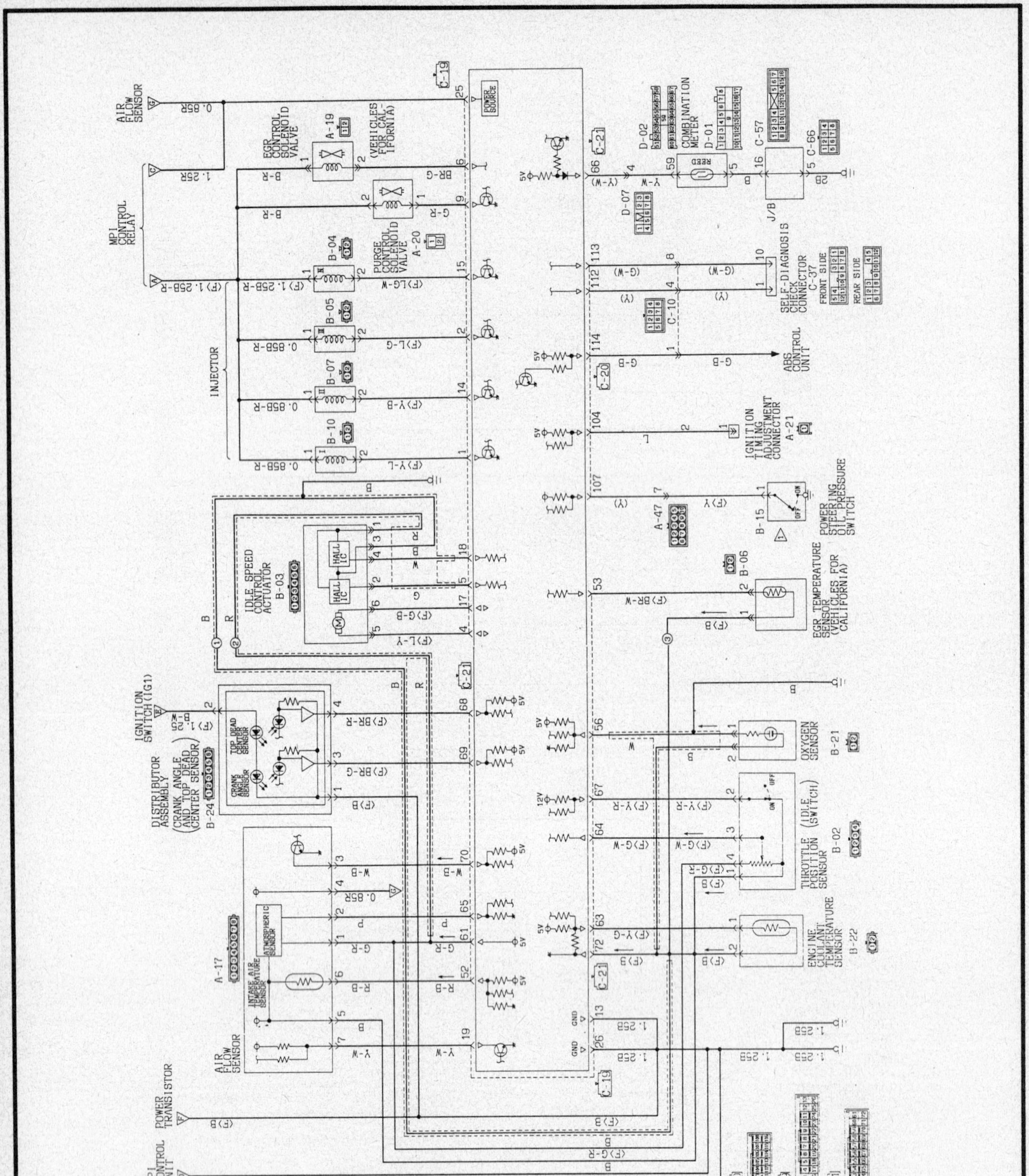

Fuel system electrical diagram, continued – Exp/LRV with 1.8L engine

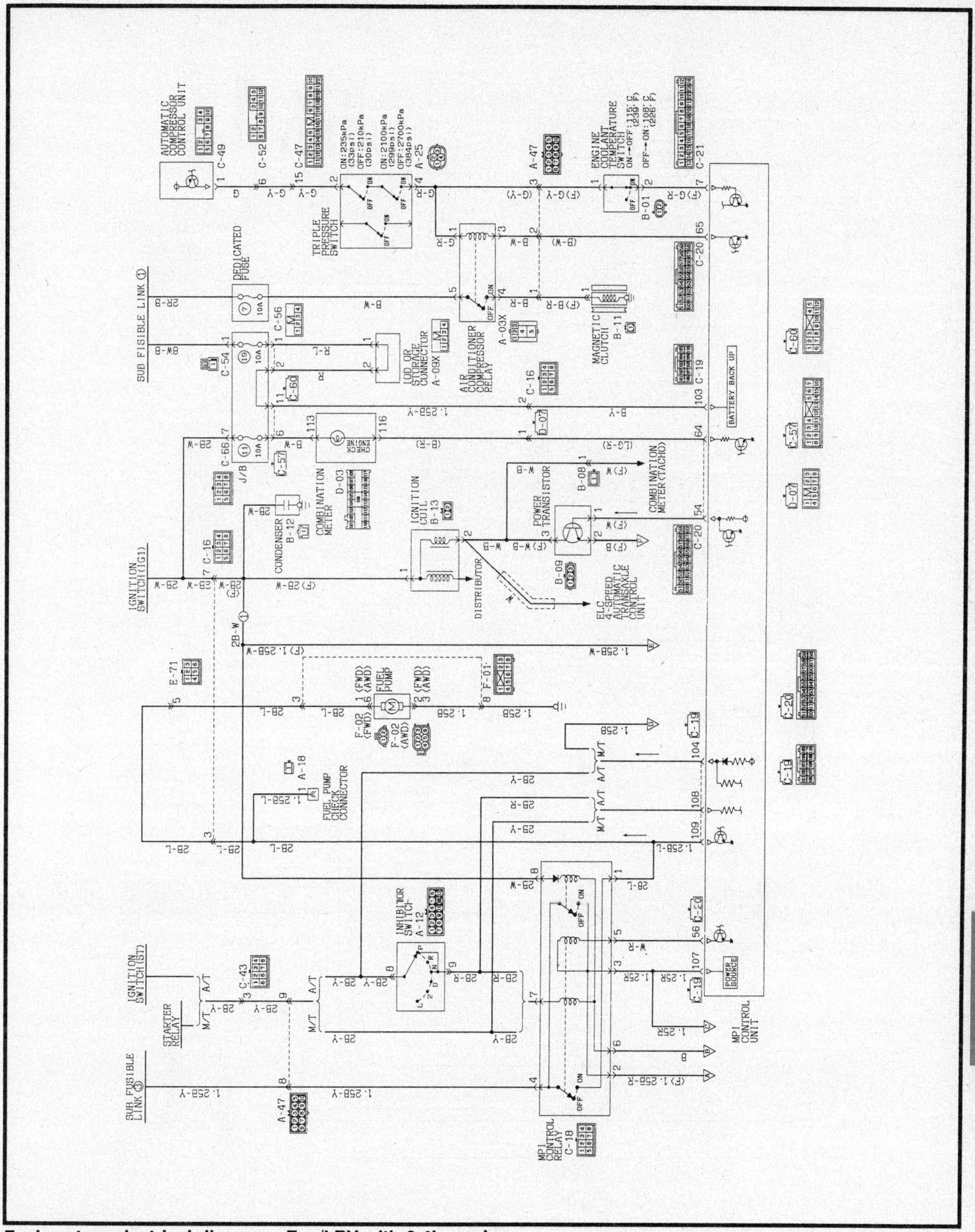

Fuel system electrical diagram – Exp/LRV with 2.4L engine

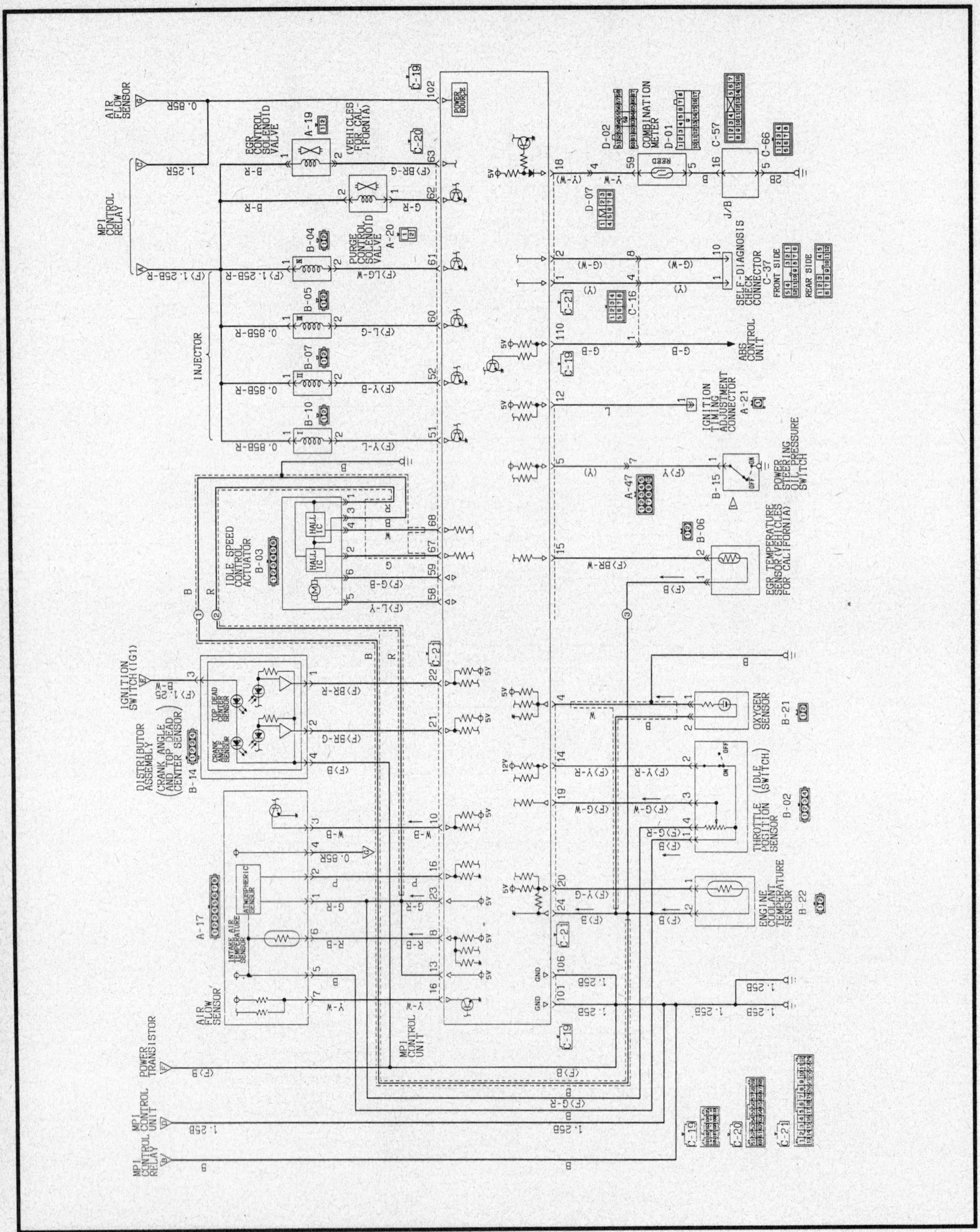

Fuel system electrical diagram, continued—Exp/LRV with 2.4L engine

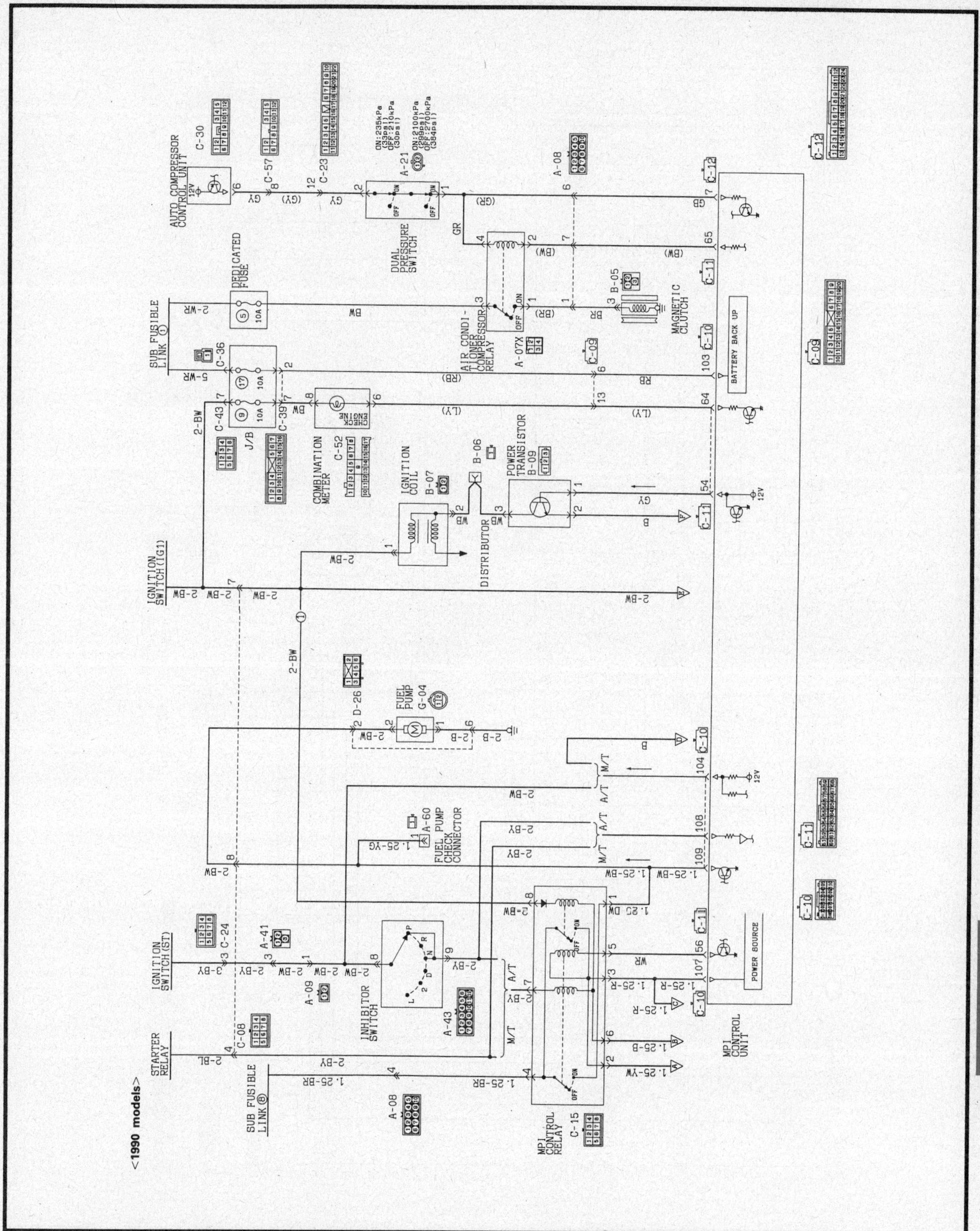

Fuel system electrical diagram—Galant

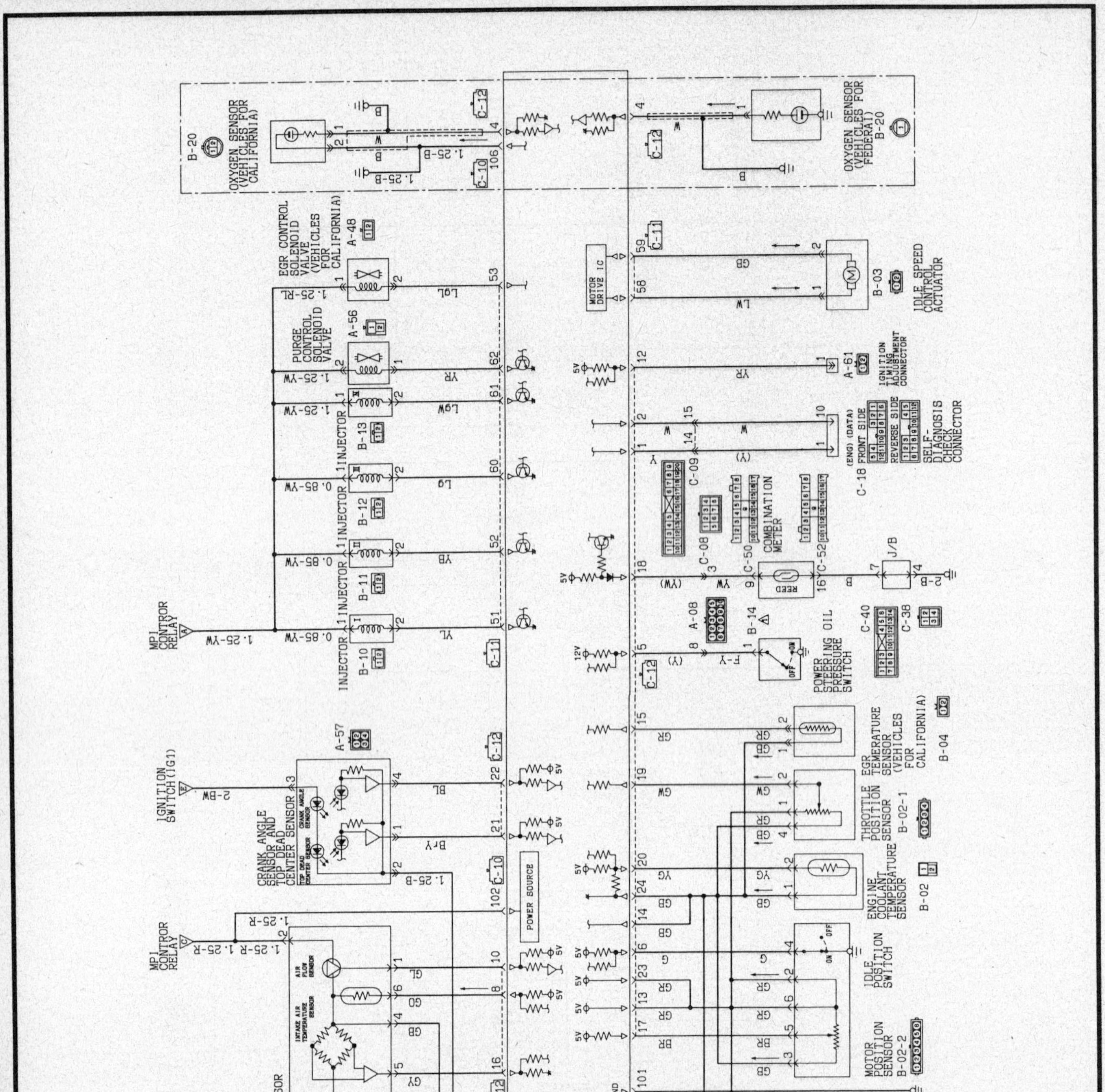

Fuel system electrical diagram, continued – Galant

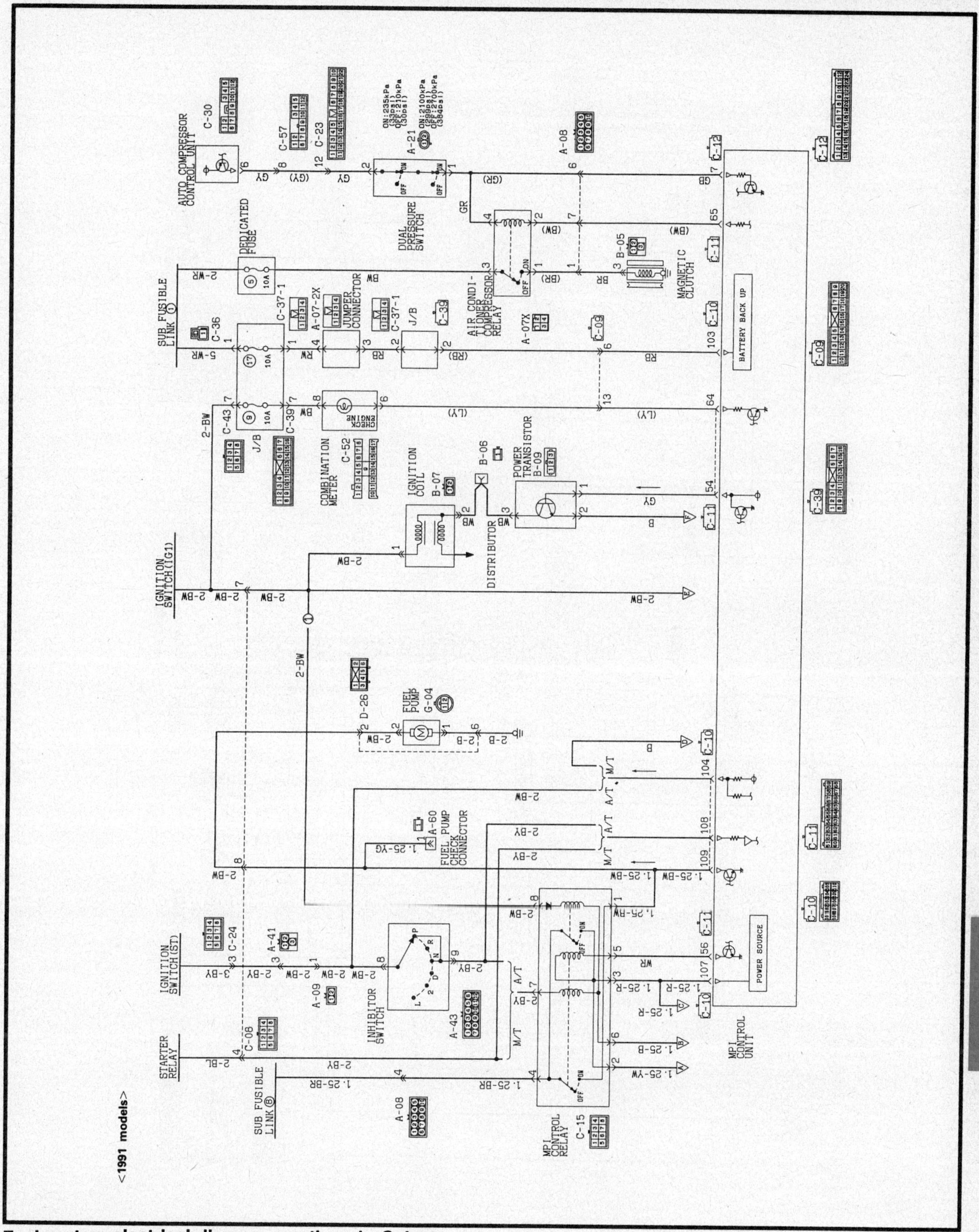

Fuel system electrical diagram, continued – Galant

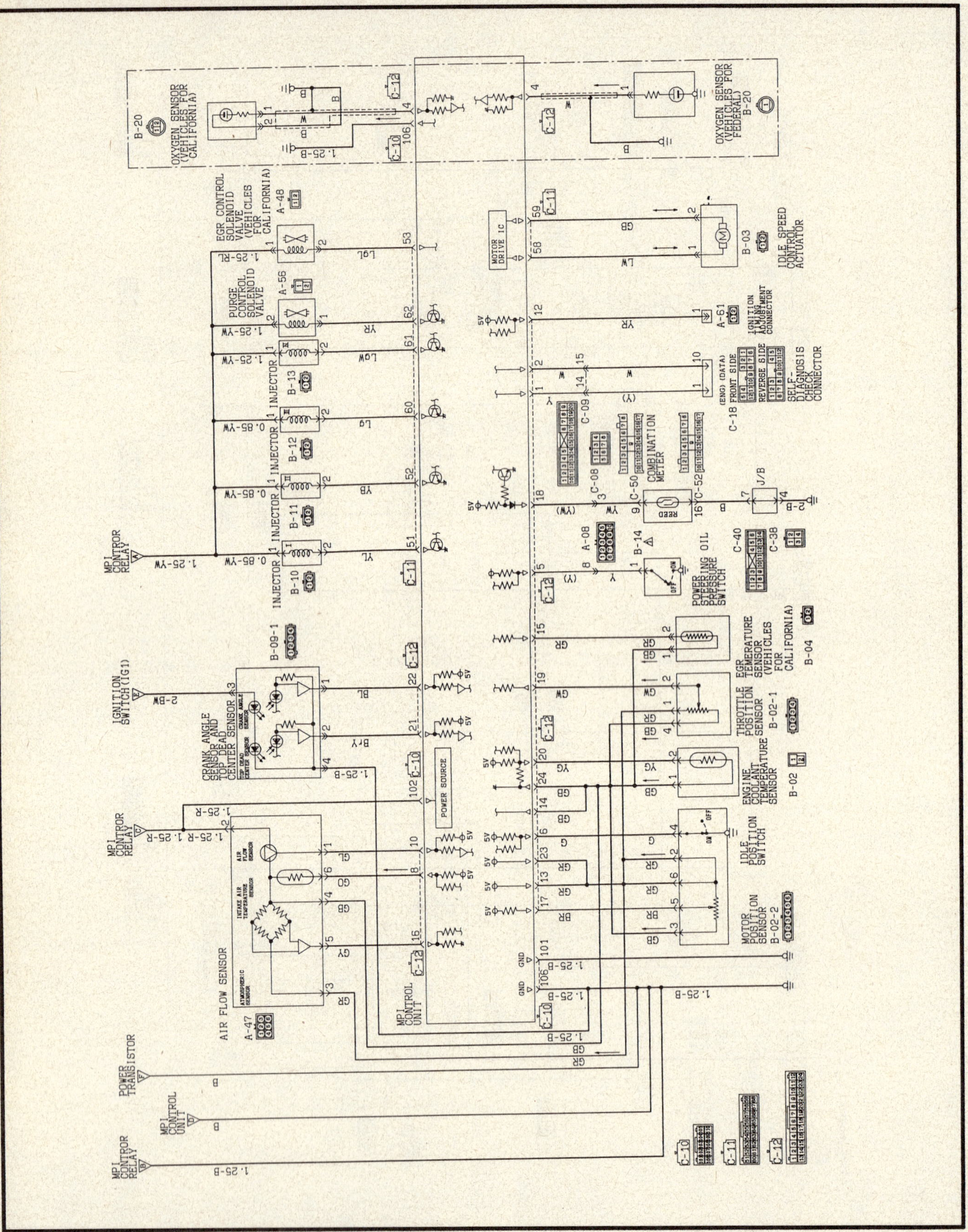

Fuel system electrical diagram, continued—Galant

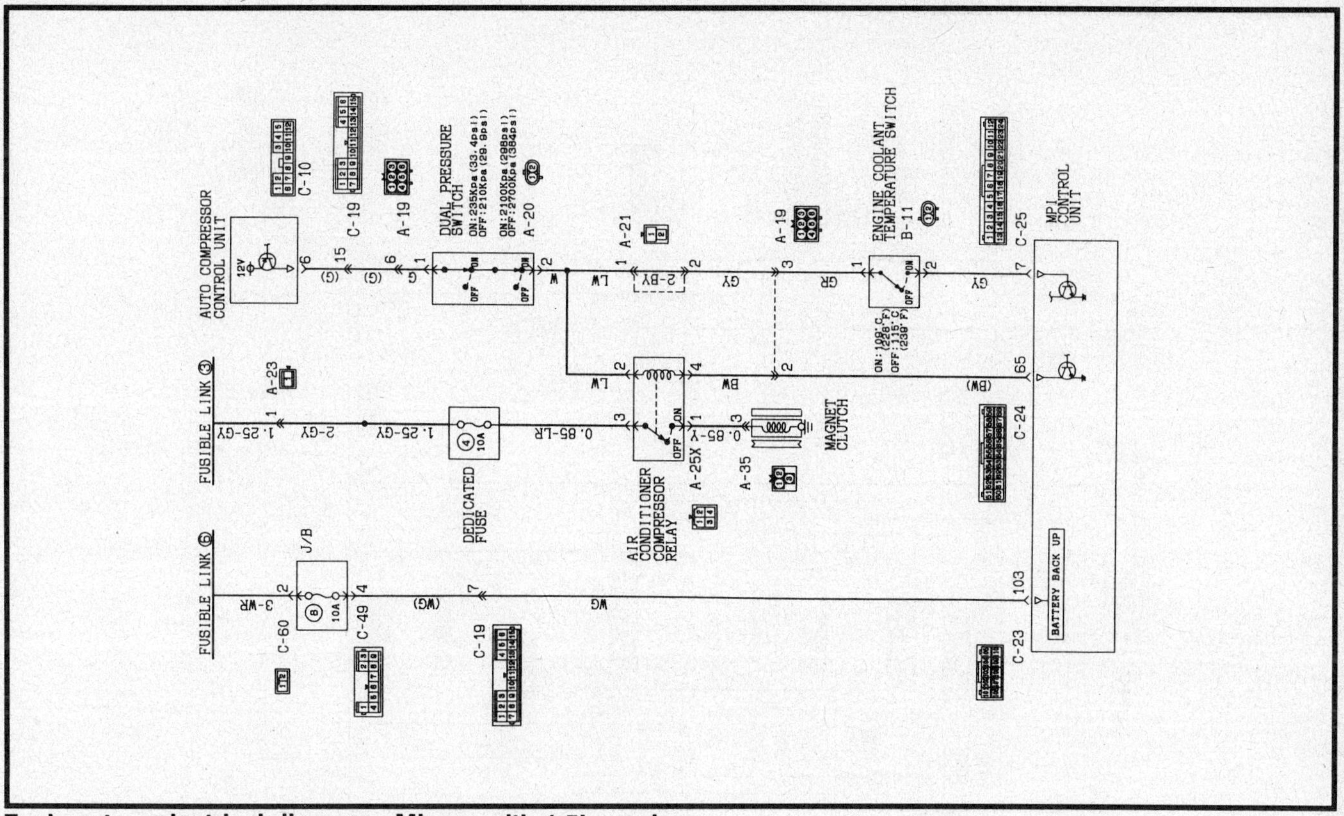

Fuel system electrical diagram – Mirage with 1.5L engine

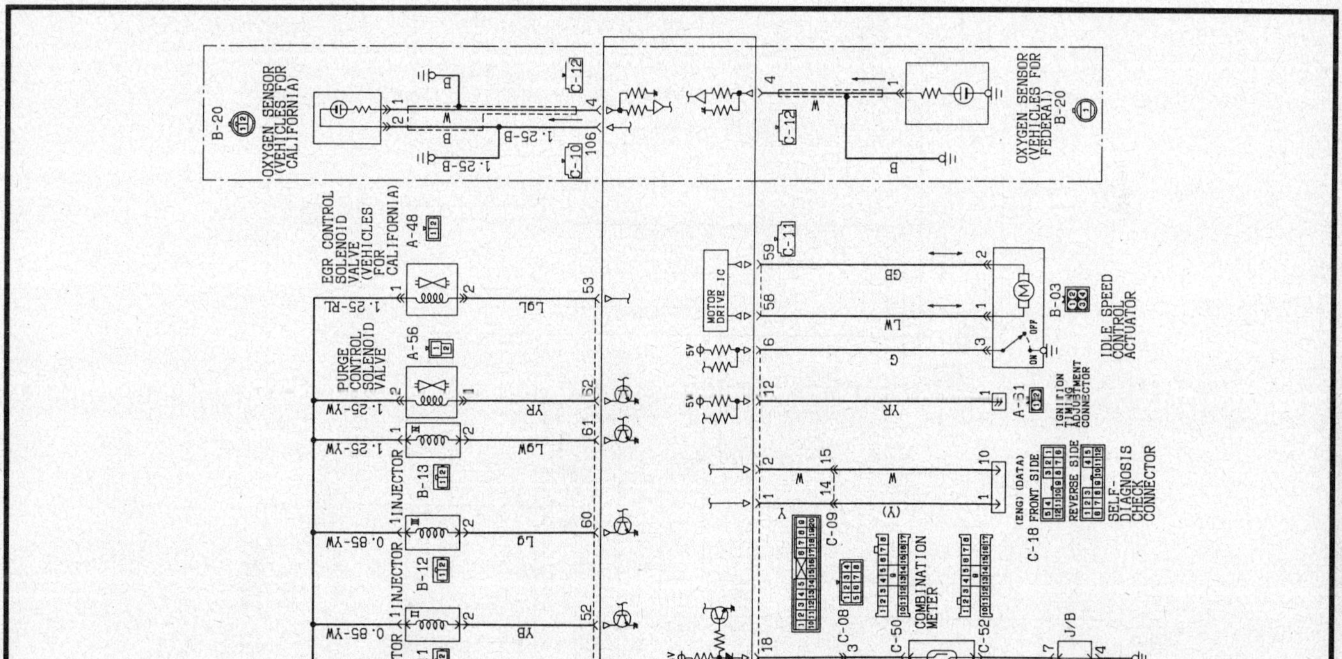

Fuel system electrical diagram, continued – Galant

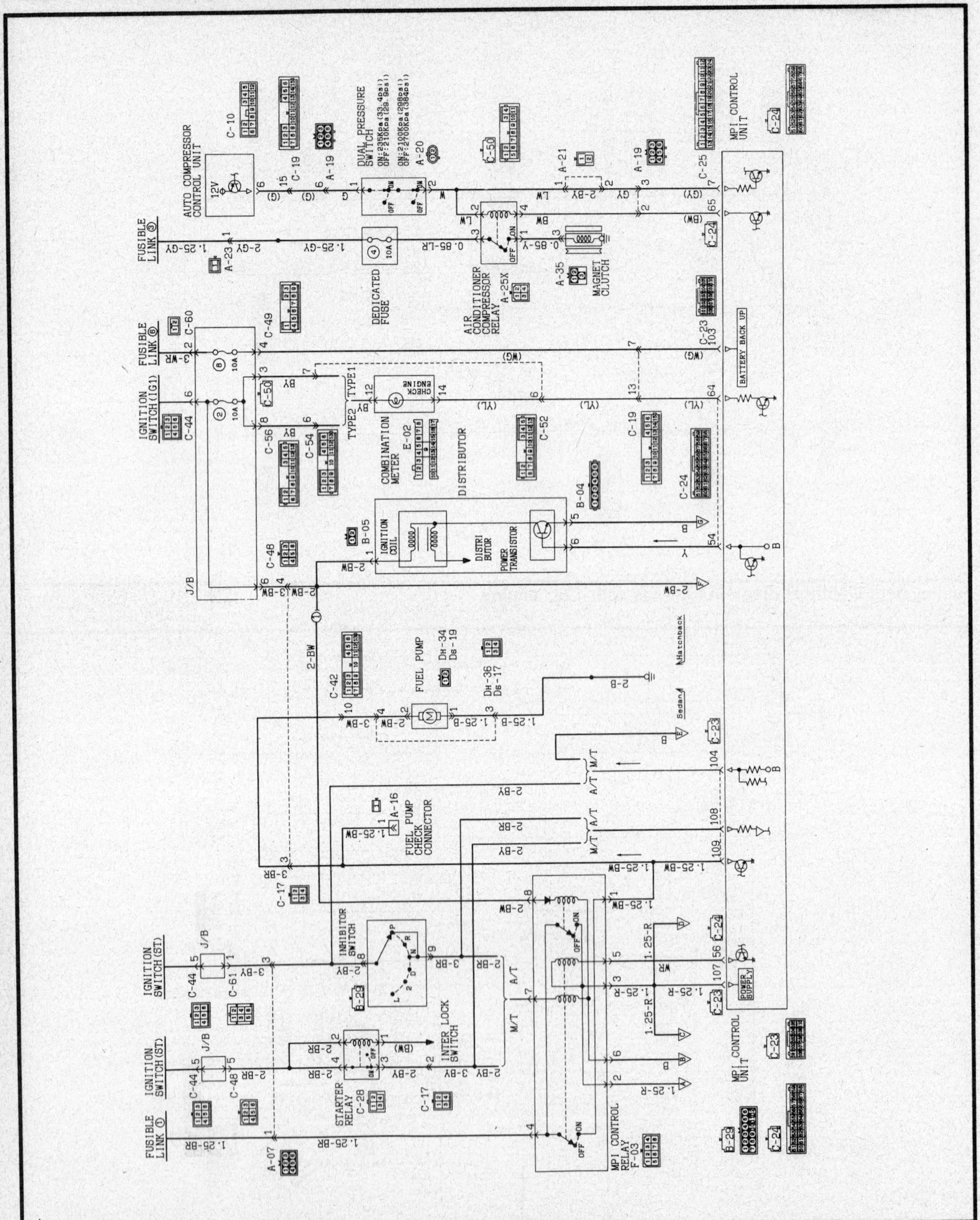

Fuel system electrical diagram, continued—Mirage with 1.5L engine

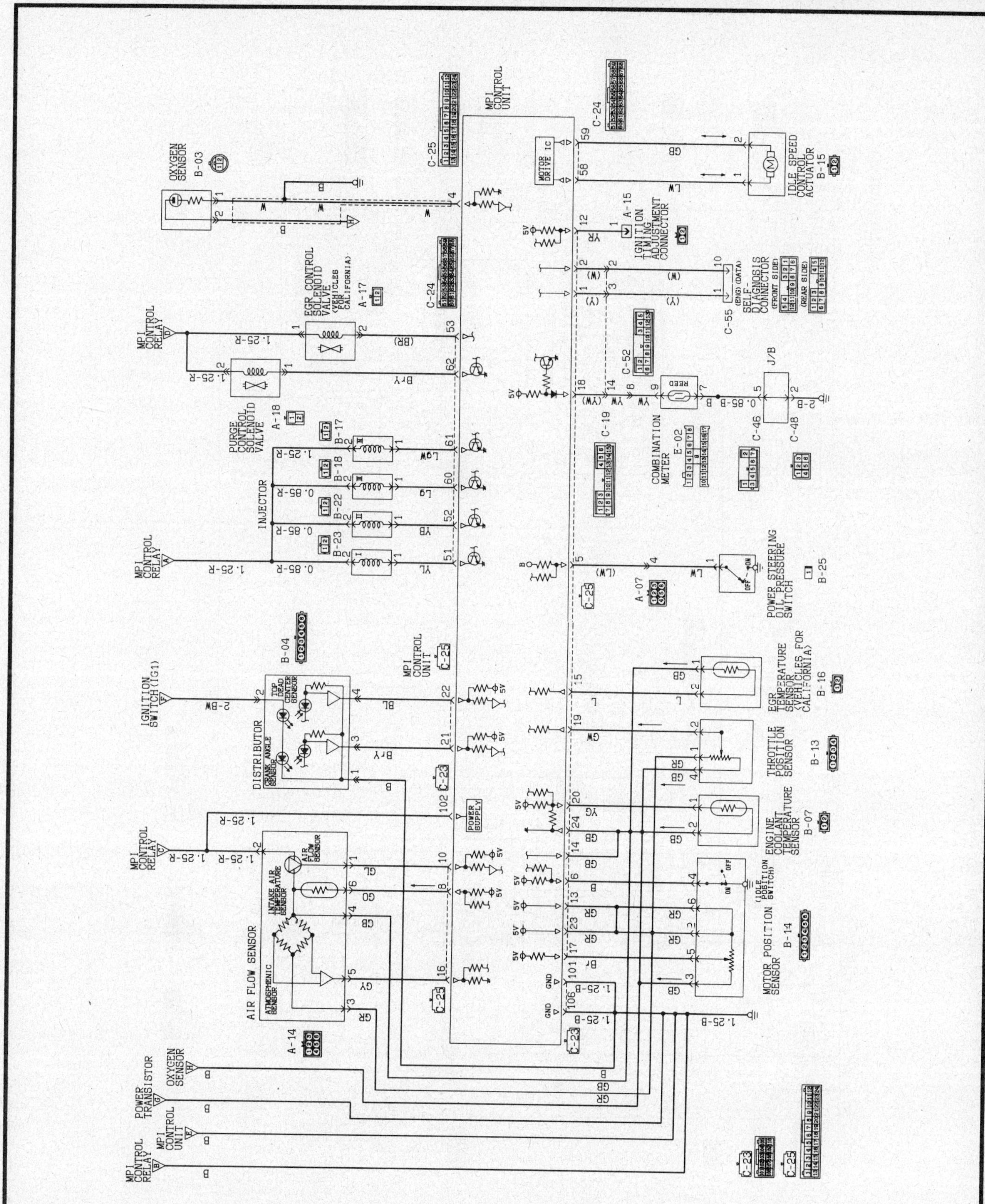

Fuel system electrical diagram, continued – Mirage with 1.5L engine

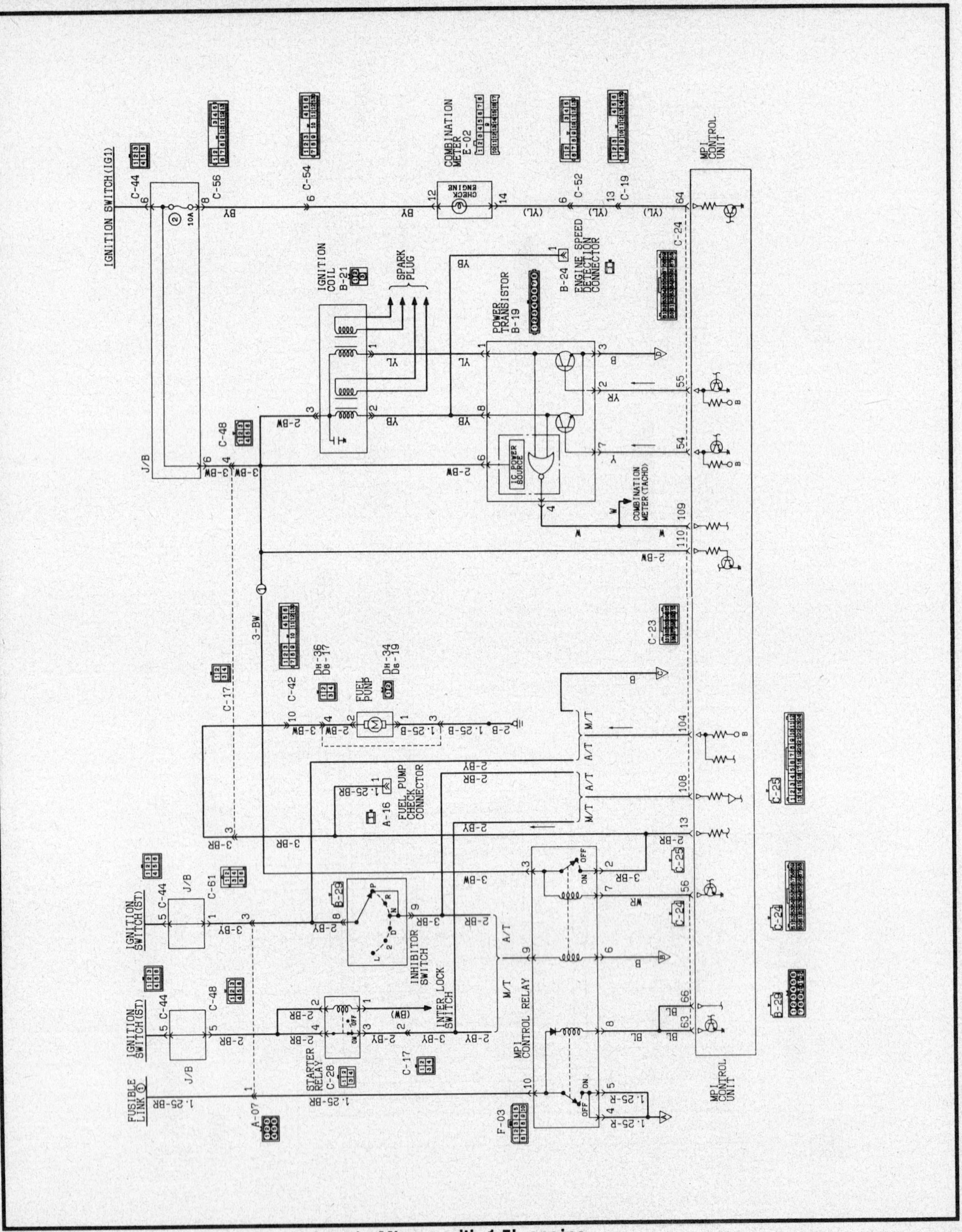

Fuel system electrical diagram, continued – Mirage with 1.5L engine

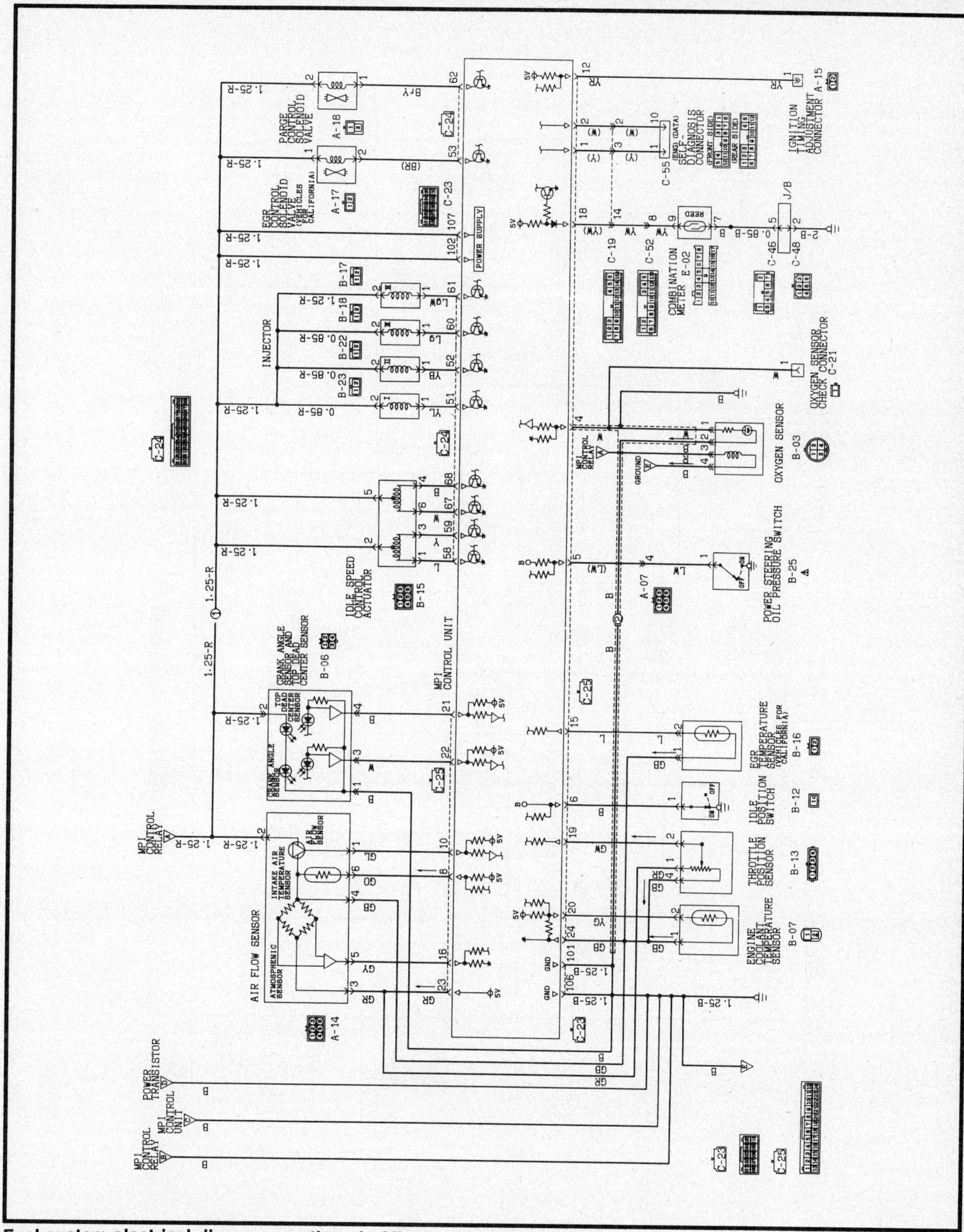

Fuel system electrical diagram, continued—Mirage with 1.6L engine

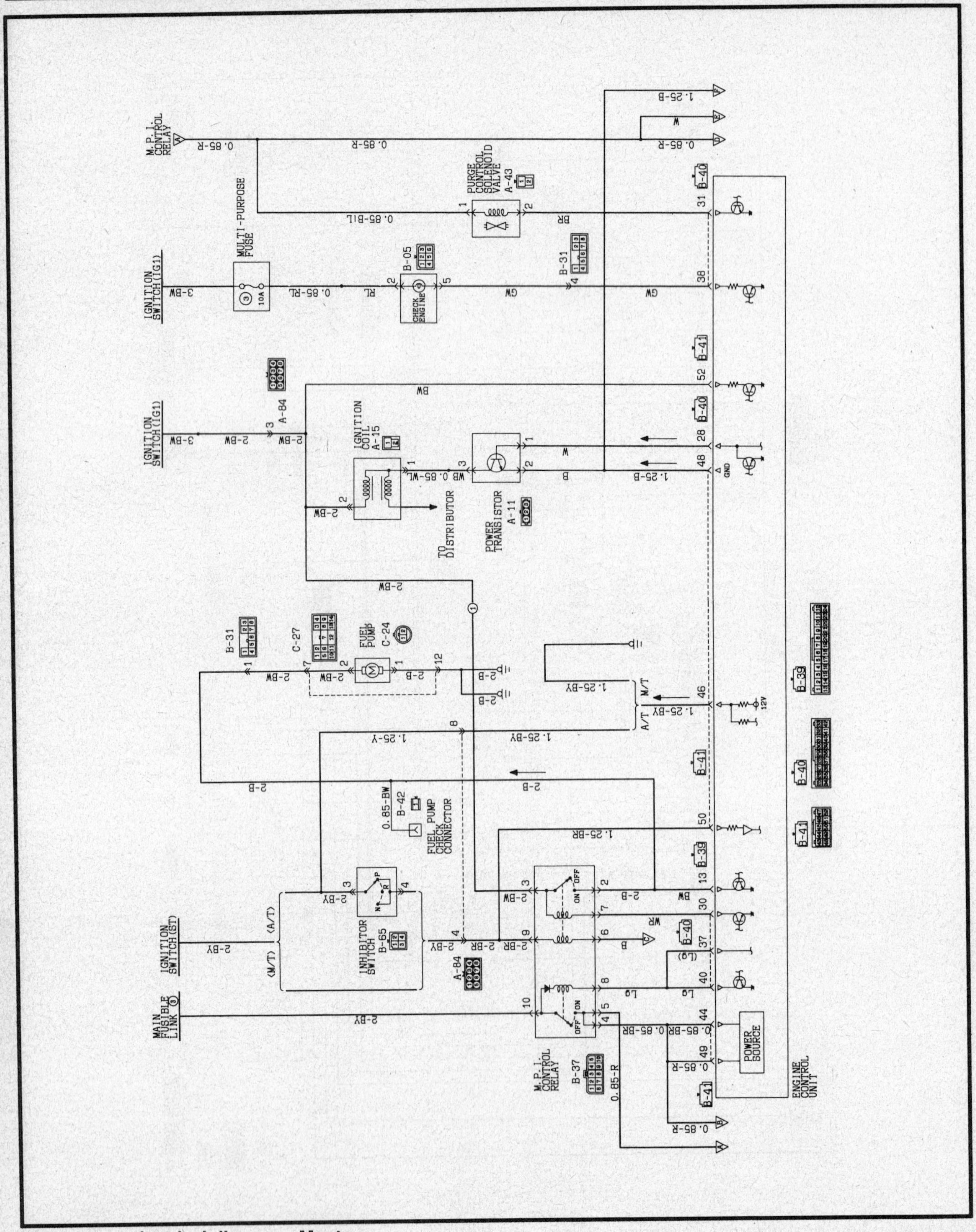

Fuel system electrical diagram—Montero

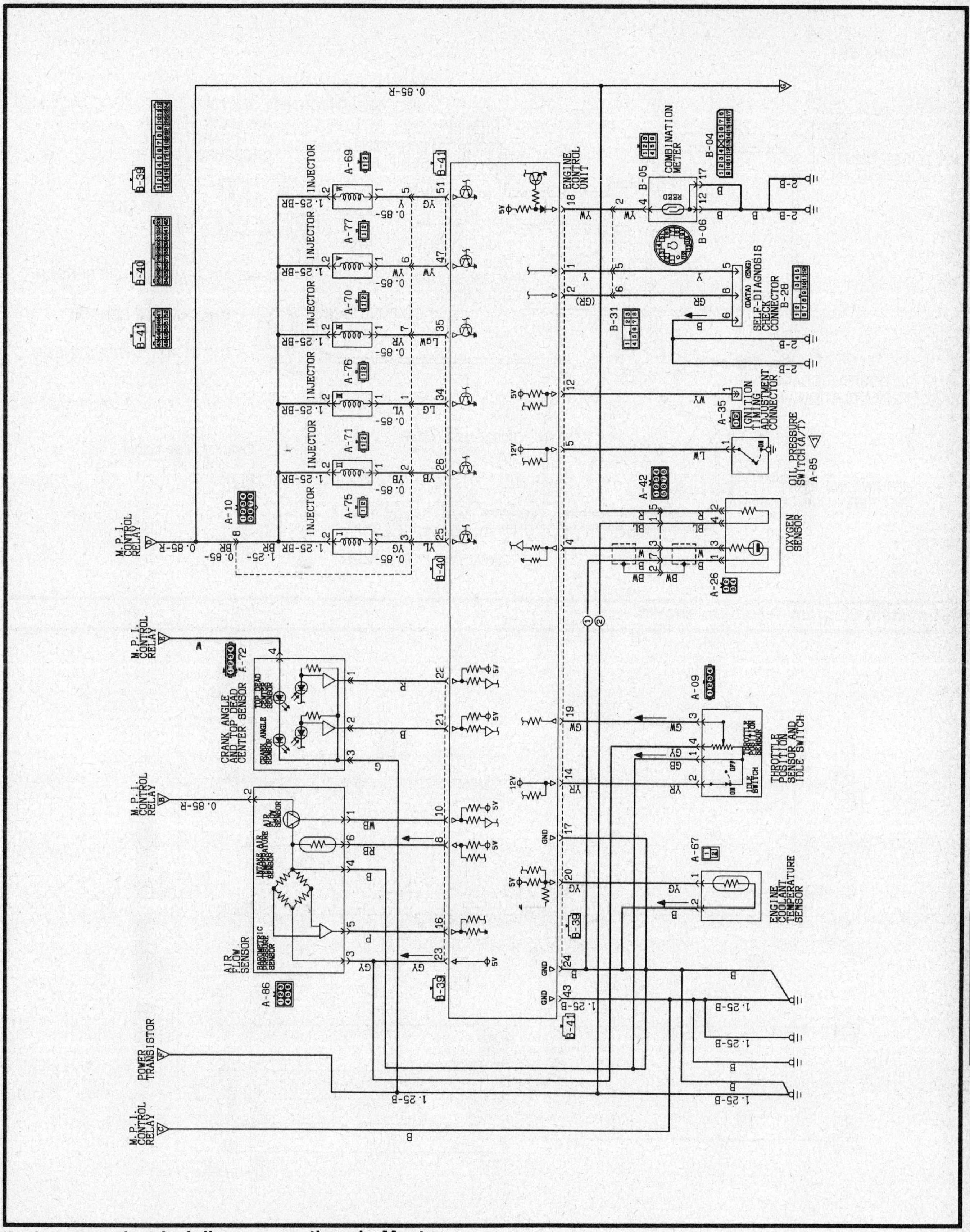

Fuel system electrical diagram, continued – Montero

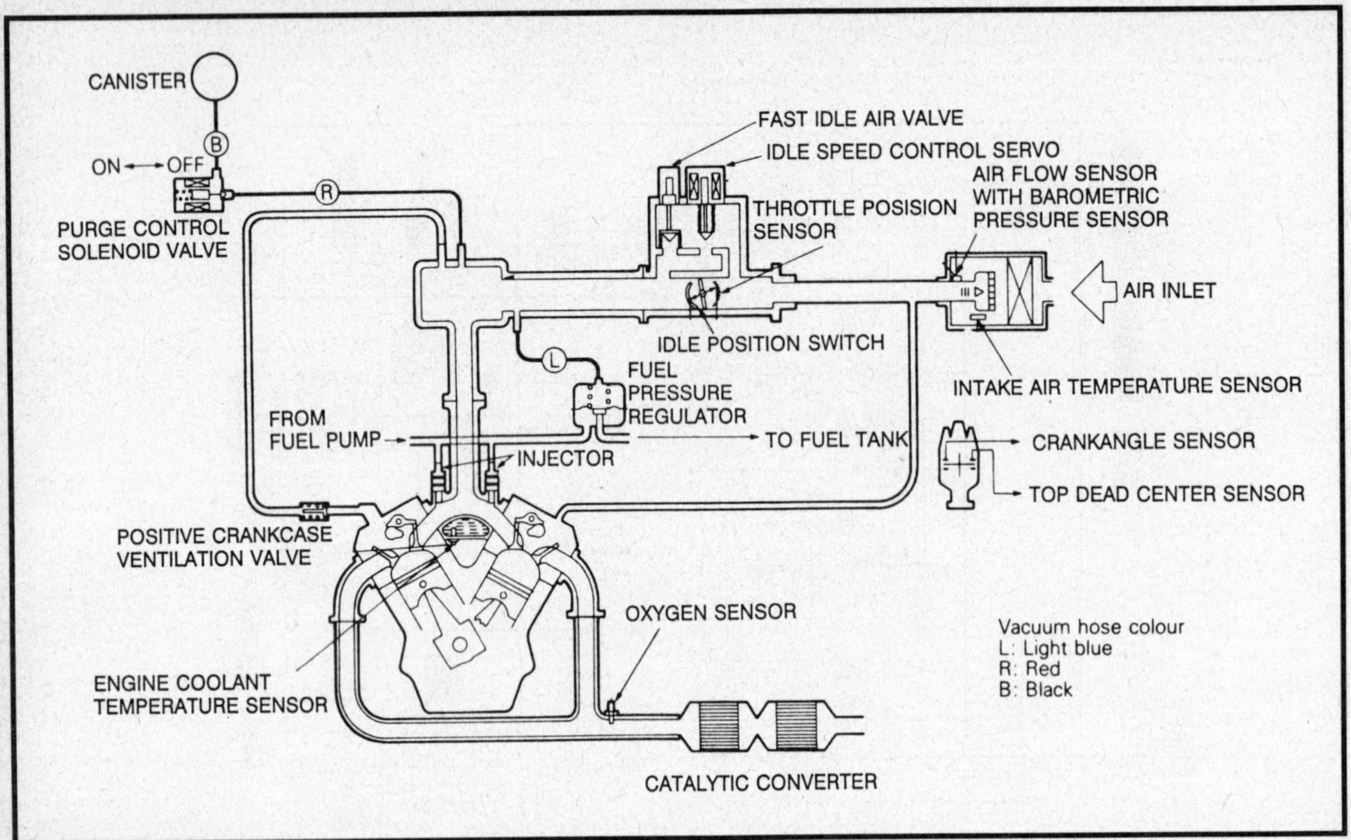

MPI system diagram—Montero shown

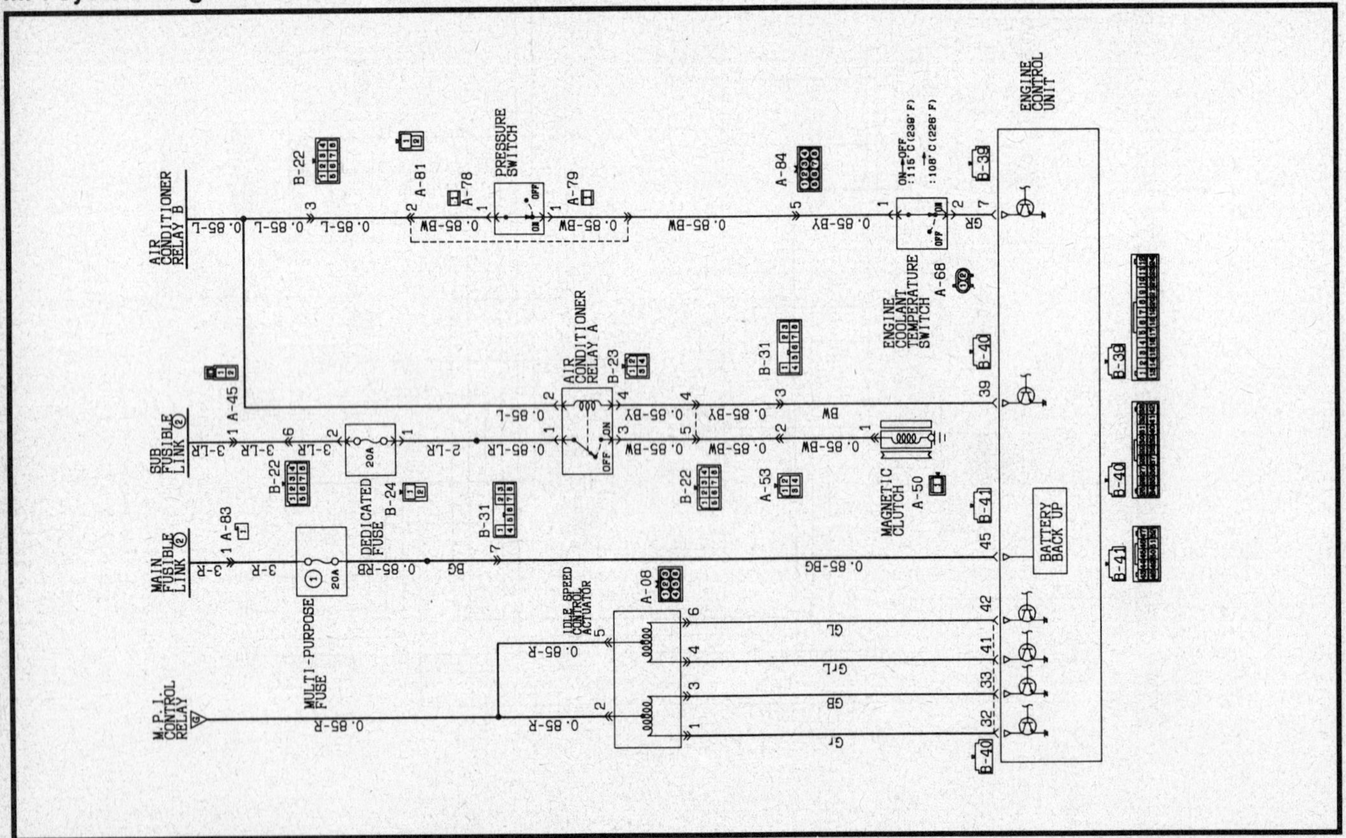

Fuel system electrical diagram, continued—Montero

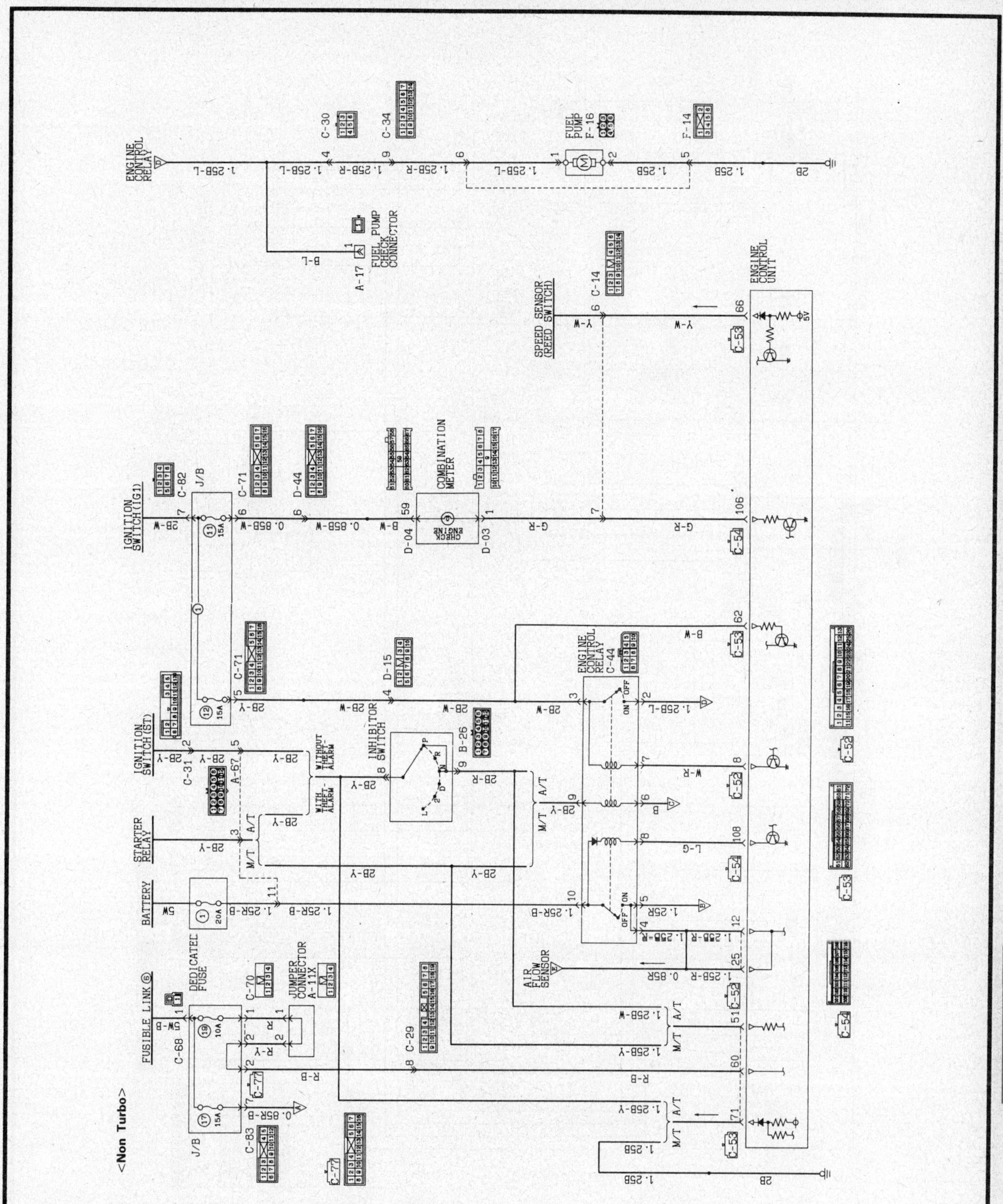

Fuel system electrical diagram—3000GT

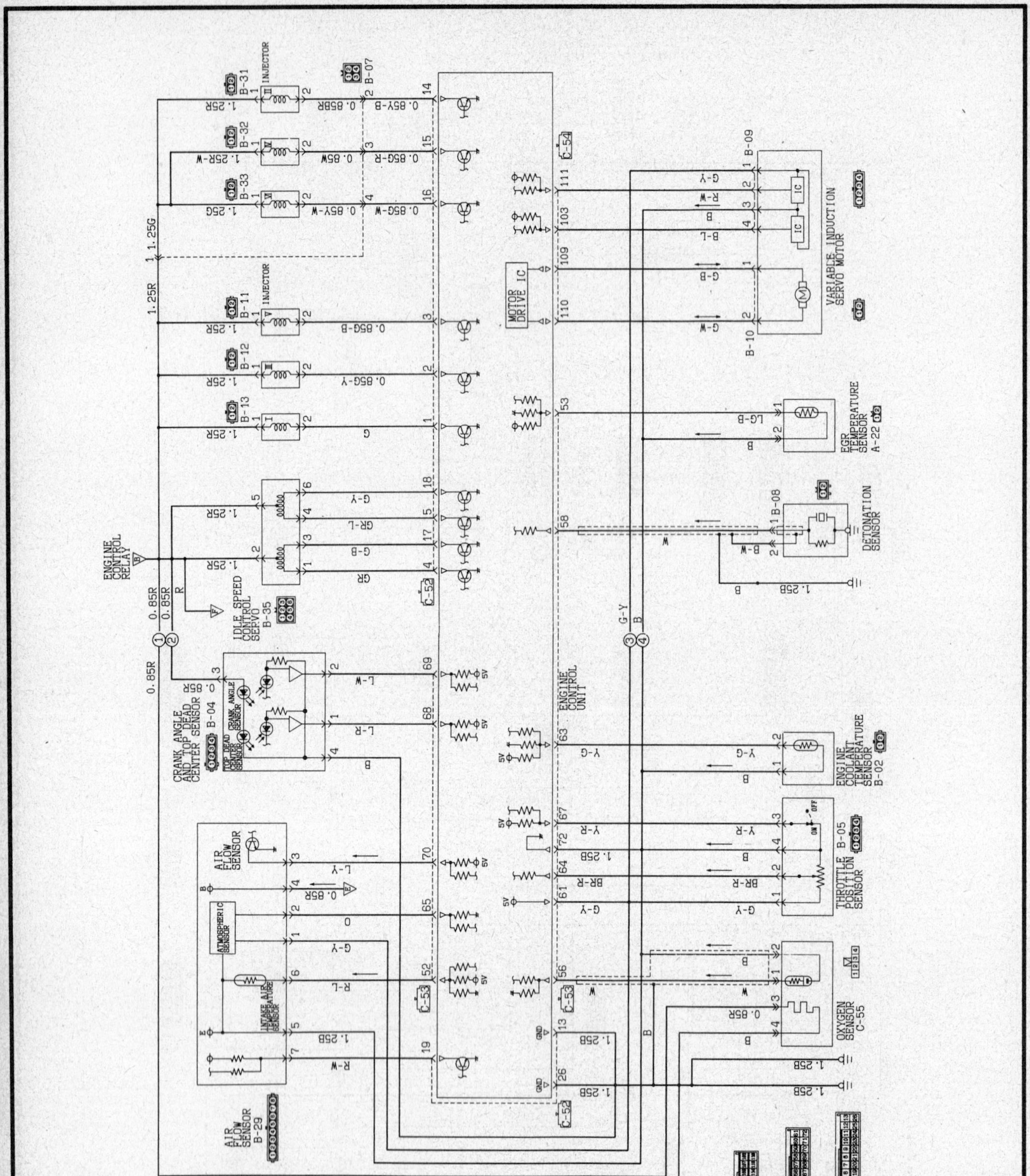

Fuel system electrical diagram, continued—3000GT

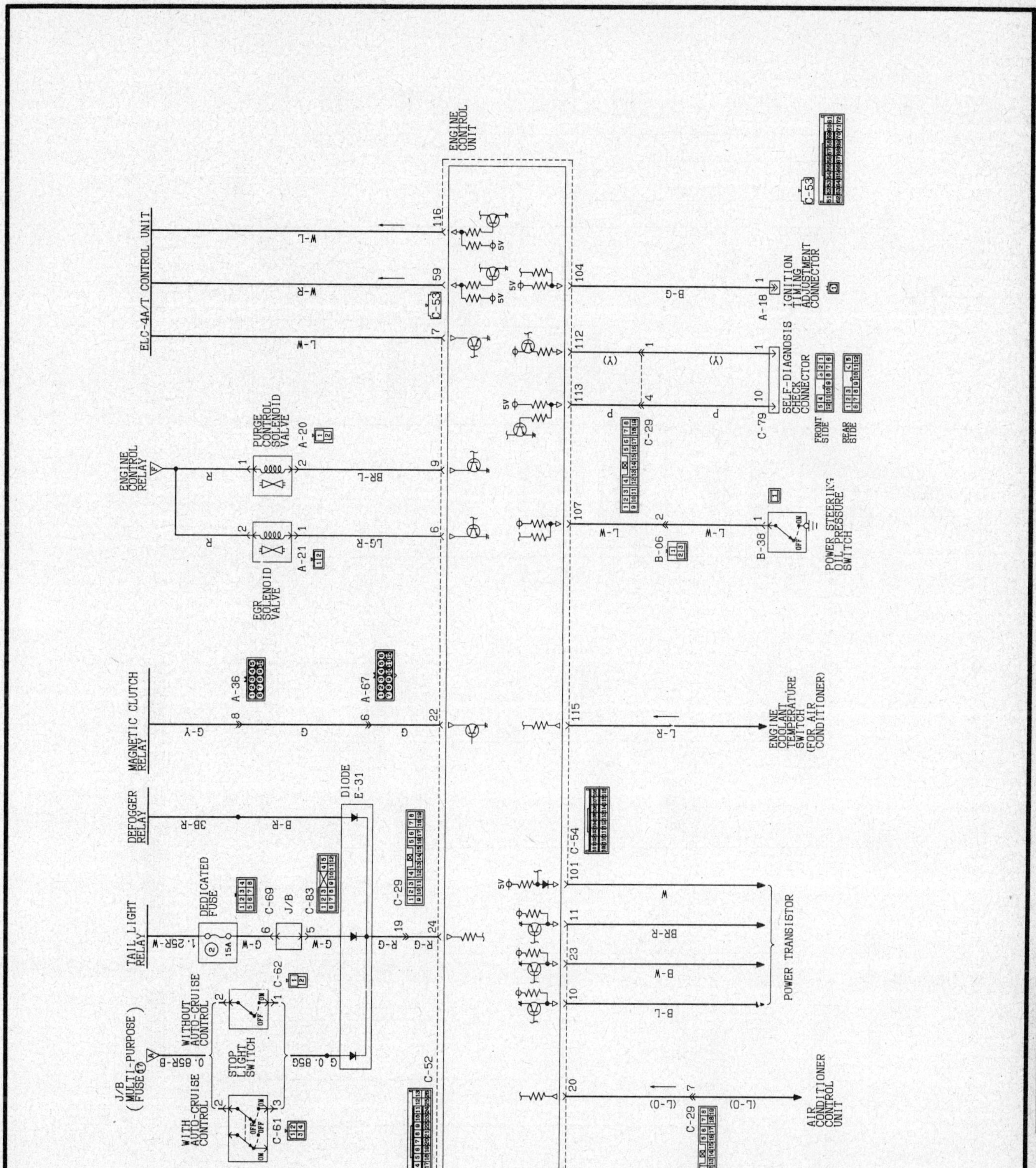

Fuel system electrical diagram, continued – 3000GT

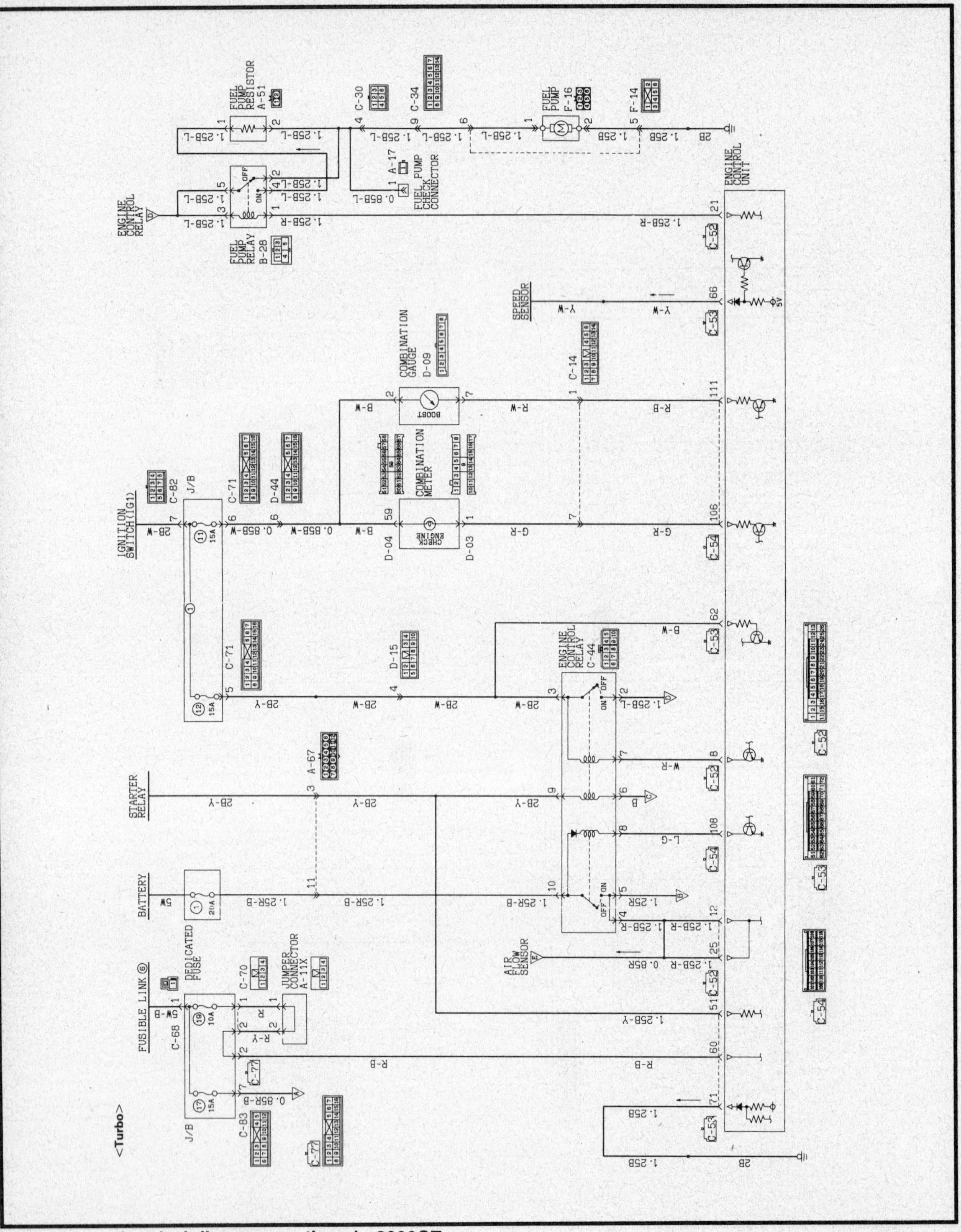

Fuel system electrical diagram, continued – 3000GT

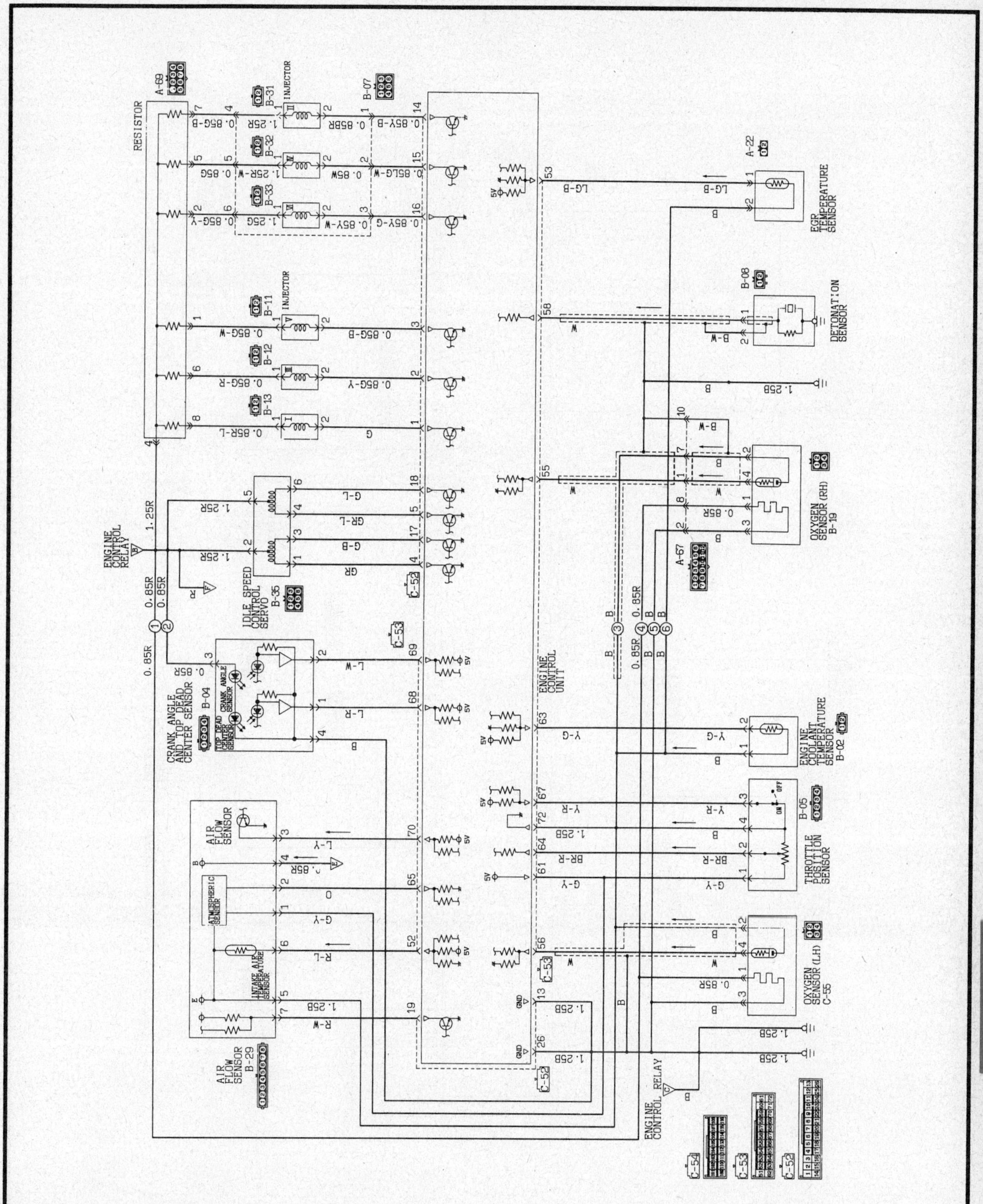

Fuel system electrical diagram, continued – 3000GT

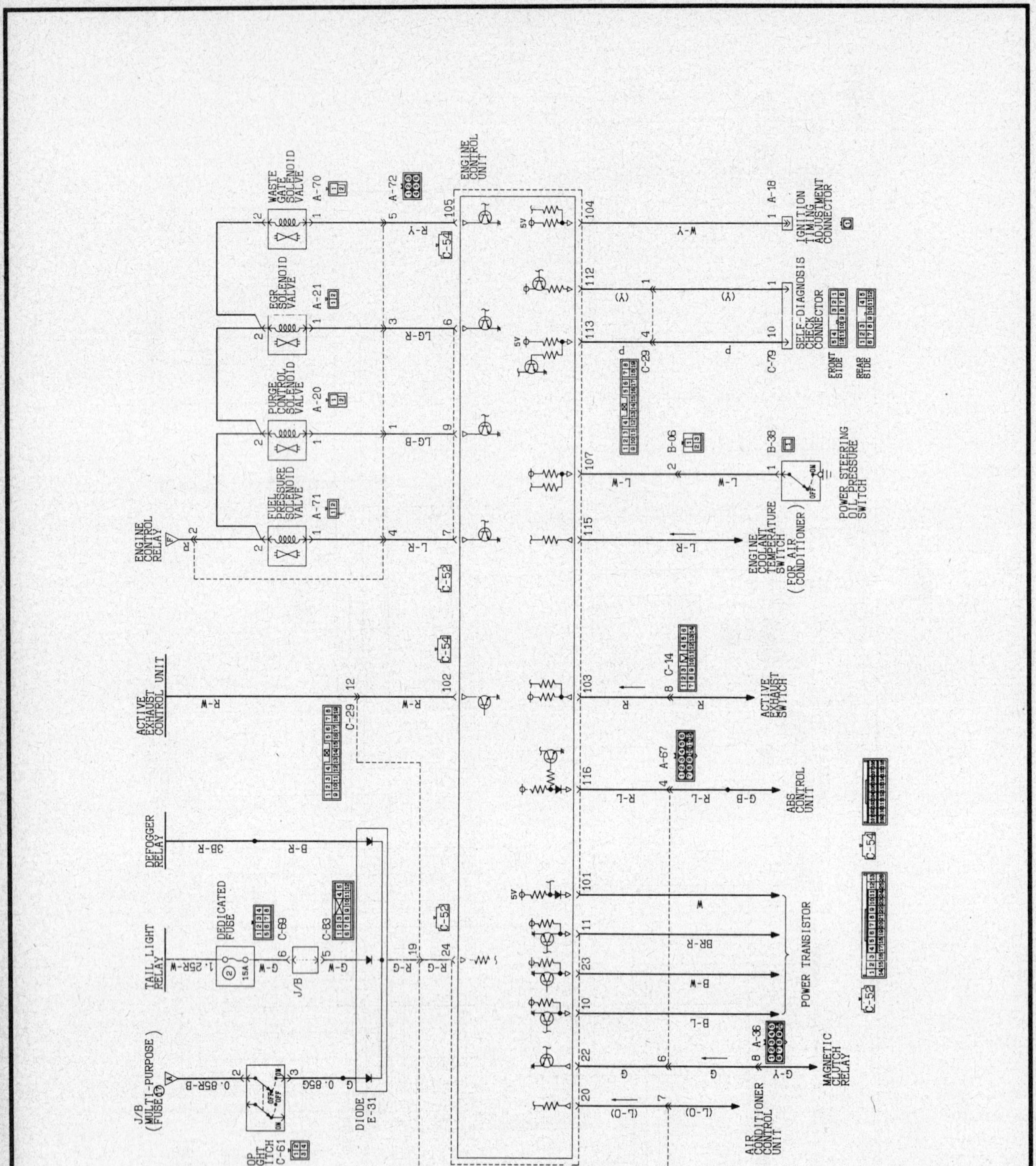

Fuel system electrical diagram, continued—3000GT

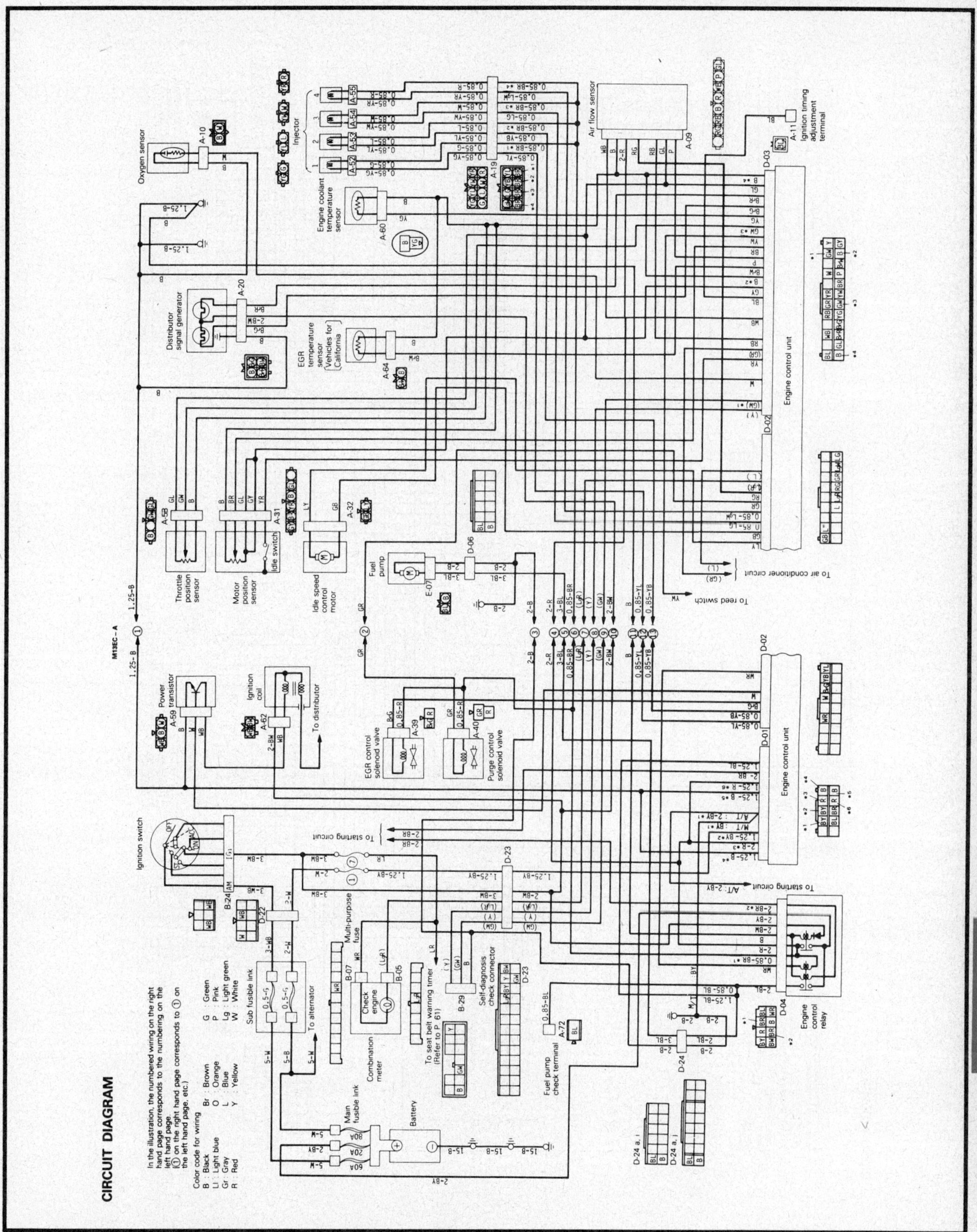

Fuel system electrical diagram—Truck with 2.4L engine

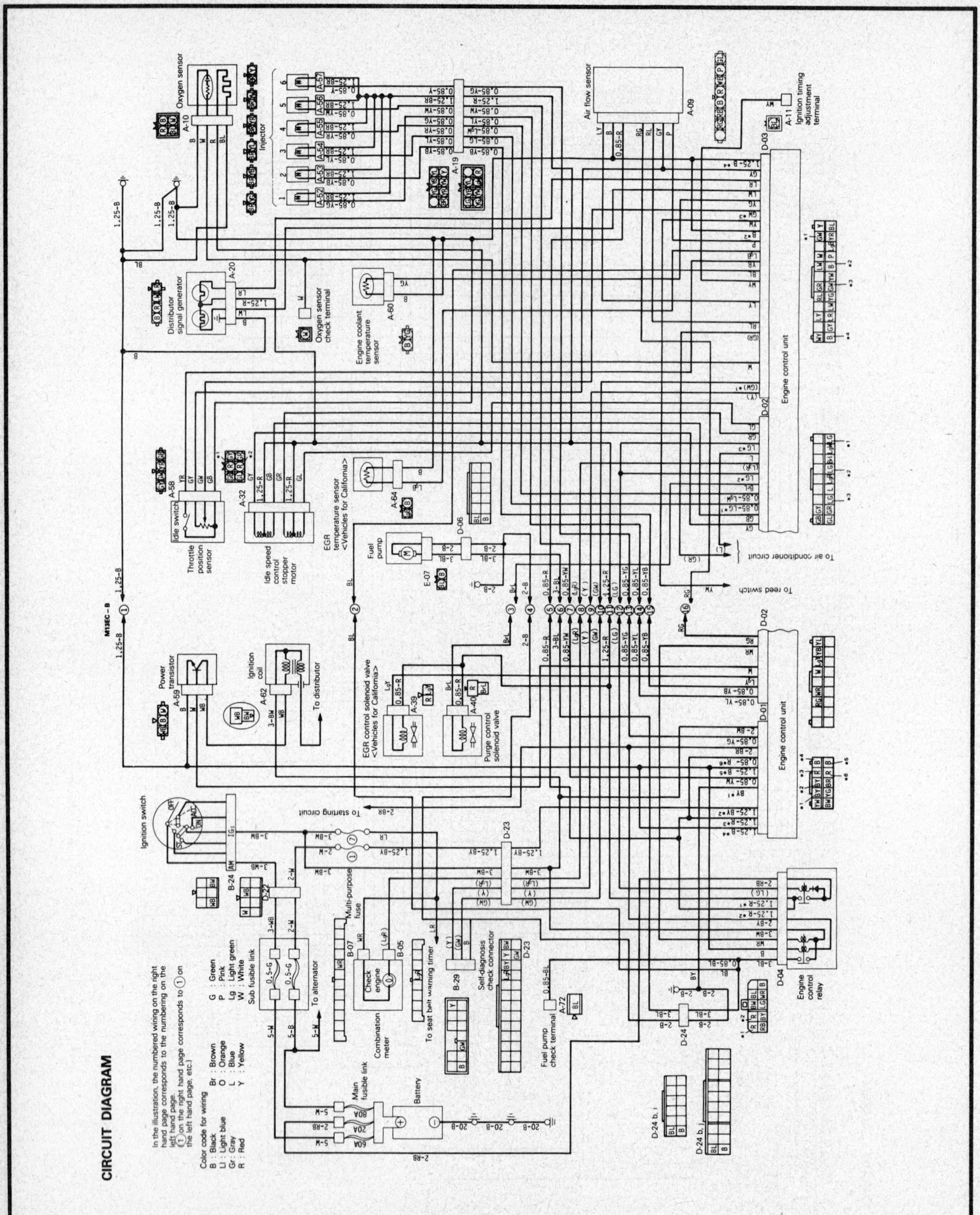

Fuel system electrical diagram, continued – Truck with 2.4L engine

Components Replacement

FUEL PRESSURE RELIEF

1. Disconnect the fuel pump harness at the rear of the tank.
2. Start engine and allow to idle.
3. After engine stalls, fuel pressure is reduced.
4. Disconnect the negative battery cable and reconnect the fuel pump wiring.
5. Remove the fuel tank cap.

NOTE: When disconnecting any fuel system hoses, always place a shop towel to prevent residual fuel from spaying.

FUEL FILTER

Removal and Installation

1. Release the fuel system pressure.
2. Turn the ignition switch **OFF**; then, disconnect the negative battery cable.

NOTE: Place a shop towel over fuel hose connections when removing, to prevent fuel spray.

3. If necessary, remove the battery and battery tray with washer tank assembly.
4. Loosen the main pipe flare nut while holding the fuel filter nut securely.
5. Remove the fuel filter mounting bolts and remove the filter from the fuel clamp or bracket, as required.

To install:

6. Temporarily position the fuel filter into the retaining clamp or bracket and; then insert the main pipe at the connector part of the filter and manually screw in the main pipe's flare nut. install the mounting bolts.
7. Holding the fuel filter nut, tighten the fuel main pipe's flare nut 22–29 ft. lbs. (30–40 Nm) and eye bolt 18–25 ft. lbs. (25–35 Nm).
8. Install the battery tray with washer tank assembly. Install the battery, if required.
9. Reconnect the negative battery cable, start the engine and check for leaks.

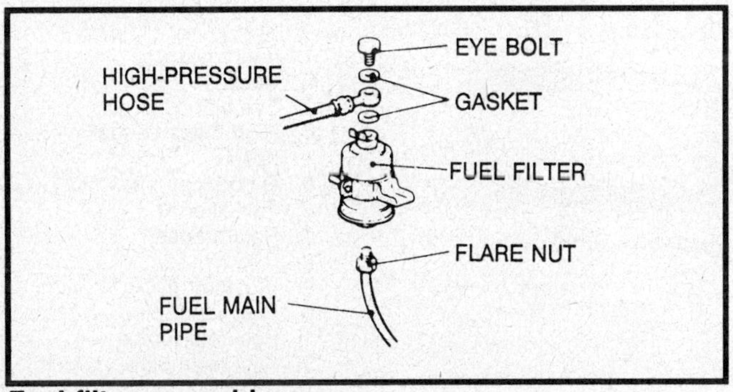

Fuel filter assembly

ENGINE COOLANT TEMPERATURE SENSOR

Removal and Installation

1. Disconnect the negative battery cable.
2. Disconnect the engine coolant temperature sensor harness connector.
3. Remove the sensor from the engine.

To install:

4. Apply sealant (3M NUT Locking No. 4171 or equivalent) to the threaded portion and install the sensor. Torque to 15–29 ft. lbs. (20–40 Nm).
5. Connect the sensor harness connector.
6. Reconnect the negative battery cable.

THROTTLE POSITION SENSOR

Removal and Installation

1. Disconnect the negative battery cable.
2. Disconnect the TPS harness connector.
3. Remove the TPS mounting screws and remove the TPS.

To install:

4. Place the TPS into position and install the mounting screws. Partially tighten the mounting screws.
5. Reconnect the TPS harness connector.
6. Connect the negative battery cable.
7. Adjust the TPS. Torque the mounting screws 1.1–1.8 ft. lbs. (1.5–2.5 Nm).

FUEL TANK AND FUEL PUMP ASSEMBLY

Removal and Installation

1. Release the fuel system pressure.
2. Turn the ignition switch **OFF**; then, disconnect the negative battery cable.
3. Raise the vehicle and support it safely.
4. Remove the drain plug and drain the fuel.
5. Disconnect the fuel filler hose, leveling hose, vapor hose, fuel return hose and pressure hose from the tank.
6. Disconnect fuel sender and fuel pump electrical harness.
7. Loosen the self-locking nuts, which holds the tank in position and remove the tank retaining straps. Remove the fuel tank.
8. If the fuel pump is to be replaced, clean the area around the pump assembly. Remove the fuel pump retaining screws and remove the fuel pump assembly from the fuel tank.
9. If the fuel gauge unit is to be replaced, clean the area around the sending unit assembly. Remove the sending unit retaining screws and remove the sending assembly from the fuel tank.

To install:

10. If removed, install the fuel pump assembly into the fuel tank. Align the 3 positioning projections of the packing with the holes in the fuel pump. When the fuel pump is installed, the holding bolt at the lower side of the fuel tank should be installed first. Use a new gasket or O-ring, as required.

NOTE: Care should be taken not to pinch the O-ring.

11. If the fuel gauge unit was removed, align the 2 positioning projections of the packing with the holes in the fuel gauge unit.

NOTE: When mounting the fuel gauge unit, incline the float at the end leftward and insert it into the fuel tank. Since the reservoir cap is provided in the fuel tank, the fuel gauge unit will contact the reservoir cap if the fuel gauge unit is inclined rightward during insertion.

12. Place the fuel tank into position. Fit the retaining straps to the fuel tank and install the self-locking nuts. Tighten the retaining straps nuts 15–22 ft. lbs. (21–31 Nm).
13. Reconnect the electrical harness connector and fuel hoses. Tighten the high pressure hose flare nut to 22–29 ft. lbs. (30–40 Nm).
14. Install the drain plug and tighten to 11–18 ft. lbs. (15–25 Nm).
15. Lower the vehicle and refill the fuel tank.
16. Reconnect the negative battery cable, start the engine and check for leaks.

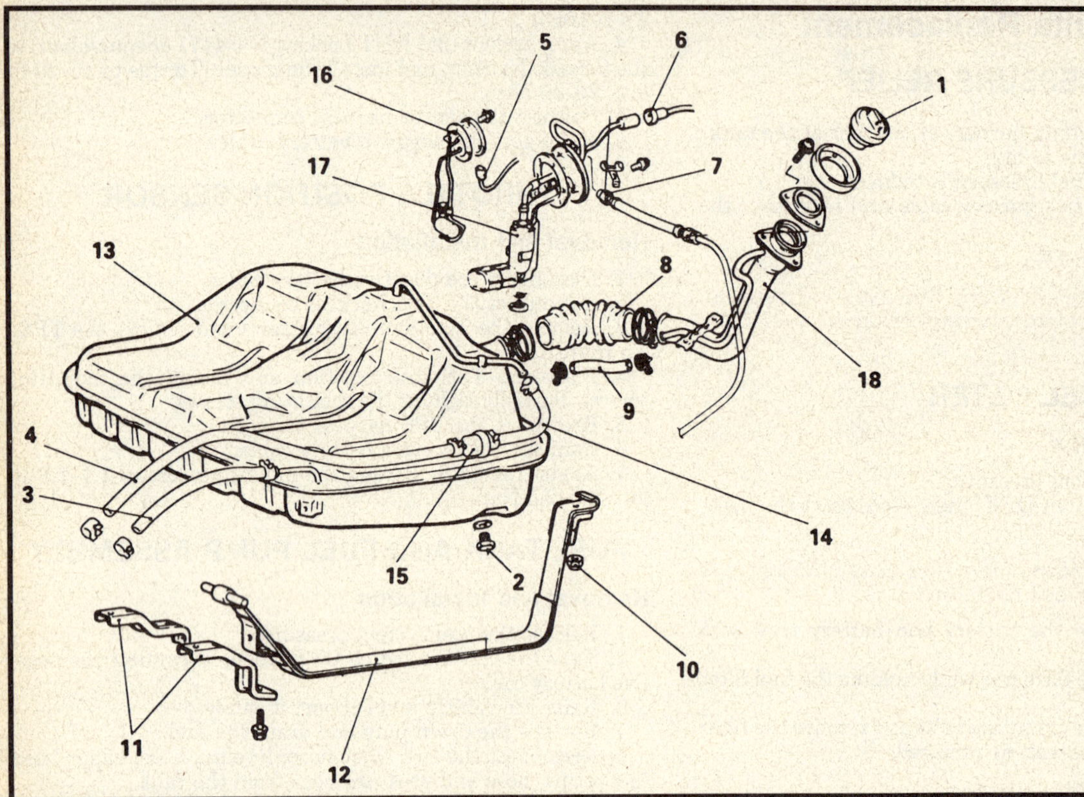

1. Fuel tank cap
2. Drain plug
3. Return hose
4. Vapor hose
5. Fuel gauge unit connector
6. Electrical fuel pump connector
7. High pressure hose to fuel pump
8. Filler hose
9. Vapor hose
10. Self-locking nut
11. Tank band strap
12. Tank band
13. Fuel tank
14. Vapor hose
15. Overfill limiter (2-way valve)
16. Fuel gauge unit
17. Electrical fuel pump
18. Fuel filler neck

Fuel tank removal and installation – typical

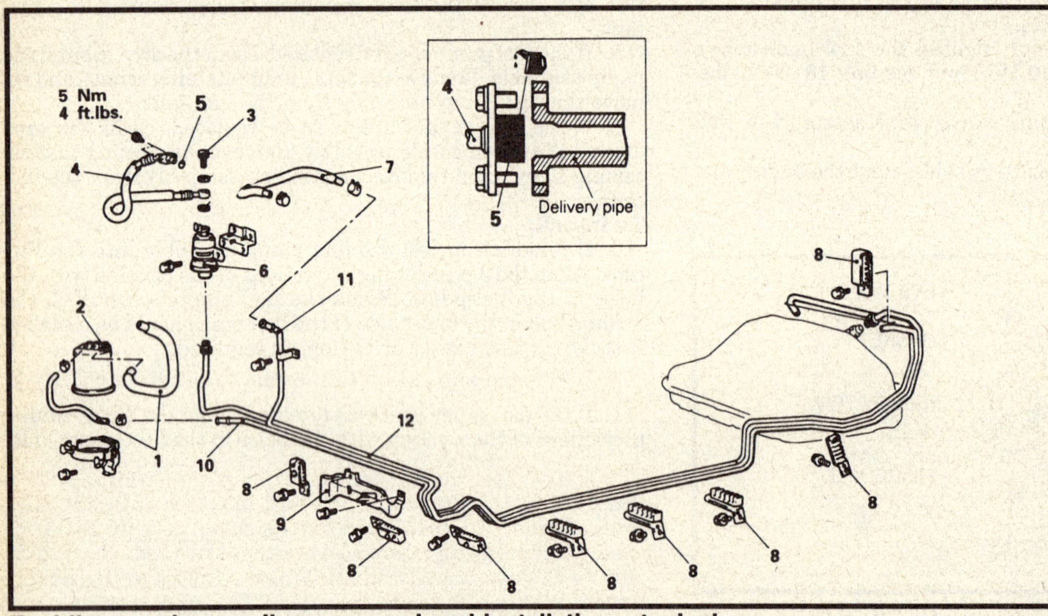

1. Vapor hose
2. Canister
3. Eye bolt
4. High pressure fuel hose
5. O-ring
6. Fuel filter
7. Return hose
8. Clip
9. Protector (DOHC)
10. Fuel vapor pipe
11. Fuel return pipe
12. Fuel main pipe

Fuel lines and vapor lines removal and installation – typical

DELIVERY PIPE AND INJECTORS

Removal and Installation

1. Relieve the fuel system pressure.
2. Turn the ignition switch **OFF**; then, disconnect the negative battery cable.
3. Disconnect the wiring harness connectors from the injectors.

NOTE: Place a shop towel over fuel hose connections when removing, to prevent fuel spray.

4. Disconnect the PCV and vacuum hoses, as required.
5. Disconnect the high pressure fuel hose and return hose connections. Cover hoses immediately to prevent dirt from entering.
6. Disconnect the vacuum hose from the pressure regulator.
7. On 2.0L engine, disconnect the accelerator cable clamp and remove the center cover.

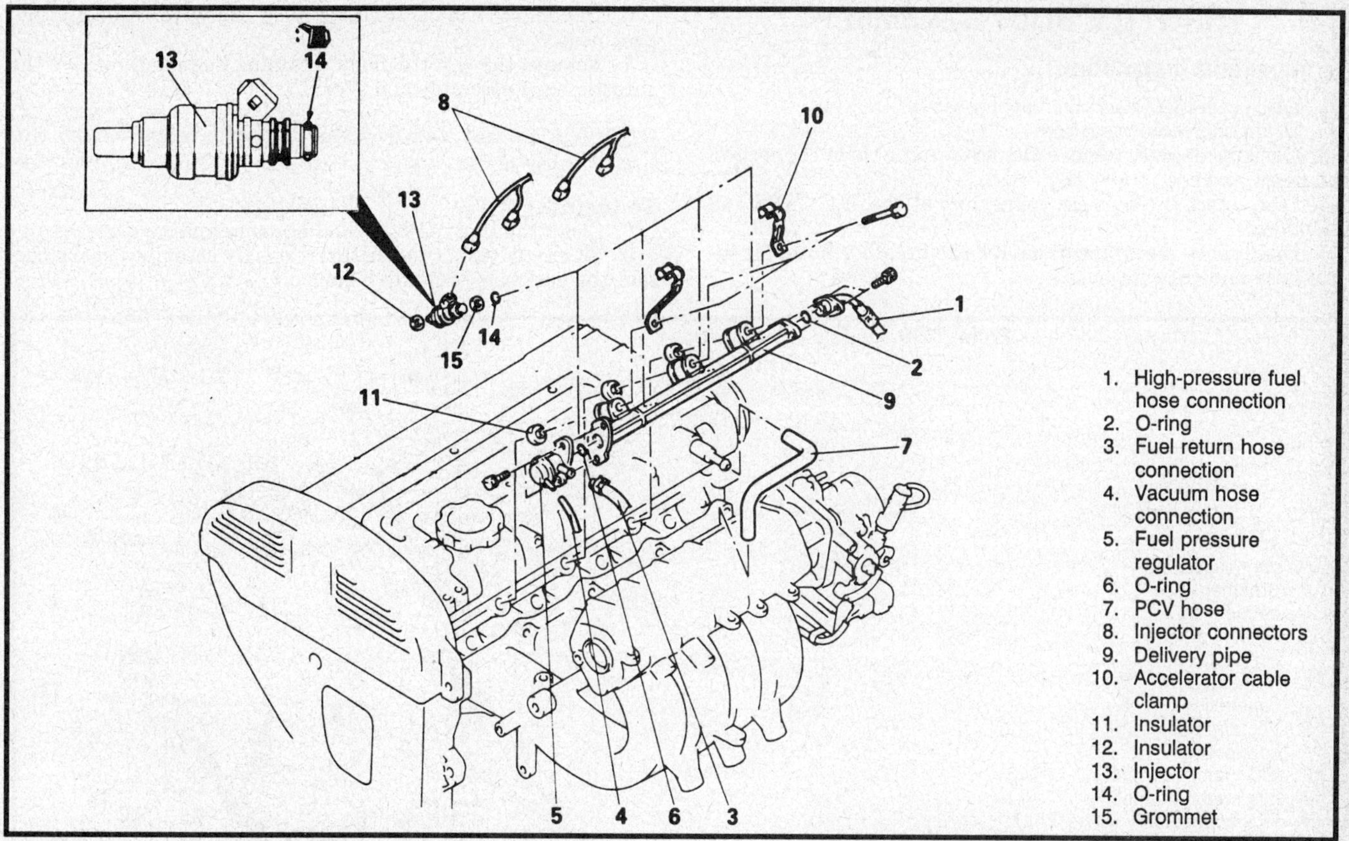

Injector removal and installation—typical 4-cylinder

8. Remove the delivery pipe retaining bolts.

NOTE: When removing the delivery pipe, be careful not to drop any injectors. Also, be aware that fuel may flow out while removing.

9. Remove the delivery pipe with the fuel injectors and pressure regulator.
10. If removal of the injector(s) are necessary, separate the injector(s) from the delivery pipe.

To install:

11. If the injector(s) were removed:
 a. Install a new grommet(s) and O-ring(s) to the injector(s).
 b. Apply a coating of solvent (spindle oil or gasoline) to the O-ring(s) of the injector(s).
 c. Install the injector to the delivery pipe by twisting to the right and left. Make certain the injector turns smoothly.

NOTE: If the injector does not turn smoothly, the O-ring may be jammed; remove the injector and re-insert it into the delivery pipe.

12. Check that the insulators are properly fitted into position. Install the delivery pipe assembly to the engine. Install the retaining bolts and tighten to 7–9 ft. lbs. (10–13 Nm).
13. Fit a new O-ring to the high pressure fuel hoses. Apply light oil or fuel to the O-ring, then insert. Be careful not to damage the O-ring during installation.
14. On 2.0L engine, install the center cover and connect the accelerator cable clamp.
15. Reconnect the wiring harness to the injectors.
16. Reconnect the vacuum lines and PCV, as required.
17. Reconnect the negative battery cable, start the engine and check for leaks.

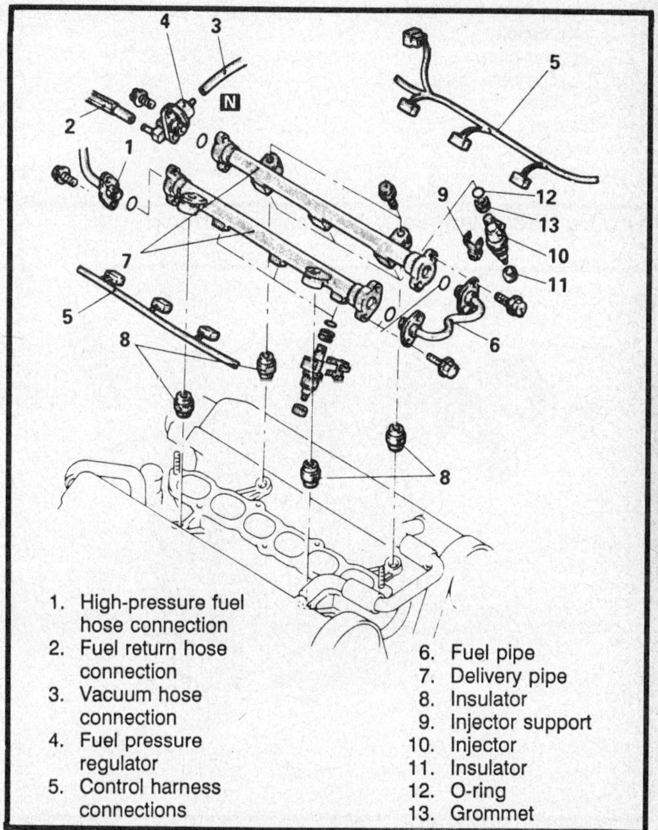

Injector removal and installation—typical 6-cylinder

THROTTLE BODY ASSEMBLY

Removal and Installation

1. Disconnect the negative battery cable.
2. Drain the cooling system.
3. On 2.0L engine, remove the accelerator cable connection and position aside.
4. Disconnect the harness connectors at the TPS, MPS and ISC motor.
5. Disconnect the connections for the breather hose, air intake hose and vacuum hose.
6. Disconnect the connections from the water hose and bypass hose.
7. Remove the throttle body retaining bolts and remove the throttle body assembly and gasket.

NOTE: Do not remove the throttle valve from the throttle body.

To install:

8. Clean the throttle body and housing mating surfaces.
9. Fit a new gasket to the throttle body assembly. Make certain the gasket is properly installed.

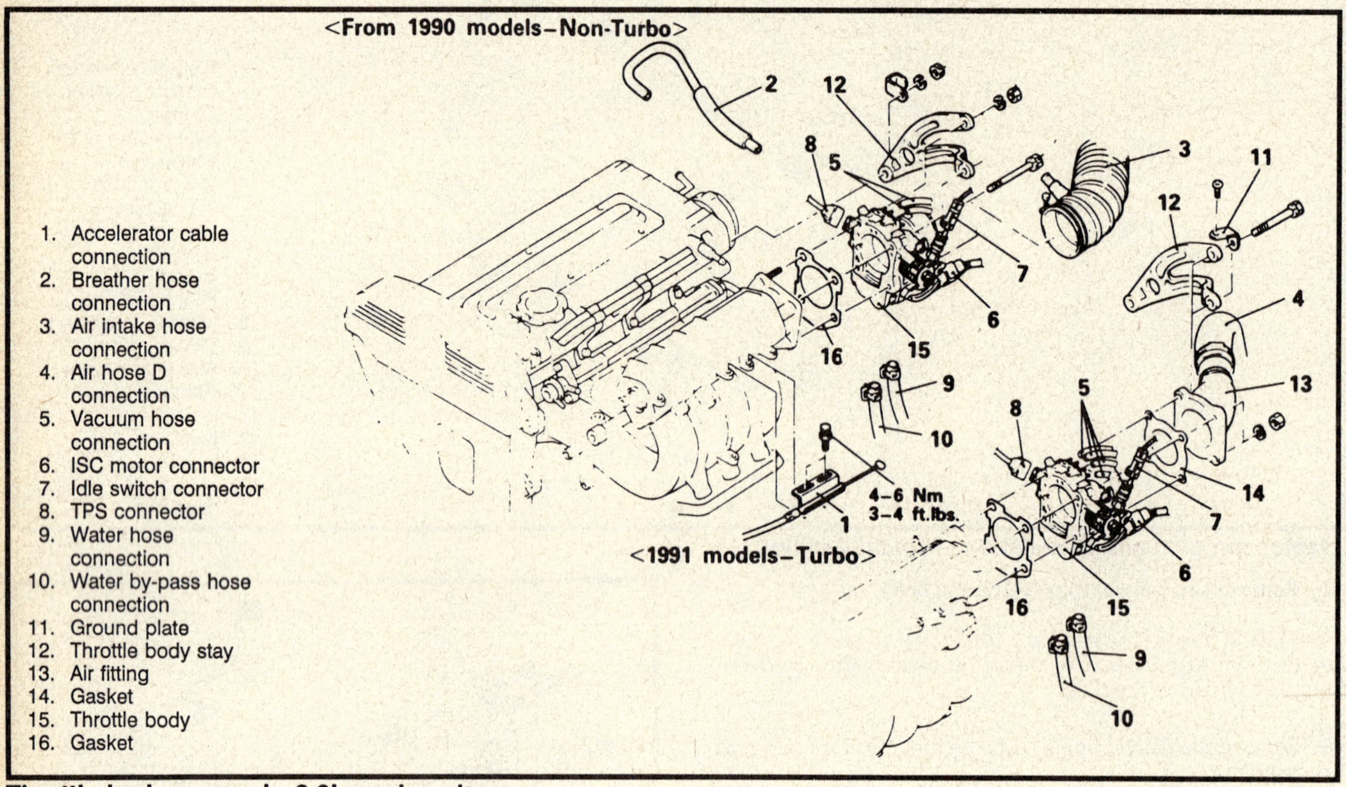

Throttle body removal–2.0L engine shown

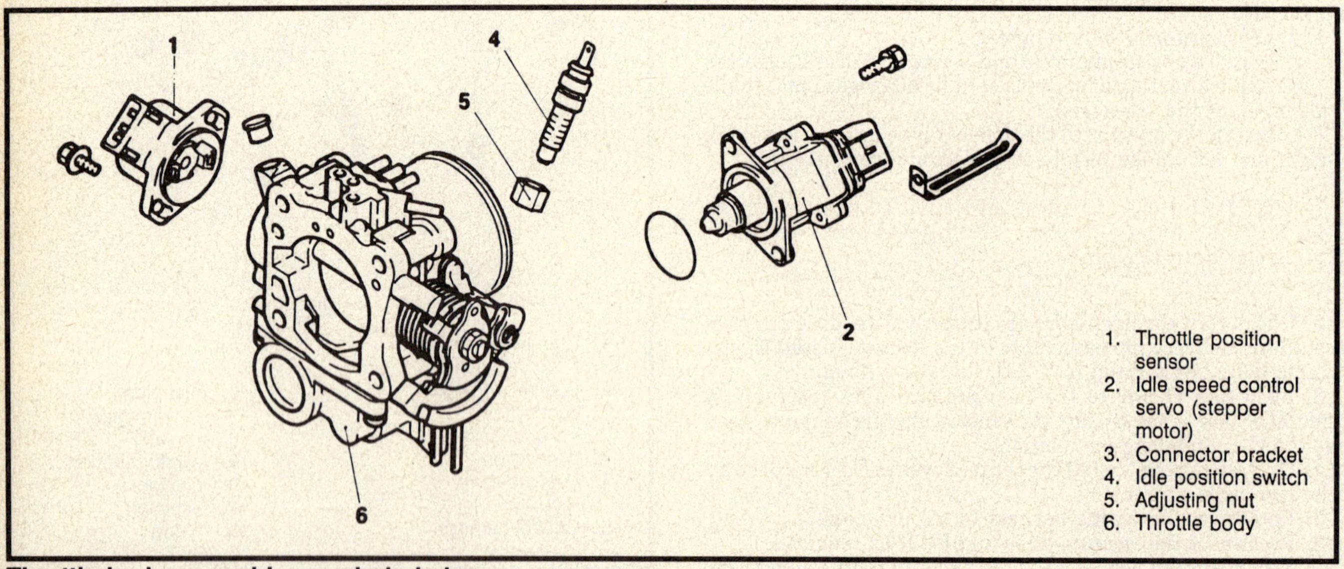

Throttle body assembly–exploded view

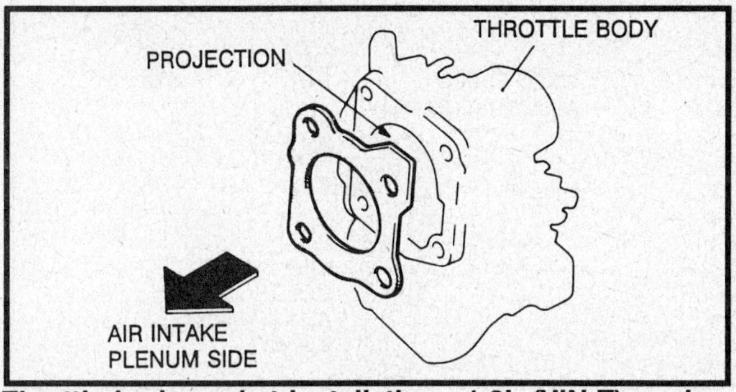

Throttle body gasket installation – 1.8L (VIN T) engine

10. Install the throttle body assembly. Tighten the retaining bolts 11–16 ft. lbs. (15-22 Nm).
11. Reconnect the water hose and by-pass hose.
12. Reconnect the breather hose, air intake hose and vacuum hose.
13. Reconnect the harness connectors at the TPS, MPS and ISC motor.
14. On 2.0L engine, connect the accelerator cable.
13. Refill the cooling system.
14. Reconnect the negative battery cable, start the engine and check for leaks.

Cleaning

1. Operate the engine until warm. Then, turn the engine **OFF**.
2. Disconnect the air intake hose from the throttle body. Check the throttle valve surface for carbon deposits, rotating the valve.
3. Plug the bypass passage entrance of the throttle body, as required.

NOTE: Do not allow any cleaning liquid to enters the bypass passage.

4. Use a suitable carbon deposit spray and remove any carbon deposit, by spraying the solvent into the valve through the throttle body intake port. Let it sit for approximately 5 minutes.
5. Start the engine. Race it several times, then idle it for approximately 1 minute. Repeat this step, if necessary.
6. Attach the air intake hose.
7. Attach the bypass hose, as required.
8. Erase the ECU diagnosis memory.
9. Adjust the basic idle speed.
10. Adjust the TPS, as required.

CURB IDLE RPM

Checking

EXCEPT 1.5L (VIN J) ENGINE

1. Operate the engine until operating temperature is reached.
2. Turn OFF all accessories. Place the gear selector in **P** (automatic transaxle) or **N** (manual transaxle).
3. Set the steering wheel in straight ahead position. Apply the parking brake and block the drive wheels.
4. Check that the electric cooling fan is OFF.
5. Stop the engine and connect a tachometer and timing light. If available, the Multi-Use Tester (MUT) (Tool MB991341 or equivalent) may be used to check engine rpm.
6. If a tachometer is used, connect the tachometer as follows:
   a. Insert a paper clip (from the harness side) into the single-pin connector, as indicated. Be careful not to disconnect the connector.
   b. Connect the tachometer to the paper clip.
   c. On 2.0L (VIN R AND U) and trucks with 3.0L engines,

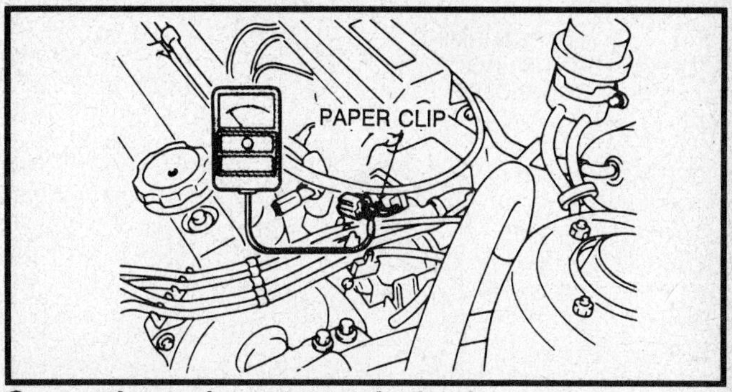

Connecting tachometer to single-pin connector, typical – 1.8L (VIN T) and 2.0L (VIN R AND U) engines shown

Self-diagnosis data-transfer switching terminal – 2.0L (VIN R AND U) engine

Connecting tachometer to single-pin connector – Trucks with 3.0L engines

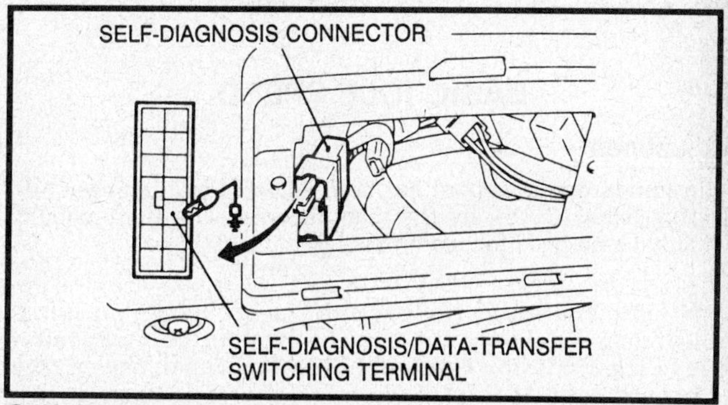

Self-diagnosis data-transfer switching terminal – Trucks with 3.0L engines

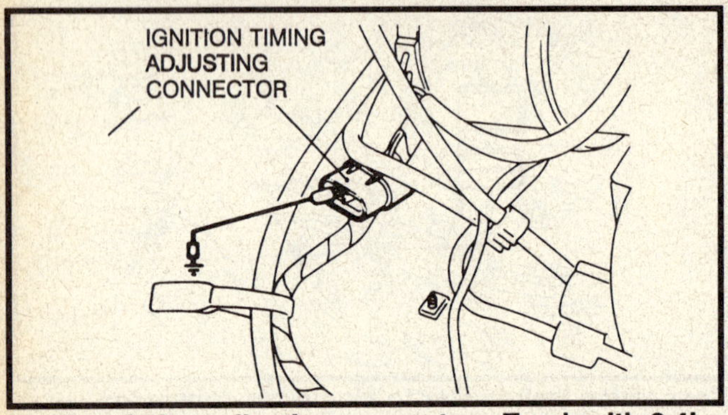

Ignition timing adjusting connector—Truck with 2.4L and 3.0L engines

use a jumper wire to ground the self-diagnosis data-transfer switching terminal of the diagnosis connector.

7. If the Multi-Use Tester (MUT) is used, connect it to the diagnosis connector.
8. Ground the ignition timing adjustment terminal.
9. Operate the engine at curb idle. Check basic ignition timing and adjust, if necessary. Timing should be between 3–7 degrees BTDC.
10. Remove the ground from the ignition timing adjustment terminal and allow the engine to idle for 2 minutes.
11. Check the curb idle rpm. Curb idle rpm should be 600–800 rpm.

NOTE: The curb idle rpm is automatically regulated by the Idle Speed Control (ISC) system.

1.5L (VIN J) ENGINE

1. Operate the engine until operating temperature is reached.
2. Turn OFF all accessories. Place the gear selector in **P** (automatic transaxle) or **N** (manual transaxle).
3. Set the steering wheel in straight ahead position. Apply the parking brake and block the drive wheels.
4. Check that the electric cooling fan is OFF.
5. Stop the engine and connect a tachometer and timing light. If available, the Multi-Use Tester (MUT) (Tool 09391-33200 or equivalent) may be used (Press Code 22 and read the idle rpm).
6. Operate the engine at curb idle.
7. Check basic ignition timing and adjust, if necessary. Timing should be 3–7 degrees BTDC.
8. Operate the engine for more than 5 seconds between 2000–3000 rpm.
9. Operate the engine at idle for 2 minutes; then check the curb idle rpm. Should be 600–800 rpm.
10. If the curb idle rpm is not within specifications, go to Idle Speed Control and TPS sensor adjustment.

BASIC IDLE SPEED

Adjustment

The standard idling speed has been adjusted by the Speed Adjusting Screw (SAS), by the manufacturer, and there should usually be no need for readjustment.

NOTE: The following procedures are based on informations available at the time of production. Always refer to the Emission Control Label, located in the engine compartment. If the information contains there-in differs from the following procedures, use those on the Emission Control Label.

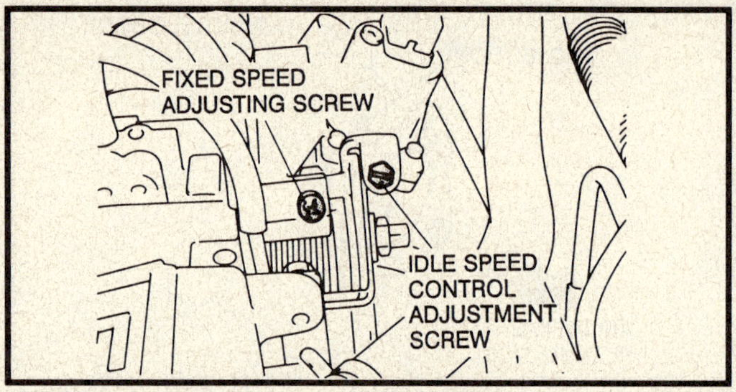

Speed Adjusting Screw (SAS) location—1.5L (VIN A), 1.8L (VIN T) and 2.0L (VIN V) engines

Speed Adjusting Screw (SAS) location—Trucks with 3.0L engine

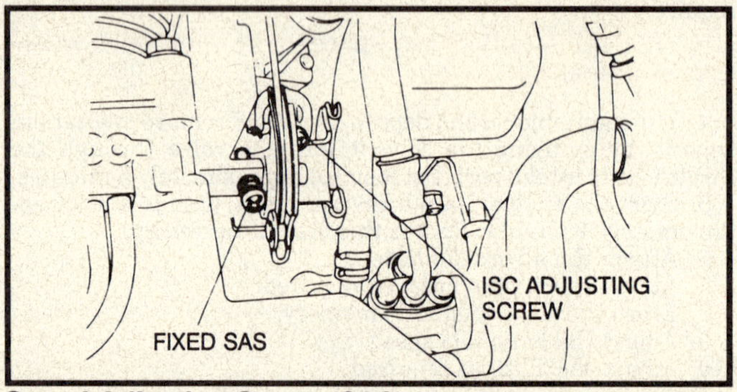

Speed Adjusting Screw (SAS) location—Trucks with 2.4L engine

1.5L (VIN A), 1.8L (VIN T), 2.0L (VIN V), 3.0L (VIN B, C AND S) AND TRUCKS WITH 2.4L ENGINES

1. Operate the engine until operating temperature is reached.
2. Turn OFF all accessories. Place the gear selector in **P** (automatic transaxle) or **N** (manual transaxle).
3. Set the steering wheel in straight ahead position. Apply the parking brake and block the drive wheels.
4. Check that the electric cooling fan is OFF.
5. Stop the engine and connect a tachometer and timing light. If available, the Multi-Use Tester (MUT) (Tool MB991341 or equivalent) may be used to check engine rpm.
6. Loosen the accelerator cable.
7. Turn the ignition switch to the **ON** position for approximately 15 seconds. Do not start the engine at this time. Check to ensure the idle speed control servo if fully retracted to the curb idle position.

NOTE: When the ignition switch is first turned ON, the ISC plunger extends to the fast idle position; then after 15 seconds, it retracts to the initial position. Initial position: the suitable idling position at which the motor position sensor output voltage is 0.9 volts.

8. Turn the ignition switch **OFF**. Disconnect the ISC servo connector allowing the ISC servo to be secure in the initial position.
9. Sufficiently loosen the fixed Speed Adjusting Screw (SAS).
10. Start the engine and let it run at idle speed.
11. Check the basic idle rpms. Basic idle speed should be 650–750 rpm

NOTE: The curb idle rpm on a new vehicle driven less than 300 miles (500 Km) may be 20–100 rpm lower than specifications, but no adjustment is necessary.
If the engine stalls or the engine speed is too low, clean the throttle valve area.

12. If the curb idle speed rpm is not within specifications, adjust the ISC adjusting screw to obtain the specified value.

NOTE: When the ISC adjustment screw is turned to make the adjustment, use a hexagonal wrench if possible. Also, in order to prevent backlash of the screw, make the final adjustment at the tightening side.

13. Tighten the fixed Speed Adjusting Screw (SAS) until the engine speed starts to increase; then loosen it until the engine speed ceases to drop (touch point) and then loosen a half turn from the touch point.
14. Turn the ignition switch **OFF** and adjust the accelerator cable.
15. Adjust the TPS.
16. Start the engine and allow it to operate for approximately 10 minutes. Check that the idling condition is normal.
17. Remove all test equipment.

1.8L (VIN D), 2.0L (VIN R AND U) AND CARS WITH 2.4L ENGINES

1. Operate the engine until operating temperature is reached.
2. Turn OFF all accessories. Place the gear selector in **P** (automatic transaxle) or **N** (manual transaxle).
3. Set the steering wheel in straight ahead position. Apply the parking brake and block the drive wheels.
4. Check that the electric cooling fan is OFF.
5. Stop the engine and connect a tachometer and timing light. If available, the Multi-Use Tester (MUT) (Tool MB991341 or equivalent) may be used.
7. If the Multi-Use Tester (MUT) is used, connect it to the diagnosis connector.

NOTE: When the multi-use tester is connected, the diagnosis control terminal/data transfer switching terminal is grounded.

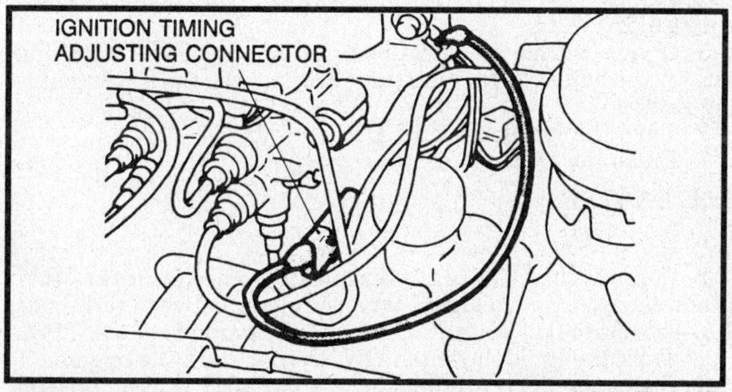

Ignition timing adjusting connector – 2.0L (VIN R AND U) engine

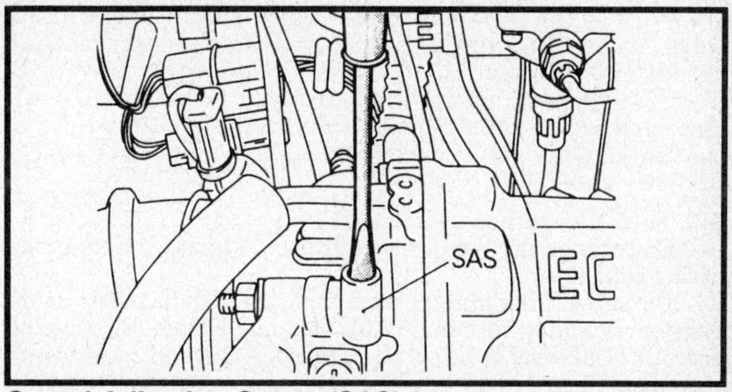

Speed Adjusting Screw (SAS) location – 2.0L (VIN R AND U) engine

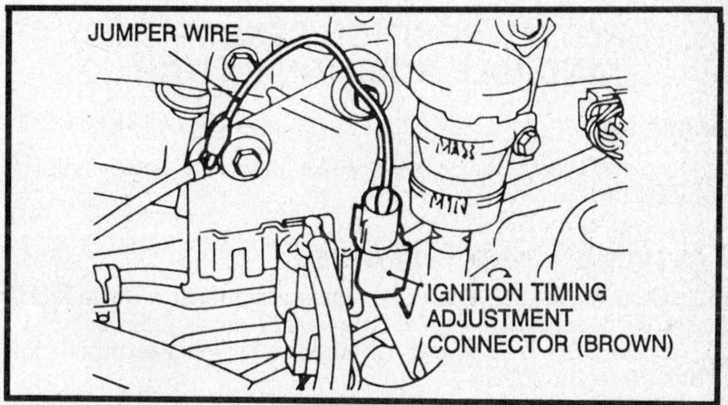

Ignition timing adjusting connector – 1.8L (VIN D) engine

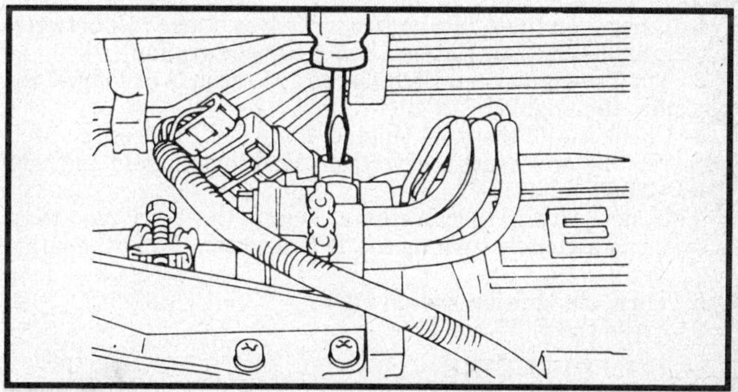

Speed Adjusting Screw (SAS) location – 1.8L (VIN D) engine

8. On 1.8L (VIN D) and 2.4L engines, remove the waterproof connector form the ignition timing adjustment connector and jumper it to ground.
9. Start the engine and let it run at idle speed.
10. Check the basic idle rpms. Basic idle speed should be 700–800 rpm.

NOTE: The curb idle rpm on a new vehicle driven less than 300 miles (500 Km) may be 20–100 rpm lower than specifications, but no adjustment is necessary.
If the engine stalls or the engine speed is too low, clean the throttle valve area.

11. If the curb idle speed rpm is not within specifications, adjust turn the Speed Adjusting Screw (SAS) to obtain the specified value.

NOTE: If the idling speed is higher than the standard value, even when the SAS is fully closed, check whether or not the idle switch (fixed SAS) has been moved. If there is no indication that the idle switch has been moved, it is possible there is leakage as a result of deterioration of the Fast Idle Air Valve (FIAV). If so, the throttle body should be replaced.

12. Turn the ignition switch **OFF**.
13. Disconnect the jumper wire from the ignition timing adjusting connector.
14. Disconnect the jumper wire from the self-diagnosis data-transfer switching terminal (2.0L engine) or diagnosis control terminal (1.8L VIN D and 2.4L engines), if the multi-use tester was not used.
15. Start the engine and allow it to operate for approximately 10 minutes. Check that the idling condition is normal.
16. Remove all test equipments.

THROTTLE POSITION SENSOR (TPS) AND IDLE POSITION SWITCH

Adjustment

Always perform the basic idle speed adjustment, before adjusting TPS.

**1.5L (VIN A) AND
2.0L (VIN R, U AND V) ENGINES**

1. Connect the MUT to the diagnostic connector. If the MUT is not used, connect a digital-type voltmeter, as follows:
   a. On 2.0L (VIN R and U) engine—between terminals 24 and 19 of the ECU.

NOTE: Do not disconnect the ECU connector from the ECU.

   b. On 1.5L and 2.0L (VIN V) engines, disconnect the TPS connector and use a jumper harness (MB991348 or equivalent) between the 2 TPS connector halves. Make test between terminals 2 (sensor output) and 4 (sensor ground).
2. Turn the ignition switch **ON** for 15 seconds or longer. Do not start the engine.
3. Check the TPS output voltage. If the MUT is used, pressed code No. 14 and read the voltage. Readings should indicate 0.480–0.520 volts.
4. If not within specifications, loosen the TPS mounting screws and adjust by turning the TPS. Tighten the TPS mounting screws.
5. Turn the ignition switch **OFF**.
6. Erase the ECU memory.

1.8L (VIN D) ENGINE

1. Disconnect the TPS harness connector.
2. Connect an ohmmeter between terminals 1 and 2 of the TPS.
3. Insert a feeler gauge 0.0256 in. (0.65mm) thick between the fixed SAS and the throttle lever.
4. Loosen the TPS mounting bolts and turn the TPS fully clockwise. Check for continuity on the ohmmeter.
5. Slowly turn the TPS counter-clockwise until the point where no continuity is indicated; then tighten the TPS mounting bolts.
6. Connect the TPS harness connector.

1.8L (VIN T) ENGINE

1. Connect the MUT to the diagnostic connector. If the MUT is not used, connect a digital-type voltmeter between terminals 24 and 19 of the ECU.

NOTE: Do not disconnect the ECU connector from the ECU.

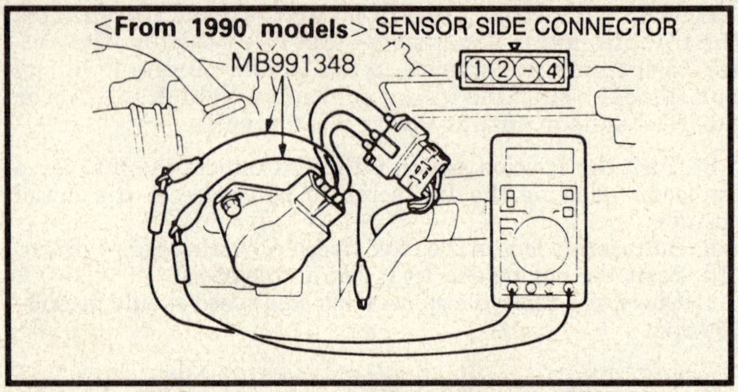

Adjusting TPS using jumper leads—1.5L (VIN A) and 2.0L (VIN V) engines

2. Turn the ignition switch **ON** for 15 seconds or longer. Do not start the engine.
3. Check the TPS output voltage. If the MUT is used, pressed code No. 14 and read the voltage. Readings should indicate 0.480–0.520 volts.
4. If not within specifications, loosen the TPS mounting screws and adjust by turning the TPS. Tighten the TPS mounting screws.
5. Turn the ignition switch **OFF**.
6. Erase the ECU memory.

2.4L ENGINE

Car

1. Disconnect the TPS harness connector.
2. Connect an ohmmeter between terminals 1 and 2 of the TPS.
3. Insert a feeler gauge 0.0256 in. (0.65mm) thick between the fixed SAS and the throttle lever.
4. Loosen the TPS mounting bolts and turn the TPS fully clockwise. Check for continuity on the ohmmeter.
5. Slowly turn the TPS counter-clockwise until the point where no continuity is indicated; then tighten the TPS mounting bolts.
6. Connect the TPS harness connector.

Truck

1. Connect the MUT to the diagnostic connector. If the MUT is not used, disconnect the TPS connector and use a jumper harness (MB991348 or equivalent) between the 2 TPS connector halves.
2. Connect a digital-type voltmeter between terminals 2 (sensor output) and 4 (sensor ground).
3. Turn the ignition switch **ON** for 15 seconds or longer. Do not start the engine.
4. Check the TPS output voltage. If the MUT is used, pressed code No. 14 and read the voltage. Readings should indicate 0.480–0.520 volts.
5. If not within specifications, loosen the TPS mounting screws and adjust by turning the TPS. Tighten the TPS mounting screws.
6. Turn the ignition switch **OFF**.
7. Erase the ECU memory.

3.0L ENGINE

Car With TCL

1. Connect the MUT to the diagnostic connector. If the MUT is not used, connect a digital-type voltmeter between terminals 2 (sensor output) and terminal 4 (sensor ground) of the TPS.
2. Turn the ignition switch **ON**. Do not start the engine.
3. Check the TPS output voltage. If the MUT is used, pressed code No. 14 and read the voltage. Readings should indicate 0.580–0.690 volts.

4. If not within specifications, remove the throttle body, loosen the TPS mounting screws and adjust by turning the TPS. Tighten the TPS mounting screws.
5. Turn the ignition switch **OFF**.
6. Erase the ECU memory.

Car Without TCL

1. Disconnect the TPS harness connector.
2. Connect an ohmmeter between terminals 3 and 4 of the TPS.
3. Insert a feeler gauge 0.0256 in. (0.65mm) thick between the fixed SAS and the throttle lever.
4. Loosen the TPS mounting bolts and turn the TPS fully clockwise. Check for continuity on the ohmmeter.
5. Slowly turn the TPS counter-clockwise until the point where no continuity is indicated; then tighten the TPS mounting bolts.
6. Connect the TPS harness connector.

Truck

1. Disconnect the TPS harness connector.
2. Connect an ohmmeter between terminals 1 and 2 of the TPS.
3. Insert a feeler gauge 0.0256 in. (0.65mm) thick between the fixed SAS and the throttle lever.
4. Loosen the TPS mounting bolts and turn the TPS fully clockwise. Check for continuity on the ohmmeter.
5. Slowly turn the TPS counter-clockwise until the point where no continuity is indicated; then tighten the TPS mounting bolts.
6. Connect the TPS harness connector.

IDLE POSITION SWITCH AND ACCELERATOR PEDAL POSITION SENSOR (APS)

Adjustment

3.0L ENGINE (VEHICLES WITH TCL)

1. Disconnect the APS connector.
2. Connect an ohmmeter, using jumper wires, between terminal 2 (idle position switch) on the APS and terminal 1 (sensor ground).
3. Insert a feeler gauge 0.020 in. (0.5mm) thick approximately 0.12 in. (3mm) deep between the accelerator lever and the throttle lever.

NOTE: If the feeler gauge is inserted deeper than 0.12 in. (3mm), the accelerator lever opening degree will be larger that set, thus resulting in improper adjustment.

4. Loosen the APS mounting bolts and turn the sensor counter-clockwise as far as it will go. There should be no continuity between terminals 1 and 2 at this time.
5. Slowly turn the APS clockwise until the point where no continuity is indicated; then tighten the APS mounting bolts.
6. Connect the APS harness connector.
7. Connect a digital voltmeter between terminal 3 (sensor output) and terminal 1 (sensor ground) of the APS.
8. Turn the ignition switch **ON**. Do not start the engine.
9. Check the APS output voltage. If the MUT is used, select item No. 11 and read the voltage. Readings should be 0.40–1.0 volts.
10. If not within specifications, check the APS and related harness.
11. Remove the feeler gauge.
12. Turn the ignition switch **OFF**.

IDLE SPEED CONTROL (ISC) AND THROTTLE POSITION SENSOR (TPS)

Adjustment

1.5L (VIN J) ENGINE

1. Operate the engine until operating temperature is reached.
2. Turn OFF all accessories. Place the gear selector in **P** (automatic transaxle) or **N** (manual transaxle).
3. Set the steering wheel in straight ahead position. Apply the parking brake and block the drive wheels.
4. Loosen the accelerator cable.
5. Connect the multi-use tester to the diagnostic connector in the fuse box. If the multi-use tester is not used, install a tachometer and a voltmeter. Insert the voltmeter probes at rear of the TPS connector. Do not disconnect the TPS connector from the throttle body.
6. Turn the ignition switch to the **ON** position for approximately 15 seconds; then check to be sure the idle speed control servo if fully retracted to the curb idle position. Do not start the engine.

NOTE: When the ignition switch is first turned ON, the ISC plunger extends to the fast idle position; then after 15 seconds, it retracts to curb idle position.

7. Turn the ignition switch **OFF**.
8. Disconnect the ISC motor connector and secure the ISC motor at the fully retracted position.
9. In order to prevent the throttle valve from sticking, open it 2 or 3 times; then release it to let it click shut, and sufficiently loosen the fixed SAS.
10. Start the engine and let it run at idle speed.
11. Check that the curb idle rpm is within specifications.
12. If the curb idle speed rpm is not within specifications, adjust the ISC adjusting screw to obtain the specified value.

NOTE: The curb idle rpm on a new vehicle driven less than 300 miles (500 Km) may be 20–100 rpm lower than specifications, but no adjustment is necessary.
If the engine stalls or the engine speed is too low, clean the throttle valve area.

13. Tighten the fixed Speed Adjusting Screw (SAS) until the engine speed starts to increase; then loosen it until the engine speed ceases to drop (touch point) and then loosen a half turn from the touch point.
14. Stop the engine.
15. Turn the ignition switch to the **ON** and check that the TPS output voltage is within specifications. Should be 0.48–0.52 volts.
16. If adjustment is necessary, loosen the TPS mounting screws and turn the TPS until the specified value is obtained. Tighten the TPS mounting screws.
17. Turn the ignition switch **OFF**.
18. Adjust the accelerator cable play: Automatic transmission – 0.12–0.2 in. (3–5mm). All others 0.04–0.08 in. (1–2mm).
19. Connect the ISC motor connector.
20. Disconnect the voltmeter, and connect the TPS connector.
21. Start the engine and check that the idle speed is correct.
22. Turn the ignition switch **OFF** and erase any fault code stored in the system memory.

IDLE POSITION SWITCH (FIXED SPEED ADJUSTING SCREW)

Adjustment

2.0L (VIN R AND U) ENGINES

NOTE: The idle position switch has been precisely adjusted by the manufacturer and its setting should not be altered. However, if the idle position switch is replaced, readjustment is necessary.

1. Loosen the acceleration cable tension.
2. Disconnect the idle switch (fixed SAS) harness connector.
3. Loosen the idle switch (fixed SAS) locknut.
4. Turn the idle switch counterclockwise until the throttle valve closes.
5. Connect an ohmmeter between the terminal of the switch and body.
6. Adjust the idle switch until continuity is indicated. Then turn the idle switch the amount of additional $^{15}/_{16}$ turns.
7. Tighten the locknut, while keeping the switch from moving. Reconnect the harness connector.
8. Adjust the acceleration cable.
9. Adjust the curb idle speed and TPS.

FIXED SPEED ADJUSTING SCREW

Adjustment

1.8L (VIN D), 2.4L (VIN W) AND 3.0L ENGINES

NOTE: The fixed SAS has been precisely adjusted by the manufacturer and its setting should not be altered. However, if for any reasons it must adjusted, us the following procedure.

1. Loosen the acceleration cable tension.
2. Back out the fixed SAS locknut and turn the fixed SAS counter-clockwise until the thorttle valve is fully closed.
3. Tighten the fixed SAS until the point where it just make contact (throttle plate just begins to open). From this point, tighten an additional 1¼ turns.
4. Hold the fixed SAS securely and tighten the locknut.
5. Adjust the accelerator cable, basic idle speed, idle position switch and TPS.

ACCELERATOR CABLE

Adjustment

3.0L (VIN B, C AND S) ENGINES

1. Check that there is no sharp bends in the cable assembly.
2. Check that the throttle lever touches the fixed SAS (stopper).
3. Check the inner cable for correct slack.
4. If there is too much slack or no slack, adjust the play as follows:
   a. Loosen the adjusting bolts on the air intake plenum and secure the outer cable so that the free play of the inner cable is 0.12–0.20 in. (3–5mm).
5. After adjustment, confirm that the throttle valve fully opens and closes.

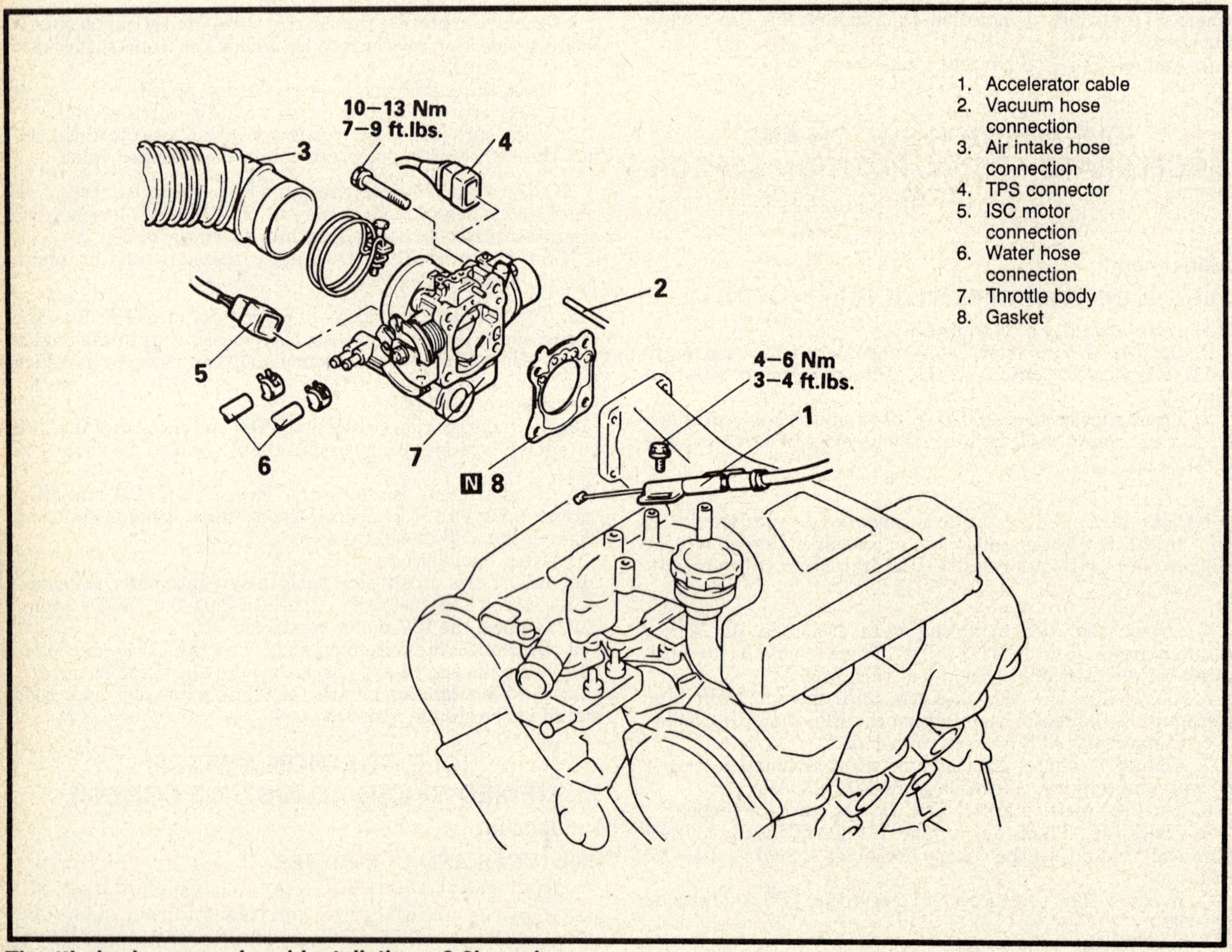

Throttle body removal and installation—3.0L engine

Feedback Carburetors

15 SECTION

HONDA CARBURETOR SYSTEM

General Information

The carburetor consists of 2 side draft carburetors, each having a variable venturi. The variable venturi carburetors allow a smooth increase of engine speed and output, due to the change in venturi area in proportion to the carburetor intake airflow rate.

CARBURETOR SYSTEMS

Automatic Choke

This system provides easy engine starting under a wide range of air temperatures. The system consists of the following:

- Choke valve and linkage
- Choke opener
- Fast idle and fast idle unloader system

The choke valve is located in the primary throttle bore of the carburetor. When the engine is not running, the choke valve angle is determined by the thermowax valve and the engine coolant temperature.

Choke Opener System

CHOKE OPENER

When the engine coolant is below 60°F (15°C), the thermovalve operates to open and manifold vacuum is bled from the choke opener. The choke opener diaphragm is retracted to an intermediate position by vacuum and spring force. When the engine coolant temperature exceeds 60°F (15°C), the thermovalve shuts off the vacuum bleed and allows the choke opener to retract fully and pull the choke valve open.

Power Valve

This system is provided to supply supplementary fuel into the primary main fuel passage when the vehicle is operated in the power mode.

In normal driving modes other than acceleration, manifold vacuum is applied on the diaphragm of the power valve and the valve is closed. When the throttle valve is suddenly opened to accelerate the vehicle, the power valve opens because of the decreased manifold vacuum and supplies additional fuel to the primary main fuel passage through the power jet, providing smooth acceleration performance.

Idle Controller

The idle controller keeps the idle speed from dropping when the compressor is running. An idle controller operated by the manifold vacuum is attached to the throttle linkage. Vacuum is controlled by an idle control solenoid valve that is energized when the A/C switch is on and compressor clutch is energized.

When the A/C is off or the compressor is not running, the idle control solenoid valve is de-energized and the vacuum in the idle controller is relieved through the air filter on the idle control solenoid valve.

Air Jet Controller

To maintain optimum air-fuel ratio, the slow air jet controls air flow into the primary jets of the carburetor throats. When the

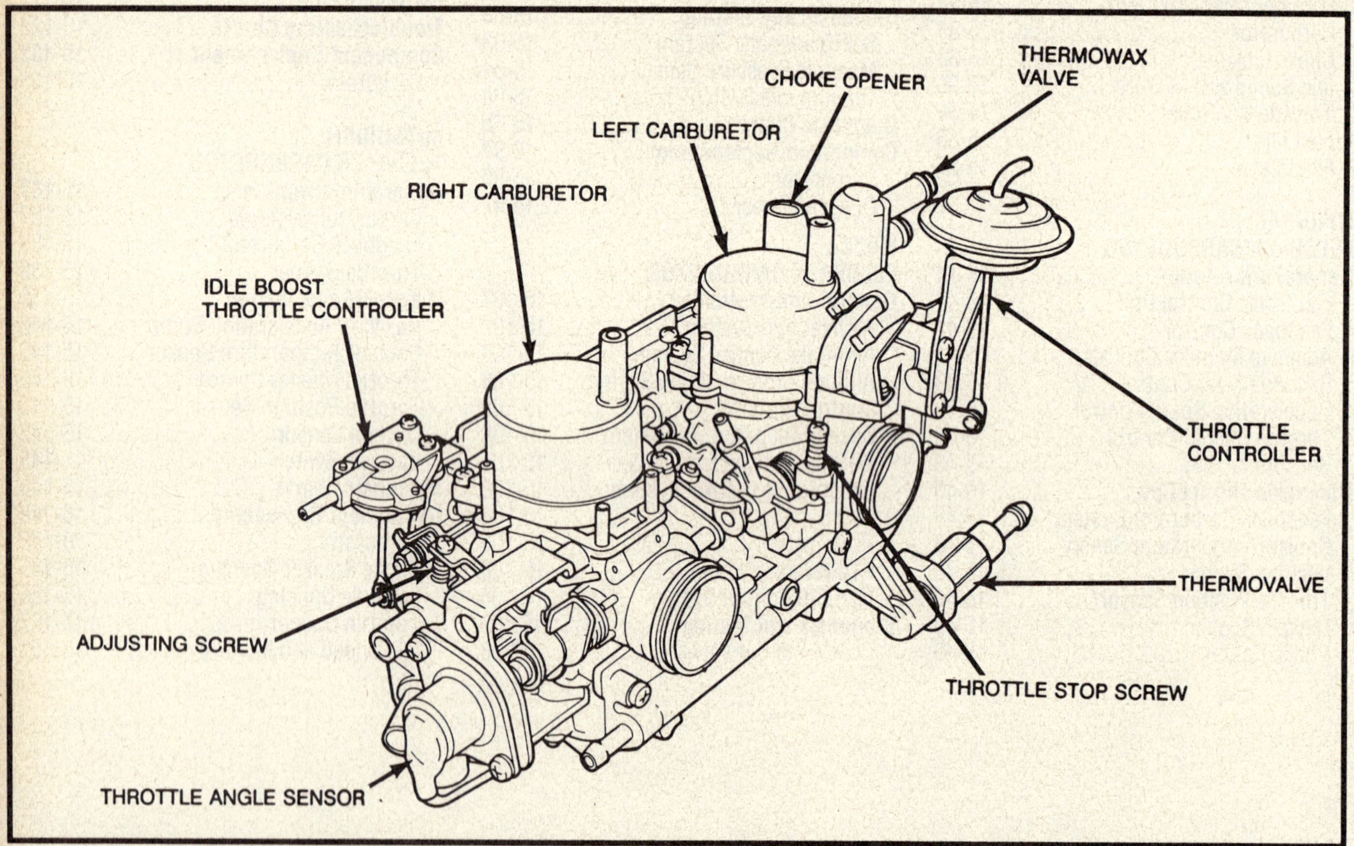

Prelude side draft carburetors

vehicle is being started or in power mode, the control unit energizer the air leak solenoid valve in the air cleaner to close the extra air passage, increasing fuel flow.

Slow Mixture Cut-off

The primary slow mixture cut-off solenoid valve is provided to cut-off the idle mixture passage upstream of the bypass port to prevent the engine from dieseling. The valve also function to cut-off fuel during deceleration.

Vacuum Piston Control System

This system allows the vacuum piston to rise when the engine is operating under heavy load. Vacuum to the piston is controlled by solenoids activated by the ECU, based on inputs from MAP, speed, rpm sensors.

FEEDBACK CONTROL

The feedback control system maintains the proper air/fuel mixture ratio by allowing air into the intake manifold, as is necessary, to adjust a temporarily fuel-rich condition. This system in made up of 4 subsystem: the air-fuel ratio control, shot of air control, deceleration air supply and hot engine start control.

Air Fuel Ratio

This system is designed to achieve a stoichimetric air-fuel ratio, needed for proper operation of the 3-way catalyst. The carburetor mixture is calibrated on the richer side. The air supply through the EACV dilutes the mixture for controlling the mixture as needed.

Shot of Air Control

This system provides air into the intake manifold to improve emissions performance and prevent afterburning due to over rich mixtures during deceleration.

The control unit receives signals from speed, coolant temperature, MAP and rpm sensors. The air shot is induced from the EACV when the manifold vacuum increases suddenly except when the vehicle is moving at very low speed with the coolant temperature below normal. The amount of air provides is based on the amount of manifold vacuum.

Deceleration Air Supply

This system is designed to improve emission performance by supplying air into the intake manifold during deceleration in relatively high engine rpm.

The control unit receives signals from speed, coolant temperature, MAP, gear position and rpm sensors. The air is induced from the EACV, based on signals from the ECU.

Hot Engine Start Control

This system is designed to provide air into the intake manifold for engine starting when the coolant temperature is very high. The EACV provides the additional air flow based on signals from the ECU.

Speed Sensor

The speed sensor sends a 4 pulse signal to the ECU for each revolution of the speedometer cable.

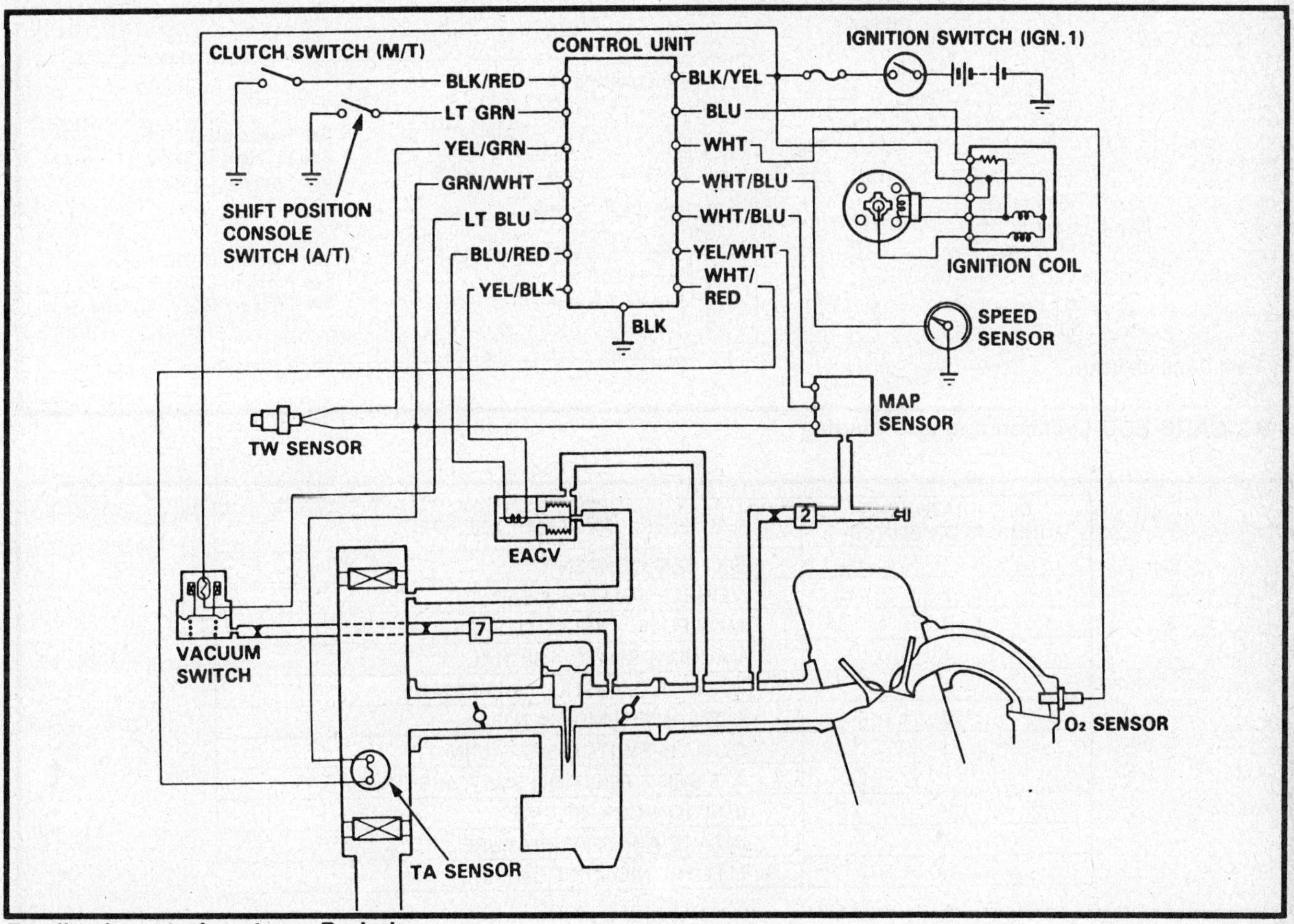

Feedback control system—Prelude

Diagnosis and Testing

FUEL PRESSURE

1. With ignition switch **OFF**, remove the dashboard under cover and pull down the fuse/relay box.
2. Remove the fuel cut-off relay from the fuse box.
3. Connect the No.1 and No. 2 terminals in the fuse box, with a jumper wire.
4. Disconnect the fuel line at the filter and connect a pressure gauge.

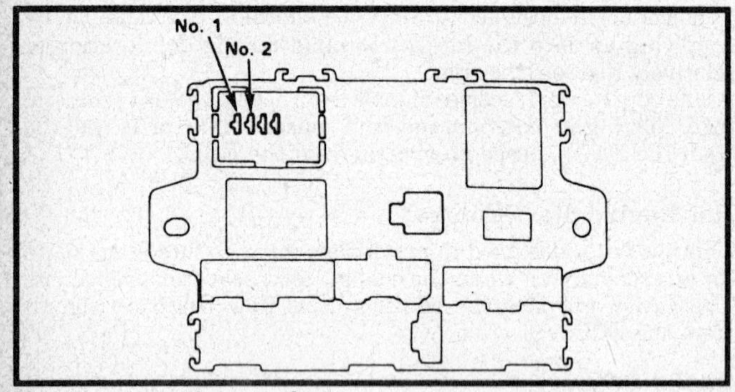

Fuel pump relay socket

5. Turn the ignition switch **ON**.
6. If the pressure is below 1.3 psi (8.8 kPa), replace the pump.
7. Remove the gauge and measure the fuel flow.
8. If flow is less than 20 oz. (600cc) in 1 minute, check for restrictions; if no restrictions, replace the pump.

SELF-DIAGNOSTIC SYSTEM

Self-Diagnosis Indicators

On Prelude, there is a single LED display. The LED will blink consecutively to indicate the trouble code. The ECU is located behind the carpet, on the passenger's side under the dash.

Sometimes the dash warning light and/or ECU LED will come ON, indicating a system problem, when, in fact, there is only a bad or intermittent electrical connection. To troubleshoot a bad connection, note the ECU LED pattern that is lit, refer to the diagnosis chart and check the connectors associated with that circuit (disconnect, clean or repair if necessary and reconnect those connections). Then, reset the ECU memory as described, restart the car and drive it for a few minutes and then recheck the car and drive it for a few minutes and then recheck the LED. If the same pattern lights up, begin system troubleshooting; if it does not light up, the problem was only a bad connection.

The memory for the PGM-FI warning light on the dashboard will be erased when the ignition switch is turned **OFF**; however, the memory for the LED display will not be canceled. Thus, the warning light will not come ON when the ignition switch is again turned **ON** unless the trouble is once more detected.

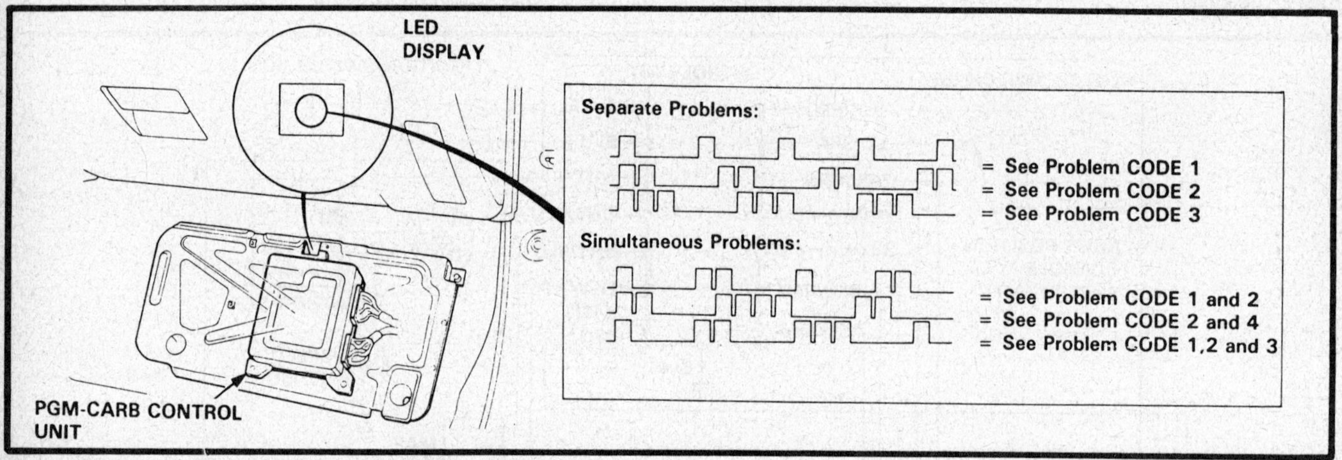

PRG-CARB ECU Location and LED display

SELF-DIAGNOSIS INDICATOR BLINKS	SYSTEM INDICATED
1	OXYGEN CONTENT
2	VEHICLE SPEED PULSER
3	MANIFOLD ABSOLUTE PRESSURE
4	VACUUM SWITCH SIGNAL
5	MANIFOLD ABSOLUTE PRESSURE
6	COOLANT TEMPERATURE
7	CLUTCH SWITCH SIGNAL (M/T)
	A/T SHIFT POSITION SWITCH SIGNAL (A/T)
8	IGNITION COIL SIGNAL
10	INTAKE AIR TEMPERATURE
14	ELECTRONIC AIR CONTROL

Self-Diagnosis Codes—Prelude

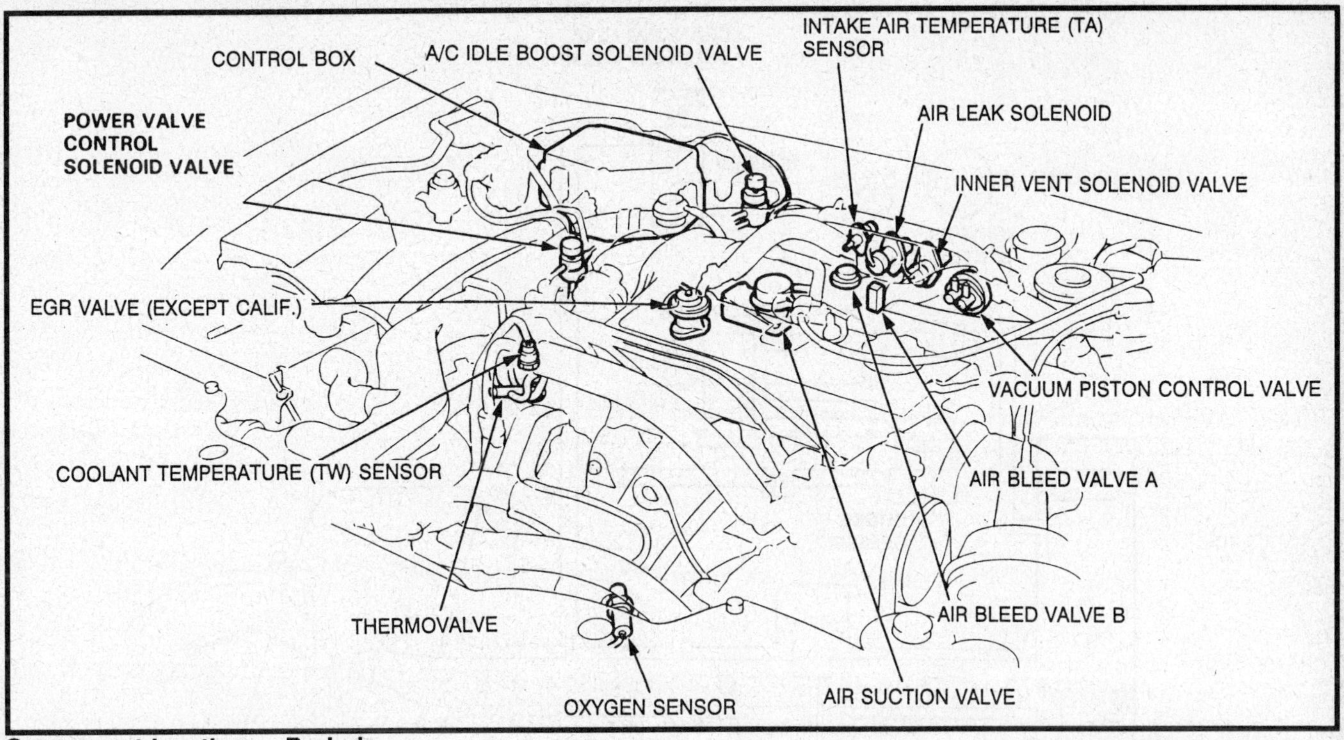

Component locations – Prelude

Troubleshooting should be done according to the LED display even if the warning light is OFF.

READING CODES

If there is a code present in the ECU memory, the LED light on the ECU will flash the code when the ignition is **ON**. Pull down the inspection flap in the carpet under the dash to expose the LED.

CLEARING CODES

To clear the trouble codes, pull the fuse marked 'EFI, ECU' for 10 seconds with the ignition switch in the **OFF** position. The fuse is located in the underhood relay box.

POWER VALVE

Testing

1. With engine coolant temperature below 69°F (21°C), disconnect hose from power valve and connect a hand pump to valve. Apply vacuum. Vacuum should remain steady.
2. If vacuum decreases, replace diaphragm.
3. Start engine and disconnect hose from power valve and connect a vacuum gauge to hose. There should be no vacuum present.
4. If vacuum is present, check hoses for proper connections or blockage. If necessary, replace the air bleed valve B and retest.
5. Allow engine to warm to normal operating temperature.
6. Disconnect the vacuum hose from power valve and connect a vacuum gauge to hose. There should be vacuum present.
7. If vacuum is not present, check hoses for proper connections or blockage. If necessary, replace the air bleed valve B and retest.

VACUUM PISTON CONTROL SYSTEM

Testing

1. Disconnect the vacuum hose from the carburetor and connect a vacuum pump.
2. Apply vacuum to the vacuum piston control valve.
3. There should be a controlled vacuum leak. If the unit does not hold any vacuum, check for broken lines and disconnections. Replace the vacuum piston control valve if no other fault is found.
4. Start engine and allow to reach normal operating temperature.
5. Apply vacumm. The valve should not hold vacuum. If it holds vacuum, disconnect vacuum hose No. 28 from the air cleaner and apply vacuum. If there is vacuum replace the vacuum piston control valve. If there is no vacuum go to "Vacuum Piston Control Solenoid Valve" chart.
6. If Step 5 did not hold vacuum, disconnect vacuum hose No. 28 and connect a vacuum gauge. Quickly raise engine speed to 3000 rpm and snap throttle closed. There should be no vacuum for a moment. If vacuum did not drop go "Vacuum Piston Control Solenoid Valve" chart.

CHOKE OPENER

Testing

1. With engine temperature below 59°F (15°C), disconnect vacuum hose No. 26 from the choke opener and connect a vacuum pump.
2. Check the vacuum with ignition switch turned to **START**.
3. There should be no vacuum; if there is vacuum, go to "Choke Opener" chart.
4. Start engine and disconnect the No. 18 vacuum hose from the choke opener and connect hand vacuum pump to the disconnected hose.
5. It should not hold vacuum. If it holds vacuum, check hoses for blockage, replace the valve if necessary.

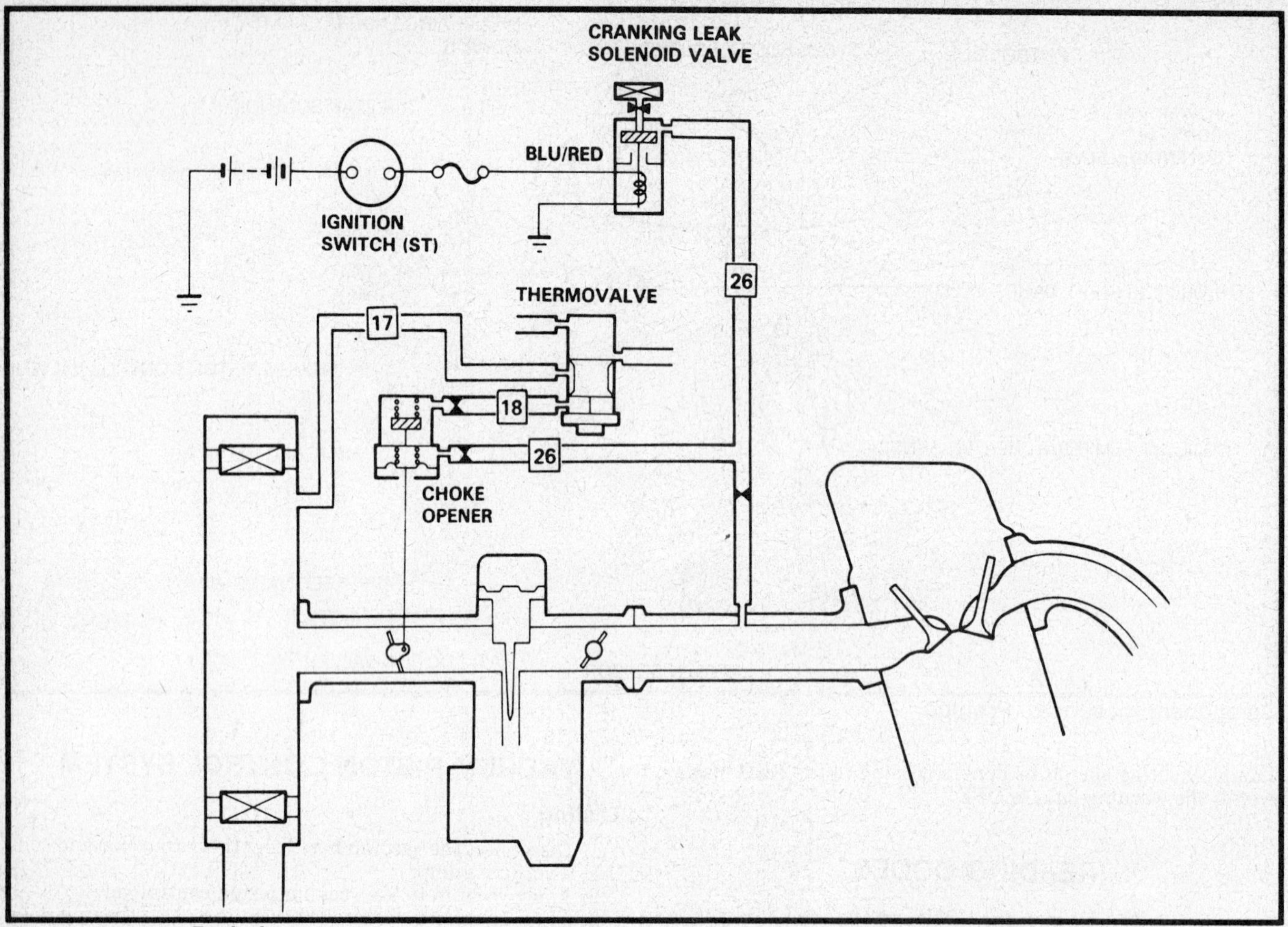

Choke opener—Prelude

6. Start engine and allow to reach normal operating temperature.
7. Disconnect the No. 18 vacuum hose from the choke opener and connect hand vacuum pump to the disconnected hose.
8. It should hold vacuum. If it does not hold vacuum, check hoses for leaks and replace the valve if necessary.

FUEL CUT-OFF RELAY

Testing

1. With ignition switch **OFF**, remove the dashboard under cover and pull down the fuse/relay box.
2. Remove the fuel cut-off relay from the fuse box.
3. Check for continuity between the No. 3 terminal and ground.
4. Attach a voltmeter across terminals No. 2 and No. 3. With ignition switch **ON**, there should be battery voltage. If no voltage, check Black/Yellow wire and No. 12 fuse.
5. Turn ignition switch **OFF**.
6. Attach voltmeter across terminals No. 4 and No. 3. With ignition switch **ON**, there should be battery voltage. If no voltage, check Blue wire to the ignition coil.
7. Turn ignition switch **OFF**.
8. Connect the No.1 and No. 2 terminals in the fuse box with a jumper wire.
9. Turn ignition switch **ON**, the fuel pump should run. If not, check the wiring to the fuel pump.
10. If any test fails and wire connections are good, replace relay and retest.

PURGE CONTROL

Testing

1. Disconnect the No. 19 vacuum hose from the purge vacuum and connect a vacuum gauge to the hose.
2. Start engine. With engine temperature below 145°F (63°C) there should be no vacuum.
3. If there is vacuum, go to "Purge Cut-Off Solenoid Valve" chart.
4. Allow engine to reach normal operating temperature. Raise engine rpm to 3500 rpm. There should be vacuum.
5. If there is no vacuum, go to "Purge Cut-Off Solenoid Valve" chart.

TWO-WAY VALVE

Testing

1. Remove the gas cap.
2. Remove the vapor line from the fuel tank and connect to T-fitting from a vacuum gauge/pump.
3. Slowly apply vacuum while watching the gauge.
4. Vacuum should stabilize at 0.2–0.6 in. Hg (5–15mm Hg); if not, install new valve and retest.
5. Change hose from vacuum to pressure fitting on hand pump. Slowly pressurize vapor line while watching gauge.
6. Pressure should stabilize at 0.4–1.4 in. Hg (10–35mm Hg). If not, install a new valve and retest.

TROUBLESHOOTING GUIDE—PRELUDE

SYMPTOM		SYSTEM: PGM-CARB CONTROL SYSTEM — PGM-CARB CONTROL UNIT	OXYGEN SENSOR	VEHICLE SPEED PULSER	MANIFOLD ABSOLUTE PRESSURE SENSOR	VACUUM SWITCH	COOLANT TEMPERATURE SENSOR	CLUTCH SWITCH SIGNAL
CHECK ENGINE WARNING LIGHT TURNS ON		□ or ☼	CHECK	CHECK	CHECK	CHECK	CHECK	CHECK
SELF-DIAGNOSIS INDICATOR (LED) BLINKS		⓪ or ✱	1	2	3 or 5	4	6	7
ENGINE WON'T START								
DIFFICULT TO START ENGINE WHEN COLD		(BU)					②	
IRREGULAR IDLING	WHEN COLD FAST IDLE OUT OF SPECIFIC	(BU)					②	
	ROUGH IDLE	(BU)			②			
	WHEN WARM RPM TOO HIGH	(BU)						
	WHEN WARM RPM TOO LOW	(BU)						
FREQUENT STALLING	WHILE WARMING UP	(BU)			②		③	
	AFTER WARMING UP	(BU)			②			
POOR PERFORMANCE	MISFIRE OR ROUGH RUNNING	(BU)	③	③	②			
	FAILS EMISSION TEST	(BU)	③		①			
	LOSS OF POWER	(BU)			③			

* If codes other than those listed above are indicated, count the number of blinks again. If the indicator is in fact blinking these codes, replace the original control unit.

(BU): When the Check Engine warning light and the self-diagnosis indicator are on, the back-up system is in operation. Substitute a know-good control unit and recheck. If the indication goes away, replace the original control unit.

SYMPTOM		PGM-CARB CONTROL SYSTEM — A/T SHIFT POSITION SIGNAL	IGNITION COIL SIGNAL	INTAKE AIR TEMPERATURE SENSOR	A/C SIGNAL	CARBURETOR	FUEL SUPPLY	AIR INTAKE	EMISSION CONTROL — ELECTRONIC AIR CONTROL VALVE	OTHER EMISSION CONTROL
CHECK ENGINE WARNING LIGHT TURNS ON		CHECK	CHECK	CHECK					CHECK	
SELF-DIAGNOSIS INDICATOR (LED) BLINKS		7	8	10					14	
ENGINE WON'T START							①			
DIFFICULT TO START ENGINE WHEN COLD						①				
IRREGULAR IDLING	WHEN COLD FAST IDLE OUT OF SPECIFIC					①				③
	ROUGH IDLE			③		①			③	③
	WHEN WARM RPM TOO HIGH	②			③	①				
	WHEN WARM RPM TOO LOW					①			②	
FREQUENT STALLING	WHILE WARMING UP					①			③	
	AFTER WARMING UP					①				
POOR PERFORMANCE	MISFIRE OR ROUGH RUNNING					①	③			
	FAILS EMISSION TEST					②		③	③	③
	LOSS OF POWER					③	②	①		③

NOTE: Across each row in the chart, the systems that could be sources of a symptom are ranked in the order they should be inspected starting with ①. Find the symptom in the left column, read across to the most likely source, then refer to the page listed at the top of that column. If inspection shows the system is OK, try the next most likely system ②, etc.

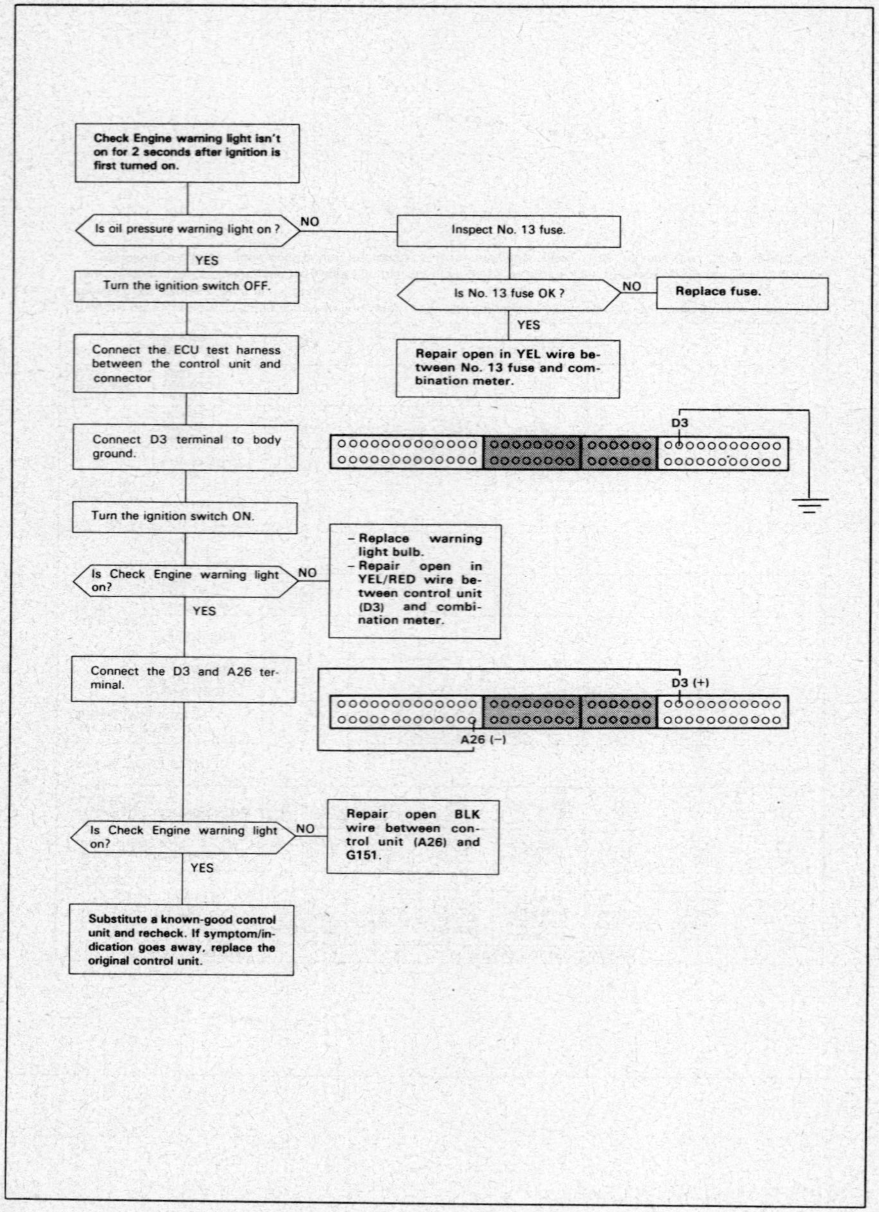
Check Engine warning light isn't on for 2 seconds after ignition is first turned on.
Is oil pressure warning light on ?
NO
Inspect No. 13 fuse.
Is No. 13 fuse OK ?
NO
Replace fuse.
YES
Repair open in YEL wire between No. 13 fuse and combination meter.
YES
Turn the ignition switch OFF.
Connect the ECU test harness between the control unit and connector
Connect D3 terminal to body ground.
D3
Turn the ignition switch ON.
Is Check Engine warning light on?
NO
– Replace warning light bulb.
– Repair open in YEL/RED wire between control unit (D3) and combination meter.
YES
Connect the D3 and A26 terminal.
D3 (+)
A26 (–)
Is Check Engine warning light on?
NO
Repair open BLK wire between control unit (A26) and G151.
YES
Substitute a known-good control unit and recheck. If symptom/indication goes away, replace the original control unit.

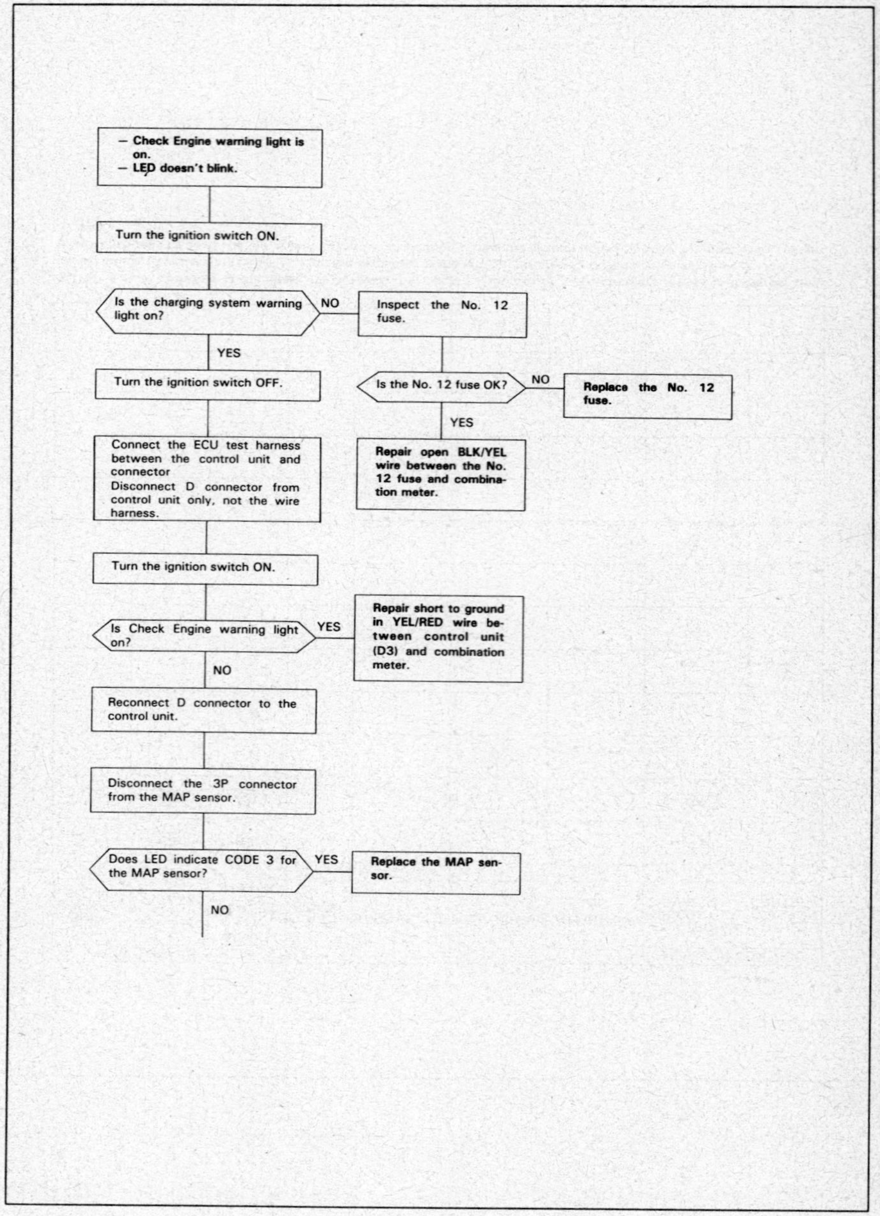
– Check Engine warning light is on.
– LED doesn't blink.
Turn the ignition switch ON.
Is the charging system warning light on?
NO
Inspect the No. 12 fuse.
Is the No. 12 fuse OK?
NO
Replace the No. 12 fuse.
YES
Repair open BLK/YEL wire between the No. 12 fuse and combination meter.
YES
Turn the ignition switch OFF.
Connect the ECU test harness between the control unit and connector
Disconnect D connector from control unit only, not the wire harness.
Turn the ignition switch ON.
Is Check Engine warning light on?
YES
Repair short to ground in YEL/RED wire between control unit (D3) and combination meter.
NO
Reconnect D connector to the control unit.
Disconnect the 3P connector from the MAP sensor.
Does LED indicate CODE 3 for the MAP sensor?
YES
Replace the MAP sensor.
NO

ECU AND CHECK LIGHT—PRELUDE

Reconnect the connector.

Disconnect the 3P connector from the throttle angle sensor.

Does Check Engine warning light remain on?
- NO → **Replace the throttle angle sensor.**
- YES ↓

Disconnect A connector from the A/T control unit.

Does Check Engine warning light remain on?
- NO → **Replace the A/T control unit.**
- YES ↓

Disconnect the 3P connector from the MAP sensor.

(For A/T model only)

Turn the ignition switch OFF.

Disconnect D connector from control unit only, not the wire harness.

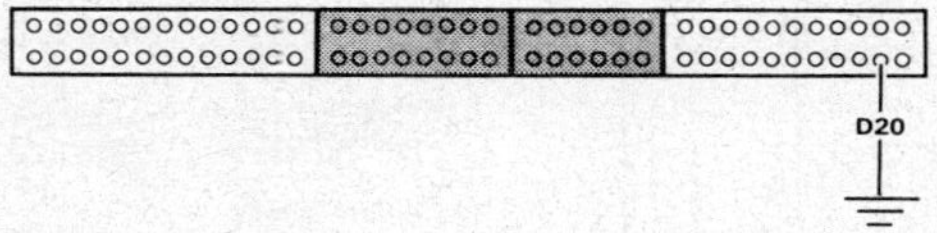

Check for continuity between the body ground and the D20 terminal.

Does continuity exist?
- YES →
  - **On M/T models, repair short to ground in YEL/WHT wire between control unit (D20) and MAP sensor.**
  - **On A/T models, repair short to ground in YEL/WHT wire between control unit (D20) and either the throttle angle sensor, the A/T control unit or the MAP sensor.**
- NO ↓

Reconnect D connector to the control unit and the 3P connector to the MAP sensor.
Also, on A/T models, reconnect the 3P connector to the throttle angle sensor and A connector to the A/T control unit.

Turn the ignition switch ON.

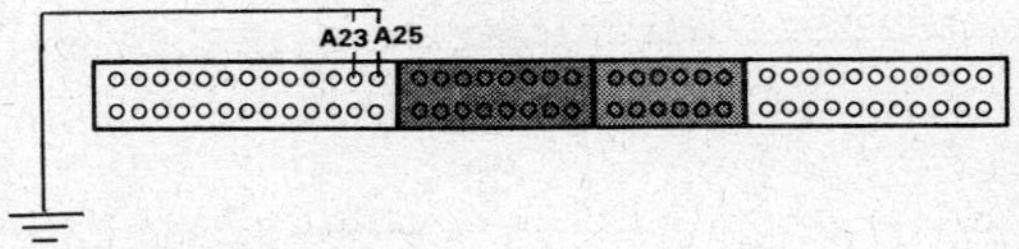

Measure voltage between A23, A25 and body ground.

Is there battery voltage?
- NO → **Repair open in BLK/YEL wire between control unit (A23, A25) and No. 12 fuse.**
- YES ↓

Substitute a known-good control unit and recheck.
If symptom/indication goes away, replace the original control unit.

OXYGEN SENSOR—CODE 1—PRELUDE

Self-diagnosis LED indicate code 1: A problem in the Oxygen (O_2) Sensor circuit.

The oxygen sensor, detects the oxygen content in the exhaust gas, and inputs the control unit. In operation, the control unit receives the signals from the sensor and varies the duration during which fuel is injected. The oxygen sensor is installed on the exhaust manifold.

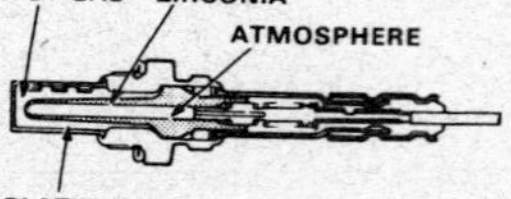

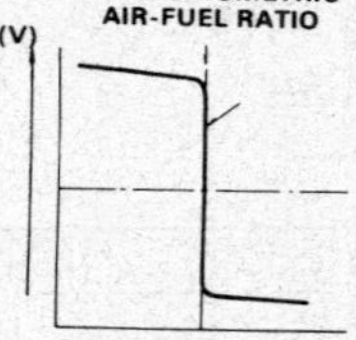

- Check Engine warning light has been reported on.
- LED indicates CODE 1.

Warm up engine to normal operating temperature (the cooling fan comes on).

Turn the ignition switch OFF.

Remove EFI, ECU fuse in the under-hood relay box for 10 seconds to reset control unit.

Start engine and allow to idle for 1 minute.

Raise engine speed to 3,000 rpm.

Remove #2 and #7 hose quickly from the vacuum hose manifold and plug the vacuum hose manifold.

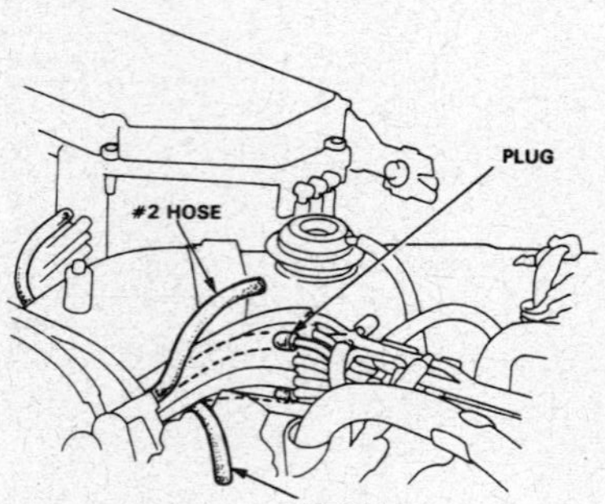

Hold engine speed steady at 3,000 rpm for 20 seconds.

Is Check Engine warning light on and does LED indicate CODE 1?

- NO: Intermittent failure, check and seat connectors at O_2 sensor, at the battery and at the control unit.
- YES: Turn the ignition switch OFF.

Disconnect the wire harness from the O_2 sensor and connect a voltmeter between the O_2 sensor connector and engine ground.

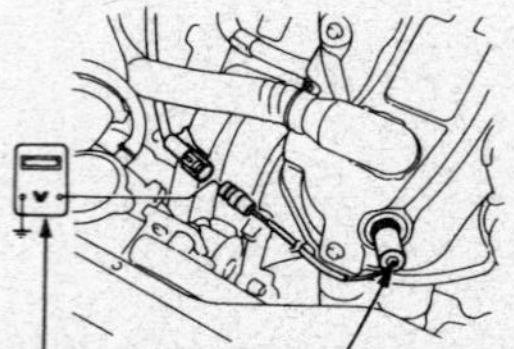

Start the engine and measure the voltage between O_2 sensor connector and engine ground.

Is voltage above 0.6 at wide open throttle and below 0.4 when the throttle is quickly released?

- NO: Replace O_2 sensor.
- YES: Turn the ignition switch OFF.

Reconnect O_2 sensor.

Connect the ECU test harness between the control unit and connector

Start the engine.

Meassure voltage between D13 (+) terminal and D21 (−) terminal.

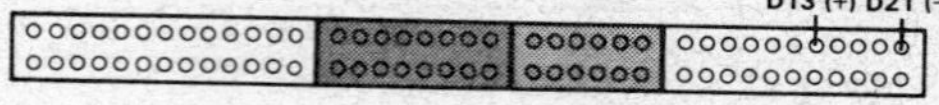

Is voltage above 0.6 V at wide open throttle and below 0.4 V when the throttle is quickly released?

- NO: Repair open or short in WHT wire between the control unit (D13) and O_2 sensor.
- YES: Substitute a known-good control unit and recheck. If symptom/indication goes away, replace the original control unit.

VEHICLE SPEED SENSOR—CODE 2—PRELUDE

Self-diagnosis LED indicates code 2: A problem in the Vehicle Speed Sensor circuit.

The signal generated by the speed pulser, which produces 4 pulses (switch closures to ground) for each revolution of the speedometer cable.

NOTE: Because it is an inter-related system, a leak in the vacuum switch or its hose may cause a "false" code 2 indication.

- Check Engine warning light has been reported on.
- LED indicates CODE 2.

Connect the ECU test harness between the control unit and connector

Block rear wheels and set the parking brake. Jack up the front of the car and support with safety stands.

WARNING Block rear wheels before jacking up front of car.

Turn the ignition switch ON.

Slowly rotate left front wheel and measure voltage between D9 (+) terminal and D21 (−) terminal.

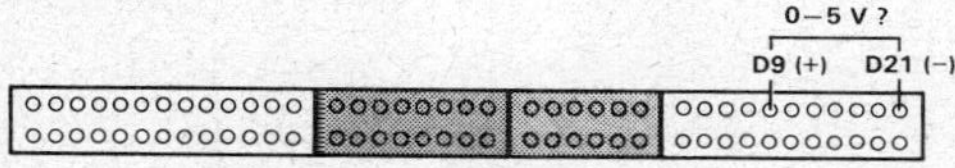

Does voltage pulse 0V and 5V?

YES:

- **Raise the engine speed to 3,000 rpm with no load, then check if the LED indicates CODE 2. If LED indicates CODE 2, repair vacuum leak in #7 hose or in vacuum switch.**
- **Substitute a known-good control unit and recheck. If prescribed voltage is now available replace the original control unit.**

NO:

Turn the ignition switch OFF.

Disconnect connector from the control unit only, not the wire harness.

Turn the ignition switch ON.

Slowly rotate left front wheel and measure voltage between D9 (+) terminal and A25 (−) terminal.

Does voltage pulse 0V and 12V ?

NO:

- **Repair open or short in WHT/BLU wire between control unit (D9) and the speed sensor.**
- **Replace speed sensor.**

YES:

Substitute a known-good control unit and recheck. If prescribed voltage is now available replace the original control unit

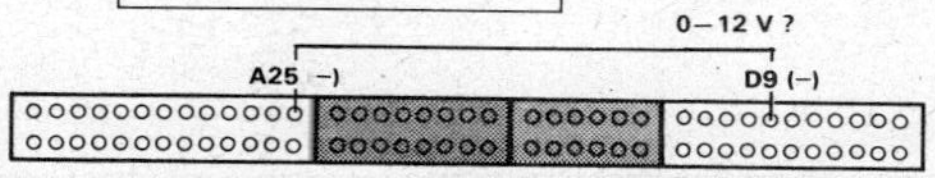

MAP SENSOR—CODE 3—PRELUDE

Self-diagnosis LED indicates code 3: Most likely an electrical problem in the Manifold Absolute Pressure (MAP) Sensor system.

Self-diagnosis LED indicates code 5: Most likely a mechanical problem (broken hose) in the Manifold Absolute Pressure (MAP) Sensor system.

The MAP sensor converts manifold absolute pressure into electrical signals and inputs the control unit.

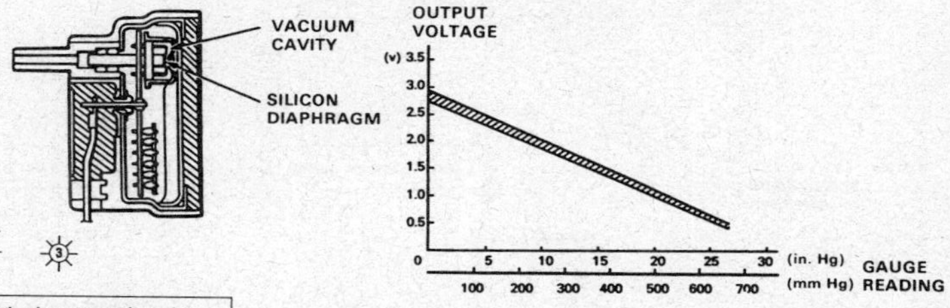

- Engine is warm and running.
- Check Engine warning light has been reported on.
- LED indicates CODE 3.

Turn the ignition switch OFF.

Remove EFI,ECU fuse in the under-hood relay box for 10 seconds to reset control unit.

Start the engine and allow to idle.

Is Check Engine warning light on and does LED indicate CODE 3 ?

NO: **Intermittent failure (test drive may be necessary.)**

YES:

Turn the ignition switch OFF.

Connect the test harness between the MAP sensor and wire harness.

Turn the ignition switch ON.

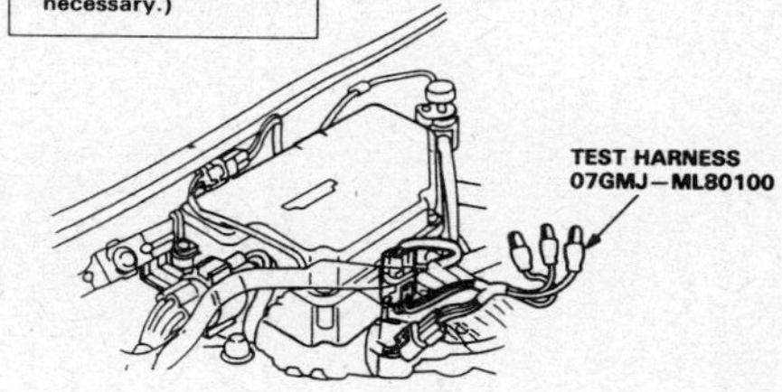

MAP SENSOR—CODE 3—PRELUDE

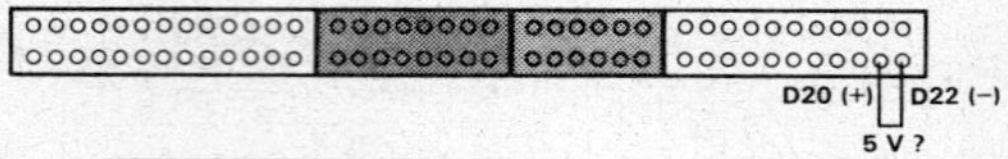

- Measure voltage between RED (+) terminal and GRN (–) terminal.
- Is there approx. 5V ?
  - NO → Measure voltage between RED (+) terminal and body ground.
    - Is there approx. 5V ?
      - YES → Repair open in GRN/WHT wire between control unit (D22) and MAP sensor.
      - NO → Turn the ignition switch OFF.
        - Connect the ECU test harness between the control unit and connector
        - Measure voltage between D20 (+) terminal and D22 (–) terminal with the ignition switch on.
        - Is there approx. 5V ?
          - YES → Repair open in YEL/WHT wire between control unit (D20) and MAP sensor.
          - NO → Substitute a known-good control unit and recheck. If prescribed voltage is now available, replace the original control unit.
  - YES ↓
- Measure voltage between WHT (+) terminal and GRN (–) terminal.
- Is there approx. 3 V ?
  - NO → – Replace MAP sensor.
    – Repair short in WHT/BLU wire between control unit (D17) and MAP sensor.
  - YES ↓
- Turn the ignition switch OFF.
- Connect the ECU test harness between the control unit and connector
- Turn the ignition switch ON.
- Measure voltage between D17 (+) terminal and D22 (–) terminal.
- Is there approx. 3V ?
  - NO → Repair open in WHT/BLU wire between control unit (D17) and MAP sensor.
  - YES → Substitute a known-good control unit and recheck. If symptom/indication goes away, replace the original control unit.

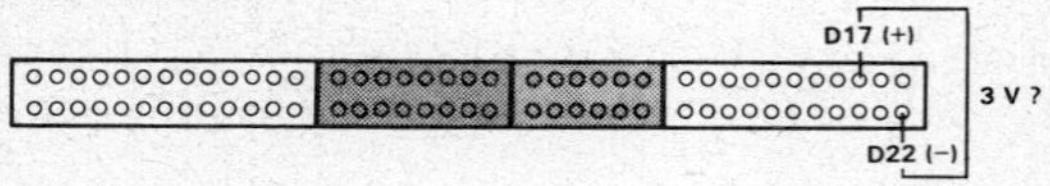

MAP SENSOR—CODE 5—PRELUDE

- Check Engine warning light has been reported on.
- LED indicates CODE 5.

1. Turn the ignition switch OFF.
2. Remove EFI.ECU fuse in the under-hood relay box for 10 seconds to reset control unit
3. Start engine and keep engine rpm at idle.
4. Is Check Engine warning light on and does LED indicate CODE 5 ?
   - NO → Intermittent failure (test drive may be necessary.)
   - YES ↓
5. Stop engine.
6. Remove #2 hose from the vacuum hose manifold and connect a T-fitting from a vacuum gauge between the vacuum hose manifold and the MAP sensor.
7. Start engine.
8. Is there vacuum ?
   - NO → Repair as necessary.
   - YES ↓
9. Connect a vacuum pump to #2 hose and apply vacuum.
10. Does it hold vacuum ?
    - NO → Connect a vacuum pump directly to the MAP sensor and apply vacuum.
      - Does it hold vacuum?
        - NO → Replace MAP sensor.
        - YES → Replace #2 hose.
    - YES

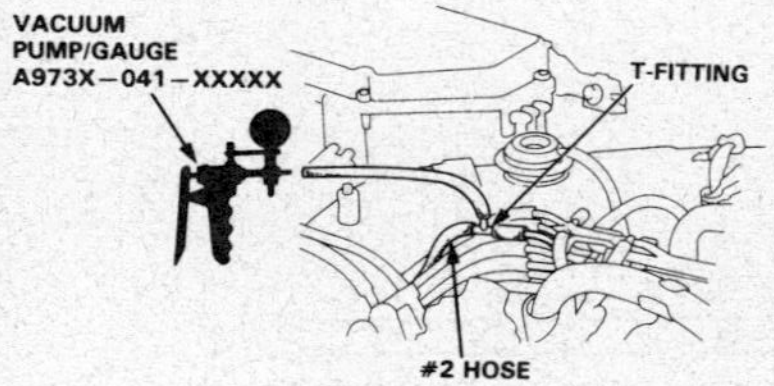

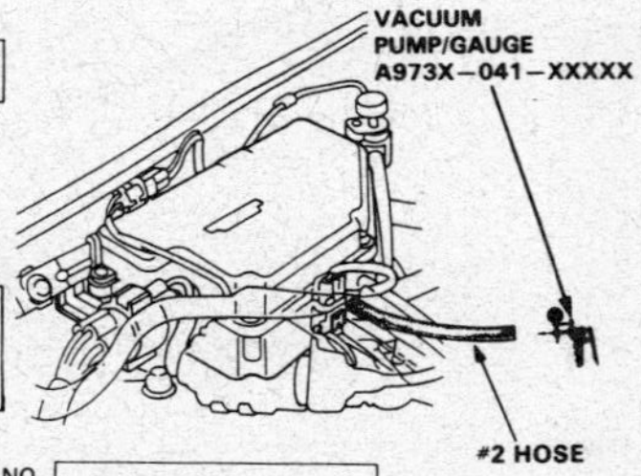

MAP SENSOR—CODE 5—PRELUDE

Stop engine.

Connect the test harness between the MAP sensor and wire harness.

Turn the ignition switch ON.

Measure voltage between WHT (+) terminal and GRN (−) terminal.

Is there approx. 3V?

- NO → **Replace MAP sensor.**
- YES → **Substitute a known-good control unit and recheck. If symptom/indication goes away, replace the original control unit.**

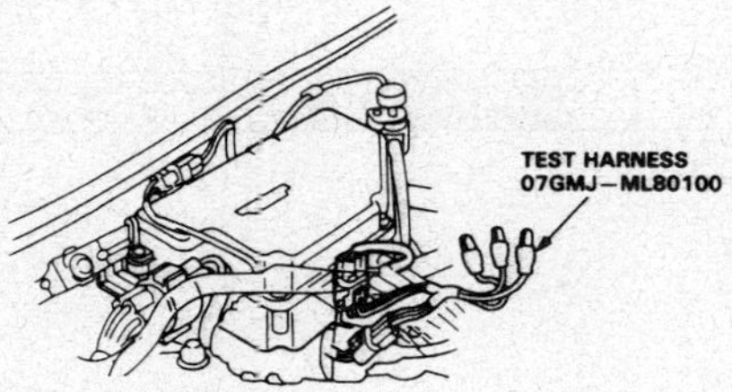

VACUUM SWITCH—CODE 4—PRELUDE

Self-diagnosis LED indicates code 4: A problem in the vacuum switch.

The vacuum switch converts carburetor ported vacuum into electrical signals and inputs the control unit.

— Engine is warm and running.
— Check Engine warning light has been reported on.
— LED indicates CODE 4.

Turn the ignition switch OFF.

Remove EFI.ECU fuse in the under-hood relay box for 10 seconds to reset control unit.

Start the engine and allow to idle for at least 30 seconds.

Is Check Engine warning light on and does LED indicate CODE 4?

- NO → **Intermittent failure. Check connectors at the control box and control unit.**
- YES ↓

Turn the ignition switch OFF.

Disconnect the 8P connector on the control box.

Measure resistance between No.5 terminal and No.6 terminal on the control box.

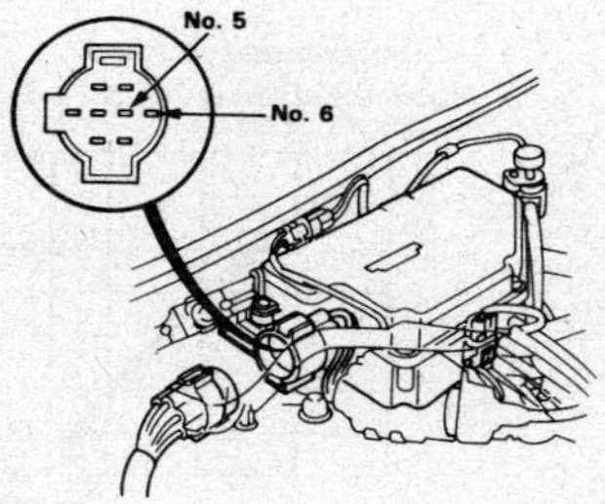

VACUUM SWITCH—CODE 4—PRELUDE

Does continuity exist?
- NO → Remove #7 hose from the vacuum hose manifold and connect a T-fitting from a vacuum gauge between the vacuum hose manifold and the vacuum switch.
  - Start engine and allow to idle.
  - Is there more than 1.2" Hg of vacuum?
    - NO → – Replace vacuum switch. – Repair open in BLK/YEL or WHT wire in the control box.
    - YES → Check routing of #7 hose.
- YES → Turn the ignition switch ON.

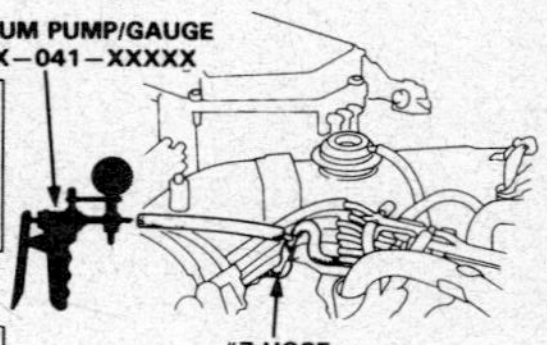

Measure voltage between BLK/YEL terminal and body ground on the wire harness.

Is there battery voltage?
- NO → Repair open in BLK/YEL wire between 8P connector and No. 12 fuse.
- YES → Turn the ignition switch OFF.

Reconnect the 8P connector to the control box.

Connect the ECU test harness between the control unit and body connector

Turn the ignition switch ON.

Measure voltage between D7 (+) terminal and D21 (−) terminal.

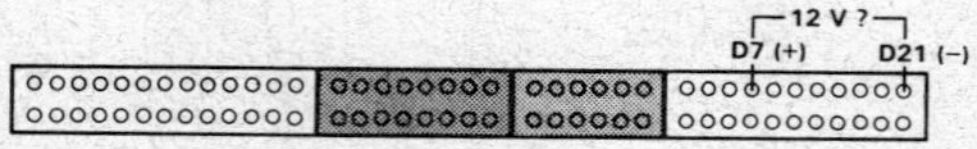

Is there battery voltage?
- NO → Repair open in LT BLU wire between control unit (D7) and vacuum switch.
- YES → Substitute a known-good control unit and recheck. If symptom/indication goes away, replace the original control unit.

COOLANT TEMPERATURE (TW) SENSOR—CODE 6 PRELUDE

Self-diagnosis LED indicates code 6: Most likely a problem in the Coolant Temperature (TW) Sensor circuit.

The TW sensor is a temperature dependant resistor (thermistor). The resistance of the thermistor decreases as the coolant temperature increases as shown below.

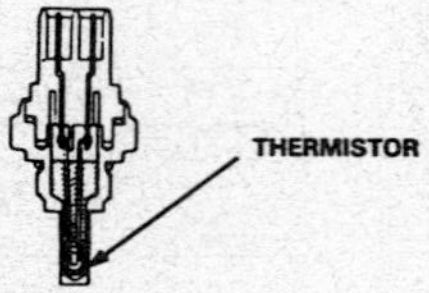

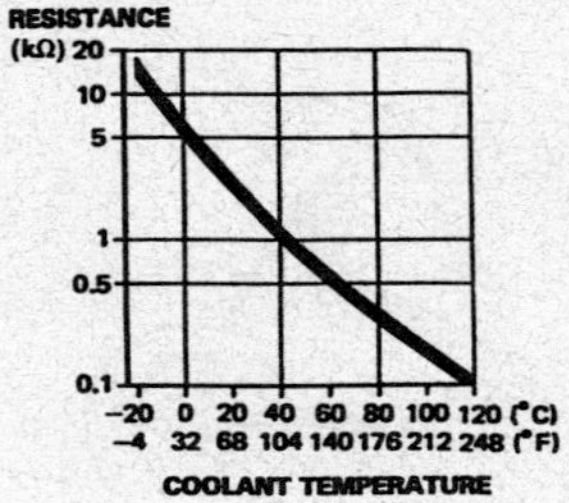

– Check Engine warning light has been reported on.
– LED indicates CODE 6.

Turn the ignition switch OFF.

Remove EFI, ECU fuse in the under-hood relay box for 10 seconds to reset control unit.

Turn the ignition switch ON.

Is Check Engine warning light on and does LED indicates CODE 6?
- NO → Intermittent failure. (test drive may be necessary.)
- YES → Warm up engine to normal operating temperature (the cooling fan comes on).

Disconnect the 2P connector on the TW sensor.

Measure resistance between the 2 terminals on the TW sensor.

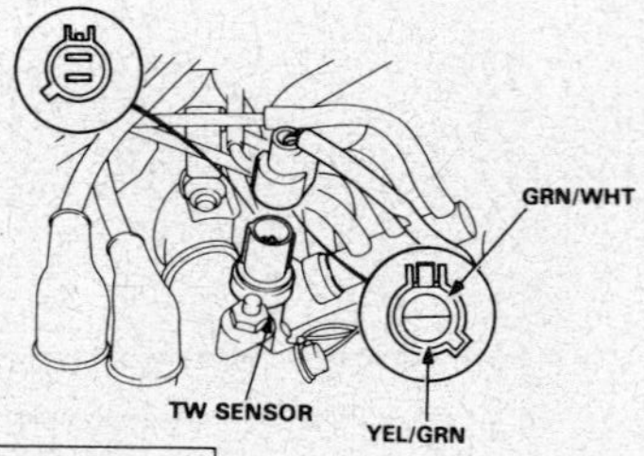

Is there 200—400Ω?
- NO → Replace TW sensor.
- YES

COOLANT TEMPERATURE (TW) SENSOR—CODE 6 PRELUDE

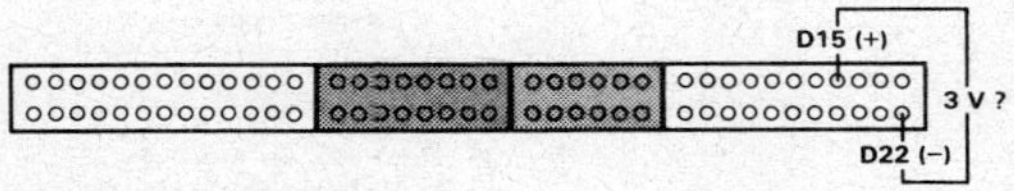

Measure voltage between YEL/GRN (+) terminal and body ground.

Is there approx. 5V ?
- NO → Turn the ignition switch OFF. → Connect the ECU test harness between the control unit and connector → Turn the ignition switch ON. → Measure voltage between D15 (+) terminal and D22 (−) terminal. → Is there approx. 5 V ?
  - YES → **Repair open in YEL/GRN wire between control unit (D15) and TW sensor.**
  - NO → Disconnect D connector from the wire harness only, not the control unit. → Measure voltage between D15 (−) terminal and D22 (−) terminal. → Is there approx. 5V ?
    - YES → **– Repair short in YEL/GRN wire between control unit (D15) and TW sensor.** **– Repair short in YEL/GRN wire between A/T control unit and TW sensor.**
    - NO → **Substitute a known-good control unit and recheck. If prescribed voltage is now available, replace the original control unit.**
- YES → Measure voltage between YEL/GRN (+) terminal and GRN/WHT (−) terminal. → Is there approx. 5V ?
  - NO → **Repair open in GRN/WHT wire between control unit (D22) and TW sensor.**
  - YES → **Substitute a known-good control unit and recheck. If symptom/indiction goes away, replace the original control unit.**

CLUTCH SWITCH SIGNAL—CODE 7—PRELUDE

(7) Self-diagnosis LED indicates code 7: A problem in the Clutch Switch circuit.

This signals the control unit when the clutch is engaged.

– Check Engine warning light has been reported on.
– LED indicates CODE 7.

Warm up engine to normal operating temperature (the cooling fan comes on).

Turn the ignition switch OFF.

Remove EFI, ECU fuse in the under-hood relay box for 10 seconds to reset control unit.

Road test with the transmission in 4th gear, accelerate to 3,000 rpm then decel to 2,000 rpm. The decel must take 5–7 seconds.

Is Check Engine warning light on and does LED indicate CODE 7?
- NO → **Intermittent failure, system is OK at this time (test drive may be necessary).**
- YES → Turn the ignition switch OFF. → Connect the ECU test harness between the control unit and connector → Turn the ignition switch ON. → Measure voltage between D8 (+) terminal and D21 (−) terminal. → Is there less than 1 V?
  - NO → Turn the ignition switch OFF.
  - YES

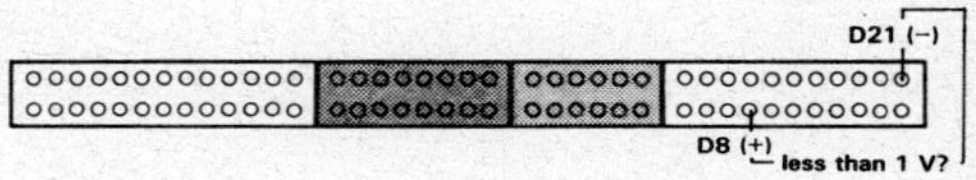

CLUTCH SWITCH SIGNAL—CODE 7—PRELUDE

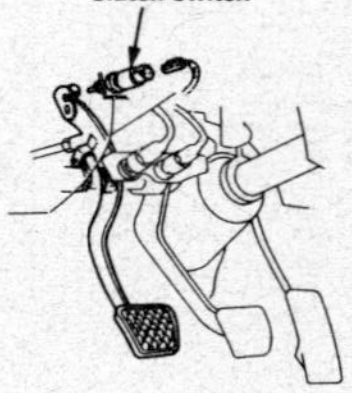

Depress the clutch pedal.

Measure voltage between D8 (+) terminal and D21 (−) terminal.

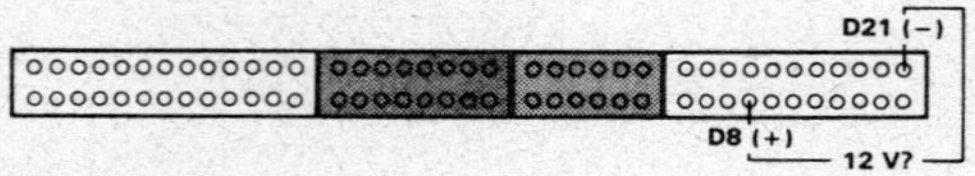

Is there battery voltage?

- YES → Substitute a known-good control unit and recheck. If symptom/indication goes away, replace the original control unit.
- NO → Turn the ignition switch OFF.

Disconnect "D" connector from main wire harness only, not the control unit.

Turn the ignition switch ON.

Is there battery voltage?

- NO → Substitute a known-good control unit and recheck. If prescribed voltage is now available, replace the original control unit.
- YES → Reconnect "D" connector to main wire harness and disconnect 2P connector on the clutch switch.

Is there battery voltage?

- YES → Replace clutch switch.
- NO → Repair short in BLK/RED, ORN wire between control unit (D8) and the clutch switch.

Disconnect the 2P connector on the clutch switch.

Connect BLK/RED terminal to BLK terminal.

Turn the ignition switch ON.

Is there less than 1 V?

- YES → Replace clutch slwtch.
- NO → Repair open in BLK/RED, ORN wire between ECU (D8) and clutch switch or BLK wire between clutch switch and G401.

AUTOMATIC TRANSMISSION SHIFT POSITION SIGNAL CODE 7—PRELUDE

(7) Self-diagnosis LED indicates code 7: A problem in the A/T Shift Position Switch circuit.

This signals the control unit when the transmission is in Neutral and Park.

- Check Engine warning light has been reported on.
- LED Indicates CODE 7.

Turn the ignition switch ON.

Observe the A/T shift indicator and select each position separately.

Does the indicator light properly ?

- NO → See A/T shift position Indicator Inspection
- YES → Warm up engine to normal operating temperature (the cooling fan comes on).

Turn the ignition switch OFF.

Remove EFI, ECU fuse in the under-hood relay box for 10 seconds to reset control unit.

Road test with the shift selector in S3, accelerate to 3,500 rpm then decel to 2,000 rpm. The decel must take 5–7 seconds.

Is Check Engine warning light on and does LED indicate CODE 7?

- NO → Intermittent failure, system is OK at this time (test drive may be necessary).
- YES → Turn the ignition switch OFF.

Connect the ECU test harness between the control unit and connector

AUTOMATIC TRANSMISSION SHIFT POSITION SIGNAL CODE 7—PRELUDE

Measure voltage between D10 (+) terminal and D21 (−) terminal in Neutral and Park position.

Is there voltage ?

- YES → Repair open in LT GRN wire between control unit (D10) and combination meter.
- NO ↓

Measure voltage between D10 (+) terminal and B8 (−) terminal in all other shift positions except Neutral and Park.

Is there battery voltage ?

- NO → Repair short in LT GRN wire between combination meter and control unit (D10).
- YES ↓

Substitute a known-good control unit and recheck.
If symptom/indication goes away, replace the original control unit.

IGNITION COIL SIGNAL—CODE 8—PRELUDE

Self-diagnosis LED indicates code 8: A problem in the ignition coil signal circuit.

- Check Engine warning light has been reported on.
- LED indicates CODE 8.

Turn the ignition switch OFF.

Remove EFI, ECU fuse in the under-hood relay box for 10 seconds to reset control unit.

Start the engine and allow to idle.

Is Check Engine warning light on and does LED indicate CODE 8 ?

- NO → Intermittent failure. Check the connectors at the ignition coil and control unit.
- YES ↓

Turn the ignition switch OFF.

Disconnect the primary terminals from the ignition coil.

Measure resistance between the B and D terminals.

Is there approx. 2000 Ω?

- NO → Replace ignition coil.
- YES ↓

Reconnect the primary terminals to the ignition coil.

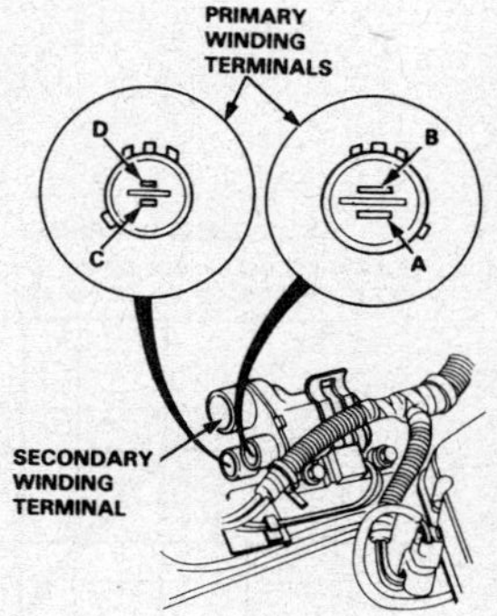

IGNITION COIL SIGNAL—CODE 8—PRELUDE

Connect the ECU test harness to main wire harness, but not the control unit

Turn the ignition switch ON.

Measure voltage between A9 (+) terminal and D21 (−) terminal.

Is there approx. 10 V?

- NO → **Repair open BLU wire between control unit (A9) and ignition coil.**
- YES → **Substitute a known-good control unit and recheck. If symptom/indication goes away, replace the original control unit.**

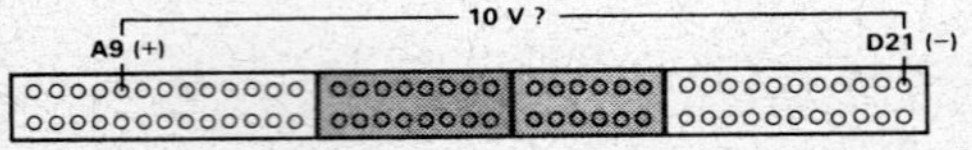

AIR TEMPERATURE (TA) SENSOR—CODE 10—PRELUDE

Self-diagnosis LED indicates code 10: Most likely a problem in the Intake Air Temperature (TA) Sensor circuit

The TA sensor is a temperature dependant resistor (thermistor). The resistance of the thermistor decreases as the intake air temperature increases as shown below.

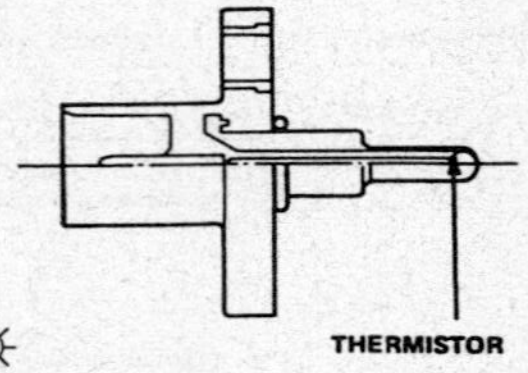

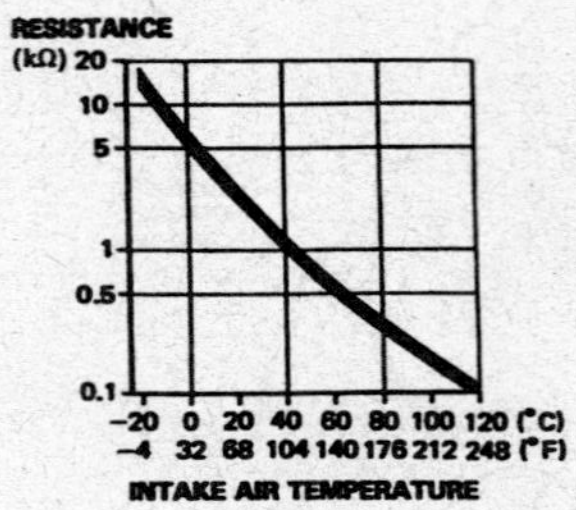

- **Check Engine warning light has been reported on.**
- **LED indicated CODE 10.**

Turn the ignition switch OFF.

Remove EFI, ECU fuse in the under-hood relay box for 10 seconds to reset control unit.

Turn the ignition switch ON.

Is Check Engine warning light on and does LED indicate CODE 10 ?

- NO → **Intermittent failure (test drive may be necessary.)**
- YES ↓

Disconnect the 2P connector on the TA sensor.

Measure resistance between the 2 terminals on the TA sensor.

Is there 1—4 kΩ ?

- NO → **Replace TA sensor.**
- YES

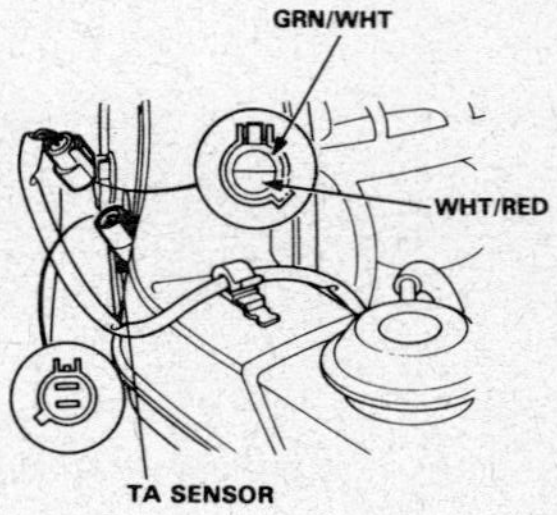

AIR TEMPERATURE (TA) SENSOR—CODE 10—PRELUDE

- Measure voltage between WHT/RED (+) terminal and body ground.
- Is there approx. 5V ?
  - NO → Turn the ignition switch OFF.
    - Connect the ECU test harness between the control unit and connector
    - Turn the ignition switch ON.
    - Measure voltage between D18 (+) terminal and D22 (−) terminal.
    - Is there approx. 5V ?
      - YES → **Repair open in WHT/RED wire between control unit (D18) and TA sensor.**
      - NO → Disconnect D connector from the wire harness only, not the control unit.
        - Measure voltage between D18 (+) terminal and D22 (−) terminal.
        - Is there approx. 5V ?
          - YES → **Repair short in WHT/RED wire between control unit (D18) and TA sensor.**
          - NO → **Substitute a known-good control unit and recheck. If prescribed voltage is now available, replace the original control unit.**
  - YES → Measure voltage between WHT/RED (+) terminal and GRN/WHT (−) terminal.
    - Is there approx. 5V ?
      - NO → **Repair open in GRN/WHT wire between control unit (D22) and TA sensor.**
      - YES → **Substitute a known-good control unit and recheck. If symptom/indication goes away, replace the original control unit.**

AIR CONDITIONING SIGNAL—PRELUDE

- **Inspection of Air Conditioning Signal.**
- Connect the ECU test harness between the control unit and connector . Disconnect "D" connector from the main wire harness only, not the control unit.
- Turn the ignition switch ON.
- Measure voltage between D6 (+) terminal and A26 (−) terminal.

- Is there battery voltage ?
  - NO → **Substitute a known-good control unit and recheck. If prescribed voltage is now available, replace the original control unit.**
  - YES → Reconnect "D" connector to the main wire harness.
- Connect A19 terminal to A26 terminal.

- Does A/C operate ?
  - NO → Connect the RED/BLU terminal of the 4P connector on the A/C clutch relay to body ground.

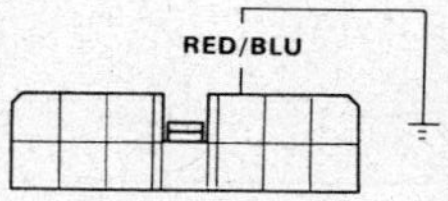

    - Does A/C operate ?
      - NO → **See Air conditioner inspection**
      - YES → **Repair open in RED/BLU wire between ECU (A19) and compressor control unit.**
  - YES → Start engine.
- Blower switch ON.
- A/C switch ON.
- Does A/C operate ?
  - NO → **See Air conditioner inspection**
  - YES → Measure voltage between D6 (+) terminal and A26 (−) terminal.

- Is voltage below 1 V ?
  - NO → **Repair open in RED/GRN wire between ECU (D6) and A/C switch.**
  - YES → **Air Conditioning Signal is OK.**

Symptom-to-Sub System Chart

NOTE:
- Across each row in the chart, the sub systems that could be sources of a symptom are ranked in the order they should be inspected, starting with ①. Find the symptom in the left column, read across to the most likely source, then refer to the page listed at the top of that column. If inspection shows the system is OK, try the next system ②, etc.
- Before starting inspection, check that other items that affect engine performance are within specification. Check the warning light and the self-diagnosis indicator, valve clearance, air cleaner, and PCV valve. In addition, check the ignition timing, function of the vacuum and centrifugal advance, and the condition of the spark plugs. If those items are all within specifications, begin with the troubleshooting listed below

SYMPTOM	SYSTEM	CARBURETOR								
		IDLE SPEED/ MIXTURE	A/C IDLE BOOST	AUTOMATIC CHOKE/ FAST IDLE SYSTEM	AIR VENT CUT-OFF (INNER VENT) SOLENOID VALVE	POWER VALVE	PRIMARY SLOW MIXTURE CUT-OFF SOLENOID VALVE	SLOW AIR JET CONTROL	VACUUM PISTON CONTROL	ACCELER-ATOR PUMP
ENGINE WON'T START					①	③	②			
DIFFICULT TO START ENGINE	WHEN COLD			①	②		③			
	WHEN WARM				②		①	③		
IRREGULAR IDLING	WHEN COLD FAST IDLE OUT OF SPECIFICATION			①			②			
	WHEN WARM RPM TOO HIGH	①	②	③						
	WHEN WARM RPM TOO LOW	①	②					③		
	ROUGH IDLE/ FLUCTUATION	①	③		②			③		
FREQUENT STALLING	WHILE WARMING UP	③		①			②			
	AFTER WARMING UP	①	③		③		②	③		
POOR PERFORMANCE	MISFIRE OR ROUGH RUNNING	③		③	①				②	
	LOSS OFF POWER				②	③			①	③
	AFTERBURN	②					①			
	HESITATION/SURGE									①

A/C IDLE BOOST SYSTEM—PRELUDE

Troubleshooting Flowchart **A/C Idle Boost System**

The A/C idle boost solenoid valve is energized when the A/C compressor is energized, applying vacuum to the diaphragm of the idle boost throttle controller. This increases the idle speed when the A/C compressor is on.

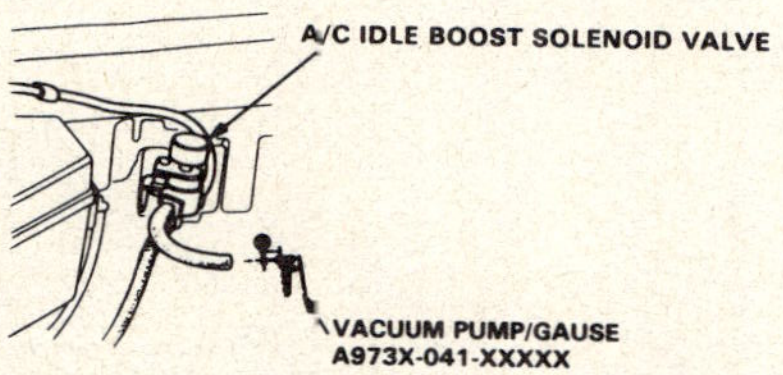

Inspection of A/C Idle Boost System

↓

Disconnect the upper vacuum hose of the solenoid valve from the idle boost throttle controller and connect a vacuum gauge.

↓

Start the engine.

↓

Is there vacuum ?
- YES → Disconnect the 2P connector near the solenoid valve. → Is there vacuum ?
  - YES → **Replace the solenoid valve.**
  - NO → **Inspect the A/C signal**
- NO ↓

Turn the A/C switch ON.

↓

Is there vacuum ?
- YES → **A/C Idle Boost solenoid valve and A/C signal are OK.**
- NO → Disconnect the 2P connector near the solenoid valve.

↓

Measure voltage between RED (+) terminal and BLK (−) terminal on the wire harness.

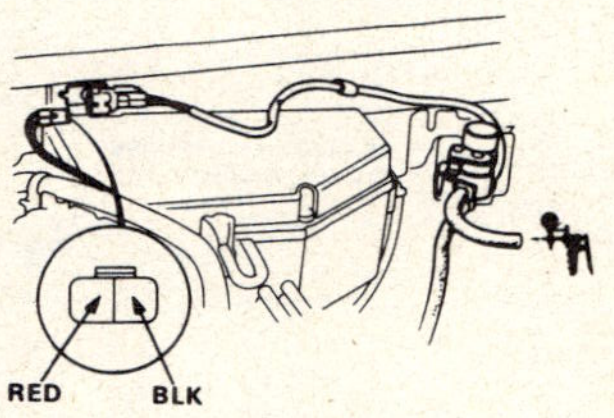

↓

Is there battery voltage ?
- NO → Measure voltage between RED (+) terminal and body ground.
- YES → Disconnect the lower vacuum hose from the solenoid valve and connect a vacuum gauge to the disconnected hose.

(cont'd)

Is there manifold vacuum?
- YES → **Replace the solenoid valve.**
- NO → **Check routing of lower hose to solenoid valve.**

Is there battery voltage ?
- YES → **Repair open in BLK wire between the solenoid valve and G203.**
- NO ↓

Turn the ignition switch OFF.

↓

Connect the ECU test harness between the control unit and the connector.

↓

Start the engine and turn the A/C switch ON.

↓

Measure voltage between A8 (+) and A26 (−).

↓

Is there battery voltage ?
- YES → **Repair open RED wire between control unit (A8) and the solenoid valve.**
- NO ↓

Turn the ignition switch OFF.

↓

Disconnect A connector from the control unit only, not the main wire harness.

↓

Check for continuity between A8 and body ground.

↓

Does continuity exist ?
- YES → **Repair short in RED wire between control unit (A8) and the solenoid valve.**
- NO → **Inspect A/C signal**
  If OK, substitute a known-good control unit and recheck. If symptom goes away, replace the original control unit.

A/C IDLE BOOST SOLENOID VALVE—PRELUDE

Troubleshooting Flowchart **A/C Idle Boost Solenoid Valve**

The A/C idle boost solenoid valve is energized when the vehicle is decelerating, applying vacuum to the diaphragm of the idle boost throttle controller for 1 second.

Inspection of A/C Idle Boost Solenoid Valve.

Warm up engine to normal operating temperature (cooling fan comes on).

Disconnect the upper vacuum hose of the solenoid valve from the idle boost throttle controller and connect a vacuum gauge.

Is there any vacuum?
- YES → Disconnect the 2P connector near the solenoid valve.
  - Is there any vacuum?
    - YES → **Replace the A/C idle boost solenoid valve.**
    - NO → **Check the warning light and the self-diagnosis indicator**
      If OK, substitute a known-good control unit and retest. If symptom goes away, replace the original control unit.
- NO → Turn the ignition switch OFF.

Disconnect the 2P connector from the solenoid valve.

Connect battery positive to terminal A and battery negative to terminal B of the connector.

Start the engine.

Is there manifold vacuum?
- NO → **Replace the A/C idle boost solenoid valve.**
- YES →
  - **A/C Idle Boost Solenoid Valve is OK.**
  - **Check open in RED wire between control unit (A8) and the solenoid valve.**
  - **Check open in BLK wire between solenoid valve and G203.**

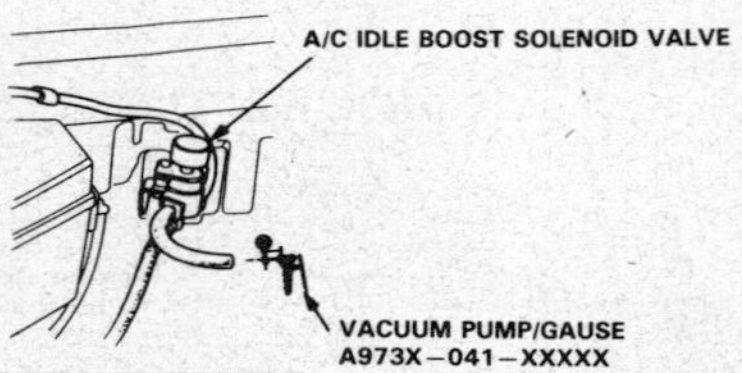

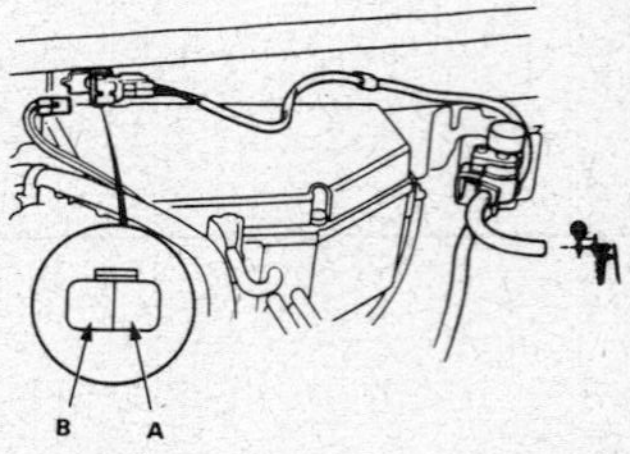

AIR LEAK SOLENOID VALVE—PRELUDE

Troubleshooting Flowchart **Air Leak Solenoid Valve**

The air leak solenoid valve is energized when the car is being started, or running in the power mode, to close the extra air passage, increasing fuel flow.

Inspection of Air Leak Solenoid Valve

Remove the air cleaner cover and filter element.

Disconnect the vacuum hose and connect a vacuum pump.

NOTE: Intake air temperature must be below 63°C (145°F).

Start the engine.

Apply 100 mmHg (4 in.Hg) vacuum to the hose.

Does solenoid valve hold vacuum?
- NO → Turn the ignition switch OFF.
  - Disconnect the GRN connector near the air cleaner.
  - Start the engine.
  - Measure voltage between GRN (+) and body ground.
  - Is there battery voltage?
    - NO → **Check the warning light and the self-diagnosis indicator . If OK, repair open in GRN wire between the solenoid valve and control unit (A12).**
    - YES → **Replace the solenoid valve.**
- YES → Warm up the engine.

NOTE: If the intake air temperature is below 63°C (145°F), warm the air bleed valve B with a dryer.

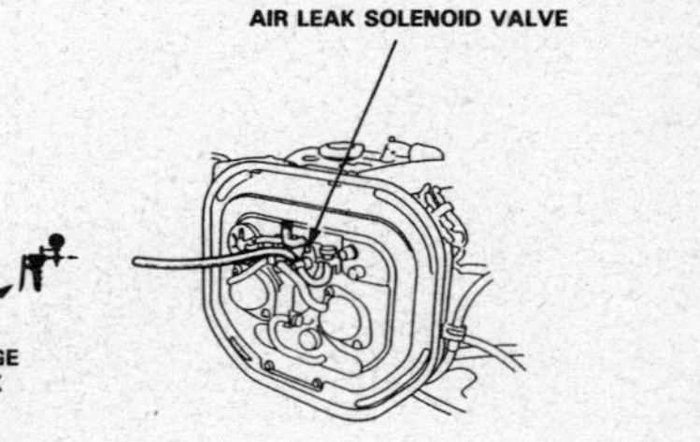

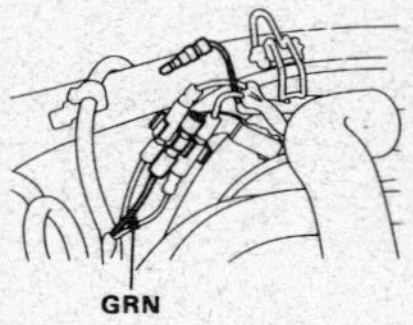

(cont'd)

AIR LEAK SOLENOID VALVE—PRELUDE

Does solenoid valve hold vacuum?
- YES → Replace the air bleed valve B.
- NO → Remove the hose from the air leak solenoid valve and cap the solenoid valve.

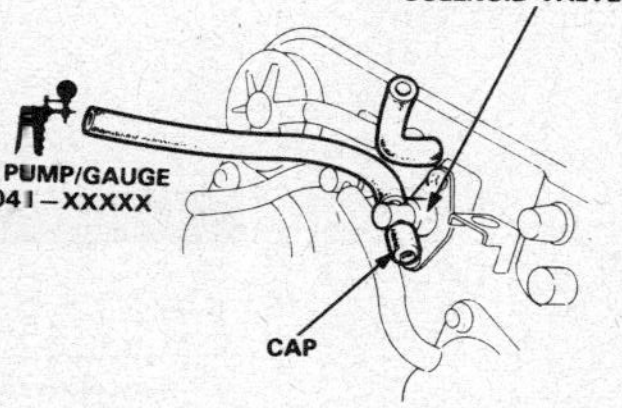

NOTE: If the intake air temperature is below 70°C (158°F), warm the TA sensor with a dryer.

Does solenoid valve hold vacuum with the engine at normal operating temperature?
- NO → Air leak solenoid valve is OK.
- YES → Turn the ignition switch OFF.

Disconnect the GRN connector near the air cleaner.

Start the engine.

Measure voltage between GRN (+) and body ground.

Is there voltage?
- YES → Check the warning light and the self-diagnosis indicator. If OK, substitute a known-good control unit and retest. If symptom goes away, replace the original control unit.
- NO → Replace the solenoid valve.

PRIMARY SLOW MIXTURE CUT-OFF SOLENOID VALVE PRELUDE

Troubleshooting Flowchart — Primary Slow Mixture Cut-off Solenoid Valve

The primary slow mixture cut-off solenoid valve is provided to cut-off the idle mixture passage between the fuel/air passage and bypass port for preventing run- on when the ignition switch is turned off.
The solenoid valve also functions to cut-off the mixture flow under deceleration.

Inspection of Primary Slow Mixture Cut-off Solenoid Valve

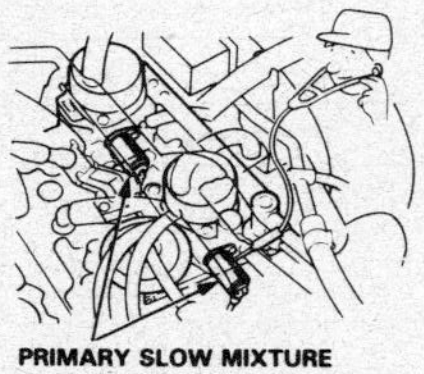

Turn the ignition switch ON.

Check the clicking sound of each solenoid valve by means of a stethoscope.

Does the solenoid valve click?
- NO → Turn the ignition switch OFF.
  - Connect the ECU test harness between the control unit and connector
  - Turn the ignition switch ON.
  - Measure voltage between A21 (+) terminal and A26 (−) terminal.
  - Is there battery voltage?
    - YES → Repair open or short in BLK/YEL wire between the solenoid valve and control unit (A21). If OK; right solenoid valve: replace the solenoid valve. left solenoid valve: inspect open in BLK wire between the solenoid valve and G151, and replace the solenoid valve.
    - NO → Check the warning light and the self-diagnosis indicator. If OK, substitute a known-good control unit and retest. If symptom goes away, replace the original control unit.
- YES → Start the engine and warm up to normal operating temperature (the cooling fan comes on).
  - Block rear wheels and set the parking brake. Jack up the front of the car and support with safety stands.
  - **⚠WARNING** Block rear wheels before jacking up front of car.
  - Place the transmission in second or "2" and accelerate, then suddenly release the throttle.

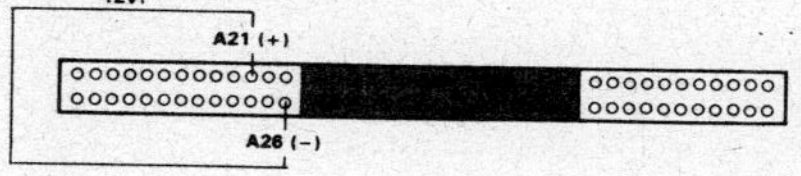

PRIMARY SLOW MIXTURE CUT-OFF SOLENOID VALVE PRELUDE

Check for the clicking sound of each solenoid valve by means of a stethoscope during deceleration above 12.5 mph and 4,000 rpm.

Does the solenoid valve click?

- YES → Solenoid valve is OK.
- NO → Turn the ignition switch OFF.
  - Connect the ECU test harness between the control unit and connector
  - Start the engine.
  - Measure voltage between A21 (+) terminal and A26 (−) terminal during deceleration above 12.5 mph and 4,000 rpm.
  - Is there no voltage for a moment?
    - YES → **Replace the solenoid valve.**
    - NO → **Check the warning light and the self-diagnosis indicator**

      If OK, substitute a known-good control unit and retest. If symptom goes away, replace the original control unit.

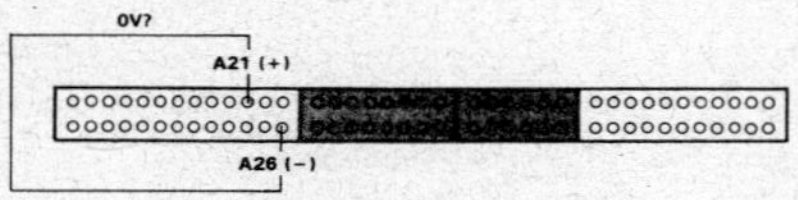

VACUUM PISTON CONTROL SOLENOID VALVE PRELUDE

Troubleshooting Flowchart Vacuum Piston Control Solenoid Valve

The vacuum piston control solenoid valve is activated by the electric current of the control unit which receives the signals of intake manifold vacuum (MAP sensor) and engine speed (ignition coil), and intake manifold vacuum is fed to the diaphragm of vacuum piston control valve through check valve C so as to open the passage from vacuum piston upper chamber to atmosphere.

Inspection of Vacuum Piston Control Solenoid Valve

- Remove the control box and open the control box lid.
- Disconnect the lower vacuum hose of the solenoid valve from the 3-way joint and connect a vacuum pump.
- Disconnect the #28 vacuum hose of the solenoid valve from the vacuum hose manifold and connect a vacuum gauge.
- Start the engine and warm up to normal operating temperature (the cooling fan comes on).
- Apply vacuum.
- Is vacuum indicated on the gauge?
  - YES
  - NO → Turn the ignition switch OFF.
    - Disconnect the 8P connector on the control box.
    - Start the engine.
    - Measure voltage between YEL (+) terminal and BLK (−) terminal on the wire harness.
    - Is there battery voltage?
      - YES → **Replace the solenoid valve.**
      - NO

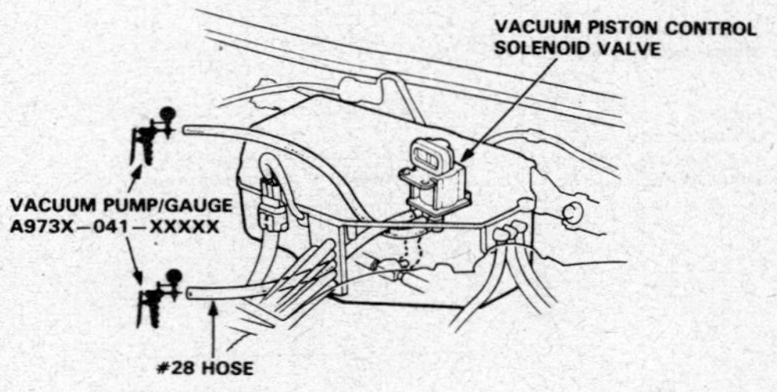

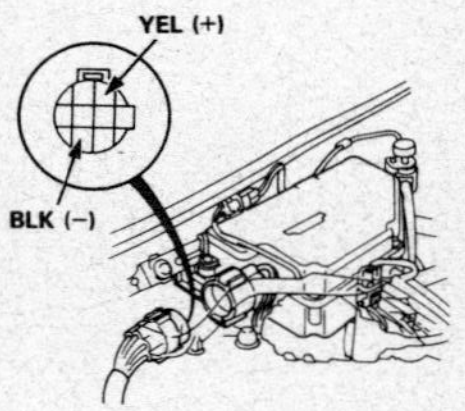

VACUUM PISTON CONTROL SOLENOID VALVE PRELUDE

Quickly raise engine speed to 3,000 rpm, then close the throttle suddenly.

Does the vacuum indication on gauge decrease for a moment?

- YES → **Vacuum piston control solenoid valve is OK.**
- NO → Turn the ignition switch OFF.

Disconnect the BP connector on the control box.

Start the engine and raise engine speed to 3,000 rpm, then close the throttle suddenly.

Measure voltage between YEL (+) terminal and BLK (–) terminal on the wire harness.

Does battery voltage decrease for a moment?

- YES → **Check the warning light and the self-diagnosis indicator If OK, substitute a known-good control unit and retest. If symptom goes away, replace the original control unit.**
- NO → **Replace the solenoid valve.**

Measure voltage between YEL (+) terminal and body ground.

Is there battery voltage ?

- YES → **Repair open BLK wire between the control box and G202.**
- NO → **Check the warning light and the self-diagnosis indicator If OK, inspect open in YEL wire between the control box and control unit (A6).**

POWER VALVE CONTROL SOLENOID VALVE—PRELUDE

Troubleshooting Flowchart **Power Valve Control Solenoid Valve**

The power valve control solenoid valve is controlled by the control unit which receives signals from the MAP sensor, TA sensor, TW sensor and ignition coil, to bleeds vacuum from the power valve diaphragm.

Inspection of Power Valve Control Solenoid Valve.

Disconnect the #14 vacuum hose of the solenoid valve from the vacuum hose manifold and connect a vacuum gauge.

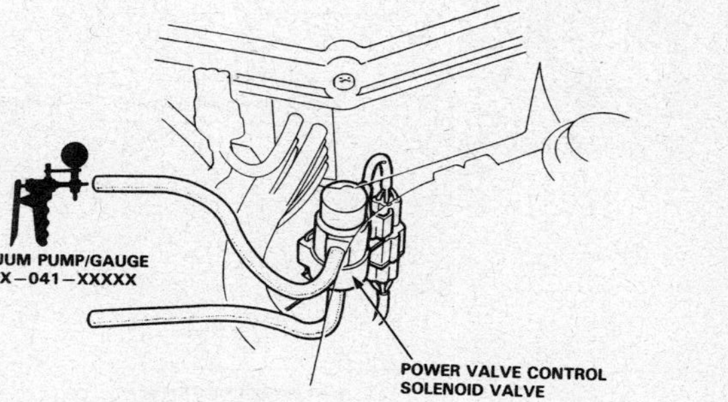

Start the engine.

NOTE: Intake air temperature must be below 15°C (59°F).

Is vacuum indicated on the gauge?

- NO → Warm up to normal operating temperature (the cooling fan comes on).
- YES → Turn the ignition switch OFF.

Disconnect the 4P connector near the solenoid valve.

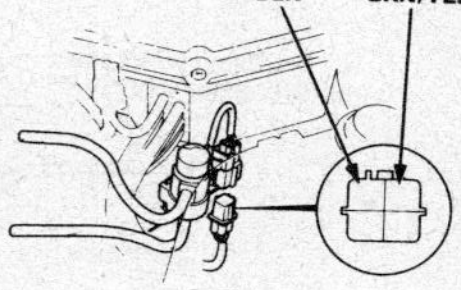

Start the engine.

Measure voltage between GRN/YEL (+) terminal and BLK (–) terminal on the wire harness.

Is there battery voltage?

- YES → **Replace the solenoid valve.**
- NO → Measure voltage between GRN/YEL (+) terminal and body ground.

Is there battery voltage?

- YES → **Repair open in BLK wire between the solenoid valve and G202.**
- NO

(cont'd)

POWER VALVE CONTROL SOLENOID VALVE—PRELUDE

Is vacuum indicated on the gauge?

- YES → **Solenoid valve is OK.**
- NO → Turn the ignition switch OFF.

Check the warning light and the self-diagnosis indicator (page 11-16). If OK, inspect open in GRN/YEL wire between the control box and control unit (A5).

1. Turn the ignition switch OFF.
2. Disconnect the 4P connector near the solenoid valve.
3. Start the engine.
4. Measure voltage between GRN/YEL (+) terminal and BLK (−) terminal on the wire harness.
5. Is there voltage?
   - NO → **Replace the solenoid valve.**
   - YES → **Check the warning light and the self-diagnosis indicator. If OK, substitute a known-good control unit and retest. If symptom goes away, replace the original control unit.**

CRANKING LEAK SOLENOID VALVE (CHOKE OPENER) PRELUDE

Troubleshooting Flowchart — Cranking Leak Solenoid Valve

The cranking leak solenoid valve bleeds vacuum from the choke opener during cranking.

1. **Inspection of Cranking Leak Solenoid Valve**
2. Disconnect #26 vacuum hose from the vacuum hose manifold.
3. Connect vacuum pump and continually apply vacuum.

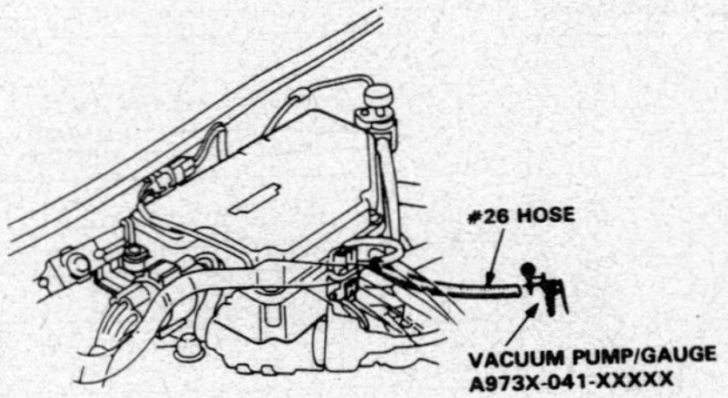

4. Does the gauge show vacuum?
   - NO → **– Check the vacuum hose for proper connection or leakage. – Replace the solenoid valve.**
   - YES → continue.
5. Does vacuum leak?
   - NO → **– Repair vacuum hose pinch or blockage. – Replace the solenoid valve.**
   - YES → continue.
6. Connect a T-fitting vacuum gauge the between #26 hose and the vacuum hose manifold.
7. Turn the ignition switch to the START position.
8. Is there less than 10'' of vacuum in the start position?
   - YES → **Cranking Leak Solenoid Valve is OK.**
   - NO → Turn the ignition switch OFF.
     - Disconnect the 8P connector on the control box.
     - Measure voltage between BLU/RED (+) terminal and BLK (−) terminal on the wire harness.

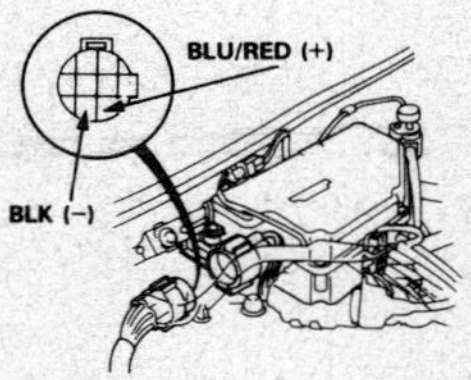

CRANKING LEAK SOLENOID VALVE (CHOKE OPENER) PRELUDE

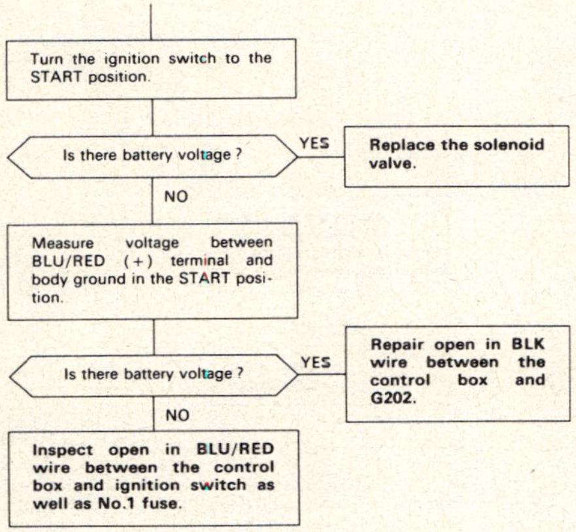

EMISSION CONTROLS TROUBLESHOOTING—PRELUDE

NOTE:

- Across each row in the chart, the sub systems that could be sources of a symptom are ranked in the order they should be inspected, starting with ①. Find the symptom in the left column, read across to the most likely source, then refer to the page listed at the top of that column. If inspection shows the system is OK, try the next system ②, etc.
- Before starting inspection, check that other items that affect engine performance are within specification. Check the warning light and the self-diagnosis indicator, valve clearance, air cleaner, and PCV valve. In addition, check the ignition timing, function of the vacuum and centrifugal advance, and the condition of the spark plugs. If those items are all within specifications, begin with the troubleshooting listed in this page.

SYMPTOM \ SYSTEM		FEEDBACK CONTROL	THROTTLE CONTROL	EGR	EVAPORATIVE CONTROL	AIR INJECTION
ENGINE WON'T START					①	
DIFFICULT TO START ENGINE	WHEN COLD	②		③	①	
	WHEN WARM	②			①	
IRREGULAR IDLING	WHEN COLD FAST IDLE OUT OF SPECIFICATION	①	③	②		
	WHEN WARM RPM TOO HIGH		①			
	WHEN WARM RPM TOO LOW	①		②		
	ROUGH IDLE/ FLUCTUATION	①		②		
FREQUENT STALLING	WHILE WARMING UP	①		②		
	AFTER WARMING UP	①		②		
POOR PERFORMANCE	MISFIRE OR ROUGH RUNNING	②		①		
	LOSS OFF POWER	②			①	
	AFTERBURN	③	②			①
	HESITATION/SURGE	①		②		

FEEDBACK SYSTEM (EACV)—PRELUDE

Troubleshooting Flowchart EACV

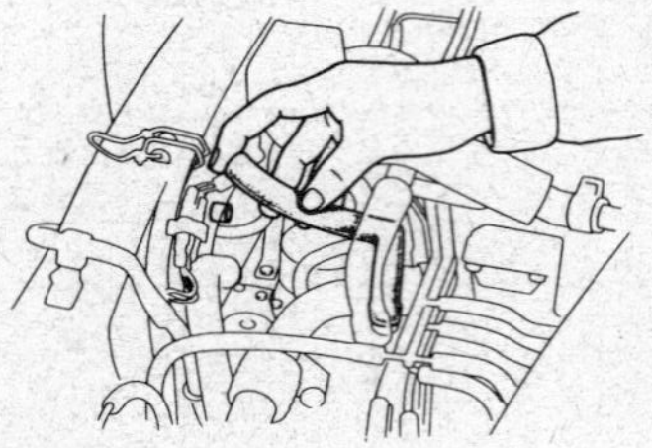

Inspection of EACV.

Start the engine and warm up to normal operating temperature (the cooling fan comes on).

Disconnect the EACV hose from the air cleaner.

Raise the engine speed to 3,500 rpm.

Is there vacuum ?
- NO → **Check the warning light and the self-diagnosis indicator If OK, replace the EACV and retest.**
- YES ↓

Raise the engine speed to 3,500 rpm, then close the throttle suddenly.

Is there vacuum ?
- NO → Check the warning light and the self-diagnosis indicator
  - Is warning light on and does LED indicate code 14?
    - YES → **Go to EACV trouble-shooting flow chart**
    - NO → **Check the hose for proper connection, cracks, brockage or disconnected hose. If OK, replace the EACV and retest.**
- YES ↓

EACV is OK.

FEEDBACK SYSTEM (EACV)—PRELUDE

(CHECK) (14) Self-diagnosis LED indicates code 14: A problem in the Electronic Air Control Valve (EACV) circuit.

The EACV opens the air passage from the air cleaner case to the intake manifold in proportion to the intensity of the electric current received from the control unit.

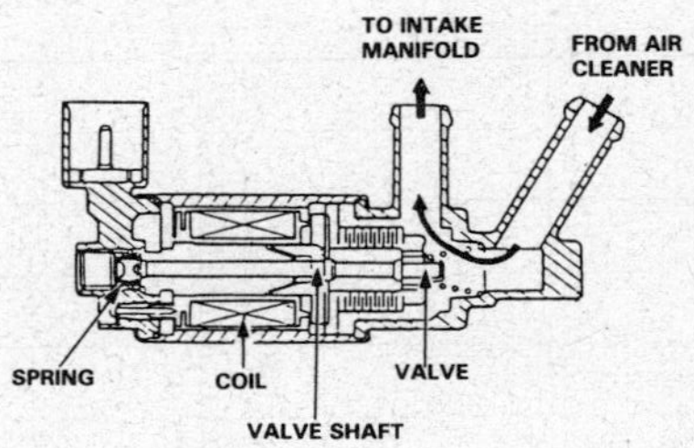

(CHECK) (14)

- **Engine is running.**
- **Check Engine warning light has been reported on.**
- **LED indicates CODE 14.**

Turn the ignition switch OFF.

Remove EFI, ECU fuse in the under-hood relay box for 10 seconds to reset control unit.

Start engine.

Raise the engine speed to 3,500 rpm, them close the throttle suddenly.

Is Check Engine warning light on and does LED indicate CODE 14?
- NO → **Intermittent failure (test driving may be necessary.)**
- YES ↓

Stop engine.

Disconnect the 2P connector on the EACV.

Measure resistance between the 2 terminals on the EACV.

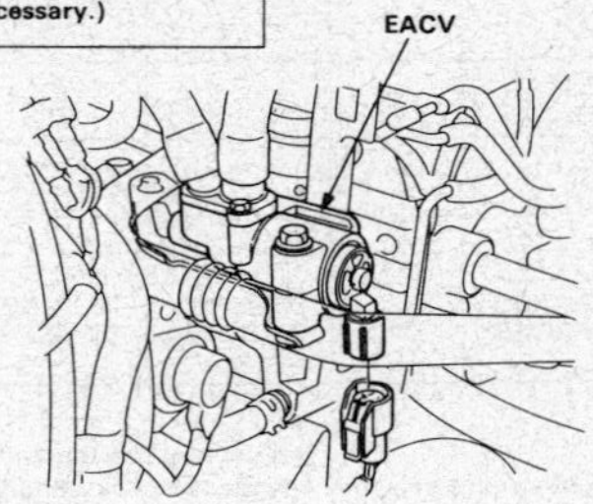

FEEDBACK SYSTEM (EACV) — PRELUDE

- Is there 8–15Ω?
  - NO → Replace EACV.
  - YES ↓
- Check for continuity to body ground on each terminal on the EACV.
- Does continuity exist?
  - YES → Replace EACV.
  - NO ↓
- Reconnect the 2P connector to EACV.
- Connect the ECU test harness to main wire harness, but not the control unit

- Check for continuity to the A24, A26 terminal and the body ground.
- Does continuity exist?
  - NO → **Repair open BLK wire between control unit (A24, A26) and G151.**
  - YES ↓

- Measure resistance between A3 terminal and A1 terminal.
- Is there 8–15 Ω?
  - NO → **Repair open in BLU/RED wire between control unit (A1) and EACV or YEL/BLK wire between control unit (A3) and EACV.**
  - YES ↓
- **Substitute a known-good control unit and recheck. If prescribed voltage is now available, replace the original control unit.**

AIR SUCTION CONTROL SYSTEM — PRELUDE

Troubleshooting Flowchart **Air Suction Control System**

The air suction control solenoid valve is controlled by the control unit which receives signals from MAP sensor, TA sensor, TW sensor, ignition coil, vacuum switch and speed sensor, to open the air suction valve.

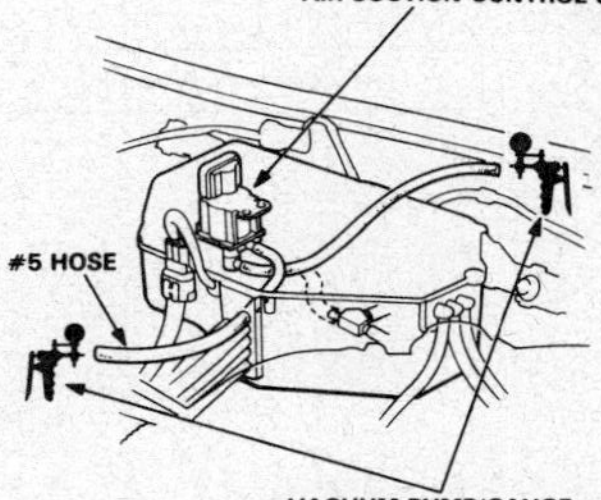

- **Inspection of Air Suction Control System**
- Remove the control box and open the control box lid.
- Disconnect the lower vacuum hose of the solenoid valve from the 3-way joint and connect a vacuum pump.
- Disconnect the #5 vacuum hose of the solenoid valve from the vacuum hose manifold and connect a vacuum gauge.
- Start the engine and warm up to normal operating temperature (the cooling fan comes on).
- Apply vacuum.
- Does solenoid valve hold vacuum?
  - YES ↓
    - Block rear wheels and set the parking brake. Jack up the front of the car and support with safety stand.
    - **⚠WARNING Block rear wheels before jacking up front of car.**
    - Place the shift or selector (lever in second or "2" and accelerate above 12.5 mph, then release the throttle.
  - NO →
    - Turn the ignition switch OFF.
    - Disconnect the 8P connector on the control box.
    - Start the engine.
    - Measure voltage between **RED** (+) terminal and **BLK** (−) terminal on the wire harness.
    - Is there voltage?
      - YES → **Check the warning light and the self-diagnosis indicator If OK, substitute a known-good control unit and retest. If symptom goes away, replace the original control unit.**
      - NO → **Replace the solenoid valve.**

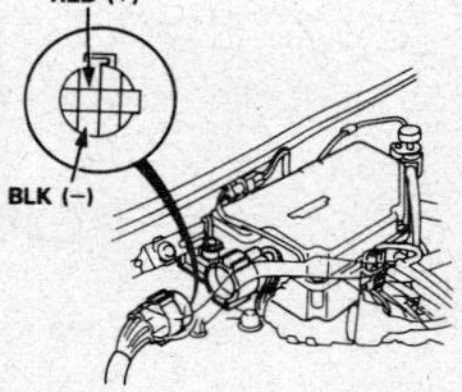

AIR SUCTION CONTROL SYSTEM—PRELUDE

- Is vacuum indicated on the gauge?
  - NO → Turn the ignition switch OFF.
    - Disconnect the 8P connector on the control box.
    - Place the shift or selector lever in second or "2" and accelerate above 12.5 mph, then release the throttle.
    - Measure voltage between RED (+) terminal and BLK (–) terminal on the wire harness.
    - Is there battery voltage?
      - YES → **Replace the solenoid valve.**
      - NO → Measure voltage between RED (+) terminal and body ground.
        - Is there battery voltage?
          - YES → **Repair open in BLK wire between the control box and G202.**
          - NO → **Check the warning light and the self-diagnosis indicator. If OK, inspect open in RED wire between the control box and control unit (A14).**
  - YES → Allow it to idle.
    - Is vacuum indicated on the gauge.
      - YES → **Replace the solenoid valve.**
      - NO → **System is OK.**

PURGE CUT-OFF SOLENOID VALVE—PRELUDE

Troubleshooting Flowchart — Purge Cut-off Solenoid Valve

When the engine coolant temperature is above the set temperature of the TW sensor, the purge cut-off solenoid valve is activated by the control unit receiving signals from each sensor.

- **Inspection of Purge Cut-off Solenoid Valve**
- Remove the control box and open the control box lid.
- CAL: Disconnect the lower vacuum hose of the solenoid valve from the 3-way joint and connect a vacuum gauge.
- Disconnect the upper vacuum hose of the solenoid valve; 49ST: from the 3-way joint. CAL: from the tube holder and connect a vacuum pump.
- Start the engine.
  NOTE: Engine coolant temperature must be below 63° C (145° F)
- Apply vacuum.
- Does solenoid valve hold vacuum?
  - YES → Turn the ignition switch OFF.
    - Disconnect the 8P connector on the control box.
    - Start the engine.
  - NO

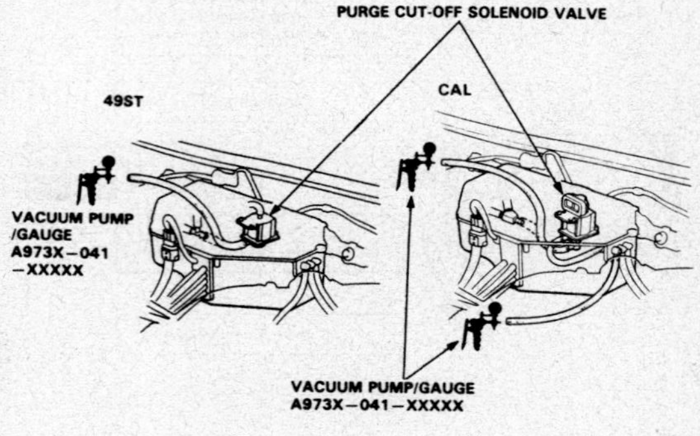

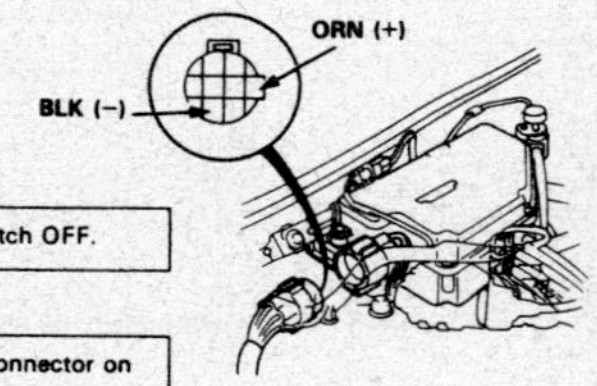

PURGE CUT-OFF SOLENOID VALVE—PRELUDE

Warm up to normal operating temperature (the cooling fan comes on).

Block rear wheels and set the parking brake. Jack up the front of the car and support with safety stands.

⚠WARNING Block rear wheels before jacking up front of car.

Put the transmission into 2nd gear, and accelerate to 20 mph.

Apply vacuum.

49ST: Does solenoid valve hold vacuum ?
CAL: Is vacuum indicated on the gauge ?

- YES → **Solenoid valve is OK.**
- NO → Stop the engine.

Disconnect the 8P connector on the control box.

Start the engine.

Put the transmission into 2nd gear, and accelerate to 20 mph.

Measure voltage between ORN (+) terminal and BLK (–) terminal on the wire harness.

Is there battery voltage ?

- YES → **Replace the solenoid valve.**
- NO → Measure voltage between ORN (+) terminal and body ground.

Is there battery voltage ?

- YES → **Repair open in BLK wire between the control box and G202.**
- NO → **Check the warning light and the self-diagnosis indicator. If OK, inspect open in ORN wire between the control box and control unit (A4).**

Measure voltage between ORN (+) terminal and BLK (–) terminal on the R engine compartment wire harness.

Is there voltage ?

- YES → **Check the warning light and the self-diagnosis indicator. If OK, substitute a known-good control unit and retest. If symptom goes away, replace the original control unit.**
- NO → **Replace the solenoid valve.**

INNER VENT SOLENOID VALVE—PRELUDE

Troubleshooting Flowchart — Inner Vent Solenoid Valve

The inner vent solenoid valve remains closed during engine cranking in order to facilitate engine startability by preventing fuel vapor in the float chamber from drawing into carburetor.

When the engine starts to run, the inner vent solenoid valve is activated to open the passage from air cleaner case.

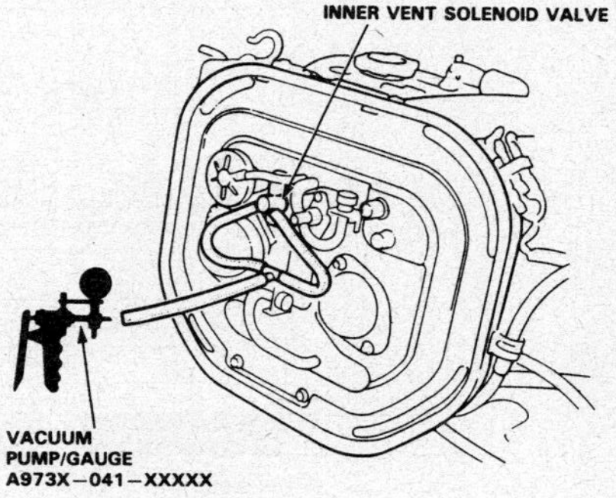

Inspection of Inner Solenoid Valve

Remove the air cleaner cover and filter element.

Disconnect two vacuum hose from the carburetor and connect a vacuum pump.

Start the engine and warm up to normal operating temperature (the cooling fan comes on).

Apply vacuum.

Does solenoid valve hold vacuum?

- YES → Turn the ignition switch OFF.
- NO → Stop the engine. → Apply vacuum.

Disconnect the GRN/RED connector and BLK connector near the air cleaner.

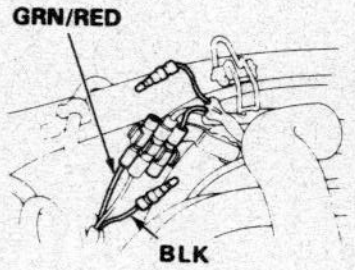

Start the engine.

Measure voltage between GRN/RED (+) terminal and BLK (–) terminal.

Is there battery voltage ?

- YES → **Replace the solenoid valve.**
- NO

AIR VENT CUT-OFF SOLENOID VALVE—PRELUDE

Troubleshooting Flowchart **Air vent Cut-off Solenoid Valve**

The air vent cut-off solenoid valve remains closed during engine cranking in order to facilitate engine startability by preventing fuel vapor in the float chamber from drawing into carburetor.

When the engine starts to run, air vent cut-off solenoid valve is activated to open the passage from air cleaner case. The opening of the air vent cut-off solenoid valve is delayed for 6 seconds when engine coolant is hot.

Inspection of Air Vent Cut-off Solenoid Valve.

Disconnect the upper hose of the solenoid valve from the air cleaner and connect a vacuum pump.

Disconnect the lower hose of the solenoid valve.

Start the engine and warm up to normal operating temperature (the cooling fan comes on).

Apply vacuum.

Does solenoid valve hold vacuum ?

YES → Turn the ignition switch OFF.

Connect the ECU test harness between the control unit and connector

Start the engine.

Measure voltage between D2 (+) terminal and A26 (−) terminal.

NO → Stop the engine.

Apply vacuum.

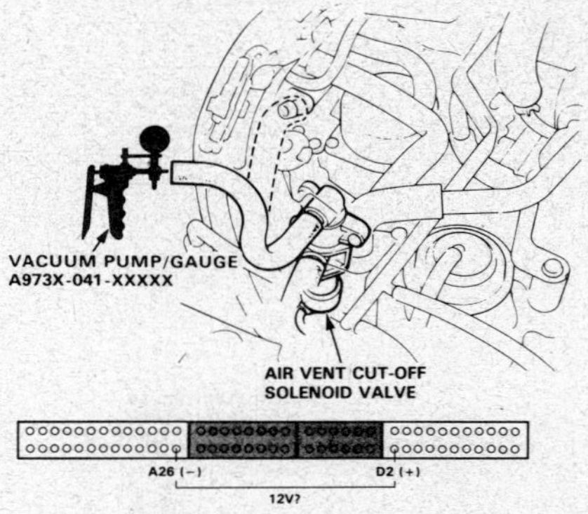

INNER VENT SOLENOID VALVE—PRELUDE

Does solenoid valve hold vacuum ?

YES → **Solenoid valve is OK.**

NO → Turn the ignition switch OFF.

Disconnect the GRN/RED connector and BLK connector near the air cleaner.

Turn the ignition switch ON.

Measure voltage between GRN/RED (+) terminal and BLK (−) terminal.

Is there voltage ?

NO → **Check the warning light and the self-diagnosis indicator. If OK, substitute a known-good control unit and retest. If symptom goes away, replace the original control unit.**

YES → **Replace the solenoid valve.**

Measure voltage between GRN/RED (+) terminal body ground.

Is there battery voltage ?

YES → **Repair open in BLK wire between the solenoid valve and G151.**

NO → **Check the warning light and the self-diagnosis indicator. If OK, inspect open in GRN/RED wire between the solenoid valve and control unit (A2).**

AIR VENT CUT-OFF SOLENOID VALVE—PRELUDE

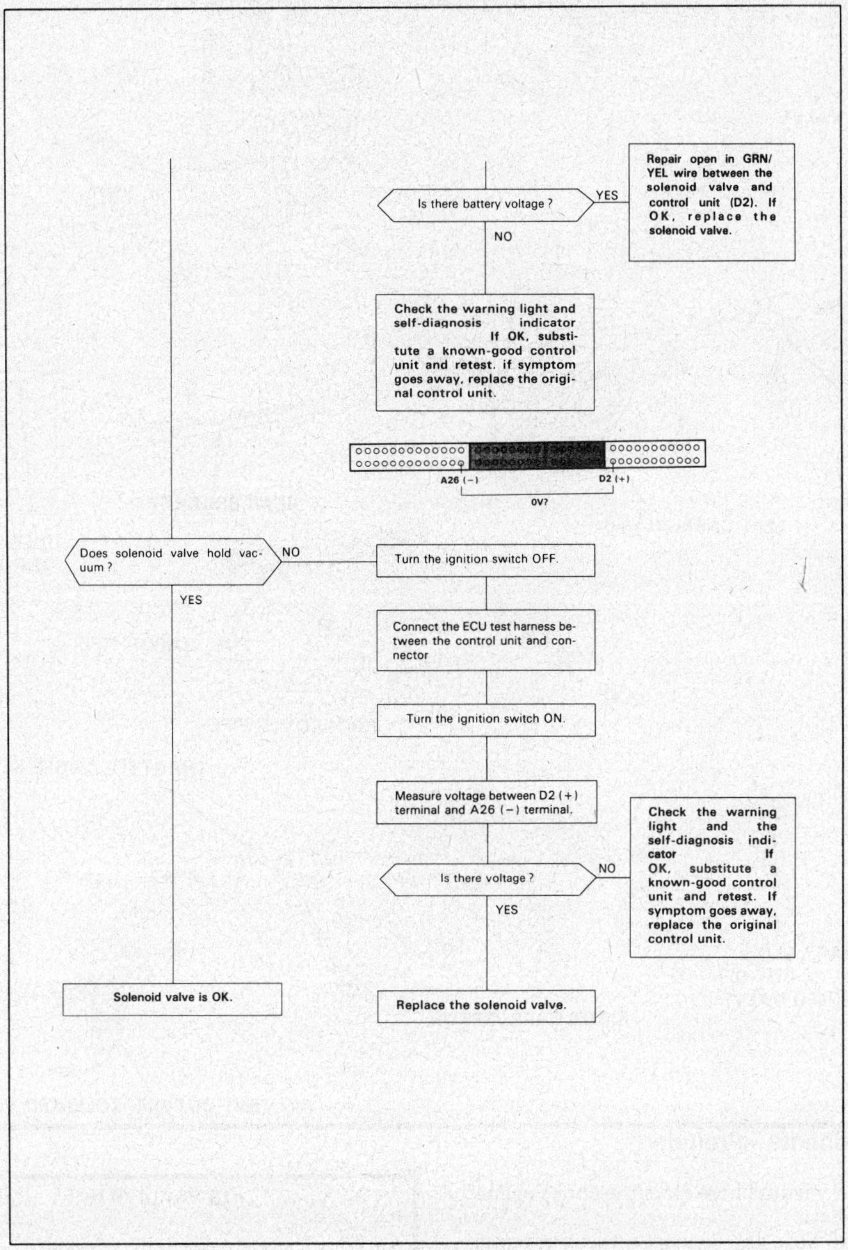

Component Replacement

CARBURETOR

Synchronization

1. Disconnect the RED braided hose (marked "PURGE") from the canister and connect a vacuum/pressure gauge to the hose.
2. Connect a tachometer to engine and start the engine.
3. Allow engine to reach normal operating temperature. Idle engine until cooling fan comes on.
4. Raise engine idle to 3000 rpm by turning the idle speed screw and note the vacuum gauge reading.
5. While the engine is running at 3000 rpm, pinch the smooth RED vacuum hose under the choke housing and note the vacuum gauge reading.
6. If the vacuum gauge reading did not change, the carburetors are adjusted properly. If the readings were different, adjust the synchronization screw until they remain the same.
7. Remove the tachometer and vacuum gauge and reconnect the purge vacuum hose.
8. Readjust idle as needed.

CHOKE LINKAGE

Adjustment

1. Remove the carburetor.

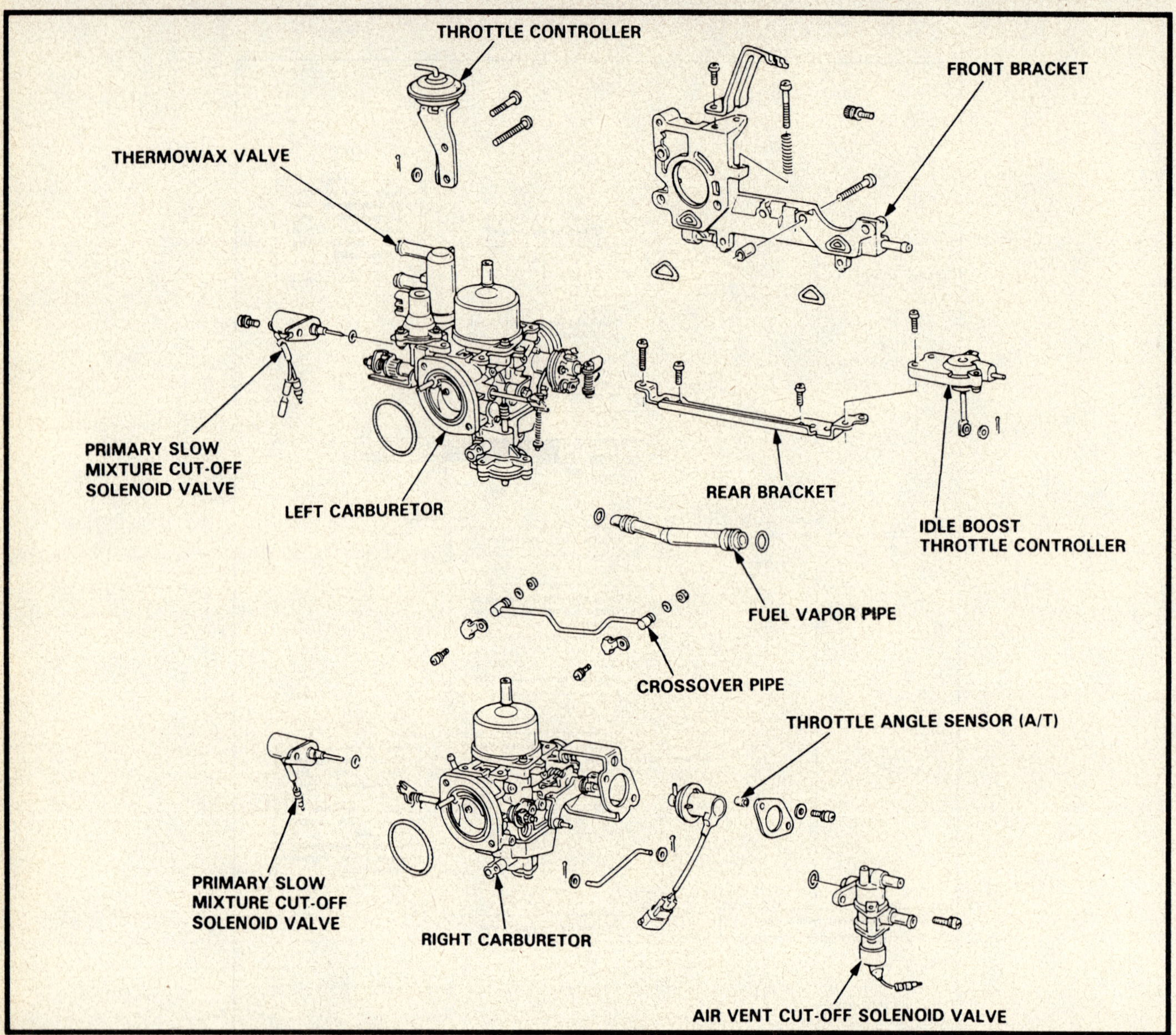

Carburetor and components—Prelude

2. Disconnect the No. 18 vacuum hose from the choke opener and leave open to atmosphere.
3. Disconnect the No. 26 hose and connect a vacuum pump.
4. Apply at least 8 in. Hg (200mm Hg).

NOTE: If vacuum drops below 8 in. Hg (200mm Hg), the bleed valve is open. Slowly re-apply vacuum until highest level is maintained with out losing vacuum. This will bring opener rod as close as possible to bleed valve.

5. Turn choke drive lever clockwise and measure clearance between choke valve and casting. First stage clearance should be 0.070–0.078 in. (1.80–1.98mm).
6. Adjust by bending tab A as needed.
7. Cap the end of vacuum hose No. 18 and apply vacuum.
8. Turn the choke drive lever clockwise until tab B seats against the stopper.
9. Measure the clearance between the choke valve and the casting.
10. The second stage clearance should be 0.168–0.176 in.

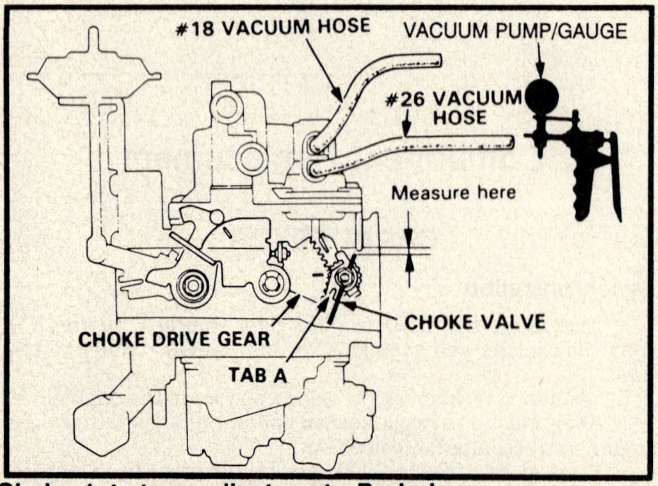

Choke 1st stage adjustment—Prelude

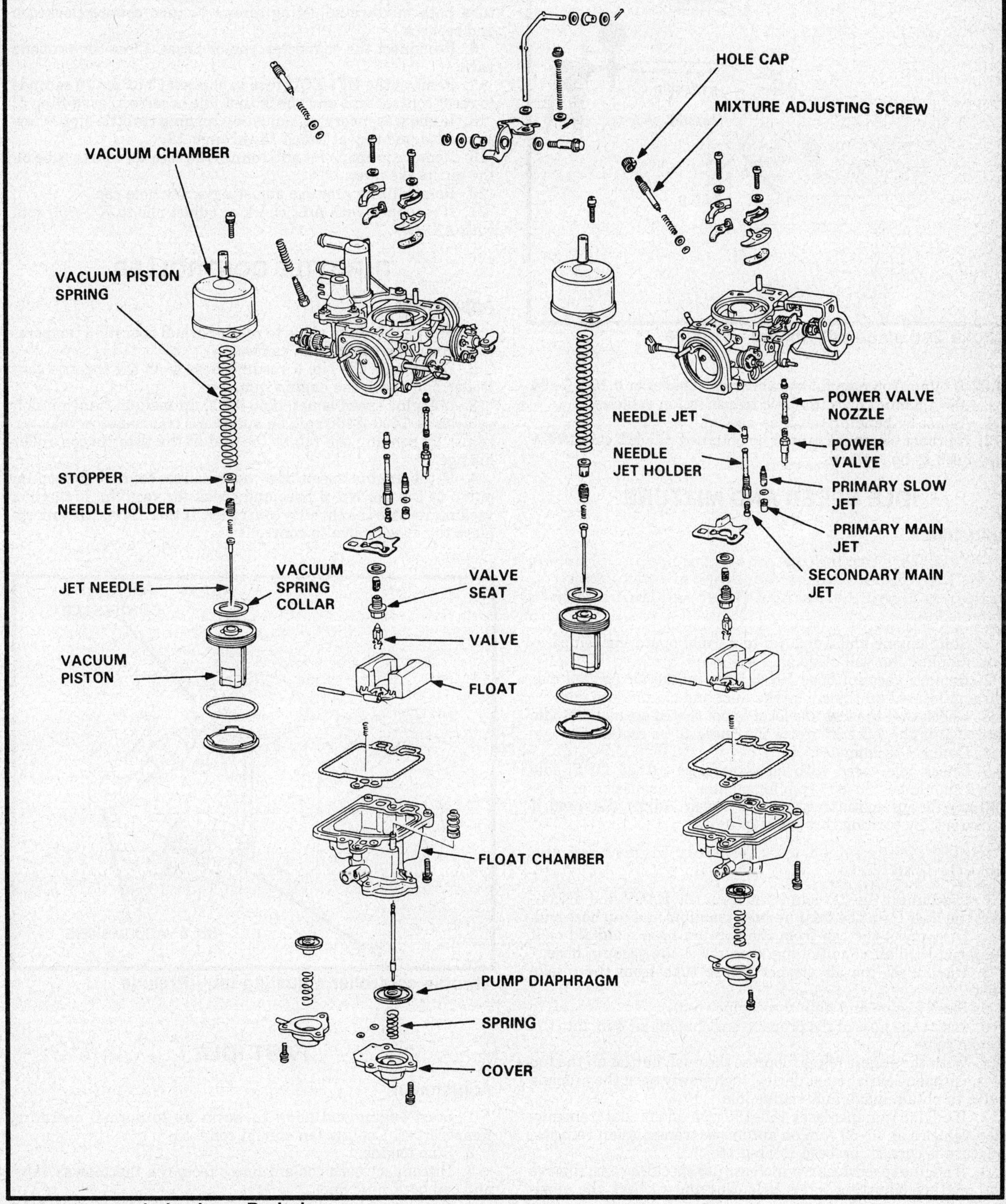

Carburetor explode view—Prelude

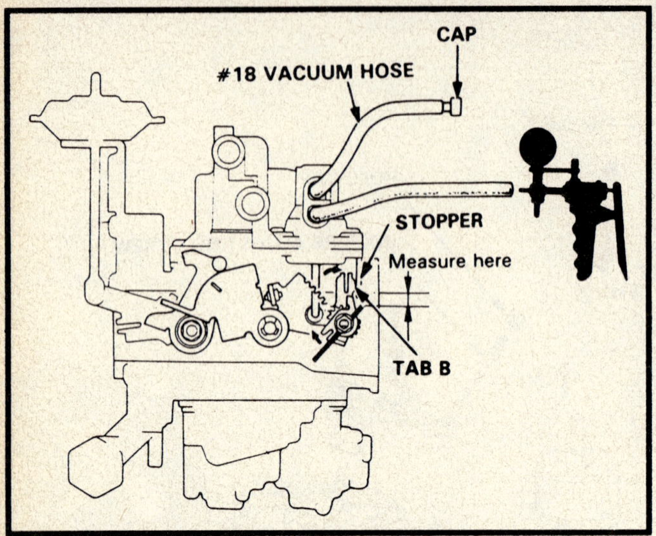

Choke 2nd stage adjustment—Prelude

(4.22–4.40mm) on manual transmission vehicles or 0.186–0.194 in. (4.65–4.83mm) on automatic transmission vehicles.

11. Adjust by bending tab B as needed.
12. If proper clearance cannot be obtained, the left carburetor may need to be replaced.

IDLE SPEED AND MIXTURE

Adjustment

NOTE: This procedure requires a propane enrichment kit. Check that carburetors are synchronized properly before making idle speed and mixture inspection.

1. Start engine and warm up to normal operating temperature. Cooling fan will come on.
2. Remove vacuum hose No. 8 from intake air control diaphragm located on the air intake duct and clamp hose end.
3. Check that the fast idle lever is not seated against fast idle cam, if it is the left carburetor may need to be replaced.
4. Connect tachometer.
5. Check idle speed with all accessories turned **OFF**. Idle speed should be 750–850 rpm for manual transmission or 700–800 rpm for automatic transmission in gear. Adjust idle speed, if necessary, by turning throttle stop screw.

NOTE: If idle speed is excessively high check the throttle controller.

6. Disconnect the 2P connector from the EACV and disconnect the hose form the vacuum hose manifold and cap hose end.
7. Disconnect the cap from the vacuum hose manifold or if equipped with air conditioning, disconnect the vacuum hose.
8. Disconnect the air cleaner intake tube from the intake dust.
9. Start engine and note the engine rpm.
10. Insert the hose of the propane enrichment kit 4 in. into the intake tube.
11. With the engine idling, depress the push button on the top of the propane enrichment device, then slowly open the propane valve to obtain maximum engine idle.
12. If engine rpm increases 150–190 rpm on manual transmission vehicles or 40–60 rpm on automatic transmission vehicles, mixture is correct, proceed to Step 16.
13. If engine speed does not increase per specifications, remove the mixture adjusting screw hole caps and recheck the maximum propane enriched rpm.
14. If the propane enriched speed is low, the mixture is rich, turn both mixture adjusting screws ¼ turn clockwise and recheck.
15. If the propane enriched speed is high, the mixture is lean, turn both mixture adjusting screws ¼ turn counterclockwise and recheck.
16. Reconnect the connecter, cap or hoses. Close the propane valve.
17. Remove the EFI-ECU fuse in the relay box for 10 seconds to reset control unit and check that idle is correct, as in Step 5.
18. If idle was incorrect, adjust by turning throttle stop screw as needed and repeat Steps 13 through 18.
19. Remove propane kit and connect air cleaner intake tube on the air intake duct.
20. Reinstall the mixture adjusting screw hole cap.
21. If equipped with A/C, check or adjust idle to 700–800 rpm with A/C **ON**.

THROTTLE CONTROLLER

Adjustment

1. Start engine and allow to reach normal operating temperature, until the cooling fan comes on.
2. Disconnect the No. 6 vacuum hose from the throttle controller and check the engine rpm.
3. If engine speed is not 1700–2700 on manual transmission vehicles or 1400–2400 rpm on automatic transmission vehicles, adjust by bending the tab at the end of the throttle controller linkage.
4. If the engine speed does not change, connect a vacuum pump to the the No. 6 hose and check for vacuum. If there is vacuum replace the throttle controller. If there is no vacuum replace the check valve to controller.

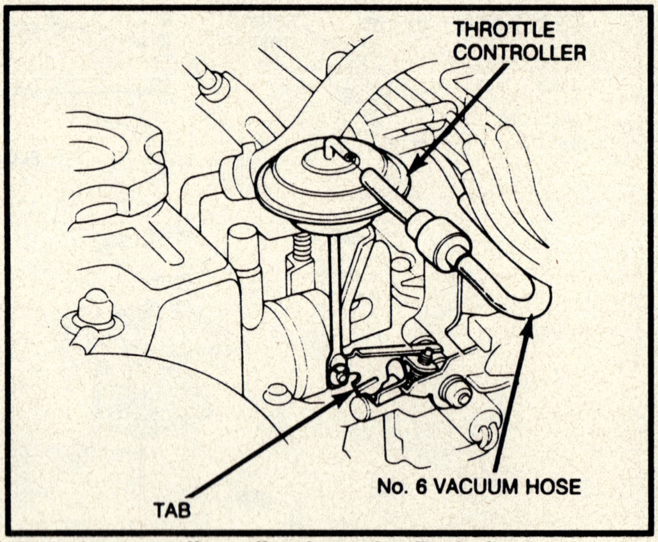

Throttle controller adjusting tab—Prelude

FAST IDLE

Adjustment

1. Start engine and allow to warm up to normal operating temperature. Cooling fan should come on.
2. Stop engine.
3. Disconnect both coolant hoses from the thermowax valve and cap both hose ends.
4. Connect a tachometer to engine and check idle speed.

5. Connect a source of cold water to thermowax valve and slowly cool down valve while checking idle changes.
6. Consult the temperature to idle speed chart and adjust fast idle screw as needed. If proper idle cannot be obtained, the left carburetor may need to be replaced.
7. Reinstall coolant hoses.
8. Start engine and allow to reach normal operating temperature.
9. Check that the fast idle lever is not against the fast idle cam. If it is, the left carburetor may need to be replaced.

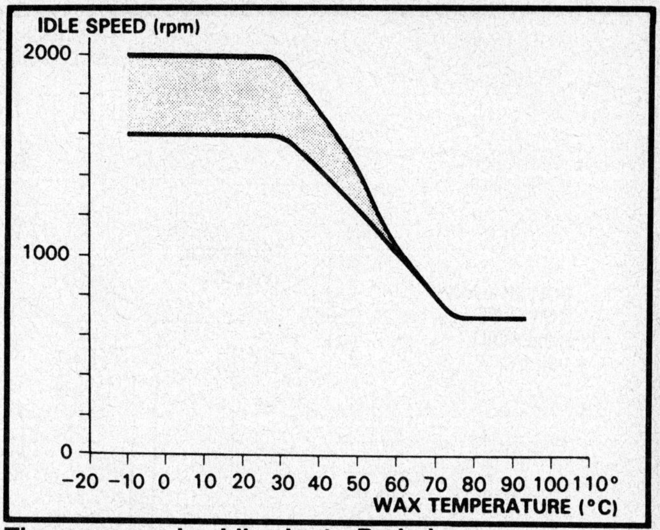

Thermowax valve idle chart—Prelude

FLOAT LEVEL

Adjustment

1. Remove carburetors.
2. Remove float chambers.
3. Using float level gauge, measure float level with float tip lightly contacting float valve and carburetor float chamber surface inclined about 30 degrees from vertical. Float level should be 0.56–0.64 in. (15–17mm). Adjust as necessary.

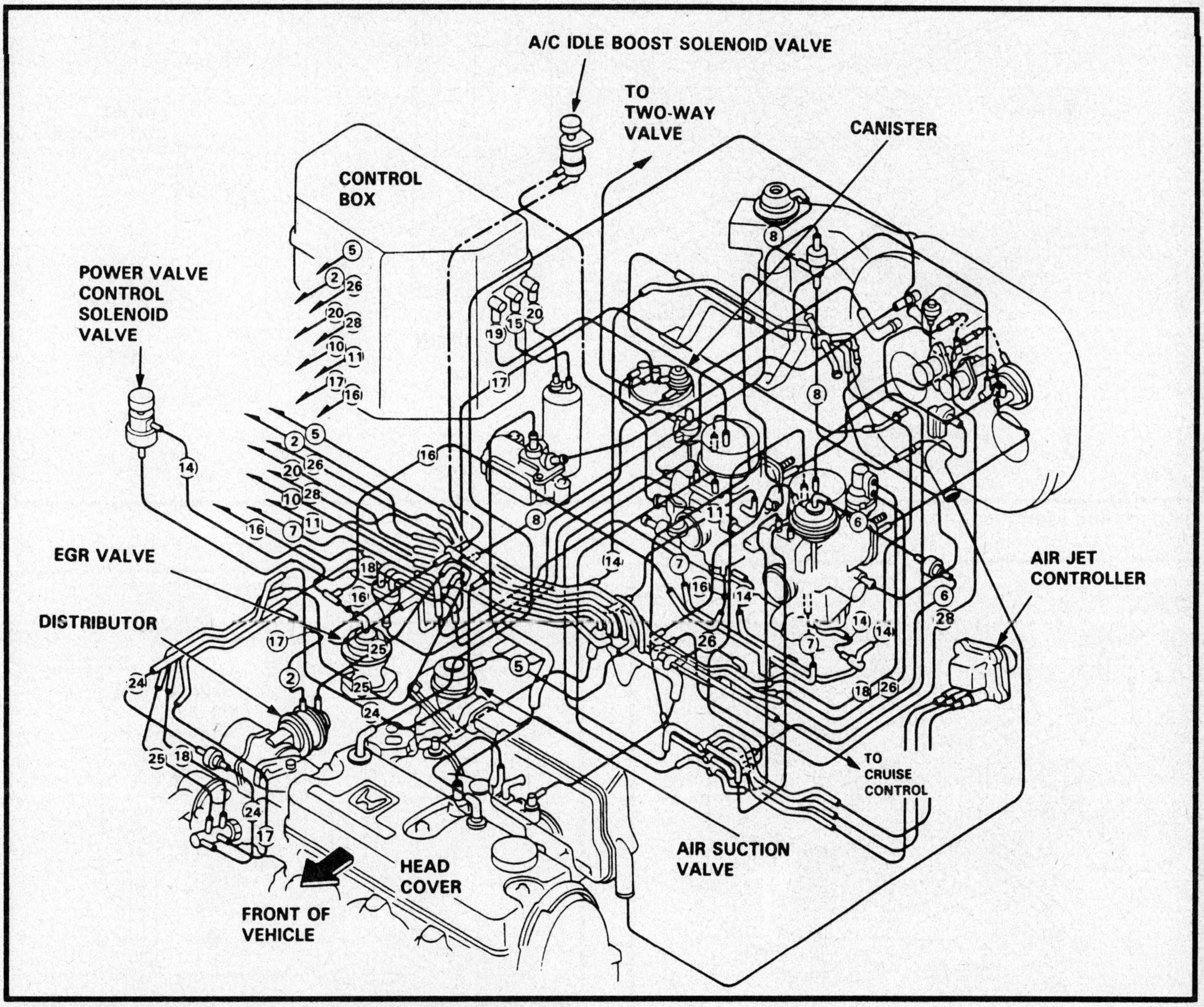

Vacuum line identification—Prelude except California

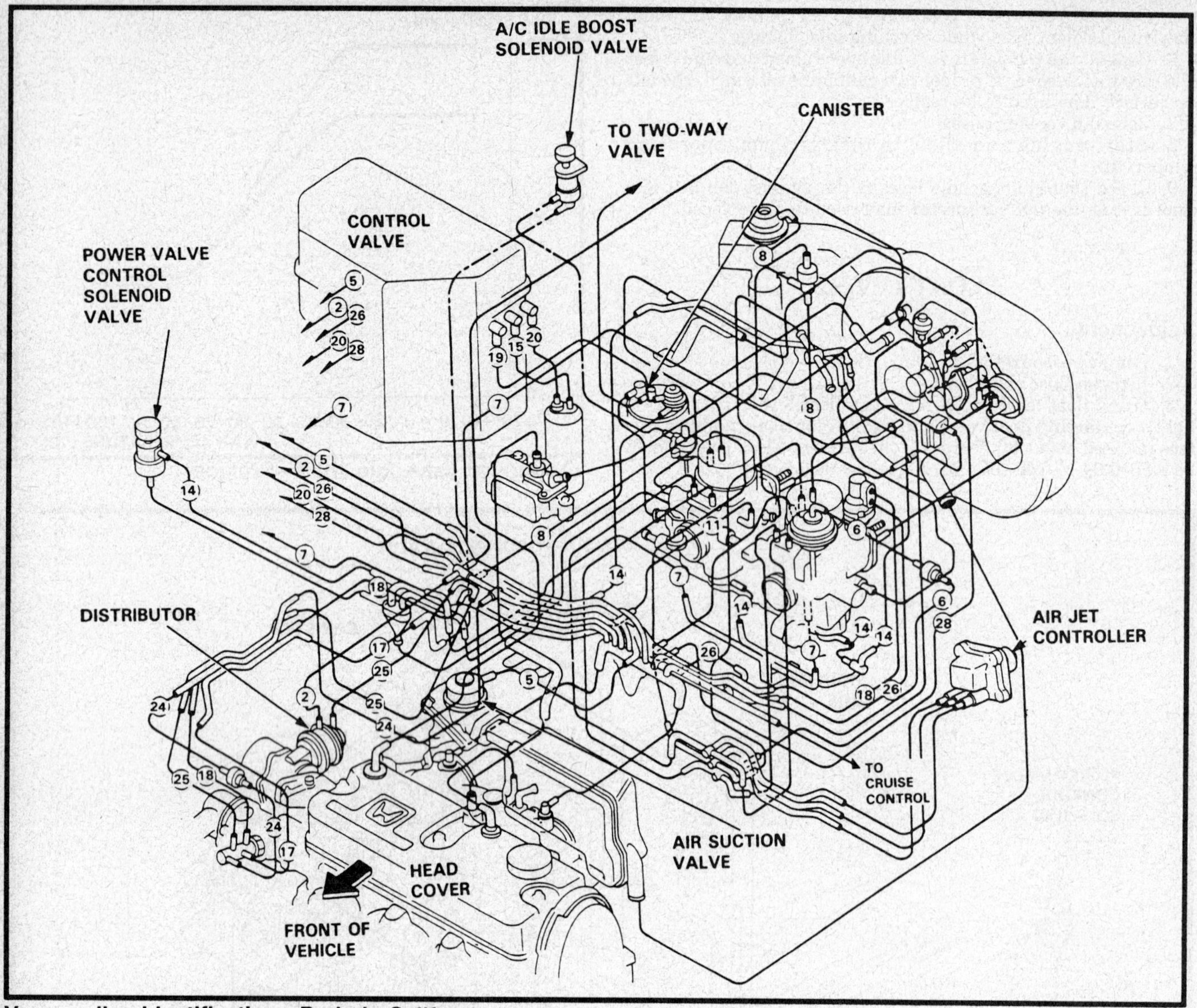

Vacuum line identification—Prelude California

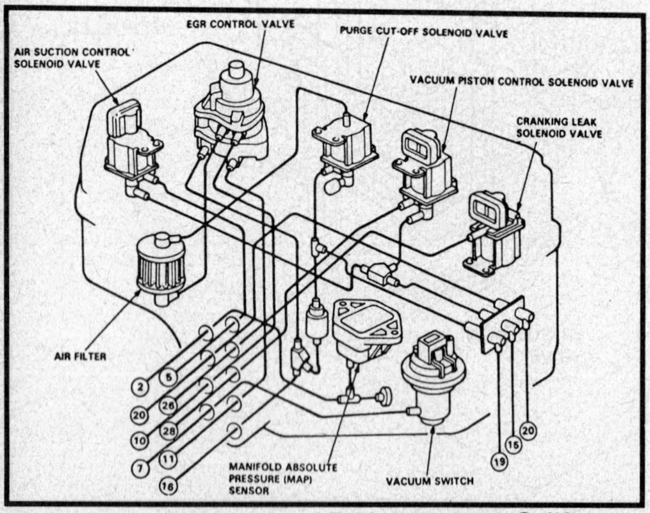

Control box components—Prelude except California

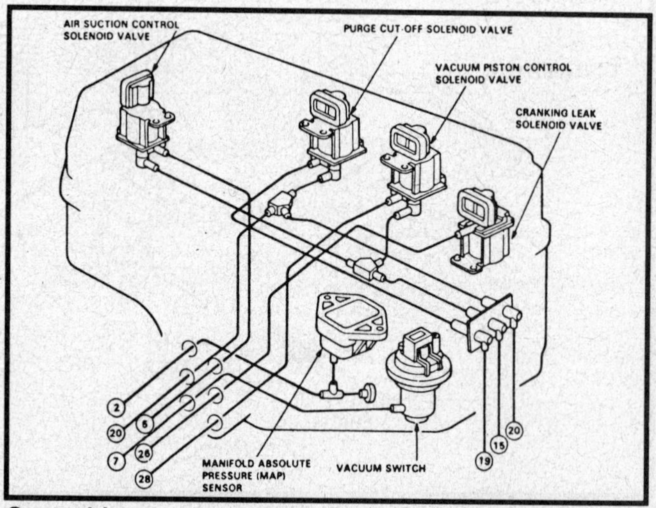

Control box components—Prelude California

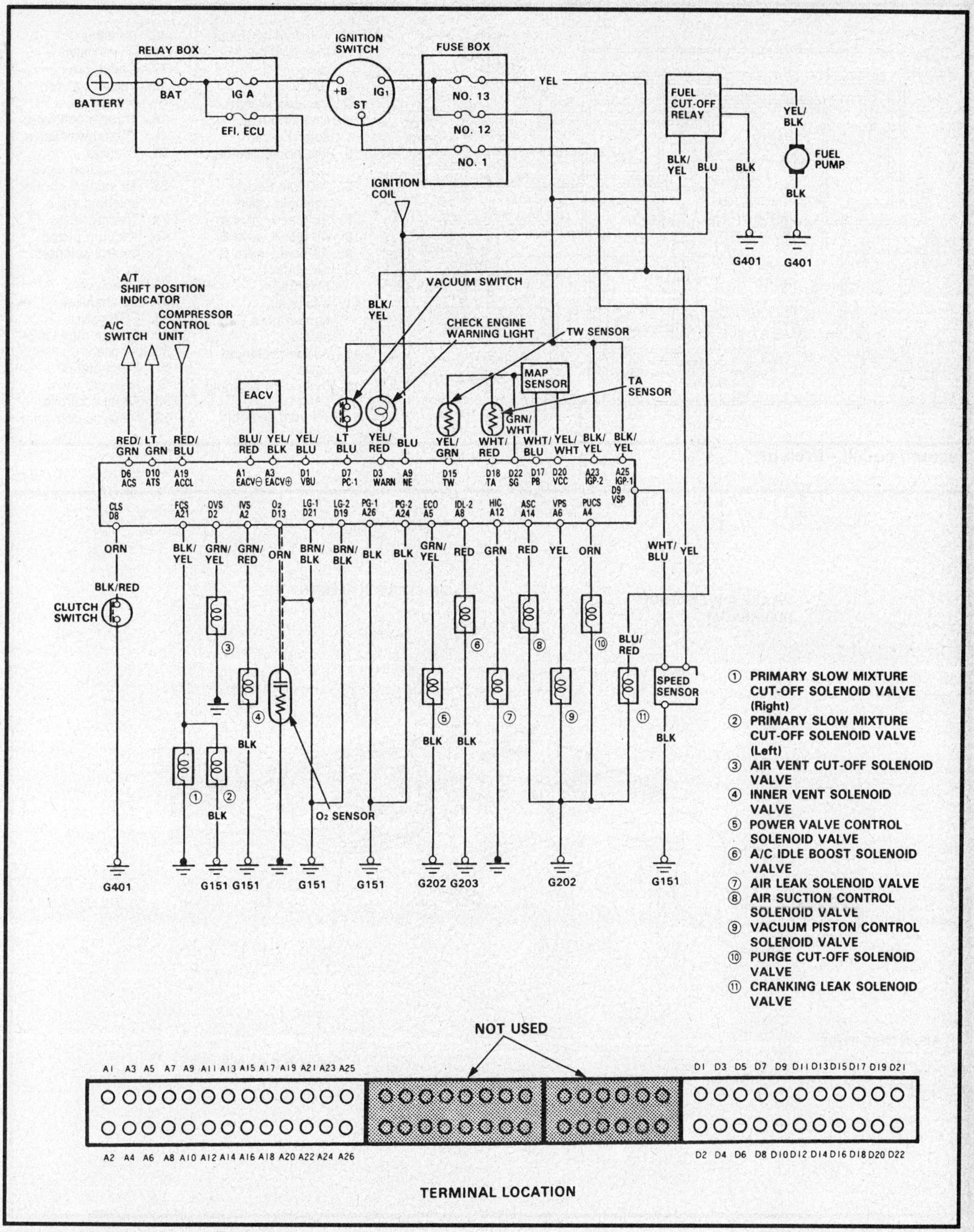

Electrical connections—Prelude

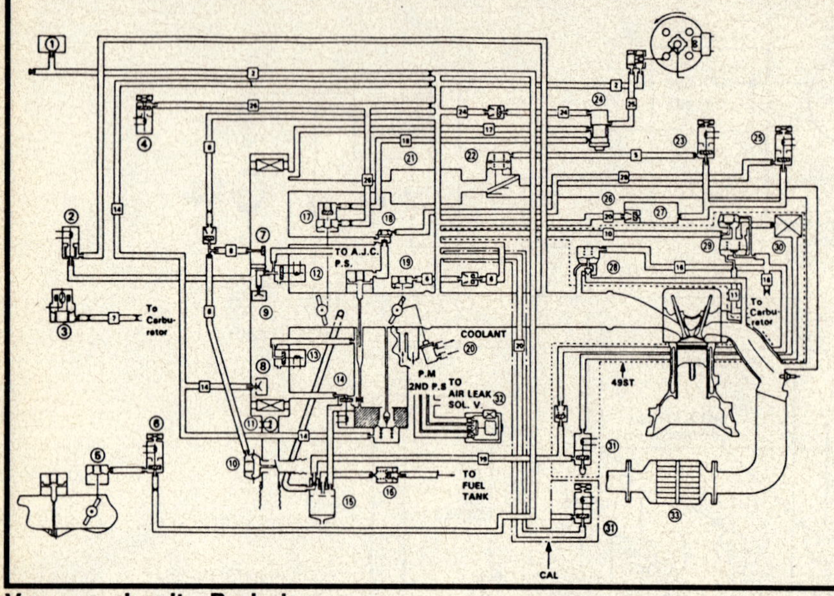

1. Manifold Absolute Pressure (MAP) sensor
2. EACV
3. Vacuum switch
4. Cranking leak solenoid valve
5. Idle boost throttle controller
6. A/C idle boost solenoid valve
7. Air bleed valve A
8. Air bleed valve B
9. Air bleed valve C
10. Air control diaphragm
11. Intake air temperature (TA) sensor
12. Air leak solenoid valve
13. Inner vent solenoid valve
14. Air vent solenoid valve
15. Canister
16. 2-way valve
17. Choke opener
18. Vacuum piston control valve
19. Throttle controller
20. Thermowax valve
21. Silencer
22. Air suction valve
23. Air suction control solenoid valve
24. Thermovalve
25. Vacuum piston control solenoid valve
26. Check valve C
27. Air chamber
28. EGR valve
29. EGR control valve
30. Air filter
31. Purge cut-off solenoid valve
32. Air jet controller
33. Catalytic converter

Vacuum circuit—Prelude

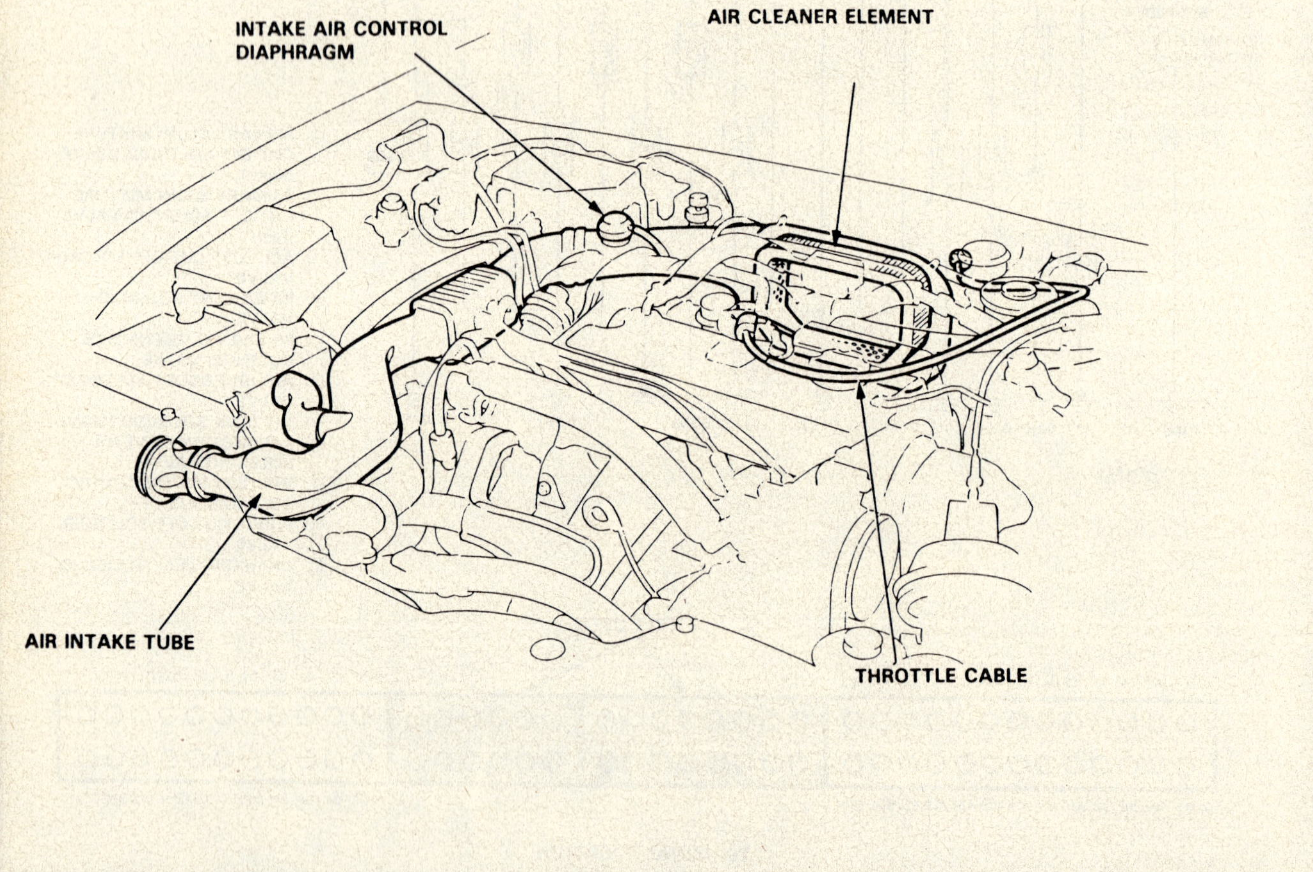

Air intake tract—Prelude

HYUNDAI FEEDBACK CARBURETOR (FBC) SYSTEM

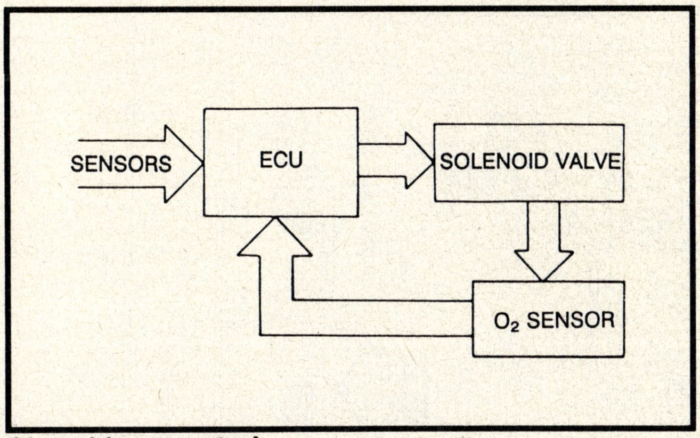

Closed loop control

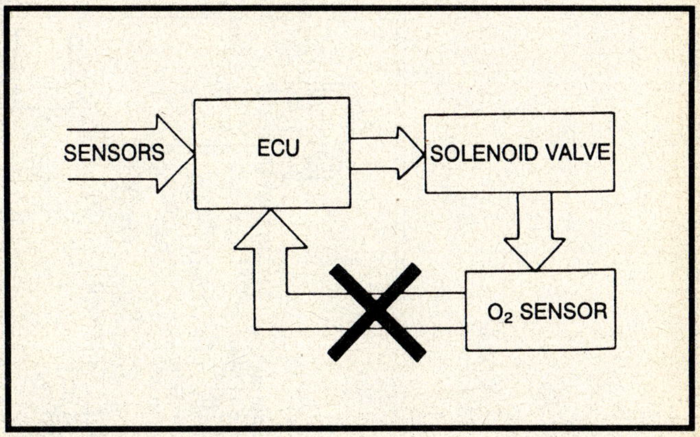

Open loop control

General Information

The feedback carburetor system provides the capability to perform closed loop fuel control. It also provides the capability to control the secondary air system, the deceleration spark control system and the throttle opener system. The basic functions of this system are depicted below. Input signals from a variety of sensors are fed to a microprocessor based Electronic Control Unit (ECU). The ECU then generates output signals for all of the controlled functions. The feedback carburetor is a 2-barrel, downdraft carburetor designed for closed loop system. When used with the closed loop system of mixture control, this carburetor includes special design features for optimum air/fuel mixtures during all ranges of engine operation. Fuel metering is accomplished through the use of 3 solenoid-operated on/off valves (jet mixture, enrichment and deceleration solenoids), adding or reducing fuel to the engine.

The activation of the on/off valve is controlled by the length of the time current is supplied to the solenoid. The solenoid operates at a fixed frequency. By varying the amount of time the solenoid is energized during each cycle (defined as duty cycle), the air/fuel mixture delivered to the engine can be precisely controlled. The duty cycle to the solenoid is controlled by the ECU in response to the signals from the exhaust oxygen sensor, throttle position sensor. Incorporated in the feedback carburetor are 8 basic systems of operation: fuel inlet, primary metering, secondary metering, accelerating pump, choke, jet mixture, enrichment and fuel cutoff.

ELECTRONIC CONTROL UNIT (ECU)

The electronic control unit is mounted in the passenger compartment and consists of a printed circuit board mounted in a protective metal box. It receives analog inputs from the sensors and converts them into digital signals. These signals and various discrete inputs are processed and used by the ECU in controlling the fuel delivery, secondary air, deceleration spark and throttle opener managements.

FEEDBACK CARBURETOR AIR/FUEL CONTROL

The feedback carburetor air/fuel ratio is controlled by the Electronic Control Unit (ECU). The ECU monitors the throttle position, engine speed, coolant temperature, intake air temperature and exhaust oxygen concentration to calculate the fuel flow required to yield the desired air/fuel ratios for all operating conditions. Closed loop control is used to adjust the fuel flow to yield a near stoichiometeric air/fuel ratio when required. The fuel flow is modified to account for special operating conditions, such as hot starts, acceleration and deceleration.

Mixture Control System

The air-fuel ratio is maintained by the ECU in either closed looped or opened loop operating modes. In opened loop operation, before the engine has reached normal operating temperature, the ECU controls fuel mixture according to preprogramed values. In closed loop operation, after the engine has warmed up, the air-fuel ratio is based on the oxygen sensor signal.

The ECU senses the oxygen sensor signal and provides feedback to the Feedback Solenoid Valve (FBSV) to maintain the proper mixture. The ECU also provides control signals to the Slow Cut Solenoid Valve (SCSV) which is kept is the wide open, or 100% duty, during this time.

When the FBSV is energized, the main fuel jet is closed, leaving the primary main jet passage as the only fuel passage. This reduces the amount of fuel, resulting in a leaner air-fuel mixture. When the FBSV is de-energized, the main fuel jet is opened to provide 2 fuel passages, including the primary main jet passage. Since this will increase the amount of fuel, richer air-fuel mixture is obtained.

With the ON/OFF operation of the SCSV, the slow fuel passage is opened and closed. The air-fuel ratio at deceleration is controlled in this manner.

ADAPTIVE MEMORY CONTROL

During the closed loop operation, the ECU controls the duty cy-

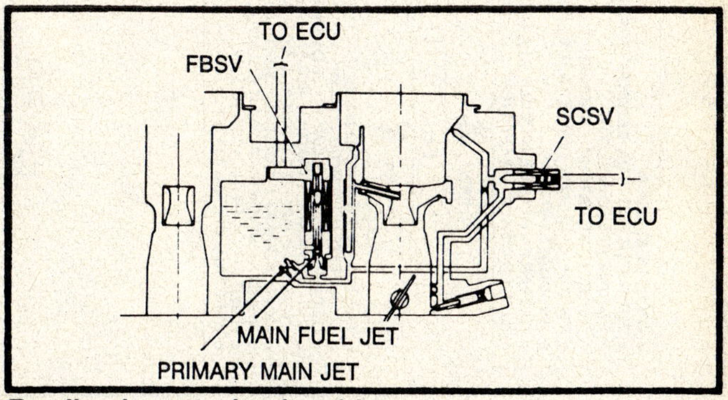

Feedback control solenoids

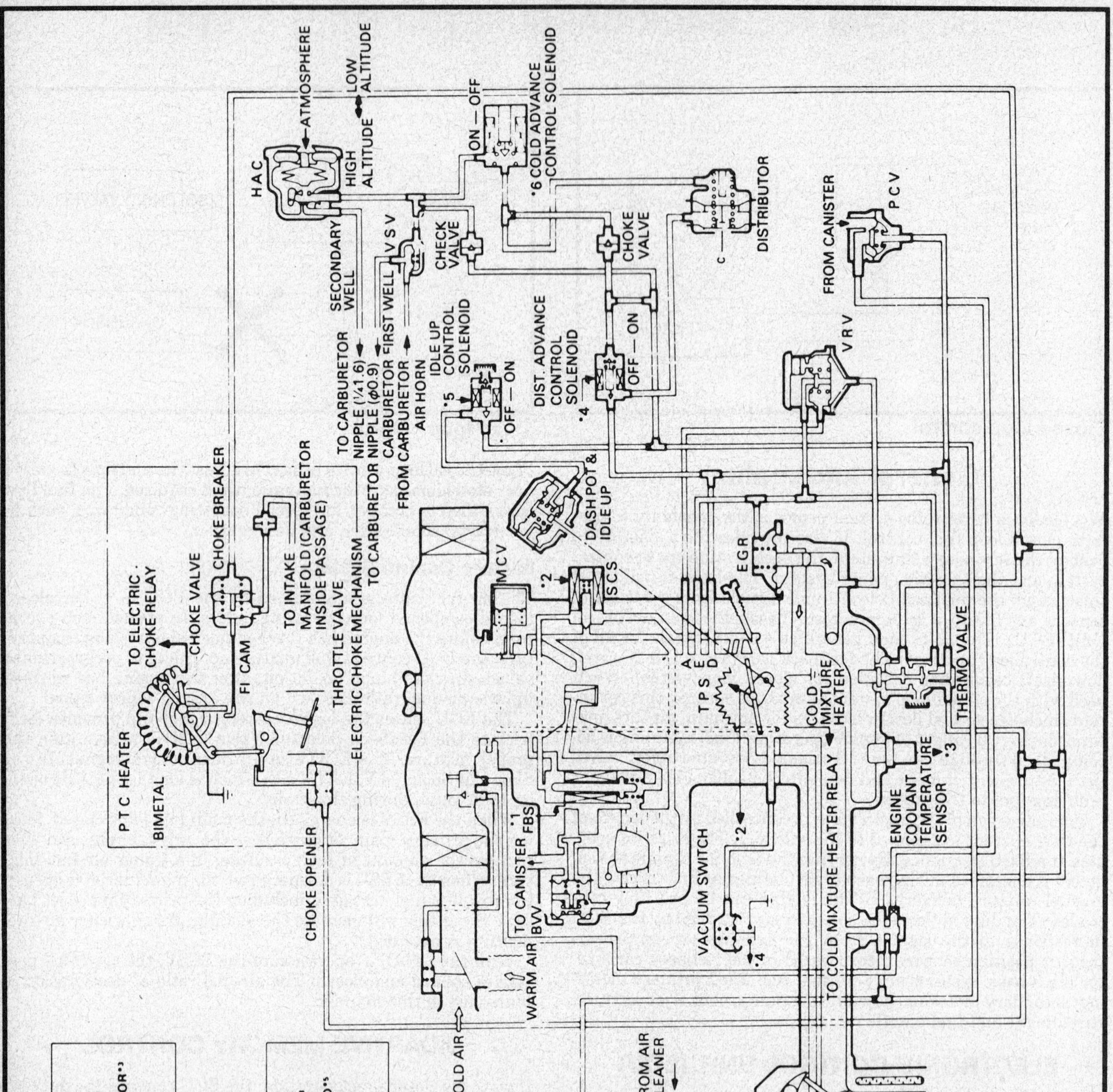

Feedback system diagram – typical

cle of the jet mixture control solenoid valve, based on the output voltage signal from the exhaust sensor. The mean values of the duty cycle is stored in a Random Access Memory (RAM) and the last ones are stored, even if the ignition switch is turned **OFF**.

SECONDARY AIR CONTROL

A solenoid is used to control the air control valve signal vacuum. The solenoid is controlled by the ECU, based on the engine speed, idle position and coolant temperature. This valve sends air to the exhaust manifold.

DECELERATION SPARK CONTROL

In order to decrease the hydrocarbon (HC) emissions emitted during vehicle deceleration, ignition timing is advanced by the solenoid operated vacuum valve on the distributor, changing the vacuum supplied to the valve from carburetor ported vacuum to intake manifold vacuum. The solenoid valve is controlled by the electronic controlled unit, based on engine speed.

THROTTLE OPENER CONTROL

When the tail switch or the blower motor switch is in the **ON** position, the engine idle speed is increased by changing the throttle valve opening. In order to decrease the hydrocarbons (HC) emissions emitted during vehicle deceleration and to improve the engine brake performance, the ECU de-energizes the solenoid valve which supplies the manifold vacuum to the throttle opener at or above the preset engine speed and then the throttle opener operation is suspended.

SENSORS

EXHAUST OXYGEN SENSOR

The oxygen sensor is mounted in the exhaust manifold. The output signal from the sensor, which varies with the oxygen content of the exhaust gas stream, is provided to the ECU for use in controlling closed loop compensation of fuel delivery.

COOLANT TEMPERATURE SENSOR

The coolant temperature sensor is installed in the intake manifold. This sensor provides data to the ECU for use in controlling fuel delivery and secondary air management.

ENGINE SPEED SENSOR

The engine speed sensor signal comes from the ignition coil. Electric signals are sent to the ECU, where the time between these pulses is used to calculate engine speed, which is used in

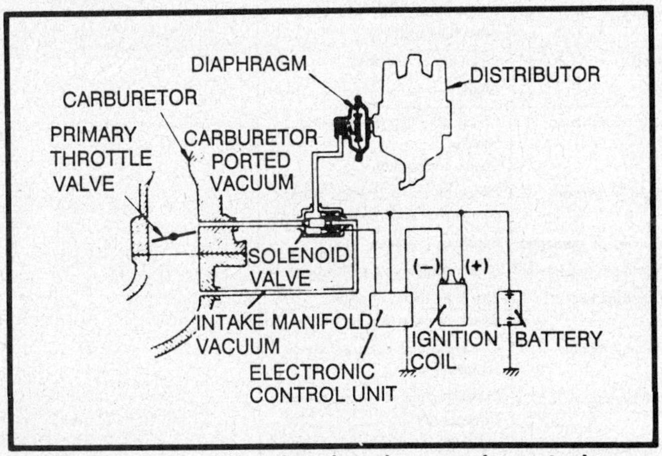

Exploded view of the deceleration spark control system

controlling fuel delivery, secondary air management, deceleration spark and throttle opener management.

THROTTLE POSITION SENSOR

This is a potentiometer mounted to the carburetor. The TPS provides throttle angle information to the ECU to be used in controlling the fuel delivery and secondary management.

VACUUM SWITCH

The switch is installed on the floor board or the fender and is turned on when the throttle valve is closed (idling) position. Information from this switch is provided to the ECU for use in controlling fuel delivery and secondary air management.

SERVICE PRECAUTIONS

- Before the battery terminals are disconnected, make sure the ignition switch is set to the **OFF** position. If the battery terminals are disconnected while the engine is running or when the ignition switch is in the **ON** position, malfunction of the computer or damage to the semi-conductors could result.
- Disconnect the battery cables before charging the battery.
- When the battery is connected, be sure not to reverse the polarity.
- Make sure that the harness connectors are securely connected. Use care not to allow the entry of water or oil into the connectors.

Diagnosis and Testing

NOTE: For Self-Diagnostic System and Accessing Trouble Code Memory, see Section 3 "SELF-DIAGNOSTIC SYSTEM".

FEEDBACK CARBURETOR SYSTEM (FBC)

If the FBC system components (sensors, carburetor control unit computer, solenoid etc.) fail, interruption of the fuel supply or failure to supply the proper amount of fuel for engine operating conditions occur. The following situations may be encountered:

1. The engine is hard to start or does not start at all.
2. Unstable idle.
3. Poor driveability.

If any of the above conditions are encountered, first perform the basic engine checks (ignition system malfunctions, incorrect engine adjustment, etc.). The feedback carburetor system can be checked by the use of a feedback carburetor checker tool and harness connector. The inspection procedure with the use of this tool is as follows:

Inspection Procedure

1. Turn the ignition switch to the **OFF** position.
2. Remove both the larger and smaller harness connectors from the carburetor control unit (ECU).
3. Set the check switch of the FBC checker to the **OFF** position.
4. Set the select switch of the FBC checker to the A position.
5. Connect the FBC harness connector to the connectors of the FBC checker, and then connect the FBC harness connector to carburetor control unit (ECU) and the harness connectors.
6. Place the FBC checker on the front passenger's seat. Perform the checks according to the "FBC SYSTEM CHECK PROCEDURES".
7. If the checker shows any departure from specifications, check the corresponding sensor and related electrical wiring, repair or replace.
8. After repair of replacement, recheck with the FBC checker to confirm that the repaired or replaced part is performing well.
9. Set the check switch of the FBC checker to the **OFF** position. Set the ignition switch to the **OFF** position.

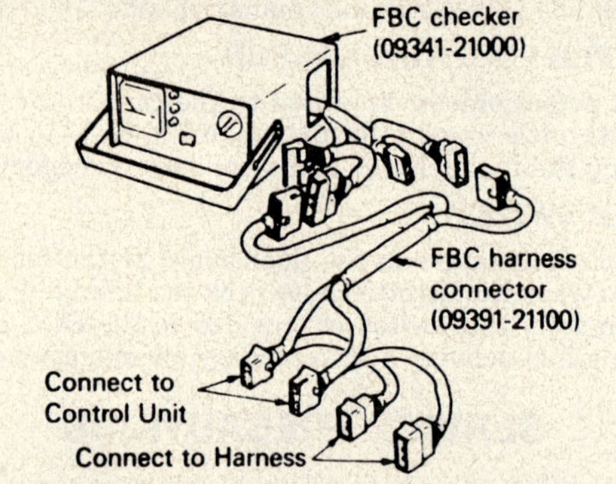

Feedback carburetor checker tool wiring harness

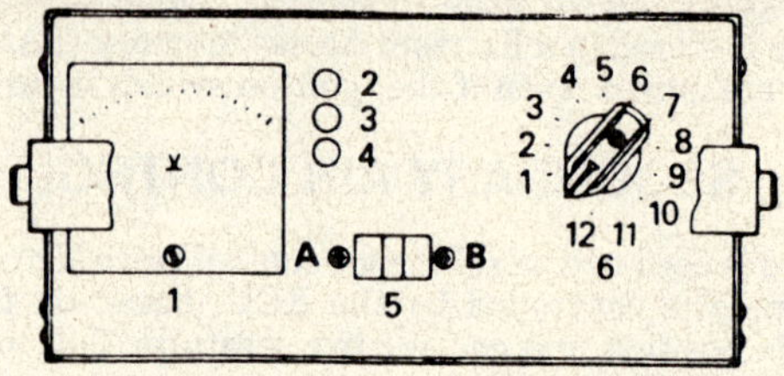

Feedback carburetor checker tool

10. Disconnect the connectors of the FBC checker and the FBC harness connector from the carburetor control unit (ECU) and the body side harness connector.
11. Connect the body side harness connector to the carburetor control unit (ECU).

NOTE: If the carburetor control unit (ECU) appears defective, perform all FBC checker tests and individual sensor tests. If everything is in proper working order and every test possible was tried, replace the carburetor control unit (ECU).

TROUBLESHOOTING THE FEEDBACK CARBURETOR SYSTEM

Symptom		Probable cause	Remedy
Engine will not start or start to hard	Carburetor	Choke valve remains open-cold engine	Clean choke bore and linkage
		Improper choke breaker operation	Check and adjust choke breaker
		Electric choke malfunction	Check electric choke body and choke valve operation
		Needle valve sticking or clogged	Repair or replace
	FBC system	Engine coolant temperature sensor malfunction	Check by using checker (Check component and replace if faulty)
		Vacuum hose disconnected or damaged	Repair or replace
		Slow-cut solenoid valve malfunction	Check component
		Feedback solenoid valve malfunction	Check component
		Vacuum switch malfunction-cold engine	Check component
		Faulty ECU	Replace
		Harness broken/short-circuited or loose connection	Repair or replace
Rough idle or engine stalls	Carburetor	Choke valve malfunction	Clean choke bore and link
		Improper fast idle-cold engine	Adjust fast idle speed
		Improper idle adjustment	Adjust idle speed
		Electric choke malfunction	Check choke body and choke valve operation
		Primary pilot jet clogged	Clean up or replace
		Dash pot malfunction	Adjust
	FBC system	Slow-cut solenoid valve malfunction	Check drive signal by using checker Check component
		Engine coolant temperature sensor malfunction	Check by using checker (Check component and replace if faulty)
		Vacuum hose disconnected or damaged	Repair or replace
		Throttle position sensor malfunction	Check component and adjust
		Engine speed sensor malfunction	Check by using checker Check harness for continuity
		Timing control system malfunction	Check system. If faulty, check components
		Throttle opener control system malfunction	Check system. If faulty, check components
		Harness broken/short-circuited or connector not connected securely	Repair or replace

TROUBLESHOOTING THE FEEDBACK CARBURETOR SYSTEM

Symptom		Probable cause	Remedy
Engine hesitates or poor acceleration	Carburetor	Acceleration pump malfunction	Clean pump discharge rate
		Choke valve remains open-cold engine	Clean choke bore and link Check choke valve operation
		Choke valve remains closed-hot engine	Clean choke bore and link Check choke valve operation
		Main jet clogged	Clean up
		Enrichment jet clogged	Clean up
		Secondary valve operation abnormal	check valve operation
	FBC system	Feedback solenoid valve malfunction	Check drive signal by using checker Check component
		Vacuum switch malfunction	Check with checker (Replace if faulty)
		Timing control system malfunction	Check system. If faulty, check components
		Engine coolant temperature sensor malfunction	Check by using checker (Check components and replace if faulty)
		Throttle position sensor malfunction	Check component and adjust
		Engine speed sensor malfunction	Check by using checker Check harnesses for continuity
		Cold mixture heater relay control system malfunction-cold engine	Check system. If faulty, check components
		Harness broken/short-circuited or connector not connected properly	Repair or replace
		Air conditioner power relay control system malfunction	Check system
Engine dieseling (runs after ignition switch is turned off)	Carburetor	Engine idle speed to high	Adjust idle speed
	FBC system	Slow-cut solenoid valve malfunction	Check component
Poor fuel mileage	Carburetor	Choke valve operation abnormal	Check valve operation
		Engine idle speed too high	Adjust idle speed
		Electric choke malfunction	Check choke body and valve operation
		Enrichment valve kept open	Repair or replace

Symptom		Probable cause	Remedy
Poor fuel mileage	FBC system	Engine coolant temperature sensor malfunction	Check by using checker (Check component and replace if faulty)
		Oxygen sensor malfunction	Check by using checker (Check component and replace if faulty)
		Timing control system malfunction	Check system. If faulty, check components
		Feedback solenoid valve malfunction	Check drive signal by using checker Check component
		Slow-cut solenoid valve malfunction	Check drive signal by using checker Check components
		Throttle position sensor malfunction	Check component and adjust
		Engine speed sensor malfunction	Check by using checker Check harness for continuity
		Harness broken/short-circuited or connector not connected securely	Repair or replace

FEEDBACK SYSTEM CHECK PROCEDURE CHART—AUTOMATIC TRANSMISSION

FBC checker setting: Select switch	FBC checker setting: Check switch	Check item	ECU Terminal# checked	Condition		Test specification
"A"	1	Power supply	A — 2	Ignition switch "LOCK" → "ON"		11V — 13V
	2	Spark advance control solenoid valve	A — 13	Idling (warm engine)		0V — 0.6V
				Start engine. Drive vehicle at a speed higher than 8 km/h (5 mph), then hold 2,000 rpm (warm engine)		13V — 15V
	3	Throttle position sensor (TPS)	B — 9	Ignition switch "LOCK"→"ON" (warm engine)	Accelerator fully closed	0.4V — 0.7V
					Accelerator fully opened	4.5V — 5.5V
	4	Engine coolant temperature sensor	B — 3	Ignition switch "LOCK" → "ON"	0°C (32°F)	3.4V — 3.6V
					20°C (68°F)	2.4V — 2.7V
					40°C (104°F)	1.5V — 1.8V
					80°C (176°C)	0.5V — 0.7V
	5					
	6	Vacuum switch for idle position	A — 6	Ignition switch "LOCK" → "ON"		9V — 13V
				Idling (warm engine)		0V — 0.6 V
	7	Idle up (throttle opener) control solenoid valve	A — 12	Idling	Power steering switch ON or lighting switch ON	0V — 0.6V
				2,000 rpm		9V — 15V
	8	Electric choke relay	A — 14	Ignition switch "LOCK"→"ON"		0V — 0.6V
				Idling		13V — 15V
	9	Air conditioner cut-off	A — 8	Ignition switch "LOCK"→"ON" and air conditioner switch "ON"*1	Accelerator fully closed	0V — 0.6V
					Accelerator fully opened	M/T 0V—0.6V
						A/T 13V—15V
	10	Power supply for sensor	B — 1	Ignition switch "LOCK" → "ON"		4.5V — 5.5V
	11					
	12	Secondary air control solenoid valve	A — 9	Idling, 70 seconds after start of warm engine		0V — 0.6V then 13V — 15V
				Quick deceleration from above 2,000 rpm to idling with gear in "N" position		Momentarily drop

*1 : ON means compressor clutch engaged

FBC Checker setting: Select switch	FBC Checker setting: Check switch	Check item	ECU Terminal# checked	Condition	Test specification
"B"	1				
	2	Feed back solenoid valve (FBSV)	A—11	Ignition switch "LOCK" → "ON"	11V — 13V
				Idling (warm engine)	2V — 12V
	3				
	4	Ignition pulse	A—15	Ignition switch "LOCK" → "START"	2V — 8V
	5				
	6				
	7	Slow cut solenoid valve	A—10	Idling	0V — 0.6V
				Quick deceleration from above 4,000 rpm to idle with transmission in position	Momentarily 13V — 15V
	8	Oxygen sensor	B — 5	Hold speed constant above 1,300 rpm, 70 seconds after start of warm engine	0V — 1V ↑ (Pulsates) ↓ 2V — 3V
	9				
	10				
	11				
	12				

FEEDBACK SYSTEM CHECK PROCEDURE CHART—MANUAL TRANSMISSION

FBC checker setting: Select switch	Check switch	Check item	ECU Terminal # checked	Condition		Test specification
"A"	1	Power supply	8 (C60-1)	Ignition switch "LOCK" → "ON"		11V—13V
	2	Spark advance control solenoid valve	3 (C60-1)	Idling (warm engine)		0V—0.6V
				Start engine. Drive vehicle at a speed higher than 8 km/h (5 mph), then hold 2,000 rpm (warm engine)		13V—15V
	3	Throttle position sensor (TPS)	3 (C60-2)	Ignition switch "LOCK" → "ON" (warm engine)	Accelerator fully closed	0.4V—0.7V
					Accelerator fully opened	4.5V—5.5V
	4	Engine coolant temperature sensor	8 (C60-2)	Ignition switch "LOCK" → "ON"	0°C (32°F)	3.4V—3.6V
					20°C (68°F)	2.4V—2.7V
					40°C (104°F)	1.5V—1.8V
					80°C (176°C)	0.5V—0.7V
	5					
	6	Vacuum switch for idle position	13 (C60-1)	Ignition switch "LOCK" → "ON"		9V—13V
				Idling (warm engine)		0V—0.6V
	7	Idle up (throttle opener) control solenoid valve	2 (C60-1)	Idling	Power steering switch ON or lighting switch ON	0V—0.6V
				2,000 rpm		9V—15V
	8	Electric choke relay	4 (C60-1)	Ignition switch "LOCK" → "ON"		0V—0.6V
				Idling		13V—15V
	9	Air conditioner cut-off	11 (C60-1)	Ignition switch "LOCK" → "ON" and air conditioner switch "ON"*1	Accelerator fully closed	0V—0.6V
					Accelerator fully opened	M/T 0V—0.6V
	10	Power supply for sensor	16 (C60-1) 17 (C60-1)	Ignition switch "LOCK" → "ON"		4.5V—5.5V
	11					
	12	Secondary air control solenoid valve	10 (C60-1)	Idling, 70 seconds after start of warm engine		0V—0.6V then 13V—15V
				Quick deceleration from above 2,000 rpm to idling with gear in "N" position		Momentarily drop

*1 : ON means compressor clutch engaged

FBC Checker setting: Select switch	Check switch	Check item	ECU Terminal # checked	Condition	Test specification
"B"	1				
	2	Feed back solenoid valve (FBSV)	1 (C60-1)	Ignition switch "LOCK" → "ON"	11V—13V
				Idling (warm engine)	2V—12V
	3				
	4	Ignition pulse	5 (C60-1)	Ignition switch "LOCK" → "START"	2V—8V
	5				
	6				
	7	Slow cut solenoid valve	9 (C60-1)	Idling	0V—0.6V
				Quick deceleration from above 4,000 rpm to idle with transaxle in position	Momentarily 13V—15V
	8	Oxygen sensor	6 (C60-2)	Hold speed constant above 1,300 rpm, 70 seconds after start of warm engine	0V—1V (Pulsates) 2V—3V
	9				
	10				
	11				
	12				

COOLANT TEMPERATURE SENSOR

Inspection

1. Disconnect the negative battery cable and drain the coolant level just below the manifold level.
2. Remove the coolant temperature sensor from the intake manifold.
3. Immerse the temperature sensing portion of the coolant temperature sensor in a container of hot water.
4. Using a suitable ohmmeter, check the resistance. The sensor should be held with its housing 0.12 in. (3mm) away from the surface of the hot water. The resistance should be as follows:
   2.45 kilo-ohms at 68°F (20°C)
   0.3 kilo-ohms at 176°F (80°C)
5. If the coolant temperature sensor has abnormally low or high resistance readings, it should be replaced.
6. When installing the sensor, be sure to coat the threads with a suitable sealant and torque it to 21 ft. lbs. (30 Nm). Fasten the harness connectors securely.

VACUUM SWITCH

Inspection

1. Disconnect the harness connector of the vacuum switch, which is usually located on the dash panel.
2. Using a suitable ohmmeter, check the continuity across the terminals.
3. Connect a suitable vacuum pump to the nipple of the switch.
4. Apply more than 12 in. Hg of vacuum to the switch. If the switch is closed (ON), the vacuum switch is good. The vacuum specifications for a properly operating switch are as follows:
   ON – more than 11.8 in. Hg
   OFF – less than 8.0 in. Hg
5. If the switch fails this inspection, replace it. If the switch is good, remove the vacuum pump, make sure the switch is open (OFF) and reconnect the harness connector.

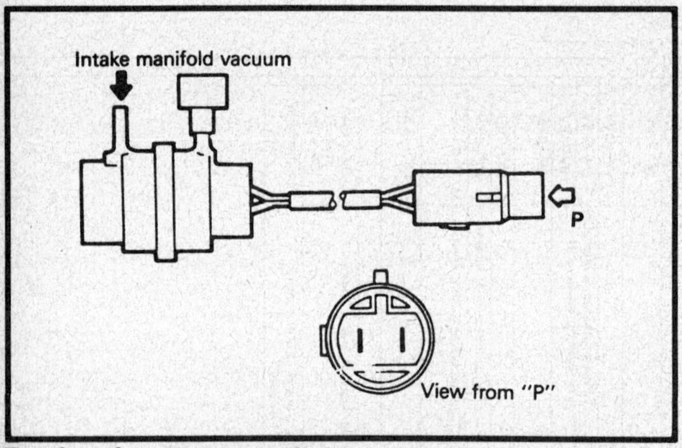

Testing the vacuum switch

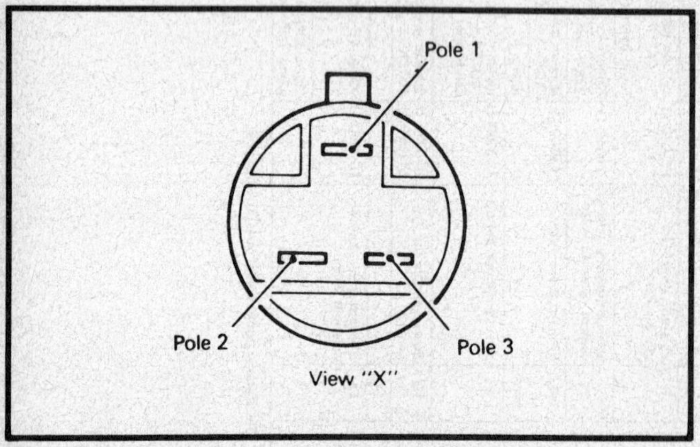

TPS connector terminal identification

THROTTLE POSITION SENSOR (TPS)

Inspection

1. Disconnect the throttle position sensor connector. Connect a suitable ohmmeter between the bottom 2 terminals (2 and 3) of the connector.
2. Check to ensure that when the throttle valve is slowly operated from the closed to wide open position, resistance changes smoothly as from approximately 3.5–6.5 kilo-ohms.
3. If the throttle position sensor fails this inspection, replace the sensor.
4. Install the new throttle position sensor assembly into the carburetor and tighten the 2 retaining screws. Reconnect and adjust the TPS.

OXYGEN SENSOR

Inspection

1. Keep the engine speed above 1400 rpm after warming it up.
2. Locate the oxygen sensor checker, which is usually located in the wiring harness along the valve cover next to the air cleaner.
3. Using a suitable voltmeter, check the voltage of the oxygen

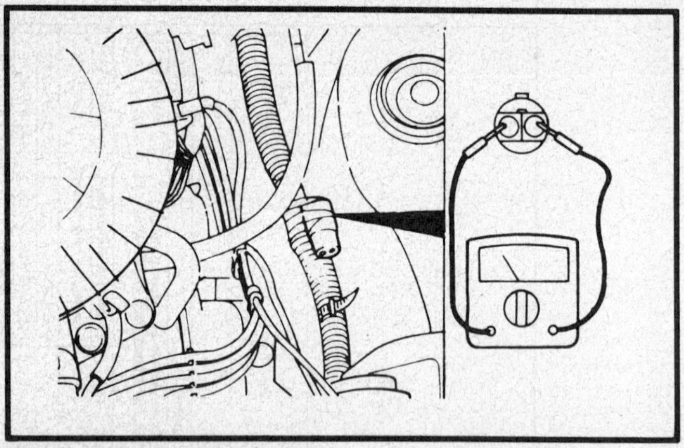

Testing the oxygen sensor

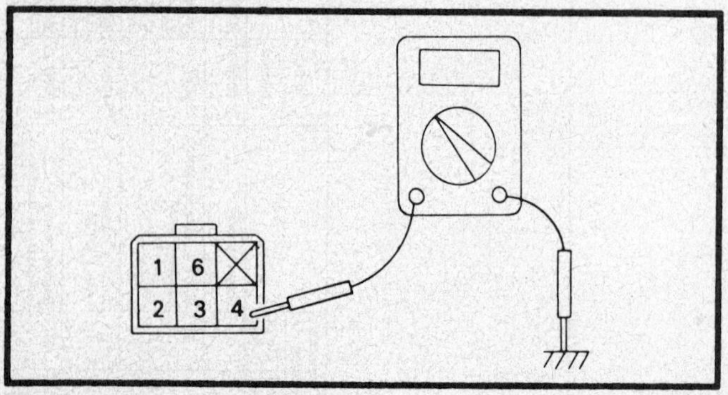

Choke heater inspection

sensor checker is 0–1 volt and 2–3 volts pulsating. after 70 seconds from the start of the engine.
4. If the voltage is not as specified, replace the oxygen sensor.

CHOKE HEATER

Inspection

1. Disconnect the electric choke heater connector and check the resistance of the heater.
2. The resistance should be approximately 6 ohms at 68°F (20°C).
3. If the resistance is a great deal out of range or the circuit is open, replace the electric choke body.

SLOW CUT SOLENOID VALVE (SCSV)

Testing

1. Disconnect the SCSV connector.
2. Using an ohmmeter, check that there is no continuity between the terminal and valve body.
3. Check that the resistance between the terminals is 48–60 ohms at 68°F (20°C).
4. Apply battery voltage directly to the SCSV terminals.
5. Check that the valve clicks.
6. If any test fails, replace the SCSV.

FEEDBACK SOLENOID VALVE (FBSV)

Testing

1. Disconnect the FBSV connector.
2. Using an ohmmeter, check that there is no continuity between the terminal and valve body.
3. Check that the resistance between the terminals is 54–66 ohms at 68°F (20°C).
4. Apply battery voltage directly to the FBSV terminals.
5. Check that the valve clicks.
6. If any test fails, replace the FBSV.

CHOKE BI-METAL ASSEMBLY

Testing

1. Disconnect the choke connector.
2. Using an ohmmeter, check the resistance of the coil.
3. The resistance should be approximately 6 ohms at 68°F (20°C).
4. If the resistance is a large amount out of range, replace the choke bi-metal assembly.

FUEL PUMP

Pressure test

1. Disconnect the hose at the carburetor and connect a fuel pressure gauge.
2. Disconnect the fuel return hose from the pump and plug the pipe.
3. Check that the pressure is 2.8–3.6 psi (19–25 kPa) with the engine idling.
4. If fuel pressure is not correct, replace the pump.

Volume test

1. Disconnect the hose at the carburetor and connect an extension hose to the fuel line.
2. Place the this hose into a measuring beaker.
3. Disconnect the fuel return hose from the pump and plug the pipe.
4. Start the engine and measure the amount of fuel pumped in less than 1 minute.

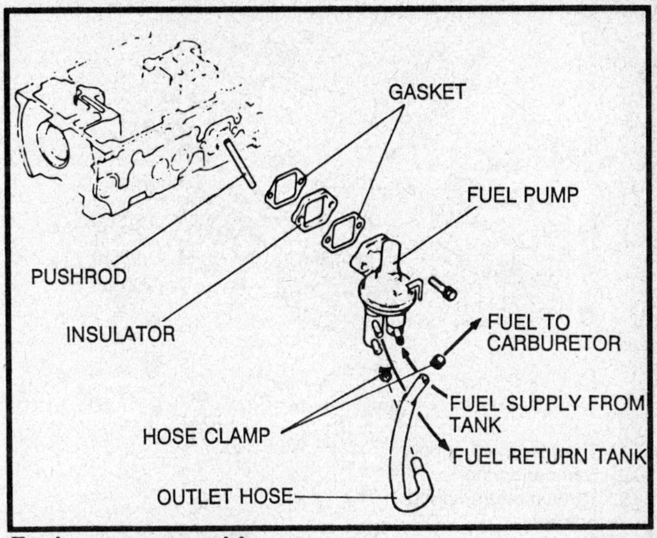

Fuel pump assembly

5. The volume should be 0.85 qts. (0.6L) per minute at 2500 rpm.
6. If the fuel volume is not correct, replace the pump.

Component Replacement

CARBURETOR

Removal and Installation

1. Disconnect the negative battery cable.
2. Remove the air cleaner assembly.
3. Disconnect the throttle cable from the carburetor.
4. Tag and disconnect the vacuum hoses from the carburetor.
5. Tag and disconnect the wiring connectors from the solenoid valves and TPS.
6. Place a container under the fuel fitting at the carburetor to collectexcess fuel and disconnect the fuel lines from the carburetor.
7. Remove the mounting bolts and carefully lift the carburetor from the engine. Take care to avoid spilling fuel from the fuel bowl.

To install:

8. Clean mating surfaces of all dirt and old gasket material.
9. Install a new gasket onto the intake manifold and carefully install the carburetor.
10. Tighten the mounting bolts alternately, a little at a time, to compress gasket evenly. Torque to 11–14 ft. lbs. (15–20 Nm).
11. Connect the throttle cable, vacuum hoses and fuel lines.
12. Check for proper choke and throttle operation.
13. Connect the negative battery cable.
14. Crank engine over several times to prime the carburetor.
15. Start engine and check for fuel leaks.
16. Make necessary mixture and idle adjustments.
17. Install the air cleaner assembly.

Overhaul

1. Remove the throttle return spring from the throttle lever and bracket.
2. Remove the dashpot rod from the free lever.
3. Remove bracket with the 2 connectors.
4. Pull the hose and delay valve off the choke breaker.
5. Remove the float chamber cover screw from the throttle body and the 4 float chamber screws from the main body. Note the size and location for reference during reassembly.
6. To remove the float chamber cover, gently pry between the enrichment cover and the cover.

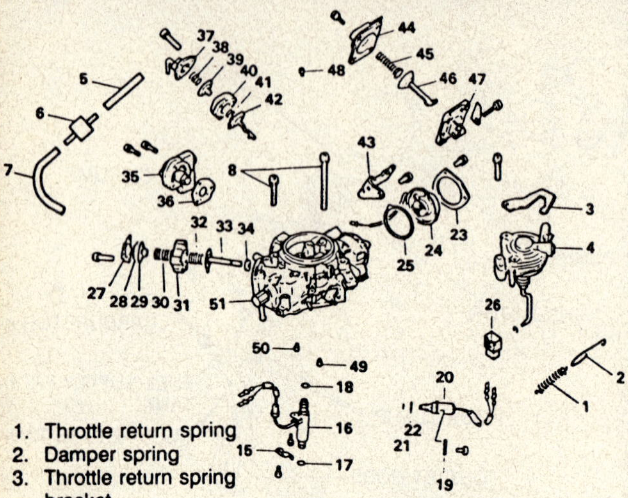

1. Throttle return spring
2. Damper spring
3. Throttle return spring bracket
4. Throttle opener dashpot
5. Hose
6. Vacuum delay valve
7. Hose
8. Screw
9. Pin
10. Float
11. Needle valve
12. Needle valve seat
13. O-Ring
14. Packing
15. Retainer
16. FBSV
17. O-Ring
18. O-Ring
19. Retainer
20. SCSV
21. O-Ring
22. O-Ring
23. Plate
24. Choke bi-metal assembly
25. Packing
26. Connector
27. Cover
28. Diaphragm
29. Spring seat
30. Spring
31. Housing
32. Spring
33. Diaphragm
34. Valve
35. MCV
36. Gasket
37. Cover
38. Spring
39. Diaphragm
40. Body
41. Spring
42. Diaphragm
43. Bracket
44. Cover
45. Spring
46. Diaphragm
47. Body
48. Primary main air jet
49. Primary pilot jet
50. Secondary pilot jet
51. Float chamber cover

Carburetor float chamber – exploded view

7. Remove the pin and then remove the float and needle valve. Do not let the float drop and take care not to damage the needle valve.

8. Gently pry the needle valve seat at both edges to remove. Use great care not to damage the float chamber cover when pushing up the needle valve seat.

9. Remove the retainer of the FBSV.

10. Using a thin blade tool, push the stopper portion of the FBSV connector and remove the 2 terminals from the connector.

NOTE: Unless the terminals are removed from the connector, the FBSV cannot be removed from the float chamber.

11. Remove the FBSV from the float chamber.

12. Remove the retainer and pull out the SCSV.

13. Using a thin blade tool, push the stopper portion of the SCSV connector and remove the 2 terminals from the connector.

14. To remove the choke assembly, grind away the rivet heads and remove the screw.

15. Remove the plate and the bi-metal choke assembly.

16. Using a thin blade tool, push the stopper portion of the choke connector and remove the terminal from the connector.

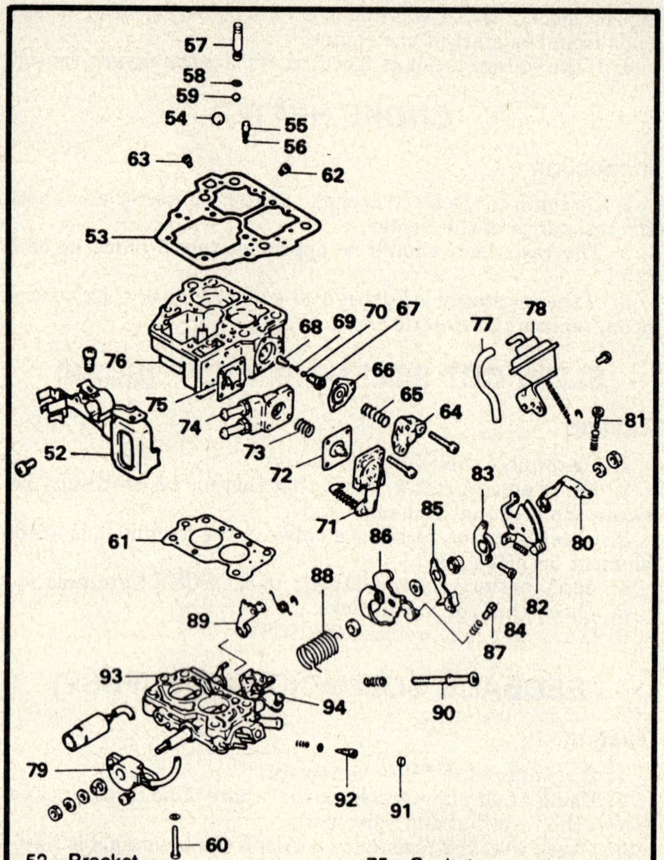

52. Bracket
53. Float chamber cover gasket
54. Steel ball
55. Weight
56. Ball
57. Plug
58. O-Ring
59. Ball
60. Screw
61. Gasket
62. Primary main jet
63. Secondary main jet
64. Cover
65. Spring
66. Diaphragm
67. Enrichment jet valve
68. Enrichment jet
69. Spring
70. Ball
71. Pump cover assembly
72. Diaphragm
73. Spring
74. Pump body
75. Gasket
76. Mixing body
77. Vacuum hose
78. Depression chamber
79. Throttle position sensor
80. Lever
81. Adjusting screw
82. Throttle lever
83. Cam follower
84. Fast idle adjusting screw
85. Free lever
86. Abutment plate
87. Idle speed adjusting screw (SAS-2)
88. Spring
89. Secondary lever
90. Idle speed adjusting screw (SAS-1)
91. Plug
92. Mixture adjusting screw (MAS)
93. Throttle body
94. Idle speed adjusting screw (SAS-3)

Carburetor main and throttle bodies – exploded view

17. Remove the 3 bowl vent valve cover screws.

18. Remove and disassemble the bowl vent valve, taking care not to misplace the springs.

19. Remove the Mixture Control Valve (MCV).

20. Remove the choke vacuum breaker cover, springs, diaphragm and linkage assembly.

21. Remove the choke bracket and choke opener.

22. Remove the pilot jet and main air jet. Use a screwdriver that is an exact fit for their slot and work carefully or damage will result.

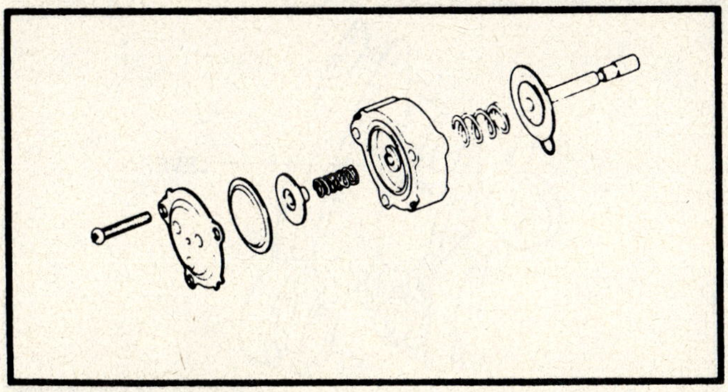

Bowl vent assembly

NOTE: Note the size symbol stamped on each jet for identification. Make certain to replace jets with size and at the correct locations.

23. Carefully lift and swing up the roll-over device weight from the bottom of the float chamber. Use a magnet to remove the roll-over steel ball.
24. Remove the accelerator pump outlet check ball and weight.
25. Separate the float chamber from the throttle body.
26. Remove the main jets. Use a screwdriver that is an exact fit for their slot and work carefully or damage will result.
27. Remove the enrichment valve cover, spring and diaphragm.
28. Loosen the enrichment jet and take out the spring and ball from the valve.
29. Remove the accelerator pump cover, diaphragm, spring and gaskets.
30. Remove the E-clip and outside washer on the secondary throttle shaft and slide the diaphragm spring link off the throttle shaft.
31. Remove the diaphragm bracket and remove the secondary throttle diaphragm.
32. Remove the nut attaching the accelerator pump lever to the throttle shaft. With an impact driver, remove the screws attaching the Throttle Position Sensor (TPS).
33. Remove the E-clip holding the TPS onto the throttle shaft.
34. For removal of plug, clamp carburetor in a vice and drill a $^5/_{64}$ in. (2mm) pilot hole in the casting surrounding the idle mixture adjusting screw (MAS), then redrill the hole to $^1/_8$ in. (3mm).
35. Insert a punch into the hole and drive out the plug.

To Reassemble:

36. Clean all parts before assembly.
37. Check for smooth operation of the throttle and choke linkage.
38. Install the TPS.
39. Install the secondary diaphragm and bracket.
40. Assemble and install the accelerator pump. Apply sealant to the screws and make certain the pump body, gasket and cover are correctly positioned.
41. Install the enrichment valve.

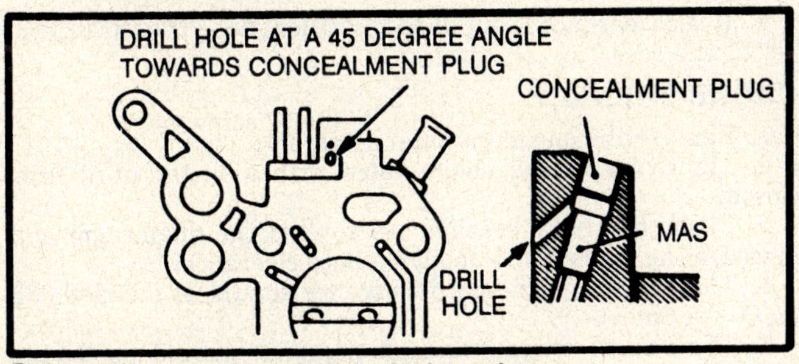

Concealment plug drill hole location

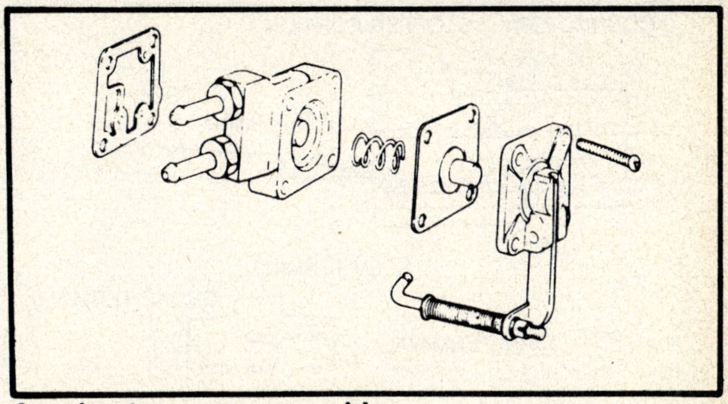

Accelerator pump assembly

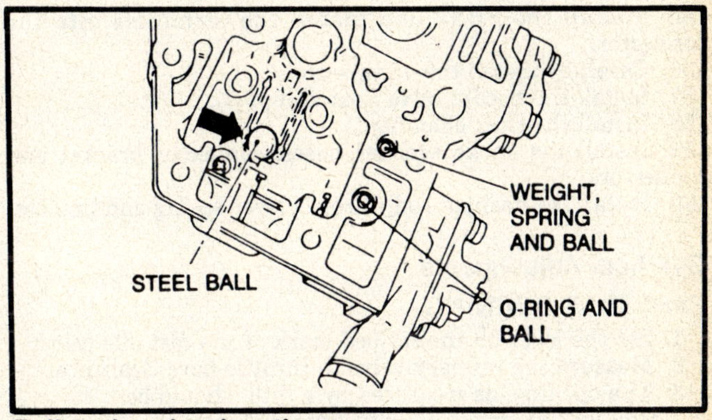

Ball and spring locations

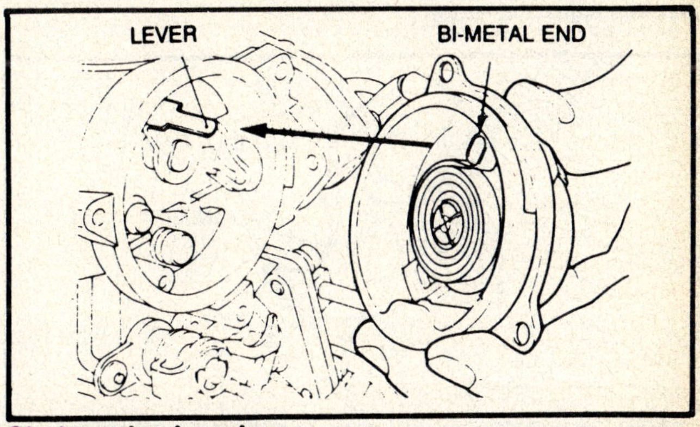

Choke valve housing

42. Install the main jets, making certain to install in the correct position.
43. Install the float chamber to the throttle body.
44. Install the accelerator pump outlet check ball and weight.
45. Install roll-over device.
46. Install the pilot jets and main air jet.
47. Install choke opener and bracket.
48. Assemble and install the choke vacuum breaker.
49. Install the MCV assembly.
50. Assemble and install the bowl vent assembly.
51. Fit the bi-metal end over the choke valve lever.
52. Install the plate and temporarily tighten the screw. Align the mating marks on the choke housing.
53. Install the rivets and the screw. Install the terminal into the correct location on the connector.
54. Install the SCSV and install the terminals into the connector.

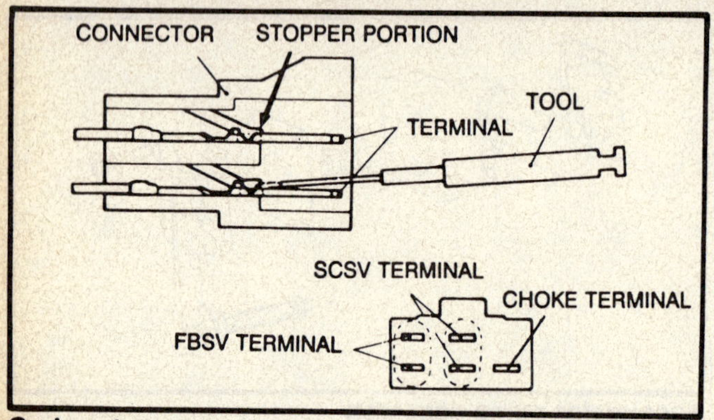

Carburetor connector terminal identification

55. Install the FBSV and install the terminals into the connector.
56. Carefully install the valve seat.
57. Install the needle valve float and pin.
58. Install the float chamber.
59. Install the choke breaker hoses, connector bracket and connectors.
60. Install the dashpot rod, throttle lever spring and bracket.

Overhaul Adjustments

FAST IDLE OPENING

1. Set the lever on the scribed mark of the fast idle cam.
2. Measure the primary valve to throttle bore clearance.
3. The opening, as measured by a drill, should be:
   Manual transmission – 0.037 in. (0.93mm).
   Automatic transmission – 0.040 in. (1.02mm).
4. If the clearance is out of specification, adjust turning the fast idle adjusting screw.

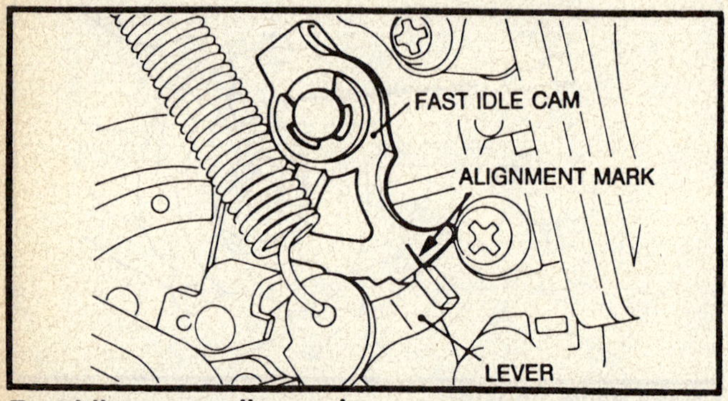

Fast idle cam scribe mark

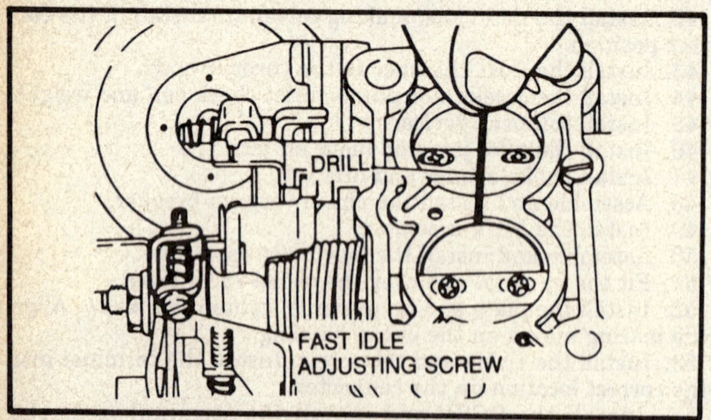

Measuring throttle bore clearance

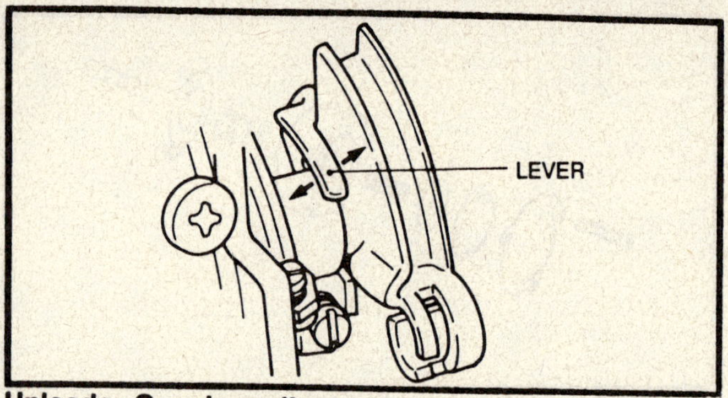

Unloader Opening adjustment

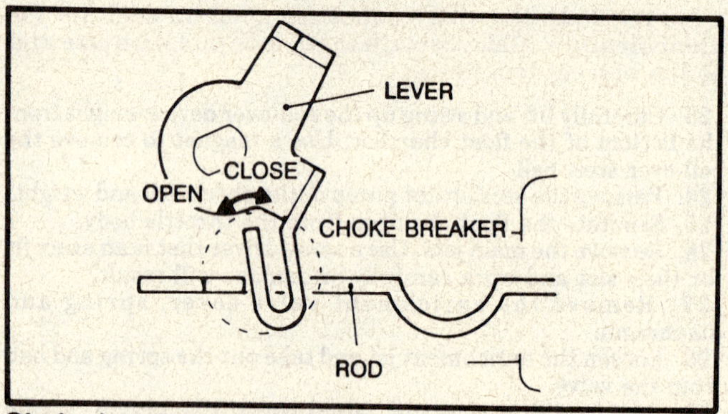

Choke breaker adjustment

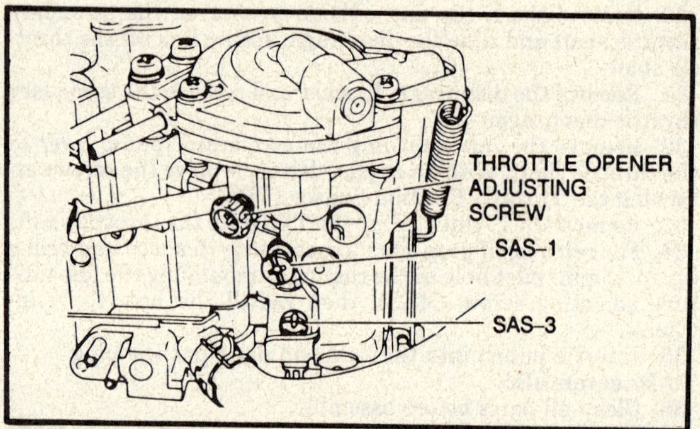

Throttle opener adjustment

UNLOADER OPENING

1. Lightly press the choke valve with a finger to fully close it.
2. Fully open the throttle valve and measure the choke valve to choke bore clearance.
3. The standard clearance should be 0.075–0.083 in. (1.9–2.1mm).
4. If the clearance is out of specification, bend the throttle lever at the adjustng tab.

CHOKE BREAKER

1. Remove the choke bi-metal assembly.
2. Lightly press the choke valve with a finger until fully closed.
3. Push the choke breaker rod toward the diaphragm and measure the choke valve to choke bore clearance.
4. The first stage standard clearance should be 0.055–0.063 in. (1.4–1.6mm).
5. The second stage standard clearance should be 0.114–0.122 in. (2.9–3.1mm).

6. If the clearance is out of specifications, adjust by bending the adjusting tab.

On-Vehicle Adjustments

IDLE SPEED

1. Bring engine to normal operating temperature.
2. Make sure electrical loads and accessories are **OFF**.
3. Place transaxle in **N** or **P**.
4. Connect a tachometer and timing light.
5. Make certain timing is 5 degrees BTDC.
6. Run engine for at least 5 seconds at 3000 rpm.
7. Allow engine to idle for 2 minutes.
8. Adjust idle speed by using the idle speed adjusting screw No. 1 (SAS–1) to obtain the following idle setting:
   1988 Excel – 650–850 rpm
   1989 Excel – 700–900 rpm
   1990 Excel – 650–750 rpm

NOTE: Do not touch the SAS–2 idle speed screw. This is preset at the factory.

IDLE MIXTURE

The idle mixture may not be adjustable on later vehicles. The mixture is set by the ECU and automatically adjusts the mixture according to the inputs from the various engine sensors.

1. Remove the carburetor from the engine and place the carburetor in a suitable fixture in order to remove the concealment plug.
2. Clamp the carburetor into a suitable vice with the idle mixture adjusting screw facing up (protect the gasket surface from the vise jaws).
3. Drill a $^5/_{64}$ in. (2mm) pilot hole in the casting surrounding the idle mixture adjusting screw, then redrill the hole ⅛ in. (3mm).
4. Insert a blunt punch into the hole and drive out the plug.
5. Reinstall the carburetor on the engine without the concealment plug.
6. With the vehicle in **P**, the drive wheels blocked and all the accessories **OFF**, connect a suitable tachometer to the engine and run the engine until it reaches normal operating temperature.
7. Turn off the engine and disconnect the negative battery cable for about 30 seconds, and reconnect the cable.
8. Disconnect the connector of the exhaust oxygen sensor, run the engine for at least 5 seconds at an engine speed of 2000–3000 rpm.
9. Allow the engine to idle for 2 minutes and set the idle CO and the engine speed to specifications (the idle CO: 0.1–0.3% at idle speed) by turning the idle speed adjusting screw and the mixture adjusting screw(s).
10. Reconnect the oxygen sensor connector, readjust the engine speed, if necessary. Shut **OFF** the engine and install the concealment plug into the hole to seal the idle mixture adjusting screw. Remove all test equipment.

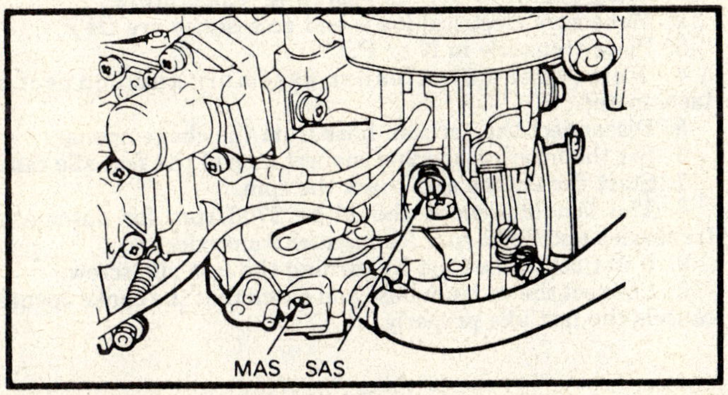

Idle Mixture and CO adjusting

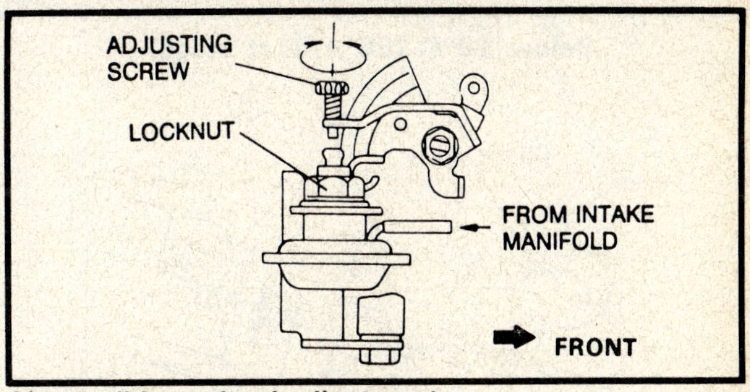

Air conditioner load adjustment

THROTTLE OPENER

1. Make certain the idle is correct.
2. By using the auxiliary lead wire, activate the idle-up control solenoid valve, apply intake manifold vacuum to the idle-up actuator and activate the idle-up actuator.
3. Open the throttle slightly to raise engine speed to 2000 rpm, then slowly close it.
4. Adjust the engine idle to the following rpm by turning the throttle opener idle screw:
   1988 Excel – 750–950 rpm
   1989 Excel – 750–950 rpm
   1990 Excel – 750–850 rpm
5. Remove the auxiliary lead wire and reconnect the idle-up solenoid valve wiring.

AIR CONDITIONING IDLE-UP

1. Allow engine to reach normal operating temperature.
2. Turn the air conditioner **ON**.

NOTE: The solenoid valve should open and intake manifold vacuum should act on the throttle opener to fully actuate it.

3. Check that the engine speed is:
   1988 Excel – 850–900 rpm
   1989 Excel – 850–900 rpm
   1990 Excel – 875–925 rpm
4. If the engine rpm is not as specified, adjust by turning the air conditioning idle adjusting screw.

THROTTLE POSITION SENSOR

1. Loosen the accelerator cable.
2. Loosen the speed adjusting screws SAS–1 and SAS–2 just enough to completely close the throttle valve completely. Record the exact number of turns loosened.

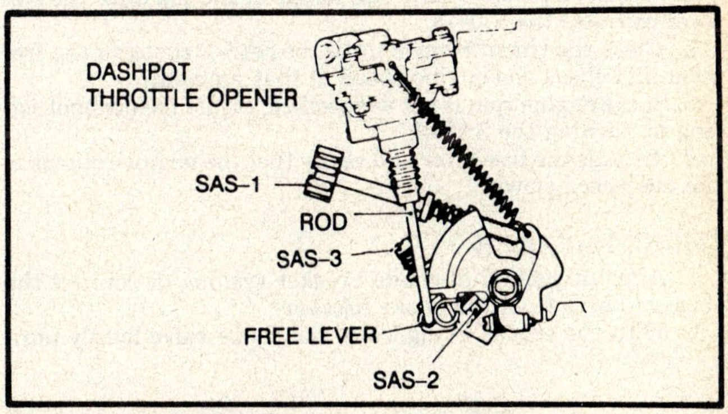

Dashpot and adjusting screw locations

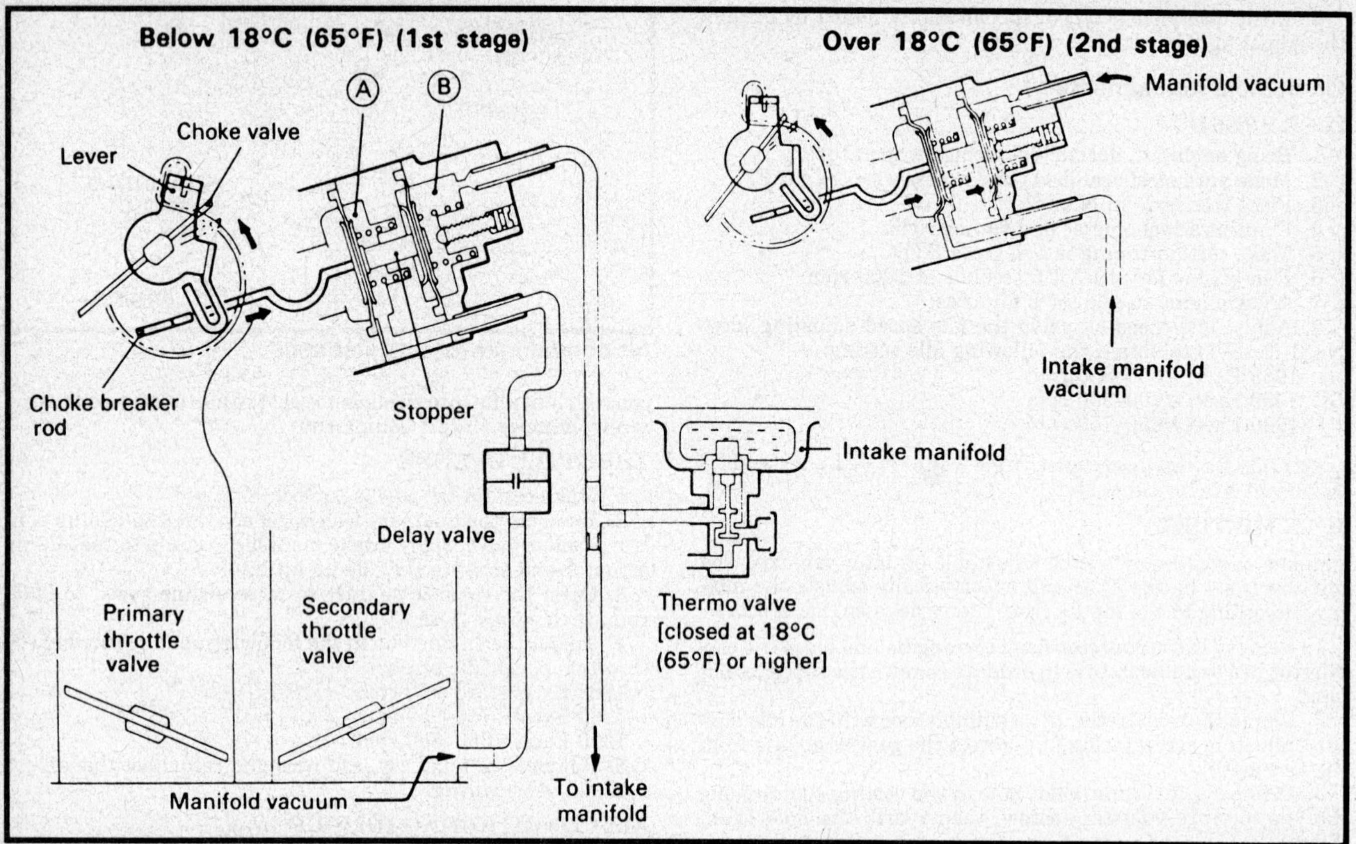

Choke breaker system inspection

3. Connect the a digital voltmeter between terminals 2 and 3 of the TPS. Do not disconnect the TPS.
4. Turn the ignition switch **ON**. The output voltage should be 0.25 volts.
5. If the voltage is not correct turn the TPS body as needed to adjust.
6. Turn the ignition switch **OFF**.
7. Tighten the SAS–1 and SAS–2 idle screw the same number of turns loosened.
8. Adjust the accelerator cable.
9. Check and adjust idle, if necessary.

DASHPOT

1. Bring engine to normal operating temperature.
2. Make sure electrical loads and accessories are **OFF**.
3. Place transaxle in **N** or **P**.
4. Open the throttle valve full stroke of the rod until the free lever contacts the SAS–3.
5. Close the throttle valve until the SAS–2 contacts the free lever and check the engine speed at that moment.
6. If the engine rpm is not as specified, adjust the dashpot setting by turning the SAS–3.
7. Release the free lever and verify that the engine returns to the idle speed slowly.

CHOKE BREAKER

1. After inspection of choke breaker system, disconnect the vacuum hoses from the choke breaker.
2. With the engine idling, close the choke valve lightly until the choke valve stops. Measure the distance between the choke valve to the choke bore.
3. The standard clearance should be 0.055–0.063 in. (1.4–1.6mm).
4. If the clearance is not correct, stop the engine, remove the choke bi-metal assembly and adjust the rod end opening to the standard clearance.
5. Reconnect the yellow striped vacuum hose and measure the choke valve to bore clearance.
6. The standard clearance should be 0.114–0.122 in. (2.0–3.1mm).
7. If not correct, adjust with the adjusting screw in the breaker diaphragm.

FAST IDLE

1. Bring engine to normal operating temperature.
2. Make sure electrical loads and accessories are **OFF**.
3. Place transaxle in **N** or **P**.
4. Remove the air cleaner assembly and connect a tachometer.
5. Disconnect the vacuum hose from the choke opener.
6. Set the lever on the 2nd highest step of the fast idle cam.
7. Start the engine and check the rpm.
8. The fast idle speed should be 2700 rpm for automatic transaxles and 2800 rpm for manual transaxles.
9. If not correct, adjust by turning the fast idle screw.
10. Connect the vacuum hose and check that the choke opener cancels the fast idle properly.

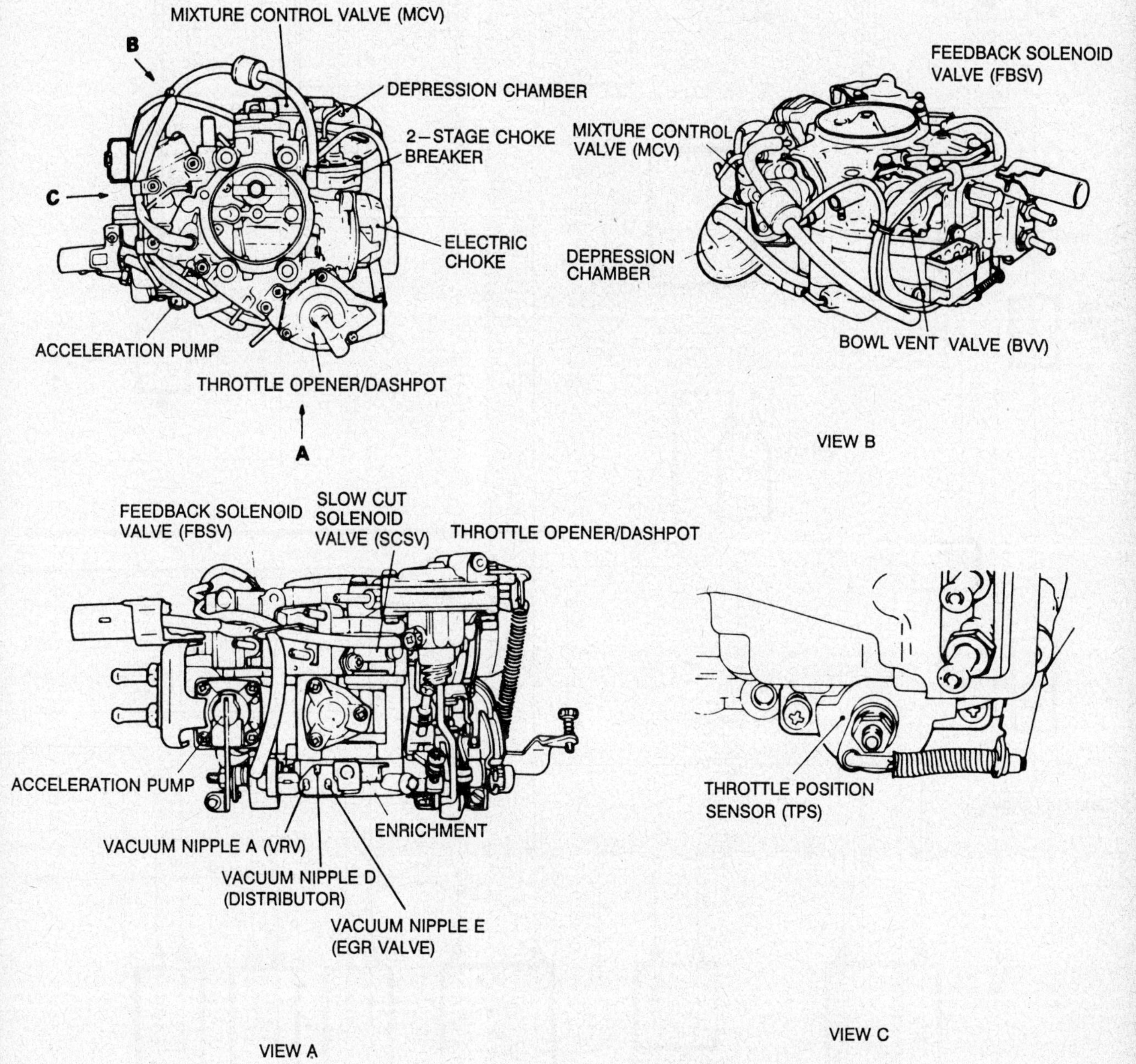

Carburetor components

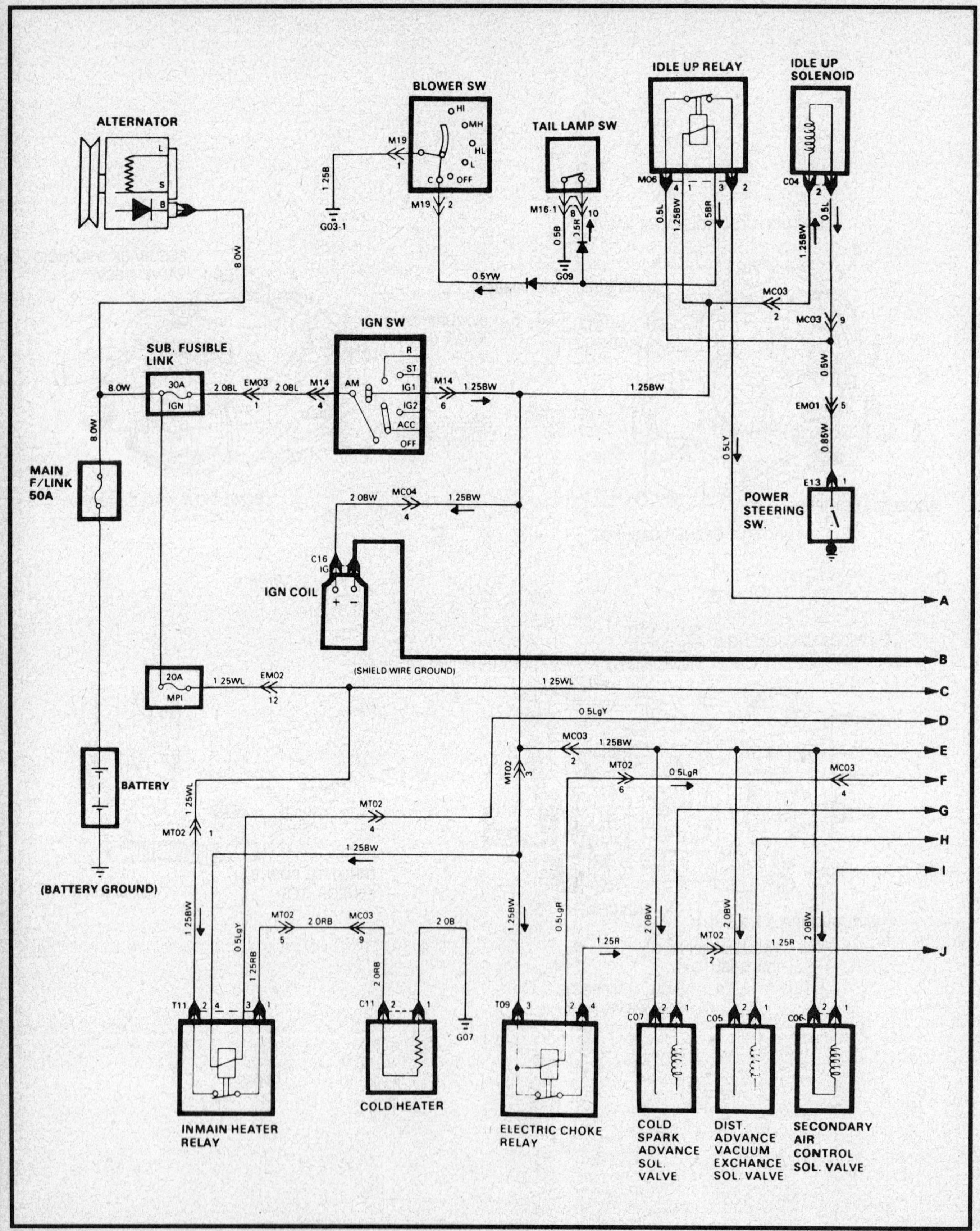

Feedback system circuit – 1990 Excel

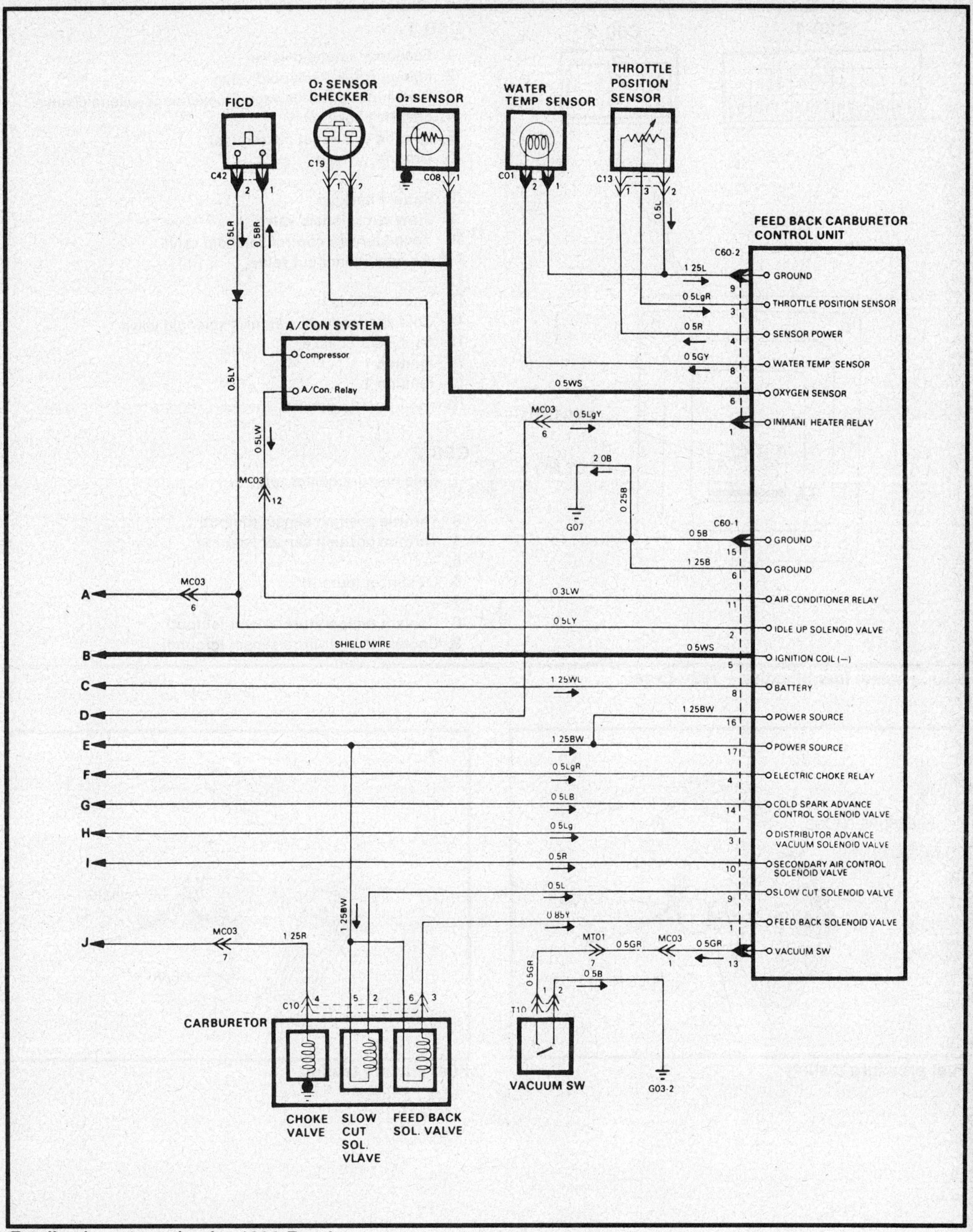

Feedback system circuit – 1990 Excel

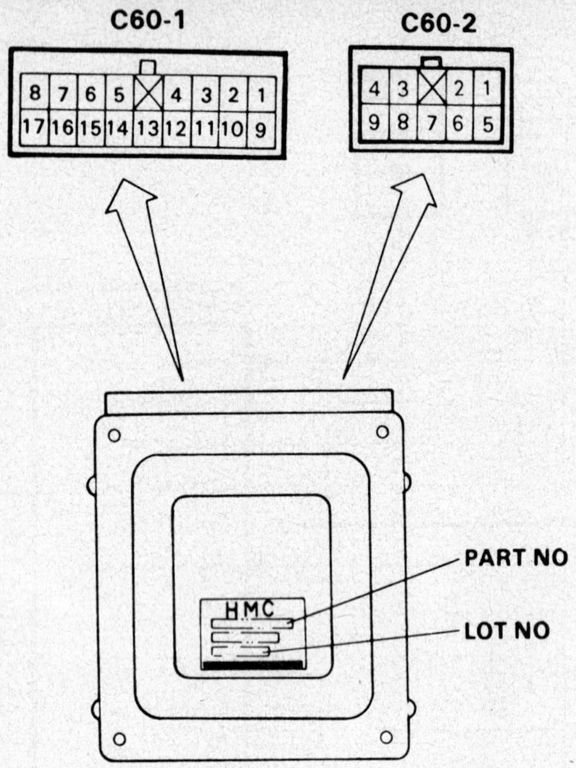

C60-1

1. Feedback solenoid valve
2. Idle up control solenoid valve
3. Distributor advance vacuum exchange solenoid valve
4. Electric choke relay
5. Ignition coil negative terminal
6. Ground
7. —
8. Battery back up
9. Slow cut solenoid valve
10. Secondary air control solenoid valve
11. Air conditioner cut relay
12. —
13. Vacuum switch
14. Cold spark advance control solenoid valve
15. O_2 sensor checker
16. Ignition 1
17. Ignition 1
18. —

C60-2

1. Cold mixture heater relay (—)
2. —
3. Throttle position sensor (output)
4. Throttle position sensor (source)
5. —
6. O_2 sensor (output)
7. —
8. Coolant temperature sensor (output)
9. Coolant temperature sensor (ground)

ECU terminal identification – 1990 Excel

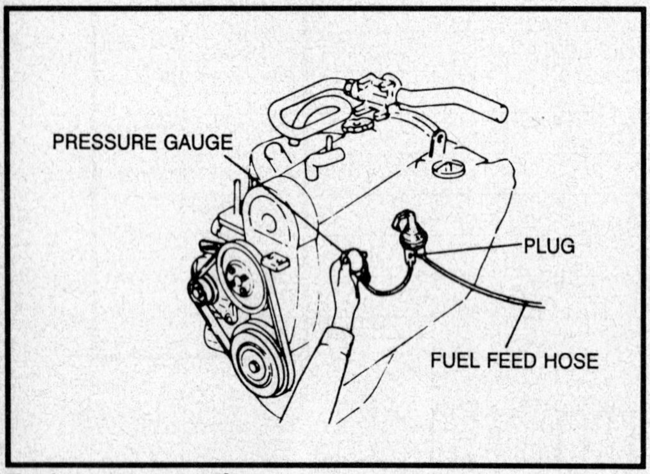

Fuel pressure testing

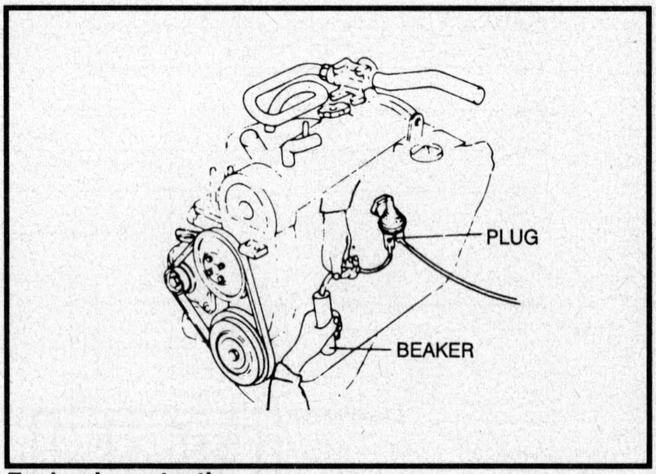

Fuel volume testing

ISUZU FEEDBACK CARBURETOR

General Information

The feedback carburetor system is used on the following models:

- Amigo with 2.3L (4ZD1) engine
- Pick-Up with 2.3L (4ZD1) engine

The FBC system includes the following sub-systems:

- Air Injection Reactor (AIR) System
- Closed Loop Emission Control System
- Coasting Fuel Cut System
- Early Fuel Evaporation System (EFE)
- Evaporative Emission Control Sytem (ESC)
- Exhaust Gas Recirculation System
- High Altitude Emission Control System
- Mixture Control Valve System
- Positive Crankcase Ventilation (PCV) System
- Thermostatically Controlled Air Cleaner System (TCA)
- 3-Way Catalytic Converter System

SYSTEM OPERATION

Air Injection Reactor System

In order to reduce HC and CO emissions, the system injects air into the exhaust ports to speed up oxidation. The air management valve switches air passage from the air pump through a vacuum management valve which is actuated by an electric signal supplied from the ECM.

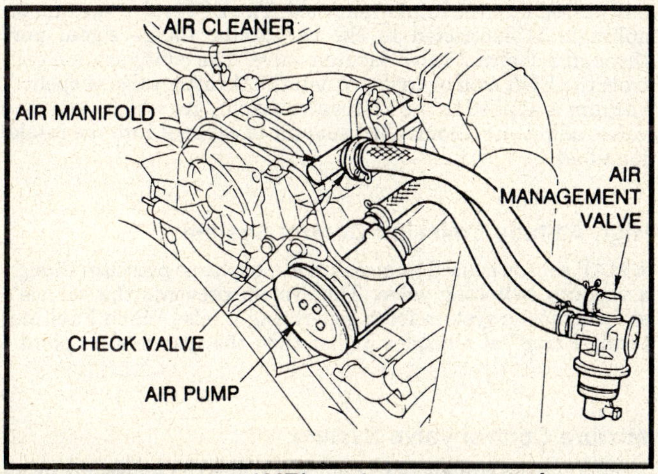

Air injection reactor (AIR) system component layout—Amigo

Closed Loop Emission Control System

The closed loop emission control system precisely controls the air/fuel ratio near the optimum mixture and allows the use of the 3-way catalyist to reduce the oxides of nitrogen and oxidize hydrocarbons and carbon monoxide. The essential components of the closed loop system are the coolant temperature sensor, oxygen sensor, electronic control module (ECM), feedback carburetor, 3-way catalytic converter, idle switch and MAP sensor, duty solenoid, fuel cut solenoid.

Coasting Fuel Cut System

While the vehicle is decelerating, this system cuts off the fuel flow in the slow circuit of the carburetor. This prevents overheating and afterburning in the exhaust system.

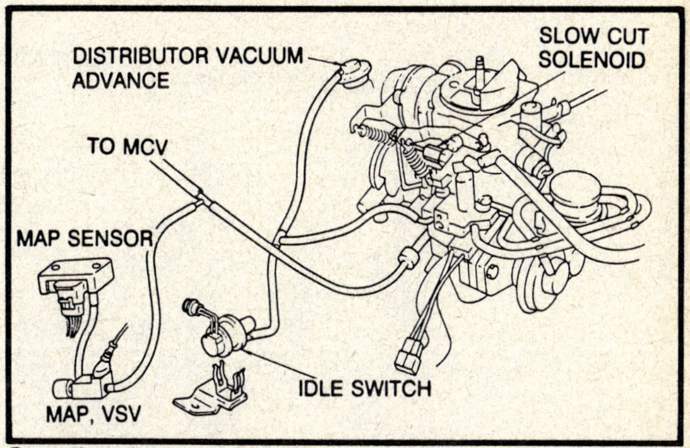

Coasting fuel cut system component layout—Amigo and Pick-Up

Early Fuel Evaporation (EFE) System

The early fuel evaporation system consists of the EFE heater, electronic control module, coolant temperature sensor and the EFE relay.

The EFE heater is equipped with a ceramic ring which is located below the primary carburetor bore. The ring is an interal part of the carburetor gasket.

The EFE heats the incoming air charge to improve atomization. This offers better cold start and drive-away performance.

The coolant temperature sensor is activated when the ignition switch is turned to the **ON** position and the engine is started. If the engine coolant temperature is below the specified value, the ECM will supply current from the battery to the coils in the EFE heater to heat the incoming fuel/air charge.

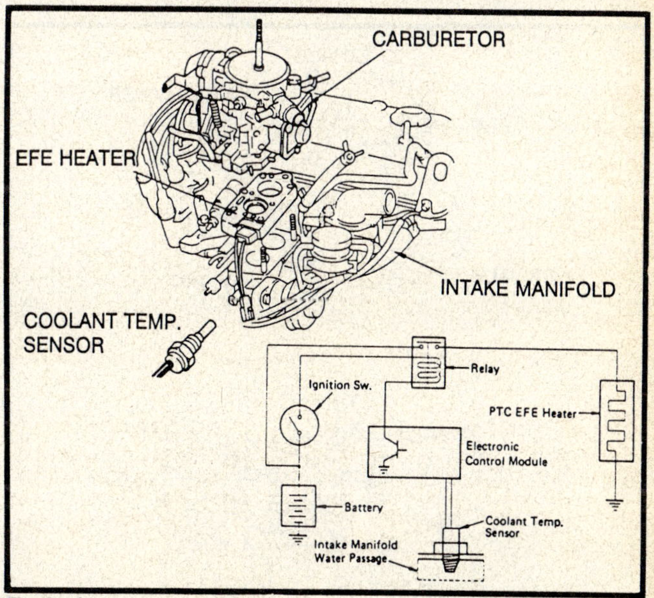

Early fuel evaporation system—Amigo and Pick-up

Evaporative Emission Control (ESC) System

The evaporative emission control system is designed to trap fuel vapors which would normally escape from the fuel tank and carburetor. This serves to limit the amount of fuel vapor released

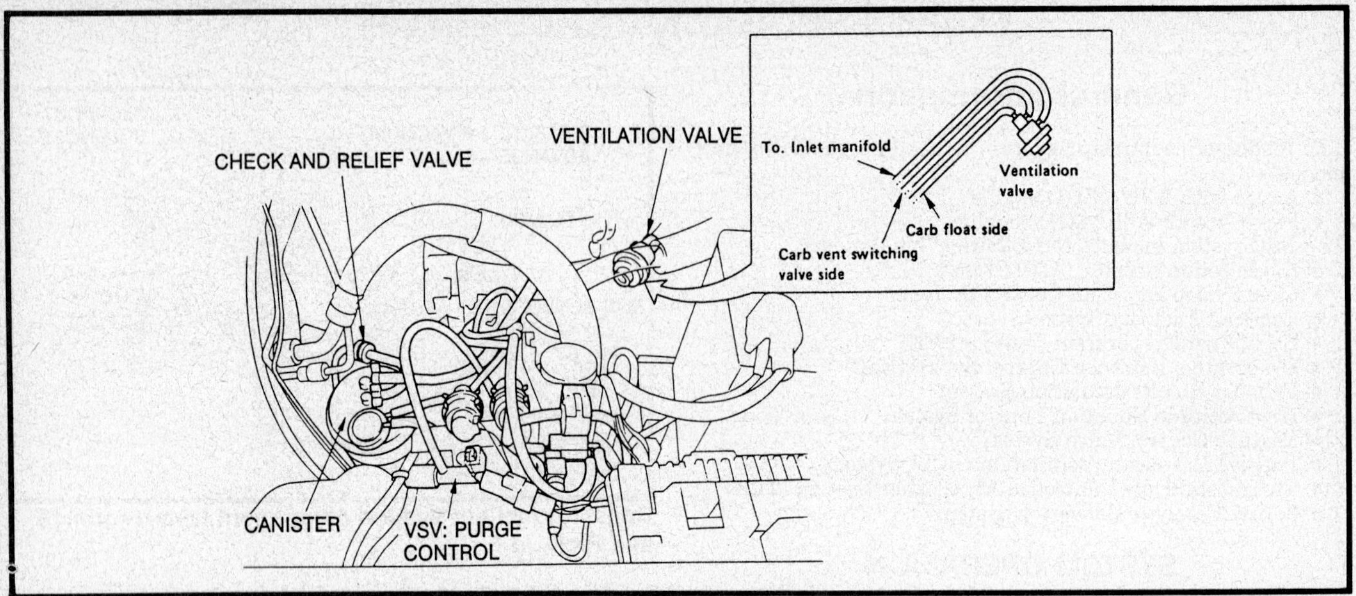

Evaporative emission control system (ECS)—Amigo and Pick-Up

to the atmosphere. The system uses a fuel vapor canister that absorbs fuel vapors and stores them until they are removed and re-burned in the engine.

Fuel vapor removal process from the carburetor uses a solenoid operated bowl vent. The non-vented fuel cap prevents the escape of vapors from the fuel tank filler. The domed fuel tank has vents with check valve assemblies located above the fuel level. The vapors are gathered and stored in the charcoal canister until the vacuum valve allows the vapors to be drawn off to the intake manifold to be burned in the engine.

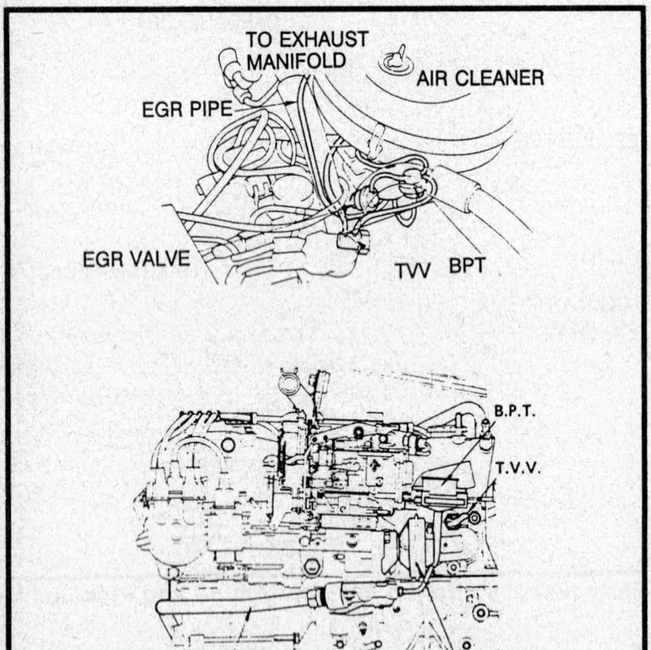

EGR system component layout– Amigo and Pick-Up

EGR System

The EGR system lowers temperatures in the combustion chamber in order to reduce nitrogen oxide emissions in the exhaust gas stream. Exhaust gases are drawn from the exhaust manifold and passed into the intake manifold. The EGR valve vacuum diaphragm is connected to the carburetor flange signal port through a thermal valve vacuum valve. The configuration controls the EGR cold override. When the throttle valve is opened, vacuum is applied to the EGR valve diaphragm which opens the valve, admitting a controlled amount of exhaust into the intake manifold.

High Altitude Emission Control System

A MAP sensor is used to measure atmospheric pressure though a vacuum switching valve. The system prevents the vehicle's self-diagnostic system from generating a false "Rich Fuel Metering Error" at altitudes higher than the system's set point.

Mixture Control Valve System

This system functions to reduce HC and CO emissions by allowing air to enter the intake manifold during vehicle deceleration.

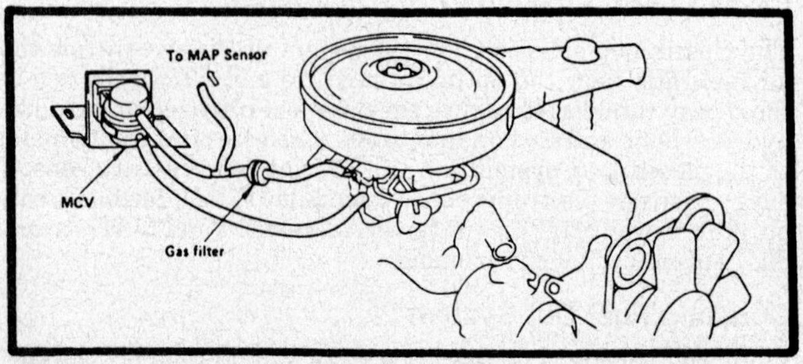

Mixture control valve system component layout—Amigo and Pick-up

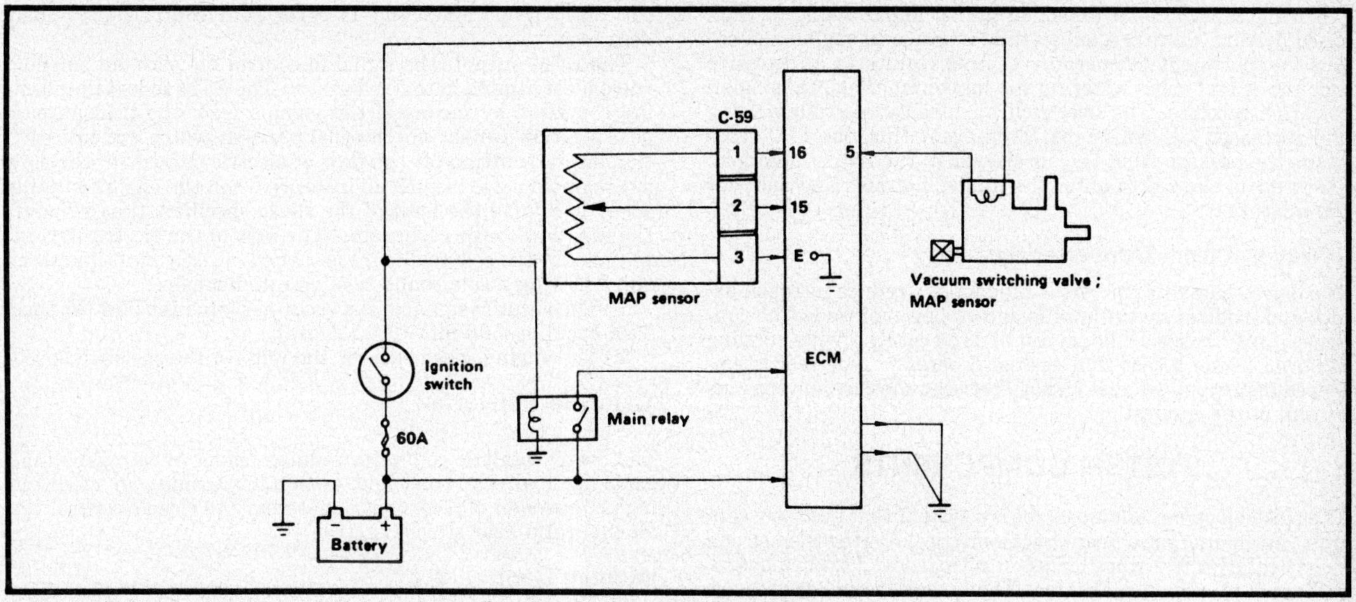

High altitude emission control system circuit—Amigo and Pick-Up

PCV System

The PCV system admits blow-by gases from the crankcase back into the intake manifold. These gases are re-burned in the combustion chamber with the fuel/air mixture.

The system is a closed type system. It includes the cylinder head cover baffle plate, PCV valve and attendant hoses and connections.

The baffle plate functions to separate the oil particles from the blow-by gas and the PCV valve controls the admission of the blow-by into the combustion chamber.

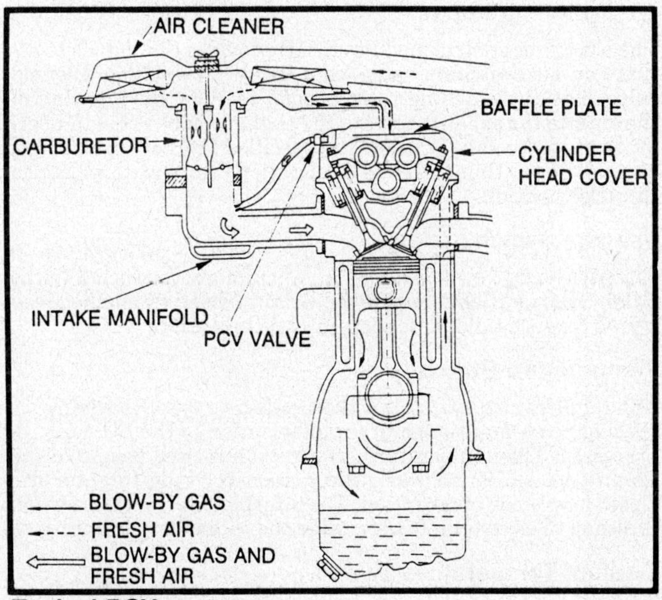

Typical PCV system

Thermostatically Controlled Air (TCA) Cleaner System

The TCA fuctions to maintain ambient air air temperature at an optimumn level so that the fuel/air ratio remains constant. This ensures fuel combustion and reduces pollutant emissions.

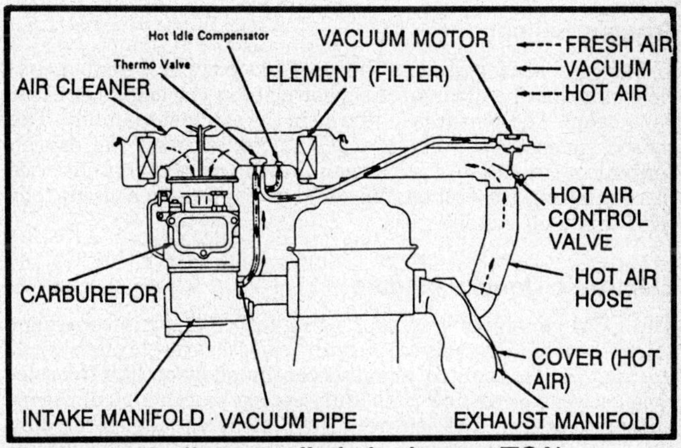

Thermostatically controlled air cleaner (TCA) system—Amigo and Pick-Up

The TCA system is mounted on the air cleaner. It consists of a vacuum motor, hot air control damper and an inlet temperature compensator (ITC) valve.

When the engine is running, there is no vacuum signal at either the vacuum motor or the ITC valve. In this condition the vacuum motor spring closes off the passage from the hot air duct.

On a cold start, the ITC valve delivers maximum vacuum to the vacuum motor which moves the hot air control damper to the fully open position. This closes the ambient air passage and opens the hot air duct. If the engine speed increases, the system vacuum level will drop allowing the diaphragm spring to overcome the vacuum force and push the hot air control damper to the fully closed position.

When the engine is running under normal conditions, the ITC valve closes the passage to the intake manifold and opens the passage to the from the air cleaner to the vacuum motor. As fresh air is fed to the vacuum motor, the diaphragm spring forces the air control valve to close off the hot air duct and open the ambient air passage.

During conditions of extended idling, hill climbing or high speed driving, there is a substantial increase in engine and engine compartment temperatures. This results in an excesive amount of fuel vapor entering the intake manifold, causing an over-rich mixture. The over-rich mixture causes rough idling and increased CO emissions. To prevent this, the ITC valve opens the passage from the air cleaner to the intake manifold. Fresh air is allowed to enter the intake manifold and lean out the mixture.

3–Way Catalytic Converter System

The 3–way catalytic converter functions to reduce nitrogen oxides and oxidizes hydrocarbons and carbon monoxide.The converter unit houses a "honeycomb" type catalyst with bonding material inside a cast iron casing. A ceramic type seal is employed to prevent exhaust leakage between the case and the outer wall of the catalyst.

SYSTEM COMPONENTS

The closed loop emission control is a system that precisely controls the air-fuel ratio near stoichiometric, allowing the use of a 3–way catalyst to reduce oxides of nitrogen and oxidize hydrocarbons and carbon monoxide. The essential components are the exhaust gas oxygen sensor, an electronic controller, a vacuum controller with a duty solenoid, a controlled air/fuel ratio carburetor and a 3–way catalytic converter.

Oxygen Sensor

The oxygen sensor used in the closed loop control system consists of a closed end zirconia sensor plate in the engine exhaust gas stream. The sensor is mounted in the exhaust manifold. The sensor generates a voltage which fluctuates with the oxygen content in the exhaust gas stream. As the oxygen content rises (lean mixture) the voltage falls, and as the oxygen content falls (rich mixture) voltage rises.

Electronic Control Module

The ECM generates a control signal to the mixture controller solenoid which controls carburetor air-fuel ratio through vacuum signals. This control signal is continuously cycling the solenoid between on and off time (duty cycle) as a function of the input voltages from the sensors. The control signal generated by the ECM is selected from 4 operational modes as follows:

1. Inhibit Mode—fixed signal to the vacuum controller solenoid.
2. Enrichment Mode—a pre-programmed duty cycle to the mixture controller solenoid.
3. Open Loop Mode—a fixed pre-programmed duty cycle to the mixture controller.
4. Closed Loop Mode—a calculated duty cycle is generated based on oxygen sensor and other sensor outputs.

During closed loop operation, the ECM monitors the voltage output of the oxygen sensor. As the sensor voltage increases and passes through the ECM threshold set point, the proportional gain immediately changes the duty cycle of the output signal. The duty cycle is further changed at a constant rate (integral gain) until the sensor input voltage decreases and passes through the ECM threshold point.

The selection of integral and proportional gain rates by the ECM is based on the engine operating conditions (idle or off-idle). condtion). At idle, different gain rates are required for optimum air/fuel ratio control than those required under partial load conditions. The ECM also stores in memory the current duty cycle being used for either idle or off–idle condition. When the ECM sees a transition from idle condition (as signaled by the vacuum switch) to off–idle, it immediately steps to the duty cycle last recorded for stoichiometric operation. From then on, (while at that engine operating condition) the system uses the basic propotional and integral gain controls previously described.

The ECM outputs the signal to control the slow cut solenoid valve incorporated in the carburetor. The ECM senses the coasting condition by means of the signals from the transmission gear position switch, clutch pedal position switch and idle position switch, cutting off the flow of electric current to the slow cut solenoid valve in such an instance when the engine running speed is beyond the limit of the speed specified, thus stopping the fuel flow to the carburetor. The flow of the electric current to the slow cut solenoid valve is turned off only at a time when the following 2 conditions exist simultaneously:

1. The vacuum signal of the vacuum switch is below the specified vacuum (308 mm Hg absolute).
2. The engine speed exceeds the limit of the specified speed.

Catalytic Converter

The 3-way catalytic converter reduces oxides of nitrogen while oxidizing hydro-carbons and carbon monoxide. To maintain high conversion effiecency it is necessary to closely control the air-fuel ratio near stoichiometric.

Vacuum Controller

The vacuum controller function is to convert the electrical signals transmitted from the electronic control module into the vacuum signals by means of the work of the vacuum control solenoid. The vacuum signals thus obtained make the fuel control actuators in the carburetor start functioning, subsequently controlling the air-fuel ratio.

This vacuum controller consists of a vacuum regulator and a vacuum control solenoid. Of these 2, the former functions to regulate the fluctuating vacuum from the intake manifold into the constant vacuum specified. The latter functions to open or close the passage from the vacuum regulator by means of the signals coming from the electronic control module. As a result, this generates a high or low level of vacuum, corresponding to the duration of the opening or closing of the passage.

Air Switching Valve

The electronic control module (ECM) provides the signals to actuate an air switching valve via a vacuum switching solenoid valve. The air switching valve is open to allow the injection of the air into the exhaust ports of the engine when (1) the temperature of the coolant is below the specified level or (2) during a given period or time after the acceleration comes up to wideopen throttle condtion.

Vacuum Switch

Electric vacuum switches sensing vacuum are connected to the ECM. These switches enable the ECM to discriminate between closed throttle (idle) and open throttle operation.

Temperature Switch

The temperature switch is a thermo-ferrite type, providing the engine coolant temperature information to the ECM microprocessor. This temperature information is used as one of the criteria to determine when the system is ready to enter the closed loop mode of operation. The information from this switch is also used as criteria to determine the secondary air injection.

Coolant Temperature Switch

This temperature sensor is mounted in the engine coolant stream. It has a high resistance (approx. 100000 ohms) when the coolant is cold and a low resistance (under 1000 ohms) when the coolant is warm. This sensor sends information on the engine temperature to the ECM which is used to vary the air/fuel ratio as the engine coolant temperature varies with the time during a cold start and to accomplish various switching functions at different temperatures on the AIR management system.

Duty Cycle in Carburetor

The duty solenoid is located in the carburetor and controls the air/fuel ratio. This solenoid is controlled by the ECM.

Barometric Switch

The barometric switch is connected to the ECM and intended to make the System Malfunction indicator lamp inoperative in the case of the "Rich Oxygen Sensor Error" alone at the prescribed altitude or higher. There is no interaction between the barometric switch and the emission control system.

Map Sensor

The MAP sensor measures pressure changes in the intake manifold with reference to zero or absolute vacuum.

The MAP sensor and the idle switch enable the ECM to discriminate between the following engine running conditions:

(1) Idling condition: Idle switch is **ON** and the MAP sensor signal is less than the specified value of 340mm Hg.

(2) Partial throttle: Idle switch is in the **OFF** position regardless of the MAP sensor input signal.

(3) Wide open throttle (WOT): Idle switch is **ON** and the MAP sensor signal is more than the specified value of 340mm Hg.

MAP VSV Control

The ECM energizes the MAP VSV for approximately 1 second under the following 3conditions to switch the MAP sensor input signal to atmospheric pressure.

(1) When the ECM is reset.

(2) When the following 3 conditions occur at the same time: engine speed is less than 200 rpm, battery voltage is more than 9.5 volts and vehicle speed is less than 1 km/h (0.8mph).

(3) When the ECM senses a high altitude condition, it energizes the MAP VSV to reach atmospheric pressure and sets the "BAROMETRIC MAP" less than 596mm Hg.

Fuel Cut Device

The fuel cutoff system consists principally of a slow cut solenoid valve that is incorporated in the carburetor, an engine speed sensor, a transmission switch, an accelerator switch, and clutch switch. This system is an auxiliary fuel system operated by the slow cut solenoid. When de-energized, the solenoid puts a plunger out to make the slow passage closed.

System Malfunction Lamp

This system utilizes a dashboard mounted malfunction indicator lamp which, for some failure modes will inform the driver of the need for unscheduled maintenance. In the event of a system malfunction the legend "CHECK ENGINE" will light and remain on as long as the fault function occurs and the engine is running. The electronic control unit incorporates a diagnostic program which will assist in diagnosing the closed loop control system faults. When activated, the diagnostic program will flash a code through the malfunction lamp which isolates the source of the system fault.

Diagnosis and Testing

NOTE: For Self-Diagnostic System and Accessing Trouble Code Memory, see Section 3 "SELF-DIAGNOSTIC SYSTEMS".

SELF-DIAGNOSTIC SYSTEMS

The diagnostic circuit check makes sure that the self-diagnostic system works, determines that the trouble codes will display and guides diagnosis to other problem areas. When the engine is running and a problem develops in the system which the self diagnosis can evaluate, the "CHECK ENGINE" light will come on and a trouble code will be stored in the ECM trouble code memory. The light will remain on with the engine running as long as there is a problem. If the problem is intermittent the "CHECK ENGINE" light will go out, but the trouble codes will be stored in the ECM trouble code memory.

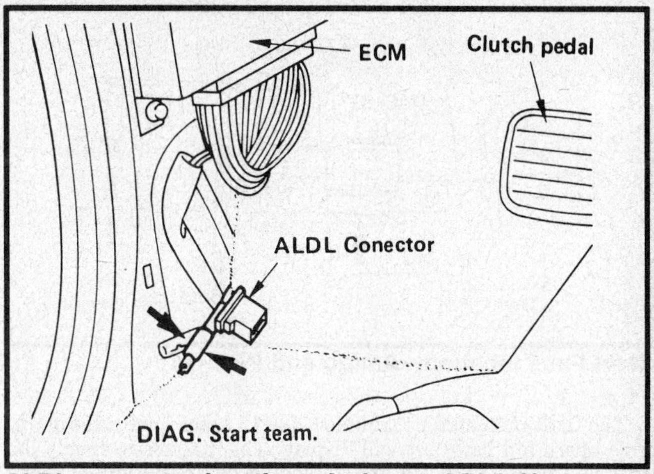

ALDL connector location—Amigo and Pick-Up

With the ignition turned on and the engine stopped, the "CHECK ENGINE" lamp should be on. This is a bulb check to indicate the light is working properly.

The trouble code "TEST" lead (white cable) and a ground lead (black cable) are branched from a harness at a distance of 8 inches from the ECM connector next to the clutch pedal. The trouble code memory is activated by placing the ignition switch in the **ON** position and engine running, connect the trouble code "TEST" lead to the ground lead.

The "CHECK ENGINE" light will begin to flash trouble Code 12. Code 12 consists of 1 flash , a short pause and then 2 more flashes. There will be a longer pause and a code 12 will repeat 2 more times. The check indicates that the self-diagnostic system is working. This cycle will repeat itself until the engine is started or the ignition switch is turned **OFF**. If more than 1 fault code is stored in the memory, the lowest number code will flash 3 imes followed by the next highest code number until all the codes have been flashed. The faults will then repeat in the same order. In most cases, codes will be checked with the engine running since no codes other than Code 12 or 51 will be present on the intial "KEY ON". Remove the ground from the test terminal before starting the engine.

A trouble code indicates a problem in a given circuit. For example, trouble Code 14 indicates a problem in the coolant sensor circuit. This includes the coolant sensor, connector harness and ECM. The procedure for finding the problem can be found by using the proper troubleshooting charts. In this case you would proceed to "TROUBLE CODE 14—SHORTED COOLANT SENSOR CIRCUIT" in the diagnostic charts provided.

When the engine is started, the "CHECK ENGINE" light will go off. If the "CHECK ENGINE" light remains on, the self-diagnostic system has detected a fault.

If a trouble code can be obtained when the "CHECK ENGINE" light is off with the engine running, the trouble code must be evaluated. A determination must be made to see if the fault is intermittent or if the engine must be at certain operating conditions to turn the "CHECK ENGINE" light ON. Faults indicated by trouble Codes 13, 31, 44 and 45 require engine operation at part throttle for up to 5 minutes after the engine has reached normal operating temperature before the "CHECK ENGINE" light will come on and store a trouble code.

Main Fuse location—Amigo and Pick-Up

The fault indicated by trouble Code 15 takes 5 minutes of engine operation before it will display. The diagnostic charts for trouble Codes 13, 31, 44 and 45 should be used if any of these trouble codes can be obtained. Be sure to remove the ground from the test terminal before starting the engine.

CLEARING THE TROUBLE CODE MEMORY

The trouble code memory is fed a continuous 12 volts even with the key in the **OFF** position. After a fault has been corrected, it will be necessary to remove the voltage for 10 seconds to clear any stored codes. The quickest way to remove the voltage is to remove the "ECM" fuse from the fuse block for 10 seconds. The voltage can also be removed by disconnecting the negative battery cable, but if the vehicle is equipped with electronic instrumentation, such as a clock and radio, they would have to be reset.

DIAGNOSTIC SYSTEM CHECK

The Diagnostic System Check is used to check the operation of the self-diagnosis system. It determines the trouble codes to display and is used as a guide to systems diagnosis.

SYSTEM PERFORMANCE CHECK

The System Performance Check is used to check the operation of the mixture control (M/C) solenoid and the main metering circuit using a tachometer, and to check the operation of the Closed Loop Emission Control System by the use of a dwell meter.

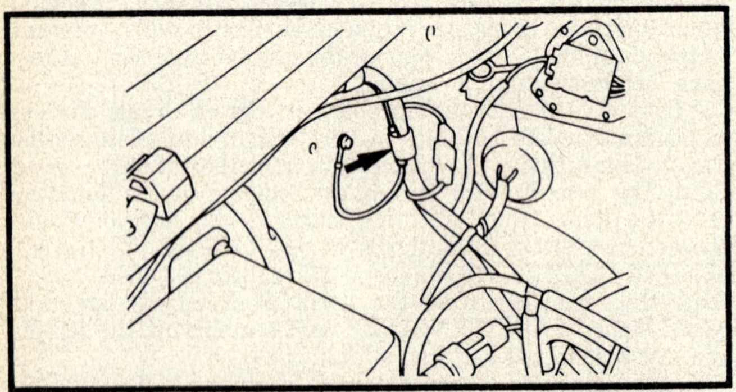

Dwell lead connector location—Amigo and Pick-Up

COOLANT TEMPERATURE SENSOR

Test

1. Drain the coolant from the radiator into a suitable container.
2. Remove the coolant temperature sensor from the intake manifold.
3. Cool the coolant temperature sensor to approximately 68°F.
4. Using a suitable ohmmeter, check to see that the resistance of the switch is approximately 2.1–2.9 kilo-ohms with the sensor at 68°F.
5. If the sensor resistance is not within specifications, replace it with as new one.
6. If the sensor resistance is within specifications, reinstall the sensor in the intake manifold, fill the radiator and start the engine so as to check for coolant leaks.

VACUUM IDLE SWITCH

Test

1. Disconnect and label the electrical connectors and vacuum hoses.
2. Using an ohmmeter, make sure there is no continuity between the positive (+) and negative (−) terminals of the switch connector.
3. Apply 79–88mm Hg of vacuum to the idle switch vacuum signal port and check for continuity across the **blue** and **black** terminals of the switch connector with the ohmmeter.
4. If there is no continuity across the switch, replace it.

MAP SENSOR

Test

1. Warm up the engine and allow to run at idle speed.
2. Disconnect vacuum hose **A** at the MAP sensor.
3. Plug the ends of the VSV and MAP sensor vacuum hoses.
4. Connnect the probes of a voltmeter across terminals **1** and **2** located at the rear face of the MAP sensor connector. The reading should be more than 2.5 volts.
5. Connect vacuum hose **A** to its original position.
6. Measure the voltage across terminals **1** and **2**. The voltage reading should be less than 2.5 volts.
7. If the voltage readings are not as specified, replace the MAP sensor.

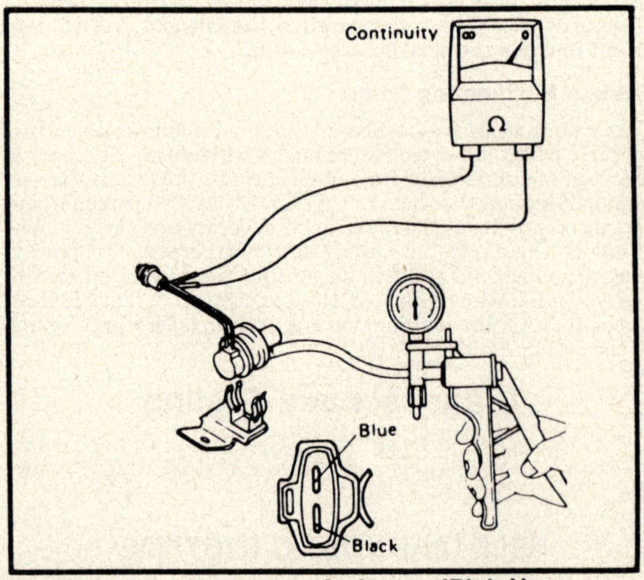

Vacuum idle switch test—Amigo andPick-Up

2.3L ENGINE – CARBURETOR TROUBLESHOOTING – PICK-UP AND AMIGO

Possible Cause of Trouble	Corrective Measure
A) Over-Flowing	
1) Poorly seated needle valve	Replace parts
2) Foreign matter on needle valve	Clean
3) Excessive clearance between float pin and hanger hole	Correct or replace
4) Fuel pump pressure excessive	Replace fuel pump
5) Incorrectly adjusted level of fuel in float chamber	Adjust
B) Fuel not Reaching Float Chamber	
1) Clogged strainer	Clean or replace
2) Restriction in fuel line other than carburetor	Check and correct
3) Defective pump system	Check and replace
C) Hard-Starting	
1) See (A)	
2) See (B)	
3) Incorrectly adjusted throttle valve opening when the choke valve is closed	Adjust
4) Excessive clearance between shaft hole and throttle valve shaft	Replace
5) Throttle valve in secondary carburetor fails to close completely	Clean and correct
D) Rought Idling	
1) See (A)	
2) See (B)	
3) See (C)-5	
4) Charburetor loose in mount	Retighten
5) Clogged slow jet	Clean
6) Clogged slow air-blead	Clean
7) Damage tip on end of idle adjust screw	Replace
8) Defective fuel cut solenoid valve	Replace
E) Fuel Consumption Excessive	
1) See (A)	
2) Clogged emulsion tube in carburetor	Clean
3) Diameter of main jets increased from wear	Replace
4) Choke valve fails to open full	Adjust
5) Incorrectly seated outlet valve	Recondition by grinding
F) Engine Lacks Power	
1) See (A)	
2) See (E)-4	
3) Cloged main jets	Clean
4) Throttle valve sticking or binding	Adjust
5) Defective fuel pump	Replace
6) See (A)-1	
7) Defective secondary valve	Clean and adjust
8) Clogged air cleaner element	Clean or replace
9) Clogged strainer	Clean or replace
10) Clogged vacuum hole	Clean

Possible Cause of Trouble	Corrective Measure
G) Flat Spot on Acceleration	
1) Secondary valve held open	Clean and adjust
2) Accelerator-pump out of specifications (delay in supply of additional fuel, amount of additional fuel is insufficient, period during which additional fuel is supplied is too short)	Replace
3) Excessive play in accelerator-pump linkage	Adjust or replace
4) Inlet and outlet valve out of specifications	Clean or replace
5) Incorrectly adjust engine idle	Adjust
6) See (D)-6	
7) See (F)-3	
8) Carburetor vent switching valve defective	Replace
H) Flat Spot on High-Speed Operation	
1) See (F)-3	
2) Clogged power Valve	Clean
3) Defective secondary valve (Leakage in diaphragm, sagged diaphragm spring, clogged vacuum port)	Replace

2.3L ENGINE – CARBURETOR SPECIFICATIONS PICK-UP AND AMIGO

Item		4ZD1
Carburetor model No.		DFP384-205
Carburetor manufacturer		Hitachi
Carburetor type		Stromberg 2-barrel, 2-stage
Fuel pressure	kg/cm^2(psi/kPa)	0.25 (3.56/24.50)
Jet orifice diameter		
Main jet primary		#89
Main jet secondary		#165
Slow jet primary		#52
Slow jet secondary		#100
Main air bleed primary		#100
Main air bleed secondary		#60
Slow air bleed primary		#150
Slow air bleed secondary		#0
Idle port diameter	mm(in)	1.7 (0.067)
Accelerator pump injector diameter	mm(in)	0.5 (0.020)
Accelerator pump delivery volume	$cm^3(in^3)$	0.8 (0.049)
Throttle valve opening angle (With the choke valve closed)	deg	15
Diaphragm spring set force	g(lb)	398 (0.88)
Diaphragm spring full stroke force	g(lb)	598 (1.32)
Primary and secondary throttle valve interlock angle	deg	47
Main nozzle primary inside diameter	mm(in)	2.6 (0.102)
Main nozzle secondary inside diameter	mm(in)	3.0 (0.118)
Throttle valve primary closing angle	deg	10
Throttle valve secondary closing angle	deg	20
Fuel pump type		Mechanical diaphragm
Fuel filter		Cartridge

2.3L ENGINE – MAIN DATA AND SPECIFICATIONS PICK-UP AND AMIGO

AIR INJECTION REACTOR SYSTEM		
Air pump data		
Construction		Positive displacement type
Number of vanes		2
EXHAUST GAS RECIRCULATION SYSTEM		
EGR valve operating negative pressure		
Starting	mmHg(inHg)	40 (1.6)
Fully opened	mmHg(inHg)	100 (3.9)
Thermal vacuum valve opening temperature	°C(°F)	Over 46 to 54 (115 to 129)
Back pressure transducer Control pressure	mm Aq(in Aq)	OFF ≧ 53.5 (2.1) ON ≦ 81.5 (3.2)
THREE-WAY CATALYTIC CONVERTER SYSTEM		
Converter size (Length x Width x Thickness)	mm(in)	431 x 228 x 89 (16.98 x 8.98 x 3.5)
Converter volume	liters(Cu.in)	2.6 (160)
Converter weight (with catalyst)	kg(lbs.)	8.2 (18)
Catalyst size	mm(in)	Pellet 3.2 (1/8) min., 6.4 (1/4 in) max.
THERMOSTATICALLY CONTROLLED AIR CLEANER SYSTEM		
Thermo sensor operating temperature		
Starting	°C(°F)	38 (100)
Fully opened	°C(°F)	44 (111)
Vacuum motor		
Opening vacuum	mmHg(inHg)	60 ± 20 (2.4 ± 0.8)
Full open vacuum	mmHg(inHg)	100 ± 20 (3.9 ± 0.8)
Idle compensator operating temperature		
Starting	°C(°F)	46 (115)
Fully opened	°C(°F)	52 (126)
EVAPORATIVE EMISSION CONTROL SYSTEM		
Check and relief valve operating pressure		
Check valve open negative pressure	mmHg(inHg)	20 to 40 (0.8 to 1.6)
Relief valve open negative pressure	mmHg(inHg)	5.0 to 20 (0.2 to 0.8)
Canister purge control valve operating pressure		
Purge control valve open	mmHg(inHg)	180 to 220 (7.1 to 8.7)

2.3L ENGINE—CLOSED LOOP EMISSION CONTROL SYSTEM—PICK-UP AND AMIGO

Vacuum switch (idle position)	Carburetor ported vacuum	Electrical signal to ECM	'ON' ; 79 — 88 mmHg ON OFF
MAP sensor	Inlet manifold air pressure or Ambient air pressure	Electrical signal to ECM	15 kPa 0.122 to 0.382V 40 kPa 1.521 to 1.688V 94 kPa 4.438 to 4.600V 102 kPa 4.859 to 5.043V
Coolant temperature sensor	Engine coolant temp.	Electrical signal to ECM	Temp. (°C) Resistance (Ω) 15.0 4450 +293 −336 128.0 80.0 +4.0 −2.3
Oxygen sensor	Oxygen content in exhaust gas stream	Voltage to ECM	'Rich' ; ≧ 800 mV 'Lean' ; ≦ 200 mV at 371°C & 450 mV BIAS
EGR gas temperature sensor (California only)	EGR gas temperature	Electrical signal to ECM	−20°C, 200°C

2.3L ENGINE—DIAGNOSTIC CIRCUIT CHECK PICK-UP AND AMIGO

Vacuum hose routing, connections and harness connector connections should be checked before this check.

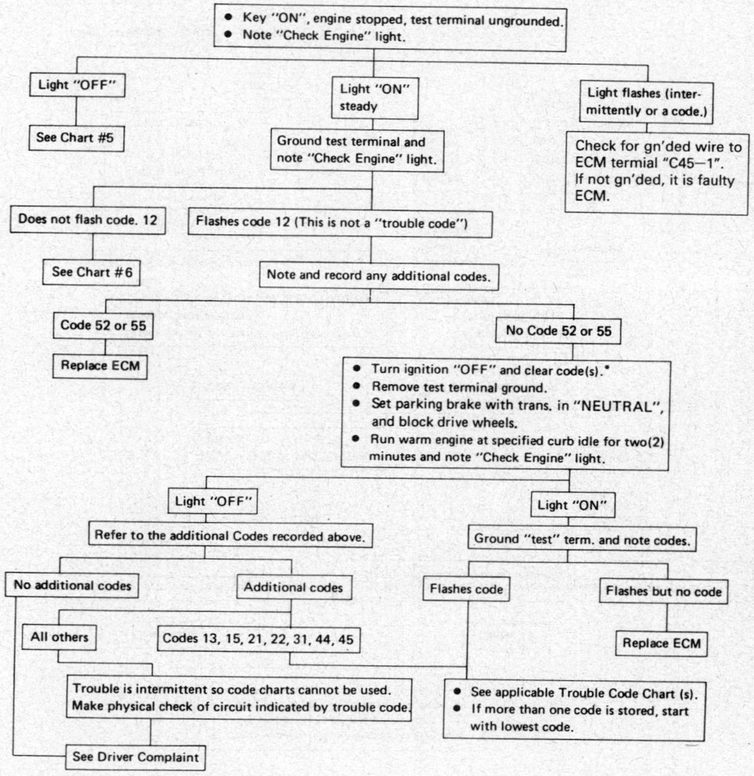

* See Code(s) Clearing Procedure

The system performance check should be performed after any repairs to the "System" have been made.

2.3L ENGINE—SYSTEM PERFORMANCE CHECK—PICK-UP AND AMIGO

The System Performance Check is used to check the operation of the duty solenoid and the main metering circuit using tachometer, and to check the operation of the Closed Loop Emission Control System by the use of a dwell meter.

SYSTEM PERFORMANCE CHECK

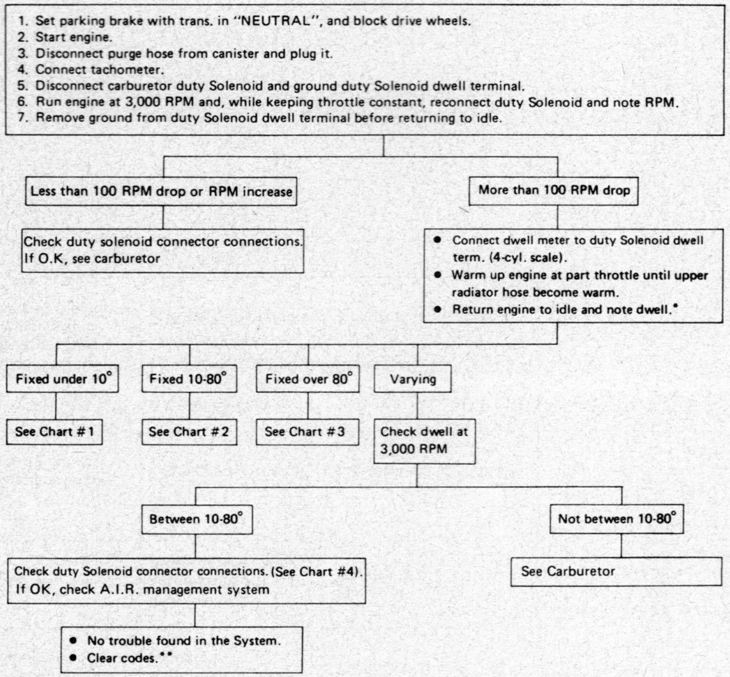

* Oxygen sensors may cool off at idle and the dwell change from varying to fixed. If this happens, running the engine at fast idle will warm it up again.

** See Code(s) Clearing Procedure

CHART #1 DWELL FIXED UNDER 10°

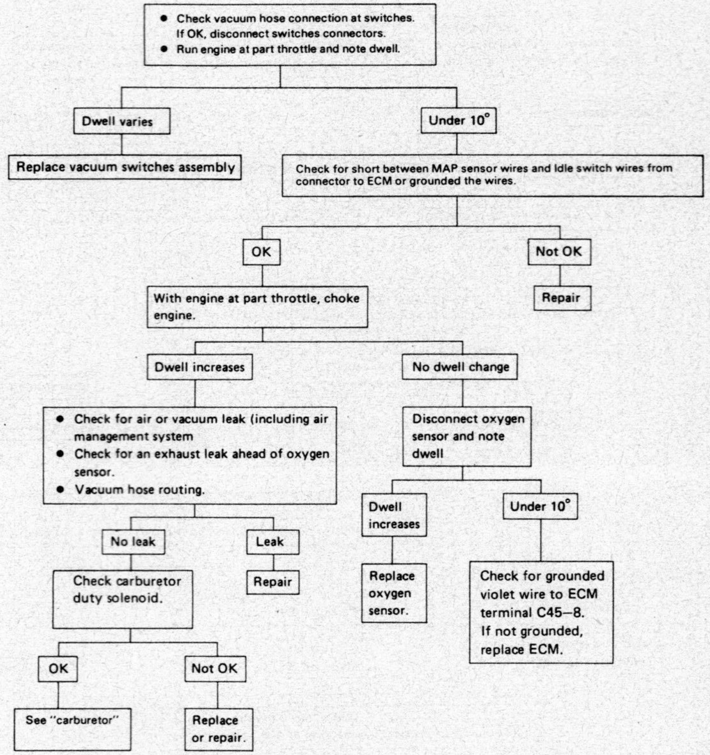

2.3L ENGINE—SYSTEM PERFORMANCE CHECK—PICK-UP AND AMIGO

CHART #2 DWELL FIXED BETWEEN 10° — 80°
Open Coolant Sensor or Oxygen Sensor Circuit

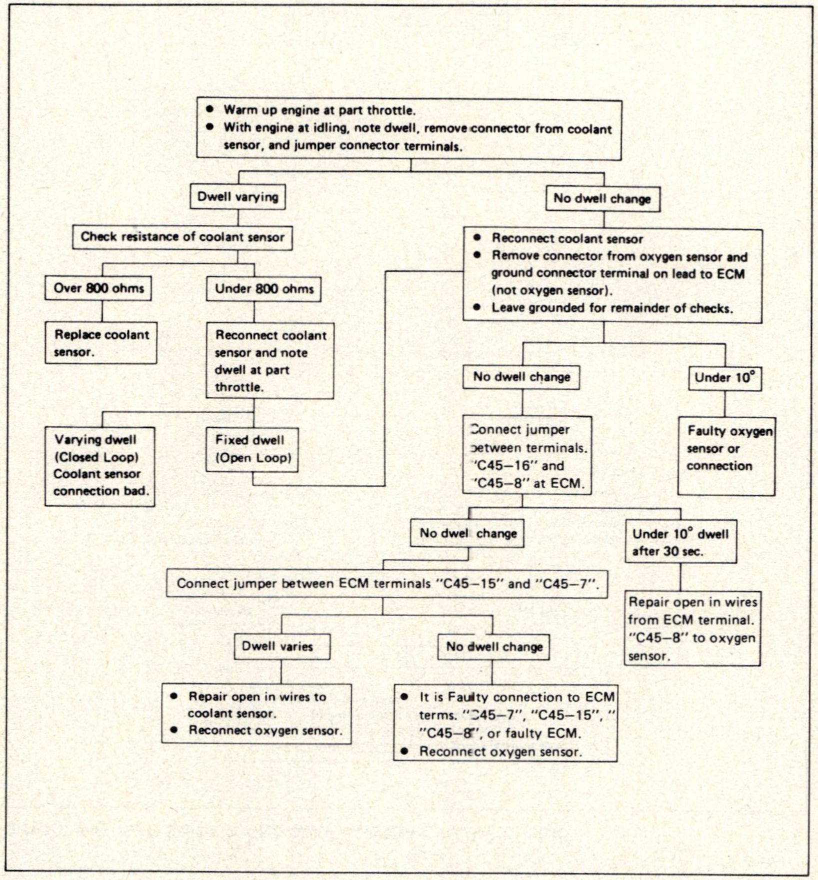

CHART #3 DWELL FIXED OVER 80°
Rich Engine Condition Indication

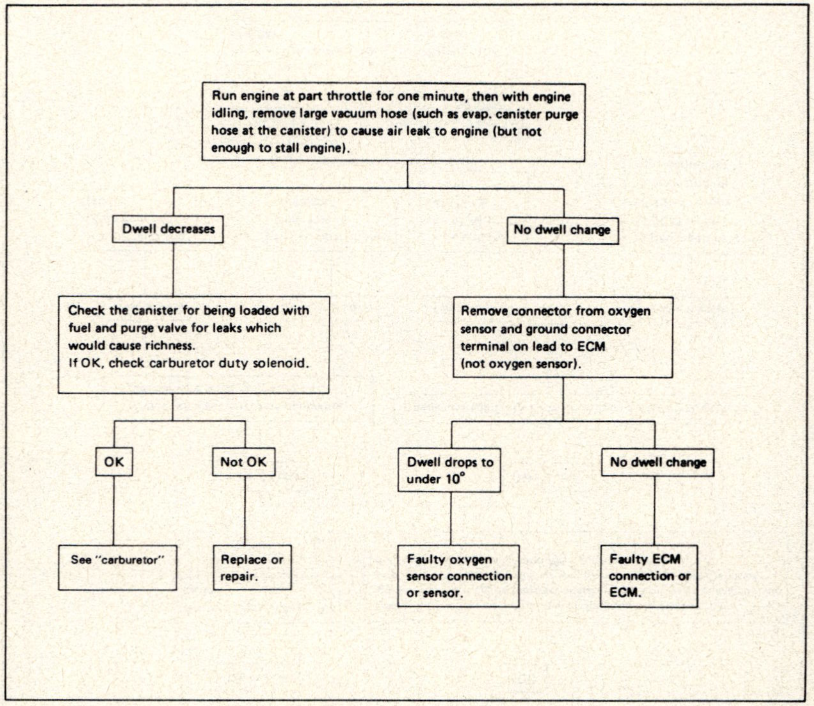

2.3L ENGINE—SYSTEM PERFORMANCE CHECK—PICK-UP AND AMIGO

CHART #4 IDLE SWITCH AND MAP SENSOR CIRCUIT CHECK

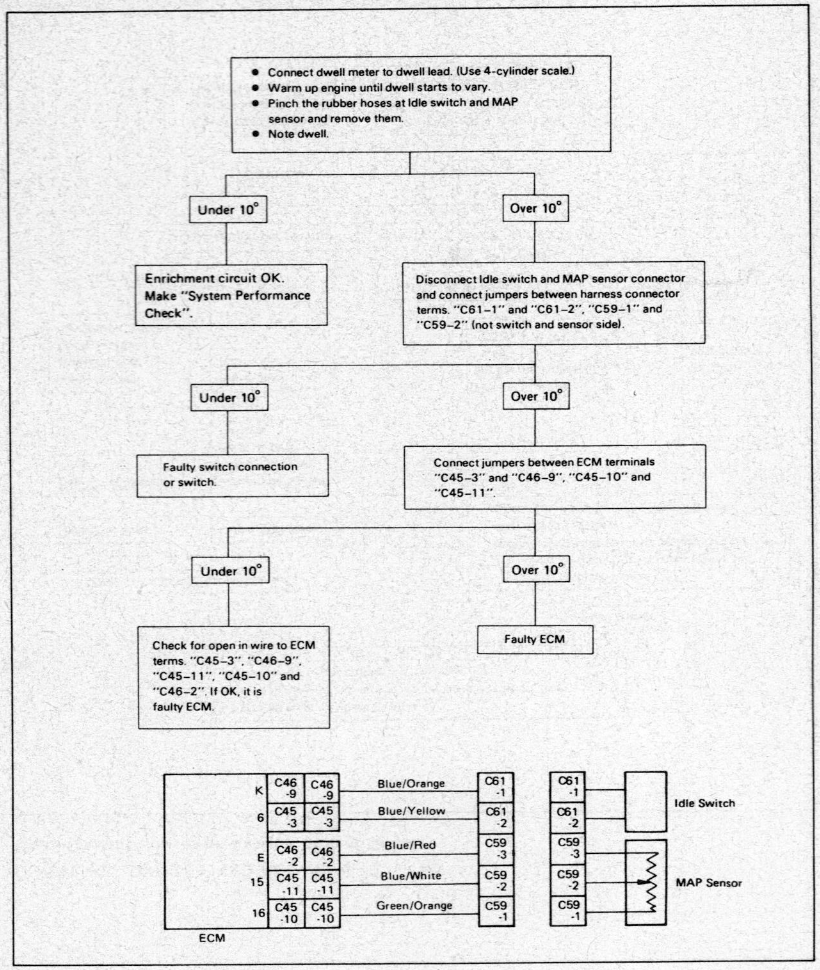

CHART #5 "CHECK ENGINE" LIGHT INOPERATIVE

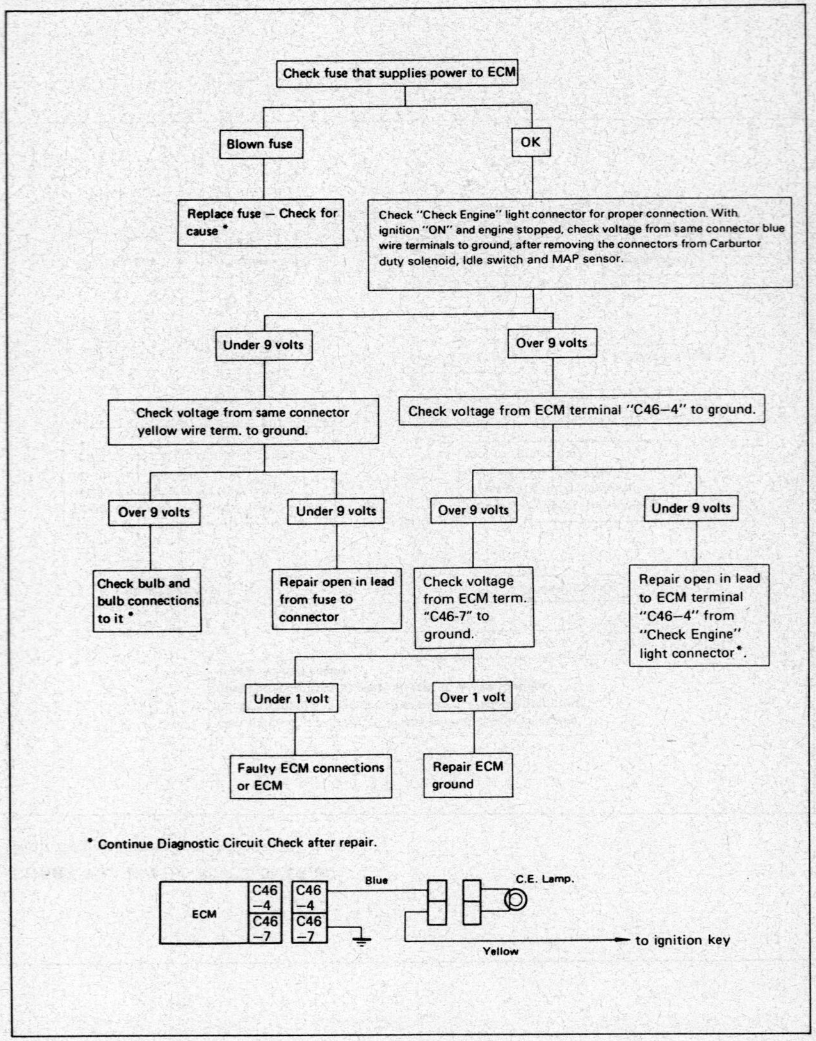

2.3L ENGINE—SYSTEM PERFORMANCE CHECK PICK-UP AND AMIGO

CHART #6 WON'T FLASH CODE 12
Check fuse that supplies power to ECM.

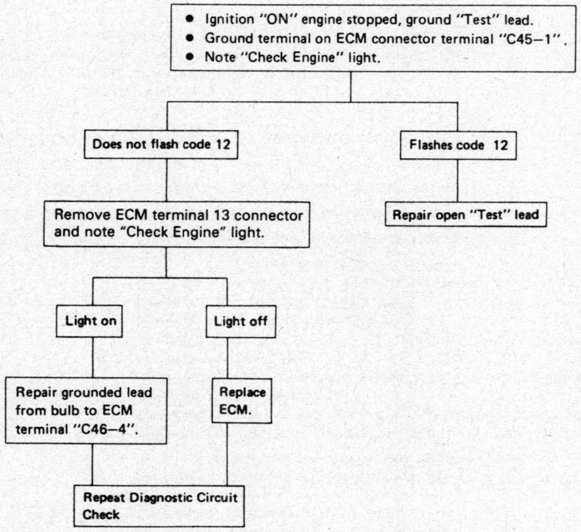

2.3L ENGINE—TROUBLE CODE IDENTIFICATION PICK-UP AND AMIGO

TROUBLE CODE IDENTIFICATION

The "CHECK ENGINE" light will only be "ON" if the malfunction exists under the conditions listed below. If the malfunction clears, the light will go out and a trouble code will be set in the ECM. Code 12 does not store in memory. If the light comes "ON" intermittently, but no code is stored, go to the "Driver Complaint" section. Any codes stored will be erased if no problem reoccurs within 50 engine starts.

The trouble codes indicate problems as follows:

TROUBLE CODE 12 No ignition reference pulses to the ECM. This code is not stored in memory and will only flash while the fault is present. This code is displayed before the engine is started.

TROUBLE CODE 13 Oxygen sensor circuit — The engine must run 1 minute at part throttle, under road load, before this code will set. This code does not set when the coolant temperature is below 70°C (158°F) and/or the time since engine start has exceeded 2 minutes.

TROUBLE CODE 14 Shorted coolant sensor circuit — The engine must run 2 minutes before this code will set.

TROUBLE CODE 15 Open coolant sensor circuit — The engine must run 5 minutes before this code will set.

TROUBLE CODE 21 Idle switch and MAP sensor failure

(1) * Vehicle speed is 0 km/h and engine speed is between 200 and 1000 rpm.
 * MAP sensor output is more than 500 mmHg abs as input signal, for 5 seconds.

(2) * Idle switch is "OFF"
 * MAP sensor output is less than 100 mmHg abs as input signal, for 5 seconds.

(3) * MAP sensor output is between 400 and 450 mmHg abs as input signal, and engine speed is more than 1500 rpm.
 * Idle switch is "ON" for more than 5 seconds.

If either condition (1), (2), (3) exists, this code will set.

TROUBLE CODE 22 Fuel cut solenoid circuit open or grounded — The engine must run under the decelerating condition over 2000 engine rpm before this code will set.

TROUBLE CODE 23 Duty solenoid circuit open or grounded.

TROUBLE CODE 25 Air management valve circuit open or grounded.

TROUBLE CODE 26 Canister V.S.V. circuit open or grounded.

TROUBLE CODE 27 Constant high voltage from V.S.V. to ECM.

TROUBLE CODE 31 No ignition reference pulses to the ECM after 10 seconds at part throttle, under load.
This code will store in memory.

TROUBLE CODE 32 EGR system failure and sensor circuit failure.

TROUBLE CODE 34 EGR gas sensor circuit failure.

2.3L ENGINE—TROUBLE CODE IDENTIFICATION PICK-UP AND AMIGO

TROUBLE CODE 44 **Fuel metering faulty lean signal**
(1) Coolant temperature is more than 158°F (70°C).
(2) Engine throttle is partially open.
(3) Engine speed is less than 2500 rpm.
(4) More than 5 seconds after fuel cut recovering.
The timer of this code will start to count when all (1) — (4) conditions exist. Then ECM judges to set code 44, to use the timer information, except when condition A or condition B exists:
Condition A "BAROMETRIC MAP" is under 596 mmHg abs.
Condition B Code 13 or 21 is stored in memory, or the coolant temperature is more than 302°F (150°C).

TROUBLE CODE 45 **Fuel metering faulty rich signal**
(1) Coolant temperature is more than 158°F (70°C).
(2) Engine throttle is partially open.
(3) Engine speed is less than 2500 rpm.
(4) Vehicle speed is between 20 and 80 km/h (12-50 mph).
(5) More than 30 seconds after starting canister purge.
The timer of this code will start to count when all (1) — (5) conditions exist. This code does not set when the high altitude condition (Above 2500 m or 8000 feet) is judged by ECM.
(MAP sensor output voltage is less than 596 mmHg abs as the input signal of atmospheric pressure.)
Then ECM judges to set code 45, to use the timer information except when condition A or condition B exists:
Condition A "BAROMETRIC MAP" is under 596 mmHg abs.
Condition B Code 13 or 21 is stored in memory, or the coolant temperature is more than 302°F (150°C).

TROUBLE CODE 51 Shorted fuel cut solenoid circuit and/or faulty ECM.
TROUBLE CODE 52 Faulty ECM — Problem is RAM in ECM.
TROUBLE CODE 53 Shorted Air management valve and/or faulty ECM.
TROUBLE CODE 54 Shorted vacuum control solenoid and/or faulty ECM.
TROUBLE CODE 55 Faulty ECM — Problem is A/D converter in ECM.

AIR : Air Injection Reactor
ECM : Electronic Control Module
WOT : Wide Open Throttle
V.S.V.: Vacuum Switching Valve

2.3L ENGINE—CONTROL MODULE TERMINAL IDENTIFICATION—PICK-UP AND AMIGO

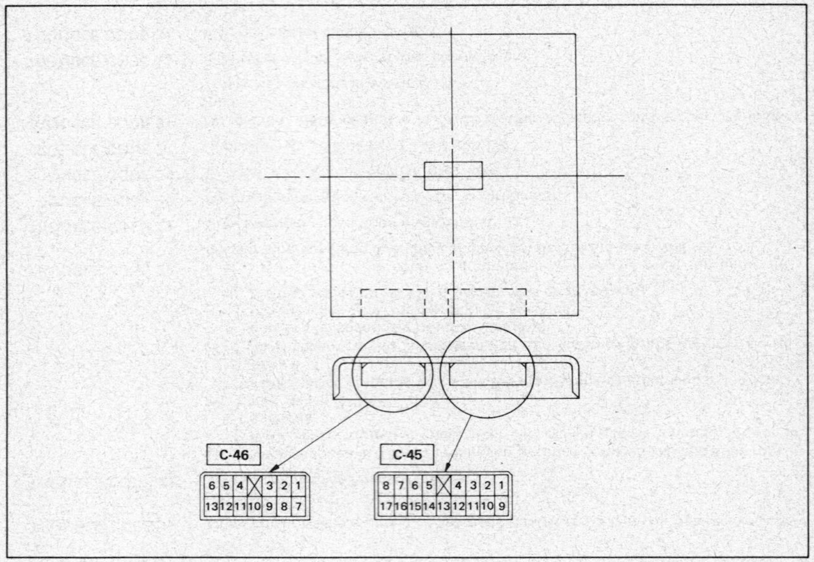

ELECTRONIC CONTROL MODULE CONNECTIONS IDENTIFICATION

Terminal C-46

Pin	Function
1	Battery + (Memory Back-up)
2	MAP Sensor Ground
3	Ignition Key +
4	System Malfunction Lamp
5	Carburetor Fuel Cut Solenoid
6	Duty Solenoid
7	Control Ground
8	Case Ground
9	Idle Position Switch Ground
10	Ignition Coil Tach. Input
11	Air Injection Reactor V.S.V. Output
12	Early Fuel Evaporation System
13	Power Ground

Terminal C-45

Pin	Function
1	ALDL Start
2	Not Used
3	Idle Switch Input
4	MAP VSV
5	Not Used
6	EGR Gas Temp. Sensor
7	Coolant Temp. Sensor Input
8	Oxygen Sensor Input
9	Not Used
10	MAP Sensor ; 5V
11	MAP Sensor ; V out
12	Canister Purge VSV
13	Inlet Air Temp. Switch Input
14	Car Speed Sensor
15	Coolant Temp. Sensor Ground
16	Oxygen Sensor Ground
17	Inlet Air Temp. Switch Ground

2.3L ENGINE—DWELL VS. DUTY AND TERMINAL IDENTIFICATION—PICK-UP AND AMIGO

RELATIONSHIP OF DWELLMETER READINGS TO FEED BACK SOLENOID CYCLING

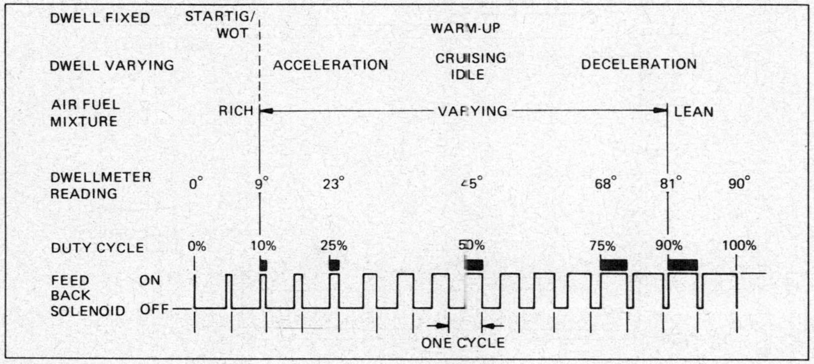

Caution:
When disconnecting or connecting the connector, turn the ignition to the "OFF" position.

When Connecting ECM Harness Connector

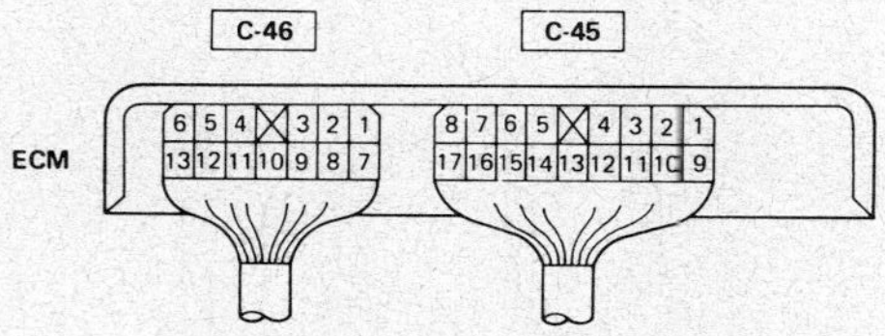

Back View of Connector

When Disconnecting ECM Harness Connector

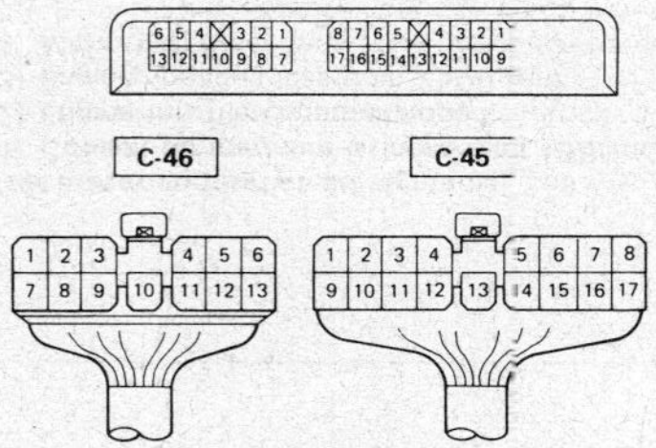

Front View of Connector

2.3L ENGINE—TROUBLESHOOTING GUIDE—CODE 12 PICK-UP AND AMIGO

TROUBLE CODE 12
NO REFERENCE PULSES TO THE ECM

① Check connections at two (2) ignition coil connectors

- Not OK → Repair
- OK →

②
- Connect voltmeter from ECM terminal "C46-10" to ground.
- Ignition "on", engine stopped and check voltage.

- OK: Over 10 volts → Faulty ECM connections or ECM.
- NG: Under 10 volts → Check for open reference lead from ignition coil to ECM.

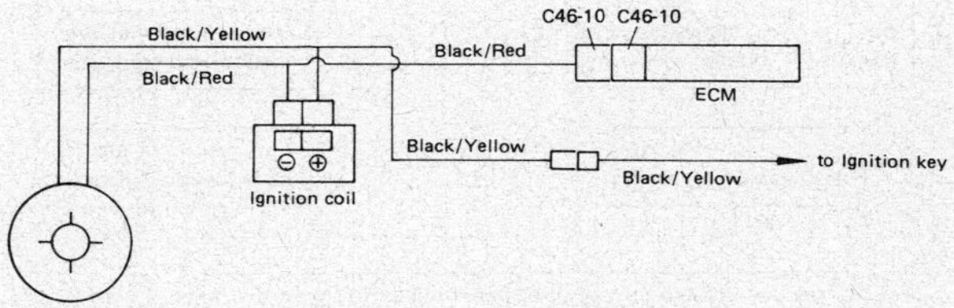

ECM : Electronic Control Module

2.3L ENGINE—TROUBLESHOOTING GUIDE—CODE 12 PICK-UP AND AMIGO

① This step checks for a poor connection at the ignition coil wire connector as being the source of no reference pulse. Check for corrosion at connector terminals not fully seated, or terminal not properly attached to the wire. The terminal must be removed from the connector and carefully inspected.

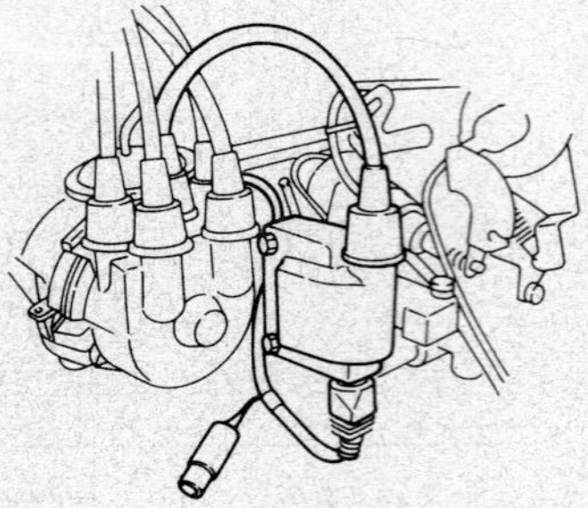

② This step determines if a reference pulse voltage is being sent to the ECM. Normal voltage should be over 10 volts. Check for good terminal contact at terminal "C46-10" of the ECM.

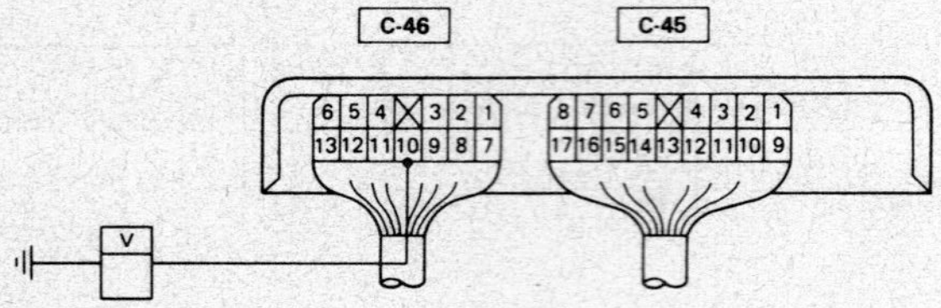

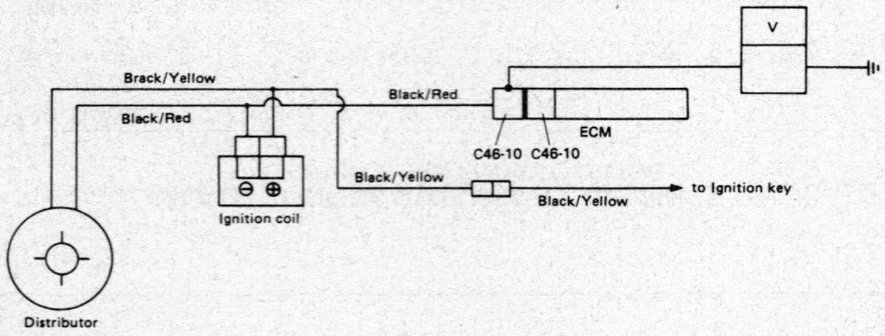

2.3L ENGINE—TROUBLESHOOTING GUIDE—CODE 13 PICK-UP AND AMIGO

TROUBLE CODE 13
OPEN OXYGEN SENSOR CIRCUIT

If 13 and 23 are displayed, go to Trouble Code 23 chart first.

①
- Connect dwell meter to dwell lead. (Use 4-cylinder scale.) Warm up engine until upper radiator hose becomes warm.
- With engine idling, disconnect oxygen sensor and ground terminal on lead to ECM (not oxygen sensor).

- OK: Under 10° dwell after 30 sec. → Faulty oxygen sensor connections or sensor.
- NG: Over 10° dwell after 30 sec. → ② Connect jumper between ECM terminals "C45-8" and "C45-16".
  - NG: Over 10° dwell after 30 sec. → ③ Check that wire ECM terminal "C45-16" (ignition OFF) is grounded.
    - Not OK → Repair
    - OK → Faulty ECM connections or ECM
  - OK: Under 10° dwell after 30 sec. → Repair open in oxygen sensor harness lead from connector to ECM.

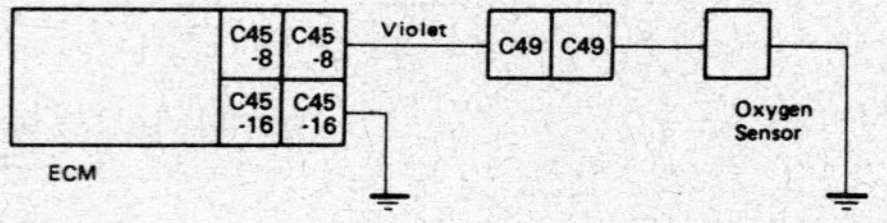

ECM : Electronic Control Module

2.3L ENGINE—TROUBLESHOOTING GUIDE—CODE 13 PICK-UP AND AMIGO

① This step tests to see if ECM is able to respond to a lean exhaust O_2 signal (low voltage).

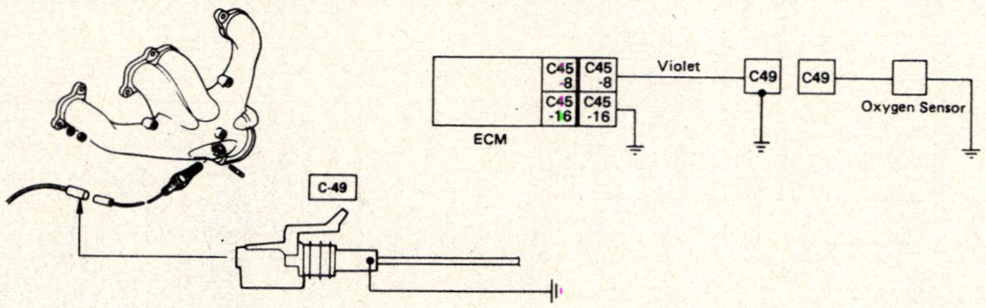

② This step determines if there is an open in the wire from the oxygen sensor connector to ECM terminal "C45-8" or if it's a faulty ECM.

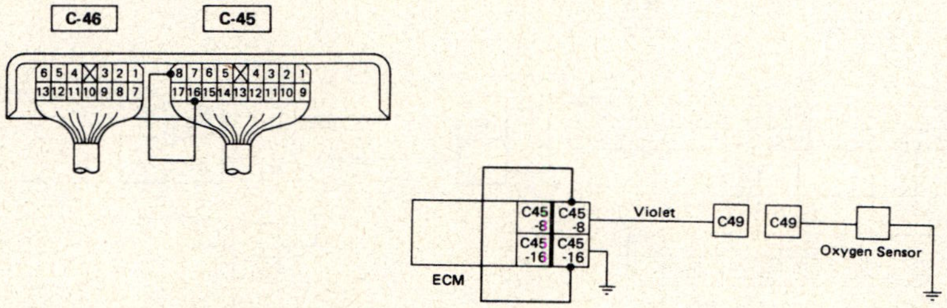

③ Check for a bad ground to ECM terminals "C45-16".

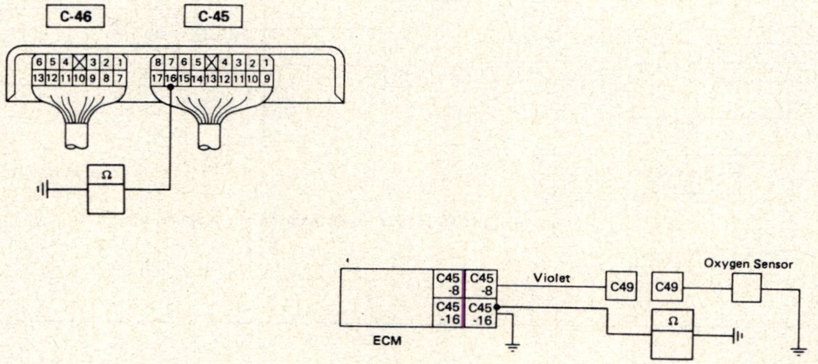

2.3L ENGINE—TROUBLESHOOTING GUIDE—CODE 14 PICK-UP AND AMIGO

TROUBLE CODE 14
SHORTED COOLANT SENSOR CIRCUIT

Check for engine overheating condition before making following test.

①
- Disconnect ECM connectors
- Check resistance between terminals "C45-7" and "C45-15" of ECM harness connector (not ECM side).

OK — Over 70 ohms

②
- Check resistance between terminal "C45-7" of ECM harness connector and ground.
- The resistance should be over 100K ohms.

OK: Faulty ECM connections or ECM.

Not OK: Check for grounded wire from sensor connector to ECM term "C45-7". If not grounded, replace coolant sensor.

NG — Under 70 ohms

③
- Disconnect coolant sensor connector.
- Check resistance between terminal "C45-7" and "C45-15" of ECM harness connector same as above.
- The resistance should be over 100K ohms.

OK: Replace coolant sensor.

Not OK: Repair short between coolant sensor wires from connector to ECM.

- Connect coolant sensor.
- Check resistance between terminal "C45-7" and "C45-15" of ECM harness connector.
- The resistance should be 100K ohms. If not over 100K ohms, replace coolant sensor.

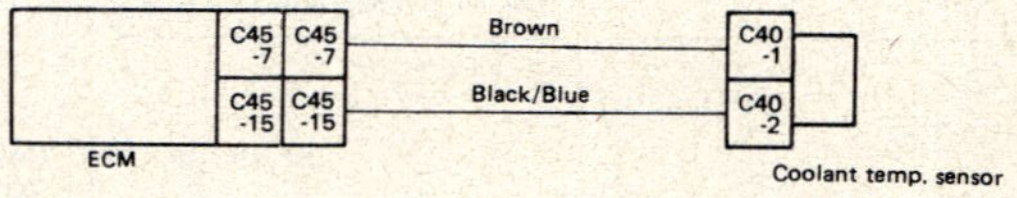

ECM : Electronic Control Module

2.3L ENGINE—TROUBLESHOOTING GUIDE—CODE 14 PICK-UP AND AMIGO

① This step determines if the coolant temp. sensor circuit is shorted.

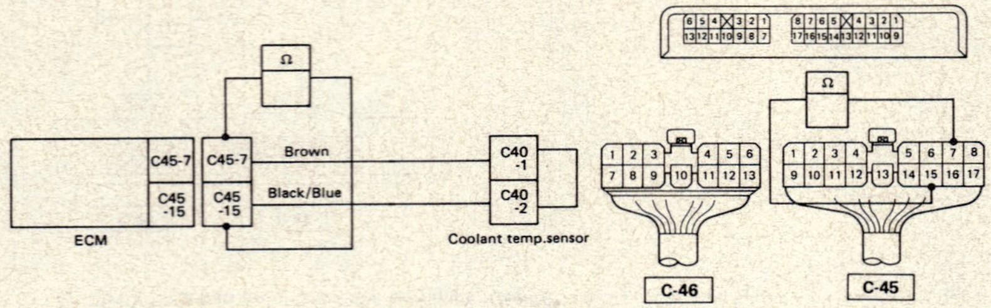

② This checks for ground in the wires, sensor to ECM and ECM to sensor.

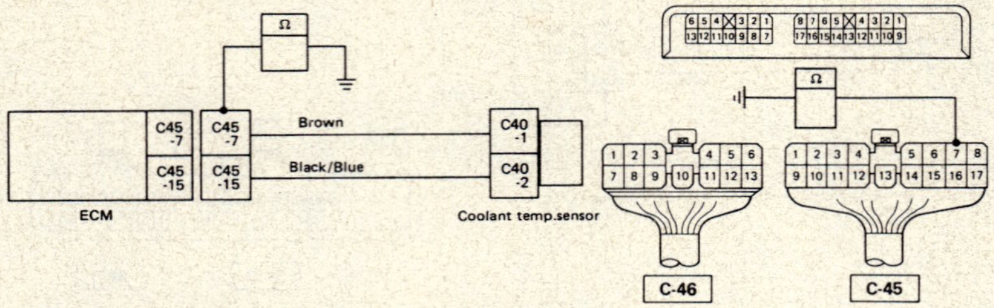

③ Check for a bad ground to ECM terminals "C45-16".

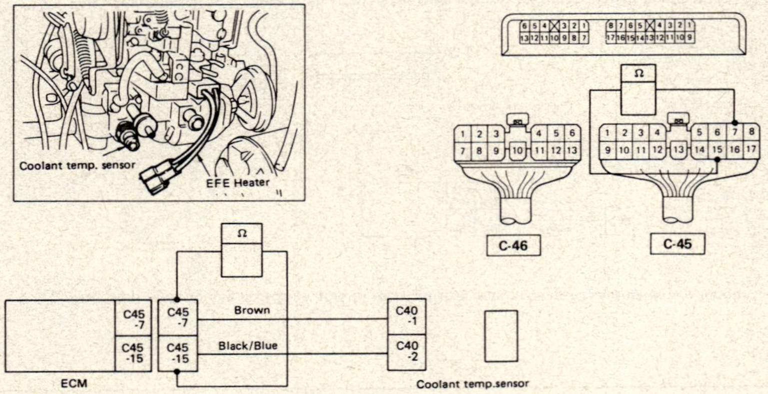

2.3L ENGINE—TROUBLESHOOTING GUIDE—CODE 15 PICK-UP AND AMIGO

TROUBLE CODE 15
OPEN COOLANT SENSOR CIRCUIT

①
- Disconnect Coolant Sensor.
- Ignition "on", engine stopped, check voltage between sensor connector terminals, should be over 2 volts.

OK → ② Check resistance of coolant sensor. Should be under 800 ohms on warm engine.

- **OK** → It is poor sensor connection or low coolant level.
- **Not OK** → Replace sensor.

Not OK → ③ Check voltage from ECM terminals "C45-7" to "C45-15", should be over 2 volts. (Sensor disconnected.)

- **Not OK** → It is faulty ECM connection or ECM.
- **OK** → Check for open in wires to ECM terminals "C45-7" and "C45-15".

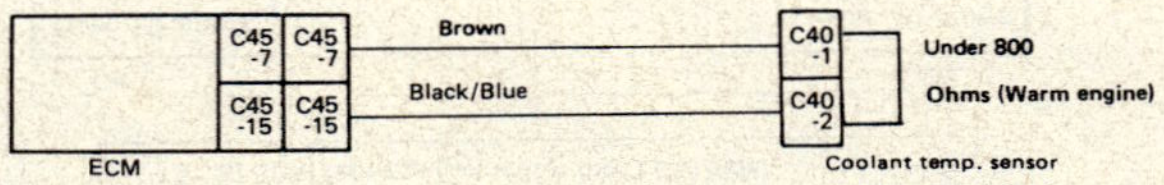

ECM : Electronic Control Module

2.3L ENGINE—TROUBLESHOOTING GUIDE—CODE 15 PICK-UP AND AMIGO

① Determine whether the low voltage at the sensor connector is due to open in the coolant sensor wires or poor terminal contact.

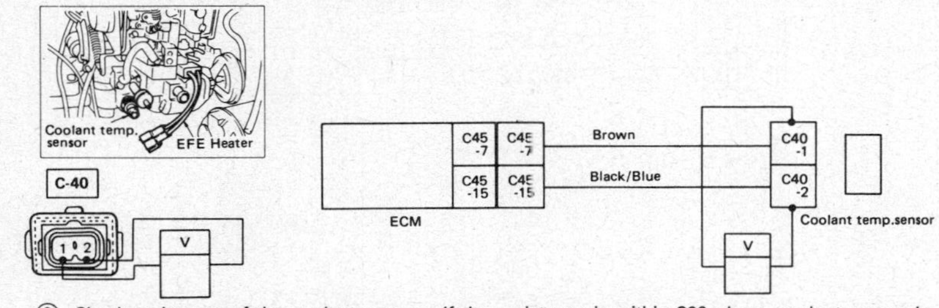

② Check resistance of the coolant sensor. If the resistance is within 800 ohms, coolant sensor is not faulty; therefore, check for corrosion at the connector or low coolant level.

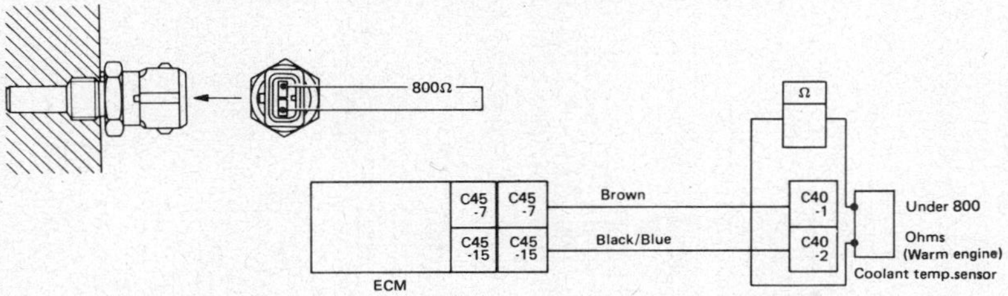

③ This step checks voltage from ECM. Voltage should be above 2 V.

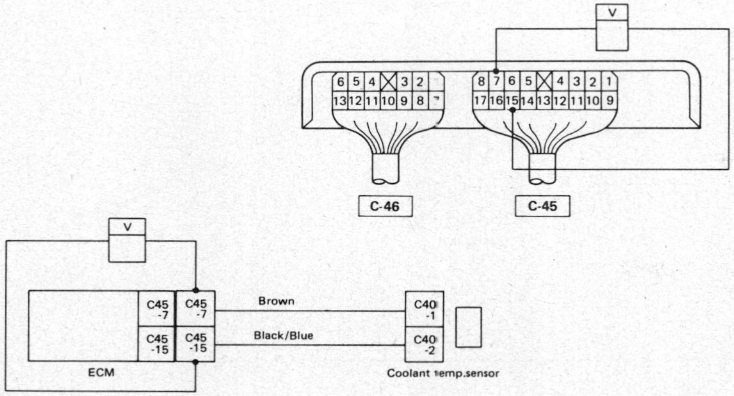

2.3L ENGINE—TROUBLESHOOTING GUIDE—CODE 21 PICK-UP AND AMIGO

TROUBLE CODE 21
IDLE SWITCH AND MAP SENSOR FAILURE

If codes 21 and 23 are displayed, go to Trouble Code 23 Chart first.

①
- Connect dwell meter to dwell lead. (Use 4-cylinder scale)
- Run engine at part throttle until dwell starts to vary.
- Disconnect vacuum switch and MAP sensor connector and connect jumpers between harness connector terms. "C61-1" and "C61-2", "C59-1" and "C59-2".
- Run engine at part throttle and note dwell.

Dwell varying

② Connect jumper between ECM terminal "C46-9" and "C45-3".

- **Dwell varying**
  - ④ Connect jumper between ECM terminal "C45-11" and "C45-10".
    - **Varying** → Faulty ECM connections or ECM.
    - **Dwell fixes to 0% after fixing to 20% for about 10 seconds. (Fixing to 20% only when engine speed is between 600 and 3400 rpm).** → Repair open in harness to terminals "C45-11" and "C45-10" of ECM.
- **Dwell fixes to 0% after fixing to 20% for about 10 seconds (Fixing to 20% only when engine speed is between 600 and 3400 rpm).** → Repair open in harness to terminals "C46-9" and "C45-3" of ECM.

Dwell fixes to 0% after fixing to 20% for about 10 seconds (Fixing to 20% only when engine speed is between 600 and 3400 rpm)

③ Disconnect jumper between harness connector terminals "C59-1" and "C59-2".

- **Dwell varying**
  - Check for short or open circuit in vacuum SW and MAP sensor. Check for ground short for MAP sensor.
    - Replace vacuum SW or MAP sensor.
    - Repair MAP sensor ground.
- **Dwell fixes to 0% after fixing to 20% for about 10 seconds. (Fixing to 20% only when engine speed is between 600 and 3400 rpm)**
  - Check for short or open from ECM harness and connector to MAP sensor harness and connector.
    - Replace MAP sensor
    - Repair: GND line (C46-2, C59-3) open output (C45-11, C59-2) or, 5 Volt line (C45-10, C59-1) short

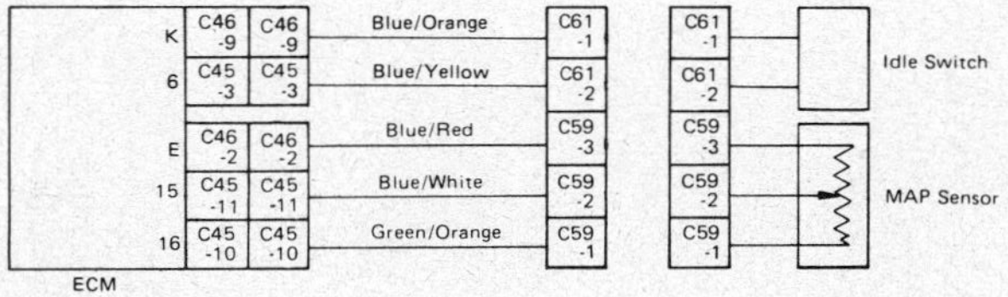

ECM : Electronic Control Module
WOT : Wide Open Throttle

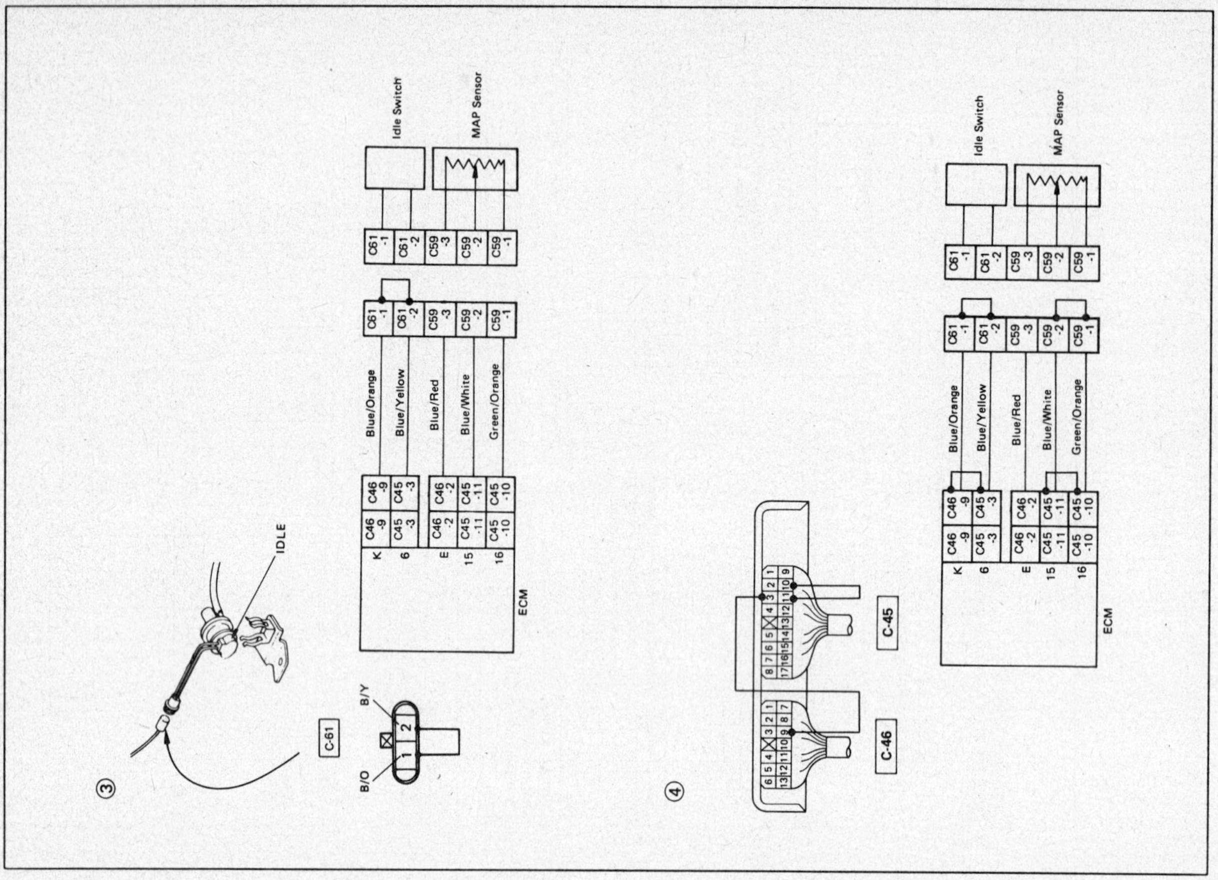
③
IDLE
C-61
B/Y
B/O
ECM
K
6
E
15
16
Blue/Orange
Blue/Yellow
Blue/Red
Blue/White
Green/Orange
Idle Switch
MAP Sensor
④
C-45
C-46

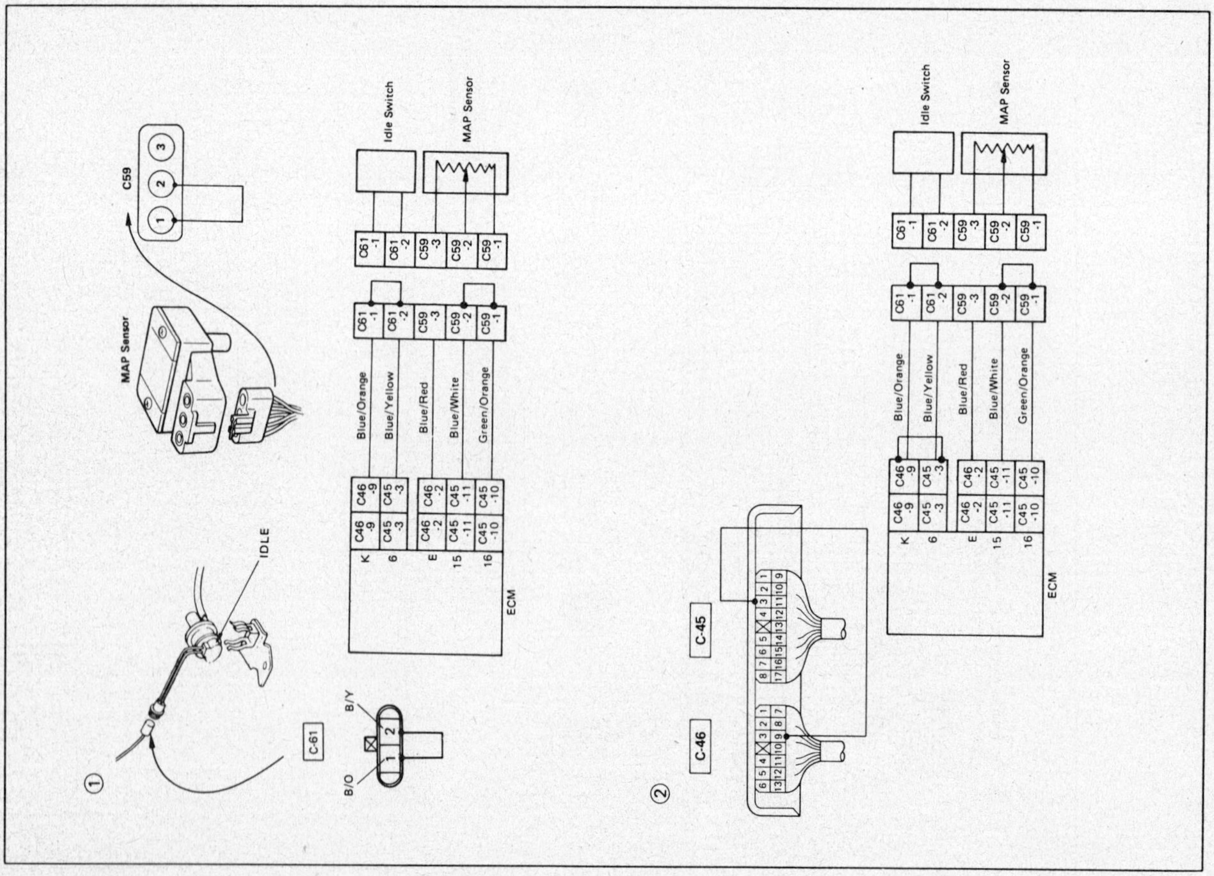
①
IDLE
MAP Sensor
C59
C-61
B/Y
B/O
ECM
K
6
E
15
16
Blue/Orange
Blue/Yellow
Blue/Red
Blue/White
Green/Orange
Idle Switch
MAP Sensor
②
C-45
C-46

2.3L ENGINE—TROUBLESHOOTING GUIDE—CODE 22,51—PICK-UP AND AMIGO

TROUBLE CODE 22, 51
FUEL CUT (F/C) SOLENOID CIRCUIT FAILURE

①
- Disconnect F/C solenoid connector
- Check F/C solenoid resistance
- The resistance should be over 15 Ω

NG: Replace F/C solenoid

OK:

Code No. 22

②
- Disconnect ECM connectors.
- Connect test light from battery ⊕ to ECM harness connector terminal "C46-5".

NG: Light "On" → Repair short to Bat. ⊕ in Light green wire to ECM term "C46-5".

OK: Light "Off" →

④
- Connect F/C solenoid connectors.
- Connect test light from ECM terminal "C46-5" to ground.
- Ignition "On" engine stopped.

Light "On" → Faulty ECM connector or ECM.

Light "Off" →

⑤
- Disconnect F/C solenoid connector.
- Connect test light from the F/C solenoid connector battery terminal White/Yellow lead to ground.
- Turn ignition "On" engine stopped.

OK: Light "On" → Repair open in wire from F/C solenoid to ECM.

NG: Light "Off" → Repair open Bat. ⊕ circuit to F/C solenoid (White/Yellow wire).

Code No. 51

③
- Disconnect ECM connectors.
- Connect test light from ground to ECM harness connector "C46-5".
- Ignition "ON" engine stopped.

Light "On" → Repair short to battery ⊕ in Light green wire to ECM term. "C46-5".

Light "Off" → Faulty ECM connector or ECM.

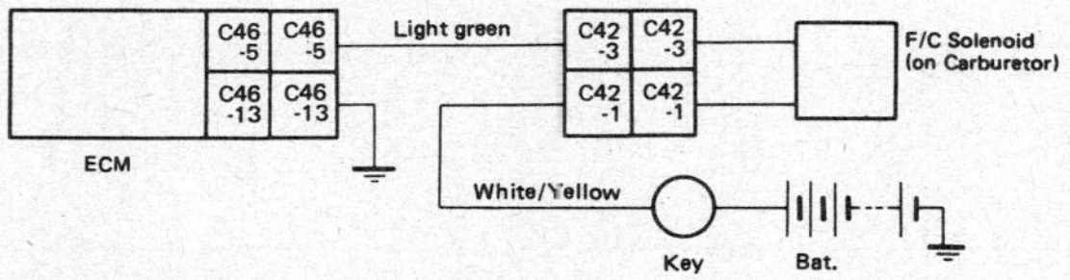

F/C : Fuel Cut
ECM : Electronic Control Module

① This step determines if the F/C solenoid is shorted. A good solenoid will have over 15 ohms resistance.

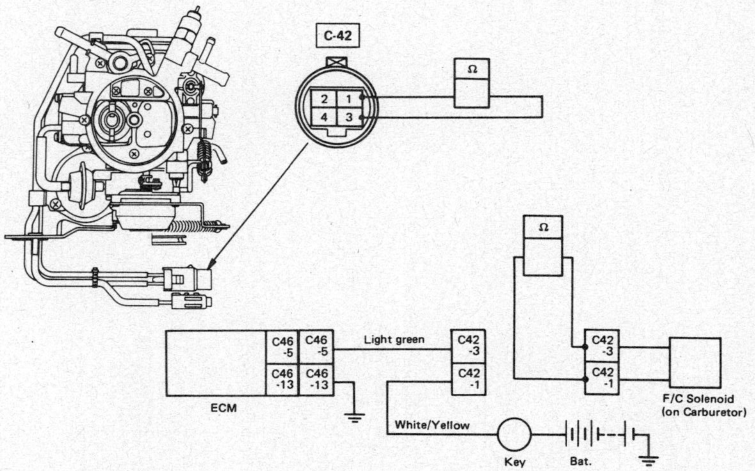

② Check for a short in the wire from V.S.V. connector to ECM terminal.

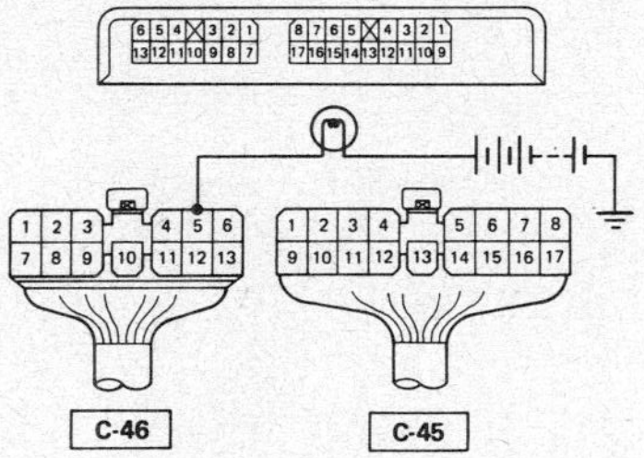

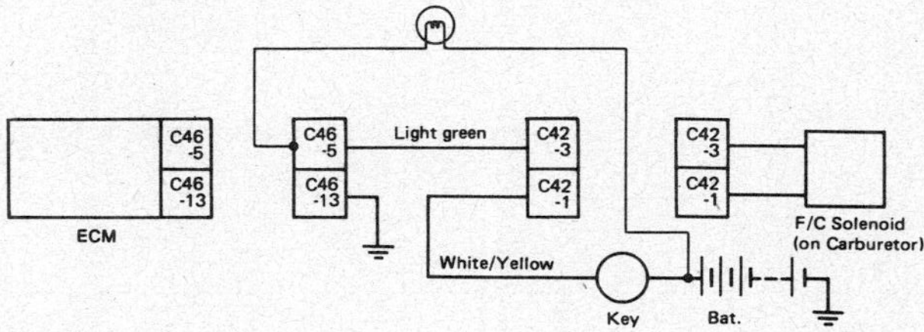

③ Check for a short in the battery (+) to ECM circuit.

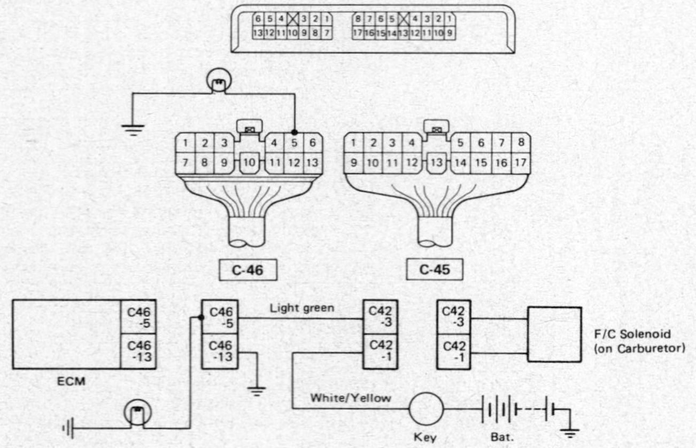

④ Check for an open in wire from battery to F/C solenoid and F/C solenoid to ECM.

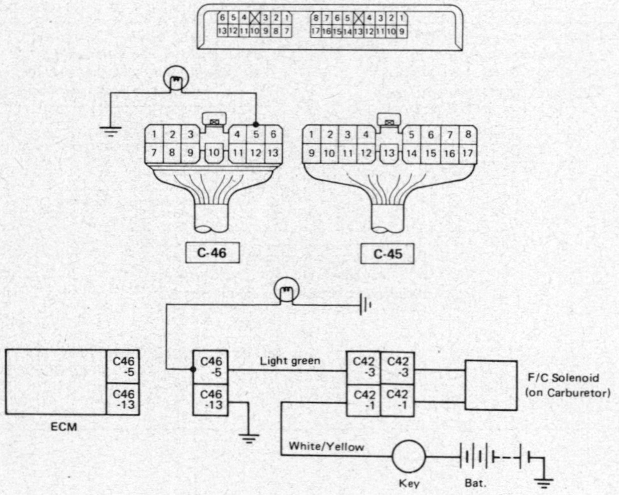

⑤ Check for an open in wire from battery to F/C solenoid.

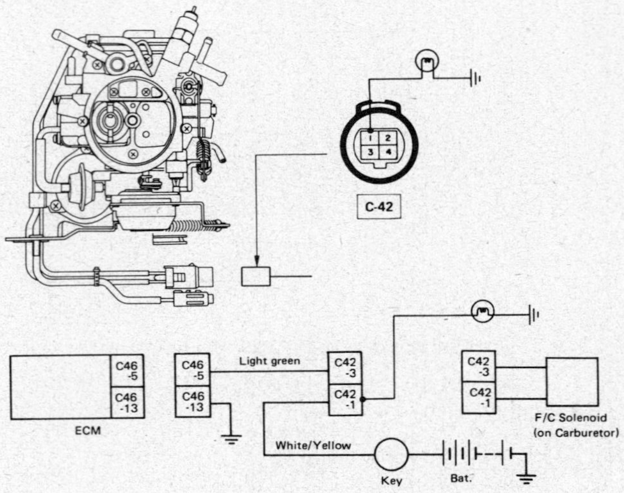

2.3L ENGINE—TROUBLESHOOTING GUIDE—CODE 23,54—PICK-UP AND AMIGO

TROUBLE CODE 23, 54
DUTY SOLENOID CIRCUIT FAILURE

Check connections at duty solenoid.

①
- Disconnect duty solenoid connector
- Check duty solenoid resistance
- The resistance should be over 20 Ω

NG: Replace duty solenoid

OK:

Code No. 23

②
- Disconnect ECM connectors.
- Connect test light from battery ⊕ to ECM harness connector terminal "C46-6".

NG — Light "On": Repair short to Bat. ⊕ in Light green wire to ECM term. "C46-6" or duty solenoid to dwell lead wire.

OK — Light "Off":

④
- Connect duty solenoid connectors.
- Connect test light from ECM terminal "C46-6" to ground.
- Ignition "On" engine stopped.

Light "On": Faulty ECM connector or ECM.

Light "Off":

⑤
- Disconnect duty solenoid connector.
- Connect test light from the duty solenoid connector battery terminal White/Yellow lead to ground.
- Turn ignition "On" engine stopped.

OK — Light "On": Repair open in wire from duty solenoid to ECM.

NG — Light "Off": Repair open Bat. ⊕ circuit to duty solenoid (White/Yellow wire).

Code No. 54

③
- Disconnect ECM connectors.
- Connect test light from ground to ECM harness connector "C46-6".
- Ignition "On" engine stopped.

Light "On": Repair short to battery ⊕ in Light green wire to ECM term. "C46-6".

Light "Off": Faulty ECM connector or ECM.

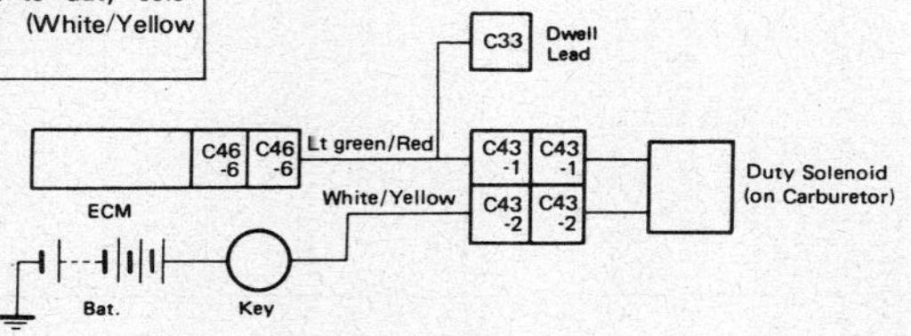

ECM : Electronic Control Module

① This step tests the resistance of the duty solenoid itself. A good solenoid will have over 20 ohms.

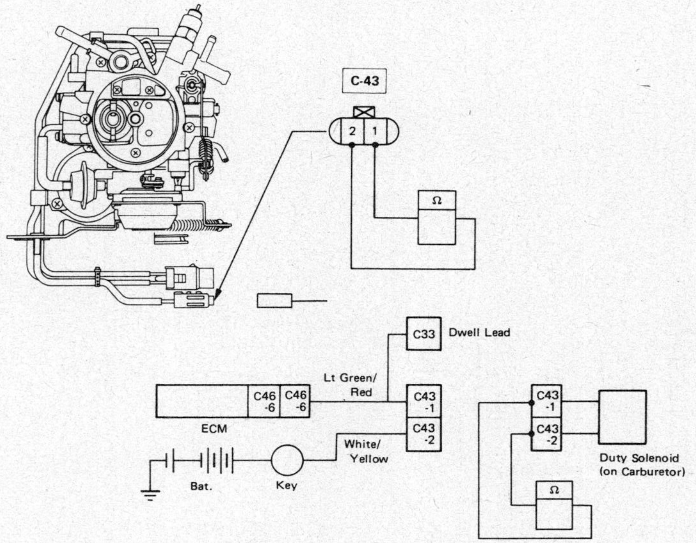

② Check for a short in the wire from duty solenoid to ECM terminal.

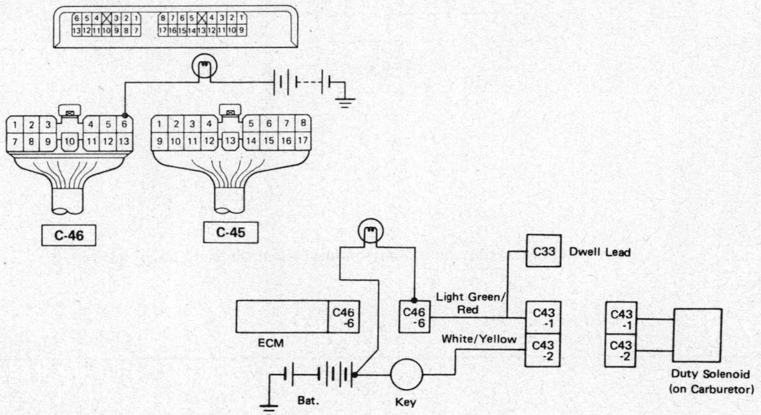

③ Check for a short in the battery (+) to ECM circuit.

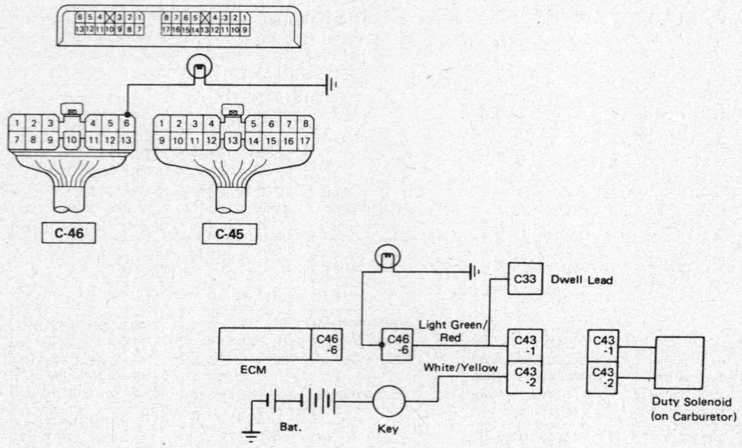

④ Check for an open in the wire from battery to duty solenoid and duty solenoid to ECM.

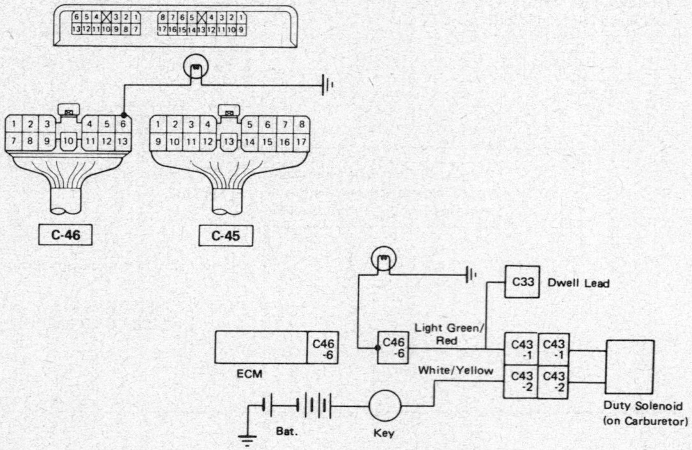

⑤ Check for an open in the wire from battery to the duty solenoid.

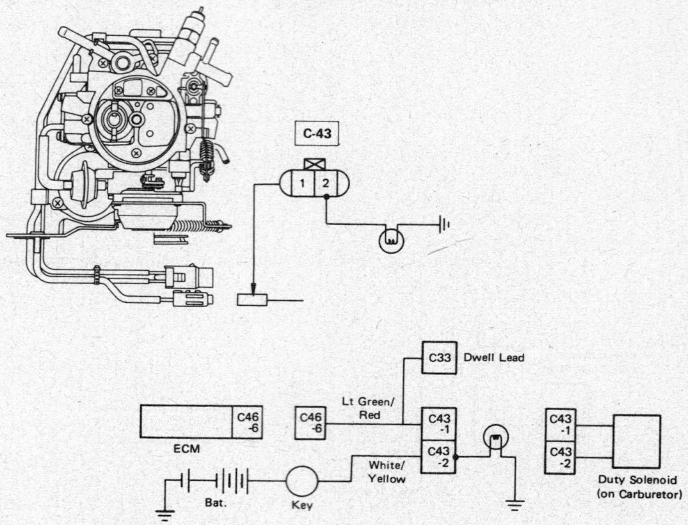

2.3L ENGINE—TROUBLESHOOTING GUIDE—CODE 25,53—PICK-UP AND AMIGO

TROUBLE CODE 25, 53
AIR MANAGEMENT VALVE CIRCUIT FAILURE

①
- Disconnect AMV connector
- Check AMV resistance
- The resistance should be over 20 Ω

NG: Replace AMV

OK:

Code No. 25

②
- Disconnect ECM connectors.
- Connect test light from battery ⊕ to ECM harness connector terminal "C46-11".

NG: Light "On" → Repair short to Bat. ⊕ in Light green wire to ECM term. "C46-11".

OK: Light "Off" →

④
- Connect AMV connectors.
- Connect test light from ECM terminal "C46-11" to ground
- Ignition "On" engine stopped.

Light "On" → Faulty ECM connector or ECM.

Light "Off" →

⑤
- Disconnect AMV connector.
- Connect test light from the AMV connector battery terminal White/Yellow lead to ground.
- Turn ignition "On" engine stopped.

OK: Light "On" → Repair open in wire from AMV to ECM.

NG: Light "Off" → Repair open Bat. + circuit to AMV (White/Yellow wire).

Code No. 53

③
- Disconnect ECM connectors.
- Connect test light from ground to ECM harness connector "C46-11".
- Ignition "On" engine stopped.

Light "On" → Repair short to battery ⊕ in Light green wire to ECM term. "C46-11".

Light "Off" → Faulty ECM connector or ECM.

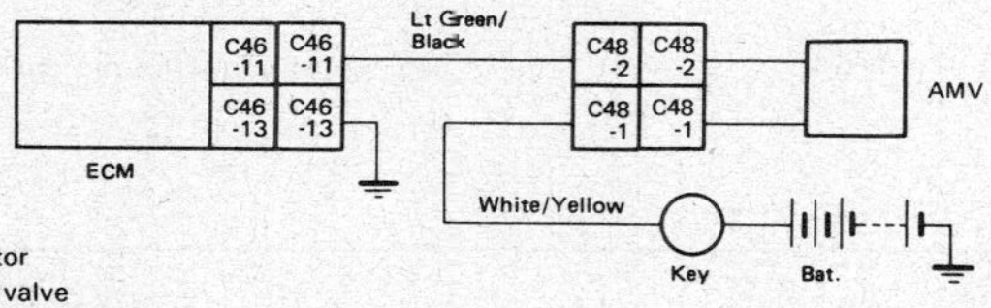

AIR : Air Ignition Reactor
AMV : Air Management valve
ECM : Electronic Control Module

① This step tests the resistance of the duty solenoid itself. A good solenoid will have over 20 ohms.

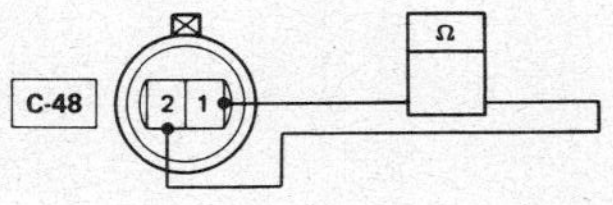

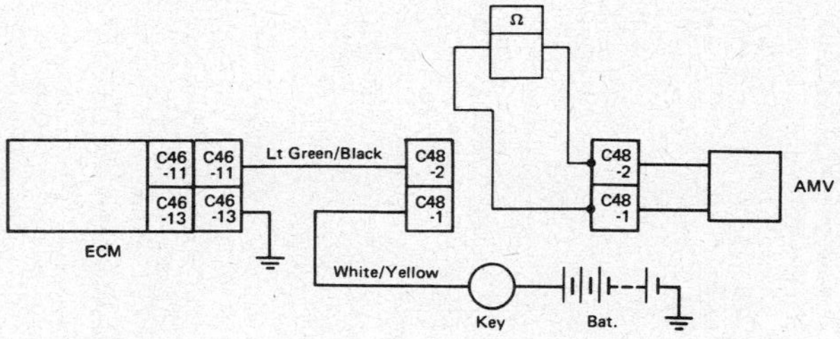

② Check for a short in the wire from AMV to ECM terminal.

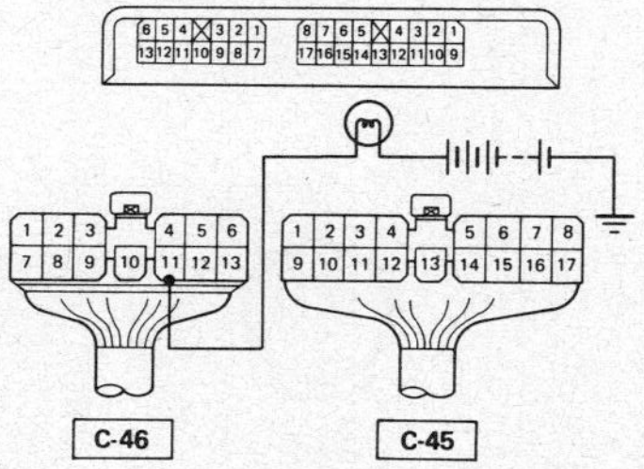

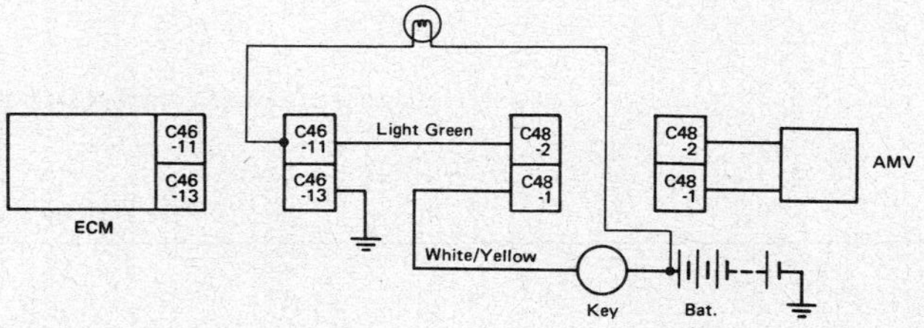

③ Check for a short in the battery (+) to ECM circuit.

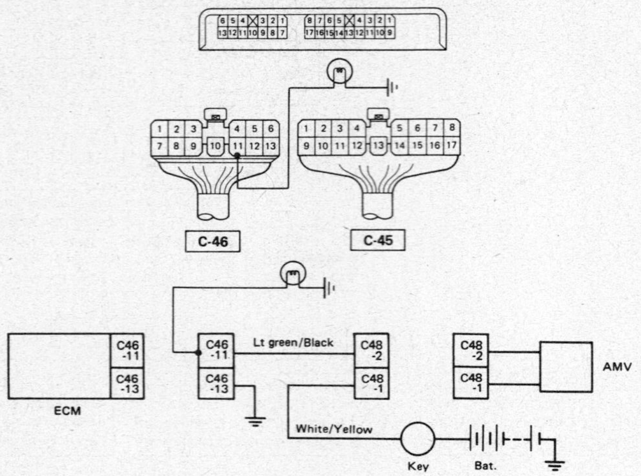

④ Check for an open in the wire from battery to AMV and AMV to ECM.

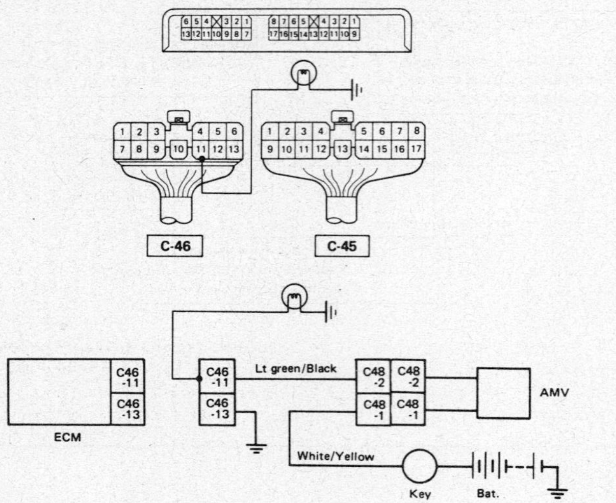

⑤ Check for an open in the wire from battery to the AMV.

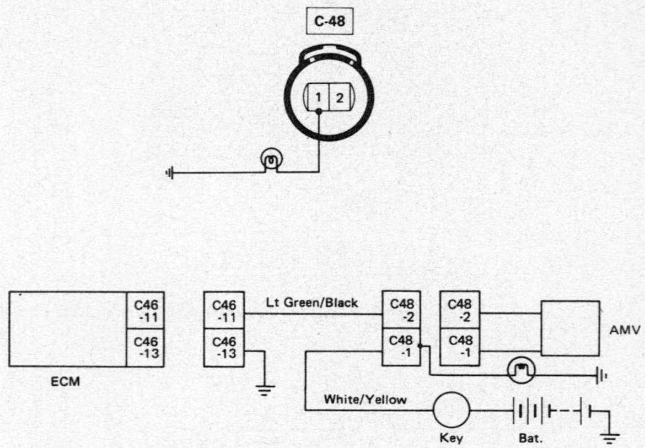

2.3L ENGINE—TROUBLESHOOTING GUIDE—CODE 26,27—PICK-UP AND AMIGO

TROUBLE CODE 26, 27
V.S.V.; CANISTER PURGE FAILURE

①
- Disconnect V.S.V. connector
- Check V.S.V. resistance
- The resistance should be over 30 Ω

NG → Replace V.S.V.

OK →

Code No. 26

②
- Disconnect ECM connectors.
- Connect test light from battery ⊕ to ECM harness connector terminal "C45-12".

NG → Light "On" → Repair short to Bat. ⊕ in Green/White wire to ECM term. "C45-12".

OK → Light "Off" →

④
- Connect V.S.V. connectors.
- Connect test light from ECM terminal "C45-12" to ground.
- Ignition "On" engine stopped.

Light "On" → Faulty ECM connector or ECM.

Light "Off" →

⑤
- Disconnect V.S.V. connector.
- Connect test light from the V.S.V. connector battery terminal White/Yellow lead to ground.
- Turn ignition "On" engine stopped.

OK → Light "On" → Repair open in wire from V.S.V. to ECM.

NG → Light "Off" → Repair open Bat. ⊕ circuit to V.S.V. (White/Yellow wire).

Code No. 27

③
- Disconnect ECM connectors.
- Connect test light from ground to ECM harness connector "C45-12".
- Ignition "On" engine stopped.

Light "On" → Repair short to battery ⊕ in Green/White wire to ECM term. "C45-12".

Light "Off" → Faulty ECM connector or ECM.

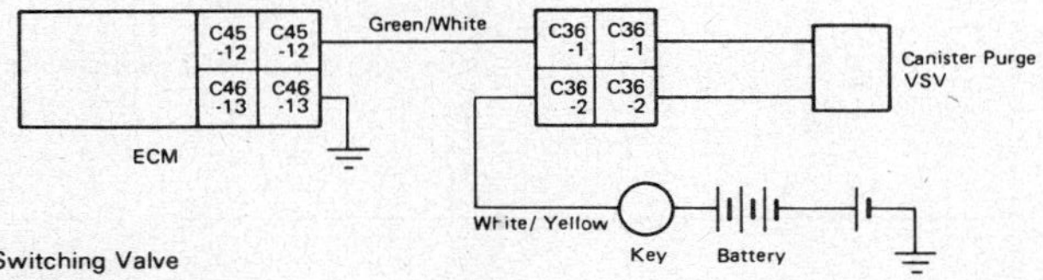

V.S.V. : Vacuum Switching Valve
ECM : Electronic Control Module

① This step tests the resistance of the V.S.V. itself. A good V.S.V. will have 30 ~ 40 ohms resistance.

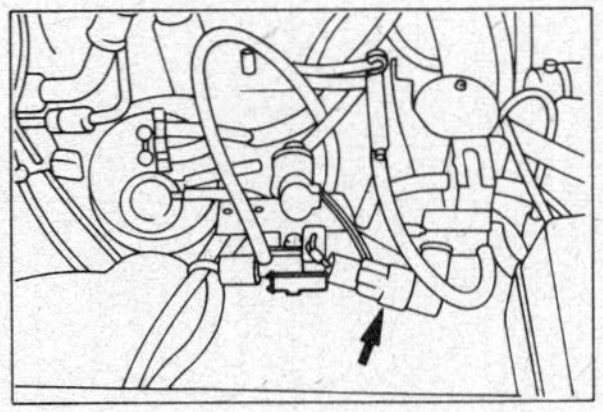

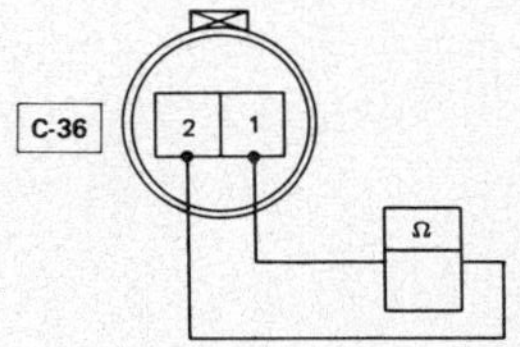

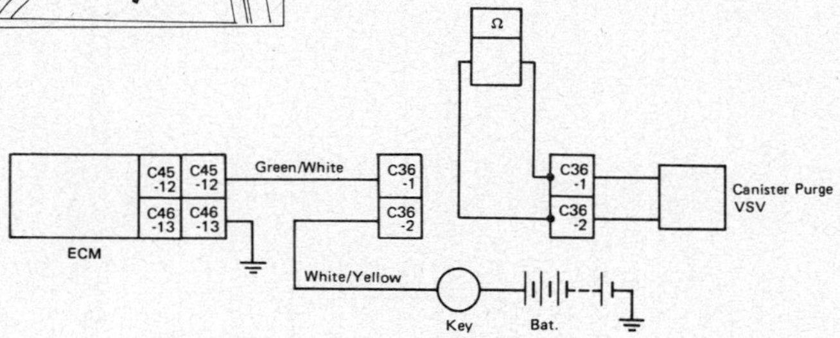

② Check for a short in the wire from V.S.V. connector to ECM terminal.

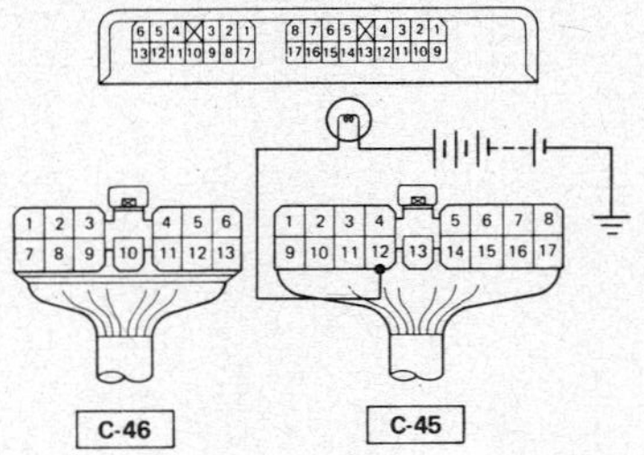

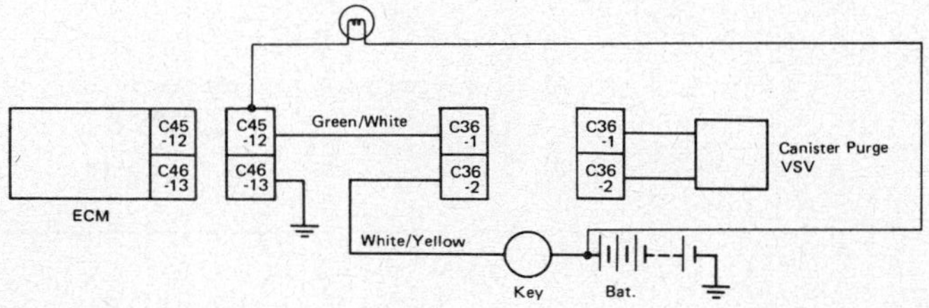

③ Check for a short in the battery (+) to ECM circuit.

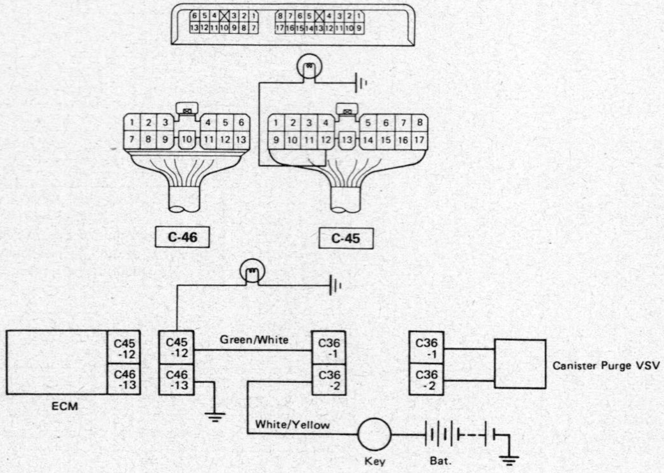

④ Check for an open in wire from battery to V.S.V. and V.S.V. to ECM.

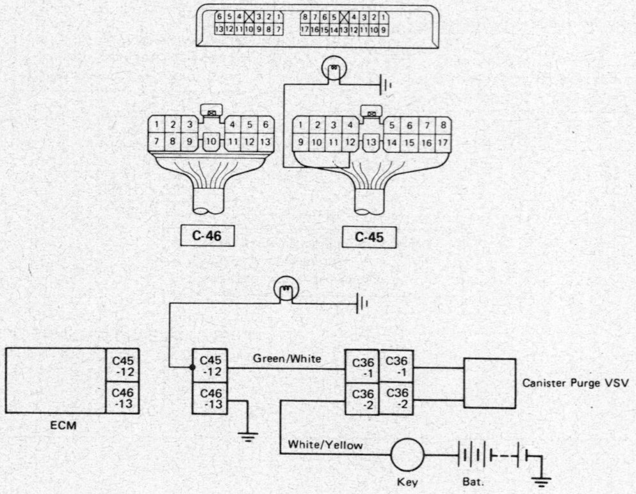

⑤ Check for an open in the wire from battery to the V.S.V.

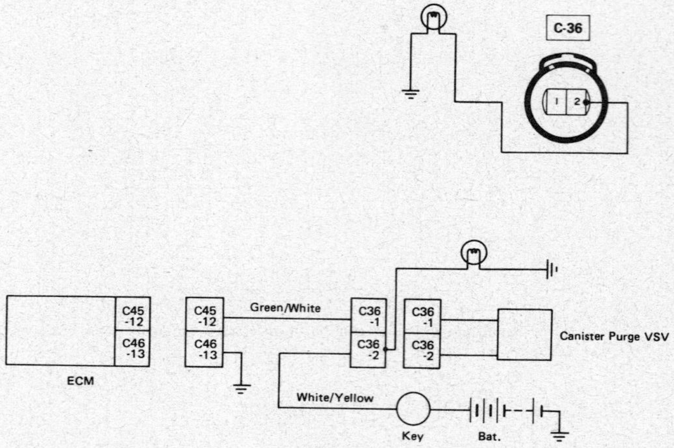

2.3L ENGINE – TROUBLESHOOTING GUIDE – CODE 31 – PICK-UP AND AMIGO

TROUBLE CODE 31
NO REFERENCE SIGNAL

① Check connections at two (2) ignition coil connectors, and four (4) terminal vacuum switches connector

- Not OK → Repair
- OK → ②
  - Connect voltmeter from ECM terminal "C46-10" to ground.
  - Ignition "on", engine stopped and check voltage.
    - OK: Over 10 volts → Faulty ECM connections or ECM.
    - NG: Under 10 volts → Check for open reference lead from ignition coil to ECM.

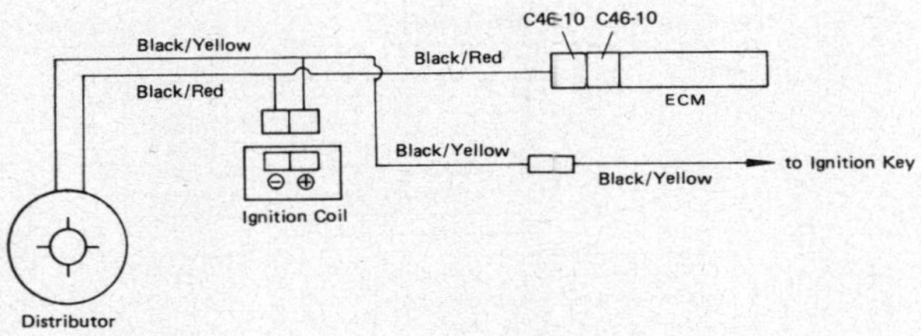

ECM : Electronic Control Module

① This step checks for a poor connection at the ignition coil wire connector as being the source of no reference pulse. Check for corrosion at connector terminals not fully seated, or terminal not properly attached to the wire. The terminal must be removed from the connector and carefully inspected.

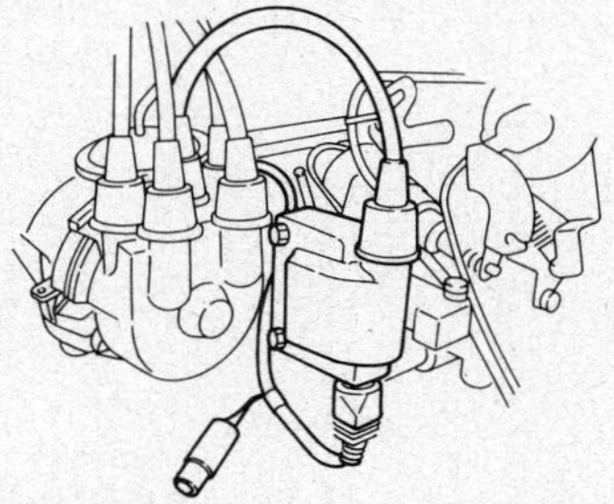

② This step determines if a reference pulse voltage is being sent to the ECM. Normal voltage should be over 10 volts. Check for good terminal contact at terminal "C46-10" of the ECM.

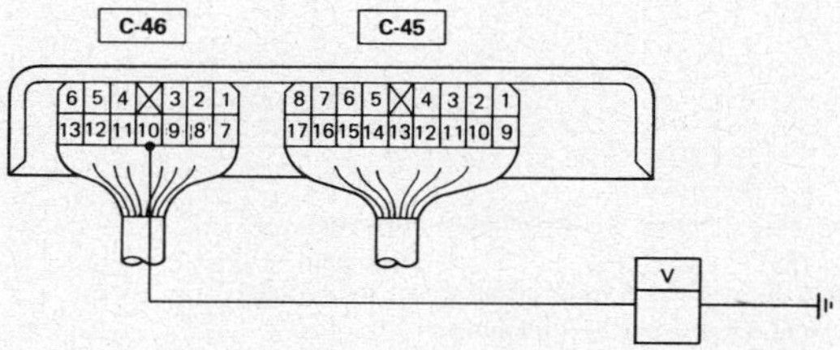

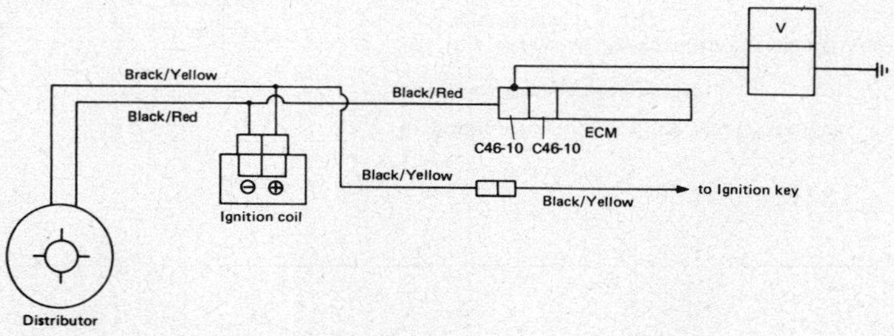

2.3L ENGINE—TROUBLESHOOTING GUIDE—CODE 32 PICK-UP AND AMIGO

TROUBLE CODE 32
EGR SYSTEM FAILURE AND
SENSOR CIRCUIT FAILURE

Signal code No. 32

- Code No. 14, 15, 21 ('91 model California only) and 34 is displayed
  - Go to Code No. 14, 15, 21 ('91 model California only) and 34
- Code No. 14, 15, 21 ('91 model California only) and 34 is not displayed
  - ① Warm up engine and run engine under the specified reappearance pattern.
    - Code No. 14, 15, 21 ('91 model California only) and 34 is displayed → Go to Code No. 14, 15, 21 ('91 model California only) and 34
    - Code No. 14, 15, 21 ('91 model California only) and 34 is not displayed
      - ② At 1/4 to 1/2 throttle, run engine with vehicle stopped, and rear wheels turning check the action of EGR valve shaft.
        - There is a lift of the EGR valve shaft
          - EGR gas passage clogging or EGR valve defective
            - Check EGR valve
              - OK: Clean EGR gas pipe and passage
              - NG: Replace EGR valve
        - There is not a lift of the EGR valve shaft
          - ③ Check that there is any defects of signal hose, such as break, clogging, deformation or leakage.
            - NG: Correct or replace the hose
            - OK: Stop engine. Set up vacuum testing tool
              - ④ • Disconnect the signal hose at CARBURETOR. • Apply vacuum to EGR valve via BPT and TVV, without via CARBURETOR. • Check the lift of EGR valve shaft.
                - OK: Clean or replace CARBURETOR
                - NG: ⑤ • Apply vacuum to EGR valve via BPT, without via TVV. • Check the lift of EGR valve shaft.
                  - OK: Replace TVV
                  - NG: ⑥ • Apply vacuum directly to EGR valve. • Check the lift of EGR valve shaft.
                    - OK: Replace BPT
                    - NG: Replace EGR valve

2.3L ENGINE—TROUBLESHOOTING GUIDE—CODE 32,34 PICK-UP AND AMIGO

①

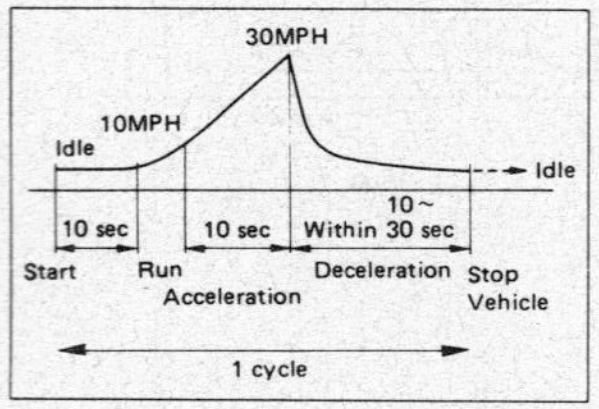

Inspection for Signal Code No. 32 and 34

Specified reappearance pattern

① Run engine at idle condition, at normal operating temperature.

② Keep at idle for more than 10 seconds, with vehicle stopped.

③ Drive vehicle at just over 10 MPH. Accelerate quickly to more than 30 MPH (must reach 30 MPH within 10 seconds of starting acceleration).

④ Within 10~30 seconds, decelerate safely to a complete stop.

⑤ Repeat above procedure ② through ④ several times.

②

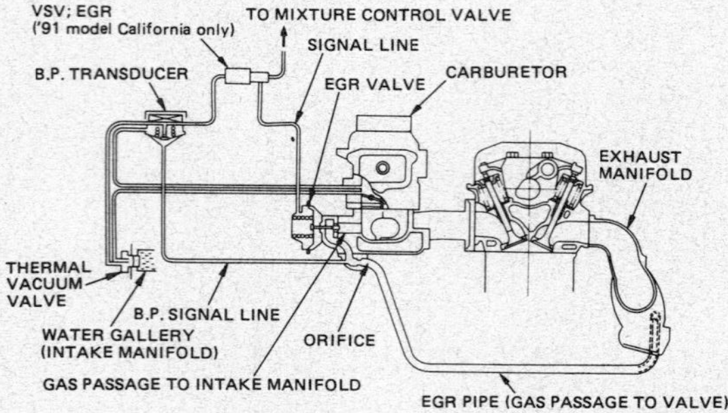

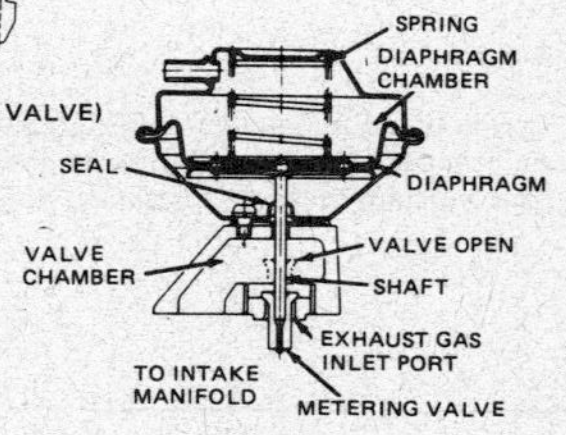

EGR VALVE

2.3L ENGINE—TROUBLESHOOTING GUIDE—CODE 32,34—PICK-UP AND AMIGO

③

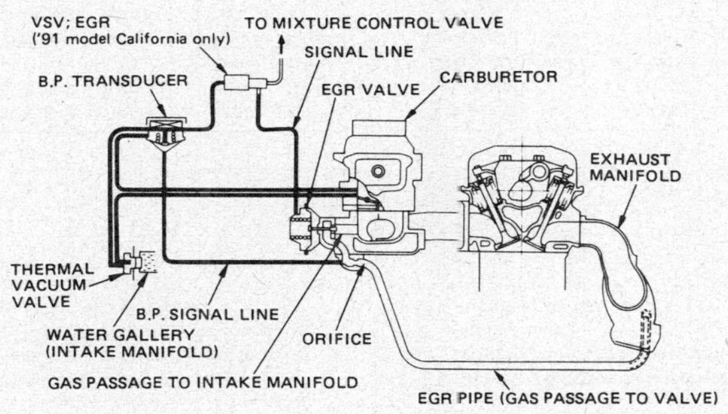

④

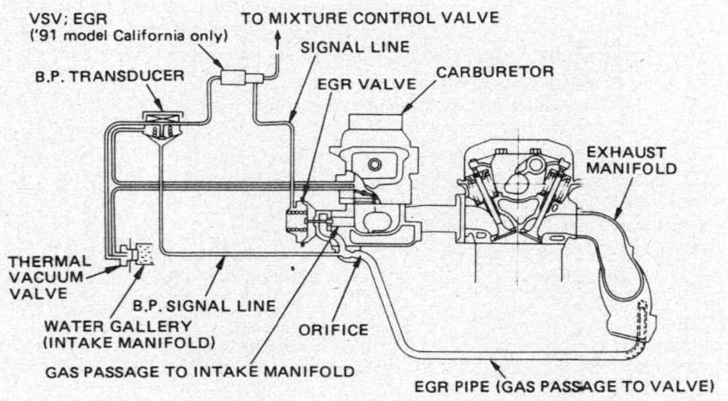

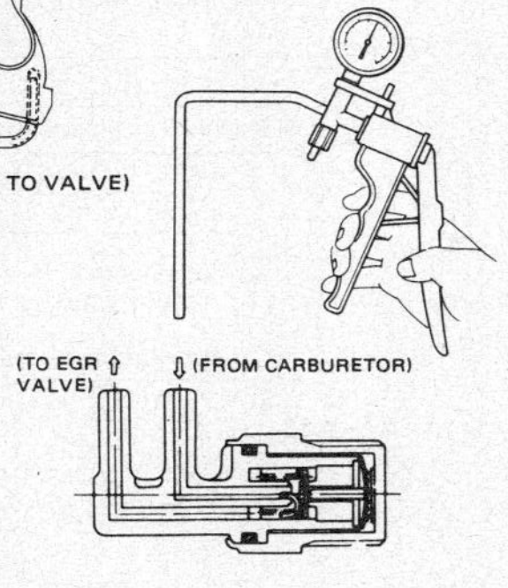

THERMAL VACUUM VALVE

⑤

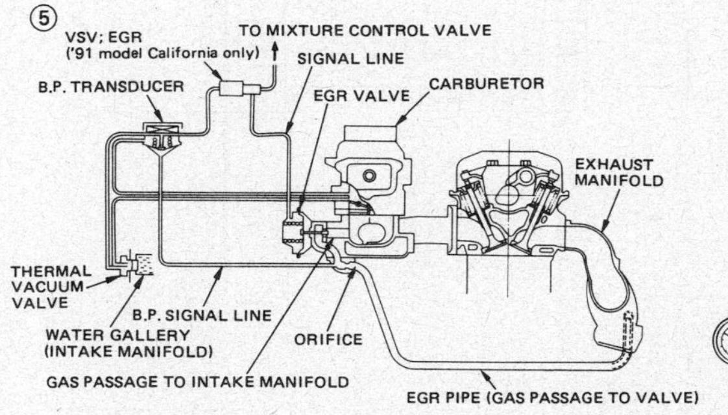

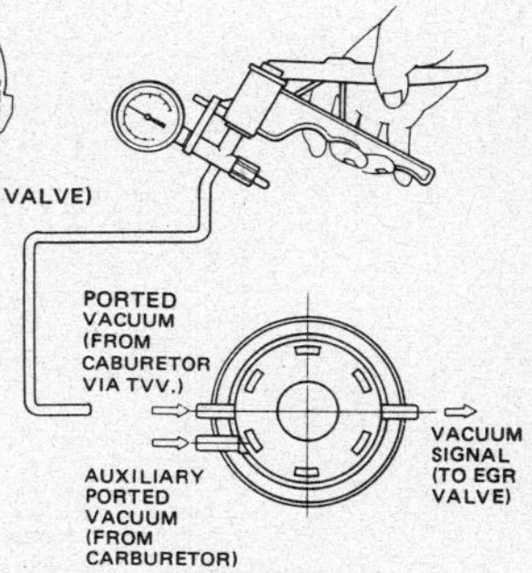

B.P. TRANSDUCER

⑥

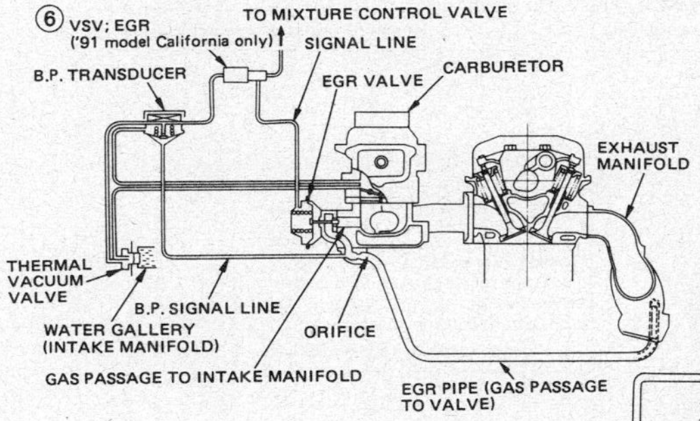

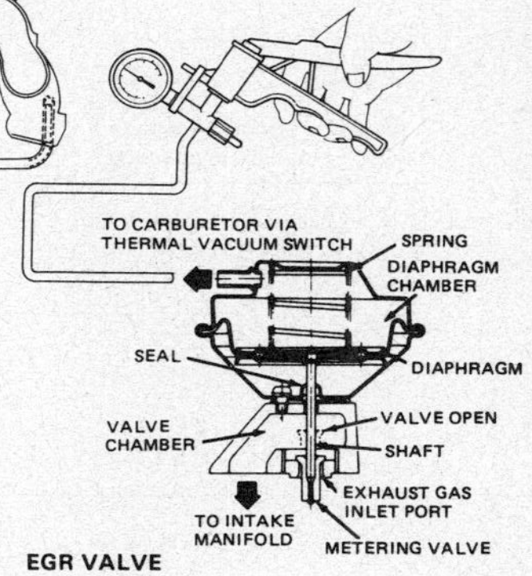

EGR VALVE

2.3L ENGINE—TROUBLESHOOTING GUIDE—CODE 34—PICK-UP AND AMIGO

TROUBLE CODE 34
EGR TEMPERATURE SENSOR OR HARNESS NG

- Run engine at idle condition at normal operating temperature more than 5 minutes.
- Then drive vehicle faster than 40 MPH, more than 10 minutes.
- "Check engine" light is either continuously or intermittently illuminated.

No code No. 34 → Refer to "INTERMITTENT"

Code No. 34 is displayed →
- Key off.
- Inspect EGR gas temp. sensor for loose fit, contamination or deformation of terminals or housings.

NG → Correct or replace

OK →

①
- Disconnect EGR temp. sensor connector.
- Check resistance from the EGR temp. sensor connector (harness side) terminal Black/Blue lead to ground.
- It should be less than 1 Ω

OK →

②
- Disconnect ECM connectors.
- Check resistance between terminal "C45-6" of ECM harness connector (not ECM side) to ground.
- It should be ∞.

OK →

④
- Remove cable side terminal of "C45-6" from ECM connector.
- Connect ECM connectors.
- Connect voltmeter from ECM terminal "C45-6" to ground.
- Ignition "On" engine stopped and check voltage.
- It should be "OV".

→ (A)

Not OK → Repair ground from ECM term. "C45-15" to sensor connector

Not OK (from ①) →

③
- Check resistance between terminal "C46-7" of ECM harness connector and ground.
- It should be less than 1 Ω.

OK →

⑤
- Disconnect ECM harness connectors.
- Check resistance between terminals "C45-15" and "C38-2".
- It should be less than 1 Ω.

OK → Faulty ECM connector or ECM

Not OK → Repair open in harness from sensor to ECM terminal "C45-15"

Not OK (from ③) → Repair open in harness from ECM terminal "C46-7" to ground

(A)

OK →

⑥
- Disconnect ECM connectors.
- Check resistance between terminal "C45-6" of ECM harness connector (not ECM side) to terminal "C38-1" of EGR gas temp. sensor connector.
- It should be less than 1 Ω.

OK →

⑦
- Remove EGR gas temp. sensor.
- Measure the resistance of EGR gas temp. sensor.
- Check the resistance table for EGR gas temperture sensor.

OK → Faulty ECM connector or ECM

Not OK → Replace EGR gas temp. sensor

Not OK (from ⑥) → Repair open in harness from ECM terminal "C45-6" to terminal "C38-1" (Yellow/Green lead)

Not OK (from (A)) → Repair short in wire of ECM terminal "C45-6" and other wires

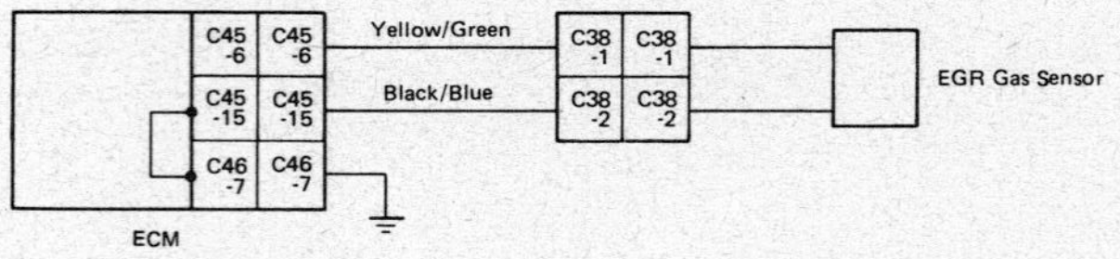

2.3L ENGINE—TROUBLESHOOTING GUIDE—CODE 34—PICK-UP AND AMIGO

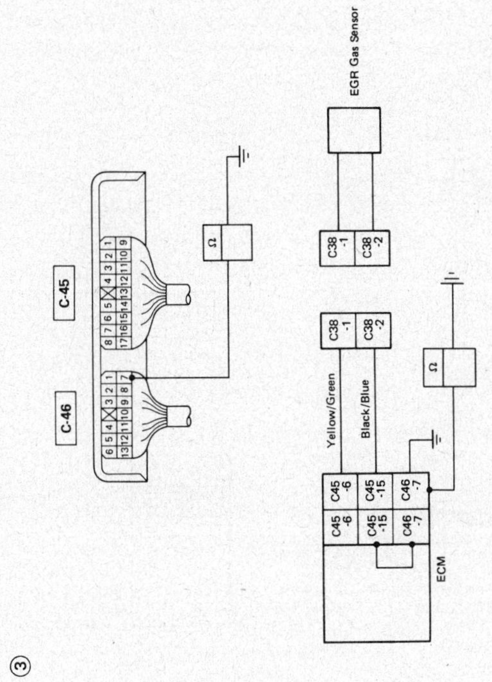

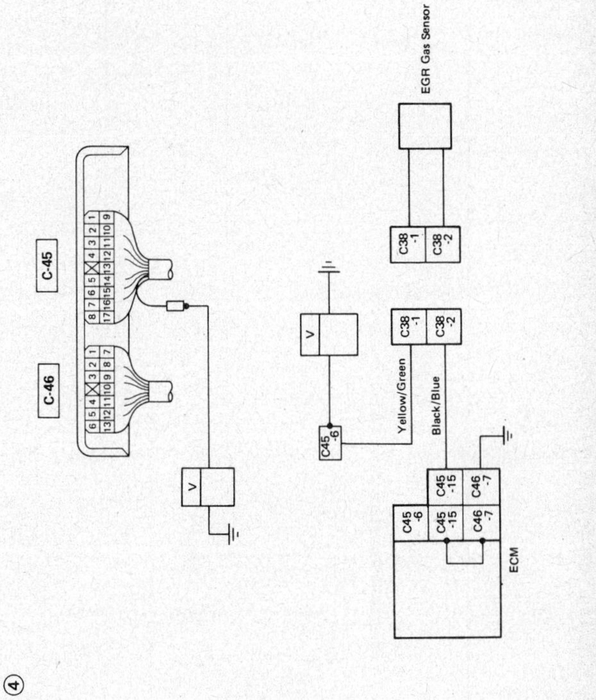

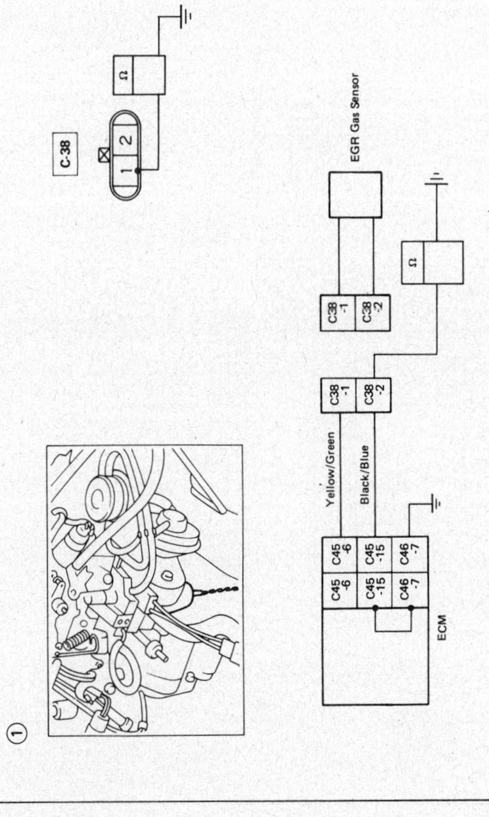

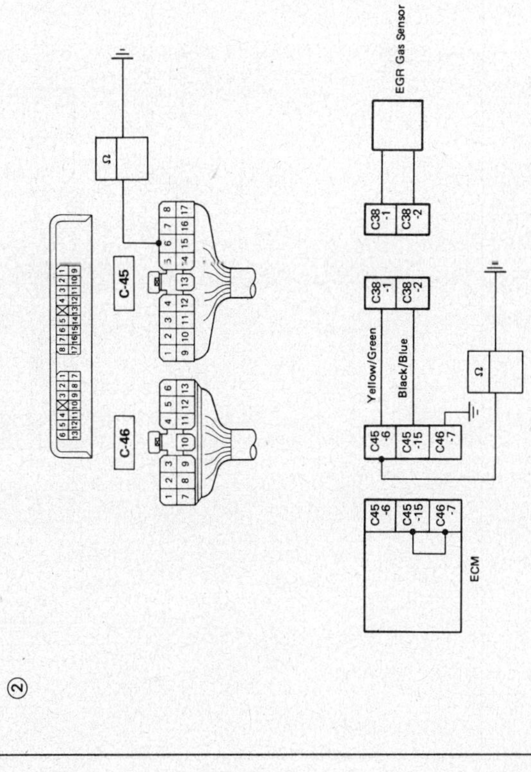

⑤

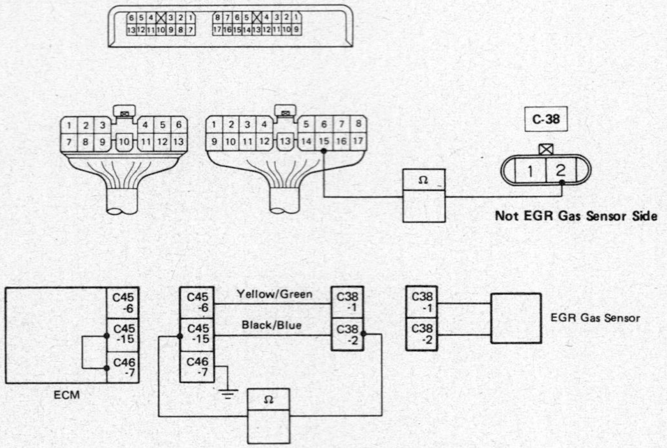

⑥

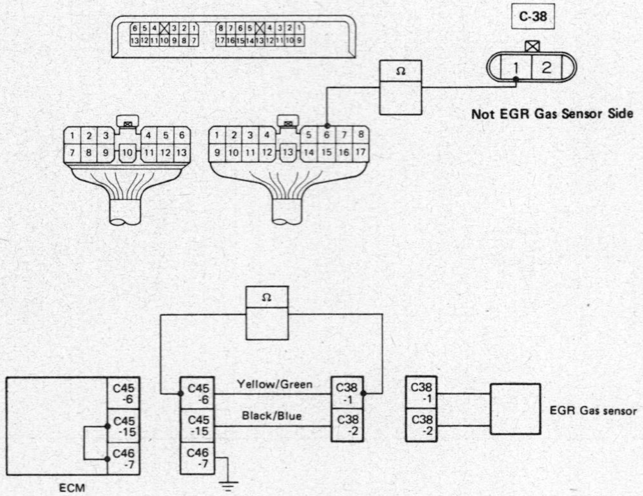

⑦ This step tests the resistance of the EGR gas sensor.

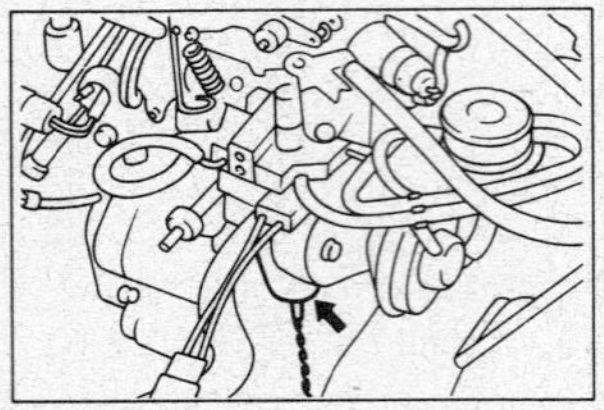

Temperature °C(°F)	Resistance KΩ
0 (32)	28
20 (68)	12.1
40 (109)	5.8
80 (176)	1.7

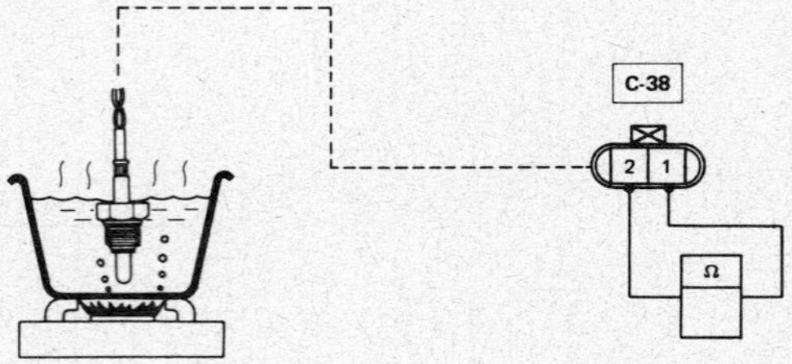

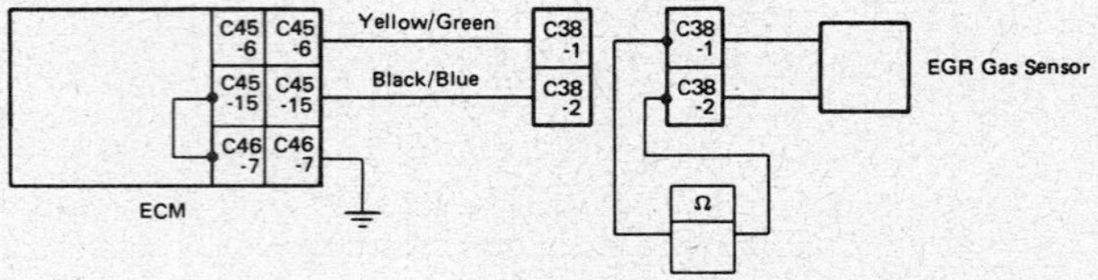

2.3L ENGINE—TROUBLESHOOTING GUIDE—CODE 44—PICK-UP AND AMIGO

TROUBLE CODE 44
OXYGEN SENSOR — LEAN INDICATION

①
- Connect dwell meter to dwell lead. (Use 4-cylinder scale.)
- Warm up engine until upper radiator hose becomes warm.
- Start engine and choke engine at part throttle.
- Note dwell.

OK — Dwell increases
- Check for air leak (including air management system
- Check for exhaust leak ahead of oxygen sensor.

  - **OK** — No leak → Check for carburetor duty solenoid
    - OK → carburetor check procedure.
    - Not OK → Replace or repair.
  - **NG** — Leak → Repair and recheck.

NG — Not dwell change

②
- Disconnect oxygen sensor.
- Connect dry battery (1.5V) + to violet wire to ECM (not oxygen sensor) and dry battery — to ground.
- Note dwell.

  - **OK** — Increases → Replace oxygen sensor.
  - **NG** — Under 10° → Check for grounded Violet wire to ECM term. "C45-8". If not grounded, replace ECM.

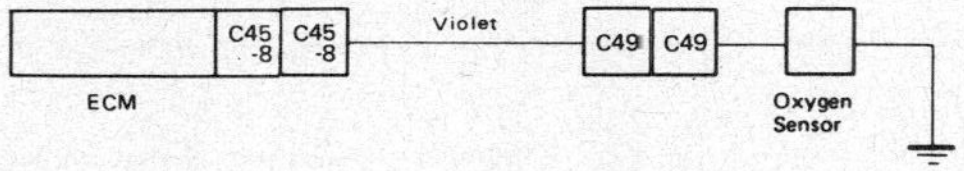

ECM : Electronic Control Module

① Determine whether or not the O_2 sensor is delivering a lean indication to the ECM unit.

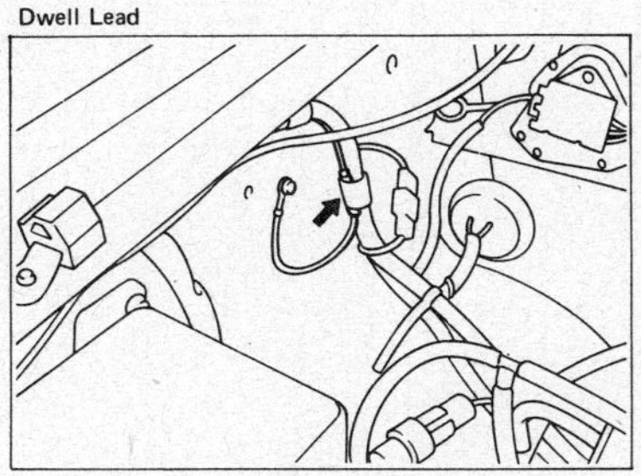

② Determine whether or not the harness from the O_2 sensor to the ECM unit is grounded.

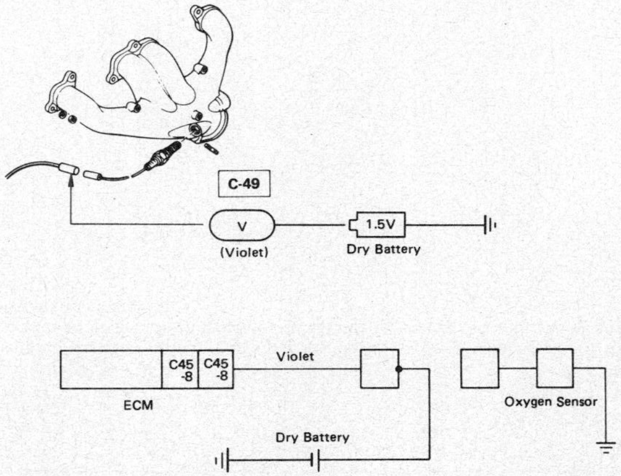

TROUBLE CODE 45
OXYGEN SENSOR — RICH INDICATION

Note: A restricted air cleaner can cause a code 45.

①
- Connect dwell mater to dwell lead.
- Start engine and run at part throttle until dwell starts to vary, then with engine idling, note dwell.
- Remove large vacuum hose (such as canister purge hose at canister) to cause an air leak (but not enough to stall engine).

OK → Dwell decreases → Check carburetor duty solenoid
- OK → carburetor check procedure.
- Not OK → Replace or repair

NO → No dwell change →
- Disconnect oxygen sensor.
- Ground connector terminal on lead to ECM (not oxygen sensor).

- OK → Dwell drops to under 10° → Faulty oxygen sensor connection or sensor
- NG → No dwell change → Faulty ECM connection or ECM

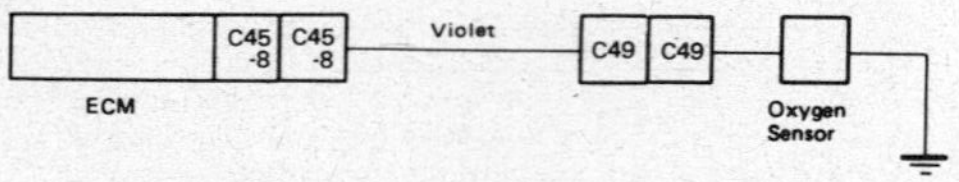

ECM : Electronic Control Module

① This step tests for a lean condition, (air leak), in the engine to see if the ECM can respond. A decrease in the dwell indicates ECM and O_2 sensor are not faulty and looks for source of constant rich condition.

Dwell Lead

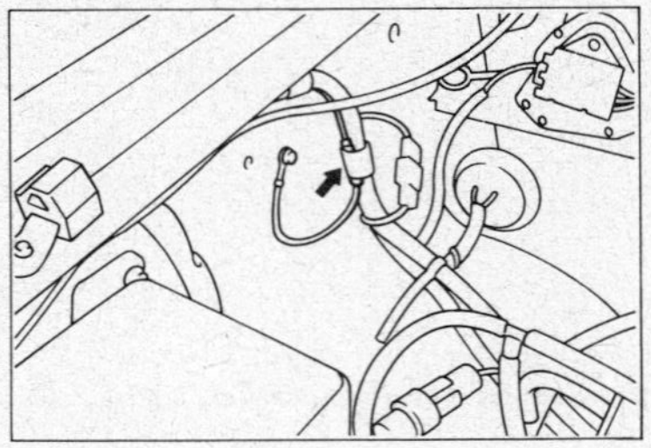

② This step tests to see if ECM is able to respond to a lean exhaust O_2 signal (low voltage).

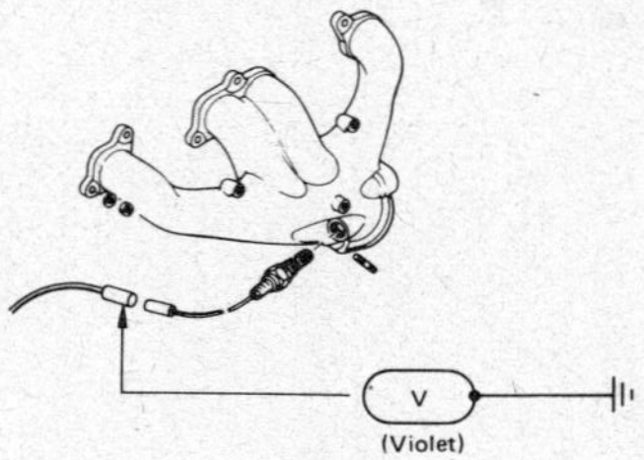

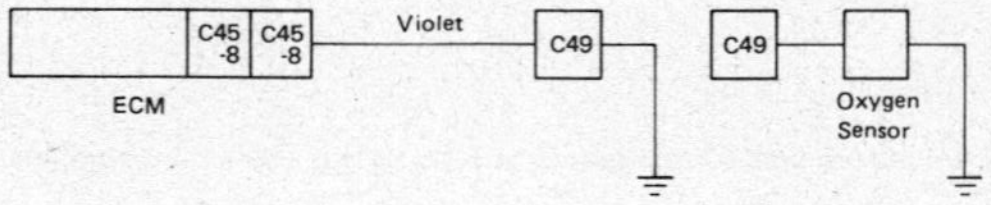

2.3L ENGINE—TROUBLESHOOTING GUIDE—CODE 52 PICK-UP AND AMIGO

TROUBLE CODE 52
RAM PROBLEM

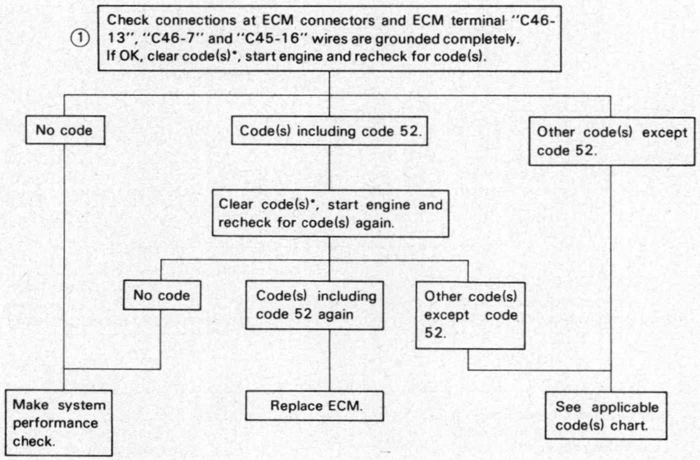

* See Code(s) Clearing Procedure

ECM : Electronic Control Module

2.3L ENGINE—TROUBLESHOOTING GUIDE—CODE 55 PICK-UP AND AMIGO

TROUBLE CODE 55
A/D CONVERTER PROBLEM

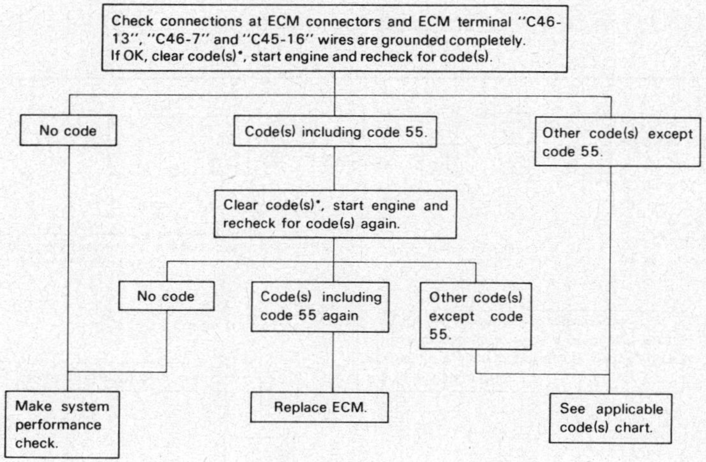

* See Code(s) Clearing Procedure

ECM : Electronic Control Module

2.3L ENGINE—TROUBLESHOOTING GUIDE PICK-UP AND AMIGO AIR MANAGEMENT

①
- Coolant at normal operating temperature.
- Start engine and run at idle, observe direction of air in the first 3 seconds. Air should go to the exhaust ports for 3 seconds and then air should go to atmosphere port. Does it?

YES:
- Air management system is operating properly.
- Check for manifold vacuum signal at valve with engine idling. Vacuum should be 10'HG (34 kPa). Repair if required.

NO:

②
- Ignition "ON" and engine stopped.
- Disconnect connector from valve solenoid and connect a test light between harness connector terminals. Is the test light "ON"?

YES:
- Check for grounded wire from solenoid to ECM.
- If not grounded, replace ECM.

NO:

③
- Ground diagnostic terminal.
- Is the test light "ON"?

YES:
- It is a faulty valve solenoid connector or solenoid. Replace AMV.

NO:

④
- Connect test light between harness terminal "A" to ground.
- Is the test light "ON"?

YES:
- Check for an open in wire from solenoid to ECM.
- If OK, check resistance of solenoid windings.
- If under 20 OHMs, replace AMV.

NO:

⑤
- Check for open fuse or open in wire to ignition.

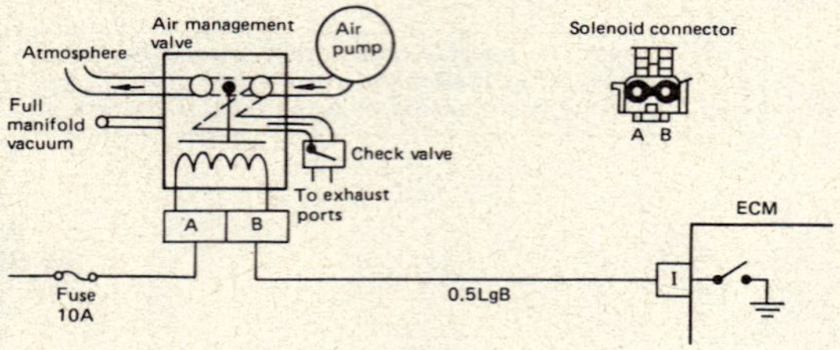

2.3L ENGINE—TROUBLESHOOTING GUIDE—IDLE SWITCH PICK-UP AND AMIGO

- Engine stopped.
- Disconnect Idle switch connector and check resistance between terminals "C61-1" and "C61-2", of the switch connector (not harness side).
- The resistance should be 0 ohm.

Not OK: Replace Switch Asm. → Make system performance check.

OK:
- Disconnect vacuum hose from the switch and connect vacuum pump.
- Applying 4 inch Hg. vacuum to switch by vacuum pump, check resistance between terminals "C61-1" and "C61-2", of the connector (not harness side) and note vacuum change.
- The resistance should be over 100K ohms.
- The vacuum should not decrease.

Not OK: Replace Switch Asm. → Make system performance check.

OK:
- Idle switch is OK.
- Check for short between Idle switch wires from connector to ECM or grounded wires. If shorted or grounded, repair wires.

→ Make system performance check.

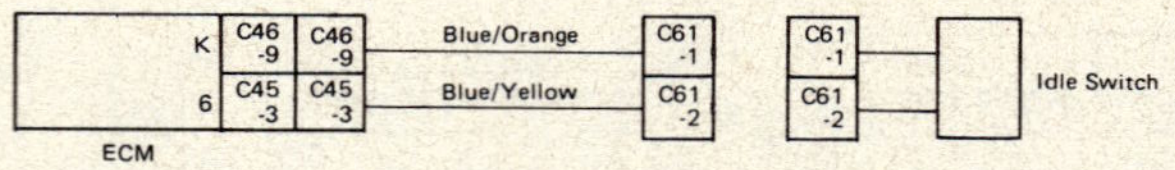

2.3L ENGINE—TROUBLESHOOTING GUIDE—DUTY SOLENOID—PICK-UP AND AMIGO

- Disconnect oxygen sensor connector and vacuum switch connector.
- Ignition "on", engine stopped.
- Check repeated clicks from duty solenoid.

No repeated clicks → Replace duty solenoid.

Repeated clicks → "carburetor"

Component Replacement

OXYGEN SENSOR

The oxygen sensor uses a permanently attached pigtail connector. The pigtail should not be removed from the oxygen sensor. Damage or removal of the pigtail or connector could affect proper operation of the oxygen sensor. Care should be taken when handling the oxygen sensor in order to preserve its effciency. The in-line electrical connector and louvered end must be kept free of grease and other contaminants. Do not use cleaning solvents of any type to clean the oxygen sensor.

Removal and Installation

The oxygen sensor may be difficult to remove when the temperature is below 120°F. Do not use excessive force. Damage to the threads in the exhaust manifold or exhaust pipe will result.

1. Start the engine and let it reach normal operating temperature. Shut off the engine.
2. Remove the sensor electrical connector and attaching hardware.
3. Using a suitable oxygen sensor socket, remove the oxygen sensor.
4. Be sure the new oxygen sensor is pre-coated with anti-seize compound. If the old sensor is to be reinstalled, it will be necessary to coat the threads of the oxygen sensor with a suitable anti-seize compound that is compatible with oxygen sensors. Do not contaminate the oxygen sensor with a noncompatible anti-seize compound.
5. Install the sensor and torque it to 31 ft. lbs. Connect the electrical connector and attaching hardware.

CARBURETOR

The carburetor used is a Stromberg manufactured by Hitachi. This carburetor is a 2- barrel, down draft type composed of a low speed (primary) side and high speed (secondary) circuits which are integrated into a single unit.

Removal and Installation

1. Disconnect the PCV hose from the cylinder head cover.
2. Disconnect the AIR hose from the air pump.
3. Disconnect the TCA hose from the air cleaner.
4. Disconnect the air hose from the slow actuator and vacuum controller. Remove the bolts attaching the air cleaner and remove the wing nut.
5. Lift the air cleaner slightly and disconnect the thermo sensor vacuum hose and air duct, then remove the air cleaner.
6. Disconnect the lead wire connector, accelerator control cable and detent cable (Auto. trans).
7. Disconnect the fuel lines at the carburetor. Disconnect the evaporative emission control system (ESC) hose from the carburetor.
8. Remove the 4 nuts and lockwashers securing the carburetor to the mainfold and remove the carburetor from the engine.
9. Installation is the reverse order of the removal procedure. Be sure that the early fuel evaporation (EFE) grid is in good working condition. Torque the carburetor attaching nuts to 12–18 ft. lbs.

Off Vehicle Carburetor Adjustments

THROTTLE VALVE CLEARANCE

1. Fully close the choke valve (labeled **1**).
2. Align fast idling screw **2** with fast idling cam **3** first step **A**.
3. Measure clearance **B** bewteen primary throttle valve **4** and the throttle valve chamber wall. The clearance should be 0.048–0.058 with a 25 degree angle on the primary throttle valve.
4. If the clearance is not within specification, adjust it using idle adjusting screw **2**.

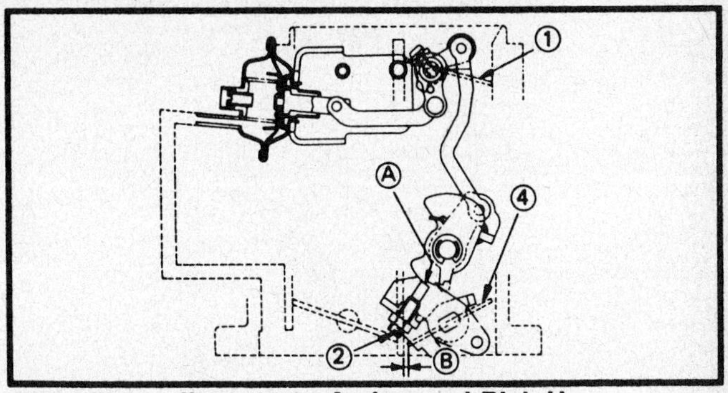

Fast idling adjustment—Amigo and Pick-Up

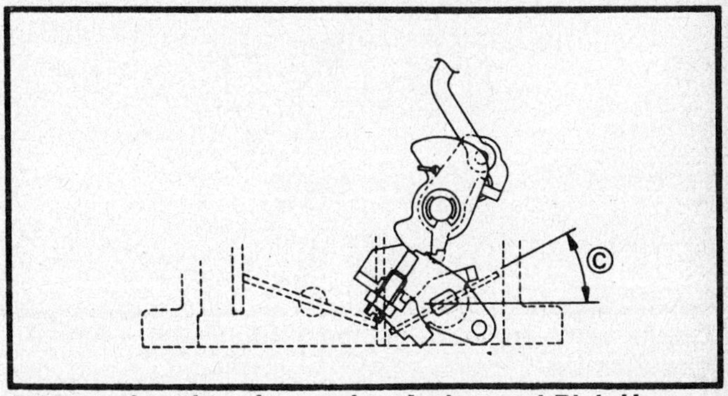

Primary throttle valve angle—Amigo and Pick-Up

PRIMARY AND SECONDARY THROTTLE VALVE INTERLOCK

1. Slowly open the primary throttle valve **1** until kick lever **2** tang **A** conatacts return plate **3**.
2. Measure clearance **B** between the primary throttle valve and the throttle valve chamber wall. The clearance should be 0.272-0.331 with a 57 degree angle on the primary throttle valve.
3. If the clearance is not as specified, adjust by carefully bending the kick lever tang.

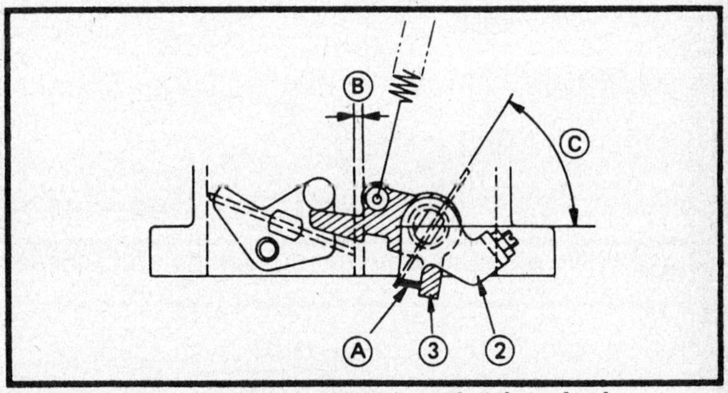

Primary and secondary throttle valve interlock adjustment—Amigo and Pick-Up

CHOKE UNLOADER

1. Fully open the primary throttle valve **1**.
2. Measure clearance **A** between choke valve **2** and the choke valve chamber wall. The clearance should be 0.105-0.129 in with a choke valve angle of 50 degrees.
3. If the clearance is not as specified, adjust by carefully bending lever tang **B**.

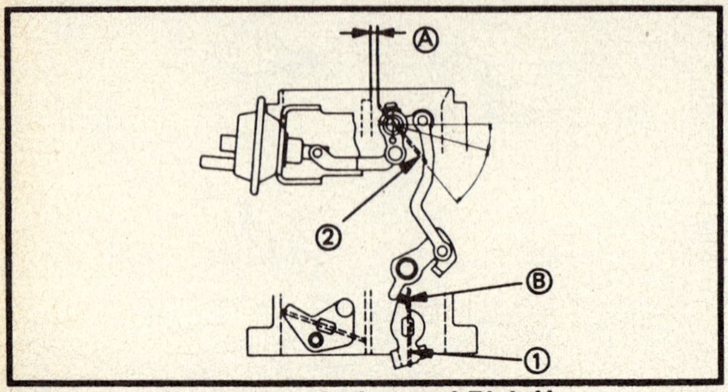

Unloader adjustment—Amigo and Pick-Up

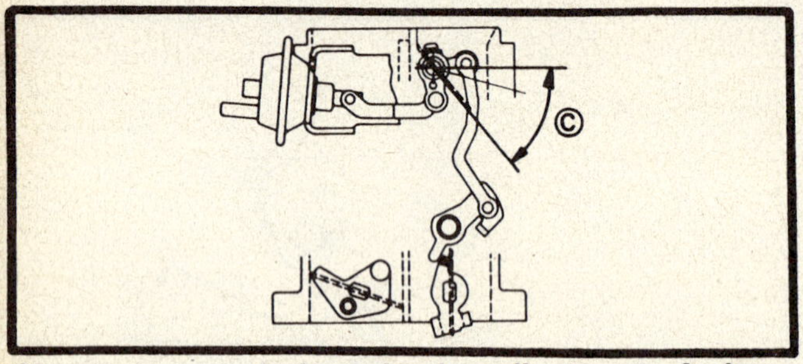

Choke valve angle for unloader adjustment—Amigo and Pick-Up

CHOKE VALVE OPENING

1. Move the tip of fast idle screw **1** until it contacts fast idling cam **2** second step **A**.
2. Measure clearance **B** between choke valve **3** and the throttle chamber wall. The clearance should be 0.031–0.050 with a choke valve angle of 30 degrees.
3. If the clearance is not as specified, adjust by carefully bending the counter lever tang.

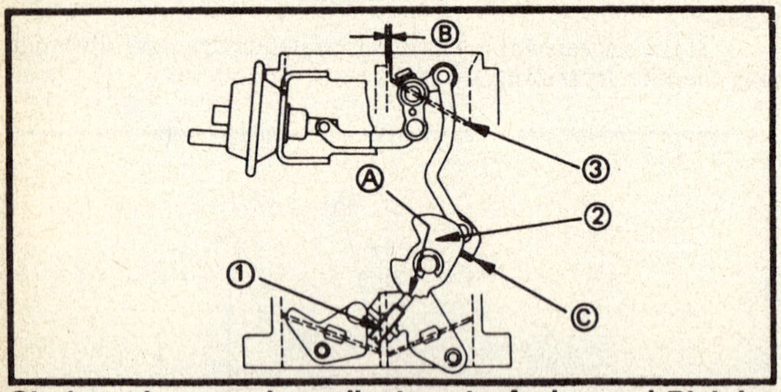

Choke valve opening adjustment—Amigo and Pick-Up

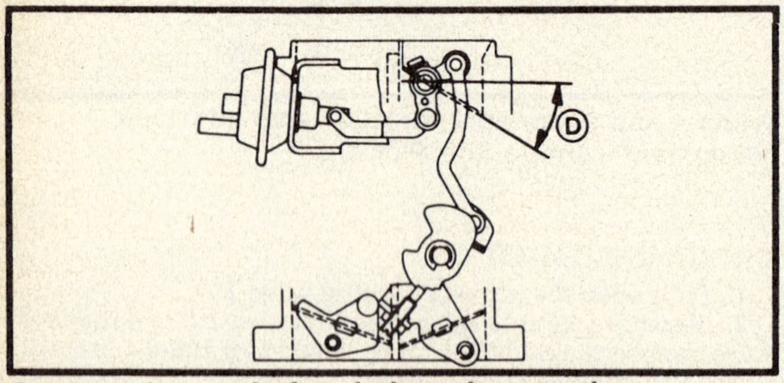

Choke valve angle for choke valve opening adjustment—Amigo and Pick-Up

FLOAT LEVEL

1. Start the engine and allow it run at idle speed.
2. Check the float chamber glass level mark **1**.
3. If the fuel level is not even with the mark, adjust it by carefully bending float seat **2**.

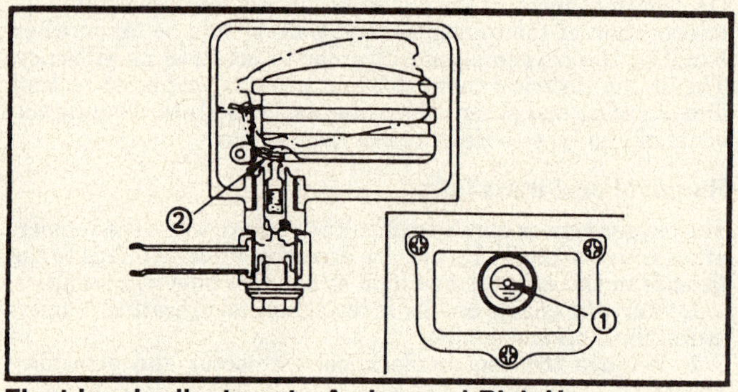

Float level adjustment—Amigo and Pick-Up

Overhaul

When disassembling or assembling the carburetor, use the proper size tools. Exercise care to guard against parts damage and keep parts free from dirt, dust and other foreign matter. Keep disassembled parts in separate groups, according to the systems from which they were removed to avoid interchanging parts. Clean all disassembled parts in carburetor cleaner and clean out the orifices with compressed air. Never use a piece of wire to clean the jets or orifices, as adjacent area may be scratched.

Wash and clean the cast parts with a soft brush in carburetor cleaner. Clean off all the carbon around the throttle valve. Wash the other parts thoroughly in carburetor cleaner. Blow all dirt and other foreign matter from the jets, fuel passages and restrictions in the body. Check to be sure that no clogging is in the air passages and fuel passages. Check for smooth operation of the throttle and choke linkage. If the operation is not smooth, clean them up. They may be replaced at this time to ensure proper carburetor performance. When replacing a main jet or a pilot jet, the old jet and the new jet must be the same size, because the jet is selected after the exact flow measurement by the factory (a number is stamped on each jet). Do not reuse old O-rings and gaskets. Replace them with new ones at every reassembly.

When reassembly of the accelerator pump is completed, fill the cylinder with fuel and check to be sure the fuel is injected smoothly. When reassembling the accelerator pump parts, be careful not to bend the piston connecting rods. When reassembling the main actuator, apply a thin film of pertroleum jelly or equivalent on the O-ring and tighten the screws carefully to prevent the O-ring from cracking.

Disassembly

1. Disconnect the accelerator pump spring. Remove the pump lever spring and then the pump arm assembly.
2. Remove the E-clip and the connecting link from the cam lever of the automatic choke.
3. Remove the 7 air horn attaching screws and lift the air horn straight up of the carburetor body.
4. Remove the float pin. Pull out the float pin and remove the float assembly. Remove and discard the air horn gasket.
5. Remove the float valve assembly, seat and packing. Remove the plunger. Remove the 2 screws holding the feedback solenoid valve to the air horn and remove the solenoid. Be sure not to tilt or bind the valve when removing.
6. Remove the choke assembly E-clip and piston link. Remove the choke breaker diaphragm unit by removing the E-clip and

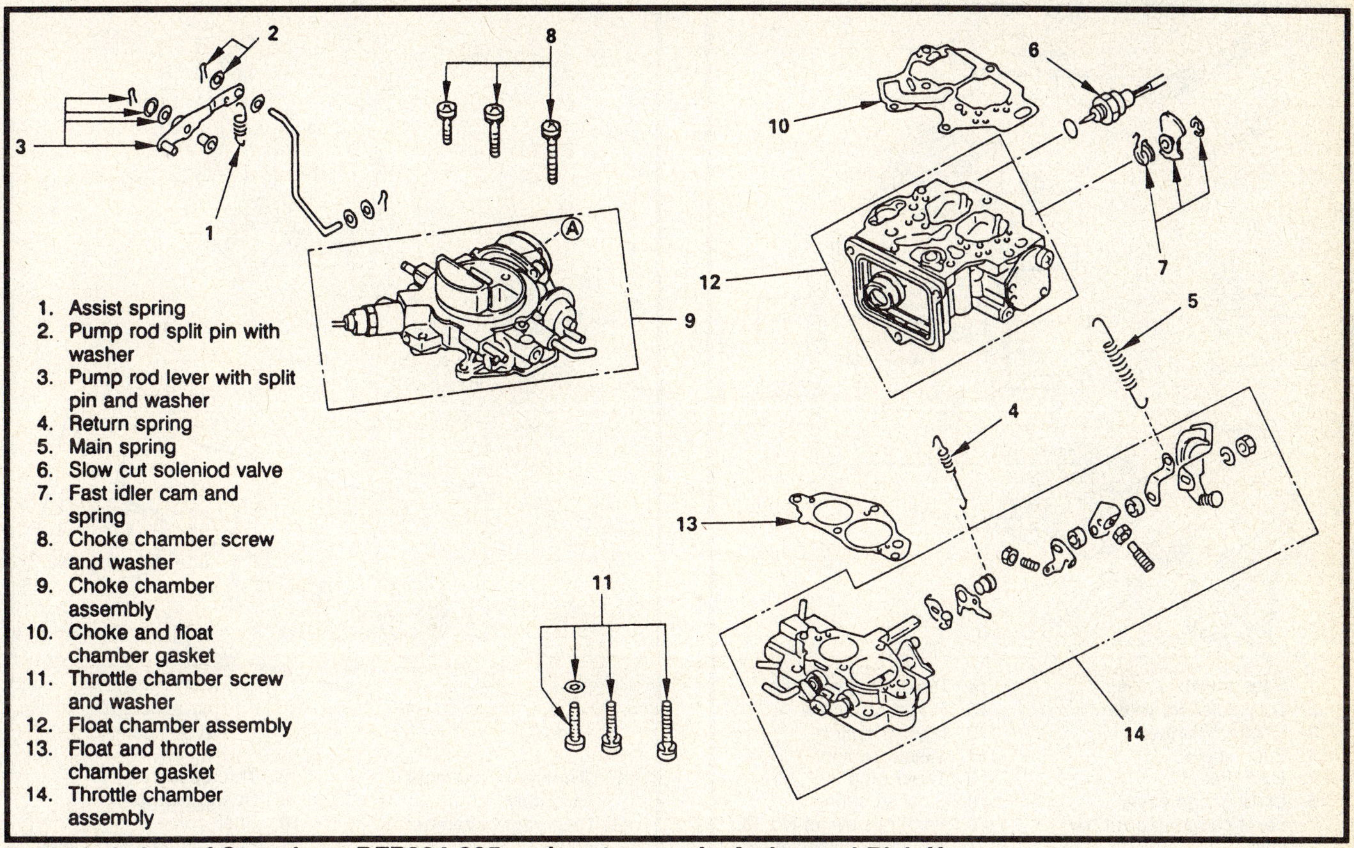

Exploded view of Stromberg DFP384-205 carburetor used—Amigo and Pick-Up

the coupling from the connecting link and loosen the 2 screws so as to remove the choke breaker diaphragm unit.

7. Remove the throttle return spring and the secondary throttle return and second lock spring.

8. Remove the E-clip and disconnect the diaphragm rod from the secondary valve shaft. Remove the slow cut solenoid valve.

9. Remove the fuel strainer clip, fuel strainer and remove the snap ring in the pump well and take out the filter and the inlet check ball. Be sure not to score the pump bore when removing the snap ring.

10. Remove the main jet plug retainer and plug. Remove the main jet with packing. Remove the primary small venturi and second small venturi assembly. Remove the main air bleed jets.

11. Remove the primary slow jet plug, remove the plug and take out the spring, pump weight and check ball.

12. Remove the secondary slow jet plug and remove the slow jet. Remove the idle up jet.

13. Remove the mixture by-pass solenoid valve. Remove the throttle position sensor retaining screws and remove the sensor.

General Inspection

Wash and clean the cast parts with a soft brush in carburetor cleaner. Clean off all the carbon around the throttle valve. Wash the other parts throughly in carburetor cleaner. Blow all dirt and other forgien matter from the jets, fuel passages and restrictions in the body. Check to be sure that no clogging is in the air passages and fuel passages. Check for smooth operation of the throttle and choke linkage. If the operation is not smooth, clean them up. They may be replaced at this time to ensure proper carburetor performance. When replacing a main jet or a pilot jet, the old jet and the new jet must be the same size, because the jet is selected after the exact flow measurement by the factory (a number is stamped on each jet). Do not reuse old O-rings and gaskets. Replace them with new ones at every reassembly.

Air Vent Solenoid Valve Inspection

1. Install the spring and the valve to the body. Be sure that the air vent solenoid body and spring are free from damage.
2. Apply a 12 volt source of power to the solenoid valve.
3. Check to see that the valve operates smoothly, if the valve does not operate properly, replace it.
4. Inspect the air horn surface for distortion, scratching and other damage, replace if necessary.

Choke Heater Inspection

1. Using a suitable ohmmeter, measure the resistance between the choke coil wire and the choke housing.
2. The resistance should be 6 ± 5 ohms at 78°F, if the resistance is not within specifications, replace the choke heater.

Slow Cut Solenoid Valve Inspection

1. Be sure that the slow cut solenoid body and spring spool are free from damage. Install the valve assembly to the body.
2. Apply a 12 volt source of power to the solenoid valve.
3. Check to see that the valve operates smoothly, if the valve does not operate properly, replace it.

Reassembly

1. Install the throttle position sensor assembly. Install the mixture by-pass solenoid valve with a new gasket. Be sure to check that each part moves smoothly.

NOTE: When the nut is to be tightened after assembly of the removed throttle lever etc., be sure to tighten the nut after turning the throttle adjusting screw.

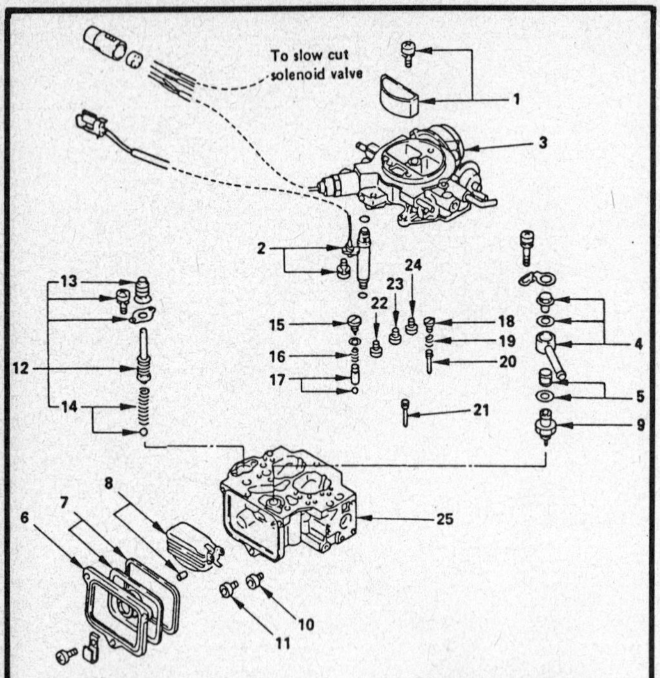

1. Vent cover
2. Duty solenoid valve
3. Choke chamber
4. Fuel nipple
5. Fuel filter
6. Level gauge cover
7. Level gauge and rubber seal
8. Float and collar
9. Needle valve
10. Secondary main jet No. 170
11. Primary main jet No. 88
12. Piston
13. Pump cover
14. Piston return spring
15. Pump set screw
16. Injector spring
17. Injector weight
18. Taper plug
19. Slow jet spring
20. Primary slow jet No. 52
21. Secondary slow jet No. 100
22. Primary main air bleed No. 100
23. Seconary main air bleed No. 60
24. Primary slow air bleed
25. Float chamber

Stromberg DFP384-205 carburetor choke chamber and float chamber assembly used—Amigo and

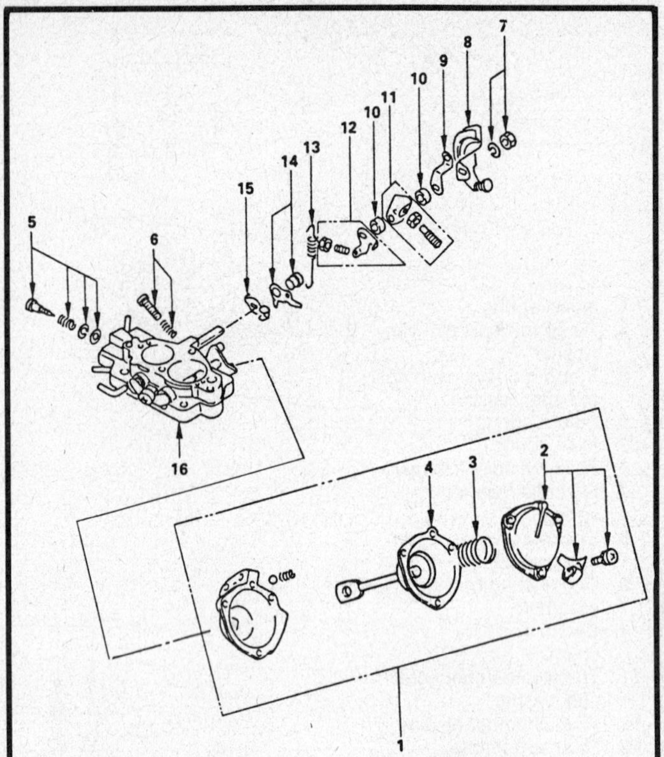

1. Diaphragm chamber assembly
2. Diaphragm chamber cover
3. Diaphragm spring
4. Diaphragm
5. Idle adjusting screw
6. Throtle adjsuting screw
7. Throttle shaft nut and washer
8. Throttle lever
9. Spring hanger
10. Shaft collar
11. Fast idler adjusting lever and screw
12. Kick lever
13. Return spring
14. Return plate and sleeve
15. Adjusting lever
16. Throttle chamber

Stromberg DFP384-205 carburetor throttle chamber assembly used—Amigo and Pick-Up

2. Install the jets as follows, be sure to assemble each jet in the proper position:
   a. Idle-up jet
   b. Secondary slow jet
   c. Primary slow jet
   d. Primary main air bleed jet
   e. Secondary main air bleed jet
   f. Secondary main jet
3. Insert the pump strainer assembly by inserting the check ball, the filter and inserting the snap ring accurately so that damage is not done to the pump bore.
4. Install the slow cut solenoid valve with a new gasket. Install the diaphragm chamber with a new gasket.
5. Place a new gasket on top of the flange and place the carburetor body on top. Install the flange screws and wire bracket and tighten the bolt.
6. Connect the diaphragm chamber rod and secondary valve shaft with the E-clip. Install the throttle return and second lock springs. Insert the accelerator pump return spring into the pump bore and install the piston from the top. Check that each part moves smoothly.
7. Install choke piston assembly along with the choke piston link. Install the choke breaker diaphragm unit and hoses. Connect the choke rod to the choke breaker diaphragm unit and secure with the E-clip.
8. Install the feedback solenoid valve as follows, pass the wires through the air horn and insert them vertically while holding the valve body. It must not be bent while inserting. Tighten the retainer screw.
9. Install the float packing and float valve. Install and tighten the needle valve seat with packing. Install the needle valve onto the valve seat.
10. Install the float and secure it with the float pin, adjust the float level as follows:
   a. Check the clearance between the float top and gasket.
   b. The float level (raised position) should be 0.59 in. (15.0mm).
   c. Bend the float tang and adjust if it is off the prescribed value. Flaws must not be provided on the contacting surface with the needle valve.
   d. Check the clearance between the float top and gasket at the lowered position of the float.
   e. The float level (lowered position) should be 1.74 in. (44.3mm).
   f. Bend the float tang and adjust if it is off the prescribed value.
11. Install the a new air horn gasket. Carefully assemble the air horn onto the carburetor body. The air horn must *not* be assembled to the body in a slanted fashion. Always be sure to install it vertically. Do not bump the float against the body. The wires must not interfere. After assembly be sure to accurately

Idle speed adjusting screw—Amigo and Pick-Up

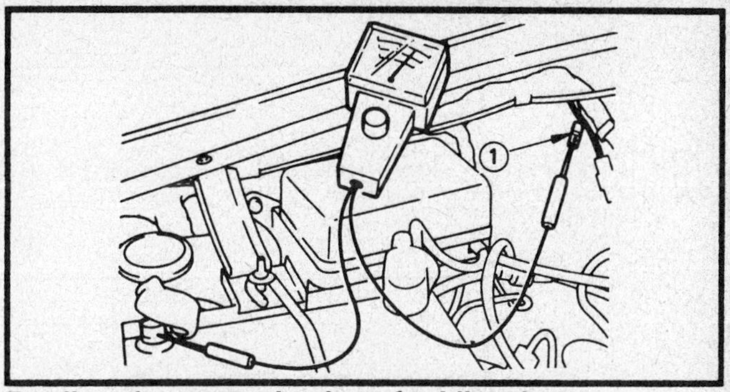

Dwell or duty meter hook-up for idle mixture adjustment—Amigo and Pick-Up

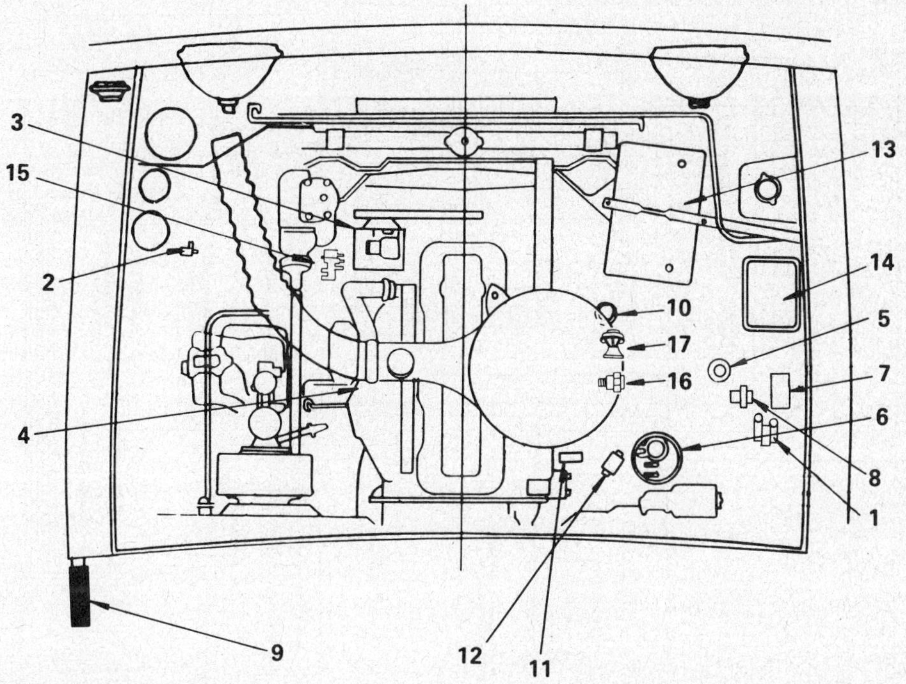

1. Vacuum switching valve;
   - Canister purge
   - Ignition
   - MAP sensor
   - EGR ('91 model California only)
2. Vacuum switching valve (A.I.R.)
3. Air pump
4. Oxygen sensor (O_2 sensor)
5. Mixture control valve (M.C.V.)
6. Canister
7. MAP sensor
8. Vacuum switch (IDLE)
9. ECM
10. Back pressure transducer
11. Ignition coil
12. Check and relief valve
13. Battery
14. Fuse and relay box
15. Air switching valve
16. Coolant temperature sensor
17. EGR valve

Component layout—Amigo and Pick-Up

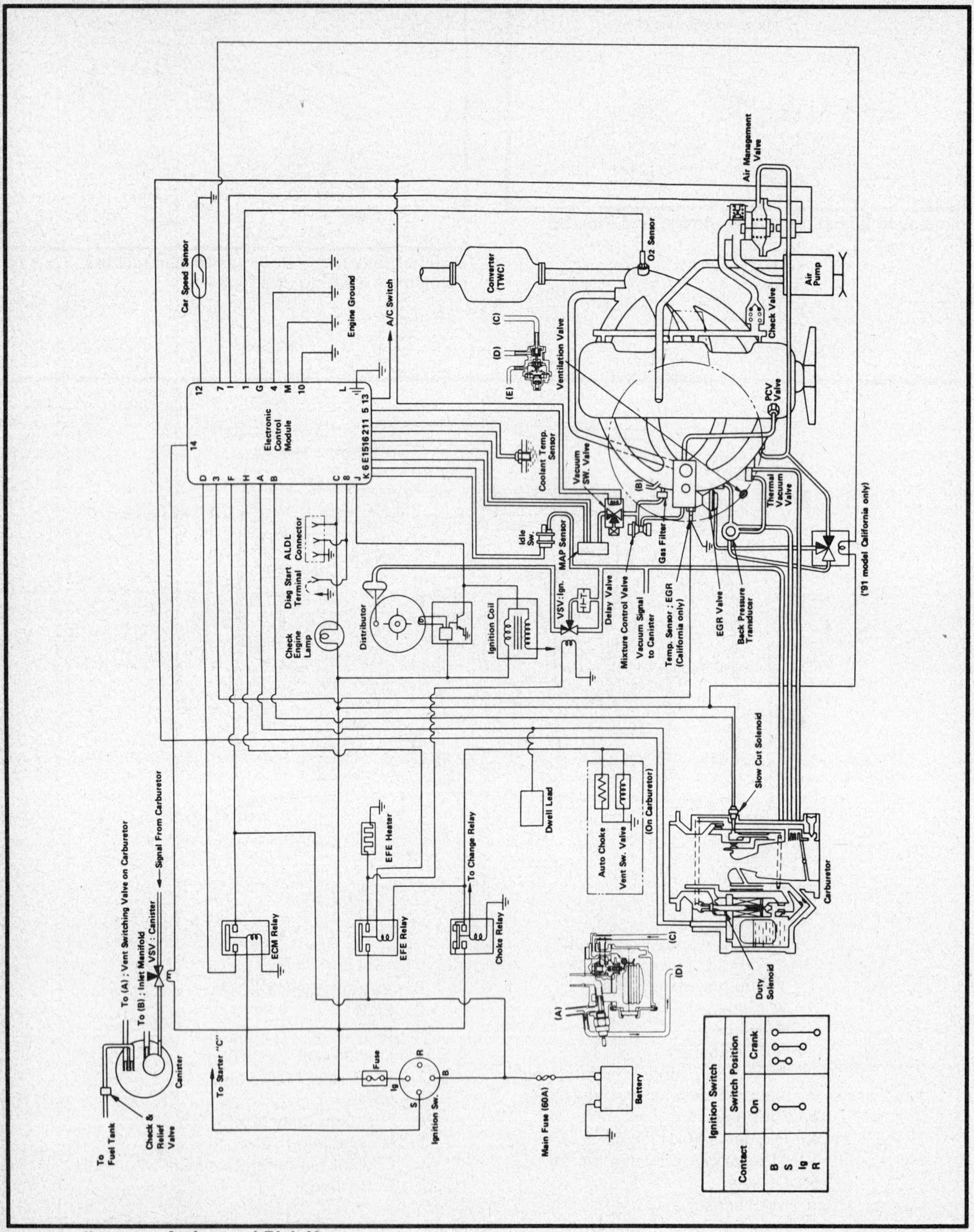

System diagram—Amigo and Pick-Up

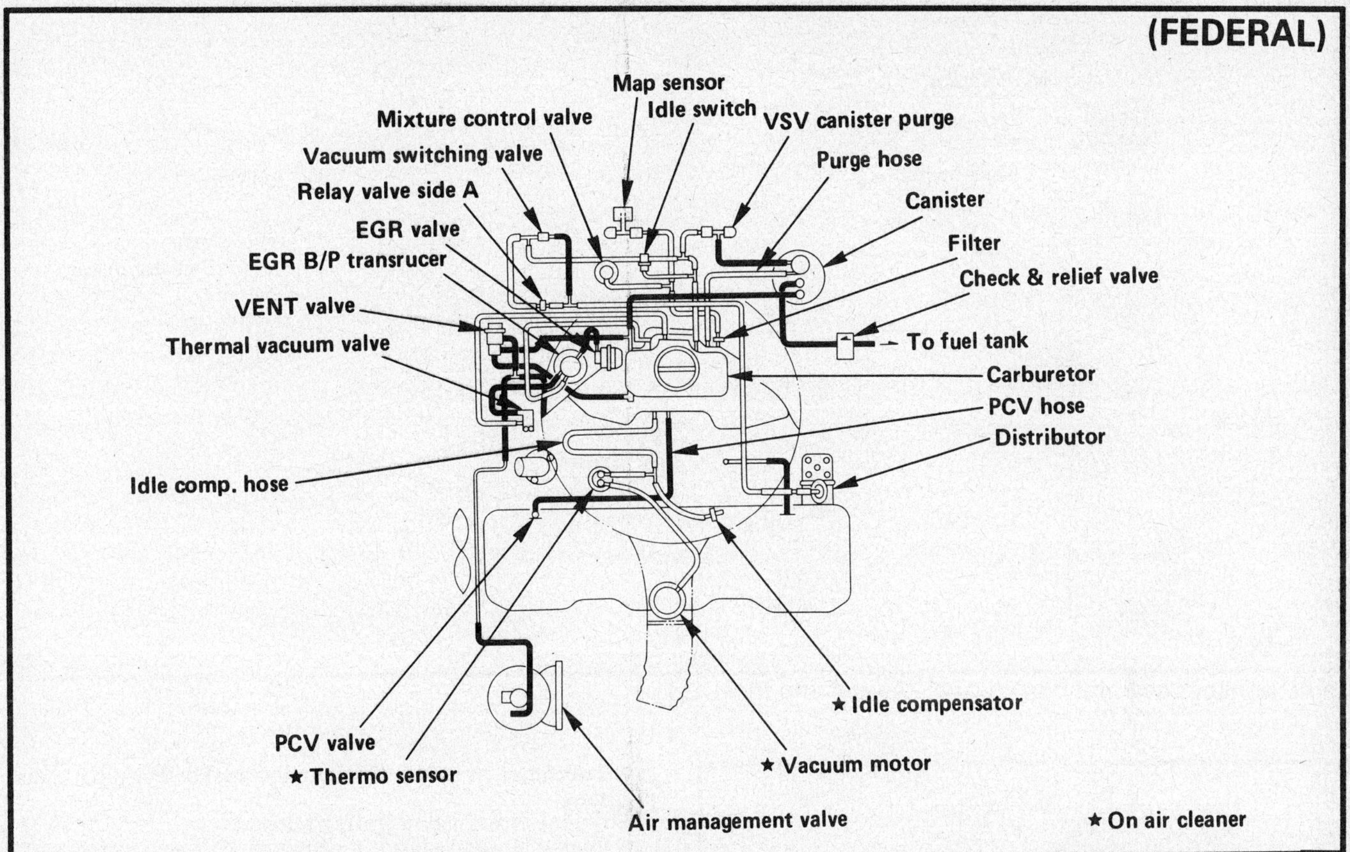

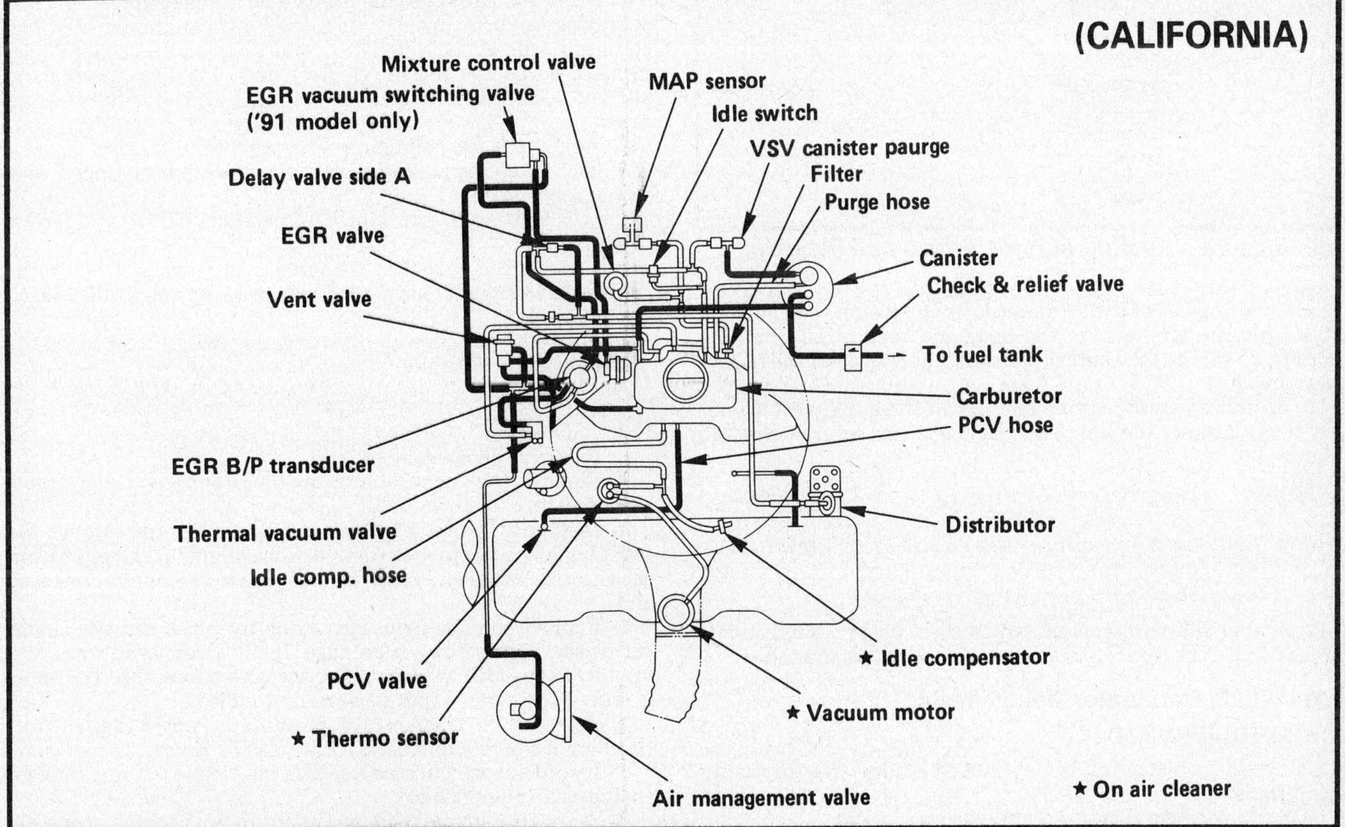

Vacuum hose routing—Amigo and Pick-Up

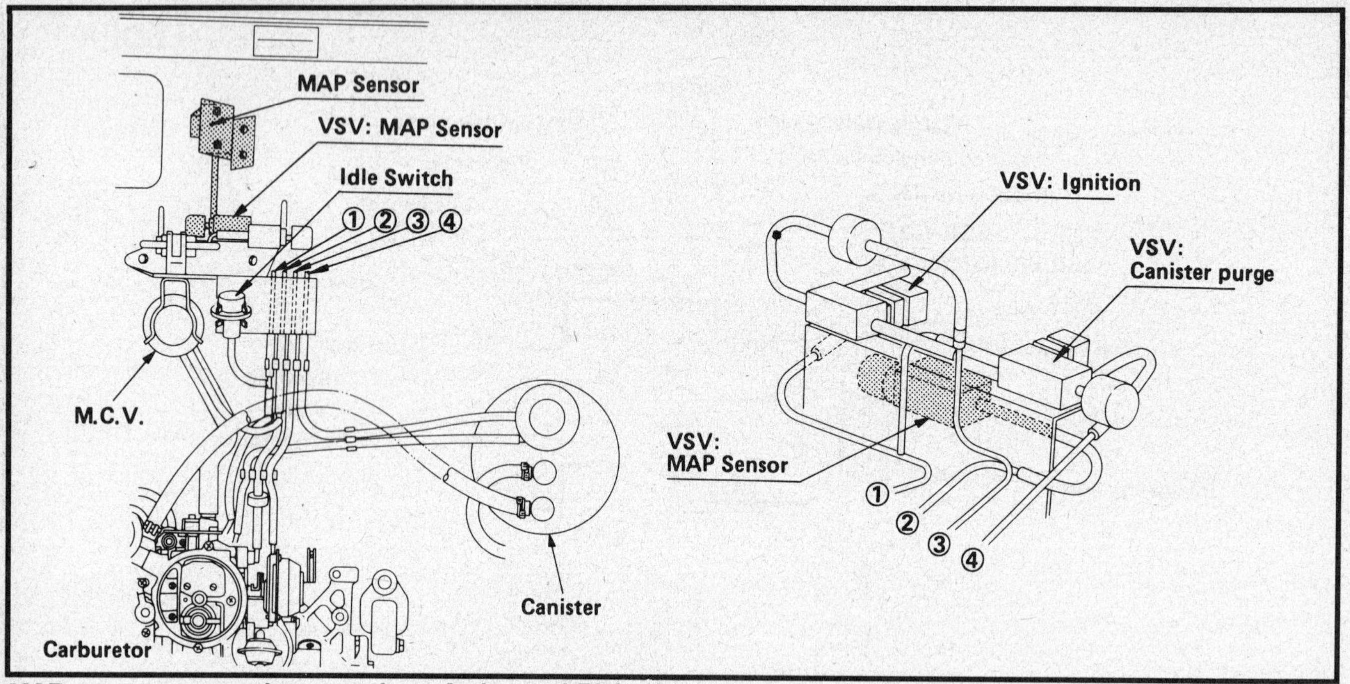

MAP sensor vacuum hose routing—Amigo and Pick- Up

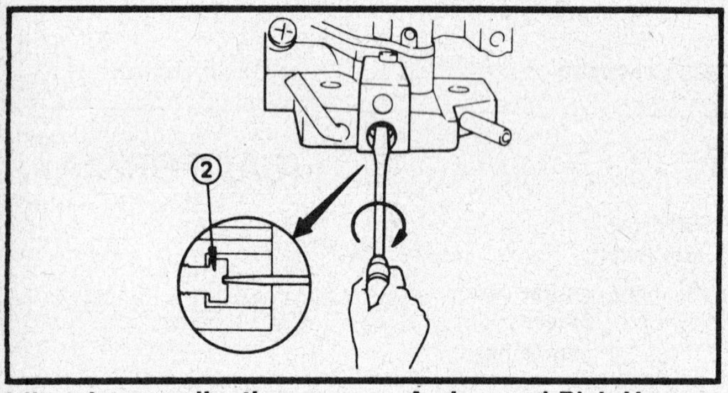

Idle mixture adjusting screw—Amigo and Pick-Up

insert the grooves of the accelerating pump piston rod and boot. Assemble the air horn to the body and install the retaining screws.Be sure to take note of the different lengths of the retaining screws.

12. Install the pump spring and install the pump arm assembly by connecting the link with the choke cam lever and install the E-clip.

NOTE: Arrange the wires of the slow cut solenoid valve on the inner side and arrange the wire of the auto choke heater on the outer side so that the wires of the auto choke heater and the slow cut solenoid valve do not interfere with the throttle return spring(s).

13. Install the wire connectors to their proper connections. Check that each part of the carburetor operates smoothly.

On Vehicle Carburetor Adjustments

CURB IDLE SPEED

Idle speed adjustment is performed under the following conditions:

- Parking brake firmly set and wheels blocked.
- All electrical accessories in the **OFF** position.
- Transmission in the **N** (Manual trans) or **P** (Auto. trans) positions.
- Engine at normal operating temperature.
- Wheels straight ahead (if equipped with power steering).

1. Make the idle speed adjustment with the engine at normal operating temperature, choke opened, air condtioner off (if so equipped), air cleaner installed and distributor vacuum line, canister purge line and the EGR vacuum line disconnected and plugged. Close the idle compensator line by bending the vacuum hose.
2. Adjust the idle adjusting screw to 900 ± 80 rpm.
3. The following procedure is if the vehicle is equipped with an air condtioner:
   a. Turn the A/C on to the **MAX** position, **COLD** and **HIGH** blower.
   b. Open the throttle to approximately 1/3 and allow the throttle to close. (This allows the speed-up solenoid to reach full travel).
   c. Adjust the speed up controller adjusting screw to set the idle at 900 ± 50 rpm.

IDLE MIXTURE

1. Remove the air cleaner.
2. Remove the carburetor from the engine and place on a work bench.
3. Drill out the idle mixture screw seal plug and remove the seal plug using the proper tool. Remove all metal shavings from the throttle flange and from the seal plug hole using compressed air.
4. Remove the idle adjusting screw from the throttle chamber. Inspect the screw for damage. If the screw head was damaged from the drill, replace the screw with a new one. The same is true if the screw taper is worn or stepped.
5. Reinstall the idle mixture screw all the way into the throttle chamber bore and then back it out 1½ turns.
6. Install the carburetor and the air cleaner. Connect a tachometer to the engine.
7. Start the engine and warm it up to normal operating temperature.

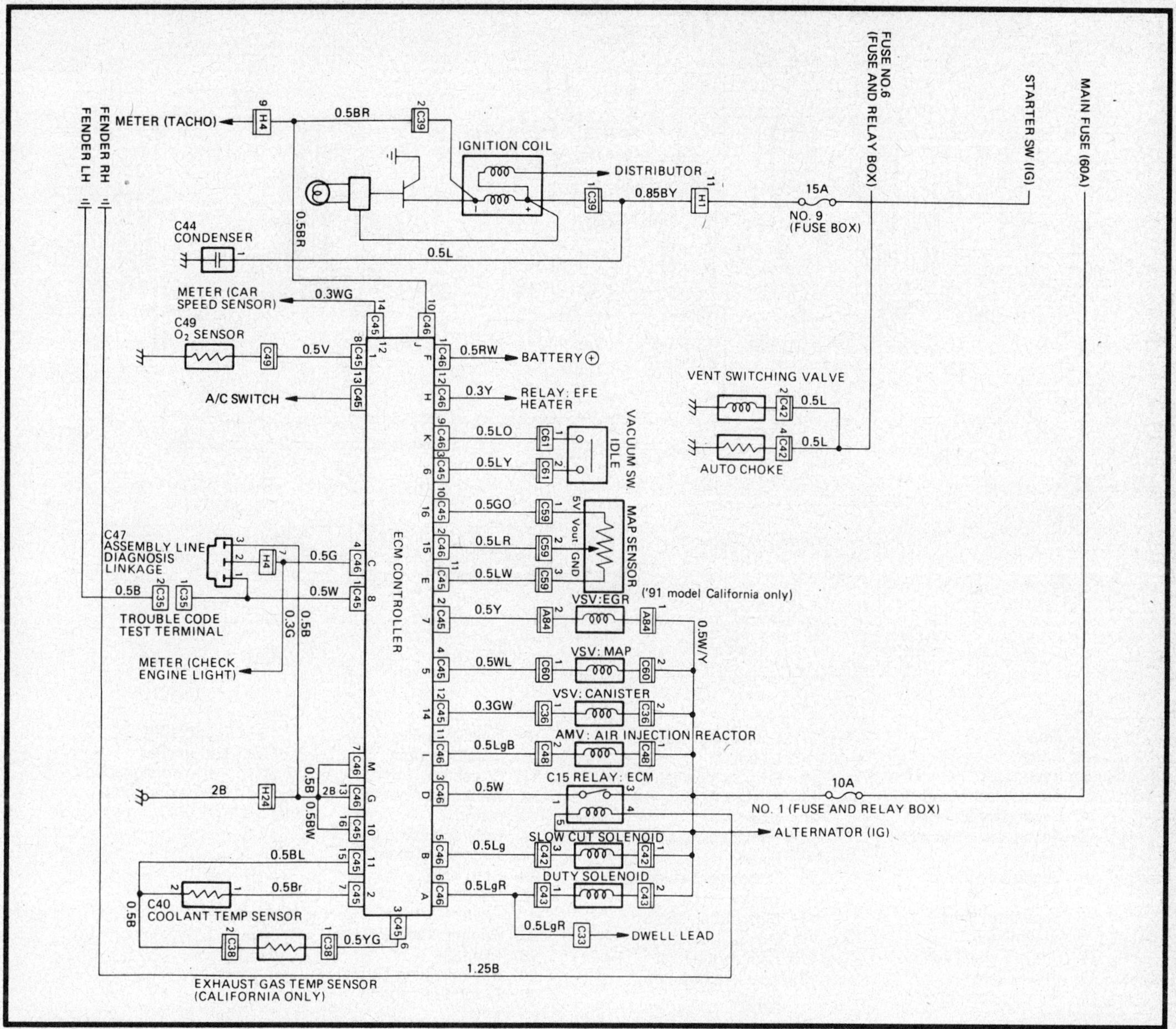

Wiring diagram—Amigo and Pick-Up

8. Make sure the choke valve is fully open. If not, open it.
9. Make sure the air conditioner (if equipped) and all electrical accessories are switched to the **OFF** position.
10. Disconnect and plug the distributor vacuum line.
11. Disconnect and plug the canister purge line.
12. Disconnect and plug the EGR vacuum line.
13. Kink the idle compensator line to close it off.
14. Adjust the idle speed to 900 rpm using the idle speed adjusting screw.
15. Connect a dwell or duty meter to the engine as follows. If using a dwell meter, set the selector switch to the "4-cylinder" scale. Connect the positive (+) side of the meter to monitor lead 1 and the negative side of the meter to a suitable engine ground. After obtaining a varying dwell or duty reading, adjust the average dwell or duty to the following specifications by turning the idle mixture screw:
    a. Average dwell using dwell meter – 36 degrees.
    b. Average dwell using duty meter – 40 degrees.
16. Adjust the idle speed to 900 rpm using the idle speed adjusting screw.
17. Remove the air cleaner.
18. Tap in a new seal plug into the throttle chamber bore so that the plug is flush with the surface of the carburetor.
19. Install the air cleaner.

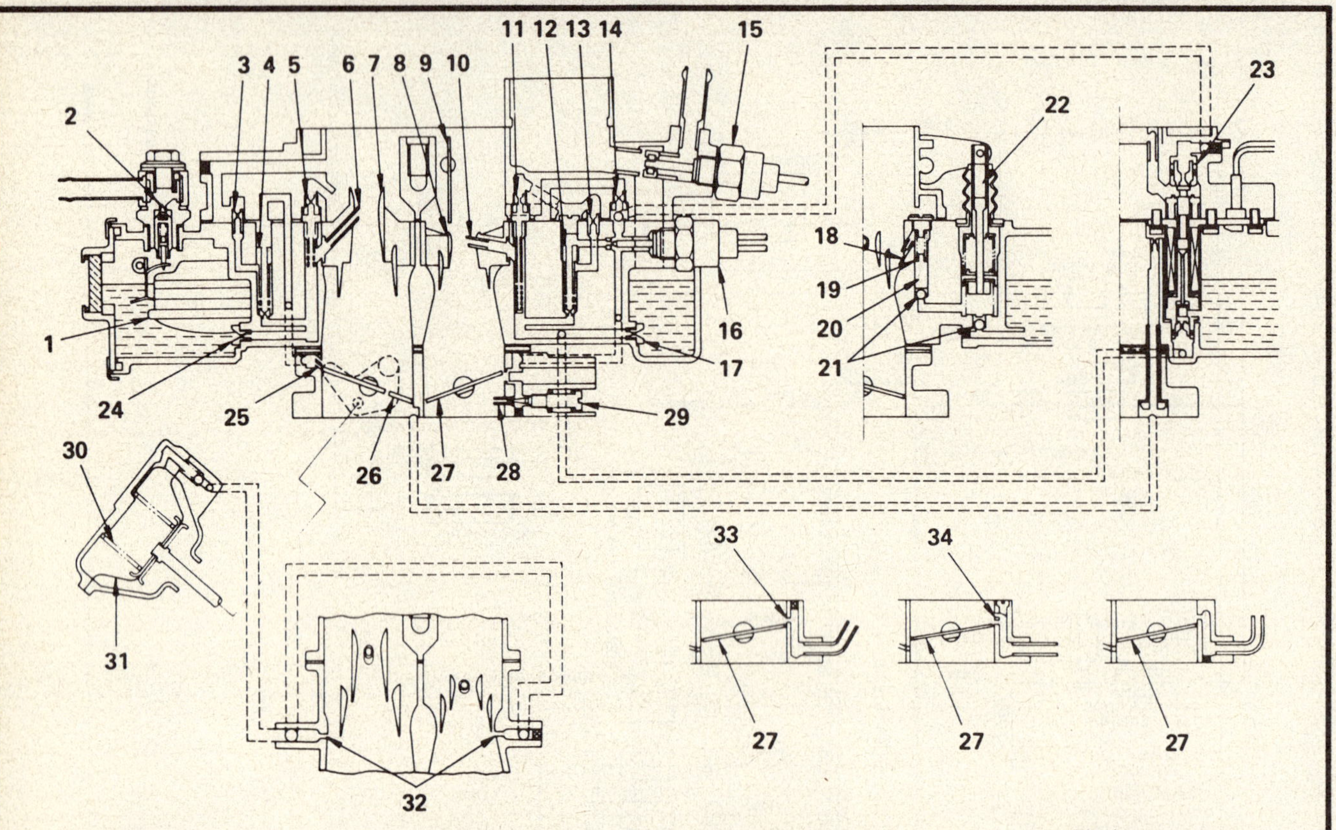

1. Float
2. Needle valve
3. Secondary economizer bleed
4. Secondary slow jet
5. Secondary main air bleed
6. Secondary main nozzle
7. Secondary small venturi
8. Primary small venturi
9. Choke valve
10. Primary main nozzle
11. Primary main air bleed
12. Primary slow jet
13. Primary economizer bleed
14. Primary slow air bleed
15. Switch vent solenoid valve
16. Slow cut solenoid valve
17. Primary main jet
18. Accelerator pump injector
19. Injector spring
20. Injector weight
21. Check ball
22. Accelerator pump piston
23. Duty solenoid valve
24. Secondary main jet
25. Step hole
26. Secondary throttle valve
27. Primary throttle valve
28. Idle hole
29. Idle adjusting screw
30. Diaphragm spring
31. Diaphragm
32. Vacuum jet
33. Advance signal port
34. EGR signal port

Carburetor components

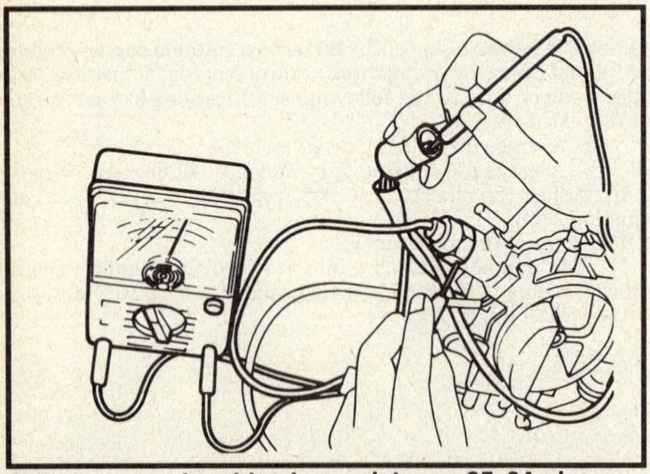

Switch vent solenoid valve resistance 25–34 ohms

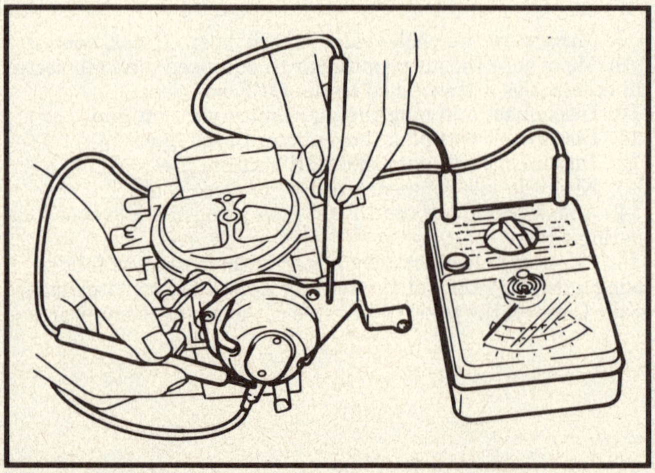

Thermostat resistance check 1.1–3.1 ohms

MAZDA B2200 FEEDBACK CARBURETOR

General Information

FEEDBACK SYSTEM

This system controls the air/fuel ratio to approximately 14.7:1 (optimum ratio), and reduces CO, HC and NOX emissions, and minimizes fuel consumption.

The Feedback System consists of an ignition coil, neutral switch, clutch switch, idle switch, water thermo sensor, water temperature switch, intake air temperature sensor, O_2 sensor, vacuum sensor, A/C switch as a sensor (input), A/F solenoid valve as an actuator (output), and an emission control unit as a processor.

The emission control unit controls the opening duration of the A/F solenoid valve to maintain the optimum air/fuel ratio. The A/F valve controls the amount of fuel to the primary main circuit through the solenoid controlled fuel jet and the air added to the primary slow circuit through the solenoid controlled air bleed.

AIR INJECTION SYSTEM

This system supplies secondary air into the exhaust system to burn CO and HC in the exhaust gas and to control the O_2 signal for the emission control unit.

This system consists of reed valves, air control valves and an ACV solenoid valve controlled by the emission control unit. Reed valve (A) supplies secondary air into the exhaust manifold when the No. 1 or No. 2 air control valve passages opens, or when both open. Reed valves (B) and (C) supply secondary air into the exhaust pipe just behind the front catalytic converter through exhaust gas pulsation.

EXHAUST GAS RECIRCULATION (EGR) SYSTEM

The EGR supplies exhaust gas into the intake manifold to reduce NOX emissions. This system consists of an EGR control

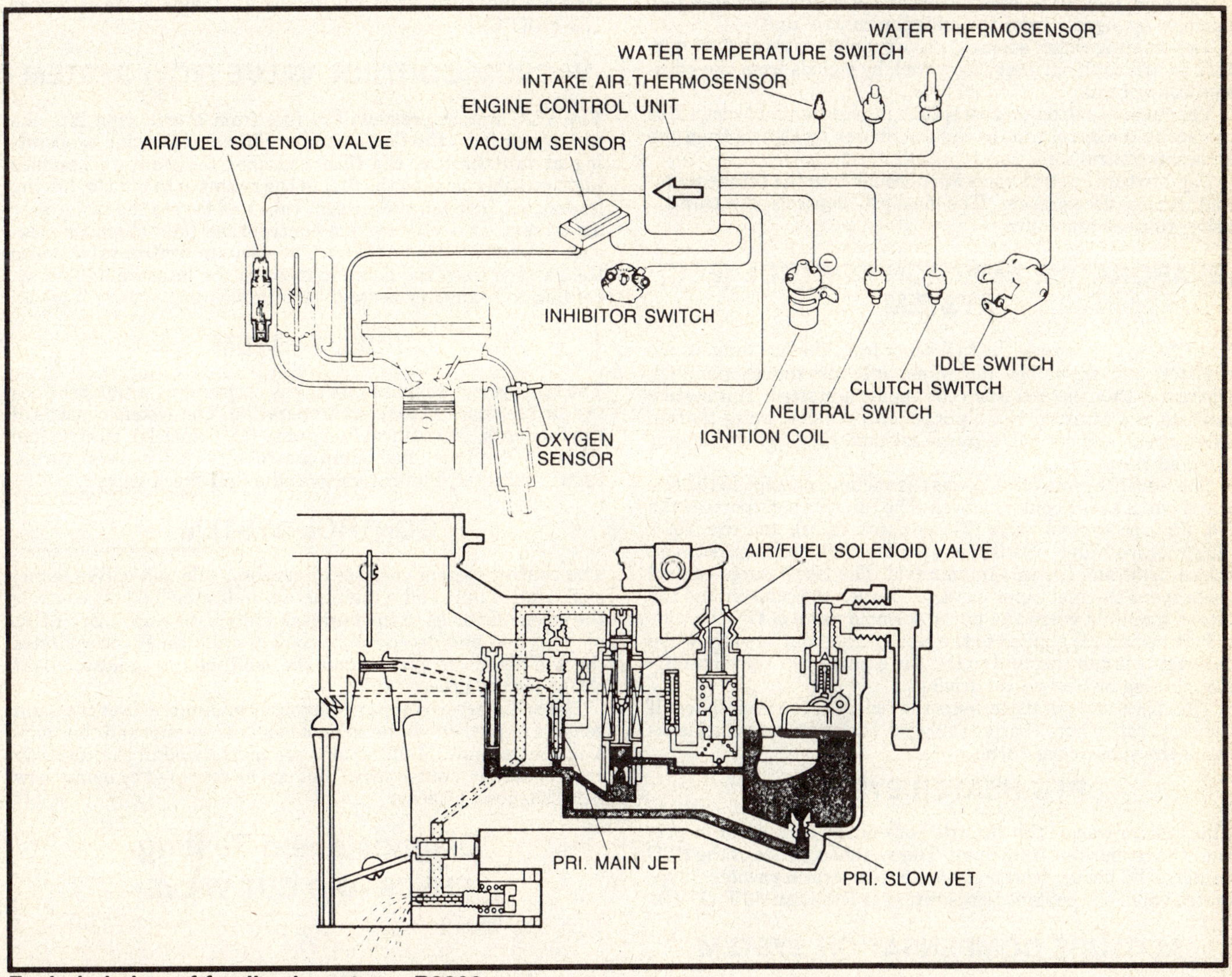

Exploded view of feedback system—B2200

valve, EGR position sensor and a duty solenoid valve controlled by the emission control unit.

The EGR control valve controls the amount of exhaust gas flowing into the intake manifold, according to the vacuum, controlled by the duty solenoid valve.

The duty solenoid valve consists of a vacuum valve and a vent valve. The vacuum valve opens the vacuum passage to the EGR control valve. The vent valve vents the vacuum from the vacuum valve, to control the vacuum according to signal from the emission control unit.

The emission control unit determines the amount of EGR gas, recirculated by the EGR position sensor on the EGR valve, and controls the opening duration of the vacuum and vent valves. The amount of exhaust gas recirculated is determined by the ignition coil signal, water thermo sensor, water temperature sensor, vacuum sensor and atmospheric pressure sensor.

DECELERATION CONTROL SYSTEM

This system control the air/fuel mixture and advances the ignition timing to reduce CO, HC and fuel consumption, and to prevent the front catalytic converter from overheating.

This system consists of a slow fuel cut solenoid valve, coasting richer solenoid valve, vacuum solenoid valve, dashpot and mixture control valve.

The slow fuel cut solenoid valve closes the primary slow fuel passage, by signals from the emission control unit.

The coasting richer solenoid valve supplies extra air/fuel mixture to add to the primary slow fuel, by signals from the emission control unit.

The vacuum solenoid valve applies intake manifold vacuum to the vacuum control unit of the distributor by signals from the emission control unit.

The mixture control valve supplies air into the intake manifold during deceleration. The dashpot, controls the throttle valve, to close gradually.

EVAPORATIVE EMISSION CONTROL (EEC) SYSTEM

The EEC system stores the fuel vapor from the fuel tank, in the canister, and draws the fuel vapor into the intake manifold, where it is then burned when the engine is started. This system consists of a canister, No. 3 purge control valve, water thermo valve, check-and-cut valve, purge solenoid valve and an air vent solenoid valve.

The water thermo valve opens the vacuum passage to the No. 1 and No. 3 purge control valves. The canister incorporates the No. 2 purge control valve (2-way check valve), and the No. 1 purge control valve, which opens the fuel vapor passage between the canister and the intake manifold. The No. 3 purge control valve opens the fuel vapor passage between the canister and the intake manifold when the purge solenoid valve is **ON**.

Port vacuum is applied to the No. 1 purge control valve while the engine is running and to the No. 3 purge control valve during running or heavy load driving.

The check-and-cut valve vents the vapors to the atmosphere if the evaporative hoses become clogged. It also prevents fuel leakage if the vehicle overturns.

PTC HEATER SYSTEM

This system warms the throttle body of the carburetor to prevent the carburetor from icing. The system consists of the PTC heater, PTC heater relay and a water temperature switch. It operates when the coolant temperature is less than 63°F (17°C).

ALTITUDE COMPENSATION SYSTEM

This system increases the amount of air to the carburetor to prevent an over-rich air/fuel ratio at high altitudes. The system consists of a high altitude compensator and the carburetor. The high altitude compensator provides additional air bleeds for the primary and secondary main fuel circuits and supplies additional air into the intake manifold.

AIR INTAKE TEMPERATURE CONTROL SYSTEM

This system controls the air intake temperature to prevent icing, and operates, depending on the air temperature around the control valve.

POSITIVE CRANKCASE VENTILATION (PCV) SYSTEM

The PCV system reburns the combustion blowby gases. The system consists of the PCV valve, which operates when the engine is running, controls the flow of blowby gases according to the intake manifold vacuum.

HOT IDLE COMPENSATION SYSTEM

This system supplies secondary air into the intake manifold to stabilize idle sped when the intake air temperature is higher 153°F (67°C).

VACUUM CONTROL VALVE (VCV) SYSTEM

The VCV system prevents the fuel from overflowing into the carburetor from the float chamber. When the engine is operating at full throttle, the float chamber temperature becomes high, possibly causing the fuel in the chamber to bubble, forcing its way out through the air vent tube and into to the carburetor air stream. The VCV system controls the float chamber pressure to prevent this bubbling. The vacuum control valve opens the passage from the float chamber to the intake manifold according to secondary venturi vacuum.

FUEL SYSTEM

The fuel system supplies fuel to the carburetor, which provides the air/fuel mixture for engine operation. The system consists of the fuel tank, mechanical fuel pump (M/T models), electric fuel pump (A/T models), fuel pump control unit (A/T models), carburetor, fuel filter, fuel cut valve and a fuel check valve

CONTROL SYSTEM

The control system consists of sensors, solenoid valves, emission control unit and a malfunction indicator light. It controls the solenoid valves in the feedback system, idle-up, EEC, EGR, air injection and deceleration control systems. It incorporates the self-diagnosis system and the malfunction indicator light (MIL) for the driver.

The self-diagnosis system diagnoses malfunctions of the main sensors (input), of all the solenoid valves (output) and the emission control unit. Malfunctions are memorized in the emission control unit as trouble codes that can be retrieved by using thew Self-Diagnosis Checker

Diagnosis and Testing

CHECK AND CUT VALVE

NOTE: This test should be performed with the valve in the horizontal position. Otherwise the weight of the valve will cause it to move out of position and close the passage.

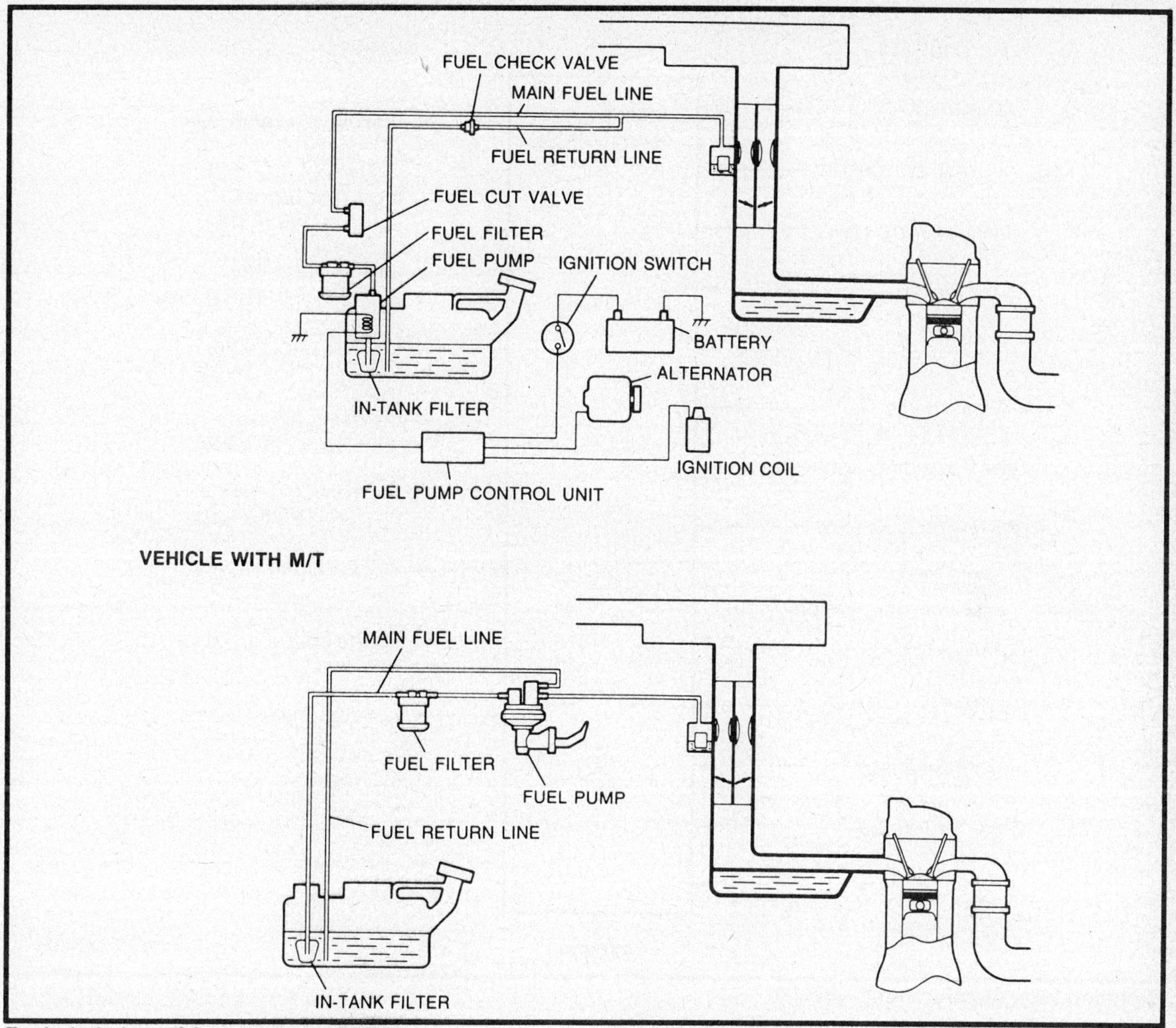

Exploded view of fuel system—B2200

Inspection

1. Remove the check and cut valve, which is located just before the fuel tank.
2. Connect a suitable pressure gauge to the passage of the fuel tank and blow through port (A) of the valve.
3. The valve should open with 0.78–1.00 psi. Remove the pressure gauge and connect it to the passage open to the atmosphere.
4. Blow through the valve from port (B) and if the valve opens at a pressure of 0.14–0.71 psi, the valve is operating properly. If the valve fails any of these tests, replace the valve.

WATER THERMO VALVE

Inspection

1. Remove the water thermo valve and immerse the valve into a container of water.
2. Heat the water gradually and observe the temperature. Blow through the valve, and if air passes through the valve at 129.2°F or higher, the valve is operating properly.

SLOW FUEL CUT SOLENOID VALVE

Inspection

1. Run the engine at idle speed and disconnect the slow fuel cut solenoid valve connector.
2. If the engine stalls, the solenoid valve is operating properly. If the engine does not stall, replace the solenoid valve.

AIR/FUEL SOLENOID VALVE

Inspection

1. Start the engine and allow it to reach normal operating temperature.
2. Install a suitable dwellmeter to the A/F check connector (BrY), and check that the dwellmeter reads 20–70 degrees.
3. Ground the A/F check connector (BrY) using a jumper

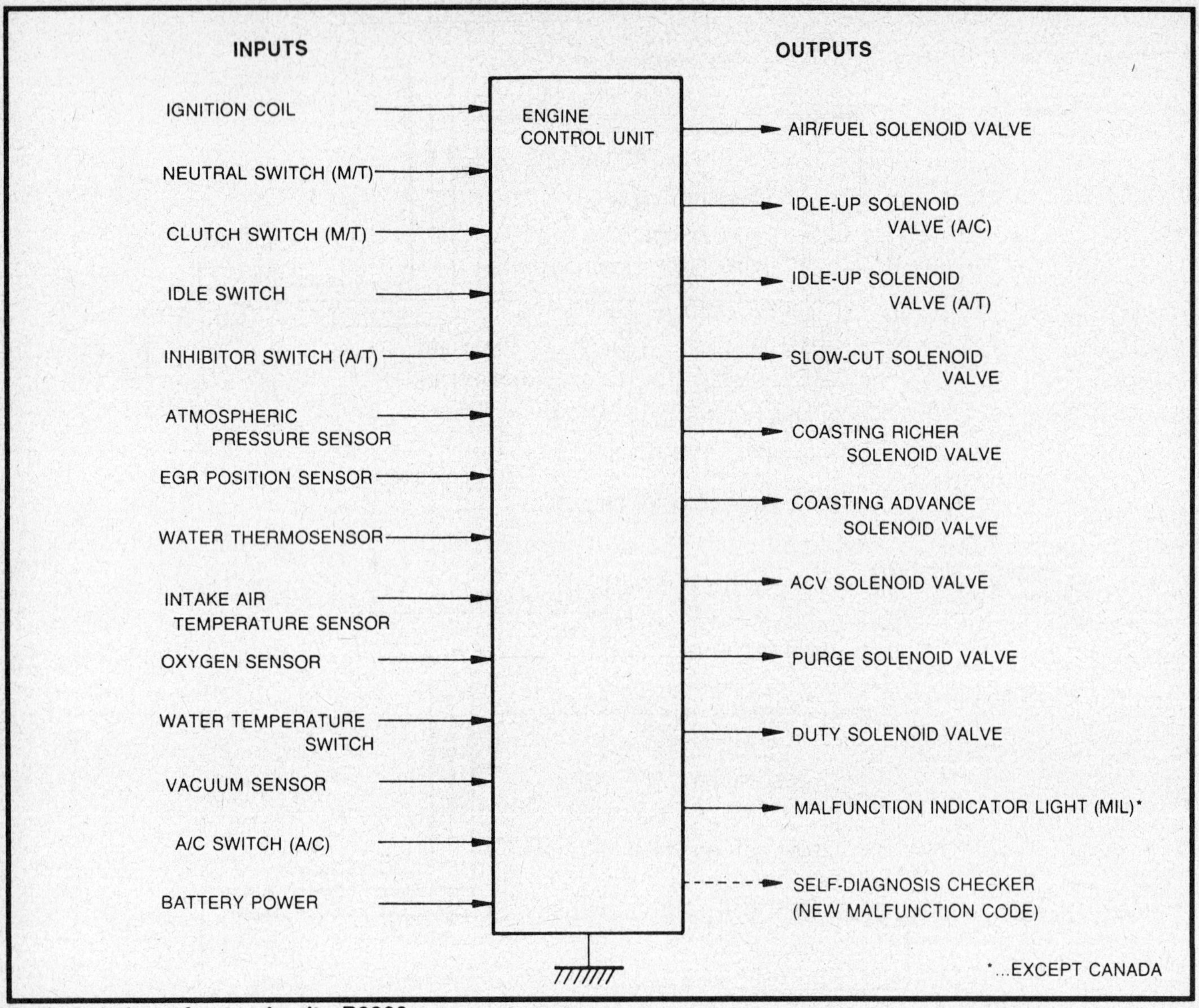

Exploded view of control unit – B2200

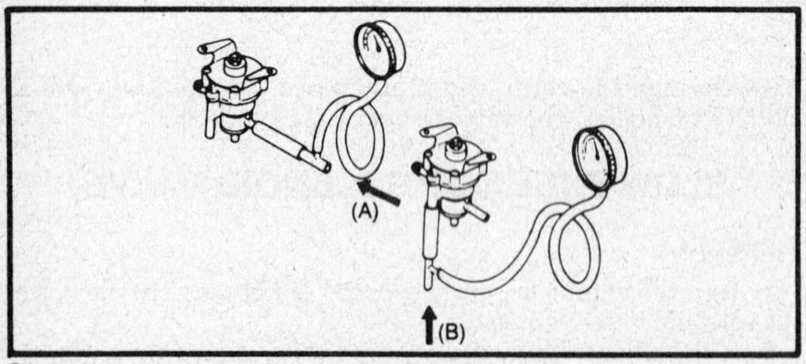

Check and cut valve test – B2200

wire, and check that the engine rpm drops. If not, clean the A/F solenoid valve and the carburetor, or replace the air horn assembly.

NOTE: Clean by using gasoline and blow it out by using compressed air. Do not submerge the unit in the cleaner. The air/fuel solenoid valve must be replaced with a new air horn. The air/fuel solenoid is not available separately.

OXYGEN SENSOR

The oxygen sensor is mounted in the exhaust manifold. The output signal from this sensor, which varies with the oxygen content of the exhaust gas, is provided to the control unit for use in controlling the closed loop compensation of fuel delivery.

Inspection

1. Start the engine and allow it to reach normal operating temperature. Stop the engine.
2. Disconnect the oxygen sensor electrical connector. Connect a suitable voltmeter between the oxygen sensor and ground.
3. Start, and run the engine up to 4000 rpm or until the voltmeter reaches approximately 0.7 volts.
4. Increase and decrease engine rpm suddenly several times. When the engine rpm is increased, the voltmeter should read 0.5–1.0 volts.
5. When the engine rpm is decreased, the voltmeter should read 0–0.4 volts.
6. If specifications are not as specified, replace the oxygen sensor.

PTC HEATER

Inspection

1. Disconnect the PTC heater connector, which is located at the base of the carburetor.
2. Connect an ohmmeter between the the PTC connector and ground and check for continuity.
3. If there is continuity at the connector, the PTC heater operating properly. If there is no continuity, replace the PTC connector.

PTC HEATER RELAY

Inspection

1. Apply battery voltage to the No. 1 terminal, and ground the No. 2 terminal of the PTC heater relay.
2. Using a suitable ohmmeter, check that there is continuity at terminals No. 3 and No. 4.
3. If the relay fails this test, replace it.

IDLE COMPENSATOR

Inspection

1. Check to see that the valve is in the closed position when the bi-metal temperature is lower than the operating temperature (145°–160°F).
2. With the valve in the closed position, apply vacuum to the vacuum hose. If excessive leakage is noted, replace the idle compensator.
3. When the bi-metal temperature exceeds 160°F (71°C), check that the valve opens. If not, replace the idle compensator.

VACUUM SOLENOID VALVE

Inspection

1. Disconnect the vacuum solenoid valve electrical connector and the vacuum hoses, then remove the valve.
2. Connect vacuum hoses to both ports on the valve. Blow through the valve from hose (A), and check that air comes out from the valve filter.
3. Apply battery voltage and ground the solenoid valve terminals using jumper wires.
4. Blow air through the valve from hose (A), and check that air comes out from port (B).
5. If the vacuum solenoid valve fails the above test, replace it.

MIXTURE CONTROL VALVE

Inspection

1. Start the engine and allow it to reach normal operating temperature.
2. Plug the intake port of the mixture control valve, and check the engine rpm does not drop.

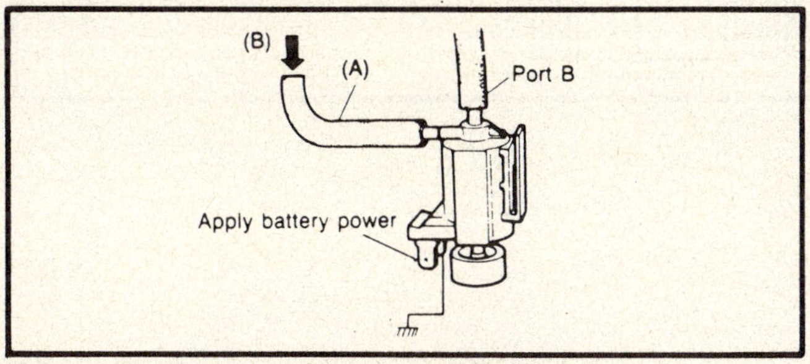

Vacuum solenoid valve test—B2200

3. Increase the engine rpm and quickly decrease. Check that air is pulled into the intake port, after the accelerator is released.

HIGH ALTITUDE COMPENSATOR

Inspection

1. Disconnect the vacuum hoses from the high altitude compensator valve.
2. Blow through each port on the high altitude compensator valve.
3. Air should pass through at a high altitude of 1640 feet and air should be blocked at a low altitude of less than 1640 feet. If not, replace the high altitude compensator valve.

TROUBLESHOOTING WITH THE SELF-DIAGNOSIS CHECKER

The self-diagnosis checker (49–H018–9A1) is used to retrieve code numbers of malfunctions which have happened and were memorized or are continuing. The malfunction is indicated by the code number and a buzzer.

The monitor lamp (green light) indicates operation of the oxygen sensor. This indicator shows a lean air/fuel mixture by the light illuminating constantly and a rich air/fuel mixture by the light not illuminating. Normal air/fuel ratio is indicated by a flashing light.

If there is more than 1 malfunction, the code numbers will display on the self-diagnosis checker 1 by 1 in numerical order. In the case of malfunctions, 09, 13, and 01, the code numbers are displayed in a order of 01, 09 and then 13.

The memory of malfunctions is canceled by disconnecting the negative battery cable for at least 5 seconds.

The emission control unit has a built in fail-safe mechanism for the main input sensors. If a malfunction occurs, the emission control unit will substitute values. This will slightly effect the driving performance, but the vehicle may still be driven.

Self-diagnosis for the output solenoid valves function within 3 seconds after turning the ignition switch to the **ON** position. It stops when the engine starts, even within 3 seconds.

The malfunction indicator light indicates a pattern the same as the buzzer of the self-diagnosis checker when the self-diagnosis check connector is grounded. When the self-diagnosis check connector is not grounded, the lamp illuminates steady while malfunction of the main input sensor occurs and goes out if the malfunction recovers. However, the malfunction code is memorized in the emission control unit.

Inspection

1. Connect the Self-Diagnosis Checker (49–H018–9A1) or equivalent to the check connector. The check connector is usually located above the right side wheel housing.
2. Set the select switch to the "A" position.
3. Ground the self-diagnosis check connector with a suitable jumper wire.
4. Turn the ignition switch to the **ON** position. Check that the number "88" flashes on the digital display and the buzzer sounds for 3 seconds after turning the ignition switch **ON**.
5. If the number "88" does not flash, check the check connector wiring.
6. If the number "88" flashes and the buzzer sounds continuously for more than 20 seconds, replace the emission control unit and perform Steps 3 and 4 again.
7. Note the code numbers and check the causes, repair as necessary. Be sure to recheck the code numbers by performing the "After Repair Procedure," after repairing.

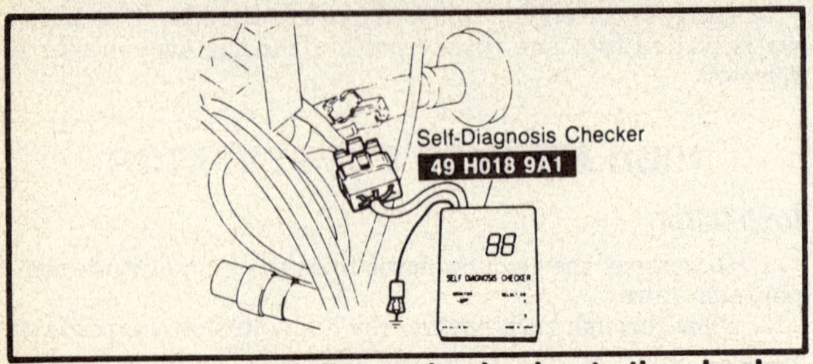

Connecting the self-diagnosis checker to the check connector – B2200

After Repair Procedure

1. Cancel the memory of malfunctions by disconnecting the battery negative cable for at least 5 seconds, then reconnect it.
2. Turn the ignition switch to the **ON** position, but do not start the engine for at least 6 seconds.
3. Start the engine and allow it to reach normal operating temperature, then run it at 2000 rpm for approximately 4 minutes.
4. Connect the Self-Diagnosis Checker (49–H018–9A1) or equivalent to the check connector. The check connector is usually located above the right side wheel housing.
5. Ground the self diagnosis check connector (YB) with a suitable jumper wire.
6. Check that there are no code numbers being displayed.

TROUBLE CODE IDENTIFICATION CHART – B2200

Code No.	Location of malfunction	Buzzer	Control unit fail-safe function
01	IG pulse circuit	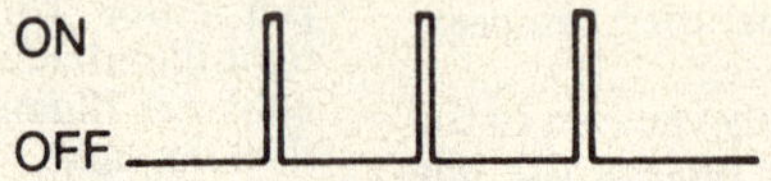	–
09	Water thermosensor or circuit		Maintains constant **80°C (176°F)** signal
13	Vacuum sensor or circuit		Holds air/fuel solenoid valve to **0%** duty and cuts off EGR
14	Atmospheric pressure sensor or circuit		Maintains constant signal of sea-level pressure
15	Oxygen sensor or circuit		Holds air/fuel solenoid valve to **20%** duty
16	EGR control system		–
	EGR position sensor or circuit		Cuts off EGR
17	Feedback system		Holds air/fuel solenoid valve to **30%** duty
18	Air/fuel solenoid valve or circuit		–

TROUBLE CODE IDENTIFICATION CHART – B2200

Code No.	Location of malfunction	Buzzer	Control unit fail-safe function
22	Slow fuel cut solenoid valve or circuit	ON / OFF	–
23	Coasting richer solenoid valve or circuit	ON / OFF	–
26	Purge solenoid valve or circuit	ON / OFF	–
28	Duty solenoid vacuum valve or circuit	ON / OFF	–
29	Duty solenoid vent valve or circuit	ON / OFF	–
30	ACV solenoid valve or circuit	ON / OFF	–
34	Idle-up solenoid valve (for A/C) or circuit	ON / OFF	–
35	Idle-up solenoid valve (for A/T) or circuit	ON / OFF	–
45	Vacuum solenoid valve or circuit	ON / OFF	–

01 → 4-second period →
09 → 4-second period →
13 → 4-second period →
Repeats above

7BU04B-135

Note

**a) If more than one malfunction occurs, the code numbers will be displayed on the Self-Diagnosis Checker one by one in numerical order.
For example, for malfunctions 09, 13, and 01, the code numbers are displayed in the order 01, 09, then 13.**

b) The memory of malfunctions is canceled when the negative battery cable is disconnected for approximately five seconds.

TROUBLE CODE DIAGNOSTIC CHART – B2200

No.17 code display (Feedback system)

- Does monitor lamp of **Self-Diagnosis Checker** illuminate at idle?
  - NO → Possible Cause: Air leak in vacuum hoses or emission components; Contaminated oxygen sensor; Clogged fuel jet(s)
  - YES ↓
- Are spark plugs clean?
  - NO → Clean or replace
  - YES ↓
- Is oxygen sensor output voltage OK?
  - NO → Possible Cause: Oxygen sensor faulty
  - YES ↓
- Is 1A terminal voltage of engine control unit OK?
  - NO → Possible Cause: Open or short circuit in wiring harness from oxygen sensor connector to engine control unit
  - YES → Perform after-repair procedure

No.18, 22, 23, 26, 28, 29, 30, 34, 35, and 45 code displays (Solenoid valves)

- Is there poor connection at connector in wiring circuit of indicated solenoid valve?
  - YES → Repair or replace
  - NO ↓
- Is signal or voltage of connector for indicated solenoid valve OK?
  - NO → Possible Cause: Open or short circuit in wiring harness of indicated solenoid valve; Engine control unit faulty
  - YES → Perform after-repair procedure

TROUBLE CODE DIAGNOSTIC CHART—B2200

No.15 code display (Oxygen sensor)

- Is oxygen sensor output voltage OK?
  - NO → Replace oxygen sensor
  - YES → Is sensitivity of oxygen sensor OK?
    - NO → Replace oxygen sensor
    - YES → Perform after-repair procedure

No.16 code display (EGR position sensor)

Note
Inspect the vacuum hose to the EGR control valve for air leakage, blockage and damage if the MIL illuminates only during cruising.

- Is there poor connection at connector?
  - YES → Repair or replace
  - NO → Is resistance in EGR position sensor OK?
    - NO → Replace EGR position sensor
    - YES → Is there **4.5—5.5V** at C terminal of EGR position sensor connector?
      - NO → Possible Cause: Open or short circuit in wiring harness from C terminal to engine control unit
      - YES → Measure 1F terminal voltage of the engine control unit
        - More than 4.8V → Possible Cause:
          Open circuit in wiring harness of ground
          Open circuit in wiring harness from A terminal to 1F terminal of engine control unit
        - 0.2V → Possible Cause:
          Short circuit in wiring harness from A terminal to 1F terminal of engine control unit
          Engine control unit faulty
        - 0.2—4.8V → Perform after-repair procedure

(continued)

- Measure 1E terminal voltage of engine control unit
  - More than 4.9V → Possible Cause:
    Open circuit of ground harness
    Open circuit of wiring harness from D terminal of vacuum sensor connector to 1E terminal of engine control unit
  - Less than 0.25V → Possible Cause:
    Short circuit of wiring harness connected to vacuum sensor
    Open circuit of wiring harness from B terminal of vacuum sensor connector to 1G terminal of engine control unit
  - 0.25—4.9V → Perform after-repair procedure

No.14 code display (Atmospheric pressure sensor)

- Is there poor connection at connector?
  - YES → Repair or replace
  - NO → Is there battery voltage at C terminal of atmospheric pressure sensor connector?
    - NO → Possible Cause:
      Open circuit in wiring harness from ignition switch to atmospheric pressure sensor
    - YES → Is there **4.5—5.5V** at B terminal of atmospheric pressure sensor connector?
      - NO → Possible Cause:
        Short circuit in wiring harness from B terminal of atmospheric pressure sensor to engine control unit
        Engine control unit faulty
      - YES → Measure 1H terminal voltage of emission control unit
        - More than 4.9V → Possible Cause:
          Open circuit of ground harness
          Open circuit of wiring harness from D terminal of atmospheric pressure sensor connector to 1H terminal of engine control unit
        - Less than 1.4V → Possible Cause:
          Short circuit in wiring harness from B terminal of atmospheric pressure sensor connector to 1G terminal of engine control unit
        - 1.4—4.9V → Perform after-repair procedure

TROUBLE CODE DIAGNOSTIC CHART—B2200

No.01 code display (IG pulse)

- Is connector connected to negative terminal of ignition coil?
  - NO → Connect connector
  - YES → Is 2A terminal voltage of engine control unit OK?
    - NO → Possible Cause: Open or short circuit in wiring harness from ignition coil to 2A terminal
    - YES → Perform after-repair procedure

No.09 code display (Water thermosensor)

- Is there poor connection at connector?
  - YES → Repair or replace
  - NO → Is resistance of water thermosensor within **0.057—84.5 kΩ** ?
    - NO → Replace water thermosensor
    - YES → Measure 1C terminal voltage of engine control unit
      - 12V → Possible Cause: Open circuit in wiring harness
      - 0V → Possible Cause: Short circuit in wiring harness from 1C terminal to water thermosensor; Engine control unit faulty

No.13 code display (Vacuum sensor)

- Is there poor connection at connector?
  - YES → Repair or replace
  - NO → Is there battery voltage at C terminal of vacuum sensor connector?
    - NO → Possible Cause: Open circuit in wiring harness from ignition switch to vacuum sensor
    - YES → Is there **4.5—5.5V** at B terminal of vacuum sensor connector?
      - NO → Possible Cause: Short circuit in wiring harness from B terminal of vacuum sensor to engine control unit; Engine control unit faulty
      - YES → **Next step**

CONTROL UNIT PIN LOCATION AND VOLTAGE TEST—B2200

Terminal	Connected to	Voltage	Condition
2E (Output)	Self-Diagnosis Checker (Digital display)	Approx. 12V	Buzzer: OFF
		Less than 1.5V	Buzzer: ON
		Code signal	When self-diagnosis check connector grounded
2F (Output)	Air/fuel (A/F) solenoid valve	Monitor reading: 1.5—3.8V (fluctuating) Actual voltage: 3.5—12V (fluctuating)	At idle
		0—14V (fluctuating or fixed)	During running
2H (Output)	Coasting richer solenoid valve	Approx. 12V	At idle
		Less than 1.5V	At 2,500—1,400 rpm with in-gear deceleration (Voltage indicated one second after conditions met)
2I (Output)	Self-Diagnosis Checker (Monitor lamp)	Less than 1.5V	Monitor lamp: ON
		Approx. 12V	Monitor lamp: OFF
2J (Output)	ACV solenoid valve	Approx. 12V	At idle
		Less than 1.5V	At 1,500 rpm or more, warmed up, no load
2K (Output)	Duty solenoid valve (Vent)	Approx. 12V	While cranking
		Approx. 12V	During warm up
		Approx. 12V	At idle
		Voltage decreases (Green and red lights flash)	During acceleration
2L (Output)	Duty solenoid valve (Vacuum)	Approx. 12V	While cranking
		Approx. 12V	During warm up
		Approx. 12V	At idle
		Voltage decreases (Green and red lights flash)	During acceleration
2M (Output)	Idle-up solenoid valve (A/C)	Less than 1.5V	At idle (A/C: ON)
		Approx. 12V	At 1,400 rpm or below (A/C: ON)
2N (Ground)	Engine ground	Less than 1.5V	—

Connectors

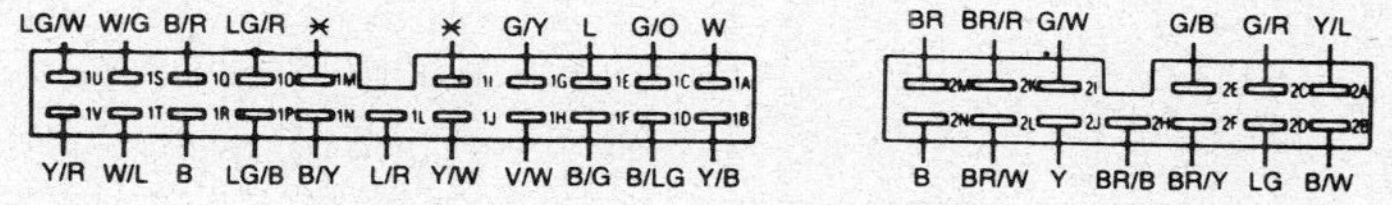

Note

a) In-gear deceleration is as follows.

Vehicle with M/T .... transmission in gear (not neutral), clutch pedal released, and throttle valve closed fully

Vehicle with A/T .... transmission in gear (not P or N) and throttle valve closed fully.

b) When inspecting "2E" terminal voltage, connect the Self-Diagnosis checker.

CONTROL UNIT PIN LOCATION AND VOLTAGE TEST—B2200

Terminal	Connected to	Voltage	Condition
1A (Input)	Oxygen sensor	0.3—0.7V	At idle
		More than 0.45V	During acceleration
		Less than 0.45V	During deceleration
1B (Input)	Self-diagnosis check connector	Approx. 12V	Check connector; Not grounded
		0V	Check connector; Grounded
1C (Input)	Water thermosensor	Approx. 0.5V	Warmed-up engine (Thermostat: Open)
1D (Ground)	Water thermosensor, EGR position sensor, Vacuum sensor, Atmospheric pressure sensor, Intake air thermosensor	Less than 1.5V	—
1E (Input)	Vacuum sensor	Approx. 1.3V	At idle
		Approx. 4.0V	Engine stopped (Atmospheric pressure)
1F (Input)	EGR position sensor	Approx. 0.7V	At idle
		0.7—4.7V	During driving
1G (Power supply)	EGR position sensor, Vacuum sensor, Atmospheric pressure sensor	4.5—5.5V	—
1H (Input)	Atmospheric pressure sensor	Approx. 4V	Sea level
1J (Input)	Intake air thermosensor	Approx. 4.1V	At 20°C (68°F)
1L (Memory power)	Battery	Approx. 12V	—
1N (Input)	Neutral and clutch switch	Approx. 12V	In gear
		Less than 1.5V	In neutral or depress clutch pedal
	Inhibitor switch	Less than 1.5V	In N or P range
		Approx. 12V	In other ranges
1O (Input)	Idle switch	Approx. 12V	At idle
		Less than 1.5V	At more than 1,200 rpm with no load
1P (Ground)	Idle switch	Less than 1.5V	—
1Q (Input)	Water temperature switch	Approx. 12V	Radiator coolant temp.: above 17°C (63°F)
		Less than 1.5V	Radiator coolant temp.: below 17°C (63°F)
1R (Ground)	Engine ground	Less than 1.5V	—
1S (Output)	Coasting advance solenoid valve	Approx. 12V	At idle
		Less than 1.5V	At 1,700—2,500 rpm during in-gear deceleration
1T (Output)	Idle-up solenoid valve (A/T)	Less than 1.5V	At less than 1,000 rpm in R, D, 2, or 1 range
		Approx. 12V	In N or P range or more than 1,100 rpm without A/C switch: ON
1U (Output)	Malfunction indicator light	Approx. 12V	light: OFF
		Less than 1.5V	light: ON
1V (Output)	Purge solenoid valve	Approx. 12V	At idle
		Less than 1.5V	At 1,400 rpm with warmed-up engine
2A (Input)	Ignition coil negative terminal	Approx. 12V	—
2B (Battery power)	Ignition switch (ON)	Approx. 12V	Ignition switch: ON
		0V	Ignition switch: OFF
2C (Input)	Air-conditioner magnetic clutch circuit	Approx. 12V	Air conditioner: ON
		0V	Air conditioner: OFF
2D (Output)	Slow fuel cut solenoid valve	Less than 15.V	Ignition switch: ON
		Less than 1.5V	At idle
		Approx. 12V	At 2,500 rpm or more during in-gear deceleration

TROUBLESHOOTING WITH THE ENGINE SIGNAL MONITOR

The Engine Signal Monitor (49–9200–162) and Adapter Harness (49–U018–001) is used to check the control unit terminal voltages. This monitor checks the individual terminal voltages through selection by the monitor switch.

Inspection

1. Start the engine and allow it to reach normal operating temperature. Stop the engine.
2. Disconnect the emission control unit connector.
3. Connect Adapter Harness (49–U018–001) between the emission control unit and the connector.
4. Connect Engine Signal Monitor (49–9200–162) to the Adapter Harness.
5. Position the select switch and the monitor switch on the Engine Signal Monitor to the terminal number to be tested. Check the voltage of each terminal.

FEEDBACK SYSTEM

System Inspections

1. Start the engine and allow it to reach normal operating temperature, and let run at idle.
2. Connect a suitable dwellmeter to the A/F check connector (BrY), and note the dwellmeter readings.
3. If dwellmeter reading 0 degrees, check the following:
   a. Ignition pulse signal for the emission control unit.
   b. Vacuum sensor.
   c. Water thermo sensor.
4. If dwellmeter reading is 27 degrees, check the following:
   a. Intake air temperature sensor.
   b. Vacuum hose routing.
   c. O_2 sensor.

5. If dwellmeter reading is 36 degrees, check the water thermo sensor.

6. If the dwellmeter reading is fluctuating out of the 20–70 degrees range, check the following:

a. Vacuum hose routing.
b. O_2 sensor.
c. A/F solenoid valve and wiring harness.
d. Clogged jets and air bleeds in the carburetor.

AIR INJECTION SYSTEM

System Inspection

1. Start the engine and allow it to reach normal operating temperature.

2. Place a thin piece of paper over the inlet port of reed valves (B) and (C). Increase engine speed to 1500 rpm and check that air is being pulled in.

3. Increase engine speed to 3000 rpm, and check that no exhaust gas is leaking from the air inlet port. If a malfunction is detected, replace the reed valve.

4. Disconnect and plug the vacuum hose from the No. 2 air control valve. Place a thin piece of paper over inlet port (D) of reed valve (A).

5. Increase engine speed to 1500 rpm, and check that air is being pulled in. If not, check the No. 1 air control valve and the reed valve.

6. Disconnect and plug the vacuum hose from the No. 1 air control valve. Connect a suitable hand operated vacuum pump to the No. 2 air control valve. Apply approximately 3.54 in. Hg (90mm Hg) to the No. 2 air control valve.

7. Place a thin piece of paper over inlet port (E) of reed valve (A). Increase engine speed to 1500 rpm, and check that air is being pulled in. If not, check the No. 2 air control valve and the reed valve.

8. Stop the engine. Disconnect the water temperature switch connector. Start the engine at let run at idle. Check that there is no vacuum at the No. 2 air control valve vacuum hose.

9. Increase engine speed to 1500 rpm and check that vacuum

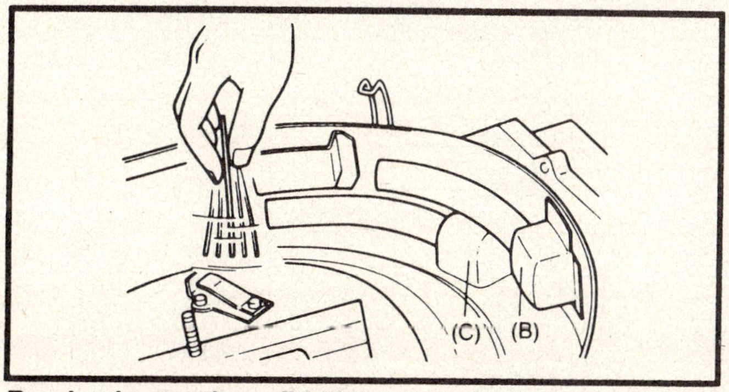

Reed valve testing – B2200

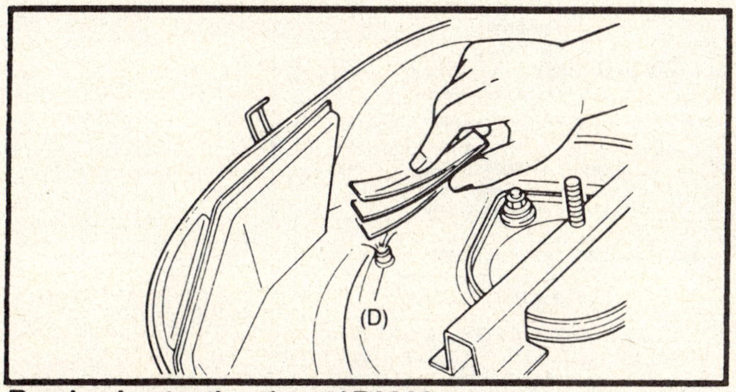

Reed valve testing (cont.)B2200

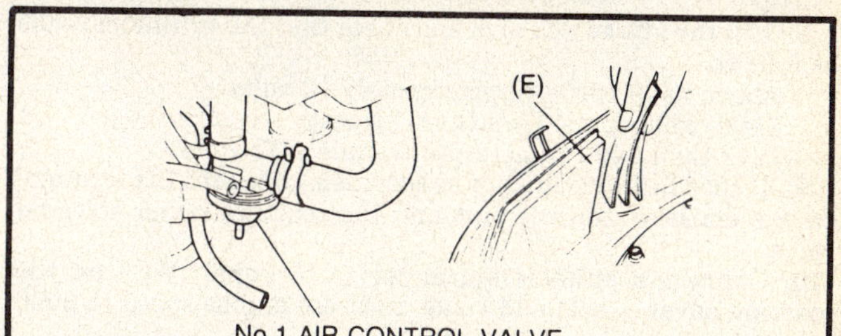

Reed valve testing (cont.)B2200

is present at the vacuum hose. If a malfunction is detected, check the ACV solenoid valve.

10. Reconnect the vacuum hoses to the No. 1 and No. 2 air control valves, then the water temperature switch connector.

EGR SYSTEM

System Inspection

1. Check the vacuum hose routing. Repair or replace if necessary.

2. Start the engine and allow it to reach normal operating temperature, and let run at idle.

3. Disconnect and plug the vacuum hose from the EGR control valve. Check that the engine operates smoothly. If not, check the EGR control valve.

4. Connect a suitable vacuum gauge to the EGR control valve hose. Check that the gauge shows no reading.

5. Rev the engine, and check that the gauge shows no vacuum. If a malfunction is detected, check the duty solenoid valve and the 2L and 2K terminals of the emission control unit.

6. Connect a suitable vacuum gauge between the duty solenoid valve and the EGR control valve. Rev the engine and note the amount of vacuum on the gauge.

7. Disconnect and plug the vacuum hose to the EGR control valve. Rev the engine and check that the gauge reading is higher than the reading obtained in Step 6. If not, check the EGR position sensor, the 1F terminal of the emission control unit and the duty solenoid valve.

8. Disconnect the water temperature switch electrical connectors, then connect a jumper wire between the connectors. With the EGR control valve hose plugged, check that the gauge shows no reading when the engine is accelerated. If a reading is obtained, check the duty solenoid valve and the 1Q terminal on the emission control unit.

DECELERATION CONTROL SYSTEM

System Inspection

1. Start the engine and allow it to reach normal operating temperature.

2. On models equipped with automatic transmissions, disconnect the inhibitor switch connector. On models equipped with manual transmissions, disconnect the neutral switch connector.

3. On all models, remove the air cleaner assembly.

4. Connect a suitable voltmeter to the F terminal (Lg) on the carburetor connector. Increase engine speed to 3000 rpm.

5. Lift the idle switch arm and check that the voltmeter reads as follows:

Above 2500 rpm – Approximately 12 volts
Below 2500 rpm – Less than 1.5 volts

6. If specifications are not as specified, check the 2D terminal of the emission control unit and the slow fuel cut solenoid valve.

7. Connect a suitable voltmeter to the (H) terminal (BrB) on the carburetor connector. Increase engine speed to 3000 rpm.

8. Lift the idle switch arm and check that the voltmeter reads as follows:
   Above 2500 rpm—Approximately 12 volts
   1400–2500 rpm—Less than 1.5 volts
   Less than 1400 rpm—Approximately 12 volts
9. If specifications are not as specified, check the 2H terminal on the emission control unit and the coasting richer solenoid valve.
10. Connect a suitable voltmeter to terminal (WG) on the coasting advance solenoid valve. Increase engine speed to 3000 rpm.
11. Lift the idle switch arm and check that the voltmeter reads as follows:
   Above 2500 rpm—Approximately 12 volts
   1700–2500 rpm—Less than 1.5 volts
   Less than 1700 rpm—Approximately 12 volts
12. If specifications are not as specified, check the 1S terminal on the emission control unit and the vacuum solenoid valve.

EVAPORATIVE EMISSION CONTROL (EEC) SYSTEM

System Inspection

1. Check the vacuum hose routing. Repair or replace if necessary.
2. Start the engine and allow it to reach normal operating temperature.
3. Disconnect the vacuum hose from the No. 1 purge control valve, and connect a suitable vacuum gauge to the hose.
4. Increase engine speed to 2500 rpm and check that the vacuum gauge reads more than 5.9 in. Hg (150mm Hg). If specifications are not as specified, check the water thermo valve. Reconnect the vacuum hose to the No. 1 purge control valve.
5. Disconnect the vacuum hose from the canister, and connect a suitable vacuum gauge to the hose. Check that vacuum is present when engine speed exceeds 1400 rpm.
6. If no vacuum exists, check the purge solenoid valve, No. 3 purge control valve and the 1V terminal on the emission control unit. Reconnect the vacuum hose to the canister.
7. Disconnect the evaporation hose from the evaporation pipe, then connect a suitable hand operated vacuum pump to the evaporation pipe.
8. Operate the vacuum pump, and check that no vacuum is held in the system. If so, check the check-and-cut valve and the evaporation pipe for clogging.

PTC HEATER SYSTEM

System Inspection

1. Disconnect the PTC heater electrical connector.
2. Connect a suitable ohmmeter between the PTC connector and ground, and check for continuity.

SPECIFICATIONS CHART—B2200

Item			Transmission type	Manual	Automatic
Fuel tank capacity	Short bed		Liter (US gal, Imp gal)	56 (14.8, 12.3)	
	Long bed		Liter (US gal, Imp gal)	66 (17.4, 14.5)	
Fuel filter	Type			Filter paper; with magnet	
Fuel pump	Type			Mechanical	Electrical
	Fuel pressure		kPa (kg/cm², psi)	26—32 (0.26—0.33, 3.7—4.7)	20—25 (0.20—0.25, 2.8—3.6)
	Flow rate		cc (cu in)/min	860 (52.5)	1,150 (70.2)
Carburetor	Type			Downdraft (2-barrel, 2-stage, auto-choke)	
	Throat diameter	Pri.	mm (in)	30 (1.181)	
		Sec.	mm (in)	34 (1.339)	
	Venturi diameter	Pri.	mm (in)	24.5 x 15 x 8 (0.965 x 0.591 x 0.315)	
		Sec.	mm (in)	31 x 10 (1.220 x 0.394)	
	Main jet	Pri.	mm (in)	1.04 (0.0409)	
		Sec.	mm (in)	1.50 (0.0591)	
	Main air bleed	Pri.	mm (in)	0.60 (0.0236)	
		Sec.	mm (in)	0.50 (0.0197)	
	Slow jet	Pri.	mm (in)	0.52 (0.0205)	
		Sec.	mm (in)	0.85 (0.0335)	
	Slow air bleed	Pri: No.1	mm (in)	0.80 (0.0315)	
		Pri: No.2	mm (in)	1.10 (0.0433)	
		Sec: No.1	mm (in)	0.80 (0.0315)	
		Sec: No.2	mm (in)	0.50 (0.0197)	
	Coasting richer jet		mm (in)	0.42 (0.0165)	
	Coasting richer air bleed	No.1	mm (in)	1.60 (0.0630)	
		No.2	mm (in)	2.60 (0.1024)	
	High-speed richer jet		mm (in)	1.80 (0.0709)	
	High-speed richer air bleed		mm (in)	1.00 (0.0394)	
	Solenoid controlled fuel jet		mm (in)	0.85 (0.0335)	
	Solenoid controlled air bleed		mm (in)	1.50 (0.0591)	
	Float level	High	mm (in)	11.6—12.6 (0.457—0.496)	10.7—11.7 (0.421—0.461)
		Low	mm (in)	46.0—47.0 (1.811—1.850)	
	Fast idle adjustment	Throttle valve clearance	mm (in)	0.84—1.04 (0.033—0.041)	
		Choke valve clearance	mm (in)	0.60—1.14 (0.024—0.045)	
	Secondary throttle valve adjustment	Throttle valve clearance	mm (in)	7.35—8.25 (0.289—0.325)	
	Unloader system adjustment	Choke valve clearance	mm (in)	2.80—3.62 (0.110—0.143)	
	Choke diaphragm adjustment	Choke valve clearance	mm (in)	1.70—2.16 (0.067—0.085)	
Air cleaner	Fresh-Hot			Bimetal, automatic	
	Element type			Wet	
Accelerator cable	Deflection		mm (in)	1—3 (0.04—0.12)	
Idle speed			rpm (in neutral or P range)	800^{+50}_{-0}	

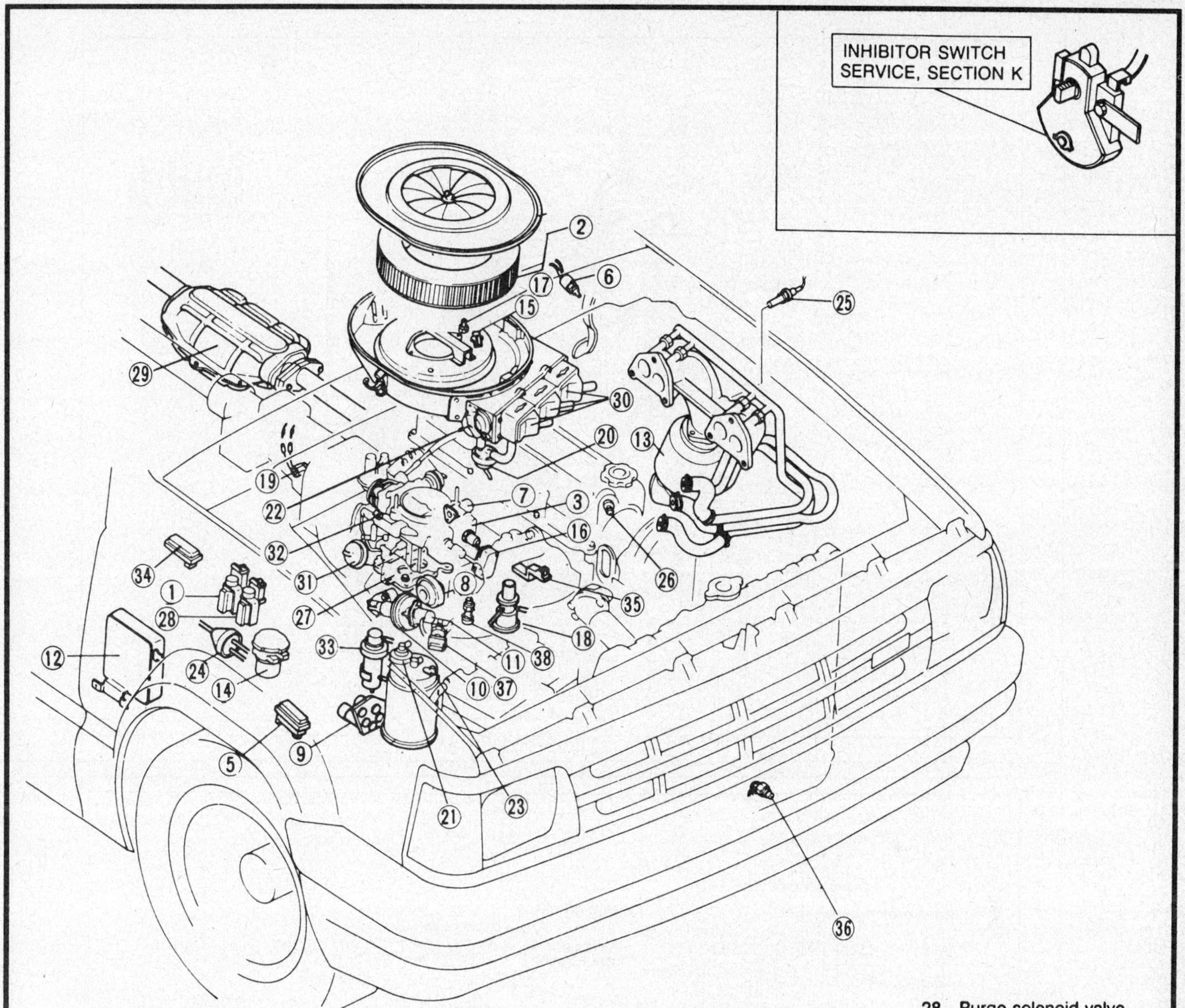

1. ACV solenoid valve
2. Air cleaner
3. Air vent solenoid valve
4. Air/fuel solenoid valve
5. Atmospheric pressure sensor
6. Clutch switch
7. Coasting richer solenoid valve
8. Dashpot
9. Duty solenoid valve
10. EGR control valve
11. EGR position sensor
12. Engine control unit
13. Front catalytic converter
14. High-altitude compensator
15. Idle compensator
16. Idle switch
17. Intake air thermosensor
18. Mixture control valve
19. Neutral switch
20. No. 1 air control valve
21. No. 1 purge control valve
22. No. 2 air control valve
23. No. 2 purge control valve
24. No. 3 purge control valve
25. Oxygen sensor
26. PCV valve
27. PTC heater
28. Purge solenoid valve
29. Rear catalytic converter
30. Reed valve
31. Servo diaphragm
32. Slow fuel cut solenoid
33. Vacuum control valve
34. Vacuum sensor
35. Vacuum solenoid valve
36. Water temperature switch
37. Water thermosensor
38. Water thermovalve

Component locations—B2200

VACUUM HOSE ROUTING—B2200

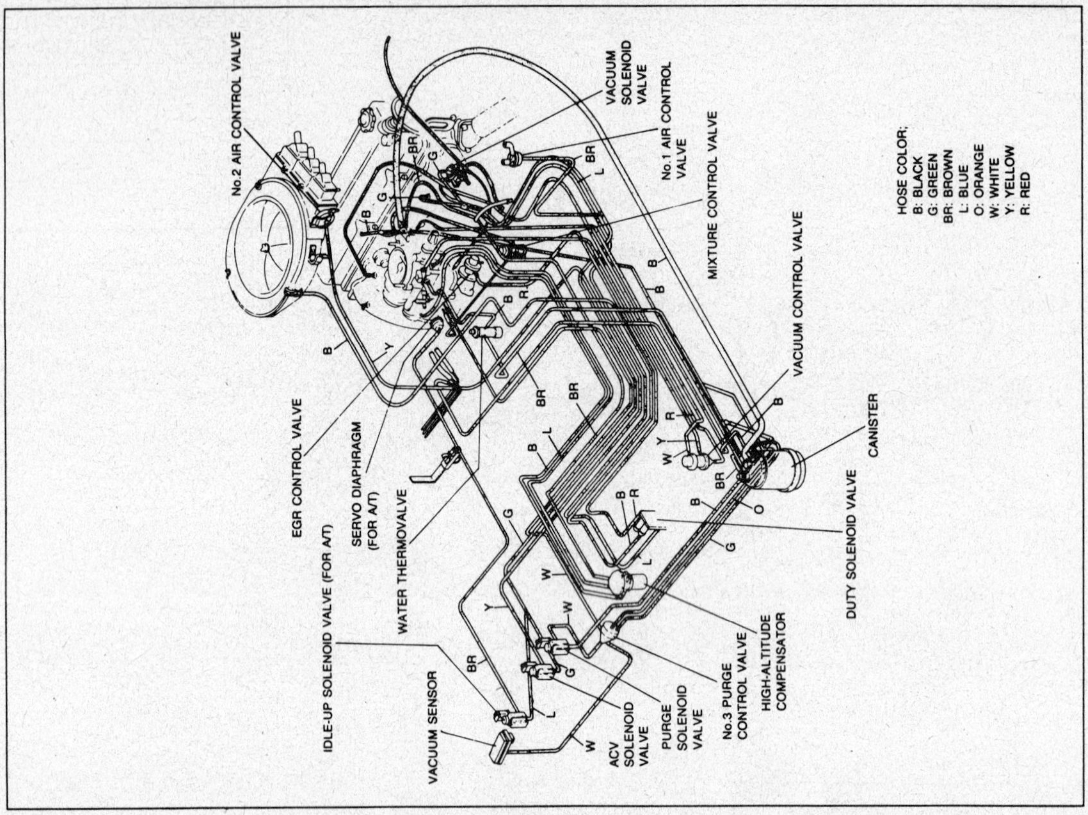

SYSTEM SCHEMATIC—B2200

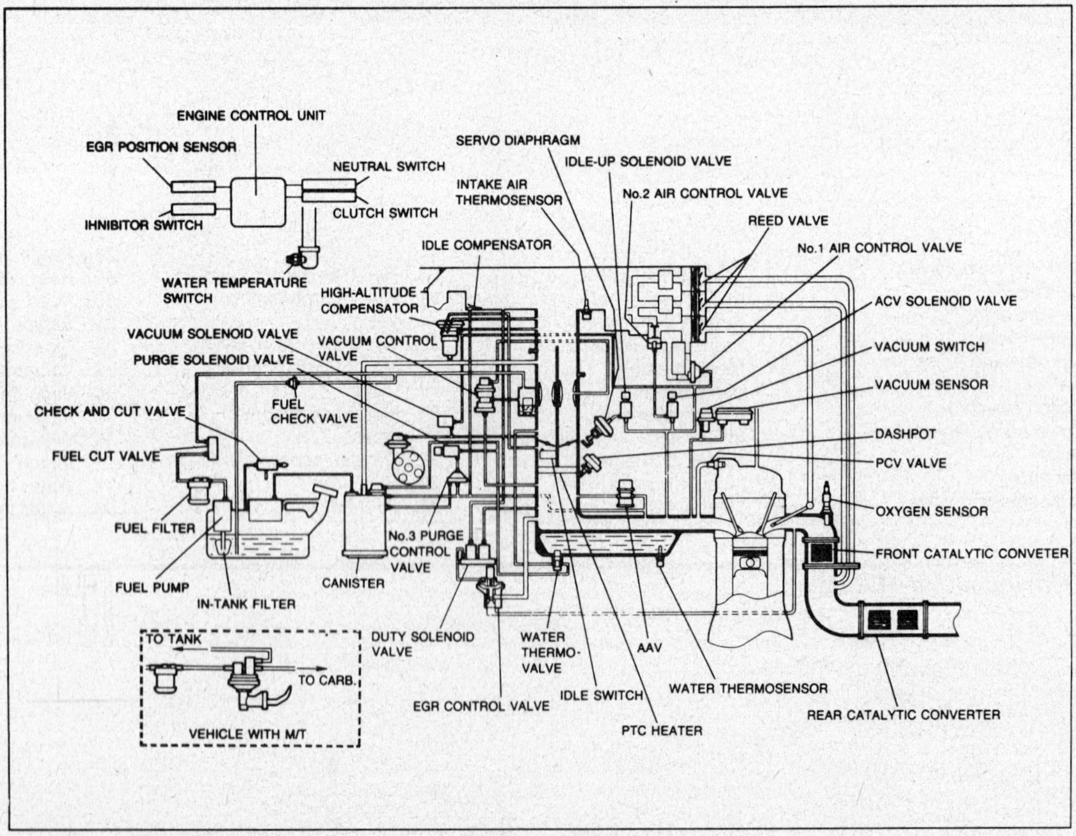

SYSTEM SCHEMATIC—B2200

For 2.2ℓ
■ ENGINE CONTROL SYSTEM ■ IGNITION SYSTEM ■ FUEL PUMP (For A/T) ■ P.T.C. HEATER

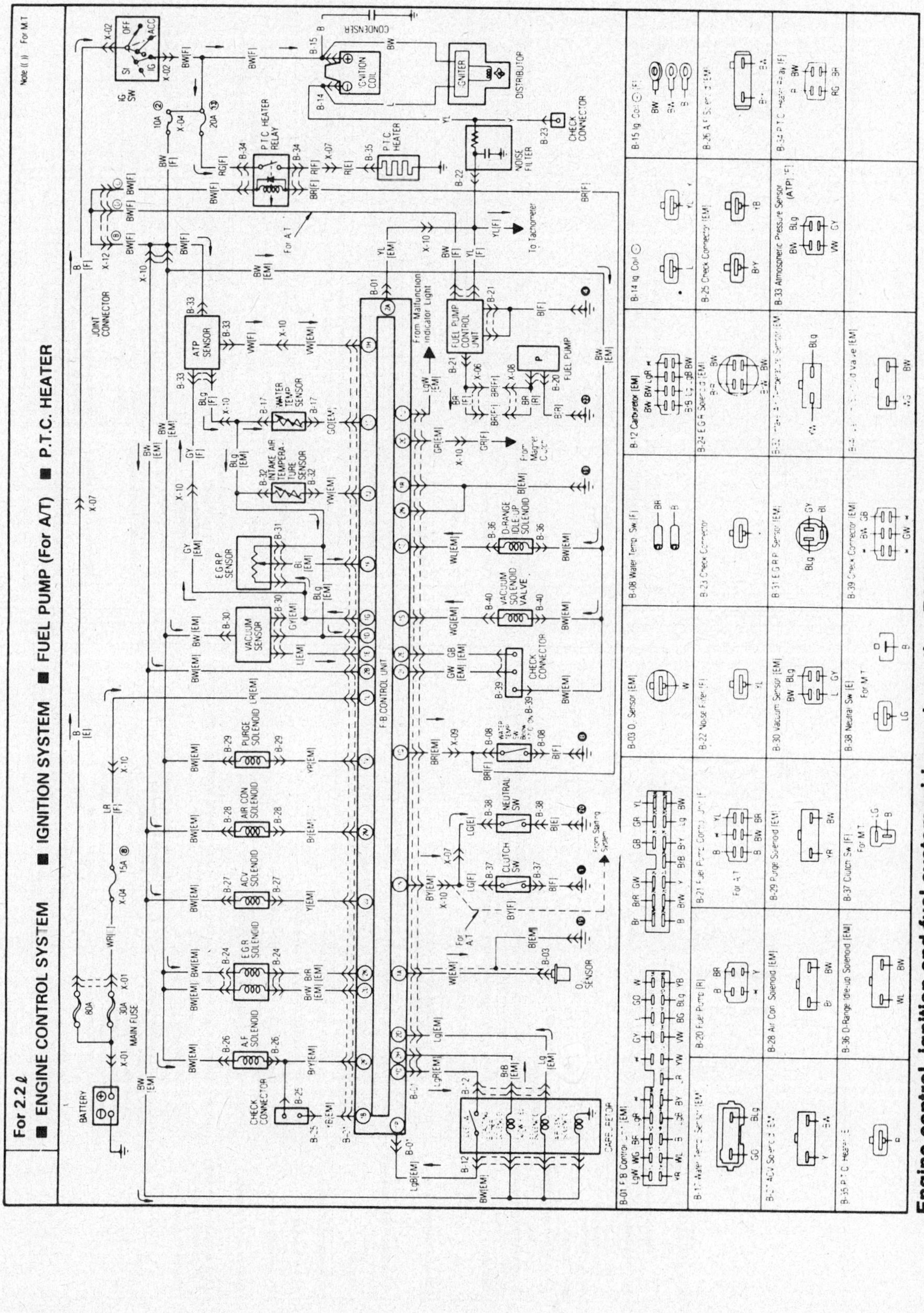

Engine control, ignition and fuel system wiring schematic—B2200

COMPONENT DESCRIPTION CHART – B2200

COMPONENT DESCRIPTIONS

Component	Function	Remarks
ACV solenoid valve	Applies vacuum to No.2 air control valve according to signal from engine control unit	
Air/fuel solenoid valve (in carburetor)	Controls air/fuel mixture according to signal from engine control unit	
Air cleaner	Filters air into carburetor	
Air vent solenoid valve	Vents float chamber to the canister while engine stopped	
Atmospheric pressure sensor	Detects atmospheric pressure (altitude); sends signal to engine control unit	Decreases amount of EGR at high altitude **(higher than 1,000 m, 3,280 ft)**
Canister	Stores gas tank and carburetor fumes while engine stopped When engine started, fumes drawn into intake manifold	
Check-and-cut valve	Vents fuel tank to atmosphere if vent line from check valve to No.2 purge control valve is clogged	
Coasting richer solenoid valve	Opens carburetor secondary stage fuel circuit during deceleration	
Dashpot	Gradually allows throttle closing during deceleration	
Duty solenoid valve **1) Vent valve** **2) Vacuum valve**	Controls vacuum to activate EGR control valve Opens vent according to signal from engine control unit Opens vacuum line according to signal from engine control unit	
EGR position sensor	Detects EGR control valve lift; sends signal to control unit	
EGR control valve (with EGR position sensor)	Introduces exhaust gas to intake manifold	Operates during acceleration and constant speed driving
Engine control unit	Detects the following: Engine speed Engine coolant temperature Intake manifold vacuum Atmospheric pressure Radiator coolant temperature Intake air temperature Oxygen concentration EGR valve lift Throttle opening In-gear condition Air conditioner ON/OFF	Ignition coil negative (—) terminal Water thermosensor Vacuum sensor Atmospheric pressure sensor Water temperature switch Intake air thermosensor Oxygen sensor EGR position sensor Idle switch Neutral and clutch switch or inhibitor switch Air conditioner

Component	Function	Remarks
Engine control unit	Controls operation of the following: Air/fuel (A/F) solenoid valve Idle-up solenoid valves Slow fuel cut solenoid valve Coasting richer solenoid valve Vacuum solenoid valve ACV solenoid valve Purge solenoid valve Duty solenoid valve	
Front catalytic converter	Reduces HC & CO by oxidation Reduces NOx	Converts into CO_2 and H_2O Honeycomb construction
Fuel check valve	Prevents leakage through fuel return line if vehicle turns over	
Fuel cut valve	Prevents leakage from main fuel line if the vehicle turns over	
Fuel filter	Filters fuel entering fuel pump and carburetor	
Fuel pump	Pumps fuel from fuel tank to carburetor	
Fuel pump cut relay	Operates fuel pump according to ignition pulse or alternator operation	
High-altitude compensator	Maintains air/fuel mixture when atmospheric pressure drops because of elevation	Adds air to air bleeds in carburetor and intake manifold Operates at altitude of 500 m (1,640 ft) or higher
Idle compensator	Keeps idle constant with temperature change	
Idle switch	Detects throttle opening	OFF at idle ON at 1,000—1,200 rpm
Idle-up solenoid valve **1) air conditioner** **2) automatic transmission**	Applies vacuum to servo diaphragm according to signal from engine control unit	
Inhibitor switch	Detects select lever position; sends signal to engine control unit	Senses transmission operating range
Intake air thermosensor	Detects intake air temperature; sends signal to engine control unit	Fixes duty of air/fuel solenoid valve at high air temperature (higher than 67°C, 153°C)
Mixture control valve	Supplies fresh air into intake manifold at first period of sudden deceleration	
Neutral and clutch switches	Detects in-gear operation and clutch engagement; sends signal to engine control unit	

COMPONENT DESCRIPTION CHART (CONT.)—B2200

Component	Function	Remarks
No.1 air control valve	Supplies secondary air to reed valve A according to intake manifold vacuum	
No.1 purge control valve	Purges fuel vapor (stored in canister) into intake manifold during running	
No.2 air control valve	Supplies secondary air to reed valve A when ACV solenoid valve is ON	
No.2 purge control valve	Pressure and vacuum valves operate in accordance with fuel tank pressure	
No.3 purge control valve	Purges fuel vapor (stored in canister) into intake manifold when purge solenoid valve is ON	
Oxygen sensor	Detects exhaust oxygen concentration; sends signal to control unit	
PTC heater	Heats throttle body of carburetor and prevents icing	
Purge solenoid valve	Applies vacuum to No.3 purge control valve according to signal from engine control unit	
Rear catalytic converter (except for Canada)	Reduces HC & CO by oxidation	Converts into CO_2 and H_2O Honeycomb construction
Reed valves **Reed valve A** **Reed valves B and C**	Supplies secondary air to exhaust manifold (valve A) Supplies secondary air to exhaust pipe just behind front catalytic converter (valves B and C)	One-way valve on air cleaner
Servo diaphragm	Opens throttle valve by vacuum from idle-up solenoid valve (for A/C and A/T)	
Slow fuel cut solenoid valve	Cuts off primary slow fuel during deceleration or when ignition switch is OFF	Improves fuel consumption and prevents run-on
Vacuum control valve	Vents float chamber to intake manifold during heavy-load driving	
Vacuum sensor	Detects intake manifold vacuum; sends signal to engine control unit	
Vacuum solenoid valve	Applies intake manifold vacuum to vacuum control unit; advances ignition timing during deceleration	
Water temperature switch	Detects radiator coolant temperature; sends signal to engine control unit	ON at **15—19°C (59—66.2°F)** or lower
Water thermosensor	Detects intake manifold coolant temperature; sends signal to engine control unit	Thermistor
Water thermovalve	Opens and closes depending on engine coolant temperature	Opens at **46—54°C (114.8—129.2°F)** or higher

VACUUM HOSE COLOR CODE IDENTIFICATION CHART—B2200

Component	Color	Connected to:
ACV solenoid valve	Yellow Green	No.2 air control valve Intake manifold
Canister	Green Black Brown Black	No.3 purge control valve Intake manifold Water thermovalve Evaporative pipe
Distributor	Black	Intake manifold
Duty solenoid valve	Red Blue Black	EGR control valve Intake manifold Air cleaner
High-altitude compensator	Blue Black Brown	Carburetor (Primary main) Carburetor (Intake manifold) Carburetor (Secondary main)
Idle compensator	Black	Intake manifold
No.1 air control valve	Brown to green	Intake manifold
No.3 purge control valve	White Orange to black Green	Purge solenoid valve Intake manifold Canister
Purge solenoid valve	White Black to brown	No.3 purge control valve Intake manifold
Vacuum control valve	Red White Yellow	Intake manifold Secondary venturi (in carburetor) Float chamber (in carburetor)
Vacuum sensor	White	Intake manifold
Vacuum solenoid valve	Green Black	Intake manifold Distributor

TROUBLESHOOTING CHART – B2200

Hard start or won't start (Cranks OK)

STEP	QUICK INSPECTION		ACTION		POSSIBLE CAUSE
1	(If trouble occurs only when engine is cold) Check if choke valve is fully closed when engine is cold	Yes	Go to Next Step		
		No	Replace automatic choke assembly		
2	Check if indicator atop battery is blue	Yes	Go to Next Step		
		No	Check if electrolyte level of battery is between upper and lower lines	Yes	Recharge battery
				No	Add distilled water
3	Check if fuel level is at specified mark on carburetor sight glass	Yes	Go to Next Step		
		No	(Higher than specified) Disassemble carburetor and check points shown		Check needle and seat for wear or rust
					Check float for damage
					Set float level
			(Lower than specified) M/T vehicle: Check for specified fuel pressure **Fuel pressure: 26—32 kPa (0.26—0.33 kg/cm², 3.7—4.7 psi)**	Yes	Set float level
				No	Replace fuel pump
			(Lower than specified) A/T vehicle: Check for fuel pump operation sound at fuel filler port [Ign ON, fuel pump control unit terminal-wire (B/R) and (B/W) jumped]	Yes	Check fuel pressure
					Set float level
				No	Check fuel pump control unit
					Replace fuel pump
4	Check for spark at disconnected high-tension lead while cranking	Yes	Go to Next Step		
		No	Check ignition system		
5	Check if spark plug condition is OK	Yes	Go to Next Step		
		No	Repair or replace spark plug(s)		
6	Check for air leakage by listening for sucking noise (Engine running)	Yes	Check points shown		Intake air system component damaged
					Vacuum hose disconnected or damaged
					Bolts or nuts loose
					Gasket damaged
		No	Go to Next Step		
7	Check for correct vacuum hose routing	Yes	Go to Next Step		
		No	Repair		

Hard start or won't start (Cranks OK) (Cont'd)

STEP	QUICK INSPECTION		ACTION		POSSIBLE CAUSE
8	Check if air cleaner element is clean	Yes	Go to Next Step		
		No	Replace		
9	Pinch PCV hose and check if condition improves	Yes	Replace PCV valve		
		No	Go to Next Step		
10	Start engine Block intake port of mixture control valve and check if engine speed drops	Yes	Replace mixture control valve		
		No	Increase engine speed and quickly decelerate Check that air is pulled into intake port for **1—2 sec** after accelerator is released	Yes	Go to Next Step
				No	Replace mixture control valve
11	Check for malfunction code with SST [Ign ON, test connector (Green: 1-pin) grounded]	Yes	Check for cause by referring to specified check sequence		
		No	Go to Next Step		
12	Disconnect and plug vacuum hose to EGR control valve and check if condition improves	Yes	Check ECU (2K), (2L) terminal voltage with SST **Voltage: 2K—Approx.12V 2L—Approx. 12V (While cranking)**	Yes	Check duty solenoid valve
				No	Check ECU (1C), (1Q) and (2A) terminal voltage with SST
		No	Check EGR control valve for operation	Yes	Go to Next Step
				No	Replace EGR control valve
13	(Only for "won't start" problem) Check if "clicking" is heard from slow fuel cut solenoid valve when ignition switch is turned OFF → ON	Yes	Go to Next Step		
		No	Check ECU (2D) Terminal voltage with SST **Voltage: Less than 1.5V (Ign ON)**	Yes	Check slow fuel cut solenoid valve
				No	Check ECU (2B) terminal voltage with SST
14	Check carburetor		Check points shown		Check jet(s) for clogging
					Check nozzle for clogging
15	Check engine condition		Check points shown		Camshaft timing
					Compression

TROUBLESHOOTING CHART – B2200

Engine stalls during warm up

STEP	QUICK INSPECTION		ACTION		POSSIBLE CAUSE
1	Check if choke valve is slightly open just after starting	Yes	Check for correct choke diaphragm adjustment	Yes	Go to Next Step
				No	Adjust
		No	Check points shown		Check choke diaphragm for damage
					Check choke diaphragm vacuum hose for disconnection or damage
2	Check if choke valve opens as engine warms up	Yes	Go to Next Step		
		No	Check voltage at choke heater (Y/L) wire **Voltage: 6—8V (At idle)**	Yes	Replace automatic choke assembly
				No	Repair or replace wiring harness
3	Check if engine stalls when throttle valve is opened slightly	Yes	Go to Next Step		
		No	Check points shown		Mixture adjustment screw adjustment
					Slow jet clogged
4	Check for air leakage by listening for sucking noise	Yes	Check points shown		Intake air system components damaged
					Vacuum hose disconnected or damaged
					Bolts or nuts loose
					Gasket damaged
		No	Go to Next Step		
5	Check for correct vacuum hose routing	Yes	Go to Next Step		
		No	Repair		
6	Pinch PCV hose and check if condition improves	Yes	Replace PCV valve		
		No	Go to Next Step		
7	Disconnect air hoses (B), (L), and (BR) from carburetor Check high-altitude compensator by blowing through each hose **500 m (1,640 ft) or higher: Air flows** **Less than 500 m (1,640 ft): Air does not flows**	Yes	Go to Next Step		
		No	Replace high-altitude compensator		

Engine stalls during warm up (Cont'd)

STEP	QUICK INSPECTION		ACTION		POSSIBLE CAUSE
8	Move control valve (for air intake temperature control system) inside air cleaner Verify that it moves freely and that spring force is felt	Yes	Go to Next Step		
		No	Replace air cleaner		
9	Check for malfunction code with SST [Ign ON, test connector (Green: 1-pin) grounded]	Yes	Check for cause by referring to specified check sequence		
		No	Go to Next Step		
10	Check switches for correct operation with SST monitor lamp [Ign ON, test connector (Green: 1-pin) grounded]	Yes	Go to Next Step		
		No	Check for cause by referring to specified check sequence		
11	Disconnect and plug vacuum hose to EGR control valve and check if condition improves	Yes	Check ECU (2K) and (2L) terminal voltage with SST **Voltage:** **2K—approx. 12V** **2L—approx. 12V** **(During warm up)**	Yes	Check duty solenoid valve
				No	Check ECU (1C), (1Q), and (2A) terminal voltage with SST
		No	Check EGR control valve for operation	Yes	Go to Next Step
				No	Replace EGR control valve
12	Check carburetor		Check points shown		Check main jet for clogging
					Check main nozzle for clogging

TROUBLESHOOTING CHART – B2200

Hard restarting when hot					
STEP	QUICK INSPECTION		ACTION		POSSIBLE CAUSE
1	Check if choke valve is fully open when engine is hot	Yes	Go to Next Step		
		No	Check voltage at choke heater (Y/L) wire **Voltage: 6—8V (At idle)**	Yes	Replace automatic choke assembly
				No	Repair or replace wiring harness
2	Check if fuel level is at specified mark on carburetor sight glass	Yes	Go to Next Step		
		No	(Higher than specified) Disassemble carburetor and check points shown		Check needle and seat for wear or rust
					Check float for damage
					Set float level
			(Lower than specified) M/T vehicle: Check for specified fuel pressure **Fuel pressure: 26—32 kPa (0.26—0.33 kg/cm², 3.7—4.7 psi)**	Yes	Set float level
				No	Replace fuel pump
			(Lower than specified) A/T vehicle: Check for fuel pump operation sound at fuel filler port [Ign ON, fuel pump control unit terminal-wire (B/R) and (B/W) jumped]	Yes	Check fuel pressure
					Set float level
				No	Check fuel pump control unit
					Replace fuel pump
3	Pinch PCV hose and check if condition improves	Yes	Replace PCV valve		
		No	Go to Next Step		
4	Check for malfunction code with SST [Ign ON, test connector (Green: 1-pin) grounded]	Yes	Check for cause by referring to specified check sequence		
		No	Go to Next Step		
5	Check switches for correct operation with SST monitor lamp [Ign ON, test connector (Green: 1-pin) grounded]	Yes	Go to Next Step		
		No	Check for cause by referring to specified check sequence		

Hard restarting when hot (Cont'd)					
STEP	QUICK INSPECTION		ACTION		POSSIBLE CAUSE
6	Check if "clicking" is heard from slow fuel cut solenoid valve when ignition switch is turned OFF → ON	Yes	Go to Next Step		
		No	Check ECU (2D) terminal voltage with SST **Voltage: Less than 1.5V (Ign ON)**	Yes	Check slow fuel cut solenoid valve
				No	Check ECU (2B) terminal voltage with SST
7	Check if "clicking" is heard from air vent solenoid valve when ignition switch is turned OFF → ON	Yes	Go to Next Step		
		No	Check for solenoid valve operation	Yes	Check wiring harness
				No	Replace solenoid valve
8	Check if idle compensator is in closed position when bimetal temperature is less than specified **Opening temperature: 63°—71°C (145°—160°F)**	Yes	Go to Next Step		
		No	Replace idle compensator		
9	Disconnect and plug vacuum hose (B) from charcoal canister and check if condition improves	Yes	Check vacuum hose routing		
		No	Go to Next Step		
10	Warm up engine and run it at idle Connect dwellmeter to check connector (White: 1-pin) and check if reading is **within 20°—70°**	Yes	Go to Next Step		
		No	**(Fixed at 0°)** Check points shown		ECU (2A) terminal voltage
					ECU (1E) terminal voltage
					ECU (1O) terminal voltage
			(Fixed at 27°) Check points shown		ECU (1J) terminal voltage
					Vacuum hose routing
					ECU (1A) terminal voltage
					Oxygen sensor sensitivity
			(Fixed at 36°) Check points shown		ECU (1C) terminal voltage
			(Fluctuating out of 20°—70° range) Check points shown		Vacuum hose routing
					ECU (1A) terminal voltage
					Oxygen sensor sensitivity
					ECU (2F) terminal voltage
					Air/fuel solenoid valve operation
					Clogged jets and air bleeds in carburetor
					Idle mixture adjustment
11	Check carburetor		Check point shown		Loose jet(s)

Note: High RVP (winter) fuel can cause vapor lock in warm weather if used.

TROUBLESHOOTING CHART—B2200

STEP	QUICK INSPECTION		ACTION		POSSIBLE CAUSE
	High idle speed after warm up				
1	Check if choke valve is fully open when engine is hot	Yes	Go to Next Step		
		No	Check voltage at choke heater (Y/L) wire **Voltage: 6—8V (At idle)**	Yes	Replace automatic choke assembly
				No	Repair or replace wiring harness
2	Check for correct accelerator cable free play **Free play: 1—3mm (0.039—0.188 in)**	Yes	Go to Next Step		
		No	Adjust		
3	Check if idle speed can be adjusted by turning TAS	Yes	Adjust idle speed		
		No	Go to Next Step		
4	Check for correct ignition timing (Vacuum hose disconnected)	Yes	Go to Next Step		
		No	Adjust ignition timing		
5	Check for malfunction code with SST [Ign ON, test connector (Green: 1-pin) grounded]	Yes	Check for cause by referring to specified check sequence		
		No	Go to Next Step		
6	Disconnect vacuum hose(s) from servo diaphragm and check if condition improves	Yes	Check ECU terminal voltage (1T), (2M) with SST **Voltage: 1T—Less than 1.5V (At less than 1,000 rpm in R, D, 2 or 1 range) Approx. 12V (In N or P range or more than 1,100 rpm without A/C switch: ON) 2M—Less than 1.5V [At idle (A/C: ON)] Approx. 12V [At 1,400 rpm or below (A/C: ON)]**	Yes	Check idle-up solenoid valve
				No	Check ECU terminal voltage (IN) and (2C) with SST
		No	Go to Next Step		
7	Check if throttle lever separates from dashpot rod **at approx. 1,900—2,100 rpm**	Yes	Go to Next Step		
		No	Adjust		
8	Check carburetor		Check point shown		Carburetor linkage

STEP	QUICK INSPECTION		ACTION		POSSIBLE CAUSE
	Engine idles roughly or stalls				
1	Check for air leakage by listening for sucking noise	Yes	Check points shown		Intake air system component damaged
					Vacuum hose disconnected or damaged
					Bolts or nuts loose
					Gasket damaged
		No	Go to Next Step		
2	Check if fuel level is at specified mark on carburetor sight glass	Yes	Go to Next Step		
		No	(Higher than specified) Disassemble carburtor and check points shown		Check needle and seat for wear or rust
					Check float for damage
					Set float level
			(Lower than specified) M/T vehicle: Check for specified fuel pressure **Fuel pressure: 26—32 kPa (0.26—0.33 kg/cm² 3.7—4.7 psi)**	Yes	Set float level
				No	Replace fuel pump
			(Lower than specified) A/T vehicle: Check for fuel pump operation sound at fuel filler port [Ign ON, fuel pump control unit terminal-wire (B/R) and (B/W) jumped]	Yes	Check fuel pressure
					Set float level
				No	Check fuel pump control unit
					Replace fuel pump
3	Disconnect high-tension lead from individual cylinders and check if condition changes	Yes	Go to Next Step		
		No	Check ignition system		Spark plug
					High-tension lead
					Distributor cap, rotor
4	Check for correct ignition timing **Ignition timing: 6° ± 1° BTDC**	Yes	Go to Next Step		
		No	Adjust ignition timing		
5	Turn throttle adjustment screw counterclockwise and check if condition improves	Yes	Adjust idle speed		
		No	Go to Next Step		
6	Pinch PCV hose and check if condition improves	Yes	Replace PCV valve		
		No	Go to Next Step		
7	Check for malfunction code with SST [Ign ON, test connector (Green: 1-pin) grounded]	Yes	Check for cause by referring to specified check sequence		
		No	Go to Next Step		
8	Check switches for correct operation with SST monitor lamp [Ign ON, test connector (Green: 1-pin) grounded]	Yes	Go to Next Step		
		No	Check for cause by referring to specified check sequence		

TROUBLESHOOTING CHART – B2200

Engine idles roughly or stalls (Cont'd)

STEP	QUICK INSPECTION		ACTION		POSSIBLE CAUSE
9	Check for correct EGR system vacuum hose routing	Yes	Go to Next Step		
		No	Repair or replace vacuum hose		
10	Disconnect and plug vacuum hose to EGR control valve and check if condition improves	Yes	Check ECU (2K) and (2L) terminal voltage with SST **Voltage:** **2K—Approx. 12V** **2L—Approx. 12V** **(At idle)**	Yes	Check duty solenoid valve
				No	Check ECU (1C), (1Q), and (2A) terminal voltage with SST
		No	Check EGR control valve for operation	Yes	Go to Next Step
				No	Replace EGR control valve
11	Check if "clicking" is heard from slow fuel cut solenoid valve when ignition switch is turned OFF → ON	Yes	Go to Next Step		
		No	Check ECU (2D) terminal voltage with SST **Voltage:** **Less than 1.5V** **(Ign ON)**	Yes	Check slow fuel cut solenoid valve
				No	Check ECU (2B) terminal voltage with SST
12	Check if idle compensator is closed when bimetal temperture is **below 63°—71°C (145°—160°F)**	Yes	Go to Next Step		
		No	Replace idle compensator		
13	Disconnect air hoses (B), (L) and (BR) from carburetor Check high-altitude compensator by blowing through each hose **500 m (1,640 ft) or higher:** **Airflows** **Less than 500 m (1,640 ft):** **Air does not flow**	Yes	Go to Next Step		
		No	Replace high-altitude compensator		
14	Check vacuum control valve				
15	Check carburetor		Check points shown		Check jet(s) for clogging
					Check carburetor fuel line for clogging
16	Check engine condition		Check point shown		Compression

Hesitation on acceleration or start-up

STEP	QUICK INSPECTION		ACTION		POSSIBLE CAUSE
1	Check if fuel level is at specified mark on sight glass	Yes	Go to Next Step		
		No	(Higher than specified) Disassemble carburetor and check points shown		Check needle and seat for wear or rust
					Check float for damage
					Set float level
			(Lower than specified) M/T vehicle: Check for specified fuel pressure **Fuel pressure:** **26—32 kPa** **(0.26—0.33 kg/cm², 3.7—4.7 psi)**	Yes	Set float level
				No	Replace fuel pump
			(Lower than specified) A/T vehicle: Check for fuel pump operation sound at fuel filler port [Ign ON, fuel pump control unit terminal-wire (B/R) and (B/W) jumped]	Yes	Check fuel pressure
					Set float level
				No	Check fuel pump control unit
					Replace fuel pump
2	Check if fuel is discharged from accelerator pump nozzle when opening throttle valve	Yes	Go to Next Step		
		No	Check if accelerator pump is damaged	Yes	Replace accelerator pump
				No	Clean carburetor fuel passages
3	Check for correct ignition timing **Ignition timing** **6° ± 1° BTDC**	Yes	Go to Next Step		
		No	Adjust ignition timing		
4	Check for correct idle speed **Idle speed** **800 $^{+50}_{-0}$ rpm (A/T: P range)**	Yes	Go to Next Step		
		No	Adjust idle speed		
5	Check for air leakage with throttle valve opened	Yes	Repair		
		No	Go to Next Step		
6	Check for malfunction code with SST [Ign ON, test connector (Green: 1-pin) grounded]	Yes	Check for cause by referring to specified check sequence		
		No	Go to Next Step		
7	Check switches for correct operation with SST monitor lamp [Ign ON, test connector (Green: 1-pin) grounded]	Yes	Go to Next Step		
		No	Check for cause by referring to specified check sequence		

TROUBLESHOOTING CHART—B2200

Hesitation on acceleration or start-up (Cont'd)

STEP	QUICK INSPECTION		ACTION		POSSIBLE CAUSE
8	Warm up engine and run it at idle	Yes	Go to Next Step		
	Connect dwellmeter to check connector (White: 1-pin) and check if dwellmeter reading is **within 20°—70°**	No	**(Fixed at 0°)** Check points shown		ECU (2A) terminal voltage
					ECU (1E) terminal voltage
					ECU (1O) terminal voltage
			(Fixed at 27°) Check points shown		ECU (1J) terminal voltage
					Vacuum hose routing
					ECU (1A) terminal voltage
					Oxygen sensor sensitivity
			(Fixed at 36°) Check points shown		ECU (1C) terminal voltage
			(Fluctuating out of 20°—70° range) Check points shown		Vacuum hose routing
					ECU (1A) terminal voltage
					Oxygen sensor sensitivity
					ECU (2F) terminal voltage
					Air/fuel solenoid valve operation
					Clogged jets and air bleeds in carburetor
					Idle mixture adjustment
9	Increase engine speed to **4,500 rpm** and check if dwellmeter indicates a **fixed 0°**	Yes	Go to Next Step		
		No	Replace Engine control unit		
10	Check for correct ignition timing advance	Yes	Go to Next Step		
		No	Insufficient centrifugal advance: Distributor malfunction		
			Insufficient vacuum advance: Check for vacuum routing	Yes	Distributor malfunction
				No	Vacuum hose
11	Disconnect and plug vacuum hose to EGR control valve and check if condition improves	Yes	Check ECU (2K) and (2L) terminal voltage with SST **Voltage: Drops from 12V and green and red lights flash (While accelerating)**	Yes	Check duty solenoid valve
				No	Check ECU (1C), (1Q), and (2A) terminal voltage with SST
		No	Check EGR control valve	Yes	Go to Next Step
				No	Replace EGR control valve
12	Check carburetor		Check point shown		Clogged primary main jet or nozzle

Lack of power

STEP	QUICK INSPECTION		ACTION		POSSIBLE CAUSE
1	Check if air cleaner element is clean	Yes	Go to Next Step		
		No	Replace air cleaner element		
2	Check if fuel level is at specified mark on carburetor sight glass	Yes	Go to Next Step		
		No	(Higher than specified) Disassemble carburetor and check points shown		Check needle and seat for wear or rust
					Check float for damage
					Set float level
			(Lower than specified) M/T vehicle: Check for specified fuel pressure **Fuel pressure: 26—32 kPa (0.26—0.33 kg/cm², 3.7—4.7 psi)**	Yes	Set float level
				No	Replace fuel pump
			(Lower than specified) A/T vehicle: Check for sound at fuel pump operation fuel filler port [Ign ON, fuel pump control unit terminal-wire (B/R) and (B/W) jumped]	Yes	Check fuel pressure
					Set float level
				No	Check fuel pump control unit
					Replace fuel pump
3	Check ignition timing **Ignition timing: 6° ± 1° BTDC**	Yes	Go to Next Step		
		No	Adjust ignition timing		
4	Check for correct ignition timing advance	Yes	Go to Next Step		
		No	Insufficient centrifugal advance: Distributor malfunction		
			Insufficient vacuum advance: Check for correct vacuum hose routing	Yes	Distributor malfunction
				No	Repair vacuum hose
5	Check if spark plug condition is OK	Yes	Go to Next Step		
		No	Repair or replace spark plug(s)		
6	Check for malfunction code with SST [Ign ON, test connector (Green: 1-pin) grounded]	Yes	Check for cause by referring to specified check sequence		
		No	Go to Next Step		

TROUBLESHOOTING CHART—B2200

Lack of power (Cont'd)

STEP	QUICK INSPECTION		ACTION		POSSIBLE CAUSE
6	Check switches for correct operation with SST monitor lamp [Ign ON, SST connector (Green: 1-pin) grounded]	Yes	Go to Next Step		
		No	Check for cause by referring to specified check sequence		
7	Disconnect and plug vacuum hose to EGR control valve and check if condition improves	Yes	Check ECU (2K) and (2L) terminal voltage with SST **Voltage: Drops from 12V and green and red lights flash (While acceleration)**	Yes	Check duty solenoid valve
				No	Check ECU (1C), (1Q), and (2A) terminal voltage with SST
		No	Check EGR control valve for operation	Yes	Go to Next Step
				No	Replace EGR control valve
8	Warm up engine and run it at idle	Yes	Go to Next Step		
	Connect dwellmeter to check connector (White: 1-pin) and check if reading is **within 20°—70°**	No	**(Fixed at 0°)** Check points shown		ECU (2A) terminal voltage
					ECU (1E) terminal voltage
					ECU (1O) terminal voltage
			(Fixed at 27°) Check points shown		ECU (1J) terminal voltage
					Vacuum hose routing
					ECU (1A) terminal voltage
					Oxygen sensor sensitivity
			(Fixed at 36°) Check points shown		ECU (1C) terminal voltage
			(Fluctuating out of 20°—70° range) Check points shown		Vacuum hose routing
					ECU (1A) terminal voltage
					Oxygen sensor sensitivity
					ECU (2F) terminal voltage
					Air/fuel solenoid valve operation
					Clogged jets and air bleeds in carburetor
					Idle mixture adjustment
9	Check engine condition		Check compression		
10	Check carburetor		Check point shown		• Clogged primary main jet or nozzle • Clogged secondary main jet or nozzle • Secondary throttle valve opening
11	Check exhaust system for clogging				

Afterburn on deceleration

STEP	QUICK INSPECTION		ACTION		POSSIBLE CAUSE
1	Check for correct ignition timing **Ignition timing: 6° ± 1° BTDC**	Yes	Go to Next Step		
		No	Adjust ignition timing		
2	Check for correct ignition timing advance	Yes	Go to Next Step		
		No	Insufficient centrifugal advance: Distributor malfunction		
			Insufficient Vacuum advance: Check for vacuum routing	Yes	Distributor malfunction
				No	Repair vacuum hose
3	Check if air cleaner element is clean	Yes	Go to Next Step		
		No	Replace		
4	Check for malfunction code with SST [Ign ON, test connector (Green: 1-pin) grounded]	Yes	Check for cause by referring to specified check sequence		
		No	Go to Next Step		
5	Check switches for correct operation with SST monitor lamp [IGN ON, Test connector (Green: 1-pin) grounded]	Yes	Go to Next Step		
		No	Check for cause by referring to specified check sequence		

TROUBLESHOOTING CHART—B2200

Afterburn on deceleration (Cont'd)

STEP	QUICK INSPECTION		ACTION		POSSIBLE CAUSE
6	Disconnect neutral switch (M/T) or inhibitor switch (A/T) connector Decelerate engine from **3,000 rpm** and check if "clicking" sound is heard from slow fuel cut solenoid valve	Yes	Go to Next Step		
		No	Check ECU (2D) terminal voltage with SST **At idle:** **Less than 1.5V** **Above 2,500 rpm during deceleration:** **Approx. 12V**	Yes	Check slow fuel cut solenoid valve
				No	Check ECU (1N), (2A), (1O), and (1P) terminal voltage with SST
7	Disconnect neutral switch (M/T) or inhibitor switch (A/T) connector Decelerate engine from **3,000 rpm** and check if "clicking" sound is heard from coasting richer solenoid valve	Yes	Go to Next Step		
		No	Check ECU (2H) terminal voltage with SST **At idle: Approx. 12V** **At 2,500—1,400 rpm during deceleration:** **Less than 1.5V**	Yes	Check coasting richer solenoid valve
				No	Check ECU (1N), (2A), (1O), and (1P) terminal voltage with SST
8	Disconnect neutral switch (M/T) or inhibitor switch (A/T) connector Decelerate engine from **3,000 rpm** and check if "clicking" is heard from coasting advance solenoid valve	Yes	Go to Next Step		
		No	Check ECU (1S) terminal voltage with SST **At idle:** **Approx. 12V** **At 2,500—1,700 rpm during deceleration:** **Below 1.5V**	Yes	Check coasting advance solenoid valve
				No	Check ECU (1N), (2A), (1O), and (1P) terminal voltage with SST
9	Start engine Block intake port of mixture control valve and check if engine speed drops	Yes	Replace mixture control valve		
		No	Increase engine speed and quickly decelerate Verify that air is pulled into intake port for **1—2 sec** after accelerator is released	Yes	Go to Next Step
				No	Replace mixture control valve
10	Check if throttle lever separates from dashpot rod **at 1,900—2,100 rpm**	Yes	Go to Next Step		
		No	Adjust dashpot		

Afterburn on deceleration (Cont'd)

STEP	QUICK INSPECTION		ACTION		POSSIBLE CAUSE
11	Place a thin paper over inlet port of reed valves (B) and (C) Increase engine speed to **1,500 rpm** and check if air is pulled in	Yes	Increase engine speed to **3,000 rpm** and check if exhaust gas leaks from air inlet port	Yes	Replace reed valve(s)
				No	Go to Next Step
		No			Replace reed valve(s)
12	Disconnect and plug vacuum hose (Y) to No.2 air control valve Place a thin paper over inlet port of reed valve (D) and increase engine speed to **1,500 rpm** and check if air is pulled in	Yes	Go to Next Step		
		No	Check No.1 air control valve for operation	Yes	Replace reed valve
				No	Replace No.1 air control valve
13	Disconnect and plug vacuum hose to No.1 air control valve Apply **90 mmHg (3.54 inHg)** of vacuum to No.2 air control valve Place a thin paper over inlet port (E) of reed valve and increase engine speed to **1,500 rpm** and check if air is pulled in	Yes	Go to Next Step		
		No	Check No.2 air control valve for operation	Yes	Check reed valve
				No	Replace No.2 air control valve
14	Disconnect water temperature switch connector and check that no vacuum exists at No.2 air control valve vacuum hose (Y)	Yes	Increase engine speed to **1,500 rpm** and check if vacuum is felt at vacuum hose	Yes	Go to Next Step
				No	Check ACV solenoid valve
		No	Check ACV solenoid valve		
15	Check engine condition		Check points shown		Check compression
					Check valve timing

TROUBLESHOOTING CHART—B2200

High fuel consumption

STEP	QUICK INSPECTION		ACTION	POSSIBLE CAUSE
1	Check other systems for proper operation • Brake • Clutch • A/T	Yes	Go to Next Step	
		No		Brake dragging
				Clutch slipping
				A/T shifting
2	Check for correct idle speed **Idle speed: 800 $^{+50}_{-0}$ rpm (A/T: P range)**	Yes	Go to next Step	
		No	Adjust	
3	Check for correct ignition timing **Ignition timing: 6° ± 1° BTDC**	Yes	Go to Next Step	
		No	Adjust	
4	Check for correct ignition timing advance	Yes	Go to Next Step	
		No	Insufficient centrifugal advance: Distributor malfunction	
			Insufficient vacuum advance: Check vacuum hose routing — Yes	Distributor malfunction
			Insufficient vacuum advance: Check vacuum hose routing — Nc	Repair or replace vacuum hose
5	Check if air cleaner element is clean	Yes	Go to Next Step	
		No	Replace	
6	Check if fuel level is at specified mark on carburetor sight glass	Yes	Go to Next Step	
		No	Adjust float level setting	
7	Check if choke valve fully opens after warm up	Yes	Go to Next Step	
		No	Replace automatic choke assembly	

High fuel consumption (Cont'd)

STEP	QUICK INSPECTION		ACTION	POSSIBLE CAUSE
8	Check if spark plug condition is OK	Yes	Go to Next Step	
		No	Repair or replace	
9	Check for malfunction code with SST [Ign ON, test connector (Green: 1-pin) grounded]	Yes	Check for cause by referring to specified check sequence	
		No	Go to Next Step	
10	Check switches for correct operation with SST monitor lamp [IGN ON, test connector (Green: 1-pin) grounded]	Yes	Go to Next Step	
		No	Check for cause by referring to specified check sequence	
11	Warm up engine and run it at idle Connect dwellmeter to check connector (White: 1-pin) and check if reading is **within 20°—70°**	Yes	Go to Next Step	
		No	**(Fixed at 0°)** Check points shown	ECU (2A) terminal voltage
				ECU (1E) terminal voltage
				ECU (1C) terminal voltage
			(Fixed at 27°) Check points shown	ECU (1J) terminal voltage
				ECU (1A) terminal voltage
				Oxygen sensor sensitivity
				Vacuum hose routing
			(Fixed at 36°) Check points shown	ECU (1C) terminal voltage
			(Fluctuating out of **20°—70° range**) Check points shown	ECU (1A) terminal voltage
				ECU (2F) terminal voltage
				Oxygen sensor sensitivity
				Air/fuel solenoid valve operation
				Vacuum hose routing
				Idle mixture adjustment
				Clogged jet(s) and air bleed(s) in carburetor
12	Check carburetor		Check point shown	Clogged or loose jet(s) and air bleed(s)

Note: Some loss of fuel economy is expected with alchol blended fuels.

Component Replacement

CARBURETOR

Removal

Remove the parts in the following order:

1. Negative battery cable
2. Air cleaner
3. Accelerator cable and cruise control cable (if equipped)
4. Fuel hoses
5. Evaporation hose
6. Vacuum hoses
7. Connector for solenoid valves
8. Connector for throttle sensor
9. Carburetor and gasket

Disassembly

Disassemble the carburetor in the numbered order as follows:

1. Accelerator pump connecting rod
2. Connect spring
3. Air vent solenoid valve
4. Choke rod
5. Air horn
6. Automatic choke assembly
7. Float
8. Needle valve assembly
9. Accelerator pump
10. Fuel bowl sight glass
11. Idle switch
12. Slow cut fuel solenoid valve
13. Coasting richer solenoid valve
14. Dashpot
15. Accelerator pump outlet check ball
16. Accelerator pump inlet check ball
17. Primary slow jet
18. Secondary slow jet
19. Primary main jet
20. Secondary main jet
21. Plug
22. Primary venturi and nozzle
23. Secondary venturi and nozzle
24. Primary main air bleed
25. Secondary main air bleed
26. Main body
27. Vacuum diaphragm
28. Throttle link
29. Throttle body
30. Mixture adjusting screw

NOTE: Further detailed disassembly procedures are as follows:

AIR HORN AND AUTOMATIC CHOKE

1. Vacuum hose
2. Accelerator pump connecting rod, spring and lever
3. Connect spring
4. Air vent solenoid valve (separate from the connector)
5. Choke rod (disconnect)
6. Air horn and automatic choke (separate from the main body)
7. Air vent solenoid valve, spring and gasket, if necessary

NEEDLE VALVE AND FLOAT

1. Float, pin and gasket
2. Needle valve assembly
3. Fuel bowl sight glass mounting screws
4. Cover, gasket, glass and rubber gasket

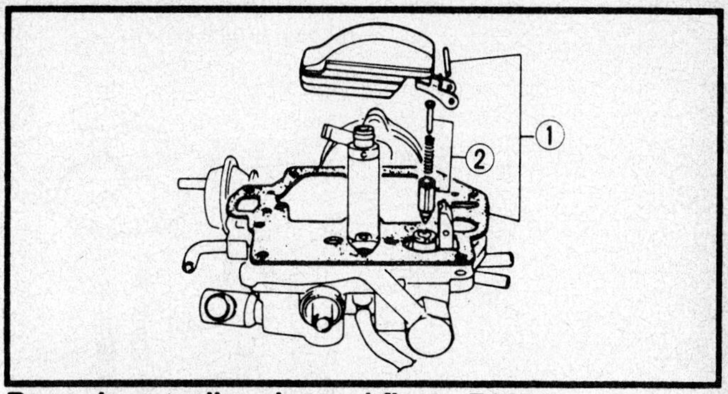

Removing needle valve and float – B2200

AIR BLEEDS AND JETS

1. Secondary slow jet
2. Secondary slow air bleed (No. 1)
3. Secondary main air bleed
4. Secondary main jet
5. Primary main air bleed
6. Slow jet and plug
7. Primary slow air bleed (No. 1)
8. Primary main jet
9. Richer air bleed (No. 2)
10. Primary slow air bleed (No. 2)
11. Richer air bleed (No. 1)
12. Richer jet

CAUTION

Note the size of all jets and air bleeds so that they a re reassembled in the correct position.

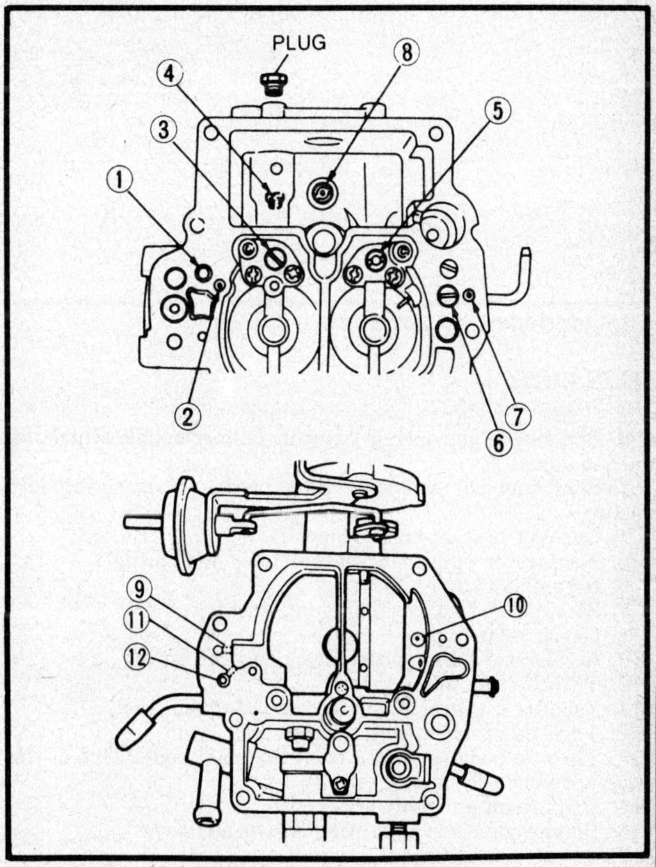

Locations of air bleeds and jets – B2200

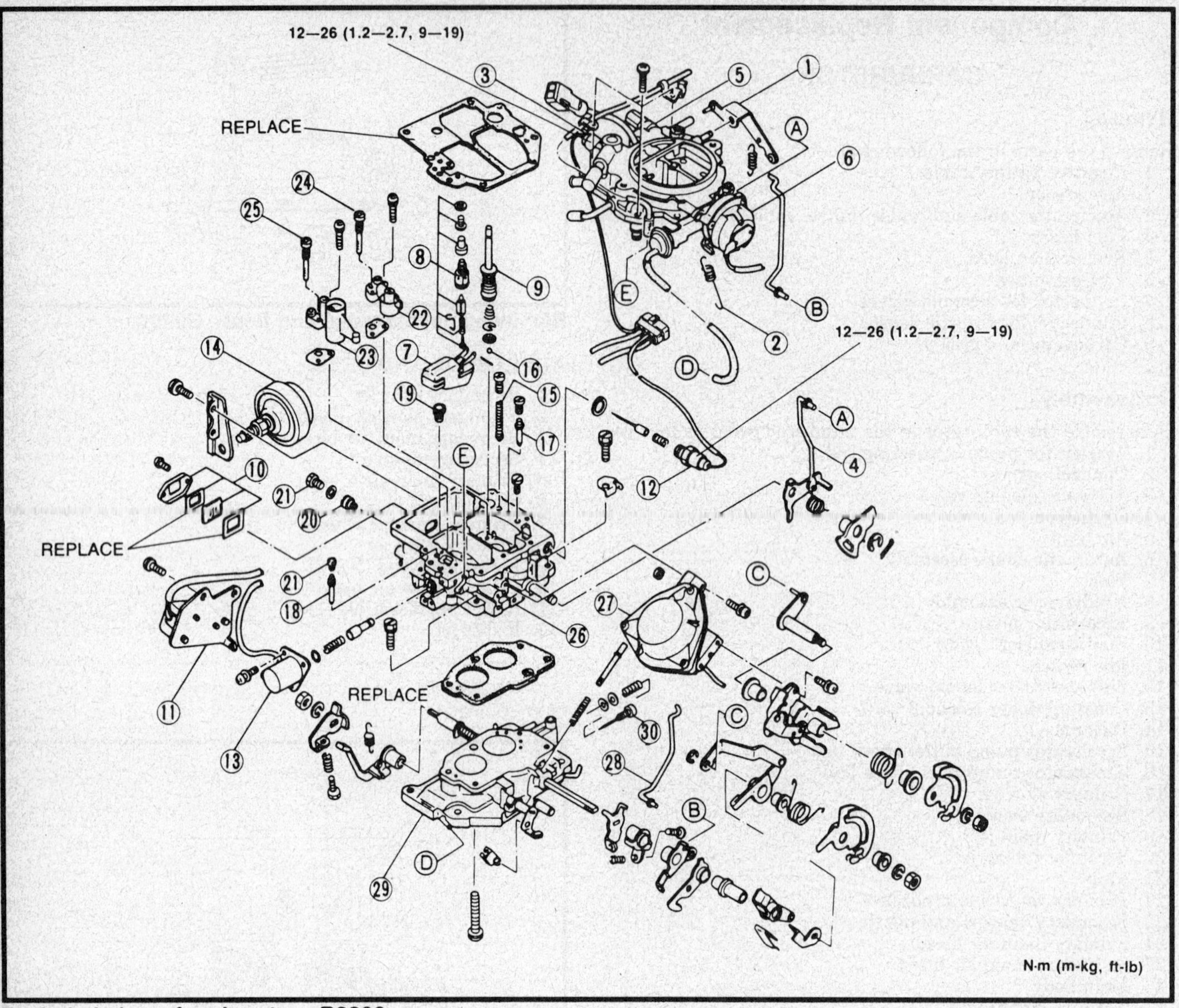

Exploded view of carburetor—B2200

MAIN BODY

1. Coasting richer solenoid valve and O-ring
2. Idle switch and spring. After installing the idle switch, be sure to adjust it.
3. Slow fuel cut solenoid valve, needle valve, spring and gasket
4. Dashpot bracket and dashpot
5. Accelerator pump plunger assembly and spring
6. Retaining clip
7. Strainer and accelerator pump inlet check ball
8. Check valve plug
9. Accelerator pump outlet check ball and spring
10. Throttle link (disconnect)
11. Vacuum diaphragm connecting rod (disconnect)
12. Throttle return spring (disconnect)
13. Throttle body (separate from the main body). One of the bolts are inside the throttle body
14. Diaphragm assembly and gasket
15. Diaphragm cover mounting screws and cover
16. Spring and diaphragm
17. Throttle lever hanger mounting screws

THROTTLE BODY

When removing the mixture adjusting screw, tap out the spring pin as shown in the figure. Do not remove the throttle valve and shaft, the venturi, the choke valve and shaft.

Inspection

Before inspection, wash all parts in carburetor cleaner and blow compressed air into the fuel passages to remove any dirt. Never use a wire to clean the jets.

1. Inspect the air horn, the main body and the throttle body for cracks or breakage.
2. Inspect the choke shaft and throttle shaft for wear. A worn throttle shaft will allow air to enter into the combustion chamber and the air/fuel mixture will become lean at low driving speeds.
3. Check the needle and seat for wear and rust.
4. Check the float for damage.
5. Examine all jets and air bleeds for clogging, clean in carburetor cleaner and blow with compressed air if necessary. Never use wire; doing so might enlarge the hole or passage, thus changing the calibration of the carburetor.

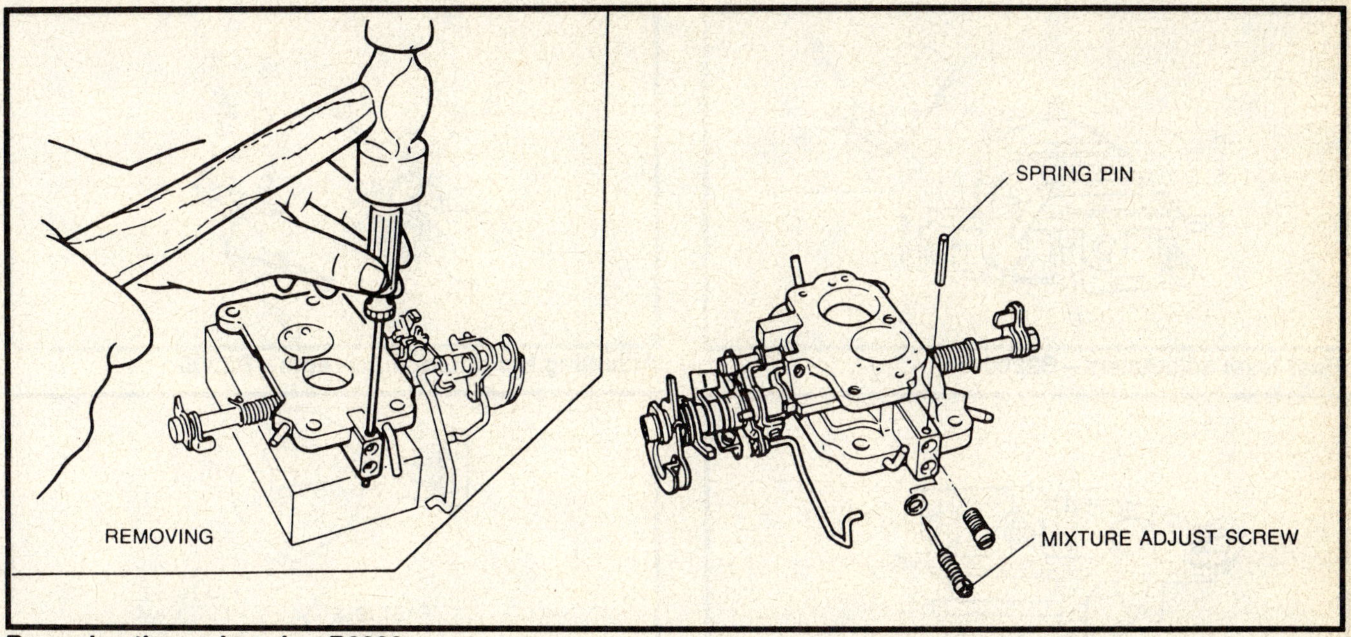

Removing the spring pin—B2200

6. Inspect the accelerator pump plunger cup. Replace the plunger if it is worn or damaged.
7. Check the diaphragm for damage.
8. Inspect the mixture adjust for burrs or ridges.
9. Check the operation of the solenoids. To check, connect the solenoid to the positive terminal of the battery and ground the body. When current is applied to the solenoid, the valve stem should be drawn into the valve body. If the valve does not operate properly, replace the solenoid.
10. Using an ohmmeter, check for continuity between the coupler and the choke heater ground. If there is no continuity, replace the choke heater.
11. To check the A/F solenoid valve, connect the terminal to the positive terminal of the battery, and ground the other terminal. Make sure that the valve is operating properly by blowing through it.
   a. When current is applied—air does not pass through the valve.
   b. When current is not applied—air passes through the valve.

NOTE: When assembling, replace the O-ring and coat it with gasoline. The A/F solenoid can not be replaced, if it is necessary to replace it, the air horn assembly must be replaced.

Assembling

Assemble the carburetor in the reverse order of disassembly, except for the following:

1. Discard the old gaskets and use new ones.
2. Make sure that all parts are in good condition and clean.
3. Both the primary and secondary venturis have there respective which are of the same shape. Therefore, during reassembly be careful not to mix up the parts.
4. Do not secure the spring pin to lock the mixture adjust screw until the idle adjustment has been completed.

Installation

Install the carburetor in the reverse order of removal. After installation, note the following:

1. Start the engine and check for leaks.
2. With the engine operating, check the fuel level. The fuel level should be at the specified mark on the sight glass.
3. Make the idle adjustment.
4. After the idle adjustment is completed, press in the spring pin.
5. Adjust the dashpot.
6. Adjust the idle switch.
7. After the idle adjustment is completed, check the fast idle speed as follows:
   a. Start and allow the engine to reach normal operating temperature.
   b. Stop the engine and remove the air cleaner.
   c. Plug the hoses of the idle compensator and the reed valves.
   d. While holding the throttle valve slightly open, push the choke to fully close it, and then release the choke valve after releasing the throttle valve.
   e. Start the engine, but do not touch the accelerator pedal.
   f. Check to see that the engine speed increases to 3000–4000 rpm. If the engine speed is not as specified, turn the fast idle adjusting screw until the correct rpm is obtained.

Adjustments

FLOAT LEVEL

This adjustment should be made without the gasket on the air horn. Before installing the air horn assembly, adjust the float level as follows:

1. Turn the air horn assembly upside-down on a stand and allow the float to lower by its own weight.
2. Measure clearance (H) between the float and the air horn. If specifications are not as specified, bend the float seat lip (a) until the proper clearance is obtained. Clearance (H) should be as follows:
   Auto Trans—0.421–0.461 in (10.7–11.7mm)
   Manual Trans—0.457–0.496 in (11.6–12.6mm).
3. Turn the air horn to the normal position and allow the float to lower by its own weight.
4. Measure clearance (L) between the bottom of the float and the air horn. If the clearance is not correct, bend the float stopper (b) until the proper clearance is obtained. Clearance (L) should be 1.811–1.850 in (46.0–47.0mm).

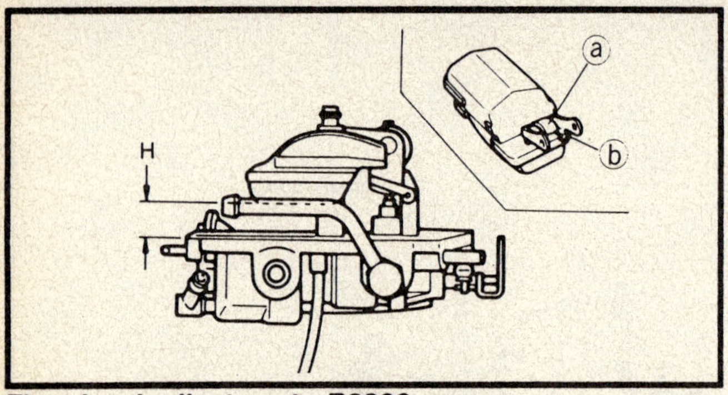

Float level adjustment – B2200

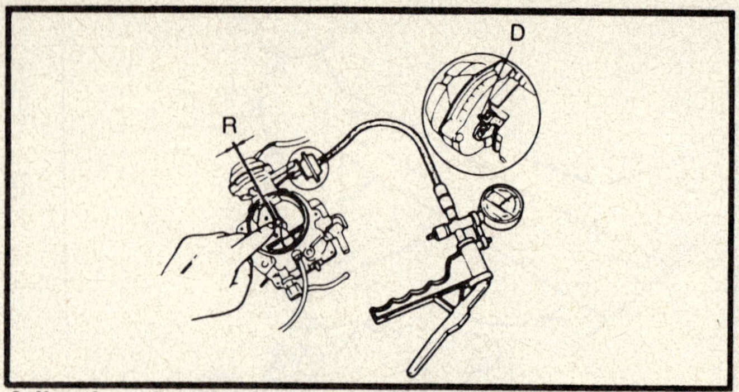

Adjusting the choke diaphragm – B2200

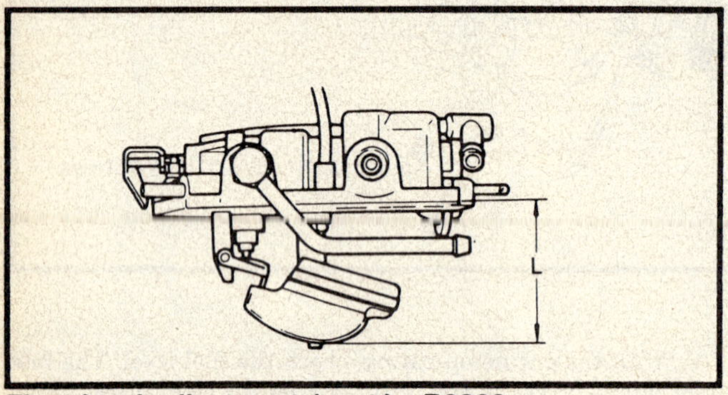

Float level adjustment (cont.) – B2200

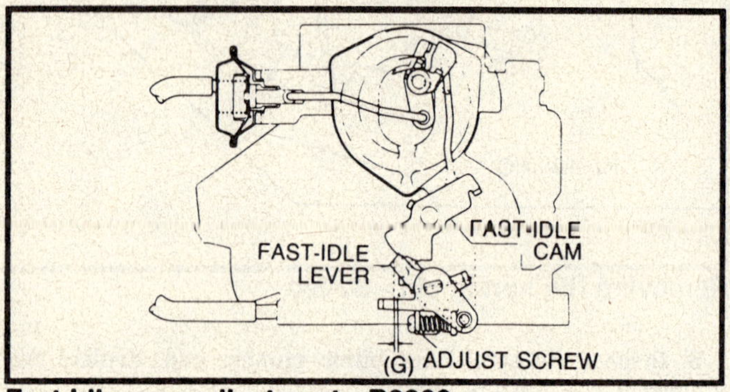

Fast idle cam adjustment – B2200

CHOKE DIAPHRAGM

1. Using a suitable hand operated vacuum pump, apply a vacuum of approximately 15.7 in. Hg (400mm Hg) to the choke diaphragm vacuum tube.
2. Push the choke valve lightly to close it, and check clearance (R). Clearance (R) should be 0.067–0.085 in (1.70–2.16mm).
3. If the clearance is not as specified, adjust it by bending the choke lever (D).

FAST IDLE CAM

1. Set the fast idle cam to the second highest position.
2. Adjust the throttle valve clearance (G) by turning the adjust screw (the clearance will become larger as the screw is turned clockwise). Throttle valve clearance should be 0.033–0.041 in. (0.84–1.04mm).

CHOKE VALVE CLEARANCE

1. Set the fast idle cam select arm to the second position.
2. Make sure that the choke valve clearance (R) is 0.024–0.045 in. (0.60–1.14mm).
3. If necessary, adjust the choke valve clearance (R) by bending the starting arm (B). If large adjustments are required, the choke rod should be bent.

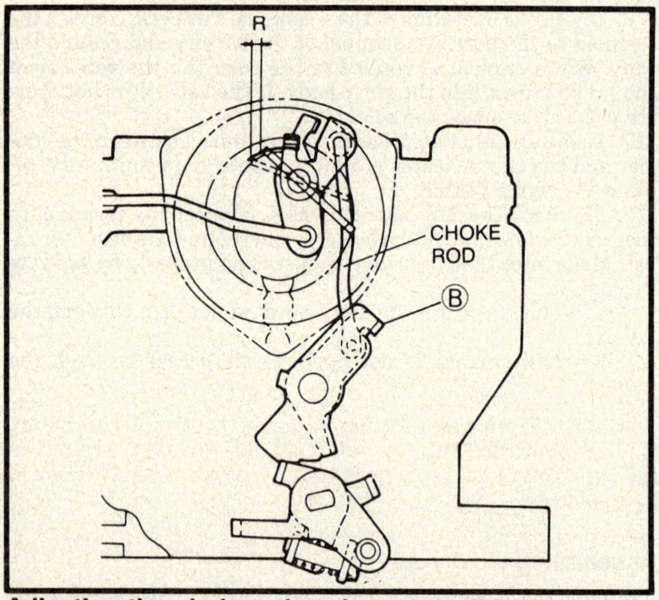

Adjusting the choke valve clearance – B2200

UNLOADER SYSTEM

1. Open the primary throttle valve all the way.
2. Measure the choke valve clearance (R). Choke valve clearance (R) should be 0.110–0.143 in. (2.80–3.62mm).
3. If the clearance obtained is not as specified, adjust by bending tab (A).

SECONDARY THROTTLE VALVE

1. The secondary throttle valve should start to open when the primary throttle valve opens 52 ± 2 degrees and should be completely open at the same time the primary throttle valve is fully open.
2. Check the clearance between the primary throttle valve and the wall of the throttle bore when the secondary throttle valve starts to open. Clearance should be 0.289–0.325 in. (7.35–8.25mm).
3. If clearance obtained is not as specified, adjust by bending tab (B).

IDLE SPEED

Check that the ignition timing, spark plugs, carburetor float level, etc. are all in normal operating condition. Turn **OFF** all lights and other unnecessary electrical loads.

1. Connect a tachometer to the engine.
2. Start and allow the engine to reach normal operating temperature. Check the idle speed and adjust it to the specification

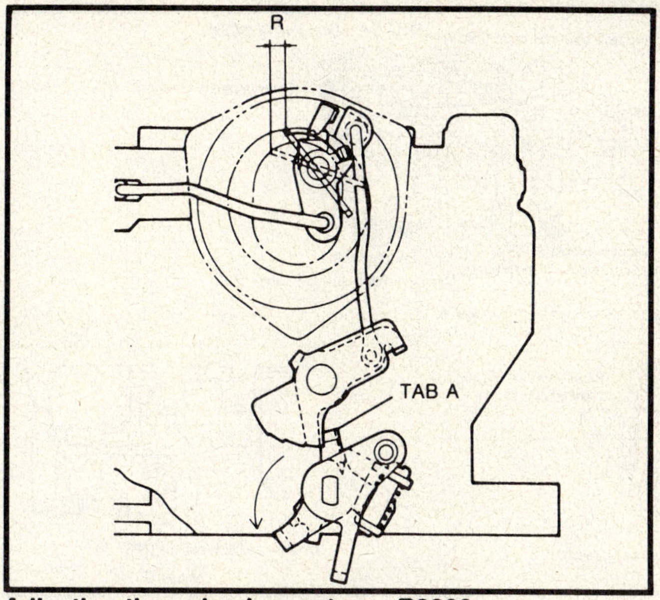

Adjusting the unloader system—B2200

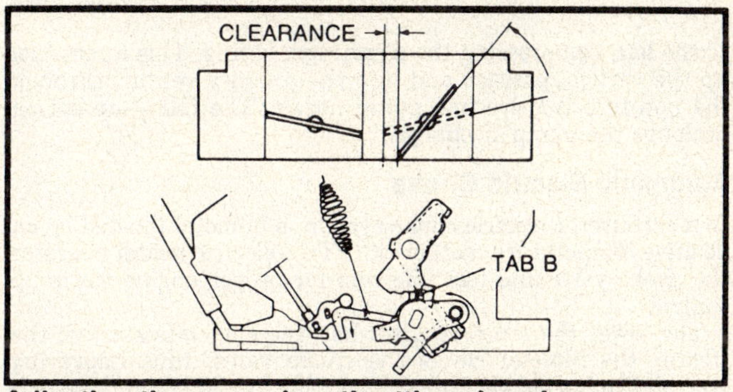

Adjusting the secondary throttle valve clearance—B2200

by turning the throttle adjusting screw. The idle speed should be 800–850 rpm in **NEURTAL** or **PARK**.

IDLE MIXTURE

1. Start the engine and allow it to reach normal operating temperature, and let the engine run at idle.
2. Connect a dwellmeter (90 degrees, 4 cylinder) to the BrY wire in the check connector of the A/F solenoid valve.
3. Check the idle mixture at specified idle speed. The idle mixture should be 20–70 degrees. If the idle mixture is not within specifications, adjust the idle mixture as follows:
   a. To adjust the idle mixture, remove the carburetor and take out the spring pin. Reinstall the carburetor.
   b. Be sure that the air cleaner is installed and the idle compensator is closed.
   c. Warm up the engine and run it at idle speed.
   d. Connect a tachometer to the engine.
   e. Connect a dwellmeter (90 degrees, 4 cylinder) to the BrY wire in the check connector of the A/F solenoid valve and read the meter. The standard value should be 27–45 degrees (at idle).
   f. If the reading obtained is not within the specified range, adjust the idle mixture by turning the mixture adjust screw.

NOTE: If the adjustment cannot be made, it is probably because of a faulty oxygen sensor, or because of either a broken wire or a short in the wiring between the oxygen sensor and the control unit.

   g. Tap in the spring pin.

DASHPOT

1. Move the throttle lever, and check that the dashpot rod comes out quickly. Release the throttle lever, and check that it returns slowly to the idle position after it has contacted the dashpot rod.
2. Start the engine and allow it to reach normal operating temperature. Stop the engine and connect a suitable tachometer to the engine.
3. Start the engine, and slowly increase engine speed. Check that the dashpot rod separates from the lever at 1900–2100 rpm.
4. If the function is not correct, loosen the lock nut and adjust by turning the dashpot.

IDLE SWITCH

The idle switch adjustment is not necessary unless it has been replaced or tampered with.

1. Start the engine and allow it to reach normal operating temperature. Stop the engine and connect a suitable tachometer.
2. Connect a voltmeter to the idle switch terminal (LgR).
3. Start the engine and increase the speed to above 2000 rpm and decelerate gradually.
4. Take a reading on the voltmeter. At idle, there should be approximately 12 volts and above 1000–1200 rpm it should be below 1.5 volts.
5. If the readings do not agree with these specifications, turn the adjusting screw to obtain the correct reading.

MITSUBISHI FEEDBACK CARBURETOR SYSTEM

General Information

All vehicles are equipped with a conventional downdraft 2-barrel compound type carburetor which incorporates automatic choke, accelerator pump, and enrichment system. In addition, a deceleration device is provided.

CARBURETOR SYSTEMS

Fuel Inlet System

The fuel inlet system constantly maintains the fuel at a prescribed level within the carburetor during operation. Fuel from the fuel tank flows through the fuel pump, fuel filter, and through a needle valve into the carburetor float chamber.

As fuel enters the float chamber, the float moves upward and closes the needle valve when the fuel is at the prescribed level. As fuel is used by the engine, the float moves down to open the needle valve, permitting fuel to flow into the float chamber.

Overfill Device

The float chamber contains a device which prevents fuel leaks in the event of vehicle roll-over. The device holds down the float to close the needle valve if the vehicle rolls-over.

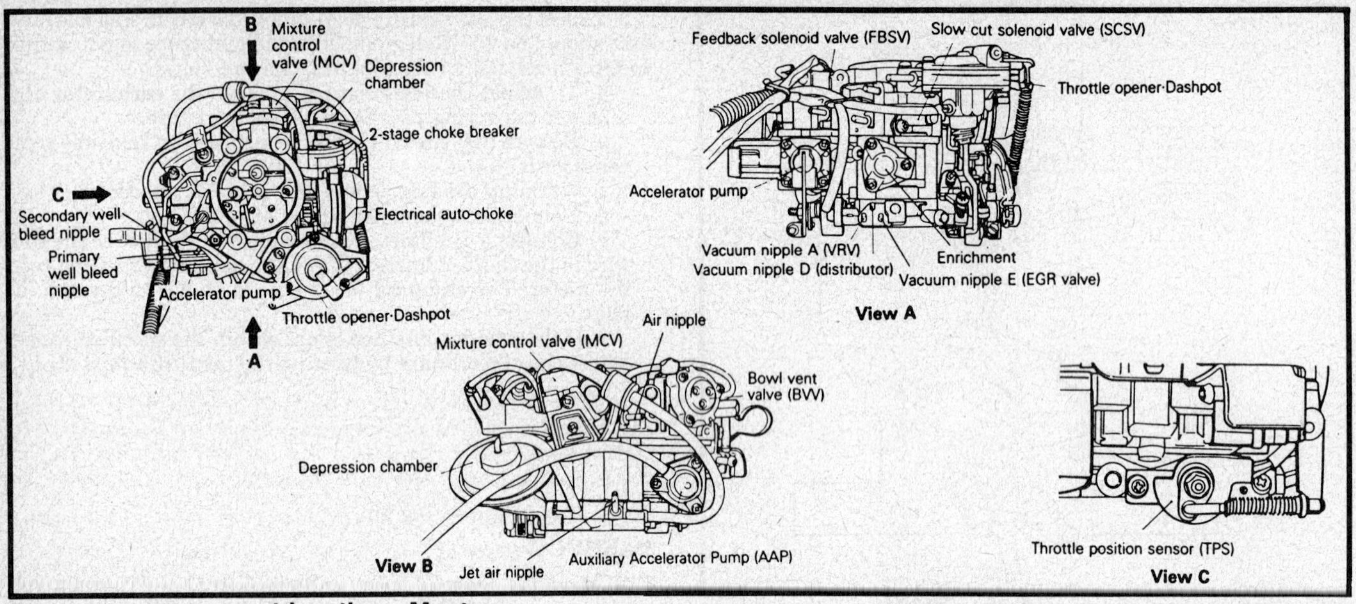

Carburetor component location – Montero

Primary Metering System

The primary metering system consists of the primary main jet, primary pilot jet, off-idle mixture adjusting screw, idle mixture adjusting screw and primary main nozzle. It provides a fuel mixture for idling, light load or cruising at part throttle, and for full throttle operation at low speeds.

Fuel Cut-Off Device

When the ignition key is turned **OFF**, the spring-loaded needle valve cuts off the fuel flow to prevent engine dieseling. Under normal engine operation, the needle valve is held open by the solenoid to provide the necessary fuel flow for engine operation.

Secondary Metering System

The secondary metering system consists of the secondary main jet, secondary pilot jet, secondary main nozzle and diaphragm chamber.

When power demand exceeds that provided by the primary barrel, the secondary throttle is brought into action by a diaphragm, which is drawn against its spring by venturi vacuum. There is a linkage between the primary and the secondary throttle shafts which delays the opening of the secondary throttle until the primary throttle has opened about half of its full travel.

Enrichment System

The carburetor contains the diaphragm enrichment system. Manifold vacuum is transmitted to the diaphragm chamber through a channel. When running at part throttle, the diaphragm chamber is drawn to the right, closing the enrichment valve. As the throttle is opened widely, the drop in manifold vacuum permits the diaphragm spring to force the diaphragm to the left and open the enrichment valve. This allows extra fuel to pass through the enrichment jet into the emulsion well, thus supplying the extra fuel required for high-load vehicle operation.

Acceleration Pump

A conventional, mechanical accelerating pump is used. It consists of the flexible diaphragm actuated by a lever, link rod and follow-up spring. The pump is charged with fuel through an inlet valve when the throttle is closed.

On opening the throttle, the spring-loaded push rod is moved to the left, compressing the diaphragm spring. This forces fuel up the vertical passage and into the primary venturi through the non-return valve and pump nozzle. The follow-up spring prolongs the pump discharge.

Automatic Electric Choke

In a carburetor electric choke system, a bi-metal (heated by an positive temperature coefficient (PTC)) electric heater operates the choke valve and fast idle cam for proper engine warm-up control.

The lower the temperature when the engine is started, the tighter the bimetal closes the choke valve, thus improving startability in cold weather.

As the choke coil is heated by the heater after start-up, the bimetal opens the choke valve gradually and pushes down the stopper lever.

Electronic Control Unit (ECU)

Based on the information from various sensors, the ECU constantly determines (computes) an optimum control for varying operating conditions and accordingly drives the output actuators.

The ECU consists of an 8-bit microprocessor, random access memory (RAM), read only memory (ROM) and input/output (I/O) interface.

Engine Coolant Temperature Sensor

The engine coolant temperature sensor installed in the coolant passage of the intake manifold is a resistor sensor. The ECU judges engine warm-up state by the sensor output voltage and makes uses this data to adjust the fuel delivery.

Throttle Position Sensor (TPS)

The TPS is a rotating variable resistor that follows the throttle shaft to sense the throttle valve angle. As the throttle shaft turns, the TPS output voltage changes and the ECU judges throttle valve opening proportional to the voltage received.

Using the TPS output signal, engine speed signal and other signals, the ECU judges the engine operating mode and controls the air-fuel ratio to optimum for that condition.

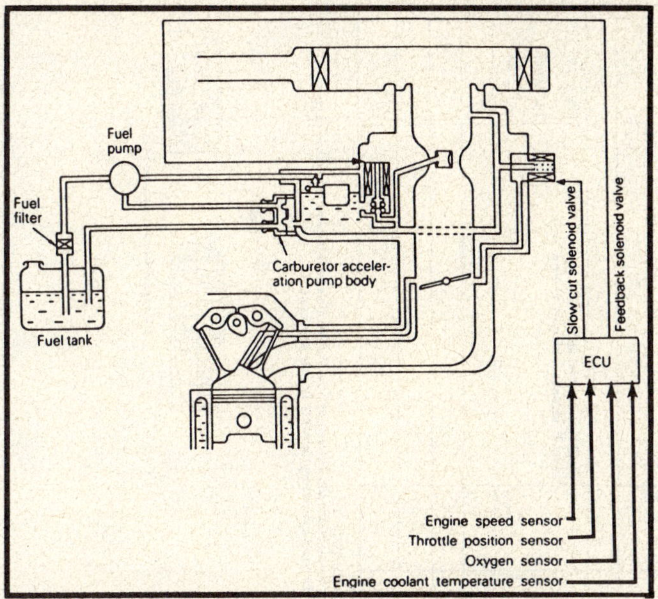

Feedback carburetor control schematic – Montero

Engine Speed Sensor (Ignition Coil Negative Terminal Voltage Sensor)

The voltage at the negative terminal of the ignition coil increases suddenly twice per crankshaft revolution. This pattern occurs synchronously with ignition timing.

By sensing this voltage change and measuring the time between peak voltages, the ECU computes the engine speed (rpm), judges the engine operating mode and adjusts the fuel flow and ignition timing.

Oxygen Sensor

The oxygen sensor installed on the exhaust manifold makes use of the principles of the solid electrolyte oxygen concentration cell. The oxygen concentration cell is characterized by sharp change of output voltage in the vicinity of the 14.7:1 air-fuel ratio.

Using such characteristics, the oxygen sensor senses the oxygen concentration in the exhaust gas and feeds it to the ECU as a voltage signal. The ECU then judges if the air-fuel ratio is rich or lean compared to the ideal 14.7:1 ratio. Based on this signal, the ECU constantly adjusts the air-fuel ratio to 14.7:1, where the emission purification rate of the 3-way catalyst is optimum. This constant monitoring and reaction is known as "closed-loop" control.

Vehicle Speed Sensor

The vehicle speed sensor uses a reed switch. The speed sensor built in the speedometer converts speedometer gear revolution (vehicle speed) into pulse signals, which are sent to the ECU.

Vacuum Switch

The vacuum switch is a contact type switch that is operated by intake manifold vacuum. When the throttle valve closes, the intake manifold vacuum acts on the vacuum switch to close its contact.

By this action, the voltage on the ECU side is grounded and the ECU senses that the throttle valve opening is at or near the idle opening.

Feedback Solenoid Valve (FBSV)

The FBSV is installed on the carburetor float chamber cover. The ECU controls the air-fuel ratio by controlling the duty cycle of the FBSV. The higher is the duty ratio, the leaner becomes the air-fuel ratio.

Slow Cut Solenoid Valve (SCSV)

The SCSV is installed on the carburetor float chamber cover. The ECU controls the carburetor slow system fuel flow by controlling the duty cycle of the SCSV.

FEEDBACK CARBURETOR OPERATION SYSTEM

Closed Loop Control (Feedback Control)

In the ordinary operation after engine warm-up, the air-fuel control is made by the ECU based on the oxygen sensor signal. The intake airflow varies with throttle position; the amount of oxygen present affects the content of the exhaust gasses and the ECU adjusts the fuel flow at the direction of the oxygen sensor.

The oxygen sensor output voltage changes sharply at the air-fuel 14.7:1 ratio. The control unit senses this oxygen sensor signal and accordingly provides feedback control of the FBSV duty ratio so that the 14.7:1 ratio (giving the best purification rate of the 3-catalyst converter) may be accurately kept. In this state, the SCSV is kept wide open (100% duty).

Open Loop Control (No Feedback Control)

During engine start, warm-up operation, high load operation and deceleration, the air-fuel ratio is in the "open-loop" control. The oxygen sensor is bypassed; the fuel metering is controlled by pre–programmed (mapped) values within the ECU. These map values are based on engine speed, throttle opening angle and engine coolant temperature, to improve startability and driveability.

During deceleration, the control on SCSV limits fuel flow for better fuel economy and for prevention of overheating of the catalyst.

When the FBSV is energized, the main fuel jet is closed to leave the primary main jet passage as the only fuel passage. This reduces the amount of fuel, resulting in leaner air-fuel mixture.

When FBSV is de-energized, the main fuel jet is opened to provide 2 fuel passages including the primary main jet passage. This increases the amount of fuel, richening the mixture. When ON-OFF operation of the SCSV occurs, the slow fuel passage is opened and closed. The air-fuel ratio during deceleration is controlled in this manner.

THROTTLE OPENER SYSTEM

For Air Conditioning

When the engine speed is below 1,200 rpm, the ECU keeps the power transistor on. When the air conditioner relay is turned on (the air compressor is driven by the engine), the throttle opener control solenoid valve is energized to introduce intake manifold vacuum to the throttle opener. The throttle opens slightly, preventing engine speed drop caused by air conditioner load.

Diagnosis and Testing

NOTE: While it is possible to test components with a voltmeter and/or an ohmmeter, complete system testing requires the use of the ECI Checker or similar diagnostic tool.

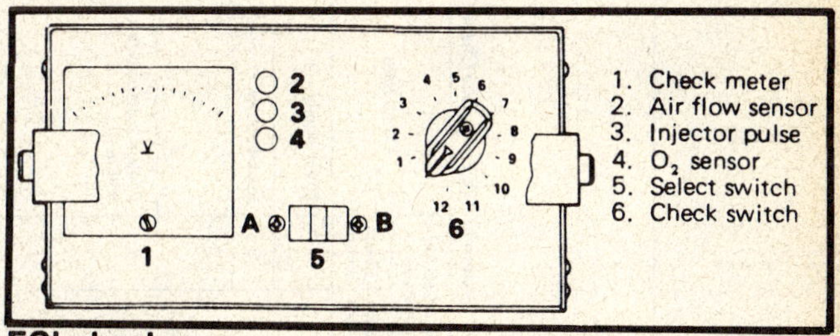

ECI checker

FBC System Check Procedure Chart for Vehicles with Manual Transaxle

ECI Checker Operation: Select Switch	ECI Checker Operation: Check Switch	Check Item	ECU Terminal # Checked	Condition		Test Specification
Set to "A"	1	Power supply	7	Ignition switch "LOCK → ON"		SV
	2	Ignition pulse	10	Idling		4V to 10V
	3	–	–	–		–
	4	Coolant temperature sensor	12	Ignition switch "LOCK → ON"	0°C (32°F)	3.4V to 3.6V
					20°C (68°F)	2.4V to 2.7V
					40°C (104°F)	1.5V to 1.8V
					80°C (176°F)	0.5V to 0.7V
	5	Power supply for sensor	3	Ignition switch "LOCK → ON"		4.5V to 5.5V
	6	Throttle position sensor (TPS)	13	Ignition switch "LOCK → ON" (warm engine)	Accelerator fully closed	0.4V to 0.7V
					Accelerator fully opened	4.5V to 5V
	7	Vacuum switch for idle position	5	Ignition switch "LOCK → ON"		9V to 13V
				Idling (warm engine)		0V to 0.6V
	8	–	–	–		–
	9	Feed back solenoid valve (FBS)	59	Ignition switch "LOCK → ON"		SV
				Idling (warm engine)		2V to 12V
	10	Slow cut-off solenoid valve (SCS)	53	Idling		0V to 0.6V
				Quick deceleration from above 4000 rpm to idling with "N" position		Momentarily SV
	11	Spark advance control solenoid valve	58	Idling (warm engine)		0V to 0.6V
				2000 rpm		SV
	12	–	–	–		–

NOTE
SV: System Voltage

FBC System Check Procedure Chart for Vehicles with Manual Transaxle

ECI Checker Operation: Select Switch	ECI Checker Operation: Check Switch	Check Item	ECU Terminal # Checked	Condition		Test Specification
Set to "B"	1	Idle up control solenoid valve	54	Idling	Air conditioner switch ON *1 or lighting switch ON	0V to 0.6V
				2000 rpm		9V to 15V
	2	A/C cutoff relay	57	Ignition switch "LOCK → ON" and A/C switch "ON" *1	Accelerator fully closed	0V to 0.6V
					Accelerator fully opened	0V to 0.6V
	3	–	–	–		–
	4	Second air control solenoid valve	55	Idling, after 70 seconds from start of warm engine		SV
				Quick deceleration from above 2000 rpm to idling with "N"		Momentarily drop
	5	–	–	–		–
	6	–	–	–		–
	7	–	–	–		–
	8	Oxygen sensor *2	1	Hold rpm constant above 1300 rpm, after 70 seconds from start of warm engine		0V to 1V ↑ (pulsates) ↓ 2V to 3V
	9	–	–	–		–
	10	–	–	–		–
	11	–	–	–		–
	12	–	–	–		–

NOTE: *1 ON means compressor clutch engaged.
*2 Because there are sometimes malfunctions and differences from the specifications in parts other than oxygen sensor, take care to inspect the parts connecting the other air-fuel ratio controls.

SV: System Voltage

ECU Terminal

View from front as installed in ECU

FBC System Check Procedure Chart for Vehicles with Automatic Transaxle

ECI Checker Operation: Select Switch	ECI Checker Operation: Check Switch	Check Item	ECU Terminal # Checked	Condition		Test Specification
Set to "A"	1	Power supply	51	Ignition switch "LOCK → ON"		SV
	2	Spark advance control solenoid valve	13	Idling (warm engine)		0V to 0.6V
				2000 rpm		13V to 15V
	3	Throttle position sensor (TPS)	55	Ignition switch "LOCK → ON" (warm engine)	Accelerator fully closed	0.4V to 0.7V
					Accelerator fully opened	4.5V to 5V
	4	ISC motor for retraction	10	Idling	Air conditioner switch ON → OFF *1	Momentarily over 6V, then 0V to 2V *2
	5	ISC motor for extension	11	Idling	Air conditioner switch ON → OFF *1	Momentarily over 6V, then 0V to 2V *2
	6	Idle position switch	1	Ignition switch "LOCK → ON" (warm engine)	Accelerator fully closed	0V to 0.6V
					Accelerator fully opened	9V to 13V
	7	ISC high motor position switch	17	Ignition switch "LOCK → ON"	After 15 seconds voltage drops	8V to 13V then 0V to 0.6V
	8	Electric choke relay	9	Ignition switch "LOCK → ON"		0V to 0.6V
				Idling		SV
	9	ISC low motor position switch	3	Ignition switch "LOCK → ON"	Voltage drops toward 0V after 15 seconds	8V to 13V, then momentarily drop
	10	Power steering switch	7	Idling		9V to 13V
				Turn steering wheel to stop		0V to 0.6V
	11	Air conditioner switch	2	Ignition switch "LOCK → ON"	Air conditioner switch OFF	0V to 0.6V
					Air conditioner switch ON *1	SV
	12	Secondary air control solenoid valve	5	Idling, 70 seconds after start of warm engine		SV
				Quick deceleration from above 2000 rpm to idling with "N" or "P" position		Momentarily drop

NOTE: *1 ON means compressor clutch engaged.
*2 Pointer indicates over 6V momentarily. If it is hard to read indication, repeat the OFF → ON or ON → OFF operation of air conditioner switch several times. If the pointer of voltmeter deflects, ISC motor is normal.

SV: System Voltage

FBC System Check Procedure for Vehicles with Automatic Transaxle

ECI Checker Operation: Select Switch	ECI Checker Operation: Check Switch	Check Item	ECU Terminal # Checked	Condition		Test Specification
Set to "B"	1	Air conditioner cutoff relay	14	Idling	Air conditioner switch OFF → ON *1	12V to 15V, then 0V to 0.6V
	2	Feed back solenoid valve (FBS)	8	Ignition switch "LOCK → ON"		SV
				Idling (warm engine)		2V to 12V
	3	–	–	–		–
	4	Ignition pulse	54	Idling		4V to 10V
	5	Coolant temperature sensor	60	Ignition switch "LOCK → ON"	0°C (32°F)	3.4V to 3.6V
					20°C (68°F)	2.4V to 2.7V
					40°C (104°F)	1.5V to 1.8V
					80°C (176°F)	0.5V to 0.7V
	6	Select lever "N" switch	6	Ignition switch "LOCK → ON"	Transaxle in "P" or "N"	0V to 0.6V
					Transaxle in "D"	6V to 13V
	7	Slow cutoff solenoid valve (SCS)	12	Idling		0V to 0.6V
				Quick deceleration from above 4000 rpm to idling with "D" position *2		Momentarily SV
	8	Oxygen sensor *3	15	Hold rpm constant above 1300, after 70 seconds from start of warm engine		0V to 1V ↑ (pulsates) ↓ 2V to 3V
	9	–	–	–		–
	10	–	–	–		–
	11	–	–	–		–
	12	–	–	–		–

NOTE: *1 ON means compressor clutch engaged.
*2 Speed sensor must show vehicle is moving.
*3 Because there are somtimes malfunctions and differences from the specifications in parts other than the oxygen sensor, take care to inspect the parts connecting the other air-fuel ratio controls.

SV: System Voltage

ECU Terminal

View from front as installed in ECU

TESTING WITH ECI CHECKER—MIRAGE/COLT AND PRECIS

FBC System Check Procedure Chart

ECI Checker Setting: Select Switch	ECI Checker Setting: Check Switch	Check Item	ECU Terminal # Checked	Condition		Test Specification
"A"	1	Power supply	2	Ignition switch "LOCK" → "ON"		SV
	2	Spark advance control solenoid valve	13	Idling (warm engine)		0V – 0.6V
				Start engine. Drive vehicle for some seconds at a speed higher than 8 km/h (5 mph), then hold 2,000 rpm (warm engine)		SV
	3	Throttle position sensor (TPS)	59	Ignition switch "LOCK" → "ON" (warm engine)	Accelerator fully closed	0.4V – 0.7V
					Accelerator fully opened	4.5V – 5.5V
	4	Engine coolant temperature sensor	53	Ignition switch "LOCK" → "ON"	0°C (32°F)	3.4V – 3.6V
					20°C (68°F)	2.4V – 2.7V
					40°C (104°F)	1.5V – 1.8V
					80°C (176°C)	0.5V – 0.7V
	5					
	6	Vacuum switch for idle position	6	Ignition switch "LOCK" → "ON"		9V – 13V
				Idling (warm engine)		0V – 0.6V
	7	Throttle opener control solenoid valve	12	Idling	Air conditioner switch ON *1 or lighting switch ON	0V – 0.6V
				2,000 rpm		9V – 15V
	8	Electric choke relay	14	Ignition switch "LOCK" → "ON"		0V – 0.6V
				Idling		SV
	9	Air conditioner cut-off relay	8	Ignition switch "LOCK" → "ON" and air conditioner switch "ON" *1	Accelerator fully closed	0V – 0.6V
					Accelerator fully opened	M/T 0V–0.6V
						A/T SV
	10	Power supply for sensor	51	Ignition switch "LOCK" → "ON"		4.5 V – 5.5V
	11	Vehicle speed sensor reed switch	56	Start engine, and drive vehicle slowly with transmission in first gear or drive range.		0V – 0.6V ↑ (pulsates) ↓ Over 2V
	12	Secondary air control solenoid valve	9	Idling, 70 seconds after start of warm engine		0V – 0.6V then SV
				Quick deceleration from above 2,000 rpm to idling with gear in "N" position		Momentarily drop

NOTE
*1 ON means compressor clutch engaged.
SV = System Voltage

FBC System Check Procedure Chart

ECI Checker Setting: Select Switch	ECI Checker Setting: Check Switch	Check Item	ECU Terminal # Checked	Condition		Test Specification
"B"	1	Shift select lever 4th or 5th switch	17	Ignition switch "LOCK" → "ON"	Transaxle in 1st	0V – 0.6V
					Transaxle in 4th or 5th	SV
	2	Feed back solenoid valve (FBS)	11	Ignition switch "LOCK" → "ON"		SV
				Idling (warm engine)		2V – 12V
	3					
	4	Ignition pulse	15	Ignition switch "LOCK → START"		2V – 8V
	5					
	6					
	7	Slow cut solenoid valve	10	Idling		0V – 0.6V
				Quick deceleration from above 4,000 rpm to idling with gear at "N" position		Momentarily SV
	8	Oxygen sensor	55	Hold speed constant above 1,300 rpm, 70 seconds after start of warm engine		0V – 1V *2 ↑ (Pulsates) 2V ↓ 3V
	9					
	10					
	11					
	12					

NOTE
*2 Failure of parts other than the oxygen sensor can also cause deviation from the specifications. Also check other parts related to air-fuel ratio control.
SV = System Voltage

Check Item	ECU Terminal # Checked	Condition		Test Specification
Cold spark advance control solenoid valve	5	Start engine. Drive vehicle for some seconds at a speed higher than 8 km/h (5 mph), then hold engine idling	Engine coolant temp. lower than 50°C (122°F)	0V – 0.6V
			Engine coolant temp. 50°C (122°F) or higher	SV
Cold mixture heater relay	57	Idling	Engine coolant temp. lower than 60°C (140°F)	0V – 0.6V
			Engine coolant temp. 60°C (140°F) or higher	SV

NOTE
SV = System Voltage

ECU terminal

View from front as installed in ECU

TESTING WITH ECI CHECKER—MONTERO

FBC SYSTEM CHECK PROCEDURE CHART (Use FBC Test Harness MD998438)

ECI Checker Operation: Select Switch	ECI Checker Operation: Check Switch	Check Item	Engine control unit Terminal No. Checked	Condition		Test Specification
Set to "A"	1	Power supply	12	Ignition switch "LOCK→ON"		SV
	2	Ignition pulse	5	Ignition switch "LOCK→START"		2V to 8V
	3	Throttle position sensor	2	Ignition switch "LOCK→ON"	Accelerator fully closed	0.4V to 0.7V
					Accelerator fully opened	4.5V to 5.5V
	4	Engine coolant temperature sensor	3	Igniton switch "LOCK→ON"	0°C (32°F)	3.4V to 3.6V
					20°C (68°F)	2.4V to 2.7V
					40°C (104°F)	1.5V to 1.8V
					80°C (176°F)	0.5V to 0.7V
	5	Power supply for sensor	8	Ignition switch "LOCK→ON"		4.5V to 5.5V
	6					
	7	Vacuum switch for idle position	10	Ignition switch "LOCK→ON"		9V to 13V
				Idling (warm engine)		0V to 0.6V
	8					
	9	Feed back solenoid valve	52	Ignition switch "LOCK→ON"		SV
				Idling (warm engine)		2V to 12V
	10	Slow cut-off solenoid valve	56	Idling		0V to 0.6V
				Quick deceleration from above 4000 rpm to idling with "N" position		Momentarily SV
	11					
	12					

NOTE
SV = System Voltage

FBC SYSTEM CHECK PROCEDURE CHART (Use FBC Test Harness MD998438)

ECI Checker Operation: Select Switch	ECI Checker Operation: Check Switch	Check Item	Engine control unit Terminal No. Checked	Condition: Test Result	Condition	Test Specification
Set to "B"	1	Idle up control solenoid valve	57	Idling	A/C switch ON*1	0V to 0.6V
				2000 rpm		9V to 15V
	2	A/C cut-off relay	54	Ignition switch "LOCK→ON" and A/C switch "ON"*1	Accelerator fully closed	0V to 0.6V
					Accelerator fully opened	0V to 0.6V
	3					
	4	Secondary air control solenoid valve	58	Idling 70 seconds after start of warm engine		0V to 0.6V then SV
				Quick deceleration from above 2000 rpm to idling with "N"		Momentarily drop
	5					
	6	Electric choke relay	59	Ignition switch "LOCK→ON"		0V to 0.6V
				Idling		13V to 15V
	7					
	8	Oxygen sensor	1	Hold rpm constant above 1300 rpm, after 70 seconds from start of warm engine		0V to 1V ↑ (pulsates) ↓ *2 2V to 3V
	9					
	10					
	11					
	12					

NOTE
*1 ON means compressor clutch engaged.
*2 Failure of parts other than the oxygen sensor can also cause deviation from the specifications. Also check other parts related to air-fuel ratio control.
SV = System voltage

Engine control unit Terminal

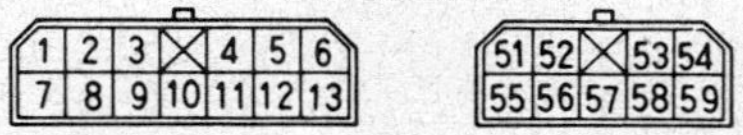

View from front as installed in Engine control unit

TESTING WITH ECI CHECKER—TRUCK

FBC SYSTEM CHECK PROCEDURE CHART
(Use FBC Test Harness MD998438)

ECI Checker Operation: Select Switch	ECI Checker Operation: Check Switch	Check Item	ECU Terminal No. Checked	Condition		Test Specification
Set to "A"	1	Power supply	7	Ignition switch "LOCK → ON"		SV
	2	Ignition pulse	10	Ignition switch "LOCK → ON"		2V to 8V
	3					
	4	Engine coolant temperature sensor	12	Ignition switch "LOCK → ON"	0 °C (32 °F)	3.4V to 3.6V
					20 °C (68 °F)	2.4V to 2.7V
					40 °C (104 °F)	1.5V to 1.8V
					80 °C (176 °F)	0.5V to 0.7V
	5	Power supply for sensor	3	Ignition switch "LOCK → ON"		4.5V to 5.5V
	6	Throttle position sensor (TPS)	13	Ignition switch "LOCK → ON" (warm engine)	Acceleratror fully closed	0.4V to 0.7V
					Accelerator fully opened	4.5V to 5.5V
	7	Vacuum switch for idle position	5	Ignition switch "LOCK → ON"		9V to 13V
				Idling (warm engine)		0V to 0.6V
	8					
	9	Feed back solenoid valve (FBS)	59	Ignition switch "LOCK → ON"		SV
				Idling (warm engine)		2V to 12V
	10	Slow cut-off solenoid valve (SCS)	53	Idling		0V to 0.6V
				Quick deceleration from above 4,000 rpm to idling with "N" position		Momentarily SV
	11					
	12					

FBC SYSTEM CHECK PROCEDURE CHART
(Use FBC Test Harness MD998438)

ECI Checker Operation: Select Switch	ECI Checker Operation: Check Switch	Check Item	ECU Terminal No. Checked	Condition		Test Specification
Set to "B"	1	Idle up control solenoid valve	54	Idling	A/C switch ON*1	0V to 0.6V
				2,000 rpm		9V to 15V
	2	A/C cut-off relay	57	Ignition switch "LOCK → ON" and A/C switch "ON"*1	Accelerator fully closed	0V to 0.6V
					Accelerator fully opened	0V to 0.6V
	3					
	4	Secondary air control solenoid valve	55	Idling, after 70 seconds from start of warm engine		0V to 0.6V then SV
				Quick deceleration from above 2,000 rpm to idling with "N"		Momentarily drop
	5					
	6					
	7					
	8	Oxygen sensor	1	Hold rpm constant above 1,300 rpm, after 70 seconds from start of warm engine		0V to 1V ↑ (pulsates) ↓ 2V to 3V*2
	9					
	10					
	11					
	12					

ECU Terminal

View from front as installed in ECU

NOTE
*1 ON means compressor clutch engaged.
*2 Failure of parts other than the oxygen sensor can also cause deviation from the specifications. Also check other parts related to air-fuel ratio control.
SV : System Voltage

For Air Conditioner Load

NOTE: Engine must be at operating temperature

1. Disconnect vacuum hose (yellow stripe) from throttle opener installed on intake manifold and connect a hand vacuum pump to nipple.
2. Check that throttle opener rod is pulled up when vacuum is applied.
3. Apply 10 in. Hg vacuum and check air tightness.
4. Start engine and close vacuum hose (yellow stripe) end with a finger to check vacuum when air conditioner switch is turned **ON** and **OFF**.

Air conditioner switch **OFF**, engine idling—vacuum absent

Air conditioner switch **ON**, engine idling—vacuum present

Air conditioner switch **ON**, engine 1200 rpm or more—vacuum absent

THROTTLE OPENER CONTROL SOLENOID VALVE

Inspection

1. Set ignition switch to **OFF**.
2. Disconnect connector from throttle opener control solenoid valve.
3. Check solenoid valve coil for continuity with an ohmmeter. Solenoid resistance should be approximately 40–48 ohms at 68°F (20°C).
4. If resistance is not approximately correct, replace solenoid valve.

THROTTLE POSITION SENSOR (TPS)

Inspection

1. Separate throttle position sensor connector.
2. Measure resistance between terminals 4 and 8. TPS resistance should be 3500–6500 ohms.
3. Connect an ohmmeter (pointer type) between terminals 8 and 3.
4. Operate throttle slowly from idle position to full open position and check that resistance makes a smooth change proportional to throttle valve opening.
5. If resistance is out of specification or fails to change smoothly, replace TPS.

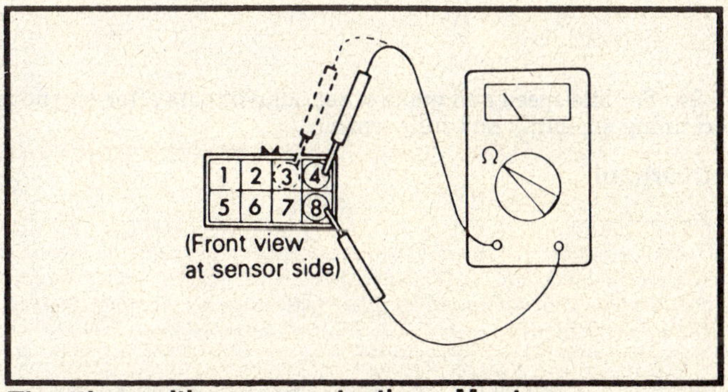

Throttle position sensor testing—Montero

COOLANT TEMPERATURE SENSOR

Inspection

1. Remove coolant temperature sensor from intake manifold.
2. With temperature sensing portion of coolant temperature sensor immersed in hot water, check resistance. Sensor should

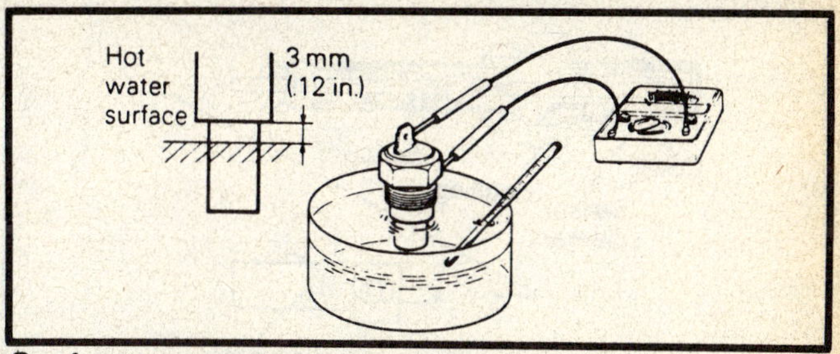

Coolant temperature sensor testing—typical

be held with its housing 3mm away from surface of hot water. Resistance should be:

5900 ohms at 32°F (0°C)
2450 ohms at 68°F (20°C)
1100 ohms at 104°F (40°C)
296 ohms at 176°F (80°C)

3. Replace sensor if not within specifications.

VACUUM SWITCH

Inspection

1. Disconnect harness connector of vacuum switch. Remove the green striped hose from the connector at the box.
2. Check continuity across terminals.
3. Connect vacuum pump to nipple.
4. Apply vacuum to the switch. When the vacuum builds above the level shown below, the switch should engage or close. Then release the vacuum while watching the gauge; the switch should open when vacuum drops below the level shown.

Switch closes when vacuum exceeds 11.8 in. Hg (40.7 kPa)
Switch opens when vacuum reduces to 7.9 in. Hg (27 kPa)

5. If switch is not within specification, replace switch.

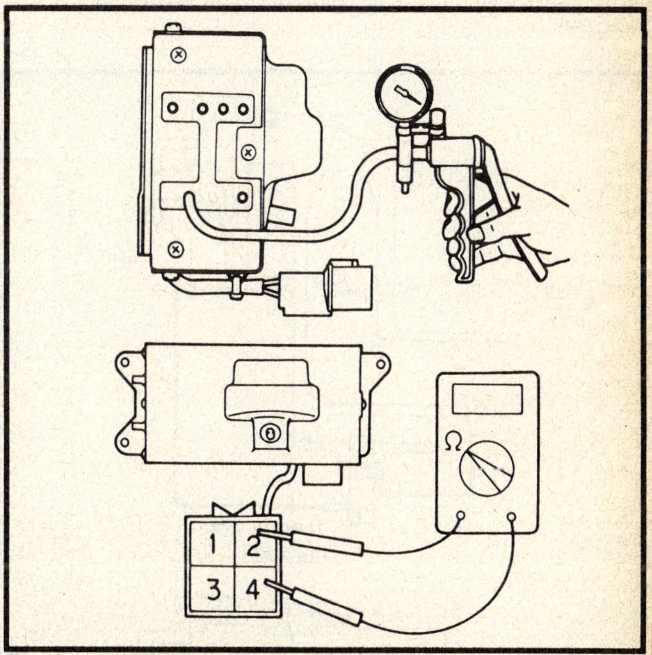

Testing vacuum switch—Montero

OXYGEN SENSOR

Inspection

NOTE: Engine must be at operating temperature and an accurate digital voltmeter should be used.

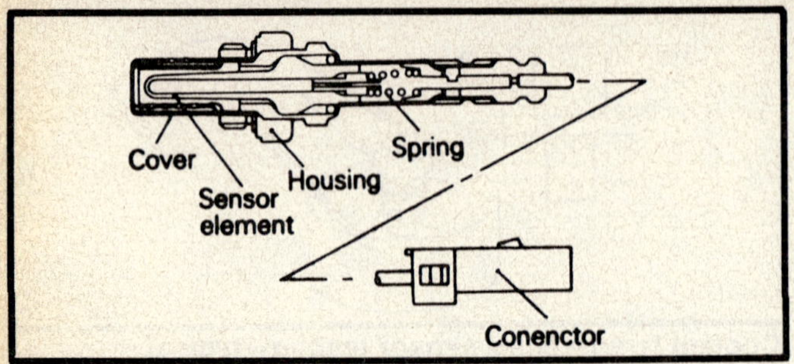

Oxygen sensor – typical

1. Separate oxygen sensor connector and connect a voltmeter to oxygen sensor connector.
2. Race engine from idle to approximately 2500 rpm; measure oxygen sensor output voltage. With a rich mixture, oxygen sensor output voltage should be approximately 1 volt. Voltage should change abruptly as engine returns to idle.

AUXILIARY ACCELERATION PUMP (AAP)

Inspection

NOTE: Engine coolant temperature must be 86°F (30°C) or lower at the start of the test procedure.

1. Remove air filter cover.
2. Start engine.
3. With engine idling, disconnect black vacuum hose from carburetor AAP. Vacuum is felt when hose end is covered by finger and fuel should spray from nozzle when hose is disconnected.
4. Warm-up engine until engine coolant temperature reaches 122°F (50°C) or higher.
5. Vacuum should not be felt at hose end when engine is fully warmed.

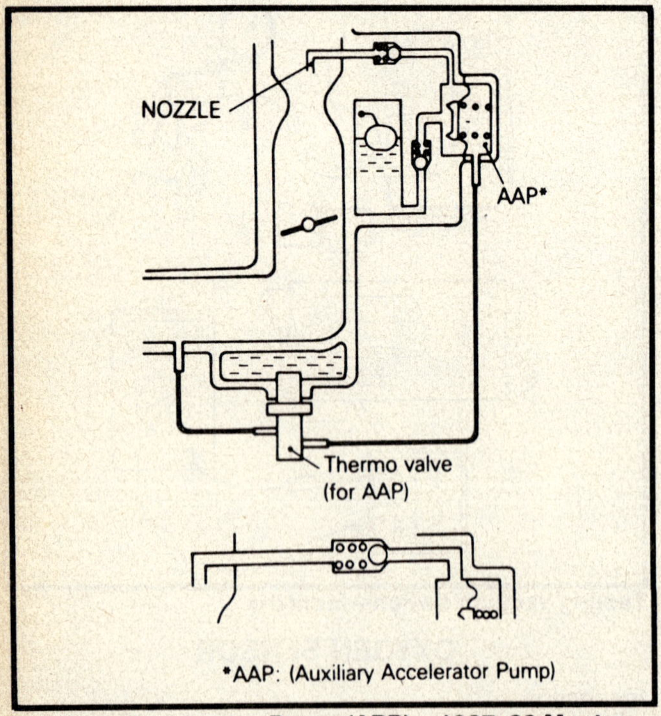

Auxiliary Accelerator Pump (APP) – 1987–88 Montero and Truck

6. If change of vacuum is incorrect, check thermo valve for AAP. If vacuum is normal but fuel is not sprayed from nozzle, clean carburetor passage.

Component Replacement

CARBURETOR

Removal and Installation

NOTE: All wires and hoses should be labeled at the time of removal.

1. Disconnect negative battery cable.
2. Drain the engine coolant to below level of intake manifold.
3. Remove air cleaner assembly.
4. Disconnect the accelerator cable.
5. Disconnect the coolant hose from carburetor.
6. Disconnect the vacuum hoses.
7. For vehicles with automatic transmissions, remove lock pin and disconnect the throttle control cable.
8. Disconnect the control harness connector.
9. Place a container under fuel hose connections. Disconnect 1 at a time and catch any spilled fuel.
10. Remove retaining bolts and lift off the engine. Keep carburetor level to avoid spillage.
11. Remove carburetor base gasket (insulator).
12. Inspect mating surfaces of the carburetor and manifold. They should be clean and free of nicks, burrs or any pieces of gasket material.

To Install:

13. Install a new carburetor gasket on manifold.
14. Locate carburetor on top of the gasket with holes lined up. Install carburetor bolts, tightening them evenly. Tighten bolts to 11–14 ft. lbs. (15–20 Nm).
15. Connect control harness connector.
16. If equipped with automatic transmission, connect throttle control cable and install locking pin securely.
17. Connect the vacuum hoses.
18. Install the coolant hose.
19. Connect the accelerator cable.
20. Reinstall the air cleaner.
21. Refill cooling system.
22. Connect the negative battery cable.
23. Start the engine and check for leaks.

CAUTION

Do NOT prime the engine by pouring fuel into the air horn of the carburetor or damage to the engine or personal injury may occur due to backfire flashback.

24. Set idle speed and make other adjustments after engine is running smoothly and fully warm.

Overhaul

CAUTION

Certain assemblies must not be disassembled or altered during overhaul. The choke plate and shaft, automatic choke linkage, inner venturi, throttle plate and shaft and fuel inlet nipple must be left alone. Damage and or reduced performance may result from tampering.

Many screws use cross-point or Phillips® heads. A proper fitting driver, which fits the head exactly, must be used.

1. Remove the carburetor. Drain remaining fuel into a container with airtight lid.
2. Remove the throttle return spring and damper spring.
3. Remove the throttle opener (automatic transmission) or dashpot (manual transmission) rod from free lever; remove opener or dashpot unit.

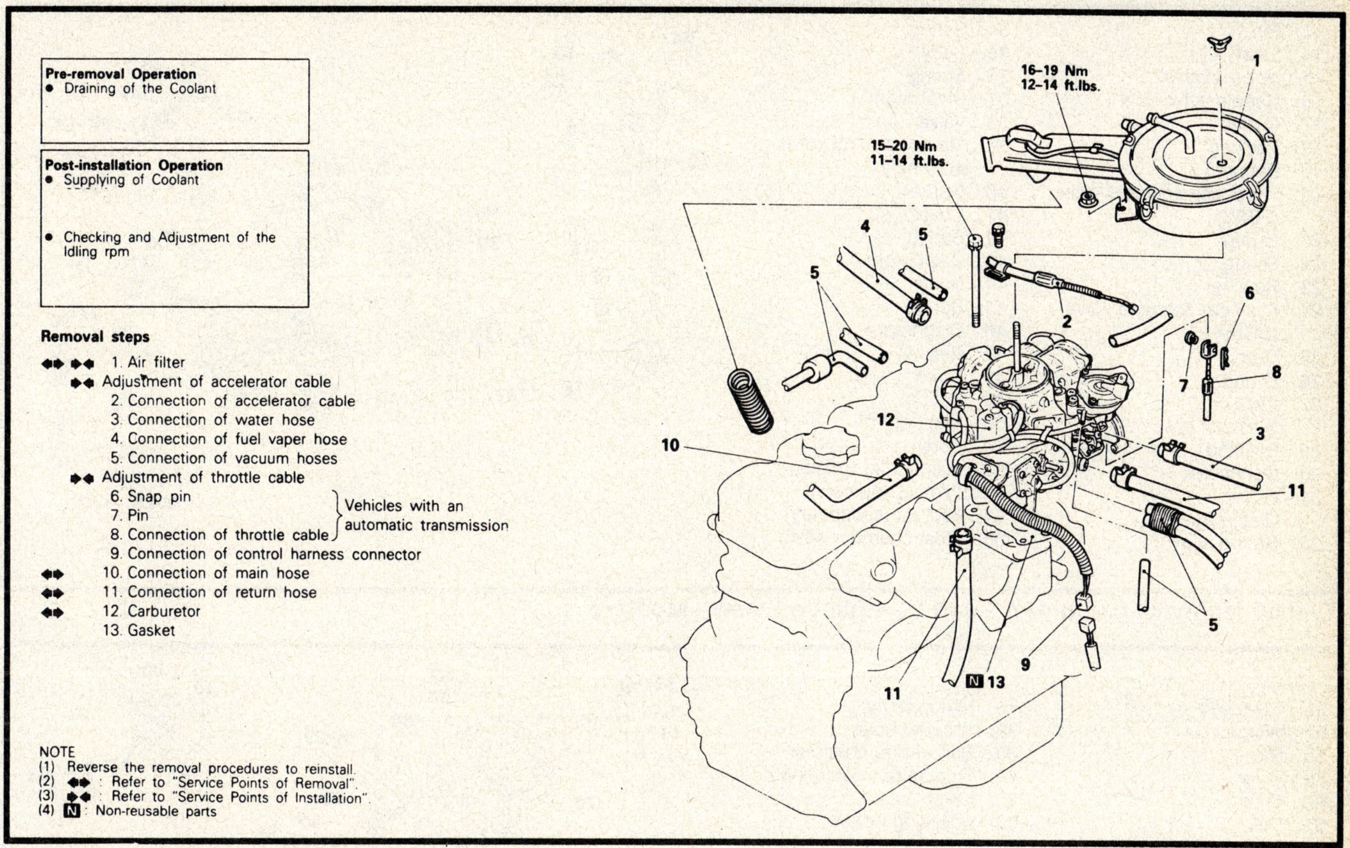

Carburetor removal and installation—Montero

4. If equipped with automatic transmission, remove idle speed control (ISC) servo.
5. Remove the connector bracket.
6. Remove the vacuum hose running from the base to the choke breaker. This vacuum line will have a delay valve in it.
7. Remove the 5 screws holding the top of the carburetor to the body and base.
8. Do not attempt to lift top by hand. Use a screwdriver blade or similar thin, flat tool. Lightly pry top upwards and lift slowly. Do not apply excessive force.
9. Remove float pivot pin and remove float. Carefully remove needle valve.

CAUTION: Do not let float drop; do not apply force to the float. The needle valve controlled by float will be damaged.

10. Valve seat may be removed by using small, flat tools to lever the seat upwards and out of position. Do not damage surrounding area.
11. Remove the electrical connector for the feedback solenoid valve (FBSV). The terminals must be removed from the connector housing before the valve can be removed. Use a thin, flat tool inserted into the connector to loosen the stopper and remove each terminal.
12. Remove the grommet from top of carburetor. Remove retainer and remove FBSV.
13. Remove the retainer and remove slow cut solenoid valve (SCSV). Hold solenoid by the body; avoid pulling on wiring.
14. Use the small screwdriver to disconnect SCSV terminals from plastic connector body. Remove the SCSV.
15. Use hand grinder or similar tool to remove heads of rivets holding the cover of choke assembly. Remove the small screw in bottom of the cover.

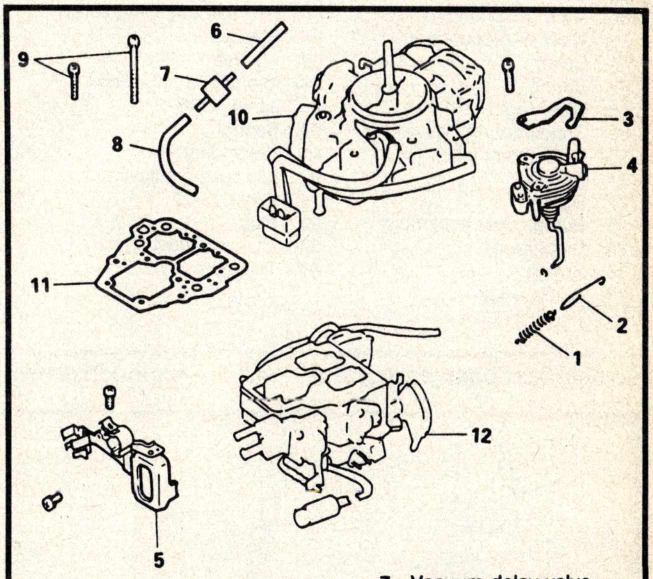

1. Throttle return spring
2. Damper spring
3. Throttle return spring bracket
4. Throttle opener/dashpot
5. Bracket
6. Hose
7. Vacuum delay valve
8. Hose
9. Screw
10. Float chamber cover (carb top)
11. Gasket
12. Mixing body (main body) and throttle body (base)

Typical feedback carburetor—part 1—exploded view—Montero

13. Pin
14. Float
15. Needle valve
16. Needle valve seat
17. O-ring
18. Packing
19. Retainer
20. Feedback solenoid valve (FBSV)
21. O-ring
22. O-ring
23. Retainer
24. Slow cut solenoid valve (SCSV)
25. O-ring
26. O-ring
27. Plate
28. Bi-metal assembly
29. Packing
30. Connector
31. Cover
32. Diaphragm
33. Spring seat
34. Spring
35. Body
36. Spring
37. Diaphragm
38. Valve
39. Mixture control valve assembly
40. Gasket
41. Cover
42. Spring
43. Diaphragm
44. Body
45. Spring
46. Diaphragm
47. Bracket
48. Cover
49. Spring
50. Diaphragm
51. Body
52. Main air jet (primary)
53. Pilot jet (primary)
54. Pilot jet (secondary)
55. Float chamber cover

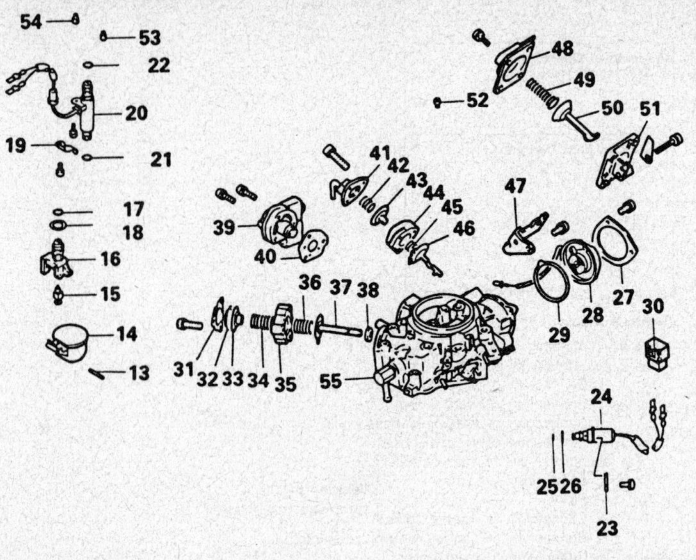

Typical feedback carburetor – part 2 – exploded view – Montero

56. Steel ball
57. Weight
58. Ball
59. Plug
60. O-ring
61. Ball
62. Screw
63. Gasket
64. Main jet (primary)
65. Main jet (secondary)
66. Cover
67. Spring
68. Diaphragm
69. Enrichment jet valve
70. Enrichment jet
71. Spring
72. Ball
73. Pump cover assembly
74. Diaphragm
75. Spring
76. Pump body
78. Mixing body
79. Vacuum hose
80. Depression chamber
81. Throttle position sensor
82. Lever
83. Adjusting screw
84. Throttle lever
85. Cam follower
86. Fast idle adjusting screw
87. Free lever
88. Abutment plate
89. Idle speed adjusting screw (SAS-2)
90. Spring
91. Secondary lever
92. Idle speed adjusting screw (SAS-1)
93. Plug
94. Mixture adjusting screw
95. Throttle body (carburetor base)

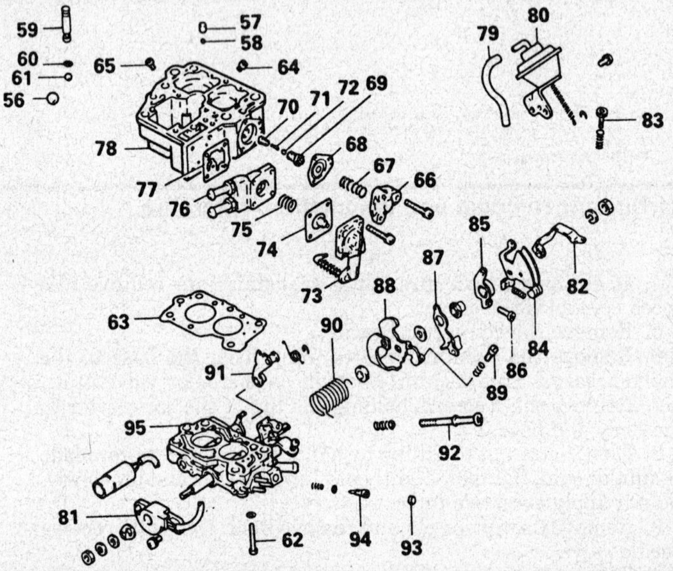

Typical feedback carburetor – part 3 – exploded view – Montero

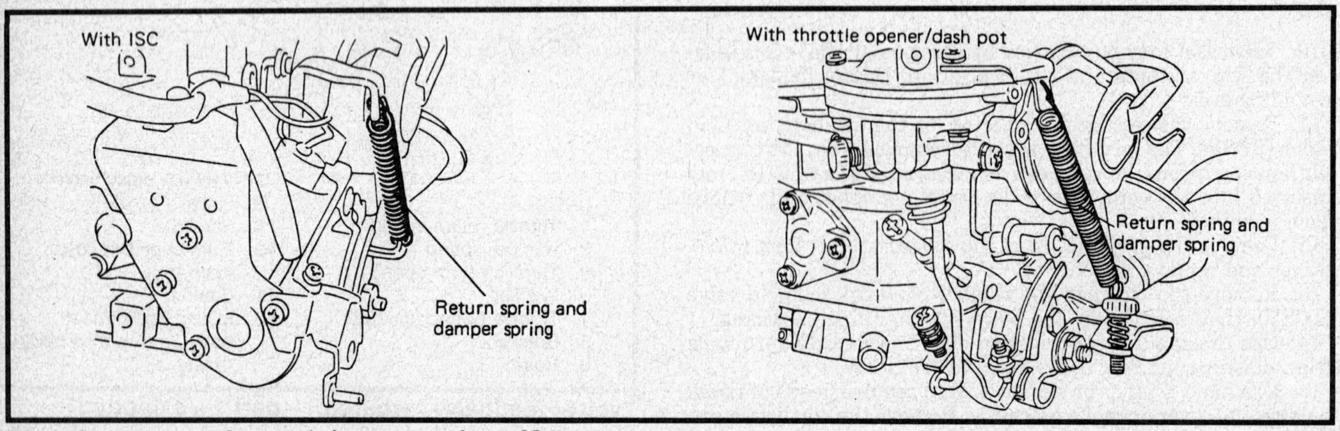

Throttle return spring and damper spring – Montero

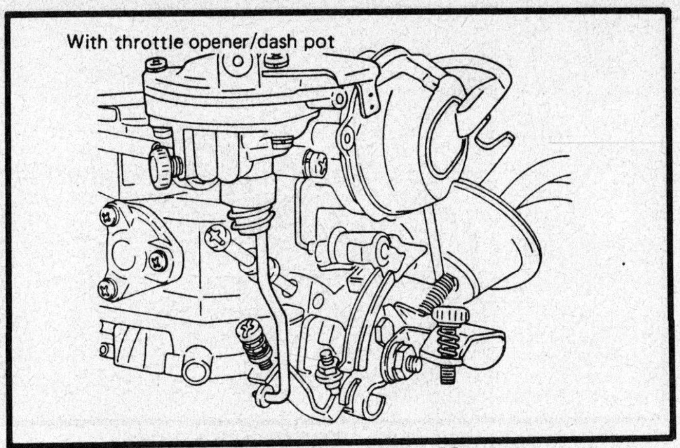

Throttle opener/dashpot location—Montero

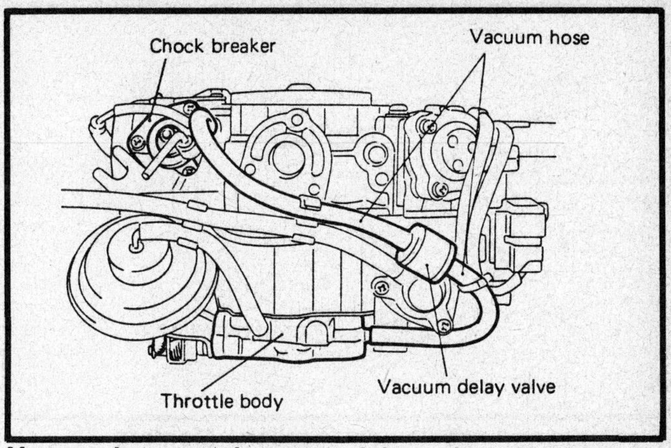

Vacuum hoses and vacuum delay valve—Montero

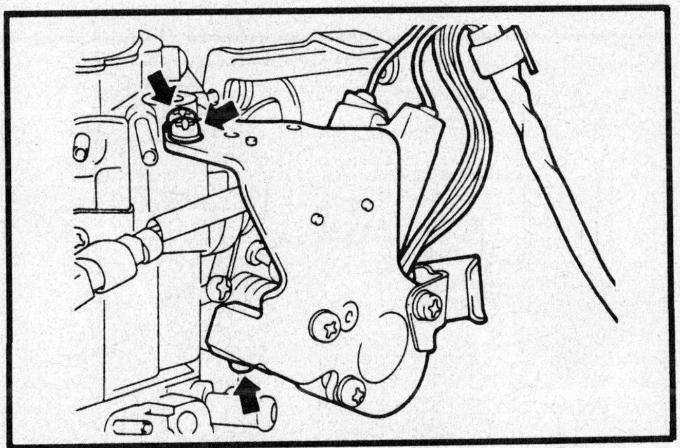
Idle speed control cover—Montero

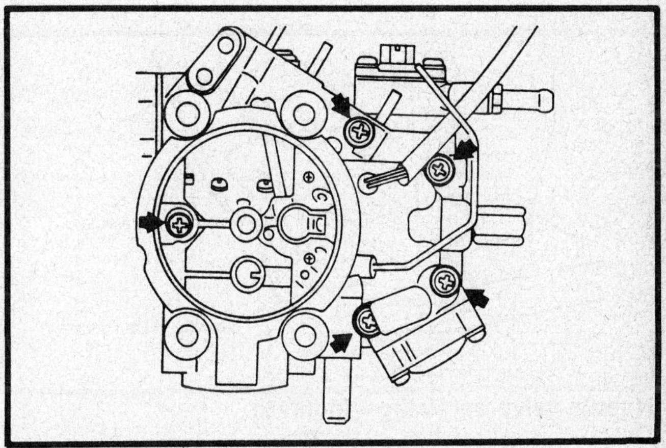
Float chamber cover screw location—Montero

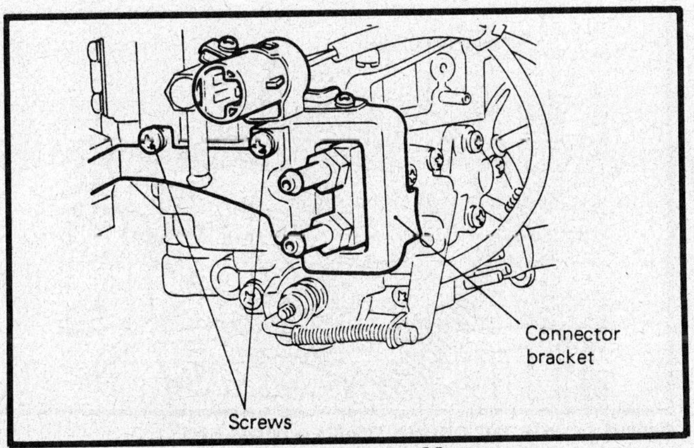

Connector bracket and screws—Montero

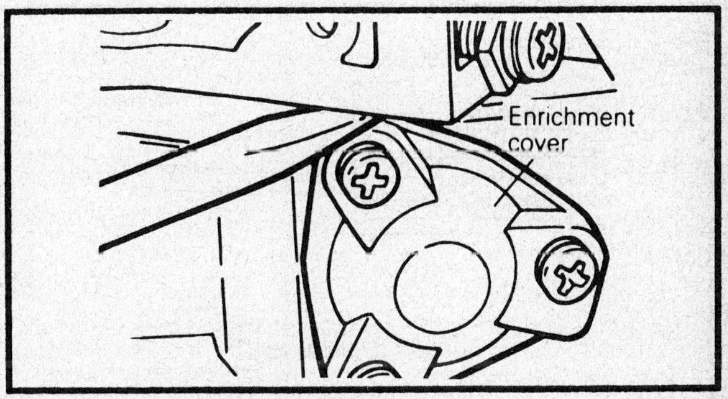

Lifting carburetor top—Montero

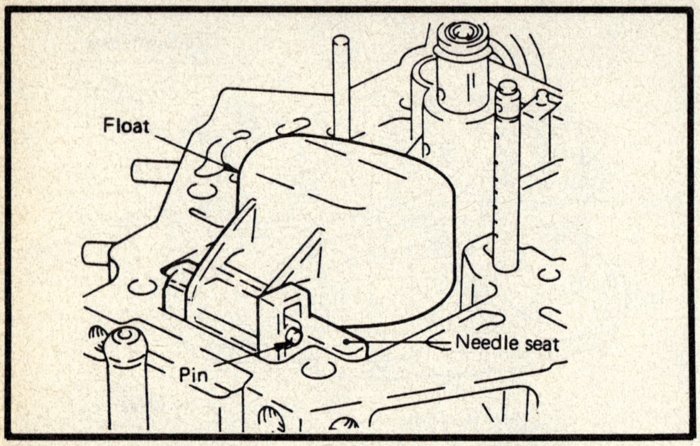

Float, pin and needle seat location—Montero

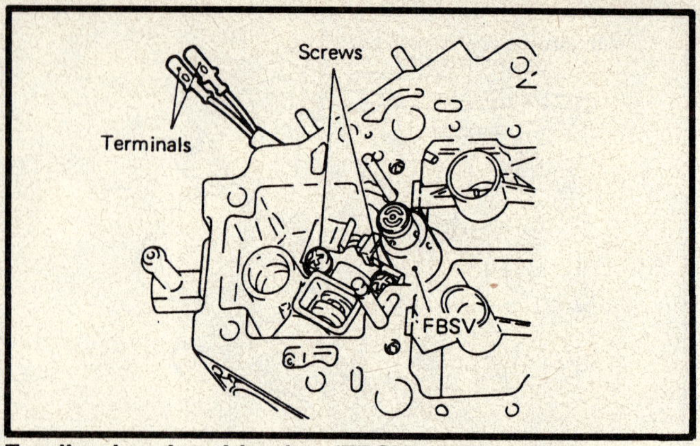

Feedback solenoid valve (FBSV) and retainer—Montero

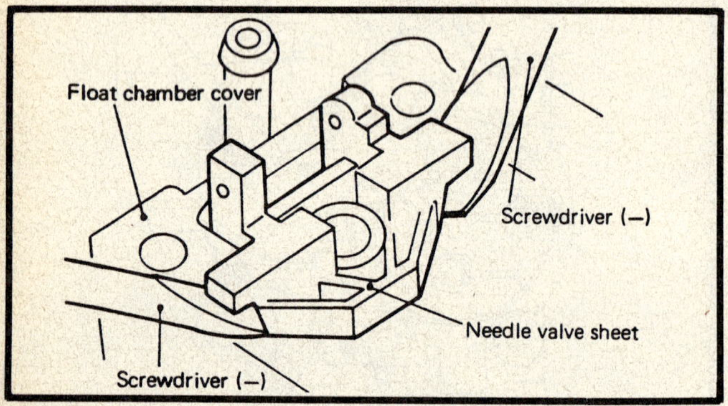

Needle valve servicing—Montero

Slow cut solenoid valve (SCSV) servicing—Montero

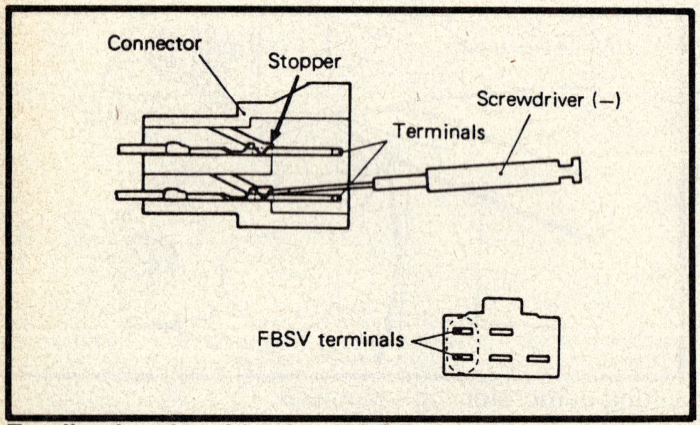

Feedback solenoid valve (FBSV) terminal removal from harness connector—Montero

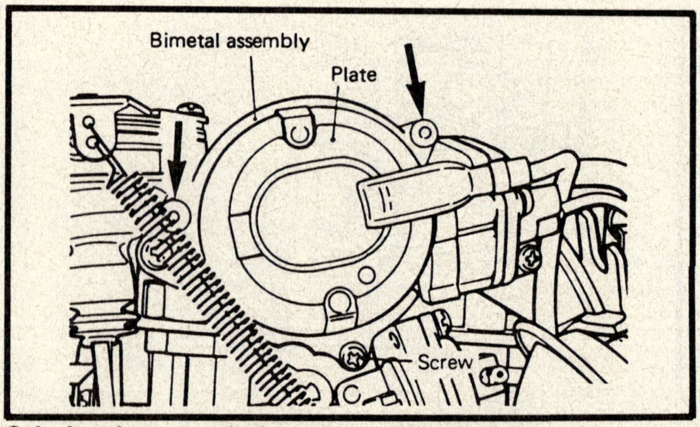

Grind points on choke cover—Montero

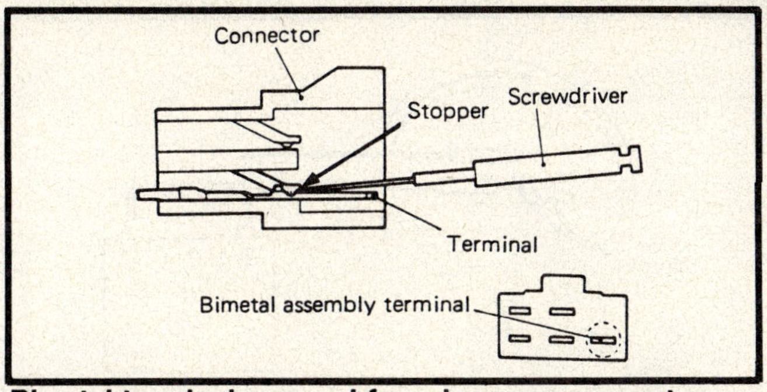

Bimetal terminal removal from harness connector–Montero

Float chamber vent valve diaphragm–Montero

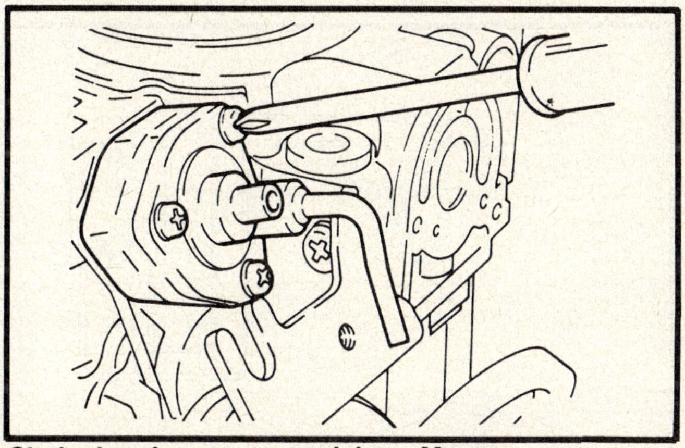

Choke breaker cover servicing–Montero

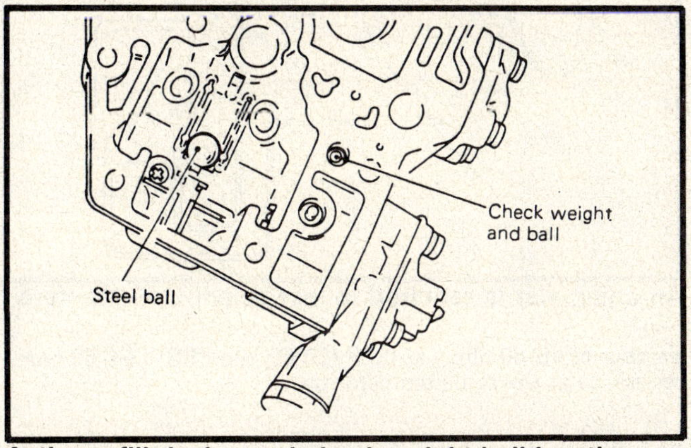

Anti-overfill device and check weight ball location–Montero

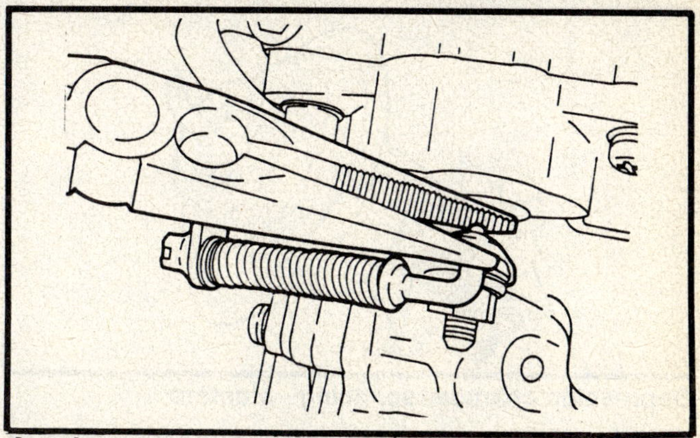

Accelerator pump rod servicing–Montero

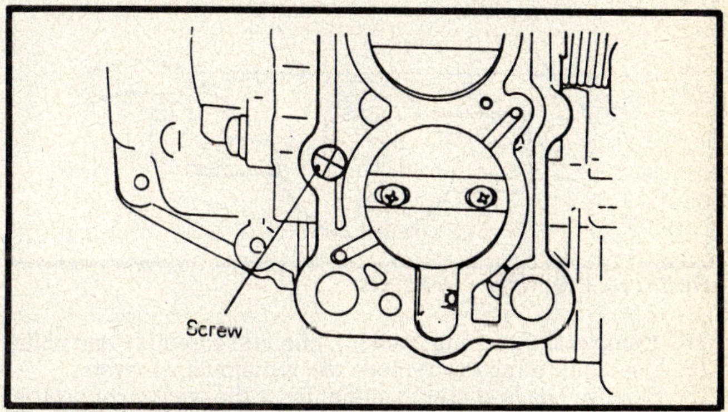

Throttle body assembly screw–Montero

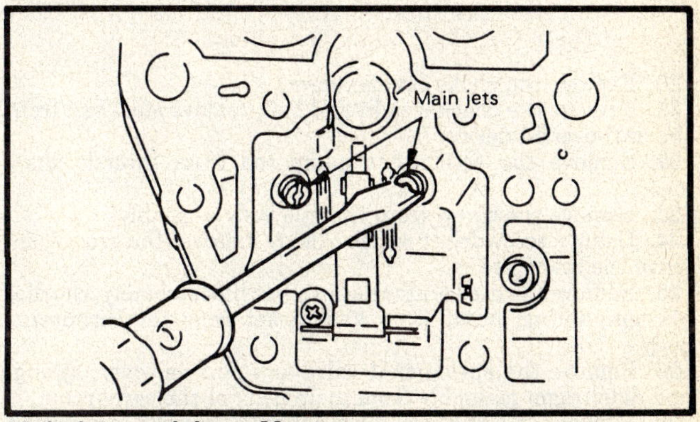

Main jet servicing–Montero

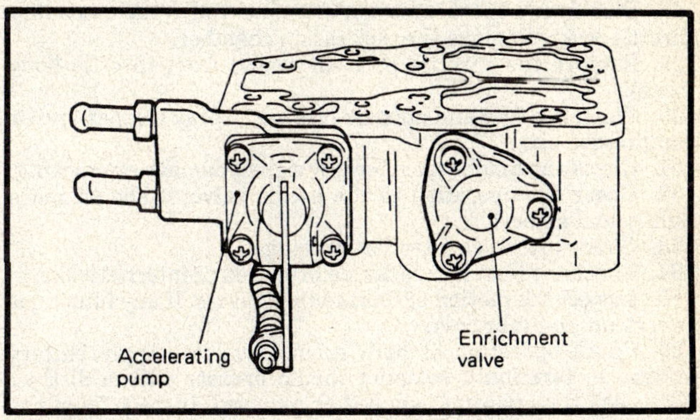

Accelerating pump and enrichment valve location–Montero

Depression chamber servicing—Montero

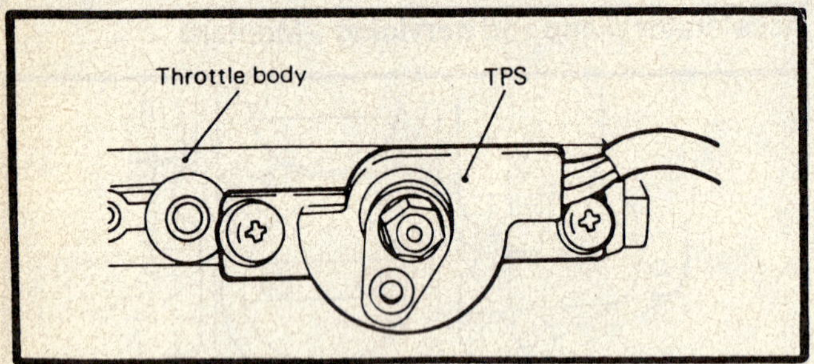

Remove the TPS—Montero

16. Remove the packing (gasket), bi-metal assembly and plate.
17. Use a pin punch to remove the remainder of rivets.
18. Remove the bi-metal terminal from the wiring connector.
19. Remove the bowl vent valve.

NOTE: There are small springs within this unit. Take note of their location; don't lose them.

20. Remove the choke breaker cover.
21. Remove the check weight and ball; remove steel ball from the anti-overfill device.
22. Remove the accelerator pump rod from throttle shaft lever.
23. Remove the screw from throttle body assembly.
24. Using a screwdriver which exactly matches the groove, remove the main jets.
25. Remove the accelerator pump cover link assembly, the diaphragm, spring, pump body and gasket from the carburetor body.
26. Remove the enrichment valve and remove cover, spring, and diaphragm assembly from main body of the carburetor.
27. Remove the vacuum hose running between depression chamber and throttle body.
28. Disconnect the depression chamber rod from secondary throttle lever. Remove the depression chamber.
29. Remove the throttle position sensor from throttle body (base).
30. Clean the fuel and air passages with spray cleaner and/or compressed air.
31. Check the diaphragms for any sign of damage or cracking.
32. Check the operation of the needle valve; it should move lightly and smoothly.
33. Check fuel inlet filter for clogging.
34. Check the float for cracks, deformation or internal leakage.
35. Inspect the motion of linkages and pivots. If any binding is felt, clean and lubricate.
36. Check operation of both solenoid valves. Apply battery voltage to terminals; solenoid should operate with a distinct click each time power is applied or removed. Inspect tip of the FBSV to insure the jet is open.
37. Use ohmmeter to check resistance. SCSV should have resistance of 48–60 ohms at 68°F (20°C); the FBSV, 54–66 ohms resistance at the same temperature.

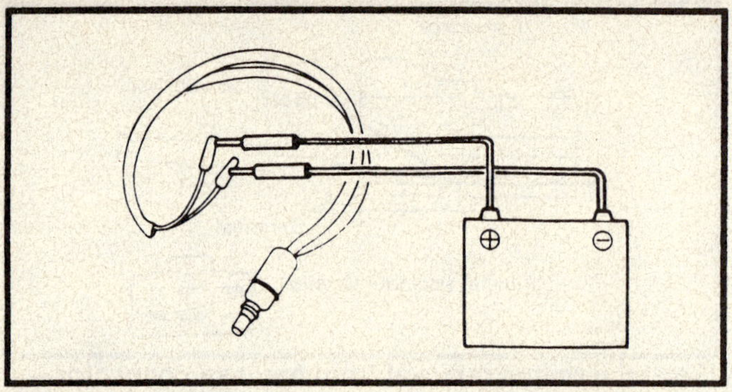

Testing solenoids; listen for click when power is applied—Montero

Solenoid tip must be clean—Montero

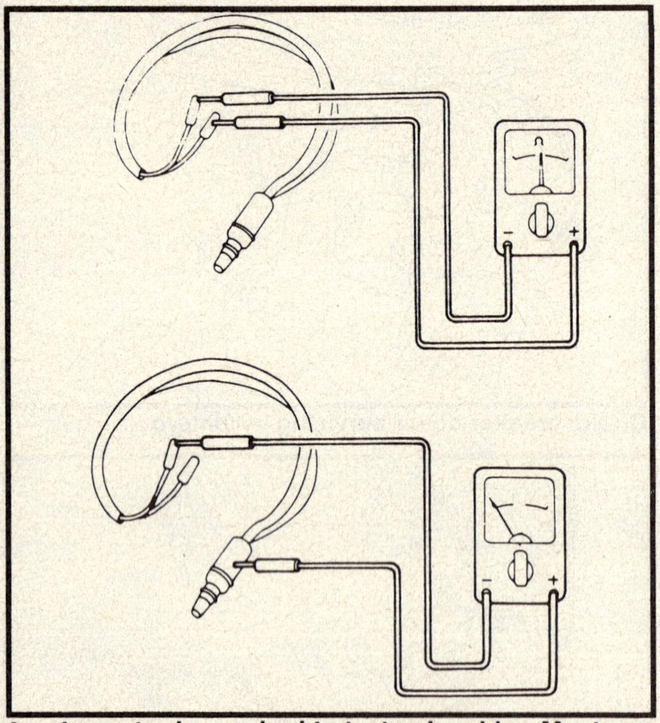

An ohmmeter is required to test solenoids—Montero

NOTE: Resistance will increase or decrease as temperature rises or falls. Make common sense allowances for the temperature in which you are testing the units.

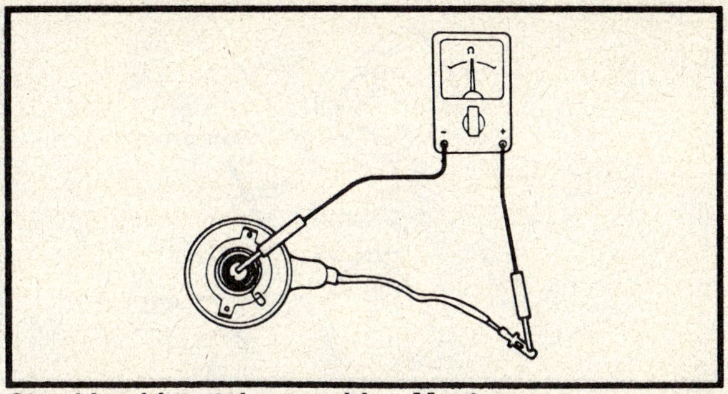

Checking bi-metal assembly—Montero

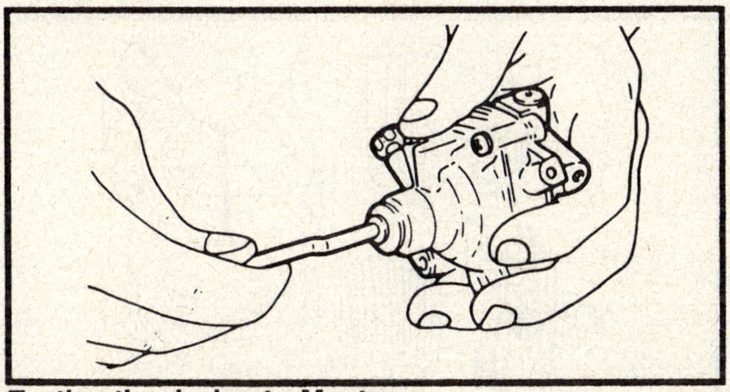

Testing the dashpot—Montero

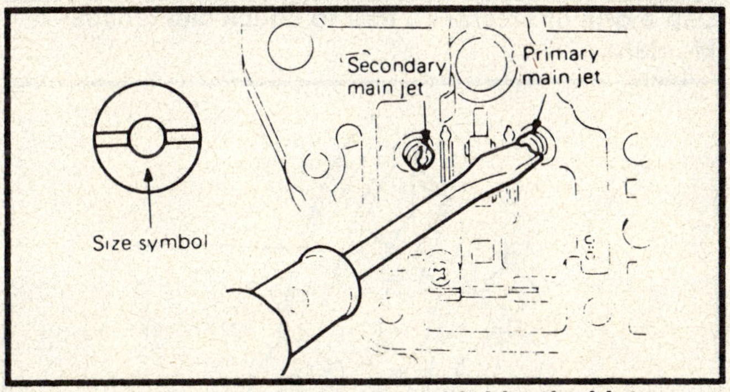

Removing jets. Always replace with identical jets—Montero

38. Hold 1 lead of ohmmeter against the case of each solenoid and touch other lead to each terminal. There should NOT be continuity between case and terminals.
39. Use ohmmeter to check bi-metal assembly. Correct resistance is approximately 6 ohms at 68°F (20°C).
40. Inspect dashpot by pulling outward on the rod. Resistance should be felt; when released, lever should return quickly.
41. Check depression chamber by pushing rod all the way into the unit and then blocking vacuum port firmly. Release the rod; if it stays in place (with the vacuum port blocked), the unit is good.
42. Before reassembly, make certain all parts are clean and dry. Any gasket or O-ring removed MUST be replaced with a new one during reinstallation.
43. If jets are to be replaced, new ones must be exact replacements. The jets have a number stamped on them for identification.
44. Install the throttle position sensor onto the throttle body.
45. Install the depression chamber. Connect chamber rod to secondary throttle lever.

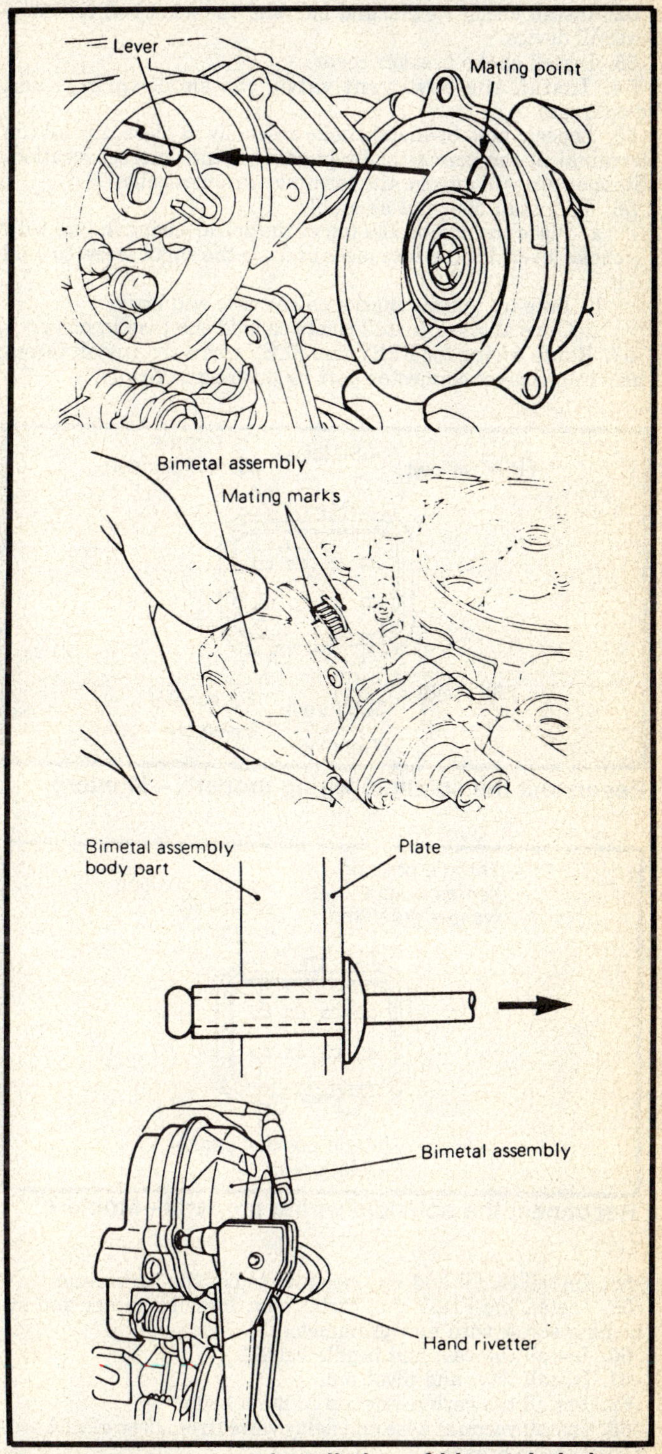

Four steps to correct installation of bi-metal choke unit—Montero

46. Install the vacuum hose from depression chamber to throttle body.
47. Assemble diaphragm, spring and cover for the enrichment valve and install (with its cover) onto the main body.
48. Assemble the accelerator pump and install.
49. Install the main jets.
50. Install the screw into the throttle body assembly.
51. Install the accelerator pump rod to throttle shaft lever.

52. Install check weight and ball and the steel ball for anti-overfill device.
53. Install choke breaker cover.
54. Install the bowl vent valve, the small springs and diaphragm.
55. Loosely hold bi-metal choke assembly in place. Install the terminal in the connector by pushing it into correct location. Stopper pin will engage the terminal automatically.
56. To install bi-metal assembly:
 a. Place the cup on the top of the spiral spring in line with choke lever. Fit cap into place and use the small screw to hold it.
 b. Line up mating marks on the case and body.
 c. Once aligned, install rivets to hold the case in place.
57. Route wiring for FBSV and SCSV correctly. Install terminals into correct connector port by pushing in firmly.

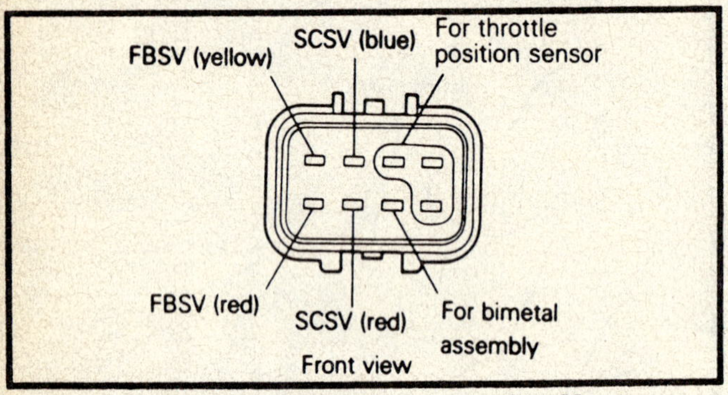

Reconnect the solenoid wiring properly—Montero

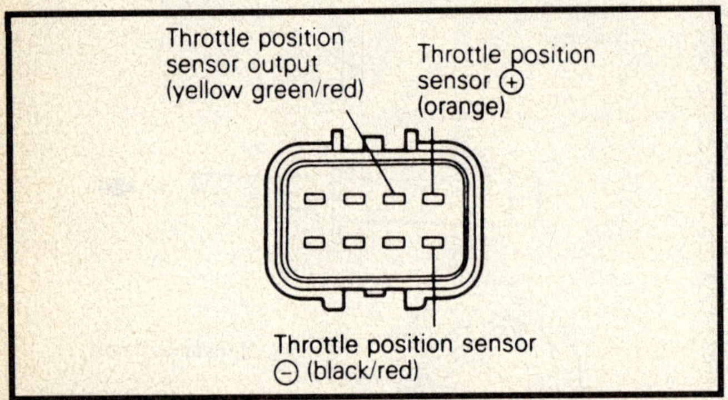

Reconnect the solenoid wiring properly—Montero

58. Install SCSV and retainer into top of the carburetor.
59. Install the FBSV into carburetor. Install retainer and attaching screw with new grommet.
60. Install the seat and needle valve.
61. Install float and pivot pin.
62. Install the carburetor top to main body.
63. Install vacuum hose and delay valve (from base of carburetor to choke breaker).
64. Install connector bracket.
65. Install the ISC servo if removed (automatic transmission only).
66. Install either dashpot or throttle opener unit and connect the rod to the free lever.
67. Install the throttle return spring and damper spring.
68. Move carburetor linkages by hand, checking that motions are smooth and there is no binding.
69. Set the high idle cam to second highest position and turn carburetor upside down. Check the clearance between primary throttle plate and throttle bore. Correct clearances are 0.031 in.

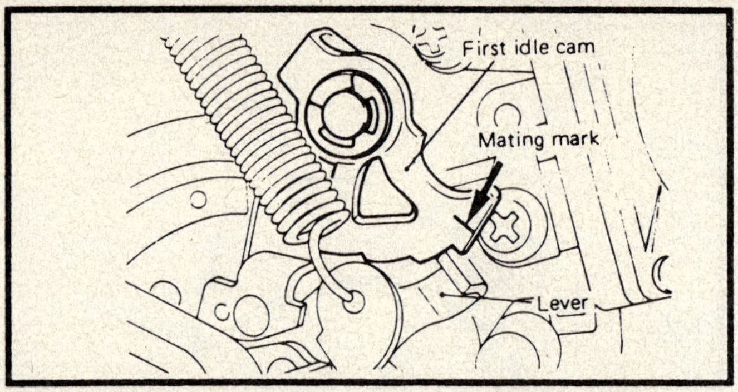

Set high idle cam before checking throttle plate clearance—Montero

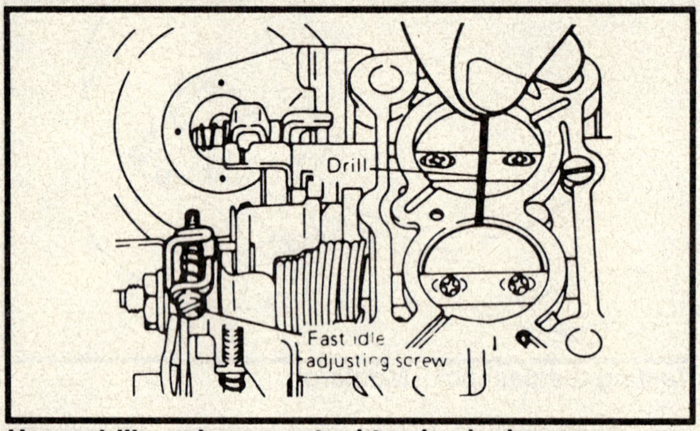

Use a drill or clearance tool to check clearances—Montero

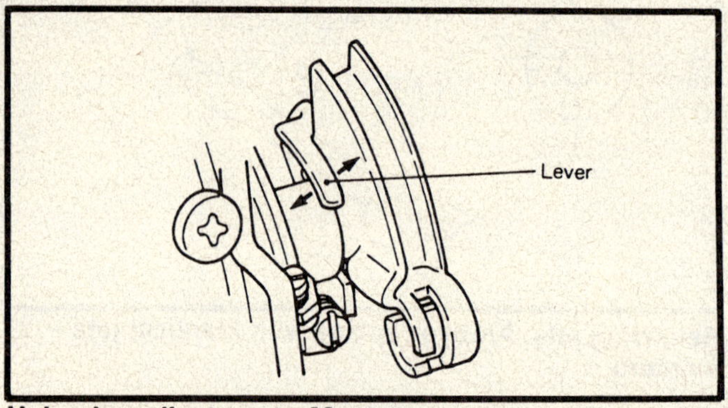

Unloader adjustment—Montero

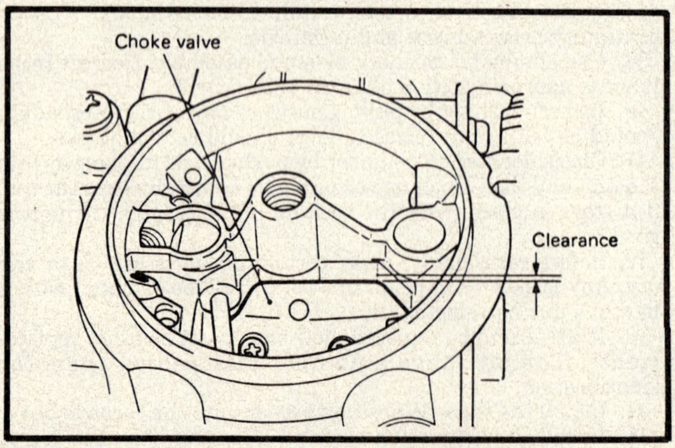

Choke valve clearance check—Montero

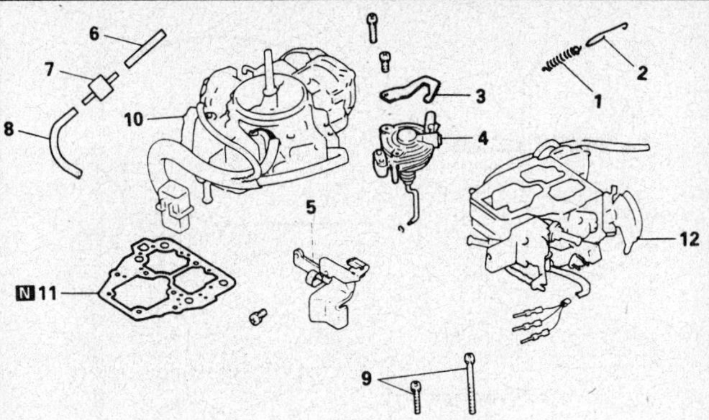

NOTE
(1) Reverse the disassembly procedures to reassemble.
(2) ◆◆ : Refer to "Service Points of Disassembly".
(3) N : Non-reusable parts

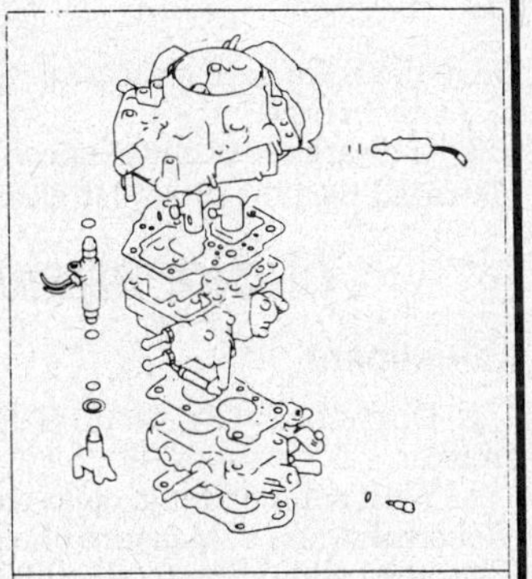

Carburetor assembly and gasket kit

Disassembly steps

1. Throttle-return spring
2. Damper spring
3. Throttle-return spring bracket
4. Throttle opener/Dash pot
5. Bracket
6. Hose
7. Vacuum-delay valve
8. Hose
9. Screw
10. ◆◆ ◆◆ Float chamber cover assembly
11. Float chamber cover gasket
12. Mixing body and throttle body assembly

Carburetor disassembly and overhaul – part 1 – Montero

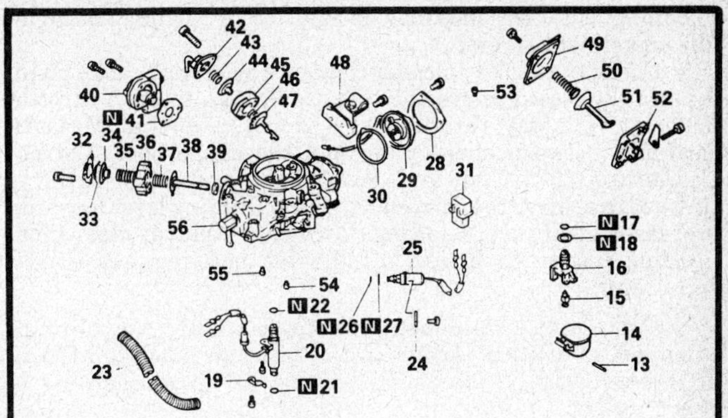

Disassembly steps

13. Pin
14. ◆◆ Float
15. ◆◆ Needle valve
16. ◆◆ Needle valve seat
17. O-ring
18. Packing
19. Retainer
20. ◆◆ ◆◆ Feedback solenoid valve (FBSV)
21. O-ring
22. O-ring
23. Tube
24. Retainer
25. ◆◆ ◆◆ Slow-cut solenoid valve (SCSV)
26. O-ring
27. O-ring
28. Plate
29. ◆◆ ◆◆ Bimetal assembly
30. Packing
31. Connector
32. Cover
33. Diaphragm
34. Spring seat
35. Spring
36. Body
37. Spring
38. Diaphragm
39. Valve
40. Mixture control valve (MCV) assembly
41. Gasket
42. Cover
43. Spring
44. Diaphragm
45. Body
46. Spring
47. Diaphragm
48. Bracket
49. Cover
50. Spring
51. Diaphragm
52. Body
53. ◆◆ ◆◆ Main air jet (primary)
54. ◆◆ ◆◆ Pilot jet (primary)
55. ◆◆ ◆◆ Pilot jet (secondary)
56. Float chamber cover

NOTE
(1) Reverse the disassembly procedures to reassemble.
(2) ◆◆ : Refer to "Service Points of Disassembly".
(3) ◆◆ : Refer to "Service Points of Reassembly".
(4) N : Non-reusable parts

Carburetor disassembly and overhaul – part 2 – Montero

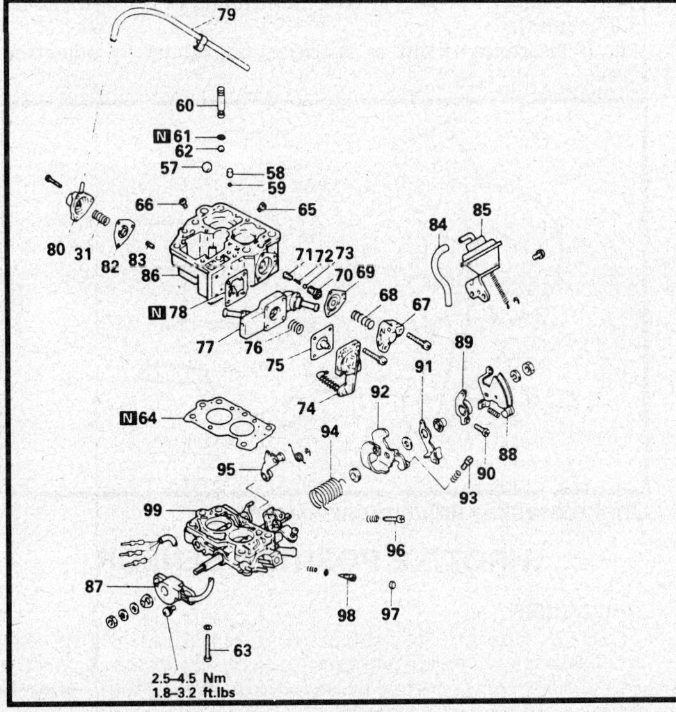

Disassembly steps

57. ◆◆ Steel ball
58. ◆◆ Weight
59. ◆◆ Ball
60. ◆◆ Plug
61. ◆◆ O-ring
62. ◆◆ Ball
63. ◆◆ Screw
64. Gasket
65. ◆◆ ◆◆ Main jet (primary)
66. ◆◆ ◆◆ Main jet (secondary)
67. Cover
68. Spring
69. Diaphragm
70. ◆◆ Enrichment jet valve
71. Enrichment jet
72. Spring
73. Ball
74. Pump cover assembly
75. Diaphragm
76. Spring
77. Pump body
78. Gasket
79. Hose
80. Auxiliary acceleration pump cover
81. Spring
82. Diaphragm
83. Check valve
84. Mixing body
85. Vacuum hose
86. Depression chamber
87. Throttle-position sensor (TPS)
88. Throttle lever
89. Cam follower
90. Fast-idle adjustment screw
91. Free lever
92. Apartment plate
93. Idle-speed adjustment screw (SAS-2)
94. Spring
95. Secondary lever
96. Idle-speed adjustment screw (SAS-1)
97. ◆◆ Plug
98. Mixture-adjustment screw (MAS)
99. Throttle body

NOTE
(1) Reverse the disassembly procedures to reassemble.
(2) ◆◆ : Refer to "Service Points of Disassembly".
(3) ◆◆ : Refer to "Service Points of Reassembly".
(4) N : Non-reusable parts

Carburetor disassembly and overhaul – part 3 – Montero

(0.80mm). The clearance may be adjusted by turning the high-idle adjustment screw.

70. Check the choke unloader clearance by setting choke plate. When fully closed, move the throttle linkage to open throttle plate(s) all the way; throttle plates should be vertical. Measure clearance between choke plate and the choke bore. Correct clearance is 0.075–0.083 in. (1.9–2.1mm). To adjust, bend throttle lever to achieve the correct clearance. Bending lever upwards increases clearance; bending downward reduces clearance. Bending too much breaks off the tab and requires a new carburetor.

NOTE: Choke unloader clearance should not change after an overhaul. Adjusting clearance greatly affects cold driveability; check clearance but do not adjust if not needed.

71. Install new base gasket and reinstall carburetor.

FAST IDLE OPENING

Adjustment

1. Set lever on second highest cam of fast idle cam.
2. Measure primary valve to throttle bore clearance. Clearance should be 0.031 in. (0.80mm).
3. If clearance is out of specification, adjust using fast idle adjusting screw.

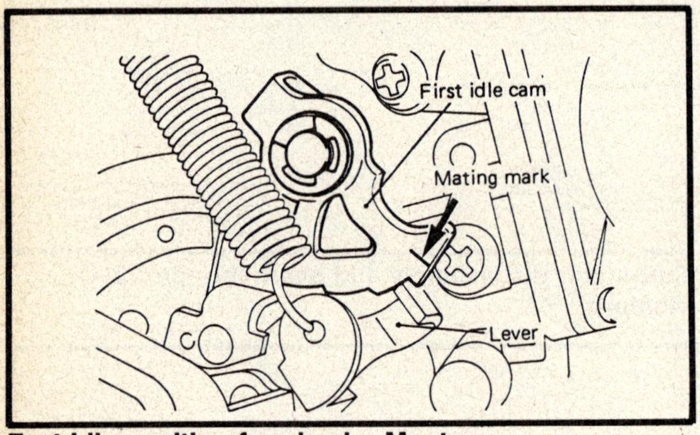

Fast idle position for check—Montero

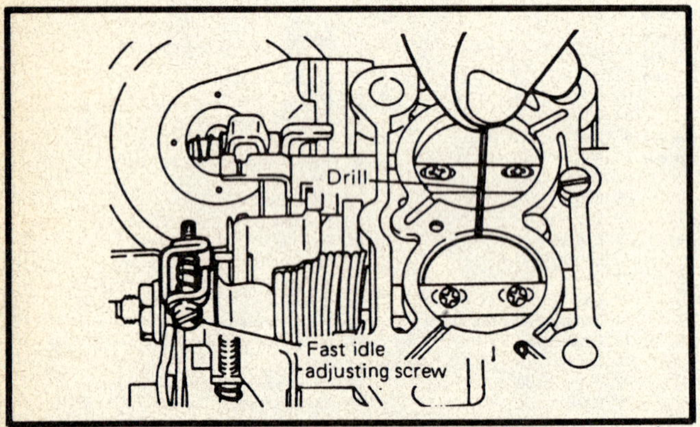

Fast idle opening check—Montero

UNLOADER OPENING

Adjustment

1. Lightly press choke valve with a finger to fully close it.
2. In this state, fully open throttle valve and measure choke

Fast idle adjustment—Montero

valve to choke bore clearance. Clearance should be 0.075–0.083 in. (1.9–2.1mm).

3. If clearance is out of specification, bend throttle lever at illustrated portion to adjust clearance to standard value.

CHOKE BREAKER OPENING

Adjustment

1. Disconnect the vacuum hose (yellow stripe) from choke breaker.
2. With engine idling, close the choke valve with a finger until choke valve stops. Measure choke valve to choke bore clearance. Clearance should be 0.091–0.980 in. (2.3–2.5mm).
3. If clearance is not as specified, stop engine, remove bimetal assembly and adjust rod end opening.

NOTE: When removing bimetal assembly, mark the electric choke body for reference.

4. Reconnect vacuum hose and measure choke valve to choke bore clearance as in Step 2. Clearance should be 0.118–0.126 in. (3.0–3.2mm).
5. If clearance is out of specification, adjust by adjusting screw.

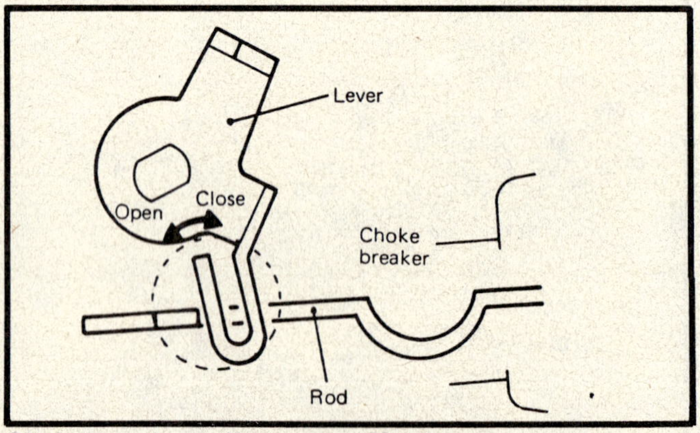

Choke breaker adjustment—Montero

THROTTLE POSITION SENSOR

Adjustment

NOTE: Coolant temperature normal, all lights and accessories OFF, transmission in N, disconnect cooling fan connector, parking brakes applied, and front wheels in straight ahead position.

1. Loosen accelerator cable.
2. Loosen speed adjusting screw No. 1 (SAS1) and No. 2 (SAS2) sufficiently to close throttle valve completely. Record number of turns loosened.

NOTE: Turning screw counterclockwise closes valve. Fast idle control should have been released (lever not resting on fast idle cam).

3. Disconnect TPS connector and connect special tool (harness connector MD998474 or equivalent) between connectors.
4. Connect an digital voltmeter between red and blue terminals 3 and 8 of special tool.
5. Turn ignition switch **ON**, engine off. Measure TPS output voltage. Output voltage should be 250 millivolt (mV).
6. If output voltage is not as specified, turn TPS body to adjust to specification.

NOTE: Turning TPS body clockwise increases output voltage.

7. Turn ignition **OFF**.
8. Tighten SAS1 and SAS2 by same number of turns as loosened in Step 2, returning them to initial state.
9. Remove voltmeter and special tool; connect TPS connector.
10. Adjust accelerator cable.
11. Start engine and check that idle speed is at specification.

THROTTLE OPENER

Adjustment

NOTE: Coolant temperature normal, all lights and accessories OFF, transmission in NEUTRAL, disconnect cooling fan connector, parking brakes applied, and front wheels in straight ahead position.

1. Make sure curb idle speed and ignition timing are within specification. Adjust if necessary.
2. Start engine.
3. Set tachometer.
4. Turn **ON** air conditioning switch.

NOTE: Solenoid valve will open and intake manifold vacuum will act on throttle opener to actuate it.

5. Check engine speed. Engine speed should be 900–950 rpm.
6. If out of specification, adjust using throttle opener (for air conditioner) adjusting screw.

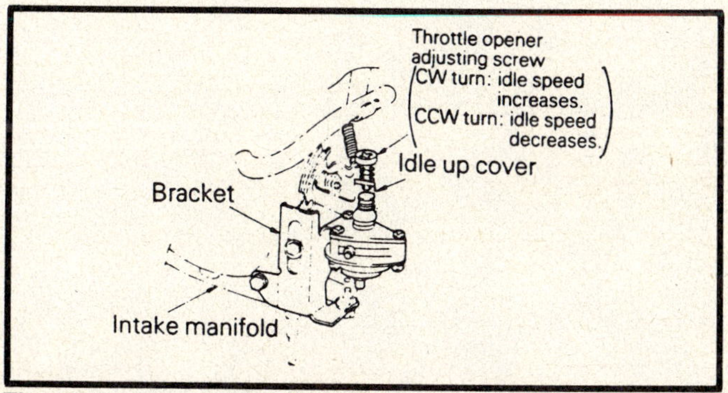

Throttle opener adjusting screw—typical

IDLE SPEED

Adjustment

NOTE: Test conditions: lights, electric cooling fan and all accessories are OFF, parking brake applied, front wheels straight ahead and transmission in NEUTRAL or PARK.

1. Run engine until coolant is at normal operating temperature.
2. Run engine for at least 15 seconds at 2000–3000 rpm.
3. Run engine at idle for 2 minutes.
4. Using a tachometer, read idling speed. Idle speed should be 800±100 rpm.
5. If idle speed is not within specified rpm, adjust speed to specification using idle speed adjusting screw No. 1 (SAS 1).

NOTE: Do not touch SAS 2. Idle speed adjusting screw (SAS 2) is preset at factory to determine relationship between throttle valve and free lever. If this setting is disturbed, throttle opener adjustment and dashpot adjustment cannot be done accurately.

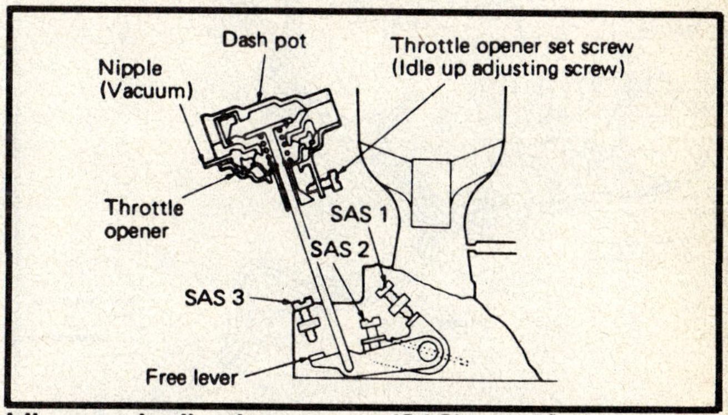

Idle speed adjusting screws (SAS)—typical

IDLE MIXTURE

Adjustment

1. Remove carburetor from engine.
2. Clamp carburetor in a vise with idle mixture adjusting screw (MAS) facing up (protect gasket surface from vise jaws).
3. Drill a $\frac{5}{64}$ in. (2mm) pilot hole in casting surrounding idle mixture adjusting screw (MAS) then redrill hole to $\frac{1}{8}$ in. (3mm).
4. Insert a blunt punch into hole and drive out plug.
5. Reinstall carburetor on engine.
6. Before adjusting, lights, electric cooling fan and all accessories must be **OFF**, transmission in **NEUTRAL** or **PARK** and ignition timing set to specification.

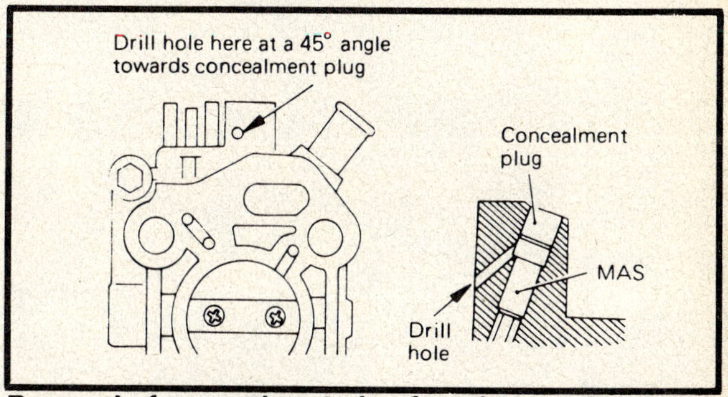

Removal of concealment plug for mixture adjusting screw (MAS)—Montero

7. Run cold engine at fast idle until coolant temperature reaches normal operating temperature.
8. Turn ignition **OFF** and disconnect cable from negative terminal of battery for about 5 seconds. Reconnect cable.
9. Disconnect exhaust oxygen sensor.
10. Run engine for more than 5 seconds at engine speed of 2000–3000 rpm.
11. Run engine at idle for 2 minutes.
12. Set idle CO and engine speed by adjusting idle speed adjusting screw (SAS) and idle mixture adjusting screw (MAS). Idle CO should be 0.1–0.3% at nominal curb idle speed.
13. Turn **OFF** the ignition switch and reconnect exhaust oxygen sensor.
14. Reset engine speed to nominal idle speed by adjusting speed adjusting screw if necessary.
15. Install concealment plug into hole to seal idle mixture adjusting screw (MAS).

VACUUM HOSE ROUTING MONTERO, TYPICAL

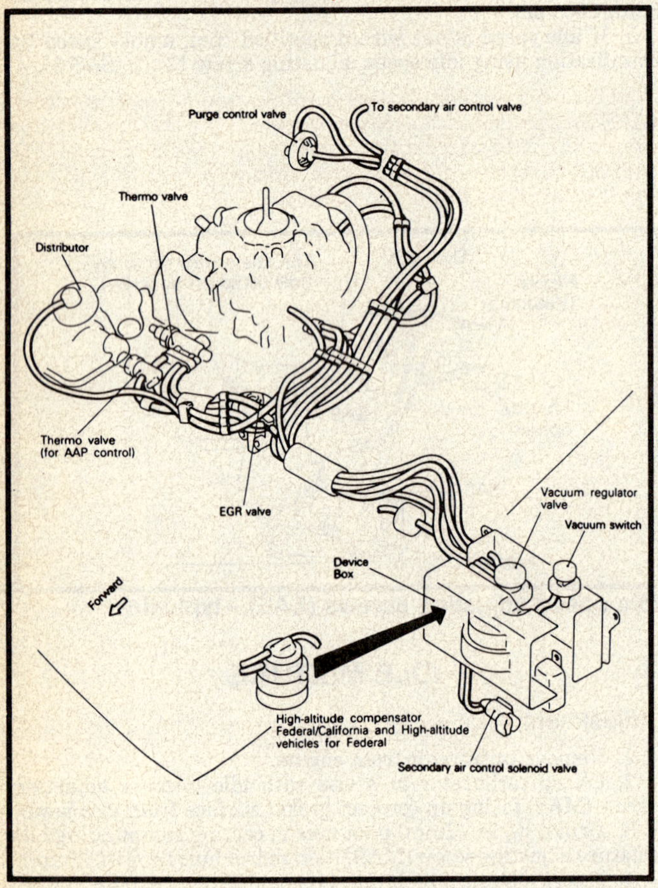

VACUUM HOSE ROUTING TRUCK, TYPICAL

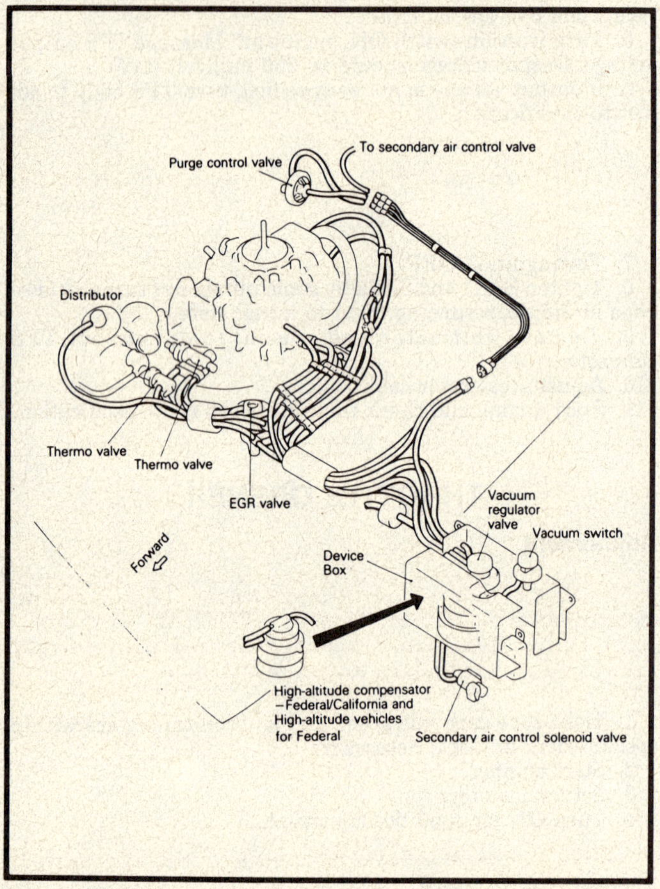

FBC SYSTEM SCHEMATIC – MONTERO AND TRUCK, FEDERAL HI ALTITUDE

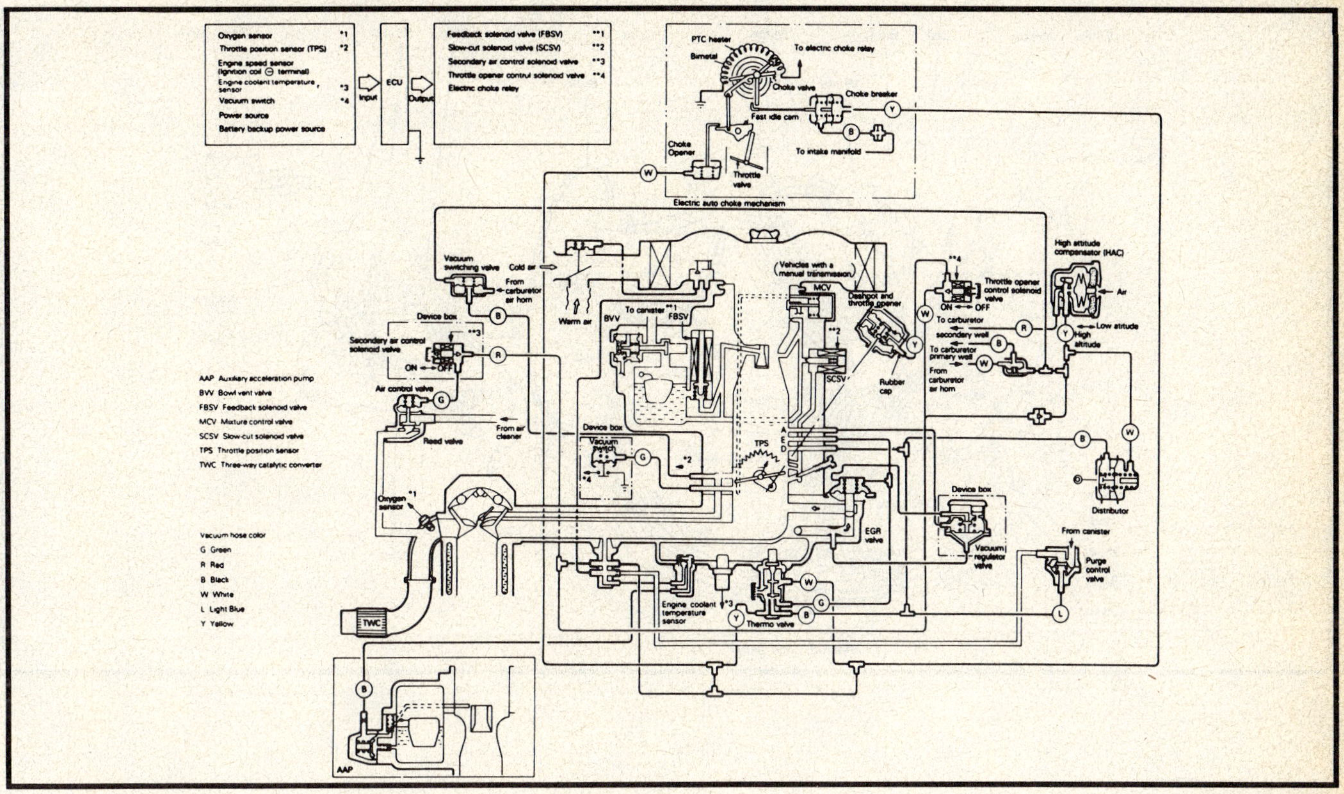

FBC SYSTEM SCHEMATIC – MONTERO AND TRUCK, CALIFORNIA

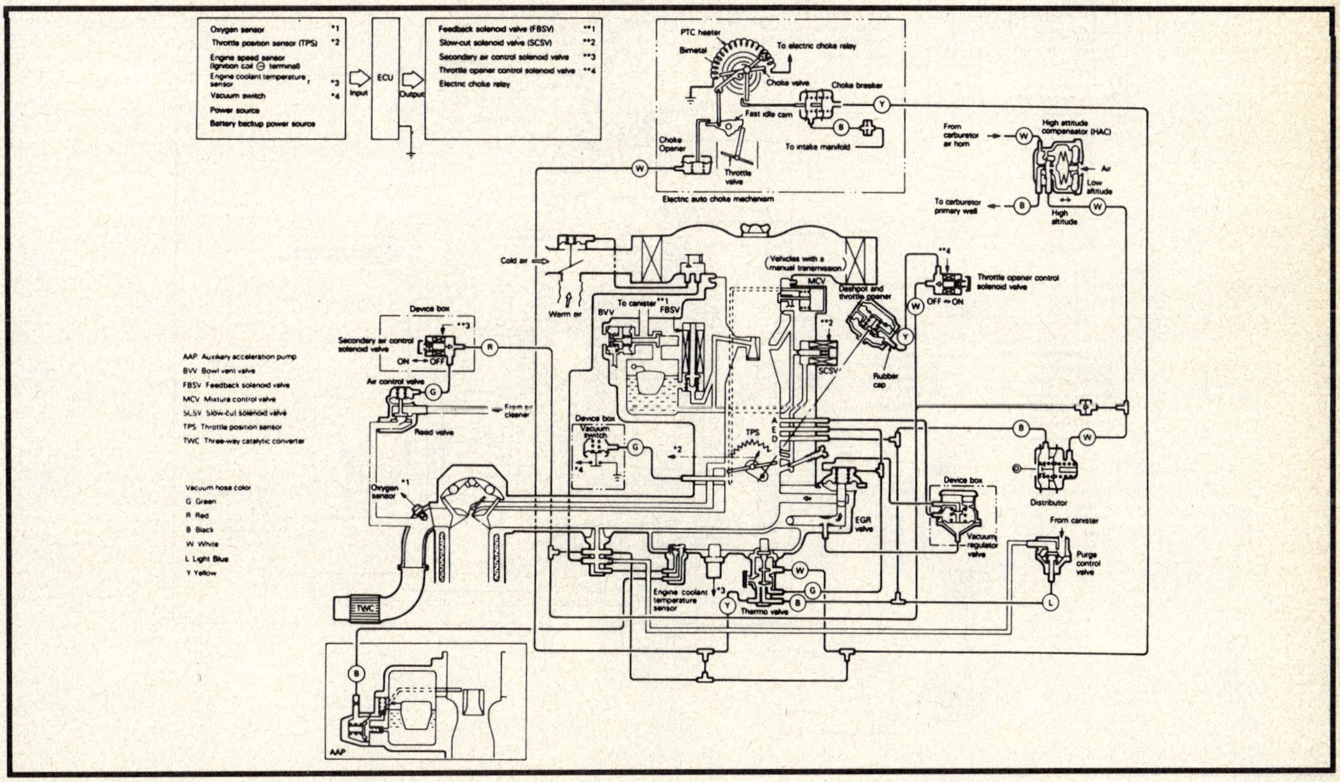

VACUUM HOSE ROUTING—PRECIS, CALIFORNIA

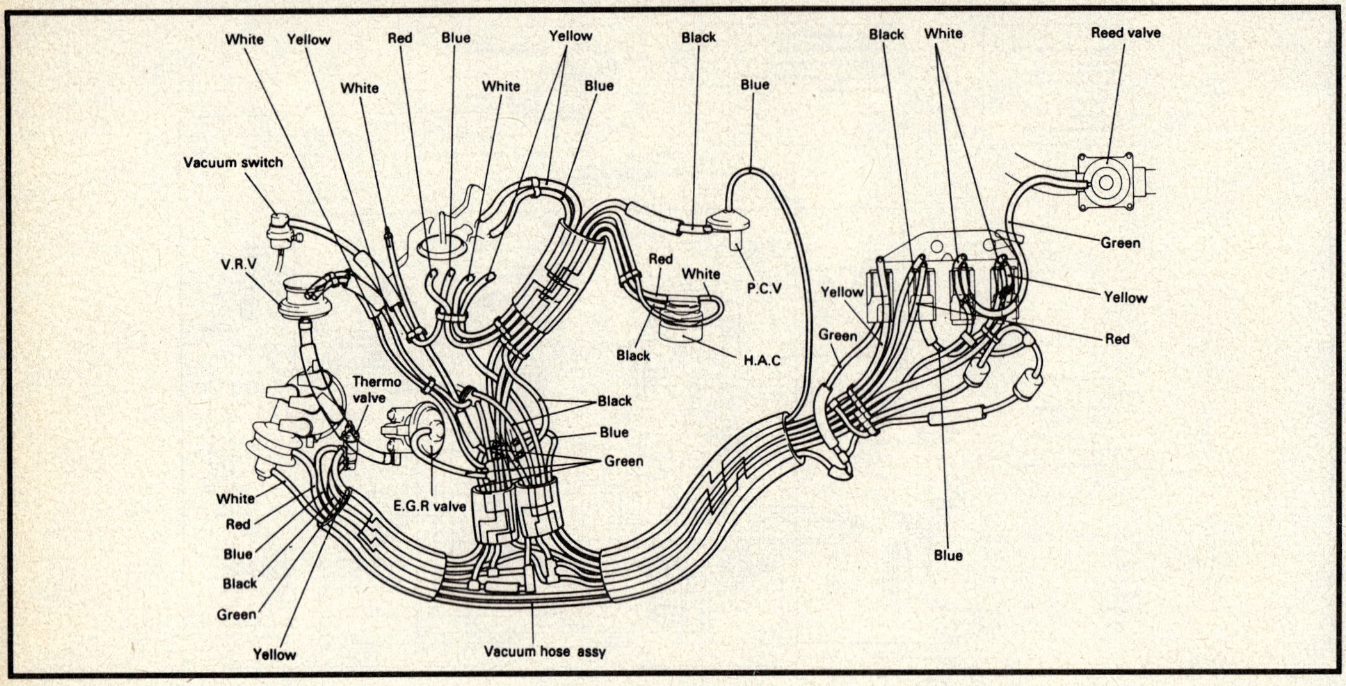

FBC SYSTEM SCHEMATIC – MONTERO AND TRUCK, FEDERAL LOW ALTITUDE

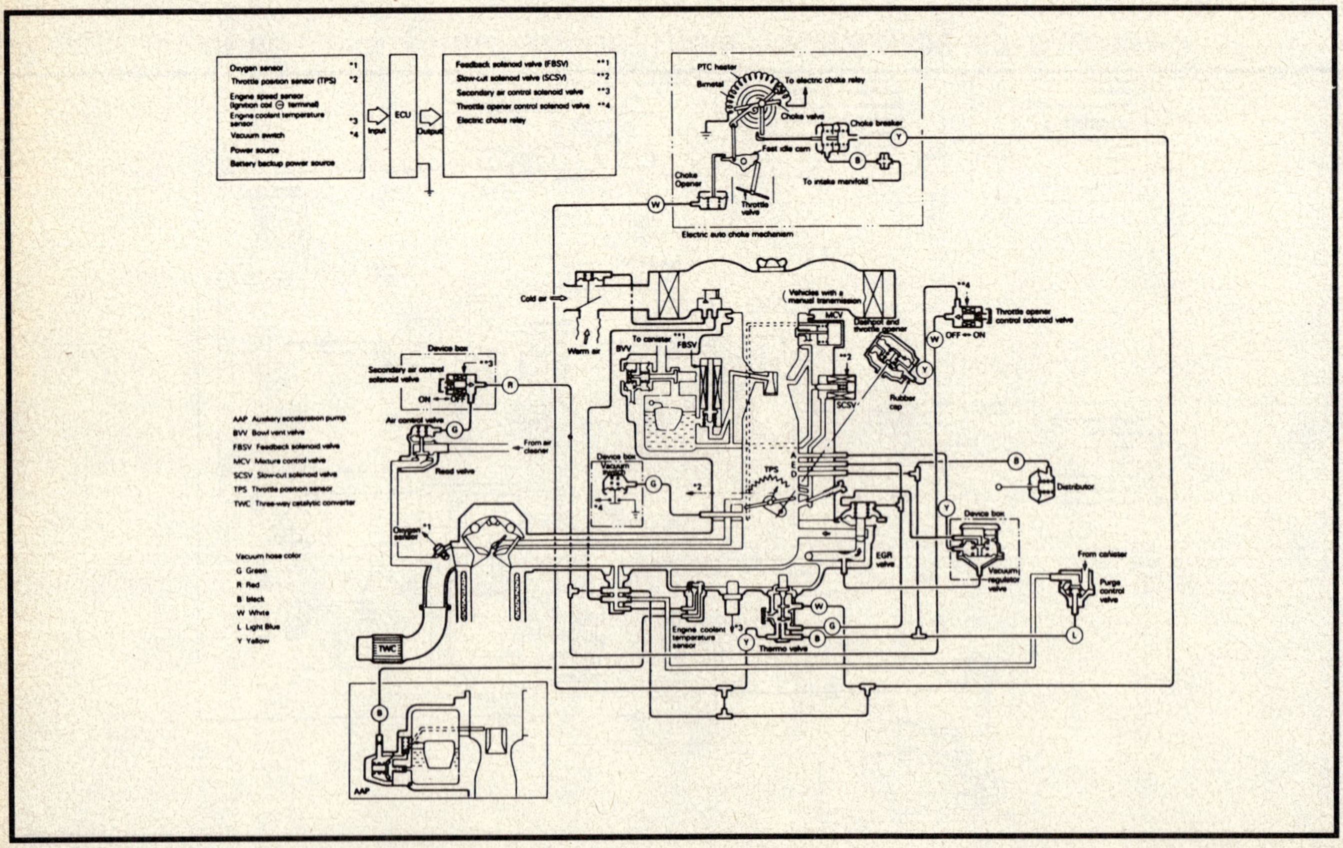

VACUUM HOSE ROUTING – MIRAGE/COLT, TYPICAL

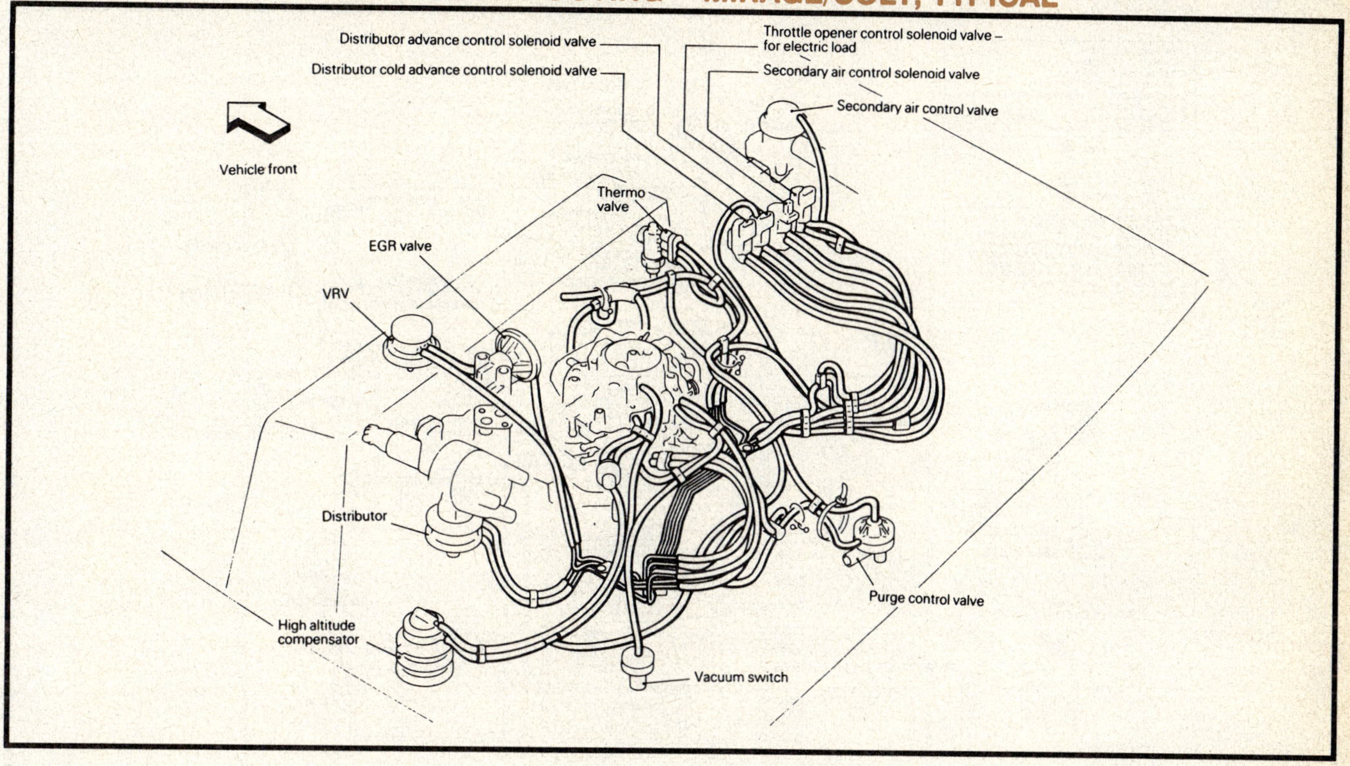

FBC SYSTEM SCHEMATIC – PRECIS, FEDERAL

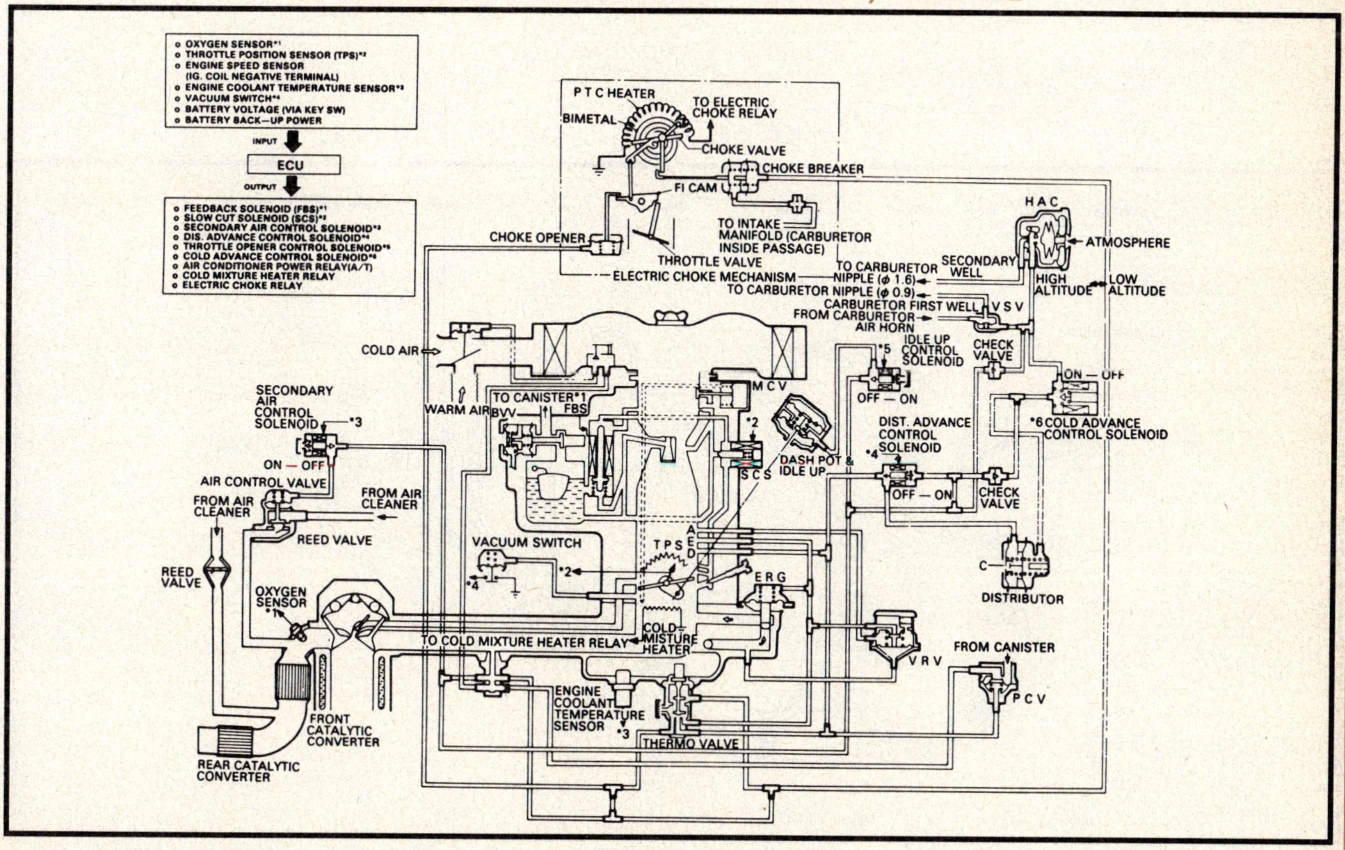

FBC SYSTEM SCHEMATIC—PRECIS, CALIFORNIA

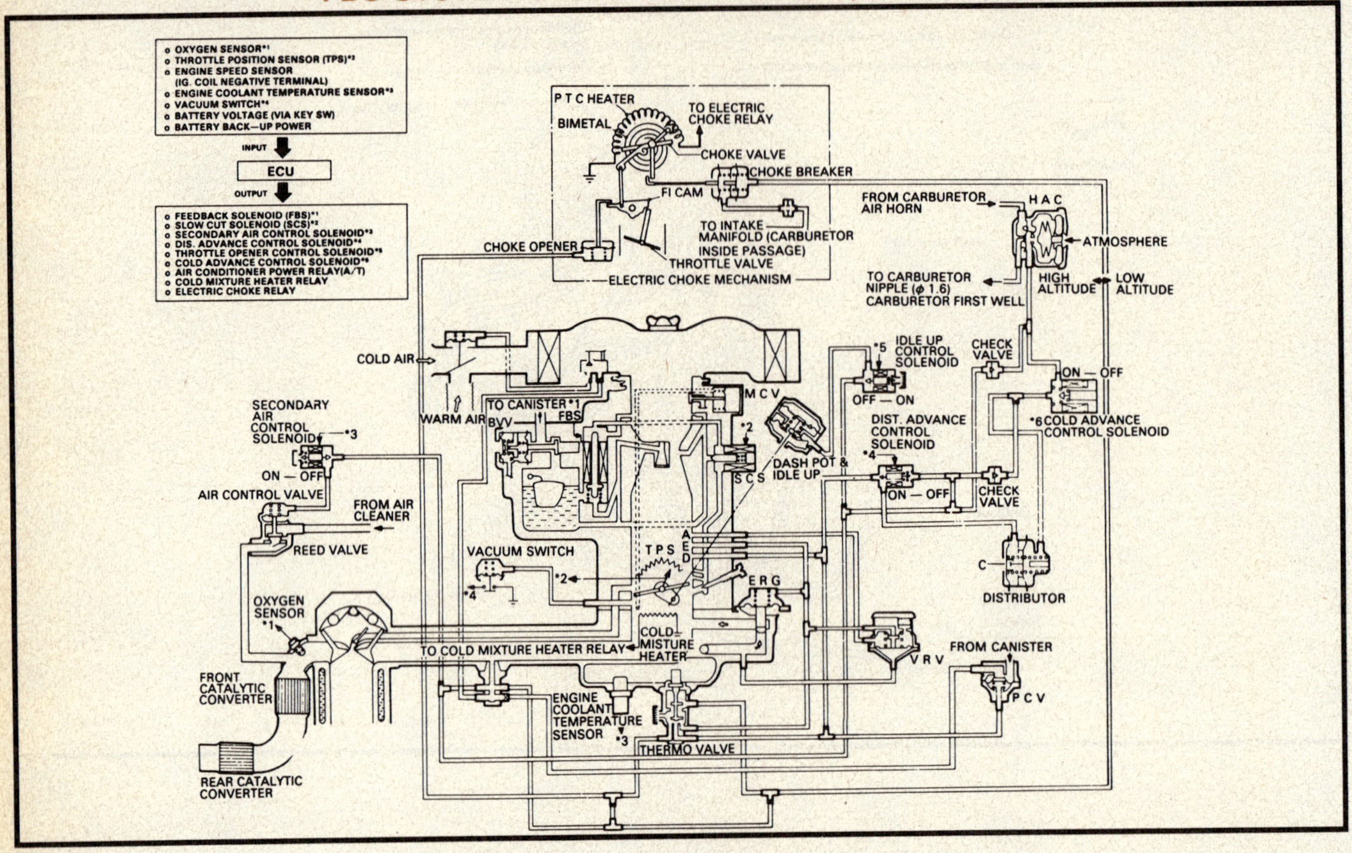

VACUUM HOSE ROUTING—PRECIS, FEDERAL

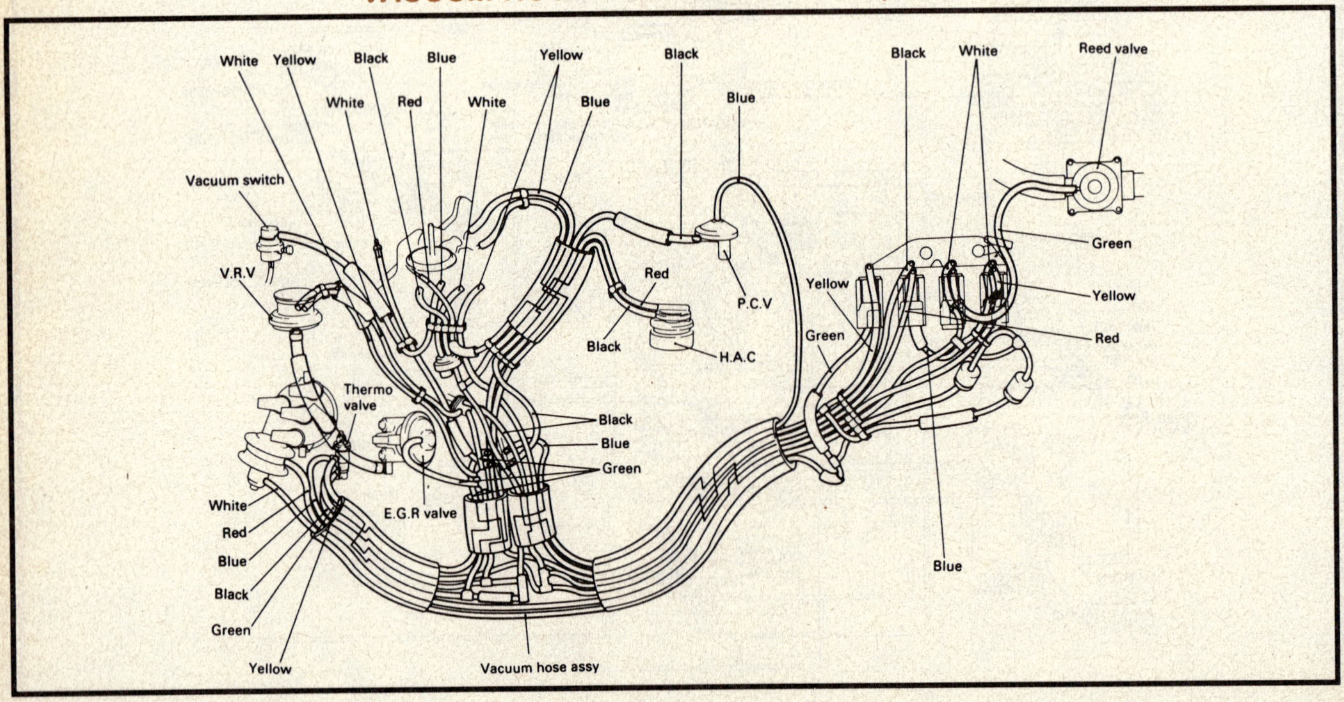

FBC SYSTEM SCHEMATIC – CORDIA/TREDIA, CALIFORNIA

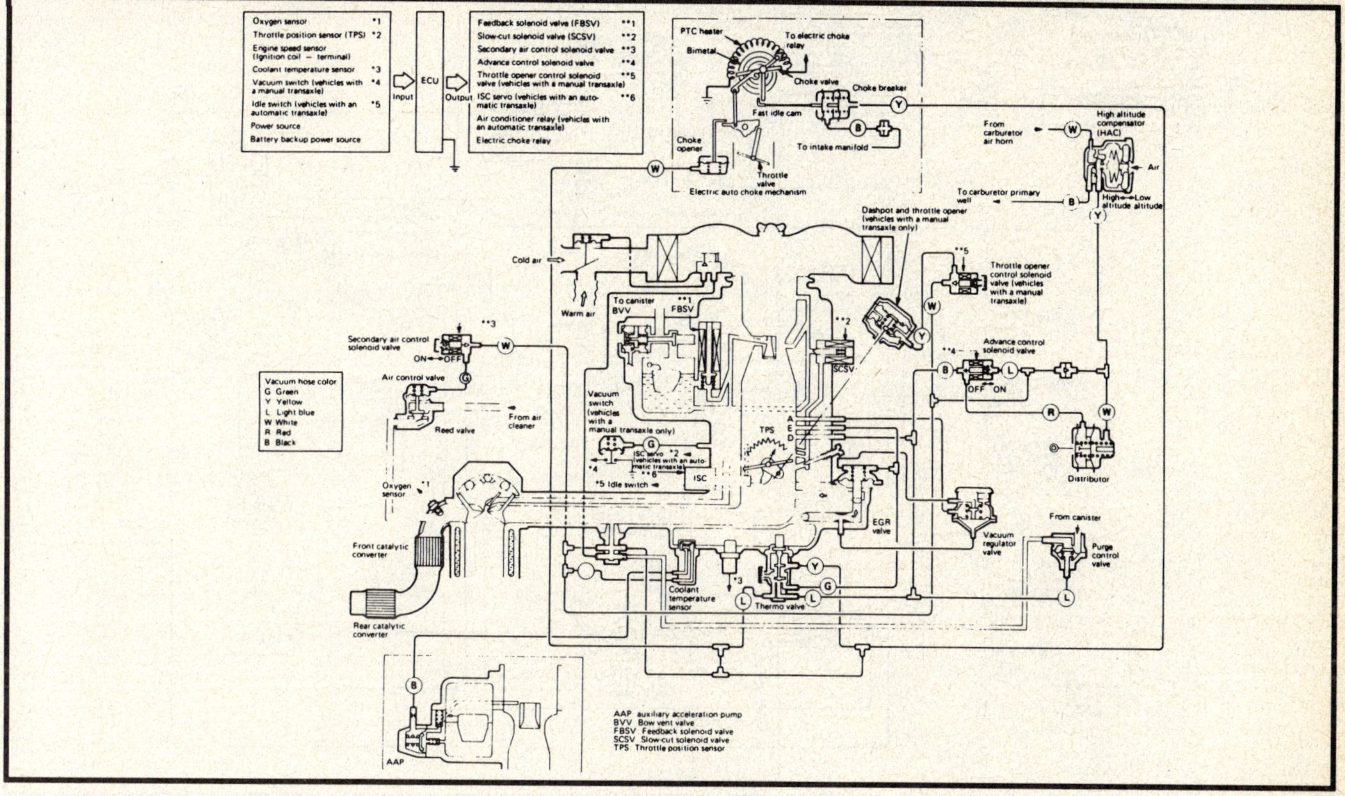

VACUUM HOSE ROUTING – CORDIA/TREDIA, TYPICAL

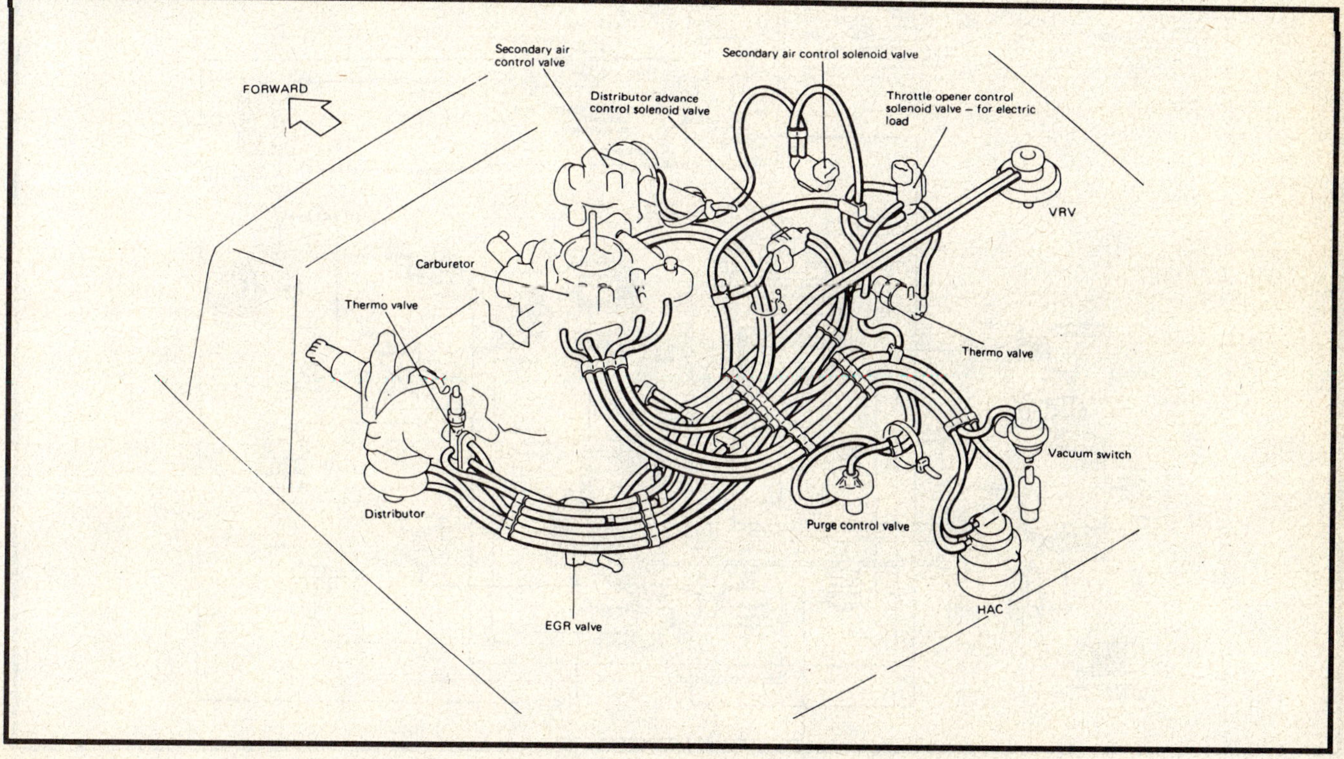

FBC SYSTEM SCHEMATIC – MIRAGE/COLT, FEDERAL

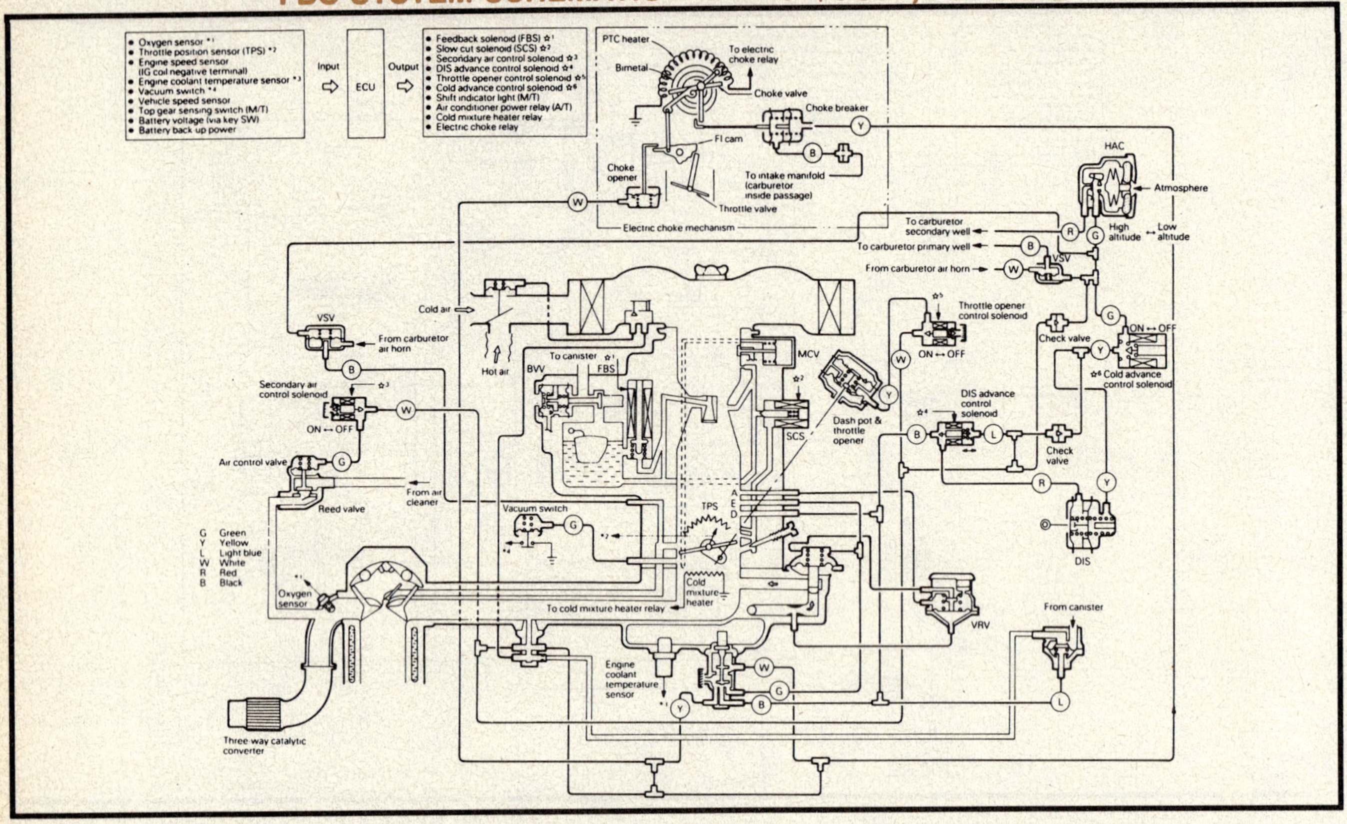

FBC SYSTEM SCHEMATIC – MIRAGE/COLT, CALIFORNIA

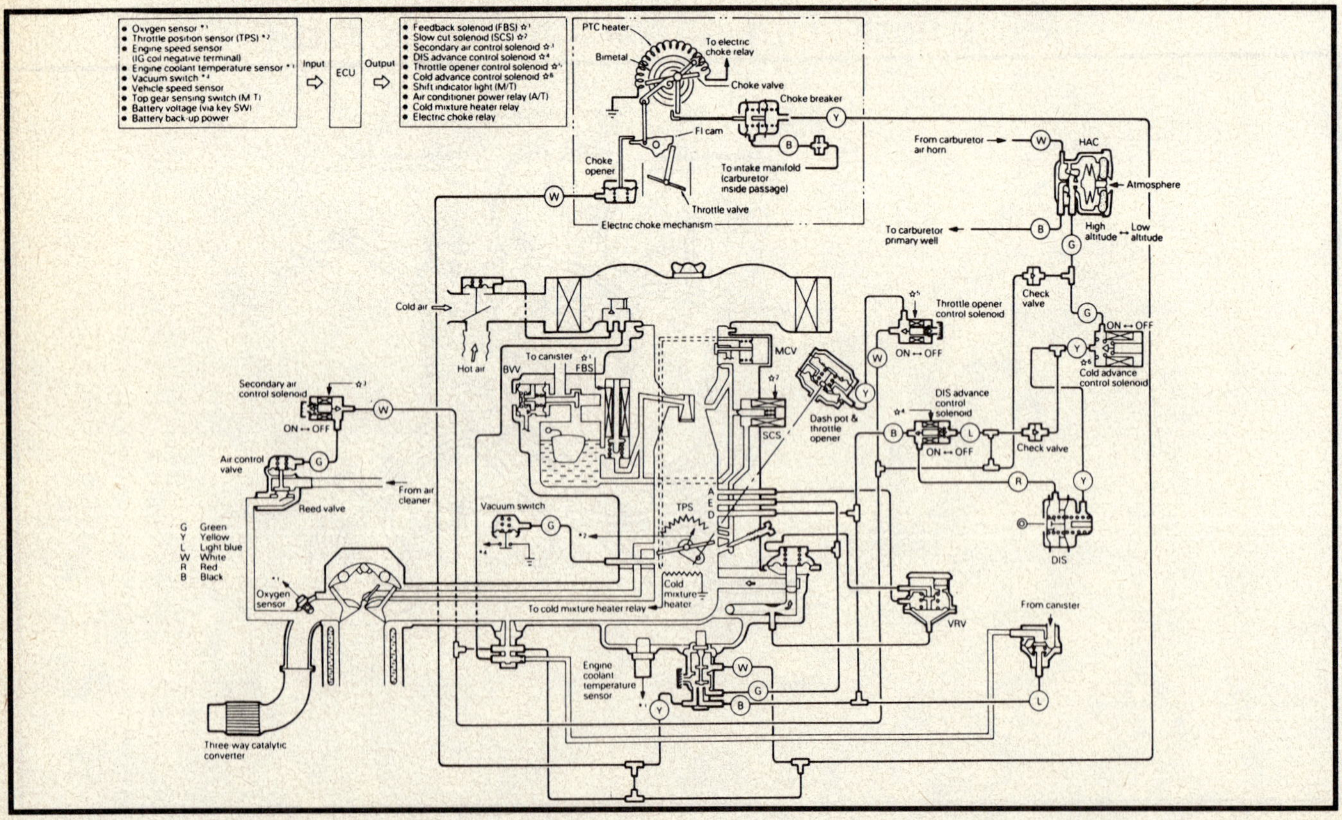

Glossary 16 SECTION

ABBREVIATIONS AND DEFINITIONS

4EAT—Electronic automatic 4 speed transaxle.

4X4L—4×4 Low input switch.

A/C DV—Air Cleaner Duct and Valve motor.

A/C P—Air Conditioning Pressure cut-out switch.

A/C—Air Conditioning.

A/F—Air/Fuel Ratio.

A/T—Automatic Transmission.

A4LD—Ford automatic 4 speed lock-up converter drive.

AAC—Auxiliary Air Control Valve.

AAV—Anti Afterburning Valve.

ABS—Anti Lock Brake system.

ACC—Air Conditioning Clutch compressor signal input to the computer relating status of the air conditioning clutch.

ACCS—Air Conditioning Cycling Switch.

ACD—Air Conditioner Demand switch.

ACT—Air Charge Temperature sensor or its signal circuit.

ACV—Air Control Valve or Thermactor Air Control Valve.

AFC—Air Flow Controlled fuel injection.

AFS—Air Flow Sensor.

AHFSS—Air conditioning and Heater Function Selection Switch input to the computer.

AI—Air Injection.

AIC—Automatic Idling Control valve.

AIR BPV—Thermactor Air Bypass Valve.

AIR GAP—The distance or space between the reluctor tooth and pick up coil.

AIR SENSOR—An air cone with a floating plate which measures air flow and determines plunger position on K-Jetronic type systems.

AIS—Air Injection System or Automatic Idle Speed circuit and/or motor.

AIV—Air Injection Valve.

ALTERNATING CURRENT (A.C.)—An electric current that is constantly changing polarity from positive to negative and back again.

AMBIENT TEMPERATURE—Temperature of the air surrounding the vehicle being serviced.

AMMETER—An electrical meter used to measure current flow (amperes) in an electrical circuit. Ammeter should be connected in series and current flowing in the circuit to be checked.

AMPERE (AMP)—The unit current flow is measured in. Amperage equals the voltage divided by the resistance.

ANTI-BFV—Anti-Backfire Valve.

AOD—Automatic Over Drive transmission.

APC—Automatic Performance Control.

ASCD—Automatic Speed Control Device, often referred to as cruise control.

ASD—Automatic Shut-Down relay driver circuit, fuel pump relay.

ATDC—After Top Dead Center.

ATM—Actuator Test Mode.

ATS—Air Temperature Sensor.

AUTOMOTIVE EMISSIONS—Gaseous and articulate compounds (hydrocarbons, nitrogen oxides and carbon monoxide) that are emitted from a vehicles crankcase, exhaust, carburetor and fuel tank.

AUXILIARY AIR REGULATOR—A rotary gate valve which stabilizes idle speed during engine warm-up.

AVOM—Analog Volt/Ohm Meter.

AWG—American Wire Gauge system.

AXOD-E—Electronic Automatic Overdrive transaxle.

AXOD—Automatic Overdrive transaxle.

BAC—Bypass Air Control system.

BAR—Unit of pressure measurement (1 bar is approximately 14.5 psi).

BASE IDLE—Idle rpm determined by throttle switch with idle speed control fully retracted.

BID—Breakerless Inductive Discharge ignition system.

BOB—Ford Breakout Box. Device which connects in series with the computer and the EEC-IV harness and permits measurements of the processor inputs and outputs also used on Mazda Navajo.

BOO—Brake On-Off input to the computer.

BOOST—Turbo charger boost solenoid or its control circuit.

BP—Barometric Pressure sensor. Used to compensate for altitude variations.

BPA—Bypass Air Valve.

BPCSV—Bypass Control Solenoid Valve.

BREAKOUT BOX—A device sometimes call pinout box, which connects in series with the computer and the harness and permits measurements of the processor inputs and outputs.

BTDC—Before Top Dead Center.

BV—Bowl Vent (carburetor).

BVT—Back-pressure Variable Transducer.

CALIBRATE—To adjust the scale of any instrument given quantitive measurements.

CANISTER PURGE SOLENOID—Electrical solenoid or its control line. Solenoid opens a valve from the fuel vapor canister line to the intake manifold when energized. This controls the flow of vapors between the carburetor bowl vent and the carbon canister.

CANISTER—A container, in a evaporative emission system, that contains charcoal to trap fuel vapors from the fuel system.

CANP—Canister Purge solenoid that opens the fuel vapor canister line to the intake manifold when energized.

CANP—Canister Purge solenoid.

CAPACITOR—A device which stores an electrical charge.

CARBON MONOXIDE—A colorless, odorless gas that is a by product of incomplete combustion of carbon. This gas is poisonous in confined spaces.

CATALYST—Special metals (i.e. platinum or palladium) within the catalytic converter that contact the hot exhaust gases and promote more complete combustion of the unburned hydrocarbons and reduction of carbon monoxide.

CATALYTIC CONVERTER—Muffler like assembly placed in the exhaust system that contains a catalyst to change hydrocarbons and carbon monoxide into water vapor and carbon dioxide.

CBD—Closed Bowl Distributor.

CC—Catalytic Converter.

CCC—Converter Clutch Control solenoid or its circuit.

CCD—Computer Controlled Dwell, used on Ford Vehicles.

CCO—Converter Clutch Override output from the computer processor to the transmission.

CCS—Coast Clutch Solenoid or its circuit.

CEC—Computerized Emission Control.

CER—Cold Enrichment Rod.

CES—Clutch Engage Switch.

CFC—Coasting Fuel Cut.

CFI—Central Fuel Injection as called throttle body injection.

CHECK ENGINE LIGHT—A dash panel light used either to aid in the identification and diagnosis of a system problems or to indicate that maintenance is required.

CHM—Cold Mixture Heater.

CID—Cylinder Identification sensor or its circuit.

CIRCUIT—The path that electricity travels in route to component and back to power source.

CIS—Continuous Injection System, Bosch K-Jetronic type system.

CLC—Converter Lock-up Clutch.

CO—Carbon monoxide.

COC—Conventional Oxidation Catalyst.

COMPUTER TIMING—The total spark advance in degrees before top dead center. Calculated by the Ford EEC-IV processor, based on sensor input.

COMPUTER—Any device capable of accepting information, comparing, adding, subtracting, multiplying, dividing and integrating this information and then supplying the results of these processes in proper form.

CONDUCTOR—Any material through which an electrical current can be transmitted easily.

CONTINUITY—Continuous or complete circuit. The type of circuit that can be checked with an ohmmeter.

CONTINUOUS SELF-TEST—A continuous test of the Ford EEC-IV system conducted whenever the vehicle is in operation.

CONVENTIONAL THEORY—The flow of current in an electrical circuit in which the direction is from positive to negative.

CP—Crankshaft Position sensor.

CPS—Crankshaft Position Sensor. Provides the ECU with engine speed and crankshaft angle (position).

CPU—Central Processing Unit.

CTS—Coolant Temperature Sensor.

CURB IDLE—Computer controlled idle rpm.

CVR—Control Vacuum Regulator.

CWM—Cold Weather Modulator.

CYL SENSOR—Crankshaft Angle Sensor.

CYLINDER IDENTIFICATION SIGNAL (CID)—A signal generated by the crankshaft timing sensor, that is used to synchronize the ignition coils, due to the fact that some models use a 2 ignition coil pack DIS system.

DFS—Decel Fuel Shut-off.

DI—Direct Ignition system, each spark plug has its own ignition coil.

DIELECTRIC SILICONE COMPOUND—Non-conducting silicone grease applied to spark plug wire boots, rotors and connectors to prevent arcing and moisture from entering a connector.

DIODE—An electrical devise that will allow current to flow in one direction only.

DIRECT CURRENT (D.C.)—An electrical current which flows in only one direction.

DIS—Distributorless Ignition System.

DOL—Data Output Link. Fuel calculation data from the EEC IV processor to the trip computer.

DRB II—Diagnostic Readout Box tester, for Chrysler system testing. Determines sensor voltage, degrees F, vacuum, rpm and mileage reading with vehicle's engine both on and off.

DSV—Deceleration Solenoid Valve.

DUAL CATALYTIC CONVERTER—Combines 2 converters in one shell. Controls NOx, HC and CO. Also called TWC.

DUAL-POINT INJECTION SYSTEM—A computer regulated system, that provides precise air/fuel ratio under all driving conditions. Same as throttle body with 2 injectors.

DV TW—Delay Valve, 2 Way.

DV—Delay Valve.

DVOM—Digital Volt/Ohm Meter.

DWELL METER—Measures the amount of time, recorded in degrees, that current passes through a closed switch.

E/L—Electrical Load control unit.

E4OD—Ford Electronic 4 speed Overdrive transmission.

EACV—Electronic Air Control Valve.

ECA—Electronic Control Assembly. Ford's engine controlling computer.

ECCS—Electronic Concentrated Control System.

ECI—Electronic Control Injection.

ECIT—Electronic Control Ignition Timing.

ECM—Electronic Control Module.

ECS—Emission Control System.

ECT—Engine Coolant Temperature sensor or its circuit, could also be Electronic Control Transmission.

ECU—ELectronic Control Unit or Engine Control Unit, To process input information to trigger the ignition control module.

EDF—Electro-Drive Fan relay or its circuit.

EEC-IV—Electronic Engine Control design 4. A computer controlled system of engine control used on Ford and some Mazda.

EEC—Evaporative Emission Control.

EEGR—Electronic Exhaust Gas Recirculation valve (Sonic).

EET—Electronic Exhaust Gas Recirculation Transducer.

EFC—Electronic Feedback Carburetor. Utilizes an electronic signal, generated by an exhaust gas oxygen sensor to precisely control the air/fuel mixture ratio in the carburetor.

EFC—Electronic Fuel Control.

EFE—Early Fuel Evaporation.

EFI—Electronic Fuel Injection. A computer controlled fuel injection system. On Ford EFI uses injectors in each intake port and CFI uses an injector in the throttle body.

EGI—Electronic Gasoline Injection.

EGO—Exhaust Gas Oxygen sensor.

EGOG—Exhaust Gas Oxygen Ground.

EGR S/O—Exhaust Gas Recirculation Shut-Off.

EGR—Exhaust Gas Recirculation system is designed to allow the flow of inert exhaust gases into the combustion chamber to cool combustion and thus reduce nitrous oxides in the exhaust.

EGRC—Exhaust Gas Recirculation Control vacuum solenoid valve or its control circuit.

EGRV—Exhaust Gas Recirculation Vent solenoid valve or its circuit.

EHC—Exhaust Heat Control vacuum solenoid or its circuit.

EICV—Electronic Idle Control Valve.

EIS—Electronic Ignition System which uses a reluctor and a pick up coil along with a module to replace the ignition points and condenser.

ELCD—Evaporative Loss Control Device.

ELECTRONIC CONTROL UNIT (ECU)—On board computer module, used to control ignition, fuel or other engine functions.

EMR—Emissions Maintenance Reminder.

EMW—Emission Maintenance Warning.

ENVIRONMENTAL PROTECTION AGENCY—Federal agency having responsibility for administering congressional programs relating to the protection of the environment.

EPA—Environmental Protection Agency.

ER—Engine running, used on some Ford system tests.

ERS—Engine RPM Sensor.

ESA—Electronic Spark Advance.

ESC—Electronic Spark Control.

ESS—Engine Speed Sensor.

EVP—EGR Valve Position sensor or its circuit.

EVR—EGR Vacuum Regulator or its circuit.

EXHAUST GAS OXYGEN SENSOR—Sensor that changes its voltage output as exhaust gas oxygen content changes as compared to the oxygen content of the atmosphere. The constantly changing electrical signal is used to control fuel mixture.

EXHAUST GAS RECIRCULATION—A procedure where a small amount of exhaust gas is readmitted to the combustion chamber to reduce peak combustion temperatures, thus reducing NOx.

FBC—FEEDBACK CARBURETOR—A system of fuel control employing a computer controlled solenoid that varies the carburetors air/fuel mixture.

FBCA—FEEDBACK CARBURETOR ACTUATOR—The computer controlled stepper motor used on Ford feedback carburetors, that varies the air/fuel mixture.

FCS—Fuel Control Solenoid.

FCV—Float Chamber Ventilation system.

FICB—Fast Idle Cam Breaker.

FIPL—Fuel Injection Pump Lever sensor or its circuit.

FMEM—Failure Mode Effects Management. Sometimes referred to limp-in mode.

FOM—Fix operating Mode, limp-in mode.

FPM—Fuel Pump Monitor. A circuit in the Ford EEC system used to monitor fuel pump operation.

FTO—Filter Tach Output. An output from the Ford DIS TFI IV module which provides a filtered ignition signal to the processor in order to control dwell.

FUEL ACCUMULATOR—Diaphragm unit which helps maintain residual fuel pressure for hot starting on CIS type fuel systems.

FUEL DISTRIBUTOR—The component which feeds fuel to the individual engine cylinders corresponding to the air flow rate metered by the air flow sensor on CIS systems.

GND, GRD or GRND—Ground. Common line leading to the negative side of the battery.

GVW—Gross Vehicle Weight.

HAI—Hot Air Intake.

HALL EFFECT PICK-UP ASSEMBLY—Used to input a signal to the electronic control unit. The system operates on the Hall Effect principle whereby a magnetic field is blocked from the pick-up by a rotating shutter assembly.

HALL EFFECT—A process where current is passed through a small slice of semi-conductor material at the same time as a magnetic field to produce a small voltage in the semi-conductor.

HBV—Heater Blower Voltage input to the ECC-IV processor reflecting heater blower voltage demand.

HC—Hydrocarbons. Any compound composed of hydrogen and carbon, such as petroleum products, that is considered a pollutant.

HCV—Exhaust Heat Control Valve.

HEDF—High-speed Electro-Drive Fan relay or its circuit.

HEGO—Heat Exhaust Gas Oxygen sensor or its circuit.

HERTZ (Hz)—The term meaning the cycles per second.

HIC—Hot Idle Compensator.

HICAS—High Capacity Active Controlled Suspension.

HLOS—Hardware Limited Operation Strategy. Certain types of computer malfunctions will place the EEC-IV system into HLOS mode. Output commands are replaced with fixed values. Sometimes referred to as limp-in mode.

HT—High Tension.

HYDROCARBON—Any compound composed of carbon and hydrogen, such as petroleum products. Excess amounts are considered undesirable contaminants.

I/O—Input/Output, for computer data transmission.

IAS—Inlet Air Solenoid valve or its circuit.

IBP—Integral Back Pressure.

IC—Integrated Circuit.

ICM—Ignition Control Module. Used by Chrysler to supply voltage for spark plug firing.

IDLE AIR STABILIZATION VALVE—Electronically controlled valve used to maintain idle speed at a predetermined level.

IDLE LIMITER—A device to control minimum and maximum idle fuel richness. The idle limiter is intended to prevent unauthorized persons from making mixture adjustments.

IDLE TRACKING SWITCH—An input device that sends a signal to the computer to indicate a closed throttle condition.

IDM—Ignition Diagnostics Monitor. A continuous monitor of the ignition input to the EEC-IV processor used to detect intermittent ignition faults.

IG—Ignition.

IGNITER—Term used by Japanese automotive and ignition manufacturers for the electronic control unit or module, sometimes spelled ignitor.

IGNITION COIL—Step-up transformer consisting of a primary and a secondary winding with an iron core. As the current flows in the primary winding stops, the magnetic field collapses across the secondary winding inducing the high secondary voltage. The coil may be oil filled or of an epoxy type design.

IMA Sensor—Idle Mixture Adjuster Sensor.

IMS—INFERRED MILEAGE SENSOR A circuit using a E-cell which deflates its state with the application of a current. As the vehicle ages the EEC-IV processor compensates for aging by changing calibration parameters.

INDUCTION—A means of transferring electrical energy in the form of a magnetic field. Principle used in the ignition coil to increase voltage.

INFINITE READING—A reading on an ohmmeter that indicates an open circuit or infinite reading.

INFINITY—An ohmmeter reading which indicates an open circuit in which no current will flow.

INJECTION VALVE—Same as injector.

INJECTOR—A solenoid or pressure-operated fuel delivery valve used for fuel injection systems.

INTERGRATED CIRCUIT (IC)—Electronic micro-circuit consisting of a semi-conductor components or elements made using thick-flim or thin-flim technology. Elements are located on a small chip made of a semi-conducting material, greatly reducing the size of the electronic control unit and allowing it to be incorporated within the distributor.

IRCM—Integrated Relay Control Module, used on some Ford systems.

ISA—Idle Speed Actuator. Extends or retracts to control engine idle speed and to set throttle stop angle during deceleration.

ISC—Idle Speed Control, this could be computer control motor, air bypass valve, or any device used to control idle rpm.

ITS—Idle Tracking Switch. An input device that sends a signal to the control module to indicate throttle position.

JSV—Jet Mixture Solenoid Valve.

KAM—Keep Alive Memory. Battery power memory locations in the computer used to store failure codes and some diagnostic parameters.

KAPWR—Keep Alive Power, used to power the KAM circuit of the processor.

KDLH—Kick-Down Low Hold.

KNOCK SENSOR—An input device that responds to spark knock, cause by over advanced ignition.

KOEO—Key On/Engine Off.

KOER—Key On/Engine Running.

KS—Knock Sensor. An input device that responds to spark knock caused by excessively advanced ignition timing.

LEAN MIXTURE—An air/fuel mixture that has excessive oxygen left after all the fuel in the combustion chamber has burned, 1 part fuel to 15 or more parts air.

LED—Light Emitting Diode.

LIMP-IN MODE—Is the attempt by the SMEC/SBEC to compensate for the failure of certain components by substituting information from other sources. Used in the Chrysler self diagnostic system.

LOAD—Any electrical device that provides resistance to current flow.

LOS—Limited Operation Strategy.

LUS—Lock-Up Solenoid.

M/C—Mixture Control.

M/T—Manual Transmission.

MAF—Mass Airflow sensor. A device used to measure the amount of intake air entering the engine on some fuel injection systems.

MAP—Manifold Absolute Pressure sensor or its circuit.

MAS—Mixture Adjust Screw.

MAT—Manifold Air Temperature.

MCT—Manifold Charge Temperature sensor.

MCV—Mixture Control Valve.

MICROPROCESSORS—A miniature computer on a silicone chip.

MIL—Malfunction Indicator Light. Check engine light.

MIRCOPROCESSOR—A miniature computer on a silicone chip.

MLP—Manual (shift) Lever Position sensor or its circuit.

MODE—An operating state i.e. closed loop vs open loop.

MODULE—Electronic control unit, amplifier or igniter of solid state or integrated design which controls the current flow in the ignition primary circuit based on input from the pick-up coil. When the module opens the primary circuit, the high secondary voltage is induced in the coil.

MONITOR BOX—An optional Ford EEC-IV test device which connects in series with the EEC-IV processor and its harness, and permits measurements in various units of the processor inputs and outputs.

MPFI—Multi-Point Fuel Injection.

MPS—Motor Position Sensor.

MS—Millisecond.

NDS—Neutral/Drive Switch.

NGS—Neutral Gear Switch or its circuit.

NITROUS OXIDES (NOx)—A compound formed during the engine's combustion process above 2500° F, when the oxygen combines with nitrogen to form nitrous oxides. This contributes to photochemical smog.

NPS—Neutral Pressure Switch or its circuit.

NTS—Negative Temperature Coefficient Resistor.

OCC—Output Cycling Check.

OCT ADJ—Octane Adjust device which modifies ignition spark.

OCT—Octane Switch.

OHM—The standard unit for measuring the resistance to current flow.

OHMMETER—The electrical meter used to measure the resistance in ohms. Self-powered and must be connected to a voltage free circuit or damage to the ohmmeter will result.

OPEN CIRCUIT—A circuit which does not provide a complete path for the flow of current.

OSC—Output State Check.

OSCILLOSCOPE—An electric testing device that shows a pattern wave form of an electrical occurrence. Used to test ignition, fuel injection, alternator and other electrical devices.

OVCV—Outer Vent Control Valve.

OVERLAY CARD—A plastic card used for the Ford Breakout or Monitor box to identify connections for each engine tested.

OXYGEN (O_2) SENSOR—Used with the feedback system to sense the presence of oxygen in the exhaust gas and signal the computer which can reference the voltage signal to an air/fuel ratio.

PA SENSOR—Atmospheric Pressure Sensor.

PCV—Positive Crankcase Ventilation. A system that controls the flow of crankshaft vapors into the engine intake manifold where they are burned in combustion rather then being discharged into the atmosphere.

PFE—Pressure Feedback EGR sensor or its circuit.

PGM-FI—Programmed Fuel Injection system.

PGM-IG—Programmed Ignition system.

PHENOMENA—Basis of symptoms; a significant occurrence.

PICK-UP COIL—Inputs signal to the electronic control unit to open the primary circuit. Consists of a fine wire coil mounted around a permanent magnet. As the reluctor's ferrous tooth passes through the magnetic field an alternating current is produced, signaling the electronic control unit. Can operate on the principle of metal detecting, magnetic induction or Hall Effect. Is also referred to as a stator or sensor.

PIP—Profile Ignition Pickup.

PORTED VACUUM SWITCH—A temperature actuated switch that changes vacuum connections when the coolant temperature changes.

POTENTIOMETER—A variable resistor used to change a voltage signal.

PRC—Pressure Regulator Control solenoid.

PRE-PUMP—In-tank fuel pump.

PRIMARY CIRCUIT—Is the low voltage side of the ignition system which consists of the ignition switch, ballast resistor or resistance wire, bypass, coil, electronic control unit and pick-up coil as well as the connecting wires and harnesses.

PROFILE IGNITION PICKUP—A hall effect vane switch that furnishes crankshaft position data to the EEC-IV processor.

PSPS—Power Steering Pressure Switch. The signal is used by computer to compensator for power steering loads.

PTC HEATER—Positive Temperature Coefficient Heater.

PULSE GENERATOR—Also called a pulse signal generator. Term used by Japanese and German automotive and ignition manufacturers to describe the pick-up and reluctor assembly. Generates an electrical pulse which triggers the electronic control unit or igniter.

PULSE WIDTH—The amount of time the control unit energizes the fuel injectors to spray fuel into the intake manifold, usually measured in millisecond.s.

PVS—Ported Vacuum Switch. A temperature-activated switch that changes vacuum connections when the coolant temperature changes.

QUICK TEST—A functional diagnostic test for Ford EEC-IV system, consisting of test hookup, key on engine off, engine running and continuous self test modes.

RAD—Radiator Temperature Switch.

RAM—Random Access Memory.

RECORDER—A device used to record the electronic signals sent to and from the engine computer. Sometimes referred to as a flight recorder, because it works similar to aircraft recording device; to store what the last functions of the engine computer that where preformed over a 1 minute or some time period.

RELAY—A switching device operated by a low current circuit, which controls the opening and closing of another higher current circuit.

RELIEF VALVE–A pressure limiting valve located in the exhaust chamber of the thermactor air pump. Its function is to relieve part of exhaust airflow if pressure exceeds a calibrated value.

RELUCTOR–Also called an armature or trigger wheel. Ferrous metal piece attached to the distributor shaft. Made up of teeth of which the number are the same as the number of engine cylinders. As the reluctor teeth pass through the pick-up magnetic field an alternating current is generated in the pick-up coil.

RESIDUAL PRESSURE–Pressure remaining in the fuel system after the engine has be shut off.

RESISTANCE–The opposition to the flow of current through a circuit or electrical device, and is measured in ohms. Resistance is equal to the voltage divided by the amperage.

RICH MIXTURE–An air/fuel mixture that has more fuel than can burn completely, 1 part fuel to 14 or less parts air.

RMS–ROOT MEAN SQUARE (effective)–The square root average of the squares of the instantaneous amplitudes taken over the duration of the pulse. Example; the wave pattern on a 110vac house receptacle would be 300vac peak-to-peak 300vac ÷ 2 × 0.70 = 105vac RMS.

ROM–Read Only Memory.

RXD–Receive data line.

SAS–Speed Adjusting Screw.

SATURATION–The state of a coil when current flow has reached the design maximum and the magnetic field has reached its maximum strength.

SBEC/SMEC–Single Board Engine Controller/Single Module Engine Controller used on Chrysler vehicles. Both regulate ignition timing, air-fuel ratio, emission control devices, cooling fan, charging system idle speed and speed control.

SBS–Supercharger Bypass Solenoid or its circuit.

SCC–Spark Control Computer.

SCSV–Slow Cut Solenoid Valve.

SDS–Service Data and Specifications.

SDV–Spark Delay Valve.

SECONDARY–The high voltage side of the ignition system, usually above 20,000 volts. The secondary includes the ignition coil, coil wire, distributor cap and rotor, spark plug wires and spark plug.

SEFI–Sequential Electronic Fuel Injection. Injectors located in intake ports that inject fuel triggered by ignition timing.

SELF-TEST–One of the 3 subsets of the Ford EEC-IV quick test modes.

SENSOR PLATE–A round plate bolted to the air flow sensor lever which floats in the stream of intake air on the CIS type systems.

SENSOR–Also called the pick-up coil or stator. See pick-up coil for definition.

SHORT CIRCUIT–An undesirable connection between a circuit and any other point.

SHUTTER–Also called the vane. Used in a Hall Effect distributor to block the magnetic field from the Hall Effect pick-up. The shutter is attached to the rotor and is grounded to the distributor shaft.

SIG RTN–Signal Return circuit for all sensors except HEGO.

SIL–Shift Indicator Light. Indicates to driver optimum time to shift gears.

SIS–Solenoid Idle Stop.

SMJ–Super Multiple Junction, the main harness connector through the bulkhead for the engine controller circuits.

SOLENOID–A wire coil with a movable core which changes position by means of electromagnetism when current flows through the coil.

SPARK DURATION–The length of time measured in milliseconds($^1/_{1000}$th second) the spark is established across the spark plug gap.

SPFI–Single Point Fuel Injection.

SPI–Single-Point Injection system, same a throttle body injection.

SPOUT–Spark output signal from the EEC-IV processor to the TFI-IV module, used to control amount of timing retard.

SQUARE WAVE–An essentially square or rectangular shaped wave. A wave that alternately assumes 1 to 2 fixed values with a negligible transition time between the 2 values.

SSI–Solid State Ignition system .

SST–Special Service Tool, or special test equipment to be used for testing or repairs.

STAR–Self Test Automatic Readout (Ford), used to access codes from the EEC-IV processor.

STATOR–Another name for a pick-coil. See pick-up coil for definition.

STI–Self Test Input (Ford) circuit in the EEC-IV systems. Used to place the computer into testing mode.

STO–Self Test Output (Ford) circuit in the EEC-IV systems. Used by the computer to send testing and fault codes to tester.

STOICHIOMETRIC–in general, the precise mixture for the most efficient process of conversion. In automotive use, a ratio of 14.7 parts air to 1 part fuel. This ratio yields the highest combustion efficiecny and therefore, the lowest emissions.

SWITCHING TRANSISTOR–Used in some electronic ignition systems, it acts as a switch for high current in response to a low voltage signal applied to the base terminal.

T.V.–Throttle Valve.

TA SENSOR–Intake Air Temperature Sensor.

TAB–Thermactor Air Bypass solenoid.

TACH INPUT–An engine rpm signal sent to the computer from the ignition coil primary circuit.

TAD—Thermactor Air Diverter solenoid.

TAS—Throttle Adjust Screw.

TBI—Throttle Body Injection (Fuel).

TCA—Thermostat Controlled Air Cleaner.

TCM—Throttle Control Module

TCP—Temperature Compensated Accelerator Pump.

TCS—Traction Control System.

TDC—Top Dead Center.

TERMINAL—A connecting point in a electrical circuit, where the circuit can be disconnected or tested.

TES—Thermal Electric Switch.

TFI—Thick Film Ignition module. Controls the coil and ignition operation on most Ford vehicles.

THERMACTOR AIR CONTROL VALVE—Combines the function of a normally closed air bypass valve and an air diverter valve in one integral valve.

THERMACTOR AIR SYSTEM—The efficiency of the catalytic converter is dependent upon temperature and the chemical makeup of the exhaust gases. These requirements are met by the thermactor air supply system.

THERMISTOR—A device that changes its resistance with temperature.

THERMO-TIME SWITCH—A switch which interupts the electrical circuit of the cold start injector based on temperature and time.

THREE-WAY CATALYST—Combines 2 converters in 1 shell. Controls NOx, HC and CO. Also called dual catalytic converter.

TIMING—Relationship between the spark plug firing and the piston position.

TK or TKS—Throttle Kicker Solenoid. An actuator moves the throttle linkage to increase idle rpm.

TKS— Throttle Kicker Solenoid, when energized, supplies manifold vacuum to the throttle kicker actuator as directed by the computer to compensate for engine loads. Also called the idle-up system.

TOT—Transmission Oil Temperature sensor.

TP or TPS—Throttle Position Sensor or its circuit. Used to signal computer the position of the throttle plates.

TRANSFER PUMP—Fuel pump located in the fuel tank, usually used with a 2 pump system.

TRANSISTOR—A semiconductor device that can control an electrical current by varying a smaller base current. This device acts like a mechanical relay with a variable resistor at the points.

TRIGGER WHEEL—See Reluctor for definition.

TTS—Transmission Temperature Switch.

TVSV—Thermostatic Vacuum Switching Valve.

TW SENSOR—Coolant Temperature Sensor.

TWC—Three-way catalyst, sometimes referred to as a dual catalytic converter. Combines 2 catalytic converters in one shell to control emissions of NOx, HC and CO.

TWSV—Three Way Solenoid Valve.

TXD—Transmitted data line.

UIM—Universal control unit.

VACUUM—A term side to describe a pressure that is less than atmospheric pressure.

VAF—Vane Air-Flow sensor or its circuit.

VAT—Vane Air-flow Temperature sensor.

VB VOLTAGE—Battery Voltage.

VBAT—Vehicle Battery voltage.

VCM—Vehicle Condition Monitor.

VCV—Vacuum Control Valve.

VECI LABEL—Vehicle Emission Control Information label, located under the hood.

VF VOLTAGE—Battery Voltage.

VIS—Variable Induction System.

VLC—Vacuum Sensor and Vacuum Line Charging solenoid valve.

VM—Vane Meter.

VOLT—The unit of electrical pressure or electromotive force.

VOLTAGE DROP—The difference in voltage between one point in a circuit and another, usually across a resistance. Voltage drop is measured in parallel with current flowing in the circuit.

VOLTMETER—An electrical meter used to measure voltage in a circuit. Voltmeters must be connected in parallel across the load or circuit.

VOM—Volt/Ohm Meter. Used to measure voltage and resistance.

VPWR—Vehicle Power supply voltage. 10-14 voltages DC.

VREF—The reference voltage or power supplied by the computer control unit to some sensors regulated at a specific voltage.

VRS—Variable Reluctance Sensor.

VSS—Vehicle Speed Sensor.

VSV—Vacuum Switching Valve.

VTV—Vacuum Transmitting Valve.